THE LIBRARY
ST. MARY'S COLLEGE OF MARYLAND
ST. MARY'S CITY, MARYLAND 20686

D1780139

DANA'S NEW MINERALOGY

DANA'S NEW MINERALOGY

The System of Mineralogy

of

James Dwight Dana and Edward Salisbury Dana

EIGHTH EDITION
ENTIRELY REWRITTEN AND GREATLY ENLARGED

RICHARD V. GAINES
EARLYSVILLE, VIRGINA

H. CATHERINE W. SKINNER
YALE UNIVERSITY

EUGENE E. FOORD
UNITES STATES GEOLOGICAL SURVEY, DENVER

BRIAN MASON
CURATOR EMERITUS, THE SMITHSONIAN INSTITUTION

ABRAHAM ROSENZWEIG
ROSENZWEIG ASSOCIATES, TAMPA

With Sections by

VANDALL T. KING
ROCHESTER ACADEMY OF SCIENCE, ROCHESTER, NEW YORK

And with Illustrations by

ERIC DOWTY
SHAPE SOFTWARE

JOHN WILEY & SONS, INC.
NEW YORK · CHICHESTER · WEINHEIM · BRISBANE · SINGAPORE · TORONTO

This text is printed on acid-free paper.

Copyright ©1997 by John Wiley & Sons, Inc.

All rights reserved. Published simultaneously in Canada.

Reproduction or translation or any part of this work beyond that permitted by Section 107 or 108 of the 1976 United States Copyright Act without the permission of the copyright owner is unlawful. Requests for permisison or further information should be addressed to the Permissions Department, John Wiley & Sons, Inc., 605 Third Avenue, New York, NY 10158-0012.

This publication is designed to provide accurate and authoritative information in regard to the subject matter covered. It is sold with the understanding that the publisher is not engaged in rendering legal, accounting, or other professional services. If legal advice or other expert assistance is required, the services of a competent professional person should be sought.

Library of Congress Cataloging-in-Publication Data
Dana, James Dwight, 1813–1895
    Dana's new mineralogy: the System of mineralogy of James Dwight
  Dana and Edward Salisbury Dana. — 8th ed., entirely rewritten and
  greatly enl. / by Richard V. Gaines . . . [et al.]; with illustrations by Eric Dowty.
      p. cm.
    Rev. ed. of: The system of mineralogy of James Dwight Dana and
  Edward Salisbury Dana. 7th ed. New York, J. Wiley & sons, inc.,
  1944- .
    Includes indexes.
    ISBN 0-471-19310-0 (alk. paper)
    1. Mineralogy. I. Dana, Edward Salisbury, 1849–1935.
II. Gaines, Richard V. III. Dana, James Dwight, 1813–1895. System
of mineralogy of James Dwight Dana and Edward Salisbury Dana.
IV. Title.
QE372.D23 1997
549–dc21   96–46965

Printed in the United States of America

10 9 8 7 6 5 4 3 2 1

Typeset by Keyword Typesetting Services Ltd, Wallington, Surrey, England.

Contents

Preface	ix
Historical Perspective	x
Acknowledgments	xii
Introduction	xv
Descriptions of Mineral Species	xviii
Abbreviations for Literature References	xxxvii

Native Elements and Alloys

Class 1 Native Elements and Alloys	1

Sulfides and Related Compounds

Class 2 Sulfides, Including Selenides and Tellurides	37
Class 3 Sulfosalts	150

Oxides

Class 4 Simple Oxides	204
Class 5 Oxides Containing Uranium, and Thorium	255
Class 6 Hydroxides and Oxides Containing Hydroxyl	266
Class 7 Multiple Oxides	292
Class 8 Multiple Oxides Containing Niobium, Tantalum, and Titanium	332

Halogenides

Class 9 Anhydrous and Hydrated Halides	373
Class 10 Oxyhalides and Hydroxyhalides	391
Class 11 Halide Complexes; Alumino-fluorides	407
Class 12 Compound Halides	419

Carbonates

Class 13 Acid Carbonates	422
Class 14 Anhydrous Carbonates	426
Class 15 Hydrated Carbonates	462
Class 16 Carbonates Containing Hydroxyl or Halogen	475
Class 17 Compound Carbonates	519

Nitrates

Class 18 Nitrates	526
Class 19 Nitrates Containing Hydroxyl or Halogen	528
Class 20 Compound Nitrates	530

Iodates

Class 21 Anhydrous and Hydrated Iodates	531
Class 22 Iodates Containing Hydroxyl or Halogen	532
Class 23 Compound Iodates	533

Borates

Class 24 Anhydrous Borates	534
Class 25 Anhydrous Borates Containing Hydroxyl or Halogen	542
Class 26 Hydrated Borates Containing Hydroxyl or Halogen	551
Class 27 Compound Borates	565

Sulfates

Class 28 Anhydrous Acid and Sulfates	568
Class 29 Hydrated Acid and Sulfates	583
Class 30 Anhydrous Sulfates Containing Hydroxyl or Halogen	623
Class 31 Hydrated Sulfates Containing Hydroxyl or Halogen	639
Class 32 Compound Sulfates	666

Selenates and Tellurates; Selenites and Tellurites

Class 33 Selenates and Tellurates	672
Class 34 Selenites, Tellurites and Sulfites	677

Chromates

Class 35 Anhydrous Chromates	692
Class 36 Compound Chromates	696

Phosphates, Aresenates, and Vanadates

Class 37 Anhydrous Acid Phosphates, etc.	697
Class 38 Anhydrous Phosphates, etc.	700
Class 39 Hydrated Acid Phosphates, etc.	735
Class 40 Hydrated Phosphates, etc.	814
Class 41 Anhydrous Phosphates, etc., Containing Hydroxyl or Halogen	815
Class 42 Hydrated Phosphates, etc., Containing Hydroxyl or Halogen	882
Class 43 Compound Phosphates, etc.	957

Antimonates, Antimonites, and Arsenites

Class 44	Antimonates	971
Class 45	Acid and Normal Antimonites and Arsenites	977
Class 46	Basic or Halogen-Containing Antimonites, Arsenites	981

Vanadium Oxysalts

Class 47	Vanadium Oxysalts	985

Molybdates and Tungstates

Class 48	Anhydrous Molybdates and Tungstates	993
Class 49	Basic and Hydrated Molybdates and Tungstates	1005

Organic Compounds

Class 50	Salts of Organic Acids	1011

Nesosilicates: Insular SiO_4

Class 51	Insular SiO_4 Groups Only	1020
Class 52	Insular SiO_4 Groups and O, OH, F, and H_2O	1061
Class 53	Insular SiO_4 Groups and Other Anions or Complex Cations	1106
Class 54	Borosilicates and Some Beryllosilicates	1116

Sorosilicates: Isolated Tetrahedral Noncyclic Groups, N>1

Class 55	Si_2O_7 Groups, Generally with No Additional Anions	1133
Class 56	Si_2O_7 with Additional O, OH, F and H_2O	1149
Class 57	Insular Si_3O_{10} and Larger Noncyclic Groups	1188
Class 58	Insular, Mixed, Single, and Larger Tetrahedral Groups	1193

Cyclosilicates

Class 59	Three-Membered Rings	1220
Class 60	Four-Membered Rings	1231
Class 61	Six-Membered Rings	1238
Class 62	Eight-MemberedRings	1267
Class 63	Condensed Rings	1269
Class 64	Rings with Other Anions and Insular Silicate Groups	1281

Inosilicates: Two-Dimensionally Infinite Silicate Units

Class 65	Single-Width Unbranched Chains, $W = 1$	1288
Class 66	Double-Width Unbranched Chains, $W = 2$	1335
Class 67	Unbranched Chains with $W > 2$	1376
Class 68	Structures with Chains of More Than One Width	1379
Class 69	Chains with Side Branches or Loops	1381

Class 70 Column or Tube Structures   1397

Phyllosilicates

Class 71 Sheets of Six-Membered Rings   1405
Class 72 Two-Dimensional Infinite Sheets with Other Than
          Six-Membered Rings   1521
Class 73 Condensed Tetrahedral Sheets   1546
Class 74 Modulated Layers   1552

Tektosilicates

Class 75 Si Tetrahedral Frameworks   1568
Class 76 Al–Si Frameworks   1595
Class 77 Zeolites   1646
Class 78 Unclassified Silicates   1702

Appendix – New Minerals for 1996   1721

Indexes

List of Mineral Names in Numerical Order   1723
List of Mineral Names in Alphabetical Order   1753
General Index   1783

Preface

Mineralogy is as old as humankind. From antiquity, people have known and highly valued minerals, with a fascination for their utility and beauty. The names of many of the 3700 minerals we know today derive from the ancient past.

However, a **systematic** approach, or a science of mineralogy, dates only from the late eighteenth and early half of the nineteenth century, when Linnaeus, better known for usurping Adam's role in naming the plants and animals, also showed an interest in classifying the minerals. His natural history collection contained some 2500 mineral specimens, and even today, individual minerals (as distinct from rocks) are referred to as **species**.

The period 1830–1859 saw a huge expansion of both popular and professional interest in natural history and great advances in the systematic classification of the many diverse parts of our environment. The first edition of Dana's **A System of Mineralogy** appeared in 1837, before **On the Origin of Species** (Darwin, 1859), a time of flux in understanding both the physical and biological history of the earth. Similarly, this eighth edition is presented against a background of changing ideas of the nature, properties, relationships, genesis, and uses of mineral species. It appears at a time when mineralogy has close affiliation with solid-state physics, materials science, and inorganic chemistry, biology, and other more wide ranging fields of science and technology.

Richard V. Gaines
H. Catherine W. Skinner
Eugene E. Foord
Brian Mason
Abraham Rosenzweig

Historical Perspective

Publication of **Dana's New Mineralogy**, the eighth Edition of **Dana's System of Mineralogy**, culminates a long series of books begun by James Dwight Dana when he wrote the first edition of **A System of Mineralogy** in 1837. Sparked by the chemical discoveries of Lavoisier and Berzelius and the crystallographic findings of Haüy, mineralogical investigators had begun to systematize the science. James D. Dana, newly appointed assistant in the chemistry laboratory at Yale University, completed his **System** at the age of 24 in 10 months. He used the Linnean principles for the classification, following and modifying those published by Friedrich Mohs. His description of individual species is basically that of Mohs' **Treatise on Mineralogy**, the English translation (1825) of Mohs' German-language book published two years earlier. The first edition of Dana's book was privately published, but the second edition (1844) and all but one other have been published by John Wiley. The **System** was the first book Wiley (then known as Wiley & Putnam) published in the sciences. Dana's mineral descriptions were considerably condensed, and he added to the localities listed for each mineral those in the United States where such were known.

The original system contained 352 minerals which can be identified with species we know today, although actual names of about 50% of them are from slightly to entirely different now; for example, gray antimony is now called stibnite, spathic iron is now known as siderite, and blue vitriol is now called chalcanthite.

The general chemistry of individual species was discussed, but no chemical formulas were presented, even though Berzelius had proposed a system of chemical notation in 1814 that introduced one- and two-letter abbreviations for the Latin names of the elements (K, Na, Fe, etc.) By the third edition of the **System** (1850), Dana adopted the Berzelian system of writing formulas, but today's reader would find them difficult to understand since the notation today was not codified until about 1880. Mineral formula presentation probably aided the systematization.

The early recognition of mineral species was entirely dependent on a knowledge of physical properties such as crystal morphology, color, luster, hardness, cleavage, streak, and specific gravity, plus simple chemical tests and blowpipe analysis. Today, the physical properties are still important, but definitive identification requires analyses developed and refined in this century such as x-ray and electron diffraction, optical and electron microscopy, and spectrographic and microprobe examination. The number of minerals we recognize as of the writing of this book has increased by a factor of 10 since 1837 and grows by about 50 new species per year. In contrast to former times, only a few milligrams of the substance is needed for complete characterization.

Dana continued to revise **A System of Mineralogy** and produced other books, many of which continue today, such as the **Manual of Mineralogy** (first edition 1848; twenty-first edition 1993), **Manual of Geology, Manual of Mineralogy and Petrography, Minerals and How to Study Them**, and other titles. Edward Salisbury Dana, the son of J.D. Dana, prepared **Dana's Textbook of Mineralogy** (first edition 1877; fourth edition 1932), as well as the sixth edition of **A System of Mineralogy** in 1892. In that edition, all of the interfacial angles from the crystal elements were recalculated, the Miller–Bravais symbols were introduced, mineral varieties were critically evaluated, and synonymy was improved and extended. The work became an international classic for over 50 years, due to the meticulous care with which it was prepared, the wonderful freedom from errors, the extraordinary detail through the author's personal knowledge of the mineralogical literature, and the inclusive scope of the work.

The desirability of having a successor to the sixth edition of the **System** became evident as early as 1915. Efforts to produce a seventh edition began in 1927 and by 1941, a partial manuscript for the first volume was nearly complete. However, World War II interrupted publication, so that the first volume of the seventh edition of the **System of Mineralogy** (elements, sulfides, sulfosalts, and oxides) by Charles Palache, Harry Berman, and Cliford Frondel, did not appear until 1944.

By then so many new minerals had been added and old ones discredited, and the methodology of mineral identification had become so much more rigorous, with new structural data leading to more scientific classification of mineral groups, that the need for completing the new edition was sorely felt. Work on volume 2 (halogenides and all oxysalts except the silicates) started in 1942, was interrupted by World War II, and was completed in 1951. Harry Berman (1902–1944) had been killed in a plane crash while on war work and Charles Palache (1869–1954) had retired, so the work was essentially completed by Clifford Frondel (1907–). Volume 3 (silica minerals) was completed by Frondel in 1962. The level of detail and complexity required by the format of the seventh edition of **System** was too daunting to attract volunteers to do the silicates, and thus the last volume, the systematic description of the most common silicate minerals, was never undertaken.

The present eighth edition, with five co-authors and a considerably less detailed format than the 7th Edition of the **System**, has been many years in preparation. During the years since it was started, the number of minerals to be covered has about doubled.

Acknowledgments

In compiling and selecting the data for this edition we have consulted the *Powder Diffraction File* published by the International Center of Diffraction Data (Newtown Square, PA 19073-3273) who have generously granted permission for reproduction of portions of their data sets; *The Mineral Data Base* by Ernest H. Nickel, CSIRO, Wembley, WA, Australia, and Monte C. Nichols published by Aleph Enterprises, Box 213, Livermore, CA 94551; and *GEOREF* the bibliographical reference compelation of the American Geological Institute, 4220 King St., Alexandria, VA, 22302-1502. However, without the contributions of colleagues and associates no effort of this magnitude could have been undertaken.

The initial stages of this project were undertaken by the late Horace Winchell of Yale University, Department of Geology and Geophysics, and Brian J. Skinner, also at Yale, was a co-author. We are grateful and wish to publically acknowledge and thank the following individuals for reviewing substantial portions of the manuscript, supplying bibliographical data on the individuals for whom minerals have been named, and offering suggestions and information on the mineral species covered in this volume.

Jay Ague: Dept. of Geology/Geophysics, Yale University.
Eugene Alexandrov: Hamden, Connecticut.
Eric Asselborn, M.D.: Montrevel-en-Bresse, Ain, France.
F. John Barlow: Appleton, Wisconsin.
Peter Bayliss: Mineralogy, Australian Museum, Sydney, NSW, Australia.
William D. Birch: Dept. of Mineralogy, Museum of Victoria, Melbourne, Australia.
Frank N. Blanchard: Dept. of Geology, University of Florida, Gainesville, Florida.
György Buda: Lorand University, Budapest, Hungary.
Fabien Cesbron: B. R. G. M., Orleans, France.
Graham Chinner: Dept. of Earth Sciences, Cambridge Univ., Cambridge, England.
Sharon Cisneros: Mineralogical Research Co., San Jose, California.
Andrew M. Clark: Dept. of Mineralogy, Natural History Museum, London, England.
Rock H. Currier: Jewel Tunnel Imports, Baldwin Park, California.
Stanley Dyl: Seaman Mineralogical Museum, Michigan College of Technology, Houghton, Michigan.

Acknowledgments

T. Scott Ercit: Canadian Museum of Nature, Ottawa, Ontario, Canada.
Richard C. Erd: U.S.G.S., Menlo Park, California.
Howard T. Evans: U.S.G.S., Reston, Virginia.
Ellen W. Faller: Dept. of Geology and Geophysics, Yale University, New Haven, Connecticut.
James A. Ferraiolo: Bowie, Maryland.
Michael Fleischer: Dept. of Mineral Sciences, Smithsonian Institution, Washington, D.C.
Claren N. ("Si") and Ann Frazier: Berkeley, California.
Carl Francis: Mineralogical Museum, Harvard University, Cambridge, Massachusetts.
Gilbert Gauthier: Maisons Laffitte, Paris, France.
Legrand A. Gould: Pebble Beach, California.
George E. Harlow: Dept. of Min. Sci., American Museum of Natural History, New York, NY.
Fritz, Fritz: Bergakademie Freiberg, Freiberg, Saxony, Germany.
Donald B. Hoover: U.S.G.S., Federal Center, Lakewood, Colorado.
John M. Hughes: Dept. of Geology, Miami University, Oxford, Ohio.
W. Barclay Kamb: Division of Geol. and Planetary Sci., Cal. Inst. of Tech., Pasadena, California.
Akira Kato: National Science Museum, Tokyo, Japan.
Petr Korbel: Dept. of Mineralogy, National Museum, Prague, Czech Republic.
Seppo Lahti: Geological Survey of Finland, Espoo, Finland.
Joseph Lhoest: Liege, Belgium.
John C. Lucking: Phoenix, Arizona.
Emil Makovicky, Inst. of Mineralogy, Univ. of Copenhagen, Copenhagen, Denmark.
Joseph A. Mandarino: Royal Ontario Museum, Toronto, Ontario.
Olaf Medenbach: Institut fur Mineralogie, Ruhr Universitat Bochum, Germany.
Duane Mellor: Geology Library, Yale University, New Haven, Connecticut.
Luiz A. D. Menezes: Belo Horizonte, Minas Gerais, Brazil.
William C. Metropolis: Mineralogical Museum, Harvard University, Cambridge, Massachusetts.
Henry O. Meyer (deceased): Dept. of Earth Sciences, Purdue Univ., West Lafayette, Indiana.
James W. Minette: U. S. Borax, Boron, California.
Paul B. Moore: Dept. of Geology, University of Chicago, Chicago, Illinois.
Margarita I. Novgorodova: Inst. Geol. of Ore Deposits, Acad. of Sciences, Moscow, Russia.
Per Nysten: University of Uppsala, Uppsala, Sweden.
Herbert Obodda: Short Hills, New Jersey.
Renato Pagano: Milan, Italy.
Ole V. Petersen: Geologisk Museum, Copenhagen, Denmark.
William W. Pinch: Rochester, New York.

Paul Powhat: Dept. of Mineral Science, Smithsonian Inst., Washington, D.C.
Alan Pring: South Australian Museum, Adelaide, SA, Australia.
Gunnar Raade: Mineralogisk-Geologisk Museum, University of Oslo, Oslo, Norway.
Richard J. Reeder: Dept. of Earth Sciences, State Univ. of N.Y., Stonybrook, N.Y.
Louise Reif: U. S. G. S., Federal Center, Lakewood, Colorado.
Charles B. Sclar: Dept. of Earth Sciences, Lehigh University, Bethlehem, Pennsylvania.
Thalassa Skinner: Cambridge, Massachusetts.
Dorian G. W. Smith: Dept. of Geology, Univ. of Alberta, Edmonton, Alberta.
Hans A. Stalder: Natural History Museum, Bern, Switzerland.
Hugo Strunz: Institute of Mineralogy, Technical University, Berlin, Germany.
Peter Susse: MINABS, Inst. of Mineralogy, Univ. of Gottingen, Gottingen, Germany.
C. Sheldon Thompson: R. T. Vanderbilt Co., Norwalk, Connecticut.
Kenneth M. Towe: Dept. of Paleobiology, Smithsonian Inst., Washington, D.C.
John J. Trelawney Esq.: Palo Alto, California.
Rudy W. Tschernich: Snohomish, Washington.
Lourens Wals: Turnhout, Belgium.
Terry C. Wallace: Dept. of Geosciences, University of Arizona, Tuscon, Arizona.
John S. White: Stewartstown, Pennsylvania.
Hans J. Wilke: Eppertshausen/Hessen, Germany.
Wendell E. Wilson: The Mineralogical Record, Tuscon, Arizona.
Barry Yampol: Oyster Bay, Long Island, New York.
Victor Yount: Warrenton, Virginia.
Josef Zemann: Institut fur Mineralogie der Universitat Wien, Austria.
Michael Zolensky: Lyndon Johnson Space Center, Houston, Texas.

We also wish to acknowledge our present editor, Philip Manor, without whom this project would never have seen the light of day. Finally, we take unto ourselves the responsibility for any errors or omissions and urge readers to write to us through John Wiley and Sons, Inc., 605 Third Avenue, New York, N.Y. 10158-0012.

Introduction: Criteria for Inclusion and Definition of a Mineral

Just as the first edition of Dana's **A System of Mineralogy**, this eighth edition also attempts to describe, catalog, and classify all known and recognized minerals reported in the literature up to December 31, 1995. It presents the results of nature's own experiments in chemical synthesis. However, since human beings are now a major factor in our global environment, some anthropogenic minerals are included, along with some data from synthetic equivalents of minerals.

CRITERIA FOR INCLUSION

The criteria for including a mineral in this book require not only that the mineral be discovered and be recognized, but also that the discovery be reported in the generally accessible literature. For minerals discovered after 1959, **recognized** means that the name of the species is approved by the Commission on New Minerals and Mineral Names (CNMMN) of the International Mineralogical Association. For minerals reported before 1959, **recognized** means generally accepted by professionals today as unique, valid species.

The International Mineralogical Association, founded in 1958 to further international cooperation in the mineralogical sciences, currently consists of representatives from 33 countries, and sponsors some 14 commissions, one of which is the Commission on New Minerals and Mineral Names (CNMMN). Proposals for a new mineral should include (1) the name and its significance; (2) geographic and geologic occurrence, source or generation of the mineral, and associated minerals, particularly those in apparent equilibrium with the proposed new mineral; (3) composition and method of analysis; (4) chemical formula, empirical and simplified; (5) crystallographic data, including single-crystal and powder diffraction results, the morphology, and crystal structure determination; (6) general appearance and physical properties; (7) the optical properties; (8) locality of the type material; and (9) relationship to other species. Acceptance as a new species requires a vote of at least 50% of the international members and an affirmative response of at least two-thirds of those voting. Once accepted, the information must be published in an appropriate journal within two years or the CNMMN would be willing to accept a proposal from anyone else. The name proposed for the mineral is voted on separately: it must not be too similar to any existing name, must be a single word, and must follow usage guidelines if it is derived from the chemical composition, physical properties, or if it is named for a country or individual. The CNMMN reviews 50 to 90 proposals a year, adding about 50 new mineral species each year.

DEFINITION OF A MINERAL

The definition of a mineral as used in this work is a pragmatic, functional one. "A mineral is an element or chemical compound that is normally crystalline and that has been formed as a result of geological processes" E. H. Nickel, CM 33:689(1995). A more formal and precise definition of what is or is not a mineral has been and continues to be debated.

The term **crystalline** in this context means that the material possesses atomic ordering on a scale that can produce an indexable diffraction pattern when the material is examined with x-rays, electrons, neutrons, or other radiation of suitable wavelength. There are, however, some naturally occurring materials that have traditionally been included among minerals even though they are noncrystalline. These include (1) a few amorphous substances that have never been crystalline to the extent of producing diffraction patterns (e.g., opal), (2) metamict materials that were crystalline at one time but whose crystallinity has been disrupted by ionizing radiation, (e.g., zircon) and (3) certain naturally occurring liquids (H_2O, as ice but not as water, and liquid mercury are both considered minerals). Petroleum and its noncrystalline bituminous forms are not regarded as minerals. The term **geological processes** in this context is expanded to include extraterrestrial substances (meterorites, moon rocks, etc.) apparently produced by processes similar to those on Earth. Secondly, deliberately synthesized materials, if substantially identical to naturally occurring minerals, may be called **synthetic equivalents**. Chemical compounds formed by the action of geological processes on anthropogenic substances have, on occasion, been accepted as minerals. Examples include the laurium minerals formed by the action of seawater on ancient metallurgical slags. However, many exotic materials are produced today with the possibility that such substances in a geological environment may have reaction products which could qualify as new minerals. Current policy of the CNMMN holds that such substances are not considered minerals.

Thirdly, strictly biogenic substances, inorganic chemical compounds produced entirely in the biological sphere such as the calcium phosphates (apatite) of bones and teeth, the carbonate species typical of mollusc shells, the iron compounds found in bacteria, and oxalate crystals in plant tissues, are all discussed in this volume as minerals because they also form under strictly geological conditions.

Apart from the named mineral species there is a rapidly growing category of potential minerals, those compounds found in minute quantities along with, or within, known minerals. The quantity is often too small to establish all the requisite characteristics to qualify as a new species, or perhaps the investigator will not deem it worthy of following the procedure to obtain a species name. At present there are about 700 of such compounds, or roughly 16% of naturally occurring mineral substances. Some are mentioned in the mineralogic literature, e.g., annual summary of compounds and new minerals in the Canadian Minerologist, but most are to be found in Chemical Abstracts.

The potential number of naturally occurring compounds that might be named as minerals is extremely large (see *Skinner and Skinner MR 11:333(1980)*) so continuity of the present rate of 50 to 60 new mineral species per annum seems assured.

Descriptions of Mineral Species: Format of Presentation

1. DANA CLASSIFICATION NUMBER

The first entry for each mineral species is a number containing four parts separated by periods. It represents a hierarchical system parallel to that of Linneaus but based on a combination of chemistry and crystal structure of the minerals. These numbers facilitate insertion and addition of a new species into a list emphasizing close chemical and structural affiliations, an advantage since each mineral is known by a different, and not necessarily related, name.

The hierarchical numbering system was developed for the seventh edition but covered only nonsilicates. That portion has been updated by James Ferraiolo (**A Systematic Classification of Nonsilicate Minerals**, Bull. Am. Mus. of Nat. Hist., Vol. 172, 1982, and Supplements) and is generally followed in this edition. The many and often very complex silicate minerals are here classified for the first time according to a structural–chemical system developed by authors Abraham Rosenzweig and Eugene Foord. The work of H. Strunz (*Mineralogische Tabellen*, 1970, Classe der Silikate, pp.359–493) was incorporated with modification. The silicates are covered in classes 51 to 78.

The first consideration in generating the numbers was division into **classes** based on composition or, in the case of silicates, on dominant structural elements. The table of contents lists and numbers each of these classes. Species containing more than one anion or anion group may be assigned to a separate class or may be assigned according to dominant chemical entity. The first number indicating **class** is followed by a number for **type**, which in some cases specifies the atomic (cation to anion, or to anion group) ratio for the mineral. This and the remaining numbers assigned to each species relate to structural similarity considerations. Mineral species with similar structures within their class and type are listed together.

2. NAME

The currently accepted name of the mineral is given next, complete with all appropriate diacritical marks, and is followed by the chemical formula. The derivation of the name, if known, opens the body of the description.

3. CHEMICAL FORMULA

All of the chemical formulas we present approach an **ideal** composition, that is, the formula includes the elements that are essential or usually found in the natural occurences of the mineral and describe the proper atomic ratios among the constituent elements and moieties. We generally follow the presentation of M. Fleischer and J. A. Mandarino, **Glossary of Mineral Species**, seventh edition (The Mineralogical Record, Tucson, Arizona, 1995), which accurately describes the stoichiometry but does not necessarily incorporate structural characteristics even though they may be known. An appropriate formula that would uniquely describe a mineral species both chemically and structurally is currently under debate in the mineralogical community [AC(A)46:1(1990)]. In some mineral presentations important relationships are indicated by the way in which the formulas are written.

3A. Isostructural Substitution

Isostructural substitution* is shown by the use of commas to separate two or more elements within parentheses. Such a presentation implies that the elements are similar enough in their structural behavior that they may occupy the same site in the crystal structure of the species. This is a common option for elements with similar charge, ionic size, and bonding character; that is, for example, more than one element can occupy a site with octahedral coordination (**C. Klein and C. Hurlbut**, John Wiley & Sons, Inc., New York 1985, p.136). The individual elements (with or without parentheses), define an **end-member** species. A multiple element presentation (within parentheses) suggests that the structure is not grossly altered by the substitution of one element by the other.

When there is a continuum between end member compositions, substitution may be complete defining a mineral series, but often there is only partial substitution.

3B. Vacancies

Vacancies or voids at certain sites in the structure are indicated in the chemical formula by the symbol $\square$. It is a useful and appropriate way to express the substitution of ions with different charges; for example, $3Mn^{2+} \Leftrightarrow 2Fe^{3+}$

---

* Isostructural substition is also known as "isomorphous" substitution. We have avoided the term "isomorphic" as it may refer to materials of the same crystal structure, or of the same morphology (habit), or both, according to context. We have also used the term "isostructural" in preference to the synonymous term "isotypic".

☐ allows the interchange of divalent and trivalent cations while preserving both the equivalence of overall charge balance and the number and disposition of sites occupied in the structures.

4. NAME DERIVATION

The body of the presentation begins with details on the name, such as the date of the initial description, sometimes the author to whom the name is attributable, and the source of the name, whether it be based on composition, physical properties, geographic discovery site, or person honored. When the name is derived from the initial occurrence (the **type locality**), that locality will be shown in italics in the subsequent section on localities. For names deriving from languages other than English, the foreign language term or, at least its meaning, is given.

At the urging of the CNMMN, an effort is being made to reduce the proliferation of species names by appending chemical or crystallographic criteria to a root name. For example, the Levinson Rule, introduced in 1966, for mineral containing rare earth isostructurals with other rare earth species appends the R.E.E. cation to a root name, the root name being the name of the first member of the series discovered: for example, tritomite-(Ce) and tritomite-(Y) are end members [Ce+Y] in the tritomite series [AM 51:152(1966)]. Chemical prefixes are also employed. However, the chemical prefix does not always mean that the two species have the same crystal structure: for example, sklodowskite and cuprosklodowskite.

5. GROUPS, DIMORPHS AND POLYMORPHS, SYNONYMS AND VARIETIES, AND POLYTYPES

The relationship between the species being described and other species follows the derivation of the name. Here, other members of an isostructural series, polymorphs, or closely related species are mentioned, with cross-references to the appropriate Dana number. When recognized, membership in a group is noted. Occasionally, the phrase "an inadequately characterized species" may occur when the species is accepted, but the data are either incomplete or seem internally contradictory.

5A. Groups

The term **group** refers to an aggregate or multiple (usually three or more) species that have identical or closely similar structure and chemistry. A group may be named when there are several end members of isostructural series (e.g., the calcite group; see 14.1.1.1) or when several species have chemical identity and are structurally related through the multiplicity of one or more lattice parameters. The term **subgroup** is used when many species exhibit the same basic structural and chemical components but there are discrete ranges for some of the chemical constituents (see the pyroxene group and subgroups).

5B. Dimorphs and Polymorphs

Some chemical compositions have more than one three-dimensional structural configuration. **Dimorphs** are two, and **polymorphs** are three or more, species with the same chemical composition but distinct crystalline structures. Each species is given a different name and number. For example, calcite $CaCO_3$ (14.1.1.1) has a polymorph aragonite (14.1.3.1), while vaterite (14.1.2.1) has yet another configuration.

5C. Synonyms and Varieties

Many minerals show variations in color or morphology usually due to trace amounts of certain elements. Especially when the mineral species is common, such as quartz (75.1.3.1), many varieties (jasper, onyx) are known. These names are **varieties** but in another sense are **synonyms** as the basic chemistry and structure are identical to those of quartz.

5D. Polytypes

Materials whose basic structural components are layers often exhibit different layer superposition or different periodicity. These variations give rise to distinct but related structures called **polytypes** rather than dimorphs or polymorphs. The micas (71.2.2.X) are sheet silicates which offer examples of stacking sequences, and different polytypes are named by appending an appropriate shorthand notation denoting the structural consideration.

6. CRYSTALLOGRAPHY

The term crystal as used in this volume implies a solid with long-range internal order detectable by diffraction techniques, whether or not the material exhibits external faces, but its unqualified use here generally implies at least a partial development of faces. The text context serves to make any necessary distinctions. A very few materials are included her that lack this long-range order (e.g., opal). Lattice constants are generally derived by X-ray diffraction studies carried out on small single crystals, but may sometimes be derived from powder diffraction of fine grained samples.

Any crystalline substance is characterised by a distinct chemical composition whose components are arranged in a distinctive three-dimensional array or structure. In the majority of minerals species, the composition is not fixed but varies within defineable limits because of substitution of ions (atoms) of similar size and coordination number. This combination of composition and structure constitutes the distinction among mineral species.

The explanation of terms in the following paragraphs are intentionally brief; their meaning and use are well developed in numerous texts, a few of which are given below; abbreviations used in the subsequent text precede each entry.

(AB) Azaroff, L. V., & M. J. Buerger (1958), **The Powder Method**, McGraw Hill, New York.

(B) Bunn, C. W. (1961), **Chemical Crystallography**, 2nd ed., Oxford Clarendon Press.

(GL) Glusker, J. P., & M. Lewis (1994), **Crystal Structure Analysis for Chemists and Biologists**, VCH, New York.

(GT) Glusker, J. P., & K. N. Trueblood (1985), **Crystal Structure Analysis: A Primer**, Oxford University Press.

(ITXC) **International Tables for X-ray Crystallography**, v. 1, Kynoch Press, Birmingham, 1969.

(KC) Klein, C., & C. S. Hurlbut, Jr. (1993), **Manual of Mineralogy**, 21st ed., John Wiley & Sons, New York.

(DT) Dana, E. S., and W. E. Ford (1932), **A Textbook of Mineralogy**, 4th ed., John Wiley & Sons, New York.

(D7) Palache, Berman & Frondel (1944), **The System of Mineralogy**, 7th ed., v.1, John Wiley & Sons, New York.

(P) Phillips, F. C., (1971), **An Introduction to Crystallography**, 4th ed., John Wiley & Sons, New York.

6A. Crystallographic Principles

1) Crystal Systems (KH, Ch. 2; P, pp. 34-51)

Crystal systems as used here are based on the symmetry of the internal structure rather than on the traditional equality/non-equality of axial ratios and interaxial angles. It follows that coincidental dimensional equality of two axes of an orthogonal set does not require that the third axis have 4-fold symmetry, but rather that the existence of a 4-fold axis requires that the remaining two axes be orthogonal and dimensionally equal.

Some confusion arises when the unique symmetry is of order three (3) or six (6). The former is a subset of the latter, and it follows that such crystals may be divided into a trigonal system and a hexagonal system, or they may be combined into a single system (hexagonal) with subdivisions. In this work, the only distinction made within the composite hexagonal systems for which rhombohedral symmetry exists.

2) Crystal Classes, Point Group Symmetry, and Symmetry Notations (KH, Ch. 2; P. pp. 338–340)

The symmetry elements defining a system may combine with additional symmetry elements to yield thirty-two combinations known as **crystal classes** or **crystallographic point groups**, those combinations constrained by the requirement of spacial repetition to symmetries of order 1, 2, 3, 4, or 6. Point groups with other than the above rotational symmetries may apply to such subjects as molecular symmetry, spectroscopic data, etc.

The **Hermann-Mauguin** notation is generally used in presentation of crystallographic data, and consists of a combination of numerals (1, 2, 3, 4, 6) denoting the order of rotational symmetry and "barred" numerals ($\bar{1}, \bar{2}, \bar{3}, \bar{4}, \bar{6}$) denoting the order of rotary-inversion symmetry. The $\bar{2}$ axis is equivalent to reflection normal to that axis, and is usually replaced by the symbol m. A slash denotes perpendicularity of a rotational element and a reflection.

The **Schoenflies** notation is commonly used by spectroscopists and for most representation of non-crystallographic symmetry.

4) Miller Indices and Zone Symbols (KH, Ch. 2)

The **Miller indices** of a crystallographic plane or face represents the coordinates of the plane-normal in reciprocal space. (*N.B.* For a face or form "indices" is often replaced by "index" without a change in meaning. The four-term Miller-Bravais indices, (hkil), are used relative to hexagonal crystals. The third term, however, is a function of the first two such that i = −(h + k).)

Zone symbols, enclosed in square brackets, [uvw], are the coordinates of a line in direct space with respect to the crystallographic axes.

5) Crystal Forms (KH, Ch. 2)

A crystal **form** is a set of faces generated from a given face by the symmetry elements present. The Miller indices of a face are enclosed in parentheses, i.e., (123), those of a form in braces, i.e., {123}.

The form {hkl} where h, k and l are positive or negative, unequal integers is the **general form** whose individual faces are acted upon by all symmetry elements present; it has the maximum possible number of faces for a form in the given point group and is distinctive for it. A face for which h, k, and l are no longer unequal or have a value of zero, may shift to a position perpendicular to a symmetry element, and that symmetry element no longer duplicates the face. The resultant **special form** will have the same number or fewer faces than the general form. The most commonly encountered forms are those with relatively simple Miller indices (corresponding to planes with a high density of lattice nodes); consequently, the correct assignment of crystal class can not always be deduced from the crystal morphology.

6) Space Groups (ITXC, v. 1: P. pp. 241–301)

In the internal crystal structure the rotation and reflection elements of the point group may combine with translational components yielding **centered cells, screw axes** and **glides**. These translational components are generate the 230 possible **space groups**.

7) Choice of Settings (ITXD, v.1, pp. 542–553)
In the mineralogical and crystallographic literature there are followed a number of different conventions, often conflicting, that dictate the assignment of unit cell edges to the axes, a, b, and c. Space groups may be set in a "standard" orientation but alternative settings are possible. The reasons for choosing an alternate setting are varied, but usually depend on the organization of data or the comparison of related structures.

8) Twinning and Epitaxy (KH, pp. 100–105; P, pp. 171–193)
Twinning is the crystallographically rational, non-parallel intergrowth of two or more individual crystals of the same species.

Epitactic relationships are those in which the structure of two species have certain features in common such that the substrate or primary phase acts as a template or nucleation site for the overgrowth or secondary phase.

6B. Order of Presentation of Crystallographic Data

1) Crystal System - Denoted by the abbreviations: TRI = triclinic; MON = monoclinic; ORTH = orthorhombic; TET = tetragonal; HEX = hexagonal including trigonal; HEX-R = hexagonal with a rhombohedral lattice; ISO = isometric.

2) Space Group and Point Group - The next entry is the space group or the possible space groups, and generally corresponds to the space group for the referenced structure determination, but may have been reset to emphasize correspondence with related species. The Hermann-Mauguin point group is nowhere stated *per se*, but an abbreviated H-M notation may be derived from the space group by dropping the lattice type symbol, deleting all subscripts from screw axis symbols, and replacing all glide symbols by an *m*. Thus, *Pnma* becomes *mmm*.

3) Lattice Constants - The unit cell edges, if known, are given in Ångstrom units, followed by the interaxial angles in decimal degrees, either as fixed value or ranges.

4) Cell Contents and Calculated Density - The cell contents, Z, is the number of formula units per unit cell as defined by the lattice constants. The calculated density in gms/cm^3, **D**, is obtained by the equation $D = (ZM)/(NV)$, where M is the formula weight, N is Avogadro's number, and V is the cell volume in cm^3. This numerical value should be close to that of the measured Specific Gravity, G, reported as a physical property of the species.

5) Powder Diffraction Pattern - (AZ) Next is the X-ray powder diffraction pattern consisting of the eight strongest lines, unindexed, arranged in decreasing order of spacing, and subscripted by the relative intensity from 10 down to 1. The pattern is preceded by the ICDD number in italics (**International Center for Diffraction Data Mineral Powder Diffraction File**), or the letters **PD** if the data was taken directly from the literature. Where appropriate, several diffraction patterns may be given. It should be kept in mind that relative intensity and even the lines included in the strongest eight may vary due to compositional variation within the species and sample preparation artifacts.

6) **Structure** - (GL, GT) A literature reference to a structure determination may appear next, and may be followed by a statement of the salient features of the structure. For important or larger mineral groups, this information will appear in the group description.

If a crystal structure drawing is present, the following criteria apply to them: (a) **Radii** - Atomic radii are nominally one-half the standard ionic radii. For labeling and other reasons, however, it is necessary in many cases to modify these radii. The minimum radius shown is 0.45 Å, meaning that many smaller cations are not shown at true relative size. Anion radii may also be modified downward. Corners of polyhedra, almost always occupied by oxygen atoms, are shown at 0.2 Å, as are unlabeled hydrogen atoms.

(b) Projection and Extent of Structure - The structures are in perspective except where noted, and the perspective distance is usually about 50 Å, although it may be less for simple structures or more for complex ones. Typically, the projection is down the shortest crystal axis, although this is by no means a rule. The drawing may show only a portion of the unit cell; constraints of space do not always allow showing the true cell repeats or symmetry.

(c) Crystal Axes - Crystal axes are shown at one-half the true relative length end to end, except where noted. Labels on the axes have a white background, and when axes point toward the observer the labels are drawn over the ends of the axes and *vice versa*. The axes are shown in true perspective for their position as shown. Thus they do not show the exact orientation of the center of the structure.

(d) Atom Labels - In most cases all atoms and polyhedra are labeled, but in complex structures this may not be true. The label refers to the predominant species in a site. Hydrogen atoms are not labeled, but may be identified by their 0.2 Å radius, that is otherwise used only for oxygen at polyhedra corners. Due to space constraints, "C" is used for the carbonate group and "W" for water. The "W" label is also used for tungsten but appears only in the anhydrous species scheelite and huebnerite.

7) **Habit and Twinning** - There follows a statement of the dominant morphological habit of the species expressed as the Miller indices of the forms present. Where morphological description differ from the structural setting, this fact is noted and a transformation matrix may be given. Furthermore, a verbal description of the gross features of the morphology (acicular, tabular, etc.) and/or a statement of the texture of crystalline aggregates. These terms are generally self-explanatory but may be found in any standard dictionary or DT, pp. 204–206. Included here is a statement of any observed twinning including the twin law and its frequency of occurrence.

7. Physical Properties

7A. Morphology and Habit

A brief description of the distinct morphology exhibited by euhedral crystals of a species follows. This includes the following information: When planar faces are observed, they are described relative to the crystal symmetry by the Miller indices. In a crystal drawing, faces are identified by letters, and possibly other symbols, that are identified in the legend to the figure in terms of Miller indices. General shapes of minerals and mineral aggregates are also described.

7B. Twinning

Twinning is the crystallographically rational intergrowth of two or more individual crystals of the same species. The intergrowth generally occurs along a layer or zone of atoms that are common to the two orientations at the boundary of the individuals, but in many cases this site has not been elucidated and only the morphological relationships of the individuals may be described.

Twinning commonly involves a 180° rotation about a zone axis or face-normal having no even-order symmetry, or a reflection across a plane that is not a mirror plane of the individual crystals. If the individual crystals are centrosymmetric, the morphological appearance of the twin displays both the rotation and reflection aspects of the twinning process. The twins may be contact or penetration twins, and may be designated as twins, trillings, four-lings, and so on, in accordance with the number of individuals observed.

Polysynthetic twinning is repeated, usually small-scale twinning that results in parallel slabs in alternating orientation. The usual boundary between the individuals is called the **composition plane**. It often but not always coincides with the twin plane. An irregular composition surface between individuals is common in penetration twins.

The description of twinning may appear in a group description when it applies to all members of a mineral group, or it may be stated as part of the habit description of an individual species. The Miller(–Bravais) index of the twin plane or the zone symbol of the twin axis is stated, as is the frequency of that type of twin. If the twinning relationship has been assigned a specific name, that is stated as the twin law.

7C. Color

Color, an important but variable mineral property, may be due to a variety of factors, including not only the presence of intrinsically colored constituents but also small amounts of foreign substances as well as physical but nonchemical factors such as crystal lattice defects. Consequently, many minerals are found in a wide variety of colors. Unlike soils, there is no widely accepted system for describing the color of minerals in standardized terms. Gemstones, especially diamonds, are of course an exception and are evaluated by color as well as other criteria, although there is no universially accepted system for describing the colors of colored gemstones.

7D. Streak

Particularly with opaque and nearly opaque minerals, color is difficult to distinguish. Producing a streak by, for example, rubbing a sample across an abrasive porcelain or glass surface may allow characteristic colors of the powdered material to be recognized.

7E. Luster

The luster of minerals depends on the appearance of its surfaces in reflected light. It is a complex property that depends on a mineral's reflectivity; index of refraction; and the propensity to cleave, giving flat or rough surfaces; and its response to polarized light. Luster is described by terms such as those listed in Table 1.

TABLE 1 LUSTER

Luster	Description	Examples
Metallic	The luster of metals, applies only to opaque materials	Gold, copper, tin
Sub-Metallic	imperfectly metallic	Columbite, wolframite
Adamantine	The luster of diamond, other materials of high index of refraction, and usually of high density; Index of refraction approximately 1.9 to 2.5.	Diamond
Vitreous	Like broken glass; Index of refraction approximately 1.3 to 1.8.	Quartz
Sub-Vitreous	Imperfectly vitreous	
Resinous	Like amber	Opal, some yellow varieties of sphalerite
Greasy	The luster of oily glass, nearly as lustrous as 'resinous'	Nepheline
Pearly	Like a pearl, or like the luster of a stack of glass plates.	Talc, brucite
Silky	Like silk, used only with fibrous minerals	Sillimanite

The degree or intensity of luster may be described by such terms as **splendent, shining, glistening,** or **glimmering. Dull** is used to describe a mineral totally lacking in luster.

7F. Cleavage and Parting

Cleavage is the planar fracture along a crystallographically rational plane that is structurally controlled. It represents a plane of weaker (but not necessarily weak) bonding within the crystal structure, and consequently can occur everywhere and anywhere within the mineral sample.

If present, cleavage is stated in terms of the indices of the form to which it is parallel, with a modifying adjective denoting the ease of propagation and/or the perfection of the cleavage surface. The self-explanatory terms are in approximate diminishing order of perfection: **perfect** or **excellent, good, fair, poor**. Other self-explanatory terms used are: **easy, difficult, interrupted**.

Parting is the development of planar fractures that tend to be widely spaced relative to cleavage and which do not occur everywhere within a specimen. Parting may be related to structural features in some degree but develops as a result of nonuniform forces such as external stress, twin boundaries, oriented mechanical inclusions, and so on, and is thus not characteristic of all samples of the species.

7G. Fracture

Fracture is the way in which crystalline substances break when not conforming to cleavage or parting. Some fracture patterns are distinctive. Chonchoidal fracture, typical of quartz, gives a fracture similar to the interior of a seashell. Fracture may also be described as hackly, jagged, fibrous, uneven, or irregular. Fibrous materials show distinctive elongate fracture.

7H. Tenacity

Tenacity depends on the nature of the chemical bonding in a mineral species and is described by terms such as those in Table 2.

TABLE 2 TENACITY

Tenacity	Description	Examples
Malleable	Can be flattened with a hammer blow without breaking or crumbling into fragments. A property only of metals.	Gold, silver
Ductile	Can be changed in shape by pressure, especially when it can be drawn out into a wire.	Gold, silver, copper
Sectile	Can be cut with a knife like cold wax, so that shavings may be turned, yet the mineral may break with a sharp blow.	Chlorargylite, chalcocite
Flexible	Bends easily, and stays bent after the pressure is removed.	Talc
Elastic	May be bent or pulled out of shape, but returns to its original form when the pressure is removed	Mica
Brittle	Separates into fragments with a sharp blow, or when cut by a knife	Characteristic of most minerals

7I. Hardness

Hardness can be related to Mohs scale, or measured quantitatively with Vickers hardness tester. Mohs scale assigns numbers to reference minerals, which correspond to simple tests (Table 3).

TABLE 3 MOHS' SCALE OF HARDNESS

Hardness	Reference Mineral	Description and Feel
1	Talc	Soft, greasy feel like talc and graphite, and flakes of the mineral will be left on the fingers
2	Gypsum	Scratched easily with the fingernail, as gypsum
3	Calcite	Just scratched by a copper coin, but not normally scratched by the fingernail
4	Fluorite	Scratched by a knife without difficulty
5	Apatite	Scratched with a knife, but with difficulty
6	Orthoclase	Not scratched with a knife, but is scratched by a file, and will scratch ordinary glass
7	Quartz	Scratches glass easily, but is scratched by topaz and a few other minerals
8	Topaz	Hard
9	Corundum	Harder
10	Diamond	The hardest known mineral

Hardness is an easily measured and valuable property of appropriately sized samples. Microscopic samples require the Vickers hardness tester, which is more quantitative. Results are presented as VHN_n, where n represents the amount of loading of the device used to test the sample.

7J. Density/Specific Gravity (KC, p203)

The density **D** (grams/cm^3) may be calculated from the lattice parameters as follows:

$$D = \frac{Z \times M}{N \times V}$$

where **Z** is the number of formula units in the unit cell, **M** the formula weight of the compound, **N** is Avogadro's number, and **V** the volume of the unit cell. It is presented with the lattice parameter information.

It may be compared with the measured specific gravity, **G** (weight of a substance relative to the weight of an equal volume of water at 4°C) which serves to confirm the number of formula units, or composition, of the unit cell. A variety of techniques have been developed to measure the specific gravity, depending on the size of the sample and its chemical composition (C. Klein and C. S. Hurlbut, John Wiley & Sons, New York 1985, p. 203).

7K. Infrared, Raman, and Mössbauer Spectroscopic Data, and Differential Thermal Analysis

Spectroscopic and differential thermal analysis data are indicated when reported in the literature and referenced.

7L. Fluorescence/Phosphorescence

Materials may luminesce on direct exposure to different wavelengths of ultraviolet light, x-rays or cathode rays, emitting light of a different and distinctive wavelength. This fluorescence is the result of the excitation of specific electrons, often from transition metal elements included in the mineral. If the luminescence persists after illumination of the sample ceases, the sample is said to be phosphorescent (see **WARR** and **ROBB**). Some minerals also react to heat by giving off visible light and are notably thermoluminescent. Fluorite, ideally CaF_2, is the classic fluorescent species. There are some famous mineral localities, such as Franklin, New Jersey, where many of the mineral species are fluorescent when exposed to UV light.

7M. Optical Properties

The optical properties presented in this work are those that can usually be determined by use of the petrographic microscope appropriately equipped for transmitted or reflected light observations, and by use of appropriate immersion media. The reader is referred to the cited references or to similar texts for additional information regarding any of the reported phenomena and the theory and practice of their determination.

7M(1) Opaque Minerals

Opaque materials are those that do not transmit visible spectrum light or do so only in the very thinnest of fragments. Details of the methods and principals involved may be found in Craig, J. R., and D. J. Vaughn, **Ore Microscopy and Ore Petrography**, John Wiley & Sons, New York, 1994, and extensive descriptions for various species may be found in Uytenbogaardt, W., and E. A. J. Burke, **Tables for Microscopic Identification of Ore Minerals**, 2nd ed., Elsevier-North Holland, New York, 1971.

The word "opaque" may also be used in the description of other species; in such cases the reference is to diaphaneity in a bulk sample.

7M(2) Non-opaque Minerals

The optical properties of transparent crystalline substances can be described in terms of the basic equations of electromagnetism as may be found in numerous text such as Slater, J. C., and H. H. Frank, **Electromagnetism**, McGraw-Hill, New York, 1947.

The optical properties of mineral species as well as other light-transmitting substances differ with direction in the crystal as a function of the symmetry of the internal structure. Detailed discussions of these optical properties can be found in any college-level text on the subject of optical crystallography (e.g. Nesse, W. D., **Introduction to Optical Mineralogy, 2nd ed.**, Oxford University Press, New York, 1991; Hartshorn, N. H., and A. Stuart, **Crystals and the Polarizing Microscope, 3rd ed.**, Edward Arnold Publishers, London, 1960; and Wahlstrom, E. E., **Optical Crystallography, 3rd ed.**, John Wiley & Sons, New York, 1960 listed above).

The optical properties of concern may be described by reference to a geometrical construct known as the **indicatrix**, generally a triaxial ellipsoid in which any axis (line passing through the origin) of the indicatrix is proportional in length to the refractive index of light vibrating in that particular direction. A section passing through the origin of the ellipsoid is an ellipse whose axial lengths correspond to the refractive indices of the two polarized rays propagated normal to the section.

A triaxial ellipsoid indicatrix has two sections that are circular; the normals to these are the two optic axes, and hence the material is said to be **biaxial**. If the indicatrix is an ellipsoid of revolution, there is but one circular section the normal to which is the optic axis, and the material is **uniaxial**. If the indicatrix is a sphere, all sections are circular and geometrically equivalent, and the material is **isotropic**.

Orthorhombic, monoclinic, or triclinic crystals are biaxial; tetragonal and hexagonal crystals are uniaxial; isometric crystals and amorphous substances are istropic. The constraints regarding orientation of optical vibration directions relative to crystallographic axes will be found in any text on optical mineralogy or optical crystallography.

Optical properties are very sensitive to mechanical stress, and the strain produced may result in observed optical character of lower symmetry than is expected for the crystal's ideal symmetry. Such situations result in an anomalous double refraction manifested as a weak (very low) birefringence.

TABLE 4. ALTERNATIVE NOTATION FOR REFRACTIVE INDICES

n_α	or	n_x	or	n_p
n_β	or	n_y	or	n_m
n_γ	or	n_z	or	n_g
		n_e	or	n_e
		n_ω	or	n_o

7M3. Order of Data Entry

a) For opaque species: Optical character, isotropic or anisotropic, and reflectivity data for various wavelengths are given.

b) For non-opaque species: The initial property reported is the optical character and sign (Biaxial+, Biaxial−, Uniaxial+, Uniaxial−, or Isotropic). In situations where either wavelength dispersion or compositional variation leads to a change in the sign, the sign may be given as $(+, -)$. This is particularly true for biaxial species with an optic axial angle close to 90 degrees.

The order of entry for remaining optical properties may vary slightly from entry to entry, but the following are included where relevant and known:

1) For biaxial species the axial angle, 2V. When the optic axial angle measured over a particular indicatrix axis varies to either side of 90°, 2V may be subscripted with an X or Z to indicate which axes is being referred to, as for example: Biax $(+,-)$ $2V_Z = 82\text{-}94°$.

2) The refractive indices either as fixed values or ranges. The indices are to be assumed as being for white light, unless a reference is made to a specific wavelength. Dispersion follows in the format $r > v$ or $r < v$ in accordance with whether the indices are greater for red or violet light, but the notation may also refer to dispersion of the optic axis or orientation of the indicatrix.

3) For biaxial species a statement of the indicatrix orientation is given. The correspondence between X, Y and Z with crystallographic axes a, b, and c for orthorhombic crystals; for monoclinic crystals, an identification of the vibration direction, X, Y, or Z coinciding with the b crystallographic axis, and a statement of the angle (extinction angle) between at least one other vibration direction and some crystallographic axis or other observable property such as cleavage (i.e., $Y = b$, $Z \wedge c = 11°$); for triclinic crystals the orientation, when known, of the indicatrix relative to the crystallographic axes is given in the form of one or more extinction angles.

4) Pleochroism/dichroism are stated if that phenomenon is recognizable. The format varies, but will consist of a statement of the color or color range associated with each vibration direction, i.e. X = colorless, Y = yellow, Z 4= pink, or as a statement of optical density, i.e., X<Y<Z. Color and optical density may be combined in the statement.

Variations in any of these properties, particularly those that are composition dependent will be found in the accompanying discussion. Other special optical properties such as pronounced changes in indicatrix orientation, and optical activity will also be found here. Optical activity, the rotation of the polarization plane of light passing through the crystal, is a phenomenon associated with crystals that lack both a center of symmetry and reflection symmetry. There are few minerals for which this rotation is large enough to be observed even in relatively thick sections.

8. COMPOSITION AND PHASE RELATIONSHIPS

Observations concerning the range in composition or substitutions of ions or elements in the formula given are discussed where appropriate, as are any unusual aspects of the composition. Full chemical analyses of the species are generally not provided. Phase relationships relative to other species, and stability fields of the species in question, with or without phase diagrams, are also given here.

9. Occurrence

The overwhelming majority of mineral species we know come from the crust of the earth. A relatively few species have been found uniquely in meteorites, and two are found only in lunar samples. With the exception of a very low content of H_2O, lunar rocks so far examined are very similar to some rock types known on earth.

The crust of the earth consists of igneous, metamorphic, and sedimentary rocks, plus in some regions vast amounts of ice, which in the form of glaciers and ice caps may be considered a monomineralic rock. Nearly 99% of the composition of these rocks consists of only nine chemical elements: oxygen, silicon, aluminum, iron, calcium, sodium, potassium, magnesium, and titanium. Another 0.8% includes chlorine, carbon, phosphorus, sulfur, manganese, and hydrogen. The remaining elements altogether constitute less than 0.3% of the crust. These rarer elements are scattered through the rocks in accessory minerals, as substitutions in more common minerals, or as localized concentrations. The latter account for the ore deposits of many of the less common elements.

The bulk mineral composition of the crust consists mostly of common minerals: quartz, feldspars, micas, amphiboles, pyroxenes, olivines, garnets, clays, calcite, dolomite, apatites, rutile, hematite, magnetite, pyrite, and others. Such minerals constitute at least 98% of the crust but probably less than 200 of the approximately 3600 recognized species. The remainder of species are found in rather special circumstances, where conditions favoring the formation of minerals containing the rarer elements, or rare combinations of the common elements, obtain. These special circumstances include principally the geologic bodies listed in Table 5.

TABLE 5 - GEOLOGIC BODIES

1. Granite pegmatites, including their oxidized and/or hydrothermally altered phases
2. Syenitic pegmatites
3. Massive sulfide deposits
4. Oxidized sulfide deposits
5. Contact metamorphic deposits
6. Carbonates
7. Evaporites
8. Volcanic exhalates
9. Special metamorphic situations, such as Franklin, NJ and Långban, Sweden
10. Sedimentary, having mineral concentrations

10. LOCALITIES

Each mineral description in this book includes a selection of the best known localities, and the number of localities listed is proportional to the general importance of the mineral. For minerals of special interest to collectors, the number of localities listed has been expanded considerably, especially for calcite and quartz. Many rare minerals are known from only one place, and for other rare minerals all known localities are listed.

The localities are usually listed in a sequence depending on geographic location (Table 6).

TABLE 6 - ORDER OF PRESENTATION OF LOCALITIES INFORMATION

1. The United States, from east to west
2. Canada
3. Mexico
4. Central America and the Caribbean
5. Greenland
6. Iceland
7. European countries, from west to east
8. Asia, including Siberia
9. China and Japan
10. India
11. South east Asia
12. Middle East
13. African Countries, from north to south
14. Oceania
15. Australia
16. Antarctica
17. South America

The localities section includes a description of the geological occurrence for important, abundant minerals as well as for rare minerals and, where possible, for the type locality.

The type locality, the locality where the mineral was first discovered, is italicized. In a very few instances there may be more than one type locality, whenever a mineral was found and studied in more than one place before being formally described and named. For some older minerals, mostly for those described before 1850, the type locality is unknown.

Localities providing economically important deposits or specimens of unusual quality may be indicated by an asterisk (*), and those for truly exceptional, museum-quality specimens by an exclamation point (!).

Abbreviations of geographic place names used in this book are given in Table 7.

Descriptions of Mineral Species

TABLE 7 - GEOGRAPHICAL ABBREVIATIONS

States, (as in postal usage)		Washington	WA
Alabama	AL	West Virginia	WV
Alaska	AK	Wisconsin	WI
Arizona	AZ	Wyoming	WY
Arkansas	AR		
California	CA	Provinces, Canada	
Colorado	CO	Alberta	AB
Connecticut	CT	British Columbia	BC
Delaware	DE	Labrador	NF
Florida	FL	(part of Newfoundland Province)	
Georgia	GA	Manitoba	MB
Hawaii	HI	New Brunswick	NB
Idaho	ID	Newfoundland	NF
Illinois	IL	Northwest Territories	NWT
Indiana	IN	Nova Scotia	NS
Iowa	IA	Ontario	ON
Kansas	KS	Quebec	QUE
Kentucky	KY	Saskatchewan	SK
Louisiana	LA	Yukon	YT
Maine	ME		
Maryland	MD	States, Australia	
Massachusetts	MA	New South Wales	NSW
Michigan	MI	Northern Territory	NT
Minnesota	MN	Queensland	Q
Mississippi	MS	South Australia	SA
Missouri	MO	Tasmania	TAS
Montana	MT	Victoria	VIC
Nebraska	NE	Western Australia	WA
Nevada	NV		
New Hampshire	NH	States, Mexico	
New Jersey	NJ	Aguascaientes	AGS
New Mexico	NM	Baja California Norte	BCN
New York	NY	Baja California Sur	BCS
North Carolina	NC	Campeche	CAMP
North Dakota	ND	Chiapas	CHIS
Ohio	OH	Chihuahua	CHIH
Oklahoma	OK	Coahuila	COAH
Oregon	OR	Colima	COL
Pennsylvania	PA	Distrito Federal	DF
Puerto Rico	PR	Durango	DGO
Rhode Isand	RI	Guanajuato	GTO
South Carolina	SC	Guerrero	GRO
South Dakota	SD	Hidalgo	HGO
Tennessee	TN	Jalisco	JAL
Texas	TX	Mexico	MEX
Utah	UT	Michoacan	MICH
Vermont	VT	Morelos	MOR
Virginia	VA	Nayarit	NAY

Nuevo Leon	NL	Distrito Federal	DF
Oaxaca	OAX	Espirito Santo	ES
Puebla	PUE	Goias	GO
Queretaro	QRO	Maranhao	MA
Qunitana Roo	QR	Mato Grosso Norte	MT
San Luis Potosi	SLP	Mato Grosso Sul	MS
Sinaloa	SIN	Minas Gerais	MG
Sonora	SON	Para	PA
Tabasco	TAB	Paraiba	PB
Tamaulipas	TMPS	Parana	PR
Tlaxcala	TLAX	Pernambuco	PE
Veracruz	VER	Piaui	PI
Yucatan	YUC	Rio de Janeiro	RJ
Zacatecas	ZAC	Rio Grande Norte	RN
		Rio Grande Sul	RS
States, Brazil		Rondonia	RO
Acre	AC	Roraima	RR
Alagoas	AL	Santa Catarina	SC
Amapa	AP	Sao Paulo	SP
Amazonas	AM	Sergipe	SE
Bahia	BA	Tocantins	TO
Ceara	CE		

11. AUTHORSHIP

The initials of the author primarily responsible for the mineral description are given in small capital letters, followed by references to the description as a whole.

References

References in this work are cited in italics in an abbreviated form, and the following list provides complete citations.

AB	J. W. Anthony, Bideaux, Bladh, and Nichols, **Handbook of Mineralogy**, Mineral Data Publishers, 1990–1995
ABC	Academia Brasileire de Ciencias
AC	Acta Crystallographica, with parts AC(A), AC(B), AC(C)
ACA	Analytica Chemica Acta
ACH	Analytical Chemistry
ACHS	Acta Chemica Scandinavica
ACP	Annals of Chemistry and Physics
ACPII	Annals of Chemistry and Physics, Series II
AGS	Acta Geologica Sinica
AGSSA	Annals of the Geological Survey of South Africa
AJC	Australian Journal of Chemistry
AJS	American Journal of Science
AJUS	Antarctic Journal of the United States
AKMG	Arkiv für Kemi, Mineralogi och Geologi
AM	American Mineralogist
AMG	Arkiv für Mineralogi och Geologi
AMS	Acta Mineralogica Sinica
ANCH	Angewandte Chemie
ANCP	Annals de Chimie (Paris)
ANE	Advances in Nuclear Engineering
ANLR	Accadèmia Nazionale dei Lincei, Classe di Sciènze Fisiche, Matemàtiche e Naturali, Atti Rendicónti
APK	A. P. Komyakov, **Mineralogy of Hyperagpaitic Alkaline Rocks**, Nauka, Moscow, 1990
AS	Archives de Sciences
AST	Atti della Accademia della Scienze di Torino
ASTR	Advances in Sciences and Technology in the USSR
AUCG	Acta Universitatis Carolinae Geologica
AUF	Der Aufschluss
AUSM	Australian Mineralogist
B&B	G. W. Brindley and G. Brown, **Clay Structures**. Monograph 5, The Mineralogical Society, 1980
BCGF	Bulletin de la Commission Geologique de Finlande

BCRPSN	Bulletin Centre Recherche Pan, Société National Pétroles d'Aquitaine
BGGU	Bulletin Grønlands Geologiske Undersögelse
BGSA	Bulletin of the Geological Society of America
BGSC	Bulletin of the Geological Society of Czechoslovakia
BGSD	Bulletin of the Geological Society of Denmark
BGSF	Bulletin of the Geological Society of Finland
BIG	Bulletin of the Institute of Geology, Sofia
BM	Bulletin de Minéralogie (pre-1978, vol. 101, Bulletin de la Société Française de Minéralogie et Crystallographie, BSFM), to 1988
BMM	Boletin de Mineralogía (Mexico)
BMNH	British Museum (Natural History)
BNSM	Bulletin of the Natural Science Museum, Tokyo, Series C
BNSMTC	Bulletin of the Natural Science Museum, Tokyo, Series C
BSASA	Bulletin of the South African Speleological Association
BSBG	Bulletin de la Société Belge de Geologie
BSCASAY	Boll. Sci. Cons. Acad. Sci. Arts RSF Yugoslavia, Section A
BSE	Boletin de la Sociedad Española de Mineralogía
BSFM	Bulletin de la Société Française de Minéralogie et Crystallographie
BTEM	Beudant, **Traite élémentaire de minéralogie, 1832**
C&S	A. J. Criddle and C. J. Stanley, **The Quantitative Data File for Ore Minerals**, Second Issue, BMNH, 1986
CA	Chemical Abstracts
CCM	Clays and Clay Minerals
CE	Chemie der Erde
CG	Chemical Geogoly
CGEM	China Gems
CIWY	Carnegie Institute of Washington Yearbook
CJC	Canadian Journal of Chemistry
CJG	Chinese Journal of Geochemistry
CJP	Czechoslovakian Journal of Physics
CLGM	A. N. Mariano, Cathodluminescence of Geological Materials
CLM	Clay Minerals
CM	Canadian Mineralogist
CMB	Clay Mineralogy Bulletin
CMG	Casopis pro Mineralogii a Geologii
CMP	Contributions to Mineralogy and Petrography (pre-1964, Beiträge zur Mineralogie un Petrographie)
CP	Contemporary Physics
CR	Comptes Rendus Académie de Sciences, Paris
CRASP	Comptes Rendus Académie de Sciences, Paris
CRC	**Handbook of Chemistry and Physics, CRC Press, Boca Raton, Florida**
CRT	Crystal Researh and Technology

CS	Clay Science
CSB	Chinese Science Bulletin
CSC	Crytal Structure Communications
CSJB	Chemical Society of Japan Bulletin
CSMQ	Colorado School of Mines Quarterly
CSR	Crystal Structure Reports
CSREM	Crystal Structures of Rare Earth Minerals
CUSGM	Clug. Univ. Studies, Series Geology Mineral
D6	**Dana's System of Mineralogy**, 6th ed. including Supplements 1, 2 and 3, John Wiley & Sons, New York, 1920
D7	**Dana's System of Mineralogy**, 7th Ed., Volumes I, II, and III, John Wiley & Sons, New York, 1944, 1952, 1960
DAN SSR	Doklady Akademii Nauk, Soyuz Sovietskikh Sotsialisticheskikh Respublik
DAN SSSR	Doklady Akademii Nauk, Soyuz Sovietskikh Sotsialisticheskikh Respublik
DANA	Doklady Academii Nauk Azerbaijan SSR
DANES	Doklady Academii Nauk USSR, Earth Sciences Section
DANESS	Doklady Academii Nauk USSR, Earth Sciences Section
DANP	Doklady Academii Nauk USSR, Physics Section
DANR	Doklady Academii Nauk Rossiyskoy
DANS	Doklady Academii Nauk USSR, also known as Akademiya Nauk USSR(SSSR), Doklady
DANT	Doklady Akademii Nauk Tadzikistan
DHZI	W. A. Deer, R. A. Howie, and J. Zussman, **Rock Forming Minerals**, R. A. Longman Inc., New York, 1962–1963
DHZ2	W. A. Deer, R. A. Howie, and J. Zussman, **Rock Forming Minerals**, 2nd ed., Longmans–Halsted Press, New York, 1978 (published in parts 1A, 1B, and 2A)
DI	**Dana's System of Mineralogy**, 7th ed., Vol. I, John Wiley & Sons, New York, 1944
DII	**Dana's System of Mineralogy**, 7th ed., Vol. II, John Wiley & Sons, New York, 1952
DIII	**Dana's System of Mineralogy**, 7th ed., Vol. III, John Wiley & Sons, New York, 1960
DJ	DeJong **Kompendium der Kristallkunde**, 1958
DM	Dana's Manual of Minerolgy, Kline and Hurlbut
DS	**(Dana's) A System of Mineralogy**, 6th ed., John Wiley & Sons, New York, 1892
DT	**Dana's Textbook of Mineralogy**, 4th ed., John Wiley & Sons, New York, 1932
EG	Economic Geology
EJM	European Journal of Mineralogy
ENM	Willard L. Roberts, Thomas J. Campbell, and George R. Rapp, Jr., **Encyclopedia of Minerals**, 2nd ed., 1990

EOB	J. Sinkankas, **Emerald and Other Beryls**, Chiton Book Co., Radnor, PA, 1981
EOS	EOS—Transactions of the American Geophysical Union
EPSL	Earth and Planetary Science Letters
ERS	Estratti dei Rendicónti della Società Italiano di Mineralogia et Petrologia
F	M. Fleischer and J. A. Mandarino, **Glossary of Mineral Species**, The Mineralogial Record, Tucson, Arizona, 1995
FELD	J. V. Smith and W. L. Brown, **Feldspar Minerals**, Springer-Verlag, New York, 1988
FF	Feldspars and Feldspathoids
FM	Forschritte der Mineralogie
G	Geochemistry (English translation of Geokhimiya)
G&G	Gems and Gemology
GAC-MAC-Abstr.	Geological Association of Canada–Mineralogical Association of Canada Abstracts
GCA	Geochimica et Cosmochimica Acta
GFF	Geologiska Föreningens i Stockholm, Förhandlingar
GG	Geologiya i Geofizika
GI	Geochemistry International (English translation)
GJJ	Geochemical Journal (Japan)
GMP	Geokhimiya Mineralogiya & Petrologiya
GRL	Geophysical Research Letters
GRM	Geologia Rudnykh Mestorozhedniye
GSAAP	Geological Society of America Abstracts and Program
GSAJ	Geological Association of Australia Journal
GSAM	Geological Society of America Memoirs
GSAPA	Geological Society of America, Program Abstract
GSASP	Geological Society of America, Special Paper
GSCB	Geological Society of Czechoslovakia Bulletin
GSCP	Geological Society of Canada Paper
GSIB	Geological Survey of Israel, Bulletin
GSICR	Geological Survey of Israel, Current Research
GSSA	Geological Survey of South Africa Bulletin
GT	Geology Today
GZ	Geologicheskii Zhurnal
H&K	C. S. Hurlbut and R. Kammerling, **Gemology**, John Wiley & Sons, Inc., New York, 1991
HBMP	Heidelberg, Beiträge zur Mineralogie und Petrologie
IANKSG	Izvestiya Academii Kazakhskoi SSR, Seriya Geologiya
IANS	Izvestia Akademii Nauk USSR
IC	Journal of Inorganic Chemistry
ICDD	International Center for Diffraction Data
IGC	International Geological Congress

IGR	International Geological Review
IJES	Israel Journal of Earth Sciences
IM	Izomorfizm Mineral
IMMA	International Mineralogical Association Meeting Abstracts
IMR	Industrial Minerals and Rocks, AIME
INDM	Indian Mineralogist
IRS	V. C. Farmer, ed. **The Intfra-red Spectra of Minerals**, Monograph 4, Mineralogical Society (England), 1974
ISM	J. W. Salisbury, L. S. Walter, N. Vergo, and D. D'Aria, **Infrared** (2.1–25µm), Johns Hopkins University Press, Baltimore, 1991
ITXRDI	N. F. M. Henry and K. Lonsdale, eds., **The International Tables for X-ray Crystallography**, Vol. 1, Kynoch Press, Birmingham, England and more recent editions
JAC	Journal of Applied Crystallography
JACES	Journal of the American Ceramics Society
JACS	Journal of the American Chemical Society
JADE	Roger Kevern, ed., **Jade**, Annes Publishing Co., London, 1991
JCG	Journal of Crystal Growth
JCH	Journal of Contaminant Hydrology
JCL	Journal of Less Common Metals
JCP	Journal of Chemical Physics
JCPDS	Joint Committee on Powder Diffraction Standards (now ICDD)
JCS	Journal of the Chemical Society (London)
JCSD	Journal of the Chemical Society, Dalton Transactions (Inorganic Chemistry)
JCSF	Journal of the Chemical Society, Faraday Transactions
JCSSJ	Journal of the Clay Science Sciety of Japan
JCUG	Journal of the Chinese University Geosciences
JG	Journal of Gemology
JGL	Journal of Glaciology
JGLBW	Jähresheft Geologie Landesamtes Baden–Würtemburg
JGR	Journal of Geophysical Research
JGSA	Journal of the Geological Society of Australia
JGSJ	Journal of the Geological Society of Japan
JGSK	Journal of the Geological Society of Korea
JINC	Journal of the International Nuclear Commission
JISI	Journal of the Iron and Steel Institute
JJMPE	Journal of the Japanese Association of Mineralogists, Petrologists, and Economic Geologists (to 1988); thereafter Journal of Mineralogy, Petrology and Economic Geology
JMC	Journal of Metal Chemistry
JMSJ	Journal of the Mineralogical Society of Japan
JP	Journal of Petrology
JPC	Journal of Physical Chemistry
JPD	Journal of Powder Diffraction
JPRC	Journal für Praktische Chimie

JRNBS	Journal of Research, National Bureau of Standards
JSS	Journal of Soil Science
JSSC	Journal of Solid State Chemistry
JVGR	Journal of Volcanic and Geothermal Research
JVS	J. V. Smith and W. L. Brown, **Feldspar Minerals**, Springer-Verlag, New York, 1988
K	Kristallografiya
KB	Kostov, I. and V. Breskovske, phosphate, arsenic and vandate minerals. Crystal Chemistry and Classification Kliment Ohridskis, Univ. Press, Sofia, 1989
KL	**Klockmann's Lehrbuch der Mineralogie**, 16th ed. 1980
KT	Kexue Tonbao
KUS	Kali and Steinsalz
KV	Kora Vyvetriviyana (Moscow)
KY	Kwangwu Yanshi
L	Lock, **The Chemical Physics of Ice**
L&L	Lazebnik and Lazebnik, **Mineralogy and Geochemistry of Ultrabasic and Basic Rocks of Yakutia**
LAP	Lapis
LC	LaChapelle, **Field Guide to Snow Crystals**
LIT	Lithos
LJ	Lapidary Journal
LPS	Lunar and Planetary Science
LS	Lunar Science
M	Meteoritics
MA	Mineralogical Abstracts
MACSC	Mineralogical Association of Canada, Short Course, Vol. 1
MAR	Marine Geology
MASLI	Memorias da Academia das Ciencias de Lisboa, Classe de Ciencias I
MD	Mineralium Deposita
MG	Modern Geology
MGR	Meddelelser om Grönland
MIN	Minerali (Vols. 1–4)
MJ	Mineralogical Journal
MJJ	Mineralogical Journal (Japan)
MK	Mir Kamenya (World of Stones)
MM	Mineralogical Magazine
MMAD	A. LaCroix, **Mineralogie de Madagascar**, 1922
MMN	P. Embry and Fuller **Manual of New Mineral Names**
MMS	H. Möenke, **Mineralspectren**, Akademie-Verlag, Berlin, P. Vol. I (1962) and Vol. II (1966)
MN	Mineral News
MP	Mineralogy and Petrology (Austria)
MPOL	Mineralogica Polonica
MPVII	Micro Probe VII

MR	Mineralogial Record
MRB	Materials Research Bulletin
MRE	Vlasov, **Mineralogy of Rare Elements**, Vol. II, 1966
MS	Mineralogicheskii Sbornik
MSARM	Mineralogical Society of America, Reviews in Mineralogy (starts with Vol. 7)
MSASC	Mineralogical Society of America, Short Course Notes Vols. 1–6
MSASP	Mineralogical Society of America, Special Paper
MSF	Materials Science Forum
MSJSP	Mineralogical Society of Japan Special Publication
MSL	Mineralogical Society of London
MSLM4	Mineralogical Society of London Monograph 4, 1974
MSUZ	Les Minéraux Secondaire d'Uranium du Zaire
MSUZC	Les Minéraux Secondaire d'Uranium du Zaire Complement
MT	J. A. Mandarino and V. Anderson, **Monteregian Treasures**, 1989
MU	Moscow University Geological Bulletin
MZ	Mineralogicheskii Zhurnal
N	D. A. C. Newman **Chemistry of Clays and Clay Minerals**, Monograph 6, Mineralogical Society, 1987
NAT	Nature
NATW	Naturwissenschaften
NBSC	National Bureau of Standards Circular
NBSM	National Bureau of Standards Monograph
NBSR	National Bureau of Standards Journal of Research
NGT	Norsk Geologisk Tidsskrift
NJBA	Neues Jahrbuch für Mineralogie (Abhandlungen)
NJBM	Neues Jahrbuch für Mineralogie Mitteilungen (Monatshefte)
NJMA	Neues Jahrbuch für Mineralogie (Abhandlungen)
NJMM	Neues Jahrbuch für Mineralogie (Monatshefte)
NR	Norelco Reporter
NW	Naturwissenschaften
NZ	G. Gottardi and E. Galli, **Natural Zeolites**, Springer-Verlag, New York, 1985
O61ZC	**Proceedings of the 6th International Zeolite Conference**, Butterworth, New York, 1984
OAQ	Anzeiger der Oesterreichischen Akademie der Wissenschaften
P&J	J. Picot and Z. Johan, **Atlas of Ore Minerals**, Elsevier, New York, 1982
P51CZ	**Proceedings of the 5th International Conference on Zeolites**
PASP	Proceedings of the Academy of Sciences, Philadelphia
PCM	Physics and Chemistry of Minerals
PD	Journal of Powder Diffraction
PGPI	Physics of Earth and Planetary Interiors
PI, PII	Partington, **History of Chemistry**, 1961
PICCM	Proceedings of the International Clay Conference, Madrid
PJA	Proceedings of the Japan Academy

PLC	Post and LaChapelle, **Glacier Ice**
PM	Periodico di Mineralogia
PNAS	Proceedings of the US National Academy of Science
PPSL	Proceedings of the Physical Society, London
PSSC	Progress in Solid State Chemistry
PSSSA	Proceedings of the Soil and Science Society of America
PT	Phase Transitions
PU	Proceedings of the Ussher Society, United Kingdom
R	P. Ramdohr, **The Ore Minerals and Their Intergrowths**, Pergamon Press, Elmsford, New York, 1980
R&M	R. V. Dietrich and B. J. Skinner, **Rocks and Rock Minerals**, John Wiley & Sons, New York, 1979
RC	Recherche Travails Chemique Pays-Bas et Belgique
REVM	Reviews in Mineralogy
RI	Revista Mineralogica Italiana
RM	Rocks and Minerals
RMS	20th Rochester Mineralogical Symposium, Rochester, New York, 1993
ROBB	M. Robbins, **Fluorescence, Gems and Minerals Under Ultraviolet Light**, Geoscience Press, Phoenix, Arizona
ROU	A. Rouse, **Garnet**, Butterworth, New York
RSIMP	Rendiconti della Società Italiana di Mineralogia e Petrologia
RSMI	Rendiconti della Società Italiana di Mineralogia e Petrologia
RZ	Referativnyi Zhurnal, Geodeziya
S	H. Strunz, **Mineralogische Tabellen**, 1970
SCI	Science [American Association for the Advancement of Science]
SEG	Southeastern Geology
SG	Sedimentary Geology
SGBA	Société Belge de Géologie, Annales
SGBB	Société Belge de Géologie, Bulletin
SGBM	Société Belge de Géologie, Memoire
SGG	Soviet Geology and Geophysics
SGS	Scientia Geologia Sinica
SGU	Sveriges Geologiska Undersökning
SIMP	Società Italiana di Mineralogia e Petrologia
SJG	Scottish Journal of Geology
SMMB	Boletin de Mineralogía (Mexico)
SMPM	Schweizerische Mineralogische und Petrographische Mitteilungen
SPC	Soviet Physics and Crystallography
SPD	Soviet Physics Doklady
SR	Wyckoff, **Structure Reports**
SRF	H. C. Skinner, M. Ross, and C. Frondel, **Asbestos and Other Fibrous Materials**, Oxford University Press, Oxford, 1988
SRNU	Scientific Reports, Niigata University, Series E. Geology
SS	Scientia Geologica Sinica

SSC	Solid State Communications
SSPN	Soil Science and Plant Nutrition (Japan)
ST	Strunz, **Mineralogische Tabellen, 1970**
SUBBSC	Studia Universitatis, Babes-Bolyai, Series Chimie
SWV	Salisbury, Walter and Vergo, **Mid-infrared Spectra of Minerals,** USGS OFR 87-263, U.S. Geological Survey, Washington, D.C., 1987
T&G	O. F. Tuttle and J. Gittins, eds., **Carbonatites**, John Wiley & Sons, Inc., New York, 1966
TA	Thermochimica Acta
TAGU	Transactions of the American Geophysical Union
TMMAN	Trudy Mineralogicheskogo Muzeya, Akademiya Nauk USSR
TMPM	Tschermaks Mineralogische und Petrographische Mitteilungen
TN	Terra Nova
TRSSA	Transactions of the Royal Society of South Australia
UCG	Universitatis Carolinae (Prague) Geologica
USGSB	U.S. Geological Survey Bulletin
USGSC	U.S. Geological Survey Circular
USGSJR	U.S. Geological Survey Journal of Research
USGSM	U.S. Geological Survey Monographs
USGSOF	U.S. Geological Survey Open File Report
USGSPP	U.S. Geological Survey Professional Paper
VLA	Vlasov, **Mineralogy of Rare Elements**, Vol. II, 1966
VMRE	Vlasov, **Mineralogy of Rare Elements**, Vol. II, 1966
W	H. Winchell, **Optical Properties of Minerals**, Academic Press, New York, 1965
WARR	T. S. Warren, S. Gleason, R. C. Bostwick, and E. R. Verbeek, **Ultraviolet Light and Fluorescent Minerals**, Thomas S. Warren, Publisher, distributed by Williams Minerals, Rio, West Virginia
WEW	Wies. Eysel. W., Min. Pet. Inst. Univ. Heidelberg
WWH	W. H. Harrity, **The Sea-Side Book**, London, 1849, pp. 2–4 as quoted in L. Barber, **The Heyday of Natural History**, 1820–1870, Doubleday and Co., New York, 1980
Z	Zeolites
ZAAC	Zeitschrift Anorganische Allgemeine Chemie
ZK	Zeitschrift für Kristallographie
ZMGP	Zentrallblatt für Mineralogie, Geologie und Paleontologie
ZPC	Zeitschrift für Physikalische Chemie
ZSRD	Zapishici Srpskoge Geoloshkogo Drushtva
ZVMO	Zapiski Vsesoyuznogo Mineralogischeskogo Obshchestvo
ZW	R. W. Tschernich, **Zeolites of the World**, Geoscience Press, Tucson, Arizona, 1992

Class 1

Native Elements and Alloys

GOLD GROUP

The members of the gold group are elements or intermetallic compounds that share the same basic cubic close-packed structure: *Fm3m*.

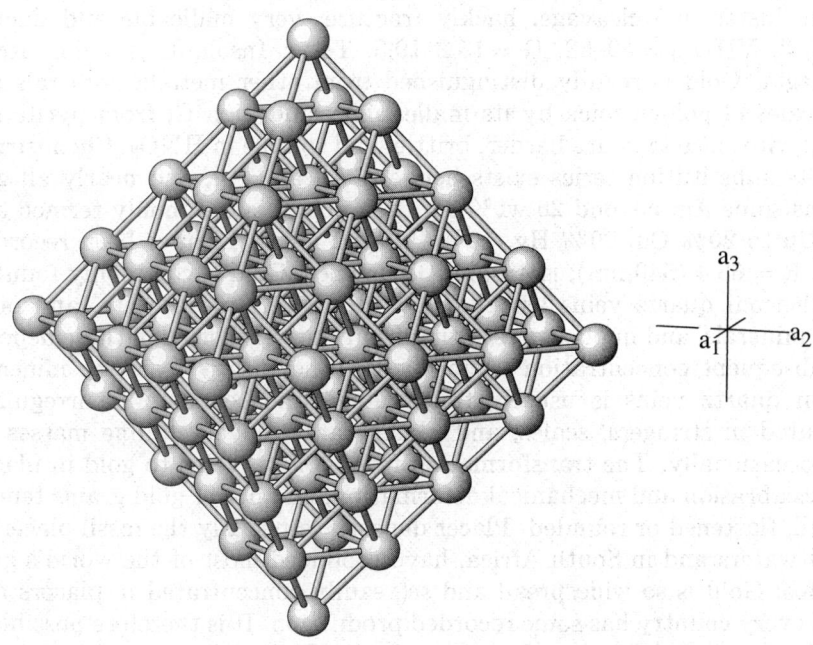

CCP: cubic closest-packed (gold, copper, platinum, nickel, taenite)
Space group Fm3m. In this view the [111] direction is vertical, showing the horizontal layers of atoms. Adopted by gold, silver, copper, platinum, and related metals.

CUBIC CLOSE-PACKED ELEMENTS

Mineral	Formula	Space Group	a	D	
Gold	Au	$Fm3m$	4.079	19.28	1.1.1.1
Silver	Ag	$Fm3m$	4.085	10.50	1.1.1.2
Copper	Cu	$Fm3m$	3.615	8.94	1.1.1.3
Lead	Pb	$Fm3m$	4.951	11.34	1.1.1.4
Aluminum	Al	$Fm3m$	4.049	2.697	1.1.1.5

Note: Taenite (1.1.11.2) and nickel (1.1.11.5) are isostructural with the gold group.

1.1.1.1 Gold Au

Known since antiquity. Gold group. ISO $Fm3m$. $a = 4.079$, $Z = 4$, $D = 19.28$. 4-784(syn): 2.36_{10} 2.04_5 1.44_3 1.23_4 1.18_1 0.936_2 0.912_2 0.833_2. **Physical properties:** octahedral and dodecahedral crystals, often in parallel groups or dendritic and arborescent forms; also massive, and as scales and rounded grains in alluvium. Golden yellow color and streak, inclining to silver-white with increasing substitution by Ag, or to orange-red with substitution by Cu; metallic luster. No cleavage, hackly fracture, very malleable and ductile. $H = 2\frac{1}{2}$–3, $VHN_{10} = 30$–58. $G = 15.2$–19.3. **Tests:** Insoluble in acids except aqua regia. Gold is readily distinguished from other metallic minerals and from scales of golden mica by its malleability and high G; from pyrite and chalcopyrite since they are harder, brittle, and soluble in HNO_3. **Chemistry:** A complete substitution series exists between Au and Ag, and nearly all gold contains some Ag; beyond 20 wt% Ag the mineral is commonly termed *electrum*. Up to 20% Cu, 20% Hg, 10% Pd, and minor Bi have been recorded. **Optics:** $R = 63.4$ (540 nm); isotropic. **Occurrence:** Gold is commonly found in hydrothermal quartz veins associated with pyrite, arsenopyrite, and other sulfide minerals, and in placers resulting from the weathering of these deposits and subsequent concentration of gold with other heavy resistant minerals. Gold in quartz veins is usually fine grained and tends to be irregularly distributed in stringers, scales, and plates; large polycrystalline masses are found occasionally. The transformation from gold in veins to gold in placers involves abrasion and mechanical deformation, and placer gold grains tend to be small, flattened or rounded. Placer deposits, especially the fossil placers of the Witwatersrand in South Africa, have produced most of the world's gold. **Localities:** Gold is so widespread and so readily concentrated in placers that almost every country has some recorded production. It is therefore possible to list only the more important localities. In the United States, gold occurs in small amounts in MD, VA, NC, SC, and GA; fine specimens in quartz were found in White Hall mines, Spotsylvania Co., and Partridge mine, Orange Co., VA, and at Dahlonega, GA; Homestake mine, Lawrence Co., SD; Silver Bow, Madison, and Broadwater Cos., MT; Leadville, Lake Co., Breckenridge(*), Summit Co., Ground Hog mine(*), Gilman Co., Silverton and Telluride, San Juan Co., CO; Bingham and Tintic, Salt Lake Co., UT; Mohave and Cochise Cos., AZ; Elmore, Idaho, and Valley Cos., ID; Goldfield, Esmeralda Co., Tonopah, Nye Co., Carlin and Cortez, Eureka Co., Winnemucca(*), Humboldt Co., NV; many localities in CA, especially in the Mother Lode belt in Mariposa, Tuolumne, Calaveras, Amador, and El Dorado Cos.—exceptional crystallized specimens from Placerville, El Dorado Co., Greenville, Plumas Co., Red Ledge mine, Nevada Co., Jamestown, Tuolumne Co., Colorado quartz mine, Mariposa Co., Bonanza mine, Trinity Co., Michigan Bluff district, Placer Co., in AK, placers at Fairbanks, Iditarod, and Nome, and quartz veins at Juneau. In Canada, Noranda district, PQ; Porcupine, Kirkland Lake, and Larder Lake deposits, ON; Atlin and Caribou, BC; and Klondike district, YT. In Mexico, Altar and Magdalena, SON; Parral, CHIH; and localities in GTO and ZAC. In Europe, minor occurrences in the Alps; major occurrences in Transylvania, Romania, especially Rosia Montana (Verespatik). In Russia, gold is found on the eastern slope of the Ural Mts. for a distance of 500 miles, from Nizhne Tagilsk in the north to Orsk in the south; in Siberia important placer deposits near Tomsk, Yeniseisk, Lena R., and upper Amur R. In India the chief producer is the Kolar field in Mysore. In Africa, Ashanti, Ghana; Kilo and Moto, Zaire; Bulawayo and Hartley, Zimbabwe; Witwatersrand, South Africa, a belt of fossil placers about 120 miles long that has been mined to depths exceeding 2 miles. Many important gold mining districts in Australia, including Hill End, Hillgrove, and Mt. Boppy, NSW; Mt. Morgan and Cooktown, Q; Bendigo and Ballarat, VIC; Kalgoorlie, Coolgardie, and Norseman, WA; Tennant Creek, NT. In New Zealand, Hauraki Peninsula, North Island, and placer deposits in Nelson, Westland, Otago, and Southland. Emperor mine, Vatukuola, Fiji. Mt. Kare, Papua New Guinea. In South America, Antioquia and Cauca, Colombia; Santa Elena, Venezuela; the coastal ranges of northern Chile and scattered localities in Peru, Bolivia, and Ecuador; Morro Velho mine, Nova Lima, MG, Brazil. BM *DI:90*, *R:321*, *P&J:281*, *ABI:189*, *MR 13:323(1982)*.

1.1.1.2 Silver Ag

Known since antiquity. Gold group. ISO $Fm3m$. $a = 4.085$, $Z = 4$, $D = 10.50$ (Silver 3C). Hexagonal polytypes 2H and 4H are known. *4-783*(syn): 2.36_{10} 2.04_4 1.45_3 1.23_3 1.18_1 0.938_2 0.914_1 0.834_1. **Physical properties:** Cubic and octahedral crystals, but commonly as reticulated, arborescent, and wiry forms; twinning common on {111}. Silver-white, usually tarnishing to gray to black, silver-white streak, metallic luster. No cleavage, hackly fracture, malleable and ductile. $H = 2\frac{1}{2}-3$, $VHN_{100} = 61-65$. $G = 10.5$. **Tests:** Soluble in HNO_3; color, malleability, and G usually diagnostic. **Chemistry:** A complete solid solution series exists between Ag and Au. Mercurian (kongsbergite, up to 37 at % Hg), cuprian, arsenian (huntilite), antimonian (animikite), and bismuthian (chilenite) varieties have been reported. **Optics:** $R = 80.8$ (540 nm); isotropic. **Occurrence:** Native silver is widely distributed in small amounts, principally in the oxidized zone of ore deposits. The larger deposits, however, are probably the result of the deposition of silver from hydrothermal solutions of primary origin. Deposits of this kind are of three main types: silver associated with various silver minerals, sulfides, and zeolites in a gangue of calcite, barite, fluorite, and quartz (Kongsberg); silver with arsenides and sulfides of nickel and cobalt in a calcite or barite gangue (Cobalt); with uraninite and nickel–cobalt minerals (Great Bear Lake, Jachymov). **Localities:** Only some of the more important localities can be given. Prospect Park, Paterson, NJ, with zeolites; Keweenaw Peninsula(*), MI, as fine crystallized aggregates and "half-breeds" (crystallized together with native copper); Butte and Elkhorn districts, MT; Georgetown(*), Crested Butte(*), Creede(*), Aspen(*), and other localities, CO; Bunker Hill, Shoshone Co., and Poorman mine, Owyhee Co., ID; Tombstone(*), Globe(*), Bisbee, and other localities, AZ; Calico district, San Bernadino Co., Cerro Gordo(*), Inyo Co., Bodie district, Mono Co., and other localities in CA. In Canada, at Cobalt(*) and Thunder Bay district, ON, Highland Bell mine(*), Beaverdell, BC, and Great Bear Lake, NWT. Many localities in Mexico, especially Batopilas(*), CHIH, Guanajuato(*), GTO, and localities in SON, DGO, and ZAC. Kongsberg(!), Norway (large specimens of crystallized and wire silver); Ste.-Marie-aux-Mines, Alsace, France; Andreasberg, Freiberg(*), and Schneeberg(*), Germany; Pribram(*), Jachymov, and Schemnitz, Czech Republic; Monte Narba, Sardinia, Italy; Smeingorsk, Semenowsky, and Altai, Siberia, Russia; Rudnui(*) and Dhezkazgan(*), Kazakhstan; Salida and Mangani, Sumatra; Broken Hill(*) and Cobar, NSW, Australia; Huancavelica, Peru; Potosi(*), Oruro(*), and other localities in Bolivia; Copiapo, Huantaya(*), Chanarcillo(*), and Chanca, Chile; Tsumeb, Namibia. BM *DI:96, R:313, P&J:71, ABI:475.*

1.1.1.3 Copper Cu

Known since antiquity. Gold group. ISO $Fm3m$. $a = 3.615$, $Z = 4$, $D = 8.94$. *4-836*(syn): 2.09_{10} 1.81_5 1.28_2 1.09_2 1.04_1 0.904_1 0.829_1 0.808_1. **Physical properties:** Cubic and dodecahedral crystals, often flattened or elongated; also arborescent, platy, and as masses weighing many tons. Light rose, tarnishing to copper-red and brown, metallic streak and luster. No cleavage, hackly fracture, highly malleable and ductile. $H = 2\frac{1}{2}–3$, $VHN_{100} = 77–99$. $G = 8.95$. **Tests:** Soluble in HNO_3; color malleability and G usually diagnostic. **Chemistry:** Usually remarkably pure, containing only traces of other elements. Arsenian copper (whitneyite) is reported with more than 11% As, but most arsenian specimens are mixtures with algodonite, so analyses are questionable. **Optics:** $R = 47.6$; isotropic. **Occurrence:** Probably all native copper is of secondary origin, formed by the reduction of copper-bearing solutions by iron minerals. Rarely an ore mineral, but large deposits of Keweenaw Peninsula, MI, are exceptions; here copper is associated with native silver, chalcocite, bornite, epidote, zeolites, and calcite, and is found as a cement in interstices of sandstone and conglomerate, filling amygdaloidal openings in interbedded lavas, and in veins traversing the country rocks. **Localities:** In traprocks and associated Triassic sandstones in MA, CT, NJ, PA, and VA. In MI a narrow belt 200 miles long in the Keweenaw Peninsula(*), which was for many years the major copper district of the world: huge masses were found there, the largest about 420 tons, and large groups of crystals. Santa Rita, Grant Co., NM; Ajo(*), Pima Co., Bisbee(*), Cochise Co., and Ray, Gila Co., AZ; Copper R. district, AK. Cap d'Or, NS, Kamloops, BC, and Upper White R., YT, Canada. Cananea, SON, and Mapimi, DGO, Mexico. Cornwall, England, in many mines. Rheinbreitenbach and Friedrichssegen, Rheinland-Pfalz, Germany. Bogoslovsk(*), Nizhne-Tagilsk, and Sverdlovsk, Ural Mts., Russia. Broken Hill(*), NSW, Mt. Isa, Q, and Wallaroo and Burra-Burra, SA, Australia; Aniseed Valley, near Nelson, New Zealand. Shaba, Zaire; Broken Hill mine, Zambia; Onganja mining district, Namibia. Andacolla, near Coquimbo, and Chuquicamata, Chile; Corocoro(*), Bolivia. **Alteration**: Alteration of native copper first produces a thin cuprite film over the surface, which gives it a characteristic copper-brown color; weathering may produce malachite, azurite, atacamite, and other secondary minerals. Copper pseudomorphs have been observed after cuprite, aragonite, calcite, azurite, chalcocite, and antlerite. BM *DI:99, R:308, P&J:138, ABI:108, MR 23(2):1(1992)*.

1.1.1.4 Lead Pb

Known since antiquity. Gold group. ISO $Fm3m$. $a = 4.951$, $Z = 4$, $D = 11.34$. $4\text{-}686$(syn): 2.86_{10} 2.48_5 1.75_3 1.49_3 1.43_1 1.14_1 1.11_1 0.837_1. Rarely as octahedral, cubic, and dodecahedral crystals; commonly as rounded masses and plates; recorded as pellets in alluvium, but probably artificial. Gray-white, tarnishing to dull lead-gray, metallic luster. No cleavage, sectile and malleable. $H = 1\frac{1}{2}$, $VHN_{100} = 5$. $G = 11.37$. $R = 48.9$ (540 nm). Franklin, NJ; Shafter district, Presidio Co., TX; Wood R. district, ID; Keno Hill, YT, Canada; Ilimaussaq, Greenland; *Långban(*), Värmland, Sweden* (crystallized masses up to 60 kg); Red Cap mine, Chillagoe, Q, Australia; El Dorado, Gran Sabana, Venezuela. BM *DI:102, R:337, P&J:304, ABI:292*.

1.1.1.5 Aluminum Al

Identified as a mineral in 1978. Gold group. ISO $Fm3m$. $a = 4.049$, $Z = 4$, $D = 2.697$. $4\text{-}787$(syn): 2.34_{10} 2.02_5 1.43_2 1.22_2 1.17_1 0.929_1 0.906_1 0.827_1. Flat platelets and scaly masses. Gray-white, metallic luster. $H = 2\text{–}3$. $G = 2.707$. Tsepochechnyi intrusive, Siberia, and Tolbachik volcano, Kamchatka, Russia. BM *DANS 243:191(1978), 313:433(1990); ZVMO 113:210(1984); AM 65:205(1980)*.

1.1.2.1 Auricupride Cu₃Au

Named in 1950 for the composition. *See* isoferroplatinum group. May contain up to 8% Pd. ISO $Pm3m$. $a = 3.753$, $Z = 1$, $D = 12.2$. $35\text{-}1357$(syn): 3.75_2 2.65_1 2.17_{10} 1.87_4 1.33_2 1.13_2 0.860_1 0.838_1. Platy aggregates and anhedral grains. Yellow with a reddish tint, metallic luster. No cleavage, ductile and malleable. $H = 3\frac{1}{2}$, $VHN_{10} = 54$. $G = 11.5$. $R = 63$ (580 nm). Occurs in serpentinites as a product of low-temperature unmixing of Au-Cu alloys. Steinmauern, Baden, Germany; Laksia and Pefkos, Cyprus; *Karabash and Talnakh deposits, Ural*

Mts., Russia; El Indio mine, east of Coquimbo, Chile. BM *FM 28:69(1950)*, *AM 62:595(1977)*, *R:334*, *ABI:36*.

1.1.2.2 Tetraauricupride CuAu

Named in 1982 for the composition and symmetry. See the tetraferroplatinum group. TET. $P4/mmm$. $a = 3.960$, $c = 3.670$, $Z = 2$, $D = 15.03$. *25-1220*(syn): 3.67_3 2.80_2 2.23_{10} 1.98_3 1.84_1 1.74_1 1.35_1 1.19_1. Microscopic anhedral grains. Golden yellow, metallic luster. Malleable. $H = 4\frac{1}{2}$, $VHN_{20} = 288$. $R = 61.2$ (546 nm); weakly anisotropic. *Sardala, Xinjiang, China*, in serpentinite. BM *AM 68:1250(1983)*, *ABI:523*.

1.1.2.3 Yuanjiangite AuSn

Named in 1994 for the locality. HEX $P6_3/mmc$. $a = 4.316$, $c = 5.510$, $Z = 2$, $D = 11.78$. *AM 80*: 3.73_3 3.09_4 2.22_{10} 2.16_6 1.55_3. Granular aggregates. Silver-white, black streak, metallic luster. Slightly ductile. $VHN_{25} = 172$–274. $G = 11.7$–11.9. $R = 74$–76 (540 nm); distinctly anisotropic, pale yellow to brown. In placers, *Yuanjiang R., near Yuanlin, Hunan Prov., China*. BM *APM 13:232(1994)*, *AM 80:1330(1995)*.

1.1.3.1 Maldonite Au₂Bi

Named in 1870 for the locality. ISO. $Fd3m$. $a = 7.971$, $Z = 8$, $D = 15.70$. *12-734*: 2.82_4 2.41_{10} 2.30_5 1.63_3 1.54_6 1.41_5 1.20_3 1.04_4. Octahedral crystals and massive, granular. Silver-white, tarnishing to copper-red and black, metallic luster. Cleavage {001}, {110}, distinct; conchoidal fracture; malleable and sectile. $H = 1\frac{1}{2}$–2, $VHN_{100} = 147$–264. $G = 15.46$. $R = 55.8$ (540 nm). *Ingram, ON, Canada*; *Salsigne deposit, Aude*, and *Scoufour, Cantal, France*; *Baita, Romania*; *Tyrnyauz district, Caucasus Mts., Russia*; *Syrymbet, Kazakhstan*; *Maldon, VIC, Australia*. BM *DI:95 R:336*, *P&J:254*, *ABI:310*.

1.1.4.1 Anyuiite Au(Pb,Sb)₂

Named in 1989 for the locality. TET $I4/mcm$. $a = 7.338$, $c = 5.658$, $Z = 4$, $D = 13.33$. *25-365*(syn): 2.84_{19} 2.48_3 2.32_5 2.24_4 1.70_2 1.54_2 1.48_2 1.32_1. Detrital grains. Silver-gray, tarnishing to dull lead-gray, metallic luster. $H = 3$, $VHN_{20} = 146$. $R = 63.9$ (540 nm); weakly anisotropic. *Bolshoi Anyui R., Kolyma region, Russia*, in gold-bearing alluvium. BM *MZ 11(4):88(1989)*, *AM 76:299(1991)*.

1.1.5.1 Zinc Zn

Known since antiquity. See the osmium group. HEX $P6_3/mmc$. $a = 2.665$, $c = 4.947$, $Z = 2$, $D = 7.14$. *4-831*(syn): 2.47_5 2.31_4 2.09_{10} 1.69_3 1.34_3 1.33_2 1.17_2 1.12_2. Microscopic grains. Blue-white, metallic luster. Cleavage {0001}, perfect; brittle. $H = 2$. $G = 6.9$–7.2. *Keno Hill, YT, Canada*, precipitated from cold brines; *Syrymbet, Kazakhstan*; *Dulcinea de Llampos mine, near Copiapó, Chile*, an oxidation product of sphalerite. Other reported occurrences doubtful and need confirmation. BM *DI:127*, *CM 6:692(1961)*, *AM 55:1019(1970)*, *ABI: 585*.

1.1.5.2 Cadmium Cd

Identified as a mineral in 1979. See the osmium group. HEX $P6/mmc$. $a = 2.979$, $c = 5.618$, $Z = 2$, $D = 8.64$. *5-674*(syn): 2.81_7 2.58_3 2.35_{10} 1.90_3 1.52_3 1.49_2 1.32_2 1.26_1. Microscopic grains. Tin-white, metallic luster. *Ust-Khannin intrusive, Vilyui R., Russia*, in heavy concentrate from a gabbro. BM *DANS 248:1426(1979)*, *AM 65:1065(1980)*, *ABI:75*.

1.1.6.1 Danbaite $CuZn_2$

Named in 1983 for the locality. ISO Space group unknown. $a = 7.762$, $Z = 12$, $D = 7.36$. *39-400*: 2.36_5 2.16_4 2.08_{10} 1.59_3 1.37_4 1.23_3 1.22_3 1.16_3. Microscopic spherulitic aggregates. Silver-white to gray-white, metallic luster. H = 4, $VHN_{20} = 234–288$. *Danba, Sichuan Prov., China*, in a Cu–Ni deposit. BM *KT 22:1383(1983)*, *AM 69:566(1984)*, *ABI: 126*.

1.1.6.2 Zhanghengite CuZn

Named in 1986 for Zhang Heng (78–139), a famous astronomer in ancient China. ISO $Pm3m$. $a = 2.95$, $Z = 1$, $D = 8.32$. *2-1231*(syn): 2.96_1 2.08_{10} 1.47_2 1.20_3 1.04_1 0.932_1 0.851_1 0.788_1. Microscopic grains. Golden yellow, bronze streak, metallic luster. No cleavage. $VHN_{10} = 140–150$. $R = 81.1$ (549 nm). Occurs in the Boxian meteorite that fell in *Boxian Co., Anhui Prov., China*, on October 20, 1977. BM *AM 75:244(1990)*.

1.1.7.1 Mercury Hg

Known since antiquity. HEX-R $R\bar{3}m$. $a = 3.46$, $c = 6.71$, $Z = 3$, $D = 14.40$. *9-253*(syn): 2.74_{10} 2.23_8 1.73_5 1.46_6 1.37_6 1.12_5 0.937_6 0.865_6. Liquid globules, solid below $-40°$. Tin-white, metallic luster. G = 13.6. Volatile, fumes highly toxic. Occurs in cinnabar or associated with it in mercury deposits. Terlingua, TX; New Almaden, Santa Clara Co., and in mercury deposits in Sonoma, Napa, and Lake Cos., CA; Almaden, Spain; Landsberg, near Obermoschel, Germany; Idrija, Slovenia; Mt. Avala, near Belgrade, Serbia. BM *DI:103*, *R:337*, *ABI:323*.

1.1.8.1 Moschallandsbergite Ag_2Hg_3

Named in 1938 for the locality. ISO $Im3m$. $a = 10.04$, $Z = 10$, $D = 13.41$. *11-67*: 2.67_4 2.36_{10} 1.97_4 1.67_4 1.45_4 1.37_7 1.28_5 1.24_6. Dodecahedral crystals commonly modified by the cube and trapezohedron, also massive, granular. Silver-white, metallic luster. Cleavage {011}, {001}, distinct; conchoidal fracture; somewhat brittle. $H = 3\frac{1}{2}$, $VHN_{100} = 153$. G = 13.5–13.7. Sala mine, Västmanland, Sweden; *Landsberg*(*) *(Moschallandsberg), near Obermoschel, Pfalz, Germany*; Chalanches mine, Allemont, Isere, France. BM *DI:103*, *R:337*, *P&J:275*, *ABI:341*.

1.1.8.2 Schachnerite $Ag_{1.1}Hg_{0.9}$

Named in 1972 for Doris Schachner (1904–1989), German mineralogist. HEX $P6_3/mmc$. $a = 2.978$, $c = 4.842$, $Z = 2$, $D = 13.4$. $27\text{-}618$(syn): 2.58_3 2.42_5 2.27_{10} 1.49_4 1.27_5 0.954_5 0.937_4 0.860_6. Crystals to 1 cm. Cream-white, metallic luster. $H = 3\frac{1}{2}$, $VHN_{100} = 148$. $G = 12\text{–}15$. $R = 72$ (589 nm); very weakly anisotropic. Sala mine, Västmanland, Sweden; *Landsberg, near Obermoschel, Pfalz, Germany*. BM *NJMA 117:1(1972), AM 58:347(1973), R:338, ABI:463*.

1.1.8.3 Paraschachnerite Ag_3Hg_2

Named in 1972 for its relationship to schachnerite. ORTH *Cmcm*. $a = 2.961$, $b = 5.13$, $c = 4.83$, $Z = 2$, $D = 12.98$. $27\text{-}617$: 2.56_3 2.40_6 2.27_{10} 1.76_3 1.48_4 1.36_5 1.26_6 0.831_4. Small pseudohexagonal crystals. Gray, metallic luster. $H = 4$, $VHN_{100} = 87$. $R = 70.9\text{–}73.7$ (589 nm); weakly anisotropic. Sala mine, Västmanland, Sweden; *Landsberg, near Obermoschel, Germany*; Kremikovci, Bulgaria. BM *NJMA 117:1(1972), AM 58:347(1973), R:338, ABI:288*.

1.1.8.4 Luanheite Ag_3Hg

Named in 1984 for the locality. HEX Space group unknown. $a = 6.61$, $c = 10.98$, $Z = 6$, $D = 12.57$. $41\text{-}1417$: 2.83_7 2.45_5 2.00_6 1.74_5 1.50_{10} 1.20_9 1.13_7 1.11_6. Microscopic grains and spherical aggregates. White, tarnishing to black, black streak, metallic luster. Malleable. $H = 2\frac{1}{2}$, $VHN = 44\text{–}75$. $G = 12.5$. $R = 64\text{–}71$ (540 nm). *Luan He R., Hebei Prov., China*, in a gold placer. BM *AM 73:192(1988), ABI:303*.

1.1.8.5 Eugenite $Ag_{11}Hg_2$

Named in 1986 for Eugen F. Stumpfl (b.1931), Austrian mineralogist. ISO $I\bar{4}3m$. $a = 16.02$, $Z = 4$, $D = 10.45$. AM 80: 2.37_{10} 2.10_8 1.46_7 1.24_7 1.19_6 1.03_5 0.950_8 0.925_8. Grains up to 4 mm. White. $VHN_{15} = 96$. $G = 10.75$. $R = 80.1$ (546 nm); isotropic. Bisbee, AZ; *Lubin mine, Poland*, in Zechstein Cu deposit. BM *AM 80:845(1995)*.

1.1.8.6 Weishanite $(Au,Ag)_3Hg_2$

Named in 1984, derivation not given. HEX $P6_3/mmc$. $a = 2.927$, $c = 4.818$, $Z = 2$, $D = 18.17$. $4\text{-}808$(syn): 2.52_5 2.40_5 2.23_{10} 1.74_6 1.46_7 1.35_8 1.24_8 1.22_7. Microscopic aggregates. Pale yellow, metallic luster. Ductile and malleable. $H = 2\frac{1}{2}$, $VHN = 51$. $R = 76.3$ (534 nm); weakly anisotropic. *Poshan mining district, Tongbai, Henan Prov., China*, in an Au–Ag deposit. BM *AM 73:196(1988), ABI:571*.

1.1.9.1 Kolymite Cu_7Hg_6

Named in 1980 for the locality. Dimorphous with belendorffite.. ISO $I\bar{4}3m$. $a = 9.418$, $Z = 4$, $D = 13.11$. $33\text{-}470$: 2.98_3 2.52_4 2.22_{10} 2.09_3 2.01_3 1.52_2 1.39_2 1.28_3. Microscopic cubo-octahedral crystals. Tin-white, tarnishing to brown-black, metallic luster. Brittle. $H = 4$, $VHN_{20} = 220\text{–}267$. $G = 13.0$. $R = 72.1$

(540 nm). Comstock Lode, NV; *Krokhalin antimony deposit, Kolyma R., Russia.* BM *ZVMO 109:206(1980), AM 66:218(1981), ABI:270, EM 1:719(1989).*

1.1.9.2 Belendorffite Cu_7Hg_6

Named in 1992 for Klaus Belendorff (b.1956), German mineral collector. Dimorphous with kolymite. HEX-R $R3m$. $a = 9.408$, $\alpha = 90.47°$, $Z = 4$, $D = 13.15$. *22-241*(syn): 6.66_5 3.31_5 2.98_8 2.97_8 2.52_{10} 2.23_{10} 2.22_{10} 2.01_5. Globule. Silver-white, tarnishing to brown-black, silver-white streak, metallic luster. $VHN_{25} = 125$. $G = 13.2$. $R = 71.9$ (540 nm); weakly anisotropic. *Landsberg, near Obermoschel, Pfalz, Germany.* BM *NJMM 21(1992), AM 77:1305(1992).*

1.1.10.1 Leadamalgam Pb_2Hg

Named in 1981 for the composition. TET $I4/mmm$. $a = 3.545$, $c = 4.525$, $Z = 0.667$ (*sic*), $D = 11.98$. *39-395*: 2.78_1 2.49_4 2.25_4 1.78_4 1.68_4 1.49_{10} 1.40_{10}. Microscopic grains. Silver-white, metallic luster. $H = 1\frac{1}{2}$, $VHN_{100} = 12$. *Shiaonanshan, Inner Mongolia, China*, in heavy concentrates from a Cu–Ni sulfide deposit. BM *AM 70:215(1985), ABI:293.*

IRON–NICKEL GROUP

The iron–nickel group consists of six nonisostructural species that reflect the various proportions of

(Fe,Ni)

that are stable at various temperatures. Fe and Ni form a complete series of solid solutions at high temperatures, but at lower temperatures the relationships are more complex. The names *kamacite* and *taenite*, corresponding to α-Fe and γ-Fe, were originally introduced to describe meteoritic forms of Fe–Ni. Both are perfectly good mineral names and are preferred to the terms *iron* and *nickel iron* used by some authorities.

On cooling through 900°, pure Fe inverts from the face-centered structure to the body-centered structure. The inversion temperature decreases with increasing Ni content. In the slowly cooled meteorites, the nickel–iron exsolves kamacite along the octahedral planes, giving rise to Widmanstätten structure typical of most iron meteorites. At about 500° kamacite containing 7.5% nickel is in equilibrium with taenite containing about 30% Ni. *GCA 52:617(1988).*

IRON–NICKEL GROUP

Mineral	Formula	Space Group	a	c	D	
Kamacite	α-(Fe,Ni)	$Im3m$	2.859		7.90	1.1.11.1
Taenite	γ-(Fe,Ni)	$Fm3m$	3.596		8.14	1.1.11.2
Tetrataenite	FeNi	$P4/mmm$	2.533	3.582	8.275	1.1.11.3
Awaruite	Ni_3Fe	$Fm3m$	3.560		8.32	1.1.11.4
Nickel	Ni	$Fm3m$	3.524		8.91	1.1.11.5
Wairauite	FeCo	$Im3m$	2.86		8.32	1.1.11.6

1.1.11.1 Kamacite α-(Fe,Ni)

Named in 1861 from the Greek for *bar*. Iron–nickel group. Contains up to 7.5% Ni. ISO $Im3m$. $a = 2.859$, $Z = 2$, $D = 7.90$. *18-645*: 3.03_1 2.95_1 2.03_{10} 1.97_6 1.48_1 1.44_3 1.17_7 1.02_1. Terrestrial kamacite occurs as disseminated grains or rounded masses; meteoritic kamacite as exsolved lamellae or in hexahedrites (5–7% Ni) as single crystals weighing many kilograms. Steel-gray, metallic luster. Cleavage {001}, hackly fracture. H = 4, $VHN_{100} = 110$–155. G = 7.84–7.92. R = 45 (546 nm). Magnetic. Easily soluble in dilute HCl. Terrestrial kamacite is rare; most occurrences are in basalts in which the iron has been reduced by the assimilation of carbonaceous material. The most important occurrence is on Disko Is., Greenland, where masses of up to 20 tons are present in basalts that have broken through coal seams. Other significant localities include Bühl, Kassel, Germany, and Putorana plateau, Taimyr Penin., Russia. Kamacite is a major component of most iron meteorites and is a minor mineral in most chondrites. It occurs as microscopic grains in some lunar rocks. BM *DI:114, R:353, P&J:161, ABI:256*.

1.1.11.2 Taenite γ-(Fe,Ni)

Named in 1861 from the Greek for *band*. Iron–nickel group. Contains 25–50% Ni. ISO $Fm3m$. $a = 3.596$, $Z = 4$, $D = 8.14$. *23.297*(syn): 2.08_{10} 1.80_8 1.27_5 1.08_8 1.04_5 0.900_3. Small grains and exsolved lamellae; also massive. Silver-white to steel-gray, metallic luster. Malleable. H = 5–$5\frac{1}{2}$, $VHN_{100} = 350$–500. G = 8.01–8.08. Magnetic. Slowly soluble in dilute HCl. Occurs in iron meteorites containing more than 7% Ni and is a major component of Ni-rich ataxites. An accessory mineral in many chondrites and in some lunar rocks. BM *DI:117, R:353, ABI:510*.

1.1.11.3 Tetrataenite FeNi

Named in 1980 for the tetragonal ordered form of taenite. Iron–nickel group. Contains 48–57% Ni. TET $P4/mmm$. $a = 2.533$, $c = 3.582$, $Z = 1$, $D = 8.275$. Microscopic grains, and rims on taenite. Cream-white, metallic luster. H = $3\frac{1}{2}$, $VHN_{25} = 170$–200. Weakly anisotropic. An accessory mineral in many

1.1.11.3 Tetrataenite

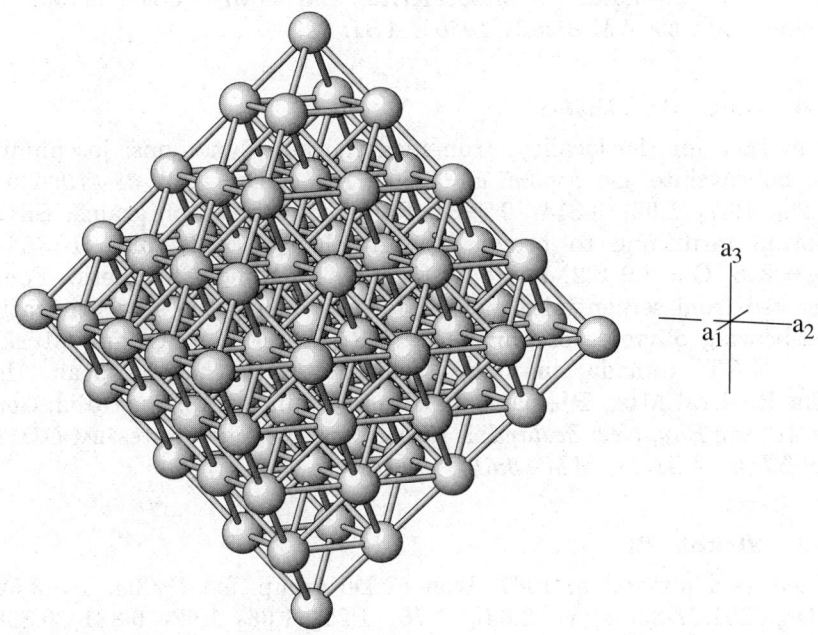

CCP: cubic closest-packed (gold, copper, platinum, nickel, taenite)
Space group Fm3m. In this view the [111] direction is vertical, showing the horizontal layers of atoms. Adopted by gold, silver, copper, platinum, and related metals.

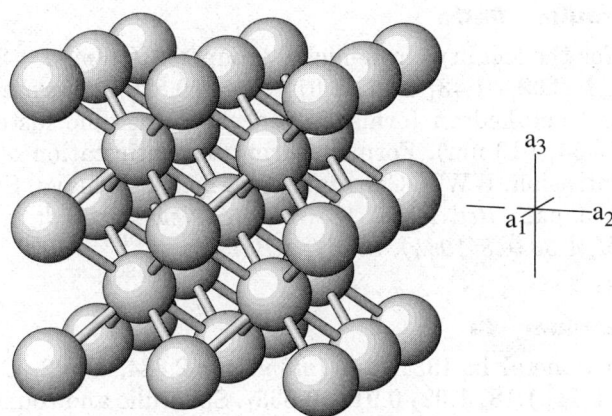

BCC: body-centered cubic (iron, kamacite)
Space group Im3m. Each atom is coordinated by eight others instead of 12 as in closest packing. Adopted by iron and related metals.

meteorites, most abundant in mesosiderites and slowly cooled chondrites; forms below 350°. BM *AM 65:624(1980), ABI: 527*.

1.1.11.4   Awaruite Ni_3Fe

Named in 1885 for the locality. Iron–nickel group. Synonyms: josephinite, souesite, bobrovskite. ISO $Fm3m$. $a = 3.560$, $Z = 4$, $D = 8.32$. *38-419*: 2.04_{10} 1.77_6 1.25_3 1.07_4 1.02_1 0.814_1 0.792_1. Rounded pebbles or grains. Silver-white, often tarnishing to brown, metallic luster. Malleable. $H = 5\frac{1}{2}$–6, $VHN_{100} = 346$. $G = 7.9$–8.2. $R = 64.2$ (540 nm). Strongly magnetic. Found in or derived from serpentinite masses; formed during the serpentinization of nickel-bearing olivine. Josephine Co., OR; Fraser R., BC, and Muskox intrusion, NWT, Canada; in many serpentinites in the European Alps; Bobrovka R., Ural Mts., Russia; Lord Brassey mine, TAS, Australia; *Gorge R., near Awarua Bay, New Zealand*. A rare accessory in meteorites. BM *DI:117, R:356, P&J:82, ABI:38, MM 43:647(1980)*.

1.1.11.5   Nickel   Ni

Recognized as a mineral in 1967. Iron–nickel group. ISO $Fm3m$. $a = 3.524$, $Z = 4$, $D = 8.91$. *4-850*(syn): 2.03_{10} 1.76_4 1.25_2 1.06_2 1.02_1 0.881_1 0.808_1, 0.788_2. Microscopic cubic grains and flakes. Silver-white, metallic luster. $H = 4\frac{1}{2}$, $VHN_{50} = 172$–184. $G = 8.90$. $R = 52.8$ (548 nm). Occurs in serpentinized basic rocks as a result of low-temperature hydrothermal activity. Grasshopper Mt., Tulameen, BC, Canada; Kaltenberg, Valais, Switzerland; *Bogota, near Canala, New Caledonia*; Mt. Clifford, near Agnew, WA, Australia; Jerry R., South Westland, New Zealand. BM *MM 40:247(1975), R:362, EG 76:1686(1981), ABI:349*.

1.1.11.6   Wairauite   FeCo

Named in 1964 for the locality. Iron–nickel group. ISO $Im3m$. $a = 2.86$, $Z = 1$, $D = 8.32$. *44-1433*: 2.02_{10} 1.43_1 1.17_3 1.01_1 0.904_2 0.825_1. Microscopic grains showing cube and octahedron forms. Silver-white, metallic luster. $H = 4\frac{1}{2}$, $VHN = 255$. $R = 54$ (510 nm). Formed during serpentinization of ultrabasic rocks. Muskox intrusion, NWT, Canada; Selva, near Poschiavo, Switzerland; Oko, near Koti, Japan; *Red Hills, Wairau R., New Zealand*; and in some meteorites. BM *MM 33:948(1964), R:357, ABI:566*.

1.1.12.1   Chromium   Cr

Recognized as a mineral in 1981. ISO $Im3m$. $a = 2.884$, $Z = 2$, $D = 7.20$. *6-694*(syn): 2.04_{10} 1.44_2 1.18_3 1.02_2 0.912_2 0.833_1. Spherulic and rounded grains. White, metallic luster. $H = 9$, $VHN = 1875$–2000. $G = 7.17$. $R = 67.9$ (540 nm). Yakutia, Russia, as inclusions in diamond; *Sichuan Prov., China*, in heavy concentrate from sand. BM *KT 26:959(1981), AM 67:854(1982), ABI:98*.

1.1.12.2 Chromferide Fe_3Cr_{1-x} ($x = 0.6$)

Named in 1986 for the composition. ISO $Pm3m$. $a = 2.859$, $Z = 1$, $D = 6.69$. *41-1466*: 2.87_2 2.02_{10} 1.66_1 1.43_8 1.28_5 1.16_{10} 1.01_7. Microscopic grains. Pale gray, metallic luster. No cleavage. $H = 4$, $VHN_{100} = 260$. $R = 55.3$ (540 nm). Magnetic. *Kumak region, Southern Ural Mts., Russia*, in quartz veins with ferchromide, graphite, and halite. BM *ZVMO 115:355(1986), AM 73:190(1988)*.

1.1.12.3 Ferchromide Cr_3Fe_{1-x} ($x = 0.6$)

Named in 1986 for the composition. ISO $Pm3m$. $a = 2.882$, $Z = 1$, $D = 6.18$. *ZVMO 115*: 2.88_1 2.04_{10} 1.66_5 1.44_6 1.29_5 1.17_9 1.02_7 0.77_8. Microscopic grains. Pale gray, metallic luster. No cleavage. $H = 6-6\frac{1}{2}$, $VHN_{100} = 900$. $R = 58.8$ (540 nm). Magnetic. *Kumak region, Southern Ural Mts., Russia*, in quartz veins with chromferide, graphite, and halite. BM *ZVMO 115:355(1986), AM 73:190(1988)*.

1.1.13.1 Tin Sn

Known since antiquity. TET $I4_1/amd$. $a = 5.831$, $c = 3.182$, $Z = 4$, $D = 7.286$. *4-673*(syn): 2.92_{10} 2.79_9 2.06_3 1.66_2 1.48_2 1.44_2 1.21_2. Rounded grains or aggregates. Tin-white, metallic luster. Ductile and malleable. $H = 2$, $VHN_{10} = 7-9$. $G = 7.31$. $R = 82-88$ (540 nm); moderately anisotropic. Nesbitt LaBine mine, Beaverlodge, SK, Canada, associated with pitchblende; Elkiaidan R., Uzbekistan; Oued Berkou, Algeria, in greisen; *Clarence R., near Oban, NSW, Australia*, in placers. BM *DI:126, R:362, ABI: 531*.

1.1.14.1 Indium In

Recognized as a mineral in 1964. TET $I4/mmm$. $a = 3.252$, $c = 4.946$, $Z = 2$, $D = 7.29$. *5-642*(syn): 2.72_{10} 2.47_2 2.30_4 1.68_2 1.62_1 1.47_2 1.40_2 1.36_1. Grains to 1 mm. Gray with yellow tint, metallic luster. $H = 3$, $VHN = 130-159$. $R = 92.3$; weakly anisotropic. *Eastern Transbaikal, Russia*, in albitized granite in association with native lead. BM *AM 52:299(1967), ABI:234*.

1.1.15.1 Cupalite (Cu,Zn)Al

Named in 1985 for the composition. ORTH Space group unknown. $a = 6.95$, $b = 4.16$, $c = 10.04$, $Z = 10$, $D = 5.18$. *39-1371*: 5.07_{10} 4.12_8 3.59_2 2.83_1 2.61_1 2.32_1 2.02_1. Microscopic grains. Yellow, metallic luster. $H = 4-4\frac{1}{2}$, $VHN_{50} = 272-318$. $R = 62.9$ (540 nm); weakly anisotropic. *Listvenitovii R., Koryak Mts., Kamchatka, Russia*, in weathered serpentinite. BM *ZVMO 114:90(1985), AM 71:1278(1986), ABI:118*.

1.1.15.2 Khatyrkite (Cu,Zn)Al$_2$

Named in 1985 for the locality. TET $I4/mcm$. $a = 6.063$, $c = 4.872$, $Z = 4$, $D = 4.357$. *25-12*(syn): 4.30_{10} 3.04_4 2.37_7 2.15_4 2.12_9 1.92_7 1.90_6 1.29_2. Prismatic crystals and microscopic grains. Yellow-gray, metallic luster. Cleavage {100}. $H = 5$, $VHN_{50} = 540$. $R = 76.1$ (540 nm); strongly anisotropic. *Kha-*

tyrka R., Koryak Mts., Kamchatka, Russia, in weathered serpentinite of the Khatirskii ultramafic zone. BM *ZVMO 114:90(1985), AM 71:1278(1986), ABI:263*.

1.1.16.1 Cohenite (Fe,Ni)$_3$C

Named in 1889 for Emil W. Cohen (1842–1905), German mineralogist and meteorite researcher. Synonym: Cementite is Fe$_3$C in steel. Contains 0.7–2.3% Ni and up to 0.3% Co. ORTH *Pnma*. $a = 5.091$, $b = 6.743$, $c = 4.526$, $Z = 4$, $D = 7.675$. *35-772*(syn): 2.39_4 2.38_4 2.11_6 2.07_7 2.03_6 2.01_{10} 1.98_5 1.85_4. Platy crystals or rounded grains. White, tarnishing to yellow, gray streak, metallic luster. Cleavage {100}, {010}, {001}; very brittle. H = 7, VHN$_{50}$ = 1100. G = 7.20–7.65. Strongly magnetic. Insoluble in dilute HCl. Disko Is., Greenland, and Putorana plateau, Taimyr Penin., Russia, in terrestrial iron; an accessory mineral in some iron meteorites and enstatite chondrites. Recorded in metallic particles in lunar rocks, probably of meteoritic origin. BM *DI:122, GCA 31:143(1967), LAP 18(4):32(1993)*.

1.1.16.2 Haxonite (Fe,Ni)$_{23}$C$_6$

Named in 1974 for Howard J. Axon (1924–1992), English metallurgist. Contains 3.5–5.6% Ni and up to 0.4% Co. ISO *Fm3m*. $a = 10.55$, $Z = 4$, $D = 7.70$. *25-405*: 2.36_{10} 2.15_{10} 1.86_6 1.76_5 1.24_6 1.22_5 1.16_4 1.06_5. Microscopic grains. Tin-white, metallic luster. H = 7, VHN = 850. A minor accessory in some iron meteorites, often in association with cohenite. BM *AM 59:209(1974)*.

1.1.17.1 Tongbaite Cr$_3$C$_2$

Named in 1983 for the locality. ORTH *Pnma*. $a = 5.57$, $b = 11.47$, $c = 2.816$, $Z = 4$, $D = 6.657$. *35-804*(syn): 2.31_{10} 2.24_6 1.95_4 1.87_4 1.82_3 1.78_3 1.70_4 1.19_7. Microscopic prismatic crystals. Pale yellow-brown, dark gray streak, metallic luster. H = $8\frac{1}{2}$, VHN$_{50}$ = 1931. R = 55 (546 nm); distinctly anisotropic. *Liu Zhuang, Tongbai Co., Henan Prov.*, and Hong district, Tibet, *China*, in ultrabasic rocks. BM *AM 70:218(1985)*.

1.1.18.1 Siderazot Fe$_5$N$_2$

Named in 1876 for the composition. HEX *P6$_3$22*. $a = 4.678$, $c = 4.369$, $Z = 1$, $D = 3.08$. *3-925*(syn): 2.34_{10} 2.19_{10} 2.06_{10} 1.59_{10} 1.34_{10} 1.24_{10} 1.15_{10} 1.13_{10}. Thin coating on lava. Tin-white, metallic luster. G = 3.147. Mt. Etna and Mt. Vesuvius, *Italy*. BM *DI:126, ZK 74:511(1930)*.

1.1.18.2 Roaldite (Fe,Ni)$_4$N

Named in 1981 for Roald Nielsen (b.1928), Danish metallurgist. Contains 5.6–6.4% Ni and 0.5% Co. ISO *Pm3m*. $a = 3.79$, $Z = 1$, $D = 7.168$. *6-627*(syn): 2.19_{10} 1.90_8 1.70_2 1.34_7 1.14_9 1.05_2 0.949_5. Micron-size platelets. Tin-white, metallic luster. H = $5\frac{1}{2}$–$6\frac{1}{2}$, VHN = 600–900. An accessory mineral in a few iron meteorites. BM *AM 66:1100(1981)*.

1.1.19.1 Osbornite TiN

Named in 1870 for George Osborne, who provided the original material. ISO $Fm3m$. $a = 4.242$, $Z = 4$, $D = 5.39$. Halite structure. $38\text{-}1420$(syn): 2.45_7 2.12_{10} 1.50_5 1.28_2 1.22_2 1.06_1 0.948_2. 0.865_1. Minute octahedra. Golden yellow, metallic luster. $H = 8\text{-}9$. $G = 5.25$. A trace constituent in meteorites, in some enstatite chondrites and achondrites. BM $DI:124$, MM $26:36(1941)$.

1.1.19.2 Khamrabaevite (Ti,V,Fe)C

Named in 1984 for Ibrahim K. Khamrabaev (b.1920), Uzbek mineralogist. ISO $Fm3m$. $a = 4.319$, $Z = 4$, $D = 4.91$. Halite structure. $32\text{-}1383$(syn): 2.50_8 2.16_{10} 1.53_6 1.31_3 1.25_2 1.08_1 0.993_1 0.968_3. Skeletal cubic crystals. Dark gray, metallic luster. No cleavage, fracture irregular. $H = 9$, $VHN_{100} = 2230\text{-}2290$. $R = 47.5$ (580 nm). *Chaktal Range, Arashan Mts., Uzbekistan, in basaltic porphyrite.* BM $ZVMO$ $113:697(1984)$, AM $70:1329(1985)$.

1.1.20.1 Carlsbergite CrN

Named in 1971 for the Carlsberg Foundation, Denmark, which supported the research. ISO $Fm3m$. $a = 4.13$, $Z = 4$, $D = 6.18$. Halite structure. $11\text{-}65$(syn): 2.39_8 2.07_{10} 1.46_8 1.25_6 1.20_3 0.950_5 0.926_6 0.846_6. Micron-size laths and irregular grains. Pale gray, metallic luster. $H > 7$, $VHN > 1000$. $G = 5.9$. $R = 41.5$ (546 nm). A minor accessory in many iron meteorites. BM AM $57:1311(1972)$.

1.1.21.1 Barringerite (Fe,Ni)$_2$P

Named in 1969 for Daniel M. Barringer (1860–1929), U.S. mining engineer, who established the meteoritic origin of Meteor Crater, Arizona. HEX $P\bar{6}2m$. $a = 5.877$, $c = 3.437$, $Z = 3$, $D = 6.91$. $33\text{-}670$(syn): 2.23_{10} 2.05_7 1.92_4 1.72_2 1.70_2 1.68_1 1.28_2 1.11_1. Micron-size grains. Tin-white, metallic luster. A trace constituent in the *Ollague pallasite*, possibly a product of artificial reheating. BM SCI $165:169(1969)$, AM $55:317(1970)$.

1.1.21.2 Schreibersite (Fe,Ni)$_3$P

Named in 1847 for Karl von Schreibers (1775–1852), Austrian meteorite researcher. Synonym: rhabdite. The Ni content ranges from 7.0 to 65.1%. TET $I\bar{4}$. $a = 9.056$, $c = 4.471$, $Z = 8$, $D = 7.13$. $19\text{-}617$(syn): 2.20_{10} 2.14_5 2.11_4 2.03_8 1.98_{10} 1.79_4 1.76_6 1.69_3. Thin plates (rhabdite) or rounded crystals. Tin-white, tarnishing to brass-yellow, gray streak, metallic luster. Cleavage {001}, perfect; very brittle. $H = 6\frac{1}{2}\text{-}7$, $VHN = 800\text{-}950$. $G = 7.0\text{-}7.3$. Strongly magnetic. An accessory mineral in most iron meteorites and many stony ones; has been observed in lunar samples, possibly of meteoritic origin. A trace constituent in terrestrial iron from Disko Is., Greenland. BM $DI:124$, $R:361$, GCA $29:513(1965)$.

1.1.22.1 Perryite $(Ni,Fe)_8(Si,P)_3$

Named in 1965 for Stuart H. Perry (1874–1957), U.S. newspaper publisher and researcher on iron meteorites. HEX-R $R3c$. $a = 6.640$, $c = 37.98$, $Z = 12$, $D = 7.63$. *17-222*(syn): 2.61_4 2.59_4 2.15_7 2.10_7 2.06_3 1.98_{10} 1.92_{10} 1.78_4. Microscopic platelets. Tin-white, metallic luster. Twinned on {0001}. Occurs in association with kamacite in the *Horse Creek iron meteorite* and in some enstatite chondrites and achondrites. BM *MM 36:850(1968), 37:905(1970); AM 77:452(1992)*.

1.1.23.1 Suessite $(Fe,Ni)_3Si$

Named in 1982 for Hans E. Suess (1909–1993), Austrian–U.S. geochemist. Contains up to 4.5% Ni. ISO $Im3m$. $a = 2.841$, $Z = 0.5$, $D = 7.083$. *35-519*(syn): 2.01_{10} 1.42_1 1.16_3. Microscopic blebs and elongated grains. Cream-white, metallic luster. $R = 51.6$ (546 nm). Magnetic. An accessory mineral in the *North Haig ureilite meteorite*, Kurama ridge, Uzbekistan. BM *AM 67:126(1982)*.

1.1.23.2 Gupeiite Fe_3Si

Named in 1984 for the Gupeikou gateway in the Great Wall of China. ISO $Fm3m$. $a = 5.670$, $Z = 4$, $D = 7.15$. *11-616*(syn): 3.25_4 2.83_4 1.99_{10} 1.70_4 1.62_2 1.41_{10} 1.15_{10} 0.995_{10}. Cores of microscopic spheres. Steel-gray, black streak, metallic luster. Brittle. $H = 5$, $VHN_{50} = 494–514$. $R = 53.5$ (546 nm). Strongly magnetic. *Yanshan area, Hebei, China*, in placers. BM *AM 71:228(1986)*.

1.1.23.3 Xifengite Fe_5Si_3

Named in 1984 for the Xifengkou gateway in the Great Wall of China. HEX $P6_3/mcm$. $a = 6.757$, $c = 9.440$, $Z = 2$, $D = 7.15$. *11-615*(syn): 2.35_5 2.18_6 2.00_{10} 1.94_8 1.92_9 1.37_6 1.27_9 1.24_8. Cores of microscopic spheres. Steel-gray, black streak, metallic luster. Brittle. $H = 5$, $VHN_{50} = 633–694$. $R = 44$ (546 nm). Strongly magnetic. *Yanshan area, Hebei, China*, in placers. BM *AM 71:228(1986)*.

PLATINUM GROUP

The members of the platinum group are elements sharing a cubic close-packed structure, space group $Fm3m$. They are isostructural among themselves and with the gold group.

CUBIC CLOSE-PACKED ELEMENTS

Mineral	Formula	Space Group	a	D	
Platinum	Pt	$Fm3m$	3.924	21.46	1.2.1.1
Iridium	Ir	$Fm3m$	3.839	22.56	1.2.1.2
Rhodium	(Rh, Pt)	$Fm3m$	3.803	12.43	1.2.1.3
Palladium	Pd	$Fm3m$	3.890	12.01	1.2.1.4

Note: Platinum, palladium, rhodium, and iridium form solid solutions among themselves. Many were separately named as minerals, but with the advent of the electron microprobe analyzer, most of the introduced names have been discarded as superfluous [*CM 29:231(1991)*].

1.2.1.1 Platinum   Pt

Named in 1748 from the Spanish *platina*, diminutive of *plata*, silver. Platinum group. Natural occurrences always alloyed with other Pt-group elements and with Fe. ISO $Fm3m$. $a = 3.924$, $Z = 4$, $D = 21.46$. *29-717*(syn): 2.24_{10} 1.94_6 1.37_6 1.17_8 0.890_9 0.867_8 0.792_{10}. Cubic crystals rare, usually as grains or flakes, rarely as sizable nuggets. Tin-white to steel-gray, dark gray to black in Fe-rich varieties; metallic luster. No cleavage, hackly fracture, malleable and ductile. $H = 4-4\frac{1}{2}$, $VHN_{100} = 297-339$. $G = 14-19$. Nonmagnetic to magnetic, magnetism increasing with Fe content. Insoluble except in hot aqua regia. Platinum is usually found in or associated with mafic and ultramafic igneous such as olivine gabbro and pyroxenite (Urals) and peridotite and dunite (Urals and Transvaal). Also found in quartz veins associated with hematite, chlorite, and pyrolusite. Commonly concentrated in placers. Platinum is widely distributed in placers throughout the world, but most commercial production is from primary deposits such as the Merensky Reef of the Bushveld Igneous Complex in South Africa and the Stillwater complex in Montana, and as a by-product from nickel ores as at Sudbury, ON, Canada; Norilsk, Russia; and Kambalda, WA, Australia. In the United States, placer deposits occur at Cape Blanco, Curry Co., OR, at several localities in CA, and at Platinum Creek, Goodnews Bay, AK. Also at Riviere-du-Loup, Beauce Co., PQ, near Edmonton, AB, Kamloops, Similkameen district, and Tulameen R., BC, Canada; Nizhni Tagil district, Ural Mts., and Konder(*), near Nelkan, Khabarovsk Oblast, Russia; near Papayan, Cauca, Colombia; Bom Sucesso, near Serro, MG, Brazil. BM *DI:106, R:340, P&J:302, ABI:410*.

1.2.1.2 Iridium   Ir

Named in 1804 because its salts are of varied colors. Platinum group. Always alloyed with other Pt-group elements; platiniridium is a rejected name for platinian iridium, osmiridium a rejected name for osmian varieties. ISO $Fm3m$. $a = 3.839$, $Z = 4$, $D = 22.56$. 6-598(syn): 2.22_{10} 1.92_5 1.36_4 1.16_5 1.11_2 0.881_4 0.859_4 0.784_5. Cubic habit, usually as rounded grains. Silver-white, metallic luster. Cleavage {001}, indistinct; hackly fracture. H = 6–7, $VHN_{100} = 867$. G = 19–21. Occurs in Pt-bearing ultrabasic rocks and in placers derived from them. Salmon Creek, Siskiyou Co., and other localities in CA; Goodnews Bay, AK; Tulameen, Spruce Creek, and Atlin, BC, Canada; Nizhni Tagil district, Ural Mts., Russia; Ava, near Mandalay, Burma; Ioma area, Papua New Guinea; Pieman R., TAS, Australia. BM *DI:110, R:339, ABI:239, CM 29:231(1991)*.

1.2.1.3 Rhodium (Rh,Pt)

Identified as a mineral in 1974. Platinum group. ISO $Fm3m$. $a = 3.803$, $Z = 4$, $D = 12.43$. 5-685(syn): 2.20_{10} 1.90_5 1.35_3 1.15_3 1.10_1 0.951_1 0.872_2 0.850_1. Microscopic subhedral grain. Silver-white, metallic luster. $H = 3\frac{1}{2}$, $VHN_{30} = 136$–194. $R = 72.6$ (546 nm). *Stillwater Complex, MT*, a single grain with composition $Rh_{0.57}Pt_{0.43}$ in heavy-mineral concentrates. BM *CM 12:399(1974), AM 61:340(1976), ABI:439*.

1.2.1.4 Palladium Pd

Named in 1803 for the asteroid Pallas, discovered in 1802. Platinum group. ISO $Fm3m$. $a = 3.890$, $Z = 4$, $D = 12.01$. 5-681(syn): 2.25_{10} 1.95_4 1.38_3 1.17_2 1.12_1 0.972_1 0.892_1 0.870_1. Rarely as octahedra; commonly in grains, sometimes with a radial fibrous texture. White to pale steel-gray, metallic luster. No cleavage, malleable and ductile. $H = 4\frac{1}{2}$–5. G = 11.9. R = 67.8 (540 nm). Easily confused with platinum but has a much lower density and is soluble in HNO_3. Occurs as a primary phase in platinum deposits and as an oxidation product of palladium-bearing sulfides. Ural Mts., Russia; Merensky Reef, Transvaal, South Africa; Choco, Cauca, Colombia; Potaro R., Guyana; Bom Sucesso, near Serro, MG, Brazil. BM *DI:109, R:338, P&J:286, ABI:376, MR 23:471(1992)*.

OSMIUM GROUP

The members of the osmium group share a hexagonal close-packed structure.

HEXAGONAL CLOSE-PACKED ELEMENTS

Mineral	Formula	Space Group	a	c	D	
Osmium	(Os,Ir,Ru)	$P6_3/mmc$	2.733	4.319	22.59	1.2.2.1
Ruthenium	(Ru,Ir,Os)	$P6_3/mmc$	2.704	4.326	12.36	1.2.2.2
Rutheniridosmine	(Ir,Os,Ru)	$P6_3/mmc$	2.726	4.326	20.6	1.2.2.3
Rhenium	Re	$P6_3/mmc$	2.760	4.458	21.0	1.2.2.4

1.2.2.2 Ruthenium

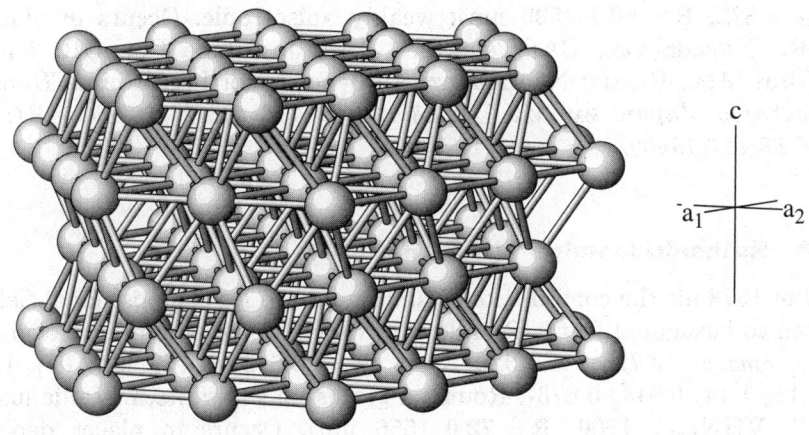

HCP: hexagonal closest-packed (osmium, iridium)

Space group P6$_3$/mmc. This shows the true symmetry with all atoms alike. No common minerals have this structure, but the arrangement is adopted by oxygen and sulfur atoms in many.

Notes:
1. Osmium and ruthenium form alloys with iridium, one of which is found as the mineral rutheniridosmine.
2. Zinc (1.1.5.1) and cadmium (1.1.5.2) are isostructural with osmium.

1.2.2.1 Osmium (Os,Ir,Ru)

Named in 1804 from the Greek for *smell*, because of the odor of its volatile oxide. Osmium group. May contain up to 50% Ir, 20% Ru, and minor amounts of the other platinum group elements. *Iridosmine, siserskite*, and *nevyanskite* are obsolete names for iridian osmium. HEX $P6_3/mmc$. $a = 2.733$, $c = 4.319$ $Z = 2$, $D = 22.59$. *6-662*(syn): 2.37_4 2.16_4 2.08_{10} 1.60_2 1.37_2 1.23_2 1.16_2 1.14_2. Crystals tabular {0001}, also as flattened grains. Tin-white to steel-gray, metallic luster. Cleavage {0001}, perfect but difficult; slightly malleable. H = 7, VHN$_{25}$ = 1206–1246. G = 17.6–21.2. R = 55 (540 nm); strongly anisotropic. Occurs with other platinoid minerals, usually in placers. American R., Sacramento Co., Salmon Creek, Siskiyou Co., and other localities in CA; Similkameen, Atlin, and Quesnel, BC, Canada; several localities in the Ekaterinburg Oblast, Ural Mts., and in Yakutsk and Transbaikalia, Russia; Anduo, Tibet, China; Guma Water, Sierra Leone; Birbir R., Ethiopia; Rustenberg area, Transvaal, South Africa; Pieman R., TAS, Australia; Lee R., Nelson, New Zealand. BM *DI:111, R:350, P&J:284, AB:369, CM 29:231(1991)*.

1.2.2.2 Ruthenium (Ru,Ir,Os)

Identified as a mineral in 1974. Osmium group. HEX $P6_3/mmc$. $a = 2.704$, $c = 4.326$, $Z = 2$, $D = 12.36$. *6-663*(syn): 2.34_4 2.14_4 2.06_{10} 1.58_3 1.35_3 1.22_3

1.14_3 1.13_2. Microscopic grains. Silver-white, metallic luster. H = $6\frac{1}{2}$, VHN_{100} = 872. R = 60.1 (530 nm); weakly anisotropic. Occurs in placers. Yuba R., Nevada Co., CA; Tennessee Pass, Josephine Co., OR; Nizhni Tagil, Ural Mts., Russia; Nuanquanzi Au deposit, northern China; *Horakanai, Hokkaido, Japan*. BM *MJJ 7:438(1974), AM 61:177(1976), ABI:454, TMPM 43:137(1990)*.

1.2.2.3 Rutheniridosmine (Ir,Os,Ru)

Named in 1973 for the composition. Osmium group. Its compositional field is restricted to hexagonal Ir–Os–Ru alloys in which Ir is the dominant element. HEX $P6_3/mmc$. a = 2.726, c = 4.326, Z = 2, D = 20.6. *41-601*: 2.36_8 2.07_{10} 1.59_4 1.36_6 1.15_5 1.14_4 0.914_4 0.870_5. Rounded grains. Silver-white, metallic luster. H = 6–7, VHN_{100} = 1100. R = 72.0 (556 nm). Occurs in placer deposits derived from ultrabasic rocks. Spruce Creek and Tulameen R. areas, BC, Canada; Ural Mts. and Yakutia, Russia; Witwatersrand, South Africa; Massif du Sud, New Caledonia; Heazlewood district, TAS, Australia; Ironstone Creek, Parapara, Nelson, New Zealand. BM *CM 12:104(1973), 29:231(1991), AM 60:946(1975), ABI:453*.

1.2.2.4 Rhenium Re

Identified as a mineral in 1976. Osmium group. HEX $P6_3/mmc$. a = 2.760, c = 4.458, Z = 2, D = 21.0. *5-702*(syn): 2.39_3 2.23_3 2.11_{10} 1.63_1 1.38_2 1.26_2 1.17_2 1.15_2. Microscopic grains. Silver-white, metallic luster. *Transbaikal, Russia*, in wolframite; Allende meteorite, in Ca,Al-rich inclusions (97% Re, 3% Ru). BM *AM 63:1283(1978)*.

1.2.3.1 Hongshiite PtCu

Named in 1974 for the locality. HEX-R $R\bar{3}m$. a = 10.703, c = 13.197, Z = 48, D = 15.75. *42-1326*: 4.35_3 2.20_{10} 1.90_8 1.35_5 1.33_5 1.15_8 1.10_3 0.856_5. Microscopic grains. Lead-gray, black streak, metallic luster. H = 4, VHN_{50} = 204–277. G = 15.00. R = 60.5–62.8 (540 nm); weakly anisotropic, red-brown to gray. Goodnews Bay, AK; *Hong district, China*; Merensky Reef, Transvaal, South Africa. BM *AM 61:185(1976), 69:411(1984), ABI:223*.

TETRAFERROPLATINUM GROUP

The five members of the tetraferroplatinum group are intermetallic compounds sharing the general formula

AB   or a multiple thereof

where
A = Pt, Pd
B = Cu, Fe, Ni, Hg

and sharing the tetragonal $P4/mmm$ structure. The structure is related to that of the isoferroplatinum group.

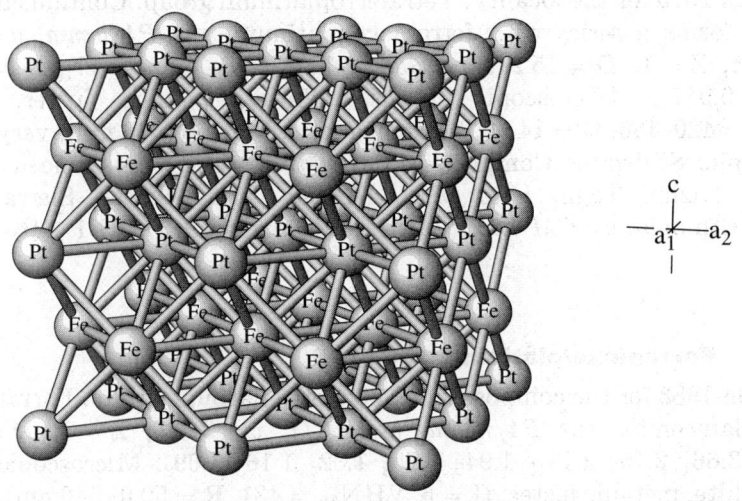

TETRAFERROPLATINUM: PtFe

TETRAFERROPLATINUM GROUP

Mineral	Formula	Space Group	a	c	D	
Tetraferroplatinum	PtFe	$P4/mmm$	3.84	3.72	15.07	1.2.4.1
Tulameenite	Pt_2FeCu	$P4/mmm$	3.902	3.585	15.21	1.2.4.2
Ferronickelplatinum	Pt_2FeNi	$P4/mmm$	3.871	3.635	15.39	1.2.4.3
Potarite	PdHg	$P4/mmm$	3.026	3.702	15.03	1.2.4.4

Note: Tetraauricupride (1.1.2.2) is isostructural with these minerals.

1.2.4.1 Tetraferroplatinum   PtFe

Named in 1975 for the composition and crystal system. Tetraferroplatinum group. Composition somewhat variable around PtFe; may contain minor amounts of Cu, Rh, Ir, Sb. TET $P4/mmm$. $a = 3.84$, $c = 3.72$, $Z = 2$,

D = 15.07. *26-1139*(syn): 2.20_{10} 1.34_5 1.16_9 1.13_6 1.10_6 1.04_5 1.02_6 0.918_6. Rounded grains or flakes. Silver-white, metallic luster. $VHN_{100} = 384$. G = 14.3. R = 61.3 (546 nm); weakly anisotropic. Ferromagnetic. Occurs in Pt-bearing placers; probably more common than reported because of confusion with ferroan platinum. *Tulameen* and Similkameen R., *BC, Canada*; Talnakh, Norilsk, Russia; Pirogues R., New Caledonia; Mooihoek and Witwatersrand, Transvaal, South Africa. BM *CM 13:117(1975), AM 61:341(1976), TMPM 36:175(1987), ABI: 525.*

1.2.4.2 Tulameenite Pt$_2$FeCu

Named in 1973 for the locality. Tetraferroplatinum group. Contains minor Ir, Ni, Sb; forms a series with ferronickelplatinum. TET $P4/mmm$. $a = 3.902$, $c = 3.585$, Z = 1, D = 15.21. *42-1359*: 2.19_{10} 1.95_5 1.32_6 1.17_{10} 1.09_9 1.02_8 0.976_6 0.941_7. Microscopic grains. White, metallic luster. H = 5, $VHN_{50} = 420$–456. G = 14.90. Ferromagnetic. R60–66 (546 nm); very weakly anisotropic. Stillwater Complex, MT; *Tulameen* and Similkameen R., *BC, Canada*; Nizhni Tagil, Ural Mts., Russia; Guma Water, Sierra Leone; Yubdo, Ethiopia. BM *CM 12:21(1973), AM 59:383(1974), ABI: 542.*

1.2.4.3 Ferronickelplatinum Pt$_2$FeNi

Named in 1983 for the composition. Tetraferroplatinum group. Forms a series with tulameenite. TET $P4/mmm$. $a = 3.871$, $c = 3.635$, Z = 1, D = 15.39. *35-702*: 3.66_1 2.75_1 2.19_{10} 1.94_5 1.70_3 1.32_4 1.16_3 1.09_2. Microscopic grains. Silver-white, metallic luster. H = 5, $VHN_{50} = 481$. R = 59.0 (540 nm); weakly anisotropic. Occurs in placers derived from ultrabasic rocks. Goodnews Bay, AK; *Koryak Mts., Kamchatka, Russia*. BM *ZVMO 112:487(1983), AM 69:1190 (1984), ABI: 155.*

1.2.4.4 Potarite PdHg

Named in 1926 for the locality. Tetraferroplatinum group. TET $P4/mmm$. $a = 3.026$, $c = 3.702$, Z = 2, D = 15.03. *13-149*(syn): 2.34_{10} 2.14_5 1.40_6 1.27_8 1.17_5 1.14_5 0.928_5 0.850_8. Small grains or nuggets. White to steel-gray, metallic luster. Brittle. H = $3\frac{1}{2}$, $VHN_{100} = 132$. G = 13.5–16.1. R = 61.0 (540 nm). Musonoi and Mindingi, Zaire; Merensky Reef, Transvaal, South Africa; *Potaro R., Guyana*, in placers. BM *DI:105, AM 45:1093(1960), R:362, ABI:418.*

ISOFERROPLATINUM GROUP

The members of the isoferroplatinum group are intermetallic compounds sharing the general chemical formula

$$A_3B$$

where

A = Pt, Pd
B = Fe, Sn, or Pb

and sharing the same cubic structure in space group $Pm3m$.

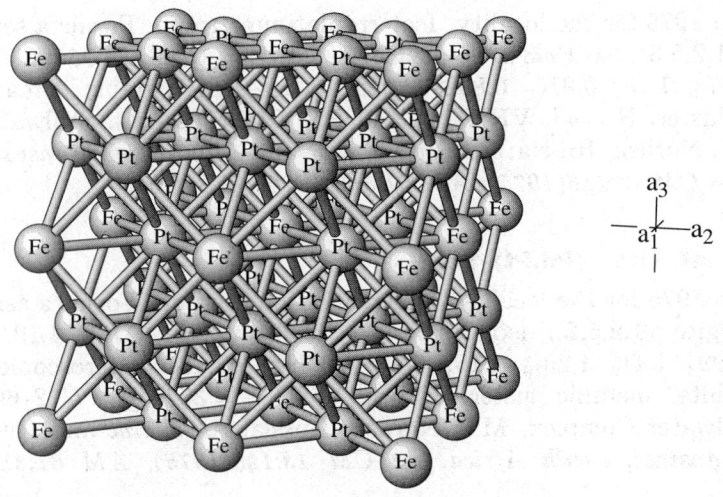

ISOFERROPLATINUM: Pt_3Fe

ISOFERROPLATINUM GROUP

Mineral	Formula	Space Group	a	D	
Isoferroplatinum	Pt_3Fe	$Pm3m$	3.866	18.42	1.2.5.1
Rustenbergite	$(Pt,Pd)_3Sn$	$Pm3m$	3.988	15.08	1.2.5.2
Atokite	$(Pd,Pt)_3Sn$	$Pm3m$	3.991	14.19	1.2.5.3
Zvyagintsevite	$(Pd,Pt,Au)_3(Pb,Sn)$	$Pm3m$	4.024	13.41	1.2.5.4
Chengdeite	Ir_3Fe	$Pm3m$	3.792	19.30	1.2.5.5

Notes:
1. Auricupride (1.1.2.1) is isostructural with these minerals.
2. See the tetraferroplatinum group for related tetragonal structures.

1.2.5.1 Isoferroplatinum Pt$_3$Fe

Named in 1975 for the composition and crystal system. Isoferroplatinum group. Contains minor amounts of Ir, Pd, Ru, and Rh. ISO $Pm3m$. $a = 3.866$, $Z = 1$, $D = 18.42$. *29-1423*: 2.22_6 1.93_{10} 1.72_2 1.57_1 1.36_7 1.17_{10} 1.11_5 1.07_1. Microscopic cubic crystals, and rounded grains and flakes. Silver-white, metallic luster. $VHN_{100} = 503$–572. Occurs in Pt-bearing placers; probably more common than reported because of confusion with ferroan platinum. Stillwater Complex, MT; Goodnews Bay, AK; *Tulameen* and Similkameen R., *BC, Canada*; Norilsk and Kondor, Russia; Witwatersrand and Merensky R., Transvaal, South Africa; Round Hill, near Orepuki, New Zealand. BM *CM 13:117(1975)*, *AM 61:341(1976)*, *ABI:243*.

1.2.5.2 Rustenbergite (Pt,Pd)$_3$Sn

Named in 1975 for the locality. Isoferroplatinum group. Forms a series with atokite (1.2.5.3). ISO $Pm3m$. $a = 3.988$, $Z = 1$, $D = 15.08$. *29-968*: 2.30_{10} 1.99_8 1.41_9 1.20_{10} 1.15_4 0.915_9 0.892_8 0.815_9. Microscopic grains. Cream-white, metallic luster. $H = 4\frac{1}{2}$, $VHN_{25} = 365$. $R = 59.8$ (546 nm). Stillwater Complex, MT; Norilsk, Russia; *Rustenberg mine* and Atok mine, *Transvaal, South Africa*. BM *CM 13:146(1975)*, *AM 61:340(1976)*, *ABI:451*.

1.2.5.3 Atokite (Pd,Pt)$_3$Sn

Named in 1975 for the locality. Isoferroplatinum group. Forms a series with rustenbergite (1.2.5.2). ISO $Pm3m$. $a = 3.991$, $Z = 1$, $D = 14.19$. *29-967*: 2.30_{10} 1.99_8 1.41_9 1.20_{10} 1.15_4 0.915_9 0.892_8 0.815_9. Microscopic grains. Cream-white, metallic luster. $H = 4\frac{1}{2}$, $VHN_{25} = 357$. $R = 59.2$–65.1 (546 nm). Stillwater Complex, MT; Norilsk, Russia; *Atok mine* and Rustenberg mine, *Transvaal, South Africa*. BM *CM 13:146(1975)*, *AM 61:340(1976)*, *ABI:35*.

1.2.5.4 Zvyagintsevite (Pd,Pt,Au)$_3$(Pb,Sn)

Named in 1966 for Orest E. Zvyagintsev (1894–1967), Russian geochemist. Isoferroplatinum group. May contain up to 4% Au. ISO $Pm3m$. $a = 4.024$, $Z = 1$, $D = 13.41$. *20-827*(syn): 2.32_{10} 2.01_8 1.42_7 1.22_9 1.16_4 1.01_3 0.923_6 0.900_5. Microscopic grains. Cream-white, metallic luster. $H = 4\frac{1}{2}$, $VHN_{15} = 241$–318. $R = 65.5$ (540 nm). Stillwater Complex, MT; *Talnakh, Norilsk, Russia*. BM *GRM 8:94(1966)*, *AM 52:299(1967)*, *R:363*, *ABI:588*.

1.2.5.5 Chengdeite Ir$_3$Fe

Named in 1995 for the locality. ISO $Pm3m$. $a = 3.792$, $Z = 1$, $D = 19.30$. *MR 27*: 2.18_8 1.89_6 1.34_7 1.26_2 1.20_1 1.14_{10} 1.09_8. Microscopic grains. Black, black streak, metallic luster. No cleavage. $H = 5$, $VHN_{50} = 452$. Strongly magnetic. In reflected light white with a yellow tint; isotropic; $R = 69.3$ (546 nm). *Luan*

R., Chengde Co., China, in placer deposits. BM *AGS 69(3):215(1995)*, *MR 27:202(1996)*, *AM 81:516(1996)*.

1.2.6.1 Paolovite   Pd$_2$Sn

Named in 1974 for the composition: palladium and olovo (Russian = tin). ORTH *Pbnm*. $a = 8.11$, $b = 5.662$, $c = 4.234$, $Z = 4$, $D = 11.32$. *26-1297*: 3.19_3 2.36_4 2.28_{10} 2.16_7 1.96_5 1.40_4 1.34_3 1.32_4. Microscopic grains. Lilac-rose, metallic luster. H = $4\frac{1}{2}$, VHN$_{50}$ = 380. R = 48–53 (540 nm). Geordie Lake, ON, Canada; *Oktyabr mine, Talnakh, Norilsk, Russia*; Atok and Western mines, Transvaal, South Africa. BM *GRM 16:98(1974)*, *AM 59:1331(1974)*; *ABI:380*, *CM 28:489(1990)*.

1.2.7.1 Taimyrite   (Pd,Cu,Pt)$_3$Sn

Named in 1982 for the locality (Taimyr Peninsula). ORTH Space group unknown. $a = 16.11$, $b = 11.27$, $c = 8.64$, $Z = 22$, $D = 10.20$. *41-1409*: 2.36_5 2.29_5 2.16_{10} 2.04_2 1.44_4 1.30_3 1.22_4 1.14_2. Microscopic grains. Bronze-gray, metallic luster. H = 5, VHN$_{50}$ = 480. R = 45–50 (550 nm); anisotropic, yellow-gray to dark gray. *Talnakh, Norilsk, Russia*. BM *ZVMO 111:78(1982)*, *AM 68:1252(1983)*, *ABI:511*.

1.2.8.1 Cabriite   Pd$_2$SnCu

Named in 1983 for Louis J. Cabri (b.1934), Canadian mineralogist. ORTH *Pmmm*. $a = 7.88$, $b = 7.88$, $c = 3.94$, $Z = 4$, $D = 10.73$. *37-418*: 2.29_{10} 2.17_9 1.84_3 1.43_3 1.23_8 1.22_4 1.18_3 1.09_2. Microscopic grains. White, metallic luster. H = 4, VHN$_{50}$ = 258–282. G = 11.1. R = 45.0–52.0 (540 nm); strongly anisotropic, gold to gray-brown. Thetford mines, PQ, Canada; *Oktyabr mine, Talnakh, Norilsk, Russia*. BM *CM 21:481(1983)*, *AM 69:1190(1984)*, *ABI:74*.

1.2.9.1 Stannopalladinite   Pd$_3$Sn$_2$

Named in 1947 for the composition. HEX *P6$_3$/mmc*. $a = 4.390$, $c = 5.655$, $Z = 1$, $D = 9.79$. *4-801*(syn): 2.27_{10} 2.20_{10} 1.58_7 1.42_5 1.19_7 0.988_6 0.944_6 0.834_7. Microscopic crystals. Pale brown-rose, metallic luster. H = $4\frac{1}{2}$–5, VHN = 387–452. R = 53–54 (540 nm); strongly anisotropic, red to blue. *Monchegorsk, Kola Penin.*, and *Talnakh, Norilsk, Russia*; Merensky Reef, Transvaal, South Africa. BM *DANS 58:1137(1947)*, *MA 10:453(1949)*, *CM 19:599(1981)*, *ABI:492*.

1.2.10.1 Niggliite   PtSn

Named in 1937 for Paul Niggli (1888–1953), Swiss mineralogist. HEX *P6$_3$/mmc*. $a = 4.100$, $c = 5.432$, $Z = 2$, $D = 13.18$. Nickeline structure. *25-*

614(syn): 3.55_6 2.97_7 2.16_{10} 2.05_8 1.49_8 1.20_9 1.13_6 0.927_6. Microscopic grains. Silver-white, metallic luster. H $= 5\frac{1}{2}$, $VHN_{25} = 590$. R $= 45$–57 (540 nm); strongly anisotropic, cream-pink to dark blue or black. Sudbury, ON, Canada; Monchegorsk, Kola Penin., and Talnakh, Norilsk, Russia; *Waterfall Gorge, Insizwa, South Africa.* BM *DI:347, MM 38:794(1972), R:363, P&J:407, ABI:352.*

1.2.11.1 Plumbopalladinite Pd_3Pb_2

Named in 1970 for the composition. HEX $P6_3/mc$. $a = 4.465$, $c = 5.709$, $Z = 1$, $D = 12.36$. 4-797(syn): 3.21_7 2.30_{10} 2.24_{10} 1.76_6 1.60_8 1.42_6 1.30_8 1.21_8. Microscopic grains. White with a rose tint, metallic luster. H $= 5$, $VHN_{50} = 407$–441. R $= 51$–54 (540 nm); strongly anisotropic, orange-brown to dark brown. Stillwater Complex, MT; *Mayak mine, Talnakh, Norilsk, Russia*; Merensky Reef, Transvaal, South Africa. BM *GRM 5:63(1970), AM 56:1121(1971), ABI:413.*

ARSENIC GROUP

Four semimetal minerals are isostructural with arsenic and crystallize in a rhombohedral lattice with space group $R\bar{3}m$. Each atom is bonded to six of its neighbors, but more closely to three than the other three. The shorter of the bonds, which are due to a degree of covalency bonding within As_4 groups, eliminates simple sphere packing and results in a puckered sheet of atoms, which is connected to the next sheet by the longer bonds. The sheets extend in the (0001) plane, resulting in perfect basal cleavage.

ARSENIC GROUP

Mineral	Formula	Space Group	a	c	D	
Arsenic	As	$R\bar{3}m$	3.76	10.56	5.78	1.3.1.1
Antimony	Sb	$R\bar{3}m$	4.307	11.273	6.70	1.3.1.2
Stibarsen	SbAs	$R\bar{3}m$	4.02	10.80	6.48	1.3.1.3
Bismuth	Bi	$R\bar{3}m$	4.546	11.860	9.75	1.3.1.4

Notes:
1. The tsumoite, tetradymite, and joséite group minerals are very similar to the arsenic group: they have the same space group, with a in the range 4.1–4.5 Å, and c a multiple of 2, 3, and 4.
2. Arsenolamprite (1.3.2.1) is an orthorhombic form of arsenic.
3. Paradocrasite (1.3.3.1) is a semimetal of antimony and arsenic but crystallizes in a monoclinic space group.

1.3.1.1 Arsenic As

Known since antiquity. Arsenic group. Usually contains small amounts of Sb in solid solution. Dimorphous with arsenolamprite. HEX-R $R\bar{3}m$. $a = 3.76$, $c = 10.56$, $Z = 6$, $D = 5.78$. 5-632(syn): 3.52_3 3.11_1 2.77_{10} 2.05_2 1.88_3 1.77_1

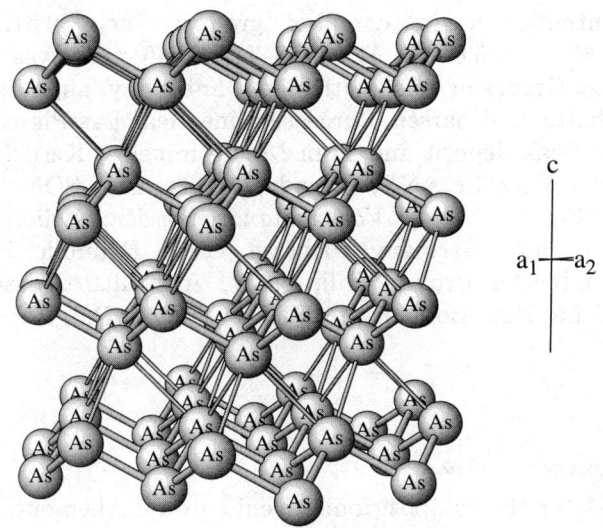

ARSENIC

Arsenic is in cubic closest-packed arrangement. However, layers are squeezed together in pairs, and the bonds between As atoms in these two layers (thick) are much shorter than those to the next layer (thin).

1.76_1 1.56_1. Crystals rare; usually massive, reniform, stalactitic, often in concentric layers. Tin-white, tarnishing to dark gray, tin-white streak, metallic luster. Cleavage {0001}, perfect; {10$\bar{1}$4}, fair; uneven fracture; brittle. H = $3\frac{1}{2}$, VHN_{100} = 72–173. G = 5.6–5.8. R = 47.4–53.2 (540 nm); distinctly anisotropic. On heating, sublimes without melting, giving dense white fumes with a garlic odor (poisonous). Widely distributed. Sterling Hill, NJ, and Washington Camp(*), Santa Cruz Co., AZ; near Port Alberni, Vancouver Is., and Watson Creek, Fraser R., BC, Canada; Ste.-Marie-aux-Mines, Haut-Rhin, France; Wolfsberg and Andreasberg(*), Harz, and Freiberg, Schneeberg, Marienberg, and Annaberg(*), Saxony, Germany; Jachymov and Pribram, Czech Republic; Sacaramb, Hunyad, and Cavnic, Romania; Akatani mine(*), Fukui Pref., Japan; Bidi, Sarawak; Forbes, NSW, Australia; Kaponga mine, near Coramandel, New Zealand; Huallapon mine, Pasta Bueno, Peru; Copiapo, Chile. BM *DI:128, R:365, P&J:76, ABI:24*.

1.3.1.2 Antimony Sb

Known since antiquity, identified as a mineral in 1748. Arsenic group. Usually contains a little As. HEX-R $R\bar{3}m$. $a = 4.307$, $c = 11.273$, $Z = 6$, $D = 6.70$. *35-732*(syn): 3.75_3 3.11_{10} 2.25_7 2.15_6 1.88_4 1.42_6 1.37_7 1.26_4. Rarely as pseudocubic or thick tabular crystals, usually massive, foliated, or granular; tin-white, gray streak, metallic luster. Cleavage {0001}, perfect and easy; less distinct on {10$\bar{1}$1}, {10$\bar{1}$4}, {11$\bar{2}$0}; twins on {10$\bar{1}$4} common, often polysynthetic or in complex groups; very brittle. H = 3–3$\frac{1}{2}$, VHN$_{100}$ = 50–69. G = 6.6–6.7. R = 70.0–71.6 (540 nm); weakly pleochroic and anisotropic. Occurs in veins with silver, antimony, and arsenic minerals, often with stibnite or stibarsen. Surcease mine, near Las Plumas, Butte Co., and Antimony Peak deposit and Tom Moore mine(*), Kern Co., CA; Lake George mine(*), York Co., NB, Canada; Moctezuma, SON, Mexico; Allemont, Isere, France; *Sala, Västmanland, Sweden*; Seinäjoki, Finland; Andreasberg(*), Harz, Germany; Pribram, Czech Republic; Sarawak, Borneo; Consols mine(*), Broken Hill, NSW, Australia; Huasco, Atacama, Chile; Quime, La Paz, Bolivia. BM *DI:132, R:373, P&J:69, ABI:16, MR 22:263(1991)*.

1.3.1.3 Stibarsen SbAs

Named in 1941 for the composition. Arsenic group. Allemontite is stibarsen intergrown with either arsenic or antimony. HEX-R $R\bar{3}m$. $a = 4.02$, $c = 10.80$, Z = 3, D = 6.48. *31-80*: 3.60$_3$ 2.92$_{10}$ 2.13$_6$ 2.01$_7$ 1.66$_4$ 1.47$_2$ 1.28$_4$ 1.01$_2$. Usually massive, reniform or mammilary, or fine granular. Tin-white, tarnishing to gray, metallic luster. Cleavage {0001}, perfect. H = 3–4. G = 5.8–6.2. R = 61.6–66.3 (540 nm). Occurs in hydrothermal veins and granite pegmatites. Ophir mine, Comstock Lode, NV; Bernic Lake, MB, and Atlin and Alder Is., BC, Canada; *Varuträsk, Västerbotten, Sweden*; Mine des Chalanches, near Allemont, France; Andreasberg, Harz, Germany; Pribram, Czech Republic; Broken Hill, NSW, Australia. BM *DI:130, AM 26:456(1941), R:371, P&J:352, ABI:496*.

1.3.1.4 Bismuth Bi

Known since the Middle Ages. Arsenic group. HEX-R $R\bar{3}m$. $a = 4.546$, $c = 11.860$, Z = 6, D = 9.75. *5-519*(syn): 3.95$_1$ 3.28$_{10}$ 2.37$_4$ 2.27$_4$ 1.97$_1$ 1.87$_2$ 1.49$_1$ 1.44$_2$. Indistinct crystals, often in parallel groupings; also arborescent, foliated, or granular. Silver-white with a reddish hue, iridescent tarnish; silver-white streak, metallic luster. Cleavage {0001}, perfect and easy; {10$\bar{1}$1}, good; {10$\bar{1}$4}, poor; sectile; brittle. H = 2–2$\frac{1}{2}$, VHN$_{100}$ = 16–18. G = 9.7–9.8. R = 62.6–63.8 (540 nm); distinctly anisotropic. Easily fusible, MP = 270°. Occurs in hydrothermal veins and in pegmatites. Monroe, CT; Chesterfield district, SC; French Creek and Las Animas mines, Boulder Co., CO; Pala, San Diego Co., CA; Cobalt, ON, and Great Bear Lake, NWT, Canada; Dolcoath and other mines, Cornwall, England; Meymac, Correze, France; Altenberg, Schneeberg(*), and Annaberg, Saxony, Germany; Jachymov, Czech Republic; Ikuno(*), Hyogo Pref., Japan; Wolfram Camp(*), Q, and Kingsgate(*), NSW, Australia; Uncia, Chorolque, Llallagua, Tazna(*), Colquechaca(*), and Potosi(*), Bolivia; Bogueiraa(*), Parelhas, RN, Brazil. BM *DI:134, R:374, P&J:91, ABI:55*.

1.3.3.1 Paradocrasite

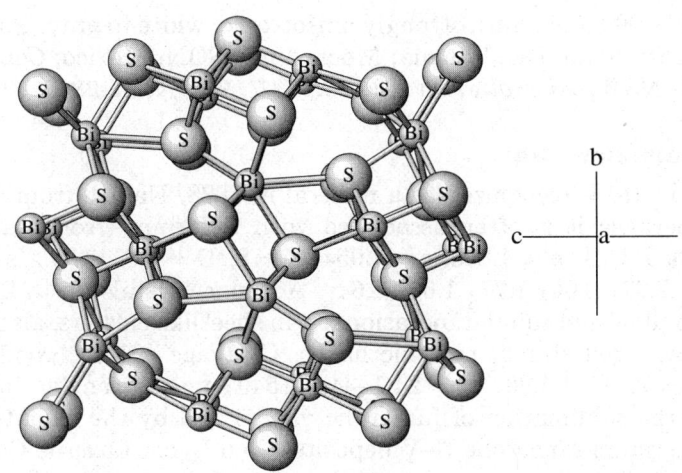

BISMUTHINITE: Bi_2S_3
Bi is in very irregular 5- and 6-coordination.

1.3.1.5 Stistaite SbSn

Named in 1970 for the composition: stibium (antimony) and stannum (tin). HEX-R $R\bar{3}m$. $a = 8.648$, $c = 10.675$, $Z = 12$, $D = 6.91$. 33-118(syn): 3.07_{10} 2.18_3 2.16_3 1.78_1 1.77_1 1.53_1 1.38_1 1.37_1. Microscopic pseudocubic crystals. Gray, metallic luster. H= 3, VHN = 103–127. G = 6.91. R = 81.6 (540 nm). *Elkiaidan R., Uzbekistan*; Rio Tamana, Colombia, in heavy-mineral concentrates from placer deposits. BM *ZVMO 99:68(1970)*, *AM 56:358(1971)*, *NJMM:117(1981)*, *ABI:501*.

1.3.2.1 Arsenolamprite As

Named in 1886 for the composition and the Greek for *brilliance*, in allusion to its luster. Dimorphous with arsenic. Contains up to 3% Bi. ORTH *Bmab*. $a = 3.65$, $b = 4.47$, $c = 11.0$, $Z = 8$, $D = 5.55$. 30-100: 5.49_1 3.47_2 2.74_{10} 2.24_2 1.88_1 1.74_3 1.69_1 1.52_1. Needles, radial aggregates of plates, or massive. Gray-white, tarnishing to black, black streak, metallic luster. Perfect cleavage in one direction. H = 2. G = 5.3–5.5. R = 44.0–49.6 (540 nm); weakly anisotropic. Ste.-Marie-aux-Mines, Alsace, France; Machenheim and Wittichen, Black Forest, and *Marienberg, Saxony, Germany*; Lengenbach, Binn, Switzerland; Jachymov, Czech Republic; Mina Alacran, Copiapó, Chile. BM *DI:130*, *CE 20:71(1959)*, *R:370*, *P&J:77*, *ABI:26*.

1.3.3.1 Paradocrasite $Sb_2(Sb, As)_2$

Named in 1971 from the Greek for *unexpected alloy*. MON *C2*. $a = 7.252$, $b = 4.172$, $c = 4.431$, $\beta = 123.13°$, $Z = 1$, $D = 6.44$. 25-48: 3.72_4 3.06_{10} 2.21_6 2.09_7 1.73_4 1.52_4 1.39_4 1.33_4. Microscopic stubby prisms. Silver-white, black streak, metallic luster. Perfect {010} parting. H = 3, VHN_{100} = 118.

$G = 6.52$. $R = 70.0$ (546 nm); strongly anisotropic, white to gray. Polysynthetically twinned. Atlin, BC, Canada; Moctezuma, SON, Mexico; *Consols mine, Broken Hill, NSW, Australia*. BM *AM 56:1127(1971), ABI:282*.

1.3.4.1 Selenium Se

Discovered in 1818, recognized as a mineral in 1828. Named from the Greek for *moon*, because it is often associated with tellurium (from Latin *tellus*, earth). HEX-R $P3_121$. $a = 4.366$, $c = 4.954$, $Z = 3$, $D = 4.81$. *6-362*(syn): 3.78_6 3.01_{10} 2.18_2 2.07_4 2.00_2 1.77_2 1.65_1 1.64_1. Acicular crystals up to 2 cm long, frequently hollow and tubular, occasionally in sheetlike clusters, also as glassy droplets. Gray, red streak, metallic luster. Cleavage $\{01\bar{1}2\}$, good; crystals flexible. $H = 2$. $G = 4.80$. $R = 24.1$–34.1 (540 nm); strongly anisotropic. Formed by the sublimation of fumarolic vapors and by the oxidation of Se-bearing material in sandstone U–V deposits. Glen Lyon, Luzerne Co., PA, on old burning coal-waste heaps, as a surface fumarolic product (!). Occurs in many U–V mines in the Colorado Plateau; United Verde mines, Jerome, AZ, in a fire zone; Kladno, Czech Republic, on burning heaps of pyritic sediments; Pacajake mine, Colquechaca, Chile, with selenide minerals. BM *DI:136, R:382, P&J339, ABI:468*.

1.3.4.2 Tellurium Te

Recognized as a mineral in 1802. Named from Latin *tellus*, earth. HEX-R $P3_121$. $a = 4.457$, $c = 5.929$, $Z = 3$, $D = 6.23$. *4-554*(syn): 3.86_2 3.23_{10} 2.35_4 2.23_3 2.09_1 1.84_2 1.62_1 1.48_1. Prismatic crystals up to 3 cm; usually massive, columnar to fine granular. Tin-white, gray streak, metallic luster. Cleavage $\{10\bar{1}0\}$, perfect; $\{0001\}$, imperfect; brittle. $H = 2$–$2\frac{1}{2}$, $VHN_{10} = 48$–66. $G = 6.1$–6.3. $R = 57.1$–68.7 (540 nm); strongly anisotropic. A hydrothermal vein mineral, associated with gold and silver tellurides. Cripple Creek, Teller Co., Magnolia, Goldhill, Ballarat, and Central districts, Boulder Co., and Vulcan(*), Gunnison Co., CO; Wilcox district, NM; Delamar(*), Lincoln Co., NV; Frood mine, Sudbury, ON, Canada; Moctezuma(*), SON, Mexico; *Facebaj, Transylvania, Romania*; Balia, Turkey; Kochbulak(*), Uzbekistan; Teine mine, near Sapporo, and Kawazu mine, near Simoda, Japan; Shimian mine, Sichuan Prov., China; Empress mine, Vatukoula, Fiji; Kalgoorlie, WA, Australia. BM *DI:138, R:383, P&J:364, ABI:516*.

1.3.5.1 Sulfur α-S

Known since antiquity. Dimorphous with rosickyite. ORTH *Fddd*. $a = 10.45$, $b = 12.84$, $c = 24.46$, $Z = 128$, $D = 2.08$. *8-247*(syn): 3.85_{10} 3.44_4 3.33_3 3.21_6 3.11_3 3.08_2 2.84_2 2.62_1. **Structure:** The atoms are bonded into puckered rings of eight atoms that can be regarded as S_8 molecules; each atom is bonded to two neighbors in its ring by homopolar bonds; the distance between S atoms in adjacent rings is much greater than the distance within the ring. **Physical properties:** Dipyramidal crystals, also thick tabular and disphenoidal; at least 56 crystal forms are known; also massive, reniform, and stalactitic.

1.3.5.2 Rosickyite

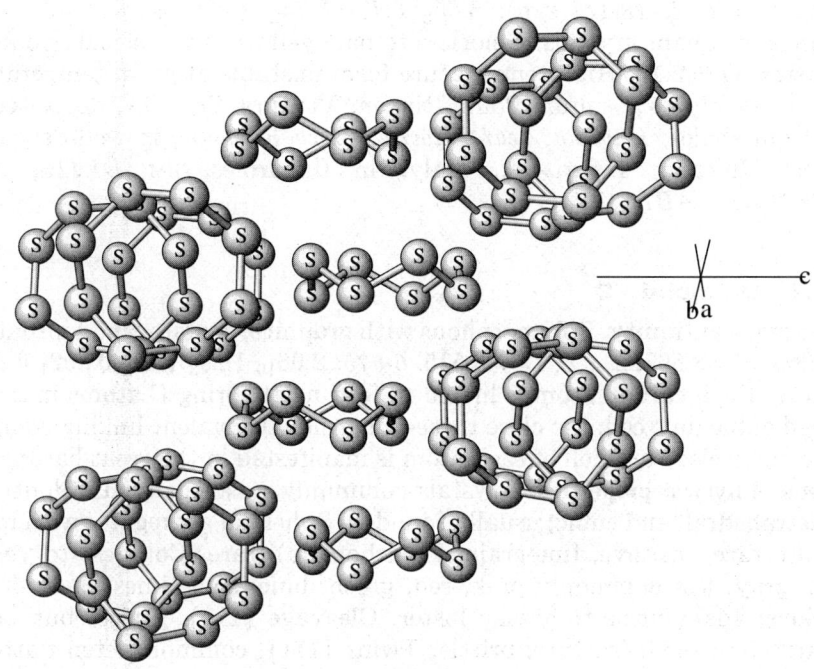

SULFUR

Each S atom is bonded to two others to form puckered eight-membered rings.

Pale yellow to yellow-brown, white streak, resinous to greasy luster. Cleavage {001}, {110}, {111}, imperfect; parting {111}; conchoidal to uneven fracture; brittle. H = $1\frac{1}{2}$–$2\frac{1}{2}$. G = 2.07. **Tests:** MP = 113°; burns in air, giving choking fumes of SO_2. **Chemistry:** Usually pure; Se and Te may be present in trace amounts. **Optics:** Biaxial (+); $N_x = 1.958$, $N_y = 2.038$, $N_z = 2.245$; XY = ab; $2V = 69°$; $r < v$. **Ocurrence:** Widely distributed in regions of recent volcanic activity, as a sublimate in fumaroles and by the oxidation of H_2S. Most commercial deposits are in salt domes, where it has been formed by the bacterial decomposition of calcium sulfate. **Localities:** LA and TX, in salt domes; Yellowstone Park, WY, in fumaroles; Sulphur Bank, Lake Co., Leviathan mine, Alpine Co., Mt. Shasta, Siskiyou Co., and other localities in CA; Mauna Loa and Kilauea, HI; San Felipe(*), BC, Mexico; Conil(*), near Cadiz, Spain; Agrigento(*), Sicily; Shor Su(*), Fergana, Uzbekistan; numerous localities in Japanese volcanic regions; White Is., New Zealand. **Uses:** The principal source of sulfur compounds, widely used in industry and agriculture. BM *DI:140, ABI:506.*

1.3.5.2 Rosickyite γ-S

Named in 1931 for Vojtech Rosicky (1880–1942), Czech mineralogist. Dimorphous with α-sulfur. MON $P2/n$. $a = 8.50$, $b = 13.16$, $c = 9.29$, $β = 124.86°$,

$Z = 32$, $D = 2.02$. *13-141*(syn): 6.65_3 3.79_1 3.74_2 3.29_{10} 3.04_1 2.49_1 1.45_2. Microscopic equant crystals. Colorless to pale yellow, white streak, adamantine luster. $G = 2.19$. High-temperature form unstable at room temperature, inverting slowly to α-sulfur. Point Rincon, Ventura Co., CA, deposited by gases from shales; *Havirna, near Letovice, Czech Republic*, in well-developed crystals; Vulcano, Lipari Is., Italy, in fumaroles. BM *DI:145, ACB 30:1396(1974), ABI:446*.

1.3.6.1 Diamond C

Known since antiquity. Polymorphous with graphite, chaoite, and lonsdaleite. ISO $Fd3m$. $a = 3.567$, $Z = 8$, $D = 3.515$. *6-675*: 2.06_{10} 1.26_3 1.08_2 0.892_1 0.818_2. **Structure:** Each carbon atom is linked to four neighboring C atoms in a face-centered cubic unit cell; the close three-dimensional covalent linking completing the outer electron shell of each atom is manifested in the great hardness of diamond. **Physical properties:** Crystals commonly octahedral, also dodecahedral, tetrahedral, and cubic; usually rounded. Spherical aggregates with radial structure rare; massive, fine-grained (carbonado), rare. Colorless to yellow, brown, gray; less commonly pink, red, green, blue; sometimes black due to inclusions; adamantine to greasy luster. Cleavage {111}, perfect but rather difficult; conchoidal fracture; brittle. Twins {111}, common, often flattened parallel to the twin plane; tetrahedral crystals twinned on {001}. $H = 10$, greater than any other substance; slightly variable with orientation. $G = 3.50$–3.53. Some specimens fluorescent and phosphorescent (usually blue) under UV and X-rays. Triboelectric; some crystals triboluminescent. **Tests:** Easily recognized by its hardness and luster. **Chemistry:** Pure C in

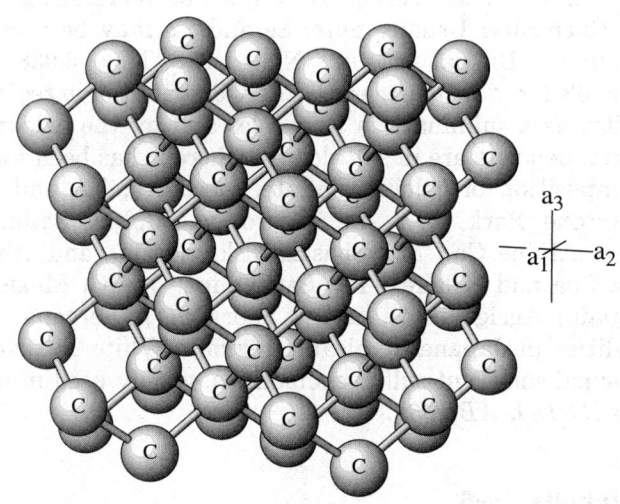

DIAMOND
Space group Fd3m. Each atom is coordinated by four others. Also adopted by silicon.

colorless material, but dark varieties yield up to 20% ash, consisting mainly of SiO_2, Fe_2O_3, MgO, Al_2O_3, CaO, TiO_2. **Optics:** Isotropic; N = 2.418; dispersion strong. **Occurrence:** Diamond is the product of deep-seated crystallization of ultrabasic igneous magmas, intruded as dikes or pipes of kimberlite or lamproite. Erosion of these primary deposits has concentrated diamonds in placers of various geologic ages; these include stream deposits, wave-concentrated deposits along present and ancient shorelines, and old conglomerates. World production of diamond in 1990 was approximately 100 million carats (equivalent to 20 metric tons). Most of this was from primary deposits, as follows (in parentheses, production in 10^3 carats). Argyle, WA, Australia (34,662); Botswana (17,352); Yakutia, Russia (15,000); South Africa (8,708); Tanzania (85). Significant producers from secondary deposits were Zaire (19,427); Brazil (1,500); Angola (1,300); Namibia (761); Ghana (637); Central African Republic (381); Venezuela (333); Guinea (135); Liberia (100); Sierra Leone (78). **Localities:** In addition to these producing localities, diamonds have been found in many places throughout the world. In the United States, alluvial diamonds have been recorded from the Appalachians in VA, WV, NC, TN, GA, AL, and in the Great Lakes region from WI, IL, IN; many diamonds have been found in placers in CA. One primary deposit is the Prairie Creek lamproite pipe near Murfreesboro, Pike Co., AR, where many thousands of stones have been collected, mostly less than 1 carat, but one weighed 40.2 carats. Small diamonds have been found in kimberlite pipes of the State Line Group, CO–WY. Numerous kimberlites, some diamondiferous, have been found in SK, AB, and NWT, Canada. India was the only source of diamonds until 1729, when the Brazilian deposits were discovered; three principal localities are known, the Golconda region in Mysore and Hyderabad, Sambalpur, and Wairaarh, near Nagpur, and the Panna district, Central Provinces. The diamonds are in secondary deposits, and current production is small, 15,000 carats in 1990. There is some production of alluvial diamonds in Borneo, 30,000 carats in 1990. Diamond-bearing deposits have been found in northern Hunan Prov., Liaoning Prov., and Shandong Prov., China; 1990 production is estimated at 1 million carats. Microscopic diamonds have been recorded from a few meteorites. **Uses:** In 1954 the General Electric Co. succeeded in synthesizing diamond; since then this material has taken a large share of the market for abrasives and other nongem uses. BM *DI:146*, *ABI:131*, *G&G 28:234(1992)*.

1.3.6.2 Graphite C

Known since antiquity. Polymorphous with diamond, chaoite, and lonsdaleite. HEX $P6_3/mmc$. $a = 2.463$, $c = 6.714$, $Z = 4$, $D = 2.26$. *23-64*: 3.36_{10} 2.03_5 1.68_8 1.23_3 1.16_5 1.12_2 0.994_4 0.829. **Structure:** Planes of atoms linked in hexagons, the planes being stacked so that half the atoms in the B layer are over atoms in the A layer, and the others are over the centers of the hexagons; the C layer is then exactly over the A layer. A rhombohedral polytype has been described in which the C layer is shifted in the same direction as

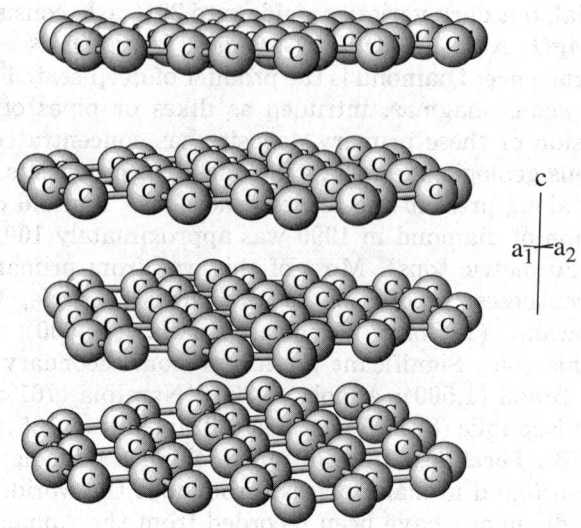

GRAPHITE
Each carbon is bonded to only three others, forming a sheet with hexagonal rings. The sheets are stacked to form rhombohedral symmetry, but they are only held together with van der Waals forces.

B, with the D layer over A. **Physical properties:** Hexagonal crystals, tabular on {0001}; also scaly, columnar, radiating, granular, compact, or earthy. Black, black streak, metallic or earthy luster. Cleavage {0001}, perfect and very easy; glides readily on {0001}; flexible; inelastic. H = 1, VHN_{10} = 7–11. G = 2.1–2.2. Conducts electricity. **Tests:** Insoluble in acids. Distinguished by color, softness (soapy feel), and black streak. **Chemistry:** Pure carbon. **Optics:** R = 7.4–20.5 (540 nm); strongly pleochroic and anisotropic. **Occurrence:** Formed by metamorphism of sedimentary carbonaceous material, by the oxidation of hydrocarbons, and by the reduction of CO and CO_2. Found as disseminated scales in igneous and metamorphic rocks, and massive in metamorphosed coal beds. Major commercial deposits are mined in China, Korea, Siberia, India, Mexico, Brazil, and Czech Republic; minor deposits in Madagascar, Turkey, Romania, Germany, Austria, and Sri Lanka. Occurs as nodules in iron meteorites. **Localities:** Some localities for good specimens are Ticonderoga, Essex Co., NY; Franklin and Sterling Hill, Sussex Co., NJ; Crestmore, Riverside Co., CA; Buckingham, PQ, and National mine, Hastings Co., ON, Canada; Morodillas mine, SON, Mexico; Egalugssuit and Disko Is., Greenland; Barrowdale, Cumbria, England; Kongsberg, Norway; St. Michel and Pargas, Finland; Passau, Bavaria, Germany; Valle Risagliardo, Piedmont, Italy; Nizhni Tunguski, Siberia, and Botogol, East Sayan, Russia; Arumanaloir, Travancore, India; Sabaragamuwa, Sri Lanka.
BM *DI:152, R:384, P&J:184, ABI:192.*

1.3.6.3 Lonsdaleite C

Named in 1967 for Kathleen Lonsdale (1903–1971), English crystallographer. Polymorphous with diamond, graphite, and chaoite; lonsdaleite has a hexagonal close-packed structure, in contrast to the cubic close-packed structure of diamond. HEX $P6_3/mmc$. $a = 2.52$, $c = 4.12$, $Z = 4$, $D = 3.52$. *19-268*(syn): 2.19_{10} 2.06_{10} 1.92_5 1.50_3 1.26_8 1.17_5 1.08_5 1.06_3. Microscopic cubes and cubo-octahedra, also polycrystalline aggregates, mixed with diamond. Pale yellow to gray, adamantine luster. $H = 3$. $G > 3.20$. $N \sim 2.41$. Associated with diamond in the *Canyon Diablo*, Goalpara, and Allan Hills 77283 meteorites; in diamond-bearing placers, Yakutia, and in soil at the Tunguska impact site, Russia. BM *NAT 214:587(1967)*, *AM 52:1579(1967)*, *MZ 7:27(1985)*, *ABI:301*.

1.3.6.4 Chaoite C

Named in 1969 for Edward C. T. Chao (b.1919), Chinese–U.S. petrologist. Polymorphous with diamond, graphite, and lonsdaleite. HEX $P6/mmm$. $a = 8.948$, $c = 14.078$, $Z = 168$, $D = 3.43$. *22-1069*: 4.47_{10} 4.26_{10} 4.12_8 3.71_4 3.22_4 3.03_6 2.55_6 2.28_6. Thin microscopic lamellae, alternating with graphite. Black. $H = 1$–2. Occurs in shock-metamorphosed graphite gneisses, and in ureilite meteorites. *Ries Crater, Bavaria, Germany*; Goalpur and Dyalpur meteorites. BM *SCI 161:363(1968)*, *AM 54:326(1969)*, *55:1067(1970)*, *ABI:93*.

1.3.7.1 Silicon Si

Identified as a mineral in 1989. ISO $Fd3m$. $a = 5.431$, $Z = 8$, $D = 2.33$. *27-1402*(syn): 3.14_{10} 1.92_6 1.64_3 1.36_1 1.25_1 1.11_1 1.05_1 0.918_1. Microscopic dodecahedral and cubo-octahedral crystals. Nuevo Potosi, Cuba, enclosed in gold grains from an ophiolite complex; Kola Penin., in gneisses from the super-deep drill hole, and Yakutia, in kimberlites, and Tolbachik, Kamchatka, in volcanic exhalations, Russia. BM *DANS 309:1182(1989)*, *AM 76:668(1991)*.

1.3.8.1 Moissanite α-SiC

Named in 1905 for Henri Moissan (1852–1907), French chemist. Carborundum is the synthetic product, widely used as an abrasive. HEX $P6_3mc$. $a = 3.081$, $c = 5.031$, $Z = 2$, $D = 3.22$ (2H polytype); 5H, 6H, 15R, 33R polytypes are known; β-SiC is isometric, $a = 4.359$. *29-1126*(syn): 2.67_8 2.52_5 2.36_{10} 1.83_2 1.54_4 1.42_3 1.31_2 1.29_1. Small hexagonal platy crystals. Colorless to blue or green, metallic luster. Cleavage {0001}, poor. $H = 9\frac{1}{2}$. $G = 3.22$. Uniaxial (+); $N_O = 2.654$, $N_E = 2.967$; weakly pleochroic in blue. Originally described from the Canyon Diablo meteorite, but probably a contaminant. Bridger, WY (β-SiC); Yakutia, Russia, in kimberlite; many localities in alluvium in the former USSR, but contamination with carborundum is a possibility; Fuxian, Liaoning, China, in kimberlite. Recorded as a rare accessory in lunar samples. BM *DI:123*, *NAT 346:352(1990)*, *MZ 12(3):17(1990)*, *AM 77:208(1992)*.

1.3.9.1 Sinoite Si$_2$N$_2$O

Named in 1964 for the composition. ORTH $Cmc2_1$. $a = 8.843$, $b = 5.473$, $c = 4.835$, $Z = 4$, $D = 2.84$. $18\text{-}1171$(syn): 4.68_8 4.44_{10} 3.36_{10} 2.60_5 2.42_8 2.38_8 1.39_8 1.37_8. Microscopic plates. Colorless, white streak, vitreous luster. $G = 2.80\text{-}2.85$. Biaxial $(-)$; $N_x = 1.740$; $N_y, N_z = 1.855$; $Z = c$; 2V small. Occurs as an accessory in some enstatite chondrites. BM *SCI 146:256(1964)*, *AM 50:521(1965)*.

1.3.10.1 Nierite Si$_3$N$_4$

Named in 1995 for Alfred O. A. Nier (1912–1994), professor of physics, University of Minnesota, a pioneer in mass spectrography. HEX-R $P31c$. $a = 7.758$, $c = 5.623$, $Z = 4$, $D = 3.184$. A hexagonal polymorph was also observed. *9-250*(syn): 4.32_5 3.88_3 3.37_3 2.89_9 2.60_8 2.55_{10} 2.32_6 2.08_5. Micron-size laths elongated in the c direction. Colorless. $H = 9$. $G_{syn} = 3.17\text{-}3.18$. Uniaxial $(-)$; $N_O = 2.03$, $N_E = 2.02$. Occurs in trace amounts in a few chondritic meteorites. BM *Meteoritics 30:387(1995)*.

Class 2

Sulfides, Including Selenides and Tellurides

2.1.1.1 Algodonite Cu₆As
Named in 1857 for the locality. HEX $P6_3/mmc$. $a = 2.600$, $c = 4.228$, $Z = 2$, $D = 8.75$. *9-429*: 2.25_2 2.11_4 1.30_2 1.19_2 1.11_2 1.09_1 1.06_1 0.837_2. Massive granular, or incrustations of minute crystals. Silver-white to steel-gray, tarnishing to dull bronze, metallic luster. No cleavage, conchoidal fracture. $H = 4$, $VHN_{100} = 245$–302. $G = 8.38$. $R = 62.8$ (540 nm); weakly anisotropic. Occurs in hydrothermal deposits, intimately associated with other copper arsenides. Numerous mines in Houghton, Keweenaw, and Baraga Cos., MI; Långban, Värmland, Sweden; Mechernich and Bruckenberg Shaft, near Zwickau, Germany; Tsumeb, Namibia; Corocoro, Bolivia; *Algodones mine, Coquimbo*, and Cerro de los Sequas, Rancagua, *Chile*. BM *DI:171, R:393, P&J:63, ABI:8*.

2.1.2.1 Horsfordite Cu₅Sb
Named in 1888 for Eben N. Horsford (1818–1893), U.S. chemist. Massive. Silver-white, easily tarnished, metallic luster. Uneven fracture, brittle. $H = 4$–5. $G = 8.81$. *Mytilene, Lesbos Is., Greece*. BM *DI:173, R:399, ABI:224*.

2.1.3.1 Telargpalite (Pd,Ag)₃Te
Named in 1974 for the composition. ISO Space group unknown. $a = 12.60$, $Z = 16$, $D = 7.378$. *26-1453*: 3.05_4 2.74_3 2.42_{10} 2.22_2 2.10_5 1.82_2 1.48_3 1.34_2. Microscopic grains. Light gray, metallic luster. $H = 2$–$2\frac{1}{2}$, $VHN_{10} = 62$. $R = 50.1$ (540 nm). Stillwater Complex, Nye Co., MT; *Oktyabr mine, Talnakh area, Norilsk*, and Lukkulaisvaara massif, Karelia, *Russia*. BM *ZVMO 103:595(1974), AM 60:489(1975), 66:1103(1981), ABI:514*.

2.1.4.1 Duranusite As₄S
Named in 1973 for the locality. ORTH Space group unknown. $a = 3.576$, $b = 6.759$, $c = 10.074$, $Z = 2$, $D = 4.53$. *25-1479*: 5.62_9 5.04_9 3.02_7 2.92_{10} 2.81_7 2.68_8 1.97_9 1.79_9. Microscopic grains and wiry aggregates. Red, metallic luster. $H = 2$, $VHN_{25} = 58$. $R = 30.4$–31.5 (540 nm). Mt. Washington Cu mine, Vancouver Is., BC, Canada; *Duranus, Alpes-Maritimes, France*. BM *BM 96:131(1973), AM 60:945(1975), ABI:141*.

2.1.5.1 Bezsmertnovite Au$_4$Cu(Te,Pb)

Named in 1979 for Marianna S. Bezsmertnaya (1915–1991) and Vladimir V. Bezsmertny (b.1912), Russian geologists. ORTH $Pmmn$. $a = 24.215$, $b = 4.025$, $c = 16.245$, $Z = 16$, $D = 15.17$. 41-579: 3.30_7 2.61_8 2.33_{10} 1.74_9 1.43_3 1.41_4 1.14_4 1.07_4. Microscopic grains. Golden yellow, metallic luster. $H = 4\frac{1}{2}$, $VHN_{20} = 353$. $R = 37.5$ (540 nm); weakly anisotropic. An undisclosed locality in *Kamchatka, Russia*. BM *DANS 249:185(1979), AM 66,878(1981), ABI:52.*

2.1.6.1 Bilibinskite Au$_3$Cu$_2$PbTe$_2$

Named in 1978 for Yuri A. Bilibin (1901–1952), Russian geologist. ISO Space group unknown. $a = 4.095$, $Z = 0.5$, $D = 14.27$. 29-544: 3.29_1 3.06_4 3.00_1 2.37_{10} 2.05_7 1.45_6 1.23_8 1.18_2. Massive. Rose-brown, brown streak, submetallic luster. No cleavage. $H = 4\frac{1}{2}$, $VHN_{20} = 329$–419. $G = 12.7$. $R = 21.5$–23.2 (540 nm). *Far Eastern region, Russia*, and Kazakhstan, in weathering zone of gold–telluride deposits. BM *ZVMO 107:310(1978), AM 64:652(1979), ABI:53.*

2.2.1.1 Dyscrasite Ag$_3$Sb

Named in 1832 from the Greek for a *bad alloy*. ORTH $Pm2m$. $a = 2.996$, $b = 5.235$, $c = 4.830$, $Z = 1$, $D = 9.76$. 10-452: 2.61_3 2.42_4 2.29_{10} 1.77_3 1.51_3 1.37_4 1.28_3 1.26_2. Usually massive, foliated, or granular; pseudohexagonal crystals rare, usually poor. Silver-white, tarnishing to lead-gray, metallic luster. Cleavage {001}, {011}, distinct; {110}, imperfect; multiple twinning common on {110}, forming hexagonal aggregates. $H = 3\frac{1}{2}$, $VHN_{100} = 153$–179. $G = 9.71$. $R = 60.1$–62.7; weakly pleochroic and anisotropic. Decomposed by HNO_3 with a white precipitate of Sb_2O_3. Occurs in a few silver deposits, and then, often abundant. Poughkeepsie Gulch, Silverton, CO, and Reese River mining district, NV; Ste.-Marie-aux-Mines, Alsace, France; Hiendelaencina, Spain; Andreasburg(*), Harz, and *Wolfach, Baden, Germany*; Pribram(*), Czech Republic; Kongsberg and Sulitjelma, Norway; Consols mine(*), Broken Hill, NSW, Australia, in large masses; Chanarcillo, Chile; Colquechaca, Bolivia. BM *DI:173, R:407, P&J:153, ABI:142.*

2.2.1.2 Allargentum Ag$_{1-x}$Sb$_x$ ($x = 0.09$–0.16)

Named in 1949 from the Greek for *another* and the Latin *argentum*, silver. HEX $P6_3/mmc$. $a = 2.945$, $c = 4.780$, $Z = 2$, $D = 10.12$. 25-54(syn): 2.55_7 2.40_7 2.25_{10} 1.74_4 1.47_4 1.35_6 1.25_5 1.23_4. Microscopic grains and lamellae, intergrown with silver. Silver-white, metallic luster. $H = 3\frac{1}{2}$, $VHN = 172$–203. $G_{syn} = 10.0$. $R = 69.5$–70.4 (540 nm); weakly anisotropic. Occurs in high-grade Ag–Sb ores. Many mines in the *Cobalt area, ON, Canada*; Wasserfall, Vosges, France; Wenzel mine, Oberwolfach, Baden, and Hartenstein, Saxony, Germany; Hällefors, Bergslagen, Sweden; Consols mine, Broken Hill, NSW, Australia. BM *FM 28:69(1950), CM 10:163(1970), BM 101:563(1978), ABI:9.*

2.2.2.1 Domeykite Cu_3As

Named in 1845 for Ignacio Domeyko (1802–1889), Chilean mineralogist. ISO $I\bar{4}3d$. $a = 9.62$, $Z = 16$, $D = 7.93$. *9-333*: 3.95_4 3.05_4 2.15_4 2.05_{10} 1.97_5 1.89_7 1.31_5 1.22_4. Massive, reniform, botryoidal. Tin-white to steel-gray, tarnishing to orange-brown, metallic luster. No cleavage, uneven fracture. $H = 3-3\frac{1}{2}$, $VHN_{100} = 213-235$. $G = 7.2-7.9$. $R = 51.9$ (540 nm). Inverts above 90° to β-domeykite (HEX, $a = 7.02$, $c = 7.25$). Soluble in HNO_3. Commonly intergrown with algodonite or β-domeykite. Franklin, NJ; Portage Lake, Houghton Co., and Mohawk and other mines in Keweenaw Co., MI; Cashin mine, Paradox Valley, CO; Michipicoten Is., Lake Superior, ON, Canada; near Helstone, Cornwall, England; Zwickau, Saxony, Germany; near Nachod, Bohemia, Czech Republic; Talmessi, Iran; Corocoro, Bolivia; Calabazo, Coquimbó, many other localities in Chile. BM *DI:172*, *EG 66:135(1971)*, *R:393*, *P&J:149*, *ABI:138*.

2.2.2.2 Kutinaite Cu_2AgAs

Named in 1970 for Jan Kutina (b.1924), Czech mineralogist. ISO Space group unknown. $a = 11.76$, $Z = 28$, $D = 8.86$. *23-957*(syn): 2.70_9 2.41_8 2.27_{10} 2.08_{10} 1.99_6 1.96_7 1.78_7 1.39_6. Microscopic grains. Silver-white, metallic luster. $H = 4\frac{1}{2}$, $VHN_{25} = 237-270$. $G = 8.38$. $R = 42.6$ (540 nm). Associated with other arsenides from *Černý Důl, Czech Republic*; Wasserfall, Vosges, France; Niederbeerbach, Odenwald, Germany; Meskani mine, Anarak district, Iran. BM *AM 55:1083(1970)*, *R:398*, *P&J:230*, *TMPM 34:167(1985)*.

2.2.2.3 Dienerite Ni_3As

Named in 1921 for Karl Diener (1862–1928), Austrian paleontologist, who discovered the mineral. ISO A single cubic crystal, about 5 mm. White, with a gray tinge, metallic luster. *Near Radstadt, Salzburg, Austria*. BM *DI:175*, *ABI:133*.

2.2.3.1 Bogdanovite $(Au,Te,Pb)_3(Cu,Fe)$

Named in 1979 for Aleksei A. Bogdanov (1907–1971), Russian geologist. ISO (pseudocubic; optics show probable ORTH symmetry) $Pm3m$. $a = 4.088$, $Z = 1$, $D = 13.72$. *34-1302*: 2.36_{10} 2.15_1 2.05_6 1.45_6 1.23_8 1.18_3 1.09_3 0.992_2. Massive. Rose-brown to bronze brown, tarnishing to blue-black, semimetallic luster. No cleavage. $H = 4\frac{1}{2}$, $VHN_{20} = 235-270$. $R = 5.6-29.5$ (540 nm); anisotropic, yellow to gold to purple. Cu and Fe vary reciprocally from $Cu_{3.4}Fe_{0.3}$ to $Cu_{0.7}Fe_{2.7}$. Bisbee, AZ; *Dalnii Vostok, Kazakhstan*, and Ago Kamchatka, Russia, in oxidation zone of Au–Te deposits. BM *AM 64:1329(1979)*, *ABI:58*, *CM 28:751(1990)*.

2.2.4.1 Atheneite $(Pd,Hg)_3As$

Named in 1974 for the Greek goddess Pallas Athene, in reference to the Pd content. HEX $P6/mmm$. $a = 6.798$, $c = 3.483$, $Z = 2$, $D = 10.16$. *26-889*: 2.42_{10}

2.25_9 2.22_6 1.87_7 1.37_8 1.30_8 1.26_8 1.03_7. Irregular alluvial grains. Cream-yellow, metallic luster. H = 5, VHN_{100} = 419–442. G = 10.2. R = 54.0–56.7 (540 nm); weakly pleochroic, distinctly anisotropic. Near Zlatoust, Ural Mts., Russia; Merensky Reef, Transvaal, South Africa; *Itabira, MG, Brazil*, in gold washings. BM *MM 39:528(1974), AM 59:1330(1974), ABI:34.*

2.2.5.1 Vincentite $(Pd,Pt)_3(As,Sb,Te)$

Named in 1974 for Ewart A. Vincent (b.1919), professor of mineralogy, Oxford University. TET $I\bar{4}$. a = 9.974, c = 4.822, Z = 8, D = 11.27. *26-1452*: 4.18_{10} 3.95_{10} 3.24_{10} 2.75_{10} 2.00_{10} 1.75_{10} 0.944_{10}. Microscopic grains. Brown-gray, metallic luster. H = 6, VHN_{15} = 494. R = 51.8–52.8 (546 nm); weakly anisotropic. Talnakh, Norilsk, Russia; *Riam Kanan R., Borneo*, in Pt–Au alluvial concentrates. BM *MM 39:525(1974), AM 59:1332(1974).*

2.2.6.1 Keithconnite $Pd_{3-x}Te$ $(x = 0.14$–$0.43)$

Named in 1979 for Keith Conn (b.1923), U.S. geologist who investigated the Stillwater Complex. HEX-R $R\bar{3}$. a = 11.45, c = 11.40, Z = 7, D = 10.16. *34-1461*: 2.26_{10} 2.22_2 2.16_9 1.40_2 1.37_2 1.32_3 1.27_2 0.791_4. Microscopic grains. Cream to gray, metallic luster. H = $4\frac{1}{2}$–5, VHN_{15} = 394–424. R = 44.4–44.9 (540 nm); moderately to strongly anisotropic. *Stillwater Complex, MT*, in heavy-mineral concentrates; Shinkolobwe, Shaba, Zaire. BM *CM 17:589 (1979), AM 66:1275(1981), ABI:259.*

2.3.1.1 Koutekite Cu_5As_2

Named in 1958 for Jaromir Koutek (1902–1983), Czech mineralogist. HEX $P6_3/mmc$. a = 11.51, c = 14.54, Z = 18, D = 8.38. *29-534*(syn): 2.45_8 2.16_5 2.09_{10} 2.02_{10} 2.00_{10} 1.79_4 1.75_4 1.15_4. Microscopic grains. Blue-gray, metallic luster. H = $3\frac{1}{2}$, VHN_{25} = 136–157. G = 8.48. R = 39.7–43.1 (540 nm); strongly anisotropic. Occurs in arsenical copper deposits. Mohawk, Keweenaw Co., MI; Daluis, Vallee du Var, and Wasserfall, Vosges, France; *Černý Důl, Czech Republic*; Talmessi and Meskani deposits, Anarak district, Iran. BM *NAT 181:1553(1958), R:399; P&J:227, ABI:274, AM 78:677(1993).*

2.3.2.1 Orcelite $Ni_{5-x}As_2$ $(x = 0.23)$

Named in 1959 for Jean Orcel (1896–1978), French mineralogist. HEX $P6_3cm$. a = 6.70, c = 12.39, Z = 6, D = 8.11. *35-493*: 3.25_2 3.09_2 2.95_4 2.14_6 1.99_{10} 1.93_{10} 1.76_6 1.64_2. Microscopic grains. Yellow-white, metallic luster. G = 7.25. R = 49.1 (540 nm); weakly pleochroic, noticeably anisotropic. Occurs as inclusions in pentlandite in ultramafic rocks. Nipissing mine, Cobalt, ON, and Table Mt., NF, Canada; Ronda massif, Malaga, Spain; Vourinos, Greece; Zolotaya Gora, Ural Mts., Russia; Beni Bousera, Morocco; *Tiebaghi massif, New Caledonia.* BM *AM 45:753(1960), BM 103:198(1980), CM 22:553(1984), ABI:363.*

2.3.3.1 Stibiopalladinite Pd$_5$Sb$_2$

Named in 1929 for the composition. HEX $P6_3/cm$. $a = 7.606$, $c = 14.21$, $Z = 6$, $D = 10.85$. *25-597*: 3.35_5 2.84_5 2.48_5 2.27_{10} 2.19_{10} 2.02_5 1.58_6 1.28_5. Massive and as minute grains. Silver-white to steel-gray, metallic luster. No cleavage, uneven fracture. H = 6, VHN$_{100}$ = 610. G = 9.5. R = 52.5–54.2 (540 nm); weakly pleochroic, distinctly anisotropic. Occurs in platinum deposits. Goodnews Bay, AK; Lac des Iles Complex, ON, Canada; Morozova mine, Norilsk, and near Zlatoust, Ural Mts., Russia; Pirogues R., New Caledonia; *Farm Tweefontein, near Potgieterstust, Transvaal, South Africa.* BM *DI:175, AM 61:1249(1976), R:414, P&J:354, ABI:497.*

2.3.3.2 Palarstanide Pd$_5$(Sn,As)$_2$

Named in 1981 for the composition. HEX $P6_322$. $a = 6.784$, $c = 14.80$, $Z = 6$, $D = 10.27$. *17-223*(syn): 2.36_4 2.33_4 2.26_2 2.20_{10} 2.11_8 1.97_3 1.95_3 1.52_3. Hexagonal prisms and rounded grains. Steel-gray, metallic luster. Cleavage {0001}, perfect. H = 5, VHN$_{50}$ = 470. R = 53.2–54.5 (540 nm); weakly anisotropic in gray. *Talnakh, Norilsk, Russia.* BM *ZVMO 110:487(1981); AM 67:858(1982), 74:1219(1989); ABI:375.*

2.3.4.1 Parkerite Ni$_3$Bi$_2$S$_2$

Named in 1937 for Robert L. Parker (1893–1973), professor of mineralogy, ETH, Zurich, Switzerland. MON $C2/m$. $a = 11.066$, $b = 8.085$, $c = 7.965$, $\beta = 134.0°$, $Z = 8$, $D = 8.53$. *26-1283*(syn): 5.73_2 4.04_3 3.98_3 2.86_7 2.84_{10} 2.34_6 2.28_2 1.99_3. Microscopic grains and irregular masses. Bronze, black streak, metallic luster. Cleavage {001}, perfect; lamellar twinning on {111}. H = 3, VHN$_{15}$ = 125. G = 8.4. R = 45.8–50.5 (540 nm); strongly anisotropic, green-gray to yellow brown. Gaspé Cu mine, PQ, and Sudbury and Cobalt-Gowganda, ON, Canada; Zinkwand mine, Styria, Austria; Karik'yavr and the Allarechensk region, Kola Penin., and Norilsk area, Ural Mts., Russia; Rakha deposit, Bihar, India; *Waterfall Gorge, Insizwa, South Africa.* BM *AM 28:343(1943), 59:296(1974); R:404; P&J:289; ABI:389.*

2.3.5.1 Shandite Pb$_2$Ni$_3$S$_2$

Named in 1950 for Samuel J. Shand (1882–1957), professor of petrology, Columbia University. HEX-R $R\bar{3}m$. $a = 5.576$, $b = 13.658$, $Z = 3$, $D = 8.87$. *26-494*: 4.55_2 3.94_7 2.79_{10} 2.28_7 1.97_4 1.76_2 1.61_2 1.39_2. Microscopic grains. Tin-white, metallic luster. Rhombohedral cleavage. H = 4. G = 8.72. Jeffrey mine, Asbestos, PQ, Canada; Isua belt, Godthåb region, Greenland, in serpentine; *Trial Harbour, TAS*, in serpentine, and Nullagine, WA, *Australia.* BM *AM 35:425(1950), 59:296(1974); R:405; ABI:473.*

2.3.5.2 Rhodplumsite Pb$_2$Rh$_3$S$_2$

Named in 1983 for the composition. HEX-R $R\bar{3}m$. $a = 5.73$, $c = 14.00$, $Z = 3$, $D = 9.85$. *35-720*: 4.64_2 4.02_3 2.86_{10} 2.44_3 2.33_6 2.01_6 1.81_3 1.22_2. Microscopic grains. Gray, metallic luster. Strongly anisotropic. *Umutnaya R., Ural Mts., Russia*, in Pt placer. BM *MZ 5:87(1983), MA 35:88(1984), ABI:441.*

2.3.6.1 Vozhminite $(Ni,Co)_4(As,Sb)S_2$

Named in 1982 for the locality. HEX Space group unknown. $a = 17.46$, $c = 7.20$, $Z = 18$, $D = 6.20$. *35-667*: 8.7_{10} 3.07_9 2.72_6 2.30_7 2.11_9 1.94_5 1.82_5 1.78_{10}. Probably massive. Yellow-brown, black streak, metallic luster. One distinct cleavage. $H = 4$–5, $VHN_{100} = 376$–480. $R = 47.5$–50.9 (540 nm). *Vozhmin serpentinite, Karelia, Russia.* BM *ZVMO 111:480(1982), AM 68:645(1983), ABI:562.*

2.4.1.1 Acanthite Ag_2S

Named in 1855 from the Greek for *thorn*, referring to the shape of the crystals. Much acanthite is paramorphic after the isometric phase *argentite* stable only above 177°. MON $P2_1/n$. $a = 4.229$, $b = 6.931$, $c = 7.862$, $\beta = 99.61°$, $Z = 4$, $D = 7.24$. *14-72*(syn): 3.08_6 2.84_7 2.61_{10} 2.58_7 2.46_7 2.44_8 2.42_6 2.38_8. Pseudocubic or pseudo-octahedral crystals up to 8 cm, also massive. Black, black streak, metallic luster. Indistinct cleavage, subconchoidal fracture, very sectile. $H = 2$–$2\frac{1}{2}$, $VHN_{50} = 21$–25. $G = 7.2$–7.4. $R = 31.2$ (540 nm); weakly anisotropic. Occurs worldwide in silver ores. The following localities have provided fine specimens: Butte, MT; Comstock Lode, Virginia City, and Tonopah, NV; Cobalt, ON, Highland Bell mine, BC, and El Bonanza mine, near Port Radium, NWT, Canada; Santa Eulalia and Batopilas, CHIH, Arizpe, SON, Zacatecas, ZAC; and Guanajuato, GTO, Mexico; Liskeard, Cornwall, England; Kongsberg, Norway; Andreasberg, Harz, and Freiberg, Saxony, Germany; Jachymov, Czech Republic; Sarrabus, Sardinia, Italy; Broken Hill, NSW, Australia; Chanarcillo, Atacama, Chile; Colquechaca, Bolivia. BM *DI:176, 191; CM 12:365(1974); R:471; P&J:72; ABI:1.*

2.4.1.2 Naumannite Ag_2Se

Named in 1845 for Karl F. Naumann (1797–1873), German mineralogist. ORTH $P2_12_12_1$. $a = 4.333$, $b = 7.062$, $c = 7.764$, $Z = 4$, $D = 7.87$. *24-1041*(syn): 2.74_2 2.67_{10} 2.58_9 2.43_3 2.25_3 2.12_3 2.08_4 2.01_3. Pseudocubic crystals, also massive and in thin plates. Black, metallic luster. Cleavage {001}, perfect; sectile and malleable. $H = 2\frac{1}{2}$, $VHN_{25} = 33$–84. $G = 7.4$–7.8. $R = 35.7$–37.3 (540 nm); distinctly anisotropic. Widely distributed in seleniferous silver ores. Silver City district, Owyhee Co., ID, L-D mine, Wenatchee, WA; Kidd Creek mine, near Timmins, ON, Canada; *Tilkerode* and Andreasberg, *Harz Mts., Germany*; Kongsberg, Norway; Sacaramb, Romania; Sanru mine, Hokkaido, Japan; Colquechaca, Bolivia; Cachueta, Mendoza, Argentina. BM *DI:179, CM12:365(1974), R:478, P&J:277, ABI:345.*

2.4.1.3 Aguilarite Ag_4SeS

Named in 1891 for Ponciano Aguilar (1853–1935), superintendent of the San Carlos mine, Guanajuato, where the mineral was first found. ORTH $P2_12_12_1$.

2.4.3.1 Uytenbogaardtite

$a = 4.33$, $b = 7.09$, $c = 7.76$, $Z = 2$, $D = 7.56$. *27-620*: 2.88_5 2.67_2 2.59_2 2.43_{10} 2.23_3 1.73_3 1.60_2 1.48_4. Skeletal pseudododecahedral crystals, also massive. Black, metallic luster. No cleavage, hackly fracture, sectile. $H = 2\frac{1}{2}$, $VHN_{25} = 23$. $G = 7.40-7.53$. $R = 31.9$ (540 nm); very weakly anisotropic. A relatively low-temperature mineral in silver deposits rich in selenium. Silver City, Owyhee Co., ID, Comstock Lode, Virginia City, NV, L-D mine, Wenatchee, WA; *San Carlos mine*(*) and other mines, *GTO, Mexico*; Sanru mine, Hokkaido, Japan; Maritoto mine, near Paeroa, New Zealand. BM *DI:178, CM 12:365(1974), P&J:60, ABI:2.*

2.4.2.1 Hessite Ag$_2$Te

Named in 1843 for Henri Hess (1802-1850), Swiss-Russian chemist of St. Petersburg. MON $P2_1/n$. $a = 8.13$, $b = 4.48$, $c = 8.09$, $\beta = 112.9°$, $Z = 4$, $D = 8.40$. *12-695*: 4.53_1 3.19_2 3.01_6 2.87_8 2.31_{10} 2.25_7 2.20_2 2.14_6. Pseudocubic crystals up to 1.7 cm, also massive. Steel-gray to lead gray, metallic luster. Cleavage {001}, indistinct; even fracture; somewhat sectile. $H = 2-3$, $VHN_{100} = 22-24$. $G = 7.9-8.4$. $R = 38.4-41.1$ (540 nm); distinctly anisotropic, dark orange to dark blue. Widely distributed in Te-rich Ag-Au ores. Mines in Boulder, Eagle, and San Juan Cos., CO; Bisbee and Tombstone, Cochise Co., AZ; Calaveras, Tuolumne, and Nevada Cos., CA; numerous localities in ON, PQ, and BC, Canada; San Sebastian, JAL, Mexico; Loch Duich, near Kyle of Lochalsh, Scotland; *Sacaramb*, Baita, and Botes, *Romania*; Savodinski mine, Altai Mts., Russia; Chugu mine, Ishikawa Pref., Japan; Kalgoorlie and Norseman, WA, Australia; Condorriaco mine, near Coquimbo, Chile. BM *DI:184, AM 34:342(1949), R:421, P&J:199, ABI:216.*

2.4.2.2 Cervelleite Ag$_4$TeS

Named in 1989 for Bernard Cervelle (b.1940), French mineralogist. ISO Space group unknown. $a = 14.03$, $Z = 24$, $D = 8.53$. *EM 1*: 6.29_8 5.74_2 5.00_{10} 4.64_4 4.29_9 3.77_6 2.60_2 2.45_2. Microscopic grains. Black, metallic luster. $VHN_{10} = 26$. $R = 34.7-38.3$ (540 nm). *La Bambolla (Mina Moctezuma) mine, Moctezuma, SON, Mexico*, rimming acanthite; Chadak, Uzbekistan. BM *EM 1:375(1989), AM 75:1431(1990).*

2.4.3.1 Uytenbogaardtite Ag$_3$AuS$_2$

Named in 1978 for Willem Uytenbogaardt (b.1918), Dutch mineralogist. TET $P4_122$. $a = 9.75$, $b = 9.85$, $Z = 8$, $D = 8.30$. *33-587*(syn): 6.98_8 4.38_6 3.09_6 2.81_6 2.73_{10} 2.61_9 2.12_8 1.99_5. Microscopic grains. Gray-white, metallic luster. $H = 2$, $VHN_{15} = 20$. $R = 30.3-34.6$ (546 nm); distinctly pleochroic and anisotropic. Occurs in hydrothermal Au-Ag quartz veins, with acanthite. New Bullfrog mine, Nye Co., and Comstock Lode, Virginia City, NV; Smeinogorski, Altai Mts., Russia; Cerotan, Java, and *Tambang Sawah, Benkoelen district, Indonesia.* BM *CM 16:651(1978), AM 65:209(1980), ABI:552.*

2.4.3.2 Fischesserite Ag₃AuSe₂

Named in 1971 for Raymond Fischesser (1911–1991), French mineralogist. ISO $I4_132$. $a = 9.967$, $Z = 8$, $D = 9.10$. *25-367*: 7.08_6 2.66_{10} 2.23_8 2.04_8 1.95_6 1.82_8 1.33_6 1.27_7. Microscopic grains. Gray-white, metallic luster. H = 2, $VHN_{10} = 47$–55. R = 29.7 (540 nm); light pink in reflected light. De Lamar mine, Owyhee Co., ID; Kidd Creek mine, near Timmins, ON, Canada; Hope's Nose, Torquay, England; *Predborice, Czech Republic*, in calcite vein; Flamenco, Atacama, Chile. BM *BM 94:381(1971), AM 57:1554(1972), P&J:163, ABI:157.*

2.4.3.3 Petzite Ag₃AuTe₂

Named in 1845 for W. Petz, who first described the mineral. ISO $I4_132$. $a = 10.38$, $Z = 8$, $D = 9.22$. *12-424*: 7.31_5 3.66_3 2.77_{10} 2.44_6 2.32_6 2.12_8 2.03_7 1.89_5. Massive, fine granular to compact. Steel-gray to iron-black, often tarnished, metallic luster. Cleavage {001}, subconchoidal fracture, slightly sectile to brittle. H = $2\frac{1}{2}$–3, $VHN_{10} = 48$. G = 8.7–9.1. R = 38.2 (540 nm); anisotropic in part. Widely distributed in Te-rich Ag–Au deposits. Mines in Boulder, Hinsdale, and Lake Cos., CO; Tuolumne and Calaveras Cos., CA; many localities in PQ and ON, Canada; Botes, *Sacaramb*, and Baia Sprie, *Romania*; Zhanatyube deposit, Kazakhstan; Kalgoorlie, WA, Australia; Emperor mine, Vatukoula, Fiji. BM *DI:186, AM 34:350(1949), R:424, P&J:298, ABI:402.*

2.4.4.1 Jalpaite Ag₃CuS₂

Named in 1858 for the locality. TET $I4_1/amd$: $a = 8.673$, $c = 11.756$, $Z = 8$, $D = 6.78$. *12-207*(syn): 6.93_4 4.32_6 2.80_{10} 2.75_{10} 2.48_6 2.43_{10} 2.35_{10} 2.12_6. Massive, granular. Lead-gray, tarnishing to black, metallic luster. Cleavage prismatic, good; sectile and malleable. H = 2–$2\frac{1}{2}$, $VHN_{25} = 29$. G = 6.7–6.8. R = 30.7–33.3 (540 nm); distinctly pleochroic and anisotropic. A low-temperature mineral in Ag deposits. Mines in Clear Creek, Mineral, Park, and Boulder Cos., CO; Mogollon, Catron Co., NM; Bisbee, AZ; *Jalpa*, ZAC, Queretaro, QRO, and La Mesa, CHIH, *Mexico*; Kongsberg, Norway; Bohutin, near Pribram, Czech Republic; Savodinski mine, Altai Mts., Russia; Sado mine, Niigata Pref., Japan; Broken Hill, NSW, Australia. BM *AM 53:1530(1968), R:486, P&J:216, ABI:246.*

2.4.5.1 Mckinstryite (Ag,Cu)₂S

Named in 1966 for Hugh E. McKinstry (1896–1961), U.S. economic geologist. ORTH *Pnam*. $a = 14.043$, $b = 15.677$, $c = 7.803$, $Z = 32$, $D = 6.57$. *19-406*: 3.51_6 3.06_6 2.86_6 2.61_{10} 2.57_4 2.41_4 2.07_7 1.95_5. Granular aggregates of microscopic crystals. Steel-gray, tarnishing to black, dark steel-gray streak, metallic luster. One poor cleavage, subconchoidal fracture. $VHN_{25} = 43$–45. G = 6.61. R = 24.0–27.9 (540 nm); strongly anisotropic, light tan to gray. A low-temperature mineral in Ag deposits. Colorado Central mine, near Georgetown, and Bulldog Mt. mine, near Creede, CO; Mogollon, Catron Co., NM; *Foster mine, Cobalt, ON*, and Echo Bay mine, Great Bear Lake, NWT,

Canada; Jalpa, ZAC, Mexico; Godejord, Grong dist., Norway; Pribram, Czech Republic; Sedmochislenitsi mine, Vratsa district, Bulgaria; Tort Kudak deposit, Kazakhstan; Sado mine, Niigata Pref., Japan; Broken Hill, NSW, Australia. BM *EG 61:1383(1966), AM 53:1530(1968), R:484, P&J:263, ABI:320*.

2.4.6.1 Stromeyerite AgCuS

Named in 1832 for Friedrich Stromeyer (1766–1835), German chemist, who first analyzed the mineral. ORTH *Cmcm*. $a = 4.06$, $b = 6.66$, $c = 7.99$, $Z = 4$, $D = 6.26$. *9-499*: 3.46_7 3.33_8 3.07_6 2.61_{10} 2.55_6 2.07_8 1.99_7 1.89_6. Pseudohexagonal prismatic crystals, also massive. Steel-gray, bluish tarnish, dark steel-gray streak, metallic luster. No cleavage, conchoidal fracture, brittle. $H = 2\frac{1}{2}$–3, $VHN_{25} = 70$–72. $G = 6.2$–6.3. $R = 27.7$–31.0 (540 nm); strongly anisotropic, blue to violet. An accessory mineral in Ag–Cu deposits. Many mines in CO; Butte, MT; Silver King and Magma mines, Superior, AZ; Cobalt-Gowganda district, ON, and Silver King mine, near Nelson, BC, Canada; Pribram, Czech Republic; Rudelstadt and Kupferberg, Silesia, Poland; Godejord, Grong Dist., Norway; Agdarinsk deposit, Azerbaijan; *Smeinogorski, Altai Mts., Russia*; Mt Lyell, TAS, and Broken Hill, NSW, Australia; Tsumeb, Namibia. BM *DI:190, EG 61:1(1966), R:481, P&J:355, ABI:502*.

2.4.6.2 Eucairite AgCuSe

Named in 1818 from the Greek for *opportunity*, because it was discovered shortly after the discovery of selenium. ORTH *Pmmn*. $a = 4.105$, $b = 4.070$, $c = 6.310$, $Z = 2$, $D = 7.89$. *25-1180*: 3.42_3 3.16_3 2.89_5 2.63_6 2.49_{10} 2.05_3 2.04_3 1.58_3. Massive, granular. Silver-white to lead-gray, metallic luster. No cleavage, somewhat sectile. $H = 2\frac{1}{2}$, $VHN_{25} = 27$–31. $G = 7.6$–7.8. $R = 37.1$–37.2 (540 nm); strongly anisotropic. Widely distributed in Cu–Se deposits. Cougar mine, San Miguel Co., CO; Kidd Creek mine, near Timmins, ON, and Martin Lake mine, near Lake Athabasca, SK, Canada; Hope's Nose, Torquay, England; *Skrikerum, near Kalmar, Sweden*; Lerbach, Harz Mts., Germany; Jachymov, Czech Republic; Kletno, Poland; Kalgoorlie, WA, Australia; Aguas Blancas, Copiapó, Chile; Sierra de Umango, La Rioja Prov., Argentina. BM *DI:183, AM 35:337(1950), R:484, P&J:158, ABI:152*.

CHALCOCITE GROUP

Copper and sulfur combine according to the formula

$$Cu_{2-x}S$$

where $0 \leq x \leq 0.6$ to give seven separately named minerals. Chalcocite, Cu_2S appears in two polymorphs. One polymorph is based on a hexagonal close-packed framework of sulfur atoms containing a complex arrangement of copper atoms, mostly in triangular coordination but some in a tetrahedral coordination. Above 103°, chalcocite appears as a hexagonal polymorph, $P6/mmc$,

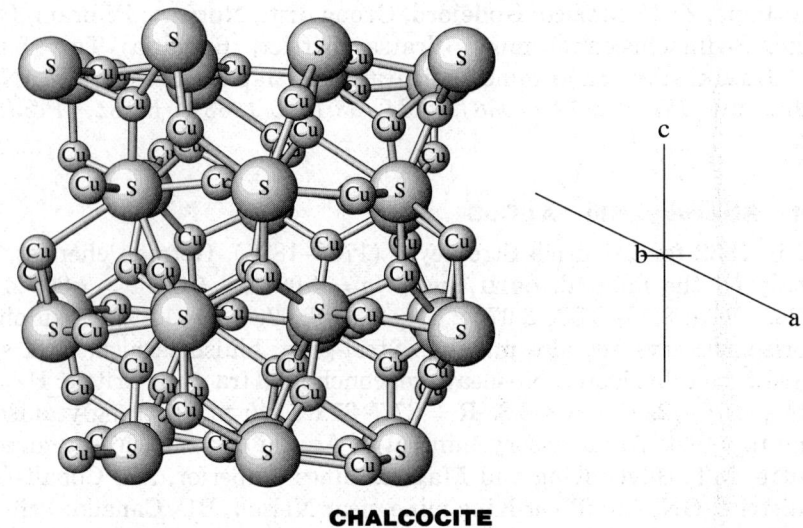

CHALCOCITE

and the lower-temperature polymorph has been characterized in the lower-symmetry space group $P2_1/c$.

COPPER SULFIDES

Mineral	Formula	Space Group	a	b	c	β	Z	
Chalcocite	Cu_2S	$P2_1/c$	15.235	11.885	13.496	116.26	48	2.4.7.1
Chalcocite*	Cu_2S	$P6/mmc$	3.95		6.75		2	2.4.7.1
Djurleite	$Cu_{1.97}S$	$P2_1/c$	26.896	15.745	13.565	90.13	128	2.4.7.2
Digenite	$Cu_{1.80}S$	$Fm3m$	5.57				4	2.4.7.3
Roxbyite	$Cu_{1.78}S$	$C2/m$	53.79	30.90	13.36	90.0	64	2.4.7.4
Anilite	$Cu_{1.75}S$	$Pnma$	7.89	7.84	11.01		16	2.4.7.5
Geerite	$Cu_{1.60}S$	$F43m$	5.410				4	2.4.7.6
Spionkopite	$Cu_{1.40}S$	$P3m1$	22.962		41.429		504	2.4.7.7

*High-temperature polymorph.
Note: See also pyrrhotite (2.8.10.1), which is commonly Fe_7S_8 and which is also based on close-packed sulfur atoms with Fe in octahedral coordination. See also the nickeline group. Pyrrhotite is a nickeline structure but with ordered vacant sites.

2.4.7.1 Chalcocite Cu_2S

Named in 1832 from the Greek for *copper*. Chalcocite group. MON $P2_1/c$. $a = 15.235$, $b = 11.885$, $c = 13.496$, $\beta = 116.26°$, $Z = 48$, $D = 5.79$; above 103° inverts to a hexagonal polymorph, $P6/mmc$; $a = 3.95$, $b = 6.75$, $Z = 2$, $D = 5.80$. *ZK 150:299(1979)*. 33-490: 3.28_4 2.73_4 2.41_5 2.40_7 2.33_3 2.21_4 1.97_7 1.88_{10}. Crystals rare, pseudohexagonal tabular or prismatic; usually massive, compact, sometimes with a sooty coating. **Structure**: Based on a hexagonal close-packed sulfur framework containing a complex arrangement of Cu atoms, mostly in triangular coordination, some in tetrahedral coordination. **Physical properties:** Black, dark gray to black streak, submetallic luster. Cleavage {110}, indistinct; conchoidal fracture; brittle to somewhat sectile. H = $2\frac{1}{2}$–3, VHN$_{100}$ = 84–87. G = 5.5–5.8. R = 31.8 (540 nm); weakly anisotropic. **Tests:** Soluble in HNO$_3$. Easily reduced to metallic copper. **Occurrence:** A common and widely distributed ore mineral of copper. The principal occurrence is in the supergene-enriched zone of sulfide deposits. Oxidizing surface waters attack the primary sulfides, such as chalcopyrite, forming soluble sulfates, which react with the primary sulfides at greater depth to precipitate chalcocite, forming a blanketlike deposit at the water-table level. This contains a considerably greater content of copper than the primary ore, and it has yielded important amounts of copper at Bingham Canyon, UT, Morenci, AZ, Ely, NV, Rio Tinto, Spain, and several other localities. Chalcocite also occurs in hydrothermal sulfide veins with bornite, enargite, chalcopyrite, and pyrite, as at Butte, MT. **Localities:** The following localities are known for fine specimens. Bristol, CT; Flambeau mine, Rush Co., WI; Butte, MT; Bisbee and Miami, AZ; Kennecott, Copper R., AK; St. Just and Redruth, Cornwall, England; Jachymov, Czech Republic; Dognacska, Romania; Bogolovsk, Ural Mts., Russia; Dzhezkazgan, Kazakhstan; Mindouli, Congo Republic; Khan mine, near Arandis, and Tsumeb, Namibia; Messina, Transvaal, and Nababiep West mine, Cape Province, South Africa. **Uses:** An important ore of copper. BM *DI:187, EG 61:1(1966), R:441, P&J:116, ABI:88.*

2.4.7.2 Djurleite Cu$_{1.97}$S

Named in 1962 for Seved Djurle (b.1928), Swedish chemist, who first synthesized the compound. Chalcocite group. MON $P2_1/c$. a = 26.896, b = 15.745, c = 13.565, β = 90.13°, Z = 128, D = 5.77. *34-660*(syn): 3.76$_2$ 3.19$_2$ 3.04$_2$ 2.65$_2$ 2.42$_2$ 2.39$_9$ 1.96$_{10}$ 1.87$_{10}$. Hexagonal and tabular crystals, usually massive, compact. Black, black streak, submetallic luster. No cleavage, twinning common on {110}, conchoidal fracture. H = $2\frac{1}{2}$–3, VHN$_{100}$ = 65–85. G = 5.7. R = 29.6–30.2 (540 nm); weakly anisotropic. A widely distributed ore mineral of copper; present in many of the large porphyry copper deposits of the western United States, as in Butte, MT; Bisbee and Globe-Miami, AZ; as crystals up to 1 cm at the benitoite mine, San Benito Co., CA; *Barranca de Cobre* and Salvadora mine, *CHIH, Mexico*; Wheal Owles, Cornwall, and Seathwaite Tarn, Cumbria, England; Bandaksli, Norway; Osarizawa and Kawazu mines, Honshu, Japan; Bagacay, Samar Is., Philippines; Mt. Gunson mine, SA, Australia; Tsumeb, Namibia. Often confused with chalcocite, with which it is closely associated. BM *AM 47:1181(1962), R:441, P&J:116, ABI:137.*

2.4.7.3 Digenite Cu$_{1.80}$S

Named in 1844 from the Greek for *two kinds*, referring to the presumed presence of both cuprous and cupric copper. Chalcocite group. Contains about 1% Fe, which may be an essential component. ISO (> 75°) $Fm3m$. a = 5.57,

$Z = 4$, $D = 5.63$. *23-692*(syn): 3.21_4 3.01_1 2.78_5 2.14_1 1.97_{10} 1.75_1 1.68_2 1.39_1. Octahedral crystals rare, usually massive. Blue to black, black streak, submetallic luster. Cleavage {111}, fracture conchoidal, brittle. $H = 2\frac{1}{2}$–3, $VHN_{100} = 86$–106. $G = 5.5$–5.7. $R = 21.6$ (540 nm); distinctly blue in polished section. Widely distributed in copper deposits. Butte(*), MT; Bisbee, Jerome, and Superior, AZ; Kennecott, AK; Cananea, SON, Mexico; Botallack mine, Cornwall, England; *Sangerhausen, Thuringia, Germany*; Tsumeb, Namibia. BM *DI:180, AM 56:1889(1971), R:441, P&J:147, ABI:134.*

2.4.7.4 Roxbyite $Cu_{1.78}S$

Named in 1988 for the locality. Chalcocite group. MON $C2/m$. $a = 53.79$, $b = 30.90$, $c = 13.36$, $\beta = 90.0°$, $Z = 64$, $D = 5.56$. *23-958*(syn): 3.35_6 3.00_5 2.86_8 2.63_7 2.37_9 1.94_{10} 1.86_9 1.68_4. Microscopic crystals and powdery aggregates. Blue-black. Cleavage {100}, poor. $H = 2\frac{1}{2}$, $VHN_{50} = 83$. $R = 24.5$–29.8 (546 nm); anisotropism very weak. *Roxby Downs, SA, Australia*; El Teniente mine, Chile. BM *MM 52:323(1988), ABI:448.*

2.4.7.5 Anilite $Cu_{1.75}S$

Named in 1969 for the locality. Chalcocite group. ORTH *Pnma*. $a = 7.89$, $b = 7.84$, $c = 11.01$, $Z = 16$, $D = 5.68$. *24-58*: 3.35_5 3.33_4 2.78_6 2.70_4 2.69_3 2.54_5 2.17_4 1.96_{10}. Prismatic or platy crystals, up to 5 mm. Blue-gray, black streak, metallic luster. $H = 3$. Yarrow Creek, AB, Canada; Neudorf, Harz, Germany; Lubin mine, Silesia, Poland; Bor, Serbia; *Ani mine, Akita, Japan*; Wallaroo, SA, Australia; Estrella mine, Atacama, Chile. BM *AM 54:1256(1969), R:441, ABI:14.*

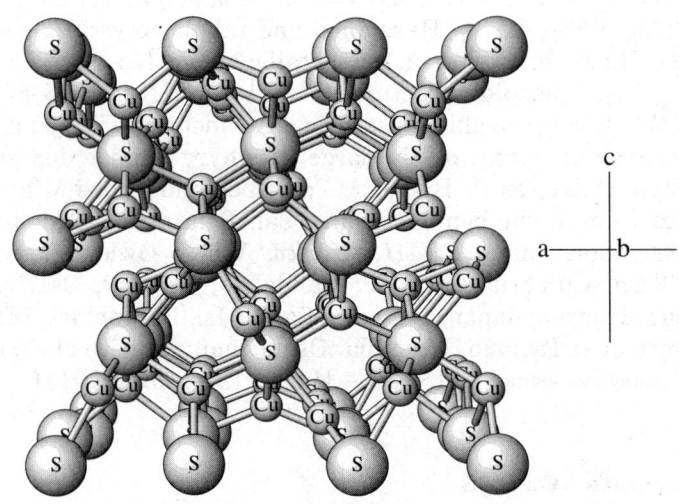

ANILITE: Cu · 7S (chalcocite group)

Sulfur atoms are in approximately body-centered cubic arrangement, with copper 3- and 4-coordinated. Not closely related to the ideal chalcocite (CuS) structure, in which S is HCP.

2.4.7.6 Geerite $Cu_{1.60}S$

Named in 1980 for Adam Geer (1895–1973), of Utica, NY, who collected the mineral. Chalcocite group. ISO $F\bar{4}3m$. $a = 5.410$, $Z = 4$, $D = 5.61$. *33-491*: 3.13_{10} 2.71_1 1.92_5 1.87_1 1.68_1 1.64_3 1.25_1 1.11_2. Microscopic platelets. Black, metallic luster. $H = 3$, $VHN_{25} = 75$–96. Moderately anisotropic. *DeKalb Twp., St. Lawrence Co., NY*; Cuchillo, Sierra Co., NM; Grube Clara, Black Forest, Germany; near Eretria, Othris Mts., Greece. BM *CM 18:519(1980), NJMM:489(1981), ABI:175.*

2.4.7.7 Spionkopite $Cu_{1.40}S$

Named in 1980 for the locality. Chalcocite group. HEX-R Space group uncertain, probably $P3m1$. $a = 22.962$, $c = 41.429$, $Z = 504$, $D = 5.13$. *36-380*: 3.28_2 3.08_9 2.78_3 2.39_2 2.30_3 1.91_{10} 1.82_3 1.62_2. Black, metallic luster. $H = 2\frac{1}{4}$, $VHN_{10} = 120$–162. $R = 19.0$–27.4 (540 nm); pleochroic in blue to blue-white [like yarrowite, formerly called "blaubleibender Covellit" (blue-remaining covellite) for this reason]. In a red-bed copper deposit at *Spionkop Creek, AB, Canada*; DeKalb Twp., NY; Bisbee, AZ; Schulenberg, Harz, Germany; near Eretria, Othris Mts., Greece. BM *CM 18:511(1980), R:676, ABI:489.*

2.4.8.1 Weissite $Cu_{2-x}Te$

Named in 1927 for Louis Weiss, owner of the Good Hope mine. HEX $P\bar{3}m1$. $a = 8.342$, $c = 21.69$, $Z = 24$, $D = 6.44$. *10-421*(syn): 3.61_7 3.24_6 2.55_3 2.17_3 2.09_{10} 2.01_7 1.80_5 1.45_3. Dark blue-black, tarnishing to deep black, black streak, metallic luster. $H = 3$, $VHN_{100} = 145$–158. $G = 6$. $R = 33.7$–34.9 (540 nm); distinctly pleochroic and anisotropic. Occurs in hydrothermal veins with other tellurides. *Good Hope and Mammoth Chimney mines, Vulcan, Gunnison Co., CO*; near Winston, Sierra Co., NM; Teine mine, Hokkaido, and Kawazu, Honshu, Japan; Kalgoorlie, WA, Australia. BM *DI:199, R:419, P&J:388, ABI:573.*

2.4.9.1 Bellidoite Cu_2Se

Named in 1975 for Eleodoro Bellido Bravo (1918–1992), director of Servicio de Geologia y Mineria, Peru. Dimorphous with berzelianite. TET $P4_2/n$. $a = 11.52$, $c = 11.74$, $Z = 32$, $D = 7.03$. *29-575*: 6.8_3 3.5_1 3.38_8 2.32_1 2.26_5 2.06_{10} 1.76_7 1.52_1. Microscopic anhedral grains. Silver-white, metallic luster. $H = 1\frac{1}{2}$–2, $VHN_{100} = 33$–41. $R = 28.5$ (589 nm); very weakly anisotropic. Occurs with other selenides in calcite at *Habri, near Tisnov, Moravia, Czech Republic.* BM *EG 70:384(1975), R:468, ABI:43.*

2.4.10.1 Berzelianite Cu_2Se

Named in 1850 for Jöns J. Berzelius (1779–1848), Swedish chemist. Dimorphous with bellidoite. ISO $Pm3m$. $a = 5.739$, $Z = 4$, $D = 6.90$. *6-680*: 3.33_9 2.03_{10} 1.73_8 1.43_3 1.32_2 1.17_4 1.01_1. Fine-grained powdery and as dendritic crusts. Lead-gray, black streak, metallic luster. No cleavage, uneven fracture, brittle. H = 2, VHN_{100} = 21–24. G = 6.71. R = 25.8 (540 nm). Occurs with other selenides in hydrothermal veins. Aurora, Mineral Co., NV; Pinky Fault uranium deposit, Lake Athabaska, SK, Canada; *Skrikerum, near Tryserum, Kalmar, Sweden*; Lerbach, Tilkerode, Zorge, and Clausthal, Harz, Germany; Habri, near Tisnov, Moravia, Czech Republic; Kalgoorlie, WA, and El Sharana, NT, Australia; Sierra de Umango, La Rioja, Argentina. BM *DI:182, AM:351(1950), R:467, P&J:88, ABI:50.*

2.4.11.1 Cuprostibite $Cu_2(Sb,Tl)$

Named in 1969 for the composition. TET $P4/nmm$. a = 3.990, c = 6.09, Z = 2, D = 8.42. *22-601*: 3.33_2 2.82_4 2.56_5 2.07_{10} 1.99_4 1.67_2 1.42_3 1.17_4. Fine-grained aggregates. Steel-gray with violet-red tint on fresh fractures, metallic luster. Cleavage in one direction, uneven fracture. H = 4, VHN_{50} = 216–249. R = 38.7–45.5 (540 nm); strongly pleochroic and anisotropic. Franklin, NJ; *Mt. Nakalak, near Narssaq, Greenland*, in vein in sodalite syenite; Långban and Långsjön, Sweden, in mineralized dolomite; Zolotaya Gora, Urals, Russia. BM *AM 55:1810(1970); NJMM:201(1982), 145(1988); ABI:123.*

2.4.12.1 Crookesite Cu_7TlSe_4

Named in 1866 for William Crookes (1832–1919), English chemist, who discovered thallium. May contain up to 5% Ag. TET $I\bar{4}$. a = 10.40, c = 3.93, Z = 2, D = 7.44. *6-280*: 5.21_1 3.30_{10} 3.01_8 2.60_{10} 2.32_3 2.11_5 1.84_4 1.78_4. Finely granular. Lead-gray, black streak, metallic luster. Two cleavages at right angles, brittle. H = $2\frac{1}{2}$–3, VHN_{100} = 92–123. G = 6.90. R = 31.5–36.5 (540 nm); distinctly anisotropic. Occurs in hydrothermal deposits with other selenides. Pinky Fault uranium deposit, Lake Athabaska, SK, Canada; *Skrikerum, near Tryserum, Kalmar, Sweden*; Bukov, near Tisnov, Czech Republic. BM *DI:183, AM 35:347(1950), R:481, P&J:135, ZK 181:241(1987).*

2.4.12.2 Sabatierite Cu_4TlSe_3

Named in 1978 for Germain Sabatier (b.1923), French mineralogist. TET $P4/mmm$. a = 3.977, c = 9.841, Z = 1, D = 7.42. *41-1458*(syn): 3.09_{10} 2.71_6 2.53_2 2.44_3 1.99_7 1.85_3 1.67_4 1.41_2. Microscopic polycrystalline aggregates. Gray, metallic luster. H = 2, VHN_{15} = 54. R = 27.9–30.6 (540 nm); strongly anisotropic. *Bukov deposit, Rožná district, Czech Republic.* BM *BM 101:557(1978), AM 64:1331(1979), P&J:410, ZK 181:241(1987), ABI: 456.*

2.4.13.1 Carlinite Tl_2S

Named in 1975 for the locality. HEX-R $R\bar{3}$. a = 12.12, c = 18.175, Z = 27, D = 8.55. *29-1344*(syn): 3.03_{10} 2.29_1 2.02_1 1.75_1 1.51_1 1.20_1 1.08_1 0.916_1. Microscopic anhedral grains. Dark gray, black streak, metallic luster. Cleavage {0001}, perfect; hackly fracture. H = 1, VHN_{50} = 23.5. G = 8.1. R = 38.8–40.1 (546 nm); moderately anisotropic. Occurs in carbonaceous limestone in the *Carlin mine, Eureka Co., NV.* BM *AM 60:559(1975), R:467, ABI:81.*

2.4.14.1 Palladoarsenide Pd_2As

Named in 1974 for the composition. MON $P2/m$. $a = 9.24$, $b = 8.47$, $c = 10.45$, $\beta = 94.00°$, $Z = 18$, $D = 10.54$. *17-227*(syn): 2.89_3 2.60_7 2.35_5 2.31_5 2.22_{10} 2.15_{10} 2.13_4 1.96_4. Microscopic grains. Steel-gray, metallic luster. Cleavage perfect in two directions, brittle. $H = 5$, $VHN_{100} = 390$. $R = 52.5$–54.8 (540 nm); moderately anisotropic. *Stillwater Complex, MT*, in heavy-mineral concentrates; *Lac des Iles Complex, ON, Canada; Oktyabr mine, Talnakh, Norilsk, Russia; Merensky Reef, Transvaal, South. Africa.* BM *ZVMO 103:104(1974), AM 60:162(1975), CM 13:321(1975), ABI:377.*

2.4.15.1 Palladobismutharsenide $Pd_2(As,Bi)$

Named in 1976 for the composition. ORTH $Pmcn$ or $P2_1cn$. $a = 7.467$, $b = 18.946$, $c = 6.797$, $Z = 20$, $D = 10.8$. *29-962*(syn): 2.59_8 2.49_9 2.38_4 2.26_4 2.23_{10} 2.20_8 2.13_5 2.08_9. Microscopic grains. Cream-white, metallic luster. $H = 6$, $VHN_{25} = 429$–483. $R = 52.1$–53.0 (546 nm); weakly to distinctly anisotropic. *Stillwater Complex, MT*, in heavy-mineral concentrates. BM *CM 14:410(1976), ABI:378.*

2.4.16.1 Majakite   PdNiAs

Named in 1976 for the locality. HEX Space group unknown. $a = 6.066$, $c = 7.20$, $Z = 6$, $D = 10.42$. *29-965*: 3.04_4 2.65_{10} 2.40_4 2.30_5 2.19_7 1.19_{10} 1.80_4 1.75_3. Microscopic grains. Gray-white, metallic luster. $H = 6$, $VHN_{50} = 520$. $G_{syn} = 9.33$. $R = 52.8$ (540 nm); weakly anisotropic. *Majak mine, Talnakh, Norilsk, Russia*, intergrown with chalcopyrite and platinum-group minerals. BM *ZVMO 105:698(1976), AM 62:1260(1977), ABI:307.*

2.4.17.1 Petrovskaite   AuAgS

Named in 1984 for Nina Petrovskaya (b.1910), Russian mineralogist. Contains up to 1.3% Se. MON Space group unknown. $a = 4.943$, $b = 6.670$, $c = 7.221$, $\beta = 95.68°$, $Z = 4$, $D = 9.44$. *38-396*: 7.25_3 3.87_3 2.77_{10} 2.63_5 2.39_4 2.25_4 1.80_3 1.47_4. Microscopic rims on gold. Dark gray to black, dark gray streak, dull metallic luster. $H = 2$–$2\frac{1}{2}$, $VHN = 40$–48. Occurs with gold in the *Maikain gold deposit, Kazakhstan.* BM *ZVMO 113:602(1984), AM 70:1331 (1985), ABI:401.*

2.4.18.1 Novakite $(Cu,Ag)_{21}As_{10}$

Named in 1961 for Jiri Novak (1902–1971), Czech mineralogist. MON Space group uncertain, probably $C2/m$. $a = 16.269$, $b = 11.711$, $c = 10.007$, $\beta = 112.7°$, $Z = 4$, $D = 8.01$. *39-370*: 3.15_6 2.52_7 2.38_5 2.00_9 1.96_8 1.91_7 1.88_{10} 1.18_8. Granular aggregates or veinlets. Steel-gray, tarnishing to black, black streak, metallic luster. $H = 3$–$3\frac{1}{2}$. $R = 50.6$–55.6 (540 nm);

anisotropic, blue-gray to light brown. *Černý Důl mine, near Berghaus, Czech Republic*, in calcite veins. BM *AM 46:885(1961)*, *TMPM 34:167(1985)*, *P&J:279*, *ABI:356*.

2.5.1.1 Umangite Cu_3Se_2

Named in 1891 for the locality. TET $P\bar{4}2_1m$. $a = 6.406$, $c = 4.282$, $Z = 2$, $D = 6.59$. *19-402*(syn): 3.55_{10} 3.20_5 3.11_6 2.26_5 2.03_4 1.91_4 1.83_8 1.78_6. Small grains. Blue-black with reddish cast, tarnishing to iridescent purple, black streak, metallic luster. Poor rectangular cleavage, uneven fracture, brittle. $H = 3$, $VHN_{100} = 88$–100. $G = 6.44$–6.49. $R = 13.6$–16.5 (540 nm); strongly pleochroic and anisotropic. Occurs with other selenides in hydrothermal veins. From Lodge Bay, Lake Athabaska to Ato Bay, Beaverlodge Lake, SK, Canada; Skrikerum, Sweden; Andreasberg, Tilkerode, Zorge, and other localities in Harz, Germany; Chaméau mine, Puy-de-Dome, France; Habri, near Tisnov, Moravia, Czech Republic; Kalgoorlie, WA, Australia; Tsumeb, Namibia; *Sierra de Umango* and *Sierra de Cachueta, La Rioja, Argentina*. BM *DI:194*, *AM 35:354(1950)*, *R:469*, *P&J:379*, *ABI:549*.

2.5.2.1 Bornite Cu_5FeS_4

Known as a mineral since 1725. Named in 1845 for Ignaz von Born (1742–1791), Austrian mineralogist. ORTH (pseudo-TET) *Pbca*. $a = 10.950$, $b = 21.862$, $c = 10.950$, $Z = 16$, $D = 5.09$; above 228° ISO *Fm3m*, $a = 5.50$, $Z = 2$. *CM 16:397(1978)*. *14-323*: 3.31_4 3.18_6 2.74_5 2.50_4 1.94_{10} 1.65_3 1.26_5 1.12_5. **Habit:** Crystals rare, pseudocubic, dodecahedral, rarely octahedral; usually massive. Twin plane {111}; often as penetration twins. **Structure:** Based on

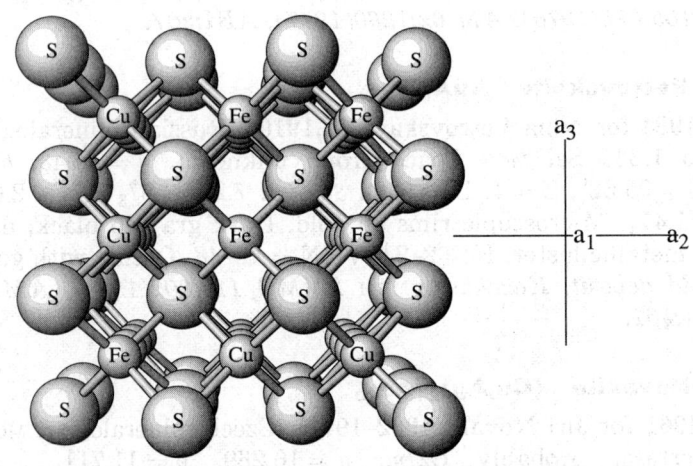

BORNITE: Cu_5FeS_4

Copper and iron are 4-coordinated by S and S is 4-coordinated by Fe and S. The Cu and Fe tetrahedra share edges, with the metals only 2.75 Å apart.

cubic close-packed sulfur framework with Cu atoms in a complex arrangement in triangular and tetrahedral coordination. **Physical properties:** Copper-red to brown on fresh surfaces, tarnishes rapidly to iridescent purple (peacock ore), gray-black streak, metallic luster. Cleavage {111} in traces, conchoidal or uneven fracture brittle. H = 3, VHN_{100} = 92. G = 5.06–5.09. R = 21.1 (540 nm); weakly pleochroic and anisotropic. **Tests:** Easily fusible to a brittle magnetic globule. Soluble in HNO_3 with the separation of sulfur. **Occurrence:** A common and widespread mineral present in many important copper deposits throughout the world. Also occurs in dikes, in basic intrusives, in contact-metamorphic deposits, in pegmatites, and in quartz veins. **Localities:** Bristol(*), CT; Butte(*), MT; many mines in AZ, including Bisbee, Globe, Ajo, Magma; many localities in CA; Kennecott, AK; Marble Bay mine, Texada Is., BC, Canada; Carn Brea(*) and other mines near Redruth, Cornwall, England; Freiberg and other localities in Saxony, Germany; near Pragratten, Austria; Talnakh, Norilsk, Russia; Dzhezkazgan(*), Kazakhstan; N'ouva mine, Talate, Morocco; Kipushi(*) and Kambove(*), Zaire; Mangula mine(*), Zimbabwe; Ookiep, Cape Prov., and Messina, Transvaal, South Africa; Mt. Lyell, TAS, Australia; Morococha, Peru; Braden mine, Chile. **Uses:** An important ore of copper. BM *DI:195, AM 63:1(1978), R:487, P&J:98, ABI:62.*

2.5.3.1 Heazlewoodite Ni$_3$S$_2$

Named in 1896 for the locality. HEX-R $R32$. $a = 5.741$, $c = 7.139$, $Z = 3$, $D = 5.87$. *8-126*: 4.11_5 2.89_9 2.38_4 2.03_5 1.83_{10} 1.66_8 1.29_2 1.08_3. Massive, fine-grained; rarely in minute crystals. Pale bronze, light bronze streak, metallic luster. H = 4, VHN_{100} = 230–254. G = 5.82. R = 57.0 (540 nm); strongly anisotropic, brown to blue-gray. Widely distributed as an accessory mineral in serpentinite. Cedar Hill quarry, Lancaster Co., PA; Josephine Co., OR; White Creek, Fresno Co., and Dorleska mine (minute crystals), Trinity Co., CA; Jeffrey mine, Asbestos, PQ, and Miles Ridge, YT, Canada; Poschiavo, Grisons, Switzerland; Cap Corse, Corsica, France; Hirt, near Friesach, Carinthia, Austria; Kop Krom mine, near Askale, Turkey; Talnakh, Norilsk, Russia; near Rabad Sefid, Iran; *Heazlewood* and Trial Harbour, *TAS, Australia*; Barberton, Transvaal, South Africa. BM *AM 32:484(1947), 62:341(1977); R:405, P&J:196; ABI:210.*

2.5.4.1 Oregonite Ni$_2$FeAs$_2$

Named in 1959 for the state of Oregon. HEX Space group unknown. $a = 6.083$, $c = 7.130$, $Z = 3$, $D = 6.92$. *13-368*: 3.57_4 2.95_4 2.31_{10} 2.12_{10} 1.99_7 1.79_7 1.76_7 1.74_7. Fine-grained pebbles with smooth brown crust. White, metallic luster. H = 5. R = 48.0 (540 nm); weakly anisotropic. *Josephine Creek, Josephine Co., OR*; Alexo mine, near Timmins, ON, Canada; near Skouriotissa, Cyprus, in serpentinite. BM *NJMM:239(1959), AM 45:1130(1960), R:400, P&J:282, ABI:364.*

2.5.5.1 Thalcusite $Tl_2Cu_3FeS_4$

Named in 1976 for the composition. TET $I4/mmm$. $a = 3.882$, $c = 13.25$, $Z = 1$, $D = 6.51$. *29-580*(syn): 3.73_6 3.31_4 2.92_{10} 2.54_9 2.19_4 1.72_8 1.61_6 1.07_5. Microscopic grains. Gray, metallic luster. $H = 2\frac{1}{2}$, $VHN_{10} = 88$. $G = 6.15$. $R = 28.4$–29.9 (540 nm); moderately anisotropic, light yellow to very dark gray. Mont Saint-Hilaire, PQ, Canada; Ilimaussaq intrusion, near Narsaq, Greenland; *Talnakh, Norilsk, Russia*; Rajpura-Dariba, Rajasthan, India. BM *ZVMO 105:202(1976)*, *AM 62:396(1977)*, *ABI:528*.

2.5.5.2 Bukovite $Tl_2Cu_3FeSe_4$

Named in 1971 for the locality. TET $I4/mmm$. $a = 3.976$, $c = 13.70$, $Z = 1$, $D = 7.45$. *25-312*: 3.43_5 3.00_{10} 2.60_9 2.26_7 1.99_7 1.77_8 1.71_5 1.66_6. Microscopic crystals and massive. Gray-brown, metallic luster. Cleavage {001}, good; {100}, imperfect. $H = 2$, $VHN_{20} = 64$. $R = 27.7$–29.4 (540 nm); weakly pleochroic. Occurs in calcite veins with other selenides at *Bukov*, Petrovice, and Predborice, *Czech Republic*. BM *BM 94:529(1971)*, *AM 57:1910(1972)*, *NJMA 138:122(1980)*, *P&J:109*.

2.5.5.3 Murunskite $K_2Cu_3FeS_4$

Named in 1981 for the locality. TET $I4/mmm$. $a = 3.88$, $c = 13.10$, $Z = 1$, $D = 3.81$. *33-1005*: 6.52_{10} 3.29_2 2.90_6 2.53_8 2.31_1 2.10_3 1.94_5 1.72_4. Microscopic grains. Copper-red to brown, iridescent tarnish, oxidizes rapidly to give a black sooty coating, metallic luster. Imperfect cleavage, brittle. $H = 3$, $VHN = 109$. $R = 18.9$ (540 nm); moderately anisotropic. Khibeny massif, Kola, and *Murun alkalic massif, Yakutia, Russia*. BM *ZVMO 110:468(1981)*, *AM 67:624(1982)*, *ABI:342*.

2.5.6.1 Argyrodite Ag_8GeS_6

Named in 1886 for the Greek for *silver-containing*. Forms a series with canfieldite (2.5.6.2). ORTH $Pna2_1$. $a = 15.149$, $b = 7.476$, $c = 10.589$, $Z = 4$, $D = 6.25$; above 223°: ISO $Im3m$, $a = 21.11$, $Z = 32$, $D = 6.49$. *14-356*: 3.14_3 3.02_{10} 2.66_4 2.44_3 2.03_3 1.86_5 1.78_2 1.17_2. Pseudocubic crystals up to 18 cm, also as radiating aggregates and botryoidal crusts. Steel-gray, gray-black streak, metallic luster. No cleavage, crystals twinned on the spinel law {111}, conchoidal or uneven fracture, brittle. $H = 2\frac{1}{2}$. $G = 6.1$–6.3. $R = 25.0$ (540 nm); weakly anisotropic. Rico, Dolores Co., CO; Vipont mine, Box Elder Co., UT; Dolly Varden mine, Alice Arm, BC, Canada; Silvermines, Co. Tipperary, Eire; *Himmelfürst* and other mines, *Freiberg, Saxony, Germany*; Fournial mine, Cantal, France; Wolyu mine, Youngdong district, South Korea; Aullagas, Colquechaca(*), Potosi(*), Porco, and other localities, Bolivia. BM *DI:356*, *NJMM:269(1978)*, *R:479*, *P&J:75*, *ABI:23*.

2.5.6.2 Canfieldite Ag_8SnS_6

Named in 1894 for Frederick A. Canfield (1849–1926), U.S. mining engineer and mineral collector. May contain up to 20% Te, 4% Se, and 2% Ge. Forms a series with argyrodite (2.5.6.1). ORTH $Pna2_1$. $a = 15.298$, $b = 7.548$, $c = 10.699$, $Z = 4$, $D = 6.31$; above 175° ISO $Im3m$. $a = 21.54$, $Z = 32$, $D = 6.24$. *38-434*(syn): 3.25_5 3.23_3 3.11_5 3.08_{10} 3.02_3 2.73_5 2.51_2 2.07_2. Pseudocubic crystals up to 1 cm, also botryoidal, massive. Steel-gray with a reddish tinge, tarnishing to black, gray-black streak, metallic luster. No cleavage, crystals twinned on the spinel law {111}, conchoidal or uneven fracture, brittle. $H = 2\frac{1}{2}$, $VHN_{50} = 109$. $G = 6.2$–6.3. $R = 25.1$ (540 nm). Leadville, Lake Co., CO; Bisbee, AZ; Himmelsfürst mine, Freiberg, Saxony, Germany; Pribram and Kutna Hora, Czech Republic; Belukhinsk deposit, Transbaikalia, Russia; Ikuno, Omidani, and Ashio mines, Japan;' Cirtan, Java; *Aullagas*(*), near *Colquechaca, Bolivia*. BM *DI:356, NJMM: 269(1978), R:479, P&J:75, ABI:79*.

2.5.7.1 Daomanite $CuPtAsS_2$

Named in 1974 for the locality. ORTH $Ama2$. $a = 5.875$, $b = 15.79$, $c = 3.761$, $Z = 4$, $D = 7.57$. *42-1327*: 8.0_8 4.71_5 3.25_7 3.06_{10} 2.91_5 2.10_8 1.87_5 1.83_8. Microscopic tabular crystals. Steel-gray with a yellow tint, metallic luster. Cleavages {100}, {001}, {010}, {110}, from most to least perfect. $H = 3\frac{1}{2}$, $VHN_{15} = 169$–175. $R = 37$–43 (544 nm); strongly pleochroic and anisotropic. *Dao* and *Ma* districts, *Yanshan, China*. BM *AM 61:184(1976), 65:408(1980); ABI:128*.

2.5.8.1 Imiterite Ag_2HgS_2

Named in 1985 for the locality. MON $P2_1/c$. $a = 4.039$, $b = 8.005$, $c = 6.580$, $\beta = 107.12°$, $Z = 2$, $D = 7.85$. *39-328*: 4.88_3 3.47_5 3.14_3 2.92_3 2.77_{10} 2.75_{10} 2.46_8 1.47_4. Anhedral grains up to 1 mm. Gray, metallic luster. $H = 2\frac{1}{2}$, $VHN_{100} = 86$. $R = 29.5$–32.1 (540 nm); strongly anisotropic, blue to red-brown. Golden Rule mine, Tuolumne Co., CA; *Imiter mine, Jbel Sarhro, Anti-Atlas, Morocco*. BM *BM 108:457(1985), AM 71:1277(1986), ABI:230, MR 21:500(1990)*.

2.5.9.1 Chvilevaite $Na(Cu,Fe,Zn)_2S_2$

Named in 1988 for T. N. Chvileva (b.1925), Russian mineralogist. HEX-R $P3m1$. $a = 3.873$, $c = 6.848$, $Z = 1$, $D = 3.94$. *42-1366*: 6.85_6 3.40_9 3.02_{10} 2.40_{10} 1.87_9 1.48_5 1.19_5. Microscopic grains. Bronze, tarnishing rapidly to a sooty black coating, metallic luster. Cleavage {001}, perfect; brittle. $H = 3$, $VHN_{20} = 110$–153. $R = 17.5$–21.6 (540 nm); strongly pleochroic and anisotropic. Occurs in sphalerite at the *Akatuya deposit, Transbaikal, Russia*. BM *ZVMO 117:204(1988), AM 74:946(1988), ABI:99*.

2.6.1.1 Dimorphite As_4S_3

Named in 1850 from the Greek for two forms, I and II. I: ORTH $Pnma$; $a = 9.07$, $b = 8.01$, $c = 10.30$, $Z = 4$, $D = 3.51$; II: ORTH $Pnma$; $a = 11.24$, $b = 9.90$, $c = 6.56$, $Z = 4$, $D = 3.60$. 26-125: 6.27_5 5.16_{10} 4.16_4 3.87_3 3.13_3 3.12_6 2.96_5 2.91_3 (I). 26-126: 5.64_3 4.89_{10} 3.91_4 3.07_3 2.94_3 2.84_3 2.14_5 1.62_5 (II). Dipyramidal crystals. Orange-yellow, yellow streak, adamantine luster. $H = 1\frac{1}{2}$. $G = 2.58$. Solfatara, Italy, in a fumarole; Alacran sulfide deposit, Copiapo, Chile. BM DI:197, NJMM:423(1972), ZK 138:161(1973), R:890, ABI:135.

JOSÉITE GROUP

The joséite group of structures are bismuth minerals corresponding to the formula

$$Bi_mX_p$$

where

Bi may be partially replaced by Pb
X = S, Se, Te
$m + p = 7$

The structures are all similar rhombohedral lattices, $R\bar{3}m$, and they are similar to the arsenic group, but with a 4c-axis repeat. AM 76:257(1991).

JOSÉITE GROUP

Mineral	Formula	Space Group	a	c	D	
Joséite	Bi_4TeS_2	$R\bar{3}m$	4.24	39.69	8.23	2.6.2.1
Joséite-B	Bi_4Te_2S	$R\bar{3}m$	4.34	40.83	8.44	2.6.2.2
Ikunolite	Bi_4S_3	$R\bar{3}m$	4.15	39.19	7.97	2.6.2.3
Laitakarite	Bi_4Se_2S	$R\bar{3}m$	4.225	39.93	8.28	2.6.2.4
Pilsenite	Bi_4Te_3	$R\bar{3}m$	4.451	41.888	8.45	2.6.2.5
Poubaite	$Bi_2PbSe_2Te_2$	$R\bar{3}m$	4.252	40.10	7.88	2.6.2.6
Rucklidgeite	$(Bi,Pb)_3Te_4$	$R\bar{3}m$	4.422	41.49	8.06	2.6.2.7

Notes:
1. Related minerals sharing the composition and the threefold-inversion axis are: protojoséite (2.6.3.1), kochkarite (2.6.3.2), aleksite (2.6.3.3), and hedleyite (2.6.3.4).
2. See also the tsumoite group for bismuth minerals with formula BiX.

2.6.2.1 Joséite Bi_4TeS_2

Named in 1853 for the locality. Joséite group. Three varieties are recognized (A,B,C), differing in Te:S ratio; joséite-C is questionable. HEX-R $R\bar{3}m$. $a = 4.24$, $c = 39.69$, $Z = 3$, $D = 8.23$. 12-735: 4.4_2 3.62_2 3.08_{10} 2.24_5 2.11_5 1.75_3 1.54_3 1.54_3. Platy crystals. Silver-white, tarnishing to lead-gray, gray streak, metallic luster. Cleavage {0001}, perfect; flexible. H = 2, VHN_{25} = 29–43. G = 8.1–8.3. R = 56.2–60.0 (540 nm); moderately anisotropic. Glacier Gulch, near Smithers, and Good Hope claim, near Hedley, BC, Canada; Carrock Fell and Coniston, Cumbria, England; Serrania de Rondo, Spain; Smolotely, Czech Republic; Bihot Mts., Romania; Stepnyak gold deposit, Kazakhstan; Koshbulak, Uzbekistan; Sosukchan deposit, Yakutia, Russia; Tsumo mine, Akita Pref., Japan; Kingsgate, NSW, Australia; *San José, MG, Brazil*. BM *DI:166; AM 34:342(1949); 56:1839(1971); R:439; P&J:367; ABI:252.*

2.6.2.2   Joséite-B Bi_4Te_2S

Named in 1949 for its relationship to joséite. Joséite group. HEX-R $R\bar{3}m$. a = 4.34, c = 40.83, Z = 3, D = 8.44. *9-435*: 4.53_2 3.16_{10} 2.30_4 2.17_5 1.95_2 1.78_3 1.57_2 1.38_2. Platy crystals. Silver-white, tarnishing to lead-gray, gray streak, metallic luster. Cleavage {0001}, perfect; flexible; inelastic. H = 2, VHN_{25} = 46. G = 8.3. R = 51.0–53.7 (546 nm); distinctly anisotropic. *Glacier Gulch, near Smithers*, and Good Hope claim, near Hedley, *BC, Canada*; Tunaberg, Sweden; Smolotely, near Pribram, Czech Republic; Stepnyak deposit, Kazakhstan; Sosukchan deposit, NE Yakutia, Russia; Tsumo mine, Akita Pref., Japan; San José, MG, Brazil. BM *AM 34:367(1949), 76:257(1991); EM 5:165(1993).*

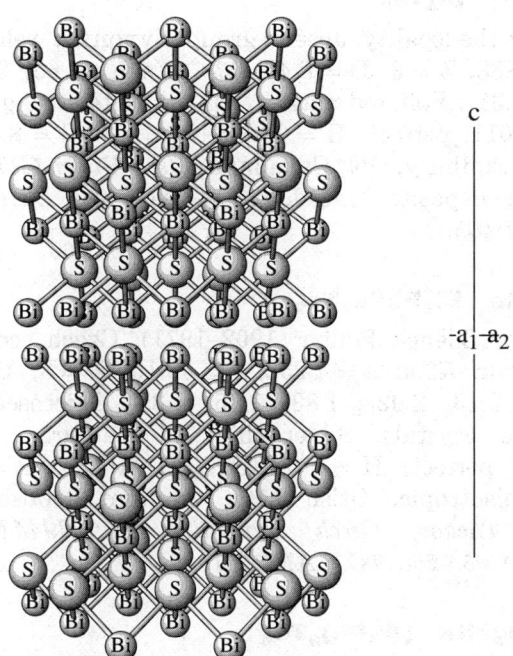

IKUNOLITE: Bi_4S_3 (joséite group)

Multiple-layer sandwich structure with the sequence Bi–S–Bi–S–Bi–S–Bi. Sulfurs are in hexagonal closest-packed arrangement.

2.6.2.3 Ikunolite Bi_4S_3

Named in 1959 for the locality. Joséite group. HEX-R $R\bar{3}m$. $a = 4.15$, $c = 39.19$, $Z = 3$, $D = 7.97$. *12-730*: 6.56_2 4.34_5 3.54_2 3.02_{10} 2.21_3 2.08_2 1.87_2 1.51_1. Foliated massive, plates to 3 cm. Lead-gray, dark gray streak, metallic luster. Cleavage {0001}, perfect; flexible. H = 2, $VHN_{100} = 45$. G = 7.8. R = 45.0–48.9 (540 nm); moderately anisotropic. *Ikuno mine, Hyogo Pref., Japan*; Kingsgate, NSW, Australia. BM *MJJ 2:397(1959); AM 45:477(1960), 47: 1431(1962); P&J:207; ABI:228.*

2.6.2.4 Laitakarite Bi_4Se_2S

Named in 1959 for Arne Laitakari (1890–1975), director of the Geological Survey of Finland. Joséite group. HEX-R $R\bar{3}m$. $a = 4.225$, $c = 39.93$, $Z = 3$, $D = 8.28$. *14-220*: 4.43_6 3.59_6 3.07_{10} 2.25_8 2.11_8 1.74_7 1.54_6 1.41_6. Foliated plates and sheets. Gray-white, metallic luster. Cleavage {0001}, perfect; flexible; inelastic. H = $2–2\frac{1}{2}$, $VHN_{100} = 54–78$. G = 8.12. R = 45.0–53.0 (540 nm); moderately anisotropic. Kidd Creek mine, near Timmins, ON, Canada; Falun mine, Sweden; *Orijärvi mine, Finland*; La Creusaz, Valais, Switzerland; Nevskoe, Magadan region, Russia; Akenobe mine, Hyogo Pref., Honshu, Japan. BM *AM 44:908(1959), 47:806(1962); P&J:232; ABI:283.*

2.6.2.5 Pilsenite Bi_4Te_3

Named in 1853 for the locality. Joséite group. Synonyn: wehrlite. HEX-R $R\bar{3}m$. $a = 4.451$, $c = 41.888$, $Z = 3$, $D = 8.45$. *33-216*(syn): 4.65_1 3.24_{10} 2.36_4 2.23_3 1.83_2 1.49_1 1.42_1 1.31_1. Foliated masses. Tin-white to steel-gray, metallic luster. Cleavage {0001}, perfect. H = 2, $VHN_{25} = 53$. G = 8.4–8.6. Sylvanite, Hidalgo Co., NM; Sudbury, ON, Canada; *Deutsch-Pilsen (Börzöny), Hungary*; Koronuda Au–Cu deposit, Macedonia, Greece. BM *DI:167, R:438, AM 69:215(1984), ABI:405.*

2.6.2.6 Poubaite $Bi_2PbSe_2Te_2$

Named in 1978 for Zdenek Pouba (1902–1971), Czech economic geologist. Joséite group. HEX-R $R\bar{3}m$. $a = 4.252$, $c = 40.10$, $Z = 3$, $D = 7.88$. *29-762*: 3.63_4 3.09_8 2.25_5 2.13_6 2.02_{10} 1.83_4 1.75_5 1.35_5. Microscopic euhedral to subhedral lathlike crystals. Silver-white to lead-gray, metallic luster. Cleavage {0001}, perfect. H = $2\frac{1}{2}$, $VHN_{25} = 74–122$. R = 48.5–53.3 (540 nm); strongly anisotropic. Otish Mts. uranium deposit, PQ, Canada; *Oldrichov, near Tachov, Czech Republic*. BM *NJMM:9(1978), AM 63:1283(1978), CM 25:625(1987), ABI:421.*

2.6.2.7 Rucklidgeite $(Bi,Pb)_3Te_4$

Named in 1977 for John C. Rucklidge (b.1938), Canadian mineralogist. Joséite group. Contains up to 3% Sb and 2% Ag. HEX-R $R\bar{3}m$. $a = 4.422$, $c = 41.49$, $Z = 3$, $D = 8.06$. *29-234*: 3.22_{10} 2.34_9 2.21_4 1.98_5 1.82_4 1.61_4 1.47_6 1.38_4. Foliated aggregates and microscopic grains. Silver-white, lead-gray streak, metallic luster. Perfect basal cleavage, brittle. H = 2, VHN_{10} = 52–63. G = 7.74. R = 62.1–65.3 (540 nm); distinctly anisotropic. *Hesperus mine, La Plata Co., CO; Campbell mine, Bisbee, AZ; Robb-Montbray mine, Montbray Twp., PQ, and Ashley deposit, Bannockburn Twp., ON, Canada; Ilomantsi, Finland; Oldrichov, near Tachov, Czech Republic; Zod, Armenia; Kochkar Au deposits, Ural Mts., Russia; Yanahara mine, Okayama Pref., Japan; Kambalda, WA, Australia.* BM *ZVMO 106:62(1977), AM 63:599(1978), ABI:450, CM 31:99(1993).*

2.6.3.1 Kochkarite Bi$_4$PbTe$_7$

Named in 1989 for the locality. HEX-R $P\bar{3}m1$. a = 4.416, c = 79.20, Z = 3, D = 7.89. *44-1439*: 3.23_{10} 2.36_6 2.21_2 2.01_3 1.98_2 1.82_2 1.61_1 1.48_2. Tabular crystals. Silver-gray, lead-gray streak, metallic luster. Perfect basal cleavage; fragments flexible, inelastic. H = 2–2$\frac{1}{2}$, VHN_{15} = 28–80. G = 7.94. R = 59.0–63.2 (540 nm); strongly anisotropic. *Kochkar deposit, southern Ural Mts., Russia.* BM *GRM 31:98(1989), AM 76:1434(1991).*

2.6.3.2 Aleksite Bi$_2$PbTe$_2$S$_2$

Named in 1978 for the locality. HEX-R $P\bar{3}m1$. a = 4.238, c = 79.76, Z = 6, D = 7.59. *29-765*: 3.63_3 3.09_{10} 2.25_4 2.12_6 1.97_3 1.35_4 1.31_4 1.21_3. Platy grains to 1 mm. Pale gray, metallic luster. Cleavage {0001}, perfect. H = 2. VHN_{20} = 40–65. R = 53.2 (540 nm); weakly anisotropic. *Near Tybo, Nye Co., NV; Barringer mine, Timmins, ON, Canada; St. David's mine, near Dolgellau, Wales; Alekseev mine, Sutemskii region, Stanovoi Mts., Russia.* BM *ZVMO 107:315(1978), AM 64:652(1979), ABI:7.*

2.6.3.3 Hedleyite Bi$_7$Te$_3$

Named in 1945 for the locality. HEX-R $R\bar{3}m$. a = 4.47, c = 119.0, Z = 6, D = 8.93. *12-719*: 3.25_{10} 2.37_5 2.24_4 1.99_3 1.85_3 1.63_4 1.48_4 1.42_3. Platy masses. Tin-white, tarnishing iron-black, gray streak, metallic luster. Cleavage {0001}, perfect; flexible lamellae. H = 2, VHN_{25} = 31–42. G = 8.91. R = 66.2–66.7 (540 nm); weakly anisotropic. *Good Hope claim and Oregon mine, near Hedley, BC, and Burwash Creek placer, Kluane Lake district, YT, Canada; Vaddas–Rieppe area, Norway; Kumbel skarn deposit, Kyrgyzstan; Ugat, Uzbekistan; Vostok-2, Maritime Territory, Russia.* BM *AM 30:644(1945), 34:364(1949); R:438; P&J:365; ABI:211.*

2.6.4.1 Genkinite (Pt,Pd)$_4$Sb$_3$

Named in 1977 for Alexandr D. Genkin (b.1919), Russian mineralogist. TET $P4_12_12$. a = 7.736, c = 24.161, Z = 8, D = 9.26. *29-133*: 3.47_3 3.02_9 2.27_{10} 2.15_4 2.12_4 1.93_6 1.91_5 1.27_4. Microscopic grains. Pale brown, metallic luster. H = 5$\frac{1}{2}$, VHN_{25} = 578–612. R = 50.8–51.1 (540 nm); moderately to strongly anisotropic. *Fox Gulch, Goodnews Bay, AK; Unst and Fetlar, Shetland Is., Scotland; Birbir R., Ethiopia; Onverwacht mine and Driekop mine, Transvaal, South Africa.* BM *CM 15:389(1977), 26:979(1988); AM 64:654(1979); P&J:404.*

2.6.5.1 Temagamite Pd_3HgTe_3

Named in 1973 for the locality. ORTH Space group unknown. $a = 11.608$, $b = 12.186$, $c = 6.793$, $Z = 6$, $D = 9.36$. 26-881(syn): 2.91_{10} 2.64_4 2.19_9 2.09_4 1.96_7 1.66_5 1.62_5 1.46_5. Microscopic grains. White, metallic luster. $H = 2\frac{1}{2}$, $VHN_{25} = 92$. $G = 9.5$. $R = 52.9$–53.9 (546 nm); weakly anisotropic. Stillwater Complex, MT; New Rambler mine, Albany Co., MT; *Temagami Cu deposit, Temagami I., ON, Canada*; Merensky Reef, Transvaal, South Africa. BM *CM 12:193(1973), AM 60:947(1975), R:440, ABI:520.*

2.6.6.1 Donharrisite $Ni_8Hg_3S_9$

Named in 1989 for Donald C. Harris (b.1936), Canadian mineralogist. MON $C2/m$. $a = 11.66$, $b = 6.91$, $c = 10.92$, $\beta = 97.43°$, $Z = 2$, $D = 5.18$. *CM 27*: 5.75_7 5.09_7 3.71_5 3.33_6 2.68_6 2.61_3 2.55_{10} 2.22_3. Microscopic micalike flakes. Brown, gray-brown streak, metallic luster. Cleavage {001}, perfect; conchoidal fracture; brittle. $H = 2$, $VHN_5 = 47$. $R = 44.2$–47.1 (540 nm); distinctly anisotropic. *Erasmus mine, Leogang, Salzburg Prov., Austria.* BM *CM 27:257(1989).*

PENTLANDITE GROUP

The pentlandite group consists of isostructural sulfides and selenides corresponding to

$$AB_8X_8$$

where

A = Ag, Cd, Mn, Fe, Ni, Co, Pb
B = Fe, Ni, Co, Cu
X = S, Se

The structures are all cubic, $Fm3m$, based on a cubic close packing of sulfur atoms with iron or nickel in a tetrahedral coordination forming a highly condensed B_8S_{14} group.

PENTLANDITE GROUP

Mineral	Formula	Space Group	a	D	
Pentlandite	$(Fe,Ni)_9S_8$	$Fm3m$	10.042	5.08	2.7.1.1
Argentopentlandite	$Ag(Fe,Ni)_8S_8$	$Fm3m$	10.50	5.05	2.7.1.2
Cobalt pentlandite	Co_9S_8	$Fm3m$	9.973	5.22	2.7.1.3
Shadlunite	$(Pb,Cd)(Fe,Cu)_8S_8$	$Fm3m$	10.91	4.72	2.7.1.4
Manganese-shadlunite	$(Mn,Pb)(Cu,Fe)_8S_8$	$Fm3m$	10.73	4.56	2.7.1.5
Geffroyite	$(Cu,Fe,Ag)_9(Se,S)_8$	$Fm3m$	10.889	5.39	2.7.1.6

2.7.1.1 Pentlandite

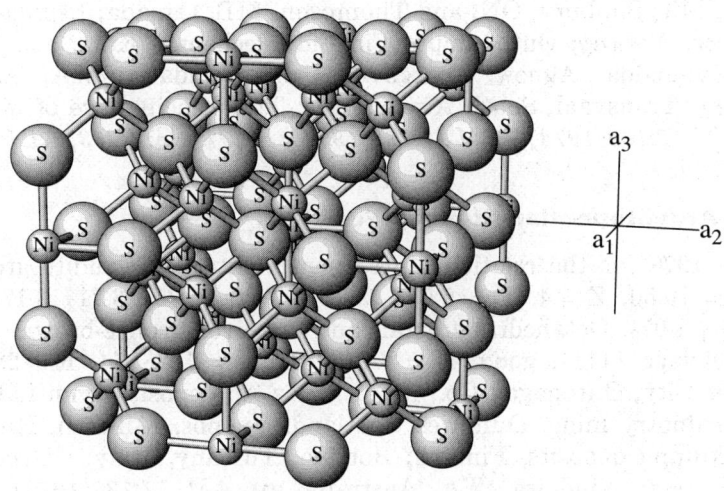

PENTLANDITE: (Ni, Fe)$_9$S$_8$

One type of Ni atom is in octahedral coordination and the other is tetrahedral. The tetrahedral Ni atoms share three edges with other tetrahedra and there is probably Ni–Ni bonding. S atoms are 4r- and 5-coordinated.

Notes:
1. Compare also with the chalcocite structures based on a hexagonal close packing of sulfur atoms.
2. Compare with the sphalerite structures based on an arrangement of cubic close-packed sulfur atoms.
3. Godlevskite (2.7.4.1) and kharaelakhite (2.7.5.1) fulfill the chemical formula, but they are reported in lower-symmetry orthorhombic space groups.

2.7.1.1 Pentlandite (Fe,Ni)$_9$S$_8$

Named in 1856 for Joseph B. Pentland (1797–1873), Irish scientist, who first noted the mineral. Pentlandite group. ISO $Fm3m$. $a = 10.042$, $Z = 4$, $D = 5.08$. *30-657*: 5.79_3 3.03_{10} 2.90_3 1.93_3 1.77_{10} 1.31_1 1.25_1 1.02_2. **Habit:** Massive, typically in granular aggregates; large grains show well-developed {111} parting planes. **Structure:** Based on cubic close-packed S atoms, with (Fe,Ni) in tetrahedral coordination forming highly condensed B$_8$S$_{14}$ groups joined at corners about an octahedrally coordinated (Fe,Ni) atom. **Physical properties:** Bronze-yellow, bronze-brown streak, metallic luster. No cleavage, conchoidal fracture. H = $3\frac{1}{2}$–4, VHN$_{100}$ = 210. G = 4.9–5.2. R = 49.2 (540 nm). Nonmagnetic. **Tests:** Insoluble in 1:1 HCl. Fuses readily to a steel-gray bead. **Chemistry:** Most analyses show Fe/Ni ratio near 1:1, but Fe ranges from 25 to 43%, and Ni from 19 to 34%; Co may be present, up to 11%. **Occurrence:** Pentlandite occurs in basic igneous rocks along with iron and nickel sulfides and arsenides. It is nearly always intimately associated with pyrrhotite. It is an accessory in some stony meteorites. **Localities:** The most abundant nickel mineral, pentlandite is widely distributed throughout the world. Lick Fork, Floyd Co., VA;

Stillwater, MT; Sudbury, ON, and Thompson, MB, Canada; *Espedalen, near Lillehammer, Norway*; Outokumpu, Finland; Pechenga, Kola, and Norilsk, Russia; Kambalda, Agnew, Spargoville, and Windaira, WA, Australia; Rustenburg, Transvaal, South Africa. Uses: The principal ore of nickel. BM *DI:242, CM 12:169(1973), MM 41:345(1977), R:497, P&J:293, ABI:396*.

2.7.1.2 Argentopentlandite Ag(Fe,Ni)$_8$S$_8$

Named in 1977 for the relationship to pentlandite. Pentlandite group. ISO $Fm3m$. $a = 10.50$, $Z = 4$, $D = 5.05$. 25-406: 6.06_2 5.25_2 3.71_2 3.17_{10} 3.02_2 2.02_4 1.86_{10} 1.07_3. Octahedral crystals and massive. Bronze-brown, metallic luster. Cleavage {111}, good. H = 3–$3\frac{1}{2}$, VHN$_{50}$ = 132–154. R = 28.8 (540 nm). Silver City, Ontonagon Co., MI; Agassiz Au deposit, Lynn Lake, MB, Canada; Talnotry mine, Dumfries, Scotland; Vuonos, Miihkali, Hietajärvi, and Outokumpu deposits, Finland; Bottino, Tuscany, Italy; *Oktyabr mine, Norilsk, Russia*; Windaira, WA, Australia. BM *AM 57:137(1972), ZVMO 106:688(1977), ABI:20*.

2.7.1.3 Cobalt Pentlandite Co$_9$S$_8$

Named in 1959 for the composition and relationship to pentlandite. Pentlandite group. Contains some Fe and Ni replacing Co. ISO $Fm3m$. $a = 9.973$, $Z = 4$, $D = 5.22$. 12-723: 5.75_6 3.01_{10} 2.88_6 2.29_5 1.92_8 1.76_{10} 1.30_5 1.02_6. Microscopic crystals and exsolved lamellae. Yellow-brown, metallic luster. Cubic cleavage. H = 4–$4\frac{1}{2}$, VHN$_{100}$ = 278–332. R = 57.0 (540 nm). Vauze mine, Noranda, PQ, and Langis mine, Cobalt, ON, Canada; Erglodd mine, central Wales; *Varislahti deposit* and Savonranta and Outokumpu deposits, *Finland*; Talmessi mine, near Anarak, Iran; Shimokaua mine, Hokkaido, Japan; Amianthus mine, Barberton, South Africa. BM *AM 44:897(1959), CM 9:597(1969), ABI:104*.

2.7.1.4 Shadlunite (Pb,Cd)(Fe,Cu)$_8$S$_8$

Named in 1973 for Tatyana Shadlun (b.1912), Russian mineralogist. Pentlandite group. ISO $Fm3m$. $a = 10.91$, $Z = 4$, $D = 4.72$. 25-1426: 5.42_3 3.84_4 3.29_{10} 3.16_2 2.11_4 1.93_9 1.67_2 1.42_2. Microscopic grains and veinlets. Yellow-gray, metallic luster. Polysynthetic twinning seen in polished section. H = $3\frac{1}{2}$–4, VHN$_{20}$ = 210. R = 24.5 (540 nm). In Cu ore, *Majak mine* and *Oktyabr mine, Norilsk, Russia*. BM *ZVMO 102:63(1973), AM 58:1114(1973), R:592, ABI:472*.

2.7.1.5 Manganese Shadlunite (Mn,Pb)(Cu,Fe)$_8$S$_8$

Named in 1973, the Mn analog of shadlunite. Pentlandite group. ISO $Fm3m$. $a = 10.73$, $Z = 4$, $D = 4.56$. 25-1425: 3.78_2 3.23_{10} 3.08_3 2.46_2 2.07_3 1.89_9 1.64_2 1.10_4. Microscopic grains and veinlets. Yellow-gray, metallic luster. Polysynthetic twinning seen in polished section. H = $3\frac{1}{2}$, VHN$_{20}$ = 195. R = 29.0

(540 nm). *Oktyabr deposit, Norilsk, Russia.* BM *ZVMO 102:63(1973), AM 58:1114 (1973), ABI:311.*

2.7.1.6 Geffroyite $(Cu,Fe,Ag)_9(Se,S)_8$

Named in 1982 for Jacques Geffroy (b.1918), French metallurgist. Pentlandite group. ISO $Fm3m$. $a = 10.889$, $Z = 4$, $D = 5.39$. *35-523*: 3.28_9 3.15_9 2.44_4 2.09_6 1.93_{10} 1.82_4 1.66_5 1.11_6. Microscopic grains. Cream-brown, metallic luster. $H = 4$, $VHN_{15} = 68–72$. $R = 30.1$ (540 nm). *Chaméane, Puy-de-Dôme, France*; San Miguel mine, Moctezuma, SON, Mexico. BM *TMPM 29:151(1982), AM 67:1074(1982), P&J:404, ABI:176.*

2.7.2.1 Mackinawite $(Fe,Ni)S_{0.9}$

Named in 1962 for the locality. May contain up to 9% Cu and 8% Co. TET $P4/nmm$. $a = 3.676$, $c = 5.032$, $Z = 2$, $D = 4.29$. *15-37*: 5.03_{10} 2.97_8 2.31_8 1.84_4 1.81_8 1.73_4 1.24_4 1.06_6. Microscopic pyramidal crystals or anhedral grains. Bronze-gray, black streak, metallic luster. Cleavage {001}, good; soft. $VHN_{15} = 63–89$. $R = 21.6–45.3$ (540 nm); strongly pleochroic and anisotropic. Widely distributed in small amounts; formed by hydrothermal activity in ore deposits, during serpentinization of ultrabasic rocks, and in the reducing environment of sediments and sedimentary rocks; an accessory in meteorites and lunar rocks. Howard Montgomery quarry, Howard Co., MD; *Mackinaw mine, Snohomish Co., WA*; Kramer borate deposit, Kern Co., CA; Muskox intrusion, NWT, Canada; Kynance Cliff, Lizard, Cornwall, England; Outokumpu mine, Finland; Talnakh area, Norilsk, Russia; Broken Hill, NSW, and Scotia nickel deposit, WA, Australia; and many other localities. BM *AM 48:511(1963), R:683, P&J:247, ABI:305.*

2.7.3.1 Yarrowite Cu_9S_8

Named in 1980 for the locality. HEX-R Space group uncertain, probably $P3m1$. $a = 3.800$, $c = 67.26$, $Z = 3$, $D = 4.89$. *CM 18:514*: 5.96_2 5.03_3 3.68_3 3.06_6 2.85_3 2.77_4 2.51_2 1.90_{10}. Black, metallic luster. Cleavage {0001}. $H = 2\frac{1}{2}$, $VHN_{15} = 93–98$. $R = 10.7–25.6$ (540 nm); pleochroic in blue to blue-white [formerly called "blaubleibender Covellit" (blue-remaining covellite) for this reason]. In red-bed copper deposit at *Yarrow Creek, AB, Canada*; High Rolls district, Otero Co., NM; Cannington Peak, Somerset, England. BM *CM 18:511(1980), R:676, ABI:583.*

2.7.4.1 Godlevskite Ni_9S_8

Named in 1969 for Mikhail N. Godlevsky (1902–1984), Russian economic geologist. May contain up to 3% Fe. ORTH $C222$. $a = 9.180$, $b = 11.263$, $c = 9.457$, $Z = 4$, $D = 5.273$. *22-1193*: 3.28_5 2.85_{10} 2.33_4 2.18_4 2.10_5 1.80_9 1.79_8 1.65_8. Microscopic grains. Yellow, metallic luster. Complexly twinned. $H = 4\frac{1}{2}–5$, $VHN_{50} = 383–415$. $R = 50.5–52.2$ (540 nm); strongly anisotropic. Near Moapa, Clark Co., NV; Orford mine, PQ, Texmont mine, south of Timmins, ON, Canada; *Zapolyarnyi mine, Norilsk, Russia*; Mt. Clifford, near Leonora, WA, Australia; Bou Azzer, Morocco; Amianthus mine, Barberton, South Africa. BM *GRM 11:115(1969), AM 55:317(1970), R:505, P&J:182, CM 26:283(1988)*.

2.7.5.1 Kharaelakhite $(Pt,Cu,Pb)_9S_8$

Named in 1985 for the locality. Contains up to 6% Fe and 5% Ni. ORTH *Pmmm*. $a = 9.713$, $b = 8.333$, $c = 14.52$, $Z = 4$, $D = 7.66$. *39-411*: 6.30_2 3.04_4 2.80_{10} 2.67_2 2.59_2 1.85_3 1.81_5 1.73_3. Thin rims on braggite–cooperite. Gray, metallic luster. $R = 37.1$–42.2 (540 nm); distinctly anisotropic. *Talnakh deposit, Kharaelakh Plateau, Norilsk, Russia*. BM *MZ 7:78(1985), AM 74:1215(1989)*.

GALENA GROUP

The galena group minerals consist of isostructural species having the chemical formula

$$AB$$

where

$A = $ Pb, Mn, Ca, Mg
$B = $ S, Se, Te

and the structure is the halite structure, $Fm3m$, a face-centered cubic close-packed association of two different elements.

GALENA GROUP

Mineral	Formula	Space Group	a	D	
Galena	PbS	$Fm3m$	5.94	7.60	2.8.1.1
Clausthalite	PbSe	$Fm3m$	6.12	8.28	2.8.1.2
Altaite	PbTe	$Fm3m$	6.44	8.31	2.8.1.3
Alabandite	MnS	$Fm3m$	5.22	4.05	2.8.1.4
Oldhamite	CaS	$Fm3m$	5.69	2.59	2.8.1.5
Niningerite	(Mg,Fe,Mn)S	$Fm3m$	5.17	2.66	2.8.1.6

Notes:
1. The halite, periclase, and galena groups are all essentially isostructural.
2. Borovskite (2.8.1.7) and crerarite (2.8.1.8) are possible members of this group.
3. See the periclase (4.2.1.1) group for other halite structures, with oxygen in place of S, Se, or Te.

2.8.1.1 Galena PbS

Name is the Latin word for *lead ore* or *dross from melted lead*. Galena group. ISO $Fm3m$. $a = 5.94$, $Z = 4$, $D = 7.60$. *5-592*(syn): 3.43_8 2.97_{10} 2.10_6 1.79_4 1.71_2 1.48_1 1.33_2 1.21_1. **Habit**: Cubes and cubo-octahedral crystals; commonly massive, coarse to very fine granular. Twinned on {111}, penetration or contact twins; on $\{114\}$, with lamellae giving diagonal striations on cleavages. **Physical properties:** Lead-gray, lead-gray streak, metallic luster. Cleavage $\{100\}$, perfect; parting or cleavage on $\{111\}$. $H = 2\frac{1}{2}$, $VHN_{100} = 79\text{--}104$. $G = 7.58$. $R = 42.4$ (540 nm). **Tests:** On heating, emits SO_2 fumes and yields a globule of metallic lead. Decomposed by HNO_3 with the formation of $PbSO_4$ and the separation of S. **Chemistry:** Usually pure PbS; the silver reported in some analyses is usually due to admixed silver minerals. Bi, Ag, Sb, Hg, and Cu have been found in amounts greater than 1 wt%. **Occurrence:** Galena is the commonest lead mineral and occurs in many types of deposits. Extensive deposits of galena (and sphalerite) occur as irregular masses in solution cavities in limestone. Hydrothermal vein deposits of galena frequently contain significant silver values. Some important deposits are of contact or regional metamorphic origin. **Localities:** Some localities for outstanding specimens are: Balmat, St. Lawrence Co., NY; Tri-State mining district, MO–KS–OK; Breckenridge, Creede, and Leadville, CO; Coeur d'Alene, ID; Admiralty Is., AK; Bathurst, NB, Kidd Creek, ON, Kimberley, BC, Pine Point, NWT, Canada; Naica and Santa Eulalia, CHIH, Mexico; Mogul mine, Tipperary, Eire; Alston Moore and Weardale, Cumberland, England; Andreasberg, Neudorf and Clausthal, Harz, and Freiberg, Saxony, Germany; Pribram, Czech Republic; Bottino, Tuscany, Italy; Trepca, Serbia; Madan, Rhodope Mts., Bulgaria; Dalnegorsk, eastern Siberia, Russia; Tsumeb, Namibia; Mt. Isa, Q, and Broken Hill, NSW, Australia; Huanzala mine, Dept. Ancash, Peru. **Uses:** The principal ore of lead. BM *DI:200, R:646, P&J:171, CM 27:363(1989), ABI:170.*

2.8.1.2 Clausthalite PbSe

Named in 1832 for the locality. Galena group. ISO $Fm3m$. $a = 6.12$, $Z = 4$, $D = 8.28$. *6-354*(syn): 3.54_3 3.06_{10} 2.17_7 1.85_2 1.77_2 1.53_1 1.37_3 1.25_2. Massive, commonly fine granular. Lead-gray, lead-gray streak, metallic luster. Cleavage $\{100\}$, good. $H = 2\frac{1}{2}\text{--}3$, $VHN_{100} = 44\text{--}49$. $G = 8.0\text{--}8.2$. $R = 51.3$ (540 nm). Widely distributed in small amounts. Rifle and Garfield mines, Garfield Co., CO; Corvusite mine, Grand Co., UT; Kidd Creek and Hemlo mines, ON, and Goldfields dist., SK, Canada; *Clausthal*, Tilkerode, Lerbach, and Zorge, *Harz, Germany*; Lasovice, Bohemia, Czech Republic; El Sharana, NT, Australia; Pacajake, Colquechaca, Bolivia; Sierra de Umango and Sierra de Cachueta, La Rioja, Argentina. BM *DI:204, AM 44:166(1959), R:659, P&J:123.*

2.8.1.3 Altaite PbTe

Named in 1845 for the locality. Galena group. ISO $Fm3m$. $a = 6.44$, $Z = 4$, $D = 8.31$. *8-28*(syn): 3.23_{10} 2.28_8 1.86_3 1.61_2 1.44_5 1.31_4 1.07_2 1.02_2. Cubic crystals rare; usually massive, fine-grained. Tin-white with a yellow tinge, tarnishing to bronze-yellow, metallic luster. Cleavage $\{100\}$, perfect. $H = 3$, $VHN_{25} = 51$. $G = 8.1\text{--}8.3$. $R = 69.7$ (540 nm). Widely distributed, usually in hydrothermal Au deposits. Kings Mt. mine, Gaston Co., NC; Red Cloud mine, Boulder Co., CO; Hilltop mine, Dona Ana Co., NM; Stanislaus mine,

Calaveras Co., CA; Mattagami Lake mine, Mattagami, PQ, Kirkland Lake mines and Sudbury, ON, Canada; Adervielle, Hautes-Pyrenees, France; Sacaramb, Stanija, and Herja, Romania; Merisi Cu deposit, Georgia; *Savodinsky mine, Zyryanovsk, Altai Mts., Kazakhstan*; Nojori mine, Japan; Kalgoorlie and Norseman, WA, Australia; Condorriaco mine, Coquimbo, Chile. BM *DI:205, AM 34:361(1949), R:661, P&J:66, ABI:11.*

2.8.1.4 Alabandite MnS

Named in 1832 for the locality. Galena group. Composition usually near MnS, but may contain up to 22% Fe and 7% Mg. ISO $Fm3m$. $a = 5.22$, $Z = 4$, $D = 4.05$. $6\text{-}518$(syn): 3.02_1 2.61_{10} 1.85_5 1.58_1 1.51_2 1.31_1 1.17_2 1.07_2. Usually massive or granular. Black, tarnishing to brown, green streak, submetallic luster. Cleavage {100}, perfect. $H = 3\frac{1}{2}\text{-}4$, $VHN_{100} = 240\text{-}251$. $G = 4.0\text{-}4.1$. $R = 22.8$ (540 nm). Occurs in low-temperature vein deposits, and as an accessory mineral in some enstatite chondrites. Schellbourne, White Pine Co., MT; Manhattan mine, Park Co., CO; Bisbee, and Lucky Cuss mine(*), Tombstone, AZ; McDame Creek, BC, Canada; Mina La Preciosa, Tlalchichuca, PUE, Mexico; Litosice, Bohemia, Czech Republic; Molinello mine, Liguria, Italy; Sacaramb and other localities, Romania; Ruen, Bulgaria; *Alabanda, Caria, Turkey*; Dastakert deposit, Armenia; Solongo and Norilsk, Russia; Noda-Tamagawa, Kaso, and other Mn mines, Honshu, Japan; Broken Hill(*), NSW, Australia; Uchucchacua mine, Peru. BM *AM 56:1269(1971), R:642, P&J:62, ABI:5.*

2.8.1.5 Oldhamite CaS

Named in 1862 for Thomas Oldham (1816–1878), director of the Geological Survey of India. Galena group. May contain up to 2% Mg and 6% Fe. ISO $Fm3m$. $a = 5.69$, $Z = 4$, $D = 2.59$. $8\text{-}464$(syn): 2.85_{10} 2.01_7 1.64_2 1.42_1 1.27_2 1.16_1 0.95_1 0.90_1. Microscopic grains. White to pale brown, white streak. Cleavage {100}, perfect. $H = 4$. $G = 2.58$. Isotropic; $N = 2.14$. Occurs in enstatite chondrites and enstatite achondrites. *Bustee meteorite, India.* BM *DI:208, R:642, GCA 46:2083(1982), ABI:360.*

2.8.1.6 Niningerite (Mg,Fe,Mn)S

Named in 1967 for Harvey H. Nininger (1887–1986), U.S. meteorite researcher. Galena group. Contains up to 37% Fe, 12% Mn, 3% Ca, and 2% Cr. ISO $Fm3m$. $a = 5.17$, $Z = 4$, $D = 2.66$ (MgS). $35\text{-}730$(syn): 3.00_1 2.60_{10} 1.84_6 1.50_2 1.30_1 1.16_1 1.06_1 0.87_1. Microscopic grains. Gray. An accessory mineral in enstatite chondrites. BM *SCI 155:451(1967), AM 56:1269(1971), R:642, GCA 46:2083(1982).*

2.8.1.7 Borovskite Pd_3SbTe_4

Named in 1973 for Igor B. Borovski (1909–1985), Russian chemist. Possibly galena group. ISO $Fm3m$. $a = 5.79$, $Z = 1$, $D = 8.12$. $26\text{-}1426$: 3.34_3 2.90_{10} 2.04_6 1.60_3 1.55_5 1.30_4 1.24_3 1.18_4. Microscopic grains. Dark gray, metallic luster. $H = 2\frac{1}{2}$, $VHN_{10} = 88$. *Khautovaarsk deposit, Karelia, Russia.* BM *ZVMO 102:427(1973), AM 59:873(1974), ABI:63.*

2.8.1.8 Crerarite $Bi_3(Pb,Pt)(S,Se)_{4-x} (x = 0.1\text{–}0.2)$

Named in 1994 for David Crerar (1945–1994), professor of geochemistry, Princeton University. Possibly galena group. ISO $Fm3m$. $a = 5.86$, $Z = 1$, $D = 7.75$. *NJMM*: 3.37_5 2.94_{10} 2.07_3 1.77_2 1.69_2 1.47_5 1.35_1 1.20_1. Microscopic grains. White-gray. $R = 50$ (540 nm). Found in amphibolite boulders, *Lac Sheen, near Belleterre, PQ, Canada.* BM *NJMM:567(1994).*

SPHALERITE GROUP

The sphalerite group consists of six minerals with the formula

$$AB$$

where

$A = $ Zn, Fe, Hg, Cd
$B = $ Se, Te

and a structure in space group $F\bar{4}3m$. The structure is based on cubic close-packed sulfur atoms, with A atoms in tetrahedral coordination.

SPHALERITE GROUP

Mineral	Formula	Space Group	a	D	
Sphalerite	(Zn,Fe)S	$F\bar{4}3m$	5.406	4.10	2.8.2.1
Stilleite	ZnSe	$F\bar{4}3m$	5.667	5.27	2.8.2.2
Metacinnabar	HgS	$F\bar{4}3m$	5.903	7.86	2.8.2.3
Tiemannite	HgSe	$F\bar{4}3m$	6.085	8.24	2.8.2.4
Coloradoite	HgTe	$F\bar{4}3m$	6.460	8.09	2.8.2.5
Hawleyite	CdS	$F\bar{4}3m$	5.818	4.87	2.8.2.6

Notes:
1. Polhemusite (2.8.3.1) is a tetragonal (Zn,Hg)S structure with a cubic pseudocell with $a = 5.33$.
2. See the pentlandite group for other structures based on closed-packed arrangements of sulfur atoms.
3. See wurtzite (2.8.7.1) for a hexagonal polymorph of ZnS.
4. See the tetrahedrite group for similar structures accommodating additional elements.

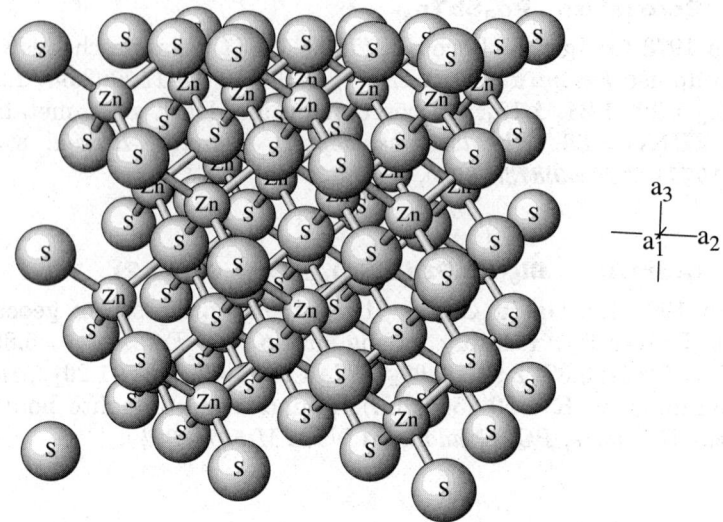

SPHALERITE: ZnS

Space group $F\bar{4}3m$. Each atom is coordinated by four of the other type. Each type of atom forms a cubic closest-packed arrangement. *See also* WURZITE.

2.8.2.1 Sphalerite (Zn,Fe)S

Named in 1847 from the Greek for *treacherous*, the mineral sometimes being mistaken for galena but yielding no lead. The old German miners' name *Blende* for sphalerite means *blind* or *deceiving*. Sphalerite group. ISO $F\bar{4}3m$. $a = 5.406$, $Z = 4$, $D = 4.10$. 5-566(syn): 3.12_{10} 2.71_1 1.91_5 1.63_3 1.35_1 1.24_1 1.10_1 1.04_1. **Habit:** Tetrahedral or dodecahedral crystals, often distorted and complex; frequently as cleavable masses, coarse to fine granular, fibrous, concretionary, or botryoidal. Twinning common on {111}, as simple or multiple contact or penetration twins. **Physical properties:** Yellow, brown, black, less commonly red or green; white or colorless when iron-free; white to yellow or brown streak, resinous to adamantine luster. Cleavage {011}, perfect. $H = 3\frac{1}{2}$–4, $VHN_{100} = 208$–224. $G = 3.9$–4.1, decreasing with Fe content. Pyroelectric. Sometimes triboluminescent. **Tests:** Soluble in HCl with evolution of H_2S. **Chemistry:** The Fe content ranges up to about 26%, corresponding to nearly 50 mol % FeS; Mn may be present up to 6%, and Cd up to 1.7%. **Optics:** $R = 16.6$ (540 nm); isotropic; $N = 2.37$–2.47, increasing with Fe content. **Occurrence:** Sphalerite has been reported in small amounts as a primary constituent of granite and in pegmatites. It is often found in contact-metamorphic deposits and in veins. Large deposits (Mississippi valley type) occur along solution channels in limestone and dolomite. It is an accessory mineral in some meteorites and lunar rocks. **Localities:** Sphalerite is the commonest zinc mineral and occurs widely throughout the world; only a few localities for fine specimens can be cited. Thomasville quarry, York Co., PA; Franklin, NJ; Elmwood mine, Carthage, TN; many localities of the Mississippi valley type in WI, IL, MO, KS, and OK; Hanover, NM; Butte, MT; Big Four mine, Kremmling, CO; Bingham and Park City, UT; Coeur d'Alene, ID; Sullivan mine, Kimberley, BC, and Watson Lake(*), YT, Canada; Cananea, SON, Santa Eulalia and Naica, CHIH, Charcas, SLP, Mexico; Alston Moor and Egremont, Cumberland, England; Picos de Europa(*), Santander, Spain; Lengenbach(*), Binn, Switzerland; Bottina, Tuscany, Italy; Rodna and Baia Sprie, Romania; Trepca, Serbia; Madan, Rhodope Mts., Bulgaria; Dalnegorsk, eastern Siberia, Russia; Chichibu mine, Saitama Pref., Japan; Mt. Isa, Q, and Broken Hill, NSW, Australia; Huaron, Casapalca, and Huancavelica, Peru. Uses: The principal ore of zinc. BM *DI:210, AC(A) 36:482(1980), R:506, P&J:95, ABI:488.*

2.8.2.2 Stilleite ZnSe

Named in 1956 for Hans Stille (1876–1966), German geologist. Sphalerite group. ISO $F\bar{4}3m$. $a = 5.667$, $Z = 4$, $D = 5.27$. *5-522*(syn): 3.27_{10} $2,00_7$ 1.71_4 1.42_1 1.30_1 1.16_2 1.09_1 0.96_1. Microscopic grains. Gray, gray streak, metallic luster. H = 5. G = 5.3. *Shinkolobwe, Shaba Prov., Zaire*; Santa Brigida mine, La Rioja Prov., Argentina. BM *AM 42:584(1957), R:520, ABI:499.*

2.8.2.3 Metacinnabar HgS

Named in 1870 for its relationship to cinnabar. Sphalerite group. Synonyms: onofrite (selenian), saukovite (cadmian). May contain up to 12% Cd, 9% Zn, 10% Se. ISO $F\bar{4}3m$. $a = 5.903$, $Z = 4$, $D = 7.86$. *22-729*: 3.41_{10} 2.95_4 2.09_8 1.78_7 1.70_2 1.36_3 1.20_3 1.14_3. Usually massive, rarely as small tetrahedral crystals. Gray-black, black streak, metallic luster. Shows lamellar twinning, subconchoidal to uneven fracture, brittle. H = 3. G = 7.5–7.7. R = 25.6 (540 nm). Widely distributed in Hg deposits. Terlingua, Brewster Co., TX; Marysvale, Piute Co., UT; Reward mines, King Co., WA; New Almaden, Santa Clara Co., New Idria, San Benito Co., Redington mine, Lake Co., and other Hg deposits in CA; Read Is., BC, Canada; San Onofre, ZAC, Almoloya, GRO, and Guadalcazar, SLP, Mexico; Leogang, Salzburg, Austria; Levigliani, Tuscany, Italy; Idria, Slovenia; Baia Sprie, Romania; Uland deposit, Gornyi Altai, Russia; Wen-Shan-Chang, Kweichow, China. BM *DI:215, AM 44:471(1959), R:521, P&J:267, ABI:327.*

2.8.2.4 Tiemannite HgSe

Named in 1855 for W. Tiemann, who discovered the mineral in 1829. Sphalerite group. ISO $F\bar{4}3m$. $a = 6.085$, $Z = 4$, $D = 8.24$. *8-469*(syn): 3.51_{10} 3.04_2 2.15_5 1.84_3 1.76_1 1.52_1 1.40_1 1.24_1. Tetrahedral crystals and massive. Dark gray to black, black streak, metallic luster. No cleavage, twinned on {111}, uneven to conchoidal fracture, brittle. H = $2\frac{1}{2}$, $VHN_5 = 22$–26. G = 8.2–8.5. R = 29.8 (540 nm). Marysvale, Piute Co., UT; New Idria, San Benito Co., and San Joaquin Ranch, Orange Co., CA; Nicholson Bay, Lake Athabaska, SK, Canada; Tilkerode, Lehrbach, Zorge, and *Clausthal, Harz, Germany*; Lasovice, Bohemia, Czech Republic; El Sharana mine, NT, Australia; Pacajake, near Colquechaca, Bolivia; Sierra de Umango, La Rioja Prov., Argentina. BM *DI:217*, *AM 35:358(1950)*, *R:523*, *P&J:370*.

2.8.2.5 Coloradoite   HgTe

Named in 1877 for the state of Colorado. Sphalerite group. ISO $F\bar{4}3m$. $a = 6.460$, $Z = 4$, $D = 8.09$. *32-665*(syn): 3.73_{10} 3.23_1 2.28_6 1.95_3 1.62_1 1.48_1 1.32_1 1.24_1. Massive, granular. Dark gray to black, black streak, metallic luster. No cleavage, brittle, friable. $H = 2\frac{1}{2}$, $VHN_{100} = 25$–28. $G = 8.0$–8.1. $R = 35.4$ (540 nm). *Keystone area, Boulder Co.*, and Good Hope mine, Gunnison Co., *CO*; Norwegian mine, Tuolumne Co., CA; Robb–Montbray mine, PQ, and mines in the Kirkland Lake area, ON, Canada; Uzel'ginsk deposit, Ural Mts., Russia; Emperor mine, Vatukoula, Fiji; Kalgoorlie, WA, Australia. BM *DI:218*, *AM 34:342(1949)*, *R:524*, *P&J:127*.

2.8.2.6 Hawleyite   CdS

Named in 1955 for James E. Hawley (1897–1965), Canadian mineralogist. Sphalerite group. Dimorphous with greenockite. ISO $F\bar{4}3m$. $a = 5.818$, $Z = 4$, $D = 4.87$. *10-454*: 3.36_{10} 2.90_4 2.06_8 1.75_6 1.45_2 1.34_3 1.19_3 1.12_3. Fine-grained powdery coatings: Orange, orange streak, earthy luster. Isotropic; $N = 1.78$. Occurs as coatings on sphalerite. Franklin, NJ; Eureka Co., NV; Crestmore, Riverside Co., CA; *Hector–Calumet mine, Galena Hill, YT, Canada*; Tynagh, Galway, Eire; Los Blancos mine, Sierra de Cartagena, Spain; Komna, Moravia, Czech Republic; Norilsk area, Ural Mts., Russia; Tui mine, Te Aroha, New Zealand; Mina Coquimbana, Atacama, Chile. BM *AM 40:555(1955)*, *NJMM:205(1971)*, *507(1974)*, *ABI:208*.

2.8.3.1 Polhemusite   (Zn,Hg)S

Named in 1978 for Clyde Polhemus Ross (1891–1965), U.S. geologist. TET $P4/n$. $a = 8.71$, $c = 14.74$, $Z = 24$, $D = 4.23$. *31-870*: 3.63_6 3.16_7 3.08_{10} 1.89_8 1.61_7 1.22_6 1.09_6 1.02_6. Microscopic tetragonal crystals. Black, resinous to adamantine luster. Knee-shaped twins with {605} as composition plane. $H = 4$, $VHN_{25} = 262$. $R = 17.1$–18.3 (546 nm); moderately to strongly anisotropic. *B and B deposit, Big Creek district, Valley Co., ID*; Getchell mine, Humboldt Co., NV. BM *AM 63:1153(1978)*, *ABI:415*.

2.8.4.1 Xingzhongite   (Ir,Cu)S

Named in 1974, presumably for the locality. ISO $Fd3m$. $a = 9.970$, $Z = 16$, $D = 7.94$. *29-551*: 5.80_8 3.00_8 2.88_5 2.49_6 1.77_{10} 1.34_4 1.21_8 1.17_5. Microscopic grains. Steel-gray, metallic luster. $H = 6$, $VHN_{50} = 753$. $R = 38.9$ (544 nm). Fox Gulch, Goodnews Bay, AK; *China*, locality not given; Tiebaghi, New Caledonia. BM *AM 61:184(1976)*, *69:412(1984)*, *CM 26:177(1988)*, *ABI:582*.

2.8.5.1 Cooperite PtS

Named in 1928 for Richard A. Cooper (1890–1972), South African metallurgist, who described the mineral. May contain up to 6% Pd. Dimorphous with braggite. TET $P4_2/mmc$. $a = 3.47$, $c = 6.10$, $Z = 2$, $D = 10.27$. *26-1302*: 3.05_2 3.02_{10} 2.45_2 1.91_3 1.75_1 1.74_1 1.51_1 1.50_2. Crystal fragments and microscopic grains. Steel-gray, metallic luster. Cleavage {011}, conchoidal fracture. H = 4–5, $VHN_{100} = 743–1018$. $G = 9.5$. $R = 39.4–47.2$ (540 nm); strongly anisotropic. Stillwater Complex, MT; in gravels, Trinity, Merced, and Stanislaus rivers, CA; Lac des Iles Complex, PG, and Tulameen district, BC, Canada; Gusevogorskii and Norilsk deposits, Russia; Round Hill, near Orepuki, New Zealand; at several localities along the *Merensky Reef, Bushveld Complex, Transvaal, South Africa*. BM *DI:258*, *AM 63:832(1978)*, *R:696*, *P&J:129*, *ABI:107*.

2.8.5.2 Vysotskite PdS

Named in 1962 for N. K. Vysotskii, Russian geologist, discoverer of platinum at Norilsk. Contains up to 17% Ni and 14% Pt. Forms a series with braggite. TET $P4_2/m$. $a = 6.371$, $c = 6.540$, $Z = 8$, $D = 6.40$. *15-151*: 2.91_{10} 2.86_{10} 2.64_7 2.61_8 2.15_6 1.86_7 1.73_6 1.72_8. Microscopic grains. Silver-white, metallic luster. $VHN_{100} = 830$. $G = 8.4$. $R = 45.1–45.8$ (540 nm); moderately anisotropic. Stillwater Complex, MT; near Moapa, Clark Co., NV; Yuba R., Nevada Co., CA; Lac des Iles Complex, ON, Canada; Konttijärvi Intrusion, Finland; *Norilsk, Russia*; Merensky Reef, Transvaal, South Africa. BM *ZVMO 91:718(1962)*; *AM 48:708(1963), 63:832(1978)*; *R:695*; *P&J:102*; *ABI:565*.

2.8.5.3 Braggite (Pt,Pd)S

Named in 1932 for W. H. (1862–1942) and W. L. (1890–1971) Bragg, pioneers in the X-ray investigation of crystals. Contains up to 5% Ni. Forms a series with vysotskite. TET $P4_2/m$. $a = 6.38$, $c = 6.59$, $Z = 8$, $D = 8.90$. *9-421*: 2.93_3 2.86_{10} 2.64_3 1.85_3 1.71_2 1.60_2 1.42_3 1.39_2. Prisms to 2 cm and rounded grains. Steel-gray, metallic luster. H = 5, $VHN_{100} = 981$. $G = 8.9$. $R = 42.6–44.2$ (540 nm); distinctly anisotropic. Stillwater Complex, MT; Norilsk, Russia; Round Hill near Orepuki, New Zealand; *Merensky Reef, Bushveld Complex, Transvaal, South Africa*. BM *DI:259*, *AM 63:832(1978)*, *R:695*, *P&J:102*, *ABI:67*.

2.8.6.1 Polarite Pd(Bi,Pb)

Named in 1969 for the locality (Polar Ural Mts.). ORTH $Ccm2_1$. $a = 7.191$, $b = 8.693$, $c = 10.681$, $Z = 16$, $D = 12.51$. *23-1298*: 3.33_2 2.65_{10} 2.50_3 2.25_5 2.16_9 1.64_5 1.40_3 1.22_3. Microscopic grains. White with yellow tint, metallic luster. H = 4, $VHN_{50} = 168–232$. $R = 59.2$ (540 nm); slightly anisotropic. Goodnews Bay, AK; *Talnakh, Norilsk, Russia*; Union mine, Merensky Reef, Transvaal, South Africa. BM *ZVMO 98:708(1969)*, *AM 55:1810(1970)*, *R:364*, *ABI:414*.

WURTZITE GROUP

Wurtzite minerals have the formula

$$AB$$

where

A = Cd, Zn
B = S, Se

and the structure is hexagonal, $P6_3mc$.

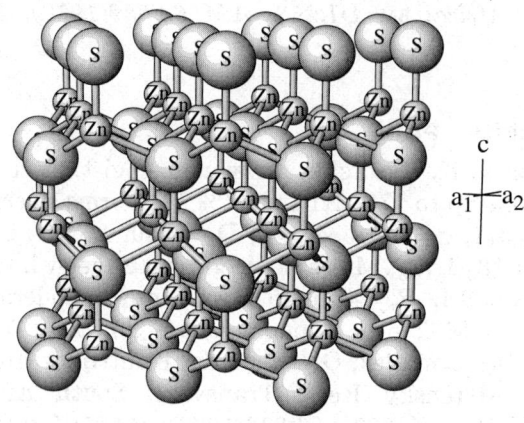

WURTZITE: ZnS

Space group $P6_3mc$. Each atom is coordinated by four of the other type. Each type of atom forms a hexagonal closest-packed arrangement.

WURTZITE GROUP

Mineral	Formula	Space Group	a	c	D	
Wurtzite	ZnS	$P6_3mc$	3.820	6.260	4.09	2.8.7.1
Greenockite	CdS	$P6_3mc$	4.136	6.713	4.82	2.8.7.2
Cadmoselite	CdSe	$P6_3mc$	4.299	7.010	5.66	2.8.7.3

Notes:
1. The common 2H polytype of wurtzite is described above. Other polytypes with larger C repeats are known.
2. See also the nickeline group, $P6_3/mmc$.

2.8.7.1 Wurtzite ZnS

Named in 1861 for Charles A. Wurtz (1817–1884), French chemist. Wurtzite group. Trimorphous with sphalerite and matraite. May contain up to 8% Fe and 4% Cd. HEX $P6_3mc$. $a = 3.820$, $c = 6.260$, $Z = 2$, $D = 4.09$ (the common 2H polytype; several polytypes with larger c-axis repeats are known). $5\text{-}492$(syn): 3.31_{10} 3.13_9 2.93_8 2.27_3 1.91_7 1.76_5 1.63_5 1.30_1. Hemimorphic pyramidal crystals up to 1.5 cm; also massive and as botryoidal banded crusts. Brown to black, brown streak, resinous to submetallic luster. Cleavage $\{11\bar{2}0\}$, easy; $\{0001\}$, difficult. $H = 3\frac{1}{2}\text{-}4$. $G = 4.0\text{-}4.1$. $R = 16.6$ (540 nm). Uniaxial (+); $N_O = 2.356$, $N_E = 2.378$. Occurs in hydrothermal veins and on concretions in sedimentary rocks. Thomaston Dam, Litchfield Co., CT; Sterling Hill mine, Sussex Co., NJ; Union Bridge quarry, Frederick, MD; Witmer(*), Allegheny Co., PA; Joplin, MO; Butte, MT; Gladenbach, Odenwald, Germany; Bleischarley mine, Brzezing, Giesche Spolka, Poland; Pribram, Czech Republic; Mezica, Slovenia; Baia Sprie, Romania; Beregovsk, Ukraine; Hercules mine, Mt. Read, TAS, Australia; Tsumeb, Namibia; Animas mine(*), Potosi, and Colquechaca, Llallagua, Oruro, and other localities, Bolivia. BM *DI:226, AM 35:29(1950), AC(B) 26:1192(1970), R:577, P&J:395, ABI:579.*

2.8.7.2 Greenockite CdS

Named in 1840 for Charles Murray Cathcart (1783–1859), Lord Greenock. Wurtzite group. Dimorphous with hawleyite. HEX $P6_3mc$. $a = 4.136$, $c = 6.713$, $Z = 2$, $D = 4.82$. $6\text{-}314$(syn): 3.58_8 3.37_6 3.16_{10} 2.45_3 2.07_6 1.90_4 1.79_2 1.76_5. Hemimorphic pyramidal crystals up to 3 cm and earthy coatings. Yellow to red color and streak, resinous to adamantine luster. Cleavage $\{11\bar{2}2\}$, distinct; $\{0001\}$, imperfect. $H = 3\text{-}3\frac{1}{2}$, $VHN_{25} = 98$. $G = 4.8\text{-}4.9$. $R = 19.6$ (540 nm). Uniaxial (+), $N_O = 2.506$, $N_E = 2.529$. Paterson(*), Passaic Co., and Franklin and Sterling Hill, Sussex Co., NJ; Arlington quarry, Leesburg, VA; Friedensville, Lehigh Co., PA; Joplin, MO; Cerro Gordo, Inyo Co., CA; *Bishopton(*), Renfrewshire, Scotland*; Pierrefitte, Hautes-Pyrenees, France; Los Blancos mine, Sierra de Cartagena, Spain; Pribram, Czech Republic; Kti-Teberda deposit, Caucasus Mts., Russia; Tsumeb, Namibia; Asunta mine(*), Potosi, and Llallagua mines, Bolivia. BM *DI:228; AM 42:184(1957), 46:1382(1961); R:582; P&J:187; ABI:194.*

2.8.7.3 Cadmoselite CdSe

Named in 1957 for the composition. Wurtzite group. HEX $P6_3mc$. $a = 4.299$, $c = 7.010$, $Z = 2$, $D = 5.66$. $8\text{-}459$(syn): 3.72_{10} 3.51_7 2.55_4 2.15_9 1.98_7 1.83_5 1.80_1 1.46_2. Microscopic hexagonal pyramids. Black, black streak, resinous to adamantine luster. Perfect cleavage, apparently prismatic. $H = 4$, $VHN = 203\text{-}222$. $G = 5.47$. *Tuva, Russia.* BM *ZVMO 86:626(1957), AM 43:623, R:582, ABI:76.*

2.8.8.1 Hypercinnabar HgS

Named in 1978 for its trimorphous relationship with cinnabar and metacinnabar. HEX Space group unknown. $a = 7.01$, $c = 14.13$, $Z = 12$, $D = 7.54$. $19\text{-}798$(syn): 3.70_7 3.43_8 3.08_{10} 2.97_6 2.80_9 2.24_6 1.98_{10} 1.89_{10}. Microscopic grains in metacinnabar. Black with purple cast, dark purple streak, adamantine luster. No cleavage. H = 3, $VHN_{25} = 51$. G = 7.43. R = 25 (540 nm); distinctly anisotropic. Manhattan, Nye Co., NV; *Mt. Diablo mine, Contra Costa Co., CA;* Gravelotte, northern Transvaal, South Africa. BM *AM 63:1143(1978), ABI:226.*

2.8.9.1 Troilite FeS

Named in 1863 for Domenico Troili (1722–1792), Italian Jesuit, who observed the mineral in the Albareto meteorite in 1766. HEX $P\bar{6}2c$. $a = 5.968$, $c = 11.740$, Z = 12, D = 4.84. *24-80*(syn): 4.73_1 2.98_6 2.94_1 2.66_4 2.09_{10} 1.72_3 1.33_1. Rounded nodules in iron meteorites and anhedral grains in stony meteorites. Bronze, gray streak, metallic luster. H = 4, $VHN_{100} = 250$. G = 4.7–4.8. An accessory mineral in most meteorites and in some lunar rocks. Terrestrial occurrences rare: Alta mine, Del Norte Co., CA; Disko Is., Greenland, with native iron in basalt. BM *DI:231, SCI 167:621(1970), GCA 39:1457(1975).*

2.8.10.1 Pyrrhotite $Fe_{1-x}S$ (x = 0–0.2)

Named in 1835 from the Greek for redness, in reference to its color. MON $A2/a$. $a = 12.811$, $b = 6.870$, $c = 11.885$, $\beta = 117.3°$, Z = 32, D = 4.62 (the commonest polytype, Fe_7S_8). *29-723*: 2.98_9 2.64_{10} 2.07_6 2.06_9 2.05_{10} 1.72_8 1.60_1 1.43_3. **Habit:** Pseudohexagonal crystals rare; usually massive, granular; twinned on $\{10\bar{1}2\}$. **Structure:** A modified nickeline structure, based on hexagonal close-packed S atoms with Fe atoms in octahedral coordination. In the monoclinic type, one in every eight Fe atoms is omitted ($x = 0.125$). **Physical properties:** Bronze to brown, tarnishing on exposure, gray-black streak, metallic luster. No cleavage, basal parting, subconchoidal to uneven fracture, brittle. H = $3\frac{1}{2}$–$4\frac{1}{2}$, $VHN_{100} = 373–409$. G = 4.6–4.7. Magnetic, varying inversely with Fe

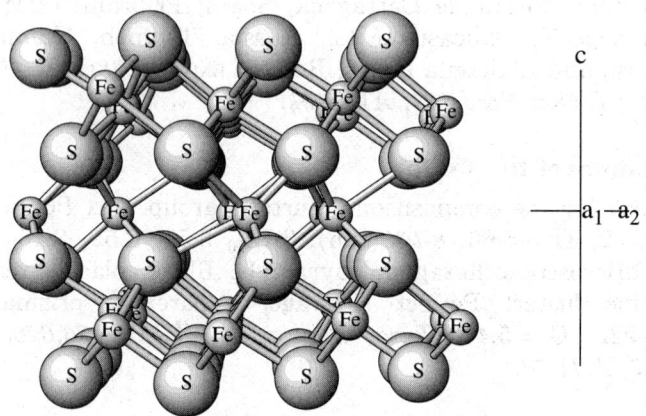

PYRRHOTITE: $Fe_7 S_8$

Sulfur is in hexagonal closest-packed arangement, as are potential iron sites (see NICCOL). Iron in pyrrhotites is always deficient from the ideal 1 : 1 ratio, and several different symmetries are shown, depending in part on the size of the deficiency.

content. **Tests:** Magnetism. Dissolves in HCl with evolution of H_2S. **Chemistry:** Usually pure iron sulfide; many analyses report minor Ni, but this is due to admixed pentlandite. **Optics:** R = 30.3–31.0 (540 nm); distinctly anisotropic. **Occurrence:** Mainly in mafic igneous rocks, frequently as magmatic segregations; also in pegmatites, in high-temperature hydrothermal and replacement veins, and as an accessory in metamorphic rocks. **Localities:** Widely distributed throughout the world. Tilly Foster mine, Brewster, NY; Franklin, NJ; Gap mine, Lancaster Co., PA; Galax, Carroll Co., VA; Ducktown, TN; Butte, MT; Crestmore, CA; Sudbury, ON, and Bluebell mine(*), Riondell, BC, Canada; Potosi and San Antonio mines(*), Santa Eulalia, CHIH, Mexico; Falun, Sweden; Outokumpu, Finland; Andreasberg, Harz, Germany; Bottino, Tuscany, Italy; Herja(*), Romania; Trepca(*), Serbia; Dalnegorsk(*), eastern Siberia, Russia; many mines in Japan; Kambalda, WA, Australia; Morro Velho(*), Nova Lima, MG, Brazil. BM *DI:231*, *EG 70:824(1975)*, *R:592*, *P&J:315*, *ABI:428*.

2.8.10.2 Smythite Fe_9S_{11}

Named in 1957 for Charles H. Smyth, Jr. (1866–1937), U.S. economic geologist. May contain minor Ni. HEX-R $R3m$. $a = 3.47$, $c = 34.50$, $Z = 3$, $D = 4.09$. *25-1182*: 11.5_{10} 5.75_2 2.99_3 2.57_3 2.27_2 2.17_2 1.99_1 1.74_2. Microscopic hexagonal plates. Bronze-yellow, dark gray streak, metallic luster. Cleavage {0001}, perfect. VHN = 388. G = 4.06. Strongly magnetic. Strongly pleochroic and anisotropic. *Bloomington, Monroe Co., IN*, inclusions in calcite crystals; near Helena, Boulder Co., MT; Boron, Kern Co., CA; Silverfields mine, Cobalt, and Nicopor mine, north of Schreiber, ON, and Bird River mines, Lac du Bonnet, MB, Canada; Clara mine, Black Forest, and Hartenstein, Saxony, Germany; Lengenbach quarry, Binn, Switzerland; Kerch Peninsula, Ukraine; Kamaishi mine, Iwate Pref., Japan; Wadi Kamal, Saudi Arabia. BM *AM 42:309(1957)*, *R:612*, *P&J:345*, *ABI:482*.

NICKELINE GROUP

The nickeline structure minerals have the general formula

$$AX$$

where

A = Ni, Co, Pt, Pd
X = As, Sb, Bi, Se, Te

and the structure is hexagonal, $P6_3/mmc$. In this structure, each X atom is surrounded by six metal atoms in the corners of a trigonal prism. Each metal atom is surrounded by six X atoms in an octahedral coordination. Even when the ratio of atoms differs considerably from 1:1, most compounds will form the structure. Nonstoichiometric compositions often result from vacant cation sites. The vacant sites may be disordered; if they are ordered, superstructures

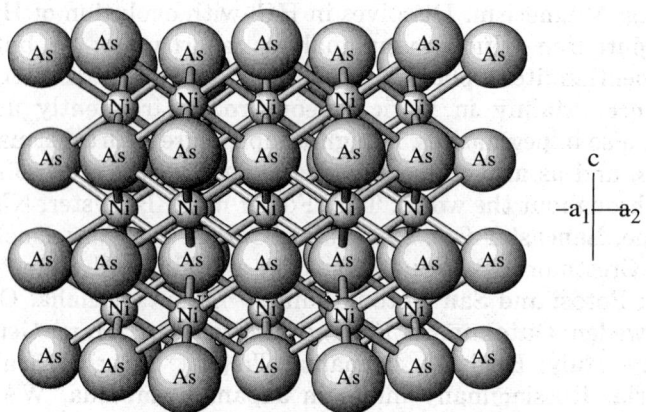

NICCOLITE (nickelite, nickeline): NiAs

Space group $P6_3/mmc$. Each atom is coordinated by six of the other type. Each is the arrangement of hexagonal closest packing. Troilite, FeS, found in extraterrestrial rocks, has essentially the same structure, and pyrrhotites, Fe1 · xS, have the same basic structure with Fe vacancies in various locations.

may be formed with lower symmetry. Pyrrhotite (2.8.10.1) is a monoclinic variation on the nickeline structure due to regular vacancies in the metal position.

NICKELINE GROUP

Mineral	Formula	Space Group	a	c	D	
Nickeline	NiAs	$P6_3/mmc$	3.609	5.019	7.84	2.8.11.1
Breithauptite	NiSb	$P6_3/mmc$	3.926	5.138	8.74	2.8.11.2
Sederholmite	NiSe	$P6_3/mmc$	3.624	5.228	7.11	2.8.11.3
Hexatestibio-panickelite	(Ni,Pd)(Sb,Te)	$P6_3/mmc$	3.98	5.35	8.94	2.8.11.4
Sudburyite	PdSb	$P6_3/mmc$	4.079	5.587	9.41	2.8.11.5
Kotulskite	Pb(Te,Bi)	$P6_3/mmc$	4.19	5.67	8.26	2.8.11.6
Sobolevskite	PdBi	$P6_3/mmc$	4.23	5.69	11.88	2.8.11.7
Stumpflite	Pt(Sb,Bi)	$P6_3/mmc$	4.175	5.504	13.52	2.8.11.8
Langisite	(Co,Ni)As	$P6_3/mmc$	3.538	5.127	8.17	2.8.11.9
Freboldite	CoSe	$P6_3/mmc$	3.632	5.301	7.56	2.8.11.10

Note: Many binary compounds of V, Cr, Mn, Fe, Co, Ni, Pd, and Rh with S, Se, Te, As, Sb, and Bi have this structure or a related one. See, for example, the wurtzite structure group.

2.8.11.1 Nickeline NiAs

Named in 1832 for the composition. Nickeline group. Synonym: niccolite. HEX $P6_3/mmc$. $a = 3.609$, $c = 5.019$, $Z = 2$, $D = 7.84$. *31-900*: 2.66_{10} 1.96_9 1.81_8 1.50_2 1.48_2 1.33_3 1.07_4 1.03_3. Crystals rare, pyramidal and often malformed; usually massive, also reticulated and arborescent. Pale copper-red, tarnishing to gray or black, black streak, metallic luster. No cleavage, may be twinned on $\{10\bar{1}1\}$ producing fourlings, conchoidal to uneven fracture, brittle. H = 5–5$\frac{1}{2}$. G = 7.78. R = 47.0–51.6 (540 nm); strongly pleochroic and anisotropic. Alters to annabergite. Soluble in aqua regia. Occurs in ore deposits with other Ni and As minerals, with pyrrhotite and chalcopyrite in ore deposits derived from norites, and in vein deposits with Co and Ag minerals. Chatham, CT; Franklin, NJ; Keweenaw Penin., MI, in Cu deposits; Copper King mine, Boulder Co., and Gem mine, Fremont Co., CO; Stanislaus mine, Calaveras Co., and Long Lake, Inyo Co., CA; Cobalt, Gowganda, Sudbury, and Thunder Bay, ON, and Great Bear Lake, NWT, Canada; Alistos, SIN, Mexico; Chalcanches, Isère, France; Andreasberg, Mansfeld, and Eisleben, Harz, and Freiberg, Annaberg, Schneeberg, and Sangerhausen, Saxony, Germany; Jachymov, Czech Republic; Schladming, Styria, Austria; Talmessi mine, near Anarak, Iran; Ait Ahmane mine, near Bou-Azzer, Morocco; Cochabamba, Cercado, Bolivia. BM *DI:236, R:615, P&J:278, ABI:350.*

2.8.11.2 Breithauptite NiSb

Named in 1845 for Johann F. A. Breithaupt (1791–1873), German mineralogist. Nickeline group. HEX $P6_3/mmc$. $a = 3.926$, $c = 5.138$, $Z = 2$, $D = 8.74$. *41-1439*: 2.84_{10} 2.06_7 1.97_7 1.61_2 1.53_3 1.42_2 1.25_2 1.07_3. Tabular crystals rare, usually massive. Light copper-red, red-brown streak, metallic luster. No cleavage, brittle. H = 5$\frac{1}{2}$. G = 7.59–8.23. R = 41.0–48.6 (540 nm); distinctly pleochroic and anisotropic. Occurs in hydrothermal veins associated with Co–Ni Ag ores. Coyote Peak, Humboldt Co., CA; many mines in the Cobalt area, ON, Canada; Sulitjelma, Norway; *Andreasberg, Harz Mts., Germany*; Sarrabus, Sardinia, Italy; Tunaberg, Saxberget, and Langsjön, Sweden; Norilsk, Russia; Broken Hill, NSW, Australia. BM *DI:238, R:623, P&J:107, ABI:69.*

2.8.11.3 Sederholmite NiSe

Named in 1964 for Jakob J. Sederholm (1863–1934), director of the Geological Survey of Finland. Nickeline group. HEX $P6_3/mmc$. $a = 3.624$, $c = 5.288$, $Z = 2$, $D = 7.11$. *18-888*: 2.70_{10} 2.02_8 1.81_6 1.54_4 1.50_4 1.35_3 1.16_3 1.08_3. Microscopic grains. Brass yellow, metallic luster. Strongly anisotropic. Occurs with other nickel selenides in calcite veins at *Kuusamo, northeastern Finland.* BM *AM 50:519(1965), R:625, P&J:339, ABI:466.*

2.8.11.4 Hexatestibiopanickelite (Ni,Pd)(Sb,Te)

Named in 1974 for the crystal system and composition. Nickeline group. HEX $P6_3/mmc$. $a = 3.98$, $c = 5.35$, $Z = 2$, $D = 8.94$. *29-932*: 2.89_{10} 2.11_8 1.99_7 1.64_4 1.58_5 1.45_3 1.34_3 1.11_6. Brown-gray, metallic luster. H = 2, VHN = 75–108. R = 58.2–62.3 (590 nm); weakly anisotropic. *Southwestern China*, in Cu–Ni sulfide deposit. BM *AM 61:182(1976), ABI:218.*

2.8.11.5 Sudburyite PdSb

Named in 1974 for the locality. Nickeline group. Contains up to 8% Ni. HEX $P6_3/mmc$. $a = 4.079$, $c = 5.587$, $Z = 2$, $D = 9.41$. $26\text{-}888$(syn): 2.98_7 2.18_{10} 2.03_7 1.49_6 1.30_5 1.20_8 1.15_6 1.08_4. Microscopic grains. White, metallic luster. $H = 4\text{--}4\frac{1}{2}$, $VHN_{25} = 281\text{--}311$. $G_{syn} = 9.37$. $R = 59.3\text{--}62.8$ (540 nm); moderately anisotropic, yellow-gray to dark gray. Blue Lake, PQ, and *Frood and Copper Cliff South mines, Sudbury, ON, Canada*, as tiny inclusions in cobaltite and maucherite, and in heavy-mineral concentrates; Danba, Sichuan Prov., China; Witwatersrand, South Africa. BM *CM 12:275(1974)*, *AM 61:178(1976)*, *P&J:411, ABI:505*.

2.8.11.6 Kotulskite Pd(Te,Bi)

Named in 1963 for Vladimir K. Kotulskii (1879–1951), Russian geologist. Nickeline group. HEX $P6_3/mmc$. $a = 4.19$, $c = 5.67$, $Z = 2$, $D = 8.26$. $15\text{-}394$: 3.05_{10} 2.24_9 2.09_9 1.53_7 1.33_7 1.24_8 1.21_7 1.17_8. Microscopic grains. Steel-gray, metallic luster. No cleavage. $VHN_{100} = 222$. $R = 59.6\text{--}67.4$ (540 nm); strongly anisotropic. An accessory mineral in many Pt deposits. Stillwater Complex, MT; New Rambler mine, Albany Co., WY; Key West mine, Clark Co., NV; Sudbury, Lac des Iles Complex, and Thierry mine, ONT, Canada; *Monchegorsk deposit, Kola*, and Norilsk, *Russia*; Shiaonanshan, Inner Mongolia, and Danba, Sichuan, China; Merensky Reef, Transvaal, South Africa. BM *ZVMO 92:33(1963)*, *AM 48:1181(1963)*, *EG 71:1159(1976)*, *R:838, P&J:406*.

2.8.11.7 Sobolevskite PdBi

Named in 1975 for Petr G. Sobolevski (1781–1841), Russian mineralogist. Nickeline group. HEX $P6_3/mmc$. $a = 4.23$, $c = 5.69$, $Z = 2$, $D = 11.88$. $29\text{-}238$: 3.07_{10} 2.26_{10} 2.11_9 1.74_4 1.69_5 1.34_4 1.24_4 1.18_5. Microscopic grains. Gray-white, metallic luster. $H = 4$, $VHN_{50} = 236$. $R = 52.4\text{--}57.7$ (540 nm); strongly anisotropic, pale cream to red-brown. *Oktyabr mine, Talnakh, Norilsk, Russia*, in chalcopyrite ore; Lubin copper mine, Zechstein, Poland. BM *ZVMO 104:568(1975)*, *AM 61:1054(1976)*, *TMPM 28:1(1981)*, *ABI:483*.

2.8.11.8 Stumpflite Pt(Sb,Bi)

Named in 1972 for Eugen F. Stumpfl (b.1931), Austrian mineralogist. Nickeline group. HEX $P6_3/mmc$. $a = 4.175$, $c = 5.504$, $Z = 2$, $D = 13.52$. $25\text{-}1482$: 3.62_6 3.03_{10} 2.19_{10} 2.09_8 1.72_4 1.51_5 1.22_4 1.15_5. Microscopic grains. Cream-white, metallic luster. $H = 5$, $VHN_{50} = 385$. $R = 59.3\text{--}63.0$ (540 nm); pleochroic and anisotropic, cream to yellow-brown. *Driekop mine, Transvaal, South Africa*, in platinum concentrates; near Nizhni Tagil, Russia. BM *BM 95:610(1972)*, *AM 59:211(1974)*, *P&J:356, ABI:503*.

2.8.11.9 Langisite (Co,Ni)As

Named in 1969 for the locality. Nickeline group. HEX $P6_3/mmc$. $a = 3.538$, $c = 5.127$, $Z = 2$, $D = 8.17$. *24-333*: 3.07_2 2.63_{10} 1.97_9 1.77_8 1.49_4 1.47_3 1.32_3 1.14_3. Microscopic grains. Pink-buff, metallic luster. $H = 6$, $VHN_{50} = 780$–857. $R = 46.1$–47.4; moderately anisotropic, blue-gray to light brown. *Langis mine, Cobalt, ON, Canada*, in Co ore. BM *CM 9:597(1969), AM 57:1910 (1972), ABI:284.*

2.8.11.10 Freboldite CoSe

Named in 1957 for Georg Frebold (b.1891), German mining engineer. Nickeline group. HEX $P6_3/mmc$. $a = 3.632$, $c = 5.301$, $Z = 2$, $D = 7.56$. *15-464*: 2.4_5 2.05_{10} 1.71_5 1.08_5 1.07_{10} 0.99_5 0.96_{10}. Microscopic grains. Copper-red, metallic luster. Temple Mt., Emery Co., UT; Pinky Fault U deposit, SK, Canada; *Trogtal quarry* and Andreasberg, *Harz*, and Hartenstein, Saxony, *Germany*. BM *NJMM:133(1955); AM 41:164(1956), 44:907(1959); R:614; ABI:161.*

2.8.12.1 Covellite CuS

Named in 1832 for Niccolo Covelli (1790–1829), who described the mineral from Vesuvius. Isostructural with klockmannite (2.8.12.2). HEX $P6_3/mmc$. $a = 3.7938$, $c = 16.341$, $Z = 6$, $D = 4.68$. *6-464*(syn): 3.22_3 3.05_7 2.81_{10} 2.72_6 1.90_9 1.74_4 1.57_2 1.56_4. Hexagonal plates, also massive or spheroidal. Based on a hexagonal framework of sulfur atoms with copper in triangular and tetrahedral coordination. Indigo-blue to black, sometimes highly iridescent in brass-yellow and deep red, lead-gray to black streak, submetallic luster inclining to resinous. Cleavage {0001}, perfect; flexible in thin leaves. $H = 1\frac{1}{2}$–2, $VHN_{100} = 128$–138. $G = 4.6$–4.7. $R = 6.8$–21.9 (540 nm); pleochroic, deep blue to blue-white; strongly anisotropic. Easily fusible; when heated burns with a blue flame and fuses to a globule. Covellite is associated with other copper minerals in the zone of secondary enrichment; it is a primary mineral in a few places. Rarely as a volcanic sublimate (*Vesuvius*). Franklin, NJ; Summitville(*), Rio Grande Co., CO; Butte(*), MT; many localities in AZ, including Bisbee, Ajo, Mammoth, Globe-Miami, and Clifton-Morenci; Darwin mines, Inyo Co., CA; Kennecott, AK; Copper Mountain mine, BC, Canada; Dillenburg, Hesse, Germany; Alghero(*), Sardinia, Italy; Bor(*), Serbia; Moonta, SA, Australia. BM *DI:248, R:676, P&J:132, ABI:112.*

2.8.12.2 Klockmannite CuSe

Named in 1928 for Friedrich Klockmann (1858–1937), German mineralogist. HEX $P6_3/mmc$. $a = 3.94$, $c = 17.25$, $Z = 6$, $D = 6.12$. Isostructural with covellite (2.8.12.1). *6-427*: 3.35_6 3.18_9 2.88_{10} 2.19_3 2.00_4 1.97_8 1.82_6 1.62_5. Granular aggregates. Slate-gray, tarnishing to blue-black, metallic luster. Perfect basal cleavage. $H = 3$, $VHN_{100} = 82$–104. $G_{syn} = 5.99$. $R = 15.8$–37.4 (540 nm); strongly anisotropic. Montreal R., ON, and Beaverlodge Lake, SK, Canada; Moctezuma, SON, Mexico; Lehrbach and Tilkerode, Harz, Germany; Skrikerum, Kalmar, Sweden; Bukov and Petrovice, Czech Republic; *Sierra de*

Umango and Sierra de Cacheuta, La Rioja, Argentina. BM *DI:251*, *AM 34:435(1949)*, *R:681*, *P&J:224*, *ABI:267*.

2.8.13.1 Vulcanite   CuTe

Named in 1961 for the locality. ORTH $Pmmn$. $a = 3.16$, $b = 4.08$, $c = 6.93$, $Z = 2$, $D = 7.11$. *22-252*(syn): 3.50_7 2.87_8 2.34_5 2.31_5 2.02_{10} 1.66_7 1.52_5 1.44_5. Microscopic prismatic grains. Light bronze to yellow-bronze, metallic luster. Two cleavages, probably pinacoidal. $H = 1$–2, $VHN_{50} = 34$–40. $R = 15.2$–67.8 (540 nm); strongly pleochroic and anisotropic. *Good Hope mine, Vulcan, Gunnison Co., CO*; Bryngovsk Au–Te deposit, Ural Mts., Russia; Iriki mine, Kagoshima Pref., Japan. BM *AM 46:258(1961)*, *R:419*, *P&J:385*, *ABI:564*.

2.8.14.1 Cinnabar   HgS

Name is from medieval Latin *cinnabaris*, probably derived from Persian *zinjifrah*, apparently meaning dragon's blood, in allusion to its red color. Trimorphous with metacinnabar and hypercinnabar. HEX-R $P3_121$. $a = 4.149$, $c = 9.595$, $Z = 3$, $D = 8.19$. *6-256*(syn): 3.36_{10} 3.17_3 2.86_9 2.07_3 1.98_4 1.77_2 1.74_3 1.68_3. **Habit**: Crystals rhombohedral or thick tabular on $\{0001\}$ to short prismatic; commonly in crystalline incrustations, granular, massive, or as an earthy coating. Twinning common with $\{0001\}$ as twin plane, simple and penetration. **Structure**: Consists of helical —Hg—S—Hg—S— chains parallel to the hexagonal c axis. **Physical properties**: Scarlet-red, darkening on exposure to light, scarlet streak, adamantine to submetallic luster. Cleavage $\{10\bar{1}0\}$, perfect; somewhat sectile. $H = 2$–$2\frac{1}{2}$, $VHN_{10} = 82$–156. $G = 8.1$–8.2. **Tests**: Heated in an open tube gives Hg vapor, which condenses to a silvery coating. **Chemistry**: Pure HgS. **Optics**: Uniaxial(+); $N_O = 2.905$, $N_E = 3.256$. $R = 28.4$–29.2 (540 nm); strongly anisotropic. **Occurrence**: Cinnabar is the most abundant mercury mineral. It occurs in veins or impregnations formed at relatively low temperatures near recent volcanic rocks or in hot-spring areas. **Localities**: Cinnabar is widely distributed and only the more significant localities can be given. Terlingua, Brewster Co., TX; Poverty Peak, Humboldt Co., and near Lovelock, Pershing Co., NV; New Idria(*), New Almaden, and other localities, CA; Almaden(*), Spain; Monte Amiata, Italy; Idria(*), Slovenia; Mt. Avala(*), Serbia; Nikitova, Ukraine; Hydercahn(*), Ferghana Basin, Uzbekistan; Tsar Tien mine(*), Hunan Prov., China; Kilkivan, near Gympie, Q, Australia; Bau(*), Sarawak; Huancavelica(*), Peru. **Uses**: The principal source of mercury and its compounds. BM *DI:251*, *R:673*, *P&J:122*, *ABI:100*.

2.8.15.1 Matraite   ZnS

Named in 1958 for the locality. Trimorphous with sphalerite and wurtzite. HEX-R $R3m$. $a = 3.8$, $c = 9.4$, $Z = 3$, $D = 4.13$. Pyramidal crystal aggregates. Yellow-brown, vitreous luster. Telluride, San Miguel Co., CO; *Matra Mts., Hungary*. BM *AM 45:1131(1960)*, *ABI:316*.

2.8.16.1 Millerite NiS

Named in 1845 for William H. Miller (1801–1880), English mineralogist, who first studied the crystals. HEX-R $R3m$. $a = 9.620$, $c = 3.149$, $Z = 9$, $D = 5.37$. 12-41: 4.81_6 2.95_4 2.78_{10} 2.51_7 2.23_6 1.86_9 1.82_5 1.74_4. Needlelike crystals, commonly in radiating groups, also massive. Brass-yellow, often with an iridescent tarnish, green-black streak, metallic luster. Cleavage $\{10\bar{1}1\}$, $\{01\bar{1}2\}$, perfect. $H = 3$–$3\frac{1}{2}$, $VHN_{20} = 179$. $G = 5.3$–5.5. $R = 52.9$–54.8 (540 nm); strongly anisotropic. Normally, a mineral of low-temperature origin, it occurs rather commonly as tufts of capillary crystals in cavities in limestone or dolostone; also as a late mineral in nickel sulfide deposits. Sterling mine(*), Antwerp, NY; Gap mine(*), Lancaster Co., PA; Keokuk(*), IA, many localities in Paleozoic limestones in the Mississippi valley; Temagami(*), ON, Marbridge mine, La Motte Twp., PQ, and Thompson mine(*), MB, Canada; Müsen(*) and Wissen(*), North Rhine-Westphalia, Germany; Kotalahti, Finland; Jachymov, Czech Republic; Kambalda and Black Swan, WA, Australia; Mina Carlote, Cuba. BM *DI:239, CM 12:248(1974), EG 69:391,1335(1974), R:626; P&J:270, ABI:333.*

2.8.16.2 Mäkinenite NiSe

Named in 1964 for Eero Mäkinen (1886–1953), Finnish geologist. HEX-R $R3m$. $a = 10.01$, $c = 3.28$, $Z = 9$, $D = 7.23$. 18-887: 4.99_6 2.88_{10} 2.63_{10} 2.33_{10} 1.95_{10} 1.81_4 1.71_4 1.67_4. Microscopic grains. Yellow, metallic luster. $H = 3$. $R = 47.4$–48.8 (540 nm); strongly pleochroic and anisotropic. Occurs with other nickel selenides in calcite veins at *Kuusamo, northeastern Finland.* BM *AM 50:519(1965), R:630, P&J:252, ABI:308.*

2.8.17.1 Ruthenarsenite RuAs

Named in 1974 for the composition. Contains minor amounts of Ir, Os, Rh, Pd, and Ni. ORTH $Pnma$. $a = 5.628$, $b = 3.239$, $c = 6.184$, $Z = 4$, $D = 9.81$. 26-947: 2.89_3 2.70_7 2.12_5 2.06_{10} 1.78_4 1.75_4 1.34_4 1.30_4. Microscopic grains. Pale brown, metallic luster. $H = 6$, $VHN_{100} = 743$–933. $R = 47.5$–49.5 (546 nm); distinctly pleochroic and anisotropic. *Papua New Guinea*, in nuggets of rutheniridosmine; Anduo, Tibet, China; Onverwacht mine, Witwatersrand, South Africa. BM *CM 12:280(1974), AM 61:177(1976), ABI:452.*

2.8.17.2 Cherepanovite RhAs

Named in 1985 for Vladimir A. Cherepanov (1927–1983), Russian mineralogist. ORTH $Pnma$. $a = 5.70$, $b = 3.59$, $c = 6.00$, $Z = 4$, $D = 9.72$. 38-407: 3.01_{10} 2.10_6 2.01_1 1.91_5 1.81_2 1.77_2 1.73_2 1.50_2 1.35_2. Microscopic grains. Gray-white, gray-black streak, metallic luster. One distinct cleavage. $H = 6$, $VHN_{50} = 726$–754. $R = 43.6$–45.5 (540 nm); weakly pleochroic and anisotropic. *Koryak–Kamchatka ophiolite belt, Russia.* BM *ZVMO 114:464(1985), AM 71:1544(1986), ABI:95.*

2.8.18.1 Modderite CoAs

Named in 1924 for the locality. Contains up to 9% Fe. ORTH $Pmcm$. $a = 3.458$, $b = 5.869$, $c = 5.292$, $Z = 4$, $D = 8.28$. 9-94(syn): 2.59_9 2.55_6 2.06_5 1.97_{10} 1.70_5 1.44_5 1.21_5 1.05_6. Microscopic grains. Black, metallic luster. $H = 4$, $VHN = 26$. *Modderfontein mine, Witwatersrand, South Africa*, in heavy-mineral concentrate; Dashkesan deposit, Azerbaijan. BM *DI:242*, *ZVMO 106:347 (1977)*, *R:625*, *ABI:334*.

2.8.19.1 Westerveldite (Fe,Ni)As

Named in 1972 for Jan Westerveld (1905–1962), Dutch mineralogist. ORTH $Pnam$. $a = 5.436$, $b = 6.024$, $c = 3.372$, $Z = 4$, $D = 7.86$. 12-799(syn): 2.64_6 2.59_{10} 2.12_2 2.08_4 2.02_4 2.00_6 1.73_3 1.69_2. Microscopic grains. Gray, metallic luster. $H = 6$–7, $VHN_{25} = 707$–798. $G = 8.13$. $R = 43.4$–47.9 (540 nm); distinctly pleochroic and anisotropic. Birchtree mine, near Thompson, MB, Canada; Ilimaussaq intrusion, near Narssaq, Greenland; *La Gallega mine, near Ojen, Malaga, Spain*; Tunaberg, Sweden; Seinäjoki, Finland. BM *AM 57:354(1972)*, *CM 12:137(1973)*, *P&J:389*, *ABI:574*.

TSUMOITE GROUP

The tsumoite group is a rhombohedral group of bismuth minerals with the general formula

$$BiX \quad \text{or a multiple thereof}$$

where

$X = S, Se, Te$

and the mineral is isostructural with tsumoite in space group $P\bar{3}m1$. The structure is similar to that of the arsenic, joséite, and tetradymite group minerals.

TSUMOITE GROUP

Mineral	Formula	Space Group	a	c	
Tsumoite	BiTe	$P\bar{3}m1$	4.422	24.050	2.8.20.1
Sulphotsumoite	Bi$_3$Te$_2$S	$P\bar{3}m1$	4.316	23.43	2.8.20.2
Nevskite	BiSe	$P\bar{3}m1$	4.18	22.8	2.8.20.3
Ingodite	Bi$_2$TeS	$P\bar{3}m1$	4.248	23.075	2.8.20.4

Note: The poorly characterized mineral platynite (2.8.21.1), (Bi,Pb)(Se,S) may belong to the group.

2.8.20.1 Tsumoite BiTe

Named in 1978 for the locality. Tsumoite group. HEX-R $P\bar{3}m1$. $a = 4.422$, $c = 24.050$, $Z = 6$, $D = 8.23$. 31-200(syn): 3.23_{10} 2.36_4 2.21_3 2.00_1 1.82_2 1.61_1

2.8.20.3 Nevskite

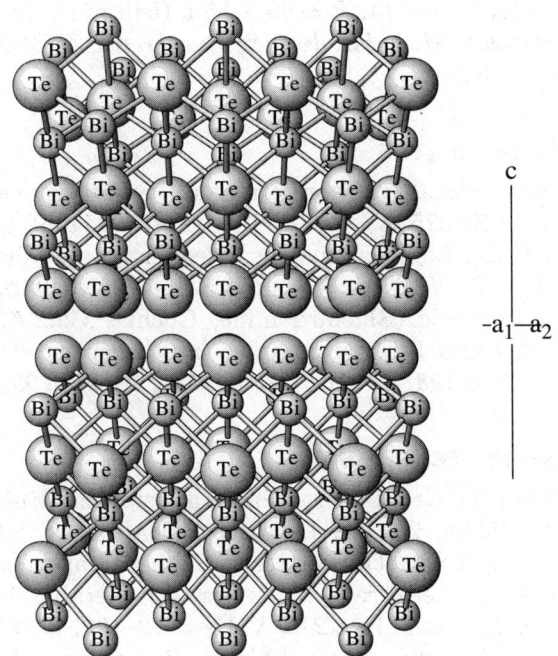

TSUMOITE: BiTe

According to the published structure, this has layers Bi–Te–Bi–Te–Bi–Te Te–Bi–Te–Bi–Te–Bi, that is, a 12-layer repeat.

1.48_1, 1.41_1. Tabular crystals. Silver-white, steel-gray streak, metallic luster. Cleavage {0001}, perfect. H = $2-2\frac{1}{2}$, VHN_{100} = 59. G = 8.16. R = 58.1–59.6 (546 nm); moderately anisotropic. Sylvanite, Hidalgo Co., NM; Tunaberg, Sweden; Tyrnauz, Caucasus Mts., and Magadan region, eastern Siberia, Russia; *Tsumo mine, Shimane Pref., Japan.* BM *AM 63:1162(1978), 76:257(1991); ABI:540; EM 5:165(1993).*

2.8.20.2 Sulphotsumoite Bi_3Te_2S

Named in 1982 for its relationship to tsumoite. Tsumoite group. HEX-R $P\bar{3}m1$. $a = 4.316$, $c = 23.43$, Z = 2, D = 8.03. *38-442*: 4.67_3 3.16_{10} 2.32_6 2.16_5 1.96_3 1.78_4 1.58_3 1.37_4. Microscopic aggregates. Gray-white, metallic luster. Perfect basal cleavage. H = 2, VHN_5 = 64–66. R = 53.6–57.3 (540 nm). *Egerlyakh deposit, Yakutia*, and Magadan region, *Russia.* BM *ZVMO 11:316(1982); AM 68:1250(1983), 76:257(1991); ABI:507.*

2.8.20.3 Nevskite   BiSe

Named in 1984 for the locality. Tsumoite group. HEX-R $P\bar{3}m1$. $a = 4.18$, $c = 22.8$, Z = 6, D = 8.31. *29-246*(syn): 22.8_2 4.56_1 3.58_2 3.06_{10} 2.24_4 2.09_3 1.90_1 1.73_2. Microscopic grains. Lead-gray, metallic luster. Perfect basal cleavage. H = 2–3, VHN = 60–114. R = 50.5–55.5 (540 nm); distinctly anisotropic. *Nevskii Sn deposit, Magadan district, Russia.* BM *ZVMO 113:351(1984), AM 70:875(1985), ABI:347.*

2.8.20.4 Ingodite Bi$_2$TeS

Named in 1981 for the locality. Tsumoite group. Synonym: grünlingite. HEX-R $P\bar{3}m1$. $a = 4.248$, $c = 23.075$, $Z = 3$, $D = 7.98$. *41-1408:* 4.59_2 3.10_{10} 2.27_5 2.13_4 1.93_3 1.65_5 1.25_3 1.22_3. Lamellar masses. Steel-gray, metallic luster. Perfect basal cleavage. H = 2, VHN$_{25}$ = 60. R = 52.4–57.8 (540 nm); moderately pleochroic and anisotropic. Bluebird mine, Cochise Co., AZ; Brandy Gill, Cumbria, England; Baita Bihor, Romania; *Ingoda deposit, Transbaikal, Russia.* BM *ZVMO 110:594(1981); AM 67:855(1982), 76:257(1991); ABI:235.*

2.8.21.1 Platynite (Bi,Pb)(Se,S)

Named in 1910 from the Greek *to broaden*, in allusion to its platy habit. HEX-R $R\bar{3}m$. $a = 8.49$, $c = 20.80$, $Z = 12$, $D = 7.45$. *ZVMO 118:* 5.45_2 3.97_{10} 3.39_2 2.97_4 2.93_2 2.60_2 2.55_2 1.98_2. Thin plates to 3 cm and aggregates of anhedral grains. Silver-gray to steel-gray, black streak, metallic luster. Cleavage {0001}, perfect; {10$\bar{1}$1}, fair. H = 2–3, VHN = 70–100. G = 7.98. R = 45–52 (540 nm). *Falun, Sweden;* Baltic Shield, Russia. BM *DI:474, NJMM 154(1963), ZVMO 118(3):22(1989), AM 76:305(1991), ABI:411.*

2.8.22.1 Realgar AsS

Named in 1747 from the Arabic for *powder of the mine*. Dimorphous with pararealgar. MON $P2_1/n$. $a = 9.324$, $b = 13.534$, $c = 6.585$, $\beta = 106.43$, $Z = 16$, $D = 3.57$. *24-77*(syn): 6.03_4 5.73_6 5.41_{10} 3.16_8 3.05_6 2.98_5 2.93_6 2.72_7. **Habit:** Crystals uncommon, generally short prismatic and striated parallel to c; usually coarse to fine granular or as an incrustation. Contact twins on {100}. **Structure:** Consists of separate cagelike As$_4$S$_4$ molecules. **Physical properties:** Red to orange-yellow color and streak, resinous to greasy luster. Cleavage {010}, good; {$\bar{1}$01}, {100}, {120} indistinct; sectile. H = 1½–2. G = 3.5–3.6. Disintegrates on long exposure to light to a red-yellow orpiment-bearing powder. **Tests:** Characteristic color and hardness; emits a garlic odor on heating. **Chemistry:** Pure AsS. **Optics:** Biaxial (−); N$_x$ = 2.538, N$_y$ = 2.684, N$_z$ = 2.704; 2V = 40°. Pleochroic, nearly colorless to pale golden yellow. **Occurrence:** As a minor constituent in hydrothermal sulfide veins with orpiment and other As minerals; also sporadically in limestones, dolostones, and claystones, and as a sublimate in fumaroles. **Localities:** Only the more significant are listed. Norris Basin, Yellowstone, WY; Mercur(*), Tooele Co., UT; Getchell mine(*), Humboldt Co., and White Caps mine(*), Nye Co., NV; Franklin(*), King Co., and Monte Cristo(*), Snohomish Co., WA; Boron, Kern Co., CA; Noche Buena, ZAC, Mexico; Matra, Corsica, France; Andreasberg, Harz, and Freiberg, Saxony, Germany; Lengenbach, Binn, Switzerland; Baia Sprie, Cavnic(*), and Sacaramb(*), Romania; Allchar(*), Macedonia;

2.8.22.4 Alacranite

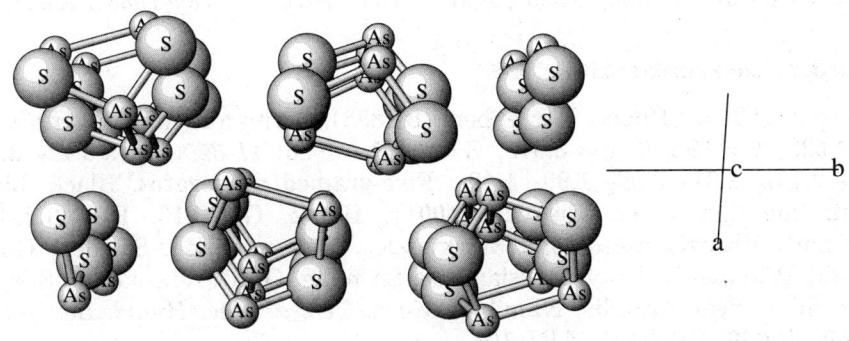

REALGAR: AsS
Each As is bonded to one other As and two S atoms, and each S to two As atoms, forming As_4S_4 molecular cages. This view shows one layer of the molecules.

Shimen(*), Hunan, China; Nishinomaki mine, Gumma Pref., Japan. BM *DI:255, AC 5:775(1952), R:889, P&J:324, ABI:436.*

2.8.22.2 Pararealgar AsS

Named in 1980 for its relationship to realgar. MON $P2_1/c$. $a = 9.909$, $b = 9.655$, $c = 8.502$, $\beta = 97.29°$, $Z = 16$, $D = 3.50$. *33-127*: 5.56_9 5.14_{10} 4.90_3 3.75_8 3.30_5 3.11_4 3.03_5 2.80_7. Powdery to granular aggregates. Yellow to orange-brown, bright yellow streak, vitreous to resinous luster. No cleavage, uneven fracture. $H = 1$–$1\frac{1}{2}$. $G = 3.52$. $N > 2.02$. Golconda mine, Humboldt Co., NV; Hemlo Au deposit, near Marathon, ON, and *Mt. Washington* and Gray Rock, *BC, Canada*; Lengenbach, Binn, Switzerland. Probably occurs with realgar at other localities. BM *CM 18:525(1980); AM 66:1277(1981), 80:400(1985); ABI:387.*

2.8.22.3 Uzonite As_4S_5

Named in 1985 for the locality. MON $P2_1/m$. $a = 7.94$, $b = 8.08$, $c = 7.10$, $\beta = 100.1°$, $Z = 2$, $D = 3.41$. *39-331*: 5.81_{10} 5.31_6 3.60_8 3.10_6 2.91_8 2.82_6 2.75_5 2.03_5. Microscopic prismatic crystals. Yellow, pearly luster. Cleavage {001}, distinct. $H = 1\frac{1}{2}$, $VHN_8 = 66$–71. $R = 19.4$–20.0 (550 nm). *Uzon caldera, Kamchatka, Russia.* BM *ZVMO 114:369(1985), AM 71:1280(1986), ABI:553.*

2.8.22.4 Alacranite As_8S_9

Named in 1986 for the locality. MON $C2/c$. $a = 9.92$, $b = 9.48$, $c = 8.91$, $\beta = 100.83°$, $Z = 2$, $D = 3.46$. *25-57*: 5.75_8 5.01_4 3.93_7 3.20_8 3.08_7 3.01_{10} 2.82_6 2.53_3. Microscopic prismatic crystals. Orange-yellow, orange-yellow streak, greasy or adamantine luster. Cleavage {100}, imperfect; conchoidal fracture. $H = 1\frac{1}{2}$, $VHN_{20} = 69$. $G = 3.43$. $R = 13.2$–14.5 (550 nm). Biaxial (+);

$N_x = 2.39$, $N_z = 2.52$. Uzon caldera, Kamchatka, Russia; *Mina Alacran*, *Chile*. BM *ZVMO 115:360(1986); AM 55:1338(1970), 73:189(1988); ABI:6.*

2.8.23.1 Herzenbergite SnS

Named in 1934 for Robert Herzenberg (b.1885), German chemist. ORTH *Pbnm*. $a = 4.328$, $b = 11.190$, $c = 3.978$, $Z = 4$, $D = 5.20$. *14-620*(syn): 3.42_1 3.24_2 2.83_3 2.79_{10} 2.31_2 2.02_1 1.99_1 1.40_7. Fine-grained aggregates. Black, black streak, metallic luster. Cleavage {001}. $H = 2$. $G = 5.16$. $R = 43.2$–49.9 (540 nm); strongly anisotropic. Järkvissle, Västernorrland, Sweden; Golub deposit, Primorskiy Kray, Russia; Hoei Sn mine, Oita Pref., Japan; Stiepelmann mine, near Arandis, Namibia; *Maria-Teresa mine, Huari, Bolivia*. BM *DI:259, R:669, P&J:405, ABI:215.*

2.8.24.1 Empressite AgTe

Named in 1914 for the locality. ORTH Space group uncertain, probably *Pmnm*. $a = 8.90$, $b = 20.07$, $c = 4.62$, $Z = 16$, $D = 7.58$. *16-412*: 4.02_4 3.81_6 3.33_6 3.18_5 2.70_{10} 2.23_8 2.14_5 2.04_5. Granular and compact masses. Pale bronze, gray-black to black streak, metallic luster. No cleavage, uneven to subconchoidal fracture, brittle. $H = 3\frac{1}{2}$, $VHN_{25} = 139$. $G = 7.5$–7.6. *Empress Josephine mine, Saguache Co.*, and Red Cloud mine, Boulder Co., *CO*; Tombstone, AZ; Grotto deposit, near Pitman, BC, Canada; Kochbulak deposit, Uzbekistan; Kawazu mine, Shizuoka Pref., Japan; Emperor mine, Vatakuola, Fiji; Kalgoorlie, WA, Australia. BM *DI:260, AM 49:325(1964), R:425, ABI:146.*

2.8.25.1 Muthmannite (Ag,Au)Te

Named in 1911 for Friedrich W. Muthmann (1861–1913), German chemist and crystallographer. ORTH Space group unknown. $a = 7.211$, $b = 4.425$, $c = 5.100$, $Z = 4$, $D = 11.43$. *42-1377*: 5.09_2 3.03_{10} 2.94_5 2.11_8 1.95_3 1.70_3 1.32_3 1.04_3. Tabular crystals. Brass-yellow, gray-white on fresh fracture, metallic luster. Perfect cleavage in zone of elongation. $H = 2\frac{1}{2}$. *Sacaramb, Romania*. BM *DI:260, ABI:343.*

CHALCOPYRITE GROUP

The chalcopyrite minerals are isostructural sulfides and selenides with the formula

$$CuBX_2$$

where

B = Fe, Ga, In
X = S, Se

and the structure is related to that of sphalerite, with a doubled c spacing.

2.9.1.1 Chalcopyrite

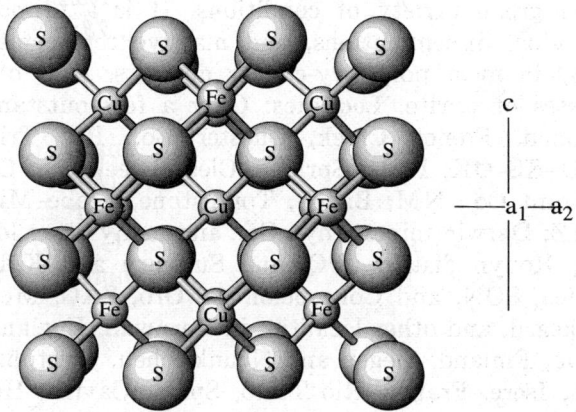

CHALCOPYRITE: CuFeS$_2$

The chalcopyrite structure can be derived from that of sphalerite by replacing alternate Zn atoms with Cu and Fe.

CHALCOPYRITE GROUP

Mineral	Formula	Space Group	a	c	D	
Chalcopyrite	CuFeS$_2$	$I\bar{4}2d$	5.289	10.423	4.18	2.9.1.1
Eskebornite	CuFeSe$_2$	$I\bar{4}2d$	5.518	11.048	5.45	2.9.1.2
Gallite	CuGaS$_2$	$I\bar{4}2d$	5.351	10.484	4.37	2.9.1.3
Roquesite	CuInS$_2$	$I\bar{4}2d$	5.523	11.141	4.74	2.9.1.4

Notes:
1. See also the chalcocite group for a discussion of other copper sulfides.
2. The stannite group minerals have essentially the same structure, adjusted for a greater diversity of elements.

2.9.1.1 Chalcopyrite CuFeS$_2$

Named in 1747 from the Greek for *brass and pyrite*. Chalcopyrite group. TET $I\bar{4}2d$. $a = 5.289$, $c = 10.423$, $Z = 4$, $D = 4.18$. *35-752*: 3.04_{10} 2.64_1 1.87_2 1.86_4 1.59_3 1.57_1 1.21_1 1.08_1. **Habit**: Equant tetrahedral crystals often modified by scalenohedral faces, up to 10 cm; often massive, compact. Twinned on {112} and {012}, penetration or cyclic. **Structure**: The structure is very similar to that of sphalerite. The ordered distribution of Cu and Fe atoms results in a tetragonal unit cell with c double the pseudocube edge. **Physical properties**: Brass-yellow, often with an iridescent tarnish, green-black streak, metallic luster. Cleavage {011}, sometimes distinct; uneven fracture, brittle. H = $3\frac{1}{2}$–4, VHN$_{100}$ = 187–203. G = 4.1–4.3. **Tests**: On charcoal fuses to a magnetic globule. Soluble in HNO$_3$ with separation of S, and gives a green solution. **Chemistry**: Usually pure CuFeS$_2$, but Se may replace S to a minor extent. **Optics**: R = 35.1–36.4 (540 nm); weakly anisotropic. **Occurrence**: The most widespread copper mineral and the most important source of that metal, formed under a great variety of conditions. It is a primary mineral in hydrothermal veins, disseminations, and massive replacements. It is the primary mineral in most porphyry-copper deposits. It is often associated with large masses of pyrite. Localities: Only a few outstanding localities can be mentioned. French Creek, Chester Co., PA; Tri-State mining district, MO–KS–OK; Idaho Springs, Clear Creek Co., CO; Santa Rita and Bayard, Grant Co., NM; Bisbee, Tombstone, Globe–Miami, and Clifton–Morenci, AZ; Darwin mines, Inyo Co., and many other localities in CA; Noranda mine, Rouyn district, PQ, and Sudbury and Kidd Creek, ON, Canada; Cananea, SON, and Concepcion del Oro, ZAC, Mexico; Redruth, Carn Brea, Liskeard, and other localities in Cornwall, England; Falun, Sweden; Outokumpu, Finland; Siegen and Neunkirchen, Westphalia, Germany; Bourg d'Oisans, Isere, France; Rio Tinto, Spain; Cavnic, Herja, and Baia Sprie, Romania; near Chenzu, Hunan, China; Ani mine, Akita, and Ashio mine, Taguchi, Japan; Nababiep mine, Cape Prov., and Messina, Transvaal, South Africa; Cobar, NSW, Moonta, SA, and Mt. Lyell, TAS, Australia; Casapalca, Peru; Braden mine, Chile. BM *DI:219, AC(B) 29:579(1973), R:526, P&J:118, ABI:89*.

2.9.1.2 Eskebornite CuFeSe$_2$

Named in 1949 for the locality. Chalcopyrite group. TET $I\bar{4}2d$. $a = 5.518$, $c = 11.048$, $Z = 4$, $D = 5.45$. *14-312*: 5.53_8 3.19_9 2.48_6 1.96_{10} 1.67_8 1.39_5 1.27_5 1.13_6. Microscopic tabular crystals. Brass-yellow, tarnishing to dark brown to black, metallic luster. Cleavage {001}, perfect. H = $3–3\frac{1}{2}$, VHN$_{100}$ = 116. G = 5.35. R = 29.4–35.0 (540 nm); markedly anisotropic. Martin Lake mine, Lake Athabaska, SK, Canada; *Eskeborn adit, Tilkerode, Harz, Germany*; Chaméane, Puy-de-Dôme, France; Nove Mesto, Moravia, Czech Republic; Habri, Peru; Sierra de Cachueta and Sierra de Umango, Argentina. BM *FM 28:69(1949), CM 10:786(1971), R:612, P&J:156, NJMM:337(1988)*.

2.9.1.3 Gallite CuGaS$_2$

Named in 1958, the first gallium mineral. Chalcopyrite group. TET $I\bar{4}2d$. $a = 5.351$, $c = 10.484$, $Z = 4$, $D = 4.37$. *25-279*(syn): 3.06_{10} 1.89_5 1.87_8 1.61_6 1.34_2 1.23_2 1.22_3. Microscopic grains and exsolution lamellae. Gray, gray streak, metallic luster. H = $3–3\frac{1}{2}$. G = 4.2. R = 19.2 (540 nm); weakly anisotropic. Pinal del Rio Prov., Cuba; Radka and Chelopech deposits, Bulgaria; Prince Leopold mine, Kipushi, Zaire; *Tsumeb, Namibia*. BM *NJMM: 241(1958), R:542, P&J:173, ABI:173*.

2.9.1.4 Roquesite CuInS$_2$

Named in 1963 for Maurice Roques (b.1911), French geologist. Chalcopyrite group. TET $I\bar{4}2d$. $a = 5.523$, $c = 11.141$, $Z = 4$, $D = 4.74$. *27-159*(syn): 4.95_1 3.20_{10} 2.79_1 2.76_1 1.96_3 1.95_1 1.68_1 1.67_1. Microscopic grains. Gray with a bluish tint, metallic luster. H = 4, VHN$_{100}$ = 241. R = 22.2–22.5 (540 nm); very weakly anisotropic. Mt. Pleasant, NB, Canada; Geevor mine, Penzance,

2.9.2.1 Stannite

Cornwall, England; Långban, Värmland, Sweden; *Charrier, Allier, France*; Syrymbet, Kazakhstan; Tosham Sn prospect, Haryana, India; Akenobe and Ikuno mines, Hyogo Pref., Japan; Ulsan mine, Korea. BM *BM 86:7(1963), AM 54:1202(1969), R:544, P&J:329, ABI:445*.

2.9.1.5 Lenaite AgFeS$_2$

Named in 1995 for the locality. TET Space group uncertain, probably $P4_2mc$. $a = 5.64$, $c = 10.34$, $Z = 4$, $D = 4.63$. $ZVMO\ 124$: 3.43_1 3.15_{10} 2.82_1 2.59_1 2.45_2 2.34_2 1.91_4 1.69_2. Microscopic granular aggregates. Creamy color in reflected light. $H = 4\frac{1}{2}$, VHN = 310. R = 32–36 (546 nm); moderately anisotropic. *Khachakchansky Ag deposit, Lena R., Russia.* BM *ZVMO 124(5):85(1995)*.

STANNITE GROUP

The stannite group includes tetragonal sulfides with the general formula

$$A_2BCS_4$$

where

A = Cu, Ag
B = Fe, Cd, Cu, Zn, Hg
C = Sn, Ge, In

and the structure is essentially identical with that of chalcopyrite.

STANNITE GROUP

Mineral	Formula	Space Group	a	c	
Stannite	Cu$_2$FeSnS$_4$	$I\bar{4}2m$	5.453	10.747	2.9.2.1
Černýite	Cu$_2$CdSnS$_4$	$I\bar{4}2m$	5.487	10.845	2.9.2.2
Briartite	Cu$_2$FeGeS$_4$	$I\bar{4}2m$	5.325	10.53	2.9.2.3
Kuramite	Cu$_2$CuSnS$_4$	$I\bar{4}2m$	5.542	10.908	2.9.2.4
Sakuraiite	Cu$_2$CuInS$_4$	$I\bar{4}2m$	5.455	10.9	2.9.2.5
Hocartite	Ag$_2$FeSnS$_4$	$I\bar{4}2m$	5.74	10.96	2.9.2.6
Pirquitasite	Ag$_2$ZnSnS$_4$	$I\bar{4}2m$	5.786	10.829	2.9.2.7
Velikite	Cu$_2$HgSnS$_4$	$I\bar{4}2m$	5.554	10.911	2.9.2.8
Kesterite	Cu$_2$ZnSnS$_4$	$I\bar{4}$	5.44	10.88	2.9.2.9
Ferrokesterite	Cu$_2$(Fe,Zn)SnS$_4$	$I\bar{4}$	5.427	10.871	2.9.2.10

2.9.2.1 Stannite Cu$_2$FeSnS$_4$

Named in 1832 from the Latin for *tin*. Stannite group. May contain up to 2% In and Zn. TET $I\bar{4}2m$. $a = 5.453$, $c = 10.747$, $Z = 2$, $D = 4.47$. *35-582*: 3.13_{10} 2.73_8 1.93_3 1.91_3 1.64_2 1.63_1 1.24_1 1.11_1. Rarely as pseudo-octahedral crystals, usually massive. Steel-gray to iron-black, black streak, metallic luster. Cleavage {110} and {001}, indistinct; uneven fracture. H = 4, VHN_{25} = 216–265. G = 4.3–4.5. R = 27.4–28.6 (540 nm); distinctly anisotropic. Occurs widely in tin-bearing veins of hydrothermal origin; an important ore mineral in some deposits. Etta mine, Keystone, SD; Lost River, Seward Penin., AK; Brunswick tin mines, NB, and Snowflake mine, BC, Canada; *Wheal Rock mine* and other mines, *Cornwall, England*; Kutna Hora, Czech Republic; Norilsk, Russia; Chenzhou, Hunan, China; Zeehan, TAS, Australia; Llallagua, Potosi, and other mines, Bolivia. BM *DI:224, CM 17:125(1979), R:548, P&J:349, ABI:490.*

2.9.2.2 Černýite Cu_2CdSnS_4

Named in 1978 for Petr Černý (b.1934), Czech–Canadian mineralogist. Stannite group. TET $I\bar{4}2m$. $a = 5.487$, $c = 10.845$, Z = 2, D = 4.95. *29-537*: 5.39_3 3.15_{10} 2.73_3 1.94_5 1.93_5 1.65_4 1.64_4 1.12_5. Microscopic grains. Steel-gray, black streak, metallic luster. H = 4, VHN_{50} = 189–210. R = 23.4–25.6 (546 nm); weakly anisotropic. *Hugo mine, Keystone, SD*; Tanco mine, MB, Canada. BM *CM 16:139, 147(1978), AM 64:653(1979), ABI:85.*

2.9.2.3 Briartite Cu_2FeGeS_4

Named in 1965 for Gaston Briart, who studied the Kipushi deposit. Stannite group. Some of the Fe is replaced by Zn, and the Ge by Ga. TET $I\bar{4}2m$. $a = 5.325$, $c = 10.53$, Z = 2, D = 4.27. *25-282*: 3.06_{10} 2.66_8 2.63_4 1.88_{10} 1.87_{10} 1.60_{10} 1.59_5 1.53_4. Microscopic grains. Gray, metallic luster. H = 4. R = 27.6 (540 nm); weakly anisotropic. *Prince Leopold mine, Kipushi, Zaire*; Tsumeb, Namibia. BM *BM 88:432(1965), AM 51:1816(1966), R:575, P&J:108, ABI:71.*

2.9.2.4 Kuramite Cu_2CuSnS_4

Named in 1979 for the locality. Stannite group. TET $I\bar{4}2m$. $a = 5.542$, $c = 10.908$, Z = 2, D = 4.56. *30-494*: 3.13_{10} 2.70_2 2.44_1 1.91_8 1.64_6 1.24_3 1.11_4 1.04_2. Microscopic grains. Gray, metallic luster. H = $4\frac{1}{2}$, VHN_{100} = 353. R = 26.5 (540 nm); distinctly anisotropic. Bisbee, AZ; *Kuramin Mts., Uzbekistan*, in gold–sulfide–quartz ores. BM *ZVMO 108:564(1979), AM 65:1067 (1980), ABI:280.*

2.9.2.5 Sakuraiite Cu_2CuInS_4

Named in 1965 for Kin-ichi Sakurai (1912–1993), Japanese mineral collector. Stannite group. Contains Zn, Fe, Ag replacing Cu, and Sn replacing In. TET $I\bar{4}2m$. $a = 5.455$, $c = 10.9$, Z = 2, D = 4.45. *21-882*: 5.47_1 3.85_1 3.15_{10} 2.73_1 2.44_1 1.93_4 1.65_2 1.57_1. Microscopic grains. Steel-gray, lead-gray streak, metallic luster. No cleavage. H = 4, VHN_{100} = 243–282. R = 22.7 (540 nm). *Ikuno mine, Hyogo Pref., Japan*. BM *AM 53:1421(1968), 73:934(1988); ABI:459.*

2.9.2.6 Hocartite Ag$_2$FeSnS$_4$

Named in 1968 for Raymond Hocart (1896–1983), French mineralogist. Stannite group. Contains up to 7% Zn replacing Fe. TET $I\bar{4}2m$. $a = 5.74$, $c = 10.96$, $Z = 2$, $D = 4.77$. *21-1337*: 5.08$_2$ 3.26$_{10}$ 2.87$_4$ 2.74$_3$ 2.03$_5$ 1.98$_8$ 1.72$_7$ 1.67$_3$. Microscopic grains. Yellow-brown, metallic luster. H = 4, VHN$_{100}$ = 209–221. R = 24.5–25.4 (540 nm); distinctly anisotropic. Dean mine, Lander Co., NV; Bisbee, AZ; *Fournial, Cantal, France; Tacama tin mine*, and Hocaya and Chocaya tin mines, *Bolivia*; Pirquitas deposit, Jujuy, Argentina. BM *BM 91:383(1968), AM 54:573(1969), R:548, P&J:202, ABI:220*.

2.9.2.7 Pirquitasite Ag$_2$ZnSnS$_4$

Named in 1982 for the locality. Stannite group. TET $I\bar{4}2m$. $a = 5.786$, $c = 10.829$, $Z = 2$, $D = 4.82$. *35-544*: 3.27$_{10}$ 2.90$_4$ 2.72$_3$ 2.05$_6$ 1.98$_8$ 1.74$_8$ 1.29$_4$ 1.17$_4$. Microscopic anhedral grains, polysynthetically twinned. Gray-brown, metallic luster. H = 4, VHN$_{25}$ = 218. R = 22.2–24.1 (540 nm); strongly anisotropic, brick-red to pale green. *Pirquitas deposit, Jujuy Prov., Argentina*. BM *BM 105:229(1982), P&J:408, AM 68:1249(1983)*.

2.9.2.8 Velikite Cu$_2$HgSnS$_4$

Named in 1977 for A. S. Velikiy (1913–1970), Russian geologist. Stannite group. TET $I\bar{4}2m$. $a = 5.554$, $c = 10.911$, $Z = 2$, $D = 5.42$. *30-494*: 3.92$_1$ 3.18$_{10}$ 2.77$_1$ 2.47$_1$ 1.96$_2$ 1.94$_3$ 1.67$_2$ 1.65$_1$. Microscopic grains. Dark gray, gray streak, metallic luster. VHN = 215–305. G = 5.59. R = 26.4 (540 nm); distinctly anisotropic. Bisbee, AZ; *Khaidarkan Sb–Hg deposit, Kyrghyzstan*. BM *DANS 300:432(1988), AM 75:933(1990)*.

2.9.2.9 Kesterite Cu$_2$ZnSnS$_4$

Named in 1958 for the locality. Stannite group. TET $I\bar{4}$. $a = 5.44$, $c = 10.88$, $Z = 2$, $D = 4.50$. *21-583*: 4.85$_2$ 3.15$_{10}$ 2.73$_3$ 1.93$_7$ 1.65$_5$ 1.25$_3$ 1.11$_4$ 1.05$_3$. Massive. Green-black, black streak, metallic luster. H = $4\frac{1}{2}$, VHN$_{100}$ = 328–348. G = 4.5–4.6. R = 23.4–24.3 (540 nm). Hugo mine, Custer Co., SD; Bisbee, AZ; Brunswick tin mine, NB, Kidd Creek mine, Timmins, ONT, Snowflake mine, Revelstoke, BC, Canada; St. Michael's Mount and Cligga mine, Cornwall, England; Chizeuil, Saone-et-Loire, France; Kirki, Thrace, Greece; *Kester deposit, Yakutia, Russia*; Oruru, Bolivia; Pirquitas deposit, Jujuy, Argentina. BM *ZVMO 88:165(1958), AM 44:1329(1959), R:560, P&J:226, ABI:261*.

2.9.2.10 Ferrokesterite Cu$_2$(Fe,Zn)SnS$_4$

Named in 1989 for its relationship to kesterite. Stannite group. TET $I\bar{4}$. $a = 5.427$, $c = 10.871$, $Z = 2$, $D = 4.51$. *31-463*: 4.86$_1$ 3.13$_{10}$ 2.71$_1$ 1.92$_5$ 1.64$_3$ 1.25$_1$ 1.11$_1$ 1.04$_1$. Massive. Steel-gray, black streak, metallic luster. Cleavage {110}, distinct. H = 4, VHN$_{100}$ = 228–255. R = 26.0–27.0 (546 nm). *Cligga mine, Cornwall, England*. BM *CM 27:673(1989)*.

2.9.3.1 Mawsonite $Cu_6Fe_2SnS_8$

Named in 1965 for Douglas Mawson (1882–1958), Australian geologist and Antarctic explorer. May contain minor Zn, Ge, As, and Se. TET $P\bar{4}m2$. $a = 7.603$, $c = 5.358$, $Z = 1$, $D = 4.65$. *30-486*: 5.38_1 4.38_1 3.10_{10} 2.87_1 2.69_1 1.90_6 1.62_2 1.10_1. Microscopic grains. Bronze, metallic luster. $H = 3\frac{1}{2}$. $R = 20.6–23.0$ (540 nm); strongly pleochroic and anisotropic. Magnetic. Usually occurs in association with bornite. Bisbee, AZ; Kidd Creek mine, Timmins, ON, and Maggie Cu deposit, BC, Canada; Chizeuil, Saone-et-Loire, France; Neves-Corvo, Portugal; Khayragatsch deposit, Kuramin Mts., Uzbekistan; Akenobe, Tada, and Ikuno mines, Hyogo Pref., and other localities in Japan; *Mt. Lyell, TAS, and Tingha, NSW, Australia*; Cholquijirca, Peru; Vila Apacheta, Bolivia; Tsumeb, Namibia; Kipushi, Shaba, Zaire. BM *AM 50:900(1965)*, *CM 17:125(1979)*, *R:575*, *P&J:262*, *ABI:319*.

2.9.3.2 Chatkalite $Cu_6FeSn_2S_8$

Named in 1981 for the locality. TET $P\bar{4}m2$. $a = 7.61$, $c = 5.373$, $Z = 1$, $D = 4.97$. *35-683*: 3.11_8 2.87_3 2.70_2 2.55_2 1.90_{10} 1.63_4 1.57_4 1.06_3. Microscopic grains. Pale rose, metallic luster. $H = 4$, $VHN_{20} = 274$. $R = 27.2–27.6$ (540 nm); weakly anisotropic. Cove Au deposit, Lander Co., NV; *Chatkalo-Kuramin Mts., Uzbekistan*. BM *MZ 3:79(1981)*, *AM 67:621(1982)*, *ABI:94*.

2.9.3.3 Stannoidite $Cu_8Fe_3Sn_2S_{12}$

Named in 1969 for its similarity to stannite. Contains up to 4% Zn. ORTH *I222*. $a = 10.76$, $b = 5.40$, $c = 16.09$, $Z = 2$, $D = 4.29$. *22-237*: 5.4_1 4.83_1 4.13_1 3.11_{10} 2.70_2 2.39_1 1.91_7 1.62_2. Massive. Brass-brown, dark brown-gray streak, metallic luster. Subconchoidal to uneven fracture. $H = 4$, $VHN_{100} = 258$. $R = 23.4–25.2$ (540 nm); strongly anisotropic, orange-red to yellow-gray. Campbell mine, Bisbee, AZ; Maggie deposit, BC, Canada; St. Michael's Mount, Cornwall, England; Långban, Värmland, Sweden; Vaultry, Haute-Vienne, and Chizeuil, Saone-et-Loire, France; Chelopech, Bulgaria; Maikan deposit, Kazakhstan; *Konjo mine, Okayama Pref.*, and other localities in *Japan*; Tingha, NSW, Australia; Colquijirca, Peru; Vila Apacheta, Bolivia. BM *AM 54:1495(1969)*, *R:562*, *P&J:350*, *ABI:491*.

2.9.4.1 Renierite $Cu_{11}GeAsFe_4S_{16}$

Named in 1948 for Armand Renier (1876–1951), Belgian geologist. May contain up to 4% Zn and 2% Ga. TET $P\bar{4}2m$. $a = 10.622$, $c = 10.551$, $Z = 2$, $D = 4.40$. *9-424*: 7.50_1 4.31_2 3.06_{10} 2.82_1 2.65_3 1.87_8 1.60_6 1.21_3. Microscopic equant crystals and granular aggregates. Bronze, black streak, metallic luster. $H = 3\frac{1}{2}$, $VHN_{100} = 328$. $G = 4.38$. Moderately magnetic. $R = 24.3–25.7$ (540 nm); strongly pleochroic and anisotropic. Originally known as "orange bornite" because of its similarity to that mineral. Inexo No. 1 mine, Jamestown, CO; Ruby Creek deposit, Brooks Range, AK; Pinar del Rio, Cuba; San Giovanni mine, Sardinia, Italy; Celopech and Radka deposits, Bulgaria; Akhtala, Armenia; Vaygach, Arkhangelsk, Russia; Shakanai mine,

Akita Pref., Japan; *Prince Leopold mine, Kipushi, Zaire*; Tsumeb, Namibia.
BM *AM 35:136(1950), 71:210(1986); R:574; P&J:325; ABI:438.*

2.9.4.2 Germanite Cu$_{26}$Ge$_4$Fe$_4$S$_{32}$

Named in 1922 for the germanium content. May contain minor amounts of Zn, As, and Ga. ISO $P\bar{4}3n$. $a = 10.586$, $Z = 1$, $D = 4.47$. *36-395*: 3.05$_{10}$ 2.65$_1$ 1.87$_7$ 1.60$_4$ 1.32$_1$ 1.21$_2$ 1.08$_2$ 1.02$_1$. Minute cubic crystals rare, usually massive. Reddish gray, tarnishing to dull brown, dark gray to black streak, metallic luster. No cleavage, brittle. H = 4. G = 4.4–4.6. R = 20.6 (540 nm). A primary source of germanium. Inexco No. 1 mine, Jamestown, CO; Ruby Creek deposit, Brooks Range, AK; Pinar del Rio, Cuba; Bancairoun mine, Alpes-Maritimes, France; Radka deposit, Bulgaria; Vaygach, Archangelsk, and Norilsk, Urals, Russia; Shakanai mine, Akita Pref., Japan; *Tsumeb, Namibia*.
BM *DI:385, EG 52:612(1954), R:572, P&J:177, AM 69:943(1984), ABI:179.*

2.9.5.1 Morozeviczite (Pb,Fe)$_3$Ge$_{1-x}$S$_4$

Named in 1975 for Josef Morozevicz (1865–1941), Polish mineralogist. Forms a series with polkovicite. ISO. $a = 10.61$, $Z = 8$, $D = 6.62$. *44-1466*: 3.08$_{10}$ 2.80$_6$ 2.15$_9$ 2.05$_6$ 1.79$_5$ 1.57$_5$ 1.47$_5$ 1.11$_5$. Massive. Brown-gray, dark gray streak, metallic luster. H = 3, VHN$_{50}$ = 119–124. R = 43.0–44.0 (535 nm). *Polkovice mine, Silesia, Poland.* BM *AM 66:437(1981), ABI:340.*

2.9.5.2 Polkovicite (Fe,Pb)$_3$(Ge,Fe)$_{1-x}$S$_4$

Named in 1975 for the locality. Forms a series with morozeviczite. ISO Space group unknown. $a = 10.61$, $Z = 8$. *39-410*: 3.56$_9$ 3.47$_{10}$ 3.08$_{10}$ 2.18$_{10}$ 2.15$_9$ 1.83$_7$ 1.67$_7$ 1.39$_7$. Massive. Brown-gray, dark gray streak, metallic luster. H = 3, VHN$_{50}$ = 119–124. R = 43.0–44.0 (535 nm); distinctly anisotropic. *Polkovice mine, Silesia, Poland.* BM *AM 66:437(1981), ABI:416.*

2.9.6.1 Hemusite Cu$_6$SnMoS$_8$

Named in 1965 after an ancient name for the Balkan Mts. ISO Space group unknown. $a = 10.82$, $Z = 4$, $D = 4.47$. *25-300*: 3.61$_2$ 3.11$_{10}$ 2.87$_2$ 2.55$_2$ 2.09$_2$ 1.92$_5$ 1.86$_3$ 1.63$_3$. Microscopic grains and aggregates. Gray, metallic luster. H = 4, VHN$_{25}$ = 166. R = 24.2 (589 nm). Chelopech, Bulgaria; Kochbulak deposit, Uzbekistan. BM *AM 56:1847(1971), ABI:213.*

2.9.6.2 Kiddcreekite Cu$_6$SnWS$_8$

Named in 1984 for the locality. ISO Space group unknown. $a = 10.856$, $Z = 4$, $D = 4.88$. *37-427*: 6.29$_{10}$ 5.41$_3$ 3.27$_3$ 3.14$_5$ 2.71$_2$ 2.49$_2$ 1.92$_6$ 1.84$_3$. Microscopic grains. Gray, metallic luster. H = $3\frac{1}{2}$, VHN$_{100}$ = 183. R = 22.5 (540 nm). Campbell mine, Bisbee, AZ; *Kidd Creek mine, near Timmins, ON, Canada*. BM *CM 22:227(1984), AM 70:437(1985), ABI:264.*

2.9.7.1 Putoranite $Cu_{1.1}Fe_{1.2}S_2$

Named in 1980 for Putoran Mts., Siberia. ISO $Pn3m$. $a = 5.30$, $Z = 2$, $D = 6.14$. $41\text{-}1404$: 3.05_{10} 2.65_2 1.87_9 1.60_6 1.33_2 1.22_3 1.08_5 1.02_2. Single crystals rare, usually as coarse-grained aggregates. Yellow, metallic luster. Polysynthetically twinned. $H = 4$, $VHN_{50} = 263$. $R = 38.0$ (540 nm). *Oktyabr mine, Talnakh, Norilsk, Russia.* BM *ZVMO 109:335(1980)*, *AM 66:638 (1981)*, *ABI:424*.

2.9.8.1 Talnakhite $Cu_9Fe_8S_{16}$

Named in 1968 for the locality. Contains up to 1% Ni. ISO $I\bar{4}3m$. $a = 10.591$, $Z = 2$, $D = 4.83$. $25\text{-}287$: 3.75_4 3.06_{10} 2.64_5 1.87_9 1.60_7 1.21_5 1.08_6 1.02_5. Microscopic laths. Brass-yellow, metallic luster. $VHN_{50} = 261\text{-}277$. $G = 4.29$. $R = 39.4$ (540 nm). Stillwater Complex, Sweetwater Co., MT; Strachimir, Bulgaria; Filizchai deposit, Azerbaijan; *Talnakh, Norilsk, Russia.* BM *ZVMO 97:63(1968)*, *AM 57:380(1972)*, *P&J:360*, *ABI:512*.

2.9.8.2 Haycockite $Cu_4Fe_5S_8$

Named in 1972 for Maurice H. Haycock (b.1900), Canadian mineralogist. ORTH $P222$. $a = 10.71$, $b = 10.71$, $c = 31.56$, $Z = 12$, $D = 4.35$. $25\text{-}289$: 3.07_{10} 2.67_4 1.89_6 1.88_8 1.61_6 1.59_4 1.21_5 1.09_6. Microscopic grains. Brass-yellow, metallic luster. $H = 4$, $VHN_{100} = 206\text{-}231$. $R = 33.3\text{-}34.5$ (540 nm); moderately anisotropic. Duluth Gabbro Complex, MN; near Madera, SON, Mexico; Krzemianka, Poland; Talnakh, Norilsk, Russia; *Mooihoek Farm, Lydenburg, Transvaal, South Africa.* BM *AM 57:689(1972)*, *AC(B) 31:2105(1975)*, *ABI:209*.

2.9.8.3 Mooihoekite $Cu_9Fe_9S_{16}$

Named in 1972 for the locality. TET $P\bar{4}2m$. $a = 10.585$, $c = 5.383$, $Z = 1$, $D = 4.37$. $25\text{-}286$(syn): 3.77_3 3.07_{10} 1.89_8 1.87_4 1.60_6 1.32_4 1.22_5 1.08_6. Microscopic grains. Brass-yellow, metallic luster. $H = 4$, $VHN_{100} = 240\text{-}246$. $G = 4.36$. $R = 35.2$ (540 nm). Duluth Gabbro Complex, MN; Talnakh, Norilsk, Russia; *Mooihoek Farm, Lydenburg, Transvaal South Africa.* BM *AM 57:689(1972)*, *CM 16:23(1978)*, *ABI:339*.

2.9.9.1 Raguinite $TlFeS_2$

Named in 1969 for Eugene Raguin (b.1900), French geologist. ORTH Space group unknown. $a = 12.40$, $b = 10.44$, $c = 5.26$, $Z = 8$, $D = 6.33$. $22\text{-}1468$: 4.70_2 4.17_6 3.35_6 2.89_{10} 2.83_2 2.64_4 2.35_3 2.07_1. Microscopic pseudohexagonal plates. Bronze, metallic luster. $G = 6.4$. $R = 25.4\text{-}32.1$ (540 nm); strongly pleochroic and anisotropic. *Allchar, Macedonia.* BM *BM 92:38,237(1969)*, *AM 54:1495,1741(1969)*, *R:562*, *P&J:319*, *ABI:430*.

2.9.10.1 Teallite PbSnS$_2$

Named in 1904 for Jethro J. H. Teall (1849–1924), director of the Geological Survey of Great Britain. ORTH *Pbnm*. $a = 4.258$, $b = 11.447$, $c = 4.103$, $Z = 2$, $D = 6.48$. *14-189*: 3.41_6 3.27_4 2.84_{10} 2.33_4 2.03_4 1.80_3 1.42_5 1.09_4. Thin tabular crystals, usually massive aggregates of thin foliae. Black, black streak, metallic luster. Cleavage {001}, perfect; flexible; inelastic. $H = 1\frac{1}{2}$. $G = 6.36$. $R = 43.4$–47.2 (540 nm); distinctly anisotropic. Ivigtut, Greenland; Sinantscha, Sikhote-Alin, and Smirnovsk, Transbaikalia, Russia; *Santa Rosa mine* and many other tin mines in *Bolivia*, sometimes as an important ore mineral. BM *DI:439, R:670, P&J:362, ABI:513*.

2.9.11.1 Rasvumite KFe$_2$S$_3$

Named in 1970 for the locality. ORTH *Cmcm*. $a = 9.035$, $b = 11.022$, $c = 5.426$, $Z = 4$, $D = 3.04$. *33-1017*: 6.98_6 5.51_{10} 3.49_2 3.40_5 2.94_1 2.91_1 2.33_1 1.78_1. Microscopic grains. Steel-gray, metallic luster. Cleavage {110}, perfect. $H = 4\frac{1}{2}$, VHN = 243–433. $G = 3.1$. Slightly magnetic. $R = 17.7$–23.8 (540 nm); strongly pleochroic and anisotropic. Point of Rocks, NM; Coyote Peak, Humboldt Co., CA; Mont Saint-Hilaire, PQ, Canada; *Rasvumchorr apatite deposits* and Kikusvumchorr apatite deposits, *Kola, Russia*. BM *ZVMO 99:712(1970); AM 64:776(1979), 65:477(1980); P&J:322; ABI:433*.

2.9.12.1 Sternbergite AgFe$_2$S$_3$

Named in 1828 for Count Caspar Sternberg (1761–1838), Czech botanist. Dimorphous with argentopyrite. ORTH *Ccmb*. $a = 6.625$, $b = 11.558$, $c = 12.750$, $Z = 8$, $D = 4.30$. *11-61*: 4.29_{10} 3.42_2 3.22_5 2.79_7 2.63_4 1.94_3 1.90_3 1.79_5. Thin pseudohexagonal plates, forming rosettes or fanlike aggregates. Brown to black, black streak, metallic luster. Cleavage {001}, perfect; flexible cleavage lamellae. $H = 1$–$1\frac{1}{2}$, VHN$_{50}$ = 31–44. $G = 4.1$–4.2. $R = 32.3$–37.2 (540 nm); distinctly pleochroic and anisotropic. Leroy mine, Cochise Co., AZ; Highland Bell mine, Beaverdell, BC, Canada; Andreasberg, Harz, and Freiberg(*), Saxony, Germany; *Jachymov*, Pribram, and in the Krusnehory Mts., *Czech Republic*; Ruen Pb–Zn deposit, Bulgaria. BM *DI:246, R:639, P&J:73, NJMM:458(1987), ABI:494*.

2.9.12.2 Picotpaulite TlFe$_2$S$_3$

Named in 1970 for Paul Picot (b.1931), French mineralogist. ORTH *C222*. $a = 5.40$, $b = 10.72$, $c = 9.04$, $Z = 4$, $D = 5.23$. *25-953*: 5.4_5 4.53_5 4.26_9 3.80_7 3.33_7 2.91_{10} 2.56_5 2.51_5. Microscopic pseudohexagonal plates. Gray, metallic luster. $H = 2$, VHN$_{15}$ = 41. $R = 24.2$–31.0 (540 nm); strongly pleochroic and anisotropic. *Allchar, Macedonia*. BM *BM 93:545(1970), AM 57:1909(1972), P&J:299, ABI:403*.

2.9.13.1 Cubanite CuFe$_2$S$_3$

Named in 1843 for the locality. Dimorphous with isocubanite. ORTH *Pcmn*.
$a = 6.46$, $b = 11.117$, $c = 6.233$, $Z = 4$, $D = 4.03$. *9-324*: 3.49_4 3.22_{10} 3.00_4 2.79_4 1.87_8 1.75_7 1.28_4 1.17_5. Thick tabular and elongated prismatic crystals, commonly twinned on {110}; usually massive. Brass- to bronze-yellow, gray-black streak, metallic luster. No cleavage, conchoidal fracture. $H = 3\frac{1}{2}$. $G = 4.0$–4.2. Strongly magnetic. $R = 36.3$–41.0 (540 nm); distinctly anisotropic. Widely distributed; occasionally an important ore mineral. Fierro, Grant Co., NM; Christmas mine, Gila Co., AZ; Mackinaw mine, Snohomish Co., WA; Landlocked Bay, Prince William Sound, AK; Chibougamau(*), PQ (exceptional crystals) and Sudbury, ON, Canada; *Barracanao, Cuba*; Kaveltorp, Örebro, Sweden; Pitkäranta, Karelia, and Norilsk, Russia; Kohmori mine, Kyoto Pref., Japan; Broken Hill, NSW, Australia; Beni Bousera, Morocco; Morro Velho mine, Nova Lima, MG, Brazil. A rare accessory in meteorites and lunar rocks. BM *DI:243*, *ZK 140:218(1974)*, *R:630*, *P&J:136*, *ABI:117*.

2.9.13.2 Argentopyrite AgFe$_2$S$_3$

Named in 1868 for its composition and similarity to pyrite. Dimorphous with sternbergite. ORTH *Pmmn*. $a = 6.64$, $b = 11.47$, $c = 6.45$, $Z = 4$, $D = 4.27$. *7-347*: 5.81_3 3.62_5 3.34_{10} 3.32_{10} 3.11_4 1.93_5 1.91_4 1.81_7. Pseudohexagonal prismatic crystals. Bronze-brown, tarnishing to lead-gray, gray streak, metallic luster. No cleavage, uneven fracture. $H = 3\frac{1}{2}$–4, $VHN_{100} = 225$–242. $G = 4.0$–4.3. $R = 26.8$–34.7 (540 nm); strongly pleochroic. Andreasberg, Harz, and Freiberg and other Saxony localities, Germany; *Jachymov*, Pribram, and Medenec, *Czech Republic*; Omidani mine, Hyogo Pref., Japan; Broken Hill, NSW, Australia; Colquechaca, Bolivia. BM *DI:248*; *AM 39:475(1954)*, *54:1198(1969)*; *R:639*; *P&J:73*; *ABI:21*.

2.9.13.3 Isocubanite CuFe$_2$S$_3$

Named in 1988 as a dimorph of cubanite, stable above 250°. ISO *Fm3m*. $a = 5.303$, $Z = 4/3$, $D = 4.075$. *27-166*: 3.05_{10} 2.64_3 1.87_8 1.59_7 1.32_4 1.21_5 1.08_6 1.02_5. Microscopic cubo-octahedral crystals. Bronze, metallic luster. $H = 3\frac{1}{2}$, $VHN_{100} = 175$. $R = 34.8$ (540 nm). Described from submarine "black smokers," East Pacific Rise and Red Sea. Recorded in some ore deposits as chalcopyrrhotite. Sudbury, ON, Canada; Tunaberg, Sweden; Talnakh deposit, Norilsk, Russia; Mooihoek Farm, Lydenburg, Transvaal, South Africa. BM *R:545*, *MM 52:509(1988)*, *AM 74:503(1989)*, *ABI:242*.

2.9.14.1 Idaite Cu$_5$FeS$_6$

Named in 1958 for the locality. HEX $P6_3/mmc$. $a = 3.90$, $c = 16.95$, $Z = 1$, $D = 4.21$. Covellite structure. *13-161*(syn): 3.14_{10} 2.82_{10} 2.70_8 1.89_{10} 1.85_{10} 1.56_{10} 1.32_8 1.08_8. Microscopic grains and exsolution lamellae. Copper-red to brown, metallic luster. $H = 2\frac{1}{2}$–$3\frac{1}{2}$. $G = 4.20$. $R = 26.1$–29.4 (540 nm); strongly anisotropic. White Canyon, San Juan Co., UT; Algoma, Jarvis Twp., ON, Canada; Radka deposit, Pazardzhik, Bulgaria; Kaikita mine,

Aomori Pref., and other localities in Japan; *Ida mine, Khan*, and *Tsumeb, Namibia*; several mines in the Copiapo district, Chile. BM *NJMM:142(1958), 241(1976); MM 43:193(1979); R:692; P&J:206; ABI:227*.

2.9.15.1 Nukundamite $(Cu,Fe)_4S_4$

Named in 1979 for the locality. HEX-R $P\bar{3}m1$. $a = 3.782$, $c = 11.187$, $Z = 1$, $D = 4.53$. *34-1409*: 3.73_1 3.27_3 3.14_{10} 2.83_7 2.80_5 1.89_6 1.85_5 1.57_3. Microscopic hexagonal crystals and massive. Copper-red, dark gray to black streak, metallic luster. Perfect basal cleavage. $H = 3$, $VHN_{20} = 103–110$. $D = 4.30$. $R = 15.8–23.7$ (540 nm); strongly pleochroic and anisotropic. *Undu mine, Nukundamu, Fiji*. BM *MM 43:193(1979), AM 66:398(1981), ABI:359*.

2.9.16.1 Mohite Cu_2SnS_3

Named in 1982 for Gunter Moh (1929–1993), German mineralogist. TRIC $P1$. $a = 6.64$, $b = 11.51$, $c = 19.93$, $\alpha = 90.0°$, $\beta = 109.75°$, $\gamma = 90.0°$, $Z = 12$, $D = 4.75$. *35-684*: 3.66_1 3.34_1 3.13_{10} 2.82_1 2.72_2 2.63_1 2.44_2 1.92_7. Microscopic grains. Gray, metallic luster. $H = 4\frac{1}{2}$, $VHN_{20} = 179$. $R = 25.5$ (540 nm); distinctly anisotropic. *April Fool mine, Lincoln Co., NV; Kochbulak deposit, Uzbekistan*. BM *ZVMO 111:110(1982), AM 68:281(1983), ABI:335*.

2.9.17.1 Caswellsilverite $NaCrS_2$

Named in 1982 for Caswell Silver (b.1916), U.S. geologist. HEX-R $R\bar{3}m$. $a = 3.541$, $c = 19.35$, $Z = 3$, $D = 3.30$. *10-292*: 2.58_{10} 1.89_9 1.76_9 1.46_8 1.29_8 1.20_8 1.13_{10} 1.04_{10}. Microscopic anhedral grains. Yellow-gray tarnishing brown, metallic luster. $H = 2$, $VHN_{15} = 17–45$. $R = 21.8–32.8$ (540 nm); strongly pleochroic and anisotropic. *Norton County enstatite achondrite* and Quizhen enstatite chondrite. BM *AM 67:132(1982), GCA 49:1781(1985), ABI:83*.

2.9.17.2 Schöllhornite $Na_{0.3}CrS_2 \cdot H_2O$

Named in 1985 for Robert Schöllhorn, German chemist. HEX-R Space group uncertain, probably $R3m$. $a = 3.32$, $c = 26.6$, $Z = 3$, $D = 2.77$. *39-322*(syn): 8.85_{10} 4.43_1 2.81_5 2.53_5 2.21_1 1.66_5. Microscopic grains. Gray, metallic luster. Perfect basal cleavage. $H = 2$, $VHN_{15} = 80$. $G_{syn} = 2.70$. $R = 16.0–19.3$ (540 nm); distinctly pleochroic and anisotropic. A weathering product of caswellsilverite in the *Norton County enstatite achondrite*. BM *AM 70:638(1985), ABI:465*.

2.9.18.1 Petrukite $(Cu,Fe,Zn)_3SnS_4$

Named in 1989 for William Petruk (b.1930), Canadian mineralogist. Contains up to 6% In and 1% Ag. ORTH $Pmn2_1$. $a = 7.667$, $b = 6.440$, $c = 6.276$, $Z = 2$, $D = 4.61$. *CM27*: 3.13_{10} 2.71_4 2.42_2 1.92_6 1.64_5 1.24_5 1.11_6 10.04_3. Microscopic grains. Brown to gray, metallic luster. Cleavage {110}, {100}, {010}, distinct; parting parallel to {001}. H = $4\frac{1}{2}$, VHN_{100} = 319. R = 27.7–27.8 (546 nm). Mt. Pleasant mine, Charlotte Co., NB, and *Herb claim, Turnagain R., BC, Canada*; Ikuno mine, Hyogo Pref., Japan. BM *CM 27:673(1989)*.

2.9.19.1 Bartonite $K_3Fe_{10}S_{14}$

Named in 1981 for Paul B. Barton (b.1930), U.S. mineralogist. TET $I4/mmm$. a = 10.424, c = 20.626, Z = 4, D = 3.286. *35-476*: 9.31_3 5.99_8 3.14_3 3.00_{10} 2.39_2 2.38_3 1.84_3 1.83_4. Brown-black masses, black streak, submetallic luster. Cleavage {112}, distinct; conchoidal fracture. H = $3\frac{1}{2}$, VHN_{15} = 94–120. G = 3.305. R = 20.9–21.8 (540 nm); distinctly anisotropic. *Coyote Peak, Humboldt Co., CA*, in an alkalic diatreme. BM *AM 66:369(1981)*, *ABI:41*.

LINNAEITE GROUP

The linnaeite group minerals are sulfides and selenides conforming to the general formula

$$A^{2+}B^{3+}X_4$$

where

A = Fe, Co, Ni, Cu, Zn
B = Fe, Co, Ni, Cr, In, Rh, Ir, Pt
X = S, Se

and the minerals are cubic, space group $Fd3m$, isostructural with one another and with the spinel group.

LINNAEITE GROUP

Mineral	Formula	Space Group	a	D	
Linnaeite	$CoCo_2S_4$	$Fd3m$	9.43	4.83	2.10.1.1
Carrollite	$CuCo_2S_4$	$Fd3m$	9.48	4.83	2.10.1.2
Fletcherite	$CuNi_2S_4$	$Fd3m$	9.520	4.76	2.10.1.3
Tyrrellite	$(Cu,Co,Ni)_3Se_4$	$Fd3m$	10.005	6.59	2.10.1.4
Bornhardtite	$CoCo_2Se_4$	$Fd3m$	10.2	6.17	2.10.1.5
Siegenite	$(Ni,Co)_3S_4$	$Fd3m$	9.387	4.89	2.10.1.6
Polydymite	$NiNi_2S_4$	$Fd3m$	9.48	4.74	2.10.1.7
Violarite	$FeNi_2Se_4$	$Fd3m$	9.46	4.72	2.10.1.8
Trüstedtite	$NiNi_2Se_4$	$Fd3m$	9.94	6.65	2.10.1.9

LINNAEITE GROUP (CONT'D)

Mineral	Formula	Space Group	a	D	
Greigite	$FeFe_2S_4$	$Fd3m$	9.876	4.08	2.10.1.10
Daubréelite	$FeCr_2S_4$	$Fd3m$	9.995	3.83	2.10.1.11
Indite	$FeIn_2S_4$	$Fd3m$	10.618	4.59	2.10.1.12
Kalininite	$ZnCr_2S_4$	$Fd3m$	9.974	4.04	2.10.1.13
Florensovite	$Cu(Cr,Sb)_2S_4$	$Fd3m$	10.005	4.28	2.10.1.14
Cuproiridisite	$CuIr_2S_4$	$Fd3m$	9.92	7.84	2.10.1.15
Cuprorhodisite	$CuRh_2S_4$	$Fd3m$	9.88	5.48	2.10.1.16
Malanite	$Cu(Pt,Ir)_2S_4$	$Fd3m$	9.92	7.92	2.10.1.17

2.10.1.1 Linnaeite $CoCo_2S_4$

Named in 1845 for Carolus Linnaeus (1707–1778), Swedish taxonomist. Linnaeite group. May contain minor amounts of Cu, Ni, and Fe. ISO $Fd3m$. $a = 9.43$, $Z = 8$, $D = 4.83$. 11-121: 3.34_4 2.83_{10} 2.36_7 1.93_3 1.82_6 1.67_8 1.09_4 1.06_4. Octahedral crystals, also massive, granular, or compact; crystals twinned on {111}. Light gray to steel-gray, gray streak, metallic luster. Cleavage {001}, imperfect, subconchoidal to uneven fracture. $H = 4\frac{1}{2}$–$5\frac{1}{2}$, $VHN_{50} = 492$. $G = 4.5$–4.8. $R = 43.6$ (540 nm). Fuses to a magnetic globule; soluble in HNO_3 with the separation of sulfur. Widely distributed in hydrothermal veins with other cobalt and nickel sulfides. Springfield and Mineral Hill mines, Carroll Co., MD; Mine La Motte, Madison Co., MO; Gladhammar, Kalmar, and *Bastnäs, Västmannland, Sweden*; Müsen, Littfeld, and Altenberg, Westphalia, Germany; Kladno, Czech Republic; Musonoi mine, Shaba, Zaire; N'Kana mine, Kitwe, Zambia. BM *DI:262, R:697, P&J:238, ABI:297.*

2.10.1.2 Carrollite $CuCo_2S_4$

Named in 1852 for the locality. Linnaeite group. May contain minor amounts of Ni and Fe. ISO $Fd3m$. $a = 9.48$, $Z = 8$, $D = 4.83$. 9-425: 3.35_4 2.86_{10} 2.37_5 1.83_6 1.67_8 1.23_3 1.19_3 1.10_3. Octahedral crystals, often twinned on {111}; also massive, granular, or compact. Silver-white to steel-gray, gray streak, metallic luster. Cleavage {011}, imperfect; subconchoidal to uneven fracture. $H = 4\frac{1}{2}$–$5\frac{1}{2}$, $VHN_{100} = 507$–586. $G = 4.5$–4.8. $R = 41.8$ (540 nm). *Patapsco mine, Carroll Co., MD*; Mine La Motte, Madison Co., MO; Gladhammar, Kalmar, Sweden; Kohlenbach mine, Westphalia, Germany; Neves-Corvo, Portugal; Chernomorets, Bulgaria; Sazare mine, Ehime Pref., Japan; Kambalda, WA, Australia; Kambove and Kolwezi mines(*), Shaba, Zaire; Rokana mine, Kitwe, Zambia; Tsumeb, Namibia; Carrizal Alto deposit, Atacama, Chile. BM *DI:262, AM 59:302(1974), R:697, P&J:238, ABI:82.*

2.10.1.3 Fletcherite $CuNi_2S_4$

Named in 1977 for the locality. Linnaeite group. Contains up to 17% Co. ISO $Fd3m$. $a = 9.520$, $Z = 8$, $D = 4.76$. *29-540*: 3.36_5 2.87_6 2.39_6 1.83_8 1.68_{10} 1.37_6 1.24_6 1.19_6. Microscopic crystals. Steel-gray, metallic luster. $H = 4\frac{1}{2}$–5, $VHN_{25} = 446$–464. $R = 43.4$ (546 nm). *Fletcher mine, Reynolds Co., MO*; New Rambler mine, Medicine Bow Mts., WY. BM *EG 72:480(1977)*, *AM 62:596(1977)*, *ABI:159*.

2.10.1.4 Tyrrellite $(Cu,Co,Ni)_3Se_4$

Named in 1952 for Joseph B. Tyrrell (1858–1957), Canadian geologist. Linnaeite group. ISO $Fd3m$. $a = 10.005$, $Z = 8$, $D = 6.59$. *8-1*: 5.78_4 3.54_4 3.02_6 2.89_7 2.50_9 1.93_6 1.77_{10} 1.51_4. Microscopic subhedral to anhedral grains. Pale bronze, black streak, metallic luster. Poor cubic cleavage, conchoidal fracture. $H = 3\frac{1}{2}$, $VHN_{100} = 343$–368. $G = 6.6$. $R = 47.3$ (540 nm). *Eagle shaft, Beaverlodge Lake, SK, Canada*; Bukov, Moravia, Czech Republic. BM *AM 37:542(1952)*, *CM 10:731(1970)*, *R:703*, *P&J:376*, *ABI:546*.

2.10.1.5 Bornhardtite $CoCo_2Se_4$

Named in 1955 for Wilhelm Bornhardt (1864–1946), German geologist. Linnaeite group. ISO $Fd3m$. $a = 10.2$, $Z = 8$, $D = 6.17$. *15-463*: 2.7_{10} 2.4_{10} 2.3_{10} 2.18_{10} 2.00_{10} 1.96_{10} 1.42_{10}. Microscopic crystals. Pink-white, metallic luster. $H = 4$. $R = 42.4$ (540 nm). *Pinky Fault deposit, SK, Canada*; Trogtal quarry and Tilkerode, Harz, Germany; Sierra de Cachueta, Mendoza Prov., Argentina. BM *NJMM:133(1955)*, *AM 41:164(1956)*, *R:702*, *P&J:97*, *ABI:61*.

2.10.1.6 Siegenite $(Ni,Co)_3S_4$

Named in 1850 for the locality. Linnaeite group. ISO $Fd3m$. $a = 9.387$, $Z = 8$, $D = 4.89$. *20-782*(syn): 3.32_4 2.83_{10} 2.35_5 1.81_4 1.66_8 1.22_2 1.08_2 1.05_3. Octahedral crystals, often twinned on $\{111\}$; also massive, granular or compact. Silver-white to steel-gray, gray streak, metallic luster. Cleavage $\{001\}$, imperfect; subconchoidal to uneven fracture. $H = 4\frac{1}{2}$–$5\frac{1}{2}$, $VHN_{100} = 459$–540. $G = 4.5$–4.8. $R = 46.2$ (540 nm). Mine la Motte, Madison Co., and Miliken mine(*), Reynolds Co., MO; Langis mine, near Cobalt, ON, Canada; *Siegen district*(*), *Westphalia, Germany*; Kladno, Czech Republic; near Brestovsko, Bosnia; Etropole and Mladenovo, Bulgaria; Kamaishi mine, Iwate Pref., Japan; Kambalda, WA, Australia; Kilembe, Uganda; Shinkolobwe mine, Shaba, Zaire. BM *DI:262*, *CM 9:597(1968)*, *R:697*, *P&J:238*; *ABI:474*.

2.10.1.7 Polydymite $NiNi_2S_4$

Named in 1876 from the Greek for *many* and *twin*, because the mineral often occurs in twinned crystals. Linnaeite group. ISO $Fd3m$. $a = 9.48$, $Z = 8$, $D = 4.74$. *8-106*: 2.85_9 2.36_9 1.82_9 1.67_{10} 1.37_6 1.23_8 1.19_7 1.10_7. Octahedral crystals, twinned on $\{111\}$; also massive, granular, or compact. Steel-gray, black streak, metallic luster. Cleavage {001}, imperfect; subconchoidal to uneven fracture. H = $4\frac{1}{2}$, VHN_{100} = 379–427. G = 4.5–5.8. R = 45.4 (540 nm). Hamilton, Hancock Co., IL; Miliken mine, Reynolds Co., MO; *Grunau mine, near Siegen, Westphalia, Germany*; Kunratice and Rozany, Czech Republic; Saint Marina, Khaskovo district, Bulgaria; Novo-Aidyrlinsk, Ural Mts., Russia; Kambalda, WA, Australia; Dry Nickel mine, Bindura, Zimbabwe. BM *DI:262, R:697, P&J:238, ABI:418*.

2.10.1.8   Violarite $FeNi_2S_4$

Named in 1924 for its color. Linnaeite group. Contains minor amounts of Co and Cu. ISO *Fd3m*. a = 9.46, Z = 8, D = 4.72. *11-95*: 3.35_3 2.85_{10} 2.36_5 1.82_6 1.67_8 1.18_4 1.12_4 1.06_5. Massive. Violet-gray, black streak, metallic luster. Cleavage {001}, perfect; brittle. H = $4\frac{1}{2}$–$5\frac{1}{2}$, VHN_{100} = 455–493. G = 4.6. R = 42.5 (540 nm). Gap mine, Lancaster Co., PA; Lick Fork deposit, Floyd Co., VA; Key West mine, Clark Co., NV; Friday mine, Julian, CA; Marbridge mine, Malartic, PQ, *Sudbury, ON*, English Lake, MB, *Canada*; Praborna mine, St. Marcel, Italy; Kambalda, WA, Australia; Dry Nickel mine, Bindura, Zimbabwe. BM *DI:262, R:697, P&J:238, ABI:560*.

2.10.1.9   Trüstedtite $NiNi_2Se_4$

Named in 1964 for Otto Trüstedt (1866–1929), Finnish mining engineer. Linnaeite group. Contains 6.4% Co. ISO *Fd3m*. a = 9.94, Z = 8, D = 6.65. *18-889*: 5.75_4 3.52_4 3.00_8 2.87_8 2.48_{10} 1.91_6 1.76_{10} 1.50_2. Microscopic euhedral crystals. Yellow, metallic luster. H = 3. R = 50.3 (540 nm). *Kuusamo, Finland*. BM *AM 50:519(1965), R:703, P&J:373, ABI:539*.

2.10.1.10   Greigite $FeFe_2S_4$

Named in 1964 for Joseph W. Greig (1895–1977), U.S. mineralogist. Linnaeite group. Synonym: melnikovite (part). ISO *Fd3m*. a = 9.876, Z = 8, D = 4.08. *16-713*: 3.50_3 2.98_{10} 2.47_6 2.02_1 1.90_3 1.75_8 1.56_1 1.51_1. Microscopic octahedral grains and powdery coatings. Black, metallic luster. H = 4, VHN_{50} = 401–423. Strongly magnetic. R = 30.7 (540 nm). Widely distributed in lacustrine clays and in hydrothermal sulfide deposits. *Boron, Kern Co.*, and Coyote Peak, Humboldt Co., *CA*; near Zacatecas, ZAC, Mexico; Treore mine, Cornwall, England; Montemesola, Taranto, Italy; Lojane deposit, Macedonia; Kerch Peninsula, Ukraine; Hanoaka mine, Akita Pref., Japan. BM *AM 49:543(1964), R:697, P&J:238, ABI:195*.

2.10.1.11   Daubréelite $FeCr_2S_4$

Named in 1876 for Gabriel A. Daubrée (1814–1896), French meteorite researcher. Linnaeite group. ISO *Fd3m*. a = 9.995, Z = 8, D = 3.83. *4-651*(syn): 3.53_8 3.01_{10} 2.50_8 2.04_6 1.92_8 1.77_{10} 1.30_8 1.25_8. Exsolution lamellae in troilite, and microscopic grains. Black, black streak, metallic luster. Distinct cleavage, brittle. H = 5, VHN_{100} = 260–303. G = 3.81. R = 33.0

(540 nm). An accessory mineral in many meteorites. BM $DI:265$, $R:703$, $ABI:129$.

2.10.1.12 Indite $FeIn_2S_4$

Named in 1963 for its indium content. Linnaeite group. ISO $Fd3m$. $a = 10.618$, $Z = 8$, $D = 4.59$. $33\text{-}613$: 3.75_4 3.20_{10} 2.65_2 2.17_1 2.04_4 1.88_5 1.62_1 1.38_1. Microscopic grains. Iron-black, metallic luster. $H = 4\frac{1}{2}$, $VHN = 309$. $R = 27.0$ (540 nm). *Dzhalindin deposit, Little Khingan Ridge, eastern Siberia, Russia.* BM $ZVMO$ $92:445(1963)$, AM $49:439(1964)$, $R:703$, $P\&J:211$, $ABI:233$.

2.10.1.13 Kalininite $ZnCr_2S_4$

Named in 1985 for Pavel V. Kalinin (1905–1981), Russian mineralogist. Linnaeite group. ISO $Fd3m$. $a = 9.974$, $Z = 8$, $D = 4.04$. $16\text{-}507$(syn): 3.53_4 3.00_8 2.50_5 1.76_{10} 1.30_2 1.25_2 1.02_5. Microscopic aggregates. Black, black streak, metallic luster. $H = 5$, $VHN_{50} = 468$. $R = 35.2$ (540 nm). *Slyudyanka complex, Baikal region, Russia.* BM $ZVMO$ $114:622(1985)$, AM $72:223(1987)$, $ABI:255$.

2.10.1.14 Florensovite $Cu(Cr,Sb)_2S_4$

Named in 1989 for Nikolai A. Florensov (1909–1986), Russian geologist. Linnaeite group. ISO $Fd3m$. $a = 10.005$, $Z = 8$, $D = 4.28$. $43\text{-}1468$: 3.53_5 3.01_{10} 2.50_8 1.92_8 1.77_{10} 1.30_6 1.16_5 1.02_8. Microscopic grains. Black, adamantine to metallic luster. $H = 5$, $VHN_{50} = 477\text{-}541$. Strongly magnetic. $R = 26.5$ (540 nm). *Slyudyanka complex, Baikal region, Russia.* BM $ZVMO$ $118:57(1989)$, AM $75:1209(1990)$.

2.10.1.15 Cuproiridisite $CuIr_2S_4$

Named in 1985 for the composition. Linnaeite group. Contains minor amounts of Pt, Rh, and Fe. ISO $Fd3m$. $a = 9.92$, $Z = 8$, $D = 7.84$. $39\text{-}329$: 3.01_{10} 2.49_7 1.92_8 1.76_9 1.29_6 1.15_5 1.11_5 1.01_8. Microscopic grains. Iron-black, metallic luster. Very brittle. $H = 5\frac{1}{2}$, $VHN_{30} = 578$. $R = 33.6\text{-}37.2$ (540 nm). *Finero, Italy; Konder massif, Yakutia, Russia.* BM $ZVMO$ $114:187(1985)$, AM $71:1277(1986)$, $ABI:120$.

2.10.1.16 Cuprorhodisite $CuRh_2S_4$

Named in 1985 for the composition. Linnaeite group. Contains minor amounts of Ir, Pt, and Fe. ISO $Fm3m$. $a = 9.88$, $Z = 8$, $D = 5.48$. $38\text{-}405$: 5.7_4 3.0_{10} 2.48_7 1.90_8 1.76_{10} 1.29_5 1.10_5 1.01_9. Microscopic grains. Iron-black, metallic luster. Very brittle. $H = 5$, $VHN_{50} = 498$. $R = 36.7\text{-}39.2$ (540 nm). *Finero, Italy; Konder massif, Yakutia, Russia.* BM $ZVMO$ $114:187(1985)$, AM $71:1277(1986)$, $ABI:122$.

2.10.1.17 Malanite $Cu(Pt,Ir)_2S_4$

Named in 1974, presumably for the locality. Linnaeite group. ISO $Fd3m$. $a = 9.92$, $Z = 8$, $D = 7.92$. *38-406*: 5.8_5 3.01_{10} 2.49_9 1.92_9 1.76_{10} 1.29_6 1.11_7 1.01_9. Microscopic grains. Silver-white, metallic luster. $VHN_{20} = 462$. Ojen, Spain; Konder massif, Yakutia, Russia; *undisclosed locality, China*; Merensky Reef, Transvaal, South Africa. BM *AM 61:185(1976), 67:1081(1982); ZVMO 114:187(1985); ABI:309.*

2.10.2.1 Wilkmanite Ni_3Se_4

Named in 1964 for Wanold W. Wilkman (1872–1937), Finnish geologist. MON $I2/m$. $a = 6.22$, $b = 3.63$, $c = 10.52$, $\beta = 90.53°$, $Z = 2$, $D = 6.88$. *18-890*: 5.25_4 2.70_{10} 2.02_{10} 2.00_8 1.82_8 1.80_{10} 1.53_6 1.50_6. Probably massive. Yellow-gray, metallic luster. Distinctly pleochroic and anisotropic. *Kuusamo, Finland*. BM *AM 50:519(1965), ABI:375.*

2.10.2.2 Brezinaite Cr_3S_4

Named in 1969 for Aristides Brezina (1848–1909), Austrian meteoriticist. MON $I2/m$. $a = 5.96$, $b = 3.425$, $c = 11.27$, $\beta = 91.53°$, $Z = 2$, $D = 4.11$. *24-310*: 5.67_7 5.23_4 2.98_7 2.64_{10} 2.61_6 2.06_7 2.03_4 1.72_7. Microscopic anhedral grains. Gray, metallic luster. $VHN = 400$. $R = 37$–40; strongly anisotropic. *Tucson iron meteorite* and New Baltimore iron meteorites. BM *AM 54:1509(1969), ABI:70.*

2.10.2.3 Heideite $(Fe,Cr)_{1+x}(Ti,Fe)S_4$ ($x = 0.15$)

Named in 1974 for Friedrich Heide (1891–1973), German geochemist. MON $I2/m$. $a = 5.929$, $b = 3.426$, $c = 11.46$, $\beta = 90.10°$, $Z = 2$, $D = 3.99$. *20-1303*(syn): 5.72_2 5.26_3 2.96_3 2.63_9 2.06_{10} 1.71_4 1.61_3 1.44_2. Microscopic anhedral grains. Black, metallic luster. $H = 3\frac{1}{2}$–$4\frac{1}{2}$. $G_{syn} = 3.94$. An accessory mineral in the *Bustee enstatite achondrite*. BM *AM 59:465(1974), ABI:212.*

2.10.3.1 Rhodostannite $Cu_2FeSn_3S_8$

Named in 1968 for its reddish color compared to stannite. TET $I4_1/a$. $a = 7.29$, $c = 10.31$, $Z = 2$, $D = 4.82$. *29-558*(syn): 5.96_5 3.65_4 3.11_9 2.58_8 1.99_7 1.83_{10} 1.35_4 1.06_4. Microscopic grains. Reddish gray, metallic luster. $H = 4$, $VHN = 243$–266. $R = 28.8$–30.4 (540 nm); distinctly anisotropic. Dean mine, Lander Co., NV; Bisbee, AZ; *Vila Apacheta, Bolivia*; Pirquitas deposit, Jujuy, Argentina. BM *MM 36:1045(1968), NJMM:166(1975), P&J:326, ABI:440.*

2.10.3.2 Toyohaite $Ag_2FeSn_3S_8$

Named in 1991 for the locality. TET $I4_1/a$. $a = 7.464$, $c = 10.80$, $Z = 2$, $D = 4.94$. *44-1440*: 6.10_3 3.77_2 3.72_4 3.21_{10} 2.64_3 2.03_4 1.88_4 1.86_3. Microcrystalline aggregates. Gray-brown, metallic luster. $R = 24.1$–26.1 (546 nm); distinctly anisotropic. *Toyoha mine, Sapporo, Japan.* BM *MJJ 15:222(1991)*, *AM 77:1117(1992)*.

2.10.4.1 Konderite $PbCu_3(Rh,Pt,Ir)_8S_{16}$

Named in 1984 for the locality. HEX $P6/m$. $a = 7.024$, $c = 16.48$, $Z = 1$, $D = 4.09$. *38–393*: 5.1_2 2.98_{10} 2.85_5 2.46_9 2.25_2 1.76_{10} 1.72_5 1.29_3. Microscopic grains. Steel-gray, metallic luster. Cleavage in two directions, brittle. $VHN_{50} = 372$–793. $R = 42.5$–44.8 (520 nm). *Konder massif, Yakutia, Russia.* BM *ZVMO 113:703(1984)*, *AM 71:229(1986)*, *ABI:271*.

2.10.4.2 Inaglyite $PbCu_3(Ir,Pt)_8S_{16}$

Named in 1984 for the locality. HEX $P6/m$. $a = 7.03$, $c = 16.44$, $Z = 1$. *37-441*: 5.7_9 2.98_{10} 2.84_6 2.44_8 2.00_4 1.90_4 1.75_7 1.70_5. Microscopic grains. Steel-gray, metallic luster. Cleavage in two directions, brittle. $VHN_{50} = 347$–726. $R = 41.6$–42.8 (520 nm). *Inagli massif, Yakutia, Russia.* BM *ZVMO 113:712(1984)*, *AM 71:228(1986)*, *ABI:231*.

2.11.1.1 Orpiment As_2S_3

Named in 1747 from the Latin *auripigmentum*, golden paint, in allusion to its color. MON $P2_1/n$. $a = 11.49$, $b = 9.59$, $c = 4.25$, $\beta = 90.45°$, $Z = 4$, $D = 3.49$. *19-84*: 4.85_{10} 4.02_5 3.22_3 2.79_3 2.72_3 2.47_4 2.09_3 1.76_4. **Habit:** Crystals uncommon; typically small, short prismatic on c with pseudo-orthorhombic appearance; commonly in foliated, columnar, or fibrous masses; also granular or powdery. **Structure:** Consists of As and S atoms bonded together in complex sheets parallel to the ac plane, giving rise to perfect {010} cleavage. **Physical**

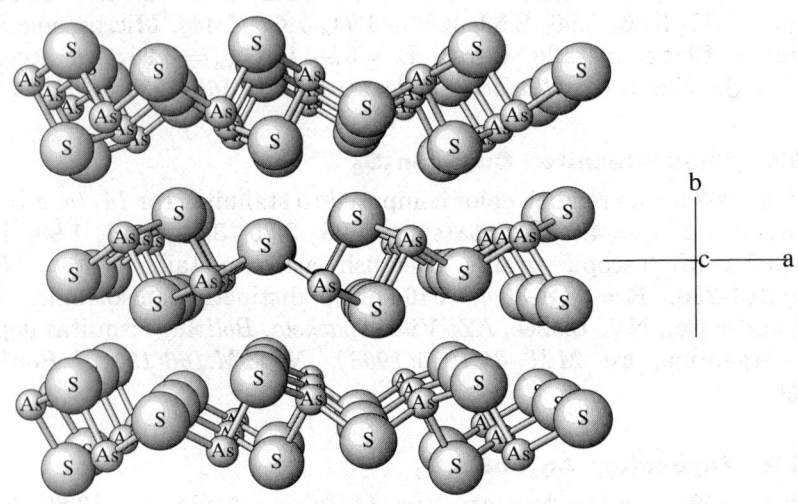

ORPIMENT: As_2S_3

As is bonded to three S and S to two As—an essentially covalent arrangement—forming corrugated sheets.

properties: Lemon yellow to brownish yellow, pale yellow streak, pearly luster on cleavage surfaces, resinous elsewhere. Cleavage {010}, perfect; lamellae flexible but not elastic; sectile. H = $1\frac{1}{2}$–2. G = 3.49. Tests: Easily recognized by its color and cleavage. Chemistry: Usually pure As_2S_3; may contain up to 1.5% Sb and 0.4% Tl. Optics: Biaxial (−); $N_x = 2.4$, $N_y = 2.81$, $N_z = 3.02$; $2V = 76°$; pleochroic in shades of yellow. R = 20.5–27.7 (540 nm). Occurrence: A low-temperature hydrothermal mineral, occurring in veins and in hot-spring deposits; also as an alteration product of realgar. Localities: Widely distributed; fine specimens from the following places: Mercur, Tooele Co., UT; Getchell mine, Humboldt Co., and White Caps mine, Nye Co., NV; Tajov, Czech Republic; Baia Sprie, Romania; Allchar, Macedonia; Jelamerk, Kurdistan, Turkey; Lukum, Georgia; Zarehchuran mine, Takab, Iran; Elbrozka, Caucasus, and Menkule, Yakutia, Russia; Guizhou, Hunan Prov., China; Jozankei, Hokkaido, Japan; Quirivilca mine, La Libertad, Peru. BM *DI:266, ZK 136:48(1972), R:890, P&J:283, ABI:366.*

2.11.1.2 Getchellite AsSbS₃

Named in 1965 for the locality. MON $P2_1/a$. $a = 11.85$, $b = 8.99$, $c = 10.16$, $\beta = 116.5°$, Z = 8, D = 4.02. *18-142*: 4.92_5 4.58_5 4.44_8 3.61_7 2.88_{10} 2.53_7 2.32_5 2.24_6. Subhedral crystals and massive. Dark blood-red, sometimes with a purple to green iridescent tarnish; orange-red streak; pearly to vitreous luster on cleavage surfaces, resinous elsewhere. Cleavage {001}, perfect; cleavage flakes flexible but inelastic; splintery fracture; sectile. H = $1\frac{1}{2}$–2. G = 3.92. R = 28.1–30.6 (540 nm). Getchell mine, Humboldt Co., NV; Zarehchuran mine, Takab, Iran; Kaidarkan, Kyrgyzstan; Gal-Khaya deposit, Yakutia, Russia; Toya mine, Hokkaido, Japan. BM *AM 50:1817(1965), AC(B) 29:2536(1973), R:704, P&J:179, ABI:182.*

STIBNITE GROUP

The stibnite group consists of antimony and bismuth sulfides and selenides corresponding to the general formula

$$A_2B_3$$

where

A = Sb, Bi
B = S, Se

and the space group is monoclinic, *Pbnm*. The structure of stibnite consists of a zigzag chain parallel to the c axis of S and Sb atoms. The long prismatic crystals are parallel to these structural chains, and the perfect {010} cleavage occurs between them.

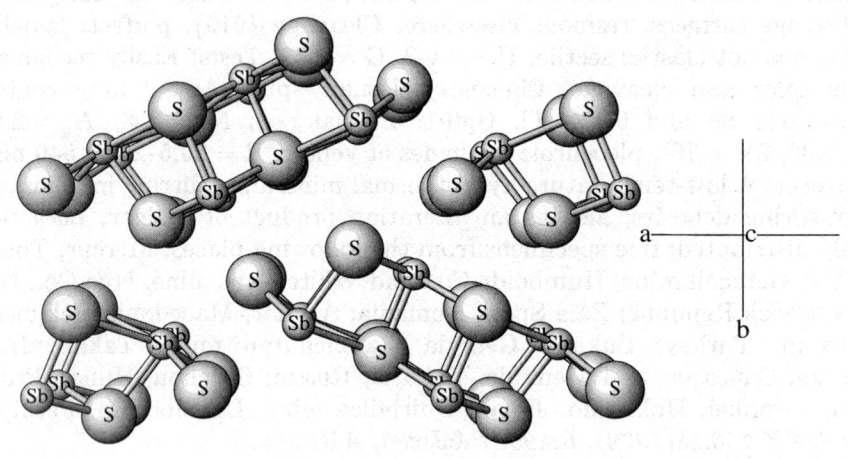

STIBNITE: Sb_2S_3

Bond lengths are highly variable. This view shows the complex chains which are isolated, ignoring bonds over 3.2 Å. With this limit, all Sb is 5-coordinated.

STIBNITE GROUP

Mineral	Formula	Space Group	a	b	c	D	
Stibnite	Sb_2S_3	Pbnm	11.229	11.310	3.839	4.63	2.11.2.1
Antimonselite	Sb_2Se_3	Pbnm	11.593	11.747	3.398	5.88	2.11.2.2
Bismuthinite	Bi_2S_3	Pbnm	11.149	11.304	3.981	6.81	2.11.2.3
Guanajuatite	Bi_2Se_3	Pbnm	11.37	11.55	4.054	7.54	2.11.2.4

2.11.2.1 Stibnite Sb_2S_3

Named in 1832 from the Latin *stibium*, an old name for the mineral. Stibnite group. ORTH *Pbnm*. $a = 11.229$, $b = 11.310$, $c = 3.839$, $Z = 4$, $D = 4.63$. 6-474(syn): 5.05_6 3.57_7 3.56_7 3.05_9 2.76_{10} 2.68_5 2.53_5 1.94_5. **Habit:** Crystals common, prismatic along c, typically striated or grooved parallel to c; often appear bent or twisted as a consequence of translation gliding; commonly in complex aggregates of acicular crystals; also in radiating or columnar masses, or granular. **Structure:** Consists of Sb and S atoms bonded together in a complex sheet structure parallel to the ac plane, giving rise to perfect {010} cleavage. **Physical properties:** Lead-gray, commonly with a black to iridescent tarnish, lead-gray streak, metallic luster. Cleavage {010}, perfect; cleavage flakes flexible, inelastic; slightly sectile. $H = 2$, $VHN_{100} = 71-86$. $G = 4.63$. **Tests:** Easily fusible. Soluble in HCl. **Chemistry:** Usually pure Sb_2S_3. **Optics:** R = 32.6–48.2 (540 nm); strongly anisotropic. **Occurrence:** Occurs most commonly in hydrothermal vein and replacement

deposits of low-temperature origin. Under surface and near-surface conditions stibnite may oxidize to white or yellow stibiconite, often as pseudomorphs. Localities: Stibnite is the commonest antimony mineral and is very widely distributed; the following localities are known for fine specimens. White Caps mine, Nye Co., NV; Antimony Peak, Kern Co., Ambrose mine, San Benito Co., Rand mines, San Bernardino Co., and many other localities in CA; Prince William mine, York Co., NB, Canada; San Martin, ZAC, Wadley mine, Wadley, SLP, Los Tejocotes, OAX, and other localities in Mexico; Wolfsberg and Andreasberg, Harz, Germany; Pribram and Kremnica, Czech Republic; Herja, Stiavnica, and Baia Sprie, Romania; Allchar, Macedonia; Nerchinsk, Siberia, Russia; Kadamdzhai, Tajikstan; Xikuandshan, Hunan, China; Ichinokawa mine, Shikoku, Japan; Kusa mine, near Bau, Sarawak; San Jose and La Salvadora mines, Oruro, Amigos mine, Potosi, and other localities in Bolivia. Uses: The principal ore of antimony.
BM *DI:270, ZK 135:308(1972), R:705, P&J:353, ABI:498.*

2.11.2.2 Antimonselite Sb_2Se_3

Named in 1993 for the composition. Stibnite group. Contains up to 4.8% Hg and 2.4% Cu. ORTH *Pbnm.* $a = 11.593$, $b = 11.747$, $c = 3.398$, $Z = 4$, $D = 5.88$. *AM 79*: 3.70_7 3.17_5 2.87_{10} 2.63_6 1.75_4. Microscopic acicular crystals and anhedral grains. Black, black streak, metallic luster. $VHN_{30} = 120$. $R = 39.0–42.0$ (546 nm). Occurs in a polymetallic deposit at *Kaiyang, Guizhou Prov., China*.
BM *AM 79:387(1994).*

2.11.2.3 Bismuthinite Bi_2S_3

Named in 1832 for the composition. Stibnite group. May contain minor amounts of Sb and Se. ORTH *Pbnm.* $a = 11.149$, $b = 11.304$, $c = 3.981$, $Z = 4$, $D = 6.81$. *17-320*(syn): 3.97_4 3.57_{10} 3.53_6 3.12_8 2.81_5 2.72_3 2.52_4 1.95_4. Prismatic to acicular crystals, elongated and striated parallel to c; usually massive with fibrous or foliated texture. Lead-gray, lead-gray streak; metallic luster. Cleavage {010}, perfect; {100}, {110}, imperfect; flexible and somewhat sectile. $H = 2$, $VHN_{50} = 170$. $G = 6.8$. $R = 37.7–50.0$ (540 nm); strongly anisotropic. Occurs in hydrothermal veins and granite pegmatites. Haddam, CT; Wickes, Jefferson Co., MT; Clear Creek mines, Chaffee Co., Breckenridge district, Summit Co., and other localities in CO; Midnight Owl mine, Yavapai Co., AZ; Pala, San Diego Co., CA; Glacier Gulch, BC, Canada; Nacozari, SON, and Guanajuato, GTO, Mexico; Carrock Fell, Cumbria, and Redruth district, Cornwall, England; Spind, Farsum, Norway; Schneeberg and Altenberg, Saxony, Germany; Bresso, Piedmont, Italy; Smolotely, Czech Republic; Moravicza and Baita, Romania; Zidarova, Bulgaria; Strezhen deposit, Altai, Russia; Horobetsu mine, Hokkaido, Japan; Kingsgate, NSW, and Mt. Biggenden mine, Q, Australia; Fefena, Madagascar; Llallagua, Potosi, and other localities in Bolivia. BM *DI:275, TMPM 14:55(1970), R:710, P&J:92, ABI:56.*

2.11.2.4 Guanajuatite Bi_2Se_3

Named in 1873 for the locality. Stibnite group. May contain up to 7% S. ORTH $Pbnm$. $a = 11.37$, $b = 11.55$, $c = 4.054$, $Z = 4$, $D = 7.54$. $10\text{-}475$: 5.16_4 3.65_9 3.19_{10} 2.88_6 2.58_5 2.31_5 1.99_7 1.77_4. Acicular crystals, elongated parallel to c; also massive, fibrous or foliated. Lead-gray, lead-gray streak, metallic luster. Cleavage {010}, distinct; {001}, indistinct; somewhat sectile. $H = 2\frac{1}{2}\text{-}3$, $VHN_{10} = 53\text{-}82$. $G = 6.3\text{-}7.0$. $R = 43.1\text{-}52.0$ (540 nm); strongly pleochroic and anisotropic. Near Salmon, Lemhi Co., ID; Thomas and Essex mines, Darwin, Inyo Co., CA; *Santa Catarina mine, Guanajuato, GTO, Mexico*; Andreasberg, Harz, Germany. BM *DI:278, R:714, P&J:188, ABI:197.*

2.11.3.1 Metastibnite Sb_2S_3

Named in 1888 as a dimorph of stibnite. May contain up to 9% Pb. Amorphous. Powdery coatings. Red, red streak, submetallic luster. $H = 2\text{-}3$. Occurs in fumaroles and as coatings on stibnite. *Steamboat Springs, Washoe Co., NV*; The Geysers, Sonoma Co., CA; Lac Nicolet mine, South Ham, PQ, Canada; Rujevac Sb–Zn–Pb deposit, Serbia; Alacran mine, Copiapó, Chile; San Jose mine, Oruro, Bolivia. BM *DI:275, AM 55:2104(1970), R:709, P&J:267, ABI:328.*

2.11.4.1 Wakabayashilite $(As,Sb)_{11}S_{18}$

Named in 1970 for Yaichiro Wakabayashi (1874–1943), Japanese mineralogist. MON $P2_1/m$. $a = 29.128$, $b = 6.480$, $c = 29.128$, $\beta = 120.0°$, $Z = 8$, $D = 4.06$. $29\text{-}1406$: 6.28_{10} 4.78_7 3.49_8 3.24_4 3.08_4 2.54_3 2.42_4 1.59_4. Prisms or fibers up to 2 cm. Golden to lemon yellow, orange-yellow streak, silky luster. Cleavage {100}, {010}, {10$\bar{1}$}, perfect. $H = 1\frac{1}{2}$. $G = 3.96$. $R = 22.0$ (540 nm); strongly pleochroic. White Caps mine(*), Manhattan, Nye Co., NV; Jas Roux, Hautes-Alpes, France; Khaidarkan(*), Kyrgyzstan; Gal-Khaya deposit, Yakutia, Russia; Shuiluo deposit, Guangxi Autonomous Region, China; *Nishinomaki mine, Gumma Pref., Japan*. BM *AM 57:1311(1972), CM 13:418(1975), R:891, P&J:386, ABI:567.*

2.11.5.1 Pääkkönenite Sb_2AsS_2

Named in 1981 for Viekko Pääkkönen (1907–1980), Finnish geologist. MON Space group unknown. $a = 5.372$, $b = 3.975$, $c = 11.41$, $\beta = 89.71°$, $Z = 2$, $D = 5.21$. $33\text{-}92$: 3.9_4 3.13_4 2.87_{10} 2.68_6 2.27_3 2.08_2 1.99_2 1.75_3. Microscopic anhedral grains. Dark gray, gray streak with brown tint, metallic luster. Cleavage in one direction, brittle. $H = 2\frac{1}{2}$, $VHN_{10} = 87$. $R = 37.2\text{-}47.8$ (540 nm); strongly anisotropic. *Kalliosalo deposit, Seinäjoki region, Finland*; Pribram, Czech Republic. BM *ZVMO 110:480(1981), AM 67:858(1982), ABI:373.*

2.11.6.1 Laphamite $As_2(Se,S)_3$

Named in 1986 for Davis M. Lapham (1931–1974), U.S. mineralogist. MON $P2_1/n$. $a = 11.86$, $b = 9.756$, $c = 4.265$, $\beta = 90.17°$, $Z = 4$, $D = 4.60$. $26\text{-}123$(syn): 4.98_8 2.95_5 2.91_{10} 2.87_9 2.82_7 1.80_6 1.77_5 1.72_7. Prismatic crystals up to 5 mm; dark red, orange-red streak, resinous luster. Cleavage {010}, perfect; flexible; inelastic; soft. G = 4.5. R = 29.5–34.0 (540 nm); moderately pleochroic and anisotropic. *Burnside, Northumberland Co., PA*, deposited from a burning coal-waste dump. BM *MM 50:279(1986)*, *AM 72:1024(1987)*, *ABI:285*.

TETRADYMITE GROUP

The tetradymite structures are rhombohedral selenides and tellurides having the chemical formula

$$A_2X_3$$

where

A = Bi or Sb
B = Te, Se, S

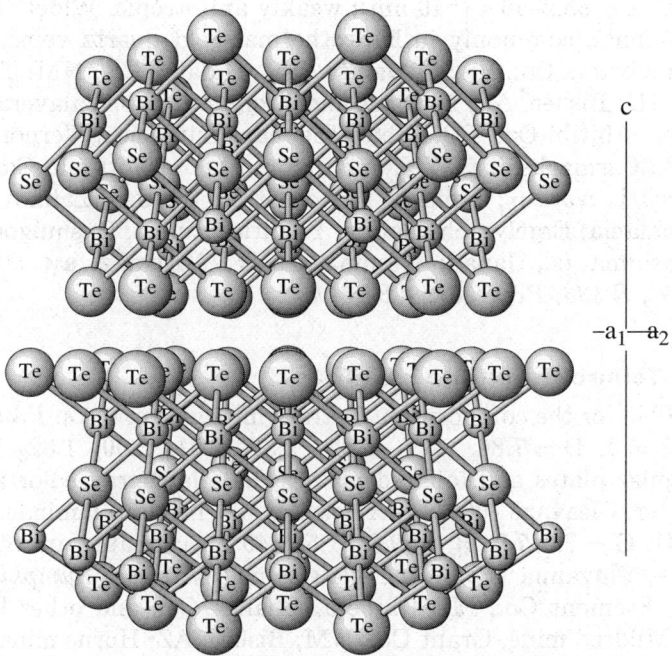

KAWAZULITE: Bi_2Te_2Se

Layer structure with the sequence Te–Bi–Se–Bi–Te. Te atoms in successive layers face each other.

and the structure is in space group $R\bar{3}m$, based on layers of successive planes composed entirely of A or X atoms [$AM\ 76:257(1991)$]. The structure is similar to that of the arsenic, joséite, and tsumoite groups.

TETRADYMITE GROUP

Mineral	Formula	Space Group	a	c	D	
Tetradymite	Bi_2Te_2S	$R\bar{3}m$	4.238	29.59	7.27	2.11.7.1
Tellurobismuthite	Bi_2Te_3	$R\bar{3}m$	4.43	29.91	7.86	2.11.7.2
Tellurantimony	Sb_2Te_3	$R\bar{3}m$	4.262	30.45	6.51	2.11.7.3
Paraguanajuatite	$Bi_2(Se,S)_3$	$R\bar{3}m$	4.140	28.64	7.68	2.11.7.4
Kawazulite	$Bi_2(Te,Se,S)_3$	$R\bar{3}m$	4.240	29.66	8.11	2.11.7.5
Skippenite	Bi_2Se_2Te	$R\bar{3}m$	4.183	29.12	7.94	2.11.7.6

2.11.7.1 Tetradymite Bi_2Te_2S

Named in 1831 for the fourfold twinned crystals. Tetradymite group. HEX-R $R\bar{3}m$. $a = 4.238$, $c = 29.59$, $Z = 3$, $D = 7.27$. *19-1330*: 3.10_{10} 2.29_{10} 2.11_8 1.97_8 1.93_8 1.64_8 1.35_8 1.30_8. Rarely as acute rhombohedral crystals, commonly as foliated to bladed masses. Steel-gray, steel-gray streak, metallic luster. Cleavage {0001}, perfect; flexible, inelastic laminae. H = $1\frac{1}{2}$, $VHN_{25} = 30$–44. G = 7.1–7.4. R = 53.2–56.4 (540 nm); weakly anisotropic. Widely distributed in small amounts, commonly in hydrothermal gold–quartz veins. Whitehall mines, Spotsylvania Co., VA; Sylvanite dist., Hidalgo Co., NM; Trail Creek, Blaine Co., ID; Bisbee, AZ; Melones and Morgan mines, Calaveras Co., CA; Eureka mine, Abitibi Co., PQ, White Elephant mine, near Vernon, BC, Discovery Fork, Carmacks dist., YT, and many other localities in Canada; *Narverud, Telemark, Norway*; Boliden, Västerbotten, Sweden; Zahkov, Slovakia; Ciclova, Romania; Ergelyatch deposit, Yakutia, Russia; Dashuigou, Sichuan, China; Tsushima Is., Japan; Nanima, NSW, Australia. BM *DI:161*, *AM 34:370(1949)*, *R:436*, *P&J:367*, *ABI:524*.

2.11.7.2 Tellurobismuthite Bi_2Te_3

Named in 1863 for the composition. Tetradymite group. HEX-R $R\bar{3}m$. $a = 4.43$, $c = 29.91$, $Z = 3$, $D = 7.86$. *8-27*: 3.23_{10} 2.36_7 2.21_5 2.00_3 1.82_3 1.61_2 1.49_3 1.41_4. Irregular plates and foliated masses. Pale lead-gray color and streak, metallic luster. Cleavage {0001}, perfect; flexible, inelastic laminae. H = $1\frac{1}{2}$–2, $VHN_{25} = 51$. G = 7.6–7.9. R = 59.6–61.5 (540 nm); weakly anisotropic. Tellurium mine, Fluvanna Co., VA; *Field's vein, Dahlonega, Lumpkin Co., GA*; Whitehorn, Fremont Co., Mt. Chipeta, Chaffee Co., and other localities in CO; Little Mildred mine, Grant Co., NM; Bisbee, AZ; Horne mine, Noranda, PQ, Lucky Jim mine, Quadra Is., BC, Canada; Boliden, Västerbotten, Sweden; Kaler deposit, Armenia; Uzelga Cu deposit, Ural Mts., Russia; Zhanatyube deposit, Kazakhstan; Oya mine, Miyage Pref., Kiura mine, Oita Pref., and other localities in Japan. BM $DI:160$, ACB $30:1307(1974)$, $R:436$, $P\&J:365$, $ABI:517$.

2.11.7.3 Tellurantimony Sb_2Te_3

Named in 1973 for the composition. Tetradymite group. HEX-R $R\bar{3}m$. $a = 4.262$, $c = 30.45$, $Z = 3$, $D = 6.51$. 15-874(syn): 3.38_1 3.16_{10} 2.35_4 2.13_3 1.77_1 1.58_1 1.47_1 1.36_1. Microscopic laths. Gray, gray streak, metallic luster. $H = 2$, $VHN_{25} = 50$. $R = 64.1$–64.3 (540 nm); moderately anisotropic. *Mattagami Lake mine, Galinee Twp., PQ, Canada*; Kobetsuzawa mine, Sapporo, Japan. BM CM $12:55(1973)$, $P\&J:363$, $ABI:515$.

2.11.7.4 Paraguanajuatite $Bi_2(Se,S)_3$

Named in 1948 as a dimorph of guanajuatite. Tetradymite group. HEX-R $R\bar{3}m$. $a = 4.140$, $c = 28.64$, $Z = 3$, $D = 7.68$. 33-124(syn): 4.78_2 3.56_2 3.04_{10} 2.24_3 2.11_1 2.07_3 1.91_1 1.71_1. Intergrown with guanajuatite. Lead-gray, lead-gray streak, metallic luster. $H = 2$, $VHN_{10} = 27$–50. $R = 51.0$–55.0 (540 nm); distinctly pleochroic and anisotropic. *Santa Catarina mine, Guanajuato, GTO, Mexico*; Falun, Sweden. BM AM $34:619(1949)$, $R:714$, $P\&J:188$, $ABI:383$.

2.11.7.5 Kawazulite $Bi_2(Te,Se,S)_3$

Named in 1970 for the locality. Tetradymite group. HEX-R $R\bar{3}m$. $a = 4.240$, $c = 29.66$, $Z = 3$, $D = 8.11$. 29-248: 4.92_4 3.64_3 3.12_{10} 2.61_2 2.31_5 2.12_5 1.76_2 1.65_1. Thin plates. Silver-white to tin-white, light steel-gray streak, metallic luster. Cleavage {0001}, perfect; flexible; inelastic. $H = 1\frac{1}{2}$, $VHN_{15} = 38$. $R = 60.0$ (547 nm); strongly anisotropic. Lone Pine mine, Grant Co., NM; Ward mine, White Pine Co., NV; Dianne Cu–U claims, Mazenod Lake, NWT, Canada; *Kawazu mine, Shizuoka Pref.*, and Suttsu mine, Hokkaido, Japan. BM AM $57:1312(1972)$, CM $19:341(1981)$, $ABI:258$, $NJMA$ $169:305(1995)$.

2.11.7.6 Skippenite Bi_2Se_2Te

Named in 1987 for George Skippen (b.1936), Canadian geologist. Tetradymite group. HEX-R $R\bar{3}m$. $a = 4.183$, $c = 29.12$, $Z = 3$, $D = 7.94$. 42-1373: 9.71_5 4.85_6 3.58_6 3.07_{10} 2.27_7 2.13_5 2.09_8 1.42_5. Massive aggregates of lamellar crystals. Steel-gray, black streak, metallic luster. Cleavage {0001}, perfect. $H = 2$, $VHN_{25} = 52$–74. $R = 49.0$–50.3 (540 nm); moderately anisotropic. *Otish Mts. uranium deposit, PQ, Canada*. BM CM $25:625(1987)$, AM $74:947(1989)$, $ABI:479$.

2.11.8.1 Montbrayite Au_2Te_3

Named in 1946 for the locality. May contain up to 3% Bi, 2% Pb, and 1% Ag and Sb. TRIC $P1$. $a = 12.11$, $b = 13.44$, $c = 10.80$, $\alpha = 104.38°$, $\beta = 97.50°$, $\gamma = 107.93°$, $Z = 12$, $D = 9.70$. 25-364(syn): 9.30_1 7.36_2 4.43_4 3.81_1 2.98_8 2.92_6 2.12_4 2.09_{10}. Microscopic grains and small masses. Yellow-white, metallic luster. One perfect cleavage. H = $2\frac{1}{2}$, VHN$_{100}$ = 238. G = 9.94. R = 63.5 (540 nm); weakly anisotropic. April Fool mine, Lincoln Co., NV; *Robb–Montbray mine, Montbray Twp., PQ, Canada*; Kalgoorlie, WA, Australia. BM *AM 31:515(1946), 57:146(1972), R:435, ABI:338.*

2.11.9.1 Ottemannite Sn_2S_3

Named in 1966 for Joachim Ottemann (b.1914), German mineralogist. ORTH *Pnam*. a = 8.864, b = 14.020, c = 3.747, Z = 4, D = 4.76. *14-619*(syn): 7.0$_4$ 5.5$_8$ 4.13$_{10}$ 3.74$_4$ 3.26$_3$ 2.75$_4$ 2.67$_5$ 2.26$_3$. Microscopic crystals. Gray, metallic luster. H = 2. G$_{syn}$ = 4.87. *Cerro Rico de Potosi*, and Maria Teresa mine, *near Huari, Bolivia*; Stiepelmann mine, Arandis, Namibia. BM *AM 51:1551(1966), CE 33:243(1974), R:704, ABI:370.*

2.11.10.1 Nagyagite $Pb_{13}Au_2Sb_3Te_6S_{16}$

Named in 1845 for the locality. ORTH (pseudo-TET). a = 8.336, b = 30.20, c = 8.288, Z = 2, D = 7.51. *8-3*(syn): 3.02$_{10}$ 2.81$_6$ 2.43$_4$ 2.08$_3$ 1.82$_3$ 1.70$_3$ 1.51$_6$ 1.46. Platy {010} with rectangular striations; also massive, granular. Blackish lead-gray color and streak, metallic luster. Cleavage {010}, perfect; {101}, good; flexible cleavage flakes. H = $1\frac{1}{2}$, VHN$_{100}$ = 60–94. G = 7.4–7.5. R = 38.8–42.8 (540 nm); distinctly anisotropic. Widely distributed in Au–Te deposits. Gold Mill, Boulder Co., and Cripple Creek, Teller Co., CO; Dorleska mine, Trinity Co., CA; Muronian mine, Moss Twp., ON, and Olive Mabel claim, Gainer Creek, BC, Canada; Schellgadener, Austria; Jilovć, Czech Republic; *Sacarimb (Nagyag), Romania*; Rendaiji mine, Shizuoka Pref., Japan; Tarua goldfield, Fiji; Sylvia mine, Tararu Creek, New Zealand; Kalgoorlie, WA, Australia; Farallon Negro, Catamarca Prov., Argentina. BM *DI:168, R:433, P&J:276, ABI:344, MM 58:479(1994).*

2.11.11.1 Buckhornite $AuPb_2BiTe_2S_3$

Named in 1992 for the locality. ORTH *Pmmm* or *P*222. a = 9.374, b = 12.326, c = 4.073, Z = 2, D = 8.35. *CM 30*: 3.74$_8$ 3.11$_8$ 2.76$_{10}$ 2.46$_8$ 2.39$_{10}$ 2.04$_6$ 1.82$_4$ 1.71$_6$. Euhedral to subhedral blades to 1.5 mm. Black, gray streak, metallic luster. One good cleavage, sectile, flexible. VHN$_{10}$ = 54. R = 41.0–47.8 (540 nm); distinctly anisotropic. *Buckhorn mine, near Jamestown, Boulder Co., CO*; Jilovć, Czech Republic; Megradzor deposit, Armenia. BM *CM 30:1039(1992), CR 318:1225(1994).*

2.11.12.1 Bowieite $(Rh,Ir,Pt)_2S_3$

Named in 1984 for Stanley H. U. Bowie (b.1917), Scottish mineralogist. ORTH *Pbcn*. a = 8.462, b = 5.985, c = 6.138, Z = 4, D = 6.45. *35-736*(syn): 3.07$_7$ 3.01$_{10}$ 2.99$_7$ 2.14$_7$ 1.76$_6$ 1.73$_8$ 1.16$_8$ 1.15$_{10}$. Microscopic anhedral grains. Gray, metallic luster. H = 6–7, VHN$_{100}$ = 1288. R = 46.2–47.8 (540 nm); weakly anisotropic. *Salmon R., Goodnews Bay, AK*; Gusevogorskii massif, Ural Mts., Russia; Gaositai, Hebei Prov., China; Pirogues R., New Caledonia. BM *CM 22:543(1984), ABI:66.*

2.11.12.2 Kashinite (Ir,Rh)$_2$S$_3$

Named in 1985 for S. A. Kashin (1900–1980), Russian geologist. ORTH *Pbcn*. $a = 8.469$, $b = 6.001$, $c = 6.143$, $Z = 4$, $D = 10.23$. *41-1416*: 3.07_8 3.01_{10} 2.99_7 2.15_5 1.74_5 1.19_6 1.14_8 1.03_5. Microscopic subhedral crystals. Gray-black, metallic luster. $H = 7\frac{1}{2}$, $VHN_{50} = 1529$. $R = 45.8–47.5$ (540 nm); distinctly anisotropic. Goodnews Bay, AK; *Nizhni Tagil, Ural Mts., Russia*, in placers. BM *ZVMO 114:617(1985)*, *AM 72:223(1987)*, *ABI:257*.

PYRITE GROUP

Many metal disulfides share the pyrite structure

$$AX_2$$

where

A = Au, Co, Cu, Fe, Mn, Ni, Os, Pd, Pt, Ru
X = As, Bi, S, Sb, Se, Te

and structure is in the cubic space group *Pa3*. In the pyrite structure, metal atoms occupy the face-centered cubic lattice positions, and the S atoms lie in dumbbell-like pairs (S_2^{2-} groups) parallel to the four cube diagonals of the lattice. Each Fe is surrounded by six S atoms in octahedral coordination. Each pair of S atoms is coordinated to six Fe atoms.

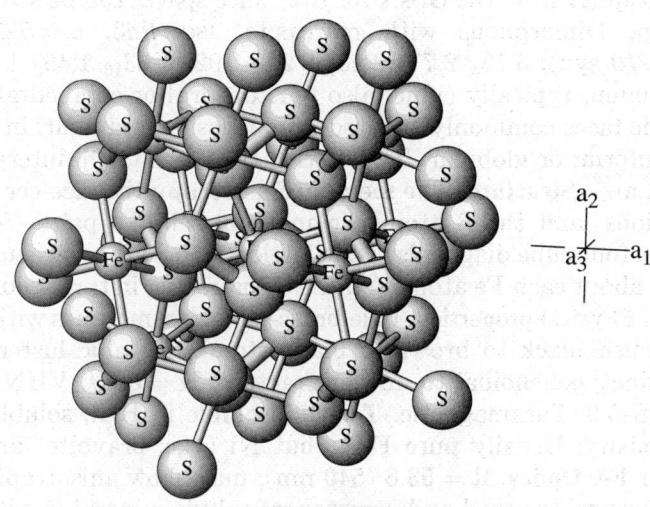

PYRITE: FeS$_2$
Fe and S$_2$ groups are in an NaCl arrangement.

PYRITE GROUP

Mineral	Formula	Space Group	a	D	
Pyrite	FeS_2	$Pa3$	5.417	5.01	2.12.1.1
Vaesite	NiS_2	$Pa3$	5.670	4.47	2.12.1.2
Cattierite	CoS_2	$Pa3$	5.534	4.82	2.12.1.3
Penroseite	$NiSe_2$	$Pa3$	5.991	6.70	2.12.1.4
Trogtalite	$CoSe_2$	$Pa3$	5.931	7.20	2.12.1.5
Villamaninite	CuS_2	$Pa3$	5.694	4.50	2.12.1.6
Fukuchilite	$(Cu,Fe)S_2$	$Pa3$	5.604	4.66	2.12.1.7
Krutaite	$CuSe_2$	$Pa3$	6.116	6.62	2.12.1.8
Hauerite	MnS_2	$Pa3$	6.100	3.48	2.12.1.9
Laurite	RuS_2	$Pa3$	5.610	6.22	2.12.1.10
Aurostibite	$AuSb_2$	$Pa3$	6.659	9.91	2.12.1.11
Krutovite	$NiAs_2$	$Pa3$	5.794	6.93	2.12.1.12
Sperrylite	$PtAs_2$	$Pa3$	5.967	10.78	2.12.1.13
Geversite	$PtSb_2$	$Pa3$	6.44	10.91	2.12.1.14
Insizwaite	$Pt(Bi,Sb)_2$	$Pa3$	6.691	13.59	2.12.1.15
Erlichmanite	OsS_2	$Pa3$	5.620	9.52	2.12.1.16
Dzarkenite	$FeSe_2$	$Pa3$	5.783	7.34	2.12.1.17
Gaotaiite	$Ir_{1-x}Te_2$	$Pa3$	6.413	10.0	2.12.1.18
Mayingite	$IrBiTe$	$Pa3$	6.502	12.77	2.12.1.19

Note: See also the cobaltite and marcasite groups.

2.12.1.1 Pyrite FeS_2

Named in antiquity from the Greek for *fire*, since sparks can be struck from it. Pyrite group. Dimorphous with marcasite. ISO $Pa3$. $a = 5.417$, $Z = 4$, $D = 5.01$. *6-710*(syn): 3.13_4 2.71_9 2.43_7 2.21_5 1.92_4 1.63_{10} 1.45_3 1.04_3. **Habit:** Crystals common, typically cubic, also pyritohedral or octahedral; pyritohedral and cubic faces commonly striated; also massive, granular, in some cases radiated, reniform, or globular. Twins with twin axis [110], interpenetrating (Iron Cross Law). **Structure:** The metal atoms occupy the face-centered cubic lattice positions, and the S atoms lie in dumbbell-like pairs (S_2^{2-} groups) parallel to the four cube diagonals of the lattice. The S atoms are in octahedral coordination about each Fe atom, and the S_2 pairs are, in turn, coordinated to six Fe atoms. **Physical properties:** Pale brass-yellow, sometimes with iridescent tarnish, greenish black to brownish black streak, metallic luster. Cleavage {001}, indistinct; conchoidal to uneven fracture. $H = 6-6\frac{1}{2}$, $VHN_{100} = 1505-1520$. $G = 4.8-5.0$. Paramagnetic. **Tests:** Insoluble in HCl, soluble in strong HNO_3. **Chemistry:** Usually pure FeS_2, but Ni (var. bravoite) and Co may substitute for Fe. **Optics:** $R = 53.8$ (540 nm); may show anisotropism. **Occurrence:** The most widespread and commonest sulfide mineral, pyrite occurs in almost all geological environments: an an accessory mineral in igneous rocks; as a magmatic segregation; in pegmatites; in contact-metamorphic deposits; in hydrothermal veins and replacement deposits; as a sublimation product; and in sediments and sedimentary rocks. **Localities:** Only a few localities for large or fine crystals can be mentioned. Chester, VT; Franklin, NJ; French Creek, Chester Co., PA; Amex mine, Boss, MO; Butte, MT; Rico and Leadville, CO; Park City and Bingham, UT; Holland mine, Santa Cruz Co., and Magma mine, Superior, AZ; Spruce Peak, King Co., WA; Prince of Wales I., AK; Nanisvik mine, NWT, Canada; Cananea, SON, Naica and Santa Eulalia, CHIH, Noche Buena and Concepcion del Oro, ZAC, Mexico; Panasquiero, Portugal; Ambasaguas and Navajun, Logroño Prov., Spain; Rio Marina, Elba, and Brosso, Piedmont, Italy; Kassandra, Greece; Kovdor, Kola, Russia; Akchitau, Kazakhstan; Mt. Stewart mine, Leadville, NSW, Australia; Quiruvilca, La Libertad, Huanzala, Huanoco, Morococha, and San Cristobal, Junin, Peru; Lapiana mine, Chicote Grande, Tacna, and Llallagua, Bolivia; Serra do Cabral, MG, Brazil. **Alteration:** Alters readily to iron sulfates and iron oxides. Pseudomorphs of brown limonite after pyrite crystals are common. Uses: A source of sulfur for the manufacture of sulfuric acid. BM *DI:281, R:791, P&J:311, ABI:426.*

2.12.1.2 Vaesite NiS$_2$

Named in 1945 for Johannes Vaes (1902–1978), Belgian mineralogist. Pyrite group. ISO $Pa3$. $a = 5.670$, $Z = 4$, $D = 4.47$. *11-99*: 2.83_{10} 2.54_4 2.32_4 2.00_5 1.71_8 1.57_3 1.09_6 1.00_4. Small octahedral and cubic crystals. Silver-gray, black streak, metallic luster. Cubic cleavage. H = 5, VHN$_{100}$ = 743–837. G = 4.45. R = 31.6 (540 nm). Miliken (Sweetwater) mine, Reynolds Co., MO; Orphan mine, Grand Canyon, AZ; Schneeberg, Nentershausen, and Iserlohn, Germany; Carmenes district, Leon Prov., Spain; St. Marina, Khaskovo district, Bulgaria; Kosaka mine, Japan; Scotia mine, Kalgoorlie, WA, Australia; *Kasompi mine* and Shinkolobwe mine, *Shaba, Zaire;* San Santiago mine, La Rioja Prov., Argentina. BM *AM 30:483(1945), R:809, P&J:105, MM 43:950(1980), ABI:554.*

2.12.1.3 Cattierite CoS$_2$

Named in 1945 for Felicien Cattier (1869–1946), chairman of the board, Union Minière du Haut-Katanga. Pyrite group. ISO $Pa3$. $a = 5.534$, $Z = 4$, $D = 4.82$. *19-362*(syn): 2.77_7 2.48_4 2.26_4 1.96_4 1.67_{10} 1.48_4 1.06_{10} 1.03_4. Cubic crystals and granular. Silver-gray with a pink tinge, black streak, metallic luster. Cubic cleavage. H = 5, VHN$_{100}$ = 1018–1114. G = 4.82. R = 36.7 (540 nm). Bald Knob, Alleghany Co., NC; Gänsberg and Hohensachsen, Black Forest, Germany; near Filipstad, Värmland, Sweden; *Shinkolobwe mine, Shaba, Zaire.* BM *AM 30:483(1945), ZK 150:163(1979), R:809, P&J:105, ABI:84.*

2.12.1.4 Penroseite NiSe$_2$

Named in 1926 for Richard A. F. Penrose (1863–1931), U.S. geologist. Pyrite group. Dimorphous with kullerudite. Contains up to 9% Co and 6% Cu. ISO $Pa3$. $a = 5.991$, $Z = 4$, $D = 6.70$. *6-507*: 3.02_3 2.68_{10} 2.45_{10} 1.81_9 1.66_3 1.60_4

1.30_2 1.15_3. Reniform masses with radiating columnar structure. Steel-gray, black streak, metallic luster. Cleavage {001}, perfect; {011}, distinct; subconchoidal fracture; brittle. H = 3, VHN_{100} = 500–583. G = 6.6–6.7. R = 41.1 (540 nm). Hope's Nose, Devon, England; Kuusamo, Finland; Tilkerode, Harz, Germany; Shinkolobwe mine, Shaba, Zaire; Pacajake mine(*), Hiaco, Bolivia. BM *DI:294; AM 35:360(1950), 74:1171(1989); R:818; P&J:293; ABI:395.*

2.12.1.5 Trogtalite $CoSe_2$

Named in 1955 for the locality. Pyrite group. Contains 9% Cu and 5% Pd. Dimorph of hastite. ISO $Pa3$. a = 5.931, Z = 4, D = 7.20. *25-253*: 2.95_7 2.64_{10} 2.42_9 1.79_8 1.64_5 1.59_7 1.30_5 1.14_6. Microscopic grains. Iron-gray, black streak, metallic luster. R = 41.3 (540 nm). *Trogtal quarry, near Lautenthal, Harz, Germany*; Musonoi mine, Shaba, Zaire. BM *NJMM:133(1955), AM 41:164(1956), R:819, P&J:372, ABI:537.*

2.12.1.6 Villamaninite CuS_2

Named in 1920 for the locality. Pyrite group. Contains considerable Ni, Co, and Fe. ISO $Pa3$. a = 5.694, Z = 4, D = 4.50. *29-556*: 3.29_2 2.85_{10} 2.55_3 2.33_3 2.01_3 1.72_4 1.58_1 1.10_2. Cubic and octahedral crystals and massive. Black, black streak, metallic luster. Perfect cubic cleavage, uneven fracture. H = $4\frac{1}{2}$, VHN_{20} = 535–710. G = 4.4–4.5. R = 23.4 (540 nm). *Providencia mine, near Villamanin, Leon Prov., Spain*; Lubin mine, near Legnica, Poland. BM *NJMM:174(1968), AM 74:1173(1989), R:816, P&J:383, ABI:558.*

2.12.1.7 Fukuchilite $(Cu,Fe)S_2$

Named in 1969 for Nobuyo Fukuchi (1877–1934), Japanese mineralogist. Pyrite group. ISO $Pa3$. a = 5.604, Z = 4, D = 4.66. *24-365*(syn): 3.24_6 2.80_{10} 2.50_6 2.29_6 1.98_6 1.69_8 1.14_6 1.08_8. Microscopic grains. Gray, gray streak, submetallic luster. H = 6. G = 4.86. *Hanawa mine, Akita Pref., Japan*. BM *MJJ 5:399(1969); AM 55:1811(1970), 74:1173(1989); ABI:167.*

2.12.1.8 Krutaite $CuSe_2$

Named in 1972 for Tomas Kruta (b.1906), Czech mineralogist. Contains minor Ni and Co. Pyrite group. ISO $Pa3$. a = 6.116, Z = 4, D = 6.62. *25-309*: 3.02_9 2.71_{10} 2.48_8 1.83_6 1.68_4 1.62_3 1.32_3 1.17_4. Microscopic subhedral crystals. Gray, gray streak, metallic luster. H = 4, VHN_{100} = 317. R = 34.9 (540 nm). *Petrovice mine, Moravia, Czech Republic*; El Dragon mine(*), Potosi, Bolivia. BM *BM 95:475(1972), AM 59:210(1974), R:820, P&J:229, ABI:277.*

2.12.1.9 Hauerite MnS$_2$

Named in 1846 for Josef von Hauer (1778–1863) and his son Franz von Hauer (1822–1899), Austrian geologists. Pyrite group. ISO $Pa3$. $a = 6.100$, $Z = 4$, $D = 3.48$. *25-549*(syn): 3.05_{10} 2.73_7 2.49_7 2.16_7 1.84_8 1.25_7 1.18_8 1.08_7. Octahedral crystals and globular aggregates. Red-brown to brown-black, red-brown streak, adamantine to submetallic luster. Perfect cubic cleavage. $H = 4$. $G = 3.46$. $R = 24.8$ (540 nm); isotropic, $N = 2.69$. Salt domes in Texas; *Kalinka, Slovakia*; Jeziorko and Grzybow, Poland; Raddusa(*), Sicily, Italy; Yazovsk and Podgornensk deposits, Aktyubinsk, Ural Mts., Russia. BM *DI:293, R:826, P&J:194, ABI:207*.

2.12.1.10 Laurite RuS$_2$

Named in 1866 for Laura R. Joy, wife of C. A. Joy (1823–1891), U.S. chemist. Pyrite group. Forms a series with erlichmanite, OsS$_2$. ISO $Pa3$. $a = 5.610$, $Z = 4$, $D = 6.22$. *19-1107*(syn): 3.24_8 2.81_{10} 2.51_2 2.29_2 1.98_6 1.69_{10} 1.62_2 1.08_4. Small cubic, octahedral, and pyritohedral crystals and rounded grains. Iron-black, dark gray streak, metallic luster. Cleavage {111}, perfect. $H = 7\frac{1}{2}$, $VHN_{25} = 2760$–2898. $G = 6.2$–7.0. $R = 40.2$ (540 nm). Stillwater Complex, MT; Yuba R., Nevada Co., and Trinity R., Trinity Co., CA; Goodnews Bay, AK; Tulameen area, BC, Canada; Konder massif, Yakutia, Russia; *Pontijn, Tanah Laut, Borneo*; Pirogues R., New Caledonia; Maud Creek, Howard R., New Zealand; Guma Water, Sierra Leone; Merensky Reef, Transvaal, South Africa. BM *DI:291, R:820, P&J:233, MM 47:465(1983), ABI:289*.

2.12.1.11 Aurostibite AuSb$_2$

Named in 1952 for the composition. Pyrite group. ISO $Pa3$. $a = 6.659$, $Z = 4$, $D = 9.91$. *8-460*(syn): 3.85_3 3.33_7 2.98_8 2.72_6 2.36_6 2.01_{10} 1.78_3 1.28_3. Microscopic grains. Gray, metallic luster. $H = 3$, $VHN = 280$–292. $G_{syn} = 9.98$. $R = 61.8$ (540 nm). Occurs in hydrothermal quartz veins. Tonawanda mine, Cadillac Twp., PQ, Chesterville mine, Larder Lake, ON, and *Giant Yellowknife mine, NWT, Canada*; Suliteljema, Norway; Krasna Hora, near Milesov, Czech Republic; Bestyube goldfield, Kazakhstan; Mobale mine, Kivu, Zaire; Sebaque area, Que-Que, Zimbabwe; Hillgrove, NSW, and Costerfield, VIC, Australia. BM *AM 32:461(1952), R:824, ABI:37, NJMM:537(1990)*.

2.12.1.12 Krutovite NiAs$_2$

Named in 1976 for Georgi A. Krutov (1902–1989), Russian mineralogist. Pyrite group. ISO $Pa3$. $a = 5.794$, $Z = 4$, $D = 6.93$. *29-928*: 2.90_6 2.59_{10} 2.37_8 2.05_5 1.75_8 1.61_5 1.55_6 1.02_6. Microscopic grains. Gray-white, metallic luster. $H = 5\frac{1}{2}$, $VHN = 630$. $R = 65.0$ (540 nm). *Geshiber vein, Jachymov, Czech Republic*; Kohmahs deposit, Tuva ASSR, Russia. BM *ZVMO 105:59(1976), AM 62:173(1977), ABI:278*.

2.12.1.13 Sperrylite $PtAs_2$

Named in 1889 for Francis L. Sperry (1861–1906), U.S. chemist, who found the mineral. Pyrite group. ISO $Pa3$. $a = 5.967$, $Z = 4$, $D = 10.78$. $9\text{-}452$: 3.43_4 2.98_6 2.11_5 1.80_{10} 1.33_4 1.22_4 1.15_7 1.05_4. Cubic and cubo-octahedral crystals, up to 3 cm. Tin-white, black streak, metallic luster. Cleavage {100}, indistinct; conchoidal fracture, brittle. $H = 6\frac{1}{2}$, $VHN_{100} = 960\text{--}1277$. $G = 10.6$. $R = 55.2$ (540 nm). The commonest Pt mineral. Rambler mine, Encampment, and Mouat mine, Stillwater, MT; *Vermilion mine*, and other mines in the *Sudbury district*, ON, Fraser R., BC, *Canada*, in placers; Hitura Cu–Ni deposit, Finland; Talnakh(*), Norilsk (centimeter-sized crystals), and in alluvial deposits, Yakutia, Russia; Kambalda Ni-mine, WA, Australia; Merensky Reef mines(*), Transvaal, South Africa. BM *DI:292*, *EG 71:1299(1976)*, *R:821*, *P&J:346*, *ABI:487*.

2.12.1.14 Geversite $PtSb_2$

Named in 1961 for Traugott W. Gevers (1900–1991), South African geologist. Pyrite group. ISO $Pa3$. $a = 6.44$, $Z = 4$, $D = 10.91$. $14\text{-}141$(syn): 2.92_3 1.94_8 1.72_5 1.31_5 1.24_{10} 1.19_6 1.17_5 1.14_8. Microscopic grains. Gray, metallic luster. $H = 4\frac{1}{2}\text{--}5$, $VHN_{50} = 726\text{--}766$. $R = 58.9$ (540 nm). Occurs in concentrates of Pt minerals. Fox Gulch, Goodnews Bay, AK; Morozova mine, Norilsk, and Vozhmin massif, Karelia, Russia; *Driekop mine, eastern Transvaal, South Africa*. BM *MM 32:833(1961)*, *ABI:183*.

2.12.1.15 Insizwaite $Pt(Bi,Sb)_2$

Named in 1972 for the locality. Pyrite group. ISO $Pa3$. $a = 6.691$, $Z = 4$, $D = 13.59$. $25\text{-}512$: 2.96_8 2.70_8 2.34_5 2.00_{10} 1.77_7 1.44_5 1.28_6 1.17_6. Microscopic grains. White, metallic luster. $H = 5$, $VHN_{25} = 488\text{--}540$. $R = 54.0$ (540 nm). Fox Gulch, Goodnews Bay, AK; Coleman mine, Sudbury, ON, Canada; Norilsk region, Russia; *Waterfall Gorge, Insizwa, South Africa*. BM *MM 38:794(1972)*, *AM 58:805(1973)*, *ABI:236*.

2.12.1.16 Erlichmanite OsS_2

Named in 1971 for Josef Erlichman (b.1935), U.S. chemist. Pyrite group. Forms a series with laurite, RuS_2. ISO $Pa3$. $a = 5.620$, $Z = 4$, $D = 9.52$. $19\text{-}882$(syn): 3.24_{10} 2.81_9 1.99_6 1.69_9 1.62_2 1.29_3 1.26_3 1.08_3. Small pyritohedral crystals and microscopic grains. Gray, metallic luster. $VHN_{100} = 1730\text{--}1950$. $G = 8.28$. $R = 42.9$ (540 nm). *MacIntosh mine, Trinity R., Humboldt Co.*, and American R., Sacramento Co., CA; Tulameen R., BC, Canada; Ural Mt. placers, and Konder massif, Yakutia, Russia; Tiebaghi, and Pirogues R., New Caledonia; Maud Creek, Howard R., New Zealand; Guma Water, Sierra Leone; Joubdo stream, Birbir R., Ethiopia; Merensky Reef, Transvaal, South Africa. BM *AM 56:1501(1971)*, *MM 47:465(1983)*, *R:821*, *ABI:149*.

2.12.1.17 Dzarkenite FeSe$_2$

Named in 1995 for the locality. Pyrite group. Dimorphous with ferroselite. ISO $Pa3$. $a = 5.783$, $Z = 4$, $D = 7.34$. ZVMO 124: 2.89_5 2.59_{10} 2.36_8 2.05_4 1.74_5 1.55_6 1.26_3 1.11_4. Microscopic octahedral crystals. Black, black streak, metallic to adamantine luster. No cleavage. H = 5. R = 42.6 (546 nm); isotropic, rosy yellow in reflected light. *Suluchekinskoye Se deposit, Dzarkenskaya depression, SE Kazakhstan.* BM *ZVMO 124(1):85(1995)*.

2.12.1.18 Gaotaiite Ir$_{1-x}$Te$_2$ (x = 0.25)

Named in 1995 for the locality. Pyrite group. ISO $Pa3$. $a = 6.413$, $Z = 4$, $D = 10.0$. AM 81: 2.86_7 2.60_6 1.93_{10} 1.71_6 1.24_8 1.13_9 1.04_8 0.978_8. Microscopic subhedral grains and veinlets. Black, black streak, metallic luster. Brittle. H = 3, VHN$_{20}$ = 117. R = 46.3 (540 nm); in reflected light white with bluish tint, isotropic to weakly anisotropic. In placer concentrates and chromite ore near the village of *Gaotai, 200 km NNE of Beijing, China*. BM *AMS 15(1):1(1995), AM 81:249(1996)*.

2.12.1.19 Mayingite IrBiTe

Named in 1995 for the locality. Pyrite group. ISO $Pa3$. $a = 6.502$, $Z = 4$, $D = 12.77$. AM 81: 2.89_7 1.96_{10} 1.74_8 1.25_8 1.21_7 1.15_7 1.05_7 0.991_7. Massive aggregates and veinlets to 1 mm. Black, black streak, metallic luster. Brittle. H = 4, VHN$_{50}$ = 178. R = 50.7 (540 nm); in reflected light white with a yellow tint, moderately anisotropic. In placer concentrates and chromite ore near *Maying, 230 km NNE of Beijing, China*. BM *AMS 15(1):5(1995), AM 81:251(1996)*.

MARCASITE GROUP

Marcasite group minerals have the general formula

$$AX_2$$

where

A = Fe, Co, Ni, Ru, Os
X = S, Se, Te, As, Sb

and the minerals crystallize in the orthorhombic space group $Pnnm$. In the marcasite structure the metal atoms occupy the lattice points of an orthorhombic body-centered lattice, and the sulfur atoms surround each metal atom in sixfold coordination. The structure is related to that of pyrite (pyrite and marcasite are dimorphous) but with the S_2^{2-} pairs lying in the a–b plane. Reference: B. J. Wuensch, Sulfide Mineralogy, *Reviews in Mineralogy*, Vol. 1, Mineralogical Society of America, Washington, DC, 1974.

2.12.2.1 Marcasite

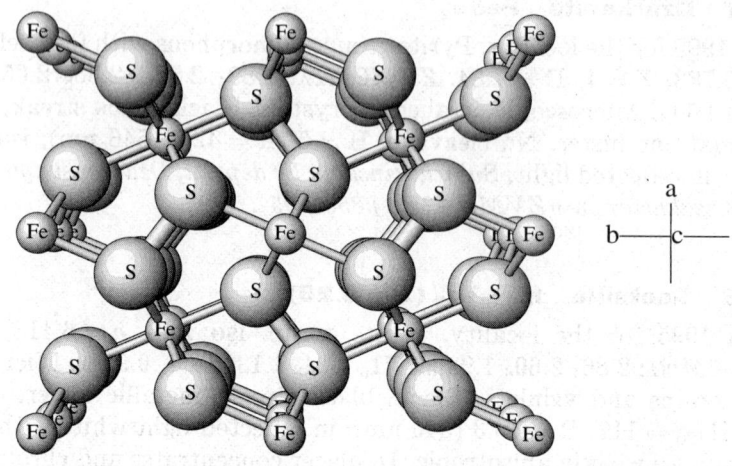

MARCASITE: FeS$_2$

Marcasite has the same octahedral coordination of iron by sulfur and disulfide ions as pyrite, but arranged differently into orthorhombic symmetry.

MARCASITE GROUP

Mineral	Formula	Space Group	a	b	c	D	
Marcasite	FeS$_2$	$Pnnm$	4.436	5.414	3.381	4.91	2.12.2.1
Ferroselite	FeSe$_2$	$Pnnm$	4.815	5.808	3.599	7.05	2.12.2.2
Frohbergite	FeTe$_2$	$Pnnm$	5.29	6.27	3.86	8.07	2.12.2.3
Hastite	CoSe$_2$	$Pnnm$	4.84	5.72	3.60	7.23	2.12.2.4
Mattagamite	CoTe$_2$	$Pnnm$	5.312	6.311	3.889	8.00	2.12.2.5
Kullerudite	NiSe$_2$	$Pnnm$	4.89	5.96	3.67	6.73	2.12.2.6
Omeiite	OsAs$_2$	$Pnnm$	5.409	6.167	3.021	11.19	2.12.2.7
Anduoite	RuAs$_2$	$Pnnm$	5.41	6.206	3.01	8.25	2.12.2.8
Löllingite	FeAs$_2$	$Pnnm$	5.300	5.983	2.882	7.47	2.12.2.9
Seinäjokite	FeSb$_2$	$Pnnm$	5.810	6.490	3.190	8.27	2.12.2.10
Safflorite	(Co,Fe)As$_2$	$Pnnm$	5.173	5.954	2.999	7.46	2.12.2.11
Rammelsbergite	NiAs$_2$	$Pnnm$	4.759	5.797	3.539	7.09	2.12.2.12
Nisbite	NiSb$_2$	$Pnnm$	5.178	6.319	3.832	8.01	2.12.2.13

Notes:
1. The structure of arsenopyrite is a derivative of the marcasite structure.
2. Many of the marcasite minerals have dimorphs in the pyrite group.
3. The stability relations between marcasite and pyrite are unclear.

2.12.2.1 Marcasite FeS$_2$

Named in 1845 from an Arabic or Moorish word applied to pyrite and similar minerals. Dimorphous with pyrite. ORTH $Pnnm$. $a = 4.436$, $b = 5.414$, $c = 3.381$, $Z = 2$, $D = 4.91$. 24-74: 3.43$_7$ 2.71$_3$ 2.69$_{10}$ 2.41$_4$ 2.31$_3$ 1.91$_3$ 1.75$_5$ 1.59$_2$. **Habit:** Crystals common, typically tabular {010}, also pyramidal;

faces commonly curved; also stalactitic, globular, or reniform, with radiating internal structure. Twinning common on {101}, producing swallowtail contact twins, and repeated to form stellate fivelings. **Structure:** Consists of S_2^{2-} pairs as in pyrite but with the dumbbell groups lying in the a–b plane in the orthorhombic structure; the Fe atoms are in octahedral coordination with S. **Physical properties:** Pale brass-yellow, tending toward light green on exposure, tin-white on fresh fracture, gray to black streak, metallic luster. Cleavage {101}, distinct; uneven fracture. H = 6–6$\frac{1}{2}$, VHN_{200} = 915–1099. G = 4.89. Tends to decompose in moist conditions to iron sulfates. **Tests:** Reacts like pyrite but is more readily decomposed. **Chemistry:** Usually pure FeS_2. **Optics:** R = 48.6–55.5 (540 nm); strongly anisotropic. **Occurrence:** Most marcasite occurs in near-surface deposits, where it has formed at low temperatures from acid solutions; pyrite, the more stable form, is deposited under conditions of higher temperature and alkalinity or low acidity. It is frequently found in sedimentary rocks and coal, often as concretions or replacing fossils. **Localities:** Only a few of the more significant can be given. Hackerville, OH; Pleasant Ridge, IN; the Pb–Zn deposits of the Mississippi Valley and the Tri-State mining district, MO–KS–OK; Temple, Bell Co., TX; Algoma mines, Wawa, ON, and Nanisivik mine, Baffin I., NWT, Canada; Santa Eulalia, CHIH, Mexico; Tavistock, Devon, and between Folkestone and Dover, Kent, England; Herne, Westphalia, Clausthal, Harz, and Freiberg, Saxony, Germany; Duchov, Vinitrov, Most, and other places in Czech Republic; Angren coalfield, Uzbekistan; Oppu mine, Akita Pref., Japan; Llallagua, Bolivia. BM *DI:311, AM 44:685(1959), R:839, P&J:257, ABI:312.*

2.12.2.2 Ferroselite FeSe$_2$

Named in 1955 for the composition. ORTH *Pnnm.* $a = 4.815$, $b = 5.808$, $c = 3.599$, Z = 2, D = 7.05. *12-291*(syn): 3.07_2 2.89_5 2.59_{10} 2.49_9 1.89_8 1.80_4 1.71_3 1.46_3. Small prismatic crystals and stellate and cruciform twins. Tin-white to steel-gray, black streak, metallic luster. H = 6–6$\frac{1}{2}$, VHN_{25} = 858–933. G = 7.20. R = 45.6–51.4 (540 nm); strongly anisotropic. An accessory mineral in Colorado Plateau uranium deposits; Kuusamo, Finland; Petrovice and Predborice deposits, Czech Republic; *Tuvinsk Autonomous Territory, Russia.* BM *DANS 105:812(1955), AM 41:671(1956), GCA 16:296(1959), R:845, P&J:162, ABI:156.*

2.12.2.3 Frohbergite FeTe$_2$

Named in 1947 for Max H. Frohberg (1901–1970), Canadian geologist. Marcasite group. ORTH *Pnnm.* $a = 5.29$, $b = 6.27$, $c = 3.86$, Z = 2, D = 8.07. *14-419*: 3.29_3 2.81_{10} 2.71_7 2.07_5 1.94_3 1.85_4 1.58_3 1.11_3. Microscopic grains. White, metallic luster. H = 3–4, VHN = 250–297. R = 50.0–52.4 (540 nm); strongly anisotropic. Gold Hill, north of San Simon, Cochise Co., AZ; *Robb–Montbray mine* and Noranda mine, *PQ*, and Lindquist Lake, BC, *Canada*; Bodennec, Finistere, France; Sacaramb and Fata Baii, Romania; Jabal Sayid, Saudi Arabia. BM *AM 32:310(1947), R:858, P&J:169, ABI:165.*

2.12.2.4 Hastite CoSe$_2$

Named in 1955 for P. F. Hast, German mining engineer. Marcasite group. Dimorphous with trogtalite. ORTH $Pnnm$. $a = 4.84$, $b = 5.72$, $c = 3.60$, $Z = 2$, $D = 7.23$. 10-408: 2.9_3 2.58_{10} 2.47_{10} 1.90_{10} 1.61_3 1.44_3 1.02_{10}. Small idiomorphic grains and radiating aggregates. Red-brown, metallic luster. $H = 6$. $R = 46.9$–52.2 (540 nm); strongly pleochroic and anisotropic. La Sal mine, Montrose Co., CO; *Trogtal quarry, near Lauenthal, Harz, Germany*; San Francisco mine, La Rioja Prov., Argentina. BM *NJMM:133(1955)*, *AM 41:164(1956)*, *R:845*, *P&J:192*, *ABI:204*.

2.12.2.5 Mattagamite CoTe$_2$

Named in 1973 for the locality. Marcasite group. ORTH $Pnnm$. $a = 5.312$, $b = 6.311$, $c = 3.889$, $Z = 2$, $D = 8.00$. 11-553(syn): 3.14_8 2.82_{10} 2.71_{10} 2.07_{10} 1.94_8 1.86_8 1.58_8 1.55_8. Microscopic grains. Gray-violet, metallic luster. $H = 4\frac{1}{2}$–$5\frac{1}{2}$, $VHN_{25} = 383$–404. $R = 47.9$–53.9 (540 nm); weakly pleochroic and anisotropic. *Mattagami Lake mine, Galinee Twp., PQ, Canada*. BM *CM 12:55(1973)*, *AM 59:382(1974)*, *P&J:260*, *ABI:317*.

2.12.2.6 Kullerudite NiSe$_2$

Named in 1964 for Gunnar Kullerud (1921–1989), Norwegian–U.S. mineralogist. Marcasite group. Dimorphous with penroseite (2.12.1.4). ORTH $Pnnm$. $a = 4.89$, $b = 5.96$, $c = 3.67$, $Z = 2$, $D = 6.73$. 18-886: 3.12_2 2.94_8 2.64_{10} 2.55_{10} 1.93_8 1.84_8 1.69_4 1.65_6. Massive, fine-grained. Lead-gray, metallic luster. *Kuusamo, Finland*. BM *AM 50:519(1965)*, *R:846*, *ABI: 279*.

2.12.2.7 Omeiite OsAs$_2$

Named in 1978 for the locality. Marcasite group. ORTH $Pnnm$. $a = 5.409$, $b = 6.167$, $c = 3.021$, $Z = 2$, $D = 11.19$. 34-336(syn): 4.07_9 3.09_3 2.69_3 2.63_{10} 2.01_4 1.93_3 1.91_6 1.70_3. Microscopic tabular crystals. Steel-gray, metallic luster. Cleavage parallel to elongation. $R = 39.2$ (546 nm); distinctly anisotropic. *Omeishan, Sichuan Prov., China*, in a Cu–Ni sulfide deposit. BM *AM 64:464(1979)*, *ABI:361*.

2.12.2.8 Anduoite RuAs$_2$

Named in 1979 for the locality. Marcasite group. Forms a series with omeiite. ORTH $Pnnm$. $a = 5.41$, $b = 6.206$, $c = 3.01$, $Z = 2$, $D = 8.25$. 33-1144: 2.00_5 1.92_{10} 1.50_9 1.21_7 1.19_7 1.13_8 1.10_9 1.08_4. Microscopic grains and granular aggregates. Lead-gray, gray-black streak, metallic luster. One cleavage, brittle. $H = 6\frac{1}{2}$–7, $VHN_{50} = 1078$. *Anduo, Tibet, China*. BM *AM 65:808(1980)*, *ABI:13*.

2.12.2.9 Löllingite FeAs$_2$

Named in 1845 for the locality. Marcasite group. Minor substitution of Co for Fe (up to 6.5%) and S (up to 6.7%) and Sb (up to 5.6%) for As has been

2.12.2.12 Rammelsbergite

recorded. ORTH $Pnnm$. $a = 5.300$, $b = 5.983$, $c = 2.882$, $Z = 2$, $D = 7.47$. *25-249*: 2.62_3 2.57_{10} 2.40_8 1.97_4 1.87_7 1.85_3 1.66_7 1.64_3. Prismatic crystals, or massive; twinned on {011}, sometimes trillings. Silver-white to steel-gray, gray-black streak, metallic luster. Cleavage {010} and {101}, sometimes distinct; uneven fracture; brittle. $H = 5-5\frac{1}{2}$, $VHN_{100} = 859-920$. $G = 7.4-7.5$. $R = 53.1-53.6$ (540 nm); strongly anisotropic. Occurs with calcite and sulfides of Fe and Cu in veins; also in pegmatites. Auburn, ME, and Center Strafford, NH, in pegmatites; Franklin and Sterling Hill, NJ, in skarn; Black Hills, SD, in many pegmatites (300-kg masses at Ingersoll mine); Gunnison Co., CO, in several mines; Pala, CA, in pegmatites; Cobalt, ON, Canada, in many mines; Andreasberg, Harz Mts., Germany; *Lölling, Carinthia, Austria*; Varuträsk, Sweden, in pegmatite; Guadalcanal, Sierra Morena, Spain; Broken Hill, NSW, Australia. BM *DI:303*, *AM 53:1856(1968)*, *R:854*, *P&J:242*, *ABI:300*.

2.12.2.10 Seinäjokite FeSb₂

Named in 1976 for the locality. Marcasite group. Contains up to 6% Ni and 8% As. ORTH $Pnnm$. $a = 5.810$, $b = 6.490$, $c = 3.190$, $Z = 2$, $D = 8.27$. *29-129*: 2.81_{10} 2.59_9 2.03_8 1.79_6 1.63_2 1.52_2 1.21_3 1.17_3. Small grains. Light gray, metallic luster. $H = 4\frac{1}{2}$, $VHN_{30} = 332$. $R = 60.8$ (540 nm). *Seinäjoki, Vaasa, Finland*, in native antimony; Ilimaussaq intrusion, near Narssaq, Greenland; Zolotaya Gora, Urals, Russia. BM *ZVMO 105:617(1976)*, *AM 62:1059(1977)*, *ABI: 467*.

2.12.2.11 Safflorite (Co,Fe)As₂

Named in 1835 from *Zaffer*, a German term for cobalt pigments. Marcasite group. Contains up to 16% Fe. ORTH $Pnnm$. $a = 5.173$, $b = 5.954$, $c = 2.999$, $Z = 2$, $D = 7.46$. *23-88*: 2.96_1 2.60_6 2.57_8 2.38_{10} 1.86_5 1.85_2 1.65_2 1.64_2. Prismatic crystals but commonly massive with radiating fibrous structure. Cyclic fiveling twins with {011} as twin plane, also cruciform twins with {101} as twin plane. Tin-white, tarnishing to dark gray, black streak, metallic luster. Cleavage {100}, distinct; uneven to conchoidal fracture. $H = 4\frac{1}{2}-5$, $VHN_{100} = 792-882$. $G = 7.1-7.4$. $R = 53.5-54.5$ (540 nm); strongly anisotropic. Quartzburg district, Grant Co., OR; Cobalt and South Lorrain, ON, Eldorado mine, Great Bear Lake, NWT, Hedley, BC, Canada; Batopilas, CHIH, Mexico; *Schneeberg* and Annaberg, *Saxony*, Bieber, Hesse, Wittichen, Baden, *Germany*; Tunaberg, Sweden; Sarrabus, Sardinia, Italy; Almalyk, Uzbekistan; Broken Hill, NSW, Australia; Bou Azzer, Morocco. BM *DI:307*, *AM 53:1856(1968)*, *R:347*, *P&J:332*, *ABI:457*.

2.12.2.12 Rammelsbergite NiAs₂

Named in 1854 for Karl F. Rammelsberg (1813–1899), German mineral chemist. Marcasite group. Trimorphous with krutovite and pararammelsbergite. ORTH $Pnnm$. $a = 4.759$, $b = 5.797$, $c = 3.539$, $Z = 2$, $D = 7.09$. *15-441*: 2.82_{10} 2.52_9 2.42_6 2.00_3 1.84_7 1.75_3 1.66_3 1.21_{10}. Prismatic crystals rare, usually massive. Tin-white with a reddish tint, black streak, metallic luster. Cleavage {101}, distinct; uneven fracture. H = $5\frac{1}{2}$–6, VHN_{100} = 630–758. G = 7.0–7.2. R = 56.0–58.2 (540 nm); strongly anisotropic. Occurs with other Ni and Co minerals in vein deposits of the Co–Ni–Ag type. Mohawk and other copper mines, Keweenaw Co., MI; Cobalt, ON, and El Dorado mine, Great Bear Lake, NWT, Canada; Schneeberg, Saxony, Richeldorf, Hesse, Eisleben, Thuringia, Germany; Lölling, Carinthia, Austria; Sarrabus, Sardinia, Italy; and a number of other worldwide occurrences. BM DI:309, R:352, P&J:320, ABI:432.

2.12.2.13 Nisbite $NiSb_2$

Named in 1970 for the composition. Marcasite group. ORTH Pnnm. a = 5.178, b = 6.319, c = 3.832, Z = 2, D = 8.01. 25-1083(syn): 3.08_5 2.76_{10} 2.69_9 2.03_8 1.84_7 1.56_6 1.17_6 1.10_6. Microscopic grains. White, metallic luster. H = 5, VHN_5 = 420–513. R = 59.4 (546 nm); weakly anisotropic. Red Lake mines, Kenora district, ON, Canada; Gruvåsen, Getön, and Tunaberg, Bergslagen, Sweden; Zolotaya Gora, Urals, and Festilvanoe mine, Magadan region, Russia. BM CM 10:232(1970), 18:165(1980); R:863; ABI:354.

COBALTITE GROUP

The cobaltite group consists of sulfides and selenides corresponding to

ABX

where

A = Co, Ni, Pt, Ir, Rh, Pd
B = As, Sb, Bi
X = S, Se, Te

and the structure is cubic, or pseudocubic, and related to that of pyrite. The structure is similar to that of pyrite but with the S_2 pairs of pyrite replaced by As–S, Sb–S, As–Se, Bi–Se, or Bi–Te pairs.

COBALTITE GROUP

Mineral	Formula	Space Group	a	D	
Cobaltite	CoAsS	$Pca2_1$	5.582	6.34	2.12.3.1
Gersdorffite	NiAsS	$P2_13$	5.688	5.98	2.12.3.2
Ullmannite	NiSbS	$P2_13$	5.886	6.85	2.12.3.3
Willyamite	CoSbS	$Pca2_1$	5.860	7.02	2.12.3.4
Tolovkite	IrSbS	$P2_13$	6.027	10.50	2.12.3.5
Platarsite	PtAsS	$Pa3$	5.790	8.41	2.12.3.6
Irarsite	IrAsS	$Pa3$	5.777	11.92	2.12.3.7
Hollingworthite	RhAsS	$Pa3$	5.769	7.91	2.12.3.8
Jolliffeite	NiAsSe	$Pa3$	5.831	7.10	2.12.3.9

2.12.3.3 Ullmannite

COBALTITE GROUP (cont'd)

Mineral	Formula	Space Group	a	D	
Padmaite	PdBiSe	$P2_13?$	6.448	9.86	2.12.3.10
Michenerite	PdBiTe	$P2_13$	6.651	10.00	2.12.3.11
Maslovite	PtBiTe	$Pa3$	6.687	11.81	2.13.3.12
Testibiopalladite	Pd(Sb,Bi)Te	$P2_13$	6.572	8.93	2.12.3.13

2.12.3.1 Cobaltite CoAsS

Named in 1832 for the composition. Cobaltite group. ORTH pseudo-ISO $Pca2_1$. $a = 5.582$, $Z = 4$, $D = 6.34$. *18-431*: 2.77_8 2.49_{10} 2.27_9 1.97_6 1.68_{10} 1.49_8 1.22_5 1.07_8. Cubic and pyritohedral crystals, often in combination; also massive, granular. Silver-white to steel-gray, gray-black streak, metallic luster. Cleavage {001}, perfect; uneven fracture. $H = 5\frac{1}{2}$, $VHN_{200} = 1171$–1346. $G = 6.3$. $R = 50.3$ (540 nm). Occurs in high-temperature hydrothermal deposits, and as veins in contact-metamorphic deposits. Cobalt district, ON, Canada; Botallack, Cornwall, England; Skutterud, near Modum, Norway; Tunaberg(*), Södermanland, and Riddarhyttan(*) and Håkanboda(*), Västmanland, Sweden; Mt. Cobalt, Q, Torrington and Broken Hill, NSW, and Bimbowrie, SA, Australia; Bou Azzer, Morocco. BM *DI:296*, *AM 67:1048(1982)*, *R:827*, *P&J:124*, *ABI:103*.

2.12.3.2 Gersdorffite NiAsS

Named in 1843 for Johann von Gersdorff (1781–1849), owner of the nickel mine at Schladming, Styria, Austria. Cobaltite group. May contain minor amounts of Co, As, Bi. ISO $P2_13$ (other polytypes are known). $a = 5.688$, $Z = 4$, $D = 5.98$. *24-519*: 4.02_2 2.84_7 2.54_{10} 2.32_6 2.01_3 1.72_6 1.58_2 1.52_2. Octahedral and pyritohedral crystals; also massive, granular. Silver-white to steel-gray, gray-black streak, metallic luster. Cleavage {001}, perfect; uneven fracture. $H = 5\frac{1}{2}$, $VHN_{100} = 657$–767. $G = 5.9$. $R = 45.0$ (540 nm). Snowbird mine, Mineral Co., MT; Sudbury and Cobalt, ON, Canada; Craignure mine, Strathclyde, Scotland; Müsen, Wissen, and Ramsbeck, Westphalia, Rammelsberg and Wolfsberg, Harz, Lobenstein, Thuringia, Germany; *Schladming, Styria, Austria*; Dobsina, Czech Republic; Ait Ahmane mine, Bou Azzer, Morocco; Cochabamba, Bolivia. BM *DI:298*, *R:833*, *P&J:178*, *CM 24:27(1986)*, *ABI:180*.

2.12.3.3 Ullmannite NiSbS

Named in 1843 for Johan C. Ullman (1771–1821), German chemist and mineralogist, who discovered the mineral. Cobaltite group. May contain up to 10% As, 5% Bi, and 3% Co. ISO $P2_13$. $a = 5.886$, $Z = 4$, $D = 6.85$. *30-861*: 4.16_3 2.94_1 2.63_{10} 2.40_6 1.77_4 1.63_2 1.57_3 1.28_1. Cubic crystals, twinned on [110]. Silver-white to steel-gray, gray-black streak, metallic luster. Cleavage

{001}, perfect; uneven fracture; brittle. H = 5–5$\frac{1}{2}$, VHN$_{100}$ = 592–627. G = 6.5–7.0. R = 46.1 (540 nm). Kerr Addison mine, Timiskaming dist., ON, and Nicholson mine, near Goldfields, SK, Canada; Settlingstones mine, Northumberland, England; Tunaberg, Sweden; Siegen, Müsen, and *Freussberg, Westphalia*, Neudorf, Harz, and Lobenstein, Thuringia, *Germany*; Ar mine, Basses-Pyrenees, France; Monte Narba and Masaloni, Sardinia, Italy; Lölling, Carinthia, Austria; Roznava, Slovakia; Talnakh, Norilsk, Russia; Tsushima Is., Japan; Broken Hill, NSW, Australia. BM *DI:301, R:837, P&J:377, CM 24:27(1986), ABI:548.*

2.12.3.4 Willyamite CoSbS

Named in 1896 for the locality. Forms a series with ullmannite. Cobaltite group. Trimorphous with costibite and paracostibite. ISO $Pca2_1$. $a = 5.860$, $Z = 4$, $D = 7.02$. *26-1106*: 4.14$_3$ 2.62$_{10}$ 2.39$_7$ 1.77$_6$ 1.63$_4$ 1.57$_5$ 1.28$_4$ 1.09$_4$. Small zoned crystals. Tin-white to steel-gray, gray-black streak, metallic luster. Perfect cubic cleavage. H = 5–5$\frac{1}{2}$. G = 6.76. Espeland, Norway; Tunaberg, Sweden; *Consols mine, Broken Hill (Willyama), NSW, Australia*. BM *DI:301, AM 56:361(1971), CM 24:27(1986), ABI:576.*

2.12.3.5 Tolovkite IrSbS

Named in 1981 for the locality. Cobaltite group. ISO $P2_13$. $a = 6.027$, $Z = 4$, $D = 10.50$. *35-656*: 3.49$_6$ 2.99$_9$ 2.13$_8$ 1.81$_{10}$ 1.23$_7$ 1.15$_9$ 1.07$_9$ 1.01$_8$. Microscopic anhedral grains. Steel-gray, metallic luster. Conchoidal fracture, brittle. H = 6, VHN$_{10}$ = 1431–1703. R = 42.9 (586 nm). Fox Gulch, Goodnews Bay, AK; Similkameen R., BC, Canada; *Ust-Belskii massif, Tolovka R., Russia*. BM *ZVMO 110:474(1981); AM 67:1076(1982), 74:1173(1989); ABI:534.*

2.12.3.6 Platarsite PtAsS

Named in 1977 for the composition. Cobaltite group. Contains up to 13% Rh, 12% Ru, and 18% Ir. ISO $Pa3$. $a = 5.790$, $Z = 4$, $D = 8.41$. *29-974*: 3.35$_8$ 2.90$_9$ 2.59$_5$ 2.36$_5$ 2.05$_6$ 1.75$_{10}$ 1.18$_5$ 1.11$_7$. Crystals to 1 mm. Gray, metallic luster. H = 7–7$\frac{1}{2}$, VHN$_{100}$ = 1177–1343. G = 8.0. R = 48.0 (540 nm). Fox Gulch, Goodnews Bay, AK; *Onverwacht mine* and Driekop, Union, and Western mines, *Transvaal, South Africa*. BM *CM 15:385(1977), 17:117(1979); P&J:408; ABI:408.*

2.12.3.7 Irarsite IrAsS

Named in 1966 for the composition. Cobaltite group. Forms a series with hollingworthite. ISO $Pa3$. $a = 5.777$, $Z = 4$, $D = 11.92$. *19-591*: 3.32$_{10}$ 2.87$_{10}$ 2.57$_8$ 2.04$_9$ 1.74$_{10}$ 1.29$_7$ 1.11$_9$ 1.02$_8$. Small grains. Iron-black, metallic luster. H = 6$\frac{1}{2}$, VHN$_{100}$ = 1546. R = 40.6 (540 nm). American R., Sacramento Co., and Stanislaus R., Stanislaus Co., CA; Fox Gulch, Goodnews Bay, AK; Werner Lake and Sudbury, ON, and Pipe mine, MB, Canada; Cliff and Harold's Grave quarries, Shetland Is.; Hitura Ni–Cu deposit, Finland; Pletene deposit, Bulgaria; Vourinos, Greece; Pirogues R., New Caledonia; *Onverwacht mine* and Driekop mine, Transvaal, South Africa. BM *ZVMO 95:700(1966)*, *AM 52:1580(1967)*, *R:874*, *P&J:212*, *ABI:237*.

2.12.3.8 Hollingworthite RhAsS

Named in 1965 for Sidney E. Hollingworth (1899–1966), English geologist. Cobaltite group. Forms a series with irarsite. ISO $Pa3$. $a = 5.769$, $Z = 4$, $D = 7.91$. *30-1037*: 3.31_{10} 2.87_{10} 2.57_8 2.35_7 2.04_9 1.74_{10} 1.29_8 1.11_9. Microscopic euhedral to anhedral grains. Gray, metallic luster. $VHN_{100} = 650$–700. R = 50.8 (540 nm). Stillwater Complex, MT; Yuba R., Nevada Co., CA; Goodnews Bay, AK; Werner Lake and Sudbury, ON, Canada; Cliff and Harold's Grave quarries, Shetland Is.; Hitura Cu–Ni deposit, Finland; Norilsk, Russia; Pirogues R., New Caledonia; *Driekop mine* and Onverwacht mine, *Transvaal, South Africa*. BM *AM 50:1068(1965)*, *R:838*, *MD 22:178(1987)*, *ABI:222*.

2.12.3.9 Jolliffeite NiAsSe

Named in 1991 for Alfred W. Jolliffe (1907–1988), Canadian geologist. Cobaltite group. ISO $Pa3$. $a = 5.831$, $Z = 4$, $D = 7.10$. *CM 29*: 2.92_5 2.60_{10} 2.38_8 2.06_3 1.76_8 1.62_4 1.56_5 1.46_1. Microscopic anhedral grains. Gray, metallic luster. R = 51.1 (540 nm). *Shirley Penin., Lake Athabasca, SK, Canada*. BM *CM 29:411(1991)*.

2.12.3.10 Padmaite PdBiSe

Named in 1991 for the locality. Cobaltite group. ISO Space group uncertain, probably $P2_13$. $a = 6.448$, $Z = 4$, $D = 9.86$. *ZVMO 120*: 2.89_{10} 2.63_9 2.28_3 1.94_9 1.72_5 1.41_3 1.38_4 1.20_3. Microscopic anhedral grains. Pale yellow, metallic luster. H = 3–4, $VHN_{20} = 260$–272. R = 48.6 (540 nm). *Padma R., Karelia, Russia*, in metashales. BM *ZVMO 120(3):85(1991)*, *AM 78:450(1993)*.

2.12.3.11 Michenerite PdBiTe

Named in 1958 for Charles E. Michener (b.1907), Canadian geologist. Cobaltite group. ISO $P2_13$. $a = 6.651$, $Z = 4$, $D = 10.00$. *25-92*(syn): 2.97_{10} 2.72_8 2.00_9 1.77_8 1.45_7 1.28_6 1.23_7 1.17_6. Microscopic grains. Silver-white to gray-white, black streak, metallic luster. H = 4, $VHN_{25} = 306$–317. G = 9.5. R = 58.2 (540 nm). A principal Pd mineral in Cu–Ni sulfide deposits. Stillwater Complex, MT; New Rambler mine, Albany Co., WY; *Sudbury, ON,* and Pipe mine, MB, *Canada*; Hitura, Finland; Monchegorsk deposit, Kola, and Oktyabr mine, Norilsk, Russia; Danba, Sichuan Prov., China; Kambalda, WA, Australia; Merensky Reef, Transvaal, South Africa. BM *CM 6:200(1958)*, *11:903(1973)*; *R:825*; *P&J:269*; *ABI:331*.

2.12.3.12 Maslovite PtBiTe

Named in 1979 for G. D. Maslov (1915–1968), Russian mineralogist. Cobaltite group. ISO $Pa3$. $a = 6.687$, $Z = 4$, $D = 11.81$. *39-1389*: 3.01_{10} 2.71_8 2.36_2

2.02_6 1.85_2 1.79_3 1.29_2 1.18_2. Microscopic grains. Gray-white, metallic luster. H = 4, VHN_{20} = 262–388. R = 56.1 (540 nm). *Oktyabr mine, Norilsk, Russia.* BM *GRM 21:94(1979); AM 65:406(1980), 74:1171(1989); ABI:314.*

2.12.3.13 Testibiopalladite Pd(Sb,Bi)Te

Named in 1974 for the composition. Cobaltite group. ISO $P2_13$. $a = 6.572$, Z = 4, D = 8.93. *29-961*: 3.28_5 2.94_{10} 2.68_8 1.98_9 1.76_7 1.27_6 1.16_6 1.07_7. Microscopic grains. Gray-white, metallic luster. Cleavage imperfect in two directions, brittle. H = $3\frac{1}{2}$–4, VHN = 165. R = 53.9 (540 nm). *Danba, Sichuan Prov., China*; Kambalda, WA, Australia. BM *AM 61:182(1976), ABI:522.*

ARSENOPYRITE GROUP

Arsenopyrite minerals are arsenides or sulfides with the general formula

$$AAs_2 \quad \text{or} \quad ABS$$

where

A = Fe, Co, Ru, Os, Ir
B = As, Sb

and a monoclinic (pseudo-orthorhombic) structure with space group $P2_1/c$. The structure is a derivative of the marcasite structure in which one-half of the S is replaced by As.

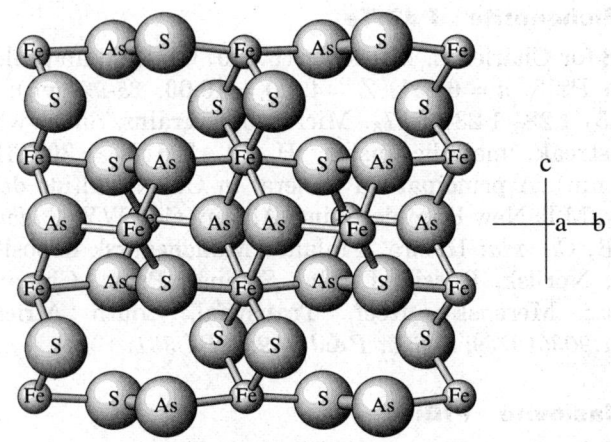

ARSENOPYRITE: FeAsS

2.12.4.2 Gudmundite

ARSENOPYRITE GROUP

Mineral	Formula	Space Group	a	b	c	β	
Arsenopyrite	FeAsS	$P2_1/c$	5.744	5.675	5.785	112.3	2.12.4.1
Gudmundite	FeSbS	$P2_1/c$	10.02	5.94	6.74	90.0	2.12.4.2
Osarsite	(Os,Ru)AsS	$P2_1/c$	5.933	5.916	6.009	112.35	2.12.4.3
Ruarsite	RuAsS	$P2_1/c$	5.945	5.920	6.025	112.92	2.12.4.4
Iridarsenite	$IrAs_2$	$P2_1/c$	6.060	6.071	6.158	113.3	2.12.4.5
Clinosafflorite	$CoAs_2$	$P2_1/c$	5.916	5.872	5.960	116.4	2.12.4.6

2.12.4.1 Arsenopyrite   FeAsS

Named in 1847 for the composition, a contraction of the older term *arsenical pyrites*. Arsenopyrite group. MON $P2_1/c$. $a = 5.744$, $b = 5.675$, $c = 5.785$, $\beta = 112.3°$, $Z = 4$, $D = 6.20$. *14-218*: 2.68_{10} 2.66_{10} 2.44_9 2.42_9 2.41_9 1.96_5 1.82_7 1.81_9. **Habit**: Pseudo-orthorhombic prismatic crystals, frequently twinned on {012} to produce cruciform twins or star-shaped trillings; also granular or compact. **Structure**: Closely related to that of marcasite; each Fe atom is surrounded by a somewhat distorted octahedron, of which one face is a triangle of three As atoms, the other a triangle of three S atoms. **Physical properties**: Silver-white to steel-gray, gray-black streak, metallic luster. Cleavage {001}, distinct; uneven fracture. $H = 5\frac{1}{2}$–6, $VHN_{100} = 1081$. $G = 6.0$–6.2. Tests: Garlic odor when ground or hammered; decomposed by HNO_3 with separation of sulfur. **Chemistry**: May contain some Co substituting for Fe. **Optics**: $R = 49.7$–52.0 (540 nm); strongly anisotropic. **Occurrence**: The most abundant arsenic mineral, it typically occurs as an early formed mineral in moderate- to high-temperature conditions. Occurs most commonly in high-temperature gold–quartz veins, in high-temperature cassiterite deposits, and in some contact-metamorphic deposits. **Localities**: Only a few localities can be mentioned: Franconia, Grafton Co., NH; Franklin, Sussex Co., NJ; Cobalt district, ON, Canada; Santa Eulalia(*) and Santa Barbara(*), CHIH, Mexico; many mines in Cornwall, England; Sala, Tunaberg, and Boliden, Sweden; Altenberg, Ehrenfriedersdorf, and Freiberg(*), Saxony, Germany; Panasqueira(*), Portugal; Trepca(*), Serbia; Ashio mine, Togichi Pref., and Ohira mine(*), Miyazaki Pref., Japan; Torrington and Broken Hill, NSW, Australia; Llallagua(*), Bolivia. BM *DI:316*, AM *46:1448(1961)*, *R:863*, *P&J:271*, *ABI:28*.

2.12.4.2 Gudmundite   FeSbS

Named in 1928 for the locality. Arsenopyrite group. MON $P2_1/c$. $a = 10.02$, $b = 5.94$, $c = 6.74$, $\beta = 90.0°$, $Z = 8$, $D = 6.94$. *8-104*: 4.10_5 3.00_4 2.93_4 2.56_{10} 1.91_8 1.62_6 1.46_6 1.41_7. Pseudo-orthorhombic prismatic crystals, twinned on {101} forming penetration and contact twins, with cruciform and butterfly shapes. Silver-white to steel-gray, gray-black streak, metallic luster. No cleavage, uneven fracture. $H = 6$, $VHN_{100} = 572$–673. $G = 6.72$. $R = 49.9$–55.3

(540 nm); distinctly pleochroic and anisotropic. Lac Nicolet mine, South Ham, PQ, Hemlo Au deposit and near Red Lake, ON, Canada; *Gudmundstorp, near Sala*, and other localities in *Sweden*; Waldsassen, Bavaria, Germany; Kalliolampi, near Nurma, Finland; Pezink, Saxova, and Vlastejovice, Czech Republic; Askot, Pithoragarh, India; Broken Hill, NSW, Australia. BM *DI:325, R:872, P&J:189, ABI:198.*

2.12.4.3 Osarsite (Os,Ru)AsS

Named in 1972 for the composition. Arsenopyrite group. May contain minor amounts of Ir, Ni, Fe, Co. MON $P2_1/c$. $a = 5.933$, $b = 5.916$, $c = 6.009$, $\beta = 112.35°$, $Z = 4$, $D = 8.44$. *25–595*: 3.79_{10} 2.96_5 2.78_6 2.74_7 2.01_6 1.89_{10} 1.87_8 1.83_6 1.70_6. Microscopic grains. Gray, metallic luster. *Gold Bluff, Humboldt Co., CA*, in placer; Vourinos, Greece; Kola Peninsula, and near Zlatoust, Ural Mts., Russia; Anduo, Tibet, China; Witwatersrand, Transvaal, South Africa. BM *AM 57:1029(1972), EG 71:1399(1976), R:875, ABI:367.*

2.12.4.4 Ruarsite RuAsS

Named in 1979 for the composition. Arsenopyrite group. MON $P2_1/c$. $a = 5.945$, $b = 5.920$, $c = 6.025$, $\beta = 112.92°$, $Z = 4$, $D = 7.08$. *27-568*(syn): 3.79_{10} 2.78_6 2.74_7 2.01_6 1.89_{10} 1.87_8 1.83_6 1.70_6. Microscopic grains. Lead-gray, gray-black streak, metallic luster. $H = 6-6\frac{1}{2}$, $VHN_{50} = 734-893$. $R = 44.1$ (546 nm); distinctly pleochroic and anisotropic. *Anduo, Tibet, China*, in heavy-mineral concentrates; Merensky Reef, Transvaal, South Africa. BM *AM 65:1068(1980), ABI:449.*

2.12.4.5 Iridarsenite IrAs$_2$

Named in 1974 for the composition. Arsenopyrite group. MON $P2_1/c$. $a = 6.060$, $b = 6.071$, $c = 6.158$, $\beta = 113.3°$, $Z = 4$, $D = 10.92$. *14-411*(syn): 3.90_{10} 3.03_4 2.83_5 2.78_5 2.05_4 1.91_5 1.88_4 1.73_5. Microscopic grains. Gray, metallic luster. $H = 5\frac{1}{2}$, $VHN_{100} = 540$. $R = 45.4-46.1$ (540 nm); weakly anisotropic. Occurs as inclusions in Os–Ir–Ru nuggets. *Fox Gulch, Goodnews Bay, AK*; near Zlatoust, Ural Mts., Russia; Tibet, China; *Papua New Guinea*; Witwatersrand and Merensky Reef, Transvaal, South Africa. BM *CM 12:280(1974), AM 61:177(1976), ABI:238.*

2.12.4.6 Clinosafflorite CoAs$_2$

Named in 1971 as the monoclinic analog of safflorite. Arsenopyrite group. MON $P2_1/c$. $a = 5.916$, $b = 5.872$, $c = 5.960$, $\beta = 116.4°$, $Z = 4$, $D = 7.48$. *14-412*(syn): 2.67_5 2.65_5 2.54_{10} 2.43_6 2.41_5 1.87_3 1.86_3 1.66_4. Massive. Tin-white, metallic luster. No cleavage. $H = 4\frac{1}{2}-5$, $VHN_{100} = 719-875$. $R = 52.0-54.7$ (540 nm). *Cobalt, ON, Canada*, in Co–Ni ores; Nord mine, Värmland, Sweden; Aghbar mine, near Bou Azzer, Morocco. BM *CM 10:877(1971), 21:129(1983); ABI:102.*

2.12.5.1 Pararammelsbergite NiAs$_2$

Named in 1940 for its polymorphism with rammelsbergite. ORTH $Pbca$. $a = 5.770$, $b = 5.838$, $c = 11.419$, $Z = 8$, $D = 7.20$. $18\text{-}876$(syn): 2.86_5 2.83_5 2.56_{10} 2.52_{10} 2.37_7 2.36_7 1.82_5 1.74_7. Crystals tabular {001}, usually massive. Tin-white, metallic luster. Cleavage {001}, perfect; uneven fracture. H = 5, VHN$_{100}$ = 681–830. G = 7.0–7.1. R = 59.9 (540 nm); strongly anisotropic, yellow, brown, gray. Occurs in Ni–Co deposits. Franklin, NJ; Cobalt, *Gowganda*, South Lorrain, and Kerr Lake, *ON*, Camsell R., NWT, *Canada*; Cerny Dul, near Berghaus, Czech Republic; Talmessi and Meskani mines, Iran; Bou Azzer, Morocco. BM *DI:310, R:862, P&J:288, TMPM 34:167(1985), ABI:386*.

2.12.5.2 Paxite CuAs$_2$

Named in 1961 from the Latin *pax*, peace. MON $P2_1/c$. $a = 5.839$, $b = 5.111$, $c = 8.084$, $\beta = 99.7°$, $Z = 4$, $D = 5.96$. $39\text{-}369$: 3.60_7 3.30_5 3.14_{10} 2.62_{10} 2.50_9 1.99_4 1.80_7 1.69_6. Massive. Steel-gray, metallic luster. Cleavage perfect in one direction. H = $3\frac{1}{2}$–4, VHN$_{25}$ = 146. G = 5.4. R = 44.2–49.5 (540 nm); strongly anisotropic, gray-green to dark brown. Černý Důl, Czech Republic, in hydrothermal calcite veins; Mohawk, Keweenaw Co., MI; Niederbeerbach, Odenwald, Germany. BM *AM 47:1484(1962), TMPM 34:167(1985), ABI:392*.

2.12.6.1 Glaucodot (Co,Fe)AsS

Named in 1849 from the Greek for *blue*, for its use in making blue glass. Dimorphous with alloclasite. ORTH $Cmmm$. $a = 6.64$, $b = 28.39$, $c = 5.64$, $Z = 24$, $D = 6.16$. $5\text{-}643$: 2.72_{10} 2.45_8 2.42_7 1.83_9 1.64_4 1.38_4 1.13_6 1.01_6. Prismatic crystals, twinned on {012}, forming cruciform penetration twins; also massive. Silver-white to gray-white, black streak, metallic luster. Cleavage {010}, perfect; {101}, distinct; uneven fracture; brittle. H = 5. G = 5.9–6.1. R = 50.2–51.1 (540 nm); distinctly pleochroic and anisotropic. Franconia, Grafton Co., NH; Standard Consolidated mine, Sumpter, Grant Co., OR; Cobalt–Gowganda district, ON, Canada; Tynebottom mine, near Alston, Cumbria, England; Skutterud, Norway; Håkansbö(*), Västmanland, and Tunaberg(*), Södermanland, Sweden; Gosenbach, Westphalia, Germany; Oravita, Romania; Dashkesan deposit, Azerbaijan; Abakan deposit, Siberia, Russia; Simokuzukawa, Japan; Carcoar, NSW, Australia; Witwatersrand, Transvaal, South Africa; *Huasco, Atacama, Chile*. BM *DI:322, R:871, P&J:181, ABI:187*.

2.12.6.2 Alloclasite (Co,Fe)AsS

Named in 1866 from the Greek for *another* and *fracture*, because of its cleavages. Dimorphous with glaucodot. MON $P2_1$. $a = 4.641$, $b = 5.606$, $c = 3.415$, $\beta = 90.03°$, $Z = 2$, $D = 6.00$. $25\text{-}246$: 3.58_3 2.80_3 2.75_{10} 2.47_9 2.40_5 1.82_7 1.71_3 1.64_3. Prismatic crystals, commonly in columnar or radiating aggregates. Steel-gray, black streak, metallic luster. Cleavage {101}, perfect; {010}, distinct; subconchoidal to uneven fracture; brittle. H = 5, VHN$_{100}$ = 818–940. G = 5.95. R = 49.3–51.5 (540 nm). Kibblehouse quarry, Perkiomenville, PA;

Silverfields mine, Cobalt, and Siscoe Metals mine, Gowganda, ON, Canada; Scar Crag mine, Keswick, England; Lautaret Pass, Hautes-Alpes, France; *Elizabeth mine, Oravita, Romania*; Dogatani mine, Japan; Mt. Isa, Q, Australia; Bou Azzer, Morocco. BM *DI:324; CM 10:838(1971), 14:561(1976); R:872; P&J:65; ABI:10.*

2.12.7.1 Costibite CoSbS

Named in 1970 for the composition. Dimorphous with paracostibite. ORTH $Pmn2_1$. $a = 3.603$, $b = 4.868$, $c = 5.838$, $Z = 2$, $D = 6.90$. *22-1082*: 4.86_5 3.08_4 2.90_6 2.60_{10} 2.50_9 1.91_8 1.80_5 1.72_4. Microscopic grains. Gray, metallic luster. $H = 6$, $VHN_{15} = 781$. Gruvåsen and Getön deposits, Hällefors, Sweden; *Consols mine, Broken Hill, NSW, Australia.* BM *AM 55:10(1970); CM 13:188(1975), 18:165(1980); R:861; ABI:111.*

2.12.7.2 Paracostibite CoSbS

Named in 1970 for its dimorphous relationship to costibite. ORTH *Pbca*. $a = 5.768$, $b = 5.949$, $c = 11.666$, $Z = 8$, $D = 7.06$. *23-1062*: 5.83_6 2.92_6 2.88_7 2.83_6 2.65_6 2.56_{10} 2.03_9 1.76_7. Microscopic anhedral and subhedral grains. White, metallic luster. $H = 6\frac{1}{2}$, $VHN_{100} = 654–763$. $G = 6.9$. $R = 48.9–49.3$ (540 nm); weakly pleochroic and anisotropic. *Trout Bay, Red Lake area, ON, Canada;* Wheal Cock mine, St. Just, Cornwall, England; Gruvåsen and Getön deposits, Hällefors, Sweden. BM *CM 10:232(1970), 13:188(1975), 18:165(1980); ABI:381.*

2.12.8.1 Lautite CuAsS

Named in 1881 for the locality. ORTH $Pna2_1$. $a = 11.350$, $b = 5.456$, $c = 3.749$, $Z = 4$, $D = 4.88$. *25-1179*: 3.11_{10} 3.09_9 2.98_2 1.91_4 1.89_2 1.87_1 1.61_1 1.60_1. Short prismatic crystals and massive. Black, black streak, metallic luster. Cleavage {001}, brittle. $H = 3–3\frac{1}{2}$, $VHN_{100} = 239–259$. $G = 4.9$. $R = 30.2–31.4$ (540 nm); weakly pleochroic and anisotropic. *Lauta, near Marienberg, Saxony,* and Niederbeerbach, Odenwald, *Germany;* Gabe Gottes mine, Rauenthal, Haut-Rhin, France. BM *DI:327, AC 19:543(1965), R:525, P&J:234, ABI:290.*

2.12.9.1. Bambollaite Cu(Se,Te)$_2$

Named in 1972 for the locality. TET $I4_1/amd$. $a = 3.865$, $c = 5.632$, $Z = 1$, $D = 4.95$. *25-313*: 3.19_{10} 1.96_7 1.93_4 1.69_3 1.65_5 1.41_1 1.27_2 1.12_2. Granular aggregates. Gray, gray streak, metallic luster. *Moctezuma mine* (also known as *La Bomballa*), *near Moctezuma, SON, Mexico.* BM *CM 11:738(1972), ABI:40.*

2.12.10.1 Molybdenite MoS$_2$

Named in 1807 from the Greek *molybdos*, lead. HEX (a $3R$ polytype is known) $P6_3mmc$. $a = 3.160$, $c = 12.295$, $Z = 2$, $D = 5.00$. *6-97*: 6.15_{10} 2.74_2 2.67_1 2.28_5 2.05_1 1.83_3 1.58_1 1.54_1. **Habit:** Hexagonal crystals, thin-to-thick tabular on

2.12.10.1 Molybdenite

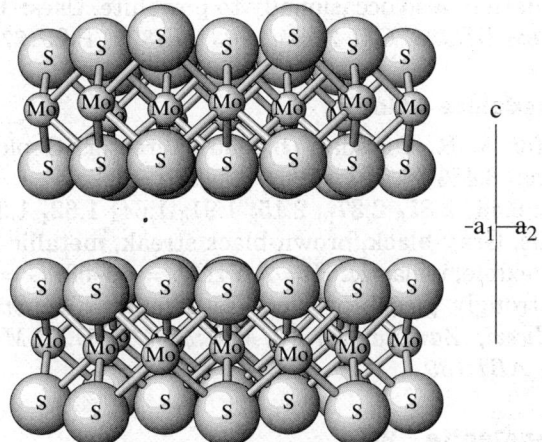

MOLYBDENITE: MoS_2

There is no strong bonding between MoS_2 sandwich layers. Mo is coordinated by six S atoms in each layer, but the sulfur atoms layers lie over one another (not in closest-packed arrangement), and the coordination polyhedron is a trigonal prism rather than an octahedron. Sulfurs are in hexagonal closest packed arrangement.

{0001}; commonly foliated, massive, or in scales. **Structure**: MoS_2 has two polytypes: molybdenite-2H (hexagonal), molybdenite-3R (rhombohedral), and polymorph jordisite (amorphous). When molybdenite is referred to with no suffix, it refers to the 2H polymorph. In this polymorph, MoS_6 trigonal prisms are linked together by sharing edges to form sheets parallel to {0001}. The sheets are linked by weak bonds between sulfurs of adjoining sheets, in hexagonal stacking sequence, resulting in perfect basal cleavage. **Physical properties**: Color and streak lead-gray with a bluish tint, metallic luster. Cleavage {0001}, perfect; flexible, inelastic lamellae. H = $1-1\frac{1}{2}$. G = 4.6–4.7. **Tests**: Decomposed by HNO_3, leaving a residue of MoO_3. **Chemistry**: Usually pure MoS_2. **Optics**: R = 21.3–45.5 (540 nm); extremely pleochroic and anisotropic. **Occurrence**: The commonest molybdenum mineral. Occurs as an accessory in some granites and pegmatites, and in high-temperature veins with scheelite, wolframite, topaz, and fluorite, and in contact-metamorphic deposits. **Localities**: Only the more significant are listed. Climax, Lake Co., CO; Childs–Aldwinkle mine(*), Pinal Co., and Santo Nino mine(*), Santa Cruz Co., AZ; Crown Point mine(*), Chelan Co., WA; Cosumnes mine(*), El Dorado Co., Pine Creek mine, Inyo Co., and other localities in CA; Alford Twp. and Temiskaming district, ON, and Con mine, Yellowknife, YT, Canada; Carrock Fell and Caldbeck Fell, Cumberland, England; Knaben(*), Vest-Agder, Norway; Altenberg, Saxony, Germany; Miass, Ural Mts., and Adunchilon Mts., Transbaikal, Russia; Tae Hwa mine(*), S. Korea; Hirase mine(*), Gifu Pref., and Busshoji mine(*), Hyogo Pref., Japan; Azegour, Morocco; Onganja mine(*), Namibia; Wolfram Camp(*), Q, Deepwater(*) and Kingsgate(*), NSW, and Moina district, TAS, Australia. **Alteration**: Frequently alters to ferrimolybdite in the zone of oxidation; also occasionally to powellite. Uses: The principal ore of molybdenum. BM $DI:328$, CM $7:524(1963)$, $R:874$, $P\&J:273$, $ABI:336$.

2.12.10.2 Drysdallite $MoSe_2$

Named in 1973 for A. R. Drysdall (b.1933), director, Geological Survey of Zambia. Contains 3.4% S. HEX $P6_3/mmc$. $a = 3.287$, $c = 12.925$, $Z = 2$, $D = 6.97$. 29-914: 6.46_8 2.85_6 2.37_{10} 2.15_3 1.91_6 1.64_4 1.62_4 1.35_2. Microscopic pyramidal crystals. Gray-black, brown-black streak, metallic luster. Cleavage {0001}, perfect; flexible; inelastic. $H = 2$, $VHN_8 = 46$–58. $G = 6.90$. $R = 21.6$–37.8 (540 nm); strongly pleochroic and anisotropic. *Kapijimpanga uranium deposit, near Solwezi, Zambia.* BM $NJMM:433(1973)$, AM $59:1139(1974)$, $R:880$, $P\&J:150$, $ABI:139$.

2.12.10.3 Tungstenite WS_2

Named in 1917 for the composition. HEX (an R polymorph is known) $P6_3/mmc$. $a = 3.154$, $c = 12.362$, $Z = 2$, $D = 7.73$. 8-237(syn): 6.18_{10} 3.09_1 2.73_3 2.67_3 2.28_4 1.83_2 1.58_2 1.53_1. Scaly or feathery aggregates. Dark lead-gray, metallic luster. Cleavage {0001}, sectile. $H = 2\frac{1}{2}$. $G = 7.4$. $R = 18.2$–36.0 (540 nm); strongly pleochroic and anisotropic. *Emma mine, Little Cottonwood district, UT*; Kidd Creek mine, near Timmins, ON, and near Chase, BC, Canada; Crevola d'Ossola, Piedmont, Italy; Lyangar, Uzbekistan; Angokitsk deposit, Buryat ASSR, and Tamvatnei deposit, Kamchatka, Russia; Tsumeb, Namibia. BM $DI:331$, $R:880$, $P\&J:374$, CM $10:729(1970)$, $ABI:543$.

2.12.11.1 Jordisite MoS_2

Named in 1909 for Eduard F. Jordis (1868–1917), colloid chemist. Dimorphous with molybdenite. Amorphous. Massive. Black, gray-black streak, submetallic luster. Sectile. Ambrosia Lake, McKinley Co., NM; Kiggins mercury mine, Clackamas Co., OR; Sun Valley mine, Coconino Co., AZ; *Himmelsfürst mine, near Freiberg, Saxony, Germany*; Bleiberg, Carinthia, Austria; Carrizal Alto, Atacama, Chile; Campo do Agostinho, Brazil. BM $DI:331$, $R:878$, AM $56:1832(1971)$, $ABI:251$.

2.12.12.1 Jeromite $As(S,Se)_2$

Named in 1928 for the locality. Amorphous. Black coatings, a sublimate from burning sulfide ores. *United Verde mine, Jerome, AZ.* BM AM $13:227(1928)$, $DI:144$, $ABI:249$.

2.12.13.1 Krennerite $AuTe_2$

Named in 1877 for Joseph Krenner (1839–1920), Hungarian mineralogist. May contain up to 6% Ag. Dimorphous with calaverite. ORTH $Pma2$. $a = 16.54$, $b = 8.82$, $c = 4.46$, $Z = 8$, $D = 8.86$. 8-20: 3.03_{10} 2.94_6 2.23_5 2.11_7 2.07_4 1.78_4 1.69_4 1.52_4. Short prismatic crystals. Silver-white to pale brass-yellow, gray streak, metallic luster. Cleavage {001}, perfect; subconchoidal to uneven fracture; brittle. H = 2–3, VHN_{100} = 166–186. G = 8.4–8.6. R = 60.4–62.4 (540 nm); strongly anisotropic. Occurs in hydrothermal Au–Te deposits. Cripple Creek, Teller Co., and Smoky Hill mine, Boulder Co., CO; Campbell mine, Bisbee, and Tombstone, Cochise Co., AZ; Bencourt mine, Louvricourt, PQ, Canada; *Sacaramb* and *Facebanya, Romania*; Zhanatyube deposit, Kazakhstan; Konstatin–Ovkoye deposit, Krasnoyarsk region, Russia; Emperor mine, Vatukoula, Fiji; Kalgoorlie and Mulgabbie, WA, Australia. BM *DI:333, R:430, P&J:228, TMPM 33:253(1984), ABI:275*.

2.12.13.2 Calaverite AuTe$_2$

Named in 1868 for the locality. May contain up to 4% Ag. Dimorphous with krennerite. MON $C2/m$. $a = 8.76$, $b = 4.410$, $c = 10.15$, $\beta = 125.2°$, $Z = 4$, $D = 9.31$. 7-344: 2.99_{10} 2.91_7 2.19_5 2.09_9 2.06_6 1.50_5 1.32_5 1.20_5. Bladed crystals, stout or slender prisms, also massive, granular. Silver-white to pale brass-yellow, yellow- to green-gray streak, metallic luster. No cleavage, subconchoidal to uneven fracture, very brittle. H = $2\frac{1}{2}$–3, VHN_{100} 197–213. G = 9.1–9.4. R = 61.3–65.9 (540 nm); weakly pleochroic and anisotropic. Widespread in Au–Te deposits. Cripple Creek(*), Teller Co., Central City, Gilpin Co., Gold Hill, Boulder Co., and other localities in CO; Spotted Horse mine, Marden, MT; *Stanislaus mine, Calaveras Co., CA*; Robb–Montbray mine, PQ, Kirkland Lake and Hemlo gold deposit, ON, Canada; Zhanatyube deposit, Kazakhstan; Koshbulak, Uzbekistan; Klyuchi, eastern Siberia, Russia; Nishizaki, Gifu Pref., and Date mine, Hokkaido, Japan; Emperor mine, Vatukoula, Fiji; Kalgoorlie, WA. Australia. BM *DI:335, R:431, P&J:111, ZK 169:227 (1984), ABI:77*.

2.12.13.3 Sylvanite (Au,Ag)$_2$Te$_4$

Named in 1835 for Transylvania, where it was first found. MON $P2/c$. $a = 8.96$, $b = 4.49$, $c = 14.62$, $\beta = 145.4°$, $Z = 2$, $D = 8.11$. 9-477: 5.09_1 3.05_{10} 2.98_2 2.25_3 2.14_5 1.99_3 1.80_2 1.52_2. Complex prismatic, tabular, or bladed crystals, also columnar to granular. Contact, lamellar, or penetration twins give arborescent forms resembling written characters (graphic tellurium). Silver-white, tarnishing to pale yellow, gray streak, metallic luster. Cleavage {010}, perfect; uneven fracture; brittle. H = 2, VHN_{100} = 154–172. G = 8.1–8.2. R = 50.6–61.1 (540 nm); strongly pleochroic and anisotropic. Occurs in minor amounts in many Au–Ag deposits. Cripple Creek(*), Teller Co., and Magnolia, Gold Hill, and Sunshine districts, Boulder Co., CO; Cornucopia district, Baker Co., OR; Melones and Stanislaus mines, Calaveras Co., CA; Dome mine, Porcupine, ON, Canada; Glava, Värmland, Sweden; Sacaramb(*) and Facebanya, Romania; Zhanatyube deposit, Kazakhstan; Emperor mine(*), Vatukuola, Fiji; Thames district, New Zealand; Kalgoorlie and Norseman, WA, Australia; Arakara goldfields, Guyana. BM *DI:338, R:426, P&J:359, TMPM 33:203(1984), ABI:509*.

2.12.13.4 Kostovite CuAuTe$_4$

Named in 1966 for Ivan Kostov (b.1913), Bulgarian mineralogist. ORTH $Pma2$. $a = 16.50$, $b = 8.84$, $c = 4.42$, $Z = 4$, $D = 7.94$. *35-521*: 3.03_{10} 2.23_8 2.10_9 1.97_7 1.77_5 1.34_6 1.06_5 1.04_5. Microscopic grains. Gray-white, metallic luster. One good cleavage, brittle. $H = 2$–$2\frac{1}{2}$, VHN = 180–186. R = 50.5–61.7 (540 nm); distinctly anisotropic. Buckeye Gulch, near Leadville, CO; Campbell mine, Bisbee, AZ; *Chelopech, Bulgaria*; Kochbulak deposit, Uzbekistan; Ashanti mine, Ghana. BM *AM 51:29(1966), R:430, ABI:272, MM 54:617(1990)*.

MELONITE GROUP

The melonite group minerals are rhombohedral sulfides, selenides, and tellurides with the general formula

$$AX_2$$

where

A = Ni, Sn, Pd, Pt, Ir
X = S, Se, Te, and minor Bi

and the space group is $P\bar{3}m1$. The minerals have a layer-lattice structure similar to that of brucite, Mg(OH)$_2$.

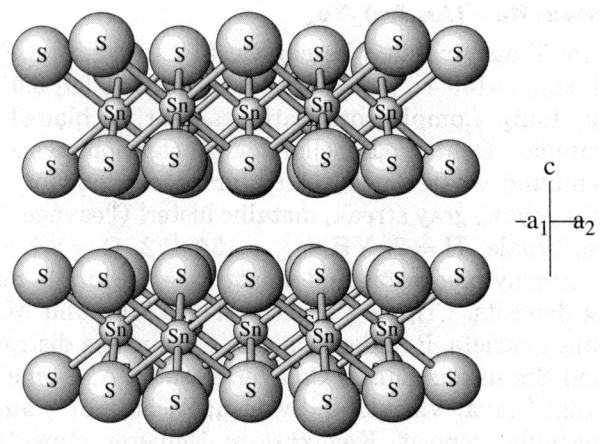

BERNDTITE: SnS$_2$

This has sandwich layers of S–Sn–S. The sulfur atoms are in hexagonal closest packing, unlike molybdenite.

MELONITE GROUP

Mineral	Formula	Space Group	a	c	D	
Melonite	$NiTe_2$	$P\bar{3}m1$	3.843	5.265	7.74	2.12.14.1
Kitkaite	NiTeSe	$P\bar{3}m1$	3.716	5.126	7.19	2.12.14.2
Moncheite	$(Pt,Pd)(Te,Bi)_2$	$P\bar{3}m1$	4.049	5.288	7.08	2.12.14.3
Merenskyite	$Pd(Te,Bi)_2$	$P\bar{3}m1$	4.036	5.132	8.29	2.12.14.4
Berndtite	SnS_2	$P\bar{3}m1$	3.639	5.868	4.46	2.12.14.5
Shuangfengite	$IrTe_2$	$P\bar{3}m1$	3.933	5.390	10.14	2.12.14.6

2.12.14.1 Melonite $NiTe_2$

Named in 1866 for the locality. Melonite group. HEX-R $P\bar{3}m1$. $a = 3.843$, $c = 5.265$, $Z = 1$, $D = 7.74$. *8-4*: 2.82_{10} 2.64_3 2.06_5 1.92_5 1.59_2 1.55_6 1.23_2 1.08_2. Hexagonal plates and small grains. Reddish white, tarnishing brown, dark gray streak, metallic luster. Cleavage {0001}, perfect. $H = 1–1\frac{1}{2}$, $VHN_{50} = 46–59$. $G = 7.4–7.7$. $R = 57.7–59.8$ (540 nm); moderately to strongly anisotropic. Cripple Creek, Teller Co., and Magnolia district, Boulder Co., CO; Bisbee, AZ; *Melones mine* and Stanislaus mine, *Calaveras Co., CA*; Robb–Montbray mine, Montbray Twp., PQ, and Sudbury and Hemlo gold deposit, ON, Canada; Monchegorsk deposit, Kola, Russia; Zhanatyube deposit, Kazakhstan; Yokozuru mine, Kyushu, Japan; Emperor mine, Vatakuola, Fiji; Kalgoorlie and Kambalda, WA, Australia; O'Okiep, Namaqualand, South Africa. BM *DI:341, R:420, P&J:265, ABI:321*.

2.12.14.2 Kitkaite NiTeSe

Named in 1965 for the locality. Melonite group. HEX-R $P\bar{3}m1$. $a = 3.716$, $c = 5.126$, $Z = 1$, $D = 7.19$. *18-896*: 5.13_2 2.73_{10} 2.57_3 2.01_5 1.86_3 1.54_2 1.51_4 1.36_1. Small platy crystals and massive. Pale yellow, metallic luster. Cleavage {0001}, good. $H = 3$, $VHN_{100} = 137$. $G = 7.22$. $R = 52.3–57.9$ (540 nm); distinctly anisotropic. *Kuusamo, Kitka R., Finland.* BM *AM 50:581(1965), R:421, P&J:223, ABI:266*.

2.12.14.3 Moncheite $(Pt,Pd)(Te,Bi)_2$

Named in 1963 for the locality. Melonite group. Forms a series with merenskyite. HEX-R $P\bar{3}m1$. $a = 4.049$, $c = 5.288$, $Z = 1$, $D = 7.08$. *15-392*: 5.32_6 2.93_{10} 2.11_8 2.02_7 1.66_6 1.58_5 1.46_7 1.28_7. Small euhedral to subhedral crystals and microscopic grains. Steel-gray, metallic luster. Basal cleavage. $H = 3\frac{1}{2}$, $VHN_{10} = 146$. $R = 56.0–56.9$ (540 nm); strongly anisotropic. Stillwater Complex, MT; New Rambler mine, Albany Co., WY; San Joaquin R., Fresno Co., CA; Sudbury and Lac des Isles Complex, ON, Canada; *Monchegorsk deposit, Monche Tundra, Kola,* and Oktyabr mine, Norilsk, *Russia*; Tao and Ma districts, China; Merensky Reef and Witwatersrand, Transvaal,

South Africa. BM *ZVMO 92:33(1963), AM 48:1181(1963), R:838, P&J:407, ABI:337.*

2.12.14.4 Merenskyite Pd(Te,Bi)$_2$

Named in 1966 for Hans Merensky (1871–1952), South African geologist. Melonite group. Forms a series with moncheite. HEX-R $P\bar{3}m1$. $a = 4.036$, $c = 5.132$, $Z = 1$, $D = 8.29$. 29-970(syn): 2.87_{10} 2.06_9 2.01_9 1.65_8 1.53_7 1.44_7 1.28_8 1.17_7 1.08_7. Microscopic grains. Tin-white, metallic luster. H = $3\frac{1}{2}$, VHN$_{10}$ = 82–128. R = 66.6–68.2 (540 nm); strongly anisotropic. Stillwater Complex, MT; New Rambler mine, Albany Co., WY; near Moapa, Clark Co., NV; Lac des Iles Complex, Temagami Lake, Sudbury, and other localities, ON, Canada; Norilsk, Russia; Shiaonanshan, Inner Mongolia, China; *Rustenburg mine, Merensky Reef,* and Messina, *Transvaal, South Africa.* BM *MM 35:815(1966), R:838, P&J:267, ABI:324.*

2.12.14.5 Berndtite SnS$_2$

Named in 1964 for Fritz Berndt, German mineralogist. Melonite group. HEX-R (a 4H polytype is known) $P\bar{3}m1$. $a = 3.639$, $c = 5.868$, $Z = 1$, $D = 4.46$. 23-677(syn): 5.89_{10} 3.16_3 2.78_6 2.16_3 1.82_3 1.74_2 1.67_1 1.53_1. Tabular hexagonal crystals. Pale yellow, yellow streak, adamantine luster. Cleavage {0001}, good. H < 2. G = 4.5. Uniaxial (–); N$_O$ > 1.93, N$_E$ = 1.705. Forestville, Northumberland Co., PA; Lagares-do-Estanho pegmatite, and Panasqueira, Portugal; Stiepelmann mine, Arandis, Namibia; *Cerro de Potosi, Bolivia.* BM *NJMM:94(1964), AM 51:1551(1966), R:880, ABI:47.*

2.12.14.6 Shuangfengite IrTe$_2$

Named in 1994 for the locality. Melonite group. HEX-R $P\bar{3}m1$. $a = 3.933$, $c = 5.390$, $Z = 1$, $D = 10.14$. AM 80: 2.85_{10} 2.10_8 1.95_6 1.58_7 1.16_6. Microscopic aggregates and veinlets. Black, black streak. Cleavage {0001}, perfect; brittle. H = 3, VHN$_{20}$ = 108. R = 40–48 (540 nm); weakly anisotropic in blue and yellow tints. In placer concentrates, *Shuangfeng, 190 km NNE of Beijing, China.* BM *AMS 14:322(1994), AM 80:1329(1995).*

2.12.15.1 Froodite PdBi$_2$

Named in 1958 for the locality. MON $C2/m$. $a = 12.75$, $b = 4.29$, $c = 5.67$, $\beta = 102.9°$, $Z = 4$, $D = 11.52$. 11-251: 2.97_7 2.77_{10} 2.48_7 2.21_7 2.14_5 1.64_6 1.56_8 1.42_6. Flat cleavage fragments and rounded grains. Gray, tarnishing rapidly, black streak, metallic luster. Cleavage {100}, perfect; {001}, less perfect; uneven fracture; brittle. H = $2\frac{1}{2}$, VHN$_{15}$ = 136. G = 12.5–12.6. R = 56.1–58.4 (540 nm); anisotropic, light to dark gray. Occurs in mill concentrates of Pb–Cu ores. Fox Gulch, Goodnews Bay, AK; *Frood mine, Sudbury, ON,* and Pipe mine, MB, *Canada;* Oktyabr mine, Talnakh area, Norilsk, Russia; Merensky Reef, Transvaal, South Africa. BM *CM 6:200(1958), 11:903(1973); R:826; ABI:166.*

2.12.15.2 Urvantsevite Pd(Bi,Pb)$_2$

Named in 1976 for Nikolai N. Urvantsev (1893–1985), Russian geologist, who discovered the Norilsk deposits. The high-temperature polymorph of froodite. TET $I4/mmm$. $a = 3.358$, $c = 12.949$, $Z = 2$, $D = 11.93$. $13\text{-}160$(syn): 3.26_7 2.65_{10} 2.37_8 2.05_4 1.91_3 1.62_5 1.42_4 1.11_4. Microscopic grains. Gray-white, metallic luster. Perfect cleavage in one direction. $H = 2$, $VHN_{10} = 47$–68. $R = 54.0$–55.2 (540 nm); weakly pleochroic and anisotropic. Tilkerode, Harz, Germany; *Majak mine, Talnakh, Norilsk, Russia*. BM *ZVMO 105:704(1976), AM 62:1260(1977), ABI:550*.

2.12.16.1 Borishanskiite Pd(As,Pb)$_2$

Named in 1975 for Serafima S. Borishanskaya (1907–1988), Russian mineralogist. Possibly identical with polarite. ORTH $Ccm2_1$. $a = 7.18$, $b = 8.62$, $c = 10.66$, $Z = 16$, $D = 10.2$. *ZVMO 104*: 2.65_{10} 2.50_6 2.25_6 2.16_9 1.68_6 1.39_6 1.17_6. Microscopic grains. Steel-gray, metallic luster. $H = 3\frac{1}{2}$–4, $VHN_{20} = 241$. $R = 57.3$–57.8 (580 nm); moderately anisotropic. *Oktyabr mine, Talnakh, Norilsk, Russia*, in Cu–Ni sulfide ores. BM *ZVMO 104:57(1975), AM 61:502(1976), ABI:60*.

2.12.17.1 Skutterudite CoAs$_{2-3}$

Named in 1845 for the locality. Synonym: smaltite (an As-deficient variety). ISO $Im3$. $a = 8.204$, $Z = 8$, $D = 6.82$. $10\text{-}328$(syn): 5.80_2 3.35_2 2.59_{10} 2.19_4 1.84_4 1.68_2 1.61_3 1.41_2 1.21_2. Cubes and cubo-octahedra, also massive; twinned on {112} to produce sixlings. Tin-white, tarnishing to gray, black streak, metallic luster. Cleavage {001}, {111}, distinct; {011}, poor; conchoidal to uneven fracture; brittle. $H = 5\frac{1}{2}$–6, $VHN_{100} = 810$–915. $G = 6.5$–6.9. $R = 54.0$ (540 nm). An important Co ore. Chatham, CT; Franklin, NJ; Silver City, NM; Horace Porter mine, Gunnison Co., CO; Cobalt(*), Gowganda, and Sudbury, ON, Hazelton, BC, and Eldorado mine, Great Bear Lake, NWT, Canada; *Skutterud, Modum, Norway*; Tunaberg(*), Sweden; Andreasberg, Harz, Wittichen, Black Forest, Schneeberg(*), Annaberg, and Freiberg, Saxony, Germany; Lölling, Austria; Khovu-Akay(*), Tuva, Russia; Bou Azzer(*), Morocco; Gashani, Zimbabwe; Punta Brava, Copiapó, Chile. BM *DI:342, CM 9:559(1971), R:881, P&J:343, ABI:480*.

2.12.17.2 Nickel-Skutterudite NiAs$_{2-3}$

Named in 1892 for its relationship to skutterudite. Synonym: chloanthite. ISO $Im3$. $a = 8.17$, $Z = 8$, $D = 6.40$. $25\text{-}566$: 5.87_4 3.39_4 2.62_{10} 2.22_7 1.86_7 1.77_4 1.70_6 1.63_8. Cubes and cubo-octahedral crystals, also massive; twinned on {112} as sixlings. Tin-white to silver gray, black streak, metallic luster. Cleavage {001} and {111}, distinct; conchoidal to uneven fracture; brittle. $H = 5\frac{1}{2}$–6, $VHN_{100} = 690$. $G = 6.5$. Franklin, NJ, *Bullards Peak, Grant Co., NM*; Cobalt, ON, Canada; Bieber, Hesse, Andreasberg, Harz Mts., Schneeberg(*) and Annaberg, Saxony, Germany; Dobsina, Czech Republic; Kovuakhsin deposit, Ural Mts., Russia. BM *DI:342, ABI:351*.

2.12.17.3 Kieftite CoSb$_3$

Named in 1994 for Cornelis Kieft (b.1924), Dutch mineralogist. ISO $Im3$. $a = 9.041$, $Z = 8$, $D = 7.63$ (Skutterudite group). $CM32$: 2.86_{10} 2.42_6 2.02_8 1.77_6 1.55_6 1.33_7 1.05_6 0.933_8. Microscopic euhedral to subhedral grains. Tin-white, gray streak, metallic luster. No cleavage, conchoidal fracture, brittle. Crystal forms {100}, {110}, {111}. $VHN_{100} = 464$. $G = 7.2$. $R = 58.7$ (546 nm); isotropic. *Tunaberg, Sweden*, in Cu–Co sulfide ores. BM *CM 32:179(1994)*.

2.13.1.1 Kermesite Sb$_2$S$_2$O

Named in 1843 from *kermes*, an old chemical name for red amorphous antimony sulfide. TRIC $P\bar{1}$. $a = 10.84$, $b = 4.07$, $c = 10.25$, $\alpha = 90.0°$, $\beta = 101.53°$, $\gamma = 90.0°$, $Z = 4$, $D = 4.85$. *30-92*: 5.30_{10} 4.32_5 4.08_6 3.14_4 2.92_5 2.70_{10} 2.50_3 2.26_1. Platy and acicular crystals, often in radiating clusters. Cherry-red, red-brown streak, adamantine to submetallic luster. Cleavage {001}, perfect; {100}, good; sectile; flexible thin splinters. $H = 1$–$1\frac{1}{2}$, $VHN_5 = 30$–90. $G = 4.68$. $R = 25.2$–30.6 (540 nm). Widely distributed as an alteration product of stibnite. Boron, Kern Co., and Kalkar quarry, Santa Cruz Co., CA; Lac Nicolet mine, South Ham Twp., PQ, and other localities in Canada; Sombrerete, ZAC, Mexico; Braunsdorf, near Freiberg, Saxony, Germany; Pernek(*), Pezinok(*), and Pribram, Czech Republic; Cetine mine, near Siena, Italy; Kadamja, Kirgizstan; Taratu Creek, Thames, New Zealand; Rivertree and Broken Hill, NSW, Australia; Globe and Phoenix mine(*), Que Que, Zimbabwe; Santa Cruz and San Francisco mines, Poopo, Bolivia. BM *DI:279, R:715, P&J:222, ABI:260*.

2.13.2.1 Sarabauite CaSb$_{10}$O$_{10}$S$_6$

Named in 1978 for the locality. MON $C2/c$. $a = 25.37$, $b = 5.654$, $c = 16.87$, $\beta = 117.58°$, $Z = 4$, $D = 4.99$. *29-293*: 4.46_3 4.23_4 3.47_8 3.22_{10} 3.18_6 3.16_5 2.82_9 2.58_4. Small tabular crystals. Carmine-red, orange streak, resinous luster. No cleavage, somewhat sectile. $H = 4$, $VHN_{20} = 272$. $G = 4.8$. Biaxial (−); 2V near 90°; pleochroic, red-yellow to brown-red. *Sarabau mine, Sarawak*. BM *AM 63:715(1978), AC(B)34:3569(1978), ABI:461*.

2.13.3.1 Cetineite (K,Na)$_{3.5}$(Sb$_2$O$_3$)$_3$(SbS$_3$) · 3H$_2$O

Named in 1987 for the locality. HEX $P6_3$. $a = 14.230$, $c = 5.579$, $Z = 2$, $D = 4.29$. *40-1458*: 12.41_8 4.67_5 4.11_6 3.58_4 3.42_4 3.00_7 2.92_{10} 2.69_6. Microscopic acicular crystals. Orange-red, orange streak, resinous luster. $H = 3$, $VHN_{20} = 127$–156. Uniaxial (+); $N > 1.74$; weakly pleochroic, orange to orange-brown. *Cetine mine, near Siena, Italy*. BM *NJMM:419(1987); AM 73:398(1988), 74:1399(1989); ABI:86*.

2.14.1.1 Valleriite $4(Fe,Cu)S \cdot 3(MgAl)(OH)_2$

Named in 1870 for Johan G. Wallerius (1709–1785), Swedish chemist and mineralogist. HEX-R $R\bar{3}m$. $a = 3.792$, $c = 34.10$, $Z = 6$, $D = 3.21$. *29-554*: 11.4_{10} 5.71_{10} 3.80_5 3.27_6 3.23_5 2.85_5 1.89_5 1.86_5. The structure consists of alternate layers of brucite-like $(Mg,Al)(OH)_2$ sheets and trigonal $(Cu,Fe)S$ sheets [double S layer with (Cu,Fe) in tetrahedral coordination]. The unit cell given corresponds to the sulfide portion of the structure. Massive, nodular, and platy. Bronze-yellow, black streak, dull metallic luster. Cleavage {0001}, perfect. $H = 1$. $G = 3.14$. $R = 10.2$–16.2 (540 nm); strongly pleochroic and anisotropic. Widely distributed in small amounts. Elizabeth mine, South Strafford, VT; Continental mine, Fierro, Grant Co., NM; Pima mine, Pima Co., and Christmas mine, Gila Co., AZ; Marbridge mine, Malartic, PQ, Sudbury, ON, and Little Chief mine, near Whitehorse, YT, Canada; *Aurora mine, Ljusnarsberg,* and *Kaveltorp, Kopparberg, Sweden*; near Pefkos, Cyprus; Monchegorsk, Kola, and Norilsk, Russia; Alnalyk, Uzbekistan; Akagane mine, Iwate Pref., Japan; Loolekop, Palabora, Mooihoek, and Onverwacht, Transvaal, South Africa; Atacocha, Peru. BM *DI:235, ZK 127:73(1968), R:683, P&J:382, ABI:555.*

2.14.2.1 Tochilinite $6Fe_{0.9}S \cdot 5(Mg,Fe)(OH)_2$

Named in 1971 for Mitrofan S. Tochilin (1910–1968), Russian scientist. MON Space group unknown. $a = 5.35$, $b = 15.54$, $c = 10.89$, $\beta = 94.81°$, $Z = 2$, $D = 3.03$. *41-590*: 10.84_4 5.41_{10} 2.63_2 2.23_5 2.05_2 1.86_4 1.75_3 1.53_2. Radiating aggregates of cylindrical acicular crystals, and as felted masses. Bronze-black, black streak, dull metallic luster. Cleavage {001}. $H = 1$–2, $VHN_5 = 15$–49. $G = 2.96$. $R = 10.2$–14.9 (546 nm); strongly pleochroic and anisotropic. Cornwall mine, Lebanon Co., PA; Jeffrey mine, Asbestos, Cross and Maxwell quarries, Wakefield, and Amos area, PQ, near Bancroft, ON, and Muskox Intrusion, NWT, Canada; Lizard ultramafic body, Cornwall, England; *Mamonovo deposit, Voronezh,* and diamond pipes, Yakutia, *Russia*; Kamaishi mine, Iwate Pref., Japan; Mt. Keith, WA, Australia; Jacupiranga mine, São Paulo, Brazil; an accessory mineral in carbonaceous chondrite meteorites. BM *ZVMO 100:477(1971); AM 57:1552(1972), 71:1201(1986); R:692; ABI:533.*

2.14.3.1 Haapalaite $2(Fe,Ni)S \cdot 1.6(Mg,Fe)(OH)_2$

Named in 1973 for Paavo Haapala (b.1906), Finnish geologist. HEX-R Space group uncertain, probably $R\bar{3}m$. $a = 3.64$, $c = 34.02$, $Z = 3$, $D = 3.57$. *26-1135*: 11.3_{10} 5.67_9 3.18_8 2.84_4 2.59_4 2.27_4 1.84_7 1.82_7. Thin scales. Bronze-red, metallic luster. $H = 1$, $VHN_3 = 9$–11. Strongly anisotropic. *Kokka serpentinite, near Outokumpu, Finland.* BM *AM 58:1111(1973), R:692, ABI:201.*

2.14.3.2 Yushkinite $V_{0.6}S_{0.6}(Mg,Al)(OH)_2$

Named in 1984 for Nikolai P. Yushkin (b.1936), Russian mineralogist. HEX-R $P\bar{3}m1$. $a = 3.20$, $c = 11.3$, $Z = 2$, $D = 3.31$. *38-432*: 11.4_3 5.68_{10} 2.76_8 2.36_2 1.60_4 1.58_6 1.53_3 1.39_3. Scaly aggregates. Pink-violet, metallic luster. $H < 1$.

$G = 2.94$. $R = 18.7$ (540 nm); strongly anisotropic. In quartz–carbonate veins. *Pay-Khoy, Silova-Yakha R., Russia.* BM *MZ 6:91(1984), AM 71:846(1986), ABI:584.*

2.14.4.1 Vyalsovite FeS · Ca(OH)$_2$ · Al(OH)$_3$

Named in 1992 for Leonid N. Vyalsov (b.1939), Russian mineralogist. ORTH Space group uncertain, probably *Cmmm*. $a = 14.20$, $b = 20.98$, $c = 5.496$, $Z = 8$, $D = 1.96$. *44-1452*: 5.40_{10} 4.76_4 2.88_2 2.16_5 1.97_3 1.89_3 1.85_3 1.68_4. Microscopic grains and veinlets. Crimson. $R = 8.2$–11.4 (540 nm); strongly anisotropic. In skarn, *Talnakh, Norilsk, Russia.* BM *AM 77:201(1992)*.

2.14.5.1 Erdite NaFeS$_2$· 2H$_2$O

Named in 1980 for Richard C. Erd (b.1924), U.S. mineralogist. MON $C2/c$. $a = 10.693$, $b = 9.115$, $c = 5.507$, $\beta = 92.17°$, $Z = 4$, $D = 2.22$. *33-1253*: 6.94_{10} 5.34_8 4.56_4 3.47_2 2.97_2 2.90_5 2.31_2 2.01_1. Small bladed crystals and granular masses. Copper-red, black streak, metallic luster. Cleavage {110}, good. $H = 2$, $VHN_{15} = 22$–67. $G = 2.30$. $R = 8.8$–20.7 (540 nm); extremely pleochroic. *Coyote Peak, Humboldt Co., CA,* in a mafic alkalic diatreme; Mont St.-Hilaire, PQ, Canada; Lovozero massif, Kola, Russia, in pegmatite. BM *AM 65:509(1980), ABI:148*.

2.14.6.1 Coyoteite NaFe$_3$S$_5$ · 2H$_2$O

Named in 1983 for the locality. TRIC $P1$ or $P\bar{1}$. $a = 7.409$, $b = 9.881$, $c = 6.441, \alpha = 100.42°$, $\beta = 104.62°$, $\gamma = 81.48°$; $Z = 2$, $D = 2.88$. *35-565*: 9.6_6 7.13_9 5.60_6 5.12_{10} 3.91_5 3.08_7 3.03_8 1.90_8. Small clots. Black, black streak, metallic luster. Cleavage {$\bar{1}11$}, perfect. $H = 1\frac{1}{2}$. $G = 2.5$–2.6. Moderately magnetic. Strongly anisotropic. *Coyote Peak, Humboldt Co., CA,* in a mafic alkalic diatreme. BM *AM 68:245(1983), ABI:113*.

2.14.7.1 Orickite CuFeS$_2$ · H$_2$O

Named in 1983 for the locality. HEX Space group unknown. $a = 3.695$, $c = 6.16$, $Z = 4$, $D = 4.21$. *37-412*: 10.78_8 6.92_3 5.41_2 4.35_{10} 3.66_9 3.60_7 3.45_5 2.52_2. Small clots. Brass-yellow, black streak, metallic luster. Cleavage {001}, good. $H = 3\frac{1}{2}$. Weakly magnetic. Strongly anisotropic. *Coyote Peak, near Orick, Humboldt Co., CA,* in a mafic alkalic diatreme. BM *AM 68:245(1983), ABI:365*.

2.15.1.1 Ardaite Pb$_{19}$Sb$_{13}$S$_{35}$Cl$_7$

Named in 1981 for the locality. MON Space group unknown. $a = 22.09$, $b = 21.11$, $c = 8.05$, $\beta = 103.02°$, $Z = 2$, $D = 6.11$. *42-585*: 3.71_2 3.39_{10} 3.36_8 3.28_2 2.87_6 2.82_4 2.78_3 2.10_4. Microscopic acicular aggregates. Green-gray, metallic luster. $R = 32.3$–35.1 (520 nm); distinctly pleochroic and anisotropic. *Madjarovo, Arda R., Bulgaria;* Dressfall mine, Gruvåsen, near Filipstad, Sweden. BM *CM 19:419(1981), MM 46:357(1982), ABI:19*.

2.15.2.1 Djerfisherite $K_6(Fe,Cu,Ni)_{25}S_{26}Cl$

Named in 1966 for D. Jerome Fisher (1896–1988), U.S. mineralogist. ISO $Pm3m$. $a = 10.41$, $Z = 1$, $D = 3.90$. *25-635*: 10.4_6 3.33_7 3.17_7 3.03_7 2.38_6 2.01_4 1.84_{10} 1.30_5. Microscopic anhedral grains. Yellow-green, submetallic luster. $H \sim 3$, $VHN_{20} = 172$. $R = 22.9$ (540 nm). *Kota–Kota enstatite chondrite and other meteorites; Coyote Peak, Humboldt, CA; Ilimaussaq Intrusion, Greenland; Lovozero and Khibina massifs, Kola, Norilsk, and Yakutian kimberlites, Russia.* BM *SCI 153:166(1966), R:1046, P&J:148, AM 64:776(1979), ABI:136*.

2.15.2.2 Thalfenisite $Tl_6(Fe,Ni,Cu)_{25}S_{26}Cl$

Named in 1979 for the composition. ISO $Pm3m$. $a = 10.29$, $Z = 1$, $D = 5.26$. *33-1401*: 4.16_5 3.42_9 3.24_7 2.96_{10} 2.35_6 2.09_3 1.97_4 1.81_7. Microscopic grains. Brown, metallic luster. $H = 3$, $VHN_{10} = 130$–164. $R = 25.6$ (560 nm). *Oktyabr mine, Norilsk, Russia.* BM *ZVMO 108:696(1979), AM 66:219(1981), ABI:529*.

2.15.2.3 Owensite $(Ba,Pb)_6(Cu,Fe,Ni)_{25}S_{27}$

Named in 1995 for DeAlton R. Owens (b.1934), Canadian mineralogist. ISO $Pm3m$. $a = 10.349$, $Z = 1$, $D = 4.78$. *CM 33*: 3.46_4 3.28_4 3.12_2 3.00_9 2.38_9 1.84_{10} 1.78_4 1.73_2. Microscopic anhedral grains. Black, black streak, metallic luster. $H = 3\frac{1}{2}$, $VHN_{10} = 137$. $R = 24.7$ (540 nm); in reflected light pale brown-gray, isotropic. *Wellgreen deposit, Kluane district, YT, Canada*, associated with magnetite, chalcopyrite, and pentlandite. BM *CM 33:665(1995)*.

2.15.3.1 Kolarite $PbTeCl_2$

Named in 1985 for the locality. ORTH Space group unknown. $a = 5.93$, $b = 3.25$, $c = 3.89$, $Z = 1$, $D = 8.99$. *39-323*: 3.91_4 3.27_{10} 2.35_5 2.00_4 1.86_3 1.79_3 1.50_3 1.35_3. Microscopic aggregates. Gray, metallic luster. $R = 27.7$–28.6 (540 nm). *Champion Reef mine, Kolar Gold Fields, Karnataka, India.* BM *CM 23:501(1985), AM 71:1545(1986), ABI:269*.

2.15.4.1 Radhakrishnaite $PbTe_3(Cl,S)_2$

Named in 1985 for B. P. Radhakrishna (b.1918), Indian mineralogist. TET Space group unknown. $a = 5.71$, $c = 3.77$, $Z = 1$, $D = 8.93$. *39-324*: 3.78_6 3.16_{10} 2.73_4 2.29_4 2.04_3 1.92_5 1.78_5 1.59_4. Microscopic aggregates. Brown-gray, metallic luster. $R = 31.1$–31.8 (540 nm); distinctly anisotropic. *Champion Reef mine, Kolar gold fields, Karnataka, India.* BM *CM 23:501(1985), AM 71:1545(1986), ABI:429*.

2.15.5.1 Perroudite $Hg_5Ag_4S_5(Cl,I,Br)_4$

Named in 1987 for Pierre Perroud (b.1943), Swiss mineralogist. ORTH $P2_12_12_1$. $a = 17.43$, $b = 12.24$, $c = 4.35$, $Z = 2$, $D = 6.60$. *42-1336*: 5.01_2 3.94_6 3.69_3 3.01_{10} 2.97_8 2.74_3 2.64_4 2.45_3. Fibrous aggregates of microscopic prismatic crystals. Red, orange-red streak, vitreous to adamantine luster. Cleavage

{100}, perfect; irregular fracture. Biaxial (+); N = 2.3–2.4; 2V = 70°; pleochroic, yellow to red-brown. *Cap-Garonne, Var, France*; Broken Hill, NSW, and Coppin Pool, WA, Australia. BM *AM 72:1251(1987), ABI:399*.

2.15.6.1 Capgarronite HgAg(Cl,Br,I)S

Named in 1992 for the locality. Synonym: tocornalite. ORTH $P2_12_12_1$. $a = 6.803$, $b = 12.87$, $c = 4.528$, $Z = 4$, $D = 6.19$. *44-1451*: 6.43_4 3.76_6 3.64_6 3.28_3 2.66_{10} 2.27_4 2.05_2 1.70_1. Small prismatic crystals. Black, gray-black streak, subadamantine to submetallic luster. $N > 2.1$; pleochroic, purple to dark brown. *Cap-Garonne, Var, France*; Broken Hill, NSW, Australia; Chanarcillo, Chile. BM *DII:25, AM 77:197(1992)*.

2.16.1.1 Mertieite-I $Pd_{11}(Sb,As)_4$

Named in 1973 for John B. Mertie (1888–1980), geologist, U.S. Geological Survey. HEX Space group unknown. $a = 15.04$, $c = 22.41$, $Z = 18$, $D = 10.6$. *25-598*: 2.28_{10} 2.23_7 2.17_9 2.02_6 1.92_6 1.20_6 1.10_6 1.09_6. Small grains. Brass-yellow, metallic luster. $VHN_{50} = 561$–593. $R = 51.3$ (546 nm); weakly pleochroic, distinctly anisotropic. *Goodnews Bay, AK*, in placer concentrates; Talnakh, Norilsk, Russia. BM *AM 58:1(1973), CM 13:321(1975), ABI:325*.

2.16.2.1 Isomertieite $Pd_{11}Sb_2As_2$

Named in 1974 for its relationship to mertieite. ISO *Fd3m*. $a = 12.283$, $Z = 8$, $D = 10.33$. *26-833*: 2.36_9 2.17_{10} 1.53_7 1.45_6 1.29_6 1.25_7 1.19_7 1.00_7. Microscopic grains. Pale yellow-white, metallic luster. $H = 5\frac{1}{2}$, $VHN_{100} = 587$–597. $R = 54.8$ (540 nm); weakly anisotropic in shades of brown. Occurs in heavy-metal concentrates with gold and other Pd minerals. Lac des Iles Complex, ON, Canada; Hope's Nose, Torquay, England; Konttijärvi intrusion, northern Finland; near Zlatoust, Ural Mts., Russia; Waiau R., Southland, New Zealand; *Itabira, MG, Brazil*. BM *MM 39:528(1974), 46:371(1982), ABI:244*.

2.16.3.1 Mertieite-II $Pd_8(Sb,As)_3$

Named in 1973 as for mertieite-I. HEX-R *R3c*. $a = 7.601$, $c = 42.904$, $Z = 12$, $D = 11.29$. *30-93*(syn): 2.48_4 2.28_9 2.19_{10} 2.02_4 1.93_4 1.49_4 1.28_5 1.27_5. Small grains. Brass-yellow, metallic luster. $H = 6$, $VHN_{50} = 561$–593. $R = 50.1$ (540 nm); distinctly anisotropic, dark blue-gray to dark brown. Occurs as small grains in precious-metal concentrates. Stillwater Complex, MT; *Goodnews Bay, AK*; Hope's Nose, Torquay, England; Oktyabr mine, Norilsk, and near Zlatoust, Ural Mts., Russia; Merensky Reef, Transvaal, South Africa. BM *AM 58:1(1973), CM 13:321(1975), MM 46:371(1982), ABI:326*.

2.16.4.1 Stillwaterite Pd_8As_3

Named in 1975 for the locality. HEX $P\bar{3}$. $a = 7.426$, $c = 10.316$, $Z = 3$, $D = 10.88$. *30-880*(syn): 2.36_{10} 2.21_5 2.14_7 2.13_8 1.99_3 1.76_4 1.36_5 1.21_4. Microscopic grains. Cream-gray, metallic luster. $H = 4\frac{1}{2}$, $VHN_{50} = 384$. $G = 10.4$.

$R = 52.5$–53.2 (546 nm); weakly anisotropic. *Stillwater Complex, MT*, in Banded and Upper zones, associated with gold and other Pd-arsenides; Roby zone, Lac des Iles Complex, ON, Canada. BM *CM 13:321(1975)*, *AM 62:1060(1977)*, *ABI:500*.

2.16.5.1 Arsenopalladinite $Pd_8(As,Sb)_3$

Named in 1974 for the composition. Contains up to 6% Sb. TRIC $P1$. $a = 7.43$, $b = 13.95$, $c = 7.35$, $\alpha = 92.88°$, $\beta = 119.5°$, $\gamma = 87.85°$, $Z = 4$, $D = 11.03$. *29-959*: 2.34_6 2.28_2 2.19_2 2.13_{10} 1.41_4 1.39_2 1.24_3 1.21_3. Rounded alluvial grains. White with a cream tint, metallic luster. $H = 4$, $VHN_{100} = 379$–449. $G = 10.4$. $R = 52.1$–53.9; strongly anisotropic, red-brown to blue-gray. Stillwater Complex, MT; Merensky Reef, Transvaal, South Africa; *Itabira, MG, Brazil*, in gold concentrates. BM *MM 39:528(1974)*, *CM 15:70(1977)*, *ABI:27*.

2.16.6.1 Telluropalladinite Pd_9Te_4

Named in 1979 for the composition. MON $P2_1/c$. $a = 7.45$, $b = 13.95$, $c = 8.82$, $\beta = 91.9°$, $Z = 4$, $D = 10.64$. *35-699*: 2.54_2 2.24_{10} 2.13_3 2.09_4 1.98_3 1.96_3 1.40_3 1.31_5. Microscopic grains. Cream-yellow, metallic luster. $H = 4$–$4\frac{1}{2}$, $VHN_{15} = 376$–399. $G_{syn} = 10.25$. $R = 50.0$ (540 nm); moderately to strongly anisotropic. *Stillwater Complex, MT*. BM *CM 17:589(1979)*, *ABI:519*.

2.16.7.1 Balkanite $Cu_9Ag_5HgS_8$

Named in 1973 for the locality. ORTH Space group uncertain, probably $P222$. $a = 10.62$, $b = 9.42$, $c = 3.92$, $Z = 1$, $D = 6.64$. *25-299*(syn): 3.14_7 3.01_8 2.98_{10} 2.70_8 2.61_9 2.55_{10} 1.99_8 1.96_9. Microscopic prismatic crystals and small grains. Steel-gray, metallic luster. No cleavage. $H = 3\frac{1}{2}$, $VHN_{100} = 79$–92. $G_{syn} = 6.32$. $R = 28.0$–34.0 (540 nm); strongly anisotropic. Manhattan, Nye Co., NV; near Agua Prieta, SON, Mexico; *Sedmochislenitsi mine, Vratsa district, Balkan Mts., Bulgaria*. BM *AM 58:11(1973)*, *ABI:39*.

2.16.7.2 Danielsite $(Cu,Ag)_{14}HgS_8$

Named in 1987 for John L. Daniels (b.1931), who collected the mineral. ORTH Space group unknown. $a = 9.644$, $b = 9.180$, $c = 18.156$, $Z = 4$, $D = 6.54$. *40-500*: 4.44_2 3.65_2 3.02_2 2.83_3 2.62_{10} 2.39_5 1.96_6 1.88_6. Microscopic grains. Gray, metallic luster. $H = 2$–$2\frac{1}{2}$, $VHN_{10} = 38$. $R = 30$–31 (546 nm); moderately anisotropic. *Coppin Pool, WA, Australia*. BM *AM 72:401(1987)*, *73:187(1988)*; *ABI:127*.

2.16.8.1 Betekhtinite $Cu_{10}(Pb,Fe)S_6$

Named in 1955 for Anatolii G. Betekhtin (1897–1962), Russian mineralogist. ORTH *Immm*. $a = 14.693$, $b = 22.72$, $c = 3.861$, $Z = 4$, $D = 6.14$. *25-1223*: 5.65_5 4.77_5 3.08_8 2.94_{10} 2.52_5 2.35_7 1.93_6 1.83_9. Prismatic crystals to 2 cm and massive. Black, black streak, metallic luster. Cleavage in three directions. $H = 3$–$3\frac{1}{2}$. $G = 5.96$–6.05. $R = 32.1$–32.5 (540 nm); strongly anisotropic. St.

Cloud mine, Sierra Co., NM; near Laird, Sutherland, Scotland; *Mansfeld Kupferschiefer, Eisleben, Germany*; Mürtschenalp, Switzerland; Radka deposit, Pazardzik, Bulgaria; Bulancak deposit, Giresun, Turkey; Dzhezkazgan(*), Kazakhstan; Yoshino mine, Yamagata Pref., Japan; Mt. Lyell, TAS, Australia; Kipushi(*), Shaba, Zaire; Tsumeb, Namibia; La Leona mine, Santa Cruz Prov., Argentina. BM *AM 41:371(1956), NJMM:121(1960), P&J:89, ABI:51.*

2.16.9.1 Larosite $(Cu,Ag)_{21}(Pb,Bi)_2S_{13}$

Named in 1972 for Fred LaRose, a discoverer of silver ore at Cobalt. ORTH Space group unknown. $a = 22.15$, $b = 24.03$, $c = 11.67$, $Z = 10$, $D = 6.18$. *25-311*: 3.21_5 2.92_9 2.85_6 2.56_4 2.50_2 2.47_6 2.16_4 1.98_{10}. Microscopic acicular crystals. Black, metallic luster. $H = 3$, $VHN_{100} = 124$. $R = 29.7-31.7$ (546 nm); moderately anisotropic. *Foster mine, Cobalt, ON, Canada.* BM *CM 11:886(1972), AM 59:382(1974), R:486, ABI:287.*

2.16.10.1 Sopcheite $Ag_4Pd_3Te_4$

Named in 1982 for the locality. ORTH Space group unknown. $a = 9.645$, $b = 7.906$, $c = 11.040$, $Z = 4$, $D = 9.95$. *35-671*: 4.12_4 3.33_{10} 2.70_5 2.56_6 2.30_5 2.15_4 2.03_4 1.81_7. Microscopic grains. Gray, metallic luster. $H = 3\frac{1}{2}$, $VHN_{10} = 134-209$. $R = 41.6-43.1$ (540 nm); weakly anisotropic. Lac des Iles Complex, and Levack West mine, Sudbury, ON, Canada; *Sopcha massif, Monchegorsk pluton, Kola, Russia.* BM *ZVMO 111:114(1982), AM 68:472(1983), CM 22:233(1984), ABI:484.*

2.16.11.1 Henryite $Cu_4Ag_3Te_4$

Named in 1983 for Norman F. M. Henry (1909–1983), English mineralogist. ISO Space group unknown. $a = 12.20$, $Z = 8$, $D = 7.96$. *37-423*: 7.03_6 3.04_8 2.35_6 2.16_{10} 1.36_6 1.25_6 1.08_8 1.02_6. Microscopic anhedral grains. Gray, metallic luster. $H = 3$, $VHN_{100} = 109-115$. $R = 33.1$ (540 nm). *Campbell mine, Bisbee, AZ;* Ashanti mine, Ghana. BM *BM 106:511(1983), AM 70:216(1985), ABI:214.*

2.16.12.1 Furutobeite $(Cu,Ag)_6PbS_4$

Named in 1981 for the locality. MON Space group uncertain, probably Cm. $a = 20.025$, $b = 3.963$, $c = 9.705$, $\beta = 101.57°$, $Z = 4$, $D = 6.74$. *35-534*: 3.43_5 2.95_9 2.78_3 2.61_5 2.55_7 2.50_{10} 2.14_4 2.09_4. Microscopic grains. Gray, metallic luster. $H = 3$, $VHN_{25} = 100-108$. $R = 34.6-36.2$ (540 nm); moderately anisotropic. Erasmus mine, Leogang, Austria; *Furutobe mine, Akita Pref., Japan;* Tsumeb, Namibia. BM *BM 104:737(1981), AM 67:1075 (1982), ABI:169.*

2.16.13.1 Stützite $Ag_{5-x}Te_3$ (x = 0.24–0.36)

Named in 1878 for Andreas Stütz (1747–1806), Austrian mineralogist. HEX $C6/mmm$. $a = 13.49$, $c = 8.48$, $Z = 7$, $D = 8.02$. *18-1187*: 3.58_4 3.53_5 3.37_4

2.16.16.1 Maucherite

3.05_6 2.63_6 2.57_7 2.55_6 2.18_{10}. Highly modified hexagonal crystals, usually massive, granular. Lead-gray, lead-gray streak, metallic luster. $H = 3\frac{1}{2}$, $VHN_{25} = 73$–90. $G = 8.00$. $R = 40.7$–40.8 (540 nm); moderately anisotropic. Empress Josephine mine, Saguache Co., and other localities in CO; Campbell mine, Bisbee, AZ; Linquist Lake, PQ, and Temagami Lake, ON, Canada; Moctezuma, SON, Mexico; Glava, Värmland, Sweden; *Sacaramb, Romania*; Kochbulak deposit, Uzbekistan; Kawazu mine, Shizuoka Pref., Japan. BM *DI:167, R:425, P&J:357, ABI:504.*

2.16.14.1 Gortdrumite $Cu_6Hg_2S_5$

Named in 1983 for the locality. Contains up to 3% Fe. ORTH Space group unknown. $a = 14.958$, $b = 7.900$, $c = 24.10$, $Z = 12$, $D = 6.80$. *34-184*: 6.03_4 4.58_{10} 3.38_7 3.08_3 3.02_3 2.88_5 2.78_5 2.67_3. Microscopic anhedral grains. Lead-gray, metallic luster. $H = 3\frac{1}{2}$–4, $VHN_{10} = 186$–230. $R = 26.7$–30.5 (540 nm); strongly anisotropic. *Gortdrum deposit, Co. Tipperary, Eire.* BM *MM 47:35(1983), AM 69:407(1984), ABI:191.*

2.16.15.1 Rickardite Cu_7Te_5

Named in 1903 for Thomas A. Rickard (1864–1953), U.S. mining engineer. ORTH Space group unknown. $a = 4.003$, $b = 19.893$, $c = 12.220$, $Z = 4$, $D = 7.39$. *26-1117*: 6.12_6 3.35_9 3.34_{10} 3.06_8 2.56_6 2.07_{10} 2.04_9 1.99_5. Massive. Purple-red tarnishing to brown, metallic luster. $H = 3\frac{1}{2}$, $VHN_{50} = 72$–85. $G = 7.54$. $R = 13.0$–18.4 (540 nm); strongly pleochroic and anisotropic. *Good Hope mine, Gunnison Co.*, and Empress Josephine mine, Saguache Co., CO; Bisbee and Tombstone, Cochise Co., AZ; Horne mine, Noranda, PQ, Canada; Sebastian mine, San Salvador; Chatkal Ridge, Kirghizstan; Darasun deposit, Transbaikalia, Russia; Teine mine, Hokkaido, and Kawazu mine, Shizuoka Pref., Japan; Kalgoorlie, WA, Australia; O'Okiep, Namaqualand, South Africa. BM *DI:198, R:416, P&J:327, ABI:442.*

2.16.15.2 Oosterboschite $(Pd,Cu)_7Se_5$

Named in 1970 for M. R. Oosterbosch (b.1908), Belgian mining engineer. ORTH Space group unknown. $a = 10.42$, $b = 10.60$, $c = 14.43$, $Z = 8$, $D = 8.48$. *24-371*: 4.48_6 2.74_7 2.65_{10} 2.60_8 2.49_5 2.24_7 1.90_7 1.85_8. Microscopic anhedral grains, polysynthetically twinned. Black, black streak, metallic luster. $H = 4\frac{1}{2}$–5, $VHN_{100} = 340$. $R = 46.0$–51.8 (540 nm); anisotropic, blue-gray to brown-gray. *Musonoi mine, Shaba, Zaire.* BM *BM 93:476(1970), AM 57:1553(1972), R:468, P&J:280, ABI:362.*

2.16.16.1 Maucherite $Ni_{11}As_8$

Named in 1913 for Wilhelm Maucher (1879–1930), German mineral dealer. TET $P4_12_12$. $a = 6.872$, $c = 21.821$, $Z = 4$, $D = 8.05$. *8-85*: 2.69_9 2.01_{10} 1.71_{10} 1.45_5 1.21_6 1.13_4 1.11_5 1.08_5. Tabular or pyramidal crystals, also massive, radiating, or granular. Gray, gray-black streak, metallic luster. No cleavage, uneven fracture. $H = 5$, $VHN_{100} = 623$–723. $G = 8.00$. $R = 48.4$–50.8 (540 nm); weakly anisotropic. Occurs in hydrothermal veins with other nickel arsenides and sulfides. Mohawk mine, Keweenaw Co., MI, Gem mine, Fremont Co., CO, Mackinaw mine, Snohomish Co., WA; Cobalt district, ON, in several mines, Jeffrey mine, Asbestos, PQ, Canada; *Eisleben, Saxony,* and Mansfeld, Thuringia, *Germany*; Los Jarales and Vimbodi, Spain; Bou Azzer, Morocco; Talmessi mine, near Anarak, Iran. BM *DI:192, AM 56:203(1973), R:400, P&J:261, ABI:318.*

2.16.17.1 Athabascaite Cu_5Se_4

Named in 1970 for the locality. ORTH Space group unknown. $a = 8.227$, $b = 11.982$, $c = 6.441$, $Z = 4$, $D = 6.63$. *21-1016*: 3.44_3 3.24_{10} 3.02_6 2.00_8 1.89_5 1.82_3 1.66_4 1.37_3. Microscopic laths and anhedral grains. $H = 2\frac{1}{2}$, $VHN_{15} = 78$. $R = 17.8$–25.5 (540 nm); strongly anisotropic. *Martin Lake mine, Uranium City, Lake Athabasca, SK, Canada*; Skrikerum, Kalmar, Sweden; Chaméane, Puy-de-Dôme, France; Petrovice and Predborice deposits, Czech Republic. BM *CM 10:207(1970), R:505, P&J:80, ABI:33.*

2.16.18.1 Penzhinite $Ag_4Au(S,Se)_4$

Named in 1984 for the locality. Contains 3% Cu. HEX $P6_322$. $a = 13.779$, $c = 16.980$, $Z = 18$, $D = 8.35$. *38-398*: 3.38_5 3.04_2 2.71_9 2.59_{10} 2.14_9 2.11_9 1.99_6 1.78_5. Microscopic grains. $R = 31.2$–34.5 (540 nm); anisotropic, cream-yellow to green-gray. *Penzhina R., Kamchatka, Russia.* BM *ZVMO 113:356(1984), AM 70:875(1985), ABI:397.*

2.16.19.1 Palladseite $Pd_{17}Se_{15}$

Named in 1977 for the composition. Contains 4% Cu and 2% Hg. ISO $Pm3m$. $a = 10.606$, $Z = 2$, $D = 8.33$. *29-1437*(syn): 3.35_5 3.20_{10} 2.84_6 2.57_7 2.50_4 2.04_8 1.88_{10} 1.77_4. Small detrital grains. White, metallic luster. $H = 4\frac{1}{2}$–5, $VHN_{100} = 390$–437. $G_{syn} = 8.30$. $R = 45.7$ (540 nm). *Itabira, MG, Brazil.* BM *MM 41:123(1977), AM 62:1059(1977), ABI:379.*

2.16.20.1 Cameronite Cu_7AgTe_{10}

Named in 1986 for Eugene N. Cameron (b.1910), U.S. economic geologist. TET $P4_2/mmc$. $a = 12.695$, $c = 42.186$, $Z = 16$, $D = 7.15$. *40-471*: 3.46_{10} 2.96_2 2.12_{10} 1.82_3 1.80_6 1.38_4 1.22_4 1.15_3. Microscopic anhedral grains. Black, black streak, metallic luster. No cleavage, subconchoidal fracture, brittle. $H = 3\frac{1}{2}$–4, $VHN_{100} = 151$–172. $R = 32.1$–33.8 (540 nm); distinctly anisotropic. *Good Hope mine, Vulcan, CO.* BM *CM 24:379(1986), AM 72:1023(1987), ABI:78.*

2.16.21.1 Patronite VS_4

Named in 1906 for Antenor Rizo-Patron (1866–1948), Peruvian engineer, who discovered the Minasragra deposit. MON $I2/a$. $a = 12.110$, $b = 10.420$, $c = 6.780$, $\beta = 100.8°$, $Z = 8$, $D = 2.83$. *14-179*: 6.3_3 5.47_6 5.15_{10} 3.92_5 3.82_4

3.02_5 2.72_4 1.98_4. Massive, finely granular. Gray-black, metallic luster. H = 2, G = 2.81. R = 20.0–32.8 (540 nm); strongly pleochroic. A major ore of vanadium. *Minasragra, near Cerro de Pasco, Peru.* BM *DI:347, R:887, P&J:290, ABI:390.*

2.16.22.1 Vasilite (Pd,Cu)$_{16}$(S,Te)$_7$

Named in 1990 for Vasil Atanasov (b.1933), Bulgarian mineralogist. ISO $I\bar{4}3m$. $a = 8.922$, Z = 2, D = 8.796. *CM 28*: 3.64_8 2.58_6 2.39_{10} 2.23_6 2.10_8 1.45_8 1.21_7 1.01_7. Small detrital grains. Gray, black streak, metallic luster. No cleavage, uneven fracture, brittle. H = $4\frac{1}{2}$–5, VHN$_{20}$ = 467–506. R = 37.7 (540 nm). In heavy-mineral concentrates, near *Novoseltsi, Bourgas region, Bulgaria.* BM *CM 28:687(1990).*

2.16.23.1 Luberoite Pt$_5$Se$_4$

Named in 1992 for the locality. MON $P2_1/c$. $a = 6.584$, $b = 4.602$, $c = 11.10$, $\beta = 101.6°$, Z = 2, D = 13.02. *31-950*(syn): 5.45_7 3.27_7 2.93_{10} 2.78_9 2.64_6 2.47_6 1.88_9 1.87_{10}. Small prismatic crystals and rounded grains. Dark bronze, metallic luster. VHN$_{15}$ = 461. G$_{syn}$ = 12.79. R = 46.3–52.8 (540 nm); strongly anisotropic. In placers, *Lubero region, Kivu Prov., Zaire.* BM *EM 4:683(1992), AM 78:450(1993).*

Class 3

Sulfosalts

COLUSITE GROUP

The colusite group consists of complex sulfosalts of copper with the general formula
$$Cu_{26}V_2B_6S_{32}$$
where

B = As, Ge, Sn, Sb

and where the space group is $P\bar{4}3n$. The structure is closely related to that of sphalerite.

COLUSITE GROUP

Mineral	Formula	Space Group	a	D	
Colusite	$Cu_{26}V_2As_6S_{32}$	$P\bar{4}3n$	10.62	4.44	3.1.1.1
Germanocolusite	$Cu_{26}V_2(Ge,As)_6S_{32}$	$P\bar{4}3n$	10.57	4.55	3.1.1.2
Nekrasovite	$Cu_{26}V_2(Sn,As)_6S_{32}$	$P\bar{4}3n$	10.73	4.62	3.1.1.3
Stibiocolusite	$Cu_{26}V_2(Sb,Sn,As)_6S_{32}$	$P\bar{4}3n$	10.71	4.66	3.1.1.4

3.1.1.1 Colusite $Cu_{26}V_2As_6S_{32}$

Named in 1933 for the locality. Colusite group. May contain minor Fe, Sb, Sn, and Ge. ISO $P\bar{4}3n$. $a = 10.62$, $Z = 1$, $D = 4.44$. 9-10: 3.07_{10} 2.65_2 1.88_6 1.60_4 1.32_2 1.22_3 1.09_3 1.02_2. Tetrahedral crystals rare; usually massive, granular. Bronze-brown, black streak, metallic luster. No cleavage, brittle. $H = 4\frac{1}{2}$, $VHN_{100} = 322$–379. $G = 4.4$–4.5. $R = 30.0$ (540 nm). *Colusa claim* and other mines in *Butte, MT*; Red Mountain Pass, Ouray Co., and Buffalo Boy mine, near Silverton, CO; Magma mine, Gila Co., and Campbell mine, Bisbee, AZ; Kidd Creek mine, near Timmins, ON, Canada; Chiseuil, Saone-et-Loire, France; Lorano, near Carrara, Italy; Bor, Serbia; Chelopech, Bulgaria; Gay Cu–Zn deposit, southern Ural Mts., Russia; Cerro de Pasco, Peru; Chuquicamata, Antofagasta, Chile. BM *DI:386, R:562, P&J:129, ABI:106, AM 79:750(1994)*.

3.1.1.2 Germanocolusite $Cu_{26}V_2(Ge, As)_6S_{32}$

Named in 1992 for its relationship to colusite. Colusite group. ISO $P\bar{4}3n$. $a = 10.57$, $Z = 1$, $D = 4.55$. *AM 79:* 3.05_{10} 2.64_4 1.87_5 1.60_3 1.32_3 1.21_3 1.08_3 1.02_3. Microscopic grains, usually in bornite. No cleavage. $VHN_{40} = 280$–370. $R = 27.3$–29.5 (546 nm); color yellow, isotropic; no internal reflections. Chelopech, Bulgaria; *Urup, northern Caucasus, Russia*; Maykain, northeastern Kazakhstan; Tsumeb, Namibia. BM *AM 79:387(1994)*.

3.1.1.3 Nekrasovite $Cu_{26}V_2(Sn, As)_6S_{32}$

Named in 1984 for Ivan Y. Nekrasov (b.1929), Russian mineralogist. Colusite group. Contains minor Fe and Sb. ISO $P\bar{4}3n$. $a = 10.73$, $Z = 1$, $D = 4.62$. *41-1410:* 3.41_1 3.09_{10} 2.68_2 1.89_8 1.62_6 1.23_5 1.10_3 1.03_3. Microscopic grains. Pale brown, metallic luster. $H = 4\frac{1}{2}$, $VHN_{20} = 314$. $R = 26.4$ (540 nm). Campbell mine, Bisbee, AZ; Neves-Corvo, Portugal; *Kairagach deposit, Kuramin Mts., Uzbekistan*. BM *MZ 6:88(1984)*, *AM 70:437(1985)*, *ABI:346*.

3.1.1.4 Stibiocolusite $Cu_{26}V_2(Sb, Sn, As)_6S_{32}$

Named in 1992 for its relationship to colusite. Colusite group. ISO $P\bar{4}3n$. $a = 10.71$, $Z = 1$, $D = 4.66$. *DANS 324:* 3.10_{10} 1.89_9 1.61_7 1.23_4 1.09_6 1.03_4. Microscopic anhedral grains. Dark gray, gray streak, metallic luster. No cleavage, conchoidal fracture. $H = 4\frac{1}{2}$, $VHN_{50} = 273$. $R = 29.5$ (580 nm). *Kairagach deposit, Kuramin Mts., Uzbekistan*, in tetrahedrite ore. BM *DANS 324:411(1992)*, *AM 79:186(1994)*.

3.1.2.1 Vinciennite $Cu_{10}Fe_4Sn(As, Sb)S_{16}$

Named in 1985 for Henri Vincienne (1898–1965), French mineralogist. TET $P4_122$. $a = 10.697$, $c = 10.697$, $Z = 2$, $D = 4.25$. *39-327:* 4.37_3 3.09_{10} 2.68_5 1.90_9 1.61_7 1.34_3 1.23_4 1.09_4. Small grains. Orange, metallic luster. Conchoidal fracture, very brittle. $H = 4$, $VHN_{100} = 321$. $R = 26.5$–27.7 (540 nm); weakly anisotropic. Maggie Cu deposit, near Ashcroft, BC, Canada; *Chizeuil, Saone-et-Loire, France*; Neves-Corvo, Portugal; Huaron, Peru. BM *BM 108:447(1985)*, *AM 71:1280(1986)*, *CM 25:227(1987)*, *ABI:559*.

3.1.3.1 Levyclaudite $Pb_8Sn_7Cu_3(Bi, Sb)_3S_{28}$

Named in 1990 for Claude Levy (b.1924), French mineralogist. MON Space group uncertain, probably $A2$. $a = 11.84$, $b = 5.825$, $c = 5.831$, $\beta = 92.6°$, $Z = 1$, $D = 6.04$. *EM 2:* 5.9_3 3.95_{10} 3.72_1 3.17_1 2.96_8 2.81_2 2.37_2 2.07_2. Microscopic lamellae in thin veinlets. Gray, black streak, metallic luster. Cleavage {100}, good. $H = 2\frac{1}{2}$–3, $VHN_{25} = 80$–87. $R = 32.5$–34.0 (546 nm); strongly anisotropic. *Aghios Philippos Pb–Zn deposit, Kirki district, Greece*. BM *EM 2:711(1990)*, *AM 76:2022(1991)*.

3.1.4.1. Cylindrite $Pb_3Sn_4FeSb_2S_{14}$

Named in 1893 for its cylindrical form. TRIC $P\bar{1}$. $a = 11.73$, $b = 5.79$, $c = 5.80$, $\alpha = 90.0°$, $\beta = 92.38°$, $\gamma = 93.87°$, $Z = 2$, $D = 5.43$. *27-246*: 3.9_9 3.85_{10} 3.47_6 3.06_7 2.89_{10} 2.85_7 2.04_7 2.03_7. Concentric tubular shells and aggregates up to 3 cm, also massive. Dark lead-gray, black streak, metallic luster. Cleavage {100}, perfect. $H = 2\frac{1}{2}$, $VHN_{100} = 54$–93. $G = 5.4$–5.5. $R = 32.0$–36.4 (540 nm); distinctly anisotropic. Smirnovsk, Transbaikalia, Russia: *Mina Santa Cruz, Poopo*, and other mines in *Bolivia*. BM *DI:482, R:748, P&J:143, ABI:124.*

3.1.4.2 Franckeite $FePb_5Sn_3Sb_2S_{14}$

Named in 1893 for Carl and Ernest Francke, German mining engineers. TRIC $P\bar{1}$. $a = 46.9$, $b = 5.82$, $c = 17.3$, $\alpha = 90.0°$, $\beta = 94.68°$, $\gamma = 90.0°$, $Z = 8$, $D = 5.88$. *15-25*: 4.30_5 3.44_{10} 3.11_5 2.91_{10} 2.86_{10} 2.82_{10} 2.36_5 2.05_8. Thin tabular crystals; usually massive, radiating or foliated. Dark gray-black color and streak; metallic luster. Cleavage {010}, perfect; flexible, inelastic cleavage flakes. $H = 2\frac{1}{2}$–3. $G = 5.9$. $R = 36.4$–38.2 (540 nm); weakly anisotropic. Thompson mine, Inyo Co., and Kalkar quarry, Santa Cruz Co., CA; Coal R., YT, Canada; Vens Haut, Cantal, France; Smirnovsk, Transbaikalia, and Sinantscha deposit, Sikote-Alin, Russia; Dachang, Guangxi Prov., China; Hoei mine, Oita Pref., Japan; Renison Bell mine, TAS, and Walla Walla mine, Rye Park, NSW, Australia; *Poopo*, Oruro, Llallagua, Huanuni, and other localities in *Bolivia*. BM *DI:448, R:747, P&J:166, ABI:160.*

3.1.4.3 Incaite $Pb_4Sn_4FeSb_2S_{13}$

Named in 1974 for the original inhabitants of the locality. TRIC Space group uncertain, probably $P\bar{1}$. $a = 17.29$, $b = 5.79$, $c = 5.83$, $\alpha = \beta = \gamma = 90.0°$, $Z = 1$, $D = 5.75$. *39-1362*: 4.31_6 3.83_4 3.43_7 3.23_4 3.12_4 2.86_{10} 2.06_4 2.03_8. Microscopic lamellae. Lead-gray, metallic luster. Cleavage {100}, perfect. $R = 29.0$–33.2 (540 nm); distinctly anisotropic. Near Mejillones, Antofagasta, Chile; *Poopo, Bolivia*. BM *NJMM:235(1974), AM 60:486(1975), R:748, ABI:232.*

3.1.4.4 Potosiite $Pb_6Sn_2FeSb_2S_{14}$

Named in 1981 for the locality. TRIC Space group uncertain, probably $P\bar{1}$. $a = 5.915$, $b = 5.938$, $c = 17.239$, $\alpha = 91.63°$, $\beta = 91.02°$, $\gamma = 90.84°$, $Z = 1$, $D = 6.20$. *35-726*: 5.76_1 4.32_3 3.45_9 2.94_1 2.88_{10} 2.16_1 2.07_1 1.92_1. Microscopic crystals in felted masses. Metallic luster. Cleavage {001}, perfect; {010}, good. $H = 2\frac{1}{2}$, $VHN_{100} = 105$. $R = 34.4$–35.1 (589 nm); moderately anisotropic. Herb claim, Turnagain R., BC, Canada; *Andacaba deposit, Potosi, Bolivia*. BM *AM 68:1249(1983), CM 24:45(1986), ABI:420.*

3.1.5.1 Miharaite $Cu_4FePbBiS_6$

Named in 1980 for the locality. ORTH $Pb2_1/m$. $a = 10.854$, $b = 11.985$, $c = 3.871$, $Z = 2$, $D = 6.06$. *33-461:* 3.25_3 3.11_3 3.03_{10} 3.00_7 2.70_3 2.64_3 2.18_5 1.94_7. Microscopic grains. Gray, metallic luster. $H = 4$, $VHN_{25} = 190$–230. $R = 31.7$–32.6 (589 nm); moderately anisotropic. Neves-Corvo, Portugal; Ulsan mine, Kyongsangnam Prov., Korea; *Mihara mine* and Imooka mine, *Okayama Pref., Japan*. BM *AM 65:784(1980), ABI:332.*

3.1.6.1 Billingsleyite $Ag_7(As, Sb)S_6$

Named in 1968 for Paul Billingsley (1887–1962), U.S. geologist, who collected the material. ISO $P2_13$. $a = 10.481$, $Z = 4$, $D = 5.90$. *21-1334*(syn): 6.06_4 3.49_4 3.16_4 3.03_{10} 2.91_5 2.80_7 2.47_7 1.85_5. Fine-grained aggregates. Dark lead-gray, metallic luster. $H = 2\frac{1}{2}$. $G = 5.92$. *North Lily mine, East Tintic district, UT*. BM *AM 53:1791(1968), R:726, ABI:54, CM 28:751(1990).*

3.1.7.1 Arsenpolybasite $(Ag, Cu)_{16}As_2S_{11}$

Named in 1963 as the As analog of polybasite. Contains up to 1% Sb. MON $C2/m$. $a = 26.08$, $b = 15.04$, $c = 23.84$, $\beta = 90°$, $Z = 16$, $D = 6.08$. *37-428* (syn): 6.38_3 6.02_3 3.68_3 3.08_4 3.01_{10} 2.09_3 1.84_6. Thin pseudohexagonal plates. Black, black streak, metallic luster. Cleavage {001}, imperfect. $H = 2$–3. $G = 6.2$. Biaxial (–); $N > 2.72$; $2V = 22°$. Creede, Mineral Co., CO; *Neuer Morgenstern mine, Freiberg, Saxony, Germany*; Quespisiza, Peru. BM *AM 48:565(1963), 52:1311(1967); CM 8:172(1965); ABI:30.*

3.1.7.2 Polybasite $(Ag, Cu)_{16}Sb_2S_{11}$

Named in 1829 from the Greek for *many* and *base*, in allusion to the many metallic bases present. Forms a series with pearceite (3.1.8.1). MON $C2/m$. $a = 26.17$, $b = 15.11$, $c = 23.89$, $\beta = 90.0°$, $Z = 16$, $D = 6.36$. *36-391:* 3.19_4 3.15_{10} 2.98_{10} 2.91_7 2.86_9 2.67_8 2.52_8 2.42_6. Pseudohexagonal plates and massive. Black, black streak, metallic luster. Cleavage {001}, imperfect; uneven fracture. $H = 2$–3. $G = 6.1$–6.2. $R = 30.8$–32.8 (540 nm); moderately anisotropic. Occurs in silver-bearing veins, occasionally an important ore mineral. Leadville, Ouray, Rico(*), and other localities, CO; Tonopah, Goldfield, and the Comstock Lode, NV; Las Chispas mine(*), Arizpe, SON, Guanajuato, GTO, and other localities, Mexico; Sabana Grande(*), Honduras; Andreasberg(*), Harz, and Freiberg(*), Saxony, Germany; Pribram, Czech Republic; Broken Hill, NSW, Australia. BM *DI:351, R:729, P&J:306, ABI:417.*

3.1.8.1 Pearceite $(Ag, Cu)_{16}As_2S_{11}$

Named in 1896 for Richard Pearce (1837–1927), U.S. chemist and metallurgist. Forms a series with polybasite (3.1.7.2). MON $C2/m$. $a = 12.64$, $b = 7.29$, $c = 11.90$, $\beta = 90.0°$, $Z = 2$, $D = 6.07$. *8-130:* 2.97_{10} 2.80_9 2.47_6 2.34_5 2.30_6 2.16_3 1.99_5 1.83_6. Pseudohexagonal plates. Black, black streak, metallic luster. No cleavage, conchoidal to irregular fracture, brittle. H = 3, VHN_{100} = 130–142. G = 6.15. R = 29.3–32.3 (540 nm); moderately anisotropic. *Molly Gibson mine, Aspen, Pitkin Co.*, and Rico, Dolores Co., CO; Neihart, Cascade Co., Flathead mine, Flathead Co., and Drumlummon mine, Marysville, MT; Lakeview district, Bonner Co., ID; Morrison and Ross mines, Holtyre, ON, and Husky mine, YT, Canada; Arizpe, SON, and Veta Rica mine, Sierra Mojada, COAH, and Guanajuato, GTO, Mexico; Sark's Hope mine, Channel Is.; Clara mine, Black Forest, Germany; Francoli, Tarragona Prov., Spain; Seikoshi, Shizuoka Pref., Japan; Arqueros, Chile. BM *DI:353, R:729, P&J:306, ABI:393.*

3.1.8.2 Antimonpearceite $(Ag,Cu)_{16}Sb_2S_{11}$

Named in 1963 as the Sb analog of pearceite. MON $C2/m$. $a = 12.81$, $b = 7.41$, $c = 11.91$, $\beta = 90.0°$, Z = 2, D = 6.40. *36-392*(syn): 3.11_7 2.96_9 2.83_{10} 2.49_6 2.38_5 2.31_4 2.18_4 1.86_5. Pseudohexagonal plates. Black, black streak, submetallic luster. No cleavage, conchoidal to irregular fracture, brittle. H = 3. G = 6.34. North Lily mine, East Tintic district, UT; *Guanajuato, GTO, Mexico*; Clara mine, Black Forest, Germany; Sarrabus, Sardinia, Italy; Seikoshi mine, Shizuoka Pref., Japan; Taltal, Chile. BM *AM 48:565(1963), CM 8:172(1965), ABI:15.*

3.1.9.1 Petrovicite $Cu_3HgPbBiSe_5$

Named in 1976 for the locality. ORTH *Pnam*. $a = 16.176$, $b = 14.684$, $c = 4.331$, Z = 4, D = 7.76. *29-567:* 3.62_7 3.55_8 3.19_8 3.12_{10} 2.96_{10} 2.72_5 2.26_5 2.11_5. Microscopic tabular crystals. Metallic luster. One imperfect cleavage. H = 3, VHN_{25} = 102. R = 45.2–46.5 (540 nm); weakly anisotropic. *Petrovice deposit, Czech Republic.* BM *BM 99:310(1976), AM 62:594(1977), P&J:297, ABI:400.*

3.1.10.1 Benleonardite $Ag_8(Sb,As)Te_2S_3$

Named in 1986 for Benjamin F. Leonard (b.1921), U.S. geologist. TET Space group unknown. $a = 6.603$, $c = 12.726$, Z = 2, D = 7.79. *40-511:* 12.7_7 3.19_3 2.94_{10} 2.86_3 2.61_4 2.33_2 2.16_4 2.12_2. Microscopic laths and anhedral grains. Black, metallic luster. H = 3, VHN_{25} = 105–125. R = 32.2–34.7 (540 nm); moderately to strongly anisotropic. *Moctezuma mine, Moctezuma, SON, Mexico*; Rotgulden, Salzburg, Austria. BM *MM 50:681(1986), AM 73:439(1988), ABI:46, LAP 18(5):18(1993).*

3.1.11.1 Aschamalmite $Pb_6Bi_2S_9$

Named in 1983 for the locality. MON $C2/m$. $a = 13.71$, $b = 4.09$, $c = 31.43$, $\beta = 91.0°$, Z = 4, D = 7.33. *35-612:* 3.53_4 3.43_{10} 3.38_9 3.00_4 2.94_6 2.93_6 2.86_5 2.07_4. Laths up to 5 cm. Lead-gray, lead-gray streak, metallic luster. Cleavage {001}, perfect. H = $3\frac{1}{2}$, VHN_{25} = 150–181. R = 43.4–46.3 (546 nm); moderately anisotropic. Granite Gap, Hidalgo Co., NM; *Ascham Alm, Untersulz-*

bachtal, Austria; Val Basso, Novara, Italy. BM *NJMM:433(1983)*, *AM 69:810(1984)*, *ABI:32*.

3.1.12.1 Tsnigriite Ag$_9$SbTe$_3$(S, Se)$_3$

Named in 1992 from the initials of the Russian name for the Central Scientific-Research Institute of Geological Prospecting in Moscow. MON $P2/m$ or Pm. $a = 8.89$, $b = 8.292$, $c = 19.50$, $\beta = 97.02°$, $Z = 4$, $D = 7.38$. *ZVMO 121*: 4.26$_5$ 3.78$_7$ 2.89$_4$ 2.85$_4$ 2.29$_4$ 2.20$_{10}$ 2.10$_4$ 2.00$_4$. Microscopic anhedral grains, gray. VHN$_{20}$ = 125. R = 32.2–36.6 (546 nm); weak bireflectance, anisotropic in brown and gray. Occurs in the Au–Ag deposit of *Vyskovol'tnoye, Tian Shan, Uzbekistan*, and in polymetallic ore of Bethumy, Rajasthan, India. BM *ZVMO 121(5):95(1992)*, *AM 79:389(1994)*.

3.2.1.1 Enargite Cu$_3$AsS$_4$

Named in 1850 from the Greek for *distinct*, referring to the cleavage. Dimorphous with luzonite. ORTH $Pnm2_1$. $a = 6.431$, $b = 7.402$, $c = 6.149$, $Z = 2$, $D = 4.47$. *35-580*(syn): 3.22$_4$ 3.21$_8$ 3.07$_9$ 2.84$_8$ 1.85$_6$ 1.73$_{10}$ 1.59$_5$ 1.26$_4$. Crystals tabular on {001} or prismatic and typically striated parallel to c; also massive and granular. Twinning common on {320}, sometimes as star-shaped cyclic trillings. Based on hexagonal close-packed S atoms with Cu and As atoms in tetrahedral coordination. Gray-black to iron-black, black streak, metallic luster. Cleavage {110}, perfect; {100}, {010}, distinct; uneven fracture; brittle. H = 3. G = 4.4–4.5. Striated crystals and cleavage are distinctive. Soluble in aqua regia. May contain up to 6% Sb and 3% Fe. R = 25.0–26.4 (540 nm); strongly anisotropic in violet-red and olive-green tones. In vein and replacement deposits, apparently formed at moderate temperatures, in some places in

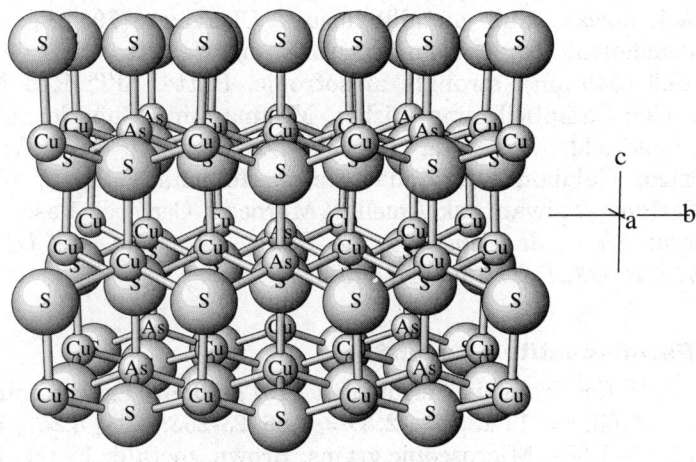

ENARGITE: Cu$_3$AsS$_4$

The enargite structure is derived from that of wurtzite by replacing one-fourth of the Zn atoms with As and three-fourths with Cu.

sufficient quantity to be an important copper ore. Butte(*), MT; Red Mountain district(*), Ouray and San Juan Cos., CO; Tintic, UT; Magma mine, Superior, AZ; Alghero and Calabona, Sardinia, Italy; Matzenkopfl, Brixlegg, Austria; Bor, Serbia; Chinkuashi mine, Keelung, Taiwan; Mancayan, Luzon, Philippines; Tsumeb, Namibia; Mina Luz Angelica, Quiruvilca(*), Morococha(*), and Cerro de Pasco(*), Peru; Sierra de Famatina, La Rioja Prov., Argentina. BM *DI:389, MD 4:72(1969), R:583; P&J:155, ABI:147.*

3.2.2.1 Luzonite Cu_3AsS_4

Named in 1874 for the locality. Dimorphous with enargite; forms a series with famatinite (3.2.2.2). TET $I\bar{4}2m$. $a = 5.290$, $c = 10.465$, $Z = 2$, $D = 4.47$. *10-450* (syn): 3.05_{10} 1.86_9 1.59_7 1.58_6 1.32_5 1.20_6 1.08_6 1.07_5. Small equant crystals twinned on {112}, usually massive, granular. Deep pink-brown, black streak, metallic luster. Cleavage {101}, good; {100}, distinct; conchoidal to uneven fracture; brittle. $H = 3\frac{1}{2}$. $G = 4.4–4.6$. $R = 24.4–27.0$ (540 nm); strongly anisotropic. Butte, MT; Summitville, Rio Grande Co., CO; Goldfield, Esmeralda Co., NV; Bisbee, AZ; Calabona, Sardinia, Italy; Chelopech, Bulgaria; Kadzaharan Cu–Mo deposit, Armenia; Teine mine, Hokkaido, Kasuga and Akeshi mines, Kagoshima Pref., and other localities in Japan; Chinkuashi mine(*), Keelung, Taiwan; *Mancayan, Luzon, Philippines*; Cerro de Pasco and Huaron, Peru; Sierra de Famatina, La Rioja Prov., Argentina. BM *AM 42:766(1957), MD 4:72(1969), R:588, P&J:246, ABI:304.*

3.2.2.2 Famatinite Cu_3SbS_4

Named in 1873 for the locality. Forms a series with luzonite (3.2.2.1). TET $I\bar{4}2m$. $a = 5.385$, $c = 10.748$, $Z = 2$, $D = 4.70$. *35-581*(syn): 3.11_{10} 2.69_2 1.90_7 1.62_4 1.35_1 1.23_2 1.20_1 1.10_3. Massive, granular, or reniform. Deep pink-brown, black streak, dull metallic luster. Cleavage {101}, good; {100}, distinct; conchoidal to uneven fracture; brittle. $H = 3\frac{1}{2}$. $G = 4.5–4.7$. $R = 23.8–25.0$ (540 nm); strongly anisotropic. Butte, MT; Red Mountain, Ouray Co., CO; Campbell mine, Bisbee, Magma mine, Superior, and Tombstone, AZ; Goldfield, Esmeralda Co., NV; Darwin, Inyo Co., CA; Cananea, SON, Mexico; Calabona, Sardinia, Italy; Hokuetsu, Japan; Chinkuashi mine(*), Keelung, Taiwan; Isk Amellal, Morocco; Cerro de Pasco and Morococha, Peru; *Sierra de Famatina, La Rioja, Argentina*. BM *DI:387, AM 42:766(1957), R:588, P&J:160, ABI:153.*

3.2.2.3 Permingeatite Cu_3SbSe_4

Named in 1971 for François Permingeat (1917–1988), French mineralogist. TET $I\bar{4}2m$. $a = 5.63$, $c = 11.23$, $Z = 2$, $D = 5.86$. *25-263:* 5.04_4 3.25_{10} 1.99_9 1.70_8 1.41_5 1.29_6 1.15_7 1.08_5. Microscopic grains. Brown, metallic luster. $H = 4–4\frac{1}{2}$, $VHN_{50} = 234$. $R = 26.2–26.3$ (540 nm); strongly anisotropic. *Predborice, Czech Republic.* BM *BM 94:162(1971), AM 57:1554(1972), R:592, P&J:295, ABI:398.*

3.2.3.1 Sulvanite Cu_3VS_4

Named in 1900 for the composition. Forms a series with arsenosulvanite (3.2.3.2). ISO $P\bar{4}3m$. $a = 5.393$, $Z = 1$, $D = 3.91$. $11\text{-}104$: 5.4_{10} 3.12_5 2.40_4 1.91_8 1.80_3 1.63_4 1.35_3 1.10_4. Cubes to 2.5 cm; usually massive. Gray, sometimes with a bronze tint, black streak, metallic luster. Perfect cubic cleavage. $H = 3\frac{1}{2}$, $VHN_{100} = 135\text{-}157$. $G = 3.9\text{-}4.0$. $R = 31.0$ (540 nm). Thorpe Hills and near Mercur, UT; Ponte Castiola, Corsica, France; near Carrara, Italy; Bor, Serbia; Pay-Khoy, northern Ural Mts., Russia; Karatau Mts., Kazakhstan; *Edelweiss mine*, south of *Burra, SA, Australia*; Kipushi, Shaba, Zaire; Tsumeb, Namibia. BM *DI:384, R:576, P&J:358, ABI:508*.

3.2.3.2 Arsenosulvanite $Cu_3(As,V)S_4$

Named in 1941 for its relationship to sulvanite (3.2.3.1), with which it forms a series. ISO $P\bar{4}3m$. $a = 5.271$, $Z = 1$, $D = 4.33$. $25\text{-}265$: 3.04_{10} 2.77_1 1.87_{10} 1.67_1 1.59_5 1.32_1 1.26_1 1.08_3. Microscopic grains. Bronze-yellow, metallic luster. $H = 3\frac{1}{2}$. $G = 4.0\text{-}4.2$. $R = 31.7$ (540 nm). Bisbee, AZ; Bor, Serbia; Kapan deposit, Armenia; Osarizawa mine, Akita Pref., Japan. BM *ZVMO 70:161(1941); AM 40:368(1955), 46:465(1961), 79:750(1994)*.

3.2.4.1 Stephanite Ag_5SbS_4

Named in 1845 for Archduke Victor Stephan (1817–1867), former mining director of Austria. ORTH $Cmc2_1$. $a = 7.793$, $b = 12.295$, $c = 8.506$, $Z = 4$, $D = 6.43$. $11\text{-}108$: 3.56_3 3.08_{10} 2.89_6 2.78_3 2.58_9 2.19_4 2.13_5 1.83_4. Short prismatic to tabular crystals; also massive. Iron-black, iron-black streak, metallic luster. Cleavage {010} and {021}, imperfect; subconchoidal to uneven fracture; brittle. Crystals twinned on {110}, often repeated to form pseudohexagonal groups. $H = 2\text{-}2\frac{1}{2}$. $G = 6.2\text{-}6.5$. $R = 27.6\text{-}29.0$ (540 nm); strongly anisotropic. Widely distributed as a late-stage mineral in hydrothermal silver deposits. Comstock Lode, NV; Cobalt, ON, and United Keno Hill mines, YT, Canada; Arizpe(*), SON, Fresnillo(*), ZAC, and Guanajuato, GTO, Mexico; Wheal Boys, Endellion, Cornwall, England; Espedalen, Norway; Andreasberg(*), Harz, and *Freiberg*(*), Schneeberg, and Marienberg, *Saxony, Germany*; Pribram(*), Hodrusa, and Jachymov(*), Czech Republic; Sarrabus(*), Sardinia, Italy; Lengenbach, Binn, Switzerland; Ozernoe deposit, Transbaikal, Russia; Altyn-Topkan, Tadjikstan; Broken Hill, NSW, Australia; Azegour, Morocco; San Cristobal, Peru; Colquechaca(*), Bolivia; Chanarcillo, Atacama, Chile. BM *DI:358, R:727, P&J:351, ABI:493*.

3.2.4.2 Selenostephanite Ag_5SbSe_4

Named in 1985 as the Se analog of stephanite. ORTH $P2_12_12_1$. $a = 7.86$, $b = 11.84$, $c = 8.92$, $Z = 4$, $D = 7.82$. $39\text{-}334$: 3.45_6 2.96_{10} 2.64_9 2.52_7 2.28_9 2.23_9 1.92_9 1.89_9. Microscopic laths. Gray, metallic luster. $H = 3$, $VHN_{20} = 95\text{-}116$. $R = 32.6\text{-}34.8$ (580 nm); moderately anisotropic. *Chukotskiy, Siberia, Russia*. BM *ZVMO 114:627(1985), AM 72:225(1987), ABI:469*.

HAUCHECORNITE GROUP

The hauchecornite group minerals are complex nickel sulfides with general formula
$$Ni_9B_2S_8 \quad \text{or a multiple thereof}$$
where
B = As, Sb, Bi, Te

and the structure is tetragonal, $P4/mmm$. The structure consists of double Ni–S ribbons linked by Bi–Ni chains, all parallel to [001].

HAUCHECORNITE GROUP

Mineral	Formula	Space Group	a	c	D	
Hauchecornite	Ni_9BiSbS_8	$P4/mmm$	7.29	5.40	6.58	3.2.5.1
Bismutohauchecornite	$Ni_9Bi_2S_8$	$P4/mmm$	7.37	5.88	6.25	3.2.5.2
Tellurohauchecornite	Ni_9BiTeS_8	$P4/mmm$	14.64	10.87	6.50	3.2.5.3
Arsenohauchecornite	$Ni_{18}Bi_3AsS_{16}$	$I4/mmm$	10.271	10.807	6.52	3.2.5.4
Tučekite	$Ni_9Sb_2S_8$	$P4/mmm$	7.174	5.402	6.14	3.2.5.5

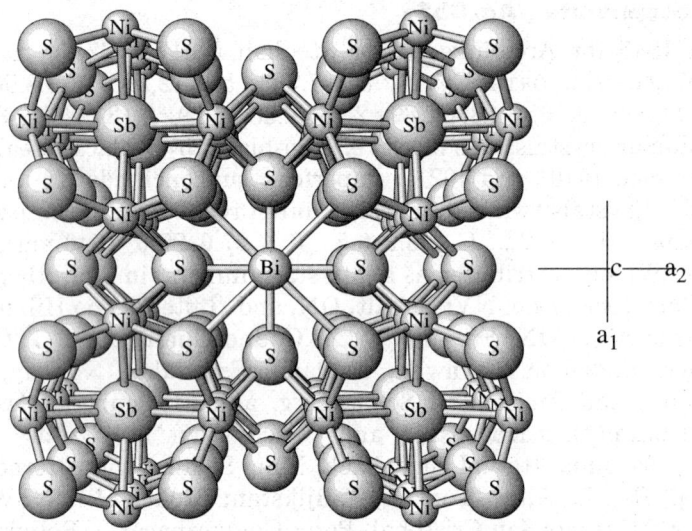

HAUCHECORNITE: Ni_9BiSbS_8

3.2.5.1 Hauchecornite Ni_9BiSbS_8

Named in 1893 for William Hauchecorn (1828–1900), German geologist. Hauchecornite group. TET $P4/mmm$. $a = 7.29$, $c = 5.40$, $Z = 1$, $D = 6.58$. *6-457*: 4.35_5 3.65_4 3.26_4 2.80_{10} 2.39_6 2.30_6 1.87_5 1.82_3. Small tabular and prismatic crystals. Pale bronze-yellow, gray-black streak, metallic luster. No cleavage, conchoidal fracture. H = 5, VHN_{50} = 447–655. G = 6.4–6.5. R = 45.2–48.2 (540 nm); distinctly anisotropic. Zimmer Lake, SK, Canada; *Friedrich mine, Hamm, Westphalia, Germany*; Talnakh, Norilsk, Russia. BM *AM 35:440(1950), R:403, P&J:193, ABI:206*.

3.2.5.2 Bismutohauchecornite $Ni_9Bi_2S_8$

Named in 1980 as the Bi analog of hauchecornite. Hauchecornite group. TET $P4/mmm$. $a = 7.37$, $c = 5.88$, $Z = 1$, $D = 6.25$. Tabular crystals, also massive. Bronze-yellow, metallic luster. VHN = 360–392. Weakly anisotropic. Near Zimmer Lake, SK, Canada; *Oktyabr mine, Norilsk, Russia*; Mihara mine, Okayama Pref., and Tsumo mine, Shimane Pref., Japan. BM *MM 43:873(1980), AM 66:436(1981), ABI:57*.

3.2.5.3 Tellurohauchecornite Ni_9BiTeS_8

Named in 1980 as the Te analog of hauchecornite. Hauchecornite group. TET $P4/mmm$. $a = 14.64$, $c = 10.87$, $Z = 8$, $D = 6.50$. *34-566*: 5.2_3 4.35_4 3.66_4 3.28_4 2.80_{10} 2.41_5 2.31_6 1.87_4. Microscopic grains. Bronze-yellow, metallic luster. $VHN_{50} = 812$–825. R = 43.9–47.7 (546 nm). *Strathcona mine, Sudbury, ON, Canada*. BM *MM 43:877(1980), AM 66:436(1981), ABI:518*.

3.2.5.4 Arsenohauchecornite $Ni_{18}Bi_3AsS_{16}$

Named in 1980 as the As analog of hauchecornite. Hauchecornite group. TET $I4/mmm$. $a = 10.271$, $c = 10.807$, $Z = 2$, $D = 6.52$. *38-346*: 4.33_7 3.63_7 3.24_7 2.77_{10} 2.38_9 2.28_8 1.85_8 1.81_7. Small tabular crystals, also massive. Bronze, gray-black streak, metallic luster. No cleavage, conchoidal fracture. H = $5\frac{1}{2}$, $VHN_{50} = 516$–655. G = 6.35. R = 46.2–47.1 (546 nm). *Vermilion mine, Sudbury, ON, Canada*; Karagaily, Kazakhstan; Tsumo mine, Shimane Pref., Japan. BM *MM 43:877(1980), AM 66:436(1981), CM 27:137(1989), ABI:25*.

3.2.5.5 Tučekite $Ni_9Sb_2S_8$

Named in 1978 for Karel Tuček (1906–1990), Czech mineralogist. Hauchecornite group. TET $P4/mmm$. $a = 7.174$, $c = 5.402$, $Z = 1$, $D = 6.14$. *29-927*: 4.33_7 3.58_5 3.21_6 2.76_{10} 2.38_8 2.28_8 1.85_8 1.79_7. Microscopic grains. Pale brass-yellow, metallic luster. No cleavage, conchoidal fracture, brittle. H = 5–6, $VHN_{20} = 718$. R = 45.4–48.0 (540 nm); strongly anisotropic. Rocheservières, Vendeé, France; Vozhmin massif, Karelia, Russia: *Kanowna* and *Whim Creek, WA, Australia*; Witwatersrand, South Africa. BM *MM 42:278(1978), AM 64:465(1979), ABI:541*.

3.2.6.1 Arcubisite Ag_6CuBiS_4

Named in 1976 for the composition. Microscopic grains; gray, metallic luster. R = 32.1–33.7 (540 nm); distinctly anisotropic. *Ivigtut, Greenland*. BM *LIT 9:253(1976), AM 63:424(1978), ABI:18*.

3.2.7.1 Chalcothallite $(Cu, Fe)_6Tl_2SbS_4$

Named in 1967 for the composition. TET $I4/mmm$. $a = 3.827$, $c = 34.280$, $Z = 2$, $D = 6.74$. $42\text{-}557$: 3.80_7 3.63_5 3.02_7 2.70_5 2.58_5 2.45_{10} 1.91_{10} 1.62_5. Lamellar aggregates to 2 cm. Lead-gray to iron-black, iridescent tarnish, black streak, metallic luster. Cleavage {001}, perfect; {100}, {010}, indistinct. H =2–$2\frac{1}{2}$, $VHN_{50} = 68$–76. $G = 6.6$. $R = 25.4$–29.1 (540 nm); distinctly anisotropic. Near *Taseq Lake, Ilimaussaq Intrusion, Greenland*. BM AM 53:1775(1968), NJMA 138:122(1980), P&J:402, ABI:91.

3.2.8.1 Chaméanite $(Cu, Fe)_4AsSe_4$

Named in 1982 for the locality. Possibly identical with mgriite (3.4.7.2). ISO $I\bar{4}mmm$. $a = 11.039$, $Z = 8$, $D = 6.17$. $35\text{-}524$: 3.19_{10} 1.95_9 1.67_8 1.38_4 1.27_6 1.13_7 1.06_5 0.975_3. Microscopic grains. Gray, metallic luster. H = 4, $VHN_{25} = 247$–292. $R = 27.1$ (540 nm). *Chaméane, Puy-de-Dôme, France*. BM TMPM 29:151(1982), AM 67:1074(1982), P&J:403, ABI:92.

3.2.9.1 Fangite Tl_3AsS_4

Named in 1993 for Jen-Ho Fang (b.1929), U.S. crystal chemist. ORTH $Pnma$. $a = 8.894$, $b = 10.855$, $c = 9.079$, $Z = 4$, $D = 6.185$. AM 78: 3.98_6 3.36_7 3.24_5 2.81_5 2.71_9 2.68_5 2.54_{10} 2.27_6. Subhedral to anhedral grains to 1 mm. Deep red to maroon, orange streak; vitreous to submetallic luster. H = 2–$2\frac{1}{2}$, $VHN_{100} = 61$. $G_{syn} = 6.20$. $R = 24$ (546 nm). BM *Mercur deposit, Tooele Co., UT*; Allchar, Macedonia. AM 78:1096(1993).

3.3.1.1 Jordanite $Pb_{14}As_6S_{23}$

Named in 1864 for H. Jordan (1808–1887), German doctor, who provided the original material. MON $P2_1/m$. $a = 8.92$, $b = 31.88$, $c = 8.457$, $\beta = 117.7°$, $Z = 2$, $D = 6.38$. $21\text{-}466$: 3.70_6 3.53_7 3.50_4 3.37_5 3.18_{10} 3.06_5 3.05_6 2.88_5. Tabular pseudohexagonal crystals to 4 cm, commonly twinned on {001}; also massive. Lead-gray, iridescent tarnish, black streak, metallic luster. Cleavage {010}, perfect; {001} parting; conchoidal fracture; brittle. H = 3, $VHN_{50} = 106$–141. $G = 6.4$. $R = 38.6$–42.2 (540 nm). Edwards mine, Balmat, NY; Keystone mine, Birmingham, PA; Zuni mine, near Silverton, CO; Penberthy Croft mine, St. Hilary, Cornwall, England; Sala, Sweden; Wiesloch, near Heidelberg, Germany; *Lengenbach, Binn, Switzerland*; Beuthen, Upper Silesia, Poland; Horni Benesov, Czech Republic; Carrara, Italy; Sacaramb, Romania; Yunosawa and Okoppe mines, Aomori Pref., Japan; Sidi Bou Aouane and Djebel Hallouf, Tunisia. BM DI:398, R:754, P&J:218, ABI:250.

3.3.1.2 Geocronite $Pb_{14}(Sb, As)_6S_{23}$

Named in 1839 from the Greek for *Earth* and *Saturn*, the alchemical names for antimony and lead. MON $P2_1/m$. $a = 8.96$, $b = 31.85$, $c = 8.48$, $\beta = 118.0°$, $Z = 2$, $D = 6.56$. $30\text{-}691$: 3.71_6 3.54_{10} 3.39_8 3.19_9 3.06_{10} 2.99_7 2.89_{10} 2.24_6. Crystals rare, usually massive, granular. Lead-gray, lead-gray streak, metallic luster. Cleavage {011}, distinct; uneven fracture. H = $2\frac{1}{2}$, VHN_{100} = 144–160. G = 6.3–6.5. R = 38.8–40.1 (540 nm); distinctly anisotropic. Inexco mine, Jamestown, CO; Park City and Tintic, UT; Darwin, Inyo Co., CA; Hemlo deposit, Thunder Bay, ON, Canada; Noche Buena, ZAC, Mexico; Kilbricken mine, Co. Clare, Eire; Treore mine, St. Treath, Cornwall, England; *Sala, Sweden*; Bayerland mine, Oberpfalz, Germany; Pietrasanta, Tuscany, and Salafossa, Tyrol, Italy; Uzunzhal, Kazakhstan; Smirnovsk, Transbaikal, Russia; Rajpura, Rajasthan, India; Huachocolpa, Peru; Xanda mine(*), Virgem de Lapa, MG, Brazil. BM *DI:395, AM 61:963(1976), R:774, P&J:175*.

3.3.2.1 Gratonite $Pb_9As_4S_{15}$

Named in 1939 for Louis C. Graton (1880–1970), U.S. economic geologist. HEX-R $R3m$. a = 17.701, c = 7.792, Z = 3, D = 6.23. *13-446:* 5.45_2 3.78_6 3.72_6 3.43_{10} 2.85_4 2.73_5 2.05_3 1.75_3. Hexagonal prisms to 15 mm, also massive. Dark lead-gray, black streak, metallic luster. No cleavage, brittle. H = $2\frac{1}{2}$, VHN_{100} = 130–146. G = 6.22. R = 34.2–34.8 (540 nm). Pinal del Rio, Cuba; Wiesloch, near Heidelberg, Germany; Blei-Scharley mine, Upper Silesia, Poland; Rio Tinto, Spain; Zvezdel, Bulgaria; *Excelsior mine(*), Cerro de Pasco, Peru*. BM *DI:397, R:757, P&J:186, ABI:193*.

3.3.3.1 Heyrovskyite $Pb_{10}AgBi_5S_{18}$

Named in 1971 for Jaroslav Heyrovsky (1890–1967), Czech chemist. ORTH *Cmcm*. a = 4.110, b = 13.600, c = 30.485, Z = 2, D = 7.11. *25-1404*(syn): 3.75_2 3.54_3 3.41_{10} 3.35_3 2.94_3 2.85_3 2.18_2 2.09_4. Acicular to prismatic crystals, to 20 mm. Tin-white, gray-black streak, metallic luster. Cleavage {001}, poor; conchoidal fracture; brittle. H = $3\frac{1}{2}$–4, VHN_{50} = 166–234. G = 7.17. R = 40.6–44.6 (540 nm); strongly anisotropic. Idarado mine, Ouray Co., CO; near Austin, Lander Co., NV; near Castlegar, BC, Canada; Clara mine, Black Forest, Germany; Rauris Goldberg, Salzburg, Austria; Furka Pass and Goppenstein, Switzerland; *near Hurky, Czech Republic*; Balikesir Balya deposit, Turkey; Spokoinoe and Shumilovsk deposits, Transbaikal, Russia; Juno mine, Tennant Creek, NT, Australia. BM *MD 6:133(1971), R:781, NJMA 127:62(1976), ABI:219, CM 29:553(1991)*.

3.3.4.1 Nuffieldite $Pb_2Cu(Pb, Bi)Bi_2S_7$

Named in 1968 for Edward W. Nuffield (b.1914), Canadian mineralogist. ORTH *Pbnm*. a = 4.61, b = 21.38, c = 4.03, Z = 4, D = 7.04. *19-675:* 4.00_9 3.66_{10} 3.54_{10} 3.16_8 2.87_3 2.54_7 1.87_6 1.35_4. Small prismatic to acicular crystals. Lead-gray, gray-black streak, metallic luster. Cleavage {001}, perfect; uneven fracture; very brittle. H = $3\frac{1}{2}$, VHN = 149–178. G = 7.01. R = 39.0–43.1 (546 nm). *Lime Creek stock, Alice Arm, BC, Canada*; Les Houches, Haute-Savoie, France; Maleevskoe, Rudny Altai, Russia; Akchatau, Kazakhstan. BM *CM 9:439(1968), 32:359(1994); ABI:358*.

3.3.5.1 Meneghinite $Pb_{13}CuSb_7S_{24}$

Named in 1852 for Giuseppi Meneghini (1811–1889), Italian mineralogist. ORTH $Pn2_1m$. $a = 11.343$, $b = 24.028$, $c = 4.126$, $Z = 1$, $D = 6.47$. *29-559:* 4.13_5 3.74_8 3.49_6 3.27_{10} 3.08_3 2.92_9 2.77_3 2.74_4. Slender prismatic crystals, also massive, fibrous to compact. Dark lead-gray, black streak, metallic luster. Cleavage {010}, perfect; conchoidal fracture; brittle. $H = 2\frac{1}{2}$. $G = 6.36$. $R = 40.0–46.0$ (540 nm); strongly anisotropic. Kalkar quarry, Santa Cruz Co., CA; Marble Lake, Barrie Twp., ON, and Bluebird–Mayflower mine, Trail Creek, BC, Canada; Pengenna mine, St. Kew, Cornwall, and Shallowford Bridge, near South Moulton, Devon, England; Vallon de Merlier, Alpes-Maritimes, and Longerey, Savoie, France; Hällefors, Sweden; Aijala, Finland; Ochsenkopf, Saxony, and Goldkronach, Bavaria, Germany; *Bottino mine, near Seravezza, Italy*; Saki and Gouskhalat, Iran; Pundung, Korea; Pingtoushan, Kiangsi, China; Broken Hill, NSW, Australia. BM *DI:402, R:772, P&J:266, ABI:322.*

TETRAHEDRITE GROUP

The tetrahedrite group is made up of isometric sulfides, selenides, and tellurides with the general formula

$$A_{12}B_4X_{13}$$

where

$A =$ Cu, Ag, Fe, Hg, Zn
$B =$ As, Sb, Te
$X =$ S, Se

and the structure is cubic, space group $I\bar{4}3m$. The structure is closely related to that of sphalerite.

Tetrahedrite Group

Mineral	Formula	Space Group	a	D	
Tetrahedrite	$(Cu,Fe)_{12}Sb_4S_{13}$	$I\bar{4}3m$	10.327	5.07	3.3.6.1
Tennantite	$(Cu,Fe)_{12}As_4S_{13}$	$I\bar{4}3m$	10.186	4.61	3.3.6.2
Freibergite	$(Ag,Cu,Fe)_{12}Sb_4S_{13}$	$I\bar{4}3m$	10.554	5.22	3.3.6.3
Hakite	$(Cu,Hg)_{12}Sb_4S_{13}$	$I\bar{4}3m$	10.88	6.30	3.3.6.4
Giraudite	$(Cu,Zn)_{12}Sb_4S_{13}$	$I\bar{4}3m$	10.578	5.75	3.3.6.5
Goldfieldite	$Cu_{12}(Te,As)_4S_{13}$	$I\bar{4}3m$	10.304	4.95	3.3.6.6
Argentotennantite	$(Ag,Cu)_{12}(As,Sb)_4S_{13}$	$I\bar{4}3m$	10.584	5.05	3.3.6.7

3.3.6.1 Tetrahedrite $(Cu,Fe)_{12}Sb_4S_{13}$

Named in 1845 for its crystal form. Tetrahedrite group. Forms a series with tennantite and freibergite. ISO $I\bar{4}3m$. $a = 10.327$, $Z = 2$, $D = 5.07$. *24-*

3.3.6.1 Tetrahedrite

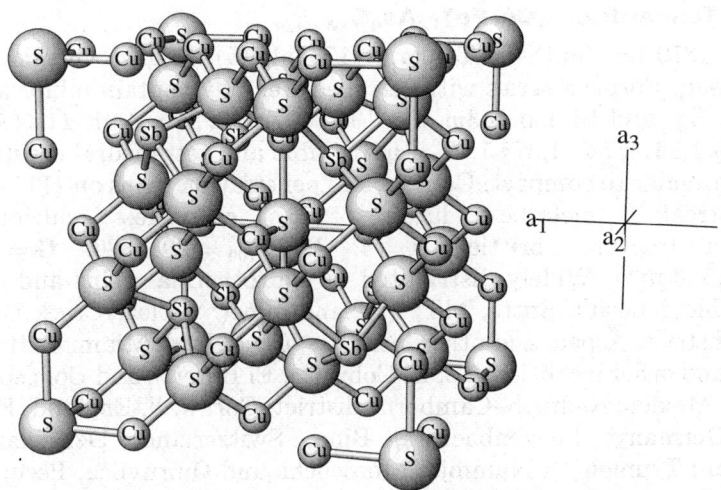

TETRAHEDRITE: $Cu_{12}Sb_4S_{13}$

Tetrahedrite has basically the sphalerite arrangement, with one-fourth of the Cu replaced by Sb. However, one-fourth of the sulfurs are missing from their normal positions, and there are a smaller number of interstitial sulfurs in the Cu–Sb layer—these are at the corners and center of the cell in this view.

1318(syn): 3.65_2 2.98_{10} 2.76_1 2.58_3 2.43_1 2.03_1 1.83_4 1.56_3. **Habit:** Crystals uncommon, typically tetrahedral; usually massive, granular to compact. Contact and penetration twins on {111}. **Structure:** Based on cubic close-packed S atoms with Cu and Fe in octahedral coordination and Sb in tetrahedral coordination. **Physical properties:** Gray to black, brown to black streak, metallic luster. No cleavage, subconchoidal to uneven fracture. H = 3–4, VHN_{100} = 312–351. G = 4.6–5.1. **Tests:** Decomposed by HNO_3 with the separation of sulfur. Crystal form and lack of cleavage typical. **Chemistry:** Forms a series with tennantite and with the Ag analog freibergite. May contain up to 8% Zn, 2% Cd, 17% Hg, and 4% Bi. **Optics:** Isotropic; N > 2.72. R = 32.8 (540 nm). **Occurrence:** Widely distributed in hydrothermal veins and contact-metamorphic deposits; probably the commonest sulfosalt mineral. **Localities:** Only the most significant are given. Butte(*), MT; Park City(*), UT; Tombstone, Globe, and other localities, AZ; Naica(*), CHIH, Noche Buena and Concepcion del Oro, ZAC, Mexico; Herodsfoot mine, Cornwall, England; Allevard(*), Isere, and Irazein, Ariege, France; Dillenberg(*), Hesse, and Freiberg, Saxony, Germany; near Brixlegg(*), Tyrol, Austria; Cavnic(*) and Botes, Romania; Tenes and Mouzaia, Algeria; Mt. Isa, Q, Broken Hill, NSW, and Renison Bell, TAS, Australia; Mercedes mine(*), Huanoco, and Casapalca, Junin, Peru; Oruro, Bolivia; El Teniente mine(*), near Santiago, Chile. **Uses:** An important copper ore, and may carry significant silver values. BM *DI:374, R:562, P&J:139, AM 73:389(1988), ABI:526.*

3.3.6.2 Tennantite (Cu, Fe)$_{12}$As$_4$S$_{13}$

Named in 1819 for Smithson Tennant (1761–1815), English chemist. Tetrahedrite group. Forms a series with tetrahedrite; may contain minor amounts of Zn, Ag, Hg, and Bi. ISO $I\bar{4}3m$. $a = 10.186$, $Z = 2$, $D = 4.61$. *11-102:* 2.94_{10} 2.55_3 2.40_2 1.80_8 1.54_5 1.27_2 1.17_3 1.04_3. Cubic and tetrahedral crystals; also massive, granular to compact. Contact and penetration twins on {111}. Black, black streak, metallic luster. No cleavage, subconchoidal to uneven fracture, brittle. $H = 4$, $VHN_{100} = 294$–380. $G = 4.6$–4.7. $R = 28.5$ (540 nm). Widely distributed in hydrothermal veins and contact-metamorphic deposits. Butte, MT; Freeland mine(*), Clear Creek Co., Central City district, Aspen, and other localities, CO; Globe–Miami district, Cerbat Mts., and other localities, AZ; El Cobre(*), El Bote(!), and Concepcion del Oro, ZAC, Mexico; Redruth–Camborne district, Cornwall, England; Freiberg, Saxony, Germany; Lengenbach(*), Binn, Switzerland; Dzhezkazgan(*), Kazakhstan; Tsumeb(*), Namibia; Morococha and Quiruvilca, Peru; Rancagua(*), Chile. BM *DI:374, R:563, P&J:139, ABI:521.*

3.3.6.3 Freibergite (Ag, Cu, Fe)$_{12}$Sb$_4$S$_{13}$

Named in 1853 for the locality. Tetrahedrite group. Forms a series with tetrahedrite. ISO $I\bar{4}3m$. $a = 10.554$, $Z = 2$, $D = 5.22$. *27-190*(syn): 3.73_4 3.05_{10} 2.64_5 2.07_4 1.86_7 1.71_4 1.59_5 1.08_4. Tetrahedral crystals and massive. Gray to black, black streak, metallic luster. No cleavage. $H = 4$, $VHN_{100} = 263$–340. $R = 32.6$ (540 nm). An important source of silver. Hi-Ho mine, Cobalt, ON, and Keno Hill area, YT, Canada; Tyndrum, Perthshire, Scotland; Sladekarr and Vena, Sweden; Espedalen, Norway; *Freiberg, Saxony, Germany*; Kutna Hora, Czech Republic; Knappenstube mine, Salzburg, Austria; Vratsa region, Bulgaria; Yukhondzha, Yakutia, Russia; Inakuraishi, Koryu, and Sanyu mines, Hokkaido, Japan; Mt. Isa, Q, Australia. BM *DI:374, MD 9:117(1974), MM 50:717(1986), ABI:162.*

3.3.6.4 Hakite (Cu, Hg)$_{12}$Sb$_4$Se$_{13}$

Named in 1971 for Jaroslav Hak (b.1931), Czech mineralogist. Tetrahedrite group. ISO $I\bar{4}3m$. $a = 10.88$, $Z = 2$, $D = 6.30$. *25-297:* 3.14_{10} 2.91_7 2.57_6 2.13_5 1.99_7 1.93_9 1.76_6 1.64_8. Microscopic anhedral grains. Gray-brown, metallic luster. $H = 4\frac{1}{2}$, $VHN_{40} = 306$. $R = 33.4$ (540 nm). Moctezuma, SON, Mexico; *Predborice* and Bukov, *Czech Republic*. BM *BM 94:45(1971), AM 57:533(1972), R:571, P&J:190, ABI:202.*

3.3.6.5 Giraudite (Cu, Zn)$_{12}$Sb$_4$Se$_{13}$

Named in 1982 for Roger Giraud (b.1936), French microprobe expert. Tetrahedrite group. ISO $I\bar{4}3m$. $a = 10.578$, $Z = 2$, $D = 5.75$. *35-525:* 3.05_{10} 2.83_3 2.50_5 2.08_4 1.93_6 1.87_9 1.71_4 1.59_7. Microscopic anhedral grains. Pale gray, metallic luster. $VHN_{25} = 233$–333. $R = 31.7$ (540 nm). *Chaméane, Puy-de-Dôme, France.* BM *TMPM 29:151(1982), AM 67:1074(1982), P&J:404, ABI:185.*

3.3.6.6 Goldfieldite $Cu_{12}(Te, As)_4S_{13}$

Named in 1909 for the locality. Tetrahedrite group. ISO $I\bar{4}3m$. $a = 10.304$, $Z = 2$, $D = 4.95$. *29-531:* 5.15_1 3.64_3 2.97_{10} 2.77_1 2.58_2 2.43_2 1.82_6 1.55_3. Massive crusts. Lead-gray to iron-black, black streak, metallic luster. No cleavage. $H = 3-3\frac{1}{2}$, $VHN_{100} = 291-342$. $R = 30.1$ (540 nm). Butte, MT; Lucky Cuss mine, Tombstone, and Bisbee, AZ; *Mohawk mine* and Claremont mine, *Goldfield, NV*; Kuramin Mts., Uzbekistan; Kawazu mine, Shizuoka Pref., Japan. BM *DI:384, R:571, ABI:190.*

3.3.6.7 Argentotennantite $(Ag, Cu)_{12}(As, Sb)_4S_{13}$

Named in 1986 as the Ag analog of tennantite. Tetrahedrite group. ISO $I\bar{4}3m$. $a = 10.584$, $Z = 2$, $D = 5.05$. *DANS 290:* 3.06_{10} 2.65_3 2.07_2 1.87_8 1.60_4 1.23_2 1.21_2 1.08_2. Microscopic grains. Gray-black, red-brown to black streak, resinous luster. Conchoidal fracture. $H = 3\frac{1}{2}$, $VHN = 285-320$. $R = 30.4$ (540 nm). *Kvartsitoviye Gorky deposit, Kazakhstan.* BM *DANS 290:206(1986), AM 73:439(1988), ABI:22.*

3.3.7.1 Lengenbachite $Pb_{37}Ag_7Cu_6As_{23}S_{78}$

Named in 1905 for the locality. TRIC $P1$ or $P\bar{1}$. $a = 36.92$, $b = 11.51$, $c = 70.20$, $\alpha = \gamma = 90.0°$, $\beta = 92.58°$, $Z = 2$, $D = 5.78$. *14-418:* 9.31_1 7.69_1 6.28_1 4.60_3 3.06_{10} 2.93_2 2.84_9 2.04_2. Thin blades to 4 cm. Lead-gray to gray-black, frequently with blue tarnish, black streak, metallic luster. Cleavage {100}, perfect; flexible; inelastic. $H = 1\frac{1}{2}-2$, $VHN = 35$. $G = 5.80-5.85$. $R = 35.9-37.4$ (540 nm); weakly anisotropic. *Lengenbach, Binn, Switzerland.* BM *DI:398, R:754, ABI:294, NJMA 166:169(1994).*

3.4.1.1 Proustite Ag_3AsS_3

Named in 1832 for Joseph L. Proust (1754–1826), French chemist. Dimorphous with xanthoconite (3.4.2.1). HEX-R $R3c$. $a = 10.79$, $c = 8.69$, $Z = 6$, $D = 5.63$. *11-470*(syn): 5.4_4 3.27_8 3.17_7 3.11_7 2.74_{10} 2.55_8 2.48_9 2.22_4. Pyramidal, rhombohedral, and scalenohedral crystals, commonly twinned on [$10\bar{1}4$] and {$10\bar{1}1$}; also massive, compact. Scarlet-red, darkening on exposure to light, vermilion streak, adamantine luster. Cleavage{$10\bar{1}1$}, distinct; conchoidal to uneven fracture; brittle. $H = 2-2\frac{1}{2}$, $VHN_{25} = 70-105$. $G = 5.6$. $R = 30.0-32.3$ (540 nm); strongly anisotropic. Widely distributed, occasionally an important silver ore. Georgetown and Red Mountain district, CO; Poorman mine, Owyhee Co., ID; Keeley mine(*), Cobalt, ON, Canada; Batopilas(*), CHIH, and Sombrerete, ZAC, Mexico; Ste.-Marie-aux-Mines(*), Haut-Rhin, France; Freiberg(*) and Niederschelma(*), Saxony, Germany; Lengenbach, Binn, Switzerland; Sarrabus(*), Sardinia, Italy; Jachymov(*) and Pribram, Czech Republic; Sacaramb, Romania; Nishizawa mine, Japan; Broken Hill(*), NSW, Australia; Chanarcillo(*), Atacama, Chile. BM *DI:366, R:783, P&J:309, ABI:423.*

3.4.1.2 Pyrargyrite

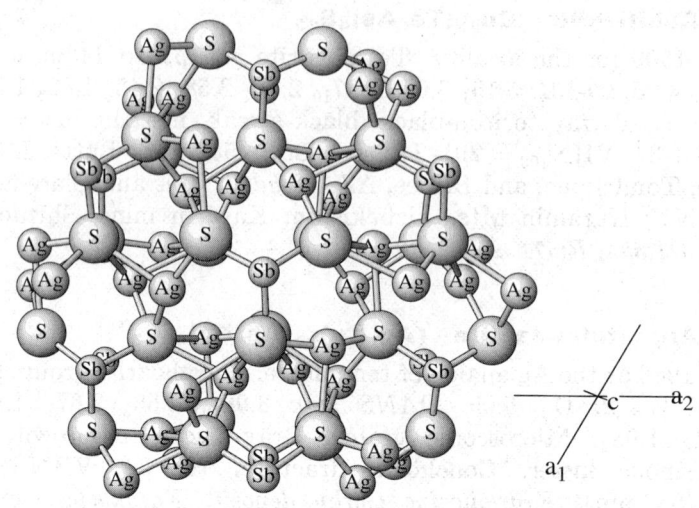

PYRARGYRITE: Ag_3SbS_3

3.4.1.2 Pyrargyrite Ag_3SbS_3

Named in 1831 from the Greek for *fire* and *silver*, referring to its color and composition. Dimorphous with pyrostilpnite (3.4.2.2). HEX-R $R3c$. $a = 11.047$, $c = 8.719$, $Z = 6$, $D = 5.86$. 21-1173(syn): 3.34_6 3.22_7 3.19_6 2.78_{10} 2.57_6 2.54_6 2.27_{22} 1.96_3. Prismatic and scalenohedral crystals; often twinned on $\{10\bar{1}4\}$ and $\{10\bar{1}1\}$; also massive, compact. Deep red, darkening on exposure to light, red streak, adamantine luster. Cleavage $\{10\bar{1}1\}$, distinct; conchoidal to uneven fracture; brittle. $H = 2\frac{1}{2}$. $G = 5.8$–5.9. $R = 28.2$–34.0 (540 nm); strongly anisotropic. An important ore of silver. Poorman mine, Owyhee Co., ID; Comstock Lode(*) and Reese R. district, NV; Cobalt(*), ON, and Beaverdell, BC, Canada; Las Chispas mine(*), Arizpe, SON, Fresnillo(*), ZAC, Guanajuato(*), GTO, and other localities, Mexico; Hiendelaencina(*), Guadalajara, Spain; Espedalen, Norway; Andreasberg(*), Harz, and Freiberg(*), Saxony, Germany; Jachymov and Pribram(*), Czech Republic; Strezhen deposit, Altai, Russia; Broken Hill(*), NSW, Australia; San Cristobal(*), Peru; Colquechaca(*), Bolivia; Chañarcillo(*), Atacama, Chile. BM *DI:362*, *NJMM:181(1966)*, *R:785*, *P&J:309*, *ABI:425*.

3.4.2.1 Xanthoconite Ag_3AsS_3

Named in 1840 from the Greek for *yellow* and *powder*, in allusion to its streak. Dimorphous with proustite (3.4.1.1). MON $C2/c$. $a = 12.00$, $b = 6.21$, $c = 17.08$, $\beta = 110.0°$, $Z = 8$, $D = 5.45$. 21-1455: 5.47_2 4.01_3 3.35_2 3.13_6 3.07_3 3.01_8 2.97_{10} 2.82_7. Tabular crystals twinned on $\{001\}$, also reniform and hemispheric aggregates. Dark red to brown, orange-yellow streak, adamantine luster. Cleavage $\{001\}$, distinct; subconchoidal fracture; brittle. $H = 2$–3, $VHN_{25} = 71$–93. $G = 5.5$–5.6. $R = 23.7$–27.3 (540 nm). Flathead mine, Sanders Co., MT; General Petite mine, Elmore Co., ID; Cobalt, ON, Canada; Batopilas, CHIH, Mexico; Ste.-Marie-aux-Mines, Haut-Rhin, France; Andreasberg, Harz, *Freiberg*, Schneeberg, and other localities in *Saxony, Germany*; Lengenbach, Binn, Switzerland; Jachymov and Pribram, Czech Republic; Baia Sprie, Romania; Nishizawa mine, Japan. BM *DI:371, R:789, P&J:396, ABI:580.*

3.4.2.2 Pyrostilpnite Ag$_3$SbS$_3$

Named in 1868 from the Greek for *fire* and *shining*, in allusion to its color and luster. Dimorphous with pyrargyrite. MON $P2_1/c$. $a = 6.84$, $b = 15.84$, $c = 6.24$, $\beta = 117.15°$, $Z = 4$, $D = 5.98$. *25-1187:* 3.33_4 3.22_{10} 3.12_9 2.86_3 2.84_7 2.64_4 2.62_5 2.42_3. Tabular crystals and sheaflike aggregates. Red, orange-yellow streak, adamantine luster. Cleavage {010}, perfect; conchoidal fracture. $H = 2$, $VHN_{100} = 107$. $G = 5.94$. Biaxial (+); N very high. Bulldog mine, Creede, CO; Silver City district, Owyhee Co., ID; Randsburg, San Bernardino Co., CA; Cobalt, ON, Canada; Hiendelaencina, Guadalajara Prov., Spain; Andreasberg, Harz, *Freiberg, Saxony, Germany*; Pribram and Trebsko, Czech Republic; Kushikino mine, Kagoshima Pref., Japan; Broken Hill, NSW, and Long Tunnel, Heazlewood, TAS, Australia; Chañarcillo, Atacama, Chile. BM *DI:369, R:789, P&J:396, ABI:427.*

3.4.3.1 Seligmanite PbCuAsS$_3$

Named in 1902 for Gustav Seligman (1849–1920), German mineral collector. Forms a series with bournonite (3.4.3.2). ORTH *Pmmn*. $a = 8.134$, $b = 8.710$, $c = 7.634$, $Z = 4$, $D = 5.43$. *11-92:* 5.72_6 4.70_4 3.85_8 2.72_{10} 2.56_5 1.91_5 1.75_7 1.41_5. Prismatic to tabular crystals, usually twinned on {110}. Black, black streak, metallic luster. Cleavage very poor on {001}, {100}, and {010}; conchoidal fracture; brittle. $H = 3$, $VHN_{100} = 168$–181. $G = 5.38$. $R = 32.4$–33.6 (540 nm); strongly anisotropic. Balmat–Edwards mine, Balmat, NY; Butte, MT; Bingham, UT; Hemlo gold deposit, near Marathon, ON, and Whisky Creek mines, near Woodcock, BC, Canada; Wiesloch, near Heidelberg, Germany; *Lengenbach, Binn, Switzerland*; Tsumeb, Namibia; Cerro de Pasco, Peru; Desierto mine, near Iquique, Chile. BM *DI:411, R:754, P&J:340, ABI:470.*

3.4.3.2 Bournonite PbCuSbS$_3$

Named in 1805 for Jacques L. de Bournon (1751–1825), French mineralogist. Forms a series with seligmanite (3.4.3.1). ORTH $Pn2_1/m$. $a = 8.168$, $b = 8.712$, $c = 7.811$, $Z = 4$, $D = 5.84$. *12-94:* 3.91_3 3.84_4 2.74_{10} 2.69_5 1.99_3 1.95_3 1.85_3 1.77_5. Crystals short prismatic to tabular, commonly twinned on {110} forming cruciform or cogwheel aggregates; also massive, granular, or compact. Steel-gray to black color and streak, metallic luster. Cleavage {010}, imperfect; subconchoidal to uneven fracture; brittle. $H = 2\frac{1}{2}$–3, $VHN_{100} = 176$–205. $G = 5.83$. $R = 34.5$–36.3 (540 nm); weakly anisotropic. Widely distributed, sometimes in fine crystals. Park City(*), UT; Mineral King mine, BC,

Canada; Naica, CHIH, and Noche Buena(*), ZAC, Mexico; Herodsfoot mine(*), Liskeard, and *Wheal Boys mine, Endellion, Cornwall, England*; Georg mine(*), near Horhausen, Westerwald, and Clausthal, Neudorf(*), and Wolfsberg(*), Harz, Germany; Pontgibaud, Puy-de-Dôme, and St. Laurent Le Minier(*), Gard, France; Rimska Jama, Bosnia; Trepca, Serbia; Rodna(*), Baia Sprie, Cavnic, and Sacaramb, Romania; Stratoni, Greece; Chichibu mine(*), Saitama, Japan; Broken Hill, NSW, Australia; Zancudo mine, near Medellin, Colombia; Quiruvilca mine(*), La Libertad, Peru; Huancacha, Colquechaca, Machacamarca(*), and Oruro(*), Bolivia. BM *DI:406, ZK 131:397(1970), R:734, P&J:101, AM 62:1097(1977), ABI:65*.

3.4.3.3 Součekite   PbCuBi(S, Se)$_3$

Named in 1979 for Frantisek Souček (1911–1989), Czech mineral collector. ORTH $Pn2_1/m$. $a = 8.153$, $b = 8.498$, $c = 8.080$, $Z = 4$, $D = 7.60$. *33-197*: 4.25_6 4.04_8 3.77_5 2.76_{10} 2.72_{10} 2.02_6 1.83_5 1.78_5. Microscopic anhedral grains and veinlets. Lead-gray, metallic luster. H = $3\frac{1}{2}$, VHN$_{25}$ = 179. Otish Mts. uranium deposit, PQ, Canada; *Oldrichov, near Tachov, Czech Republic*. BM *NJMM:289(1979), AM 65:209(1980), CM 25:625(1987), ABI:486*.

3.4.4.1 Lapieite   CuNiSbS$_3$

Named in 1984 for the locality. ORTH $P2_12_12_1$. $a = 7.422$, $b = 12.508$, $c = 4.900$, $Z = 4$, $D = 4.97$. *37-436*: 3.18_9 2.96_{10} 2.77_2 2.64_2 2.23_2 1.86_6 1.84_9 1.60_3. Microscopic subhedral grains. Gray, metallic luster. Cleavage {100}, fair. H = $4\frac{1}{2}$–5. *Lapie R., St. Cyr Ranges, YT, Canada*. BM *CM 22:561(1984), AM 70:1329(1985), ABI:286*.

3.4.4.2 Mückeite   CuNiBiS$_3$

Named in 1989 for Arno Mücke (b.1937), German mineralogist. ORTH $P2_12_12_1$. $a = 7.509$, $b = 12.551$, $c = 4.877$, $Z = 4$, $D = 6.07$. *NJMM*: 3.18_8 2.98_{10} 2.90_4 2.23_3 2.16_3 2.09_6 1.86_5 1.84_7. Tabular euhedral or subhedral crystals to 1 mm. Pale gray, gray-black streak, metallic luster. Cleavage {010}, very good; {001}, good. H = $3\frac{1}{2}$, VHN = 136–165. G = 5.88. R = 34.1–39.0 (546 nm); moderately anisotropic. *Grüne Au mine, Schutzbach, near Siegen, Germany*. BM *NJMM:193(1989), AM 75:708(1990)*.

AIKINITE GROUP

The members of the aikinite group are isotypic Cu–Pb–Bi sulfosalts and have closely related orthorhombic structures derived from that of bismuthinite, Bi$_2$S$_3$, *Pbnm*, $a = 11.23$, $b = 11.27$, $c = 3.91$. *NJMM:56(1990)*. The general formula is

$$Pb_xCu_xBi_yS_n$$

3.4.5.1 Aikinite

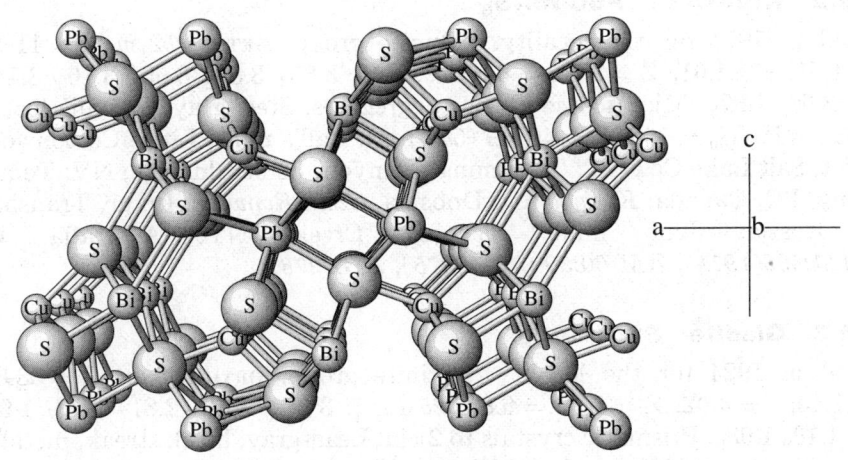

AIKINITE: PbCuBiS$_3$
All coordinations are irregular: Pb 5, Cu 4, and Bi 5 by sulfur.

AIKINITE GROUP

Mineral	Formula	Space Group	a	b	c	D	
Aikinite	PbCuBiS$_3$	Pbnm	11.32	11.64	4.04	7.19	3.4.5.1
Krupkaite	PbCuBi$_3$S$_6$	Pb2$_1$m	11.15	11.51	4.01	6.98	3.4.5.2
Gladite	PbCuBi$_5$S$_9$	Pbnm	33.66	11.45	4.02	6.88	3.4.5.3
Hammarite	Pb$_2$Cu$_2$Bi$_4$S$_9$	Pbnm	33.45	11.58	4.01	7.05	3.4.5.4
Friedrichite	Pb$_5$Cu$_5$Bi$_7$S$_{18}$	Pb2$_1$m	33.84	11.65	4.01	7.13	3.4.5.5
Pekoite	PbCuBi$_{11}$S$_{18}$	Pb2$_1$m	33.50	11.32	3.99	6.80	3.4.5.6
Lindströmite	Pb$_3$Cu$_3$Bi$_7$S$_{15}$	Pbnm	56.07	11.57	4.01	7.03	3.4.5.7

3.4.5.1 Aikinite PbCuBiS$_3$

Named in 1843 for Arthur Aikin (1773–1854), English geologist. Aikinite group. ORTH *Pbnm*. $a = 11.32$, $b = 11.64$, $c = 4.04$, $Z = 4$, $D = 7.19$. *25-310:* 4.06$_5$ 3.67$_{10}$ 3.62$_5$ 3.59$_9$ 3.18$_8$ 3.16$_7$ 2.86$_9$ 2.68$_5$. Prismatic to acicular crystals, also massive. Black, black streak, metallic luster. Cleavage {010}, indistinct; uneven fracture. H = 2–2$\frac{1}{2}$. G = 7.07. R = 46.3–47.3 (540 nm); distinctly anisotropic. Widely distributed in small amounts. Chantilly quarry, Loudoun Co., VA: Ouray, San Juan Co., and Sunnyside mine, Eureka Co., CO; Sells mine, Alta, Salt Lake Co., UT; Bisbee, AZ; near Cucomungo Spring and Sylvania Mts., NV; Owen Lake, BC, Canada; Taxco and Huitzuco, GRO, Mexico; Gardette mine, near Bourg d'Oisans, France; Dobsina, Czech Republic; Pilica, Poland; Zidarova, Bulgaria; Kirki mine, Greece; *Beresovsk, Ural Mts.*, and Inkur, Transbaikal, *Russia*; Karagaily, Kazakhstan; Huancavelica, Peru. BM *DI:412, R:738, P&J:61, AM 61:15(1976), ABI:3*.

3.4.5.2 Krupkaite $PbCuBi_3S_6$

Named in 1974 for the locality. Aikinite group. ORTH $Pb2_1m$. $a = 11.15$, $b = 11.51$, $c = 4.01$, $Z = 2$, $D = 6.98$. *29-563:* 3.65_3 3.59_2 3.55_3 3.16_3 3.14_{10} 2.84_4 2.66_3 1.97_3. Microscopic fibrous aggregates. Steel-gray, metallic luster. $H = 3\frac{1}{2}$, $VHN_{100} = 165$. $R = 46$–49 (546 nm). Ball's mine, Little Cottonwood district, Salt Lake Co., UT; Cucomungo Canyon, Esmeralda Co., NV; Temiskaming, PQ, Canada; *Krupka* and Dobsina, *Czech Republic*; Inkur, Transbaikal, Russia; Juno mine, Tennant Creek, NT, Australia. BM *NJMM:533(1974), AM 60:300,737(1975), ABI:276.*

3.4.5.3 Gladite $PbCuBi_5S_9$

Named in 1924 for the locality. Aikinite group. ORTH *Pbnm*. $a = 33.66$, $b = 11.45$, $c = 4.02$, $Z = 4$, $D = 6.88$. *25-1422:* 3.56_{10} 3.10_7 2.81_7 1.95_7 1.92_7 1.31_9 1.10_9 1.08_9. Prismatic crystals to 2 cm. Lead-gray, black streak, metallic luster. Cleavage {010}, good; {100}, fair. $H = 2$–3, $VHN_{100} = 166$. $G = 6.96$. Comstock mine, Dos Cabezas, Cochise Co., AZ; Tanco pegmatite, Bernic Lake, MB, Canada; Bleka, Telemark, Norway; *Gladhammar, Kalmar, Sweden*; Krupka, Czech Republic; Baita Bihor, Romania. BM *DI:483, AKMG 4:377(1968), ABI:186.*

3.4.5.4 Hammarite $Pb_2Cu_2Bi_4S_9$

Named in 1924 for the locality. Aikinite group. ORTH *Pbnm*. $a = 33.45$, $b = 11.58$, $c = 4.01$, $Z = 4$, $D = 7.05$. *30-179:* 4.03_6 3.65_{10} 3.57_{10} 3.16_8 3.14_{10} 2.85_{10} 2.67_5 2.57_5. Prismatic or acicular crystals. Steel-gray with red tint, black streak, metallic luster. Cleavage {010}, good; flat conchoidal fracture. $H = 3$–4, $VHN_{100} = 243$. $G = 6.73$. $R = 39.3$–47.5 (540 nm). Victoria district, Dona Ana Co., NM; *Gladhammar, Kalmar, Sweden*; Baita, Romania; Inkur deposit, Transbaikal, Russia. BM *DI:442, P&J:191, AM 61:15(1976), ABI:203.*

3.4.5.5 Friedrichite $Pb_5Cu_5Bi_7S_{18}$

Named in 1978 for Otmar M. Friedrich (1902–1991), Austrian mineralogist. Aikinite group. ORTH $Pb2_1m$. $a = 33.84$, $b = 11.65$, $c = 4.01$, $Z = 2$, $D = 7.13$. *29-561:* 4.05_4 3.64_{10} 3.58_{10} 3.16_{10} 2.85_8 2.68_3 2.58_4 1.98_4. Prismatic crystals to 1.5 mm, or granular aggregates. Black, metallic luster. $H = 4$, $VHN_{50} = 224$. $G = 6.98$. $R = 40.1$–46.2 (546 nm); distinctly anisotropic. Fremont mine, east of Hachita, Grant Co., NM; near Panguitch, Garfield Co., UT; *Sedl, east of the Habachtal, Salzburg, Austria*; Inkur, Transbaikal, Russia. BM *CM 16:127(1978), AM 64:654(1979), ABI:164, NJMM:56(1990).*

3.4.5.6 Pekoite $PbCuBi_{11}S_{18}$

Named in 1976 for the locality. Aikinite group. ORTH $Pb2_1m$. $a = 33.50$, $b = 11.32$, $c = 3.99$, $Z = 2$, $D = 6.80$. *29-560:* 3.62_3 3.59_2 3.56_3 3.14_{10} 3.03_4 2.55_2 2.33_2 1.97_3. Prismatic crystals to 2 mm. Lead-gray, lead-gray streak, metallic luster. Cleavage {010}, good. Comstock mine, Dos Cabezas, Cochise Co., AZ; Germania Consolidated mine, Fruitland, Stevens Co., WA; Tanco pegmatite, Bernic Lake, MB, Canada; Baita, Romania; near Narechen, Rhodope Mts., Bulgaria; Kochbulak deposit, Uzbekistan; *Peko mine, Tennant Creek, NT, Australia*. BM *CM 14:322,578(1976), ABI:394*.

3.4.5.7 Lindströmite $Pb_3Cu_3Bi_7S_{15}$

Named in 1924 for Gustav Lindström (1838–1916), Swedish mineral chemist. Aikinite group. ORTH *Pbnm*. $a = 56.07$, $b = 11.57$, $c = 4.01$, $Z = 4$, $D = 7.03$. *42-543:* 4.03_5 3.65_{10} 3.58_6 3.56_8 3.16_7 3.14_9 2.84_9 2.66_4. Striated prismatic crystals to 1 cm. Lead-gray, black streak, metallic luster. Cleavage {100} and {010}, good. $H = 3-3\frac{1}{2}$. $G = 7.01$. $R = 41.1$–45.4 (540 nm). Beaver Mts., Beaver Co., UT; Manhattan, Nye Co., NV; Silver Miller mine, Cobalt, ON, Canada; Yecora, near Iglesia, SON, Mexico; *Gladhammar, Kalmar, Sweden*; Beresovsk district, Ural Mts., and Inkur, Transbaikal, Russia. BM *DI:459, P&J:237, CM 15:527(1977), ABI:296*.

3.4.6.1 Marrite $PbAgAsS_3$

Named in 1904 for John E. Marr (1857–1933), English geologist. MON $P2_1/a$. $a = 7.291$, $b = 12.68$, $c = 5.998$, $\beta = 91.22°$, $Z = 4$, $D = 5.82$. *21-1338:* 3.45_{10} 3.15_3 3.00_7 2.91_4 2.75_{10} 2.05_5 2.01_4 1.00_4. Equant to tabular crystals. Lead-gray to gray-black, brown-black streak, metallic luster. No cleavage, conchoidal fracture, brittle. $H = 3$, VHN = 168. $R = 31.5$–34.0 (546 nm); distinctly anisotropic. *Lengenbach, Binn, Switzerland*. BM *DI:487, SMPM 45:711(1965), R:750, ABI:313*.

3.4.6.2 Freieslebenite $PbAgSbS_3$

Named in 1845 for Johann K. Freiesleben (1774–1846), German mining commissioner. MON $P2_1/a$. $a = 7.54$, $b = 12.82$, $c = 5.89$, $\beta = 92.23°$, $Z = 4$, $D = 6.22$. *10-468:* 3.48_8 3.25_3 2.98_7 2.83_{10} 2.08_4 2.01_3 1.78_5 1.73_3. Prismatic striated crystals to 2 cm. Lead-gray to black, gray streak, metallic luster. Cleavage {110}, imperfect; subconchoidal to uneven fracture. $H = 2-2\frac{1}{2}$. $G = 6.2$. $R = 37.0$–42.3 (540 nm); weakly anisotropic. Garfield mine, Gunnison Co., CO; Castle Dome Mts., Yuma Co., AZ; Cobalt, ON, and Mt. Nansen mines, YT, Canada; Pontgibaud, Puy-de-Dôme, Ussel, Correze, and Vialas, Lozère, France; Hiendelaencina(*), Guadalajara Prov., Spain; Hällefors, Sweden; *Himmelfurst mine(*), Freiberg*, and *Marienberg, Saxony, Germany*; Pribram, Czech Republic; Dossena, Bergamo, Italy; Baia Sprie(*), Romania; Bakadzhik deposit, Bulgaria; Oruro(*), Bolivia. BM *DI:416, R:744, P&J:168, ABI:163*.

3.4.7.2 Mgriite Cu_3AsSe_3

Named in 1982 from the initial letters of the Moscow Geological Exploration Institute. Possibly identical with chaméanite (3.2.8.1). ISO $Fd3m$. $a = 5.530$, $Z = 1$, $D = 5.87$. *35-675:* 3.18_{10} 1.95_{10} 1.67_5 1.38_3 1.27_4 1.13_4 1.06_3 0.978_4.

Massive. Gray, metallic luster. No cleavage. H = $4\frac{1}{2}$, $VHN_{20} = 287$–379. R = 26.9 (540 nm). *Erzgebirge, Saxony, Germany.* BM *ŽVMO 111:215(1982), AM 68:280(1983), ABI:329.*

3.4.8.1 Wittichenite Cu_3BiS_3

Named in 1853 for the locality. ORTH $P2_12_12_1$. $a = 7.68$, $b = 10.33$, $c = 6.70$, $Z = 4$, $D = 6.20$. *9-488:* 4.55_4 3.83_3 3.19_3 3.08_8 2.85_{10} 2.66_4 2.39_3 1.90_3 1.82_3. Acicular crystals and massive. Gray to black, black streak, metallic luster. Conchoidal fracture, brittle. H = 3, $VHN_{100} = 170$–187. G = 6.0–6.2. R = 33.5–35.2 (540 nm); weakly anisotropic. Widely distributed in small amounts. Fairfax quarry, Centerville, VA; Butte, MT; Bisbee, AZ; Cobalt, ON, and Maid of Erin mine, Rainy Hollow district, BC, Canada; Seathwaite Tarn, Cumbria, England; Tjøstulflaten, Telemark, Norway; *Wittichen, Black Forest, Germany*; Mangualde, Portugal; Musomiste, Bulgaria; Akavan deposit, Armenia; Chumauk deposit, Uzbekistan; Strezhan deposit, Transbaikal, Russia; Mt. Gunson mine, Pernatty Lagoon, SA, Australia; Colquijirca, Peru. BM *DI:373, R:721, P&J:391, ABI:577.*

3.4.8.2 Skinnerite Cu_3SbS_3

Named in 1974 for Brian J. Skinner (b.1928), Australian–U.S. economic geologist. MON $P2_1/c$. $a = 7.815$, $b = 10.252$, $c = 13.270$, $\beta = 90.35°$, $Z = 8$, $D = 5.11$. *26-1110*(syn): 3.91_8 3.21_7 3.19_7 3.05_7 2.83_{10} 2.63_9 2.62_9 1.81_9. Microscopic grains. Gray, metallic luster. H = 3, $VHN_{50} = 148$–166. Distinctly anisotropic. Belmont, Nye Co., NV; *Ilímaussaq Intrusion, Greenland*; Grube Clara, Black Forest, Germany; Kosice, Czech Republic. BM *AM 59:889(1974), ABI:478.*

3.4.9.1 Ellisite Tl_3AsS_3

Named in 1979 for Albert J. Ellis (b.1929), New Zealand geochemist. HEX-R $R3m$. $a = 12.324$, $c = 9.647$, $Z = 7$, $D = 7.18$. *35-723:* 5.33_4 3.56_2 3.21_6 2.85_1 2.76_1 2.67_{10} 2.33_3 1.78_2. Rhombohedral crystals to 1.3 mm and small masses. Dark gray, pale brown streak, metallic luster. Good rhombohedral cleavage, hackly fracture. H = 2, $VHN_{50} = 36$–44. $G_{syn} = 7.10$. R = 29.2 (546 nm); strongly anisotropic. *Carlin gold deposit, Eureka Co., NV.* BM *AM 64:701(1979), ABI:144.*

3.4.10.1 Christite $TlHgAsS_3$

Named in 1977 for Charles L. Christ (1916–1980), U.S. mineralogist. MON $P2_1/n$. $a = 6.113$, $b = 16.188$, $c = 6.111$, $\beta = 96.71°$, $Z = 4$, $D = 6.37$. *29-1337*(syn): 4.04_5 3.63_9 3.49_6 3.37_3 2.98_{10} 2.80_3 2.70_5 2.21_4. Subhedral grains to 1 mm. Orange to deep red, orange streak, adamantine luster. Cleavage {010}, perfect; {100}, {001}, {$\bar{1}$01}, good. H = 1–2, $VHN_{10} = 28$–35. $G_{syn} = 6.2$. *Carlin gold deposit, Eureka Co., NV*; Allchar, Macedonia. BM *AM 62:421(1977), ABI:96.*

3.4.10.2 Laffittite AgHgAsS$_3$

Named in 1974 for Pierre Laffitte (b.1925), French geologist. MON Aa.
$a = 7.732$, $b = 11.285$, $c = 6.643$, $\beta = 115.16°$, $Z = 4$, $D = 6.07$. *35-566:* 5.30$_1$
3.63$_2$ 3.51$_{10}$ 3.20$_8$ 3.02$_7$ 2.69$_6$ 2.14$_2$ 1.90$_3$. Anhedral grains to 1 mm. Dark red, red-brown streak, adamantine luster. H = 3, VHN$_{25}$ = 92–138. G$_{syn}$ = 6.11. R = 30.7–32.5 (540 nm); strongly anisotropic. Getchell mine, Humboldt Co., NV; *Jas Roux mine, Hautes-Alpes, France*; Chauvai, Kyrgyzstan. BM *BM 97:48(1974), AM 68:235(1983), R:783, P&J:231, ABI:282.*

3.4.11.1 Routhierite TlCuHg$_2$As$_2$S$_6$

Named in 1974 for Pierre Routhier (b.1916), French economic geologist. TET
I4mm. $a = 9.977$, $c = 11.290$, $Z = 4$, $D = 6.81$. *29-1338:* 4.15$_{10}$ 3.53$_8$ 2.99$_{10}$ 2.50$_9$ 1.87$_7$ 1.83$_6$ 1.76$_7$ 1.52$_6$. Anhedral grains and veinlets. Violet-red, metallic luster. Two perpendicular cleavages. H = $3\frac{1}{2}$, VHN$_{25}$ = 148. R = 28.2–29.3 (540 nm); weakly anisotropic. Hemlo gold deposit, near Marathon, ON, Canada; *Jas Roux mine, Hautes-Alpes, France.* BM *BM 97:48(1974); AM 60:947(1975), 75:935(1990); P&J:330; ABI:447.*

3.4.11.2 Stalderite TlCu(Zn, Fe, Hg)$_2$As$_2$S$_6$

Named in 1993 for Hans-Anton Stalder (b.1925), Swiss mineralogist. TET
I$\bar{4}$2m. $a = 9.865$, $c = 10.938$, $Z = 4$, $D = 4.97$. *SMPM 75:* 4.09$_5$ 3.42$_3$ 2.94$_{10}$ 2.65$_2$ 2.54$_2$ 2.44$_4$ 1.81$_2$ 1.74$_2$. Microscopic pseudoisometric crystals. Black, red-brown streak, metallic luster. No cleavage; uneven to conchoidal fracture. H = $3\frac{1}{2}$–4, VHN$_{10}$ = 135. In reflected light gray-white, very weakly anisotropic, R = 27–28 (540 nm). *Lengenbach, Binn, Switzerland.* BM LAP *18(10):40(1993), AM 78:845(1993), SEPM 75:337(1995).*

3.4.12.1 Samsonite Ag$_4$MnSb$_2$S$_6$

Named in 1910 for the locality. MON $P2_1/n$. $a = 10.92$, $b = 8.082$, $c = 6.682$, $\beta = 92.79°$, $Z = 2$, $D = 5.51$. *11-74:* 6.4$_3$ 5.54$_3$ 3.20$_{10}$ 3.01$_9$ 2.86$_5$ 2.59$_6$ 2.51$_4$ 2.43$_5$. Prismatic crystals to 1 cm. Steel-black, dark red streak, metallic luster. Conchoidal fracture, brittle. H = $2\frac{1}{2}$, VHN$_{100}$ = 187–212. G = 5.51. R = 30.0–31.2 (540 nm); weakly anisotropic. Silver Miller mine(*), Cobalt, ON, Canada; *Samson vein(*), Andreasberg, Harz, Germany.* BM *DI:393, R:790, P&J:334, ABI:460.*

3.4.13.1 Nowackiite Cu$_6$Zn$_3$As$_4$S$_{12}$

Named in 1965 for Werner Nowacki (1909–1989), Swiss mineralogist. HEX-R
R3. $a = 13.44$, $c = 9.17$, $Z = 3$, $D = 4.38$. *25-323:* 3.13$_6$ 1.89$_7$ 1.61$_6$ 1.22$_5$ 1.09$_8$ 1.04$_2$ 1.02$_{10}$. Small equant crystals. Dark gray to black, metallic luster. *Lengenbach, Binn, Switzerland.* BM *AM 51:532(1966), ZK 124:352(1967), R:571, ABI:357.*

3.4.13.2 Aktashite $Cu_6Hg_3As_4S_{12}$

Named in 1968 for the locality. Forms a series with gruzdevite (3.4.13.3). HEX-R $R3$. $a = 13.72$, $c = 9.32$, $Z = 3$, $D = 5.47$. *25-298:* 4.04_2 3.10_{10} 2.69_7 1.90_{10} 1.62_{10} 1.55_2 1.35_3 1.24_4. Microscopic euhedral to anhedral grains. Gray-black, black streak, metallic luster. Conchoidal to uneven fracture, brittle. $H = 4\frac{1}{2}$, $VHN_{50} = 300-346$. $G = 5.5$. Hemlo gold deposit, near Marathon, ON, Canada; Moctezuma, SON, Mexico; Jas Roux mine, Hautes-Alpes, France; *Gal-Khaya deposit, Yakutia, and Aktash deposit, Altai, Russia;* Chauvai, Kyrgyzstan. BM *AM 56:358(1971), 58:562(1973); DANS 251:96(1980); ABI:4.*

3.4.13.3 Gruzdevite $Cu_6Hg_3Sb_4S_{12}$

Named in 1981 for V. S. Gruzdev (1938–1977), Russian mineralogist. Forms a series with aktashite (3.4.13.2). HEX-R $R3$. $a = 13.902$, $c = 9.432$, $Z = 3$, $D = 5.85$. *35-659:* 3.16_{10} 2.12_3 1.93_9 1.65_8 1.36_3 1.25_5 1.11_6 1.05_5. Possibly massive. Gray-black, metallic luster. $H = 4$, $VHN_{30} = 295$. $R = 32.8$ (540 nm); weakly anisotropic. Moctezuma, SON, Mexico; *Chauvai deposit, Kyrgyzstan.* BM *DANS 261:971(1981), AM 67:885(1982), ABI:196.*

3.4.14.1 Galkhaite $(Cs, Tl)(Hg, Cu)_6As_4S_{12}$

Named in 1972 for the locality. ISO $I\bar{4}3m$. $a = 10.365$, $Z = 2$, $D = 5.36$. *35-533:* 7.33_6 4.23_9 2.99_{10} 2.77_8 2.59_4 1.89_2 1.83_5 1.56_3. Cubic crystals to 1 cm, also massive. Dark orange-red, orange-yellow streak, vitreous to adamantine luster. No cleavage; conchoidal to uneven fracture, brittle. $H = 3$, $VHN_{100} = 186-202$. $G = 5.4-5.7$. $R = 24.6$ (540 nm). Getchell mine, Humboldt Co., and Carlin mine, Elko Co., NV; Hemlo gold deposit, near Marathon, ON, Canada; *Gal-Khaya, Yakutia, Russia;* Khaidarkan, Kyrgyzstan. BM *DANS 205:1194(1972), AM 59:208(1974), CM 20:575 (1982), ABI:172.*

LILLIANITE GROUP

The lillianite group minerals are sulfosalts with the general formula

$$A_mB_nS_6 \quad \text{or a multiple thereof}$$

where

$A = Pb, Ag, Mn$
$B = Sb, Bi$

The structures of these rare minerals are complex (see figure). *NJMA 165:331(1993).*

3.4.15.1 Lillianite

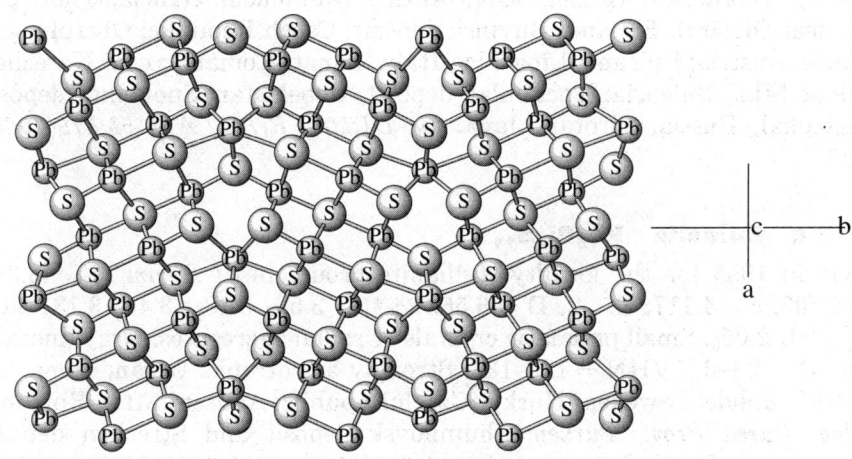

LILLIANITE: $Pb_3Bi_2S_6$

The Pb–Bi sites (Pb and Bi are indistinguishable by X-ray methods) are irregular and may be anywhere from 2- to 8-coordinated, depending on the length cutoff. Pb and Bi must be at least partially disordered and could be completely disordered.

LILLIANITE GROUP

Mineral	Formula	Space Group	a	b	c	D	
Lillianite	$Pb_3Bi_2S_6$	$Bbmm$	13.522	20.608	4.112	7.14	3.4.15.1
Bursaite	$Pb_5Bi_4S_{11}$	$Bbmm$	13.399	20.505	4.117	6.56	3.4.15.2
Gustavite	$PbAgBi_3S_6$	$Bbmm$	13.510	20.169	4.092	7.01	3.4.15.3
Andorite	$PbAgSb_3S_6$	$Pmma$	12.994	19.148	4.303	5.41	3.4.15.4
Uchucchacuaite	$Pb_3AgMnSb_5S_{12}$	$Pmmm$	12.67	19.32	4.38	5.57	3.4.15.5
Ramdohrite	$Pb_6Ag_3Sb_{11}S_{24}$	$Pmma$	12.99	19.21	4.29	5.59	3.4.15.6
Roschinite	$Pb_2Ag_5Sb_{13}S_{24}$	$Pnma$	12.946	19.048	4.236	5.36	3.4.15.7
Fizelyite	$Pb_{14}Ag_5Sb_{21}S_{48}$	$Pnmm$	13.167	19.246	8.700	5.24	3.4.15.8

3.4.15.1 Lillianite $Pb_3Bi_2S_6$

Named in 1889 for the Lillian mine, Leadville, CO, the discredited type locality. Lillianite group. Dimorphous with xilingolite. ORTH $Bbmm$. $a = 13.522$, $b = 20.608$, $c = 4.112$, $Z = 4$, $D = 7.14$. $29\text{-}763$(syn): 3.52_{10} 3.42_7 3.01_6 2.91_8 2.78_6 2.16_5 2.07_6 1.78_7. Prismatic crystals rare; usually granular

massive or radiating fibers. Steel-gray, black streak, metallic luster. Cleavage
{100}, good; conchoidal fracture. H = 2–3. G = 7.0–7.2. R = 51.5 (540 nm);
distinctly anisotropic. Near Minersville, Beaver Co., UT; Twin Lakes, Fresno
Co., CA; Yecora, near Iglesia, SON, Mexico; Montenem, Galicia, Spain; Jilijärvi, near Orijärvi, Finland; Buvinsk deposit, Czech Republic; Oberpinzgau,
Salzburg, Austria; Vulcano, Lipari Is., Italy; Banat, Romania; near Narechen,
Rhodope Mts., Bulgaria; Kochbulak deposit, Uzbekistan; Spokoinoe deposit,
Transbaikal, Russia; Cirotan, Java. BM $DI:404$, $R:781$, AM $54:579(1969)$,
$ABI:295$.

3.4.15.2 Bursaite $Pb_5Bi_4S_{11}$

Named in 1955 for the locality. Lillianite group. ORTH $Bbmm$. $a = 13.399$,
$b = 20.505$, $c = 4.117$, $Z = 2$, $D = 6.56$. $25\text{-}431$: 3.56_6 3.48_{10} 3.40_9 3.13_7 3.00_5
2.92_4 2.90_4 2.05_6. Small prismatic crystals or radial aggregates. Gray, metallic
luster. H = $2\frac{1}{2}$–$3\frac{1}{2}$, VHN = 126–134. Strongly anisotropic. Organ, Dona Ana
Co., NM; Boliden, Sweden; Hurky, Czech Republic; Apuseni Mts., Romania;
Uludag, Bursa Prov., Turkey; Shumilovsk deposit and Strezhen deposit,
Transbaikal, Russia. BM AM $41:671(1956)$, $57:328(1972)$; $NJMA$
$158:293(1988)$; $ABI:73$.

3.4.15.3 Gustavite $PbAgBi_3S_6$

Named in 1970 for Gustav Hageman (1842–1916), Danish chemical engineer.
Lillianite group. ORTH $Bbmm$. $a = 13.510$, $b = 20.169$, $c = 4.092$, $Z = 4$,
$D = 7.01$. $24\text{-}143$: 3.98_5 3.64_8 3.40_8 3.38_8 3.36_{10} 3.00_{10} 2.90_{10} 2.75_8. Tabular
crystals to 2 mm. Gray, metallic luster. Two distinct cleavages. H = $3\frac{1}{2}$,
$VHN_{100} = 175$–218. R = 42–46 (546 nm); distinctly anisotropic. Silver Bell
mine, Red Mt., Ouray Co., and Old Lout mine, San Juan Co., CO; South
Mt., Owyhee Co., ID; Darwin, Inyo Co., and Randsburg, San Bernardino
Co., CA; Tanco pegmatite, Bernic Lake, MB, and Terra mine, Camsell R.,
NWT, Canada; *Ivigtut, Greenland*; Rotgülden, Salzburg, Austria; Monteneme
deposit, Galicia, Spain; Kti-Teberda deposit, Caucasus Mts., Russia. BM CM
$10:173(1970)$, $13:411(1975)$, $ABI:200$.

3.4.15.4 Andorite $PbAgSb_3S_6$

Named in 1894 for Andor von Semsey (1833–1923), Hungarian mineral
collector. Lillianite group. ORTH $Pmma$. $a = 12.994$, $b = 19.148$, $c = 4.303$,
$Z = 4$, $D = 5.41$. $35\text{-}596$: 3.76_6 3.44_7 3.30_{10} 3.02_6 2.91_9 2.75_7 2.28_6 2.07_5. Tabular and prismatic crystals to 3 cm, also massive. Dark gray, black streak,
metallic luster. No cleavage, smooth conchoidal fracture, brittle. H = 3–$3\frac{1}{2}$.
G = 5.35. R = 36.3–40.4 (540 nm); weakly anisotropic. Keyser mine, Morey
district, Nye Co., NV; Bear Basin, King Co., WA; Thompson mine, Darwin,
Inyo Co., CA; near Takla Lake, BC, and near Nansen Creek, YT, Canada;
Les Farges mine(*), near Ussel, Correze, and Bournac, Montagne Noire,
France; Pribram, Czech Republic; *Baia Sprie*(*), *Romania*; Llallagua,

Itos(*), and San Jose(*) mines, Oruro, and Tatasi mine, Potosi, Bolivia. BM
DI:457, R:742, P&J:68, ABI:12.

Quatrandorite and Senandorite
Incompletely characterized minerals occurring in microscopic intergrowths
with andorite. BM NJMM:175(1984), AM 70:219(1985), ZK 180:141(1987).

3.4.15.5 Uchucchacuaite $Pb_3AgMnSb_5S_{12}$

Named in 1984 for the locality. Lillianite group. ORTH $Pmmm$. $a = 12.67$,
$b = 19.32$, $c = 4.38$, $Z = 2$, $D = 5.57$. 38-369: 3.80_3 3.49_3 3.30_{10} 2.90_8 2.75_3
2.29_1 2.19_1 2.08_3. Subhedral to anhedral grains. Lead-gray, metallic luster.
$H = 3\frac{1}{2}$, $VHN_{100} = 168$. $R = 34.1$–43.3 (540 nm). *Uchuc–Chacua deposit, Cajatambo Prov., Peru.* BM BM 107:$597(1984)$, AM 70:$1332(1985)$, ABI:547.

3.4.15.6 Ramdohrite $Pb_6Ag_3Sb_{11}S_{24}$

Named in 1930 for Paul Ramdohr (1890–1985), German mineralogist. Lillianite group. ORTH $Pmma$. $a = 12.99$, $b = 19.21$, $c = 4.29$, $Z = 2$, $D = 5.59$. 25-459: 3.82_2 3.48_3 3.32_{10} 3.04_3 2.94_6 2.78_5 2.21_5 2.09_2. Long prismatic crystals to 1 cm, twinned on {010}. Gray-black, gray-black streak, metallic luster. Uneven fracture, brittle. $H = 2$, $VHN_{20} = 206$. $G = 5.43$. $R = 36.0$–40.9 (540 nm); moderately anisotropic. Bear Basin, King Co., WA; Round Valley mine, Bishop Creek, Inyo Co., CA; Alyaskitovoye deposit, Yakutia, Russia; *Colorado vein, Chocaya mine, Potosí, Bolivia.* BM DI:450, R:742, $P\&J$:164, AM 76:$2020(1991)$, ABI:431.

3.4.15.7 Roshchinite $Pb_2Ag_5Sb_{13}S_{24}$

Named in 1990 for Yuri V. Roshchin (1934–1979), Russian geologist. Lillianite group. ORTH $Pnma$. $a = 12.946$, $b = 19.048$, $c = 4.236$, $Z = 2$, $D = 5.36$. 42-1422: 3.24_6 2.85_8 2.70_6 2.10_3 2.05_3 1.99_3 1.87_5 1.77_{10}. Short prismatic crystals to 4 mm. Silver-gray to lead-gray, pale steel-gray streak, metallic luster. Very brittle. $VHN_{15} = 95$. $G = 5.27$. $R = 35.7$–39.7 (560 nm); distinctly anisotropic. *Kvartsitovyye Gorki deposit, Kazakhstan.* BM $DANS$ 312:$197(1990)$, AM 77:$450(1992)$.

3.4.15.8 Fizelyite $Pb_{14}Ag_5Sb_{21}S_{48}$

Named in 1926 for Sandor Fizely (1856–1918), Hungarian mining engineer, who discovered the mineral. Lillianite group. ORTH $Pnmm$. $a = 13.167$, $b = 19.246$, $c = 8.700$, $Z = 2$, $D = 5.24$. 23-753: 3.79_3 3.47_5 3.32_{10} 3.29_6 2.94_6 2.88_3 2.78_3 2.18_3. Small striated prisms, twinned on {010}. Steel-gray to dark lead-gray, dark gray streak, metallic luster. Cleavage {010}, very brittle. $H = 2$. $G = 5.56$. $R = 38.0$–44.9 (540 nm); distinctly anisotropic. Morey, Nye Co., NV; Les Farges mine, Corrèze, and Bournac, Montagne Noire, France; Pribram, Czech Republic; *Herja* and *Baita, Romania*; Inakuraishi mine, Hokkaido, Japan. BM $DI:450$, $R:742$, $P\&J:164$, $NJMM:175(1984)$, $ABI:158$.

3.4.16.1 Xilingolite $Pb_3Bi_2S_6$

Named in 1982 for the locality. Dimorphous with lillianite (3.4.15.1). MON $C2/m$. $a = 13.65$, $b = 4.078$, $c = 20.68$, $\beta = 93.0°$, $Z = 4$, $D = 7.07$. $22\text{-}651$(syn): 3.72_3 3.38_{10} 3.32_3 3.08_3 2.92_4 2.84_5 2.17_5 2.08_4. Prismatic crystals to 8 mm, twinned on {001}. Lead-gray, gray streak, metallic luster. $H = 3$, $VHN = 103$. $G = 7.08$. $R = 41.8$–44.5 (546 nm); distinctly anisotropic. *Chaobuleng district, Xilingola League, Inner Mongolia, China.* BM $AM\ 69:409(1984)$, $ABI:581$.

3.4.17.1 Kirkiite $Pb_{10}Bi_3As_3S_{19}$

Named in 1985 for the locality. HEX $P6_322$. $a = 8.69$, $c = 26.06$, $Z = 2$, $D = 6.88$. $39\text{-}339$: 3.65_5 3.48_6 3.26_{10} 3.07_7 2.85_6 2.19_5 2.00_4 1.82_5. Microscopic euhedral to subhedral crystals. Gray, metallic luster. $H = 3$, $VHN_{100} = 150$. Weakly anisotropic. *Aghios Philippos deposit, near Kirki, Greece.* BM $BM\ 108:667(1985)$, $AM\ 71:1278(1986)$, $ABI:265$.

3.4.18.1 Erniggliite $Tl_2SnAs_2S_6$

Named in 1992 for Ernst Niggli (b.1917), Swiss mineralogist. HEX-R $P\bar{3}$. $a = 6.680$, $c = 7.164$, $Z = 1$, $D = 5.24$. $SMPM\ 72$: 4.51_4 3.34_{10} 3.06_5 3.03_6 2.68_5 2.25_4 1.93_4 1.87_4. Microscopic hexagonal crystals. Steel-gray to black, red-black streak, metallic luster. Cleavage {0001}, good. $H = 2$–3, $VHN_{10} = 48$. $R = 26.8$–27.7 (543 nm). *Lengenbach, Binn, Switzerland.* BM $SMPM\ 72:293(1992)$.

3.5.1.1 Neyite $Pb_7(Cu, Ag)_2Bi_6S_{17}$

Named in 1969 for Charles S. Ney (1918–1975), Canadian geologist. MON $C2/m$. $a = 37.5$, $b = 4.07$, $c = 41.6$, $\beta = 96.80°$, $Z = 8$, $D = 7.11$. $23\text{-}1156$: 3.72_{10} 3.51_{10} 3.39_3 2.92_{10} 2.27_5 2.08_4 2.04_6 1.46_4. Prismatic to bladed crystals. Lead-gray, metallic luster. Conchoidal fracture, brittle. $H = 2\frac{1}{2}$. $G = 7.02$. Moderately anisotropic. *Johnny Lyon Hills, near Benson, Cochise Co., AZ; Lime Creek deposit, Kitsault, near Alice Arm, BC, Canada.* BM $CM\ 10:90(1969)$, $AM\ 55:1444(1970)$, $ABI:348$.

3.5.2.1 Boulangerite $Pb_5Sb_4S_{11}$

Named in 1837 for Charles L. Boulanger (1810–1849), French mining engineer. MON $P2_1/a$. $a = 21.56$, $b = 23.51$, $c = 8.09$, $\beta = 100.7°$, $Z = 8$, $D = 6.19$. $18\text{-}688$: 3.91_2 3.73_{10} 3.67_2 3.43_2 3.22_5 3.03_4 2.82_4 2.69_3. Prismatic to needlelike crystals, sometimes forming rings, and as fibrous masses. Lead-gray, brown to brown-gray streak, metallic luster. Cleavage {100}, good; flexible in thin crystals; brittle. $H = 2\frac{1}{2}$–3. $G = 6.0$–6.3. $R = 38.0$–40.3 (540 nm); distinctly anisotropic. Widely distributed in small amounts.

Augusta Mt., Gunnison Co., CO; Iron Mountain mine, Mineral Co., MT; Coeur d'Alene, ID; Echo district, Union Co., NV; Cleveland mine, Stevens Co., WA; Madoc, ON, Canada; Noche Buena, ZAC, Mexico; Sala, Västmannland, and Boliden, Västerbotten, Sweden; Wolfsberg, Harz, and Waldsassen, Bavaria, Germany; *Molières, Gard, France*; Pribram, Czech Republic; Bottino, Tuscany, Italy; Trepca, Serbia; Ciprovici, Bulgaria; Broken Hill, NSW, Australia; Llallagua, Bolivia. BM *DI:420, R:770, P&J:99, ABI:64*.

3.5.2.2 Falkmanite
Probably identical with boulangerite (3.5.2.1). BM *NJMM:498(1989)*.

3.5.3.1 Sterryite $Pb_{10}Ag_2(Sb,As)_{12}S_{29}$
Named in 1967 for T. Sterry Hunt (1826–1892), Canadian mineralogist. ORTH *Pbam*. $a = 28.48$, $b = 42.62$, $c = 4.090$, $Z = 4$, $D = 5.91$. *20-562*: 4.14_5 3.68_9 3.54_6 3.26_{10} 2.97_6 2.84_7 2.35_6 2.05_6. Fibrous bundles and anhedral grains. Black, black streak, metallic luster. Cleavage parallel to c, perfect. $H = 3\frac{1}{2}$. Strongly pleochroic. *Madoc, ON, Canada*. BM *CM 9:191(1967), MR 13:95(1982), ABI:495*.

3.5.4.1 Diaphorite $Pb_2Ag_3Sb_3S_8$
Named in 1871 for the Greek for *difference*; resembling but distinct from freieslebenite. MON $C2_1/a$. $a = 15.849$, $b = 32.084$, $c = 5.901$, $\beta = 90.17°$, $Z = 8$, $D = 6.02$. *9-126*: 3.30_{10} 2.95_4 2.89_2 2.81_8 2.03_4 2.00_2 1.76_3 1.70_3. Prismatic crystals. Steel-gray, metallic luster. No cleavage, subconchoidal to uneven fracture, brittle. $H = 2\frac{1}{2}$–3, $VHN_{100} = 219$. $G = 6.04$. $R = 37.3$–40.7 (540 nm); strongly anisotropic. Wood River district, Blaine Co., ID; near Morey, Nye Co., NV; Lake Chelan district, Chelan Co., WA; Catorce, SLP, Mexico; Fournial, Cantal, and Pontgibaud, Puy-de-Dôme, France; Freiberg, Saxony, Germany; *Pribram* and Kutna Hora, *Czech Republic*; Baia Sprie, Romania; Mangazeika and Bulatsk deposits, Yakutia, Russia; Zancudo, Colombia; Machacamarca, Bolivia. BM *DI:414, R:745, P&J:146, ABI:132*.

3.5.5.1 Schirmerite $Pb_{3-6}Ag_3Bi_{7-9}S_{18}$
Named in 1874 for J. H. L. Schirmer, superintendent of the U.S. Mint, Denver, CO. ORTH *Bbmm*. $a = 13.45$, $b = 44.4$, $c = 4.02$, $Z = 8$, $D = 7.58$. *25-760*: 3.21_9 2.92_9 2.88_7 2.84_7 2.79_7 2.38_7 2.28_7 2.04_{10}. Lath-shaped crystals, also massive, granular. Lead-gray to iron-black, metallic luster. No cleavage, uneven fracture, brittle. $H = 2$, VHN = 150–206. $G = 6.74$. $R = 42.4$–47.7 (540 nm); distinctly anisotropic. *Treasury mine, Park Co*., Old Lout mine, near Ouray, Lake City, Hinsdale Co., and Magnolia district, Boulder Co., *CO*; Darwin, Inyo Co., CA; Corrie Buie, Perthshire, Scotland; Montenem, Galicia, Spain; Zambaraks deposit, eastern Karamazar, Tajikstan. BM *DI:424, R:746, P&J:338, NJMA 131:56(1977), ABI:464*.

3.5.6.1 Ourayite $Pb_4Ag_3Bi_5S_{13}$

Named in 1977 for the locality. ORTH $Bbmm$. $a = 13.457$, $b = 44.042$, $c = 4.100$, $Z = 4$, $D = 7.11$. *33-1188:* 3.43_{10} 3.33_6 2.96_9 2.85_6 2.09_9 2.04_7 1.79_7 1.32_5. Microscopic laths. Black, metallic luster. Strongly anisotropic. *Old Lout mine, Ouray, CO*; Pitiquito, SON, Mexico; Ivigtut, Greenland. BM *MJMA 131:56(1977), AM 64:243(1979), CM 22:565(1984), ABI:371*.

3.5.7.1 Madocite $Pb_{17}(Sb, As)_{16}S_{41}$

Named in 1967 for the locality. ORTH $Pba2$. $a = 27.2$, $b = 34.1$, $c = 4.06$, $Z = 2$, $D = 5.98$. *20-567:* 3.87_5 3.67_7 3.40_{10} 3.36_9 3.11_4 3.03_4 2.93_6 2.72_8. Small elongate grains. Gray-black, gray-black streak, metallic luster. Cleavage {010}, perfect; conchoidal fracture. $H = 3$, $VHN_{50} = 155$. $R = 42.3$ (546 nm); strongly anisotropic. *Madoc, ON, Canada*; Jas Roux, Hautes-Alpes, France; Boliden, Västerbotten, Sweden; Novoye, Khaidarkan, Kirgizstan. BM *CM 9:7(1967), AM 53:1421(1968), MR 13:94(1982), ABI:306*.

3.5.8.1 Wallisite $PbTl(Cu, Ag)As_2S_5$

Named in 1965 for the locality. TRIC $P\bar{1}$. $a = 7.980$, $b = 8.979$, $c = 7.761$, $\alpha = 106.11°$, $\beta = 114.48°$, $\gamma = 65.48°$, $Z = 2$, $D = 5.71$. *27-279:* 3.72_4 3.62_4 3.50_3 3.33_8 3.31_{10} 2.86_8 2.81_7 2.64_5. Crystals to 1 mm, usually massive. Lead-gray, metallic luster. Cleavage {001}, good. *Lengenbach, Binn, Canton Wallis, Switzerland*. BM *AM 54:1497(1969), ZK 127:349(1968), ABI:568*.

3.5.8.2 Hatchite $PbTl(Ag, Cu)As_2S_5$

Named in 1912 for Frederick H. Hatch (1864–1932), English geologist. TRIC $P\bar{1}$. $a = 8.038$, $b = 9.167$, $c = 7.807$, $\alpha = 105.23°$, $\beta = 113.62°$, $\gamma = 64.82°$, $Z = 2$, $D = 5.81$. *25-463:* 4.54_3 4.28_3 3.75_2 3.68_2 3.57_4 3.40_8 3.31_{10} 2.86_9. Equant crystals to 1 mm. Lead-gray to gray-black, chocolate-brown streak, metallic luster. *Lengenbach, Binn, Switzerland*. BM *DI:487, SMPM 45:702(1965), ZK 125:249(1967), ABI:205*.

3.5.9.1 Cosalite $Pb_2Bi_2S_5$

Named in 1868 for the locality. ORTH $Pbnm$. $a = 19.09$, $b = 23.89$, $c = 4.058$, $Z = 8$, $D = 7.13$. *13-502:* 4.45_1 4.10_1 3.96_1 3.44_{10} 3.37_3 2.96_2 2.81_3 1.91_2. Prismatic crystals, often acicular; usually massive or as fibrous aggregates. Steel-gray to lead-gray, black streak, metallic luster. No cleavage, uneven fracture. $H = 2\frac{1}{2}$–3. $G = 6.9$–7.0. $R = 41.5$–44.5 (540 nm); weakly anisotropic. Widely distributed in small amounts. Red Mountain district, San Juan Co., CO; Magma mine, Superior, and Landsman Camp, Gila Co., AZ; Deer Park, Stevens Co., WA; Darwin, Inyo Co., CA; Boston Creek, McElry Twp., and Cobalt, ON, Canada; *Nuestra Senora mine, Cosala, SIN, Mexico*; Carrock Fell mine, Cumbria, England; Bjelkes mine, Nordmark, and Boliden, Västerbotten, Sweden; Iilijärvi, Finland; Hurky and Zlata Hory, Czech Republic; Vaskö, Hungary; Baita and Moravicza, Romania; Trepca, Serbia;

Goce-Delav, Bulgaria; Kara Oba(*), Kazakhstan; Ustarasai, Uzbekistan; Strezhen deposit, Altai, Russia; near Oberon, NSW, Australia; Amparindravato, Madagascar. BM *DI:445*, *R:778*, *P&J:131*, *ABI:110*.

3.5.9.2 Veenite $Pb_2(Sb, As)_2S_5$

Named in 1967 for R. W. van der Veen (1883–1925), Dutch economic geologist and metallographer. ORTH $P2_1cn$. $a = 4.22$, $b = 26.2$, $c = 7.90$, $Z = 4$, $D = 5.96$. *20-560:* 3.81_{10} 3.42_8 3.26_8 3.23_5 3.03_9 2.93_5 2.83_5 2.76_7. Flattened prismatic crystals; also massive, granular. Steel-gray, black streak, metallic luster. No cleavage, conchoidal fracture, very brittle. $H = 3\frac{1}{2}$, $VHN_{100} = 155$. $G = 5.92$. Moderately anisotropic. *Madoc, ON, Canada*; Huachocolpa, Huancavelica, Peru. BM *CM 9:7(1967)*, *AM 53:1422(1968)*, *ABI:556*.

3.5.9.3 Dufrénoysite $Pb_2As_2S_5$

Named in 1845 for Pierre A. Dufrénoy (1792–1857), French mineralogist. MON $P2_1/m$. $a = 8.41$, $b = 25.85$, $c = 7.88$, $\beta = 90.5°$, $Z = 8$, $D = 5.62$. *10-453:* 3.74_{10} 3.56_5 3.40_5 3.21_6 3.00_9 2.90_5 2.70_8 2.36_6. Tabular crystals to 3 cm. Lead-gray, chocolate-brown streak, metallic luster. No cleavage, parting {010}, conchoidal fracture. $H = 3$, $VHN_{100} = 135$–146. $G = 5.60$. $R = 36.3$–39.6 (540 nm); strongly anisotropic. Silver Star claims, near Fruitland, Stevens Co., WA; Hemlo gold deposit, near Marathon, ON, Canada; *Lengenbach and Reckibach, Binn, Switzerland*. BM *DI:442*, *SMPM 45:732(1965)*, *R:752*, *P&J:151*, *ABI:140*.

3.5.10.1 Owyheeite $Pb_{10}Ag_3Sb_{11}S_{28}$

Named in 1921 for the locality. ORTH $Pnam$. $a = 22.82$, $b = 27.20$, $c = 8.19$, $Z = 4$, $D = 6.44$. *5-510:* 3.49_7 3.37_4 3.25_{10} 2.90_5 2.84_6 2.23_5 2.13_3 2.05_6. Small acicular crystals, also massive to coarsely fibrous. Gray to black, red-brown streak, metallic luster. Cleavage {001}, very brittle. $H = 2\frac{1}{2}$, $VHN_{100} = 107$. $G = 6.2$–6.5. $R = 38.0$–44.6 (540 nm); strongly anisotropic. Domingo and Garfield mines, Gunnison Co., CO; *Poorman mine, Silver City district, Owyhee Co., ID*; Kaiser Tunnel, Morey, Nye Co., NV; Alma property and Rambler mine, Slocan mining district, BC, Canada; near Yecora, SON, Mexico; Bourneix, Haute-Vienne, France; Kutna Hora, Czech Republic; Rajpura–Dariba, Rajasthan, India; Rivertree, NSW, and Meerschaum mine, near Omeo, VIC, Australia; Roc-Blanc, Morocco; Marmato, Colombia. BM *DI:423*, *AM 34:398(1949)*, *R:745*, *P&J:285*, *TMPM 32:271(1984)*, *ABI:372*.

3.5.11.1 Imhofite $Tl_6As_{15}S_{25}$

Named in 1965 for Josef Imhof (1902–1969), Swiss mineral collector. MON $P2_1/n$. $a = 8.755$, $b = 24.425$, $c = 5.739$, $\beta = 108.28°$, $Z = 1$, $D = 4.39$. *30-1343:* 3.94_9 3.62_7 3.56_8 3.25_6 3.16_7 2.85_8 2.68_6 2.67_{10}. Microscopic plates and aggregates. Copper-red, adamantine luster. $H = 2$. Strongly anisotropic.

Lengenbach, Binn, Switzerland. BM $AM\ 51:531(1966)$, $ZK\ 144:323(1976)$, $ABI:229$.

3.5.12.1 Izoklakeite $Pb_{26}(Ag, Cu, Fe)_2(Sb, Bi)_{20}S_{57}$

Named in 1986 for the locality. ORTH $Pnnm$. $a = 33.88$, $b = 38.02$, $c = 4.070$, $Z = 2$, $D = 6.51$. $36\text{-}415$: 3.77_3 3.40_{10} 3.31_4 3.16_3 2.88_4 2.15_6 2.04_4 1.75_4. Acicular aggregates to 1 mm. Lead-gray, gray-black streak, metallic luster. Cleavage {001}, good; conchoidal fracture; brittle. $H = 4$, $VHN_{50} = 150\text{--}212$. $G = 6.47$. $R = 40.6\text{--}45.2$ (540 nm); distinctly anisotropic. *Izok Lake, NWT, Canada*; Vena mine, Bergslagen, Sweden; Turtschi, Binn, Switzerland. BM $CM\ 24:1(1986)$, $AM\ 72:821(1987)$, $ABI:245$.

3.5.12.2 Giessenite $Pb_{26}(Cu, Ag, Fe)_2(Bi, Sb)_{20}S_{57}$

Named in 1963 for the locality. MON $P2_1/n$. $a = 34.34$, $b = 38.05$, $c = 4.06$, $\beta = 90.33°$, $Z = 2$, $D = 6.93$. $38\text{-}445$: 3.91_6 3.79_6 3.65_6 3.61_6 3.43_{10} 3.40_9 2.03_7 1.75_6. Microscopic acicular crystals. Gray-black, gray-black streak, metallic luster. $H = 2\frac{1}{2}\text{--}3$, $VHN = 65$. $R = 34.0$ (549 nm); weakly anisotropic. Bjorkåsen deposit, Ofoten, Norway; Vena mine, Bergslagen, Sweden; *Turtschi, Giessen, near Binn, Switzerland*; Otome mine, Yamanashi Pref., Japan. BM $SMPM\ 43:471(1963)$, $CM\ 24:19(1986)$, $ABI:184$.

3.5.13.1 Jaskólskiite $Pb_{2-x}Cu_x(Sb, Bi)_{2-x}S_5$ ($x = 0.2$)

Named in 1984 for Stanislaw Jaskólski (1896–1981), Polish mineralogist. ORTH $Pbnm$. $a = 11.331$, $b = 19.871$, $c = 4.100$, $Z = 4$, $D = 6.50$. $38\text{-}363$: 3.78_4 3.71_{10} 3.60_5 3.33_6 2.97_8 2.76_6 2.75_5 2.05_5. Small aggregates. Lead-gray, dark gray streak, metallic luster. $H = 4$, $VHN_{100} = 165\text{--}179$. $R = 36.5\text{--}42.0$ (540 nm); strongly anisotropic. *Vena mine, Bergslagen, Sweden*; Izok Lake, NWT, Canada. BM $CM\ 22:481(1984)$, $ZK\ 171:179(1985)$, $ABI:248$.

3.5.14.1 Zoubekite $Pb_4AgSb_4S_{10}$

Named in 1986 for Vladimir Zoubek (b.1903), Czech geologist. ORTH Space group unknown. $a = 18.698$, $b = 6.492$, $c = 4.577$, $Z = 1$, $D = 5.21$. $39\text{-}342$: 3.67_4 3.25_4 3.07_5 2.39_6 2.22_8 1.96_5 1.80_{10} 1.33_8. Microscopic laths. Steel-gray, black streak, metallic luster. No cleavage, uneven fracture. $H = 3\frac{1}{2}$, $VHN_{15} = 154\text{--}170$. $R = 37.6\text{--}43.1$ (540 nm); strongly anisotropic. *Pribram, Czech Republic*. BM $NJMM:1(1986)$, $AM\ 72:227(1987)$, $ABI:587$.

3.5.15.1 Watanabeite $Cu_4(As, Sb)_2S_5$

Named in 1993 for Takeo Watanabe (1907–1986), Japanese mineralogist. ORTH Space group unknown. $a = 14.51$, $b = 13.30$, $c = 17.96$, $Z = 16$, $D = 4.66$. MM57: 4.43_1 3.36_1 3.00_{10} 2.59_2 2.24_1 1.98_1 1.83_4 1.56_2. Microscopic anhedral grains. Silvery lead-gray, lead-gray streak, metallic luster. No cleavage, uneven fracture, brittle. $H = 4\frac{1}{2}$, $VHN_{100} = 253\text{--}306$. $G = 4.66$.

$R = 31.1$–32.0 (546 nm); weakly anisotropic. *Teine mine, Sapporo, Hokkaido, Japan*, in a hydrothermal vein. BM *MM 57:643(1993)*.

3.6.1.1 Proudite $Pb_{15}Cu_2Bi_{19}(S,Se)_{44}$

Named in 1976 for J. S. Proud (b.1907), Australian mining director. MON $C2/m$. $a = 31.96$, $b = 4.12$, $c = 36.69$, $\beta = 109.52°$, $Z = 2$, $D = 7.08$. *30-491:* 3.90_4 3.83_5 3.49_7 3.45_4 3.22_5 2.96_{10} 2.91_4 2.07_6. Acicular crystals or irregular laths. Silver-gray, metallic luster. Two good cleavages at 40°. $VHN_{50} = 38$–57. Strongly anisotropic. Janos, CHIH, Mexico; *Juno mine, Tennant Creek, NT, Australia*. BM *AM 61:839(1976)*, *ABI:422*.

3.6.2.1 Eskimoite $Pb_{10}Ag_7Bi_{15}S_{36}$

Named in 1977 for the Eskimos, the natives of Greenland. MON *Am*. $a = 30.194$, $b = 4.100$, $c = 13.459$, $\beta = 93.35°$, $Z = 1$, $D = 7.10$. *33-1187:* 3.53_3 3.36_{10} 2.96_5 2.87_6 2.08_4 2.05_5 1.75_5 1.67_4. Small lamellar grains, twinned on {010}. Black, metallic luster. $H = 4$, $VHN_{50} = 162$–223. $R = 43.7$–46.5 (546 nm); distinctly anisotropic. Flathead mine, Niarada, MT; Manhattan, Nye Co., NV; *Ivigtut, Greenland*; Montenem, Galicia, Spain; Rauris Goldberg, Salzburg, Austria; Kochbulak deposit, Uzbekistan. BM *NJMA 131:56(1977)*, *AM 64:243(1979)*, *ABI:151*.

3.6..3.1 Treasurite $Pb_6Ag_7Bi_{15}S_{32}$

Named in 1977 for the locality. MON $B2/m$. $a = 26.538$, $b = 4.092$, $c = 13.349$, $\beta = 92.77°$, $Z = 1$, $D = 7.06$. *33-1189:* 3.63_5 3.49_{10} 3.22_8 2.93_5 2.86_5 2.82_5 1.99_6 1.96_6. Small granular aggregates. Black, metallic luster. *Treasury mine, Geneva district, CO*; 10 km SW of Tyrone, Grant Co., NM; Kochbulak deposit, Uzbekistan. BM *NJMA 131:56(1977)*, *AM 64:243(1979)*, *ABI:535*.

3.6.4.1 Playfairite $Pb_{16}Sb_{18}S_{43}$

Named in 1967 for John Playfair (1748–1819), Scottish scientist. MON $P2_1/m$. $a = 45.12$, $b = 4.139$, $c = 21.45$, $\beta = 92.53°$, $Z = 4$, $D = 5.64$. *20-563:* 3.98_4 3.49_4 3.39_{10} 3.32_{10} 2.97_4 2.79_7 2.09_6 1.77_4. Small tabular crystals. Lead-gray to black, black streak, metallic luster. Cleavage {100}, perfect. $H = 3\frac{1}{2}$, $VHN_{50} = 154$. $R = 35.4$–39.2 (589 nm); strongly pleochroic. *Madoc, ON, Canada*; Novoye, Khaidarkan, Kyrgyzstan. BM *CM 9:191(1967)*, *AM 53:1424(1968)*, *ABI:412*.

3.6.5.1 Cannizzarite $Pb_4Bi_6S_{13}$

Named in 1925 for Stanislao Cannizzaro (1826–1910), Italian chemist. MON $P2_1/m$. $a = 4.13$, $b = 4.09$, $c = 15.48$, $\beta = 98.56°$, $Z = 1$, $D = 6.95$. *9-357:* 3.82_{10} 3.49_3 3.01_6 2.87_5 2.68_6 2.22_5 2.03_5 1.91_4. Thin laths and felted masses. Pale gray, metallic luster. $H = 3$, $VHN_{20} = 132$. $G = 6.7$. $R = 49.2$–51.6 (540 nm); strongly anisotropic. Landsman Camp, Graham Co., AZ; Santa Maria, Grisons, and Goppenstein, Valais, Switzerland; *Vulcano, Lipari Is., Italy*; Shumilkovsk deposit, Transbaikal, and Vysokogorsk deposit, Far Eastern Region, Russia. BM $DI:472$; AM $38:536(1953)$, $72:229(1987)$; $ABI:80$.

3.6.5.2 Wittite $Pb_3Bi_4(S,Se)_9$

Named in 1924 for Th. Witt (1847–1908), Swedish mining engineer. MON $P2/m$. $a = 4.19$, $b = 4.08$, $c = 15.56$, $\beta = 101.4°$, $Z = 1$, $D = 6.91$. 25-460: 3.63_3 3.38_{10} 2.99_6 2.90_9 2.75_3 2.13_4 2.04_8 1.75_3. Massive. Lead-gray, black streak, metallic luster. One good cleavage. $H = 2$–$2\frac{1}{2}$, $VHN_{25} = 45$–63. $G = 7.12$. $R = 42.4$–44.0 (540 nm); distinctly anisotropic. Middlemarch Canyon, Cochise Co., AZ; Falun, Sweden; Nevskoe deposit, Magadan district, Russia. BM $DI:451$, $P\&J:392$, AM $65:789(1980)$, $ABI:578$, $TMPM$ $46:137(1992)$.

3.6.6.1 Launayite $Pb_{22}Sb_{26}S_{61}$

Named in 1967 for Louis de Launay (1860–1938), French economic geologist. MON $C2/m$. $a = 42.58$, $b = 4.015$, $c = 32.16$, $\beta = 102.00°$, $Z = 2$, $D = 5.98$. 20-568: 4.17_8 3.97_3 3.45_{10} 3.40_6 2.92_8 2.84_5 2.75_5 2.01_7. Microscopic grains. Lead-gray, black streak, metallic luster. Cleavage {100}, {001}, perfect. $H = 3\frac{1}{2}$, $VHN = 179$. $R = 36.9$–43.8 (546 nm); strongly anisotropic. Madoc, ON, Canada. BM CM $9:191(1967)$, AM $53:1423(1968)$, $ABI:288$.

3.6.7.1 Jamesonite $Pb_4FeSb_6S_{14}$

Named in 1825 for Robert Jameson (1774–1854), Scottish mineralogist. Forms a series with benavidesite. MON $P2_1/a$. $a = 15.65$, $b = 19.03$, $c = 4.03$, $\beta = 91.8°$, $Z = 2$, $D = 5.71$. 13-461: 4.06_3 3.82_3 3.70_4 3.43_{10} 3.08_3 2.81_4 2.71_4 1.91_3. Acicular crystals, as radiating and fibrous masses. Lead-gray, lead-gray streak, metallic luster. Cleavage {001}, good; brittle. $H = 2\frac{1}{2}$, $VHN_{100} = 66$–86. $G = 5.5$–5.7. $R = 34.0$–39.3 (540 nm); strongly anisotropic. Widely distributed in small amounts. Slate Creek, Shoshone Co., ID; St. Paul mine, Monashee Mt., BC, Canada; Noche Buena(*) and Noria mine, ZAC, Mexico; Endellion, Cornwall, England; Sala, Västmannland, Sweden; Clausthal, Harz, and Freiberg, Saxony, Germany; Pribram, Czech Republic; Aranyidka and Baia Sprie, Romania; Trepca(*), Serbia; Balkumeisk deposit, Yakutia, Russia; Dachang, Guangxi, China; Chichibu mine, Saitama, Japan; Rivertree, NSW, Australia; Oruro(*), Bolivia. BM $DI:451$, $R:765$, $P\&J:217$, CM $25:667(1987)$, $ABI:247$.

3.6.7.2 Benavidesite $Pb_4(Mn,Fe)Sb_6S_{14}$

Named in 1982 for Alberto Benavides (b.1920), Peruvian mining engineer. Forms a series with jamesonite. MON $P2_1/a$. $a = 15.74$, $b = 19.14$, $c = 4.06$, $\beta = 91.50°$, $Z = 2$, $D = 5.60$. 35-546: 4.10_3 3.85_2 3.45_{10} 3.17_2 3.10_2 2.96_2 2.83_4 2.74_3. Microscopic acicular crystals and rounded grains, also massive. Lead-gray, dark gray-brown streak, metallic luster. $H = 2\frac{1}{2}$, $VHN_{15} = 97$. $R = 39.0$–42.4 (540 nm); strongly anisotropic. Sätra mine, near Doverstorp, Södermanland, Sweden; Dachang, Guangxi, China; Uchucchacua, Cajatambo

Prov., Peru. BM *BM 105:166(1982), AM 68:280(1983), CM 25:667(1987), ABI:44.*

3.6.8.1 Dadsonite $Pb_{21}Sb_{23}S_{55}Cl$

Named in 1969 for Alexander S. Dadson (1906–1958), Canadian mineralogist. MON Space group unknown. $a = 19.05$, $b = 4.11$, $c = 17.33$, $\beta = 96.33°$, $Z = 0.5$, $D = 5.51$. *21-942:* 4.10_4 3.78_7 3.62_6 3.38_{10} 2.84_7 2.79_6 2.22_4 2.07_4. Minute fibrous crystals. Lead-gray, black streak, metallic luster. $H = 2\frac{1}{2}$, $VHN_{15} = 226$–279. $G = 5.68$. $R = 39.1$–43.9 (540 nm); distinctly anisotropic. Red Bird mine, Pershing Co., NV; Madoc, ON, and *Giant property, Yellowknife, NWT, Canada*; Saint-Pons, Provence, France; Wolfsberg, Germany. BM *MM 37:437(1969), CM 17:601(1979), ABI:125.*

3.6.9.1 Parajamesonite $Pb_4FeSb_6S_{14}$

Named in 1947 as a dimorph of jamesonite. Probably ORTH Space group unknown. *41-1401:* 4.67_4 4.21_{10} 3.78_6 3.29_1 3.01_1 2.49_2 2.23_2 2.02_2. Columnar crystals to 8 mm. Gray, metallic luster. $G = 5.48$. *Herja, Romania.* BM *SMPM 27:183(1947), AM 34:133(1949), ABI:384.*

3.6.10.1 Eclarite $Pb_9(Cu, Fe)Bi_{12}S_{28}$

Named in 1984 for Eberhard Clar (b.1904), Austrian mineralogist. ORTH *Pnma*. $a = 54.76$, $b = 4.030$, $c = 22.75$, $Z = 4$, $D = 6.88$. *35-627:* 3.49_4 3.41_{10} 3.01_6 2.89_7 2.74_4 2.14_5 2.04_5 2.01_8. Fan-shaped aggregates of acicular crystals to 1.5 cm, and granular. Tin-white, metallic luster. Cleavage {010}, distinct. $H = 2\frac{1}{2}$, $VHN_{50} = 87$–191. $G = 6.85$. $R = 42.9$–47.2 (546 nm); distinctly anisotropic. *Bärenbad, west of Hollersbachtal, Salzburg, Austria.* BM *TMPM 32:103,259(1984), AM 70:215(1985), ABI:143.*

3.6.11.1 Vikingite $Pb_8Ag_5Bi_{13}S_{30}$

Named in 1977 for the Vikings, who settled Greenland in the tenth century. MON $C2/m$. $a = 13.603$, $b = 25.248$, $c = 4.112$, $\beta = 95.55°$, $Z = 1$, $D = 6.94$. *43-690:* 3.62_4 3.40_{10} 3.05_4 2.91_5 2.10_4 2.06_8 1.88_4 1.75_9. Microscopic lamellar grains. Gray, metallic luster. $H = 3\frac{1}{2}$, $VHN_{50} = 185$. $R = 43.1$–43.8 (546 nm); distinctly anisotropic. Apache Hills, near Hachita, Grant Co., NM; South Mountain, Owyhee Co., ID; Gabbs, Nye Co., NV; *Ivigtut, Greenland*; La Roche Balue, France; Kochbulak deposit, Uzbekistan. BM *NJMA 131:56(1977), AM 64:243(1979), ABI: 557, NJMM:454(1992).*

3.6.12.1 Sorbyite $Pb_{19}(Sb, As)_{20}S_{49}$

Named in 1967 for Henry C. Sorby (1826–1908), English chemist and the founder of metallography. MON $C2/m$. $a = 45.15$, $b = 4.146$, $c = 26.53$, $\beta = 113.4°$, $Z = 2$, $D = 5.52$. *20-564:* 4.13_6 4.02_4 3.64_3 3.44_{10} 3.38_9 3.04_4 2.96_6 2.10_5. Microscopic needles and prismatic crystals. Black, black streak, metallic luster. Cleavage {001}, perfect. $H = 3\frac{1}{2}$, $VHN_{50} = 175$. $R = 37$–43

(546 nm); strongly anisotropic. Near Candelaria, Mineral Co., NV; *Madoc, ON, Canada*; Novoye, Khaidarkan, Kyrgyzstan. BM *CM 9:191(1967), MR 13:93(1982), ABI:485.*

3.6.13.1 Baumhauerite $Pb_3As_4S_9$

Named in 1902 for Heinrich A. Baumhauer (1848–1926), German mineralogist. TRIC $P1$. $a = 22.80$, $b = 8.357$, $c = 7.89$, $\alpha = 90.05°$, $\beta = 97.26°$, $\gamma = 89.88°$, $Z = 4$, $D = 5.33$. A monoclinic polytype, baumhauerite-2A, is known. *12-281:* 4.15_{10} 3.59_8 3.47_8 3.26_6 3.02_6 2.96_9 2.77_8 2.72_6. Short prismatic crystals to 25 mm and granular aggregates. Lead-gray to steel-gray, chocolate-brown streak, metallic luster. Cleavage {100}, good; conchoidal fracture. $H = 3$, $VHN_{50} = 179$. $G = 5.33$. $R = 35.1$–39.2 (540 nm); strongly anisotropic. Sterling Hill, NJ; Zuni mine, Silverton, CO; Madoc, ON, Canada; *Lengenbach, Binn, Switzerland*; Novoye, Kaidarkan, Kyrgyzstan. BM *DI:460, P&J:85, ABI:42, AM 75:915(1990).*

3.6.14.1 Sinnerite $Cu_6As_4S_9$

Named in 1964 for Rudolph von Sinner (1890–1961), Swiss scientist. TRIC $P1$. $a = 9.06$, $b = 9.83$, $c = 9.08$, $\alpha = 90.00°$, $\beta = 109.50°$, $\gamma = 107.80°$, $Z = 2$, $D = 4.47$. *25-264:* 3.34_2 3.02_{10} 2.61_4 1.85_8 1.58_7 1.56_2 1.31_2 1.21_3. Prismatic crystals 1–2 mm. Steel-gray, metallic luster. VHN = 357–390. $R = 29.5$–31.5 (546 nm); distinctly anisotropic. *Lengenbach, Binn, Switzerland.* BM *SMPM 44:5(1964), 45:691(1965), AM 57:824(1972), ABI:477.*

3.6.15.1 Berryite $Pb_3(Ag,Cu)_5Bi_7S_{16}$

Named in 1966 for Leonard G. Berry (1914–1982), Canadian mineralogist. MON $P2_1/m$. $a = 12.707$, $b = 4.021$, $c = 28.92$, $\beta = 102.6°$, $Z = 2$, $D = 6.87$. *19-703:* 3.74_6 3.51_{10} 3.50_9 3.26_6 3.06_5 2.91_{10} 2.88_7 2.82_7. Lathlike aggregates to 1 mm; also massive, granular. Blue-gray, metallic luster. $H = 3$, $VHN_{100} = 131$–152. $G = 6.7$. $R = 38.6$–47.0 (546 nm); distinctly anisotropic. *Missouri mine, Park Co.*, and Mike mine, San Juan Co., *CO*; near Owen Lake, south of Houston, BC, Canada; Ivigtut, Greenland; Nordmark, Värmland, Sweden; Grube Clara, Black Forest, Germany; Kochbulak deposit, Uzbekistan; Adrasman, Kaptar–Hana, and Tary Ekan deposits, Karamazar, Tajikstan. BM *CM 8:407(1966), 11:1016(1973); P&J:86; ABI:48.*

3.6.16.1 Robinsonite $Pb_4(Sb,Bi)_6S_{13}$

Named in 1952 for Stephen C. Robinson (1911–1981), Canadian mineralogist. MON $I2/m$. $a = 23.698$, $b = 3.980$, $c = 24.466$, $\beta = 93.9°$, $Z = 4$, $D = 5.7$. *35-513:* 4.06_5 3.97_5 3.81_4 3.44_9 3.41_{10} 3.21_5 3.04_4 2.67_5. Striated prismatic crystals, also massive, fibrous to compact. Bluish lead-gray, metallic luster. Uneven fracture, brittle. $H = 2\frac{1}{2}$–3. $G = 5.64$–5.75. Strongly anisotropic. *Red Bird mine, Pershing Co.*, and Silver Coin mine, Humboldt Co., *NV*; Madoc, ON, and Dodger mine, Salmo, BC, Canada; Vall de Ribes, eastern Pyrenees, Spain;

Male Zelezne, Czech Republic. BM *AM 37:438(1952)*, *R:764*, *CM 20:97(1982)*, *ABI:443*.

3.6.17.1 Liveingite $Pb_9As_{13}S_{28}$

Named in 1901 for George D. Liveing (1827–1924), English chemist. Synonym: Rathite-II. MON $P2_1/m$. $a = 7.944$, $b = 70.84$, $c = 8.335$, $\beta = 90.13°$, $Z = 4$, $D = 5.29$. *31-678*: 4.11_9 3.70_7 3.42_9 2.99_{10} 2.35_9 2.09_9 1.93_8 1.92_8. Prismatic crystals to 4 cm. Lead-gray to gray-black, chocolate brown streak, dull metallic luster. Cleavage {100}, perfect. H = 3, $VHN_{100} = 169$–180. G = 5.3. R = 35.6–40.1 (540 nm); distinctly anisotropic. *Lengenbach, Binn, Switzerland*. BM *DI:462*, *AM 54:1498(1969)*, *R:751*, *ABI:298*.

3.6.18.1 Weibullite $Pb_6Bi_8(S, Se)_{18}$

Named in 1910 for Mats Weibull (1856–1923), Swedish mineralogist, who first described it. ORTH *Pnma*. $a = 53.68$, $b = 4.11$, $c = 15.40$, $Z = 4$, $D = 6.73$. *29-761*: 3.85_{10} 3.64_5 3.27_8 3.08_9 3.02_7 2.81_7 2.28_6 2.14_6. Indistinct prismatic crystals; usually massive, prismatic to fibrous, or foliated. Steel-gray, dark gray streak, metallic luster. One good cleavage, very brittle. H = 2–3, $VHN_{100} = 136$–156. G = 6.97. R = 40.0–46.6 (540 nm); strongly anisotropic. *Falun, Sweden*; Nevskoe deposit, Magadan district, Russia. BM *DI:473*, *R:778*, *P&J:387*, *ABI:570*, *CM 18:1(1980)*.

3.6.19.1 Kobellite $Pb_{22}Cu_4(Bi, Sb)_{30}S_{69}$

Named in 1839 for Franz von Kobell (1803–1882), German mineralogist. Forms a series with tintinaite. ORTH *Pnmm*. $a = 22.62$, $b = 34.08$, $c = 4.02$, $Z = 1$, $D = 6.51$. *8-122*: 3.93_3 3.51_{10} 3.38_{10} 3.11_3 2.84_5 2.13_6 2.03_4 1.74_4. Massive, fibrous or radiated; also granular. Lead-gray, black streak, metallic luster. Cleavage {010}, good. H = $2\frac{1}{2}$. G = 6.48. R = 41.9–46.4 (540 nm); distinctly anisotropic. Superior quarry, Raleigh, Wake Co., NC; Silver Bell mine, Red Mountain, San Juan Co., and Leadville, Lake Co., CO; Dodger mine, Salmo, BC, and Tintina mines, YT, Canada; *Vena mine, near Askersund, Sweden*; Rammelsberg, Harz, Germany; Smolotely, near Pribram, Czech Republic; Ciclova, Romania; Ustarasai deposit, Uzbekistan; Zeehan, TAS, Australia. BM *DI:447*, *R:780*, *P&J:225*, *ABI:268*.

3.6.19.2 Tintinaite $Pb_{22}Cu_4(Sb, Bi)_{30}S_{69}$

Named in 1968 for the locality. Forms a series with kobellite. ORTH *Pnnm*. $a = 22.234$, $b = 34.158$, $c = 4.049$, $Z = 1$, $D = 5.52$. *20-565*: 3.96_4 3.72_3 3.52_8 3.40_{10} 3.27_4 2.87_5 2.83_4 2.71_7. Blades and small masses to 2 mm. Lead-gray, black streak, metallic luster. Cleavage {010}, good. $VHN_{50} = 151$. G = 5.48. *Tintina mines, YT*, and Deer Park mine, Rossland, BC, *Canada*; Boliden, Västerbotten, Sweden. BM *CM 9:371(1968)*, *22:219(1984)*; *AM 54:573(1969)*, *70:441(1985)*; *ABI:532*.

FÜLÖPPITE GROUP

The fülöppite minerals are a group of monoclinic Pb–Sb sulfosalts with the general formula

$$Pb_{3+2n}Sb_8S_{15+2n}$$

where

$n = 0-3$

and the space group is $C2/c$. The fülöppite group minerals are rare, and their structures complex (see figure).

FÜLÖPPITE GROUP

Mineral	Formula	Space Group	a	b	c	β	
Fülöppite	$Pb_3Sb_8S_{15}$	$C2/c$	13.44	11.73	16.93	94.71	3.6.20.1
Plagionite	$Pb_5Sb_8S_{17}$	$C2/c$	13.47	11.82	19.99	107.33	3.6.20.2
Heteromorphite	$Pb_7Sb_8S_{19}$	$C2/c$	13.63	11.94	21.29	90.92	3.6.20.3
Semseyite	$Pb_9Sb_8S_{21}$	$C2/c$	13.64	11.96	24.46	105.87	3.6.20.4
Rayite	$Pb_8(Ag,Tl)_2Sb_8S_{21}$	$C2/c$	13.60	11.96	24.49	103.94	3.6.20.5

3.6.20.1 Fülöppite $Pb_3Sb_8S_{15}$

Named in 1929 for Bela Fülöpp, Hungarian mineral collector. Fülöppite group. MON $C2/c$. $a = 13.44$, $b = 11.73$, $c = 16.93$, $\beta = 94.71°$, $Z = 4$, $D = 5.19$. *22-648*: 3.89_{10} 3.63_5 3.38_5 3.23_7 3.20_9 2.92_8 2.82_8 2.75_7. Short prismatic and pyramidal crystals to 2 mm. Gray-black, gray-black streak, metallic luster. $H = 2\frac{1}{2}$, $VHN_{100} = 140-157$. $G = 5.22$. $R = 31.9-40.1$ (540 nm); distinctly anisotropic. Wet Swine Gill, Caldbeck fells, Cumbria, England; Auliac, Cantal, Bodennec, Finistère, and Bournac, Montagne Noire, France; Baia Mare, Romania. BM *DI:463, R:762, P&J:170, ABI:168*.

3.6.20.2 Plagionite $Pb_5Sb_8S_{17}$

Named in 1833 from the Greek for *oblique*, in reference to the obliquity of the crystals. Fülöppite group. MON $C2/c$. $a = 13.47$, $b = 11.82$, $c = 19.99$, $\beta = 107.33°$, $Z = 4$, $D = 5.59$. *22-1129*: 5.87_4 3.87_8 3.77_7 3.61_5 3.26_9 3.21_{10} 2.91_9 2.62_6. Tabular striated crystals to 1 cm; also massive, granular to compact. Dark lead-gray color and streak, metallic luster. Cleavage {112}, very good; conchoidal to uneven fracture; brittle. $H = 2\frac{1}{2}$, $VHN_{100} = 150$. $G = 5.5-5.6$. $R = 34.0-42.5$ (540 nm); distinctly anisotropic. Porter claim, Carbon Hill, Wheaton R. District, YT, Canada; Leyraux, Cantal, Chazelles, Haute-Loire, and Bournac, Montagne Noire, France; *Wolfsberg*(*), *Harz, Germany*; Azatec deposit, Armenia; Balkumeisk deposit, Yakutia, Russia; Kochbulak(*), Uzbekistan; San José mine, Oruro, Bolivia. BM *DI:464, R:762, P&J:301, ZK 139:351(1974), ABI:407*.

3.6.20.4 Semseyite

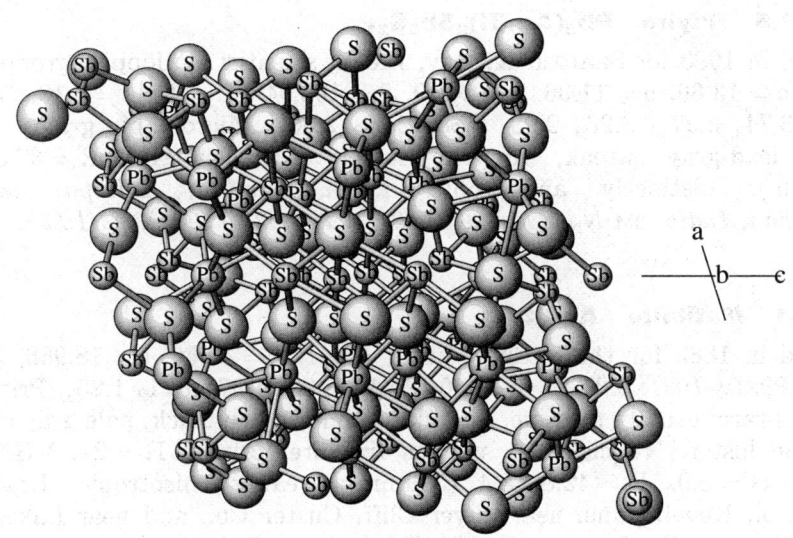

PLAGIONITE: $Pb_5Sb_8S_{17}$
Coordination is irregular and for Sb, sometimes asymmetric, typical for sulfosalts.

3.6.20.3 Heteromorphite $Pb_7Sb_8S_{19}$

Named in 1876 from the Greek for *different* and *form*, in allusion to the difference from a proposed dimorphous mineral. Fülöppite group. MON $C2/c$. $a = 13.63$, $b = 11.94$, $c = 21.29$, $\beta = 90.92°$, $Z = 4$, $D = 5.82$. *30-693:* 3.86_8 3.82_7 3.41_9 3.31_9 3.30_9 3.26_8 3.10_8 2.89_{10}. Pyramidal crystals and massive. Black, black streak, metallic luster. Cleavage {112}, good; uneven fracture; brittle. $H = 2\frac{1}{2}$–3. $G = 5.6$–5.7. *Arnsberg, Westphalia,* and Wolfsberg, Harz, *Germany;* Pribram, Czech Republic; Kara Kamar deposit, Tajikstan; Yecora, SON, Mexico. BM *DI:465, R:763, ZK 151:193(1980), ABI:217.*

3.6.20.4 Semseyite $Pb_9Sb_8S_{21}$

Named in 1881 for Andor von Semsey (1833-1923), Hungarian mineral collector. Fülöppite group. MON $C2/c$. $a = 13.64$, $b = 11.96$, $c = 24.46$, $\beta = 105.87°$, $Z = 4$, $D = 6.08$. *22-1130:* 4.22_4 3.88_6 3.81_9 3.35_7 3.26_{10} 2.95_9 2.86_8 2.15_5. Tabular and prismatic crystals, sometimes grouped in rosettes. Black, black streak, metallic luster. Cleavage {112}, perfect. $H = 2\frac{1}{2}$. $G = 6.03$. $R = 36.0$–43.3 (540 nm); strongly anisotropic. Brobdignag prospect, near Silverton, CO; Madoc, ON, Canada; Louisa mine, Eskdale, Dumfriesshire, Scotland; Bwlch mine, near Conway, Wales; Les Anglais mine, Cantal, Montlucan and Bournac, Montagne Noire, France; Wolfsberg, Harz, Germany; Milesov, Krasna Hora, and Bohutin, Czech Republic; Baia Sprie, Herja, and Rodna, Romania; Madan, Bulgaria; Julcani district, Peru; San José mine, Oruro, Bolivia. BM *DI:466, R:764, P&J:341, ABI:471.*

3.6.20.5 Rayite $Pb_8(Ag,Tl)_2Sb_8S_{21}$

Named in 1983 for Santosh K. Ray, Indian scientist. Fülöppite group. MON $C2/c$. $a = 13.60$, $b = 11.96$, $c = 24.49$, $\beta = 103.94°$, $Z = 4$, $D = 6.13$. *35-594:* 3.90_3 3.74_3 3.37_{10} 3.26_9 2.98_5 2.88_2 2.29_2 2.06_3. Microscopic grains. Lead-gray, lead-gray streak, metallic luster. No cleavage. $R = 37.5–39.6$ (540 nm); distinctly anisotropic. *Rajpura–Dariba, Udaipur district, Rajasthan, India.* BM *NJMM:296(1983), AM 69:211(1984), ABI:435.*

3.7.1.1 Matildite $AgBiS_2$

Named in 1883 for the locality. HEX-R $P\bar{3}m1$. $a = 4.066$, $c = 18.958$, $Z = 3$, $D = 6.99$. *24-1031:* 6.31_3 3.45_2 3.30_8 2.83_{10} 2.03_5 1.97_6 1.71_3 1.25_3. Prismatic crystals rare; usually massive, granular. Iron-gray to black, pale gray streak, metallic luster. No cleavage, uneven fracture, brittle. $H = 2\frac{1}{2}$, $VHN_{100} = 72–85$. $G = 6.9$. $R = 42.8–48.4$ (540 nm); weakly anisotropic. Leadville, Lake Co., Revell mine, near Silver Cliff, Custer Co., and near Lake City, Hinsdale Co., CO; Bisbee, AZ; Mayflower mine, Boise Basin area, Ada Co., ID; Darwin mine, Inyo Co., CA; Cobalt, ON, Glacier Gulch, BC, and Camsell R., Great Bear Lake, NWT, Canada; Schapbach, Black Forest, Germany; Panasqueira, Portugal; Bustarviejo, near Madrid, Spain; *Matilda mine, near Morococha, Peru.* BM *DI:429, R:665, P&J:366, ABI:315.*

3.7.1.2 Bohdanowiczite $AgBiSe_2$

Named in 1967 for Karol Bohdanowicz (1864–1947), Polish economic geologist. HEX-R $P\bar{3}m1$. $a = 4.205$, $c = 19.650$, $Z = 3$, $D = 7.87$. *29-1441:* 3.57_3 3.42_6 2.93_{10} 2.10_8 2.04_8 1.71_5 1.46_5 1.29_4. Microscopic anhedral grains. Gray, metallic luster. $H = 3$, $VHN = 63–96$. $R = 49.8–51.5$ (546 nm); distinctly anisotropic. Kidd Creek mine, near Timmins, ON, Canada; near Vanos, CHIH, Mexico; Frederik VII mine, near Julianehaab, Greenland; *Kletno, Sudetes Mts., Poland.* BM *MM 43:131(1979), AM 64:1333(1979), CM 18:353(1980), ABI:59.*

3.7.1.3 Volynskite $AgBiTe_2$

Named in 1965 for I. S. Volynskii (1900–1962), Russian mineralogist. HEX-R $P\bar{3}m1$. $a = 4.468$, $c = 20.755$, $Z = 3$, $D = 7.943$. *18-1173:* 3.21_8 3.09_{10} 2.33_3 2.21_5 2.15_3 1.82_3 1.61_2 1.55_2. Microscopic grains. Gray, metallic luster. $H = 2–2\frac{1}{2}$, $VHN_{25} = 42–66$. $R = 53.0–54.6$ (540 nm); weakly pleochroic and anisotropic. Campbell mine, Bisbee, AZ; Ashley deposit, Bannockburn Twp., ON, Canada; Ivigtut, Greenland; Glava, Värmland, Sweden; Ilomantsi, Finland; *Armenia,* undisclosed locality; Zhana–Tyube Au–Te deposit, Kazakhstan; Kochbulak deposit, Uzbekistan; Champion Reef mine, Kolar, India; Yokozuru mine, Kyushu, Japan; Kambalda, WA, Australia. BM *AM 51:531(1966), R:429, CM 21:137(1983), ABI:561.*

3.7.2.1 Trechmannite AgAsS$_2$

Named in 1905 for Charles O. Trechmann (1851–1917), English mineralogist. Dimorphous with smithite. HEX-R $R\bar{3}m$. $a = 13.98$, $c = 9.12$, $Z = 18$, $D = 4.78$. *16-700:* 7.0_6 4.26_6 3.64_6 3.15_8 3.04_6 2.70_{10} 1.94_7 1.89_8. Small equant crystals. Cinnabar-red color and streak, adamantine luster. Cleavage {$10\bar{1}1$}, good; {0001}, fair; conchoidal fracture; brittle. $H = 1\frac{1}{2}$–2, $VHN_{25} = 97$–100. $R = 23.4$–29.9 (540 nm); moderately anisotropic. Niederbeerbach, Odenwald, Germany; *Lengenbach, Binn, Switzerland*. BM *DI:432, SMPM 45:695(1965), ZK 129:163(1969), ABI:536.*

3.7.3.1 Smithite AgAsS$_2$

Named in 1905 for G. F. Herbert Smith (1872–1953), English mineralogist. Dimorphous with trechmannite. MON $A2/a$. $a = 17.23$, $b = 7.78$, $c = 15.19$, $\beta = 101.2°$, $Z = 24$, $D = 4.93$. *13-136*(syn): 3.21_8 2.82_{10} 2.72_6 1.98_3 1.95_5 1.70_4 1.66_4 1.61_4. Platy pseudohexagonal and equant crystals to 3 mm. Cinnabar-red color and streak, adamantine luster. Cleavage {100}, perfect; conchoidal fracture; brittle. $H = 1\frac{1}{2}$–2. $G = 4.88$. $R = 35.0$–36.6 (540 nm). Silver mines, Co. Tipperary, Eire; Jas Roux, Hautes-Alpes, France; Wiesloch, Black Forest, Germany; *Lengenbach, Binn, Switzerland*. BM *DI:430, R:726, P&J:344, ABI:481.*

3.7.3.2 Miargyrite AgSbS$_2$

Named in 1829 from the Greek for *less* and *silver*, as it contains less silver than other red silver sulfosalts. MON $A2/a$. $a = 13.324$, $b = 4.410$, $c = 12.881$, $\beta = 98.48°$, $Z = 8$, $D = 5.25$. *19-1137*(syn): 3.44_8 3.19_3 3.10_3 2.89_{10} 2.75_7 2.01_4 1.97_3 1.91_2. Thick tabular crystals to 1 cm, often deeply striated; also massive. Steel-gray to iron-black, cherry-red streak, adamantine to metallic luster. Cleavage {010}, imperfect; subconchoidal fracture. $H = 2\frac{1}{2}$. $G = 5.2$–5.3. $R = 27.4$–40.0 (540 nm); strongly anisotropic. Widely distributed in small amounts, sometimes an important ore mineral. Silver City and Flint districts, Owyhee Co., ID; Randsburg district(*), San Bernardino Co., CA; Sombrerete and Veta Grande, ZAC, and Catorce, SLP, Mexico; Andreasberg(*) and Clausthal, Harz, and *Braunsdorf, near Freiberg, Saxony, Germany*; Pribram and Kutna Hora, Czech Republic; Hiendelaencina, Guadalajara Prov., Spain; Strezhen deposit, Altai, Russia; Huancavelica, Peru; Tres Puntas, Atacama, and Huantajaya, Tarapaca, Chile; Colquechaca, Potosi, and Pulcayo, Bolivia. BM *DI:424, R:662, P&J:268, ABI:330.*

3.7.4.1 Aramayoite Ag(Sb,Bi)S$_2$

Named in 1926 for Felix Aramayo (1846–1929), Bolivian mine director. TRIC $P\bar{1}$. $a = 7.74$, $b = 8.84$, $c = 8.32$, $\alpha = 100.7°$, $\beta = 90.00°$, $\gamma = 103.9°$, $Z = 6$, $D = 5.65$. *4-696:* 3.44_2 3.22_4 3.16_1 2.82_{10} 2.05_2 1.94_3 1.71_2 1.40_2. Platy crystals. Iron-black, red-brown to black streak, metallic luster. Cleavage {010}, perfect; {100}, fair; {001}, poor; flexible cleavage flakes; inelastic; sectile. $H = 2\frac{1}{2}$. $G = 5.60$. $R = 39.0$–42.2 (540 nm); strongly anisotropic. Niarada,

Flathead Co., MT; Monts de Blond and Le Bourneix, Hautes-Vienne, and Ste.-Marie-aux-Mines, Haut-Rhin, France; Herminia, Julcani district, and San Genaro mine, Huancavelica, Peru; *Animas mine, Chocaya, Bolivia*; El Indio mine, east of Coquimbo, Chile. BM *DI:427, R:664, P&J:70, ABI:17.*

3.7.5.1 Chalcostibite CuSbS$_2$

Named in 1847 for the composition. ORTH *Pnam.* $a = 6.016$, $b = 14.505$, $c = 3.800$, $Z = 4$, $D = 5.00$. *24-347:* 3.11_{10} 3.09_9 2.99_9 2.97_9 2.30_7 2.11_6 1.89_6 1.75_8. Bladed crystals to 4 cm, also massive. Lead-gray to iron-gray, gray streak, metallic luster. Cleavage {010}, perfect; {001}, {100} less so; subconchoidal fracture; brittle. $H = 3–4$, $VHN_{50} = 226–279$. $G = 4.95$. $R = 37.5–43.7$ (540 nm); distinctly anisotropic. Mt. Washington mine, Vancouver Is., BC, and Porter property, Carbon Hill, YT, Canada; Nacozari and Moctezuma, SON, Mexico; *Wolfsberg, Harz, Germany*; Milesov and Krasna Hora, Czech Republic; St.-Pons(*), Provence, France; Capileira, Granada Prov., Spain; Tereksai, Kyrghyzstan; Darasun deposit, Yakutia, Russia; Macayan, Philippines; Rar el Anz(*), east of Casablanca, Morocco; Colquechaca, Unica, Oruro, and other localities, Bolivia. BM *DI:433, R:717, P&J:119, ABI:90.*

3.7.5.2 Emplectite CuBiS$_2$

Named in 1853 from the Greek for *entwined* or *interwoven*, in allusion to its sometimes intimate association with quartz. ORTH *Pnam.* $a = 6.137$, $b = 14.541$, $c = 3.898$, $Z = 4$, $D = 6.43$. *10-474:* 7.38_5 3.23_9 3.13_7 3.05_{10} 2.17_4 1.86_3 1.80_3 1.66_3. Striated prismatic crystals, flattened on {010}. Tin-white to gray, metallic luster. Cleavage {010}, perfect; {001}, less so; conchoidal to uneven fracture; brittle. $H = 2$, $VHN_{100} = 181$. $G = 6.38$. $R = 35.9–37.0$ (540 nm); strongly anisotropic. Bisbee, AZ; El Cobre mine, Concepcion del Oro, ZAC, Mexico; Corrie Buie, Perthshire, Scotland; Åmdal mines, Telemarken, Norway; *Johanngeorgenstadt* and other localities, *Saxony, Germany*; Krupka and Cinvec, Czech Republic; Zidarova, Bulgaria; Ankavan deposit, Armenia; Khayragatsch deposit, Kuramin Mts., Uzbekistan; Akenobe mine, Hyogo Pref., and Hade mine, Okayama Pref., Japan; Cerro Blanco, near Copiapó, Chile. BM *DI:435, R:719, P&J:154, ABI:145.*

3.7.6.1 Lorandite TlAsS$_2$

Named in 1894 for Eötvös Lorand (1848–1919), Hungarian physicist. MON $P2_1/a$. $a = 12.29$, $b = 11.34$, $c = 6.14$, $\beta = 104.6°$, $Z = 8$, $D = 5.51$. *19-1331:* 3.68_5 3.59_6 3.58_{10} 3.18_2 3.00_4 2.88_6 2.75_3 2.63_3. Tabular, prismatic, or pyramidal crystals to 2 mm. Red tarnishing to dark gray, cherry-red streak, adamantine to metallic luster. Cleavage {100}, perfect; {201}, very good; {001}, good; flexible cleavage lamellae. $H = 2–2\frac{1}{2}$. $G = 5.53$. $R = 27.0–29.3$ (540 nm); strongly anisotropic. Rambler mine, near Encampment, Albany Co., WY; Carlin gold deposit, Eureka Co., NV; Lengenbach, Binn, Switzerland; *Allchar, Macedonia*; Dzhizhikrut deposit, Tajikstan. BM *DI:437, R:732, P&J:244, ABI:302*.

3.7.7.1 Weissbergite TlSbS$_2$

Named in 1978 for Byron G. Weissberg (b.1930), New Zealand chemist. TRIC P1. $a = 11.8$, $b = 6.4$, $c = 6.1$, $\alpha = 109.9°$, $\beta = 81.8°$, $\gamma = 105.4°$, $Z = 4$, $D = 6.22$. *29-1331*(syn): 3.75_1 3.71_1 3.65_1 3.58_1 2.97_{10} 2.84_1 2.35_1 1.49_1. Microscopic anhedral grains. Steel-gray, dark gray streak, metallic luster. Cleavage: one perfect, two excellent, one good. $H = 1\frac{1}{2}$. $G = 5.79$. $R = 35.0$–36.8 (546 nm); strongly anisotropic. Lookout Pass, Tooele Co., UT; *Carlin gold mine, Eureka Co., NV*; Allchar, Macedonia. BM *AM 63:720(1978), ABI:572, MR 24:446(1993)*.

3.7.8.1 Sartorite PbAs$_2$S$_4$

Named in 1868 for Sartorius von Waltershausen (1809–1876), German mineralogist. Synonyms: scleroclase, skleroklas. MON *Pbnm*. $a = 19.62$, $b = 7.89$, $c = 4.19$, $\beta = 90°$, $Z = 4$, $D = 5.13$. *11-76*: 9.8_4 4.13_5 3.52_9 3.46_8 2.95_9 2.75_{10} 2.33_7 2.10_4. Prismatic crystals to 10 cm. Steel-gray, chocolate-brown streak, metallic luster. Cleavage {100}, fair; conchoidal fracture; very brittle. $H = 3$, VHN = 196. $G = 5.05$. $R = 34.6$–38.6 (540 nm); weakly anisotropic. Zuni mine, San Juan Co., CO: *Lengenbach*(*), *Binn, Switzerland*; Pitone quarry, near Seravezza, Tuscany, Italy. BM *DI:478, SMPM 45:714(1965), R:750, P&J:335, ABI:462*.

3.7.8.2 Guettardite Pb(Sb,As)$_2$S$_4$

Named in 1967 for Jean Guettard (1715–1786), French geologist. MON $P2_1/a$. $a = 20.095$, $b = 7.946$, $c = 8.783$, $\beta = 101.2°$, $Z = 8$, $D = 5.49$. *20-561*: 4.19_5 3.90_5 3.52_{10} 2.80_9 2.67_5 2.65_5 2.33_4 2.14_4. Minute acicular crystals and anhedral grains. Gray-black, metallic luster. Cleavage {001}, perfect; conchoidal fracture; very brittle. $H = 3\frac{1}{2}$, VHN$_{50}$ = 180–197. $R = 36.1$–41.2 (546 nm). Brobdignag mine, San Juan Co., CO. *Madoc, ON, Canada*; Jas Roux deposits, Hautes-Alpes, France; Pitone quarry, near Seravezza, Tuscany, Italy; Novoye, Khaidarkan, Kyrgyzstan. BM *CM 9:191(1967), 18:13(1980); ABI:199*.

3.7.8.3 Twinnite Pb(Sb,As)$_2$S$_4$

Named in 1957 for Robert M. Thompson (1918–1967), Canadian mineralogist; Thompson means *son of Thomas*, which is Aramaic for *twin*. TRIC P$\bar{1}$. $a = 19.583$, $b = 7.981$, $c = 8.564$, $\alpha = 89.77°$, $\beta = 89.62°$, $\gamma = 90.27°$, $Z = 8$, $D = 5.32$. *20-559*: 4.18_5 3.91_5 3.51_{10} 2.78_7 2.69_5 2.65_5 2.34_8 2.15_5. Microscopic grains. Black, black streak, metallic luster. Cleavage {100}, perfect; polysynthetically twinned on {100}. $H = 3\frac{1}{2}$, VHN$_{50}$ = 147. $R = 35.0$–42.0 (540 nm); strongly pleochroic. Hemlo gold deposit, near Marathon, and *Madoc, ON, Canada*; Jas Roux, Hautes-Alpes, France; Rujevac, Yugoslavia;

Novoye, Khaidarkan, Kyrgyzstan. BM *CM 9:191(1967)*, *AM 53:1424(1968)*, *ABI:545*.

3.7.9.1 Galenobismutite PbBi$_2$S$_4$

Named in 1878 for its composition. ORTH *Pnam*. $a = 11.840$, $b = 14.553$, $c = 4.081$, $Z = 4$, $D = 7.12$. *20-571:* 3.47_{10} 3.41_4 3.36_4 3.02_2 2.90_4 2.20_3 2.05_5 2.04_5. Acicular and platy crystals, commonly massive, columnar to fibrous, or compact. Tin-white to lead-gray, black streak, metallic luster. Cleavage {110}, good; flexible cleavage plates. $H = 2\frac{1}{2}-3\frac{1}{2}$, $VHN_{100} = 142-194$. $G = 6.9-7.0$. $R = 45.9-48.8$ (540 nm); strongly anisotropic. Belzazzar mine, Boise Co., ID; Hatfield mine, Okanogan Co., and Germania mine, Stevens Co., WA; Siscoe mine, Gowganda, ON, Cariboo mine, Barkerville, BC, and Dublin Gulch, YT, Canada; Corrie Buie, Perthshire, Scotland; Gupworthy mine, Somerset, England; *Ko mine, Nordmark*, Gladhammar, and Falun(*), *Sweden*; Pechtelsgrun, Saxony, Germany; Oberpinzgau, Salzburg, Austria; Karaoba, Kazakhstan; Chatkal Mts., Uzbekistan; Strezhen deposit, Altai, Russia; Chenzhou, Hunan, China; Tae Hwa mine(*), South Korea; Kingsgate, NSW, and Mt. Farrell, TAS, Australia; Spekboom Valley, Transvaal, South Africa; Cerro Bonete, Bolivia. BM *DI:471, R:777, P&J:172, ABI:171*.

3.7.9.2 Sakharovaite Pb(Bi,Sb)$_2$S$_4$

Named in 1955 for Marina S. Sakharova (b.1917), Russian mineralogist. ORTH Space group unknown. *25-1398:* 3.41_{10} 3.09_4 2.81_4 2.72_6 2.30_4 2.24_4 2.03_6 1.91_4. Acicular crystals in radiating aggregates to 1 cm. Lead-gray, metallic luster. One perfect cleavage. Spissko–Gemer deposits, Czech Republic; *Ustarasaisk deposit, Tian Shan Mts., Kazakhstan*. BM *AM 41:814(1956), 45:1134(1960); ABI:458*.

3.7.9.3 Berthierite FeSb$_2$S$_4$

Named in 1827 for Pierre Berthier (1782–1861), French chemist. ORTH *Pnam*. $a = 11.383$, $b = 14.127$, $c = 3.755$, $Z = 4$, $D = 4.70$. *24-509*(syn): 4.35_4 3.66_9 3.63_5 3.18_8 3.00_5 2.86_8 2.62_9 2.60_{10}. Prismatic crystals to 1 cm, also massive, prismatic to fibrous and granular. Dark steel-gray, often with iridescent tarnish, dark brown-gray streak, metallic luster. Prismatic, indistinct cleavage. $H = 2-3$, $VHN_{100} = 201$. $G = 4.64$. $R = 30.9-40.8$ (540 nm); strongly anisotropic. Widely distributed. Boron, Kern Co., CA; San Martin, DGO, and Concepcion del Oro, ZAC, Mexico; Pontgibaud, Puy-de-Dôme, Valcros, Provence, *Chazelles, Auvergne*, and Charbes, Alsace, *France*; Braunsdorf, near Freiberg, Saxony, Germany; Pribram and Kutna Hora, Czech Republic; Niccioleta mine, Tuscany, Italy; Herja(*) and Baia Sprie(*), Romania; Ribnova, Rhodope Mts., Bulgaria; Belukha, Transbaikal, Russia; Nakaze mine, Hyugo-ken, Honshu, Japan. BM *DI:481, R:724, P&J:87, ABI:49*.

3.7.9.4 Garavellite FeSbBiS$_4$

Named in 1979 for Carlo Garavelli (b.1929), Italian mineralogist. ORTH $Pnam$. $a = 11.439$, $b = 14.093$, $c = 3.754$, $Z = 4$, $D = 5.65$. *33-641:* 14.0$_6$ 7.08$_6$ 3.62$_{10}$ 3.20$_{10}$ 2.98$_8$ 2.89$_8$ 2.63$_{10}$ 2.51$_{10}$. Microscopic anhedral grains. Gray, metallic luster. $H = 4$, VHN$_{50}$ = 212–222. R = 32.8–34.7 (546 nm); strongly anisotropic. *Valle del Frigido, near Massa, Tuscany, Italy.* BM *MM 43:99(1979), AM 64:1329(1979), ABI:174.*

3.7.10.1 Simonite TlHgAs$_3$S$_6$

Named in 1982 for Simon Engel, son of Peter Engel (b.1942), Swiss crystallographer. MON $P2_1/n$. $a = 5.935$, $b = 11.387$, $c = 15.87$, $\beta = 90.4°$, $Z = 4$, $D = 5.09$. *44-1460:* 9.30$_2$ 5.37$_1$ 4.38$_2$ 3.63$_{10}$ 3.05$_1$ 2.84$_5$ 2.77$_1$ 2.73$_2$. Microscopic crystals. Pale red. *Allchar, Macedonia.* BM *ZK 161:159(1982), AM 69:211(1984), ABI:476, MR 24:446(1993).*

3.7.10.2 Edenharterite TlPbAs$_3$S$_6$

Named in 1992 for Andreas Edenharter (b.1933), Swiss crystal chemist. Contains up to 3% Sb. ORTH $Fdd2$. $a = 15.465$, $b = 47.507$, $c = 5.843$, $Z = 16$, $D = 5.09$. *EM 4:* 3.80$_{10}$ 3.76$_2$ 3.39$_2$ 3.12$_2$ 2.90$_2$ 2.77$_4$ 2.73$_4$ 2.66$_2$. Parallel intergrowths of lathlike crystals to 2 mm. Brown-black to black, raspberry-red streak. Cleavage {100}, distinct; uneven fracture. $H = 2\frac{1}{2}$–3, VHN$_{10}$ = 95–101. R = 28.5–31.5 (546 nm); very weakly anisotropic. *Lengenbach, Binn, Switzerland.* BM *EJM 4:1265(1992).*

3.7.11.1 Livingstonite HgSb$_4$S$_8$

Named in 1874 for David Livingstone (1813–1873), Scottish missionary and African explorer. MON $A2/a$. $a = 30.25$, $b = 4.00$, $c = 21.48$, $\beta = 104.20°$, $Z = 8$, $D = 4.98$. *25-555*(syn): 10.46$_3$ 5.20$_{10}$ 3.81$_5$ 3.75$_2$ 3.47$_8$ 2.85$_6$ 2.59$_4$ 2.27$_6$. Prismatic crystals to 12 cm, often in radiating aggregates. Black, red streak, adamantine to metallic luster. Cleavage {001}, perfect; {010}, {100}, poor; flexible. $H = 2$, VHN$_{100}$ = 96–125. G = 4.8–5.0. R = 30.4–40.5 (540 nm); strongly anisotropic. *Huitzuco*(*), *GRO*, and Guadalcazar, SLP, *Mexico*; Pedrosa del Rey, Leon, Spain; Khaidarkan, Kyrgyzstan; Matsuo mine, Iwate Pref., Japan. BM *DI:485, R:733, P&J:241, ABI:299.*

3.7.12.1 Rathite Pb$_3$As$_5$S$_{10}$

Named in 1896 for Gerhard von Rath (1830–1888), German mineralogist. May contain up to 4% Tl. MON $P2_1/n$. $a = 25.16$, $b = 7.94$, $c = 8.47$, $\beta = 100.5°$, $Z = 4$, $D = 5.31$. *9-426:* 4.19$_6$ 3.60$_8$ 3.39$_7$ 2.97$_6$ 2.87$_7$ 2.75$_{10}$ 2.22$_5$ 2.12$_5$. Prismatic crystals, polysynthetically twinned on {100}. Lead-gray, chocolate-brown streak, metallic luster. Cleavage {100}, perfect; parting on {010}; subconchoidal fracture. $H = 3$, VHN = 161. G = 5.37. R = 33.8–39.2 (540 nm); strongly anisotropic. *Lengenbach, Binn, Switzerland.* BM *DI:455, SMPM 45:726(1965), R:752, P&J:323, ABI:434.*

3.7.13.1 Nordströmite $Pb_3CuBi_7Se_4S_{10}$

Named in 1980 for T. Nordström (1843–1920), Swedish mining engineer. MON $P2_1/m$. $a = 17.97$, $b = 4.11$, $c = 17.62$, $\beta = 94.3°$, $Z = 2$, $D = 7.12$. *35-466:* 3.88_3 3.58_5 3.48_3 3.43_3 3.07_{10} 3.01_3 2.90_3 2.24_7. Microscopic fibrous masses. Lead-gray, metallic luster. One good cleavage. H = $2-2\frac{1}{2}$, $VHN_{25} = 130-143$. Johnny Lyon Hills, north of Benson, Cochise Co., AZ; *Falun, Sweden.* BM *AM 65:789(1980), CM 18:343(1980), ABI:355.*

3.7.14.1 Junoite $Pb_3Cu_2Bi_8(S,Se)_{16}$

Named in 1975 for the locality. MON $C2/m$. $a = 26.66$, $b = 4.06$, $c = 17.03$, $\beta = 127.20°$, $Z = 2$, $D = 6.77$. *29-564:* 3.90_8 3.55_{10} 3.39_4 3.23_5 2.97_5 2.92_7 2.21_4 2.07_4. Microscopic tabular crystals and anhedral grains. Lead-gray, metallic luster. One good cleavage. H = 3–4, VHN = 114–213. R = 41–49 (546 nm); strongly anisotropic. Near Potts, Lander Co., NV; Kidd Creek mine, Timmins, ON, Canada; Kochbulak deposit, Uzbekistan; *Juno mine, Tennant Creek, NT, Australia.* BM *AM 60:548(1975), CM 18:353(1980), ABI:254.*

3.7.15.1 Vrbaite $Hg_3Tl_4Sb_2As_8S_{20}$

Named in 1912 for Karl Vrba (1845–1922), Czech mineralogist. ORTH $C2ca$. $a = 13.399$, $b = 23.389$, $c = 11.287$, $Z = 4$, $D = 5.40$. *20-1264:* 5.18_6 4.31_5 4.04_{10} 3.33_8 3.15_6 3.08_4 2.57_8 2.29_6. Small tabular or pyramidal crystals. Dark gray-black, red-yellow streak, submetallic to metallic luster. Cleavage {010}, good; uneven to conchoidal fracture; brittle. H = $3\frac{1}{2}$. G = 5.30. R = 29.6–32.5 (540 nm); distinctly anisotropic. *Allchar, Macedonia.* BM *DI:484, R:732, P&J:384, ABI:563.*

3.7.16.1 Rohaite Cu_5TlSbS_2

Named in 1978 for John Rose-Hansen (b.1937), Danish mineralogist. ORTH Space group uncertain, probably $Pmmm$. $a = 7.602$, $b = 3.801$, $c = 20.986$, $Z = 4$, $D = 7.752$. *39-412:* 3.80_9 3.08_{10} 2.62_5 2.61_5 2.58_5 2.39_{10} 2.16_5 1.90_9. Microscopic euhedral and subhedral crystals in aggregates to 5 mm. Gray, metallic luster. H = 3, $VHN_{25} = 94$. R = 23.9–35.4 (540 nm); strongly anisotropic. *Ilimaussaq intrusion, Greenland.* BM *AM 65:208(1980), NJMA 138:122(1980), ABI:444.*

3.7.17.1 Dervillite Ag_2AsS_2

Named in 1941 for Henri Derville, French paleontologist. MON $P2/a$. $a = 6.833$, $b = 12.932$, $c = 9.638$, $\beta = 99.55°$, $Z = 8$, $D = 5.62$. *35-628:* 3.25_3 3.17_2 3.08_{10} 3.02_8 2.84_5 2.74_2 2.66_3 2.07_2. Microscopic crystals. Gray, metallic luster. H = 1, $VHN_{100} = 19.5$. R = 22.2 (540 nm); moderately anisotropic. *Ste.-Marie-aux-Mines, Haut-Rhin, France.* BM *BM 106:519(1983), AM 68:1041(1983), ABI:130.*

3.7.18.1 Watkinsonite $Cu_2PbBi_4(Se,S)_8$

Named in 1987 for David H. Watkinson (b.1937), Canadian mineralogist. MON Space group uncertain, probably Pm. $a = 12.921$, $b = 3.997$, $c = 14.989$, $\beta = 109.2°$, $Z = 2$, $D = 7.82$. *42-1375*: 3.57_9 2.98_{10} 2.93_{10} 2.41_7 2.25_6 2.14_7 2.07_7 1.48_7. Anhedral grains to 3 mm. Black, metallic luster. No cleavage; conchoidal fracture. $H = 3\frac{1}{2}$, $VHN_{25} = 155–186$. $R = 48.0–48.6$ (540 nm); moderately anisotropic. Otish Mts. uranium deposit, PQ, Canada. BM *CM 25:625(1987)*, *AM 74:948(1989)*, *ABI:569*.

3.8.1.1 Zinkenite $Pb_9Sb_{22}S_{42}$

Named in 1826 for J. K. L. Zinken (1798–1862), German mineralogist. HEX $P6_3$. $a = 44.15$, $c = 8.62$, $Z = 8$, $D = 5.34$. *7-334*: 3.95_1 3.45_{10} 3.02_2 2.80_4 2.13_2 2.06_2 1.99_3 1.83_3. Acicular crystals in columnar and radiating fibrous aggregates; also massive. Steel-gray, steel-gray streak, metallic luster. Cleavage $\{11\bar{2}0\}$, indistinct. $H = 3–3\frac{1}{2}$, $VHN_{100} = 185$. $G = 5.2–5.3$. $R = 37.9–42.5$ (540 nm); distinctly anisotropic. Brobdignag prospect, near Silverton, CO; Morey, Nye Co., and Eureka, Eureka Co., NV; Bonanza Creek, Bridge R. district, BC, and Yellowknife Bay area, NWT, Canada; Grainsgill, Carrock Fell, Cumbria, England; Saint-Pons(*), Haute-Provence, and Pontgibaud, Puy-de-Dôme, France; *Wolfsberg(*)*, *Harz*, and Aldersbach, Bavaria, Germany; Sacaramb and Baia Mare, Romania; Ziddi deposit, Tajikstan; Magnet mine, Dundas, TAS, Australia; San José, Itos, and other mines, Oruro(*), Bolivia. BM *DI:476*, *R:761*, *P&J:398*, *ABI:586*.

3.8.2.1 Cuprobismutite $Cu_{10}Bi_{12}S_{23}$

Named in 1884 for the composition. MON $C2/m$. $a = 17.628$, $b = 3.911$, $c = 15.190$, $\beta = 101.18°$, $Z = 1$, $D = 6.24$. *29-536*(syn): 6.27_2 3.62_5 3.45_3 3.22_5 3.08_{10} 2.85_3 2.72_6 1.95_3. Prismatic crystals and thin blades, also massive. Gray, metallic luster. $VHN_{100} = 159$. $G = 6.31$. $R = 36.8–41.3$ (540 nm); distinctly anisotropic. Missouri mine, Hall's Valley, Park Co., and Silver Cliff, Custer Co., CO; Fairfax quarry, Centerville, VA; Marysvale, Piute Co., UT; Baicolliou, Switzerland; Krupka, Czech Republic. BM *DI:437*; *AM 37:447(1952)*, *58:967(1973)*; *R:718*; *P&J:141*; *ABI:119*.

3.8.3.1 Tvalchrelidzeite $Hg_{12}(Sb,As)_8S_{15}$

Named in 1975 for Georgy A. Tvalchrelidze (b.1915), Georgian mineralogist. MON Space group unknown. $a = 11.51$, $b = 4.39$, $c = 15.62$, $\beta = 92.14°$, $Z = 1$, $D = 7.74$. *29-904*: 3.49_{10} 3.29_7 3.19_3 2.92_{10} 2.89_{10} 2.08_{10} 2.03_4 1.75_3. Granular aggregates. Lead-gray, red-black streak, adamantine luster. One perfect cleavage. $H = 3\frac{1}{2}$, $VHN_{100} = 172$. $G = 7.38$. $R = 36.7–39.5$ (540 nm); distinctly anisotropic. Getchell mine, Humboldt Co., NV; Hemlo gold deposit, near Marathon, ON, Canada; *Gomi deposit, Georgia*; Tyute deposit, Altai Mts., Russia. BM *DANS 225:911(1975)*, *AM 62:174(1977)*, *ABI:544*.

3.8.4.1   Hodrushite $Cu_8Bi_{12}S_{22}$

Named in 1970 for the locality. MON $A2/m$. $a = 27.205$, $b = 3.927$, $c = 17.575$, $\beta = 92.15°$, $Z = 2$, $D = 6.45$. *25-267:* 3.62_8 3.48_6 3.22_7 3.10_{10} 2.72_8 2.55_6 1.72_6 1.45_5. Crystals rare, usually as anhedral grains and fine-grained aggregates. Steel-gray, metallic luster. No distinct cleavage, very brittle. $H = 3\frac{1}{2}$, $VHN = 187-213$. $G = 6.35$. $R = 37.4-42.8$ (540 nm); weakly anisotropic. Pioche, Lincoln Co., NV; Bisbee, AZ; *Banska Hodrusa, near Banka Stiavnica, Czech Republic.* BM *MM 37:641(1970), R:723, ABI:221.*

3.8.4.2   Paderaite $Cu_6Pb_2AgBi_{11}S_{22}$

Named in 1985 for Karel Padera (b.1923), Czech mineralogist. MON $P2_1/m$. $a = 28.44$, $b = 3.95$, $c = 17.55$, $\beta = 106.1°$, $Z = 2$, $D = 6.91$. *42-618:* 6.21_3 3.63_7 3.21_6 3.18_3 3.06_{10} 2.85_4 2.66_4 2.19_3. Microscopic intergrowths with bismuthinite. Steel-gray, metallic luster. Moderately anisotropic. *Baita, Romania.* BM *NJMM:557(1985), CM 24:513(1986), ABI:374.*

3.8.6.1   Hutchinsonite $(Pb,Tl)_2As_5S_9$

Named in 1904 for Arthur Hutchinson (1866–1937), English mineralogist. ORTH *Pbca*. $a = 10.80$, $b = 35.35$, $c = 8.16$, $Z = 8$, $D = 4.58$. *8-124:* 5.40_2 4.44_5 3.78_7 3.05_6 2.74_{10} 2.38_3 2.22_3 1.90_3. Prismatic and platy crystals to 6 mm. Dark red to black, red streak, adamantine to submetallic luster. Cleavage {010}, good; conchoidal fracture; brittle. $H = 1\frac{1}{2}-2$, $VHN_{100} = 170$. $G = 4.6-4.7$. $R = 30.0-31.0$ (546 nm); anisotropic. Wiesloch, Black Forest, Germany; *Lengenbach(*), Binn, Switzerland;* Quiruvilca(*), Peru. BM *DI:468, SMPM 45:707(1965), R:740, P&J:205, ABI:225.*

3.8.7.1   Pierrotite $Tl_2Sb_6As_4S_{16}$

Named in 1970 for Roland Pierrot (b.1930), French mineralogist. ORTH $Pna2_1$. $a = 38.746$, $b = 8.816$, $c = 7.989$, $Z = 4$, $D = 4.75$. *25-938:* 3.63_8 3.59_{10} 3.49_9 2.84_8 2.70_9 2.52_6 2.47_6 2.35_8. Polycrystalline aggregates. Gray-black, metallic luster. $H = 3\frac{1}{2}$. $G = 4.97$. $R = 31.3-35.0$ (540 nm); distinctly anisotropic. *Jas Roux, Hautes-Alpes, France.* BM *BM 93:66(1970); AM 57:1909(1972), 70:220(1985); ABI:404.*

3.8.8.1   Gerstleyite $Na_2(Sb,As)_8S_{13} \cdot 2H_2O$

Named in 1956 for J. M. Gerstley (b.1907), President, Pacific Coast Borax Company. MON *Cm*. $a = 9.911$, $b = 23.05$, $c = 7.097$, $\beta = 127.85°$, $Z = 2$, $D = 3.53$. *11-367:* 11.9_{10} 5.64_7 4.03_7 3.56_3 3.05_9 2.81_5 2.74_4 1.93_4. Platy-fibrous spherules in clay. Cinnabar-red color and streak, weakly adamantine luster. Cleavage {010}, {100}, perfect; {001}, poor. $H = 2\frac{1}{2}$. $G = 3.62$. Biaxial; $N > 2.01$; strongly birefringent; pleochroic. X, salmon-red; Y,Z, deep blood-red. *Baker mine, Boron, Kern Co., CA.* BM *AM 41:839(1956), ABI:181.*

3.8.9.1 Gillulyite $Tl_2(As,Sb)_8S_{13}$

Named in 1991 for James C. Gilluly (1896–1980), U.S. geologist. MON $P2/n$. $a = 9.584$, $b = 5.679$, $c = 21.501$, $\beta = 100.07°$, $Z = 2$, $D = 4.14$. *AM 76:* 4.14_6 3.87_6 3.63_9 3.08_{10} 2.81_{10} 2.72_6 2.50_7 1.76_7. Microscopic crystals to 1 mm, and cleavable masses to 2 cm. Deep red, brick-red streak, submetallic luster. Cleavage {001}, perfect. $H = 2–2\frac{1}{2}$, $VHN_{15} = 87–132$. $G = 4.02$. $R = 24.1–28.0$ (546 nm); distinctly anisotropic. *Mercur gold deposit, Tooele Co., UT.* BM *AM 76:653(1991), 78:1099(1993), 80:394(1995).*

PAVONITE GROUP

The pavonite group consists of structurally related bismuth sulfosalts with the general formula

$$A_xBi_yS_z$$

where

A = Ag, Cu, Pb

and the structure is monoclinic in space group $C2/m$. The pavonite minerals are rare (see figure).

PAVONITE GROUP

Mineral	Formula	Space Group	a	b	c	β	
Pavonite	$AgBi_3S_5$	$C2/m$	13.35	4.03	16.34	94.5	3.8.10.1
Makovickyite	$Ag_{1.5}Bi_{5.5}S_9$	$C2/m$	13.83	4.04	14.72	97.5	3.8.10.2
Benjaminite	$Ag_3Bi_7S_{12}$	$C2/m$	13.25	4.05	20.25	103.14	3.8.10.3
Mummeite	$Ag_2CuPbBi_6S_{13}$	$C2/m$	13.47	4.06	21.63	92.9	3.8.10.4
Borodaevite	$(Ag,Pb)_6Bi_9S_{17}$	$C2/m$	13.515	4.098	26.00	93.00	3.8.10.5
Cupropavonite	$Cu_2AgPbBi_5S_{10}$	$C2/m$	13.45	4.02	33.06	93.50	3.8.10.6

3.8.10.1 Pavonite $AgBi_3S_5$

Named in 1954 from the Latin *pavo*, peacock, for Martin A. Peacock (1898–1950), Canadian mineralogist. Pavonite group. Contains minor Pb and Cu. MON $C2/m$. $a = 13.35$, $b = 4.03$, $c = 16.34$, $\beta = 94.5°$, $Z = 4$, $D = 6.80$. *29-1138:* 5.34_5 3.59_9 3.46_{10} 3.37_7 3.32_7 2.96_7 2.85_{10} 2.01_7. Tiny bladed crystals and massive. Tin-white to lead-gray, metallic luster. Indistinct cleavage. $H = 2$, $VHN_{50} = 188$. $G_{syn} = 6.8$. $R = 44.0–48.4$ (540 nm); strongly anisotropic. Silver Bell mine, Red Mountain, Ouray Co., and Poughkeepsie Gulch, San Juan Co., CO; Apache Hills mine, near Hachita, Grant Co., NM; Manhattan, Nye Co., and Pioche, Lincoln Co., NV; Keeley mine, South Lorrain Twp., ON, Canada; Panasqueira, Portugal; Banat, Romania; *Porvenir mine, Cerro Bonete, Potosi, Bolivia.* BM *AM 39:409(1954), CM 15:339(1977), R:716, P&J:291, ABI:391.*

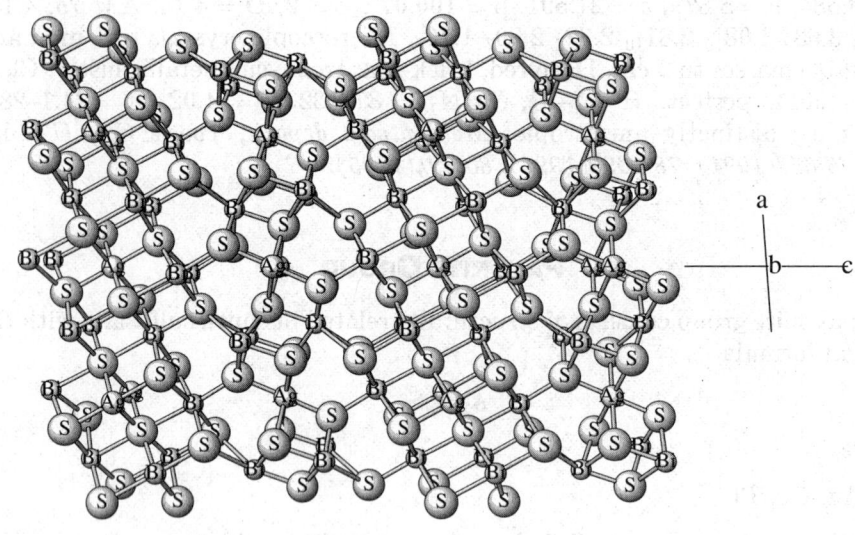

PAVONITE: AgBi$_3$S$_5$

3.8.10.2 Makovickyite Ag$_{1.5}$Bi$_{5.5}$S$_9$

Named in 1994 for Emil Makovicky (b.1941), Slovak–Danish mineralogist. Pavonite group. MON $C2/m$. $a = 13.83$, $b = 4.04$, $c = 14.72$, $\beta = 97.5°$, $Z = 2$, $D = 6.70$. NJMA: 3.63_5 3.49_5 3.42_1 2.97_3 2.85_{10} 2.27_4 2.12_3 2.01_3. Anhedral grains to 2 mm. Gray, metallic luster. No apparent cleavage. VHN$_{50}$ = 210–221. R = 43.2 (546 nm). *Felbertal, Austria; Baita Bihorului, Romania.* BM *NJMA 168:147(1994).*

3.8.10.3 Benjaminite Ag$_3$Bi$_7$S$_{12}$

Named in 1924 for Marcus Benjamin (1857–1932), editor, U.S. National Museum. Pavonite group. May contain minor Cu and Pb. MON $C2/m$. $a = 13.25$, $b = 4.05$, $c = 20.25$, $\beta = 103.14°$, $Z = 2$, $D = 6.68$. 29-577: 3.54_8 3.43_8 3.30_5 2.94_5 2.85_{10} 2.81_6 2.02_6 2.01_6. Laths to 6 mm and massive. Lead-gray, lead-gray streak, metallic luster. One fair cleavage. H = $3\frac{1}{2}$, VHN$_{50}$ = 186–232. R = 44.0–48.6 (540 nm); strongly anisotropic. *Alaska mine, Poughkeepsie Gulch, near Ouray, CO; Outlaw mine,* 12 miles north of *Manhattan,* and at Round Mt., *Nye Co., NV;* Canadian Keely mine, Cobalt, ON, and Terra Company mine, Camsell R., NWT, Canada; Adrasman deposit, Kuramin Mts., Kyrgyzstan; AW mine, near Tenterfield, NSW, Australia; Porvenir mine, Cerro Bonete, Potosi, Bolivia. BM *DI:441, AM 38:550(1953), CM 13:402(1975), ABI:45, MR 22:171(1991).*

3.8.10.4 Mummeite $Ag_2CuPbBi_6S_{13}$

Named in 1986 for William G. Mumme (b.1936), Australian mineralogist. Pavonite group. MON $C2/m$. $a = 13.47$, $b = 4.06$, $c = 21.63$, $\beta = 92.9°$, $Z = 1$, $D = 6.64$. *NJMM:* 5.55_3 3.53_5 3.43_5 3.34_3 2.86_{10} 2.02_{10} 1.65_4 1.29_4. Short columnar crystals to 1 mm. Gray, metallic luster. No cleavage. $VHN_{50} = 203$. $R = 42.8–47.0$ (540 nm); distinctly anisotropic. *Alaska mine, Poughkeepsie Gulch, Silverton, CO.* BM *NJMM:555(1992)*.

3.8.10.5 Borodaevite $(Ag, Pb)_6Bi_9S_{17}$

Named in 1992 for Yuri S. Borodaev (b.1923), Russian mineralogist. Pavonite group. Contains up to 6% Sb. MON $C2/m$ or Cm. $a = 13.515$, $b = 4.098$, $c = 26.00$, $\beta = 93.00°$, $Z = 2$, $D = 6.95$. *ZVMO 121:* 3.49_8 3.37_9 3.24_9 2.82_{10} 2.01_7 1.99_8. Millimeter-size platy crystals and anhedral grains. Gray, black streak, metallic luster. Occurs in an Sn–W deposit at *Alaskitivoje, near Ust-Nera, Yakutia, Russia.* BM *ZVMO 121(4):115(1992), MA 45:111(1994)*.

3.8.10.6 Cupropavonite $Cu_2AgPbBi_5S_{10}$

Named in 1979 for its similarity to pavonite. Pavonite group. MON $C2/m$. $a = 13.45$, $b = 4.02$, $c = 33.06$, $\beta = 93.50°$, $Z = 4$, $D = 7.04$. *42-552:* 5.36_7 3.60_8 3.47_{10} 3.00_7 2.97_7 2.86_9 2.84_7 2.02_7. Lamellar intergrowths with pavonite. Tin-white to lead-gray, metallic luster. $R = 43.0–48.2$ (540 nm); distinctly anisotropic. *Alaska mine, Poughkeepsie Gulch, San Juan Co.*, and *Missouri mine, Hall's Valley, Park Co., CO; Campbell mine, Bisbee, AZ; April Fool mine, Delamar, Lincoln Co., NV.* BM *BM 102:351(1979), AM 65:206(1980), CM 18:181(1980), ABI:121*.

3.8.11.1 Ustarasite $Pb(Bi, Sb)_6S_{10}$

Named in 1955 for the locality. Space group unknown. *25-429:* 3.89_5 3.53_{10} 3.08_7 2.51_7 1.92_7 1.73_7 1.48_{10} 1.14_7. Prismatic crystals. Silver-gray to gray, metallic luster. One perfect cleavage. $H = 2\frac{1}{2}$. Strongly anisotropic. *Montenem, Galicia, Spain; Ustarasaisk deposit, western Tian-Shan, Kazakhstan.* BM *AM 41:814(1956), ABI:551*.

3.8.12.1 Chabournéite $(Tl, Pb)_{21}(Sb, As)_{91}S_{147}$

Named in 1981 for the locality. TRIC $P1$. $a = 16.346$, $b = 42.602$, $c = 8.534$, $\alpha = 95.86°$, $\beta = 86.91°$, $\gamma = 96.88°$, $Z = 1$, $D = 5.12$. *35-474:* 3.93_7 3.57_{10} 3.36_7 2.85_7 2.81_8 2.71_7 2.35_7 2.14_9. Microscopic crystals. Black, red-brown streak, greasy to submetallic luster. Conchoidal fracture. $H = 3$, $VHN_{100} = 168$. $G = 5.10$. $R = 30.8–36.6$ (540 nm); strongly anisotropic. *Jas Roux deposit, near Chabournéou Glacier, Hautes-Alpes, France; Toya mine, Abuta, Hokkaido, Japan.* BM *BM 104:10(1981), AM 67:621(1982), ABI:87*.

3.8.13.1 Rebulite $Tl_5Sb_5As_8S_{22}$

Named in 1982, derivation not given. MON $P2_1/c$. $a = 17.405$, $b = 7.358$, $c = 31.88$, $\beta = 104.90°$, $Z = 4$, $D = 4.94$. 44-1461: 4.13_4 3.99_6 3.86_3 3.46_3 3.36_9 3.08_{10} 2.93_7 2.17_7. Crystals to 3 mm. Dark gray, red-brown streak, submetallic luster. $G = 4.81$. Allchar, Macedonia. BM ZK 160:109(1982), AM 68:644(1983), ABI:437, MR 24:446(1993).

3.8.13.2 Jankovićite $Tl_5Sb_9(As,Sb)_4S_{22}$

Named in 1995 for Slobodan Janković, Serbian mineralogist. TRIC $P\bar{1}$. $a = 7.393$, $b = 8.707$, $c = 17.584$, $\alpha = 103.81°$, $\beta = 91.79°$, $\gamma = 109.50°$, $Z = 1$, $D = 5.08$. TMPM 53: 3.46_{10} 3.39_6 3.18_5 3.08_7 2.80_4 2.58_3 2.29_6 1.74_4. Microscopic tabular grains. Black, brown-violet streak, metallic luster. Cleavage {100}, subconchoidal fracture, brittle. $H = 2$, $VHN_{50} = 98$. $R = 32.2$–35.1 (540 nm); in reflected light white to blue-white, with dark red internal reflections; distinctly anisotropic. Allchar, Macedonia. BM TMPM 53:125(1995), EJM 6:479(1995).

3.8.14.1 Parapierrotite $Tl(Sb,As)_5S_8$

Named in 1975 for its relation to pierrotite. MON $P2_1/n$. $a = 8.02$, $b = 19.35$, $c = 9.03$, $\beta = 91.97°$, $Z = 4$, $D = 5.04$. 29-1330: 4.15_9 3.93_6 3.70_9 3.60_7 3.49_{10} 2.91_9 2.83_{10} 2.36_9. Prismatic crystals to 3 mm. Black, submetallic luster. $H = 2\frac{1}{2}$–3, $VHN = 76$. $G = 5.07$. $R = 29.1$–39.7 (540 nm); distinctly anisotropic. Near Lookout Pass, Tooele Co., UT; Hemlo gold deposit, near Marathon, ON, Canada; Jas Roux, Hautes-Alpes, France; Allchar, Macedonia. BM TMPM 22:200(1975), AM 61:504(1976), P&J:287, ABI:385.

3.8.14.2 Bernardite $Tl(As,Sb)_5S_8$

Named in 1989 for Jan Bernard (b.1928), Czech mineralogist. MON $P2_1/c$. $a = 15.647$, $b = 8.038$, $c = 10.750$, $\beta = 91.27°$, $Z = 4$, $D = 4.34$. MM 53: 4.46_7 4.28_7 4.09_6 3.78_6 3.72_6 3.07_7 3.06_{10} 2.68_6. Thick tabular crystals to 4 mm. Black, deep red streak, metallic luster. No cleavage, subconchoidal to uneven fracture. $H = 2$. $G = 4.51$. $R = 25.0$–26.7 (540 nm). Lengenbach, Binn, Switzerland; Allchar, Macedonia. BM MM 53:531(1989), AM 75:1209(1990).

3.8.15.1 Vaughanite $TlHgSb_4S_7$

Named in 1989 for David J. Vaughan (b.1946), English mineralogist. TRIC Space group uncertain, probably $P1$. $a = 9.012$, $b = 13.223$, $c = 5.906$, $\alpha = 93.27°$, $\beta = 95.05°$, $\gamma = 109.16°$, $Z = 2$, $D = 5.56$. MM 53: 12.5_2 8.42_2 4.34_3 4.20_{10} 3.55_2 3.31_6 2.75_4 2.32_3. Microscopic anhedral grains. Black, black streak, metallic luster. No cleavage, brittle. $H = 3$–$3\frac{1}{2}$, $VHN_{25} = 100$–115. $R = 32.1$–39.4 (540 nm); moderate to strong anisotropism. Hemlo gold deposit, near Marathon, ON, Canada. BM MM 53:79(1989), AM 75:710(1990).

3.9.1.1 Criddleite $TlAg_2Au_3Sb_{10}S_{10}$

Named in 1988 for Alan J. Criddle (b.1944), English mineralogist. MON Space group uncertain, probably $A2/m$. $a = 19.96$, $b = 8.057$, $c = 7.809$, $\beta = 92.08°$, $Z = 2$, $D = 6.57$. $MM\ 52$(syn): 5.63_9 5.00_4 3.91_5 2.86_7 2.81_{10} 2.02_6 1.96_7 1.40_4. Microscopic lathlike or anhedral grains. Black, black streak, metallic luster. $H = 3-3\frac{1}{2}$, $VHN_{25} = 94-129$. $G_{syn} = 6.86$. $R = 36.7-38.4$ (540 nm); distinctly anisotropic. *Hemlo gold deposit, near Marathon, ON, Canada*; Viges, Creuse, France. BM *MM 52:691(1988)*, *AM 75:706(1990)*, *ABI:114*, *CM 28:739(1990)*.

Class 4

Simple Oxides

4.1.1.1 Cuprite Cu$_2$O

Named in 1845 from Latin *cuprum*, copper, in allusion to its composition. ISO $Pn3m$. $a = 4.270$, $Z = 2$, $D = 6.100$. *NBSC 539(2):23(1953)*. *5-667*(syn): 3.02_1 2.47_{10} 2.14_4 1.51_3 1.29_2 1.23_1 0.980_1 0.955_1. **Structure:** In the structure of cuprite the oxygens form a body-centered cubic framework. Within each cube of this network there are four copper atoms, which occupy the centers of smaller cubes generated by dividing the primary cube into eight different cubes; only half of these smaller cubes are occupied, distributed alternately in the larger cube and the copper atoms forming the corners of a tetrahedron. *AC(A) 46:271(1990)*. **Habit:** Crystals octahedral or cubic; rarely dodecahedral; sometimes highly modified; also massive or earthy. In the variety *chalcotrichite* crystals are greatly elongated [001] into capillary shapes. **Physical properties:** Dark red to cochineal red, sometimes almost black. Chalcotrichite is bright red. Brownish-red, shining streak; adamantine, submetallic

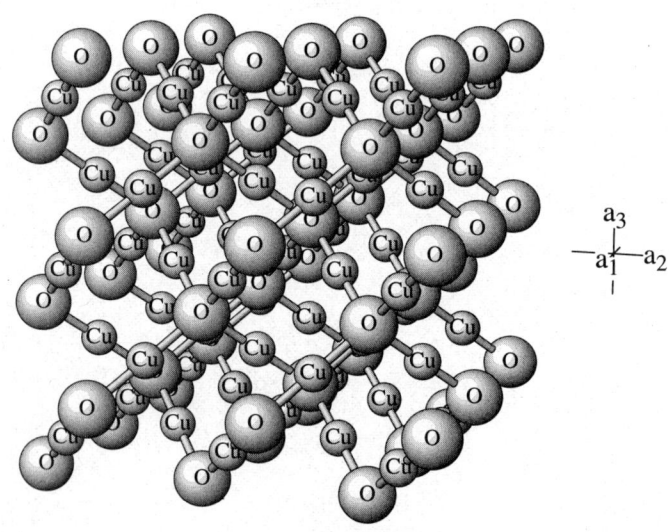

CUPRITE: Cu$_2$O

In cuprite each large atom (O) is coordinated to four small atoms (Cu) and each small atom to two large ones.

luster. Cleavage {111}, interrupted; {001}, rare; conchoidal to uneven fracture; brittle. H = $3\frac{1}{2}$–4. G = 6.14. Fusible with difficulty. **Tests:** Soluble in hot, concentrated HCl, and the solution when diluted gives a heavy white precipitate of Cu_2Cl. Soluble in NH_4OH and in concentrated solutions of HNO_3, H_2SO_4, and NaOH. **Optics:** Isotropic; N = 2.849. Usually exhibits anomalous birefringence and pleochroism. In RL, bluish white with strong internal reflections. R = 31.1 (470 nm), 26.0 (546 nm), 24.6 (589 nm), 23.5 (650 nm). *P&J:140*. **Occurrence:** Widespread in the oxidized zone of copper deposits, usually associated with native copper, malachite, azurite, tenorite, chrysocolla, and limonite. **Localities:** In PA at Cornwall, Lebanon Co. Common in several AZ mines, especially at Bisbee(!!), where it was found in large transparent crystals; also at Morenci, Ray, and other mines(!). Santa Rita district, NM. Chessy, France(*), in octahedral or dodecahedral crystals, often pseudomorphed by malachite and azurite. Fine crystallized specimens in Cornwall, England(!) at Redruth, Liskeard, Gwennap, St. Just, and especially the Phoenix mine, Linkinhorne, to 30 mm. In the Ural Mts., Russia, at Bogoslovsk and Ekaterinburg. In Namibia at Tsumeb(!) and at the Emke mine, Onganga(!!) in transparent crystals up to 8 cm. In fine crystals at the Mashamba West mine, Zaire(!) and at Likasi(!). In Australia at Broken Hill, NSW(!), and at Burra Burra, SA. Corocoró, Bolivia. Andacollo, Chile, in fine crystals(!). **Use:** An ore of copper. RG *DI:491*.

4.1.2.1 Ice H_2O

From Middle English *is* or *iis*, related to Dutch *ijs* and German *eis*. HEX $P6_3mmc$. $a = 4.523$, $c = 7.367$, Z = 4, D = 0.9167 at 0°. *16-687:* 3.90_{10} 3.66_{10} 3.40_8 2.67_4 2.25_9 2.07_6 1.92_5 1.52_2. Data at 90° K. **Habit:** Commonly (as snow) in hexagonal plates, six-rayed stars, and branched stars with dendritic or fern-frond-like arms, all flattened (0001), and in simple and complex combinations of all of these, of great variety and beauty; and in hexagonal prismatic columns and needles elongated [0001], and in combinations of these with the preceding: for example, a prismatic column capped by hexagonal plate or star at both ends, forming a capped column, a tsuzumi type of snow crystal (named after the shape of the Japanese tomtom). Crystal habit is controlled by air temperature and humidity in the clouds at the time the snow crystal grows. Size is generally in the range 0.05 to 7 mm. As frost, in the form of dendritic or feathery aggregates; as skeletal hexagonal prisms with hopper-shaped faces, and as hollow, cup-shaped, or scroll-shaped crystals, the faces being coarsely ribbed and stepped parallel to $\langle 1000 \rangle$ and [0001]. Also as needles $\langle 1000 \rangle$ formed by rapid crystallization on the surface of freezing water. As rounded polycrystalline bodies with concentric structure (as in hailstones and icicles). Often massive. Massive polycrystalline ice such as glacier ice contains crystal grains typically in the size range 1 mm to as much as $\frac{1}{2}$ m, generally increasing in size with age. *LC 8:101, L:173, CP 23:3(1992)*. A generalized type of twinning has been inferred to explain 12-branched snow crystals and certain crystal-orientation relations observed in glacier ice. *JXLG 28:21(1975), JGL 21:607(1978).* **Structure:** Ice has a structure built of discrete water molecules little altered in size or shape from their configuration in water vapor but linked together in a three-dimensional framework by hydrogen bonds. The oxygen atoms of the water molecules are in a hexagonal arrangement like the Si atoms in β-tridymite. Every oxygen is linked tetrahedrally to four other oxygens at 2.76 Å by hydrogen bonds, in which the H atom lies very near each line connecting adjacent O atoms, nearer to one O (0.96 Å) than the other (1.80 Å). At any moment only two H atoms (protons) are situated close to any one O atom, forming a discrete H_2O molecule with O—H distances 0.96 Å and with a H—O—H angle 104.50°. The protons hop back and forth within each bond and also from bond to bond, forming a proton-disordered but still hydrogen-bonded structure. *PPSL 45:768(1933), F 23:147.*

In addition to ordinary ice I, there are at least nine other polymorphs, designated ice II, ice III, and so on. All of these except one are formed under high pressure, are considerably denser than ice I ($D \geq 1.14$) and revert spontaneously to ice I when the high pressure is released. They all have structures in which the water molecules are tetrahedrally hydrogen-bonded to one another. For example, the structure of ice III is analogous to that of keatite (a name given to a synthetic tetragonal polymorph of SiO_2) in the same way that ice I is analogous to β-tridymite. *AC(B) 41(3)169(1985), F:49.*
Physical properties: Colorless; in large transparent masses very pale blue; usually white from included gas bubbles or flaws; white streak, vitreous luster. Cleavage (0001), perfect but difficult to achieve; conchoidal fracture; brittle at low temperatures (less so near the melting point) or on short time scales (like a hammer blow), but when loaded slowly over long periods of time, ice flows in a quasi-viscous manner, as in glacier flow. Single crystals deform readily by dislocation gliding on (0001); deformation of polycrystalline ice (such as glacier ice) is slower and involves several other mechanisms in addition. *F:185.* The hardness of ice is very temperature dependent; at $-44°$ the hardness of ice is about 4 (like fluorite); that is why snow crunches in very cold weather. With a temperature of $-78°$ (dry ice temperature) the hardness becomes ≈ 6 (like feldspar). At such temperatures, which are attained in some arctic or mountainous regions, blowing snow can abrade and significantly erode many rocks. *AJS 238:61(1939).* Static friction between ice and the running surfaces of skis, sled runners, and so on, is low at temperatures above about $-10°$, but at low temperatures the friction increases and skis do not slide freely. Electrical conductivity is low but appreciable, due to hopping of protons in the ice structure; hence ice has been called a protonic semiconductor. The dielectric constant is very high (≈ 100), for similar reasons. *F:198.* Ice melts to water above 0°, with properties as follows: Color, when pure, similar to that of ice. Density increases 9% on melting, a very unusual property. As a function of temperature, the density reaches a maximum, $G = 0.999973$, at 4°. This property results in water at 4° sinking to the bottom of lakes and rivers, displacing less dense water from below. This, in effect, maintains a reservoir of slightly warmer water at depth, which, in

4.1.2.1 Ice

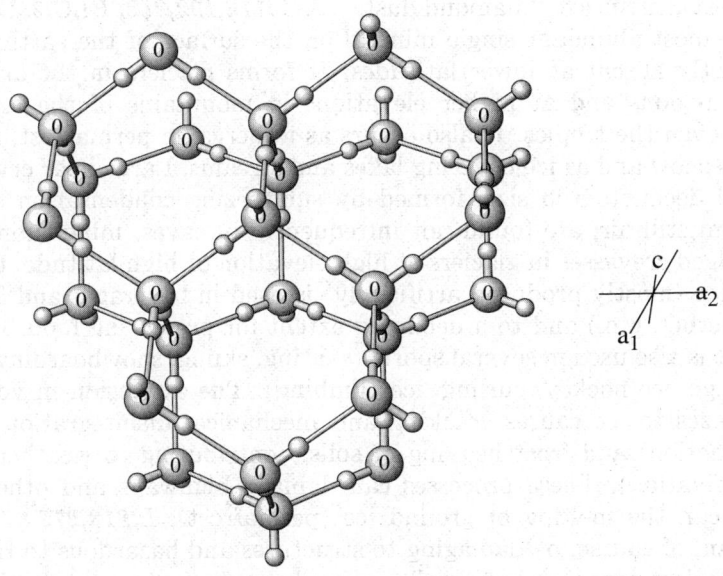

ICE: H$_2$O

The positions of the oxygen atoms in hexagonal ice are similar to the positions of silicon atoms in tridymite—basically a hexagonal closest-packed array. The water molecules are linked to each other by hydrogen bonds. Since the space-group symmetry of the oxygen positions is trigonal, the orientation of these bonds must be disordered over the long range to result in this symmetry.

combination with the fact that ice floats, constrains such water bodies to freeze from the top down, slowly, and retards their freezing to the bottom. Water is nearly unique among chemical compounds in this property. The melting point of ice is lowered by the addition of electrolytes, such as NaCl, CaCl$_2$, and so on. Compared to other liquids, the specific heat, heat of fusion, heat of vaporization, surface tension, and dielectric constant of water are abnormally high. **Chemistry:** Most water-soluble substances such as salts enter only very sparingly into the ice structure. A water solution of salts, such as seawater, will produce relatively pure ice upon freezing, most of the dissolved salts remaining behind in a residual brine. In the Greenland and Antarctic ice sheets, below a depth of about 1,000 m, the air that had been included as bubbles in the ice at shallower depths reacts with some of the ice to form a transparent solid hydrate (N$_2$,O$_2$) · mH$_2$O (with $m \approx 6$) of cubic ;structure ($a = 12.0$ Å) for which the name craigite has been proposed. *SCI 165:489(1969)*. (This icelike mineral is also called clathrate hydrate.) As a result, the bubbles in the ice at these depths disappear and the mass becomes almost perfectly transparent. The clathrate hydrate structure is analogous to that of melanophlogite in the same way that ice I is analogous to β-tridymite. *SCI 148:232(1965)*. **Optics:** Uniaxial (+); N$_O$ = 1.3091, N$_E$ = 1.3105 (Na). **Varieties:** Snow, corn snow (spring snow), firn (névé), glacier ice, graupel, hail, frost, hoar, rime, frazil, anchor ice, grease ice, pancake ice, sea ice, shore-fast ice, river ice, lake ice, candle ice, icicle, icing, aufeis (naled), ground ice (permafrost), cirrus ice, "diamond dust,"... *L:112,192,242, PLC:2*. **Localities:** Ice is the most abundant single mineral on the surface of the earth, despite being mostly absent at lower latitudes. It forms glaciers in the arctic and antarctic regions and at higher elevations in mountains of the temperate zone and even the tropics. It also occurs as icebergs, in permafrost, and seasonally as snow and as ice covering lakes and streams. Large hoar crystals up to several decimeters in size, formed by subfreezing condensation of water vapor from still air, are found not infrequently in caves, mine tunnels, and snow-bridged crevasses in glaciers at high elevation or high latitude. **Uses and hazards:** Ice (mostly produced artificially) is used in beverages and food (ice cream, sherbet, etc.) and to a declining extent for preserving food by refrigeration. It is also used in several sports (skating, skiing, snowboarding, tobogganing, luge, ice hockey, curling, ice climbing). The expansion in volume as water freezes to ice causes cracking and mechanical disintegration of rock (congelifraction) and frost heaving of soils, contributing to weathering and erosion (nivation). These processes can damage highways and other structures, as can the melting of ground ice (permafrost). *L:213,272,374*. Snow and ice can, of course, be damaging to structures and hazardous to transportation. Glaciers pose rare but locally severe hazards, and travel on glaciers is hazardous because of crevasses, which are tensile fractures in deforming ice. RG Reviewed and expanded by Barclay Kamb. *DI:494, LC(1969), L(1990), PLC(1971)*.

PERICLASE GROUP

The periclase structure minerals are anhydrous oxides with the formula

$$AO$$

where

A = Mg, Ni, Mn, Cd, Ca, Fe, or Ti

and the structure is the halite structure, face-centered cubic space group $Fm3m$.

PERICLASE GROUP

Mineral	Formula	Space Group	a	D	
Periclase	MgO	$Fm3m$	4.211	3.581	4.2.1.1
Bunsenite	NiO	$Fm3m$	4.177	6.806	4.2.1.2
Manganosite	MnO	$Fm3m$	4.445	5.365	4.2.1.3
Monteponite	CdO	$Fm3m$	4.695	8.24	4.2.1.4
Lime	CaO	$Fm3m$	4.8105	3.345	4.2.1.5
Wüstite	FeO	$Fm3m$	4.307	5.97	4.2.1.6
Hongquiite	TiO	$Fm3m$	4.293	5.36	4.2.1.7

4.2.1.3 Manganosite

Notes:
1. See the galena group and the halite group, which are essentially isostructural with the periclase group.
2. Periclase minerals occur in high-temperature metamorphic environments, and their formation depends on the nearly complete absence of water. In the presence of water, they alter readily, chiefly to the hydroxide. Periclase minerals are rare in the near-surface environment.

4.2.1.1 Periclase MgO

Named in 1840 from the Greek for *around* and *fracture*, alluding to the perfect cubic cleavage. Periclase group. ISO $Fm3m$. $a = 4.211$, $Z = 4$, $D = 3.581$. *NBSC 539(1):37(1953)*. *4-829*(syn): 2.43_1 2.11_{10} 1.49_5 1.27_1 1.22_1 1.05_1 0.942_2 0.860_2. Usually as irregular rounded grains; rarely as octahedrons, cubes, or cubo-octahedrons. Colorless, white, yellow, green, black (due to impurities); white streak; vitreous luster. Cleavage {001}, perfect; {111}, imperfect; parting {011} on glide planes. $H = 5\frac{1}{2}$. $G_{syn} = 3.56$. Easily soluble in HCl and HNO_3. Finely powdered periclase gives an alkaline reaction with water [$MgO + H_2O \rightarrow Mg(OH)_2$]. Most periclase is relatively pure MgO. However, material from Mt. Somma, Italy, has 6% FeO, and that from Nordmark, Sweden, has 9% MnO and 2.5% ZnO, the manganese being due to oriented inclusions of manganosite. Isotropic; $N = 1.732$ (Na). Periclase is usually formed in high-temperature metamorphic deposits derived from dolomitic marble. Found at Crestmore, CA, in part altered to brucite, associated with chondrodite, wilkeite, and spinel. Organ Mts., NM, with forsterite in dolomite xenoliths. Nordmark, Sweden, with hausmannite in dolomitic marble; also at Langban. *Monte Somma, Vesuvius, Italy*, in ejected blocks of limestone and at Teulada, Sardinia. RG *DI:499, AM 61:266(1976)*.

4.2.1.2 Bunsenite NiO

Named in 1868 for Robert Wilhelm Bunsen (1811–1899), German chemist, inventor of spectroscopy, who also gave his name to a type of gas burner. Periclase group. ISO $Fm3m$. $a = 4.177$, $D = 6.806$. *NBSC 539(1):47(1953)*. *4-835*(syn): 2.41_9 2.09_{10} 1.48_6 1.26_2 1.21_1 1.04_8 0.934_2 0.853_2. Octahedral crystals, sometimes modified by {110} or {011}. Dark pistachio green, brownish black streak, vitreous luster. $H = 5\frac{1}{2}$. $G_{nat} = 6.90$, $G_{syn} = 6.7$. Difficultly soluble in acids. Isotropic; $N = 2.37$ (Li). Found in the oxidized portion of a nickel–uranium vein at *Johanngeorgenstadt, Saxony, Germany*, with nickel and cobalt arsenates. RG *DI:500*.

4.2.1.3 Manganosite MnO

Named in 1874 for the composition. Periclase group. ISO $Fm3m$. $a = 4.445$, $Z = 4$, $D = 5.365$. *NBSC 539(5):45(1955)*. *7-230*(syn): 2.57_6 2.22_{10} 1.57_6 1.34_2 1.28_1 1.11_1 1.02_1 0.994_2. Rarely octahedral, sometimes modified by {001} or {011}; usually in irregular grains or masses, or exsolved laminae in zincite. Emerald green, turning black upon exposure, brown streak, vitreous luster. Cleavage {001}, fair. $H = 5\frac{1}{2}$. $G_{syn} = 5.0-5.4$, $G_{nat} = 5.364$. Dissolves with difficulty in strong HCl or HNO_3 to a colorless solution. Isotropic; $N = 2.19$. In RL, isotropic gray, with emerald green internal reflections. $R = 14.7$ (470 nm), 13.9 (546 nm), 13.5 (589 nm), 13.6 (650 nm). *P&J:256*. From *Franklin, NJ*, in masses and very rare crystals associated with franklinite, willemite, calcite, and zincite. Långban, Värmland, Sweden, and at Nordmark, with pyrochroite, hausmannite, periclase, and garnet in dolomite. RG *DI:501, BM 92:500(1969)*.

4.2.1.4 Monteponite CdO

Named in 1946 for the locality. Periclase group. Synonym: cadmium oxide. ISO $Fm3m$. $a = 4.695$, $Z = 4$, $D = 8.24$. *NBSC 539(2):27(1953)*. *5-640*(syn): 2.71_{10} 2.35_9 1.66_4 1.42_3 1.36_1 1.08_1 1.05_1 0.958_1. Octahedral; as coatings; pulverulent. Gray to black, metallic luster. Cleavage {111}. $H = 3$. $G_{syn} = 8.1$–8.2. Soluble in dilute acids. Isotropic; $N = 2.49$ (Li). In transmitted light, red to orange-brown. From *Genarutta, near Iglesias, Monte Poni, Sardinia, Italy*, as a coating over hemimorphite. RG *DI:502, EG 41:761(1946)*.

4.2.1.5 Lime CaO

Named in 1935 from an old English word for the substance, related to Dutch *lijm* and Latin *limus*, mud. Periclase group. ISO $Fm3m$. $a = 4.8105$, $Z = 4$, $D = 3.345$. *NBSM 25(14):49(1977)*. *43-1001*(syn): 2.78_4 2.41_{10} 1.701_5 1.450_1 $1,389_1$ 1.203_1 1.076_1 0.982_1. Colorless. Cleavage {011}, perfect; parting {011}. $H = 3\frac{1}{2}$. $G_{syn} = 3.3$. Isotropic; $N = 1.838$. Hydrolyzes in water to $Ca(OH)_2$; rapidly alters in air to portlandite and then more slowly to calcite. Found at *Vesuvius, Italy*, as powdery intergranular films in blocks of limestone that were enveloped by lava. RG *DI:503, NBSC 539(1):43(1953)*.

4.2.1.6 Wüstite FeO

Named in 1927 for Ewald Wüst, German geologist. Periclase group. ISO $Fm3m$. $a = 4.307$, $Z = 4$, $D = 5.97$. *AC 13:140(1960)*. *6-615:* 2.49_8 2.15_{10} 1.52_6 1.30_3 1.24_2 1.08_2 0.988_1 0.963_2. Minute octahedral crystals. Gray, metallic luster. $G = 5.74$. Strongly magnetic. Found at *Scharnhaüsen, Württemberg, Germany*; Jharia coal field, Bihar, India, in a natural coke; in the fusion crusts of iron meteorites. RG *DI:504, DANS 168:1390(1960), MM 35:664(1965), NJMM:65(1971)*.

4.2.1.7 Hongquiite TiO

Named in 1976 for the locality. Periclase group. ISO $Fm3m$. $a = 4.293$, $Z = 4$, $D = 5.36$. *29-1361:* 2.48_5 2.14_{10} 1.53_7 1.30_5 1.08_8 0.985_1 0.960_2. Perfect cubo-octahedral crystals. Bright white, metallic luster. Brittle. $VHN_{100} = 710$. Nonmagnetic. Isotropic. In reflected light, white with a pinkish tint. Red internal reflections. $R = 28.6$ (466 nm), 27.5 (544 nm), 35.8 (589 nm), 32.6 (656 nm). Found in platinum ores associated with high-temperature rocks of the garnet–hornblende–pyroxenite type in the *Tao district, Hongqui, China*. RG *AM 61:184(1976)*.

4.2.2.1 Zincite (Zn, Mn)O

Named in 1845 for the composition. HEX $C6mc$. $a = 3.249$, $c = 5.205$, $Z = 2$, $D = 5.680$. *NBSC 539(2):25(1953)*. *36-1451*(syn): 2.81_6 2.60_4 2.48_{10} 1.911_2 1.625_3 1.477_3 1.378_2 1.358_1. Crystals rare, hemimorphic pyramidal with large negative base $\{000\bar{1}\}$; usually massive, as irregular grains and rounded masses. Red to orange, rarely yellow; subadamantine luster. Cleavage $\{10\bar{1}0\}$, perfect but difficult; parting $\{000\underline{1}\}$, sometimes distinct; conchoidal fracture; brittle. $G = 5.64$–5.68. Artificial material fluoresces green to yellow under LWUV; natural material is rarely or only faintly fluorescent, possibly due to quenching by 3–6% manganese impurity. Uniaxial $(+)$; $N_O = 2.013$, $N_E = 2.029$ (Na). Nonpleochroic. In reflected light light rose-brown. $R = 12.5$ (470 nm), 12.0 (546 nm), 11.8 (589 nm), 11.5 (650 nm). *P&J:397*. Found at *Franklin* and *Sterling Hill, NJ*(!), where zincite constituted about 1% of the ore, associated with calcite, willemite, franklinite, and other minerals. Crystals only in secondary veins or fractures. Also reported from Olkusz, Kielce, Poland; Bottino, near Saravezza, Tuscany, Italy; Paterna, Almeria, Spain; the Heazlewood mine, TAS, Australia. Many specimens of alleged naturally crystallized zincite are actually smelter products and not natural. RG *DI:504*.

4.2.2.2 Bromellite BeO

Named in 1825 after Magnus von Bromell (1679–1731), Swedish physician and mineralogist. HEX $P6_3mc$. $a = 2.698$, $c = 4.379$, $Z = 2$, $D = 3.009$. *NBSM 25(21):38(1985)*. *43-1000*(syn): 2.34_9 2.19_6 2.061_{10} 1.598_2 1.349_3 1.238_3 1.149_2 1.129_1. Structure: *SR 53A:123(1986)*. Prismatic, hemimorphic crystals with dominant $\{0001\}$ and opposite pyramidal $\{10\bar{1}1\}$ termination. White, vitreous luster. Cleavage $\{10\bar{1}0\}$, perfect; $\{0001\}$, good. $H = 9$. $G = 3.017$. Normally not fluorescent, but bromellite from Norway is fluorescent white under SWUV. Uniaxial $(+)$; $N_O = 1.719$, $N_E = 1.733$. Reported in skarn S of Sierra Blanca, Hudspeth Co., TX. *Långban, Värmland, Sweden(!)* in a calcite–hematite skarn associated with swedenborgite, richterite, and manganophyllite; at the Saga I quarry, Tvedalen, Norway; near Ekaterinburg, Ural Mts., Russia, in an emerald mine, associated with phenacite(*). A large polycrystalline mass from this mine weighed 5 kg and was surrounded by a phenacite crust in talc schist. RG *DI:506*.

4.2.3.1 Tenorite CuO

Named in 1841 for Michele Tenore (1781–1861), Italian botanist, University of Naples; melacon is from the Greek for *black* and *dust*. Synonym: melaconite, for a massive, pitchy variety, often admixed with other minerals. MON $C2/c$. $a = 4.684$, $b = 3.425$, $c = 5.129$, $\beta = 99.47°$, $Z = 4$, $D = 6.51$. *NBSC 539(1):49(1953)*. *41-254*(syn): 2.531_6 2.524_{10} 2.31_{10} 1.867_3 1.581_1 1.506_2 1.409_1 1.375_1. Crystals rare and minute; as thin twinned aggregates and laths $\{100\}$ and elongated $[001]$; commonly as scales or pulverulent; twinning common on $\{011\}$ producing featherlike forms and dendritic patterns. Gray-black, metallic luster. Cleavages in zones [011] and [0$\bar{1}$1]; conchoidal to uneven fracture; brittle. Flexible and elastic thin scales. H = $3\frac{1}{2}$. G_{nat} = 6.4, G_{syn} = 6.45. Infusible. Easily soluble in dilute HCl or HNO_3. In RL, gray-white with a brownish tint; strongly pleochroic, very strongly anisotropic. R = 30.7–20.3 (470 nm), 30.1–19.6 (546 nm), 29.5–19.5 (589 nm), 29.0–19.4 (650 nm). *P&J:366*. Formerly common in the oxidized portion of copper deposits, but less abundant than cuprite. Often intergrown with chrysocolla. Present to some degree in virtually all oxidized copper deposits. Copper Harbor, Keweenaw Point, MI. In AZ abundant at Bisbee; also at Globe and Morenci. Bingham Canyon, UT; Butte, MT. In Spain at Rio Tinto, Huelva. On lava at *Vesuvius, Italy*. Tsumeb, Namibia. Brittle, lustrous masses at Kakanda and elsewhere in Zaire. Chuquicamata, Chile. An ore of copper. RG *DI:507*.

4.2.4.1 Litharge γ-PbO

Named in 1917 from a Greek word given by Dioscorides to a material obtained in the process of separating lead from silver by pyrometallurgy, the material being lead oxide. Dimorph: massicot, β-PbO. TET $P4/nmm$. a = 3.976, c = 5.023, Z = 2, D = 9.355. *NBSC 539(2):30(1953)*. *5-561*(syn): 3.12_{10} 2.81_6 2.51_2 1.99_1 1.87_4 1.68_2 1.56_1 1.54_1. As alteration crusts on massicot; dense lamellar cleavages. Red, greasy to dull luster. Cleavage {110}. H = 2. G = 9.14. Easily fusible. Soluble in HCl and HNO_3, slowly soluble in alkalis. Uniaxial (−); N_O = 2.665, N_E = 2.535 (Li). *Cucamonga Peak, San Bernardino Co.*, and Ft. Tejon, Kern Co., CA. Mineral Hill district, Hailey, Blaine Co., ID, with native lead and leadhillite. Tintic district, UT. Långban, Värmland, Sweden(!). Zashuran River, Kurdistan, Azerbaijan, with massicot, orpiment, realgar, and cerussite. RG *DI:514*, *NJMA 94:1187(1960)*.

4.2.5.1 Romarchite SnO

Named in 1971 for the initials of the Royal Ontario Museum (ROM), where the mineral was identified, and *archaeology*. TET $P4/nmm$. a = 3.802, c = 4.836, Z = 2, D = 6.398. *NBSC 539(4):28(1955)*. *6-395*(syn): 2.99_{10} 2.69_4 2.42_1 1.90_1 1.80_3 1.60_3 1.49_1 1.48_1. Black. Found as small crystals as an oxidation product on tin cooking utensils submerged since 1830 in the Winnipeg River at *Boundary Falls, MB, Canada*, associated with hydroromarchite. Abundant in the U.S. Virgin Islands in a vein with cassiterite, native tin, copper, and lead, and various lead–tin oxides and hydroxides. RG *CM 10:916(1971)*.

4.2.6.1 Montroydite HgO

Named in 1903 for Montroyd Sharp, one of the owners of the mine in Texas where the mineral was found. ORTH *Pnma*. a = 6.608, b = 5.518, c = 3.519, Z = 4, D = 11.209. *NBSC 539(9):39(1960)*. *37-1469*(syn): 2.97_{10} 2.84_7 2.76_5 2.41_6 1.82_4 1.63_1 1.50_2 1.44_1. Crystals long prismatic [001] or equant, often highly modified. As twisted, wormlike aggregates; massive, powdery. Red to reddish brown, yellow-brown streak, vitreous to subadamantine luster.

Cleavage {010}, perfect; crystals glide easily, with T {010}; sectile; very flexible, but not elastic. H = $2\frac{1}{2}$. G = 11.3. In the closed tube volatilizes completely, giving a sublimate of native Hg in the upper part of the tube. Soluble in cold HCl or HNO_3, and in alkali chloride solutions. Biaxial (+); XYZ = abc; N_x = 2.37, N_y = 2.5, N_z = 2.65 (Li); 2V large. From *Terlingua, Brewster Co., TX*, in veins with calcite, terlinguaite, mercury, and eglestonite. Two miles W of Redwood City, San Mateo Co., CA, with eglestonite, calomel, mercury, and cinnabar, and at the Socrates mine, Sonoma Co. Formerly abundant at Huahuaxtla, GRO, Mexico, with a similar mineral suite. Moschellandsberg, Germany, with moschellandsbergite and cinnabar. Guanxi, China(!), in acicular crystals to 10 mm. RG *DI:511, ACHS 10:852(1966)*.

4.2.7.1 Massicot β-PbO

Named in 1841 for a French name for lead oxide, related to the Spanish *mazacote*, of eventual Arabic origin. Dimorph: litharge (γPbO). ORTH *Pbcm*. a = 5.891, b = 5.489, c = 4.755, Z = 4, D = 9.642. *NBSC 539(2):32(1953)*. 38-1477(syn): 3.07_{10} 2.95_2 2.75_2 2.38_2 1.849_1 1.796_1 1.723_2 1.640_2. Massive, earthy to scaly. Dull sulfur-yellow, sometimes reddish; lighter yellow streak; greasy to dull luster. Cleavage {100} and {010}, flexible but not elastic. H = 2. G_{syn} = 9.56. Biaxial (+); probably Y = a; N_x = 2.51, N_y = 2.61, N_z = 2.71 (Li), 2V = 90°; strong dispersion. Y, light sulfur-yellow; Z, deep yellow. Massicot is a product of oxidation of galena and found as a sublimation product in fumaroles. It is associated with cerussite and other secondary lead minerals and with limonite and antimony oxides. Found at Austinville, Wythe Co., VA; Leadville and Creede, CO; Redemption mine, Esmeralda Co., NV. Popocatepetl, Ixtacihuatl, and Perote volcanos, Mexico. *Stolberg, near Aachen, Germany*. In Sardinia, Italy, in the Oreddo Valley with cerussite, caledonite, and other lead minerals. In the Marico district, Transvaal, South Africa, with minium, vanadinite, and wad. Caracoles, Bolivia. Melrose, NSW, also Dundas, TAS, Australia. RG *DI:516, NJMA 94:1187(1960)*.

CORUNDUM–HEMATITE GROUP

Corundum minerals have the chemical formula

$$A_2O_3$$

where

A = Al, Fe^{3+}, Cr, or V

and are isostructural in the rhombohedral space group $R\bar{3}c$. The structure consists of an arrangement of oxygen atoms in approximately hexagonal close packing. Between oxygen layers there are sites for cations octahedrally coordinated by six oxygen ions, but only two-thirds of the available positions

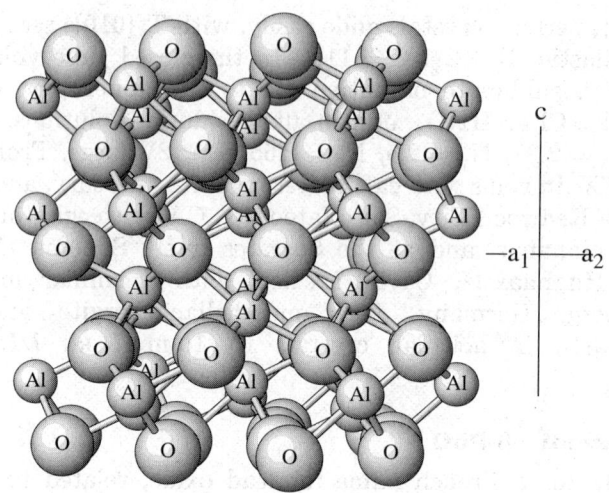

CORUNDUM: Al_2O_3

Al atoms occupy two-thirds of the octahedral interstices in a hexagonal-closest-packed array of oxygens. Because the octahedra share faces in pairs, they are considerably distorted.

are filled. Groups of three oxygen ions form a common face to two neighboring octahedra, and thus the groups are linked to a pair of A ions. *DHZ II:536(1992), AC(A)46:271(1990)*.

CORUNDUM–HEMATITE GROUP

Mineral	Formula	Space Group	a	c	D	
Corundum	Al_2O_3	$R\bar{3}c$	4.758	12.991	3.987	4.3.1.1
Hematite	Fe_2O_3	$R\bar{3}c$	5.036	13.749	5.270	4.3.1.2
Eskolaite	Cr_2O_3	$R\bar{3}c$	4.954	13.584	5.245	4.3.1.3
Karelianite	V_2O_3	$R\bar{3}c$	4.954	14.008	5.016	4.3.1.4

Notes:
1. See also the ilmenite structures, which are lower-symmetry versions of the corundum structure that accommodate different A and B metals in place of corundum's two A metals.
2. Corundum minerals are typical of high-temperature environments, but hematite also forms extensively in soils.
3. Crystals are all particularly hard.

4.3.1.1 Corundum Al_2O_3

The name *corundum* is probably derived from the Tamil name of the mineral, *kuruntam*, and eventually from the Sanskrit for *ruby*. Corundum–hematite group. HEX-R $R\bar{3}c$. $a = 4.758$, $c = 12.991$, $Z = 6$, $D = 3.987$. *NBSC 539(2):20(1953)*. PD 43-1484(syn): 3.48_7 2.55_{10} 2.38_4 2.09_{10} 1.740_5 1.602_{10} 1.405_4 1.374_6. **Habit:** Crystals commonly prismatic, often barrel-shaped, modified by $\{0001\}$, $\{11\bar{2}1\}$, $\{22\bar{4}1\}$, $\{22\bar{4}3\}$, also $\{11\bar{2}0\}$; also tabular $\{0001\}$, or rhombohedral with $\{10\bar{1}1\}$ and $\{0001\}$; striae on $\{0001\}$ parallel to $[01\bar{1}0]$.

4.3.1.1 Corundum

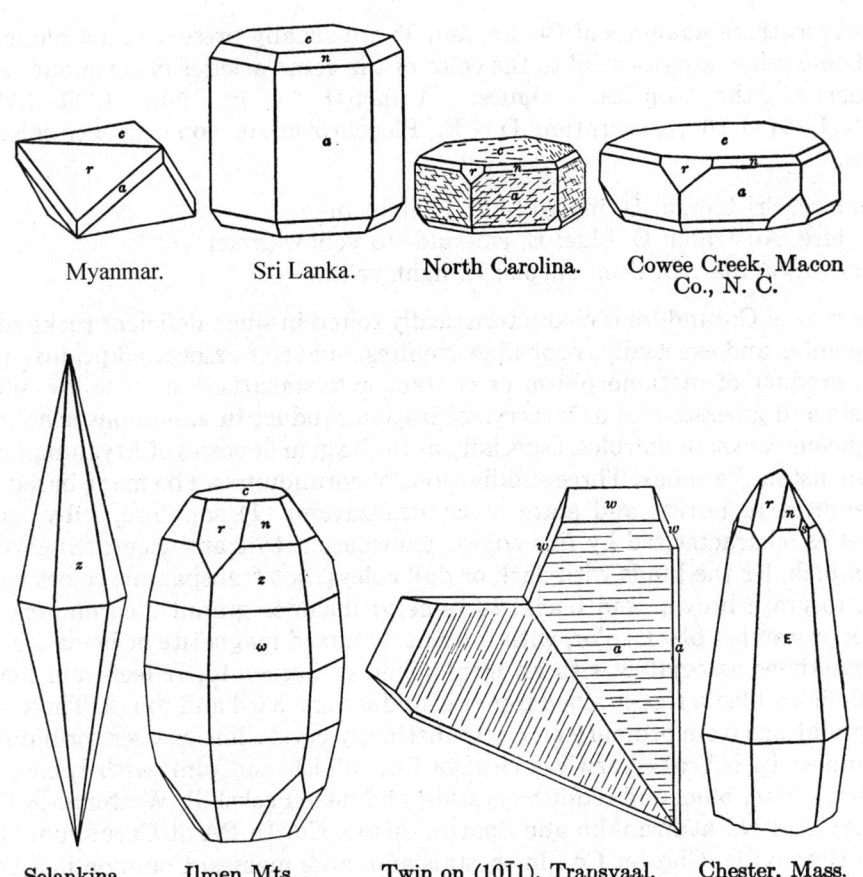

CORUNDUM: Al$_2$O$_3$

Forms: c{0001}, r{10$\bar{1}$1}, a{11$\bar{2}$0}, n{22$\bar{4}$3}, z{22$\bar{4}$1}, w{11$\bar{2}$1}, e{33$\bar{6}$1}

Large blocks showing parting on rhombohedron and base, due to exsolution of boehmite, frequently observed. Massive, granular (emery). Twinning common on {10$\bar{1}$1}, usually lamellar. Rarely penetration or arrowhead twins; also on {0001}, less common. **Physical properties:** Corundum can be found in white, gray, blue-gray, several shades of blue (sapphire), blood-red (ruby); also yellow, orange, pink, violet, green, black. For a discussion of the causes of color of corundum gems and other minerals and gems, see *G&G XXIV*:81(1988). Emery is usually black due to admixed fine magnetite. Uncolored streak, vitreous luster. No cleavage; uneven to conchoidal fracture; brittle, but very tough when compact (emery). H = 9; Mohs' scale defines hardness 9 as the hardness of corundum. G = 4.0. Rubies are fluorescent and phosphorescent in LWUV, a rich red color. Some sapphires are weakly fluorescent, pink. Insoluble in acids. Infusible. **Chemistry:** Substitution of other elements for Al is low; Fe to a maximum of 1%, and a trace of Ti.

However, trace amounts of Cr, Fe, and Ti are usually present and are important since they are essential to the color of the gem varieties of corundum and influence the optics. **Optics:** Uniaxial $(-)$; $N_O = 1.765\text{--}1.776$, $N_E = 1.757\text{--}1.767$; absorption $O > E$. Pleochroism in some deeper-colored gems:

Sapphire, Sri Lanka: O, indigo blue; E, light blue
Sapphire, Australia: O, blue; E, emerald- to yellow-green
Ruby, Myanmar: O, deep purple; E, light yellow

Occurrence: Corundum is characteristically found in silica-deficient rocks such as syenites and especially, nepheline syenites and their associated pegmatites; as a product of metamorphism or contact metasomatism in some low-silica schists and gneisses and as a recrystallization product in aluminous xenoliths in igneous rocks; in marbles, especially as in the gem deposits of Myanmar and Afghanistan. **Varieties:** Three subdivisions of corundum can be made based on differences in purity and state of crystallization: (1) sapphire, ruby: gem varieties, characterized by fine colors, transparent to translucent; (2) corundum: includes the kinds with dark or dull colors, not transparent; colors light blue to gray, brown, and black; (3) emery: includes granular corundum, of black or grayish-black color, with intimately mixed magnetite or hematite. It is sometimes associated with an iron spinel, or hercynite. It feels and looks much like a black, fine-grained iron ore, and is very hard and tough. There are all gradations from ordinary emery to distinctly crystalline coarser corundum. **Localities:** In NY at Warwick, Orange Co., bluish and pink with spinel; at Amity, white, blue, and reddish crystals; and near Peekskill, Westchester Co. (emery). In NJ at Franklin and Sparta, Sussex Co. In PA at Corundum Hill near Unionville, Chester Co., in crystals and large masses. Common in a belt from VA through western NC and SC to GA, especially at Franklin, Macon Co., NC(*), and at Laurel Creek, GA. In ONT, Canada, corundum is found in syenites and anorthosites in Hastings Co., at South Burgess in Leeds Co. in red and blue crystals, and many places in Renfrew Co. Arnes, Norway(!) in fine crystals. Bellerberg, Eifel, Germany, in volcanic rock. Ticino, Switzerland, in dolomite. Island of Naxos and nearby islands, Greece (emery), as well as in Asia Minor and Smyrna, Turkey. Zlatoust and Miask, Ilmenskie Mts., southern Ural Mts., Russia(!). Fine red (ruby) crystals are found in gneiss in the Mysore district, India(!). Large opaque crystals come from Steinkopf, Namaqualand, Cape Prov., South Africa, and an important deposit with giant crystals from Zoutpansberg, Transvaal. In Brazil, large corundum crystals come from near Formoso, Goias(*). **Gemmology:** Ruby and sapphire are two of the four stones traditionally considered as precious (as opposed to semiprecious). Corundum can be found in any color and makes fine gems in shades of red, blue, pink, green, yellow, orange, and purple. Only those of a rich red color are called ruby; the pure blue stones are called *sapphire*, and all the other colors can be called *fancy sapphire*, although some go by specialized names for particular shades, such as padparasha sapphire, for an orange-pink color. Star rubies and sapphires owe their asterism to submiscroscopic inclusions of rods (crystals) of rutile which have exsolved oriented parallel to the a crystallographic axes. The star effect is produced with a source of light, such as the sun, incident along the direction of the c axis; such stones are always cut *en cabochon* with the base normal to [0001]. The color of ruby is due to Cr, and of sapphire of various colors mainly to Fe^{3+} and Fe^{2+}. Sapphires and rubies are among the most commonly manufactured stones; several different processes having been developed for doing so (including the star stones). The discrimination of the best synthetics from natural stones may be extremely difficult and depends in part on the study of minute inclusions under a microscope. Also, natural stones of poor or uneven color can be greatly improved by heating and other processes. **Gem localities:** Near Helena, MT, gem sapphires are found in alluvials along and near the Missouri River, especially at Yogo Gulch. They are derived from a lamprophyre dike that cuts limestone. Alluvial sapphire was studied and described from Espaly, Artois, France, about 1715. Fine blue and fancy sapphires are found in Sri Lanka in the Ratnapura and Rakwana districts, in alluvial deposits, associated with pale rubies, chrysoberyl cat's eye, alexandrite, taaffeite, and other gems. They are formed in schists and other metamorphic rocks. Rubies are found in Afghanistan at Jagdalak, 50 km E of Kabul, in limestone. The best rubies and many sapphires come from the Mogok area in upper Mandalay, Myanmar, in a region encompassing 70 km^2. The rubies occur *in situ* in limestone and also in the soil of the hillsides and the gem-bearing gravels of the Irrawaddy R., associated with gem spinel and, rarely, peridot. Superb "electric blue" sapphires are found in Kashmir, India, at Soomjam, Padar, at 4,500 m elevation, where metamorphics, including marble beds, are intruded by desilicated pegmatites. Gem rubies and sapphires are found SE of Bangkok, Thailand, and near Battambang, Cambodia. In Madagascar, equant rubies and some sapphires are abundant in decomposed mica schists at Vatondrangy SE of Antsirabé; these are generally opaque, but tabular rubies, including some gems, are found near Ampanihy in the southern part of the country. In Tanzania important deposits of gem sapphires and fancy sapphires in many pastel hues are found in the Umba region near Arusha, and bright red opaque ruby in green chromian zoisite is mined in the Matubitu Mts. In Australia very dark blue, almost black gem sapphires are mined at Anakie, Q, and some fine golden-colored stones have been found there also. **Synthesis:** Synthetic corundum abrasives are manufactured on a large scale by the electric furnace treatment of highly purified $Al(OH)_3$ from bauxite. **Uses:** The principal use of corundum is as an abrasive, and in this application it is normally called *alumina*. It is used as a gem, and synthetic crystals are used in bearings for watches and fine instruments, and for lasers. RG *DI:520; ZK 117:235(1962), 120:342(1964).*

4.3.1.2 Hematite Fe$_2$O$_3$

Named in antiquity from the Greek *haimatitis*, blood-red, in allusion to the color of the powdered mineral; "concreted blood" (Theophrastus). Corundum–hematite group. Synonyms: specularite, oligiste. HEX-R $R\bar{3}c$. $a = 5.036$,

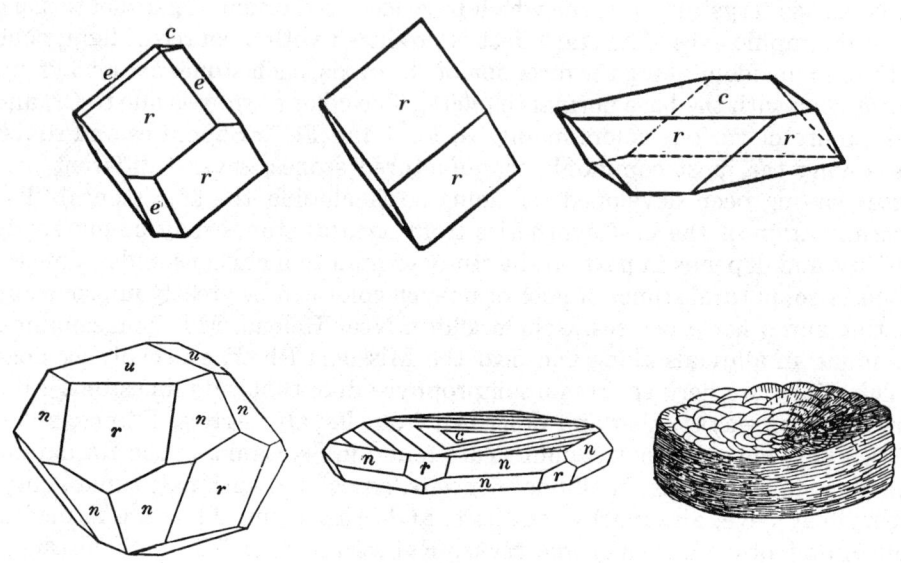

HEMATITE: Fe$_2$O$_3$
Forms: $c\{0001\}$, $r\{10\bar{1}1\}$, $n\{2\bar{2}43\}$, $u\{10\bar{1}4\}$

$c = 13.749$, $D = 5.270$. *NBSM 25(18):37(1981)*. *NBSN*(syn): 3.68_3 2.70_{10} 2.52_7 2.21_2 1.84_4 1.69_5 1.49_3 1.45_3. **Structure:** *AM 51:123(1966)*. **Habit:** Thick to thin tabular $\{0001\}$, often as subparallel growths on $\{0001\}$ or as rosettes ("iron roses"). Rhombohedral $\{10\bar{1}1\}$ to produce pseudocubic crystals; rarely prismatic [0001]; rarely scalenohedral. Sometimes micaceous to platy. Also compact columnar, or fibrous, frequently radiating. In reniform masses with smooth fracture (kidney ore) and in botryoidal and stalactitic shapes. Commonly earthy, ocherous, or frequently admixed with clay or other impurities. Also granular, friable to compact, concretionary, oolitic. Twinning on $\{0001\}$ as penetration twins, or with $\{10\bar{1}0\}$ as composition plane. On $\{10\bar{1}1\}$, usually lamellar, and showing twin striae on base and rhombohedron. **Physical properties:** Steel gray to black crystals; earthy and compact fine-grained material, dull to bright red. In thin splinters or scales deep blood-red by transmitted light or internal reflections. Cherry-red to reddish-brown streak; metallic luster in crystals, earthy when fine grained. No cleavage, subconchoidal to uneven fracture, parting on $\{0001\}$ and $\{10\bar{1}1\}$, brittle in crystals, soft and unctuous in earthy varieties, elastic in thin laminae. H = 5–6. G = 5.26. Nonmagnetic. Infusible. Soluble in concentrated HCl. Becomes magnetic when heated in reducing flame. **Chemistry:** In the hematite cell there is very little substitution of other elements. Al and Ti may approach 1%; Cr and V much less. Earthy varieties may contain H$_2$O and/or admixed minerals such as goethite and other hydrous iron oxides. **Optics:** Uniaxial $(-)$; N$_O$ = 3.22, N$_E$ = 2.94 (Na); strong dispersion; absorption O > E. R = 31.7–

4.3.1.2 Hematite

28.1 (470 nm), 30.0–26.5 (546 nm), 28.95–25.15 (589 nm), 26.8–23.3 (650 nm). *C&S:157, BM(NH)(1985)*. **Occurrence:** Most large hematite ore bodies are of sedimentary origin, bedded, or are metamorphosed sedimentary deposits like the *itabirites* of Brazil, which are thick sequences over vast areas of alternating thin beds of hematite and silica. In places, most of the silica has been removed by subsequent solution; high-grade hematite ore results. Much of the original deposition is believed to have been caused by bacterial action on dilute iron-bearing lake waters, precipitating hydrated iron oxides which were subsequently dehydrated to hematite. Contact metasomatism by intrusive igneous bodies, mostly granites, has produced high-grade hematite or, often, mainly magnetite. Hematite is also occasionally found as a sublimate on lava as a result of volcanic activity and fumaroles. Hematite is ubiquitous in soils, either derived directly from the decomposing underlying rocks or as a result of alteration of pyrite, magnetite, goethite, siderite, or iron silicates. It accounts for the red color of soils throughout the world, especially in tropical regions, and of many Triassic sedimentary deposits. **Localities:** Major iron deposits, predominantly hematite, are found around the shores of Lake Superior in MN and WI in the Iron Ranges, and in the southern Appalachians, especially AL; in Ukraine, China, and India; in Australia, especially in WA; in Africa, especially Liberia; in Venezuela in the Orinoco region, and in Brazil in the states of MG and PA. Fine crystallized or botryoidal hematite specimens are found in so many places throughout the world that only the most important can be listed.

Hematite crystals are found at several localities in northern NY, especially in St. Lawrence Co.; at Franklin, NJ, in coarse crystalline masses showing rhombohedral parting. In Quartzsite(*), AZ, as brilliant tabular crystals associated with quartz; in the Thomas Range, UT, with topaz in lithophysae in rhyolite. In Mexico in the mines of Concepcion del Oro, ZAC. Beautiful crystals of hematite have long come from the Rio Marina, Elba(!), Italy, and also from Vesuvius, Etna, and Stromboli volcanos. In Switzerland, fine rosettes of crystals are found at Passo di Lucendro, St. Gothard, Ticino(!) and also in plates with rutile at Cavradi, Tavetsch, Grisons, and at Alp Lercheltini, Binntal, Wallis(!). Splendid botryoidal specimens come from Ulverstone(!), Lancashire, England ("kidney ore"). In France, fine crystals are found at Framont, Alsace, and in Puy-de-Dôme(!) on Mt. Dor. From Saurüssel, Zillertal, Tyrol, Austria(!) in large roses; in Saxony, Germany, from Altenberg and Johanngeorgenstadt. At Dognáska, Romania(!), in superb twins. In Russia, from the district of Ekaterinburg, Ural Mts.; and at Korshunovskoye, W of Lake Baikal, Siberia(!) magnificent druses of complex crystals. In volcanic rocks, in beautiful crystals, on Ascension Island(*), South Atlantic Ocean. Nador, Morocco, and in South Africa in fine crystals from the Wessels mine, Cape Prov.(*). In Brazil there are many fine localities. In MG at the Miguel Burnier(!) iron mines and at Dom Bosco, Ouro Preto district, large rosettes of thick crystals are found much like the Swiss "iron roses." In the latter locality the hematite is associated with gem topaz in quartz veins. At the Casa de Pedra iron mine,

Congonhas(!), exceptional well-faceted crystals are found, having sharp faces and brilliant luster, ranging up to 15 cm in size, of a variety of habits from tabular to equant to scalenohedral, and often twinned under a variety of twin laws. At the magnesite deposit of Serra das Eguas(!), Brumado, BA, brilliant crystals and groups are found associated with magnesite. Also in BA in Novo Horizonte(*), crystallized hematite is often associated with epitaxial growths of golden colored rutile, forming stars of great beauty often embedded in clear quartz crystals. Uses: The most important source of iron. Hundreds of millions of tons are mined annually for the iron and steel industries worldwide. Ground hematite is also widely used as a red pigment and as a polishing compound (rouge). Compact black hematite or crystals are sometimes cut and faceted as a semiprecious gemstone. RG *DI:527, AM 47:1332(1962)*.

4.3.1.3 Eskolaite Cr_2O_3

Named in 1958 for Pentti Eskola (1883–1964), Finnish geologist, University of Helsinki. Corundum–hematite group. HEX-R $R\bar{3}c$. $a = 4.954$, $c = 13.584$, $Z = 6$, $D = 5.245$. *NBSC 539(5):22(1955)*. *38-1479*(syn): 3.63_7 2.67_{10} 2.48_9 2.18_4 1.815_4 1.672_9 1.465_3 1.432_4. Crystals prismatic, less commonly platy, with $\{0001\}$ and $\{10\underline{1}0\}$. Lustrous black, light green streak, vitreous luster. No cleavage. $H = 9$. $G = 5.18$. Nonmagnetic. Infusible. Insoluble in acids or alkalis. Chemical analysis showed 4.5% V_2O_3 and 0.5% Fe_2O_3 in isomorphous substitution in the mineral, giving the formula $(Cr_{1.90}V_{0.09}Fe_{0.01})O_3$. Pleochroic; O, emerald green; E, olive green. Absorption very strong. In RL, gray, with emerald green internal reflections. $R = 22.0$ (470 nm), 20.8 (546 nm), 20.0 (589 nm), 19.6 (650 nm). *P&J:157*. Found at the *Outokumpu copper mine, Finland*(!), in crystals to 10 mm associated with pyrrhotite, chalcopyrite, pentlandite, uvarovite, chromian spinel, quartz, calcite, and other minerals in a chromian tremolite skarn. As rolled pebbles from the Merume River, Guyana ("merumite"), intergrown with bracewellite, guyanaite, mcconnellite, and grimaldiite. As rounded crystals in stream gravels from the Serra das Almas, Parimirim, BA, Brazil, associated with bahianite and cassiterite. RG *AM 43:1098(1958), USGSPP 887:9(1976)*.

4.3.1.4 Karelianite V_2O_3

Named in 1963 for Karelia, the region in Finland where the Outokumpu mine is situated. Corundum–hematite group. HEX-R $R\bar{3}c$. $a = 4.954$, $c = 14.008$, $Z = 6$, $D = 5.016$. *NBSM 25(20):108(1984)*. *34-187*(syn): 3.66_7 2.71_{10} 2.48_8 2.19_4 1.83_3 1.70_9 1.47_2 1.43_3. Disseminated grains intergrown with sulfides. Black, black streak. Conchoidal fracture. $H = 8-9$. $G = 4.87$. In polished section, etched by 1:1 HNO_3; not affected by other reagents. Analyzed mineral contained about 4% Fe_2O_3, trace Mn, as isomorphous substitution. In RL, brownish olive-gray. No internal reflections. Weakly pleochroic, strongly anisotropic. $R = 15.0$ (470 nm), 16.9 (546 nm), 17.8 (589 nm), 18.9 (650 nm). *P&J:220*. From the *Outokumpu copper mine, Finland*, in veins in skarn, associated with pyrrhotite, chalcopyrite, chrome spinel, and coulsonite. RG *AM 48:33(1963)*.

4.3.2.1 Akdalaite $(Al_2O_3)_4 \cdot H_2O$

Named in 1970 for the Kazakh name for the locality. HEX $P6_1$ or $P6_122$. $a = 12.87$, $c = 14.97$, $Z = 18$, $D = 3.673$. *25-17*: 3.24_7 2.49_9 2.35_9 2.32_8 2.11_{10} 1.86_7 1.42_{10} 1.39_{10}. Accumulations of tabular and elongated tabular crystals showing hexagonal outline. White, vitreous–porcellanous luster. No cleavage, irregular fracture, brittle. $H = 7\frac{1}{4}$. $G = 3.68$. Uniaxial $(-)$; $N_O = 1.747$, $N_E = 1.741$; extinction parallel to elongation. Found in the *Solvech fluorite deposit, Karaganda region, Kazakhstan*, in veinlets cutting fluorite–magnetite–diopside–vesuvianite–andradite skarn and amesite–fluorite–muscovite rock. RG *AM 56:635(1971), ZVMO 99:333(1970), IGR 13:675(1971)*.

4.3.2.2 Ferrihydrite $Fe_5O_6(OH)_3 \cdot 3H_2O$; FeO(OH); (?)

Named in 1973 from the composition. HEX Space group unknown. $a = 5.08$, $c = 9.4$, $Z = 1$. *CCM 36:111(1988)*. *29-712*: 2.50_{10} 2.21_8 1.96_8 1.72_5 1.51_7 1.48_8. *AM 75:437(1988)*: 2.52_5 2.23_4 1.98_3 1.72_3 1.51_5 1.46_9. (Powder lines are few and broad, probably due to the colloidal particle size and poor structural organization of the mineral.) Yellow brown ochers to dark brown masses, earthy luster. No cleavage. $G_{syn} = 3.96$. Soft. Highly soluble in ammonium oxalate. Easily synthesized by slow hydrolysis of ferric salts at pH between 3 and 9.5. The presence of iron-fixing bacteria greatly increases the yield. The exact composition is uncertain, being variously given as FeO(OH); $Fe_5O_6(OH)_3 \cdot 3H_2O$; $5Fe_2O_3 \cdot 9H_2O$; $Fe_2O_3 \cdot 2FeO(OH) \cdot 2.6H_2O$, and others, probably due to the loosely held H_2O and hence variability of analytical results on various samples examined. Ferrihydrite, a primary precipitate of iron-fixing bacteria or of the natural hydrolysis of iron salts in solution, is probably a precursor of goethite(limonite); feroxyhite; and many other iron minerals, including some that form large iron deposits. From Kazakhstan; from various localities in Finland; near Hannover, Germany. Widely distributed elsewhere in the world, but seldom recognized or characterized. RG *AM 65:1044(1980), 60:485(1975); IGR 16,1131(1974); PICCM:333(1973)*.

PEROVSKITE GROUP

The perovskite group minerals have the general formula

$$ABO_3$$

where

A = Ca, Na, Ce, and rarely Sr
B = Ti or Nb.

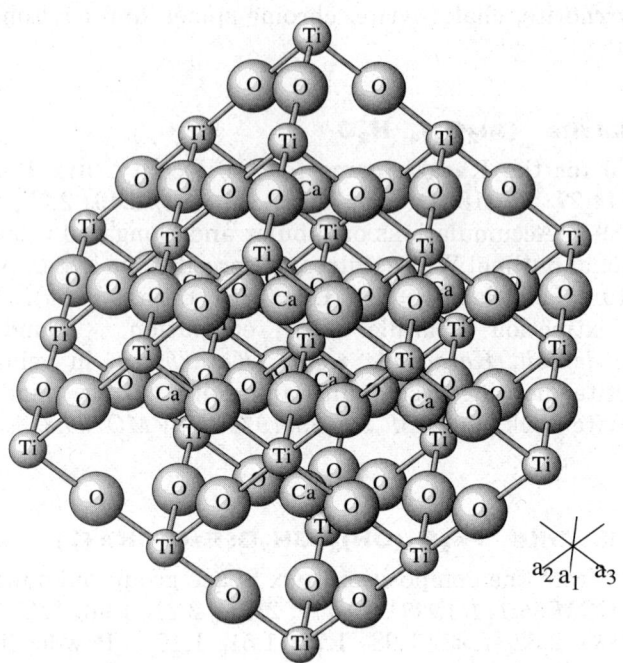

PEROVSKITE: CaTiO$_3$

Space group Pm3m. This is an idealized structure; perovskite itself has lower symmetry. Ca atoms replace O atoms in a CCP arrangement and Ti atoms are in selected octahedral interstices. Ca–O bonds are not shown.

The ideal perovskite structure is cubic, with a approximately 3.8 Å. The B ions are surrounded by regular octahedra of oxygen that share corners to form a three-dimensional framework. The A ions occupy large holes between the octahedra, and each is surrounded by 12 oxygens. The actual structure is related to the ideal cube by a shear on {010}, leaving $b \approx 7.64$ Å and a rhombus-shaped a–c face with $\beta = 90.80°$. *DHZ II:556, AC C39:1323(1983)*.

PEROVSKITE GROUP

Mineral	Formula	Space Group	a	b	c	D	
Perovskite	CaTiO$_3$	Pnma	5.441	7.644	5.381	4.036	4.3.3.1
Latrappite	(Ca,Na)(Nb,Ti)O$_3$	Pcmn	5.448	7.777	5.553	4.40	4.3.3.2
Loparite-(Ce)	(Ce,Na,Ca)(Ti,Nb)O$_3$	Pm3m	3.889			5.25	4.3.3.3
Lueshite	β-NaNbO$_3$	Pbma	5.569	15.523	5.505	4.575	4.3.3.4
Tausonite	SrTiO$_3$	Pm3m	3.905			4.85	4.3.3.5

4.3.3.1 Perovskite CaTiO$_3$

Named in 1839 after L. A. Perovski (1792–1856), Russian mineralogist, of St. Petersburg. Perovskite group. ORTH *Pnma*. $a = 5.441$, $b = 7.644$, $c = 5.381$, $Z = 4$, $D = 4.036$. *NBSM 25(9):17(1971)*. PD *NBSM*(syn): 3.82_1 2.72_4 2.70_{10} 1.91_5 1.567_1 1.563_2 1.557_3 1.35_1. **Habit**: Crystals cubic, sometimes highly modified, but the faces often irregularly distributed. As cubo-octahedra or octahedra, especially in cerian or niobian varieties. Rarely massive granular, or reniform. Twinning on {111}; also as penetration or lamellar twins. **Physical properties**: Black, brown, yellow, with adamantine luster, or metallic when black; gray to colorless streak. Cleavage {001}, imperfect; subconchoidal to uneven fracture. $H = 5\frac{1}{2}$. $G = 4.01$. Nonfluorescent. Decomposed by hot concentrated H_2SO_4 and by HF. Infusible. **Chemistry**: Wide compositional variation is the rule, as Nb substitutes for Ti up to 44.9 wt % Nb_2O_5 (Nb/Ti = 2:5); Ce for Ca at least up to Ce/Ca = 4:7, and Na for Ca to at least Na/Ca = 1:4. Early analyses showing even higher levels of substitution were not certainly of perovskite but may have been of loparite-(Ce) or latrappite. The unit cell parameters of perovskite increase with Nb content, up to $a = 5.45$, $b = 7.78$, $c = 5.55$ with concomitant X-ray changes. **Optics**: Biaxial (+); XYZ = *acb*; $N = 2.34$; $2V = \pm 90°$; r > v; absorption Z > X. In RL, isotropic gray, with yellow to orange-brown internal reflections. $R = 17.8$ (470 nm), 16.7 (546 nm), 16.3 (589 nm), 16.0 (650 nm). *P&J:296*. **Occurrence**: Perovskite occurs both as an accessory mineral of the magmatic stage and as a late-stage deuteric mineral in basic igneous rocks, especially carbonatites and related rocks. Also in basic pegmatites associated with such rocks and in metamorphosed limestones at the contact with alkalic or basic intrusives. Also found in chlorite or talc schists. Perovskite is thought to be a major constituent of the upper mantle portion of the earth. **Localities**: Niobian perovskite is found in limestone metamorphosed at the contact with alkaline rocks at Magnet Cove, AR(*). Abundant as an accessory in carbonatite at Powderhorn, Gunnison Co., CO. In nepheline syenite pegmatite, Bearpaw Mts., MT. Cerian perovskite is found in alkalic pegmatite at Moose Creek, near Leanchoil, BC, Canada. From the Gardiner complex, Kangerlussuaq Fjord, eastern Greenland, in large black pseudo-octahedral crystals up to 8 cm(!!). In Switzerland in crystals and reniform masses in talc schist near the Findelen glacier, Zermatt, Valais(*). In Italy at the Wildkreuzjoch, Pfitschtal, Trentino(!); also in Piedmont at Emarese, Valle d'Aosta, and so on. At Schelingen in the Kaiserstuhl, Baden, Germany, in contact metamorphic limestone, and as an accessory mineral in alkalic–basic intrusives of this district. In chlorite schist, *Zlatoust district, Ural Mts., Russia*(!). In Brazil, in magnetite-rich carbonatite at Catalao, GO, and near Bagagem, MG. RG *DI:730, CM 7:683(1963)*.

4.3.3.2 Latrappite (Ca,Na)(Nb,Ti)O$_3$

Named in 1964 for La Trappe, PQ, Canada, a village close to where the mineral was found. Perovskite group. ORTH *Pcmn*. $a = 5.448$, $b = 7.777$,

$c = 5.553$, $Z = 4$, $D = 4.40$. CM 8:121(1964). 16-694: 3.89_8 2.77_3 2.74_{10} 1.94_6 1.74_1 1.60_1 1.58_3 1.37_1. Small black cubes, gray streak, submetallic luster. Cleavage {001}, imperfect; subconchoidal fracture. $H = 5\frac{1}{2}$. $G = 4.40$. Latrappite has Nb/Ti > 1:1. Analyzed mineral shows 43.9% Nb_2O_5, Nb/Ti = 5:2. A nearly complete solid solution series exists between perovskite, latrappite, and lueshite. From *Oka, QUE, Canada*, in a pyrochlore-rich carbonatite, in white calcite associated with diopside, biotite, and apatite. Also from Vogtsburg, Kaiserstuhl, Germany, and Yva Prov., Sri Lanka. RG CM 7:683(1963), AM 49:819(1964).

4.3.3.3 Loparite-(Ce) (Ce, Na, Ca)(Ti, Nb)O_3

Named in 1922 after the Russian name for the Lapp inhabitants of the Kola Peninsula. Perovskite group. ISO $Pm3m$. $a = 3.889$, $Z = 0.3$, $D = 5.25$. DANS 140:1090(1961). 35-618: 2.75_{10} 2.24_2 1.94_4 1.59_5 1.37_3 1.23_2 1.04_3 0.917_2. As cubes. Interpenetration twins on {111}. Loparite is defined as that portion of the solid-solution series in which REE/Ca > 1:1. Adequate studies of analyzed materials are lacking. Some weak extra lines in the powder diagram suggest that the true symmetry is other than isometric. The value for Z is unsatisfactory; the indicated cell volume based on $a = 3.89$ Å is almost exactly one-fourth of the cell volume based on an orthorhombic perovskite cell. Needs further study. Black, brownish streak, metallic luster. No cleavage, conchoidal fracture. $H = 5\frac{1}{2}$. $G = 4.86$. Isotropic; $N = 2.33$. Translucent red in fine chips. From Buer, Bjorkedal, Possgrunn, southern Norway; *Khibina Tundra, Kola Peninsula, Russia*; also from an unspecified location in Siberia in albitized and greisenized granites. RG CMP:84-365(1983), DANS 140:1407(1961), MA 16:455.

4.3.3.4 Lueshite β-NaNbO_3

Names in 1959 for the locality. Perovskite group. Polymorph: natroniobite. ORTH $Pbma$. $a = 5.569$, $b = 15.523$, $c = 5.505$, $Z = 8$, $D = 4.575$. NBSM 25(18):64(1981). 19-1221(syn): 3.9_9 2.76_{10} 1.949_7 1.746_4 1.594_7 1.379_3 1.235_2 1.044_2. Black pseudocubes with faces lightly striated and gray streak. Cleavage {001}, imperfect, easy. $H = 5\frac{1}{2}$. $G = 4.44$. Biaxial; $N = 2.30$. Powderhorn, CO, in carbonatite. Mt. St.-Hilaire, PQ, Canada. Igdlunguaq, Greenland, in nepheline syenite, associated with neptunite, analcime, and altered eudialyte. Originally found in the Ilimaussaq igneous complex, Greenland, and named igdloite; later found in carbonatite at *Lueshe, 150 km N of Goma, Zaire*, as incrustations on a compact yellow magnesian–calcian mica, at the contact with a cancrinite syenite, and renamed lueshite. RG AM 44:1327(1959), 46:1004(1961).

4.3.3.5 Tausonite SrTiO_3

Named in 1984 for L. V. Tauson (1917–1989), Russian geochemist. Perovskite group. ISO $Pm3m$. $a = 3.905$, $Z = 1$, $D = 4.85$. 35-734(syn): 3.90_1 2.76_{10} 2.25_3 1.95_5 1.59_4 1.38_3 1.24_2 1.04_2. As cubes and cubo-octahedrons; also irregular grains. Tausonite has cell contents close to loparite and is probably a member of the perovskite group. A mineral intermediate between loparite and tausonite has been described from Brazil and Paraguay in carbonatites. Red, reddish brown, gray; adamantine luster. Conchoidal fracture, brittle. H = 6–6$\frac{1}{2}$. G = 4.88. Insoluble in dilute acids. Isotropic, sometimes slightly anisotropic. From alkalic rocks of the *Murunskii massif, Aldan, Siberia, Russia*. RG *AM 70:218(1985), ZVMO 113:86(1984).*

4.3.4.1 Natroniobite NaNbO$_3$

Named in 1960 for the composition. Polymorph: lueshite. MON Space group unknown. *26-1380:* 3.79$_4$ 3.06$_9$ 2.97$_{10}$ 1.89$_5$ 1.72$_6$ 1.60$_8$ 1.59$_4$ 1.54$_4$. Irregular grains and fine-grained aggregates. Yellowish, brownish, blackish. H = 5$\frac{1}{2}$–6. G = 4.40. Biaxial (+); X $\wedge$ c = 10–15°; N$_x$ = 2.10–2.13, N$_y$ = 2.19–2.21, N$_z$ = 2.21–2.24; 2V = 10–30°; r < v. From the *Lesnaya Baraka* and *Sallanlatvi massifs, Kola Peninsula, Russia*, in dolomitic carbonatites with apatite, phlogopite, perovskite, and pyrochlore. RG *AM 47:1483(1962), ZVMO 92:190(1962).*

ILMENITE GROUP

The ilmenite group is isostructural within itself and nearly isostructural with the corundum group, and conforms to the formula

$$ATiO_3$$

where

A = Fe^{2+}, Mg, Mn, Zn, Fe^{3+}, and rarely Sb^{5+}

and a structure in space group $R\bar{3}$. The structure has a rhombohedrally centered lattice, similar to that of the corundum–hematite group but with Ti and another element replacing the A$_2$ of the corundum structure. Along the threefold rotoinversion axis of ilmenite pairs of Ti^{4+} ions alternate with pairs of Fe^{2+} ions, forming cation layers perpendicular to that threefold axis. Each cation layer is a mixture of Fe^{2+} and Ti^{4+}.

ILMENITE GROUP

Mineral	Formula	Space Group	a	c	D	
Ilmenite	FeTiO$_3$	$R\bar{3}$	5.088	14.093	4.784	4.3.5.1
Geikielite	MgTiO$_3$	$R\bar{3}$	5.054	13.898	3.895	4.3.5.2
Pyrophanite	MnTiO$_3$	$R\bar{3}$	5.140	14.290	4.596	4.3.5.3
Ecandrewsite	(Zn, Fe, Mn)TiO$_3$	$R\bar{3}$	5.090	14.036	4.99	4.3.5.4
Melanostibite	Mn(Fe^{3+}, Sb^{5+})O$_3$	$R\bar{3}$	5.226	14.325	5.63	4.3.5.5

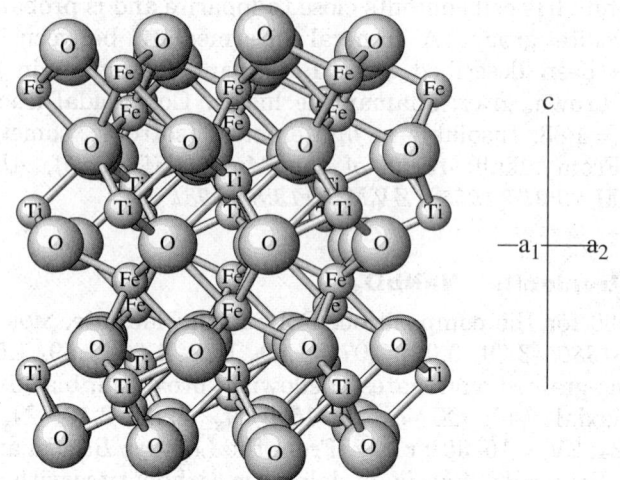

ILMENITE: FeTiO₃

The structure is similar to hematite or corundum, but the pairs of octahedra that share faces are occupied by Fe + Ti.

While the minerals of this group are very similar structurally and in cell parameters to those of the corundum–hematite group, there are important differences. There is nearly complete solid solution between Fe and Mn, and between Fe and Mg, and at least partial solution between Zn and Fe, Mn. In contrast, in the corundum group the minerals are virtually end members with exceedingly small solid solutions of other elements. *DHZ II:543(1992), MJJ 14:179(1989).*

Note: Ilmenite is an abundant mineral; all the others are rare.

4.3.5.1 Ilmenite FeTiO₃

Named in 1827 after the locality in the Ilmenskie Mts., Russia. Ilmenite group. HEX-R $R\bar{3}$. $a = 5.088$, $c = 14.093$, $Z = 6$, $D = 4.784$. *NBSM 25(15):34(1978)*. *29-733*(syn): 3.74_3 2.75_{10} 2.54_7 2.24_3 1.868_4 1.726_6 1.506_3 1.469_4. **Habit**: Crystals usually thick tabular {0001}; sometimes thin laminae; also acute rhombohedral. Compact, massive; as disseminated grains, or as a heavy constituent in sand. Twinning on {10$\bar{1}$1}, lamellar. **Physical properties**: Black, black streak, metallic to submetallic luster. No cleavage, parting {0001} and {10$\bar{1}$1}, conchoidal fracture. H = 5–6. G = 4.72. Nonmagnetic or only slightly magnetic, with the latter likely to represent higher Fe$_e$O$_3$ content in the structure, or microscopic inclusions of magnetite. **Tests**: Slowly soluble in hot 1:1 HCl. Infusible. **Optics**: Uniaxial (−); birefringence very strong; absorption E > O, faint. Transmits only red light. In polished section, grayish white. VHN$_{100}$ = 566–698. Anisotropy strong; bireflectance weak. R = 19.7–16.9 (470 nm), 19.2–16.4 (546 nm), 19.6–17.0 (589 nm), 20.2–18.0 (650 nm). *C&S:166, BM(NH)(1985)*. **Chemistry**: Extensive solid solution exists between ilmenite and geikielite (to 70 mol % MgTiO₃) and between ilmenite and pyrophanite (to 64 mol % MnTiO₃). But the majority of terrestrial ilmenites contain Fe₂O₃ up to 6% maximum and only minor amounts of MgO and

4.3.5.2 Geikielite

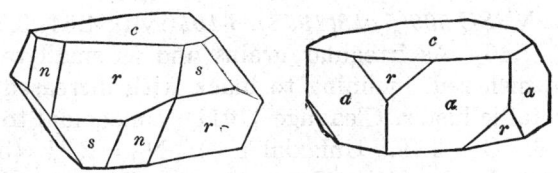

ILMENITE: FeTiO$_3$
Forms: c{0001}, r{10$\bar{1}$1}, n{22$\bar{4}$3},

MnO. Exceptions are the ilmenites in kimberlites and ultrabasic xenoliths, which frequently contain major amounts of MgTiO$_3$; the pyrophanite molecule may become important in carbonatites and certain differentiated acid rocks. Above 6%, Fe$_2$O$_3$ is believed to be due to admixed or exsolved magnetite or hematite. **Occurrence:** Ilmenite occurs principally as a common accessory mineral in igneous rocks such as gabbros, diorites, and anorthosites, as veins and disseminated deposits, sometimes of large extent. Massive accumulations may be found as dikes cutting anorthosite. Widespread in granitic rocks, as a minor accessory. In some pegmatites, especially those of intermediate or alkalic composition. Heavy-mineral concentrates derived from beach sands may contain major ilmenite, associated with magnetite, rutile, zircon, and other dense minerals, which are recovered from such deposits in Florida, India, Australia, and Brazil. **Localities:** Crystals are found in pegmatites at Chester and at Quincy, MA. At Sanford Lake, Tahawus, NY, a large deposit is mined associated with magnetite and garnet in anorthosite(*), and at Warwick, Amity, and Monroe in metamorphosed limestone associated with spinel, chondrodite, and rutile. In VA at Roseland in a large dike with apatite and rutile ("nelsonite" rock). Iron Mt., WY, in a large ilmenite–magnetite dike in anorthosite. Summit, OR. In PQ, Canada, at St. Urbain, Charlevoix, as veinlets disseminated in anorthosite. Bancroft, ON, in large crystals(*). In Norway at Kragerö, as large crystals in veins in diorite(!). Northern Sweden at Routivare, associated with pyrrhotite in altered gabbro. Crystals are found at Alp Uercheltini, Binnatal, Switzerland, and exquisite ilmenite roses at Maderanertal, Uri(!). In *Russia* in the *Ilmenskie Mts., southern Ural Mts.*(!), in crystals to 20 cm. Large beach sand deposits are found at Travancore, southern India. At the Ambatofotsikely pegmatite, Betafo, Madagascar, in crystals to 20 cm(!). In Australia, at Woodstock, WA (manganoan ilmenite). **Uses:** Ilmenite is a major source of TiO$_2$, which is used as a pigment in paints and enamels and as a raw material in the manufacture of titanium. RG *DI:534*, *DHZII 543(1992)*.

4.3.5.2 Geikielite MgTiO$_3$

Named in 1892 for Archibald Geikie (1835–1924), director of the Geological Survey of Great Britain. Ilmenite group. HEX-R $R\bar{3}$. $a = 5.054$, $c = 13.898$,

$Z = 6$, $D = 3.895$. *NBSC 539(5):43(1955)*. *6-494*(syn): 4.64_3 3.70_5 2.72_{10} 2.53_6 2.22_7 1.85_4 1.71_6 1.46_4. As irregular grains and as small tabular crystals. Ruby-red to brownish red, inclining to black with increased Fe^{2+} content; metallic to submetallic luster. Cleavage $\{10\bar{1}1\}$, conchoidal to subconchoidal fracture. $H = 5-6$. $G = 3.79$. Uniaxial $(-)$; $N_O = 2.31$ (brownish red), $N_E = 1.95$ (pinkish). In CA at the Jensen quarry, Riverside Co., in disseminated crystals or veinlets in marbles associated with spinel, diopside, fosterite, pyrite, and pyrrhotite, in crystals close to end-member composition; and in magnesian marbles in the Sta. Lucia Mts., Monterrey Co. From *Sri Lanka* as pebbles in gem-bearing gravels. RG *AM 44:879(1959), 34:835(1949); DI:535.*

4.3.5.3 Pyrophanite MnTiO₃

Named in 1890 from the Greek for *fire*, and *to appear*, in allusion to its red color. Ilmenite group. HEX-R $R\bar{3}$. $a = 5.140$, $c = 14.290$, $Z = 6$, $D = 4.596$. *NBSM 25(15):42(1978)*. *29-902*(syn): 3.78_3 2.79_{10} 2.57_7 2.26_3 1.89_4 1.75_4 1.52_4 1.48_3. Fine grained, scaly. Deep blood red, metallic to submetallic luster. Cleavage $\{02\bar{2}1\}$, perfect; $\{10\bar{1}2\}$, less so; conchoidal to subconchoidal fracture. $H = 5-6$. $G = 4.54$. Uniaxial $(-)$; $N_O = 2.481$ (yellowish red), $N_E = 2.21$ (Na). In RL, gray. Reddish internal reflections. Weakly pleochroic, moderately anisotropic. $R = 22-18$ (470 nm), 19–16 (546 nm); 18–14 (589 nm). From the Sterling Hill mine, Ogdensburg, NJ, associated with pyroxene, gahnite, calcite, and biotite. *Harstig mine, Värmland, Sweden*, associated with ganophyllite, garnet, and manganophyllite, in cavities later filled with calcite. At Queluz, SW of Ouro Preto, MG, Brazil. RG *DI:535, CM 23:491(1985)*.

4.3.5.4 Ecandrewsite (Zn,Fe,Mn)TiO₃

Named in 1988 for E. C. Andrews, pioneering geologist in the Broken Hill region of New South Wales. Ilmenite group. HEX-R $R\bar{3}$. $a = 5.090$, $c = 14.036$, $Z = 6$, $D = 4.99$. *26-1500*(syn): 3.72_2 2.73_{10} 2.54_8 2.23_2 1.860_4 1.713_4 1.500_3 1.466_3. Irregular grains; also euhedral tabular crystals. Although ecandrewsite is a member of the ilmenite group, zinc substitution is virtually absent in other members of the group. Dark brown to black, dark brown to black streak, submetallic luster. No cleavage observed. $H = 5\frac{1}{2}$, $VHN_{100} = 500-600$. In RL, grayish white with a pinkish tinge. Weakly pleochroic, strongly anisotropic. $R = 19.5-17.5$ (470 nm), 19.3–17.4 (546 nm), 19.2–17.3 (589 nm), 18.9–17.1 (650 nm). San Valentin mine, La Union district, Sierra de Cartagena, Spain, in oxidized ore intergrown with zincian ilmenite. From the *Melbourne-Rockwell mine, northern Broken Hill, NSW, Australia*, with almandine-spessartine, ferroan gahnite, and rutile in quartz-rich metasediments. RG *MM 52:237(1988), AM 74:501(1989).*

4.3.5.5 Melanostibite Mn(Fe³⁺,Sb⁵⁺)O₃

Named in 1892 in allusion to its color and antimony content. Ilmenite group. HEX-R $R\bar{3}$. $a = 5.226$, $c = 14.325$, $Z = 6$, $D = 5.63$. *AM 53:1104(1968)*. *20-699:*

4.78_5 4.31_4 3.82_6 2.81_{10} 2.61_8 2.29_4 1.92_5 1.77_8. Foliated masses; also tiny striated crystals with prism and pyramid. Twinning on {0001} by rotation. Black, cherry-red streak. Cleavage {0001}, perfect. H = 4. Soluble in 1:1 HCl. Weakly magnetic after ignition. Analytical data show that the Fe/Sb ratio is close to 1:1. Although melanostibite contains no titanium, the Fe^{3+} and Sb^{5+} is considered to be the equivalent of Ti^{4+}. This, combined with the similar unit cell dimensions and resulting formula, puts the mineral within the ilmenite group. Opaque. From the *Sjö mine, Grythytte parish, Örebro, Sweden*, in veinlets in dolomite. RG *DII:1041, AMG 4:449(1967)*.

4.3.5.6 Briziite NaSbO$_3$

Named in 1994 for G. Brizzi (1936–1992), Italian amateur mineralogist who collected the original specimens. Ilmenite group. HEX $R\bar{3}$. $a = 5.301$, $c = 15.932$, Z = 6, D = 4.95. *PD:* 5.30_5 3.00_5 2.65_7 2.37_7 1.874_{10} 1.471_7 1.185_5. Crystals minute, platy to thin tabular, hexagonal in outline, to 0.2 mm. Aggregates of crystals are pale pink to yellow. Cleavage {001}, perfect; flexible. $VHN_{15} = 57$. G = 4.8. Nonfluorescent. Uniaxial (−); $N_E = 1.631$, $N_O = 1.84_{calc}$. Found at the *Cetine antimony mine, Siena, Tuscany, Italy*, encrusting weathered waste material and slag. RG *EJM 6:667(1994)*.

4.3.6.1 Macedonite PbTiO$_3$

Named in 1971 for the locality, in former Yugoslavia. TET $P4/mmm$. $a = 3.889$, $c = 4.209$, Z = 1, D = 8.09. *6-452*(syn): 4.15_3 3.90_5 2.84_{10} 2.76_5 2.30_4 1.95_3 1.66_2 1.61_4. *AM*(nat): 3.84_6 2.84_{10} 2.73_8 2.28_7 1.95_7 1.60_9 1.37_5 1.23_5. Irregular grains and rare stout prismatic crystals with pseudocubic habit. Black, vitreous luster. Cleavage not observed. H = $5\frac{1}{2}$. G = 7.82. Soluble in hot HCl, HNO$_3$, and H$_2$SO$_4$. In RL, grayish white. Internal reflections yellow-brown and dark brown. Pleochroism not observed; very weakly anisotropic, almost isotropic. From *Crni Kamen, near Prilep, Macedonia*, in amazonite-rich quartz syenite veins, associated with rutile, zircon, and barite. RG *AM 56:387(1971), LIT 4:101(1971)*.

4.3.7.1 Maghemite γ-Fe$_{2.67}$O$_4$

Named in 1927 from the first syllables of *magnetite* and *hematite*, in allusion to the magnetic properties and composition. Synonym: γ-Fe$_2$O$_3$. ISO $P4_132$. $a = 8.351$, Z = 8, D = 4.860. *24-81*(low temp.)(syn): 2.95_3 2.78_2 2.52_{10} 2.08_2 1.70_1 1.61_3 1.48_5 1.09_2. *25-1402*(high temp.)(syn): 2.95_3 2.51_{10} 2.09_2 1.70_1 1.60_2 1.47_4 1.27_1 1.09_1. Massive. Brown, brown streak. H = 5. Highly magnetic, about the same as magnetite. The composition is essentially Fe$_2$O$_3$, but structural considerations show that the formula should be written Fe$_{2.67}$O$_4$. Usually impure, admixed with silica, gibbsite, kaolin, goethite, and other iron oxides and hydroxides. Isotropic; N > 2.40. In RL, white to grayish blue, isotropic. Reflectance: > magnetite, < hematite. Formed by slow oxidation at low temperatures of preexisting iron minerals such as siderite, magnetite, or lepidocrocite. Common in gossans, also in laterites, and very widely distributed throughout the world in rocks of this type, especially in tropic regions. Common in the oxidized outcrops of magnetite deposits. First recognized in a gossan at *Iron Mt., Shasta Co., CA.* Durant, OK, as rounded ocherous masses. Windpass mine, BC, Canada. Katzenbuchel, Baden-Württemberg, Germany. In the Bushveld complex, South Africa. Near Sydney, NSW, Australia, in laterites. RG *DI:708, AM 55:925(1970), PD 3:104(1988), CMP 69:249(1979).*

4.3.7.2 Bixbyite (Mn, Fe)$_2$O$_3$

Named in 1897 for Maynard Bixby (1853–1935) of Salt Lake City, UT, miner and mineral collector, discoverer of red beryl in Utah. Synonym (in part): partridgeite (=Mn$_2$O$_3$), which was named in 1943 for F. C. Partridge (1903–1939), senior mineralogist, Geological Survey of South Africa. The name *partridgeite*, a virtually Fe-free, orthorhombic variety of bixbyite, has never been submitted to the IMA, CNMMN (International Mineralogical Association, Committee on New Minerals and Mineral Names), as it was found before the IMA came into existence; however, the structure is distinct from that of bixbyite. ISO $Ia3$. $a = 9.415$, $Z = 16$, $D = 5.026$. *NBSC 539(9):37(1960). 24-508*(syn Mn$_2$O$_3$): 3.84_2 2.719_{10} 2.35_1 2.01_1 1.846_1 1.664_3 $1.452_{<1}$ 1.420_1. Unit cell for Mn$_2$O$_3$: ORTH $Pcab$. $a = 9.416$, $b = 9.424$, $c = 9.405$, $Z = 16$, $D = 5.031$. *NBSM 25(11):95(1974)*. As cubes, sometimes modified by {112}; penetration twinning on {111}. Structure: *AC(B)27:821(1971)*. Black, black streak, metallic to submetallic luster. Cleavage {111} in traces, irregular fracture. $H = 6$–$6\frac{1}{2}$. $G = 4.945$. Difficultly fusible. Slowly soluble in 1:1 HCl, with evolution of Cl$_2$. In RL, gray, weakly anisotropic. $R = 22.9$ (470 nm), 23.6 (546 nm), 23.1 (589 nm), 22.3 (650 nm). *P&J:94*. The amount of Fe that can enter the bixbyite composition varies with the temperature of formation of the mineral, from about 1 to 60% Fe$_2$O$_3$. Bixbyite from the type locality in Utah, a high-temperature environment, contains about 1:1 Fe$_2$O$_3$/Mn$_2$O$_3$, or usually an excess of Fe$_2$O$_3$. From 1 to 60% the structure is isometric and a solid-solution series exists. Below 1% Fe$_2$O$_3$ the structure becomes orthorhombic, pseudoisometric. Such material, found only in Postmasburg and Mamatwan, Cape Prov., South Africa, [partridgeite] and exhibits strong anisotropism in polished section. Synthetic Mn$_2$O$_3$ also has this structure. *EG 78:1111(1983)*. Most minerals that have been identified as bixbyite have Mn$_2$O$_3$ > Fe$_2$O$_3$; outside the type locality, hardly any bixbyite is Fe dominant. For those that are Mn dominant and still have the isometric structure, the name *manganbixbyite* would be appropriate, while for the Fe-dominant variety, *bixbyite* would be retained. This question is being submitted to the IMA, CNMMN. From on and near *Topaz Mt., Thomas Range, Juab Co., UT*(!), in lithophysae in rhyolite, associated with topaz, spessartine, red beryl, and hematite; also with fluorite. Black Range tin district, NM, in rhyolite (manganbixbyite). Barranca del Cobre, Chih., Mexico. Långban, Värmland, Sweden, in metamorphosed manganese ores (manganbixbyite); Ultevis, Sweden (bixbyite); Ribes, Girona, Spain;

Sitapár, Central Provs., India, with braunite and hollandite (manganbixbyite). Postmasburg, Cape Prov., South Africa (manganbixbyite). Valle de las Plumas, Patagonia, Argentina (manganbixbyite). RG *DI:550*, *ZK 129:360(1969)*, *NJMM 1973:426*, *AM 74:1325(1989)*.

4.3.8.1 Avicennite Tl_2O_3

Named in 1958 for Abu Ali Ibn Sina ("Avicenna") (980–1037), Arab physician and scientist, who lived in Bukhara. ISO $Ia3$. $a = 10.543$, $Z = 16$, $D = 10.354$. *NBSM 25(16):77(1979)*. *NBSM*(syn): 4.30_1 3.04_{10} 2.64_4 2.48_1 2.07_1 1.86_3 1.59_3 1.55_1. As small grayish-black cubes, grayish-black streak, metallic luster. Indistinct cleavage, uneven fracture, very brittle. Dissolves with difficulty in concentrated acids. From near *Dzhuzumli, Mt. Zirabulaksk, Bukhara, Uzbekistan*, in a hematite–calcite vein cutting banded, silicified limestones. Vizarron de Marques, Cadereyta, QRO, Mexico, as a coating on quartz and hematite. RG *AM 44:1324(1959)*, *MA 72:3282(1972)*.

4.3.9.1 Arsenolite As_2O_3

Named in 1854 for the composition. Dimorph: claudetite. ISO $Fd3m$. $a = 11.074$, $Z = 16$, $D = 3.865$. *NBSC 539(1):51(1953)*. *36-1490*(syn): 6.40_4 3.20_{10} 2.77_3 2.54_3 2.13_1 1.959_2 1.670_1 1.552_1. Octahedral; also botryoidal or stalactitic; earthy. White, yellowish, reddish; vitreous to silky luster. Cleavage {111}; conchoidal fracture. $H = 1\frac{1}{2}$. $G = 3.87$. In closed-tube sublimes, condensing above in small octahedra. Slightly soluble in hot water. Tastes astringent, sweetish (**poisonous**). Isotropic; $N = 1.755$ (Na). Formed from the oxidation of arsenic and arsenic sulfides; also formed as a sublimate in mine fires and burning coal seams. In NV from the Ophir mine, as an alteration of native As, and in fine crystals from the White Caps mine, Manhattan(!), from the burning of the shaft. In CA at the Armagosa mine, San Bernardino Co., and as an alteration product of enargite at the Exchequer and Monitor mines, Alpine Co. In Canada at Watson Creek, BC, and near Lake Wanapitei, Sudbury district, ON. *Andreasberg, Harz, Germany*, as an alteration of smaltite and other arsenides and as an alteration of arsenic at the Lauta mine near Marienberg, Saxony. In France from burning coal dumps at Sorbier, near St. Étienne, Loire, and as a direct oxidation product on native As at Giftgrube vein, Rauenthal, Ste.-Marie-aux-Mines, Haute-Rhin. At Sondalo, Sondrio, and Borgofranco in Torino, Italy. Morococha, Peru, with realgar and orpiment. RG *DI:543*.

4.3.9.2 Sénarmontite Sb_2O_3

Named in 1851 for Henri de Sénarmont (1808–1862), professor of mineralogy at the School of Mines in Paris, who first described the species. Dimorph: valentinite. ISO $Fd3m$. $a = 11.152$, $Z = 16$, $D = 5.583$. *NBSC 539(3):31(1954)*. *42-1466*(syn): 6.44_2 3.22_{10} 2.79_3 2.56_1 1.971_3 1.681_3 1.610_1 1.279_1. Structure: *AC(B) 31:2016(1975)*. Octahedral. Crystals; also granular, massive, and as crusts. Above 460° senarmontite is isometric, but at ordinary temperatures becomes triclinic, pseudoisometric by twinning. *BM 57:90(1934)*. Colorless, grayish white, resinous to subadamantine luster. Cleavage {111} in traces, uneven fracture, brittle. H = 2–2$\frac{1}{2}$. G = 5.50. Isotropic; N = 2.087 (Na). Birefringence anomalous, strong, often arranged in zones or segments due to twinning. In Canada with valentinite and kermesite as an alteration of stibnite and antimony at South Ham, Wolfe Co., PQ. At Pernek, Slovakia, with kermesite. Nieddoris, Sardinia, Italy. Anzat-le-Luguet, Puy de Dôme, France. Niccioleta, Grosseto, northern Italy. Abundant and well crystallized at *Hamimat, Constantine, Algeria*(!), with kermesite, valentinite, and other minerals as an alteration of stibnite in veins in a bituminous marly sediment. Also common as an artificial product in old antimony smelters. RG *DI:544*.

4.3.10.1 Claudetite As$_2$O$_3$

Named in 1868 for Frederick Claudet, French chemist, who first described the natural mineral. Dimorph: arsenolite. MON $P2_1/n$. $a = 5.339$, $b = 12.984$, $c = 4.541$, $\beta = 94.27°$, Z = 4, D = 4.186. *NBSM 25(3):9(1964)*. *NBSM*(syn): 4.92$_3$ 3.45$_5$ 3.36$_2$ 3.33$_2$ 3.25$_{10}$ 2.77$_4$ 2.64$_2$ 2.26$_3$. Crystals thin plates on {010}, elongated [001], with {111} and {$\bar{1}$11} prominent, resembling gypsum. Twinning on {100} as penetration or contact twins. Colorless to white; vitreous luster, pearly on cleavage. Cleavage {010}, perfect; fibrous fracture, parallel to {110}; very flexible. H = 2$\frac{1}{2}$. G = 4.15. In closed-tube sublimes, condensing above as small octahedrons of arsenolite. Soluble in hot alkali solutions (**highly poisonous**). Biaxial (+); X $\wedge$ c = 84°; Y = b; Z $\wedge$ c = 6°; N$_x$ = 1.871, N$_y$ = 1.92, N$_z$ = 2.01 (Na); r < v. Occurs as a product of oxidation of orpiment, realgar, and other arsenic minerals. Also formed in some mine fires and as a condensate from gases emanating from burning coal seams. Imperial Co., CA, about 50 km N of Yuma, AZ, as an alteration product of realgar, and in Trinity Co. at Island Mt. as crusts in the gossan of a pyrrhotite deposit. *Santo Domingo mine, Portugal*. At Calañas, Andalusia, Spain. With native sulfur and orpiment at the La Salle mine, Decazeville, Aveyron, France. Well crystallized at Smolnik, Slovakia(!), as a sublimation product. RG *DI:545*, *AM 36:833(1951)*, *SR 41A:213(1975)*.

4.3.10.2 Bismite Bi$_2$O$_3$

Named in 1868 for the composition. MON $P2_1c$. $a = 5.850$, $b = 8.170$, $c = 7.512$, $\beta = 112.98°$, Z = 4, D = 9.36. *WEW(1989)*. *41-1449*(syn): 3.31$_4$ 3.26$_{10}$ 2.71$_5$ 2.69$_3$ 2.56$_3$ 1.96$_5$ 1.75$_5$ 1.51$_3$. Massive, granular to earthy. Yellow to grayish green, subadamantine to earthy luster. Uneven fracture. H = 4$\frac{1}{2}$. G = 8.6–9.2. Biaxial; N > 2.42. In pegmatite at Rincon, San Diego Co., CA, as microcrystals associated with bismutite and pucherite. Meymac, Corrèze, France, with bismutite. *Calavi, Potosi, Bolivia*, as an alteration crust on bismuth. Numerous other localities have been reported, such as Petaca, NM; Schneeberg, Germany; Cornwall, England; Mt. McDonald, TAS, Australia, and elsewhere, but the minerals are not certainly of this species. RG *DI:599*, *AM 28:521(1943)*, *NBSM 25(3):17(1964)*.

4.3.11.1 Valentinite Sb_2O_3

Named in 1845 after Basilius Valentinus, probably a mythical monk and alchemist of the late Middle Ages. Many works published under his name ("a literary forgery") were on the subject of antimony and its compounds. He is alleged to have been born in 1394, but no solid evidence of his life has been found; much of what was attributed to him had been published by others at an earlier date. *PII:183*. Dimorph: senarmontite. ORTH *Pccn*. $a = 4.914$, $b = 12.468$, $c = 5.421$, $Z = 4$, $D = 5.832$. *NBSC 539(10):6(1960)*. *11-689*(syn): 4.57_2 3.49_3 3.17_2 3.14_{10} 3.12_8 2.65_1 1.93_1 1.80_2. Crystals prismatic either [001] or [100]; also tabular {010}. Often in stellate or fan-shaped groups and as aggregates of thin plates. Also massive, lamellar, or granular. Colorless, white, yellowish, reddish; adamantine luster, pearly on cleavage. Cleavage {110}, perfect; {010}, imperfect; brittle. $H = 2\frac{1}{2}$–3. $G = 5.83$. Easily fusible in closed tube. Soluble in HCl. Colored brown and slowly dissolved by $(NH_4)_2S$ solution. Biaxial $(-)$; $XYZ = abc$; $N_x = 2.18$, $N_y = 2.35$, $N_z = 2.35$ (Na); 2V very small; $r < v$, marked. From the Ochoco district, Crook Co., OR(!), in good crystals. Picahotes mine, San Benito Co., CA, with stibnite. In Canada at South Ham, Wolfe Co., PQ, and Prince William, NB. In Mexico at Sombrerete, ZAC, and Arechuybo, CHIH. *Les Chalanches, near Allemont, Isère, France*, as an oxidation product of antimony and allemontite. Braunsdorf, Germany. Příbram, Bohemia, Czech Republic(!), crystals 10 to 15 mm in galena. Le Cetine and Pereta, Tuscany, Italy. Djebel Senza mine, Constantine, Algeria. In crystals to 2 cm at Huanuni and Tatasi(!), Bolivia. Hargreaves, Ouro Preto district, MG, Brazil. RG *DI:547*.

4.3.12.1 Sillénite $Bi_{12}SiO_{20}$

Named in 1943 for Lars Gunnar Sillén (1916–1970) of Stockholm, Sweden, who contributed much to the knowledge of the polymorphs of Bi_2O_3. ISO *I23*. $a = 10.110$, $Z = 2$, $D = 9.18$. *MJJ 15(8):343(1991)*. *37-485*(syn): 3.57_2 3.20_{10} 2.92_2 2.70_7 1.73_4 1.68_2 1.64_3 1.49_2. As cubes; massive; earthy. Yellow-green, olive-green, reddish brown; adamantine to dull luster. $VHN_{50} = 345$–386. $G = 9.16$. Isotropic; $N > 2.50$. From *Durango, Mexico*, associated with bismutite; also at Navojoa, SON. In cubes up to 5 mm from Fuka, Okayama Pref., Japan(!), on calcite in a gehlenite–spurrite skarn. In pegmatite at Monapo, Mozambique. RG *DI:601*, *AM 28:525(1943)*.

4.3.13.1 Sphaerobismoite Bi_2O_3

Named in 1995 by Walenta, for the composition and the form of the aggregate. Dimorphous with bismite. TET $P4_22_12$ or $P4_2/n$. $a = 8.08$. $c = 6.46$, $Z = 4$, $D = 7.17$. *AUF 45:245(1995)*. *PD*: 5.73_7 3.44_5 3.16_{10} 3.01_4 2.82_3 2.56_4 2.02_5 1.90_{26}. Forms spherulitic aggregates of minute tabular crystals. Green, yellowish; white streak, adamantine luster. Translucent. Nonfluorescent. No discernible cleavage, conchoidal fracture, brittle. $H = 4$. Uniaxial $(+)$, $N_O = 2.13$, $N_E = 2.18$ (Na). Nonpleochroic. Upon heating at 600° for 1 hour, converts to delta Bi_2O_3. Soluble in 1:1 HCl and 1:1 HNO_3. Found in the zone of oxidation at *Neubulach* and *Wittichen, Black Forest, Germany*, associated with bismutite, malachite, mixite, chalcopyrite, and emplectite. RG *AUF 213(1979)*.

RUTILE GROUP

The rutile group comprises nine oxides having the general formula

$$AO_2$$

where

$A =$ Ti, Mn, Sn, Pb, Ge, Nb, Fe^{3+}, Fe^{2+}, Sb^{5+}

and a tetragonal structure in space group $P4_2/mnm$. The dominant structural feature of the rutile group consists of chains of edge-sharing TiO_6 octahedra that run parallel to c. The octahedra share two opposite edges to form straight chains, which are, in turn, linked by corners to form the framework. The oxygen atoms are coordinated to three titanium atoms in a nearly planar configuration, forming nearly equilateral triangles. *CM 17:77(1979)*.

The formula AO_2 can accommodate several types of structures, with the result that several of these minerals have one or more naturally occurring polymorphs. With the exception of rutile, members of the group show little miscibility, probably due to considerable difference in the ionic radii of the A ions.

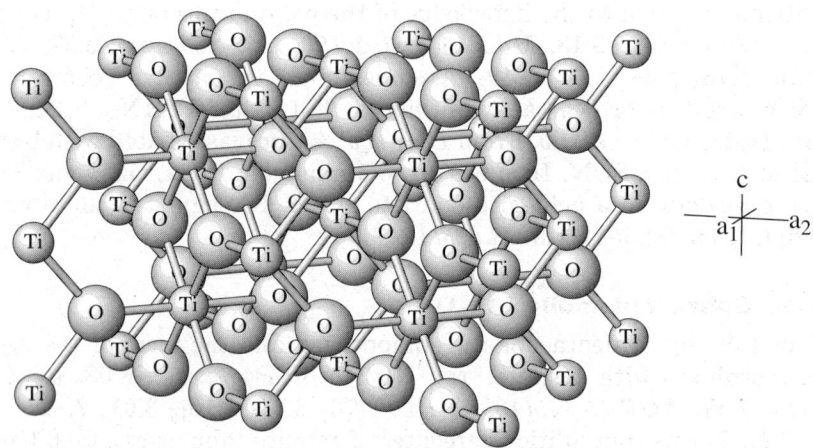

RUTILE: TiO_2

Space group $P4_2/mmm$. TiO_6 octahedra share edges to form chains in the c-axis direction, and the chains are also cross-linked in the same way.

4.4.1.1 Rutile

RUTILE GROUP

Mineral	Formula	Space Group	a	c	D	
Rutile	TiO_2	$P4_2/mnm$	4.593	2.959	4.250	4.4.1.1
Ilmenorutile	$(Ti, Nb, Ta, Fe^{3+}, Fe^{2+})O_2$	$P4_2/mnm$	4.62	2.99	4.3–5.0	4.4.1.2
Struverite	$(Ti, Ta, Nb, Fe^{3+}, Fe^{2+})O_2$	$P4_2/mnm$	4.645	2.999	5.0–5.7	4.4.1.3
Pyrolusite	MnO_2	$P4_2/mnm$	4.400	2.874	5.189	4.4.1.4
Cassiterite	SnO_2	$P4_2/mnm$	4.738	3.188	6.995	4.4.1.5
Plattnerite	PbO_2	$P4_2/mnm$	4.952	3.386	9.563	4.4.1.6
Argutite	GeO_2	$P4_2/mnm$	4.396	2.863	6.28	4.4.1.7
Squawcreekite	$(Fe^{3+}, Sb^{5+}, Sn, Ti)O_2$	$P4_2/mnm$	4.667	3.101	6.02–6.06	4.4.1.8
Stishovite	SiO_2	$P4_2/mnm$	4.173–4.1791	2.6655–2.6659	4.29	4.4.1.9

4.4.1.1 Rutile TiO_2

Named in 1803 from the Latin *rutilus*, red, in allusion to the usual color. Rutile group. Polymorphs: anatase, brookite. TET $P4_2/mnm$, $a = 4.593$, $c = 2.959$, $Z = 2$, $D = 4.250$. *NBSM 25(7):83(1969)*. *NBSM*(syn): 3.25_{10} 2.49_5 2.19_3 1.69_6 1.62_2 1.36_2 1.35_1 0.820_1. **Habit**: Commonly prismatic, often slender to acicular [001]. Prism zone vertically striated. Usually terminated by {101} or {111}. Rarely pyramidal or equant. Also granular, massive, coarse to fine. Twinning on {011} common. Often geniculated, or contact twins of varied habit, sometimes sixlings or eightlings occasionally forming circular groupings, doughnut-shaped. Also reticulated groups ("sagenite"). Epitaxial intergrowths of rutile with hematite, magnetite, ilmenite, brookite, and anatase are common, and oriented microscopic needles of rutile are found in corundum, phlogopite, and quartz, giving rise to asteriated material such as star rubies or star quartz. **Physical properties**: Red, brown, yellow, to black in varieties high in Fe, Nb, or Ta; pale brown to yellow streak; adamantine–metallic luster. Cleavage {110}, distinct; {100}, less so; parting on {092} and {011} due to twin gliding; conchoidal to uneven fracture. $H = 6-6\frac{1}{2}$. $G = 4.23$. Nonfluorescent. **Tests**: Insoluble in acids. Infusible. Decomposed by fusion with alkali or alkali carbonate. **Chemistry**: High amounts of Fe^{3+}, Fe^{2+}, Nb, and Ta may be present, as well as minor amounts of Sn, Cr, and V. The substitution of Nb^{5+} and/or Ta^{5+} is coupled with Fe^{3+} and Fe^{2+} in charge compensation; these varieties are black and have densities ranging up to 5.7, depending on content of Nb, Ta, and Fe. High-Nb rutile is called niobian rutile or *ilmenorutile* (4.4.1.2), and high-Ta rutile is tantalian rutile or *strüverite* (4.4.1.3). These two mineral names are considered as distinct species names in this book, depending on the content of Nb and Ta, but many consider them to be merely varieties of rutile. *NJMA 101:142(1964)*. **Optics**: Uniaxial (+); $N_O = 2.612$, $N_E = 2.889$ (Na). Dichroism O, distinct, in red;

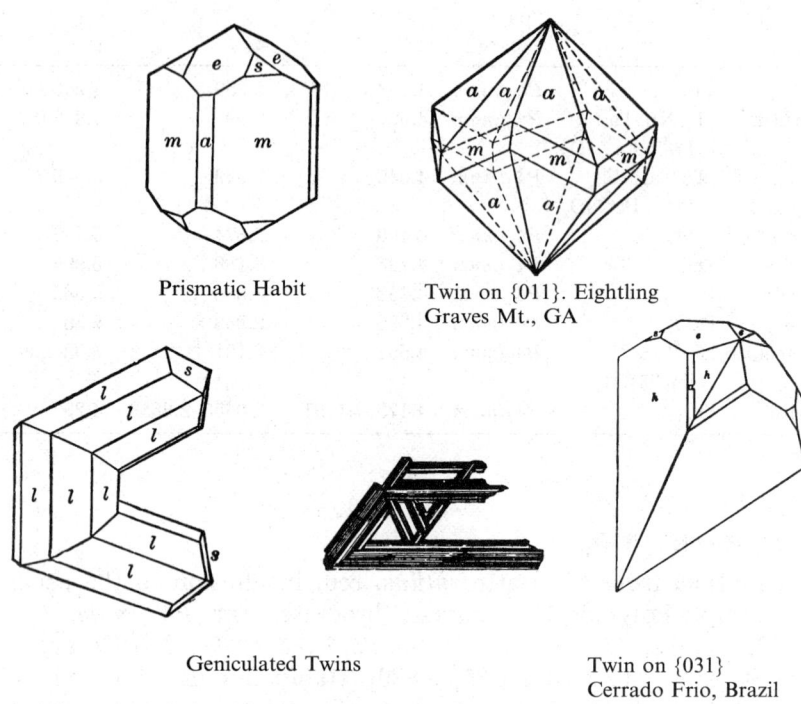

Prismatic Habit

Twin on {011}. Eightling Graves Mt., GA

Geniculated Twins

Twin on {031} Cerrado Frio, Brazil

RUTILE: TiO$_2$
Forms: a{100}, c{001}, m{110}, e{101}, l{310}, h{210}

E, brown, yellow, or green. Absorption E > O. Often anomalously biaxial. In RL pale gray, pleochroic, abundant internal reflections of varying shades of red-brown, yellow to colorless, strong anisotropism. R = 25.3–22.0 (470 nm), 23.9–20.5 (546 nm), 23.3–19.9 (589 nm), 22.7–19.2 (650 nm). *P&J:331.* **Occurrence:** Rutile is a common accessory mineral in metamorphic rocks, especially gneisses and schists; in granites, syenites, anorthosites, and plutonic rocks rich in hornblende. It is also common in aluminous rocks and in rocks rich in apatite, such as nelsonites; in quartzite with kyanite, pyrophyllite, or lazulite. Also common in pegmatites and in alpine cleft-type veins and fissures. **Localities:** Graves Mt., GA(!), in large, equant crystals and twins up to 10 cm and more, associated with kyanite, pyrophyllite, and lazulite in quartzite; Hiddenite, Alexander Co., NC(!), in fine crystals in pegmatite with muscovite, dolomite, apatite, beryl, and green spodumene (hiddenite); also in NC in Cleveland Co. N of Boiling Springs. Parksburg, Chester Co., PA(!), in multiple twins, sometimes doughnut-shaped. Roseland, Nelson Co., VA, with ilmenite and apatite ("nelsonite") in anorthosite. Magnet Cove, AR(*), in crystals and twins and in part pseudomorphous after large crystals of brookite. In large sharp crystals in pyrophyllite with andalusite at the Champion mine, White Mts., Mono Co., CA(!); also at the Jensen quarry, Riverside. In Norway in large amounts in apatite veins in Fogne and in the region of Gjerstadvand and Vegaardsheien; in Sweden at Horrsjöberg, Värmland, with lazulite, pyrophyllite, and kyanite in quartzite. Finely crystallized at many localities in the alpine regions of Switzerland, France, Italy, and Austria, as drusy crystals in crevices and vein deposits mostly in metamorphic rocks, with adularia, albite, hematite, anatase, brookite, titanite, mica, and so on; especially in Switzerland at Alp Lercheltini, Binnatal, Wallis(!); Cavradi, Tavetsch, Grisons, epitaxial with hematite; Lukmanierschlucht, Tavetsch, as sagenite; Pfitsher-Joch, Val di Passa, Tyrol, Italy, large fibrous crystals; Saurüssel, Zillertal, Austria. Near Ambatofinandrahana, Madagascar. In Brazil in fine V-twins in the Ouro Preto district, MG(!), associated with precious topaz in veins. Brumado, BA(!), in fine crystals in a large magnesite deposit. Novo Horizonte, BA(!), as acicular, golden-colored crystals ("Venus hair") embedded in clear quartz crystals ("rutilated quartz"). This rutile sometimes occurs in epitaxial six-rayed stellate groups growing from a hematite crystal nucleus. Uses: Used for making titanium metal and for coating welding rods. RG *DI:554*.

4.4.1.2   Ilmenorutile $(Ti, Nb, Ta, Fe^{3+}, Fe^{2+})O_2$

Named in 1854 for the original locality in the Ilmenskie Mts., Russia, and its similarity to rutile. Rutile group. TET $P4_2mnm$. $a = 4.62$, $c = 2.99$, $Z = 2$, $D = 4.3$–5.0. *16-934:* 3.28_{10} 2.51_9 2.32_3 2.20_5 1.70_{10} 1.64_6 1.47_3 1.37_4. PD for TiO_2, 59.5%; Nb_2O_5, 28.75%; Ta_2O_5, 1.65% from Vezna, Moravia, Czech Republic. Crystals equant to short prismatic, often twinned on {011}. Black to brownish red, grayish or greenish-black streak, submetallic luster. $H = 6$–6.5. $G = 4.3$–5.0. Slightly magnetic; magnetic susceptibility increases with iron content. Nb, coupled with Ta, Fe^{2+}, and/or Fe^{3+} enter the composition of rutile up to a maximum of about 33% Nb_2O_5 or higher amounts of $(Nb_2O_5 + Ta_2O_5)$. Much natural rutile contains at least 1–2% Nb_2O_5. The point at which the Nb_2O_5 content justifies the name of ilmenorutile rather than niobian rutile is arbitrary and not fixed; some feel that with >10% Nb_2O_5, the term *ilmenorutile* can be used. Ilmenorutile also makes a complete series with strüverite (4.4.1.3). Common in pegmatites. Also in some albitites and apogranites. Niobian rutile and/or ilmenorutile are common at Magnet Cove, AR. In pegmatites at Evje, Iceland, and Tvedstrand, southern Norway. Val Vigesso, Piedmont, Italy. In pegmatite near *Miask, Ilmenskie Mts., Russia*. Near Ambatofinandrahana, Madagascar; also at Ampangabe in crystals to 13 cm and at Fefena in rose quartz. Near Mokota, Zimbabwe. Araçuai, (!) MG, Brazil. RG *DI:558, AM 44:620(1959), NJMA 101:142(1964)*.

4.4.1.3   Strüverite $(Ti, Ta, Nb, Fe^{3+}, Fe^{2+})O_2$

Named in 1907 for Giovanni Strüver (1842–1915), professor of mineralogy at the University of Rome. Rutile group. TET $P4_2mnm$. $a = 4.645$, $c = 2.999$, $Z = 2$, $D = 5.0$–5.7. *17-543:* 3.28_{10} 2.52_9 2.32_4 2.21_3 1.71_{10} 1.64_6 1.47_3 1.38_4.

PD: for TiO_2, 47.8; Ta_2O_5, 38.7; Nb_2O_5, 6.2; FeO, 7.3; Etta mine, SD. Crystals prismatic to blocky; twinning on {011}. Black, gray-black; black streak; metallic luster. H = 6–6.5. G = 5.0–5.7. Magnetic susceptibility slight, but increases with content of Fe^{3+}. Traces of Ta are common in rutile and a complete series exists up to about 41% Ta_2O_5. There is coupled substitution of Ta^{5+}, Fe^{3+}, and/or Fe^{2+} for Ti^{4+}; the solid-solution composition may be considered to be a mixture of TiO_2, Fe^{2+}, Ta_2O_6, and $Fe^{2+}TaO_4$; $Fe^{3+}NbO_4$ should also be added, since there is a complete solid-solution series between strüverite and ilmenorutile. The Committee of New Minerals and New Mineral Names of the International Mineralogical Association has not yet ruled on the validity of the names *strüverite* and *ilmenorutile*, as opposed to *tantalian rutile* and *niobian rutile*; in any event, any line of demarcation drawn between tantalian rutile and strüverite, or their niobian analogs, would be arbitrary and subject to disagreement. The basis of the controversy lies in the fact that even with high percentages of Nb and Ta, there is a higher mol % of TiO_2 in these minerals than Nb or Ta or Nb + Ta; there is no basic structural change, and the powder diffraction patterns differ to only a small degree from pure rutile. A little Mn and Sn also substitute in the composition. From the Etta mine, near Keystone, SD. Also in pegmatite near *Craveggia, Val Vigesso, Piedmont, Italy*(*). Near Fefena and Ampangabe, Madagascar(!). In alluvium near Salak North, Malaysia, where much strüverite is recovered by ocean bottom dredging as a tantalum ore. From Madagascar at Tsaramanga, Tongafeno, Ankazobe, Beltsiry, and other localities. In fine crystals at Barra do Cuiete, Conselheira Pena, MG(!!), and at Parelhas, RGN(!), Brazil. Used as an important ore of Ta. RG *DI:558*, *AM 49:792(1964)*, *NJMA 101:142(1964)*.

4.4.1.4 Pyrolusite MnO_2

Named in 1827 from the Greek for *fire* and *to wash*, because it is used to decolorize the brown and green tints of glass. Rutile group. Synonym: polianite. Polymorphs: ramsdellite, akhtenskite. TET $P4_2/mnm$. $a = 4.400$, $c = 2.874$, Z = 2, D = 5.189. *NBSM 25(10):39(1972)*. 24-735(syn): 3.11_{10} 2.41_6 2.11_2 1.62_6 1.56_1 1.306_2 1.304_2 1.000_1. **Habit:** Crystals uncommon, long to short prismatic [001] or equant. In druses of small crystals, and as paramorphs or overgrowths over crystals of manganite. Usually columnar or fibrous, botryoidal, granular massive, earthy. As dendritic growths on fracture surfaces or enclosed in chalcedony (moss agate). The term *polianite* refers to crystallized pyrolusite, which originally was thought to be a distinct mineral. **Structure:** MnO_2 exists in at least three natural polymorphs: pyrolusite, ramsdellite, and akhtenskite. A fourth probable polymorph, β-pyrolusite, is also believed to exist on the basis of powder-diffraction and other data. All of these polymorphs share one dimension of the unit cell near 2.87, and a second dimension near 4.44. Two are orthorhombic (β-pyrolusite and ramsdellite), one tetragonal (pyrolusite) and one hexagonal (akhtenskite). Differentiation of these polymorphs is difficult and usually requires careful investigation using EM. *AC(B) 32,2200(1976)*, *AM 64,1199(1979)*. **Physical properties:** Gray or black, black or bluish-black streak, metallic luster. Cleavage {110}, perfect; uneven fracture; brittle. H = 2 (earthy), $6\frac{1}{2}$ (crystals), 2–6 (massive). G = 5.0. Nonmagnetic. **Tests:** Soluble in concentrated HCl with evolution of Cl_2. Massive varieties usually yield a little water in the closed tube. **Chemistry:** Most pyrolusite that is not mixed with other Mn phases or impurities is very pure MnO_2. Small amounts of iron and water are often present, up to about 1%; the presence of other elements such as Ba usually indicates an admixture of other minerals. **Optics:** In RL, cream-white. Very weakly pleochroic, distinctly anisotropic. R = 31.4–18.9 (470 nm), 31.3–18.7 (546 nm), 30.6–18.5 (589 nm), 29.2–18.2 (650 nm). *P&J:313*. **Occurrence:** Pyrolusite is one of the most common manganese minerals. It is apparently always formed under oxidizing conditions and high pH. The chief types of occurrence are as bog, lacustrine, or shallow marine deposits; in the oxidized zone of manganiferous ore deposits and rocks, and as deposits formed by circulating meteoric waters. Colloidal processes undoubtedly play a major role in the transportation and deposition of pyrolusite. Bacterial action is also important. Associated minerals are other manganese and iron oxides and hydrated oxides, including hausmannite, braunite, manganite, psilomelane, nsutite, marokite, hematite, goethite, and "limonite." Pyrolusite is also deposited as nodules on the deep-sea floor, probably related to bacterial action and/or spreading centers between plates. Vast tonnages of these are known to cover millions of square kilometers of the Pacific and other ocean floors. **Localities:** Pyrolusite is found in small bodies of manganese ores in the Piedmont plateau and with the residual iron ores of the Paleozoic sedimentary rocks, Appalachian region. With the hematite ores of the Lake Superior region, as at Negaunee, MI. It has been mined at Batesville, Independence Co., AR. Was found at Leadville, CO, and at Butte, MT, as a result of oxidation and leaching of manganiferous carbonate ores. Fine crystals near Bathurst, Gloucester Co., NB, Canada(!). Horni Blatnà, Czech Republic(!), in large crystals (polianite). Large deposits are mined in Russia and especially in Kutasi, Georgia. Deposits formed by the weathering of rocks containing manganese silicates are mined extensively in India, and similar deposits are found in Ghana. Also found in large deposits in Brazil in MG and SP. **Uses:** Used as a principal source of manganese in the manufacture of steel; as a decolorizer in the manufacturing of glass; in electric batteries; as an oxidizing agent in the chemical industry, and as a coloring material in bricks, pottery, and so on. RG *DI:562*.

4.4.1.5 Cassiterite SnO_2

Named in 1832 from the Greek *kassiteros*, tin. Rutile group. TET $P4_2/mnm$. $a = 4.738$, $c = 3.188$, Z = 2, D = 6.995. *NBSC 539(1):54(1953)*. *14-567*(nat): 3.33_8 2.60_8 1.75_{10} 1.42_4 1.40_4 1.14_3 1.082_4 1.076_4. *21-1250*(syn): 3.35_{10} 2.64_8 2.37_2 1.77_6 1.68_2 1.50_1 1.44_2 1.42_2. **Habit:** Untwinned crystals are the exception, usually short prismatic [001] with {110} and {100} prominent; sometimes

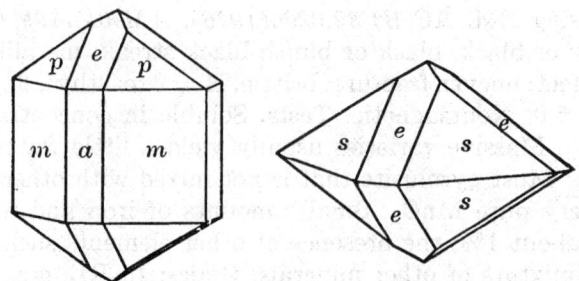

Prismatic and Low Pyramidal Habits

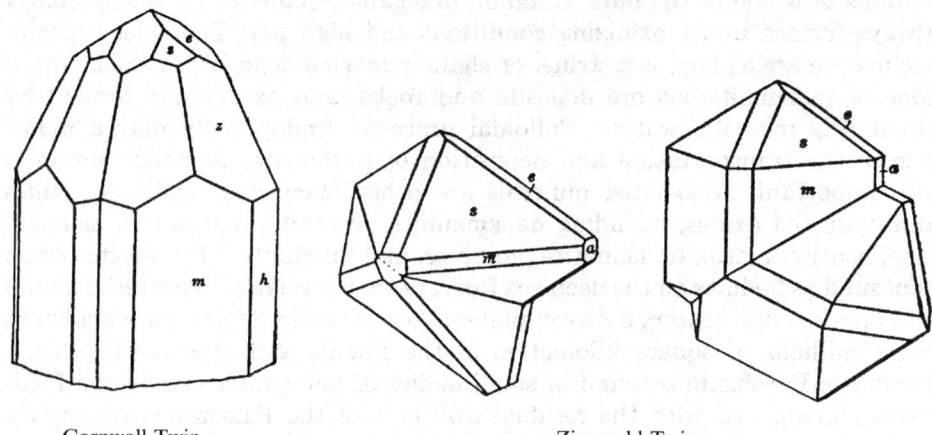

Cornwall Twin · Zinnwald Twins

CASSITERITE: SnO_2

Forms: $a\{100\}$, $c\{001\}$, $e\{011\}$, $h\{120\}$, $m\{110\}$, $s\{111\}$, $z\{321\}$

long prismatic with acute terminations due to the development of a steep pyramid (needle tin). Also massive, radial-fibrous botryoidal crusts or concretionary masses (wood tin); granular. Twinning very common on {011}, both contact and penetration; often repeated, producing complex forms; sometimes stellate fivelings. Twins also reported on {031}. **Physical properties:** Black, brown, yellow-gray, rarely red, colorless, or white. The color is often in zones parallel to crystal surfaces. White, grayish, brownish streak; adamantine to metallic luster in black varieties, inclining to greasy on fracture surfaces. Cleavage {100}, imperfect; {110}, indistinct; subconchoidal to uneven fracture; brittle. H = 6–7. G = 6.99. Nonfluorescent. Nonmagnetic. **Diagnostic tests:** Slowly attacked by acids; decomposed by fusion with alkalis. Fragments of cassiterite when placed in dilute HCl or H_2SO_4 with metallic zinc become coated with a gray deposit of metallic tin which turns bright on friction. This specific test for cassiterite is used for identifying the mineral when admixed with other black phases, such as in tantalite concentrates. **Chemistry:** Fe^{3+} is usually present in substitution for Sn, to

4.4.1.5 Cassiterite

about Fe/Sn = 1:6, but iron-free cassiterite also occurs; the black color of some cassiterites is related to the iron content. Ta and Nb may also be present in small amounts, in cassiterite derived from pegmatites, and may reach 8% Ta_2O_5. Wood tin may contain unusually high amounts of several elements substituting in the cell, up to (wt %) 10% FeO; 3.5% ZnO; 3.1% SiO_2; 2.4% In_2O_3; 1.5% As_2O_3; and smaller amounts of WO_3, Al_2O_3, and Sb_2O_5, the sum total of such impurities reaching as much as 13% in reported analyses. **Optics:** Uniaxial (+); $N_O = 2.001$, $N_E = 2.098$. Dichroism usually weak or absent, but may be strong. Absorption E > O. In RL, light gray; very weakly pleochroic, strongly anisotropic. R = 12.6–11.4 (470 nm), 12.1–11.0 (546 nm), 12.0–10.9 (589 nm), 12.0–10.7 (650 nm). *P&J:114*. **Varieties:** *Wood tin* is a type of fine-grained botryoidal cassiterite found in shallow high-temperature deposits. It has concentric structure and is radiated fibrous internally, so that broken pieces with brownish color often have the appearance of wood. The structure of wood tin can accommodate significant amounts of certain other elements, including Sb, As, In, and Ge. **Occurrence:** Cassiterite is characteristic of deep-seated high-temperature veins, and of association with highly acid igneous rocks such as granites and their derivatives. Important placer deposits are derived from the weathering and erosion of stockworks of cassiterite-bearing veins in granite. Also, large tonnages of cassiterite are found disseminated as small grains in albitites and apogranites, often accompanied by niobium–tantalum minerals, zircon, and xenotime. Cassiterite is also found in some pegmatites, associated with columbite–tantalite, lepidolite, albite, topaz. It is associated with certain rhyolites, although these do not contribute a significant tonnage to total production. Vein-type deposits are important in Cornwall, England; central Europe; and especially Bolivia. Associated minerals are wolframite, tourmaline, topaz, fluorite, and arsenopyrite, and to a lesser extent pyrrhotite, pyrite, chalcopyrite, galena, sphalerite, and several tin-bearing sulfides, such as stannite and teallite. In rhyolites and related extrusive rocks, it is found as wood tin, sometimes associated with rare antimonates and arsenates, such as ordoñezite and durangite. **Localities:** Very little cassiterite has been mined in the United States, and most U.S. occurrences are in pegmatites. In New England at Norway, Hebron, and elsewhere in Oxford Co., ME; also in NH, MA, and CT. Large crystals have been found at the Herbb No. 2 pegmatite, Powhatan Co., VA(!), associated with wodginite and topaz. Harney peak, near Custer, SD. In rhyolite in the Black Range, Sierra and Catron Cos., NM. Veins in the York district, Seward Peninsula, AK, have associated Be mineralization. Fine cross tin is found in rhyolite in several areas in Mexico, especially at Sierra Fria, Rincon de Romos, AGS, but also in GTO, SLP, ZAC, and DGO. In Cornwall, England, veins were mined by Celtic peoples at the time of the Phoenicians; fine crystals were found at St. Agnes, Camborne, St. Just, St. Austell. Panasqueira, Portugal(*) to 3 cm, often twinned, sometimes complex "cross-bit" twins. In Bohemia in veins at Horni Slavoc and Cinovec, often well crystallized(!), and in the neighboring district of Saxony, Germany, at Altenberg and Ehrenfriedersdorf. In Italy at San Piero in Campo, Elba(*), at several of the mines. In crystals to 10 cm at the Iultin deposit(!), eastern Chukotka, Siberia, Russia, also at Merek(!), far eastern Siberia. Bain Mod, Hentiy district, Outer Mongolia(*), crystals to 6.5 cm. Nuristan, Laghman, Afghanistan(!), sharp crystals and twins to 10 cm. Stak Nala, Gilgit Div., Pakistan(*). Important deposits, mostly alluvial, occur in the Malay Peninsula and on the islands of Billiton and Banka off the coast of Sumatra, Indonesia; also in Thailand, where it is associated with tantalite. Leiyang, Hunan, China. Tae Wha, N of Chungju, Korea(!), with scheelite and wolframite. Fine crystals at Bisesero and Gifurve and other pegmatites in Rwanda(!) and adjacent regions; also disseminated in an enormous pegmatite at Manono, Zaire. Abundant in NSW, Australia over an area of 20,000 km^2 in the New England range, with fine crystals coming from Emmaville, Glen Innes, and Elsinore(!). Important deposits are found in Bolivia at Llallagua(!), Oruro, and many other places; these are high-temperature vein deposits, many with complex mineralogy. At Viloco (formerly called "Araca")(!) one vein has produced magnificent druses of large, honey-colored twin crystals. In Brazil, fine bipyramidal crystals have been found with topaz and aquamarine in pegmatite in MG at the Fazenda de Funil, Sta. Maria de Itabira(!); Galileia(!), Lavra Toa, Sao José da Safira(!), and other places. Important alluvial deposits are found in RO derived from stockworks of veins in granite, and in Pittinga, AM, a large decomposed apogranite is mined containing cassiterite, columbite, zircon, and xenotime. Use: The principal and practically the only ore of tin. RG *DI:574, NMGS(1986)*.

4.4.1.6 Plattnerite α-PbO$_2$

Named in 1845 for K. F. Plattner (1800–1858), professor of metallurgy and assaying at Freiburg, Sachsen. Rutile group. Polymorph: scrutinyite. TET $P4_2/mnm$. $a = 4.952$, $c = 3.386$, $Z = 2$, $D = 9.563$. *MR 1:75(1970)*. Structure: *AC(B) 36:2394(1980)*. *41-1492*: 3.5_{10} 2.80_9 2.48_2 1.856_5 1.753_1 1.567_1 1.525_1 1.485_1. Prismatic [001]. Usually in nodular or botryoidal masses, dense to fibrous. Twinning on {011}, as contact or penetration twins. Black; brownish black when massive; chestnut brown streak; metallic, adamantine luster, tarnishing and becoming dull on exposure. No cleavage; fracture of masses, small conchoidal to fibrous; brittle. $H = 5\frac{1}{2}$. $G = 9.56$. Easily fusible. Decomposed to red-lead and oxygen when moderately heated. Soluble in HCl with evolution of Cl$_2$. Difficultly soluble in HNO$_3$ or H$_2$SO$_4$ with evolution of oxygen. Uniaxial; $N = 2.30 \pm 0.05$. In RL, gray-white with red-brown internal reflections. Pleochroic and anisotropic. $R = 21.1$–19.8 (470 nm), 19.4–19.9 (546 nm), 18.4–18.9 (589 nm), 16.7–17.1 (650 nm). *P&J:303*. From the Morning mine, Mullan, Shoshone Co., ID, in large botryoidal masses, in part fibrous; also in the Gilmore district, Lehmi Co. Ojuela mine, Mapimi, DGO, Mexico(*), as druses of small crystals on limonite. *Leadhills, Scotland,* with cerussite, leadhillite, and pyromorphite. Zandjir mines, Yezd, Iran. Tsumeb, Namibia, with minium and massicot as coatings. RG *DI:581*.

4.4.1.7 Argutite GeO$_2$

Named in 1983 for the locality. Rutile group. TET $P4_2/mnm$. $a = 4.396$, $c = 2.863$, $Z = 2$, $D = 6.28$. *AM 69:406(1984)*. *35-729*(syn): 3.11_{10} 2.40_6 2.20_2 2.11_1 1.62_5 1.55_2 1.304_2 1.300_2. As minute prismatic crystals embedded in sphalerite. Colorless; in RL, light gray. Anisotropy distinct. $R = 11.6$ (470 nm), 10.97 (546 nm), 10.93 (589 nm), 10.93 (650 nm). From a zinc deposit at the *Argut Plane, Haute Garonne, Pyrenees, France*, associated with sphalerite, cassiterite, siderite, and rarely briartite. RG *TMPM 31:97(1983), AM 69:406(1984)*.

4.4.1.8 Squawcreekite (Fe^{3+}, Sb^{5+}, Sn, Ti)O$_2$

Names in 1991 for the locality. Rutile group. TET $P4_2/mnm$, $a = 4.667$, $c = 3.101$, $Z = 2$, $D = 6.02$–6.06. *NJMM*: 3.30_{10} 2.58_5 2.33_2 2.26_1 2.09_1 1.73_4 1.65_1 1.55_1. Microscopic prismatic crystals. Yellow-brown, pale yellow-brown streak, adamantine luster. Cleavage {100}, imperfect; conchoidal to subconchoidal fracture. $H = 6$–$6\frac{1}{2}$. Uniaxial (+); $N_O = 2.18$, $N_E = 2.28$; r < v, weak. From the *Squaw Creek tin prospect, Catron Co., NM*. Esperanza tin mine, near Madera, DGO, Mexico. RG *NJMM:363(1991)*.

4.4.1.9 Stishovite SiO$_2$

Named in 1962 by Chao et al. for Sergei Mikhailovich Stishov (b.1937), Russian crystallographer, who first synthesized the mineral in 1961. Rutile group. Polymorphous with quartz, tridymite, cristobalite, coesite, and lutecite. TET $P4_2/mnm$. $a = 4.1773$–4.1791, $c = 2.6655$–2.6659, $Z = 2$, $D = 4.29$. *PD 45-1374*(nat): 2.96_{10} 2.25_2 1.98_4 1.87_1 1.53_4 1.48_1 1.33_1 1.235_1. Isostructural with rutile. Si is in octahedral (sixfold) coordination, compared to the tetrahedral coordination characteristic of all other Si minerals. *JSSC 47:185(1983)*. Raman data indicate that the mineral may be triclinic with lattice vectors three times those of the rutile-type cell. Crystals are elongated on c; natural crystals are usually submicron in size, but synthetic crystals are as much as 230 microns in length. Colorless to white, vitreous luster. $H = 2080$ kg/mm^2 parallel to elongation and 1700 kg/mm^2 perpendicular to elongation. $G = 4.28$–4.35. Insoluble in common acids and only very slowly soluble in HF. Separated from other silica polymorphs by HF treatment. IR: *AM 75:951(1990)*. Raman spectra: *CA 95:135975(1981)*. Essentially pure SiO$_2$. Uniaxial (+); syn: $N_O = 1.800$, $N_e = 1.845$; nat: N about 1.77–1.78. Positive elongation; colorless. The mineral is metastable and reverts to quartz after heating at 230° for 3 days; converts to cristobalite when heated in air to 900°. Formed from quartz through high transient shock pressure and high temperature, accompanying the impact of a large meteorite. Occurs with coesite and quartz in sandstone at *Meteor crater, Coconino Co., Arizona* (Canyon Diablo meteorite). Also occurs with coesite in the Ries crater, in granite and pumiceous tuff, Bavaria, Germany. Popigai impact structure, Siberia, Russia; Lonar crater, Deccan Plateau, India. Vredefort Dome (astrobleme), South Africa, in pseudotachylite veins. RG *JGR 67:419(1962), PCM 13:146(1986), Am 75:739(1990), AC(B) 47:561(1991)*.

4.4.2.1 Varlamoffite $(Sn, Fe)(O, OH)_2$

Named in 1946 for Nicolas Varlamoff (1910–1976), Belgian geologist, who collected the first specimens of the mineral. TET $P4_2/mnm$. $a = 4.667$, $c = 3.096$, $Z = 1$, $D = 3.26$; $D = 5.14$ after heating. *MA 83:3628*. PD: $3.30_7(b)$ $2.58_7(b)$ $1.41_2(b)$. PD after heating to 1000°: 3.33_{10} 2.63_8 1.76_7 1.67_4. In the natural state, nearly amorphous, but becomes somewhat crystalline upon heating. Earthy, claylike. Yellow to orange-red, yellow to reddish orange streak, vitreous to greasy luster. $G = 3.21$–3.36. Nonfluorescent. Average $N = 1.81$. Soluble with difficulty in acids. Is not reduced to tin when treated with dilute acid with powdered zinc (as contrasted with cassiterite). Some Fe is invariably present, and lower percentages of other elements, such as Al and Si, may be found. Published analyses show some variation, and the formula is somewhat uncertain. Cligga mine, Cornwall, England. Sungei Lah, Chenderiang, Perak, Malaysia, on cassiterite. From a greisenized Nb–Ta-bearing stock in Guanxi, China, with cassiterite. Sardine mine, near Ewan, Kangaroo Hills district, Q, Australia, in the oxidized portion of a stannite–cassiterite vein. *Kalima, Maniema, Zaire*. RG *AM 34:618(1949); MM 35:622(1965), 37:624(1970); ZVMO 99:232(1970)*.

4.4.3.1 Downeyite SeO_2

Named in 1977 for Wayne F. Downey, Jr. of Harrisburg, PA, who collected and preserved the first specimens. TET $P4_2/mbc$. $a = 8.364$, $c = 5.064$, $Z = 8$, $D = 4.161$. *NBSM 25(7):60(1969). 22-1314:* 4.18_5 3.74_6 3.23_5 3.01_{10} 2.53_2 1.93_2 1.88_1 1.83_1. Crystals prismatic [001] to acicular, in tufts. Colorless, adamantine luster. Nonfluorescent. Extremely hygroscopic. Crystals of downeyite in air absorb moisture and soon dissolve to a droplet of selenious acid. Stable in air only above 200°, or in a dry atmosphere such as one would obtain in a sealed desiccator. Uniaxial (+); $N > 1.80$. Moderate bireflectance. From *near Forestville and near Glen Lyon, Luzerne Co., PA*, as a product of burning coal-waste piles, where it deposits as a crystalline condensate close to vents of hot gas. RG *AM 62:316(1977)*.

4.4.3.2 Paratellurite TeO_2

Named in 1960 from the Greek *para* and *tellurite*, with which it forms a dimorphous pair. TET $P4_32_12$. $a = 4.810$, $c = 7.613$, $Z = 4$, $D = 6.018$. *NBSC 539(10):55(1960). 42-1365:* 3.40_{10} 3.10_1 2.98_{10} 2.41_2 1.872_5 1.700_1 1.661_2 1.493_1. Usually fine grained, massive. Rare crystals are short prismatic to equant, usually somewhat rounded. Also commonly pseudomorphous after tellurite crystals. Grayish white, honey-colored, black; white streak; resinous to waxy luster. No cleavage. $H = 1$. $G = 5.60$. Nonfluorescent. Uniaxial;

N > 2.05. From the *Moctezuma mine, Moctezuma, SON, Mexico*(*), where it forms a black oxidation rind around masses of native tellurium. The black color is due to included finely divided tellurium. Also as waxy, honey-colored coatings on fractures in native Te, and as pseudomorphs after later tellurite crystals at some distance from the tellurium masses. RG *AM 45:1272(1960), SPC 32:354(1987)*.

4.4.4.1 Anatase TiO$_2$

Named in 1801 for the Greek for *extension*, in allusion to the greater length of the common pyramidal faces compared to those of other tetragonal minerals. Synonym: octahedrite, named because of the common pseudo-octahedral habit. Polymorphs: rutile, brookite. TET $I4_1/amd$. $a = 3.785$, $c = 9.514$, Z = 4, D = 3.893. *NBSM 25(7):82(1969)*. Structure: *ZK 136:273(1972)*. 21-1272(syn): 3.52_{10} 2.43_1 2.38_2 2.33_1 1.89_3 1.70_2 1.67_2 1.48_1. Commonly acute pyramidal {011}, less commonly tabular or obtuse pyramidal, with {017} or {013}; often highly modified; twinning on {112} rare. Brown, indigo-blue, green, gray, black; colorless to pale yellow streak; adamantine luster, metallic–adamantine in dark varieties. Cleavage {001}, {011}, perfect; subconchoidal fracture; brittle. H = $5\frac{1}{2}$–6. G = 3.90. Insoluble in acids; infusible; decomposed by fusion with KHSO$_4$. Uniaxial (−); N$_O$ = 2.561, N$_E$ = 2.488. Absorption E > O or O > E. Often biaxial with small 2V, especially in deeply colored varieties. Substitution of other elements into the anatase structure is not found to a significant extent as in the case of rutile, although up to 1% or more of Fe, Sn, Nb, and/or Ta are known. Anatase is typically found in veins or crevices of the alpine type ("alpine clefts") in gneiss or schist, where it is derived by leaching from the surrounding rock by hydrothermal solutions. Associated minerals are quartz, brookite, adularia, hematite, chlorite, titanite, and others. From MA with brookite and titanite near diabase dikes at Somer-

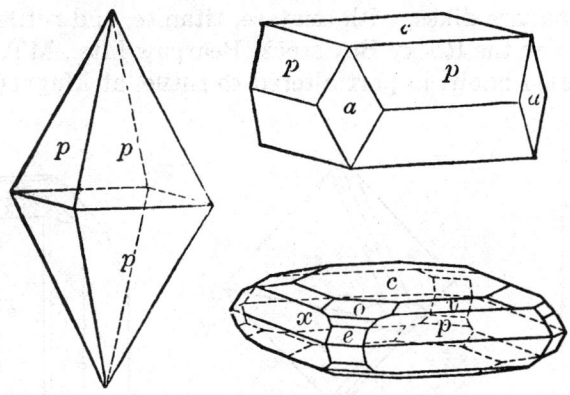

ANATASE: TiO$_2$
Forms: p{111}, e{101}, v{117}

ville and in pockets in pegmatite at Quincy with aegirine, fluorite, and ilmenite. Arvon, Buckingham Co., VA, in joint planes in slate with quartz and pyrite. In fine blue crystals in a diorite dike on Beaver Creek, Gunnison Co., CO(!). With brookite on quartz at Placerville, CA. In many localities in Switzerland, especially superb large complex yellow crystals at Alp Lercheltini, Binntal, Wallis(!) and black octahedra (2 cm) at Crapteig and Crappasusta(!), Thusis. In Italy at Monte Cervandone, Ossola(!). In France at *Maronne, le Bourg d'Oisans, Isère*. In large, brilliant blue-black crystals on quartz crystals from Tysse, Hardangervidda, southern Norway(!). In MG, Brazil, as alluvial rolled pebbles ("favas") at Diamantina, and at Minas Novas(!) in fine crystals on smoky quartz. RG *DI:583*.

4.4.5.1 Brookite TiO$_2$

Named in 1825 for Henry James Brooke (1771–1857), English crystallographer and mineralogist. Polymorphs: rutile, anatase. ORTH *Pbca*. $a = 9.174$, $b = 5.456$, $c = 5.138$, $Z = 8$, $D = 4.120$. *NBSM 25(3):57(1964)*. Structure: *CM 17:77(1979)*. 29-1360: 3.51_{10} 3.47_8 2.90_9 2.48_3 2.41_2 1.89_3 1.69_2 1.66_3. Usually tabular $\{010\}$ and elongated $[001]$; also prismatic $[001]$, with $\{120\}$ prominent; rarely tabular $\{001\}$; also pseudohexagonal with $\{120\}$ and $\{111\}$ equally developed. Various shades of brown, to black with adamantine–metallic luster; uncolored to grayish to yellowish streak. Cleavage $\{120\}$, indistinct; subconchoidal fracture; brittle. $H = 5\frac{1}{2}$–6. $G = 4.14$. Insoluble in acids. Small amounts of Fe^{3+} are usually present, and a niobian variety is known. Biaxial (+); $Z = b$; $N_x = 2.583$, $N_y = 2.584$, $N_z = 2.700$ (Na); 2V small. Weakly pleochroic, in yellowish, reddish, and brownish tints; very strong dispersion. Absorption $Z > Y > X$. Optical properties vary widely with wavelength and temperature. Brookite is found principally in metamorphic rocks in fissures and open cavities in veins of the Alpine cleft type, with anatase, rutile, titanite, albite, chlorite, and so on. Found at the Ellenville lead mine, Ulster Co., NY, on quartz with galena and chalcopyrite. Somerville, MA, in quartz–calcite veinlets in diabase dikes, with anatase, titanite, and rutile. In nepheline syenite pegmatite in the Rocky Boy stock, Bearpaw Mts., MT. In large thick crystals with varied habit, in part altered to rutile, at Magnet Cove, AR(!).

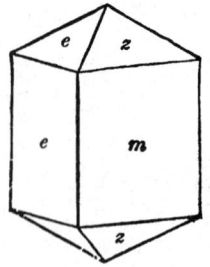 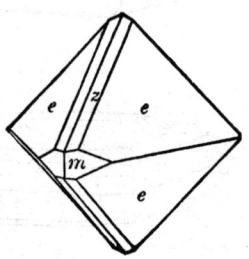 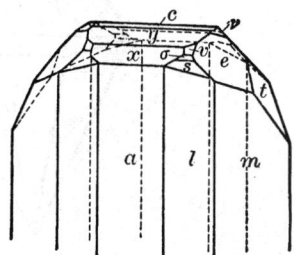

BROOKITE: TiO$_2$
Forms: $z\{112\}$, $m\{110\}$, $e\{122\}$

Prenteg, Tremadog, Gwynedd, Wales(!). Tête Noir(!) and Finhaut, Mt. Blanc massif, Wallis, Switzerland, in crystals >40 mm; also Rieder Tobel(!), Maderanertal, Uri, in tabular crystals. In Austria superb crystals >50 mm at Frossnitzalp(!), Tyrol. In *France* at *Le Plan du Lac, St. Christophe-en-Oisans, Isère*(!). In volcanic tuff at Biancavilla near Mt. Etna, Italy. In Russia, fine crystals in the Atliansk placers near Miask, Ural Mts.(!), and in the Murunskiy alkalic massif, NE of Lake Baikal, Siberia(!). In the diamond placers of MG and BA, Brazil, as pebbles ("favas"). RG *DI:589, DHZII:554(1992)*.

4.4.6.1 Tellurite TeO$_2$

Named in 1849 for the composition. Dimorph: paratellurite. ORTH *Pbca*. $a = 5.607$, $b = 12.034$, $c = 5.463$, $Z = 8$, $D = 5.75$. *NBSC 539(9):57(1960)*. *9-433*: 3.72_9 3.28_{10} 3.01_5 2.80_3 2.73_5 2.30_3 2.05_2 1.93_3. Crystals prismatic to acicular [001], usually tapered, and as thin laths {010}. Also grouped in tufts. Structure: *ZK 124:228(1967)*. Yellow, orange, colorless, white; white streak, subadamantine luster. Cleavage {010}, perfect; flexible. H = 2. G = 5.90. Easily soluble in acids and alkalis. When heated in closed tube fuses to a brown droplet and sublimes. Biaxial (+); XYZ = *bac*; $N_x = 2.00$, $N_y = 2.18$, $N_z = 2.35$ (Li); 2V near 90°; r < v, moderate. Found in the zone of oxidation of mines containing native Te or tellurides. Keystone and other mines, Magnolia district, Boulder Co., CO; also at Cripple Creek with sonoraite. Moctezuma(!) and San Miguel mines, Moctezuma, SON, Mexico, in crystals up to 3 cm and groups, in part altered to paratellurite, with tellurium, emmonsite, zemannite, denningite, spiroffite, and other tellurium oxysalts. *Zalathna and Facebaia, Transylvania, Roumania*. Nicolaevsky mine, western Altai, Kazakhstan. With tetradymite and nagyagite at the Rendaizi mine, Idu, Japan. RG *DI:593*.

4.4.6.2 Scrutinyite PbO$_2$

Named in 1988 to reflect the close scrutiny and sustained examination required to determine the physical properties with a very limited amount of sample material. Polymorph: plattnerite. ORTH *Pbcn*. $a = 4.971$, $b = 5.956$, $c = 5.438$, $Z = 4$, $D = 9.867$. *SR 16:224(1952)*. *45-1416*: 3.82_4 3.12_{10} 2.72_5 2.61_3 2.48_3 1.840_8 1.635_4 1.527_5. Crystals minute, platy {100}. Dark reddish brown, dark brown streak, submetallic luster. Cleavage {100}, perfect; {010}, imperfect. Nonfluorescent. In RL, gray-white, reddish-brown internal reflections. Bireflectance very weak. Pleochroism not seen. Weakly anisotropic, with a bluish color. R = 17.9–18.8 (546 nm), 17.0–17.9 (589 nm). From the *Sunshine No.1 mine, Hansonburg district, Bingham, NM*, associated with plattnerite and murdochite on fluorite and quartz. Ojuela mine, Mapimi, DGO, Mexico, with plattnerite. RG *CM 26:905(1988), AM 75:709(1990)*.

4.4.7.1 Ramsdellite γ-MnO$_2$

Named in 1943 for Lewis S. Ramsdell (1895–1975), mineralogist, University of Michigan. Polymorphs: pyrolusite, akhtenskite. ORTH *Pnma*. $a = 9.32$,

$b = 4.46$, $c = 2.850$, $Z = 4$, $D = 4.87$. SR 15:184(1951). 43-1455: 4.06_{10} 2.55_4 2.43_3 2.34_3 2.14_2 1.903_2 1.655_2 1.616_1. Crystals stout prismatic [001], usually pseudomorphous after groutite. Gray, black; black streak; metallic to dull luster. Cleavages on three pinacoids and a prism. $H = 2$–4. $G = 4.83$. In RL, yellowish white, pleochroic, strongly anisotropic; high reflectance. From the Monroe–Tener mine, Chisholm, MN. *Lake Valley, Sierra Co., NM.* East River, Pictou Co., NS, Canada. Gavilan mine, BC, Mexico. Bülten-Adenstedt, Niedersachsen, Germany, in crystals to 10 mm. Horni Blatnà and Prebuz, Bohemia, Czech Republic. Kodjas Karil mine, Roumelia, Turkey. Dongri Buzurg, India. RG *ACHS 3:163(1949); AM 47:47(1962), 64:1199(1979).*

4.4.8.1 Nsutite Mn(O, OH)$_2$

Named in 1962 for the locality. HEX Space group unknown. $a = 9.65$, $c = 4.43$, $Z = 12$, $D = 4.86$. AC 12:341(1959). 17-510: $4.00_{10}(b)$ 2.42_7 2.33_7 2.13_5 1.64_{10} 1.60_5 $1.48_2(b)$ $1.37_4(b)$. *14-614*(manganoan nsutite): 4.10_{10} 2.66_6 2.45_6 2.36_6 2.16_8 2.13_6 1.67_8 1.39_6. Massive, earthy. Gray brown; black streak; dull luster. $H = 6.5$–8.5. $G = 4.5$. The formula may be written $Mn^{4+}_{1-x}Mn^{2+}_{x}O_{2-2x}(OH)_{2x}$ where $x = 0.06$–0.07; for manganoan nsutite $x = 0.16$, and some of the Mn may be in the Mn^{3+} form. Analyzed nsutite from Ghana, for which the strongest X-ray diffraction line is 1.64, gave 90.6% MnO_2 and 2.6% MnO. Another variety, for which the strongest X-ray line is 1.67, gave 74.6% MnO_2 and 11.0% MnO. These varieties are referred to as nsutite and manganoan nsutite, respectively; the observed shift in X-ray diffraction lines is due to an expansion of the unit cell caused by an increased amount of lower-valent (larger) Mn ions. In reflected light, slightly creamy white. Weakly pleochroic, distinctly anisotropic. For nsutite: $VHN_{300} = 1150$; $R = 29.4$ (green), 28.5 (orange), 27.2 (red). For manganoan: $VHN_{300} = 900$; $R = 22.1$ (green), 16.0 (orange), 14.2 (red). Nsutite is a widespread mineral, which has been observed in most of the world's major manganese deposits and probably remains unrecognized in others. Manganoan nsutite typically is derived from the oxidation of manganese carbonate minerals such as rhodochrosite or kutnohorite, and nsutite is formed by the heating or further oxidation of manganoan nsutite. Phillipsburg, MT. Molango, HGO, Mexico. Tartana, Greece. Bülten-Adenstedt, Niedersachsen, Germany. Mysore, India. Busuango Island, Philippines. *Nsuta, Ghana.* In Australia at Mossman, Q, and in Barabba, NSW. In Brazil at Conselheiro Lafaiette and in the Quadrilatero Ferrifero, MG. Used as an important ore of manganese. RG *AM 47:246(1962), 50:170(1965).*

4.4.9.1 Vernadite (Mn^{4+}, Fe, Ca, Na)(O, OH)$_2 \cdot n$H$_2$O

Named in 1944 for Vladimir Ivanovich Vernadskii (1863–1945), Russian naturalist and geochemist. TET $I4/m$. $a = 9.866$, $c = 2.844$, $Z = 4$, $D = 2.95$. 15-604: 6.81_3 3.11_6 2.39_{10} 2.15_6 1.83_4 1.65_3 1.54_4 1.42_4. Massive, fine grained; as crusts. The name vernadite is given to poorly crystalline supergene hydrated Mn^{4+} oxides giving diffraction reflections at 2.4 and 1.4 Å and is distinguished from birnessite by absence of reflections at 7.0–7.2 and 3.5–3.6 Å. It is said to be structurally similar to feroxyhite (Δ-FeOOH). Black, chocolate-brown streak, dull luster, reddish to chocolate brown when dispersed. Vernadite occurs as a weathering product of Mn^{2+}-containing oxides, carbonates, and silicates. It occurs in marine and freshwater iron-manganese crusts and concretions, often in relict bacterial forms. Originally described from an unspecified locality in Russia. Said to be widespread as a superficial oxidation product of manganese oxides. RG *AM 31:85(1946), 64:1334(1979); MD 14:249(1979); IGR 22:58(1980)*.

4.4.10.1 Akhtenskite ϵ-MnO$_2$

Named in 1979 for the locality. HEX $P6_3/mmc$. $a = 2.80$, $c = 4.45$, $Z = 1$, $D = 4.78$. 30-820(syn): 2.42_7 2.13_4 1.64_{10} 1.40_4 1.21_1 1.19_1 1.07_1 0.847_1. Small platy particles. Dark gray, black. G = 4.00. *Aktenskii limonite deposit, southern Ural Mts., Russia*, derived from oxidation of Mn-bearing siderite in the absence of Mn-oxidizing microorganisms. It is associated with cryptomelane, nsutite, pyrolusite, and todorokite. Also from the Gumeshevskii deposit, Ural Mts., and from Urbanus, Eisengans, Germany. RG *AM 68:473(1983)*.

4.4.11.1 Paramontroseite VO$_2$

Named in 1955 for the paramorphic relationship with the host phase, montroseite. ORTH *Pnma*. $a = 4.89$, $b = 9.39$, $c = 2.93$, $Z = 4$, $D = 4.18$. 25-1003: 4.35_4 3.39_{10} 2.65_5 2.48_3 2.21_4 2.18_3 1.61_3 1.43_4. Found as an integral alteration product of crystals of montroseite, which occurs in bladed masses. Physical properties are those of montroseite, with which paramontroseite is invariably intimately intermixed. Black, black streak, submetallic luster. Cleavage {010}, {110}, good. G = 4.0. Slightly magnetic. From the *Bitter Creek and other mines, Uravan mineral belt, Paradox Valley, Montrose Co., CO*. RG *AM 40:861(1955)*.

4.4.12.1 Cerianite-(Ce) CeO$_2$

Named in 1955 for the composition. ISO $Fm3m$. $a = 5.411$, $Z = 4$, $D = 7.215$. *NBSC 539(1):56(1953)*. 43-1002: 3.12_{10} 2.71_3 1.913_5 1.632_3 1.241_1 1.210_1 1.105_1 1.041_1. Small octahedrons; also, massive, powdery. Isostructural with uraninite and thorianite. Greenish amber, gray-green; subadamantine luster. ThO_2 may substitute for CeO_2 in the cell, as well as other rare earths and minor amounts of Nb_2O_5, Ta_2O_5, and ZrO_2. From mineral claims in *Lackner Twp., Sudbury mining district, ON, Canada*, in carbonatite rock in a nephelinized hybrid gneiss. *Cañon Colorado, Sierra de Bermejillo, DGO, Mexico. Morro de Ferro, Poços de Caldas, MG, Brazil*. RG *NBSM 20:38(1983), AM 40:560(1955)*.

4.4.13.1 Tazheranite $(Zr, Ti, Ca)O_2$

Named in 1969 for the locality. The cubic modification of ZrO_2. ISO $Fm3m$, $F43$, or $F43c$. $a = 5.108$, $Z = 4$, $D = 4.83$. AM $55:318(1970)$. 22-540: 2.94_{10} 2.55_6 1.80_{10} 1.54_{10} 1.47_4 1.17_5 1.04_5 0.983_5. Irregular grains, and rounded to thick tabular isometric crystals; yellowish orange, reddish orange, to cherry-red; adamantine to greasy luster. No cleavage; brittle. $H = 7\frac{1}{2}$. $G = 5.01$. Nonfluorescent. Nonmagnetic. Isotropic; $N = 2.35$. From the *Tazheran alkalic massif, W of Lake Baikal, Siberia, Russia*, within bands of periclase–brucite marbles, with calcite, dolomite, spinel, forsterite, pyrrhotite, melilite, clinohumite, ludwigite, magnesioferrite, calzirtite, baddeleyite, geikielite, perovskite, rutile, and zircon. RG $DANS$ $186:142$, $186:917(1969)$; SPC $14:922(1970)$.

4.4.13.2 Srilankite $(Ti, Zr)O_2$

Named in 1983 for the country of origin. ORTH $Pbcn$. $a = 4.706$, $b = 5.553$, $c = 5.024$, $Z = 4$, $D = 4.765$. ZK $164:59(1983)$. 21-1236: 3.5_6 2.84_{10} 2.12_6 1.996_6 1.697_6 1.663_{10} 1.646_{10} 1.480_6. As rolled pebbles. Black, submetallic to adamantine luster. One indistinct cleavage; conchoidal fracture; brittle. $H = 6.5$. Optically biaxial; $2V = 16°$. Bireflectance high; translucent, blue. $R = 19.4$–18.6 (470 nm), 18.4–16.0 (546 nm), 18.5–16.4 (589 nm), 18.7–17.0 (650 nm). From *Rakwana, Sabaragamuva, Sri Lanka*, in gem gravels, with zirconolite, baddeleyite, geikielite, spinel, and perovskite, as <1-mm inclusions in these minerals. RG $NJMM$ $1983:151$, AM $69:212(1984)$.

4.4.14.1 Baddeleyite ZrO_2

Named in 1892 for Joseph Baddeley who brought the original specimens from Sri Lanka. MON $P2_1/c$. $a = 5.145$, $b = 5.207$, $c = 5.311$, $\beta = 99.23°$, $Z = 4$, $D = 5.836$. AC $18:983(1965)$. 37-1484: 3.70_1 3.17_{10} 2.84_7 2.62_2 2.54_1 1.848_2 1.819_2 1.804_1. Crystals short to long prismatic [001], usually flattened {100}; sometimes radially fibrous or in botryoidal masses. Twinning common, {100}, often polysynthetic; also on {110} and more rarely on {201}. Black, reddish, yellow, green, colorless; vitreous to greasy luster. Cleavage {001}, nearly perfect; subconchoidal to uneven fracture; brittle. $H = 6\frac{1}{2}$. $G = 5.82$. Biaxial; $X \wedge c = 12.8°$; $Y = b$; $N_x = 2.136$, $N_y = 2.236$, $N_z = 2.243$; $r > v$. X, dark brown; Y, brown; Z, light brown. From near Bozeman, MT, in a corundum syenite. Mt. Somma, Italy, in druses in a sanidinite rock, with fluorite, nepheline, pyrochlore, and allanite. As rounded crystals in gem gravels at *Ralkwana* and *Balangoda, Sri Lanka*, with zircon, tourmaline, spinel, and several rare-earth minerals. In Phalaborwa, Transvaal, South Africa, in a large carbonatite, with magnetite, diopside, apatite, and phlogopite, in fine black crystals up to 3×8 cm(!). In jacupirangite at Jacupiranga, SP, Brazil, with ilmenite, zirkelite, magnetite, apatite, perovskite, and other minerals, and at Poços de Caldas, MG, in botryoidal masses associated with alkalic rocks in a carbonatite(*). As an accessory in some lunar rocks. RG $DI:608$, AM $40:275(1955)$, AC $12:507(1959)$.

4.4.15.1 Tugarinovite MoO$_2$

Named in 1980 for Aleksei Ivanovich Tugarinov (1917–1977), geochemist, Vernadskii Institute, Moscow, whose wife, V. G. Kruglova, discovered the mineral and named it. MON $P2_1/c$. $a = 5.59$, $b = 4.82$, $c = 5.51$, $\beta = 119.53°$, $Z = 4$, $D = 6.58$. AM 66:438(1981). 32-671(syn): 3.42_{10} 2.44_3 2.43_7 2.40_4 1.725_3 1.723_4 1.711_4 1.709_4. Crystals 0.5 to 1.5 mm, prismatic or thick tabular. Dark lilac brown, greenish-gray streak, metallic to greasy luster. H = $4\frac{1}{2}$. Insoluble in hot HCl or H$_2$SO$_4$, slowly soluble in boiling concentrated HNO$_3$. Biaxial. Pleochroism pale yellowish to bluish olive-brown. Strong bireflectance. R = 20.3–14.7 (546 nm). From *eastern Siberia, Russia*, in quartz metasomites, associated with uraninite, molybdenite, zircon, and galena. RG ZVMO 109:465(1980), DANS 264:689(1982).

4.4.16.1 Cervantite Sb$_2$O$_4$

Named in 1850 for the locality. ORTH $Pna2_1$. $a = 5.436$, $b = 4.810$, $c = 11.76$, $Z = 4$, $D = 6.641$. NBSC 539(10):8(1960). 11-694: 3.45_3 3.07_{10} 2.94_4 2.65_2 2.40_2 1.86_2 1.78_2 1.72_2. Fine-grained, earthy; rarely minute acicular. Yellow, reddish white; greasy to earthy luster. H = 4–5. G = 6.5. Nonfluorescent. Biaxial; N$_x$ = 2.0, N$_z$ = 2.1; strong bireflectance; positive elongation. Commonly found as pseudomorphs after stibnite and after various Sb sulfosalts. Emma mine, Salt Lake Co., UT. Wood River district, Blaine Co., ID. San Emidio mine, Kern Co., CA. In Mexico from the Altar district, SON; from Zacualpan, MEX; and from the San José mine, (!) Catorce, SLP, in large, sharp pseudomorphs after stibnite. *Cervantes, Galicia, Spain*. Pereta, Tuscany, Italy. Allemont, Isère, France. Cornwall, England. Siegen region, Nordheim-Westfalen, Germany. With valentinite at Semsa, Algeria, and at Djebel Tayta, near Guelma. In Australia in the Wiluna district, WA, and the Heazlewood district, TAS. From many localities in Peru and Bolivia. RG DI:595, AM 47:1221(1962), NJMM 1962:93.

4.4.17.1 Lenoblite V$_2$O$_4$·2H$_2$O

Named in 1970 for André Lenoble, French mineralogist–geologist. ORTH $P2_12_12_1$. $a = 2.995$, $b = 4.820$, $c = 14.800$. 23-727: 7.46_7 4.59_{10} 4.05_4 3.72_9 2.80_4 2.53_8 2.28_5 2.20_5. Azure blue, altering to greenish. From the uranium deposit at *Mine Oklo, Mounana* (near Franceville), *Gabon*, associated with duttonite. Also from Kyzyl-Kum, Uzbekistan, with other V minerals. RG BM 93:235(1970), AM 56:635(1971), ZVMO 100:488(1971).

4.5.1.1 Molybdite MoO$_3$

Named in 1963 for the composition. ORTH $Pnma$. $a = 3.962$, $b = 13.858$, $c = 3.697$, $Z = 4$, $D = 4.709$. NBSC 539(3):30(1954). 5-508: 6.93_3 3.81_8 3.46_6 3.26_{10} 2.70_2 2.66_4 2.31_3 1.85_2. Flat needles or thin plates, forming irregular aggregates, flattened on {010}. Light greenish yellow to nearly colorless, adamantine luster. Cleavage {100}, perfect; {001}, distinct; flexible. Biaxial (+); XYZ = bac; N > 2.0; 2V large; very high birefringence; parallel extinction; positive elongation. From *Krupka, northwestern Bohemia, Czech Republic*, associated with molybdenite and betpakdalite, in cavities in topaz–quartz greisen. This mineral is easily confused with ferrimolybdite, which is the normal alteration product of molybdenite and looks almost identical. RG *AM 49:1497(1964), NBSM 20:118(1984)*.

4.5.2.1 Tungstite $WO_3 \cdot H_2O$

Named in 1868 for the composition. ORTH *Pnma*. $a = 5.249$, $b = 10.711$, $c = 5.133$, $Z = 4$, $D = 5.75$. *CM 22:681(1984)*. *43-679:* 5.36_8 3.46_{10} 2.62_4 2.56_5 1.851_4 1.831_4 1.731_5 1.634_4. Massive, pulverulent to earthy; as microscopic platy crystals. Yellow to yellowish green, resinous luster. Cleavage {001}, perfect. $H = 2.5$. $G = 5.52$. Soluble in alkalis but not in acids. Biaxial $(-)$; X = acute bisectrix; $N_x = 1.82$, $N_z = 2.04$; $2V = 27°$; X, colorless; Y,Z, deep yellow; birefringence = 0.22; absorption $Z > Y > X$. Forms as an alteration product of wolframite, ferberite, and scheelite. From *Lane's mine, Monroe, CT*. As an alteration of huebnerite at Osceola, NV. Blue Wing district, Lehmi Co., ID. Kootenay Belle mine, Salmo, BC, Canada. Drakewall mine, Callington, Cornwall, England. Calacalani, Oruro, and at Tazna, Bolivia. RG *DI:605, AM 29:194(1944)*.

4.5.3.1 Meymacite $WO_3 \cdot 2H_2O$

Named in 1961 for the locality. Originally called ferritungstite. *PD:* 3.81 (diffuse). Massive, earthy; yellow-brown, resinous–vitreous luster. Conchoidal fracture. $G = 4.0$. Isotropic; $N > 2.0$. From *Meymac, Corrèze, France*, as an alteration product of scheelite. A highly questionable species. RG *BM 88:613(1965), AM 53:1065(1968)*.

4.5.4.1 Alumotungstite $(W, Al)(O, OH)_3$ (?)

Named in 1971 for the composition. ISO Space group unknown. $a = 10.20$, $Z = 8$. Crystals octahedrons, minute; yellow. Isotropic; $N > 1.90$. Originally reported as the aluminian analog of ferritungstite. From the *Kramat Pulai mine, Kinta district, Perak, Malaysia*, associated with yttrotungstite-(Y). Gifurve mine, Rwanda, associated with ferritungstite; similarly at the Nyamulilo and the Kirwa mines, Uganda. RG *MR 12:81(1981), MM 38:261(1971)*.

4.5.5.1 Ferritungstite $(K, Ca)_{0.2}(W, Fe)_2(O, OH)_6 \cdot H_2O$

Named in 1911 for the composition. ISO $Fd3m$. $a = 10.352$, $Z = 1$, $D = 4.75$. *GAC/MAC 13:A37(1988)*. *11-331:* 5.94_{10} 3.10_9 2.97_{10} 2.57_6 1.82_8 1.55_7 0.990_6 0.870_6. As octahedrons, sometimes platy, representing cyclic twins (sixlings); occasionally fibrous. Ocherous yellow, earthy luster. $G = 5.02$. Isotropic; $N = 2.09$–2.15. Usually formed as an alteration product of scheelite. *Germania tungsten mine, Stevens Co., WA*. Nevada scheelite mine, Mineral Co., NV. La Bertrande, Haute Vienne, France. Borralha mine, Minho, Portugal.

Nyakabingu, Rwanda. Namulilo mine, Uganda. RG *DII:1093*, *AM 42:83(1957)*, *MR 12:81(1981)*.

4.5.6.1 Sidwillite $MoO_3 \cdot 2H_2O$

Named in 1985 for Sidney A. Williams (b.1933), American mineralogist, petrologist, and economic geologist. MON $P2_1/c$. $a = 10.618$, $b = 13.825$, $c = 10.482$, $\beta = 91.63°$, $Z = 16$, $D = 3.11$. *BM 108:813(1985)*. *39-363:* 6.94_8 3.79_5 3.68_6 3.46_3 3.32_{10} 3.25_9 2.66_5 2.62_4. Tabular crystals; bright yellow, resinous luster. Cleavage {010}, perfect. $H = 2\frac{1}{2}$. $G = 3.12$. Biaxial $(-)$; $X = b$; $Z \wedge a = 47°$; $N_x = 1.70$, $N_y = 2.21$, $N_z = 2.38$; $2V = 48°$. X, colorless; Y,Z, yellow. From *Lake Como, CO*, as an oxidation product of jordisite in a quartz vein, associated with pyrite. RG *AM 71:1546(1986)*.

4.6.1.1 Shcherbinaite V_2O_5

Named in 1972 for Vladimir V. Shcherbina (1907–1978), Russian geochemist. ORTH *Pmmn*. $a = 11.51$, $b = 3.559$, $c = 4.371$, $Z = 2$, $D = 3.372$. *NBSC 539(8):66(1959)*. *41-1426:* 5.77_3 4.38_{10} 4.09_3 3.41_8 2.88_5 2.76_2 2.61_3 1.919_1. Fibrous aggregates of acicular crystals. Yellow-green; vitreous luster. Brittle. $H = 3-3\frac{1}{2}$. $G = 3.2$. Readily soluble in HCl and HNO_3. Biaxial; $N \approx 2.42$. From fumaroles in the cupola of the *Bezymyanni volcano, Kamchatka, Siberia, Russia*. RG *DANS 193:135(1970)*, *ZVMO 101:464(1972)*, *AM 58:560(1973)*.

4.6.2.1 Navajoite $V_2O_5 \cdot 3H_2O$ or $(V, Fe^{3+})_{10}O_{24} \cdot 12H_2O$

Named in 1955 for the Navajo Indians, on whose reservation the mineral occurs. MON $C2/m$. $a = 34.92$, $b = 3.597$, $c = 11.79$, $\beta = 97.0°$, $Z = 1$, $D = 2.53$. *AM 75:508(1990)*. *45-1401:* 17.4_1 11.8_{10} 10.5_1 5.79_1 3.41_2 3.18_1 2.90_1 1.992_1. Fibrous. Dark brown, brown streak, silky to adamantine luster. $H < 2$. $G = 2.56$. Biaxial $(-)$; Z parallel to fiber length; $N_x = 1.905$, $N_y = 2.02$, $N_z = 2.025$; parallel to extinction. X,Y, yellowish brown; Z, dark brown. From the *Monument No.2 Mine, Monument Valley, Apache Co., AZ*, associated with corvusite, tyuyamunite, rauvite, hewettite, steigerite, and limonite. RG AM 40:207(1955), 44:332(1959).

4.6.3.1 Ilsemannite $Mo_3O_8 \cdot nH_2O$ (?)

Named in 1871 for J. C. Ilsemann (1727–1822), mining commissioner at Clausthal, Harz, who had studied molybdenite. *21-574:* 4.23_5 3.36_{10} 3.14_2 2.72_4 2.44_3 2.22_3 1.93_3 1.64_5. Earthy; as thin crusts and stains. Blue-black, earthy luster. Peptized by water, giving a blue solution or sol. Near *Ouray, UT*, with halotrichite and molybdite in sandstone. *Gibson, Shasta Co., CA*, as an alteration product of molybdenite. *Bleiberg, Carinthia, Austria*, with wulfenite, barite, and gypsum. In sandstone in the *Hlatimbe valley, Natal, South Africa*. *Bamford, Q, Australia*, staining molybdenite-bearing quartz. RG *DI:603, AM 36:609(1951)*.

4.6.4.1 Paramelaconite $Cu_2^{2+}Cu_2^{1+}O_3$

Named in 1891 for the Greek *para*, near, and *melaconite*, because it is near melaconite in composition. TET $I4_1amd$. $a = 5.837$, $c = 9.932$, $Z = 4$, $D = 5.93$. $AC(B)$ 34:22(1978). 33-480: 3.17_{10} 2.92_4 2.48_1 1.89_2 1.88_1 1.73_2 1.59_1 1.54_1. Crystals stout prismatic [001]; {010} striated [100]. Black, brownish-black streak, metallic/adamantine luster. No cleavage, conchoidal fracture. $H = 4\frac{1}{2}$. $G = 5.9$. Readily soluble in cold dilute mineral acids and in dilute NH_4OH and NH_4Cl solutions. In RL, white, with a pinkish-brown tint. Pleochroism weak, anisotropism strong. From the *Copper Queen mine, Bisbee, AZ(!)*, in fine crystals, associated with cuprite, goethite, tenorite, connellite, and malachite. Algomah mine, Ontonagon Co., MI. RG *DI:510*, *AM 63:180(1978)*.

4.6.5.1 Murdochite $PbCu_6O_{8-x}(Cl, Br)_{2x}$

Named in 1955 for Joseph Murdoch (1890–1973), U.S. mineralogist, University of California, Los Angeles. ISO $Fm3m$. $a = 9.224$, $Z = 4$, $D = 6.06$. $AC(C)$ 39:1143(1983). 7-28: 5.30_{10} 3.25_8 2.78_8 2.66_{10} 2.30_9 2.11_9 1.63_{10} 1.56_8. As minute cubes, also as octahedra; black, black streak. $H = 4$. $G = 6.47$. Insoluble in cold dilute HNO_3, but dissolves with effervescence upon warming the acid. *Mammoth–St. Anthony mine, Tiger, AZ*, with wulfenite. Mex-Tex mine, Hansonburg district, NM. RG *AM 69:815(1984)*, *NJMM 1972:104*.

4.6.6.1 Tantite Ta_2O_5

Named in 1983 for the composition. TRIC $P1$. $a = 3.80$, $b = 3.79$, $c = 35.74$, $\alpha = 90.92°$, $\beta = 90.18°$, $\gamma = 90.0°$, $Z = 6$, $D = 8.45$. *MZ 5(3):90(1983)*. 21-1198(syn): 3.78_6 3.77_7 3.76_5 3.36_8 3.35_7 2.98_{10} 2.46_7 2.45_8. As grains and microscopic veinlets; colorless, transparent; adamantine luster. No cleavage. $H = 7$. Nonfluorescent. Weak blue cathodoluminescence. In RL, grayish white; strongly anisotropic. $R = 14.2–17.5$, depending on orientation and wavelength. From granite pegmatites in the *Kola region, Russia*, associated with microlite, stibiotantalite, and holtite. RG *AM 69:1193(1984)*.

4.6.7.1 Nolanite $(V, Fe, Ti)_{10}O_{14}(OH)_2$

Named in 1957 for Thomas B. Nolan (1901–1992), geologist and director of the U.S. Geological Survey. HEX $P6_3mc$. $a = 5.890$, $c = 9.255$, $Z = 1$, $D = 4.60$. AC 11:703(1958). 19-640: 4.62_6 3.41_8 2.63_{10} 2.47_{10} 2.44_8 2.22_6 1.96_6 1.46_8. Thick tabular crystals; black, brownish-black streak. $H = 5–5\frac{1}{2}$. $G = 4.65$. In RL, dark brown and deep blue internal reflections. From the *Eldorado uranium mine, Beaverlodge region, Lake Athabaska, SK, Canada*, in dolomite, with pitchblende, calcite, and sulfides. Kalgoorlie, WA, Australia, with quartz, gold, and various tellurides. RG *AM 52:734(1967)*, *68:833(1983)*.

Class 5

Oxides Containing Uranium and Thorium

5.1.1.1 Uraninite UO$_2$

Named in 1792 for the composition. Uranium was named for the planet Uranus (named for the Greek god), which had been discovered shortly before the element was discovered. Two names, *uraninite* and *pitchblende*, are in common use for this mineral. Uraninite is the correct name for the species, but pitchblende is a rather sharply defined colloform variety of uraninite. It differs from ordinary uraninite in the crystal size, habit, and mode of occurrence. The particle size of pitchblende varies from about 10^{-2} to 10^{-6} mm. ISO $Fm3m$. $a = 5.468$, $Z = 4$, $D = 10.968$. *NBSC 539(2):33(1953)*. *41-1422*: 3.15_{10} 2.73_5 1.933_5 1.647_5 1.367_1 1.254_2 1.222_2 1.116_2. *USGSB 1064*(syn): 3.14_{10} 1.93_8 1.65_9 1.25_6 1.22_6 1.12_6 1.05_7 0.924_7. **Habit:** Octahedral crystals; less commonly cubes, or combinations of these. Twinning on {111}, rare. Pitchblende is massive, botryoidal, or reniform; banded layers. **Structure:** The crystal structure of uraninite is analogous to that of fluorite, CaF$_2$. The U^{4+} ions are each surrounded by eight O^{2-} ions at the corners of a cube, and each O^{2-} ion by four U^{4+} ions at the corners of an octahedron. **Physical properties:** Black, brownish black, dark gray, greenish; streak brownish-black, grayish, olive-green; semimetallic (especially in well-crystallized material) to submetallic luster, pitchy, dull. No cleavage, uneven to conchoidal fracture, brittle. $H = 5-6$. $G = 7.5-9.7$. Nonfluorescent. Highly radioactive. Nonmagnetic. **Tests:** Soluble in HF, HNO$_3$, H$_2$SO$_4$, but only very slowly attacked by HCl. Infusible. **Chemistry:** Uraninite, ideally UO$_2$, is always partially oxidized, with part of the U^{4+} converted to U^{6+}. Natural material, especially the colloform varieties (pitchblende), approach U$_3$O$_8$ for this reason. In crystallized material Th is usually present, and a complete solid-solution series exists between UO$_2$ and ThO$_2$, although in natural material ThO$_2$ seldom exceeds 14%. Rare earths are also commonly present in small amounts; total REO is normally less than 12%. Pb is also invariably present, from a trace to about 20%; this is believed to be derived mostly from the radioactive decay of the uranium. The unit cell edge of uraninite can vary from 5.37 to 5.55, depending on the state of oxidation; the more highly oxidized it is, the smaller the cell edge. The cell edge also increases linearly with increasing content of Th. He and possibly Ar are usually present, up to 2.6 wt%. **Optics:** Transparent in thin splinters, usually greenish, yellowish, or deep brown in color. Refractive index is high. In polished section, light gray with a brownish tint. Dark brown internal reflections. Isotropic. R = 16.7 (470 nm), 16.0 (546 nm), 15.8 (589 nm), 16.0 (650 nm). *P&J:381*. **Occurrence:** Uraninite is found in pegmatites, usually in crystals, associated with monazite, Nb–Ta minerals, zircon, fluorite, and so on. When in hydrothermal veins, the normal mode of occurrence is as pitchblende; such deposits account for the bulk of uranium production. In high-temperature veins it is associated with cassiterite, arsenopyrite, and Co–Ni–Bi–As minerals; in moderate-temperature hydrothermal veins, with galena and other sulfides; also found as impregnations in sandstone, often associated with V, Cu, and other ores. **Localities:** Topsham, ME(!) in fine cubo-octahedral crystals. Ruggles pegmatite, Grafton, NH, in dendritic masses of crystals associated with curite, uranophane, and so on; fine crystals at Spruce Pine(!) and other localities in western NC, also pegmatitic; in veins in Gilpin Co. and Jefferson Co., CO, with sulfides; at Baringer Hill, Llano Co., TX, in pegmatite with gadolinite and Nb–Ta minerals; in sandstone and conglomerate near Moab, UT, and near Grants, NM. Eldorado mine, Great Bear Lake, NWT, Canada, in veins with native silver and Co and Ni mineralization; also at Lake Athabasca, SK, and in very sharp crystals to 10 cm in Cardiff Twp, ONT. Placeres de Guadalupe, CHIH, Mexico(*), with gold. Cornwall, England(!), botryoidal masses in tin-bearing veins. Fuente-Ovejuna pegmatite, Sierra Albarrana, Cordoba, Spain(!), crystals to 40 kg. Château Lambert, Vosges, France; also at Chaméane, Puy de Dôme; Barzot, Saone-et-Loire, and le Retail, Vendée. In *Bohemia, Czech Republic,* at *Jáchymov* and *Horni Slavkov* (pitchblende) in veins; Schneeberg and Johanngeorgenstadt, Sachsen, Germany. Ajmer, Rajputana, India, in pegmatite. Kotoge and Masaki, Fukuoka Pref., Japan, in pegmatite. Azgour, Morocco (pitchblende). Shinkolobwe and elsewhere in Shaba, Zaire, as replacement bodies of crystallized uraninite (crystals to 30 mm) with oxidized uranium minerals and copper ores. Morogoro, Tanzania. El Sharana, NT, Australia. **Uses:** The principal ore of uranium. In early times used for coloring glass yellow-green. Now used for generating nuclear power, atomic bombs, and so on. RG *DI:611, USGSB 1064:11(1958), EG 56:241(1961), MR 5:79(1974).*

5.1.1.2 Thorianite ThO$_2$

Named in 1904 for the composition. ISO $Fm3m$. $a = 5.600$, $Z = 4$, $D = 9.991$. *NBSC 539(1):57(1953). 42-1462:* 3.23_{10} 2.80_4 1.979_5 1.688_5 1.284_2 1.252_1 1.077_2 0.946_2. Usually as cubes, sometimes modified by $\{111\}$ or $\{113\}$. Penetration twins on $\{111\}$, common. Black; dark gray; brownish black, with gray to greenish-gray streak; horny to submetallic luster. Cleavage $\{001\}$, poor; uneven to subconchoidal fracture; brittle. H = 5–6. G = 9.7–9.8. Nonfluorescent. Radioactive. Diamagnetic. Infusible; at high temperatures emits a strong white glow. Thorianite forms a complete solid-solution series with uraninite. It also forms a partial series with cerianite, CeO_2. Uranium apparently is always present in moderate to large amounts; varieties with Th/U ≈ 1:1 are called uranothorianite. Pb is also invariably present and is probably of radiogenic origin. The cell edge of thorianite varies from 5.60 to 5.50, decreasing with increasing content of U, Pb, or Ce, and with oxidation of the U^{4+} content to U^{6+}. Isotropic; N ≈ 2.20 to 2.35. In RL, gray, isotropic, with reddish-brown internal reflections. R = 16.4 (470 nm), 15.3 (546 nm), 15.1 (589 nm), 14.8 (650 nm). *P&J:369.* Thorianite is a primary mineral found chiefly in pegmatites or as an accessory mineral in granites. Known occurrences are principally in alluvial deposits derived from these rocks or placers. It is also occasionally found in calcite–fluorite–apatite veins that are transitional to pegmatites. From Easton, PA, in serpentine formed by the alteration of magnesium silicates at the contact of limestone with pegmatite. In black sands in the Missouri River near Helena, MT, and similarly in the Wiseman and Nixon Fork districts, AK. In *Sri Lanka* from the *Balangoda district*, in placers associated with thorite, allanite, baddeleyite, geikielite, and cassiterite. In Madagascar in pyroxenites near Betroka(!), in fine crystals up to 15 cm (uranothorianite) with spinel, phlogopite, diopside; also at Betanimena, Morafeno, and Amboanemba. RG *DI:620, USGSB 1064:47(1958)*.

5.2.1.1   Metaschoepite $UO_3 \cdot < 2H_2O$

5.2.1.2   Paraschoepite $UO_3 \cdot 2H_2O$

5.2.1.3   Schoepite $UO_3 \cdot 2H_2O$

Named in 1923 (schoepite); 1947 (para-); 1960 (meta-); for Alfred Schoep (1881–1966), Belgian mineralogist, who studied the mineralogy of uranium extensively. *Meta*, from the Greek indicating a lower hydrate than schoepite; *para*, from the Greek indicating a close relationship to schoepite. ORTH *Pbna*. $a = 13.99$, $b = 16.72$, $c = 14.73$, Z = 32, D = 4.69 (metaschoepite). ORTH *Pbca*. $a = 14.12$, $b = 16.83$, $c = 15.22$, Z = 32, D = 4.83 (paraschoepite). ORTH *Pbca*. $a = 14.33$, $b = 16.79$, $c = 14.73$, Z = 32, D = 4.83 (schoepite). M: *43-364*: 7.33_{10} 3.67_3 3.58_4 3.49_2 3.48_1 3.22_5 3.16_3 2.56_2. P: *23-1461*: 7.53_{10} 3.56_{10} 3.32_5 3.22_{10} 3.17_5 2.01_3 1.78_4 1.73_3. S: *13-407*: 7.28_{10} 5.08_7 3.66_1 3.51_1 3.44_2 3.22_1 2.89_1 2.54_1. $UO_2(OH)_2$: 5.09_{10} 3.45_3 3.39_2 2.89_1 2.54_1 2.48_1 2.02_1 1.77_1. Crystals tabular on {001}, elongated [010], with pseudohexagonal aspect; also massive, as crusts; twinning not observed. Schoepite, the highest hydrate of the three closely related minerals, dehydrates spontaneously at room temperature to metaschoepite and paraschoepite. Any given sample of schoepite is likely to consist of a mixture of two or even three of these compounds. Crushing or grinding of the crystals also causes dehydration, which may progress as far as $UO_2(OH)_2$, a compound not yet found in nature. The physical properties of these minerals are closely similar, where it has been possible to distinguish between them. Diagnosis of the compounds is best done through powder diffraction data (where it must be kept in mind that any given unknown of the group is almost certainly a mixture of unknown proportions), or through their optical properties. Amber (schoepite); yellow (meta- and paraschoepite); yellow streak. Cleavage {001}, perfect; {010}, fair; brittle. H = $2\frac{1}{2}$. G = 4.8. Fluorescent pale green under LWUV and SWUV. Biaxial (−); $\hat{X}Y\hat{X} = cba$; metaschoepite: $N_x = 1.695$, $N_y = 1.730$, $N_z = 1.750$; paraschoepite: $N_x = 1.700$, $N_y = 1.750$, $N_z = 1.770$; schoepite: $N_x = 1.70$, $N_y = 1.720$, $N_z = 1.735$. Schoepite, and by inference, metaschoepite and paraschoepite as its normal alteration products, have been found at Beryl Mt., Acworth, NH, as pseudomorphs after uraninite. Several mines on the Colorado Plateau, including the Monument No. 2 mine, Apache Co., AZ, and the Happy Jack mine, White Canyon, San Juan Co., UT. From Germany at Wölsendorf, Bavaria, with ianthinite and becquerelite, and at Menzenschwand, Black Forest. Margnac II, Haute-Vienne, France, and at Mas d'Alary, Hérault. *Kasolo, Shaba, Zaire*, and also at Musonoi; relatively abundant at Shinkolobwe, Zaire, with other secondary uranium minerals. RG *DI:627, USGSB 1064:72(1958), AM 45:1026(1960), IANS Ser. Geol.:140(1979), JINC 30:1745(1968)*.

5.2.2.1 Masuyite $Pb_3U_8^{6+}O_{27} \cdot 10H_2O$

Named in 1947 for Gustave Masuy (1905–1945), Belgian geologist, shot by the German occupiers at the end of World War II. ORTH *Pban*. $a = 41.93$, $b = 24.22$, $c = 42.61$, $Z = 432$, $D = 5.338$. *AM 45:1026(1960)*. 13-408: 7.08_{10} 3.56_3 3.52_7 3.48_2 3.16_1 3.12_5 2.01_2 1.95_2. Crystals tabular on {001}, with pseudohexagonal aspect; invariably twinned, both on {110} and {130}. Reddish orange, brownish orange; adamantine luster. Cleavage {001}, perfect; {010}, fair. G = 5.08. Nonfluorescent. Biaxial (−); $XYZ = cba$; $N_y = 1.895$, $N_z = 1.915$ (Na); $2V \approx 50°$. X,Y, golden yellow. Lodive, Hérault, France. *Shinkolobwe, Shaba, Zaire*, minute crystals lining cavities in uraninite, and also at Kamoto. RG *USGSB 1064:78(1958), ZK 113:132(1960)*.

5.3.1.1 Studtite $UO_4 \cdot 4H_2O$

Named in 1947 for Franz Edward Studt, Belgian prospector and geologist, author of the first geological map (1913) of Shaba, Zaire. MON *C2, Cm*, or *C2/m*. $a = 11.85$, $b = 6.80$, $c = 4.25$, $\beta = 93.85°$, $Z = 2$, $D = 3.64$. *AM 59:166(1974)*. 16-206(syn): 5.88_{10} 4.23_1 3.55_1 3.49_2 3.40_2 3.39_3 2.94_1 2.65_1. Thin crusts composed of minute fibers; as masses of radial-fibrous tufts of acicular crystals on renardite; yellow. No cleavage. $G_{syn} = 3.58$. Soft; nonfluorescent. Biaxial (+); $N_x = 1.537$ (nearly colorless); $N_z = 1.680$; 2V small; parallel to elongation. From Menzenschwand, Black Forest, Germany, with billietite, uranophane, and rutherfordine on barite, quartz, and limonite. Davignac, Corrèze, France. Menzenschwand, Black Forest, Germany. In *Zaire* at *Shinkolobwe, Shaba*, with uranophane and rutherfordine, and at Kobokobo, Kivu. RG *BSBG 70:B212(1947), AM 33:384(1948)*.

5.3.1.2 Metastudtite $UO_4 \cdot 2H_2O$

Named in 1983 from Greek *meta* and *studtite*, indicating a lower hydrate than studtite. ORTH *Immm*. $a = 6.51$, $b = 8.78$, $c = 4.21$, $Z = 2$, $D = 4.67$. *35-571:* 5.22_{10} 4.38_5 3.79_5 3.54_8 3.21_5 2.76_3 2.67_2 2.47_2. Silky fibers; radiated nodules. Pale yellow, yellowish streak, silky luster. Nonfluorescent. Soluble in hot HCl without effervescence. Biaxial (+); $Z = c$; $N_x = 1.640$, $N_y = 1.658$, $N_z = 1.760$; $2V = 48°$; no pleochroism. From *Shinkolobwe, Shaba, Zaire*, on dolomite with rutherfordine, becquerelite, masuyite, kasolite, and wolsendorfite; also on massive uraninite. RG *AM 68:456(1983)*.

5.3.2.1 Vandenbrandeite $Cu(UO_2)(OH)_4$

Named in 1932 for Pierre van den Brande (1896–1957), Belgian geologist, who discovered the ore deposit at Kalongwe, Zaire. TRIC $P\bar{1}$. $a = 7.86$, $b = 5.44$, $c = 6.10$, $\alpha = 91.86°$, $\beta = 102.0°$, $\gamma = 89.62°$, $Z = 2$, $D = 5.26$. *USGSB 1064:100(1958)*. *4-340:* 5.06_4 4.29_{10} 2.92_8 2.56_4 2.09_3 1.85_3 1.47_3 1.32_3. *8-325*(syn): 5.26_9 4.44_{10} 3.85_4 2.97_8 2.59_7 2.20_5 2.03_6 1.64_5. Crystals tabular {001}, somewhat elongated; also scalelike. Dark green to blackish green, vitreous luster. Cleavage {110}, perfect. $H = 4$. $G = 5.03$. Nonfluorescent. Biaxial (−); $Z \wedge \text{elongation} = 40°$; $N_x = 1.765$, $N_y = 1.792$, $N_z = 1.80$; $2V \approx 90°$; strong dispersion. X, blue-green; Z, yellow-green. Rabejac, Lodève, Herault, France, on pitchblende. In *Zaire* at *Kalongwe, Shaba*, with kasolite, uranophane, sklodowskite, torbernite, and malachite; also from Luiswishi(!) doubly terminated crystals to 1 cm, and at Musonoi(!), abundant brilliant crystals. RG *DI:632, CSC 6:53(1977)*.

5.4.1.1 Clarkeite $(Na, Ca, Pb)_2U_2(O, OH)_7$

Named in 1931 for Frank Wigglesworth Clarke (1847–1931), American geochemist, U.S. Geological Survey. *8-315:* 5.77_8 3.34_9 3.17_{10} 2.69_5 1.97_7 1.86_6 1.76_5 1.64_5. Massive, fine grained; brown; dark reddish brown, yellowish-brown streak, waxy luster. Conchoidal to splintery fracture. $H = 4-4\frac{1}{2}$. $G = 6.29$. Nonfluorescent. Readily soluble in acids. Biaxial (−); $N_x = 1.997$, $N_y = 2.098$, $N_z = 2.108$; $2V = 30°-50°$; $r < v$, weak; weakly pleochrosic. From the *Deer Park mine, Spruce Pine, Mitchell Co., NC*. Ajmer district, Rajputana, India, associated with uraninite, fourmarierite, and uranophane in pegmatite. RG *DI:624, USGSB 1064:95(1958), AM 41:127(1956)*.

5.4.2.1 Calciouranoite $(Ca, Ba, Pb)U_2O_7 \cdot 5H_2O$

Named in 1974 for the composition. Amorphous. *41-1433:* 4.0_5 3.41_{10} 3.05_{10} 2.68_5 2.00_5 (diffraction diagram made after heating). Brown, orange. $G = 4.62$. From the oxidation zone of a U–Mo deposit, *Russia*, associated with other uranium oxides. RG *ZVMO 103:108(1974), IGR 16:1255(1974), AM 60:161(1975)*.

5.4.2.2 Bauranoite $BaU_2O_7 \cdot 4-5H_2O$

Named in 1973 for the composition. *25-1469:* 3.41_7 3.09_{10} 1.98_6 1.948_6 1.908_6 1.723_6 1.667_6 1.325_5. Dense, fine-grained aggregates. Reddish brown. H = 5. G = 5.3. Biaxial (−); $N_x = 1.911–1.916$, $N_z = 1.920–1.932$; 2V = 81°. From a U–Mo deposit in *Russia*, replacing uraninite, replaced by uranophane. RG *ZVMO 102:75(1973), AM 58:1111(1973)*.

5.4.3.1 Metacalciouranoite $(Ca, Na, Ba)U_2O_7 \cdot 2H_2O$

Named in 1973 from the Greek *meta* and *calciouranoite*, indicating a lower hydrate. Space group unknown. *25-1451:* 3.43_6 3.09_{10} 1.99_4 1.94_4 1.92_3 1.73_3 1.68_7 1.28_7. Dense, fine-grained aggregates. Orange. G = 4.90. Biaxial (−); $N_x = 1.897$, $N_y = 1.911$, $N_z = 1.932$; 2V = 81°. From the oxidation zone of a U–Mo deposit in *Russia*, replacing pitchblende and being replaced by uranophane. RG *ZVMO 102:75(1973), AM 58:1111(1973)*.

5.4.3.2 Wölsendorfite $(Pb, Ba, Ca)U_2O_7 \cdot 2H_2O$

Named in 1957 for the locality. ORTH Space group unknown. $a = 11.95$, $b = 13.99$, $c = 7.02$, Z = 6, D = 6.82. *AM 42:919(1957). 29-786:* 6.91_4 3.46_8 3.09_{10} 2.44_3 1.993_3 1.913_3 1.725_4 1.676_4. *MM 46:131(1982):* 7.03_{10} 6.03_2 3.50_8 3.47_5 3.13_5 3.10_6 2.47_3 2.01_3. Fine, platy aggregates. Orange-red, adamantine to vitreous luster. Cleavage {001}, good. H = 5. G = 6.8. Biaxial; $N_x = 2.05$, $N_z = 2.09$. Eldorado mine, Great Bear Lake, NWT, Canada, on pitchblende. *Wölsendorf, Bavaria, Germany*, in fissures in fluorite, and at Menzenschwand, Black Forest. In France at Kerségalec-en-Lignol, Vendée, with pitchblende; Margnac II, Haute-Vienne; and La Faye, Grury, Sâone et Loire. Shinkolobwe, Shaba, Zaire, with secondary uranium minerals. RG *BM 105:606(1982), ZVMO 103:718(1974)*.

5.5.1.1 Agrinierite $(K_2, Ca, Sr)U_3O_{10} \cdot 4H_2O$

Named in 1972 for Henri Agrinier (1928–1971), French engineer, Mineralogy Laboratory, Commission de Energie Atomique. ORTH *Cmmm*. $a = 14.04$, $b = 24.07$, $c = 14.13$, Z = 16, D = 5.62. *25-630:* 7.08_{10} 6.05_7 3.52_8 3.49_9 3.15_9 3.13_{10} 2.79_6 2.02_8. Crystals tabular {001}, with pseudohexagonal section; sector twinning shown. Orange. Cleavage {001}, good. G = 5.7. Biaxial (−); XZ = *ab*; $N_y = 2.01$, $N_z = 2.06$ (Na); 2V = 55°. From the *Margnac uranium deposit, Haute Vienne, Massif Central, France*, with uranophane in small cavities in "gummite" (a massive orange-red alteration product of pitchblende). RG *MM 38:781(1972), AM 58:805(1973), BM 105:606(1982)*.

5.5.2.1 Rameauite $K_2CaU_6O_{20} \cdot 9H_2O$

Named in 1972 for Jacques Rameau (1926–1960), French prospector, who discovered the deposit where the mineral occurs. MON *C2/c* or *Cc*. $a = 13.97$, $b = 14.26$, $c = 14.22$, $\beta = 121.02°$, Z = 4, D = 5.55. *AM 58:805(1973). 25-631:* 7.12_{10} 3.57_9 3.50_{10} 3.47_9 3.19_9 3.14_{10} 3.12_{10} 2.50_7. Crystals prismatic, flattened

{010}, elongated [001], with pseudohexagonal section; twinning on {100}. Orange. Cleavage {010}, good. G = 5.6. Biaxial; $N_y = 1.95$, $N_z = 1.97$ (Na); 2V = 32°. *Margnac uranium deposit, Haute Vienne, Massif Central, France*, with calcite and uranophane on pitchblende. RG *MM 38:781(1972)*, *BM 105:606(1982)*.

5.5.3.1 Protasite Ba(UO$_2$)$_3$O$_3$(OH)$_2$ · 3H$_2$O

Named in 1986 for Jean Protas (b.1931), University of Nancy, France. MON Pn. $a = 12.295$, $b = 7.221$, $c = 6.956$, $\beta = 90°24'$, $Z = 2$, $D = 5.827$. *AM 72:1230(1987)*. *39-349*: 7.06_5 6.00_2 3.58_3 3.14_{10} 3.11_3 1.98_2 1.75_2 1.60_2. Pseudohexagonal plates, flattened {010}; sector twinning observed. Bright orange, subadamantine luster. Cleavage {010}, good; hackly fracture; brittle. Biaxial (−); $N_y = 1.79$–1.83, $N_z = 1.79$–1.83; high birefringence; 2V = 60–65°. No pleochroism. From *Shinkolobwe, Shaba, Zaire*, associated with uranophane and uraninite. RG *AM 72:224(1987)*.

5.6.1.1 Ianthinite U^{4+}U$_5^{6+}$O$_{17}$ · 10H$_2$O

Named in 1926 for the Greek for *violet*, in allusion to its color. ORTH Space group unknown. $a = 11.52$, $b = 7.15$, $c = 30.38$, $Z = 4$, $D = 4.99$. *AM 44:1103(1959)*. *12-272*: 7.63_{10} 3.81_8 3.59_6 3.35_6 3.24_8 2.61_4 2.53_4 2.03_4. Small plates or laths flattened on {001} and elongated along the b axis; also thick tabular {001}. Violet-black, submetallic luster. Cleavage {001}, perfect. H = 2–3. G = 5.16. Nonfluorescent. Biaxial (−); XYZ = cba; $N_x = 1.674$ (colorless); $N_y = 1.90$ (violet); $N_z = 1.92$ (dark violet). 2V small. Ianthinite slowly alters in air, first to a violet-brown compound, and finally to a yellow substance ("epi-ianthinite"), which is schoepite probably mixed with other uranium compounds. In Germany at Wölsendorf, Bavaria, and at Menzenschwand, Baden-Württemberg. In France at Bigay, Lachaux, Puy de Dôme, and Les Bois-Noirs, St.-Priest-la-Prugne, Loire. From *Shinkolobwe, Shaba, Zaire*. RG *DI:633*, *USGSB 1064:56(1958)*.

5.7.1.1 Compreignacite K$_2$U$_6$O$_{19}$ · 11H$_2$O

Named in 1964 for the locality. ORTH $Pnnm$. $a = 7.16$, $b = 12.16$, $c = 14.88$, $Z = 2$, $D = 5.13$. *AM 50:807(1965)*. *17-167*: 7.40_{10} 3.70_6 3.58_6 3.53_8 3.34_6 3.19_8 2.55_6 2.32_5. *33-1049*(syn): 7.46_{10} 3.71_8 3.58_4 3.53_6 3.23_7 3.20_9 2.56_4 2.03_4. Small crystals flattened on {001}; twins on {110}, common. Yellow. Cleavage {001}, perfect. G = 5.03. Biaxial (−); XYZ = cab; $N_x < 1.790$, $N_y = 1.798$, $N_z = 1.802$; 2V = 10°–15°; no dispersion. X, colorless; Y,Z, yellow. From the *Margnac mine, Compreignac, Haut Vienne, France*, and in Cornwall, England, at West Wheal Owles, St. Just. RG *BM 87:365(1964)*, *105:606(1982)*.

5.7.1.2 Becquerelite Ca(UO$_2$)$_6$O$_4$(OH)$_6$ · 8H$_2$O

Named in 1922 for Antoine Henri Becquerel (1852–1908), French physicist, who discovered radioactivity. ORTH $Pn2_1a$. $a = 13.838$, $b = 12.378$, $c = 14.924$, $Z = 4$, $D = 5.119$. *AM 72:1230(1987)*. *13-405*: 7.44_{10} 4.68_1 3.73_3 3.54_2 3.20_3

2.57_1 2.04_1 1.94_1. 29-389(syn): 7.49_{10} 3.74_4 3.55_4 3.45_2 3.21_8 3.16_3 2.57_3 2.04_3. Usually prismatic, elongated [010]; occasionally tabular on {001}; massive; twinning not observed. Lemon yellow, amber, brownish yellow; yellow streak; adamantine to greasy luster. Cleavage {001}, perfect; also {101}. H = $2\frac{1}{2}$. G = 5.2. Nonfluorescent. Biaxial (−); XYZ = cab; $N_x = 1.730$, $N_y = 1.825$, $N_z = 1.830$ (Na); 2V ≈ 30°. X, pale yellow; Y,Z, deep yellow. From the Monument No. 2 mine, Apache Co., AZ, and elsewhere on the Colorado Plateau in small quantities. Eldorado mine, Great Bear Lake, NWT, Canada. In Germany at Wölsendorf, Bavaria, with schoepite and uraninite in a fluorite deposit, and at Ellwieler, Pfalz. Margnac II mine, Haute-Vienne, France, and at Bigay-Lachaux, Puy de Dôme. *Kasolo* and Shinkolobwe, *Shaba, Zaire*(!), as an alteration product of uraninite, associated with schoepite, fourmarierite, curite, soddyite, and other secondary uranium minerals. Bemasoandro, Malakialina, Madagascar. RG *DI:625, USGSB 1064:62(1958), AM 45:1026(1960)*.

5.7.1.3 Billietite $Ba(UO_2)_6O_4(OH)_6 \cdot 8H_2O$

Named in 1947 for Valère Billiet (1903–1945), Belgian crystallographer, University of Ghent. ORTH $Pnb2_1$. a = 12.072, b = 30.167, c = 7.145, Z = 4, D = 5.093. *AM 72:1230(1987)*. 29-208(syn): 7.53_8 3.76_5 3.50_7 3.18_{10} 2.56_3 2.05_3 2.04_5 1.95_4. Tabular on {001}, with a pseudohexagonal aspect; usually twins on {110}. Deep golden yellow; adamantine luster. Cleavage {001}, perfect. G = 5.28. Biaxial (−); XYZ = cab; $N_x = 1.76$, $N_y = 1.800$, $N_z = 1.805$ (Na); 2V = 37°. X, pale greenish yellow; Y, greenish yellow; Z, deep yellow. Kruth, and Margnac II, Haute-Vienne, France, common as crusts on gummite. Menzenschwand, Schwarzwald, Germany. In *Shaba, Zaire*, at *Shinkolobwe*, associated with fourmarierite, uranophane, and torbernite, and at Swambo and Luiswishi. RG *USGSB 1064:68(1958), ZK 113:132(1960), BM 105:606(1982)*.

5.8.1.1 Vandendriesscheite $PbU_7O_{22} \cdot 12H_2O$

5.8.1.2 Metavandendriesscheite $PbU_7O_{22} \cdot nH_2O$ (n < 12)

Named in 1947 for Adrien Vandendriessche (1914–1940), Belgian mineralogist and geologist. Metav- (1960) indicates a lower state of hydration than V-. ORTH $Pmma$ or $P2_1ma$. a = 14.07, b = 40.85 (V.), b = 41.31 (MV.), c = 43.33, Z = 36. *AM 45:1026(1960)*. 13-117: 7.25_{10} 4.53_1 3.61_{10} 3.53_3 3.17_8 2.52_3 2.03_2 1.99_4. Prismatic, elongated [100], {010} striated parallel to [100]; also tabular on {001}, elongated [100], with pseudohexagonal aspect; twinning not observed. Yellowish orange, orange; adamantine luster. Cleavage {001}, perfect. G = 5.45. Nonfluorescent. Biaxial (−); XYZ = cba; $N_x = 1.780$, $N_y = 1.850$, $N_z = 1.860$ (Na); 2V = 60°. X, colorless; Y,Z, golden yellow. Vandendriesscheite crystals have been shown to have varying proportions of the normal and meta phases, corresponding to varying degrees of partial dehydration which takes place in a normal atmosphere. Differentiating between the phases, even using X-ray powder diffraction, is not an easy matter. The properties of the two phases are otherwise closely similar. From Newry, ME. Palermo and Ruggles pegmatites, NH. Spruce Pine district, Mitchell Co., NC. Eldorado mine, Great Bear Lake, NWT, Canada, with fourmarierite on altered uraninite. Limouzat quarry, Les Bois-Noirs, Loire, France, common as an orange crust near pitchblende. Kalongwe and *Shinkolobwe, Shaba, Zaire*, in the orange-red oxidic zone of alteration ("gummite" zone) of uraninite. RG *BSBG 70:B212(1947), USGSB 1064:81(1958), AM 45:1059(1960)(Meta-)*.

5.9.1.1 Uranosphaerite $Bi_2U_2O_9 \cdot 3H_2O$

Named in 1873 for the composition *uranium* and the Greek for *sphere*, in allusion to its globular forms. Probably ORTH Space group unknown. *8-321*: 5.25_6 4.37_4 3.87_7 3.47_6 3.16_{10} 1.97_4 1.90_5 1.83_8. In hemispherical forms, radial-fibrous, made up of acutely terminated crystals elongated [001]. Brick-red; orange-yellow, yellow streak, greasy to dull luster. Cleavage {001}. H = 2–3. G = 6.36. Biaxial (+); XYZ = *abc*; $N_x = 1.959$, $N_y = 1.981$, $N_z = 2.060$; 2V = 56°; r < v, strong; no pleochroism. From the *Weisser Hirsch, mine, Neustadtl, near Schneeberg, Sachsen, Germany*, with walpurgite, uranospinite, troegerite, zeunerite, erythrite, and cobalt oxides. Kersegalec-en-Lignol, Vendée, France, in red masses. RG *DI:631, AM 42:905(1957)*.

5.9.2.1 Fourmarierite $PbU_4^{6+}O_{13} \cdot 4H_2O$

Named in 1924 for Paul Fourmarier (1877–1970), Belgian geologist. ORTH $Bb2_1m$. a = 13.986, b = 16.400, c = 14.293, Z = 8, D = 5.978. *AM 72:229 (1987). 13-116*: 7.20_{10} 4.36_1 3.58_5 3.55_2 3.18_5 3.14_1 2.52_1 1.98_1. Crystals tabular {001}, elongated [010], with pseudohexagonal aspect; also acicular in orange tufts. Reddish orange; brown, subadamantine luster. Cleavage {001}, perfect; {100}, fair. H = 3–4. G = 5.74. Nonfluorescent. Easily soluble in acids to a yellow solution. Biaxial (−); XYZ = *cab*; $N_x = 1.863$, $N_y = 1.885$, $N_z = 1.890$ (Na); 2V = 50°. Y, pale amber yellow; Z, amber yellow. From the Spruce Pine district, Mitchell Co., NC, as an alteration of uraninite in pegmatites; Palermo mine, near North Groton, NH. Monument No. 2 mine, Apache Co., AZ, in oxidized, mineralized fossil wood. Eldorado mine, Great Bear Lake, NWT, Canada, on altered uraninite. Ajmer district, Rajputana, India. *Shinkolobwe mine, Shaba, Zaire*(!), with kasolite, torbernite, and other secondary uranium minerals. Mounana, Gabon. RG *DI:628, USGSB 1064:87(1958), AM 45:1026(1960)*.

5.9.3.1 Curite $Pb_2U_5^{6+}O_{17} \cdot 4H_2O$

Named in 1921 for Pierre Curie (1859–1906), French physicist known for his research on radioactivity. Exact formula uncertain. ORTH *Pnam*. a = 12.513, b = 13.002, c = 8.373, Z = 1, D = 7.435. *SR 48A:246(1981). 14-267*: 6.23_{10} 3.96_7 3.52_4 3.14_8 3.06_6 2.53_4 2.09_4 1.85_4. Crystals prismatic [001], striated [001]; usually massive, sugary, or earthy. Orange-red, orange streak, adamantine luster. Cleavage {100}; brittle. H = 4–5. G = 7.37. Nonfluorescent. Easily soluble in dilute acids. Biaxial (−); XYZ = bac; N_x = 2.06, N_y = 2.11, N_z = 2.15; 2V large; r > v, strong. X, pale yellow; Y, light red-orange; Z, dark red-orange. Curite is a secondary mineral formed with other uranium oxysalts during the alteration of uraninite. La Crouzille, Puy de Dôme, France, as a component of "gummite." Ellweiler, Pfalz, Germany. *Shinkolobwe mine*, and elsewhere in *Shaba, Zaire*(!), associated with fourmarierite, schoepite, vandendriesscheite, soddyite, kasolite, dewindtite, and torbernite; also in Kivu at Kobokobo. In Madagascar at Bemasoandro. RG *DI:629, USGSB 1064:92(1958), TMPM 26:279(1979)*.

5.9.4.1 Mourite $U^{4+}Mo_5^{6+}O_{12}(OH)_{10}$

Named in 1962 for the composition, *molybdenum* and *uranium*. MON Pc or $P2/c$. a = 24.443, b = 7.182, c = 9.901, β = 102.22°, Z = 4, D = 4.22. *AM 56:163(1971). 24-1359:* 5.97_{10} 3.30_2 3.23_2 3.18_2 3.13_2 2.98_2 2.89_4 2.49_2. Fan-shaped, radiating-fibrous, or vermiform aggregates of plates with rectangular outline, a few microns across; fibers are elongated [010]. Violet, violet-blue streak, adamantine luster. Cleavage {100}, pronounced. H = 3. G = 4.17. Slowly soluble in hot concentrated HCl; dissolves in concentrated HNO_3 with separation of molybdic acid. Insoluble in H_2SO_4. N > 1.790. Pleochroism pale greenish violet to dark blue-violet. $Z \wedge c$ = 10°; positive elongation; strong anisotropism; extinction parallel to fiber axis. From the Boso–Hackney prospect, Tordillo hill, 20 km SW of Falls City, Karnes Co., TX. From a U–Mo deposit in *Russia*, associated with iriginite, powellite, sodium–uranospinite, uranophane, tyuyamunite, goethite, jarosite, and kaolinite. RG *ZVMO 91:67(1962), AM 47:1217(1962)*.

5.9.5.1 Richetite $PbU_4O_{13} \cdot 4H_2O$

Named in 1947 for Emile Richet (1884–1938), chief geologist, Union Minière du Haut Katanga. TRIC $P1$ or $P\bar{1}$. a = 20.81, b = 12.06, c = 16.30, α = 103.8°, β = 115.1°, γ = 90.4°, Z = 9, D = 6.02. *AM 70:1335(1985). 38-370:* 7.12_8 3.55_6 3.47_6 3.12_{10} 2.49_4 2.01_3 1.96_4 1.75_3. Pseudohexagonal tablets, flattened on {010}; twinning gives hexagonal aspect. Black, adamantine luster. Cleavage {010}, perfect; [010], fair. Biaxial (−); X parallel to c; $Y \wedge [110]$ = 85°, $Z \wedge [110]$ = 5°; N_x ≈ 1.9, N_y ≈ 2.0, N_z ≈ 2.0; 2V large. X, pale brown to colorless; Y,Z, dull brown. *Shinkolobwe mine, Shaba, Zaire*, with uranophane. RG *BSBG 70:B212(1947), USGSB 1064:91(1958), BM 107:581(1984)*.

5.9.6.1 Sayrite $Pb_2(UO_2)_5O_6(OH)_2 \cdot 4H_2O$

Named in 1983 for David Sayre (b.1924), American crystallographer. MON $P2_1c$. a = 10.704, b = 6.960, c = 14.533, β = 116.82°, Z = 2, D = 6.76. *BM 106:299(1983). 35-601:* 7.01_8 5.67_5 3.51_9 3.43_4 3.11_{10} 3.05_7 1.96_5 1.92_5. Crystals elongated [010], flattened {$\bar{1}$02}. Yellow-orange, reddish orange. Cleavage

5.9.6.1 Sayrite

{$\bar{1}02$}, distinct. Biaxial (−); X ≈ a, Y = b, Z ≈ c; N_y ≈ 1.94, N_z ≈ 1.95; 2V large. Y, pale yellow; Z, amber. From the *Shinkolobwe mine, Shaba, Zaire*, associated with uranophane, becquerelite, masuyite, and richetite. RG *AM* 69:568(1984).

Class 6

Hydroxides and Oxides Containing Hydroxyl

DIASPORE GROUP

The diaspore group consists of five minerals corresponding to the formula

$$AO(OH)$$

where

A = Al, Fe, Mn, (V, Fe), or Cr

and the structure is orthorhombic, *Pnma* or *Pnmd*.

The structure of diaspore is based on layers of oxygen atoms, the sequence of which is that of hexagonal close packing. Aluminum ions occupy octahedrally coordinated sites between layers in such a way as to form strips of octahedra, the direction of which defines the *c* parameter of the unit cell. The strips have the width of two octahedra and yield an orthorhombic unit cell in which *a* is twice the distance between oxygen layers. Manganese may substitute for Al and Fe up to about 6% in diaspore and goethite; and Fe for Al and V in diaspore and montroseite to a similar extent. In groutite little Fe substitution is reported; but 5.7% Sb was found replacing manganese in groutite from Franklin, NJ. *DHZ II:574(1952), PCM 5:179(1979)*.

DIASPORE GROUP

Mineral	Formula	Space Group	a	b	c	D	
Diaspore	AlO(OH)	*Pnma*	4.396	9.426	2.844	3.380	6.1.1.1
Goethite	FeO(OH)	*Pnma*	4.62	9.95	3.01	4.26	6.1.1.2
Groutite	α-MnO(OH)	*Pnmd*	4.560	10.700	2.870	4.171	6.1.1.3
Montroseite	(V,Fe)O(OH)	*Pnma*	4.54	9.97	3.03	4.11	6.1.1.4
Bracewellite	CrO(OH)	*Pnma*	4.492	9.860	2.974	4.286	6.1.1.5

6.1.1.1 Diaspore AlO(OH)

Named in 1801 from the Greek *dias*, to scatter, for the unusual decrepitation before the blowpipe. Diaspore group. Dimorph: böhmite. ORTH *Pnma*. $a = 4.396$, $b = 9.426$, $c = 2.844$, $Z = 4$, $D = 3.380$. *NBSC 539(3):41(1954)*. 5-355: 3.99_{10} 2.56_3 2.32_6 2.13_5 2.08_5 1.63_4 1.48_2 1.38_2. **Habit:** Crystals

6.1.1.1 Diaspore

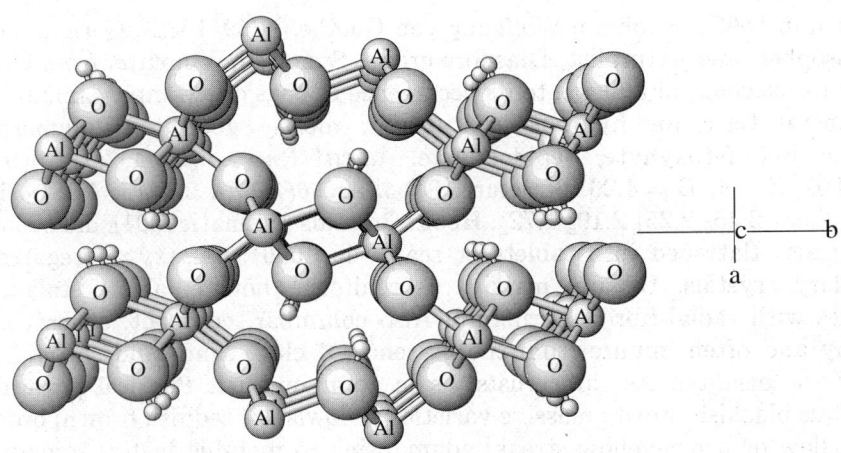

DIASPORE: AlOOH

Like boehmite, diaspore has chains of Al octahedra, but they are linked along their edges to form a framework. The edge-sharing chains are double, unlike those in manganite, MnOOH, which are single. Oxygen atoms are hexagonal closest packed, and H atoms occupy the interstitial positions not occupied by Al.

commonly thin platy {010}, elongated [001]; sometimes prismatic or acicular; rarely tabular {100}. Also massive, foliated, and in thin scales. Disseminated. Twinning on {061} or on {021} to produce pseudohexagonal aggregates and fine V-twins. **Physical properties**: White, grayish white, colorless; also greenish gray, hair brown, yellowish, lilac, pink; sometimes violet-blue in one direction, reddish plum in another, and asparagus green in a third; rose-red to dark red in the manganoan variety. Vitreous to brilliant luster, pearly on cleavage. Cleavage {010}, perfect: {110}, less so; conchoidal fracture; very brittle. $H = 6\frac{1}{2}$-7. $G = 3.4$. **Tests**: Not attacked by acids, but soluble in H_2SO_4 after ignition. **Chemistry**: Mn^{3+} substitutes for Al up to about 5% Mn_2O_3; also Fe^{3+} substitutes for Al. **Optics**: Biaxial (+); $XYZ = cba$; $N_x = 1.702$, $N_y = 1.722$, $N_z = 1.750$; $2V = 84°$; $r < v$, weak; absorption $X < Y < Z$. **Localities**: Chester, MA(!), as large plates and crystals with emery. In cavities in corundum in the Culsagee mine, Franklin, Mason Co., NC(*). Near Laws at White Mountain, CA, with andalusite, corundum, pyrophyllite, and lazulite in quartz–mica schist. Widespread in the bauxites and aluminous clays of AR and MO. Banská Belá, Slovakia. Jordanów, Silesia, Poland. *Mramorskoi, Kossoibrod, Ural Mountains, Russia*(!), in fine crystals in emery-bearing chlorite schists. Mugla, eastern Anatolia, Turkey(!!) (manganiferous diaspore), in fine crystals often of gem quality and from which many stones have been cut. Common in the bauxites of Hungary and France. Shokozan, Bungo, Japan, with alunite, pyrophyllite, and kaolinite in a hydrothermally altered rock. **Use**: An ore of aluminum when it is a constituent of bauxite. RG *DI:675*.

6.1.1.2 Goethite FeO(OH)

Named in 1806 for Johann Wolfgang von Goethe (1749–1832), German poet, philosopher, and naturalist. Diaspore group. Synonym: limonite, from Greek word for *meadow*, in allusion to its occurrence in bogs or swamps. Limonite is a general term for hydrous iron oxides, mostly goethite. Polymorphs: akaganéite, feroxyhyte, lepidocrocite. ORTH *Pnma*. $a = 4.62$, $b = 9.95$, $c = 3.01$, $Z = 4$, $D = 4.26$. Structure: *SR 33A:266(1968)*. *29-713:* 4.98_1 4.18_{10} 2.69_4 2.58_1 2.45_5 2.25_1 2.19_2 1.72_2. **Habit:** Crystals prismatic [001] and striated [001]; also flattened into tablets or scales on {010}; velvety aggregates of capillary crystals. Usually massive as reniform, botryoidal, or stalactitic masses with radial-fibrous structure. Also columnar, compact, ocherous, or earthy and often impure from the presence of clay, sand, and so on. Also oolitic or pisolitic. As thin crusts, as in common rust. **Physical properties:** Crystals blackish brown; massive varieties yellowish or reddish brown; brownish-yellow or ocher-yellow streak; adamantine to metallic luster, sometimes dull, earthy; also silky. Cleavage {010}, perfect; {100}, less so; uneven fracture; brittle. $G = 4.28$. **Tests:** In CT gives off water and is converted into hematite. Soluble in HCl. Most varieties test positive for Mn. Nonmagnetic. **Optics:** Biaxial $(-)$; $XYZ = bca$, $N_x = 2.260$, $N_y = 2.393$, $N_z = 2.398$ (Na); 2V small to medium; $r > v$, strong; absorption $Z > Y > X$; in polished section gray. Slightly pleochroic, strongly anisotropic. X, clear yellow; Y, brownish yellow; Z, orange-yellow. $R = 16, 18$ (546 nm); 17, 20 (470 nm); 15, 17 (589 nm); 14, 16 (650 nm). **Chemistry:** Mn is often present substituting for Fe, in amounts up to 5% Mn_2O_3. **Occurrence:** Formed under oxidizing conditions as a weathering product of ferriferous minerals, especially siderite, pyrite, magnetite, and glauconite. Also as a direct inorganic or biogenic precipitate from marine or meteoric waters and widespread as a deposit in bogs and springs ("bog iron ore"). Forms the weathered capping or gossan of sulfide veins. As a principal component of laterites, which are formed in countries with warm, humid climates where rock decay has progressed without interruption for a long time. Rarely as a low-temperature hydrothermal vein product. Also the principal component of common rust everywhere where iron and steel objects have been subject to corrosion. **Localities:** Goethite is a very common mineral nearly everywhere, and only a few localities can be mentioned. Formerly common in the Lake Superior hematite deposits, notable specimens having been found in MI at the Jackson mine, Negaunee(!), and the Superior mine near Marquette. Well crystallized in sheaflike groups in pockets in pegmatite in the Pikes Peak granite near Florissant, CO(!) and fine pseudomorphs of "limonite" (goethite) after pyrite from Pelican Point, Great Salt Lake, UT(!). As residual brown iron ores in the Appalachian region, mainly in AL, GA, VA, and TN. Fine, large crystals were found in Cornwall, England, at the Restormel mine, Lostwithiel, and the Botallack mine, St. Just(!). In France as thin acicular crystals to 30 mm at Chaillac, Vienne. In Germany, numerous localities in Nordrhein–Westfalen, Rhine, and Nassau, especially at Siegen, Eiserfeld Horhausen, and Oberstein.

Widespread in residual iron ore deposits in Ukraine; and in Russia acicular crystals with amethyst from Lake Onega(*), and at Bakal, southern Ural Mts.(!). As fine large pseudomorphs after pyrite and sometimes siderite crystals from near Diamantina, MG, Brazil(*). Uses: An important ore of iron; also used as pigment in ochers. RG *DI:680; AM 32:659(1947), 44:254(1959); GSAM 85:187(1962).*

6.1.1.3 Groutite α-MnO(OH)

Named in 1945 for Frank Fitch Grout (1880–1958), American petrologist, University of Minnesota. Diaspore group. Polymorphs: manganite, feitknechtite. ORTH *Pnmd.* $a = 4.560$, $b = 10.700$, $c = 2.870$, $Z = 4$, $D = 4.171$. *NBSM 25(11):97(1974).* Structure: *AC(B) 24:1233(1968). 12-733*(syn): 4.20_{10} 2.81_3 2.67_3 2.53_1 2.37_5 2.31_2 1.69_3 1.61_2. Crystals lens-shaped or wedge-shaped, flattened {100}; also acicular [001], flattened {010}, and elongated [001]. Black, dark brown streak, submetallic/adamantine luster. Cleavage {010}, perfect; {100}, fair; brittle. $H = 3\frac{1}{2}$. $G = 4.144$. Nonfluorescent. In CT gives off water. Biaxial (+); $XYZ = cba$; $N_x = 2.1$–2.2, $N_z = 2.1$–2.2; 2V medium. X, yellowish brown; Z, dark brown to purplish black. From *Sagamore open pit, Cuyuna Range, MN,* and other mines in the district. Talcville, near Gouverneur, NY, with calcite in vugs in talc. Franklin, Sussex Co., NJ, in franklinite and andradite, an antimonian groutite with 7.6% Sb_2O_5. From Germany at the Bülten–Adenstedt mine, near Peine, Niedersachsen, in fine radial aggregates making crystallized balls. RG *AM 32:654(1947), 52:858(1967).*

6.1.1.4 Montroseite (V, Fe)O(OH)

Named in 1953 for the locality. Diaspore group. ORTH *Pnma.* $a = 4.54$, $b = 9.97$, $c = 3.03$, $Z = 4$, $D = 4.11$. *AM 40:861(1955). 11-152*: 4.31_{10} 3.38_8 2.64_{10} 2.50_8 2.42_6 2.22_8 1.97_6 1.51_8. Crystals microscopic, bladed, elongated [001]. Black, black streak, submetallic luster. Cleavage {010} and {110}, good; brittle. $G = 4.0$. Weakly magnetic. Opaque. *Bitter Creek mine, Paradox Valley, Montrose Co., CO,* interstitial to sand grains in a sandstone. Also found in other nearby mines: the Matchless mine, Mesa Co.; the Juniper mine, Grand Co., and in UT at Temple Mt., Emery Co. RG *AM 38:1235(1953).*

6.1.1.5 Bracewellite CrO(OH)

Named in 1967 for Smith Bracewell (1899–1979), director, British Guiana Geological Survey, who first noted the mineral assemblage in which the mineral occurs. Diaspore group. Polymorphs: grimaldiite, guyanaite. ORTH *Pnma.* $a = 4.492$, $b = 9.860$, $c = 2.974$, $Z = 4$, $D = 4.286$. *25-1497*: 4.93_5 4.09_{10} 2.65_8 2.55_7 2.47_5 2.41_7 2.16_7 1.30_6. As masses and grains intergrown with other Cr oxides. Black, deep red; dark brown streak. $G = 4.47$. From the *Merume River, Mazaruni district, Guyana,* as black alluvial pebbles consisting of one or more of the minerals bracewellite, eskolaite, grimaldiite, guyanaite, and mcconnellite. The crude pebbles are locally called "merumite." RG *USGSPP 887:13(1976)*, *AM 62:593(1977)*.

6.1.2.1 Boehmite AlO(OH)

Named in 1927 for Johannes Böhm (1857–1938), German geologist and paleontologist, who first recognized the species. Dimorph: diaspore. ORTH $Cmcm$. $a = 2.868$, $b = 12.227$, $c = 3.700$, $Z = 4$, $D = 3.070$. *NBSC 539(3):38(1954)*. 21-1307(syn): 6.11_{10} 3.16_6 2.35_5 1.86_3 1.85_2 1.66_1 1.45_2 1.31_2. Microscopic lenticular crystals tabular {001}. Usually disseminated or as pisolitic aggregates. Structure: *CLM 29:435(1981)*. White to gray. Cleavage {010}. $G = 3.03$. Biaxial $(-)$; $YZ = cb$; $N \approx 1.64$. Widely distributed in bauxite and at times forms the principal constituent of the rock. Also may exsolve from corundum and is responsible for the rhombohedral and basal parting reported in that mineral. Linwood–Barton district, GA. Gánt and Baratka, Hungary. Stangenrod, Harz, Germany. Mt. Maggiore, Italy. Ayrshire, Scotland. In France at Recoux, Var, Péreille, and Cardarcet, Ariège. RG *DI:645*, *CCM 27:81(1979)*.

6.1.2.2 Lepidocrocite γ-FeO(OH)

Named in 1813 from the Greek for *scale* and *thread* or *fiber*, in allusion to the scaly or feathery habit. Polymorphs: goethite, akaganéite, feroxyhyte. ORTH $Cmc2_1$. $a = 3.08$, $b = 12.50$, $c = 3.07$, $Z = 4$, $D = 3.960$. **Structure:** The structure of lepidocrocite is similar to that of boehmite and consists of iron-centered oxygen octahedra linked together in chains parallel to X by sharing diagonally opposite edges. The chains make up double sheets of octahedra, the repeat distance of which defines the a parameter of the unit cell. The oxygens within the double octahedral layers are in cubic close-packed relationship. *DHZ II:580,576*, *SR 44A:224(1979)*. 44-1415(syn): 6.26_6 3.29_{10} 2.47_8 2.43_3 2.36_4 1.936_7 1.535_3 1.525_3. **Habit:** Scales flattened {010}, sometimes aggregated into palmate or plumose groups; loose rosettes; massive, bladed to fibrous, or micaceous. Fibrous varieties elongated [100]. **Physical properties:** Red, reddish brown; dull-orange streak; submetallic luster. Cleavage {010}, perfect; {100}, less perfect; {001}, good; brittle. $H = 5$. $G = 4.09$. **Optics:** Biaxial $(-)$; $XYZ = bca$; $N_x = 1.94$, $N_y = 2.20$, $N_z = 2.51$; $2V = 83°$; slight dispersion; absorption $X < Y < Z$. X, yellow; Y, dark red-orange; Z, darker red-orange. In RL, gray-white. Strongly pleochroic, strongly anisotropic. $R = 18.4$, 11.9 (470 nm); 16.8, 10.9 (546 nm); 15.9, 10.7 (589 nm); 15.2, 10.4 (650 nm). *P&J:236*. **Occurrence:** Lepidocrocite is formed under the same conditions as goethite, and the two minerals are often found together in the same deposit. **Localities:** From the Lake Superior iron region, MN and MI; Easton, PA; Iron Mt. mine, Shasta Co., CA. Well crystallized at the Eisenzeche mine, Eiserfeld, Siegen, Germany. Rancié en Sem, near Vicdessos, Ariège, France. Near Trosna and other localities, Ukraine. Zlaté Hory, Czech Republic. Sometimes found as acicular inclusions in amethystine quartz, in the Alps, Brazil, and so on. (The mineral from Eiserfeld was originally described and named *goethite*, which then was shown to exist in many other places. Later, it was proved that the Eiserfeld material was actually distinct from the common goethite and was identical to lepidocrocite as described from Zlaté Hory. The name *goethite*, however, has been retained for the commoner of the two polymorphs.) RG *DI:642, GSAM 85:185(1962), CLM 14:285(1979)*.

6.1.2.3 Guyanaite CrO(OH)

Named in 1967 for the country of origin. Polymorphs: bracewellite, grimaldiite. ORTH $Pnnm$. $a = 4.857$, $b = 4.295$, $c = 2.958$, $Z = 2$, $D = 4.574$. AM *62:593(1977)*. 20-312(syn): 3.22_{10} 2.53_2 2.43_8 2.15_1 1.72_3 1.64_5 1.61_2 1.42_2. As discrete grains intergrown with other components of "merumite." Reddish brown, golden brown, dark green. $G = 4.53$. From the *Merume River, Mazaruni District, Guyana*, as a constituent of alluvial pebbles locally called *merumite*, which consist of an intergrowth of eskolaite, bracewellite, guyanaite, grimaldiite, and mcconnellite. RG *USGSPP 887:10(1976)*.

6.1.3.1 Manganite γ-MnO(OH)

Named in 1827 for the composition. Polymorphs: groutite, feitknechtite. MON $P2_1/c$. $a = 8.98$, $b = 5.28$, $c = 5.71$, $β = 90°$, $Z = 8$, $D = 4.38$. ZK *118:303(1963)*. **Structure:** The manganese atoms occupy the centers of O—OH octahedra, and the octahedra share opposite edges and form chains parallel to c. The octahedra are somewhat distorted, the Mn—O distances along b being greater than the MN—O and Mn—OH distances in the rectangle in the plane of the shared edges. Pairs of OH groups in the a direction form one side of these rectangles, opposite pairs of O atoms forming the other side, thus alternating along the chain. The OH groups of one ribbon bind themselves to the O atoms of the adjacent ribbon to form an (010) sheet corrugated in the direction [001]. The outstanding excellence of the (010) cleavage over the other cleavages in the prismatic zone is thus produced. ZK *95:163(1936), 118:303(1963)*. 41-1379: 3.41_{10} 2.64_2 2.52_1 2.417_2 2.414_2 1.782_2 1.672_2 1.670_2. **Habit:** Crystals pseudo-orthorhombic, short to long prismatic [001], often terminated by {001} alone. Prismatic faces deeply striated [001]. Crystals often grouped in bundles or markedly composite subparallel [001]. Rarely massive, stalactitic. Twins on {011}, both as contact and penetration twins, often repeated; lamellar twinning on {100} invariably present. **Physical properties:** Dark steel gray, black; reddish-brown to black streak, submetallic luster. Cleavage {010}, perfect; {110} and {001}, less so; uneven fracture; brittle. $H = 4$. $G = 4.33$. **Tests:** Infusible. Soluble in HCl with evolution of Cl. **Optics:** Biaxial (+); $XYZ = abc$; $N_x = 2.25$, $N_y = 2.25$, $N_z = 2.53$; 2V small; $r > v$, very strong; absorption X and $Y > Z$. X, reddish brown; Z, red brown. In RL, gray-white with a brownish tint. Internal reflections blood red. Weakly pleochroic. $R = 23.9$, 17.9 (470 nm); 22.0, 16.9 (546 nm); 21.0, 16.4 (589 nm); 20.2, 15.9 (650 nm). *P&J:255*. **Occurrence:** Manganite occurs as a low-temperature hydrothermal vein mineral with barite, calcite, siderite, braunite, and hausmannite. As replacement deposits formed by meteoric waters, with pyrolusite, psilomelane, and goethite. In residual clays. **Localities:** Jackson mine, Negaunee, MI, and elsewhere in the Lake Superior iron district. Well crystallized at Powell's Fort, near Woodstock, Shenandoah Co., VA(!). Livermore–Tesla district, Alameda Co., CA. Lake Valley district, Sierra Co., NM. Abundant and finely crystallized at *Ilfeld, Harz, Germany*(!!), in veins in porphyrite with barite and calcite; also at Ilmenau, Thuringia, in part pseudomorphous after calcite. In France at Las Cabesses and Rancié, Ariège. Botallack mine, St. Just, Cornwall, and at Egremont, Cumberland, England. Bölet, Sweden. N'Chwaning No. II mine, Kalahari manganese field, Kuruman, Cape, South Africa(!). RG *DI:646, AM 50:1296(1965)*.

6.1.4.1 Heterogenite-2H   CoO(OH)

6.1.4.2 Heterogenite-3R   CoO(OH)

Named in 1872 (3R) and 1973 (2H) from the Greek for *other* and *kind*, in reference to its difference in composition from other minerals that it otherwise resembles. 3R: HEX $P6_3/mmc$. $a = 2.855$, $c = 8.805$, $Z = 2$, $D = 4.912$. *NJMM:215(1967)*. Heterogenite-3R is isostructural with grimaldiite. 2H: HEX-R $R\bar{3}m$. $a = 2.855$, $c = 13.156$, $Z = 3$, $D = 4.931$. *MM 39:152(1973)*. *26-1107*(2H-nat): 4.39_{10} 2.47_7 2.38_5 2.16_8 1.64_8 1.43_4 1.26_5 1.24_9. *7-169*(3R-syn): 4.38_{10} 2.43_1 2.31_9 1.98_1 1.80_8 1.50_2 1.43_3 1.36_3. Botryoidal; reniform or globules; porous aggregates; crystalline masses; with crystals prismatic and {0001} well developed. Twinning common in crystals. The two polytypes of heterogenite have similar physical properties and can be distinguished only through their powder patterns. Crystals lustrous black; other forms black, blackish brown, reddish brown; dark brown streak; metallic to vitreous luster (crystals), to dull luster. Cleavage {0001}, good; conchoidal fracture; somewhat malleable. H = 4–5, crystals; H = 3–4, massive forms. G = 4.0–4.72. A small amount of Ni is usually present, up to about 2%. Copper is also often detected but is believed usually to represent admixed impurities. In RL, white to gray. $N \approx 2.7$. Bireflectance 0.07. Strongly anisotropic. $R = 16.5$–23.5 (589 nm). From *Mindingi, Shaba, Zaire* (heterogenite 2H), where it is abundant and used as an ore of cobalt. *Schneeberg, Saxony, Germany* (heterogenite 3R), as an alteration product of smaltite, with pharmacosiderite and calcite. RG *DI:652*.

6.1.4.3 Feitknechtite   β-MnO(OH)

Named in 1965 for Walter Feitknecht (1899–1975), Swiss, professor of chemistry, University of Bern, who first synthesized the compound. Polymorphs: groutite, manganite. Probably HEX Space group unknown. $a = 8.6$, $c = 9.30$. *18-804*(syn): 4.62_{10} 2.64_5 2.36_2 1.96_1 1.55_1 1.50_1. Crystals are thin hexagonal plates. This mineral was indexed on a tetragonal cell; however, this is inconsistent with SEM photographs of synthetically prepared material, which show clear-cut hexagonal habit. Frondel described "hydrohausmannite" from Franklin, NJ, and Sweden in 1953. This mineral was later shown to be a mixture; one of the components of this mixture, β-MnO(OH), was subsequently named feitknechtite. Black, brownish black; brown streak. G = 3.80. Uniaxial (−); $N_O = 2.055$, $N_E = 1.95$. From *Franklin, Sussex Co., NJ*, as an alteration product of pyrochroite and lining solution cavities in a matrix of calcite, sussexite, and zincite. Långban and Pajsberg, Sweden, in altered pyrochroite. RG *AM 38:761(1953), 50:1296(1965)*.

6.1.4.4 Feroxyhyte Δ-FeO(OH)

Named in 1976 for the composition. Polymorphs: goethite, lepidocrocite, akaganéite. HEX Space group unknown. $a = 2.93$, $c = 4.60$, $Z = 1$, $D = 4.20$. *AM 62:1057(1977). 13-87*: 4.61_2 2.55_{10} 2.26_{10} 1.69_{10} 1.47_{10} 1.27_2 1.22_2 1.10_2. Massive; as nodules and concretions. Δ-FeO(OH) consists of two phases, both hexagonal, one ordered and magnetic, the other, feroxyhyte, disordered or only slightly ordered and nonmagnetic. Alters spontaneously to goethite upon exposure to air. From *Pacific Ocean* Fe–Mn nodules, and from the *Baltic, White*, and *Kara seas*. RG *IGR 19:873(1977), CCM 28:272(1980)*.

6.1.5.1 Grimaldiite CrO(OH)

Named in 1967 for Frank S. Grimaldi (1915–1985), former chief chemist, U.S. Geological Survey. Polymorphs: bracewellite, guyanaite. HEX-R $R\bar{3}m$. $a = 2.973$, $c = 13.392$, $Z = 3$, $D = 5.61$. *AM 62:593(1977). 9-331*: 4.44_{10} 2.53_1 2.41_3 2.23_1 1.86_3 1.54_1 1.49_2 1.41_1. Tabular rhombohedral crystals; scallop-shaped minute platy aggregates. Structure: *AC 10:423(1957)*. Deep red; pinkish brown (Bolivia). G = 5.49. *Merume River, Mazaruni District, Guyana*, as a constituent of black alluvial pebbles locally known as "merumite," consisting of intergrowths of eskolaite, bracewellite, guyanaite, mcconnellite, and grimaldiite. Hiaca mine, Colquechaca, Bolivia, with penroseite in barite. RG *USGSPP 887:17(1976), MM 48:560(1984)*.

6.1.6.1 Akaganéite β-FeO(OH, Cl)

Named in 1961 for the locality. Polymorphs: goethite, lepidocrocite, feroxyhyte. MON $I2/m$. $a = 10.600$, $b = 3.034$, $c = 10.513$, $\beta = 90.23°$, $Z = 8$, $D = 3.6$. *AM 76:272. 34-1266*(syn): 7.47_4 5.28_3 3.33_{10} 2.63_3 2.55_6 2.30_4 1.95_2 1.64_4. Massive; as crusts; similar to "limonite." The structure consists of double chains of edge-sharing $FeO_3(OH)_3$ octahedra extending along the c axis. The octahedra comprising the double chains share corners with adjacent chains to form tunnels approximately 5×5 Å in size. Each tunnel is composed of adjoining cavities formed by eight OH groups. Infrared studies indicate that at least half of these groups are replaced by Cl–OH in nonstoichimetric proportions. *MM 33:270(1962), 32:545(1960)*. Brown to dark brown, brown streak, dull to earthy luster. G = 3.0. Other physical properties similar to "limonite" (goethite). From the *Akagané mine, Iwate, Japan*, as an oxidation product of pyrrhotite, associated with goethite. As an important corrosion crust on some meteorites. A major Fe-oxide component in soils and geothermal brines. A corrosion product of some steels. RG *NJMA 113:29(1970), GCA 41:539(1977), CLM 14:273(1979), AC(A) 35:197(1979).*

BRUCITE GROUP

The brucite minerals are isostructural simple hydroxides corresponding to the formula

$$A(OH)_2$$

where

A = Mg, Fe^{2+}, Mn, Ca, or Ni

and which crystallize with rhombohedral symmetry in space group $P\bar{3}m1$.

Brucite has a layered structure in which each layer has two sheets of (OH) parallel to the basal plane, with a sheet of Mg ions between them, each MG lying between six (OH). The hydroxyl groups are hexagonal close-packed ABABA, with each (OH) linked to three Mg on one side and fitted into three (OH) of the next layer.

This arrangement of the structure leads to polarization of the (OH) group, with the result that the layers of brucite are held together not only by van der Waals forces but also by a type of hydrogen bond. *DHZ 5:89, SR 32A:242(1967), ZK Supp. 4:316(1991).*

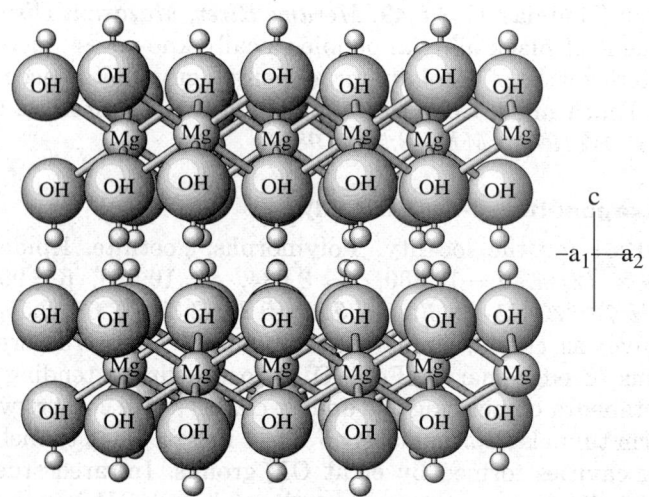

BRUCITE: Mg(OH)$_2$
Two close-packed layers of hydroxyl with Mg in all octahedral interstices. Hydrogen bonds are the only links between sandwiches. Symmetry requires hydrogen atoms to be equidistant from three oxygen atoms to which they might be bonded—in reality, they may be disorderd. *See also* GIBBSITE.

6.2.1.2 Amakinite

BRUCITE GROUP

Mineral	Formula	Space Group	a	c	D	
Brucite	$Mg(OH)_2$	$P\bar{3}m1$	3.147	4.769	2.368	6.2.1.1
Amakinite	$(Fe, Mg)(OH)_2$	$P\bar{3}m1$	6.918	14.52	2.73	6.2.1.2
Pyrochroite	$Mn(OH)_2$	$P\bar{3}m1$	3.323	4.738	3.268	6.2.1.3
Portlandite	$Ca(OH)_2$	$P\bar{3}m1$	3.593	4.909	2.241	6.2.1.4
Theophrastite	$Ni(OH)_2$	$P\bar{3}m1$	3.131	4.608	3.95	6.2.1.5

Note: See the melonite (2.12.14.1) group for structurally similar sulfides, selenides, and tellurides.

6.2.1.1 Brucite $Mg(OH)_2$

Named in 1824 for Archibald Bruce (1777–1818), American mineralogist, who first described the species. Brucite group. HEX-R $P\bar{3}m1$. $a = 3.147$, $c = 4.769$, $Z = 1$, $D = 2.368$. *NBSC 539(6):30(1956)*. *44-1482*(syn): 4.79_5 2.72_1 2.37_{10} 1.796_3 1.572_3 1.493_1 1.374_1 1.309_1. Crystals tabular {0001}, often subparallel aggregates of plates. Massive, foliated; fibrous; rarely fine granular. White, pale green, gray, blue; manganoan varieties yellow to red; white streak; vitreous to waxy luster, pearly on cleavages. Cleavage {0001}, perfect; flexible foliae; sectile. $H = 2\frac{1}{2}$. $G = 2.39$. Pyroelectric. Infusible; ignited material reacts alkaline. Easily soluble in acids. Uniaxial (+); $N_O = 1.561$, $N_E = 1.581$; manganoan brucite: $N_O = 1.59$, $N_E = 1.60$. Typically formed as a low-temperature hydrothermal vein mineral in serpentine and chloritic or dolomitic schist. Also as an alteration product of periclase in crystalline limestones. *Hoboken, Hudson Co., NJ*, with hydromagnesite, artinite, and dolomite in serpentine. Well crystallized from the Tilly Foster iron mine, Brewster, Putnam Co., NY. The Wood mine, Texas, Lancaster Co., PA(!), in large fine crystals with chromite in serpentine. Lodi district, NV, in a large deposit with hydromagnesite. Crestmore, Riverside Co., CA, as an alteration of periclase in marble. In Italy in periclase limestone blocks thrown up by Vesuvius. In Sweden as an alteration of periclase in Nordmark, Långban, and elsewhere. Asbest, near Ekaterinburg, Ural Mts., Russia(!), in fine transparent crystals to 5 cm; also in the Kempirsai massif, southern Ural Mts.(!). Phalaborwa, Transvaal, South Africa, in a large carbonatite. RG *DI:636, AM 48:86(1963)*.

6.2.1.2 Amakinite $(Fe, Mg)(OH)_2$

Named in 1962 for the Amakin expedition, which prospected the diamond pipes of Yakutia. Brucite group. HEX-R $P\bar{3}m1$. $a = 6.918$, $c = 14.52$, $Z = 12$, $D = 2.73$. *15-125*: 5.49_7 2.80_8 2.30_{10} 1.73_9 1.55_7 1.53_8 1.39_7 1.27_7. Crystals rhombohedra; also as irregular grains. Pale green to yellow green, when fresh. Rough cleavage observed occasionally. $H = 3\frac{1}{2}–4$. $G = 2.98$. Weakly magnetic. In CT decrepitates, giving off water, and turns brown to black. Soluble in HCl. Uniaxial (+); $N_O = 1.707$, $N_E = 1.722$. Alters rapidly in air, with formation of brown crusts of hydrous ferric oxide. From the *Lucky East-*

ern kimberlite pipe, *Yakutia, Siberia, Russia*. RG *ZVMO 91:72(1962)*, *AM 47:1218(1962)*.

6.2.1.3 Pyrochroite Mn(OH)$_2$

Named in 1864 from the Greek *puro*, fire, and *khroma*, color, because of the change in color upon ignition. Brucite group. HEX-R $P\bar{3}m1$. $a = 3.323$, $c = 4.738$, $Z = 1$, $D = 3.268$. *MJJ 3:30(1960)*. *18-787*(syn): 4.72_9 2.87_4 2.45_{10} 1.83_6 1.66_4 1.57_3 1.38_2 1.37_2. Crystals tabular {0001}; also rhombohedral with {10$\bar{1}$2} or {10$\bar{1}$1}; rarely prismatic [0001]. Usually foliated massive and as thin veinlets. Colorless to pale greenish or bluish, but upon exposure soon becomes bronze-brown and then black; pearly luster on cleavages. Cleavage {0001}, perfect; thin flakes flexible. $H = 2\frac{1}{2}$. $G = 3.25$. Easily soluble in dilute HCl. In CT becomes verdigris green, and finally brownish black. Mg substitutes for Mn up to at least Mn/Mg = 5:1. Uniaxial (−); $N_O = 1.723$ (brown), $N_E = 1.681$ (lighter brown); absorption O > E. From Franklin and Sterling Hill, Sussex Co., NJ, in secondary hydrothermal veinlets cutting franklinite ore, with rhodochrosite, calcite, willemite, gageite, chlorophoenicite, and hodgkinsonite. From Långban and *Persberg, Vårmland, Sweden*. In crystal druses in limonite near Prijedor, Bosnia. Noda–Tamagawa mine, Japan. Kuruman manganese district, Cape Prov., South Africa. RG *DI:639*, *AM 50:1296(1965)*.

6.2.1.4 Portlandite Ca(OH)$_2$

Named in 1933 because Ca(OH)$_2$ is a common product of hydration of portland cement. Brucite group. HEX-R $P\bar{3}m1$. $a = 3.593$, $c = 4.909$, $Z = 1$, $D = 2.241$. *NBSC 539(1):58(1953)*. *44-1481*(syn): 4.92_7 3.11_3 2.63_{10} 1.928_3 1.795_3 1.69_1 1.48_1 1.45_1. Structure: *AC 14:950(1961)*. Hexagonal plates {0001}. Colorless, pearly luster on cleavage. Cleavage {0001}, perfect; flexible; sectile. $H = 2$. $G = 2.23$. Soluble in water, giving alkaline solution. Easily soluble in dilute HCl. Uniaxial (−); $N_O = 1.574$, $N_E = 1.547$. From *Scawt Hill, Co. Antrim, Northern Ireland*, with afwillite and calcite in larnite–spurrite contact rocks. Vesuvius, Italy, in fumaroles. In good crystals in the "mottled" zone complex, Hatrurim region, Israel, with fluorapatite and ettringite. RG *DI:641*, *AM 48:925(1963)*.

6.2.1.5 Theophrastite Ni(OH)$_2$

Named in 1981 for Theophrastos (372–287 B.C.), the first Greek mineralogist. Brucite group. HEX-R $P\bar{3}m1$. $a = 3.131$, $c = 4.608$, $Z = 1$, $D = 3.95$. *AM 66:1021(1981)*. *14-117*: 4.61_{10} 2.71_5 2.33_{10} 1.75_4 1.56_3 1.48_2 1.33_1 1.30_1. Microacicular crystals forming layers like velvet, aligned perpendicular to the surface of other minerals that are coated by it. Emerald green, pale green streak, vitreous luster. Cleavage {0001}, perfect; conchoidal fracture. $H = 3\frac{1}{2}$. $G = 4.00$. Soluble in acids. Uniaxial (+); $N_O \approx 1.759$, $N_E \approx 1.760$. Birefringence very weak. From *Vermion, Macedonia, Greece*, in a magnetite–chromite mine, with vesuvianite, chlorite, andradite, Ni-bearing serpentine minerals, and chlorite. RG *MM 46:1(1982)*.

6.2.2.1 Behoite Be(OH)$_2$

Named in 1970 for the composition, Be + HO (= OH). Dimorph: clinobehoite. ORTH $P2_12_12_1$. $a = 4.639$, $b = 7.071$, $c = 4.548$, Z = 4, D = 1.92. *39-1411*(syn): 3.88_9 3.83_6 2.95_3 2.79_3 2.39_{10} 2.20_2 1.983_2 1.962_2. Crystals equant, showing combinations of {011} with {110}, giving a pseudo-octahedral habit. Colorless; pastel shades of brown, red, or yellow due to impurities; vitreous luster. No cleavage; conchoidal fracture. H = 4. G = 1.92. Nonfluorescent. Biaxial (−); XYZ = bac; $N_x = 1.533$, $N_y = 1.544$, $N_z = 1.548$; 2V = 82°; r < v, strong. From *Rode Ranch pegmatite, Llano Co., TX*, as an alteration product of gadolinite, associated with bastnaesite and tengerite. With bertrandite and fluorite at Sierra Blanca, Hudspeth Co., TX. Mt. St.-Hilaire, QUE, Canada, as radialfibrous spherules and coating serandite(!). RG *AM 55:1(1970)*, *PD 4:50(1989)*.

6.2.3.1 Clinobehoite Be(OH)$_2$

Named in 1989 for the monoclinic symmetry and its chemical identity with behoite. Dimorph: behoite. MON $P2_1$. $a = 11.020$, $b = 4.746$, $c = 8.646$, β = 98.93°, Z = 12, D = 1.92. PD 5.43_8 3.98_6 3.76_7 3.61_9 3.16_7 2.71_{10} 2.44_6 2.31_7. Platy, cuneiform crystals in radial aggregates. Structure: *ZK 185:612(1988)*. Colorless, white; vitreous to pearly luster. Cleavage {100}, perfect; {010}, {001}, imperfect; brittle. H = 2–3. G = 1.93. Soluble in HCl. Biaxial (−); X ≈ c, Y = b; Z ∧ a ≈ 9°; $N_x = 1.539$, $N_y = 1.544$, $N_z = 1.548$; 2V = 80°. Dispersion strong. From *Mursinsk, Ural Mts., Russia*, in hydrothermally altered zones of desilicated pegmatites, associated with bavenite, bityite, analcime, phillipsite, and albite. RG *MZ 11(5):88(1989)*.

6.2.4.1 Spertiniite Cu(OH)$_2$

Named in 1981 for Francesco Spertini (b.1937), chief geologist, Jeffrey mine. ORTH $Cmc2$. $a = 2.947$, $b = 10.593$, $c = 5.256$, Z = 4, D = 3.94. *CM 19:337(1981)*. 35-505: 5.29_8 3.73_9 2.63_{10} 2.50_6 2.36_5 2.27_7 1.93_3 1.72_7. Crystals microscopic, tabular, grouped in botryoidal or radial aggregates. Structure: *AC(C) 46:2279(1980)*. Blue, blue-green; vitreous luster. Cleavage not observed. G = 3.93. Soft. Nonfluorescent. Soluble in acids, insoluble in cold water. Biaxial; $N_x = 1.720$ (colorless), $N_z > 1.800$ (dark blue); parallel extinction; positive elongation. From the *Jeffrey mine, Shipton Twp., Richmond Co., QUE, Canada*, in a rodingite dike, with diopside, grossular, vesuvianite, chalcocite, and atacamite. RG *AM 67:860(1982)*.

6.2.5.1 Calumetite Cu(OH, Cl)$_2$ · 2H2O

Named in 1963 for the locality. ORTH *15-669*: 7.50_{10} 3.76_5 3.42_3 3.30_3 3.02_6 2.48_8 2.34_3 1.99_3. Spherules, and sheaves of scaly crystals, with {001} and {110}. Azure blue, powder blue; pearly luster. Cleavage {001}, perfect; brittle.

H = 2. Insoluble in water or NH_4OH; soluble in cold, dilute acids. Biaxial $(-)$; XYZ = cab; $N_x = 1.666$, $N_y = 1.690$, $N_z = 1.690$; 2V = 2°; Z > Y > X. X, blue. From the *Centennial mine, Calumet, Houghton Co., MI*, associated with copper, malachite, cuprite, atacamite, paratacamite, buttgenbachite, and other minerals. RG *AM 48:614(1963)*.

6.2.6.1 Anthonyite $Cu(OH, Cl)_2 \cdot 3H_2O$

Named in 1963 for John W. Anthony (1920–1992), mineralogist, University of Arizona. MON Space group unknown. β = 112.63°. *15-670:* 5.84_{10} 5.17_4 4.14_7 3.99_6 3.44_6 3.18_4 3.07_4 2.87_6. Crystals prismatic, elongated [001]. Lavender. Cleavage {100}, perfect; sectile; bends readily on {100} cleavage. H = 2. Insoluble in water; soluble in cold, dilute acids. Dehydrates rather rapidly in air. Biaxial $(-)$; Y = b; Z $\wedge$ c = 13°; $N_x = 1.526$, $N_y, N_z = 1.602$; absorption Z = Y > X. X, lavender; Z, deep smoky blue. From the *Centennial mine, Calumet, Houghton Co., MI*, associated with copper, cuprite, paratacamite, and an unidentified copper chloride. RG *AM 48:614(1963)*.

6.2.7.1 Duttonite $V^{4+}O(OH)_2$

Named in 1957 for Clarence E. Dutton (1841–1912), geologist, U.S. Geological Survey, one of the first to work in the Colorado Plateau region. MON $C2/c$. $a = 8.80$, $b = 3.95$, $c = 5.96$, β = 90.67°, Z = 4, D = 3.24. Structure: *AC(A) 11:56(1958)*. *10-377:* 4.40_{10} 3.61_8 3.29_1 2.64_1 2.48_2 2.45_2 1.97_2 1.84_2. Crusts of six-sided platy crystals, flattened {001}, pseudo-orthorhombic. Pale brown. Cleavage {100}, pronounced. H = $2\frac{1}{2}$. G ≈ 3.1. Biaxial $(+)$; XYZ = acb; $N_x = 1.810$, $N_y = 1.900$, $N_z > 2.01$; 2V ≈ 60°; dispersion r < v, moderate. X, pale pinkish brown; Y, pale yellow-brown; Z, pale brown. From the *Peanut mine, Bull canyon mining district, about 24 km W of Naturita, Montrose Co., CO*, with melanovanadite and selenium. RG *AM 42:455(1957)*, *BM 93:235(1970)*.

6.2.8.1 Tivanite $TiVO_3(OH)$

Named in 1981 for the composition. MON $P2_1c$. $a = 7.494$, $b = 4.552$, $c = 10.005$, β = 129.78°, Z = 4, D = 4.17. *35-494:* 3.92_2 3.88_{10} 2.80_9 2.64_7 2.54_2 1.69_3 1.68_3 1.67_2. Irregular grains. Black, black streak, metallic to submetallic luster. VHN_{50} = 650 kg/mm². In RL, dark brown–deep blue. No internal reflections. Pleochroism not discernible; weakly to moderately anisotropic. R = 16.8 (470 nm), 17.4 (546 nm), 17.7 (589 nm). From *Kalgoorlie, WA, Australia* in the "green leader" mineralized zone, associated with V-bearing sericite, nolanite, and tomichite. RG *AM 66:866(1981)*.

6.2.9.1 Sweetite $Zn(OH)_2$

Named in 1984 for Jessie M. Sweet (1901–1979), curator, mineral collection, British Museum (Natural History). Polymorphs: wülfingite, ashoverite. TET Space group unknown. $a = 8.222$, $c = 14.34$, Z = 20, D = 3.41. *AM 70:438(1985)*. *38-356:* 4.53_4 3.57_6 2.92_{10} 2.71_2 2.26_2 1.84_1 1.76_2 1.56_1. Tetragonal bipyramids, probably {112}, with pseudo-octahedral aspect. White, vitreous luster. G = 3.33. Soluble in dilute HCl with slight effervescence. Uniaxial (−); $N_O = 1.636$, $N_E = 1.628$. Northwest of *Milltown, near Ashover, Derbyshire, England*, in fluorite veins in a limestone quarry, on colorless fluorite, associated with barite and ashoverite. RG *MM 48:267(1984)*.

6.2.10.1 Wülfingite Zn(OH)₂

Named in 1985 for Ernst Anton Wülfing (1860–1930), professor of mineralogy and petrography, Heidelberg University. Polymorphs: sweetite, ashoverite. ORTH $P2_12_12_1$. $a = 8.490$, $b = 5.162$, $c = 4.197$, $Z = 4$, $D = 2.50$. *AM 73:196(1988)*. 38-385(syn): 4.41_{10} 4.25_8 3.28_{10} 3.21_8 2.73_8 2.46_6 2.28_8 2.22_8. As fine grained incrustations; crystals minute, acicular. White to colorless, white streak, waxy luster. No cleavage, conchoidal fracture. H = 3. Soluble in dilute acids. Biaxial (−); $N_x = 1.572$, $N_y = 1.578$, $N_z = 1.580$ (Na); $2V = 60°$; r > v, extreme. From *Richelsdorfer Hütte, Richelsdorf, Hessen, Germany*, as a weathering product of zinc-bearing slag, associated with simonkolleite, hydrocerussite, diaboleite, zincite, and hydrozincite. Also found at Milltown, Ashover, Derbyshire, England, associated with sweetite and ashoverite on fluorite. RG *NJMM:145(1985)*.

6.2.11.1 Ashoverite Zn(OH)₂

Named in 1988 for the locality. Polymorphs: sweetite, wülfingite. TET Space group uncertain, probably $I4_1/amd$. $a = 6.83$, $c = 33.35$, $Z = 32$, $D = 3.44$. *MM 52:699(1988)*. 41-1359: 3.33_1 3.03_1 2.90_{10} 2.32_1 1.82_3 1.70_2 1.47_1 1.01_1. Crystals are square plates. Colorless to milky white, white streak, vitreous to dull luster. Cleavage {001}, perfect. G = 3.3. Fluoresces bluish white in SWUV. Uniaxial (+); $N_O = 1.629$, $N_E = 1.639$. From *Milltown, near Ashover, Derbyshire, England*, in a vein in a limestone quarry, on fluorite associated with sweetite, wülfingite, and γ-Zn(OH)₂. RG *MM 52:699(1988), AM 75:431(1990)*.

6.2.12.1 Paraotwayite Ni(OH)₂₋ₓ(SO₄, CO₃)₀.₅ₓ

Named in 1987 for its close physical and chemical resemblance to otwayite. MON Pm. $a = 7.89$, $b = 2.96$, $c = 13.63$, $\beta = 91.1°$, $Z = 6$, $D = 3.520$. *CM 25:409(1987)*. 41-1424: 7.95_2 6.81_{10} 5.08_8 3.86_5 3.39_2 2.95_4 2.24_8 1.97_3. Parallel to subparallel fibers; as cross-fiber veinlets. Emerald green, white streak, silky luster. No cleavage, brittle. $VHN_{20} = 223$ kg/mm². G = 3.30. Nonfluorescent. Insoluble in cold HCl or HNO₃. Biaxial; $N_x = 1.655$, $N_z = 1.705$; parallel extinction. Fibers are length-slow. From the *Otway nickel deposit, Nullagine, Pilbara region, WA, Australia*, in veinlets with dolomite and gaspeite, cutting nodules of nickeloan chrysotile with millerite and polydymite. RG *AM 73:1496(1988)*.

6.2.13.1 Cianciulliite MnMg$_2$Zn$_2$(OH)$_{10}$·2–4H$_2$O

Named in 1991 for John Cianciulli, mineral collector of Sussex, NJ. MON C2/m. $a = 15.47$, $b = 6.369$, $c = 5.576$, $\beta = 101.28°$, Z = 2, D = 2.87. *44-1407*: 7.61$_{10}$ 5.47$_2$ 4.17$_2$ 3.96$_5$ 3.45$_3$ 3.00$_4$ 2.75$_6$ 2.67$_3$. Crystals tabular {100}. Dark reddish brown; adamantine to vitreous luster, dull to pearly on {100}. Cleavage {100}, perfect; flexible. H ≈ 2. Nonfluorescent. In polished section gray; anisotropism weak; bireflectance weak; nonpleochroic; internal reflection orange-red. R = 8.5 (470 nm); 8.1 (546 nm), 8.0 (589 nm), 8.0 (650 nm). From the *Franklin mine, Franklin, Sussex Co., NJ*, with willemite and cahnite. RG *AM 76:1789(1991)*.

6.3.1.1 Gibbsite γ-Al(OH)$_3$

Named in 1822 for George Gibbs (1777–1834), original owner of the Gibbs mineral collection acquired by Yale College early in the nineteenth century. Polymorphs: bayerite, doyleite, nordstrandite. MON $P2_1/c$. $a = 8.684$, $b = 5.078$, $c = 9.736$, $\beta = 94.53°$, Z = 8, D = 2.421. *ZK 139:129(1974)*. *33-18*(syn): 4.85$_{10}$ 4.37$_7$ 4.32$_5$ 3.31$_3$ 2.45$_4$ 2.39$_6$ 2.05$_4$ 1.80$_3$. **Habit**: Crystals tabular {001}, with a hexagonal aspect. As lamellar-radiate spheroidal concretions. Massive; stalactitic; compact/earthy. Twinning about [130] as twin axis very common; on {001}, common; on {100} and {110}, rare. Usually, the {001} twin law occurs in combination with other laws. **Structure**: The fundamental unit of structure is a layer of Al ions sandwiched between two sheets of close-packed hydroxyl ions. Two out of three of the octahedrally coordinated sites between the oxygen layers are occupied by cations. The layers may be regarded as built of octahedra linked laterally by sharing edges; the network so described may be described as an orthogonal (pseudohexagonal) cell. The oxygen layers are in the sequence ABBAABB; thus in the ideal structure of gibbsite, oxygens at the bottom of one layer lie directly above oxygens at the top of the layer below. The structure of gibbsite is in fact somewhat distorted from the ideal; the Al–O octahedra are not regular but have their shared edges longer than the remainder, and the successive layers of octahedra are slightly displaced in the a direction, resulting in a monoclinic cell. *DHZ 5:93, ZK Supp. 4:316(1991)*. **Physical properties**: White, grayish, greenish, reddish white; reddish or yellow when impure; vitreous to pearly luster. Cleavage {001}, perfect; tough. H = $2\frac{1}{2}$–$3\frac{1}{2}$. G = 2.40. Nonfluorescent. **Tests**: Infusible; in CT yields water. Slowly soluble in concentrated H$_2$SO$_4$; readily soluble in hot alkali solutions; gives an argillaceous odor when breathed on. **Optics**: Biaxial (+); X = b, Y ∧ c = 69°, Z ∧ c = −21°; N$_x$ = 1.568, N$_y$ = 1.568, N$_z$ = 1.587; 2V = 0°; variable dispersion, positive or negative elongation. Inclined extinction when fibrous. **Occurrence**: Most gibbsite is a secondary product resulting from the weathering, leaching, and alteration of aluminous minerals. It is an important constituent of many bauxites and may be the chief aluminum mineral present. *Bauxite* is defined as an aluminous lateritic rock in which aluminum hydroxides, amorphous or crystalline, predominate over other lateritic constituents. Gibbsite is also formed as a low-temperature

6.3.3.1 Nordstrandite

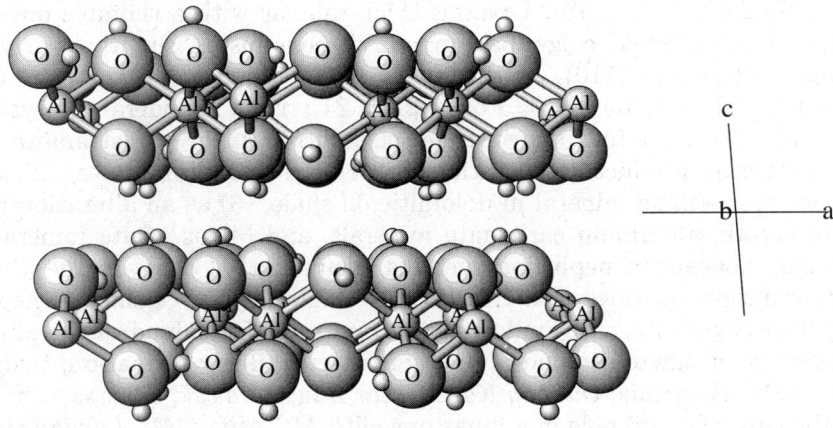

GIBBSITE: Al(OH)$_3$
Al atoms occupy two-thirds of the octahedra interstices between two closest-packed hydroxide layers. Sandwiches are held together with hydrogen bonds. Note that some hydrogen bonds are within layers, according to this structure determination. *Compare* BRUCITE.

hydrothermal mineral in veins or cavities in alkaline or other igneous rocks. **Localities:** From *Richmond, Berkshire Co., MA*, as radial-fibrous crusts in a limonite deposit. Also crystallized near Slatoust in the Schischimsk Mts., Ural Mts., Russia(!), and at Salka, southern Ural Mts.(!); Eikaholm and Lille-Arö, Langesundfjord, Norway(!); Serro, MG, Brazil(*). Important bauxite deposits consisting largely of gibbsite are in AR; Vogelsberg, Hesse, Germany; France; Hungary; in Madras and Maharashtra, India; Ghana; Ouro Preto, MG, Brazil; Suriname; Guyana; Jamaica; Australia; and several other countries throughout the world. Use: The principal ore mineral of aluminum. RG *DI:663, AM 55:43(1970), AC(B) 32:1719(1976)*.

6.3.2.1 Bayerite α-Al(OH)$_3$

Named in 1928 for Karl J. Bayer, German metallurgist. Polymorphs: gibbsite, nordstrandite, doyleite. MON $P2_1/c$. $a = 5.062$, $b = 8.671$, $c = 4.713$, $\beta = 90.45°$, $Z = 4$, $D = 3.06$. Structure: *ZK 125:317(1967)*. 20-11(syn): 4.71_9 4.35_7 3.20_3 2.22_{10} 1.72_4 1.60_1 1.55_1 1.46_1. Finely fibrous. Colorless, silky luster. $G = 3.05$. From the *Hatrurim formation, Israel*, with calcite, portlandite, and ettringite. RG *MM 33:723(1963), AM 55:43(1970)*.

6.3.3.1 Nordstrandite β-Al(OH)$_3$

Named in 1958 for R. A. van Nordstrand (b.1917), Sinclair Research Laboratories, who first synthesized the compound. Polymorphs: gibbsite, doyleite, bayerite TRIC $P\bar{1}$. $a = 6.148$, $b = 6.936$, $c = 5.074$, $\alpha = 95.77°$, $\beta = 99.07°$,

$\gamma = 83.30°$, $Z = 4$, $D = 2.454$. *CM 20:77(1982)*. *24-6*(syn): 4.79_{10} 4.32_3 4.21_2 2.39_3 2.27_3 2.02_3 1.90_2 1.78_1. Crystals thick tabular with a rhombic outline. Colorless to coral pink, beige, pale green; vitreous luster, slightly pearly on cleavage. Cleavage {110}, perfect. $G = 2.43$. Biaxial (+); $Z \wedge c = 32°$; $N_x = 1.580$, $N_y = 1.583$, $N_z = 1.602$; $2V = 24°$; $r < v$, moderate; negative elongation. There are four known types of occurrences for nordstrandite: (1) as a weathering product of bauxitic soils derived from limestones, (2) as a vein- or fissure-filling mineral in dolomitic oil shale, (3) as an alteration product of certain aluminum carbonate minerals, and (4) as a late mineral in pegmatitic pockets in nepheline syenites. Found in oil shale in the Green River formation, northwestern CO. Mt. St.-Hilaire, QUE, Canada, in nepheline syenite pegmatite, and similarly at Narssarssuk, Greenland. As an alteration product of dawsonite and alumohydrocalcite, Berry formation, Sydney Basin, NSW, Australia; *Gunony Kapor, near Bau, Sarawak, Malaysia*, in soil from the edge of a sinkhole in a limestone cliff; *Mt. Alifan–Mt. Lamlan ridge, Guam*, in secondary solution cavities in limestone, near the contact with deeply weathered basalt flows. RG *NJMA 109:185(1968)*; *AM 55:43(1970)*, *60:285(1975)*.

6.3.4.1 Doyleite Al(OH)$_3$

Named in 1985 for E. J. Doyle (b.1905), Ottawa, Ontario, who collected the first specimens. Polymorphs: gibbsite, bayerite, nordstrandite. TRIC $P1$ or $P\bar{1}$. $a = 5.002$, $b = 5.175$, $c = 4.980$, $\alpha = 97.5°$, $\beta = 118.6°$, $\gamma = 104.73°$, $Z = 2$, $D = 2.482$. *38-376*: 4.79_{10} 4.30_2 4.18_2 4.09_2 2.36_4 1.97_3 1.86_3 1.84_3. Rosettes of platy crystals; also finely granular, botryoidal, or smooth porcelainlike crusts. Colorless, also white to bluish white; white streak; vitreous luster, also dull. Cleavage {010}, perfect; {100}, poor; crystal plates slightly flexible but not elastic. $H = 2\frac{1}{2}$–3. $G = 2.48$. Nonfluorescent. Insoluble in cold mineral acids. Biaxial (+); $N_x = 1.545$, $N_y = 1.553$, $N_z = 1.566$ (Na); $2V = 77°$; $r > v$, strong. From *Mt. St.-Hilaire, QUE, Canada*, in vugs in nepheline syenite pegmatite, associated with albite, pyrite, and calcite. Also from the Francon quarry, Montreal, QUE, in vugs in silicocarbonate sills cutting Ordovician limestones, with weloganite, calcite, and quartz. RG *CM 23:21(1985)*.

6.3.5.1 Dzhalindite In(OH)$_3$

Named in 1963 for the locality. ISO $Im3$. $a = 7.95$, $Z = 8$, $D = 4.34$. *16-161*(syn): 4.42_4 3.99_{10} 2.82_9 1.78_9 1.63_8 1.40_3 1.20_3 1.07_3. Yellow-brown grains. Sn, Fe, and Zn may substitute to a limited extent in the formula. In polished section, dark gray. Isotropic; $N = 1.725$. $R = 8.2$ (589 nm). From the Mt. Pleasant sulfide–cassiterite body, SW of Frederickton, NB, Canada, in fractures in sphalerite, with calcite, galena, and quartz. *Dzhalindin ore deposit, Little Khingan ridge, far eastern Siberia, Russia*, as an oxidation product of indite, in cracks in cassiterite. RG *ZVMO 92:445(1963)*, *AM 49:439(1964)*, *CM 10:781(1971)*.

6.3.5.2 Söhngeite Ga(OH)$_3$

Named in 1965 for Adolf Paul Gerhard Söhnge (b.1913), South African geologist; chief geologist, Tsumeb Corp., Namibia. ISO $Im3$. $a = 7.47$, $Z = 8$, $D = 3.847$. *AM 51:1815(1966)*. *18-532:* 3.74_{10} 2.63_6 2.36_3 2.15_4 1.87_4 1.67_7 1.53_6 1.24_3. Crystalline aggregates. Light brown. $H = 4–4\frac{1}{2}$. $G = 3.84$. Isotropic; $N = 1.736$. From *Tsumeb, Namibia*, on corroded germanite containing gallite. RG *NW 52:493(1965)*.

6.3.5.3 Bernalite Fe(OH)$_3 \cdot n$H$_2$O

Named in 1993 for John Desmond Bernal (1901–1971), British crystallographer and historian of science. Hydrated iron hydroxide, with n ranging from 0.0 to <0.25. ORTH $Immm$. $a = 7.544$, $b = 7.560$, $c = 7.558$ (pseudocubic), $Z = 8$, $D = 3.23$. *AM 78:827(1993)*. The structure is not related to that of other iron hydroxides or gibbsite but belongs to the stottite or schoenfliesite groups, consisting of a three-dimensional network of corner-connected Fe(OH)$_6$ octahedra. 3.78_{10} 2.68_2 2.39_2 2.19_1 2.02_1 1.892_1 1.692_2 1.545_1. Crystals are pseudo-octahedra, up to 3 mm. Yellow-green to dark bottle green, apple green streak, vitreous to adamantine luster. No cleavage; conchoidal to uneven fracture; brittle. $H = 4$. $G = 3.32$. Biaxial; $N = 1.92–1.94$; other optical properties could not be determined due to micropolysynthetic twinning. Bernalite was found on museum specimens collected long ago from the upper levels of the open pit at the *Proprietary mine, Broken Hill, NSW, Australia*, associated with coronadite and goethite. RG *NW 79:509(1992)*.

WICKMANITE GROUP

The wickmanite minerals are basically tin hexahydroxides, with some substitution, leading to group of minerals of the general formula

$$AB(OH)_6$$

where

A = Mn, Mg, Fe^{2+}, Zn, Ca, Na
B = usually Sn, but also Ge and Sb^{5+}

and a structure that may be cubic, trigonal, or tetragonal.

In the structure, A and B atoms form a NaCl-like structure, with the (OH) groups octahedrally coordinating the cations, giving an infinite, corner-sharing three-dimensional framework of coordinating octahedra. This normally gives an isometric structure; however, in five members of this group the polyhedra connected along the c axis are alternately rotated slightly along opposite senses of rotation, which lowers the symmetry to tetragonal. The hexagonal–rhombohedral structure of burtite is closely related, but it has not been studied in detail. *AM 73:657(1988)*.

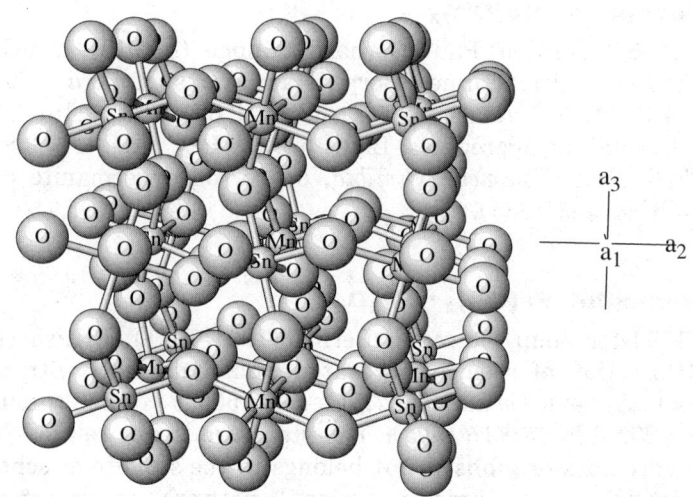

WICKMANITE: MnSn(OH)$_6$
Unlike most other oxides and hydroxides, Mn and Sn octahedra in wickmanite share only corners, not edges. Mn and Sn apparently alternate strictly.

WICKMANITE GROUP

Mineral	Formula	Space Group	a	c	D	
Wickmanite	MnSn(OH)$_6$	$Pn3m$	7.873		3.82	6.3.6.1
Schoenfliesite	MgSn(OH)$_6$	$Pn3$	7.759		3.483	6.3.6.2
Natanite	FeSn(OH)$_6$	$Pn3m$	7.69		4.035	6.3.6.3
Vismirnovite	ZnSn(OH)$_6$	$Pn3m$	7.72		4.073	6.3.6.4
Burtite	CaSn(OH)$_6$	$R\bar{3}$	11.49	14.08	3.22	6.3.6.5
Mushistonite	(Cu, Zn, Fe)Sn(OH)$_6$	$Pn3m$	7.735		4.08	6.3.6.6
Stottite	FeGe(OH)$_6$	$P4_2/n$	7.594	7.488	3.545	6.3.7.1
Tetrawickmanite	MnSn(OH)$_6$	$P4_2/n$	7.787	7.797	3.79	6.3.7.2
Jeanbandyite	(Fe, Mn)Sn(OH)$_6$	$P4_2/n$	7.648	7.648		6.3.7.3
Mopungite	NaSb(OH)$_6$	$P4_2/n$	7.994	7.859	3.264	6.3.7.4

6.3.6.1 Wickmanite MnSn(OH)$_6$

Named in 1967 for Franz-Erik Wickman (b.1915), Swedish mineralogist, who has studied the mineralogy of Långban extensively. Wickmanite group. Polymorph: tetrawickmanite. ISO $Pn3m$. $a = 7.873$, $Z = 4$, $D = 3.82$. *AM* 53:1063(1968). Structure: *AC* 13:601(1960). 20-727: 4.55$_4$ 3.93$_{10}$ 2.78$_6$ 2.37$_3$ 1.97$_4$ 1.76$_7$ 1.61$_5$ 1.05$_4$. Crystallizes as octahedra. Brownish yellow, greenish yellow, orange. No cleavage. H = 3–4. G = 3.89. Slowly soluble in common acids. Isotropic; N = 1.705. From Whealcock Zawn, Botallack, Cornwall, England, with axinite and grossular. *Långban, Värmland, Sweden.* Tvedalen,

Norway, in a nepheline syenite dyke. Pitkäranta mining district, Lake Ladoga, Karelia, Russia, from the dump of the Toivo shaft, in an altered garnet skarn. Tsumeb mine, Namibia (germanian wickmanite). Llallagua, Bolivia, in cavities in stannite. RG *AMG 4:395(1967)*, *MM 53:388(1989)*.

6.3.6.2 Schoenfliesite MgSn(OH)$_6$

Named in 1971 for Arthur Moritz Schoenflies (1853–1928), professor of mathematics, University of Frankfurt. Wickmanite group. ISO $Pn3$. $a = 7.759$, $Z = 4$, $D = 3.483$. *AM 57:1557(1972)*. *9-27*: 4.48_4 3.87_{10} 2.74_4 2.33_2 1.94_2 1.78_2 1.58_2 1.49_2. Small grains; prismatic crystals and fibers; minute cubo-octahedral crystals. Colorless. No cleavage. H = 4. G > 3.32. Soluble in HCl; slowly soluble in NH$_4$OH. Forms a solid-solution series with wickmanite, MnSn(OH)$_6$. Isotropic; N = 1.667. From *Brooks Mountain, Seward Peninsula, AK*, associated with hulsite, goethite, magnetite, fluorite, and hematite. Pitkäranta mining district, Lake Ladoga, Karelia, Russia, in chlorite–serpentine rock, with chondrodite, diopside, fluorite, calcite, dolomite, and magnetite. RG *ZK 134:116(1971)*, *CM 15:437(1977)*.

6.3.6.3 Natanite FeSn(OH)$_6$

Named in 1981 for Natan Il'ich Ginzburg (1917–1985), Russian mineralogist and geochemist, expert on pegmatites and the mineralogy of Be, Nb, and Ta. Wickmanite group. ISO $Pn3m$. $a = 7.69$, $Z = 4$, $D = 4.035$. *AM 67:1077(1982)*. *33-687*: 3.73_9 2.71_7 2.22_5 1.92_3 1.71_{10} 1.56_7 1.28_3 1.03_4. *31-653*(syn): 4.53_2 3.92_{10} 2.77_5 2.26_1 1.96_2 1.75_4 1.60_3 1.30_1. Greenish brown, vitreous luster. No cleavage. H = $4\frac{1}{2}$. Soluble in dilute HCl, insoluble in concentrated Na$_2$CO$_3$. Isotropic; N = 1.755. From the *Trudov* and *Mushiston deposits, Tadzhikistan*. RG *ZVMO 110:492(1981)*.

6.3.6.4 Vismirnovite ZnSn(OH)$_6$

Named in 1981 for Vladimir Ivanovich Smirnov (1910–1988), Russian economic geologist, Moscow State University. Wickmanite group. ISO $Pn3m$. $a = 7.72$, $Z = 4$, $D = 4.073$. *AM 67:1077(1982)*. *33-1376*: 3.84_{10} 2.22_7 1.73_9 1.57_8 1.29_4 1.16_4 1.12_5 1.03_6. Pale yellow, vitreous luster. No cleavage. H = 3.9. Soluble in dilute HCl, insoluble in concentrated Na$_2$CO$_3$. Isotropic; N = 1.735. From the *Trudov* and *Mushiston deposits, Tadzhikistan*, as an oxidation product of stannite, with varlamoffite, goethite, malachite, azurite, and so on. RG *ZVMO 110:492(1981)*.

6.3.6.5 Burtite CaSn(OH)$_6$

Named in 1981 for Donald M. Burt (b.1943), mineralogist and economic geologist; professor, Arizona State University. Wickmanite group. HEX-R (pseudo-ISO) Space group uncertain, probably $R\bar{3}$. $a = 11.49$, $c = 14.08$, $Z = 12$, $D = 3.22$. (Pseudo-ISO cell: $Pn3$; $a = 8.128$, $Z = 4$.) *9-30*(syn): 4.69_5 4.06_{10} 2.87_4 2.57_1 2.45_1 2.34_2 1.81_2 1.66_2. Crystals are pseudo-octahedrons, superficially altered to varlamoffite. Colorless, vitreous luster. Cleavage

{0001}, good cubic cleavage; very brittle. H = 3. G = 3.28. Nonfluorescent. Slowly soluble in cold dilute HCl. Uniaxial (+); $N_O = 1.633$, $N_E = 1.633$ (Na). Birefringence <0.0003. Nearly isotropic. Optical data demonstrate that the true symmetry is HEX-R, pseudo cubic. From a garnetite tin skarn located in the cliffs of the W bank of the *Beht R.* (33°31'26" lat., 5°49'50" W. long.), ≈ 40 km SW of *Meknès, central Morocco*, associated with wickmanite, stokesite, datolite, and varlamoffite. RG *CM 19:397(1981), AM 67:854(1982)*.

6.3.6.6 Mushistonite (Cu, Zn, Fe)Sn(OH)$_6$

Named in 1984 for the locality. Wickmanite group. ISO $Pn3m$. $a = 7.735$ (Zn > Fe), $a = 7.705$ (Fe > Zn), Z = 4, D = 4.08. *AM 70:1331(1985)*. *ZVMO* (Zn > Fe): 3.88_{10} 2.74_5 2.45_1 2.33_1 2.23_2 1.93_2 1.73_3 1.58_2; (Fe > Zn): 3.84_{10} 2.73_5 1.72_2. CuSn(OH)$_6$: *20-369*(syn): 4.48_5 4.05_6 3.79_{10} 2.77_6 2.24_6 1.714_6 1.694_6 1.562_6. Fine grained, earthy aggregates. Brownish green. H = 4–4$\frac{1}{2}$. Soluble in dilute HCl. From the *Mushiston tin deposit, Tadzhikistan*, in the oxidation zone, replacing stannite, associated with varlamoffite. RG *ZVMO 113:612(1984), MR 17:383(1986)*.

6.3.7.1 Stottite FeGe(OH)$_6$

Named in 1958 for Charles E. Stott (1896–1978), U.S. geologist and general director, Tsumeb mine, Namibia. Wickmanite group. TET $P4_2/n$. $a = 7.594$, $c = 7.488$, Z = 4, D = 3.545. *AM 73:657(1988)*. Structure: *AC 14:205. 11-161:* 3.77_{10} 2.66_8 2.17_6 1.89_6 1.69_7 1.67_6 1.54_6 1.25_6. Crystals tetragonal bipyramidal, up to 10 mm in size, pseudo-octahedral with {111} dominant. Brown, gray-white streak, greasy luster. Cleavage {100}, {010}, good; {001}, fair. G = 3.60. Zinc may replace Fe^{2+}, up to nearly as much mol% Zn as Fe^{2+}. Uniaxial (−); $N_O = 1.7375$, $N_E = 1.728$; r > v, weak. Usually anomalously biaxial. From *Tsumeb, Namibia*, in the deep oxidized orebody, formed by the action of circulating waters on renierite and germanite. RG *NJMM 1958:85, AM 43:1006(1958)*.

6.3.7.2 Tetrawickmanite MnSn(OH)$_6$

Named in 1973 for its relationship to wickmanite but with tetragonal symmetry. Wickmanite group. Polymorph: wickmanite. TET $P4_2/n$. $a = 7.787$, $c = 7.797$, Z = 4, D = 3.79. *MR 4:24(1973)*. *25-553:* 4.52_3 3.94_{10} 3.88_3 2.77_9 2.26_3 1.76_5 1.75_3 1.61_4. Crystals pyramidal {112} with {001} and minor {100}; also pseudocubic (Långban); twinning not observed. Honey yellow, brownish orange. No cleavage. G = 3.65. Uniaxial (−); $N_O = 1.724$, $N_E = 1.720$. From the *Foote mine, Kings Mountain, Cleveland Co., NC*, associated with albite, bavenite, siderite, and eakerite. Also from Långban, Värmland, Sweden. RG *AM 58:966(1973)*.

6.3.7.3 Jeanbandyite (Fe, Mn)Sn(OH)$_6$

Named in 1982 for Jean Arney Bandy (1900–1991), widow of Mark C. Bandy, who collected the original jeanbandyite specimens at Llallagua and whose superb mineral collection she donated to the Los Angeles County Museum of Natural History. Wickmanite group. TET $P4_2/n$. $a = 7.648$, $c = 7.648$. *35-663:* 4.41_2 3.83_{10} 2.71_7 2.21_2 1.91_2 1.71_6 1.56_4 1.28_2. As crudely pseudo-octahedral crystal aggregates in epitactic growth upon the vertices of octahedral crystals of wickmanite. Forms observed {111}, {100}, and {001}. Brownish orange, buff streak, vitreous to subadamantine luster. Cleavage {001}, {100}, fair. $H = 3\frac{1}{2}$. $G = 3.81$. Slowly soluble in cold 1:1 HCl. Uniaxial $(-)$; $N_O = 1.837$, $N_E = 1.833$. From *Llallagua, Potosi, Bolivia*, in the Contacto vein and elsewhere in the mine, in vugs in association with fluorapatite, stannite, and other minerals. Santa Eulalia, CHIH, Mexico, also as an epitaxial overgrowth on wickmanite. RG *MR 13:235(1982), AM 68:471(1983).*

6.3.7.4 Mopungite $NaSb^{5+}(OH)_6$

Named in 1985 for the locality. Wickmanite group. TET $P4_2/n$. $a = 7.994$, $c = 7.859$, $Z = 4$, $D = 3.264$. *MR 16:73(1985).* *38-411:* 4.58_{10} 3.99_8 3.93_4 1.629_5 1.325_5 1.265_5 1.205_7 1.192_5. Crystals pseudocubic; as crusts. Colorless, white; vitreous luster. No cleavage. $H = 3$. $G = 3.21$. Nonfluorescent. Soluble in hot but not in cold water. Decomposed by HCl and HNO_3; soluble in cold dilute tartaric acid. Uniaxial $(-)$; $N_O = 1.614$, $N_E = 1.605$. From *Mopung Hills, Churchill Co., NV*, associated with stibiconite, senarmontite, romeite, and tripuhyite, all pseudomorphous after stibnite, and with selenium and sulfur. RG *AM 70:1330(1985).*

6.3.8.1 Jamborite $(Ni^{2+},Ni^{3+},Fe)(OH)_2(OH,S,H_2O)$

Named in 1973 for John L. Jambor (b.1936), mineralogist, Geological Survey of Canada. HEX Space group unknown. $a = 3.07$, $c = 23.3$, $Z = 3$, $D = 2.69$. *25-1363:* 7.78_{10} 3.89_4 2.59_6 2.32_1 1.98_1 1.53_5 1.50_3 1.32_1. *27-340*(syn): 11.0_{10} 7.68_{10} 3.79_6 2.59_6 2.31_6 1.95_6 1.54_6 1.51_6. Microscopic, fibrous, or lamellar crystals. Green. $G = 2.67$. Uniaxial $(-)$; $N_O = 1.607$, $N_E = 1.602$ (Na). From Hall's Gap, KY, as alteration of millerite. From *Italy* at *Montescuto Ragazza near Bologna and* at *Castelluccio di Moscheda near Modena*, in cavities in ophiolitic rock, with calcite, dolomite, and quartz, pseudomorphous after millerite. RG *AM 58:835(1973), MM 44:339(1981).*

6.3.9.1 Bottinoite $Ni[Sb^{5+}(OH)_6]_2 \cdot 6H_2O$

Named in 1992 for the locality. HEX-R $P31m$, $P312$, or $P\bar{3}1m$. $a = 16.026$, $c = 9.795$, $Z = 6$, $D = 2.81$. *AM 77:1301(1992).* 4.88_5 4.62_{10} 2.34_8 2.16_5 2.09_6 1.806_7 1.751_6 1.648_5. Roselike aggregates of small tabular crystals. Light blue-green, very light blue streak, vitreous luster. Conchoidal fracture; brittle. $H = 3$. $G = 2.83$. Nonfluorescent. Uniaxial $(+)$; $N_O = 1.600$, $N_E = 1.605$. Occurs on fractures in schist with calcite, siderite, quartz, and sulfides, perched on ullmanite crystals, at the *Bottino mine, Alpe Apuani, Italy*. RG

6.4.1.1 Lithiophorite $A_{14}Li_6(OH)_{42}Mn_3^{2+}Mn_{18}^{4+}O_{42}$

Named in 1870 from *lithium* and the Greek for *to bear*, in allusion to the presence of lithium in its composition. HEX $P3_1$. $a = 13.37$, $c = 28.20$, $Z = 3$, $D = 3.376$. *41-1378:* 9.43_7 4.71_{10} 3.14_1 2.37_2 1.880_1 1.453_1 1.447_1 1.436_1. Structure: *AM 67:817(1982)*. As fine scales; also compact; botryoidal. Black, blue-black; blackish-gray streak; metallic to dull luster. $H = 2$–3. $G = 3.14$–3.36. In RL, white, with very strong pleochroism and anisotropism, no internal reflections. $R = 20.2$–9.7 (546 nm). Although lithium is part of the name of the mineral, Li_2O is highly variable in content, ranging from 0.1 to 3.3%; it is possibly not an essential component. Any of the elements Co, Ni, Cu, Zn, and/or Pb are also usually present in significant amounts up to a maximum of about 5%, but no one of these is essential. Lithiophorite is widespread in soils and in the weathering zones of a variety of ore deposits. It is probably one of the major components of the poorly defined manganese oxide *wad (DI:566)*, in association with asbolan, vernadite, birnessite, cryptomelane, and so on. In contrast to some of these minerals, lithiophorite is readily identifiable with a crystalline structure that produces excellent X-ray diffraction patterns. From Charlottesville and elsewhere in VA, coating fractures in quartz veins. Lecht mines, Tomintoul, Banffshire, Scotland, with cryptomelane. *Schneeberg, Sachsen, Germany*. In Australia at Groote Eyelandt, Kalgoorlie, and Pilbara, WA, and at Rockhampton and Gympie areas, Q. Itatiaia mine, Conselheira Pena, MG, Brazil, in brilliant botryoidal masses in pegmatite(!). Used as an ore of manganese. RG *DI:569, AM 52:1545(1967), MM 37:618(1970), AUM 2:25(1987)*.

6.4.1.2 Quenselite $PbMn^{3+}O_2(OH)$

Named in 1926 for Percy Dudgeon Quensel (1881–1966), Swedish mineralogist and petrologist, University of Stockholm. MON $P2/c$. $a = 5.61$, $b = 5.70$, $c = 9.15$, $\beta = 93.0°$, $Z = 4$, $D = 7.07$. *GSAM 85:200(1962). 23-351:* 3.68_7 3.60_7 3.04_{10} 2.95_6 2.72_8 2.44_4 2.08_4 1.99_4. Crystals tabular {010}, striated on {011} parallel to [100]. Structure: *ZK 134:321*. Pitch black, dark brown to gray streak, metallic to adamantine luster. Cleavage {001}, perfect; micaceous; thin laminae flexible. $H = 2\frac{1}{2}$. $G = 6.84$. Soluble in dilute HCl, with evolution of Cl. Biaxial (+); $X = b$, $Y \approx c$; $N \approx 2.30$; pleochroism deep brown; absorption $Z > X$. From *Långban, Värmland, Sweden*, in the Amerika stope, with calcite and barite in crevices cutting hausmannite–hematite–braunite ore. RG *DI:729*.

6.4.2.1 Hydroromarchite $Sn_3O_2(OH)_2$

Named in 1971 for being a hydrate and for its relationship to romarchite. TET $P4/mnc$ or $P4nc$. $a = 7.98$, $c = 9.17$, $Z = 4$, $D = 4.80$. *AM 58:552(1973). 25-1303:* 3.50_{10} 3.26_5 2.96_8 2.77_9 2.48_4 1.92_5 1.90_5 1.77_4. Crusts of white crystals. Found as a corrosion product on tin cooking utensils found in the remains of a sunken canoe lost 150 years earlier near *Boundary Falls, Winnipeg River, ONT, Canada*. RG *CM 10:916(1971)*.

6.4.3.1 Häggite $V_2O_2(OH)_3$

Named in 1958 for Gunnar Hägg (1903–1986), Swedish chemist, University of Uppsala. MON $C2/m$. $a = 12.17$, $b = 2.99$, $c = 4.83$, $\beta = 98.25°$, $Z = 2$, $D = 3.53$. *AM 45:1144(1960)*. *29-1380:* 4.80_{10} 4.05_5 3.51_1 3.02_3 2.44_3 1.96_2 1.82_1 1.49_1. Crystals acicular. Black. From *Carlile, Crook Co., WY*. Also from Kara Tau Range, Kazakhstan. RG *AM 65:210(1980)*, *ZK 116:482(1961)*.

6.4.4.1 Hydrocalumite $Ca_2Al(OH)_6[Cl_{1-x}(OH)_x] \cdot 3H_2O$

Named in 1934 in allusion to the composition. Formula also written $Ca_4Al_2(OH)_{12}(Cl, CO_3, OH, H_2O)_{2.5} \cdot 4H_2O$. *NJMM 1980:322*. MON $P2/c$. $a = 10.020$, $b = 11.501$, $c = 16.286$, $\beta = 104.22°$, $Z = 2$, $D = 2.15$. *42-558:* 7.9_{10} 3.95_9 3.86_3 3.75_2 2.88_5 2.53_2 2.37_3 2.29_3. Massive. Colorless, light green; vitreous luster, pearly on cleavage. Cleavage {001}, perfect. $H = 3$. $G = 2.15$. Strongly pyroelectric. Biaxial $(-)$; $X \wedge c = <3°$; $Y = b$; $N_x = 1.535$, $N_y = 1.553$, $N_z = 1.557$; $2V = 24°$. From *Scawt Hill, County Antrim, Northern Ireland*, in vugs in a larnite rock, with afwillite, portlandite, and ettringite. Boisséjour, Puy de Dôme, France, with ettringite. Montalto di Castro, Viterbo, Italy. RG *DI:667*, *BM 86:149(1963)*, *AM 74:1403(1989)*.

6.4.5.1 Iowaite $Mg_4Fe^{3+}(OH)_{10}Cl \cdot 3H_2O$

Named in 1967 for the locality. HEX Space group unknown. $a = 3.119$, $c = 24.52$, $Z = 0.6$, $D = 2.11$. Structure: *AM 54:296(1969)*. *20-500:* 8.11_{10} 4.05_4 2.64_2 2.36_3 2.02_2 1.56_1 1.53_1 1.45_1. Crystals platy {0001}. Bluish green to pale green; alters in air to whitish green with a rusty red overtone, due to loss of zeolitic water; white streak; greasy luster. Cleavage {0001}, perfect. $H = 1\frac{1}{2}$. $G = 2.11$. Insoluble in water; readily soluble in mineral acids. The amount of zeolitic water is somewhat variable, depending on atmospheric humidity and other factors; it is all driven off at a temperature of 280°. Uniaxial $(-)$; $N_O = 1.543$, $N_E = 1.533$; crystals length-fast. From *Sioux Co., IA*, in veinlets in serpentinite, encountered in drill core at a depth of 300–450 m, associated with chrysotile, dolomite, brucite, calcite, magnesite, and pyrite. Also from the Komsomolskiy mine, Talnakh, Siberia, Russia(!), in crystals to 2.5 cm, with valeriite, száibelyite, magnetite, brucite, and serpentine in metamorphosed dolomites associated with porphyry copper deposits. Phalaborwa, Transvaal, South Africa(!), in sharp crystals to 1 cm. RG *AM 52:1261(1967)*, *MA 77:851*.

6.4.6.1 Meixnerite $Mg_6Al_2(OH)_{18} \cdot 4H_2O$

Named in 1975 for Heinrich Herman Meixner (1908–1981), Austrian mineralogist, University of Salzburg. HEX-R $R\bar{3}m$. $a = 3.046$, $c = 22.93$, $Z = 3/8$, $D = 1.953$. *38-478:* 7.64_6 3.82_7 2.57_6 2.29_7 1.941_{10} 1.523_8 1.494_6 1.268_5. Crystals tabular. Structurally related to hydrotalcite. Colorless. Cleavage {0001}, perfect. $H = 2$. $G \approx 1.9$. Uniaxial; $N = 1.517$; bireflectance: 0.016. From *Ybbs Persenberg, Austria*, as a secondary mineral with talc and later aragonite in cracks of a pyrope–serpentine rock. RG *TMPM 22:79(1975)*, *AM 61:176(1976)*.

6.4.7.1 Janggunite $Mn^{4+}_{5-x}(Mn^{2+}, Fe^{3+})_{1+x}O_8(OH)_6; X=0.2$

Named in 1977 for the locality. ORTH Space group unknown. $a = 9.324$, $b = 14.05$, $c = 7.956$, $Z = 4$, $D = 3.58$. *AM 63:794(1978)*. *29-889:* 9.34_{10} 7.09_{10} 4.62_8 4.17_8 3.55_{10} 3.10_{10} 2.47_8 1.53_8. Radiating groups of flakes; colloform bands, and dendritic masses. Black, brownish-black to dark brown streak, dull luster. Cleavage in one direction; very fragile. $H = 2–3$. $G = 3.59$. In RL, grayish white to gray. No internal reflections; strongly anisotropic, yellowish brown to gray. $R = 13, 15$ (546 nm). From the *Janggun mine, Bonghwa, Korea*, with calcite, nsutite, and todorokite in the zone of supergene manganese oxides. RG *MM 41:519(1977)*, *AM 63:426(1978)*.

6.4.8.1 Kimrobinsonite $Ta(OH)_3(O, CO_3)$

Named in 1985 for Kim Robinson (b.1951), Australian geologist of Perth, WA, who found the first specimen. ISO Space group unknown. $a = 3.812$, $Z = 1$, $D = 6.87$. *CM 23:573(1985)*. *38-461:* 3.81_{10} 2.97_1 2.70_7 2.20_2 1.91_3 1.70_5 1.56_4 1.27_2. Cryptocrystalline, earthy. White; white streak; dull, chalky luster. Friable; soft. Isotropic; $N \gg 2$ (2.23_{calc}). From *Mt. Holland, WA, Australia*, in a deeply weathered pegmatite, intergrown with cesstibitantite, in an apparent pseudomorph after an unknown precursor Ta–Sb mineral, associated with montmorillonite, Li-muscovite, and tourmaline. RG *AM 72:1025(1987)*.

6.4.9.1 Asbolane $(Co, Ni)_{1-y}(Mn^{4+}O_2)_{2-x}(OH)_{2-2y+2} \cdot nH_2O$

Named in 1847 from the Greek for *to soil like soot*. Synonym: cobaltian wad. HEX Space group unknown. $a = 2.823$, $c = 9.34$. *42-1319:* 9.6_{10} 4.83_{10} 3.23_1 2.45_4 2.38_1 1.810_1 1.540_5 1.420_5. Compact to earthy; massive; sometimes botryoidal. *Wad* is a general term covering soft, impure oxides of manganese. Asbolan is a cobaltian wad with a well-defined structure type; it may also contain significant amounts of Ni, Fe, Cu, V, Ca, and Ba. *AM 67:417(1982)*. Black; bluish or brownish black with streak like the color, dull luster. Usually very soft; also hard, to $6\frac{1}{2}$. In RL, white to brownish; pleochroism and anisotropism very strong; no internal reflections. $R = 25.5–13.2$ (546 nm). *P&J:78*. From the Blackbird district, Lehmi Co., ID: Mine La Motte, MO (nickelian); Silver Bluff, SC; Schneeberg region, Sachsen, Germany; Markirch, Alsace, France; Capo Calamita, Elba, Italy; Lipov deposit, middle Ural Mts., Russia; and at Nizhni Tagilsk, Siberia; Shaba cobalt district, Zaire; abundant in New Caledonia with garnierite as residual deposits on weathered peridotite containing disseminated Co and Ni minerals. RG *DI:566,568; ZVMO 116:210(1987); CR 307:155(1988)*.

6.4.10.1 Schwertmannite $Fe^{3+}_{16}O_{16}(OH)_{12}(SO_4)_2$

Named in 1994 for Udo Schwertmann (b.1927), German soil scientist, Technical University, Munich. TET $P4/m$. $a = 10.66$, $c = 6.04$, $Z = 1$, $D = 3.77–$

6.4.10.1 Schwertmannite

3.99. *MM 58:641(1994)*. *PD* (all lines broad): 4.86_4 3.39_5 2.55_{10} 2.28_2 1.95_1 1.66_2 1.51_2 1.46_2. Schwertmannite is a common compound but difficult to characterize and study, recognized so far in at least 40 localities worldwide. It is poorly crystalline, very fine grained, the individual crystallites measuring only a few nanometers in size. Its proposed structure is akin to that of akaganéite [FeO(OH, Cl)], with sulfate instead of chloride as a stabilizing element in the tunnel cavities. The OH/SO_4 ratio appears to be nonstoichiometric, and the formula might be written $Fe^{3+}O(OH, SO_4)$, in which case it could be considered to be a polymorph of goethite, lepidocrocite, feroxyhite, and akaganéite. It has a high specific surface area, in the range 100 to 200 m^2/g. Most specimens consist of spherical to ellipsoidal particles (aggregates) that are 200 to 500 nm in diameter. Needlelike structures with average widths and thicknesses of 2 to 4 nm and lengths of 60 to 90 nm radiate from the particle surfaces and give the material a pin-cushion morphology. It is very soluble in cold 5 *M* HCl or in ammonium oxalate at pH 3.0. Upon heating, it converts to hematite with an intermediate phase of $Fe_2(SO_4)_3$. It was first recognized in the *Pyhäsalmi sulfide mine, Oulu Prov., Finland*. The sample was collected as an ocherous, secondary precipitate forming a crust on stones inundated by acidic drainage from a mound of concentrate. It usually forms as a result of rapid oxidation of Fe^{2+} in acid, sulfate-rich effluents, produced from decomposition of primary sulfides, forming what had previously been described as "amorphous" ferric hydroxide. Associated minerals may be jarosite, natrojarosite, goethite, and ferrihydrite. Other recognized occurrences have been noted in Europe, North America, and Australia. Most are related to mining activity or other results of human activity, but one in the Zillertal Alps, Switzerland, at 2600 m elevation, is the purely natural result of the oxidation of a pyritic schist. RG *AM 80:847(1995)*.

Class 7

Multiple Oxides

7.1.1.1 Delafossite $Cu^{1+}Fe^{3+}O_2$

Named in 1873 for Gabriel Delafosse (1796–1878), French mineralogist. HEX-R $R\bar{3}m$. $a = 3.035$, $c = 17.166$, $Z = 3$, $D = 5.507$. Structure: *SR 37A:262(1971)*. *39-246*: 2.86_6 2.51_{10} 2.24_2 1.662_2 1.518_2 1.437_1 1.431_1 1.341_1. Tabular to equant crystals, with {0001} and {10$\bar{1}$1}; as botryoidal crusts; contact twins on {0001}. Black, black streak, metallic luster. Cleavage {10$\bar{1}$1}, imperfect; brittle. H = $5\frac{1}{2}$. G = 5.41. Weakly magnetic. Soluble in HCl and H_2SO_4 but insoluble in HNO_3. Easily fusible. Becomes magnetic on heating. RL, brownish white. Distinctly pleochroic; O, light golden brown; E, darker rose brown; anisotropism strong with characteristic green hues. R = 22.7–19.0 (470 nm), 23.3–19.1 (546 nm), 23.0–19.0 (589 nm), 23.6–20.3 (650 nm). *P&J:145*. From Bisbee, AZ, as crystals and botryoidal crusts with hematite, cuprite, copper, and tenorite. Pope–Shenon copper mine, Salmon, ID. Copreasa mine, Sahuaripa, SON, Mexico. Cartagena mine, Pedroso, Sevilla, Spain. *Ekaterinburg, Ural Mts., Russia*. RG *DI:674; AM 31:539(1946), 53:1779(1968)*.

7.1.1.2 Mcconnellite $CuCrO_2$

Named in 1976 for R. B. McConnell (1903–1986), British geologist, former director, British Guiana Geological Survey. HEX-R $R\bar{3}m$. $a = 2.983$, $c = 17.160$, $Z = 3$, $D = 5.61$. *39-247*(syn): 2.85_6 2.55_1 2.47_{10} 2.21_3 1.646_2 1.488_4 1.426_2 1.319_2. Tabular rhombohedral crystals. Dark red. G = 5.49. From the *Merume River, Mazaruni district, Guyana*, as alluvial pebbles consisting of mcconnellite intergrown with other chromium oxides: grimaldiite, eskolaite, bracewellite, and/or guyanaite. RG *USGSPP 887:17(1976), AM 62:593(1977), JACS 77:896(1955)*.

7.1.2.1 Crednerite $CuMnO_2$

Named in 1849 for Karl F. H. Credner (1809–1876), German mining geologist and mineralogist. MON $C2/m$. $a = 5.530$, $b = 2.884$, $c = 5.898$, $\beta = 104.6°$, $Z = 2$, $D = 5.49$. *MA 15:261*. *18-448*: 2.85_8 2.71_{10} 2.42_{10} 2.24_6 1.61_6 1.56_8 1.27_5 1.09_5. Crystallizes as six-sided plates, in radiating, hemispherical, or spherulitic groupings; also earthy coatings. Twinning on {41$\bar{1}$}, giving a pseudohexagonal aspect. Metallic, bright luster. Cleavage {001},{100}, perfect; {111}, good. H = 4. G = 5.38. Soluble in HCl with evolution of Cl; insoluble in HNO_3. RL creamy white; pleochroism yellowish to gray-brown; strongly anisotropic. Polysynthetic twinning sometimes visible. R = 33.3–25.5 (470 nm), 34.2–23.7 (546 nm), 33.4–23.2 (589 nm), 31.6–22.6 (650 nm). *P&J:134*. *Friedrichsroda, Thüringen, Germany*, intergrown with psilomelane and hausmannite, associated with malachite, barite, calcite, and wad. Higher Pitts, Mendip Hills, Somerset, England, with hydrocerussite and malachite, and at Merehead quarry, near Shepton Mallet(!), in crystal roses to 20 mm. Tachgagalt and Idikel, Morocco. RG *DI:722, AM 51:1819(1966)*.

SPINEL GROUP

The spinel group comprises 22 minerals with the general formula

$$A^{2+}B_2^{3+}O_4$$

where

A = Mg, Mn, Fe, Zn, Co, Cu, or Ge
B = Al, Fe^{3+}, Cr, or V

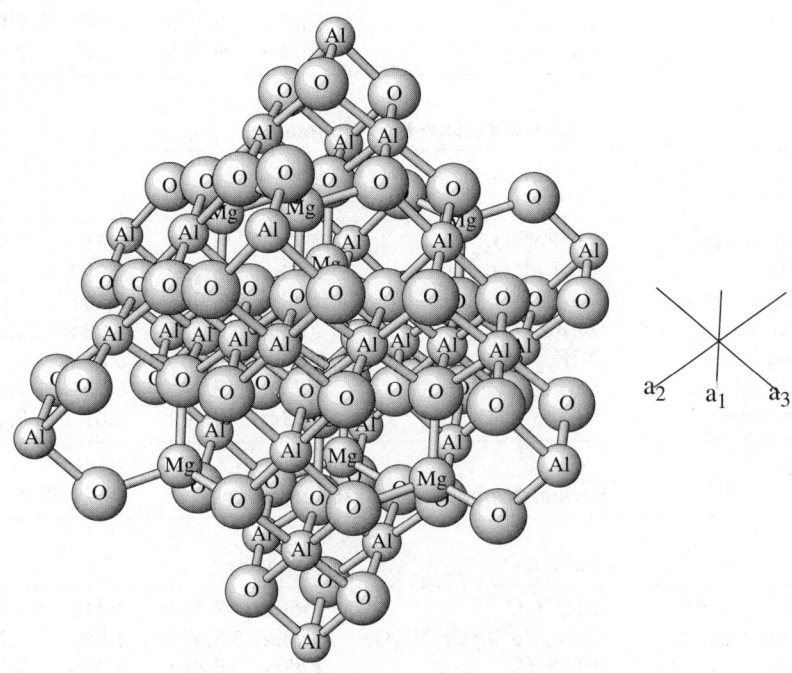

SPINEL: $MgAl_2O_4$

Space group *Fd3m*. This ball-and-stick model has the [111] direction vertical to show the approximately CCP oxygens (horizontal). Mg atoms are in tetrahedral interstices, and Al in octahedral interstices.

and an isometric structure in space group $Fd3m$.

There are 32 oxygen ions and 24 cations in the unit cell. Eight of the cations are in tetrahedral coordination in the A site. The remaining 16 cations are in octahedral coordination in the B site.

Perpendicular to the triad axis, layers of oxygen atoms alternate with layers of cations. The cation layers in which all the cations are in octahedral coordination alternate with other layers in which the cations are distributed among the A and B positions in the proportion A/B=2:1.

Spinels occur in two structural types, normal and inverse, that differ in their distribution of cations among the A and B positions. Writing the general formula $R_8^{2+}R_{16}^{3+}O_{32}$, the two distributions are:

Normal: $8R^{2+}$ in the A site, $16 R^{3+}$ in the B site
Inverse: $8R^{3+}$ in the A site, $8R^{2+}$ plus $8R^{3+}$ in B

Subgroups are defined in terms of the B position.

SPINEL GROUP: ALUMINUM SUBGROUP

Mineral	Formula	Space Group	a	D	
Spinel	$MgAl_2O_4$	$Fd3m$	8.083	3.578	7.2.1.1
Galaxite	$MnAl_2O_4$	$Fd3m$	8.258	4.077	7.2.1.2
Hercynite	$FeAl_2O_4$	$Fd3m$	8.153	4.260	7.2.1.3
Gahnite	$ZnAl_2O_4$	$Fd3m$	8.085	4.607	7.2.1.4

SPINEL GROUP: IRON SUBGROUP

Mineral	Formula	Space Group	a	D	
Magnesioferrite	$MgFe_2^{3+}O_4$	$Fd3m$	8.383	4.51	7.2.2.1
Jacobsite	$MnFe_2^{3+}O_4$	$Fd3m$	8.499	4.989	7.2.2.2
Magnetite	$Fe^{2+}Fe_2^{3+}O_4$	$Fd3m$	8.396	5.197	7.2.2.3
Franklinite	$ZnFe_2^{3+}O_4$	$Fd3m$	8.441	5.234	7.2.2.4
Trevorite	$NiFe_2^{3+}O_4$	$Fd3m$	8.339	5.368	7.2.2.5
Cuprospinel	$CuFe_2^{3+}O_4$	$Fd3m$	8.369	5.251	7.2.2.6
Brunogeierite	$Ge^{2+}Fe_2^{3+}O_4$	$Fd3m$	8.409	5.51	7.2.2.7

SPINEL GROUP: CHROMIUM SUBGROUP

Mineral	Formula	Space Group	a	D	
Magnesiochromite	$MgCr_2O_4$	$Fd3m$	8.333	4.414	7.2.3.1
Manganochromite	$(Mn, Fe^{2+})(Cr, V)_2O_4$	$Fd3m$	8.47	4.88	7.2.3.2
Chromite	$FeCr_2O_4$	$Fd3m$	8.379	5.055	7.2.3.3
Nichromite	$(Ni, Co)(Cr, Fe^{3+})_2O_4$	$Fd3m$	8.316	5.24	7.2.3.4
Cochromite	$(Co, Ni)(Cr, Al)_2O_4$	$Fd3m$	8.292	5.01	7.2.3.5
Zincochromite	$ZnCr_2O_4$	$Fd3m$	8.352	5.434	7.2.3.6

7.2.1.1 Spinel

SPINEL GROUP: VANADIUM SUBGROUP

Mineral	Formula	Space Group	a	D	
Vuorelainenite	$(Mn, Fe^{2+})(V, Cr)_2O_4$	$Fd3m$	8.48	4.64	7.2.4.1
Coulsonite	$Fe^{2+}V_2O_4$	$Fd3m$	8.297	5.15	7.2.4.2
Magnesiocoulsonite	MgV_2O_4	$Fd3m$	8.385	4.31	7.2.4.3

SPINEL GROUP: TITANIUM SUBGROUP

Mineral	Formula	Space Group	a	D	
Qandilite	Mg_2TiO_4	$Fd3m$	8.403	4.04	7.2.5.1
Ulvospinel	$Fe_2^{2+}TiO_4$	$Fd3m$	8.535	4.777	7.2.5.2

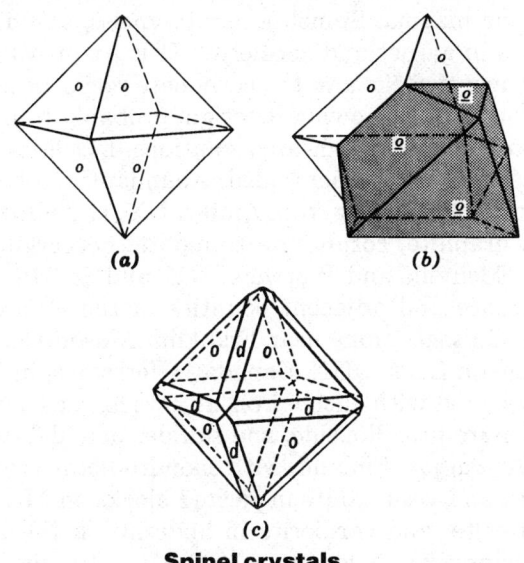

Spinel crystals

7.2.1.1 Spinel MgAl$_2$O$_4$

Named in 1546, possibly from the Latin *spinella*, little thorn, in allusion to the spine-shaped octahedral crystals. Spinel group. ISO $Fd3m$. $a = 8.083$, $Z = 8$, $D = 3.578$. *NBSM 25(9):25(1971)*. *NBSM*(syn): 4.66_4 2.86_4 2.44_{10} 2.02_7 1.56_5 1.43_6 1.052_1 0.825_2. Structure: *AC(A) 36:122(1980), (B) 40:96(1984)*. **Habit:** Crystals are usually octahedrons, rarely modified by {110} or {100}. Twinning common, on {111} (the *spinel law*), with twinned aggregates often flattened parallel to the composition plane, and forming triangular tablets. **Physical properties:** Green, black, brown, red, blue, and so on; white streak, vitreous luster. No cleavage; parting {111}, indistinct; conchoidal to uneven fracture. $H = 7\frac{1}{2}$–8. $G = 3.56$. Nonfluorescent, except for some gem varieties, especially the red ones. Nonmagnetic. **Optics:** Isotropic; $N = 1.718$. **Chemistry:** Virtually complete solid-solution substitution is possible in this AB_2O_4 mineral, in which in the A position Fe^{2+}, Ni, Co, Mn, and Zn may substitute for Mg, and in the B position Fe^{3+} and Cr may substitute for the Al. Other elements that more rarely enter into this structure are Cu and Ge in the A position and V in the B position. **Occurrence:** Spinel occurs embedded in granular limestone or dolomite as a metamorphic mineral, frequently associated with chondrodite. Spinel (common spinel; also *pleonaste, picotite,* and chromite) occurs as an accessory constituent in many basic rocks, especially those of the peridotite group; its formation is the result of the crystallization of a magma very low in silica, high in magnesia, and containing alumina. Since, as in many peridotites, alkalis are absent, feldspars cannot form, and the Al_2O_3 present must form either spinel or corundum. The variety *pleonaste* occurs in rocks of this type that are high in iron content and it is commonly associated with magnetite, the two minerals being the first to crystallize from the magma. Spinel occurs in gneiss, amphibolite, and so on, often as inclusions in garnet and cordierite. It is frequently found in contact zones between eruptive rocks and the carbonate rocks, occurring often in the cavities of the igneous rocks, owing its origin probably to pneumatolytic conditions or to deposits from superheated solutions. **Localities:** Only a few of the many known localities can be mentioned. Abundant in a belt of Precambrian crystalline limestones stretching from Amity, NY, to Andover, NJ, associated with phlogopite, graphite, corundum, tremolite, norbergite, dravite, and so on, especially at Edenville and Warwick, NY, and Sparta and Franklin, NJ. Also in St. Lawrence and adjacent counties in the Grenville marble; with corundum at the Culsagee mine near Franklin, Mason Co., NC. The Parker mine, Notre Dame du Laus, QUE, Canada(!), ferroan spinel. In Greenland at Disko Island associated with native iron. In Sweden at Kaveltorp, Nya-Kopparberg, with chondrodite, diopside, and sulfides in a high-temperature replacement deposit. At Pargas, Finland, with chondrodite in crystalline limestone. In Italy with mica and vesuvianite in ejected blocks on Mt. Somma, and with sillimanite, andalusite, and cordierite in andesite on the Lipari Islands. At Nikolae-Maximilianovskaya, southern Ural Mts., Russia(!), in superb large crystals; also in Aldan, Yakutia, Siberia(!), crystals to 30 cm. Near Betroka, Madagascar(!), large, sharp, complex gray-green crystals, some the size of a human head, with phlogopite and diopside. **Gemology:** Only Mg–Al spinel is of significance as a gemstone. It occurs in a wide range of colors, red, pink, orange, blue, purple, and violet being the most common. Although some fine blue stones are known, especially from Myanmar, it is mostly the red stones that are seen in the jewelry trade; the similarity of these stones to ruby has given rise to names such as *balas ruby* and *ruby spinel*. To add to the confusion, fine red spinels accompany ruby in the gem gravels of Myanmar and elsewhere. Spinel may also exhibit color change, appearing grayish blue in daylight and violet to purple in incandescent light.

The distinction of spinel from ruby is relatively easy, based mostly on the lower refractive index and specific gravity of spinel and its being optically isotropic. Both stones, of red color, are strongly fluorescent in LWUV. The red color of both stones is due to a trace of Cr.

Spinel is produced, along with ruby, sapphire, peridot, and other gems, from the gem gravels of the Mogok region in Myanmar; from the gem gravels of Thailand; and especially from those of Sri Lanka, where it is accompanied by sapphire of many colors, chrysoberyl (including cat's eye and alexandrite), and several rarer gemstones. It is also found in association with corundum in the alluvial deposits of the Umba River, Tanzania, and fine pink stones come from the Pamir Mountains of Tadzhikistan.

Synthetic spinel has been made by the Verneuil process as boules of various colors. In the blue color, this material is often fluorescent, weak to strong red in LWUV and chalky blue in SWUV, whereas natural blue spinel is not fluorescent. RG *DI:689, GSAM 85:191(1962), SPC 13:599(1968)*.

7.2.1.2 Galaxite $MnAl_2O_4$

Named in 1932 from the plant, *galax*, abundant in Alleghany Co., NC, where the mineral was found, and the nearby town of Galax, VA. Spinel group. ISO $Fd3m$. $a = 8.258$, $Z = 8$, $D = 4.077$. *NBSC 539(9):35(1960)*. 29-880(syn): 2.90_7 2.47_{10} 2.05_3 1.58_5 1.45_4 1.07_1 0.837_2 0.793_2. Massive or small grains. Black or mahogany red, red-brown streak, vitreous luster. $H = 7\frac{1}{2}$. $G = 4.234$. Isotropic; $N = 1.923$. From the *Bald Knob mine, Alleghany Co., NC*, associated with jacobsite, alleghanyite, and calcite. RG *DI:689, ZAAC 415:69(1975), AM 68:449(1983)*.

7.2.1.3 Hercynite $FeAl_2O_4$

Named in 1847 for the Latin name for the Bohemian forest *Silva Hercynia*, where the mineral was first found. Spinel group. ISO $Fd3m$. $a = 8.153$, $Z = 8$, $D = 4.260$. *NBSM 25(19):48(1982)*. 34-192(syn): 2.88_6 2.46_{10} 2.04_2 1.66_2 1.57_4 1.44_4 1.06_1 0.832_1. Massive; as rounded black grains. Dark gray-green to leek-green streak, vitreous luster. $H = 7\frac{1}{2}$–8. $G = 3.93$. Isotropic; $N = 1.800 \pm 0.005$. Near Whittles, Pittsylvania Co., VA, with hogbomite in emery. About 5 km NE of Colton, NY, in a pelitic Grenville paragneiss, with sillimanite, ilmenite, and rutile. Sweetwater Wash in the Old Woman Mts., 60 km SW of Needles, CA, with ilmenite, anorthite, staurolite, and corundum. Near Cuauhtemoc, CHIH, Mexico. Načetín, near Pobezovice, Bohemia, Czech Republic. Le Prese, Veltlin, Switzerland, in gabbro with corundum and sillimanite. Near Erode, Coimbatore, Madras, India. In cassiterite placers near Moorina, TAS, Australia. RG *DI:689, AM 64:736(1979)*.

7.2.1.4 Gahnite $ZnAl_2O_4$

Named in 1807 for Johann Gottlieb Gahn (1745–1818), Swedish chemist and mineralogist, discoverer of manganese. Spinel group. ISO $Fd3m$. $a = 8.085$, $Z = 8$, $D = 4.607$. *NBSC 539(2):38(1953)*. Structure: *ZK 120:476(1974)*. 5-669(syn): 2.86_8 2.44_{10} 1.86_1 1.65_2 1.56_4 1.43_4 1.05_1 0.825_1. As octahedral crystals; also irregular grains, massive. Dark green; yellow; brown, gray streak, vitreous luster. Conchoidal fracture; brittle. H = $7\frac{1}{2}$–8. G = 4.57. Isotropic; N = 1.775–1.805. Found in crystalline schists; in granite pegmatites, especially those rich in Li; in contact metamorphosed limestones; in high-temperature replacement ore deposits in schists or marbles or quartzose rocks. From Franklin and Sterling Hill, NJ, in the crystalline limestone wallrock adjacent to the orebody and occasionally in the ore itself, in crystals that may reach 12 cm on edge. In CO as an abundant gangue mineral in the Sedalia copper mine, Salida, and well crystallized at Cotopaxi, Chaffee Co. In granite pegmatites at Topsham, ME; Haddam, CT; Mitchell Co., NC. *Nafversberg and Eric Matts mines, Falun, Sweden*. Träskböle, Perniö parish, Finland, in pegmatite. Greenbushes, WA, Australia, in a large Li–Ta pegmatite. In Brazil in Li–Ta pegmatites at Alto do Giz, Equador, and at the Ermo mine, Carnauba dos Dantas, RGN, and in MG at several Li pegmatites, sometimes in gem-clear nodules. Ambatofotsikely, Madagascar, in pegmatite. RG *DI:689*.

7.2.2.1 Magnesioferrite $MgFe_2^{3+}O_4$

Named in 1859 for the composition. Spinel group. ISO $Fd3m$. $a = 8.383$, $Z = 8$, $D = 4.51$. *GSAM 85:192(1962)*. 17-464(syn): 2.96_4 2.53_{10} 2.09_3 1.71_1 1.61_3 1.48_4 1.09_1 0.855_1. Rarely crystallized as octahedrons. Usually granular, massive. Black, black streak, with metallic to submetallic luster. Octahedral parting; brittle. H = 6–$6\frac{1}{2}$. G = 4.6. Strongly magnetic. Isotropic; N = 2.38. Found in metamorphosed dolomites, and in fumaroles, where it appears to have formed by the reaction at high temperature of steam and ferric chloride with magnesian material; also in some high-grade metamorphic environments. Sterling Hill, NJ; Magnet Cove, AR. Långban, Värmland, Sweden, occasionally with associated bromellite. From the fumaroles of 1855, Fosse de Cancherone, Mt. Somma, Vesuvius, Italy; also on Mts. Etna and Stromboli, and on the Lipari Is. In France on Puy de la Tache, Mt. Doré, Puy de Dôme. RG *DI:701, AMG 5:1(1969)*.

7.2.2.2 Jacobsite $MnFe_2^{3+}O_4$

Named in 1869 for the original locality in Jakobsberg, Sweden. Spinel group. Polymorph: iwakiite (7.2.6.2). ISO $Fd3m$. $a = 8.499$, $Z = 8$, $D = 4.989$. *NBSC 539(9):36(1960)*. 10-319(syn): 4.91_2 3.01_3 2.56_{10} 2.12_2 1.73_2 1.64_3 1.50_4 1.11_3. Usually massive. Black, brownish-black streak, metallic–submetallic luster. No cleavage; brittle. H = $5\frac{1}{2}$–6. G = 4.76. Highly magnetic. Isotropic; N = ±2.3. In RL, gray, isotropic. R = 18.3 (470 nm), 17.7 (546 nm), 17.3 (589 nm), 16.7 (650 nm). *P&J:215*. From the Bald Knob mine, Alleghany Co., NC, with galaxite, alleghanyite, and calcite. *Jakobsberg mine, Värmland, Sweden*, in crystalline limestone with native copper, and at Långban with tephroite and calcite. Debarstica, near Tatar-Pazardzik, Bulgaria. Manganese Hill, 40 km N of Eilat, southern Negev, Israel. RG *DI:698, AM 45:734(1960)*.

7.2.2.3 Magnetite $Fe^{2+}Fe_2^{3+}O_4$

Named in 1845, said to be derived from Magnes, a Greek shepherd, who first discovered the mineral on Mt. Ida by noting that the nails of his shoes and the iron ferrule of his staff clung to a rock. Spinel group. ISO $Fd3m$. $a = 8.396$, $Z = 8$, $D = 5.197$. *NBSM 25(5):31(1967)*. Structure: *AC(C) 40:1491(1984)*. *19-629*(syn): 2.97_3 2.53_{10} 2.10_2 1.71_1 1.62_3 1.48_4 1.28_1 1.09_1. **Habit:** Usually octahedral, sometimes dodecahedral; sometimes complex and highly modified, but {001} form rare. Also granular; massive; as black sands. Twinning on {111} common (spinel twins); sometimes polysynthetic twinning lamellae producing striations on an octahedral face. **Physical properties:** Iron black, black streak, metallic to submetallic luster. Parting octahedral, often highly developed; brittle. $H = 5\frac{1}{2}$–6. $G = 5.175$. Strongly magnetic; some varieties possessing polarity ("lodestone"). **Tests:** Soluble in HCl; the solution reacts for both ferrous and ferric iron. When heated to 550°, loses its magnetism. **Chemistry:** Ti may be present up to about 6%. Fe^{2+} sometimes partly replaced by Mg and more rarely Ni and/or Mn. Fe^{3+} may be replaced by Al (up to 1.5%), Cr, and V. **Optics:** Isotropic; $N = 2.42$ (Na). In RL, brownish gray. $VHN_{100} = 737$. $R = 20.0$ (470 nm), 19.9 (546 nm), 20.0 (589 nm), 19.95 (650 nm). *C&S:218*. **Occurrence and Paragenesis:** Magnetite is one of the most abundant and widespread of oxides. It is found in diverse geological environments and in some deposits in sufficient abundance to constitute an important iron ore. Following are the principal types of occurrence: (1) as a magmatic segregation deposit, with apatite and pyroxenes; (2) as an accessory mineral in acid igneous rocks, widespread but in minor amounts; (3) in calcic igneous rocks such as gabbros and anorthosites, often associated with ilmenite; (4) in many carbonatites, often associated with apatite; (5) in metasomatic rocks associated with limestones, with garnet, diopside, epidote, hematite, pyrite, and other sulfides; (6) in chlorite schists with pyrite; (7) as an accessory in granite pegmatites; (8) in sulfide vein deposits of relatively high temperature; (9) as a product of fumarolic activity; (10) as a detrital mineral in marine or fluvatile deposits, often in large amounts; (11) rarely, as an oxidation product in the zone of weathering. **Alteration:** Magnetite is often pseudomorphously altered to hematite ("martite") or to goethite. **Localities:** Major orebodies of magnetite are found in NY, especially in Warren, Clinton, and Essex Cos. The Tilly Foster mine at Brewster, Putnam Co., NY(!), produced fine crystals, often associated with chondrodite. At the Z.C.A. No. 4 mine, Balmat, St. Lawrence Co.(!), fine bright cubes of magnetite are found with anhydrite, halite, colorless sphalerite, and talc. In PA a major orebody at Cornwall, Lebanon Co.(!), produced fine dodecahedrons, and the French Creek mine, Chester Co.(!), produced brilliant octahedral crystals. In Chester, VT(*), in chlorite schist, well crystallized with pyrite crystals. Magnet Cove, near Hot Springs, AR, both in crystals and as lodestone. In Millard Co., UT(*), well crystallized and as martite. In Mexico at Cerro del Mercado, DGO.(!), in part as martite pseudomorphs. What may be the largest known magnetite orebodies are mined in Sweden at Kiruna and Gellivare. These are magmatic segregation deposits. Also found at Persberg and Nordmark, Värmland, the latter mine furnishing splendid dodecahedral crystals(!). At Alp Lercheltini, Binntal, Valais, Switzerland(!), in fine lustrous crystals in Alpine clefts. At Traversella, Piedmont, Italy, in a contact of limestones and diorite. In simple and twinned octahedra in the Zillertal, Tyrol, Austria(*). Contact metamorphic deposits have long been mined in the Banat district, Romania; at Moravita(!) large dodecahedral crystals have been found. Magnitogorsk, Ural Mts., Russia, in a large deposit; at Kashnakar, near Zlatoust, southern Ural Mts.(!), in sharp 2.5-cm octahedra modified by dodecahedra; in cubic crystals at Tamvatnei, northeast Siberia(!). Large deposits are found in South Africa, as magmatic segregations in the Bushveld complex and elsewhere. In Brazil at Brumado, BA(!), in fine octahedra in magnesite; and also well crystallized in the Jacupiranga carbonatite, SP(!). The most powerful natural magnets (lodestone) are found in Siberia, Russia; in the Harz Mts., Germany; and on Mt. Elba, Italy. **Uses:** An important ore of iron. Also, when finely ground and mixed with water, used to make heavy media used in mineral separation processes. RG *DI:698*.

7.2.2.4 Franklinite $ZnFe_2O_4$

Named in 1819 for the locality, and for Benjamin Franklin (1706–1790) after whom the town was named. Spinel group. ISO $Fd3m$. $a = 8.441$, $Z = 8$, $D = 5.324$. *NBSM 25(9):60(1971) 22-1012*(syn): 2.98_4 2.54_{10} 2.11_2 1.72_1 1.62_3 1.49_4 1.29_1 1.10_1. Crystallizes as octahedrons, rarely modified by dodecahedron; crystals usually have rounded edges; also granular, massive. Black, reddish brown to black streak, metallic to submetallic luster. Conchoidal to uneven fracture; parting on $\{111\}$; brittle. $H = 5\frac{1}{2}-6\frac{1}{2}$. $G = 5.2$. Slightly magnetic. Infusible. Soluble in HCl, sometimes with evolution of a small amount of Cl_2. No franklinites with only Zn and Fe^{3+} have been found. Normally, much of the Zn is replaced by Mn^{2+}, and a portion of the Fe^{3+} is replaced by Mn^{3+}. Some of the Zn may also be replaced by Fe^{2+}. Isotropic, $N = 2.36$. In RL, gray-white, isotropic. $R = 19.2$ (470 nm), 18.5 (546 nm), 17.6 (589 nm), 16.7 (650 nm). *P&J:167*. Once abundant at the *Franklin and Sterling Hill mines, Franklin and Ogdensburg, NJ*, where it constituted an important ore of zinc, accompanied by calcite, willemite, zincite, and about 330 other minerals. RG *DI:698; GSAM 85:193(1962); AM 50:1672(1965), 64:599(1979); LIT 5:69(1972)*.

7.2.2.5 Trevorite $NiFe_2^{3+}O_4$

Named in 1921 for Tudor Gruffydd Trevor (1865–1958), Welsh–South African geologist, mining inspector for Pretoria district, Transvaal. Spinel group. ISO $Fd3m$. $a = 8.339$, $Z = 8$, $D = 5.368$. *NBSC 539(10):44(1960). 44-1485*(syn): 4.81_1 2.95_3 2.51_{10} 2.41_1 2.09_2 1.702_1 1.605_3 1.474_3. Usually massive; rarely as octahedral crystals. Black, dusky brown streak, submetallic luster. $H = 6-6\frac{1}{2}$. $G = 5.33$. Strongly magnetic. The composition varies from nearly pure $NiFe_2O_4$ to $\sim(Ni_6Fe_4)Fe_2O_4$, with traces of Co and Mg also substituting in the Ni position. Isotropic; $N = \sim 2.3$. $VHN_{50} = 937 \pm 10$ kg/mm^2. $R = 24.1$ (546 nm), 24.0 (589 nm). From 3 km W of the Scotia talc mine, *Bon Accord, Barberton, Transvaal, South Africa*, associated with nickel–olivine, nickel–serpentine, and nickel–chlorite. RG *DI:698; AM 54:1204(1969), 57:1524(1972)*.

7.2.2.6 Cuprospinel $CuFe_2^{3+}O_4$

Named in 1973 for the composition and its identity as a member of the spinel group. Spinel group. ISO $Fd3m$. $a = 8.369$, $Z = 8$, $D = 5.251$. *25-283:* 4.79_3 2.96_5 2.52_{10} 2.10_3 1.61_4 1.48_6 1.27_2 1.09_3. Irregular black grains, black streak. $H = 6\frac{1}{2}$. $G = 5.40$. In RL, gray, similar to magnetite, but without the pinkish tint. Isotropic. $VHN_{100} = 985$ kg/mm^2. $R = 22.7$ (470 nm), 21.7 (546 nm), 21.0 (589 nm), 20.0 (650 nm). From the *Consolidated Rambler mine, near Baie Verte, NFD, Canada*, as a product of a burning ore dump, intergrown with hematite and other spinels. RG *CM 11:1003(1973), AM 59:381(1974)*.

7.2.2.7 Brunogeierite $Ge^{2+}Fe_2^{3+}O_4$

Named in 1972 for Bruno H. Geier (b.1902), chief mineralogist, Tsumeb Corp. Spinel group. ISO $Fd3m$. $a = 8.409$, $Z = 8$, $D = 5.51$. *NJMM:263(1972)*. *25-359:* 2.97_5 2.53_{10} 2.10_2 1.716_3 1.618_4 1.486_8 1.095_3 0.858_3. As thin gray crusts with metallic luster. $H = \sim 5-5\frac{1}{2}$. Ferromagnetic. RL isotropic, gray, reflectance lower than that of magnetite. From *Tsumeb, Namibia*, as crusts on tennantite which encloses renierite, and surrounded by stottite. RG *AM 58:348(1973)*.

7.2.3.1 Magnesiochromite $MgCr_2O_4$

Named in 1868 for the composition and its identity as a member of the spinel group. Spinel group. Synonym: chrompicotite. ISO $Fd3m$. $a = 8.333$, $Z = 8$, $D = 4.414$. *NBSC 539(9):34(1960)*. *10-351*(syn): 4.81_6 2.95_1 2.51_{10} 2.41_1 2.08_5 1.60_4 1.47_5 1.41_1. Massive, granular. Black, brown streak, metallic luster. No cleavage, uneven fracture, brittle. $H = 5\frac{1}{2}$. $G = 4.2$. Sometimes feebly magnetic. Some Fe^{2+} is invariably substituted for Mg, and a complete series exists between $MgCr_2O_4$ and $FeCr_2O_4$. In RL, gray. From *Scottie Creek, Lillooet district, BC, Canada*. In the *Schwarzenberg, near Tampadel, Silesia, Germany*, in serpentine. Ferdinandovo, Bulgaria. Mt. Djeti, Togo (ferrian). Dun Mt., New Zealand, in dunite. RG *DI:709*.

7.2.3.2 Manganochromite $(Mn, Fe^{2+})(Cr, V)_2O_4$

Named in 1978 for its composition and identity as a member of the chromite sub-group. Spinel group. ISO $Fd3m$. $a = 8.47$, $Z = 8$, $D = 4.88$. *31-630:* 4.89_3 2.99_3 2.55_{10} 2.44_3 2.12_1 1.73_3 1.63_{10} 1.50_{10}. As euhedral to irregular embedded grains. Black. $H = 6\frac{1}{2}$. V_2O_3 substituting for Cr_2O_3 is present up to 27.6%. Similarly, FeO substitutes for MnO up to 10%. A small amount of ZnO also substitutes for MnO. In RL, isotropic, brownish gray. $R = 16.8$ (470 nm), 16.5 (546 nm), 16.5 (589 nm), 16.3 (650 nm). From the *Nairne deposit, Bru-*

kunga, E of Adelaide, SA, Australia, associated with pyrrhotite and diopside, in pyritic beds of the Kanmantoo metasediments. RG *AM 63:1166(1978)*.

7.2.3.3 Chromite $FeCr_2O_4$

Named in 1845 from the composition. Spinel group. Donathite, a supposed tetragonal polymorph of chromite, has been discredited. *AM 77:1120(1992)*. ISO $Fd3m$. $a = 8.379$, $Z = 8$, $D = 5.055$. *NBSM 25(19):50(1992)*. *GSAM 85:195(1962)*(nat): 4.80_4 2.95_2 2.51_{10} 2.08_5 1.60_6 1.47_8 1.08_9 1.04_4. *34-140*(syn): 4.84_1 2.96_3 2.53_{10} 2.09_2 1.61_4 1.48_5 1.09_1 0.855_1. Crystals usually octahedral, extremely uncommon; usually massive or granular. Black, brown streak, metallic luster. Uneven fracture; brittle. $H = 5\frac{1}{2}$. $G = 4.7$. Sometimes feebly magnetic. Isotropic; $N = 2.08$–2.16. In RL, grayish white, with a brownish tint. Isotropic. $R = 12.7$ (470 nm), 12.0 (546 nm), 11.8 (589 nm), 11.6 (650 nm). *P&J:120*. MgO usually replaces part of the FeO and a complete series exists between chromite and magnesiochromite. Al_2O_3 may replace part of the Cr_2O_3, as does Fe_2O_3; these may result in serious reduction of the Cr_2O_3 content in chromite ore. Chromite deposits are invariably associated with ultramafic ferromagnesian rocks, especially serpentines, peridotites, and dunites. In these chromite may be present as magmatic segregations of large size, or more commonly disseminated as an accessory mineral. Weathering and decomposition of these rocks may give rise to alluvial deposits of chromite. Wood mine, near Texas, Lancaster Co., PA. In the Webster–Addie ultramafic ring, Jackson Co., NC. In CA in Shasta and Siskiyou Cos. and in the Great Serpentine Dike in Eldorado and Placer Cos. In the serpentine areas of QUE, Canada. Abundant in good crystals to 10 mm at Haroldswick and Swinaness, Unst Island, Shetlands, Scotland. *Gassin, Var, France*. Ramberget, near Hestmandö, Norway. Nizhne–Tagilsk district, Perm, eastern Ural Mts., Russia, associated with platinum. Important deposits are mined in Turkey, at Antioch, Smyrna, and Brussa. In the Zhob valley igneous complex, western Pakistan. In the serpentine of the Great Dike, Zimbabwe. From the Bushveld Complex, South Africa. Tiebaghi, Thio, New Caledonia, in octahedrons >50 mm. Also in the Philippines, Cuba, and elsewhere. The principal, and only, ore of chromium. RG *DI:709; AM 38:1134(1053), 45:579(1960)*.

7.2.3.4 Nichromite $(Ni, Co)(Cr, Fe^{3+})_2O_4$

Named in 1978 for the composition. Spinel group. ISO $Fd3m$. $a = 8.316$, $Z = 8$, $D = 5.24$. *23-1271*(syn): 4.9_2 2.95_5 2.51_{10} 2.08_2 1.70_2 1.60_3 1.47_5 1.32_1. Minute dark green to black grains, greenish-gray streak, metallic luster. Conchoidal fracture. In RL, isotropic. From the *Bon Accord nickel deposit, Barberton, Transvaal, South Africa*, associated with trevorite, liebenbergite, and bunsenite. RG *BRGMII 3:225(1978), AM 65:811(1980)*.

7.2.3.5 Cochromite (Co,Ni)(Cr,Al)$_2$O$_4$

Named in 1978 for the composition, and its being a member of the chromite subgroup. Spinel group. ISO $Fd3m$. $a = 8.292$, $Z = 8$, $D = 5.01$. *AM 65:811(1980)*. *22-1084*(syn): 4.82_1 2.95_3 2.51_{10} 2.08_2 1.60_3 1.47_4 1.08_1 0.850_1. As minute grains. Black, greenish-gray streak, metallic luster. Conchoidal fracture. In RL, isotropic, gray. $R = 14.3$ (470 nm), 13.7 (546 nm), 14.0 (589 nm), 13.9 (650 nm). $VHN_{50} = 1218$ kg/mm^2. From *Bon Accord*, Barberton district, Transvaal, South Africa, associated with chromite and trevorite. RG *NBSM 9:21(1971)*.

7.2.3.6 Zincochromite ZnCr$_2$O$_4$

Named in 1987 for the composition. Spinel group. ISO $Fd3m$. $a = 8.352$, $Z = 8$, $D = 5.434$. *ZVMO 116:367(1987)*. *22-1107*: 2.95_5 2.51_{10} 2.08_2 1.700_1 1.603_4 1.472_4 1.270_1 1.084_1. Minute octahedral crystals. Brownish black, brown streak, semimetallic luster. No cleavage. $H = 5\frac{1}{2}$. Weakly paramagnetic. In RL, brownish gray with brown internal reflections, isotropic. $R = 12.2$ (470 nm), 11.8 (546 nm), 11.6 (589 nm), 11.6 (650 nm). From the *Onega trough, southern Karelia, Russia*, with quartz and amorphous Cr–V–Fe oxides and hydroxides. RG *AM 73:932(1988)*.

7.2.4.1 Vuorelainenite (Mn,Fe^{2+})(V,Cr)$_2$O$_4$

Named in 1982 for Yrjo Vuorelainen (1922–1988), Finnish mineralogist with the Outokumpu Co., who investigated and described many new species from Finland. Spinel group. ISO $Fd3m$. $a = 8.48$, $Z = 8$, $D = 4.64$. *35-550*: 4.90_2 2.99_4 2.56_{10} 2.12_3 1.73_2 1.63_8 1.50_8 1.10_2. As black euhedral grains, submetallic luster. In RL, isotropic, brownish gray. $VHN_{50} = 900$ kg/mm^2. $R = 15.7$ (470 nm), 15.4 (546 nm), 15.6 (589 nm), 16.2 (650 nm). From the *Stäta (Doverstorp) pyrite deposit, Bergslagen, central Sweden*, associated with pyrrhotite, rutile, pyrophanite, and other minerals. RG *CM 20:281(1982), AM 68:472(1983)*.

7.2.4.2 Coulsonite Fe^{2+}V$_2$O$_4$

Named in 1937 for Arthur Lennox Coulson (b.1898), geologist, Geological Survey of India. Spinel group. ISO $Fd3m$. $a = 8.297$, $Z = 8$, $D = 5.15$. *AM 47:1284(1962)*. *15-122*: 4.79_3 2.93_6 2.50_{10} 2.07_8 1.69_3 1.60_9 1.47_9 1.27_2. Irregular to subhedral bluish-gray grains, dark brown to black streak, metallic luster. $H = 4\frac{1}{2}$–5. $G = 5.2$. Slightly magnetic. In RL, isotropic, bluish gray. $R = 23.5$ (green). From the Buena Vista Hills, 32 km SE of Lovelock, NV, in metamorphosed magnetite-rich hornblende andesite. *Singhbum, Bihar, India*, associated with titaniferous magnetite ores. RG *AM 48:948(1963)*.

7.2.4.3 Magnesiocoulsonite MgV$_2$O$_4$

Named in 1995 by Reznitsky et al. Spinel group. The magnesium analog of coulsonite. Analogous to chromite–magnesiochromite. Forms a series with magnesiochromite. ISO $Fd3m$. $a = 8.385$, $Z = 8$, $D = 4.31$. P.D. 4.84_9 2.95_3 2.52_{10} 2.09_8 1.612_8 1.482_9 1.092_7 1.048_5. *ZVMO 124:91(1995)*. Grains are equant and have an octahedral habit. Black, black streak, metallic luster; opaque. Uneven fracture; brittle. $H = 6\frac{1}{2}$. Insoluble in HNO_3. In RL, light gray; isotropic; no bireflectance; nonpleochroic. $R = 14.0$ (470 nm), 13.7 (546 nm), 13.7 (589 nm), 13.7 (650 nm). Found in the Slyuyaanka parametamorphic complex, western Pribaikal. Russia, in quartz–diopside rocks, associated with tremolite, calcite, pyrite, kalininite, florensovite, goldmanite, kammererite, and karelianite. RG

7.2.5.1 Qandilite Mg$_2$TiO$_4$

Named in 1985 for the rocks of the Qandil group, where the mineral was found. Spinel group. ISO $Fd3m$. $a = 8.403$, $Z = 8$, $D = 4.04$. *AM 73:930(1988)*. *25-1157*(syn): 4.88_5 2.99_2 2.54_{10} 2.11_6 1.722_1 1.624_3 1.492_5 1.099_1. Small grains with euhedral outline. Structure inverse spinel type, with Ti^{4+} in octahedral sites and Mg in both tetrahedral and octahedral sites in a ratio of 2:1. *AC(B) 45:542(1989)*. Iron black, black streak, metallic luster. Cleavage {111}, perfect; brittle. $H = 7$. $G = 4.03$. Strongly magnetic, similar to magnetite. Soluble in hot 1:1 HCl. Although the theoretical endmember composition is Mg_2TiO_4, natural qandilite is a solid solution mixture of 64% Mg_2TiO_4, 23% $FeFe_2O_4$, and 13% $MgFe_2O_4$. In RL, isotropic, pinkish gray. $VHN_{100} = 998$ kg/mm^2. $R = 13.4$ (470 nm), 13.1 (546 nm), 13.0 (589 nm), 13.0 (650 nm). From the Kangerdlugssuaq region, East Greenland, in a periclase–forsterite marble in contact with an alkalic ultramafic intrusion. *Quandil rocks, Dupezeh Mt., Hero Town, Qala-Dizeh region, Iraq*, in a forsterite skarn, associated with serpentinized olivine, calcite, perovskite, and spinel. RG *MM 45:135(1982)*.

7.2.5.2 Ulvöspinel Fe$_2^{2+}$TiO$_4$

Named in 1943 for the locality and its identity as a member of the group. Spinel group. ISO $Fd3m$. $a = 8.535$, $Z = 8$, $D = 4.777$. *NBSM 25(20):61(1984)*. *34-177*(syn): 4.92_1 3.02_3 2.57_{10} 2.13_2 1.74_1 1.64_3 1.51_4 0.871_1. The structure is that of an inverse spinel, with Ti^{4+} in octahedral sites and Fe^{2+} in both octahedral and tetrahedral sites in a ratio of 1:2. *JAC 11:121(1978)*. As minute grains. Metallic luster. In RL, isotropic, brown. $R = 17.0$ (470 nm), 17.6 (546 nm), 18.4 (589 nm), 19.1 (650 nm). *P&J:378*. From the Aiyansh basalt lava flow, Nass River valley, and the Itcha Mt. hawaiite lava flow, BC, Canada. *Ulvö, Angermanland archipegalo, northern Sweden*. RG *AM 40:138(1955); NJMA 94:993(1960), 120:31(1974); CM 18:339(1980)*.

7.2.6.1 Kusachiite CuBi$_2$O$_4$

Named in 1995 for Isao Kusachi (b.1942), Japanese mineralogist, Okayama University, Fuka, Japan. TET $P4/ncc$. $a = 8.511$, $c = 5.823$, $Z = 4$, $D = 8.64$. *PD*: 4.26_2 3.19_{10} 2.91_2 2.70_2 2.40_1 1.947_2 1.728_1 1.652_1. Prismatic crystals to 0.5 mm and as platy grains in globular aggregates. Black, metallic luster.

Cleavage {110}, perfect. H = $4\frac{1}{2}$, VHN_{25} = 292–357. G = 8.5. Readily soluble in HCl. Opaque; in RL, gray with deep red to dark brown internal reflections, weak bireflectance; pleochroic brownish gray to gray. R = 21.1–19.0 (482 nm), 20.2–18.0 (545 nm), 19.7–17.6 (589 nm), 19.5–17.3 (659 nm). From *Fuka, Okayama Pref., Japan*, on calcite crystals in a cavity in a calcite vein at a gehlenite–spurrite skarn/limestone contact, associated with henmilite, sillenite, bakerite, tenorite, bultfonteinite, apophyllite, cuspidine, and thaumasite. RG *MM 59:545(1995)*, *AM 81:517(1996)*.

7.2.6.2 Iwakiite $Mn^{2+}(Fe^{3+},Mn^{3+})_2O_4$

Named in 1979 for the locality. Synonym: α-vredenbergite. Dimorphous with jacobsite. TET $P4_2/nnm$. $a = 8.519$, $c = 8.540$, $Z = 8$, $D = 4.88$. *MJJ 9:383(1979)*. Structure: *ZK 185:605(1988)*. 38-430: 4.93_3 3.02_4 2.57_{10} 2.135_2 2.13_2 1.64_3 1.509_4 1.506_4. Small grains. Greenish black, black streak, metallic luster. No cleavage. H = $6-6\frac{1}{2}$, $VHN_{100} = 726$. G = 4.85. Strongly magnetic. In RL, olive-gray, moderately anisotropic. R = 19.9 (470 nm), 20.8 (546 nm), 20.6 (589 nm), 19.8 (650 nm). From the *Gozaisho mine, Iwaki, Japan*, in manganese ore, associated with rhodonite, quartz, braunite, hematite, rhodochrosite, and spessartine. RG *AM 65:406(1980)*.

7.2.7.1 Hausmannite $Mn^{2+}Mn_2^{3+}O_4$

Named in 1827 for Johann F. L. Hausmann (1782–1859), professor of mineralogy, University of Göttingen. TET $I4_1/amd$. $a = 5.762$, $c = 9.470$, $Z = 4$, $D = 4.834$. SR *54A:141(1987)*. 24-734: 4.92_3 3.09_4 2.77_9 2.49_{10} 2.46_2 1.80_3 1.58_3 1.54_5. Crystals pseudo-octahedral {011}; also stout prismatic; granular, massive. Twin plane {112}, often repeated as fivelings. Brownish black, chestnut-brown streak, submetallic luster. Cleavage {001}, nearly perfect; uneven fracture; brittle. H = $5\frac{1}{2}$. G = 4.84. Slightly paramagnetic. Soluble in hot HCl, with evolution of Cl_2. Small amounts of Zn and Fe may substitute for the Mn. Uniaxial (−); $N_O = 2.46$, $N_E = 2.15$ (Li); nonpleochroic, deep reddish brown. In RL, grayish white, distinctly anisotropic, red internal reflections. R = 19.5–16.5 (470 nm), 18.1–15.5 (546 nm), 17.6–14.9 (589 nm), 17.1–14.4 (650 nm). *P&J:195*. Franklin, NJ, with zinc ore. Batesville district, AR, with pyrolusite and "psilomelane." Prefumo Canyon district, San Luis Obispo Co., CA. Långban, Värmland, Sweden(!), in fine crystals, and at Jacobsberg in granular limestone with garnet, magnetite, hematite, manganophyllite, and jacobsite. *Ilfeld, Harz*, and at Öhrenstock near Ilmenau, Thuringia(!), *Germany*. Lower Averser valley, Graubunden, Switzerland. Aosta valley, Torino, Italy. Miguel Burnier, MG, Brazil(!), in fine crystals. Kuruman manganese district, Cape Prov., South Africa(!), in fine crystals to 4 cm. RG *DI:712*, *AM 50:1305(1965)*.

7.2.7.2 Hetaerolite $ZnMn_2^{3+}O_4$

Named in 1877 from the Greek *hetaros*, companion, in allusion to the association with chalcophanite in some specimens. TET $I4_1/amd$. $a = 5.720$,

$c = 9.245$, $Z = 4$, $D = 5.252$. *NBSM 25(10):61(19??)*. *24-1133:* 3.05_5 2.86_2 2.72_7 2.47_{10} 2.02_2 1.56_3 1.52_4 1.43_2. Crystals usually pseudo-octahedral; also massive. Twin plane {112}, as fivelings. Black, dark brown streak; submetallic shining luster. Cleavage {001}, indistinct; uneven fracture; brittle. H = 6. G = 5.18. In RL, gray, weakly pleochroic; anisotropic, red internal reflections. R = 20.6–16.2 (470 nm), 18.3–14.6 (546 nm), 17.7–14.1 (589 nm), 17.1–13.7 (650 nm). *P&J:200. Franklin* and *Sterling Hill* mines, *Sussex Co., NJ*, associated with calcite, willemite, manganosite, chalcophanite, gahnite, zincite, and other minerals. RG *DI:715, AM 50:1672(1965), MA 74:3413(1974)*.

7.2.7.3 Hydrohetaerolite $Zn_2Mn_4^{3+}O_8 \cdot H_2O$

Named in 1929 because the composition is that of hetaerolite plus water. TET $I4_1/amd$. $a = 5.73$, $c = 9.00$, $Z = 4$, $D = 4.77$. *AM 41:268(1956)*. *9-459:* 3.02_7 2.87_4 2.66_8 2.47_{10} 1.57_5 1.51_7 1.43_4 1.26_3. Massive; as fibrous crusts with a botryoidal surface. Dark brown to brownish black, dark brown streak, submetallic luster. Cleavage parallel to elongation of fibers. H = 5–6. G = 4.64. Soluble in HCl, giving off Cl_2. Uniaxial (−); $N_O = 2.26$, $N_E = 2.10$. In RL, creamy gray. From the *Wolftone mine, Leadville, Lake Co., CO*, associated with chalcophanite, smithsonite, and hemimorphite in oxidized ore. Sterling Hill mine, Ogdensburg, NJ. RG *DI:717, AM 27:48(1942)*.

7.2.8.1 Minium $Pb^{4+}Pb_2^{2+}O_4$

Named in antiquity from the Latin *minium*, derived from the Iberian word for cinnabar, the Romans getting their cinnabar from Spain. Although originally applied to cinnabar, the name is now used for red lead oxide. TET $P4_2/mbc$. $a = 8.815$, $c = 6.565$, $Z = 4$, $D = 8.924$. *NBSC 539(8):32(1959)*. Structure: *SR 41A:212(1975)*. *41-1493*(syn): 6.23_2 3.38_{10} 3.12_2 2.90_5 2.79_4 2.63_3 1.91_2 1.76_3. Massive, earthy, or pulverulent; as microscopic scales. Red to brownish red, orange-yellow streak, greasy luster. H = 2.5. G = 9.05. Soluble in excess 1:1 HCl, with evolution of Cl_2. Decomposed by HNO_3, with brown residue of PbO_2. On heating, color darkens and becomes black, but original color is restored on cooling. Uniaxial; N ∼ 2.42. Pleochroism: O, deep reddish brown; E, nearly colorless; negative elongation. Abnormal green interference colors. From Austin's mine, Wythe Co., VA. Jay Gould mine, Alturas Co., ID, with native lead and galena. In oxidized ore in several of the mines at Leadville, CO, with galena, cerussite, and iron oxides. Godiva mine, Tintic, UT. In Mexico at the Santa Fé mine, Bolaños, DGO, with massicot and cerussite, and at Zimapan, HGO. Leadhills, Scotland, and at Weardale, Yorkshire, England. Långban, Värmland, Sweden, with native lead. Badenweiler, Baden, and Bleiberg, Baden-Württemberg, Germany. Mine des Molérats, St.-Prix, Saône-et-Loire, France. Sarrabus, Sardinia, Italy. Zmeyewskaja-Gora, Altai Mts., Kazakhstan. Broken Hill, NSW, Australia, as the product of a mine fire. RG *DI:517, NJMA 94:1187(1960)*.

7.2.9.1 Chrysoberyl $BeAl_2O_4$

Named in 1790 from the Greek for *golden* and *beryl* (a misnomer), in allusion to the common color. ORTH *Pnma*. $a = 9.404$, $b = 5.476$, $c = 4.427$, $Z = 4$, $D = 3.699$. *NBSC 539(9):10(1960)*. **Structure**: Chrysoberyl is isostructural with olivine, $(Mg, Fe)Si_2O_4$, and is the hexagonal close-packed analog of the cubic close-packed spinel structure. The oxygens form a distorted hexagonal close-packed array in which one-eighth of the tetrahedral interstices are occupied by beryllium in tetrahedral coordination, and one-half of the octahedral sites are filled by aluminum in octahedral coordination. The beryllium–oxygen tetrahedra point alternately either way along both the X and Y directions. The Al atoms occupy two sets of nonequivalent lattice positions; half are located at centers of symmetry and half on reflection planes, the former having as nearest neighbors two oxygens from two adjacent tetrahedra, and the latter two oxygens from one adjacent tetrahedron. *AM 48:804(1963)*. *45-1445*: 3.23_4 2.56_7 2.26_4 2.09_8 1.618_{10} 1.613_9 1.465_4 1.361_6. **Habit**: Crystals usually tabular {001}; sometimes stout prismatic [100]. Striated on {001} parallel to [100]. Twinning common, with twin plane {130}. Twinned crystals are generally flattened perpendicular to the composition plane. Simple, somewhat heart-shaped flattened twins are perhaps the most common form of occurrence of two individuals. Also common are contact and penetration twins, often repeated and forming pseudohexagonal crystals as viewed along [001], with or without reentrant angles. Such crystals are called *trillings*. **Physical properties**: Green, yellow, greenish brown; sometimes green or blue in daylight and red or brownish pink in incandescent light (alexandrite); vitreous luster. Cleavage {110}, distinct; {010}, imperfect; uneven to conchoidal fracture; brittle. $H = 8\frac{1}{2}$. $G = 3.75$. Nonfluorescent, except for alexandrite, which is fluorescent weak to moderate red in both LWUV and SWUV. Nonmagnetic. Insoluble in acids. **Chemistry**: Fe^{3+} may substitute for Al up to about 6% Fe_2O_3. **Optics**: Biaxial (+); $XYZ = cba$; $N_x = 1.746$, $N_y = 1.748$, $N_z = 1.756$; $2V = 70°$; $r > v$. X, pale pink; Y, pale yellow; Z, pale green. **Occurrence**: Normally found in pegmatites; rarely in some fluorine-rich veins. Alexandrite is usually found in mica schist. **Localities**: At many localities in ME, especially Stoneham, Greenwood, Topsham(*), Haddam, CT(*), with columbite, tourmaline, gahnite, beryl. Drew Hill, Golden Gate Canyon, Jefferson Co., CO(!), in large crystals and twins to 10 cm. Saetersdalen, Norway. Mourne Mts., Northern Ireland. In dolomite with corundum at Campolungo, near St. Gothard, Switzerland. At the Malysheva mine, middle Ural Mts., Russia, in fine nongem alexandrite twins. Hyderabad region, India(*), twins to 5 cm. In schist on the Novello claims, Victoria, Zimbabwe(!) (nongem alexandrite), as pseudohexagonal trillings. In fine flattened crystals and trillings N of Lake Alaotra, Madagascar(!). At numerous places in Brazil, especially in ES in superb trillings to 6 cm at Itaguaçu(!), and at Collatina(!) in magnificent V-twins and single crystals to over 40 cm; in BA fine trillings of nongem alexandrite in schist at Carnaiba(*), associated with emerald; at Itamaraju(!) in large sharp crystals and twins; Municipo de Prado(!), fine trillings; from Jaqueto, a 24-kg or larger V-twin; Medeiros Neto(!) 5-cm crystals; Teixeira de Freitas, gemmy flat twins and trillings to 5 cm. In MG at Minas Novas in fine small single crystals and twins, and at Americana, Mun. do Teófilo Otoni. **Gemology:** Chrysoberyl occurs in three distinct gem varieties, two of which, if of fine quality, are among the most expensive, rare, and sought-after gems:

1. Common gem chrysoberyl occurs in shades of light yellow, green, and brown. It is colored by ferric iron. Density is 3.71–3.73; mean index of refraction 1.75, with birefringence 0.009; these values also hold for the other gem varieties of chrysoberyl. The principal sources of gem material are Sri Lanka and Brazil.

2. *Alexandrite*, which at the original locality in Russia is light grass green in daylight and raspberry-red in artificial light. It was named after the czarevitch, later Czar Alexander II of Russia, because it was supposedly first found on the day set apart for celebrating his majority, in 1833. Most alexandrite from Brazil is greenish blue in daylight and deep red in artificial light; some rare stones are deep red in any kind of light and show no color change. From Sri Lanka, alexandrite is slightly yellowish green in daylight and pinkish brown in artificial light. Still other color changes are known, and, in effect, any chrysoberyl that shows any change whatever is often locally called alexandrite, but few of them make attractive gems. The ones that do show a change also vary greatly in the ease with which the change can be demonstrated, depending on the stone and the quality of the light sources used. Larger stones (above 10 ct) that have good quality of change tend to be very dark. The really good, attractive alexandrites are extremely rare and command a high price. Only the Russian stones and a very few from Brazil fit in this category. The colors of alexandrite are caused by a trace of Cr replacing the Al plus variable trace amounts of Fe^{3+}. The most desirable colors have the lowest Fe content. The principal sources are in mica schist with phenacite and emerald near the Takowaja River, 90 km E of Ekaterinburg, Ural Mts., Russia(!); in the gem gravels of Sri Lanka; and near Hematita, MG, Brazil(!).

3. *Cat's eye* is a chrysoberyl that is milky or opalescent in which myriad minute acicular inclusions parallel to the c axis of the crystal, in a properly oriented, cabochon-cut stone, reflect a band of light at right angles to the direction of the inclusions, reminiscent of the eye of a cat when a flashlight is directed at it. Good cat's eyes are usually honey yellow or greenish yellow, although alexandrite cat's eyes are known. Fine cat's eyes may reach up to 100 ct and command high prices, comparable to fine sapphires. The best cat's eyes come from Sri Lanka(!); from near Trivandrum, Kerala, India(!); and from Brazil at Padre Paraiso(!) or near Minas Novas, MG.

Synthetics: Synthetic alexandrite has been made by both the Czochralski and the flux-grown methods; their detection requires microscopic examination of minute inclusions usually present. They are also more strongly fluorescent than natural alexandrite. Synthetic corundum with a purplish-blue daylight color and fine red artificial light color has long been available as an alexandrite "look-alike"; it can easily be identified by the difference in refractive index from true alexandrite. RG $H\&K(1991)$, AM $24{:}267(1939)$, $DI{:}718$, $NJMA$ $134{:}117(1979)$, PCM $4{:}426(1987)$.

7.2.10.1 Marokite $CaMn_2^{3+}O_4$

Named in 1963 for the country of origin. ORTH $Pmab$ or $P2_1ab$. $a = 9.71$, $b = 10.03$, $c = 3.162$, $Z = 4$, $D = 4.63$. BM $86{:}359(1963)$. Structure: BM $89{:}318(1966)$. 16-709: 2.87_6 2.71_{10} 2.56_6 2.29_8 2.22_{10} 2.07_8 1.63_6 1.58_6. Platy crystals flattened {010}; no twinning. Black, reddish-brown streak, bright shiny luster. Cleavage {100}, perfect; {001}, good. $G = 4.64$. Biaxial $(-)$; $X = c$; $N_x = 2.07$, $N_y \sim 2.42$, $N_z \sim 2.42$; $2V = 20$–$25°$. X, safflower red; Y,Z, opaque. Plane of optic axes is {100}. In RL, brownish gray with red internal reflections; strongly anisotropic; pleochroic, yellowish gray to gray-brown. $R = 19.7$–17.4 (470 nm), 18.3–16.0 (546 nm), 18.7–16.0 (589 nm), 17.9–15.5 (650 nm). $P\&J{:}258$. From *Tachgagalt, Ouarzazate, Morocco*, associated with calcite, barite, hausmannite, braunite, crednerite, and polianite. Black Rock mine, northwestern Cape Prov., South Africa. RG AM $49{:}817(1964)$, $53{:}495(1968)$.

7.2.11.1 Taaffeite $BeMg_3Al_8O_{16}$

Named in 1951 for Edward Charles Richard Taaffe (1898–1967), Bohemian–Irish gemologist, of Dublin, who discovered the species in a cut stone. HEX $P6_3mc$. $a = 5.687$, $c = 18.34$, $Z = 2$, $D = 3.579$. $NJMM{:}$ $393(1983)$. 35-701: 4.58_5 4.34_4 2.60_6 2.42_{10} 2.38_4 2.04_6 1.47_5 1.42_5. Hexagonal, stout prismatic crystals. Pale mauve; green, vitreous luster. $H = 8$. $G = 3.613$. Fluoresces green under UV radiation. Uniaxial $(-)$; $N_O = 1.723$, $N_E = 1.718$ (Na). From the gem gravels of *Sri Lanka*. Nanling district, Hsianghualing range, Hunan, China, with chrysoberyl and phenakite. Mt. Painter, SA, Australia, with spinel and högbomite, in high Mg–Al-bearing schists. Used to a limited extent as a gem. RG AM $37{:}360(1952)$, $69{:}215(1984)$; MM $43{:}575(1980)$.

7.2.12.1 Musgravite $(Mg,Fe,Zn)_2BeAl_6O_{12}$

Named in 1967 for the locality. HEX-R $R\bar{3}m$. $a = 5.675$, $c = 41.096$, $Z = 6$, $D = 3.69$. $NJMA$ $146{:}15(1983)$. 34-191: 4.57_4 4.21_3 2.66_6 2.41_{10} 2.27_4 2.05_7 1.487_4 1.419_8. Platy masses, flattened {001}. Blackish green, vitreous luster. $H = 8$. $G = 3.68$. From a pegmatite near zircon point, 2 km S of McIntyre Island, Enderby Land, Antarctica, with sapphirine, surinamite, sillimanite, chrysoberyl, garnet, and rutile. *NNE of Ernabella Mission, Musgrave Range, SA, Australia*. RG MM $36{:}305(1967)$, AM $66{:}1028(1981)$.

7.2.12.2 Pehrmanite $Be(Fe,Zn,Mg)_2Al_6O_{12}$

Named in 1981 for Gunnar Pehrman (1895–1980), professor emeritus, University of Åbo, Turku, Finland. HEX-R $R32$ or $R\bar{3}m$. $a = 5.70$, $c = 41.16$, $Z = 6$, $D = 4.07$. 35-503: 3.15_3 2.86_8 2.67_4 2.53_3 2.42_{10} 2.06_4 1.492_3 1.426_5. Subhedral hexagonal tabular crystals, and as epitaxial overgrowths on nigerite. Light gray-green, vitreous luster. No cleavage, very brittle. H = $8-8\frac{1}{2}$, $VHN_{100} = 1700$. Uniaxial $(-)$; $N_O = 1.79$; parallel extinction; positive elongation. O, pale greenish; E, pale grayish brown. R = 8.3 (470 nm), 8.1 (546 nm), 8.0 (589 nm), 7.9 (650 nm). From the wall zone of the *Rosendal pegmatite, Kemiö Island, southwestern Finland*, embedded in albite, with quartz, sillimanite, nigerite, muscovite, spessartine, chlorite, biotite, epidote, allanite, and calcite. RG *CM 19:311(1981), AM 67:859(1982), NJMA 146:15(1983)*.

7.2.13.1 Filipstadite $(Mn, Mg)_2(Sb^{5+}_{0.5}Fe^{3+}_{0.5})O_4$

Named in 1988 for the town of Filipstad, near which Långban is located. ORTH, pseudo-ISO Space group unknown. $a = 8.640$ (cubic subcell), Z = 8, D = 4.9. *AM 73:413(1988)*. PD: 4.97_2 3.05_3 2.56_{10} 2.16_4 1.66_6 1.53_7 1.32_2 1.13_3. Euhedral equant microcrystals. Black, brown streak, metallic luster. No cleavage, conchoidal fracture, brittle. $VHN_{100} = 831$. In RL, gray, with rare, red internal reflections, weakly anisotropic in shades of dark brownish gray. From *Långban, near Filipstad, Värmland, Sweden*, intergrown with ingersonite and jacobsite. RG

7.2.14.1 Yafsoanite $Ca_3Te_2Zn_3O_{12}$

Named in 1982 for the Russian acronym for Yakut Filial, Siberian Branch, Academy of Sciences, USSR ("Yafsoan" in Russian). ISO $Ia3d$. $a = 12.632$, Z = 8, D = 5.54. *ZVMO 111:118(1982), MP 40:111(1989)*. 35-661: 4.47_3 3.16_8 2.82_9 2.58_{10} 1.751_3 1.683_{10} 1.577_3 1.024_5. Yafsoanite, rather than being a tellurite or tellurate, has been shown structurally to be a garnet-type oxide with atoms Ca, Te^{6+}, and Zn coordinated to eight, six, and four oxygen atoms, respectively. This accounts for its hardness, nearly 6, far harder than any of the known tellurites or tellurates. Crystals show dodecahedron, cube, and octahedron, also as radiating concentric material. Light to dark brown, also white; vitreous luster. No cleavage. Soluble in acids, unreacted upon by KOH or $FeCl_2$ solution. Isotropic; N = 1.800. Found at the Empire mine, Tombstone, AZ, in colorless dodecahedrons on dugganite; originally from the *Kuranakh gold deposit, central Aldan, Yakutia, Siberia, Russia*, associated with dugganite and desclozite. RG *AM 68:282(1983), 75:937(1990)*.

7.3.1.1 Welinite $Mn_6(W^{6+}, Mg)_2Si_2(O, OH)_{14}$

Named in 1967 for Eric Welin (b.1923), Swedish mineralogist and geochronologist. HEX $P3$. $a = 8.155$, $c = 4.785$, Z = 1, D = 4.41. Structure: *SR 34A:387(1969). AM 53:1064(1968)*. 20-1389: 7.00_5 4.07_6 3.95_3 3.10_8 2.84_4 2.33_8 1.78_{10} 1.54_6. Fissure fillings and embedded crystal sections. Red-brown to reddish black, resinous luster. Cleavage {0001}, poor; brittle. H = 4. G = 4.47. Uniaxial (+); $N_O = 1.864$, $N_E = 1.88$. *Långban, Värmland, Sweden*, associated with calcite, barite, sarkinite, and adelite filling fissures in hausmannite ore. RG *AMG 4:459(1968), AM 71:1522(1986)*.

7.3.1.2 Franciscanite $Mn_6V_2Si_2(O,OH)_{14}$

Named in 1986 for the Franciscan complex, the formation in California where it was found. HEX $P3$. $a = 8.148$, $c = 4.804$, $Z = 1$, $D = 3.93$. AM 71:1522(1986). Structure: NJMM:493(1986). 40-487: 3.97_3 3.11_9 2.84_9 2.67_4 2.33_{10} 1.79_7 1.54_5 1.37_3. Glassy, irregular grains. Red to brownish red, brownish-red streak, vitreous luster. No cleavage, uneven fracture. H = 4. G = 4.1. Weakly magnetic, possibly due to microscopic inclusions of a magnetic mineral. Uniaxial (+); $N_O = 1.859$, $N_E = 1.876$; intense absorption; E ≫ O. O, wine-red; E, dark red to black. From the *Pennsylvania mine, San Antonio Valley, Santa Clara Co., CA*, associated with sonolite, hausmannite, and gageite, in chert. RG

7.3.1.3 Örebroite $Mn_6(Fe,Sb)_2Si_2(O,OH)_{14}$

Named in 1986 for the locality. HEX $P3$. $a = 8.183$, $c = 4.756$, $Z = 1$, $D = 4.77$. AM 71:1522(1986). 40-486: 4.08_3 3.94_2 3.10_{10} 2.84_5 2.68_4 2.33_7 1.78_9 1.55_5. Massive, polycrystalline. Dark brown, reddish-brown streak, vitreous luster. No cleavage, irregular fracture. H = 4. Slightly magnetic, possibly due to inclusions of a magnetic mineral. Uniaxial (+); $N_O = 1.857$, $N_E = 1.875$. From *Sjögruvan, Grythyttan, Örebro, Sweden*, associated with calcite or with hausmannite and dolomite. RG

7.3.2.1 Grossite $CaAl_4O_7$

Named in 1994 for Shulamit Gross (b.1923), Israeli mineralogist, Geological Survey of Israel, who first found the mineral. MON $C2/c$. $a = 12.94$, $b = 8.910$, $c = 5.446$, $\beta = 107.0°$, $Z = 4$, $D = 2.88$. EJM 6:591(1994). PD: 4.46_4 3.61_1 3.51_{10} 2.88_1 2.60_4 2.54_1 2.44_2 1.764_2. Prismatic, subhedral, microscopic crystals. White to colorless, white streak, vitreous luster, transparent. Biaxial (+); $N_x = 1.618$, $N_y = 1.618$, $N_z = 1.652$; $2V = 12°$. Nonfluorescent. Meteoritic grossite is related to high-temperature processes in the early solar system and has been found in certain meteorites associated with Ca,Al-rich inclusions (*Acfer-182; ALH 85085*, etc.), where it may be a major component, with perovskite, hibonite, melilite, and spinel. Also formed by high-temperature metamorphism of some limestones, associated with brownmillerite and mayenite in the *Hatrurim formation of Israel*. RG AM 80:630(1995).

7.4.1.1 Hibonite $(Ca,Ce)(Al,Ti,Mg)_{12}O_{19}$

Named in 1956 for Paul Hibon, French prospector, who discovered the mineral. HEX $P6_3/mmc$. $a = 5.61$, $c = 22.16$, $Z = 2$, $D = 3.84$. Structure: NJMA 109:192(1968). 38-470(syn): 2.78_5 2.62_9 2.48_{10} 2.29_3 2.11_8 2.01_6 1.573_3 1.534_6. Hexagonal prismatic crystals, flattened {0001}, or with steep pyramids; {0001} face commonly divided into six sectors. Black to brownish black. Cleavage {0001}, easy, parting on {10$\bar{1}$0}, subconchoidal fracture. H = $7\frac{1}{2}$–8. G = 3.84. Slowly soluble in mixed 1:1 $H_2SO_4 + H_3PO_4$. Uniaxial (−); $N_O = 1.807$, $N_E = 1.79$. O, gray, E, gray. Mrassu R., Siberia, Russia(!). *Esiva, Andranondambo, Madagascar*(!), in alluvial deposits; also at Andakato and Besakoa, in metamorphosed limestone rich in calcic plagioclase, with corundum, spinel, and thorianite. In the Allende and some other meteorites. RG *AM 42:119(1957): GCA 44:685(1980), 46:575(1982).*

7.4.1.2 Yimengite $K(Cr, Ti, Fe, Mg)_{12}O_{19}$

Named in 1983 for the locality. HEX $P6_3/mmc$. $a = 5.875$, $c = 22.940$, $Z = 2$, $D = 4.35$. *AM 70:218(1985): 37-480:* 2.78_9 2.63_{10} 2.45_4 2.24_5 2.13_4 1.67_6 1.62_9 1.48_8. Irregular grains, platy to tabular {0001}. Black, brown streak, metallic luster. Cleavage {0001}, perfect; {10$\bar{1}$1}, fair. $H = 4$. $G = 4.34$. Insoluble in dilute mineral acids. In RL, gray to grayish white, weakly pleochroic. $R = 12.7$, 17.7 (402 nm); 11.3, 18.2 (438 nm); 15.1, 16.0 (498 nm); 15.7, 16.2 (546 nm); 13.0, 16.5 (589 nm); 13.5, 16.7 (624 nm); 12.6, 15.5 (641 nm). *Yimengshan area, Shandong, China*, in kimberlite dikes, with olivine, phlogopite, pyrope, chromite, ilmenite, chrome-diopside, apatite, zircon, moissanite, and perovskite. Guaniamo, Bolivar Prov., Venezuela, in a kimberlite sill. RG *MM 53:305(1989).*

7.4.1.3 Hawthorneite $Ba(Ti_3Cr_4Fe_2^{2+}Fe_2^{3+}Mg)O_{19}$

Named in 1989 for John Barry Hawthorne (b.1934), chief geologist, DeBeers Consolidated Mines, South Africa; an expert on kimberlites and diamonds. HEX $P6_3/mmc$. $a = 5.871$, $c = 23.06$, $Z = 2$, $D = 5.02$. PD: 4.65_2 3.84_2 2.94_4 2.77_8 2.62_{10} 2.41_5 2.23_3 2.12_2. Minute grains. Black, metallic luster. Uniaxial $(-)$; $N_{O\,calc} = 2.39$, $N_{E\,calc} = 2.33$. In RL, gray with no internal reflections, moderately bireflectant, nonpleochroic. Moderately anisotropic, brown to gray. $R = 18.3$, 17.1 (470 nm); 17.5, 16.4 (546 nm); 17.3, 16.3 (589 nm); 17.05, 16.0 (650 nm). From the *Bultfontein mine, Kimberley, Cape Prov., South Africa*, in a harzburgite xenolith, associated with lindsleyite, chromian spinel, chromian rutile, and magnesian-chromian rutile. RG *AM 74:668(1989).*

7.4.2.1 Magnetoplumbite $Pb(Fe^{3+}, Mn^{3+})_{12}O_{19}$

Named in 1925 to indicate the magnetic properties and lead content of the mineral. HEX $P6_3/mmc$. $a = 5.873$, $c = 23.007$, $Z = 2$, $D = 5.71$. *AM 74:1186(1989). 43-666:* 3.91_4 2.97_6 2.81_{10} 2.65_9 2.45_6 2.26_5 1.618_6 1.487_8. Steep pyramidal crystals, doubly terminated. Gray-black, dark brown streak, metallic luster. Cleavage {0001}, perfect. $H = 6$. $G = 5.52$. Strongly magnetic. Slowly soluble in 1:1 HCl, with slight evolution of Cl_2. In RL, gray with a brown tint; weakly anisotropic. $R = 25.8$–24.0 (470 nm), 24.5–22.9 (546 nm), 23.7–22.2 (589 nm), 22.7–21.2 (650 nm). *P&J:251. Långban, Värmland, Sweden*, associated with manganophylite and kentrolite. RG *DI:728, AM: 36:512(1951), NJMM:141(1980).*

7.5.1.1 Braunite I $3Mn_2^{3+}O_3 \cdot Mn^{2+}SiO_3$

Named in 1831 for Kammerrat Wilhelm Braun (1790–1872) of Gotha, Germany, who was active in mineralogical and geognostical matters. TET $I4_1/acd$. $a = 9.408$, $c = 18.668$, $Z = 8$, $D = 4.860$. *AM 61:1226(1976).* **Structure:** The

7.5.1.1 Braunite I

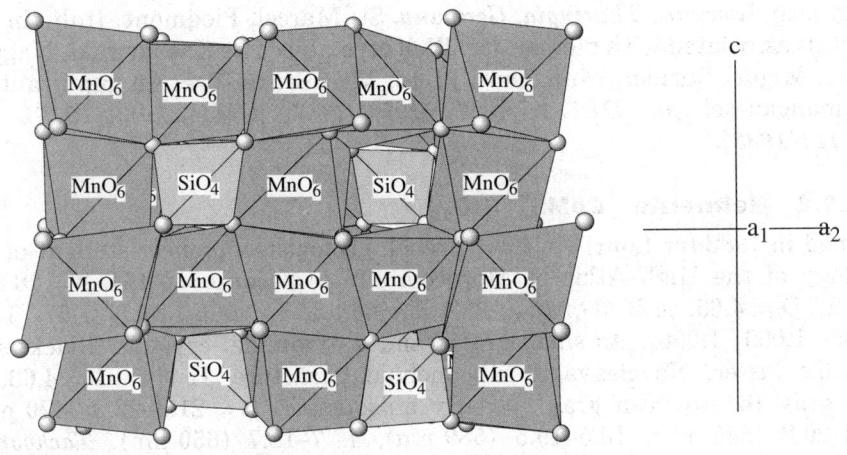

BRAUNITE: Mn_7SiO_{12}

The structure can be described as a three-dimensional framework of intersecting chains of Mn^{3+} octahedra, with Si (4-coordinated) and Mn^{2+} (8-coordinated) occupying channels.

underlying features of the braunite structure are two three-dimensional octahedral arrangements of cations around vacant sites. These cation arrangements must not be confused with the coordination of oxygen atoms around each of these cations. One such arrangement is referred to as the $(Mn^{2+}Mn_4^{3+}Si)$ arrangement, and the other as the (Mn_6^{3+}) arrangement. Certain Mn and Si atoms are surrounded by six each, of the $(Mn^{2+}Mn_4^{3+}Si)$ arrangements, and three other Mn atoms are surrounded by four each of the $(Mn^{2+}Mn_4^{3+}Si)$ and two of the (Mn_6^{3+}) arrangements. One-seventh of the manganese atoms are divalent, and six-sevenths are trivalent. The Mn^{2+} ions have distorted cubic coordination with the surrounding oxygens; the Mn^{3+} atoms have octahedral coordination, and the Si atoms have tetrahedral coordination. *AM 60:1098(1975)*.
41-1367: 3.49_1 2.71_{10} 2.36_1 2.14_1 1.665_1 1.659_4 1.420_2 1.411_1. **Habit:** Pyramidal crystals, {011}, {131}; striated on {001}, {201} parallel to {010}; also granular, massive; twinning on {112}. **Physical properties:** Brownish black to steel gray, black to gray streak, submetallic luster. Cleavage {112}, perfect; uneven to subconchoidal fracture; brittle. $H = 6-6\frac{1}{2}$. $G = 4.8$. Weakly magnetic. **Tests:** Soluble in HCl with evolution of Cl_2. Leaves a residue of gelatinous silica. **Chemistry:** Ba is often present, in amounts up to several percent. Some Fe substitutes for Mn. **Optics:** In RL, weakly anisotropic. $R = 22.7-21.3$ (470 nm), $21.1-19.5$ (546 nm), $20.5-18.9$ (589 nm), $19.7-18.4$ (650 nm). *P&J:104*. **Localities:** Spiller manganese mine, NE of Mason Co., TX, as lenses in quartzite. Autlán, JAL, Mexico, in a major manganese deposit. Nombre de Dios, Panama, with pyrolusite and "psilomelane." Lång-

ban, Värmland, and Jakobsberg, Nordmark, Sweden. *Ohrenstock* and Elgersberg, near *Ilmenau, Thuringia, Germany*. St. Marcel, Piedmont, Italy, in fine crystals associated with piemontite. Well crystallized at Kacharwaki, Nagpur, India. Miguel Burnier, Mun. Ouro Preto, MG, Brazil. Use: An important ore of manganese. RG *DI:551, AM 52:20(1967), CMP 49:21(1975), EG 78:1111(1983)*.

7.5.1.2 Neltnerite $CaMn_6^{3+}SiO_{12}$

Named in 1982 for Louis Neltner, French geologist, a pioneer student of the geology of the High Atlas of Morocco. TET $I4_1/acd$. $a = 9.464$, $c = 18.854$, $Z = 8$, $D = 4.65$. *AM 68:282(1983)*. 35-666: 2.73_{10} 2.36_3 2.17_3 1.67_8 1.43_6 1.085_5 1.059_4 1.056_4. As small grains and dipyramidal crystals. Black, submetallic luster. No cleavage, subconchoidal fracture. H = 6. G = 4.63. In RL, gray to brownish gray. Weakly anisotropic. R = 21.3–22.3 (470 nm), 19.2–20.2 (546 nm), 18.5–19.5 (589 nm), 17.7–18.7 (650 nm). *Tachgagalt, Morocco*, with braunite I, braunite II, marokite, and crednerite. RG *BM 105:161(1982)*.

7.5.1.3 Braunite II $7(Mn,Fe)_2O_3 \cdot CaSiO_3$

Named in 1967 as in braunite I (7.5.1.1). TET $I4_1/acd$. $a = 9.431$, $c = 37.774$, $Z = 8$, $D = 4.85$. *AM 65:756(1980)*. 41-1368: 3.68_1 2.72_{10} 2.361_1 2.354_2 2.08_1 1.666_2 1.420_1 1.361_1. Braunite II is compositionally distinct from braunite I and represents an ordered version of braunite I with an approximately double-size cell. Black, submetallic luster. H = 6. G = 4.75. Nonmagnetic. In RL, yellow-brown with properties like braunite. From the *Black Rock and Wessels mines, Kalahari Manganese district, Cape Prov., South Africa*. Tachgagalt, Anti-Atlas, Morocco, with neltnernite, marokite, hausmannite, crednerite, henritermierite, gaudefroyite, and calcite. Used as an ore of manganese. RG *EG 78:1111(1983), AM 71:1543(1986)*.

7.5.1.4 Abswurmbachite $Cu^{2+}Mn_6^{3+}[O_8/SiO_4]$

Named in 1991 for Irmgard Abs-Wurmbach (b.1938), German crystal chemist, who studied braunite. TET $I4_1/acd$. $a = 9.406$, $c = 18.546$, $Z = 8$, $D = 4.96$. *NJMA 163:117(1991)*. 41-576: 3.48_1 2.70_{10} 2.35_2 2.13_2 1.663_1 1.651_3 1.416_1 1.402_1. Minute fibrous to prismatic, also equant, grains. Black, brownish-black streak, metallic luster. Brittle. $VHN_{25} = 920$. In RL, gray, with weak anisotropy, no internal reflections, nonpleochroic. R = 20.8–21.2 (470 nm), 19.6–20.0 (546 nm), 19.2–19.7 (589 nm), 18.7–19.2 (650 nm). *Mount Ochi, near Karystos, Evvia, Greece*, in low-grade, high-pressure metamorphosed quartzites, with tenorite, shattuckite, piemontite, sursassite, ardennite, rutile, and hollandite. RG *AM 77:670(1992)*.

7.5.2.1 Painite $CaZrB(Al_9O_{18})$

Named in 1957 for A. C. D. Pain (d.1971), British gem collector, who first noticed the mineral. HEX $P6_3$. $a = 8.715$, $c = 8.472$, $Z = 2$, $D = 3.996$. *AM*

7.6.1.1 Cesárolite

61:88(1976). *10-405:* 7.63_5 5.76_{10} 3.70_8 2.52_{10} 2.37_8 2.01_8 1.73_7 1.42_7. Hexagonal prismatic crystal, pseudo-orthorhombic. Garnet red, vitreous luster. H = 8. G = 4.01. Insoluble in acids. Uniaxial (−); $N_O = 1.816$, $N_E = 1.787$. O, pale brownish orange; E, ruby red. Near Franklin, NJ, in marble. From *Ohngaing, Mogok district, Sagaing, Myanmar*, where a single crystal was recovered in 1952 from gem concentrates. Additional crystals have subsequently been found there. RG *MM 31:420(1957)*, *42:518(1978)*, *50:267(1986)*; *AM 42:580(1957)*.

7.5.3.1 Birnessite $(Na, Ca, K)_x(Mn^{4+}, Mn^{3+})_2O_4 \cdot 1.5H_2O$

Named in 1956 for the locality. MON $C2/m$. $a = 5.174$, $b = 2.850$, $c = 7.336$, $\beta = 103.18°$, Z = 1, D probably = 3.4. *AM 75:477(1990)*. *43-1456*(syn): 7.14_{10} 3.57_3 2.52_1 2.43_1 2.22_1 2.15_1 1.823_1 1.476_1. Massive, earthy, as pseudomorphs. Dark brown to black. H = 1.5. G = 3.0. Uniaxial (−); $N_O \sim 1.73$, $N_E \sim 1.69$. In RL, grayish white, strongly pleochroic and anisotropic. R = 26.1–10.5 (470 nm), 23.1–9.8 (546 nm), 21.9–9.6 (589 nm), 20.8–9.4 (650 nm). *P&J:90*. Found in a wide variety of geological settings. It is a major Mn phase in many soils; an important component in desert varnishes; common in manganese nodules from the ocean floor; and it is also found as an alteration product of Mn-rich ore deposits. Franklin, Sussex Co., NJ, as a weathering product of franklinite. Cummington, MA, as an oxidation product of rhodonite, rhodochrosite, tephroite, spessartine, alleghanyite, and manganoan dolomite. Material abundant at Mt. St.-Hilaire, QUE, Canada, often pseudomorphous after serandite, has erroneously been called birnessite, but is something different as it contains no Na. *Birness, Aberdeenshire, Scotland*, as grains <8 mm cementing gravel in a road cut 30 km N of Aberdeen. RG *MM 31:283(1956)*; *AM 45:871(1960)*, *69:814(1984)*; *JGSK 16:105(1980)*.

7.5.4.1 Zenzénite $Pb_3(Fe^{3+}, Mn^{3+})_4Mn_3^{4+}O_{15}$

Named in 1991 for Nils Zenzén (1883–1959), Swedish mineralogist and historian of science. HEX $P6_3/mcm$. $a = 10.008$, $c = 13.672$, Z = 4, D = 6.83. *CM 29:350(1991)*. *45-1345:* 3.42_5 3.18_8 2.83_7 2.66_{10} 2.37_6 2.10_4 2.07_3 1.687_8. Subhedral to euhedral grains, tabular {0001}. Black. Cleavage {0001}, distinct. H = $5\frac{1}{2}$. Weakly paramagnetic. In RL, bright white with no internal reflections, nonpleochroic. Weak bireflectance. R = 31.0–26.1 (470 nm), 29.5–25.0 (546 nm), 28.5–24.4 (589 nm), 27.3–23.5 (650 nm). From *Långban, Värmland, Sweden*, in carbonate–phyllosilicate skarn, associated with magnesioferrite, jacobsite, hausmannite, and dolomite. RG *CM 29:347(1991)*.

7.6.1.1 Cesárolite $PbH_2Mn_3^{4+}O_8$

Named in 1920 for Giuseppe R. P. Cesáro (1849–1939), professor of mineralogy and crystallography, University of Liége, Belgium. HEX-R Space group unknown. $a = 2.811$, $c = 20.386$. *IMA ABST:334(1994)*. *14-489:* 3.40_4 3.21_1 2.27_1 2.19_{10} 2.09_8 1.76_5 1.56_5 1.41_3. Friable masses, botryoidal crusts. Steel gray, submetallic to dull luster. H = $4\frac{1}{2}$. G = 5.29. Soluble in HCl with evolution of Cl_2. From *Sidi Amor Ben Salem, Tunisia*, in cavities in galena. RG *DI:744, GSAM 85:206(1962), CE 27:258(1967)*.

7.6.2.1 Bartelkeite $PbFe^{2+}Ge_3O_8$

Named in 1981 for Wolfgang Bartelke, mineral collector and specialist in Tsumeb minerals. MON $P2_1$ or $P2_1/m$. $a = 5.431$, $b = 13.689$, $c = 5.892$, $\beta = 111.48°$, $Z = 2$, $D = 4.97$. *AM 67:413(1982)*. 35-508: 5.08_5 4.74_6 4.42_8 3.50_5 2.91_{10} 2.87_8 2.75_7 2.20_7. Crystals tabular $\{10\bar{1}\}$ or acicular [101]. Colorless or pale green. Cleavage $\{101\}$, distinct. $H = 4$. Soluble in hot HCl. Biaxial $(-)$; $Z = b$; $N_x = 1.885$, $N_y = 1.910$, $N_z = 1.913$; $2V = \sim 35°$; $r < v$. *Tsumeb, Namibia*, in primary ore, in cavities containing germanite, renierite, tennantite, and galena. RG *CE 40:201(1981)*.

7.6.3.1 Kamiokite $Fe_2^{2+}Mo_3^{4+}O_8$

Named in 1985 for the locality. HEX $P6_3mc$. $a = 5.782$, $c = 10.053$, $Z = 2$, $D = 6.02$. *MJJ 12:393(1985)*. Structure: *AC(C) 42:9(1986)*. 36-526(syn): 5.02_{10} 3.54_{10} 2.78_3 2.50_8 2.43_8 2.00_4 1.65_3 1.57_4. Minute euhedral grains, thick hexagonal crystals. Black, black streak, metallic to submetallic luster. Cleavage $\{0001\}$, perfect; even fracture. $H = 4\frac{1}{2}$. $G = 5.96$. In RL, gray, no internal reflections, distinctly pleochroic, strongly anisotropic. $R = 24.0-17.6$ (470 nm), 22.9–16.5 (546 nm), 22.4–16.0 (589 nm), 22.3–15.8 (650 nm). *P&J:219*. From the Mohawk and Ahmeek mines, MI, as inclusions in domeykite and algodonite. *Kamioka mine, Gifu Pref., Japan*, in quartz veinlets containing molybdenite, K-feldspar, fluorite, ilmenite, and scheelite. RG *AM 68:1038(1983), 73:191(1988); TMPM 35:67(1986)*.

7.7.1.1 Pseudobrookite $Fe_2^{3+}TiO_5$

Named in 1878 from the Greek *pseudo*, false, and *brookite*, because the mineral at one time was mistakenly thought to be brookite. ORTH $Bbmm$. $a = 9.797$, $b = 9.981$, $c = 3.730$, $Z = 4$, $D = 4.36$. *ZK 73:97(1930)*. Structure: *SR 51A:208(1984)*. 41-1432(syn): 4.90_4, 3.49_{10} 2.75_8 2.46_2 2.41_2 1.97_2 1.87_2 1.54_2. Usually tabular $\{100\}$ and elongated [001], sometimes long prismatic [001] or acicular. Black, dark reddish brown; reddish brown to ochre-yellow streak; metallic to adamantine luster, greasy on fracture surfaces. Cleavage $\{010\}$, distinct; uneven to subconchoidal fracture. $H = 6$. $G = 4.39$. Soluble in hot 1:1 HCl or H_2SO_4, decomposed by HF, infusible. Biaxial (+); $XYZ = bca$; $N_x = 2.38$, $N_y = 2.39$, $N_z = 2.42$ (Li); $2V = 50°$; $r < v$; absorption, in brown, $X < Y < Z$. Usually formed as a pneumatolytic or fumarolic product in volcanic igneous rocks, or as a reaction product of xenoliths in such rocks; usually associated with tridymite, hematite, magnetite, sanidine, apatite, and rutile. At Topax Mt. and elsewhere in the Thomas Range, Juab Co., UT(!), in lithophysae in rhyolite, with topaz, bixbyite, hematite, red beryl, and ilmenite. Red Cone, Crater Lake, OR(!), in cavities in basalt with hypersthene and apatite. In France, in cavities in andesite and trachyte at Riveau Grande, Mt. Doré, Puy de Dôme, with tridymite, hypersthene, and sanidine. Hessenbrucker Hammer, Hessen, Germany, as a product of fumarolic action on basalt. On lava from the 1872 eruption of Vesuvius, Italy, with hematite, magnetite, and sellaite. In *Uroiu, Transylvania, Romania*(!), in cavities in andesite, with tridymite, hypersthene, and garnet. As a pneumatolytic product on basalt on the Island of Réunion, Indian Ocean. RG *DI:736, GSAM 85:204(1962), AM 73:1377(1988)*.

7.7.1.2 Armalcolite FeMgTi$_4$O$_{10}$

Named in 1970 for Neil A. Armstrong (b.1930), Edwin E. Aldrin, Jr. (b.1930), and Michael Collins (b.1930), the three astronauts on whose voyage to the moon the samples containing the mineral were collected. ORTH *Bbmm*. $a = 9.770$, $b = 9.983$, $c = 3.739$, $Z = 2$, $D = 4.00$. *AM 62:913(1977). 41-1444*(syn): 4.89$_4$ 3.49$_{10}$ 2.75$_7$ 2.46$_2$ 2.41$_2$ 1.97$_2$ 1.87$_3$ 1.55$_2$. Small grains, commonly with rectangular outline. In RL, gray, with no internal reflections, pleochroic, strongly anisotropic, pale gray to dark bluish gray. In armalcolite the middle range of a binary solid-solution series has been named; it is more usual to name the end members. In synthetic material a complete series is known to exist between FeTi$_2$O$_5$ (pseudobrookite) and MgTi$_2$O$_5$. From *Tranquillity Base, Moon*, in rocks collected on the *Apollo XI* mission, associated with ilmenite. In lamproites from Smoky Butte, Garfield Co., MT. Kuonamski region, Yakutia, Russia, in picrite porphyrites. Jagersfontein mine, Orange Free State, South Africa, in kimberlite, associated with olivine, perovskite, ilmenite, and mathiasite. RG *AM 55:2136 (1970), 60:566(1975), 73:1377(1988); GCA Supp. 1, 1:55(1970); EPSL 29:91(1976); MZ 2:87 (1980); CMP 77:307(1981)*.

7.7.2.1 Berdesinskiite V$_2^{3+}$TiO$_5$

Named in 1981 for Waldemar Berdesinski (1911–1990), German mineralogist, University of Heidelberg. MON $P2/c$ or Pc. $a = 10.11$, $b = 5.084$, $c = 7.03$, $\beta = 111.47°$, $Z = 4$, $D = 4.536$. *AM 67:1074(1982). 37-463:* 4.72$_8$ 2.90$_{10}$ 2.68$_{10}$ 2.54$_8$ 2.45$_{10}$ 1.73$_{10}$ 1.65$_{10}$ 1.45$_{10}$. Minute grains. Black, metallic luster. H = 6–6$\frac{1}{2}$. In RL, reddish brown to yellow-brown, weakly pleochroic and anisotropic. R = 18.1–18.7 (470 nm), 19.6–20.5 (546 nm), 20.1–21.2 (589 nm), 20.6–21.6 (650 nm). *Lasamba Hill, Kwale district, Voi, Kenya*, in a kornerupine deposit in weathered gneiss, associated with schreyerite, tourmaline, and kornerupine. RG *NJMM:110(1983), AM 68:1038(1983)*.

7.8.1.1 Todorokite (Na,Ca,K)$_x$(Mn^{4+},Mn^{3+})$_6$O$_{12}$ · 3–4.5H$_2$O

Named in 1934 for the locality. MON $P2/m$. $a = 9.764$, $b = 2.842$, $c = 9.551$, $\beta = 94.14°$, Z = 1, D = 3.65. *AM 73:861(1988). 38-475:* 9.56$_{10}$ 4.86$_2$ 4.77$_2$ 2.44$_2$ 2.40$_4$ 2.39$_3$ 2.355$_2$ 2.345$_2$. **Habit:** Spongy banded and reniform aggregates composed of minute lathlike crystals. **Structure:** Todorokite consists predominantly of triple chains of edge-sharing (Mn^{4+}O$_6$) octahedra that corner-share to form large roughly square tunnels parallel to b that measure three octahedra on a side. These tunnels accommodate water molecules and cations (Na,

Ca, K, etc.). The octahedra at the edges of the triple chains are larger than those in the middle and therefore likely accommodate the larger cations (i.e., Mg, Mn^{3+}, Cu^{2+}, Co, Ni, etc.). This structure is similar to that of hollandite and romanechite, but with larger tunnels. *AM 73:861(1988), 68:972(1983).* **Physical properties:** Black–brown, metallic luster. Cleavages {100} and {010}, perfect. $H = 1\frac{1}{2}$. $G = 3.67$. **Tests:** Soluble in HCl with evolution of Cl_2. **Optics:** $Y = b$; $Ns > 1.74$; brown; absorption $Z > X$. In RL, gray-white, very strongly pleochroic and anisotropic. $R = 21.3$–11.4 (470 nm), 17.7–9.9 (546 nm), 16.6–9.4 (589 nm), 15.7–9.0 (650 nm). *P&J:371.* **Occurrence:** Although todorokite has been demonstrated to be a common product of oxidation and leaching of primary manganese carbonate and silicate minerals, its major occurrence is as the dominant Mn^{4+} phase of deep-sea ferromanganese nodules. **Localities:** Sterling Hill mine, Sussex Co., NJ. Charco Rodondo and many other mines, Oriente Province, Cuba, with pyrolusite, cryptomelane, manganite, and "psilomelane"; Romanêche, France; Farragudo, Portugal; Huttenberg, Carinthia, Austria; *Todoroki mine, Hokkaido, Japan*, as an alteration product of inesite and rhodochrosite; in Australia from the Gladstone area, Q, and the Kalgoorlie and Pilbara areas, WA. N'Chwaning mine and other mines in the Kuruman manganese belt, northern Cape Prov., South Africa. **Uses:** An ore of manganese. As a component of deep-sea manganese nodules, todorokite is not only significant as a potential source of manganese, but also as a source of Cu, Co, and Ni, which are taken up in appreciable

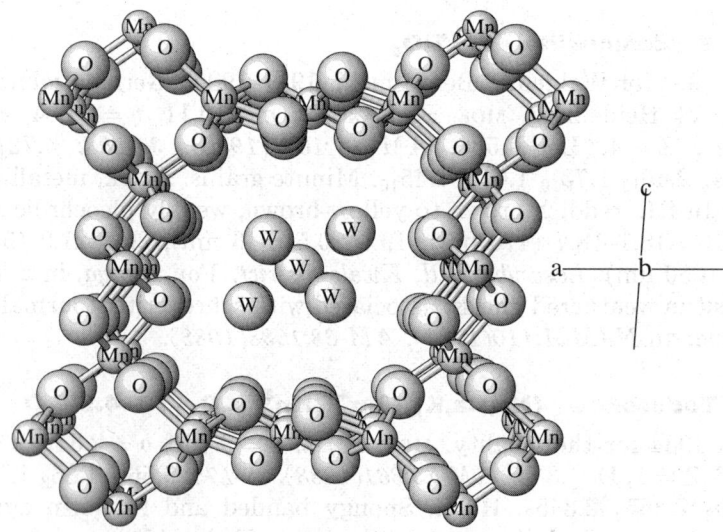

TODOROKITE: $NaMn_6O_{12} \cdot 3H_2O$ (variable)

Octahedral chains of Mn octahedra are linked sideways to form large channels. Water molecules and Na and other large cations occupy the centers of the channels. The channel atoms are disordered over the positions shown.

amounts up to several percent in some of these nodules. RG *DI:573; AM 45:1167(1960), 64:1333(1979); MM 36:757(1968), 50:336(1986).*

7.8.1.2 Woodruffite $(Zn,Mn^{2+})_2Mn_5^{4+}O_{12} \cdot 4H_2O$

Named in 1953 for Samuel Woodruff, late-nineteenth-century miner and mineral collector at Franklin, who collected and preserved many of the finest specimens found at that locality. TET Space group unknown. $a = 8.42$, $c = 9.28$, $Z = 2$, $D = 3.98$. *MM 33:507(1963). 16-338:* 9.34_5 4.66_{10} 4.08_3 2.66_6 2.48_6 1.86_4 1.69_4 1.48_5. Botryoidal masses, also pulverulent. Iron-black, chocolate brown when pulverulent; brownish streak; dull luster. Smooth conchoidal fracture. $H = 4\frac{1}{2}$. $G = 4.01$. Sterling Hill mine, Ogdensburg, Sussex Co., NJ, in oxidized material derived from the orebody outcrop. Sandur mine, Mysore, India, associated with cryptomelane, lithiophorite, pyrolusite, ramsdellite, jacobsite, manganite, and braunite. RG *AM 38:761(1953).*

7.8.2.1 Chalcophanite $(Zn,Fe^{2+})Mn_3^{4+}O_7 \cdot 3H_2O$

Named in 1875 from the Greek for *copper* and *to appear*, in allusion to the change in color on ignition. HEX-R $R\bar{3}$. $a = 7.533$, $c = 20.794$, $Z = 6$, $D = 3.86$. *AM 73:1401(1988).* Structure: *AC 8:165(1955). 45-1320:* 6.93_{10} 4.07_3 3.51_2 2.55_2 2.45_2 2.23_4, 1.897_2 1.590_2. Crystals tabular {0001}, sometimes with {10$\bar{1}$1} equally developed giving an equant pseudo-octahedral habit. Commonly as drusy, botryoidal, or stalactitic crusts; massive; granular; or platy; fibrous. Iron black to bluish black, chocolate-brown streak, metallic luster. Cleavage {0001}, perfect; flexible in thin plates. $H = 2\frac{1}{2}$. $G = 4.00$. Uniaxial (−); $N_O \gg 2.72$, $N_E \sim 2.72$. O, nearly opaque; E, deep red. In RL, white, with deep-red internal reflections, very strongly pleochroic and extremely highly anisotropic. $R = 33.0$–10.2 (470 nm), 27.7–9.6 (546 nm), 26.0–9.3 (589 nm), 24.6–9.1 (650 nm). *P&J:117.* A common weathering product in many Mn-bearing base metal deposits. Sterling Hill mine, Ogdensburg, NJ; Bisbee, AZ; Wolftone mine, Leadville, CO, with limonite, smithsonite, and hemimorphite. Groot Eyelandt, NT; Buchan, VIC; and Kambalda, WA, Australia. RG *DI:739; GSAM 85:206(1962); MM 44:109(1981), 49:752(1985).*

7.8.2.2 Aurorite $(Mn^{2+},Ag,Ca)Mn_3^{4+}O_7 \cdot 3H_2O$

Named in 1967 for the locality. TRIC $P\bar{1}$. $a = 7.53$, $b = 7.53$, $c = 8.20$, $\alpha = 90°$, $\beta = 117.20°$, $\gamma = 120°$, $Z = 2$, $D = 3.88$. *19-88:* 6.94_{10} 4.06_5 3.46_7 2.54_5 2.45_4 2.23_5 1.56_5 1.43_5. Small irregular masses and platy or scaly grains. $H < 3$. In RL, white to gray, with no internal reflections and strongly pleochroic; strongly anisotropic, cream white to medium gray. Aurora mine, Treasure Hill, Hamilton, NV, associated with todorokite, cryptomelane, pyrolusite, quartz, and manganoan calcite. RG *EG 62:186(1967), AM 52:1581(1967).*

7.8.2.3 Jianshuiite (Mg, Mn)Mn$_3^{4+}$O$_7 \cdot$ 3H$_2$O

Named in 1992 for the locality. TRIC $P\bar{1}$. $a = 7.534$, $b = 7.525$, $c = 8.204$, $\alpha = 89.7°$, $\beta = 7.53°$, $\gamma = 120.0°$, $Z = 2$, $D = 3.598$. *AM 79:185(1994)*. PD: 6.97$_{10}$ 5.54$_1$ 4.09$_1$ 3.48$_1$ 2.45$_1$ 2.23$_2$. Massive; as soft, porous aggregates of microscopic grains. Brown to brownish black, opaque. Soft. G = 3.60. In RL, grayish white, distinctly anisotropic, brown internal reflectance. R = 23.0 (470 nm), 19.9 (546 nm), 19.1 (589 nm), 18.6 (650 nm). A hypogene mineral associated with other hydrated manganese oxides in Mn ore near *Lu village, Jianshui Co., Yunnan, China.* RG *AMS 12:69(1992)*.

7.8.2.4 Ernienickelite NiMn$_3^{4+}$O$_7 \cdot$ 3H$_2$O

Named in 1994 for Ernest (Ernie) Henry Nickel (b.1925), Canadian–Australian mineralogist, CSIRO (Commonwealth Scientific and Industrial Research Organization) and IMA (International Mineralogical Association). HEX-R $R\bar{3}$ or $R3$. $a = 7.514$, $c = 20.52$, $Z = 6$, $D = 3.83$. *CM 32:333(1994)*. PD: 6.84$_{10}$ 4.01$_2$ 3.44$_1$ 2.75$_1$ 2.53$_1$ 2.22$_3$ 1.884$_2$ 1.575$_2$. Crystallizes in very small, almost circular, thin plates up to 0.5 mm × 0.5 mm × 0.02 mm, and rosettes of plates. Opaque, nearly black, with a red-brown cast; yellow-brown streak; submetallic to vitreous luster. Cleavage {001}, perfect; splintery fracture; brittle. H = 2. G = 3.84. Uniaxial (−); N$_O$ > 2.00 m, N$_E$ = 1.97. Minor Mg and Co substitute for Ni in the formula. Found in cavities in chalcedony in a lateritized dunite, associated with goethite, magnesite, nimite, nontronite at the *SM7 open pit N of Siberia, Kalgoorlie district, WA, Australia.* RG

7.8.3.1 Lazarenkoite CaFe^{3+}As$_3^{3+}$O$_7 \cdot$ 3H$_2$O

Named in 1981 for Eugeni Konstantinovich Lazarenko (1912–1979), mineralogist; academician, Academy of Science, Ukraine. ORTH Space group unknown. $a = 21.80$, $b = 12.64$, $c = 8.40$, $Z = 10$, $D = 3.59$. *35-669:* 11.2$_9$ 8.40$_{10}$ 6.55$_8$ 4.66$_9$ 3.31$_4$ 3.20$_4$ 3.14$_4$ 3.01$_3$. Fibrous crystal incrustations. Bright orange, resinous/silky luster. H = 1. G = 3.45. IR data show negative for (As^{5+}O$_4$)$^{3-}$ or (As^{3+}O$_3$)$^{3-}$. Insoluble in water, readily soluble in acids. Biaxial(−); N$_x$ = 1.820, N$_y$ = 1.920, N$_z$ = 1.955; 2V = 30°; positive elongation. X, pale yellow; Y, pale brown; Z, rose brown. *Western Siberia, Russia,* as an oxidation product on skutterudite–loellingite ores. RG *MZ 3(3):92(1981), AM 67:415(1982), CA 100:37241p(1984)*.

CRYPTOMELANE GROUP

The cryptomelane group of minerals consists of hard black, fine-grained, and visually indistinguishable minerals corresponding to the formula

$$A(B_1B_2)_8O_{16}$$

where

7.9.1.2 Cryptomelane

A = Ba, K, Pb, or (Na,K)
$B_1 = Mn^{4+}$
$B_2 = Mn^{2+}$

and the structure is that of hollandite, monoclinic but pseudotetragonal. The structure is a distinctive tunnel structure consisting of a framework of octahedrally coordinated metal ions. The octahedra are lined by edge sharing to form double columns parallel to the tunnel axis. The double strings in turn share corners to form a three-dimensional framework containing square tunnels which run through the structure. The tunnels, two octahedra wide, accommodate the K, Na, Ba, and Pb ions and H_2O when present. *AC(B) 36:1056(1982)*.

CRYPTOMELANE GROUP

Mineral	Formula	Space Group	a	b	c	β	
Hollandite	$Ba(Mn^{4+}, Mn^{2+})_8O_{16}$	$I2/m$	10.026	2.878	9.729	91.03°	7.9.1.1
Cryptomelane	$K(Mn^{4+}, Mn^{2+})_8O_{16}$	$I2/m$	9.956	2.871	9.706	90.95°	7.9.1.2
Manjiroite*	$(Na, K)(Mn^{4+}, Mn^{2+})_8O_{16}$	$I4/m$	9.916		2.864		7.9.1.3
Coronadite	$Pb(Mn^{4+}, Mn^{2+})_8O_{16}$	$I2/m$	9.938	2.868	9.834	90.38°	7.9.1.4

*Cell constants given for the tetragonal cell.

Notes:
1. Space group originally thought to be tetragonal, but detailed examination of powder data indicates a monoclinic cell.
2. The group constitutes a large portion of the material formerly called *psilomelane*.

7.9.1.1 Hollandite $Ba(Mn^{4+},Mn^{2+})_8O_{16}$

Named in 1906 for Thomas H. Holland (1868–1947), British geologist and director, Geological Survey of India. Cryptomelane group. MON (pseudo-TET) $I2/m$. $a = 10.026$, $b = 2.878$, $c = 9.729$, $\beta = 91.03°$, $Z = 1$, $D = 4.83$. *38-476:* 4.86_2 3.52_2 3.46_2 3.17_4 3.14_9 3.10_{10} 3.07_3 2.41_4. Short prismatic crystals, terminated by a flat pyramid; also massive, fibrous. Silvery gray to black, black streak, metallic to shining luster. Prismatic, distinct cleavage; brittle. H = 6 on crystal faces, less on fracture surfaces. G = 4.95. In RL, white, weakly pleochroic, strongly anisotropic. R = 36.2–30.1 (440 nm), 32.8–27.6 (546 nm), 31.2–26.5 (589 nm), 29.8–25.7 (650 nm). *P&J:204*. Ultevis, Sweden. *Kajlidongri, Jhabua state, India*, as rough crystals in quartz veins traversing a manganese orebody; also at Sitapar, Chhindwara district, and elsewhere in India. Ambatomiady, Madagascar. Nuba Mt., Kordofan Province, Sudan, intergrown with pyrolusite. RG *DI:743*, *AM 64:1199(1979)*, *NJMM: 30(1986)*, *MJJ 13:119(1986)*.

7.9.1.2 Cryptomelane $K(Mn^{4+},Mn^{2+})_8O_{16}$

Named in 1942 from the Greek for *hidden* and *black*, as the identity of this common black manganese mineral had been lost within the group of such minerals. Cryptomelane group. MON (pseudo-TET) $I2/m$. $a = 9.956$, $b = 2.871$, $c = 9.706$, $\beta = 90.95°$, $Z = 1$, $D = 4.398$. $AC(B)$ $38:1056(1982)$. 44-1386: 7.01_3 6.89_3 4.85_3 3.12_5 3.09_5 2.41_4 2.40_{10} 2.39_3. Botryoidal, massive, fine grained to radial fibrous. Gray to black, submetallic to earthy luster. $G = 4.3$. In RL, grayish white. R values close to those for hollandite. Cryptomelane is the commonest of the hard, black, fine-grained manganese oxides formerly called "psilomelane"; a few of many localities are: Mena, AR; Tombstone, Cochise Co., AZ; Deming, NM; Philipsburg, MT; *Chindwara, India*. RG *AM* 27:607(1942), MM 32:166(1959), CMP 55:191(1976).

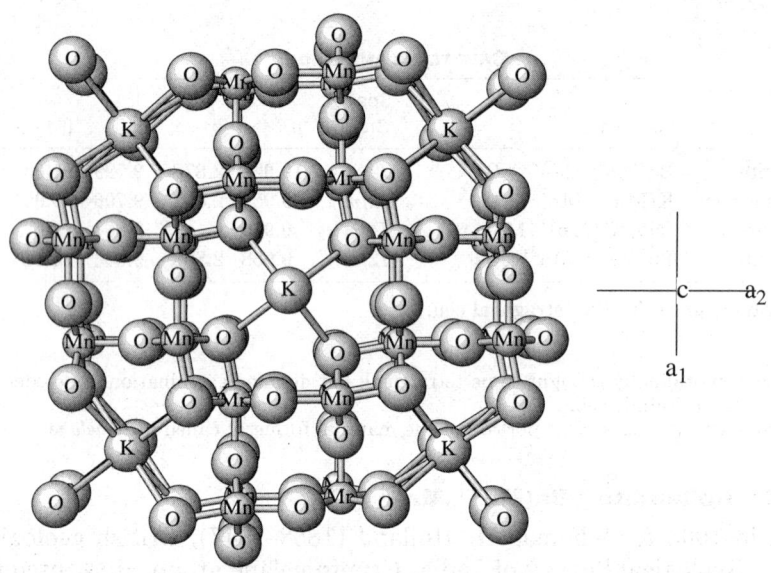

CRYPTOMELANE: KMn$_8$O$_{16}$

Chains of Mn octahedra link sideways to four square columns, which are linked to each other at their edges. K atoms occupy "tunnels" between the columns.

7.9.1.3 Manjiroite (Na,K)(Mn^{4+},Mn^{2+})$_8$O$_{16}$

Named in 1967 for Manjiro Watanbabe (1891–1980), Japanese mining geologist and professor, Tohoku University. Cryptomelane group. TET $I4/m$. $a = 9.916$, $c = 2.864$, $Z = 1$, $D = 4.41$. *AM* 53:2103(1968). 21-1153: 7.02_{10} 4.94_8 3.14_9 2.41_{10} 2.16_7 1.84_5 1.55_5 1.43_4. Compact masses. Dark brownish gray, brownish-black streak, dull luster. No cleavage, conchoidal fracture. VHN = 181. $G = 4.29$. In RL, anisotropic, weakly pleochroic. *Kohare mine, Iwate Pref., Japan*, and in several other nearby deposits, with pyrolusite, nsutite, birnessite, cryptomelane, and goethite. RG.

7.9.1.4 Coronadite Pb(Mn^{4+},Mn^{2+})$_8$O$_{16}$

Named in 1904 for Francisco Vasquez de Coronado (ca. 1500-1554), Spanish explorer of the American southwest, and for the locality, the Coronado vein. Cryptomelane group. MON (pseudo-TET) ($I2/m$. $a = 9.938$, $b = 2.868$, $c = 9.834$, $\beta = 90.38°$, $Z = 1$, $D = 5.35$. *AM 74:913(1989)*. *41-596:* 6.99$_1$ 3.49$_3$ 3.12$_{10}$ 2.40$_2$ 2.21$_2$ 2.16$_1$ 1.837$_1$ 1.548$_1$. Massive, radial fibrous. Dark gray to black, brownish-black streak, submetallic to dull luster. H = 4$\frac{1}{2}$-5. G = 5.44. In RL, white, strongly anisotropic. *Horse Shoe Shaft, Coronado vein, Clifton-Morenci district, AZ*; Bou Tazoult, Imini, Morocco. Broken Hill, NSW, Australia. RG *AM 27:48(1942), DI:742*.

7.9.2.1 Romanechite (Ba,H$_2$O)(Mn^{4+},Mn^{3+})$_5$O$_{10}$

Named in 1910 for the locality at Romanèche, Saône-et-Loire, France. Synonym: "psilomelane." *Psilomelane* is from the Greek for *smooth* and *black*, in allusion to its appearance. *Psilomelane* was abandoned as the name of a specific mineral in the 1970s by decision of the IMA Commission on New Minerals and New Mineral Names. The reason was that since the term was introduced in 1827 to refer to a hard black manganese oxide mineral from Schneeberg, Saxony, it was subsequently applied to all such minerals, which by their nature are difficult to distinguish from each other. It is now known that these minerals constitute a group of six or more distinct species. In 1910, Lacroix proposed the name *romanechite* for a mineral identical to the Schneeberg material but from Romanèche, France, and this name has been used in France since that time. It was felt that the name *psilomelane* has been used for so many different minerals that it has ceased to be a well-defined term. Therefore, romanechite will be used in its stead for the mineral (Ba,H$_2$O)$_2$Mn$_5$O$_{10}$, although the type locality for this mineral continues to be its original source (under a different name) at Schneeberg. *Psilomelane* will probably continue to be used as a convenient term for hard black manganese minerals whose true identity has not been established. MON $C2/m$. $a = 13.929$, $b = 2.846$, $c = 9.678$, $\beta = 92.65°$, $Z = 2$, $D = 4.74$. *AM 73:1155(1988)*. *14-627:* 6.96$_6$ 3.48$_6$ 2.88$_4$ 2.41$_{10}$ 2.37$_5$ 2.26$_4$ 2.19$_9$ 1.824$_4$. Massive, often botryoidal; also earthy and pulverulent. Iron black to gray; brownish-black to black streak, shining; submetallic to dull luster. H = 5–6. G = 4.71. In RL, gray-white, pleochroic, moderately reflective, strongly anisotropic. R = 35.0–28.0 (470 nm), 32.0–26.0 (546 nm), 30.3–25.1 (589 nm), 28.8–24.1 (650 nm). *P&J:328,308*. Mayfield prospect, Chispa, Jeff Davis Co., TX; Hoggett manganese group, Hidalgo Co., NM; Sodaville, NV; Tolbard mine, Paymaster district, Imperial Co., CA. Talamantes mine, Parral, CHIH, Mexico. Restormel mine, Lostwithiel, Cornwall, England; *Romanèche-Thorins, near Mâcon, Sâone-et-Loire, France;* Schneeberg, Saxony, Germany; Tekrasni, Singhbum district, Central Prov., India; Pilbara district, Australia. RG *DI:668, AM 45:176(1960), MA 87:3126(1987), IANS:68(1983)*.

7.9.3.1 Freudenbergite $Na_2Fe_2^{3+}Ti_6O_{16}$

Named in 1961 for Wilhelm Freudenberg (b.1881), German geologist, who studied the rocks of Odenwald, Germany, where the mineral was found. MON $C2/m$. $a = 12.267$, $b = 3.823$, $c = 6.483$, $\beta = 107.10°$, $Z = 4$, $D = 3.97$. AM 46:65(1961). 17-531: 6.32_5 3.63_{10} 3.10_8 3.02_8 2.73_8 2.71_8 2.07_8 2.05_8. Isostructural with cryptomelane group. AC(B) 34:255(1978). Small grains. Blackish. Cleavage {001}, {010}, good. $G = 4.3$. *Mickelsberg, Katzenbuckel, Odenwald, Germany,* with sanidine, orthopyroxene, apatite, amphibole, and feldspar. Liberia, in garnet–clinopyroxene–plagioclase xenoliths in kimberlite, associated with rutile, ilmenite, perovskite, and titanite. This freudenbergite is the Fe^{2+} analog, with composition $Na_2Fe^{2+}Ti_7O_{16}$. RG *NJMM:12(1961), ZK 132:157(1970), MA 85:788(1985)*.

7.9.4.1 Priderite $(K,Ba)(Ti,Fe^{3+})_8O_{16}$

Named in 1951 for Rex Tregilgas Prider (b.1910), geologist, University of Western Australia. TET $I4/m$. $a = 10.139$, $c = 2.966$, $Z = 1$, $D = 3.92$. MM 36:867(1968). 6-296: 7.14_7 5.05_7 3.20_{10} 2.47_6 2.23_4 1.891_6 1.687_4 1.588_5. Rectangular prismatic crystals. Black; reddish, gray streak, adamantine luster. Cleavage {001}, perfect; {010}, fair. $G = 3.86$. Weakly paramagnetic. Uniaxial (+), $N_O > 2.10$. O, deep reddish brown; E, deep reddish brown to black. *Walgidee Hills, Fitzroy Basin, Kimberley, WA, Australia,* as a characteristic accessory of all the leucite-bearing rocks in the area. RG *AM 36:793(1951), AUSM 1:298(1985)*.

7.9.4.2 Ankangite $Ba(Ti,V)_8O_{16}$

Named in 1989 for the locality. TET $I4/m$. $a = 10.139$, $c = 2.961$, $Z = 1$, $D = 4.389$. 45-1393: 3.59_3 3.21_{10} 2.48_2 2.27_2 2.23_2 1.891_1 1.690_1 1.589_2. Prismatic crystals, showing {100} and {110}. Black, grayish-black streak, vitreous to adamantine luster. Uneven fracture, brittle. $VHN_{50} = 874$. $G = 4.44$. In RL, rosy gray-white, anisotropic, weakly to distinctly bireflectant, no internal reflections. $R = 20.1, 12.8$ (589 nm). From *Ankang Co., Shaanxi Prov., China,* in a quartz vein, with barite, barytocalcite, roscoelite, and diopside. RG *AM 76:2020(1991)*.

7.9.5.1 Mannardite $BaTi_6(V,Cr)_2O_{16} \cdot H_2O$

Named in 1986 for George William Mannard (1932–1982), Canadian mineralogist and president, Kidd Creek Mines Ltd., long interested in the mineral deposits of southern BC. TET $I4_1/a$. $a = 14.356$, $c = 5.911$, $Z = 4$, $D = 4.28$. CM 24:55(1986). 42-615: 3.20_{10} 2.47_7 2.26_3 2.22_5 1.89_7 1.69_5 1.59_8 1.39_4. Prismatic, elongated crystals, with {110} and {001}; {110} striated [001]. Jet black; appearance similar to rutile, with white to grayish streak and adamantine luster. Cleavage {001}, prominent; uneven to subconchoidal fracture; very brittle. $H = 7$. $G = 4.12$. Uniaxial (+); $N_O = 2.26$, $N_E = 2.42$ (calc). In RL, pale reddish brown, no internal reflections; distinctly pleochroic, light to dark brown; strongly anisotropic. $R = 15.7–16.8$ (470 nm), 15.2–17.5 (546 nm), 15.25–18.05 (589 nm), 15.3–18.55 (650 nm). *Rough claims, Sifton Pass, Kechica River, BC, Canada*, in quartz–carbonate veins, with barytocalcite, norsethite, quartz, barite, and sulvanite. Brunswick No. 12 orebody, Bathurst, NB, Canada. RG *AM 73:193(1988)*.

7.9.5.2 Redledgeite $BaTi_6Cr_2^{3+}O_{16} \cdot H_2O$

Named in 1961 for the locality. This mineral was originally described in 1928 with a seriously incorrect analysis and named "chromrutile." *AM 13:69(1928)*. The same material was reexamined by Strunz and, on the basis of new data, renamed *redledgeite*, since it was not a variety of rutile. *NJMM:107(1961)*. TET $I4_1/a$. $a = 14.320$, $c = 5.893$, $Z = 4$, $D = 4.413$. *CM 24:55(1986)*. *39-352*: 7.10_2 3.57_5 3.20_{10} 2.83_3 2.47_6 2.22_4 1.89_5 1.58_6 1.39_5. Redledgeite is isostructural with mannardite, and the powder patterns are almost identical. The weak line 7.10_2 is, however, diagnostic for redledgeite, as it is absent from the mannardite pattern. Prismatic or equant bipyramidal crystals. Black; another type is pale green; adamantine luster. $H = 6\frac{1}{2}$. $G = 3.72$. *Red Ledge mine, Washington district, Nevada Co., CA*, with kammererite and chromite. RG *AM 46:1201(1961)*, *NJMM:116(1963)*.

7.10.1.1 Ranciéite $(Ca,Mn) Mn_4^{4+}O_9 \cdot 3H_2O$

Named in 1859 for the locality. HEX Space group unknown. $a = 2.86$, $c = 7.50$. *22-718*: 7.49_{10} 3.74_1 2.46_1 2.34_1 2.06_1 1.76_1 1.43_1. Massive; as stalactites with colloform, banded texture. Black; brownish; violet, metallic luster. $G = 3.2$. Oriente Prov., Cuba. *Rancié, near Vicdessos, Ariège, France*, in cavities in limonite, as radiated groups of cleavable lamellae; also at the Filhols mine, Pyrénées-Orientales. Monte Gelato, Mazzano, Romano, Italy, in lake deposited tuffs, associated with halloysite. RG *DI:572*, *BM 92:191(1969)*, *AM 54:1741(1969)*, *MM 50:111(1986)*.

7.10.1.2 Takanelite $(Mn,Ca)Mn_4^{4+}O_9 \cdot H_2O$

Named in 1971 for Katsutoshi Takane (1899–1945), professor of mineralogy, Tohoku University, Sendai, Japan. HEX Space group unknown. $a = 8.68$, $c = 9.00$, $Z = 3$, $D = 3.78$. *25-164*: 7.57_{10} 4.43_2 3.77_3 2.46_2 2.35_2 2.07_2 1.42_2. Massive; in irregular nodules. Gray; black, brownish-black streak, submetallic to dull luster. No cleavage. $VHN_{100} = 480$. $G = 3.41$. In RL, yellowish gray, with weak yellowish-white to light yellowish-gray pleochroism, and moderate anisotropy, yellowish gray to light brownish gray. Takanelite is the Mn^{2+} analog of ranciéite. *Nomura mine, Ehime, Japan*, in the oxidation zone of a braunite–rhodochrosite–caryopilite deposit, intimately intergrown with braunite, halloysite, goethite, and quartz. RG *AM 56:1487 (1971)*, *69:814 (1984)*; *JGSK 76:105(1980)*.

7.10.2.1 Karibibite $Fe_2^{3+}As_4^{3+}(O,OH)_9$

Named in 1973 for the locality. ORTH Space group unknown. $a = 27.91$, $b = 6.53$, $c = 7.20$, $Z = 6$, $D = 4.04$. *25-1405*: 6.35_4 5.35_3 3.50_5 3.18_8 3.09_7

2.80_8 2.67_4 2.38_{10}. Spindle-shaped bundles of fibers. Brownish yellow. Very soft. G = 4.07. Fluoresces yellow in SWUV light. Paramagnetic. Readily soluble in dilute acids or in dilute alkali hydroxide solutions. Biaxial (−); $N_x = 1.96$, $N_z > 2.10$. X, straw yellow; Z, light brownish yellow. *Karibib lithium pegmatite, Namibia*, in vugs in loellingite, associated with eosphorite, scorodite, and quartz. Lavra do Enio (a pegmatite), Galileia, MG, Brazil, after löllingite. RG *LIT 6:265(1973), AM 59:382(1974), NJMA 138:94(1980)*.

7.10.3.1 Otjisumeite $PbGe_4O_9$

Named in 1981 for the Herrero name for the locality, Tsumeb. TRIC. $P1$ or $P\bar{1}$. $a = 6.945$, $b = 6.958$, $c = 9.279$, $\alpha = 102.56°$, $\beta = 103.03°$, $\gamma = 114.46°$, Z = 2, D = 5.77. *35-473:* 5.87_3 4.20_4 3.41_5 2.95_{10} 2.41_3 2.22_3 1.85_4 1.78_4. Pseudohexagonal crystals, elongated along c axis. White or colorless, subadamantine luster. Cleavage {001} in traces. H = 3. Biaxial; X $\wedge$ c = 3–5°; $N_x = 1.920$, $N_z = 1.943$; 2V = 20°. *Tsumeb mine, Namibia*, in small cavities in tennantite–germanite–renierite ore, associated with schaurteite, chalcocite, siderite, and gypsum. RG *NJMM:49(1981), AM 72:1026(1987)*.

7.11.1.1 Muskoxite $Mg_7Fe_4^{3+}O_{13} \cdot 10H_2O$

Named in 1969 for the igneous complex (the Muskox intrusion) in which the mineral occurs. Probably pseudo HEX-R Space group unknown. $a = 3.07$, $c = 4.6$. *22-709:* 4.61_8 4.44_2 4.12_3 2.66_3 2.31_{10} 1.75_6 1.54_4 1.46_2. Aggregates of small crystals, also anhedral grains. Dark reddish brown, light orange-brown streak, vitreous luster. Cleavage {0001}, perfect. H = 3. G = 3.15. Biaxial (−); N = 1.80–1.81; 2V = 10–40°. *Muskox intrusion, Coppermine River area, NWT, Canada*, in drill core, associated with coalingite, serpentine, and ill-defined mixtures of Mg–Fe–Mn oxides. RG *AM 54:684(1969)*.

7.11.2.1 Brownmillerite $Ca_2(Al,Fe)_2O_5$

Named in 1964 for Lorrin Thomas Brownmiller, chief chemist of the Alpha Portland Cement Co., Easton, PA, who was among the first to prepare and describe the artificial compound, which is found in cement clinker. ORTH Pcmm. $a = 5.567$, $b = 14.521$, $c = 5.349$, Z = 4, D = 3.732. Structure: *AC(B) 27:2311 (1971)*. *30-226:* 7.25_5 2.78_3 2.67_4 2.64_{10} 2.58_2 2.05_4 1.93_4 1.81_5. Minute square platelets. Reddish brown. G = 3.76. Biaxial (−); $N_x < 2.02$, $N_y, N_z > 2.02$. X and Y are in plane of platelets; diagonal extinction. X, yellowish brown; Y,Z, reddish brown. *Ettringer, Bellerberg, Mayen, Laacher See area, Eifel, Rhineland–Pfalz, Germany*, in metamorphosed marly limestone inclusions in effusive volcanic rocks, associated with wollastonite, gehlenite, larnite, diopside, mayenite, spinel, portlandite, and other minerals. From the "mottled zone" of the Hatrurim formation, Israel–Jordan. RG *NJMM:22 (1964), AM 50:2106(1965)*.

7.11.2.2 Srebrodol'skite Ca$_2$Fe$_2^{3+}$O$_5$

Named in 1985 for Boris Ivanovich Srebrodol'ski (b.1927), Ukranian mineralogist, Lvov State University. ORTH *Pnma*. $a = 5.420$, $b = 14.752$, $c = 5.594$, $Z = 4$, $D = 4.03$. *ZVMO 114:195(1985)*. *38-408:* 7.38$_{10}$ 3.69$_{10}$ 2.80$_6$ 2.71$_6$ 2.68$_{10}$ 2.08$_4$ 1.95$_7$ 1.84$_{10}$. Minute grains and crystals, flattened {010}. Black, brownish red; grayish-brown streak; adamantine–metallic luster. Cleavage {hol}, good. $H = 5.5$. $G = 4.04$. Nonfluorescent. Weakly magnetic. Soluble in HCl. Biaxial $(-)$; X parallel to [010]; $N_x \sim 2.24$, $N_y \sim 2.25$, $N_z \sim 2.27$. *Kopeysk, Chelyabinsk coal basin, Ural Mts., Russia*, in a coal seam in pieces of petrified wood thermally metamorphosed by burning coal, which encloses earthy masses of portlandite, carbonates, and srebrodol'skite. RG *AM 71:1279(1986)*.

7.11.3.1 Mayenite Ca$_{12}$Al$_{14}$O$_{33}$

Named in 1964 for the locality. ISO $I\bar{4}3d$. $a = 11.982$, $Z = 2$, $D = 2.68$. *NBSC 539(9):20(1960)*. *9-413*(syn): 4.89$_{10}$ 3.00$_5$ 2.68$_{10}$ 2.45$_5$ 2.19$_4$ 1.95$_3$ 1.66$_3$ 1.60$_3$. Minute rounded grains. Colorless. $G = 2.85$. Isotropic; $N = 1.643$. *Ettringer, Bellerberg, Mayen, Eifel, Rhineland–Pfalz, Germany*, in thermally metamorphosed inclusions of marly limestone in effusive volcanic rocks, associated with brownmillerite, wollastonite, diopside, portlandite, ettringite, and other minerals. Mayenite is also a common component of portland cement clinker. RG *AM 50:2106(1965)*.

7.11.4.1 Hematophanite Pb$_4$Fe$_3^{3+}$O$_8$(OH,Cl)

Named in 1928 from the Greek *haimato*, blood, and *phaneros*, visible, in allusion to the blood-red color in transmitted light. TET *P4mmm*. $a = 3.910$, $c = 15.287$, $Z = 1$, $D = 8.11$. Structure: *SR 48A:182(1981)*. *27-271:* 3.9$_6$ 3.77$_4$ 2.76$_{10}$ 2.71$_{10}$ 2.24$_4$ 1.953$_4$ 1.590$_6$ 1.566$_4$. Thin tablets {001}, in part in lamellar aggregates. Deep red-brown, yellowish-red streak, submetallic luster. Cleavage {001}, very good. $H = 2-3$. $G = 7.70$. Easily soluble in dilute HCl or in HNO$_3$. Uniaxial $(-)$. Low birefringence, blood red by transmitted light. *Jakobsberg mine, Värmland, Sweden*, with jacobsite, plumboferrite, andradite, copper, cuprite, and cerussite in calcite. RG *DI:728, AM 56:625(1971)*.

7.11.5.1 Plumboferrite PbFe$_4^{3+}$O$_7$

Named in 1881 for the composition. HEX-R *P*312. $a = 11.88$, $c = 47.23$, $Z = 42$, $D = 6.55$. *GSAM 85:198(1962)*. *22-656:* 3.88$_4$ 2.96$_8$ 2.81$_9$ 2.64$_{10}$ 1.842$_4$ 1.680$_6$ 1.645$_6$ 1.484$_6$. Thick tabular crystals; in cleavable masses. Black, red streak. Cleavage {0001}. $H = 5$. $G = 6.07$. Easily soluble in HCl with evolution of Cl$_2$ and residue of lead chloride. In RL, gray, with a brownish tint. Weakly anisotropic. $R = 27.2-26.0$ (470 nm), $25.3-24.8$ (546 nm), $24.3-23.9$ (589 nm), $22.8-22.4$ (650 nm). *P&J:305*. *Jakobsberg mine, Nordmark, Värmland, Sweden*; also from the Sjö mine in Örebro. RG *DI:726*.

7.11.6.1 Carboirite $Fe^{2+}Al_2GeO_5(OH)_2$

Named in 1983 for the locality. TRIC $P\bar{1}$. $a = 9.513$, $b = 5.569$, $c = 9.296$, $\alpha = 96.05°$, $\beta = 101.31°$, $\gamma = 89.27°$, $Z = 4$, $D = 3.95$. 35-586: 4.79_5 4.53_{10} 3.02_8 2.73_9 2.48_9 1.85_6 1.61_7 1.38_6. Crystallizes in pseudohexagonal plates. Green, vitreous luster. $H = 6$. Biaxial (+); $N_x = 1.731$, $N_y = 1.735$, $N_z = 1.740$; $2V = 7°$. X, bluish green; Y, light blue; Z, colorless to yellow. *Carboire, Marimana Dome, Ariège, Pyrenees, France*, in a zinc deposit in sphalerite with Ge-bearing quartz. RG *TMPM 31:97(1983), AM 69:406(1984)*.

7.11.7.1 Högbomite 4H, 5H, 6H, 15H $(Mg,Fe)_{1.4}Ti_{0.3}Al_4O_8$

7.11.7.2 Högbomite 8H $(Al,Fe^{2+},Fe^{3+},Mg,Ti,Zn)_{11}O_{15}(OH)$

7.11.7.3 Högbomite 15R, 18R, 24R $(Mg,Fe)_{1.4}Ti_{0.3}Al_4O_8$

Named in 1916 for Arvid Gustaf Högbom (1857–1940), Swedish geologist, University of Uppsala. The diffraction patterns of all högbomites studied have been found to exhibit either hexagonal or rhombohedral symmetry. The dimensions of the unit cell are in each instance $a = 5.72$ Å and $c =$ close to some multiple of 4.6 Å. The crystallography of högbomite can thus be described in terms of a series of polytypes of constant a, to each of which a symbol of the form of nH or nR can be assigned, where H and R indicate hexagonal and rhombohedral lattice types and n is given by $c = 4.6$ nÅ. The polytypes arise by variation of the stacking sequence of approximately close-packed oxygen layers, with interstitial cations on fourfold and sixfold sites. The formulas may be represented by $R^{2+}_{1.0-1.6}Ti^{4+}_{0.2-0.4}R^{3+}_{3.7-4.3}(OH)^-_{0-0.4}$ where $R^{2+} = $ Fe,Mg,Zn and $R^{3+} = $ Al,Fe. The polytypes so far observed are 4H, 5H, 6H, 8H, 15H, 15R, 18R, and 24R. The unit cell constants for each of these polytypes are:

4H: $P6_3mc$; $a = 5.72$, $c = 18.35$, $Z = 4$, $D = 3.72$. *MM 33:563(1963)*.
5H: $P\bar{6}2m$; $a = 5.72$, $c = 23.0$, $Z = 5$, $D = 3.73$. *MM 33:563(1963)*.
6H: $P6_3mc$; $a = 5.72$, $c = 27.5$, $Z = 6$, $D = 3.75$. *MM 33:563(1963)*.
8H: $P6_3mc$; $a = 5.734$, $c = 18.389$, $Z = 1$, $D = 4.04$. *AM 67:373(1982)*.
15H: $a = 5.72$, $c = 69.0$, $Z = 15$, $D = 3.73$. *MM 33:563(1963)*.
15R: $R32$ or $R\bar{3}m$; $a = 5.72$, $c = 69.0$, $Z = 15$, $D = 3.73$. *MM 33:563(1963)*.
18R: $R32$ or $R\bar{3}m$; $a = 5.738$, $c = 83.4$, $Z = 18$, $D = 3.68$. *MM 33:563(1963)*.
24R: $R32$ or $R\bar{3}m$; $a = 5.7$, $c = 55.8$, $Z = 24$. *NJMM:401(1990)*.

There is no obvious correlation between högbomite polytypes and composition, optics, or other physical properties, and even powder data are not entirely reliable to differentiate the polytypes. Positive identification can only be made through determination of the unit cell through single-crystal X-ray identification. Several polytypes are often found in a single deposit, and even a single högbomite crystal may consist of several polytypes. While certain polytypes tend to be concentrated in particular deposits, the location alone is not a reliable means of identification. Insufficient powder diffraction patterns of högbomite of known polytype have been made to be sure if any

lines or patterns are diagnostic of polytype. Observed differences may well be caused by högbomite's substantial variations in chemical composition.
5H—*16-336:* 4.60_4 2.86_4 2.49_6 2.43_{10} 2.08_6 1.978_6 1.436_6 1.426_8.
8H—*MM 41:342(1977):* 2.94_2 2.86_3 2.60_5 2.43_{10} 2.05_2 1.60_2 1.47_3 1.43_5.
18R—*16-167:* 2.93_4 2.87_6 2.69_5 2.55_5 2.44_{10} 2.42_4 2.08_4 1.434_6.
Crystals thick to thin tabular {0001}; as minute grains. Twinning on {0001}, sometimes repeated. Black, grey streak, metallic–adamantine luster. Cleavage {0001}, imperfect; conchoidal fracture; brittle. H = $6\frac{1}{2}$. G = 3.81. Weakly magnetic. Uniaxial (−). O, dark golden brown, E, light golden brown. In RL, gray, weakly pleochroic, distinctly anisotropic. Sometimes exhibits lamellar twinning. Localities: (4H), (5H) Perseus claim, Routevaara, near Kvikkjokk, Lapland, Sweden; (6H) near Peekskill, Westchester Co., NY, with emery; (8H) Johannsen's mine, NNE of Alice Springs, Strangeways Range, NT, Australia; (15H) Castor claim, Routevaara, Lapland, Sweden; (15R) Toombeola, Galway Co., Eire; (18R) Dentz farm, Letaba district, Transvaal, South Africa; (24R) Mautia Hill, Morogoro area, Tanzania. Also (unspecified polytypes) near Whittles, Pittsylvania Co., VA, with hercynite, magnetite, and corundum (emery); Bathurst Twp., Lanark Co., ONT, Canada, with corundum pseudomorphs after spinel. RG *DI:723; AM 37:600(1952), 76:1734(1991); NJMM:373(1977).*

7.11.8.1   Nigerite 6H $(Zn,Mg,Fe^{2+})(Sn,Zn)_2(Al,Fe^{3+})_{12}O_{22}(OH)_2$

7.11.8.2   Nigerite 24R $(Fe^{2+},Zn)_4Sn_2(Al,Fe^{3+})_{15}O_{30}(OH)_2$

Named in 1947 for Nigeria, the country of origin. Two polytypes of nigerite are known, 6H and 24R. In this respect it is similar to högbomite, with which it is structurally related. Nigerite 6H: HEX $P\bar{3}m$ or $P\bar{3}m1$. a = 5.72, c = 13.86, Z = 2, D = 4.47. *MM 28:129(1947).* Nigerite 24R: HEX-R $R\bar{3}m$. a = 5.730, c = 55.60, Z = 3, D = 4.42. *AM 64:1255(1979).* (6H) *26-1391:* 4.65_9 4.00_9 2.83_{10} 2.42_{10} 1.643_9 1.544_{10} 1.426_9 1.412_9. (24R) *38-436:* 4.52_4 4.20_4 2.86_5 2.84_7 2.435_7 2.430_{10} 2.42_5 1.43_4. Crystals are hexagonal plates, also anhedral grains. Light to dark brown, adamantine luster. Brittle. H = 8–9. G = 4.51. Weakly magnetic. Insoluble in acids. Uniaxial (+); N_O = 1.80, N_E = 1.81. Sometimes optically negative. In pegmatites of Seixoso-Macieira, northern Portugal. (6H) *Egbe district, Kabba Prov., Nigeria,* in quartz–sillimanite rocks closely associated with cassiterite-bearing granite pegmatites, associated with sillimanite, andalusite, gahnite, cassiterite, columbite, chrysoberyl, and apatite. The nigerite often forms epitaxial overgrowths on gahnite, {0001} of nigerite on {111} of gahnite. (24R) *Mt. Garnet, Q, Australia.* Jornal area, Amapa, Brazil, in quartz–mica veins with cassiterite, andalusite, tantalite, tourmaline, and chrysoberyl. Venus Creek tantalite placers, Sinnamary R. area, French Guiana. RG *AM 33:98(1948), 52:864(1967); MM 39:837(1974).*

7.11.9.1 Pengzhizhongite-6H $(Mg,Zn,Fe,Al)_4(Sn,Fe)_2Al_{10}O_{22}(OH)_2$

7.11.9.2 Pengzhizhongite-24R $(Mg,Zn,Fe,Al)_4(Sn,Fe)_2Al_{10}O_{22}(OH)_2$
Named in 1989 for Peng Zhizhong, late Chinese mineralogist. The Mg analog of nigerite. HEX, HEX-R Space group unknown. $a = 5.692$, $c = 13.78$, $Z = 1$, $D = 4.16$. PD: 2.85_9 2.42_{10} 1.639_3 1.545_4 1.414_5 1.050_3. Minute tabular crystals. Light yellowish brown, light yellow, to colorless; white streak; vitreous luster. $H = 8$. $G = 4.22$. From high-temperature metasomatic tungsten ore in the Anhua area, Hunan, China, associated with scheelite, magnetite, rutile, nigerite, cassiterite, zircon, and taaffeite. RG *AMS 9:20(1989)*, *AM 76:1730(1991)*.

7.11.10.1 Kleberite $FeTi_6O_{13} \cdot 4H_2O$ (?)
Named in 1978 for Will Kleber (1906–1970), Humboldt University, Berlin. Data unsatisfactory; a doubtful species. Probably HEX Space group uncertain, probably $P6_3mcm$. $a = 4.94$, $c = 4.59$. PD: 2.46_3 2.16_6 1.67_{10} 1.42_6. Small grains. Dark brown to black. $H = 4-4\frac{1}{2}$. $G = 3.27$. Uniaxial $(-)$ or biaxial $(-)$; $N \gg 1.80$. Bireflectance $= 0.04$–0.05. From *near Königshain and near Borna, western Weissel basin, Germany*, in the heavy-mineral fraction from Tertiary sediments. RG *AM 64:655(1979)*.

7.11.11.1 Mongshanite $(Mg,Fe^{2+},Cr)_2(Ti,Zr)_5O_{12}$
Named in 1982. Data inadequate; a doubtful species. HEX Space group unknown. *AM 73:441(1988)*. PD: 3.44_4, 2.89_{10} 2.26_7 2.14_7 1.80_5 1.60_5 1.45_7 0.904_7. As inclusions in ilmenite. Physical properties similar to ilmenite. *Yimengshan area (?), Shandong, China*, in kimberlite dikes. *MA 84M:4292(1984)*.

7.11.12.1 Diaoyudaoite $NaAl_{11}O_{17}$
Named in 1986 for the locality. HEX $P6_3/mmc$. $a = 5.602$, $c = 22.626$, $Z = 2$, $D = 3.21$. *AM 75:240(1990)*. *45-1451*: 11.3_{10} 5.62_4 2.70_7 2.26_5 2.24_6 2.05_6 1.430_6 1.404_8. Minute thin tabular crystals. Colorless to light green, vitreous luster. Well-developed cleavage, conchoidal fracture. $H = 7\frac{1}{2}$. $G = 3.30$. Uniaxial $(-)$; $N_O = 1.688$, $N_E = 1.663$. From seafloor muds near *Diaoyu Dao Island, N of Taiwan*, at 1,500 m depth, where it is abundant in the heavy-mineral fraction, associated with hornblende, epidote, dolomite, chlorite, and biotite, and containing inclusions of native Cr (?). *AMS 6(3):224(1986)*.

7.11.13.1 Chiluite $Bi_6Te_2^{6+}Mo_2^{6+}O_{21}$
Named in 1989 for the locality. HEX $P6_322$, $P6_3m$, or $P6_3$. $a = 8.970$, $c = 12.207$, $Z = 1$, $D = 3.67$. *45-1369*: 3.3_{10} 3.05_9 2.88_4 2.51_5 2.06_5 1.905_3 1.790_3 1.655_5. Plumose replacements, veinlets, irregular to hexagonal grains. Yellow. $H = 3$. $G = 3.65$. Uniaxial $(-)$; $N_O = 2.4$, $N_E = 2.3$; parallel extinction; negative elongation. In RL, gray, weakly anisotropic, no bireflectance.

R = 19.5–18.4 (470 nm), 18.3–17.3 (546 nm), 17.8–16.75 (589 nm), 17.3–16.45 (650 nm). *Chilu, Fujian, China*, in a molybdenite deposit, associated with bismuthinite, molybdenite, joseite, koechlinite, and cassiterite. RG *AM 76:666(1991)*.

7.11.14.1 Rilandite (Cr,Al)$_6$SiO$_{11}$ · 5H$_2$O

Named in 1933 for James L. Riland (1857–1938?), newspaper publisher of Meeker, CO, on whose claim the mineral was found. Symmetry and space group unknown. Not amorphous, but powder diffraction pattern shows broad lines only. Massive, pitchy, with an angular, platy habit. Dark brownish black, grayish-brown streak, dull to brilliant luster. Conchoidal fracture, very brittle. H = 2–3. Needs further study. From carnotite claims *20 km ENE of Meeker, CO*, in sandstone, associated with fossil wood. RG *AM 18:195(1933)*.

7.11.15.1 Lindqvistite Pb$_2$(Mn^{2+},Mg)Fe$^{3+}_{16}$O$_{27}$

Named in 1993 for Bengt Lindqvist (b.1927), Swedish mineralogist and petrologist, and curator, Swedish Natural History Museum. HEX $P6_3/mmc$. $a = 5.951$, $c = 33.358$, $Z = 2$, $D = 5.76$. *AM 78:1304(1993)*. PD: 4.17$_6$ 3.33$_4$ 3.01$_6$ 2.98$_7$ 2.80$_{10}$ 2.78$_5$ 2.62$_{10}$ 2.61$_9$. Found in subhedral tabular crystals (max. 5 mm) flattened on (0001). Black, brownish-black streak, submetallic luster. Cleavage (0001), perfect, brittle. H = 6. Slowly soluble in cold 3:1 HCl, insoluble in HNO$_3$ or H$_2$SO$_4$. In RL, gray with a faint brownish tint, weakly bireflectant, moderately anisotropic, no internal reflections. R = 23.6 (470 nm), 22.8 (546 nm), 22.2 (589 nm), 21.4 (650 nm). Found at the *Jakobsberg mine, Filipstad, Sweden*, associated with hematite, jacobsite, plumboferrite, andradite, and other minerals in skarn. RG

===== Class 8 =====

Multiple Oxides Containing Niobium, Tantalum, and Titanium

8.1.1.1 Fergusonite-(Y) $YNbO_4$

Named in 1827 for Robert Ferguson (1767–1840), Scottish advocate, politician, and mineralogist. TET $I4_1/a$. $a = 5.16$, $c = 10.89$, $Z = 4$, $D = 5.38$ (for pure $YNbO_4$). Fergusonite is usually metamict. $9\text{-}443$(heated): 3.12_{10} 3.01_2 2.96_9 2.74_4 2.64_2 2.53_1 1.90_5 1.86_3. Crystals prismatic [001] to pyramidal, sometimes with {001} prominent; as irregular masses or grains. Black to brownish black, gray, yellow, brown; brown, yellow-brown, greenish-gray streak; vitreous–submetallic luster. Cleavage {111} in traces, subconchoidal fracture, brittle. H = $5\frac{1}{2}\text{-}6\frac{1}{2}$. G = 4.2–5.7. Fergusonite takes up water in the process of metamictization; the D and H decrease with hydration. It is essentially an oxide of Y, RE, niobium, and tantalum, with the type formula ABO_4, where A = Y, Er, Ce, La, Nd, Di, U, Zr, Th, Ca, Fe^{2+}. B = Nb, Ta, Ti, Sn, and W. A complete series apparently exists between the predominantly niobian fergusonite and the tantalian formanite, although the Ta-rich members are far scarcer. Water is reported in nearly all analyses but is undoubtedly due to alteration. The nearly invariable presence of at least traces of U and Th result in most fergusonites being metamict. Isotropic (when metamict); N = 2.06–2.19, depending on composition. Spruce Pine, Mitchell Co., NC, in pegmatites with allanite; abundant at Barringer Hill, Llano Co., TX, with gadolinite and other rare-earth minerals. Murchison Twp., ONT, Canada, with euxenite. *Qeqertaussaq, Julienhaab district, Greenland.* Iveland district, Norway, in pegmatites with gadolinite, thalenite, and so on, and in Hundholmen with allanite, yttrofluorite, xenotime, and gadolinite; in Russia in the Ilmenskie Mts. at the Blum mine, with samarskite and ilmenorutile. Ytterby, Sweden, in pegmatite with xenotime, cyrtolite, and biotite. In Sri Lanka in gem gravels; Hakatamura and Hagata, Iyo, Japan, in pegmatite; Morogoro, Tanzania. In Madagascar S of Lake Itasy; at Ambatofotsikely; and elsewhere. RG *DI:757; AM 40:805(1955), 46:1516(1961), 52:1536(1967).*

8.1.1.2 Formanite-Y $YTaO_4$

Named in 1944 for Francis Gloster Forman (1904–1980), government geologist, WA, Australia. TET $I4_1/a$. $a = 7.75$, $c = 11.41$, $Z = 4$, $D = 7.03$ (data from synthetic $YTaO_4$). $26\text{-}1478$(heated): 3.13_{10} 2.93_9 2.72_6 1.90_7 1.64_7 1.57_7 1.22_6 1.19_6. As anhedral pebbles, also tabular crystals. Black, pale yellow streak, vitreous to submetallic luster. H = 6. G = 6.24. Isotropic (metamict); N = 2.14. Siberia, Russia, in albitized pegmatite with monazite and gadolinite. *Cooglegong, WA, Australia*, in placers with cassiterite, monazite, euxenite, and gadolinite. Used as an ore of tantalum. RG *DI:757, DANS 154:86(1964)*.

8.1.1.3 Fergusonite-(Ce)   (Ce,Nd,La)NbO$_4$·0.3H$_2$O

Named in 1986 for the composition, and as a member of the fergusonite group. TET Space group unknown. $a = 5.17$, $c = 5.30$ (heated material). Irregular poikiloblasts and rare prismatic dipyramidal crystals. Dark red to black. G = 5.48. Uniaxial (−); $N_O = 2.30$, $N_E = 2.20$. *Novopoltavsk massif, Ukraine*, in carbonatite, associated with magnetite, columbite, allanite, bastnaesite, parisite, apatite, uranpyrochlore, aeschynite, fersmite, and monazite. RG *AM 74:946(1989)*.

8.1.1.4 Fergusonite-(Nd)   (Nd,Ce)(Nb,Ti)O$_4$

Named in 1987 for the composition and as a member of the fergusonite group. Space group unknown. Granular, as equant grains. Dark brown, adamantine luster. From *Bayan Obo, Inner Mongolia, China*, in a rare-earth deposit, in veinlets of aegirine. RG *AM 74:946(1989)*.

8.1.2.1 β-Fergusonite-(Y)   YNbO$_4$

Named in 1961 from the symmetry, and the composition as a Y-dominant member of the fergusonite group. MON $C2/c$. $a = 7.037$, $b = 10.945$, $c = 5.298$, $\beta = 134.07°$, Z = 4, D = 5.57. *23-1486*(syn): 3.12_{10} 2.96_9 2.73_5 2.64_5 1.90_7 1.86_6 1.76_4 1.63_4. Light yellow. G = 5.65. *Central-eastern Russia*, in the apical parts of microcline granite stocks. Karasugama mine, Fukushima, Fukushima Pref., Japan. RG *AM 46:1516(1961)*.

8.1.2.2 β-Fergusonite-(Ce)   (Ce,La,Nd,Th)NbO$_4$

Named in 1973 for the symmetry, composition, and its being a member of the fergusonite group. MON $C2/c$. $a = 5.19$, $b = 11.34$, $c = 5.48$, $\beta = 84.95°$, Z = 4. *AM 60:485(1975)*. PD: 3.21_{10} 3.07_{10} 1.96_8 1.91_7 1.71_6 1.67_6 1.61_6. *33-332*(syn): 3.23_{10} 3.06_{10} 2.85_3 2.76_2 2.57_2 1.98_3 1.91_2 1.70_2. Well-formed crystals, frequently terminated by bipyramids. Red; reddish brown, vitreous to greasy luster. G = 5.34. From *northern China*, in a skarn deposit, associated with diopside, phlogopite, and Ce-apatite. RG *AM 62:397(1977)*.

8.1.2.3 β-Fergusonite-(Nd)   (Nd,Ce)NbO$_4$

Named in 1984 for the symmetry, composition, and its being a member of the fergusonite group. MON (heat-treated material) $I2$. $a = 5.07$, $b = 5.62$, $c = 5.41$, $\beta = 93°$, Z = 2, D = 6.45. *32-680*: 3.21_8 3.04_{10} 2.82_8 2.73_3 2.57_4 1.899_3 1.799_3 1.603_4. Irregular grains, rarely prismatic crystals. Red, brownish red, yellowish brown; pale brown streak, vitreous to greasy luster. No cleavage, subconchoidal fracture. Moderately paramagnetic. Isotropic (metamict); $N = 2.0$. *Bayan Obo, Inner Mongolia, China*, in an Nb–rare earth deposit in proterozoic riebeckite–ferroan dolomite, associated with biotite, magnetite, monazite, bastnaesite, aeschynite-(Nd), ilmenite, and fergusonite-(Ce). RG *AM 69:406(1984)*.

8.1.3.1 Yttrotantalite-(Y) $(Y,U,Fe^{2+})(Ta,Nb)O_4$

Named in 1802 from the composition. Dimorphous with formanite. ORTH Space group unknown. Metamict. Prismatic [001] with {110} and {010}; also tabular {010}. Black; brown, gray streak, submetallic to vitreous/greasy luster. Cleavage {010}, indistinct, conchoidal fracture. $H = 5-5\frac{1}{2}$. $G = 5.7$. Isotropic; $N = 2.15$. Rare metals mine, Aquarius Range, Mojave Co., AZ. In *Sweden* at Finbo and Brodbo, near Falun, with albite, garnet, and topaz; and at *Ytterby* with fergusonite-(Y). Norway near Berg, in Råde, in pegmatite, and at Hattevik E of Dillingö. RG *DI:763*.

8.1.3.2 Yttrocolumbite-(Y) $(Y,U,Fe)(Nb,Ta)O_4$

Named in 1937 from the composition. Inadequately characterized to be certain that the reported material is the Nb-dominant analog of yttrotantalite. Polymorphous with fergusonite-(Y) and β-fergusonite-(Y). ORTH Space group unknown. Metamict. Black, brilliant luster. $H = 6$. $G = 5.49$. From *Mozambique*. RG *MASLI:369(1937), AM 25:155(1940)*.

8.1.4.1 Ishikawaite $(U,Y,Fe)(Nb,Ta)O_4$

Named in 1922 for the locality. More study needed, possibly is uranpyrochlore. Probably ORTH Space group unknown. Metamict. Tabular crystals. Black, dark brown streak, waxy luster. Conchoidal fracture. $H = 5-6$. $G = 6.3$. From the *Ishikawa district, Fukushima, Iwaki Pref., Japan*, in pegmatite, with samarskite. RG *DI:766*.

8.1.5.1 Loranskite-(Y) $(Y,Ce,Ca)(Zr,Ta)_2O_6$ (?)

Named in 1899 for Apollonie Mikhailovich Loranski (1847–1917), inspector and teacher of the Mining Institute, St. Petersburg, Russia. An ill-defined mineral of uncertain formula and identity. Probably ORTH Space group unknown. Metamict. Crystals resemble samarskite. Black, greenish-gray streak, submetallic luster. Conchoidal fracture, brittle. $H = 5$. $G = 4.4$. *Impilaks, Lake Ladoga, Finland*, in pegmatite, with "wiikite" (possibly euxenite). RG *DI:767*.

8.1.6.1 Stibiocolumbite $SbNbO_4$

Named in 1915 for the composition. ORTH $Pc2_1n$. $a = 5.559$, $b = 11.79$, $c = 4.929$, $Z = 4$, $D = 5.728$. *AM 48:1348(1963)*. 16-907(syn): 4.54_1 3.52_2 3.13_{10} 2.95_3 2.78_1 2.70_1 2.46_1 1.74_2. Crystals prismatic [001]; {010} and

{110} striated parallel to [001]; hemimorphic; twinning common, polysynthetic, twin axis [001], composition plane {010}. Brown, reddish yellow, greenish yellow; pale yellow to yellow-brown streak; resinous to adamantine luster. Cleavage {010}, distinct; {100}, indistinct; subconchoidal to granular fracture; brittle. H = $5\frac{1}{2}$. G = 5.98. Markedly pyroelectric. The Nb-dominant members of the stibiocolumbite–stibiotantalite group are decidedly rare compared to the Ta-rich members, and the nearly pure Nb end member has not been found. Biaxial (+); XYZ = abc; $N_x = 2.40$, $N_y = 2.42$, $N_z = 2.46$ (Na); 2V = 73°. In pegmatite at the *Himalaya mine, Mesa Grande, San Diego Co., CA*, in fine crystals associated with pink beryl, lepidolite, and rubellite. Varuträsk, Sweden, in pegmatite. RG *DI:767*.

8.1.6.2 Stibiotantalite SbTaO$_4$

Named in 1893 for the composition. ORTH $Pc2_1n$. a = 5.535, b = 11.814, c = 4.911, Z = 4, D = 7.583(syn). *AM 48:1348(1963)*. *16-908:* 3.51_4 3.12_{10} 2.95_4 2.692_2 2.687_2 2.456_1 1.888_2 1.735_2. Isostructural with stibiocolumbite (8.1.6.1). Crystals prismatic, tabular [001]; also equant. Hemimorphic. Polysynthetic twinning common, twin axis [001], composition plane {010}. Brown, reddish yellow, greenish yellow; resinous to adamantine luster. G = 7.34. Biaxial (+); $N_x = 2.374$, $N_y = 2.404$, $N_z = 2.457$ (Na). Himalaya mine, Mesa Grande, San Diego Co., CA. Muiane pegmatite, Alta Ligonha, Mozambique, in fine, large crystals up to 10 cm, with rubellite, morganite, and manganotantalite. Benson No. 3 mine, Mtoko, northeastern Zimbabwe, with simpsonite, microlite, and manganotantalite. Abundant at *Greenbushes, WA, Australia*, with other tantalum minerals in a large lithium pegmatite. Used as an ore of tantalum. RG *DI:767, AM 42:178(1957)*.

8.1.6.3 Bismutotantalite Bi(Ta,Nb)O$_4$

Named in 1929 from the composition. ORTH *Pcmn*. a = 5.633, b = 11.763, c = 4.957, Z = 4, D = 8.76. *AM 48:1348(1963)*. Isostructural with bismutocolumbite (8.1.6.4). *16-909*(syn): 3.54_2 3.14_{10} 2.94_3 2.48_2 2.03_1 1.90_2 1.73_3 1.53_2. Crystals stout prismatic [001]. Black to light brown, black to yellowish-brown streak, submetallic to adamantine luster. Cleavages {101} and {010}, good; subconchoidal fracture. H = 5. G = 8.4–8.8. Not pyroelectric. Insoluble in acids, including HF. Biaxial (+); $N_x = 2.39$, $N_y = 2.40$, $N_z = 2.43$; 2V = 80°; r < v; parallel extinction. *Gamba Hill, Busiro Co., Uganda*, in pegmatite with schorl, muscovite, and cassiterite. In Brazil in pure alluvial pebbles and boulders (some of which exceeded 25 kg) at Saco da Onça, Carnauba dos Dantas, RGN, and at the Alto Manoel Baldoino, Junco, PA, in fine large crystals with bismutomicrolite. RG *DI:769, NJMM:241(1955), AM 42:178(1957)*.

8.1.6.4 Bismutocolumbite Bi(Nb,Ta)O$_4$

Named in 1992 for the composition. ORTH *Pcmn*. a = 5.677, b = 11.731, c = 4.992, Z = 4, D_{meas} = 7.56, D_{calc} = 7.66. Synthetic BiNbO$_3$ has a = 4.980,

$b = 11.709$, $c = 5.675$ *(ICDD 16-295)*. *45-1372:* 3.16_{10} 2.93_9 2.84_5 2.50_5 1.899_4 1.871_4 1.769_5 1.734_8. *PD*(syn): 3.74_2 3.15_{10} 2.92_4 2.84_3 2.55_2 2.04_3 1.90_3 1.73_4. Isostructural with bismutotantalite but not with stibiocolumbite–stibiotantalite. Prismatic crystals as much as 2 mm long by 1 mm across (Malkhan) and as much as 4 cm long from the Pala district. Black, yellow streak, submetallic luster, transparent in small (<0.03 mm) fragments. Polysynthetic twinning present. Nonfluorescent. $H = 5\frac{1}{2}$. Cleavage {001}, perfect; conchoidal fracture; brittle. Material from the Malkhan district contains only 0.28% wt% Sb_2O_3 and traces of Sn, Ti, and Pb. Material from Pala, CA, contains major Sb, as well as Bi, Nb, and Ta. Biaxial (+); XYZ = acb; $N_x = 2.38$, $N_y = 2.42$, $N_z = 2.47$; $2V_{z\,calc} = 85°$. In miarolytic cavities within extremely fractionated granitic pegmatites. Associated minerals include elbaite, albite, quartz, and minor rare minerals. *Danburitovaya mine, Malkhan Range, central Transbaikalia, Russia.* Occurs as portions of compositionally zoned crystals (including stibiocolumbite and stibiotantalite) from the Stewart mine, Pala district, CA. RG *ZVMO 121:130(1992)*.

8.1.7.1 Alumotantite $AlTaO_4$

Named in 1981 for the composition. ORTH *Pbcn*. $a = 4.473$, $b = 11.308$, $c = 4.775$, $Z = 4$, $D = 7.47$. *CM 30:653(1992)*. *36-440:* 5.66_5 3.64_7 3.13_{10} 2.89_8 2.44_5 1.84_5 1.65_5 1.07_5. Colorless, adamantine luster. No cleavage. Strongly luminescent, bright blue in cathode rays. Nonfluorescent in UV. $VHN_{100} = 1650–1690$. *Kola Penin., Russia,* in albitized areas of granite pegmatites, forming rims around simpsonite and also around natrotantite. Bikita pegmatite, Zimbabwe, replacing and overgrowing simpsonite. Alto do Giz, Equador, RGN, Brazil, replacing simpsonite. RG *ZVMO 110:338(1981)*, *AM 67:413(1982)*, *IGR 24:835(1982)*.

WODGINITE GROUP

The wodginite group includes four isostructural minerals with the general formula

$$ABC_2O_8$$

where

$A = Mn^{2+}, Fe^{2+}, Li, \square$
$B = Sn^{4+}, Ti, Fe^{3+}, Ta$
$C = Ta, Nb$

and a common structure is found in $C2/c$.

The principal structure element consists of octahedra forming α-PbO_2-like zigzag chains via sharing of polyhedron edges. Oxygen atoms form approximately hexagonal close-packed layers. All of the cations in wodginite are in octahedral coordination. They occupy one-half of the octahedral interstices at each level in the close-packed array of anions. There are three cationic sites, called A, B, and C. A and B lie at the same level, and the C sites lie at a

8.1.8.1 Wodginite

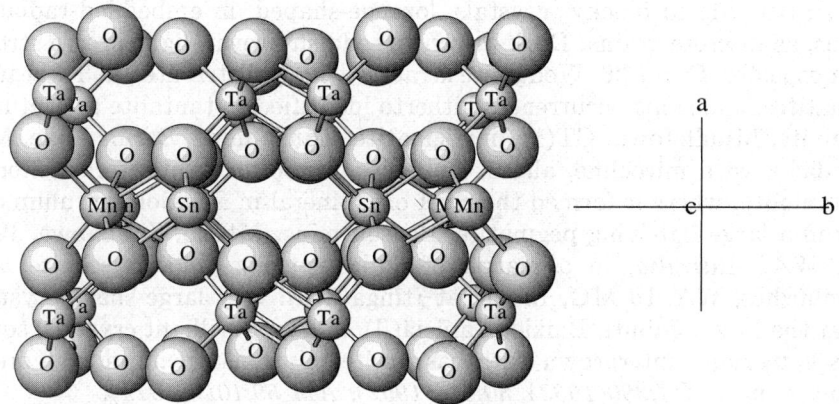

WODGINITE: MnSnTa$_2$O$_8$
Like wolframite (huebnerite), wodginite contains hexagonal closest-packed layers of oxygen stacked on (100), with all cations in octahedral interstices. All-Ta layers alternate with Mn-Sn layers. Double chains of Ta-Ta and Mn-Sn edge-sharing octahedra are parallel to the c axis.

different level. These layers alternate along a. Chains are interconnected by corner sharing of linkages between levels. The A and B polyhedra alternate along the zigzag chains of their level; adjacent levels consist only of chains of C polyhedra. Ideally, the A site contains only Mn, the B site Sn, and the C site Ta, resulting in the ideal formula MnSnTa$_2$O$_8$.

There is some solid solution among the various members, but all the members of the group except wodginite are very rare, and insufficient analytical data exist to say how extensive this is. Fe, Ti, and Li are usually present to some degree in all wodginites. Vacancies are common in the A site and probably in the B site. *CM 30:597(1992)*.

WODGINITE GROUP

Mineral	Formula	Space Group	a	b	c	β	D	
Wodginite	MnSnTa$_2$O$_8$	$C2/c$	9.489	11.429	5.105	91.1	7.69	8.1.8.1
Ferrowodginite	Fe^{2+}SnTa$_2$O$_8$	$C2/c$	9.415	11.442	5.103	91.8	7.02	8.1.8.2
Titanowodginite	MnTiTa$_2$O$_8$	$C2/c$	9.466	11.431	5.126	90.32	6.89	8.1.8.3
Lithiowodginite	LiTa$_3$O$_8$	$C2/c$	9.441	11.516	5.062	91.07	7.74	8.1.8.4

Note: Partial ordering is common and results in suppression of some lines of the powder pattern, which may be alleviated by heating the sample.

8.1.8.1 Wodginite MnSnTa$_2$O$_8$

Named in 1963 for the locality. Wodginite group. MON $C2/c$. $a = 9.489$, $b = 11.429$, $c = 5.105$, $\beta = 91.1°$, $Z = 4$, $D = 7.69$. *CM 14:550(1976)*. Structure: *CM 30:597(1992)*. 29-901: 3.67$_7$ 3.00$_{10}$ 2.95$_7$ 2.87$_3$ 2.55$_2$ 2.50$_3$ 1.83$_1$

1.77_3. Prismatic to blocky crystals, lozenge-shaped, in embedded radiating groups, as discrete grains. Dark brown to reddish brown, light brown streak, resinous luster. G = 7.36. Wodginite is rather widely distributed in Ta-bearing pegmatites, and some occurrences hitherto identified as tantalite are actually wodginite. Middletown, CT(*), in pegmatite. Tanco mine, Bernic Lake, MB, Canada(!), with mirocline, albite, lepidolite, microlite, tantalite, cassiterite, and tapiolite, where it formed the chief ore mineral in a major tantulum orebody in a large flat-lying pegmatite. Benson mine, Mtoko, Zimbabwe. *Wodgina, WA, Australia*, in a matrix of albite, quartz, and muscovite; also Greenbushes, WA. In MG, Brazil, at Itinga(!!) in very large sharp crystals, and at the Lavra Jabuti, Baixio, Galileia(!!), in large, brilliant crystals, sometimes epitaxially intergrown with cassiterite. Used as an important ore of tantalum. RG *CM 7:390(1963), 30:613(1992); AM 59:1040(1974)*.

8.1.8.2 Ferrowodginite $Fe^{2+}SnTa_2O_8$

Named in 1992 for the composition. Wodginite group. MON $C2/c$. $a = 9.415$, $b = 11.442$, $c = 5.103$, $\beta = 90.8°$, $Z = 4$, $D = 7.02$. *CM 30:633(1992)*. PD: 4.16_5 3.64_2 2.97_{10} 2.86_2 2.55_3 2.49_4 2.35_2 1.455_3. Minute irregular grains embedded in cassiterite. Dark brown to black, a dark brown streak, vitreous luster. No cleavage, irregular fracture. $H = 5\frac{1}{2}$. Anisotropic; $N > 2.0$. From a pegmatite near *Sukula, Tammela–Somero pegmatite province, southwestern Finland*, with cassiterite, tapiolite, and bismuth. RG

8.1.8.3 Titanowodginite $MnTiTa_2O_8$

Named in 1992 for the composition. Wodginite group. MON $C2/c$. $a = 9.466$, $b = 11.431$, $c = 5.126$, $\beta = 90.32°$, $Z = 4$, $D = 6.89$. *CM 30:637(1992)*. PD: 3.64_5 2.98_{10} 2.97_{10} 2.50_4 1.77_2 1.720_2 1.715_2 1.455_2. Euhedral bipyramidal crystals. Dark brown to black, dark brown streak, vitreous luster. No cleavage, irregular fracture, brittle. $H = 5\frac{1}{2}$. $G = 6.86$. Nonfluorescent. Biaxial (+); $N > 2.0$. From the *Tanco mine, Bernic Lake, MB, Canada*, in the saccharoidal albite unit, with manganocolumbite, microlite, quartz, and beryl. RG

8.1.8.4 Lithiowodginite $LiTa_3O_8$

Named in 1990 for the composition. Wodginite group. MON $C2/c$. $a = 9.441$, $b = 11.516$, $c = 5.062$, $\beta = 91.07°$, $Z = 4$, $D = 7.74$. *MZ 12:94(1990)*. 36-1480(syn): 4.18_3 3.65_6 2.98_{10} 2.94_9 2.53_4 2.50_5 1.779_4 1.706_4. Forms the central portion of wodginite crystals. Polysynthetically twinned. Dark pink to red, pale pink streak, adamantine luster. No cleavage, uneven fracture. H = 5–6. G = 7.5. Nonluminescent under UV or cathode rays. Tanco pegmatite, Bernic Lake, MB, Canada, in compact crystalline aggregates to 1 cm, associated with simpsonite. From albitic zones in pegmatite, *Kalba, eastern Kazakhstan*, associated with wodginite, irtyshite, simpsonite, and ixiolite. Greenbushes, WA, Australia. RG *AM 76:667(1991)*.

8.1.9.1 Liandratite $U^{6+}(Nb,Ta)_2O_8$

Named in 1978 for Georges Liandrat and his wife, of Samoëns, France, who had prospected extensively in Madagascar. HEX $P\bar{3}1m$. $a = 6.36$, $c = 4.01$, $Z = 1$, $D = 6.87$. 29-1435(metamict, heated): 4.01_8 3.18_{10} 2.49_4 2.00_1 1.84_3 1.69_2 1.67_1 1.20_2. As rough crystals, liandratite forming a 1- to 2-mm epitaxial crust on petschekite crystals. Yellow to yellow brown, yellow-white streak, vitreous luster. Conchoidal fracture. $H = 3\frac{1}{2}$. $G = 6.8$. Isotropic; $N = 1.83$. *Antsaoka I pegmatite, Berere region, near Tsaratanana, Madagascar*, associated with petschekite, struverite, monazite, ilmenite, garnet, and tourmaline. Gruta de Generosa, Sabinopolis, MG, Brazil. RG *AM 63:941(1978)*.

8.1.9.2 Petscheckite $U^{4+}Fe^{2+}(Nb,Ta)_2O_8$

Named in 1978 for Eckhard Petsch (b.1939), of Idar-Oberstein, Germany, who was a noteworthy prospector in Madagascar. Petscheckite progressively alters into partially iron-leached and oxidized forms which are designated "oxypetscheckite" and "hydroxypetscheckite." Only the altered compounds have been found; the original, source material, petscheckite, is hypothesized on theoretical grounds. These compounds have the respective general formulas $U_3^{4+}Fe_2^{3+}(Nb, Ta)_6O_{24}$ and $U^{4+}Fe_{1/3}^{3+}\square_{2/3}(Nb, Ta)_2O_7(OH)$. HEX $P\bar{3}1m$. $a = 6.42$, $c = 4.02$, $Z = 1$, $D = 6.99$ (oxypetscheckite). *AM 63:941(1978)*. 29-1426(heated): 4.02_{10} 3.21_{10} 2.51_8 2.01_2 1.86_2 1.85_3 1.70_4 1.49_2. Metamict, hydroxypetscheckite decomposes on heating and the cell constants are unknown. Tabular crystals, hexagonal shape. Black, brownish-black streak, submetallic luster. $H = 5$. $G = 7$. *Antsaoka I pegmatite, Berere district, Madagascar*, associated with columbite, struverite, ilmenite, monazite, and liandratite, the latter forming a thin epitaxial coating on the petscheckite crystals. RG

8.1.10.1 Ixiolite $(Ta,Nb,Mn,Sn,Fe^{2+})O_2$

Named in 1857 for Ixion, a mythological character related to Tantalus (ixiolite is mineralogically related to tantalite). ORTH *Pbcn*. $a = 4.785$, $b = 5.758$, $c = 5.160$, $D = 6.93$. *CM 14:540(1976)*. 15-733: 3.65_3 2.98_{10} 2.57_1 2.51_2 1.77_1 1.75_2 1.72_2 1.46_3. Rectangular prismatic crystals, with {001}, {010}, {100}, {130}, {112}; also, irregular grains, anhedral masses. The structure of ixiolite can be regarded as a disordered columbite–tantalite structure in which the a axis is one-third the length of the columbite a axis, and the various metal ions are more or less randomly distributed between the layers of oxygen atoms. The unit cell contains M_4O_8, in which M is Ta, Nb, Mn, Sn, and Fe^{2+}. While Ta and Mn are usually dominant, the other metals can vary over rather wide limits. Blackish gray; dark brown, grayish-brown streak, submetallic luster. Uneven to subconchoidal fracture. $H = 6-6\frac{1}{2}$. $G = 7.1$. In addition to the metals listed in the formula, certain ixiolites may also have significant quantities of Sc^{3+}, Ti, Fe^{3+}, W, U, and/or Nb > Ta. Helen Beryl mine, Custer, SD. Tanco mine, Bernic Lake, MAN, Canada, with wodginite, tantalite, microlite, cassiterite, tapiolite, and albite in a large Li/Cs-rich pegmatite. *Skogböle, Kimito, Finland*, in pegmatite with tapiolite. Betanimena, Madagascar (scandian). Muiane, Alto Ligonha, Mozambique (scandian). Araçuai, MG, Brazil. RG *DI:778; AM 48:961(1963), 59:1020(1974); IMA Abst.:265(1986)*.

8.1.11.1 Samarskite-(Y)   (Y,Fe,U)(Nb,Ta)O$_4$

Named in 1847 for Vasilii Erafovich Samarski-Bykhovets (1803–1870) of the Russian Corps of Mining Engineers. ORTH *Pbcn*. $a = 5.740$, $b = 4.774$, $c = 5.068$. Always metamict. *39-361*(heated): 3.72_2 3.09_{10} 2.96_9 2.81_2 2.61_3 2.48_2 1.912_3 1.854_2. Crystals prismatic [001]; also tabular {100} and {010}, usually rough; often massive, and then not easily identified; often found in close association with columbite crystals. Black, sometimes with a brownish tint; externally often brown or yellowish brown due to alteration; dark reddish brown to black streak; vitreous to resinous luster; crystals externally dull. Cleavage {010}, indistinct(?); conchoidal fracture; brittle. $H = 5$–6. $G = 5.69$. Isotropic; $N = 2.20 \pm 0.05$. Found in granite pegmatites, especially those related to orogen-related granites having high Y and RE contents. Wiseman's mica mine, Mitchell Co., NC; also at Spruce Pine, associated with columbite and uranmicrolite. Pelton's quarry, Portland, CT. Topsham, ME, with monazite. Ohio City, CO, in Li-rich pegmatite. Maisonneuve, Berthier Co., QUE, Canada, with beryl and fergusonite. Anneröd, near Moss, Norway, with columbite in pegmatite. *Miask, Ural Mts., Russia*(!), with columbite, aeschynite, monazite, and garnet. Antanamalaza, Madagascar(!), with almandine and magnetite. In Brazil at Divino de Ubá(!), MG, and also near Sabinopolis(!), in parallel growth with columbite. RG *DI:797, DANS 160:693(1965), AM 70:856(1985)*.

8.1.12.1 Qitianlingite   Fe$_2^{2+}$Nb$_2$W^{6+}O$_{10}$

Named in 1985 for the locality. ORTH *Pbcn*. $a = 23.706$, $b = 5.723$, $c = 5.045$, $Z = 4$, $D = 6.42$. *AM 73:1497(1988)*. Qitianlingite is a polysome of wolframite and columbite (i.e., the structure can be described as containing alternate slices of columbite and wolframite structures). Each slice has both the structure and the composition of one of the two parental minerals. Structure: *IMA Abst.:261(1986)*. *41-1415:* 3.65_7 2.96_{10} 2.87_6 2.52_7 2.49_4 1.77_8 1.39_6 1.19_6. Tabular crystals, flattened {001}. Black, dark brown streak, metallic–semimetallic luster. Cleavage {100}, perfect. VHN$_{50} = 520$–580. Moderately magnetic. Insoluble in dilute HCl. In RL, yellowish white, weakly pleochroic and anisotropic. $R = 12.0$–14.0 (546 nm). *Qitian Ling, Hunan, China*, in a pegmatite with zinnwaldite, cassiterite, wolframite, and "wolframixiolite." RG

PYROCHLORE GROUP

The pyrochlore group consists of 22 isometric minerals with the general formula

$$A_{1-2}B_2O_6(O, OH, F) \cdot nH_2O$$

where

A = Ca, K, Ba, Y, Ce, Pb, U, Sr, Cs, Na, Sb^{3+}, Bi, and/or Th
B = Nb, Ta, Ti, Sn, Fe, W

The pyrochlore structure can be derived from the fluorite structure by removing part of the anions so that half the cubes are changed to flattened octahedra. The A-site cations are arranged in cubes, and the B-site cations are arranged in flattened octahedra.

The cations A and B form a face-centered cubic array (as do the Ca atoms in fluorite) and the anions are located in the tetrahedral interstices of the cationic array. The face-centered A and B cations are ordered in alternate {110} rows in every other {001} plane and in alternate {110} rows in the other {001} planes. This results in three kinds of tetrahedral interstices for anions: 48 f positions having 2A and 2B nearest neighbors; 8 a positions having four B near neighbors; and 8 b positions having four A near neighbors. In the pyrochlore structure the 8 a positions are vacant: $A_2B_2O_7$ versus $4CaF_2$. The four B cations that neighbor such an anion vacancy tend to be electrostatically shielded from one another by a displacement x of each 48 f anion from the center of its tetrahedral interstice toward its two neighbor B cations. *VLA 2:499(1966), AC(B) 35:1450(1979)*. The pyrochlore group is subdivided into four subgroups. *AC(B) 35:1450(1979), JSSC:123(1978)*.

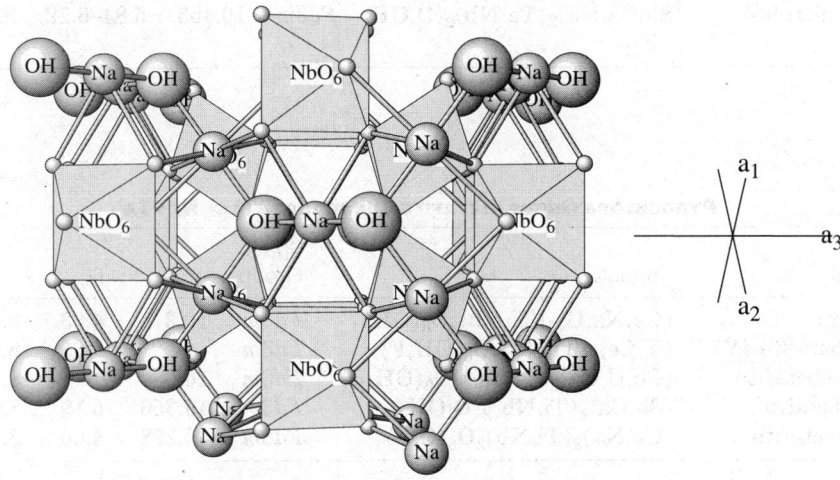

PYROCHLORE: CaNaNb$_2$O$_6$(OH,F)

NbO$_6$ octahedra share corners to form an open framework with channels in the ⟨110⟩ directions, which are occupied by the Ca and Na atoms (disordered over a single site) and OH,F.

Pyrochlore Group–Pyrochlore Subgroup: $Nb > Ta; (Nb+Ta) > 2Ti$

Mineral	Formula	Space Group	a	D	
Pyrochlore	$(Na,Ca)_2Nb_2O_6(OH,F) \cdot nH_2O$	$Fd3m$	10.43	4.26	8.2.1.1
Kalipyrochlore	$(K,Sr)_{2-x}Nb_2O_6(O,OH) \cdot nH_2O$	$Fd3m$	10.604	3.42	8.2.1.2
Bariopyrochlore	$(Ba,Sr)_2(Nb,Ti)_2(O,OH)_7$	$Fd3m$	10.562	4.01	8.2.1.3
Yttropyrochlore-(Y)	$(Y,Na,Ca,U)_{1-2}(Nb,Ta)_2(O,OH)_7$	$Fd3m$	10.0–10.34	4.17	8.2.1.4
Ceriopyrochlore-(Ce)	$(Ce,Ca,Y)_2(Nb,Ta)_2O_6(OH,F)$	$Fd3m$		4.25	8.2.1.5
Plumbopyrochlore	$(Pb,Y,U,Ca)_{2-x}Nb_2(O,OH)_7$	$Fd3m$	10.534	6.75	8.2.1.6
Uranpyrochlore	$(U,Ca,Ce)_2(Nb,Ta)_2O_6(OH)$	$Fd3m$	10.333		8.2.1.7
Strontiopyrochlore	$SrNb_2(O,OH)_7$	$Fd3m$	10.53	3.99	8.2.1.8

Pyrochlore Group–Microlite Subgroup: $Ta > Nb; (Ta+Nb) > 2Ti$

Mineral	Formula	Space Group	a	D	
Microlite	$(Na,Ca)_2Ta_2O_6(O,OH,F)$	$Fd3m$	10.40	6.34	8.2.2.1
Bariomicrolite	$(Ba,U,Pb)_2(Ta,Nb)_2(O,OH)_7$	$Fd3m$	10.570	5.60	8.2.2.2
Plumbomicrolite	$(Pb,Na,Ca)_2(Ta,Nb)_2O_6(O,OH,F)$	$Fd3m$	10.56	7.2	8.2.2.3
Uranmicrolite	$(U,Ca)_2(Ta,Nb)_2O_6(OH)$	$Fd3m$	10.40	6.45	8.2.2.4
Bismutomicrolite	$(Bi,Ca)(Ta,Nb)_2O_6(OH)$	$Fd3m$	10.485	6.83	8.2.2.5
Stannomicrolite	$(Sn^{2+}Fe^{2+}Mn^{2+})_2(Ta,Nb,Sn^{4+})_2(O,OH)_7$	$Fd3m$	10.57	7.56	8.2.2.6
Stibiomicrolite	$(Sb,Ca,Na)_2(Ta,Nb)_2(O,OH,F)_7$	$Fd3m$	10.455	5.84–6.22	8.2.2.7

Pyrochlore Group–Betafite Subgroup: $2Ti > Nb+Ta$

Mineral	Formula	Space Group	a	D	
Betafite	$(Ca,Na,U)_2(Ti,Nb)_2O_6(OH)$	$Fd3m$	10.3	4.63	8.2.3.1
Yttrobetafite-(Y)	$(Y,Ce)_2(Ti,Nb)_2O_6(OH,F)$	$Fd3m$		5.05	8.2.3.2
Plumbobetafite	$(Pb,U,Ca)_2(Ti,Nb)_2O_6(OH,F)$	$Fd3m$	10.33		8.2.3.3
Stibiobetafite	$(Sb,Ca)_2(Ti,Nb)_2(O,OH)_7$	$Fd3m$	10.356	5.19	8.2.3.4
Calciobetafite	$(Ca,Na)_2(Ti,Nb)_2O_6(O,F)$	$Fd3m$	10.298	4.60	8.2.3.5

8.2.1.1 Pyrochlore

PYROCHLORE GROUP–CESSTIBTANTITE SUBGROUP: ABC_4O_{12}; A = Na, Cs; B = Bi, Sb^{3+}; C = Ta, Nb

Mineral	Formula	Space Group	a	D	
Cesstibtantite	$Cs_{0.31}(Sb,Na,Pb,Ni)_{0.91}(Ta,Nb)_2(O,OH,F)_6(OH,F)_{0.69}$	$Fd3m$	10.515	6.49	8.2.4.1
Natrobistantite	$(Na,Cs)Bi(Ta,Nb,Sb)_4(O,OH)_{12}$	$Fd3m$	10.502	6.32	8.2.4.2

Notes:
1. Cesplumtantite 8.7.11.1, although tetragonal and having a slightly different formula, has a powder pattern very similar to the pyrochlores and might also be considered a member of the pyrochlore group. Further work on this mineral would be indicated.
2. The stibconite structures are very similar.
3. Because of the complex chemistry of these minerals, there was much confusion in their classification and a large number of mineral names were in use to describe the members, depending upon the perceived chemistry. The present nomenclature and classification follows the 1977 report of the IMA committee formed to consider the matter in *AM 62:403(1977)*.
4. The formulas shown here are simplified; in actuality, many other elements can be found in small amounts in naturally occurring material, among them U and Th. These radioactive elements render many pyrochlore minerals metamict, and the diffraction data and cell constants can be determined only after heating the material under investigation.

8.2.1.1 Pyrochlore $(Na,Ca)_2Nb_2O_6(OH,F) \cdot nH_2O$

Named in 1826 from the Greek for *fire* and *green*, because some specimens of the mineral turn green after ignition. Pyrochlore group. ISO $Fd3m$. $a = 10.34$–10.43, $Z = 8$, $D = 4.26$. *CM 6:610(1961). 13-254:* 5.98_2 3.13_2 3.00_{10} 2.60_2 1.84_6 1.57_5 1.50_1 1.19_2. Usually octahedral crystals, sometimes modified by cube faces; as anhedral grains; spinel-law twins common. Brown, black, yellowish brown; light brown to yellowish-brown streak; vitreous to resinous luster. Cleavage {111}, rarely distinct; subconchoidal to uneven fracture; brittle. $H = 5$–$5\frac{1}{2}$. $G = 4.45$. Nonfluorescent in UV light. Infusible. May turn yellowish green or other color upon ignition. Insoluble in HCl, slowly decomposed by concentrated H_2SO_4. Isotropic; $N = 1.96$. In RL, isotropic. Strong brown to yellow internal reflections. $R = 11.0$ (470 nm), 10.7 (546 nm), 10.5 (589 nm), 10.4 (650 nm). Pyrochlore is abundant in some carbonatites and is typically found in nepheline syenite rocks and their derivatives. It is also found in pegmatites, particularly those derived from syenitic magmas. St. Peter's Dome area(!), El Paso Co., CO, in pegmatite. In Canada at Oka, QUE, in carbonatite, with latrappite, niocalite, and so on; Verity carbonatite, Blue River, BC(!). In *Norway* at Fredricksvärn and Laurvik in nepheline syenite pegmatites with zircon, polymignite, and cerian apatite. Alnö region, Sweden, in limestone in contact with nepheline syenite intrusive, with perovskite, zircon, apatite, and magnetite. Miask, Ilmen Mts., Russia, in nepheline syenite with aeschynite and zircon. Los archipelago, Conakry, Guinea, in nepheline syenite, with villiaumite, astrophyllite, fluorite, catapleite, and lavenite. The most important ore of niobium. The biggest deposits are in carbonatites, which in some instances may carry up to 4% Nb_2O_5 contained in one of the pyrochlore group minerals; the particular mineral varies from one carbonatite to another, but bariopyrochlore and pyrochlore are among the most important. RG $DI:748$, AM $62:403(1977)$.

8.2.1.2 Kalipyrochlore $(K,Sr)_{2-x}Nb_2O_6(O,OH) \cdot nH_2O$

Named in 1978 for the composition, with K (kalium) dominant in the A position. Pyrochlore group. ISO $Fd3m$. $a = 10.604$, $Z = 8$, $D = 3.42$. CM $32:415(1994)$. 30-991: 6.16_8 3.21_7 3.07_{10} 2.05_2 1.88_4 1.79_2 1.60_3 1.22_1. As octahedral crystals. Greenish; grayish, white streak, vitreous luster. $H = 4$–$4\frac{1}{2}$. $G = 3.44$. Isotropic; $N = 1.93$–1.99. Formed as a result of leaching pyrochlore by potassium-bearing solutions in the zone of weathering. The crystals are zoned, having a core of normal pyrochlore. *Goma, Lueshe, Kivu, Zaire*, in alluvial deposits and weathered soils derived from the Lueshe carbonatite. RG AM $63:528(1978)$.

8.2.1.3 Bariopyrochlore $(Ba,Sr)_2(Nb,Ti)_2(O,OH)_7$

Named in 1977 for the composition. Pyrochlore group. Formerly called "pandaite," for the locality in Tanzania. ISO $Fd3m$. $a = 10.562$, $Z = 8$, $D = 4.01$. AM $44:1324(1959)$. 12-285: 6.09_8 3.18_5 3.04_{10} 1.86_7 1.59_7 1.21_5 1.18_5 1.016_6. Small euhedral crystals, $\{111\}$ modified by $\{100\}$. Yellowish gray, tan, olive-gray. Cleavage $\{111\}$, poor; conchoidal fracture. $H = 4\frac{1}{2}$–5. $G = 4.0$. Usually radioactive. Isotropic; $N = 2.07$–2.10. In RL (white light), 13.2%. *Panda Hill, near Mbeya, Tanzania*, in biotite-rich altered carbonatite. Araxa, MG, and Catalão, GO, Brazil, in carbonatites. The Araxa deposit is the world's largest and richest niobium deposit. RG MM $32:10(1959)$, AM $62:403(1977)$.

8.2.1.4 Yttropyrochlore-(Y) $(Y,Na,Ca,U)_{1-2}(Nb,Ta)_2(O,OH)_7$

Named in 1977 for the composition. Pyrochlore group. Formerly called "obruchevite" for the Russian geologist, V. A. Obruchev (1863–1956). ISO $Fd3m$. $a = 10.0$–10.34, $Z = 8$, $D = 4.17$. AM $43:797(1958)$. 18-765: 5.50_2 4.43_2 3.09_3 3.02_{10} 2.97_2 2.95_{10} 2.80_2 2.66_2. 25-1015(heated): 3.45_4 3.15_5 2.98_{10} 2.58_3 2.50_3 1.70_9 1.55_5 1.49_7. Dense masses without crystal form. Chocolate brown, vitreous to adamantine luster. Conchoidal fracture. $H = 4\frac{1}{2}$–5. $G = 3.7$. Isotropic; $N = 1.830$–1.835. *Alakurtti region, Karelia, Russia*, in pegmatites, associated with allanite, garnet, columbite, and fergusonite. RG AM $62:403(1977)$.

8.2.1.5 Ceriopyrochlore-(Ce) $(Ce,Ca,Y)_2(Nb,Ta)_2O_6(OH, F)$

Named in 1977 for the composition. Pyrochlore group. Formerly called "marignacite," for the French chemist, G. de Marignac (1817–1894). ISO $Fd3m$. $Z = 8$, $D = 4.25$. 41-1385: 3.7_3 3.3_6 3.0_{10} 2.56_3 2.46_3 1.82_4 1.53_3 1.51_3. Forms octahedral crystals. Light to dark brown, light brown to yellowish-brown streak, resinous luster. Conchoidal fracture, brittle. $H = 5$–$5\frac{1}{2}$. $G = 4.13$. Isotropic. Sometimes anomalously birefringent. *Wausau, Marathon Co., WI*, an alkalic pegmatite, with acmite, lepidomelane, rutile, fluorite, and various Zr minerals. Schelingen, Kaiserstuhl, Germany. Nei Monggol, China. RG DI:755, AM 62:403(1977).

8.2.1.6 Plumbopyrochlore (Pb, Y, U, Ca)$_{2-x}$Nb$_2$(O, OH)$_7$

Named in 1966 for the composition. Pyrochlore group. ISO $Fd3m$. $a = 10.534$, $Z = 8$, $D = 6.75$. 25-453: 3.02_{10} 2.62_6 1.86_9 1.58_9 1.21_7 1.18_6 1.08_8 1.02_8. As grains, and zoned octahedral crystals. Brown, greenish yellow, red. $G = 5.04$. Isotropic; $N = 2.08$. *Ural Mts., Russia*, in metasomatically altered granitic rocks, with microcline, albite, and aegirine; Keivy, Kola Penin.(!), in large crystals. RG AM 55:1068(1970), 62:403(1977).

8.2.1.7 Uranpyrochlore (U,Ca,Ce)$_2$(Nb,Ta)$_2$O$_6$(OH)

Named in 1977 for the composition. Formerly called "hatchettolite," for Charles Hatchett (1765–1847), English chemist, who discovered the element niobium. Pyrochlore group. ISO $Fd3m$. $a = 10.333$, $Z = 8$. MM 53:257(1989). 29-1411: 7.3_4 3.01_{10} 2.61_3 1.84_8 1.72_1 1.58_6 1.20_2 1.17_2. Octahedrons, modified by {001} and {113}; also massive. Black, red, red-brown; yellowish brown, adamantine to resinous luster. Subconchoidal fracture, brittle. $H = 4$. $G = 4.5$–4.9. Isotropic; $N = 1.98$. From *mica mines, Mitchell Co., NC*, with samarskite. Harding pegmatite, Dixon, NM. Woodcox mine, Hybla, ONT, Canada, in pegmatite with columbite, cyrtolite, titanite, amazonite. Ndale, Ft. Portal area, Uganda, in a calcite-bearing tuff related to a recent carbonatite intrusion, associated with pyrochlore, zircon, titanite, monazite, magnetite, biotite, and other minerals. RG $DI:754$, AM 62:403(1977).

8.2.1.8 Strontiopyrochlore SrNb$_2$(O,OH)$_7$

Named in 1986 for the composition. Pyrochlore group. ISO $Fd3m$. $a = 10.53$, $Z = 8$, $D = 3.99$. AM 73:930(1988). Octahedral crystals. Pale yellow. $VHN = 246$. $G = 3.80$. Isotropic; $N = 2.08$. *Yenisei Ridge, Russia*, in dolomitic carbonatites. RG $DANS$ 290:1212(1986).

8.2.2.1 Microlite (Na, Ca)$_2$Ta$_2$O$_6$(O,OH,F)

Named in 1835 from the Greek for *small*, in allusion to the minute size of the crystals from the original locality. Pyrochlore group. ISO $Fd3m$. $a = 10.40$, $Z = 8$, $D = 6.34$. 3-1139: 3.00_7 1.84_8 1.57_{10} 1.36_6 1.19_7 1.16_6 1.060_6 1.000_7. Crystallizes as octahedrons, sometimes modified by {011}, {001}, or {113}; also as anhedral grains. Yellow, brown, green, reddish, black; pale yellowish or brownish streak; resinous to vitreous luster. Cleavage {111}, difficult; subconchoidal to uneven fracture; brittle. $H = 5$–$5\frac{1}{2}$. $G = 6.42$. Nonfluorescent. Isotropic; $N = 2.023$. Microlite is found typically in pegmatites, especially those rich in albite, or in lepidolite and other lithium minerals. Also disseminated as small grains in some albitites and apogranites. *Chesterfield, Hampshire Co., MA*, in lithium pegmatite with albite, elbaite, spodumene, columbite, and cassiterite. Topsham, ME, with albite, topaz, elbaite, and lepidolite. In CT in pegmatites at Haddam, Middletown, and Portland. Amelia Court House, VA(!!), in large crystals in cleavelandite with manganotantalite, columbite, and amazonite. Harding mine, Dixon, NM(*), in considerable amount embedded in pink lithian muscovite, with beryl, albite, and uranpyrochlore. Tanco mine, Bernic Lake, MB, Canada, in albite with wodginite and other tantalum minerals in a very large lithium pegmatite. Also Mattawa, ONT(!), as brown octahedra to 1 cm. Narssarssuk, Greenland. Varuträsk, Sweden, with stibiotantalite in pegmatite. Skogböle, Finland, with ixiolite. Échassières, Massif Centrale, France, in kaolinized apogranite. Fine crystals near Shingus, Gilgit, Pakistan. Yi Chun mine, Jiangxi Prov., China, in a lithian albitite. Jooste mine, Karibib, Namibia. Greenbushes and Wodgina, WA, Australia, in pegmatites. In Brazil at many localities, especially São João del Rei, MG(!); Xanda mine, Araçuai, MG(!); at Alto do Giz, Equador, RGN, with manganotantalite and simpsonite. An important ore of tantalum. RG *DI:748, AM 62:403(1977)*.

8.2.2.2 Bariomicrolite $(Ba,U,Pb)_2(Ta,Nb)_2(O,OH)_7$

Named in 1977 for the composition, being a member of the microlite subgroup. Pyrochlore group. Formerly called "rijkeboerite," for A. Rijkeboer, Dutch chemist. ISO $Fd3m$. $a = 10.570$, $Z = 8$, $D = 5.60$. *AM 48:1415(1963)*. *16-616*: 6.04_8 3.18_7 3.03_{10} 2.63_6 1.87_6 1.78_4 1.59_5 1.08_4. Crystallizes in octahedrons. Pink, reddish, or yellowish brown, white or colorless. $H = 4\frac{1}{2}$–5. $G = 5.7$. $R_{mean} = 13.2$. From a pegmatite near *Chi Chico, São João del Rei, MG, Brazil*, associated with cassiterite. RG *AM 62:403(1977)*.

8.2.2.3 Plumbomicrolite $(Pb,Na,Ca)_2(Ta,Nb)_2O_6(O,OH,F)$

Named in 1961 for the composition as a member of the microlite subgroup. Pyrochlore group. ISO $Fd3m$. $a = 10.56$, $Z = 8$, $D = 7.2$ (for Ta/Nb = 10:1). *AM 47:1220(1962)*. *44-670*: 2.99_{10} 2.60_6 1.85_9 1.58_{10} 1.209_8 1.178_8 1.078_8 1.017_9. Crystalline masses and rare octahedra. Greenish yellow; orange, greasy luster. $H = 6$. $G = 6.5$–7.2. Insoluble in acids. Keivy, Kola Penin., Russia(!), as large yellow crystals (often octahedra) frequently in excess of 3 cm. *Kivu, Zaire*, in alluvium derived from nearby spodumene pegmatites. RG *AM 62:403(1977)*.

8.2.2.4 Uranmicrolite $(U,Ca)_2(Ta,Nb)_2O_6(OH)$

Named in 1957 for the composition, as a member of the microlite subgroup. Pyrochlore group. Formerly called "djalmaite," for Djalma Guimaraes (1895–1973), Brazilian mineralogist and petrologist. ISO. $Fd3m$. $a = 10.40$, $Z = 8$, $D = 6.45$. *43-693*: 2.95_{10} 2.55_8 1.818_{10} 1.552_{10} 1.183_8 1.154_8 1.054_8 0.995_8. Crystallizes in octahedra, modified by {113} and {001}. Yellowish brown; greenish brown; brownish black, light yellow streak, greasy luster. No cleavage, irregular fracture. $H = 5\frac{1}{2}$. $G = 5.75$–5.88. Highly radioactive. Isotropic; $N = 1.97$. In pegmatite from *Fazenda Posse, Brejaúba district, Municipo de Conceição de Serro, MG, Brazil*, with columbite, magnetite, samarskite, garnet, beryl, tourmaline, and bismuth minerals; also Parelhas, RGN, Brazil, in fine crystals. RG $DI:805$; $AM\ 26:347(1941)$, $62:403(1977)$; $ABC\ 21(4)$: $337(1949)$.

8.2.2.5 Bismutomicrolite $(Bi,Ca)(Ta,Nb)_2O_6(OH)$

From the composition and as a member of the microlite subgroup. Pyrochlore group. Formerly called "westgrenite," for Arne Westgren (1889–1975), Swedish chemist and crystallographer, who synthesized $BiTa_2O_6F$. ISO $Fd3m$. $a = 10.485$, $Z = 8$, $D = 6.83$. $AM\ 48:215(1963)$. 26-1042: 5.94_6 3.14_8 3.00_{10} 2.60_6 2.00_6 1.84_9 1.76_6 1.57_9. Crystallizes as octahedra, also massive, in veinlets. Yellow; pink; brown, resinous to dull luster. Uneven fracture. $H = 5$. $G = 6.5$. Infusible; insoluble in acids. Isotropic; $n = > 2.00$. *Wampewo Hill, Busiro Co., Buganda, Uganda*, as veinlets in bismutotantalite. In Brazil at the Alto Manoel Baldoino, Junco, PB, in masses with bismutotantalite; Ermo mine, Carnauba dos Dantas, RGN, in small octahedral crystals, with spessartine, gahnite, euclase, bertrandite, and tantalite. RG $AM\ 62:403(1977)$.

8.2.2.6 Stannomicrolite $(Sn^{2+},Fe^{2+},Mn^{2+})_2(Ta,Nb,Sn^{4+})_2(O,OH)_7$

Named in 1977 from the composition and as a member of the microlite subgroup. Pyrochlore group. Formerly called "sukulaite" for the locality. ISO $Fd3m$. $a = 10.57$, $Z = 8$, $D = 7.56$. $AM\ 53:2103(1968)$. 23-1441: 3.05_{10} 2.64_6 1.87_8 1.59_8 1.52_4 1.21_4 1.18_4 1.02_4. Grains and narrow rims surrounding ferrowodginite. Yellowish brown. In RL, gray with a reddish tint; reddish-brown internal reflections. R value > cassiterite. *Sukula, Tammela, Finland*. RG $BCGF\ 229:173(1967)$, $AM\ 62:403(1977)$, $NJMM:249(1987)$.

8.2.2.7 Stibiomicrolite $(Sb,Ca,Na)_2(Ta,Nb)_2(O,OH,F)_7$

Named in 1938 for the composition and as a member of the microlite subgroup. Pyrochlore group. ISO $Fd3m$. $a = 10.455$, $Z = 8$, $D = 5.84$–6.22. $AM\ 73: 1499(1988)$. 40-516: 6.03_8 3.14_9 3.01_{10} 2.61_7 2.01_5 1.85_7 1.58_8 1.36_4. Microscopic veinlets replacing stibiotantalite. Greenish-white; white, white streak. $H = <$stibiotantalite. Isotropic; $N = >1.90$; 2.07–2.16 (calc.). In RL, gray, darker than tantalite. *Varuträsk, Sweden*, in pegmatite with albite, lepidolite, lithiophilite, alkali beryl, cassiterite, columbite–tantalite, and allemontite, as veinlets replacing stibiotantalite, associated with native Sb in the veinlets. RG $GFF\ 60:216(1938)$, $109:105(1987)$; $DI:757$.

8.2.3.1 Betafite $(Ca,Na,U)_2(Ti,Nb)_2O_6(OH)$

Named in 1912 for the locality. Pyrochlore group. In part, once named "samiresite," a plumboan betafite (Lacroix). ISO $Fd3m$. $a = 10.3$, $Z = 8$, $D = 4.63$. Metamict. Physical properties and degree of hydration are variable, due to the variable degree of alteration of betafite from different sources. 13-197: 2.96_{10} 2.56_2 1.814_5 1.546_4 1.482_2 1.284_1 1.179_2 1.148_2. Known only as crystals, which are octahedral, usually modified by {011}, more rarely {113} and {001}. Greenish brown; yellow; dark brown; black, waxy to vitreous to submetallic luster. Conchoidal to uneven fracture; brittle. $H = 4$–$5\frac{1}{2}$. Strongly radioactive. Nonfluorescent. Isotropic; N = 1.92. Brown Derby pegmatite, Ohio City, Gunnison Co., CO. Silver Crater mine, Bancroft, ONT, Canada(!), in fine crystals. Mina El Muerto, Telixtlahuaca, OAX, Mexico(!). In numerous places in *Madagascar*, in fine crystals up to 15 cm, especially W of *Betafo*(!); Antanifotsy(*); *Ambatolampikely*(!); E of Ampangabé(!); Samiresy ("samiresite"); and at Tongafeno with beryl, columbite, and zircon. RG *DI:803, CM 6:610(1961), AM 46:1519(1961)*.

8.2.3.2 Yttrobetafite-(Y) $(Y,Ce)_2(Ti,Nb)_2O_6(OH,F)$

Named in 1962 from the composition and as a member of the betafite subgroup. Pyrochlore group. ISO $Fd3m$. Z = 8, D = 5.05. Metamict. *AM 49:440(1964). 41-1434*(heated to 900°): 3.26_8 2.99_7 2.50_6 1.76_2 1.69_{10} 1.63_3 1.18_6 1.10_3. Greenish, greasy to dull luster. G = 3.65–4.90. *Alakurtti region, northwestern Karelia, Russia*, in quartz–plagioclase–microcline pegmatite. RG *AM 62:403(1977)*.

8.2.3.3 Plumbobetafite $(Pb,U,Ca)(Ti,Nb)_2O_6(OH,F)$

Named in 1969 for the composition and as a member of the betafite subgroup. Pyrochlore group. ISO $Fd3m$. $a = 10.33$, Z = 8. Metamict. Rounded grains, crude octahedral crystals. Yellowish, adamantine luster. G = 4.64. *Burpala massif, N of Lake Baikal, Siberia, Russia*, in a microcline–albite–quartz–riebeckite–aegirine dyke cutting nepheline syenite. RG *AM 55:1068(1970), 62:403(1977)*.

8.2.3.4 Stibiobetafite $(Sb,Ca)_2(Ti,Nb)_2(O,OH)_7$

Named in 1979 for the composition and as a member of the betafite subgroup. Pyrochlore group. ISO $Fd3m$. $a = 10.356$, Z = 8, D = 5.19. *CM 17:583(1979). 37-695*(heated): 5.94_3 2.99_{10} 2.58_4 1.83_6 1.56_5 1.50_3 1.19_3 1.16_3. Anhedral veinlets and replacement aggregates, also poorly developed octahedra. Dark brown; brownish black, pale brown streak, vitreous luster. Brittle. H ~ 5. G = 5.30. Isotropic; N_{calc} = >1.78, 2.2. *Věžná, Moravia, Czech Republic*, as a replacement of columbite–ixiolite, associated with niobian rutile, cassiterite, and metamict zircon, in a sodic pegmatite. RG *AM 66:1278(1980)*.

8.2.3.5 Calciobetafite $(Ca,Na)_2(Ti,Nb)_2(O,OH)_7$

Named in 1983 for its composition and as a member of the betafite subgroup. Pyrochlore group. ISO $Fd3m$. $a = 10.298$, Z = 8, D = 4.60. *AM 68:262(1983). 42-2*: 2.97_{10} 2.57_3 1.82_5 1.55_4 1.49_1 1.18_2 1.15_1 1.05_1. Octahedral crystals. Reddish brown. G = 4.81. Isotropic. Oka, QUE, Canada, in carbonatite. *Campi Flegrei, Monte di Procida, Campania, Italy*, in a sanidinite, associated with zirconolite 3O and zirconolite 3T. RG

8.2.4.1 Cesstibtantite (Cs,Na)Sb^{3+}Ta$_4$O$_{12}$

Named in 1982 for the composition. Pyrochlore group. ISO $Fd3m$. $a = 10.515$, $Z = 8$, $D = 6.49$. *MP (1993). 35-672:* 3.17_9 3.04_{10} 2.02_8 1.860_{10} 1.587_{10} 1.370_9 1.017_9 1.012_{10}. Small grains. Colorless; gray; yellow-orange; black, vitreous to adamantine luster. Brittle. $H = 5$. $G = 6.5$. Isotropic; $N > 1.8$. The formula for cesstibtantite from the type locality is $Cs_{0.31}(Sb,Na,Pb,Bi)_{0.91}$ $(Ta,Nb)_2(O,OH,F)_6(OH,F)_{0.69}$. From the Tanco mine, the formula was $(Cs,K)_{0.23}(Na,Sb,Pb,Ca,Bi)_{1.06}(Ta,Nb)_2(O,OH,F)_6(OH,F)_{0.55}$. Cesstibtantite differs from the normal pyrochlores in that it contains significant amounts of very large cations such as Cs. As these cations are too large for the conventional eight-coordinated A site, they occupy the 18-coordinated ϕ site. The formula can be written $A_{2-m}B_2O'_6(\phi,A')_{1-n}$, where $A = $ Na,Sb,Pb,Bi,Ca,Sn; $B = $ Ta,Nb; $O' = $ O,OH,F; $\phi = $ OH,F; and $A' = $ Cs,K. It is intermediate between the normal pyrochlore (with only monovalent anions at the ϕ site), and inverse pyrochlore structures (with only large cations at the ϕ site). Tanco mine, Bernic Lake, MB, Canada, with wodginite, simpsonite, and microlite. *Leshaia, Kola Penin., Russia*, in albitized zones of a pegmatite, as aggregates along simpsonite–stibiotantalite grain boundaries. RG *IGR 24:843(1982), AM 67:413(1982), MP 48:235(1993).*

8.2.4.2 Natrobistantite (Na,Cs)Bi(Ta,Nb,Sb)$_4$(O,OH)$_{12}$

Named in 1983 for the composition. Pyrochlore group. ISO $Fd3m$. $a = 10.502$, $Z = 4$, $D = 6.32$. *AM 69:407(1984)*. Possibly isostructural with bismutomicrolite, or a Cs-bearing variety of that mineral. *35-706:* 6.07_6 3.17_5 3.03_{10} 2.63_8 1.86_9 1.58_9 1.21_5 1.17_5. Octahedral crystals; aggregates and grains. Blue-green, yellow-green, colorless. $G = 6.2$. Colorless grains fluoresce red-orange in UV light; colored grains do not fluoresce. Isotropic. $R = 16.7$ (486 nm), 17.0 (589 nm), 15.7 (656 nm). *Kyokbogor, China*, in granite pegmatite, formed by replacement of bismutotantalite and replaced in turn by microlite; associated with bismuth oxides, pucherite, and clinobisvanite. RG *MZ 5(2):82(1983).*

8.2.5.1 Zirkelite (Ti,Ca,Zr,Fe^{3+},Ce)O$_{2-x}$

Named in 1960 for Ferdinand Zirkel (1838–1912), German petrographer, University of Leipzig. Metamict. Probably ISO Space group unknown. $a = 5.02$. *38-450:* 2.9_{10} 2.52_4 1.78_7 1.52_6 1.46_2 1.26_2 1.16_3 1.13_2. Irregular segregations and grains. Iron black, light brown streak, adamantine to greasy luster. $H = 6$. $G = 4.37$. Isotropic; $N = 2.5$. Reddish brown in thin section. *Aldan, Arbarastakh massif, Kola Penin., Russia*, in coarse-grained micaceous pyroxenite, in a group of Aldan ultrabasic rocks. Jacupiranga, SP, Brazil, in carbonatite with baddeleyite, perovskite, and so on. RG *DI:740, AM 68:262(1983), MM 53:565(1989).*

8.2.5.2 Zirconolite-3O (Ca,Fe,Y,Th)$_2$Fe(Ti,Nb)$_3$Zr$_2$O$_{14}$

Named in 1828 for the composition. ORTH *Cmca*. $a = 10.148$, $b = 14.147$, $c = 7.278$, $Z = 4$. Zirconolite-3O is the nonmetamict orthorhombic polytype (possibly a polymorph) of $CaZrTi_2O_7$. Nomenclature of this group was revised in 1989. *MM 53:565(1989), AM 68:272(1983)*. PD: 2.97_{10} 2.57_3 1.82_5 1.55_4 1.49_1 1.18_2 1.15_1 1.05_1. Prismatic crystals, elongated [100]. Dark red. G = 4.8. Anisotropic, parallel extinction, weakly pleochroic. *Fredricksvärn, Norway. Monte di Procida, Campi Flegrei, Campania, Italy*, in sanidinite, associated with calciobetafite and zirconolite-3T. RG *DI:764; AM 42:581(1957), 60:341(1975)*.

8.2.5.3 Zirconolite-2M $CaZrTi_2O_7$

Named in 1989 for the composition. Initially described as "zirkelite." See zirkelite (8.2.5.1). Renamed by the IMA. *MM 53:565(1989)*. MON $C2/c$. $a = 12.445$, $b = 7.288$, $c = 11.487$, $\beta = 100.38°$, $Z = 8$, D = 4.9. Structure: *AC(B) 37:306(1981)*. 34-167: 2.93_{10} 2.82_4 2.513_3 2.507_3 1.806_1 1.797_2 1.754_2 1.749_2. Brown, black. G = 4.2. *Kaiserstuhl, Germany*, in carbonatite. RG *AM 67:615(1982), T&G:183-204*.

8.2.5.4 Zirconolite-3T $CaZrTi_2O_7$

Named in 1989 for the composition. Formerly called "zirkelite" (8.2.5.1). Nomenclature revised by the IMA in 1989. *MM 53:566(1989)*. HEX-R $P3_12$. $a = 7.287$, $c = 16.886$, $Z = 3$, D = 4.88. *AM 68:266(1983)*. 15-12: 2.98_{10} 2.84_2 2.53_3 2.30_1 1.82_5 1.75_3 1.525_1 1.510_1. Crystals rough, hexagonal, platy {0001}; small grains. Black; brownish black, reddish-brown streak, resinous to submetallic luster. Splintery fracture, brittle. H = $5\frac{1}{2}$. G = 4.72. Uniaxial. *Monte di Procida, Campi Flegrei, Campania, Italy*, with zirconolite-3O and calciobetafite in sanidinite; *Walaweduwa, Bambarabotuwa district, Sabaragamuwa, Sri Lanka*. RG *MM 16:309(1913), DI:741*.

8.2.5.5 Zirconolite $(Ca,Fe^{2+},RE,Zr,Th)(Zr,Nb,Ti,Ta,Fe^{3+})Ti_2O_7$

Named in 1989 for the composition. Originally called "polymignite." Metamict. In 1989, the IMA, CNMMN, approved the name *zirconolite* (without suffix) for minerals having the composition ABX_2O_7 (ideally, $CaZrTi_2O_7$) which are noncrystalline or of undetermined polytypoid. Metamict "polymignite" is zirconolite. *MM 53:566(1989)*. Pseudomorphous crystals, when present, are prismatic [001], striated [001], somewhat flattened {100}; massive; as anhedral grains. Black, dark brown streak, submetallic–metallic luster. Conchoidal fracture. H = $6\frac{1}{2}$. G = 4.80. Nonfluorescent. Infusible. Decomposed by concentrated H_2SO_4. Isotropic; N ≈ 2.22. In *Norway* at *Fredriksvärn* in pegmatite with magnetite, nepheline, zircon, and pyrochlore; also on the *Island of Svenor* and at *Bommestad, Langesundfjord*. RG *DI:764*.

8.2.6.1 Scheteligite $(Ca,Y,Sb)_2(Ti,Ta,Nb,W)_2O_6(OH)$ (?)

Named in 1937 for Jacob Schetelig (1875–1935), Norwegian mineralogist, director of the Oslo Mineralogical Museum. A questionable mineral. Probably ORTH Space group unknown. Metamict. Rough, probably orthorhombic crystals. Black, grayish to pale yellow streak, brilliant luster. Conchoidal fracture.

$H = 5\frac{1}{2}$. $G = 4.74$. *Torvelona, Iveland, Norway*, associated with tourmaline, bismuth, euxenite, thortveitite, monazite, and beryl. RG *AM 23:293(1938)*, DI:757.

8.2.7.1 Orthobrannerite $U^{6+}U^{4+}Ti_4O_{12}(OH)_2$

Named in 1978 for the orthorhombic symmetry and similarity to brannerite. Metamict. ORTH (after heating) $P2_122$. $a = 7.37$, $b = 11.67$, $c = 6.33$, $Z = 2$, $D = 5.46$. *33-1434*: 4.87_7 3.89_8 3.17_{10} 2.45_9 2.29_5 1.82_4 1.66_9 1.20_6. Prismatic crystals, with {001}, {120}, {021}, {140}. Striated [001]. Black, dark brown to black streak, adamantine luster. Conchoidal fracture. $VHN_{100} = 525$. $G = 5.46$. Nonfluorescent. Strongly radioactive. Unattacked by H_2SO_4 or HCl, dissolves in $H_2SO_4 + H_3PO_4$. Isotropic; $N = 2.33$. In RL, grayish white, with reddish-brown internal reflections. *Yunnan, China*, in a biotite pyroxene syenite, and in *Sichuan* in an alkalic lamprophyre. RG *AM 64:656(1979)*.

8.2.8.1 Parabariomicrolite $BaTa_4O_{10}(OH)_2 \cdot 2H_2O$

Named in 1986 for the similarity to bariomicrolite but having a different symmetry. HEX-R $R\bar{3}m$. $a = 7.429$, $c = 18.505$, $Z = 3$, $D = 5.97$. *CM 24:655(1986)*. *39-401*: 6.18_5 3.17_6 3.08_4 3.04_{10} 2.64_5 1.88_4 1.86_4 1.59_4. As irregular replacements of microlite. White; pale pink, white streak, vitreous or pearly luster. Cleavage {001}, {101}, good; brittle. $H = 4$. Nonfluorescent. Anisotropic; $N > 2.0$. *Alto do Giz, Equador, RGN, Brazil*, associated with simpsonite, microlite, manganotantalite, tapiolite, natrotantite, alumotantite, stibiotantalite, beryl, spodumene, and petalite. RG *AM 73:194(1988)*.

8.3.1.1 Ferrotapiolite $(Fe,Mn)Ta_2O_6$

Originally called "tapiolite," named in 1863 for *Tapio*, god of the forest in Finnish mythology. Renamed *ferrotapiolite* in 1983 for the composition. Synonym: tapiolite. TET $P4_2/mnm$. $a = 4.762$, $c = 9.272$, $Z = 2$, $D = 8.17$. *23-1124*: 4.22_3 3.36_{10} 2.58_9 2.38_3 1.746_7 1.680_3 1.502_2 1.407_3. Short prismatic or equant crystals, with {001}, {100}, {110}, and {113} well developed. Twinning on {013}, very common. Twins often show one or more curved faces, which are very characteristic. Tapiolite exists in an ordered and a disordered phase, the ordered phase having a trirutile structure and the disordered one a rutile-type structure. The relationship to rutile is shown by direct correspondence of the a cell dimension, but in ordered tapiolite the c dimension is tripled. Black, cinnamon brown to brownish-black streak, submetallic to subadamantine luster. No cleavage, uneven to subconchoidal fracture. $H = 6-6\frac{1}{2}$. $G = 7.90$. Ferrotapiolite is always Ta-rich, and the solubility of $FeNb_2O_6$ in the structure is limited to about 20%. Ferrotapiolite is also nearly always Fe-rich, although a small amount of Mn is usually present. Only one or two known examples show $Mn > Fe^{2+}$ and then only to a limited degree, so that manganotapiolite is always an intermediate member of the series $FeTa_2O_6$–$MnTa_2O_6$. Uniaxial (+); $N_O = 2.27$, $N_E = 2.42$ (Li). O, pale reddish brown; E, nearly opaque.

Twin lamellae on {013}. In ME at Topsham, Sagadahoc Co., and Paris, Oxford Co. Old Mike mine, 6 km N of Custer, Custer Co., SD. Tanco mine, Bernic Lake, MB, Canada, with wodginite, tantalite, ixiolite, and microlite. *Kulmala pegmatite, Sukula, Tammela, Finland*, and also at Skogböle, Kimito. Chanteloube, Limoges, France. Muiane mine, Alto Ligonha, Mozambique, in fine crystals up to 8 kg. Greenbushes, and at Strelley, Pilbara, WA, Australia. In Brazil at many localities, especially at the Alto Boqueirão, and at Alto dos Preas, Parelhas, RGN; Alto Marimbondo, Carnauba dos Dantas, RGN; in MG at the Barra do Cuiete, Mun. Conselheira Pena, and at the Amuro mine, near São José da Safira. RG *DI:775, MM 42:477(1978), BGSF 55:101(1983)*.

8.3.1.2 Manganotapiolite (Mn,Fe)(Ta,Nb)$_2$O$_6$

Named in 1983 for the composition. Dimorphous with manganotantalite. TET $P4_2/mnm$. $a = 4.762$, $c = 9.272$, $Z = 2$, $D = 7.72$. *35-626*: 4.24_4 3.37_{10} 2.59_9 2.38_6 1.75_9 1.68_6 1.50_4 1.41_4. Prismatic, minute crystals. Dark brown, dark brown streak. $VHN_{100} = 711$. End-member tetragonal Mn(Ta,Nb)$_2$O$_6$ has not been found; Mn > Fe members are very rare, the maximum Mn/Fe ratio so far recognized is Mn$_{59}$Fe$_{32}$. In RL, gray, with strong red-brown internal reflections; weakly pleochroic. R = 14.7–14.1 (546 nm). *Viitaniemi, Eräjärvi, Orivesi, Finland*, in a lithium pegmatite, associated with ferrotapiolite. RG *BGSF 55:101(1983), AM 70:217(1985)*.

COLUMBITE GROUP

The columbite group of minerals consists of five isostructural minerals having the general formula

$$AB_2O_6$$

where

A = Fe, Mn, Mg
B = Nb, Ta

and an orthorhombic structure in space group *Pbcn*.

In the structure of the columbite minerals, the oxygen anions are in pseudohexagonal close packing parallel to (100), and the metal cations occupy one-half of the available pseudo-octahedral interstices. In the idealized structure AB$_2$O$_6$: (Fe,Mn)(Nb,Ta)$_2$O$_6$, the two different cations (Fe,Mn) and (Nb,Ta) are ordered into different octahedral sites in the required ratio of 1:2 so that they also occur in layers parallel to (100) with layers of (Fe,Mn) cations at $x = 0$ and $x = \frac{1}{2}$ and of (Nb,Ta) cations at $x = \frac{1}{6}, \frac{1}{3}, \frac{2}{3}$, and $\frac{5}{6}$. In terms of the linking of cation-oxygen octahedra, the structure can be regarded as chains of staggered octahedra running parallel to z with one chain containing, in the ideal structure, only (Fe,Mn) or (Nb,Ta) cations in the centers of the octahedra, and with one layer of (Fe,Mn) chains followed by two layers of (Nb,Ta)

Columbite Group

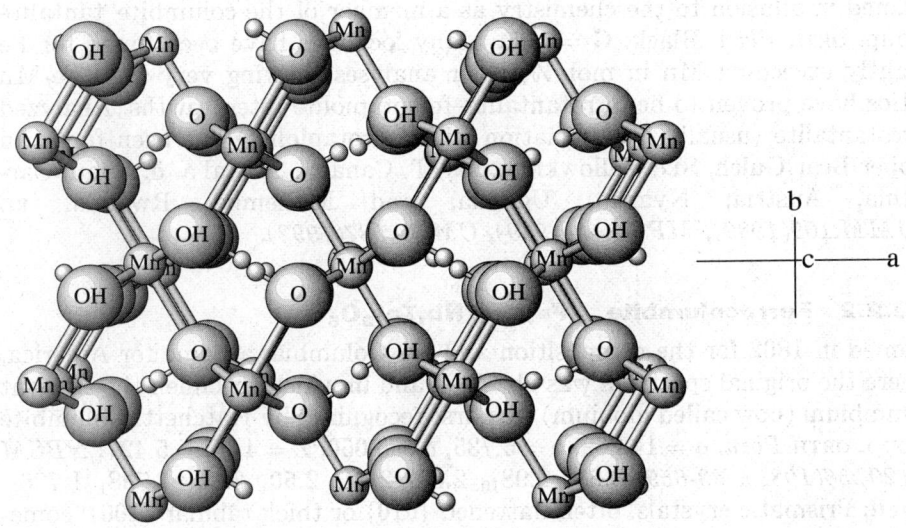

COLUMBITE

chains. In any one chain, one octahedron shares one edge with the octahedron above and one with the octahedron below it along z, and any one chain is linked to its neighboring parallel chains through certain octahedral corners (oxygens). This octahedral linking is such that each oxygen is coordinated by three cations. *CM 14:540(1976)*.

COLUMBITE GROUP

Mineral	Formula	Space Group	a	b	c	D	
Ferrotantalite*	$(Fe,Mn)(Ta,Nb)_2O_6$	Pbcn					8.3.2.1
Ferrocolumbite	$(Fe,Mn)(Nb,Ta)_2O_6$	Pbcn	14.269	5.735	5.050	5.27	8.3.2.2
Manganotantalite	$(Mn,Fe)(Ta,Nb)_2O_6$	Pbcn	14.440	5.766	5.093	7.95	8.3.2.3
Manganocolumbite	$(Mn,Fe)(Nb,Ta)_2O_6$	Pbcn	14.433	5.764	5.083	5.28	8.3.2.4
Magnocolumbite	$(Mg,Fe,Mn)(Nb,Ta)_2O_6$	Pbcn	14.193	5.700	5.032	5.23	8.3.2.5

*End member not observed in nature.
A nearly complete solid solution series exists between the first four members of the group, the end members of all of them except ferrotantalite having been observed in nature. The end member $FeTa_2O_6$ has the tetragonal or tapiolite structure. Most tantalites tend to be Mn dominant, and those in which Fe > Mn are rare. Most of the Fe,Mn columbites are also intermediate members, with the Fe-dominant members being somewhat more abundant. The Mg-dominant member requires special geological circumstances for its formation and hence is very rare; Mg does not enter into the composition of the other members to a significant degree. Minor quantities of Fe^{3+}, Sc, Ti, Sn, and W are also commonly present.
Note: These minerals are economically important, especially for their tantalum content.

8.3.2.1 Ferrotantalite (Fe,Mn)(Ta,Nb)$_2$O$_6$

Named in allusion to the chemistry as a member of the columbite–tantalite group. ORTH *Pbcn*. Black. G = 7.95. Many localities have been found for Fe slightly exceeding Mn in mol %; older analyses showing very high Fe/Mn ratios have proven to be ferrotantalite–ferrotapiolite intergrowths. Analyzed ferrotantalite (usually in association with ferrotapiolite) has been found in Upper Bear Gulch, SD; Yellowknife, NWT, Canada; Spittal a. d. Drau, Carinthia, Austria; Nyanga, Uganda; and Muhembe, Rwanda. RG *NJMM:109(1989)*, *MP 41:53(1989)*, *CM 30:587(1992)*.

8.3.2.2 Ferrocolumbite (Fe,Mn)(Nb,Ta)$_2$O$_6$

Named in 1802 for the composition and for Columbia, a name for America, where the original specimen was obtained and in which specimen the element columbium (now called niobium) was first recognized by Hatchett. Columbite group. ORTH *Pbcn*. $a = 14.269$, $b = 5.735$, $c = 5.050$, $Z = 4$, $D = 5.427$. *NBSM 25(20):56(1984)*. 33-659: 3.68$_4$ 2.98$_{10}$ 2.87$_1$ 2.54$_1$ 2.50$_1$ 2.39$_1$ 1.738$_1$ 1.726$_2$. **Habit:** Prismatic crystals, often flattened {010} or thick tabular {100}; sometimes nearly equant; massive. Twin plane {201}, common. Usually contact twins, flattened, heart-shaped. **Physical properties:** Black, brownish-black streak, submetallic luster. Cleavage {010}, distinct; {100}, less so; subconchoidal to uneven fracture; brittle. H = 6. G = 5.20–6.76 (depending on Nb/Ta ratio). Paramagnetic. **Tests:** Soluble in HF, partially soluble in hot concentrated H$_2$SO$_4$, insoluble in HCl or HNO$_3$. In distinguishing between columbites and tantalites, density is the most useful determinant. Between 5.20 and 6.76, Nb > Ta (mol %) and the mineral is in the columbite group. Between 6.76 and 7.95, Ta > Nb and it is a tantalite. For (Fe,Mn)(Nb,Ta)$_2$O$_6$, where Fe/Mn = 3:1 and Nb/Ta = 1:1, the mineral would contain 31.3% Nb$_2$O$_5$ and 51.9% Ta$_2$O$_5$. Above 31.3%, Nb$_2$O$_5$ is therefore in the columbite range, and above 51.9%, Ta$_2$O$_5$ is in the tantalite range. Proportions of Fe/Mn other than 3:1 would make only a small difference in these numbers. **Optics:** Biaxial (−); N ≈ 2.40; Z > X; strong absorption; weakly anisotropic. **Occurrence:** Ferrocolumbite and manganocolumbite are typical of granitic pegmatites and other environments where Nb > Ta. They may also rarely be found in alkalic pegmatites; as an alteration product of pyrochlore and other Nb minerals; and as disseminations in albitites and apogranites, associated with zircon, xenotime, cassiterite, and ilmenorutile. Ferrocolumbite is the commonest member of the columbite–tantalite group. **Localities:** Near *New London*, CT, and elsewhere in CT, as at Haddam, Portland, and Branchville. Numerous localities in ME, NH, and MA. Amelia Courthouse, VA; in SD at the Etta, Ingersoll, Peerless, Old Mike, and other mines. Ivigtut, Greenland, in the cryolite mine, nearly pure FeNb$_2$O$_6$. Many localities in Europe, especially in Norway, Sweden, Finland, Germany, Italy, and France. Important alluvial deposits with associated cassiterite are found near Jos, Nigeria. In Madagascar in very large crystals at Ambatofotsikely, and at Ampangabé, Morafeno, Samiresy, and Befanamo. Also common in the pegmatite regions of Zimbabwe, Zaire, Uganda, and South Africa. Many localities in Brazil: fine large crystals are found in MG at the Golconda district and at São José da Safira; and in AM in the large Pittinga alluvial deposit derived from a weathered apogranite. Use: An important ore of tantalum and niobium. RG *DI:780, NJMA 106:1(1966), NJMM:372(1985).*

8.3.2.3 Manganotantalite (Mn,Fe)(Ta,Nb)$_2$O$_6$

Named in 1887 for the mythical Tantalus, in allusion to the tantalizing difficulties encountered in making a solution of the mineral in acids, preparatory to analysis. *Mangano* for the Mn-dominant chemistry. Dimorphous with manganotapiolite. Columbite group. ORTH *Pbcn.* $a = 14.440$, $b = 5.766$, $c = 5.093$, $Z = 4$, $D = 7.95$. *NJMM:372(1985). DANS 148:97(1963).* PD(nat): 3.62_9 2.93_{10} 1.89_7 1.71_8 1.52_8 1.46_8 1.44_8 0.988_8. 33-909(syn): 7.22_3 3.70_5 3.62_1 2.99_{10} 2.55_1 2.51_2 2.41_1 1.74_2. **Habit:** Crystals prismatic [001], usually flattened {010} or {100}; also equant. Twin plane {201}, common. Usually heart-shaped contact twins. **Physical properties:** Black, brownish black, orange-red; brownish-black to orange streak; semimetallic luster. Cleavage {010}, distinct; {100}, less so; subconchoidal to uneven fracture; brittle. $H = 6$–$6\frac{1}{2}$. $G = 6.76$–7.95. Nonfluorescent. Paramagnetic. **Tests:** Partially decomposed by hot H_2SO_4, insoluble in HCl or HNO_3, soluble in HF. **Chemistry:** Most tantalites have Mn > Fe, and pure end member $MnTa_2O_6$ is known from a number of localities. When Fe^{2+} is very low, less than 1%, the mineral is orange-red in color, sometimes transparent. Nb is present in all proportions up to Ta/Nb = 1:1, but the high-Nb members tend to be Mn poor. **Optics:** Biaxial (+); $N \approx 2.12$; $2V = 34°$; r < v, moderate; strong absorption. In RL, red-brown internal reflections. **Localities:** Manganotantalite is the most abundant mineral that is Ta-dominant. Mt. Apatite and Paris, ME; Portland, CT; Amelia Courthouse, VA(!), with microlite in albite; Etta mine, Keystone, and Helen Beryl mine, Custer, SD. Pala, San Diego Co., CA. Tanco mine, Bernic Lake, MB, Canada(!), with wodginite and microlite. *Utö, Sweden.* In the tin placers of Malaysia and Thailand. Kokoto Hai mine, Xinjiang, China(!). Shingus, Gilgit, Pakistan(!). Bikita and Benson mines, Zimbabwe. Alto Ligonha, Mozambique, at the Morrua(!), Muiane(!), and Marropino mines, sometimes in large crystals and in transparent sharp red twins. In Brazil in RGN, at the Alto do Giz, Equador(!), and at various mines in Parelhas and Carnauba dos Dantas. Greenbushes, WA, Australia; also at Wodgina and Cooglegong. Use: The principal ore of tantalum. RG *DI:783, NJMA 106:1(1966), NJMM:372(1985).*

8.3.2.4 Manganocolumbite (Mn,Fe)(Nb,Ta)$_2$O$_6$

Named in 1892 for the composition. Columbite group. ORTH *Pbcn.* $a = 14.433$, $b = 5.764$, $c = 5.083$, $Z = 4$, $D = 5.28$. *NJMM: 372(1985).* 33-899(syn): 7.21_1 3.70_3 2.99_{10} 2.88_1 2.54_1 2.51_1 1.75_1 1.74_1. 45-1360: 3.65_8 2.97_{10} 2.86_3 2.50_2 1.771_1 1.737_2 1.721_2 1.454_3. Habit as with ferrocolumbite. Black, brownish black. Synthetic end member $MnNb_2O_6$ is red, just like natural end member manganotantalite. $G = 5.20$–6.76. Other properties as with ferrocolumbite.

Manganocolumbite is nearly as common as ferrocolumbite, but very few occurrences approach the $MnNb_2O_6$ end member. Manganocolumbite can be distinguished from ferrocolumbite only by chemical analysis. Most samples in collections have never been analyzed for the Mn/Fe ratio and are merely labeled "columbite." No natural $MnNb_2O_6$ end members have been observed iron-free enough to be red, but synthetic $MnTa_2O_6$ is red. Old Mike mine, Custer Co., and Elk Creek, Pennington Co., SD. San Diego mine, San Diego Co., CA. Londonderry mine, Coolgardie, WA, Australia. RG *DI:780, AM 63:1166(1978)*.

8.3.2.5 Magnocolumbite   (Mg,Fe,Mn)(Nb,Ta)$_2$O$_6$

Named in 1963 from the composition. Columbite group. ORTH *Pbcn*. $a = 14.193$, $b = 5.700$, $c = 5.032$, $Z = 4$, $D = 5.23$. *33-875*(syn): 7.11_3 3.64_6 3.55_2 2.95_{10} 2.85_1 1.724_1 1.712_2 1.446_1. Small prismatic crystals. Black to brownish black, dark brown streak, semimetallic luster. Cleavage {010} and {100}; uneven fracture. $H = 6\frac{1}{2}$. $G = 5.25$. Biaxial $(-)$; $XY = cb$; $N_x = 2.33$ (brownish yellow), $N_z = 2.40$ (brownish red); $2V \approx 80°$. Magnocolumbite forms under the very unusual conditions in pegmatite genesis of Mg saturation and virtual absence of Fe. *Kukh-i-la, southwestern Pamir Mts., Tadzhikistan*, in pegmatite cutting dolomitic marbles, which have been partially assimilated forming cordierite, dravite, spinel, andalusite, and kyanite. It is commonly intergrown with ilmenorutile. Also from eastern Siberia in a similar environment. RG *AM 48:1182(1963), ZVMO 96:720(1967), MA 69:582(1969)*.

8.3.3.1 Fersmite   (Ca,Ce,Na)(Nb,Ti)$_2$(O,OH,F)$_6$

Named in 1946 for Alexandr Evgenievich Fersman (1883–1945), Russian mineralogist and geochemist. Usually metamict. ORTH *Pbcn*. $a = 14.926$, $b = 5.752$, $c = 5.204$, $Z = 4$, $D = 4.78$. Structure: *AM 55:90(1970)*. Fersmite is essentially a distorted Ca analog of columbite. *39-1392:* 3.77_2 3.74_2 3.06_{10} 2.69_2 1.886_1 1.805_3 1.773_2 1.529_2. Short prismatic crystals [001]. Massive, as irregular grains. Brown; black, grayish-brown streak, resinous luster. No cleavage, conchoidal to uneven fracture, brittle. $H = 4\frac{1}{2}$. $G = 4.79$. Biaxial $(+)$; $N_x = 2.07$, $N_y = 2.08$, $N_z = 2.19$ (Li); $2V = 20$–$25°$. X, pale greenish; Y, yellow to colorless; Z, dark greenish yellow to olive yellow. From the Dark Star claim, Ravalli Co., MT, in marble, intergrown with ferrocolumbite, associated with monazite, ancylite, barite, quartz, and apatite. In BC, Canada, disseminated in large amounts in the Aley carbonatite E of Williston Lake, probably pseudomorphous after pyrochlore. *Vishnevye Mts., Russia*, as an alteration product of pyrochlore, intergrown with ferrocolumbite. Madagascar, at Anjanabonoina in large (>2 kg) complex lustrous crystals. RG *AM 32:373(1947)*.

8.3.4.1 Brannerite   (U,Ca,Ce)(Ti,Fe)$_2$O$_6$

Named in 1920 for John Casper Branner (1850–1922), American geologist. Metamict. MON $C2/m$. $a = 9.812$, $b = 3.770$, $c = 6.925$, $\beta = 118.97°$, $Z = 2$, $D = 6.37$. Structure: *AC 21:974(1966)*. *12-477*(syn): 6.04_4 4.74_{10} 3.44_{10}

3.35_{10} 3.02_4 2.90_4 2.77_3 2.04_4. Indistinct prismatic crystals, usually as rounded or irregular grains. Black, altered material brownish olive-green, yellow brown, yellow; dark greenish-brown to yellowish-brown streak; vitreous luster when fresh, resinous when altered. Conchoidal fracture. H = $4\frac{1}{2}$–$5\frac{1}{2}$. G = 4.2–5.4. Nonfluorescent. Strongly radioactive. Decomposed by hot concentrated H_2SO_4. Isotropic; N = 2.26 (Na). *Kelly Gulch, Custer Co., ID*, as alluvial grains. California vein, Mt. Antero, Chaffec Co, CO(!) in crystals to 1.5 cm. Dean's mine, Coleville, Mono Co., CA. In a quartz pebble conglomerate at the Pronto mine, Blind River district, ONT, Canada, on the northern shore of Lake Huron. In pegmatite at Fuenteovejuna, Cordoba, Spain(!), crystals to 30 cm. In Switzerland abundant at Lodrino (Iragna), Tessin. In Australia at Crocker's Well, Manna Hill, SA, in quartz veins with rutile, xenotime, apatite, and zircon; and with uraninite in calcite veins in the Nichols Nob copper mine, Flinders Range, SA. RG *DI:774; CM 6:483(1960), 20:271(1982); NJMA 106:1(1966); IANS:63(1982)*.

8.3.4.2 Thorutite (Th,U,Ca)Ti$_2$(O,OH)$_6$

Named in 1958 for the composition. Metamict. MON $C2/m$. a = 9.822, b = 3.824, c = 7.036, β = 118.83°, Z = 2, D = 6.08. *19-1351*(syn): 6.16_5 4.77_{10} 4.30_5 3.49_{10} 3.40_8 2.93_5 2.57_5 2.49_5. Short prismatic crystals. Black, pale brown streak, resinous luster. Conchoidal fracture. G = 6.0. Radioactive. Turns golden when heated in CT and gives off water. Isotropic; N > 2.1. From an unspecified locality in *Russia* (?) in a syenite massif in veins of microcline and sericitized nepheline, associated with thorite, zircon, and calcite. RG *ZVMO 87:201(1958); AM 43:1007(1958), 48:1419(1963)*.

8.3.5.1 Lucasite-(Ce) CeTi$_2$(O,OH)$_6$

Named in 1987 for Hans Lucas, Australian geologist, who discovered the mineral. MON $C2/c$. a = 5.178, b = 8.756, c = 9.768, β = 93.52°, Z = 4, D = 5.00. *AM 72:1006(1987)*. *40-1455*: 3.38_{10} 3.26_6 3.20_8 2.58_7 2.54_4 2.23_5 2.03_4 1.83_6. Crystal grains striated [010], and aggregates. Light to dark brown; gray, white streak, resinous luster. Cleavage {001}, prominent; conchoidal fracture; brittle. VHN_{10} = 824. Biaxial; N_{calc} = 2.32; 2V large; extreme birefringence. In RL, gray, with brown internal reflections. R = 16.6 (470 nm), 15.6 (546 nm), 15.3 (589 nm), 15.0 (650 nm). From the *Argyle AK1 mine, Kimberley, WA, Australia*, in a diamondiferous olivine lamproite tuff, associated with anatase, almandine, chromian pyrope, ilmenite, barite, chromite, and chromian diopside. RG

8.3.6.1 Aeschynite-(Ce) (Ce,Ca,Fe)(Ti,Nb)$_2$(O,OH)$_6$

Named in 1828 from the Greek for *shame*, in allusion to the inability of chemical science, at the time of its discovery, to separate some of its constituents. Synonym: eschynite.Metamict. ORTH *Pmnb*. a = 7.538, b = 10.958, c = 5.396, Z = 4. *NBSM 25(3):24(1964)*. Structure: *SR 27:535(1962)*. *15-864*(syn): 5.48_3 4.43_3 3.11_4 3.02_8 2.98_{10} 2.70_3 2.04_3 1.96_3. Rough prismatic crystals, massive.

Dark brown, brown to nearly black streak, submetallic to resinous luster. H = 5–6. G = 4.20–5.34. Biaxial (−); N > 2.0; 2V moderate. Bireflectance = 0.10–0.15. Dark Star claim, Ravalli Co., MT, in carbonatitic rocks, associated with allanite, apatite, monazite, fersmite, magnetite, pyrite, and columbite. *Miask, Ilmen Mts., Russia*, in nepheline syenite. Hytterö, Norway. RG *DI:793, AM 46:1436(1961), DANS 142:181(1962).*

8.3.6.2 Niobo-aeschynite-(Ce) $(Ce,Ca,Th)(Nb,Ti)_2(O,OH)_6$

Named in 1960 for the composition. ORTH *Pbnm.* $a = 5.396$, $b = 11.085$, $c = 7.585$, $Z = 4$. *AM 60:309(1975). 29-311:* 5.54_2 3.45_2 3.13_4 3.05_7 2.98_{10} 2.83_2 2.70_2 2.04_2. Small water-worn grains, some with prismatic habit. Red-brown, dark brown, black; orange-brown to light reddish-brown streak; resinous luster. Conchoidal fracture. H = 4–5, $5\frac{1}{2}$. G = 5.04. Paramagnetic. Biaxial (−); XYZ = bca; $N_x = 2.27$, $N_y = 2.32$, $N_z = 2.34$; 2V = 85°. X,Y,Z, reddish brown. In placer concentrates from the Tofty tin belt, Hot Springs district, central AK. *Vishnevye Gory, Ural Mts., Russia*, in quartz–arfvedsonite veinlets cutting fenites. RG *AM 47:417(1962).*

8.3.6.3 Aeschynite-(Y) $(Y,Ca,Fe)(Ti,Nb)_2(O,OH)_6$

Named in 1966 for the composition. See aeschynite-Ce (8.3.6.1). ORTH *Pbnm.* $a = 5.18$, $b = 10.92$, $c = 7.50$, $Z = 4$, $D = 5.13$. *20-1401:* 3.07_3 2.99_7 2.91_{10} 2.77_2 2.59_2 1.85_2 1.57_3 1.51_3. Black; brown; yellow, white to pale yellow streak, submetallic–resinous–pearly luster. Cleavage {100}, {010}, {001}, perfect; uneven fracture; brittle. H = $4\frac{1}{2}$. G = 4.95. Moderately paramagnetic. Insoluble in HCl, soluble in HF or in a fusion with $KHSO_4$. Biaxial (+); $N_x = 2.19$, $N_y = 2.21$, $N_z = 2.24$; 2V = 70°; nonpleochroic. In *Norway* at Urstad, Hitterö, and abundantly at Kåbuland, Iveland Parish. Nonmetamict aeschynite is fairly common as small crystals in Alpine cleft environment, as for example at Wanni-Gletscher, Binntal, Wallis, Switzerland; at Lohningerbruch, Rauris, Austria; and Val d'Ossola, Italy. Miask, Ilmen Mts., Russia. In southern China in a Y,Re-rich muscovite granite, with xenotime, monazite, zircon, fergusonite, thorite, gadolinite, chernovite, and columbite. In Madagascar at Tongafeno, Ambohitromby, and elsewhere. RG *DI:793; AM 51:152(1966), 61:178(1976)*

8.3.6.4 Tantalaeschynite-(Y) $(Y,Ce,Ca)(Ta,Ti)_2O_6$

Named in 1974 for the composition. Metamict. ORTH *Pbnm.* $a = 5.34$, $b = 10.97$, $c = 7.38$, $Z = 4$, $D = 6.39$. *MM 39:571(1974). 26-1*(heated): 3.71_3 3.00_{10} 2.94_{10} 2.65_5 2.21_3 1.86_4 1.70_5 1.58_7. Prismatic, equant crystals, with {010}, {001}, {110}, and {130}. Brownish black; black, light yellowish-brown streak, resinous luster. No cleavage, conchoidal fracture, brittle. $VHN_{100} = 665$, H = $5\frac{1}{2}$–6. G = 5.7–6.1. Isotropic; yellow to orange pleochroism. In RL, dark gray, abundant white to dark red internal reflections. R = 14.5 (470 nm), 14.2 (546 nm), 14.0 (589 nm), 14.0 (650 nm). *Raposa pegmatite,*

São José do Sabugí, PB, Brazil, with biotite, ferrocolumbite, and beryl. Phulord alluvial deposit, Thailand. RG *AM 59:1331(1974)*.

8.3.6.5 Aeschynite-(Nd) (Nd,Ce)(Ti,Nb)$_2$(O,OH)$_6$

Named in 1982 for the composition. Metamict. Space group unknown. *34-480* (heated): 545_5 3.00_9 2.93_{10} 2.64_5 2.26_5 1.95_6 1.57_7 1.53_6. Tabular and prismatic crystals, equigranular clusters. Dark to light brown, light brown streak, adamantine luster. Conchoidal fracture, brittle. H = 5–6. G = 4.8. Nonmagnetic, radioactive. Readily soluble in warm HCl, H_2SO_4, and H_3PO_4; with difficulty in HNO_3. Biaxial; N = 2.2–2.4; 2V = 80°. *Bayan Obo, Inner Mongolia, China*, in veins in slate and metamorphosed dolomite, associated with aegirine, riebeckite, barite, fluorite, and phlogopite. RG *AM 69:565(1984)*.

8.3.7.1 Vigezzite (Ca,Ce)(Nb,Ta,Ti)$_2$O$_6$

Named in 1979 for the locality. ORTH *Pmnb*. $a = 7.559$, $b = 11.025$, $c = 5.360$, Z = 4, D = 5.54. *MM 43:459(1979)*. *34-1316*: 4.82_9 3.78_8 3.04_{10} 2.97_{10} 2.68_4 2.37_4 1.71_6 1.60_7. Crystals flat prismatic, {001} striated [001]. Orange-yellow. Cleavage {001}, perfect; brittle. H = $4\frac{1}{2}$–5. Nonfluorescent. Biaxial; $N_x = 2.14$, $N_z = 2.315$ (Na); strong birefringence; extreme dispersion; positive elongation. *Alpe Rosso, and Pizzo Marcio, near Orcesco, Valle Vigezzo, Novara, Piemonte, Italy*, in albitic rock, with pyrochlore, columbite, fersmite, bavenite, and emerald. RG *AM 65:812(1980)*.

8.3.7.2 Rynersonite Ca(Ta,Nb)$_2$O$_6$

Named in 1978 for the Rynerson family, San Diego Co., CA. Members of the family have operated mines in the Himalaya pegmatite–aplite dike system since 1900. ORTH *Pmnb*. $a = 7.505$, $b = 11.063$, $c = 5.370$, Z = 4, D = 6.394. *AM 63:709(1978)*. *39-1430*(syn): 4.84_{10} 3.75_8 3.04_{10} 2.69_4 2.36_5 1.714_3 1.596_3. Masses of lath-like crystals. Reddish pink, white streak, earthy luster. Uneven fracture. G = 6.40. Nonfluorescent. Biaxial (+); XYZ = *cba*; N > 2.05 (straw yellow); birefringence 0.14. *Himalaya mine, Mesa Grande district, San Diego Co., CA*, as an alteration product of stibiocolumbite–stibiotantalite crystals, associated with fersmite and antimonian microlite. Wampewo pegmatite, Mengo district, Uganda, with microlite in pseudomorphs after bismutotantalite. RG *SMPM 59:15(1979)*.

8.3.8.1 Polycrase-(Y) (Y,Ca,Ce,U,TH)(Ti,Nb,Ta)$_2$O$_6$

Named in 1870 from the Greek for *many* and *mixture*, in allusion to the many rare elements it contains. Metamict. ORTH *Pbcn*. $a = 14.62$, $b = 5.55$, $c = 5.19$, Z = 4, D = 4.30–5.87. *VLA:479*(heated to 800°): 2.98_7 2.90_{10} 1.85_6 1.67_5 1.57_6 1.50_5 1.19_5 1.07_6. Crystals stoutly prismatic [001], sometimes flattened {010}; also massive. Twinning common on {201}, flattened {010} and striated [001]. Black; yellowish, grayish, or reddish-brown streak; submetallic luster. Subconchoidal to conchoidal fracture. H = $5\frac{1}{2}$–$6\frac{1}{2}$. G = 5.0–5.9. The G decreases with alteration and increases after ignition. It also varies significantly with increase or decrease of the Nb/Ta ratio. Radioactive. Crystal chemical relationships in the polycrase–euxenite group and the identification of its members are complicated by most of them being metamict and by different degrees of hypogene alteration and weathering effects. Deficiencies in the A site are common, coupled with the introduction and removal of various cations. Consequently, heating experiments often fail to restore the original structure of the euxenite-type minerals because of these secondary deviations from the original stoichiometry. Isotropic; N ≈ 2.25. Polycrase–euxenite are commonly found in K-rich, granite pegmatites, associated with monazite, beryl, columbite, magnetite, biotite, ilmenite, xenotime, gadolinite, allanite, and garnet. They are less common in albite–O-rich pegmatites or Li-rich pegmatites. Davis farm, near Zirconia, Henderson Co., NC. Baringer Hill, Llano Co., TX. In placers near Pioneerville, Boise Co., ID. *Rasvåg, Hitterö, Norway*, and scores of other localities in Norway. Lön Jonköping, Sweden. In Madagascar well crystallized from numerous localities, in potash-rich pegmatites with beryl, columbite, monazite betafite, as at Ankazobe; Ranomafana; Ampangabe; Samiresy. In Brazil in MG at Gruta de Generosa, Sabinopolis; Pomba; Fazenda Santa Clara, and so on. RG *DI:787*.

8.3.8.2   Euxenite-(Y) $(Y,Ca,Ce)(Nb,Ta,Ti)_2O_6$

Named in 1870 from the Greek for *friendly to strangers, hospitable*, in allusion to the rare elements that it contains. Metamict. See also Polycrase-(Y) (8.3.8.1). ORTH *Pbcn*. $a = 14.643$, $b = 5.553$, $c = 5.195$, $Z = 4$. *ZK 152:69(1980)*. *5-603*(heated): 3.66_3 2.98_{10} 2.43_3 1.82_4 1.77_3 1.72_4 1.64_3 1.49_4. Physical properties and localities for euxenite are listed together with those for polycrase, with the understanding that positive identification of the species is not possible in most cases without chemical analysis. These minerals are common worldwide and only a few of the known occurrences can be mentioned. Morton, Delaware Co., PA; Trout Creek Pass, Chaffee Co., CO; in ONT, Canada, at the J. G. Gole quarry, Murchison Twp. in lustrous euhedral crystals to 4 cm, associated with fergusonite-(Y); also in Sabine and Mattawan Twps., Nipissing district. Pegmatite del Bosco, Arvogno, Val di Crana, Val Vigezzo (Novara), Italy. *Jolster, Söndfjord, Norway*, as well as on Hitterö and Kragerö Is. Huntila, Pitkäranta, Finland. Fine crystals from Madagascar at numerous localities, especially Ambatofotsikely and on Mt. Vohambohitra. RG *DI:787*; *AM 35:386(1950)*, *50:2084(1965)*; *NJMA 103:1(1965)*; *CM 15:92(1977)*.

8.3.8.3   Tanteuxenite-(Y) $(Y,Ca,Ce)(Ta,Nb,Ti)_2(O,OH)_6$

Named in 1928 for the composition. Metamict. ORTH *Pcan*. $a = 5.18$, $b = 14.57$, $c = 5.52$, $Z = 4$. *31-1434*(heated): 3.12_{10} 2.87_7 2.67_3 2.52_3 2.28_1 2.15_1. Indistinct tabular crystals. Brownish black, resinous luster. Subconchoidal fracture. H = 5–6. G = 5.4–5.9. Isotropic. Mattawa, ONT, Canada, as euhedral crystals to 5 mm. Piano de Lavonchio, Craveggia, Val Vigezzo,

Piedmont, Italy. *Cooglegong, Pilbara, WA, Australia*, in tin placers. Liha, Kivu, Zaire. RG *DI:791, USGSB 1036:111(1958), MM 32:308(1959)*.

8.3.8.4 Yttrocrasite-(Y) $(Y,Th,U)Ti_2(O,OH)_6$

Named in 1906 from the Greek for *mixture*, because it is a mixture of yttrium with many other elements. Metamict. Probably ORTH Space group unknown. A doubtful species. May be identical to polycrase. As rough crystals. Black, resinous to pitchy luster. Conchoidal fracture. $H = 5\frac{1}{2}$–6. $G = 4.80$. Radioactive. Isotropic; $N \approx 2.12$–2.15. From *Burnet Co., TX*, 5 km E of the Baringer Hill pegmatite in Llano Co., TX. RG *DI:793, AM 67:156(1982)*.

8.3.8.5 Uranopolycrase $(U,Y)(Ti,Nb,Ta)_2O_6$

Named for the composition, the uranium-dominant analog of polycrase-Y. ORTH *Pbcn*. $a = 14.51$, $b = 5.558$, $c = 5.173$, $Z = 4$. *EJM 5:1161(1993)*. *PD*(heated): 3.21_1 2.99_{10} 2.78_3 2.51_1 1.90_5 1.86_1 1.77_4 1.48_4. Crystals prismatic, tabular {100}. Brown–red, opaque, brownish streak, adamantine luster. Cleavage {100}, good. In RL, pale gray with bluish tones; anisotropism and bireflectance not observed. Weak internal reflections. $VHN_{20} = 659$. Found in pegmatite at the *Fonte del Prete quarry, southern Piero in Campo, Elba, Italy*, as inclusions in beryl crystals with stilbite. Associated minerals are quartz, orthoclase, albite, lepidolite, and elbaite. RG

8.3.9.1 Kassite $CaTi_2O_4(OH)_2$

Named in 1965 for Kicolai Grigorevich Kassin (1885–1949), Russian geologist and academician, discoverer of the Afrikanda massif on the Kola Peninsula. ORTH Space group uncertain, probably *Ammm*. $a = 8.99$, $b = 9.55$, $c = 5.26$, $Z = 2$, $D = 3.28$. *AM 52:559(1967)*. *39-357*: 7.85_{10} 3.59_3 3.26_7 2.99_4 2.79_3 2.57_7 1.910_8 1.898_6. Crystals flattened on {010}, forming six-sided plates. As rosettes of plates and as spherules. Twinning on {101} and {181}, very common. Pale yellow, adamantine luster. Cleavage {010}, perfect; {101}, distinct; very brittle. $H = 5$. $G = 3.28$. Insoluble in acids. Biaxial $(-)$; $XZ = ab$; $N_x = 1.95$, $N_y = 2.13$, $N_z = 2.21$; $2V = 58°$; $r > v$, very strong. Josephine Creek, OR. Possibly from the Diamond Jo quarry, Magnet Cove alkalic complex, Hot Springs Co., AR, in a nepheline syenite pegmatite. *Afrikanda massif, Kola Penin., Russia*, in miarolytic cavities of alkalic pegmatites. associated with cafetite and titanite, and replacing ilmenite. Henan Prov., China. RG *ZVMO 106:114(1977); AM 71:1045(1986), 76:283(1991)*.

8.3.10.1 Thoreaulite $Sn^{2+}(Ta,Nb)_2O_6$

Named in 1933 for Jacques Thoreau (1886–1971), University of Louvain, Belgium. MON $C2/c$. $a = 17.140$, $b = 4.865$, $c = 5.548$, $\beta = 91.0°$, $Z = 4$, $D = 8.27$. *AM 55:367(1970)*. *44-1411*: 8.59_2 3.57_1 3.10_4 3.07_4 2.86_{10} 1.853_2 1.697_2 1.683_3. Crystals prismatic [001]. Brown, yellow streak with brownish tint, resinous to adamantine luster. Cleavage {100}, perfect; {011}, imperfect. $H = 6$. $G = 7.6$–7.9. Biaxial $(+)$; $Y = b$; $Z \wedge c = 27°$; $N \approx 2.38$; $2V = 25°$; birefringence = 0.039. Ungursai, eastern Kazakhstan. From *Zaire* at *Manono, Shaba*, with cassiterite in pegmatite; also at Maniema and at Kubitaka, Punia. Lavra Urubú, Itinga, MG, Brazil, in a lithium pegmatite with lepidolite, rubellite, and petalite. RG *DI:802, AM 59:1026(1974), SPD 20:528(1975), CM 26:889(1988)*.

8.3.10.2 Foordite $(Sn^{2+},Pb)(Nb,Ta)_2O_6$

Named in 1988 for Eugene E. Foord (b.1946), mineralogist and geologist, U.S. Geological Survey, specialist in Nb and Ta minerals. Isostructural with thoreaulite, 8.3.10.1. MON $C2/c$. $a = 17.093$, $b = 4.877$, $c = 5.558$, $\beta = 90.85°$, $Z = 4$, $D = 6.66$. Structure: *CM 26:889(1988)*. *45-1350:* 3.59_4 3.10_5 3.07_5 2.85_{10} 2.78_2 1.853_3 1.694_5 1.681_5. Found as alluvial pebbles. Brownish yellow, greenish tint, yellowish-white streak, vitreous to adamantine luster. Cleavage {100}, perfect; {011}, poor. $H = 6$. $G = 6.73$. Biaxial (+); $N_{calc} = 2.294$; 2V moderate. Birefringence > 0.10. From *Lutsiro, Sebeya River, Rwanda*, and *Punia, Zaire*. RG.

8.3.11.1 Changbaiite $PbNb_2O_6$

Named in 1978 for the locality. HEX-R $R3m$. $a = 10.499$, $c = 11.553$, $Z = 9$, $D = 6.51$. *AM 64:242(1979)*. *29-780*(syn): 5.25_2 3.11_8 3.03_{10} 2.17_3 1.926_1 1.764_4 1.625_2. Small tabular crystals, and as spherules. Principal forms {0001}, {01$\bar{1}$1}; {10$\bar{1}$1}, {11$\bar{2}$0}. Colorless, white, pale brown, pale yellow-green; white streak; adamantine to pearly luster. Cleavage {0001}, perfect; {10$\bar{1}$1}, distinct; hackly to conchoidal fracture; brittle. $H = 5\frac{1}{2}$. $G = 6.48$. Nonfluorescent. Nonmagnetic. Insoluble in HCl, HNO_3, and H_2SO_4; slightly soluble in warm H_3PO_4. Uniaxial (−); $N_O = 2.485$, $N_E = 2.476$; r > v. R = 15.9 (546 nm). *Changbai Mt., Tonghua, Jilin, China*, in kaolin veinlets and kaolin-filled cavities. RG

8.3.12.1 Kobeite-(Y) $(Y,U)(Ti,Nb)_2(O,OH)_6$

Named in 1950 for the locality. May be identical with polycrase. Needs further study. Metamict. ORTH Space group unknown. $a = 14.758$, $b = 5.753$, $c = 4.985$, $Z = 4$. *AM 42:742(1957)*. *PD*(heated): 2.92_{10} 2.78_3 2.50_4 2.29_2 1.81_7 1.74_3 1.54_2 1.52_3. *44-1449:* 5.42_2 4.39_1 3.96_1 3.11_1 2.94_{10} 2.75_3 1.987_2 1.916_2. Irregular grains. Dark brown, black, resinous to vitreous luster. Brittle. $G = 5.0$. Strongly radioactive. Isotropic; $N = 2.205$. *Shiraishi, Kobe-Mura, Kyoto Pref., Japan*, in a pegmatite, associated with parisite, zircon, monazite, tscheffkinite, xenotime, and biotite. Also from a cobble found in the Paringa R., South Westland, New Zealand. RG *AM 36:925(1951), MJJ 3:139(1961)*.

8.4.1.1 Schreyerite $V_2Ti_3O_9$

Named in 1976 for Werner Schreyer (b.1930), Ruhr University, Bochum, Germany. Polymorph: kyzylkumite. MON Space group unknown. $a = 7.06$, $b = 5.01$, $c = 18.74$, $\beta = 119.4°$, $Z = 4$, $D = 4.480$. *30-1429:* 4.08_5 3.38_6 2.87_8 2.74_{10} 2.43_2. Minute grains embedded in rutile; lamellar twinning prevalent.

Black. VHN = 1150. Insoluble in acids. In RL, reddish brown, moderately anisotropic, yellow-brown to reddish-brown pleochroism. N ≈ 2.7. *Kwale District, S of Voi, Kenya*, in vanadium-enriched gneiss, associated with kyanite, sillimanite, rutile, kornerupine, biotite, tourmaline, diopside, and epidote. RG *NW 63:293(1976); AM 62:395(1977), 63:1182(1978); NJMM:110(1983)*.

8.4.1.2 Olkhonskite $(Cr,V)_2Ti_3O_9$

Named in 1994 for the locality. See schreyerite (8.4.1.1). MON Space group unknown. $a = 7.03$, $b = 5.02$, $c = 18.83$, $\beta = 119.6°$, $Z = 4$, $D = 4.48$. PD: 2.88_7 2.75_{10} 2.43_7 1.426_7 1.386_{10}. Forms platy inclusions 1–20 mm wide and 50–150 mm long in rutile. Black, metallic luster. Knoop hardness = 1412 Kp/mm². In RL, light gray, noticeably bireflectant, weakly pleochroic, weakly anisotropic. R = 21.1–18.1 (470 μm), 19.9–18.5 (546 μm), 19.8–18.4 (589 μm), 20.0–18.6 (650 μm). Found in rutile in quartzite schists of the Olkhon series on the western shore of *Lake Baikal, Russia, near Olkhon Is.*, associated with schreyerite, eskolaite, karelianite, and other Cr–V–Ti oxides. RG *ZVMO 123(4):98(1994), AM 81:251*.

8.4.1.3 Kyzylkumite $V_2^{3+}Ti_3O_9$

Named in 1981 for the locality. Polymorph: schreyerite. MON Space group unknown. $a = 33.80$, $b = 4.578$, $c = 19.99$, $\beta = 93.4°$, $Z = 18$, $D = 3.77$. *AM 67:855(1982)*. 35-486: 3.70_8 2.92_{10} 2.60_5 2.20_4 2.19_5 1.69_5 1.68_6 1.65_5. Prismatic crystals and grains. Black, vitreous to resinous luster. G = 3.75. Insoluble in acids. *Kyzyl-Kum, Uzbekistan*, in siliceous schists, associated with chlorite, pyrite, and rutile. RG *ZVMO 110:607(1981), IGR 24:740(1982)*.

8.4.2.1 Pseudorutile $Fe_2^{3+}Ti_3O_{8.33}(OH)_{1.33} \cdot nH_2O$

Named in 1966 for the composition, as an intermediate product between ilmenite and rutile. HEX $P6_322$. $a = 14.375$, $c = 4.615$, $Z = 6$, $D = 4.817$. Structure: *AM 60:898(1975)*. 29-1494: 3.67_4 3.50_{10} 3.23_3 2.66_9 2.51_8 1.69_7 1.66_3 1.48_3. Microscopic grains and inclusions averaging <20 nm. G = 4.20. Paramagnetic. Pseudorutile is a very common and widespread alteration product of ilmenite and is present in ilmenite placers and beach sands throughout the world. Near *Hackberry, AZ; FL; NJ; India; Brazil; Singkep Is., Sumatra, Indonesia; South Neptune Is., SA, Australia*; as a submicroscopic intergrowth with ilmenite and rutile in all alluvial ilmenites. RG *AM 35:117(1950), 52:299(1967), 68:985(1983); NAT 211:179(1966); MM 43:659(1980)*.

CRICHTONITE GROUP

The chrichtonite group minerals have the general formula

$$AB_{21}O_{38}$$

where

A = a large cation = Sr, Na, Pb, Ca, Ba, K, and U + REE
B = a small cation, dominantly Ti, Cr, Fe, Mg, Zn, and Nb

The atomic arrangement of the crichtonite group is based on a close-packed anion framework with a nine-layer stacking sequence, in which the A atoms occupy one of the anion sites in the cubic layers. The other metal atoms are ordered into both tetrahedral and octahedral interstices between the anion layers. Two types of metal-atom polyhedral arrangements occur, between pairs of hexagonally stacked anion layers, and between hexagonal and cubic anion layers, h-M-h and h-M-c, respectively. Between hexagonally stacked anion layers, two types of metal atoms, M(1) and M(3), occupy octahedral sites only. The M(3)O$_6$ octahedra, articulate by edge sharing to form interconnected 12-member hexagonal rings, with an isolated M(1)O$_6$ octahedron at the center of each ring. The h-M-c metal-atom layers, two per unit cell, contain metals in both octahedral and tetrahedral sites. The octahedrally connected atoms, M(4) and M(5), join by edge sharing into six-membered

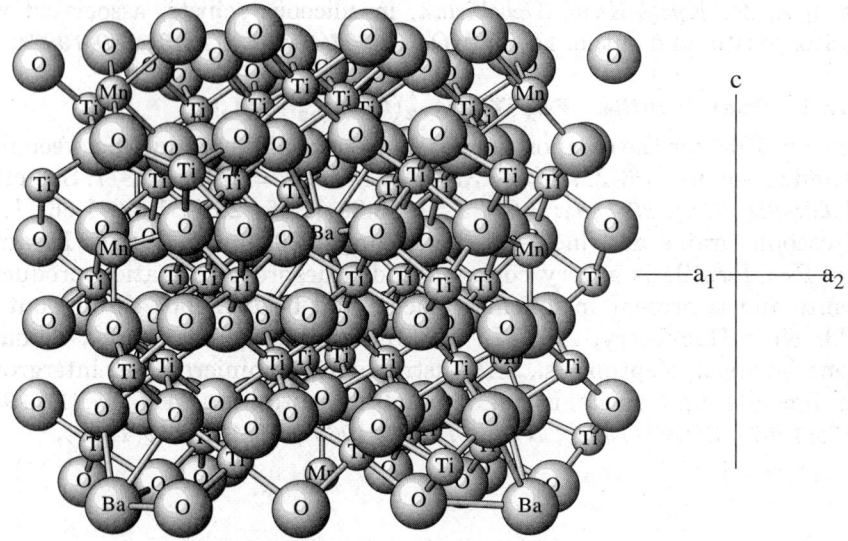

LINDSLEYITE: Ba(Ti,Mn)$_{21}$O$_{38}$

This structure is related to perovsite in that it is based on cubic-closest-packed oxygens, with large cations, Ba in this case, replacing some oxygens, and smaller cations, mostly Ti, in octahedral interstices. However, there are also some small cations, Mn in this case, in tetrahedral coordination, and the proportion of large cations is much smaller than in perovskite.

hexagonal rings that are interlinked via corner sharing with basal oxygens of the $M(2)O_4$ tetrahedra. The cubic-stacked anion layers have A cations ordered into one of the available 13 anion sites per unit cell. *AM 61:1203(1976)*.

CRICHTONITE GROUP

Mineral	Formula	Space Group	a	c	D	
Landauite	$NaMnZn_2(Ti,Fe)_6Ti_{12}O_{38}$	$R\bar{3}$	10.366	20.77	4.46	8.5.1.1
Loveringite	$(Ca,Ce,La)(Ti,Fe,Cr)_{21}O_{38}$	$R\bar{3}$	10.337	20.676	4.415	8.5.1.2
Crichtonite	$(Sr,La,Ce)(Ti,Fe^{3+}Fe^{2+},Mn)_{21}O_{38}$	$R\bar{3}$	10.374	20.746	4.54	8.5.1.3
Senaite	$Pb(Ti,Fe,Mn)_{21}O_{38}$	$R\bar{3}$	10.393	20.811	4.59	8.5.1.4
Davidite-(La)	$(La,Ce,Ca)(Y,U)(Ti,Fe)_{20}O_{38}$	$R\bar{3}$	10.376	20.910	4.72	8.5.1.5
Davidite-(Ce)	$(Ce,La)(Y,U)(Ti,Fe)_{20}O_{38}$	$R\bar{3}$	10.376	20.910	4.72	8.5.1.6
Mathiasite	$(K,Ca,Sr)(Ti,Cr,Fe,Mg)_{21}O_{38}$	$R\bar{3}$	10.36	20.65	4.39	8.5.1.7
Lindsleyite	$(Ba,Sr)(Ti,Cr,Fe,Mg)_{21}O_{38}$	$R\bar{3}$	10.37	20.52	4.63	8.5.1.8

8.5.1.1 Landauite $NaMnZn_2(Ti,Fe)_6Ti_{12}O_{38}$

Named in 1966 for Lev D. Landau (1908–1968), Russian physicist, and academician, Moscow University. Crichtonite group. HEX-R $R\bar{3}$. $a = 10.366$, $c = 20.77$, $Z = 3$, $D = 4.46$. *CM 16:63(1978)*. 18-672: 3.36_6 3.02_6 2.83_{10} 2.21_7 2.11_9 1.78_8 1.58_8 1.43_8. Fine-grained aggregates of irregular grains. Black, gray streak, semimetallic luster. No cleavage, conchoidal fracture. $H = 7\frac{1}{2}$. $G = 4.42$. Dissolves with difficulty in a mixture of HNO_3 and H_2SO_4. Biaxial (−); $N_x = 2.373$, $N_z = 2.388$; $2V = 60°$; $r > v$, weak; absorption $Z > Y > X$. *Burpala alkalic complex, northern Baikal region, Siberia, Russia*, in albite veinlets in quartz syenite and in syenitic pegmatites, associated with albite, polylithionite, brookite, strontian chabazite, monazite, and bastnaesite. RG *DANS 166:1420(1966), 261:741(1981); AM 51:1546(1966)*.

8.5.1.2 Loveringite $(Ca,Ce,La)(Ti,Fe,Cr)_{21}O_{38}$

Named for John Francis Lovering (b.1930), Australian geochemist, University of Melbourne. Crichtonite group. HEX-R $R\bar{3}$. $a = 10.337$, $c = 20.676$, $Z = 3$, $D = 4.415$. *NBSM 25(16):106(1979)*. Structure: *AM 63:28(1978)*. 30-263: 3.38_8 3.04_7 2.87_{10} 2.83_7 2.47_5 2.24_6 2.13_6 1.59_6. Irregular grains, acicular crystals. Black, iron-gray streak, metallic–submetallic luster. In RL, white. $VHN_{10} = 440$. *Jimberlana intrusion, near Norseman, WA, Australia*, associated with quartz-K feldspar intergrowths and phlogopite. RG *MM 42:187(1978), CM 17:635(1979), AM 68:474(1983)*.

8.5.1.3 Crichtonite (Sr,La,Ce)(Ti,Fe^{3+},Fe^{2+},Mn)$_{21}$O$_{38}$

Named in 1813 for Alexander Crichton (1763–1856), Scottish physician and mineral collector. Crichtonite group. Crichtonite was long thought to be identical with ilmenite, but in 1953 was shown to have distinct chemistry and unit cell. HEX-R $R\bar{3}$. $a = 10.374$, $c = 20.746$, $Z = 3$, $D = 4.54$. *22-1121:* 3.39_{10} 2.88_{10} 2.84_8 2.24_8 2.13_{10} 1.796_8 1.596_{10} 1.440_{10}. Rare crystals are steep rhombohedra terminated by {0001}, more commonly in rose type groups. Black, metallic, brilliant luster. Cleavage {0001}, perfect; conchoidal fracture; brittle. H = 4.5. G = 4.46. *Plan du Lac, St. Christophe en d'Oisans, Isère, France*, associated with anatase, quartz, chlorite, hematite, and feldspar. Selva, Tavetsch, Switzerland, in crystals to 25 mm; also Wanigletscher, Binntal, Wallis. In Italy from Pizzo Cervandone, Ossola, crystals to 10 mm. RG *DI:537; AM 38:734(1953), 55:534(1970); MM 37:349(1969)*.

8.5.1.4 Senaite Pb(Ti,Fe,Mn)$_{21}$O$_{38}$

Named in 1898 for Joaquim da Costa Sena, Esola de Minas, Ouro Preto, Brazil. Crichtonite group. HEX-R $R\bar{3}$. $a = 10.393$, $c = 20.811$, $Z = 3$, $D = 4.59$. *45-1388:* 3.40_{10} 3.06_5 2.99_5 2.89_{10} 2.85_5 2.47_5 2.25_5 1.602_5. Rhombohedral, tabular, crystals with {0001}, {10$\bar{1}$1}, {20$\bar{2}$1}, {40$\bar{4}$1}. Black, brownish-black streak, submetallic luster, conchoidal fracture. H = 6+. G = 5.30. Uniaxial (−); N ≈ 2.50 (Li); low to moderate bireflectance. From *Dattos, Curralinho*, and other localities near *Diamantina, MG, Brazil*, in diamond-bearing stream placers, and at the Fazenda Guariba(!) in the same district as crystals in fractures in quartzite. RG *DI:541; AM 53:869(1968), 61:1203(1976)*.

8.5.1.5 Davidite-(La) (La,Ce,Ca)(Y,U)(Ti,Fe)$_{20}$O$_{38}$

Named in 1906 for Tenatt William Edgeworth David (1858–1934), Australian geologist. Crichtonite group. Metamict. The original davidite. HEX-R $R\bar{3}$. $a = 10.376$, $c = 20.910$, $Z = 3$, $D = 4.72$. *13-505:* 3.42_7 3.07_5 3.00_4 2.90_{10} 2.85_6 2.48_4 2.25_5 2.14_4. Massive, as anhedral grains. Black, vitreous luster. G = 4.42. Strongly radioactive. Davidite is a fairly common mineral in uranium-rich granite pegmatites and in syenitic rocks; also in pegmatite-like phases of calcite–apatite–diopside–titanite rocks of the Grenville type. Pandora prospect, Quijotos Mts., Pima Co., AZ. McLean Point, NS, Canada, in sandstone with other uranium minerals. Nuevo Continente mine, Mun. Nacozari, SON, Mexico, with autunite. Tete district, Mozambique, in large crystals in a scapolite–calcite–dolomite rock, and in epidiorite with rutile, titanite, magnetite, ilmenite, apatite, tourmaline, and molybdenite. *Radium Hill mine, Olary Prov., SA, Australia*, associated with rutile, ilmenite, and carnotite; also at Billeroo and Crocker's Well, SA. RG *DI:542, AM 46:700(1961), CM 23:495(1986)*.

8.5.1.6 Davidite-(Ce) (Ce,La)(Y,U)(Ti,Fe)$_{20}$O$_{38}$

Named in 1960 for the composition. Crichtonite group. Metamict. HEX-R $R\bar{3}$. $a = 10.376$, $c = 20.910$, $Z = 3$, $D = 4.72$. *8-291:* 3.37_7 3.21_7 2.86_8 2.23_8 1.70_{10}

1.69_{10} 1.59_8 1.435_9. Crystals tabular $\{0001\}$, or pyramidal with ditrigonal pyramids dominant; also cuboidal; commonly massive. Black; dark brown; reddish, grayish-black to dark brown streak, vitreous to submetallic, shiny, luster. Conchoidal to uneven fracture, brittle. H = 6. G = 4.4–4.9. Strongly radioactive. Slowly attacked by HF and strong H_2SO_4 or HNO_3. Isotropic; $N \approx 2.3$. In RL, light gray. Faraday mine, Bancroft, ONT, Canada, and at the Baska mine, Kotelmach Lake, Foster Lakes area, SK. *Tuftane, Iveland, Norway*, in pegmatite with thortveitite, gadolinite, and euxenite. RG *DI:542, USGSB 1064:337(1958), NGT 40:277(1960), AM 51:152(1966)*.

8.5.1.7 Mathiasite $(K,Ca,Sr)(Ti,Cr,Fe,Mg)_{21}O_{38}$

Named in 1983 for Frances Celia Morna Mathias (b.1913), British–South African geochemist, University of Capetown, South Africa. Crichtonite group. HEX-R $R\bar{3}$. $a = 10.36$, $c = 20.65$, $Z = 3$, $D = 4.39$. *37-420*: 4.14_5 3.39_6 2.99_8 2.88_8 2.25_9 2.14_{10} 1.79_7 1.44_{10}. Black, metallic luster. Conchoidal fracture. $VHN_{100} = 1505$. In RL, tan, weakly pleochroic, weakly to moderately anisotropic. *Jagersfontein mine, Orange Free State, South Africa*, in kimberlite, associated with olivine, diopside, phlogopite, chromian ilmenorutile, Mg–Cr spinel, ilmenite, perovskite, and titanite. Also from the Bultfontein mine, and at Kolonkwanen in the MacKenzie's Post district, near the border with Botswana, Cape Prov. RG *AM 68:494(1983)*.

8.5.1.8 Lindsleyite $(Ba,Sr)(Ti,Cr,Fe,Mg)_{21}O_{38}$

Named in 1983 for Donald H. Lindsley (b.1934), mineralogist, Department of Earth Sciences, State University of New York at Stony Brook. Crichtonite group. HEX-R $R\bar{3}$. $a = 10.37$, $c = 20.52$, $Z = 3$, $D = 4.63$. *37-421*: 3.04_4 2.87_7 2.83_7 2.13_{10} 1.80_{10} 1.59_{10} 1.50_5 1.44_{10}. Small grains and aggregates. Black, metallic luster. Conchoidal fracture. $VHN_{100} = 1505$. In RL, tan with no internal reflections, and weakly pleochroic, pale tan to brown; weakly to moderately anisotropic. *DeBeers mine, Kimberley, Cape Prov., South Africa*, in kimberlite, in peridotite nodules, associated with phlogopite, K-richterite, Cr-spinel, ilmenite, and Cr-diopside. Also found at the Bultfontein mine. RG *AM 68:494(1983)*.

8.6.1.1 Franconite $Na_2Nb_4O_{11} \cdot 9H_2O$

Named in 1984 for the locality. MON Space group unknown. $a = 22.22$, $b = 12.86$, $c = 6.36$, $\beta = 92.23°$, $Z = 4$, $D = 2.71$. *CM 22:239(1984)*. *38-357*: 11.0_{10} 6.06_4 5.55_7 4.73_6 4.21_5 3.21_5 3.18_6 2.63_5. Aggregates of small globules consisting of radiating, bladed crystals. White, white streak, vitreous to silky luster. No cleavage, basal parting. G = 2.72. Nonfluorescent. Difficultly soluble in HCl. Biaxial (−); $N_x = 1.72$, $N_y = 1.78$, $N_z = 1.79$; $2V \approx 35°$; Y perpendicular to elongation; Z parallel to elongation. *Francon Quarry, Montreal Is., QUE, Canada*, in a dawsonite-bearing still in vugs associated with weloganite, calcite, and quartz. RG *AM 70:436(1985), DANS 305:700(1989)*.

8.6.1.2 Hochelagaite $(Ca,Na,Sr)Nb_4O_{11} \cdot 8H_2O$

Named in 1986 for Hochelaga, the original (Native American) name for Montreal, where the mineral was found. MON Space group unknown. $a = 19.98$, $b = 12.88$, $c = 6.446$, $\beta = 93.42°$, $Z = 4$, $D = 2.85$. *CM 24:449(1986)*. *36-418:* 10.0_{10} 6.18_2 5.39_5 4.96_5 3.21_7 3.12_8 2.80_4 1.98_3. Polycrystalline globules consisting of microscopic radiating blades. White, white streak, vitreous luster. No cleavage, brittle. $H = 4$. $G = 2.9$. Nonfluorescent. Hochelagaite is similar in structure to franconite, and the chemical composition differs only in the presence of Ca instead of Na, and a small difference in the H_2O content. Despite being closely related, however, it is doubtful that they have identical structures or that a complete solid-solution series exists between them. Biaxial $(-)$; $N_x = 1.72$, $N_y = 1.81$, $N_z = 1.82$; $2V \approx 35°$; Y perpendicular to blades; Z parallel to elongation. *Francon quarry, St. Michel district, Montreal Is., QUE, Canada*, in vugs in a dawsonite-bearing still, associated with franconite, weloganite, calcite, and quartz. Also from Mt. St.-Hilaire, QUE. RG *AM 72:1024(1987)*.

8.6.2.1 Calciotantite $Ca(Ta,Nb)_4O_{11}$

Named in 1982 for the composition. HEX-R $P6_322$. $a = 6.245$, $c = 12.323$, $Z = 2$, $D = 7.46$. *15-679*(syn): 6.15_6 5.37_6 3.06_8 3.00_{10} 2.77_{10} 2.47_8 1.92_6 1.55_8. *MZ 4(3):75(1982)*. 6.11_5 3.12_7 3.02_{10} 2.78_9 2.47_6 1.80_7 1.55_7 1.51_{10}. Square to rectangular to hexagonal crystals as inclusions in microlite. Colorless, adamantine luster. No cleavage. $VHN_{40} = 970$. Nonfluorescent in UV light, luminesces blue in cathode rays. $R = 16.9$–16.3 (486 nm), 17.0–16.6 (551 nm), 17.1–16.7 (589 nm), 18.2–18.0 (656 nm). *Kola Penin., Russia*, in granite pegmatites, with albite, spodumene, scheelite, apatite, wodginite, and microlite. RG *AM 68:471(1983)*, *SPC 33:498(1988)*.

8.6.2.2 Natrotantite $Na_2Ta_4O_{11}$

Named in 1981 for the composition. HEX-R $R\bar{3}c$. $a = 6.209$, $c = 36.609$, $Z = 6$, $D = 7.706$. Structure: *BM 108:541(1985)*. *38-463:* 5.15_3 3.05_5 3.01_{10} 2.77_8 2.47_3 1.79_{25} 1.70_{14} 1.547_5. Colorless; yellowish, adamantine luster. No cleavage, uneven fracture. $VHN_{40} = 1270$. Nonfluorescent in UV light, strongly luminescent yellow-green in cathode rays. In RL, strongly pleochroic and strongly anisotropic. $R = 15.0$–13.6 (470 nm), 12.4–11.9 (589 nm), 12.0–12.0 (650 nm). *Kola Penin., Russia*, in albitized areas of granite pegmatites, intergrown with microlite and rimming simpsonite. Alto do Giz, Equador, RGN, Brazil, as inclusions in simpsonite, with microlite and manganotantalite. RG *ZVMO 110:338(1981)*, *AM 67:413(1982)*, *IGR 24:835(1982)*.

8.6.2.3 Irtyshite $Na_2(Ta,Nb)_4O_{11}$

Named in 1985 for the locality. HEX $P6_3/m$. $a = 6.231$, $c = 36.77$, $Z = 6$, $D = 7.03$. *38-391:* 6.13_7 5.18_6 3.08_9 3.03_{10} 2.78_9 2.48_6 1.93_4 1.80_4. Microscopic veinlets and irregular grains. Colorless, adamantine luster. No cleavage, uneven fracture, brittle. In RL, bireflectant, nonpleochroic. *Irtysh River area, eastern Kazakhstan,* from granite pegmatites, in altered thoreaulite, lithiotantite, and ixiolite. RG *AM 71:1545(1986).*

8.7.1.1 Murataite $(Y,Na)_6(Zn,Fe)_5(Ti,Nb)_{12}O_{29}(O,F)_{10}F_4$

Named in 1974 for Kiguma Jack Murata (b.1909), geochemist, U.S. Geological Survey. ISO $F\bar{4}3m$. $a = 14.863$, $Z = 4$, $D = 4.64$. *CM (1995). 26-1383:* 8.51_3 2.86_{10} 2.47_6 2.14_2 1.75_8 1.49_8 1.43_5 1.14_5. Embedded grains with cubic outline. Black, submetallic luster. No cleavage, conchoidal fracture. $H = 6-6\frac{1}{2}$, $VHN_{100} = 827$. $G = 4.69$. Isotropic; $N \approx 2.13$. In RL, gray, with orange internal reflections, slightly pleochroic. $R = 13.6$ (546 nm). *St. Peter's Dome, El Paso Co., CO,* in a pegmatite associated with riebeckite, quartz, astrophyllite, aegirine, and zircon. Burpala massif, northern Baikal region, Siberia, Russia, in rare-metal alkali syenite pegmatites, associated with landauite and brookite. RG *AM 59:172(1974), MA 34:722(1983), CM 33:1223(1995).*

8.7.2.1 Uhligite $Ca_3(Ti,Al,Zr)_9O_{20}$

Named in 1909 for Alfred Louis Johannes Uhlig (1883–1919), German geologist, leader of the African expedition in which the material was first collected. May possibly be perovskite. Needs further investigation. ISO or pseudo-ISO Space group unknown. $a = 7.654$. Black, gray to brownish-gray streak, metallic luster. Cleavage {001}, imperfect; conchoidal fracture. $H = 5\frac{1}{2}$. $G = 4.15$. Exhibits an internal structure of biaxial lamellae, similar to those in perovskite. *Lake Magadi, Rift Valley, Kenya,* in nepheline syenite. RG *DI:735.*

8.7.3.1 Cafetite $Ca(Fe^{3+},Al)_2Ti_4O_{12} \cdot 4H_2O$

Named in 1959 for the composition. Probably ORTH Space group unknown. $a = 31.34$, $b = 12.12$, $c = 4.96$, $Z = 6$, $D = 3.19$. *AM 45:476(1960). 20-243:* 7.2_3 5.2_3 4.78_5 3.64_4 3.31_{10} 2.29_4 1.764_{10} 1.504_4. Radial-fibrous or tangled aggregates of acicular crystals. Pale yellow; colorless, adamantine luster. Flexible fibers, brittle. $H = 4-5$. $G = 3.28$. Biaxial $(-)$; $Z \wedge c = 2-4°$; $N_x = 1.95$, $N_y = 2.08$, $N_z = 2.11$; the optics are inconsistent with ORTH symmetry; cafetite may be monoclinic. *Afrikanda massif, Kola Penin., Russia,* in miarolitic cavities in pegmatites cutting pyroxenite, associated with kassite. RG *ZVMO 88:444(1959); AM 71:1045(1986), 76:283(1991).*

8.7.4.1 Calzirtite $CaZr_3TiO_9$

Named in 1961 for the composition. TET $I4_1/acd$. $a = 15.094$, $c = 10.043$, $Z = 8$, $D = 4.94$. Structure: *AM 71:815(1986). 15-121:* 2.95_{10} 2.55_6 1.80_{10} 1.54_9 1.17_7 1.14_5 0.981_6 0.860_6. Tetragonal prismatic, bipyramidal crystals. Light to dark brown, brownish green, nearly black, wine-red, yellow-green; brown streak; semimetallic to adamantine luster, greasy on fracture surfaces. No cleavage, conchoidal fracture, brittle. $H = 6-7$. $G = 5.01$. Nonfluorescent in UV light, luminesces red in cathode rays. Uniaxial $(+)$; $N_O = 2.22$, $N_E = 2.26$. In RL, light gray, with strong reddish brown internal reflections. $R = 15.0-16.4$ (546 nm). *Meimecha Kotui, eastern Siberia, Russia,* in metasomatic calcite–forsterite–magnetite rocks of an alkalic ultrabasic massif; also from the Guly massif; from carbonatites in the Kola Penin., and elsewhere. In Uganda in residual soils from the Bukusu carbonatite complex, with perovskite, ilmenite, zircon, and baddeleyite. In Brazil from the Tapira carbonatite, Sacramento, MG, and from the Jacupiranga carbonatite, SP. RG *AM 46:1515(1961), 52:1880(1967); MM 35:544(1965), 36:778(1968)*.

8.7.5.1 Simpsonite $Al_4(Ta,Nb)_3O_{13}(OH)$

Named in 1938 for Edward Sidney Simpson (1875–1939), government mineralogist of Western Australia. HEX *P*3. $a = 7.38$, $c = 4.51$, $Z = 1$, $D = 6.83$. Structure: *SR 27:540(1962)*. *15-705*(syn): 2.84_9 2.12_7 1.64_{10} 1.39_{10} 1.28_6 1.22_5 1.05_5 1.02_9. Hexagonal, tabular, or equant crystals; rarely prismatic; prism zone striated [0001]. White, colorless, dark yellow, yellow-brown; adamantine to greasy luster. No cleavage, conchoidal fracture. $H = 7\frac{1}{2}$. $G = 6.81$. Strongly fluorescent bluish white in SWUV. Insoluble in acids. Uniaxial $(-)$; $N_O = 2.045$, $N_E = 2.005$. Tanco mine, Bernic Lake, MB, Canada, with microlite, wodginite, cesstibtantite, and tapiolite. Northern Kola Penin., Russia, with cleavelandite, cesium beryl, spodumene, montebrasite, pollucite, petalite, eucryptite, elbaite, manganotantalite, microlite, and stibiotantalite. In Zimbabwe at the Benson No. 3 pegmatite, Mtoko, with stibiotantalite, and in the Bikita pegmatite. *Tabba Tabba, WA, Australia*, in pegmatite. In Brazil from the Alto do Giz, Equador, RGN, in fine crystals up to 3 cm, associated with manganotantalite, microlite, and gahnite; also at the nearby Onça mine with tapiolite and microlite. RG *AM 25:313(1940), DI:771, MM 33:458(1963)*.

8.7.6.1 Rankamaite $(Na,K)_{1-x}(Ta,Nb,Al)_4(O,OH)_{11}$

Named in 1969 for Kalervo Rankama (b.1913), Finnish geochemist, University of Helsinki. ORTH *Pban*. $a = 17.184$, $b = 17.689$, $c = 3.942$, $Z = 6$, $D = 5.84$. T. S. Ercit, Ph.D. thesis, University of Manitoba (1986). *25-9*: 4.11_3 3.94_4 3.47_5 3.38_6 3.01_8 2.97_{10} 2.79_3 1.74_3. Massive, felted acicular crystals. White to creamy white, white streak, dull, chalky luster. $H = 3–4$. $G = 5.5$. Nonfluorescent. Biaxial; $N > 2.1$; $Z = c$; positive elongation. Tanco mine, Bernic Lake, MB, Canada, a mineral intermediate between rankamaite and sosedkoite. Near *Mumba, Kivu, Zaire*, in alluvial concentrates, with simpsonite, cassiterite, manganotantalite, and muscovite, probably derived from the alteration of simpsonite in a lithium pegmatite. RG *AM 55:1814(1970), BGSF 41:47(1979), DANS 264:442(1982)*.

8.7.6.2 Sosedkoite $(K,Na)_{1-x}(Ta,Nb,Al)_4(O,OH)_{11}$

Named in 1982 for A. F. Sosedko (1901–1957), Russian mineralogist. ORTH *Pban*. $a = 17.25$, $b = 17.73$, $c = 3.95$, $Z = 6$, $D = 6.90$. T. S. Ercit, Ph.D. thesis, University of Manitoba (1986). *35-687*: 6.1_5 3.95_{10} 3.47_5 3.03_9 2.79_5 2.38_5 2.28_4 1.97_6. Acicular crystals. Colorless, adamantine luster. No cleavage. $VHN_{20} = 830$. Nonfluorescent, luminesces weak blue under cathode rays. Tanco mine, Bernic Lake, MB, Canada, a mineral intermediate between sosedkoite and rankamaite. *Kola Penin., Russia*, in granite pegmatites, in microlite and cesstibtantite, and along grain boundaries of these minerals with simpsonite and stibiotantalite. RG *DANS 264:133;442(1982), AM 68:644(1983)*.

8.7.7.1 Lithiotantite Li(Ta,Nb)$_3$O$_8$

Named in 1983 for the composition. MON $P2_1/c$. $a = 7.444$, $b = 5.044$, $c = 15.255$, $\beta = 107.18°$, $Z = 4$, $D = 7.08$. *35-718:* 6.03_7 4.13_5 2.96_{10} 2.49_5 1.90_5 1.77_6 1.72_8 1.45_8. Equant crystals, poorly formed. Colorless, stained pink by cassiterite inclusions; adamantine luster. No cleavage, conchoidal to uneven fracture, brittle. $H = 6-6\frac{1}{2}$. $G = 7.0$. Nonfluorescent, luminesces weak yellow-green in cathode rays. $N > 1.9$; very strong dispersion. In RL, grayish white, nonpleochroic, weakly anisotropic. *Eastern Kazakhstan*, in edge zones of microcline–albite pegmatites, with cassiterite, rankamaite, and thoreaulite. RG *MZ 5(1):91(1983), AM 69:1191(1984)*.

8.7.8.1 Belyankinite Ca$_{1-2}$(Ti,Zr,Nb)$_5$O$_{12}$ · 9H$_2$O (?)

Named in 1950 for Dmitri Stepanovich Belyankin (1876–1953), Russian mineralogist and petrographer. ORTH or MON Space group unknown. Amorphous to X-rays. *38-455*(heated): 3.21_8 2.90_3 2.48_6 2.18_3 1.69_{10} 1.63_2 1.36_5 1.26_2. Massive. Light yellow, brownish yellow; vitreous to oily luster, pearly on cleavage. Cleavage perfect in one direction; uneven fracture. $H = 2-3$. $G = 2.4$. Completely soluble in hot HCl, infusible. Biaxial $(-)$; $N_x = 1.740$, N_y, $N_z = 1.775-1.780$. X, brown; Y,Z, light brown to yellow-brown. *Kola Penin., Russia*, in nepheline syenite pegmatite, with microcline, nepheline, aegirine, and zeolites. RG *DANS 71:925(1950), AM 37:882(1952)*.

8.7.8.2 Manganbelyankinite (Mn,Ca)$_2$(Ti,Nb)$_5$O$_{12}$ · 9H$_2$O

Named in 1957 for the composition and relation to belyankinite. Amorphous. *38-454*(heated): 3.21_5 2.89_5 2.48_8 2.17_3 1.88_3 1.69_{10}. Brownish black, resinous luster. Biaxial $(-)$; $2V = 29°$. *Mt. Kedykverpakh, Lovozero massif, Kola Penin., Russia*, in pegmatite. RG

8.7.8.3 Gerasimovskite (Mn,Ca)(Nb,Ti)$_5$O$_{12}$ · 9H$_2$O

Named in 1957 for Vasili Ivanovich Gerasimovskii (1907–1979), Russian mineralogist and geochemist; discoverer of 10 new minerals, and of the industrial loparite deposit on the Lovozero massif. AMOR *AM 43:1221(1958)*. *38-456:* 3.7_8 3.18_6 2.10_4 1.89_{10} 1.64_2. Platy masses. Brown, gray; pearly luster. Perfect cleavage in one direction. $H = 2$. $G = 2.55$. Biaxial $(-)$; $N_x \approx 1.74$, $N_y \approx 1.81$, $N_z \approx 1.81$; $2V = 18°$; parallel extinction; positive elongation. *Lovozero massif, Kola Penin., Russia*, on Mt. Punkuruaiv, Mt. Nepkha, and Mt. Alluaiv in alkaline ussingite-bearing pegmatites. RG

8.7.9.1 Jeppeite (K,Ba)$_2$(Ti,Fe)$_6$O$_{13}$

Named in 1984 for John Frederick Biccard Jeppe (b.1920), Australian geologist of Nedlands, WA, who discovered the mineral. MON $C2/m$. $a = 15.543$, $b = 3.837$, $c = 9.123$, $\beta = 99.25°$, $Z = 2$, $D = 3.98$. *MM 48:263(1984). 37-426:* 4.50$_4$ 3.07$_{10}$ 2.99$_{10}$ 2.81$_{10}$ 2.09$_6$ 2.07$_6$ 1.92$_8$ 1.41$_5$. Finely prismatic to acicular aggregates. Black; brown, pale brown streak, submetallic luster. Cleavage {100}, {20$\bar{1}$}, perfect; brittle. H = 5–6, G = 3.94. Biaxial (+); YZ = bc; X ∧ $a = 10°$; N$_x$ = 2.13, N$_y$ = 2.21, N$_z$ = 2.35. X, blue; Y, dark greenish brown; Z, brown. In RL, weak bireflectance. *Walgidee Hills, Fitzroy basin, Kimberley, WA, Australia*, in lamproite, associated with priderite, richterite, scherbakovite, wadeite, perovskite, and apatite, in a celadonite–chlorite matrix. RG *AM 70:872(1985)*.

8.7.10.1 Zimbabweite Na(Pb,Na,K)$_2$(Ta,Nb,Ti)$_4$As$_4$O$_{18}$

Named in 1986 for the country of origin. ORTH $Ccmb$. $a = 12.245$, $b = 15.287$, $c = 8.684$, $Z = 4$, $D = 5.97$. Structure: *AM 73:1186(1988). 39-356:* 3.82$_6$ 3.24$_5$ 3.20$_{10}$ 3.03$_6$ 2.99$_7$ 2.88$_7$ 2.55$_5$ 1.913$_5$. Crystals are euhedral square tablets showing {010}, {201}, {100}, and {111}. Yellow-brown, white streak, adamantine luster. Cleavage {010}, perfect; brittle. H = 5. G = 6.20. Nonfluorescent. Nonmagnetic. Insoluble in common acids or bases. Biaxial (+); XYZ = cba; N > 2.10; 2V = 80°; r < v, strong. X, pale yellow brown, Y, light reddish brown, Z, reddish brown. *St. Ann's mine, Karoi district, SE of Miami, Zimbabwe*, in kaolinized pegmatite. RG *BM 109:331(1986)*.

8.7.11.1 Cesplumtantite (Cs,Na,Ca)$_2$(Pb,Sb,Sn)$_3$Ta$_8$O$_{24}$

Named in 1986 for the composition. Pyrochlore group. TET Space group unknown. $a = 13.552$, $c = 6.445$, $Z = 2$, $D = 6.87$. *MZ 8(5):92(1986). 40-488:* 6.11$_5$ 3.19$_5$ 3.05$_{10}$ 2.65$_5$ 2.04$_5$ 1.87$_7$ 1.59$_7$ 1.18$_5$. Microscopic veinlets, showing complex polysynthetic twinning. Colorless, white streak, adamantine luster. No cleavage. VHN$_{40}$ = 1240. In RL, light gray, with no internal reflections, weakly bireflectant, strongly anisotropic. *Manono, Shaba, Zaire*, as veinlets in thoreaulite, associated with lithiotantite, cassiterite, calciotantite, and microlite. RG *AM 74:501(1989)*.

8.7.12.1 Koragoite (Mn^{2+},Mn^{3+})$_3$(Nb,Ta)$_3$(Nb,Mn^{2+})$_2$W$_2$O$_{20}$

Named in 1995. MON $P2_1$. $a = 24.73$, $b = 5.056$, $c = 5.760$, $\beta = 103.5°$, $Z = 2$, $D = 6.11$. *PD:* 6.0$_5$ 3.74$_8$ 3.69$_8$ 2.98$_{10}$ 1.783$_5$ 1.744$_6$ 1.732$_7$ 1.456$_5$. Platy crystals. Found in pegmatite in *Pamir, Tadzhikistan*. An inadequately described mineral. RG *K 40(3):469(1995), AM 81:250(1996)*.

Class 9

Anhydrous and Hydrated Halides

HALITE GROUP

The halite group minerals are isostructural and have general formula

$$AX$$

where

A = Na, K, or Li
X = Cl, F

and the structure is in the cubic space group $Fm3m$.

The structure of halite was the first structure established through the use of X-rays, by W. L. Bragg in 1914. If ions of one type are taken as the corners and face centers of the cell, those of opposite sign lie at the midpoints of cell edges and at the cube center. Each Na atom is thus surrounded by six Cl atoms, and each Cl atom by six Na atoms. The Na and Cl atoms are thus arranged alternately in rows parallel to the edges of the cell. *DHZ 5:367.*

HALITE GROUP

Mineral	Formula	Space Group	a	D	
Halite	NaCl	$Fm3m$	5.640	2.164	9.1.1.1
Sylvite	KCl	$Fm3m$	6.93	1.987	9.1.1.2
Villiaumite	NaF	$Fm3m$	4.634	2.799	9.1.1.3
Carobbiite	KF	$Fm3m$	5.347	2.524	9.1.1.4
Griceite	LiF	$Fm3m$	4.027	2.638	9.1.1.5

Notes:
1. The galena and the periclase groups are both isostructural with halite.
2. Osbornite (1.1.19.1), khamrabaevite (1.1.19.2), and carlsbergite (1.1.20.1) are all isostructural with the halite group.

9.1.1.1 Halite NaCl

Named in antiquity from the Greek *hals*, salt. Halite group. Synonyms: salt, rock salt. ISO $Fm3m$. $a = 5.640$, $Z = 4$, $D = 2.164$. *NBSC 539(2):41(1953).* 5-628(syn): 3.26_1 2.82_{10} 1.99_6 1.63_2 1.41_1 1.26_1 1.15_1 0.892_1. **Habit:** Usually cubes, rarely with {111}. Sometimes forms hopper-shaped crystals. Massive,

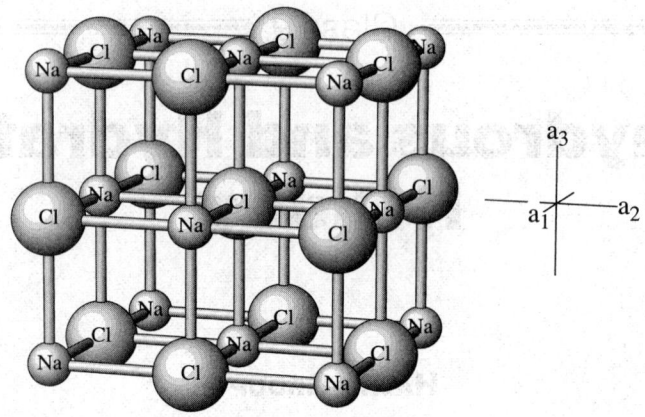

HALITE: NaCl
Space group *Fm3m*. Each atom is coordinated by six of the other kind.

bedded. Twinning on {111}, rare. **Physical properties:** Colorless, white, reddish, blue, purple. The red color may be due to occluded hematite; the bluish and purple colors are due to defects in the structure and disappear if the mineral is dissolved and recrystallized. White streak, vitreous luster. Cleavage {001}, perfect, easy; conchoidal fracture; brittle. H = 2. G = 2.168. Nonfluorescent under UV radiation, but occasionally foreign material included in the crystal will fluoresce, giving the impression that the halite is fluorescent. **Tests:** Soluble in water. With concentrated H_2SO_4 gives HCl. Tastes salty. Fuses at 804°, often with decrepitation. **Optics:** Isotropic; N = 1.544. **Occurrence:** Halite occurs in deposits of several types. Quantitatively most important are the evaporites, in which halite forms bedded deposits reaching thicknesses of 1000 m and more. Also found in salt domes, which have been derived from deep beds of rock salt that have been plastically deformed. It is a common constituent of salt lakes and playas and may be found crystallizing in these waters near the shore. It crystallizes from seawater in lagoons; halite derived from the evaporation of sea spray is carried into the upper atmosphere in minute crystals, where it assists in nucleating clouds and fog. Often found in volcanic fumaroles; also found as crystals in the fluid inclusions in many minerals. **Localities:** Halite is widely distributed throughout the world in sedimentary deposits of all ages. Important deposits have been mined for many years in upper Silurian sediments in central and western NY. The world's largest rock salt mine is near Retsof, NY. Some is mined by underground methods, but much is also recovered through brine wells. Louann salt formation, AL. In LA, TX, and elsewhere along the Gulf coast important deposits are known, many of them in salt domes; also in AZ in the Verde Valley, Yavapai Co., in fine clear masses. In CA it is associated with borate deposits in San Bernardino Co. In UT, salt is obtained from the waters of Great Salt Lake. In Canada in ONT in Bruce, Huron, and Lambton Cos.; in SK in extensive beds, and in NF. Famous deposits are found in Germany at Stassfurt, Saxony, and in Leopoldshall; also at Heringen–Werra (Hessen) fine transparent deep blue cleavages are found. Rhythmically banded clear, lilac, pale blue fibrous halite comes from Wittolshein, Alsace, France; also in Landes and in Lorraine. Large deposits are known in Galicia, Poland, at Kalusz, Bochnia, and Wieliczka; at the latter mine wonderful cubic crystals up to 50 cm on edge were found, the finest in the world. Salzburg, Austria, at Hallein, and in the Tyrol near Innsbruck. Crystallized halite is found in Sicily, Italy, at Girgenti and other localities, and vast beds of it are known to underlie much of the Mediterranean Sea. In Spain near Barcelona; in Cheshire in England. Deposits are known at numerous places in Russia and in China. In Pakistan in the Salt Range. In Africa, in Algeria and Ethiopia. Salar de Uyuni, Bolivia, crystal groups beneath the crust of an enormous salt flat. Uses: The principal uses for salt are for culinary purposes and for food preservation; as a basis for the manufacture of hydrochloric acid and most other chlorine compounds; for the manufacture of sodium, sodium compounds, and glass; and for melting ice and snow on highways, and so on. RG DII:4.

9.1.1.2 Sylvite KCl

Named in 1832 for the old chemical name for the substance *sal digestivus Sylvii*, or digestive salt, of Francis Sylvius de le Boë (1614–1672), Dutch physician and chemist of Leyden. Halite group. ISO $Fm3m$. $a = 6.93$, $Z = 4$, $D = 1.987$. NBSC 539(1):65(1953). 41-1476: 3.15_{10} 2.23_4 1.817_1 1.573_1 1.407_1 1.284_1 1.049_1 0.995_1. Commonly in cubes or in cubes modified by octahedrons; also granular crystalline masses intermixed with halite and other salts; compact. Colorless to white, translucent to transparent; other colors due to impurities; vitreous luster. Cleavage {100}, perfect; uneven fracture. $H = 2$. $G = 1.97$–1.99. Soluble in water. Tastes somewhat salty. Isotropic; $N = 1.490$. Sylvite is formed like halite from the fractional crystallization of seawater. It forms sylvite-rich layers in many salt deposits but is much less abundant than halite. Louann salt formation, Alabama. Important deposits occur in the Permian basin of NM and TX, particularly in the region of Carlsbad, NM. In Canada vast deposits are being mined in SK. At Stassfurt and Leopoldshall, northern Germany, and in the southern Harz region; associated with sylvite and halite are many other chlorides and sulfates. Kalusz, Galicia, Poland. *Vesuvius, Italy*, in fumaroles, and at Mt. Etna, Sicily. From the nitrate beds in the desert of Tarapacá, Chile, and adjacent regions of Peru. Used as a raw material in the manufacture of potassium and potassium compounds; an essential ingredient in fertilizers, for which use millions of tons are mined annually. RG DII:7, MM 29:669(1951).

9.1.1.3 Villiaumite NaF

Named in 1908 for Mr. Villiaume, French explorer, in whose collection of rocks from Guinea the mineral was first found. Halite group. ISO $Fm3m$. $a = 4.634$, $Z = 4$, $D = 2.799$. NBSC 539(1):63(1953). 36-1455: 2.68_1 2.32_{10} 1.64_4 1.40_1

1.34_1 1.16_1 1.04_1 0.946_1. Crystals rare, as cubes sometimes modified by octahedron; as cleavable masses; granular. Carmine red, artificial NaF is colorless, white streak, vitreous luster. Cleavage {001}, perfect. H = $2-2\frac{1}{2}$. G = 2.79. Soluble in water to form a colorless solution; upon recrystallization, the crystals are colorless. Very poisonous. When heated to 300°, becomes colorless. Isotropic; N = 1.326. Natural material is anomalously uniaxial (−); weakly birefringent, strongly dichroic. O, pink to carmine; E, yellow. Villiaumite is formed in alkalic rocks such as nepheline syenites, usually in pegmatitic phases. Point of Rocks, Colfax Co., NM. Mt. St.-Hilaire, QUE, Canada, as cleavable masses and small crystals. Ilimaussaq intrusion, Julienhaab district, southwestern Greenland; Kola Penin., Russia; *Rouma Island, Los Archipelago, Guinea.* Poços de Caldas, MG, Brazil, in small crystals. RG *DII:10, NJMM:111(1981)*.

9.1.1.4 Carobbiite KF

Named in 1956 for Guido Carobbi (1900–1983), professor of mineralogy at the University of Florence. Halite group. ISO $Fm3m$. $a = 5.347$, Z = 4, D = 2.54. *NBSC 539(1):64(1953)*. *36-1458*(syn): 3.09_2 2.67_{10} 1.89_5 1.61_1 1.54_1 1.34_1 1.20_1 1.09_1. Crystallizes as small cubes. Colorless, vitreous luster. Cleavage {001}, perfect. G = 2.52. Isotropic; N = 1.362. From *Vesuvius, Italy*, in cavities in lava. RG *AM 42:117(1957)*.

9.1.1.5 Griceite LiF

Named in 1989 for Joel D. Grice (b.1946), curator of minerals, Canadian Museum of Nature, Ottawa, Canada. Halite group. ISO $Fm3m$. $a = 4.027$, Z = 4, D = 2.638. Structure: *CM 27:125(1989)*. *45-1460*(syn): 2.32_9 2.01_{10} 1.424_5 1.213_1 1.163_1 0.924_1 0.900_1 0.822_1. As millimeter-size aggregates; powdery coatings; translucent irregular grains in villiaumite. White, dull to vitreous luster. Cleavage {001}, good; uneven to conchoidal fracture. H = $4\frac{1}{2}$. G = 2.62. Some specimens fluoresce light yellow in SWUV. Insoluble in H_2O. Not attacked by 30% HCl. Isotropic; N = 1.392. *Poudrette quarry, Mt. St.-Hilaire, QUE, Canada*, in a sodalite inclusion in nepheline syenite, associated with ussingite, eudyalite, sphalerite, serandite, and lovozerite; also as inclusions in villiaumite or pseudomorphs of it. RG

9.1.2.1 Hydrohalite $NaCl \cdot 2H_2O$

Named in 1847 for the water content and similarity to halite in composition. MON $P2_1/c$. $a = 6.331$, $b = 10.118$, $c = 6.503$, $\beta = 114.4°$, Z = 4, D = 1.654. *NJMM:325(1972)*. Structure: *AC(B) 30:2363(1974)*. *29-1197*: 3.87_8 3.82_9 3.63_5 2.98_{10} 2.88_6 2.67_{10} 2.52_{10} 2.24_9. Stable only in very cold concentrated NaCl solutions at temperatures below −5°. Colorless. G = 1.63. First observed during the winter at a salt works in *Salzburg, Germany*. Said to have been found just below the surface, encrusting rocks in a fjord on the coast of Greenland. From Lake Bonney, Taylor Valley, Victoria Land, Antarctica. RG *DII:15, AJUS 10:179(1975)*.

9.1.3.1 Sal Ammoniac NH$_4$Cl

Named in 1546 from the medieval Latin for salt of Ammon, used by early writers for common rock salt from near the Oracle of Ammon in Egypt. Later, the name was applied to the present compound, which was manufactured in Egypt. ISO $Pm3m$. $a = 3.876$, $Z = 1$, $D = 1.527$. NBSC 539(1):59(1953). 7-7(syn): 3.87_3 2.74_{10} 2.24_1 1.94_1 1.73_1 1.58_3 1.37_1 1.23_1. Crystals usually trapezohedral {112}. May be modified by {113}; rarely shows gyroidal forms. Frequently, skeletal and in parallel growth. Twinning on {111}. Colorless to white, yellowish brown when mixed with FeCl$_3$, vitreous luster. Cleavage {111}, imperfect; conchoidal fracture; somewhat plastic, not easily powdered. $G = 1.532$. Isotropic; $N = 1.639$. Weakly birefringent after deformation. Found chiefly at volcanic fumaroles, especially where lava has flowed over soil or vegetation. Also produced by condensation of gases from burning coal seams and in guano deposits. Fissure Springs, Death Valley, and at Mud Volcanos, Niland, CA. The volcanos of Kilauea, Pelée, Vesuvius, Etna, Stromboli, and Vulcano are noteworthy sources. In large crystal groups from burning coal dumps at St. Etienne, Loire, France. In coal seams at Newcastle, England, and elsewhere. Ravat, Tadzhikistan, nice crystals to 50 mm. From the guano deposits of Chile and Peru. RG DII:15.

9.1.4.1 Chlorargyrite AgCl

Named in 1875 for the composition: chlorine and silver (argyros). Synonyms: cerargyrite, horn silver. ISO $Fm3m$. $a = 5.549$, $Z = 4$, $D = 5.571$. NBSC 539(4):44(1955). 31-1238(syn): 3.20_5 2.77_{10} 1.96_5 1.67_2 1.60_2 1.39_1 1.24_1 1.13_1. Crystals usually cubes, often modified by {111} and {011}. Crystal faces usually curved and poorly formed. As crusts and waxy coatings, hornlike masses, dendritic, columnar, or stalactitic. Twinning on {111}. Colorless when fresh (AgCl) to yellowish or green with increasing ratio of AgBr. Becomes violet brown to gray on exposure to light, the Br-rich varieties showing less tendency to change. Resinous to adamantine luster. Uneven fracture, sectile, plastic, tough. $H = 2\frac{1}{2}$. $G = 5.55-6.00$. Insoluble in water and in dilute HCl but soluble in NH$_4$OH, KCN, and Na$_2$S$_2$O$_3$ solutions. A fragment of chlorargyrite placed on a strip of zinc and moistened with a drop of water swells up, turns black, and is partly reduced to metallic silver, which shows its metallic luster upon being pressed with the point of a knife. Fuses in the flame of a candle. "Horn silver" minerals can be distinguished by their waxy luster and plasticity. Isotropic; $N = 2.07-2.16$. A complete solid-solution series exists between chlorargyrite and bromargyrite; the intermediate compositions, with AgCl/AgBr close to 1:1, is known as *embolite*. Limited solid solution also exists between Ag(Cl,Br) and AgI, especially toward the Br-rich portion of the series, the maximum, observed ratio in natural material being about 3.5:1. I-rich varieties, which are known as *iodobromite*, are yellowish to orange. Formed in the oxidized zone of silver-bearing deposits, especially in arid regions, associated with native silver, cerussite, malachite, limonite, and so on. Chlorargyrite was formerly abundant in many of the Western silver camps, especially Leadville, CO; Tombstone, AZ; Rand district, CA; Rochester, Bullfrog, and Chloride districts, NV; Frisco, UT; Lake Valley, NM; Poorman's Lode, ID. Once common in the silver mines of Germany, at Johanngeorgenstadt, Freiberg, and Schneeberg in Saxony, and Andreasberg, Harz. Jáchymov, Bohemia, Czech Republic. Altai Mts., Kazakhstan. Also formerly in large amounts at Chañarcillo in Atacama, and in Antofagasta and Tarapacá, in Chile. Potosi, Bolivia. Very abundant in the early days of mining at Broken Hill, NSW, Australia(!). Formerly an important ore of silver. RG *DII:11*.

9.1.4.2 Bromargyrite Ag(Br,Cl)

Named in 1841 for the composition. ISO $Fm3m$. $a = 5.775$, $Z = 4$, $D = 6.48$. NBSC *539(4):46(1955)*. *6-438*(syn): 3.33_1 2.89_{10} 2.04_6 1.67_2 1.44_1 1.33_1 1.29_1 1.18_1. Crystallizes as cubes, sometimes with $\{111\}$; crystal faces usually rounded; also massive, hornlike. Yellowish green to green, shining streak, waxy to adamantine luster. Very plastic. $H = 2\frac{1}{2}$. $G = 6.00$–6.47. Isotropic; $N = 2.16$–2.25. For chemistry, tests, and occurrence, see chlorargyrite (19.1.4.1). In AZ at Tombstone and at Bisbee. Many other occurrences of bromargyrite have been reported for the western United States, but the exact identity of the material is uncertain. In Mexico in the San Onofre and Plateros districts, ZAC, and at Parilla, DGO. Dernbach, Nassau, Germany; Huelgoat, Finistère, France. Donetsk basin, Ukraine. Chkalov, Russia. Formerly abundant at Chañarcillo, Chile(!), with silver and iodargyrite; also once abundant at Broken Hill, NSW(!), Australia. RG *DII:11*, AM *47:982(1962)*.

9.1.5.1 Iodargyrite AgI

Named in 1854 for the composition. HEX $P6_3mc$. $a = 4.592$, $c = 7.510$, $Z = 2$, $D = 5.683$. NBSC *539(8):51(1959)*. *9-374*(syn): 3.98_6 3.75_{10} 3.51_4 2.73_2 2.30_9 2.12_3 1.96_5 1.33_1. Crystals prismatic [0001] or tabular $\{0001\}$, usually showing hemimorphic character; also in rosettes, massive, or scaly. May form paramorphs after miersite, on $\{30\bar{3}4\}$. Colorless, changing to pale yellow on exposure to light; also dark yellow, greenish yellow, brownish; yellow, shining streak; resinous to adamantine luster; pearly on cleavage. Cleavage $\{0001\}$, perfect; conchoidal fracture; sectile; flexible. $H = 1\frac{1}{2}$. $G = 5.69$. Uniaxial $(+)$; $N_O = 2.21$, $N_E = 2.22$; strong dispersion; abnormal green interference colors. From the oxidized zone of silver deposits in arid regions. Tonopah, NV, in fine crystals; also at Majuba Hill. Commonwealth mine, Pearce Hills, Cochise Co., AZ. *Albarradón, near Mazapil, ZAC, Mexico.* Chañarcillo, Atacama, Chile. Broken Hill, NSW, Australia, in fine specimens. RG *DII:22*.

9.1.6.1 Tocornalite   Silver mercury iodide: exact formula uncertain

Named in 1867 for S. F. Tocornal, rector of the University of Santiago, Chile. A questionable species. Space group unknown. *PD:* 6.37_7 3.76_9 3.61_9 2.64_{10} 2.25_7 2.04_5. AM *58:348(1973)*. Massive, granular. Light to bright yellow; on exposure to light turns green, then black; dull luster. Soft. *Chañarcillo, Atacama, Chile*, with other halides of silver and mercury. Proprietary mine, Broken Hill, NSW, Australia, with embolite; also with iodargyrite and cinnabar. RG *DII:25*.

9.1.7.1 Nantokite CuCl

Named in 1867 for the locality. ISO $Fd3m$. $a = 5.416$, $Z = 4$, $D = 4.138$. *NBSC 539(4):35(1955)*. *6-344*(syn): 3.13_{10} 2.71_1 1.92_6 1.63_3 1.35_1 1.24_1 1.11_1 1.04_1. Massive, granular. Colorless or white, altering to greenish or grayish; white streak; adamantine luster. Cleavage {001}, conchoidal fracture. $H = 2\frac{1}{2}$. $G = 4.14$. Easily soluble in acids and NH_4OH. Easily fusible. Isotropic; $N = 1.93$. Parrsboro area, NS, Canada, as seawater corrosion on copper in basalt joints exposed on beaches. Mapimi, DGO, Mexico(!), in fine blades to 20 mm. *Nantoko, Copiapó, Chile*, with oxidized copper minerals. Broken Hill, NSW, Australia. RG *DII:18*.

9.1.7.2 Miersite (Ag,Cu)I

Named in 1898 for Henry Alexander Miers (1848–1942), English mineralogist, Oxford University. ISO $F\bar{4}3m$. $a = 6.495$, $Z = 4$, $D = 5.686$. *NBSC 539(9):48(1960)*. *2-499*: 3.72_6 3.23_{10} 2.28_8 1.95_8 1.49_2 1.32_2 1.29_2. Tetrahedral crystals; as crusts; twinning on {111}. Canary yellow, yellow streak, adamantine luster. Cleavage {011}, perfect. $G = 5.65$. Insoluble in dilute HNO_3, with dilute H_2SO_4 and zinc decomposes to metallic Ag plus Cu. Fuses easily to a red liquid. Isotropic; $N = 2.20$. Cu may replace Ag up to 20 mol %. Forms a partial series with marshite. Mildren and Steppe Claim, Pima Co., AZ, as overgrowths on iodargyrite. *Broken Hill, NSW, Australia*. RG *DII:19, AM 20:585(1935)*.

9.1.7.3 Marshite CuI

Named in 1892 for C. W. Marsh, Australian mineral collector, who first described the mineral. ISO $F\bar{4}3m$. $a = 6.051$, $Z = 4$, $D = 5.710$. *NBSC 539(4):38(1955)*. *6-246*(syn): 3.49_{10} 3.03_1 2.14_6 1.82_3 1.51_1 1.39_1 1.24_1 1.16_1. Crystals tetrahedral, also as crusts. Twinning on {111}, sometimes multiple. Colorless to pale yellow, becoming brownish red on exposure to light; yellow streak; adamantine luster. Cleavage {011}, perfect; conchoidal fracture; brittle. $H = 2\frac{1}{2}$. $G = 5.68$. Turns black in dilute HNO_3. Isotropic; $N = 2.35$. *Broken Hill, NSW, Australia*, with oxidized copper minerals. Chuquicamata, Chile. RG *DII:20*.

9.1.8.1 Calomel HgCl

Named in 1776 for the Greek for *beautiful* and *honey*, in allusion to the sweet taste; or for *beautiful* and *black*, in reference to black mercury sulfide, to which the name was first applied. TET $I4mm$. $a = 4.480$, $c = 10.906$, $Z = 2$, $D = 7.162$. *NBSM 25(13):30(1976)*. Structure: *ZK 187:305(1989)*. *26-312*: 4.15_8 3.17_{10} 2.73_3 2.24_2 2.07_4 1.970_2 1.962_3 1.732_1. Variable habit; often tabular $\{001\}$; prismatic $[001]$; pyramidal; equant, especially in complex twins. As drusy crusts, massive, earthy. Twinning on $\{112\}$, contact or penetration. Colorless, white, grayish, yellowish, brown; color deepens on exposure to light; adamantine luster. Cleavage $\{110\}$, good; $\{011\}$, imperfect; conchoidal fracture; plastic; sectile. $H = 1\frac{1}{2}$. $G = 7.15$. Fluoresces brick red in SWUV. Insoluble in water, soluble in aqua regia. Blackens when treated with alkalis. Tastes sweet. In CT sublimes, condensing in upper part of tube. Uniaxial $(+)$; $N_O = 1.973$, $N_E = 2.656$; occasionally, weakly pleochroic. A secondary mineral, found in the oxidized zone of mercury-bearing deposits, usually associated with native Hg and cinnabar. Terlingua, TX(!), in complex crystals. Howard Co., AR. Sunflower district, Maricopa, Co., AZ. In Mexico at Huahuaxtla, GRO, and at El Doctor, QRO, as an alteration of selenian metacinnabar. *Moschellandsberg, Bavaria, Germany.* Idria, Italy. Almaden, Spain. Avala, Serbia(!), fine crystals to 10 mm. Used as an ore of mercury. RG *DII:25*.

9.1.8.2 Kuzminite $Hg_2(Br,Cl)_2$

Named for Alexei Michailovich Kuzmin (1891–1980), Russian mineralogist, Tomsk Polytechnical Institute, western Siberia. TET $I4/mmm$. $a = 4.597$, $c = 11.034$, $Z = 2$, $D = 7.64$. *AM 73:192(1988)*. *40-514:* 4.26_2 3.25_4 2.76_2 2.30_1 2.10_2 1.989_1 1.768_1 1.453_{10}. Irregular grains and powdery masses. Colorless, bluish gray, or dull white, sometimes with brown tint; white to yellowish streak. Cleavage $\{100\}$, good; uneven fracture. $H = 2$. Soluble in warm HCl and aqua regia. Darkens in KOH solution. Uniaxial $(+)$; $N > 2.00$. Slightly pleochroic, strongly birefringent. From the *Kadyrel mercury mine, Tuva, Siberia, Russia,* in calcite veins, with calomel, eglestonite, lavrentievite, corderoite, and mercury. RG *ZVMO 115:595(1986)*.

9.1.8.3 Moschelite Hg_2I_2

Named in 1989 for the locality. TET $I4/mmm$. $a = 4.920$, $c = 11.600$, $Z = 2$, $D = 7.75$. *NJMM:524(1989)*. *6-245:* 4.54_3 3.49_{10} 2.91_3 2.47_3 2.23_5 2.10_2 1.881_2 1600_1. Fine crystalline crusts and small irregular plates. Lemon yellow, which changes to dark olive-green upon exposure to light; brown streak; adamantine luster. Imperfect cleavage, conchoidal fracture, sectile. $H = 1$–2. From the dump at the *Backofen mine, Moschel landsberg district, Obermoschel, Rhineland Pfalz, Germany,* associated with cinnabar, mercury, calomel, and terlinguaite. RG

9.2.1.1 Fluorite CaF_2

Named in antiquity from the Latin *to flow*, since it melts more easily than other minerals with which it was confused. Synonym: fluorspar. ISO $Fm3m$. $a = 5.463$, $Z = 4$, $D = 3.181$. *NBSC 539(1):69(1953)*. *35-816*(syn): 3.16_9 1.932_{10} 1.647_3 1.366_1 1.253_1 1.115_2 1.051_1 0.864_1. **Habit:** Crystals common, usually cubes $\{001\}$; less often octahedrons $\{111\}$; rarely $\{011\}$; combinations of the above forms are fairly common, with small modifying faces $\{0kl\}$ very common, especially $\{013\}$ and $\{012\}$. Cube corners often modified by $\{113\}$

9.2.1.1 Fluorite

and {124}. {001} may be partly or wholly replaced by a vicinal form {0kl}. Faces {001} and {0kl} are usually smooth and lustrous; {111} is rough, dull, and may be made up of many tiny pits and growths with flat sides parallel to {001}. Large crystals are commonly composite, showing coarse textured mosaic structure. Frequently massive, coarse to fine granular; occasionally earthy; rarely columnar, fibrous, or globular. Twins on {111}, usually as interpenetrating cubes. Overgrowths of fluorite on the corners, or surrounding an earlier-formed crystal of different color, habit, and so on, are well known. **Structure:** The structure of fluorite was one of the earliest (1914) to be investigated by X-rays and serves as a type for several fluorides, chlorides, oxides, and intermetallic compounds. Ca ions are arranged on a cubic face-centered lattice, while each fluorine ion is at the center of one of the smaller cubes obtained by dividing the unit cube into eight parts. Each calcium is therefore coordinated by eight fluorine ions, while each fluorine ion is surrounded by four calcium ions arranged at the corners of a regular tetrahedron. Other minerals with the fluorite structure include uraninite, cerianite, and thorianite. Some oxides, selenides, and tellurides of formula A_2X have the "antifluorite" structure in which the metals occupy the positions of F in fluorite, and the nonmetals the positions of Ca. *DHZII 5:347.* **Physical properties:** Commonly colorless, white, and shades of purple, green, blue, and yellow. Pink and red are rare. Colorless and perfectly transparent when pure (used in optics). Some crystals are green by transmitted light but appear blue by reflected light, due to fluorescence. Crystals may show more or less sharp zones of coloration, either forming phantom crystals inside or following the loci of certain faces or edges as they develop in the growth of the crystal. Color may be lightened or lost in many cases by heating. Proximity of uraninite or other radioactive minerals to fluorite in the deposit may result in a deep purple to nearly black color. For a discussion of the causes of color in fluorite and other minerals and gems, reference is made to *G&G XXIV:81(1988).* White streak, vitreous luster; may be dull in massive varieties. Cleavage {111}, perfect and very easy; flat conchoidal, splintery, or uneven fracture; brittle. H = 4 (Mohs' scale defines hardness 4 as the hardness of fluorite). G = 3.18; G may increase to as much as 3.6 in yttrian or cerian fluorite. Heating may cause thermoluminescence; a variety producing green thermoluminescence is called *cholorphane*. Occasionally, triboluminescent. Some varieties are strikingly fluorescent blue under LWUV, an effect related to the presence of trace amounts of erbium. Other colors may be caused by other activators. The term *fluorescence* was derived from the observed glow of certain English fluorites when exposed to daylight containing UV light. Electrically nonconductive. Diamagnetic. **Tests:** Decomposed by H_2SO_4 with evolution of HF (**poison!**) and with difficulty in hot HCl. In closed tube (CT) decrepitates and usually luminesces. **Chemistry:** Y and/or Ce may substitute for Ca to the extent of $Y + Ce/Ca = 1:5$; other rare earths are common in spectrographic amounts. Radiation such as gamma rays from included or nearby radioactive atoms releases atoms of Ca and F in fluorite, causing the very dark violet color. Such material smells of HF or F_2 while being powdered

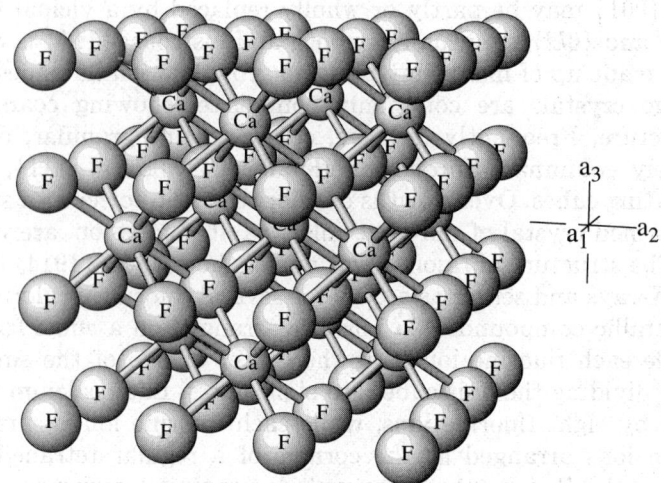

FLUORITE: CaF$_2$

Space group Fm3m. Ca atoms are coordinated by eight F atoms, and F atoms are coordinated by four Ca atoms.

(*antozonite*, also "stink floss"), and turns white and transparent when heated. **Optics:** Isotropic; N = 1.4338 (Na). For yttrian varieties, N is up to 1.457. Birefringence absent or weak; common in cleaved, cut, pressed, or otherwise deformed grains. **Occurrence:** A common mineral, fluorite occurs (1) as a vein mineral, either as the chief constituent or with metal ores, especially of Pb and Ag, and usually with quartz, barite, calcite, dolomite; (2) in cavities in secondary carbonate rocks; (3) in replacement deposits associated with igneous contacts; in tactites; (4) in high-temperature pneumatolytic deposits, such as pegmatites and occasionally in cassiterite veins; (5) uncommon as a late hydrothermal product in cavities and joints in granite and its alkalic relatives, sometimes in very fine crystals in alpine-type veins; (6) in veinlike bodies cutting syenite, with calcite, apatite, uraninite, hornblende, biotite; (7) as a hot spring deposit. **Alteration:** Pseudomorphs by substitution of quartz and several other minerals after fluorite are common. Fluorite also forms pseudomorphs after calcite, barite, and galena. **Localities:** Only a few of the many known deposits can be listed. Westmoreland, NH(!) in large transparent green octahedral crystals; Trumbull, CT (chlorophane); Clay Center(*) and Tiffin, OH, with celestite; fine, large (up to 20 cm) multicolored crystals from Rosiclare, IL(!) and adjacent regions of IL and KY; Ft. Wayne, IN(*), in a limestone quarry; Elmwood mine, near Carthage, TN(!) superb cubic crystals reaching over 50 cm, with barite, sphalerite, calcite, and so on; Mt. Antero and Wagon Wheel Gap, CO(*); Hansonburg District, Socorro Co.(*), and fine purple octahedrons from Pine Canyon, Grant Co.(!), NM. In Canada at Madoc, ON in fine crystals(!), and at Wilberforce and elsewhere a dark purple variety associated with U mineralization; at the Rock Candy mine near Grand Forks, BC(!). In Mexico at Naica, CHIH(!) in fine green, colorless, and pale violet complex crystals; Ivigtut, Greenland. Several localities in England are famous for magnificent crystal groups, as Alston Moor and Cleator Moor, Cumberland(!); Weardale and Durham, Yorkshire(!); Castleton, Derbyshire(!), where a beautiful blue–purple banded massive variety called "blue john" is fashioned into jewelry and objects d'art; from Cu and Sn veins in Cornwall(!) and in Tavistock, Devon. At Kongsberg, Norway(!), colorless, green, violet, and blue cubes and octahedrons, sometimes associated with native silver. In Germany, widespread in Saxony and at several places in the Harz Mts. and in Baden; Wölsendorf, Bavaria, is especially known for the deep purple, nearly opaque variety antozonite. Cínovec, Bohemia, Czech Republic. Alpine-type veins in Göschener Alp, Aar massif, Switzerland(!) and Argentiéres, Mt. Blanc massif, France(!), in beautiful pink octahedrons. Fine blue cubes come from Le Beix, Puy de Dôme, France(!). Sarnthal, Tyrol, Austria(*). In Spain in fine purple modified cubes at La Collada and Berbes, Asturias(!). At Dal'negorsk, Primorskiy Krai, Siberia, Russia(!), magnificent large completely colorless to light green or blue cubic crystals associated with calcite, danburite, sulfides, and so on; in Kazakhstan at Kara Oba, superb cubes and octahedrons, mostly green or purple, with wolframite, pyrite, quartz, cosalite, bertrandite, and at Akchatau(!), purple octahedrons with bertrandite and apatite. In Pakistan in the Tormiq area(!) large pink and greenish cubo-octahedrons formed in alpine-type deposits, and at Nagar, near Aliabad, Hunza Valley(!), superb large cubes and octahedrons, some of exceptional deep pink color associated with gemmy blue beryl crystals and muscovite. Chiang Mai province, Thailand(!), small colorless to light violet transparent cubes coating large stibnite crystals. Obira mine, Kyushu, Japan(!), superb pink octahedrons and greenish cubes with quartz, arsenopyrite, and so on. Zeerust district, Transvaal, South Africa. Okarusu, Orongo Mts., Namibia(*). Mt. Bischof, TAS; and Goulburn Co. and near Deepwater, NSW, Australia. At Huanzala, Peru(!), superb, exceptionally sharp rich pink octahedrons, some with a greenish central core, usually associated with pyrite and sphalerite. Yttrian fluorite occurs in the Franklin marble, Sussex Co., NJ; in Orange Co., NY; Lake George and elsewhere, CO; Hundholmen, Norway; Falun (also cerian), Sweden. Uses: Fluorite is the primary source of all fluorine chemicals, including Freon and other fluorocarbons. It is an essential ingredient in the manufacture of steel, abou 7 kg being required per ton. It is also used in the ceramic industry as an ingredient in glazes. Annual production amounts to several million tons. A major part of world output is exported from several localities in China, including some beautiful banded material for decor applications. Other major production comes from mines in NL, SLP, COAH, and GTO in Mexico. RG *DII:29; AM 37:916(1952), 45:855(1960), 52:1003(1967); USGSPP 650:D69.*

9.2.1.2 Frankdicksonite BaF_2

Named in 1974 for Frank W. Dickson (b.1922), American geochemist, Stanford University. ISO $Fm3m$. $a = 6.200$, $Z = 4$, $D = 4.886$. *NBSC 539(1):70(1953).* 4-452(syn): 3.58_{10} 3.10_3 2.19_8 1.87_5 1.55_1 1.42_1 1.39_1 1.27_1.

Minute cubes. Colorless, vitreous luster. Cleavage {111}, perfect. $H = 2\frac{1}{2}$. Somewhat soluble in water. Istropic; $N = 1.472$. Said to have been found at the *Carlin gold mine, Eureka Co., NV*, embedded in quartz veinlets along a contact between latite and silicified limestone; no matrix specimens are known. RG *AM 59:885(1974)*.

9.2.1.3 Tveitite-(Y) $Ca_{1-x}(Y,RE)_xF_{2-x}(x \approx 0.3)$

Named in 1977 for John Peder Tveit (1909–1978), Norwegian miner, who found it in his quarry. MON Space group unknown. $a = 3.924$, $b = 3.893$, $c = 5.525$, $\beta = 90.27°$. *AM 62:1060(1977)*. *29-364:* 3.22_4 3.18_{10} 2.76_5 1.963_5 1.949_7 1.676_4 1.664_6 1.265_8. White to pale yellow, greasy luster, fluorescent faint yellow-orange in SWUV. Biaxial (−); $N_x = 1.476$, $N_y = 1.479$, $N_z = 1.481$; $2V = 34°$. Shows complex polysynthetic twinning. Barringer Hill, TX. In pegmatite at *Hoydalen, Telemark, Norway*, with amazonite, beryl, gadolinite, and monazite. RG *LIT 10:81(1977)*.

9.2.2.1 Sellaite MgF_2

Named in 1868 for Quintino Sella (1827–1884), Italian mining engineer and mineralogist. TET $P4_2mnm$. $a = 4.623$, $c = 3.052$, $Z = 2$, $D = 3.172$. *NBSC 539(4):33(1955)*. *41-1443:* 3.27_{10} 2.55_2 2.23_8 2.07_2 1.711_6 1.634_2 1.526_1 1.375_2. Crystals acicular to stout prismatic or nearly equant, fibrous aggregates. Colorless to white, buff, rarely black; vitreous luster. Cleavage {010} and {110}; conchoidal fracture. $H = 5$. $G = 3.15$. Nonfluorescent, except sellaite from Brazil, which fluoresces blue in SWUV. Slightly soluble in water; decomposed by concentrated H_2SO_4. Uniaxial (+); $N_O = 1.378$, $N_E = 1.390$ (Na). *Moutiers, Savoie, France*, in bituminous dolomite/anhydrite rock in a glacial moraine. Bleichrode, Harz, Germany. In Italy at Carrara in cavities in marble; also at Vesuvius and at Etna, Sicily. Brock's Creek, NT, Australia. In Brazil in a magnesite mine at Brumado, BA(!), in transparent crystals of great perfection up to 10×3 cm, in vugs, sometimes twinned (V-twins), associated with magnesite and quartz. RG *DII:37, NJMA 136:10(1979)*.

9.2.3.1 Lawrencite $FeCl_2$

Named in 1877 for John Lawrence Smith (1818–1883), American chemist, mineralogist, and student of meteorites. HEX-R $R\bar{3}m$. $a = 3.59$, $c = 17.56$, $Z = 3$, $D = 3.22$. *1-1106*(syn): 5.90_6 3.07_3 2.54_{10} 1.95_1 1.80_6 1.72_1 1.47_2 1.14_2. Massive, in veinlets. White, altering in air to green, brown; vitreous luster. Cleavage {0001}, perfect. $G = 3.16$. Soft. Uniaxial (−); $N = 1.567$; weak birefringence. Found in fissures in iron meteorites and in native iron from Ovifak, Greenland. Reported as a sublimation product on Vesuvius, Italy. RG *DII:40*.

9.2.3.2 Scacchite $MnCl_2$

Named in 1869 for Arcangelo Scacchi (1810–1894), Italian mineralogist, University of Naples. HEX-R $R\bar{3}m$. $a = 3.706$, $c = 17.569$, $Z = 3$, $D = 3.000$. *NBSM 25(8):43(1970)*. *22-720:* 5.85_{10} 3.16_3 2.59_8 1.98_1 1.85_3 1.81_3 1.71_1

1.51_1. Crusts and efflorescences. Colorless when fresh, becoming rose-red and dirty red-brown upon exposure. Cleavage {0001}, perfect. Soft. G = 2.98. Soluble in water, deliquescent, easily fusible. Uniaxial (−); $N_O = 1.708$; $N_E = 1.622$. *Vesuvius, Italy*, as a deliquescent salt in fumaroles, associated with halite, sylvite, lawrencite, and other salts. RG *DII:40*.

9.2.3.3 Chloromagnesite MgCl₂

Named in 1873 for the composition. The existence of the mineral chloromagnesite is questioned by some because its deliquescent character would make its existence as a solid in nature improbable, as well as making its recovery and study difficult. The material studied was prepared synthetically and examined in a special dry atmosphere. HEX-R $R\bar{3}m$. $a = 3.632$, $c = 17.795$, Z = 3, D = 2.339. *NBSM 25(11):94(1974)*. *37-774*(syn): 5.89_{10} 2.94_5 2.57_1 1.809_1 1.485_1 1.474_1 1.178_1 1.145_1. Colorless. G = 2.325. Deliquescent, very soluble in water. Uniaxial (−); $N_O = 1.675$, $N_E = 1.59$. *Vesuvius, Campania, Italy*, as a sublimate in fumaroles, associated with halite and sylvite. RG *DII:41, JAC 17:483(1984)*.

9.2.4.1 Rokühnite (Fe,Mg)Cl₂ · 2H₂O

Named in 1980 for Robert A. Kühn (b.1911), professor of mineralogy, Kaliforschungsinstitut, Hannover, Germany. MON $C2/m$. $a = 7.352$, $b = 8.561$, $c = 3.637$, $\beta = 98.06°$, Z = 2, D = 2.385. *AM 66:219(1981)*. *25-1040*(syn): 5.55_{10} 4.28_8 2.88_5 2.77_5 2.76_9 2.40_5 2.14_4 2.09_5. Microscopic crystals, tabular {110} or as short flexible fibers. Colorless. Cleavage {110}, very good; {010}, good. G = 2.35. Soluble in water. Rapidly hydrates in air to FeCl₂ · 4H₂O. Biaxial (+); $X = b$; $Z \wedge c = -49°$; $N_x = 1.605$, $N_y = 1.633$, $N_z = 1.703$; $2V = 64°$; $r \leq v$. Found in intergrowths of carnallite, rinneite, and halite at the *Salzdetfurth* and *Sigfried–Geisen mines* in the *Zechstein Basin, Germany*. RG *NJMM 1980:125*.

9.2.5.1 Sinjarite CaCl₂ · 2H₂O

Named in 1980 for the locality. TET Space group unknown. $a = 7.21$, $c = 5.86$, Z = 2, D = 1.96. *AM 65:1069(1980)*. *1-989*: 6.10_4 4.35_5 3.06_7 2.83_{10} 2.52_2 2.35_2 2.16_2 2.12_6. Massive, granular. Prismatic crystals, [001]. Pale pink, white streak, vitreous to resinous luster. Prismatic cleavage, well developed. $H = 1\frac{1}{2}$. G = 1.66. Soluble in water; very deliquescent. Melts at 172° in its water of crystallization. Uniaxial; N = 1.54; elongated crystals give parallel extinction. Found in detritus in a wadi (intermittent stream) cutting across the Sinjar anticline, *Sinjar, W of Mosul, Iraq*. RG *MM 43:643(1980)*.

9.2.6.1 Antarcticite CaCl₂ · 6H₂O

Named in 1965 for the locality. HEX $P321$. $a = 7.876$, $c = 3.955$, Z = 1, D = 1.712. *NBSM 25(12):16(1975)*. *26-1053*: 6.80_4 3.93_9 3.42_7 2.79_8 2.58_6 2.27_6 2.16_{10} 1.71_3. Acicular to columnar crystals. Colorless or white. Cleavage {0001}, perfect; {1010}, nearly perfect; brittle. H = 2–3. G = 1.71. Soluble in water; deliquescent; temperature sensitive, and melts easily in its own water of crystallization. Uniaxial $(-)$; $N_O = 1.550$, $N_E = 1.492$. Bristol dry lakes, San Bernardino Co., CA. *Don Juan pond*, a saline pool in *Victoria Land, Antarctica*. RG *SCI 149:975(1965), AM 54:1018(1969)*.

9.2.7.1 Cotunnite PbCl$_2$

Named in 1825 for Domenico Cotugno (Cotunnius) (1736–1822), Italian physician and anatomist, University of Naples. ORTH *Pnam*. $a = 7.622$, $b = 9.045$, $c = 4.535$, $Z = 4$, $D = 5.908$. *NBSM 25(12):23(1975)*. Structure: *SPC 21:38(1976)*. *26-1150*(syn): 4.06_4 3.89_8 3.81_4 3.58_{10} 2.78_6 2.51_5 2.15_4 2.10_4. Crystals flattened {010} and elongated [001]; also massive, granular. Colorless to white; yellowish; greenish, adamantine to silky or pearly luster. Cleavage {010}, perfect; subconchoidal fracture; sectile. $H = 2\frac{1}{2}$. $G = 5.80$. Slightly soluble in water; more so in sodium acetate solution. Easily fusible. Biaxial $(+)$; $XYZ = bac$; $N_x = 2.199$, $N_y = 2.217$, $N_z = 2.260$ (Na); $2V = 67°$. Usually forms as an alteration of galena under arid, saline conditions. Bentley district, Mojave Co., AZ. *Vesuvius, Italy*, in fumaroles. From several localities in northern Chile and southern Peru. RG *DII:42, MA 69:1075(1969)*.

9.2.7.2 Hydrophilite CaCl$_2$

Named in 1813 from the Greek for *water* and *friend*, in allusion to its extreme hygroscopic properties. A questionable species, for the same reasons as stated under chloromagnesite, MgCl$_2$. *MM 43:682(1980)*. ORTH *Pnnm*. $a = 6.261$, $b = 6.429$, $c = 4.167$, $Z = 2$, $D = 2.175$. *NBSM 25(11):18(1974)*. *24-223*(syn): 4.48_9 3.05_{10} 2.86_4 2.36_3 2.33_5 2.24_3 2.08_2 1.91_3. Massive, as crusts. Twinning polysynthetic, complex. Colorless. Prismatic cleavage, perfect. Biaxial $(+)$; $N_x = 1.600$, $N_y = 1.605$, $N_z = 1.613$; 2V moderate. *Lüneberg, Hannover, Germany*, with anhydrite, gypsum, and halite. Mt. Etna, Sicily, Italy, from fumaroles. RG *DII:41*.

9.2.8.1 Eriochalcite CuCl$_2$ · 2H$_2$O

Named in 1884 for the Greek for *wool* and *copper*, because it is a fibrous copper mineral. ORTH *Pbmn*. $a = 7.416$, $b = 8.093$, $c = 3.749$, $Z = 2$, $D = 2.516$. *NBSM 25(18):33(1981)*. Structure: *ZK 189:13(1989)*. *33-451*(syn): 5.47_{10} 4.05_6 3.35_2 3.09_4 2.64_8 2.21_3 2.02_2 1.60_2. Lichenlike aggregates of crystals elongated [001]; also wool-like. Twinning on {021}. Bluish green to greenish blue, vitreous luster. Cleavage {110}, perfect; {001}, good; conchoidal fracture. $H = 2\frac{1}{2}$. $G = 2.47$. Easily soluble in water; easily fusible, yielding water. Biaxial $(+)$; $XYZ = bca$; $N_x = 1.646$, $N_y = 1.685$, $N_z = 1.745$ (Na); $2V = 75°$; $r < v$, strong. X, pale green; Y, pale olive-green; Z, pale blue. Found in fumaroles during the 1869 eruption of *Vesuvius, Italy*. In Chile at the Quetena mine, Antofagasta, with atacamite and bandylite. RG *DII:44*.

9.2.8.2 Tolbachite CuCl$_2$

Named in 1983 for the locality. MON $C2/m$. $a = 6.89$, $b = 3.31$, $c = 6.82$, $\beta = 122.3°$, $Z = 2$, $D = 3.42$. AM 69:408(1984). 34-198(syn): 5.78_{10} 3.45_2 2.92_1 2.89_1 2.37_1 1.87_1 1.81_1 1.65_1. Fibrous, mosslike. Brown to golden brown. Cleavage {001}, perfect. Readily soluble in water. In air alters quickly to eriochalcite, CuCl$_2 \cdot$ 2H$_2$O. Biaxial, pleochroism pale greenish to dark brown; high birefringence; positive elongation. Index of refraction could not be measured because the mineral reacts with immersion media. *Tolbachik volcano, Kamchatka, Russia*, in fumaroles associated with melanothallite, dolerophanite, tenorite, and euchlorine. RG *DANS 270:415(1983)*.

9.2.9.1 Bischofite MgCl$_2 \cdot$ 6H$_2$O

Named in 1877 for Gustav Bischof (1792–1870), German mineral chemist and geologist. MON $C2/m$. $a = 9.871$, $b = 7.113$, $c = 6.079$, $\beta = 93.73°$, $Z = 2$, $D = 1.585$. NBSM 25(11):37(1974). 25-515(syn): 4.10_{10} 3.96_3 2.98_4 2.88_7 2.74_4 2.73_6 2.64_9 1.85_4. Crystals short prismatic [001]; granular, foliated, sometimes fibrous. Twinning polysynthetic, due to pressure. Colorless to white, vitreous luster. Conchoidal to uneven fracture. $H = 1$–2. $G = 1.604$. Soluble in water, very deliquescent, easily fusible. Tastes bitter, stinging. Biaxial (+); $X = b$; $Y \wedge c = 9.5°$; $N_x = 1.498$, $N_y = 1.505$, $N_z = 1.525$ (Na); $2V = 79°$; $r > v$. Found in saline deposits at *Stassfurt* and *Leopoldshall, Germany*, associated with carnallite and kieserite. RG *DII:46*.

9.2.9.2 Nickelbischofite NiCl$_2 \cdot$ 6H$_2$O

Named in 1979 as the nickel analog of bischofite. MON $I2/m$. $a = 8.786$, $b = 7.076$, $c = 6.625$, $\beta = 97.22°$, $Z = 2$, $D = 1.932$. NBSM 25(11):42(1974). 25-1044(syn): 5.59_{10} 5.50_4 4.82_4 2.93_4 2.75_3 2.69_2 2.41_2 2.18_2. Prismatic crystals, with {110}, {001}, and {100}; as powdery coatings; crusts. Emerald green, very pale green to white streak, vitreous luster. Cleavage {001}, perfect; subconchoidal fracture; brittle. $H = 1\frac{1}{2}$. $G = 1.929$. Deliquescent, readily soluble in water. Biaxial (+); $X \wedge c = 8°$; $Y = b$; $N_x = 1.589$, $N_y = 1.617$, $N_z = 1.644$; $2V = 87°$; absorption $c > a$. X, greenish yellow; Y, pale green; Z, green. Material from Oxford, TX, has been presumed artificial. Found in drill core from the *Dumont ultramafic body, near Amos, QUE, Canada*. In sublimates from fumaroles on *Mt. Shirane, Gumma Pref., Japan*. RG *CM 17:107(1979)*, AM 67:156(1982).

9.2.10.1 Laurelite Pb(F,Cl,OH)$_2$

Named in 1989 for the locality. HEX $P6$, $P\bar{6}$, or $P6/m$. $a = 10.252$, $c = 3.973$, $Z = 6$, $D = 6.52$. AM 74:927(1989). 45-1457: 3.33_{10} 3.13_4 2.95_4 2.09_4 1.934_6 1.812_4 1.709_4 1.282_4. Acicular crystals, twinned by rotation on [0001]. Colorless, white streak, silky luster. Cleavage {001}, imperfect; conchoidal fracture; brittle. $H = 2$. $G = 6.2$. Nonfluorescent. Slowly soluble in water, rapidly so in 1:1 HCl. Uniaxial (+); $N_O = 1.903$, $N_E = 1.946$. *Grand Reef mine, Laurel*

Canyon, Graham Co., AZ, in a vug surrounded by galena and fluorite and associated with grandreefite, pseudograndreefite, and aravaipaite. RG

9.2.11.1 Matlockite PbFCl

Named in 1852 for the locality. TET $P4/nmm$. $a = 4.110$, $c = 7.233$, $Z = 2$, $D = 7.111$. NBSM 25(13):25(1976). 26-311(syn): 3.62_4 3.57_{10} 2.91_5 2.72_4 2.27_4 2.06_2 1.79_2 1.78_3. Crystals usually tabular {001}, also short pyramidal; as hemispherical rosettes; massive; lamellar. Colorless to yellow or pale amber, greenish; adamantine luster, pearly on {001}. Cleavage {001}, perfect; subconchoidal fracture. $H = 2\frac{1}{2}$–3. $G = 7.12$. Soluble in HNO_3, easily fusible. Uniaxial (−); $N_O = 2.145$, $N_E = 2.006$ (Na). Mammoth mine, Tiger, Pinal Co., AZ, in cleavages to 8 cm across. *Bage mine, Bole Hill, Cromford, near Matlock, Derbyshire, England*(!), associated with phosgenite. Sainte Lucie, Lozére, France, on galena. Laurium, Greece, in altered slag. Challocolla, Tarapacá, Chile, with pseudoboleite. RG *DII:59*.

9.2.11.2 Rorisite CaFCl

Named in 1990 from the Latin *roris*, dew, alluding to the transparent droplets that cover the mineral in moist air. TET $P4/nmm$. $a = 3.890$, $c = 6.810$, $Z = 2$, $D = 2.94$. AM 76:1731(1991). 24-185(syn): 2.75_5 2.56_{10} 2.14_5 1.963_2 1.945_4 1.562_2 1.550_3 1.376_1. Small tabular crystals flattened {001}. Colorless, white streak, vitreous luster. Cleavages {001}, {110}, perfect; conchoidal fracture; brittle. $H = 2$. $G = 2.78$. Soluble in water; in moist air the colorless plates become turbid and coated with drops of $CaCl_2$; the turbid plates consist of fine grained fluorite. Uniaxial (−); $N_O = 1.668$, $N_E = 1.635$. Fluoresces weak violet in SWUV. *Kopeysk, Chelyabinsk coal basin, Ural Mts., Russia*, in carbonatized wood fragments in old burnt dumps, associated with fluorite, periclase, and troilite. RG *ZVMO 119(3):73(1990)*.

9.3.1.1 Molysite FeCl₃

Named in 1868 from the Greek for *stain*, in allusion to its staining of lavas on which it occurs. HEX-R $R\bar{3}$. $a = 6.065$, $c = 17.42$, $Z = 6$, $D = 2.912$. JAC 22:173(1989). 1-1059(syn): 5.90_3 4.79_1 2.68_{10} 2.08_4 1.75_3 1.67_1 1.63_2 1.46_1. Crystals tabular {0001}, massive, as coatings. Yellow to brownish red, purple-red. Crystals are green by reflected light. Cleavage {0001}, perfect. $G = 2.90$. Soft. Very deliquescent, hydrolyzed by water to give a ferric hydrate, fuses at 282°. Uniaxial (−); very strongly birefringent. *Vesuvius, Italy*, as a sublimate in fumaroles; also on Etna and Vulcano. From a fumarole in Iceland. RG *DII:47*.

9.3.2.1 Hydromolysite FeCl₃ · 6H₂O

Named in 1965 for the composition, the hydrated analog of molysite. MON $C2/m$. $a = 11.834$, $b = 7.029$, $c = 5.952$, $\beta = 100.47°$, $Z = 2$, $D = 1.844$. NBSM 25(17):40(1980). 33-645(syn): 6.03_1 5.87_8 5.82_{10} 4.41_1 3.16_2 2.91_1 2.75_1 2.44_1. As small crystals. Orange-red. Somewhat deliquescent. Biaxial;

N ≈ 1.78. *Rio Marina, Elba, Italy,* in a pyrite mine, where seawater has reacted with oxidizing pyrite, giving rise to incrustations containing gypsum, copiapite, halite, and hydromolysite. RG *AM 44:908(1959), 51:1551(1966); ZVMO 94:189(1965).*

9.3.3.1 Chloraluminite $AlCl_3 \cdot 6H_2O$

Named in 1874 for the composition. HEX Space group unknown. $a = 11.831$, $c = 11.910$, $Z = 6$, $D = 1.666$. *NBSC 539(7):3(1957). 8-453*(syn): 5.95_3 5.14_3 3.89_4 3.68_4 3.30_{10} 3.25_6 2.76_4 2.31_5. Crystalline crusts and stalactites, also small rhombohedral crystals. Colorless to white or yellowish. Uniaxial (−); $N_O = 1.560$, $N_E = 1.507$. Found in fumaroles on *Vesuvius, Italy*, during the 1872 and 1906 eruptions. RG *DII:50, AC(B) 27:1069(1971).*

9.3.4.1 Fluocerite-(Ce) $(Ce,La)F_3$

Named in 1845 for the composition. Synonym: tysonite. Originally called fluocerite. HEX $P6_3/mcm$. $a = 7.130$, $c = 7.298$, $Z = 6$, $D = 6.12$. *MM 47:41(1983)*. Structure: *SPC 33:105(1988). 38-452:* 3.64_8 3.57_8 3.20_8 2.06_9 2.01_9 1.792_9 1.731_8 1.325_7. Prismatic crystals, rolled alluvial pebbles, irregular nodules. Pale yellow to straw yellow, vitreous to resinous luster. Cleavage {0001}, distinct; {11$\bar{2}$0}, indistinct; subconchoidal to uneven fracture; splintery; brittle. H = 4–5. G = 5.90. Soluble in H_2SO_4, insoluble in HCl or HNO_3. Uniaxial (−); $N_O = 1.613$, $N_E = 1.609$. Bethlehem, NH. Little Patsy quarry, South Platte district, Jefferson Co., CO. Snowbird mine, MT. In pegmatite at *Finbo and Brodbo, Dalarne, Sweden*. Hammarøj, Nordland, Norway. From albite pegmatites near Odegi, Afu Hills, Nigeria. RG *DII:48, MR 21:429(1990).*

9.3.4.2 Fluocerite-(La) $(La,Ce)F_3$

Named in 1969 for the composition. Formerly called fluocerite. HEX Space group unknown. $a = 4.135$, $c = 7.295$, $Z = 6$, $D = 5.94$. *AM 69:566(1984). 32-483:* 3.68_5 3.60_4 3.23_{10} 2.07_6 2.02_7 1.807_4 1.745_3 1.336_2. Platy to tabular crystals. Pale greenish yellow, vitreous luster. Pyramidal, imperfect cleavage; conchoidal fracture. VHN = 390 kg/mm². G = 5.93. Uniaxial (−); $N_O = 1.609$, $N_E = 1.603$. *Central Kazakhstan*, in hydrothermal quartz veins in granite. RG

9.3.5.1 Gananite α-BiF_3

Named in 1984 for the locality. ISO $P\bar{4}3m$. $a = 5.825$, $Z = 4$, $D = 8.928$. *AM 73:1494(1988). 11-10*(syn): 3.38_{10} 2.93_8 2.07_{10} 1.76_9 1.34_7 1.31_6 1.20_7 0.99_7. Irregular aggregates of grains. Brown to black; greenish black, dark gray streak, resinous to semimetallic luster. No cleavage, uneven fracture, brittle. H = 3–3½. Soluble in HCl and HNO_3. In RL, isotropic, gray. R = 12.3–10.1 (435 nm), 12.7–11.1 (546 nm), 12.9–12.6 (589 nm), 13.3–11.7 (650 nm). *Laikeng mining district, Ganan area, Jiangxi, China*, with bismuth, bismuthinite, pyrite, and chalcopyrite in wolframite-bearing quartz veins. RG

9.3.6.1 Rosenbergite $AlF[F_{0.5}(H_2O)_{0.5}]_4 \cdot H_2O$ or $AlF_3 \cdot 3H_2O$

Named in 1993 for Philip E. Rosenberg (b.1931), U.S. geochemist, Washington State University. TET $P4/n$. $a = 7.715$, $c = 3.648$, $Z = 2$, $D = 2.111$. *EJM 5:1167(1993)*. PD(Cetine mine): 5.47_{10} 2.44_7 2.03_7 1.775_8 1.725_9 1.388_7 1.306_7 1.127_7. *9-108*(syn): 5.45_{10} 3.86_6 3.33_2 2.73_3 2.44_6 2.03_1 1.728_4. Tetragonal prisms, ≤ 0.25 mm, in small aggregates. Colorless, transparent, vitreous luster. Cleavage {001}, good. $VHN_{15} = 103$. $G = 2.10$. Nonfluorescent. Uniaxial $(-)$, $N_O = 1.427$, $N_E = 1.403$. Found at the *Cetine mine, Tuscany, Italy*, in cavities in highly silicified limestone, associated with gypsum, fluorite, elpasolite, ralstonite, and onoratoite, and as a sublimate from fumaroles in the crater of *Mt. Erebus, Ross Island, Antarctica*, associated with gypsum and other phases that were unidentified. RG *AM 73:855(1988), 79:765(1994)*.

Class 10

Oxyhalides and Hydroxyhalides

10.1.1.1 Atacamite $Cu_2(OH)_3Cl$

Named in 1801 for the locality. Polymorphs: paratacamite, botallackite. ORTH *Pnam.* $a = 6.030$, $b = 6.865$, $c = 9.120$, $Z = 4$, $D = 3.756$. *USGSPP 800B:119(1972)*. Structure: *AC(C) 42:1277(1986)*. 25-269: 5.48_{10} 5.03_7 2.84_5 2.78_5 2.76_6 2.28_7 2.27_5 1.61_6. Crystals slender prismatic [001]; also, tabular {010}; rarely pseudo-octahedral. In crystalline aggregates, massive, granular, or fibrous. Twinning on {110}, forming doublets or triplets; also several other twin laws. Bright green, dark emerald green; blackish green, apple green streak, adamantine to vitreous luster. Cleavage {010}, perfect; {101}, fair; conchoidal fracture; brittle. $H = 3-3\frac{1}{2}$. $G = 3.77$. Easily soluble in acids, easily fusible, in CT gives off water and forms a gray sublimate. Biaxial $(-)$; XYZ= *bac*; $N_x = 1.831$, $N_y = 1.861$, $N_z = 1.880$; $2V = 74°$; $r < v$, strong. X, pale green; Y, yellow green; Z, grass green. Atacamite is a secondary mineral found in the oxidized zone of copper deposits, chiefly in arid regions. It often alters pseudomorphously to malachite. Sparingly at Bisbee and at Jerome, AZ. Tintic, UT. Majuba Hill, NV. Boleo, BCS, Mexico. Vesuvius, Italy, in fumaroles. In Australia superb crystals to 80 mm at the New Cornwall mine, near Kedina, NSW(!). *Atacama desert*, northern *Chile*(!), where it has been noted in several mines, in fine crystal groups and rosettes of blades to 25 mm, and at Chuquicamata, Antofagasta, where it was an important ore mineral. Formed as an alteration product on ancient and modern bronzes and copper objects. One of the corrosion products common on the Statue of Liberty. RG *DII:69*.

10.1.1.2 Hibbingite $\gamma\text{-}Fe_2^{2+}(OH)_3Cl$

Named in 1994 for the city of Hibbing, Minnesota. Isostructural with atacamite. ORTH *Pnam.* $a = 6.31$, $b = 9.20$, $c = 7.10$, $Z = 4$, $D = 3.04$. *AM 79:555(1994)*. PD: 5.62_4 5.16_2 2.86_6 2.33_{10} 2.12_5 1.87_2 1.745_2 1.652_5. As microveinlets in serpentinite. Colorless to pale green, turning yellowish with oxidation. Soft. Soluble in water and ethanol. Upon protracted exposure to air, alters to akaganeite. Biaxial; $N = 1.6-1.7$; moderately birefringent, fast length. Found in drill core of partially serpentinized troctolitic or peridotitic rocks near the base of the *Duluth complex, MN*, associated with magnetite and serpentine, and in the deep copper zone in the footwall of the *Strathcona mine,*

Sudbury district, ON, Canada, as fracture fillings in chalcopyrite, pentlandite, and cubanite. RG *CM 27:311(1989)*.

10.1.2.1 Paratacamite $Cu_2(OH)_3Cl$

Named in 1873 in allusion to the dimorphous relationship to atacamite. Polymorphs: atacamite, botallackite. HEX-R $R\bar{3}$. $a = 13.654$, $c = 14.041$, $Z = 24$, $D = 3.754$. *JAC 4:530(1971)*. Structure: *AC(B) 31:183(1975)*. 25-1427: 5.50_7 5.44_{10} 2.90_3 2.78_5 2.76_6 2.74_6 2.27_8 1.71_5. Rhombohedral crystals; also granular, massive, and in fibrous or spherulitic crusts. Twinning on $\{10\bar{1}1\}$, sometimes polysynthetic. Green to greenish black, green streak, vitreous luster. Cleavage $\{10\bar{1}1\}$, good; conchoidal to uneven fracture. $G = 3.74$. Tests same as for atacamite. Uniaxial (+); $N_O = 1.843$, $N_E = 1.848$; nonpleochroic; sometimes anomalously biaxial. Common through alteration of copper minerals by seawater under cold conditions. Bisbee, AZ. Levant and Cligga Head mines, St. Just, Cornwall, England. *Generosa* and *Herminia mines, Sierra Gorda, Chile*. Broken Hill, NSW, and in fine crystals at the Nangaroo mine, Murrin Murrin, WA, Australia. RG *DII:74, NJMM:335(1972)*.

10.1.3.1 Botallackite $Cu_2(OH)_3Cl$

Named in 1865 for the locality. Polymorphs: atacamite, paratacamite. MON $P2_1/m$. $a = 5.717$, $b = 6.126$, $c = 5.636$, $\beta = 93.07°$, $Z = 1$, $D = 3.598$. Structure: *MM 49:87(1985)*. 8-88: 5.66_{10} 2.84_4 2.68_3 2.57_7 2.40_8 1.93_3 1.53_4 1.25_3. Crusts of tiny bladed crystals. Pale bluish green. Cleavage in one direction. $G \approx 3.6$. Tests same as for atacamite. Biaxial (+); X perpendicular to cleavage; $N_x = 1.775$, $N_y = 1.800$, $N_z = 1.846$; $2V = 78°$; $r > v$, strong. *Botallack mine, St. Just, Cornwall, England*(!), and the nearby Levant mine; also at the Cligga Head mine. Identified as an alteration of ancient bronze objects. RG *DII:76, MM 29:34(1958)*.

10.1.4.1 Kempite $Mn^{2+}(OH)_3Cl$

Named in 1924 for James Furman Kemp (1859–1926), American mining geologist and professor at Columbia University. ORTH *Pnam*. $a = 6.49$, $b = 9.52$, $c = 7.12$, $Z = 4$, $D = 2.96$. 25-1158: 5.70_8 5.36_7 2.97_7 2.90_2 2.84_2 2.39_{10} 2.12_2 1.69_4. Crystals prismatic [001]. Emerald green. $H = 3\frac{1}{2}$. $G = 2.94$. Soluble in dilute acids. In CT turns black and gives off acid water. Biaxial (−); XYZ= *cba*; $N_x = 1.684$, $N_y = 1.695$, $N_z = 1.698$; $2V = 55°$. *Alum Rock Park, E of San Jose, Santa Clara Co., CA*, in a boulder of manganese minerals, which has since been completely collected and carried away, associated with pyrochroite, hausmannite, and rhodochrosite. RG *DII:73, AJS 8:145(1924)*.

10.1.5.1 Korshunovskite $Mg_2Cl(OH)_3 \cdot 4H_2O$

Named in 1982 for the locality. TRIC $P1$ or $P\bar{1}$. $a = 8.64$, $b = 6.25$, $c = 7.42$, $\alpha = 101.40°$, $\beta = 103.90°$, $\gamma = 72.70°$, $Z = 2$, $D = 1.798$. *AM 68:643(1983)*. 36-388: 8.04_{10} 4.03_7 3.84_8 2.87_6 2.70_6 2.44_9 2.19_5 1.857_7. Elongated prismatic grains. Colorless. $H = 2$. $G = 1.787$. Slowly soluble in water, rapidly so in weak acids. Biaxial (−); $N_x = 1.516$, $N_y = 1.538$, $N_z = 1.547$; 2V = 62°; negative elongation; extinction 6–8° to the elongation. *Korshunov iron ore deposit, Irkutsk, Siberia, Russia*, in drill core at a depth of 770 m, as veinlets 1–2 mm wide in dolomitic marble, associated with ekaterinite and shabynite. RG *ZVMO 111:324(1982)*.

10.2.1.1 Zavaritskite BiOF

Named in 1962 for Aleksandr Nicolaevich Zavaritskii (1884–1952), Russian petrographer. TET $P4/nmm$. $a = 3.75$, $c = 6.23$, $Z = 2$, $D = 9.21$. *AM 48:210(1963)*. 22-114: 3.11_4 2.65_{10} 2.39_4 2.02_6 1.873_{10} 1.793_4 1.618_{10} 1.325_6. Gray, translucent in thin edges, semimetallic to greasy luster. G = 9.0. Uniaxial; $N_O = 2.213$; weakly birefringent. *Sherlovy Gory deposits, eastern Transbaikal, Siberia, Russia*, psuedomorphous after bismuthinite. *Egbushu mine, Hirugawa Mura, Ena Gun, Gifu, Japan.* RG *DANS 146:680(1962)*.

10.2.1.2 Bismoclite BiOCl

Named in 1935 for the composition. TET $P4/nmm$. $a = 3.891$, $c = 7.369$, $Z = 2$, $D = 7.752$. *NBSC 539(4):54(1955)*. 6-249(syn): 7.38_4 3.44_{10} 2.75_8 2.68_{10} 2.21_3 1.95_4 1.83_3 1.57_3. Minute scaly crystals, massive, earthy, columnar. Creamy white, grayish, yellowish, brown; greasy to silky luster; also dull, earthy. Cleavage {001}, perfect; plastic. H = 2–2½. G = 7.72. Soluble in acids, but BiOCl reprecipitates upon considerable dilution. Melts at red heat and on cooling solidifies to a yellow mass. In CT yields some water, then sublimes to yellow droplets. Substitution of (OH) for Cl is apparently only partial, and does not extend to BiO(OH). Uniaxial (−); $N_O = 2.15$. Strongly birefringent. Formed by the oxidation of bismuth or bismuthinite, usually under arid conditions. *Eagle and Blue Bell mines, Tintic district, UT. Grizzly Bear mine, Goldfield, NV. Jackal's Water, NNE of Steinkopf, Namaqualand, Cape Prov., South Africa; Bygoo, NSW, Australia; Manoel Baldoino mine, Junco, PB, Brazil.* RG *DII:60, CM 8:390(1965)*.

10.2.1.3 Daubréeite BiO(OH,Cl)

Named in 1876 for Gabriel Auguste Daubrée (1814–1896), French mineralogist and geologist. TET $P4/nmm$. $a = 3.86$, $c = 7.41$, $Z = 2$, $D = 7.56$ for OH/Cl = 1:1. 43-674: 7.35_5 3.70_1 2.72_8 2.66_{10} 2.46_1 1.673_{10} 1.563_8 1.259_5. Found only massive, compact, earthy, columnar to platy, fibrous. Creamy white, grayish, yellowish brown; greasy to silky luster, dull, pearly on cleavage surfaces. Cleavage {001}, perfect; very plastic. H = 2–2½. G = 6.5. In the series BiO(OH)–BiOCl there is only partial substitution of (OH) for Cl, and BiO(OH) is not known. Daubréeite refers to material with OH/Cl ≥ 1:1. Uniaxial (−); $N_O = 1.91$; weakly birefringent. *Constancia mine, Tazna, Bolivia*, mixed with clay. RG *DII:60, MM 24:49(1935), ST:168*.

10.2.2.1 Laurionite PbCl(OH)

Named in 1887 for the locality. Dimorphous with paralaurionite. ORTH $Pnma$. $a = 9.699$, $b = 4.020$, $c = 7.111$, $Z = 4$, $D = 6.212$. Structure: ZK $141:246(1975)$. $31\text{-}680$(syn): 5.70_3 4.01_{10} 3.56_4 3.33_3 3.30_6 2.87_3 2.51_3 2.30_5. Crystals elongated {010} or tabular {100}. Colorless to white; adamantine luster, pearly on {100}. Cleavage {101}, distinct; not brittle. $H = 3\text{-}3\frac{1}{2}$. $G = 6.24$. Biaxial $(-)$; XYZ= cab, $N_x = 2.077$, $N_y = 2.116$, $N_z = 2.158$ (Na); $2V = 88°$. Laurium (Laurion), Greece, in ancient lead slags exposed to the action of seawater, associated with phosgenite, paralaurionite, fiedlerite, and other secondary lead minerals. Wheal Rose, near Sithney, Cornwall, England. RG $DII:62$.

10.2.3.1 Paralaurionite PbCl(OH)

Named in 1899 from the Greek *para* and *laurionite*, a dimorphous pair. MON $C2/m$. $a = 10.79$, $b = 3.98$, $c = 7.19$, $\beta = 117.22°$, $Z = 4$, $D = 6.28$. MM $29:341(1950)$. $16\text{-}158$: 5.14_{10} 3.49_6 3.21_{10} 2.98_7 2.51_9 2.44_6 2.01_6 1.70_6. Crystals thin tabular {100}, or lathlike by elongation [001]. Twinning on {100} common, simulating orthorhombic symmetry. Colorless to white, also sulfur-yellow, pale greenish, or violet; subadamantine luster. Cleavage {001}, perfect; crystals bend readily about [010] due to translation gliding; not brittle. $G = 6.15$. Soft. Soluble in HNO_3, easily fusible. Biaxial $(-)$; $Y = b$; $Z \wedge c = 25°$; $N_x = 2.05$, $N_y = 2.15$, $N_z = 2.20$; $2V = 68°$; $r < v$, strong. Mammoth mine, Tiger, AZ; Wheal Rose, near Sithney, Cornwall; and common in large cleavage masses in hydrocerussite at Shepton Mallet, Mendip Hills, Somerset, England; Laurium, Greece, in ancient lead slags altered by seawater. In large bladed gray crystals with leadhillite in Toussit, Oujda, northern Morocco(!); San Rafael mine, Cerro Gordo, Chile. RG $DII:64$, GSC $60(4):58(1960)$.

10.2.4.1 Blixite $Pb_2Cl(O,OH)_2$

Named in 1958 for Ragner Blix (b.1898), chemist, Swedish Museum of Natural History. ORTH Space group unknown. $a = 5.832$, $b = 5.694$, $c = 25.47$, $Z = 8$. AM $45:908(1960)$. $12\text{-}542$: 3.88_8 3.18_5 2.93_{10} 2.83_6 2.15_5 2.12_6 2.04_6 1.66_8. Thin crystalline coatings. Dirty yellow to yellowish green, pale yellow streak, vitreous luster. Distinct cleavage in one direction. $H = 3$. $G = 7.35$. Biaxial $(+)$; $N_x = 2.05$, $N_z = 2.20$; $2V = 80°$. Långban, Värmland, Sweden, in the *Amerika stope*, in fissures in dolomite; also in fractures in braunite–hausmannite ore. RG AMG $2:411(1958)$.

10.2.5.1 Perite $PbBiO_2Cl$

Named in 1960 for Per Adolf Geijer (1886–1976), Swedish geologist. ORTH $Bmmb$. $a = 5.627$, $b = 5.575$, $c = 12.425$, $Z = 4$, $D = 8.24$. AM $46:765(1961)$. $13\text{-}352$: 3.77_8 2.86_{10} 2.78_7 2.07_7 1.97_7 1.66_7 1.62_9 1.25_8. Crystals tabular {010}. Sulfur yellow, adamantine luster. Cleavage {010}, distinct. $H = 3$. $G = 8.16$. Easily soluble in dilute acids. Biaxial; $N > 2.4$. Blue Bell claims, Baker, CA.

Långban, Värmland, Sweden, associated with hausmannite and calcite; also from Glen Florrie, WA, Australia. RG *AMG 2:565(1960)*.

10.2.5.2 Nadorite PbSbO$_2$Cl

Named in 1870 for the locality. ORTH *P4nc*. $a = 5.60$, $b = 5.44$, $c = 12.22$, $Z = 4$, $D = 7.05$. *17-469:* 3.71$_3$ 2.80$_{10}$ 2.70$_3$ 2.06$_3$ 2.03$_3$ 1.95$_3$ 1.62$_3$ 1.59$_3$. Crystals thin tabular {010} or prismatic [100]; in subparallel groups. Twins on {101}. Yellow to smoky brown, white to yellowish streak, resinous to adamantine luster. Cleavage {010}, perfect. H = $3\frac{1}{2}$–4. G = 7.02. Biaxial (+); XYZ= *bca*; N$_x$ = 2.30, N$_y$ = 2.35, N$_z$ = 2.40 (Li); 2V = 85°; r > v, strong. Commonly partially altered to bindheimite. Långban, Värmland, Sweden; the Bodannan mine, St. Endellion, Cornwall, England, as an alteration of jamesonite. *Djebel Nador, Constantine, Algeria*; Touissit, Oujda, northern Morocco, with anglesite(!). RG *DII:1039, PM 52:335(1973)*.

10.2.6.1 Thorikosite Pb$_3$(Sb,As)O$_3$(OH)Cl$_2$

Named in 1985 after the ancient town of Thorikos, near which the Laurium mines are located. TET *I4/mmm*. $a = 3.919$, $c = 12.854$, $Z = 1$, $D = 7.24$. *AM 70:845(1985). 38-403:* 6.39$_3$ 3.75$_7$ 3.20$_3$ 2.89$_{10}$ 2.78$_5$ 2.10$_3$ 1.96$_2$ 1.62$_3$. Crystals prismatic, flattened {001}. Light yellow, vitreous luster. Cleavage {001}, perfect; very brittle. Found in vugs in ancient lead slags at *Laurium, Greece*, associated with paralaurionite, hydrocerussite, sphalerite, and calcite. RG

10.2.7.1 Asisite Pb$_7$SiO$_8$Cl$_2$

Named in 1988 for the Asis farm, where the Kombat mine (source of the mineral) is located. TET *I4/mmm*. $a = 3.897$, $c = 22.81$, $Z = 1$, $D = 8.041$. *AM 73:643(1988). 45-1429:* 3.83$_3$ 3.46$_4$ 2.96$_{10}$ 2.75$_6$ 1.984$_4$ 1.757$_4$ 1.627$_6$ 1.602$_5$. Platy crystals to 0.5 mm. Yellow to yellow-green, adamantine luster. Cleavage {001}, perfect. H = $3\frac{1}{2}$. Uniaxial (−); N$_O$ = 2.39 N$_E$ = 2.32 (Na) (calc); r < v. In RL, gray, with distinct internal reflections; nonpleochroic; weakly anisotropic; R = 18.4, 6.0 (470 nm); 17.2, 5.3 (546 nm); 16.8, 5.1 (589 nm); 16.3, 4.9 (650 nm). Kombat mine, Namibia, with barite in interstices of coarsely crystalline hematophanite. RG

10.2.8.1 Zharchikhite AlF(OH)$_2$

Named in 1988 for the locality. MON *P2$_1$/a*. $a = 5.164$, $b = 7.843$, $c = 5.179$, $\beta = 116.25°$, $Z = 4$, $D = 2.82$. *AM 74:504(1989). 45-1430:* 3.98$_{10}$ 2.92$_8$ 2.31$_7$ 1.926$_5$ 1.833$_9$ 1.788$_5$ 1.737$_7$ 1.289$_7$. Prismatic crystals to 2.5 mm, elongated [001]; drusy aggregates; massive, fine-grained. Colorless to white, vitreous luster. Cleavage {010}, perfect; very brittle. H = 4.5. G = 2.81. Insoluble in acids, slowly soluble in NaOH solution. Biaxial (−); Y = *b*; Z ∧ *c* = 43°. N$_x$ = 1.532, N$_y$ = 1.552, N$_z$ = 1.567 (Na); 2V = 80°; inclined dispersion, r > v. *Zharchinskoye molybdenum deposit, Zabaykalye, Transbaikal, Russia*, in a fault breccia, assoiated with prosopite, ralstonite, gearksutite, barite, and siderite. RG *ZVMO 117:79(1988)*.

10.2.9.1 Pinalite $Pb_3(WO_4)OCl_2$

Named in 1989 for the county of origin. ORTH $A2aa$ or $Amaa$. $a = 11.073$, $b = 13.086$, $c = 5.624$, $Z = 4$, $D = 7.78$. AM $74:934(1989)$. 6.52_3 3.78_9 3.28_4 2.93_{10} 2.81_4 2.77_3 2.12_3 1.642_5. Acicular bladed crystals elongated [001], flattened {010}. Bright yellow, adamantine luster. No cleavage, brittle. Nonfluorescent. Biaxial (+); $XYZ = bca$; $N_x = 2.490$, $N_y = 2.495$, $N_z = 2.505$ (calc); nonpleochroic; $2V = 70°$; $r > v$, moderate. From the *Mammoth mine, Pinal Co., AZ*, in solution cavities in quartz, with leadhillite, cerussite, matlockite, and so on. RG

10.3.1.1 Mendipite $Pb_3Cl_2O_2$

Named in 1839 for the locality. ORTH $Pnma$. $a = 11.87$, $b = 5.806$, $c = 9.48$, $Z = 4$, $D = 7.30$. Structure: BM $94:323(1971)$. 23-332: 7.40_6 3.78_6 3.51_7 3.08_6 3.04_8 2.90_6 2.78_{10} 2.64_9. 25-448(syn): 3.52_6 3.05_{10} 2.79_{10} 2.64_9 2.10_6 1.65_6 1.64_6 1.63_6. Fibrous or columnar masses, often radiated. Colorless to white or gray, often tinged with pink or other colors; pearly to silky luster on cleavages, resinous to adamantine across fractures. Cleavage {110}, perfect; {100}, {010}, fair; conchoidal to uneven fracture. $H = 2\frac{1}{2}$. $G = 7.24$. Soluble in dilute HNO_3; in CT becomes yellow and decrepitates. $PbCl_2$ then sublimes, and the residue melts to a yellow liquid (PbO). Biaxial (+); $XYZ = abc$; $N_x = 2.24$, $N_y = 2.27$, $N_z = 2.31$; $2V_{calc} = 83°$; $r < v$, very strong. Near *Churchill* and at Higher Pitts, *Mendip Hills, Somerset, England*, and in some quantity at the Foster Yeoman quarry, Shepton Mallet. Also at the Kunibert mine, Brilon, Westphalia, Germany. RG $DII:56$, BM $94:108(1971)$.

10.3.2.1 Fiedlerite $Pb_3(OH)_2Cl_4$

Named in 1887 for Carl Gustav Fiedler (1791–1853), Saxon commissioner of mines, who directed an exploratory expedition to Laurium, Greece, where the mineral was found. MON $P2_1/a$. $a = 16.62$, $b = 8.02$, $c = 7.20$, $\beta = 102.20°$, $Z = 4$, $D = 5.64$. BM $87:125(1964)$. 15-59: 3.89_{10} 3.54_5 3.33_5 3.22_5 2.81_8 2.55_{10} 2.01_6 1.63_3. Lathlike crystals, tabular {100} and elongated [010]. Twinning on {100}. Colorless to white, adamantine luster. Cleavage {100}, good. $H = 3\frac{1}{2}$. $G = 5.88$. Slowly soluble in HNO_3; in CT decrepitates, loses water, and fuses, leaving a sublimate of $PbCl_2$. Biaxial (−); $Z = b$; $Y \wedge c = -34°$; $N_x = 1.98$, $N_y = 2.04$, $N_z = 2.10$; $2V = 85°$; $r < v$, perceptible. *Laurium, Greece*, formed by action of seawater on ancient lead slags. RG $DII:67$.

10.3.3.1 Corderoite $Hg_3S_2Cl_2$

Named in 1974 for the locality. Dimorphous with lavrentievite. ISO $I2_13$. $a = 8.949$, $Z = 4$, $D = 6.827$. AM $59:652(1974)$. Structure: $AC(B)$ $24:156(1968)$. 39-1386: 6.3_4 3.66_{10} 2.83_6 2.58_7 2.39_3 2.24_3 2.11_3 1.757_6. Claylike, as small masses or replacements. Grain size submicroscopic (<2 μm). Light orange-pink, but becomes light gray to black upon exposure to sunlight or artificial light; dull luster. $H = 2$. $G = 6.895$. Isotropic; $N > 2.50$. *Cordero mine, W of McDermitt, Humboldt Co., NV*, in late Miocene playa sediments, with cinnabar, montmorillonite, quartz, and cristobalite. Kadyrel mercury mine, Tuva, Siberia, Russia. The corderoite replaces cinnabar in all cases. RG *JCG 1:271(1967)*.

10.3.4.1 Lavrentievite Hg$_3$S$_2$(Cl,Br)$_2$

10.3.4.2 Arzakite Hg$_3$S$_2$(Br,Cl)$_2$

Named in 1984 for Michail Alexeevich Lavrentiev (1900–1980), Russian mathematician and academician, founder of the Siberian Academy of Science. Arzakite named in 1984 for the locality. Lavrentievite is dimorphous with corderoite (10.3.3.1). MON or TRIC $P2/m$, $P2$, or Pm. Lavrentievite: $a = 8.94$, $b = 5.194$, $c = 18.33$, $\beta = 92.73°$, $Z = 5$, $D = 7.26$–7.50. Arzakite: $a = 8.99$, $b = 5.24$, $c = 18.45$, $\beta = 92.46°$, $Z = 5$, $D = 7.64$. *AM 70:873(1985)*. *41-606* (lavrentievite): 5.00_3 3.96_4 3.38_5 3.01_6 2.61_{10} 2.29_4 2.20_4 1.587_5. *28-658*(arzakite)(syn): 5.10_3 4.06_8 3.43_3 3.07_6 3.06_6 2.65_{10} 2.32_5 1.618_3. Small grains. $H = 2$–$2\frac{1}{2}$. $G = 7.4$. Both are found at the Kadyrel and *Arzak mercury deposits, Tuva, Siberia, Russia*, with cinnabar and corderoite. Lavrentievite is also found at the Kadyrelski ore showing, with calomel, eglestonite, and mercury. RG *DANS 290:177(1986)*.

10.3.5.1 Grechishchevite Hg$_3$S$_2$(Br,Cl,I)$_2$

Named in 1990 for Oleg Konstantinovich Grechishchev, Russian engineering geologist, long-time student of mercury ores from Tuva, Siberia. TET $P4_2m$, $P\bar{4}m2$, $P4mm$ or $P4mmm$. $a = 13.225$, $c = 8.685$, $Z = 8$, $D = 7.23$ (for Hg$_3$S$_2$(Br$_{1.0}$Cl$_{0.5}$I$_{0.5}$). *AM 76:1729(1991)*. *45-1418*: $4,18_3$ 3.95_6 3.12_2 3.02_7 2.65_{10} 2.60_4 2.34_3 1.682_4. Minute intergrown crystals and grains. Bright to dark orange; reddish orange, deep yellow streak, vitreous–adamantine luster. One cleavage, prismatic; brittle. $H = 2\frac{1}{2}$. Unaltered in dilute acids. Blackens slowly upon immersion in 40% KOH solution. Uniaxial (+); N > 2. In RL, grayish white with strong orange internal reflections; distinctly pleochroic. $R = 17.3, 19.8$ (546 nm). *Arzak and Kadyrel mercury deposits, Tuva, Siberia, Russia*, in the oxidized zone on fractures, associated with bromian calomel, eglestonite, lavrentievite, arzakite, corderoite, and mercury. RG

10.3.6.1 Radtkeite Hg$_3$S$_2$I

Named in 1991 for Arthur S. Radtke (b.1936), geologist, U.S. Geological Survey. ORTH $Fmmm$, $F222$, or $Fmm2$. $a = 16.85$, $b = 20.27$, $c = 9.133$, $Z = 16$, $D = 7.05$. *Am 76:1715(1991)*. *44-1408*(syn): 3.89_5 3.73_2 3.38_2 2.71_4 2.65_9 2.54_{10} 1.827_2 1.697_2. Submicroscopic grains and minute prismatic crystals. Bright yellow-orange to orange, yellow-orange streak, adamantine luster. Cleavage {010}, perfect; {001}, {100}, very good; hackly to conchoidal fracture. $H = 2$–3. Biaxial (+); N>2.0; $2V \approx 35°$. *McDermitt mercury mine, Humboldt Co., NV*, associated with quartz, cinnabar, and corderoite. RG

10.4.1.1 Penfieldite $Pb_2Cl_3(OH)$

Named in 1892 for Samuel L. Penfield (1856–1905), American mineralogist and mineral chemist. HEX $P6/m$. $a = 11.28$, $c = 48.65$, $Z = 36$, $D = 6.00$. *BM 91:407(1968)*. *22-384:* 5.70_4 3.73_{10} 3.31_6 3.14_8 2.74_6 2.56_6 2.27_6 2.15_4. Crystals prismatic [0001], with $\{10\bar{1}1\}$ or steep pyramids; also tabular $\{0001\}$; twins common. Colorless to white, adamantine to greasy luster. Cleavage $\{0001\}$, distinct; not brittle. $G = 5.82$–5.86. Decomposed by water, soluble in dilute HNO_3. Uniaxial (+); $N_O = 2.13$, $N_E = 2.21$. *Laurium, Greece*, in ancient lead slags exposed to the action of seawater. Sierra Gorda, Antofagasta, Chile, with pseudoboleite. RG *DII:66*.

10.4.2.1 Terlinguaite $Hg^{1+}Hg^{2+}ClO$

Named in 1900 for the locality. MON $C2/c$. $a = 19.53$, $b = 5.92$, $c = 9.48$, $\beta = 143.80°$, $Z = 8$, $D = 9.31$. Structure: *AC 9:956(1956)*. *25-559*(syn): 5.76_8 4.34_6 4.17_8 3.68_6 3.28_8 2.96_8 2.82_8 2.51_{10}. Complex crystals, usually prismatic [010], flattened $\{001\}$, or equidimensional; usually rounded and sometimes in parallel growth; frequently powdery, or as aggregates. Sulfur-yellow to brown, becoming olive-green on exposure to light; yellow streak; brilliant, adamantine luster. Cleavage $\{\bar{1}01\}$, perfect. $H = 2\frac{1}{2}$. $G = 8.725$. Biaxial (−); $Y \wedge c = 83°$; $N_x = 2.35$, $N_y = 2.64$, $N_z = 2.66$; $2V = 20°$; $r < v$, extreme. *Terlingua, Brewster Co., TX*, with calomel, eglestonite, kleinite, montroydite, mercury, and cinnabar. Cahill mine, Humboldt Co., NV. RG *DII:52*.

10.4.3.1 Kleinite $Hg_2N(Cl,SO_4) \cdot nH_2O$

Named in 1905 for Carl Klein (1842–1907), German mineralogist, University of Berlin. HEX $P6_3/mmc$. $a = 13.56$, $c = 11.13$, $Z = 18$. Structure: *SR 12:219(1949)*. Kleinite has hexagonal symmetry above 130°; below that the symmetry is possibly triclinic. *39-389:* 5.19_3 4.04_2 2.93_{10} 2.89_7 2.79_9 2.59_{10} 2.02_2 1.690_3. Crystals prismatic [0001] with light vertical striations on prism; also equant. Yellow to orange, yellow streak, adamantine to greasy luster. Cleavage $\{0001\}$, easy; $\{10\bar{1}1\}$, imperfect. $H = 3\frac{1}{2}$–4. $G = 8.0$. Soluble in warm HCl or HNO_3 without deposition of calomel. Since the Cl^-, SO_4^{-2}, and H_2O are somewhat variable, it is probable that these anions and water occupy open channelways as in mosesite. Biaxial (−); $N_x = 2.16$, $N_y = 2.18$; 2V small; $r < v$, very strong. *Terlingua, Brewster Co., TX*, with terlinguaite, montroydite, and other mercury oxysalts. RG *DII:87, AM 17:541(1932)*.

10.4.4.1 Melanothallite Cu_2OCl_2

Named in 1870 from the Greek for *black* and *green* (the color of a young twig), alluding to the black color changing to green upon oxidation. ORTH *Fddd*. $a = 9.595$, $b = 9.693$, $c = 7.461$, $Z = 8$, $D = 4.08$. *AM 68:852(1983)*. *35-679:* 5.04_{10} 2.95_5 2.82_2 2.52_8 2.42_2 2.18_2 1.93_2 1.47_2. Crystals are thin plates. Black to bluish black, deep brown in fine fragments, vitreous luster. Prismatic cleavage, perfect; brittle. Alters rapidly in air to a green material. Partly dissolved by water, completely so by warm acids. Originally found among sublimates on *Vesuvius, Italy*, associated with eriochalcite, chalcocyanite, euchlorine, and dolerophanite. Later found as a result of the fissure eruption of Tolbachin volcano, Kamchatka Penin., Siberia, Russia, in 1975–1976, with the same associated minerals. RG *DII:44*, *ZVMO 111:562(1982)*.

10.5.1.1 Cadwaladerite $Al(OH)_2Cl \cdot 4H_2O$

Named in 1941 for Charles Meigs Biddle Cadwalader (1885–1959), American museum administrator, Academy of Natural Sciences, Philadelphia. An inadequately characterized mineral. Amorphous. Small masses and grains. Lemon yellow, transparent, vitreous luster. Conchoidal fracture. G = 1.66. Isotropic; $N \approx 1.513$. *Cerro Pintado, Tarapacá, Chile*, in a sulfate deposit. RG *DII:77*.

10.5.2.1 Poyarkovite Hg_3ClO

Named in 1981 for Vladimir Erastovich Poyarkov, Russian geologist, well-known investigator of mercury and antimony deposits. MON Space group uncertain, probably $C2/m$. $a = 18.82$, $b = 9.02$, $c = 16.79$, $\beta = 112.40°$, $Z = 24$, $D = 9.88$. *AM 67:860(1982)*. 33-920: 4.20_4 3.09_5 2.96_4 2.83_{10} 2.74_8 2.60_6 1.88_4 1.80_7. Irregular grains and aggregates. Deep raspberry- to cherry-red, turning darker and finally black on exposure; red streak; vitreous to adamantine luster. Irregular to conchoidal fracture, very brittle. $H = 2-2\frac{1}{2}$. $G = 9.50$. Instantly blackened by KOH solution. Decomposed by HNO_3, insoluble in HCl. Biaxial; $N > 2.0$. Pleochroism deep red to brownish red. In RL, strongly anisotropic; colors may vary with orientation, but always include azure to blue. Janos, CHIH, Mexico. *Khaidarkan deposit, Siberia, Russia*, in close contact with eglestonite and calomel. RG *ZVMO 110:501(1981)*.

10.5.3.1 Pinchite $Hg_5O_4Cl_2$

Named in 1974 for William W. Pinch (b.1940), American mineral collector, who first recognized the species. ORTH *Ibam*. $a = 11.54$, $b = 6.08$, $c = 11.64$, $Z = 4$, $D = 9.26$. *CM 12:417(1974)*. 26-1267: 3.94_6 3.26_4 2.92_5 2.84_{10} 2.70_8 2.17_2 1.79_2 1.64_2. Euhedral crystals up to 1 mm, with tabular $\{001\}$ habit. Dark brown to black, reddish-brown streak, adamantine luster. No cleavage. Soft. $G = 9.5$. Biaxial; $N > 2.20$; pleochroic, red to almost opaque; strongly birefringent. *Terlingua, Brewster Co., TX*, with terlinguaite and montroydite. RG

10.5.4.1 Eglestonite $Hg_6^{1+}Cl_3O(OH)$

Named in 1903 for Thomas Egleston (1832–1900), American mineralogist and metallurgist, Columbia University. ISO $Ia3d$. $a = 16.036$, $Z = 16$, $D = 8.65$. *AM 62:396(1977)*. Structure: *AM 77:839(1992)*. 29-909: 4.01_4 3.27_{10} 2.60_1 2.54_5 2.32_1 1.89_6 1.709_1 1.336_1. Crystals dodecahedral, commonly modified by cube, usually rounded; also as a thin earthy film coating cinnabar. Yellow to brown, becoming darker and finally black on exposure to light; adamantine to resinous luster. Uneven to conchoidal fracture; brittle. $H = 2\frac{1}{2}$. $G = 8.45$. In

NH$_4$OH turns black immediately. Decomposed by acids with separation of calomel. Isotropic; N = 2.49. *Terlingua, Brewster Co., TX*; McDermitt, NV; in CA at Kings mine, King's Co.; Socrates mine, Sonoma Co., and New Idria mine, San Benito Co. Also Pike Co. and elsewhere in AR; Almaden, Ciudad Real, Spain; Arzak and Kadyrel mercury deposits, Tuva, Siberia, Russia; Monarch mercury mine, Transvaal, South Africa. RG *DII:51*

10.5.4.2 Kadyrelite Hg$_4^{1+}$(Br,Cl)$_2$O

Named in 1987 for the locality. ISO *Ia3d*. $a = 16.22$, $Z = 24$, $D = 8.79$. *AM 74:503(1989)*. *42-1371*: 4.06$_3$ 2.63$_2$ 2.57$_8$ 2.34$_2$ 1.912$_{10}$ 1.731$_2$ 1.350$_2$ 1.102$_2$. Small grains and powdery coatings. Bright to dull orange, orange-yellow streak, vitreous to adamantine luster. No cleavage, conchoidal to uneven fracture; brittle. H = 2$\frac{1}{2}$–3. Turns black in 40% KOH solution; in HCl slowly becomes dark gray; readily soluble in 1:1 HNO$_3$, leaving a cottonlike white residue. Isotropic; N > 2.0. In RL, grayish white; intense orange internal reflections; no bireflectance. Found in cavities in carbonate veins, associated with calomel, eglestonite, corderoite, and other mercury minerals at the *Kadyrelski mercury deposit, Tuvinskaya, Armenia*. RG *ZVMO 116:733(1987), AM 77:839(1992)*.

10.5.5.1 Comancheite Hg$_{13}^{2+}$(Cl, Br)$_8$O$_9$

Named in 1981 for the Comanche Indians, who were the first miners in the Terlingua district, using cinnabar for war paint. ORTH *Pnnm* or *Pnn2*. $a = 18.407$, $b = 21.641$, $c = 6.677$, $Z = 4$, $D = 8.0$. *35-510*: 5.68$_7$ 5.42$_6$ 3.82$_3$ 3.10$_4$ 2.88$_8$ 2.71$_5$ 2.67$_{10}$ 2.45$_5$. Tiny crystalline masses or stellate acicular crystals. Often mistaken for montroydite. Red; crystals, orange-yellow to yellow resinous luster, vitreous crystals. Cleavage {001}, {110}, fair; brittle. H = 2. G = 7.7. Unaffected by cold concentrated acids; in 40% KOH solution turns dull orange-brown. Biaxial; N$_x$ = 1.78, N$_z$ = 1.79; strong absorption; parallel extinction; length-fast. Indices lower than predicted by the Gladstone–Dale law. From the *Mariposa mine, Terlingua district, Brewster Co., TX*, encrusting calcite. RG *CM 19:393(1981), AM 67:622(1982)*.

10.5.6.1 Claringbullite Cu$_4$Cl(OH)$_7$ · nH$_2$O

Named in 1977 for Frank Claringbull (1911–1990), British X-ray crystallographer, keeper of minerals and director of the British Museum (Natural History). HEX Space group uncertain, probably $P6_3$. $a = 6.671$, $c = 9.183$, $Z = 2$, $D = 3.99$. *29-539*: 5.75$_{10}$ 4.89$_8$ 4.58$_8$ 2.89$_7$ 2.70$_{10}$ 2.45$_9$ 1.80$_7$ 1.67$_7$. Crystals platy {0001}. Blue to blue-green. Cleavage {0001}, perfect; {10$\bar{1}$1}, {11$\bar{2}$0}, distinct. Soft. G = 3.9. Uniaxial (−); N$_O$ = 1.782, N$_E$ = 1.780. Bisbee, AZ. *Nchanga open-pit mine, Zambia*, with cuprite. M'sesa mine, Kambowe, Shaba, Zaire. RG *MM 41:433(1977), AM 63:793(1978)*.

10.5.7.1 Onoratoite $Sb_8O_{11}Cl_2$

Named in 1968 for Ettore Onorato (1899–1971), Italian mineralogist, University of Rome. MON $C2/m$. $a = 19.047$, $b = 4.053$, $c = 10.318$, $\beta = 110.25°$, $Z = 2$, $D = 5.425$. Structure: $AC(C)$ $40{:}1506(1984)$. 21-52: 4.39_7 3.19_{10} 3.04_5 2.82_5 2.68_6 2.60_3 2.54_3 1.81_3. Acicular to bladed crystals, elongated [010], flattened {001}; fibrous [010]. White. $G = 5.3$. Soluble in cold concentrated HCl, and NaOH solution. In CT sublimes, giving oily droplets of $SbCl_3$ and minute octahedrons of Sb_2O_3. Biaxial (−); $X \wedge c = 12°$, $Y \wedge a = 8°$, $Z \wedge b = 0$–$14°$; $N_y = 2.18$–2.23, $N_z = 2.23$–2.26. Bireflectance = 0.24. *Cetine di Cotorniano, Rosia, Siena, Italy*, as an alteration product of stibnite. RG *MM 36:1037(1968), MR 8:285(1977), AM 70:440(1985)*.

10.5.8.1 Simonkolleite $Zn_5(OH)_8Cl_2 \cdot H_2O$

Named in 1985 for Werner Simon (b.1939) and Kurt Kolle (b.1949), mineral collectors from Cornberg, Germany, who submitted the first samples for investigation. HEX-R $R\bar{3}m$. $a = 6.334$, $c = 23.58$, $Z = 3$, $D = 3.35$. *NJMM:145(1985)*. 7-155: 7.87_{10} 3.58_3 3.17_4 2.94_3 2.73_6 2.67_7 2.37_4 1.59_4. Tabular to lamellar hexagonal crystals to 1 mm. Colorless, white streak, vitreous luster. Cleavage {0001}, perfect. $H = 1\frac{1}{2}$. $G = 3.20$. Soluble in dilute acids. Uniaxial (+); $N_O = 1.657$, $N_E = 1.700$. Found as a weathering product of Zn-bearing slags at the *Richelsdorf foundry, Hesse, Germany*. RG *AM 73:194(1988)*.

10.5.9.1 Abhurite $Sn_3O(OH)_2Cl_2$

Named in 1985 for the locality. HEX-R $R\bar{3}m$, $R3m$, or $R32$. $a = 10.017$, $c = 44.014$, $Z = 21$, $D = 4.34$. 39-314: 4.14_5 3.63_4 3.40_5 3.27_4 3.24_4 2.89_7 2.82_5 2.53_{10}. Platy crystals, occasionally twinned {0001}. White to tan, colorless in crystals. Very fragile. $G = 4.29$. Dissolves rapidly in HNO_3 and slowly in HCl. Uniaxial (+); $N \approx 2.06$. Negligible birefringence. Found on corroded tin ingots recovered from an ancient wreck in the *Sharm Abhur*, a cove in the Red Sea, about 30 km N of *Jiddah, Saudi Arabia*. RG *CM 23:233(1985)*.

10.5.10.1 Damaraite $Pb_4O_3Cl_2$

Named in 1990 for the Damara sequence, the dolostones that host the deposit at the type locality. ORTH $Pma2$, $Pmam$, or $P2_1am$. $a = 15.104$, $b = 6.891$, $c = 5.806$, $Z = 3$, $D = 7.84$. *MM 54:593(1990)*. 6-405(syn): 6.5_6 3.75_8 2.92_{10} 2.77_7 2.12_7 1.965_7 1.686_7 1.629_7. Subhedral grains <0.2 mm. Colorless, adamantine luster. Cleavage {010}, good; subconchoidal to irregular fracture. $H = 3$, $VHN_{50} = 150$. In RL, gray to bluish gray, faintly pleochroic, weakly anisotropic. From the *Kombat mine, 49 km S of Tsumeb, Namibia*, in veins of barite–calcite–jacobsite–hematite, associated with hausmannite, hematophanite, and copper. RG *AM 77:670(1992)*.

10.5.11.1 Parkinsonite $(Pb,Mo,\square)_8O_8Cl_2$

Named in 1994 for Reginald F. D. Parkinson, English mineral collector from Somerset, U.K. TET $I4/mmm$. $a = 3.992$, $c = 22.514$, $Z = 1$, $D = 7.39$. MM $58:59(1994)$. PD: 5.63_9 3.75_2 3.53_1 2.99_3 2.818_{10} 2.25_3 1.994_1 1.876_1. Small cleavage plates and crystalline patches. Bright red to purplish red, translucent, scarlet streak, adamantine luster. Cleavage {001}, perfect; sectile. $H = 2-2\frac{1}{2}$. $G_{syn} = 7.32$. Nonfluorescent. In RL, gray, moderately bireflectant, weakly pleochroic, strong red internal reflections. $R = 22.0$ (470 nm), 20.5 (546 nm), 19.6 (589 nm), 18.8 (650 nm). Found at *Merehead Quarry, Cranmore, Somerset, England*, in fractures and cavities in carbonate vugs in veins of manganese and iron oxides, associated with mendipite, diaboleite, cerussite, and chloroxiphite. RG

10.5.12.1 Lorettoite $\alpha\text{-}Pb_7O_6Cl_2$

Named in 1916 for the original locality. ORTH (pseudo-TET). $a = 5.49$, $b = 22.9$, $c = 5.50$, $Z = 2$, $D = 7.77$. 6-393: 3.86_6 3.49_6 2.98_{10} 2.77_8 2.31_6 2.25_6 1.993_7 1.642_7. Composite bladelike microcrystals forming wartlike aggregates. Light yellow to reddish yellow, yellow streak, adamantine luster. Cleavage parallel to blades, perfect. $H = 2\frac{1}{2}-3$. $G = 7.39$. Uniaxial $(-)$; $N_O = 2.40$, $N_E = 2.37$ (Li). Originally described from Loretto, TN, but this material was later shown to be an artifact. AM $64:1303(1979)$. More recently the same compound was found in conjunction with the ancient lead slags at *Laurium, Greece*, with hydrocerussite. Since the various laurium compounds have long been accepted as valid minerals despite their having resulted from human activity, lorettoite may eventually also be considered as valid, although the IMA Commission on New Minerals and New Mineral Names has not ruled on the matter. RG LAP $11:25(1986)$, $13:11(1988)$; MM $56:53(1992)$.

10.5.13.1 Hanawaltite $Hg_6^{1+}Hg^{2+}O_3Cl_2$

Named in 1996 by Roberts et al. in honor of J. D. (Don) Hanawalt (1903–1987), who was a pioneer in the field of X-ray powder diffraction. ORTH $Pbma$. $a = 11.790$, $b = 13.881$, $c = 6.450$, $Z = 4$, $D = 9.51$. $CM:34(1996)$. PD: 5.25_8 4.35_4 3.16_6 3.05_{10} 2.95_7 2.68_5 2.41_5 1.747_4. Crystals are subhedral to anhedral, platy to somewhat bladed, max. 0.3 mm × 0.3 mm. Black to very dark brown, black to dark red-brown streak, metallic luster, nonfluorescent. Cleavage {001}, good; uneven fracture, brittle. $H < 5$. In polished section white with dark red internal reflections; pleochroic, white to blue-white; strongly bireflectant. $R = 22.8$, 29.6 (470 nm); 20.7, 25.7 (546 nm); 20.15, 24.35 (589 nm); 20.2, 22.9 (650 nm). Found in a prospect pit near the abandoned *Clear Creek mercury mine, New Idria district, San Benito Co., CA*, associated with calomel, mercury, cinnabar, montroydite, and quartz. RG PD $11:45(1996)$.

10.6.1.1 Diaboleite $Pb_2CuCl_2(OH)_4$

Named in 1923 from the Greek for *apart* or *distinct from* and the mineral boleite. TET $P4mm$. $a = 5.870$, $c = 5.494$, $Z = 1$, $D = 5.410$. Structure: ZK

$134{:}69(1971)$. $21\text{-}468$: 5.50_{10} 3.28_9 2.95_8 2.57_9 2.28_{10} 2.06_8 1.75_9 1.53_9. Crystals tabular $\{001\}$; also subparallel aggregates of thin plates. Deep blue, pale blue streak. Cleavage $\{001\}$, perfect; conchoidal fracture. $H = 2\frac{1}{2}$. $G = 5.42$. Tests similar to chloroxiphite (10.6.4.1). Uniaxial $(-)$; $N_O = 1.98$, $N_E = 1.85$; absorption $O > E$. Distinct crystals(!) at the *Mammoth mine, Tiger, AZ*, in the Collins veins. Mendip Hills, England, in mendipite, with chloroxiphite, and hydrocerussite, in oxidized iron and manganese ores. RG *DII:82, MM 36:933(1968), NJMM:116(1970)*.

10.6.2.1 Koenenite $Na_4Mg_4Cl_{12} \cdot Mg_5Al_4(OH)_{22}$

Named in 1902 for Adolf von Koenen (1837–1915), German geologist of Göttingen, who first found the mineral. HEX-R $R\bar{3}m$. $a = 34.72$, $c = 32.64$, $Z = 3$, $D = 2.162$. Structure: *ZK 126:7(1968)*. $43\text{-}1490, 43\text{-}684$: 10.8_3 5.43_{10} 2.82_3 2.57_4 2.33_3 2.28_3 2.04_5 1.99_4. Crystals acute scalenohedra or rhombohedra; as crusts. Colorless to yellow, or deep red due to included scales of hematite; pearly luster on cleavage. Cleavage $\{0001\}$, perfect; flexible. $H = 1\frac{1}{2}$. $G = 1.98$. Uniaxial $(+)$; $N_O = 1.52$, $N_E = 1.55$. O, red brown; E, colorless. From the potash mine *Justus I near Volpriehausen*, and elsewhere in *Hannover, Germany*. RG *DII:86, NJMM:161(1966)*.

10.6.3.1 Yedlinite $Pb_6CrCl_6(OH,O)_8$

Named in 1974 for Neal Yedlin (1908–1977), American amateur mineralogist and micromounter, who first observed the species. HEX-R $R\bar{3}$. $a = 12.868$, $c = 9.821$, $Z = 3$, $D = 5.80$. Structure: *AM 59:1160(1974)*. $27\text{-}269$: 6.44_3 4.51_7 3.88_3 2.95_{10} 2.62_7 2.47_3 2.30_2 1.79_2. Prismatic crystals with $\{11\bar{2}0\}$, $\{1\bar{1}01\}$, $\{0001\}$, and other forms. Red-violet, white streak, vitreous luster. Cleavage $\{11\bar{2}0\}$, distinct; somewhat sectile. $H = 2\frac{1}{2}$. $G = 5.85$. Uniaxial $(-)$; $N_O = 2.125$, $N_E = 2.059$. O, pale cobalt blue; E, lavender. *Mammoth mine, Tiger, AZ*, associated with diaboleite, wulfenite, dioptase, phosgenite, and wherryite. RG *AM 59:1157(1974)*.

10.6.4.1 Chloroxiphite $Pb_3CuCl_2(OH)_2O_2$

Named in 1923 from the Greek for *green* and *blade* or *straight sword*, in allusion to the color and habit of the crystals. MON $P2_1/m$. $a = 10.458$, $b = 5.750$, $c = 6.693$, $\beta = 97.78°$, $Z = 2$, $D = 6.84$. Structure: *MM 41:357(1977)*. $33\text{-}467$: 10.4_7 5.96_3 3.89_5 3.85_5 2.91_4 2.87_{10} 2.83_8 2.69_4. Bladed crystals, rarely terminated, elongated [010] and flattened $\{\bar{1}01\}$, in subparallel groupings. Dull olive-green, similar to epidote; pale greenish-yellow streak; resinous to adamantine luster. Cleavage $\{\bar{1}01\}$, perfect; $\{100\}$, distinct; very brittle. $H = 2\frac{1}{2}$. $G = 6.76\text{-}6.93$. Soluble in HNO_3. In CT breaks into cleavage fragments, evolves water followed by vaporization of $PbCl_2$. The residue melts to a brown liquid, which on cooling becomes a bright green glass. Biaxial $(-)$; $Z = b$; X perpendicular to $\{\bar{1}01\}$; $N_x = 2.16$, $N_y = 2.24$, $N_z = 2.25$; $r > v$, medium to strong. X,Y, brown; Z, green. *Higher Pitts, Mendip Hills, Somerset, England*, embedded in mendipite. RG *DII:84, MM 43:901(1980)*.

10.6.5.1 Zirklerite (Fe,Mg)$_9$Al$_4$Cl$_{18}$(OH)$_{12}$ · 14H$_2$O (?)

Named in 1928 after Zirkler, director of the Aschersleben potash works, Germany. An inadequately characterized mineral. HEX-R Space group unknown. Light gray. Rhombohedral cleavage. H = $3\frac{1}{2}$. G = 2.6. Soluble in water with separation of iron and aluminum hydroxides. Uniaxial (+); N ≈ 1.552. *Adolfsglück mine, Hope, Hannover, Germany*, as the chief constituent of a rock containing halite, clay, anhydrite, quartz, and minor rinneite. RG *DII:87*.

10.6.6.1 Boleite Cu$_{24}$Pb$_{26}$Ag$_9$Cl$_{62}$(OH)$_{48}$

Named in 1891 for the locality. ISO $Pm3m$. $a = 15.29$, Z = 1. Structure: *SR 39A:195(1973)*. 27-1206: 5.09$_3$ 4.41$_{10}$ 3.82$_8$ 3.51$_4$ 2.70$_6$ 2.55$_4$ 2.33$_4$ 1.99$_4$. Crystals are cubes, cubo-octahedrons, or rarely dodecahedral. Deep prussian blue; blue streak with a greenish tint; vitreous luster, pearly on cleavages. Cleavage {001}, perfect; {101}, good. H = $3-3\frac{1}{2}$. G = 5.05. Soluble in HNO$_3$; in CT, loses water and fuses. Isotropic; N = 2.05. *Mammoth–St. Anthony mine, Tiger, AZ*; near *Phillipsburg, MT*, in the foundations of an old silver mill, the result of chemical spills in the mill; *Boleo, near Santa Rosalia, BCS, Mexico*, with and on anglesite, paratacamite, pseudoboleite, and cumengéite in white clay; at the *South mine, Broken Hill, NSW, Australia*; *Challocollo, Cerro Gordo, Tarapacá, Chile*. Used as an ore of silver. RG *DII:78*, *JSSC 6:86(1973)*, *MR 5:280(1974)*.

10.6.7.1 Cumengéite Cu$_{20}$Pb$_{21}$Cl$_{42}$(OH)$_{40}$

Named in 1893 for Edouard Cumengé (1828–1902), French mining engineer. Synonym: cumengite. TET $I4/mmm$. $a = 15.065$, $c = 24.436$, Z = 2, D = 4.66. Structure: *MM 50:157(1986)*. 27-174: 12.3$_7$ 4.83$_9$ 3.98$_7$ 3.74$_7$ 3.22$_7$ 2.66$_8$ 2.37$_{10}$ 2.01$_8$. Octahedral or cubo-octahedral crystals, with nearly equal development of forms; also found as parallel overgrowths on crystals of boleite, as in the case of pseudoboleite. Indigo blue, sky-blue streak, vitreous luster (not pearly on cleavages). Cleavage {101}, good; {110}, distinct; {001}, poor. H = $2\frac{1}{2}$. G = 4.67. Soluble in HNO$_3$. Uniaxial (−); N$_O$ = 2.026, N$_E$ = 1.965. O, dark blue; E, pure blue. From *Boleo, near Santa Rosalia, BCS, Mexico*. *Gunver Head, Padstow, Cornwall, England*; also at *Newport Beach, Falmouth*. *Christian Levin coal mine, Essen, Ruhr, Germany*. Also found in the *Laurium slags, Greece*. RG *DII:79*, *NJMM:116(1970)*, *MR 5:280(1974)*.

10.6.8.1 Pseudoboleite Pb$_{31}$Cu$_{24}$Cl$_{62}$(OH)$_{48}$

Named in 1895 from the Greek *pseudo*, false, and *boleite*, a mineral it resembles. TET $I4/mmm$. $a = 15.24$, $c = 30.74$, Z = 2, D = 5.07. *NJMM:113(1992)*. 22-470: 4.43$_{10}$ 3.83$_{10}$ 3.51$_4$ 2.71$_{10}$ 2.55$_6$ 2.39$_6$ 2.33$_7$ 1.99$_6$. Observed only in epitaxial overgrowths on boleite, with the {001} faces of the two minerals in common. Euhedral crystals, due to reentrant angles along the cube edges of boleite, resemble twins. Indigo blue; vitreous luster, pearly on cleavages. Cleavage {001}, perfect; {101}, nearly so. H = $2\frac{1}{2}$. G = 4.85. Uniaxial (−); N$_O$ = 2.03, N$_E$ = 2.00. *Mammoth–St. Anthony mine, Tiger, Pinal*

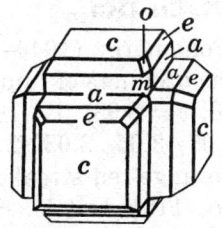

PSEUDOBOLEITE: $Pb_{31}Cu_{24}Cl_{62}(OH)_{48}$
Forms: c{001}, a{010}, m{110}, e{011}, o{112}

Co., and Daisy shaft, Banner mine, Pima Co., AZ, with connellite, gerhardtite, and atacamite. *Boleo, near Santa Rosalia, BCS, Mexico*(!), with boleite and cumengéite. RG *DII:80, MR 5:280(1974)*.

10.6.9.1 Bideauxite $Pb_2AgCl_3(F,OH)_2$

Named in 1970 for Richard A. Bideaux (b.1935), American mineralogist, who first noted the mineral. ISO $Fd3m$. $a = 14.132$, $Z = 16$, $D = 6.256$. 25-461: 4.26_7 4.08_7 3.53_9 3.24_7 2.72_{10} 2.50_9 1.98_5 1.84_6. Forms small crystals and masses. Colorless, turning pale lavender on exposure to light; adamantine luster. No cleavage, conchoidal fracture, brittle to sectile. $H = 3$. $G = 6.27$. Isotropic; $N = 2.18$–2.19. *Mammoth–St. Anthony mine, Tiger, AZ*, where it envelops and replaces boleite, associated with leadhillite, anglesite, matlockite, and cerussite. RG *MM 37:637(1970), AM 57:1003(1972)*.

10.6.10.1 Chlormagaluminite $(Mg,Fe^{2+})_4Al_2(OH)_{12}(Cl_2,CO_3) \cdot 2H_2O$

Named in 1982 for the composition. Originally called "chlor-manasseite." HEX $P6/mcm$, $P6cm$, or $P\bar{6}c2$. $a = 5.29$, $c = 15.46$, $Z = 1$, $D = 2.06$. *AM 68:849(1983)*. 38-446: 7.67_{10} 3.86_8 2.60_8 2.34_9 2.17_9 1.839_{10} 1.555_8 1.526_9. Aggregates of hexagonal dipyramids. Colorless to yellow-brown. Cleavage {0001}, perfect. $G = 2.0$. Uniaxial (+); $N_O = 1.540$, $N_E = 1.560$. Found in skarns from the *Kapaev explosion pipe, Angara R., southern Siberian Platform, Russia*, associated with magnetite and chlorite. RG *ZVMO 111:121(1982)*.

10.6.11.1 Kelyanite $Hg_{16}^{1+}Hg_{20}^{2+}Sb_3(Cl,Br)_9O_{28}$

Named in 1982 for the locality. MON Space group uncertain, probably $C2/m$. $a = 23.50$, $b = 13.62$, $c = 10.31$, $\beta = 97°$, $Z = 2$, $D = 8.51$. *AM 68:1248(1983)*. 35-714: 3.78_6 3.30_{10} 3.24_5 2.72_6 2.53_6 2.36_5 1.95_5 1.66_5. Irregular grains. Reddish brown. No cleavage. $G = 8.57$. In RL, grayish white, with red internal reflections, weakly bireflectant, anisotropic. $R = 20.5$–19.9 (460 nm), 18.9–18.1 (546 nm), 17.8–17.1 (620 nm). From the oxidation zone of the stibnite–cinnabar ores of the *Kelyan deposit, Buryatia, Ural Mts., Russia*, associated with calomel, eglestonite, Sb oxides, Hg, and shakovite. RG *ZVMO 111:330(1982)*.

10.6.12.1 Ponomarevite $K_4Cu_4OCl_{10}$

Named in 1988 for Vasili V. Ponomarev (1940–1976), Russian chemist and vulcanologist, investigator of the products of fumarolic activity at Tolbachik. MON $C2/c$. $a = 14.73$, $b = 14.86$, $c = 8.93$, $\beta = 104.9°$, $Z = 4$, $D = 2.72$. *AM 75:709(1990)*. *45-1417*: 7.31_8 6.07_7 3.65_6 3.05_6 2.80_{10} 2.79_8 2.47_7 2.31_6. Fine grained to glassy crusts. Red, orange-red streak, vitreous to resinous luster. Cleavage {001}, imperfect; also {110}; brittle. $H = 2\frac{1}{2}$. $G = 2.78$. Soluble in cold water; unstable in air, altering slowly to mitscherlichite. Biaxial $(-)$; $N_x = 1.686$, $N_y = 1.718$, $N_z = 1.720$. Found as a product of fumarolic activity of the Tolbachik main fracture eruption, Kamchatka, Siberia, Russia, with halite, sylvite, tenorite, tolbachite, dolerophanite, and piypite. RG *DANS 300:1197(1988)*.

Class 11

Halide Complexes; Alumino-Fluorides

11.1.1.1   Neighborite   NaMgF$_3$
Named in 1961 for Frank Neighbor (b.1906), American petroleum geologist, Sun Oil Co., who observed and preserved the first specimens. ORTH *Pnma*. $a = 5.363$, $b = 7.676$, $c = 5.503$, $Z = 4$, $D = 3.06$. *AM 46:379(1961)*. 13-303: 3.83_4 2.71_5 2.30_3 2.23_2 2.20_1 1.92_{10} 1.58_1 1.56_3. Octahedral crystals, showing polysynthetic twinning; also as rounded grains. Colorless to pink; also brown when in alkalic syenitic rocks; vitreous luster. Uneven fracture. $H = 4\frac{1}{2}$. $G = 3.03$. Insoluble in water, soluble in acids. Nearly isotropic; $N = 1.364$. Birefringence 0.003. *South Ouray, Uintah Co., UT*, from drill cores penetrating the Green River Formation, associated with burbankite, nahcolite, barytocalcite, garrelsite, wurtzite, quartz, and dolomite. Mt. St.-Hilaire, QUE, Canada. Gjerdingen, Nordmark, Norway. RG *AM 56:1519(1971)*, *DANS 210:666(1973)*.

11.1.2.1   Carnallite   KMgCl$_3$ · 6H$_2$O
Named in 1856 for Rudolph von Carnall (1804–1874), Prussian mining engineer. ORTH *Pnna*. $a = 16.154$, $b = 22.508$, $c = 9.575$, $Z = 12$, $D = 1.590$. *NBSM 25(8):50(1970)*. The structural framework of carnallite is made up of KCl$_6$ octahedra in a very open network with interstitial Mg(H$_2$O)$_6$ octahedra. There are two nonequivalent KCl$_6$ octahedra, both distorted, one more than the other. Two-thirds of these octahedra share faces (have three common corners), and the other one-third share corners with these and with each other. The large openings in this resulting framework accommodate the unusual Mg octahedrally coordinated with neutral water molecules. These isolated chargeless ions are not bonded to each other or to the rest of the structure, but occupy fixed positions in the framework. The cation and water coordination around Cl is octahedral. There are five different chlorine anions, and each is coordinated to two potassium cations and to four water molecules (transmitting the charge of magnesium cations). *AM 70:1309(1985)*.   24-869(syn): 4.69_4 3.75_5 3.60_7 3.56_6 3.32_{10} 3.04_7 2.93_{10} 2.35_5. Crystals thick tabular {001}, pseudohexagonal, pyramidal; usually massive, granular. Colorless to white, often reddish due to enclosed oriented scales of hematite; rarely yellow or blue; greasy luster, dull to shining. No cleavage, conchoidal fracture. $H = 2\frac{1}{2}$. $G = 1.602$. Soluble in water. Tastes bitter. Deliquescent. In the closed

tube, fuses easily. Biaxial (+); XYZ= cba; $N_x = 1.468$, $N_y = 1.47$, $N_z = 1.495$; $2V_{calc} = 66°$; r < v. Carnallite typically forms in marine evaporite salt beds, in part by reaction of preexisting saline minerals with liquor rich in potash. From the Permian salt basin of TX and NM, especially in the mines at Carlsbad, NM. *Stassfurt district, Germany.* Besenrode, upper Silesia, and Kalusz, Galicia, Poland. Barcelona and Lerida Provs., Spain. Ozinski, Saratov, Russia, with other potash salts. Ethiopia, on the plains of Danakil. Carnallite is an important source of potash for fertilizers and other purposes. RG *DII:92, MM 29:667(1951), NJMM:100(1973).*

11.1.3.1 Chlorocalcite $KCaCl_3$

Named in 1872 for the composition, which was originally thought to be chloride of calcium alone. Although chlorocalcite is highly deliquescent, rendering its recovery and study difficult, it has been observed in enough places that there is little doubt about its existence. ORTH Space group unknown. $a = 7.551$, $b = 10.42$, $c = 7.251$, $Z = 4$, $D = 2.155$. *NBSM 25(7):36(1969)*. 21-1170(syn): 5.23_3 3.70_3 3.14_4 3.12_3 3.06_2 2.61_{10} 2.38_2 2.11_3. Forms cubelike crystals, sometimes with octahedral- or dodecahedral-like truncations; also prismatic or tabular. White, sometimes stained violet. Cubelike cleavage, with one direction better than the other two. $H = 2\frac{1}{2}$–3. Biaxial (−); $N = 1.568$; very low birefringence; shows polysynthetic twinning. Strongly hygroscopic. Tastes bitter. *Mt. Vesuvius, Campania, Italy,* as a volcanic sublimate, associated with halite, sylvite, and hematite. Desdemona potash mine, Leinetal, Prussia, Germany. RG *DII:91.*

11.2.1.1 Pseudocotunnite K_2PbCl_4 (?)

Named in 1873 from the Greek *pseudo,* false, and *cotunnite,* a mineral it resembles in some respects. ORTH *Pnma.* $a = 11.80$, $b = 5.77$, $c = 9.82$, $Z = 4$, $D = 4.25$. Platy crystals, or acicular [001]; as warty aggregates of crusts. Colorless to white, yellow, or greenish yellow; dull luster. Biaxial; $N \approx 2.0$; strongly birefringent. Soluble in warm water, easily fusible. *Vesuvius, Italy,* in fumaroles, associated with cotunnite. RG *DII:96, ST:168.*

11.2.2.1 Avogadrite $(K,Cs)BF_4$

Named in 1926 for Amadeo Avogadro (1776–1856), Italian physicist, University of Turin. Barite group [see barite (28.3.1.1)]. ORTH *Pnma.* $a = 8.659$, $b = 5.480$, $c = 7.030$, $Z = 4$, $D = 2.507$–3.0, depending on Cs content. Structure: *AC(B) 25:2161(1969).* 16-378(syn): 3.51_5 3.41_{10} 3.26_8 3.06_8 2.80_4 2.23_3 2.09_4 2.07_6. Minute crystals, tabular {001} or elongated [010] or [100]; as dense crusts. Colorless to white. $G = 2.6$. Biaxial (−); $N_x = 1.324$, $N_y = 1.3245$, $N_z = 1.325$; 2V large, 90°(calc). Soluble in water, easily fusible. Tastes bitter. *Vesuvius, Italy,* in fumaroles. RG *DII:97, MA 11:535(1952).*

11.2.3.1 Ferruccite   NaBF$_4$

Named in 1933 for Ferruccio Zambonini (1880–1932), Italian mineralogist. ORTH *Cmcm*. $a = 6.837$, $b = 6.262$, $c = 6.792$, $Z = 4$, $D = 2.507$. Structure: *AC(B) 24:1703(1968)*. *11-671*(syn): 3.82_2 3.41_9 3.39_{10} 2.84_3 2.31_4 2.14_2 2.03_2 2.00_2. Minute crystals, tabular {001} or {010}. Colorless to white. Cleavage {100}, {010}, and {001}. $H = 3$. $G = 2.49$. Biaxial (+); XYZ= *cba*; $N_x = 1.301$, $N_y = 1.301$, $N_z = 1.307$; $2V = 11°$. Soluble in water, giving a bitter, acid taste; easily fusible. *Vesuvius, Italy,* in fumaroles. RG *DII:98*.

11.2.4.1 Barberiite   NH$_4$BF$_4$

Named in 1994 for Franco Barberi, Italian volcanologist, University of Pisa. ORTH *Pnma*. $a = 9.062$, $b = 5.673$, $c = 7.267$, $Z = 4$, $D = 1.90$. *AM 79:381(1994)*. *15-745*(syn): 4.53_6 4.48_{10} 3.84_5 3.64_5 3.54_9 3.19_9 2.90_6 2.54_5. Minute crystals, tabular to platy on {001}, forming globular aggregates. Colorless, transparent to translucent, white streak, vitreous luster. Cleavage {100}, perfect; {010}, {001}, good. VHN$_{25}$ = 14.2 mg/mm^2. $G = 1.89$. Very unstable in air. Biaxial; $N_{\text{mean calc}} = 1.308$; $2V = \pm 90°$. Found in fumaroles at the *Fossa crater on Vulcano, Aeolian Islands, Italy,* where it formed at a temperature of $\leq 600°$, associated with sulfur, salammoniac, sassolite, malladrite, realgar, and various rare sulfides. RG

11.3.1.1 Douglasite   K$_2$Fe^{2+}Cl$_4$ · 2H$_2$O

Named in 1879 for the locality at Douglashall, Germany. MON Space group unknown. $\beta = 104.76°$. *41-1358:* 5.07_5 4.16_6 4.03_6 3.67_8 3.08_{10} 2.97_5 2.37_4 1.822_4. Coarsely granular. Light green, altering to brownish red; vitreous luster. Uniaxial (+); $N_O = 1.488$, $N_E = 1.500$. *Douglasshall, NW of Stassfurt, Saxony, Germany,* associated with carnallite, sylvite, and halite. RG *DII:100*.

11.3.2.1 Mitscherlichite   K$_2$CuCl$_4$ · 2H$_2$O

Named in 1925 for Eilhard Mitscherlich (1794–1863), German crystallographer and chemist, who first prepared the compound. He also was the discoverer of isomorphism, in 1819. TET $P4_2/mnm$. $a = 7.454$, $c = 7.909$, $Z = 2$, $D = 2.416$. *NBSM 25(9):34(1971)*. *23-478*(syn): 5.42_7 3.96_2 3.16_5 3.07_3 2.71_{10} 2.64_{10} 1.98_3 1.58_3. Crystals pyramidal {011} or short prismatic [001], twinning on {011}. Greenish blue, vitreous luster. $H = 2\frac{1}{2}$. $G = 2.42$. Uniaxial (–); $N_O = 1.638$, $N_E = 1.613$. O, sky blue; E, grass green. Soluble in water. *Vesuvius, Italy,* as stalactites on the floor of the crater. RG *DII:100*.

11.4.1.1 Erythrosiderite   K$_2$Fe^{3+}Cl$_5$ · H$_2$O

Named in 1872 for the Greek for *red* and *iron*. ORTH *Pnma*. $a = 13.585$, $b = 9.706$, $c = 7.018$, $Z = 4$, $D = 2.364$. *NBSM 25(14):27(1977)*. *29-1004*(syn): 5.68_4 5.57_4 4.88_1 2.99_2 $2,84_2$ 2.78_{10} 2.44_3 2.43_4. Tabular or pseudo-octahedral crystals. Ruby-red to brownish red, vitreous luster. Cleavage {210} and {011}, perfect. $G = 2.37$. Slightly deliquescent. Biaxial (+);

XYZ= abc; $N_x = 1.715$, $N_y = 1.75$, $N_z = 1.80$; $2V = 62°$; $r > v$, strong. *Vesuvius, Italy*, in fumaroles, associated with hematite. Stassfurt, Germany, as an efflorescence on rinneite. RG *DII:101*.

11.4.1.2 Kremersite $(NH_4,K)_2Fe^{3+}Cl_5 \cdot H_2O$

Named in 1853 for Peter Kremers (b.1827), German chemist. ORTH *Pnma*. $a = 13.657$, $b = 9.808$, $c = 7.028$, $Z = 4$, $D = 2.175$. *NBSM 25(14):8(1977)*. *28-734*(syn): 5.72_8 5.61_6 4.90_4 2.86_3 2.80_{10} 2.79_6 2.78_6 2.45_7. Pseudo-octahedral crystals with {011} and {210} in equal development. Ruby-red to brownish red, vitreous luster. Cleavage {210} and {011}, perfect. $G = 2.00$. Biaxial (+). Soluble in water, very deliquescent. In natural kremersite the NH_4 has replaced only a little over half of the K, and the name refers to compounds where $NH_4/K > 1$. Al can substitute for some of the Fe. *Vesuvius, Italy*, in fumaroles. RG *DII:101*.

11.5.1.1. Hieratite K_2SiF_6

Named in 1882 for the locality in the crater of Vulcano (Greek name, Hiera). ISO *Fm3m*. $a = 8.133$, $Z = 4$, $D = 2.719$. *NBSC 539(5):50(1955)*. *7-217*(syn): 4.70_{10} 2.88_7 2.35_7 2.03_5 1.66_1 1.56_1 1.44_2 1.29_1. Octahedral or cubo-octahedral crystals; as stalactitic concretions. Colorless to white or gray, vitreous luster. Cleavage {111}, perfect. $H = 2\frac{1}{2}$. $G = 2.668$. Isotropic; $N = 1.340$ (Na). Soluble in water; in CT sublimes without residue. *Vulcano, Lipari Is., Sicily, Italy*, in fumaroles; also on Vesuvius. RG *DII:103*.

11.5.1.2 Cryptohalite $(NH_4)_2SiF_6$

Named in 1873 from the Greek for *concealed* and *salt*, in allusion to its intimate mixture with other salts at the original locality. Dimorphous with bararite (11.5.2.2). ISO *Fm3m*. $a = 8.395$, $Z = 4$, $D = 2.000$. *NBSC 539(5):5(1955)*. *44-1484*(syn): 4.84_{10} 4.20_1 2.97_1 2.53_1 2.42_2 2.10_1 1.615_1 1.484_1. Massive or arborescent; mammillary. Colorless to white or gray, vitreous luster. Cleavage {111}, perfect. $H = 2\frac{1}{2}$. $G = 2.00$. Isotropic; $N = 1.369$ (Na). Soluble in water, in CT sublimes without residue. *Vesuvius, Italy*, in fumaroles. Kladno, Bohemia, Czech Republic, in burning coal-waste heaps. Barari, Jharia coal field, India, in crevices as a sublimate over a burning coal seam. RG *DII:104*.

11.5.2.1 Malladrite Na_2SiF_6

Named in 1926 for Alessandro Malladra (1868–1945), Italian volcanologist, director of the Vesuvius Observatory. HEX-R *P321*. $a = 8.866$, $c = 5.043$, $Z = 3$, $D = 2.729$. *NBSM 25(16):68(1979)*. Structure: *AC 17:1408(1964)*. *33-1280*(syn): 4.43_{10} 4.21_9 3.33_9 3.06_5 2.28_9 1.80_5 1.66_2 1.59_2. Crystals prismatic [001]; as crusts. Pale rose to colorless. $G = 2.755$. Uniaxial (−); $N_O = 1.313$, $N_E = 1.309$. Slightly soluble in water, more so in hot water. *Vesuvius, Italy*, in fumaroles. RG *DII:105*.

11.5.2.2 Bararite $(NH_4)_2SiF_6$

Named in 1951 for the locality. Dimorphous with cryptohalite (11.5.1.2). HEX-R $P\bar{3}m1$. $a = 5.77$, $c = 4.78$, $Z = 1$, $D = 2.144$. 44-1424: 4.99_{10} 4.77_2 3.45_3 2.39_1 2.21_4 1.753_1 1.724_1 1.479_1. Crystals tabular {0001}; also arborescent, mammillary. White to colorless, vitreous luster. Cleavage {0001}, perfect. $H = 2\frac{1}{2}$. $G = 2.15$. Uniaxial $(-)$; $N_O = 1.406$, $N_E = 1.391$ (Na). Soluble in water; in the closed tube sublimes without residue. Vesuvius, Italy, in fumaroles. *Barari, Jharia coalfield, Bengal, India*, as a crusted sublimate above a burning coal seam. RG *DII:106*, AM *37:361(1952)*.

11.5.3.1 Rinneite $K_3NaFe^{2+}Cl_6$

Named in 1909 for Freidrich Wilhelm Berthold Rinne (1863–1933), German crystallographer and petrographer. HEX-R $R\bar{3}c$. $a = 12.024$, $c = 13.852$, $Z = 6$, $D = 2.348$. Structure: *SR 49A:115(1982)*. *20-925*: 6.00_6 5.75_4 3.65_2 3.46_3 2.81_4 2.65_4 2.59_5 2.51_{10}. Massive, granular. Colorless when pure and fresh; usually rose, violet, or yellow, becoming brown on exposure; brilliant luster, sometimes silky. Cleavage {11$\bar{2}$0}, good; splintery, conchoidal fracture. $H = 3$. $G = 2.35$. Uniaxial $(+)$; $N_O = 1.589$, $N_E = 1.589$ (Na); birefringence 0.0008. Soluble in water, tastes astringent, in CT easily fusible. Formed as a secondary product in marine evaporite salt deposits. *Wolkramshausen, near Nordhausen, Saxony, Germany*, and at several localities in Hannover. RG *DII:107*.

11.5.4.1 Chlormanganokalite K_4MnCl_6

Named in 1906 for the composition. HEX-R $R\bar{3}c$. $a = 11.94$, $c = 14.81$, $Z = 6$, $D = 2.310$. *3.856*(syn): 5.90_5 4.90_2 3.55_5 3.45_2 2.82_3 2.69_8 2.55_{10} 2.50_2. Crystals rhombohedrons, sometimes arranged in parallel groups. Pale yellow to canary yellow, vitreous luster. Conchoidal fracture, brittle. $H = 2.5$. $G = 2.31$. Uniaxial $(+)$; $N \approx 1.59$; low birefringence. *Vesuvius, Italy*, with halite, sylvite, and hematite crystals, lining cavities in blocks of scoria. RG *DII:109*, ST:*163*.

11.5.5.1 Tachyhydrite $Mg_2CaCl_6 \cdot 12H_2O$

Named in 1856 from the Greek for *quick water*, in allusion to its ready deliquescence. HEX-R $R\bar{3}$. $a = 10.136$, $c = 17.318$, $Z = 3$, $D = 1.673$. Structure: *AC(B) 36:2736(1980)*. *34-788*: 5.77_4 5.07_1 3.81_1 3.10_1 2.88_{10} 2.61_3 1.983_2 1.443_3. Massive. Colorless to yellow, vitreous luster. Cleavage {10$\bar{1}$1}, perfect. $H = 2$. $G = 1.667$. Uniaxial $(-)$; $N_O = 1.520$, $N_E = 1.512$ (Na). Very deliquescent, tastes sharp and bitter. *Stassfurt* and elsewhere in *Saxony, Germany*, a rare constituent of marine evaporites. RG *DII:95*.

11.5.6.1 Gagarinite-(Y) $NaCaY(F,Cl)_6$

Named in 1961 for Yuri Alekseevich Gagarin (1934–1968), Russian cosmonaut, first person to travel in space. HEX-R $P\bar{3}$. $a = 5.99$, $c = 3.53$, $Z = 1$, $D = 4.02$. *DANS 149:672(1963)*. *15-69*: 5.24_3 3.00_5 2.92_5 2.28_2 2.09_9 1.96_2 1.72_{10} 1.13_3. Cryptocrystalline, massive. Creamy to yellowish or rosy, white streak, dull to vitreous luster. Cleavage {0001}, fair; brittle. $H = 4\frac{1}{2}$. $G = 4.21$. Slightly magnetic, similar to glaucophane. Uniaxial $(+)$; $N_O = 1.472$, $N_E = 1.492$; sometimes anomalously biaxial, with 2V to 20°. Soluble in acids, partially decomposed by H_2O. Found in alkaline rocks. Washington Pass, WA. Langesundfjord, Norway. *Kazakhstan*. RG *DANS 141:954(1961)*, *AM 47:805(1962)*.

11.5.7.1 Zajacite-(Ce) Na(Ce,REE,Ca)$_2$F$_6$

Named in 1995 for Ihor Stephan Zajac (b.1935), Canadian geologist, who led the exploration group that discovered the Strange Lake, Canada, deposit. HEX-R $P\bar{3}$. $a = 6.099$, $c = 11.064$, $Z = 3$, $D = 4.55$. PD: 5.29_7 3.04_{10} 2.35_3 2.15_7 1.757_8 1.361_3 1.152_4 0.919_4. Forms anhedral grains to 2 mm. Colorless to pale pink, white streak, vitreous luster, nonfluorescent. No cleavage, conchoidal fracture. $H = 3\frac{1}{2}$. $G = 4.44$. Uniaxial $(+)$; $N_E = 1.483$, $N_O = 1.503$ (Na). From the peralkaline granite intrusive complex at *Strange Lake, near Schefferville, QUE/NF border, Canada*, associated with bastnaesite-(Ce) and vlasovite. RG *CM 34:(1996)*.

11.6.1.1 Cryolite Na$_3$AlF$_6$

Named in 1799 from the Greek for *ice* and *stone*, in allusion to the icelike appearance of the mineral. MON $P2_1/n$. $a = 5.402$, $b = 5.596$, $c = 7.756$, $\beta = 90.28°$, $Z = 2$, $D = 2.963$. Structure: *CM 13:37(1975)*. *22-772*(syn): 4.54_6 4.44_4 3.89_7 2.80_3 2.75_{10} 2.34_3 1.94_{10} 1.57_4. Rare, equant, pseudocubic crystals; usually massive, coarsely granular; twinning common, by several different laws, often repeated or polysynthetic. Colorless to white; also brownish, reddish, brick-red, rarely black; white streak; vitreous to greasy luster. No cleavage; uneven fracture; parting on {100} and {110}, producing cuboidal forms; brittle. $H = 2\frac{1}{2}$. $G = 2.97$. Weakly thermoluminescent. Biaxial $(+)$; $Y = b$; $Z \wedge c = -44°$; $N_x = 1.3376$, $N_y = 1.3377$, $N_z = 1.3387$ (Na); $2V = 43°$; $r < v$, horizontal. Small clear fragments, when placed in water, become nearly invisible. Easily soluble in $AlCl_3$ solution, soluble in H_2SO_4 with evolution of HF (**poison!**). Fusible in small fragments in a candle flame. Alters by weathering to a series of secondary fluorides, including pachnolite, thomsenolite, gearksutite, prosopite, cryolithionite, weberite, jarlite, and others. St. Peter's Dome, Pikes Peak area, CO; Morefield mine, Amelia, VA; Francon quarry, Montreal, QUE, Canada, in small yellow crystals; also Mt. St.-Hilaire. Cryolite was the principal mineral in an extraordinary large pegmatite mass at *Ivigtut, Greenland*. This body, which was mined more or less continuously for over 100 years, also contained siderite, microcline, quartz, fluorite, chiolite, topaz, and minor sulfides and oxides, in addition to other rare and unusual fluorides, many formed by alteration. Miask, Ural Mts., Russia. Molten cryolite is used as a solvent for bauxite in the electrolysis of aluminum oxide to form aluminum. Since the depletion of the Ivigtut deposit, all cryolite used today is synthetic. RG *DII:110*, *AM 35:149(1950)*, *MA 78:877(1978)*, *MR 24:31(1993)*.

11.6.2.1 Elpasolite K$_2$NaAlF$_6$

Named in 1883 for the locality in El Paso Co., CO. ISO $Fm3m$. $a = 8.122$, $Z = 4$, $D = 3.002$. *NBSM 25(9):43(1971)*. Structure: *NJMM:481(1987)*. *22-1235*(syn): 4.69$_2$ 2.87$_{10}$ 2.35$_8$ 2.03$_9$ 1.66$_3$ 1.44$_4$ 1.28$_1$ 1.09$_1$. Massive, as indistinct cubo-octahedral crystals. Colorless to white, vitreous to greasy luster; also purple. No cleavage, uneven fracture. H = 2$\frac{1}{2}$. G = 2.99. Isotropic; N = 1.376. Morefield pegmatite, Amelia, VA, in purple masses, with chiolite, cryolite, and prosopite. *St. Peter's Dome, Pikes Peak region, El Paso Co., CO*, lining cavities in massive pachnolite in a cryolite pegmatite. RG *DII:114*.

11.6.3.1 Colquiriite LiCaAlF$_6$

Named in 1980 for the locality. HEX-R $P\bar{3}1c$. $a = 5.02$, $c = 9.67$, $Z = 2$, $D = 2.95$. *43-1481*: 3.96$_6$ 3.22$_{10}$ 2.22$_9$ 2.11$_3$ 1.976$_4$ 1.735$_8$ 1.551$_4$ 1.445$_4$. Anhedral grains to 1 cm. White. No cleavage, conchoidal fracture. H = 4. G = 2.94. Nonfluorescent. Uniaxial (−); N$_O$ = 1.388, N$_E$ = 1.385. *Colquiri cassiterite deposit, Bolivia*, intergrown with ralstonite and gearksutite. RG *TMPM 27:275(1980), AM 66:879(1981)*.

11.6.4.1 Cryolithionite Na$_3$Li$_3$Al$_2$F$_{12}$

Named in 1904 for the mineral cryolite, which it resembles, plus the additional lithium content. ISO $Ia3d$. $a = 12.125$, $Z = 8$, $D = 2.770$. *NBSM 25(9):23(1971)*. *22-416*: 4.28$_{10}$ 3.03$_6$ 2.71$_4$ 2.38$_4$ 2.21$_5$ 1.97$_6$ 1.92$_2$ 1.62$_3$. Dodecahedral crystals, massive, as irregular grains. Colorless to white, vitreous luster. Cleavage {011}, distinct; uneven fracture. H = 2$\frac{1}{2}$–3. G = 2.77. Isotropic; N = 1.3395 (Na). *Ivigtut, Greenland*, intimately associated with cryolite. Also with cryolite at Miask, Ural Mts., Russia. RG *DII:99, AM 56:18(1971), MR 24:33(1993)*.

11.6.5.1 Pachnolite NaCaAlF$_6$ · H$_2$O

Named in 1863 from the Greek for *frost* and *stone*, in allusion to its appearance. Dimorphous with thomsenolite (11.6.6.1). MON $F2/d$. $a = 12.117$, $b = 10.414$, $c = 15.680$, $\beta = 90.36°$, $Z = 16$, $D = 2.983$. Structure: *CM 21:561(1983)*. *5-356*: 3.95$_{10}$ 3.26$_1$ 3.02$_2$ 2.92$_1$ 2.79$_7$ 2.40$_1$ 2.16$_5$ 1.97$_9$. Prismatic crystals, bounded by {110} and {111}, always twinned [001]. White, vitreous luster. Cleavage {001}, indistinct; uneven fracture. H = 3. G = 2.98. Biaxial (+); X = b; Z $\wedge$ c = 69°; N$_x$ = 1.411, N$_y$ = 1.413, N$_z$ = 1.420; 2V = 76°; r < v, weak; strong horizontal dispersion. Soluble in H$_2$SO$_4$, easily fusible; in CT decrepitates, yields acid water, etches the glass, and leaves a white coating of silica. Morefield mine, Amelia, VA. St. Peter's Dome, El Paso Co., CO, with thomsenolite, elpasolite, and other fluorides. *Ivigtut, Greenland*, as an alteration product of cryolite. Gjerdingen, Norway. Hagendorf Süd, Germany. RG *DII:114, MR 24:34(1993)*.

11.6.6.1 Thomsenolite $NaCaAlF_6 \cdot H_2O$

Named in 1868 for Hans Peter Jorgen Julius Thomsen (1826–1909), Danish physical chemist and founder of the Greenland cryolite industry. Dimorphous with pachnolite (11.6.5.1). MON $P2_1/c$. $a = 5.583$, $b = 5.508$, $c = 16.127$, $\beta = 96.43°$, $Z = 4$, $D = 2.974$. *NBSM 25(8):132(1970)*. Structure: *AC 23:162(1967)*. *5-343:* 4.02_{10} 3.43_1 2.92_5 2.16_2 2.00_8 1.96_9 1.76_3 1.64_2. Crystals long prismatic {110} and {001}; occasionally, pseudocubic; also massive. White to colorless or pale pink, sometimes stained with iron oxides; vitreous luster, pearly on {001}. Cleavage {001}, perfect; {110}, distinct; uneven fracture; brittle. $H = 2$. $G = 2.98$. Biaxial $(-)$; $Z = b$; $X \wedge c = -52°$; $N_x = 1.407$, $N_y = 1.414$, $N_z = 1.415$; $2V = 50°$; $r < v$, weak. Easily soluble in H_2SO_4; in CT decrepitates, yields acid water, and etches the glass. St. Peter's Dome, Pikes Peak region, El Paso Co., CO. In rhyolite at the Spider claim, Juab Co., UT, associated wtih drusy tridymite and sodium boltwoodite. *Ivigtut, Greenland*, with pachnolite, ralstonite, and fluorite. Miask, Ural Mts., Russia. RG *DII:116, MR 24:34(1993)*.

11.6.7.1 Carlhintzeite $Ca_2AlF_7 \cdot H_2O$

Named in 1979 for Carl Hintze (1851–1916), German mineralogist, professor of mineralogy at the University of Breslau and compiler of the *Handbuch der Mineralogie*. TRIC (pseudo-MON $C\bar{1}$ or $C1$. $a = 9.48$, $b = 6.98$, $c = 9.30$, $\alpha = 91.13°$, $\beta = 104.85°$, $\gamma = 90.00°$, $Z = 4$, $D = 2.89$. *CM 17:103(1979)*. *33-249:* 4.56_7 3.69_6 3.48_{10} 2.85_4 2.28_3 2.24_3 2.08_3 1.46_4. Crystals prismatic [010], always twinned about [010], with forms {100}, {001}, and {110}; sometimes in radiating druses resembling pectolite. White to colorless, vitreous luster. $G = 2.86$. Biaxial $(+)$; $X = b$; $Z \wedge c = 10°$; $N_x = 1.411$, $N_y = 1.416$, $N_z = 1.422$; $2V = 77°$. *Hagendorf, Germany*, in pegmatite, associated with rockbridgeite, pyrite, strengite, and apatite, formed from the hydrothermal alteration of primary triphylite. RG *AM 65:205(1980)*.

11.6.8.1 Gearksutite $CaAl(F,OH)_5 \cdot H_2O$

Named in 1868 from the Greek *Ge*, Earth, because it is usually earthy, and *arksutite*, from the locality near Arksut fjord. Pseudo-TET (possibly MON) Space group unknown. *5-283:* 4.55_{10} 3.34_7 3.15_8 2.28_8 2.15_6 2.07_4 1.93_8 1.74_6. Massive or earthy, chalky, consisting of minute needles. White, dull luster. $H = 2$. $G = 2.77$. Biaxial $(-)$; $X = b$; $Z \wedge$ elongation; $N_x = 1.448$, $N_y = 1.454$, $N_z = 1.456$; $2V_{calc} = 60°$. In CO at St. Peter's Dome, Pikes Peak area, in a cryolite pegmatite, and Wagon Wheel Gap, in wall-rock alteration of a fluorite–barite vein in rhyolite. Grand Reef mine, Graham Co., AZ. Sierra Blanca Peaks, Hudspeth Co., TX. *Ivigtut, Greenland*, as an alteration product of cryolite. Miask, Ural Mts., Russia. RG *DII:119, AM 34:383(1949)*.

11.6.9.1 Prosopite $CaAl_2(F,OH)_8$

Named in 1853 from the Greek for *mask*, in allusion to the deceptive character of the mineral. MON $C2/c$. $a = 6.705$, $b = 11.158$, $c = 7.364$, $\beta = 95°$, $Z = 4$,

$D = 2.898$. *NJMM:329(1986)*. 5-307: 4.35_{10} 3.24_4 3.07_5 2.31_4 2.13_6 1.91_5 1.84_6 1.81_6. Crytals tabular {010}; also massive, granular to powdery. Colorless to white or grayish, also lavender; vitreous to greasy luster. Cleavage {111}, perfect; uneven to conchoidal fracture. $H = 4\frac{1}{2}$. $G = 2.89$. Biaxial (+); $Y = b$; $Z \wedge c = -35°$; $N_x = 1.501$, $N_y = 1.503$, $N_z = 1.510$ (Na); $2V = 63°$; $r > v$, strong. Soluble in H_2SO_4, infusible. Morefield mine, Amelia, VA, in fine-grained lavender to gray masses, associated wtih topaz, fluorite, albite, and kaolinite. St. Peter's Dome, El Paso Co., CO. Ivigtut, Greenland. *Altenberg, Sachsen, Germany*, in greisen, partly altered to kaolin and fluorite. In the tin veins of Schlaggenwald, Bohemia, Czech Republic. Mt. Bischoff, TAS, Australia. RG *DII:121, AM 34:383(1949), DANS 190:665(1970)*.

11.6.10.1 Jarlite NaSr$_3$Al$_3$(F,OH)$_{16}$

Named in 1933 for Carl Frederick Jarl (1872–1951), Danish chemist, owner of the Danish cryolite industry, 1912–1943. MON $C2/m$. $a = 15.942$, $b = 10.821$, $c = 7.241$, $\beta = 101.82°$, $Z = 4$, $D = 4.06$. Structure: *CM 21:553(1983)*. 5-594: 3.63_4 3.45_4 3.19_9 3.11_6 2.98_{10} 2.15_7 2.00_4 1.82_6. Minute crystals, tabular {100}; often in sheaflike aggregates; also massive. Colorless to white; gray; greenish gray, vitreous luster. $H = 4-4\frac{1}{2}$. $G = 3.78-3.93$. Biaxial (±); $Y = b$; $X \wedge c = -6°$; $Z \wedge c = 84°$; $2V \approx 90°$. Some Ca, Mg, and Ba can substitute for Sr. Soluble in AlCl$_3$ solution; easily fusible, with effervescence. *Ivigtut, Greenland*, with cryolite, thomsenolite, and chiolite. RG *DII:118, AM 34:383(1949), NJMM:543(1985), MR 24:35(1993)*.

11.6.10.2 Calcjarlite Na(Ca,Sr)$_3$Al$_3$(F,OH)$_{16}$

Named in 1973 for the composition, calcium-dominant jarlite. MON $C2/m$. $a = 16.19$, $b = 9.868$, $c = 7.157$, $\beta = 99.2°$, $Z = 4$, $D = 3.32$. 29-1195: 3.51_5 3.44_3 3.16_6 3.04_7 2.96_{10} 2.68_8 2.23_4 2.15_4. Massive. White, vitreous luster. No cleavage, uneven fracture. $H = 4$. $G = 3.51$. Biaxial (+); $Z \wedge c = 10-15°$; $N_x = 1.425$, $N_y = 1.428$, $N_z = 1.432$; $2V = 72°$. *Yenisei region, Siberia, Russia*, with fluorite and usovite. RG *ZVMO 99:458(1970), AM 59:873(1974)*.

11.6.11.1 Chiolite Na$_5$Al$_3$F$_{14}$

Named in 1846 from the Greek for *snow*, in allusion to its appearance. TET $P4/mnc$. $a = 7.014$, $c = 10.400$, $Z = 2$, $D = 2.998$. *NBSM 25(16):63(1979)*. 30-1144(syn): 5.81_2 5.20_3 2.91_{10} 2.33_5 2.00_3 1.79_3 1.75_2 1.55_3. Crystals rare, dipyramidal {011}; usually massive, granular; twinning on {011}, sometimes distorted into prismatic shapes. Colorless to white; vitreous luster, pearly on cleavage. Cleavage {001}, perfect; {011}, distinct. $H = 3\frac{1}{2}-4$. $G = 3.00$. Soluble in H_2SO_4 and in AlCl$_3$ solution, easily fusible. Uniaxial (−); $N_O = 1.343$, $N_E = 1.350$. In crystals at the Morefield mine, Amelia, VA, associated with elpasolite, cryolite, and prosopite. Ivigtut, Greenland, in small masses. *Miask, Ilmen Mts., Ural Mts., Russia*, in a cryolite pegmatite, with topaz, phenacite, fluorite, and other fluorides. RG *DII:123, BGSD 26:95(1977)*.

11.6.12.1 Ralstonite $Na_xMg_xAl_{2-x}(F,OH)_6 \cdot H_2O$

Named in 1871 for J. Grier Ralston of Norristown, PA (1815–1880), minister, educator, and amateur scientist, who first observed the mineral. ISO $Fd3m$. $a = 9.91$, $Z = 8$, $D = 2.78$. Structure: $NJMM:97(1984)$. 45-1331: 5.74_{10} 2.99_6 2.86_5 2.28_1 2.03_1 1.903_2 1.755_3 1.497_1. Crystallizes as octahedra; more rarely as cubes, or the two in combination. Colorless to white, vitreous luster. Cleavage $\{111\}$, imperfect; uneven fracture. $H = 4\frac{1}{2}$. $G = 2.50$–2.68. Isotropic; $N = 1.37$–1.43. Decomposed by H_2SO_4 with evolution of HF (**poison!**); in closed tube whitens, yields acid water, then a copious white sublimate that etches the tube. The composition varies rather widely, even in a single crystal, from near $NaMgAl(F,OH)_6 \cdot H_2O$ to $Na_{0.5}Mg_{0.5}Al_{1.5}(F,OH)_6 \cdot H_2O$ and F/OH from 2:1 to 11:1. The hypothetical Al-rich end member is basic aluminum fluoride, $Al_2(F,OH)_6 \cdot H_2O$, but natural material with this composition has not been found. St Peter's Dome, El Paso Co., CO; *Ivigtut, Greenland*; also in Kazakhstan, and in Kamchatka, Siberia, Russia. RG *DII:126, AM 50:1851(1965), MM 54:599(1990)*.

11.6.13.1 Weberite Na_2MgAlF_7

Named in 1938 for Theobald C. F. Weber (1823–1886), one of the founders of the cryolite industry in Denmark. ORTH *Imma*. $a = 7.060$, $b = 10.000$, $c = 7.303$, $Z = 4$, $D = 2.948$. Structure: $TMPM$ $25:57(1978)$. 5-733: 5.90_5 5.06_6 2.96_9 2.90_9 2.30_5 1.78_{10} 1.55_5 0.882_6. Irregular grains and masses. Light gray, white streak, vitreous luster. Cleavage $\{101\}$, poor; $\{010\}$, indistinct; uneven fracture. $H = 3\frac{1}{2}$. $G = 2.96$. Biaxial $(+)$; XYZ= abc; $N_x = 1.346$, $N_y = 1.348$, $N_z = 1.350$; $2V = 83°$. Soluble in $AlCl_3$ solution; in closed tube gives acid vapors. St. Peter's Dome, El Paso Co., CO. *Ivigtut, Greenland*, with cryolite and other fluorides. RG *DII:127, AM 34:383(1949)*.

11.6.14.1 Usovite $Ba_2CaMgAl_2F_{14}$

Named in 1967 for Mikhail Antonovich Usov (1883–1939), Russian geologist. MON $C2/c$ or Cc. $a = 13.565$, $b = 5.200$, $c = 14.557$, $\beta = 91.35°$, $Z = 4$, $D = 4.26$. 19-1391: 3.41_{10} 2.04_7 1.205_9 1.132_8 1.093_8 1.025_8 1.012_8 1.005_8. Irregular grains, or indistinct elongated platy forms. Brown, white streak, vitreous to greasy luster. Cleavage in one direction, perfect; irregular fracture. $H = 3\frac{1}{2}$. $G = 4.18$. Biaxial $(+)$; Y parallel to cleavage; $N_x = 1.441$, $N_y = 1.442$, $N_z = 1.444$; $2V = 70°$; absorption $X = Y > Z$. X,Y, brownish yellow; Z, pale yellow. *Upper Noiby River area, Yenisei region, Siberia, Russia*, in a fluorite vein. RG *AM 52:1582(1967), 60:739(1975)*.

11.6.15.1 Yaroslavite $Ca_3Al_2F_{10}(OH)_2 \cdot H_2O$

Named in 1966 for the locality. ORTH Space group unknown. $a = 8.76$, $b = 5.54$, $c = 4.52$, $Z = 4$, $D = 3.15$. 18-272: 4.51_6 3.66_7 3.45_{10} 2.83_5 2.26_5 2.23_8 1.839_6 1.47_6. Radial-fibrous or spherical growths. White, vitreous luster. Pinacoidal cleavage, irregular fracture. $H = 4$. $G = 3.09$. Slightly fluorescent pale violet in SWUV. Biaxial $(+)$; $N_x = 1.413$, $N_z = 1.423$; $2V = 74°$; negative elongation. *Yaroslav, Siberia, Russia*, in cavities in sellaite in banded fluorite–sellaite–tourmaline ores. RG *ZVMO 95:39(1966), AM 51:1546(1966)*.

11.6.16.1 Tikhonenkovite SrAl(OH)F$_4$ · H$_2$O

Named in 1964 for Igor Petrovich Tikhonenkov (1927–1961), Russian student of alkaline rocks and minerals. Dimorphous with acuminite (11.6.17.1). MON Space group unknown. $a = 8.73$, $b = 10.62$, $c = 5.02$, $\beta = 102.43°$, $Z = 4$, $D = 3.30$. *17-501:* 4.89_{10} 4.44_5 3.64_9 3.33_5 3.27_8 3.14_5 2.30_6 2.10_7. Prismatic crystals. Colorless to slightly pink, vitreous luster. Cleavage {001}, perfect; conchoidal to uneven fracture. $H = 3\frac{1}{2}$. $G = 3.26$. Biaxial $(-)$; $N_x = 1.452$, $N_y = 1.456$, $N_z = 1.458$; $2V = 70°$. *Karasug, Tannu-Ola range, Tuva, Russia*, in fissures and druses in iron ores, with gearksutite, fluorite, celestite, strontianite, and barite. RG *DANS 156:104,345(1964), 174:193(1967); AM 49:1774(1964)*.

11.6.17.1 Acuminite SrAlF$_4$(OH) · H$_2$O

Named in 1987 from the Latin for *spearhead*, in allusion to the crystal shape. Dimorphous wtih tikhonenkovite (11.6.16.1). MON $C2/c$. $a = 13.223$, $b = 5.175$, $c = 14.251$, $\beta = 111.6°$, $Z = 8$, $D = 3.305$. Structure: *ZK 194:221(1991)*. PD: 4.77_{10} 4.71_{10} 3.50_{10} 3.35_{10} 3.31_8 3.29_8 2.07_9 2.03_8. Aggregates of crystals shaped like spear points and about 1 mm long, with {110} and {111}; twins on {100}. Colorless, vitreous luster. Cleavage {001}, perfect. $H = 3\frac{1}{2}$. $G = 3.295$. Nonfluorescent. Biaxial $(+)$; $X = b$; $Y \wedge c = 15°$; $N_x = 1.451$, $N_y = 1.453$, $N_z = 1.463$; $2V = 46\text{–}57°$; $r > v$, very strong. *Ivigtut, Greenland*, with celestite, fluorite, jarlite, thomsenolite, and other fluorides. RG *NJMM 1987:502, AM 73:1492(1988), MR 24:36(1993)*.

11.6.18.1 Artroeite PbAlF$_3$(OH)$_2$

Named in 1995 for Arthur (Art) Roe (1912–1993), U.S. chemist, professor, University of North Carolina, micromounter, and mineralogist. TRIC $P\bar{1}$. $a = 6.259$, $b = 6.791$, $c = 5.053$, $\alpha = 90.92°$, $\beta = 107.45°$, $\gamma = 104.45°$, $Z = 2$, $D = 5.47$. *AM 80:179(1995)*. PD: 4.42_0 4.05_4 3.27_4 3.22_4 2.60_7 2.19_7 2.03_5 2.015_4. Minute crystals, bladed, flattened on {10$\bar{1}$}, usually twinned on [010]. Colorless, white streak, vitreous luster. Cleavage {100}, perfect; {010}, good; conchoidal fracture; brittle. $H = 2\frac{1}{2}$. $G = 5.36$. Biaxial $(-)$; $N_x = 1.629$, $N_y = 1.682$, $N_z = 1.691$; $r > v$, strong. Found in a single 15 mm × 5 mm vug surrounded by galena and fluorite, at the *Grand Reef mine*, near *Klondyke, Graham Co., AZ*. RG

11.6.19.1 Aravaipaite Pb$_3$AlF$_9$ · H$_2$O

Named in 1989 for the locality in the Aravaipa mining district, AZ. TRIC $P1$ or $P\bar{1}$. $a = 5.842$, $b = 25.20$, $c = 5.652$, $\alpha = 93.83°$, $\beta = 90.13°$, $\gamma = 85.28°$, $Z = 4$, $D = 6.37$. *45-1458:* 12.5_8 3.65_7 3.50_6 3.33_6 3.13_{10} 2.92_4 2.82_4 2.03_4. Crystals platy {010}, polysynthetically twinned. Colorless, white streak, vitreous to pearly luster. Cleavage {010}, perfect; {100}, {001}, {101}, {10$\bar{1}$}, good; irregular fracture; flexible. H = 2. No fluorescence. Biaxial (−); $X \wedge \phi = 67°$; $Z \wedge \theta = 76°$; $N_x = 1.678$, $N_y = 1.690$, $N_z = 1.694$; $2V = 70°$; $r < v$, strong. Formed by the reaction of supergene solutions on galena and fluorite. *Grand Reef mine, Laurel Canyon, Graham Co., AZ*, in a vug surrounded by layers of quartz, fluorite, and galena. RG *AM 74:927(1989)*.

11.6.20.1 Bøgvadite Na$_2$SrBa$_2$Al$_4$F$_{20}$

Named in 1988 for Richard Bøgvad (d.1952), Danish, former chief geologist for the cryolite mining company. ORTH *Pnmn* or *Pn2n*. $a = 7.110$, $b = 19.907$, $c = 5.347$, $Z = 2$, $D = 3.898$. *BGSD 37:21(1988)*. *45-1419:* 9.97$_4$ 6.69$_4$ 3.24$_{10}$ 3.19$_5$ 2.92$_5$ 2.67$_4$ 2.38$_4$ 2.12$_5$. Blocky crystals, showing {110}, {010}, and {012}. Colorless, white streak, vitreous luster. Uneven fracture. VHN$_{25}$ = 300. G = 3.85. Nonfluorescent. Biaxial (−); XYZ= *cab*; OAP = {010}; $N_x = 1.433$, $N_y = 1.436$, $N_z = 1.439$ (Na); $2V = 87°$. *Ivigtut, Greenland*, with jarlite, ralstonite, barite, muscovite, and quartz. RG *AM 76:1728(1991), MR 24:36(1993)*.

11.6.21.1 Karasugite SrCaAl[F,(OH)]$_7$

Named in 1994 for the locality. MON $P2_1/c$. $a = 8.215$, $b = 11.989$, $c = 6.076$, $\beta = 96.22°$, $Z = 2$, $D = 3.206$. *NJMM:209(1994)*. PD: 6.76$_7$ 4.25$_{10}$ 3.64$_8$ 3.15$_7$ 3.06$_8$ 3.03$_7$ 2.84$_7$ 2.13$_8$. Rosettes and fan-shaped aggregates of microcrystals, bladed on {100}, elongated [001], twinned with (100) as both twin and composition plane. Colorless, transparent, white streak, vitreous luster. Cleavage {100}, perfect; splintery fracture; brittle. Nonfluorescent. Biaxial (+); Y = *b*, $X \wedge c = 7°$; $N_x = 1.424$, $N_y = 1.432$, $N_z = 1.442$; $2V_{meas} = 94.5°$. From the *Karasug* Fe–REE–barite–fluorite deposit, in the *western part of the Tannu-Ola range, N of Karasug in southern Siberia, Russia*, in fissures cutting limonite–hematite ores, with "limonite," tikhonenkovite, and gearksutite. RG *AM 80:185(1995)*.

Class 12

Compound Halides

12.1.1.1 Stenonite $(Sr,Ba,Na)_2Al(CO_3)F_5$

Named in 1962 for Nicholas Stenonis (Niels Steensen) (1638–1686), Danish anatomist, discoverer of the law of constancy of interfacial angles. MON $P2_1m$. $a = 5.450$, $b = 8.704$, $c = 13.150$, $\beta = 98.20°$, $Z = 4$, $D = 3.847$. Structure: *CM 22:245(1984)*. *15-366:* 3.39_{10} 2.24_6 2.21_6 2.17_{10} 2.14_6 1.93_8 1.87_6 1.79_6. Small masses and grains. Colorless to white, vitreous luster. Cleavage {001}, {120}, and {$\bar{1}$20}, good. $H = 3\frac{1}{2}$. $G = 3.86$. Biaxial (−); $X = b$; $Z \wedge c = 32°$; $N_x = 1.452$, $N_y = 1.527$, $N_z = 1.538$; $2V = 43°$. *Ivigtut, Greenland*, associated with jarlite, cryolite, fluorite, and other fluorides. RG *AM 48:1178(1963)*.

12.1.2.1 Grandreefite $Pb_2SO_4F_2$

Named in 1989 for the locality. MON $A2/a$. $a = 8.667$, $b = 4.442$, $c = 14.242$, $\beta = 107.41°$, $Z = 4$, $D = 6.96$. Structure: *AM 76:278(1991)*. *45-1455:* 3.16_{10} 3.12_9 3.08_8 2.28_5 1.824_5 1.806_6 1.780_4 1.370_5. Crystals prismatic [001], with {120}, {130}, {293}, and {101}. Colorless, white streak, subadamantine luster. No cleavage, conchoidal fracture, brittle. $H = 2\frac{1}{2}$. $G = 7.0$. Nonfluorescent. Biaxial (+); $N_x = 1.872$, $N_y = 1.873$, $N_z = 1.897$. 2V small; $2V_{calc} = 23°$; $r < v$, weak. Coarse polysynthetic twinning observed in polarized light. Decomposed by cold water, slowly soluble in high-index immersion fluids. *Grand Reef mine, Graham Co., AZ*, in a vug surrounded by fluorite and galena, believed to have been formed by reaction of supergene fluids with these minerals. RG *AM 74:927(1989)*.

12.1.3.1 Pseudograndreefite $Pb_6SO_4F_{10}$

Named in 1989 for its close similarity to grandreefite in its composition, physical properties, and structure. ORTH $F222$. $a = 8.518$, $b = 19.574$, $c = 8.493$, $Z = 4$, $D = 7.08$. *45-1456:* 3.89_2 3.20_{10} 3.00_3 2.21_3 2.12_2 1.900_2 1.779_7 1.364_3. Square crystals (pseudotetragonal), tabular on {010}. Colorless, white streak, subadamantine luster. No cleavage, conchoidal fracture, brittle. $H = 2\frac{1}{2}$. $G = 7.0$. Nonfluorescent. Biaxial (+); XYZ= *cab*; $N_x = 1.864$, $N_y = 1.865$, $N_z = 1.873$; $2V = 30°$; $r > v$, strong. Decomposed in cold water, slowly attacked by high-index optical immersion fluids. *Grand Reef mine, Graham Co., AZ*, associated with grandreefite, laurelite, aravaipaite, galena, and fluorite. RG *AM 74:927(1989)*.

12.1.4.1 Creedite $Ca_3Al_2SO_4(OH)_2F_8 \cdot 2H_2O$

Named in 1916 for the locality. MON $C2/c$. $a = 13.936$, $b = 8.606$, $c = 9.985$, $\beta = 94.4°$, $Z = 4$, $D = 2.738$. Structure: *NJMM:69(1983)*. *8-72*: 7.30_9 6.90_9 5.79_8 3.92_9 3.67_7 3.48_{10} 3.07_8 2.16_9. Crystals short prismatic to acicular [001], with {111}, {110}, and/or {$\bar{1}$11} dominant; also {001}; commonly in radiating aggregates or drusy masses. Colorless to white; rarely purple; vitreous luster. Cleavage {100}, perfect; conchoidal fracture; brittle. $H = 4$. $G = 2.71$–2.73. Biaxial $(-)$; $Y = b$; $Z \wedge c = 42°$; $N_x = 1.463$, $N_y = 1.476$, $N_z = 1.484$ (Na); $2V = 64°$; $r > v$, strong. Slowly soluble in acids. *Wagon Wheel Gap, Creede quadrangle, Mineral Co., CO*(*), in a fluorite–barite vein. Grand Reef mine, Graham Co., AZ. Granite, Nye Co., NV. Santa Eulalia, CHIH, Mexico(!), in fine purple crystals and groups. Aktschatau, central Kazakhstan(!), in large well-crystallized violet groups. Crystals to 2.5 cm at the Colquiri mine, La Paz Dept., Bolivia. RG *DII:129, AM 37:787(1952)*.

12.1.5.1 Chukhrovite-(Y) $Ca_3(Y,Ce)Al_2(SO_4)F_{13} \cdot 10H_2O$

12.1.5.2 Chukhrovite-(Ce) $Ca_3(Ce,Nd)Al_2(SO_4)F_{13} \cdot 10H_2O$

Named in 1960 for Fedor Vasilevich Chukhrov (1908–1988), Russian mineralogist, academician, and director, Institute of Geology of Ore Deposits, Moscow. Chukhrovite-(Y): ISO $Fd3$. $a = 16.710$, $Z = 8$, $D = 2.274$ (Y/Ce = 1:1), $D = 2.35$ (Y). Chukhrovite-(Ce): ISO $FD\#$. $a = 16.80$, $Z = 8$. Structure: *AM 66:392(1981)*. *14-61*(chukhrovite-Y): 3.26_9 2.84_8 2.57_9 2.25_9 2.19_{10} 1.83_{10} 1.68_8 1.51_8. *43-667*(chukhrovite-Ce): 9.75_{10} 5.93_8 4.20_5 3.22_7 2.56_6 2.24_5 2.17_6 1.824_5. Crystallizes as cubo-octahedrons. White to colorless, vitreous to pearly luster. Cleavage {111}, indistinct; irregular fracture; brittle. $H = 3$. $G = 2.27$–2.40. Isotropic; $N = 1.440$–1.483. Chukhrovite-(Y) was found in the oxidized zone of the *Kara Oba molybdenum–tungsten deposit, central Kazakhstan*, associated with creedite, gearksutite, halloysite, fluorite, and other minerals. Chukhrovite-(Ce) was found at the *Clara mine, Oberwolfach, middle Black Forest, Germany*. RG *CE 38:331(1979); AM 65:1065(1980), 45:1132(1960); ZVMO 102:201(1973)*.

12.1.6.1 Boggildite $Na_2Sr_2Al_2(PO_4)F_9$

Named in 1954 for Ove Balthasar Boggild (1872–1956), Danish geologist and mineralogist, University of Copenhagen; author of *Mineralogia Groenlandia* (1905) and *The Mineralogy of Greenland* (1953). MON, pseudo-ORTH $P2_1/c$. $a = 5.251$, $b = 10.464$, $c = 18.577$, $\beta = 107.32°$, $Z = 4$, $D = 3.692$. Structure: *CM 20:263(1982)*. *14-417*: 3.96_7 3.89_8 3.16_{10} 3.13_7 2.88_5 2.87_5 2.75_5 2.63_7. Massive, with twinning lamellae. Flesh-red. $H = 4$–5. $G = 3.66$. $Z = b$; $X \wedge a = 36°$; $N_x = 1.462$, $N_y = 1.466$, $N_z = 1.469$; $2V = 80°$. *Ivigtut, Greenland*, with black cryolite, siderite, sphalerite, and fluorite near a greisen contact with the cryolite orebody. RG *AM 39:848(1954), 41:959(1956)*.

12.1.7.1 Barstowite $Pb_4(CO_3)Cl_6 \cdot H_2O$

Named in 1991 for Richard William Barstow (1947–1982), Cornish mineral dealer. MON $P2_1$ or $P2_1/m$. $a = 4.218$, $b = 9.180$, $c = 16.673$, $\beta = 91.48°$, $Z = 2$, $D = 5.77$. PD: 4.16_5 4.02_{10} 3.79_3 3.05_3 2.48_3 2.38_6 2.30_8 2.11_4. Prismatic crystals (microscopic) with rectangular outline. Colorless to white, adamantine luster. Prismatic, imperfect cleavage; uneven fracture; brittle. $H = 3$. $G = 5.71$. In polished section gray to dark gray; moderately anisotropic; colorless internal reflections. R = (air): 13.5, 14.0 (470 nm); 12.9, 13.4 (546 nm); 12.7, 13.1 (589 nm); 12.4, 12.8 (650 nm). *Bounds Cliff, St. Endellion, Cornwall, England*, in a quartz–dolomite vein containing Pb, Sb, and Cu, just above sea level, with barstowite, phosgenite, and other minerals, formed by the action of seawater on the exposed ore. RG *MM 55:121(1991)*, *AM 77:670(1992)*.

12.1.8.1 Arzrunite $Cu_4Pb_2SO_4(OH)_4Cl_6 \cdot 2H_2O$

Named in 1899 after Andreas Arzruni (1847–1898), German mineralogist, University of Aachen, who first recognized the mineral. Probably ORTH Space group unknown. Prismatic crystals, with {110}, {010}, and {001}. Blue to bluish green. Biaxial, parallel extinction, optic axis perpendicular to {110}. Strongly pleochroic, blue to nearly colorless. *Buena Esperanza mine, Challacolla, Tarapacá, Chile*, with lanarkite. RG *DII:130*.

Class 13

Acid Carbonates

13.1.1.1 Nahcolite NaHCO₃

Named in 1929 after the chemical formula. Synonym: sodium bicarbonate. MON $P2_1/n$. $a = 7.475$, $b = 9.686$, $c = 3.481$, $\beta = 93.38°$, $Z = 4$, $D = 2.16$. *JCP 1:634(1933)*. Structure consists of infinite chains of cross-linked HCO_3^- ions associated with Na, forming layers of HCO_3, and Na, nearly parallel to $c(101)$. *15-700*(syn): 3.48_3 3.06_4 2.96_7 2.94_{10} 2.60_{10} 2.68_3 2.21_4 2.04_3. Crystals prismatic [001], with {110}, {010} dominant, terminated by {101}; often as aggregates or friable porous masses. Colorless, white to gray, or buff; colorless streak; transparent; vitreous to pearly luster. Cleavage {101}, perfect; {111}, good; {100}, distinct; brittle; contact and penetration twins common. H = 2.5. G = 2.21. IR. *MSLM 4:261(1974)*. Biaxial (−); Y = b; X ∧ c = 27°; $N_x = 1.377$, $N_y = 1.503$, $N_z = 1.583$; 2V = 75°; r < v, weak. *AM 25:769(1940)*. Natural material contains <0.5% Ca, Fe, Al. Easily soluble in H_2O, forming alkaline solution which on heating gives off CO_2; also soluble in glycerine. Lacustrine evaporite mineral. Large deposits are found in central Searles Lake, San Bernardino, CA, in thin beds associated with gaylussite, thenardite, burkeite, northupite, borax, and halite. Concretions up to 1.5 m diameter in oil shales of Green River formation (Eocene), Anvil Pts., Rifle, CO. Admixed with trona, thermonatrite as efflorescence in old Roman conduit at hot springs, Naples, Italy. In dry Little Lake Mogadi, 40 km south of Lake Mogadi, Kenya, as fibrous pseudomorphs after gaylussite, and as alteration

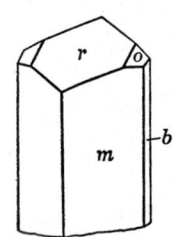

Searles Lake Prismatic Habit
Forms: a{100}, b{010}, m{110}, o{111}, r{101}

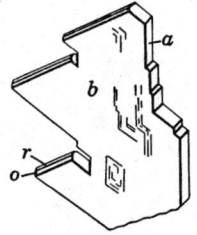

Searles Lake Twin on {101}

Nahcolite

rims around thermonatrite. Medical use: to attain alkalinity; a major source of sodium carbonate in the manufacture of soaps. HCWS $DII:134$.

13.1.2.1 Kalicinite KHCO$_3$

Named in 1898 after the composition, potassium = kalium. MON $P2_1/a$. $a = 15.173$, $b = 5.628$, $c = 3.711$, $\beta = 104.63°$, $Z = 4$ $D = 2.15$. $AC(B)$ $30:1155(1974)$. 12-292(syn): 7.34_2 3.67_{10} 2.97_4 2.86_9 2.82_8 2.63_9 2.38_4 1.832_3. Fine-crystalline aggregates; synthetic material short prismatic crystals. Colorless to white–yellow, transparent. Cleavage {100}, {001}, {101}. Soft. $G_{syn} = 2.17$. Biaxial (−); $X \wedge c = 30°$; $N_x = 1.380$, $N_y = 1.482$, $N_z = 1.578$. $2V = 81.5°$. Soluble in H_2O. Found as a result of decomposition under a dead tree, *Chypis, Canton Wallis, Switzerland*, possibly in association with trona in Hungary. HCWS $DII:136$.

13.1.3.1 Teschemacherite (NH$_4$)HCO$_3$

Named in 1868 by English chemist Frederick Edward Teschemacher (1791–1863), who described the species. ORTH $Pccn$. $a = 7.225$, $b = 10.709$, $c = 8.746$, $Z = 8$, $D = 1.51$. $TMPM\ 29:67(1981)$. 9-415(syn): 5.34_6 4.05_4 3.62_5 3.20_2 3.07_3 3.01_4 3.00_{10} 2.16_3. See also 44-1483. Compact crystalline masses; synthetic crystals short prisms [001]. Colorless, or white–yellow, transparent. Cleavage {110}, perfect; brittle. $H = 1\frac{1}{2}$. $G = 1.57$. Biaxial (−); $N_x = 1.4227$, $N_y = 1.5358$, $N_z = 1.5545$; $2V = 24°$; $r < v$, weak. $AJS\ 269:97(1971)$. Decomposes in moist air, soluble in H_2O, effervesces with dilute acids. Synthesized. In guano deposits: Saldanha Bay, Cape Prov., South Africa; on Chicha and Guanape Is., Peru. Deposited at the capped wellhead for geothermal waters at Broadlands, New Zealand. HCWS $DII:137$, $AM\ 57:1304(1972)$.

13.1.4.1 Trona Na$_3$CO$_3$(HCO$_3$) · 2H$_2$O

Named in 1773 for the reduced form of the word for native salt in Arabic, *natrun*. MON $C2/c$(syn). $a = 20.362$, $b = 3.481$, $c = 10.291$, $\beta = 106.48°$, $Z = 4$, $D = 2.147$. Structure: $AC(B)\ 38:2874(1982)$. Planar $(CO_3)^{2-}$ groups, somewhat distorted, are linked by H^+ and coordinated octahedrally with Na. 29-1447: 9.77_5 4.89_6 3.20_2 3.07_8 2.65_{10} 2.44_3 2.25_3 2.03_3. Rare crystals elongated [010], flattened {001}; {$h0l$} faces striated parallel to [010]; often fibrous. Colorless to gray or yellow–white, translucent, usually columnar or massive, vitreous luster. Cleavage {100}, perfect, uneven columnar fracture. $H = 2\frac{1}{2}$–3. $G = 2.11$. IR. $MSLM\ 4:263(1974)$. Biaxial (−); $Y \wedge c = -7°$; $N_x = 1.418$, $N_y = 1.492$, $N_z = 1.543$; $2V = 75.8°$. Not altered by exposure to dry air. Alkaline taste. Crystals of trona may be obtained by reacting $NaHCO_3$ with $Na_2CO_3 · 10H_2O$ in sodium chloride solution at temperatures up to 195°. Lacustrine evaporitic mineral; also as efflorescence on soils in arid regions. Commonly associated minerals are natron, thermonatrite, nahcolite, halite, glauberite, thenardite, mirabilite, and gypsum. Trona has wide occurrence in the western United States, especially in soda lakes near Laramie, and with shortite, northupite, bradleyite, and pirssonite in Green River oil shales of

Sweetwater Co., WY; present in large quantities in NV at Double Springs Marsh near Fallon. Trona occurs with borax, hanksite, thenardite, and other salts at Searles Lake, San Bernardino Co., at Borax Lake, Lake Co., and as layers along the shores of Owens Lake, Inyo Co., CA.; at shallow depths in Lake Goodenough, BC, Canada; in the subsurface residual brines from former alkaline lakes, Lake Texcoco, Mexico. The oldest known deposits of trona are in the Nile Valley near Memphis, Egypt. It has been found at the eastern Sahara Oasis of Bilma, Libya; also found in the alkali deserts of Hungary, Mongolia, and Tibet; as efflorescences near Lake Chad, Chad; at Lake Magadi, Kenya; and at Lake Nyasa, Tanzania. HCWS *DII:138*.

13.1.5.1 Nesquehonite $Mg(HCO_3)(OH) \cdot 2H_2O$

Named in 1890 after the locality. MON $P2_1/n$. $a = 7.705$, $b = 5.367$, $c = 12.121$, $\beta = 90.6°$, $Z = 4$, $D = 1.856$. Structure: *AC(B) 28:1031(1972)*. *20-669*(syn): 6.48_{10} 4.91_1 3.85_8 3.23_2 3.03_3 2.62_6 2.34_2 1.925_2. Radiating tufts of acicular crystals, also botryoidal. Colorless to white, transparent to translucent, vitreous luster. Some felted aggregates are pseudomorphs after lansfordite. Cleavage {110}, perfect; {001}, less so; splintery to fibrous fracture. $H = 2\frac{1}{2}$. $G = 1.852$. Biaxial (−); $N_{x\,calc} = 1.417$, $N_y = 1.503$, $N_z = 1.527$; $2V = 53°$; $r < v$. Easily soluble in dilute acids with effervescence; soluble in H_2O, especially in the presence of CO_2. Like lansfordite, appears to be a recent product formed under atmospheric conditions. Found in a coal mine at *Nesquehoning, near Lansford, Carbon Co., PA*, where it formed the base of stalactites and was associated with lansfordite; also lining crevices in sepentinites, especially noted at Franscia asbestos mine, Val Lanterna, Lombardy, Italy. Described from a mineral springs deposit Rohitsch–Sauerbrunn, Austria. HCWS *DII:225*.

13.1.6.1 Wegscheiderite $Na_5CO_3(HCO_3)_3$

Named in 1962 after R. Wegscheider (b.1913), who synthesized the compound. TRIC $P1$. $a = 10.04$, $b = 15.56$, $c = 3.466$, $\alpha = 91.92°$, $\beta = 95.82°$, $\gamma = 108.67°$, $Z = 2$, $D = 2.334$. *15-653*: 3.68_6 2.95_{10} 2.91_3 2.83_4 2.80_5 2.66_4 2.64_6 2.21_6. Crystals acicular to bladed and fibrous up to 5 cm in length with no well-developed faces. Colorless when pure, red-brown with 4% organic inclusions, vitreous luster. Prismatic cleavage, subconchoidal fracture. Non-piezoelectric. $H = 2\frac{1}{2}-3$. $G = 2.34$. Biaxial (−); $N_x = 1.433$, $N_y = 1.519$, $N_z = 1.528$; $2V = 32.4°$; strongly birefringent. Soluble in cold H_2O, decomposes rapidly in common acids releasing CO_2. A lacustrine mineral deposited in strata up to depths of 645 m in *Perkins Well No. 1, in the Wilkins Peak member of Green River Shales, northeastern Sweetwater Co., WY*. Trona is replaced by wegscheiderite, which is replaced by halite. HCWS *AM 48:400(1963)*.

13.1.7.1 Sergeevite $Ca_2Mg_{11}(CO_3)_9(HCO_3)_4(OH)_4 \cdot 6H_2O$

Named in 1980 for Evengi M. Sergeev (b.1914), engineering geologist, Moscow University. Compare with huntite (14.4.3.1). HEX-R $R32$. $a = 19.01$,

$c = 7.82$, $Z = 3$, $D = 2.64$. *AM 66:1100(1981)*. *41-1403:* 7.14_3 3.58_3 3.37_2 2.87_3 2.82_{10} 2.68_1 1.965_3 1.755_2. IR shows CO_3^{-2}, $(HCO_3)^-$, OH and H_2O. White, very fine-grained, dull aggregates with uneven to conchoidal fracture. $H = 3\frac{1}{2}$. $G = 2.2$–2.35. $N_{av} = 1.581$, birefringence ≈ 0.010. Decomposed by cold $1:10$ HCl. Weathering product of pyroxene, garnet skarn, Tyrnyauz, Caucasus, Russia. HCWS *ZVMO 109:217(1980)*.

13.1.8.1 Barentsite $Na_7Al(CO_3)_2(HCO_3)_2F_4$

Named in 1983 to honor Dutch sailor Willem Barents and the sea that washes the Kola Peninsula. TRIC $P1$ or $P\bar{1}$. $a = 6.472$, $b = 6.735$, $c = 3.806$, $\alpha = 92.5°$, $\beta = 97.33°$, $\gamma = 119.32°$, $Z = 1$, $D = 2.55$. *AM 69:565(1984)*. *35-693:* 2.89_9 2.78_{10} 2.66_{10} 2.54_3 2.32_5 2.17_7 1.870_4 1.670_3. Crystals up to 3–5 mm. Colorless, water-clear, vitreous luster, but pearly on faces. Cleavage {001}, {100}, perfect; brittle. $H = 3$. $G = 2.56$. IR, DTA. Biaxial ($-$); X near (001); $X \wedge a \approx 45°$; $N_x = 1.358$, $N_y = 1.479$, $N_z = 1.530$; $2V = 62°$; $r < v$, weak. In air becomes covered with white dull powder, effervesces in HCl. Described from drill core 600 m deep in *Mt. Restinyon, Kola Penin., Russia*, where it occurs with trona and natrite in the interstices of hydrothermal veinlets composed of shortite, albite, and natrolite that cut pegmatitic rock. HCWS *ZVMO 112:474(1983)*.

Class 14

Anhydrous Carbonates

The anhydrous carbonates fall into two divisions based on their crystallization character. One division consists of minerals with hexagonal/rhombohedral form and structure, similar to calcite, the most common carbonate mineral species. The other division includes minerals which are often pseudo-hexagonal, such as those of the aragonite group. A subset under each of these divisions are the double carbonates such as dolomite, $CaMg(CO_3)_2$, and alstonite, $BaCa(CO_3)_2$.

The structures of all these minerals are dominated by the trigonal carbonate $(CO_3)^{2-}$ group whose C-O distances, 1.28A ±0.02A, and O-C-O angles, 120° ± 2°, are effectively constant for all carbonates whose structures have been determined *FM 58:95 (1981)*. The group, planar in the hexagonal/rhombohedral anhydrous carbonates becomes slightly puckered in the orthorhombic/monoclinic forms.

CALCITE GROUP

The calcite group minerals are carbonates conforming to the formula

$$ACO_3$$

where

A = Ca, Mg, Mn, Fe, Co, Ni, Zn, or Cd

and the minerals crystallize in the hexagonal–rhombohedral space group $R\bar{3}c$.

The eight known endmembers (Table) typically show at least some solid solution (substitution) of several different cations. For example, there is complete miscibility for the cation pairs Ca–Cd, Mg–Co, Mg–Mn, Fe–Mg, and Fe–Mn, and limited miscibility for Ca–Mg, Ca–Fe, Ca–Mn, Ca–Co, Ca–Ni, and Ni–Mg. *MSASC 11:11(1983)*, *AM 51:677(1966)*, *DII:151–178*.

The carbonate group, effectively an oxygen-bounded equilateral triangle with carbon at the center, are planar, have trigonal symmetry, and form layers alternating with layers of the divalent metal ions in octahedral coordination with six oxygens from six carbonate groups. All layers are perpendicular to the c axis, with carbonate triangles of successive layers pointing in opposite directions.

Calcite Group

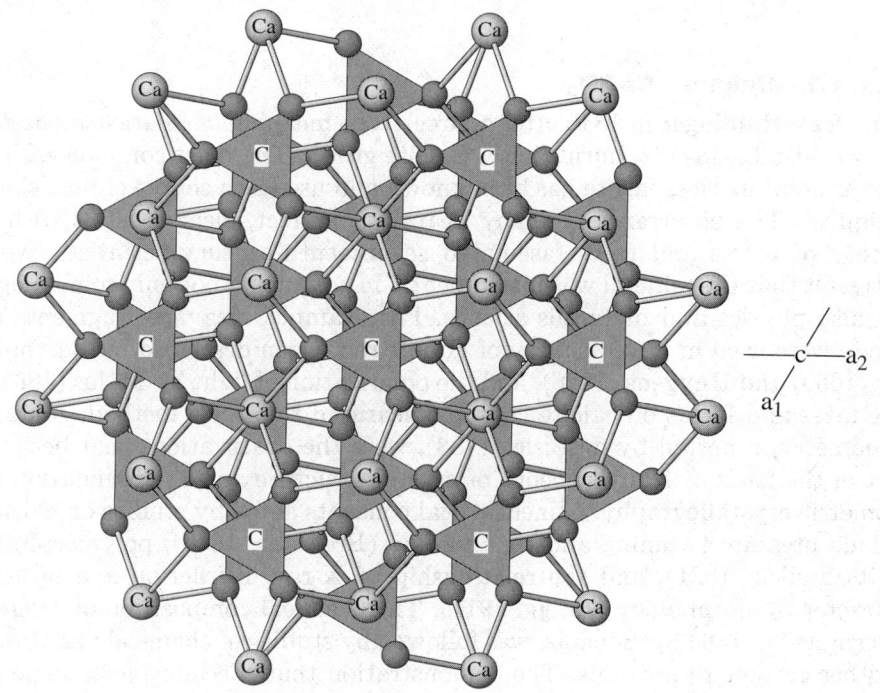

CALCITE: CaCo₃

Projection down the unique c axis of calcite shows the basic hexagonal features of this structure dominated by planar $(CO_3)^{2-}$ groups. However, the appearance belies the fact that the CO_3 groups are actually rhombohedral, and form layers alternating with layers of divalent metal ions perpendicular to c. Every other layer of CO_3 groups in the sequence Ca-CO_3-Ca-CO_3 point in opposite directions. The divalent cations in the calcite group minerals have octahedral coordination with six oxygens from six different CO_3 groups. Compare with aragonite (14.1.3.1), and the ordered double carbonate dolomite (14.2.1.1.).

CALCITE GROUP

Mineral	Formula	Space Group	a	c	D	
Calcite	$CaCO_3$	$R\bar{3}c$	4.9896	17.061	2.71	14.1.1.1
Magnesite	$MgCO_3$	$R\bar{3}c$	4.63328	15.0129	3.00	14.1.1.2
Siderite	$FeCO_3$	$R\bar{3}c$	4.6916	15.3796	3.936	14.1.1.3
Rhodochrosite	$MnCO_3$	$R\bar{3}c$	4.7682	15.6354	3.72	14.1.1.4
Sphaerocobaltite	$CoCO_3$	$R\bar{3}c$	4.6581	14.958	4.208	14.1.1.5
Smithsonite	$ZnCO_3$	$R\bar{3}c$	4.656	15.027	4.43	14.1.1.6
Otavite	$CdCO_3$	$R\bar{3}c$	4.923	16.287	5.03	14.1.1.7
Gaspéite	$NiCO_3$	$R\bar{3}c$	4.621	14.93	4.35	14.1.1.8

MSASC 11:11(1983), AM51:677(1966), DII 151-178, ASTR(1992)

Notes:
1. There is virtually no substitution for the carbonate group in the calcite group structures.
2. Carbonate groups are largely responsible for the characteristically large birefringence of the group *MSASC 11:49(1983)*.
3. See aragonite (14.1.3.1), the orthorhombic polymorph of calcium carbonate, and the dolomite (14.2.1.1) group, ordered double carbonates.
4. The aragonite group minerals crystalize with larger A cations in a different structure.
5. See vaterite (14.1.2.1) a hexagonal polymorph of calcium carbonate.
6. See the dolomite group minerals crystalize in space group $R\bar{3}$.

14.1.1.1 Calcite CaCO$_3$

Named by Haidinger in 1845 after a Greek root meaning *to reduce to a powder by heat* (also Latin *calx*, burnt lime). Calcite group. The most common carbonate mineral species, calcite has been known and used as a source of lime since antiquity. The occurrence of many beautiful, perfect, clear crystals with a variety of forms and faces fascinated seventeenth-century scientists, with the result that the mineral was instrumental in advancing not only mineralogy but also physics and materials science. For example, cleavage fragments of calcite were used in the discovery of double refraction described by Bartholinus (1669) and Huygens (1678), and the polarization of light by Malus (1808). The faces and forms on calcite crystals measured by a new method, contact goniometry, reported by de Lisle (1783), were the observations that became part of the basis of Hauy's theory of crystal structure and the foundation of geometric crystallography. Mineralogical concepts aided by studies on calcite include pressure twinning and twin gliding (Brewster, 1815), polymorphism (Mitscherlich, 1821), and the relationship of X-ray diffraction and optical character to morphology (Bragg, 1914). The chemical composition of calcite, determined in 1804 by Bucholz, was followed by studies of chemical variations in other carbonate minerals. The demonstration that the interfacial angle of the crystal forms varied with composition for isostructural compounds, such as those in the calcite group, established the concept of an isostructural series of minerals forms are isomorphic. The polymorphic relationship of calcite to aragonite was recognized by Hauy and Mitscherlich (1821), who postulated the structural relationship, but it was Bragg (1924) using X-ray diffraction who confirmed the actual structural differences. Calcite could be called a "mineralogists' mineral."

Calcite, the dominant member of the calcite group (see group description), has had numerous names over its long history, such as *calcspar*, *kalkspar*, *kalkstein*, *doppelstein*, *Iceland spar*, *lublinite*, and many other varietal names: *satin spar*, *slate spar*, *cave pearls*, *stinkcalc*, *sand calcite*, depending on the habit, inclusions, or environment of formation. **Physical properties:** HEX-R $R\bar{3}c$. $a = 4.9896$, $c = 17.061$, $Z = 6$, $D = 2.71$ *ZK 156:233(1981)*. *5-586*(syn): 3.86_1 3.035_{10} 2.495_1 2.285_2 1.913_2 1.875_2 1.604_1. *24-27*(calc): 3.852_3 3.030_{10} 2.284_2 2.094_3 1.907_2 1.8726_3 1.6040_2. See also *43-697*, magnesian calcite. **Structure:** Calcite (Fig. 14.1.1.1A) is the thermodynamically stable form of CaCO$_3$ at the surface of the earth. Figure 14.1.1.1B illustrates the relationship between the true unit cell of calcite, the commonly observed cleavage rhomb, and the hexagonal prism. There are two low-temperature polymorphs of CaCO$_3$: aragonite (14.1.3.1) and vaterite (14.1.2.1). A slightly different structure, calcite II, forms via displacive transition at 15 kb and room temperature [*AC(B) 31:343(1975)*], and another, calcite III, forms above 20 kb [*AM 75:801(1990)*]. Calcite IV and V, higher-temperature forms, probably with rotational anion disorder, have not yet been fully described. *MSASC 11:191(1983)*. **Habit:** Hundreds of different crystal faces and forms have been described for calcite, and a few of the common habits are illustrated in

14.1.1.1 Calcite

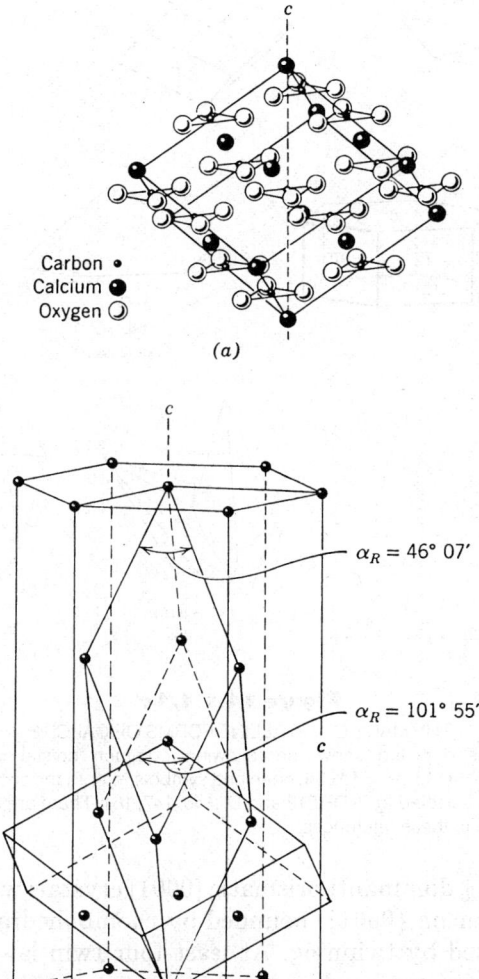

Figures 14.1.1.1a and 14.1.1.1b
Calcite Rhomb and Unit Cell Relationships (a) The arrangement of Ca^{2+} and CO_3^{2-} groups relative to the cleavage rhomb of calcite. (b) The relationship between the steep, true rhombohedral unit cell ($\alpha = 46.11°$) of calcite to the cleavage rhomb ($\alpha = 101.91°$) and the hexagonal cell in which $c/a = 3.42$ (AJS 254:65 (1956) MSASC 11:1 (1983) DM 20:329 Fig.10.2).

Fig. 14.1.1.1C. Note that the forms have been given individual letters (see *DII:143* for a complete list). Miller indices, which identify the forms, may refer to the cleavage rhomb or to the hexagonal cell with $c/a = 0.854$. The typical rhomb, and the scalenohedral forms, may be modified by an amazing variety of faces (e.g., the Lake Superior crystal on which $\{18.0.\overline{18}.1\}$ has been described). Commonly observed calcite crystal forms: obtuse to acute rhombohedrons $\{02\overline{2}1\}$, $\{40\overline{4}1\}$, $\{05\overline{5}4\}$, $\{03\overline{3}2\}$ alone or in combination; scaleno-

14.1.1.1 Calcite

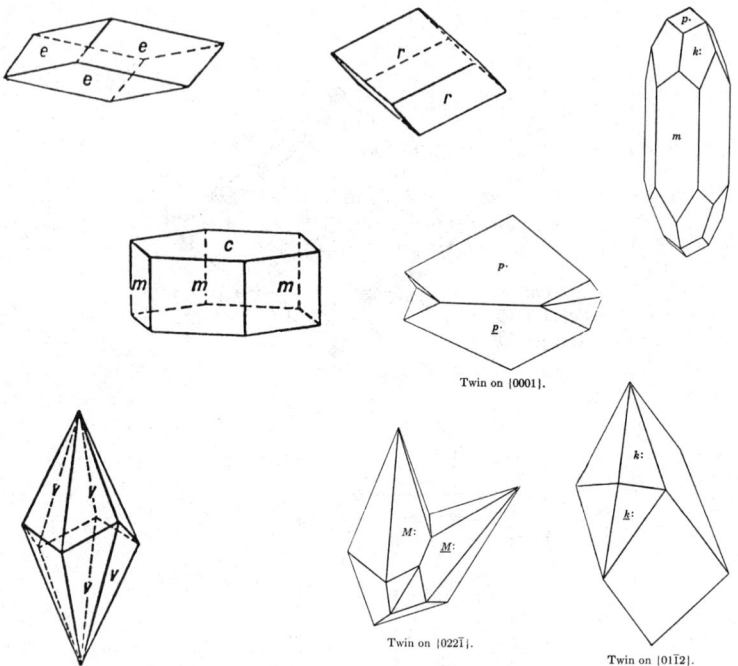

Figure 14.1.1.1c
THE MANY CRYSTALLINE FORMS OF CALCITE
The crystal morphologies depicted show the following common crystal faces: c = 0001 e = 01$\bar{1}$2 k := 21$\bar{3}$1 m = 10$\bar{1}$0 M := 74$\overline{11}$3 p· = 10$\bar{1}$1. Underlined symbols refer to the second or twin crystal face in the pair. These forms were selected from DF:512 and DII:146, 147, 161. There are many additional morphologies and faces presented in these references.

hedrons with {21$\bar{3}$1} dominant; prismatic [0001] crystals with {10$\bar{1}$0}, {11$\bar{2}$0}; thin-to-thick tabular on {0001}, bounded by rhombohedral faces, and a variety of shapes created by twinning. At least four twin laws have been determined. The most common is when the twin plane and the composition plane are {01$\bar{1}$2}. A repeated effect, produced by pressure, this twinning law is especially encountered in marble. Also common is twinning on {0001} with {0001} as the composition surface, which produces reentrant angles (see Fig. 14.1.1.1C). Twins with {10$\bar{1}$1} or {02$\bar{2}$1} as twin plane and composition surface are uncommon but may result in heart-shaped crystals ("butterfly twins") or distorted forms in some localities. Calcite also is found as fibers, both coarse and fine, sometimes lamellar, and may be granular and compact, forming multiple aggregates or earthy (chalk). Nodular (oölitic) and stalactitic forms are often observed.

Colorless and transparent to white when pure, natural samples are usually white but do occur in virtually every color, including blue and black, reflecting, at least in part, the isomorphous substitution of other, mostly divalent, elements defect color centers within the crystals, or admixtures of other minerals. Streak, usually white, will be influenced by admixtures. Luster

14.1.1.1 Calcite

vitreous, sometimes pearly to iridescent on cleavage surfaces. Cleavage $\{10\bar{1}1\}$, perfect; ready parting along twin lamellae $\{01\bar{1}2\}$, $\{0001\}$; brittle. $H = 3$, variable on different faces or planes and directions within the faces or planes. Mohs' hardness $H = 3$ is defined as the hardness of calcite. $G \approx 2.96$. Natural and artificial etch figures conform with HEX-R symmetry $\bar{3}2/m$. Specimens may be fluorescent, or phosphorescent, on excitation by UV, X-ray, cathode radiation, or sunlight; occasionally, thermoluminescent *JP(C) 17:2027(1984)*. Reddish-yellow fluorescent Mn-containing (up to 3.5% $MnCO_3$) calcites are common at Franklin and Sterling Hill, NJ, and Långban, Sweden. IR *(SWV:54)* and Raman spectra have been measured for all members of the calcite group [*AM 62:36(1976)*] as well as on calcite II [*AM 75:801(1990)*] and the Mg-containing biogenic forms [*AM 76:641(1991)*]. **Optics:** Uniaxial $(-)$; $N_O = 1.658$, $N_E = 1.486$. In transmitted light calcite is colorless, occasionally abnormally biaxial (2V up to 30°) as a result of mechanically induced strain. Variations in N value with temperature, composition, and wavelength have been summarized. *DII:151*. **Chemistry:** The most common, and perhaps the most important, solid substitution in nature considering the widespread occurrence of limestones, marls, dolomitic limestones, and marbles is that of Mg in calcite. The chemical variations have been used to interpret the environments that existed during the formation of these sedimentary and metamorphic rocks. Phase relations determined for the binary system $CaCO_3$–$MgCO_3$ show two solid-solution series separated by the double carbonate ordered phase dolomite $CaMg(CO_3)_2$ (14.2.1.1): calcite–dolomite, and dolomite–magnesite (14.1.1.2). Biologically mediated metastable formation and the geological persistence of magnesian calcites (Mg content may be up to 20% $MgCO_3$) is widespread and well documented. Calcite is easily produced via precipitation from solution below 30° unless sulfates or salts containing Mg, Pb, Sr, or other large cations are present. Single crystals may be grown via slow diffusion of a soluble carbonate solution into a Ca-containing gel with the crystal habit dependent on associated other ions or molecules. The kinetics of precipitation and dissolution have been reviewed *MSASC 11:227(1983)*. Calcite is usually detected, and discriminated from dolomite or magnesite, by fizzing on contact with cold dilute (1:10) HCl. **Occurrence:** Perhaps the most common occurrences of calcite are the massive and thick sedimentary rocks in which calcite is usually predominant: limestone, chalk, marl, oolitic limestone, tufa, travertine, calc–sinter. The grain size of calcite in these rocks ranges from submicroscopic to coarse, depending on the depositional environment and subsequent metamorphism, or diagenesis. Other low-temperature environments in which calcite forms include earthy crusts and efflorescences on soils, the intricate formations (stalactites) in calcareous springs or streams in caves, as cementing material in sedimentary rocks, and as veins deposited from lime-bearing carbonated surface waters which are found in all types of rock. Calcite is a common gangue mineral and is a minor but widespread constituent in all types of rock alteration. The unusually large accumulations of sedimentary strata representing quiet-water calcite deposits are often lamellar, porous, concretionary, and fossiliferous and may have been deposited in the sea or by lakes and springs. Freshwater calcite deposits occur on all continents. Coralline and shelly deposits characteristic of invertebrate organisms that inhabit the seas are usually calcitic, but may be aragonitic, indicating a specific invertebrate species or distinctive temperature of deposition. The crystalline habit of calcite in hydrothermal deposits has also been used to indicate temperature or other conditions of deposition. Often the petrifying substance of fossil organic materials (i.e., wood), calcite is usually the replacement for the initial depositional biomineral aragonite. It is also replaced by, or replaces, other mineral species (i.e., there are pseudomorphs of quartz, chalcedony, or iron oxides after calcite, and calcite pseudomorphous after gypsum, barite, and fluorite). In all low-temperature deposits calcite is normally finely crystalline. Aggregates of calcite crystals are common, as are clusters that reflect oriented overgrowth of secondary calcite (epitaxy). Platy aggregates, lamellar aggregates and soft, scaly, often powdery forms, as well as fibrous calcite with a silky luster, have been described. The mineral may exhibit a range of minor and trace elements which, together with C and O isotopic data, have been used to determine the conditions and time of formation of these deposits. **Localities.** There is an almost endless number of localities for crystallized calcite worldwide, the following are some of the most noteworthy: 1. *Euhedral or single-crystal localities.* The greatest source of optical calcite (one crystal measured 7 m × 2.5 m) was the cavernous basalt at Helgustader, Eskifjord, Iceland (iceland spar). From NY as fine crystals to 5 cm from Rossie, in large pink twinned crystals, suitable for faceting flawless rainbow-hued gemstones to several thousand carats, from Balmat, St. Lawrence Co. In NJ in basalt quarries associated with zeolites from Great Notch; at Upper New Street and Prospect Park quarries, West Paterson, with prehnite and zeolites; red-fluorescent calcite with green-fluorescent willemite and yellow-fluorescent esperite and about 250 other minerals from Franklin and Sterling Hill, Sussex Co. Baseball-sized crystals with pyrite inclusions forming a three-rayed star from Staunton, VA. The Elmwood and Gordonsville mines near Carthage, TN, in magnificent transparent amber, large twinned (0001) crystals with sphalerite, fluorite, and barite. In IN, from North Vernon, Jennings Co., fine pale-brown crystals of low prismatic hexagonal form terminated by low rhombohedrons, with a nearly spherical shape. In OH at White Rock quarry in Clay Center, as yellow rhombs of calcite on white blades of celestite. Splendid white to clear crystals up to 15 cm enclosing brilliant native copper in mines on the Keweenaw Peninsula, MI, notably from the Calumet, Quincy, and other mines. The Sweetwater and Buick mines, Reynolds Co., MO, produce crystals over 60 cm associated with large galena crystals; also giant "dogtooth" yellow crystals, sometimes with a lavender core, and "fishtail" twins from Joplin and nearby Treece, KS, and Picher, OK, in the Tri-State zinc mining district, associated with sphalerite, galena, dolomite, and chalcopyrite. Sand calcite crystals with 70% sand included from Rattlesnake Butte in the badlands of Washington Co., SD. Near Terlingua, TX, calcite that fluoresces brilliant red or yellow and phosphoresces bright blue, depending on the material and the UV wavelength used. At the Iceberg claim near the Harding mine, Taos Co., NM, fine optical calcite in flawless rhombs to 20 cm. In AZ at Ray with cuprite inclusions; at Bisbee with malachite inclusions; and at the Cyprus mine near Baghdad clear twins of many kinds, resembling English material. In Canada at Chibogamou, QUE, gray scalenohedra with cubanite phantoms; from St. Clothilde as yellow modified rhombs on dolomite. In Mexico many localities: in CHIH the Buena Tierra mine with red phantoms; the Potosi mine with clear to red crystals on stalactites; at Areponapuchic as large V-twins, some over 11 kg; N of Chihuahua city the "cocoanut" geodes with many types of calcite associated with crystallized quartz, including smoky and amethyst types. In DGO large boxwork clusters of thin snow-white plates to over 1 m across ("angel wing") from La Amparo mine, San Antonio; this area has produced much optical iceland spar. In SLP the San Sebastian mine at Charcas has tabular crystals to 18 cm across in red-fluorescent groups to several meters across, associated with danburite, datolite, and stilbite. In Guanajuato, GTO, the Valenciana, San Juan de Rayas, and Peregrina mines in crystals to 40 cm associated with amethyst and dolomite with complex, light purple, interpenetrating twins. In ZAC the El Cobre mine has green crystals with malachite and azurite. In Norway as white rhombs with silver at Kongsberg. In Sweden the Malmberget mine, Lappland, as clear orange V-twins; also from Långban. Beautiful brilliant crystals of many forms, often twinned, occur in England from the following classic localities: Alston Moor, Egremont; Frizington, Cumbria; Weardale, Durham; the Stank mine near Furness in Lancashire; Wheal Ray and elsewhere in Cornwall. In Germany from Andreasburg, and as heart-shaped twins from the Elbingerode, Harz; at Freiburg and Schneeburg, Saxony. At Pribram, Czech Republic, in thin, flowerlike formations. Felsobanya, Romania, has rhombs, balls, and stalactites; at Trepca, Serbia, as white rhombs with sulfides. In France from Pau, Bearn as clear brilliant crystals in geodes, and from the famous Fontainebleu deposit near Paris rhombs and large rounded flowing shapes of sand calcite of great beauty and complexity. In Spain at La Collada and La Viesca, large white scalenohedra on blue fluorite, and yellow crystals on dolomite from the zinc mines at Reocin. In Kazakhstan as beautiful clear yellow (faceting quality) twins over 5 cm from Sarbay–Sokolov, and from Dzezkazgan associated with chalcocite, bornite, and silver. One of the world's outstanding localities is Dal'negorsk, Primorski Kraj, far eastern Russia, with many associations and habits. Hunan, China, has provided many large white to clear yellow and gray phantom crystals with realgar and orpiment. In India from vugs in the Deccan traps, clear crystals with okenite, prehnite, apophyllite, and zeolites from Bombay Island; also from Nasik, Maharashtra, with stilbite and powellite. Mashamba West, Shaba, Zaire, has fine clear and pink cobaltian calcite associated with malachite. One of the finest calcite localities is Tsumeb, Namibia, where crystals are associated with dioptase, malachite, and other minerals; also at Onganja, rhombs and dogtooth crystals with cuprite. In Australia large crystals and iceland spar were found at the Garibaldi mine, Lionville, NSW. In Peru white manganoan calcite crystals, sometimes with sulfides, from Huanzala and Pachapqui, Huanaco; also at Casapalca, Junin. At Irai, RGS, Brazil, and neighboring Artigas, Uruguay, white crystals of many forms occur inside amethyst geodes in basalt.

2. *Massive varieties.* Chalks, such as the White Cliffs of Dover, England, and limestone strata in the rock sequences on every continent, though predominantly calcitic, often contain other carbonate species, clays, and quartz. Marbles, equally globally distributed, modulate calcite dominance with the metamorphic formation of dolomite from hydrothermal activity. Vein calcites are usually described from mining localities, where the associations may be with massive sulfides or native metals such as copper and silver. Colored calcites, indicating some solid solution of Mn, Fe, Zn, Co, Pb, Sr, Mg, or Ba, are often described as one of the gangue minerals in well-known mines, such as Långban, Sweden. Calcite hotsprings have been described at Hammam Meskoutine, Algeria, and at the Kennardt district, Cape Prov., South Africa. Tufa deposits are prominent in northwestern NV at the shores of former Lake Lahontan, and in Mono Lake, CA. The tufa contains particularly interesting pseudomorphs that have occasionally been given separate names, such as *barleycorn* described originally from Obersdorf near Sangerhausen, Thuringia, Germany; *jarrowite*, prismatic pyramidal and laterally grooved or rough, brown/red-brown crystals, originally described from the river clays of Jarrow on Tyne, Durham, England; or *glendonites*, widely distributed in the Permo-Carboniferous mudstones of NSW and Q, Australia (perhaps after glauberite). Travertine outcrops found near Rome, Italy, have been used as building materials for millennia. Stalactites and stalagmites, many still growing in numerous caverns, are well-known tourist attractions, such as in Mammoth Cave, KY and Guilin, China. Fine-grained but well-crystallized calcite is the hallmark of invertebrate species from modern to ancient forms. However, calcitic species may be replaced by other minerals, such as pyrite, in black shales. HCWS *DF 511, DII:141, MSASC:11(1983)*.

14.1.1.2 Magnesite Mg(CO)$_3$

Named in 1807 for the composition. Calcite group. See also the Ca and Mg double carbonates dolomite, $CaMg(CO_3)_2$ (14.2.1.1), and huntite, $(Mg_3Ca)(CO_3)_4$ (14.4.3.1). HEX-R $R\bar{3}c$. $a = 4.633$, $c = 15.013$, $Z = 6$, $D = 3.00$. *ZK 156:233(1981)*. 8-479(syn): 2.74_{10} 2.50_2 2.10_5 1.939_1 1.700_4 1.354_1 1.338_1 0.9134_1. Distinct colorless and transparent crystals of magnesite, either rhombohedral or prismatic, are rare; usually massive, coarse to fine-grained, or compact and porcelaneous, chalky, lamellar, or fibrous. Normally white or light gray, with coloration influenced by the amount of cation solid solution (e.g., the ferroan forms of magnesite are often yellow or brownish, becoming brown on exposure, and the manganoan forms grade into pink). White streak. Cleavage $\{10\bar{1}1\}$, perfect; conchoidal fracture. $H = 3\frac{1}{2}-4\frac{1}{2}$. $G = 3.0$, variable with level of cation substitution. IR. *MSLM 4:239(1974)*.

Magnesite is normally nonfluorescent, but some fine-grained varieties fluoresce greenish to bluish white under LWUV; may also be phosphorescent in UV or triboluminescent. Uniaxial $(-)$; $N_O = 1.700$, $N_E = 1.509$ (pure $MgCO_3$). Slightly soluble in cold HCl, soluble with effervescence in warm acids. Virtually no Ca is found in natural or synthetically produced magnesite, but the biologically mediated formation of magnesian calcites is well documented. *MSASC 11:97(1983)*. Cryptocrystalline magnesite deposits, often with admixed opaline silica, appear to be the result of the alteration of magnesian-rich rocks such as serpentinites, peridotites, or dunites by carbonate-containing waters. Magnesite, usually in association with dolomite, occurs as the result of the replacement of calcitic rocks, but primary sedimentary magnesite has been described. Magnesite has been discovered in cavities in volcanoes, oceanic salt deposits, igneous rocks, and meteorites. Magnesite is associated with serpentinites and with dolomites in deposits on the east and west coasts of the United States and from similar localities in Canada (e.g., Grenville, Argenteuill Co., QUE). Large bodies of sedimentary magnesite occur in the Muddy Valley district, Clark Co., NV; hydrothermal magnesite in dolomite was described from the Paradise Range, NV. Magnesite is found in chlorite schists in the Pfitschtal and at Greiner in Zillertal, as well as in rock salt at Hall, Tyrol, Austria, which may be a sedimentary precipitate. Notable deposits of magnesite are recorded from Radenthein, Carinthia, Austria; from Snarum in Buskerud, Norway; on the island of Euboea, Greece; near Salem, Madras, India; as veinlets in gabbro-amphibolites, Saryku-Boldy, Kazakhstan; as coarse crystals stratified with dolomite and marble in Shenking Prov. and on the Liaodong Pensinsula, Liaoning, near Mukden, northeastern China. Magnesite has been described as precipitating from brackish lagoons along with magnesian calcite, hydromagnesite, and dolomite in the Coorong, southeastern SA, Australia. Large clear magnesite crystals and twins are obtained from a major magnesite deposit at Brumado, BA, Brazil(!), associated with dolomite, hematite, uvite, talc, sellaite, and many other minerals. Magnesite is one of the sources of the magnesia in the manufacture of chemical compounds such as epsom salts. HCWS *DII:162, MSASC:11(1983)*.

14.1.1.3 Siderite FeCO₃

Named in 1845 after the Greek for *iron*. Calcite group. *Chalybite* has been used in England. HEX-R $R\bar{3}c$. $a = 4.6916$, $c = 15.3796$, $Z = 6$, $D = 3.936$. *ZK 156:233(1981)*. *29-696*: 3.59_3 2.80_{10} 2.35_2 2.13_2 1.965_2 1.738_3 1.732_4 1.506_1. Rhombohedral crystals with $\{10\bar{1}1\}$ or $\{01\bar{1}2\}$ often curved or composite; also tabular or prismatic or scalenohedral. Often massive, coarse to fine granular, botryoidal, and globular with fibrous structure. Twinning uncommon. Translation gliding $\{0001\}$, $\{10\bar{1}1\}$. Yellowish–grayish brown to red brown, dark brown color usually due to oxidation, also shades of gray to pale green, occasionally white; white streak, vitreous or silky luster; translucent. $H = 4$. $G = 3.96$; decreases with substitution of Mn, Mg, and Ca for Fe. IR *MSLM*

4:239(1974), Mossbauer *NJMM:101(1975)*. Uniaxial (−); colorless to yellow/yellow-brown; $N_O = 1.875$, $N_E = 1.633$ (pure).

Studies in the ternary system $CaCO_3$–$MgCO_3$–$FeCO_3$ confirm a limited solution of Ca in siderite, a somewhat larger amount of Fe in solution in calcite, and a complete solid solution Fe–Mg between magnesite and siderite, but an Fe analog of dolomite, $CaFe(CO_3)_2$, has not been synthesized, nor does it exist in nature *MSASC 11:66(1983)*. The phase ankerite, or ferroan dolomite, is discussed in 14.2.1.2. A complete Fe–Mn solid solution from siderite to rhodochrosite also exists.

Siderite occurs primarily in bedded biosedimentary deposits with shale, clay, and organics (coal), where it is often massive and fine grained, or concretionary. These morphologies indicate that siderite was probably biogenically generated under low-oxygen, low-Eh conditions. Also found in metamorphosed sedimentary deposits with larger grain size, in pegmatites, as a gangue mineral in hydrothermal vein deposits, and in nepheline syenite pegmatites. A large vein deposit of siderite mined by the original colonists is still accessible at Mine Hill, Roxbury, CT(*); formerly mined with limonite, "black-band" ore, clay stone, and coal measures in eastern PA, IL, ID, OH, KY, and WV; in cretaceous clays of coastal plain, MD. Prominent in the iron ranges of Northern Peninsula, MI, and as an important gangue mineral in lead silver mines of Coeur d'Alene and other districts in ID, and in Bisbee district, AZ. Globular fibrous siderite is found in diabase in Weehawken, NJ, and Spokane, WA. Canadian occurrences: Slocan district, BC; Keno City, YT; and at the Acadia mines, Londonderry, NS; fine crystals to 15 cm at Mt. St.-Hilaire(!), QUE, Canada. Siderite is associated with cryolite at Ivigtut(*), Greenland. Well-crystallized siderite was found in the tin veins of Cornwall, especially at Camborne, Redruth, St. Austell, and at Tavistock, Devonshire; mined with the black-band ores in Somerset, Durham, Northumberland, and Yorkshire, England. Important replacement and metamorphic deposits are found in Erzberg near Eisenerz, Styria, and at Húttenberg-Lölling, Carinthia, Austria; in the silver veins at Freiberg, in the Harz Mts., in the Siegen region of Westphalia, and in Kamsdorf, Thuringia, Germany; finely crystalline siderite in the Piedmont, Italy; at Isère, France; in Val Tavetsch, Switzerland. Siderite was also a common constituent of the clay iron ores near Radom, central Poland, and of the black-band ores in the Kertsch and Taman peninsulas, Russia. At Tatasi, Bolivia; also at the Morro Velho gold mine(!), Nova Lima, and in large perfect crystals up to 15 cm in alpine cleft-type vugs in gneiss on Ibituruna Mt.(!), Governador Valadare, MG, Brazil. HCWS *DII:166*, *MSASC:11(1983)*.

14.1.1.4 Rhodochrosite $MnCO_3$

Named in 1800 after the (rose) color. Calcite group. HEX-R $R\bar{3}c$. $a = 4.7682$, $c = 15.6354$, $Z = 6$, $D = 3.72$ *MSASC 11:11(1983)*. 44-1472(syn): 3.67_3 2.85_{10} 2.40_2 2.178_2 2.005_2 1.773_2 1.766_3 1.538_1. Distinct crystals infrequent but rhombohedral forms often rounded, rarely scalenohedral, thick tabular, pris-

14.1.1.4 Rhodochrosite

matic. Usually massive, coarsely granular, incrusting, botryoidal and globular aggregates that are concentrically banded and radially fibrous to columnar. In parallel growth with dolomite, and with $Li(NO_3)_{syn}$. $\{01\bar{1}2\}$ twinning rare. Various shades of pink to red, also yellow-gray, beige to brown; white streak; vitreous to pearly luster, subtransparent to translucent. Cleavage $\{10\bar{1}1\}$, perfect; sometimes $\{01\bar{1}2\}$ parting; translation gliding T$\{0001\}$, t$[10\bar{1}0]$. $H = 3\frac{1}{2}$–4. $G = 3.70$ (pure). IR. *MSLM 4:239(1974)*. Uniaxial ($-$); colorless to pale rose; $N_O = 1.816$, $N_E = 1.597$; variations with Ca, Fe, and Mg substitution. Deep-red varieties may be faintly dichroic, O > E. The $CaCO_3$–$MnCO_3$ system resembles the $CaCO_3$–$MgCO_3$ system with the double carbonate phase kutnahorite, $CaMn(CO_3)_2$ (14.2.1.3), separating two solid solution series. *MSASC 11:60(1983)*. Rhodochrosite is a common primary gangue mineral in moderate- to low-temperature hydrothermal veins with Ag, Pb, Zn, Cu ores containing manganite, albabandite, tetrahedrite, and sphalerite, along with calcite, siderite, dolomite, fluorite, barite, and quartz; also high-temperature metasomatic deposits with rhodonite, friedelite, garnet, braunite, alabandite, hausmannite, and tephroite; and in pegmatites, especially with lithiophilite, also as a secondary mineral in deposits of iron and manganese oxyhydroxides and oxides. Occurrences found in the pegmatites of Poland, ME; at Franklin, Sussex Co., NJ; in veins in Batesville, AR; in CO as rose-red crystals in John Reed mine(*), Alicante, the Climax mine(!), and the Matchless mine(!), near Leadville, Lake Co.; at the Moose mine(*), Central City, Gilpin Co.; at the Sweet Home mine(!), near Alma, Park Co., in clear simple rhombs to 10 cm on quartz druses with fluorite; at the American Tunnel mine(!), Gladstone, San Juan Co., in rhombs to 5 cm on white quartz. Rhodochrosite is abundant at Butte and at Phillipsburg, MT; at the silver mines near Austin, Tonopah, and other districts in NV; at Park City, Marysvale, and Bingham, UT; and in the Tesla district, Alameda Co., with bementite in southern Trinity Co., CA. In Canada at Mt. St.-Hilaire(!), QUE, in color-zoned (orange-red with a red-brown exterior) rhombs to 10 cm, usually twinned, associated with aegirine, albite, leucophanite, microcline, polylithionite, and sphalerite; brownish to pinkish rhombs. In Mexico at the Potosi mine(!), Sta. Eulalia, CHI, in rhombs and scalenohedra to 5 cm and botryoids of pink to blood-red color, often gem quality, associated with fluorite, hubnerite, helvite, kutnahorite, and so on. Rhodochrosite is found in western Merionethshire, England; with rhodonite, friedelite at Vielle Aure, Hautes-Pyrenees, in France; in the Ardennes, Belgium; at Waldalgesheim, Germany, in spheres to 8 mm and fine microcrystals at Grube Wolf, Herdorf, Siegerland. A well-known European locality is Kapnik, Romania; also Ljubija, Slovenia; bladed rhodochrosites are found on sphalerite crystals from Trepca(!), Serbia; in the Polunchony district, northern Ural Mts., and Krasnoyarsk, Siberia, Russia. In Japan there are beautiful banded and botryoidal formations that contain rhombohedral crystals in a druse of rhodochrosite associated with pyrite, sphalerite at the Inakuraishi(*) and Yakuno(*) mines, Hokkaido, where the ore consists of rhodochrosite, alabandite, sphalerite, galena, pyrite, marcasite, pyrrhotite, tetrahedrite–tennantite, pyrargyrite, also at the Oppu mine, Aomori Pref. At Mounana, Gabon; as clear red scalenohedral crystals at N'Chwaning mine(!), Kurumen Mt. district, and in banded stalactitic deposits at Hotazel(!), Kalahari manganese field in the Cape Prov., South Africa; at Burra, SA, and at Broken Hill, NSW, Australia. Exceptional large red rhombs to 15 cm have been found at Huallapon mine(!), Pasto Bueno, Ancash Prov., Peru; at la Rioja, Capillitas bed(!), in Andalgala Co., Catamarca, Argentina, rhodochrosite is found in caverns as large, deep-pink, banded stalactites from which polished decorative objects are made. HCWS *DII:174*, *MSASC 11(1983)*.

14.1.1.5 Sphaerocobaltite $CoCO_3$

Named in 1877 for composition and form. Calcite group. Also called cobaltocalcite. HEX-R $R\bar{3}c$. $a = 4.6581$, $c = 14.958$, $Z = 6$, $D = 4.208$. *AM 46:1283(1961)*. *11-692*(syn): 3.55_4 2.74_{10} 2.33_2 2.11_2 1.948_2 1.702_3 1.697_3 1.495_1. Crystals are rare; usually as small spherical masses or crusts with concentric and radiating structure. Rose-red, altering to gray, brown, or velvet black; peach-blossom pink streak; translucent. Cleavage $\{10\bar{1}1\}$. H = 4. G = 4.13. IR. *MSLM 4:239(1974)*. Uniaxial (−); $N_O = 1.855$, $N_E = 1.60$; dichroic, violet-red to rose-red. Small amounts of Ni, Fe, and Ca substitute for Co. Found at Boleo, BCS, Mexico; *Schneeberg, Saxony, Germany*, with erythrite and annabergite in Co–Ni veins; in dolomite at Étoile du Congo mine(!), Shaba, Zaire; Jervois Range, NT, Australia. HCWS *DII:175*.

14.1.1.6 Smithsonite $ZnCO_3$

Named in 1832 after James Smithson (1754–1829), whose generous bequest facilitated the founding of the Smithsonian Institution in Washington, DC. Calcite group. HEX-R $R\bar{3}c$. $a = 4.653$, $c = 15.026$, $Z = 6$, $D = 4.43$. *MSASC 11:11(1983)*. *8-449*: 3.55_5 2.75_{10} 2.33_3 2.11_2 1.946_3 1.776_1 1.703_5 1.515_1. Rarely well crystallized; rhombohedral faces generally curved, rough, and composite; ordinarily botryoidal, reniform, stalactitic; encrusting; coarsely granular to compact; massive, earthy, and friable. As porous to cellular or cavernous masses with drusy surfaces. Grayish white to dark gray, greenish or brownish white, also green to apple-green, blue, green-blue, yellow, pink, brown, or white; rarely colorless and transparent; translucent; white streak. Cleavage $\{10\bar{1}1\}$, not perfect; translation gliding T$\{0001\}$, $t[10\bar{1}1]$; uneven to conchoidal fracture; brittle. H = 4–4$\frac{1}{2}$. G = 4.2 (pure). IR. *MSLM 4:239(1974)*. May fluoresce greenish or bluish white in UV. Uniaxial (−); $N_O = 1.848$, $N_E = 1.621$ (Na). Synthetic studies (> 600°C) showed no evidence of a double carbonate, but see minrecordite, $CaZn(CO_3)_2$ (14.2.1.4). Limited amounts of Fe, Co, Cu, Mn, Cd, Mg, and Pb are often found in natural smithsonite. Soluble in acids with effervescence. Varieties with a range of colors (e.g., ferroan, manganoan) are well known. A common mineral but not yet described as a primary hydrothermal phase, smithsonite is normally found in the oxidized zone of ore deposits or as a replacement of adjacent calcareous rocks, where it probably derived from the alteration of sphalerite. Often found in association with hemimorphite, cerussite, malachite, azurite, anglesite, pyromorphite, mimetite, aurichalcite, willemite, and hydrozincite. Smithsonite is found at Friedensville, Lehigh Co., PA; Mineral Point, Iowa Co., WI; Joplin and Granby districts, MO; in large deposits at Monarch and Leadville, CO; Tintic and Ophir districts, UT; Elkhorn district, MT; as fine green to blue botryoidal masses at Kelly mine(!), Magdalena Co., NM; Wood River district, ID; in the 79 mine(*), Gila Co., the Mammoth St. Anthony mine(*), Tiger, Pinal Co., micromount samples were obtained at the Old Yuma mine, Pima Co., AZ; at Bisbee, Cochise Co.; and in the Cerro Gordo district, Inyo Co., CA. In Mexico as pink, botryoidal masses with balls to 5 cm at Choix, SIN, and at the San Antonio mine, CHIH, as stalactites and botryoidal masses to 25 cm in gray, green, yellow, and bright orange (Cd). Abundant in the lead–zinc deposits of the Iglesias district in Sardinia, Italy; in the Aachen district, Rhineland, Germany; at Tarnowskie and Bytom, Silesia, Poland; a cobaltiferous variety at Stoima Planina, Bulgaria: in a variety of colors at Laurium mines(*), Attica, Greece; at Broken Hill Mine, Zambia, both botryoidal and as crystals; fine rhombohedral and scalenohedral colorless, white, pink, bright green, blue, yellow, and in-between-shade crystals up to several centimeters were found at Tsumeb(!), Namibia, associated with cerussite, mimetite, duftite, willemite, and rosasite; at Proprietary mine(!), Broken Hill, NSW, Australia, in botryoidal and mammillary forms as well as crystals. HCWS *DII:176*, *MSASC:11(1983)*.

14.1.1.7 Otavite $CdCO_3$

Named in 1906 for the Otavi district, Namibia, where Tsumeb is located. Calcite group. HEX-R $R\bar{3}c$. $a = 4.923$, $c = 16.287$, $Z = 6$, $D = 5.03$. *MSASC 11:11(1983)*. Structure: *DANS 245:1099(1979)*. *42-1342*(syn): 3.78_7 2.95_{10} 2.46_3 2.07_2 1.891_1 1.839_2 1.826_3 1.583_2. See also *8-456*(syn). Minute rhombohedral crystals $\{10\bar{1}1\}$. Crusts of white to yellow-brown or reddish, adamantine luster. Not brittle. $H = 4$. $G = 4.96$. Uniaxial $(-)$; $N_O = 1.851$, $N_E = 1.612$. A secondary mineral associated with smithsonite, cerussite, olivenite, pyromorphite, azurite, and malachite from *Tsumeb, Namibia*. HCWS *DII:181*.

14.1.1.8 Gaspéite $NiCO_3$

Named in 1966 for the locality, Gaspé Peninsula, QUE, Canada. Calcite group. HEX-R $R\bar{3}c$. $a = 4.621$, $c = 14.93$, $Z = 6$, $D = 4.35$ ($\approx 10\%$ Mg gaspeite). *12-771*(syn): 3.51_5 2.71_{10} 2.30_3 2.09_4 1.926_3 1.755_2 1.681_5 1.673_4. See also *23-437*, $(Ni, Mg)CO_3$. Rhombohedral crystals up to 0.5 mm length. Light green, yellow-green streak, vitreous to dull luster. Good rhombohedral cleavage $(10\bar{1}1)$, uneven fracture. $H = 4\frac{1}{2}$–5. $G = 3.71$. IR, DTA magnesian gaspéite. Uniaxial $(-)$; light green; $N_O = 1.83$, $N_E = 1.61$ for magnesian–gaspeite. Gaspéite occurs as a 60-cm wide light green vein, essentially pure (minor serpentine and chrome spinel), in varicolored siliceous dolomite in *Lemieux*

Twp., Gaspé Nord Co., QUE, Canada, in association with crystalline millerite, niccolite, annabergite, gersdorfite, and magnesite; at Kambalda, WA, Australia. HCWS *AM 51:677(1966)*.

14.1.2.1 Vaterite $CaCO_3$

Named in 1911 after Heinrich Vater (1859–1930), mineralogist and chemist of Tharandt, Saxony, Germany. Polymorphs: calcite (14.1.1.1) and aragonite (14.1.3.1). HEX $P6_3/mmc$. $a = 7.135$, $c = 16.98$, $Z = 12$, $D = 2.645$. *MM 32:535(1960)*. *33-268*: 4.22_3 3.57_6 3.29_{10} 2.73_9 2.11_2 2.06_5 1.854_3 1.820_7. See also *13-192*(syn). Often spherulitic, radially fibrous, admixed with aragonite or calcite. Hexagonal plates or lens-shaped skeletal crystals resembling snowflakes were grown artificially. White, or colorless, transparent. $G = 2.645$. IR, Raman. *MSLM 4:247(1974)*. Uniaxial (+); $N_O = 1.550$, $N_E = 1.650$. Synthesized from hydrogels containing KCO_3 at 5°. Inverts to aragonite or calcite in boiling water, and to calcite if water contains NaCl. Dry crystals are stable in air but change to calcite at 440°. Vaterite has been found as rims around amorphous Ca_3SiO_5 in boreholes of Arco No. 1, Susi, Ulster Co., NY; as pseudomorphs after larnite in *Ballycraigy, Larne, Northern Ireland*; in the Hatrurim formation, Israel; in the nacre of mollusks. HCWS *DII:181, ZK 128:183(1969)*.

14.1.2.2 Gregoryite $(Na_2,K_2,Ca, \square)CO_3$

Named in 1981 for J. W. Gregory (1864–1932), English geologist. HEX $P6_3mc$. $a = 5.215$, $c = 6.584$, $Z = 2$, $D = 2.269$. Phase stable $> 490°$; X-ray data at 565°. *25-815*(syn): 3.72_3 3.29_4 2.66_{10} 2.61_8 2.14_4 1.863_3 1.707_1 1.506_1. Pattern probably also contains nyerereite (14.3.4.1). Equigranular, rounded, brownish, in a matrix containing bladed crystals of nyerereite and opaque grains of alabandite, MnS. Phase plots close to the solubility limit of $CaCO_3$ in $(NaK)_2(CO_3)$. Intimately associated with nyerereite in the *carbonate lavas* that emanated from *Oldoinyo Lengai volcano, Tanzania*. HCWS *AM 66:879(1981), LIT 13:23(1980)*.

ARAGONITE GROUP

The aragonite mineral group is similar to the calcite mineral group, both having the chemical formula

$$ACO_3$$

but in aragonite

$$A = Ca, Sr, Ba, Pb$$

and these species crystallize in the orthorhombic system, space group $Pmcn$.

The structure [*AM 56:758(1956)*] consists of alternating layers of cations and trigonal carbonate groups perpendicular to the c axis, with the triangles pointing in opposite directions in alternate layers, similar to the minerals of the calcite group. The larger size and coordination number of the cations in arago-

Aragonite Group

nites group leads to an orthorhombic structure type. The carbon of the aragonite CO_3^{2-} is slightly out of the plane (0.026 Å in aragonite, less in strontianite and witherite) slightly displaced along the b axis (0.20 Å in aragonite) from one carbonate layer to the next along c, and the plane defined by the three oxygens is tilted about 2.5° from (001). Further, the divalent cations are not strictly planar. Alternate atoms vary, each being ±0.05 Å out of the plane and coordinated with a total of nine oxygens: six bonds are made with two (edge) oxygens on three separate carbonate groups, and three bonds with single (corner) oxygens of three other carbonate groups. The result is a distorted polygon in which the distances to corner oxygens are shorter than those with the six edge oxygens (2.41–2.45 Å versus 2.52–2.66 Å for aragonite). Because the cations are pseudohexagonally arranged within (001), the disposition of the slightly puckered atoms produces an ABAB layered sequence along c, and the stacking order of the alternating cation and anion planes along c becomes $AC_1BC_2AC_1BC_2$. The result is an orthorhombic rather than hexagonal–rhombohedral space group. *MSASC 11:146(1983)*.

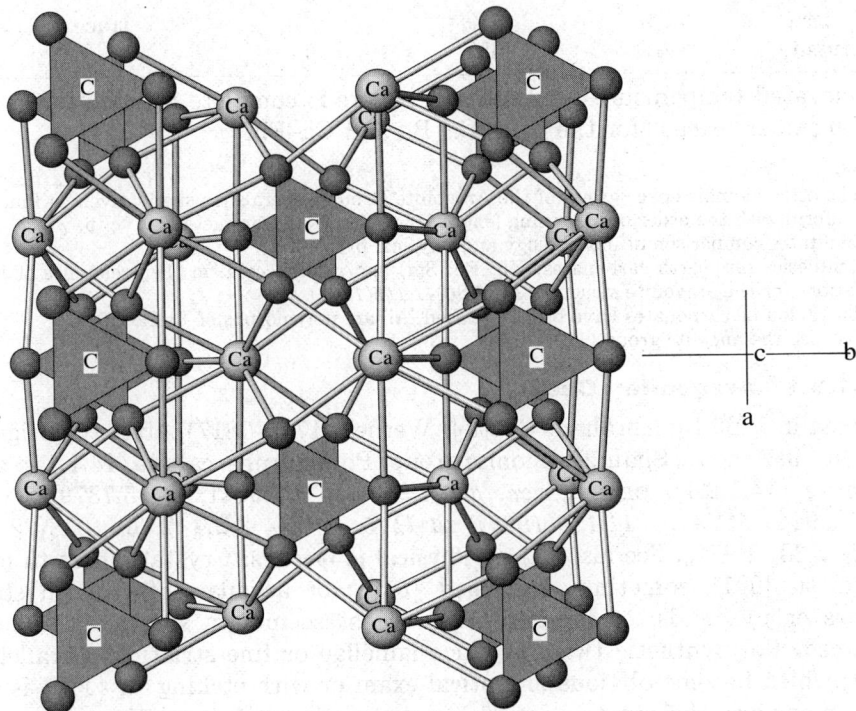

ARAGONITE: CaCO₃

The aragonite structure is similar to calcite (14.1.1.1) with alternating layers of CO_3 pointing in opposite directions but larger size cations take on nine-fold coordination with the oxygens of the CO_3 groups. This c axis projection illustrates the different polygonal arrangement for calcium. Minerals in the group normally have a prismatic habit and orthorhombic, pseudo-hexagonal symmetry. Aragonite, for example, has a prism angle of 62–63° and repeated twinning on {110} results in pseudohexagonal aggregates. Compare with ancylite (16b.1.1.1).

Aragonite Group*

Mineral	Formula	Space Group	a	b	c	D	
Aragonite	$CaCO_3$	$Pmcn$	4.9598	7.9641	5.7379	2.944	14.1.3.1
Witherite	$BaCO_3$	$Pmcn$	5.3126	8.8958	6.4284	4.29	14.1.3.2
Strontianite	$SrCO_3$	$Pmcn$	5.090	8.358	5.997	3.785	14.1.3.3
Cerussite	$PbCO_3$	$Pmcn$	5.1800	8.492	6.134	6.582	14.1.3.4

*For data from synthetic samples, see MSASC 11:146(1993) table 1a).

There is limited isomorphous substitution detected between members of the group. MSASC 11:159(1983).

Maximum Miscibility in the Aragonite Group (mol %)

	Ca	Sr	Ba	Pb
Aragonite	—	14%	0	2%
Strontianite	27%	—	3%	0
Witherite	0.5%	8%	—	trace
Cerussite	3%	3%	0	—

At elevated temperatures and pressures there is complete miscibility for the end members except for the pairs Ca–Ba and Ca–Pb.

Notes:
1. The orthorhombic space group of the Aragonite Group minerals is usually given as Pmcn to conform with the order of increasing length for the unit cell parameters: a <c<b, which facilitates comparison of morphology and physical properties.
2. Synthetics rare-earth carbonates (Yb, Eu, Sm) and radium confirm the preference of large cations for the aragonite structure MSASC 11:145(1983).
3. Three double carbonates have been described. All are polymorphs of $BaCa(CO_3)_2$.
4. See also the ancylite group (16b.1.1.1).

14.1.3.1 Aragonite $CaCO_3$

Named in 1790 by Abraham Gottlob Werner (1749–1817), after the original locality in Aragon, Spain. Aragonite group. Polymorphs: calcite (14.1.1.1) and vaterite (14.1.2.1). ORTH $Pmcn$. $a = 4.9598$, $b = 7.9641$, $c = 5.7379$, $Z = 4$, $D = 2.944$. MSASC 11:146(1983). 41-1475: 3.40_{10} 3.27_5 2.70_6 2.48_4 2.37_5 2.34_3 2.33_3 1.977_6. See also 5-453. **Physical properties:** Crystals short to long prismatic [001], sometimes flattened {010}; or acicular, often with steep domes or pyramids; or tabular {001}. Most seemingly single crystals are twinned. Polysynthetic twins produce lamellae or fine striations parallel to [001] which become obvious on optical exam or with etching. If {110} is the twin plane and the composition plane, the result is the formation of pseudo-hexagonal aggregates of contact and penetration types (Fig. 14.1.3.1A). Aragonite is also found as stalactites; as columnar, radiating, or stellate aggregates; as lamellar crusts composed of fibers; as acicular crystals. Colorless to white, gray, pale to deep blue, green, rose-red or violet; uncolored streak; transparent to translucent; vitreous to resinous luster. Cleavage {010}, distinct; subconchoidal fracture; translation gliding T{010}, t[100],

14.1.3.1 Aragonite

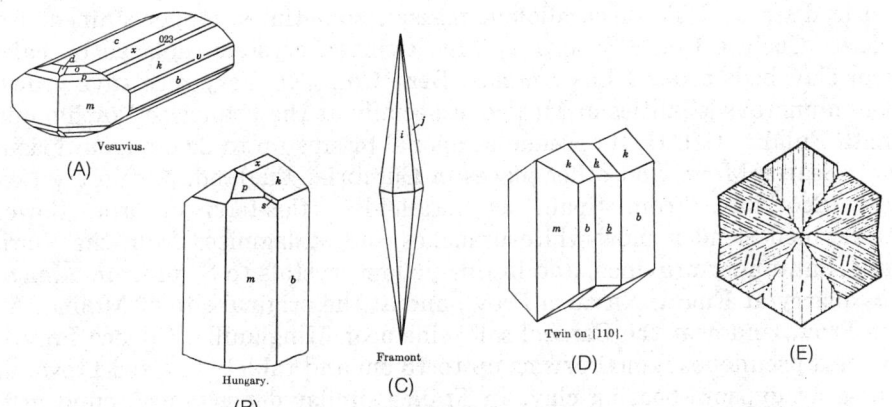

Figure 14.1.3.1a
CRYSTAL FORMS OF ARAGONITE
(A) Prismatic [001], (B) Prismatic {110}, {010}, (C) Acicular with steep pyramids, (D) twin on {110}, (E) Pseudohexagonal aggregate of contact and penetration twins. b = (010), c = (001), d = (102), i = (021), j = (0, 10, 1), k = (011), m = (110), p = (111), s = (121), v = (031), x = (012).

and twin gliding; brittle. $H = 3\frac{1}{2}$–4. $G = 2.947$. IR, Raman *MSLM 4:242(1974)*, *SWVD:36(1991)*. Fluorescent under electron, X-ray, UV, usually pale rose, yellow, bluish, with greenish phosphorescence in LWUV, yellowish in SWUV. Biaxial (−); $N_x = 1.531$, $N_y = 1.681$, $N_z = 1.685$; $2V_{calc} = 18.25°$; $r < v$; optics variable with cationic substitution. **Chemistry:** Some minor substitution of Sr, Pb, and Zn are usual with the varieties exhibiting zoning. Limited F substitution detected in inorganic marine aragonites and Sr, Na, and Mg in aragonitic shell materials. Soluble with dilute acids, with effervescence. Easily synthesized from concentrated solutions of calcium salts, especially in the presence of the larger atoms named above, at relatively higher temperature than calcite, or with more rapid precipitation from highly concentrated reactants. **Occurrence:** Aragonite, the most common orthorhombic anhydrous carbonate and one of the three polymorphs of $CaCO_3$, is unstable relative to calcite at surficial temperature and pressure; paramorphs of calcite after aragonite are common. However, the mineral is often found in stalactites and stalagmites; as pisolites, sinters, and massive lamellar deposits at geysers and hot springs along with sulfates; as sea-bottom oolites; with siderite in iron deposits, with calcite and dolomite and other magnesian minerals in altered serpentines, or as the replacement mineral in sedimentary, metamorphic, and basic igneous rocks; in ore deposits, where it forms from low-temperature and pressure aqueous solutions; as the carbonate phase in invertebrate skeletons and the sediments derived from them. Aragonite may be a primary phase in high-pressure metamorphic rocks. **Localities:** Good crystals come from many localities, but North American specimens are uncommon. Fine crystals are found in numerous caves along the Appalachian mountain chain, and with calcite, from the Magdalena district, Socorro Co., NM. A zincian variety occurs with secondary zinc ores from Leadville, CO; in the Tintic district, UT; as coralloidal masses, sometimes copper-stained, from Bisbee, Cochise Co., AZ; also as large twinned crystals replaced by calcite from clay beds around Las Animas, Bent Co., CO. Very nice cave growths from numerous localities in Mexico, especially at the Francisco Portillo mine, Santa Eulalia, CHIH. As pseudohexagonal prisms up to 20 cm from Frizington, Cleator Moor, and other places in Cumbria, England, and in the Leadhills, Scotland. From Spain as coralloidal (flos-ferri or iron flowers) intertwined slender snow-white branches and stalagmites from the Florida mine, Santander province, and in fine yellow crystals to 8 cm from a magnesite quarry at Eugui, Navarra Prov., and at the original site of Molina, Aragon Prov., and near the Caracol salt mine near Minglanilla, Cuenca Prov., as twinned pseudohexagonal prisms up to 15 cm and tablets in a red-brown and blue-gray gypsum-bearing clay. In France similar deposits are found in the region of Dax, Pouillon, and Bastennes; also occurring in volcanic rocks of the Auvergne, Puy de Dôme. From Germany in basaltic rock at Kaiserstuhl, Baden, and as pseudomorphs after gypsum (schaumkalk) at Wiederstadt in the Harz. A plumbian variety (tarnowitzite) occurs in dolomitic lead ore from Tarnowitz, Silesia, Poland. From Austria as flos-ferri at Erzberg near Eizsenerz and Huttenberg, Lolling, also as crystals from Werfen and Leogang, Salzburg. Beautiful orthorhombic clear yellow crystals (suitable for faceting) to over 10 cm from veins traversing basalt near Horschenz (Bilin), Czech Republic, and as pseudohexagonal twins from Herrengrund, and as the finest white elongated prisms reaching 30 cm from Podrechany, Hungary. Numerous localities from the sulfur mines of Sicily, Italy, particularly from Racamulto, Cianciana, and Agrigento as fine pseudohexagonal prisms to 10 cm. Fine flos-ferri from the old iron, lead, and silver mines around Laurion, in southern Greece. Also found with the geyser deposits and sulfurous hot pools at Osorezan; Tottoku dist.; at Kuriyama Yuzawa, Shimotsuke Pr., at the Takasecawa R., Shinano Pr.; as nodular aggregates in clay and gypsum at Matsushiro, Iwami Pr., Japan. From Morocco in bursts of crystals similar to the Spanish pseudohexagonal twins from Tezouta, near Sefrou and as bluegreen and olive green botryoidal crusts on galena, and as plumbian variety white pseudohexagonal prisms to 7 cm from Touissit mine, south of Oujda; as clear, white, and blue needlelike and white elongated pseudohexagonal prisms of plumbian variety crystals from Tsumeb, Namibia. Pseudomorphs of copper after aragonite are found at Corocoro, Bolivia. HCWS *DII:191, MSASC 11:195(1983)*.

14.1.3.2 Witherite BaCO$_3$

Named in 1790 after William Withering (1741–1799), English physician and mineralogist, who called attention to the mineral. Aragonite group. See alstonite (14.2.5.1), paralstonite (14.2.2.2), and barytocalcite (14.2.6.1), all polymorphs of the double carbonate BaCa(CO$_3$)$_2$. ORTH *Pmcn*. $a = 5.3126$, $b = 8.8958$, $c = 6.4284$, $Z = 4$, $D = 4.29$. *MSASC 11:146(1983)*. *45-1471*:

3.72_{10} 3.66_5 3.22_2 2.63_3 2.59_4 2.15_5 2.02_4 1.941_3. See also *44-1487*(syn). Crystals always repetitively twinned on {110}, resulting in pseudohexagonal dipyramids, also short prismatic [001], tabular to lenticular. Morphology often complex, faces rough, horizontally striated; also globular, botryoidal, granular, coarse fibrous. Colorless to milky white-gray, also tinted yellow, brown, or green; white streak; transparent to translucent. Cleavage {010}, distinct; {110}, imperfect; uneven fracture. H = 3–3$\frac{1}{2}$. G = 4.291. IR, Raman. *MSLM 4:242(1974)*. Fluorescent and phosphorescent in X-ray, electron beams, UV. Thermoluminescent. Biaxial (−); XYZ= *cba*; $N_x = 1.529$, $N_y = 1.676$, $N_z = 1.677$ (Na); 2V = 16°; r > v, weak. Minor amounts of Sr and Ca detected in analyses. Soluble in dilute HCl. Witherite is the most common barium mineral after barite, although a rare mineral. It occurs in low-temperature hydrothermal veins with barite, alstonite, calcite, galena, and other sulfides. Fine crystals at Minerva mine, Cave in Rock, Hondin Co., IL(!); at El Potal, Mariposa Co., CA. Important commercial vein deposits with barite occur in northern England, particularly *Alston Moor, Cumberland*; in Lamothe coal measures and at Brioude, Haute-Loire, France; in barite veins in Arkhyz and Djalankol deposits, North Caucasus, Kopets Mts., Karakala region, Turkmenistan; as radial aggregates at Tsuaki silver mine Ugo Pref., Japan. HCWS *DII:196*, *MSASC 11:145(1983)*.

14.1.3.3 Strontianite SrCO$_3$

Named in 1791 after the town in Scotland where the mineral was found. Aragonite group. The element Sr was discovered in this material (1790). ORTH *Pmcn*. a = 5.090, b = 8.358, c = 5.997, Z = 4, D = 3.785. *MSASC 11:146(1983)*. *5-418*(syn): 3.54_{10} 3.45_7 2.48_3 2.46_4 2.45_3 2.05_5 1.986_3 1.905_4. See also *44-1421*, calcian strontianite. Crystals short prismatic [001], often acicular, or acute spear-shaped. High calcian varieties often show steep pyramidal forms; often pseudohexagonal. {110} and {010} striated horizontally, {*hhl*} and {0*kl*} sometimes rounded; massive, columnar to fibrous; granular; rounded masses. Twinning common, twin plane {110}, usually contact rarely penetration twins; repeated twinning gives trillings; polysynthetic twinning results in obvious twin lamellae. Colorless to gray, yellowish, greenish, yellow-brown to reddish; transparent to translucent. Cleavage {110}, nearly perfect; uneven to subconchoidal fracture; brittle. H = 3$\frac{1}{2}$. G = 3.76 (pure). IR, Raman. *MSLM 4:242(1974)*. Fluorescent and phosphorescent in X-ray, electron beams, and UV, sometimes thermoluminescent. Biaxial (−); XYZ= *cba*; $N_x = 1.520$, $N_y = 1.667$, $N_z = 1.669$ (Na); $2V_{calc} = 7.1°$; r < v, weak. Solid solutions of Ca in strontianite up to 27 mol %, Ba up to 3.3%. Soluble in dilute HCl. Strontianite is a low-temperature hydrothermal mineral associated with barite, celestite, and calcite, often as veins in limestone or marble; occasionally, part of the gangue with sulfide minerals; also forms geodes, concretionary forms in clays. Geodes and veins with celestite and calcite in limestones of Schoharie Co., NY; with aragonite in limestones in Mifflin Co., PA; also with celestite in calcite cap rock of salt domes, Gulf coast LA and

TX; large deposits in Strontium Hills, N of Barstow, San Bernardino Co., CA. In Canada vein deposits in limestones at Nepean, Carleton Co., ONT, and in Cariboo district, BC. Commercially important deposits in marls of Westphalia, Germany, principally near Hamm and Münster, also with metal veins in the Harz and near Freiberg, Saxony; also with zeolites at Oberschaffhausen in the Kaiserstuhl, Baden. Original description from veins in a gneiss at *Strontian, Argyllshire, Scotland*. With celestite in a lead–silver deposit in Sierra Mojada district, COAH, Mexico; with celestite, gypsum, and phosphatic nodules in clay at Trichy, India. HCWS *DII:196, MSASC 11:158(1983)*.

14.1.3.4 Cerussite $PbCO_3$

Artificial lead carbonate was written about as early as 400 B.C. by the Greeks, named after *cerussa* (Latin, *white lead*). Aragonite group. ORTH *Pmcn*. $a = 5.1800$, $b = 8.492$, $c = 6.134$, $Z = 4$, $D = 6.582$. *MSASC 11:146(1983)*. 5-417(syn): 3.59_{10} 3.50_4 3.07_2 2.52_2 2.49_3 2.08_3 1.933_2 1.859_2. **Physical and chemical properties**: Extremely varied crystal forms, including tabular {010}, elongated [001] or [100], equant, dipyramidal, pseudohexagonal, rarely acicular, thin tabular {001}. {010} and {0kl} striated on [100]; commonly groups in clusters due to almost universal twinning on {110}, {130}. Colorless to white, gray, smoky, sometimes dark gray to black, due to included particles of sulfides, manganese oxides, or carbonaceous materials; also blue to green with copper impurity; colorless to white streak; adamantine to vitreous, resinous, and pearly luster, sometimes submetallic if dark due to surface films; transparent to translucent. Cleavage {110}, {201}, distinct; conchoidal fracture; brittle. $H = 3-3\frac{1}{2}$. $G = 6.55$. IR, Raman. *MSLM 4:242(1974)*. Fluorescent yellow with X-rays, LWUV. Biaxial (−); $N_x = 1.804$, $N_y = 2.077$, $N_z = 2.079$ (Na); $2V_{calc} = 9°$; $r > v$, strong. Soluble with effervescence in 1:10 HNO_3. Natural cerussite shows little cation substitution. **Occurrence**: Cerussite is a secondary Pb mineral, an alteration product of galena, PbS, and anglesite, $PbSO_4$. It is the corrosion product of metallic lead or lead alloys and has been known since Roman times. A common lead mineral, cerussite is found in the upper oxidized portions of ore deposits with iron oxides, anglesite, pyromorphite, malachite, azurite, smithsonite, and other lead, zinc, or copper secondary minerals. **Localities**: Fine specimens were found at Wheatley Mines, Phoenixville, PA(!); Austinville, VA; Joplin, MO; Mineral Point in southern WI; abundant at Leadville, CO(*); Elkhorn and Wickes districts, Jefferson Co., MT; an important ore mineral in silver–lead districts of Shoshone and Coeur d'Alene, ID. Twinned crystals from Stevenson–Bennet mine(!), Organ district, Dona Ana Co., NM, and in the Magdalena district(!) and other places; fine twin specimens and reticulated masses with wulfenite and other minerals at the Mammoth–St. Anthony mine(!), Tiger, Pinal Co.; in the Warren district(*); the Hilltop mine(!), Cochise Co.; the 79 mine(!), Gila Co.; Flux mine(!), Santa Cruz Co.; Red Cloud mine(!), Yuma Co., AZ; also occurs in the Eureka and Goodsprings districts, NV; at the Modoc mine(!), Monster mine(!), and the Cerro Gordo mine, Inyo Co., as well as others in southern CA. In Mexico at the Bilbao mine(*), ZAC, and in CHIH at Santa Eulalia, and as pseudomorphs after anglesite at Mina Erupcion(!), Los Lamentos. In Canada at Salmo, BC, in the Moyie and H.B. mines. Abundant in Spain in Sierra de Cartagena, Murcia, and in Linares regions of Andalusia; fine crystals were found with leadhillite, caledonite, and linarite at Leadhills(!), Scotland; at Mies, Bohemia, Czech Republic; at Rézbánya, Comitat Bihar, Hungary; with anglesite and phosgenite in lead mines of Monteponi(!) and Montevecchio(!), Sardinia, Italy; V-shaped twins occur at the Touissit Mine(!), Oujda district, and crystals associated with barite are found at Mibladen, Sefron region, Morocco; exceptional large transparent twin crystals at the Kombat mine(!), Klein Otavi, and crystals, reticulated twins, of exceptional size, variety, and perfection are associated with azurite, malachite, smithsonite, and other minerals at the Tsumeb mine(!), Namibia; at the Broken Hill mine, Zambia; occurs with zinc phosphates at Mindouli, Zaire. At the Proprietary mine(!), Broken Hill, NSW, Australia, the major portion of the oxidized zone was pure cerussite, and magnificent crystals and reticulated groups of large size were obtained, and also in TAS, the Mt. Bischoff(!) and Magnet(!) mines near Waratah produced bright yellow crystals of chromian cerussite and the Comet mine(!) near Dundas gave spectacular specimens; in New Zealand the Tui mine(!) on Mt. Te-Aroha; at the San José mine(*), Oruro Prov., Bolivia. Cyclic twinned crystals and flat crystals up to 5 cm length were found at the Boquira mine(!), BA, Brazil, and the Santana Nova prospect(*) near Lagerado, SP, gave crystals and twins to 10 cm. HCWS *DII:200, MSASC 11:174(1983)*.

14.1.4.1 Rutherfordine (UO_2)CO_3

Named in 1906 for New Zealand–born, British atomic physicist and Nobel laureate Ernest Rutherford (1871–1937). ORTH *Pmmn*. $a = 4.845$, $b = 9.205$, $c = 4.296$, $Z = 2$, $D = 5.72$. *SCI 121:472(1955)*. 11-263: 4.61_{10} 4.30_7 3.92_3 3.23_4 2.64_3 2.31_2 2.06_2 1.926_2. Lathlike crystals up to 3 mm length with [001] elongation, large (100), (010); striated on (100) parallel to [001]. Pale yellow to amber. Cleavage parallel to (010), perfect; (001) fair. Biaxial (+); XYZ= bca (c elongation); $N_x = 1.715$, $N_y = 1.730$, $N_z = 1.795$; $2V_{calc} = 53°$. *AM 41:844(1957)*. The positive sign, which distinguishes this carbonate mineral, is probably due to the disposition of linear $(O-U-O)^{+2}$ ions perpendicular to the layers of $(CO_3)^{2-}$ groups which parallel (010) in the structure. Soluble in acids. Found as pseudomorphs after uraninite in pegmatites in *Uruguru Mts., Morogoro district, Tanzania*; as radiating strawlike clusters on uraninite with masuyite from Shinkolobwe, Shaba, Zaire; El Sharana, NT, Australia. HCWS *DII:247*.

14.1.5.1 Widenmannite $Pb_2(UO_2)(CO_3)_3$

Named in 1961 after Bergrat J. F. Widenmann (1764–1798), who first discovered uranium containing micas in the Schwarzwald. ORTH *Pnnm, Pnm2$_1$*, or *P22$_1$2$_1$*. $a = 8.99$, $b = 9.36$, $c = 4.95$, $Z = 2$, $D = 6.89$. *SMPM 56:167(1976)*.

27-281: 4.16$_{10}$ 3.34$_7$ 3.19$_8$ 2.57$_4$ 2.34$_{10}$ 1.911$_5$ 1.869$_5$ 1.473$_5$. Fragile, tabular (100), yellow crystals. Pearly to silky luster. Cleavage {100}, perfect. Not fluorescent in UV. Biaxial (−); XYZ= *bac*; N$_x$ = 1.803, N$_y$ = 1.905, N$_z$ = 1.945; 2V = 63°. Soluble in 1:10 HNO$_3$. Occurs as tufts or radiating bundles of lathlike crystals in cavities associated with compact quartz, cerussite, hügelite, hallimondite, and altered galena in the *Michael mine, Weiler bei Lehr, Schwarzwald, Germany*. HCWS *AM 47:415(1961)*.

14.1.6.1 Natrite Na$_2$CO$_3$

Named in 1983 for the composition. Zabuyelite (14.1.6.2) is the Li analog; see also thermonatrite (15.1.1.1). MON *C2* or *Cm*. $a = 8.906$, $b = 5.238$, $c = 6.045$, $\beta = 101°$, Z = 4, D = 2.55. *ZVMO 111:220(1983)*. *37-451:* 2.96$_{10}$ 2.62$_3$ 2.60$_4$ 2.55$_6$ 2.37$_7$ 2.25$_4$ 2.19$_2$ 2.18$_4$. Massive, pseudohexagonal. Gray-white to colorless, dull luster. Cleavage {001}, perfect; (100) (110), less perfect. H = 3½, G = 2.54. Soluble in water. DTA, DGA data. IR confirms absence of water in compound. Biaxial (−); Y = *b*; N$_x$ = 1.410, N$_y$ = 1.535, N$_z$ = 1.543; OAP (010); 2V = 28°. Occurs in deep workings and drill cores in mines of Kola Penin., Khibina massif, and with foyaite(?) on mine dumps on *Mt. Karnasurt, Lovozero massif, Kola Penin., Russia*. HCWS *MA:755(1983)*.

14.1.6.2 Zabuyelite Li$_2$CO$_3$

Named in 1978 for the locality. MON *C2/c*. $a = 8.356$, $b = 4.964$, $c = 6.185$, $\beta = 114.6°$, Z = 4, D = 2.10. *AM 75:243(1990)*. *22-1141*(syn): 4.16$_9$ 3.80$_2$ 3.03$_3$ 2.92$_8$ 2.81$_{10}$ 2.63$_3$ 2.49$_2$ 2.43$_4$. Prismatic crystals 1.5–7 μm long embedded in rock salt layers with gaylussite and northupite, or as doubly terminated (pyramids) prisms up to 1.2 mm long from periphery of Li-rich salt lake. Colorless, transparent, vitreous luster. Cleavage {100}, perfect; {011}, moderate; brittle; twinned (100). Not fluorescent. G = 2.09. Biaxial (−); Y = *b*; Z∧*c* = 10°; N$_x$ = 1.429, N$_y$ = 1.567, N$_z$ = 1.574; 2V = 25°. Slightly soluble in water, effervesces in 1:10 HCl. Found at *Zabuye Salt Lake, Nagri, Tibet, China*. CS *MM 54:671(1990)*.

DOLOMITE GROUP

The dolomite group, four double carbonates of divalent cations, has the general formula

$$AB(CO_3)_2$$

where

A = Ca
B = Mg, (Fe^{2+},Mg), Mn^{2+}, Zn

and where the space group is $R\bar{3}$.

The dolomite group minerals show similarities with the calcite group. (CO$_3$)$^{2-}$ groups in planar layers are perpendicular to the *c* axis and alternate with layers of cations. The two types of cations are segregated into self-contained layers giving rise to the ordered sequence A–(CO$_3$)–B–CO$_3$–A–(CO$_3$)–B(CO$_3$)... along c. The ordering influences the cation-oxygen bond lengths, and results in a lowering of the space group symmetry from R$\bar{3}$c in calcite to R$\bar{3}$ in dolomite.

DOLOMITE GROUP

Mineral	Formula	Space Group	a	c	D	
Dolomite	CaMg(CO$_3$)$_2$	R$\bar{3}$	4.8069	16.0034	2.85	14.2.1.1
Ankerite	Ca(Fe^{2+},Mg, Mn^{2+})(CO$_3$)$_2$	R$\bar{3}$	4.830	16.167	3.00–3.10	14.2.1.2
Kutnohorite	Ca(Mn^{2+},Mg, Fe^{2+})(CO$_3$)$_2$	R$\bar{3}$	4.915	16.639	3.12	14.2.1.3
Minrecordite	CaZn(CO$_3$)$_2$	R$\bar{3}$	4.8183	16.0295	3.44	14.2.1.4

Notes:
1. CaFe(CO$_3$)$_2$, is not known experimentally nor in nature.
2. Substitution and disordering between the A and B cation positions is common with a concomitant reduction of stability of the ordered structure relative to single cation rhombohedral end members. This is especially true for Fe-containing systems.
3. Microscale compositional (and structural) variations are evidenced in the saddle-shaped habit of some dolomite group minerals (R. Faux, pers. com., 1996).

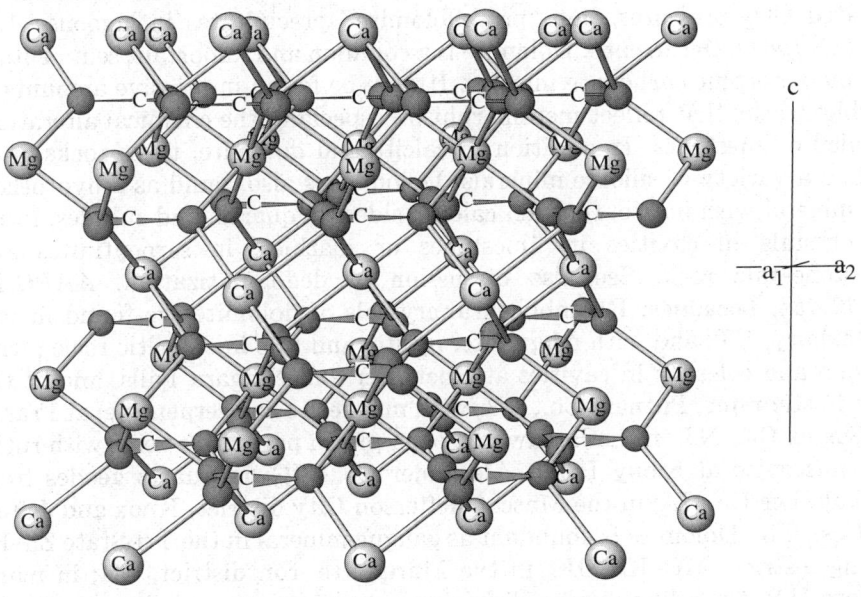

DOLOMITE: CaMg(CO$_3$)$_2$

This projection down a* of dolomite, a rhombohedral double carbonate, is virtually identical to adleite, but with two ordered and distinct cation layers representing the 50:50 mole proportions of A and B cations. The space group changes to R$\bar{3}$.

14.2.1.1 Dolomite CaMg(CO$_3$)$_2$

Named in 1791 after Déodat Guy Silvain Tancrède Gratet de Dolomieu (1750–1801), French engineer and mineralogist. Dolomite group. HEX-R $R\bar{3}$. $a = 4.8069$, $c = 16.0034$, Z = 3, D = 2.85. *MSACS 11:22(1983)*. *36-426*: 3.70_1 2.89_{10} 2.67_1 2.40_1 2.19_2 2.02_1 1.805_1 1.787_1. See also *11-78*. **Physical properties:** Commonly rhombohedral $\{10\bar{1}1\}$, or $\{40\bar{4}1\}$ dominant; prismatic $\{11\bar{2}0\}$ terminated by rhombohedrons; tabular $\{0001\}$ with $\{11\bar{2}0\}$. $\{10\bar{1}1\}$ often striated horizontally or curved, the "saddle" forms. Twinning with $\{0001\}$, $\{10\bar{1}0\}$, or $\{11\bar{2}0\}$ twin planes common. Also massive, coarse to fine granular; fibrous or pisolitic. Colorless to white, grayish, greenish, becoming more brown with increase of Fe^{2+}, or reddish with Mn^{2+}; streak same colors; vitreous to pearly luster, sometimes porcelaineous or opaline; subtranslucent. Cleavage $\{10\bar{1}1\}$, perfect; conchoidal fracture; brittle; twin gliding $\{02\bar{2}1\}$; translation gliding T$\{0001\}$, t$[10\bar{1}0]$. H = $3\tfrac{1}{2}$–4. G = 2.85, values increase when Fe or Mn substitutes for Mg. Uniaxial (–); N$_O$ = 1.679, N$_E$ = 1.500 (pure). IR, Raman. *MSLM 4:254(1974)*, *ISM:82,84,86(1991)*. Some varieties fluoresce in UV; triboluminescent. **Chemistry:** Limited substitution in the cation positions; ferroan, cobaltian, plumbian, and manganoan dolomite with distinct coloration have been described. The maximum Ca detected in dolomite can be expressed as Ca$_{57}$Mg$_{43}$. Differences in the unit cell parameters between low- and high-temperature dolomite add to the complexity of interpretation of cation ordering. *MSASC 11:37(1983)*. Synthesis of dolomite at ordinary temperatures and pressures has not been demonstrated except at elevated CO$_2$ pressures, but "protodolomite" precipitates in lagoons. *AJS 261:449(1963)*. **Occurrence:** Dolomite is a common and important sedimentary and metamorphic carbonate mineral. It may be found in massive amounts in marbles, rocks that reflect metamorphism, especially the chemical alteration, of calcitic limestones. In addition to calcite and dolomite, these rocks often contain a variety of silicate minerals. Dolomite is also found as a hydrothermal mineral with barite, fluorite, calcite, siderite, quartz, and sulfides; forming crystals in cavities in limestones or marbles; in serpentinites and soapstone–talc rock. See also discussion of dedolomitization. *AAPG(B) 12:762–764*. **Localities:** Rhombohedral crystals of dolomite are found in talc at Roxbury, VT, and with magnesite, pyrite, and talc in chloritic rocks; with gypsum and celestite in cavities at Rochester and Niagara Falls, and at the Tilly Foster mine, Putnam Co., NY, with magnetite and serpentine; at Franklin, Sussex Co., NJ, in the Buckwheat open pit; in pegmatite veins with rutile and muscovite at Stony Point, Alexander Co., NC; in quartz geodes from Keokuk, Lee Co., IA; in the Mascot–Jefferson City district, Knox and Jefferson Cos., TN. Dolomite is abundant as gangue mineral in the Tri-State Zn–Pb mining district, MO–KS–OK; in the Marquette iron district, MI; in many western U.S./Canadian mining districts as a gangue mineral (i.e., at Climax mine, Lake Co., and Camp Bird mine, Ouray Co., CO); ferroan dolomite is found at the Vekol mine, Casa Grande, AZ; at Guanajuato silver mines, GTO, Mexico, and at Santa Eulalia(*), CHIH, in crystals to 5 cm. Sharp, clear, transparent individuals to 10 cm and crystalline groups were found in a quarry at Eugui, Navarre, Spain(!); in England in the metallic veins of Cornwall; in France at Markirch, Alsace; in the German mines of Freiberg, Saxony. Fine crystals are found in Brosso(!), Traversella in the Piedmont, and at Pfitschtal in Trentino, Tuscany, while massive dolomite strata gave the name to the Dolomites, a mountain chain in northern Italy; notable crystals occur in dolomite rock at Lengenbach quarry(!), Binnenthal, Valais, and in rock salt at Bex, Switzerland; clear crystals are found near Djelfa(*), Algeria; and Co, Mn, Pb, and Zn varieties are found at the Tsumeb mine(!), Tsumeb, Namibia. A relatively homogeneous gray dolostone comprised almost entirely of interlocking rhombohedral dolomite crystals is found in Minowa, Kuzuu, Ashio Mts., Tochigi, Japan; dolomite is also found in the Ibi district, Gifu Pref.; at the Proprietary mine, Broken Hill, NSW, and in the modern sediments of the Coorong, southeastern SA, Australia; in Brazil at the Pedra Preta mine(!), Brumado district, BA, in sharp yellowish crystals and twins to 20 cm associated with crystalline magnesite and other minerals; ferroan dolomite at the Morro Velho gold mine(!), Nova Lima, MG, Brazil. Dolomite has been reported in carbonaceous chondrites. HCWS *DII:208*.

14.2.1.2 Ankerite Ca(Fe^{2+},Mg,Mn^{2+})(CO$_3$)$_2$

Named in 1825 after Mathias Joseph Anker (1771–1843), Styrian (Austria) mineralogist. Dolomite group. The pure iron analog of dolomite does not exist naturally or synthetically; up to 70 mol % CaFe(CO$_3$)$_2$ is soluble in the dolomite structure with high Fe (>17.5 mol % Fe) containing specimens called ankerite, the others ferroan dolomite. Data that follow are for Ca[Mg$_{0.27}$Fe$_{0.63}$Ca$_{0.05}$Mn$_{0.05}$](CO$_3$)$_2$. HEX-R $R\bar{3}$. $a = 4.830$, $c = 16.167$, $Z = 3$, $D = 3.00$–3.10. *MSASC 11:24(1983)*. *41-586*: 3.71_1 2.91_{10} 2.41_1 2.20_1 2.02_1 1.818_1 1.797_3 1.795_3. See also *33-282*. Similar to dolomite with rhombic crystalline forms, also massive fine to coarse granular; twinning on {0001}, {10$\bar{1}$0}, {11$\bar{2}$0}. Colorless, white to grays, browns, darker on weathering; translucent, vitreous to pearly luster. Cleavage {10$\bar{1}$1}, perfect; subconchoidal fracture. $H = 3\frac{1}{2}$–4. $G = 2.87$. IR. *AM 45:311(1960)*. Fluorescent in UV, triboluminescent. Uniaxial $(-)$; $N_O = 1.764$, $N_E \approx 1.52$. *DII:211*. Fe and Mg contents are highly correlated, while few ankerites show Ca deficiencies. Ferroan dolomite is commonly found in similar occurrences as dolomite. High-Fe-containing species do not occur on a rock-forming scale but are associated with iron ores, such as with hematite at Antwerp Jefferson Co., NY; at Jeffrey quarry, Little Rock, AR; Snowbird mine, Mineral Co., MO; at Coeur d'Alene region in sulfide veins, Shoshone Co., ID; King Co., Spruce Peak, WA; in Nova Scotia in ores at Londonderry, Colchester Co., NB; also at Erzberg near Eisenerz, Austria, and with siderite at Csetnek, Com Gömör, and at Pribam, Bohemia, Czech Republic; at Tsumeb mine, Namibia; at the Mamatwan and Boltsburn mines, Durham, Cape Prov., South Africa; in Ogatatown, Oita Pref., Japan; at the Muzo emerald mine, Boyacá, Colombia. HCWS *DII:216, MSASC 11:28(1983)*.

14.2.1.3 Kutnohorite $Ca(Mn^{2+},Mg,Fe^{2+})(CO_3)_2$

Named in 1901 after the locality. Dolomite group. HEX-R $R\bar{3}$. $a = 4.915$, $c = 16.639$, $Z = 3$, $D = 3.12$. *JPD 3:172(1988)*. *11-345:* 3.75_2 2.94_{10} 2.44_1 2.23_2 2.04_2 1.876_1 1.837_3 1.814_3. Sample from Franklin, NJ, which contains Mn, minor Mg and <0.5% Fe. See also *19-234*, calcian kutnohorite, and *43-695*, magnesian kutnohorite. Granular, massive. White to pale rose, translucent. Cleavage $\{10\bar{1}1\}$, perfect; subconchoidal fracture; brittle. H = $3\frac{1}{2}$–4. G = 3.12. IR. *MSLM 4:253(1974)*. Uniaxial (–); $N_O = 1.727$, $N_E = 1.535$ (Franklin, NJ). Ca deficiencies of up to 10% $CaCO_3$ as well as excess Ca have been reported. Mn content varies from 38 to 84 at %. Typical sample gives diffuse diffraction patterns, which suggest microscale composition and structure inhomogeneities. Occurs at Franklin and Sterling Hill mines, Ogdensburg, NJ; Camp Bird mine, Ouray, CO; at Providencia, Mexico; at *Kutná Hora, central Bohemia, Czech Republic*; at Ryûjima mine, Nagano Pref., and Hoei mine, Oita Pref., Japan; at Broken Hill, NSW, Australia. HCWS *DII:217*; *AM 40:748(1953), 52:1751(1967)*; *AMG 4:187(1966)*; *MSASC 11:31(1983)*.

14.2.1.4 Minrecordite $CaZn(CO_3)_2$

Named in 1982 after the journal *Mineralogical Record*. Dolomite group. HEX-R $R\bar{3}$. $a = 4.8183$, $c = 16.0295$, $Z = 3$, $D = 3.44$. *35-667:* 4.04_1 3.70_2 2.89_{10} 2.41_3 2.20_2 2.02_2 1.806_4 1.789_4. Rhombs $\{10\bar{1}4\}$, up to 0.5 mm. Milky white, pearly luster. Often saddle-shaped, with $\{10\bar{1}4\}$ cleavage. H = $3\frac{1}{2}$. G = 3.45. Uniaxial (–); $N_O = 1.750$, $N_E = 1.550$. No fluorescence with UV. Found on dioptase, and with zincian dolomite, duftite, and malachite needles in the oxidation zone of *Tsumeb mine, Tsumeb, Namibia*. Mg-rich minrecordite (larger unit cell, lower indices) is also found at this locality. These colorless, glassy, transparent crystals contain ochre inclusions and are associated with cerussite and calcite. HCWS *MR 13:131(1982)*.

14.2.2.1 Norsethite $BaMg(CO_3)_2$

Named in 1960 for Keith Norseth (1927–1991), U.S. engineering geologist at Westvaco mine, west of Green River, WY. HEX-R $R32$. $a = 5.02$, $c = 16.75$, $Z = 3$, $D = 3.84$. *AM 46:420(1961)*. *12-530:* 4.21_3 3.86_4 3.02_{10} 2.66_4 2.51_4 2.10_4 1.931_4 1.864_4. Structurally similar to dolomite group. Platy $\{0001\}$ rhombs which may show $\{11\bar{2}0\}$, $\{10\bar{1}0\}$, and $\{10\bar{1}1\}$. Clear to milky white, vitreous to pearly luster. Good rhombohedral cleavage, hackly fracture. H = $3\frac{1}{2}$. G = 3.84. Uniaxial (–); $N_O = 1.694$, $N_E = 1.519$. Found as rhombohedral crystals below the main trona bed in black dolomitic oil shale at Westvaco mine, Sweetwater Co., WY, associated with shortite, labuntsovite, searlsite, loughlinite, and pyrite, also with northupite in gray shale. HCWS *MA 14:343(1960)*.

14.2.2.2 Paralstonite BaCa(CO$_3$)$_2$

Named in 1979 after alstonite. A trimorph of BaCa(CO$_3$)$_2$. See alstonite (14.2.5.1), barytocalcite (14.2.6.1); olekminskite (14.2.2.3), SrSr(CO$_3$)$_2$, forms a solid-solution series with paralstonite. HEX-R $P321$. $a = 8.692$, $c = 6.148$, $Z = 3$, $D = 3.62$. *AM 64:1332(1979)*. *33-178:* 6.15$_2$ 4.35$_1$ 3.55$_{10}$ 2.85$_1$ 2.51$_7$ 2.05$_2$ 1.943$_2$ 1.853$_2$. Structure similar to other double carbonates. *MSASC 11:156(1983)*. Pyramidal crystals up to 1 mm. Colorless to smoky white and gray masses, vitreous luster. H = 4–4$\frac{1}{2}$. G = 3.75 (high, due to barite impurity). Uniaxial (–); N$_O$ = 1.672, N$_E$ = 1.527. Dissolves with effervescence in dilute HCl. Occurs as coatings on barite at Cave-in-Rock district, IL, with yellow calcite, gray alstonite, purple fluorite, and dark brown sphalerite. HCWS *GSCP 79-1c:99(1979)*.

14.2.2.3 Olekminskite Sr(Sr,Ca,Ba)(CO$_3$)$_2$

Named in 1991 for the city Olekminsk, administration center of the Murunskij alkaline complex, Aldan Shield, Siberia. Forms a series with paralstonite (14.2.2.2). HEX-R $P321$. $a = 8.66$, $c = 6.08$, $Z = 3$, $D = 3.65$. *AM 78:451(1993)*. *44-1422:* 3.50$_{10}$ 2.49$_9$ 2.03$_9$ 1.928$_6$ 1.837$_6$ 1.581$_6$ 1.443$_6$ 1.305$_7$. Needlelike crystals up to 0.010 mm thick, round or hexagonal in cross section, form spherulitic aggregates 0.20–0.30 mm in diameter. Crystal centers are enriched in Ca and Ba, while the rims have high Sr. Transparent, vitreous luster. Brittle. H = 3. G = 3.70. Uniaxial (–); N$_O$ = 1.670, N$_E$ = 1.527; negative elongation. Analyses show virtually the complete range between SrSrCO$_3$ and paralstonite, CaBa(CO$_3$)$_2$; La and Ce in minor or trace amounts. Found in barytocalcite–quartz veins 5–10 cm wide that cut intrusive breccias accompanying the alkaline Kedrovyji massif, close to the Murunskij alkaline complex, *Aldan Shield, Siberia.* HCWS *ZVMO 120:89(1991)*.

14.2.3.1 Benstonite Ba$_6$Ca$_6$Mg(CO$_3$)$_{13}$

Named in 1961 for Orlando J. Benston (1901–1966), metallurgist at National Lead Co., Malvern, AR, who developed the method of flotation for barite. HEX-R $R\overline{3}$. $a = 18.28$, $c = 8.67$, $Z = 3$, $D = 3.648$. *AM 47:585(1962)*. *14-637:* 4.19$_2$ 3.92$_4$ 3.60$_1$ 3.35$_1$ 3.09$_{10}$ 2.54$_3$ 2.13$_3$ 1.905$_2$. Massive. White to ivory, white streak. Good rhombohedral cleavage {31$\overline{4}$2}. H = 3–4. G = 3.596. Uniaxial (–); N$_O$ = 1.690, N$_E$ = 1.527. Fluoresces red under LWUV or SWUV. Soluble in HCl. Found at the barite mine in Hot Spring Co., AR, associated with milky quartz, barite, and calcite. HCWS *NW 48:550(1961)*.

14.2.4.1 Ewaldite Ba(Ca,Y,Na,K)(CO$_3$)$_2$

Named in 1971 for Paul P. Ewald (1888–1985), German crystallographer, founder of *Zeitschrift für Kristallographie* and *Acta Crystallographica*, professor at Brooklyn Polytechnical Institute, NY. HEX $P6_3mc$. $a = 5.318$, $c = 12.837$, $Z = 2$, $D = 3.37$. *ZVMO 121:56(1992)*. Structure: *TMPM 15:201(1971)*. *43-678:* 6.39$_{10}$ 4.58$_2$ 4.31$_6$ 3.72$_4$ 3.20$_1$ 2.62$_1$ 2.44$_1$ 1.714$_1$. Tiny, <1 μm, untwinned flattened hexagonal dipyramids or prisms with {0001} terminations. Blue-green. G = 3.25. Uniaxial (−); $N_O = 1.646$, $N_E \approx 1.572$. O, dark blue-green; E, pale yellow-green. From *Green River formation, Sweetwater Co., WY*, trona mine and exploratory drill holes, crystals intimately intergrown with black mckelveyite (15.3.4.2), and associated with trona, labuntsovite, leusospherite, searlsite, and other minerals; Y-rich variety in carbonatites of Vuorijärvi ultrabasic massif, and (Ce + Nd)-rich varieties in hydrothermal sites in the Khibiny alkaline massif, Russia. HCWS *AM 56:2156(1971)*, *TMPM 15:185(1971)*.

14.2.5.1 Alstonite BaCa(CO₃)₂

Named in 1841 after locality. Trimorph. See paralstonite (14.2.2.2) and barytocalcite (14.2.6.1). TRIC $P1$ or $P\bar{1}$. $a = 17.38$, $b = 14.40$, $c = 6.123$, $\alpha = 90.35°$, $\beta = 90.12°$, $\gamma = 120.08°$, $Z = 12$, $D = 3.605$. Alstonite appears to have a superstructure based on paralstonite without long-range order of the metal cations or the CO₃ groups. *MSASC 11:156(1983)*. *27-32:* 5.65_1 3.55_{10} 2.84_1 2.51_5 2.18_1 2.05_3 1.943_2 1.846_2. Crystals steep pseudodihexagonal pyramids, faces striated horizontally, twinned. White, gray, creamy to reddish (bleached on exposure); white streak; vitreous luster. Cleavage {110}, imperfect; uneven fracture. H = 4–4½. G = 3.70. IR. *MSLM 4:258(1974)*. Biaxial (−); XYZ= cab; $N_x = 1.526$, $N_y = 1.671$, $N_z = 1.672$; $2V = 6°$; r > v, weak. Weakly fluorescent yellow in LWUV. Sr (and Mn) solid solution noted in studies of CaCO₃–SrCO₃–BaCO₃–PbCO₃ system. *MSASC 11:145(1983)*. Found at Minerva mine, Cave-in-Rock, IL; in low-temperature hydrothermal deposits at *Brownley Hill lead mine, Alston, Cumberland*, associated with calcite, barite, and witherite, also at the *Fallowfield lead mine near Hexham, Northumberland*, and in veins at New Brancepeth near Durham, England. HCWS *DII:218*.

14.2.6.1 Barytocalcite BaCa(CO₃)₂

Named in 1824 in allusion to the composition. MON $P2_1/m$. $a = 8.092$, $b = 5.2344$, $c = 6.544$, $\beta = 106.05°$, $Z = 2$, $D = 3.65$. *NBSR 75A:197(1971)*. Trimorph with alstonite (14.2.5.1) and paralstonite (14.2.2.2). *15-285:* 4.34_3 4.02_4 3.14_9 3.13_{10} 2.51_2 2.38_2 2.16_3 2.01_2. Structure similar to dolomite but with two crystallographically distinct CO₃ groups. *MSASC 11:156(1983)*. Short to long prismatic [101], [001] or equant; {100} striated [001], other faces striated [101]; massive. Often oriented growths on barite, or calcite on barytocalcite. Colorless to white, grayish, greenish, yellowish; white streak; vitreous to resinous luster; transparent to translucent. Cleavage {210}, perfect; {001}, imperfect; uneven fracture; brittle. H = 4. G = 3.66. IR. *MSLM 4:258(1974)*. Weakly fluorescent in UV. Biaxial (−); $Z = b$; $X \wedge c = 64°$; $Y \wedge c = 26°$; $N_x = 1.525$, $N_y = 1.684$, $N_z = 1.686$; $2V = 15°$; r > v, slight. Found as crystals up to 5 cm long, and also massive, in veins in limestones at *Blagill mine(!), Alston Moor, Cumberland, England*; at Långban, Värmland, Sweden, with hedyphane and hausmannite; in barite ore of Tolecheinski deposit, Krasnoyark region, Siberia, Russia. HCWS *DII:220*.

14.3.1.1 Bütschliite $K_2Ca(CO_3)_2$

Named in 1947 for Johann Adam Otto Bütschli (1848–1920), professor of zoology, Heidelberg, Germany, who studied the double carbonates of K and Ca. Dimorph with fairchildite (14.3.3.1). HEX-R $R\bar{3}m$. $a = 5.38$, $c = 18.12$, $Z = 3$, $D = 2.61$. *AM 59:353(1974)*. Structure similar to eitelite (14.3.2.1). *AM 58:211(1973)*. 25-626(syn): 3.02_3 2.86_{10} 2.69_1 2.07_1 1.690_2 1.641_2 1.044_2 0.933_2. Microscopic, barrel-shaped crystals showing $\{10\bar{1}1\}$, $\{01\bar{1}2\}$, $\{0001\}$, $\{11\bar{2}0\}$, $\{11\bar{2}2\}$; cleavage $\{0001\}$. IR. *MSLM 4:256(1974)*. Attacked by moisture, producing a $CaCO_3$ coating that protects bütschliite stability. Fairchildite is the high-temperature ($> 500°$) modification. Occurs in the clinkers of melted wood ash found in partly burned trees, mainly hemlock and fir, in the forests of the western United States. HCWS *DII:231, AM 32:607(1947)*.

14.3.2.1 Eitelite $Na_2Mg(CO_3)_2$

Named in 1955 after Wilhelm Hermann Julius Eitel (1891–1979), chemist, and founder and director of the Institute for Silicate Research, University of Toledo, Ohio. HEX-R $R\bar{3}$. $a = 4.942$, $c = 16.406$, $Z = 3$, $D = 2.74$. *AM 58:211(1973)*. 24-1227(syn): 3.79_2 2.73_3 2.60_{10} 2.47_3 2.25_2 2.07_2 1.898_2 1.531_1. Colorless rhombohedral crystals a few millimeters in diameter with rounded edges; synthetics show $\{0001\}$, $\{10\bar{1}1\}$, $\{01\bar{1}2\}$; good cleavage parallel to $\{0001\}$. $H = 3\frac{1}{2}$. $G = 3.737$. IR. *MSLM 4:256(1974)*. Uniaxial $(-)$; $N_O = 1.605$, $N_E = 1.450$. Found in drill-hole samples from the Green River formation, Uinta basin, UT, associated with reedmergnerite, leucosphenite, shortite, searlesite, and crocidolite; abundant in core from *Mapco Shrine Hospital No. 1 well, Duchesne Co., UT*, into brown marlstone, associated with trona, nahcolite, shortite, and northupite; at Malha Crater lake, Sudan. HCWS *AM 40:356(1955)*.

14.3.3.1 Fairchildite $K_2Ca(CO_3)_2$

Named in 1949 after John Gifford Fairchild (1882–1965), analytical chemist at the U.S. Geological Survey. Dimorph with bütschliite (14.3.1.1). *AM 59:353(1974)*. See also natrofairchildite (14.3.5.1). HEX $P6_3/mmc$. $a = 5.272$, $c = 13.280$, $Z = 2$, $D = 2.441$. 21-1287(syn): 6.67_1 3.19_{10} 2.70_3 2.65_7 2.23_2 2.17_2 2.04_1 1.891_1. Microscopic clear, colorless hexagonal blades parallel to $\{0001\}$ with good $\{0001\}$ cleavage. IR. *MSLM 4:256(1974)*. Uniaxial $(-)$; $N_O = 1.533$, $N_E = 1.498$. Soluble in H_2O, decomposes. P, Al, Fe, and Mn are minor contaminants. Higher temperature form than bütschliite. Fairchildite blades are enclosed in dense, light gray bluish, or gray, friable matrix with a white porcelaneous crust that also contains bütschliite and calcite. The samples are fused wood-ash clinkers that occur in partly burned fir and hemlock trees in many national parks and forests in the western United States. HCWS *DII:222, BGSA 57:1218(1947)*.

14.3.3.2 Zemkorite $Na_2Ca(CO_3)_2$

Named in 1988 for the Institute of the Earth's Crust (Russian, *zemnoy kory* = earths crust), Russian Academy of Science, Siberian Branch, where the mineral was studied. Trimorph with nyerereite (14.3.4.1), natrofairchildite (14.3.5.1). HEX $P6_3/mmc$. $a = 10.06$, $c = 12.72$, $Z = 8$, $D = 2.47$. *41-1440:* 6.36_9 4.36_{10} 3.18_6 3.04_{10} 2.52_{10} 2.18_7 2.06_{10} 1.797_8. Tabular grains, 0.1–0.5 mm across, show no faces; also as fan-shaped aggregates 3–4 mm across. Colorless, transparent, vitreous to pearly luster. Perfect cleavage parallel to tabular direction, no twinning, brittle. H = 2. G = 2.46. Uniaxial (−); $N_O = 1.535$, $N_E = 1.513$. Analyses also showed K and trace of Al. Found in drill cores at 400–450 m depth into the eastern body of the *Udachnaya pipe, Daldyn kimberlite field, Yakutiya, Russia*. The mineral occurs as fracture fillings in unaltered kimberlite, along contacts between groundmass and olivine xenoliths and xenocryts, or in the interstices. Associated with shortite and rarely halite. HCWS *AM 75:933(1990).*

14.3.4.1 Nyerereite $Na_2Ca(CO_3)_2$

Named in 1963 after the president of Tanzania, Julius K. Nyerere (b.1922). Dimorph of zemkorite (14.3.3.2), possibly also of natrofairchildite (14.3.5.1). ORTH $Cmc2$. $a = 5.044$, $b = 8.809$, $c = 12.743$, $Z = 4$, $D = 2.417$. *AM 63:600(1978). 33-1221:* 6.38_9 4.39_9 3.05_{10} 2.58_8 2.54_9 2.20_8 2.16_9 2.07_9. Colorless pseudohexagonal platy crystals up to 0.5 mm across, invariably twinned. G = 2.54. Biaxial (−); XYZ = cab; $N_x = 1.511$, $N_y = 1.533$, $N_z = 1.534$; 2V = 29°. Analyses also show Sr, Ba, K, SO_3, Cl, and F. Inverts to hexagonal cell at 292°; synthetic material has $a = 20.3$, $c = 12.4$. A major constituent of the 1960–1965 "carbonate" lavas from *Mt. Oldoinyo Lengai, Tanzania*. HCWS *AM 60:487(1975).*

14.3.5.1 Natrofairchildite $Na_2Ca(CO_3)_2$

Named in 1971 as the Na analog of fairchildite (14.3.3.1). Possibly trimorphous with nyerereite (14.3.4.1) and zemkorite (14.3.3.2). HEX $P6_3/mmc$. $a = 5.291$, $c = 13.218$, $Z = 2$, $D = 2.14$. *25-804:* 6.71_6 4.50_3 3.18_{10} 2.67_4 2.64_9 2.20_6 2.10_5 1.891_6. Single platy crystals 1–2 mm without faces, or fan-shaped aggregates. White, vitreous luster. One (probably basal) cleavage, polysynthetic twinning parallel to cleavage. H = $2\frac{1}{2}$. Uniaxial (−); $N_O = 1.459$, $N_E = 1.525$. Slightly luminescent orange in UV. Analyses show only 1.4% K, lower amounts of Sr and Ba, traces of Mg, Fe, Mn, La, and Y; decomposes in air. Found in burbankite–calcite veins at depths greater than 70 m, *Vuori-Jarvi massif, Kola Penin., Russia*. HCWS *AM 60:488(1975).*

14.4.1.1 Shortite $Na_2Ca_2(CO_3)_3$

Named in 1940 after Maxwell Naylor Short (1889–1952), professor of mineralogy, University of Arizona. ORTH $C2mm$. $a = 4.961$, $b = 11.03$, $c = 7.12$, $Z = 2$, $D = 2.621$. Structure: *AMG 1:95(1949). 21-1348:* 5.99_5 5.52_7 4.96_7 3.83_5 2.56_{10} 2.48_5 2.18_7 1.996_6. Wedge-shaped crystals, tabular parallel to {100}, equant or short prismatic {100}; {011} often striated [100] in combination with $\{00\bar{1}\}$. Colorless to pale yellow, transparent. Distinct cleavage parallel to $\{010\}$, conchoidal fracture. $H = 3$. $G = 2.6$ (probably impure). IR. *MSLM 4:256*. Fluoresces amber color in UV. Biaxial $(-)$; $XYZ = cab$; $N_x = 1.531$, $N_y = 1.555$, $N_z = 1.570$; $2V = 75°$; $r < v$, moderate. Decomposes in H_2O giving $CaCO_3$ ppt. Occurs with calcite and pyrite as single crystals in inch-wide bands in the montmorillonite clay associated with trona in *Green River formation, WY*, and UT; in oil-well cores, Sweetwater and Uintah Cos., UT; in kimberlite at Upper Canada gold mine, Ontario, Canada; minute inclusions in apatite from Tororo carbonatite complex, eastern Uganda. HCWS *AM 24:514(1939)*, *DII:222*, *USGSPP:B1(1969)*.

14.4.2.1 Sahamalite-(Ce) $MgCe_2(CO_3)_4$

Named in 1953 after Thure Georg Sahama (1910–1983), petrologist, University of Helsinki, Finland. MON. $P2_1/a$. $a = 1.679$, $b = 16.21$, $c = 4.63$, $\beta = 106.7°$, $Z = 2$, $D = 4.30$. *AM 38:941(1953)*. *6-189:* 5.33_7 4.26_7 3.90_{10} 3.65_{10} 3.03_7 2.87_{10} 2.12_7 2.10_7. Crystals up to 0.2 mm tabular parallel to $\{\bar{2}01\}$, other forms $\{1\bar{3}1\}$, $\{120\}$, $\{110\}$, $\{010\}$, $\{011\}$. Colorless. Poor cleavage parallel to $\{010\}$. $G = 4.30$. Biaxial $(-)$; Y parallel to b; $Z \wedge c = 29°$; $N_x = 1.679$, $N_y = 1.776$, $N_z = 1.807$; $2V = 57°$; $r < v$. Minor amounts of Fe^{2+} substitute for Mg, and La and Nd for Ce. Insoluble in boiling HCl, soluble boiling $3N$ HNO_3. Occurs in barite–dolomite rock at *Mountain Pass, San Bernardino Co., CA*, associated with parisite and altered bastnäsite. HCWS *AM 38:941(1953)*.

14.4.3.1 Huntite $Mg_3Ca(CO_3)_4$

Named in 1955 after Walter Frederick Hunt (1882–1975), professor of mineralogy, University of Michigan. HEX-R. $R32$. $a = 9.505$, $c = 7.821$, $Z = 3$, $D = 2.87$. *AM 46:1304(1961)*. Structure: *AC 15:238(1962)*. *14-409:* 2.89_2 2.83_{10} 2.60_1 2.43_1 1.991_1 1.972_3 1.765_2 1.757_2. Compact, chalklike with submicroscopic (<1 μm) platelets or fibers. White, earthy; subconchoidal fracture of mass. Soft. $G = 2.696$. DTA, IR. *MSLM 4:255(1974)*. $N_{av} = 1.617$. Effervesces in 1:1 HCl, HNO_3, H_2SO_4. Occurs in caves where rocks are magnesium-rich and as a near-surface weathering product of rocks containing brucite, dolomite, magnesite, and serpentines; as a diagenetic product in recent sediments. Found at *Gabbs, NM*; Carlsbad Caverns, NM; Crestmore, CA; Grotte de la Clamouse, Herault region, France; Dorog, Hungary; Kurgashinkin, Uzbekistan; aggregates of several kilograms in serpentinites, Yazd, Iran; Tea Tree Gully, SA, Australia. HCWS *AM 38:4(1953)*.

BURBANKITE GROUP

The burbankite group of REE-containing anhydrous carbonate minerals conform to the general formula

$$A_3B_3(CO_3)_5$$

where

A = Na, Ca, □
B = Ca, Sr, Ba or rare-earth elements

and crystallized in layered structure with hexagonal or monoclinic pseudo-hexagonal symmetry.

The five minerals so far determined as belonging in the group show Na as essential with series between the remaining cationic species. As described for other layered carbonates, there is evidence of supercells.

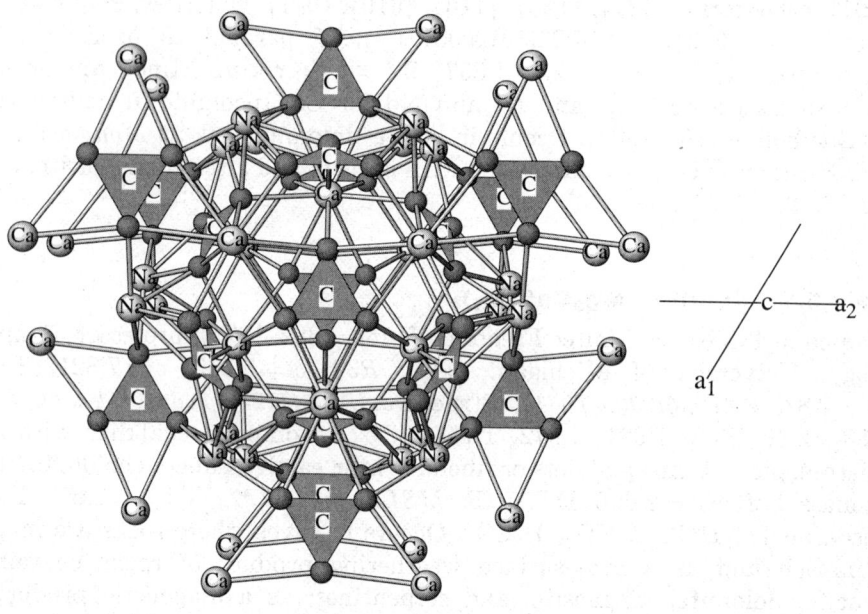

BURBANKITE: $Na_3Ca_3(CO_3)_5$

In this c axis projection of burbankite [simplified to $Na_3Ce_3(CO_3)_5$)] three distinct crystallographic sites for CO_3 groups are visible: one group at the center of the projection is on a mirror plane perpendicular to 'c' with the C_1 atom having 6_3 or rhombohedral symmetry; the second group, almost parallel to 'c' can only be seen end-on, while the third CO_3 group (also perpendicular to the c axis) appears as an outer ring. All three CO_3 groups are effectively planar. The A site cations (usually occupied by Na or Ca with a vacancy to balance any charge requirements) have eight-fold coordination. The B site cations (REE, Ba) have ten-fold coordination. Petersenite has tenfold cation coordination for both A and B sites *NJMM:161 (1985), SPC 13:192 (1968)*.

14.4.4.2 Khanneshite

BURBANKITE GROUP

Mineral	Formula	Space Group	a	b	c	β	D	
Burbankite	$(Na,Ca,\square)_3(REE, Ba, Sr)_3(CO_3)_5$	$P6_3mc$	10.514		6.520		3.5	14.4.4.1
Khanneshite	$(Na,Ca,\square)_3(Ba, REE, Sr)_3(CO_3)_5$	$P6_3mc$	10.65		6.58		3.94	14.4.4.2
Calcioburbankite	$Na_3(Ca,REE,Sr)_3(CO_3)_5$	$P6_3mc$, $P6_3mmc$, or $P\bar{6}2c$	10.447		6.318		3.46	14.4.4.3
Remondite-(Ce)	$Na_3(Ce,La,Ca,Na,Sr)_3(CO_3)_5$	$P2_1$	10.444	6.313	10.445	119.86	3.46	14.4.5.1
Petersenite-(Ce)	$Na_4(REE)_2(CO_3)_5$	$P2_1$	20.872	6.367	10.601	120.50	3.67	14.4.5.2

14.4.4.1 Burbankite $(Na,Ca,\square)_3(Sr,Ba,Ce)_3(CO_3)_5$

Named in 1955 after Wilbur Swett Burbank (1898–1975) of the U.S. Geological Survey. Burbankite group. HEX $P6_3mc$. $a = 10.514$, $c = 6.520$, $Z = 2$, $D = 3.5$. *CM* 12:342(1974). Structure: *NJMM*:161(1985), *AC(C)* 45:185(1989). 26-1374(syn): 5.24_5 3.71_6 3.03_8 2.74_4 2.63_{10} 2.62_{10} 2.14_6 1.982_4. Slender hexagonal prisms terminated by {0001}; as radiating aggregates or cleavable masses up to 3 cm across; microbotryoidal crusts. Colorless, pale yellow to pinkish. Cleavage {1000}, distinct to imperfect. H ≈ $3\frac{1}{2}$. G = 3.5. Uniaxial (−); $N_O = 1.616$, $N_E = 1.597$. May have up to 13 wt % Ba or up to 15 wt % RE (Ce, Y, La, Pr, Nd, Sm, Dy, Yb). Soluble in dilute HCl with effervescence. Found as yellow crystals intimately intergrown with ancylite and other rare-earth carbonates in veins in shonkinites(!), *Big Sandy Creek, Hill Co., Bearpaw Mts., MT*; as crusts or in veins containing much organic material and RE = 22%; associated with mckelveyite–ewaldite in the Wilkins Peak member of the Green River formation, Sweetwater Co., WY; as dihexagonal prismatic crystals embedded in analcime in pegmatite dikes(!) in Mt. St.-Hilaire, QUE, Canada, associated with microcline, aegirine, calcite, siderite, and ancylite; with kamerinite in the Nlende deposit, Kribi, Cameroon. HCWS *AM* 38:1169(1953), 62:158(1977).

14.4.4.2 Khanneshite $(Na,Ca)_3(Ba,Sr,Ce)_3(CO_3)_5$

Named in 1982 for the locality. Burbankite group. Possibly forms a series with burbankite (14.4.4.1) and remondite-(Ce) (14.4.5.1). HEX $P6_3mc$. $a = 10.65$, $c = 6.58$, $Z = 2$, $D = 3.94$. *AM* 68:1249(1983). 35-700: 9.25_3 5.34_3 3.78_5 3.29_3 3.08_6 2.66_{10} 2.19_6 2.09_4. Prismatic crystals 5–10 mm long, 2–3 mm wide. Pale yellow. Indistinct cleavage parallel elongation, rough transverse parting, brittle. Soft. G = 3.8–3.9. DTA. Uniaxial (−); $N_O = 1.623$–1.620, $N_E = 1.610$–1.609. Chemical analyses show total REE oxides >8 wt %: La, Ce, Pr, Nd, Sm, Eu, Gd, Dy, Ho, Er, Tm, Yb, Y detected. Readily dissolved by HCl. Commonly altered and replaced by barite. Disseminated in fine-grained carbonatite associated with dolomite, calkinsite, carbocernaite, barite, and chlorite at *Khanneshin, Afghanistan*. HCWS *ZVMO 111:321(1982)*.

14.4.4.3 Calcioburbankite $Na_3(Ca,REE,Sr)_3(CO_3)_5$

Named in 1995 for the resemblance to burbankite (14.4.1.1). Burbankite group. HEX $P6_3mc$, $P6_3mmc$, or $P\bar{6}2c$. $a = 10.447$, $c = 6.318$, $Z = 2$, $D = 3.46$. *CM 33:1231(1995)*. PD: 5.20_5 3.68_5 2.601_{10} 2.63_5 2.180_5 2.130_5 2.050_5. Crystals elongate parallel to [001] to 2 cm in length with deep orange interior color, also equant to 0.5 mm showing {100}, {001}, and silky white exterior with light pink interior; vitreous luster. Cleavage {100}, indistinct. H = 3–4. G = 3.45. Uniaxial (−); $N_O = 1.636$, $N_E = 1.631$; nonpleochroic. Microprobe analysis (wt %): Na_2O, 15.17; CaO, 11.8; BaO, 0.46; SrO, 7.65; La_2O_3, 9.30; Ce_2O_3, 14.38; Pr_2O_3, 1.26; Nd_2O_3, 3.76; Sm_2O_3, 0.04; CO_2 (calc), 35.13. Dissolves readily in 1:10 HCl. There are two different associations for this mineral at *Mt. St.-Hilaire, QUE, Canada:* (1) with ancylite-(Ce), calcite, chlorite-group minerals, donnayite-Y, fluorapatite, natrolite, pyrite, rhodochrosite, and rutile perhaps formed in an associated vent [*MR 21:310,368(1990)*], and (2) as subhedral crystals in a fragment of a marble xenolith where shortite is the predominant mineral phase along with aegirine, calcite, fluorite, galena, leucophanite, mangan-neptunite, microcline, molybdenite-2H and 3R, narsarsukite, pictolite, schairerite, sodalite, sphalerite, thermonatite, and titanite. HCWS

14.4.5.1 Remondite-(Ce) $Na_3(Ce,La,Ca,Na,Sr)_3(CO_3)_5$

Named in 1988 for Guy Remond (b.1935), physicist and research mineralogist at Bureau de Recherches Géologiques et Minières, Orleans, France. Burbankite group. MON $P2_1$. $a = 10.444$, $b = 6.313$, $c = 10.445$, $\beta = 119.86°$, $Z = 2$, $D = 3.46$. *CR 307:915(1988)*, *AC(C) 45:185(1989)*. PD: 5.17_4 3.01_4 3.006_7 2.62_5 2.61_7 2.59_{10} 2.13_7 2.127_5. *AM 75:433(1990)*. Massive, pseudohexagonal. Red-orange, vitreous luster, transparent. Imperfect cleavage, conchoidal fracture. H = 3–3½. G = 3.43. TGA. Biaxial (+); $N_x = 1.632$, $N_y = 1.633$, $N_z = 1.638$; $2V = 40°$. REE dominate, in addition to Ce minor amounts of La, Nd, Y, and trace Pr, Sm, Eu, Gd, Dy, Ho, Er, Yb, and Lu. Soluble in HCl, HNO_3 with effervescence. Found as massive fillings in veinlets in nepheline syenite orthogneiss at *Ebounja, near Kribi, Cameroon*, associated with nepheline, monazite, pyrochlore, betafite, bastnaesite, cancrinite, sodalite, and calcite. HCWS *MA 89:3585(1989)*.

14.4.5.2 Petersenite-(Ce) $Na_4(REE)_2(CO_3)_5$

Named in 1994 for Ole Valdemar Petersen (b.1939), mineralogist and curator at the Paterson Mineral Museum in Copenhagen, Denmark, for his contributions to the mineralogy and genesis of alkaline rocks. Burbankite group. MON $P2_1$. $a = 20.872$, $b = 6.367$, $c = 10.601$, $\beta = 120.5°$, $Z = 4$, $D = 3.67$. Structure: *CM 32:405(1994)*. Layered structure similar to burbankite but with a supercell that illustrates changes to 10- rather than eightfold coordinated cations.

PD: 9.13_3 5.22_5 4.13_3 3.70_4 2.61_{10} 2.15_3 1.921_3. Diffraction maxima that distinguish petersenite from other members of the group are at 6.84, 5.47, 4.13, and 3.18 Å. Acicular crystals up to 7 mm long, some showing striated prisms. Gray with pink, mauve, or yellow tinge; vitreous luster; translucent to transparent. Conchoidal fracture, brittle. H ≈ 3. G = 3.69. Biaxial; XZ = ba, Y ∧ c = 30°; N_x = 1.623, N_y = 1.636, N_z = 1.649; 2V = 89.7°; no dispersion, no absorption. EPMA (wt %): Ce_2O_3, 23.7; NaO, 17.4; La_2O_3, 14.49; Nd_2O_3, 5.8; Pr_2O_3, 2.0; SrO, 1.7; CaO, 1.3; Sm_2O_3, 0.6; BaO, 0.3; CO_2 (calc), 32.9. Occurs at several sites within the *Poudrette quarry, Mt. St.-Hilaire, QUE, Canada*, at the contact with the hornfels pegmatite phase of the nepheline–sodalite syenite intrusive complex: yellow-tinged crystals are found embedded in trona, and light mauve crystals in a soda-rich inclusion associated with microcline, albite, aegirine, and schomyokite-(Y). HCWS

14.4.6.1 Carbocernaite (Ca,Na)(Sr,Ce)(CO$_3$)$_2$

Named in 1961 for the composition carbon–cerium–natrium. See remondite-(Ce) (14.4.5.1) and burbankite (14.4.4.1). ORTH $Pmc2$. a = 5.214, b = 6.430, c = 7.301, Z = 2, D = 3.60. *AM 68:1251(1983)*. 25-175: 6.39_3 5.19_4 4.81_5 4.05_5 3.53_4 2.99_{10} 2.93_3 2.01_5. Platy on {100}; crystals also show {010}, {001}, {021}, {540}, and {210}. Colorless, white, turns yellowish, brownish, rose when altered; vitreous luster on crystal faces, greasy on fracture surfaces. Cleavage {100}, {021}, {010}, poor; brittle. H = 3. G = 3.53. Biaxial (−); XYZ= bac; N_x = 1.569, N_y = 1.679, N_z = 1.708; 2V = 52°; r > v, marked. Mineral contains high amounts of REE (wt %): 15% CaO, 5% Na_2O, 12% SrO, 3% BaO, 26% REE oxides (La, Ce, Nd, Sm, Y). Canadian sample contains F, no Na or Ba, more Sr. Soluble in dilute HCl. Found at Sturgeon Narrows, Thunder Bay, ONT, Canada, in dolomite–calcite–carbonatite veins in pyroxenites and ijolites of the *Vuoriyärvi massif, Kola Penin., Russia*, with chlorite, ankerite, and other minerals, and in cavities associated with alstonite, anatase, quartz, and zeolites; also at Phan Si Pan Reso, North Vietnam. HCWS *AM 46:1202(1961), CM 11:812(1972)*.

Class 15

Hydrated Carbonates

15.1.1.1 Thermonatrite $Na_2CO_3 \cdot H_2O$

Named in 1845 because it results from drying out of natron. ORTH $P2_1ab$. $a = 10.72$, $b = 5.249$, $c = 6.469$, $Z = 4$, $D_{syn} = 2.259$. Structure: *AC(B) 31:890(1975)*. *8-448*(syn): 5.35_2 5.24_2 2.77_{10} 2.75_6 2.68_6 2.48_3 2.37_6 2.01_3. Crystals rare, thin tabular parallel to {001} or {010} prismatic parallel to a; usually encrusting, an efflorescence. Colorless, white, grayish, yellowish; transparent; vitreous luster. Cleavage parallel to {100}, difficult. H = 1. G = 2.26. IR. *MSLM 4:262(1974)*. Biaxial (−); XYZ = bca; $N_x = 1.420$, $N_y = 1.506$, $N_z = 1.524$; 2V = 48°; r < v, weak. Loses H_2O easily with small amount of heat but does not dissolve in solutions containing other salts (K_2CO_3, $NaHCO_3$, NaCl) until 112°; soluble in hot water giving alkaline solution. As an efflorescence on soils in arid regions; deposits from saline lakes, usually with trona and natron. Found in Death Valley, Inyo Co., and San Joaquin Valley; with sborgite at Twenty Mule Team Canyon, and with halite at Borax Lake, Lake Co.; probably also at Owens, Searles, and other saline lakes, CA; identified from the fumaroles, Mt. Vesuvius, Italy; bedded deposit in Gorodki oil field, Kama region, Russia; deserts in other parts of the world, such as Sudan and Egypt. HCWS *DII:224*.

15.1.2.1 Natron $Na_2CO_3 \cdot 10H_2O$

Named in antiquity after *neter*, Hebraic for *alkaline salts of all varieties*; in the seventeenth century, *natron* (as distinct from *nitre*=saltpeter or potassium carbonates) was applied only to sodium carbonates. Easily confused with trona (13.1.4.1) or thermonatrite (15.1.1.1). MON *Cc*. $a = 12.83$, $b = 9.026$, $c = 13.44$, $\beta = 123.0°$, $Z = 4$, $D = 1.46$. Structure: *AC(B) 25:2656(1969)*. Basically a distorted NaCl structure in which two Na in octahedral coordination form $[Na_2(H_2O)_{10}]^{2+}$ units by sharing an edge. The units form a three-dimensional framework with spaces in which disordered but effectively planar $(CO_3)^{2-}$ groups fit. The CO_3 groups are H-bonded to the surrounding H_2O molecules. *15-800*(syn): 5.37_5 4.50_2 3.94_3 3.45_3 3.04_{10} 3.02_7 2.89_6 2.38_2. Fine-grained, granular, or columnar crusts or friable coatings. Synthetic crystals may show $\{001\}$, $\{010\}$, $\{100\}$, $\{110\}$, $\{011\}$, $\{\bar{1}01\}$, $\{\bar{1}12\}$ with tendency to tabular parallel to $\{010\}$; twins on $\{001\}$. Colorless, white, gray, or yellow, if impure; vitreous luster. Distinct cleavage parallel to $\{001\}$, imperfect parallel to $\{010\}$; conchoidal fracture; brittle. $H = 1$. $G = 1.478$. IR. *MSLM 4:262(1974)*. Biaxial $(-)$; X parallel to b; $Z \wedge c \approx 41°$; $N_x = 1.405$, $N_y = 1.425$, $N_z = 1.440$; $2V_{calc} = 82°$; $r > v$. Decomposes with slight heating to thermonatrite (15.1.1.1) and H_2O, soluble in H_2O (0.40 g/g at 30°) forming alkaline solution, effervesces with acids. Forms from H_2O solution below 32°, but when other salts are present the temperature limit may be lower. Occurs in soda lakes at Ragtown, NV; Clinton, Lilloet district, BC; Debreczin and Szegedin regions, Hungary; in salt lakes of eastern Gobi desert, Mongolia, China, usually associated with thermonatrite, trona, gaylussite, and calcite. HCWS *DII:230:304*.

15.1.3.1 Monohydrocalcite $CaCO_3 \cdot H_2O$

Named in 1964 from the composition. HEX. $P3_112$. $a = 10.566$, $c = 7.573$, $Z = 9$, $D = 2.38$. *AM 60:669(1975)*. *29-306*: 5.28_4 4.33_{10} 3.08_8 2.83_5 2.38_4 2.17_6 1.945_3 1.931_6. Cryptocrystalline, earthy. White to gray. Uniaxial $(-)$; $N_O = 1.590$, $N_E = 1.543$. *NW 46:553(1959)*. Synthesized at pH>8 and in the presence of Mg; there appears to be a relationship with algal or bacterial matter in the natural occurrences. Found in a polymetallic vein of the Vrancice deposit near Pribram, Czech Republic; in the bottom sediments of *Lake Issyk-Kul, Kirgizia, Russia*; in the beach rocks surrounding Lake Fellmongery, Robe, SA, Australia; as a precipitate in the air scrubbers of air-conditioning plants; also in otoliths and bladderstones. HCWS *AM 49:1151(1964), 62:273(1977)*.

15.1.4.1 Ikaite $CaCO_3 \cdot 6H_2O$

Named in 1963 for the locality. MON $C2/c$ or Cc. $a = 8.87$, $b = 8.23$, $c = 11.02$, $\beta = 110.2°$, $Z = 4$, $D = 1.8$. *AM 49:439(1964)*. *37-416*(syn): 5.85_3 5.17_{10} 4.16_3 2.80_5 2.64_9 2.63_7 2.61_3 2.46_3. Stout prismatic to tabular $\{001\}$ crystals. Chalky white. $G = 1.77$. IR. *MSLM 4:264(1974)*. Biaxial $(-)$; $N_x = 1.455$, $N_y = 1.538$, $N_z = 1.545$; $2V = 45°$; distinct dispersion. Dehydrates spontaneously above 8° to calcite. Natural material very close to synthetic. *AJS 41:473(1916)*. Found below the water surface at the inner portion of *Ika Fjord, Ivigtut, Greenland*, in stalagmite-like aggregates. HCWS *DII:228*.

15.1.5.1 Barringtonite $MgCO_3 \cdot 2H_2O$

Named in 1965 for the locality. TRIC $P1$ or $P\bar{1}$. $a = 9.156$, $b = 6.202$, $c = 6.092$, $\alpha = 94.0°$, $\beta = 95.5°$, $\gamma = 108.7°$, $Z = 4$, $D = 2.83$. *AM 50:2103(1965)*. *18-768*: 8.68_{10} 6.09_8 5.82_8 3.09_{10} 3.02_6 2.94_{10} 2.50_8 2.31_8. Nodules of radiating fibers averaging 0.01 mm × 0.03 mm in size. Colorless. Cleavage $\{001\}$, $\{100\}$, $\{010\}$, distinct. Biaxial $(+)$; $N_x = 1.458$, $N_y = 1.473$, $N_z = 1.501$; $2V_{calc} = 73.7°$; length slow. Found with nesquehonite on the surfaces of olivine basalt under Rainbow Falls, Sempill Creek, *Barrington Tops, NSW, Australia.* HCWS *MM 34:370(1965)*.

15.1.6.1 Lansfordite $MgCO_3 \cdot 5H_2O$

Named in 1888 for the locality. MON. $P2_1/m$. $a = 12.48$, $b = 7.55$, $c = 7.34$, $\beta = 101.7°$, $Z = 4$, $D = 1.693$. *MM 46:453(1982)*. *35-680*(syn): 7.18_3 5.24_3 5.11_3 4.58_{10} 3.24_6 3.01_3 2.84_{10} 1.710_4. Stalactitic, short prismatic crystals [001]. Colorless to white; translucent; vitreous luster resembling paraffin when fresh, but rapidly alters, becoming dull and opaque. Cleavage {001}, perfect; {100}, distinct. $H = 2\frac{1}{2}$. $G = 1.70$. IR. *MSLM 4:264(1974)*. Biaxial (+); XZ= bc; Y perpendicular to {100}; $N_x = 1.456$, $N_y = 1.469$, $N_z = 1.508$; $2V = 59.5°$ (Italy). Alters to nesquehonite; soluble, effervescent in 1:10 acids. Loses all H_2O at 110°. Precipitated with nesquehonite when alkaline carbonates added to magnesian chloride solution in the cold. Found as small stalactites attached to the roof of an anthracite mine in *Nesquehoning, Lansford, Carbon Co., PA*, associated with nesquehonite; also as small crystals in hydromagnesite deposit at Atlin, BC, Canada; with nesquehonite in serpentinous rock, Cogne, Val d'Aosta, Piedmont, Italy. HCWS *DII:228*.

15.1.7.1 Hellyerite $NiCO_3 \cdot 6H_2O$

Named in 1958 after Henry Hellyer (b.1826), first surveyor general of Van Diemen's Land, TAS, Australia. MON $C2/c$. $a = 10.77$, $b = 7.30$, $c = 18.68$, $\beta = 94.0°$, $Z = 8$, $D = 2.06$. *12-276*: 9.4_{10} 6.06_{10} 3.65_7 3.40_6 3.11_4 2.78_4 2.38_5 2.31_4. See also *24-523*(calc). Fine-grained coatings [in part on zaratite (16.1.2.1)], in shear planes in serpentinites. Pale blue, vitreous luster. Lamellar twinning parallel to perfect cleavage, two less perfect cleavages nearly 90° from first make angle of 112°. $H = 2\frac{1}{2}$. $G = 1.97$. Biaxial (−); X = Y < Z; $N_x = 1.455$, $N_y = 1.503$, $N_z = 1.549$; $2V = 85°$. Y, pale greenish blue, is normal to perfect cleavages, OAD nearly normal to cleavages. Readily decomposes to basic carbonate without much change in X-ray pattern. Found associated with zaratite at *Old Lord Brassey nickel mine, Heazlewood, TAS, Australia*. HCWS *AM 44:533(1959)*.

15.1.8.1 Joliotite $(UO_2)CO_3 \cdot nH_2O$ ($n \approx 2$)

Named in 1976 for J. Frederic Joliot (1900–1958), French physicist, and I. Joliot-Curie. ORTH *P222*, *Pmm2*, or *Pmmm*. $a = 8.16$, $b = 10.35$, $c = 6.32$, $Z = 4$, $D = 4.55$. *MA 28:208(1977)*. *29-1378*: 8.09_{10} 4.48_2 4.10_5 3.42_9 3.18_8 2.11_3 2.04_2 1.882_4. Spherulites of scaly crystals {100}, {010}, {011}. Lemon yellow, yellow streak, translucent. Cleavage parallel to (100). $H = 1-2$. $G = 4.04$. Weakly fluorescent. Biaxial (−); $N_x = 1.600$, $N_y = 1.644$, $N_z = 1.644$, 2V small. Z, yellow. Dehydration leads to contraction of a axis in unit cell. Occurs as crusts on "limonite" and barite at uranium deposit, *Menzenschwald, Schwarzwald, Germany*, associated with rutherfordine, studtite, and billietite. HCWS *SMPM 56:167(1976)*.

15.2.1.1 Pirssonite $Na_2Ca(CO_3)_2 \cdot 2H_2O$

Named in 1896 after Louis Valentine Pirsson (1860–1919), petrographer, Yale University. Suolunite is the Si analog. ORTH. $Fdd2$. $a = 11.32$, $b = 20.06$, $c = 6.00$, $Z = 8$, $D = 2.367$. *USGSPP 405:1(1962)*, Structure: *AC 23:763(1967)*. 22-476: 5.13_7 4.93_6 3.15_5 2.89_5 2.65_9 2.57_8 2.51_{10} 2.02_5. See also 24-1065(syn). Short prismatic [001] with variable upper terminations, also tabular {010}, or pyramidal; crystals notably hemimorphic with only {11$\bar{1}$} on one end, {131} more obvious than {111} on other end, also {010}, {110} occasionally {311}. Colorless to white, grayish due to inclusions; vitreous luster; transparent. Conchoidal fracture, brittle. H = 3–3$\frac{1}{2}$. G = 2.352. IR. *MSLM 4:270(1974)*. Pyroelectric parallel to [001]. Biaxial (+); $N_x = 1.504$, $N_y = 1.510$, $N_z = 1.575$ (Na); 2V = 31.3°, increases with T; r < v. Soluble, effervescent in cold 1:10 acids. From drill cores with shortite, bradleyite, northupite, gaylussite, and trona near *Green River, Sweetwater Co., WY*; as euhedral crystals in mud layers and consolidated strata *Searles Lake, San Bernardino Co.(!), CA*, and with northupite, tychite, gaylussite, and other minerals in salt lakes such as Borax Lake; in the Khibina alkaline massif, Kola Penin., Russia; with trona and thenardite at Otjiwalundo Salt Pan, Namibia. HCWS *DII:233*.

15.2.2.1 Gaylussite $Na_2Ca(CO_3)_2 \cdot 5H_2O$

Named in 1826 after Joseph Louis Gay-Lussac (1778–1850), French chemist and physicist. MON $C2/c$ or Cc. $a = 11.589$, $b = 7.779$, $c = 11.207$, $\beta = 102.0°$, $D = 1.989$. *AM 52:1570(1967)*. 21-343: 6.41_{10} 4.50_5 4.43_4 3.21_6 2.73_9 2.70_5 2.64_{10} 2.51_5. Crystals elongated parallel to a, flattened or wedgelike with {110}/{001} dominant, faces usually rough, {011} striated parallel to [1$\bar{1}$1]; {110}, {011}, {001}, {$\bar{1}$12} common. Colorless to white with yellow or gray tinge; streak uncolored; vitreous luster; transparent to translucent. Perfect cleavage parallel to {110}, difficult parallel to {001}, conchoidal fracture, brittle. H = 3–3$\frac{1}{2}$. G = 2.367. IR. *MSLM 4:270(1974)*. Biaxial (−); $N_x = 1.443$, $N_y = 1.516$, $N_z = 1.523$; 2V = 34° (Na); r < v, strong, crossed. Effloresces slowly in dry air. Soluble, effervescent in acids; slightly soluble in H_2O, giving $CaCO_3$ ppt., readily alters to calcite. Usually found in soda lakes with natron, thermonatrite, trona, pirssonite, calcite, and borax. Calcite pseudomorphs after gaylussite have been described. Found at sinks in Carson desert near Ragtown, NV; with northupite and pirssonite in trona at Borax Lake, Lake Co.; Mono Lake, Mono Co., Searles Lake, San Bernardino Co., CA. Crystals formed in a clay bed adjacent an alkali lake at *Lagunillas, Merida, Venezuela*; also in the Taboos-nor (salt lake) in eastern Gobi desert, Mongolia, China. HCWS *DII:235*.

15.2.3.1 Chalconatronite $Na_2Cu(CO_3)_2 \cdot 3H_2O$

Named in 1955 after the composition: copper and natron (soda). MON $P2_1/n$. $a = 13.72$, $b = 6.12$, $c = 9.70$, $\beta = 91.3°$, $Z = 4$, $D = 2.31$. *SCI 122:75(1955)*. 22-1458(syn): 7.82_5 6.90_{10} 5.18_7 4.18_8 3.68_9 2.89_6 2.85_7 2.53_6. Fine-grained incrustation on Egyptian bronze objects, presumably buried for centuries with owners' bodies. Greenish-blue crystals are platy or lath shaped, pseudohexagonal. Soft. G = 2.27. Biaxial (+); Y parallel to elongation; $N_x = 1.483$, $N_y = 1.530$, $N_z = 1.576$; 2V large. X, colorless; Y, pale blue; Z, blue. Decomposes in water, soluble with effervescence in cold acids. At the time this mineral was described there was a debate whether such occurrences should be given "mineral" designation. It has since been found at Grube Glückrad, near Oberschulenberg, Hartz Mts. Germany; and at Carr Boyd nickel mine 50 km north of Kalgoorlie, WA, Australia. HCWS

15.2.4.1 Baylissite $K_2Mg(CO_3)_2 \cdot 4H_2O$

Named in 1976 after Noel Stanley Bayliss, chemist, University of Western Australia, Nedlands, WA, Australia. MON $P2_1/a$. $a = 12.37$, $b = 6.24$, $c = 6.86$, $\beta = 114.5°$, $Z = 2$, $D = 2.03$. *SMPM 567:187(1976)*. *29-1017*(syn): 6.31_8 4.19_7 3.13_8 3.01_{10} 2.98_9 2.56_8 2.46_8 2.06_8. Colorless fine-grained anhedral aggregates and crusts. No cleavage, conchoidal fracture. H = 2–3. G = 2.01. Biaxial (−); $N_x = 1.465$, $N_y = 1.485$, $N_z = 1.535$; 2V = 64°. Synthetic compound known. Decomposes in H_2O and at high humidity. Found in the cable tunnel of Gersternegg/Sommerlock, in the *Grimsel Area, Canton Bern, Switzerland*, where mineralized fissures intersect Grimsel granodiorite or aplite granite; associated with quartz, feldspar, grimselite, calcite, monohydrocalcite, and schröckingerite. HCWS

15.2.5.1 Andersonite $Na_2Ca(UO_2)(CO_3)_3 \cdot 6H_2O$

Named in 1948 for Charles Alfred Anderson (1902–1990), U.S. Geological Survey. HEX-R $R\bar{3}$. $a = 18.009$, $c = 23.838$, $Z = 18$, $D = 2.875$. *NJMM:488 (1987)*. *20-1092*: 13.0_{10} 7.93_{10} 5.67_{10} 5.23_8 5.14_8 4.31_6 3.68_{10} 3.13_6. Structure: *AC(B) 37:1946(1981)*. Rhombohedral crystals up to 1 cm, also tiny pseudocubic crystals; thick crusts and veinlets. Bright yellow-green, transparent to translucent, vitreous to pearly luster. $H = 2\frac{1}{2}$. G = 2.8. IR. *MSLM 4:273(1974)*. Fluoresces bright yellow green in UV. Uniaxial (+); $N_O = 1.520$, $N_E = 1.540$. *DII:239*. Soluble in water; crystals can be obtained by evaporating a solution of K_2CO_3 and the nitrates of Na, Ca, and uranium. A secondary mineral found as efflorescences on mine walls or in veins associated with U deposits. Occurs with liebigite and carnotite in Jim Thorpe, PA; at Atomic King No. 2 mine(!) in thick crusts, crystals up to $\frac{1}{2}$ in., and at Cane Creek Wash(!) near Moab, San Juan Co., Skinny No. 1 mine, southeast of Thompsons, Grand Co., UT; large crystals encrusting sandstone in Ambrosia Lakes district, McKinley Co., NM; with bayleyite, swartzite, schröckingerite, and gypsum at *Hillside mine, Bagdad, Yavapai Co., AZ*; on tunnel walls, Geevor mine, Cornwall, England. HCWS *AM 34:274(1949)*, *USGSB 1064:115(1958)*.

15.2.6.1 Grimselite $K_3Na(UO_2)(CO_3)_3 \cdot H_2O$

Named in 1972 for the locality. HEX $P\bar{6}2c$. $a = 9.30$, $c = 8.26$, $Z = 2$, $D = 3.27$. *SMPM 52:93(1972)*. *25-679:* 8.06_8 5.78_{10} 4.66_7 4.14_6 3.63_8 3.08_8 2.67_7 2.01_7. Crusts of fine-grained aggregates, mostly anhedral; crystals showing $\{10\bar{1}0\}$, $\{10\bar{1}1\}$, and $\{0001\}$. Yellow, pale yellow streak. No distinct cleavage, conchoidal fracture, brittle. $H = 2-2\frac{1}{2}$. $G = 3.30$. Nonfluorescent. Uniaxial $(-)$; $N_O = 1.601$, $N_E = 1.480$. O-yellow; E-colorless. Analysis shows small amounts of Al, Ca, and Mg, traces of Cu, Fe, Si, and Mn. Has been synthesized from solutions of K_2CO_3, Na_2CO_3, and uranyl acetate; H_2O determined by loss on heating, DTA. Found as crusts in cable tunnel between Gerstemmegg and Sommerloch, *Grimsel area, Aar massif, Oberhasli, Switzerland*; also on fissures and veins in aplite granite, associated with schröckingerite, monohydrocalcite, and other carbonates. HCWS *AM 58:139(1973)*.

15.3.1.1 Zellerite $Ca(UO_2)(CO_3)_2 \cdot 5H_2O$

Named in 1963 for Howard D. Zeller (b.1922), U.S. Geological Survey, who found the mineral. See metazellerite (15.3.1.2). ORTH $Pmmm$ or $Pmn2_1$. $a = 11.220$, $b = 19.252$, $c = 4.933$, $Z = 4$, $D = 3.24$. *AM 51:1567(1966)*. *19-257:* 9.66_{10} 7.33_2 5.59_4 4.85_5 4.41_3 3.65_4 2.95_1 2.69_1. Hairlike, pin-cushion clumps of needlelike crystals, individual crystal faces not discernible. Light yellow. No cleavage observed. $H \approx 2$. $G = 3.25$. Weak patchy fluorescence in UV. Biaxial $(-)$; $Z = c$; $N_x = 1.536$, $N_y = 1.559$, $N_z = 1.697$; $2V = 30-45°$, dependent on hydration; $r < v$. X,Y, colorless; Z, light yellow. Zellerite is metastable relative to metazellerite at higher temperatures or lower humidities. Minor amounts of REE, K, Na, and Si in analysis. A secondary mineral that forms as clumps or small veins in oxidized uranium ores usually under relatively high CO_2 and pH > 7. Found along with liebigite and gypsum, "limonite," and partially altered iron sulfides, on fine-grained mixture of uraninite, coffinite, and iron sulfides in intergranular areas in arkose at *Lucky Mc mine, Freemont Co.*, and Pat No. 8 mine, Powder River Basin, *WY*. HCWS

15.3.1.2 Metazellerite $Ca(UO_2)(CO_3)_2 \cdot 3H_2O$

Named in 1966 for the relationship to zellerite. ORTH $Pbn2_1$ or $Pbnm$. $a = 9.718$, $b = 18.226$, $c = 4.965$, $Z = 4$, $D = 3.41$. *AM 51:1567(1966)*. *19-258*(syn): 9.1_{10} 4.70_4 4.55_2 4.41_2 4.30_4 3.98_2 3.79_5 3.17_2. Intimately associated wtih zellerite (15.3.1.1), this mineral is the trihydrate, or more dehydrated version, and is similar in physical and chemical properties to zellerite with chalky yellow color and dull luster. Biaxial$_{mean} = 1.626$. Formation sensitive to pCO_2, H_2O. Occurs with zellerite in oxidized zones at *Lucky Mc mine, Gas Hills, WY*, associated with liebigite, gypsum, "limonite," partially altered iron sulfides, and fine-grained mixture of uraninite and coffinite in intergranular areas of arkose. HCWS

15.3.2.1 Liebigite $Ca_2(UO)_2(CO_3)_3 \cdot 11H_2O$

Named in 1848 after German chemist Justus van Liebig (1803–1873). ORTH $Pba2$. $a = 16.703$, $b = 17.513$, $c = 13.741$, $Z = 8$, $D = 2.43$. *USGSB 1064:108(1958)*. *11-296:* 8.68_9 6.81_{10} 4.55_6 3.33_5 3.31_5 3.10_6 2.15_4 1.998_5. Crystals short prismatic [001], or equant, usually indistinct with rounded edges and convex or vicinal faces; commonly granular or scaly aggregates, thin crusts, botryoidal. Green to yellow-green, vitreous luster, pearly on cleavage; transparent to translucent. Cleavage {100}. $H = 2\frac{1}{2}$. $G = 2.41$. Fluorescent green in UV. Biaxial (+); $N_x = 1.497$, $N_y = 1.502$, $N_z = 1.539$; $2V = 40°$; $r > v$, moderate. X, colorless; Y,Z, pale yellow green. Soluble with effervescence in H_2SO_4 leaving a white deposit, solution is green. A secondary mineral in the oxidized portions of uranium deposits associated with other uranium-containing carbonate and hydrated carbonate minerals and β-uranophane. Found at Jim Thorpe mine, PA; Schwartzwalder mine, Jefferson Co., CO; Lucky Mc mine, Freemont Co., Silver Cliff mine at Lusk, and Pumpkin Buttes, WY; Mi Vida mine, Moab in San Juan Co., Black Apes mine in Grand Co., UT; Midnite mine, Stevens Co., WA; Nicholson mine, Goldfields, SAS, Canada; at Wheal Basset, Cornwall, England; as efflorescence on mine walls Joachymov, Bohemia, Czech Republic; at Schmiedeberg, Eisleben, and Schneeberg, Germany; at Stripa, Sweden; at *Edirne, near Adrianople, Turkey.* HCWS *DII:241.*

15.3.3.1 Bayleyite $Mg_2(UO_2)(CO_3)_3 \cdot 18H_2O$

Named in 1948 after William Shirley Bayley (1861–1943), mineralogist and economic geologist, University of Illinois, Urbana, IL. MON $P2_1/c$. $a = 26.65$, $b = 15.3$, $c = 6.53$, $β = 93.1°$, $Z = 4$, $D = 2.06$. *AM 36:1(1951)*. Syn: $P2_1/a$. $a = 26.560$, $b = 15.25$, $c = 6.505$, $β = 92.9°$. Structure: *TMPM 35:133(1986)*. H-bonded pseudohexagonal framework composed of two types of Mg—O octahedra: $Mg_1(H_2O)_6$ and two $Mg_3(H_2O)_6$, and two $UO_2(CO_3)_3$ groups, that have channels parallel to c occupied by another Mg—O octahedral group: $Mg_2(H_2O)_6$ and 12 H_2O molecules. *4-130:* 13.1_9 7.66_{10} 6.53_4 5.85_4 5.08_4 3.83_6 2.69_5 2.21_5. Short prisms [001] to acicular, bladed crystals in crusts and radiating groups. Synthetic material has beautiful prism [001] crystals, showing {100}, {110}, {210}, {001}, {401}, {021}, {211}, {111}, {$\bar{3}$11}. Light yellow, vitreous luster, transparent, becoming dull on dehydration. Conchoidal fracture. $G = 2.05$. DTA (syn). Feebly fluorescent in UV. Biaxial (−); $N_x = 1.455$, $N_y = 1.490$, $N_z = 1.500$; $2V_{calc} = 30°$; inclined extinction. Syn: $Y = b$; $X \wedge c = 11°$; $N_x = 1.453$, $N_y = 1.498$, $N_z = 1.499$; $V = 16°$. On exposure to air rapidly dehydrates to phase with $N_x = 1.502$, $N_z = 1.551$ and strong green fluorescence. Soluble in H_2O. Widespread in Colorado Plateau region, bayleyite is found in crusts on sandstone in the Ambrosia Lakes district, McKinley Co., NM; Pumpkin Buttes area of Powder River Basin, WY; at Southwest mine, Bisbee, Cochise Co., and with schröckingerite, andersonite, swartzite, and gypsum as coatings on mine walls, *Hillside mine, Yavapai Co., AZ;* as an efflorescence with epsomite and gypsum in a tunnel at Azegour,

Morocco. HCWS *DII:237*, *USGSB 1064:112(1958)*, *USGSPP 400-B:B440(1966)*.

15.3.3.2 Swartzite $CaMg(UO_2)(CO_3)_3 \cdot 12H_2O$

Named in 1951 after Charles K. Swartz (1861–1949), geologist and mineralogist, Johns Hopkins University. MON $P2_1/m$. $a = 11.12$, $b = 14.72$, $c = 6.47$, $\beta = 99.5°$, $Z = 2$, $D = 2.32$. *USGSB 1064:117(1958)*. *4-111*: 8.76_{10} 7.31_9 5.50_{10} 4.82_8 3.66_7 2.91_8 2.06_8 1.71_8. Tiny green prismatic [010] crystals in clusters or in crusts that turn dull yellowish white on dehydration. $G = 2.3$. Fluoresces bright green in UV. Biaxial (−); $N_x = 1.465$, $N_y = 1.51$, $N_z = 1.540$; $2V_{calc} = 40°$. X, colorless; Y,Z, yellow. Soluble in H_2O. Found as efflorescence on mine walls at the *Hillside mine, Yavapai Co., AZ*, associated with gypsum, bayleyite, andersonite, and schröckingerite; at Coral claim, Elk Ridge, San Juan Co., UT. HCWS *DII:238*.

15.3.4.1 Donnayite-(Y) $NaCaSr_3Y(CO_3)_6 \cdot 3H_2O$

Named in 1978 after Joseph Désiré Hubert Donnay (1902–1994), Belgian–American–Canadian crystallographer and mineralogist, and Gabrielle (Hamburger) Donnay (1920–1987), American–Canadian mineralogist, McGill University. Related to weloganite (15.3.4.3) and mckelveyite (15.3.4.2); see also ewaldite (14.2.4.1). TRIC *P*1. $a = 9.000$, $b = 8.999$, $c = 6.793$, $\alpha = 102.77°$, $\beta = 116.28°$, $\gamma = 59.99°$, $Z = 1$, $D = 3.266$. *CM 16:335(1978)*. *29-1445*: 6.10_4 4.37_7 3.21_3 2.84_{10} 2.60_4 2.04_3 2.02_3 1.978_3. Crystals all less than 2 mm in size may be platy, tabular, saucer- or barrel-shaped, and columnar; some are hemimorphic; terminations "capped" by ewaldite or mckelveyite. Colorless, white, yellow, gray; white streak; vitreous luster. $H = 3$. $G = 3.30$. Biaxial (−); $N_x = 1.551–1.561$, $N_y = 1.646$, $N_z = 1.652$; $2V = 0–20°$. Found in nepheline syenite at *Mt. St.-Hilaire, QUE, Canada*, associated with microcline, analcime, natrolite, aegirine, arfedsonite, siderite, rhodochrosite, ancylite, and others. HCWS

15.3.4.2 Mckelveyite-(Y) $(Ba,Sr)(Na,Ca,Y,REE)(CO_3)_2 \cdot 3–5H_2O$

Named in 1965 after Vincent E. McKelvey (1916–1985), director of the U.S. Geological Survey. See donnayite (15.3.4.1) and weloganite (15.3.4.3). TRIC *P*1. $a = 9.142$, $b = 9.141$, $c = 7.000$, $\alpha = 102.51°$, $\beta = 115.67°$, $\gamma = 59.99°$, $Z = 1$. *ZVMO 119:76(1990)*. PD: 6.36_5 4.46_9 4.12_5 3.29_5 2.92_{10} 2.64_8 2.26_6 2.02_7. See also *18-901*. Hemimorphic crystals and aggregates of plates < 0.5 mm (Green River); barrel-shaped pseudohexagonal crystals to 3 cm with sector zoning (Khibiny), disk-shaped pseudorhombohedral crystals to 3 mm across (Khibiny), platy aggregates (Salanlatvi, Vuoriyarvi). Greenish or yellowish gray, white, reddish brown, brown, lime green, or black with amorphous carbonaceous inclusions (Green R.); translucent to opaque, colorless in thin section. No cleavage, uneven fracture. $H = 3.5–4$. $G = 3.47$. IR, DTA (Russian samples). All Russian samples appear uniaxial (−); $N_O = 1.649$, $N_E = 1.550$ (Sallanlatvi); $N_O = 1.652–1.658$; $N_E = 1.553–1.554$ (Khibiny).

Microprobe analyses on Russian samples show Y, La, Ce, Sm, Eu, Gd, Tb, Ho, Er, and Yb; CO_2 and H_2O for formula calculated. Found in drill core from four subsurface localities in the *Green River formation, Sweetwater Co., WY*, where it forms syntactic intergrowths with ewaldite associated with labuntsovite, ewaldite, leucosphenite, searlesite, and other minerals. In Khibiny massif mckelveyite is occasionally zoned with Sr enrichment outward [toward donnayite-(Y)] and occurs in aegirine–feldspar–natrolite–calcite veins in the syenite, in siderite–ankerite–natrolite veins in carbonatized foyaite, and in the cataclastic albitite at the contact zone. In the Sallanlatvi and Vuoriyarvi massifs mckelveyite is found in solution cavities in carbonatite. Associated minerals are sphalerite, galena, pyrite, barite, strontionite, fluorite, ancylite, ewaldite, cordylite, donnayite, vinogradovite, catapleiite, epididymite, zhonghuacerite, orthoclase, zircon, and burbankite. HCWS *AM 78:236(1993), 50:593(1965), 52:860(1967); TMPM 15:185(1971)*.

15.3.4.3 Mckelveyite-(Nd) $(Ba,Sr)(Ca,Na,Nd,REE)(CO_3)_2 \cdot 3\text{--}10\ H_2O$

Named in 1993 for its composition and relationship to Mckelveyite-(Y). The Nd endmember in series with mckelveyite-(Y) (15.3.4.2). TRIC. $a = 9.183$, $b = 9.182$, $c = 7.042$, $\alpha = 102.52°$, $\beta = 115.66°$, $\gamma = 59.99°$, $Z = 1$ *ZVMO 119:76(1993)*. PD 6.39_5 4.49_9 4.15_5 2.93_{10} 2.66_8 2.28_6 2.03_7 1.964_6. Single disk-shaped pseudorhombohedral crystals to 1 mm or in parallel growth with mckelveyite-Y, platy aggregates. No cleavage, uneven fracture. H = 3–$3\frac{1}{2}$. Two varieties:1) pale green, transparent, colorless in thin section, Biaxial (–); $N_x 1.556$, $N_y 1.650$, $N_z 1.653$, $2V = 10\text{--}13°$, 2) yellow-brown, opaque, colorless in thin section, Uniaxial (–); $N_o 1.651$, $N_e 1.556$. The latter variety shows a higher amount of Ba, and Sr, and an absence of Na not compensated by Ca. Microprobe analyses: Sr, Ba, Na or ■, Ca, Y, La, Ce, Pr, Nd, Sm, Eu, Gd, Dy. CO_2 and H_2O calculated. Both varieties are described from the solution cavities in the carbonatites at the *Vuoriyarvi massif, Kola Penin., Russia* associated with chalcopyrite, sphalerite, galena, pyrite, barite, strontionite, chlorite, ancylite, ewaldite, cordylite, carbonate spatite, alstonite, carbocernaite. HCWS

15.3.4.4 Weloganite $Na_2Sr_3Zr(CO_3)_6 \cdot 3H_2O$

Named in 1968 after William E. Logan (1798–1875), first director of the Canadian Geologic Survey (1842–1870). See also donnayite (15.3.4.1) and mckelveyite-(Y) (15.3.4.2). TRIC P1. $a = 8.988$, $b = 8.988$, $c = 6.730$, $\alpha = 102.84°$, $\beta = 116.42°$, $\gamma = 59.99°$, $Z = 1$, $D = 3.21$. *CM 13:22(1975). 27-790:* 6.02_5 4.36_8 2.82_{10} 2.59_7 2.38_6 2.23_7 2.01_7 1.966_7. Crystals 2 mm to 3 cm, elongated parallel to c, pseudohexagonal in outline, pyramidal terminations with pedion, prism faces heavily striated and grooved parallel to the base, indicating parallel growth; crystals often vary in cross section, giving a hemimorphic aspect, and hourglass or other odd shapes; also massive. White, lemon-yellow, amber, thin fragments colorless and transparent, larger crystals may be zoned shades of yellow on basal sections; white streak; vitreous luster. Perfect basal cleavage, conchoidal fracture. H = $3\frac{1}{2}$. G = 3.22. Biaxial (−); N_x = 1.558, N_y = 1.646, N_z = 1.648; 2V ≈ 15°. Soluble with effervescence in dilute HCl. From the *St. Michel quarry, Montreal Is., QUE, Canada*, where an alkalic sill intrudes an Ordovician limestone, associated with calcite, quartz, dawsonite, dresserite, and other minerals. HCWS *CM 9:468(1968)*.

15.3.5.1 Voglite $Ca_2Cu(UO_2)(CO_3)_4 \cdot 6H_2O$

Named in 1853 after Josef Florian Vogl, Austrian mining official and mineralogist, who described the compound and published on uranium mines and minerals of Jachymov, Czech Republic. MON $P2_1$. a = 25.97, b = 24.50, c = 10.70, β = 104°, Z = 16, D = 3.06. *33-274:* 12.6_{10} 9.65_8 8.70_{10} 7.10_6 6.13_4 4.65_3 4.32_5 3.94_3. Rhomboidal scales flattened on {010} or coatings with plane angles of 78° and 102° striated on flat surface parallel to an edge. Emerald to grass-green, pale green streak, pearly luster on cleavages. Cleavage {010}, perfect; lamellar twinning on {100}. Biaxial (+); N_x = 1.541, N_y = 1.547, N_z = 1.564; 2V = 60°; r < v, strong. X, blue-green; Y,Z, pale yellow. Soluble with effervescence in acids. A secondary mineral found encrusting sandstone at Frey Point, and at White Canyon No. 1 mine, San Juan Co., UT; Red Mesa Trading Post, Navajo Co., AZ; with liebigite as an alteration product of uraninite at *Elias mine, Jachymov, Bohemia, Czech Republic*. HCWS *DII:237*, USGSB *1064:126(1958)*.

15.3.6.1 Fontanite $Ca(UO_2)_3(CO_3)_4 \cdot 3H_2O$

Named in 1992 for François Fontan, mineralogist, University of Paul-Sabatier, Toulouse, France. ORTH *Pmnm, Pmn2_1, $P2_1$nm*. a = 15.337, b = 17.051, c = 6.931, Z = 4, D = 4.19. *EJM 4:1271(1992)*. PD: 8.55_{10} 6.94_5 4.11_6 3.72_6 3.46_5 3.21_4 2.77_7 2.52_2. Radiating aggregates to 4 mm diameter of rectangular laths up to 1.2 mm × 0.15 mm × 0.05 mm, elongate [001], platy (010) showing {100}, {010}. Bright yellow, white streak, transparent, vitreous luster. Cleavage {010}, {001}, {100}. H = 3. G = >4.10. Fluoresces light green in LWUV. Biaxial (−); XYZ= *bac*; $N_{x\ calc}$ = 1.603, N_y = 1.690, N_z = 1.710; 2V = 49°; r > v, weak; nonpleochroic. Occurs in the alteration zone of the *Rabejec deposit* (7 km SSE of Lodève), *Hérault, France*, with billietite and uranophane, where the primary minerals are pitchblende, uraniferous carbon, and coffinite. HCWS

15.4.1.1 Calkinsite-(Ce) $Ce_2(CO_3)_3 \cdot 4H_2O$

Named in 1953 after Frank Cathcart Calkins (1878–1974) of the U.S. Geological Survey. ORTH $P2_122_1$. a = 9.57, b = 12.65, c = 8.94, Z = 4, D = 3.27. *AM 38:1169(1953)*. *6-76:* 6.54_{10} 4.78_4 4.49_4 3.27_5 2.93_3 2.13_3 2.12_3 2.07_3. Thin tabular crystals parallel to {010} with {010} dominant and also {122}, {132}, {100}, {001}, {102} with maximum length 1 mm; twins {101}. Pale yellow. Perfect cleavage parallel to {010}, distinct parallel to {101}, sometimes parallel to {001}. H ≈ $2\frac{1}{2}$. G = 3.28. Biaxial (−); N_x = 1.569, N_y = 1.657, N_z = 1.686; 2V = 54°; r < v. X = *b*, yellow; Y = *c*, colorless;

$Z = a$. Occurs with barite in vugs of weathered burbankite and ancylite veins, also showing lanthanite and goethite, at head of *Big Sandy Creek, SE Hill Co., MT*; at Khanneshin, Afghanistan. HCWS

15.4.2.1 Lanthanite-(La) $La_2(CO_3)_3 \cdot 8H_2O$

Named in 1845 for the composition. See other end members of the solid-solution series lanthanite-(Ce) (15.4.2.3), lanthanite-(Nd) (15.4.2.2), and possibly calkinsite (15.4.1.1). ORTH. *Pbnb*. $a = 9.504$, $b = 16.943$, $c = 8.937$, $Z = 4$, $D = 2.78$ for $LaCe(CO_3)_3 \cdot 8H_2O$. Structure: *AM 62:142(1977)*. Two types of REE—O polyhedra in which the cations are in 10-fold coordination: REE_1—O bonds with six O from CO_3 groups and four O from H_2O, while REE_2—O bonds with eight O from CO_3 groups and two O from H_2O, form layers parallel to the *ac* plane. The layers are connected only by H bonds. Lathlike, platy to thick tabular crystals show twinning on {101}, also scaly, granular to earthy. Colorless, white, yellowish, pinkish; pearly luster; transparent. Micaceous cleavage on {010}, not brittle. $H = 2\frac{1}{2}$–3. $G = 2.69$–2.73. Biaxial (−); $N_x = 1.52$, $N_y = 1.587$, $N_z = 1.613$; $2V \approx 63°$ (Na); $r < v$, weak. Other rare earths are present. Soluble in acids, decomposed into white powder by boiling H_2O. Alteration product in magnetite deposit, Sanford mine, Moriah, Essex Co., NY; rare mineral in oxidized zinc ores, Bethlehem, Lehigh Co., PA; with allanite in pegmatite at Baringer Hill, Llano Co., TX; coats cerite at *Bastnaes mine, near Riddarhyttan, Västmanland, Sweden*; found at Grimsel Pass, Switzerland, and at Curitiba, Paraná, Brazil. HCWS *DII:241*.

15.4.2.2 Lanthanite-(Nd) $Nd_2(CO_3)_3 \cdot 8H_2O$

Named in 1980 for the composition. Nd end member of the series. ORTH *Pbnb*. $a = 9.476$, $b = 16.940$, $c = 8.942$, $Z = 4$, $D = 2.82$. *BM 102:342(1979)*. *42-593*: 8.50_{10} 4.74_5 4.47_6 4.14_3 3.95_3 3.25_6 3.04_6 2.58_3. Orthorhombic crystals < 2 mm long flattened on [010] show many other faces with striations often seen on (010). Bright pink, vitreous luster. Cleavage {010}, perfect; {101}, good; twinned on {101}. $H = 2\frac{1}{2}$–3. Nonfluorescent. Biaxial (−); $XYZ = bca$; $N_x = 1.532$, $N_y = 1.590$, $N_z = 1.614$; $2V = -61°$. Decomposes with effervescence in dilute HCl. Analysis shows slightly higher amounts of Nd over La plus > 4 wt % Pr_2O_3, Sm_2O_3, and minor amounts of Gd, Eu, Dy, Y, Ce, and Th. Occurs in recent silty, arenaceous clays and carbonate sediments near *Curitiba, Paraná, Brazil*. HCWS *GSCP 80-1C:141(1980)*.

15.4.2.3 Lanthanite-(Ce) $Ce_2(CO_3)_3 \cdot 8H_2O$

Named in 1985 for the composition. Ce end member of the series. ORTH *Pbnb*. $a = 9.482$, $b = 16.938$, $c = 8.965$, $Z = 4$, $D = 2.79$. *AM 70:411(1985)*. *38-377*: 8.47_{10} 4.75_7 4.46_6 4.21_4 4.13_4 3.94_4 3.26_7 3.03_7. Colorless transparent plates {010} with ragged edges are sectile; white streak; vitreous luster. Cleavage {010}. $H = 2\frac{1}{2}$. $G = 2.76$. Nonfluorescent. Biaxial (−); $XYZ = bca$; $N_x = 1.532$, $N_y = 1.594$, $N_z = 1.616$; $2V = 60°$. Ce-dominant element followed by La, Nd with less than 2.5 wt % Sm, Gd, and Y. Dissolves in dilute acids with effervescence, yielding a gel-like ppt. of La hydroxides. A secondary mineral occurring in oxidized copper ore at Britannia mine, Snowdonia, northern Wales, in plates covered by radiating tufts of malachite, associated with brochantite, posnjakite, and chalcoalumite. HCWS

15.4.3.1 Tengerite-(Y) $Y_2(CO_3)_3 \cdot 2$–$3H_2O$

Named in 1838 after C. Tenger, Swedish geologist. See lokkaite (15.4.4.1) and kimuraite (15.4.5.1). ORTH $Pb2_1m$. $a = 6.078$, $b = 9.157$, $c = 15.114$, $Z = 4$, $D = 3.11$. Structure: *AM 78:425(1993)*. Corrugated sheets of Y—O in ninefold coordination and trigonal planar CO_3 groups are connected by other CO_3 groups to form a three-dimensional framework. *27-91:* 7.54_6 5.60_6 4.55_7 3.87_{10} 3.55_6 2.95_7 2.53_5 1.970_5. See also *24-1419*. Pulverulent or fibrous, white, coatings (originally on gadolinite). Most specimens are intimate mixture of at least five different rare-earth carbonate minerals: lanthanite, lokkaite-(Y), kimuraite-(Y), bastnäsite-(Ce), and tengerite-(Y). IR. Biaxial; $N_x = 1.587$, $N_z = 1.616$. In addition to Y, natural samples contain La, Ce, Pr, Nd, Sm, Gd, Tb, Dy, Er, Ho, Tm, Yb, Lu, No, and Ca. Found at Luster pegmatite, southern Platte district, Jefferson Co., CO; at Clear Creek pegmatite, Burnet and Llano Cos., TX; Evans–Lou mine QUE, Canada; Ytterby, Sweden; Rosås, Norway; Pyörönmaa, Finland; and Iisaka, Japan. HCWS *DII:275*.

15.4.4.1 Lokkaite-(Y) $CaY_4(CO_3)_7 \cdot OH_2O$

Named in 1970 for Lauri Lokka (1885–1966), mineralogist and chief chemist, Geological Survey of Finland. See tengerite (15.4.3.1). ORTH. $Pb2m$. $a = 39.35$, $b = 6.104$, $c = 9.26$, $Z = 4$, $D = 2.92$. Supercell detected in structure refinement of this mineral. *AM 71:1028(1986)*. See kimuraite (15.4.5.1). PD(calc): 19.67_3 9.84_5 6.56_7 5.83_5 4.63_6 3.93_5 3.83_{10} 2.95_4. *25-170:* 19.60_4 9.77_5 6.51_6 5.79_5 4.59_8 3.90_6 3.81_{10} 2.93_4 (Finland). Isolated disklike radial aggregates, 0.5 to 1 mm in diameter, composed of white fibers. Brittle. IR, DTA. Biaxial $(-)$; $Z = c$; $N_x = 1.569$, $N_y = 1.592$, $N_z = 1.620$. Alteration product of rare-earth minerals. Found at Evans–Lou mine, QUE, Canada; in fissures of albite in *Pyörönmaa pegmatite, Kangasala, Finland*; in fissures of alkali olivine basalt, Kirigo, Saga Pref., Japan. HCWS *AM 56:1838(1971)*.

15.4.5.1 Kimuraite-(Y) $CaY_2(CO_3)_4 \cdot 6H_2O$

Named in 1986 for Kenjiro Kimura (b.1896), geochemist, University of Tokyo, Japan. ORTH $Imm2$, $Immm$, $I222$, or $I2_12_12_1$. $a = 9.25$, $b = 23.98$, $c = 6.04$, $Z = 4$, $D = 2.98$. *AM 71:1028(1986)*. *40-473:* 12.06_{10} 6.02_4 5.93_2 4.87_1 4.64_2 4.01_2 3.76_3 2.93_1. Spherulitic aggregates of scales 4 cm long × 2 cm wide, individual crystals are tabular. Light purple to pinkish white, vitreous to silky on perfect cleavage planes {010}. $H = 2\frac{1}{2}$. $G = 2.6$ (porous nature accounts for the discrepancy with D). IR. Reddish-purple and purple strong fluorescence in UV. Biaxial $(-)$; XYZ= abc; $N_x = 1.548$, $N_y = 1.612$, $N_z = 1.626$; $2V = 70°$; $r < v$, weak. Microprobe analysis shows Y is 75% of

REE, dominant others Nd, Gd, Dy, Er, Sm, less of La, Ce, Eu, Tb, Ho, Tm, Yb, and Lu. Soluble with effervescence in 1:1 HCl. A secondary mineral found in fissures up to 10 cm wide that parallel the flow structures in an alkali olivine basalt, *Kirigo, Hizen, Higashi Matsuura, Saga Pref., Japan.* HCWS

15.4.6.1 Tuliokite $Na_6BaTh(CO_3)_6 \cdot 6H_2O$

Named in 1990 for the locality. HEX-R $R\bar{3}$. $a = 14.175$, $c = 8.605$, $Z = 3$, $D = 3.25$. *AM 76:668(1991)*. *42-1395:* 7.03_9 4.07_6 3.51_5 3.15_8 2.67_9 2.35_{10} 2.04_6 1.959_7. Prismatic crystals up to 4 mm, and also rhombohedral crystals. Light or dark gray, color dependent on amount of included organic material; vitreous luster. No cleavage, brittle. $H = 3-4$. $G = 3.15$. Nonfluorescent. IR. *AM 77:209(1992)*. Uniaxial $(-)$; $N_O = 1.574$, $N_E = 1.587$ (598 nm). Easily soluble in 1:10 HCl with evolution of CO_2. The H_2O content of the mineral is in some doubt, as single-crystal analyses gave the ideal formula with six H_2O, while microprobe/penfield H_2O analyses suggest eight. Occurs overgrowing sidorenkite, vinogradovite, and villiaumite in cavities between altered microcline crystals in 2- to 10-cm-thick pegmatite vein, or as rhombohedral crystals on pirssonite and shortite in 5- to 7-cm-thick natrolite–aegirine–microcline veinlets with trona, thermonatrite, natron, and villiaumite filling the interstices. Both occurrences in the nepheline syenites of *Mt. Kukisvumchorr*, at *Tuliok R., Khibiny massif, Kola Penin., Russia.* HCWS

15.4.7.1 Shomiokite-(Y) $Na_3(Y,Dy)(CO_3)_3 \cdot 3H_2O$

Named in 1992 for the river close to the discovery site. ORTH P. $a = 10.136$, $b = 17.348$, $c = 5.970$, $Z = 4$, $D = 2.59$. *ZVMO 121:129(1992)*. *PD:* 6.53_6 5.05_5 4.85_7 2.86_7 2.23_5 2.08_{10} Irregular grains and short prismatic pseudohexagonal crystals up to 1–2 mm in diameter, rosettelike aggregates up to 3 mm. Faces {010}, {110}, {011}, dull and uneven. Colorless, waxy to silky luster, translucent to transparent in thin pieces. Cleavage {110}, perfect; parting {001} forming needles parallel to c. $H = 2-3$. $G = 2.52$. IR. Biaxial $(+)$; XYZ= bca; $N_x = 1.528$, $N_y = 1.529$, $N_z = 1.531$; $2V = 49°$. Soluble with effervescence in 1:10 acids; becomes turbid and dehydrates on heating. Microprobe analysis: 21.93% Na_2O, 21.52% Y_2O_3, 5.03% Dy_2O_3, 1.80% Gd_2O_3, 1.24% Er_2O_3, 0.46% Tb_2O_3, 0.63% H_2O_3, 0.09% Ce_2O_3, 0.01% CaO. Occurs among the interstices of potassium feldspar crystals in pegmatites of the *Alluaiv Mt., Lovozero massif, Kola Penin., Russia*, associated with albite, cancrinite, kogarite, villiaumite, neighborite, and siderenkite. HCWS *AM 79:765(1994)*.

Class 16

Carbonates Containing Hydroxyl or Halogen

BASTNÄSITE/SYNCHYSITE/PARASITE GROUPS

Several rare-earth-element carbonate minerals conform to the general formula

$$REE(OH,F):(CO_3):Ca(CO_3)$$

where

REE = Ce, La, Nd, Y mainly, but Pr, Sm, Gd, etc. are also usually detected.

The minerals form layered structures and several subsets based on the ratios of the three formula components have been determined. These subsets can be described:

1:1:0 = bastnäsites
1:1:1 = synchysites
2:2:1 = parisites
3:3:2 = röentgenite-(Ce)

The minerals crystallize in hexagonal, rhombohedral, or with pseudohexagonal space groups in the orthorhombic and monoclinic systems.

Note that none contain molecular water, and thorbasnäsite, in which molecular water has been detected is, in spite of the name, placed in the ancylite group (16b.1.1.1).

The synchysites, parisites and röntgenite-(Ce) have additional cation-CO_3 layers. The stacked layering in the synchysites, for example, is slightly oblique to c resulting in monoclinic symmetry (see Figure). All these minerals show pronounced subcells due to the layering, and with different layer repeats and cation substitutions there are often supercells, or polytypes (see parisite-(Ce) 16a.1.5.1). Cation segregation during crystallization results in syntactic intergrowths which may interfere with definitive identification. However, it is important to note that there is not a continuous series between the one and two cation subsets of the REE-CO_3-F(OH) minerals i.e. between bastnäsites and synchysites. Further, all these minerals show large positive birefringence. Compare Ba-containing REE-Co_3-F species that show negative birefringence: cordylite (16a.1.6.1), huanghoite (16a.1.7.1), zhonghuacernite (16a.1.8.1) and cebaite (16a.1.91).

Bastnäsite Subgroup—REE, F)/CO_3 = 1:1

Mineral	Formula	Space Group	a	b	c	D	
Bastnäsite-(Ce)	Ce(CO_3)F	$P\bar{6}2c$	7.16		9.79	5.02	16a.1.1.1
Bastnäsite-(La)*	La(CO_3)F	$P\bar{6}2c$	7.118		9.762	4.89	16a.1.1.2
Bastnäsite-(Y)	Y(CO_3)F	$P\bar{6}2c$	6.57		9.48		16a.1.1.3
Hydroxylbastnäsite-(Ce)	Ce(CO_3)(OH)	$P\bar{6}2c$	7.23		9.98	4.79	16a.1.2.1
Hydroxylbastnäsite-(La)	La(CO_3)(OH)	$C222_1$	21.891	12.639	10.047	4.64	16a.1.2.2
Hydroxylbastnäsite-(Nd)	Nd(CO_3)(OH)	$P\bar{6}2c$	7.191		9.921	4.89	16a.1.2.3

AM 38:932(1953), 78:415(1993) *No data exist for the La end member. See 16a.1.1.2.

Synchysite Subgroup—(CeF)/(CO_3)/[Ca(CO_3)] = 1:1:1

Mineral	Formula	Space Group	a	b	c	β	Z	D	
Synchysite-(Ce)	CaCe(CO_3)$_2$F	$C2/c$	12.329	7.110	18.741	102.68	12	3.99	16a.1.3.1
Synchysite-(Y)	CaY(CO_3)$_2$F	$C2/c$	12.039	6.960	18.436	102.45	12	3.9	16a.1.3.2
Synchysite-(Nd)†	CaNd(CO_3)$_2$F	[orth.]	[4.039]	[6.984]	[9.045]		[2]	[4.14]	16a.1.3.3

†Needs Space Group re-examination.

Parisite Subgroup—(CeF)/(CO_3)/[Ca(CO_3)] = 2:2:1

Mineral	Formula	Space Group	a	c	Z	D	
Parisite-(Ce)	Ca(Ce,La)$_2$(CO_3)F_2	$R3$	7.125	14.018	3	4.38	16a.1.4.1
Parisite-(Nd)	CaNd$_2$(CO_3)$_3F_2$	$R3$			3		16a.1.4.2

NJMM 192(1992).

Röntgenite-(Ce)—(CeF)/(CO_3)/[Ca(CO_3)] = 3:3:2

Mineral	Formula	Space Group	a	c	
Röntgenite-(Ce)	Ca$_2$Ce$_3$(CO_3)$_5F_3$	$R3$	7.13	69.4	16a.1.5.1

16a.1.1.1 Bastnäsite-(Ce) Ce(CO_3)F

Named in 1841 after the locality. Bastnäsite group. HEX $P\bar{6}2c$. $a = 7.16$, $c = 9.79$, $Z = 6$, $D = 5.02$. *AM 38:932(1953)*. *11-340:* 4.88_4 3.56_7 2.88_{10} 2.06_4 2.02_4 1.898_4 1.674_2 1.573_4. Stacked thin plates, anhedral masses or granular. Crystals tabular {0001}, usually with {10$\bar{1}$0} alone, also deeply grooved horizontally due to oscillatory combination of {10$\bar{1}$1} and {10$\bar{1}$0}. Often intergrown with parisite, röntgenite, and/or synchysite; epitaxial or alteration pseudomorphs after fluocerite. Waxy yellow to reddish brown;

16a.1.1.1 Bastnäsite-(Ce)

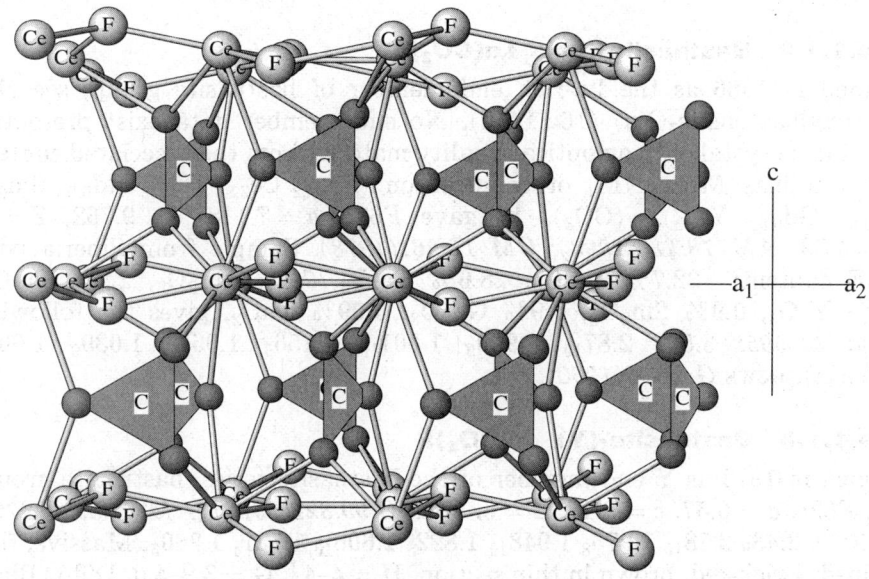

BASTNÄSITE: Ce(CO^3)F

The projection down a_2^*, perpendicular to (010) of the bastnäsite structure shows alternating layers along the c axis of planar CO_3 groups that have one edge parallel to c, and layers of REE-F groups. The REE in the bastnäsites are in 9 fold coordination with 6 oxygens and 3 F *AM 78:415 (1993)*. In the hydroxy-bastnäsites Oh replaces F.

vitreous to greasy luster, pearly on parting surfaces; transparent to translucent. Cleavage {10$\bar{1}$0}, indistinct; distinct to perfect parting on {0001}; uneven fracture; brittle. H = 4–4$\frac{1}{2}$. G = 4.9–5.2. IR shows no OH or H_2O. *GI 3:444(1966)*. Uniaxial (+); N_O = 1.722, N_E = 1.823; faintly pleochroic; E > O. Strongly piezoelectric. Soluble in strong hot acids. Found in contact or alteration zones in alkalic rocks, especially alkalic pegmatites, and in hydrothermal deposits related to alkali syenites. Occurs at a contact zone at Jamestown, Boulder Co., with cerite, fluorite, allanite, and toernebohmite; and with fluorcerite in pegmatites in the Pikes Peak granite at Cheyenne Mt., El Paso Co., CO; in great abundance associated with other rare-earth minerals and barite in dolomite breccias associated with syenitic intrusions at Mountain Pass, San Bernardino Co., CA. Originally described from *Bastnäs, Riddarhyttan district, Västermanland, Sweden,* associated with allanite, cerite, and fluorcerite in narrow bands in a contact-metamorphic amphibole skarn; also as an alteration of fluorocerite at Finbo near Fahlun, Sweden; in the Kychtym district in the Ural Mts., with allanite, cerite, britholite, and toernebohmite, and as pebbles in the gold placers of the Motchaline-log tributary of the Borzovka R., Russia; in the contact zone of an alkali granite–pegmatite with glauconite and tscheffkinite in Torendrika–Ifasina, Madagascar; in millions of tons at Bayan Obo, Inner Mongolia, China; also from the Potgietersrust tin field, Transvaal, South Africa. HCWS *DII:289*.

16a.1.1.2 Bastnäsite-(La)   La(CO$_3$)F

Named in 1966 as the La-rich end member of bastnäsite group; see also hydroxylbastnäsite-(La) (16a.1.2.2). No end member data exist presently, but single-crystal data on optical-quality material from the brecciated contact line, Gallinas Mts., NM, of composition (La$_{2.22}$ Ce$_{2.88}$ Pr$_{0.22}$ Nd$_{0.58}$ Sm$_{0.04}$ Eu$_{0.01}$ Gd$_{0.02}$ Y$_{0.03}$)$_6$ (CO$_3$)$_6$ F$_6$ gave $P\bar{6}2c$; $a = 7.118$, $c = 9.762$, $Z = 6$, $D = 4.89$. *AM 78:415(1993); CM 16:361(1978)*. Sample from Siberia with REE content: 32.7% La$_2$O$_3$, 28.6% Ce$_2$O$_3$, 7.5% Nd$_2$O$_3$, 2.2% Pr$_2$O$_3$, 1.7% Y$_2$O$_3$, 0.9% Sm$_2$O$_3$, 0.9% Gd$_2$O$_3$, 0.09% ThO$_2$, gives the following data: *41-595:* 3.55$_8$ 2.87$_{10}$ 1.900$_9$ 1.301$_{10}$ 1.156$_9$ 1.068$_9$ 1.039$_9$ 1.006$_9$ (Siberia). HCWS *G 6:596(1960)*.

16a.1.1.3 Bastnäsite-(Y)   Y(CO$_3$)F

Named in 1970 as Y end member of the bastnäsite series, bastnäsite group. HEX $P\bar{6}2c$. $a = 6.57$, $c = 9.48$, $Z = 6$. *ZVMO 99:328(1970), AM 57:594(1972)*. *25-1009:* 3.43$_8$ 2.78$_{10}$ 1.976$_8$ 1.948$_{10}$ 1.822$_8$ 1.608$_{10}$ 1.521$_4$ 1.260$_5$. Massive, fine grained. Brick-red, brown in thin section. H = 4–4$\frac{1}{2}$. G = 3.9–4.0. DTA. Biaxial (+); highly birefringent. Analysis shows Y is 40% of total rare-earth content; Dy, 11%; Er, 6.8%; Ce, 7.0%; Na, 6.2%. Occurs as pseudomorphs up to 8 cm in length after gagarinite in microcline pegmatite, Kazakhstan. HCWS *CM 16:361(1978)*.

16a.1.2.1 Hydroxylbastnäsite-(Ce)   Ce(CO$_3$)(OH)

Named in 1964 as the hydroxyl analog of bastnäsite-(Ce). Bastnäsite group. HEX $P\bar{6}2c$. $a = 7.23$, $c = 9.98$, $Z = 6$, $D = 4.79$. *AM 50:805(1965)*. *17-503:* 3.59$_9$ 2.92$_{10}$ 2.09$_9$ 2.05$_8$ 1.923$_9$ 1.698$_7$ 1.460$_5$ 1.319$_7$. *32-189*(syn): 4.99$_7$ 3.62$_9$ 2.93$_{10}$ 2.49$_2$ 2.09$_3$ 2.05$_5$ 1.926$_3$ 1.700$_2$. Reniform aggregates in cavities. Waxy yellow to dark brown or colorless, vitreous to greasy luster. Cleavage {11$\bar{2}$0}, imperfect. H $\approx$ 4. G = 4.475. IR shows OH takes the place of F in the structure. *GI 3:444(1966)*. Uniaxial (+); N$_O$ = 1.760, N$_E$ = 1.870. Analysis: 27% La$_2$O$_3$, 37.5% Ce$_2$O$_3$, 4% H$_2$O, 1.1% F. Dissolves in cold HCl or H$_2$SO$_4$. Occurs at an unspecified Russian locality where alkali–ultrabasic rocks have late carbonatite veins, associated with barite, strontianite, monazite, ancylite, altered burbankite, pyrite, and other sulfides. cs

16a.1.2.2 Hydroxylbastnäsite-(La)   La(CO$_3$)(OH)

Named in 1983 to conform with bastnäsite group designations. Bastnäsite group. ORTH $C222_1$. $a = 21.891$, $b = 12.639$, $c = 10.047$, $Z = 36$, $D = 4.64$. *JSSC 12:115(1975)*. *29-744*(syn): 5.02$_8$ 3.65$_8$ 2.95$_{10}$ 2.51$_6$ 2.11$_6$ 2.07$_8$ 1.943$_8$ 1.824$_6$ (end member with ideal composition). La/Nd appears to be a useful indicator of paragenetic type [*GCA 33:725(1969) 29:755(1965)*]. Described as fine-grained fracture fillings from lower portions of karstic bauxites in Hungary and Yugoslavia associated with bastnäsite-(Ce), bastnäsite-(La), hydroxylbastnäsite-(Nd), synchysite-(Nd), monazite-(Nd), and neodymian goyazite. HCWS *DANS 159:93(1983); AM 71:1277(1986)*.

16a.1.2.3 Hydroxylbastnäsite-(Nd) Nd(CO$_3$)(OH)

Named in 1985 as the Nd end member in the hydroxylbastnäsite series in the bastnäsite group. HEX $P\bar{6}2c$. $a = 7.191$, $c = 9.921$, $Z = 6$, $D = 4.89$. *MM 49:717(1985)*. *38-400:* 4.95$_9$ 3.60$_8$ 2.91$_{10}$ 2.48$_2$ 2.08$_3$ 2.04$_5$ 1.914$_3$ 1.690$_2$. Irregular aggregates of plates 100–200 μm in diameter. Whitish, white streak, dull. Probably parting on {001}. H = 1–2. Nonfluorescent. TG/DTG, IR. *GI 3:444(1966)*. Uniaxial (+); $N_O = 1.715$, $N_E = 1.81$. 27% La$_2$O$_3$, 31.5% Nd$_2$O$_3$, 4.4% Sm$_2$O$_3$. Calculation brings H$_2$O% > F. Occurs in the red karstic bauxites of Niksic, Montenegro, Yugoslavia. HCWS *AM 38:932(1953)*, *51:152(1966)*, *NJMM 289(1985)*.

16a.1.3.1 Synchysite-(Ce) CaCe(CO$_3$)$_2$F

Named in 1901 for "confounding," having been mistaken for parisite (16a.1.5.1). Synchysite group. MON $C2/c$. $a = 12.329$, $b = 7.110$, $c = 18.741$, $\beta = 102.68°$, $Z = 12$, $D = 3.99$. *CM 32:865(1994)*. Pronounced pseudocell $a' = 7.11$, $c' = 18.285$. *18-284:* 9.1$_6$ 4.53$_5$ 3.55$_{10}$ 3.32$_4$ 2.80$_{10}$ 2.06$_5$ 1.934$_5$ 1.873$_4$. See also *44-1431*. Pseudorhombohedral with acute rhombs and small {0001}, or thick tabular {0001}, occasionally hemimorphic. Crystals of the two habits intergrown and may be separate species. {0001} smooth and brilliant, lateral faces striated horizontally. Waxy yellow to brown, vitreous to greasy luster, translucent. Twinning {0001} common, also lamellar. Cleavage or parting on {0001}, subconchoidal to splintery fracture, brittle. H = $4\frac{1}{2}$. G = 3.90. Uniaxial (+); $N_O = 1.674$, $N_E = 1.770$ (Na). Crystals alter, becoming grayish white with reduced hardness and loss of transparency and luster, probably forming bastnäsite or calcite. Altered core in some crystals may have dark transparent terminations (possibly a parisite) that are anhydrous. Readily soluble in acids. Found with parisite in Quincy, MA; at the Scrub Oaks mine, Dover, NJ, intimately intergrown with quartz, xenotime, bastnäsite, leucoxene, and hematite; also from Henry pegmatite, Cotopaxi, CO, associated with cenosite; with gadolinite, aadularia, and xenotime in alpine vein, Piz Blas, Val Naps, Graudunden, Switzerland; in pegmatitic segregations in syenite at *Narssarssuk, Julianehaab district, Greenland*. HCWS *DII:287*, *AM 38:392(1953)*.

16a.1.3.2 Synchysite-(Y) CaY(CO$_3$)$_2$F

Named in 1966 for the composition. Synchysite group. Synonym: doverite. *AM 47:332(1962)*. MON $C2/c$. $a = 12.039$, $b = 6.960$, $c = 18.436$, $\beta = 102.45°$, $Z = 12$, $D = 3.9$. *CM 32:865(1994)*. *44-1431:* 8.99$_{10}$ 4.50$_5$ 3.48$_4$ 2.75$_{10}$ 2.01$_5$ 1.831$_5$ 1.623$_4$ 1.376$_3$ (Dover). See also *14-570*. Massive, fine-grained compact. Reddish-brown, brownish streak, dull luster. H = $4\frac{1}{2}$–$6\frac{1}{2}$. G = 3.89–4.11. *AM 45:94(1960)*. Biaxial; N = 1.643–1.73. Analysis: 35% Y, 12% Ce, 24% Ca, 5% F, on material that contained some xenotime impurities. Found at *Scrub Oaks mine, Dover, Morris Co., NJ*; Henry pegmatite, Cotopaxi, CO. CS *AM 51:152(1966)*.

16a.1.3.3 Synchysite-(Nd) CaNd(CO$_3$)$_2$F

Named in 1978 as the Nd end member of the series with ORTH Space group unknown. $a = 4.039$, $b = 6.984$, $c = 9.045$, $Z = 2$, $D = 4.14$. *AM 64:658(1979)*, however, a supercell HEX $P3_1$ with $c = 54$ Å was described *NJMM 201(1983)* and most recently a high resolution single crystal analysis of synchysite-(Ce) has shown the spacegroup to be $C2/c$ monoclinic, pseudohexagonal (see 16a.1.3.1) *CM 33:865(1994)*. A reanalysis for this Nd end member is needed. *35-589:* 9.04$_6$ 4.52$_{10}$ 3.50$_2$ 3.25$_2$ 3.02$_1$ 2.77$_4$ 2.26$_2$ 1.898$_4$. Fine aggregates up to 100 μ in diameter, light grey-blue, streak white, dull. No cleavage. H = 1. Frequent inclusions and intergrown. Nonfluorescent. Biaxial (+); N$_x$ = 1.61, N$_y$ = 1.66, N$_z$ = 1.74, negative elongation; parallel to extinction; pleochroic light violet to colorless. Electron microprobe data on three samples gave the following (average) formula: Ca$_{1.10}$(Nd$_{0.34}$La$_{0.27}$Y$_{0.12}$Pr$_{0.10}$Sm$_{0.06}$Gd$_{0.05}$Ce$_{0.04}$Dy$_{0.02}$) (CO$_3$)$_{1.90}$F$_{0.90}$. Occurs as microcavity fillings in sandstone-type uranium deposits at *Stráz block, Holicky deposit, Ceská Lepa, Bohemia, Czech Republic*, associated with florencite-(La), sphalerite, manganoan siderite, pyrite, quartz (chalcedony), and kaolinite; at Pyrörönmaa pegmatite, Kangasela, southwestern Finland; in limestones at the base of the Gerbnik bauxite deposit, Yugoslavia. HCWS *AM 47:337(1962)*, *ZVMO 111:233(1978)*.

16a.1.4.1 Huanghoite-(Ce) BaCe(CO$_3$)$_2$F

Named in 1961 for the locality near the Yellow River (Huang Ho), China. Synchysite group. HEX-R $R\bar{3}m$. $a = 5.072$, $c = 38.46$, $Z = 6$, $D = 4.84$. Structure: *NJMM:163(1993)*. *15-286:* 3.91$_7$ 3.21$_{10}$ 2.50$_7$ 2.01$_9$ 1.937$_{10}$ 1.616$_7$ 1.557$_7$ 1.325$_7$. Plates up to 10 cm × 5 cm. Honey yellow to yellowish green, greasy luster, translucent. Cleavage {0001}, distinct; irregular fracture. H ≈ 4.7. G = 4.51–4.67. Uniaxial (−); N$_O$ = 1.768, N$_E$ = 1.603; weakly pleochroic, green-yellow. Vigorously dissolved by HCl. Occurs in the calcite veins distributed in hydrothermal deposits, especially those in granite–syenites, or in hydrothermally altered dolomites, associated with aegirine, fluorite, magnetite, monazite, hematite, bastnäsite, parisite, and aeschynite at *Bayan Obo*, 150 km north of the *Yellow River, Inner Mongolia, China*. HCWS *SS 10:1007(1961)*, *ZVMO 92:199(1963)*, *AM 48:1179(1963)*.

16a.1.5.1 Parisite-(Ce) Ca(Ce,La)$_2$(CO$_3$)$_3$F$_2$

Named in 1835 for J. J. Paris, mine proprietor at Muzo, north of Bogota, Colombia. HEX-R $R3$. $a = 7.125$, $c = 14.018$, $Z = 3$, $D = 4.38$. *AM 38:932(1953)*. Several supercell polytypes have been described: (3H): $a = 7.110$, $c = 83.83$ [*NJMM:192(1992)*]; (4H): $a = 7.12$, $c = 56.4$ [*SGS 20(1993)*]; (6R, 42R): $c = 59.346$; (48R): $c = 67.824$; (16H): $c = 22.608$ *AMS 13:331(1993)*. The different lattices all show a subcell with $a' = 4.11$, $c' = 4.70$. *42-1332:* 14.03$_4$ 4.68$_4$ 3.57$_{10}$ 2.84$_{10}$ 2.06$_8$ 1.938$_6$ 1.882$_5$ 1.658$_5$. Rhombs or prismatic crystals to 4 cm with acute double hexagonal pyramids, lateral faces striated or deeply grooved. {0001} common. Some crystals show zones or cores of different color or optical properties representing intergrowths with synchysite or other members of this group of minerals. Brown, yellow to wax-yellow; vitreous to resinous luster, {0001} parting pearly; transparent to translucent. Cleavage {0001}, distinct; subconchoidal to splintery fracture; brittle. H = $4\frac{1}{2}$. G = 4.36. IR data: *MSLM* *4:272*(1974). Uniaxial (+); N_O = 1.676, N_E = 1.757 (Na). O, light yellow; E, golden yellow. Soluble in hot, strong acids. May alter by hydration. Occurs at Ballou and Fallon Bros. quarries, Quincy, MA, with aegirine, microcline, riebeckite, fluorite, ilmenite, and anatase in pegmatite pipes in granite; with pyrite in an altered trachyte near Pyrites, Ravalli Co., MT; at Lassur, France; with eudidymite from an alkali pegmatite in nepheline syenite on Övre-Arö Is., Langesund Fjord, Norway; in riebekite-granite veinlets Montorfano, Lombardy, Italy; with bastnäsite and tscheffkinite from Ifasina, Torendrika, Madagascar; at the emerald deposits at *Muzo, Boyacá*(!), 100 km N of *Bogota, Colombia*, with pyrite, marcasite, cerian dolomite, albite, calcite, quartz, gypsum, and beryl (emerald) in pockets and veins in intensely folded carbonaceous shale beds. HCWS *DII:283(1952), AM 38:932(1953)*.

16a.1.5.2   Parisite-(Nd) $CaNd_2(CO_3)_3F_2$

Named in 1986 as the Nd-rich end member of the parisite series. Synchysite group. HEX-R *R*3. No unit cell data but assumed similar in all respects with parisite-Ce, as "both have the same diffraction pattern." *AM 73:1496(1988)*. Large yellowish-brown grains in a quartz matrix, vitreous luster. Conchoidal fracture. H = 4–5. G = 4.2–4.5. Uniaxial (+); N_O = 1.679, N_E = 1.754. REE (average of two determinations): 9.3% La, 35.9% Ce, 7.7% Pr, 40.7% Nd, 4.5% Sm, 0.4% Eu, 0.95% Gd, 0.25% Dy, 0.1% Yb, 0.25% Y. Occurs in a quartz matrix with hematite, or calcite, barite, sodic pyroxene, and sodic amphibole in the *Bayan Obo iron–rare earth deposit, Inner Mongolia, China*. HCWS

16a.1.6.1   Röntgenite-(Ce) $Ca_2Ce_3(CO_3)_5F_3$

Named in 1953 after William Conrad von Röntgen (1845–1923), German physicist and discoverer of X-rays. Synchysite group. HEX-R *R*3. a = 7.13, c = 69.4, Z = 9, D = 4.19. *AM 38:868,932(1953)*. Pronounced supercell similar to other Ce–F-containing carbonates; see synchysite-(Ce) (16a.1.3.1). *35-579:* 23.1_{10} 11.6_7 7.7_1 4.63_4 3.57_5 2.82_5 2.06_2 1.940_2. Polycrystalline, similar to parisite, bastnäsite, and synchysite. Crystals show single terminations $\{10\bar{1}1\}$, $\{01\bar{1}2\}$, striated horizontally, truncated with $\{0001\}$; twinning uncommon. Waxy yellow to brown, transparent to translucent. No cleavage observed. H ≈ $4\frac{1}{2}$. Occurs as minute crystals intergrown with bastnäsite minerals at *Narssarssuk, Greenland*. HCWS

16a.1.7.1   Cordylite-(Ce) $BaCe_2(CO_3)_3F_2$

Named in 1898 for the Greek for *club*, in allusion to the shape of the crystals. Synchysite group. HEX $P6_3/mmc$. a = 5.098, c = 23.050, Z = 2, D = 3.97. *CM 13:93(1975). 27-34:* 4.34_8 4.13_6 3.84_8 3.51_9 3.19_{10} 2.55_8 2.12_7 2.04_8. Tiny, hexagonal prisms parallel to c, often with enlarged pyramidal terminations that give a scepter-like appearance due to the interaction of several different pyramidal forms. Colorless, waxy yellow to ocherous yellow; greasy to adamantine luster, pearly on {0001}, dull with surface alterations. Basal sections may show concentric hexagonal zones varying from colorless to yellow, clear to cloudy, also radial striations. Distinct cleavage parallel to {0001}, conchoidal fracture, brittle. H = $4\frac{1}{2}$. G = 4.10. Uniaxial (−); N_O = 1.764, N_E = 1.577. O, green-yellow; E, brown-yellow. Easily soluble in acids. Found in pegmatite veins in nepheline–syenite *Narssarssuk, Julianehaab district, Greenland*, associated with aegirite, ancylite, synchysite, and neptunite; at Bayan Obo, Inner Mongolia, China. HCWS *DII:285*.

16a.1.8.1 Zhonghuacerite-(Ce) $Ba_2Ce(CO_3)_3F$

Named in 1981 to honor the source and composition; the name means *a cerium mineral from China*. HEX-R Space group unknown. a = 5.07, c = 9.82, Z = 1, D = 4.66. *AM 67:1078(1982)*. *39-396:* 3.92_8 3.42_4 3.22_{10} 2.51_6 2.12_3 2.10_{10} 1.979_{10} 1.638_3 1.572_3. Powder pattern close to huanghoite (16a.1.4.1), relationship to cebaite (16a.1.9.1) not clear. Aggregates of small yellow grains, vitreous to resinous luster, transparent. Uniaxial (−); N_O = 1.745, N_E = 1.565. Contains 45% BaO, 10.5% Ce_2O_3, 12% $(La,Nd)_2O_3$, with some Sr, Fe, Mg, Al, and Th. Dissolves readily in acids. Found in a metamorphosed dolomite at *Bayan Obo, Inner Mongolia, China*, associated with cebaite, ferrodolomite, phlogopite, riebeckite, magnetite, and quartz. HCWS

16a.1.9.1 Cebaite-(Ce) $Ba_3Ce_2(CO_3)_5F_2$

Discovered in 1973, later named for the composition. MON $C2/m$ or $C2$. a = 21.426, b = 5.035, c = 13.240, β = 94.6°, Z = 4, D = 4.81. *AM 60:738(1975), 70:214(1985)*. Pronounced rhombohedral subcell a' = 5.11, c' = 9.8, similar to huanghoite-(Ce) (16a.1.4.1). *43-691:* 3.96_6 3.35_1 3.22_{10} 2.53_3 2.13_4 1.995_5 1.355_2 1.260_2. Granular aggregates and tabular grains. Orange-yellow to waxy yellow; orange-yellow to gray-white streak; vitreous, waxy luster. Probably cleavage on {30$\bar{1}$}, uneven fracture. H = $4\frac{1}{2}$–5. G = 4.66–4.31. IR, DTA. Biaxial (−); N_x = 1.598–1.604, N_y = 1.735, N_z = 1.740–1.748; 2V ≈ 5°. A Nd-rich form has been reported. *AM 73:1493(1988)*. Occurs in Fe–Nb rare-earth deposit at *East mine, Bayan Obo, Inner Mongolia, China*, associated with acmite, fluorite, barite, cordylite, huanghoite, and baiyuneboite-(Ce). HCWS

16a.2.1.1 Azurite $Cu_3(CO_3)_2(OH)_2$

Named in antiquity after the Persian *lazhward*, blue color. MON $C2_1/c$. a = 5.00, b = 5.85, c = 10.35, β = 92.33°, Z = 2, D = 3.834. *AC 11:866(1958)*. Figure 16a.2.1.1 and Table 16a.2.1.1 compare structural elements of azurite (Cu/CO_3 = 3:2) with malachite (2:1). *11-682:* 5.15_6 5.08_3 3.67_5 3.52_{10} 2.54_3 2.51_4 2.29_4 2.22_7. **Physical properties:** Crystals with varied forms (over 45 well-established, and over 100 rare or vicinal, faces have been described) are common; often tabular {001}, less common {102} or {$\bar{1}$02}; or prismatic [001] or [010]; sometimes equant or rhombohedral, often composite. Faces commonly wavy with striations on {001} parallel to a, on {100} parallel to b. Twins are rare across {$\bar{1}$01}, {$\bar{1}$02}, or {001}. May also be massive, stalactitic, or botryoidal. Azure blue to very dark blue; blue streak; vitreous to dull and earthy luster, earthy types light blue; transparent to translucent. Cleavage {011}, perfect; {100}, fair; conchoidal fracture; brittle. H = $3\frac{1}{2}$–4. G = 3.77. IR. *MSLM 4:268(1974)*. Biaxial (+); $Z \wedge c = 12.6°$, X parallel to b; $N_x = 1.730$, $N_y = 1.758$, $N_z = 1.838$; $2V = 67°$; $r > v$, strong; pleochroic, blue with absorption $Z > Y > X$. **Chemical properties:** Soluble, effervescent in dilute acids, decomposes in boiling H_2O. Most specimens quite pure; National Museum specimen No. R8545 contains less than 0.1% Zn, 0.01% Si, Al, and Fe. **Occurrence:** A secondary mineral found in the oxidized portions of copper deposits, azurite is formed by the action of carbonated waters on copper-containing minerals or by Cu-containing solutions ($CuSO_4$, $CuCl_2$) with limestones. Usually in close association with, but less common than, malachite (16a.3.2.1), azurite often alters to malachite, giving fibrous pseudomorphs (Chessy, Tsumeb). Also associated with "limonite," wad, cuprite, tenorite, chrysocolla, chalcocite, calcite, native Cu, and other Cu-bearing minerals. Pseudomorphs after cuprite, cerussite, and tetrahedrite. **Localities:** Hundreds of localities have been reported. A selected list for fine crystals follows: hollow paramorphs of azurite after gypsum to 25 cm(!) are found at Apex Mine, St. George, Washington Co., UT; in AZ crystals to 8 cm completely or partially pseudomorphed by malachite were common at the Copper Queen mine(!), Warren Mining district, Cochise Co., Bisbee(!) [see *MR 12(3):259(1981)*]; and in Gila Co., the Payson district; and in Tiger, Pinal Co., at the Mammoth St. Anthony mine(!) and at the Silver King mine(*), Pioneer district; and at the New Cornelia mine(*), Ajo, Pima Co.; and at the Copper Mt. prospect, Anaconda district. In NM at the Georgetown District(*), Grant Co.; at the Kelly mine and elsewhere in the Magdalena District(*), Socorro Co. In Mexico at the San Eligio mine, Aranzazu, the El Cobre mine, Concepcion del Oro, and the San Carlos mine(!), Mazapil, all in ZAC; in CHIH at Santa Eulalia(!), and Sierra Rica. In Cornwall, England, at Wheal Phoenix mine, near Liskeard; at Chessy(!) near Lyon, Rhône, France, where finely crystallized material is known as "Chessylite"; in Poland at Montes Chauves, near Kielce; at Rosas(*), and near Alghero, Sardinia, Italy; in Romania at Kisbanya and at Moldava, Banat district; from the Laurium mine, Greece; at Beresov(!), Tomsk and Solotuschinsk(*) in the Altai Mts., Siberia, Russia; large spherical aggregates to 8 cm were found at Shilu(!), Guangdong Prov., China; at Bou Beker, Bou Skour(!), Kerouchen, and at the Touissit mine(!) Oujda, Morocco, where azurite–malachite crystals up to several inches were found. The Tsumeb mine, Namibia, was the world's premier source for fine azurite, both in quantity, variety of crystal form, perfection, and beauty; lustrous single azurite crystals up to 25 cm long as well as magnificent groups of large crystals(!) were associated with olivenite, mimetite, mottramite, anglesite, and cerussite; also many partly or wholly converted pseudomorphs to fibrous malachite. In Australia at the Proprietary mine(!), Broken Hill, the Gladstone mine(*), Cobar district, NSW; and at the Girofla(*) and Muldiva(*) mines, Chillagoe district, Q; and in SA at Moonta(*), near Walleroo, and at the Burra mine(!), near Burra Burra, in the St. Dominick mine, Arkaroola, where spherical aggregates and rosettes as

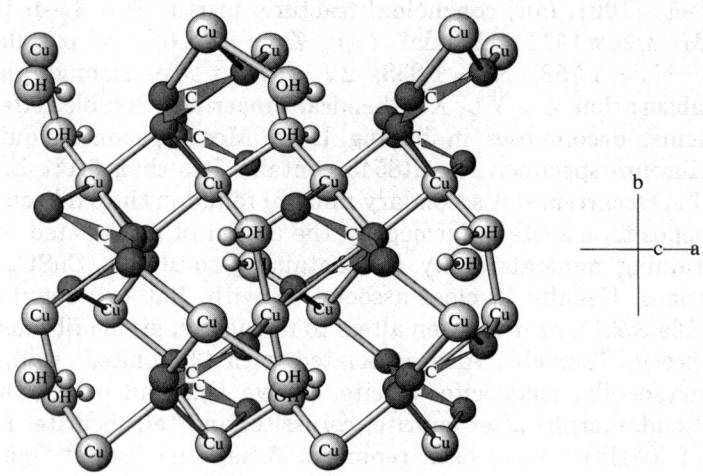

AZURITE: $CO_3(CO_3)_2(OH)_2$

A hydrated carbonate structure this c axis projection shows trigonal planar CO_3 groups almost edge on with some, but not obvious layering, and two distinct copper atom sites. Each metal ion is coordinated with two (OH) and two (CO_3) oxygen atoms approximating a square configuration, with two longer Cu-O bonds. Compare with malachite, 16a.3.2.1.

Azurite, $Cu_3(CO_3)_2(OH)_2$, $Cu:CO_3$ 3:2, a blue copper carbonate often occurs in close association with green malachite (16a.3.2.1), $Cu_2(CO)_3(OH)_2$, with $Cu:CO_3$ 2:1. Both copper carbonates crystalize with similar monoclinic space groups, but malachite has a more open structure. (See malachite 16.3.2.1)

TABLE 16.2.1.1 COMPARISON OF AZURITE AND MALACHITE

	Azurite	Malachite
Space group	$P2_1/c$	$P2_1/a$
Bond distances (in Å)		
Cu_1—2(O)	1.88	1.95 –
Cu_1—2(OH)	1.98	– 2.00
Cu_1—2(O, OH)	2.38	2.71
Cu_2—2(O)	1.92 –	1.91 –
Cu_2–2 (OH)	– 2.04	– 2.07
Cu_2–2(O, OH)	2.38	≈ 2.41
C – O	1.24 –	
	– 1.30	

References: Azurite *AC* 11:866 (1958), *NATW* 45:208 (1958), Malachite *AC* 4:200 (1951)

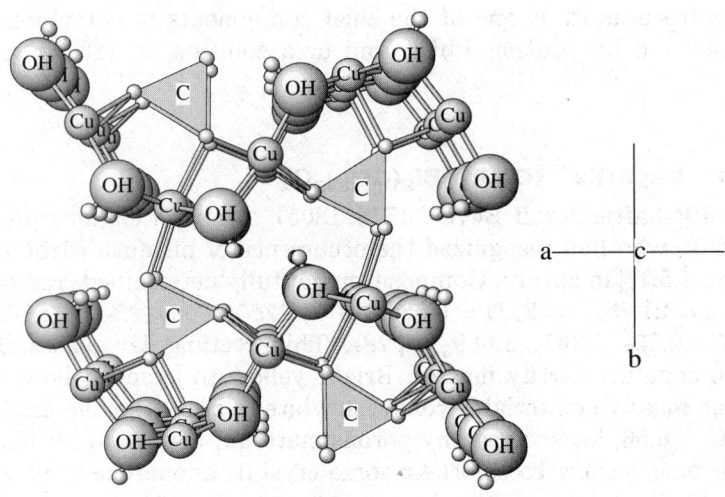

MALACHITE: $CO_2(CO_3)(OH)_4$

Also a c axis projection the layering of the trigonal, virtually planar, CO_3 groups in malachite shows them perpendicular to the c axis, parallel to (001). The two Cu ions positions in malachite have relatively longer Cu-O and Cu-(OH) bonds (TABLE 16.2.1.1) resulting in a different space group and larger size unit cell than azurite (16.2.1.1).

well as single crystals were found; in Chile at Coquimbo. See *R&M 66:24(1991)* for worldwide azurite localities. An ore of copper, azurite was also used as the blue pigment in ancient art and in Europe from the fifteenth to seventeenth centuries. In some instances it has altered to green malachite. HCWS *AM 35:985(1950)*, *DII:264*.

16a.2.2.1 Hydrocerussite $Pb_3(CO_3)_2(OH)_2$

Named in 1877 in allusion to the composition. HEX Space group unknown. $a = 5.24$, $c = 23.74$, $Z = 3$, $D = 6.84$. *TMPM 3:298(1953)*. 13-131(syn): 4.47_6 4.25_6 3.61_9 3.29_9 2.62_{10} 2.49_3 2.23_5 1.696_4. Crystals thin scales with hexagonal outline and flattened {0001}, also thick tabular {0001}, or steep pyramidal {$h0\bar{h}l$}. Colorless to white, sometimes faintly greenish; adamantine luster, pearly on {0001}; transparent to translucent. Cleavage {0001}, perfect; brittle. H = $3\frac{1}{2}$. G = 6.8. IR. *MSLM 4:268(1974)*. Uniaxial (−); $N_O = 2.09$, $N_E = 1.94$. Small amounts of Cl substitute for (OH). A secondary mineral found with other lead-bearing minerals: leadhillite, matlockite, cerussite, mendipite, and paralaurionite. Large crystals were described from the Mammoth Mine, Tiger, AZ; in Scotland in cavities in galena at Wanlockhead, Dumfries, and from Leadhills, Lanark; abundant at Mendip Hills, Somerset, England; at *Långban, Sweden*, as an alteration product of native lead; an alteration product of ancient lead slags, Laurium, Greece; at Rederovski mine, Altai Mts.,

Russia. Hydrocerussite is one of the chief components of commercial white lead, synthesized by heating $PbCl_2$ and urea solution at 180° in air. HCWS *DII:271*.

16a.2.3.1 Beyerite $(Ca,Pb)Bi_2(CO_3)_2O_2$

Named in 1943 after Adolf Beyer (1743–1805), a mining engineer in Schneeberg, Saxony, who had recognized the occurrence of bismuth carbonate [bismutite (16a.3.5.1)] in nature. Composition not fully determined. TET $I4/mmm$. $a = 3.784$, $c = 21.76$, $Z = 2$, $D = 6.58$. *AM 54:1720(1969)*. *22-1067:* 10.9_9 3.35_8 2.85_{10} 2.67_7 2.15_7 1.907_5 1.689_6 1.578_7. Thin rectangular plates flattened {001}, also compact earthy masses. Bright yellow to lemon-yellow, vitreous luster; when massive material is yellowish white to gray. Conchoidal fracture. $H = 2$–3. $G = 6.56$, lower in earthy porous material. Uniaxial $(-)$; $N_O = 2.13$, $N_E = 1.99$; pale, yellow to colorless; some crystals anomalously biaxial with small 2V. Easily soluble with effervescence in acids. A secondary mineral associated with bismuth-containing minerals. Found massive with bismutite as alteration of primary bismuth minerals in Mica Lode and School Section pegmatites in Eight Mile Park district, near Carson City, Fremont Co., and Meyers Ranch pegmatite, Park Co., CO; as compact gray masses in Stewart Mine, San Diego Co., Pala, CA, yellow platy crystals over bismutite and chalcedonic quartz *Schneeberg, Saxony, Germany*; crystals at Bisundi, Rajasthan, India, with massicot, litharge, calcite, and bismutite. HCWS *DII:282*.

ROSASITE GROUP

The rosasite group minerals are copper-containing carbonates corresponding to

$$(AB)_2(CO_3)Z_2$$

where

$A = Cu$
$B = Zn, Ni, Co, Mg$
$Z = (OH)$

The structure of these minerals [*CM 19:315(1981)*] appears to be similar to malachite (16a.3.2.1), but because of the difficulty in obtaining good (untwinned) samples, there is the possibility that the minerals may actually be triclinic with small angular deviations from monoclinicity as determined for kolwezite. Glaukospherite is pseudo-orthorhombic.

16a.3.1.2 Glaukospherite

ROSASITE GROUP

Mineral	Formula	Space Group	a	b	c	α	β	γ	D	
Rosasite	(CuZn)(CO$_3$)(OH)$_2$	$P2_1/m$ or $P2/a$	9.366	12.116	3.127		90.06		3.8	16a.3.1.1
Glaukospherite	(Cu,Ni)$_2$(CO$_3$)(OH)$_2$	$P2_1/a$	9.35	11.97	3.13		96		4.217	16a.3.1.2
Kolwezite	(Cu,Co)$_2$(CO$_3$)(OH)$_2$	$P1$ or $P\bar{1}$	9.50	12.15	3.189	93.32	90.74	91.47	3.94	16a.3.1.3
Zincrosasite	(Zn,Cu)$_2$(CO$_3$)(OH)$_2$									16a.3.1.4
Mcguinnessite	(Mg,Cu)$_2$(CO$_3$)(OH)$_2$	$P*/a$	9.398	12.011	3.379		93.28		3.076	16a.3.1.5

16a.3.1.1 Rosasite CuZn(CO$_3$)(OH)$_2$

Named in 1908 after the locality, a mine in Sardinia. Rosasite group. MON $P2_1/m$ or $P2_1$. $a = 9.366$, $b = 12.116$, $c = 3.127$, $\beta = 90.06°$, $Z = 4$, $D = 3.8$. CM 19:315(1981). MON $P2_1/a$. $a = 12.88$, $b = 9.36$, $c = 3.164$, $\beta = 110.42°$, $Z = 4$, $D = 4.15$. PD 1:56(1986). 36-1475: 6.05$_9$ 5.07$_9$ 3.69$_9$ 2.96$_7$ 2.60$_{10}$ 2.53$_5$ 2.50$_4$ 2.34$_4$. Mammillary, spherulitic or botryoidal crusts of rectangular platy or lathlike fibers (to 0.5 mm), elongated on c and invariably twinned with rotation about a^*. Green to bluish green and sky blue. Cleavage in two directions at right angles, brittle. H ≈ $4\frac{1}{2}$. G = 4.0–4.2. Biaxial (−); XYZ = ca^*b; $N_x = 1.673$, $N_y = 1.796$, $N_z = 1.811$; 2V = 33°; strong absorption light to dark emerald green; no dispersion. Analysis: Cu/Zn = 32.8:23.7 but variable even along the length of fibers. Found at Jack Pot Claim near Wellington, Majuba Hill, NV; Tombstone district, Cochise Co., AZ; with aurichalcite at Kelly, NM; with smithsonite and hemimorphite at Cerro Gordo, Inyo Co., CA; described from *Rosas mine, Sulcis, Sardinia*, where it is associated with brochantite, malachite, aurichalcite, greenockite, and siderite; at Bakara, Bulgaria; also from Tsumeb, Namibia, with malachite, smithsonite, and hydrozincite. HCWS *DII:251*.

16a.3.1.2 Glaukospherite (Cu,Ni)$_2$(CO$_3$)(OH)$_2$

Named in 1974 for the color and the habit from the Greek for *blue* and *sphere*. Rosasite group. MON Pseudo-orthorhombic. $P2_1/a$. $a = 9.35$, $b = 11.97$, $c = 3.13$, $\beta = 96°$, $Z = 4$, $D = 4.217$ (Zaire). CM 19:315(1981). 27-178: 5.96$_2$ 5.04$_3$ 3.68$_7$ 3.02$_2$ 2.95$_2$ 2.59$_{10}$ 2.52$_4$ 2.12$_3$. Fibrous green spherulites, fibers elongated parallel to c, dull to subvitreous and silky luster. Good cleavage parallel to fiber elongation. H = 3–4. G = 3.78. Biaxial (−); $N_x = 1.69$, $N_y = 1.83$, $N_z = 1.83$. Microprobe analysis shows 41.57% CuO, 25.22% NiO, 1.23% MgO, trace Ca and Zn. Found at Kambalda, Windarra, Scotia, *Carr Boyd Rocks, Widgiemooltha, and St. Ives, WA, Australia*, associated with goethite, quartz, paratacamite, gypsum, nickeloan malachite, and nickeloan magnesite; Kasompi, Zaire. HCWS *MM 39:737(1974)*.

16a.3.1.3 Kolwezite $(Cu,Co)_2(CO_3)(OH)_2$

Named in 1980 for the mine locality. Rosasite group. TRIC. $P1$ or $P\bar{1}$. $a = 9.50$, $b = 12.15$, $c = 3.189$, $\alpha = 93.32°$, $\beta = 90.74°$, $\gamma = 91.47°$, $Z = 4$, $D = 3.94$. *CM 19:315(1981)*. *29-1416:* 6.04_8 5.08_8 3.70_{10} 3.02_4 2.96_5 2.59_8 2.54_5 2.34_4. Nodules (1–10 mm) or crystalline crusts. Black to beige. H $\approx$ 4. G = 3.97. DTA and IR show similarity with rosasite and glaukospherite. *BM 103:179(1980)*. $N_{x'} = 1.688$, $N_{z'} = 1.90$. Analysis showed Cu/Co = 2:1. Found as nodules (1–10 mm) and thin microcrystalline crusts in the oxidized zone of a Cu–Co deposit, *Kolwezi–Kamoto–Musonoi, southern Shaba, Zaire*. HCWS *AM 65:1067(1980)*.

16a.3.1.4 Zincrosasite $(Zn,Cu)_2(CO_3)(OH)_2$

Named in 1959 for the composition and as a possible member of the rosasite group with Zn > Cu. Botryoidal and mammillary crusts with radial-fibrous structure, light blue to whitish with cleavage in two directions. H $\approx$ 3. Described from *Tsumeb, Namibia*. HCWS *FM 37:87(1959), AM 44:1323(1959)*.

16a.3.1.5 Mcguinnessite $(Mg,Cu)_2(CO_3)(OH)_2$

Named in 1981 for Albert L. McGuinness (1926–1990), mineral dealer, San Mateo, CA. Rosasite group. See also pokrovskite (16a.3.2.3). MON $P2/a$ or $P2_1/a$. $a = 9.398$; $b = 12.011$; $c = 3.379$; $\beta = 93.28°$; D = 3.076 (Mg-rich), 3.234 (Cu-rich). *MR 12:143(1981)*. Structure very similar to malachite but with only one octahedral cation group: sheets parallel to (102) are made up of edge sharing $Mg(OH)_5(CO_3)$ octahedra. *35-481:* 7.40_1 6.02_{10} 5.05_2 3.69_7 3.00_1 2.53_3 2.33_2 2.14_3. Spherules 0.1–2 mm in diameter composed of light blue-green fibers with Cu-rich cores darker blue, and Mg-rich rims pale blue-green to nearly white; vitreous to silky luster. Brittle, inelastic. H = $2\frac{1}{2}$. G = 3.02 (Mg-rich), 3.22 (Cu-rich). DTA data. Biaxial (−); $X \wedge c = 11°$; $N_x = 1.602$, $N_y = 1.730$, $N_z = 1.732$ (Na); negative elongation; maximum extinction; weakly pleochroic. X, very pale green; Y,Z, light blue-green. Occurs in serpentinized peridotite at *Red Mt., Mendocino Co., CA*, associated with vuagnatite, goethite, malachite, azurite, and chrysocolla; at Gabbs, NV; in Mexico at El Calvario, Municipio de Malege, BCS; and at Carr Boyd Rocks mine, WA, Australia. HCWS *AM 66:1276(1981)*.

16a.3.2.1 Malachite $Cu_2(CO_3)(OH)_2$

Named in antiquity after the Greek *mallows*, in allusion to the color. MON $P2_1/a$. $a = 9.48$, $b = 12.03$, $c = 3.21$, $\beta = 98°$, $Z = 4$, $D = 4.00$. *AC 4:200(1951)*. *10-399*(syn): 5.99_6 5.06_8 3.69_9 2.86_{10} 2.82_4 2.78_5 2.52_6 2.46_4. *41-1390*(syn): 5.97_8 5.04_{10} 3.69_{10} 2.99_2 2.86_7 2.78_3 2.52_6 2.13_2. **Structure:** see azurite (16a.2.1.1) **Physical properties:** Crystals are rare, usually short or long prismatic [001], or needles, and are often grouped in tufts and rosettes. Untwinned crystals virtually unknown, twin plane {100} sometimes penetration or polysynthetic twinning, axis [201]. Commonly massive, incrusting with the surface mammillary, botryoidal; internally fibrous, usually radial on

16a.3.2.1 Malachite

coarse to fine scale, showing color banding. Delicate fibrous aggregates, sheaf-like or compact; concentrically banded stalactites, sometimes alternating with azurite. Bright green with crystals deeper shades even blackish; pale green streak; adamantine luster in crystals, fibrous varieties silky or velvety, often dull and earthy; translucent to opaque. Cleavage $\{\bar{2}01\}$, perfect; $\{010\}$, fair; subconchoidal fracture of compact masses. H = $3\frac{1}{2}$–4. G = 4.05 (crystals), to 3.6 in massive fibrous material. IR. *MSLM 4:268(1974), GI 3:444(1966)*. Biaxial (−); X ∧ c = 23.5°; $N_x = 1.655$, $N_y = 1.875$, $N_z = 1.909$; 2V = 43°; r < v, strong. X, nearly colorless; Y, yellowish green; Z, deep green. **Chemical properties:** Some analyses contain traces of Zn, but higher contents probably mean that the sample is one of the rosasite group. Water and CO_2 lost at ≈ 315°, a lower temperature than azurite, leaving tenorite. Easily soluble in dilute acids. **Occurrence:** Malachite is the most common secondary mineral of copper and virtually every oxidized zone of Cu ore deposits will show some of this mineral. Typically found with azurite, and more abundant than that species, it is also associated with goethite–hematite, cuprite, tenorite, *wad*, calcite, chalcedony, chrysocolla, and other secondary minerals. **Localities:** Only the most important localities for specimens will be mentioned. Malachite was found in the copper mines of Ducktown, TN, and Copper Co., MI, and is widespread in NM at the Fiero-Hanover district(*) and the Santa Rita district, Grant Co.; also in the Magdalena district(*), Socorro Co. In NV in the Good Springs district. Acicular crystals were found in AZ at the Copper Queen mine(!) and other mines in Bisbee district(!), Cochise Co., where it is often alternatively banded with azurite or pseudomorphous after azurite crystals; in Gila Co., the Inspiration mine(!) has fine banded masses of alternating malachite, chrysocolla, and chalcedony; in Pima Co. at the New Cornelia open pit(*) concentric structures with azurite and quartz; in Yavapi Co., at the Yeager mine, Black Hills district, AZ. In Mexico at the El Cobre mine(!), Concepcion del Oro, ZAC, malachite pseudomorphic after azurite crystals to 12 cm. In France at Chessy(!) near Lyon, often pseudomorphous after cuprite, and with secondary arsenates and phosphates in crevices in Keuper sandstone at Cap Garonne, near Hyère, Maures; in the Rhineland, Germany, at Betzdorf; at Rudabaya and Moldawa, Banat region, Romania; in Russia at Ekaterinburg(!), in the Ural Mts., and large amounts of beautifully banded malachite came from the Demidoff mine(!) at Nizhni–Tagil(!), where one banded mass contained over 1 million kilograms of the pure mineral was mined in the early part of the nineteenth century. Cut into slabs, these pieces were made into tabletops, pillars, vases, and other decorative pieces and may be seen in many world-famous museums (e.g., Uffizi, and St. Peters, Italy); in China in the Hubei and Hunan provinces and at Shilu(*), Guandong. Malachite was called "Shilu" in ancient China, where it was highly prized. In Morocco at the Aouli mine, Aouli, at Bou Skour and at the Touissit mine(!), Oujda; cobaltian varieties and stalactites to 90 cm form at Kolwesi(!), Shaba, Zaire, while fine crystals to 10 mm are found at Kambove(*); in Zambia at Bwana Mkubwa; at Tsumeb(!), Namibia, where it is the most abundant secondary Cu mineral. Malachite forms small needles and crusts with native Cu and cuprite, thick banded masses, stalactites, and pseudomorphs after azurite. *MR 8(3)(1977)*. In Australia in NT at Rum Jungle; in SA at Arkaroola, at Wallaroo, and in the Burra district, especially at the Burra mine(!), where large masses of interbanded azurite–malachite similar to the Russian material at Demidoff were found; also at Broken Hill(!) (plumbian and zincian varieties) and at Coba, NSW, Australia; in Brazil at Morro do Bule, Ouro Preto district, fine pseudocubic crystals to 6 mm are found in a limestone quarry. As a secondary ore of Cu, malachite was mined at Bisbee, Tsumeb, and elsewhere, but probably the most prominent use, especially where banding was beautifully displayed, was for decorative pieces. Large ornamental objects and elegant jewelry continue to bring high prices. HCWS *DII:255*.

16a.3.2.2 Nullaginite $Ni_2(CO_3)(OH)_2$

Named in 1981 for the locality. MON $P2_1$ or $P2_1/m$. $a = 9.23$, $b = 12.001$, $c = 3.091$, $\beta = 90.48°$, $Z = 4$, $D = 4.07$. *CM 19:315(1981)*. Structurally similar to malachite. *35-501:* 7.30_3 5.04_3 4.62_4 3.66_4 2.58_{10} 2.56_9 1.545_3 1.541_3. Nodular bright green grains up to 2 mm or cross-fiber veinlets in matrix of chlorite and Ni-rich serpentine (pecoraite). $G = 3.56$. Biaxial; $X \wedge c = 6°$; $N = 1.67$–1.78; slightly pleochroic; maximum absorption parallel to c. Found as a secondary mineral in the sepentinized peridotites of the Otway Belt, *Nullagine district, WA, Australia*, in association with millerite, polydymite, pecoraite, gaspéite, otwayite, parkerite, shandite, and breithauptite. HCWS

16a.3.2.3 Pokrovskite $Mg_2(CO_3)(OH)_2 \cdot 0.5H_2O$

Named in 1984 for Pavel Vladimirovich Pokrovskii (1912–1979), mineralogist from the Ural Mts., Russia. MON $P2_1/a$. $a = 9.43$, $b = 12.27$, $c = 3.395$, $\beta = 96.6°$, $Z = 4$, $D = 2.58$. *AM 70:217(1985)*. Probably isostructural with malachite (16a.3.2.1); see also the rosasite group. *37-454:* 6.10_7 5.20_4 4.70_7 3.73_7 2.60_{10} 2.17_9 1.561_7 1.385_5. Prismatic to acicular crystals in spherulitic aggregates. White with a flesh tint, white streak, dull luster, opaque because of microinclusions. $H \approx 3$. $G = 2.51$, crystals; 2.27, aggregates. DTA, IR. Biaxial $(-)$; $X \wedge c = 8°$; $N_x = 1.537$, $N_{y\,calc} = 1.611$, $N_z = 1.619$; negative elongation. Insoluble in H_2O, dissolves in 1:10 HCl, turns brown when heated. Original reports of CaO in the mineral probably due to admixed dolomite. IR shows both (OH) and H_2O present. Vein mineral in ultramafic rocks. Occurs in the *Zlatogorsk intrusive, central Kazakhstan*, within dunite and associated with dolomite, magnesite, and a sjögrenite-like mineral. HCWS

16a.3.3.1 Georgeite $Cu_2CO_3(OH)_2$

Named in 1979 for George Herbert Payne (1912–1989), past chief of the mineral division, Western Australian Government Chemical Laboratories. Amorphous analog of malachite (16a.3.2.1). *AM 77:212(1992)*. Light blue, pale blue streak, vitreous to earthy luster, transparent to subopaque aggregates. Soft. $G = 2.55$. Isotropic; $N_{av} = 1.593$ (Na). Forms thin coatings associated with malachite and chalconatronite on partly weathered tremolitic rock that contains disseminated copper and iron sulfides. Analysis of synthetic material compared to georgeite from *Carr Boyd mine, 100 km N of Kalgoorlie, WA, Australia*, and Britannia mine, Wales, showed them to be identical. hcws

16a.3.4.1 Phosgenite $Pb_2(CO_3)Cl_2$

Named in 1841 for the composition *phosgen*, a name for carbonyl chloride, $COCl_2$, as the mineral contains these elements. TET $P4/mbm$. $a = 8.112$, $c = 8.814$, $Z = 4$, $D = 6.136$. *MM 31:883(1958)*. *12-218:* 4.40_6 4.04_5 3.61_8 2.79_{10} 2.56_8 2.21_4 1.800_4 1.426_4. Short prismatic [001]; rarely thick tabular {001}, usually terminated with large {001}, sometimes {111} dominant; also massive, granular. Yellowish white to yellowish brown, smoky brown, also colorless, white, rose, grayish, and greenish; white streak; adamantine luster; transparent to translucent. Cleavage {001}, {110}, distinct; {010}, less distinct; conchoidal fracture; sectile. Translation gliding T{110}, t[001] crystals easily flexed about directions perpendicular [001]. Percussion figure on {001} with rays parallel to [110]. $H = 2$–3 varying with direction. $G = 6.133$. IR. *MSLM 4:272(1974)*. Weakly fluorescent yellow in LWUV, X-rays. Uniaxial (+); $N_O = 2.118$, $N_E = 2.144$ (Na). O, reddish; E, greenish in thick sections. Sometimes weakly biaxial especially after mechanical deformation. Soluble with effervescence in dilute HNO_3. A secondary mineral formed under surface conditions from primary lead minerals such as galena; commonly associated with cerussite, forms epitaxial growths, and alters providing many pseudomorphs of cerussite after phosgenite. Found in Southampton lead mine, Hampshire Co., MA; abundant large masses in part altered to cerussite at Terrible mine, Isle, Custer Co., CO; with diaboleite at Mammoth mine, Pinal Co., AZ; in Organ district, Dona Ana Co., NM; Silver Sprout mine, Inyo Co., CA; with matlockite at Matlock and in Cromford, Derbyshire, England; crystals up to 5 in. were found at Monteponi and Montevecchio(!) near Iglesias and at Gibbas(!) near Cagliari, Sardinia, Italy; at Laurium, with laurionite and other lead minerals from the reaction of seawater on old lead slags, Greece; in fine crystals at Tsumeb(!), Namibia; at the Comet mine, Dundas, TAS, and Broken Hill, NSW, Australia. hcws *DII:258*.

16a.3.5.1 Bismutite $Bi_2(CO_3)O_2$

Named in 1841 in allusion to the composition. TET $I4/mmm$. $a = 3.870$, $c = 13.697$, $Z = 2$, $D = 8.28$. *BGSF 40:145(1960)*. *41-1488*(syn): 6.84_3 3.72_4 2.95_{10} 2.73_4 2.14_3 1.933_2 1.750_3 1.617_4. Dense earthy masses, opaline crusts, and radial-fibrous crusts or spheroidal aggregates, rarely scaly to lamellar aggregates, may appear prismatic or take on other pseudomorphous habits. Straw-yellow to brown-yellow, greenish gray, green, brown to black, especially if pseudomorph after bismuthinite; gray streak; vitreous to pearly luster in fibrous or scaly forms, dull to earthy; transparent in small grains. Cleavage {001}. $H = 2\frac{1}{2}$–$3\frac{1}{2}$, variable with state of aggregation. $G = 6.7$–7.4. $N_{av} = 2.12$–2.30 on cryptocrystalline material; moderately birefringent; fibers show parallel to extinction; positive elongation. Cu, Fe, and Ca reported in analyses may be admixture. Effervesces in acids. A secondary mineral usually an alteration of bismuthinite, or other primary Bi-containing species, in veins of Co–Ni–Ag–Bi-containing minerals, or in pegmatites with Bi minerals. Found in pegmatites at Portland, CT; Casher's Valley, Jackson Co., NC; at Leadville, Salida, Chaffee Co., and at Telluride, San Miguel Co., CO; in the Petaca district, Rio Arriba Co., and San Andreas Mts., Otero Co., and the Organ district, Hidalgo Co., and at Harding pegmatite, Dixon, NM; in the Tintic district, UT, with bismuth arsenates and bismoclite; at Bagdad copper mine W of Hillside Tavapai Co., also Hualpai Mts. E of Yucca, Mohave Co., NV; at Pala, San Diego Co., CA; at Durango in the El Carmen mine and at Guanajuato, GTO, Mexico; in Cornwall, England; Meymac, Corrèze, France; *Ullersreuth, Voigtland, Germany*; as an alteration of aikinite in Berezov, Ural Mts., Russia; also in pegmatites in Ampangabe, Madagascar; and at Boksput, in Gordonia, Cape Prov., South Africa; in the Sn–sulfide veins, Tazna district, Bolivia; at São Jose de Bryamba, MG, Brazil; and in the pegmatite pipes at Kingsgate, NSW, Australia. HCWS *DII:259.*

16a.3.6.1   Brenkite $Ca_2(CO_3)F_2$

Named in 1978 for the locality in Germany. ORTH. *Pbcn.* $a = 7.650$, $b = 7.550$, $c = 6.548$, $Z = 4$, $D = 3.126$. *NJMM:325(1978). 29-322:* 3.28_6 3.03_8 2.79_{10} 2.49_4 2.27_6 1.888_5 1.835_8 1.790_6. Radiating aggregates of lathlike crystals up to 1 mm with {010}, {102}, {021}. Colorless, white streak. No distinct cleavage. $H = 5$. $G = 3.10$. IR. Biaxial $(-)$; $XY = ca$; $N_x = 1.525$, $N_y = 1.590$, $N_z = 1.593$; $2V = 26–28°$. Dissolves with effervescence in 1:10 HCl. Found at *Brenk, Eifel, Germany.* HCWS *AM 64:241(1979).*

16a.3.7.1.   Kettnerite $CaBi(CO_3)OF$

Named in 1956 after Radim Kettner (1891–1968), professor of geology, Charles University, Prague, Czech Republic. TET $P4/nmm$. $a = 3.79$, $c = 13.59$, $Z = 2$, $D = 5.85$. *AM 43:385(1958). 38-462:* 13.5_{10} 6.78_3 3.66_4 3.31_4 2.91_9 2.10_5 1.725_6 1.588_4. See also *25-126*. Small (0.2–0.3 mm) plates, {001} dominant, also {111}. Brown, brownish yellow to lemon yellow. $G = 5.80$. Biaxial $(-)$; $N > 2.05$. In cavities on quartz–fluorite in quartz veins cutting pegmatites that contain plagioclase. Occurs near *Krupke (Graupen), northwestern Bohemia, Czech Republic,* associated with fluorite, bismuth, bismuthinite, hematite, and altered topaz. CS *CMG 1:195(1956).*

16a.3.8.1   Dawsonite $NaAl(CO_3)(OH)_2$

Named in 1874 after John William Dawson (1820–1899), Canadian geologist, principal of McGill University. ORTH. *Imam.* $a = 6.71$, $b = 10.411$, $c = 5.58$, $Z = 4$, $D = 2.434$. Structure: *CM 9:51(1967). 45-1359:* 5.67_{10} 3.38_1 2.79_5 2.61_2 2.51_1 2.15_1 1.993_1 1.730_1. Thin encrustations or rosettes of bladed to acicular crystals, and as tufts of fine needles. Colorless to white, colorless streak, vitreous to silky luster in fine aggregates, transparent. Cleavage {110}, perfect. $H = 3$. $G = 2.44$. IR. *MSLM 4:270(1974).* Biaxial $(-)$; $XY = ac$; $N_x = 1.462$,

$N_y = 1.537$, $N_z = 1.589$ (Na); $2V = 76.75°$; $r < v$. A low-temperature hydrothermal mineral from decomposing aluminous silicates. Found as coatings on joint surfaces of felspathic dike in *Trenton Limestone, McGill University environs, Montreal, QUE, Canada*, associated with calcite, dolomite, pyrite, galena, and wad. In veins associated with late Tertiary basalts in Sollingen, Germany; also at Pian Castagnaio and Santa Fiora near Mte. Amiata, Siena, Tuscany, Italy, associated with calcite, dolomite, pyrite, fluorite, and cinnabar; in Komana, Drin Valley, northern Albania; with barite east of Tenès, Alger, Algeria. HCWS *DII:278, RZ 9:B196(1985)*.

16a.3.9.1 Northupite $Na_3Mg(CO_3)_2Cl$

Named in 1895 after C. H. Northup (b.1861), grocer, of San Jose, CA, who found the first specimen. ISO $Fd3$ or $Fd3m$. $a = 13.98$, $Z = 16$, $D = 2.40$. *CR 193:1421(1931)*. *19-1213:* 8.08_5 4.96_3 3.21_4 2.86_3 2.70_8 2.47_{10} 2.37_3 2.11_6. Octahedral crystals {111} which sometimes contain symmetrical inclusions of clay. Colorless to pale yellow, gray, or brown due to included clay and organic matter; vitreous luster; transparent. Conchoidal fracture, brittle. $H = 3\frac{1}{2}$. $G = 2.38$. IR. *MSLM 4:272(1974)*. Isotropic, $N = 1.5144$ (Na); anomalous bireflectance in sectors. Crystals embedded in clay sediments, *Searles Lake, San Bernardino Co., CA*, associated with tychite and pirssonite; with shortite, bradleyite, trona, pirssonite, and gaylussite in drill cores in the oil shales, Green R., Sweetwater Co., WY. CS *DII:278*.

16a.4.1.1 Hydrozincite $Zn_5(CO_3)_2(OH)_6$

Named in 1853 after the composition. MON $C2/m$. $a = 13.58$, $b = 6.28$, $c = 5.41$, $\beta = 95.51°$, $Z = 2$, $D = 3.97$. *CM 8:385(1965)*. Structure: *AC 17:1051(1964)*. *19-1458:* 6.77_{10} 3.66_4 3.14_5 2.85_3 2.72_6 2.48_7 1.915_3 1.688_4. Massive, earthy, porous to compact, and as encrustations sometimes concentrically banded with fine, fibrous, radial structure; dense agatelike masses; stalactitic, reniform, or pisolitic. Lath or bladed crystals flattened {100} and elongated [001] often tapering to sharp point, and projecting from massive aggregates. Intimate twinning observed on single-crystal examination for structure determinations. Pure white to gray, yellowish, brownish pinkish, pale lavender; dull to shining streak; dull to earthy luster; pearly on crystals. Cleavage {100}, perfect; brittle. $H = 2-2\frac{1}{2}$. $G = 4.0$, 3.5–3.8 in massive material. IR. *MSLM 4:268*. Fluoresces pale blue to lilac in UV. Biaxial (−); $X = b$; $Z \wedge c \approx 40°$; $N_x = 1.640$, $N_y = 1.736$, $N_z = 1.750$; $2V = 40°$; $r < v$, strong. Although compositionally similar, hydrozincite does not contain Cu in significant amounts compared with aurichalcite (16a.4.2.1). A secondary mineral in the oxidized zone of ore deposits containing Zn minerals such as sphalerite. Easily soluble in acids. Found at Friedensville, Lehigh Co., PA; Linden, Iowa Co., WI; as concentric, contorted laminae and botryoidal crusts in Marion Co., AR; in Joplin, MO; with hemimorphite and smithsonite at Galena district, Cherokee Co., KS; in Magdalena district, Socorro Co., NM; Tintic district, Juab Co., UT; abundant at Goodsprings, NV; with hemimorphite and willemite in the Cerro Gordo mine, Inyo Co., CA; large quantities in Dolores mine, Udias V. Santander Prov., Spain; at Bleiberg, Carinthia, Austria; Raibl, Yugoslavia; Auronzo in Venetia and Mt. Malfidano, Iglesias, Sardinia, Italy; at Yaaz, Iran; at Bou Thaleb, Algeria; with cerussite, "limonite," and sphalerite at Narlarla, western Kimberley district, WA, Australia. HCWS $DII:248$.

16a.4.2.1   Aurichalcite $(Zn,Cu)_5(CO_3)_2(OH)_6$

Named in eighteenth century after the Greek for *mountain brass*. ORTH $P2_1/m$. $a = 13.82$, $b = 6.419$, $c = 5.29$, $\beta = 101.04°$, $Z = 2$, $D = 3.93$. $AC(B)$ $50:673(1994)$. 17-743: 6.78_{10} 3.68_7 3.25_3 2.89_4 2.81_3 2.72_4 2.61_8 1.656_4. Crystals elongated [001], forming tufted, feathery druses like velvet, rarely columnar, laminated, or granular. {100} zone striations parallel to (001), (010), producing a gridlike appearance corresponding to cleavage or twin directions. Pale green to green-blue and sky blue, silky to pearly luster, transparent. Cleavage {010}, perfect; fragile. H = 1–2. G = 3.64. IR. *MSLM 4:268(1974)*. Aurichalcite crystals from Mapimi, DGO, Mexico, show Zn/Cu = 2.4, compared to the virtually absent Cu in hydrozincite. Soluble in acids and in ammonia. A secondary mineral in oxidized copper and zinc ore deposits. Forms encrustations with malachite, azurite, cuprite, smithsonite, hemimorphite, hydrozincite, and rosasite, in mine areas at Bisbee, Cochise Co., AZ; Cottonwood Canyon, Salt Lake Co., and in Tintic district, Juab Co., UT; Magdalena district, Socorro Co., NM; Goodsprings, NV; Cerro Gordo and Defiance mines, Darwin district, Inyo Co., CA; Cigarrero mine, Almdoya, CHIH, and Mina Ojuela, Mapimi, DGO, Mexico; Cumberland, England; Leadhills, Lanark, Scotland; Chessy, near Lyon, Rhône, France; Campiglia, Tuscany, Italy; Monteponi near Iglesias, Sardinia, Italy; Laurium, Greece; Moravicza and Rézbánya, Romania; near Loktevsk on the Alei R. in the Altai district, Tomsk, Russia; in Pref. Nagato, Japan; at Mindouli, Zaire; at Tsumeb, Namibia; at Broken Hill, NSW, Australia. HCWS $DII:250$, $DS:298(1892)$, CM $8:385(1965)$.

16a.5.1.1   Plumbonacrite $Pb_{10}(CO_3)_6O(OH)_6$ III

Name proposed in 1889 for a mineral from Wanlockhead, Scotland $MM 8:200$ (1988) but incompletely characterized. See hydrocerrusite 16a.2.2.1 HEX $a = 9.076$, $c = 24.96$ $JINCh$ $28:2507(1966)$ PD: 4.26_8 3.357_7 2.953_4 2.619_{10} 2.235_4 1.699_5. This is a second lead carbonate which may be intimately mixed with hydrocerrusite in synthetic materials. Whether or not it occurs naturally has not yet been determined AM $52:563(1967)$.

16a.5.2.1   Tunisite $NaCa_2Al_4(CO_3)_4(OH)_8Cl$

Named in 1969 for the country of origin. TET $P4nmm$. $a = 11.22$, $c = 6.582$, $Z = 2$, $D = 2.48$. AM $54:1(1969)$. 27-1001: 5.62_{10} 5.07_6 3.55_8 3.29_7 3.13_6 2.75_7 2.59_9 2.56_7. Crystalline aggregates, usually < 0.2 mm, tabular on (001), with occasional crystals up to 8 mm that display (110), (100), (101). Colorless. Cleavage {001}, {110}, perfect. H = $4\frac{1}{2}$. G = 2.51. IR DTA. Uniaxial (+);

$N_O = 1.573$, $N_E = 1.599$. Easily soluble in HCl. Late hydrothermal mineral filling cavities created in veins composed mainly of white calcite; may cover portions of the calcite. Found on the dumps of the Pb–Zn ore deposit of *Sakiet Sidi Yousseff, Tunisia*. The calcite veins that contain the mineral reach a thickness of 2 m in the area of the ore consisting of galena, sphalerite, marcasite, and pyrite. HCWS *AM 69:2389(1969)*.

16a.5.3.1 Loseyite $Mn_4Zn_3(CO_3)_2(OH)_{10}$

Named in 1929 after Samuel R. Losey (b.1833) of Franklin, NJ, a mineral collector. MON $A2/a$. $a = 16.23$, $b = 5.51$, $c = 14.95$, $\beta = 95.37°$, $Z = 4$, $D = 3.37$. Structure: *AC(B) 37:1323(1981)*. 17-206: 7.49_8 3.80_9 3.68_{10} 3.54_6 2.79_6 2.77_6 2.63_{10} 2.54_7. Subparallel aggregates and radiating bundles of lath-like crystals elongated [010]. Bluish white, brownish, transparent. Cleavage not observed. H ≈ 3. G = 3.27. Biaxial (+); $Y = b$; $N_x = 1.637$, $N_y = 1.648$, $N_z = 1.676$; $2V = 64°$; $r > v$, weak. Small amounts of Mg in isomorphous substitution. In small veinlets in massive ore at *Franklin, Sussex Co., NJ*, with altered pyrochroite, sussexite, chlorophoenicite, and calcite. HCWS *DII:245*.

16a.5.3.2 Sclarite $(Zn,Mg,Mn^{2+})_4Zn_3(CO_3)_2(OH)_{10}$

Named in 1989 for Charles B. Sclar (b. 1925), professor of mineralogy, Lehigh University, PA. MON $A2/a$. $a = 16.110$, $b = 5.432$, $c = 15.041$, $\beta = 95.49°$, $Z = 4$, $D = 3.547$. Isostructural with loseyite (16a.5.3.1). Zn in both octahedral and tetrahedral coordination. PD: 7.50_{10} 5.23_2 4.82_2 3.75_4 3.63_5 3.53_4 2.62_5 2.50_4. Clusters of 0.2-mm crystals form 1.5-mm coarse-surfaced spherules; platy on {001}, elongate parallel to [010], minor {100}. Clear, colorless, transparent crystals appear grayish white in aggregates; white streak; vitreous luster. No cleavage or parting observed, brittle. H = 3–4. G = 3.51. No fluorescence in LWUV or SWUV. Biaxial (+); $Y = b$; $X \wedge c = +49°$; $N_x = 1.648$, $N_y = 1.664$, $N_z = 1.702$; $2V = 63.4°$; $r > v$, strong. Analyses showed Mg, Mn, and trace Fe in addition to Zn. A secondary mineral found in an intensely altered specimen from *Franklin mine, Franklin, Sussex Co., NJ*, associated with willemite, secondary zincite, another zinc-carbonate mineral (unnamed), rhodocrosite, gageite, chlorophoenicite, and leucophoenicite. HCWS *AM 74:1355(1989)*.

16a.5.4.1 Sabinaite $(Na,Ca)_4Zr_2Ti(CO_3)_4O_4$

Named in 1980 for Ann Phyllis Sabina Stenson (b. 1930), mineralogist, Geological Survey of Canada. MON. $C2/c$. $a = 10.171$, $b = 6.623$, $c = 17.976$, $\beta = 94.32°$, $Z = 4$, $D = 3.48$. *AM 71:231(1986)*. 36-414: 8.96_{10} 4.48_3 3.25_5 2.99_5 2.73_3 2.29_4 2.02_5 1.795_4. Fine-grained white, powdery coatings and chalky aggregates; max. dimensions 0.01 mm × 0.001 mm. St. Hilaire crystals tabular on (001) and elongated on a; forms {001}, {010}, {100}, and {110}. Cleavages {001}, perfect; {010}, distinct. DTA. Biaxial (+); $Y = b$. $X \wedge c = 13°$ in obtuse β; $N_x = 1.720$, $N_y = 1.79$, $N_z = 1.90$; $2V = 82°$ (Na); $r > v$. Found in a dawsonite-rich sill associated with calcite, quartz, weloganite, and cryolite on *St. Michel, Montreal Is.*, and at Mt. St.-Hilaire, QUE, Canada. HCWS *AM 66:1277(1981)*, *CM 23:17(1985)*.

16a.5.5.1 Rouvilleite $Na_3Ca_2(CO_3)_3F$

Named in 1991 for the locality. MON Cc or $C2/c$. $a = 8.043$, $b = 15.812$, $c = 7.030$, $\beta = 101.16°$, $Z = 4$, $D = 2.69$. *44-1471*: 7.0_2 3.52_5 2.92_3 2.90_3 2.86_{10} 2.78_1 2.69_1 1.737_2; also *CM 29:107(1991)*: 7.08_8 2.94_7 2.90_{10} 2.71_9 2.637_6 2.04_7 1.87_8 1.75_6. Irregular crystalline masses up to 3 mm across in cavities; rare euhedral crystals show {010}, {110}, {100}, {001}, but other forms displayed are inconsistent with space group determination. Tan to colorless, white streak, vitreous to slightly waxy luster, transparent. Cleavage {001}, good; {010}, imperfect; uneven fracture; brittle. $H = 3$. $G = 2.67$. IR. Not fluorescent. Biaxial $(-)$; $Y = b$; $X \wedge c = 6°$ in acute angle β; $N_x = 1.472$, $N_y = 1.562$, $N_z = 1.569$ (Na); $2V = 25-30°$; nonpleochroic. Strongly effervescent in cold 1:1 HCl, insoluble in H_2O. As described contains $> 6.8\%$ MnO. Mn-bearing and Mn-free rouvilleite have been synthesized. A secondary mineral in cavities in sodalite syenite inclusion in nepheline syenite at the *Poudrette quarry, Rouville Co., Mt. St-Hilaire, QUE, Canada*. Mineral occurs with villiaumite and kupletskite. HCWS *AM 76:2023(1991)*.

ANCYLITE GROUP

The four end members of the ancylite group, rare-earth-element-containing hydrated hydroxycarbonates, have the general formula

$$(Sr,Ca,Pb)(REE)(CO_3)_2(OH) \cdot H_2O$$

and crystallize with orthorhombic or monoclinic symmetry. The structure [*AM 60:280(1975)*] resembles that of aragonite (14.1.3.1). In spite of the introduction of (OH), H_2O and large REE cations *AM60:280(1975)*. There are relative shifts and an increase in the angle of tilt of the carbonate ions in the carbonate layers. Compare also with the bastnäsite/synchysite group.

ANCYLITE GROUP

Mineral	Formula	Space Group	a	b	c	β	D	
Ancylite-(Ce)	$SrCe(CO_3)_2(OH)$ $\cdot H_2O$	$Pmcm$	5.062	8.586	7.303		3.939	16b.1.1.1
Calcioancylite-(Ce)	$CaCe(CO_3)_2(OH)$ $\cdot H_2O$	$Pmmm$	7.24	8.49	5.02			16b.1.1.2
Calcioancylite-(Nd)	$CaNd(CO_3)_2(OH)$ $\cdot H_2O$	Pm	7.212	4.967	8.468	90.04°	4.08	16b.1.1.3
Gysinite(Nd)	$PbNd(CO_3)_2(OH)$ $\cdot H_2O$	$Pmcn$	5.04	8.50	7.25		4.82	16b.1.1.4

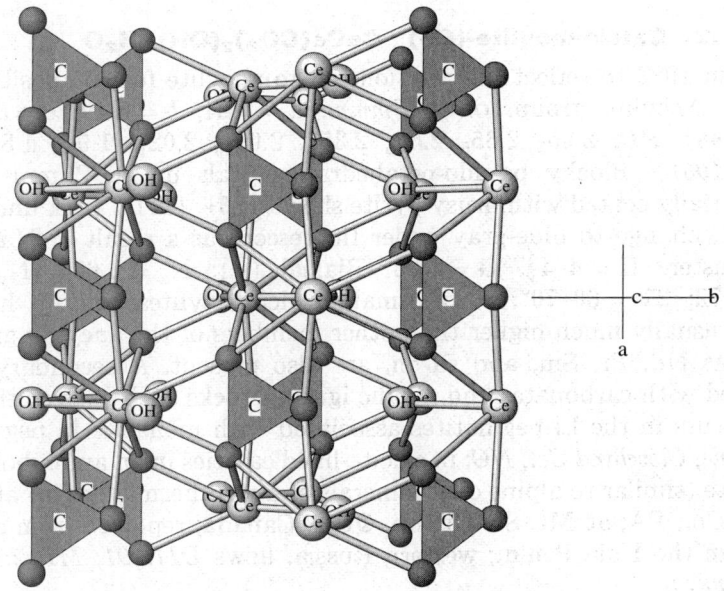

ANCYLITE: La$_2$(CO$_3$)$_2$(OH)$_2$

This c axis projection of anyclite [(simplified to Ce(CO$_3$)$_2$(OH)·H$_2$O)] shows direct similarities to the aragonite (14.1.3.1) structure. The large REE and Sr ions are in tenfold coordination with OH and O of the (CO$_3$) groups. Note only OH is depicted in this diagram as the structure analysis did not discriminate between (OH) and H$_2$O in the calculation of bonding to O AM 60:280 (1975).

16b.1.1.1 Ancylite-(Ce) SrCe(CO$_3$)$_2$(OH) · H$_2$O

Named in 1899 for Greek for *curved* in allusion to the rounded and distorted character of the crystals. Ancylite group. ORTH *Pmcn*. $a = 5.062$, $b = 8.586$, $c = 7.303$, $Z = 2$, $D = 3.939$. Structure: *AM 60:280(1975)*. *29-384:* 5.57_5 4.34_{10} 3.71_{10} 2.96_{10} 2.35_8 2.09_7 2.02_7 1.953_5. Pseudo-octahedral crystals with faces curved and often dull, also short prismatic [001], and as groups and clusters of small rounded crystals. Pale yellow with tinge of orange, yellow-brown to brown, gray; white streak; vitreous luster on faces, greasy on fracture surfaces; translucent to subtranslucent. No cleavage, splintery fracture, tough, brittle. $H = 4$–$4\frac{1}{2}$. $G = 3.95$. Biaxial (−); XYZ= *abc*; $N_x = 1.625$, $N_y = 1.700$, $N_z = 1.735$; 2V medium large; colorless, sometimes cloudy with inclusions (Narssarssuk). Occurs in hydrothermal veins associated with felsic alkalic rocks, or in pegmatites in alkaline rocks and in carbonatites. Found in veins in the Bearpaw Mts. as fine grains intimately intergrown with calcite, quartz, and barite, also in masses several feet across in Sheep Creek vein, Ravalli Co., MT; in the carbonatite, Mt. St.-Hilaire, QUE, Canada; in druses in pegmatite veinlets in nepheline–syenite at Narssarssuk in the *Julianejaab district, Greenland*, associated with aegirite, albite, microcline, zircon, synchisite, cordylite, and eudidymite; in Langesundfjord district, Norway; in the Khibina tundra, Kola Penin., Russia. HCWS *DII:291*.

16b.1.1.2 Calcioancylite-(Ce) CaCe(CO$_3$)$_2$(OH)·H$_2$O

Named in 1922 to reflect the relationship to ancylite for compositions with Ca > Sr. Ancylite group. ORTH $Pmmm$. $a = 7.24$, $b = 8.49$, $c = 5.02$. CM $8{:}398(1968)$. PD: 2.95_5 2.65_5 2.53_5 2.35_{10} 2.09_{10} 2.02_{10} 1.95_9 1.85_6. AM $46{:}1424(1961)$. Blocky pseudo-octahedral crystals up to 2 mm (Cornog, PA), partially coated with drusy pyrite show {120}, {111}. Pink under ordinary light, change to blue-gray under fluorescent as a result of Nd content; frosty luster. H = 4–4$\frac{1}{2}$. G = 4.30. Biaxial (−); $N_x = 1.654$, $N_y = 1.733$, $N_z = 1.772$; 2V = 60–70°; r < v, small. Calcioancylites contain higher Ca than Sr, usually much higher than other members of the ancylite group. La, as well as Nd, Pr, Sm, and so on, are also present. A secondary mineral associated with carbonates and alkalic igneous rocks such as syenitic pegmatites. Occurs in the Li-pegmatites associated with monazite in pegmatite at *Foote mine, Cleveland Co., NC*; in calcite-lined cavities in an amphibole gneiss–migmatite (similar to alpine cleft minerals of hydrothermal origin) at Cornog, Chester Co., PA; at Mt. St.-Hilaire, QUE, Canada; reported from a granitic boulder in the Kola Penin., western Russia. HCWS $DII{:}291$; MR $2{:}18(1971)$, $18{:}203(1987)$.

16b.1.1.3 Calcioancylite-(Nd) CaNd(CO$_3$)$_2$(OH)·H$_2$O

Named in 1991 to show the association to calcioancylite-(Ce) and the dominant element, Nd. Ancylite group. MON Pm. Note that the space group is different from ancylite-(ce). $a = 7.212$, $b = 4.976$, $c = 8.468$, $\beta = 90.04°$, Z = 1, D = 4.08. Structure: EJM $2{:}413(1990)$. 42-1424: 5.49_6 4.26_{10} 3.67_7 2.91_9 2.63_4 2.32_7 2.05_8 1.979_7. Pine cone aggregates up to 1 mm long composed of individual dipyramidal pseudoorthorhombic crystals. Pale pink, white streak, vitreous luster. Irregular fracture, brittle. H = 4–4$\frac{1}{2}$. G = 4.02. Nonfluorescent. Biaxial (−); Z = c; $N_x = 1.660$, $N_y = 1.725$, $N_z = 1.761$; 2V = 70°; r < v, weak; nonpleochroic. Dominant Nd, with Ce, Sm, Gd, Y, Pr, La, Dy, Tm also detected on analysis. Found on a pink orthoclase in a miarolitic cavity in granite in a quarry near *Baveno, Piedmont, Italy*. HCWS AM $76{:}1729(1991)$.

16b.1.1.4 Gysinite-(Nd) PbNd(CO$_3$)$_2$(OH)·H$_2$O

Named in 1985 after Marcel Gysin (1891–1974), mineralogist, University of Geneva, Switzerland. Ancylite group. ORTH $Pmcn$. $a = 5.04$, $b = 8.50$, $c = 7.25$, Z = 2, D = 4.82. 39-364: 5.54_5 4.33_{10} 3.68_{10} 3.34_4 2.95_5 2.34_5 2.07_6 2.02_5. Isolated pseudo-octahedral crystals with {111} and {110} up to 1 mm in length elongated on b; aggregates and penetration twins with twin plane (100). Light pink to reddish pink, white-light pink streak, vitreous to greasy luster, translucent. No cleavage, brittle. No fluorescence with LWUV or SWUV. Biaxial (−); XYZ= cab; $N_x = 1.745$, $N_y = 1.805$, $N_z = 1.840$; 2V = 70°; r < v. Probe analysis showed Pb, La, Nd, and no other rare-earth elements. H$_2$O confirmed by TGA. Soluble with effervescence in cold 1:10 HCl. Occurs in the uraninite-containing ore deposits at *Shinkolobwe, Shaba, Zaire*, associated with schuilingite, malachite, cerussite, talc–chlorite, bornite, wulfenite, kasolite, native gold, and garnet. HCWS *AM 70:1314(1985)*.

16b.1.2.1 Thorbastnäsite Th(Ca,Ce)(CO$_3$)$_2$F$_2$ · 3H$_2$O

Named in 1965 for the compositional and crystal structural resemblance to the bastnäsite group (16a.1.2.1). HEX $P\bar{6}2c$. $a = 6.99$, $c = 9.71$, $D = 5.70$. *AM 50:1505(1965)*. *18-1362*: 3.54$_8$ 2.85$_{10}$ 2.03$_{10}$ 1.870$_7$ 1.656$_5$ 1.279$_6$ 1.165$_5$ 1.134$_5$. Cryptocrystalline, locally subparallel aggregates, brown. $G = 4.04$ (discrepancy ascribed to "molecular water"). DTA and IR data similar to parisite and bastnäsite, but there is molecular water. Uniaxial (+); $N = 1.670$–1.678. Pb also present; Fe, Al, and Si in analyses are admixtures of other minerals. An accessory mineral in iron-rich saccaroidal albitites, may be a pseudomorph after thorite, associated with hydrous iron oxides, quartz, and albite. Found in "exocontact alkaline metasomatic rocks" of an unnamed nepheline syenite intrusion, *eastern Siberia, Russia*, replacing ferrian thorite. HCWS *ZVMO 94:105(1965)*, *MA 17:398(1965)*.

16b.1.3.1 Indigirite Mg$_2$Al$_2$(CO$_3$)$_4$(OH)$_2$ · 15H$_2$O

Named in 1971 after the locality. Space group unknown. Cell data (partially dehydrated under EM) = 6.23, parallel to elongation; 3.16, perpendicular to elongation. *ZVMO 100:178(1971)*. *25-510*: 7.62$_9$ 5.80$_{10}$ 5.24$_9$ 4.56$_8$ 3.65$_4$ 2.70$_9$ 2.60$_9$ 1.625$_5$. Radiating aggregates (rosettes) of very fine flexible fibers, needles, or plates up to 1 mm in length, flexible. Snow white, vitreous to silky luster. H $\approx$ 2. $G = 1.6$. IR, DTA. Uniaxial; $N_O = 1.472$, $N_E = 1.502$; positive elongation parallel to extinction. Insoluble in H$_2$O, alcohol, and ammonia; dissolved by acids, KOH solution. Found in the oxidation zone of the *Sarylakh Au–Sb deposit, Indigirka R., northeastern Yakutia, Russia*, associated with gypsum, amorphous Fe-oxides, quartz, and gibbsite. HCWS *AM 57:326(1972)*.

16b.1.4.1 Schuilingite-(Nd) PbCuNd(CO$_3$)$_3$(OH) · 1.5H$_2$O

Named in 1947 after H. J. Schuiling, Belgian geologist to Union Minière du Haut-Katanga, Zaire. ORTH *Pmcn*. $a = 7.43$, $b = 18.89$, $c = 6.40$, $Z = 2$, $D = 4.59$ (Schinkolobwe). *SMPM 63:1(1983)*. *41-583*: 9.46$_{10}$ 6.06$_9$ 4.80$_5$ 4.72$_6$ 4.49$_9$ 3.85$_{10}$ 2.93$_5$ 2.64$_4$. See also *35-545*. Crusts of minute acicular crystals elongated [001] with prism and terminal domes. Azure blue, turquoise; adamantine luster; translucent. Cleavage {110}, perfect; {100}, poor. H = $3\frac{1}{2}$. $G = 4.50$. Biaxial (−); XYZ = acb; $N_x = 1.730$, $N_y = 1.770$, $N_z = 1.795$; 2V = 75°; weakly pleochroic (Schinkolobwe). Found at *Kalompe, and Schinkolobwe, Shaba, Zaire*, associated with cerussite, calcite, pyrite, and so on. HCWS *DII:252*, *BM 105:225(1982)*, *AM 69:1196(1984)*.

16b.1.5.1 Shabaite-(Nd) CaNd$_2$(UO$_2$)(CO$_3$)$_4$(OH)$_2$ · 6H$_2$O

Named in 1989 for the locality. MON. $P2$, Pm, or $P2/m$. $a = 9.208$, $b = 32.09$, $c = 8.335$, β = 90.3°, $Z = 5$, $D = 3.23$. *EJM 1:85(1989)*. PD: 15.9$_{10}$ 7.31$_7$ 4.58$_6$ 4.17$_7$ 4.13$_7$ 4.01$_3$ 3.79$_3$ 3.09$_6$. Micaceous flakes and rosettes up to 5 mm in diameter, tabular on {010}, elongate [100], twinned {001}; pale greenish yellow, translucent to opaque, pearly. $H = 2\frac{1}{2}$. $G = 3.13$. Biaxial $(-)$; $Y = b$; $Z \wedge a = 3-4°$; $N_x = 1.534$, $N_y = 1.590$, $N_z = 1.600$; $2V = 44°$; nonpleochroic. Electron microprobe analyses showed 0.38% Nd, 0.15% Sm, 0.21% Y, 0.07% Dy, 0.07% Pr, 0.04% La, 0.02% Ce. Soluble in HCl. Occurs with uraninite, kamotoite-(Y), schuilingite-(Nd), and uranophane in Cu–Co deposit, *Kamoto East, 5 km W of Kolwézi, Shaba, Zaire*. HCWS *AM 75:433(1990)*.

16b.1.6.1 Astrocyanite-(Ce) $Cu_2Ce_2(UO_2)(CO_3)_5(OH)_2 \cdot 1.5H_2O$

Named in 1990 for its starlike morphology and the Greek *cyan*, blue. HEX $P6/mmm$. $a = 14.96$, $c = 26.86$, $Z = 12$, $D = 3.95$. *EJM 2:407(1990)*. 42-1396: 13.3_4 6.73_{10} 4.30_5 4.16_6 3.72_9 2.49_4 2.15_4 2.07_4. Isolated tablets {001}, or aggregates to millimeter size, rosettes. Pale blue, bright blue, and blue-green; individuals show vitreous luster, but earthy luster in rosettes; translucent to opaque. Cleavage {001}. $H = 2-3$. $G = 3.80$. Uniaxial $(-)$; $N_O = 1.638$, $N_E = 1.688$; N_E normal to {001}; pleochroic, bright blue to almost colorless. Microprobe analyses show Cu/U = 2, Ce and Nd predominant REE, La, Pr, Sm, and Y also present. Effervescent in HCl 1:10. Oxidation product of uraninite. Found in *East Kamoto Cu–Co deposit, southern Shaba, Zaire*, associated with uranophane, kamotoite-(Y), francoisite-(Nd), shabaite-(Nd), schuilingite, and masuyite. HCWS *AM 76:665(1991)*.

16b.1.7.1 Kamphaugite-(Y) $Ca_2(Y,REE)_2(CO_3)_4(OH)_2 \cdot 3H_2O$

Named in 1993 after Erling Kamphaug (b. 1931), Norwegian mineral collector who provided a specimen for study. Related to tengerite (15.4.3.1), lokkaite (15.4.4.1), kimuraite (15.4.5.1). TET $P4_12_12_1$. $a = 7.434$, $c = 21.793$, $Z = 4$, $D = 3.24$. Structure: *EJM 5:679(1993)*, *5:685(1993)*. PD: 6.14_{10} 4.38_8 3.52_6 2.83_5 2.72_2 2.63_9 2.45_2 1.891_3. Crystals roughly square in outline up to 1 mm across, platy on {001}; as divergent, roselike spherules and aggregates; each platelet is a subparallel group. White to colorless, some spherules pale yellow or pale brown; white streak; vitreous luster; transparent. No cleavage, uneven fracture. $H = 2-3$. $G = 3.19$. Nonfluorescent. Uniaxial $(+)$; $N_O = 1.627$, $N_E = 1.663$; $2V \approx 15°$; anomalously biaxial. Effervesces in 1:10 acids. Norwegian samples contain appreciable REE especially Dy. Low-temperature late-stage secondary mineral. Occurs as white crusts at Evans Lou pegmatite, QUE, Canada; at Tordal, Telemark, and at the Tangen granite pegmatite quarry, near Kragerys, as rosettes along cracks in feldspar and quartz; an alteration product of kuliokite-(Y), and in a magnetite and helvite contact skarn at *Hortekollen, Hoydalen, Oslo region, Norway*; replaces gagarinite, Kazakhstan; as white spheroids on opal from a quartz and barite vein, Transvaal, South Africa. HCWS *AM 79:387(1993)*.

16b.2.1.1 Dundasite PbAl$_2$(CO$_3$)$_2$(OH)$_4 \cdot$ H$_2$O

Named in 1893 for the locality in TAS, Australia. ORTH. *Pbmm*. $a = 9.08$, $b = 16.37$, $c = 5.62$, $Z = 4$, $D = 3.573$. Structure: *MM 33:565(1972)*. *21-936*: 7.91$_{10}$ 4.63$_5$ 3.60$_8$ 3.23$_5$ 3.09$_6$ 2.97$_4$ 2.67$_5$ 2.64$_4$. See also *33-731*(calc). Spheroidal aggregates with radiating needles parallel to c and as matted or felty crusts. White, vitreous or silky luster, transparent. Cleavage {010}, perfect. $H = 2$. $G = 3.25$. DTA, IR. *AM 35:985(1950)*. Biaxial (−); XYZ = abc; $N_x = 1.603$, $N_y = 1.716$, $N_z = 1.742$; 2V ≈ −50°. Effervescent in acids. Secondary mineral in the oxidation zone of lead deposits. Found with cerussite and allophane at mines near Trefriw, Carnarvonshire, Wales; at Clements lead mine near Maam, County Galway, Ireland; with cerussite, hemimorphite, and greenockite at the Mill Close mine, Wensley, Derbyshire, Wheal Rose, Sithney, and with allophane at Port Quin antimony mine, Cornwall, England; with crocoite and pyromorphite and iron oxyhydroxides in gossan in the *Adelaide Prop. mine, Dundas*, with cerussite and gibbsite at *Hercules mine, Mt. Read, TAS, Australia*. HCWS *DII:279*, *BSFM:83(1960)*.

16b.2.1.2 Dresserite Ba$_2$Al$_4$(CO$_3$)$_4$(OH)$_8 \cdot$ 3H$_2$O

Named in 1970 for John Alexander Dresser (1866–1954), Canadian geologist, an authority on asbestos deposits and for his contributions to the geology of the Monteregian Hills. Ba analog of dundasite (16b.2.1.1). ORTH *Pbnm*. $a = 9.27$, $b = 16.83$, $c = 5.63$, $Z = 2$, $D = 3.06$. *CM 10:84(1970)*. *20-617*: 8.09$_{10}$ 6.23$_6$ 4.68$_3$ 4.63$_2$ 3.66$_5$ 3.17$_3$ 2.73$_4$ 2.67$_3$. Spherulitic up to 3 mm in diameter; radiating blades elongated [001], flattened parallel to {010}. White, white streak, vitreous to silky luster. Biaxial (−); $N_x = 1.518$, $N_z = 1.601$; 2V = 30–40°; parallel extinction; length-slow. Dissolves in HCl with effervescence. From *St. Michel, Montreal Island, QUE, Canada*, in cavities of an alkali sill intruding Ordovician limestones associated with weloganite, plagioclase, quartz, dawsonite, and others. HCWS

16b.2.1.3 Strontiodresserite SrAl$_2$(CO$_3$)$_2$(OH)$_4 \cdot$ H$_2$O

Named in 1977 from the composition and analogy with dresserite. Sr analog of dresserite (16b.2.1.2). ORTH *Pbnm*. $a = 9.168$, $b = 16.037$, $c = 5.598$, $Z = 4$, $D = 2.73$. *GSCP 78:180(1978)*. *29-1295*: 7.93$_{10}$ 5.99$_6$ 3.55$_4$ 4.39$_8$ 3.00$_7$ 2.64$_6$ 2.05$_4$ 1.732$_5$. "Atoll-like" coatings of lathlike 0.1 mm × 0.01 mm × 0.001 mm crystals. White, silky luster. $G = 2.71$. Biaxial (−); $N_x = 1.510$, $N_y = 1.583$, $N_z = 1.595$; 2V = 42.5°. Complete solid solution between strontiodresserite and dundasite. Occurs as secondary mineral in silicocarbonatite sill at *Francon Quarry, St. Michel, Montreal Is., QUE, Canada*. HCWS *CM 15:405(1977)*.

16b.2.2.1 Hydrodresserite BaAl$_2$(CO$_3$)$_2$(OH)$_4 \cdot$ 3H$_2$O

Named in 1977 for its relation to dresserite as it gradually dehydrates to that phase. Ba analog of alumohydrocalcite (16b.2.3.1), structurally distinct. TRIC $P1$ or $P\bar{1}$. $a = 9.79$, $b = 10.42$, $c = 5.62$, $\alpha = 96.05°$, $\beta = 92.20°$, $\gamma = 115.71°$,

$Z = 2$, $D = 2.81$. *AM 64:654(1979)*. *29-145:* 9.30_3 8.75_4 8.52_{10} 5.32_3 4.26_5 3.62_3 3.42_7 3.10_6. Spheres and hemispheres up to 2 mm in diameter of radiating fibrous crystals elongated on $\{001\}$ exhibiting $\{010\}$, $\{\bar{2}10\}$, $\{102\}$. Colorless, white; white streak; vitreous luster. $H = 3$–4. $G = 2.80$. DTA. Biaxial $(-)$; $N_x = 1.502$, $N_y = 1.594$, $N_z = 1.595$; $2V = 17°$. Loses H_2O readily even at room temperature, breaking down to dresserite or another carbonate. Found in an alkalic sill *Francon quarry Montreal Is., QUE, Canada*. HCWS *CM 15:399(1977)*.

16b.2.3.1 Alumohydrocalcite   CaAl$_2$(CO$_3$)$_2$(OH)$_4$ · 3H$_2$O

Named in 1926 for the composition. TRIC $P\bar{1}$. $a = 6.498$, $b = 14.457$, $c = 5.678$, $\alpha = 95.83°$, $\beta = 93.23°$, $\gamma = 82.24°$, $Z = 2$, $D = 2.21$. *NJMM:630(1969)*. *42-592:* 7.15_8 6.44_8 6.18_{10} 5.12_4 3.23_3 3.22_5 2.85_4 2.54_3. See also *22-138*. Chalky masses of fibrous crystals. White, pale blue, violet, gray. Cleavage $\{100\}$, perfect; $\{110\}$, imperfect; brittle. $H = 2\frac{1}{2}$. $G = 2.23$. Biaxial $(-)$; $Z \wedge c = 6°$; $N_x = 1.502$, $N_y = 1.562$, $N_z = 1.585$; $2V = 64°$; extinction (Pakistan). *AM 63:795(1978)*. Found at Mount Hamilton, Santa Clara Co., CA; in a dolomite quarry near Bergisch–Gladbach, Germany; Visé, Belgium; Cr-bearing form at Nowej Rudy, Silesia, Germany; originally described as an alteration product of allophane in the *Khakassy district, Siberia, Russia*, in association with volbothite, malachite, and so on; at Chitral, Northwest Frontier Prov., Pakistan, associated with aragonite, quartz, and dickite; in Osaka Pref., Japan. HCWS *DII:280*.

16b.2.3.2 Para-alumohydrocalcite   CaAl$_2$(CO$_3$)$_2$(OH)$_4$ · 6H$_2$O

Named in 1977 for its relation to alumohydrocalcite, differing in H_2O content. *ZVMO 106:336(1977)*. See alumohydrocalcite (16b.2.3.1). *30-222:* 7.90_{10} 6.20_3 4.04_2 3.90_3 3.32_3 3.03_2 2.68_3 2.64_3. Radiating aggregates of fibrous crystals, white but colorless in thin section with extinction 10° from the fiber length. $H = 1\frac{1}{2}$–2. $G = 2.0$. DTA. Biaxial $(-)$; $N_x = 1.473$, $N_z = 1.502$; $2V = 69°$; positive elongation. HCl only slightly effective in dissolution. Found with gypsum and calcite in the oxidized zone of the sulfur deposits of *Vodinsk, Ukraine*, and *Gaurdok, Turkmenistan*. HCWS *AM 63:795(1978)*.

16b.2.4.1 Bijvoetite-(Y)   (Y,Dy)$_2$(UO$_2$)$_4$(CO$_3$)$_4$(OH)$_6$ · 11H$_2$O

Named in 1982 to honor Dutch crystallographer Johannes Martin Bijvoet (1892–1980). ORTH $C2ma$ or $Cmma$. $a = 21.22$, $b = 45.30$, $c = 13.38$, $Z = 16$, $D = 3.907$. *CM 20:231(1982)*. *35-551:* 8.61_8 6.70_{10} 4.16_6 3.52_5 3.36_5 3.00_6 2.14_4 1.763_4. Minute tabular $\{001\}$ crystals elongated along $[110]$; forms $\{001\}$, $\{110\}$, $\{130\}$, $\{010\}$. Cleavage $\{001\}$, perfect. Sulfur yellow, vitreous luster, transparent. $G = 3.95$. Nonfluorescent under SWUV or LWUV. Biaxial $(+)$; $XYZ = cab$; $N_x = 1.600$, $N_y = 1.650$, $N_z = 1.722$; $2V = 84°$. X, colorless; Y, pale yellow; Z, dark yellow. Microprobe analysis also showed Gd and Tb. Occurs with lepersonite and hydrated uranium oxides near primary uraninite in lower portion of oxidation zone in *Shinkolobwe, Zaire*. HCWS *AM 68:11248(1983)*.

16b.3.1.1 Artinite $Mg_2(CO_3)(OH)_2 \cdot 3H_2O$

Named in 1902 after Ettore Artini (1866–1928), mineralogist at the University of Milan. MON $C2$ or $C2/m$. $a = 16.56$, $b = 3.15$, $c = 6.22$, $\beta = 99.16°$, $Z = 2$, $D = 2.02$. Structure: *AC 5:286(1952)*. *6-484:* 8.18_3 5.37_8 3.69_5 2.74_{10} 2.67_2 2.27_2 2.21_4 1.918_2. Crusts of acicular crystals elongated [010]; botryoidal masses of silky fibers; spherical aggregates of radiating fibers and cross-fiber veinlets. White; white streak; single crystals, vitreous fibers silky; transparent. Cleavage {100}, perfect; {001}, good; brittle. $H = 2\frac{1}{2}$. $G = 2.02$. Biaxial (−); Y parallel to b; $Z \wedge c = 30°$; $N_x = 1.488$, $N_y = 1.534$, $N_z = 1.556$; $2V = 70°$; optic axis perpendicular to {100}. Easily soluble with effervescence in cold acids. DTA. Low-temperature hydrothermal mineral in veins and crusts in serpentinized ultrabasic rocks associated with hydromagnesite, brucite, aragonite, calcite, dolomite, magnesite, pyroaurite, and crysotile. Found in asbestos deposits at Hoboken, NJ, and on Staten Island, NY, with brucite and hydromagnesite; at Luning, NV, as cross fibers veinlets with hydromagnesite in massive brucite; with pyroaurite along joint planes in serpentinite, Eden Mills, VT; in asbestos deposits *Val Brutta, and Francia in the Val Lanterna, Lombardy, Italy*, and in the Val di Lanzo, Val d'Aosta, and near Torre Santa Maria, Val Malenco; with hydromagnesite at Kraubat Styria, Austria; at Lojane, southern Serbia; Yavoritsy, Bulgaria. HCWS *DII:263*.

16b.3.2.1 Otwayite $Ni_2(CO_3)(OH)_2 \cdot H_2O$

Named in 1977 after Charles Albert Otway (b.1922), miner and prospector of Gosnells, WA, Australia. ORTH Space group unknown. $a = 10.18$, $b = 27.4$, $c = 3.22$, $Z = 8$, $D = 3.346$. *29-868:* 6.84_{10} 5.67_8 3.02_5 2.74_6 2.53_5 2.37_4 2.24_5 1.732_4. Structure probably related to aurichalcite (16a.4.2.1). Fibrous, with individuals ≈ 0.05 μm in diameter and several 100 μm in length, form rosette aggregates. Bright green, silky luster. VHN = 247 (5 g load). $G = 3.41$. Biaxial; $N_{\text{parallel to } c} = 1.65$, $N_{\text{perpendicular to } c} = 1.72$; pleochroic, pale green to darker green perpendicular to length; parallel to extinction; length fast. Dissolves slowly in cold 1:1 HCl at room temperature, with effervescence at 100°. Found within narrow (1-mm-wide) veinlets, probably late-stage fracture fillings, in nickeloan serpentinites that contain millerite, polydymite, and apatite in the *Nullagine region of WA, Australia*. The fibers, oriented normal to the veinlet walls, are associated with magnesite, gaspéite, and pecoraite. HCWS *AM 62:999(1977)*.

16b.3.3.1 Kamotoite-(Y) $Y_2(UO_2)_4(CO_3)_3(OH)_8 \cdot 10\text{-}11H_2O$

Named in 1986 for the locality. MON $P2_1/m$. $a = 21.22$, $b = 12.93$, $c = 12.39$, $\beta = 115.3°$, $Z = 4$, $D = 3.94$. *BM 109:643(1986)*. *40-482:* 8.49_8 6.48_{10} 3.93_4 3.49_4 3.05_6 2.76_4 2.13_4 1.749_4. Crusts of elongate [100] blades up to 5 mm with sharp pointed ends; crystals show {001}, {100}, {$\bar{1}$01}. Bright yellow, vitreous luster, transparent to translucent. Twinned on (001). Cleavages {010}, possible; {001}, easy. G = 3.93. Nonfluorescent LWUV or SWUV. Biaxial (−); $N_x = 1.604$, $N_y = 1.667$, $N_z = 1.731$; 2V = 87°; positive elongation; absorb Z > Y > X; no dispersion. X = b, colorless; Y ∧ c = 25°, pale yellow-green; Z = a, bright yellow. Microprobe analyses showed also Nd, Sm, Gd, and Dy. Occurs on uraninite in *Kamoto, southern Shaba, Zaire*. HCWS *AM 73:191(1988)*.

16b.4.1.1 Zaratite $Ni_3(CO_3)(OH)_4 \cdot 4H_2O$

Named in 1851 for Antonio Gil y Zarate (1793–1861), Spanish dramatist, general director of public education, and counselor of state. A mixture of amorphous and fibrous particles rather than a discrete mineral species. *MM 33:663(1963)*. *16-164*: 8.93_7 5.07_{10} 2.73_5 2.45_7 2.00_4 1.91_2 1.74_1 1.55_2. Crusts, small stalactitic or minute mammillary, microcrystalline, also massive, compact, and as thin films on millerite. Emerald-green, paler streak, vitreous to greasy luster, translucent to transparent. Conchoidal fracture of aggregates, brittle. H = $3\frac{1}{2}$. G = 2.57. IR. *MSLM 4:268(1974)*. N = 1.56–1.61; fibers parallel to extinction; positive elongation; pleochroic. Length, emerald-green; perpendicular, yellow-green. Soluble with effervescence in 1 : 10 HCl. Synthesized by action of NiCl on nesquehonite. Associated with dolomite, aragonite, calcite, hydromagnesite, brucite, chromite, pentlandite, and pyrrhotite. Alteration product of millerite and meteoritic iron. Found with chromite at Low's mine and Wood's mine, Lancaster Co., and at West Nottingham, Chester Co., PA; in serpentinites in Swinaness, Unst, Shetland Is., England; as an incrustation on magnetite with NiS at *Cape Ortegal, Galicia, Spain*; with millerite and chromite near Kraubat, Styria, Austria; as the alteration of millerite in geodes and sideritic septaria at Rapice, Dubi, Bohemia, Czech Republic; at Lilaz, Cogne, Piedmont, Italy, in serpentinites with aragonite and nesquehonite; in the ultrabasic rocks containing Ni deposits of Khalilova massif near Orsk in the southern Ural Mts., Russia; in Broken Hill, NSW, and as an alteration product of heazlewoodite (Ni_3S_2), TAS, Australia; at Dun Mountain, Nelson, New Zealand. HCWS *DII:245*.

16b.4.2.1 Defernite $Ca_6(CO_3)_2(OH)_7Cl \cdot nH_2O$

Named in 1980 after Jacques Deferne (b.1934), Curator of Mineralogy, Museum d'Histoire Naturelle, Geneva, Switzerland. ORTH $Pna2_1$ or $Pnam$. $a = 17.860$, $b = 22.775$, $c = 3.658$, Z = 8, D = 2.42. *BM 103:185(1980)*. *42-1324*: 11.38_{10} 8.32_9 5.68_3 3.07_4 2.99_4 2.89_4 1.889_3 1.814_3. See also *35-488*. Fan-shaped aggregates 100–200 μm in size, crystals showing {001}, {100}, {010}, {101}, {201}. Colorless, vitreous luster. Cleavage {010}, perfect; {100}, distinct. G = 2.5. Biaxial (−); XYZ = cba; $N_x = 1.546$, $N_y = 1.572$, $N_z = 1.576$; 2V = 42°; r < v. Dissolves in cold HCl with effervescence. Occurs as secondary mineral in skarn contact of granite with Cretaceous limestone near *Güneyce–Ikizere, Trabson region, Turkey*, associated with vesuvianite, wollastonite, andradite, diopside, calcite, rustumite, spurrite, and hillebrandite. HCWS *AM 65:1066(1980)*.

16b.4.2.2 Holdawayite $Mn_6(CO_3)_2(OH)_7(Cl,OH)$

Named in 1988 after Michael J. Holdaway (b.1936), petrologist, Southern Methodist University, Dallas, TX. MON $C2/m$. $a = 23.437$, $b = 3.314$, $c = 16.618$, $\beta = 111.15°$, $Z = 4$, $D = 3.24$. *AM 73:632(1988)*. PD: 10.93_{10} 7.77_4 5.46_8 3.88_7 2.93_4 2.69_6 2.59_5 1.789_7. Subhedral grains up to 2 cm in diameter or fibrous aggregates. Pink, light pink streak, silky to vitreous luster, translucent, may be covered with black sooty coating. Cleavage {100}, perfect; irregular fracture, brittle. H ≈ 3. G = 3.19. Nonfluorescent. Effervesces weakly in 1:10 HCl. Biaxial (−); X = b; Z ∧ c = 45° in obtuse β; $N_x = 1.644$, $N_y = 1.719$, $N_z = 1.721$; 2V = 12° (Na); r < v. Occurs as coarse veins cutting silicate-facies Mn-bearing rocks. Found at *Kombat mine, 37 km E of Otavi and 49 km S of Tsumeb, Namibia*, as an anastomosing mesh of veinlets 1–8 mm thick in a hematite–carbonate matrix, and also as patches and blebs 1–4 cm in diameter. The mineral was restricted to the footwall zone of a complexly deformed body of intercalated manganese and iron ores, and occasionally formed up to 40% of the ore. HCWS

16b.4.3.1 Claraite $(Cu,Zn)_3(CO_3)(OH)_4 \cdot 4H_2O$

Named in 1982 for the locality. An inadequately characterized mineral. TRIC (pseudo-HEX) $P1$ or $P\bar{1}$. $a = 26.22$, $c = 21.56$, $Z = 66$, $D = 3.34$. *CE 41:97(1982)*. 35-549: 13.47_{10} 7.84_9 5.69_3 5.17_6 3.65_8 3.24_5 2.96_6 2.72_4. Radiating fibrous spherulitic aggregates up to 0.2 mm [on barite, *AM 62:175(1977)*]; blue-green with perfect {10$\bar{1}$0} cleavage. H ≈ 2. G = 3.35. Biaxial (−); Z ∧ c = 30°; $N_O = 1.645$, $N_E = 1.751$; r > v, strong. X, colorless; Z, pale green to bluish green. Dissolves in 1:10 HNO$_3$ with effervescence. Associated with malachite, azurite, and olivenite at *Clara Mine, Black Forest, Germany*. HCWS *AM 68:471(1983)*.

16b.4.4.1 Peterbaylissite $Hg_2(CO_3)(OH) \cdot 2H_2O$

Named in 1995 for Peter Bayliss, University of Calgary, Alberta, Canada, for his contributions to structural and experimental mineralogy. ORTH *Pcab*. $a = 11.130$, $b = 11.139$, $c = 10.725$, $Z = 8$, $D = 7.14$. Structure: *CM 33:47(1995)*. Pseudotetragonal. A unique mineral occurrence of Hg$^+$ ion. There are four cation sites, three occupied by Hg, the fourth C as (CO$_3$). Puckered hexagonal sheets with Hg along the edges of the hexagons and CO$_3$ and H$_2$O reinforcing bonding between sheets. PD: 6.35_4 4.95_4 4.84_5 2.969_7 2.786_7 2.648_{10} 2.419_6 1.580_2. Clusters and isolated wedge-shaped crystals elongate [001], also platy. Black to dark red-brown, dark brown to black streak, submetallic to adamantine luster, opaque. Resembles specular hematite and magnesiochromite, which also occur at New Idria. Brittle, irregular fracture. H < 5. Nonfluorescent. IR confirmed presence of H$_2$O, (OH), and CO$_3$. $N_{av\ calc} = 2.10$ (589 nm); in reflected light, gray with blue tinge, nonpleochroic, weakly to moderately bireflectant, internal reflections from yellow-white to orange, weakly anisotropic, dull dark gray and brown. R_1 % (air), R_2 % (oil): 11.4, 12.15; 2.02, 2.35 (470 nm); 10.95, 11.6; 1.88, 2.15 (546); 10.9, 11.5; 1.85, 2.11 (589); 10.7, 11.2; 1.82, 2.04 (650). Becomes transparent without effervescence in 1:1 HCl. Electron microprobe analysis is consistent with composition. Crystals up to 0.2 mm long occur on ferroan magnesite and quartz on a single specimen from the former *Clear Creek mercury mine, New Idria district, San Benito Co., CA*. HCWS

16b.5.1.1 Callaghanite $Cu_2Mg_2(CO_3)(OH)_6 \cdot 2H_2O$

Named in 1954 after Eugene Callaghan (1904–1990), director of the New Mexico Bureau of Mines. MON $C2/c$. $a = 10.006$, $b = 11.752$, $c = 8.213$, $\beta = 107.4°$, $Z = 4$, $D = 2.65$. *AM 58:551(1973)*. *11-332:* 7.45_{10} 6.17_{10} 4.80_6 3.87_9 3.72_6 3.18_9 2.67_6 2.30_7. Encrustations, tiny disseminated monoclinic crystals showing {111}, {$\bar{1}$11}, {221}, {$\bar{2}$21}, and veinlets. Blue, white streak, vitreous luster, transparent. Cleavage {111}, {$\bar{1}$11}, perfect; irregular fracture; brittle. $H = 3–3\frac{1}{2}$. $G = 2.71$. Biaxial (−); $N_x = 1.559$, $N_y = 1.653$, $N_z = 1.680$; $2V = 55°$. Occurs in magnesite and dolomite rock in Gabbs Refractory, Inc. mine near *Gabbs, Nye Co., NV*. HCWS *AM 39:316,630(1954)*.

16b.5.2.1 Urancalcarite $Ca(UO_2)_3(CO_3)(OH)_6 \cdot 3H_2O$

Named in 1984 for the composition. ORTH $Pbnm$ or $Pbn2_1$. $a = 15.42$, $b = 16.08$, $c = 6.970$, $Z = 4$, $D = 4.10$. *BM 107:21(1984)*. *38-350:* 8.06_{10} 7.00_3 4.02_8 3.49_7 3.28_3 3.19_5 2.83_4 2.10_5. Acicular crystals in radiating aggregates to 4 mm in diameter. Bright yellow. Crystals, elongated on [001], flattened on (100), show {100}, {010}, and {001}. No cleavage. $H \approx 2$–3. $G = 4.03$. Biaxial (−); $XYZ = abc$; $N_x = 1.660$, $N_y = 1.712$, $N_z = 1.736$; $2V = 67°$. Found on uraninite associated with uranophane, masuyite, and wyartite in *Shinkolobwe, Shaba, Zaire*. HCWS *AM 70:438(1985)*.

16b.5.3.1 Montroyalite $Sr_4Al_8(CO_3)_3(OH,F)_{26} \cdot 10–11H_2O$

Named in 1986 after Monterregian Hill (Mont Royal), a landmark in Montreal, QUE, Canada. Probably TRIC $a' = 7.14$, $c' = 6.55$. *CM 24:455(1986)*. Resembles aluminohydrocalcite. *40-470:* 6.57_{10} 4.00_5 3.28_6 3.19_5 2.86_4 2.55_4 2.48_4 2.36_5. Fibers, lath-shaped 20 μm long, 5 μm wide, radiate from millimeter-size white translucent hemispheres. White, white streak, waxy luster. Polysynthetic twinning noted with EM. No visible cleavage, uneven to splintery fracture, brittle. $H = 3\frac{1}{2}$. $G = 2.667$. TGA/EGA. Biaxial (−); $N_x = 1.51$, $N_y = 1.530$, $N_z = 1.545$; $2V = 80°$; Y parallel to length; X,Z approximately 45° to plane of lath. Found in cavities in a silicocarbonatite sill exposed at *Francon quarry, Montreal, QUE, Canada*, associated with albite, quartz, strontiodresserite, calcite, dawsonite, ankerite, and fluorite. HCWS

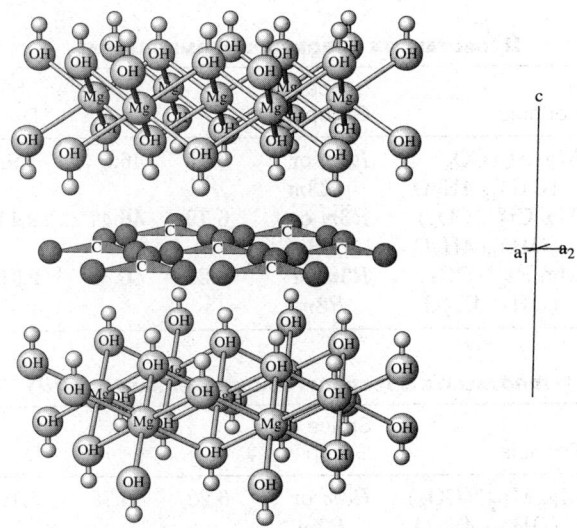

SJÖGRENITE: $Mg_6Al_2(CO_3)(OH)_{16} \cdot 4H_2O$

An a* projection of the structure (simplified) showing two brucite-like layers and an intercalated layer of CO_3 groups. Note neither the differently charged cations in the brucite-like layer, nor the disorder in the CO_3 layer that also includes H_2O are depicted.

SJÖGRENITE–HYDROTALCITE GROUP

The sjögrenite–hydrotalcite minerals conform to the formula

$$A_6B_2(CO_3)(OH)_{16} \cdot 4H_2O$$

where
A = Mg, Ni
B = Al, Cr^{3+}, Fe^{3+}, Mn^{3+}, Co^{3+}

and crystallize with hexagonal or hexagonal–rhombohedral space groups. The general structure [CCM 35:401(1982)] is composed of two types of layers; a positively charged brucite-type layer [$(A^{2+}_{1-x}B^{3+}_x)(OH)_2$] that is intercalated 1:1 with a negatively charged layer that contains $(CO_3)^{2-}$ groups and molecular water. The trigonal planar $(CO_3)^{2+}$ groups are usually highly disordered, possibly missing, or may be replaced by other anion groups. The H_2O–CO_3 interlayer contributes only to the 00l and 10l diffraction maxima. Subgroups result from the different stacking of the layers.

SJÖGRENITE SUBGROUP: HEXAGONAL

Mineral	Formula	Space Group	a	c	D	
Manasseite	$Mg_6Al_2(CO_3)(OH)_{16} \cdot 4H_2O$	$P6_3/mmc$	6.12	15.324	2.02	16b.6.1.1
Barbertonite	$Mg_6Cr^{3+}_2(CO_3)(OH)_{16} \cdot 4H_2O$	$P6_3/mmc$	6.20	15.6	2.11	16b.6.1.2
Sjögrenite	$Mg_6Fe^{3+}_2(CO_3)(OH)_{16} \cdot 4H_2O$	$P6_3/mmc$	6.22	15.61	2.097	16b.6.1.3

Hydrotalcite Subgroup: Rhombohedral

Mineral	Formula	Space Group	a	c	D	
Hydrotalcite	$Mg_6Al_2(CO_3)$ $(OH)_{16} \cdot 4H_2O$	$R\bar{3}m$ or $R3m$	6.15	46.5	1.929	16b.6.2.1
Stichtite	$Mg_6Cr_2^{3+}(CO_3)$ $(OH)_{16} \cdot 4H_2O$	$R\bar{3}m$ or $R3m$	6.19	46.47	2.11	16b.6.2.2
Pyroaurite	$Mg_6Fe_2^{3+}(CO_3)$ $(OH_{16} \cdot 4H_2O$	$R\bar{3}m$ or $R3m$	6.219	46.83	2.10	16b.6.2.3

Hydrotalcite Subgroup: Rhombohedral (cont'd)

Mineral	Formula	Space Group	a	c	D	
Desautelsite	$Mg_6Mn_2^{3+}(CO_3)$ $(OH)_{16} \cdot 4H_2O$	$R\bar{3}m$ or $R3m$	6.23	46.78	2.10	16b.6.2.4
Reevesite	$Ni_6Fe_2^{3+}(CO_3)$ $(OH)_{16} \cdot 4H_2O$	$R\bar{3}m$ or $R3m$	6.164	45.54	2.89	16b.6.3.1
Takovite	$Ni_6Al_2(CO_3)$ $(OH)_{16} \cdot 4H_2O$	$R\bar{3}m$ or $R3m$	6.04	45.16	2.79	16b.6.3.2
Comblainite	$Ni_6^{2+}Co_2^{3+}(CO_3)$ $(OH)_{16} \cdot 4H_2O$	$R\bar{3}m$ or $R3m$	6.08	45.58	3.16	16b.6.3.3

Note: See honessite and hydrohonessite for SO_4 analogs of reevesite.

Some end-member compositions have both hexagonal and rhombohedral polytypes; that is, they are polymorphs. For example, pyroaurite (hydrotalcite group) is the rhombohedral variation of sjögrenite. Both minerals are composed of $[Mg_6Al_2(OH)_{16}]^{2+}$ layers alternating with $[CO_3 \cdot 4H_2O]^{2-}$ interlayers. Pyroaurite has a different stacking order than sjögrenite but the two layers are identical.

There are many layered structures with similar compositions. Variations include: (1) cationic substitution within the brucitelike layers, (2) anionic substitution in the interlayers, (3) variations in the periodicity of layers, and (4) combinations of the several types of layers leading to mixed-layer minerals. *CCM 35:401(1987)*.

Mineral species with predominantly $R^{3+}B$ (*i.e.*, $Al(OH)_3$ instead of mg plus Al, or $(SO_4)^{2-}$ and Cl^- in addition to $(CO_3)^{2-}$ in the interlayer, or mixed layers and multiple periodicities are reported in classes 16b.7 or class 17). More are anticipated.

16b.6.1.1 Manasseite $Mg_6Al_2(CO_3)(OH)_{16} \cdot 4H_2O$

Named in 1941 after Ernesto Manasse (1875–1922), mineralogist of Florence, Italy. Sjögrenite group, aluminum end member. Dimorphous with hydrotalcite (16b.6.2.1). HEX $P6_3/mmc$. $a = 6.12$, $c = 15.324$, $Z = 1$, $D = 2.02$.

$NJMM:161(1966)$. 14-525: 7.67_{10} 2.60_5 2.49_3 2.34_4 2.17_4 2.00_4 1.84_6 1.52_3. Massive with a foliated or contorted structure, lamellar on {0001}; pyramidal prismatic crystals with triangular cross section, faces striated parallel to cleavage. White, bluish, grayish, or brownish white; transparent luster; greasy feel. Cleavage {0001}, perfect; laminae flexible but not elastic; crushes to talclike powder. H = 2. G = 2.05. Uniaxial(−); $N_O = 1.524$, $N_E = 1.510$. Occurs in association with hydrotalcite, dolomite, and serpentine minerals in the serpentinites at Amity, Orange Co., NY; at Mt. St.-Hilaire, QUE, Canada; at Nordmark, Snarum, and *Kongsberg, Norway*; in iron ore skarns of Kapaev explosive pipe, southern Siberian platform, Russia. HCWS *DI:658*, *ZVMO 113:47(1984)*, *CCM 35:401(1987)*.

16b.6.1.2   Barbertonite $Mg_6Cr_2^{3+}(CO_3)(OH)_{16} \cdot 4H_2O$

Named in 1940 for the locality. Sjögrenite group. Dimorphous with stichtite (16b.6.2.2). HEX $P6_3/mmc$. $a = 6.20$, $c = 15.6$, Z = 1, D = 2.11. *CCM 35:401(1987)*. Contorted lamellar-fibrous masses and cross-fiber veinlets. Microcrystals stacked about [0001]. Rose-pink to violet. Cleavage {0001}, perfect. H = $1\frac{1}{2}$-2. G = 2.10. Uniaxial (−); $N_O = 1.557$, $N_E = 1.529$. Found at Cunningsburgh, Scotland, at *Barberton, Transvaal, South Africa*, and at Dundas, TAS, Australia, in association with stichtite. HCWS *AM 26:196,295(1941)*, *MM 39:377(1973)*.

16b.6.1.3   Sjögrenite $Mg_6Fe_2^{3+}(CO_3)(OH)_{16} \cdot 4H_2O$

Named in 1940 after Sten Naders Hjalmar Sjögren (1856–1922), mineralogist, National Museum, Stockholm, Sweden. Sjögrenite group, iron end member. Dimorphous with pyroaurite (16b.6.2.3). HEX $P6_3/mmc$. $a = 6.22$, $c = 15.61$, Z = 1, D = 2.097. *NJMM 161(1966)*. 24-1091: 7.8_{10} 3.9_3 2.66_1 2.55_1 2.39_1 2.22_1 1.872_1 1.581_1. Thin plates {0001}, hexagonal outline, with striated and indistinct faces; plates flexible but not elastic. Yellowish or brownish; waxy to vitreous luster, pearly on cleavages; transparent. H = $2\frac{1}{2}$. G = 2.11. IR. *MSLM 4:271(1974)*. Uniaxial (−); $N_O = 1.573$, $N_E = 1.559$. Pale yellow to colorless. Soluble with effervescence in 1 : 10 acids. Epitactic overgrowths on pyroaurite crystals from Sterling Hill mine, Ogdensburg, Sussex Co., NJ; with calcite and pyroaurite as hydrothermal product at *Långban, Sweden*; at Yoshikawa, Aichi Pref., Japan. HCWS *AM 26:196,295(1941)*, *DI:659*.

16b.6.2.1   Hydrotalcite $Mg_6Al_2(CO_3)(OH)_{16} \cdot 4H_2O$

Named for the resemblance to talc in its physical properties and its water content. Sjögrenite group, aluminum end member of the hydrotalcite subgroup. Dimorphous with manasseite (16b.6.1.1). HEX-R $R\bar{3}m$ or $R3m$. $a = 6.15$, $c = 46.5$, Z = 3, D = 1.929. Structure: *NJMM 544(1969)*, *CCM 35:401(1987)*. 41-1428: 16.12_6 7.84_1 4.34_{10} 3.91_8 2.61_3 2.52_1 2.20_1 1.862_2. 22-700(syn): 7.84_{10} 3.90_6 2.60_4 2.33_3 1.990_3 1.541_4 1.498_3 1.265_1. Massive with a foliated or contorted lamellar structure on {0001}, lamellar-fibrous or in indistinct plates {0001}. White, sometimes with a brownish tint; white streak;

pearly to waxy luster; transparent. Cleavage {0001}, perfect; laminae flexible but not elastic; crushes to talc-like powder. H = 2. G = 2.06. Uniaxial (−); $N_O = 1.511$, $N_E = 1.495$; occasionally biaxial due to strain; nonpleochroic. Soluble with effervescence in 1:10 HCl. IR. *MSLM 4:271(1974)*. An alteration product of spinel at Somerville, St. Lawrence Co., and at Amity, Orange Co., also in close association with manasseite at Rossie, St. Lawrence Co., and at Antwerp and Oxbow, Jefferson Co., NY; at Vernon, Sussex Co., NJ; at Nordmark, and at *Snarum, Norway*, in an intimate mixture with manasseite as foliated masses in serpentinites; with brucite, pyroaurite, antigorite, and calcite in an asbestos mine, Bashenowo, Ural Mts., Russia; Jianshibao asbestos mine, Sichuan, China. HCWS *DI:653, AM 53:1057(1968)*.

16b.6.2.2 Stichtite $Mg_6Cr_2(CO_3)(OH)_{16} \cdot 4H_2O$

Named in 1910 after Robert Sticht (1856–1922), general manager of Mount Lyell Mining and Railway Co., TAS, Australia. Sjögrenite group, chromium end member in the hydrotalcite subgroup. Dimorphous with barbertonite (16b.6.1.2). HEX-R $R\bar{3}m$ or $R3m$. $a = 6.19$, $c = 46.47$, $Z = 3$, $D = 2.11$. *MM 39:377(1973)*. *14-330*: 7.8_{10} 4.30_1 3.91_9 2.60_4 2.32_3 1.97_3 1.540_2 1.510_2. Massive, in matted and contorted aggregates of plates or fibers and as cross-fiber veinlets; in micaceous scales. Intense lilac to rose-pink; pale to white streak; waxy to greasy, faintly pearly luster; greasy feel; transparent. Cleavage {0001}, perfect; laminae flexible but not elastic. H = $1\frac{1}{2}$–2. G = 2.16. IR. *MSLM 4:271(1974)*. Uniaxial (−); $N_O = 1.545$, $N_E = 1.518$; sometimes biaxial; pleochroic, dark rose pink to lilac, light rose-pink to lilac; positive elongation. Found as blebs or veinlets in serpentinites at Megantic mine, Black Lake, QUE, Canada; at Kaapsche Hoop, Barberton district, Transvaal, South Africa; closely associated with chromite, barbertonite, and antigorite at *Dundas, TAS, Australia*. HCWS *DI:655*.

16b.6.2.3 Pyroaurite $Mg_6Fe_2^{3+}(CO_3)(OH)_{16} \cdot 4H_2O$

Named in 1865 in allusion to the golden yellow color that appears on heating to relatively low temperatures. Sjögrenite group, hydrotalcite subgroup. Dimorphous with sjögrenite (16b.6.1.3). HEX-R $R\bar{3}m$ or $R3m$. $a = 6.219$, $c = 46.83$, $Z = 3$, $D = 2.10$. Structure: *AC(B) 24:972(1968)* and also *CCM 35:401(1987)*. *25-521*: 7.77_{10} 3.89_8 2.62_5 2.33_5 1.97_5 1.76_1 1.55_2 1.52_2. Tabular {0001} with lateral faces striated and indistinct, crushes to talc-like powder, also thick tabular {0001} with {01$\bar{1}$5} alone or with {20$\bar{2}$5}. Yellowish to brownish white, also green, colorless; white streak; waxy to vitreous, pearly luster; transparent. H = $2\frac{1}{2}$. G = 2.12. Uniaxial (−); $N_O = 1.564$, $N_E = 1.543$; pale yellow, brownish, reddish to colorless; occasionally biaxial with small 2V. A secondary mineral in low-temperature hydrothermal veins and serpentinites. Found at Sterling Hill, Sussex Co., NJ; at Cornwall, Lebanon Co., and Wood's chrome mine, Lancaster Co., PA; at Blue Mont, MD, as a coating on serpentine; in dolomitic rock at Rutherglen, ONT, Canada; in *Långban, Sweden*, as a parallel mixture of sjögrenite and pyroaurite intergrown with calcite in veins; at Kraubat, Styria, Austria, associated with brucite, aragonite, calcite, hydromagnesite, artinite, and magnesite in serpentine; and at Val Malenco, Lombardy, and Val Ramazzo, Liguria, Italy. HCWS *DI:656*.

16b.6.2.4 Desautelsite $Mg_6Mn_2^{3+}(CO_3)(OH)_{16} \cdot 4H_2O$

Named in 1979 after Paul Desautels (1920–1991), curator of minerals and gems, National Museum of Natural History, Smithsonian Institution, Washington, DC. Sjögrenite group, hydrotalcite subgroup. HEX-R $R\bar{3}m$ or $R3m$. $a = 6.23$, $c = 46.78$, $Z = 3$, $D = 2.10$. *AM 64:127(1979). 33-869:* 7.76_{10} 3.88_6 2.62_5 2.33_6 1.981_6 1.768_2 1.670_2 1.527_3. See also *44-1446*(syn). Euhedral, simple, hexagonal crystals tabular on [0001] with only {0001} and {10$\bar{1}$0}; usually less than 0.2 mm in diameter. Bright orange, light orange streak. Cleavage {0001}, perfect. H ≈ 2. G = 2.13. Does not fluoresce in UV. Uniaxial (−); $N_O = 1.569$, $N_E = 1.547$, strongly pleochroic. O, deep orange; E, light orange. Readily soluble in fumes of HCl which distinguishes it from pyroaurite, which requires liquid dissolution. Desautelsite solution is deep brown in 1:1 HCl, while pyroaurite is light greenish yellow. Occurs embedded in colorless brucite at *Cedar Hill quarry, Lancaster Co., PA*; abundant in two localities near New Idria, San Benito Co., CA, associated with the highly sheared serpentinized masses that contain talc, antigorite, and chrysotile, where desautelsite forms an orange crust on serpentine fragments and in veins with artinite (Aker's claim). HCWS

16b.6.3.1 Reevesite $Ni_6Fe_2^{3+}(CO_3)(OH)_{16} \cdot 4H_2O$

Named in 1967 for Frank Reeves (1896–1986), U.S. geologist, who discovered the Wolf Creek meteorite crater, WA, Australia. Sjögrenite group, hydrotalcite subgroup; see honessite and hydrohonessite, SO_4 equivalents of reevesite. HEX-R $R\bar{3}m$ or $R3m$. $a = 6.164$, $c = 45.54$, $Z = 3$, $D = 2.89$. *AM 56:1077(1973). 26-1286:* 7.6_{10} 3.8_5 2.60_5 2.30_4 1.947_4 1.735_2 1.541_3 1.510_3. Hexagonal platelets up to 100 μm in diameter. Greenish or canary yellow. No cleavage. G = 2.87. IR. Uniaxial (−); $N_O = 1.735$, $N_E = 1.65$ (Wolf Creek samples); $N_O = 1.72$, $N_E = 1.63$ (Barberton samples). Occurs as hexagonal plates in druses with violarite and millerite in the Ni ore of the Bon Accord area, Barberton Mt. Land, South Africa, associated with minute quartz crystals, thin films of opaline silica, and goethite stains; also as alteration product in cracks and crevices of weathered *Wolf Creek meteorite, WA, Australia*, and in many weathered meteorites. HCWS *AM 52:1190(1967)*.

16b.6.3.2 Takovite $Ni_6Al_2(CO_3)(OH)_{16} \cdot 4H_2O$

Named in 1957 for the locality. Sjögrenite group, hydrotalcite subgroup. HEX-R $R\bar{3}m$ or $R3m$. $a = 6.04$, $c = 45.16$, $Z = 3$, $D = 2.79$. *AM 62:458(1977). 15-87:* 7.57_{10} 3.77_9 2.60_4 2.55_9 2.38_5 2.26_8 1.917_9 1.708_7. Poorly crystalline massive, plates about 1 μm. Bluish green. Uniaxial; $N_O = 1.605$, $N_E = 1.598$; parallel extinction; positive elongation. DTA, IR. Secondary crust, or fracture filling; most data derived from synthetic preparations and by analogy with Mg, Al hydroxycarbonates as impure material analyses give variable Ni : Al, CO_3, and H_2O. Zincian variety: "eardleyite." *AM 62:449(1977)*. Occurs intimately mixed with other minerals, especially silicates; for example, at Var, France, takovite is associated with boehmite and kaolinite or as a thin green coating on bauxite; with gibbsite, allophane at the contact of limestone and metamorphosed serpentinite at *Takovo, Serbia;* from the Perseverence Mine, Agnew, WA, Australia. HCWS *MA 13:624(1958), AM 57:1559(1972)*.

16b.6.3.3 Comblainite $Ni_6^{2+}Co_2^{3+}(CO_3)(OH)_{16} \cdot 4H_2O$

Named in 1980 for Gordon Comblain (b.1920) of the Musée Royal de l'Afrique Central, Tervuren, Belgium, who discovered the mineral. Sjögrenite group, hydrotalcite subgroup. HEX-R $R\bar{3}m$ or $R3m$. $a = 6.08$, $c = 45.58$, $Z = 3$, $D = 3.16$. *BM 103:113(1980)*. 33-429: 7.64_{10} 3.81_5 2.61_1 2.57_7 2.28_5 1.934_4 1.519_2 1.489_3. Cryptocrystalline crusts, turquoise-blue, on altered uraninite. TGA, IR, ESCA used to determine Ni and Co valences. Uniaxial $(-)$; $N_O = 1.690$, $N_E = 1.684$; nonpleochroic, yellow-green. Found in the uranium deposits at *Shinkolobwe, Zaire*, associated with becquerelite, curite, rutherfordine, and heterogenite. HCWS *AM 65:1065(1980)*.

16b.6.4.1 Quintinite-2H $Mg_4Al_2(OH)_{12}CO_3 \cdot 3H_2O$

Named in 1995 for Quintin Wight (b.1935), Scottish–Canadian amateur mineralogist, micromounter, and author, for his contributions to the study of the minerals of Mt. St.-Hilaire. HEX $P6_322$. $a = 10.571$, $c = 15.139$, $Z = 4$, $D = 2.14$. PD: 7.57_{10} 3.78_{10} 2.60_2 2.50_2 2.34_2 2.17_2 1.991_2 1.825_2. Crystals equant to 5 mm or prismatic to 3 mm, hexagonal, trapezohedral. Orange-brown, pale brown; white streak; vitreous luster, transparent. Cleavage {0001}, perfect; brittle. H = 2. G = 2.14. Uniaxial $(+)$; $N_O = 1.533$, $N_E = 1.533$. O, yellow; E, lighter yellow. From the *Jacupiranga carbonatite, Jacupiranga, SP, Brazil*, in beautiful sharp crystals and groups, which were at first mistakenly believed to be manasseite, associated with dolomite, magnetite, fluorapatite, phlogopite, and pyrrhotite. HCWS *CM:34(1996)*.

16b.6.4.2 Quintinite-3T $Mg_4Al_2(OH)_{12}(CO_3) \cdot 3H_2O$

Named as for quintinite-2H. HEX-R $P3_112$ or $P3_212$. $a = 10.571$, $c = 22.71$, $Z = 6$, $D = 2.14$. PD: 7.57_{10} 3.78_9 2.57_4 2.28_4 1.932_4 1.524_2 1.493_2. Crystals hexagonal, tabular, to 1 mm, sometimes so thin as to appear micaceous; also as rosettes. Yellow to pale yellow or colorless, white streak, vitreous luster, transparent. Cleavage {0001}, perfect; brittle. H $\approx$ 2. G = 2.14. Uniaxial $(\pm)$; $N_O = 1.533$, $N_E = 1.533$. From *Mt. St.-Hilaire, QUE, Canada*, in nepheline syenite, mainly in silicate and carbonate cavities and pegmatites, with microcline, analcime, albite, tetranatrolite, aegirine, siderite, donnayite, kupletskite, ancylite-Ce, and so on. HCWS *CM (in press)(1996)*.

16b.6.4.3 Caresite $Fe_4^{2+}Al_2(CO_3)(OH)_{12} \cdot 3H_2O$

Named for Steve Cares (b.1909), machinist, and Janet Cares (b.1921), chemist, amateur mineralogists of Sudbury, MA, who found the mineral. HEX $P3_112$ or $P3_212$. $a = 10.805$, $c = 22.48$, $Z = 6$, $D = 2.59$. PD: 7.49_{10} 3.75_5 2.63_4 2.31_5 1.948_4 1.558_2 1.526_2. Tabular to prismatic crystals, barrel-shaped, to 0.5 mm. Green-brown with black coating of smectite or chamosite, white streak, vitreous luster, transparent. Cleavage {001}, perfect; brittle. H ≈ 2. $G = 2.59$. Uniaxial (−); $N_O = 1.599$, $N_E = 1.570$. O, pale green; E, green. Mt. St.-Hilaire, QUE, Canada, mainly in silicate and carbonate cavities and nepheline syenite pegmatites, with microcline, analcime, analcite, smectite, and aegirine. HCWS *CM (in press)(1996)*.

16b.6.4.4 Charmarite-2H $Mn_4^{2+}Al_2(CO_3)(OH)_{12} \cdot 3H_2O$

16b.6.4.5 Charmarite-3T $Mn_4^{2+}Al_2(CO_3)(OH)_{12} \cdot 3H_2O$

Named in 1995 for Charles (b.1917) and Marcel (b.1918) Weber, amateur mineralogists of Guilford, CT, who contributed extensively to the knowledge of Mt. St.-Hilaire and also found the mineral. Charmarite-2H: HEX $P6_322$. $a = 10.985$, $c = 15.10$, $Z = 4$, $D = 2.50$. PD: 7.53_{10} 3.77_6 2.58_5 2.22_4 1.856_4 1.586_2 1.552_4. Hexagonal, tabular crystals, to 0.5 mm. Orange-brown, pale brown, pale blue, colorless, transparent. $H = 2$. $G = 2.47$. Uniaxial (−); $N_O = 1.587$, $N_E = 1.547$. Charmarite-3T: HEX-R $P3_112$ or $P3_212$. $a = 10.985$, $c = 22.63$, $Z = 6$, $D = 2.50$. PD: 7.55_{10} 3.77_9 2.67_7 2.35_7 1.973_6 1.586_3 1.662_3. Hexagonal prisms, may be fluted with star-shaped cross sections. May be tipped by places of C-2H. Orange brown, pale brown, vitreous, transparent. Uniaxial (−); $N_O = 1.587$. Both polytypes from *Mt. St.-Hilaire, QUE, Canada*, in silicate and carbonate cavities and in pegmatites in nepheline syenite with microcline, analcime, albite, aegirine, astrophyllite, siderite, rhodochrosite, calcite, and others. HCWS *CM (in press) (1996)*.

16b.7.1.1 Hydromagnesite $Mg_5(CO_3)_4(OH)_2 \cdot 4H_2O$

Named in 1835 for the composition. MON $P2_1/c$. $a = 10.11$, $b = 8.94$, $c = 8.38$, $\beta = 114.5°$, $Z = 2$, $D = 2.25$. *MR 4:18(1973)*. Pseudo-ORTH. Structure: *AC(B) 33:1273(1968)*. 25-513: 9.20_4 6.40_4 5.79_{10} 4.19_3 3.32_3 3.14_2 2.90_8 2.69_3. Tufted acicular crystals parallel to c flattened parallel to {100}, also massive and loose-granular to chalky; usually multiply twinned across {100}. Colorless or white; vitreous luster in larger grains, also earthy, silky, pearly. Perfect cleavage parallel to {010}; parting parallel to {100}; brittle. $H = 3.5$. $G = 2.24$. Biaxial (+); Z parallel to b; $X \wedge c = 47°$; $N_x = 1.523$, $N_y = 1.527$, $N_z = 1.545$; 2V moderate. Effervesces in acids. Found in veinlets in serpentines and altered magnesium-rich rocks with opal, calcite, dolomite, magnesite, artinite, deweylite, and pyroaurite. Occurs in old chrome mines, such as Wood's mine, near the town of Texas, and Cedar Hill quarry, State Line district, southern Lancaster Co., PA; in serpentine rock at Sulfur Creek area, Colusa Co., and at New Idria district, CA; with magnesite in valley bottoms in the Caribou, Atlin, and Kamloops districts, BC, Canada; as an alteration product of brucite in marble-serpentinites, Kraubat, Styria, Austria; in chromite mines in Serbia; as the alteration product in the periclase marbles at Predazzo, southern Tyrol, and at Viu, Lombardy, Italy; at Shabani, Belingwe district, Zimbabwe; at São Felix do Minato mine, BA, Brazil. HCWS *DII:271*, *AM 39:24(1954)*.

16b.7.1.2 Widgiemoolthaite $(Ni,Mg)_5(CO_3)_4(OH)_2 \cdot 4-5H_2O$

Named in 1993 for the locality. Ni analog of hydromagnesite (16b.7.1.1). MON Space group unknown. $a = 10.06$, $b = 8.75$, $c = 8.32$, $\beta = 114.3°$, $Z = 2$, $D = 3.24$. *AM 78:819(1993)*. Pattern diffuse; fiber bundle analysis shows b axis ≈ 8.8 Å and superlattice 62 Å, approximately seven hydromagnesite layers. Pseudo-orthorhombic, may be twinned. PD: 9.7_1 6.30_5 5.75_{10} 4.36_4 4.14_3 2.87_4 2.45_2 2.12_3. Isolated spheroids to 2 mm in diameter of radiating fibers and up to 1 μm long. Grassy green when massive to blue green; pale blue-green streak; silky luster. Fibrous cleavage, brittle. $G = 3.13$. Biaxial (+); $N_{min} = 1.630$ parallel to length, fast ($= b$); $N_{max} = 1.640$ perpendicular to length; 2V high. Reacts weakly with 1:1 HCl, not at all with 1:10 HCl, 1:1 H_2SO_4. Found along with several other unnamed Ni minerals and gaspéite, carrboydite, pecoraite, glaukosphaerite, hydrohonessite, nepouite, kambaldite, nullaginite, otwayite, reevesite, retgersite, takovite, anabergite, and nickeloan paratacamite, in a weathered stockpile at the *132 North nickel mine, 4 km SW of Wudgiemooltha, 80 km S of Kalgoorlie, WA, Australia*. HCWS

16b.7.2.1 Dypingite $Mg_5(CO_3)_4(OH)_2 \cdot 5H_2O$

Named in 1970 for the locality. Mineral is part of a series of hydrates; see giorgiosite (16b.7.2.2), a higher hydrate, and hydromagnesite (16b.7.1.1), the lower and most stable hydrate. MON $a = 32.6$, $b = 23.8$, $c = 9.45$, $Z = 8$, $D = 2.15$. *23-1218* (Dypingdal, $5H_2O$): 15.2_3 10.6_{10} 6.34_6 5.86_9 4.20_3 3.16_4 3.07_4 2.93_4. *29-857* (Japan, $8H_2O$): 33.2_6 16.4_4 10.80_6 7.35_3 6.45_6 5.89_{10} 4.21_4 2.93_6. Globular aggregates with radiating structure, fibrous parallel to c. White to colorless, pearly luster. Fluorescent, light blue; phosphorescent, yellow-green. IR, DTA. Biaxial (+); $N_x = 1.508$, $N_y = 1.510$, $N_z = 1.516$; elongation parallel to Y. Effervescent in cold 1:10 HCl. Dehydrates to hydromagnesite on heating to 150°. Synthesized during production of high-purity magnesia from magnesite; supercell obvious in synthetic preparation. *MM 48:437(1984)*. Found at Sterling Hill mine, Ogdensburg, Sussex Co., NJ; at New Idria district and San Raphael district, Emery Co., CA; in *Dypingdal serpentine-magnesite deposit, Snarum, Norway*, as coatings on fracture surfaces on serpentines or hydrotalcite; at Yoshikawa, Aichi Pref., Japan. HCWS *AM 55:1457(1970)*.

16b.7.2.2 Giorgiosite $Mg_5(CO_3)_4(OH)_2 \cdot 6H_2O$

Named in 1905 for one of the cones, Giorgios, created during the eruption at Santorin islands in 1866. See also dypingite (16b.7.2.1). *29-858*(syn): 11.8_{10} 5.85_4 4.44_6 3.38_7 3.28_7 2.92_6 2.62_4 1.439_7. Fibrous spherulitic crusts of fine needles or globular aggregates. White. G = 2.155–2.185. DTA. Biaxial; $N_x = 1.498$–1.501, $N_z = 1.508$–1.513; parallel extinction, positive elongation. Synthesized in aqueous environment, temperature $\approx 60°$; phase is a metastable form between nesquehonite and hydromagnesite and more hydrated than dypingite. Occurs on the lava at *Alphroëssa, Santorin, Greece.* HCWS *DII:274, AM 55:1457(1970), NJMM:196(1975)*.

16b.7.3.1 Rabbittite $Ca_3Mg_3(UO_2)_2(CO_3)_6(OH)_4 \cdot 18H_2O$

Named in 1955 for John Charles Rabbitt (1907–1957), geologist with the U.S. Geological Survey. MON $a = 32.6$, $b = 23.8$, $c = 9.45$, $\beta = 90°$, $Z = 8$, $D = 2.69$. *AM 40:201(1955)*. 7-365: 11.3_5 8.24_{10} 7.79_8 5.83_5 5.72_5 4.81_5 4.71_7 4.37_8. Fibrous, acicular parallel to c, silky masses. Pale green. Two perfect cleavages parallel to fiber length, also perpendicular (probably {001}), possibly twin gliding as crystals easily deformed on grinding. $H = 2+$. $G \approx 2.6$. Biaxial (+); Y parallel to b; $Z \wedge c \approx 15°$; $N_x = 1.502$, $N_y = 1.508$, $N_z = 1.525$; $2V \approx 20°$. Soluble in HCl, slowly in H_2O. Found as efflorescence on walls of *Lucky Strike No. 2 mine, Emery Co., UT.* HCWS

16b.7.4.1 Wyartite $Ca_3U^{4+}(UO_2)_6(CO_3)_2(OH)_{18} \cdot 3$–$5H_2O$

Named in 1958 after Jean Wyart (1902–1992), mineralogist at the Sorbonne, Paris. ORTH $P2_12_12_1$. $a = 11.25$, $b = 7.08$, $c = 20.98$, $Z = 2$, $D = 4.81$. *AM 45:200(1960)*. 12-635: 10.3_{10} 5.19_3 4.26_1 3.55_1 3.47_1 3.36_1 3.28_1 3.01_1. Platy {001} small crystals elongated on [010] with {110}. The {001} forms strongly striated parallel to [010]. Black to violet-black, alters greenish; brownish-violet streak; dull, vitreous luster on cleavages, opaque to translucent. Cleavage {001}, perfect; also {010}. $H = 3$–4. $G = 4.69$. DTA. Biaxial $(-)$; $XYZ = cba$; $N_x = 1.89$, $N_z = 1.91$; $2V = 48°$; negative elongation; strongly pleochroic. X, gray; Y, violet; Z, lavender-blue. Sample alters on standing or in X-ray beam (see *12-636*). A product of the alteration of red rims surrounding uraninite at *Shinkolobwe, Shaba, Zaire.* HCWS *AM 44:908(1959), BM 82:80(1959)*.

16b.7.5.1 Brugnatellite $Mg_6Fe^{3+}(CO_3)(OH)_{13} \cdot 4H_2O$

Named in 1909 after Luigi Brugnatelli (1859–1928), mineralogist, University of Pavia, Italy. HEX-R $P6_3/mmc$. $a = 5.48$, $c = 16.00$, $Z = 1$, $D = 2.21$. *14-365*: 7.93_{10} 3.96_9 2.63_5 2.35_6 2.00_7 1.56_4 1.53_4 1.30_2. Massive in lamellar or foliated masses of small flakes flattened on {0001}; flakes may have three- or six-sided outline with striations intersecting at $60°$. Flesh-pink to yellowish or brownish white, white streak, pearly luster, transparent. Cleavage {0001}, perfect; crushes to talc-like powder. $H \approx 2$. $G = 2.14$. Uniaxial $(-)$; $N_O = 1.533$, $N_E = 1.510$; pleochroic. O, yellowish red; E, colorless. Found at *York R., Bancroft, ONT, Canada*; as crusts and coatings along cracks in serpentinites associated with artinite, hydromagnesite, pyroaurite, magnesite, chrysotile, aragonite, and brucite at *Ciappanico in Torre Sita Maria, and Sandrio, Val Malenco, Italy*, and other localities close by, such as at *Viu, Val di Lanzo*, and at Cogne, Val d'Aosta, Piedmont, and as an alteration product of brucite at Monte Ramazzo, Liguria. HCWS $DI:660$, AM $14:42(1929)$.

16b.7.6.1 Coalingite $Mg_{10}Fe_2^{3+}(CO_3)(OH)_{24} \cdot 2H_2O$

Named in 1965 after the locality. HEX-R $R\bar{3}m$. $a = 6.24$, $c = 74.8$, $Z = 4$, $D = 2.26$. AM $50:1893(1965)$. see Pyroaurite (16b.6.2.3) Structure: MM $38:286(1971)$. Layered structure composed of two brucite-like layers and one disordered CO_3 and H_2O layer repeated along c. Mg and Fe have random distributions in the brucitelike layers. Different multiples of the brucitelike layer and other stacking disorders are very probable; resembles sjögrenite structure. 26-1217: 13.4_4 6.05_5 4.20_8 2.67_3 2.34_{10} 1.884_3 1.712_3 1.558_5. 34-182(calc): 12.4_{10} 6.23_1 4.16_5 3.12_1 2.67_1 2.54_1 2.12_1 1.899_1. Platy (0001), also equant grains, 0.1 to 0.2 mm; reddish brown, resinous. $H = 1$–2. $G = 2.32$. DTA, TGA, IR. Uniaxial $(-)$; $N_O = 1.588$–1.594, $N_E = 1.560$–1.564; nonpleochroic; some strained plates show undulant extinction, anomalous biaxial (to 20°). Elongate grains oriented normal to plates are length-slow, striated parallel to length, and markedly pleochroic, from deep golden brown to colorless, absorption $O > E$, strong birefringence. Variations may be due to differences in stacking of plates or hydration level. Formed by the weathering of Fe-rich brucite in serpentinites. From the *New Idria serpentinite, Coalinga, Fresno and San Benito Cos., CA*, intimately intermixed with chrysotile or pyroaurite and associated with artinite and hydromagnesite; at Muskox intrusion, Canada. HCWS

16b.7.7.1 Carbonate-Cyanotrichite $Cu_4Al_2(CO_3,SO_4)(OH)_{12} \cdot 2H_2O$

Named in 1963 after the composition and structural similarity to cyanotrichite. ORTH Space group unknown. 16-365: 10.1_{10} 5.63_5 5.03_6 4.21_{10} 3.33_6 2.77_5 2.51_5 2.01_6. Fibrous aggregates, plates up to 5 mm long. Pale blue to azure, silky luster. $G = 2.66$. Biaxial $(+)$; $N_x = 1.616$, $N_z = 1.677$; $2V = 55$–$60°$; $r > v$, strong; negative elongation; parallel extinction; strongly pleochroic. X, colorless; Z, bright blue. Soluble in acids with effervescence. From Chester, Hampton Co., MA; Cole shaft and Lavender pit, Bisbee, Cochise Co., AZ; at *Balasauskandyk, in NW Kara-Tau area, Kazakhstan*, in cleavage surfaces and cavities of weathered vandiferous (Middle Cambrian) shales with gypsum, variscite, volborthite, malachite, aurichalcite, and azurite. HCWS $ZVMO$ $92:458(1963)$, AM $49:441(1964)$.

16b.7.8.1 Scarbroite $Al_5(CO_3)(OH)_{13} \cdot 5H_2O$

Named in 1960 after the locality. See hydroscarbroite (16b.7.9.1) and sjögrenite–hydrotalcite group (16b.6.1.1). TRIC $P1$ or $P\bar{1}$. $a = 9.94$, $b = 14.88$, $c = 26.47$, $\alpha = 98.7°$, $\beta = 96.5°$, $\gamma = 89.0°$, $Z = 4$, $D = 2.03$. Structure: MM $32:363(1960)$, $43:615(1980)$. Layers: two gibbsite, $Al(OH)_3$, and one layer containing H_2O–(CO_3) are stacked along c. 12-627: 8.66_{10} 8.34_4 5.99_5 5.63_4 4.91_4 4.46_4 4.33_5 3.72_6. See also 31-18. Compact thin rhombic plates about 1 μm in size. White. $H = 1$–2. $G = 1.85$–2.05. $N_{av} = 1.509$. Transforms at 130°

to metascarbroite, hydrates to hydroscarbroite on immersion in H_2O. Found at *South Bay, Scarborough, England*, in vertical fissures in sandstone, and probably in the bauxite deposits at Carev Most near Niksic, Montenegro, Yugoslavia. HCWS *AM 43:384(1958), 50:1504(1965); MM 32:353(1960)*.

16b.7.9.1 Hydroscarbroite $Al_{14}(CO_3)_3(OH)_{36} \cdot nH_2O$

Named in 1960 for the relationship to scarbroite. An inadequately characterized mineral. *42-588*(Scarborough): 9.0_{10} 6.77_3 6.69_3 5.72_3 4.71_5 4.39_5 4.07_2 1.167_5. Produced when scarbroite stands in humid air (never completely hydrated). Dehydrates to scarbroite in air at 40° or above. Found with scarbroite at *South Bay, Scarborough, Yorkshire, England*. HCWS *MM 32:353(1960)*.

16b.7.10.1 Sharpite $Ca_2(UO_2)_6(CO_3)_5(OH)_4 \cdot 6H_2O$

Named in 1938 after R. R. Sharp (1881–1956), English engineer, prospector, and soldier, who discovered the Shinkolobwe uranium deposit in 1915. ORTH. $a = 21.99$, $b = 15.63$, $c = 4.487$, $Z = 2$, $D = 4.61$. *NJMM:109(1984)*. *12-164*: 11.0_4 5.34_4 4.49_{10} 4.30_4 3.93_6 3.18_4 2.99_6 2.45_6. Crusts and aggregates of fine radiating fibers. Light yellowish green to gray-white. $H = 2\frac{1}{2}$. $G > 4.45$. IR, TGA. Biaxial (+); $N_x = 1.633$, $N_z \approx 1.72$; needles parallel extinction; Z = elongation; nonpleochroic; aggregates show pleochroic. X, brownish; Z, greenish yellow. Found with uranophane, curite, vandenbrandite, and becquerelite at *Shinkolobwe, Shaba, Zaire*. HCWS *DII:275, AM 70:220(1985)*.

16b.7.11.1 Albrechtschraufite $MgCa_4[(UO_2)(CO_3)_3]_2F_2 \cdot 17H_2O$

Named in 1984. TRIC $P\bar{1}$. $a = 13.562$, $b = 13.406$, $c = 11.636$, $\alpha = 115.72°$, $\beta = 107.66°$, $\gamma = 92.86°$, $Z = 2$. *AC(A) 40:C-247(1984)*. A layered structure of $MgCaF_2[(UO_2)(CO_3)_3] \cdot 8H_2O$ interconnected into a framework by $[(UO_2)(CO_3)_3]$ units, one Ca, five Ca-bonded water molecules, and four lattice water molecules. Found on a specimen of schröeckingerite from *Joachymov, Bohemia, Czech Republic*. HCWS

16b.7.12.1 Kambaldaite $Na_2Ni_8(CO_3)_6(OH)_6 \cdot 6H_2O$

Named in 1985 for the locality. HEX $P6_3$. $a = 10.340$, $c = 6.097$, $Z = 1$, $D = 3.193$. Structure: *AM 70:419(1985)*. *39-316*: 9.03_{10} 4.49_9 3.61_4 2.68_4 2.58_4 2.52_4 2.26_4 1.970_3. Needles and prismatic crystals elongated in the c direction and as cryptocrystalline masses in veins, layers, and concretions up to 2 mm. Hexagonal prisms terminated {0001}, occasional pyramids; zoned with clear core, transparent margins giving a "fish-eye" appearance. Emerald green; pale green streak; silky luster on fracture, chalky when massive. $G = 3.18$. Uniaxial (+); $N_O = 1.65$, $N_E = 1.69$. O, light green; E, emerald green. Found together with gaspéite and aragonite on fracture surfaces in altered primary Ni-sulfide ores now goethitic in the *Otter shoot, Kambalda, WA, Australia*, associated with reevesite, pyrite, pentlandite, pyrrhotite, and millerite. HCWS

16b.7.13.1 Roubaultite $Cu_2(UO_2)_3(CO_3)_2O_2(OH)_2 \cdot 4H_2O$

Named in 1970 after Marcel Roubault (b.1905), geologist, University of Nancy, France. TRIC $P\bar{1}$. $a = 7.767$, $b = 6.924$, $c = 7.850$, $\alpha = 92.16°$, $\beta = 90.89°$, $\gamma = 93.48°$, $Z = 1$, $D = 4.71$. $AC(C)$ $41:654(1985)$. 25-318: 7.74_9 6.88_8 5.55_{10} 4.42_5 3.50_7 3.45_8 3.23_8 3.18_8. Platy (100) crystals in rosettes to 3 mm in diameter. Grass- to apple-green, brilliant to slightly greasy luster. $H \approx 3$. DTA. Biaxial (+); $N_x = 1.700$, $N_y = 1.800$, $N_z = 1.82$–1.84; $2V_{calc} = 55°$; pleochroic. X,Y, colorless; Z, yellow green. From the oxidation zone of the U deposit, *Shinkolobwe, Shaba, Zaire*, on pitchblende associated with becquerelite, vandenbrandite, soddyite, and cuprosklodowskite. cs *BM 93:550(1970)*, *AM 57:1912(1972)*.

16b.7.14.1 Znucalite $Zn_{12}(UO_2)Ca(CO_3)_3(OH)_{22} \cdot 4H_2O$

Named in 1990 for the composition: Zn, U, Ca. TRIC $P1$ or $P\bar{1}$. $a = 12.692$, $b = 25.096$, $c = 11.685$, $\alpha = 89.08°$, $\beta = 91.79°$, $\gamma = 90.37°$, $Z = 4$, $D = 3.09$. *NJMM:393(1990)*. PD: 25.10_6 6.14_8 3.16_7 2.73_9 2.71_{10} 2.68_6 1.582_8 1.556_7. Porous coatings of saucer-shaped aggregates up to 10 cm^2, sometimes consisting of plates 15 μm × 8 μm × 0.4 μm. White, light yellow, grayish yellow; silky luster; translucent. Cleavage {010}, perfect. $G = 3.01$. Fluoresces yellow green in UV. IR, TGA. Biaxial (−); $N_x = 1.563$, $N_y = 1.621$, $N_z = 1.621$; $2V_{calc} \approx 0°$; negative elongation; extinction $\approx 9°$. Soluble in acids with effervescence. An oxidation product of the U dump material at the *Lill mine*, which operated in the second half of the nineteenth century *at Pribram, central Bohemia, Czech Republic*; occurs with gypsum, hydrozincite, aragonite, sphalerite, galena, and pyrite. HCWS *AM 76:1732(1991)*.

16b.7.15.1 Szymanskiite $Hg_{16}Ni_6(CO_3)_{12}(OH)_{12}(H_3O)_8 \cdot 3H_2O$

Named in 1990 after Jan T. Szymanski (b.1938), crystallographer, CANMET, Ottawa, Canada, who solved the crystal structure. HEX $P6_3$. $a = 17.415$, $c = 6.011$, $Z = 1$, $D = 4.86$. Structure: *CM 28:703(1990)*, *28:709(1990)*. 44-1395: 14.9_{10} 5.60_{10} 3.30_8 3.21_5 2.70_6 2.67_6 2.48_5 1.751_5. Acicular or prismatic crystals with HEX cross section, up to 0.4 mm long, elongate [0001], major $\{10\bar{1}0\}$, minor {0001}, striations parallel to [0001] on $\{10\bar{1}0\}$. Light blue-gray, pale blue streak, vitreous luster, transparent. Cleavage $\{10\bar{1}0\}$, poor; irregular to conchoidal fracture, brittle. Light sensitive, nonfluorescent. IR. Uniaxial (−); $N_O = 1.795$, $N_E = 1.786$; pleochroic. O, yellowish green; E, bluish green. Whitens but no effervescence in 1:10 HCl. Mineral includes structural water (OH), and H_2O. Crystal sprays of millimeter size in vugs and cavities, also as disseminated crystals within massive quartz matrix that also includes chalcedony, opal, ferroan magnesite, goethite, chromite, and minor chlorite and dolomite. Associated with cinnebar, montroydite, native mercury, edgarbaileyite, and millerite at *Clear Creek HG mine, New Idria district, San Benito Co., CA*. HCWS *AM 76:1731(1991)*.

Class 17

Compound Carbonates

17.1.1.1 Tychite Na$_6$Mg$_2$(CO$_3$)$_4$(SO$_4$)

Named in 1905 for the Greek for *luck* or *chance* in allusion to the fact that of about 5000 northupite crystals, the first and one of the last 10 turned out to be tychite. ISO *Fd*3. $a = 13.90$, $Z = 8$, $D = 2.59$. Structure: *AC(B) 36:1332(1980)*. *22-479:* 4.19_8 3.19_3 2.67_{10} 2.46_4 2.35_2 2.01_2 1.736_2 1.605_3. Octahedral crystals. White, vitreous luster, transparent. Conchoidal fracture. $H = 4$. $G = 2.456$. Isotropic; $N = 1.508$ (Na), colorless. A chromate analog has been prepared. Occurs with northupite, as, for example, in cuttings from a borehole in a clay bed, *Borax Lake, CA*, also associated with gaylussite, thenardite, pirssonite, and schairerite; in Green River formation, WY, CO, UT; at Lake Kahve, Uganda. HCWS *DII:294*.

17.1.1.2 Ferrotychite Na$_6$Fe$_2$(CO$_3$)$_4$(SO$_4$)

Named in 1981 for the composition and analogy to tychite. ISO *Fd*3. $a = 13.962$, $Z = 8$, $D = 2.794$. *ZVMO 110:600(1981)*. *35-707:* 4.18_9 2.68_{10} 2.47_8 2.36_4 2.10_3 1.958_4 1.614_6 1.351_4. Small grains (0.5–1 mm) with golden-brown surface film alteration product. Fresh surfaces colorless to light yellow, vitreous luster. Conchoidal fracture. $H = 4$. Strongly magnetic. Isotropic; $N = 1.550$. Found in a drill core at 539 m in southeastern part of *Khibina alkalic massif, Kola Penin., Russia*. HCWS *AM 67:622(1982)*.

17.1.1.3 Manganotychite Na$_6$(Mn,Fe,Mg)$_2$(CO$_3$)$_4$(SO$_4$)

Named in 1990 for the composition and resemblance to tychite (17.1.1.1) and ferrotychite (17.1.1.2). ISO *Fd*3. $a = 13.995$, $Z = 8$, $D = 2.79$. *ZVMO 119:46(1990)*. *43-1482:* 4.22_8 2.70_{10} 2.47_7 2.37_3 2.11_2 1.959_3 1.616_3 1.605_3. Irregular grains up to 1 cm across and massive aggregates to 5 cm. Pale pink, dull vitreous luster. Conchoidal fracture. $H = 4$. $G = 2.70$. Weakly magnetic. IR, DTA. $N = 1.544$. Dissolves slightly in cold H$_2$O, developing a brown coating, effervesces in 1:10 acids. Found intimately mixed with shortite in the core of hyperalkalic pegmatite veins at *Alluaiv Mt., Lovozero massif, Kola Penin., Russia*, associated with potassium feldspar, cancrinite, aegirine, villiaumite, cryolite, kogarkoite, trona, sidorenkite, and other minerals. HCWS *AM 77:448(1992)*.

17.1.2.1 Leadhillite $Pb_4(CO_3)_2(SO_4)(OH)_2$

Named in 1832 after the locality. Polymorphous with susannite (17.1.3.1) and macphersonite (17.1.4.1). MON $P2_1/a$. $a = 9.09$, $b = 11.57$, $c = 20.74$, $\beta = 90.5°$, $Z = 8$, $D = 6.58$. *AM 35:985(1950)*. 35-617: 11.6_6 6.78_3^* 3.57_{10} 2.94_7 2.89_6 2.62_6 2.11_6 2.06_6 1.554_6 Note: * line distinguishes leadhillite from susannite. Crystals usually pseudohexagonal thin to thick tabular {001} with hexagonal outline, several rhombohedral and pyramidal forms common, also prismatic [001] or equant, massive, granular. [101] when developed shows striations or may be curved. Crystals commonly twinned, twin plane {140}, also contact twins of aragonite type, penetration twins, lamellar twins with composition plane parallel to {$\bar{1}42$} or {340}. Other twin laws give rise to pseudohexagonal groupings. Colorless to white, gray, pale green, pale bluish green, pale blue or yellowish; resinous to adamantine luster, pearly on {001}; transparent to translucent. Cleavage {001}, perfect, easy. Translation gliding on {001}, also twin gliding. Yellowish fluorescence in UV. DTA, IR. *MSLM 4:273,427(1974)*. Biaxial $(-)$; $Z = b$; $X \wedge c = -5\frac{1}{2}°$; $N_x = 1.87$, $N_y = 2.00$, $N_z = 2.01$; $r < v$, strong. Soluble with effervescence in HNO_3, leaving lead sulfate. Exfoliates in hot H_2O. A secondary mineral found in the oxidized zone of lead deposits associated with cerussite, anglesite, lanarkite, caledonite, linarite, and pyromorphite. Pseudomorphs after galena, calcite; calcite and cerussite may form pseudomorphs after leadhillite. Occurs at Loudsville mine, Hampshire Co., MA; found at Joplin, MO; at Lookout and Caledonia mines, Shoshone Co., ID; with lanarkite at Leadville, CO; at Campbell and Cole shafts(!), Bisbee, Cochise Co., Grand Reef mine, Graham Co., Mammoth–St. Anthony mine(!), Tiger, Pinal Co., AZ; in the Tintic district(!), UT; Blue Bell mine, Baker, CA; from the Wanlockhead district and especially from *Susanna mine(!), Leadhills, Lanarkshire, Scotland*; at Red Gill, Cumberland, and Matlock, Derbyshire, England; at Nantycagal (Eaglebrook), Pyfed, Wales; at Mala-Calzetta lead mine near Iglesias and Tiny mine in Oriddo V., Sardinia, Italy; at Nertchinsk and Beresovak, Ural Mts., Russia; from the Toroku mine, Miyazaki Pref., Japan; at Djebel Ressas, Tunisia; at Tsumeb mine(!), Namibia; at Dundas and Victoria Magnet mine, Whyte R., TAS, Australia. HCWS *DII:295*.

17.1.3.1 Susannite $Pb_4(CO_3)_2(SO_4)(OH)_2$

Named in 1845 after the locality. Polymorphous with leadhillite (17.1.2.1) and macphersonite (17.1.4.1). HEX-R $R\bar{3}$. $a = 9.072$, $c = 11.539$, $Z = 3$, $D = 6.534$. *MM 49:759(1985)*. 39-372: 11.5_2 6.52_1^* 4.54_7 3.57_{10} 2.94_8 2.62_8 2.32_5 2.11_7 2.07_5. Note: * line distinguishes susannite from leadhillite. Acute rhombohedral crystals with {0001}, hexagonal prisms. Colorless to greenish or yellowish. IR. Uniaxial; see leadhillite (17.1.2.1). Found at Mammoth–St. Anthony mine, Tiger, Pinal Co., AZ; at Tintic district, UT; at *Susanna mine, Leadhills,*

Scotland; at Moldawa, Hungary; at Nertschinsk, Siberia, Russia. HCWS
DII:298, *CM 10:141(1972)*, *MM 48:295(1984)*.

17.1.4.1 Macphersonite $Pb_4(CO_3)_2(SO_4)(OH)_2$

Named in 1985 for Harry Gordon Macpherson (b.1925), mineralogist at Royal Scottish Museum, Edinburgh, Scotland. Polymorph of leadhillite (17.1.2.1) and susannite (17.1.3.1). ORTH *Pcab*. $a = 10.37$, $b = 23.10$, $c = 9.35$, $Z = 8$, $D = 6.6$. *MM 48:277(1984)*. *38-354*: 4.53_3 3.27_5 3.23_{10} 2.65_9 2.60_3 2.31_3 2.18_3 2.03_3. Tabular on b, and dumbbell-shaped composite crystals, occasionally pseudohexagonal. White, pale amber, colorless in thin flakes; resinous to adamantine luster. Polysynthetic and contact twins. Cleavage $\{010\}$, perfect; uneven fracture. $H = 2\frac{1}{2}$–3. $G = 6.5$. Strong yellow fluorescence in UV (Scottish samples). IR. Biaxial $(-)$; $XYZ = bca$; $N_x = 1.87$, $N_y = 2.00$, $N_z = 2.01$; $2V = 35°$; $r > v$. Effervesces in acids with precipitant $PbSO_4$. Occurs at *Leadhills, Lanarkshire, Scotland*, and at Argentolle mine St.-Prix, Saône-et-Loire, France, associated with leadhillite, susannite, cerussite, caledonite, pyromorphite, and scotlandite. HCWS *MM 48:295(1984)*.

17.1.5.1 Schröckingerite $NaCa_3(UO_2)(CO_3)_3(SO_4)F \cdot 10H_2O$

Named in 1783 after Julius Freiherr Schröckinger von Neudenberg (1814–1882), who found and described the U occurrence at Jachymov, Bohemia, Czech Republic. TRIC $P\bar{1}$. $a = 9.634$, $b = 9.635$, $c = 14.391$, $\alpha = 91.4°$, $\beta = 92.3°$, $\gamma = 120.3°$, $Z = 2$, $D = 2.56$. Structure: *TMPM 35:1(1986)*. *8-397*(syn): 14.3_3 8.48_7 7.26_{10} 5.42_2 4.80_8 3.36_2 2.88_7 2.39_2. Clusters or globular aggregates of flexible, pseudohexagonal scales up to 1 mm × 0.02 mm, flattened $\{001\}$; crusts, pisolitic; subparallel intergrowths, (001) twinning. Greenish yellow; vitreous luster, pearly on $\{001\}$, transparent. Cleavage $\{001\}$, perfect. $H = 2\frac{1}{2}$. $G = 2.51$. Strongly fluorescent yellow green in UV. DTA. Biaxial $(-)$; $XYZ = cba$; $N_x = 1.489$, $N_y = 1.539$, $N_z = 1.541$; $2V = 25°$; $r < v$; pleochroic. X, colorless; Y, pale yellow; Z, pale green. Extinctions are not parallel to faces, about 1–2° off. Loses six H_2O by 75°. Dehydration in stages does not interrupt basic structure. Found as small concretions in gypsiferous clay and in arkose near Wamsutter, Sweetwater Co., WY; at Shinarump No. 3 mine(!), Seven Mile Canyon, Moab, UT; as efflorescence on walls of Hillside mine, Yavapai Co., AZ, associated with bayleyite, swartzite, andersonite, and gypsum; as an alteration product of uraninite at *Jachymov, Bohemia, Czech Republic*; at Johanngeorgenstadt, Saxony, Germany; at Radhausberg, Bad Gastein, Austria; at La Soberania mine, San Isidro, Mendosa Prov., Argentina. HCWS *DII:236*, *USGSB 1064:121(1958)*, *AM 44:1020(1959)*.

17.1.6.1 Nasledovite $PbMn_3Al_4(CO_3)_4(SO_4)O_5 \cdot 5H_2O$

Named in 1975 after B. N. Nasledov (1891–1942) of Russia. *25-438*: 3.26_{10} 2.98_4 2.85_5 2.03_6 2.02_6 1.749_4 1.592_3 1.462_6. Oolitic up to 3 mm, made of radiating fibers. Snow-white on fresh surfaces but usually covered with dark red/brown film, silky luster. $H = 2$. $G = 3.07$. $N_{av} = 1.591$, fibers give wavy extinction. Found as fissure fillings along with cerussite in sooty pyrolusite and Fe oxides in polymetallic ore zone in granodiorite at *Sardab, E of Altyn-Topkansk ore field, Kuraminsk Mts., Kazakhstan.* HCWS *DAN-U 45:13(1958), AM 44:1325(1959).*

17.1.7.1 Motukoreaite $Na_2Mg_{38}Al_{24}(CO_3)_{13}(SO_4)_8(OH)_{108} \cdot 56H_2O$

Named in 1977 for the Maori name of the locality: Motukorea, meaning "island of the cormorants." HEX-R $R\bar{3}m$. $a = 9.172$, $c = 33.51$, $Z = 1$, $D = 1.90$ AM 74:1054(1989). Structure: *NJMM: 263(1986)*. Similar to pyroaurite and sjögrenite with two layers: one brucite-like, the other containing Na, SO_4, CO_3, and H_2O connected by OH—O hydrogen bonds. *41-1380:* 11.3_{10} 5.59_4 4.58_3 3.72_7 2.58_4 2.39_3 2.16_2 1.924_3. Fibrous radiating aggregates of hexagonal plates. White. $H = 1-1\frac{1}{2}$. $G = 1.48-1.53$. Nearly isotropic; $N_{av} = 1.51$. Samples show variable H_2O, as well as CO_3 and SO_4. It is suggested that the composition of the basaltic glass (lava) and seawater influence the final composition and formation of the mineral. Found as an alteration product in vesicles and filling veinlets as claylike cement in beach rock and basaltic volcanic tuffs on flanks of small extinct basaltic cone at *Brown's Island, (Motukorea) within Waitemata Harbor, New Zealand*; as the rim cement in volcanic breccias in weathered submarine environments. HCWS *AM 63:598(1978), MM 41:389(1977), 43:337(1979).*

17.1.8.1 Canavesite $Mg_2(CO_3)(HBO_3) \cdot 5H_2O$

Named in 1978 for the locality. MON $P2/m$. $a = 23.49$, $b = 6.164$, $c = 21.91$, $\beta = 114.91°$, $Z = 12$, $D = 1.79$. *CM 16:69(1978). 29-1431:* 9.54_{10} 8.12_4 7.80_2 7.53_2 4.56_2 3.69_1 3.33_1 3.11_2. Fibers in aggregates, rosettes; fibers elongated in the direction [010]; with several {h0l} cleavages. White, vitreous luster. $G = 1.80$. IR, DTA, TG. Biaxial (+); $Z = b$; $N_x = 1.485$, $N_y = 1.494$, $N_z = 1.505$; 2V large. Occurs in fractures in ludwigite and magnetite skarns at the *Vola Gera tunnel, Brosso iron mine, Canavese district, Piedmont, Italy.* HCWS

17.1.9.1 Harkerite $Ca_{24}Mg_8[AlSi_4(O,OH)_{16}]_2(BO_3)_8(CO_3)_8(H_2O,Cl)$

Named in 1948 after Alfred Harker (1859–1939), petrologist, Cambridge University, England. Compare with sakhaite, a borate mineral. Both minerals show marked pseudosymmetry in cubic space group $Fd3m$. *DANS 236:1203(1977).* HEX-R $R\bar{3}m$. $a = 18.131$, $\alpha = 33.46°$, $Z = 2$, $D = 3.00$ (rhombohedral cell). Structure: *AM 62:263(1977)*. Ca (in 8- or 10-fold coordination) is in incomplete cubic closest packing with O atoms, Mg octahedral coordination; structural disorder results from mutual replacement of (Si, Al) O_4–CO_3–BO_3–H_2O–OH groups. *10-465:* 5.22_7 3.39_6 3.01_6 2.84_6 2.61_{10} 2.13_8 1.84_9 1.51_7. Simple octahedral crystals. Colorless, vitreous luster. No cleavage. $G = 2.959$. IR. Frequently anomalously biaxial; $N_{av} = 1.653$. Readily alters to calcite; dissolves with effervescence in acids. Some substitution or disorder of the aluminosilicate, borate, and carbonate groups, OH and H_2O, will influence the crystalline modification. Occurs at contacts (scarns) of dolomitic limestones with granites; found in association with monticellite, calcite, bornite, chalcocite, magnetite, and diopside in *Camas Malag, Bradford area, Isle of Skye, Scotland*; Alban Hills, Italy; Balkash region, Kazakhstan, Russia; Dadang tin–iron ore deposit, China. HCWS *MM 29:621(1951)*, *AM 51:1820(1966)*.

17.1.10.1 Daqingshanite-(Ce)
$(Sr,Ca,Ba)_3(Ce,La)(CO_3)_{3-x}(PO_4)(OH,F)_x$

Named in 1983 for a mountain, Daqingshan, near the Bayan Obo iron ore deposit, China. HEX-R $R3m$, $R\bar{3}m$, or $R32$. $a = 10.058$, $c = 9.225$, $Z = 3$, $D = 3.71$. *G(C) 2:180(1983)*. *37-455*: 3.95_6 3.16_{10} 2.52_7 2.11_5 2.04_6 1.941_6 1.895_4 1.620_4. Rhombohedral crystals (0.05 mm) with rounded edges. Pale yellow, white streak, greasy-glassy luster. Cleavage $\{10\bar{1}1\}$, perfect; conchoidal fracture. $VHN_{20} = 335$. $G = 3.81$. Nonfluorescent. IR. Uniaxial (−); $N_O = 1.708$, $N_E = 1.609$; colorless. Soluble in 1:10 HCl. Found as monomineralic aggregates in veins cutting dolomite in footwall of *Bayan Obo Fe–Nb–REE deposits, Inner Mongolia, China*, with benstonite, huntite, strontianite, pyrite, phlogopite, and monazite. HCWS *AM 69:811(1984)*.

17.1.11.1 Tundrite-(Ce) $Na_3Ce_4(Ti,Nb)_2(CO_3)_3(SiO_4)_2O_4(OH) \cdot 2H_2O$

Named in 1963 after the locality, the Lovozero tundra. Solid-solution series with tundrite-(Nd) (17.1.11.2). TRIC $P1$. $a = 7.57$, $b = 13.98$, $c = 5.03$, $\alpha = 101.5°$, $\beta = 70.4°$, $\gamma = 101.5°$, $Z = 2$, $D = 3.96$. *SPD 14:304(1969)*. *25-1188*: 13.56_{10} 6.78_5 4.52_1 4.15_1 3.69_1 3.62_1 3.54_1 3.46_2. Fine acicular crystals up to 5 mm long with (210) (001) (100) (010); spherules to 15 mm in diameter; radial aggregates to 0.5 cm of flattened prismatic crystals (Khibiny). Brownish yellow, golden green; silky or vitreous luster. Cleavage parallel to (010), distinct; other cleavage perpendicular to c and 86° to (010); brittle. H ≈ 3. $G = 3.70$. DTA, TGA, IR. Biaxial (+); $Z \wedge c = 4°$; $N_x = 1.743$, $N_y = 1.80$, $N_z = 1.88$; $2V = 76°$; pleochroic. X, pale yellow; Z, greenish yellow. Khibiny sample which contains 10% CO_2, 3.75% H_2O, no SO_4 shows $N_x = 1.761$, $N_z = 1.880$, $2V \approx 90°$, Z parallel to elongation. REE variations. Alters to rhabdophane. Occurs in nepheline syenite pegmatites at *Mt. Nepkha, Lovozero tundra, Kola Penin., Russia*, associated with aegirine, lamprophyllite, and ramsayite; in Khibiny massif in a greenstone contact with nepheline syenite associated with aegirine, rinkite, apatite, and fluorite; in the Ilimaussaq alkaline intrusive, southern Greenland. HCWS *AM 50:2097(1965)*, *59:633(1971)*; *SPC 21:399(1976)*; *DANS 211:128(1973)* (Engl. transl.).

17.1.11.2 Tundrite-(Nd) $Na_3Nd_4(Ti,Nb)_2(CO_3)_3(SiO_4)_2O_4(OH) \cdot 2H_2O$

Named in 1968 for the composition and similarity to tundrite-(Ce); end member of the Nd–Ce series. TRIC $P1$ by analogy. No data available specifically on high-Nd-containing mineral. Found at the *Ilimaussaq intrusive, Greenland*;

Khibiny massif, Russia. HCWS *AM 53:1780(1968), 59:633(1974); DANS 211:128(1973)* (Engl. transl.).

17.1.12.1 Lepersonnite-(Gd)
$CaGd_2(UO_2)_{24}(CO_3)_8(SiO_4)_4(OH)_{24} \cdot 48H_2O$

Named in 1982 for Jacques Lepersonne (b.1909), honorary head of the Department of Geology and Mineralogy, Musée Royale de l'Afrique Centrale, Brussels, Belgium. ORTH *Pnnn* or *Pnn*2. $a = 16.23$, $b = 38.74$, $c = 11.73$, $Z = 2$, $D = 4.01$. *CM 20:231(1982). 35-552:* 11.1_1 8.15_{10} 6.46_1 4.06_2 3.65_7 3.21_5 3.00_1 2.86_4. Crusts and spherules (to 5 mm) of radiating acicular crystals. Bright yellow, transparent, translucent. Nonfluorescent. $G = 3.97$. Biaxial $(-)$; $Y = c$; $N_x = 1.638$, $N_y = 1.666$, $N_z = 1.682$; $2V_{calc} = 73°$; pleochroic. X, pale yellow; Y,Z, bright yellow. Microprobe analysis: 22.79% SiO_2, 76.14% UO_3, 2.09% Gd_2O_3, 1.07% Dy_2O_3, 0.41% Y_2O_3, 0.09% Tb_2O_3, 0.45% CaO, 4.02% CO_2 (by chromatography), 12.12% H_2O (TGA). Found in a black crust in the lower portion of the oxidation zone at *Shinkolobwe, Shaba, Zaire*, with bijvoetite, sklodowskite, uranophane, becquerelite, rutherfordine, and uraninite. HCWS *AM 68:1248(1983)*.

17.1.13.1 Qilianshanite $NaHCO_3 \cdot H_3BO_3 \cdot 2H_2O$

Named in 1993 for the locality composition approximate. MON *C*2. $a = 16.119$, $b = 6.928$, $c = 6.730$, $\beta = 100.46°$, $Z = 4$, $D = 1.639$. *AMS 13:97(1993). PD:* 6.36_3 4.20_1 3.46_{10} 3.17_6 2.61_1 1.731_2. Aggregates to 4 mm, tabular or prismatic crystals to 2 mm; colorless, white streak, vitreous luster, transparent. Cleavages {100}, {010}, perfect. $H = 2$. $G = 1.706$. Polysynthetically twinned. DTA. Biaxial $(-)$; $X = b$; $Z \wedge c = 9°$; $N_{x\,calc} = 1.351$, $N_y = 1.459$, $N_z = 1.486$; $2V = 50°$. Soluble in H_2O, with effervescence in 1:10 HCl. Occurs at the *Juhongtu boron deposit, SW Qilian Mt. system, Oinghai Prov., China*, associated with quartz, calcite, tincalconite, and nahcolite. HCWS *AM 79:765(1994)*.

17.1.14.1 Mineevite-(Y)
$Na_{25}Ba(Y,Gd,Dy)_2(CO_3)_{11}(HCO_3)_4(SO_4)_2F_2Cl$

Named in 1992 for D. A. Mineev (1935–1992), Russian mineralogist. Chemical formula not consistent with analysis. HEX $P6_3/M$. $a = 8.811$, $c = 37.03$, $Z = 2$, $D = 2.84$. *ZVMO 121:138(1992). PD:* 2.83_{10} 2.66_5 2.53_7 2.27_9 1.660_5. Irregular grains 0.5–1.0 cm in diameter. Colorless to pale green or yellow green, vitreous to pearly luster (on cleavage), transparent. Cleavage {0001}, perfect; steplike fracture, brittle. $H = 4$. $G = 2.85$. Fluoresces weakly yellow green in UV. IR, DTA. Uniaxial $(-)$; $N_O = 1.536$, $N_E = 1.510$. Stable in water, readily decomposed in 1:10 acids with effervescence. Microprobe analysis: 36.9% Na_2O, 7.3% BaO, 5.9% Y_2O_3, 2.3% Gd_2O_3, 1.80% Dy_2O_3, 7.6% SO_3, 1.7% Cl, 1.8% F, 0.95% Sm_2O_3; remainder less than 0.6%: Ce, Nd, Tb, Ho, Er, Yb, CO_2 30.8; 1.66% H_2O (by difference and not accounted for in formula). Occurs in the interstices of potassium feldspar crystals in the pegmatites of the Alluaiv Mts., Lovozero massif, Kola Penin., Russia, in association with nahcolite, trona, thermonatrite, sidorenkite, neighborite, aegirine, albite, sphalerite, manganotychite, and others. HCWS

17.1.15 Brianyoungite $Zn_{12}(CO_3,SO_4)(OH)_4$

Named in 1993 for Brian Young (b.1947), field geologist with the British Geological Survey, who collected the specimen. ORTH/MON Space group unknown. $a = 15.724$, $b = 6.256$, $c = 5.427$, $\beta \approx 90°$, $Z = 4$, $D = 4.11$. *MM* 57:665(1993). Structurally similar to hydrozincite. *PD:* 15.4_{10} 7.88_{10} 5.25_2 2.71_4 2.58_2 2.40_2 1.565_3. Rosettes of <100 μm composed of thin blades (≈ 1–2 μm) elongate [010] that taper to a sharp point. White, vitreous luster, transparent. Cleavage {100}, perfect; {001}, possible. $G = 3.93$–4.09. Nonfluorescent. TGA, IR. Biaxial (−); $N_O = 1.635$, $N_E = 1.650$; straight extinction. Readily soluble with effervescence in 1:10 acids. Analysis: 9.90% CO_2, 6.62% SO_3. IR: (OH) not H_2O. Occurs with gypsum on rubbly limestone in oxidized zone of *Brownley Hill mine, Nenthead, Cumbria, England*, and on specimens from Smallcleugh mine, Nenthead; at the Bastenberg mine, Ramsbeck, Germany; at Vieille, Montagne, Hollogne, Belgium. HCWS *AM* 79:1009(1994).

Class 18

Nitrates

18.1.1.1 Nitratine NaNO$_3$

Synonym: soda-niter. Named in 1845 for the composition. HEX-R $R\bar{3}c$. $a = 5.070$, $c = 16.829$, $Z = 6$, $D = 2.26$. Calcite structure. $7\text{-}271$(syn): 3.89_1 3.03_{10} 2.81_2 2.53_1 2.31_3 2.13_1 1.90_2 1.88_1. Rhombohedral crystals; usually massive, granular. Colorless or white, except when tinted by impurities; white streak; vitreous luster. Cleavage $\{01\bar{1}2\}$, perfect; conchoidal fracture; rather sectile. $H = 1\frac{1}{2}\text{-}2$. $G = 2.24\text{-}2.29$. Soluble in H$_2$O. Cooling taste. Uniaxial $(-)$; $N_O = 1.587$, $N_E = 1.336$. This water-soluble salt commonly occurs as a surface impregnation or efflorescence in arid areas. It is typically associated with gypsum, halite, and other nitrates and sulfates. The only significant deposits occur in a narrow belt about 450 miles long in the deserts of northern Chile, along the virtually rainless eastern slopes of the coast ranges. These deposits, which consist of a near-surface layer from a few inches to a few feet in thickness, and which contain numerous other salts, have supplied important amounts of nitrate for the world's demands. The nitrate rock (known as caliche) also contains iodates, which are the only mineral source of iodine; the iodine is a valuable by-product of nitrate production. In 1988 Chile produced some 750,000 tons of nitrates and 4,000 tons of iodine. Aqua Fria Mt., Brewster Co., TX, veinlets in trachyte; Hidalgo Co., NM; near Homedale, Owyhee Co., UT, in veinlets in rhyolite; Niter Buttes, SE of Lovelock, NV; Armagosa R. and Death Valley, Inyo and San Bernardino Cos., and near Tulare, Tulare Co., CA. Small deposits of nitratine similar to those in Chile occur in the deserts of Bolivia, Peru, North Africa, Egypt, Russia, Kazakhstan, and India. BM *DII:300*, AM *55:1500(1970)*, CG *67:85(1988)*.

18.1.2.1 Niter KNO$_3$

Name derived from Hebraic *neter*, used in ancient times for alkaline salts extracted by water from vegetable ashes. Synonym: saltpeter. ORTH *Pmcn*. $a = 5.414$, $b = 9.164$, $c = 6.431$, $Z = 4$, $D = 2.11$. Aragonite structure. $5\text{-}377$(syn): 4.66_2 3.78_{10} 3.73_6 3.03_6 2.76_3 2.71_2 2.66_4 2.65_6. Thin crusts, or massive, granular, or columnar. Colorless to white or gray, white streak, vitreous luster. Cleavage {001}, nearly perfect; {010}, good. H = 2. G = 2.10. Soluble in water. Tastes saline and cooling. Biaxial (−); $N_x = 1.332$, $N_y = 1.504$, $N_z = 1.504$; 2V = 7°; r < v, weak. Niter is a minor constituent in most deposits of nitratine. It is also found in some limestone caves, as in VA, TN, KY, AL, and OH, from which it was mined for the manufacture of gunpowder in the War of 1812 and the Civil War. A massive granular deposit of more than 125 tons was mined from a cave near Lava Station, Socorro Co., NM; North Table Butte, Leucite Hills, WY; Death Valley, CA. BM *DII:303*, *CG 67:85(1988)*.

18.2.1.1 Nitrobarite Ba(NO₃)₂

Named in 1882 for the composition. ISO $P2_13$. $a = 8.118$, Z = 4, D = 3.24. *24-53*(syn): 4.69_{10} 4.06_4 3.32_1 2.87_3 2.45_8 2.34_5 1.86_2 1.82_2. Octahedral crystals, twinned on {111} (spinel law). Colorless, white streak, vitreous luster. No cleavage. H = 3. G = 3.25. Isotropic; N = 1.571. Easily soluble in water. Known from one specimen from Chile, presumably from the nitrate deposits. BM *DII:305*.

18.2.2.1 Nitrocalcite Ca(NO₃)₂·4H₂O

Named in 1835 for the composition. MON $P2_1/n$. $a = 14.48$, $b = 9.160$, $c = 6.285$, β = 98.4°, Z = 4, D = 1.90. *26-1406*(syn): 7.72_7 5.42_7 5.14_{10} 4.36_6 3.58_7 3.12_6 2.81_6 2.31_4. Efflorescent silken tufts and massive. Colorless or white, white streak, vitreous luster. Soft. G = 1.90. Biaxial (−); $N_x = 1.465$, $N_y = 1.498$, $N_z = 1.504$; 2V = 50°. Soluble in water. Tastes sharp and bitter. Occurs as an efflorescence in limestone caves or on calcareous rocks and soils. Mammoth Cave and other caves, KY, Wyandotte Cave, IN, Death Valley, CA; Kuramin Mts., Kyrgyzstan. BM *DII:306*, *ZVMO 86:403(1957)*, *AC(B) 33:1861(1977)*.

18.2.3.1 Nitromagnesite Mg(NO₃)₂·6H₂O

Named in 1835 for the composition. MON $P2_1/c$. $a = 6.194$, $b = 12.71$, $c = 6.600$, β = 93.0°, Z = 2, D = 1.64. *14-101*(syn): 5.84_5 4.43_6 4.15_4 3.30_{10} 3.19_4 2.93_8 2.69_4 2.08_3. Efflorescences. Colorless or white, white streak, vitreous luster. Cleavage {110}, perfect. G = 1.58. Biaxial (−); $N_x = 1.34$, $N_y = 1.506$, $N_z = 1.504$; 2V = 5°; r < v. Soluble in water. Bitter taste. Associated with nitrocalcite in limestone caves, but definitely identified only from Madison Co., KY. BM *DII:307*, *AC 14:1296(1961)*.

Class 19

Nitrates Containing Hydroxyl or Halogen

19.1.1.1 Gerhardtite $Cu_2(NO_3)(OH)_3$

Named in 1885 for Charles F. Gerhardt (1816–1856), who first prepared the artificial compound. ORTH $P2_12_12_1$. $a = 5.592$, $b = 6.075$, $c = 13.812$, $Z = 4$, $D = 3.40$. *14-687:* 6.91_{10} 3.45_6 2.80_6 2.67_6 2.62_8 2.60_7 2.31_8 1.92_6. Thick tabular crystals. Emerald green to dark green, light green streak. Cleavage {001}, perfect; {100}, good. $H = 2$. $G = 3.40$–3.43. Biaxial (+); $N_x = 1.703$, $N_y = 1.713$, $N_z = 1.722$; 2V large; r < v, very strong; pleochroic. X,Y, green; Z, blue. Insoluble in water, soluble in dilute acid. *United Verde mine, Jerome, AZ*, in cavities in cuprite; Great Australia Mine, Cloncurry, Q, Australia; Likasi, Shaba, Zaire. BM *DII:308, ZK 116:210(1961)*.

19.1.2.1 Buttgenbachite $Cu_{19}Cl_4(NO_3)_2(OH)_{32} \cdot 2H_2O$

Named in 1925 for Henri J. Buttgenbach (1874–1964), Belgian mineralogist. HEX $P6_3/mmc$. $a = 15.82$, $c = 9.14$, $Z = 2$, $D = 3.44$. *8-136:* 13.7_{10} 7.95_{10} 3.27_8 2.75_9 2.61_6 2.51_8 2.30_{10} 1.62_9. Minute flat needles. Azure-blue, blue streak, vitreous luster. $H = 3$. $G = 3.42$. Uniaxial (+); $N_O = 1.738$, $N_E = 1.752$; nonpleochroic, blue. Mildren and Steppe claims, Pima Co., AZ; Great Australia mine, Cloncurry, Q, Australia; *Likasi, Shaba, Zaire*, in cavity in cuprite. BM *DII:572, MM 39:264(1973)*.

19.1.3.1 Sveite $KAl_7(NO_3)_4Cl_2(OH)_{16} \cdot 8H_2O$

Named in 1980 for Sociedad Venezolana de Espeleologia. MON Space group unknown. $a = 10.89$, $b = 13.04$, $c = 30.71$, $\beta = 92.1°$, $Z = 6$, $D = 2.20$. *33-986:* 10.2_{10} 6.17_2 6.00_4 4.21_2 3.69_4 3.27_2 2.75_2 2.44_6. Flaky aggregates. White. Cleavage {001}, perfect. Soft. $G = 2.0$. Biaxial (+); $N_x = 1.501$, $N_y = 1.503$, $N_z = 1.535$; 2V small. *Autana Cave, Amazonas Terr., Venezuela*, as crusts and efflorescences on the walls. BM *AM 67:1076(1982)*.

19.1.4.1 Mbobomkulite $(Ni,Cu)Al_4(NO_3SO_4)_2(OH)_{12} \cdot 3H_2O$

Named in 1980 for the locality. MON $P2_1$. $a = 10.171$, $b = 8.865$, $c = 17.145$, $\beta = 95.37°$, $Z = 4$, $D = 2.34$. *35-696:* 8.55_{10} 7.87_2 4.55_2 4.27_4 3.18_2 3.05_2 2.51_2 2.00_2. Powdery nodules. Sky blue. $G = 2.30$. Cleavage {001}, perfect. $N_x = 1.515$, $N_z = 1.585$. *Mbobo Mkulu cave, Nelspruit district, South Africa*, as friable nodules in allophane. BM *AM 67:415(1982)*.

19.1.4.2 Hydrombobomkulite $(Ni,Cu)Al_4(NO_3SO_4)_2(OH)_{12} \cdot 14H_2O$

MON Space group unknown. $a = 10.145$, $b = 17.155$, $c = 20.870$, $\beta = 90.55°$, $Z = 8$. *35-697:* 10.45_{10} 6.23_1 5.23_5 4.90_1 3.49_3 3.13_1 2.79_1 2.49_2. Blue nodules. Dehydrates in the atmosphere in a few hours to mbobomkulite. BM *AM 67:415(1982)*.

19.1.5.1 Likasite $CuNO_3(OH)_5 \cdot 2H_2O$

Named in 1955 for the locality. ORTH *Pcmn*. $a = 5.830$, $b = 6.775$, $c = 21.711$, $Z = 4$, $D = 2.89$. *30-497:* 10.8_8 5.75_{10} 3.23_6 3.18_6 2.72_7 2.66_6 2.52_7 2.17_7. Tabular crystals and massive. Blue. Cleavage {001}, perfect. $G = 2.97$. Biaxial (−); $N_x = 1.615$, $N_z = 1.685$; pleochroic. X, green-blue; Z, pale blue. *Likasi mine, Shaba, Zaire.* BM *BM78:84(1955), NJMM: 101(1986)*.

Class 20

Compound Nitrates

20.1.1.1 Darapskite Na$_3$(NO$_3$)(SO$_4$)·H$_2$O

Named in 1891 for Ludwig Darapsky (b.1857), Chilean chemist and mineralogist. MON $P2_1/m$. $a = 10.571$, $b = 6.917$, $c = 5.189$, $Z = 2$, $D = 2.20$. *23-1408*(syn): 10.3_{10} 4.13_2 3.53_2 3.46_4 3.27_2 2.87_3 2.86_3 2.59_3. Tabular crystals, platy masses, and granular; crystals twinned on {100}, frequently polysynthetic. Colorless, white streak, vitreous luster. Cleavage {100}, perfect; {010}, good; uneven fracture; brittle. $H = 2\frac{1}{2}$. $G = 2.20$. Biaxial $(-)$; $N_x = 1.388$, $N_y = 1.479$, $N_z = 1.486$; 2V small; r > v, strong. A widespread mineral in the Chilean nitrate deposits; also Chuquicamata Cu deposit, Chile, in veins; Flower Cave, Big Bend National Park, TX, and Death Valley, CA, associated with nitratine and niter; Roberts Massif, Shackleton Glacier, Antarctica, in saline arid soil. BM *DII:309, AM 55:1500(1970), MM 41:548(1977), CG 67:85(1988)*.

═══════════════════════════ Class 21 ═══════════════════════════

Anhydrous and Hydrated Iodates

21.1.1.1 Lautarite Ca(IO$_3$)$_2$

Named in 1891 for the locality. MON $P2_1/n$ or $P2_1/c$. $a = 7.148$, $b = 11.304$, $c = 7.280$, $\beta = 106.36°$, $Z = 4$, $D = 4.588$. Structure: $AC(B)$ 34:$84(1978)$. 28-221(syn): 4.36_5 3.49_6 3.43_8 3.34_4 3.32_3 3.16_{10} 2.84_3 2.01_3. Crystals short prismatic [001]; in radial or stellate aggregates. Colorless or yellowish. Cleavage {011}, good. H = $3\frac{1}{2}$–4. G = 4.50. Slightly soluble in water, soluble in HCl with evolution of Cl$_2$. Biaxial (+); Y = b; X $\wedge$ c = 25°; $N_x = 1.792$, $N_y = 1.840$, $N_z = 1.888$; 2V $\approx$ 90°; r > v, moderate. *Pampa del Pique III, Oficina Lautaro, Atacama desert, Antofagasta, Chile*, as crystals coating fractures, or embedded in gypsum bands in the nitrate deposits. RG *DII:312*.

21.1.2.1 Brüggenite Ca(IO$_3$)$_2$·H$_2$O

Named in 1971 for Juan Brüggen M. (1887–1953), Chilean geologist. MON $P2_1/c$. $a = 8.505$, $b = 10.00$, $c = 7.498$, $\beta = 95.25°$, $Z = 4$, $D = 4.267$. AM 57:$1911(1972)$. 26-1405: 4.24_8 3.74_6 3.50_6 3.24_9 3.05_{10} 2.52_6 1.777_6 1.748_6. Anhedral columnar crystals. Colorless to bright yellow, vitreous luster. Conchoidal fracture, brittle, H = $3\frac{1}{2}$. G = 4.24. Biaxial (–); Z = b; X $\wedge$ a = 9°; $N_x = 1.773$, $N_y = 1.797$, $N_z = 1.814$; 2V = 88°; r > v, moderate to strong; negative elongation. *Pampa del Pique III, Oficina Lautaro, Antofagasta, Chile*, in columnar crystals intergrown with nitratine. RG

21.1.3.1 Bellingerite Cu$_3$(IO$_3$)$_6$·2H$_2$O

Named in 1940 for Herman C. Bellinger (1867–1940), American metallurgist, manager of the copper mine at Chuquicamata. TRIC $P\bar{1}$. $a = 7.256$, $b = 7.950$, $c = 7.856$, $\alpha = 105.01°$, $\beta = 92.95°$, $\gamma = 96.95°$, $Z = 1$, $D = 4.932$. Structure: $AC(B)$ 30:$965(1974)$. 19-393: 3.82_7 3.72_{10} 3.57_5 3.35_9 3.25_5 3.17_9 3.13_6 3.07_6. Crystals prismatic [001], somewhat tabular {100}; twinning on {$\bar{1}$01}, with or without reentrant angles. Light green. Subconchoidal fracture, brittle. H = 4. G = 4.89. Slightly soluble in hot water, easily so in HCl. In CT gives water with purple fumes of iodine which crystallizes on walls of tube. Biaxial (+); $N_x = 1.890$, $N_y = 1.90$, $N_z = 1.99$; 2V medium; r > v, strong; absorption Z > X, Y. XY, light bluish green; Z, blue-green. *Chuquicamata, Antofagasta, Chile*, as isolated crystals associated with leightonite and gypsum. RG *DII:313*.

= Class 22 =

Iodates Containing Hydroxyl or Halogen

22.1.1.1 Salesite $Cu(IO_3)(OH)$

Named in 1939 for Reno H. Sales (1876–1969), chief geologist, Anaconda Copper Mining Co. ORTH *Pnma*. $a = 10.794$, $b = 6.708$, $c = 4.781$, $Z = 4$, $D = 4.900$. *AM 63:172(1978)*. *22-236*: 5.38_4 4.37_{10} 3.66_7 2.70_5 2.65_3 2.39_6 1.89_3 1.79_6. Crystals stout prismatic [001] with pyramidal terminations. Bluish green, vitreous luster. Cleavage {110}, perfect. $H = 3$. $G = 4.77$. Insoluble in H_2O, easily soluble in HNO_3; in CT, gives water and copious purple fumes of iodine, which crystallizes on the walls of the tube. Biaxial (−); XYZ= *acb*; $N_x = 1.786$, $N_y = 2.070$, $N_z = 2.075$; $2V = 0$–$5°$; $r > v$, extreme. X, colorless; Y, light bluish green; Z, bluish green. *Chuquicamata, Antofagasta, Chile*, in oxidized copper ore. RG *AM 24:388(1939)*, *DII:315*, *AC 15:1105(1962)*.

22.1.2.1 Seeligerite $Pb_3(IO_3)OCl_3$

Named in 1971 for Erich Seeliger, German mineralogist, Technical University, Berlin. ORTH (pseudo-TET) $C222_1$. $a = 7.964$, $b = 7.964$, $c = 27.288$, $Z = 8$, $D = 7.052$. *AM 57:327(1972)*. *25-450*: 3.65_8 3.21_{10} 2.81_6 2.78_8 1.99_8 1.75_8 1.69_8 1.62_9. Crystallizes in thin plates. Bright yellow. Cleavage {001}, perfect; {110}, good. $G_{syn} = 6.83$. Biaxial (−); $N_x = 2.12$, $N_y = 2.32$, $N_z = 2.32$; $2V = 4°$. *Santa Ana mine, Caracoles, Sierra Gorda, Chile*, associated with schwartzembergite, paralaurionite, and boleite. RG *NJMM:210(1971)*.

22.1.3.1 Schwartzembergite $Pb_6(IO_3)_2O_3Cl_4$

Named in 1864 for Schwartzemberg, assayer, Copiapó, Chile. ORTH (pseudo-TET) $P4_2/nnm$. $Z = 1$, $D = 7.09$. *AM 55:1814(1970)*. *24-572*: 3.78_6 2.88_{10} 2.81_8 2.09_7 1.986_6 1.758_5 1.677_5 1.633_9. Rounded, flat-pyramidal crystals, earthy crusts and masses. Yellow, reddish brown; straw-yellow streak; adamantine luster. Cleavage {001}, distinct. $H = 2$–$2\frac{1}{2}$. $G = 7.39$. Soluble in HCl with evolution of Cl_2, easily fusible. Biaxial (−); $N_x = 2.25$, $N_y = 2.35$, $N_z = 2.36$ (Li); 2V small; dispersion not noticeable. *Cachinal, Atacama desert, Chile*, with cerussite and anglesite on galena. Also from the San Rafael mine, Sierra Gorda, Caracoles district, with paralaurionite, and elsewhere in northern Chile. RG *DII:317*, *NJMM:467(1970)*.

Class 23

Compound Iodates

23.1.1.1 Dietzeite $Ca_2H_2O(IO_3)_2(CrO_4)$

Named in 1894 for August Dietze (d. 1893), German chemist, who first described the mineral. MON $P2_1/c$. $a = 10.118$, $b = 7.238$, $c = 13.965$, $\beta = 106.37°$, $Z = 4$, $D = 3.822$. Structure: *CM 31:313(1993)*. *25-132*: 3.61_9 3.48_{10} 3.13_{10} 2.98_6 2.88_6 1.870_6 1.741_5 1.711_5. Crystals tabular, {100}; elongated, [001]; fibrous crusts; columnar. Deep golden yellow. Cleavage {100}, poor; conchoidal fracture. $H = 3\frac{1}{2}$. $G = 3.62$. Soluble in hot water; solution deposits $Ca(IO_3)_2 \cdot 6H_2O$ upon cooling. Biaxial $(-)$; $Y = b$; $Z \wedge c = 6°$; $N_x = 1.825$, $N_y = 1.842$, $N_z = 1.857$; $2V = 86°$; $r < v$, very strong, inclined. *Atacama desert, Antofagasta*, and in the Oficina Maria Elena, near Tocopilla, *Chile*, with lopezite, tarapacaite, and ulexite in nitrate rock. RG *DII:318*.

23.1.2.1 Fuenzalidaite $K_6(Na,K)_4Na_6Mg_{10}(SO_4)_{12}(IO_3)_{12} \cdot 12H_2O$

Named in 1994 for Humberto Fuenzalida (1904–1966), Chilean geologist, who planned and directed the first school of geology in Chile at the University of Chile. HEX-R $P\bar{3}c1$. $a = 9.464$, $c = 27.336$, $Z = 1$, $D = 3.284$. *AM 79:1003(1994)*. PD: 13.67_5 7.05_4 3.93_{10} 3.52_2 3.10_2 3.02_4 2.68_3 2.33_2. Crystals are thin hexagonal plates on (0001), < 0.2 mm. Pale yellow, vitreous luster, transparent. Brittle. $H \approx 2–3$. Slowly soluble in water. Uniaxial $(-)$; $N_O = 1.622$, $N_E = 1.615$. In analyzed material the cation "M" (Na,K) ratio is $(Na_{2.8}K_{1.2})$ and in the anions $(S_{0.95}Se_{0.05})$. Found in the *Chilean nitrate fields* in veins and veinlets of *caliche blanco* (white nitrate ore) about 1 km S of the former plant site of *Oficina Santa Luisa*. RG

23.1.2.2 Carlosruizite $K_6(Na,K)_4Na_6Mg_{10}(SeO_4)_{12}(IO_3)_{12} \cdot 12H_2O$

Named in 1994 for Carlos Ruiz F. (b.1916), Chilean geologist, Instituto de Investigaciones Geológicas. HEX-R $P\bar{3}c1$. $a = 9.590$, $c = 27.56$, $Z = 1$, $D = 3.400$. *AM 79:1003(1994)*. PD: 13.75_3 7.10_2 3.97_2 3.56_{10} 3.08_3 3.06_4 2.72_4 2.36_1. Crystals hexagonal, thin plates on (0001), diameter <0.2 mm. Pale yellow, vitreous luster, transparent. Brittle. $H \approx 2–3$. Slowly soluble in water. Uniaxial $(-)$; $N_O = 1.655$, $N_E = 1.642$. In analyzed material the (Na,K) ratio is $(Na_{2.6}K_{1.4})$ and in the anions, $(Se_{0.54}S_{0.36}Cr_{0.10})$. Found in the *Chilean nitrate fields* in material collected long ago, from an unknown locality but probably near *Zapiga*, in samples of iquiqueite obtained by leaching *caliche amarillo* (yellow nitrate ore). RG

===== Class 24 =====

Anhydrous Borates

24.1.1.1 Sinhalite MgAlBO$_4$

Named in 1952 for *Sinhala*, the Sanskrit name for Sri Lanka. Contains up to 2% Fe$_2$O$_3$. ORTH *Pbnm*. $a = 4.332$, $b = 9.882$, $c = 5.681$, $Z = 4$, $D = 3.45$. Olivine structure. *34-157*: 4.94$_4$ 3.96$_3$ 3.25$_7$ 2.62$_6$ 2.38$_4$ 2.31$_5$ 2.14$_{10}$ 1.62$_9$. Waterworn pebbles to 2 cm and massive. Colorless, yellow, pink, brown; white streak; vitreous luster. No cleavage, conchoidal fracture. H = 7. G = 3.49. Biaxial ($-$); N$_x$ = 1.669, N$_y$ = 1.698, N$_z$ = 1.706; 2V = 55°; iron-rich varieties pleochroic in brown. Bodnar quarry, near Hamburg, Sussex Co., NJ; Johnsburg, Warren Co., NY; Bancroft area, ONT, Canada; southern Yakutia, Russia; *Ratnapura area, Sri Lanka*, in gem gravels; Mogok, Burma; Tanzania. BM *MM 29:841(1952), 35:196(1965); MR 13:226(1982); EJM 6:313(1994)*.

24.1.2.1 Behierite (Ta,Nb)BO$_4$

Named in 1962 for Jean Behier (1903-1965), French mineralogist. TET *I4$_1$/amd*. $a = 6.206$, $c = 5.472$, $Z = 4$, $D = 8.02$. Zircon structure. *7-131*: 4.1$_6$ 3.1$_8$ 2.48$_5$ 2.33$_6$ 1.94$_6$ 1.75$_4$ 1.64$_6$ 1.59$_{10}$. Octahedral crystals to 7 mm. Pink-gray, white streak, adamantine luster. Cleavage {110}, {010}, distinct; subconchoidal fracture. H = 7–7$\frac{1}{2}$. G = 7.86. Uniaxial (+); N > 2.0; strongly birefringent. *Manjaka, Madagascar*, in pegmatite. BM *AM 47:414(1962)*.

LUDWIGITE GROUP

The ludwigite minerals are borates with the general formula

$$A_2XO_2(BO_3)$$

where

A = Mg, Fe^{2+}, Ni
X = Fe^{3+}, Al, Ti, Mg, Mn^{3+}, Sb^{5+}

and a structure in space group *Pbam*. The structure of these rare minerals is complex and cannot be described adequately in a brief paragraph.

24.2.1.1 Ludwigite

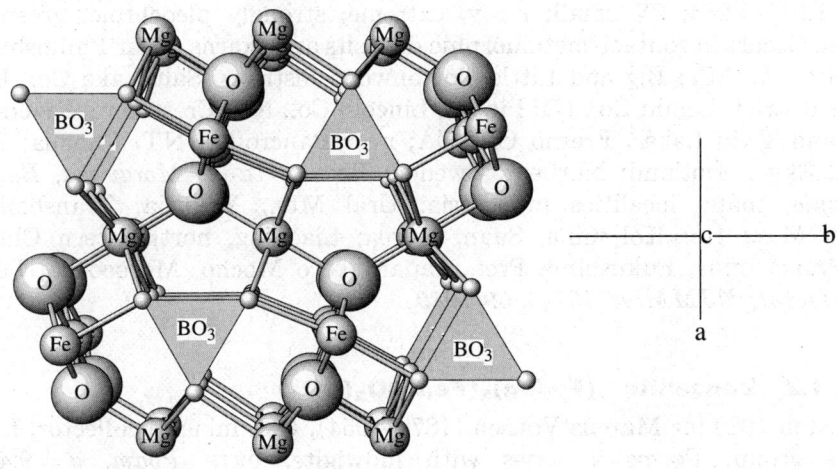

LUDWIGITE: $Mg_2AlO_2(BO_3)$
Layers of Al and BO_3 triangles alternate with layers containing Mg and O. Al and Mg are both octahedral. The packing is efficient and the density is high.

LUDWIGITE GROUP

Mineral	Formula	Space Group	a	b	c	D	
Ludwigite	$(Mg,Fe)_2(Fe,Al)O_2(BO_3)$	Pbam	9.26	12.26	3.05	3.75	24.2.1.1
Vonsenite	$(Fe,Mg)_2(Fe,Al)O_2(BO_3)$	Pbam	9.463	12.305	3.073	4.81	24.2.1.2
Azoproite	$(Mg,Fe)_2(Fe,Ti,Mg)O_2(BO_3)$	Pbam	9.279	12.232	3.001	3.62	24.2.1.3
Bonaccordite	$Ni_2FeO_2(BO_3)$	Pbam	9.213	12.229	3.001	5.19	24.2.1.4
Chestermanite	$Mg_2(Fe,Mg,Al,Sb)O_2(BO_3)$	Pbam	18.525	12.272	3.022	3.65	24.2.1.5
Fredrikssonite	$Mg_2MnO_2(BO_3)$	Pbam	9.18	12.555	2.954	3.80	24.2.1.6

24.2.1.1 Ludwigite $(Mg,Fe)_2(Fe,Al)O_2(BO_3)$

Named in 1874 for Ernst Ludwig (1842–1915), Austrian chemist. Ludwigite group. Forms a series with vonsenite. ORTH *Pbam*. $a = 9.26$, $b = 12.26$, $c = 3.05$, $Z = 4$, $D = 3.75$. *15-797*: 5.12_{10} 2.99_3 2.55_7 2.52_7 2.17_4 2.12_2 2.03_5 1.90_3. Prismatic crystals, radiating plates, and massive, fibrous or granular.

Dark green to brown and black; green-black to black streak; submetallic luster, silky in fibrous material. No cleavage. H = 5. G = 3.82–4.32, increasing with Fe content. Biaxial (+); $N_x = 1.805$–1.850, $N_y = 1.805$–1.865, $N_z = 1.915$–1.985; 2V small; r > v, extreme; strongly pleochroic, green to brown. Occurs in contact-metamorphic deposits and skarns. Near Philipsburg, Granite Co., MT; Big and Little Cottonwood districts, Salt Lake Co., UT; Texas district, Lemhi Co., ID; Pioche, Lincoln Co., NV; Crestmore, Riverside Co., and Twin Lakes, Fresno Co., CA; near Bancroft, ONT, Canada; Kilbride, Skye, Scotland; Norberg, Sweden; Brosso, Italy; *Moravicza, Banat, Romania*; many localities in Russia: Ural Mts., Yakutia, Transbaikal, Sayan Mts.; Hol Kol mine, Suan, Korea; Liaoning, northeastern China; Haneyama mine, Fukushima Pref., Japan; Toro Mocho, Morococha, Peru. BM *DII:321*; *NJMM:95(1974), 69(1989)*.

24.2.1.2 Vonsenite $(Fe,Mg)_2(Fe,Al)O_2(BO_3)$

Named in 1920 for Marcus Vonsen (1879–1954), U.S. mineral collector. Ludwigite group. Forms a series with ludwigite. ORTH *Pbam*. $a = 9.463$, $b = 12.305$, $c = 3.073$, $Z = 4$, $D = 4.81$. *25-395*: 5.15_3 4.73_1 2.84_1 2.58_{10} 1.94_1 1.60_1 1.54_2 1.40_2. Short prisms and massive, granular. Black, brown-black streak, metallic luster. No cleavage. H = 5. G = 4.3–4.8. Jayville, St. Lawrence Co., NY; *North Hill quarry, Riverside, CA*; Brooks Mt., Seward Peninsula, AK (*pageite*); Christophe mine, Breitenbrunn, Saxony, Germany; Badajoz, Spain; Capo di Bove (*breislakite*) and Mt. Vesuvius, Italy; Kamaishi mine, Iwate Pref., Japan. BM *DII:322*; *AM 46:786(1961)*; *NJMM:95(1974), 69(1989)*.

24.2.1.3 Azoproite $(Mg,Fe)_2(Fe,Ti,Mg)O_2(BO_3)$

Named in 1970 for AZOPRO, acronym for International Geological Association for Study of Deep Zones of the Earth's Crust. Ludwigite group. ORTH *Pbam*. $a = 9.279$, $b = 12.232$, $c = 3.001$, $Z = 4$, $D = 3.62$. *25-523*: 5.07_8 2.99_4 2.52_{10} 2.32_3 2.16_6 2.11_5 2.02_6 1.90_5. Prismatic crystals to 2 cm. Black, green-black streak, adamantine luster. Cleavage (possibly parting) {010}, good; {001}, less good; conchoidal fracture; brittle. H = $5\frac{1}{2}$. G = 3.63. Biaxial (+); $N_x = 1.799$, $N_y = 1.822$, $N_z = 1.855$; 2V = 80°; strongly pleochroic. X, pale green; Y, dark green; Z, red-brown. *Tazheranskii massif, Baikal, Russia*, in skarn. BM *ZVMO 99:225(1970)*, *AM 56:360(1971)*.

24.2.1.4 Bonaccordite $Ni_2FeO_2(BO_3)$

Named in 1974 for the locality. Ludwigite group. ORTH *Pbam*. $a = 9.213$, $b = 12.229$, $c = 3.001$, $Z = 4$, $D = 5.19$. *29-330*: 5.1_5 4.61_4 2.55_{10} 2.51_{10} 2.33_4 2.11_3 2.03_4 1.90_5. Microscopic prisms and radiating rosettes. Red-brown. H = 7. Biaxial; $N_{calc.\ mean} \sim 2.2$. *Bon Accord area, Barberton, Transvaal, South Africa*. BM *AM 61:502(1976)*.

24.2.1.5 Chestermanite $Mg_2(Fe,Mg,Al,Sb)O_2(BO_3)$

Named in 1988 for Charles W. Chesterman (1913–1991), U.S. geologist, who discovered the mineral. Ludwigite group. ORTH $Pbam$. $a = 18.525$, $b = 12.272$, $c = 3.022$, $Z = 8$, $D = 3.65$. *CM 26:* 5.11_{10} 3.07_2 2.75_3 2.56_9 2.48_2 2.17_6 2.00_3 1.71_2. Rare prismatic crystals to 2 mm, usually fibrous or asbestiform. Gray-green to black, gray-green streak, vitreous to silky luster. No cleavage, conchoidal to uneven fracture. $H = 6$. $G = 3.72$. Biaxial (+); $N_x = 1.753$, $N_y = 1.763$, $N_z = 1.791$; $2V = 63°$; pleochroic. X, dark blue-green; Y, green; Z, pale yellow-brown. *Twin Lakes region, Fresno Co., CA.* BM *CM 26:911(1988)*.

24.2.1.6 Fredrikssonite $Mg_2MnO_2(BO_3)$

Named in 1983 for Kurt A. Fredriksson (b.1926), Swedish–U.S. geochemist. Ludwigite group. Polymorphous with pinakiolite, orthopinakiolite, and takéuchiite. ORTH $Pbam$. $a = 9.18$, $b = 12.555$, $c = 2.954$, $Z = 4$, $D = 3.80$. *38-412:* 5.16_8 2.59_{10} 2.49_9 2.20_3 2.01_5 1.90_2 1.57_3 1.51_4. Small prismatic crystals. Red-brown, yellow to brown streak, vitreous luster. Poor cleavage. $H = 6$. $G = 3.84$. Biaxial (+); $N_x = 1.82$, $N_y = 1.86$, $N_z = 1.99$; $2V = 51°$; strongly pleochroic. X, golden brown; Z, dark red-brown to black. *Långban, Värmland, Sweden*, in dolomite. BM *AM 71:227(1986), CM 32:397(1994)*.

24.2.2.1 Warwickite $(Mg,Fe)_3TiO_2(BO_3)_2$

Named in 1838 for the locality. May contain up to 13% Cr_2O_3 and 2% Al_2O_3. ORTH $Pnam$. $a = 9.236$, $b = 9.444$, $c = 3.001$, $Z = 2$, $D = 3.40$. *12-171:* 6.68_4 4.25_4 2.99_4 2.58_{10} 2.10_4 1.60_5 1.55_4 1.50_5. Prismatic crystals to 3 cm. Dark brown to black, blue-black streak, pearly to adamantine luster. Cleavage {100}, perfect; uneven fracture; brittle. $H = 3\frac{1}{2}$–4. $G = 3.35$. Biaxial (+); $N_x = 1.772$–1.808, $N_y = 1.773$–1.813, $N_z = 1.792$–1.827; 2V small and variable; pleochroic in shades of brown. Near *Warwick, Orange Co., NY*, in marble; Twin Lakes area, Fresno Co., CA; Bancroft area, ONT, Canada; Jumilla, Murcia Prov., Spain; Taiga deposit, southern Yakutia, Russia; Hol Kol mine, Korea. BM *DII:326; AM 59:985(1974), 76:1380(1991)*.

24.2.2.2 Yuanfulite $(Fe,Mg,Al,Ti)_4O_2(BO_3)_2$

Named in 1994 for Yuan Fuli (1893–1987), Chinese geologist. ORTH $Pnam$. $a = 9.258$, $b = 9.351$, $c = 3.081$, $Z = 2$, $D = 3.80$. *AM 81:* 6.56_2 4.18_4 2.96_3 2.57_{10} 2.09_2 1.59_2 1.55_2. Prismatic crystals to 1 mm. Black, brown-black streak, adamantine to submetallic luster. Cleavage {100}, perfect. $H = 5$–6. $VHN_{50} = 843$. In RL, pale gray with red-brown internal reflections, weakly anisotropic; $R = 9.7$ (540 nm). In metamorphosed marble in the *Zhuanmiao deposit, Liaoning Prov., China*. BM *APM 13:328(1994), AM 81:252(1996), MR 27:204(1996)*.

24.2.3.1 Hulsite $(Fe,Mg)_2(Fe,Sn)O_2(BO_3)$

Named in 1908 for Alfred Hulse Brooks (1871–1924), U.S. geologist. MON $P2/m$. $a = 10.695$, $b = 3.102$, $c = 5.431$, $\beta = 94.21°$, $Z = 2$, $D = 4.54$. 17-511: 5.35_7 4.99_4 2.66_{10} 2.59_7 2.49_4 2.37_2 2.35_2 2.04_5. Small rectangular crystals and tabular masses. Black, black streak, vitreous to submetallic luster. Cleavage {110}, good. $H = 6$. $G = 4.57$. Brooks Mt., Seward Peninsula, AK. BM DII:326, NJMM:95(1974), AM 61:116(1976).

24.2.3.2 Magnesiohulsite $(Mg,Fe)_2(Mg,Fe,Sn)O_2(BO_3)$

Named in 1985 for the composition. MON $P2/m$. $a = 10.70$, $b = 3.06$, $c = 5.45$, $\beta = 94.6°$, $Z = 2$, $D = 4.15$. 41-1374: 5.32_7 2.67_{10} 2.56_9 2.47_8 2.35_5 2.13_6 2.02_8 1.97_6. Aggregates of microscopic needles. Black, dark brown streak, submetallic luster. $H = 6$. $G = 4.18$. Biaxial; $N_x = 1.88$, $N_z = 1.95$; strongly pleochroic. X, dark green; Z, dark brown. Qiliping, Changning Co., Hunan, China, in skarn. BM AM 73:929(1988).

24.2.4.1 Pinakiolite $Mg_2MnO_2(BO_3)$

Named in 1890 from the Greek for *small tablet*, in allusion to the morphology. MON $C2/m$. $a = 21.79$, $b = 5.977$, $c = 5.341$, $\beta = 95.63°$, $Z = 8$, $D = 4.09$. 36-413: 5.42_8 2.70_9 2.51_{10} 2.16_4 1.99_6 1.62_4 1.52_5 1.50_4. Thin rectangular tablets. Black, gray-brown streak, metallic luster. Cleavage {010}, good; very brittle. $H = 6$. $G = 3.88$. Biaxial (−); $N_x = 1.908$, $N_y = 2.05$, $N_z = 2.065$; $2V = 32°$; $r < v$; pleochroic. X, red-brown; Y, nearly opaque; Z, yellow-red. Långban, Värmland, Sweden, in dolomite. BM DII:324, AM 59:985(1974).

24.2.5.1 Orthopinakiolite $Mg_2MnO_2(BO_3)$

Named in 1960 as a dimorph of pinakiolite (24.2.4.1). ORTH $Pnnm$. $a = 18.357$, $b = 12.591$, $c = 6.068$, $Z = 16$, $D = 4.06$. 33-656: 5.96_2 5.19_7 3.03_3 2.80_3 2.60_{10} 2.53_9 2.38_2 2.21_3. Acicular crystals. Black, gray-brown streak, metallic luster. $H = 6$. $G = 4.03$. Långban, Värmland, Sweden, in dolomite. BM AKMG 2:551(1960), AM 46:768(1961), CM 16:475(1978).

24.2.6.1 Takéuchiite $Mg_2MnO_2(BO_3)$

Named in 1980 for Yoshio Takéuchi (b.1924), Japanese mineralogist. ORTH $Pnnm$. $a = 27.585$, $b = 12.561$, $c = 6.027$, $Z = 24$, $D = 3.93$. 33-864: 5.20_9 3.02_7 2.73_7 2.60_9 2.53_6 2.21_7 2.04_8 1.51_{10}. Acicular crystals to 8 mm. Black, brown streak, metallic luster. Uneven fracture. $H = 6$. Långban, Värmland, Sweden, in dolomite. BM AM 65:1130(1980), ZK 181:135(1987).

24.2.7.1 Blatterite $(Mn,Mg)_2(Mn,Sb)O_2(BO_3)$

Named in 1988 for Fritz Blatter (b.1943), German mineral collector, who discovered the mineral. ORTH $Pnnm$ or $Pnn2$. $a = 37.693$, $b = 12.620$, $c = 6.254$, $Z = 32$, $D = 4.35$. NJMM: 5.24_5 2.72_4 2.63_3 2.60_{10} 2.52_3 2.09_4 1.56_3 1.54_3. Prismatic crystals to 5 mm. Black, brown streak, submetallic to metallic luster. Cleavage {001}, perfect. H = 6. G = 4.7. Biaxial (−); $N_x = 1.91$, $N_y = 1.97$, $N_z = 2.00$. *Kittel mine, Värmland, Sweden.* BM *NJMM:121(1988), AM 74:1399(1989).*

24.3.1.1 Sassolite H$_3$BO$_3$

Named in 1800 for the locality. TRIC $P\bar{1}$. $a = 7.039$, $b = 7.053$, $c = 6.578$, $\alpha = 92.58°$, $\beta = 101.17°$, $\gamma = 119.83°$, $Z = 4$, $D = 1.50$. *25-97:* 6.06$_5$ 6.03$_5$ 5.91$_{10}$ 2.96$_5$ 2.92$_5$ 2.84$_7$ 2.26$_6$ 2.23$_5$. Tabular pseudohexagonal crystals. White to gray, white streak, pearly luster. Cleavage {001}, perfect. Acid taste, slightly saline and bitter. H = 1. G = 1.48. Biaxial (−); $N_x = 1.340$, $N_y = 1.456$, $N_z = 1.459$; 2V = 17°. Wichita Mts., Comanche Co., OK; in borate deposits at Boron and Death Valley, and at Sulphur Bank and the Geysers, Lake Co., CA; *Sasso, Tuscany,* deposited by hot-spring vapors, and in fumaroles at Vulcano, Vesuvius, and Solfatara, *Italy;* Bezymyanni volcano, Kamchatka, Russia; Iwo-dake volcano and other localities in Japan. BM *DI:662, AM 42:56(1957).*

24.3.2.1 Kotoite Mg$_3$(BO$_3$)$_2$

Named in 1939 for Bundjiro Koto (1856–1935), Japanese geologist. ORTH *Pnmn.* $a = 5.401$, $b = 8.422$, $c = 4.507$, $Z = 2$, $D = 3.09$. *38-1475*(syn): 3.46$_2$ 2.67$_{10}$ 2.32$_3$ 2.23$_7$ 2.18$_8$ 1.73$_5$ 1.67$_5$ 1.52$_3$. Massive, granular. Colorless or white, white streak, vitreous luster. Cleavage {110}, perfect; parting {101}. H = 6½. G = 3.10. Biaxial (+); $N_x = 1.652$, $N_y = 1.653$, $N_z = 1.673$; 2V = 21°; r > v. Big Cottonwood mine, Brighton, UT; Jumbo Mt., Snohomish Co., WA; Baita Bihor, Romania; Korotov deposit, Transbaikal, Russia; *Hol Kol mine, Suan, Korea,* in contact-metamorphosed dolomite; Neichi mine, Iwate Pref., Japan. BM *DII:328.*

24.3.2.2 Jimboite Mn$_3$(BO$_3$)$_2$

Named in 1963 for Kotora Jimbo (1867–1924), founder of the Mineralogical Institute, University of Tokyo. ORTH *Pmmn.* $a = 5.663$, $b = 8.714$, $c = 4.651$, $Z = 2$, $D = 4.09$. *19-781:* 4.1$_2$ 2.78$_{10}$ 2.59$_5$ 2.46$_2$ 2.33$_4$ 1.80$_3$ 1.73$_4$ 1.53$_2$. Probably massive. Pale purple-brown, white streak, vitreous luster. Cleavage {110}, perfect; parting {101}. H = 5½. G = 3.98. Biaxial (+); XY= *ab*; $N_x = 1.792$, $N_y = 1.794$, $N_z = 1.821$; 2V = 35°; r > v. *Kaso mine, Tochigi Pref., Japan.* BM *AM:1416(1963).*

24.3.3.1 Nordenskiöldine CaSn(BO$_3$)$_2$

Named for Nils A. E. Nordenskiöld (1832–1901), Swedish mineralogist and Arctic explorer. HEX-R $R\bar{3}$. $a = 4.857$, $c = 16.066$, $Z = 3$, $D = 4.22$. Isostructural with dolomite. *36-394:* 5.36$_8$ 4.07$_7$ 3.73$_8$ 2.91$_{10}$ 2.43$_6$ 2.21$_5$ 2.04$_6$ 1.80$_6$. Tabular crystals and subparallel growths. Colorless or yellow, white streak, vitreous luster. Cleavage {0001}, perfect; {10$\bar{1}$1}, indistinct; conchoidal fracture; brittle. H = 5½–6. G = 4.20. Uniaxial (−); $N_O = 1.778$, $N_E = 1.660$. Brooks Mt., AK; Seagull batholith, YT, Canada; *Arö, Langesundsfjord, Nor-*

way, in syenite pegmatite; Uchkosilkon tin deposit, eastern Kirgizia; Geiju tin deposit, Yunnan, China; Arandis tin deposit, Namibia. BM *DII:332*, *NJMM:111(1986)*, *CM 32:81(1994)*.

24.3.3.2   Tusionite $MnSn(BO_3)_2$

Named in 1983 for the locality. HEX-R $R\bar{3}$. $a = 4.787$, $c = 15.30$, $Z = 3$, $D = 4.78$. Dolomite structure. *35-727*: 5.12_4 4.02_4 3.65_9 2.82_{10} 2.39_6 2.17_4 2.00_6 1.74_8. Small tabular crystals, and lamellar intergrowths to 1.5 cm. Colorless to yellow-brown, white streak, vitreous luster. Cleavage {0001}, perfect. $H = 6$. $G = 4.73$. Uniaxial (−); $N_O = 1.854$, $N_E = 1.757$. Thomas Mt., Riverside Co., CA; Recice, Czech Republic; *Tusion R., Pamir Mts., Tajikistan*, in granite pegmatite. BM *DANS 272:1449(1983)*, *AM 69:1193(1984)*, *CM 32:903(1994)*.

24.3.4.1   Peprossiite-(Ce) $(Ce,La)Al_2(BO_3)_3$

Named in 1993 for Giuseppe Rossi (1938–1989), Italian mineralogist. HEX $P\bar{6}2m$. $a = 4.610$, $c = 9.358$, $Z = 1$, $D = 3.47$. *EM 5*: 3.67_{10} 3.04_{10} 2.46_8 2.31_5 2.02_5 1.95_5 1.86_5 1.84_5. Hexagonal plates to 2 mm. Pale yellow, white streak, vitreous luster. Cleavage {0001}, perfect; {11$\bar{2}$0}, distinct. $H = 2$. $G = 3.45$. Uniaxial (+); $N_O = 1.703$, $N_E = 1.711$. *Sacrofano volcanic center, Campagnano, Latium, Italy*, in pyroclastics. BM *EM 5:53(1993)*.

24.3.5.1   Takedaite $Ca_3(BO_3)_2$

Named in 1995 for Hiroshi Takeda (b.1934), Japanese mineralogist. HEX-R $R\bar{3}c$. $a = 8.638$, $c = 11.850$, $Z = 6$, $D = 3.11$. *MM 59*: 4.65_1 2.92_{10} 2.76_6 2.49_4 2.16_2 2.04_2 1.98_2 1.90_8. Microscopic granular crystals. White or pale gray, vitreous luster. $H = 4\frac{1}{2}$. $G = 3.10$. Uniaxial (−); $N_O = 1.726$, $N_E = 1.630$. Occurs in limestone skarn at *Fuka, Okayama Pref., Japan*. BM *MM 59:549(1995)*.

24.4.1.1   Suanite $Mg_2B_2O_5$

Named in 1953 for the locality. MON $P2_1/a$. $a = 12.31$, $b = 3.120$, $c = 9.205$, $\beta = 104.3°$, $Z = 4$, $D = 2.91$. *16-168*: 4.47_3 2.98_2 2.82_5 2.64_2 2.56_{10} 2.01_6 1.92_3 1.56_2. Possibly massive. White, white streak, pearly to silky luster. $H = 5\frac{1}{2}$. $G = 2.91$. Biaxial (−); $X = b$; $N_x = 1.596$, $N_y = 1.639$, $N_z = 1.670$; $2V = 70°$; $r > v$, weak. *Hol Kol mine, Suan Co., Korea*; Aldan, Yakutia, Russia; Liaoning, northeastern China; Miyako, Japan. BM *MJJ 1:54(1953)*, *AM 40:941(1995)*.

24.4.2.1   Kurchatovite $Ca(Mg,Mn,Fe)B_2O_5$

Named in 1966 for Igor V. Kurchatov (1903–1960), Russian physicist. Dimorphous with clinokurchatovite. ORTH $P2_12_12_1$. $a = 11.15$, $b = 36.4$, $c = 5.55$, $Z = 24$, $D = 3.03$. *19-648*: 2.89_4 2.78_{10} 2.67_8 2.26_7 2.01_7 1.92_9 1.63_6 1.23_8. Massive. Pale gray, white streak, vitreous luster. One perfect cleavage. $H = 4\frac{1}{2}$.

$G = 3.02$. Biaxial $(-)$; $XZ = bc$; $N_x = 1.635$, $N_y = 1.681$, $N_z = 1.698$; $2V = 66°$; $r > v$. Luminesces bright violet in LWUV. *Solongo, Transbaikalia, Russia*, in skarn. BM *ZVMO 95:203(1966), AM 51:1817(1996)*.

24.4.3.1 Clinokurchatovite Ca(Mg,Mn,Fe)B$_2$O$_5$

Named in 1983 as the dimorph of kurchatovite. MON $P2_1/b$. $a = 12.19$, $b = 10.95$, $c = 5.59$, $\beta = 102.0°$, $Z = 8$, $D = 3.27$. *36-389:* 3.09_8 2.80_{10} 2.59_5 2.03_5 1.99_4 1.72_4 1.57_4 1.24_5. Polysynthetically twinned crystals to 2 mm. White, white streak, vitreous luster. $H = 4\frac{1}{2}$. $G = 3.08$. Biaxial $(-)$; $N_x = 1.644$, $N_y = 1.675$, $N_z = 1.704$; $2V = 82-88°$. *Sayak-IV and other deposits, Ural Mts., Russia*, in skarn. BM *ZVMO 112:483(1983), AM 69:810(1984)*.

24.5.1.1 Metaborite HBO$_2$

Named in 1964 for the composition. ISO $P\bar{4}3n$. $a = 8.885$, $Z = 24$, $D = 2.49$. *15-868* (syn): 6.28_4 4.44_{10} 3.97_4 3.63_5 2.81_4 2.37_3 2.09_3 1.94_9 Crystal aggregates showing tetrahedron faces. Colorless to brown, white streak, vitreous luster. $H = 5$. $G = 2.47$. Isotropic; $N = 1.618$. Slowly soluble in water. In a salt deposit, probably in *Kazakhstan*. BM *ZVMO 93:329(1964), AM 50:261(1965)*.

24.5.2.1 Calciborite CaB$_2$O$_4$

Named in 1956 for the composition. ORTH *Pccn*. $a = 8.38$, $b = 13.81$, $c = 5.00$, $Z = 8$, $D = 2.89$. *27-67:* 7.1_6 3.81_6 3.57_8 3.44_{10} 2.66_6 2.31_6 1.98_7 1.87_7. Radial aggregates. White, white streak, vitreous luster. No cleavage, conchoidal to uneven fracture. $H = 3\frac{1}{2}$. $G = 2.88$. Biaxial $(-)$; $N_x = 1.595$, $N_y = 1.654$, $N_z = 1.670$; $2V = 54°$. Insoluble in water, readily soluble in acids. *Novofrolovsk deposit, Ural Mts., Russia*, in contact-metamorphosed limestone. BM *ZVMO 85:76(1956), AM 41:815(1956)*.

24.5.3.1 Johachidolite CaAlB$_3$O$_7$

Named in 1942 for the locality. ORTH *Cmma*. $a = 7.968$, $b = 11.724$, $c = 4.374$, $Z = 4$, $D = 3.44$. *29-280:* 3.99_2 3.51_2 2.63_{10} 2.43_3 2.00_3 1.96_5 1.82_3 1.75_2. Equant anhedral grains to 1 mm. Colorless, white streak, vitreous luster. No cleavage, subconchoidal fracture. $H = 7\frac{1}{2}$. $G = 3.37$. Biaxial $(+)$; XYZ= *acb*; $N_x = 1.712$, $N_y = 1.717$, $N_z = 1.726$; $2V = 70°$; $r > v$, strong. Fluoresces pale blue in SWUV. *Johachido, Kisshu Co., North Korea*. BM *DII:384, AM 62:327(1977)*.

24.5.4.1 Diomignite Li$_2$B$_4$O$_7$

Named in 1987 from the Greek for *divine mix*. TET $I4_1cd$. $a = 9.470$, $c = 10.279$, $Z = 8$, $D = 2.437$. *18-717*(syn): 4.07_{10} 3.91_2 3.50_5 2.66_6 2.59_4 2.24_2 2.09_2 2.05_4. Microscopic euhedral to anhedral crystals in fluid inclusions in spodumene. Uniaxial $(-)$; $N_O = 1.605$, $N_{E\,syn} = 1.560$. *Tanco pegmatite, Bernic Lake, MB, Canada*. BM *CM 25:173(1987)*.

Class 25

Anhydrous Borates Containing Hydroxyl or Halogen

25.1.1.1 Hambergite $Be_2(BO_3)(OH,F)$

Named in 1890 for Axel Hamberg (1863–1933), Swedish mineralogist. A complete series exists between OH and F end members. ORTH $Pbca$. $a = 9.78$, $b = 12.23$, $c = 4.43$, $Z = 8$, $D = 2.37$. *17-475:* 4.52_4 3.81_{10} 3.16_5 3.13_9 2.40_4 2.21_4 2.13_4 2.09_4. Prismatic crystals to 3.5 cm, showing reticulated twinning on {110}. Colorless or white, white streak, vitreous luster. Cleavage {010}, perfect; {100}, good; brittle. $H = 7\frac{1}{2}$. $G = 2.35$–2.37. Biaxial (+); $XY = ab$; $N_x = 1.543$–1.560, $N_y = 1.580$–1.591, $N_z = 1.617$–1.631; $2V = 84$–$90°$; $r > v$, weak. Occurs in granite and syenite pegmatites. Little Three and Himalaya mines, San Diego Co., CA; *Helgaråen, Langesundsfjord, Norway*; Elba, Italy; Susice, Bohemia, and Ctidruzice, Moravia, Czech Republic; Burpala pluton, north Baikal, Russia; Turakuloma Range(*), East Pamirs, Tadjikistan; Drot(*), Gilgit, Pakistan; Betafo(*) and Mt. Bity(*), Madagascar. BM *DII:370, AM 50:85(1965)*.

25.1.2.1 Fluoborite $Mg_3(BO_3)(F,OH)_3$

Named in 1926 for the composition. HEX $P6_3/m$. $a = 8.827$, $c = 3.085$, $Z = 2$, $D = 3.01$. *15-667:* 7.72_6 4.46_8 2.58_6 2.42_{10} 2.14_8 2.13_8 1.82_6 1.48_6. Hexagonal prisms to 2 cm, also fibrous, radiating. Colorless or white, white streak, vitreous luster. Cleavage {0001}, good. $H = 3\frac{1}{2}$. $G = 2.85$–2.99, increasing with F content. Uniaxial (−); $N_O = 1.525$–1.570, $N_E = 1.504$–1.534, increasing with OH content. Fluoresces cream in SWUV. Sparta, Hamberg, and Sterling Hill, NJ; Crestmore, Riverside Co., Amboy, San Bernardino Co., and Twin Lakes area, Fresno Co., CA; Bancroft, ONT, Canada; Broadford, Skye, Scotland; Querigut, France; *Tallgruvan, Norberg, Sweden*; Nocera, Italy; Pitkäranta, Karelia, and other localities in Russia; Hol Kol mine, Suan, South Korea; Seilibin, Malaya. Usually in skarn deposits. BM *DII:369, AM 48:678(1963), TMPM 21:94(1974), MR 13:223(1982)*.

25.1.3.1 Frolovite $Ca[B(OH)_4]_2$

Named in 1957 for the locality. TRIC $P\bar{1}$. $a = 7.774$, $b = 5.680$, $c = 8.136$, $\alpha = 113.15°$, $\beta = 101.67°$, $\gamma = 107.87°$, $Z = 2$, $D = 2.17$. *13-453:* 6.08_{10} 3.86_9 3.47_8 2.65_6 2.52_7 2.36_8 2.33_7 2.04_6. Veinlets and irregular masses, replacing calciborite. Gray-white, white streak, dull luster. Brittle. $H = 3\frac{1}{2}$. $G = 2.14$.

Biaxial (+); $N_x = 1.563$, $N_y = 1.572$, $N_z = 1.586$; $2V = 70°$. *Novo-Frolovsk, Turinsk region, Ural Mts., Russia,* in a skarn deposit. BM *ZVMO 86:622(1957), AM 43:385(1958), DANS 202:78(1972).*

25.1.4.1 Teepleite $Na_2B(OH)_4Cl$

Named in 1939 for John E. Teeple (1874–1931), U.S. chemist, who studied the Searles Lake deposit. TET $P4/nmm$. $a = 7.26$, $c = 4.85$, $Z = 2$, $D = 2.08$. *11-12:* 2.90_8 2.70_{10} 2.26_7 2.02_9 1.82_7 1.66_8 1.54_7 1.40_7. Tabular euhedral crystals to 5 cm. Colorless or white, white streak, vitreous to greasy luster. Alters rapidly to tincalconite on exposure. Irregular to subconchoidal fracture, very brittle. $H = 3–3\frac{1}{2}$. $G = 2.076$. Uniaxial (−); $N_O = 1.519$, $N_E = 1.503$. *Borax Lake, Lake Co.,* and *Searles Lake, San Bernardino Co., CA.* BM *DII:372, MR 6:80(1975), AC(B) 38:82(1982).*

25.1.4.2 Bandylite $CuB(OH)_4Cl$

Named in 1938 for Mark C. Bandy (1900–1963), U.S. mineralogist, who discovered the mineral. TET $P4/n$. $a = 6.14$, $c = 5.58$, $Z = 2$, $D = 2.81$. *12-631:* 5.59_{10} 4.35_6 4.13_5 3.08_8 2.54_8 1.95_7 1.66_6 1.46_5. Thick tabular or equant crystals. Blue, pale blue streak, vitreous luster. Cleavage {001}, perfect. $H = 2\frac{1}{2}$. $G = 2.81$. Uniaxial (−); $N_O = 1.691$, $N_E = 1.641$; pleochroic. O, deep blue; E, pale yellow-green. *Mina Quetena, near Calama, Antofagasta, Chile.* BM *DII:373, AM 44:875(1959).*

25.1.5.1 Vimsite $Ca[BO(OH)_2]_2$

Named in 1968 for Vsesuosny Institut Mineralogo, Moscow. MON $B2/b$. $a = 10.026$, $b = 4.440$, $c = 9.558$, $\beta = 91.31°$, $Z = 4$, $D = 2.56$. *21-134:* 6.87_6 3.72_7 3.48_{10} 3.04_6 2.61_7 2.55_6 2.50_6 2.22_8. Elongated crystals to 2 mm. Colorless, white streak, vitreous luster. Cleavage perfect along elongation. $H = 4$. $G = 2.54$. Biaxial (−); $N_x = 1.585$, $N_y = 1.614$, $N_z = 1.614$; $2V = 28°$. *Solongo, Transbaikalia, Russia,* in skarn. BM *DANS 182:821,1402(1968), AM 54:1219(1969).*

25.1.6.1 Olshanskyite $Ca_3(OH)_2[B(OH)_4]_4$

Named in 1969 for Yakov I. Olshansky (1912–1958), Russian geochemist. TRIC $P1$ or $P\bar{1}$. $a = 9.991$, $b = 14.740$, $c = 7.975$, $\alpha = 94.53°$, $\beta = 69.08°$, $\gamma = 112.44°$, $Z = 3$, $D = 2.31$. *22-144:* 7.61_5 6.78_5 4.50_3 3.35_5 3.05_8 2.81_{10} 2.57_4 2.02_4. Fibrous aggregates of microscopic twinned crystals. Colorless, white streak, vitreous luster. $H = 4$. $G = 2.23$. Biaxial (−); $N_x = 1.557$, $N_y = 1.568$, $N_z = 1.570$; $2V = 54°$; $r > v$, weak. *Eastern Siberia, Russia,* in skarn; *Fuka, Okayama Pref., Japan.* BM *DANS 184:1398(1969), AM 54:1737(1969), MM 58:279(1994).*

25.2.1.1 Sussexite (Mn,Mg)BO$_2$(OH)

Named in 1868 for the locality. Forms a series with szaibelyite (25.2.1.2). MON $P2_1/a$. $a = 12.832$, $b = 10.700$, $c = 3.283$, $\beta = 94.67°$, $Z = 8$, $D = 3.30$ (Mn end member). *13-599:* 6.32_{10} 3.43_4 3.32_4 3.11_4 2.74_8 2.48_8 2.26_4 1.79_6. Felted fibrous masses. White to pink, white streak, silky luster. $H = 3-3\frac{1}{2}$. $G = 3.0-3.3$, decreasing with Mg content. Biaxial $(-)$; $N_x = 1.63-1.67$, $N_y = 1.69-1.73$, $N_z = 1.69-1.73$; $2V \sim 25°$; $r > v$, moderate; weakly pleochroic, colorless to pink. *Franklin* and *Sterling Hill, Sussex Co., NJ;* Chicagon mine, Iron Co., MI; Gonzen, Switzerland; Val Graveglia, Italy; Matsuo mine, Kochi Pref., Japan; Hotazel, Cape Prov., South Africa. Occurs in metamorphosed Mn deposits. BM *DII:375*.

25.2.1.2 Szaibelyite (Mg,Mn)BO$_2$(OH)

Named in 1861 for Stephan Szaibely (1777–1855), Hungarian mine surveyor, who first collected the mineral. Forms a series with sussexite. MON (an ORTH dimorph is known) $P2_1/a$. $a = 12.577$, $b = 10.393$, $c = 3.139$, $\beta = 95.88°$, $Z = 8$, $D = 2.65$ (Mg end member). *29-864:* 6.2_{10} 5.17_5 3.24_4 3.02_4 2.66_8 2.42_8 2.20_8 2.08_6. Powdery or fibrous. White, white streak, silky luster in fibrous varieties. $H = 3-3\frac{1}{2}$. $G = 2.6-2.9$, increasing with Mn content. Biaxial $(-)$; $N_x = 1.57-1.62$, $N_y = 1.65-1.69$, $N_z = 1.65-1.69$; $2V \sim 25°$; $r > v$. Widely distributed: in metamorphosed limestone and dolomite, in skarns, on sepentinite, and in salt deposits. Clarke Co., AL, in salt cores; near Pioche, Lincoln Co., NV; Stinson Beach, Marin Co., CA; Penobsquis and Salt Springs, NB, and Douglas Lake, BC, Canada; Kilchrist and Kilbride, Skye, Scotland; Snarum, Norway; Tallgruvan, Norberg, Sweden; Aschersleben, Saxony, Germany; Brosso, Ivrea, Italy; Klodowa, Poland; *Baita Bihor, Romania*; Inder, Kazakhstan; Ural Mts. and eastern Siberia, Russia; Hol Kol mine, Suan, South Korea; Fengchen, Liaoning Prov., China (mined as a source of boron compounds); Bou Azzer, Morocco. BM *DII:375, AM 60:273(1975), CM 19:291(1981)*.

25.2.2.1 Sibirskite CaHBO$_3$

Named in 1962 for the locality. MON Space group unknown. $a = 3.67$, $b = 4.89$, $c = 19.30$. *15-282:* 3.74_4 3.29_6 2.93_{10} 2.58_{10} 2.20_4 2.05_4 1.93_4 1.88_6. Anhedral grains and powdery aggregates. Gray, white streak, vitreous luster. Biaxial $(-)$; $N_x = 1.555$, $N_y = 1.643$, $N_z = 1.658$; $2V = 43°$. *Solongo, Transbaikalia, Russia*, in skarn. BM *ZVMO 91:455(1962), AM 48:433(1963), DANS 216:166(1974)*.

25.2.3.1 Pinnoite MgB$_2$O(OH)$_6$

Named in 1884 for chief councillor of Mines Pinno, of Halle, Germany. TET $P4_2$. $a = 7.62$, $c = 8.19$, $Z = 4$, $D = 2.29$. *25-1119:* 5.39_{10} 3.61_4 3.14_4 2.31_7 2.25_3 2.05_5 1.91_3 1.57_2. Radiating nodules or granular. Yellow to green, pale yellow streak, vitreous luster. Uneven fracture. $H = 3\frac{1}{2}$. $G = 2.27$. Uniaxial $(-)$; $N_O = 1.565$, $N_E = 1.575$. Eagle Borax works, Death Valley, CA; *Stassfurt* and Aschersleben, *Saxony, Germany*, in salt deposits; Inder, Kazakhstan; Qinghai–Xizaig Plateau, China, in saline lakes. BM *DII:334*.

25.3.1.1 Ameghinite NaB$_3$O$_3$(OH)$_4$

Named in 1967 for Florentino (1854–1911) and Carlos (1865–1936) Ameghino, pioneer Argentine geologists. MON $C2/c$. $a = 18.428$, $b = 9.882$, $c = 6.326$, $\beta = 104.3°$, $Z = 8$, $D = 2.04$. *20-1081:* 4.95$_1$ 3.35$_2$ 3.15$_8$ 3.06$_{10}$ 2.91$_2$ 2.70$_1$ 2.66$_1$ 2.55$_3$. Nodular masses in borax. Colorless, white streak, vitreous luster. Cleavage {100}, very good; {001}, {010} poor. Conchoidal fracture, very brittle. $H = 2\frac{1}{2}$. $G = 2.03$. Biaxial $(-)$; $Z = b$; $X \wedge c = 9°$; $N_x = 1.429$, $N_y = 1.528$, $N_z = 1.538$; $2V = 33°$; $r < v$, weak. Fluoresces pale blue in UV. *Tincalayu borax deposit, Salta, Argentina*. BM *AM 52:935(1967), 60:879(1975)*.

25.3.2.1 Solongoite Ca$_2$B$_3$O$_4$(OH)$_4$Cl

Named in 1974 for the locality. MON $P2_1/c$. $a = 7.93$, $b = 7.26$, $c = 12.54$, $\beta = 94.00°$, $Z = 4$, $D = 2.53$. *26-1051:* 7.84$_9$ 3.40$_6$ 2.74$_7$ 2.61$_8$ 2.54$_8$ 2.20$_{10}$ 1.91$_8$ 1.73$_9$. Microscopic prismatic crystals. Colorless, white streak, vitreous luster. $H = 3\frac{1}{2}$. $G = 2.51$. Biaxial $(+)$; $Z \wedge c = 25°$; $N_x = 1.510$, $N_y = 1.510$, $N_z = 1.545$; $2V$ very small; $r > v$, very weak. *Solongo, Transbaikalia, Russia*. BM *ZVMO 93:117(1974), AM 60:162(1975)*.

25.3.3.1 Fabianite CaB$_3$O$_5$(OH)

Named in 1962 for Hans-Joachim Fabian, German geologist. MON $P2_1/a$. $a = 6.593$, $b = 10.488$, $c = 6.365$, $\beta = 113.38°$, $Z = 4$, $D = 2.79$. *15-631:* 3.96$_6$ 3.27$_{10}$ 3.03$_9$ 2.92$_9$ 2.62$_5$ 2.24$_4$ 2.07$_6$ 2.03$_6$. Prismatic crystals to 25 mm. Colorless, white streak, vitreous luster. Cleavage {011}. $H = 6$. $G = 2.796$. Biaxial $(-)$; $XZ = ba$; $X \wedge c = 23°$; $N_x = 1.608$, $N_y = 1.637$, $N_z = 1.650$; $2V = 65°$; $r < v$, weak. Fluoresces yellow-brown. *Rehden, near Diepholz, Germany*, in drill core from salt deposit. BM *NW 49:230(1962), AM 48:212(1963), CM 10:108(1969), ZK 132:241(1970)*.

25.3.4.1 Uralborite CaB$_2$O$_2$(OH)$_4$

Named in 1961 for the composition and locality. MON $P2_1/n$. $a = 6.881$, $b = 12.305$, $c = 9.791$, $\beta = 97.46°$, $Z = 8$, $D = 2.61$. *14-272:* 7.61$_{10}$ 6.18$_6$ 4.81$_5$ 3.42$_6$ 2.97$_9$ 2.13$_{10}$ 1.83$_5$ 1.41$_8$. Prismatic crystals to 7 mm. Colorless, white streak, vitreous luster. Distinct cleavage, parallel to elongation. $H = 4$. $G = 2.60$. Biaxial $(+)$; $N_x = 1.604$, $N_y = 1.609$, $N_z = 1.615$; $2V = 85°$; $r > v$, strong. Fluoresces violet in LWUV. *Solongo, Transbaikalia, Russia*, in skarn. BM *ZVMO 90:673(1961), AM 47:1482(1962), DANS 189:532(1969)*.

25.3.5.1 Howlite Ca$_2$SiB$_5$O$_9$(OH)$_5$

Named in 1868 for Henry How (1828–1879), Canadian chemist, who discovered the mineral. MON $P2_1/c$. $a = 12.820$, $b = 9.351$, $c = 8.608$, $\beta = 104.84°$, $Z = 4$, $D = 2.61$. *26-1404:* 12.4$_3$ 6.21$_{10}$ 4.12$_2$ 3.89$_9$ 3.09$_7$ 3.01$_4$ 2.93$_3$ 2.89$_3$. Tabular crystals rare, usually as nodular masses. White, white streak, subvitreous luster. $H = 3\frac{1}{2}$. $G = 2.58$. Biaxial $(-)$; $X = b$; $Z \wedge a = 55°$; $N_x = 1.582$, $N_y = 1.595$, $N_z = 1.607$; $2V = 70-80°$. Lang, Los Angeles Co., Furnace Creek, Inyo Co., Boron, Kern Co., and Calico Mts., San Bernardino Co., CA; Flat Bay, NF, near *Windsor*, Iona(*), and other localities in *NS, Canada*; Magdalena, SON, Mexico; Rehden borehole, Emsland, Germany; Jarando, Yugoslavia; Susurluk, Turkey; southern Ural Mts., Russia. BM *DII:362*; *AM 55:716(1970), 73:1138(1988)*.

25.4.1.1 Roweite $Ca_2Mn_2B_4O_7(OH)_6$

Named in 1937 for George Rowe (1868–1947), mine captain at Franklin. ORTH *Pbam*. $a = 9.057$, $b = 13.357$, $c = 8.289$, $Z = 4$, $D = 2.933$. *26-1065:* 3.97_{10} 3.06_3 2.60_7 2.26_3 2.18_5 2.13_4 1.71_3 1.64_3. Subhedral platy crystals to 5 mm. Amber to brown, white streak, vitreous luster. Cleavage {100}, fair. $H = 4\frac{1}{2}$. $G = 2.935$. Biaxial $(-)$; $N_x = 1.646$, $N_y = 1.658$, $N_z = 1.660$; $2V = 28°$; $r < v$, strong; moderately pleochroic, colorless to yellow-brown. *Franklin, Sussex Co., NJ*; Solongo, Transbaikalia, Russia. BM *DII:377, AM 59:60,66(1974)*.

25.4.1.2 Federovskite $Ca_2(Mg,Mn)_2B_4O_7(OH)_6$

Named in 1976 for Nikolai M. Federov (1886–1956), Russian mineralogist. Forms a series with roweite (25.4.1.1). ORTH *Pbam*. $a = 8.96$, $b = 13.15$, $c = 8.15$, $Z = 4$, $D = 2.82$. *29-347:* 3.92_{10} 3.02_7 2.59_{10} 2.28_6 2.16_6 2.12_7 1.91_6 1.69_7. Fibrous veinlets. Brown, pale streak, vitreous luster. Cleavage {100}, perfect. $H = 4\frac{1}{2}$. $G = 2.65–2.90$, increasing with Mn content. Biaxial $(-)$; $XZ = ca$; $N_x = 1.619–1.643$, $N_y = 1.627–1.653$, $N_z = 1.629–1.654$; $2V = 40°$; $r < v$, very strong; pleochroic, colorless to bright yellow. *Solongo, Transbaikalia, Russia*, in skarn. BM *ZVMO 105:71(1976), AM 62:173(1977)*.

25.5.1.1 Priceite $Ca_4B_{10}O_{19} \cdot 7H_2O$

Named in 1873 for Thomas Price (b.1837), San Francisco metallurgist, who first analyzed the mineral. TRIC Space group unknown. *9-147:* 10.9_{10} 5.99_2 5.46_8 4.29_4 3.63_{10} 3.49_7 2.72_6 2.18_5. Nodular chalky masses. White, white streak, dull luster. $H = 3-3\frac{1}{2}$. $G = 2.42$. Biaxial $(-)$; $N_x = 1.573$, $N_y = 1.594$, $N_z = 1.597$; $2V = 32-42°$; $r < v$, strong. Near *Chetco, Curry Co., OR*; Furnace Creek, Inyo Co., CA; Wentworth and Hillsborough, NB, Canada; Meldon, Devon, England; Pohla, Erzgebirge, Germany; Radotin, Czech Republic; Kirka, Turkey; Inder, Kazakhstan. BM *DII:341, AM 41:689(1956), CM 19:291(1981)*.

BORACITE GROUP

The boracite minerals are borates with the general formula

$$X_3B_7O_{13}Cl$$

where

$$X = Mg, Fe^{2+}, Mn^{2+}$$

and the structure is either an orthorhombic, pseudocubic structure or a rhombohedral structure.

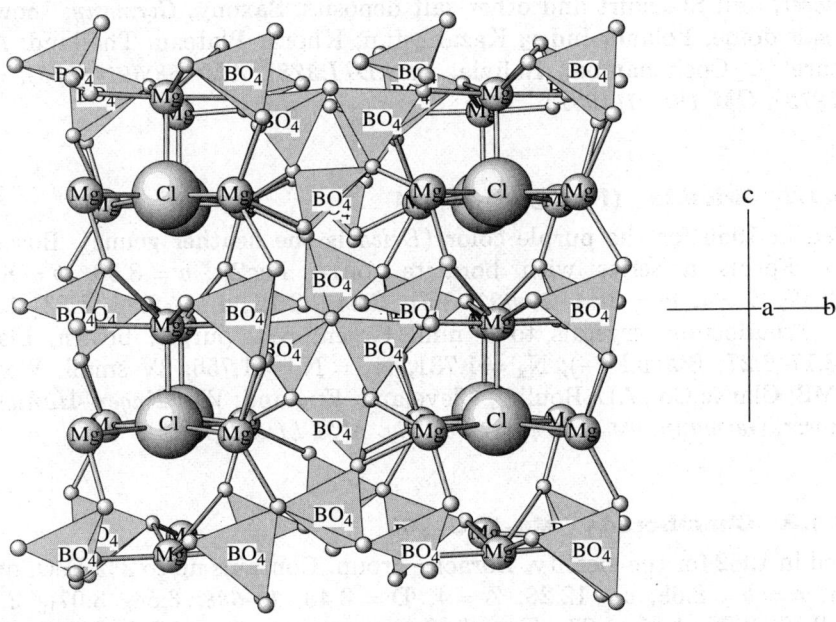

BORACITE: $Mg_3B_7O_{13}$

The high-temperature cubic form of boracite has an open three-dimensional framework of BO_4 tetrahedra—this includes some linkages of three tetrahedra to a single oxygen, which is not seen in silicates. Mg and Cl occupy large interstices. In the pictured low-temperature orthorhombic form (a displacive transition) some B–O linkages are shifted, resulting in a complex sheet structure.

BORACITE GROUP

Mineral	Formula	Space Group	a	b	c		D	
Boracite	$(Mg,Fe)_3B_7O_{13}Cl$	$Pca2_1$	8.550	8.550	12.091	2.945	25.6.1.1
Ericaite	$(Fe,Mg)_3B_7O_{13}Cl$	$Pca2_1$	8.58	8.69	12.17	3.44	25.6.1.2
Chambersite	$Mn_3B_7O_{13}Cl$	$Pca2_1$	8.68	8.68	12.26	3.48	25.6.1.3
Congolite	$Fe_3B_7O_{13}Cl$	$R3c$	8.612		21.065	3.44	25.6.2.1
Trembathite	$(Mg,Fe)_3B_7O_{13}Cl$	$R3c$	8.574		20.99	3.14	25.6.2.2

25.6.1.1 Boracite $(Mg,Fe)_3B_7O_{13}Cl$

Named in 1789 for the composition. Boracite group. Forms a series with ericaite. ORTH (ISO above 265°) $Pca2_1$. $a = b = 8.550$, $c = 12.091$, $Z = 4$, $D = 2.945$. 5-710: 3.04_7 2.72_5 2.14_4 2.06_{10} 1.83_3 1.76_5 1.36_4 1.24_5. Pseudocubic crystals. White to gray, green in ferroan varieties; white streak; vitreous luster. No cleavage, conchoidal to uneven fracture. $H = 7-7\frac{1}{2}$. $G = 2.95-3.10$, increasing with Fe content. Strongly piezoelectric and pyroelectric. Biaxial (+); $XY = ca$; $N_x = 1.658$, $N_y = 1.662$, $N_z = 1.668$; $2V = 82°$. A common accessory in salt deposits. Choctaw salt dome, Iberville Parish, LA; Clarke Co., AL; Wayne Co., MS; Wichita Mts., OK; Penobsquis and Salt Springs, NB, Canada; Aislaby(*), Yorkshire, England; Luneville, France; *Lüneberg(*), Hannover,* and Stassfurt and other salt deposits, Saxony, *Germany*; Inowroclow salt dome, Poland; Inder, Kazakhstan; Khorat Plateau, Thailand; Alto Chapare(*), Cochabamba, Bolivia. BM *DII:378, AM 58:691(1973), ZK 138(1973), CM 19:291(1981).*

25.6.1.2 Ericaite $(Fe,Mg)_3B_7O_{13}Cl$

Named in 1950 for the purple color (*Erica* is the heather genus). Boracite group. Forms a series with boracite. ORTH $Pca2_1$. $a = 8.58$, $b = 8.69$, $c = 12.17$, $Z = 4$, $D = 3.44$. 29-697: 3.51_4 3.04_{10} 2.72_6 2.15_5 2.06_7 1.83_4 1.47_2 1.24_3. Pseudocubic crystals to 4 mm. Green, red, purple, brown, black. $G = 3.17-3.27$. Biaxial (−); $N_x = 1.731$, $N_y = N_z = 1.755$, 2V small. Wayne Co., MS; Clarke Co., AL; Boulby, Cleveland, England; *Wathlingen–Hänigsen, Hannover, Germany*. BM *AM 41:372(1956), MM 41:404(1977).*

25.6.1.3 Chambersite $Mn_3B_7O_{13}Cl$

Named in 1962 for the locality. Boracite group. Contains up to 2% FeO. ORTH $Pca2_1$. $a = b = 8.68$, $c = 12.26$, $Z = 4$, $D = 3.48$. 14-638: 3.54_5 3.07_{10} 2.74_6 2.50_3 2.17_5 2.08_6 1.85_5 1.77_3. Tetrahedral crystals to 2 cm. Colorless to deep purple, white streak, vitreous luster. No cleavage, subconchoidal to uneven fracture. $H = 7$. $G = 3.49$. Biaxial (+); $N_x = 1.732$, $N_y = 1.737$, $N_z = 1.744$; $2V = 83°$. Venice salt dome, Plaquemines Co., LA; *Barber's Hill salt dome, Chambers Co., TX*; Pomyarka, Carpathian Mts., Ukraine; Hebei Prov., China. BM *AM 47:665(1962), NJMM:69(1982).*

25.6.2.1 Congolite $Fe_3B_7O_{13}Cl$

Named in 1972 for the country of origin. Boracite group. Dimorphous with ericaite (25.6.1.2). HEX-R $R3c$. $a = 8.612$, $c = 21.065$, $Z = 6$, $D = 3.44$. 25-2: 3.05_8 2.73_{10} 2.16_6 2.15_6 2.06_8 1.84_7 1.83_7 1.76_6. Fine-grained. Pale red, white streak, vitreous luster. Uniaxial (−); $N_O = 1.755$, $N_E = 1.731$. Bischofferode, Hannover, Germany; *Brazzaville, Congo*, insoluble residue from drill core. BM *AM 57:1315(1972), ZK 138:64(1973).*

25.6.2.2 Trembathite (Mg,Fe)$_3$B$_7$O$_{13}$Cl

Named in 1992 for Lowell T. Trembath (1936–1993), Canadian mineralogist. Boracite group. Dimorphous with boracite. HEX-R $R3c$. $a = 8.574$, $c = 20.99$, $Z = 6$, $D = 3.14$. *CM 30:* 3.50_3 3.03_{10} 2.71_7 2.48_2 2.14_4 2.05_7 1.83_3 1.75_2. Rhombohedra to 2 mm. Colorless to pale blue, white streak, vitreous luster. No cleavage, conchoidal fracture. H = 6–8. G = 2.84–3.34. Uniaxial (−); $N_O = 1.684$, $N_E = 1.668$. *Salt Springs potash deposit, Sussex, NB, Canada.* BM *CM 30:445(1992)*.

25.6.3.1 Strontioborite SrB$_8$O$_{11}$(OH)$_4$

Named in 1960 for the composition. MON $P2_1$. $a = 9.909$, $b = 8.130$, $c = 7.623$, $\beta = 108.4°$, $Z = 2$, $D = 2.38$. *18-1285:* 7.2_{10} 5.4_7 4.09_9 3.52_7 3.29_8 3.07_7 2.61_7 2.33_6. Platy crystals to 2 mm. Colorless, white streak, vitreous luster. G = 2.40. Biaxial (+); $N_x = 1.470$, $N_y = 1.510$, $N_z = 1.579$; 2V near 90°. *Caspian region, Russia,* in rock salt. BM *DANS 135:173(1960)*, *AM 46:768(1961)*.

25.7.1.1 Preobrazhenskite Mg$_3$B$_{11}$O$_{15}$(OH)$_9$

Named in 1956 for Pavla I. Preobrazhensk (1874–1944), Russian geologist. ORTH $Pbcn$. $a = 16.33$, $b = 9.16$, $c = 10.59$, $Z = 4$, $D = 2.45$. *31-788:* 5.28_{10} 3.79_{10} 3.23_9 3.19_9 2.82_9 2.77_9 2.64_9 2.13_{10}. Tabular crystals. Colorless to lemon yellow or gray, white streak, vitreous luster. Very friable. H = $4\frac{1}{2}$–5. Biaxial (+); $N_x = N_y = 1.570$, $N_z = 1.595$; 2V very small. *Inder(*), Kazakhstan,* in salt deposits. BM *DANS 11:1087(1956)*, *AM 42:704(1957)*, *CM 32:387(1994)*.

25.8.1.1 Jeremejevite Al$_6$(BO$_3$)$_5$F$_3$

Named in 1883 for Pavel V. Jeremejev (1830–1899), Russian mineralogist. HEX $P6_3/m$. $a = 8.558$, $c = 8.183$, $Z = 2$, $D = 3.28$. *35-504:* 4.28_{10} 3.79_6 2.80_3 2.47_3 2.20_4 2.06_5 1.68_3 1.39_3. Hexagonal prisms to several centimeters. Colorless to pale yellow-brown or blue, white streak, vitreous luster. No cleavage, conchoidal fracture. H = $7\frac{1}{2}$. G = 3.29. Uniaxial (−), in part anomalously biaxial; $N_O = 1.647$, $N_E = 1.637$. *Mt. Soktuj, S of Nerchinsk, Transbaikal, Russia;* Emmelberg, near Üdersdorf, Eifel, Germany; Usakos and Cape Cross, Namibia. BM *DII:330, CM 19:303(1981), ZK 165:255(1983)*.

25.8.2.1 Rhodizite (K,Cs)Al$_4$Be$_4$(B,Be)$_{12}$O$_{28}$

Named in 1834 for the Greek for *rose-colored,* because it tinges the blowpipe flame red. ISO $P\bar{4}3m$. $a = 7.317$, $Z = 1$, $D = 3.345$. *18-327:* 7.31_2 3.27_5 2.98_{10} 2.44_5 2.21_2 2.11_4 1.96_2 1.78_3. Dodecahedral or tetrahedral crystals. Colorless to white or yellow, white streak, vitreous to adamantine luster. Cleavage {111}, {$\bar{1}$11}, difficult; conchoidal fracture. H = $8\frac{1}{2}$. G = 3.36. Isotropic;

N = 1.693. Pyroelectric and piezoelectric. *Sarapulsk and Schaitansk, near Mursinsk, Ural Mts., Russia*; Ambatofinandrahana, Manjaka, and Antanddrokomby, Madagascar. BM *DII:329*, *TMPM 10:409(1965)*, *ZK 125:423(1967)*, *MM 50:163(1986)*.

Class 26

Hydrated Borates Containing Hydroxyl or Halogen

26.1.1.1   Berborite $Be_2(BO_3)(OH) \cdot H_2O$

Named in 1967 for the composition. HEX-R $P3$ (1T polytype; 2T and 2H polytypes are known). $a = 4.434$, $c = 5.334$, $Z = 1$, $D = 2.05$. *22-107:* 5.34_{10} 3.84_1 3.12_{10} 2.67_6 2.22_2 2.05_3 1.39_2 1.28_1. Crystals to 5 mm. Colorless, white streak, vitreous luster. Cleavage {0001}, perfect. H = 3. G = 2.03–2.06. Uniaxial (−); $N_O = 1.580$, $N_E = 1.485$. Tvedalen district, Langesundsfjord, Norway, in nepheline syenite pegmatites; *Pitkaranta, Karelia, Russia*, in skarn. BM *DANS 174:189(1967), AM 53:348(1968), NJMA 162:101(1990)*.

26.1.2.1   Wightmanite $Mg_5(BO_3)O(OH)_5 \cdot 2H_2O$

Named in 1962 for R. H. Wightman (1902–1969), Director of Exploration, Riverside Cement Co. MON $I2/m$. $a = 13.46$, $b = 3.102$, $c = 18.17$, $\beta = 91.60°$, $Z = 2$, $D = 2.58$. *14-640:* 10.7_{10} 9.07_{10} 4.74_2 3.03_3 2.67_3 2.45_2 2.35_2 2.12_2. Pseudohexagonal prisms. Colorless, white streak, vitreous luster. Cleavage {$\bar{1}01$}, perfect. H = $5\frac{1}{2}$. G = 2.59. Biaxial (−); $N_x = 1.585$, $N_y = 1.601$, $N_z = 1.602$; 2V = 29°; r < v, strong. *Crestmore, Riverside Co.*, and Twin Lakes area, Fresno Co., CA. BM *AM 47:718(1962), 59:985(1974)*.

26.1.3.1   Shabynite $Mg_5(BO_3)Cl_2(OH)_5 \cdot 4H_2O$

Named in 1980 for Leonid I. Shabynin (b.1909), Russian geologist. Possibly MON Space group unknown. *41-1405:* 9.27_{10} 5.47_6 4.70_4 4.21_5 3.69_7 2.44_8 2.36_8 1.80_5. Finely fibrous. White, white streak, silky luster. H = 3. G = 2.32. Biaxial (−); $N_x = 1.543$, $N_y = 1.571$, $N_z = 1.577$; 2V = 49°; straight extinction. *Korshunov skarn–magnetite deposit, Baikalia, Russia.* BM *ZVMO 109:569(1980), AM 66:1101(1981)*.

26.1.4.1   Hexahydroborite $Ca[B(OH)_4]_2 \cdot 2H_2O$

Named in 1976 for the composition. MON $P2/a$. $a = 8.006$, $b = 8.012$, $c = 6.649$, $\beta = 104.21°$, $Z = 2$, $D = 1.88$. *30-236:* 7.76_{10} 6.64_4 3.95_2 3.35_8 2.52_4 2.47_3 2.30_2 1.97_2. Microscopic prismatic crystals. Colorless, white streak, vitreous luster. H = $2\frac{1}{2}$. G = 1.87. Biaxial (+); $X \wedge c = 14°$; $N_x = 1.498$, $N_y = 1.503$, $N_z = 1.510$; 2V = 83°; r > v, strong. *Solongo, Transbaikalia, Russia.* BM *DANS 228:1337(1976), AM 62:1259(1977)*.

26.1.5.1 Henmilite $Ca_2Cu[B(OH)_4]_2(OH)_4$

Named in 1986 for Kitinosuke Henmi (b.1919), Japanese mineralogist. TRIC $P\bar{1}$. $a = 5.762$, $b = 7.977$, $c = 5.649$, $\alpha = 109.61°$, $\beta = 91.47°$, $\gamma = 83.69°$, $Z = 1$, $D = 2.52$. *40-481:* 5.25_{10} 4.35_8 3.93_5 3.71_8 3.30_8 2.49_8 2.43_8 2.35_8. Microscopic crystals and anhedral masses. Blue-violet, pale violet streak, vitreous luster. $H = 2$. Biaxial $(-)$; $N_x = 1.585$, $N_y = 1.608$, $N_z = 1.615$; $2V = 58°$; strongly pleochroic. X, pale pink; Y, pale purple; Z, pale blue. *Fuka mine, Okayama Pref., Japan*, in borate veins cutting marble. BM *AM 71:1234(1986)*.

26.2.1.1 Pentahydroborite $CaB_2O(OH)_6 \cdot 2H_2O$

Named in 1961 for the composition. TRIC $P\bar{1}$. $a = 7.845$, $b = 6.525$, $c = 8.124$, $\alpha = 111.62°$, $\beta = 11.19°$, $\gamma = 73.44°$, $Z = 2$, $D = 2.14$. *14-339:* 7.04_{10} 5.86_4 3.54_8 3.20_5 2.99_9 2.88_6 2.49_5 1.94_8. Small grains. Colorless, white streak, vitreous luster. No cleavage. $H = 2\frac{1}{2}$. $G = 2.03$. Biaxial $(+)$; $N_x = 1.531$, $N_y = 1.536$, $N_z = 1.544$; $2V = 73°$. *Solongo, Transbaikalia, Russia,* in skarn; Fuka mine, Okayama Pref., Japan. BM *ZVMO 90:673(1961); AM 47:1482(1962), 71:1234(1986)*.

26.3.1.1 Inyoite $CaB_3O_3(OH)_5 \cdot 4H_2O$

Named in 1916 for the locality. MON $P2_1/a$. $a = 10.621$, $b = 12.066$, $c = 8.408$, $\beta = 114.02°$, $Z = 4$, $D = 1.87$. *37-1459:* 7.67_{10} 7.57_2 3.84_2 3.45_2 3.37_2 2.80_2 2.78_2 2.55_2. Short prismatic or tabular crystals, coarse spherulitic aggregates, and massive. Colorless or white, white streak, vitreous luster. Cleavage {001}, good; irregular fracture. $H = 2$. $G = 1.875$. Biaxial $(-)$; $Y = b$; $X \wedge c = 36°$; $N_x = 1.490$, $N_y = 1.505$, $N_z = 1.516$; $2V = 86°$; $r < v$, weak. Soluble in hot water and dilute acids. *Monte Blanco, Inyo Co.,* and Boron, Kern Co., CA; Wentworth and Hillsborough, NB, Canada; Donets basin, Ukraine; Kirka, Turkey; Inder(*), Kazakhstan; Qinghai–Xizaig Plateau, China; Laguna Salinas, Peru; Monte Azul mine, Sijes, and Tincalayu, Salta Prov., Argentina. Occurs in sedimentary borate deposits. BM *DII:358, AM 38:912(1953), CM 19:291(1981), SR 49A:239(1982)*.

26.3.1.2 Inderborite $CaMg[B_3O_3(OH)_5]_2 \cdot 6H_2O$

Named in 1941 for the locality and composition. MON $C2/c$. $a = 12.137$, $b = 7.433$, $c = 19.234$, $\beta = 90.29°$, $Z = 4$, $D = 1.919$. *12-70:* 9.58_2 6.34_4 6.07_{10} 4.51_2 3.28_3 3.20_3 3.03_2 3.01_2. Crystals to 2 cm and massive. Colorless or white, white streak, vitreous luster. Cleavage {100}, good; conchoidal fracture. $H = 3\frac{1}{2}$. $G = 2.00$. Biaxial $(-)$; $Z = b$; $X \wedge c = 2°$; $N_x = 1.483$, $N_y = 1.512$, $N_z = 1.530$; $2V = 77°$. Slowly soluble in water. Furnace Creek, Inyo Co., CA; Sarikaya, Kirka Co., Turkey; *Inder, Kazakhstan*; Sijes, Salta Prov., Argentina. BM *DII:355, AC 21:A61(1966), CM 32:533(1994)*.

26.3.1.3 Inderite $MgB_3O_3(OH)_5 \cdot 5H_2O$

Named in 1937 for the locality. MON $P2_1/a$. $a = 12.026$, $b = 13.116$, $c = 6.821$, $\beta = 104.49°$, $Z = 4$, $D = 1.784$. *36-423:* 6.61_4 5.84_4 5.72_6 5.05_{10} 3.36_7 3.28_4 2.66_4 2.56_4. Tabular or acicular crystals, and nodular aggregates. Colorless or white, white streak, vitreous to pearly luster. Cleavage {010}, perfect; {110}, good. $H = 2\frac{1}{2}$. $G = 1.80$. Biaxial (+); $Z \wedge c = 9°$; $N_x = 1.488$, $N_y = 1.491$, $N_z = 1.505$; $2V = 37°$; $r > v$, weak. Boron(*), Kern Co., and Furnace Creek, Inyo Co., CA; Kirka, Turkey; *Inder*(*), *Kazakhstan*; Qinghai–Xizaig Plateau, China. BM *DII:360*, *AM 45:732(1960)*, *AC(B) 32:1329(1976)*.

26.3.2.1 Meyerhofferite $CaB_3O_3(OH)_5 \cdot H_2O$

Named in 1914 for Wilhelm Meyerhoffer (1864–1906), German chemist, who synthesized the mineral. TRIC $P\bar{1}$. $a = 6.63$, $b = 8.35$, $c = 6.46$, $\alpha = 90.55°$, $\beta = 102.28°$, $\gamma = 87.2°$, $Z = 2$, $D = 2.13$. *12-411:* 8.31_{10} 6.47_{10} 4.97_3 4.15_2 3.29_1 3.15_2 3.07_2 2.50_2. Radiating prismatic crystals, and as pseudomorphs after inyoite. Colorless or white, white streak, vitreous to silky luster. Cleavage {010}, perfect. $H = 2$. $G = 2.12$. Biaxial (−); $N_x = 1.500$, $N_y = 1.535$, $N_z = 1.560$; $2V = 78°$; $r > v$. *Monte Blanco*(*) and Gower Gulch, Furnace Creek, *Inyo Co.*, and Boron, Kern Co., *CA*; Kirka, Turkey; Inder, Kazakhstan; Sijes, Salta Prov., Argentina. BM *DII:356*, *ZK 114:321(1960)*.

26.3.3.1 Kurnakovite $MgB_3O_3(OH)_5 \cdot 5H_2O$

Named in 1940 for Nikolai S. Kurnakov (1860–1941), Russian mineralogist. TRIC $P\bar{1}$. $a = 8.344$, $b = 10.59$, $c = 6.443$, $\alpha = 99.0°$, $\beta = 108.9°$, $\gamma = 105.5°$, $Z = 2$, $D = 1.855$. *29-856:* 7.22_{10} 5.01_8 4.90_9 4.21_8 3.48_7 2.68_8 2.53_7 2.48_7. Crystals to 30 cm, and granular aggregates. Colorless or white, white streak, vitreous luster. Cleavage {010}, perfect. $H = 3$. $G = 1.847$. Biaxial (−); $Z \wedge c = 22°$; $N_x = 1.489$, $N_y = 1.510$, $N_z = 1.525$; $2V = 80°$; $r > v$. Boron(*), Kern Co., and Furnace Creek area, Inyo Co., CA; Kirka, Turkey; *Inder, Kazakhstan*; Qinghai–Xizaig Plateau, China; Tincalayu(*), Salta Prov., Argentina. BM *DII:360*, *AC(B) 30:2194(1974)*.

26.3.4.1 Hydrochlorborite $Ca_2B_4O_4(OH)_7Cl \cdot 7H_2O$

Named in 1965 for the composition. MON $I2/a$. $a = 22.783$, $b = 8.745$, $c = 17.066$, $\beta = 96.71°$, $Z = 8$, $D = 1.841$. *29-312:* 8.48_{10} 8.18_1 6.01_3 4.14_1 3.82_1 3.70_1 3.60_1 2.80_1. Platy crystals or dense masses. Colorless or white, white streak, vitreous luster. Cleavage {001}, good. Brittle. $H = 2\frac{1}{2}$. $G = 1.825$. Biaxial (+); $Y = b$; $X \wedge c = 25°$; $N_x = 1.499$, $N_y = 1.502$, $N_z = 1.521$; $2V = 45°$; $r < v$. *China*; Salar Carcote, Antofagasta, Chile. BM *SS 14:943(1965)*; *AM 62:147(1977)*, *63:814(1978)*.

26.3.5.1 Colemanite $CaB_3O_4(OH)_3 \cdot H_2O$

Named in 1884 for William T. Coleman (1824–1893), founder of the California borate industry. MON $P2_1/a$. $a = 8.726$, $b = 11.253$, $c = 6.098$, $\beta = 110.12°$,

$Z = 4$, $D = 2.43$. *33-267:* 5.63_7 4.01_6 3.85_8 3.29_8 3.14_{10} 2.89_6 2.14_6 2.01_8. Complex crystals to several cm, also massive, granular to compact, and as pseudomorphs after inyoite. Colorless to white or gray, white streak, vitreous to adamantine luster. Cleavage {010}, perfect; {001}, distinct; uneven to subconchoidal fracture. $H = 4\frac{1}{2}$. $G = 2.42$. Biaxial (+); $X = b$; $Y \wedge c = 6°$; $N_x = 1.586$, $N_y = 1.592$, $N_z = 1.614$; $2V = 55°$; $r > v$, weak. Pyroelectric and ferroelectric below $-6°$. Soluble in hot HCl. Muddy Mts., Clark Co., NV; Lila B(*), Biddy McCarthy(*), and other mines in the *Furnace Creek area, Inyo Co.*, Boron, Kern Co., Borate, San Bernardino Co., Tick Canyon, Los Angeles Co., and other localities in CA; Penobsquis and Salt Springs, NB, Canada; Kirka, Turkey; Inder(*), Kazakhstan; Sijes, Salta Prov., Argentina. An important ore of boron. BM *DII:349*, *AM 38:411(1953)*, *AC 11:761(1958)*.

26.3.6.1 Hydroboracite $CaMg[B_3O_4(OH)_3]_2 \cdot 3H_2O$

Named in 1834 for the composition. MON $P2/c$. $a = 11.769$, $b = 6.684$, $c = 8.235$, $\beta = 102.59°$, $Z = 2$, $D = 2.17$. *35-646:* 6.69_9 5.78_{10} 4.47_1 4.36_1 3.32_7 2.44_1 2.23_1 1.91_2. Radiating aggregates of prismatic to acicular crystals, also fibrous to compact. Colorless or white, white streak, vitreous or silky luster. Cleavage {010}, perfect. $H = 5-6$. $G = 2.15$. Biaxial (+); $Y = b$; $X \wedge c = 33°$; $N_x = 1.520$, $N_y = 1.534$, $N_z = 1.569$; $2V = 60-66°$; $r < v$. Thompson mine(*) and other localities in the Furnace Creek area, Inyo Co., and Boron(*), Kern Co. (a single mass weighing more than a ton and as good crystals in vugs), CA; Penobsquis and Salt Springs, NB, Canada; Woffleben, Germany; Kirka, Turkey; *Caucasus Mts., Russia*; Inder(*), Kazakhstan; Qinghai–Xizaig Plateau, China; in nitrate deposits, northern Chile; Pastos Grandes and Sijes, Salta Prov., Argentina. BM *DII:353*; *MR 4:21(1973)*, *9:379(1978)*; *CM 16:75(1978)*.

26.3.7.1 Nifontovite $Ca_3[B_3O_3(OH)_6]_2 \cdot 2H_2O$

Named in 1961 for Roman V. Nifontov (1901–1960), Russian geologist. MON $C2/c$. $a = 13.089$, $b = 9.498$, $c = 13.566$, $\beta = 119.58°$, $Z = 4$, $D = 2.36$. *27-68:* 7.28_{10} 5.52_6 3.84_8 2.95_6 2.75_7 2.24_8 2.21_7 2.06_8. Colorless, white streak, vitreous luster. Cleavage poor parallel to elongation. $H = 3\frac{1}{2}$. $G = 2.36$. Biaxial (+); $N_x = 1.575$, $N_y = 1.578$, $N_z = 1.584$; $2V = 76°$; $r > v$, strong. Fluoresces violet in LWUV. *Novofrolovo mine, Ural Mts., Russia*, in skarn; Fuka, Okayama Pref., Japan. BM *DANS 139:188(1961)*, *AM 47:172(1962)*, *MM 58:279(1994)*.

26.4.1.1 Borax $Na_2B_4O_5(OH)_4 \cdot 8H_2O$

Named in 1753 from the Arabic for white. MON $A2/a$. $a = 12.219$, $b = 10.665$, $c = 11.884$, $\beta = 106.64°$, $Z = 4$, $D = 1.71$. Contains anionic complex groups $B_4O_5(OH)_4^{2-}$ made up of two triangle and two tetrahedral groups, one corner of each tetrahedron and each triangle is occupied by (OH). All the oxygen atoms are shared by two tetrahedrons or by a tetrahedron and a triangle. *AC(B) 34:3502(1978)*. *33-1215:* 5.84_4 5.69_5 4.86_8 3.94_5 2.87_7 2.83_6 2.58_{10} 2.57_9. Short prismatic crystals [001] and often somewhat tabular {100}; prominent zones [001] and [110]. Colorless; in dry air dehydrates to the trihydrate, tincalconite, and becomes chalky; white streak; vitreous to resinous luster. Cleavage {100}, perfect; {110}, less so; conchoidal fracture; brittle. $H = 2-2\frac{1}{2}$. $G = 1.715$. Soluble in water. Tastes sweetish alkaline. Fuses easily to a colorless glass. Biaxial $(-)$; $X = b$; $Z \wedge c = 55°$; $N_x = 1.447$, $N_y = 1.469$, $N_z = 1.472$; $2V = 40°$; $r > v$, strong. Borax is an evaporite mineral associated with halite, sulfates, carbonates, and other borates in alkaline lakes and playas. Originally brought to Europe from salt lakes in Kashmir and Tibet. Borax Lake(*), Esmeralda Co., NV; Furnace Creek district, Inyo Co., Borax Lake, Lake Co., Searles Lake(*), San Bernardino Co., Boron(*), Kern Co., CA; Kirka, Turkey; Inder, Kazakhstan; Lake Ma-Pin-Montalei, Tibet; Ladakh, Kashmir; Salar Cauchari(*), Jujuy Prov., Argentina. An important source of boron compounds. BM *DII:339*, *MR 6:74,84(1975)*.

26.4.2.1   Tincalconite $Na_2B_4O_5(OH)_4 \cdot 3H_2O$

Named in 1878 from tincal, the Sanskrit for borax. HEX-R $R32$. $a = 11.12$, $c = 21.20$, $Z = 9$, $D = 1.91$. *7-277*: 8.75_6 4.71_3 4.38_9 3.44_6 2.92_{10} 2.59_3 2.26_3 2.19_4. Pseudo-octahedral crystals rare; usually as a powder or pseudomorphs, the dehydration product of borax. White, white streak, earthy or vitreous luster. $G = 1.88$. Uniaxial $(-)$; $N_O = 1.461$, $N_E = 1.474$. Localities as for borax; crystals from Searles Lake, San Bernardino Co., and Boron(*), Kern Co., CA; Larderello(*), Italy; Kirka, Turkey. BM *DII:337*; *AM 33:472(1948)*, *58:523(1973)*; *PM 43:583(1973)*.

26.4.3.1   Hungchaoite $MgB_4O_5(OH)_4 \cdot 7H_2O$

Named in 1964 for Chang Hung-Chao (1877–1951), Chinese mineralogist. TRIC $P\bar{1}$. $a = 8.811$, $b = 10.644$, $c = 7.888$, $\alpha = 103.38°$, $\beta = 108.58°$, $\gamma = 97.15°$, $Z = 2$, $D = 1.705$. *34-1288*: 7.14_6 6.71_{10} 5.85_6 5.36_9 4.09_8 4.08_7 3.36_9 2.91_8. Tabular or equant crystals to 0.5 mm, and as nodules and crusts. Colorless or white, white streak, vitreous luster. Cleavage $\{\bar{1}11\}$, $\{\bar{1}\bar{1}1\}$, good; {010}, imperfect. $H = 2\frac{1}{2}$. $G = 1.706$. Biaxial $(-)$; $X \wedge a = 28°$; $Z \wedge b = 37°$; $N_x = 1.445$, $N_y = 1.485$, $N_z = 1.490$; $2V = 15°$; $r < v$, distinct. Furnace Creek district, Inyo Co., CA; *Qinghai–Xizaig Plateau, China*, in saline lakes. BM *SS 13:525(1964)*; *AM 50:262(1965)*, *62:1135(1977)*, *64:369(1979)*.

26.4.4.1   Halurgite $Mg_2[B_4O_5(OH)_4]_2 \cdot H_2O$

Named in 1962 for the Institute of Halurgy, Moscow. MON Space group uncertain, probably $P2/c$. $a = 13.25$, $b = 7.60$, $c = 13.20$, $\beta = 92.2°$, $Z = 4$, $D = 2.25$. *15-180*: 6.76_5 4.81_9 3.87_{10} 3.69_5 3.57_6 3.29_{10} 2.64_7 2.16_9. Microscopic platy crystals or fine-grained masses. White, white streak, vitreous luster. $H = 2\frac{1}{2}-3$. $G = 2.19$. Biaxial $(+)$; $X = b$; $N_x = 1.532$, $N_y = 1.545$, $N_z = 1.572$; $2V = 70°$. *Kungur salt deposits, Ural Mts., Russia.* BM *DANS 143:693(1962)*, *AM 47:1217(1962)*.

26.4.5.1 Kernite $Na_2B_4O_6(OH)_2 \cdot 3H_2O$

Named in 1927 for the locality. MON $P2_1/c$. $a = 7.016$, $b = 9.152$, $c = 15.678$, $\beta = 108.86°$, $Z = 4$, $D = 1.905$. 25-1322: 7.40_{10} 6.64_9 3.90_2 3.71_2 3.25_3 3.13_3 2.88_3 2.47_3. Large equant crystals and cleavage masses several meters thick. Colorless, surface usually white from a coating of tincalconite; white streak; vitreous luster. Cleavage {100}, {001}, perfect; {$\bar{1}$02}, fair. $H = 2\frac{1}{2}$. $G = 1.906$. Biaxial (−); $Z = b$; $X \wedge c = 71°$; $N_x = 1.454$, $N_y = 1.473$, $N_z = 1.488$; $2V = 80°$; $r > v$, weak. Boron(*), Kern Co., CA, in clay-shales, apparently a somewhat altered buried lake deposit. A large proportion of boron production in the United States is obtained from this deposit; Furnace Creek district, Inyo Co., CA; Kirka, Turkey; Tincalayu, Salta Prov., Argentina. BM DII:335, AM 58:21,308(1973).

26.5.1.1 Sborgite $NaB_5O_6(OH)_4 \cdot 3H_2O$

Named in 1957 for Umberto Sborgi (1883–1955), Italian mineralogist. MON $P2/c$. $a = 11.119$, $b = 16.474$, $c = 13.576$, $\beta = 112.83°$, $Z = 8$, $D = 1.72$. 24-1056: 6.86_4 4.60_{10} 4.30_3 3.56_4 3.52_4 3.30_8 3.20_8 2.57_4. White, white streak, earthy luster. $H = 3\frac{1}{2}$. $G = 1.713$. Biaxial (+); $N_x = 1.431$, $N_y = 1.438$, $N_z = 1.507$; $2V = 35°$; $r > v$, weak. Furnace Creek area, Inyo Co., CA, as an alteration product of colemanite and priceite; Larderello, Tuscany, Italy, as incrustations around hot springs. BM AM 43:378(1958), AC(B) 28:3559(1972).

26.5.2.1 Santite $KB_5O_6(OH)_4 \cdot 2H_2O$

Named in 1970 for Giorgi Santi (1746–1823), Tuscan naturalist. ORTH $Aba2$. $a = 11.065$, $b = 11.171$, $c = 9.054$, $Z = 4$, $D = 1.74$. 25-624 (syn): 5.93_2 5.60_7 3.52_9 3.36_{10} 3.28_2 2.77_5 2.52_1 2.18_2 Microscopic anhedral grains. Colorless, white streak, vitreous luster. Cleavage {010}, perfect; {100}, distinct. $H = 2\frac{1}{2}$. $G = 1.74$. Biaxial (+); $XY = cb$; $N_x = 1.422$, $N_y = 1.435$, $N_z = 1.480$; $2V = 70°$. Larderello, Tuscany, Italy. BM CMP 27:159(1970), AM 56:636(1971).

26.5.3.1 Ammonioborite $(NH_4)_3B_{15}O_{20}(OH)_8 \cdot 4H_2O$

Named in 1931 for the composition. MON $C2/c$. $a = 25.27$, $b = 9.65$, $c = 11.56$, $\beta = 94.29°$, $Z = 4$, $D = 1.80$. 12-637: 8.98_4 5.70_6 5.44_3 4.82_2 3.16_{10} 3.09_{10} 3.01_5 2.88_6. Minute platy crystals or fine-grained granular masses. White, white streak, dull luster. $G = 1.76$. Biaxial (+); $Y = b$; $Z \wedge c = 7°$; $N_x = 1.470$, $N_y = 1.487$, $N_z = 1.540$; $2V = 59°$; $r < v$, weak. Larderello, Tuscany, Italy. BM DII:366, AM 44:1150(1959), SCI 171:377(1971).

26.5.4.1 Larderellite $NH_4B_5O_7(OH)_2 \cdot H_2O$

Named in 1854 for Francesco de Larderell (1848–1925), a proprietor of the Tuscan borate industry. MON $P2_1/a$. $a = 11.63$, $b = 7.615$, $c = 9.447$, $\beta = 96.75°$, $Z = 4$, $D = 1.89$. 12-633: 9.45_5 5.44_7 5.12_5 4.70_{10} 3.14_4 2.96_7 2.92_{10} 2.89_{10}. Microscopic plates. White, white streak, dull luster. Cleavage

{001}, {010}, perfect. Biaxial (+); X = b; Z ∧ c = 30°; $N_x = 1.493$, $N_y = 1.509$, $N_z = 1.561$; 2V = 58°; r < v. *Larderello, Tuscany, Italy.* BM *DII:365, AM 45:1087(1960), AC(B) 25:2264(1969).*

26.5.5.1   Ezcurrite $Na_2B_5O_7(OH)_3 \cdot 2H_2O$

Named in 1957 for Juan M. de Ezcurra (1900–1970), manager of the Tincalayu borate company. TRIC $P\bar{1}$. a = 8.598, b = 9.570, c = 6.576, α = 102.75°, β = 107.50°, γ = 71.50°, Z = 2, D = 2.05. *26-1370:* 8.99_2 6.93_{10} 4.49_1 4.20_1 3.30_1 3.13_3 3.09_4 3.08_4. Cleavable masses with bladed fibrous structure up to 7 cm. Colorless, white streak, vitreous to silky luster. Cleavage {110}, perfect; {010}, good; {100}, fair. H = 3–3½. G = 2.05. Biaxial (−); $N_x = 1.468$, $N_y = 1.507$, $N_z = 1.529$; 2V = 74°; r > v. *Tincalayu borate mine, Salta Prov., Argentina.* BM *EG 42:426(1957); AM 52:1048(1967), 58:110(1973).*

26.5.6.1   Nasinite $Na_2B_5O_8(OH) \cdot 2H_2O$

Named in 1961 for Raffaello Nasini (1854–1931), Italian chemist. ORTH $Pna2_1$. a = 12.015, b = 6.518, c = 11.173, Z = 4, D = 2.13. *29-1180:* 5.99_9 5.59_6 5.27_{10} 4.09_3 3.03_5 3.00_6 2.82_3 2.79_3. Earthy masses. Orange-yellow to pale gray, white streak. G = 2.12. Biaxial (−); Y = b; Z ∧ a = 7°; $N_x = 1.494$, $N_y = 1.512$, $N_z = 1.524$; 2V = 67°. *Larderello, Tuscany, Italy.* BM *AM 48:709(1963), AC(B) 31:2405(1975).*

26.5.7.1   Biringuccite $Na_2B_5O_8(OH) \cdot H_2O$

Named in 1961 for Vannoccio Biringuccio (1480–1537), Italian alchemist. MON $P2_1/c$. a = 11.196, b = 6.561, c = 20.757, β = 93.89°, Z = 8, D = 2.32. *35-421:* 10.3_8 5.18_4 3.45_{10} 3.05_6 3.03_6 2.85_4 2.68_3. 2.59_5. Microscopic pseudohexagonal plates or needles. Colorless or white, white streak. Cleavage {001}, {100}, good. G = 2.30. Biaxial (−); Y = b; Z ∧ a = 5°; $N_x = 1.496$, $N_y = 1.539$, $N_z = 1.557$; 2V = 63°. *Larderello, Tuscany, Italy.* BM *AM 48:709(1963), 59:1005(1974).*

26.5.8.1   Gowerite $CaB_6O_8(OH)_4 \cdot 3H_2O$

Named in 1959 for Harrison P. Gower (1890–1967), U.S. Borax & Chemical Corp. MON $P2_1/a$. a = 12.882, b = 16.630, c = 6.558, β = 121.62°, Z = 4, D = 2.003. *12-528:* 9.2_3 8.2_{10} 4.09_2 3.94_1 3.84_1 3.19_5 2.96_1 1.90_2. Globular clusters of long prismatic needles. Colorless or white, white streak, vitreous luster. Cleavage {001}, distinct. H = 3. G = 2.00. Biaxial (+); Y = b; Z ∧ c = 26°; $N_x = 1.485$, $N_y = 1.502$, $N_z = 1.550$; 2V = 61°; r > v, weak. *Furnace Creek district, Inyo Co., CA,* from the weathering of colemanite and priceite; N'Chwaning II mine, Kalahari manganese field, South Africa; Surire Salar, Arica Prov., Chile; Sijes, Salta Prov., Argentina. BM *AM 44:911(1959), 57:381(1972); MR 22:285(1991).*

26.5.9.1 Veatchite $Sr_2[B_5O_8(OH)]_2 \cdot B(OH)_3 \cdot H_2O$

Named in 1938 for John A. Veatch (1808–1870), U.S. physician and geologist, who first detected borates in mineral waters of California. MON Aa (a $P2_1$ polytype is also known). $a = 20.060$, $b = 11.738$, $c = 6.652$, $\beta = 92.10°$, $Z = 4$, $D = 2.664$. *12-712:* 10.5_{10} 5.64_1 5.12_1 3.47_2 3.32_4 2.87_1 2.60_3 2.40_1. Platy crystals and as cross-fiber veins. Colorless or white, white streak, vitreous to pearly luster. Cleavage {010}, perfect. H = 2. G = 2.66. Biaxial (+); $Z = b$; $X \wedge c = 52°$; $N_x = 1.551$, $N_y = 1.553$, $N_z = 1.621$; $2V = 37°$; $r > v$, weak. *Sterling borax mine, Tick Canyon, Los Angeles Co.*, and Billy mine, Death Valley, Inyo Co., CA; Penobsquis, NB, Canada; Aislaby, Yorkshire, England; Reyershausen, near Göttingen, Germany; Nepa salt deposit, Baikalia, Russia; Inder, Kazakhstan. BM *DII:348*, *AM 56:1934(1971)*.

26.5.10.1 Veatchite-A $Sr_2[B_5O_8(OH)]_2 \cdot B(OH)_3 \cdot H_2O$

Named in 1979 as a dimorph of veatchite. TRIC $A1$ or $A\bar{1}$. $a = 20.80$, $b = 11.72$, $c = 6.63$, $\alpha = 90°$, $\beta = 90.80°$, $\gamma = 91.95°$, $Z = 4$, $D = 2.77$. *30-1284:* 10.4_{10} 4.09_6 3.45_8 3.32_{10} 2.84_8 2.59_{10} 1.79_8 1.75_8. Nodules with euhedral crystals on the surface. Colorless or white, white streak, vitreous luster. Cleavage {100}, perfect; {011}, {01$\bar{1}$}, good. H = 2. G = 2.73. Biaxial (+); XZ = cb; $N_x = 1.549$, $N_y = 1.551$, $N_z = 1.621$; $2V = 25°$; $r < v$, strong. *Emet borate deposit, Kutahya, Turkey.* BM *AM 64:362(1979)*, *MD 19:217(1984)*.

26.5.11.1 Ulexite $NaCaB_5O_6(OH)_6 \cdot 5H_2O$

Named in 1850 for George L. Ulex (d.1883), German chemist, who determined its composition. TRIC $P\bar{1}$. $a = 8.816$, $b = 12.870$, $c = 6.678$, $\alpha = 90.36°$, $\beta = 109.05°$, $\gamma = 104.98°$, $Z = 2$, $D = 1.955$. *12-419:* 12.2_{10} 8.03_2 7.75_8 6.00_3 4.33_2 4.16_3 3.10_2 3.01_2. Nodules, sometimes loose-textured (cotton-balls), fibrous veins, rarely as needlelike crystals. Colorless or white, white streak, vitreous or silky luster. Cleavage {$\bar{1}$20}, perfect; {1$\bar{1}$0}, good. H = $2\frac{1}{2}$. G = 1.955. Biaxial (+); $X = b$; $Y \wedge c = 20°$; $N_x = 1.493$, $N_y = 1.505$, $N_z = 1.526$; $2V = 75°$. Decomposed by hot water. Occurs in salt playas and in sedimentary borate deposits. Columbus, Rhodes, and Teels Marshes, NV; Furnace Creek district, Inyo Co., Boron, Kern Co., Sterling borax mine, Los Angeles Co., Calico Mts., San Bernardino Co., CA; Fischell's Brook and Flat Bay, NFD, and Wentworth, Hillsborough and other localities in NB, Canada; Niederellenbach, Hesse, Germany; Donets Basin, Ukraine; Kirka, Turkey; Inder, Kazakhstan; Arequipa, Peru; *Iquique, Tarapaca, Chile*; salars in Salta and Jujuy Prov., Argentina. BM *DII:345*; *AM 44:712(1959)*, *63:160(1978)*; *CM 19:291(1981)*.

26.5.12.1 Probertite $NaCaB_5O_7(OH)_4 \cdot 3H_2O$

Named in 1929 for Frank H. Probert (1876–1940), University of California, who discovered the mineral. MON $P2_1/a$. $a = 13.43$, $b = 12.57$, $c = 6.59$, $\beta = 100.25°$, $Z = 4$, $D = 2.13$. *12-420:* 9.12_{10} 6.62_2 3.52_2 2.94_2 2.88_2 2.81_4 2.17_2 1.99_2. Acicular or platy crystals in rosettes or radial groups, also massive. Colorless or white, white streak, vitreous luster. Cleavage {110}, perfect. $H = 3\frac{1}{2}$. $G = 2.14$. Biaxial (+); $Y = b$; $Z \wedge c = 12°$; $N_x = 1.515$, $N_y = 1.525$, $N_z = 1.544$; $2V = 73°$; $r > v$. Partly decomposed by hot water. Southard, Blaine Co., OK; *Boron, Kern Co.*, Furnace Creek district, Inyo Co., Sterling borax mine, Los Angeles Co., Calico Mts., San Bernardino Co., *CA*; Bakhmut Basin, Ukraine; Tincalayu, Salta Prov., Argentina. BM *DII:343, AM 44:712(1959), AC(B) 38:3072(1982)*.

26.5.12.2 Tuzlaite $NaCaB_5O_8(OH)_2 \cdot 3H_2O$

Named in 1994 for the locality. MON $P2_1/c$. $a = 6.506$, $b = 13.280$, $c = 11.462$, $\beta = 92.97°$, $Z = 4$, $D = 2.23$. *AM 79:* 8.64_{10} 6.62_3 5.71_1 5.23_1 4.18_2 2.87_3 2.59_2 2.41_1. Prismatic crystals to 0.5 mm, and veinlets. Colorless or white, white streak, silky to pearly luster. Cleavage {001}, perfect. $H = 2-3$. $G = 2.21$. Biaxial (+); $Y = b$; $Z \wedge a = 14°$; $N_x = 1.532$, $N_y = 1.544$, $N_z = 1.561$; $2V = 82°$. In dolomitic marl at the *Tuzla salt mine, Bosnia*. BM *AM 79:562(1994)*.

26.5.13.1 Kaliborite $HKMg_2B_{12}O_{16}(OH)_{10} \cdot 4H_2O$

Named in 1889 for the composition. MON $C2/c$. $a = 18.93$, $b = 8.62$, $c = 14.97$, $\beta = 99.9°$, $Z = 4$, $D = 2.110$. *18-669:* 7.22_{10} 3.84_5 3.77_5 3.36_5 3.10_7 2.49_7 2.35_5 2.07_5. Small crystals, also massive granular. Colorless or white, white streak, vitreous luster. Cleavage {001}, {$\bar{1}$01}, perfect; {100}, good. $H = 4$. $G = 2.116$. Biaxial (+); $Y = b$; $Z \wedge c = 65°$; $N_x = 1.508$, $N_y = 1.526$, $N_z = 1.550$; $2V = 81°$. Eagle Borax works, Furnace Creek, Inyo Co., CA; *Schmidtmannshall* and Leopoldshall, *Saxony, Germany*; Sallent, Spain; Monte Sambuco, Sicily; Inder(*), Kazakhstan. BM *DII:367, AM 50:1079(1965), SR 31A:172(1966)*.

26.5.14.1 Hilgardite $Ca_2B_5O_9Cl \cdot H_2O$

Named for Eugene W. Hilgard (1833–1916), German–U.S. geologist, one of the first to describe the Louisiana salt deposits. Hilgardite occurs in three polytypes, designated 4M, 1A, 3A (M, MON; A, TRIC; 1, 3, 4 are the number of formula units in the unit cell). Data are for the 4M polytype. The 3A polytype has been known as *parahilgardite*. MON *Aa*. $a = 11.438$, $b = 11.318$, $c = 6.318$, $\beta = 90.06°$, $Z = 4$, $D = 2.69$. *11-404:* 5.72_4 5.67_4 2.86_{10} 2.84_{10} 2.76_6 2.55_6 2.11_8 1.99_8. Tabular hemimorphic crystals to 2 cm. Colorless or white, white streak, vitreous luster. Cleavage {010}, {100}, perfect. $H = 5$. $G = 2.71$. Biaxial (+); $Y = b$; $Z \wedge c = 2°$; $N_x = 1.630$, $N_y = 1.636$, $N_z = 1.664$; $2V = 35°$. *Choctaw salt dome, Iberville Parish, LA*, in insoluble residues of rock salt; Esmeralda Co., NV; Penobsquis, Sussex, and Salt Springs, NB, Canada; Boulby salt mine, Yorkshire, England; Reyershausen, Germany; Angara, Irkutsk region, Russia; Inder, Kazakhstan. BM *DII:382; AM 64:187(1979), 68:604(1983), 70:636(1985)*.

26.5.15.1 Tyretskite $Ca_2B_5O_9(OH) \cdot H_2O$

Named in 1964 for the locality. TRIC $P\bar{1}$. $a = 6.44$, $b = 6.45$, $c = 6.41$, $\alpha = 118.23°$, $\beta = 119.75°$, $\gamma = 73.50°$, $Z = 1$, $D = 2.59$. *26-2:* 5.82_3 3.23_6 2.93_{10} 2.86_{10} 2.80_7 2.14_9 2.06_8 1.85_9. Small platy crystals and radiating fibrous aggregates. White to brown, white streak, vitreous luster. $G = 2.189$. Biaxial (+); $N_x = 1.637$, $N_y = 1.642$, $N_z = 1.670$; $2V = 46°$. *Tyret, eastern Siberia, Russia,* in drill core from salt deposit; Inder, Kazakhstan. BM *MA 17:500(1966); AM 53:2084(1968), 70:636(1985).*

26.5.16.1 Volkovskite $KCa_4B_{22}O_{32}(OH)_{10}Cl \cdot 4H_2O$

Named in 1966 for A. I. Volkovskaya, Russian mineralogist, who found the mineral. TRIC $P1$. $a = 6.575$, $b = 23.921$, $c = 6.522$, $\alpha = 90.58°$, $\beta = 119.10°$, $\gamma = 95.56°$, $Z = 1$, $D = 2.28$. *18-1460:* 8.1_{10} 6.03_6 5.42_6 3.28_9 2.81_7 2.63_8 2.15_7 1.98_7. Thin plates to 1.5 mm. Colorless to pink, white streak, vitreous luster. Cleavage $\{010\}$, perfect; $\{001\}$, good. Brittle. $H = 2\frac{1}{2}$. $G = 2.27$. Biaxial (+); $Y = b$; $Z \wedge a = 31°$; $N_x = 1.536$, $N_y = 1.539$, $N_z = 1.603$; $2V = 24°$. Clarke Co., AL; Sussex, NB, Canada; *Nepa deposit, eastern Siberia, Russia.* BM *ZVMO 95:45(1966), AM 51:1550(1966), CM 28:351(1990).*

26.5.17.1 Pringleite $Ca_9B_{26}O_{34}(OH)_{24}Cl_4 \cdot 13H_2O$

Named in 1993 for Gordon J. Pringle, Canadian mineralogist. Dimorphous with ruitenbergite. TRIC $P1$. $a = 12.759$, $b = 13.060$, $c = 9.733$, $\alpha = 102.14°$, $\beta = 102.03°$, $\gamma = 85.68°$, $Z = 1$, $D = 2.11$. *CM 31:* 9.21_7 7.69_{10} 5.74_6 4.63_4 3.85_4 2.80_3 2.20_3 2.06_3. Subhedral platy or prismatic crystals to 2 mm. Colorless or very pale yellow, white streak, vitreous luster. Cleavage $\{110\}$, good. Brittle. $H = 3-4$. $G = 2.22$. Biaxial (+); $X = c$; $Y \wedge a = 40°$; $N_x = 1.537$, $N_y = 1.548$, $N_z = 1.570$; $2V = 77°$; $r < v$, strong. *Sussex, NB, Canada,* in salt deposit. BM *CM 31:795(1993), 32:1(1994).*

26.5.17.2 Ruitenbergite $Ca_9B_{26}O_{34}(OH)_{24}Cl_4 \cdot 13H_2O$

Named in 1993 for Arie A. Ruitenberg, Canadian geologist. Dimorphous with pringleite (26.5.17.1). MON $P2_1$. $a = 19.88$, $b = 9.715$, $c = 17.551$, $\beta = 114.85°$, $Z = 2$, $D = 2.13$. *CM 31:* 9.03_6 8.56_{10} 6.62_7 6.14_3 5.12_3 3.79_3 3.49_3 2.89_3. Anhedral 7 mm × 4 mm grain. Colorless, white streak, vitreous luster. Cleavage $\{100\}$, good. $H = 3-4$. Biaxial (+); $X = b$; $Y \wedge a = 25°$; $N_x = 1.542$, $N_y = 1.545$, $N_z = 1.565$; $2V = 47°$. *Sussex, NB, Canada,* in salt deposit. BM *CM 31:795(1993), 32:1(1994).*

26.6.1.1 Rivadavite $Na_6Mg[B_6O_7(OH)_6]_4 \cdot 10H_2O$

Named in 1967 for Bernardino Rivadavia (1780–1845), first president of Argentina. MON $P2_1/c$. $a = 15.870$, $b = 8.010$, $c = 22.256$, $\beta = 116.43°$, $Z = 2$, $D = 1.910$. *19-1211:* 14.2_{10} 7.59_{10} 6.23_5 5.34_5 3.25_7 2.95_7 2.85_5 2.45_3. Nodular masses, aggregates of platy crystals to 3 mm. Colorless or white, white streak, vitreous luster. Cleavage $\{100\}$, $\{\bar{1}01\}$, perfect; $\{010\}$, poor. $H = 3\frac{1}{2}$.

26.6.5.1 Aristarainite

$G = 1.905$. Biaxial (+); $Y = b$; $Z \wedge a = 32°$; $N_x = 1.470$, $N_y = 1.481$, $N_z = 1.497$; $2V = 80°$. Furnace Creek district, Inyo Co., CA; *Tincalayu borate deposit, Salta Prov., Argentina*. BM *AM 52:326(1967), NW 60:350(1973)*.

26.6.2.1 McAllisterite $Mg_2[B_6O_7(OH)_6]_2 \cdot 9H_2O$

Named in 1965 for James F. McAllister (b.1911), U.S. geologist, who discovered the mineral. Dimorphous with admontite (26.6.3.1). HEX-R $R3c$. $a = 11.546$, $c = 35.562$, $Z = 6$, $D = 1.864$. *18-767:* 8.72_{10} 6.65_3 5.77_5 4.36_4 4.06_5 3.35_5 3.26_5 2.81_5. Equant rhombohedral crystals to 1 mm, and efflorescences. White to amber, white streak, vitreous luster. Cleavage {0001}, {01$\bar{1}$2}, good. $H = 2\frac{1}{2}$. $G = 1.868$. Uniaxial (−); $N_O = 1.507$, $N_E = 1.465$. Sterling Hill, NJ; *Furnace Creek district, Inyo Co., CA*; Qinghai–Xizaig Plateau, China; Tincalayu borate deposit, Salta Prov., Argentina. BM *AM 50:629(1965), 52:1776(1967); SR 41A:421(1975)*.

26.6.3.1 Admontite $Mg_2[B_6O_7(OH)_6]_2 \cdot 9H_2O$

Named in 1979 for the locality. Dimorphous with mcallisterite (26.6.2.1). MON $P2_1/c$. $a = 12.68$, $b = 10.07$, $c = 11.32$, $\beta = 109.68°$, $Z = 2$, $D = 1.875$. *34-1438:* 12.1_9 7.60_{10} 5.72_4 5.29_7 3.93_8 3.09_6 2.68_9 2.00_2. Microscopic platy crystals. Colorless, white streak, vitreous luster. No cleavage, conchoidal fracture. $H = 2$–3. $G = 1.82$. Biaxial (−); $X \wedge c = 45°$; $N_x = 1.442$, $N_y = 1.500$, $N_z = 1.504$; $2V = 30°$; $r < v$. *Near Admont, Styria, Austria*, in a gypsum deposit. BM *TMPM 26:69(1979), AM 65:205(1980)*.

26.6.4.1 Aksaite $MgB_6O_7(OH)_6 \cdot 2H_2O$

Named in 1962 for the locality. ORTH *Pbca*. $a = 12.540$, $b = 24.327$, $c = 7.480$, $Z = 8$, $D = 1.975$. *15-654:* 6.4_{10} 6.3_4 6.1_5 4.72_5 4.37_3 3.59_4 3.19_5 3.11_5. Elongated crystals to 7 mm. Colorless to pale gray, white streak, vitreous luster. Cleavage {100}, {010}. $H = 2\frac{1}{2}$. $G = 1.99$. Biaxial (−); $XZ = cb$; $N_x = 1.472$, $N_y = 1.503$, $N_z = 1.526$; $2V = 80°$. *Ak-Sai, Kazakhstan*, in salt deposit. BM *ZVMO 91:447(1962); AM 48:209,930(1963), 56:1553(1971)*.

26.6.5.1 Aristarainite $Na_2Mg[B_6O_8(OH)_4]_2 \cdot 4H_2O$

Named in 1974 for Lorenzo F. Aristarain (b.1926), Argentine mineralogist. MON $P2_1/a$. $a = 18.886$, $b = 7.521$, $c = 7.815$, $\beta = 97.72°$, $Z = 2$, $D = 2.102$. *26-1379:* 7.74_{10} 5.40_1 3.87_1 3.78_1 3.12_1 3.04_1 2.58_2 2.40_1. Platy crystals to 1 mm. Colorless, white streak, vitreous luster. Cleavage {001}, {100}, perfect; {110}, fair. $H = 3\frac{1}{2}$. $G = 2.027$. Biaxial (−); $Y \wedge c = 38°$, $Z \wedge a = 46°$; $N_x = 1.484$, $N_y = 1.498$, $N_z = 1.523$; $2V = 70°$; $r > v$, weak. Fluoresces cream-white in SWUV. *Tincalayu borate deposit, Salta Prov., Argentina*. BM *AM 59:647(1974), 62:979(1977)*.

26.6.6.1 Nobleite $CaB_6O_9(OH)_2 \cdot 3H_2O$

Named in 1961 for Levi F. Noble (1882–1965), U.S. geologist. MON $P2_1/a$. $a = 14.56$, $b = 8.016$, $c = 9.838$, $\beta = 111.75°$, $Z = 4$, $D = 2.098$. *13-243:* 6.79_{10} 5.18_1 3.94_1 3.39_3 3.12_1 2.57_1 2.31_1 2.10_1. Platy crystals to 3 mm and fine-grained aggregates. Colorless or white, white streak, vitreous luster. Cleavage {100}, perfect; {001}, indistinct; sectile and flexible crystals; inelastic. H = 3. G = 2.09. Biaxial (+); Y = b; Z ∧ c = 7°; $N_x = 1.500$, $N_y = 1.520$, $N_z = 1.554$; 2V = 76°; r > v, weak. *Furnace Creek district, Inyo Co., CA*; Sijes, Salta Prov., Argentina. BM *AM 46:560(1961)*.

26.6.6.2 Tunellite $SrB_6O_9(OH)_2 \cdot 3H_2O$

Named in 1961 for George Tunell (1900–1996), U.S. mineralogist. Contains up to 15% BaO. MON $P2_1/a$. $a = 14.390$, $b = 8.213$, $c = 9.934$, $\beta = 114.03°$, $Z = 4$, $D = 2.381$. *14-616:* 6.97_3 6.57_{10} 6.21_2 5.14_3 4.53_2 3.87_2 3.52_2 3.05_2. Prismatic and tabular crystals to 9 cm and nodules of radiating prisms. Colorless, white streak, subvitreous to pearly luster. Cleavage {100}, perfect; {001}, distinct. H = $2\frac{1}{2}$. G = 2.40–2.46, increasing with Ba content. Biaxial (+); Y = b; Z ∧ c = 5°; $N_x = 1.519$, $N_y = 1.534$, $N_z = 1.569$; 2V = 68°; r > v, weak. *Furnace Creek district, Inyo Co., Boron(*), Kern Co., CA*; Kirka, Turkey. BM *AM 47:416(1962), 49:1549(1964)*.

26.6.7.1 Ginorite $Ca_2B_{14}O_{20}(OH)_6 \cdot 5H_2O$

Named in 1934 for Piero Ginori Conti (1865–1939), a leader in the development of the Tuscan borax industry. MON $P2_1/c$. $a = 13.375$, $b = 14.368$, $c = 12.261$, $\beta = 101.2°$, $Z = 4$, $D = 2.14$. *8-116:* 7.18_{10} 5.36_8 4.68_4 3.90_5 3.57_6 3.28_6 3.12_5 2.09_8. Platy crystals and dense masses. Colorless or white, white streak, vitreous luster. Cleavage {010}, good. H = $3\frac{1}{2}$. G = 2.09. Biaxial (+); Y = b; Z ∧ c = 39°; $N_x = 1.517$, $N_y = 1.525$, $N_z = 1.579$; 2V = 42°. Furnace Creek district, Inyo Co., CA; Windsor, NS, Canada; *Sasso Pisano, Tuscany, Italy*; Caspian Sea region, Russia. BM *DII:364, MM 29:955(1952), CM 19:291(1981)*.

26.6.7.2 Strontioginorite $SrCaB_{14}O_{20}(OH)_6 \cdot 5H_2O$

Named in 1959 for the composition and relationship to ginorite. MON $P2_1/a$. $a = 12.850$, $b = 14.48$, $c = 12.845$, $\beta = 101.6°$, $Z = 4$, $D = 2.265$. *13-137:* 7.25_{10} 5.40_6 4.75_4 3.92_6 3.34_6 3.15_4 2.10_8 1.19_6. Tabular crystals to 3 mm. Colorless, white streak, silky luster. Cleavage {010}, perfect; {001}, good. H = 3. G = 2.25. Biaxial (+); Y = b; Z ∧ c = 40°; $N_x = 1.512$, $N_y = 1.524$, $N_z = 1.577$; 2V = 52°. Furnace Creek district, Inyo Co., CA; *Reyershausen, near Göttingen, Germany*; Inder, Kazakhstan. BM *CMP 6:366(1959); AM 45:478(1960), 55:1911(1970); USGSJR 2:699(1974)*.

26.7.1.1 Korzhinskite $CaB_2O_4 \cdot H_2O$

Named in 1963 for Dimitri S. Korzhinsky (1899–1985), Russian geochemist. Probably ORTH Space group unknown. *16-366:* 3.11_7 2.81_7 2.21_3 2.02_{10} 1.93_3 1.91_5 1.76_4 1.37_2. Lamellar aggregates of prismatic crystals. Colorless, white streak. Cleavage in direction of elongation. Biaxial (+); $N_x = 1.642$, $N_y = 1.647$, $N_z = 1.672$; $2V = 44°$; straight extinction. *Novofrolovo mine, Turlinsk deposit, Ural Mts., Russia*, in skarn. BM *ZVMO 92:555(1963)*, *AM 49:441(1964)*.

26.7.2.1 Chelkarite $CaMgB_2O_4Cl_2 \cdot 7H_2O$

Named in 1968 for the locality. ORTH *Pbca*. $a = 13.69$, $b = 20.84$, $c = 8.26$, $Z = 10$, $D = 2.44$. *27-72:* 10.4_9 6.68_5 4.96_7 3.53_{10} 2.75_4 2.58_4 2.21_8 2.03_7. Prismatic crystals to 1.5 cm. Colorless to pale pink, white streak, vitreous luster. $G = 2.21$. Biaxial (+); $N_x = 1.520$, $N_z = 1.558$. *Chelkar, Kazakhstan*, in salt deposit. BM *AM 56:1122(1971)*.

26.7.3.1 Ekaterinite $Ca_2B_4O_7(Cl,OH)_2 \cdot 2H_2O$

Named in 1980 for Ekaterina V. Rozhkova (1898–1979), Russian mineralogist. HEX *P6*. $a = 11.86$, $c = 23.88$, $D = 2.23$. *33-270:* 11.9_7 2.51_9 2.31_{10} 2.09_{10} 2.05_{10} 1.92_{10} 1.83_8 1.28_9. Foliated aggregates of hexagonal crystals. White, white streak, pearly luster. $H = 1$. $G = 2.44$. Biaxial (−); $N_x = 1.574$, $N_y = N_z = 1.577$; 2V very small. *Korshunov skarn iron-ore deposit, Lower Ilim region, Irkutsk district, Russia*. BM *ZVMO 109:469(1980)*, *AM 66:437(1981)*.

26.7.4.1 Satimolite $KNa_2Al_4(B_2O_5)_3Cl_3 \cdot 13H_2O$

Named in 1969, origin not given. ORTH Space group unknown. $a = 12.62$, $b = 18.64$, $c = 6.97$, $Z = 2$, $D = 1.70$. *25-1350:* 9.5_9 6.3_9 4.01_9 3.51_8 3.20_{10} 2.44_8 1.97_8 1.94_7. Rounded nodules. White, white streak. $G = 2.1$. Biaxial (−); $N_x = 1.535$, $N_y = 1.552$, $N_z = 1.553$; 2V very small. *Chelkar, Kazakhstan*, in salt deposit. BM *AM 55:1069(1970)*.

26.7.5.1 Tertschite $Ca_4B_{10}O_{19} \cdot 20H_2O$

Named in 1953 for Hermann Tertsch (1880–1962), Austrian mineralogist. Probably MON Space group unknown. *AM-39:* 3.12 2.83 2.35 2.16 2.02 1.93 (strongest diffraction lines). Fibrous masses. White, white streak, silky luster. Biaxial; $Z \wedge c = 30°$; $N_x = 1.502$, $N_z = 1.517$. Fluoresces blue-violet in UV. *Kurtpinari mine, Faras, Turkey*. BM *CMP 3:443(1953)*, *AM 39:849(1954)*.

26.7.6.1 Braitschite-(Ce) $(Ca,Na_2)_7(Ce,La)_2B_{22}O_{43} \cdot 7H_2O$

Named in 1968 for Otto Braitsch (1921–1966), German mineralogist. HEX Space group unknown. $a = 12.156$, $c = 7.377$, $Z = 1$, $D = 2.84$. *21-158:* 10.5_6 4.28_{10} 3.17_5 3.16_4 3.02_9 2.81_6 2.14_3 1.91_4. Microscopic hexagonal plates and as nodules. Colorless, white, and pink, white streak, vitreous luster. $G = 2.903$.

Uniaxial (+); $N_O = 1.646$, $N_E = 1.647$. *Cane Creek potash mine, near Moab, Grand Co., UT.* BM *AM 53:1081(1968)*.

26.7.7.1 Wardsmithite $Ca_5Mg(B_4O_7)_6 \cdot 30H_2O$

Named in 1970 for Ward C. Smith (b.1906), U.S. geologist. Probably HEX Space group unknown. *23-120:* 13.5_{10} 12.3_6 7.43_2 6.73_2 6.12_6 4.72_4 3.36_5 2.74_3. Microscopic hexagonal plates and fine-grained nodules. Colorless or white, white streak, vitreous luster. Cleavage {0001}, good. $H = 2\frac{1}{2}$. $G = 1.88$. Uniaxial (−); $N_O = 1.490$, $N_E = 1.476$. *Furnace Creek district, Inyo Co., and Boron, Kern Co., CA.* BM *AM 53:349(1970)*.

26.7.8.1 Studenitsite $NaCa_2B_9O_{14}(OH)_4 \cdot 2H_2O$

Named in 1995 for the Studenitsa monastery, near the type locality. MON $P2_1/c$. $a = 11.499$, $b = 12.588$, $c = 10.530$, $\beta = 99.42°$, $Z = 4$, $D = 2.34$. *ZVMO 124:* 5.41_7 5.20_6 4.20_6 3.35_9 3.27_6 3.04_{10} 2.21_6 1.96_4. Wedge-shaped crystals to 5 mm. Colorless to pale yellow, white streak, vitreous luster. No cleavage. $H = 5\frac{1}{2}$. $G = 2.29$. Biaxial (+); $ZX = ab$; $Y \wedge c = 10°$; $N_x = 1.532$, $N_y = 1.538$, $N_z = 1.564$; $2V = 54°$; $r > v$. Insoluble in water, soluble in dilute HCl. Occurs in a borate lens in sedimentary clays in *Pobrdzhsk Potok and Piskan deposits on the Ibar R., southern Serbia.* BM *ZVMO 124(3):57(1995)*.

= Class 27 =

Compound Borates

27.1.1.1 Carboborite $Ca_2Mg[B(OH)_4]_2(CO_3)_2 \cdot 4H_2O$

Named in 1964 for the composition. MON $P2_1/n$. $a = 11.011$, $b = 6.674$, $c = 10.692$, $\beta = 116.64°$, $Z = 2$, $D = 2.15$. *17-529:* 5.63_{10} 4.86_5 4.32_{10} 3.20_5 3.14_8 2.73_7 2.44_6 2.16_8. Crystals resembling steep rhombohedrons to 3 mm. Colorless, white streak, vitreous luster. Cleavage {100}, perfect; {$\bar{1}11$}, distinct. $H = 2$. $G = 2.12$. Biaxial (−); $Y = b$; $Z \wedge c = 12°$; $N_x = 1.507$, $N_y = 1.546$, $N_z = 1.569$; $2V = 75°$. Fluoresces white in UV. Furnace Creek area, Inyo Co., CA; *Qinghai–Xizaig Plateau, China*, in saline lakes. BM *SS 13:813(1964), AM 50:262(1965), BM 104:578(1981)*.

27.1.2.1 Gaudefroyite $Ca_4Mn_3(BO_3)_3(CO_3)O_3$

Named in 1964 for Christophe Gaudefroy (b.1878), French mineralogist. HEX $P6_3$. $a = 10.61$, $c = 5.879$, $Z = 2$, $D = 3.44$. *17-154:* 9.1_8 4.89_6 4.54_8 2.95_{10} 2.69_8 2.62_{10} 2.46_{10} 2.31_6. Hexagonal prisms to 5 cm. Black, brown streak, submetallic luster. Conchoidal fracture. $H = 6$. $G = 3.44$. Uniaxial (+); $N_O = 1.81$, $N_E = 2.02$. *Tachgagalt manganese mine, near Ouarzazarte, Morocco*; Wesels and N'Chwaning mines, Cape Prov., South Africa. BM *BM 87:216(1964), AM 50:806(1965), MR 22:285(1991), NJMM:385(1993)*.

27.1.3.1 Borcarite $Ca_4Mg[B_4O_6(OH)_6](CO_3)_2$

Named in 1965 for the composition. MON $C2/m$. $a = 17.840$, $b = 8.380$, $c = 4.445$, $\beta = 102.04°$, $Z = 2$, $D = 2.790$. *22-532:* 2.90_8 2.72_7 2.66_{10} 2.38_5 2.26_6 2.21_6 1.88_9 1.70_6. Massive. Blue to green, white streak, vitreous to pearly luster. Cleavage {100}, {110}, perfect. $H = 4$. $G = 2.77$. Biaxial (−); $Z = b$; $Y \wedge c = 28°$; $N_x = 1.590$, $N_y = 1.651$, $N_z = 1.657$; $2V = 30°$; $r < v$. *Solongo, Transbaikalia, Russia*, in skarn. BM *ZVMO 94:180(1965), AM 50:2097(1965), DANS 225:823(1976), MM 59:297(1995)*.

27.1.4.1 Sakhaite $Ca_3Mg(BO_3)_2(CO_3) \cdot 0.36H_2O$

Named in 1966 for the locality, *Sakha* being the name for Siberia in the Yakut language. ISO $F4_132$. $a = 14.685$, $Z = 16$, $D = 2.79$. *19-1112:* 8.4_1 5.16_2 3.35_1 2.81_1 2.58_{10} 2.11_6 1.83_2 1.49_1. Octahedral crystals to 8 mm and massive. White to gray, white streak, vitreous luster. No cleavage. $H = 5$. $G = 2.78–2.88$. Isotropic; $N = 1.638–1.642$. *Solongo, Transbaikalia, Russia*, in skarn; Kombat mine, near Tsumeb, Namibia. BM *ZVMO 95:193(1966)*, *AM 51:1817(1966)*, *DANS 239:1103(1978)*, *MM 54:105(1990)*.

27.1.5.1 Sulfoborite $Mg_3[B(OH)_4]_2(SO_4)(OH,F)_2$

Named in 1893 for the composition. ORTH *Pnma*. $a = 10.132$, $b = 12.537$, $c = 7.775$, $Z = 4$, $D = 2.37$. *14-639:* 5.55_4 3.47_{10} 3.13_8 3.09_{10} 2.97_7 2.43_5 2.05_{10} 1.90_9. Prismatic crystals. Colorless, white streak, vitreous luster. Cleavage {110}, good; {001}, fair. $H = 4-4\frac{1}{2}$. $G = 2.38-2.45$. Biaxial $(-)$; $XY = ab$; $N_x = 1.522$, $N_y = 1.540$, $N_z = 1.552$; $2V = 70-88°$. Occurs in salt deposits. Yorkshire, England; Wittmar, Brunswick, and *Westeregeln, Saxony, Germany*; Inder(*), Kazakhstan. BM *DII:387*, *MM 34:460(1965)*, *DANS 228:1076(1976)*, *AM 68:255(1983)*.

27.1.6.1 Teruggite $Ca_4Mg[AsB_6O_{11}(OH)_6]_2 \cdot 14H_2O$

Named in 1968 for Mario E. Teruggi (b.1919), Argentine geologist. MON $P2_1/a$. $a = 15.675$, $b = 19.920$, $c = 6.255$, $\beta = 99.33°$, $Z = 2$, $D = 2.19$. *21-150:* 12.1_{10} 9.98_2 8.37_2 4.65_2 3.85_2 3.58_2 2.79_3. Rounded nodules of microscopic acicular crystals. Colorless or white, white streak, vitreous luster. Cleavage {001}, good; {110}, fair. $H = 2\frac{1}{2}$. $G = 2.20$. Biaxial $(+)$; $Z = b$; $X \wedge c = 26°$; $N_x = 1.526$, $N_y = 1.528$, $N_z = 1.551$; $2V = 33°$; $r > v$, weak. Emet mine, Kütahya, Turkey, *Loma Blanca deposit, Jujuy Prov., Argentina*. BM *AM 53:1815(1968)*, *58:1034(1973)*.

27.1.7.1 Garrelsite $Ba_3NaSi_2B_7O_{16}(OH)_4$

Named in 1955 for Robert M. Garrels (1916–1988), U.S. geochemist. MON $C2/c$. $a = 14.639$, $b = 8.466$, $c = 14.438$, $\beta = 114.21°$, $Z = 4$, $D = 3.88$. *26-1369:* 6.13_3 4.23_3 3.94_3 3.64_8 3.05_{10} 2.87_5 2.76_5 2.03_6. Bipyramidal crystals to 3 mm. Colorless, white streak, vitreous luster. $H \sim 6$. $G = 3.68$. Biaxial $(-)$; $Z = b$; $N_x = 1.620$, $N_y = 1.633$, $N_z = 1.640$; $2V = 55°$. *South Ouray No. 1 well, Uintah Co., UT*, in bore core from dolomitic shale; Boron, Kern Co., and Searles Lake, San Bernardino Co., CA. BM *BGSA 66:1597(1955)*, *NW 60:349(1973)*, *USGSJR 2:213(1974)*.

27.1.8.1 Iquiqueite $Na_4K_3Mg(CrO_4)B_{24}O_{39}(OH) \cdot 12H_2O$

Named in 1986 for the locality. HEX $P\overline{3}1c$. $a = 11.637$, $c = 30.158$, $Z = 3$, $D = 2.07$. *39-358:* 10.1_9 6.04_9 3.28_9 3.22_9 3.02_{10} 2.91_8 2.89_8 2.86_{10}. Microscopic hexagonal plates and crystal aggregates. Yellow, yellow streak, vitreous luster. Cleavage {0001}, perfect; {10$\overline{1}$0}, imperfect. $H = 2$. $G = 2.05$. Uniaxial $(-)$; $N_O = 1.502$, $N_E = 1.447$. *Iquique, Tarapaca Prov., Chile*, disseminated in nitrate ore. BM *AM 71:830(1986)*.

27.1.9.1 Moydite-(Y) $YB(OH)_4CO_3$

Named in 1986 for Louis Moyd (b.1916), Canadian mineralogist. ORTH *Pbca*. $a = 9.080$, $b = 12.222$, $c = 8.911$, $Z = 8$, $D = 3.01$. *40-508:* 6.11_{10} 4.50_9 3.18_7

3.05_3 2.82_5 2.75_3 2.53_4 1.86_6. Platy crystals to 1 mm. Yellow, white streak, vitreous luster. Cleavage {010}, good; {101}, poor. Soft. G = 3.13. Biaxial (−); XY = ab; N_x = 1.588, N_y = 1.681, N_z = 1.690; 2V = 32°; r > v. *Evans-Lou granite pegmatite, near Wakefield, QUE, Canada.* BM *CM 24:665(1986)*.

27.1.10.1 Wiserite $(Mn,Mg)_{14}(B_2O_5)_4(SiO_4)(OH,Cl)_8$

Named in 1845 for David F. Wiser (1802–1878), Swiss mineralogist. TET $P4/n$. $a = 20.192$, $c = 3.281$, Z = 2, D = 3.57. *13-593:* 14.2_{10} 6.40_6 5.08_4 4.78_4 3.36_6 2.87_6 2.53_8 2.35_6. Prismatic crystals to 1 cm or fibrous. White to pink or brown, white streak, vitreous or silky luster. H = $2\frac{1}{2}$. G = 3.54. Uniaxial (−); N_O = 1.751, N_E = 1.700; weakly pleochroic, colorless to brown. *Gonzen, Switzerland;* Japan (17 localities); Kombat mine, near Tsumeb, Namibia. BM *DII:245, SMPM 39:85(1959), AM 74:1351,1374(1989)*.

Class 28

Anhydrous Acid and Sulfates

28.1.1.1   Mercallite $KHSO_4$

Named in 1935 for Giuseppe Mercalli (1850–1914), Italian geologist, director of the Vesuvius Observatory. ORTH $Pbca$. $a = 8.412$, $b = 9.800$, $c = 19.957$, $Z = 16$. Structure: $AC(B)$ $31:302(1977)$. $11\text{-}649$(syn): 3.87_7 3.84_{10} 3.52_9 3.41_9 3.26_9 3.03_7 2.74_3 2.37_3. Stalactites composed of minute tabular crystals. Colorless, also sky blue due to admixed copper salt; vitreous luster. No cleavage. $G = 2.33$. Biaxial (+); XYZ $= bca$; $N_x = 1.445$, $N_y = 1.460$, $N_z = 1.491$; $2V = 56°$. Found as stalactites in fumaroles in the crater of *Vesuvius, Italy*, admixed with misenite, hieratite, and carobbiite. RG *DII:395*.

28.1.2.1   Misenite $K_8H_6(SO_4)_7$ (?)

Named in 1849 for the locality. MON Space group unknown. $a = 23.00$, $b = 7.104$, $c = 8.710$, $\beta = 102.66°$. $41\text{-}1363$: 4.92_4 4.83_4 4.35_2 4.24_{10} 3.95_3 3.65_6 3.49_3 2.89_3. Fibrous masses and aggregates of needle- or lathlike crystals, elongated [100] and flattened {001}. Colorless to grayish white, pearly or silky luster. Cleavage {010}, distinct. $G = 2.32$. Biaxial (+); $Z = b$; $X \wedge c = 29°$; $N_x = 1.475$, $N_y = 1.480$, $N_z = 1.487$; $2V = 81°$. Found with potassium alum, alunogen, and possibly arcanite as a transient efflorescence in a fumarole on *Cape Miseno*, near *Naples, Italy*. RG *DII:396*.

28.1.3.1   Letovicite $(NH_4)_3H(SO_4)_2$

Named in 1932 for the locality. TRIC $P1$ or $P\bar{1}$. $a = 5.87$, $b = 10.17$, $c = 8.27$, $\alpha = 101.10°$, $\beta = 111.10°$, $\gamma = 89.90°$, $Z = 2$, $D = 1.82$. $35\text{-}1500$: 4.98_8 4.95_8 4.65_4 3.81_3 3.77_6 3.39_{10} 3.36_4 2.93_8. Minute pseudohexagonal plates on {001}; granular masses; lamellar twinning common. Colorless. Cleavage {001}, distinct; uneven fracture. $G = 1.83$. Biaxial (−); $Z = b$; $X \wedge c = -12°$; $N_x = 1.501$, $N_y = 1.516$, $N_z = 1.525$ (Na); $2V = 75°$. Formed during the burning of waste coal heaps at *Letovice, Moravia, Czech Republic*, and at Kladno, Bohemia, Czech Republic of similar origin. RG *DII:397*, MA *73:4375*.

28.2.1.1   Mascagnite $(NH_4)_2SO_4$

Named in 1779 for Paolo Mascagni (1752–1815), Italian anatomist, University of Siena, who first described the natural salt. ORTH $Pnma$. $a = 7.78$, $b = 10.36$, $c = 5.99$, $Z = 4$, $D = 1.765$. $10\text{-}343$: 5.22_3 4.39_7 4.33_{10} 3.89_4 3.14_3

3.12_3 3.06_6 3.00_3. Crystals equant to short prismatic [100]; usually as stalactites or crusts; twinning on {011}. Colorless to gray or yellow, vitreous to dull luster. Cleavage {100}, good; uneven fracture; slightly sectile. $H = 2-2\frac{1}{2}$. $G = 1.768$. Biaxial (+); $XYZ = cba$; $N_x = 1.521$, $N_y = 1.532$, $N_z = 1.539$ (Na); $2V = 52°$; $r > v$, weak. A complete solid-solution series exists between mascagnite and arcanite, K_2SO_4. A compound in this series containing about 16 mol % $(NH_4)_2SO_4$ has been named taylorite, and was formed as a leachate of guano; see arcanite (28.2.1.2). Soluble in water; tastes sharp and bitter; readily fusible. Occurs as a sublimation product in fumaroles and from burning coal seams. The Geysers, Sonoma Co., CA, with boussingaultite and ammonioalunite. Commentry, France, from burning coal seams; *Vesuvius*, and Etna, *Italy*, in fumaroles; also from the solfatara of Tuscany; Nyamlagira volcano, Zaire. RG *DII:398*, *AM 36:591(1951)*, *PM 54:32(1985)*.

28.2.1.2 Arcanite K_2SO_4

Named in antiquity from the mediaeval Latin alchemical name for the compound *arcanum duplicatum*, double secret. Synonym: taylorite, ammonian arcanite, named in 1892 for W. J. Taylor (1833–1864), Philadelphia mineral chemist. ORTH *Pnma*. $a = 7.476$, $b = 10.071$, $c = 5.763$, $Z = 4$, $D = 2.70$. *AC(B) 28:2845(1972)*. 24-703: 4.16_4 3.74_2 3.00_9 2.90_{10} 2.89_8 2.42_3 2.23_2 2.08_3. Crystals are thin tablets which are cyclic twins of six individuals bounded by {100}, {110}, and {211}; as crusts, lumps, or concentric with crystalline structure. Colorless to white. Cleavage {010} and {100}, good. $G = 2.66$. Biaxial (+); $XYZ = cab$; $N_x = 1.494$, $N_y = 1.495$, $N_z = 1.497$ (Na); $2V = 67°$; $r > v$, moderate. A complete solid-solution series exists between arcanite and mascagnite; taylorite contains only about 16 mol % $(NH_4)_2SO_4$, and the name should be dropped. Found in a pine railroad tie in the *Santa Ana mine, Trabuco Canyon, Orange Co., CA*. Taylorite was described from the guano beds of the Chincha Is., Peru. RG *DII:399*, *AM 36:592(1951)*, *NJMM:75(1973)*.

28.2.2.1 Aphthitalite $(K,Na)_3Na(SO_4)_2$

Named "aphthalose" in 1813 from the Greek for *unalterable* and *salt*, in allusion to its stability in air. Synonym: *glaserite*, after Christoph Glaser, seventeenth-century German chemist. HEX-R $P\bar{3}m1$. $a = 5.680$, $c = 7.309$, $Z = 1$, $D = 2.697$. *AC(B) 36:919(1980)*. 20-928: 4.09_3 3.67_2 2.94_8 2.84_{10} 2.44_2 2.33_1 2.04_5 1.66_1. Crystals tabular {0001}; as bladed aggregates, crusts, or massive; twins on {0001} or {11$\bar{2}$0}. Colorless to white; rarely pale shades of gray, blue, green, red due to impurities; white streak; vitreous to resinous luster. Cleavage {10$\bar{1}$1}, fair; {0001}, poor; conchoidal fracture. $G = 2.66-2.71$. Uniaxial (+); $N_O = 1.490$, $N_E = 1.496$ (for 68 mol % K_2SO_4). K_2SO_4 can vary in the range 44 to 75%. Soluble in water, with a saline, bitter taste. Formed in volcanoes as incrustations in fumaroles, and in lacustrine salt deposits. Carlsbad potash district, NM; Searles lake, San Bernardino Co., CA; Kilauea, HI. *Vesuvius*, and Etna, *Italy*. Stassfurt salt deposits, Germany, especially at

Douglashall. Kalusz, Galicia, Ukraine, as crystals embedded in picromerite. RG $DII:400$, $NJMM:75(1973)$, PM $54:34(1985)$.

28.2.3.1 Thenardite Na_2SO_4

Named in 1826 for Louis Jacques Thenard (1777–1857), French chemist, University of Paris. Five different polymorphs of Na_2SO_4 are known, but thenardite is the only naturally occurring one. ORTH $Fddd$. $a = 9.829$, $b = 12.302$, $c = 5.868$, $Z = 8$, $D = 2.659$. CM $13:186(1975)$. $37-1465$: 4.66_7 3.84_2 3.18_5 3.08_6 2.78_{10} 2.65_5 2.33_3 1.87_4. Crystals dipyramidal {111}; also tabular {010}. Common forms {010}, {011}, {101}, {111}, {311}. Large crystals common. Also as crusts and efflorescences. Twins on {110} and {011}. Colorless to white; also light shades of gray, yellow, brown, red; vitreous to resinous luster. Cleavage {010}, perfect; {101}, fair; uneven to hackly fracture. $H = 2\frac{1}{2}$–3. $G = 2.664$. Thenardite from Sodaville, Mineral Co., NV, fluoresces bright white under LWUV or SWUV, after which it shows greenish-white phosphorescence. Biaxial (+); $XYZ = cba$; $N_x = 1.464$, $N_y = 1.474$, $N_z = 1.485$ (Na); $2V = 82.5°$. Analyses conform closely to the formula. Traces of K, Mg, Ca, and Cl are probably due to admixture. Soluble in water. Tastes salty. Salt lakes and playas containing Na_2SO_4 are found in arid parts of most continents. Thenardite, along with other sulfates, carbonates, and halides, is deposited from these waters. Also found as an efflorescence in soils in alkali regions, especially in caliche, and as an incrustation on lavas and around fumaroles. In CA at Searles Lake, Death Valley, Inyo Co., and at Salton Sinks, Imperial Co. In AZ as extensive beds near Camp Verde, Yavapai Co., Esmeralda Co., NV. In the sodium sulfate lakes of western Canada. First reported from *Lago Espartinus*, near *Aranjuez, Madrid Province, Spain*, where extensive bedded deposits are found. Vesuvius and Etna, Italy. In Russia in the Volga region and in Central Asia. At the Natron Lakes NW of Cairo, Egypt, and in the Sudan. At several places in Chile and in the arid regions of the west coast of South America. Ross Island, Antarctica. RG $DII:404$, BM $95:529(1972)$, $NJMM:408(1978)$.

28.2.4.1 Gianellaite $Hg_4(SO_4)N_2$

Named in 1977 for Vincent Paul Gianella (1886–1983), Mackay School of Mines, University of Nevada. ISO $F\bar{4}3m$. $a = 9.521$, $Z = 4$, $D = 7.13$. $NJMM:119(1977)$. $29-907$: 5.51_8 2.87_{10} 2.74_{10} 2.37_7 2.18_6 1.83_6 1.68_7 1.43_7. Rosettes of flattened subhedral crystals, rarely as euhedral distorted octahedra. Straw yellow. $H = 3$. $G = 7.19$. Isotropic; $N = 2.085$. Not attacked by H_2SO_4 or concentrated HNO_3, decomposed by concentrated HCl at room temperature. Darkens when heated above 130°, and when exposed to light for several weeks. Turns white and volatilizes above 400°. *Mariposa mine, Terlingua, Brewster Co., TX*, associated with terlinguaite, calomel, montroydite, mercury, and cinnabar on fracture surfaces in limestone. RG AM $62:1057(1977)$.

BARITE GROUP

The barite group consists of seven isostructural minerals that contain a tetrahedral group, usually sulfate, and a variety of other elements including most often barium or lead. The formula is

$$ATO_4$$

or a multiple thereof where

A = Pb, Ba, Sr, K, Cs
TO_4 = a tetrahedral group = SO_4, CrO_4, SeO_4, BF_4, $[GeO_2(OH)_2]$

In $BaSO_4$, for example, the sulfate ions are regular tetrahedra arranged with the sulfur and two oxygen atoms lying on a mirror plane of the structure. The other two oxygen atoms are equidistant above and below the plane. The Ba ions also lie on this same mirror plane and are in 12-fold coordination with oxygens belonging to seven different sulfate groups.

Strontium may substitute for Ba, and a complete solid-solution series may exist between barite and celestine. However, varieties of barite containing more than a few percent Sr are rare. Ca may also substitute for Ba up to a maximum ratio Ca/Ba = 1:12. Pb rarely substitutes for Ba but may reach a Pb/Ba ratio of 1:4. A complete series barite–anglesite has not been found in nature. Rarely (Musonoi, Zaire) Se may substitute for some of the S.

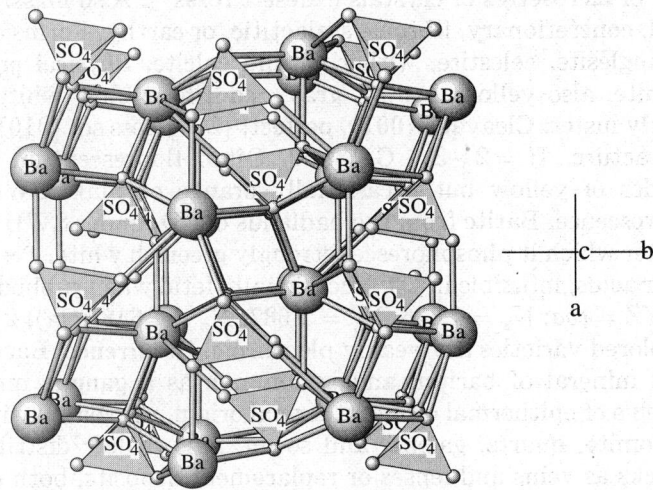

BARITE: $BaSO_4$
Perspective view down c axis.

Barite Group

Mineral	Formula	Space Group	a	b	c	D	
Avogadrite	(K,Cs)BF$_4$	*Pnma*	4.659	5.480	7.030	2.6	11.2.2.1
Barite	BaSO$_4$	*Pnma*	8.884	5.456	7.157	4.50	28.3.1.1
Celestine	SrSO$_4$	*Pnma*	8.371	5.355	6.870	3.982	28.3.1.2
Anglesite	PbSO$_4$	*Pnma*	8.482	5.398	6.959	6.36	28.3.1.3
Itolite	Pb$_3$[GeO$_2$(OH)$_2$](SO$_4$)$_2$	*Pnma*	8.47	5.38	6.94	6.67	30.2.6.1
Olsacherite	Pb$_2$(SO$_4$)(SeO$_4$)	$P222_1$	8.42	10.96	7.00	6.55	32.1.3.1
Hashemite	Ba(Cr,S)O$_4$	*Pnma*	9.103	5.528	7.331	4.52	35.3.3.1

Note: Anhydrite, CaSO$_4$ is structurally distinct from barite.

28.3.1.1 Barite BaSO$_4$

Named in 1640 from the Greek for *weight*. Barite group. Synonym: barytes, from the Greek for *heavy*. ORTH *Pnma*. $a = 8.884$, $b = 5.456$, $c = 7.157$, $Z = 4$, $D = 4.50$. *CM 15:522(1977)*. *24-1035*: 3.90_5 3.45_{10} 3.32_7 3.10_{10} 2.84_5 2.73_5 2.12_8 2.11_8. **Habit:** Often well crystallized, usually thin to thick tabular {001}, bounded by {210} alone or in combination with {101}, {011}, or other forms. Also flattened {001}, and elongated to prismatic [010] or [100]. Rarely prismatic [001]; equant. Other common forms are {211}, {010}, {100}, {201}, and {102}. Often forms aggregates of tabular crystals whose edges project into crestlike forms, or as rosettes of crystals ("desert roses"). Also massive, compact, laminated, concretionary, fibrous, stalactitic, or earthy. Forms epitaxial growths with anglesite, celestite, witherite, and calcite. **Physical properties:** Colorless or white; also yellow, brown, gray, reddish or blue; white streak; vitreous to pearly luster. Cleavage {001}, perfect; {210}, less so; {010}, imperfect; uneven fracture. $H = 2\frac{1}{2}-3\frac{1}{2}$. $G = 4.50$. Often fluorescent in LWUV, usually in shades of yellow but occasionally orange or pink; SWUV also may excite fluorescence. Barite from the badlands of SD, under SWUV, fluoresces yellow, after which it phosphoresces strongly greenish white. **Tests:** Insoluble in water or acids, infusible in CT, occasionally fetid when rubbed. **Optics:** Biaxial (+); XYZ = *cba*; $N_x = 1.636$, $N_y = 1.637$, $N_z = 1.648$ (Na); $2V = 37°$; r < v, weak. Colored varieties are weakly pleochroic. **Occurrence:** Barite is the most abundant mineral of barium and is common as a gangue mineral in metalliferous veins of epithermal or mesothermal origin, associated with fluorite, calcite, dolomite, quartz, galena, and so on. Also widely distributed in sedimentary rocks as veins and lenses or replacement deposits, both of hypogene and supergene origin. Also found in important amounts in residual deposits resulting from the weathering of limestones or dolomites. Occasionally, an important constituent of carbonatites. **Localities:** An early location of large crystals was in Cheshire, CT, while polished slabs were made from massive deposits in Jefferson Co., NY. Noteworthy residual deposits are found in southern Appalachia, especially GA, and in MO and AR. Beautiful yellow-brown crystals(!) line cavities in septaria near Elk Creek in the badlands of

28.3.1.1 Barite

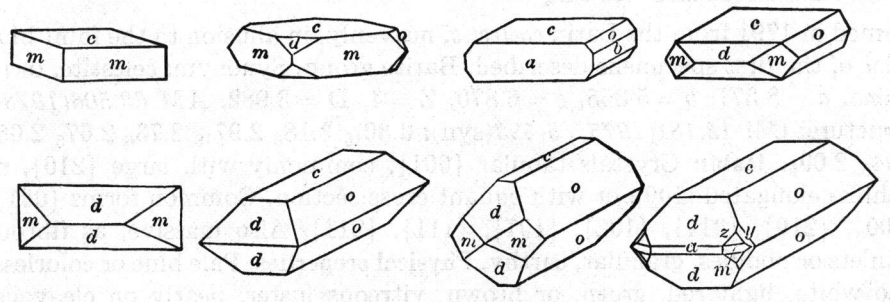

Barite
Forms: m{110}, c{001}, d{102}, o{011}, z{111}

SD. "Desert roses" and concretionary crystals containing much included sand occur at Norman in central OK. In Co as fine blue crystals(!) with calcite in veins in shale at Stoneham in Logan Co. and as brilliant, clear colorless or pale violet crystals from concretions in the Book Cliffs, Mesa Co.(!). Important commercial deposits are found near Battle Mt., Lander Co., NV, and in southeastern Tulare Co., CA. Fine crystals are found with copper ore in deep levels of the Magma mine, Superior, AZ(*). In Canada, barite is mined in NS, and good crystals are found in ONT at Galetta, and in QUE. Fine crystals(*) are found with fluorite at the Rock Candy mine near Grand Forks, BC. In Mexico large deposits are mined, especially in the states of MICH, PUE, COAH, and NL. Many localities for fine crystals are in England, especially at Alston Moor(!); Frizington(!); at the Pallaflat, Mowbray, and Dalmellington mines, Frizington, Cumbria(!), in superb blue, yellow, or amber elongated and flat large crystals and groups; and in the veins of Cornwall(*). Found at many places in Germany, often in fine crystallized specimens, as at Beihilfe mine, Halsbrücke, Saxony, and at Pöhla mine, Crottendorf. In France at La Cô-d'Abo, Puy de Dôme, in very large flawless amber prismatic crystals, and at Le Maine mine, near Autun, Saône et Loire. Moscona mine, Villabona, Oviedo, Asturias, Spain, in blue prismatic crystals, and large deposits are known in Castile and Andalusia. Silius, Cagliari, Sardinia, in large yellow and brown crystals. Baia Sprie and Cavnik, Maramures, Romania(!), associated with stibnite veins in nice, flawless, blue or red (inclusions of realgar) flat crystals. Plumbian varieties are found on Taiwan and at Shibukuro, Japan. In India from Madras in large, clear, colorless cleavages. Morocco at Taouz, Sahara, in extremely large (> 1.0 m) white elongated crystals and at M'Rit, Midelt, in colorless tabular crystals. Mashamba West, Zaire, in flat gemmy orange crystals. From Brazil in carbonatite at Araxa, MG, in beautiful green crystal groups(*); also near Joao Pessoa, PB, in roseate groups in limestone fissures. Uses: The principal use, accounting for hundreds of thousands of tons annually, is for making high-density muds for oil well drilling; also used for pigments. RG *DII:408; AM 52:1877(1967), 63:506(1978); NBSM 25(10):12(1974).*

28.3.1.2 Celestine SrSO$_4$

Named in 1791 from the Latin *coelestis*, heavenly, in allusion to the faint blue color of the first specimens described. Barite group. Synonym: celestite. ORTH *Pnma*. $a = 8.371$, $b = 5.355$, $c = 6.870$, $Z = 4$, $D = 3.982$. *AM 63:506(1978)*. Structure: *CM 13:181(1975)*. *5-593*(syn): 3.30_{10} 3.18_6 2.97_{10} 2.73_6 2.67_5 2.05_6 2.04_6 2.00_5. **Habit:** Crystals tabular {001}, commonly with large {210}, or lathlike elongated [100], or with equant cross section. Common forms {001}, {100}, {210}, {011}, {102}, {101}, {111}, {211}. Also massive; as fibrous veinlets or nodules, granular, earthy. **Physical properties:** Pale blue or colorless; also white, light red, green, or brown, vitreous luster, pearly on cleavage. Cleavage {001}, perfect; {210}, good; uneven fracture. $H = 3-3\frac{1}{2}$. $G = 3.97$. Some celestines fluoresce white under LWUV, and to a lesser extent under SWUV. **Optics:** Biaxial (+); $XYZ = cba$; $N_x = 1.622$, $N_y = 1.624$, $N_z = 1.631$ (Na); $2V = 50°$; $r < v$; pleochroic if colored. **Chemistry:** Ba substitutes for Sr and a probable complete series extends to barite. Some Ca may replace Sr, with a maximum Ca/Sr ratio about 1:11. Normally, Ba and Ca are present only up to few mol %. Very slightly soluble in water. Slowly soluble in concentrated acids or alkali carbonate solutions. Difficulty fusible. **Occurrence:** Celestine may be directly deposited from seawater or precipitated from percolating surface waters in veins and fissures, presumably leached from nearby sedimentary rocks. It also forms occasionally in hydrothermal deposits. It is common as a minor constituent of evaporites, and disseminated in shales, limestone, dolomite, and sandstone. **Localities:** At the lead mine at Rossie, NY; in *PA* at *Bellwood, Blair Co.*, in blue fibrous layers. In OH at Clay Center(*) and at Put-in-Bay on South Bass Is., Lake Erie, in a large vug in dolomite(!). In CA in fine clear crystals(*) with colemanite at Death Valley, Inyo Co., and in massive beds 3–6 m thick near Ludlow. Dundas, ONT, Canada, in pale blue crystals. In Mexico in large blue crystals to 40 cm(*) near Matehuala, SLP, and elsewhere in the region in commercial deposits. In England near Bristol, Gloucester, in good crystals(*). Found at numerous places in Germany, Austria, and the Slovakia. In the amygdules of a volcanic rock at Montecchio Maggiore, Italy, and in splendid groups of crystals at Girgenti with sulfur, aragonite, and gypsum and at the Floristella mine, Caltanisetta, Sicily(!), blue crystals > 100 mm on corroded sulfur. Near Sakoany, SW of Majunga, Madagascar, as large, brilliant, often clear blue crystals(!) in

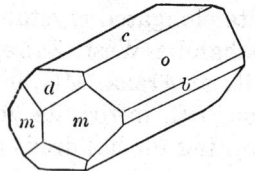
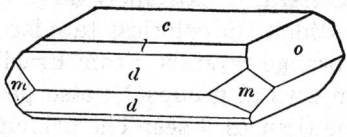

Celestine
Forms: *m*{110}, *c*{001}, *l*{104}, *d*{102}, *o*{011}

large geodes in early Tertiary marls. RG *DII:415, AM 46:189(1961), ZK 121:204(1965).*

28.3.1.3 Anglesite PbSO$_4$

Named in 1832 for the locality. Barite group. ORTH *Pnma.* $a = 8.482$, $b = 5.398$, $c = 6.959$, $Z = 4$, $D = 6.36$. Structure: *AM 63:506(1978)*. 36-1461(syn): 4.27$_5$ 4.24$_5$ 3.33$_7$ 3.22$_5$ 3.01$_9$ 2.07$_{10}$ 2.032$_6$ 2.027$_5$. Variable habit; thin to thick tabular {001}, often with {210}, {101}, and rhomboidal outline; also prismatic [001], [100], or [010] in combination with {210}, {011}, {102}; equant or pyramidal {111} or {211}. {100} and {210} often striated [001]. Usually massive, granular to compact, often with concentric banding and enclosing an unaltered core of galena. May form epitaxial overgrowths on galena or barite. Colorless to white, often tinged with gray, yellow, green, or blue; colorless streak; adamantine to resinous luster, or vitreous. Cleavage {001}, good; {210}, distinct; {001} in traces; conchoidal fracture; brittle. $H = 2\frac{1}{2}$-3. $G = 6.38$. Often fluoresces yellow under LWUV radiation. Biaxial (+); $XYZ = cba$; $N_x = 1.877$, $N_y = 1.883$, $N_z = 1.894$ (Na); $2V = 75°$. A small amount of substitution of Ba for Pb has been reported. Slowly soluble in HNO$_3$, easily so in ammonium acetate solution. Fragments placed in (NH$_4$)$_2$CO$_3$ solution are coated with white PbCO$_3$, and in (NH$_4$)$_2$S solution turn black and lustrous. Easily fusible. Anglesite is normally formed from the oxidation of galena and is common in the upper portion of lead deposits. Cavities in altering galena are often lined with fine crystals of anglesite, sometimes accompanied by sulfur crystals. Anglesite forms pseudomorphs after galena, and in turn readily alters to cerussite, which may vein massive anglesite or form crusts on it. Formerly found well crystallized(*) at the Wheatley mine, Chester Co., PA. In UT in the Tintic district, Juab Co. Fine specimens come from the Coeur d'Alene district(*), ID; also noted in AZ, NV, CO, and CA. In Mexico abundant in massive form at Sierra Mojada, COAH, associated with boleite at Boleo, BCS, and elsewhere in the country. In *Wales* at *Pary's mine* on the *Island* of *Anglesey*. In England at Matlock and Cromford, Derbyshire(*); in Scotland at Leadhills, Lanark. Found in Germany in the lead mines near Siegen, Westphalia; at Bleiberg, Carinthia, Austria; in France at Huelgoat, Brittany. In Italy on the island of Sardinia at Monte-

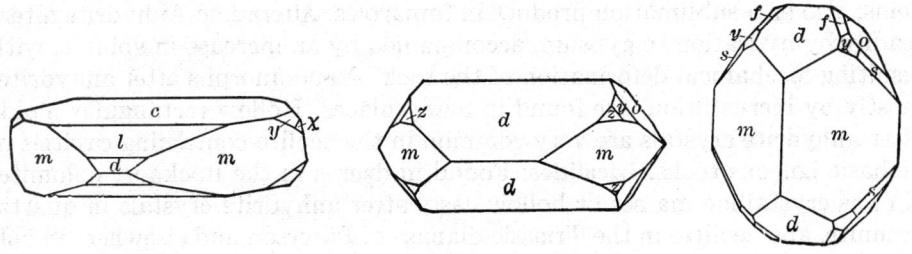

Anglesite
Forms: *m*{110}, *c*{001}, *l*{104}, *d*{102}, *o*{011}

vecchio and finely crystallized at Monteponi(!). In Africa crystals of exceptional size (to 80 cm), beauty, and transparency at Touissit, Morocco(!), with cerussite and phosgenite; at Sidi-Amor-Ben-Salem, Tunisia, and at Tsumeb, Namibia(!). In Australia at Broken Hill, NSW, and at Dundas, TAS. RG *DII:420*.

28.3.2.1 Anhydrite CaSO$_4$

Named in 1795 for the Greek term for *without water*, in allusion to its composition. ORTH *Cmcm*. $a = 6.993$, $b = 6.995$, $c = 6.245$, $Z = 4$, $D = 2.96$. *CM 13:289(1975)*. *37-1496*(syn): 3.50_{10} 2.85_3 2.33_2 2.21_3 2.09_1 1.87_2 1.75_1 1.65_1. **Structure:** The unit cell has two edges approximately equal but the structure is not pseudotetragonal. It is very different from that of barite but more reminiscent of zircon. S atoms, which are at the centers of regular tetrahedra of O's, and Ca atoms, lie on the lines of intersection of two sets of mirror planes. Planes containing evenly spaced Ca and SO$_4$ ions lie parallel to {100} and {010}, whereas layering is not so well defined parallel to {001}. This is in accord with the better cleavage on {100} and {010} than on {001}. The succession of Ca and SO$_4$ ions in the Z direction can be regarded as a chainlike feature, which also occurs in the structure of gypsum. *DHZ* 2nd ed.: *617(1992)*, *AC(B) 31:2164(1965)*. **Habit:** Crystals uncommon, usually equant or thick tabular on {010}. Usually massive, coarsely crystalline, showing pseudocubic cleavage; granular, or fibrous. Twinning on {011} or {120}. **Physical properties:** Colorless to white; also pale shades of blue or violet; rose, mauve, brownish, reddish, gray; white streak; vitreous to pearly luster. Cleavage {010}, perfect; {100}, nearly so; {001}, good to imperfect; uneven to splintery fracture. $H = 3\frac{1}{2}$. $G = 2.98$. Nonfluorescent. **Optics:** Biaxial (+); $XYZ = bac$; $N_x = 1.570$, $N_y = 1.575$, $N_z = 1.614$ (Na); $2V = 43.7°$; absorption $Z > Y > X$. **Chemistry:** The composition of anhydrite apparently varies only slightly from the theoretical formula through isomorphous substitution for Ca. Sr and Ba have been reported in amounts less than 1%. **Test:** Soluble in acids. **Occurrence:** Anhydrite is one of the most important minerals of the marine evaporites, forming beds that may reach a thickness of hundreds of meters. It precipitates from concentrated saline solutions at temperatures above 42°. It is often associated with gypsum, at the expense of which it can be formed. Occasionally found as a gangue mineral in hydrothermal veins; also as a sublimation product in fumaroles. **Alteration:** Anhydrite alters readily by hydration to gypsum, accompanied by an increase in volume, with resulting mechanical deformation of the rock. Pseudomorphs after anhydrite, mostly by incrustation, are found in many places. Hollow rectangular molds after anhydrite crystals are very common in the zeolite-containing cavities in diabasic igneous rocks. **Localities:** Found in druses in the Lockport dolomite, NY; as crystalline masses or hollow casts after anhydrite crystals in quartz, prehnite, and zeolites in the Triassic diabase of Paterson and elsewhere in NJ. From the Carlsbad salt beds of the Permian basin, TX–NM, and in the cap rock of salt domes in TX and LA. In Canada in bedded deposits in NS and on

28.3.3.1 Chalcocyanite

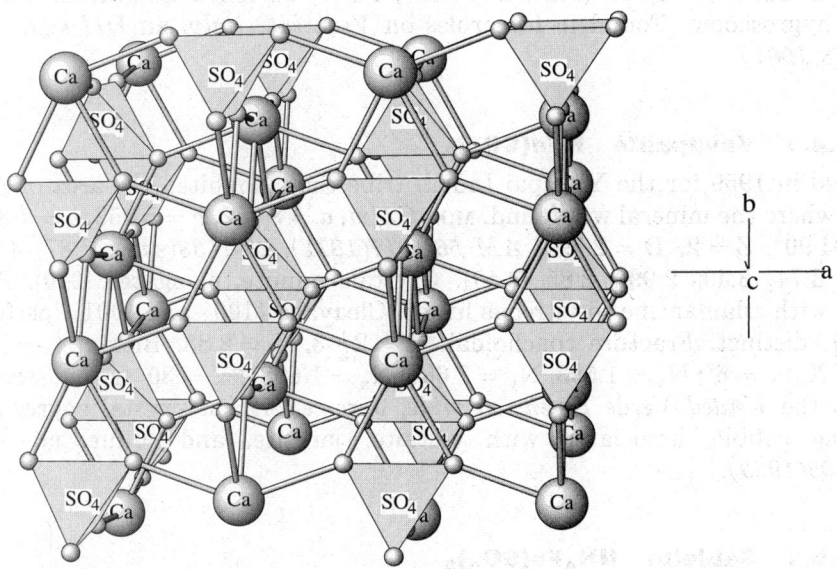

ANHYDRITE: CaSO$_4$

Ca atoms and SO$_4$ tetrahedra alternate in all three directions, as in halite, but the sulfate groups are considerably flattened in the a directions and are off center. The structure is basically the same as that of zircon and xenotime.

Calumet Island, QUE. In Mexico in orebodies at Nacozari, SON, and at Naica, CHIH, in fine blue crystals(!). Important deposits in the Stassfurt region, Germany. *Hall, near Innsbruck, Tyrol, Austria*. In Switzerland at the Simplon tunnel, Valois (km 7.25 and 9.4–9.6 from northern Portal) in fine transparent lavender crystals(!); also at many places in Swiss Triassic deposits, superb crystals to 10 cm, and nice flat twins in the clefts. In France at numerous places in the rock salt and gypsum deposits of the Pyrenees; from fumaroles at Vesuvius, Italy; from the salt beds of Krakow, Poland. In the ore at the Hanaoka and Kano mines, Ugo, Japan. With calcite, apatite, titanite, and diopside in Grenville-type deposits near Betroka, Madagascar. RG *DII:424, AC 16:767(1969), NBSC 539(4):65(1955)*.

28.3.3.1 Chalcocyanite CuSO$_4$

Named in 1873 from the Greek for *copper* and *azure blue*, in allusion to the composition, and color change when hydrated. Synonym: chalcokyanite. Originally called "hydrocyanite." ORTH *Pmna*. $a = 8.409$, $b = 6.709$, $c = 4.833$, $Z = 4$, $D = 3.89$. *MPA 39:201(1988)*. 15-775(syn): 4.19_8 3.55_{10} 2.61_{10} 2.42_5 1.96_2 1.77_3 1.43_{2} 1.429_2. Crystals tabular {100}. Colorless when pure; normally pale green, brownish, or yellowish; also sky blue, due to partial hydration. No cleavage. H = $3\frac{1}{2}$. G = 3.65. Biaxial (−); XYZ = abc; N_x = 1.724, N_y = 1.733, N_z = 1.739 (Na); 2V = 80°; r > v, extreme. Soluble in water; very hygroscopic. Found in fumaroles on *Vesuvius, Italy*. RG *DII:429, AM 46:758(1961)*.

28.3.4.1 Yavapaiite $KFe(SO_4)_2$

Named in 1959 for the Yavapai Indian tribe that inhabited the area of Arizona where the mineral was found. MON $C2/m$. a = 8.152, b = 5.153, c = 7.877, β = 94.90°, Z = 2, D = 2.891. *AM 56:1917(1971)*. *29-1438*(syn): 7.87_7 4.07_5 3.89_7 3.74_7 3.49_6 2.99_{10} 2.85_5 2.40_6. Crystals minute, elongated [010]. Pale pink, with adamantine to vitreous luster. Cleavage {100} and {001}, perfect; {110}, distinct. Fracture conchoidal. H = $2\frac{1}{2}$–3. G = 2.88. Biaxial (−); XZ = ab; X ∧ c = 6°; N_x = 1.593, N_y = 1.684, N_z = 1.698; 2V = 30°; r > v, strong. From the *United Verde Extension mine, Jerome, AZ*, as crystal aggregates coating rubble, associated with voltaite, jarosite, and sulfur. RG *AM 44:1105(1959)*.

28.3.5.1 Sabieite $NH_4Fe(SO_4)_2$

Named in 1983 for the locality. HEX-R $P321$. a = 4.82, c = 8.17, Z = 1, D = 2.361. *AM 71:229(1986)*. *24-44*: 8.3_7 3.74_{10} 2.95_5 2.42_3 2.31_1 1.871_1 1.861_1 1.397_1. Microscopic thin platelets, powdery. White. Parallel extinction. Derived from the dehydration of lonecreekite. From the *Lone Creek cavern, near Sabie, Transvaal, South Africa*. RG *AGSSA 17:29(1983)*.

28.3.5.2 Godovikovite $NH_4(Al,Fe)(SO_4)_2$

Named in 1988 for Alexandr A. Godovikov (1927–1995), Russian mineralogist, director of the Fersman Museum, Moscow. HEX-R $P321$. a = 4.75, c = 8.30, Z = 1, D = 2.52. *ZVMO 117:208(1988)*. *23-1*: 8.28_6 3.68_{10} 2.92_3 2.76_1 2.37_2 1.849_1 1.839_1 1.525_1. Compact or porous chalky aggregates of fine hairs. White, dull luster. Uneven fracture. H = 2. G = 2.53. Uniaxial (+); N_O = 1.572, N_E = 1.581. From burning dumps of coal mines near *Kopeysk, Chelyabinsk coal basin, southern Ural Mts., Russia*. RG *AM 75:241(1990)*.

28.4.1.1 Vanthoffite $Na_6Mg(SO_4)_4$

Named in 1902 for Jacobus Hendricus van't Hoff (1852–1911), Dutch physical chemist. MON $P2_1/c$. a = 9.797, b = 9.217, c = 8.199, β = 113.30°, Z = 2, D = 2.67. *AC 17:1613(1964)*. *29-1240*(syn): 4.03_{10} 3.92_7 3.43_{10} 3.11_8 3.06_5 2.91_7 2.85_6 2.83_6. Massive, bedded. Colorless, vitreous to pearly luster. Uneven to flat conchoidal fracture; friable. H = $3\frac{1}{2}$. G = 2.694. Biaxial (−); N_x = 1.485, N_y = 1.488, N_z = 1.489; 2V = 84°; r < v, weak. Formed in marine evaporites. *Wilhelmshall, Stassfurt region, Germany; Hall, Tyrol, Austria*. RG *DII:430*.

28.4.2.1 Glauberite Na$_2$Ca(SO$_4$)$_2$

Named in 1808 in allusion to its similarity in composition to Glauber's salt (Na$_2$SO$_4$ · 10H$_2$O), which was named for Johann Rudolph Glauber (1603–1668), German alchemist. MON $C2/c$. $a = 10.129$, $b = 8.306$, $c = 8.533$, $\beta = 112.20°$, $Z = 4$, $D = 2.78$. *AM 52:1272(1967)*. **Structure:** Ca atoms bind eight O's lying at the corners of a distorted square antiprism, its bases being nearly parallel to (100). Na atoms bind seven O's arranged at the corners of a polyhedron having one face centered. Each Na polyhedron shares three edges with the surrounding Na polyhedra, forming three types of zigzag chains: one along the c axial direction, another along the ab diagonal direction, and the third along the ac diagonal. The Ca polyhedra, connected to each other by one edge, are placed along the voids of the Na–O framework, parallel to the c axis. Structurally weak points are localized along layers formed by the Na atoms; these layers are parallel to (001), and along them the mineral shows the main cleavage {001}. As could be surmised from its high density, the structure of glauberite is very compact. *ZK 122:175(1965)*. *19-1187*(syn): 4.38$_5$ 3.18$_8$ 3.13$_{10}$ 3.11$_8$ 2.86$_5$ 2.81$_7$ 2.68$_6$ 1.98$_6$. **Habit:** Variable crystal habit: prismatic [001]; tabular {001}; dipyramidal through combination of {111} and {110}. Common forms {001}, {100}, {110}, and {111}. **Physical properties:** Usually

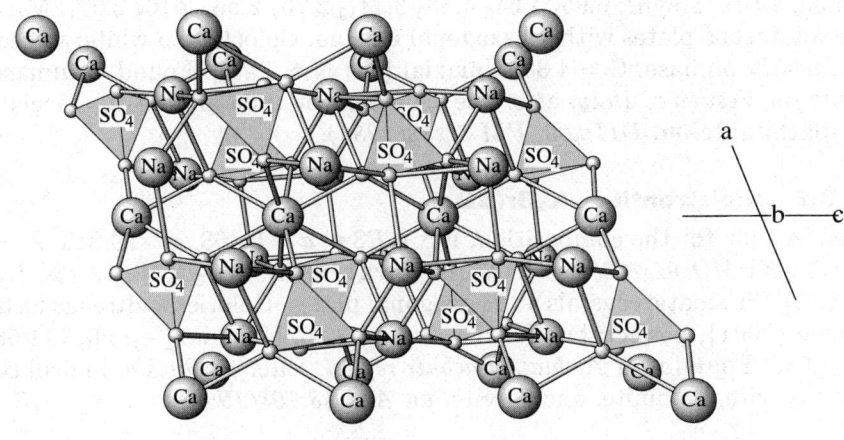

GLAUBERITE: Na$_2$Ca(SO$_4$)$_2$

Ca atoms are roughly in planes (parallel to (100); Na atoms in planes parallel to (001). Cleavage breaks weak Na–O bonds on (001).

gray or yellowish, sometimes colorless or reddish; white streak; vitreous to waxy luster, pearly on {001} cleavage. Cleavage {001}, perfect; {110}, indistinct; conchoidal fracture. H = $2\frac{1}{2}$–3. G = 2.75–2.85. **Tests:** Whitens in water due to leaching of Na_2SO_4 and deposition of gypsum. Soluble in HCl. Tastes slightly saline. Fusible to a white enamel. **Chemistry:** Analyses of natural glauberites deviate hardly at all from the stoichiometric quantities of Na, Ca, SO_4, and H_2O. **Optics:** Biaxial (−); Z = b; Y∧c = 12°; N_x = 1.515, N_y = 1.535, N_z = 1.536; 2V = 7°; r > v, strong, horizontal. **Occurrence:** Glauberite is formed under a variety of conditions: as a constituent of both marine and lacustrine evaporites, in cavities in basaltic extrusive rocks, in fumaroles, and with nitrate or borate deposits under extremely arid conditions. **Localities:** Abundant as molds and casts of prehnite and quartz after glauberite in cavities in basalt at Paterson and elsewhere in NJ. From the Verde Valley salt beds, Yavapai Co., AZ. In CA at Searles Lake and in the Saline Valley clays, Inyo Co. In the upper Silurian beds at Gypsumville, MB, Canada. First recognized as crystals in halite at *Villarubia, Toledo, Spain*. From the Triassic salt deposits of Lorraine, France. In Germany in the Stassfurt potash region, often in fine crystals(!). In fumarole deposits at Vulcano in the Lipari Islands, Italy. From the Permian salt deposits at Kairovka and elsewhere in southern Russia. The Mayo mine, Salt Range, Pakistan. In the Atacama desert, Chile. RG *DII:431*.

28.4.3.1 Palmierite $K_2Pb(SO_4)_2$

Named in 1907 for Luigi Palmieri (1807–1896), Italian scientist and director of the observatory on Vesuvius. HEX-R $R\bar{3}m$. a = 5.584, c = 20.677, Z = 3, D = 4.35. *29-1015*(syn): 6.95_4 4.64_3 4.33_5 3.14_{10} 2.75_7 2.56_4 2.16_3 2.07_4. Microscopic micaceous plates with a hexagonal outline. Colorless to white; vitreous luster, pearly on base. G = 4.33. Uniaxial (−); N ≈ 1.712. Found in fumarole deposits on *Vesuvius, Italy*, after the eruptions of 1906 and 1919, associated with aphthitalite. RG *DII:403, PM 54:34(1985)*.

28.4.3.2 Kalistrontite $K_2Sr(SO_4)_2$

Named in 1962 for the composition. HEX-R $R\bar{3}m$. a = 5.463, c = 20.843, Z = 3, D = 3.32. *ZVMO 91:712(1962)*. *29-1049*(syn): 6.94_2 4.31_2 3.13_{10} 2.73_7 2.15_2 2.06_3 1.91_2. Prismatic crystals and hexagonal plates. Colorless, vitreous luster. Cleavage {0001}, perfect; brittle. H = 2. G = 3.20. Uniaxial (−); N_O = 1.569, N_E = 1.549. Found near *Alshtan, Bachkir region, southern Russia*, in drill core with anhydrite, dolomite, and sylvite. RG *AM 48:708(1963)*.

28.4.3.3 Mereiterite $K_2Fe(SO_4)_2 \cdot 4H_2O$

Named in 1995 for Kurt Helmut Mereiter (b.1945), Austrian mineralogist, University of Vienna. Compare leonite (29.3.3.3). MON $C2/m$. a = 11.844, b = 9.556, c = 9.947, Z = 4, D = 2.358. PD: 4.78_3 3.51_5 3.44_{10} 3.33_5 3.05_3 2.41_3 2.39_5. Subhedral crystals up to 10 mm × 8 mm × 6 mm, embedded in gypsum. Pale yellow, vitreous to greasy luster, transparent. No cleavage, conchoidal fracture, brittle. H = $2\frac{1}{2}$–3. G = 2.36. Nonfluorescent. Readily soluble in water. Biaxial (+); X = [010]; Z $\wedge$ c $\approx$ 20°; N_x = 1.497, N_y = 1.501, N_z = 1.509; 2V = 71°; nonpleochroic; r > v, weak. From the *Hilarion adit, Agios Konstandinos, Laurium mining district, Greece*, formed by the oxidation of primary sulfides. RG *EJM 7:559(1995), AM 81:251(1996)*.

28.4.4.1 Langbeinite $K_2Mg_2(SO_4)_3$

Named in 1891 for A. Langbein, German chemist, Leopoldshall. ISO $P2_13$. a = 9.920, Z = 4, D = 2.824. *NJMM:182(1979)*. 19-974(syn): 4.05_3 3.14_{10} 2.99_2 2.75_2 2.65_4 2.41_1 1.95_1 1.61_1. Rarely in crystals; as nodules or grains. Colorless or pale colors, vitreous luster. No cleavage, conchoidal fracture. H = $3\frac{1}{2}$–4. G = 2.83. Reported to be Piezoelectric. (*Bond, Bell Syst. Tech. J. 22:145(1943)*). Isotropic; N = 1.533–1.535. Slightly soluble in water, easily fusible, turns white on slight heating. Found in marine evaporites. May be pseudomorphously replaced by kainite. Abundant in salt beds 24 km E of Carlsbad, NM, where some beds up to 2 m thick contain up to 90% langbeinite, associated with halite and sylvite. Partly in crystals up to 2 cm. At *Wilhelmshall, Stassfurt region, northern Germany*, also in Solvayhall and elsewhere. At Hall, Tyrol, Austria. Stebnik, Galicia, Poland. Mayo mine, Salt Range, Pakistan. Mined on a large scale as a source of potash. RG *DII:434, PM 54:1(1985)*.

28.4.4.2 Manganolangbeinite $K_2Mn_2(SO_4)_3$

Named in 1924 for its composition and relationship to langbeinite. ISO $P2_13$. a = 10.114, Z = 4, D = 3.02. *NBSM 6:43(1968)*. 20-909(syn): 5.84_1 4.13_1 3.20_{10} 3.05_2 2.70_5 2.06_1 1.98_1 1.64_2. Crystallizes as small tetrahedra. Rose-red, vitreous luster. No cleavage. H = $3\frac{1}{2}$–4. G = 3.02. Piezoelectric. Isotropic; N = 1.572. Found in a lava cavern at *Vesuvius, Italy*. RG *DII:435*.

28.4.4.3 Efremovite $(NH_4)_2Mg_2(SO_4)_3$

Named in 1989 for Ivan Antonovich Yefremov (1907–1972), Russian geologist and science fiction writer. ISO $P2_13$. a = 9.99, Z = 4, D = 2.52. 42-1432: 5.76_2 4.08_7 3.53_1 3.15_{10} 3.01_3 2.77_1 2.67_3 1.737_1. Aggregates of equant grains. Colorless to gray, vitreous to dull luster. No cleavage, uneven fracture. H = 2. Isotropic; N = 1.550. Soluble in water; in CT, evolves ammonia. Spontaneously hydrates to boussingaultite in air over several days. From burning coal dumps in the Chelyabinsk coal basin, southern Ural Mts., Russia. RG *AM 76:299(1991)*.

28.4.5.1 Millosevichite $(Al,Fe)_2(SO_4)_3$

Named in 1913 after F. Millosevich (1875–1942), Italian mineralogist, University of Rome. HEX-R $R\bar{3}$. a = 8.055, c = 21.191, Z = 6, D = 2.86. *NBSM 15:8(1978)*. 42-1410(syn): 5.82_4 4.22_2 4.03_1 3.53_4 3.50_{10} 2.92_2 2.66_2 2.62_1. Granular masses which show small clear crystals on their surfaces. Clear red to brick red, also violet-blue; vitreous luster. H = $1\frac{1}{2}$. Nearly isotropic;

$N \approx 1.573$. From *Vulcano, Lipari Islands, Italy*, in the Alum Grotto. In the Lvovsko-Volynskii Bolynskir coal basin, Ukraine, Russia, in fractures in rocks from which H_2S and SO_2 are escaping. RG *DII:539*.

28.4.5.2 Mikasaite $(Fe^{3+},Al)_2(SO_4)_3$

Named in 1994 for the locality. The Fe^{3+} analog of millosevichite (28.4.5.1). HEX-R $R\bar{3}$. $a = 8.14$, $c = 21.99$, $Z = 6$. PD: 5.99_3 4.35_2 3.56_{10} 2.97_2 2.72_2 2.64_1 2.35_1 2.24_1, in good agreement with *JCPDS 33-679* [synthetic $Fe_2(SO_4)_3$]. Forms aggregates of porous material made up of hollow spherical crystallites to 100 μm in diameter. White to light brown with a similar streak. Soft. Highly deliquescent. Uniaxial (+); $N_x = 1.504$, $N_y = 1.518$. Found as a sublimate in fissures with escaping high-temperature gas over a burning coal seam in *Ikushunbetsu, Mikasa City, Hokkaido, Japan*. RG *MM 58:649(1994), AM 80:846(1995)*.

28.4.6.1 Klyuchevskite $K_3Cu_3(Fe^{3+},Al)O_2(SO_4)$

Named in 1989 for the locality, the Klyuchevsk group of volcanoes. MON 12 $a = 18.412$, $b = 4.944$, $c = 18.640$, $\beta = 101.5°$, $Z = 4$, $D = 2.98$. *41-1463* 9.17_9 9.03_{10} 7.20_4 4.50_3 3.76_6 3.68_2 3.45_2 3.41_3. Crystals acicular, elongate [010], to 0.1mm, as unoriented aggregates, dark green to olive green, light green streak, semimetallic luster, cleavage perfect {h01}, $H = 2\frac{1}{2}$, $G = 3.00$–3.15. Soluble in cold water and in weak acids, hydrates slowly in air to a white powder. Biaxial +, $N_\alpha 1.549$ (olive), $N_\beta 1.550$ (green), $N_\gamma 1.680$ (dark olive), $2V_{calc} 11°$, $Y = b$. Found in cavities and fissures in massive sublimates at the *Great Tolbachik Fissure Eruption, Kamchatka, Russia*, associated with kamchatkite, ponomarevite, and hematite. *ZVMO 118(1):65(1989) AM 75:1210(1990) MM 56:411(1992) AM 78:454(1993)*.

28.4.6.2 Alumoklyuchevskite $K_3Cu_3(Al,Fe^{3+})O_2(SO_4)_4$

Named in 1995 as the Al-dominant analog of klyuchevskite. MON *I2*. $a = 18.423$, $b = 5.139$, $c = 18.690$, $Z = 4$, $D = 2.95$. PD: 9.15_8 9.04_{10} 7.20_5 4.51_2 3.78_4 3.76_3 3.42_2 2.79_2. Acicular, prismatic crystals, to 1 mm, in aggregates. Transparent, dark green. Cleavage (h0l), perfect. $H = 2$. $G = 3.1$. Biaxial (+); $N_x = 1.542$, $N_y = 1.548$, $N_z = 1.641$; $2V = 30°$. X, light green; Y, grayish green; Z, dark green. Soluble in water, hydrates slowly in air to a white powder. Found in fumaroles in the *Main Tolbachik fracture eruption, Kamchatka, Russia*, associated with fedotovite, tenorite, langbeinite, and lammerite. RG *ZVMO 124(1):95(1995)*.

Class 29

Hydrated Acid and Sulfates

29.1.1.1 Rhomboclase HFe(SO$_4$)$_2$ · 4H$_2$O

Named in 1891 in allusion to the basal cleavage and crystal shape. ORTH *Pnma*. $a = 9.724$, $b = 18.333$, $c = 5.421$, $Z = 4$, $D = 2.206$. *TMPM 21:216(1974)*. *27-245:* 9.15$_4$ 4.74$_4$ 4.29$_4$ 4.21$_4$ 4.05$_5$ 3.55$_4$ 3.33$_5$ 3.11$_{10}$. Crystals thin tabular {001}, also as stalactites with radiating bladed structure. Colorless to white, gray, or pale yellow, or tinted blue or green; vitreous to pearly luster. Cleavage {001}, perfect; {110}, good; conchoidal to fibrous fracture; cleavage folia are flexible. $H = 2$. $G = 2.23$. Biaxial (+); XYZ = cab; N$_x$ = 1.534, N$_y$ = 1.553, N$_z$ = 1.638 (Na); 2V = 27°. Slowly soluble in water, easily in acids. From *Smolnik, Slovakia*, associated with other sulfates in altering pyritic ores. Esperanza mine, Cerro de Pasco, Peru, as incrustations in abandoned workings. Alcaparrosa, near Cerritos Bayos, Chile, with szomolnokite and roemerite. Thin platy crystals to 2 cm(!) with melanterite and copiapite, Socavon mine, Oruro, Bolivia. RG *DII:436, TMPM 16:155(1970)*.

29.1.2.1 Matteuccite NaHSO$_4$ · H$_2$O

Named in 1952 for Vittorio Matteucci (1862–1909), Italian vulcanologist. MON *Cc*. $a = 7.81$, $b = 7.82$, $c = 8.03$, $\beta = 117.48°$, $Z = 4$, $D = 2.108$. *22-1379*(syn): 5.18$_2$ 5.11$_4$ 3.63$_3$ 3.56$_{10}$ 3.46$_7$ 3.42$_6$ 3.38$_2$ 2.76$_2$. Mixed with other salts in fine-grained stalactites. Colorless, vitreous luster. Biaxial (+); 2V = 86°. Found as a third component in stalactites in fumaroles in the 1933 eruption of *Vesuvius, Italy*, mixed with mercallite and ralstonite. The identity of this component with synthetic NaHSO$_4$ · H$_2$O is probable but not positive. RG *AM 39:848(1954), AST 109:532(1975)*.

29.1.3.1 Monsmedite H$_8$K$_2$Tl$_2$(SO$_4$)$_8$ · 11H$_2$O

Named in 1968 for the Latin *Mons Medius*, for Baia Sprie, Romania, the locality for the mineral. ORTH Space group unknown. *41-1448:* 5.49$_1$ 3.56$_4$ 3.40$_{10}$ 3.06$_1$ 2.86$_4$ 2.32$_4$ 2.08$_4$ 1.57$_4$. Cubelike crystals, with {111} and {110}. Dark green to black, nearly opaque in large crystals, pitchy luster. Prismatic and pyramidal cleavage. H > 2. $G = 3.0$. Biaxial (−); N ≈ 1.608; birefringence = 0.011; r > v. Found in geodes with marcasite and barite on kaolinite and limonite at *Baia Sprie, Baia Mare region, Romania*, in the oxidation zone of a silver-bearing lead–zinc deposit. RG *AM 53:2104(1968), 54:1496(1969)*.

29.2.1.1 Lecontite $NH_4Na(SO_4) \cdot 2H_2O$

Named in 1858 for John L. LeConte (1825–1883), American entomologist, who discovered the mineral. ORTH $P2_12_12_1$. $a = 8.216$, $b = 12.854$, $c = 6.232$, $Z = 4$, $D = 1.747$. Structure: *AC 22:683(1967)*. *15-283*(syn): 5.07_{10} 4.65_6 3.94_5 3.80_{10} 3.45_6 3.32_7 3.03_9 2.69_8. *15.370*(nat): 5.06_8 4.98_5 4.64_{10} 3.91_5 3.80_5 3.32_5 3.04_9, 1.99_7. Prismatic or lathlike crystals. Colorless, vitreous luster. Biaxial (−); $XYZ = acb$; $N_x = 1.440$, $N_y = 1.452$, $N_z = 1.453$; $r < v$, strong. Soluble in water. Tastes saline, bitter. Found embedded in bat guano in *Las Peidras cave*, near Camayagua, Honduras. RG *DII:438*; *AM 36:596(1951)*, *48:180(1963)*.

29.2.2.1 Mirabilite $Na_2SO_4 \cdot 10H_2O$

Named in 1845 from the Latin *sal mirabile*, wonderful salt, suggested by J. R. Glauber (1603–1668), who was surprised at its formation during an experiment. Synonym: Glauber's salt. MON $P2_1/c$. $a = 11.51$, $b = 10.37$, $c = 12.85$, $\beta = 107.80°$, $Z = 4$, $D = 1.465$. Structure: *AC(B) 34:3502(1978)*. *11-647*: 5.49_{10} 4.77_5 3.83_4 3.26_6 3.21_8 3.11_6 2.80_3 2.52_4. Crystals short prismatic to acicular [001] or [010]; also tabular or lathlike; commonly massive, as crusts, or stalactitic. Twins by interpenetration on {001}; also on {100}. Colorless to white, white streak, vitreous luster. Conchoidal fracture. $H = 1\frac{1}{2}$–2. $G = 1.490$. Biaxial (−); $X = b$, $Z \wedge c = 31°$; $N_x = 1.396$, $N_y = 1.410$, $N_z = 1.419$ (Na); $2V = 76°$; $r < v$, strong, crossed. Soluble in water. Tastes cool, slightly saline, and bitter. In CT, loses water and melts. Loses its water in dry air and falls to a loose powder. Formed in saline lakes, playas, and springs, and as an efflorescence on alkali soils. In CA at several places, especially at Mono Lake, Searles Lake, and Soda Lake, San Luis Obispo Co. In UT along the southern shore of Great Salt Lake, where it crystallizes during winter due to decreased solubility at low temperatures. In Canada in alkalic lakes in SK. Ischl, Hallstatt, and elsewhere in Austria; Ebro, Spain; Montedoro, Girgenti, Sicily, Italy. Salar de Pintados, Tarapacá, Chile, as ice-clear anhedral crystals. RG *DII:439*, *MR 1:12(1970)*.

29.3.1.1 Syngenite $K_2Ca(SO_4)_2 \cdot H_2O$

Named in 1872 from the Greek meaning *related*, because of its chemical resemblance to polyhalite. MON $P2_1/m$. $a = 9.777$, $b = 7.147$, $c = 6.250$, $\beta = 104.02°$, $Z = 2$, $D = 2.574$. *NBSM 25:14(1977)*. Structure: *SPC 23:271(1978)*. *28-739*: 9.49_4 5.71_6 4.62_4 3.35_4 3.17_8 2.86_{10} 2.83_5 2.74_6. In crystalline crusts and lamellar aggregates; crystals tabular {100} to prismatic [001], often rich in forms. Contact twins on {100}. Colorless, white, or faintly yellow; vitreous luster. Cleavage {110} and {100}, perfect; {010} distinct; conchoidal fracture. $H = 2\frac{1}{2}$. $G = 2.58$–2.60. Biaxial (−); $Z = b$; $X \wedge c = -2°$; $N_x = 1.501$, $N_y = 1.517$, $N_z = 1.518$ (Na); $2V = 28°$; $r < v$, strong. Does not alter in dry air. Partially dissolves in water with separation of gypsum. Easily fusible. Formed in marine salt deposits or in fumaroles. In the crater of Haleakala,

Maui, Hawaii. *Kalusz, Galicia, Poland.* Glück Auf, Germany. In lava at Vesuvius, Italy. RG *DII:442*, *ZK 124:398(1967)*, *SPC 20:773(1975)*.

29.3.1.2 Koktaite (NH$_4$)$_2$Ca(SO$_4$)$_2$ · H$_2$O

Named in 1948 for Jaroslav Kokta (1904–1970), Czech chemist, who analyzed the artificial compound. MON $P2_1/m$. $a = 6.324$, $b = 7.129$, $c = 10.098$, $\beta = 102.58°$. $Z = 2$, $D = 2.140$. *11-475:* 9.83$_{10}$ 5.83$_2$ 4.96$_4$ 4.08$_1$ 3.56$_2$ 3.30$_7$ 3.00$_2$ 2.89$_2$. Acicular or fibrous crystals, with {100}, {110}, {001}, {011}, {101}; twins on {100} frequent. Colorless to white. No cleavage. G = 2.09. Soluble in water with separation of gypsum, easily fusible. Biaxial (−); Y = b; Z ∧ c = 2°; N$_x$ = 1.524, N$_y$ = 1.532, N$_z$ = 1.536; 2V = 72°. Found as pseudomorphs after gypsum with mascagnite and ammonia alum on the waste heaps of a lignite mine at *Zeravice, near Kyjov in southeastern Moravia, Czech Republic*. RG *DII:444*, *AM 34:618(1949)*, *PM 54:29(1985)*.

29.3.2.1 Kröhnkite Na$_2$Cu(SO$_4$)$_2$ · 2H$_2$O

Named in 1876 for B. Kröhnke, who first analyzed the mineral. MON $P2_1/c$. $a = 5.807$, $b = 12.656$, $c = 5.517$, $\beta = 108.19°$, $Z = 2$, $D = 2.913$. *25-826:* 6.32$_{10}$ 4.15$_3$ 3.71$_4$ 3.35$_3$ 3.28$_8$ 3.10$_4$ 2.93$_7$ 2.76$_5$. Structure: *AC(B) 31:1753(1975)*. Short prismatic crystals; also pseudo-octahedral; commonly as crusts or cross-fiber veinlets; massive; granular. Twins on {101}, sometimes heart-shaped. Sky blue to greenish blue, vitreous luster. Cleavage {010}, perfect; {$\bar{1}$01}, poor; conchoidal fracture. H = 2$\frac{1}{2}$–3. G = 2.90. Soluble in water, easily fusible, in CT gives water. Biaxial (−); Y = b; X ∧ c = 48°; N$_x$ = 1.544, N$_y$ = 1.578, N$_z$ = 1.601 (Na); 2V = 79°; r < v, weak, inclined. Formerly an abundant ore mineral at *Chuquicamata, Antofagasta, Chile*, where it was associated with chalcanthite, natrochalcite, antlerite, and other soluble copper salts. Also at the Salvador mine, Quetana, and elsewhere in the Atacama desert region. RG *DII:444*, *PM 23:223(1954)*.

29.3.3.1 Blödite Na$_2$Mg(SO$_4$)$_2$ · 4H$_2$O

Named in 1821 for Carl August Bloede (1773–1820), German chemist. Synonym: bloedite. MON $P2_1/a$. $a = 11.128$, $b = 8.246$, $c = 5.543$, $\beta = 100.87°$, $Z = 2$, $D = 2.218$. *NBSM 25(6):43(1968)*. Structure: *CM 23:669(1985)*. *19-1215:* 4.56$_{10}$ 3.29$_{10}$ 3.25$_{10}$ 2.97$_4$ 2.73$_4$ 2.72$_4$ 2.65$_4$ 2.64$_4$. Crystals short prismatic [001], often highly modified; also massive, granular to compact. Colorless; sometimes greenish, reddish, gray due to inclusions; vitreous luster. No cleavage, conchoidal fracture, brittle. H = 2$\frac{1}{2}$–3. G = 2.25. Soluble in water. Tastes saline, bitter. Biaxial (−); Y = b; X ∧ c = 37°; N$_x$ = 1.483, N$_y$ = 1.486, N$_z$ = 1.487; 2V = 71°. Formed in both marine and lacustrine salt deposits, and with nitrates in extremely arid regions. Soda Lake, San Luis Obispo Co., CA. Stassfurt region, Germany, and at the *Ischl deposit* and Hallstatt deposit in *Austria*. Kalusz, Galicia, Poland. In salt lakes near Korduansk, Russia. Chuquicamata, Chile, and elsewhere in the Atacama desert. RG *DII:447*.

29.3.3.2 Nickelblödite Na$_2$(Ni,Mg)(SO$_4$)$_2$ · 4H$_2$O

Named in 1977 from the composition and relationship to blödite. MON $P2_1/c$. $a = 10.87$, $b = 8.07$, $c = 5.46$, $\beta = 100.72°$, $Z = 2$, $D = 2.455$. *MM 41:37(1977)*. *29-1253*: 4.47$_9$ 4.19$_7$ 3.72$_6$ 3.22$_{10}$ 3.19$_8$ 2.90$_5$ 2.59$_6$ 2.24$_5$. Efflorescences; also minute tabular crystallites. Light green. VHN$_5$ = 104–139. G = 2.43. Biaxial (−); N$_x$ = 1.513, N$_y$ = 1.518, N$_z$ = 1.520; 2V = 60–70°. Found as efflorescences encrusting ores containing violarite and pyrite at *Durkin nickel mine, Kambalda, WA, Australia*, and at *Carr Boyd Rocks, WA*. RG *AM 62:1059(1977)*.

29.3.3.3 Leonite K$_2$Mg(SO$_4$)$_2$ · 4H$_2$O

Named in 1896 for Leon Strippelmann, director of the salt works at Westerregeln, Germany. See mereiterite (29.3.3.4). MON $C2/m$. $a = 11.769$, $b = 9.539$, $c = 9.889$, $\beta = 95.30°$, $Z = 4$, $D = 2.203$. Structure: *ZK 173:75(1985)*. *21-995*: 5.88$_2$ 3.49$_4$ 3.42$_{10}$ 3.31$_4$ 3.04$_5$ 2.88$_3$ 2.38$_5$ 2.17$_2$. Crystals tabular {100} and elongated [001], usually anhedral, intergrown with other minerals; colorless to yellowish, waxy to vitreous luster. Conchoidal fracture. H = 2$\frac{1}{2}$–3. G = 2.201. Easily soluble in water. In CT, loses water and fuses. In air alters and becomes coated with a white crust of picromerite. Tastes bitter. Biaxial (+); Y = b; Z ∧ a = small; N$_x$ = 1.479, N$_y$ = 1.482, N$_z$ = 1.487; 2V ≈ 90°. Formed in marine salt deposits. Found in drill cores from wells in Eddy and Lea Cos., MN. *Westerregeln*, Leopoldshall, and elsewhere in the *German* potash districts. RG *DII:450*, *USGSPP 550B:125(1966)*, *MA 74:3459*.

29.3.4.1 Wattevillite Na$_2$Ca(SO$_4$)$_2$ · 4H$_2$O (?)

Named in 1879 for Oscar de Watteville of Paris, France. A questionable mineral species. ORTH or MON Space group unknown. *41-1360*: 5.44$_5$ 5.10$_4$ 4.39$_{10}$ 4.20$_6$ 4.03$_4$ 2.95$_5$ 2.89$_5$ 2.67$_4$. Aggregates of minute acicular crystals, in part twinned. Snow-white, silky luster. G = 1.81. Biaxial (−); N$_x$ = 1.435, N$_y$ = 1.455, N$_z$ = 1.459; 2V = 48°; no perceptible dispersion. *Einigkeit mine, Bauersberg, near Bischofsheim, Bavaria, Germany*, associated with other sulfates in pyritic lignite. RG *DII:452*.

29.3.5.1 Konyaite Na$_2$Mg(SO$_4$)$_2$ · 5H$_2$O

Named in 1982 for the locality. MON $P2_1/c$. $a = 5.784$, $b = 24.026$, $c = 8.066$, $\beta = 95.37°$, $D = 2.097$. *AM 67:1035(1982)*. *35-649*(syn): 12.0$_{10}$ 4.55$_6$ 4.20$_2$ 4.18$_2$ 4.00$_7$ 3.96$_1$ 2.81$_1$ 2.73$_1$. Efflorescences consisting of aggregates of minute crystals; colorless to white. H = 2$\frac{1}{2}$. G = 2.09. Soluble in water. Biaxial (−); X = b; Z ∧ c = 70°; N$_x$ = 1.464, N$_y$ = 11.468, N$_z$ = 1.474; 2V = 74°. *Great Konya Basin, Central Anatolia, Turkey*, as salt efflorescences on saline soils, associated with gypsum, hexahydrite, loeweite; halite, and blödite. RG *AM 74:1382(1989)*.

PICROMERITE GROUP

The picromerite group consists of five isostructural sulfate hexahydrates of general formula

$$A_2B(SO_4)_2 \cdot 6H_2O$$

where

A = K, NH_4
B = Mg, Cu, Fe, or Ni

and the space group is monoclinic. *CSMQ 1:1(1972), AC 17:1478(1964)*.

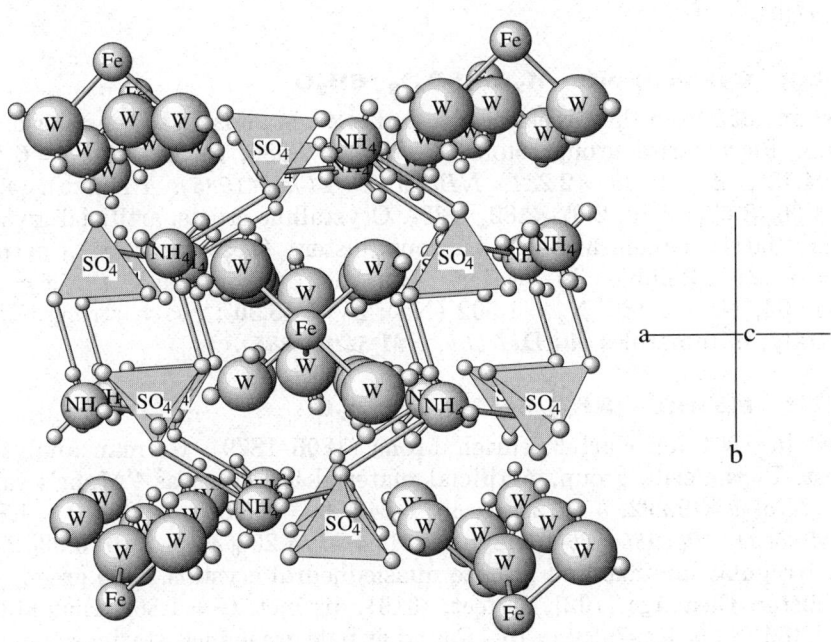

MOHRITE: $(NH_4)_2Fe(SO_4)_2 \cdot 6H_2O$ (picromerite group)

NH_4(K) and sulfate are mutually eight-coordinated and form a loose framework. Fe)$H_2O)_6$ octahedra are in interstices. Structural formula could be $NH_4)_2(SO_4)_2 \cdot Fe(H_2O)_6$.

PICROMERITE GROUP

Mineral	Formula	Space Group	a	b	c	β	D	
Picromerite	$K_2Mg(SO_4)_2 \cdot 6H_2O$	$P2_1/a$	9.096	12.254	6.128	104.78°	2.025	29.3.6.1
Cyanochroite	$K_2Cu(SO_4)_2 \cdot 6H_2O$	$P2_1/a$	9.080	12.123	6.164	104.43°	2.234	29.3.6.2
Mohrite	$(NH_4)_2Fe(SO_4)_2 \cdot 6H_2O$	$P2_1/a$	9.292	12.601	6.249	106.80°	1.859	29.3.7.1
Boussingaultite	$(NH_4)_2Mg(SO_4)_2 \cdot 6H_2O$	$P2_1/a$	9.325	12.605	6.208	107.10°	1.717	29.3.7.2
Nickel-boussingaultite	$(NH_4)_2Ni(SO_4)_2 \cdot 6H_2O$	$P2_1/c$	9.241	12.544	6.243	106.97°	1.85	29.3.7.3

29.3.6.1 Picromerite $K_2Mg(SO_4)_2 \cdot 6H_2O$

Named in 1855 for the Greek for *bitter* and *part*, in allusion to the magnesium content, salts of which are usually bitter. Picromerite group. MON $P2_1\backslash a$. $a = 9.096$, $b = 12.254$, $c = 6.128$, $\beta = 104.78°$, $Z = 2$, $D = 2.025$. *NBSM 25(8):54(1970)*. *21-1400*(syn): 4.16_9 4.06_{10} 3.71_{10} 3.16_4 3.06_7 2.96_6 2.81_4 2.39_5. Short prismatic crystals, as crusts, massive. Colorless to white; also yellowish, reddish, gray, due to admixture; vitreous luster. Cleavage $\{\bar{2}01\}$, perfect. $H = 2\frac{1}{2}$. $G = 2.028$. Soluble in water, easily fusible. Biaxial (+); $Y = b$; $X \wedge a = -1°$; $N_x = 1.460$, $N_y = 1.462$, $N_z = 1.472$ (Na); $2V = 48°$; $r > v$, weak. From fumarole deposits on *Vesuvius, Italy*. Stassfurt salt deposits, Germany; Kalusz, Galicia, Poland. RG *DII:453*, *NW 44:350(1957)*, *CE 24:94(1965)*

29.3.6.2 Cyanochroite $K_2Cu(SO_4)_2 \cdot 6H_2O$

Named in 1855 from the Greek for *blue* and *color*, in allusion to the color of the mineral. Picromerite group. MON $P2_1/a$. $a = 9.080$, $b = 12.123$, $c = 6.164$, $\beta = 104.43°$, $Z = 2$, $D = 2.234$. *NBSM 25(21):99(1985)*. *PD*(syn): 4.25_3 4.18_9 4.06_8 3.67_{10} 2.99_4 2.97_7 2.82_4 2.38_5. Crystalline crusts, artificial crystals tabular $\{001\}$. Greenish blue, vitreous luster. Cleavage $\{\bar{2}01\}$, perfect. $G_{syn} = 2.224$. Soluble in water. Biaxial (+); $Y = b$; $X \wedge c = 18°$; $N_x = 1.484$, $N_y = 1.486$, $N_z = 1.502$ (Na); $2V = 46.30°$; $r < v$, strong. Vesuvius, Italy, in fumaroles. RG *DII:454*, *PM 54:1(1985)*.

29.3.7.1 Mohrite $(NH_4)_2Fe(SO_4)_2 \cdot 6H_2O$

Named in 1964 for Karl Friedrich Mohr (1806–1879), German analytical chemist. Picromerite group. Artificial material is known as "Mohr's salt." MON $P2_1/a$. $a = 9.292$, $b = 12.601$, $c = 6.249$, $\beta = 106.80°$, $Z = 2$, $D = 1.859$. *NSBM 25(21):10(1985)*. *35-764*(syn): 5.41_4 4.27_4 4.20_{10} 4.16_4 3.80_8 3.03_5 2.83_3 2.46_3. Irregular laminae and minute quasieuhedral crystals. Pale green, vitreous luster. Cleavage $\{\bar{1}02\}$, perfect; $\{010\}$, distinct. $G = 1.862$. Biaxial (+); $N_x = 1.486$, $N_z = 1.497$; $2V = 75°$. Found as pale green incrustations from the boriferous soffoni of *Travale, Val di Cecina, Tuscany, Italy*. RG *AM 50:805(1965)*.

29.3.7.2 Boussingaultite $(NH_4)_2Mg(SO_4)_2 \cdot 6H_2O$

Named in 1864 after the French chemist Jean-Baptiste Boussingault (1802–1887). Picromerite group. MON $P2_1/a$. $a = 9.325$, $b = 12.605$, $c = 6.208$, $\beta = 107.10°$, $Z = 2$, $D = 1.717$. *NBSM 25(21):14(1985)*. Structure: *ZK 117:344(1962)*. *35-771*(syn): 5.38_6 4.26_6 4.21_{10} 4.14_3 3.80_9 3.15_5 3.01_3 2.78_3. Usually massive, as crusts or stalactites; crystals short prismatic $\{001\}$. Colorless to yellowish pink. Cleavage $\{\bar{2}01\}$, perfect. $H = 2$. $G_{syn} = 1.722$. Soluble in water. Biaxial (+); $Y = b$; $Z \wedge c = 12°$; $N_x = 1.470$, $N_y = 1.472$, $N_z = 1.479$ (Na); $2V = 51.17°$; $r > v$, perceptible. Near Mahanoy City, Schuylkill Co., PA, as small crystals in cavities in a waste coal heap; the Geysers, Sonoma

Co., CA, and in sandstone at South Mountain, Ventura Co. From the boric acid fumaroles of *Travale, Tuscany, Italy.* RG *DII:455.*

29.3.7.3 Nickel-Boussingaultite $(NH_4)_2Ni(SO_4)_2 \cdot 6H_2O$

Named in 1976 after the composition, the nickel analog of boussingaultite. Picromerite group. MON $P2_1/c$. $a = 9.241$, $b = 12.544$, $c = 6.243$, $\beta = 106.97°$, $Z = 2$, $D = 1.85$. *AC 17:1478(1964)*. *31-62:* 6.24_3 5.39_4 4.24_4 4.17_{10} 4.15_7 3.76_9 3.04_3 3.03_4. Thin films and crusts composed of tiny prismatic grains. Greenish blue. Imperfect cleavage. H = $2\frac{1}{2}$. Soluble in water. Biaxial (+); X $\wedge$ c = 0–4°; $N_x = 1.490$ (bluish), $N_y = 1.494$, $N_z = 1.501$ (yellow). From the *Norilskiye nickel deposit, Siberia, Russia,* in the oxidation zone in old workings near or on old timbers, resembling coatings of chrysocolla. RG *ZVMO 105:710(1976), AM 71:1545(1976), AC(B) 35:155(1979).*

29.3.8.1 Mosesite $Hg_2N(Cl,SO_4,MoO_4) \cdot H_2O$

Named in 1910 for Alfred J. Moses (1859–1920), professor of mineralogy, Columbia University, who first described several of the mercury minerals from Terlingua. ISO $F\bar{4}3m$. $a = 9.524$, $Z = 8$, $D = 7.53$. *AM 38:1225(1953)*. *39-388:* 5.51_8 2.88_9 2.76_{10} 2.38_4 2.19_4 1.834_3 1.685_4 1.436_3. The structure of mosesite consists of a cubic cristobalite-like configuration of Hg_2N^+ groups. This structure has open spaces occupied by H_2O and exchangeable ions such as SO_4^{2-}, Cl^-, AsO_4^{3-}, and CO_3^{2-}. Mosesite is isostructural with and essentially identical to $Hg_2NOH \cdot H_2O$, "Millon's base," and the presence of Cl^- and other ions is dependent entirely on ion exchange with whatever is present in the ore-forming solutions. Crystals are octahedrons or cubes, or a combination of both. Twinning on {111}. Yellow, becoming light olive-green on long exposure to light, very pale yellow streak, adamantine luster. Cleavage {111}, imperfect; uneven fracture; brittle. H = $3\frac{1}{2}$. G = 7.72. In CT turns dark brown to black and partially volatilizes, depositing mercury and calomel on the walls of the tube. Partially soluble in cold HCl. Isotropic; N = 2.065. *Terlingua, Brewster Co., TX,* associated with montroydite. Clack quicksilver mine, Fitting district, Pershing Co., NV. Huahuaxtla, GRO, Mexico, with terlinguaite, montroydite, eglestonite, and calomel. Several localities in Russia. RG *DII:89.*

29.4.1.1 Hydroglauberite $Na_{10}Ca_3(SO_4)_8 \cdot 6H_2O$

Named in 1969 for the composition, rather similar to glauberite, but hydrated. Probably ORTH Space group unknown. *24-1071:* 9.20_9 4.60_8 4.20_6 3.52_6 3.08_{10} 2.90_7 2.78_9 2.25_6. Dense masses consisting of microscopic fibers. Snow white, silky luster. Cleavages: one good, one distinct, one poor. Soluble in water, with separation of gypsum. Tastes salty, slightly bitter. Biaxial (−); $N_x = 1.488$, $N_z = 1.500$; parallel extinction, positive elongation. Kara-Kalpakii, Uzbekistan, as an alteration product of glauberite, associated with halite, astrakhanite, mirabilite, and polyhalite. RG *ZVMO 98:58(1969), AM 55:321(1970).*

29.4.2.1 Eugsterite $Na_4Ca(SO_4)_3 \cdot 2H_2O$

Named in 1981 for Hans P. Eugster (1925–1987), Swiss–American mineralogist, Johns Hopkins University, awarded the Roebling Medal, 1983. MON Space group unknown. $\beta = 116°$. *35-487*(syn): 9.20_4 5.50_6 4.50_3 3.45_3 3.43_{10} 2.94_2 2.76_3 2.75_5. Clusters of thin fibers, as efflorescences. Colorless. Very soft. $G \approx 2.50$. Soluble in water. Biaxial; $Y \wedge c = 27°$; $N_x = 1.492$, $N_z = 1.496$. Found as surface efflorescences on soils at *Sindo and Luanda, near Lake Victoria, Kenya*, and in the *Konya Basin, Central Anatolian Plateau, Turkey*. RG *AM 66:632(1981)*.

29.4.3.1 Löweite $Na_{12}Mg_7(SO_4)_{13} \cdot 15H_2O$

Named in 1847 for Alexander Loewe (1808–1895), Austrian chemist. Synonym: loeweite. HEX-R $R\bar{3}$. $a = 18.866$, $c = 13.434$, $Z = 3$, $D = 2.364$. *NBSM 25(14):35(1977)*. *29-1241*(syn): 10.4_{10} 5.45_3 4.29_7 4.05_6 3.46_5 3.28_5 3.18_4 2.70_5. Anhedral grains embedded in other sulfates. Colorless, also reddish yellow due to included material; vitreous luster. Conchoidal fracture. $H = 2\frac{1}{2}$–3. $G = 2.36$. Soluble in water. Tastes slightly bitter. Uniaxial $(-)$; $N_O = 1.490$, $N_E = 1.471$. Formed only in marine evaporites. With anhydrite at *Ischl* and Hallstatt, *Austria*; also at Hall, Tyrol, and in the Stassfurt area, Germany. RG *DI:446, CMP 6:201(1959), AM 55:378(1970)*.

29.4.4.1 Ferrinatrite $Na_3Fe^{3+}(SO_4)_3 \cdot 3H_2O$

Named in 1889 for the composition. HEX-R $P\bar{3}$. $a = 15.566$, $c = 8.69$, $Z = 6$, $D = 2.55$. *MM 41:375(1977)*. *41-609*: 7.78_{10} 4.39_5 3.89_2 3.43_7 3.30_5 2.94_6 2.91_9 2.83_6. Crystals short prismatic [0001], also cleavable masses or fibrous aggregates, cryptocrystalline. Colorless, white, greenish, bluish green, pale amethystine; vitreous luster. Cleavage $\{10\bar{1}0\}$, perfect; $(11\bar{2}0)$, fair; splintery fracture; brittle. $H = 2\frac{1}{2}$. $G = 2.55$–2.61. Uniaxial $(+)$; $N_O = 1.558$, $N_E = 1.613$ (Na). Alters to sideronatrite on standing in moist air. Vesuvius, Italy, in fumaroles, with aphthitalite and alum. *Mina La Compania, S of Sierra Gorda, Atacama desert, Chile*; also at Alcaparossa, Cerritos Bayos, and at Chuquicamata. RG *DII:456, TMPM 23:317(1976)*.

29.4.5.1 Polyhalite $K_2MgCa_2(SO_4)_4 \cdot 2H_2O$

Named in 1818 from the Greek for *many* and *salt*, in allusion to the several component salts present. TRIC $P1$ or $P\bar{1}$. $a = 11.69$, $b = 16.33$, $c = 7.60$, $\alpha = 91.60°$, $\beta = 90°$, $\gamma = 91.90°$, $Z = 4$, $D = 2.76$. *TMPM 14:75(1970)*. *21-982*: 6.00_1 5.93_1 3.18_{10} 2.95_2 2.91_3 2.90_1 2.89_3 2.85_3. Crystals rare, tabular $\{010\}$ or elongated [001]; usually massive or fibrous; twins on $\{010\}$ and $\{100\}$, often polysynthetic. Colorless or white, occasionally colored by included matter; vitreous to resinous luster. Cleavage $\{10\bar{1}\}$, perfect; parting $\{010\}$. $H = 3\frac{1}{2}$. $G = 2.78$. Decomposed by water with separation of gypsum. Biaxial $(-)$; $N_x = 1.547$, $N_y = 1.560$, $N_z = 1.567$; $2V = 62°$. Widely formed as a constituent of marine evaporites, associated with halite and anhydrite. Found in the TX–NM salt deposits and in Grand Co., UT. *Ischl, upper Aus-*

tria; also at Hallstatt and Hall. Abundant disseminated in halite in the Stassfurt district, Germany. Also from France, Poland, and England; very abundant in Kazakhstan. RG *DII:458, MM 29:667(1951)*.

29.4.5.2 Leightonite $K_2CuCa_2(SO_4)_4 \cdot 2H_2O$

Named in 1938 for Tomás Leighton Donoso (1896–1967), Chilean mining engineer and mineralogist, University of Santiago. ORTH *Fmmm*. $a = 11.67$, $b = 16.52$, $c = 7.492$, $Z = 4$, $D = 2.953$. *CM 7:272(1962)*. *15-128*: 3.18_6 2.90_{10} 2.51_1 2.40_1 2.22_2 1.78_3 1.46_1 1.047_1. Crystallizes as blades or laths elongated [001] and flattened {100}, rarely equant, also as cross-fiber veinlets. Pale watery blue to greenish blue, vitreous luster. No cleavage. $H = 3$. $G = 2.95$. Biaxial $(-)$; $XYZ = bca$; $N_x = 1.578$, $N_y = 1.587$, $N_z = 1.595$ (Na); $2V = 86°$; $r > v$, fairly strong. *Chuquicamata, Chile*, in open crevices, associated with atacamite and kröhnkite. RG *DII:461*.

29.4.6.1 Metavoltine $K_4Na_4(Fe^{2+},Zn)Fe_6^{3+}(SO_4)_{12}O_2 \cdot 20H_2O$

Named in 1883 from the Greek for *with* and voltine, because it was found associated with voltaite at the original locality. HEX-R $P\bar{3}$. $a = 9.545$, $c = 18.09$, $Z = 1$, $D = 2.435$. *IMA ABST:80(1994)*. *29-1043*: 18.2_8 9.1_{10} 3.76_2 3.29_3 3.07_2 2.90_2 2.60_2 2.02_4. Minute crystals tabular {0001} with prominent {10$\bar{1}$1}; fine granular, scaly, incrusting. Orange brown and yellowish brown to greenish brown, resinous luster. Cleavage {0001}, perfect. $H = 2\frac{1}{2}$. $G = 2.5$. Partially soluble in water, decomposed by dilute acids. Uniaxial $(-)$; $N_O = 1.588$–1.595, $N_E = 1.573$–1.581. O, dark yellowish brown; E, light yellow to pale greenish yellow. Near *Borate, Calico Hills, San Bernardino Co., CA*. In fumaroles on Vesuvius, Italy, and at the solfatara of Pozzuoli, near Naples. Near *Madeni Zakh, Iran*, as an alteration of pyritic trachyte. At several localities in Chile, with copiapite, alunogen, and other sulfates at Chuquicamata, Sierra Gorda, Quetena, Alcaparrosa, and La Compania. RG *DII:619, TMPM 23:155(1976)*.

29.4.7.1 Görgeyite $K_2Ca_5(SO_4)_6 \cdot H_2O$

Named in 1953 for Rolf von Görgey (1886–1915), Austrian mineralogist, student of the petrography of Austrian salt deposits. MON $C2/c$. $a = 17.47$, $b = 6.83$, $c = 18.22$, $\beta = 113.30°$, $Z = 4$, $D = 2.887$. *NJMM:126(1965)*. *18-997*: 5.60_1 3.16_7 3.03_2 3.01_{10} 2.82_4 2.76_3 1.90_2 1.84_3. Crystals thin tabular {100}. Colorless to yellowish, vitreous luster. Cleavage {100}, distinct; splintery fracture. $H = 3\frac{1}{2}$. $G = 2.75$. Insoluble in hot water. Biaxial $(+)$; $N_x = 1.560$, $N_y = 1.569$, $N_z = 1.584$; $2V = 79°$. From the *Leopold horizon, Ischl salt deposit, Austria*, with glauberite, halite, and polyhalite. RG *AM 39:403(1954), NJMM:26(1965), ZK 151:49(1980)*.

29.5.1.1 Krausite $KFe^{3+}(SO_4)_2 \cdot H_2O$

Named in 1931 for Edward Henry Kraus (1875–1973), American mineralogist, University of Michigan. MON $P2_1/m$. $a = 7.91$, $b = 5.15$, $c = 8.99$,

$\beta = 102.75°$, $Z = 2$, $D = 2.839$. AM $50{:}1929(1965)$. $18{-}1028$: 6.59_7 4.40_8 4.26_5 3.69_7 3.09_{10} 2.77_4 2.58_4 2.55_4. Crystals short prismatic [001], also equant to tabular {001}. Pale lemon yellow, white streak, vitreous luster. Cleavage {001}, perfect; {100}, good. $H = 2\frac{1}{2}$. $G = 2.84$. Slowly decomposed by water, soluble in HCl, in CT yields acid water. Biaxial (+); $Z = b$; $Y \wedge c = -35°$; $N_x = 1.588$, $N_y = 1.650$, $N_z = 1.722$; 2V large. X, colorless; Y,Z, pale yellow. *Borate, Calico Hills, San Bernardino Co., CA.* Santa Maria mine, Velardeña, DGO, Mexico, in drusy crusts in limestone. RG $DII{:}462$, AM $50{:}504(1965)$.

29.5.2.1 Goldichite $KFe^{3+}(SO_4)_2 \cdot 4H_2O$

Named in 1955 for Samuel Stephen Goldich (b.1909), American mineralogist and petrologist, University of Minnesota. MON $P2_1/c$. $a = 10.387$, $b = 10.486$, $c = 9.086$, $\beta = 101.68°$, $Z = 4$, $D = 2.461$. AM $56{:}1919(1971)$. $11{-}428$: 10.3_8 7.35_9 6.85_7 4.00_6 3.40_6 3.25_5 3.07_{10} 2.66_6. Crystals short prismatic [001], as radiating clusters, fine grained incrustations. Pale yellow-green, lavender in artificial light; vitreous luster. Cleavage {100}, perfect. $H = 2\frac{1}{2}$. $G = 2.43$. Slightly soluble in cold water, easily in hot; soluble in dilute acids. Biaxial (+); $X = b$; $Z \wedge c = 11°$; $N_x = 1.582$, $N_y = 1.602$, $N_z = 1.629$; 2V $= 82°$. *Dexter No. 7 mine, Calf Mesa, San Rafael Swell, UT*, associated with other sulfates in a pyritic uranium mine. Gansu Prov., China, in crystals to several centimeters. Santa Barbara mine, Jujuy, Argentina. RG AM $40{:}469(1955)$, $50{:}505(1965)$.

29.5.3.1 Tamarugite $NaAl(SO_4)_2 \cdot 6H_2O$

Named in 1889 for the locality. MON $P2_1/a$. $a = 7.353$, $b = 25.225$, $c = 6.10$, $\beta = 95.20°$, $Z = 4$, $D = 2.066$. AM $54{:}19(1969)$. $19{-}1186$: 12.5_2 4.22_{10} 4.21_8 4.16_1 3.96_3 3.65_6 3.15_2 2.90_3. Crystals tabular {010}, also short prismatic [001], as fibrous or granular masses. Colorless, vitreous luster. Cleavage {010}, perfect. $H = 3$. $G = 2.06$. Soluble in water. Tastes sweetish, astringent. Biaxial (+); $Y = b$; $X \wedge c = 4°$; $N_x = 1.484$, $N_y = 1.486$, $N_z = 1.497$ (Na); 2V $= 60°$. In MO near Eureka, St. Louis Co., and at Fulton, Calloway Co., altering from mendozite. St. Bartholomew Is., West Indies. At the Skouriotissa mine, Cyprus. Miseno near Naples, Italy. *Tamarugal Pampa, Cerro Pintado, Tarapacá, Chile.* Anglesea, VIC, Australia. RG $DII{:}466$, AM $51{:}1805(1966)$, MM $40{:}642(1976)$, $NJMM{:}171(1987)$.

29.5.3.2 Amarillite $NaFe^{3+}(SO_4)_2 \cdot 6H_2O$

Named in 1933 for the locality. MON $C2/c$. $a = 8.419$, $b = 10.841$, $c = 12.472$, $\beta = 95.48°$, $D = 2.222$. AM $77{:}212(1992)$. Crystals equant, thick tabular {001}; rarely prismatic [001]. Pale yellow with a greenish tint; vitreous, brilliant luster. Cleavage {110}, good; conchoidal fracture. $H = 2\frac{1}{2}{-}3$. $G = 2.19$. Soluble in water, tastes astringent. Biaxial (+); $Y = b$; $X \wedge c = -39°$; $Z \wedge c = 51°$; $N_x = 1.532$, $N_y = 1.555$, $N_z = 1.591$; 2V large; r < v. *Tierra Amarilla, near Copiapó, Chile*, as veinlets in massive coquimbite. RG $DII{:}468$.

29.5.4.1 Mendozite NaAl(SO$_4$)$_2$ · 11H$_2$O

Named in 1868 for the locality. MON $C2/c$. $a = 21.75$, $b = 9.11$, $c = 8.30$, $\beta = 92.47°$, $Z = 4$, $D = 1.781$. *AM 57:1081(1972)*. *22-475*(syn): 10.9$_3$ 4.76$_9$ 4.61$_3$ 4.58$_7$ 4.15$_5$ 3.60$_4$ 3.50$_{10}$ 2.64$_4$. Fibrous masses. Colorless to white. Cleavage {100}, good. H = 3. G = 1.730–1.765. Biaxial (−); X = b; Y ∧ $c = 30°$; $N_x = 1.449$, $N_y = 1.461$, $N_z = 1.463$; 2V = 56°; distinct, crossed dispersion. Alters to tamarugite upon exposure to air. Shimaré, Idzumo, Japan. *San Juan, Mendoza, Argentina.* Many other occurrences of hydrated sodium aluminum sulfates are known which are suspected to contain mendozite, without definite proof that this phase is present. These include places in MO, CA, and UT. RG *DII:469*.

29.5.4.2 Kalinite KAl(SO$_4$)$_2$ · 11H$_2$O

Named in 1868 for the composition: contains potassium (kalium). MON $C2/c$. $a = 19.92$, $b = 9.27$, $c = 8.304$, $\beta = 98.79°$, $Z = 4$, $D = 2.00$. *ICDD Grant-in-Aid (1988)*. *41-1362:* 4.94$_6$ 4.80$_{10}$ 4.60$_5$ 4.31$_9$ 4.12$_7$ 3.49$_8$ 2.68$_3$ 1.873$_3$. A fibrous, birefringent type of alum distinct from isometric potassium alum. Fibrous. Biaxial (−), or uniaxial; Y ∧ $c = 13°$; $N_x = 1.430$, $N_y = 1.452$, $N_z = 1.458$; 2V = 52°. San Bernardino Co., CA (uniaxial); Mt. Wingen, Australia; with jarosite from Quetena, Chile. RG *DII:471*.

29.5.5.1 Potassium alum KAl(SO$_4$)$_2$ · 12H$_2$O

Named in 1875 for the composition. Synonym: potash alum. ISO $Pa3$. $a = 12.157$, $Z = 4$, $D = 1.753$. *NBSC 539(6):36(1956)*. Structure: *AC 22:793(1967)*. *7-17*(syn): 5.44$_4$ 4.96$_2$ 4.30$_{10}$ 4.05$_5$ 3.25$_6$ 3.04$_3$ 2.95$_2$ 2.79$_4$. Octahedral crystals; as cubes when crystallized from alkaline solutions; also massive, granular, stalactitic, as coatings. Twins on {111}, rare. Colorless; vitreous luster. Cleavage {111}, poor; conchoidal fracture. H = 2–2$\frac{1}{2}$. G = 1.757. Soluble in water; in CT, dissolves in its own water of crystallization. Tastes sweetish and astringent. Isotropic; N = 1.456; anomalous birefringence, frequent. Alum Cave, Sevier Co., TN. Near Silver Peak, Esmeralda Co., NV. The Geysers, Sonoma Co., and at the Sulphur Bank mercury mine, Lake Co., CA. Numerous places in Germany and France in brown coal deposits. Similarly in Spain and England. Chuquicamata, Chile. RG *DII:472*.

29.5.5.2 Sodium alum NaAl(SO$_4$)$_2$ · 12H$_2$O

Named for the composition. Synonym: soda alum. ISO $Pa3$. $a = 12.214$, $Z = 4$, $D = 1.670$. *NBSM 25(15):68(1978)*. *29-1167*(syn): 7.05$_1$ 4.31$_{10}$ 3.53$_1$ 3.26$_1$ 2.96$_4$ 2.49$_1$ 1.908$_1$ 1.632$_1$. Structure: *AC 22:182(1967)*. Artificial crystals are octahedral. Colorless, vitreous luster. Conchoidal fracture. H = 3. G = 1.67. Soluble in water, fuses in its water of crystallization at about 63°. Isotropic; N = 1.439. K does not substitute for Na to any significant extent in natural material. A number of occurrences of sodium alum in nature have been reported, but none can be referred with certainty to this species. See mendozite (29.5.4.1). RG *DII:474*.

29.5.5.3 Tschermigite $NH_4Al(SO_4)_2 \cdot 12H_2O$

Named in 1853 for the locality. Synonym: ammonia alum. ISO $Pa3$. $a = 12.240$, $Z = 4$, $D = 1.642$. NBSC $539(6):3(1956)$. Structure: ZK $157:147(1982)$. 7-22(syn): 7.07_6 5.48_5 5.00_4 4.33_{10} 4.08_8 3.69_4 3.27_8 2.81_4. Fibrous and columnar masses, efflorescences; artificial crystals are octahedral. Colorless to white, vitreous or silky luster. Conchoidal fracture. $H = 1\frac{1}{2}$. $G = 1.645$. Melts at 93° in its water of crystallization. Tastes sweetish, astringent. Isotropic; $N = 1.459$ (Na). Natural material conforms closely to the formula. In artificial material K can substitute for (NH_4) and a complete solid-solution series can be prepared. Formed chiefly in lignite or brown coal beds; in burning coal seams or coal waste heaps; in fumaroles. The Geysers, Sonoma Co., CA, with boussingaultite, mascagnite, and voltaite. Wamsutter, WY, with gypsum and ammoniojarosite in bituminous shale. *Tschermig, Bohemia, Czech Republic*, in brown coal. In the solfatara of Pozzuoli at Naples and in the crater of Etna, Italy. With sulfur and mascagnite at Nyamlagira volcano, Zaire. RG *DII:475, NJMM:171(1987)*.

29.5.5.4 Lonecreekite $NH_4Fe^{3+}(SO_4)_2 \cdot 12H_2O$

Named in 1983 for the locality. ISO $Pa3$, $a = 12.302$, $Z = 4$, $D = 1.691$. AM $71:229(1986)$. 7-5: 7.1_6 5.51_3 4.34_{10} 4.10_6 3.71_5 3.29_7 3.08_4 2.08_3. Colorless, vitreous luster. $G = 1.693$. Isotropic; $N = 1.483$. Dehydrates to sabieite. *Lone Creek Fall cave, near Sabie, eastern Transvaal, South Africa*, where oxidizing pyrite reacts with ammonia produced by the decay of organic matter. RG *AGSSA 17:29(1985), AM 71:229(1986)*.

29.6.1.1 Bassanite $2CaSO_4 \cdot H_2O$

Named in 1910 for Francesco Bassani (1853–1916), Italian paleontologist, University of Naples. Synonym: plaster of Paris. ORTH Space group unknown. $a = 12.031$, $b = 12.695$, $c = 6.934$, $Z = 6$, $D = 2.731$. NBSM $25(18):22(1981)$. 41-224(syn): 6.01_8 3.47_5 3.01_{10} 2.80_9 1.849_2 1.845_3 1.693_2 1.665_2. Chalky, as microscopic needles in parallel arrangement, sometimes pseudomorphous after gypsum. Snow white. $G = 2.69$–2.76. Dehydrates at about 130°, somewhat unstable, rehydrates gradually to gypsum in a normal slightly humid atmosphere. Uniaxial or biaxial (+); $N_x = 1.550$, $N_z = 1.577$ (Na); 2V small; positive elongation. Found in drill core at Darby dry lake and at a dry lake near Ballarat, southeastern CA, associated with gypsum, celestite, and clay. *Vesuvius, Italy*, in cavities in leucite–tephrite blocks, 1906 eruption, and with gibbsite in fumaroles, 1911 eruption. RG *DII:476; AM 38:1266(1953), 69:910(1984); MM 35:354(1965)*.

KIESERITE GROUP

The kieserite group consists of six isostructural sulfate monohydrates of the general formula

$$ASO_4 \cdot H_2O$$

where

A = Mg, Fe, Mn, Cu, Zn, Ni

and the structure is monoclinic, isostructural with kieserite. The structure of these very rare minerals is described in *NJMA 157:121(1987)*.

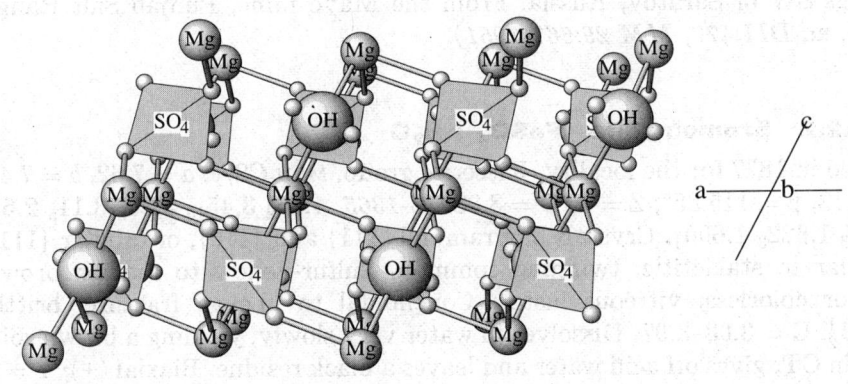

KIESERITE: $Mg(SO_4) \cdot H_2O$

Mg and SO_4/H_2O alternate in a layer structure on (001). Mg is octahedrally coordinated, to four sulfate oxygens and two H_2O.

KIESERITE GROUP

Mineral	Formula	Space Group	a	b	c	β	D	
Kieserite	$MgSO_4 \cdot H_2O$	A2/a	7.511	7.611	6.921	116.17°	2.589	29.6.2.1
Szomolnokite	$FeSO_4 \cdot H_2O$	C2/c	7.62	7.47	7.12	115.85°	3.08	29.6.2.2
Szmilkite	$MnSO_4 \cdot H_2O$	A2/a	7.766	7.666	7.120	115.85°	2.943	29.6.2.3
Poitevinite	$(Cu,Fe,Zn)SO_4 \cdot H_2O$	C2/c	7.48	7.42	7.05	114.67°	3.30	29.6.2.4
Gunningite	$(Zn,Mn)SO_4 \cdot H_2O$	A2/a	7.508	7.587	6.936	116.25°	3.364	29.6.2.5
Dwornikite	$(Ni,Fe)SO_4 \cdot H_2O$	P2/c	6.839	7.582	7.474	117.85°	3.346	29.6.2.6

29.6.2.1 Kieserite $MgSO_4 \cdot H_2O$

Named in 1860 for D. G. Kieser (1779–1862), German, president of the Jena Academy. Kieserite group. MON $A2/a$. $a = 7.511$, $b = 7.611$, $c = 6.921$, $\beta = 116.17°$, $Z = 4$, $D = 2.589$. NBSM 25(16):46(1979). 33-882(syn): 4.82_8 3.41_{10} 3.35_7 3.31_7 3.05_4 2.56_4 2.52_3 2.05_2. Massive granular, intergrown with other salts; crystals rare, bipyramidal {$\bar{1}11$} and {110}. Twinning polysynthetic; also contact on {001}. Colorless, grayish white, or yellowish; vitreous luster. Cleavage {110} and {111}, perfect; friable to firm. $H = 3\frac{1}{2}$. $G = 2.571$. Dissolves slowly in water, easily fusible. Biaxial (+); $Y = b$; $Z \wedge c = -76.5°$; $N_x = 1.520$, $N_y = 1.533$, $N_z = 1.584$ (Na); $2V = 55°$; $r > v$, moderate. Formed principally in marine evaporites. Common in the Permian salt deposits of west TX and NM, with polyhalite, anhydrite, and halite. *Stassfurt* and other places in the potash deposits of *Germany*. Good crystals were found at Hildesheim, Wathlingen, and Westerregeln; also at Hallstatt, upper Austria. Kalusz, Galicia, Poland. Kieserite constitutes up to 99% of salt deposits penetrated by borings SW of Saratov, Russia. From the Mayo mine, Punjab Salt Range, India. RG *DII:477, MM 29:667(1951)*.

29.6.2.2 Szomolnokite $FeSO_4 \cdot H_2O$

Named in 1877 for the locality. Kieserite group. MON $C2/c$. $a = 7.62$, $b = 7.47$, $c = 7.12$, $\beta = 115.85°$; $Z = 4$, $D = 3.08$. 45-1365: 4.86_2 3.45_{10} 3.30_3 3.11_5 2.53_4 1.998_3 1.622_3 1.600_4. Crystals bipyramidal ($\bar{1}11$} and {110}, or tabular {$\bar{1}11$}; globular to stalactitic; twinning common. Sulfur-yellow to reddish brown; blue or colorless, vitreous luster. Conchoidal to uneven fracture, brittle. $H = 2\frac{1}{2}$. $G = 3.03$–3.07. Dissolves in water very slowly, yielding a brown solution; in CT, gives off acid water and leaves a black residue. Biaxial (+); $Y = b$; $X \wedge c = -26°$; $N_x = 1.591$, $N_y = 1.623$, $N_z = 1.663$; $2V = 80°$; $r > v$, strong. Commonly associated with oxidizing pyrite. At the Tintic Standard mine, Dividend, UT. *Szomolnok, Slovakia*, where it was found with rhomboclase; also near Skrivan, Bohemia, Czech Republic. Alcaparrosa, Chile, with roemerite and rhomboclase. With copiapite at the Sta. Elena mine, La Alcaparrosa, San Juan, Argentina. RG *DII:479, CM 11:958(1973), PM 54:32(1985)*.

29.6.2.3 Szmikite $MnSO_4 \cdot H_2O$

Named in 1877 for Ignaz Szmik (1815–1881), Hungarian mining official at Felsobanya. Kieserite group. MON $A2/a$. $a = 7.766$, $b = 7.666$, $c = 7.120$, $\beta = 115.85°$, $Z = 4$, $D = 2.943$. NBSM 25(16):49(1979). 33-906(syn): 4.92_5 4.86_3 3.51_{10} 3.45_2 3.36_3 3.14_4 2.61_2 2.58_3. Stalactitic masses with a botryoidal surface. Dirty white to reddish, rose-red. Splintery to earthy fracture. $H = 1\frac{1}{2}$. $G = 3.15$. Biaxial (+); $Z = b$; $N_x = 1.562$, $N_y = 1.595$, $N_z = 1.632$; $2V \approx 90°$. *Baia Sprie* (formerly *Felsöbánya*), Romania, as an efflorescence. RG *DII:481, PM 54:32(1985)*.

29.6.2.4 Poitevinite (Cu,Fe,Zn)SO$_4$ · H$_2$O

Named in 1964 for Theophile Eugene Poitevin (1888–1978), Canadian mineralogist, Geological Survey of Canada. Kieserite group. MON $C2/c$. $a = 7.48$, $b = 7.42$, $c = 7.05$, $\beta = 114.67°$, $Z = 4$, $D = 3.30$. *CM 8:109(1965)*. *15-120*: 4.85$_4$ 4.72$_5$ 3.46$_{10}$ 3.34$_2$ 3.26$_2$ 3.08$_8$ 2.51$_4$ 1.67$_3$. Powdery to vermiform, very fine grained. Salmon colored. H = 3–3$\frac{1}{2}$. G = 3.30. Biaxial: $N_x = 1.610$, $N_z = 1.636$; 2V large. The natural material from Lillooet, BC, has a Cu/Fe/Zn ratio of 100:92:16. *Hat Creek, Bonaparte River, Lillooet, BC, Canada*, associated with bonattite. RG *AM 50:263(1965)*.

29.6.2.5 Gunningite (Zn,Mn)SO$_4$ · H$_2$O

Named in 1962 for Henry Cecil Gunning (1901–1991), geologist, Geological Survey of Canada. Kieserite group. MON $A2/a$. $a = 7.508$, $b = 7.587$, $c = 6.936$, $\beta = 116.25°$, $Z = 4$, $D = 3.364$. *NBSM 25(19):86(1982)*. Structure: *CM 7:209(1962)*. *33-1476*: 4.81$_4$ 4.76$_4$ 3.40$_{10}$ 3.35$_2$ 3.31$_3$ 3.06$_5$ 2.52$_5$ 2.19$_2$. Fine-grained efflorescences. White, vitreous luster. H = 2$\frac{1}{2}$. G$_{syn}$ = 3.321. Easily soluble in cold water. Biaxial; $N_x = 1.577$, $N_z = 1.630$; birefringence = 0.06. *Comstock–Keno mine, Keno Hill, YT, Canada*, and at several other mines in the district, in association with oxidizing sphalerite, arsenopyrite, pharmacosiderite, jarosite, and gypsum. RG *AM 47:1218(1962)*.

29.6.2.6 Dwornikite (Ni,Fe)SO$_4$ · H$_2$O

Named in 1982 for Edward J. Dwornik, geologist, U.S. Geological Survey. Kieserite group. MON $P2/c$. $a = 6.839$, $b = 7.582$, $c = 7.474$, $\beta = 117.85°$, $Z = 4$, $D = 3.346$. *AM 68:642(1983)*. *21-974*: 4.74$_5$ 4.73$_7$ 3.33$_{10}$ 3.31$_3$ 3.29$_4$ 3.02$_5$ 2.49$_4$ 2.33$_2$. Aggregates of very fine-grained particles. Snow white with a faint greenish tinge. Other physical properties not measurable; probably close to those of szomolnokite. Biaxial; $N_{mean} = 1.63$. *Minasragra, Peru*, with patronite, various sulfates, and bitumen. RG *MM 46:351(1982)*.

GYPSUM GROUP

The gypsum group is comprised of five isostructural minerals, including a variety of anions. The general formula is

$$AXO_4 \cdot 2H_2O$$

where

A = Ca, Nd, (Y,Er)
XO$_4$ = SO$_4$, HPO$_4$, HAsO$_4$, PO$_4$

and the structure is a layered monoclinic structure in space group $C2/c$ or Cc. In the gypsum structure, there are two sheets consisting of $(SO_4)^{2-}$ ions bound to one another by Ca ions. The two sheets are separated by layers of water molecules. Each Ca is coordinated by six O's of SO$_4$ groups and by two water molecules. The water molecules lie near the centers of triangles formed by two oxygens (of sulfate groups) to which they are hydrogen bonded and by a Ca ion. Each water molecule links a Ca both to an oxygen in the same double sheet and one in a neighboring sheet. *AC(B) 38:1056(1982)*.

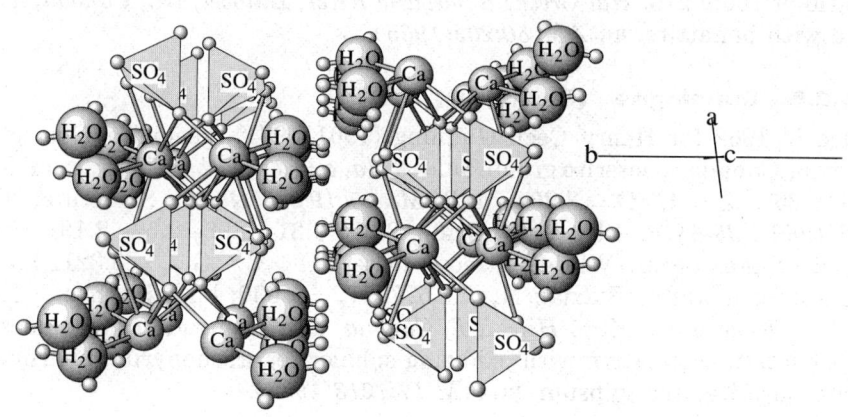

GYPSUM: CaSO₄ · 2H₂O

The (010) cleavage passes through the double layer of water molecules, where the structure is held together with hydrogen bonds at most.

GYPSUM GROUP

Mineral	Formula	Space Group	a	b	c	β	D	
Gypsum	CaSO$_4$ · 2H$_2$O	C2/c	5.678	15.20	6.52	118.43°	2.308	29.6.3.1
Brushite	CaHPO$_4$ · 2H$_2$O	C2/c	6.361	15.191	5.814	118.40°	2.318	39.1.1.1
Pharmacolite	CaHAsO$_4$ · 2H$_2$O	P2/m	5.975	15.434	6.280	114.80°	2.731	39.1.1.2
Churchite-(Nd)	NdPO$_4$ · 2H$_2$O	C2/c	6.19	15.14	5.61	115.10°	3.84	40.0.7.1
Churchite-(Y)	(Y,Er)PO$_4$ · 2H$_2$O	C2/c	5.61	15.14	6.19	115.30°	3.07	40.4.6.1

Note: Perfect (010) cleavage is due to cleavage along the sheets of water molecules in the structure.

29.6.3.1 Gypsum CaSO₄ · 2H₂O

Named in antiquity from the Greek *gypsos*, plaster, which was applied equally to gypsum, dehydrated gypsum, lime, and similar materials. Gypsum group. Synonyms: gips, selenite. MON $C2/c$. $a = 5.678$, $b = 15.20$, $c = 6.52$,

29.6.3.1 Gypsum

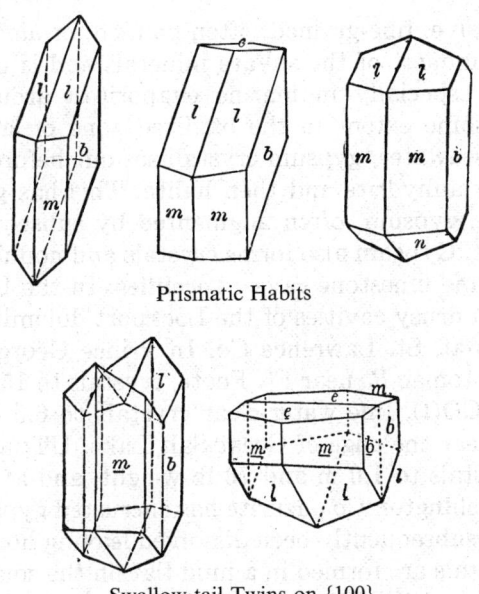

Prismatic Habits

Swallow-tail Twins on {100}

GYPSUM CaSO₄·2H₂O

Forms: b{010}, l{111}, m{110}, e{103}

β = 118.43°, Z = 4, D = 2.308. AC(B) 38:1074(1982). 33-311(syn): 7.63_{10} 4.28_{10} 3.80_2 3.07_8 2.87_5 2.69_4 2.22_2 2.09_3. **Habit**: Crystals tabular {010} with {$\bar{1}$11} and {120}. Also prismatic [001], stout to acicular, the prism zone often striated [001]. Sometimes lenticular by rounding of {$\bar{1}$11} and {$\bar{1}$03}. Crystals often have warped or curved surfaces, bent, or twisted. Rosettelike aggregates of lenticular crystals common. Also granular; massive, bedded, fibrous veinlets, or concretions. Twins common on {100} ("swallowtail twins"); on {$\bar{1}$01} as contact twins, butterfly- or heart-shaped. **Physical properties**: Colorless to white; yellowish, brownish due to impurities; white streak; subvitreous luster, pearly on cleavages. Cleavage {010}, perfect and easy, almost micaceous; {100} and {011}, distinct; cleavage fragments have rhombic form with angles of 66° and 114°; flexible, but not elastic. H = 2 (Mohs' scale defines hardness 2 as the hardness of gypsum). G = 2.317. Rarely fluorescent, or when caused by included fluorescent matter. **Tests**: Very slightly soluble in water, soluble in HCl, in CT loses water and becomes opaque, white. **Chemistry**: Only trace amounts of Ba, Sr, and occasionally Mg can substitute for the Ca, and gypsum adheres very closely to its theoretical formula. When heated above 70°, it slowly converts to the hemihydrate (bassanite); rapidly above 90°. At temperatures above 250° gypsum converts to anhydrite. **Optics**: Biaxial (+); Y = b; Z ∧ c = 51.87°; N_x = 1.521, N_y = 1.523, N_z = 1.530 (Na); 2V = 58°; r > v, strongly inclined. **Varieties**: (1) crystallized, or *selenite*—distinct crystals or broad folia, transparent; (2) fibrous, or *satin spar*—aggregates with a parallel fibrous structure, often translucent with the pearly opalescence of moonstone; (3) massive, fine-grained, often banded, or *alabaster*. **Occurrence:** Gypsum is the commonest of the sulfate minerals and is of extremely widespread occurrence, especially in marine evaporites; around solfatara and fumaroles; and to some extent in the oxidized zone of sulfide deposits. In the evaporation of seawater, gypsum crystallizes out before other sulfates or halides, followed by anhydrite and then halite. This has given rise to large sedimentary beds of gypsum, often augmented by subsequent hydration of associated anhydrite. Gypsum also forms crystals and nodules in certain clays and marls and in some limestone caves. **Localities:** In the United States, fine crystals are found in drusy cavities of the Lockport dolomite in Niagara Co., NY(*), and at Balmat, St. Lawrence Co. In Prince Georges Co., MD, in a road cut near the Potomac R. near Ft. Foote, crystals to 15 cm. At the Camp Bird mine, Ouray, CO(!), fine water-clear crystals to 6.5 cm. Large curved crystals are found near the edge of Great Salt Lake, UT; also at Hanksville, Wayne Co. (*), crystals to 1.0 m and 70 lb weight, and at the Apex mine(!) near St. George, Washington Co., azurite has encrusted gypsum crystals to 20 cm; the gypsum has subsequently been dissolved leaving hollow tubes of azurite. In OK, fine crystals are formed in a mud flat on the southwestern edge of the Salt Plains National Wildlife Refuge(*). Crystals form at shallow depth by precipitation from saline groundwater in response to weather conditions; most have zoned reddish inclusions in the form of an hourglass. In NM the sand in the White Sands National Monument is composed of gypsum; the dunes, covering about 560 km^2, constitute one of the world's greatest deposits. In AZ at the San Xavier mine, Pima Co.; in the Warren district, Cochise Co., excellent examples of the curved "ram's horn" variety. Commercial deposits of gypsum are exploited in many states and provinces of Canada. Near Winnipeg, MB, fine radiating aggregates of crystals forming balls up to 10 cm are found along the Red River floodway(!); also at Hines Creek, AB; Swift Current, SK, and at Endicott Pit No. 2, Endicott, NB. In Mexico, in a lead–zinc mine at Naica, CHIH, crystals are formed in clefts in limestone, attaining 2.0 m in length and are sharp and transparent (the "Cave of the Swords")(!). Similar large crystals are also found at the San Antonio mine, Santa Eulalia, CHIH(!). At the Bex mine, Vaud, Switzerland(!), in saline clays. Fine crystals are found at the sulfur mines of the Caltanisseta district, Sicily, Italy(!), associated with celestite and aragonite; the variety alabaster is mined extensively in Tuscany, and superb crystals are found in the pyrite mines there; also in Tuscany at the Cararra marble quarries, fine limpid crystals to 10 cm. Common in the clays at Cormeilles-en-Parisis near Paris, France(*) (hence "plaster of paris") and at Caresse, Pyrenées-Atlantique. In the salt mines of Stassfurt, Germany, and of Austria. The Machow sulfur mine, Tarnobrzeg, Poland, contains embedded gypsum crystals to 4.0 m. In Spain crystals of exceptional perfection, transparency, and brilliance to 10 cm are found in alabaster quarries near Fuentes de Ebro, SE of Zaragoza(!), and at El Gorguel, La Union, Murcia. In Ukraine at the Optimisticheskaya and Kristalnaya Caves, Tarnopol region. Gaurdak, Turkmenistan(!), fine transparent crystals. At Juayma, Ras Tanura, Saudi Arabia. In Africa pink sand "roses" are found in the Sahara desert; at the N'Chwaning mine(*), Kuruman district, South Africa. In Australia at the Broken Hill mine, NSW, and at Swan Hill, VIC. In Chile at the Teniente mine, enormous fine crystals (up to 4.0 m × 0.6 m in diameter) were found in 1974 in an open cavity of 300 m^3, as well as in similar smaller cavities; these were formed in the last phase of the primary mineralization. In Australia, at the Mt. Elliott mine, Q(!), large clear crystals to 1.0 m, some containing beautiful arborescent crystals of native Cu; from the Walaroo mine, York Peninsula, SA(!), fine crystals to 25 cm, some enclosing dolomite and chalcopyrite; also, in mud at the bottom of Lake Gilles, 30 km NW of Iron Knob, SA(!), fine transparent crystals and twins of a variety of habits; and at Broken Hill, NSW, and Swan Hill, VIC. In Brazil around the hills W of Iraí, RGS(*), large crystals (up to 20 kg) are found in amygdaloidal cavities associated with quartz and amethyst. Uses: Gypsum, upon moderate heating, alters to the hemihydrate (bassanite) normally known as plaster of paris. This material is used in huge amounts in home construction, the manufacture of wallboard, and similar products. Alabaster is widely used like marble for architectural purposes; the great pyramid of Cheops in Egypt was originally faced entirely with alabaster, which during the Middle Ages was all removed (stolen) for home construction. RG *DII:482*, *JSSC 85:23(1990)*.

29.6.4.1 Sanderite MnSO$_4$ · 2H$_2$O

Named for Bruno Sander (1884–1979), Austrian geologist, Alte Universität, Innsbruck. An inadequately characterized mineral. *20-689*(syn): 5.51$_3$ 4.42$_8$ 4.25$_9$ 3.92$_4$ 3.74$_4$ 3.25$_3$ 3.10$_{10}$ 2.73$_5$. Found as efflorescences on kieserite at the *Hansa I shaft, Empelde*, and the *Niedersachsen shaft, near Wathlingen, Austria*. RG *NJMM:28(1952)*, *AM 37:1072(1952)*.

29.6.5.1 Bonattite CuSO$_4$ · 3H$_2$O

Named in 1957 for Stefano Bonatti (1902–1968), Italian petrologist, Pisa. MON *Cc*. $a = 5.592$, $b = 13.029$, $c = 7.341$, $\beta = 97.03°$, z = 4, d = 2.68. *AC(B) 24:508(1968)*. *22-249*: 5.11$_7$ 4.42$_{10}$ 3.65$_6$ 3.42$_5$ 3.25$_6$ 3.19$_4$ 3.01$_6$ 2.50$_5$. Concretions composed of minute individual crystals; vermiform masses. Pale blue. G = 2.663. Biaxial (+); N$_x$ = 1.554, N$_y$ = 1.577, N$_{z\,syn}$ = 1.618. Avoca claim, Bonaparte R., near Lillooet, BC, Canada. *Cape Calamita, Elba, Italy*, as a secondary mineral in pyrite deposits. RG *RSIMP 13:268(1957)*; *AM 43:180(1958)*, *47:1223(1962)*; *CM 7:245(1962)*.

ROZENITE GROUP

The rozenite group consists of five minerals that are isostructural and that conform to

$$ASO_4 \cdot 4H_2O$$

where

A = Fe, Mg, Mn, Co, or Zn

and the structure is monoclinic, space group $P2_1/c$. The structure of these very rare minerals is described in *AC 15:815(1962)*.

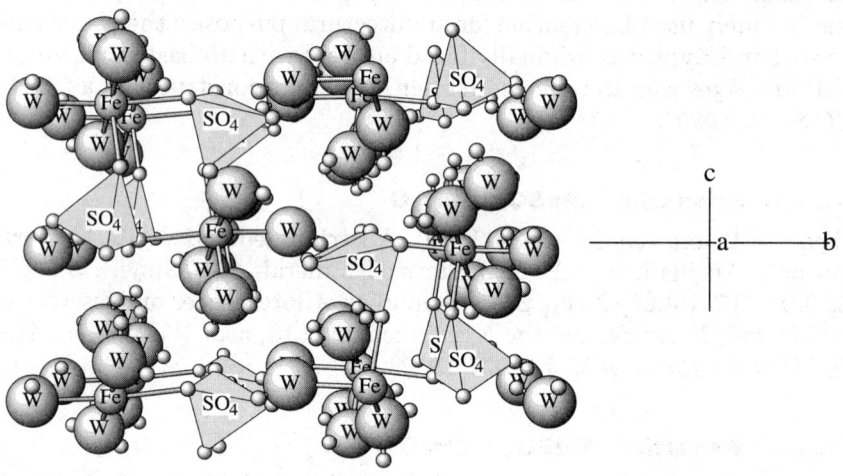

ROZENITE: $FeSO_4 \cdot 4H_2O$

This structure is composed of molecules containing two Fe octahedra and two sulfate tetrahedra, sharing orners.

ROZENITE GROUP

Mineral	Formula	Space Group	a	b	c	β	D	
Rosenite	$FeSO_4 \cdot 4H_2O$	$P2_1/c$	5.80	13.65	7.98	90.43°	2.28	29.6.6.1
Starkeyite	$MgSO_4 \cdot 4H_2O$	$P2_1/c$	5.92	13.60	7.91	90.85°	2.007	29.6.6.2
Ilesite	$MnSO_4 \cdot 4H_2O$	$P2_1/c$	5.94	13.76	8.01	90.80°	2.26	29.6.6.3
Aplowite	$(Co,Mn,Ni)SO_4 \cdot 4H_2O$	$P2_1/c$	5.94	13.56	7.90	90.50°	2.36	29.6.6.4
Boyleite	$ZnSO_4 \cdot 4H_2O$	$P2_1/c$	5.95	13.60	7.96	90.30°	2.34	29.6.6.5

29.6.6.1 Rozenite FeSO$_4$ · 4H$_2$O

Named in 1960 for Zygmunt Rozen (1874–1936), Polish mineralogist, Mining Academy, Krakov. Rozenite group. MON $P2_1/c$. $a = 5.80$, $b = 13.65$, $c = 7.98$, $\beta = 90.43°$, $Z = 4$, $D = 2.28$. AC 15:815(1962). 19-632(syn): 6.88$_3$ 5.48$_8$ 4.49$_{10}$ 3.99$_7$ 3.40$_4$ 3.23$_3$ 2.98$_4$ 2.96$_3$. Efflorescences and coatings, mixed with other sulfates, fine grained, crystalline. White or colorless. $G = 2.195$. Soluble in water. Biaxial $(-)$; $N_x = 1.527$, $N_z = 1.543$; $2V_{calc} = 82°$. Dolliver State Park, near Ft. Dodge, IA, on outcrops of the "copperas beds," with halotrichite and szomolnokite. *Stazic pyrite mine, Rudki*, and on pyritic gneiss on the slopes of *Ornak, western High Tatra, Poland*. RG AM 46:242(1961); CM 7:751(1963), 11:958(1973); MM 51:176(1987).

29.6.6.2 Starkeyite MgSO$_4$ · 4H$_2$O

Named in 1956 for the locality. Rozenite group. MON $P2_1/c$. $a = 5.92$, $b = 13.60$, $c = 7.91$, $\beta = 90.85°$, $Z = 4$, $D = 2.007$. 24-720: 6.83$_5$ 5.43$_8$ 4.70$_4$ 4.46$_{10}$ 3.95$_7$ 3.40$_5$ 3.22$_4$ 2.95$_6$. Powdery efflorescences. White to yellowish; greenish. Soluble in water. Biaxial $(-)$; $N \approx 1.496$; 2V large; r > v. Probably a complete solid-solution series exists between starkeyite and rozenite; the California material is ferroan starkeyite. *Starkey mine, Madison Co., MO*, on altered pyrite and marcasite in a dolomite rock. Alta troilite mine, Del Norte Co., CA. Hansa shaft, Empelde, Germany. RG AM 41:662(1956), AC 17:863(1963), CM 12:229(1973), MR 6:144(1975).

29.6.6.3 Ilesite MnSO$_4$ · 4H$_2$O

Named in 1881 for Malvern W. Iles (1853–1890), American metallurgist, Denver, CO. Rozenite group. MON $P2_1/c$. $a = 5.94$, $b = 13.76$, $c = 8.01$, $\beta = 90.80°$, $Z = 4$, $D = 2.26$. 32-651: 5.50$_9$ 4.55$_{10}$ 4.01$_9$ 3.46$_8$ 3.28$_8$ 3.01$_7$ 2.99$_9$ 2.58$_7$. Loosely adherent crystalline aggregates, prismatic. Clear green, becoming white on partial dehydration. $G_{syn} = 2.25$. Soluble in water. Biaxial $(-)$; $N_x = 1.511$, $N_y = 1.519$, $N_z = 1.521$; $2V_{calc} = 53°$. The original material contained Mn/Zn/Fe = 5.4:1.3:1. From several sulfide veins at the head of *Hall valley, Park Co., CO*. RG DII:486.

29.6.6.4 Aplowite (Co,Mn,Ni)SO$_4$ · 4H$_2$O

Named in 1965 for Albert Peter Low (1861–1942), Canadian geologist, former director of the Geological Survey of Canada. Rozenite group. MON $P2_1/c$. $a = 5.94$, $b = 13.56$, $c = 7.90$, $\beta = 90.50°$, $Z = 4$, $D = 2.36$. CM 8:166(1965). 16-488: 6.82$_4$ 5.44$_9$ 4.46$_{10}$ 3.94$_5$ 3.39$_4$ 3.21$_3$ 2.94$_3$ 2.56$_2$. Efflorescences and coatings, overlain by moorhouseite. Bright pink, white streak, vitreous luster. $H = 3$. $G = 2.33$. Biaxial $(-)$; $N_x = 1.528$, $N_z = 1.536$; 2V medium. Natural aplowite gave the ratio Co/Mn/Ni/Cu/Fe/Zn = 100:50:45:3:2:2. *Magnet Cove Barium Corp. mine, Walton, NS, Canada*, as an oxidation product on specimens of barite–siderite containing various sulfides, including cobaltian pyrite. RG AM 50:809(1965).

29.6.6.5 Boyleite $ZnSO_4 \cdot 4H_2O$

Named in 1978 for Robert William Boyle (b.1920), geochemist, Geological Survey of Canada. Rozenite group. MON $P2_1/c$. $a = 5.95$, $b = 13.60$, $c = 7.96$, $\beta = 90.30°$, $Z = 4$, $D = 2.34$. CE 37:73(1978). 31-818: 6.85_8 5.46_{10} 4.47_{10} 3.96_8 3.39_7 3.20_5 2.95_7 2.55_5. Earthy crusts, reniform masses. White. No cleavage, uneven fracture. $H \approx 2$. Soluble in water. Biaxial $(-)$; $N_x = 1.522$, $N_z = 1.536$; $2V \approx 70°$. Analyzed material contained $(Zn_{0.84}, Mg_{0.16})SO_4 \cdot 4H_2O$. In dry air dehydrates to gunningite. Keno Hill, Yukon, Canada. *Kropbach quarry, southern Black Forest, Germany*, formed from the decomposition of sphalerite, associated with gypsum. RG AM 64:241,464(1979).

CHALCANTHITE GROUP

The chalcanthite group consists of four isostructural minerals, sulfate pentahydrates, with the general formula

$$ASO_4 \cdot 5H_2O$$

where

A = Cu, (Fe, Cu), Mg, Mn

and the structure is in the triclinic space group $P\bar{1}$. The structure of these rare minerals is described in AC(B) 28:1448(1972). Some solid solution exists between the various members, especially between chalcanthite, siderotil, and pentahydrite. However, it is doubtful whether end-member $FeSO_4 \cdot 5H_2O$ exists, and the usual dehydration product of the higher hydrates is rozenite, $FeSO_4 \cdot 4H_2O$.

CHALCANTHITE GROUP

Mineral	Formula	Space Group	a	b	c	α	β	γ	D	
Chalcanthite	$CuSO_4 \cdot 5H_2O$	$P\bar{1}$	6.110	10.673	5.95	97.6	107.2	77.55	2.282	29.6.7.1
Siderotil	$(Fe,Cu)SO_4 \cdot 5H_2O$	$P\bar{1}$	6.26	10.63	6.06	92.1	110.2	77.1	2.212	29.6.7.2
Pentahydrite	$MgSO_4 \cdot 5H_2O$	$P\bar{1}$	6.31	10.57	6.03	81.1	109.8	105.1	1.93	29.6.7.3
Jokokuite	$MnSO_4 \cdot 5H_2O$	$P\bar{1}$	6.37	10.77	6.13	98.8	110.0	77.8	2.094	29.6.7.4

Note: The chalcanthite group minerals are all water soluble.

29.6.7.1 Chalcanthite

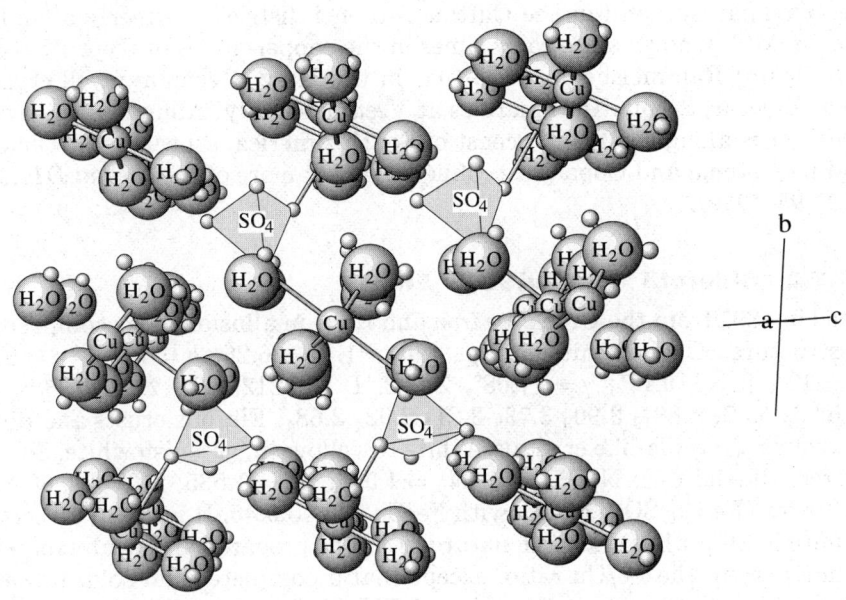

CHALCANTHITE: $CuSO_4 \cdot 5H_2O$

Each Cu octahedron has two sulfate oxygens and four H_2O. The octahedra and sulfate tetrahedra from corner-sharing chains parallel to (110)

29.6.7.1 Chalcanthite $CuSO_4 \cdot 5H_2O$

Named in antiquity from "chalcanthum," its old name, meaning *flowers of copper*. Chalcanthite group. TRIC $P\bar{1}$. $a = 6.110$, $b = 10.673$, $c = 5.95$, $\alpha = 97.58°$, $\beta = 107.17°$, $\gamma = 77.55°$, $Z = 2$, $D = 2.282$. *11-646*(syn): 5.48_6 4.73_{10} 3.99_6 3.71_9 3.30_6 2.82_4 2.75_5 2.66_4. Crystals short prismatic [001], less commonly thick tabular $\{\bar{1}\bar{1}1\}$. Natural crystals very rare, artificial crystals very easily made. Commonly stalactitic; as veinlets, massive, granular. Cruciform twins rare, with {010} in common. Sky blue of different shades, sometimes a little greenish; colorless streak; vitreous luster. Cleavage $\{1\bar{1}0\}$, imperfect; conchoidal fracture. $H = 2\frac{1}{2}$. $G = 2.286$. Soluble in water. Tastes metallic and nauseous. In CT, loses water and turns whitish but does not fuse. Biaxial $(-)$; $X \wedge \phi = 169°$, $X \wedge \sigma = 76°$; $Y \wedge \phi = 76°$, $Y \wedge \sigma = 78°$; $Z \wedge \emptyset = -57°$, $Z \wedge \sigma = 18°$; $N_x = 1.514$, $N_y = 1.537$, $N_z = 1.543$ (Na); $2V = 56°$; $r < v$. Mg, Mn, Zn, and Fe can all substitute for Cu in the cell to a minor degree, but in nature only Fe substitution is found to any extent, up to about 2%. Chalcanthite is a common mineral in the oxide zone of copper deposits and in arid regions may constitute an important ore mineral. Copper sulfate is common dissolved in mine waters, from which it may crystallize, forming crusts and stalactites on the timbers and walls of mines. Bluestone mine,

Yerington district, NV; widespread in AZ, especially at the United Verde mine, Yavapai Co., and in the Clifton–Morenci district. Formerly abundant at Butte, MT. Known since early times in the copper mines of Cyprus; at Rio Tinto, Spain; Rammelsberg and Goslar in the Harz, Germany; well crystallized at Zajecar, Serbia. In fumaroles at Vesuvius, Italy. Abundant at several copper mines along the Pacific coast of South America, especially at Chuquicamata, Quetena, and Copaquire, Chile. Used as an ore of copper. RG *DII:488*, *AM 37:95(1952)*.

29.6.7.2 Siderotil (Fe,Cu)SO$_4$ · 5H$_2$O

Named in 1891 from the Greek for *iron* and *fiber*, in allusion to its composition and structure. Chalcanthite group. TRIC $P\bar{1}$. $a = 6.26$, $b = 10.63$, $c = 6.06$, $\alpha = 92.13°$, $\beta = 110.17°$, $\gamma = 77.08°$, $Z = 2$, $D = 2.212$. *CM 7:751(1963)*. 22-357: 5.73_5 5.57_6 4.89_{10} 3.90_3 3.73_8 3.21_4 2.92_4 2.68_4. Fibrous crusts and divergent groups of needle-like crystals. White to yellowish, greenish white. Soluble in water. Biaxial (−); $N_x = 1.515$, $N_y = 1.525$, $N_z = 1.535$; $2V = 50$–$60°$. All data are for (Fe,Cu)SO$_4$ · 5H$_2$O, with Fe/Cu = 0.55:0.45. It is not at all certain that pure FeSO$_4$ · 5H$_2$O exists in nature or can be prepared artificially, or what the limits are in the Fe/Cu ratio, except that a complete solid-solution series exists for high-Cu members. Ducktown, TN; Yerington, NV; Bingham Canyon, UT; Mt. Diablo mine, Contra Costa Co., CA; *Idria, Gorizia, Italy*, with melanterite. Because of the fine-grained nature of most of the suspected occurrences and difficulties of analysis, the exact nature of some of the occurrences cited is uncertain. RG *DII:491*.

29.6.7.3 Pentahydrite MgSO$_4$ · 5H$_2$O

Named in 1951 for the composition (five waters of crystallization). Charcanthite group. TRIC $P\bar{1}$. $a = 6.31$, $b = 10.57$, $c = 6.03$, $\alpha = 81.12°$, $\beta = 109.82°$, $\gamma = 105.08°$, $Z = 2$, $D = 1.93$. 25-532(syn): 5.15_3 4.93_{10} 3.65_3 3.26_4 2.95_3 2.79_3 2.68_3 2.00_4. Granular crusts, massive; artificial crystals elongated [001]. Colorless to white. No cleavage. $G = 1.718$. Soluble in water. Biaxial (−); $N_x = 1.495$, $N_y = 1.512$, $N_z = 1.518$; $2V = 55°$; $r < v$. Reported from several localities, but usually as a very impure salt with other metals replacing part of the Mg. *The Geysers, Sonoma Co., CA*, as a pseudomorphous dehydration product of epsomite. Comstock Lode, NV. Copaquire, Tarapacá, Chile. RG *DII:492, AM 36:641(1951)*.

29.6.7.4 Jokokuite MnSO$_4$ · 5H$_2$O

Named in 1978 for the locality. Chalcanthite group. TRIC $P\bar{1}$. $a = 6.37$, $b = 10.77$, $c = 6.13$, $\alpha = 98.77°$, $\beta = 109.97°$, $\gamma = 77.83°$, $Z = 2$, $D = 2.094$. *AM 64:655(1979)*. 31-836: 5.84_{10} 5.66_6 4.98_6 4.39_3 3.28_3 2.73_7 2.29_3 1.62_3. As stalactites. Pale pink, vitreous luster. No cleavage. $H = 2\frac{1}{2}$. $G = 2.03$. Soluble in water. Biaxial (−); $N_x = 1.498$, $N_y = 1.510$, $N_z = 1.517$; $2V = 70$–$80°$; weak dispersion. *Jokoku manganese mine, southwestern Hokkaido, Japan*, with gypsum, szmikite, ilesite, rozenite, siderotil, ferrohexahydrite, malladrite, melanterite, and goslarite. RG *MJJ 9:28(1978)*.

HEXAHYDRITE GROUP

The hexahydrite group minerals are isostructural hexahydrates of sulfates corresponding to

$$ASO_4 \cdot 6H_2O$$

where

A = Mg, Zn, Fe, Co, Ni, or Mn

and the minerals crystallize with monoclinic symmetry in space group $C2/c$. The structure of these rare minerals is described in *AC 17:235(1964)*.

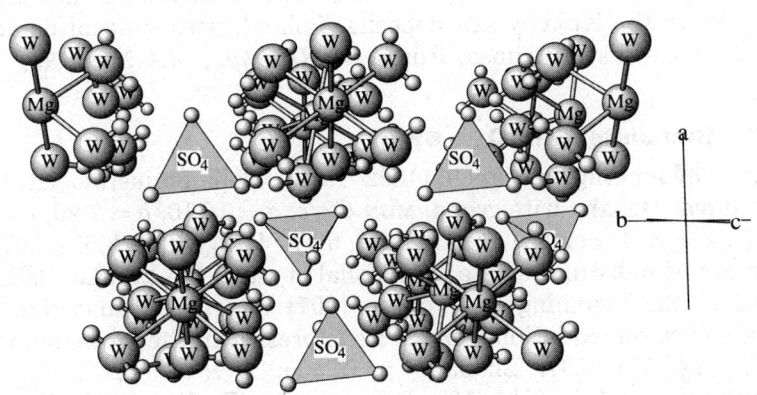

HEXAHYDRITE: $MgSO_4 \cdot 6H_2O$

Independent $Mg(H_2O)_6$ otahedra and sulfate groups, in layers on (001). The structural formula could be $Mg(H_2O)_6(SO_4)$.

HEXAHYDRITE STRUCTURE

Mineral	Formula	Space Group	a	b	c	β	D	
Hexahydrite	$MgSO_4 \cdot 6H_2O$	C2/c	10.11	7.21	24.41	98.30°	1.745	29.6.8.1
Bianchite	$ZnSO_4 \cdot 6H_2O$	C2/c	10.10	7.20	24.49	98.27°	2.031	29.6.8.2
Ferrohexahydrite	$FeSO_4 \cdot 6H_2O$	C2/c	10.08	7.28	24.59	98.37°	1.934	29.6.8.3
Nickel-hexahydrite	$NiSO_4 \cdot 6H_2O$	C2/c	9.895	7.241	24.188	98.42°	2.037	29.6.8.4
Moorhouseite	$(Co,Ni,Mn)SO_4 \cdot 6H_2O$	C2/c	10.03	7.23	24.26	98.37°	2.01	29.6.8.5
Chvaleticeite	$(Mn,Mg)SO_4 \cdot 6H_2O$	C2/c	10.05	7.24	24.3	98.00°	1.84	29.6.8.6

29.6.8.1   Hexahydrite $MgSO_4 \cdot 6H_2O$

Named in 1911 for the composition: six waters of crystallization. Hexahydrite group. MON $C2/c$. $a = 10.11$, $b = 7.21$, $c = 24.41$, $β = 98.30°$, $Z = 8$, $D = 1.745$. *24-719:* 5.45_5 5.10_5 4.88_3 4.39_{10} 4.16_4 4.04_5 2.94_3 2.90_3. Coarse columnar to delicately fibrous; crystals rare, thick tabular {001}. Colorless to white, pearly to vitreous luster. Cleavage {100}, perfect. $G = 1.757$. Soluble in water. Tastes bitter, salty. Biaxial (−); $Y = b$; $X \wedge c = -25°$; $N_x = 1.426$, $N_y = 1.453$, $N_z = 1.456$; $2V = 38°$. Formed as a dehydration product of epsomite. Oroville, WA, as efflorescence on epsomite. *Bonaparte River, BC, Canada*. Pseudomorphous after epsomite at Kelcany, Moravia, and at Kladno, Bohemia, Czech Republic. From the Kujawy salt deposits, Poland, with syngenite and other sulfates. Saki salt lakes, Crimea, Russia. RG *DII:494, MA 74:3459.*

29.6.8.2   Bianchite $ZnSO_4 \cdot 6H_2O$

Named in 1930 for Angelo Bianchi (1892–1970), Italian mineralogist, University of Padova. Hexahydrite group. MON $C2/c$. $a = 10.10$, $b = 7.20$, $c = 24.49$, $β = 98.27°$, $Z = 8$, $D = 2.031$. *12-16:* 5.85_5 5.47_7 4.42_{10} 4.03_9 3.61_6 2.97_8 2.91_8 2.27_6. Crusts of indistinct crystals; artificial crystals are tabular {001}, with {110} and {112}. Twinning common on {001} in artificial material. White, becoming yellow on oxidation if much Fe is present in the molecule; vitreous luster. $H = 2\frac{1}{2}$. $G = 2.07$. Biaxial (−); $Y = b$; $X \wedge c = -26°$; $N_x = 1.465$, $N_y = 1.494$, $N_z = 1.495$; $2V = 10°$. Data are for Zn/Fe = 2:1. Fe and Cu may substitute for some of the Zn, up to at least Zn/Fe = 2:1. From the walls of mines at *Raibl, Predil, Julian Alps, Italy*. Borieva mine, Madan region, Bulgaria. RG *DII:495, MA 75:1368, PM 54:9(1985).*

29.6.8.3   Ferrohexahydrite $FeSO_4 \cdot 6H_2O$

Named in 1962 for the composition. Hexahydrite group. MON $C2/c$. $a = 10.08$, $b = 7.28$, $c = 24.59$, $β = 98.37°$, $Z = 8$, $D = 1.934$. *ZVMO 91:490(1962). 15-393:* 4.89_6 4.43_{10} 2.97_7 2.93_7 2.80_5 2.03_6 1.88_6 1.86_6. Capillary or fibrous crystals. Colorless. Biaxial; $N_x = 1.468$, $N_z = 1.498$. *Northeastern Tataria, Russia*, in a core from lower Carboniferous deposits, possibly resulting from alteration of melanterite. RG *AM 48:433(1963).*

29.6.8.4 Nickel-Hexahydrite NiSO$_4$ · 6H$_2$O

Named in 1965 for the composition. Hexahydrite group. Dimorphous with retgersite (29.6.9.1). MON $C2/c$. $a = 9.895$, $b = 7.241$, $c = 24.188$, $\beta = 98.42°$, $Z = 8$, $D = 2.037$. *MM 39:246(1973). 33-955:* 5.42_2 4.90_5 4.78_2 4.37_{10} 4.10_2 4.00_6 2.92_3 2.89_4. Fine-grained crusts. Bluish green, vitreous luster. Cleavage {010}, perfect; {100}, fair. Biaxial (−); Z ∧ c = 45°; N$_x$ = 1.469, N$_z$ = 1.494. Considerable Mg and Fe may substitute for Ni in the structure. *Severnaya mine, Norilsk, Siberia, Russia. Noddy's Creek, TAS, Australia.* RG *ZVMO 94:534(1965), AM 51:529(1966), AC 12:72(1965).*

29.6.8.5 Moorhouseite (Co,Ni,Mn)SO$_4$ · 6H$_2$O

Named in 1965 for Walter Wilson Moorhouse (1913–1969), Canadian geologist, University of Toronto. Hexahydrite group. MON $C2/c$. $a = 10.03$, $b = 7.23$, $c = 24.26$, $\beta = 98.37°$, $Z = 8$, $D = 2.01$. *CM 8:166(1965).* Structure: *AC(C) 44:599(1988). 16-304:* 5.85_3 5.45_3 4.38_{10} 4.14_2 4.02_6 3.58_2 2.92_2 2.90_2. Crusts and efflorescences. Light pink, vitreous luster. Conchoidal fracture. $H = 2\frac{1}{2}$. $G = 1.97$. Soluble in water. Biaxial (−); Y = b; N$_x$ = 1.470, N$_z$ = 1.496; 2V ≈ 20°; X >> Z. Natural moorhouseite had Co/Ni/Mn/Cu/Fe/Zn = 100:45:21:9:6:1. *Magnet Cove Barium Corp. mine, Walton, NS, Canada,* encrusting sulfide-bearing barite–siderite rock. RG *AM 50:808(1965).*

29.6.8.6 Chvaleticeite (Mn,Mg)SO$_4$ · 6H$_2$O

Named in 1986 for the locality. Hexahydrite group. MON $C2/c$. $a = 10.05$, $b = 7.24$, $c = 24.3$, $\beta = 98.00°$, $Z = 8$, $D = 1.84$. *NJMM:121(1986). 42-614:* 5.45_8 4.91_{10} 4.47_8 3.98_8 3.69_5 3.42_7 3.25_8 2.97_7. Tough, fine-grained aggregates; loose coatings. White; also pinkish, yellowish. $H = 1\frac{1}{2}$. $G = 1.84$. Soluble in water. N$_x$ = 1.457, N$_z$ = 1.506. *Chvaletice, Bohemia, Czech Republic,* on pyrite–manganese ores, associated with melanterite, epsomite, and other sulfates. RG *AM 72:1023(1987).*

29.6.9.1 Retgersite NiSO$_4$ · 6H$_2$O

Named in 1949 for Jan Willem Retgers (1856–1896), Dutch chemical crystallographer. Dimorphous with nickel–hexahydrite (29.6.8.4). TET $P4_12_12$. $a = 6.782$, $c = 18.28$, $Z = 4$, $D = 2.075$. *NBSC 539(7):36(1957). 8-470*(syn): 4.64_2 4.57_4 4.25_{10} 3.39_1 2.96_2 2.72_2 2.57_1 2.33_1. Fibrous crusts and veinlets, elongated [001]. Crystals rare, thick tabular {001}, grading to short prismatic. Deep emerald green with a tinge of blue, greenish white streak, vitreous luster. Cleavage {001}, perfect; {110}, poor. Subconchoidal to uneven fracture; brittle. $H = 2\frac{1}{2}$. $G = 2.04$. Soluble in water. Tastes slightly bitter and metallic. Uniaxial (−); N$_O$ = 1.511, N$_E$ = 1.487. *Gap nickel mine, Lancaster Co., PA. Cottonwood Canyon, Churchill Co., NV,* as an alteration of nickeline. *Lichtenberg, Bavaria,* and *Lobenstein, Thuringia, Germany. Allarechensk deposit, Kola Penin., Russia,* on oxidized ores or as veinlets in cracks in ore. *Minasragra, Peru,* in crystals with morenosite and minasragrite. RG *DII:497, MA 19:143(1968).*

MELANTERITE GROUP

The melanterite group minerals are sulfate heptahydrates with the general formula

$$ASO_4 \cdot 7H_2O$$

with

A = Fe, Cu, Zn, Co, Mn

and a monoclinic structure in space group $P2_1/c$. The melanterite group minerals are believed to be isostructural, although in two cases cell data are incomplete. *AC 17:1167(1964)*.

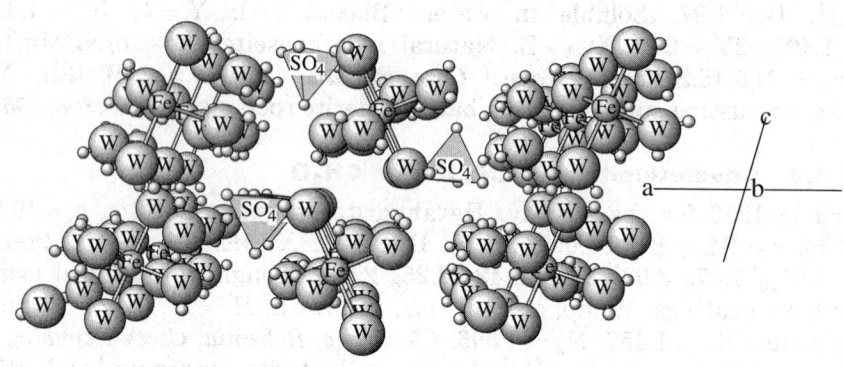

MELANTERITE: $FeSO_4 \cdot 7H_2O$

Completely independent (except for H bonds), $Fe)(H_2O)_6$ octahedra and sulfate groups alternate in layers on (100). There is one free H_2O.

MELANTERITE GROUP

Mineral	Formula	Space Group	a	b	c	β	D	
Melanterite	$FeSO_4 \cdot 7H_2O$	$P2_1/c$	14.077	6.509	11.054	105.60°	1.893	29.6.10.1
Boothite	$CuSO_4 \cdot 7H_2O$							29.6.10.2
Zinc-melanterite	$(Zn,Cu)SO_4 \cdot 7H_2O$							29.6.10.3
Bieberite	$CoSO_4 \cdot 7H_2O$	$P2_1/c$	14.13	6.55	11.00	105.40°	1.842	29.6.10.4
Mallardite	$MnSO_4 \cdot 7H_2O$	$P2_1/c$	14.15	6.50	11.06	105.60°	1.838	29.6.10.5

29.6.10.1 Melanterite FeSO$_4$ · 7H$_2$O

Named in 1850 from the Greek word for *copperas*, ferrous sulfate. Melanterite group. MON $P2_1/c$. $a = 14.077$, $b = 6.509$, $c = 11.054$, $\beta = 105.60°$, $Z = 4$, $D = 1.893$. *NBSM 25(8):38(1970)*. *22-633:* 5.49_1 4.90_{10} 4.87_5 4.03_1 3.78_6 3.73_2 3.29_2 3.21_1. Usually in stalactitic or concretionary forms; as fibrous or capillary aggregates, massive, pulverulent. Crystals rare, equant to short prismatic [001]; also pseudo-octahedral. Colorless to white or green; also greenish blue or blue with increasing substitution of Cu for Fe; upon exposure to dry air becomes yellowish white and opaque; vitreous luster. Cleavage {001}, perfect; {110}, distinct; conchoidal fracture; brittle. H = 2. G = 1.89. Soluble in water; in CT, yields water and on strong heating, SO$_2$ and SO$_3$. Tastes sweetish, astringent, and metallic. Biaxial (±); Y = b; Z $\wedge$ c = 61°; N$_x$ = 1.471, N$_y$ = 1.478, N$_z$ = 1.484 (Na); 2V large; r > v weak, inclined. A number of divalent metals, especially Cu, Mg, Zn, Co, Ni, and Mn, may substitute in part for the Fe, giving a partial solid-solution series to the respective end-member heptahydrates, especially Cu. The name *pisanite* is applied to members containing (Mg + Fe)/Cu < 5:1 and >2:1, the maximum observed in nature. *Kirovite* is a Mg-rich member having Mg/(Fe + Cu) > 1:5 and < 1:1. Melanterite and its varieties are secondary products formed by the oxidation of pyrite, marcasite, and other iron sulfides. They occur typically as beardlike efflorescences on the walls and timbers of mine workings; in the oxide zone of pyritic deposits, especially in arid regions; and in sheltered crevices in pyritic sedimentary and metamorphic rocks; also in coal and lignite deposits, where usually derived from marcasite. Associated minerals may include epsomite, chalcanthite, pickeringite, halotrichite, alunogen, alums, and gypsum. Melanterite alters in air by dehydration to siderotil or any of the other iron sulfate hydrates, depending on temperature and humidity conditions. Found at many U.S. mines; the copper deposits of Ducktown, TN; Butte, MT; Bingham Canyon, UT; Comstock Lode, NV; and the Cu deposits of Shasta and Trinity Cos., CA. Pisanite at Rammelsberg, Harz, Germany; Rio Tinto, Spain; Falun, Sweden; Quetena mine, Calama, Chile. Kirovite at the Kalata mine, Kirovgrad, Ural Mts., Russia; Idria, Italy; and Szomolnok, Czech Republic. Melanterite is used in the manufacture of fertilizers and inks and in water purification. RG *DII:499*.

29.6.10.2 Boothite CuSO$_4$ · 7H$_2$O

Named in 1903 for Edward Booth (1857–1917), American chemist, University of California. Melanterite group. MON Space group unknown. Usually massive with a crystalline structure, or fibrous. Crystals rare. Blue, paler than chalcanthite; vitreous to silky or pearly luster. Cleavage {001}, imperfect. H = 2–2½. G ≈ 2.1. Biaxial; X near c; Y = b; N$_x$ = 1.47, N$_y$ = 1.48, N$_z$ = 1.49; 2V large. Boothite is apparently separated by a gap in the solid-solution series from cuprian melanterite. *Alma pyrite mine, near Leona Heights, Alameda Co., CA*, with chalcanthite and melanterite. Also from Sain-Bel, near Lyon, Rhône, France. RG *DII:504*.

29.6.10.3 Zinc-melanterite (Zn,Cu)SO$_4$ · 7H$_2$O

Named in 1920 from the composition and relationship to melanterite. Melanterite group. MON Space group unknown. Massive, columnar, consisting of microscopic rodlike crystals. Pale greenish blue, vitreous luster. H ≈ 2. G = 2.02. Biaxial (+); Y ∧ elongation = 34°; N$_x$ = 1.479, N$_y$ = 1.483, N$_z$ = 1.488; 2V ≈ 90°; slight dispersion. The pure end member, ZnSO$_4$ · 7H$_2$O, is not stable in artificial systems. The name zinc–melanterite is applied to compositions where Zn > (Cu,Fe, etc.). Found in considerable amounts as an oxidation product of pyrite–chalcopyrite–sphalerite ore at the *Good Hope and Vulcan mines, Gunnison Co., CO*. RG *DII:508*.

29.6.10.4 Bieberite CoSO$_4$ · 7H$_2$O

Named in 1845 for the locality. Melanterite group. MON $P2_1/c$. $a = 14.13$, $b = 6.55$, $c = 11.00$, $\beta = 105.40°$, $Z = 4$, $D = 1.842$. *16-487:* 5.41$_1$ 4.87$_{10}$ 4.82$_6$ 4.02$_1$ 3.76$_8$ 3.71$_2$ 3.25$_1$ 2.73$_3$. Crusts and stalactites. Artificial crystals similar to melanterite. Rose-red or flesh-red, vitreous luster. Cleavage {001}, perfect; {110}, fair. H ≈ 2. G = 1.96. Soluble in water, in CT gives off water. Biaxial (+); Y = b; Z ∧ c = 29°; N$_x$ = 1.475, N$_y$ = 1.482, N$_z$ = 1.489 (Na); 2V$_{artif}$ = 88°. In natural material Mg may substitute for some of the Co, up to Mg/Co = 1:2.8. In artificial material a complete series exists between the Co and Fe analogs; Cu substitutes for Co in part. Formed from the oxidation of sulfide-bearing ores containing cobalt. Island Mt., Trinity Co., CA. *Bieber, Hesse, Germany*, and at *Siegen, Westphalia*, in the cobalt veins. *Chalanches, Isère, France. Tres Puntos, near Copiapó, Chile. Lomagundi, Zimbabwe*. RG *DII:505, PM 54:1(1985)*.

29.6.10.5 Mallardite MnSO$_4$ · 7H$_2$O

Named in 1879 for Ernest Mallard (1833-1894), French crystallographer. Melanterite group. MON $P2_1/c$. $a = 14.15$, $b = 6.50$, $c = 11.06$, $\beta = 105.60°$, $Z = 4$, $D = 1.838$. *MA 33:82–4639(1982)*. *33-905:* 5.49$_7$ 4.92$_{10}$ 4.88$_6$ 4.04$_3$ 3.79$_4$ 3.26$_4$ 3.13$_3$ 2.76$_4$. Fibrous masses and crusts, artificial crystals are tabular. Pale rose to colorless, vitreous luster. Cleavage {001}, good. H ≈ 2. G = 1.846$_{artif}$. Soluble in water; in CT loses water, and on strong heating affords SO$_3$ and leaves a brown residue. Biaxial (+); Y = b; Z ∧ c = 43°; N$_x$ = 1.462, N$_y$ = 1.465, N$_z$ = 1.474; 2V large. *Lucky Boy mine, Butterfield Canyon, Salt Lake Co., UT. Bayard area, Central district, Grant Co., NM* (cuprian and zincian mallardite). *Jokoku mine, Hokkaido, Japan*. RG *DII:507, BM 2:117(1879)*.

29.6.11.1 Epsomite MgSO$_4$ · 7H$_2$O

Named in 1824 for the locality. Synonym: Epsom salt. ORTH $P2_12_12_1$. $a = 11.86$, $b = 11.99$, $c = 6.858$, $Z = 4$, $D = 1.678$. *NBSC 539(7):30(1957)*. Structure: *AC 17:1361(1964)*. *36-419:* 5.98$_3$ 5.34$_3$ 5.31$_2$ 4.22$_{10}$ 4.20$_8$ 2.88$_2$ 2.67$_2$ 2.66$_3$. Crystals rare. Usually fibrous to acicular crusts, elongated [001]; botryoidal masses, stalactitic. Colorless or white, also pink (with Co) or green (with Ni); vitreous luster, also silky or earthy. Cleavage {010}, perfect; {101}, distinct; conchoidal fracture. H = 2–2$\frac{1}{2}$. G = 1.677. Soluble in water. Tastes bitter. Biaxial (–); XYZ = acb; N_x = 1.432, N_y = 1.455, N_z = 1.461; 2V = 52°; r < v, weak. Epsomite forms complete series with morenosite and goslarite. Fe, Mn, Cu, and Co can also substitute to a limited extent, up to Fe/Mg = 1:5 and Mn/Mg = 1:10. The name *epsomite* refers to Mg-dominant compounds. Epsomite commonly forms as crusts or delicate fibrous efflorescences on the walls of mine workings in coal or metal deposits, on the walls or floors of limestone caves, and as efflorescences in sheltered places on outcrops of dolomitic or calcareous sedimentary rocks. Also found in salt lakes and oceanic salt deposits. It dehydrates partially at room temperature in dry air, losing one H_2O. The change is reversible. Found in limestone caves in KY, TN, and IN. In salt lakes in Albany, Catron, and Natron Cos., WY. Carlsbad potash deposit, NM. Widespread in playas in NV; near Oroville, WA, in salt lakes with mirabilite. Near Ashcroft, BC, Canada, in lake deposits. *Epsom, Surrey, England*, in mineral waters and as deposits therefrom. Idria, Italy, in silky fibers at the mercury deposits. From mineral springs at Saidschitz and Sedlitz, Bohemia, Czech Republic, and in burning coal heaps near Kladno. Stassfurt, Germany, massive in thin layers with carnallite in the salt deposits. Used as a source of magnesium in industry; also used in medicine. RG *DII:509, JCS:816(1973)*.

29.6.11.2 Goslarite $ZnSO_4 \cdot 7H_2O$

Named in 1847 for the locality. ORTH $P2_12_12_1$. a = 11.799, b = 12.050, c = 6.822, Z = 4, D = 1.972. *NBSC 539(8):71(1958)*. *9-395:* 5.36_8 5.29_4 4.46_3 4.21_{10} 4.18_5 3.46_3 2.87_4 2.65_3. Efflorescent crusts; stalactites; massive, granular, or fibrous. Colorless to white, inclining to brownish, pinkish, or greenish in material containing Fe, Mn, or Cu; vitreous luster, silky if fibrous. Cleavage {010}, perfect. H = 2–2$\frac{1}{2}$. G = 1.978. Soluble in water; in CT, yields water. Biaxial (–); XYZ = bca; N_x = 1.457, N_y = 1.480, N_z = 1.484 (Na); 2V = 46°; r < v, weak. Formed from the oxidation of sphalerite in sulfide orebodies, occurring as an efflorescence on timbers and on the walls of mine passages. Webb City, MO (ferroan goslarite); Galena, KS (cuprian goslarite); Gagnon mine, Butte, MT; Bingham Canyon, UT; Comstock Lode, NV. *Rammelsberg mine, near Goslar, Harz, Germany*; also Freiberg, Saxony. Falun, Kopparberg, Sweden. Ceilmes, Herault, France, green incrustations on gossan. Almagrera mine, Huelva, Spain. Cerro de Pasco, Peru. La Alcaparrosa, San Juan, Argentina. RG *DII:513*.

29.6.11.3 Morenosite $NiSO_4 \cdot 7H_2O$

Named in 1851 for Antonio Moreno Ruiz (1796–1852), Spanish pharmacist and chemist, Academia de Ciencias Naturales, Madrid. ORTH $P2_12_12_1$. a = 11.86, b = 12.08, c = 6.81, Z = 4, D = 1.93. *1-403*(syn): 5.30_6 4.45_1

4.20_{10} 3.75_2 3.45_2 2.85_3 2.65_2 2.19_1. Efflorescent crusts of indistinct crystals and fibers; stalactitic. Apple green to greenish white, vitreous luster. Cleavage {010}, distinct; conchoidal fracture. H = $2-2\frac{1}{2}$. G = 1.953. Soluble in water; in CT, loses water, swells, and finally turns yellow. Tastes astringent. Biaxial $(-)$; XYZ = acb; N_x = 1.469, N_y = 1.489, N_z = 1.492; 2V = 42°; r > v, weak. Forms a complete solid-solution series with epsomite. Fe may substitute for some Ni forms by oxidation of nickel-bearing sulfides under atmospheric conditions. Dehydrates easily to retgersite. Gap nickel mine, Lancaster Co., PA. Abundant in the Sudbury nickel district, ONT, Canada, and at the Marbridge mine, LaMotte Twp., QUE, coating millerite. *Cap Hortegal, Galicia, Spain.* St. Patrick's copper mine, Avoca, Co. Wicklow, Eire. Jachymov, Bohemia, Czech Republic. Val Malenco, Lombardy, Italy. Minasragra, Peru, with retgersite. RG *DII:516, MM 33:1110(1964)*.

29.6.12.1 Minasragrite $(V^{4+}O)SO_4 \cdot 5H_2O$

Named in 1915 for the locality. MON $P2_1/a$. a = 12.947, b = 9.748, c = 7.005, β = 110.93°, Z = 4, D = 2.036. *AM 58:531(1973)*. *26-1393*: 6.05_2 5.43_6 5.14_{10} 3.91_7 3.83_4 3.73_2 3.66_3 3.51_2. Delicate efflorescences composed of minute crystals; granular, mammillary, or spherulitic. Blue, vitreous luster. No cleavage, conchoidal fracture. H = $1\frac{1}{2}-2$. G = 2.03. Easily soluble in cold water, in CT melts. Biaxial $(-)$; X = b; Z $\wedge$ c small; N_x = 1.513, N_y = 1.536, N_z = 1.545; absorption X > Y > Z. *Minasragra, Peru*, encrusting patronite. RG *DII:437*.

29.6.13.1 Stanleyite $(V^{4+}O)SO_4 \cdot 6H_2O$

Named in 1982 for Henry Morton Stanley (1841-1904), British journalist and explorer. ORTH Space group unknown. a = 12.12, b = 9.71, c = 14.92, Z = 8, D = 2.01. *MM 45:163(1982)*. *35-557*: 4.98_9 4.69_8 4.41_6 4.20_{10} 3.81_6 3.73_6 3.09_2 2.74_2. Fine-grained efflorescences consisting of minute crystallites. Aquamarine blue to cobalt blue, white streak. No cleavage. H = $1-1\frac{1}{2}$. G = 1.95. Nonfluorescent. Very soluble in cold water, slowly soluble in alcohol. Biaxial (+); N_x = 1.505, N_y = 1.519, N_z = 1.533; 2V large; absorption X = Y > Z. *Minasragra, Peru*, associated with patronite and pyrite. RG *AM 68:644(1983)*.

29.7.1.1 Ransomite $CuFe_2^{3+}(SO_4)_4 \cdot 6H_2O$

Named in 1928 for Frederick Leslie Ransome (1868–1935), U.S. mining geologist. MON $P2_1/c$. a = 4.811, b = 16.217, c = 10.403, β = 93.02°, Z = 2, D = 2.735. *AM 55:729(1970)*. Crusts and radiating tufts of needlelike crystals elongated [001]. Bright sky blue; vitreous luster, pearly on cleavage. Cleavage {001}, perfect. H = $2\frac{1}{2}$. G = 2.632. Biaxial (+); N_x = 1.631, N_y = 1.643, N_z = 1.695; 2V = 53°. Al substitutes for Fe^{3+} up to Al/Fe = 1:10. Found with other hydrated sulfates as the result of a fire in the *United Verde mine, Jerome, AZ*. RG *DII:519*.

29.7.2.1 Römerite $Fe^{2+}Fe_2^{3+}(SO_4)_4 \cdot 14H_2O$

Named in 1858 for Friedrich Adolph Römer (1809–1869), German geologist. TRIC $P\bar{1}$. $a = 6.48$, $b = 15.26$, $c = 6.27$, $\alpha = 90.07°$, $\beta = 100.80°$, $\gamma = 85.47°$, $Z = 1$, $D = 2.199$. *CM 6:348(1959)*. *34-144:* 5.06_4 4.81_{10} 4.25_2 4.12_2 4.05_7 3.98_3 3.17_2 3.05_2. Crystals cuboidal, thick tabular {010} with large {001} and prominent [001] zone. Crystal aggregates; stalactitic; granular. Rust-brown to yellow, also violet-brown; oily to vitreous luster. Cleavage {010}, perfect; {001}, good; uneven fracture. $H = 3–3\frac{1}{2}$. $G = 2.174$. Biaxial $(-)$; $N_x = 1.519–1.524$, $N_y = 1.570–1.571$, $N_z = 1.580–1.583$; $2V = 51–45°$; $r > v$, very strong, crossed. X, reddish yellow; Y, pale yellow; Z, yellow-brown. Island Mt., Trinity Co., CA. Copper Queen mine, Bisbee, AZ. *Rammelsberg, Harz, Germany*, associated with copiapite. Blyava sulfate deposits, southern Ural Mts., Russia. Madeni Zakh, Iran, with voltaite. Tierra Amarilla, Alcaparrosa, Chile. RG *DII:520, AM 55:78(1970)*.

29.7.2.2 Lishizhenite $ZnFe_2^{3+}(SO_4)_4 \cdot 14H_2O$

Named in 1990 for Li Shizhen (1518–1593), a famous Chinese pharmacologist. TRIC $P\bar{1}$. $a = 6.477$, $b = 15.298$, $c = 6.309$, $\alpha = 90.12°$, $\beta = 101.07°$, $\gamma = 93.58°$, $Z = 1$, $D = 2.201$. Structure: *AM 76:670(1991)*. *45-1394:* 5.07_7 4.79_{10} 4.11_3 4.06_7 3.98_4 3.51_8 3.05_2 2.86_4. Crystal aggregates and clusters; veinlets; massive. Crystals elongated [100], tabular prismatic and showing {010}, {100}, and {001}. Light violet, deeper in clusters; vitreous luster. Cleavage {010}, nearly perfect; {001}, poor. $H = 3\frac{1}{2}$. $G = 2.206$. Biaxial $(-)$; Optic Axis Plane $\wedge$ {010} $= 86°$; $N_x = 1.522$, $N_y = 1.568$, $N_z = 1.578$; $2V = 47°$. $r > v$, strong. *Xitieshan, Qinghai, China*, in the oxidation zone of a Pb–Zn deposit in cavities in anhydrite, associated with roemerite, copiapite, sulfur, gypsum, pyrite, and quartz. RG *AM 76:2022(1991)*.

HALOTRICHITE GROUP

The halotrichite group consists of isostructural hydrated sulfates corresponding to

$$AB_2(SO_4)_4 \cdot 22H_2O$$

where

$A = Mg, Fe^{2+}, Mn^{2+}$, or Zn
$B = Al, Fe^{3+}$, or Cr

and the structure is monoclinic in $P2_1/c$, $P2/m$, or $P2$. The structure of these rare minerals is described in *MM 40:599(1976)*.

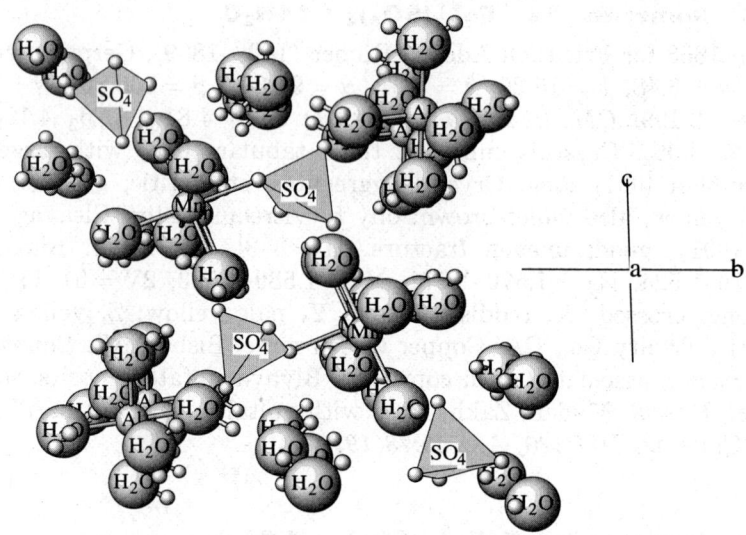

APJOHNITE: MnAl$_2$(SO$_4$)$_4$ · 22H$_2$O (halotrichite group)

This structure consists largely of independent radicals—there is only one bond from Mn to one of the four sulfate groups. Mn and Al are octahedrally coordinated by H$_2$O except for the one sulfate oxygen on Mn. The structural formula could be written 2[Al(H$_2$O)$_6$] · Mn(H$_2$O)$_5$(SO)$_4$ · 3(SO$_4$) · 9H$_2$O.

HALOTRICHITE GROUP

Mineral	Formula	Space Group	a	b	c	β	D	
Pickeringite	MgAl$_2$(SO$_4$)$_4$ · 22H$_2$O	$P2/m$	20.8	24.2	6.17	95.°	1.84	29.7.3.1
Halotrichite	Fe^{2+}Al$_2$(SO$_4$)$_4$ · 22H$_2$O	$P2/m$	20.51	24.29	6.179	101.°	1.95	29.7.3.2
Apjohnite	Mn^{2+}Al$_2$(SO$_4$)$_4$ · 22H$_2$O	$P2_1/c$	6.198	24.347	21.266	100.47°	1.87	29.7.3.3
Dietrichite	ZnAl$_2$(SO$_4$)$_4$ · 22H$_2$O	$P2_1/c$	6.240	24.434	21.379	100.60°	1.86	29.7.3.4
Bilinite	Fe^{2+}Fe$_2^{3+}$(SO$_4$)$_4$ · 22H$_2$O	$P2_1c$	6.208	24.333	21.255	100.18°	1.993	29.7.3.5
Redingtonite	(Fe^{2+},Mg)(Cr,Al)$_2$(SO$_4$)$_4$ 22H$_2$O							29.7.3.6
Wupatkiite	(Co,Mg)Al$_2$(SO$_4$)$_4$ · 22H$_2$O	$P2_1c$	6.189	24.234	21.204	100.33°	1.87	29.7.3.7

Note: A complete solid-solution series exists between pickeringite and halotrichite, and it is probable that this series extends to apjohnite.

29.7.3.1 Pickeringite MgAl$_2$(SO$_4$)$_4$ · 22H$_2$O

Named in 1844 for John Pickering (1777–1846), president, American Academy of Sciences. Halotrichite group. MON $P2/m$. $a = 20.8$, $b = 24.2$, $c = 6.17$, $\beta = 95°$, $Z = 4$, $D = 1.84$. *12-299:* 6.08$_2$ 4.97$_2$ 4.82$_{10}$ 4.32$_3$ 4.18$_2$ 4.12$_3$ 3.79$_3$ 3.51$_9$. Crystals acicular [001]; matted aggregates; tufted, spheroidal; as incrustations. Colorless to white, also yellowish, reddish; vitreous luster. Cleavage {010}, poor; conchoidal fracture; brittle. H = $1\frac{1}{2}$. G = 1.76. Soluble in water; in CT, fuses easily in its water of crystallization. Tastes astringent. Biaxial (−); Y = b; Z ∧ $c = 36°$; N$_x$ = 1.475, N$_y$ = 1.480, N$_z$ = 1.483; 2V = 60°. The

Geysers, Sonoma Co., CA. Near Tucumcari, Quay Co., NM. Fallon, Churchill Co., NV. Alum Point, Salt Lake Co., UT (manganoan). Elba, Italy, in the iron mines. Wetzelstein near Saalfeld, Germany. Terlano, Tyrol (manganoan), and Dienten, Salzburg (ferroan), Austria. Abundant in the deserts of the west coast of South America: *Cerros Pintados, near Iquique, Tarapacá, Chile*, and at Chuquicamata and Quetena. RG *DII:523, BGSA 74:709(1963)*.

29.7.3.2 Halotrichite $Fe^{2+}Al_2(SO_4)_4 \cdot 22H_2O$

Named in 1839 from the Latin *halotrichum*, which in turn is for the older German name *haarsalz*, hair salts, in allusion to its crystal habit. Halotrichite group. MON $P2/m$. $a = 20.51$, $b = 24.29$, $c = 6.179$, $\beta = 101°$, $Z = 4$, $D = 1.95$. *39-1387:* 15.8_2 6.02_4 4.93_2 4.78_{10} 4.29_3 3.95_2 3.75_3 3.48_8. Crystals acicular [001], similar to pickeringite. Colorless to white, yellowish, greenish; vitreous luster. Cleavage {010}, poor; conchoidal fracture; brittle. $H = 1\frac{1}{2}$. $G = 1.89$. Soluble in water giving an acid solution; in CT, melts in its water of crystallization. Biaxial $(-)$; $Y = b$; $Z \wedge c = 38°$; $N_x = 1.480$, $N_y = 1.486$, $N_z = 1.490$ (Na); $2V = 35°$. Probably forms a complete solid-solution series with pickeringite. However, most analyses of natural material are near the respective end members. Halotrichite and pickeringite are formed as a result of weathering of pyritic aluminous rocks and accumulate as efflorescences in sheltered places. Also observed as hairlike efflorescences in mine workings, especially coal mines. From outcrops of the "copperas" beds, Dolliver State Park, Webster Co., IA. Alum Mt. district, Grant Co., NM. In CA in the sulfate deposit near Calico Hills, San Bernardino Co., and near hot springs on Lassen Peak, Shasta Co. Glace Bay coal mines, NS, Canada. On brown coal at the Wohlforth mine, Hesse–Nassau, Germany. Idria mines, Italy, and at the solfatara of Pozzuoli near Naples. Fahlun, Sweden. The solfatara of Krisuvig, Iceland. Urumiya, Iran. Common in the Chilean desert regions. RG *DII:523, CM 10:958(1973)*.

29.7.3.3 Apjohnite $Mn^{2+}Al_2(SO_4)_4 \cdot 22H_2O$

Named in 1838 for James Apjohn (1796–1886), professor of chemistry and mineralogy, Trinity College, Dublin. Halotrichite group. MON $P2_1/c$. $a = 6.198$, $b = 24.347$, $c = 21.266$, $\beta = 100.47°$, $D = 1.87$. *MM 40:599(1976)*. *29-886:* 6.07_2 4.82_9 4.34_3 4.32_3 4.13_2 3.97_3 3.79_4 3.52_{10}. Masses and crusts composed of fibrous or acicular crystals; also asbestiform. Colorless to white; also tinted rose, pale green, or yellow; silky luster. $H = 1\frac{1}{2}$. $G = 1.78$. Soluble in water. Probably a complete series exists to pickeringite. Large amounts were found at Alum Cave, Sevier Co., TN, under an overhanging cliff at the headwaters of Little Pigeon creek. Szomolnok, Czech Republic. *Delagoa Bay, Mozambique*. RG *DII:527*.

29.7.3.4 Dietrichite $ZnAl_2(SO_4)_4 \cdot 22H_2O$

Named in 1878 for Gustav Heinrich Dietrich, Bohemian analytical chemist, Pribram, Czech Republic, who analyzed the mineral. He was said to have been among the first to be able to distinguish Fe^{2+} from Fe^{3+} in analysis. Halotrichite group. MON $P2_1/c$. $a = 6.240$, $b = 24.434$, $c = 21.379$, $\beta = 100.10°$, $Z = 4$, $D = 1.86$. *25-1173:* 4.99_3 4.84_4 4.33_{10} 4.20_2 4.15_3 3.52_{10} 2.74_2 1.88_3. Fibrous tufted aggregates; efflorescences and incrustations. Dirty white to brownish yellow, silky luster. $H = 2$. Biaxial (+); $X = b$, $Z \wedge c = 29°$; $N_x = 1.475$, $N_y = 1.480$, $N_z = 1.488$; 2V large. Analyzed material contains substantial Fe and Mn substituting for the Zn. *Felsobanya, Hungary*, as a recent efflorescence in mine workings. RG *DII:528*.

29.7.3.5   Bilinite $Fe^{2+}Fe_2^{3+}(SO_4)_4 \cdot 22H_2O$

Named in 1913 for the locality. Halotrichite group. MON $P2_1/c$. $a = 6.208$, $b = 24.333$, $c = 21.255$, $\beta = 100.30°$, $Z = 4$, $D = 1.993$. *25-1153:* 5.48_3 4.96_3 4.84_5 4.31_{10} 4.10_4 3.51_{10} 3.29_3 2.89_3. Radial-fibrous aggregates. White to yellowish. $H = 2$. $G = 1.875$. $N \approx 1.500$; inclined extinction, about 37°. *Schwaz, near Bilina, Bohemia, Czech Republic*. RG *DII:529*.

29.7.3.6   Redingtonite $(Fe^{2+},Mg)(Cr,Al)_2(SO_4)_4 \cdot 22H_2O$

An inadequately characterized mineral. Halotrichite group. Named in 1888 for the locality. MON Space group unknown. Massive, with fine parallel-fibrous structure. White; purple on fractures across fiber length; silky luster. $G = 1.761$. Inclined extinction, feeble birefringence. *Redington mercury mine, Knoxville, Napa Co., CA*. RG *DII:529, MR 2:152(1971)*.

29.7.3.7   Wupatkiite $(Co,Mg)Al_2(SO_4)_4 \cdot 22H_2O$

Named in 1995 for the prehistoric native American dwelling, Wupatki, near the discovery site of the mineral. Halotrichite group. MON $P2_1/c$. $a = 6.189$, $b = 24.234$, $c = 21.204$, $\beta = 100.33°$, $Z = 4$, $D = 1.87$. *PD:* 6.03_2 4.79_{10} 4.30_3 4.16_2 4.11_2 3.95_3 3.77_3 3.49_9. Forms cross-fiber veinlets with fibers 5–10 µm in diameter but 8 mm long. Empire rose colored, white streak, silky luster. Cleavage oblique to length of fibers. $H = 1\frac{1}{2}$. $G = 1.92$. Biaxial (−); $Z \wedge$ fiber length, to 12°; $N_x = 1.484$, $N_y = 1.477$. Found *12 km ESE of Gray Mountain, Cameron uranium district, Coconino Co., AZ*, as a postmine oxidation product at a U–Co–Ni–Mo prospect in the Chinle formation, associated with pickeringite and moorhouseite. RG *MM 59:553(1995), AM 81:518(1996)*.

29.8.1.1   Lausenite $Fe_2^{3+}(SO_4)_3 \cdot 6H_2O$

Named in 1928 for Carl Lausen, U.S. mining engineer, who first described the species. MON Space group unknown. *DANS 219:125(1974)*. *PD:* 5.55_4 3.52_{10} 3.38_{10} 3.03_5 2.85_5 2.52_4 2.28_4 2.07_6. Lumpy aggregates of minute fibers. White, silky luster. Biaxial (−); $X \wedge c = 27°$; $N_x = 1.598$, $N_y = 1.628$, $N_z = 1.654$; 2V large. *United Verde mine, Jerome, AZ*, with copiapite and other sulfates, formed as a result of a mine fire. RG *DII:530*.

29.8.2.1 Kornelite $Fe_2^{3+}(SO_4)_3 \cdot 7H_2O$

Named in 1888 after Kornel Hlavacsek (1835–1914), Hungarian mining engineer at the Szomolnok copper mine, where the mineral was found. MON $P2_1/c$. $a = 14.30$, $b = 20.12$, $c = 5.425$, $\beta = 96.80°$, D = 2.254. AM 58:535(1973). 44-1426: 10.1_4 7.11_6 6.70_{10} 3.55_8 3.50_5 3.36_5 3.20_5 2.67_5. Crystals lathlike {010} or acicular [001]; as crusts, tufted aggregates, or globular masses with radial-fibrous structure. Pale rose-pink to violet, silky luster. Cleavage {010}. G = 2.306. Soluble in water. Biaxial (+); Z = b; X $\wedge$ c = 20°; $N_x = 1.572$, $N_y = 1.586$, $N_z = 1.640$; 2V = 49–62°; r > v, perceptible. Tintic Standard mine, Dividend, UT; Copper Queen mine, Bisbee, AZ, with coquimbite, roemerite, voltaite, and rhomboclase. Smolník (Szomolnok), Slovakia (formerly Hungary), with voltaite and coquimbite. RG DII:530, BM 87:125(1964).

29.8.3.1 Coquimbite $Fe_2^{3+}(SO_4)_3 \cdot 9H_2O$

Named in 1841 for the locality. Dimorphous with paracoquimbite. HEX-R $P\bar{3}1m$. $a = 10.922$, $c = 17.084$, Z = 4, D = 2.137. AM 55:1534(1970). 44-1425: 8.28_8 5.47_6 4.60_6 3.64_6 3.50_5 3.36_{10} 2.77_5 2.76_8. Crystals short prismatic [0001] with {10$\bar{1}$1} and {11$\bar{2}$0}; pyramidal {10$\bar{1}$1}; also massive, granular. Pale violet to deep amethystine, also yellowish or greenish; vitreous luster. Cleavage {10$\bar{1}$1}, imperfect; {10$\bar{1}$0}, difficult. H = $2\frac{1}{2}$. G = 2.11. Soluble in cold water and in acids. If a water solution is heated, hydrous ferric oxide precipitates. Tastes astringent. Uniaxial (+); $N_O = 1.539$, $N_E = 1.548$. Al substitutes for Fe^{3+} up to at least Al/Fe = 1:1.49. San Rafael Swell, Emery Co., UT, in uranium mines. United Verde mine, Jerome, AZ, in the fire zone. Near Borate, San Bernardino Co., CA, in the sulfate deposit. Concepcion mine, Huelva, Spain. Skouriotissa, Cyprus. Abundant in *Chile*, especially in *Coquimbo Prov.* and at Tierra Amarilla, Quetena, Chuquicamata, and Alcaparrosa. RG DII:532, NJMM:89(1974).

29.8.4.1 Paracoquimbite $Fe_2^{3+}(SO_4)_3 \cdot 9H_2O$

Dimorphous with coquimbite. Named in 1935 from the Greek *para* and coquimbite; the minerals are polytipic. HEX-R $R\bar{3}$. $a = 10.93$, $c = 51.30$, Z = 12, D = 2.115. AM 56:1567(1971). 27-254: 8.88_{10} 8.55_2 7.62_4 5.47_3 4.71_2 4.61_3 3.37_3 2.76_1. Rhombohedral crystals, with large {01$\bar{1}$2}; also equant pseudocubic with {0001} and {01$\bar{1}$2}; or prismatic; massive, granular. Pale violet, vitreous luster. Cleavage {01$\bar{1}$2} and {10$\bar{1}$4}, imperfect. H = $2\frac{1}{2}$. G = 2.11. Soluble in water. Tastes astringent. *Tierra Amarilla*, Alcaparrosa, and Quetena, *Chile*, with coquimbite and other sulfates. RG DII:534, NJMM:89(1974).

29.8.5.1 Quenstedtite $Fe_2^{3+}(SO_4)_3 \cdot 11H_2O$

Named in 1889 for Friedrich August Quenstedt (1809–1889), German mineralogist. TRIC $P\bar{1}$. $a = 6.184$, $b = 23.60$, $c = 6.539$, $\alpha = 94.18°$, $\beta = 101.73°$, $\gamma = 96.27°$, Z = 2, D = 2.145. AM 59:582(1974). 28-496: 5.84_7 5.64_5 5.38_3 5.08_5 4.48_3 4.22_9 4.11_{10} 3.82_5. Crystals tabular {010} or short prismatic [100]; aggregates of minute crystals. Pale violet to reddish violet. Cleavage {010}, perfect; {100}, good. H = $2\frac{1}{2}$. G = 2.147. Readily soluble in water. Biaxial (+); X ∧ ɸ = −43°, X ∧ σ = 45°, Y ∧ ɸ = 128°, Y ∧ σ = 43°, Z ∧ ɸ = −138°, Z ∧ σ = 88°; N_x = 1.547, N_y = 1.566, N_z = 1.594 (Na); 2V = 70°; r < v, strong. *Tierra Amarilla, near Copiapó, Chile*, and at *Alcaparrosa near Calama*. RG *DII:535*.

29.8.6.1 Alunogen $Al_2(SO_4)_3 \cdot 17H_2O$

Named in 1832 from the Greek *to make alum*. TRIC $P\bar{1}$. a = 7.420, b = 26.97, c = 6.062, α = 89.95°, β = 97.57°, γ = 91.88°, Z = 2, D = 1.791. *AM 61:311(1976)*. 26-1010: 13.5_6 4.49_{10} 4.39_8 4.33_8 3.97_8 3.90_5 3.68_5 3.02_4. Delicate fibrous masses or crusts; efflorescences. Crystals small and rare, prismatic [001], or {010} with a six-sided outline; twinning on {010}. Colorless to white, or tinged by impurities; vitreous to silky luster. Cleavage {010}, perfect; also {100}, {$\bar{3}13$}. H = $1\frac{1}{2}$-2. G = 1.77. Easily soluble in water; in CT, gives water, and at a higher temperature, sulfuric acid; melts at about 114°. Tastes acid and sharp, like alum. Biaxial (+); X ≈ b; Z ∧ c = 42°; N_x = 1.459, N_y = 1.463, N_z = 1.473; 2V = 31°. Formed as an efflorescence in coal formations, and especially in arid regions, as a result of oxidation of pyrite or other sulfides in orebodies, giving rise to sulfuric acid, which reacts with aluminous minerals in wall rock. Also formed in fumaroles and solfatara. Abundant at *Smoky Mt., Jackson Co., NC*. In *NM* extensive deposits in the *Alum Mt. district, Grant Co.*, as crusts 1.0 m or more thick on cliffs; also in *Sandoval Co.* as a thick bed. In *CA* around the *Mt. Lassen hot springs, Shasta Co.*, and at *the Geysers, Sonoma Co*. In *BC, Canada*, near *Vernon* with epsomite in veins. *Kolosoruk, Bohemia, Czech Republic*, in brown coal, and in the opal mines of *Cervenica (Opalbanya), Slovakia*. In fumaroles on *Mt. Vesuvius, Italy*; in the solfatara of *Pozzuoli*; at *Miseno* in the sulfur caves; and at *Libiola, Sestri Levante, Genova*(*). At numerous places in Chile and Peru, as in the *Cerro Pintado near Iquique* and at *Francisco de Vergara, Antofagasta Prov*. RG *DII:537, AM 49:1763(1964), BM 96:385(1973), TMPM 21:164(1974)*.

29.8.7.1 Meta-alunogen $Al_2(SO_4)_3 \cdot 14H_2O$

Named in 1942 from the Greek *meta* and alunogen, indicating a lower hydrate than alunogen. An inadequately characterized mineral. ORTH Space group unknown. a = 12.25, b = 13.95, c = 15.95. 22-23(syn): 12.2_3 6.11_1 4.21_1 4.07_{10} 3.99_1 3.86_1 3.02_1. White, waxy or pearly luster. Biaxial (+); N_x = 1.469, N_y = 1.473, N_z = 1.491; 2V large. *Francisco de Vergara, Antofagasta, Chile*, as a dehydration product of alunogen. RG *DII:539, AM 28:61(1943)*.

29.9.1.1 Voltaite $K_2Fe_5^{2+}Fe_4^{3+}(SO_4)_{12} \cdot 18H_2O$

Named in 1841 for Alessandro Volta (1745–1827), Italian physicist. ISO $Fd3c$. a = 27.254, Z = 16, D = 2.662. 20-1388: 9.63_4 6.80_3 5.55_6 3.54_8 3.40_{10} 3.03_5 2.85_4 2.08_3. Crystals form cubes, octahedrons, and combinations of these;

rarely dodecahedral; also massive, granular. Greenish black to black; dark oil-green, grayish-green streak, resinous luster. No cleavage, conchoidal fracture. H = 3. G = 2.7. Isotropic; N = 1.593–1.608. Weak anomalous birefringence common. Al and Mg may replace a portion of the Fe^{3+} and Fe^{2+}, resulting in lower refractive indices. United Verde mine, Jerome, AZ, in the fire zone; also at Bisbee at the Copper Queen mine. At the sulfate deposit near Borate, San Bernardino Co., CA. The solfatara of *Pozzuoli near Naples, Italy*, as well as in fumaroles on Vesuvius and in the sulfur cave of Miseno. The Rammelsberg mine, near Goslar, Harz, Germany. In the pyrite mines of Huelva, Spain. With metavoltine and roemerite at Madeni Zakh, Iran. Quetena, Chuquicamata, and Alcaparrosa, Chile. RG *DII:464, NW:670(1970), TMPM 18:185(1972).*

29.9.1.2 Zincovoltaite $K_2Zn_5Fe_4^{3+}(SO_4)_{12} \cdot 18H_2O$

Named in 1987 as the zincian analog of voltaite. ISO $Fd3c$. $a = 27.18$, $Z = 16$, $D = 2.767$. *AM 75:244(1990)*. PD: 5.53_4 4.24_3 3.54_7 3.39_{10} 3.13_4 3.03_3 2.84_3. Single grains and granular aggregates; also veinlets cutting pyrite. Crystals isometric with {111} and {100}; also {110} and {211}. Greenish black to oil green, with gray-green streak and pitchy to resinous luster, vitreous on crystal faces. No cleavage, conchoidal fracture, brittle. H = 3. G = 2.756. Isotropic; N = 1.605 (Na); sometimes anomalously uniaxial. From a lead–zinc mine at Xitieshan, Qinghai, China, associated with roemerite, melanterite, and gypsum. RG

29.9.2.1 Zircosulfate $(Zr, Ti)(SO_4)_2 \cdot 4H_2O$

Named in 1965 for the composition. ORTH $Fddd$. $a = 25.92$, $b = 11.62$, $c = 5.532$, $Z = 8$, $D = 2.833$. *AM 51:529(1966)*. 8-495(syn): 6.49_5 4.90_3 4.32_{10} 3.47_4 2.98_9 2.90_2 2.33_3 2.13_2. Compact, powdery masses. Colorless to white, dull luster. H = $2\frac{1}{2}$–3. G = 2.85. Easily soluble in cold water. Biaxial (+); $N_x = 1.620$, $N_y = 1.644$, $N_z = 1.674$; 2V = 75°. Found in a cavity in nepheline syenite pegmatite in the Korgeredabin alkalic massif, southeastern Tuva, Russia, and probably resulted from the dissolution of preexisting eudialyte by sulfuric acid solutions. RG *ZVMO 94:530(1965)*.

29.9.3.1 Bazhenovite $CaS_5 \cdot CaS_2O_3 \cdot 6Ca(OH)_2 \cdot 20H_2O$

Named in 1987 for A. G. Bazhenov (petrographer) and L. F. Bazhenova (analytical chemist), Russian scientists. A questionable "mineral." Chemical classification uncertain. MON $P2_1/c$. $a = 8.45$, $b = 17.47$, $c = 8.24$, $\beta = 119.50°$, $Z = 1$, $D = 1.845$. *AM 74:500(1989)*. Crystal structure is layered, with $Ca(OH)_2$, polysulfide, and water-bearing layers parallel to {010}. *42-1361*: 8.76_{10} 4.39_{10} 2.91_6 2.81_5 2.62_5 2.28_5 1.996_7. Granular aggregates of blades flattened {010} and elongated [001]. Orange-yellow, light yellow streak, vitreous to pearly luster. Cleavage {010}, very good; brittle, but elastic in thin leaves. $H = 2$. $G = 1.83$. Biaxial (+); $XY = ba$; $Z \wedge c = 30°$; $N_x = 1.595$, $N_y = 1.619$, $N_z = 1.697$; $2V = 60°$; absorption $X > Z > Y$. X, deep yellow-green; Y, greenish yellow; Z, pale greenish yellow. *Chelyabinsk coal basin, southern Ural Mts., Russia*, from old burning coal dumps, associated with native (?) iron, sulfur, oldhamite, troilite, pyrrhotite, fluorite, and periclase in altered pyritized siderite fragments. RG *ZVMO 116:737(1987)*.

= Class 30 =

Anhydrous Sulfates Containing Hydroxyl or Halogen

30.1.1.1 Sundiusite $Pb_{10}(SO_4)(Cl_2)_8$

Named in 1980 for Nils Sundius (1886–1976), Swedish mineralogist. MON $C2$, Cm, or $C2/m$. $a = 24.67$, $b = 3.781$, $c = 11.881$, $\beta = 100.07°$, $Z = 2$, $D = 7.20$. AM 65:506(1980). 33-761: 6.10_3 3.91_2 3.74_3 3.43_2 3.10_6 3.04_6 2.98_{10} 2.74_8. Plumose aggregates of crystals. White to colorless, adamantine luster. Cleavage {100}, perfect; brittle. H = 3. G = 7.0. Not fluorescent. Biaxial (+); N > 2.10; maximum birefringence = 0.07; cleavage laths length-slow. *Amerika stope, Långban, Värmland, Sweden.* RG

30.1.2.1 Elyite $Pb_4Cu(SO_4)(OH)_8$

Named in 1972 for John Ely, an important figure in the early mining history of eastern Nevada. MON $P2_1/a$. $a = 14.248$, $b = 5.768$, $c = 7.309$, $\beta = 100.43°$, $Z = 2$, $D = 6.32$. AM 57:364(1972). 25-293: 7.19_{10} 3.34_3 3.14_3 3.08_3 3.05_3 3.00_8 2.88_4. Prismatic to fibrous crystals in radiating sprays and tufts; crystals elongate [010], tabular {001}; twinning common, on {001}. Violet, pale violet to white streak, silky luster. Cleavage {001}, good; sectile. H = 2. G ≈ 6.0. Nonfluorescent. Insoluble in water, turns white in dilute HNO_3 or HCl. Biaxial (−); $Y = b$; $X \wedge c = 45°$; $N_x = 1.990$, $N_y = 1.993$, $N_z = 1.994$ (Na); 2V = 68°; noticeable dispersion; absorption Z > Y > X. *Silver King Mine, Ward, NV*, in voids in oxidizing Pb–Cu–Zn sulfide ore, associated with earthy secondary galena. *Tsumeb, Namibia*. Common as a smelter product in slags. RG

30.1.3.1 Brochantite $Cu_4(SO_4)(OH)_6$

Named in 1824 for A. J. M. Brochant de Villiers (1772–1840), French geologist and mineralogist. MON $P2_1/a$. $a = 13.08$, $b = 9.85$, $c = 6.02$, $\beta = 103.37°$, $Z = 4$, $D = 3.98$. Structure: SR 23:451(1959). 43-1458: 6.38_8 5.35_7 3.90_{10} 3.19_5 2.92_2 2.68_5 2.52_8 1.740_2. Crystals stout prismatic to acicular [001]; also elongated [010] or tabular {001}; massive, granular. Twinning common on {100}, giving a pseudo-orthorhombic appearance. Emerald green to blackish green, pale green streak, vitreous luster, pearly on cleavage. Cleavage {100}, perfect; uneven to conchoidal fracture. H = $3\frac{1}{2}$–4. G = 3.97. Soluble in acids; in CT, yields water. Biaxial (−); $X \approx a$; $Y = b$; $Z \approx c$; $N_x = 1.728$, $N_y = 1.771$, $N_z = 1.800$ (Na); 2V = 77°; r < v, medium. X, bluish; Y, green. Formed in

623

the oxide zone of copper deposits under conditions of low acidity. Tintic district, Juab Co., UT, and also at the Apex mine, Washington Co. Organ district, Dona Ana Co., NM. In AZ at the Mammoth–St. Anthony mine, Pinal Co.(*), with linarite; in Cochise Co. at the Shattuck mine(!) in centrimetric square prisms with malachite, and at Bisbee(!). In England at Roughten Gill, Cumberland. Nassau, Hesse–Nassau, Germany. Rosas, Sardinia, Italy. Rio Tinto, Huelva, Spain. *Gumeshevsk, near Ekaterinburg, Ural Mts., Russia*. Brilliant crystals at Aïn Barbar, Constantine, Algeria(*), and in the mines at Bou Beker on the Algerian border near Oujda, Morocco. Shaba, Zaire. Spectacular specimens from Tsumeb, Namibia(!). Chuquicamata, and in the Atacama desert region, Chile. Broken Hill, NSW, Australia. Used as an ore of copper. RG *DII:541*.

30.1.5.1 Klebelsbergite $Sb_4O_4(SO_4)(OH)_2$

Named in 1929 for Kuno Klebelsberg (1875–1932), Hungarian educator. ORTH $Pca2$. $a = 5.766$, $b = 11.274$, $c = 14.887$, $Z = 4$, $D = 4.69$. *AM 65:931(1980)*. *35-464:* 6.22_8 3.89_8 3.55_8 3.15_8 3.13_{10} 2.83_8 2.44_8 1.81_8. Tufts of minute acicular crystals elongated [001] and flattened {010}. Pale yellow to orange-yellow, yellow streak, vitreous luster. No cleavage. $VHN_{10} = 200$. $G = 4.62$. Nonfluorescent. Slowly soluble in cold concentrated HCl. In CT, fuses and gives acid water. Biaxial $(-)$; $N \approx 1.95$; $2V \approx 70°$; nonpleochroic. *Baia Sprie* (formerly *Felsöbánya*), *Maramures, Romania*, on stibnite. Pereta mine, Tuscany, Italy(!), in crystals to 2 cm; also at La Cetine. RG *DII:583, AM 65:499(1980), NJMM:223(1980)*.

30.1.6.1 Kogarkoite $Na_3(SO_4)F$

Named in 1973 for Lia Nikolaevna Kogarko (b.1936), Russian geochemist, who first noticed the mineral in Russia. MON (pseudorhombohedral) Space group unknown. $a = 18.074$, $b = 6.958$, $c = 11.443$, $\beta = 107.71°$, $Z = 12$, $D = 2.679$. *MM 43:753(1980)*. *25-827:* 4.29_5 3.80_7 3.48_{10} 2.99_9 2.72_9 2.71_9 2.56_7 2.14_9. Pseudohexagonal crystals; also as chalky incrustations; twinning on {0001}. Pale blue, also white or colorless; vitreous luster. $H = 3\frac{1}{2}$. $G = 2.676$. Biaxial $(+)$; pseudouniaxial $(+)$; $N_x = 1.439$, $N_y = 1.439$, $N_z = 1.442$; 2V near $0°$. *Hortense Hot Springs, Chalk Creek, Chaffee Co., CO*, as a sublimate from steam issuing from the springs. Lovozero Massif, Kola Penin., Russia, in crystals up to 2 cm in nepheline syenite pegmatite, associated with villiaumite, nepheline, aegirine, ramsayite, and apatite. RG *AM 58:116(1973)*.

30.1.7.1 Sulphohalite $Na_6(SO_4)_2FCl$

Named in 1888 for the composition. ISO $Fm3m$. $a = 10.068$, $Z = 4$, $D = 2.503$. *AM 56:177(1971)*. *15-668:* 5.81_2 3.56_{10} 3.03_4 2.91_8 2.51_6 2.25_3 1.94_5 1.78_7. Usually dodecahedral; also octahedral, or combinations of {100}, {111}, and {110}. Colorless to pale greenish yellow, or gray, vitreous to greasy luster. Conchoidal fracture. $H = 3\frac{1}{2}$. $G = 2.505$. Slowly soluble in cold water, easily fusible. Tastes weakly saline. Isotropic; $N = 1.455$. From the saline beds of *Searles Lake, San Bernardino Co., CA*, with hanksite. Crystals attain 3 cm. Abundant at Otjiwalundo salt pan, 250 km N of Otavi, Namibia, with trona, thenardite, and pirssonite. RG *DII:548, MM 40:131(1975)*

30.1.8.1 Galeite $Na_{15}(SO_4)_5F_4Cl$

Named in 1955 for William Alexander Gale (b.1898), American physical chemist, American Potash and Chemical Corp. HEX-R $P31m$. $a = 12.97$, $c = 13.955$, $Z = 3$, $D = 2.610$. *MM 40:357(1975)*. 15-651: 3.68_7 3.52_8 3.04_4 2.97_6 2.79_{10} 2.55_7 1.85_4 1.76_6. Crystals hexagonal, barrel-shaped or tabular. Common forms are $\{11\bar{2}0\}$, $\{11\bar{2}1\}$, $\{2\bar{1}\bar{1}1\}$, and $\{0001\}$. Colorless. $G = 2.605$. Uniaxial (+); $N_O = 1.447$, $N_E = 1.449$. *Searles Lake, San Bernardino Co., CA*, found in a drill core that penetrated a lower salt bed, associated with gaylussite and northupite. RG *AM 41:672(1956), 48:485(1963), 56:174(1971)*.

30.1.9.1 Schairerite $Na_{21}(SO_4)_7F_6Cl$

Named in 1931 for John Frank Schairer (1904–1970), American physical chemist, Geophysical Laboratory, Washington, DC. HEX-R $P31m$. $a = 12.197$, $c = 19.259$, $Z = 3$, $D = 2.627$. *MM 40:131(1975)*. 16-165: 4.44_2 3.79_7 3.52_8 3.01_6 2.76_{10} 2.58_6 2.41_2 1.76_7. Rhombohedral crystals with $\{03\bar{3}2\}$ and $\{10\bar{1}1\}$ prominent; artificial crystals are tabular $\{0001\}$; twinning $\{0001\}$, common. Colorless, vitreous luster. Conchoidal fracture, brittle. $H = 3\frac{1}{2}$. $G = 2.616$. Slowly soluble in water, easily fusible. Uniaxial (+); $N_O = 1.440$, $N_E = 1.445$. Found in a drill core in a lower salt bed at *Searles Lake, San Bernardino Co., CA*, with gaylussite, tychite, pirssonite, hanksite, and other salts. RG *DII:547, AM 56:174(1971)*.

30.1.10.1 D'Ansite $Na_{21}Mg(SO_4)_{10}Cl_3$

Named in 1958 for Jean D'Ans (1881–1969), German chemist, Technical University, Berlin. ISO $I\bar{4}3m$. $a = 15.913$, $Z = 4$, $D = 2.66$. 41-1473(syn): 4.26_3 3.98_3 3.40_{10} 3.12_6 2.82_{10} 2.58_4 2.52_3 1.717_3. Synthetic crystals tetrahedral with $\{211\}$ dominant; also $\{2\bar{1}1\}$ and $\{110\}$. Colorless. $G = 2.655$. Easily fusible. Isotropic; $N = 1.489$ (Na). *Hall, near Innsbruck, Tyrol, Austria*, enclosed in bloedite, associated with vanthoffite. RG *NW 45:362(1958), AM 43:1221(1958), NJMM:152(1958), ZK 152:83(1980)*.

30.1.11.1 Chlorothionite $K_2Cu(SO_4)Cl_2$

Named in 1872 for the composition: chlorine and sulfur (Greek *theion*). ORTH $Pnma$. $a = 7.732$, $b = 6.078$, $c = 16.292$, $Z = 4$, $D = 2.678$. 29-998: 5.69_3 3.26_3 3.15_2 3.04_{10} 2.85_4 2.49_2 2.19_7 1.99_2. Crystalline incrustations. Bright blue. $H = 2\frac{1}{2}$. $G_{syn} = 2.67$. Soluble in water. Biaxial (+); 2V large; $r > v$. Found in fumaroles during the 1872 eruption of *Vesuvius, Italy*. RG *DII:547, ZK 144:226(1976)*.

30.1.12.1 Antlerite $Cu_3(SO_4)(OH)_4$

Named in 1889 for the locality. ORTH *Pnam*. $a = 8.244$, $b = 11.987$, $c = 6.043$, $Z = 4$, $D = 3.93$. *CM 27:205(1989)*. 7-407(syn): 6.01_3 5.40_3 4.86_{10} 3.60_8 2.68_8 2.57_9 2.50_3 2.13_2. Crystals thick tabular {001}, also equant or short prismatic [001], as cross-fiber veinlets, aggregates of acicular crystals, granular. Emerald green to blackish green, also light green; very similar to brochantite; pale green streak; vitreous luster. Cleavage {010], perfect; {100}, poor; brittle. $H = 3\frac{1}{2}$. $G = 3.88$. Soluble in dilute H_2SO_4, fusible. Biaxial (+); XYZ = *bac*; $N_x = 1.726$, $N_y = 1.738$, $N_z = 1.789$; $2V = 53°$. X, yellow-green; Y, blue-green; Z, green. Antlerite is a secondary mineral found in the oxidized zone of copper deposits in arid regions, associated with atacamite and other copper sulfates. *Antler mine, Hualpai Mts., Mojave Co., AZ*; also from Bisbee, Cochise Co. Northern light mine, Black Mt., NV. In the Sierra Mojada, COAH, Mexico. Torre Capdulla, Lerida, Italy. Abundant at the great Chuquicamata copper deposit, Antofagasta, Chile, where for many years it was the principal ore mineral in the oxidized zone, associated with kröhnkite, natrochalcite, atacamite, and brochantite. RG *DII:544, MJJ 3:223(1961), NAT 197:70(1963)*.

30.1.13.1 Schuetteite $Hg_3SO_4O_2$

Named in 1959 for Curt Nicolaus Schuette (1895–1975), American mining engineer and geologist, specialist in mercury deposits. HEX-R $P3_121$. $a = 7.07$, $c = 10.05$, $Z = 3$, $D = 8.36$. *AM 44:1026(1959)*. 12-724(syn): 5.19_2 3.34_7 2.91_{10} 2.60_6 2.30_2 1.93_3 1.76_2 1.56_2. Thin crusts and films. Bright canary yellow, yellow streak. Friable. $H \approx 3$. $G_{syn} = 8.18$. The color does not darken on exposure to sunlight, which distinguishes schuetteite from the yellow mercury oxychlorides. Uniaxial (−); $N > 2.10$; pleochroic, greenish yellow to orange-yellow; moderately to highly birefringent. Formed either by the oxidation of cinnabar ore exposed to sunlight, or in the bricks and remains of old mercury furnaces and condensers. It is not known to form underground under natural circumstances. *Silver Cloud mine, Ivanho district, N of Battle Mountain, NV*, and at at least five other NV mercury mines; the Opalite mine, southeastern Oregon; the Idaho Almaden mine, E of Weiser, ID; the Sulphur Bank mine, Lake Co., and elsewhere in CA; at the dismantled Chisos mine furnace, Brewster Co., TX. RG

30.1.14.1 Mammothite $Cu_4Pb_6AlSb^{5+}O_2(SO_4)_2(OH)_{16}Cl_4$

Named in 1985 for the locality. MON $C2/m$. $a = 18.933$, $b = 7.331$, $c = 11.532$, $\beta = 112.43°$, $Z = 2$, $D = 5.25$. *38-389*: 10.4_6 6.67_9 6.08_6 5.24_3 4.72_8 3.59_3 3.05_9 2.90_{10}. Microscopic crystals, tabular to prismatic; also acicular. Bright blue to greenish blue, pale blue, vitreous luster. Cleavage {010}, distinct; very brittle. $H = 3$. $G = >4.2$. Nonfluorescent. Biaxial (+); $N_x = 1.868$, $N_y = 1.892$, $N_z = 1.928$; $2V = 80°$; moderately pleochroic; absorption $Z < Y \leq X$. *Mammoth vein, Tiger, AZ*, associated with anglesite and phosgenite, and from

Laurium, Attika, Greece, with cerussite and phosgenite. RG MR 16:117(1985), TMPM 34:279(1985), AM 71:1548(1986).

30.1.15.1 Ye'elimite $Ca_4Al_6(SO_4)O_{12}$

Named in 1983 for the locality. ISO $I4_132$. $a = 18.392$, D = 2.61. 33-256(syn): 4.92_1 3.76_{10} 3.25_1 2.91_1 2.65_3 2.46_1 2.17_2 1.62_1. Colorless. G = 2.61. Isotropic; N = 1.568. Formed as a result of high-temperature metamorphism of calcareous rocks. *Hatrurim basin, W of the Dead Sea, Israel, near Har Ye'elim and Nahal Ye'elim*, a hill and a wadi. From the Hatrurim formation in cobbles in a pseudoconglomerate, with larnite, brownmillerite, and ellestadite. RG *JACH 12:413(1962)*, *GSICR:1(1983)*, *AM 72:226(1987)*.

30.1.16.1 Chenite $Pb_4Cu(SO_4)_2(OH)_6$

Named in 1986 for Tsong T. Chen (b.1942), Canadian mineralogist. TRIC $P1$. $a = 5.791$, $b = 7.940$, $c = 7.976$, $\alpha = 112.01°$, $\beta = 97.73°$, $\gamma = 100.45°$, Z = 1, D = 6.044. 45-1379: 5.57_5 4.32_7 3.61_{10} 3.57_8 3.43_7 3.39_7 3.30_5 3.01_5. Minute crystals, prismatic to equant; twinning rare. Sky blue, pale blue streak, vitreous to resinous luster. Cleavage {100}, good. H = $2\frac{1}{2}$. G = 5.98. Nonfluorescent. Biaxial; $N_x = 1.871$, $N_y = 1.909$, $N_z = 1.927$ (Na); 2V = 67°; r > v, very strong; weakly pleochroic. *Susanna mine, Leadhills, Scotland*, on partially oxidized galena and chalcopyrite, associated with caledonite, linarite, and leadhillite. Also in cavities in slag from the old Glengaber smelter at Wanlockhead, Scotland. RG *MM 50:129(1986)*, *AM 72:222(1987)*, *NJMM:259(1988)*.

30.1.17.1 Nabokoite $KCu_7Te^{4+}O_4(SO_4)_5Cl$

Named in 1987 for Sof'ya I. Naboko (b.1909), Russian vulcanologist. Needs further study and substantiation. TET $P4/ncc$. $a = 9.833$, $c = 20.591$, Z = 4, D = 3.974. *ZVMO 116:358(1987)*. 42-1337: 10.4_{10} 4.57_4 3.56_4 3.42_6 2.88_5 2.44_7 1.972_4 1.775_4. Zoned crystals, thin tabular {001}. Light yellow-brown to honey-brown zones alternating with dark brown zones (atlasovite), vitreous luster. Cleavage {001}, perfect. H = 2–$2\frac{1}{2}$. G = 4.18. Insoluble in water, soluble in HCl or HNO_3, reacts with KOH. Uniaxial (−); $N_O = 1.778$, $N_E = 1.773$. Found as a sublimate on porous basalt near a fumarole formed during the *Great Clefted Tolbachik eruption, Kamchatka, Russia*, associated with anglesite, chalcocyanite, and dolerophanite. RG *AM 73:929(1988)*, *TMPM 38:291(1988)*.

30.1.17.2 Atlasovite $Cu_6Fe^{3+}BiO_4(SO_4)_5 \cdot KCl$

Named in 1987 for Vladimir V. Atlasov (1661–1711), Russian explorer. Because of wide variations in the analytical data given, the formula and unit cell are open to question and reinterpretation. TET $P4/ncc$. $a = 9.86$, $c = 20.58$, Z = 4, D = 4.12. *ZVMO 116:358(1987)*. The unit cell was derived from the powder data by analogy with nabokoite. 42-1338: 10.8_9 10.4_{10} 4.57_5 3.57_4 3.43_7 2.89_7 2.64_4 2.45_8. Crystals tabular {001}, zoned. Dark brown, light brown streak, vitreous luster. Cleavage {001}, perfect. H = 2–$2\frac{1}{2}$. G = 4.20.

Tests same as with nabokoite. Uniaxial $(-)$; $N_O = 1.783$, $N_E = 1.776$. O, red-brown; E, yellowish. Atlasovite and nabokoite form separate zones in tabular tetragonal crystals, in which Fe^{3+} and Bi^{3+} substitute for Te and Cu in an incompletely understood manner. Found as zones in crystals that were a sublimate from the *Great Clefted Tolbachik volcanic eruption, Kamchatka, Russia*. RG *AM* 73:927(1988).

30.1.18.1 Coquandite $Sb_6O_8(SO_4) \cdot H_2O$

Named in 1992 for Henri Coquand (1811–1881), French scientist, who studied the Sb deposits of Tuscany. TRIC $P\bar{1}$. $a = 11.434$, $b = 29.77$, $c = 11.314$, $\alpha = 91.07°$, $\beta = 119.24°$, $\gamma = 92.82°$, $Z = 12$, $D = 5.78$. *MM* 56:599(1992). PD: 14.84_5 9.27_4 8.01_3 6.81_7 3.30_9 3.20_4 3.09_{10} 2.84_3. White powdery masses and thin crusts, spheroidal knobs of feathery aggregates of silky fibers, flexible lamellar crystals, minute tabular crystals, elongate [001], flattened (010). White to colorless, white streak, adamantine luster, transparent. Flexible, no cleavage. Biaxial $(+)$; $N_{mean} = 2.08$; $2V = >>60°$; length-slow; nonpleochroic; parallel extinction on {010}; axial plane parallel to [001]; polysynthetically twinned on (010), nonfluorescent. Found at the Lucky Knock mine, Tonasket, Okanogan Co., WA; at the *Pereta stibnite mine, Tuscany, Italy*, where a stibnite vein with quartzite gangue fills a fracture in limestone. The coquandite is embedded between the needles of columnar stibnite, where it is associated with peretaite and klebelsbergite. Also at the Cetine mine, Tuscany. RG *AM* 78:845(1993).

30.2.1.1 Lanarkite $Pb_2(SO_4)O$

Named in 1832 for the locality. MON $C2/m$. $a = 13.769$, $b = 5.698$, $c = 7.079$, $Z = 4$, $D = 7.00$. *37-516*(syn)(calc): 6.37_2 4.43_2 3.34_{10} 2.963_8 2.958_2 2.87_2 2.85_4 2.05_2. Crystals elongated [010], twinning rare, polysynthetic. Gray to greenish white, pale yellow; adamantine to resinous luster, pearly on cleavages. Cleavage $\{\bar{2}01\}$, perfect; $\{\bar{4}01\}$, $\{201\}$, imperfect. Thin flexible laminae. $H = 2-2\frac{1}{2}$. $G = 6.92$. Fluoresces yellow in LWUV and in X-rays. Soluble in KOH and in warm HNO_3. Biaxial $(-)$; $Y = b$; $Z \wedge c = 30°$; $N_x = 1.928$, $N_y = 2.007$, $N_z = 2.036$; $2V = 60°$; $r > v$, strong, inclined. *Susanna mine, Leadhills, Lanarkshire, Scotland*, with leadhillite, cerussite, and caledonite. Tanne, Harz Mts., Germany. Laquorre mine near Aulus, Ariège, France. Challocolla, Atacama, Chile. RG *DII:550, AC 4:471(1951), ZK 132:99(1970)*.

30.2.2.1 Dolerophanite $Cu_2(SO_4)O$

Named in 1873 from the Greek for *fallacious appearance*, because its appearance does not suggest its composition. MON $C2/m$. $a = 9.370$, $b = 6.319$, $c = 7.639$, $\beta = 122.33°$, $Z = 4$, $D = 4.16$. Structure: *AC 16:1009(1963)*. *13-189*: 6.44_5 4.76_1 3.62_{10} 2.78_2 2.62_4 2.55_2 2.26_3 2.03_1. Tabular crystals, elongated [010]; prism faces usually striated [010], other faces smooth and lustrous. Chestnut brown to dark brown and nearly black, yellowish to brown streak. Cleavage $\{\bar{1}01\}$, perfect. $H = 3$. $G = 4.17$. Easily soluble in HNO_3, decomposed by water, giving a blue solution. Biaxial (+); $Y = b$; $Z \wedge c = -10°$; $N_x = 1.715$, $N_y = 1.820$, $N_z = 1.880$; $2V = 85°$; $r > v$, very strong, crossed. X, deep brown; Y, brownish yellow; Z, lemon yellow. Found as a sublimate formed during the eruption of *Vesuvius, Italy*, in 1868, associated with chalcocyanite, euchlorine, and eriochalcite. RG *DII:551, AM 46:147(1961)*

30.2.3.1 Linarite PbCu(SO₄)(OH)₂

Named in 1839 for the locality. MON $P2_1/m$. $a = 9.691$, $b = 5.650$, $c = 4.687$, $\beta = 102.65°$, $Z = 2$, $D = 5.315$. *NBSM 25(16):34(1979)*. Structure: *AC 14:747(1961)*. 30-493: 4.85_4 4.52_6 3.63_3 3.56_6 3.15_{10} 3.11_4 2.71_3 2.59_3. Crystals elongated [010], or tabular {$\bar{1}01$} and {001}; as crusts or aggregates; twinning on {100}, fairly common. Deep azure blue, similar to azurite; pale blue streak, vitreous to subadamantine luster. Conchoidal fracture, brittle. $H = 2\frac{1}{2}$. $G = 5.35$. Soluble in HNO_3 with separation of $PbSO_4$; easily fusible; in CT decrepitates, darkens, and loses water. Biaxial (−); $Z = b$; $X \wedge c = -24°$; $N_x = 1.809$, $N_y = 1.838$, $N_z = 1.859$ (Na); $2V = 80°$; $r < v$, strong. X, pale blue; Y, clear blue; Z, prussian blue. Mammoth mine(!), Tiger, AZ, and in superb crystals at the Grand Reef mine(!!) in Laurel Canyon, Graham Co. Tintic and Eureka districts, UT. Cerro Gordo mine, Inyo Co., CA. At the Blanchard mine(*), Bingham, NM, with galena and fluorite. Beaver Mt., Slocan District, BC, Canada. *Linares, Jaen, Spain.* San Giovanni and Arenas, Sardinia, Italy. Leadhills, Lanarkshire, Scotland, and at Red Gill and other mines in Cumberland, England(!), sometimes in fine crystals. In Germany at Badenweiler and elsewhere; in Russia at Nerchinsk, and near Beresovsk, Ural Mts. Abundant at Tsumeb, Namibia(*). Broken Hill, NSW, Australia. At the Ortiz mine, Catamarca, Argentina, and at several places in Chile and Peru. RG *DII:553, MJJ 3:282(1961), MA 87:3984(1987)*.

ALUNITE GROUP

The alunite minerals are isostructural and conform to the formula

$$AB_3^{3+}(TO_4)_2(OH)_6$$

where

$A = K, Na, Tl, H_3O, NH_4, Ag, Pb, Ca, Ba$
$B = Al, Fe^{3+}$, or Cu
$T = S$ (and also minor P, As, V, Si)

and crystallize in the rhombohedral system, space group $R\bar{3}m$.

In this structure, Al^{3+} ions are at the maximum separation from each other permitted by the crystal symmetry, and are near the maximum possible distance from the S^{6+} that still permits an Al–(SO₄) distance of 1.936 Å. The Al is in octahedral coordination surrounded by a nearly regular octahedron made up of four (OH)⁻ groups and two O's of two different (SO₄) ions. The sulfur atoms lie on the trigonal axis and are surrounded by three oxygen atoms which form the base of a tetrahedron and one apical oxygen at a shorter bond length which also lies on the threefold axis. The coordination number of K is six, with six O's from six separate (SO_4) groups at 2.82 Å forming flattened octahedra. The basal oxygen atoms of the (SO_4) tetrahedra are surrounded by the three (OH) groups. Large variations of the ionic radii of the ion in the A position are allowed by the structure, ranging from $r = 0.95$ Å for Na to $r = 1.33$ Å for K, which is considerably greater than usually found in an isomorphous series. On the other hand, the $(SO_4)^{2-}$, $(PO_4)^{3-}$, and $(AsO_4)^{3-}$ differ but slightly in dimensions, making possible the phosphate–phosphate, sulfate–phosphate, or sulfate–arsenates of the related beudantite group and crandallite that are isostructural with the alunites. *AC 18:249(1965)*.

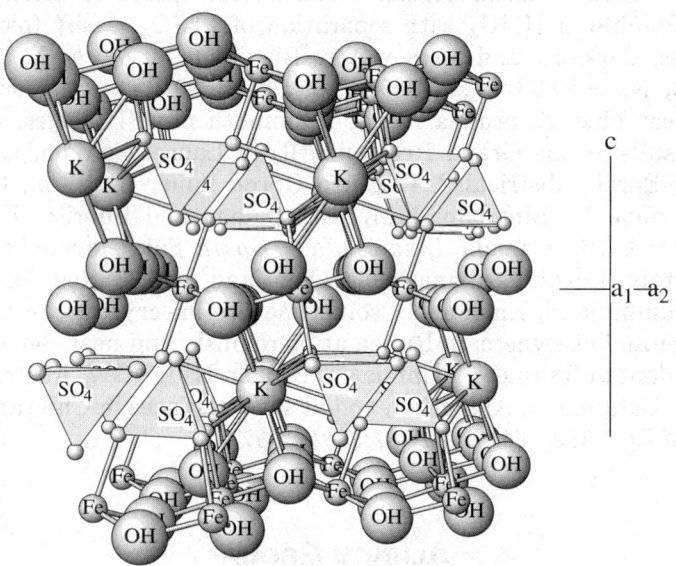

JAROSITE: $KFe_3(SO_4)_2(OH)_6$

Jarosite, which is isostrutural with alunite, in a sense consists of alternating layers of potassium sulfate and layers of iron (aluminium) hydroxide. However, hydroxide atoms form only four of the six corners of the Fe(Al)–O octahedra and the other two are sulfate oxygens. Layers are also tied together with K–OH bonds. *See also* TURQUOISE *under Phosphates, Arsenates, and Vanadates.*

ALUNITE GROUP, ALUNITE SUBGROUP: B = AL

Mineral	Formula	Space Group	a	c	D	
Alunite	$KAl_3(SO_4)_2(OH)_6$	$R\bar{3}m$	6.970	17.27	2.839	30.2.4.1
Natroalunite	$NaAl_3(SO_4)_2(OH)_6$	$R\bar{3}m$	6.990	16.905	2.819	30.2.4.2
Schlossmacherite	$(H_3O,Ca)Al_3(SO_4)_2(OH)_6$	$R\bar{3}m$	6.998	16.67	3.00	30.2.4.3
Osarizawaite	$PbCuAl_2(SO_4)_2(OH)_6$	$R\bar{3}m$	7.075	17.248	4.20	30.2.4.4

30.2.4.2 Natroalunite

ALUNITE GROUP, ALUNITE SUBGROUP: B = AL (CONT'D)

Mineral	Formula	Space Group	a	c	D	
Minamiite	$(Na,K,Ca,\square)Al_3(SO_4)_2(OH)_6$	$R\bar{3}m$	6.98	33.490	2.81	30.2.4.5
Ammonioalunite	$NH_4Al_3(SO_4)_2(OH)_6$	$R\bar{3}m$	7.013	17.855	2.58	30.2.4.6
Walthierite	$Ba_{0.5}\square_{0.5}Al_3(SO_4)_2(OH)_6$	$R\bar{3}m$	6.992	34.433	3.02	30.2.4.7
Huangite	$Ca_{0.5}\square_{0.5}Al_3(SO_4)_2(OH)_6$	$R\bar{3}m$	6.983	33.517	2.80	30.2.4.8

ALUNITE GROUP, JAROSITE SUBGROUP: B = Fe_3^{3+}

Mineral	Formula	Space Group	a	c	D	
Jarosite	$KFe_3^{3+}(SO_4)_2(OH)_6$	$R\bar{3}m$	7.304	17.268	3.25	30.2.5.1
Natrojarosite	$NaFe_3^{3+}(SO_4)_2(OH)_6$	$R\bar{3}m$	7.327	16.634	3.29	30.2.5.2
Hydronium jarosite	$(H_3O)Fe_3^{3+}(SO_4)_2(OH)_6$	$R\bar{3}m$	7.17	16.6	3.01	30.2.5.3
Ammoniojarosite	$(NH)_4Fe_3^{3+}(SO_4)_2(OH)_6$	$R\bar{3}m$	7.327	17.500	3.112	30.2.5.4
Argentojarosite	$AgFe_3^{3+}(SO_4)_2(OH)_6$	$R\bar{3}m$	7.347	16.580	3.66	30.2.5.5
Plumbojarosite*	$Pb[Fe_3^{3+}(SO_4)_2(OH)_6]_2$	$R\bar{3}m$	7.305	33.675	3.71	30.2.5.6
Beaverite	$PbCu(Fe^{3+},Al_2)(SO_4)_2(OH)_6$	$R\bar{3}m$	7.256	34.30	4.31	30.2.5.7
Dorallcharite	$Tl_{0.8}K_{0.2}Fe_3^{3+}(SO_4)_2(OH)_6$	$R\gamma m$	7.330	17.633	3.85	30.2.5.8

*Half the available A positions are occupied by Pb ions, and the others remain vacant to provide valence compensation.

Note: The alunite groups are isostructural with crandallite and with the beudantite group.

30.2.4.1 Alunite $KAl_3(SO_4)_2(OH)_6$

30.2.4.2 Natroalunite $NaAl_3(SO_4)_2(OH)_6$

Named in 1824 (alunite) and 1911 (natroalunite) for the composition. *Alunite* is a contraction of an earlier term, "aluminilite." Alunite group. HEX-R $R\bar{3}m$. $a = 6.970$ (alunite), 6.990 (natroalunite); $c = 17.27$ (alunite), 16.905 (natroalunite); $Z = 3$; $D = 2.839$ (alunite), 2.819 (natroalunite). *14-136*(alunite): 5.77_3 4.96_6 2.99_{10} 2.89_{10} 2.29_8 1.93_7 1.90_3 1.50_4. *41-1467*(natroalunite): 4.90_7 3.49_3 2.98_7 2.96_{10} 2.93_2 2.22_4 1.895_4 1.745_3. **Habit:** Massive, granular to dense; as rocky masses admixed with quartz, kaolin, or other minerals. Rare crystals usually minute pseudocubic rhombohedrons $\{01\bar{1}2\}$; fibrous or columnar. **Physical properties:** White; also grayish, yellowish, reddish, reddish brown; dull to vitreous luster, pearly on $\{0001\}$. Cleavage $\{0001\}$, distinct; $\{01\bar{1}2\}$, poor; conchoidal to splintery fracture; brittle. $H = 3\frac{1}{2}-4$. $G = 2.6-2.9$. Not fluorescent, strongly pyroelectric. **Tests:** Slowly soluble in dilute H_2SO_4, nearly insoluble in HCl and HNO_3, infusible. **Optics:** Uniaxial (+); $N_O = 1.572$, $N_E = 1.592$ (Na)(alunite); $N_O = 1.568$, $N_E = 1.58$ (Na)(natroalunite). **Chemistry:** Na can substitute for K in alunite up to at least Na/K = 9:2, but the pure Na end member is not known in nature. Na-bearing alunites beyond trace amounts are relatively uncommon. The name *natroalunite* is given to that part of the series with Na > K. **Occurrence:** Alunite and natroalunite are widespread in rocks that have been altered by solfataric action, or "alunitized." In some places they result from the action of sulfuric acid derived from the oxidation of pyrite or other sulfides on nearby rocks. Alunitization may be accompanied by kaolinization and silicification. **Localities:** Many deposits are found in the western United States, as at Red Mt., Hinsdale Co., and at the head of South River, Mineral Co., CO; in altered volcanic rocks of the Goldfield district, NV. Ubiquitous around the hot springs of Lassen Volcanic National Park, CA. Commercially important deposits occur near Marysvale, Piute Co., UT. From the solfatara deposits of *Tolfa, Civita Veccia, Italy,* which have been the basis of an alum industry since the fifteenth century. Near Almeria, Spain; Beregovo, Ukraine; Fanshan, southwestern China; near Bulla Delah, NSW, Australia. Natroalunite is much less common than alunite. At the *National Belle mine, Red Mt., Silverton, CO.* Sugarloaf Butte, near Quartzite, AZ; Funeral Range, Death Valley, CA; Kalgoorlie, WA, Australia; Salamanca, Chile; Ukusu, Izu Prov., Japan. **Uses:** Alunite is mined extensively, and is used for the manufacture of alum, potash, and alumina. RG *DII:556; AM 47:127(1962), 65:953(1980), 77:1092(1992); GSPP 1076A:2(1978); NJMM:534(1982).*

30.2.4.3 Schlossmacherite $(H_3O,Ca)Al_3(SO_4,AsO_4)_2(OH)_6$

Named in 1980 for Karl Schlossmacher (b.1887), mineralogist and honorary president of the German Gemological Society. Alunite group. $R\bar{3}m$. $a = 6.998$, $c = 16.67$, D = 3.00. *NJMM:215(1980). 33-973:* 4.92_7 3.51_4 2.98_5 2.96_{10} 2.22_3 1.90_4 1.75_3 1.46_2. The composition of analyzed material places this mineral intermediate between the alunite and beudantite groups. Since the ratio of $(SO_4)^{2-}$ to $(AsO_4)^{3-}$ is nearly 2:1, it is considered to be part of the alunite group, but some authorities differ on this. *Emma Luisa mine, Guanaco, Chile,* associated with ceruleite and chenevixite. RG *AM 65:1069(1980).*

30.2.4.4 Osarizawaite $PbCuAl_2(SO_4)_2(OH)_6$

Named in 1961 for the locality. Alunite group. $R\bar{3}m$. $a = 7.075$, $c = 17.248$, D = 4.20. *AM 47:1216(1962). 15-178:* 5.75_8 3.52_6 3.00_{10} 2.87_5 2.28_5 2.23_4 1.92_5 1.50_5. Fine grained, earthy; as powdery crusts; rarely as microscopic ring crystals. Greenish yellow, dull luster. G = 3.89–4.02. Insoluble in water and HNO_3, decomposed by boiling concentrated HCl or H_2SO_4. Uniaxial (+); $N_O = 1.714$, $N_E = 1.731$; sometimes anomalously biaxial (+). Osarizawaite is the aluminian analog of beaverite. From the *Osarizawa mine, Akita Pref., Japan.* Mt. Edgar Pastoral Station, ESE of Marble Bar, Pilbara district, and from Whim Creek, WA, Australia. RG *MJJ 3:181(1961); NJMM:39(1977), 401(1980); AM 65:1287(1980).*

30.2.4.5 Minamiite (Na,K,Ca,☐)Al$_3$(SO$_4$)$_2$(OH)$_6$

Named in 1982 for A. E. Minami (1899-1977), Japanese geochemist. Alunite group. $R\bar{3}m$. $a = 6.98$, $c = 33.490$, $D = 2.81$. *AM 67:114(1982)*. *34-143*: 4.90$_9$ 3.49$_3$ 2.98$_5$ 2.96$_{10}$ 2.22$_4$ 1.90$_4$ 1.75$_3$ 1.46$_2$. Crystals are hexagonal tablets up to 0.2 mm maximum. White to nearly colorless; vitreous luster, pearly on cleavage. Cleavage {0001}, perfect. H = 3.5. G = 2.8. Nonfluorescent. Uniaxial (+); N$_O$ = 1.580, N$_E$ = 1.604. Found in hydrothermally altered andesite at *Okumanza, Mt. Shirane, Gumma Pref., Japan*, intimately mixed with quartz and alunite. RG

30.2.4.6 Ammonioalunite (NH$_4$)Al$_3$(SO$_4$)$_2$(OH)$_6$

Named in 1988 for the composition. Alunite group. $R\bar{3}m$. $a = 7.013$, $c = 17.855$, $D = 2.58$. *AM 73:145(1988)*. *42-1334*: 5.02$_8$ 3.51$_2$ 3.02$_{10}$ 2.99$_3$ 1.977$_2$ 1.915$_4$ 1.752$_2$ 1.536$_2$. Massive, fine grained; microcrystals (<20 μm) euhedral with characteristic rhombohedral morphology. Grayish white, white streak, vitreous luster. H = 2–3. G ≈ 2.4. Nonfluorescent. Nearly insoluble in cold dilute HCl, slowly soluble in cold dilute H$_2$SO$_4$, decomposes in 1.0 *N* KOH. Uniaxial (+); N$_O$ = 1.590, N$_E$ = 1.602. "Ammonioalunite" refers to alunites NH$_4^+$ > Na$^+$ or K$^+$. A complete solid-solution series exists between these three elements, but NH$_4^+$-rich members are rare. The maximum observed NH$_4^+$ content was 0.92 mol %. Found in alteration zones, usually near acidic hot springs, with temperatures below 100°, and with abundant NH$_4^+$ and SO$_4^{2-}$ but few K$^+$ ions. *The Geysers, Sonoma Co., CA*, with opal and quartz. Also from the Ivanhoe deposit, a fossil hot spring associated with Au–Hg mineralization, Elko Co., NV. RG

30.2.4.7 Walthierite Ba$_{0.5}$☐$_{0.5}$Al$_3$(SO$_4$)$_2$(OH)$_6$

Named for Thomas Nash Walthier (1923–1990), U.S. economic geologist, St. Joseph Minerals Co., who played a significant role in exploration of El Indio deposit and the Tambo district. Alunite group. HEX-R $R\bar{3}m$. $a = 6.992$, $c = 34.443$, $Z = 6$, $D = 3.02$. *AM 77:1275(1992)*. PD: 5.73$_5$ 4.97$_5$ 3.49$_6$ 2.98$_{10}$ 2.28$_8$ 1.909$_7$ 1.747$_6$ 1.503$_4$. Subhedral to anhedral crystals, 5 to 100 μm in length in individual grains and as aggregates. White to yellowish, white streak, vitreous luster. Cleavage {001}, perfect. Uniaxial (+); N$_O$ = 1.588, N$_E$ = 1.604; negative elongation, nonpleochroic. Found in the *Tambo mining district, Chile, in the Reina vein about 7 km SE of the El Indio gold mine*, where walthierite is intimately associated with alunite, barite, quartz, pyrite, and jarosite. RG

30.2.4.8 Huangite Ca$_{0.5}$☐$_{0.5}$Al$_3$(SO$_4$)$_2$(OH)$_6$

Named in 1992 for Yunhui Huang (b.1926), Chinese mineralogist, who studied the minerals of the Hsianghualing Range, Hunan. Alunite group. HEX-R $R\bar{3}m$. $a = 6.983$, $c = 33.517$, $Z = 6$, $D = 2.80$. *AM 77:1275(1992)*. PD: 4.91$_8$ 3.49$_3$ 2.97$_{10}$ 2.46$_4$ 2.23$_5$ 1.899$_4$ 1.745$_4$ 1.375$_4$. Huangite forms anhedral to subhedral crystals typically 3–10 μm in diameter. White to yellowish, white streak, vitreous luster. Cleavage {001}, perfect. Uniaxial (+), negative elongation, nonpleochroic. Because of small grain size, N values could not be measured. Found at the *El Indio* Au–Ag–Cu deposit in the *Campana B vein, Coquimbo region, Chile*, associated with kaolinite and pyrite. Other associated minerals are enargite, minamiite, and woodhouseite. RG

30.2.5.1 Jarosite $KFe_3^{3+}(SO_4)_2(OH)_6$
30.2.5.2 Natrojarosite $NaFe_3^{3+}(SO_4)_2(OH)_6$
30.2.5.3 Hydronium jarosite $(H_3O)Fe_3^{3+}(SO_4)_2(OH)_6$

Jarosite named in 1852 for the locality, natrojarosite named in 1902 for the composition, hydronium jarosite named in 1960 for the composition. Alunite group. Synonym: hydronium jarosite was formerly called carphosiderite. HEX-R $R\bar{3}m$. $a = 7.304$ (jarosite), 7.327 (natrojarosite), 7.17 (hydronium jarosite); $c = 17.268$ (jarosite), 16.634 (natrojarosite), 16.6 (hydronium jarosite); $Z = 3$; $D = 3.25$ (jarosite), 3.29 (natrojarosite), 3.01 (hydronium jarosite). *AM 50:1595(1965)*. *22-827*(syn)(jarosite): 5.93_5 5.09_7 3.65_4 3.11_8 3.08_{10} 2.29_4 1.98_5 1.83_5. *36-425*(syn)(natrojarosite): 5.95_2 5.59_3 5.06_8 3.12_9 3.07_{10} 2.24_3 1.98_3 1.83_3. *36-427*(syn)(hydronium jarosite): 5.93_2 5.08_6 3.66_2 3.12_7 3.08_{10} 2.27_2 1.98_{13} 1.83_{13}. **Habit**: Minute crystals, usually pseudocubic {01$\bar{1}$3} or tabular {0001}. Granular massive; as crusts or coatings; fibrous, nodular, concretionary, pulverulent, or earthy. **Physical properties**: Ocherous, amber-yellow, or dark brown; pale yellow streak; sometimes glistening; subadamantine to vitreous luster, resinous on fractures. Cleavage {0001}, distinct; uneven to conchoidal fracture; brittle. $H = 2\frac{1}{2}-3\frac{1}{2}$. $G = 2.9-3.26$. Strongly pyroelectric. **Tests**: In CT affords acid water, insoluble in water, soluble in HCl. **Optics**: Uniaxial (−); jarosite: $N_O = 1.820$ (reddish brown), $N_E = 1.715$ (colorless); natrojarosite: $N_O = 1.832$ (pale yellow), $N_E = 1.750$ (colorless); hydronium jarosite: $N_O = 1.816$ (deep yellow), $N_E = 1.728$ (pale yellow). Usually anomalously biaxial with very small 2V. **Chemistry**: Al may substitute for Fe and a complete solid-solution series between alunite and jarosite probably exists, but intermediate members are rare. Na substitutes for K to at least Na/K = 1:2.4, but it is doubtful if a complete series exists between jarosite and natrojarosite. Similarly, there is at least partial substitution of K for Na in natrojarosite. Al also substitutes for Fe, and the material from Kopec, Czech Republic, has about equal Fe and Al, but the amount of Al entering jarosite is normally small. **Occurrence**: Jarosite is widespread as crusts and coatings in the oxidized zone of sulfide deposits and in cracks in the adjoining rocks. It is formed by reaction of H_2SO_4 derived from the oxidation of pyrite with the gangue and wall rock. It is usually accompanied by goethite (limonite), from which it may be difficult to distinguish. **Localities**: Jarosite is abundant in association with mineral deposits of the western United States and is probably present in nearly every major copper, lead, zinc, silver, gold, and iron mine, but is seldom seen in material that is noteworthy in any way. From the *Barranco Jaroso, Sierra Almagrera, Spain*; in Germany, France, Italy, Great Britain, and so on; in Chile at Chuquicamata and elsewhere; abundant in Australia, and throughout the world. Natrojarosite is less abundant than jarosite but it is also widespread. At many places in the United States, including the eastern side of *Soda Springs Valley, Esmeralda Co., NV*; at Santa Eulalia, CHIH, Mexico; from Bohemia, Czech Republic, as an alteration of pyritic shales; in cracks in Eocene clay in Denmark. Hydronium jarosite was found at the *Upernivik district, Langø, Greenland* (carphosiderite); also at the *Staszic mine, Góry Swiętokrzyskie (Holy Cross Mt.), Poland*. Probably many low-alkali jarosite minerals which have been investigated would prove to be hydronium jarosite if reanalyzed in the light of present knowledge. RG Jarosite: *DII:560, NJMM:406(1976), MJJ 8:419 (1977), AM 72:178(1987); natrojarosite: DII:563, NJMM:406(1976);* hydronium jarosite: *DII:566, AM 46:243(1961)*.

30.2.5.4 Ammoniojarosite $(NH_4)Fe_3^{3+}(SO_4)_2(OH)_6$

Named in 1927 for the composition. Alunite group. $R\bar{3}m$. $a = 7.327$, $c = 17.500$, $D = 3.112$. *JAC 6:490(1973)*. 26-1014(syn): 5.81_1 5.12_6 3.11_{10} 2.91_1 2.57_1 2.32_2 1.99_3 1.83_3. Small lumps and nodules composed of microscopic tabular grains. Pale ocherous yellow; dull, waxy to earthy luster. Uniaxial $(-)$; $N_O = 1.800$, $N_E = 1.750$. Found with tschermigite, epsomite, and jarosite in a lignitic shale on the *western side of the Kaibab fault, UT*. In a similar environment in the Wasatch formation, Buffalo, WY. With buddingtonite at the Sulfur Bank mercury mine, Lake Co., CA. RG *DII:562, CM 20:91(1982)*.

30.2.5.5 Argentojarosite $AgFe_3^{3+}(SO_4)_2(OH)_6$

Named in 1923 for the composition. Alunite group. $R\bar{3}m$. $a = 7.347$, $c = 16.580$, $D = 3.66$. *AM 58:936(1973)*. 41-1398(syn): 5.96_5 3.68_2 3.06_{10} 2.77_4 2.53_3 2.22_6 1.838_3 1.471_3. Fine-grained masses and coatings consisting of microscopic platy hexagonal crystals. Yellow to brown, yellow streak, brilliant luster. Cleavage {0001}. $G = 3.62$. Uniaxial $(-)$; $N_O = 1.889$–1.890 (yellow), $N_E = 1.789$–1.790 (pale yellow). *Tintic Standard mine, Tintic, UT*, with anglesite, barite, and quartz. RG *DII:565*

30.2.5.6 Plumbojarosite $Pb[Fe_3^{3+}(SO_4)_2(OH)_6]_2$

Named in 1902 for the composition. Alunite group. $R\bar{3}m$. $a = 7.305$, $c = 33.675$, $D = 3.71$. *CM 23:659(1985)*. The presence of the divalent cation Pb^{2+} results in a doubling of the c axis of the hexagonal cell when compared with jarosite, with half of the Pb^{2+} sites being vacant in an ordered arrangement. This also results in other minor compensating differences from the conventional jarosite structure. 39-1353: 3.06_{10} 2.81_5 2.53_3 2.25_8 1.975_3 1.827_3 1.532_3 1.487_4. Crusts or masses consisting of microscopic hexagonal plates; also pulverulent, earthy, or ocherous. Golden brown to dark brown, dull to glistening or silky luster. Cleavage {10$\bar{1}$4}, fair. Soft. $G = 3.665$. Uniaxial $(-)$; $N_O = 1.875$, $N_E = 1.786$. O, yellow-brown; E, nearly colorless. Common in the oxidized zone of lead mines in arid regions. *Cook's Peak district,*

Luna Co., NM; Tintic district, UT; Tombstone, AZ; Boss mine, Clark Co., NV. Bolkardag, Anatolia, Turkey. Carguaicollo tin deposits, Potosi, Bolivia. RG *DII:568*, *AM 51:443(1966)*, *CM 23:47(1985)*.

30.2.5.7 Beaverite   PbCu(Fe^{3+},Al)$_2$(SO$_4$)$_2$(OH)$_6$

Named in 1911 for the locality. Alunite group. $R\bar{3}m$. $a = 7.256$, $c = 34.30$, $D = 4.31$. *CM 23:49(1985)*. Structural considerations show that Pb^{2+} occupies the position of the alkali ion in the jarosite structure, whereas Cu^{2+} substitutes for one of the Fe^{3+} in the sheet units. The c dimension of the lattice is doubled as in plumbojarosite. *17-476*: 5.85_{10} 3.60_4 3.31_4 3.03_{10} 2.52_3 2.28_5 1.95_3 1.80_3. Earthy and friable masses consisting of microscopic hexagonal plates. Canary yellow, dull luster. $G = 4.36$. Soluble in hot HCl, insoluble in water. Uniaxial $(-)$; $N \approx 1.85$; strongly birefringent. Al may substitute for some of the Fe^{3+}, but pure ferrian beaverite has been found in Zaire. Occurs in the oxidized portion of Pb–Cu deposits in arid regions. *Horn Silver mine, Frisco, Beaver Co., UT. Kipushi, Shaba, Zaire.* RG *DII:568*, *AM 47:1085(1962)*.

30.2.5.8 Dorallcharite   Tl$_{0.8}$K$_{0.2}$Fe$_3^{3+}$(SO$_4$)$_2$(OH)$_6$

Named in 1994 for the locality and the golden yellow color (*doré* in French). Alunite group. HEX-R $R\gamma m$. $a = 7.330$, $c = 17.663$, $Z = 3$, $D = 3.85$. *EJM 6:255(1984)*. PD: 5.95_9 3.67_3 3.11_{10} 2.99_2 2.58_2 1.991_3 1.833_2. Fine-grained, earthy masses consisting of microscopic pseudocubic to pseudo-octahedral crystals (< 0.01 mm) which are probably rhombohedra. Golden yellow, yellow streak. Cleavage {0001}, distinct. Colorless in transmitted light. Uniaxial $(-)$; $N_O = 1.822$, $N_E = 1.768$; nonpleochroic. Found as an oxidation product at *Allchar, Macedonia*, a deposit known for its As–Tl mineralization and profusion of rare sulfosalts. RG *AM 80:184(1995)*.

30.2.6.1 Itoite   Pb$_3$[GeO$_2$(OH)$_2$](SO$_4$)$_2$

Named in 1960 for Tei-Ichi Ito (1898–1980), mineralogist and crystallographer, University of Tokyo, Japan. Barite group; see barite (28.3.1.1). ORTH *Pnma*. $a = 8.47$, $b = 5.38$, $c = 6.94$, $Z = 4$, $D = 6.67$. *12-641*: 4.24_8 3.79_6 3.60_6 3.33_9 3.21_7 3.00_9 2.07_{10} 2.03_8. Silky fibers. White, silky luster. Soft. Ns$_{mean} = 1.84$–1.85. *Tsumeb, Namibia*, as an alteration product of anglesite, associated with fleischerite, cerussite, and mimetite. RG *NJMM:132(1960)*, *AM 45:1313(1960)*.

30.2.7.1 Piypite   K$_2$Cu$_2$(SO$_4$)$_2$O · (Na,Cu)Cl

Named in 1984 for Boris I. Piyp (1906–1966), Russian geologist and director, Institute of Volcanology, Petropoavlovsk-Kamchatskii. TET *I4mmm*. $a = 13.60$, $c = 4.98$, $Z = 2$, $D = 3.0$. *AM 75:1215(1990)*. *37-464*: 9.63_{10} 6.79_4 4.31_2 3.21_1 3.04_7 3.01_3 2.67_2 1.92_2. Mosslike aggregates of acicular or columnar crystals. Emerald-green, dark green to black; yellowish-green streak; vitreous luster. Perfect cleavage parallel to elongation. $H = 2\frac{1}{2}$. $G = 3.10$. Unstable in air, decomposed by water, soluble in acids. Uniaxial (+); $N_O = 1.583$, $N_E = 1.695$. O, pale green; E, deep green. Found in 500° fumarole incrustations of *Great Tolbachik fissure extrusion, Kamchatka, Russia*, associated with aphthitalite, dolerophanite, euchlorine, chalcocyanite, and tenorite. RG *DANS 275:714(1984)*, *AM 70:437(1985)*.

30.2.8.1 Kamchatkite $KCu_3(SO_4)_2OCl$

Named in 1988 for the locality. ORTH $Pna2_1$. $a = 9.741$, $b = 12.858$, $c = 7.001$, $Z = 4$, $D = 3.58$. *41-1394:* 7.76_{10} 3.50_7 3.22_3 2.90_2 2.68_3 2.59_4 2.28_2 2.27_2. Prismatic crystals, elongate [001] with rhombic cross section. Greenish to yellowish brown, yellow streak, vitreous luster. Cleavage {011}, {100}, perfect. $H = 3\frac{1}{2}$. $G = 3.48$. Soluble in water and in weak acid solutions, hygroscopic. Biaxial (+); $XYZ = cab$; $N_x = 1.695$, $N_y = 1.718$, $N_z = 1.759$; $2V = 75°$; nonpleochroic; no dispersion; negative elongation. Found as a fumarolic product of the *Great Tolbachik fissure eruption, Kamchatka, Russia*, with ponomarevite, hematite, and klyuchevskite. RG *ZVMO 117:459(1988)*, *AM 75:1210(1990)*.

30.2.9.1 Cannonite $Bi_2O(OH)_2(SO_4)$

Named in 1992 for Benjamin Bartlett Cannon V (b.1950), U.S. amateur mineralogist, who first recognized the mineral and donated specimens for investigation. MON $P2_1/c$. $a = 7.700$, $b = 13.839$, $c = 5.686$, $\beta = 109.11°$, $Z = 4$, $D = 6.515$. *MM 56:505(1992)*. *45-1439:* 3.64_6 3.51_4 3.47_6 3.21_{10} 2.92_7 2.78_5 1.984_9 1.561_4. Intergrown crystalline aggregates of euhedral equant to prismatic (< 0.2 mm) crystals. Colorless, white streak, adamantine luster, transparent. No apparent cleavage, uneven to conchoidal fracture, brittle. $H = 4$. Nonfluorescent, biaxial. In RL, weakly bireflectant gray to darker gray, weakly anisotropic. Nonpleochroic, strong internal reflections. $N_{calc} = 1.91–1.99$ (589 nm). From the *Tunnel Extension mine, Ohio mining district, Marysvale, UT*, in cavities in quartz gangue, associated with cuprobismutite, bismuthinite, and covelline. RG *AM 78:845(1993)*.

30.3.1.1 Euchlorine $NaKCu_3O(SO_4)_3$

Named in 1869 from the Greek for *pale green*, in allusion to its color. MON $C2/c$. $a = 18.41$, $b = 9.43$, $c = 14.21$, $\beta = 113.7°$, $Z = 8$, $D = 3.27$. *NJMM:541(1989)*. *45-1332:* 8.44_{10} 6.61_2 3.48_3 3.24_3 2.85_4 2.84_4 2.82_5 2.54_5. Rectangular crystal tablets, also incrustations. Emerald-green. Cleavage in two directions. $G = 3.10–3.27$. Partly soluble in water. Biaxial (+); $N_x = 1.580$, $N_y = 1.605$, $N_z = 1.644$; $2V$ large. X, pale grass-green; Y, grass-green; Z, bright yellow-green. A fumarolic product from the 1868 eruption of *Vesuvius, Italy*, associated with dolerophanite, eriochalcite, and chalcocyanite. RG *DII:570*, *AM 75:1214(1990)*, *NJMA 161:241(1990)*.

30.3.2.1 Caracolite $Na_3Pb_2(SO_4)_3Cl$

Named in 1886 for the locality. HEX $P6_3/m$. $a = 9.81$, $c = 7.14$, $Z = 2$, $D = 4.50$. *NJMM:284(1967)*. *25-706*: 8.45_3 4.03_3 3.55_6 2.93_{10} 2.88_5 2.13_4 1.91_6 1.87_2. Incrustations and imperfect crystals; twinning polysynthetic, complex. Colorless, also grayish, or stained green; vitreous luster. No cleavage. $H = 4\frac{1}{2}$. $G \approx 5.1$. Partially decomposed in water, soluble in ammonium acetate solution or hot KOH solution. Biaxial $(-)$; $N_x = 1.743$, $N_y = 1.754$, $N_z = 1.764$; 2V very large; $r > v$, strong. *Mina Beatriz, near Caracoles, Sierra Gorda, Atacama, Chile*, with bindheimite, anglesite, and galena. RG *DII:546, PM 54:34(1985)*.

30.3.3.1 Cesanite $Ca_{1+x}Na_{4-x}(SO_4)_3(OH)_x$

Named in 1981 for the locality. Isotypic with apatite. HEX $P6_3/m$. $a = 9.446$, $c = 6.895$, $Z = 2$, $D = 2.831$. *MM 47:59(1983)*. *35-506*: 4.72_5 3.90_7 3.45_9 2.82_6 2.73_{10} 2.64_5 2.27_5 1.84_7. Medium to coarse granular; subhedral crystals prismatic, somewhat elongated parallel to c; occasionally twinned. Colorless to white, greasy to silky luster. Cleavage $\{0001\}$, pronounced. $H = 2$–3. $G = 2.786$. Nearly insoluble in water, slightly soluble in hot or cold acids, completely soluble in aqua regia. Uniaxial $(-)$; $N_O = 1.570$, $N_E = 1.564$. From the *Cesano I geothermal well*, on the edge of *Baccano Caldera, Cesano, Italy*, with gypsum, anhydrite, aphthitalite, gorgeyite, and kalistrontite, in veins and cavity fillings. RG *MM 44:269(1981), AM 67:621(1982)*.

30.3.4.1 Fedotovite $K_2Cu_3(SO_4)_3O$

Named in 1988 for S. A. Fedotov (b.1931), Russian geologist. MON $P2_1/c$. $a = 19.06$, $b = 9.47$, $c = 14.18$, $\beta = 112.37°$, $Z = 8$, $D = 3.17$. *AM 75:240(1990)*. *45-1405*: 8.83_{10} 6.59_1 6.54_1 4.54_1 4.41_1 4.21_1 2.94_2 2.84_1. Crystals flattened $\{100\}$, somewhat irregular; micaceous; as crusts. Emerald- to grass-green, light grass-green streak, silky to vitreous luster. Cleavage $\{100\}$, perfect. $H = 2\frac{1}{2}$. $G = 3.205$. Biaxial $(+)$; $Z = b$; $Y \wedge c \approx 0°$; $N_x = 1.577$, $N_y = 1.594$, $N_z = 1.633$; $2V = 68°$; absorption $Z > Y$. X, greenish blue; Y,Z, yellow-green. From *Great Tolbachik fissure eruption, Kamchatka, Russia*, with dolerophanite, tolbachite, chalcocyanite, piypite, melanothallite, and tenorite. RG *DANS 299:961(1988), MM 52:724(1988)*.

30.4.1.1 Klyuchevskite $K_3Cu_3Fe^{3+}O_2(SO_4)_4$

Named in 1989 for the locality. MON $I2/m$, Im, or $I2$. $a = 18.41$, $b = 4.94$, $c = 18.64$, $\beta = 101.50°$, $Z = 4$, $D = 3.02$. *ZVMO 118:65(1989)*. *41-1463*: 9.17_9 9.03_{10} 7.20_4 4.50_3 3.76_6 3.68_2 3.45_2 3.41_3. Unoriented aggregates of acicular crystals, elongated $[010]$. Dark green to olive-green, light green streak, semimetallic luster. Cleavage $\{h0l\}$, perfect. $G = 3.00$–3.15. Soluble in water and weak acids. Biaxial $(+)$; $Y = b$; $N_x = 1.549$, $N_y = 1.550$, $N_z = 1.680$; $2V = 11°$. X, olive-green; Y, green; Z, dark olive-green. From fumaroles at the *Great Tolbachik fissure eruption, Kamchatka, Russia*, with kamchatkite, ponomarevite, and hematite. RG *AM 75:1210(1990)*.

Class 31

Hydrated Sulfates Containing Hydroxyl or Halogen

31.1.1.1 Connellite $Cu_{19}Cl_4(SO_4)(OH_4)_{32} \cdot 3H_2O$

Named in 1850 for Arthur Connell (1794–1863), Scottish chemist. HEX $P\bar{6}2c$. $a = 15.78$, $c = 9.10$, $Z = 2$, $D = 3.46$. Structure: *AM 57:426(1972)*. *35-538*: 13.7_8 7.89_{10} 5.46_2 5.17_3 3.22_2 2.73_4 2.49_3 2.28_4. Crystals acicular [0001], striated [0001], in radiating groups. Azure blue, pale greenish-blue streak, vitreous luster. $H = 3$. $G = 3.36$. Insoluble in water; soluble in acids and in NH_4OH; easily fusible; in CT, affords acid water. Uniaxial (+); $N_O = 1.724$–1.738, $N_E = 1.746$–1.758; nonpleochroic. (NO_3) may substitute for up to 20% of the $(SO_4)^{2-}$ in the structure. Connellite is isostructural with buttgenbachite $Cu_{19}Cl_4(NO_3)(OH)_{32} \cdot 2H_2O$. *MM 39:266(1973)*. Bisbee, AZ, at Copper Queen(!) and several other mines, associated with cuprite, malachite, azurite, and spangolite. Grand Central mine, Tintic, UT. *Wheal Gorland, Gwennap, Cornwall, England*(!), in blue masses of felted crystals; Arenas, Sardinia, Italy; Mouzaïa, Algeria; Likasi, Shaba, Zaire(!). RG *DII:572; MM 29:280(1950), 54:425(1990)*.

31.1.2.1 Shigaite $Mn_7Al_4(SO_4)_2(OH)_{22} \cdot 8H_2O$

Named in 1985 for the locality. HEX-R $R3$ or $R\bar{3}$. $a = 9.51$, $c = 32.83$, $Z = 3$, $D = 2.35$. *AM 71:1546(1986)*. Structurally related to mooreite, lawsonbauerite, and torreyite. *38-428*: 10.9_{10} 5.47_9 4.36_4 3.66_8 2.66_4 2.46_9 2.20_6 1.94_6. Small tabular hexagonal crystals. Light yellow, very light yellow to white streak. Cleavage {0001}, perfect; moderately flexible. $H = 2$. $G = 2.32$. Nonfluorescent. Uniaxial (−); $N_O = 1.546$; absorption O > E. O, medium yellow; E, very light yellow. *Loi mine, Shiga Pref., Japan*, in manganosite–rhodochrosite–sonolite ore, in veinlets with pyrochroite, jacobsite, hausmannite, and galaxite. RG *NJMM:453(1985)*.

31.1.3.1 Mooreite $(Mg,Zn,Mn)_{15}(SO_4)_2(OH)_{26} \cdot 8H_2O$

Named in 1929 for Gideon E. Moore (1842–1895), U.S. chemist, an early investigator of the ores of Franklin and Sterling Hill, NJ. MON $P2_1/a$. $a = 11.147$, $b = 20.350$, $c = 8.202$, $\beta = 92.7°$, $Z = 2°$, $D = 2.54$. Structure: *AC(B) 36:1304(1980)*. Structurally related to shigaite, lawsonbauerite, and torreyite. *43-696*: 10.1_8 5.08_{10} 3.45_5 2.67_4 2.38_8 1.732_5 1.619_6 1.546_3. Crystals tabular to platy {010}. Colorless, vitreous luster. Cleavage {010}, perfect.

$H = 3$. $G = 2.47$. Biaxial $(-)$; $X = b$; $Z \wedge c = 44°$; $N_x = 1.533$, $N_y = 1.545$, $N_z = 1.547$; $2V = 50°$; $r > v$, perceptible. *Sterling Hill mine, Ogdensburg, NJ*, with rhodochrosite, torreyite, zincite, fluoborite, and pyrochroite in veinlets cutting zinc ore. RG *DII:574; AM 54:973(1969), 68:474(1983)*.

31.1.4.1 Torreyite $(Mg,Mn)_9Zn_4(SO_4)_2(OH)_{22} \cdot 8H_2O$

Named in 1949 from John Torrey (1796–1873), American naturalist, who early studied the Franklin minerals. MON $P2_1/c$. $a = 16.486$, $b = 9.292$, $c = 10.619$, $\beta = 95.42°$, $Z = 2$, $D = 2.65$. *AM 64:949(1979). 33-874:* 10.2_{10} 6.10_3 5.16_5 4.52_2 3.84_4 3.47_2 2.73_4 1.57_5. Massive, granular. White to colorless, dull to vitreous luster. Cleavage {010}, good; not brittle. $H = 3$. $G = 2.665$. Soluble in acids. Biaxial $(-)$; $N_x = 1.570$, $N_y = 1.584$, $N_z = 1.585$; $2V = 40°$. *Sterling Hill mine, Ogdensburg, NJ*, with altered pyrochroite, mooreite, and fluoborite, in veinlets cutting willemite–franklinite ore. RG *DII:575; AM 34:589(1949), 67:1029(1982)*.

31.1.4.2 Lawsonbauerite $(Mg,Mn)_9Zn_4(SO_4)_2(OH)_{22} \cdot 8H_2O$

Named in 1979 for Lawson H. Bauer (1889–1954), American mineral chemist at Franklin, NJ. MON $P2_1/c$. $a = 10.50$, $b = 9.64$, $c = 16.41$, $\beta = 95.22°$, $Z = 2$, $D = 2.92$. *AM 67:1029(1982). 33-873:* 10.5_{10} 6.24_3 5.24_6 4.77_2 3.90_5 3.33_3 2.77_4 1.587_5. Crystals minute, prismatic, bladed, flattened {001}, and elongate [010]. Colorless to white, dull luster. No cleavage, even fracture, brittle. $H = 4\frac{1}{2}$. $G = 2.87$. Nonfluorescent. Biaxial $(-)$; Y parallel to b; $Z \wedge c = 7°$; $N_x = 1.590$, $N_y = 1.608$, $N_z = 1.611$; $2V = 42°$; $r > v$, strong. *Sterling Hill mine, Ogdensburg, NJ*, intimately associated with sussexite and calcite, also with zincite and franklinite. RG *AM 64:949(1979), 67:1029(1982)*.

31.1.5.1 Spangolite $Cu_6Al(SO_4)Cl(OH)_{12} \cdot 3H_2O$

Named in 1890 for Norman Spang (1842–1922), U.S. mineral collector, who supplied the original specimens for study. HEX-R $C3c$. $a = 8.245$, $c = 14.34$, $Z = 2$, $D = 3.14$. *AM 34:181(1949)*. Structure: *NJMM: 349(1992). 5-142:* 7.10_{10} 3.59_8 2.66_4 2.54_7 2.36_5 2.17_4 1.98_6 1.80_7. Hemimorphic crystals; however, often holohedral in aspect; commonly tabular {0001}, less often short prismatic. Twinning rare, on {0001}, the acute antilogous poles of two individuals joined together. Dark green, bluish green, emerald-green; pale green streak; vitreous luster. Cleavage {0001}, perfect; also on {10$\bar{1}$1}, distinct; conchoidal fracture; brittle. $H = 2$ on {0001}, $H = 3$ on inclined faces. $G = 3.14$. Pyroelectric. Uniaxial $(-)$; $N_O = 1.681$–1.686, $N_E = 1.627$–1.638. O, weak, green; E, bluish green. In *AZ* at *Tombstone*, also at the Copper Queen mine, Bisbee(!), in crystals up to 3 cm, and at the Metcalf mine, Clifton–Morenci. Majuba Hill, NV; Grand Central mine, Tintic, UT; Mex-Tex mine, Bingham, NM(*). Wheal Gorland, Gwennap, Cornwall, England(!). Arenas, Sardinia, Italy. St. Rome-de-Tarn, Aveyron, and Vezzani, Corsica, France. RG *DII:576*.

31.1.6.1 Schulenbergite $(Cu,Zn)_7(SO_4,CO_3)_2(OH)_{10} \cdot 3H_2O$

Named in 1984 for the locality. HEX-R $P3$ or $P\bar{3}$. $a = 8.249$, $c = 7.183$, $Z = 1$, $D = 3.38$. *38-349:* 7.19_{10} 3.58_4 3.21_3 2.70_8 2.53_8 2.16_3 1.791_3 1.559_3. Crystals thin tabular {0001}, microscopic, often in rosettes. Light blue-green, pale blue-green streak, pearly luster. $H = 2$. $G = 3.28$. Nonfluorescent. Uniaxial $(-)$; $N_O = 1.640$, $N_E = 1.623$; faintly dichroic. From the dumps of the *Glücksrad mine, near Oberschulenberg, Harz, Germany.* RG *NJMM:17(1984), AM 70:438(1985).*

31.2.1.1 Cyanotrichite $Cu_4Al_2SO_4(OH)_{12} \cdot 2H_2O$

Named in 1839 from the Greek for *blue* and *hair*. ORTH Space group unknown. $a = 10.16$, $b = 12.61$, $c = 2.90$, $Z = 1$, $D = 2.88$. *11-131:* 10.2_{10} 5.47_5 5.26_8 3.88_9 3.38_4 3.04_4 2.39_4 2.03_4. Plushlike coatings of minute, acicular, radial-fibrous crystals. Bright azure blue, pale blue streak, silky luster. $G = 2.74$–2.95. Soluble in acids. Biaxial $(+)$; X perpendicular to laths; Z parallel to elongation; $N_x = 1.588$ (nearly colorless), $N_y = 1.617$, $N_z = 1.655$; $2V = 82°$; $r < v$, strong. Y, pale blue; Z, bright blue. Grandview mine, Grand Canyon, Coconino Co., AZ; American Eagle mine, Tintic district, UT; Cap Garonne, Var, France, with chalcophyllite and parnauite; Leadhills, Scotland; Laurium, Greece; *Moldova, Banat, Romania.* Mednorudyansk, near Nizhnii Tagil, Russia. RG *DII:578.*

31.2.2.1 Woodwardite $Cu_4Al_2(SO_4)(OH)_{12} \cdot z\text{-}4H_2O$

Named in 1866 for Samuel P. Woodward (1821–1865), English naturalist and geologist. An inadequately characterized mineral. HEX-R Space group unknown. $a = 3.00$, $c = 27.3$. *39-726:* 10.9_{10} 5.46_6 3.66_5 2.61_4 2.45_2 1.535_1. Minute botryoidal concretions with a fibrous structure. Greenish blue to turquoise-blue. $G = 2.38$. Biaxial $(+)$; $N_x = 1.552$, $N_y = 1.555$, $N_z = 1.565$. *Cornwall, England.* Klausen, Trentino, Italy. Mine du Chateau, Urbeis, Bas Rhin, France, as a postmine product. RG *DII:580, MM 40:644(1976), AM 62:599(1977).*

31.2.3.1 Zincaluminite $Zn_6Al_6(SO_4)_2(OH)_{26} \cdot 5H_2O$ (?)

Named in 1881 for the composition. A questionable, inadequately characterized mineral. ORTH or HEX Space group unknown. *41-1361:* 11.1_5 7.07_4 4.83_{10} 4.35_3 3.55_2 2.71_2 2.45_3 2.40_3. Crusts of minute crystals with a hexagonal outline. White to pale blue. $H = 2\frac{1}{2}$–3. $G = 2.26$. Soluble in acids and alkalis; in CT, affords slightly alkaline water. Uniaxial $(-)$; $N_O - 1.534$, $N_E = 1.514$. Laurium, Greece, with smithsonite, serpierite, and other secondary minerals. RG *DII:579.*

31.2.4.1 Guarinoite $(Zn,Co,Ni)_6(SO_4)(OH,Cl)_{10} \cdot 5H_2O$

Named in 1993 for André Guarino, French mineral collector. HEX $P6_3$, $P6_3/m$, or $P6_322$. $a = 8.344$, $c = 21.59$, $Z = 3$, $D = 2.77$. *AS 46:37(1993). PD:* 10.8_{10}

3.30_9 2.73_6 2.56_5 2.35_4 1.575_3. Rounded aggregates of thin hexagonal crystals to 0.04 mm × 0.2 mm. Bright to deep pink, light pink streak, vitreous to pearly luster, transparent. Cleavage {001}, perfect; irregular fracture. Soft. Soluble in HCl. Uniaxial (−); $N_O = 1.584$, $N_E = 1.544$; strongly pleochroic. Nonfluorescent. O, pink; E, light pink. From the *Cap Garonne mine, Var, France*, as a secondary mineral in the zone of oxidation, with anglesite, antlerite, cerussite, and so on. RG *AM 78:1314(1993)*.

31.2.5.1 Theresemagnanite (Co,Zn,Ni)$_6$(SO$_4$)(OH,Cl)$_{10}$ · 8H$_2$O

Named in 1993 for Thérèse Magnan for her contributions to knowledge about the Cap Garonne mine. HEX $P6_3$. $a = 8.363$, $c = 26.18$, $Z = 3$, $D = 2.48$. *AS46:37(1993)*. PD: 13.1_{10} 3.52_3 2.99_3 2.68_4 2.53_9. Thin platy crystals, tabular {0001}, making up radiating spherules to 0.2 mm. Pink to light pink, streak light pink, pearly luster, transparent. Cleavage {0001}, perfect; irregular fracture. Soft. G = 2.52. Soluble in HCl. Nonfluorescent. Uniaxial (−); $N_O = 1.568$, $N_E = 1.542$. From the *Cap Garonne mine, Var, France*, as secondary, alteration minerals on quartz, associated with anglesite, antlerite, ktenasite, cerussite, brochantite, covellite, tennantite, and gersdorffite. RG *AM 78:1314(1993)*.

31.2.6.1 Uranopilite (UO$_2$)$_6$(SO$_4$)(OH)$_{10}$ · 12H$_2$O

Named in 1882 for uranium and the Greek for *felt*, alluding to the composition and habit. Probably MON Space group unknown. $c = 8.91$. *8-443:* 9.18_8 7.12_{10} 5.51_4 4.28_8 3.65_5 3.31_4 2.99_3 2.90_3. Microscopic needles and laths forming velvety incrustations and globular or reniform masses. Lemon to golden yellow, silky luster. Cleavage {010}, perfect. G = 3.7–4.0. Fluoresces bright greenish yellow in LWUV or SWUV. Insoluble in water, soluble in dilute acids. Biaxial (+); $X = b$; $Y \wedge c = 18°$; $N_x = 1.623$, $N_y = 1.625$, $N_z = 1.634$ (Na); 2V large; r < v, extreme; also, r > v. A secondary mineral found on altering uraninite. Goldfields, SK, Canada; also Eldorado mine, Great Bear Lake, and Hottah Lake, NWT, Johanngeorgenstadt, Saxony, Germany. *Jáchymov, Bohemia, Czech Republic*; Urgeriça, Portugal; Shinkolobwe, Shaba, Zaire. RG *AM 37:394(1952), 37:950(1952); USGSB 1064:135(1958)*.

31.2.7.1 Meta-Uranopilite (UO$_2$)$_6$(SO$_4$)(OH)$_{10}$ · 5H$_2$O

Named in 1952 as a lower hydrate than uranopilite. Probably ORTH Space group unknown. *18-309:* 8.65_{10} 5.53_8 5.08_4 4.36_4 3.67_4 3.57_{10} 3.31_9 3.00_5. Minute needles and laths. Yellow, grayish, brown, or green; fluoresces yellow-green. Biaxial (−); XYZ =bca; $N_x = 1.72$, $N_y = 1.76$, $N_z = 1.76$. *Jáchymov, Bohemia, Czech Republic*, and probably at other localities where uranopilite has been found. RG *DII:582, AM 37:950(1952), USGSB 1064:140(1958)*.

31.3.1.1 Chalcoalumite CuAl$_4$(SO$_4$)(OH)$_{12}$ · 3H$_2$O

Named in 1925 in allusion to the composition. MON $P2_1$. $a = 17.090$, $b = 8.915$, $c = 10.221$, $\beta = 95.88°$, $Z = 4$, $D = 2.25$. *MR 2:126(1971)*. *25-1430:* 8.50_{10}

7.90_2 6.71_1 5.10_1 4.79_2 4.25_9 4.18_3 2.52_1. Botryoidal crusts of matted fibers or laths; also microcrystals flattened {100}; as small spherules. Twinning common, on {100} and {010}. Turquoise green to pale blue, white streak, dull to vitreous luster. Cleavage {100}, perfect; somewhat sectile. H = $2\frac{1}{2}$. G = 2.29. Slowly soluble in dilute acids, difficultly fusible. Biaxial (+); $N_x = 1.523$, $N_y = 1.525$, $N_z = 1.532$; 2V small; r > v, strong. Alters to cuprian gibbsite. *Holbrook shaft, Copper Queen mine, Bisbee, AZ*(!), as thin crusts and as platy crystals in vugs, associated with goethite, cuprite, azurite, and malachite; also in the Lavender Pit(!). RG *DII:580*.

31.3.1.2   Nickelalumite $(Ni,Cu)Al_4[(SO_4),(NO_3)](OH)_{12} \cdot 3H_2O$

Named in 1980 in allusion to the composition. MON $P2_1$. a = 10.175, b = 8.860, c = 17.174, β = 95.95°, Z = 4, D = 2.28. *AM 67:416(1982)*. *35-698*: 8.54_{10} 7.88_2 4.78_1 4.58_1 4.27_6 2.51_2 2.29_2 2.00_2. Coatings and crusts, also nodules. Pale blue. G = 2.24. Biaxial; X ∧ c = 30°; $N_x = 1.532$, $N_z = 1.543$. *Mbobo Mkulu cave, eastern Transvaal, South Africa*, as crusts on gypsum crystals. RG

31.3.2.1   Wermlandite $CaMg_7(Al,Fe)_2(SO_4)_2(OH)_{18} \cdot 12H_2O$

Named in 1971 for the locality. HEX-R $P\bar{c}1$. a = 9.303, c = 22.57, Z = 2, D = 1.96. *ZK 168:133(1984)*. *25-153*: 11.2_7 7.98_{10} 5.62_4 4.63_5 3.89_4 2.61_4 1.54_4 1.51_3. Thin hexagonal plates {0001}. Pale greenish gray. Cleavage {0001}, perfect; folia are flexible as in mica, but inelastic. H = $1\frac{1}{2}$. G = 1.932. Uniaxial (−); $N_O = 1.493$, $N_E = 1.482$; sometimes anomalously biaxial with small 2V. *Långban, Värmland, Sweden*, associated with calcite and magnetite. RG *LIT 4:213(1971)*, *AM 57:327(1972)*.

31.4.1.1   Posnjakite $Cu_4(SO_4)(OH)_6 \cdot 2H_2O$

Named in 1967 for E. W. Posnjak (1888–1949), U.S. geochemist, investigator of the system $CuO \cdot SO_3 \cdot H_2O$. MON Pc. a = 10.578, b = 6.345, c = 7.863, β = 117.98°, Z = 2, D = 3.351. *ZK 149:249(1979)*. *43-670*: 7.00_{10} 3.50_7 3.35_5 2.70_8 2.43_9 2.34_6 1.873_6 1.550_8. Grains, small tabular crystals, and fissure fillings. Blue to dark blue, bluish streak, vitreous luster. H = 2–3. G = 3.32. Insoluble in NH_4OH, in CT gives off water and melts to a dark enamel. Biaxial (−); $N_x = 1.625$, $N_y = 1.680$, $N_z = 1.706$; 2V = 57°; absorption Y ≫ Z > X. X, bluish to colorless; Y, blue to dark blue; Z, greenish blue. *Murray mine, Georgetown*, and the *Yellow Pine mine, Crisman, CO*. *Drakewalls mine, Gunnislake, Cornwall, England*, in rich blue incrustations; also the *Fowley Consols mine*. *La Petite Verriére, Les Ardillats, Rhône, France*. From quartz–tungsten veins of the *Nura–Tadinsk, central Kazakhstan*, with aurichalcite and other minerals associated with oxidizing chalcopyrite. *Piesky, near Banská Bystrica, Slovakia*. RG *AM 52:1582(1967), 68:1251(1983); NJMM: 16(1982)*.

31.4.2.1 Wroewolfeite $Cu_4(SO_4)(OH)_6 \cdot 2H_2O$

Named in 1975 for Caleb Wroe Wolfe (1908–1980), U.S. crystallographer, Boston University. Dimorphous with langite (31.4.3.1). MON Pc. $a = 6.045$, $b = 5.646$, $c = 14.337$, $\beta = 93.38°$, $Z = 2$, $D = 3.32$. AM 70:1050(1985). 27-1133: 7.15_{10} 3.58_7 2.77_2 2.63_4 2.43_2 2.38_2 2.28_2 2.00_3. Crusts of small crystals, twinning on {001}. Deep greenish blue, light blue streak, vitreous luster. Cleavage {010}, {100}, {001}, perfect. $H = 2\frac{1}{2}$. $G = 3.27$. Biaxial $(-)$; $N_x = 1.637$, $N_y = 1.682$, $N_z = 1.694$; $2V = 53°$; absorption $Y > Z > X$. X, light blue; Y, deep greenish blue; Z, greenish blue. From the oxidation zone of the old lead mine, *Loudville, MA*, with chalcocite, covellite, and langite. Ladywell mine, Shelve, Shropshire, England; Nant-y-Cagal mine, Ceulanywaesmawr, Cardigan, Wales; West Blackcraig mine, Kirkcudbright, Scotland. RG *MM 40:1(1975)*, *AM 61:179(1976)*.

31.4.3.1 Langite $Cu_4(SO_4)(OH)_6 \cdot 2H_2O$

Named in 1864 for Victor von Lang (1838–1921), Austrian physical crystallographer, University of Vienna. Dimorphous with wroewolfeite (31.4.2.1). MON Pc. $a = 7.118$, $b = 6.034$, $c = 11.209$, $\beta = 90.01°$, $Z = 2$, $D = 3.37$. BM 107:641(1984). Structure: AC(C) 40:1309(1984). 43-671: 7.07_{10} 3.57_9 2.65_7 2.62_7 2.50_8 2.21_6 2.13_9 1.773_7. Crystals small equant or elongated [100]; as scales, crusts; also earthy. Twinning common on {110}, at times multiple, resembling aragonite. Fine blue to greenish blue, vitreous to silky luster. Cleavage {001} and {010}. $H = 2\frac{1}{2}$–3. $G = 3.28$. Insoluble in water, easily soluble in dilute acids or NH_4OH. Biaxial $(-)$; $XYZ = cba$; $N_x = 1.641$, $N_y = 1.690$, $N_z = 1.712$; $2V = 66°$. X, light yellowish green; Y, blue-green; Z, sky blue. Caroline tunnel of the Ward mine, Ely, NV. Eschach, Styria, Austria. Viel Salm, Belgium. *St. Blazey and St. Just, Cornwall*; also Copper Hill mine, Redruth, and East Pool and Dolcoath, Camborne, Cornwall, *England*. Nice rippled crusts at the Allegmenes mine, Castlehawn, Beachhaven Co., Cork Ireland(*). St.-Rome-de-Tarn, Aveyron, France(*), crystals to 5 mm on barite. Mollau, Haut Rhin, France. RG *DII:583*, *AM 49:1143(1964)*, *NJMM:16(1982)*, *BM 81:257(1975)*.

31.4.4.1 Felsöbanyaite $Al_4(SO_4)(OH)_{10} \cdot 5H_2O$

Named in 1853 for the locality. HEX Space group unknown. $a = 22.556$, $c = 19.032$, $Z = 24$, $D = 2.21$. 25-1491: 5.96_3 4.79_{10} 4.64_{10} 3.67_4 2.71_4 2.45_4 2.27_5 2.19_3. Globular, radial aggregates of lamellar crystals. Colorless, vitreous luster. Cleavage {001}, perfect; also {100}, {010}. $H = 1\frac{1}{2}$. $G = 2.33$. Uniaxial $(+)$; $N_O = 1.516$, $N_E = 1.533$. *Baia Sprie, Maramures, Romania* (formerly known as *Felsöbánya*, in *Hungary*). RG *DII:585*, *MM 29:8(1950)*, *AM 50:812(1965)*.

31.4.5.1 Basaluminite $Al_4(SO_4)(OH)_{10} \cdot 4H_2O$

Named in 1948 from the composition (a basic sulfate of aluminum). MON Space group unknown. $a = 14.857$, $b = 10.011$, $c = 11.086$, $\beta = 122.28°$,

$Z = 4$, $D = 2.211$. *MM 43:931(1980). 42-556:* 9.39_{10} 4.73_6 3.69_5 2.46_4 2.28_4 2.20_4 1.887_4 1.438_6. Compact masses. White, dull luster. Conchoidal fracture, not plastic. $G = 2.10$. In CT yields water. $N \approx 1.520$; $N_{calc} = 1.539$ (Na); negative elongation. Basaluminite is probably always formed by the dehydration of hydrobasaluminite under weathering conditions. Above the Fleming coal in the Cabaniss formation, Crawford Co., KS; also from IN and TN. *Irchester, Northamptonshire, and Clifton Hill, Brighton, Sussex, England*; also in the Oxford clay, Chickerell, Dorset. Epernay, Marne, France. RG *DII:586, AM 53:722(1968), CM 9:644(1969), MM 37:29(1969).*

31.4.6.1 Hydrobasaluminite $Al_4(SO_4)(OH)_{10} \cdot 15H_2O$

Named in 1948 for the composition and relation to basaluminite. Probably the same as "winebergite" (discredited). MON Space group unknown. $a = 14.911$, $b = 9.993$, $c = 13.640$, $\beta = 112.24°$, $Z = 4$, $D = 2.277$. *8-76:* 12.6_{10} 6.18_7 5.29_7 4.70_7 4.23_6 4.00_6 3.73_7 3.07_7. Claylike. White to light yellowish brown, dull luster. Plastic. Dehydrates rapidly in air to form basaluminite. Since hydrobasaluminite is apparently always the precursor of basaluminite, the localities for hydrobasaluminite are the same as those given for basaluminite. It is a widespread mineral but difficult to recognize because of its claylike properties. KS, IN, TN; *Irchester*, and elsewhere in *England.* RG *AM 33:747(1948), 53:773(1968), 54:1363(1969); MM 43:931(1980).*

31.4.7.1 Namuwite $(Zn,Cu)_4(SO_4)(OH)_6 \cdot 4H_2O$

Named in 1982 for the country of origin, and the National Museum of Wales. HEX Space group unknown. $a = 8.29$, $c = 10.50$, $Z = 2$, $D = 2.81$. *35-528:* 10.6_{10} 5.31_2 4.15_3 2.71_4 2.63_4 2.41_2 1.57_3 1.55_2. Fine grained encrustations consisting of minute hexagonal plates. Pale sea-green, pale green streak, pearly luster. Cleavage {0001}, perfect. $H = 2$. $G = 2.77$. Uniaxial; $N = 1.577$; extremely low birefringence. *Averllyn lead mine, Llanrwst mining district, northern Wales*, encrusting hydrozincite. RG *MM 46:51(1982), AM 68:281(1983).*

31.4.8.1 Glaucocerinite $(Zn,Cu)_5Al_3(SO_4)_{1.5}(OH)_{16} \cdot 9H_2O$

Named in 1932 from the Greek for *blue* and *waxlike.* HEX-R Space group unknown. $a = 3.057$, $c = 32.52$ (pseudocell), $D = 2.33$. *MM 49:583(1985). 39-338:* 10.9_{10} 5.45_9 3.63_8 2.62_6 2.46_6 2.23_5 1.98_5 1.54_4. Warty masses with radial-fibrous structure. Sky blue to turquoise-blue, waxy luster. $H = 1$. $G = 2.40$. Uniaxial; $N_{mean} = 1.542$; sometimes anomalously biaxial; $N_x = 1.540$, $N_y = 1.554$, $N_z = 1.562$; $2V = 60°$; positive elongation; parallel to extinction. *Laurium, Greece*, with smithsonite, adamite, gypsum, and malachite. RG *DII:574, AM 72:1028(1987).*

31.4.9.1 Ramsbeckite $(Cu,Zn)_{15}(SO_4)_4(OH)_{22} \cdot 6H_2O$

Named in 1985 for the locality. MON $P2_1/c$. $a = 16.088$, $b = 15.576$, $c = 7.102$, $\beta = 90.22°$, $Z = 2$, $D = 3.41$. *NJMM:38(1988). 39-365:* 7.09_{10} 3.55_3 2.69_3 2.68_2 2.52_8 2.51_2 2.15_2 1.776_2. Equant to tabular, rhomb-shaped crystals. Green, light green streak, vitreous luster. Cleavage {001}, perfect; conchoidal fracture. H = 3.5. G = 3.39. Nonfluorescent. Soluble in dilute acids. Biaxial (−); Y = b; X ∧ c = 5°, Z ∧ a = 5°; N_x = 1.635, N_y = 1.675, N_z = 1.680; 2V = 37°; r > v. X,Y,Z, bright blue. Ecton mine, near Audubon, Montgomery Co., PA, associated with anglesite, linarite, and posnjakite. *Bastenberg mine, near Ramsbeck, Germany,* and elsewhere in old German mine dumps. RG *NJMM:550(1985), AM 72:225(1987), MR 18:131(1987).*

31.5.1.1 Vonbezingite $Ca_6Cu_3(SO_4)_3(OH)_{12} \cdot 2H_2O$

Named for Karl Ludwig ("Ludi") von Bezing (b.1945), Austrian–South African radiologist and amateur mineralogist. MON $P2_1/c$. a = 15.122, b = 14.358, c = 22.063, β = 108.68°, Z = 8, D = 2.81. *AM 77:1292(1992).* PD: 3.39_{10} 3.37_6 3.20_5 3.19_7 3.12_9 3.10_6 2.769_4 2.763_4. Euhedral crystals, elongated [001] with maximum length 10 mm. Dark blue, similar to azurite; light blue streak; vitreous luster on fracture surfaces, subvitreous on crystal faces. Macroscopically the mineral is almost indistinguishable from azurite, from which it could be distinguished mainly by the dull, earthy luster of its crystal faces. No cleavage, subconchoidal fracture, brittle. H = 4. G = 2.82. Nonfluorescent. Biaxial (−); X = b; Y ∧ a = 30.2°, Z ∧ c = 11.5°. N_x = 1.590, N_y = 1.610, N_z = 1.619; 2V = 65°; r > v, moderate, dispersion strong. X, dark blue; Y, gray blue; Z, light blue. The mineral was found in a single solution cavity at the *Wessels mine, Kalahari manganese field, northwestern Cape Prov., South Africa,* associated with yellowish sturmanite crystals, bultfonteinite, gypsum, calcite, barite, azurite, and others. RG *MR 22:279(1991).*

31.6.1.1 Devilline $CaCu_4(SO_4)_2(OH)_6 \cdot 3H_2O$

Named in 1864 for H. E. Sainte-Claire Deville (1818–1881), French chemist. MON $P2_1/c$. a = 20.870, b = 6.135, c = 22.191, β = 102.73°, Z = 8, D = 3.06. Structure: *AC(B) 28:1182(1972). 35-561:* 10.2_{10} 5.10_{10} 3.75_4 3.39_7 3.21_4 3.19_4 2.52_2 2.10_4. Thin six-sided plates, flattened {001} and striated {010}; crusts or rosettes of crystals; twinning on {010}. Dark emerald green to bluish green, light green streak, vitreous luster, pearly on {001}. Cleavage {001}, perfect; {110}, {101}, {10$\bar{1}$}, distinct; not very brittle. H = $2\frac{1}{2}$. G = 3.13. Insoluble in water and concentrated H_2SO_4, soluble in HNO_3, in CT affords water. Biaxial (−); X ≈ c, Z = b; N_x = 1.585, N_y = 1.649, N_z = 1.660; 2V = 39° ± 2°; r < v, marked. X, very pale green; Y, green; Z, turquoise-green. Ecton copper mine, Montgomery Co., PA. *Cornwall, England.* Vezzani, Corsica, France, in stalactites with spangolite. Španiá Dolina, N of Banská Bystrica, Slovakia(!), in platy rosettes to 4 cm with gypsum, azurite and malachite. RG *DII:590, AM 54:329(1969), NJMM:79(1983).*

31.6.1.2 Lautenthalite $PbCu_4[(OH)_6|(SO_4)_2] \cdot 3H_2O$

Named in 1993 for the locality. The Pb analog of devilline. MON $P2_1/c$. a = 21.642, b = 6.040, c = 22.544, β = 108.2°, Z = 8, D = 3.84. *NJMM:401(1993).* PD: 5.14_{10} 4.53_6 3.40_8 2.63_5 2.53_4 2.21_4. Crystals tabular

(100) and elongate [010] to 0.5 mm, and as sheaflike aggregates, usually twinned on (100). Bright blue, white streak, vitreous luster, transparent. Cleavages {010} and {100}. Soluble in HCl, leaving a residue of $PbSO_4$. Biaxial $(-)$; $Z = b$; $Y \wedge c = 4°$; $N_x = 1.659$, $N_y = 1.703$, $N_z = 1.732$; $2V = 79°$; $r > v$, distinct; faintly pleochroic. Found in a Ag-rich galena specimen from *Lautenthal, Harz, Germany*, on anglesite, accompanied by serpierite and devilline. Also from Öblarn, Steiermark, Austria. RG *AM 79:571(1994)*.

31.6.2.1 Serpierite $Ca(Cu,Zn)_4(SO_4)_2(OH)_6 \cdot 3H_2O$

Named for Giovanni Battista Serpieri (1832–1897), Italian mining entrepreneur, founder of the Montecatini Company, and developer of the Laurium mines, Greece. Dimorphous with orthoserpierite. MON $C2/c$. $a = 22.186$, $b = 6.250$, $c = 21.853$, $\beta = 113.37°$, $Z = 8$, $D = 3.08$. Structure: $AC(B)$ *24:1214(1968)*. *22-148:* 10.2_{10} 5.09_8 3.39_8 2.71_6 2.65_6 2.44_6 2.17_6 1.57_6. Crusts and tufted aggregates of tiny lathlike crystals, elongated [100] and flattened {001}; botryoidal. Sky blue; vitreous luster, pearly on cleavage. Cleavage {001}, perfect. $G = 3.07$. Soluble in acids. Biaxial $(-)$; $YZ = bc$; $X \wedge a = 24°$; $N_x = 1.584$, $N_y = 1.642$, $N_z = 1.647$; $2V = 35° \pm 2°$; $r > v$, strong. X, pale green; Y,Z, bluish green. Otoño mine, Altar district, SON, Mexico. *Laurium, Greece*, with smithsonite. Ross Island, Killarney, Eire. Akchagyl, Kazakhstan, with linarite and cyanotrichite. RG *DII:592, AM 54:328(1969)*.

31.6.3.1 Ktenasite $ZnCu_4(SO_4)_2(OH)_6 \cdot 6H_2O$

Named in 1950 for Constantine A. Ktenas (d.1935), Greek mineralogist. MON $P2_1/c$. $a = 5.589$, $b = 6.166$, $c = 23.751$, $\beta = 95.55°$, $Z = 2$, $D = 2.93$. *29-591:* 11.8_{10} 5.93_9 4.85_9 2.96_5 2.79_6 2.69_6 2.66_5 2.58_7. Tabular crystals to 1.01 mm. Blue-green, vitreous luster. $H = 2$–2.5. $G = 2.98$. Soluble in dilute acids and ammonia. Biaxial $(-)$; $Z = b$; $N_x = 1.511$, $N_y = 1.613$, $N_z = 1.623$; $2V = 51°$. *Laurium, Greece*, with glaucokerinite and serpierite in smithsonite. RG *AM 36:381(1951), ZK 147:129(1978)*.

31.6.4.1 Peretaite $CaSb_4^{3+}O_4(SO_4)_2(OH)_2 \cdot 2H_2O$

Named in 1980 for the locality. MON $C2/c$. $a = 24.665$, $b = 5.601$, $c = 10.185$, $\beta = 95.98°$, $Z = 4$, $D = 4.06$. *AM 65:940(1980)*. *33-264:* 12.2_{10} 6.12_2 5.45_2 3.10_3 3.06_7 2.53_2 2.45_3 2.07_2. Crystals tabular {100}, elongated [001], always twinned on {100}. Colorless, vitreous luster. Cleavage {100}, perfect. $H = 3\frac{1}{2}$. $G = 4.0$. Biaxial $(+)$; $Y \wedge b = 28°$; Z parallel to c; $N_y = 1.841$, $N_z = 1.935$; 2V very large. *Pereta, Tuscany, Italy*, in an antimony-bearing vein in silicified limestone, in vugs, associated with klebelsbergite, valentinite, kermesite, sulfur, and stibnite. Also La Cetine mine. RG.

31.6.6.1 Campigliaite $Mn^{2+}Cu_4(SO_4)_2(OH)_6 \cdot 4H_2O$

Named in 1982 for the locality. MON $C2$. $a = 21.707$, $b = 6.098$, $c = 11.245$, $\beta = 100.3°$, $Z = 4$, $D = 3.06$. *AM 67:388(1982)*. *35-529:* 10.7_{10} 5.34_6 3.56_5 2.77_1 2.67_1 2.57_1 1.78_1 1.52_1. Bladelike crystals, flattened {100} and elongated

[010]. Light blue, vitreous luster. Cleavage {100}, perfect. G = 3.00. Slowly soluble in acetic acid. Biaxial (−); X ≈ a, Y ≈ c, Z = b; $N_x = 1.589$, $N_y = 1.645$, $N_z = 1.659$; nonpleochroic; 2V = 52°; r < v, weak. *Temperino mine, Campiglia Marittima, Tuscany, Italy*, in vugs in an ilvaite-rich skarn, associated with gypsum, brochantite, and antlerite. RG

31.6.7.1 Orthoserpierite $Ca(Cu,Zn)_4(SO_4)_2(OH)_6 \cdot 3H_2O$

Named in 1985 from the relationship to serpierite, with which it is dimorphous. ORTH $Pca2_1$. a = 22.10, b = 6.20, c = 20.39, Z = 8, D = 3.07. *39-346*: 10.2_{10} 5.10_9 3.40_9 3.18_5 2.61_5 2.56_5 2.51_4 2.38_6. Masses and fibrous crusts; crystals flattened {001} and elongated [010]. Sky blue, light green streak, vitreous luster. Splintery fracture. G = 3.00. Fluorescent mauve in LWUV and SWUV. Biaxial (−); XYZ = cab; $N_x = 1.586$, $N_y = 1.645$, $N_z = 1.650$; 2V = 32°; r > v, distinct. X, nearly colorless, Y,Z, pale green. *Chessy copper mine, near Lyon, France*, with gypsum, devilline, and calcite in brecciated argillite, as a postdump product. RG *SMPM 65:1(1985), AM 72:1026(1987)*.

31.6.8.1 Rabejacite $Ca(UO_2)_4(SO_4)_2(OH)_6 \cdot 6H_2O$

Named in 1993 for the locality. ORTH Space group unknown. a = 8.73, b = 17.09, c = 15.72, Z = 4, D = 4.31. *EJM 5:873(1993)*. PD: 7.90_{10} 3.98_4 3.49_8 3.38_7. Acicular crystals, flattened tablets on {001}; or rounded nodules. Amber yellow, vitreous luster, transparent to translucent. No cleavage. H = 3. G = 4.1. Fluorescent light yellow under SWUV and LWUV. Biaxial (−); X = c; Y and Z in the plane of the tablets; $N_{x\,calc} = 1.617$, $N_y = 1.710$, $N_z = 1.758$; 2V = 68°; r > v, weak; strongly pleochroic from pale yellow (Y) to sulfur-yellow (Z). Found at *Rabejac and Mas d'Alary village a few kilometers from Lodève, France*, accompanied by gypsum and other secondary U minerals as an alteration product of pitchblende. RG *AM 79:572(1994)*.

31.7.1.1 Kainite $K_4Mg_4(SO_4)_4Cl_4 \cdot 11H_2O$

Named in 1865 from the Greek for *recent*, in allusion to its recent (secondary) formation. MON $C2/m$. a = 19.72, b = 16.23, c = 9.53, β = 94.92°, Z = 4, D = 2.176. *AM 57:1325(1972)*. *25-1237*: 8.12_7 7.77_8 7.37_{10} 4.62_4 3.08_9 3.05_4 3.03_7 2.95_3. Massive, granular; as thick tabular crystals flattened {100}, often rich in forms. Colorless, also other colors due to included foreign matter; vitreous luster. Cleavage {001}, perfect; smooth to splintery fracture; brittle. H = $2\frac{1}{2}$–3. G = 2.15. Tastes salty and bitter. Biaxial (−); Y = b; Z ∧ c = 13°; $N_x = 1.494$, $N_y = 1.505$, $N_z = 1.516$; 2V = 90°; r > v, very weak. Formed only in marine evaporite potash deposits, often as a result of subsequent metamorphism. *Eddy Co., NM*, in the Permian potash deposits, associated with sylvite and langbeinite. *Stassfurt* and many other *northern German potash deposits*, associated with sylvite, halite, carnallite, kieserite, and other salt minerals. *Kalusz, Galicia, Poland*. Used as an ore of potash and magnesium. RG *DII:594*.

31.7.2.1 Uklonskovite NaMg(SO$_4$)F · 2H$_2$O

Named in 1964 for A. S. Uklonskii (1888–1972), mineralogist, Uzbekistan. MON $P2_1/m$. $a = 7.202$, $b = 7.214$, $c = 5.734$, $\beta = 113.14°$, $Z = 2$, $D = 2.414$. *AM 71:1282(1986)*. *39-320*(syn): 6.61$_8$ 5.27$_6$ 4.88$_3$ 3.51$_{10}$ 3.31$_5$ 3.15$_6$ 3.01$_5$ 2.97$_5$ 1.70$_6$. Minute, prismatic crystals, flattened {010}, elongated [010]. Colorless, vitreous luster. $G = 2.42$. Insoluble in water, soluble in cold dilute HCl. Biaxial (+); $N_x = 1.476$, $N_z = 1.500$. From cavities in clays covering a saline layer in Neogene rocks of *Kara-Kalpakii, lower Amu-Darya river, Kara Tau, Kazakhstan*, associated with glaserite and polyhalite. Also from the Cetine mine, Tuscany, Italy. RG *DANS 158:116(1964)*, *AM 50:520(1965)*, *BM 108:133(1985)*.

31.7.3.1 Clinoungemachite Na$_9$K$_3$Fe^{3+}(SO$_4$)$_6$(OH)$_3$ · 9H$_2$O (?)

Named in 1938 for its similarity to ungemachite except for its crystal system. Physical and chemical properties presumed to be similar to ungemachite. Needs confirmation. MON (pseudo-HEX-R) Space group unknown. $\beta = 110.67°$. Crystals thick tabular {001}. Indistinguishable in appearance from ungemachite except through measurement of individual crystals. Colorless to yellowish. Soluble in weak HCl. *Chuquicamata, Chile*, with sideronatrite in massive altered iron sulfates. RG *DII:597*.

31.7.4.1 Aluminite Al$_2$(SO$_4$)(OH)$_4$ · 7H$_2$O

Named in 1807 for the composition. MON $P2_1/c$. $a = 7.440$, $b = 15.583$, $c = 11.700$, $\beta = 110.18°$, $Z = 4$, $D = 1.794$. Structure: *AC(B) 34:2407(1978)*. *30-44*: 8.98$_{10}$ 7.79$_9$ 6.37$_4$ 5.49$_3$ 5.01$_4$ 4.70$_5$ 3.74$_4$ 3.71$_3$. Reniform or nodular masses consisting of minute fibers. White, dull luster. Earthy fracture, friable. $H = 1–2$. $G = 1.7$. Soluble in acids, in CT affords much water. Biaxial (+); $X =$ elongation; $N_x = 1.459$, $N_y = 1.464$, $N_z = 1.470$; $2V \approx 90°$. Formed in recent or Tertiary clays, marls, and lignites by the action of sulfate solutions derived from the alteration of pyrite or marcasite, on aluminous silicates. Joplin, MO, coating limestone. At the railroad crossing of the Green River, UT; *Halle, Saxony, Germany*. Épernay, Marne, France, and at Auteuil near Paris. Vesuvius, Italy, with halite and sylvite. In chalk at New Haven, Sussex, England. Salt Range, Punjab, Pakistan. RG *DII:600*, *ZK 151:141(1980)*.

31.7.5.1 Meta-Aluminite Al$_2$(SO$_4$)(OH)$_4$ · 5H$_2$O

Named in 1968 for the similarity to aluminite, but a lower hydrate. MON Space group unknown. $a = 7.930$, $b = 16.879$, $c = 7.353$, $\beta = 106.73°$, $Z = 4$, $D = 2.17$. *ZK 151:141(1980)*. *32-27*(syn): 8.46$_6$ 4.52$_{10}$ 4.39$_9$ 3.74$_4$ 3.62$_5$ 3.54$_6$ 2.71$_4$ 2.27$_4$. Nodular microcrystalline aggregates and concretions. Snow white, silky luster. Soft. $G = 2.18$. Biaxial (−); $Z = b$; $N_x = 1.497$, $N_y = 1.512$, $N_z = 1.513$; 2V small. Upon heating to 55° loses some water and alters to aluminite. Otherwise stable under normal conditions. *Fumerole uranium mine, Temple Mt., San Rafael Swell, Emery Co., UT*. RG *AM 53:717(1968)*.

31.7.6.1 Despujolsite $Ca_3Mn^{4+}(SO_4)_2(OH)_6 \cdot 3H_2O$

Named for Pierre Despujols, founder of the Service Geologique, Morocco. HEX $P\bar{6}2c$. $a = 8.56$, $c = 10.76$, $Z = 2$, $D = 2.54$. *BM 91:43(1968)*. *20-226:* 7.40_6 4.26_8 3.49_4 3.34_{10} 2.57_6 2.24_4 2.13_8 2.03_6. Hexagonal prismatic crystals, in aggregates. Lemon yellow, vitreous luster. Conchoidal fracture, brittle. $H = 2\frac{1}{2}$. $G = 2.46$. Dissolves in HCl with release of chlorine; turns brown, then black, with dilute HNO_3; insoluble in cold acetic acid. Uniaxial (+); $N_O = 1.656$, $N_E = 1.682$. O, paler yellow; E, pale yellow. *Tachgagalt, Morocco*, associated with gaudefroyite. RG *AM 54:326(1969)*.

31.7.6.2 Schaurteite $Ca_3Ge^{4+}(SO_4)_2(OH)_6 \cdot 3H_2O$

Named in 1965 for Werner T. Schaurte (1893–1978), South African scientist. HEX $P\bar{6}2c$. $a = 8.525$, $c = 10.803$, $Z = 2$, $D = 2.64$. *AM 53:507(1968)*. *19-225:* 7.40_5 4.26_7 3.49_5 3.34_{10} 2.58_5 2.24_5 2.21_5 2.13_6. Minute acicular to fibrous crystals. White, silky luster. $G = 2.65$. Uniaxial (+); $N_O = 1.569$, $N_E = 1.581$. *Tsumeb, Namibia*, as an alteration product of Ge-bearing ores. RG *NJMA 123:160(1975)*.

31.7.6.3 Fleischerite $Pb_3Ge^{4+}(SO_4)_2(OH)_6 \cdot 3H_2O$

Named in 1960 for Michael Fleischer (b.1908), American geochemist, U.S. Geological Survey. HEX $P\bar{6}2c$. $a = 8.867$, $c = 10.875$, $Z = 2$, $D = 4.674$. *SR 41A:345(1975)*. *29-771:* 7.70_5 6.27_7 3.62_{10} 3.44_8 2.64_9 1.89_9 1.72_6 1.68_6. Aggregates of fibrous crystals. White to pale rose, silky luster. Soft. $G = 4.55$. Nonfluorescent. Becomes rose-violet when irradiated with X-rays. Uniaxial (+); $N_O = 1.747$, $N_E = 1.776$. *Tsumeb mine, Namibia*, associated with cerussite, mimetite, and altered tennantite, in the upper oxidation zone. RG *AM 45:1313(1960)*, *NJMM:132(1960)*, *NJMA 123:160(1975)*.

31.8.1.1 Natrochalcite $NaCu_2(SO_4)_2(OH) \cdot H_2O$

Named in 1908 for the composition. MON Space group unknown. $a = 8.812$, $b = 6.188$, $c = 7.510$, $\beta = 118.7°$, $Z = 2$, $D = 3.487$. Structure: *ZK 187:239(1989)*. *45-1364:* 6.57_8 4.82_2 3.46_8 3.20_8 2.80_{10} 2.53_7 2.31_4 1.682_2. Crystals pyramidal {111}, with {110} well developed; cross-fiber veinlets. Bright emerald green, greenish-white steak, vitreous luster. Cleavage {001}, perfect. $H = 4\frac{1}{2}$. $G = 3.49$. Slowly soluble in water, easily so in acids; in CT, affords acid water and fuses to a dark enamel. Biaxial (+); $Y = b$; $X \wedge c = -12°$; $N_x = 1.649$, $N_y = 1.655$, $N_z = 1.714$ (Na); $2V = 37°$; $r < v$, strong. *Chuquicamata, Antofagasta, Chile*, associated with kroehnkite, antlerite, bloedite, atacamite, and other sulfates. Used as an ore of copper at Chuquicamata. RG *DII:602*, *DANS 123:78(1958)*, *BGSA 74:709(1963)*.

31.8.2.1 Johannite $Cu(UO_2)_2(SO_4)_2(OH)_2 \cdot 8H_2O$

Named in 1830 after Archduke Johann (1782–1859) of Austria, founder of the Styrian Landesmuseum in Graz. TRIC $P\bar{1}$ or $P1$. $a = 8.903$, $b = 9.499$,

$c = 6.812$, $\alpha = 109.87°$, $\beta = 112.01°$, $\gamma = 100.4°$, $Z = 1$, $D = 3.44$. *TMPM 30:47(1982)*. *17-530*: 7.73_{10} 6.16_9 5.59_4 4.38_6 3.87_7 3.41_8 3.13_7 3.04_7. Crystals prismatic [001] and thick tabular {100}; as spheroidal aggregates and coatings. Emerald-green to apple-green, pale green streak, vitreous luster. Cleavage {100}, good; not brittle. $H = 2\frac{1}{2}$. $G = 3.32$. Decomposed by water, soluble in acids. Tastes bitter. Biaxial (±); $X \wedge \phi = -101°$, $X \wedge \sigma = 85°$; $Y \wedge \phi = 37° =$, $Y \wedge \sigma = 8°$; $Z \wedge \phi = 169°$, $Z \wedge \sigma = 85°$; $N_x = 1.572$, $N_y = 1.595$, $N_z = 1.614$; $2V \approx 90°$; $r > v$, strong (Joach.); $r < v$, strong (CO). Formed by the oxidation of uraninite usually in the presence of gypsum. Central City District, Gilpin Co., CO. *Jáchymov, Bohemia, Czech Republic.* Johanngeorgenstadt, Saxony, Germany, Mine du Limouzat, St.-Priest-la-Prugne, Loire, France. Cornwall, England. RG *DII:606*, *USGSB 1064:130(1958)*, *AM 68:851(1983)*, *NJMA 159:297(1988)*.

31.8.3.1 Sideronatrite $Na_2Fe^{3+}(SO_4)_2(OH) \cdot 3H_2O$

Named in 1878 for the composition. ORTH *Pnn2* or *Pnnm*. $a = 7.27$, $b = 20.50$, $c = 7.15$, $Z = 4$, $D = 2.276$. *17-156*: 10.21_{10} 6.78_4 5.86_3 3.58_4 3.38_6 3.01_8 2.68_6 1.754_3. Minute needles elongated [001], nodular masses, fibrous crusts, earthy. Yellow to pale yellow, orange, yellow-brown; pale yellow streak. Cleavage {100}, perfect. $H = 1\frac{1}{2}-2\frac{1}{2}$. $G = 2.28$. Insoluble in cold water, decomposed in boiling water, with separation of ferric oxide. Biaxial (+); XYZ= *abc*; $N_x = 1.508$, $N_y = 1.525$, $N_z = 1.586$; $2V = 58° \pm 5°$; $r > v$, strong. X, nearly colorless; Y, pale amber-yellow; Z, pale yellow. With melanterite on the Urus plateau, Cheleken Is., Caspian Sea, Turkmenistan. *San Simon mine, Huantajaya, Tarapacá, Chile*, and elsewhere in the arid regions of Chile. With voltaite at Potosi, Bolivia. RG *DII:604*, *TMPM 28:315(1981)*, *PM 54:15(1985)*.

31.8.4.1 Metasideronatrite $Na_2Fe^{3+}(SO_4)_2(OH) \cdot H_2O$

Named in 1938 for the similarity to sideronatrite but being a lower hydrate. ORTH *Pbnm* or *Pbn2₁*. $a = 7.36$, $b = 64.08$, $c = 7.08$, $Z = 16$, $D = 2.76$. *AM 58:1080(1973)*. *29-1219*: 8.05_9 6.68_7 3.99_3 3.68_{10} 3.49_4 3.15_3 2.75_5 2.67_5. Crystals rare, prismatic [001]; usually as coarse to fine crystal aggregates. Golden yellow to straw yellow, silky luster. Cleavage {100}, {010}, perfect; {001}, good; fibrous fracture. $H = 2\frac{1}{2}$. $G = 2.68$. Insoluble in cold water, decomposes in boiling water, soluble in acids, in CT yields abundant water. Biaxial (+); XYZ= *abc*; $N_x = 1.543$, $N_y = 1.575$, $N_z = 1.634$; $2V = 60°$; $r > v$, strong. X, colorless; Y, light yellow; Z, brownish yellow. *Chuquicamata, Antofagasta, Chile*, with metavoltine and other secondary sulfates. RG *DII:603*, *NJMM 1982:248*.

31.9.1.1 Butlerite $Fe^{3+}(SO_4)(OH) \cdot 2H_2O$

Named in 1928 for Gurdon Montague Butler (1881–1961), American mining geologist, University of Arizona. Dimorphous with parabutlerite (31.9.2.1). MON $P2_1/m$. $a = 6.50$, $b = 7.37$, $c = 5.84$, $\beta = 108.38°$, $Z = 2$, $D = 2.564$. *NBSM 25(10):95(1972)*. *25-409*: 4.99_{10} 4.74_1 3.60_1 3.24_1 3.17_5 3.08_1 2.50_1

1.844_1. Crystals usually tabular {001} or {100} with a tendency for elongation [010]; also pseudo-octahedral. Twinning on {$\bar{1}$05}, very common. Deep orange, pale yellow streak, vitreous luster. Cleavage {100}, perfect. $H = 2\frac{1}{2}$. $G = 2.55$. Biaxial ($\pm$); $Z = b$; $X \wedge c = -18°$; $N_x = 1.593$–1.604, $N_y = 1.665$–1.674, $N_z = 1.741$–1.731; 2V large. X, colorless; Y, faint yellow; Z, light yellow. With copiapite and other sulfates at the *United Verde mine, Jerome, AZ*, formed under fumarolic conditions after a mine fire. Dexter No. 7 mine, Calf Mesa, San Rafael Swell, Emery Co., UT. Santa Elena mine, La Alcaparrosa, Argentina. Chuquicamata, Chile. RG *DII:608, AM 40:477(1955)*.

31.9.2.1   Parabutlerite $Fe^{3+}(SO_4)(OH) \cdot 2H_2O$

Named in 1938 from the Greek *para* and butlerite, a dimorphous pair. ORTH *Pmnb* or *Pmma*. $a = 20.13$, $b = 7.38$, $c = 7.22$, $Z = 8$, $D = 2.47$. *BM 93:185(1970)*. *16-939*: 5.85_6 4.99_{10} 4.07_2 3.60_4 3.11_{10} 2.91_2 2.50_4 1.84_3. Crystals prismatic [001], striated [001]; twinning on {142}, rare. Light orange to light orange-brown, vitreous luster. Cleavage {110}, poor; conchoidal fracture; brittle. $H = 2\frac{1}{2}$. $G = 2.55$. Insoluble in water; soluble in acids; in CT, yields acid water. Biaxial (+); XYZ= *bca*; $N_x = 1.589$–1.598, $N_y = 1.660$–1.663, $N_z = 1.750$–1.737 (Na); 2V = 87°; r > v, moderate. X, colorless; Y, faint yellow; Z, light yellow. *Alcaparrosa, near Cerritos Bayos, Antofagasta, Chile*, in a bed as an alteration product of copiapite. Santa Elena mine, La Alcaparrosa, San Juan, Argentina, with butlerite, fibroferrite, and other sulfates. RG *DII:610*.

31.9.3.1   Amarantite $Fe_2^{3+}(SO_4)_2O \cdot 7H_2O$

Named in 1888 from the Greek *amaranthos*, in allusion to its red color. TRIC $P\bar{1}$. $a = 8.976$, $b = 11.678$, $c = 6.698$, $\alpha = 95.65°$, $\beta = 90.37°$, $c = 97.2°$, $Z = 2$, $D = 2.14$. *ZK 127:261(1968)*. *17-158*: 11.3_{10} 8.69_{10} 5.16_4 4.98_4 3.57_8 3.41_4 3.11_6 3.05_8. Crystals elongated [001] or flattened {100}; striated [001]; aggregates of acicular crystals, radiating, columnar, or bladed. Amaranth-red; orange-red, lemon yellow streak, vitreous luster. Cleavage {010}, {100}, perfect; brittle. $H = 2\frac{1}{2}$. $G = 2.2$. Decomposed by water, soluble in HCl. Biaxial (−); $X \wedge \phi = 82°$, $X \wedge \sigma = 72°$; $Y \wedge \phi = 178°$, $Y \wedge \sigma = 68°$; $Z \wedge \phi = -44°$, $Z \wedge \sigma = 29°$; $N_x = 1.516$, $N_y = 1.598$, $N_z = 1.621$ (Na); 2V = 30°; r < v, horizontal. X, colorless; Y, pale yellow; Z, reddish brown. Santa Maria Mts., Riverside Co., CA. *Caracoles, Sierra Gorda, Chile*; also from Copiapó, Quetena, Alcaparrosa, and Chuquicamata. RG *DII:611, MM 42:144(1978)*.

31.9.4.1   Hohmannite $Fe_2^{3+}(SO_4)_2O \cdot 8H_2O$

Named in 1888 for Thomas Hohmann, Chilean mining engineer, who discovered the mineral. TRIC $P1$ or $P\bar{1}$. $a = 9.148$, $b = 10.922$, $c = 7.183$, $\alpha = 90.28°$, $\beta = 90.78°$, $\gamma = 107.37°$, $Z = 2$, $D = 2.250$. *MM 42:144(1978)*. *17-155*: 10.4_6 8.69_8 7.92_{10} 5.31_4 3.95_3 3.46_6 3.26_3 3.12_4. Crystals short prismatic [001]; as granular aggregates. Chestnut-brown to orange and reddish, orange-yellow streak, vitreous, brilliant luster. Cleavage {010}, perfect; {110} and {1$\bar{1}$0}, fair. H = 3. G = 2.255. Insoluble in cold water, decomposed by hot water; soluble in HCl; in CT affords acid water. Biaxial (−); $N_x = 1.559$ (pale yellow), $N_y = 1.643$ (pale greenish yellow), $N_z = 1.655$ (dark greenish yellow) (Na); $2V = 40°$; r > v, extreme. Upon exposure rapidly dehydrates and crumbles to metahohmannite. Redington mine, Knoxville, Napa Co., CA, with sulfur and cinnabar. *Sierra Gorda, near Copiapó, Chile*; also at Quetena and Chuquicamata. *DII:613*.

31.9.5.1 Metahohmannite $Fe_2^{3+}(SO_4)_2(OH)_2 \cdot 3H_2O$ (?)

Named in 1938 for the similarity to hohmannite, with a lower state of hydration. Probably TRIC $P1$ or $P\bar{1}$. *39-379*: 9.7_6 7.18_{10} 7.04_6 4.27_7 3.28_7 2.96_7 2.70_7 2.49_7. Pulverulent masses. Orange. Biaxial (+); $N_x = 1.709$, $N_y = 1.718$, $N_z = 1.734$. X, pale yellow; Y, reddish yellow; Z, reddish brown. Formed rapidly from the dehydration of hohmannite. *Chuquicamata*, Quetena, and Alcaparrosa, *Chile*. RG *DII:608, MM 42:M11(1978)*.

31.9.6.1 Botryogen $Fe^{3+}Mg(SO_4)_2(OH) \cdot 7H_2O$

Name derived in 1828 from the Greek for *bear* and *a bunch of grapes*, in allusion to the botryoidal masses originally found at Falun. MON $P2_1/n$ or $P2_1/c$. $a = 10.526$, $b = 18.872$, $c = 7.136$, $\beta = 100.13°$, $Z = 4$, $D = 2.23$. Structure: *AC(B) 24:760(1968)*. *17-157*: 8.87_{10} 6.29_6 5.34_4 5.11_6 4.07_4 3.74_4 3.50_4 3.00_8. Crystals prismatic [001], striated [001], with {101} large; mostly botryoidal, reniform, or radiating globular aggregates. Light to dark orange-red, ocher-yellow streak, vitreous luster. Cleavage {010}, perfect; {110}, good; conchoidal fracture; brittle. $H = 2-2\frac{1}{2}$. $G = 2.19$. Partly soluble in boiling water, leaving an ocherous residue; soluble in HCl. Biaxial (+); $X = b$; $Z \wedge c = 12°$; $N_x = 1.523$, $N_y = 1.530$, $N_z = 1.582$; $2V = 42°$; r > v, strong. X, colorless to pale brown; Y, cinnamon-brown; Z, golden yellow. Cornwall, Lebanon Co., PA. Redington mine, Knoxville, Napa Co., CA. The copper mine at *Falun, Sweden*. Rammelsberg, Harz, Germany. Miniera di Terranera, Rio Marina, Elba, Italy(*), with pickeringite and copapite. Madeni Zakh, Iran. Abundant at Chuquicamata, Quetena, and Alcaparrosa, Chile. Santa Elena mine, La Alcaparrosa, Argentina. RG *DII:617*.

31.9.6.2 Zincobotryogen $Fe^{3+}Zn(SO_4)_2(OH) \cdot 7H_2O$

Named in 1964 as the zincian analog of botryogen. MON $P2_1/n$. $a = 10.523$, $b = 17.841$, $c = 7.137$, $\beta = 100.12°$, $Z = 4$, $D = 2.237$. *NBSM 25(20):67(1984)*. *34-186*: 8.96_{10} 6.35_6 5.16_8 4.34_4 4.10_5 3.22_5 3.03_6 2.76_5. Prismatic crystals and radiating crystalline aggregates. Bright orange-red, vitreous to greasy luster. $H = 2\frac{1}{2}$. $G = 2.201$. Biaxial (+); $N_x = 1.542$, $N_y = 1.551$, $N_z = 1.587$; $2V = 54°$; r > v; strongly pleochroic; negative elongation. Rammelsberg, Harz, Germany. *Xitieshan*, in the arid Qaidam (Chaidamu) basin, Qinghai, China, associated with pickeringite. RG *AM 46:1517(1955), 49:1776(1964); IGR 10:917(1968)*.

31.9.6.3 Xitieshanite $Fe^{3+}(SO_4)Cl \cdot 6H_2O$

Named in 1983 for the locality. MON $P2_1/c$. $a = 14.102$, $b = 6.908$, $c = 10.673$, $\beta = 111.27°$, $Z = 4$, $D = 2.025$. *AM 74:1404(1989)*. *35-719:* 6.67_6 6.09_5 5.69_5 4.96_{10} 4.81_{10} 4.21_5 3.90_9 3.25_4. Rhombic, rectangular crystals, or as massive aggregates. Bright yellowish green, yellow streak. Imperfect cleavage, uneven to conchoidal fracture. H = $2\frac{1}{2}$. G = 1.99. Soluble in cold water and in dilute acids. Biaxial (−); $N_x = 1.536$, $N_y = 1.570$, $N_z = 1.628$; 2V = 77°; r > v. X, colorless to pale yellow; Y, pale yellow; Z, light greenish yellow. From the oxidation zone of the *Xitieshan lead–zinc mine, Qaidam basin, Qinghai, China*, with copiapite, roemerite, and other sulfates. RG *AM 69:1194(1984)*, *KT 33:502(1988)*, *SS(Geol):106(1989)*.

31.9.7.1 Guildite $CuFe^{3+}(SO_4)_2(OH) \cdot 4H_2O$

Named in 1928 for Frank Nelson Guild (1870–1939), geologist, professor of mineralogy, University of Arizona. MON $P2_1/m$. $a = 9.786$, $b = 7.134$, $c = 7.263$, $\beta = 105.28°$, $Z = 2$, $D = 2.717$. *AM 63:478(1978)*. *23-217:* 9.46_4 6.50_1 5.00_3 3.61_2 3.14_{10} 2.91_1 2.36_1 2.08_1. Short prismatic, pseudocubic crystals. Brown, pale canary yellow streak, vitreous luster. Cleavage {001}, {100}, perfect; conchoidal fracture; brittle. H = $2\frac{1}{2}$. G = 2.695. Biaxial (+); $N_x = 1.623$, $N_y = 1.630$, $N_z = 1.684$; 2V small. X,Y, pale yellow; Z, greenish yellow. Formed with coquimbite and other sulfates as a result of the mine fire at the *United Verde mine, Jerome, AZ*. RG *DII:619*, *AM 55:502(1970)*.

31.9.7.2 Chaidamuite $ZnFe^{3+}(SO_4)_2(OH) \cdot 4H_2O$

Named in 1986 for the locality. MON $P2_1/m$. $a = 9.759$, $b = 7.134$, $c = 7.335$, $\beta = 106.20°$, $Z = 2$, $D = 2.72$. *AM 73:1493(1988)*. *41-1375:* 9.40_8 5.03_7 5.00_8 3.64_7 3.57_4 3.12_{10} 3.09_{10} 2.05_4. Crystals thick tabular {001} or short prismatic [001], usually as disseminated grains and granular aggregates. Brown to yellowish brown, pale yellow streak, vitreous luster. Cleavage {001}, {100}, perfect; conchoidal fracture. H = $2\frac{1}{2}$–3. G = 2.722. Nonfluorescent. Insoluble in water, slightly soluble in dilute HCl. Biaxial (+); X = b; Y ∧ c = 12°, Z ∧ a = 28°; $N_x = 1.688$, $N_y = 1.640$, $N_z = 1.632$; 2V = 44°; r < v. X, colorless to light yellow; Y, light yellow; Z, brownish yellow. *Xitieshan lead–zinc mine*, in the arid *Chaidamu (Qaidam) basin, Qinghai, China*, associated with zincobotryogen, coquimbite, copiapite, butlerite, and other sulfates. RG

31.9.8.1 Aubertite $CuAl(SO_4)_2Cl \cdot 14H_2O$

Named in 1978 for J. Aubert (b.1929), French geophysicist, who collected the mineral. TRIC $P\bar{1}$. $a = 6.288$, $b = 13.239$, $c = 6.284$, $\alpha = 91.87°$, $\beta = 94.67°$, $\gamma = 82.45°$, $Z = 1$, $D = 1.83$. Structure: *AC(B) 35:2499(1979)*. *33-447:* 6.25_5 5.59_4 4.83_4 4.50_{10} 4.25_7 3.95_6 3.69_4 3.13_4. Crusts of corroded grains. Azure blue. Cleavage {010}, perfect. G = 1.815. Biaxial (−); $N_x = 1.462$, $N_y = 1.482$, $N_z = 1.495$ (Na); 2V = 71°; r > v, moderate; optic axis nearly perpendicular to {010}. *Quetena, Antofagasta, Chile*, with copiapite, amarantite, parabutlerite, and hohmannite. RG BM 102:348(1979), AM 65:205(1980).

31.9.8.2 Svyazhinite MgAl(SO$_4$)$_2$F · 14H$_2$O

Named for Nikolai Vasilevich Svyazhin (1927–1967), Russian mineralogist. TRIC $P\bar{1}$ or $P1$. $a = 6.217$, $b = 13.306$, $c = 6.255$, $\alpha = 90.05°$, $\beta = 93.30°$, $\gamma = 82.03°$, $Z = 1$, $D = 1.69$. AM 70:877(1985). 37-443: 5.68$_7$ 4.91$_{10}$ 4.40$_5$ 4.26$_4$ 4.15$_5$ 3.39$_4$ 2.84$_5$ 2.82$_5$. Crystals tabular $\{0\bar{1}4\}$, with complex morphology. Usually as nodules. Twinning on $\{0\bar{1}4\}$. Yellowish. Biaxial (−); N$_x$ = 1.423, N$_y$ = 1.439, N$_z$ = 1.444; 2V small; 2V$_{calc}$ = 57°. Found in fractures cutting pyroxene-amphibole fenites. *Ilmeniskie Mts., Ural Mts., Russia*, associated with pyrite and fluorite. ZVMO 113:347(1984).

31.9.8.3 Magnesioaubertite (Mg,Cu)Al(SO$_4$)$_2$Cl · 14H$_2$O

Named in 1988 for the composition and relationship to aubertite. TRIC $P\bar{1}$. $a = 6.31$, $b = 13.20$, $c = 6.29$, $\alpha = 91.70°$, $\beta = 94.50°$, $\gamma = 82.60°$, $Z = 1$, $D = 1.78$. AM 75:1433(1990). 45-1403: 6.26$_5$ 5.61$_7$ 4.81$_5$ 4.50$_{10}$ 4.25$_6$ 3.97$_8$ 3.68$_6$ 3.12$_5$. Polycrystalline aggregates of grains. Sky blue, white streak, glassy luster. Cleavage $\{010\}$, good. H = 2–3. G = 1.80. Biaxial (−); X ∧ $\{010\}$ = 17°; N$_x$ = 1.466, N$_y$ = 1.481, N$_z$ = 1.488; 2V = 112°; r > v. X, colorless; Y,Z, light blue. *Grotta de Faraglione, Porte de Levante, Vulcano island (Lipari group), N of Sicily, Italy*, in crystal aggregates enclosed in pickeringite on alunogen, with metasideronatrite, sulfur, aluminocopiapite, and metavoltine. RG AUF 39:97(1988).

31.9.9.1 Wilcoxite MgAl(SO$_4$)$_2$F · 18H$_2$O

Named in 1983 for William Wilcox, who discovered the mining district in 1879. He was killed by Apaches in 1880. TRIC $P1$ or $P\bar{1}$. $a = 14.90$, $b = 6.65$, $c = 6.77$, $\alpha = 117.43°$, $\beta = 100.58°$, $\gamma = 89.17°$, $Z = 1$, $D = 1.67$. 35-575: 5.88$_4$ 5.65$_9$ 4.91$_{10}$ 4.37$_6$ 4.12$_4$ 3.38$_5$ 2.98$_4$ 2.83$_4$. Distinct triclinic crystals with $\{010\}$, $\{110\}$, $\{100\}$, $\{\bar{1}10\}$, and $\{0\bar{1}1\}$. Colorless to white, vitreous luster. H = 2. G = 1.58. Soluble in cold water and dilute acids; in CT, dissolves in its water of crystallization. Biaxial (−); N$_x$ = 1.424, N$_y$ = 1.436, N$_z$ = 1.438; 2V = 48°; imperceptible dispersion. *Lone Pine mine, Wilcox district, Catron Co., NM*, forming crusts and efflorescences, associated with khademite, gypsum, and lannonite. RG MM 47:37(1983), AM 69:408(1984).

31.9.10.1 Jurbanite Al(SO$_4$)(OH) · 5H$_2$O

Named in 1976 for Joseph John Urban (b.1915), mineral collector, Tucson, AZ, who first called attention to the species. Dimorphous with rostite. MON $P2_1/n$ or $P2_1/c$. $a = 8.396$, $b = 12.479$, $c = 8.155$, $\beta = 101.92°$, $Z = 4$, $D = 1.828$. ZK 173:33(1985). 17-388: 6.75$_{10}$ 5.73$_{10}$ 4.48$_{10}$ 4.00$_{10}$ 3.99$_{10}$ 3.90$_{10}$ 3.71$_{10}$ 2.60$_{10}$. Crystals prismatic [001] to equant, with $\{011\}$ and $\{110\}$. Colorless, vitreous luster. No cleavage, brittle. H = $2\frac{1}{2}$. G = 1.786. Soluble in water. Biaxial (−); Y = b; Z ∧ $a = -5°$; N$_x$ = 1.459, N$_y$ = 1.473, N$_z$ = 1.483; 2V = 80°. *San Manuel mine, Pinal Co., AZ*, on the 2075 level, as a postmine stalactitic assemblage with epsomite, hexahydrite, pickeringite, and other sulfates. La Cetine mine, Tuscany, Italy, with gypsum and halotrichite. RG *AM 61:1(1976)*.

31.9.11.1 Khademite $Al(SO_4)F \cdot 5H_2O$

Named in 1973 for N. Khadem (b.1910), director, Geological Survey of Iran. ORTH *Pcab*. $a = 11.181$, $b = 13.048$, $c = 10.885$, $Z = 8$, $D = 1.942$. *AM 66:1102(1981)*. *26-1011*: 4.25_{10} 4.18_7 3.90_5 3.35_2 2.74_2 2.73_6 2.50_1 2.03_2. Crystals tabular {001}, with {010}, {110}, {111}, and {313}; in crusts. Colorless. No cleavage. $G = 1.925$. Biaxial (−); $YZ = ba$; $N_x = 1.44$, $N_y = 1.460$, $N_z = 1.487$; $2V = 68°$. Lone Pine mine, Catron Co., NM, associated with wilcoxite, lannonite, and gypsum as a postmine efflorescence. *Saghand, Iran*, associated with copiapite, butlerite, parabutlerite, and other sulfates. RG *AM 60:486(1975)*, *BM 104:19(1981)*, *MM 52:133(1988)*.

31.9.11.2 Rostite $AlSO_4(OH,F) \cdot 5H_2O$

Named in 1979 for Rudolph Rost (b.1912), Czech mineralogist, who originally described the species but thought it to be "lapparentite." It has been redefined as a new species. *NJMM:193(1979)*. Dimorphous with jurbanite. ORTH *Pcab*. $a = 11.175$, $b = 13.043$, $c = 10.878$, $Z = 8$, $D = 1.961$. *NJMM:476(1988)*. *41-1382*: 4.26_{10} 4.19_6 3.91_5 3.35_1 3.27_1 2.73_1 2.50_1 2.03_1. Chalky masses. White. $G = 1.892$. Biaxial (+); Z perpendicular to flattening; Y bisects acute angle of crystals; $N_x = 1.461$, $N_y = 1.470$, $N_z = 1.484$. From the burning waste heaps of the *Schoeller coal mine, Libušín, Kladno district, Bohemia, Czech Republic*, with copiapite and ammonia alum. RG *DII:601*, *AM 64:1331(1979)*, *MM 52:133(1988)*.

31.9.12.1 Fibroferrite $Fe^{3+}(SO_4)(OH) \cdot 5H_2O$

Named in 1833 for its fibrous nature and composition. HEX-R $R\bar{3}$. $a = 20.177$, $c = 7.656$, $Z = 18$, $D = 1.996$. *MA 34M:1237(1983)*. *38-481*: 12.1_{10} 6.98_4 4.57_5 4.07_4 3.44_2 3.35_3 2.99_3 2.78_3. Fine-fibrous crusts and masses; radial fibrous, botryoidal. Fibers elongated [0001]. Pale yellow to nearly white; pale greenish shades, silky to pearly luster. Cleavage {001}, perfect. $H = 2\frac{1}{2}$. $G = 1.95$. Decomposed by water, giving an acid solution and a brown precipitate of ferric hydroxide. Uniaxial (+); $N_O = 1.513$, $N_E = 1.571$. O, colorless; E, pale amber yellow. Commonly formed from the oxidation of bodies of pyrite, pyrrhotite, or marcasite. Dehydrates partially at room temperature so that the exact state of hydration of a given sample may be uncertain. Genette Mt., AZ. From the pyrrhotite deposit, Island Mt., Trinity Co., CA; also in the Calico Hills near Borate, San Bernardino Co., with krausite, coquimbite, and other sulfates. Pallières mines, Gard, France; Pöham, Salzburg, Austria; Skouriotissa, Cyprus. *Tierra Amarilla, near Copiapó, Chile*; also abundant at Chuquicamata. Santa Elena mine, La Alcaparrosa, Argentina. Yetar Spring, near Chidlows, WA, Australia. RG *DII:614*, *TMPM 28:17(1981)*, *NJMM:171(1987)*.

31.9.13.1 Slavíkite NaMg$_2$Fe$_5^{3+}$(SO$_4$)$_7$(OH)$_6$ · 33H$_2$O

Named in 1926 for František Slavík (1876–1957), Czech mineralogist and professor, Charles University, Prague. HEX-R $R\bar{3}$. $a = 12.20$, $c = 35.13$, $Z = 3$, $D = 1.90$. *SR 41A:351(1975)*. *20-679:* 11.7$_8$ 9.04$_{10}$ 5.83$_8$ 5.41$_8$ 4.21$_8$ 3.47$_8$ 2.95$_8$ 2.70$_8$. Crystals minute, tabular {0001}, with prominent {10$\bar{1}$1}; granular; scaly; as crusts. Greenish yellow, vitreous luster. G = 1.89. Uniaxial (−); N$_O$ = 1.533–1.530, N$_E$ = 1.497–1.506. O, lemon yellow; E, nearly colorless. Formed as a weathering product of pyritic shales at *Valachov Hill, near Skřivany, Bohemia, Czech Republic*, with halotrichite, pickeringite, and gypsum. Pöham, Salzburg, Austria. Santa Elena mine, La Alcaparrosa, San Juan, Argentina, with fibroferrite. RG *DII:621; BM 87:622(1964); NJMM:93(1973), 27(1975)*.

31.9.14.1 Lannonite HCa$_4$Mg$_2$Al$_4$(SO$_4$)$_8$F$_9$ · 32H$_2$O

Named in 1983 for Dan Lannon, who in 1893 staked important claims in the Wilcox district. TET Space group unknown. $a = 6.84$, $c = 28.01$, $Z = 1$, $D = 2.322$. *AM 69:407(1984)*. *35-576:* 14.0$_{10}$ 4.84$_7$ 4.67$_4$ 3.98$_5$ 3.46$_7$ 3.33$_5$ 2.91$_4$ 2.76$_4$. Nodules consisting of microscopic square plates. Chalky white, dull luster. H = 2. G = 2.22. Insoluble in water; dissolves in cold, dilute acids; in CT yields water and HF. Uniaxial (+); N$_O$ = 1.460, N$_E$ = 1.478. *Lone Pine mine, Wilcox district, Catron Co., NM*, with wilcoxite, khademite, and gypsum, as a postmine efflorescence. RG *MM 47:37(1983)*.

31.10.1.1 Carrboydite (Ni,Cu)$_{14}$Al$_9$(SO$_4$,CO$_3$)$_6$(OH)$_{43}$ · 7H$_2$O

Named in 1976 for the locality. HEX Space group unknown. $a = 9.14$, $c = 10.34$, $Z = 4$, $D = 2.692$. *AM 61:366(1976)*. *29-926:* 10.5$_{10}$ 5.25$_8$ 3.48$_6$ 2.62$_4$ 2.55$_7$ 2.36$_4$ 2.07$_4$ 1.51$_6$. Platy crystallites forming nodules. Yellowish green to blue-green. G = 2.50. Uniaxial (−); N$_O$ = 1.56, N$_E$ = 1.54. The formula is based on a brucite-type interlayer structure. *MM 44:337(1981)*. Found in surface material at the nickel mine at *Carr Boyd Rocks, WA, Australia*. RG

ETTRINGITE GROUP

The ettringite group minerals are basically isostructural and conform to chemistry

$$Ca_6X_2Y(O,OH)_{12} \cdot 26H_2O \quad \text{or a submultiple thereof}$$

where

X = Al, Cr, Fe^{3+}, Mn^{2+}, Si
Y = (SO$_4$,CO$_3$)$_3$ or (SO$_4$)$_2$B(OH)$_4$

and the structure is hexagonal or rhombohedral. Although there is considerable variation in chemistry, and to a lesser extent in cell parameters, the structures are all very similar. For details of the structure, see *AC(B) 26:386(1970), 27:594(1971), 25:1943(1969)*.

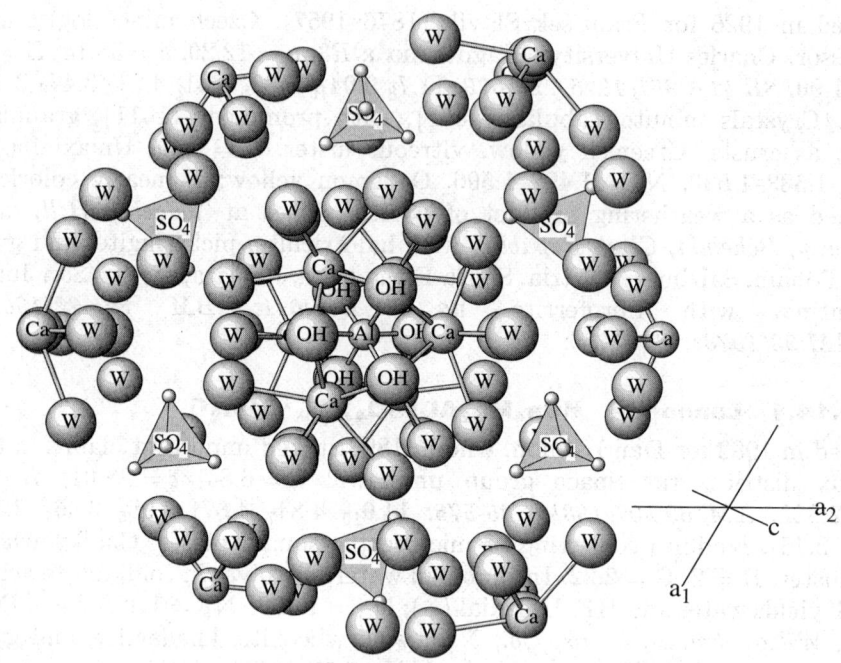

ETTRINGITE: $Ca_6Al_2(SO_4)_3(OH)_{12} \cdot 26H_2O$

This structure has columns consisting of octahedral Al alternating with triplets of eight-coordinated Ca. The ligands are (OH) or H_2O. Sulfate ions are independent, lying between columns with some free H_2O.

ETTRINGITE GROUP

Mineral	Formula	Space Group	a	c	D	
Ettringite	$Ca_6Al_2(SO_4)_3$ $(OH)_{12} \cdot 26H_2O$	$P31c$	11.23	21.44	1.754	31.10.2.1
Bentorite	$Ca_6(Cr,Al)_2(SO_4)_3$ $(OH)_{12} \cdot 26H_2O$	$P6_3/mmc$	22.35	21.41	2.021	31.10.2.2
Charlesite	$Ca_6Al_2(SO_4)_2[B(OH)_4](OH,O)_{12}$ $\cdot 26H_2O$	$P31c$	11.16	21.21	1.79	32.4.4.1
Sturmanite	$Ca_6Fe_2^{3+}(SO_4)_{2.4}[B(OH_4)_{1.2}(OH)_{12}$ $\cdot 25H_2O$	$P31c$	11.16	21.79	1.855	32.4.4.2
Jouravskite	$Ca_3(SO_4)(CO_3)$ $[Mn^{4+}(OH)_6] \cdot 12H_2O$	$P6_3$	11.06	10.50	1.93	32.4.4.3
Thaumasite	$Ca_3(SO_4)(CO_3)$ $[Si(OH)_6] \cdot 12H_2O$	$P6_3$	11.06	10.50	1.93	32.4.4.4

31.10.2.1 Ettringite $Ca_6Al_2(SO_4)_3(OH)_{12} \cdot 26H_2O$

Named in 1874 for the locality. Ettringite group. HEX-R $P31c$. $a = 11.23$, $c = 21.44$, Z = 2, D = 1.754. *NBSC 539(8):3(1959)*. *41-1451*(syn): 9.72_{10} 5.61_8 4.69_2 3.87_3 3.48_2 3.24_2 2.77_3 2.56_3. Crystals prismatic [0001], often without terminal faces, or with $\{10\bar{1}2\}$. Colorless to white, yellow to light brown. Cleavage $\{10\bar{1}0\}$, perfect. H = $2-2\frac{1}{2}$. G = 1.77. Partly decomposed by water, giving an alkaline solution; easily soluble in dilute acids. Uniaxial (−); $N_O = 1.491$, $N_E = 1.470$ (Franklin); $N_O = 1.466$, $N_E = 1.462$ (Scawt Hill). Franklin, Sussex Co., NJ, with andradite, manganophyllite, and other minerals. Lucky Cuss mine, Tombstone, AZ, as an alteration of Ca–Al silicates. *Near Ettringen, between Mayen and the Lacker See, Rhine, Germany*, in cavities in metamorphosed limestone inclusions in leucite–nepheline tephrite. Scawt Hill, Co. Antrim, Eire. Boisséjour, near Clermont-Ferrand, Puy de Dôme, France(*). N'Chwaning mine, Kuruman manganese district, Cape Prov., South Africa(!), in superb crystals and groups to 20 cm. RG *DII:589, AM 45:1137(1960), MM 39:377(1973)*.

31.10.2.2 Bentorite $Ca_6(Cr,Al)_2(SO_4)_3(OH)_{12} \cdot 26H_2O$

Named in 1980 for Y. K. Bentor, American geologist, University of California, San Diego. Ettringite group. HEX $P6_3/mmc$. $a = 22.35$, $c = 21.41$, Z = 8, D = 2.021. *AM 66:637(1981)*. *33-248*: 9.66_{10} 5.59_4 3.89_1 3.60_1 3.23_1 2.77_1 2.21_1 1.94_2. Minute hexagonal prisms with $\{10\bar{1}1\}$ $\{10\bar{1}0\}$, twinning on $\{10\bar{1}0\}$. Bright violet, very pale violet streak, vitreous luster. Cleavage $\{10\bar{1}0\}$, perfect; $\{0001\}$, distinct. H = 2. G = 2.025. Uniaxial (+); $N_O = 1.478$, $N_E = 1.484$; absorption E > O; positive elongation. O, nearly colorless; E, pale violet. *Hatrurim formation, southern Israel*, as a fracture filling cutting calcite–spurrite marble, with brownmillerite, mayenite, and other phases. RG *IJES 29:81(1980)*.

31.10.3.1 Zaherite $Al_{12}(SO_4)_5(OH)_{26} \cdot 20H_2O$

Named for M. A. Zaher, Geological Survey of Bangladesh, who discovered the mineral. TRIC $P1$ or $P\bar{1}$. $a = 18.475$, $b = 19.454$, $c = 3.771$, Z = 1, D = 2.006. *MM 49:145(1985)*. *38-379*: 18.1_{10} 9.56_1 9.08_1 4.82_1 4.61_1 4.56_1 4.44_1 3.22_1. Densely packed aggregates of small grains; in pure veinlets, micro- to cryptocrystalline. Chalk white to light bluish green, pearly to earthy luster. Well-developed cleavage in one direction. H = $3\frac{1}{2}$. G = 2.007–2.011. Biaxial (+); $N_x = 1.498$, $N_z = 1.499$ (Na); 2V moderate. *Salt Range, Pakistan*, in massive kaolinite–boehmite rock, associated with aluminite. In a sillimanite quarry 65 km W of Pofadder, Bushmanland, Cape Prov., South Africa. RG *AM 62:1125(1977), MM 48:131(1984)*.

ZIPPEITE GROUP

The zippeite group minerals are poorly characterized and probably a structurally heterogeneous group with the general formula

$$A_x(UO_2)_6(SO_4)_3(OH)_{10} \cdot nH_2O$$

where

A = K, Na (and $x = 4$, $n = 4$), or
A = Mg, Ni, Co, Zn (and $x = 2$, $n = 16$)

CM 14:429(1976).

ZIPPEITE GROUP

Mineral	Class	
Zippeite	Monoclinic	31.10.4.1
Sodium zippeite	Orthorhombic (?)	31.10.4.2
Magnesium zippeite	Orthorhombic (?)	31.10.4.3
Nickel zippeite	Orthorhombic (?)	31.10.4.4
Zinc zippeite	Orthorhombic (?)	31.10.4.5

Note: The minerals are all very fine grained, and single-crystal work on natural material has not been possible.

31.10.4.1 Zippeite $K_4(UO_2)_6(SO_4)_3(OH)_{10} \cdot 4H_2O$

Named for Franz Xaver Maxmillian Zippe (1791–1863), Austrian mineralogist. MON $C2/m$. $a = 8.81$, $b = 14.13$, $c = 8.85$, $\beta = 104.25°$, $Z = 2$, $D = 3.68$. *AM 37:394(1952)*. 29-1062: 7.06_{10} 5.45_3 3.66_3 3.50_9 3.12_8 2.87_4 2.65_4 2.22_4. Crystals (syn) are flat plates {010} with elongate, doubly terminated, rhombic outline. Twinning on {001}. Naturally, the mineral forms fine-grained pulverulent crusts, and is rather common underground in uranium mines but rare in collections. Orange-yellow. $H = 2$. $G = 3.66$. Fluorescent bright yellow, LWUV or SWUV. Highly radioactive. Fairly soluble in acid waters. Biaxial $(-)$; $X = b$; $Z \wedge a = 3°$; $N_x = 1.655$, $N_y = 1.717$, $N_z = 1.765$. X, colorless; Y, pale yellow; Z, yellow. The cell parameters and optics herein given for zippeite were determined for crystals synthetically grown whose identity as zippeite is not certain, since no analysis was made for K and the formula assumed at that time (1951) was incorrect. Nevertheless, based on powder data and other factors, it is considered probable that this synthetic material was zippeite. Gilpin Co., CO; Eldorado mine, Great Bear Lake, NWT, Canada; La Crouzille, Haute-Vienne, and Ronchamp, Haute-Saône, France. *Jáchymov (St. Joachimsthal), Bohemia, Czech Republic*; Cornwall, England. RG *MM 43:539(1979)*.

31.10.4.2 Sodium Zippeite Na$_4$(UO$_2$)$_6$(SO$_4$)$_3$(OH)$_{10}$ · 4H$_2$O

Named in 1976 from the composition. Probably ORTH Space group unknown. *29-1285*(syn): 7.34$_{10}$ 3.75$_1$ 3.66$_5$ 3.49$_4$ 3.15$_4$ 2.86$_2$ 2.12$_1$ 1.75$_1$. Velvety crusts of minute crystals; earthy. Yellow, dull luster. Cleavage parallel to flattening of grains, perfect. G > 3.3. Fluoresces bright yellow under LWUV or SWUV. Optically, the crystal grains conform to orthorhombic symmetry. Biaxial (−); N$_x$ = 1.63; the acute bisectrix X is perpendicular to the plane of flattening and cleavage, {010}, with Z parallel to the elongation or c axis; 2V moderate to large; pleochroic. X, colorless to pale yellow; Y, yellow; Z, dark yellow to golden yellow. Twinning is common. Sodium zippeite is the commonest member of the group. From many mines on the Colorado Plateau, including the Happy Jack mine, White Canyon, San Juan Co., UT; Delta mine, San Rafael Swell, Emery Co., UT; Fruita, CO; Jáchymov, Bohemia, Czech Republic. RG *CM 14:429(1976)*.

31.10.4.3 Magnesium Zippeite Mg$_2$(UO$_2$)$_6$(SO$_4$)$_3$(OH)$_{10}$ · 16H$_2$O

Named in 1976 for the composition. Probably ORTH Space group unknown. *29-876*: 9.7$_1$ 7.2$_8$ 3.58$_{10}$ 3.48$_8$ 3.11$_6$ 2.88$_2$ 2.74$_3$ 2.52$_2$. Orange-yellow. Physical properties like sodium zippeite. N$_x$ = 1.70–1.75; other optics like those of sodium zippeite. A complete solid-solution series probably exists between the Mg, Co, and Ni members. *Lucky Strike No. 2 mine, Emery Co., UT.* RG *CM 14:429(1976), MM 43:539(1979)*.

31.10.4.4 Nickel Zippeite Ni$_2$(UO$_2$)$_6$(SO$_4$)$_3$(OH)$_{10}$ · 16H$_2$O

Named in 1976 for the composition. Probably ORTH Space group unknown. *29-1434*(syn): 7.21$_{10}$ 3.94$_1$ 3.59$_5$ 3.47$_2$ 3.12$_3$ 2.65$_1$ 2.49$_1$ 1.96$_1$. Habit similar to sodium zippeite. Dark yellow to tan. Physical properties and optics similar to magnesium zippeite. *Happy Jack mine, White Canyon, San Juan Co., UT*, with sodium zippeite, uranopilite, johannite, and zeunerite. Eldorado mine, Great Bear Lake, NWT, Canada. Jáchymov, Bohemia, Czech Republic. RG *CM 14:426(1976), MM 43:539(1979)*.

31.10.4.5 Zinc Zippeite Zn$_2$(UO$_2$)$_6$(SO$_4$)$_3$(OH)$_{10}$ · 16H$_2$O

Named in 1976 for the composition. Probably ORTH Space group unknown. *29-1395*(syn): 9.62$_3$ 7.08$_{10}$ 4.84$_1$ 3.54$_5$ 3.44$_4$ 3.10$_3$ 2.47$_2$ 1.94$_2$. Habit, color, and physical properties similar to sodium zippeite. Optics similar to magnesium zippeite. *Hillside mine, Bagdad, Yavapai Co., AZ*, associated with sodium zippeite as yellow, orange, and reddish-brown coatings on quartzose ore containing disseminated sulfides and uraninite. RG *CM 14:428(1976), MM 43:539(1979)*.

31.10.4.6 Cobalt Zippeite Co$_2$(UO$_2$)$_6$(SO$_4$)$_3$(OH)$_{10}$ · 16H$_2$O

Named in 1976 for the composition. Probably ORTH Space group unknown. *29-520*(syn): 7.21$_{10}$ 3.94$_1$ 3.59$_5$ 3.47$_2$ 3.12$_3$ 2.65$_1$ 2.49$_1$ 1.96$_1$. Velvety crusts of minute crystals, botryoidal coatings, earthy. Color, physical properties, and optics similar to those of nickel zippeite. *Happy Jack mine, White Canyon, San Juan Co., UT.* RG *CM 14:429(1976), MM 43:539(1979).*

COPIAITE GROUP

The copiapite minerals have the general formula

$$AFe_4^{3+}(SO_4)_6(OH)_2 \cdot 20H_2O$$

where

A = either divalent ions Fe, Mg, Cu, Ca, or Zn, or the trivalent elements Fe^{3+} or Al and the space group is $P\bar{1}$. The structure of these very rare minerals is described in *AM 58:314(1973)*.

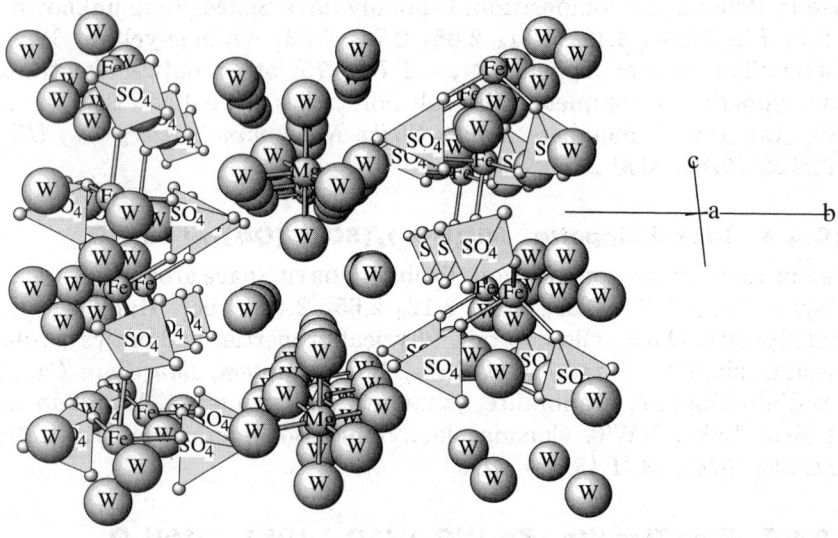

COPIAPITE (magnesio): $MgF_4^{3+}(SO_4)_6(OH)_2 \cdot 20H_2O$
Layers of $Mg(H_2O)_6$ octahedra alternating with layers containing octahedra Fe^{3+} and sulfate on (010). Layer sequence is ABA ABA . . . , where A is Fe-sulfate and B is $Mg(H_2O)_6$. Only hydrogen bonds join layers.

31.10.5.2 Magnesiocopiapite

COPIAPITE GROUP

Mineral	Formula	Space Group	a	b	c	α	β	γ	D	
Copiapite	$Fe^{2+}Fe_4^{3+}(SO_4)_6$ $(OH)_2 \cdot 20H_2O$	$P\bar{1}$	7.337	18.76	7.379	91.28	102.11°	98.57	2.118	31.10.5.1
Magnesiocopiapite	$MgFe_4^{3+}(SO_4)_6$ $(OH)_2 \cdot 20H_2O$	$P\bar{1}$	7.342	18.818	7.389	91.27	102.09°	98.51	2.05	31.10.5.2
Cuprocopiapite	$CuFe_4^{3+}(SO_4)_6$ $(OH)_2 \cdot 20H_2O$	$P\bar{1}$	7.31	18.15	7.25	92.30	102.18°	100.24	2.26	31.10.5.3
Ferricopiapite	$(Fe^{3+},Al)_{2/3}Fe_4^{3+}$ $(SO_4)_6$ $(OH)_2 \cdot 20H_2O$	$P\bar{1}$	7.390	18.213	7.290	93.42	102.12°	99.18	2.123	31.10.5.4
Calciocopiapite	$CaFe_4^{3+}(SO_4)_6$ $(OH)_2 \cdot 20H_2O$	$P\bar{1}$	7.44	18.79	7.22	85.18	104.42°	102.12	2.159	31.10.5.5
Zincocopiapite	$ZnFe_4^{3+}(SO_4)_6(OH)_2 \cdot 20H_2O$	$P\bar{1}$	7.33	18.72	7.35	91.30	102.06°	98.42	2.09	31.10.5.6
Aluminocopiapite	$(Al,Mg)_{2/3}Fe_4^{3+}$ $(SO_4)_6(OH)_2 \cdot 20H_2O$	$P\bar{1}$	7.30	18.80	7.31	91.30	102.18°	98.42	2.07	31.10.5.7

31.10.5.1 Copiapite $Fe^{2+}Fe_4^{3+}(SO_4)_6(OH)_2 \cdot 20H_2O$

Named in 1833 for the locality. Copiapite group. TRI $P\bar{1}$. $a = 7.337$, $b = 18.76$, $c = 7.379$, $\alpha = 91.47°$, $\beta = 102.18°$, $\gamma = 98.95°$, $Z = 1$, $D = 2.118$. *NJMM:158(1983)*. 35-583: 18.4_{10} 9.23_{10} 6.15_6 5.57_7 5.32_3 4.20_4 3.58_6 3.49_6. Crystals tabular {010}, sometimes highly modified; in loose aggregations of scales, granular or incrusting. Sulfur-yellow to orange; when massive, greenish yellow to olive-green; pearly luster on {010}. Cleavage {010}, perfect; {$\bar{1}01$}, imperfect. H = $2\frac{1}{2}$–3. G = 2.1. Easily soluble in water, forming a yellow solution which becomes turbid on heating; in CT affords abundant water. Biaxial (+); X perpendicular to {010}; Y = {$\bar{1}01$}; Z = {101}; $N_x = 1.509$, $N_y = 1.532$, $N_z = 1.577$; 2V = 73°; r > v, strong. X, greenish yellow to yellow; Y, yellow to colorless; Z, sulfur-yellow to yellow-green. In CA at the Redington mine, Knoxville, Napa Co.; at Sulfur Bank hot springs, Lake Co.; at the Island Mt. copper mine, Trinity Co. Comstock Lode, NV. Copper Queen mine, Bisbee, AZ. Goslar, Harz, Germany. Falun, Sweden. Vigneria and Capo Calamita, Elba, Italy. Common in coal mines in France, especially near Commentry, Plateau Central. *Tierra Amarilla, near Copiapó, Chile*, with melanterite, fibroferrite, alunogen, halotrichite, and other sulfates; also at Alcaparrosa, Quetena, and Chuquicamata. RG *DII:623*, *ZK 135:34(1972)*, *AM 58:314(1973)*, *CM 23:53(1985)*.

31.10.5.2 Magnesiocopiapite $MgFe_4^{3+}(SO_4)_6(OH)_2 \cdot 20H_2O$

Named in 1938 for the composition and relationship to copiapite. Copiapite group. TRI $P\bar{1}$. $a = 7.342$, $b = 18.818$, $c = 7.389$, $\alpha = 91.45°$, $\beta = 102.15°$, $\gamma = 98.85°$, $Z = 1$, $D = 2.05$. *ZK 135:34(1972)*. 42-599: 18.6_9 9.3_{10} 6.19_5 5.60_8 4.21_4 4.03_3 3.59_5 3.51_4. Habit, physical properties, tests as for copiapite. G = 2.04. Biaxial (+); $N_x = 1.507$, $N_y = 1.529$, $N_z = 1.576$; 2V = 67°. With amarantite in the *Santa Maria Mts., near Blythe, Riverside Co., CA*. Laird Post, BC, Canada. Falun, Sweden. RG *DII:623*, *CM 23:53(1985)*.

31.10.5.3 Cuprocopiapite $CuFe_4^{3+}(SO_4)_6(OH)_2 \cdot 20H_2O$

Named in 1938 for the composition and relation to copiapite. Copiapite group.
TRI $P\bar{1}$. $a = 7.31$, $b = 18.15$, $c = 7.25$, $\alpha = 92.5°$, $\beta = 102.3°$, $\gamma = 100.4°$, $Z = 1$, $D = 2.26$. CM 23:55(1985). 19-394: 8.81_8 6.32_4 5.82_5 3.90_3 3.76_3 3.56_{10} 3.29_3 3.08_4. Small crystals, rounded masses. Yellowish green. $G = 2.154$. Biaxial (+); $N_x = 1.558$, $N_y = 1.575$, $N_z = 1.620$; $2V = 63°$. *Chuquicamata, Antofagasta, Chile.* RG AM 23:737(1938), DII:623.

31.10.5.4 Ferricopiapite $(Fe^{3+},Al)_{2/3}Fe_4^{3+}(SO_4)_6(OH)_2 \cdot 20H_2O$

Named in 1938 for the composition and relation to copiapite. Copiapite group.
TRI $P\bar{1}$. $a = 7.390$, $b = 18.213$, $c = 7.290$, $\alpha = 93.7°$, $\beta = 102.2°$, $\gamma = 99.3°$, $Z = 1$, $D = 2.123$. AM 58:314(1973). 29-714: 18.4_7 9.06_{10} 6.04_7 5.58_8 5.39_5 4.03_5 3.58_8 3.53_7. Yellow. $H = 2\frac{1}{2}$–3. $G = 2.1$. Biaxial (+); $N_x = 1.531$, $N_y = 1.546$, $N_z = 1.597$; $2V = 52°$. *Coso Hot Spring, Inyo Co., CA. Tierra Amarilla, Copiapó, Chile.* RG DII:625, CM 23:53(1985).

31.10.5.5 Calciocopiapite $CaFe_4^{3+}(SO_4)_6(OH)_2 \cdot 20H_2O$

Named in 1960 for the composition and relation to copiapite. Copiapite group.
TRI $P\bar{1}$. $a = 7.44$, $b = 18.79$, $c = 7.22$, $\alpha = 85.3°$, $\beta = 104.7°$, $\gamma = 102.2°$, $Z = 1$, $D = 2.159$. CM 23:55(1985). 27-77: 3.46_9 3.11_{10} 3.03_{10} 2.81_{10} 1.86_{10} 1.30_9 1.27_9 1.24_9. The lack of powder lines of high value, normal to other members of the copiapite group, suggests that such lines may have been obscured during the examination of calciocopiapite; further study is needed. Powdery crusts. Grayish to brownish yellow. *Dashkesan, middle Caucasus, Azerbaijan*, as a supergene product of weathering of magnetite ores containing pyrite and calcite. RG AM 47:807(1962).

31.10.5.6 Zincocopiapite $ZnFe_4^{3+}(SO_4)_6(OH)_2 \cdot 20H_2O$

Named in 1964 for the composition and relation to copiapite. Copiapite group.
TRIC $P\bar{1}$. $a = 7.33$, $b = 18.72$, $c = 7.35$, $\alpha = 91.5°$, $\beta = 102.1°$, $\gamma = 98.7°$, $Z = 1$, $D = 2.09$. AM 49:1777(1964). 45-1323: 18.2_{10} 9.06_9 6.55_3 6.06_5 5.71_4 5.20_3 4.87_3 3.57_5. Compact, massive aggregates of crystals. Yellowish green, vitreous luster. $H = 2$. $G = 2.181$. Biaxial (+); $N_x = 1.534$, $N_y = 1.554$, $N_z = 1.586$; $2V = 78°$; $r > v$. *Xitieshan lead–zinc mine, Qaidam Basin, Qinghai Prov., China*, with copiapite, halotrichite, coquimbite, roemerite, sideronatrite, and melanterite. RG AGS 44:100(1964), IGR 10:917(1968), CM 23:53(1985).

31.10.5.7 Aluminocopiapite $(Al,Mg)_{2/3}Fe_4^{3+}(SO_4)_6(OH)_2 \cdot 20H_2O$

Named in 1947 for the composition and relation to copiapite. Copiapite group.
TRI $P\bar{1}$. $a = 7.30$, $b = 18.80$, $c = 7.31$, $\alpha = 91.5°$, $\beta = 102.3°$, $\gamma = 98.7°$, $Z = 1$, $D = 2.07$. CM 23:55(1985). 20-659: 18.1_8 9.2_{10} 6.17_7 5.58_8 5.32_3 4.68_3 3.58_5 3.50_5. Minute scales and earthy blooms. Lemon yellow, pearly luster on {010}. Cleavage {010}, perfect; {$\bar{1}$01}, imperfect. $H = 2\frac{1}{2}$–3. Soluble in water; in CT,

gives acid water. Biaxial (+); $N_y = 1.535$, $N_z = 1.585$; 2V moderate; $r > v$, strong. Y, colorless; Z, greenish yellow. *Island Mt., Trinity Co., CA.* Northern bank of Mosquito Fork of the Forty Mile R., AK. RG *DII:625, AM 52:1220(1967)*.

31.10.6.1 Honessite $Ni_6Fe_2^{3+}(OH)_{16}(SO_4) \cdot 4H_2O$

Named in 1959 for Arthur P. Honess (1886–1942), Pennsylvania State University. HEX-R $R\bar{3}m$ or $R3m$. $a = 3.083$, $c = 26.71$. *MM 44:340(1981)*. *42-573:* 8.84_{10} 4.43_4 2.65_1 2.62_2 2.39_1 2.09_1 1.54_2 1.52_2 Fine-grained coatings and masses with a microfibrous structure. Orange to yellow, reddish, green; dull luster. Soft. Slowly soluble in weak, cold acids; rapidly so in concentrated acids. Uniaxial; $N = 1.615$–1.635. Honessite is derived from the oxidation of millerite and other primary nickel minerals. *Linden, Iowa Co., WI,* as coatings in the oxidized portion of lead deposits containing traces of nickel; also at Dodgeville, Shullsburg, and Galena. Hagdale, Unst, Shetland, Scotland, on nickeliferous chromite, associated with reevesite. RG *AM 44:995(1959), BM 103:170(1980)*.

31.10.7.1 Hydrohonessite $Ni_6Fe_2^{3+}(SO_4)(OH)_{16} \cdot 7H_2O$

Named in 1981 to indicate its relationship to honessite. HEX Space group unknown. $a = 3.09$, $c = 10.80$. *36-382:* 11.0_{10} 5.56_5 3.68_4 2.71_3 2.60_2 2.39_2 2.15_1 1.91_1. Thin incrustations of tiny hexagonal plates. Bright yellow. Soluble in dilute acids. Uniaxial (−); $N_O = 1.63$, $N_E = 1.59$. *Otter Shoot nickel mine, Kambalda, WA, Australia;* also from the Carr Boyd mine, Kambalda. Unst, Shetland, Scotland. RG *MM 44:333(1981), AM 67:623(1982)*.

31.10.8.1 Clairite $(NH_4)_2Fe_3^{3+}(SO_4)_4(OH)_3 \cdot 3H_2O$

Named in 1983 for Claire Martini, South African, wife of the describer of the species, Jacques Edouard Martini. TRIC $P1$ or $P\bar{1}$. $a = 9.368$, $b = 9.150$, $c = 52.610$, $\alpha = 88.15°$, $\beta = 90°$, $\gamma = 118.37°$, $Z = 8$, $D = 2.32$. *39-1388:* 17.5_{10} 8.78_{10} 8.23_2 7.76_1 7.31_1 3.42_2 3.28_3 3.04_2. Yellow. Cleavage {001}, perfect. $G = 2.31$. Biaxial; X parallel to c; $N_x = 1.595$, $N_z = 1.607$. X, pale yellow; Z, dark yellow. Formed by reaction of the oxidation products of pyrite with ammonia produced by the decay of organic matter. *Lone Creek Fall cave, near Sabri, eastern Transvaal, South Africa.* RG *AM 71:229(1986)*.

31.10.9.1 Caminite $MgSO_4 \cdot xMg(OH)_2 \cdot yH_2O$

Named in 1983 from the Latin *caminus*, chimney, in allusion to its mode of occurrence. TET $I4_1/amd$. $a = 5.239$, $c = 12.988$, $Z = 4$, $D = 2.145$. *AM 71:819(1986). 39-359:* 3.35_{10} 3.22_8 2.06_2 2.04_2 1.87_5 1.85_2 1.67_2 1.62_3. Colorless or white. Cleavage {001}, good. $H = 2\frac{1}{2}$. $G = 2.5$. Uniaxial (−); $N_O = 1.534$, $N_E = 1.532$. Caminite is precipitated from seawater as part of the structure of "black smoker" chimneys on the *East Pacific Rise, 21° N latitude, 2600 m depth.* It is associated with anhydrite of similar origin and various sulfides. RG *GCA 47:2053(1983)*.

Class 32
Compound Sulfates

32.1.1.1 Burkeite Na$_6$(SO$_4$)$_2$CO$_3$
Named in 1921 for William Edmund Burke (1880–1966), chemical engineer, American Potash and Chemical Co., who discovered the artificial salt. ORTH $Pmnm$. $a = 5.167$, $b = 9.215$, $c = 7.055$, $Z = 2$, $D = 2.61$. *NBSM 25(11):52(1974)*. *24-1134*: 3.85$_4$ 3.80$_8$ 3.53$_8$ 2.80$_{10}$ 2.78$_6$ 2.64$_8$ 2.58$_8$ 1.93$_3$. Crystals tabular {100}, reticular aggregates of platy crystals, massive, as nodules; twinning common as X-shaped interpenetration twins, on {110}. White to pale buff or grayish; vitreous to greasy luster, like cryolite. Conchoidal fracture, brittle. H = $3\frac{1}{2}$. G = 2.57. Soluble in cold water, easily fusible. Biaxial (−); XYZ= cab; N$_x$ = 1.448, N$_y$ = 1.489, N$_z$ = 1.493; 2V = 34°; $r > v$, distinct. In artificial material (SO$_4$) and (CO$_3$) can substitute mutually over a wide range. *Searles Lake, San Bernardino Co., CA*, in drill cuttings from a clay stratum, associated with gaylussite, trona, sulfohalite, borax, northupite, and other salts. Konya basin, central Anatolia, Turkey, with halite and trona. Eastern shore of Lake Turkmana, northern Kenya. RG *DII:633, MM 43:341(1979), NJMM:203(1988)*.

32.1.2.1 Hectorfloresite Na$_9$(SO$_4$)$_4$(IO$_3$)
Named in 1989 for Hector Flores W. (1906–1984), Chilean geologist and professor, University of Chile. MON Space group unknown. $a = 18.775$, $b = 6.936$, $c = 14.239$, β = 108.92°, Z = 4, D = 2.90. *AM 74:1207(1989)*. PD: 6.17$_1$ 4.69$_2$ 3.88$_{10}$ 2.79$_3$ 2.70$_8$ 1.94$_2$ 1.68$_1$ 1.56$_1$. PD(syn): 4.68$_3$ 3.88$_6$ 2.79$_2$ 2.70$_{10}$ 2.37$_1$ 2.34$_1$ 2.16$_1$ 1.94$_1$. Minute, prismatic crystals, elongated [010], with multiple twinning to give a pseudohexagonal morphology. White. No cleavage, conchoidal fracture, brittle. H = 2. G = 2.80. Biaxial (−); N$_x$ = 1.493, N$_y$ = 1.521, N$_z$ = 1.523 (Na); 2V = 26°; r < v, faint. *Alianza nitrate mine, Oficina Victoria, Tarapacá, Chile*, associated with darapskite, nitratine, bloedite, halite, and glauberite. RG

32.1.3.1 Olsacherite Pb$_2$(SO$_4$)(SeO$_4$)
Named in 1969 for Juan A. Olsacher (1903–1964), Argentine mineralogist, University of Cordoba. Barite group. See barite (28.3.1.1). ORTH $P222_1$. $a = 8.42$, $b = 10.96$, $c = 7.00$, Z = 4, D = 6.55. *AM 54:1523(1969)*. Isostructural with barite. *22-1135*: 4.32$_4$ 3.85$_5$ 3.34$_{10}$ 3.23$_{10}$ 3.02$_{10}$ 2.79$_6$ 2.09$_7$ 2.04$_4$.

Acicular crystals, terminated by steep pyramids, elongated [010]. Colorless, vitreous luster. Cleavage {101}, good; {010}, fair; very brittle. H = 3–3$\frac{1}{2}$. Biaxial (−); XYZ= acb; $N_x = 1.935$, $N_y = 1.966$, $N_z = 1.983$ (Na); 2V = 80°; no dispersion. Tests with synthetic lead selenate–sulfates indicate that (SO_4) and (SeO_4) can mutually substitute to a limited but unquantified extent. *Pacajake mine, Hiaco, Bolivia*, associated with cerussite, anglesite, wulfenite, chalcomenite, ahlfeldite, and selenium; also from the El Dragon mine, Potosi. Cachueta, Mendoza, Argentina. RG

32.2.1.1 Rapidcreekite $Ca_2(SO_4)(CO_3) \cdot 4H_2O$

Named in 1986 for the locality. ORTH $Pbcn$. $a = 15.49$, $b = 19.18$, $c = 6.157$, Z = 8, D = 2.239. *CM 24:51(1986)*. *39-351:* 7.78_{10} 4.31_7 3.88_7 3.11_8 2.92_5 2.80_6 2.56_5 1.90_5. Acicular crystals, as sprays and crystalline coatings. White, vitreous luster. Cleavage {010}, perfect; {100}, good; splintery fracture; brittle. H = 2. G = 2.21. Nonfluorescent. Biaxial (+); XYZ= cab; $N_x = 1.516$, $N_y = 1.518$, $N_z = 1.531$; 2V = 45°. *Rapid Creek, Big Fish river area, YT, Canada*, along the valley of a tributary (*Crescent Creek*), with gypsum and aragonite on joint surfaces of siderite iron formation. RG *AM 72:224(1987)*.

32.2.2.1 Humberstonite $Na_7K_3Mg_2(SO_4)_6(NO_3)_2 \cdot 6H_2O$

Named in 1970 for James Thomas Humberstone (1850–1939), who worked on the saline minerals of the Chilean nitrate deposits. HEX-R R. $a = 10.900$, $c = 24.410$, Z = 3, D = 2.252. *AM 55:1518(1970)*. *21-682:* 8.80_4 8.14_6 7.47_3 4.53_4 4.07_3 3.39_{10} 2.72_7 2.58_4. Minute crystals, platy {0001}, in aggregates. Colorless, vitreous luster. Cleavage {0001}, perfect; irregular fracture; brittle. H = 2$\frac{1}{2}$. G = 2.252. Nonfluorescent. Soluble in water, insoluble in acetone or alcohol. Uniaxial (−); $N_O = 1.474$, $N_E = 1.436$. *Taltal nitrate district, Atacama desert, Chile*, at several places, associated with bloedite and nitratine. RG *KUS 5:190(1969), AM 55:1072(1970)*.

32.2.3.1 Ungemachite $K_3Na_8Fe^{3+}(SO_4)_6(NO_3)_2 \cdot 6H_2O$

Named in 1936 for Henri-León Ungemach (1880–1936), Belgian crystallographer. HEX-R R. $a = 10.898$, $c = 24.989$, Z = 3, D = 2.259. *AM 71:826(1986)*. *20-1326:* 8.82_3 8.33_6 7.55_3 3.43_{10} 2.96_2 2.72_6 2.59_4 1.88_3. Crystals thick tabular {0001} with numerous rhombohedral forms. Colorless to pale yellow, vitreous luster. Cleavage {0001}, perfect; irregular fracture; brittle. H = 2$\frac{1}{2}$. G = 2.287. Easily soluble in dilute HCl. Uniaxial (−); $N_O = 1.502$, $N_E = 1.449$. *Chuquicamata, Antofagasta, Chile*, as vein fillings and in cavities in jarosite and metasideronatrite, associated with metavoltine, fiberferrite, and sodium alum. RG *DII:596*.

32.3.1.1 Hanksite $KNa_{22}(SO_4)_9(CO_3)_2Cl$

Named in 1885 for Henry G. Hanks (1826–1907), a state mineralogist of California. HEX $P6_3/m$. $a = 10.465$, $c = 21.191$, Z = 2, D = 2.585. *AM 58:799(1973)*. Structure: *AC(B) 28:3614(1972)*. *25-1348:* 3.81_{10} 3.53_8 3.43_6

2.93_3 2.87_2 2.79_7 2.62_5 2.45_2. Hexagonal prismatic crystals {100}, with {102} as quartzoids; also tabular {0001}, often of large size; as interpenetrating groups. Colorless, sometimes slightly yellow or gray due to included clay; vitreous to dull luster. Cleavage {0001}, good; uneven fracture; brittle. H = $3-3\frac{1}{2}$. G = 2.562. Weakly fluorescent pale yellow in LWUV. Soluble in water, effervesces weakly in dilute acids, easily fusible. Tastes saline. Uniaxial $(-)$; $N_O = 1.481$, $N_E = 1.461$. Abundant at *Searles Lake, San Bernardino Co., CA*, associated with halite, borax, trona, and aphthitalite. Also reported from Death Valley. RG *DII:628*.

32.3.2.1 Caledonite $Cu_2Pb_5(SO_4)_3(CO_3)(OH)_6$

Named in 1932 from *Caledonia* (an ancient name for Scotland), because it was found in Scotland. ORTH $Pmn2_1$. $a = 20.089$, $b = 7.146$, $c = 6.560$, $Z = 2$, $D = 5.689$. Structure: *AC(B) 29:1986(1973)*. *29-565*: 4.69_6 3.15_4 3.14_{10} 3.03_6 2.75_5 2.27_3 2.23_4 1.86_6. Prismatic crystals, elongated [001]; often striated and with vicinal faces in [001] zone. Deep verdigris green or bluish green, resinous luster. Cleavage {010}, perfect; {100}, {101}, incomplete; uneven fracture; brittle. H = $2\frac{1}{2}$-3. G = 5.76. Dissolves with effervescence in nitric acid, leaving a residue of lead sulfate; in CT decrepitates and affords water. Biaxial $(-)$; XYZ= *cab*; $N_x = 1.818$, $N_y = 1.866$, $N_z = 1.909$; weakly pleochroic; $2V = 85°$; r < v, faint. Found in small amounts in the oxidized zone of copper–lead deposits, with cerussite, anglesite, linarite, leadhillite, malachite, azurite, and other sulfates–carbonates. Mammoth mine(!), Tiger, AZ; Organ District, Dona Ana Co., NM; Cerro Gordo, Inyo Co., CA; Talisman mine, Beaver Creek, UT; *Leadhills*(!), *Lanarkshire, Scotland*. Red Gill, Cumberland, England. Malacalzetta, Iglesias, and Tiny Arenas, Sardinia, Italy. Preobrazhenskyi mine, Berezovsk, Ural Mts., Russia. Tchah Mille, Anarak, Iran(!), crystals to 1 cm. Challocollo, Sierra Gorda, Atacama, Chile. RG *DII:630, NW 57:127(1970), MM 40:536(1976)*.

32.3.3.1 Wherryite $CuPb_4O(SO_4)_2(CO_3)(OH,Cl)_2$

Named in 1950 for Edgar T. Wherry (1885–1982), American mineralogist and plant ecologist. MON $P2_1/m$. $a = 20.82$, $b = 5.79$, $c = 9.17$, $\beta = 91.28°$, $Z = 4$, $D = 7.22$. *AM 55:505(1970)*. *23-218*: 4.78_4 4.57_4 3.22_2 3.16_{10} 3.06_5 2.90_3 2.75_5 1.88_4. Usually massive, fine granular; minute crystals, equant to acicular. Light apple-green to yellowish green, light yellow. G = 6.45. Slowly soluble in cold HCl or HNO_3. Biaxial $(-)$; $N_x = 1.942$, $N_y = 2.010$, $N_z = 2.024$ (Na); $2V = 50°$. *Mammoth mine, Tiger, Pinal Co., AZ*, with leadhillite, paralaurionite, diaboleite, chrysocolla, and cerussite. RG *AM 35:93(1950), DII:632, BM 74:528(1951)*.

32.3.4.1 Hauckite $Fe_3^{3+}(Mg,Mn)_{24}Zn_{18}(SO_4)_4(CO_3)_2(OH)_{81}$ (?)

Named in 1980 for Richard Hauck (b.1935), New Jersey mineral collector. HEX $P6/mmm$. $a = 9.17$, $c = 30.21$, $Z = 1$, $D = 3.10$. *33-872*: 7.80_9 5.02_5 4.57_6 3.96_{10} 3.78_5 3.02_6 1.75_3 1.58_2. Crystals simple hexagonal plates {0001}, with

{010}. Bright orange to light yellow, vitreous to slightly pearly luster. Cleavage {0001}, perfect and easy; brittle. H = 2–3. G = 3.02. Nonfluorescent. Uniaxial (+); $N_O = 1.630$, $N_E = 1.638$. O, golden brown; E, pale yellow. *Sterling Hill mine, Ogdensburg, NJ*, associated with phlogopite and mooreite on a calcite–serpentine matrix. RG *AM 65:192(1980)*.

32.3.5.1 Heidornite $Na_2Ca_3(SO_4)_2[B_5O_8(OH)_2]Cl$

Named in 1956 for F. Heidorn, German geologist, who studied the Zechstein formation. MON $C2/c$. $a = 10.21$, $b = 7.84$, $c = 18.79$, $\beta = 93.5°$, $Z = 4$, $D = 2.698$. *NJMM:157(1967)*. 25-805: 3.77_8 3.11_{10} 2.96_7 2.90_6 2.81_6 2.74_8 2.54_6 1.98_6. Prismatic, spearlike crystals, with {110}, and {11$\bar{1}$} and {11$\bar{2}$}. Colorless, vitreous luster. Cleavage {001}, perfect. H = 4–5. G = 2.753. Biaxial (+); $Y = b$; $Z \wedge a = 23°$; $N_x = 1.579$, $N_y = 1.588$, $N_z = 1.604$ (Na); $2V = 74°$; $r < v$. Found associated with glauberite in a cavity in drill core from 1,968 m depth in the upper anhydrite of the *Zechstein formation*, on the *German–Dutch border, NE of Nordhorn*. RG *AM 42:120(1957)*.

32.4.1.1 Nakauriite $(Mn,Ni,Cu)_8(SO_4)_4(CO_3)(OH)_6 \cdot 48H_2O$

Named in 1976 for the locality. ORTH Space group unknown. $a = 14.585$, $b = 11.47$, $c = 16.22$, $Z = 2$, $D = 2.35$. *AM 62:594(1977)*. 28-538: 7.31_{10} 7.08_1 4.84_1 3.94_1 3.65_2 2.40_1 2.37_2 1.91_2. Aggregates of minute acicular to fibrous crystals. Sky blue. G = 2.39. Dissolves with effervescence in dilute acids. Biaxial (−); $N_x = 1.585$, $N_y = 1.604$, $N_z = 1.612$ (Na); $2V = 65°$; parallel to extinction; positive elongation. X, colorless; Y, light greenish blue; Z, pale sky blue. *Nakauri, Achi Pref., Japan*, as veins cutting massive serpentinite, associated with chrysotile, magnetite, pyroaurite, and artinite. RG *AM 67:156(1982)*.

32.4.2.1 Tatarskite $Ca_6Mg_2(SO_4)_2(CO_3)_2Cl_4(OH)_4 \cdot 7H_2O$

Named in 1963 for Vitalii Borisovich Tatarskii (b.1907), Russian mineralogist, Leningrad University. Probably ORTH Space group unknown. 15-785: 5.34_8 2.97_{10} 2.92_7 2.63_9 2.52_6 2.31_6 2.00_8 1.59_7. Coarsely crystalline masses. Colorless; vitreous luster, pearly on cleavage. Two pinacoidal cleavages, fair. H = $2\frac{1}{2}$. G = 2.341. Soluble in boiling water or cold dilute HCl. Biaxial (−); $N_x = 1.567$, $N_y = 1.654$, $N_z = 1.722$; $2V = 83°$; parallel to extinction; positive elongation. From drill core in the *Caspian Depression, Kazakhstan*, at depths of 850–900 m in anhydrite rock, associated with halite, bischofite, magnesite, hilgardite, and strontiohilgardite. RG *ZVMO 92:697(1964)*, *AM 49:1151(1964)*.

32.4.3.1 Mountkeithite $(Mg,Ni)_{11}(Fe^{3+},Cr)_3(SO_4,CO_3)_{3.5}(OH)_{24} \cdot 11H_2O$

Named in 1981 for the locality. HEX Space group unknown. $a = 10.698$, $c = 22.545$, $Z = 2$, $D = 1.95$. *MM 44:345(1981)*. 41-578: 11.30_{10} 5.63_8 4.63_2 3.76_6 2.65_4 2.55_4 1.55_5 1.51_3. Flaky grains and crusts; scales; rosettes. Pale pink to white, pearly luster. G = 2.12. Uniaxial (−); $N_O = 1.52$, $N_E = 1.51$. *Mount Keith nickel deposit, 400 km NNW of Kalgoorlie, WA, Australia*, associated with stichtite, hexahydrite, and morenosite on a bleached serpentinite matrix. RG

32.4.4.1 Charlesite $Ca_6Al_2(SO_4)_2[B(OH)_4](OH,O)_{12} \cdot 26H_2O$

Named in 1983 in honor of Charles Palache (1869–1954), professor of mineralogy, Harvard University. Ettringite group. See ettringite (31.10.2.1). HEX-R $P31c$. $a = 11.16$, $c = 21.21$, $Z = 2$, $D = 1.79$. *35-606:* 9.70_{10} 5.58_8 4.65_4 3.85_8 2.75_7 2.54_7 2.19_7 2.13_5. Crystals are hexagonal bipyramids with {104} only. Colorless to white, vitreous luster. Cleavage {101}, perfect; brittle. $H = 2\frac{1}{2}$. $G = 1.77$. Weakly fluorescent in SWUV, light violet or light green. Uniaxial (−); $N_O = 1.492$, $N_E = 1.475$. *Franklin, Sussex Co., NJ*, on the 800 level close to the Palmer shaft, associated with willemite, rhodonite, ganophyllite, pectolite, and grossular. RG *AM 68:1033(1983)*.

32.4.4.2 Sturmanite $Ca_6Fe_2^{3+}(SO_4)_{2.4}[B(OH)_4]_{1.2}(OH)_{12} \cdot 25H_2O$

Named in 1983 for B. Darko Sturman (b.1937), Canadian mineralogist, Royal Ontario Museum, Toronto. Ettringite group. See ettringite (31.10.2.1). HEX-R $P31c$. $a = 11.16$, $c = 21.79$, $Z = 2$, $D = 1.855$. *35-637:* 9.67_{10} 5.58_7 4.75_2 3.89_7 2.77_5 2.58_6 2.22_5 2.16_4. Crystals are simple hexagonal bipyramids with {104} and traces of {114}. Bright yellow, vitreous luster. Cleavage {100}, perfect; brittle. $H = 2\frac{1}{2}$. $G = 1.847$. Nonfluorescent. Uniaxial (+); $N_O = 1.500$, $N_E = 1.505$. O, pale green; E, pale yellowish green. Some crystals zoned, with cores uniaxial (−); $N_O = 1.499$, $N_E = 1.497$. In the structure of sturmanite there is some substitution of $[B(OH)_4]^-$ radicals for $(SO_4)^{2-}$ so that the ratio of (SO_4) to $[B(OH)_4]$ is not exactly 2:1, thereby preserving charge balance. *Black Rock mine, Kuruman district, Cape Prov., South Africa*, associated with barite and hematite. RG *CM 21:705(1983)*.

32.4.4.3 Jouravskite $Ca_3(SO_4)(CO_3)[Mn^{4+}(OH)_6] \cdot 12H_2O$

Named in 1965 for Georges Jouravsky (1896–1964), chief geologist, Division de la Géologie du Maroc. Ettringite group. See ettringite (31.10.2.1). HEX $P6_3$. $a = 11.06$, $c = 10.50$, $Z = 2$, $D = 1.93$. The Mn^{4+} is in six-coordination with hydroxyl ions, as in the case of Si^{4+} in thaumasite (32.4.4.4). *AC(B) 25:1943(1969)*. *18-668:* 9.60_{10} 5.53_6 3.81_6 3.54_4 3.42_5 2.73_5 2.52_5 2.17_5. Small groups of grains occurring as spots on other manganese minerals. Greenish yellow to greenish orange. Cleavage {100}, good. $H = 2\frac{1}{2}$. $G = 1.95$. Uniaxial (−); $N_O = 1.556$, $N_E = 1.540$; interference colors abnormal, green and russet. *Tachgagalt No. 2 vein, Anti Atlas, Morocco*, associated with calcite and manganite on a matrix of other manganese minerals. RG *BM 88:254(1965), AM 50:2102(1965)*.

32.4.4.4 Thaumasite Ca$_3$(SO$_4$)(CO$_3$)[Si(OH)$_6$] · 12H$_2$O

Named in 1878 from the Greek for *astonishment*, in allusion to its remarkable composition, consisting of silicate, carbonate, and sulfate components. Ettringite group. See ettringite (31.10.2.1). HEX $P6_3$. $a = 11.030$, $c = 10.396$, $Z = 2$, $D = 1.876$. *NJMM:60(1983)*. The silicon atom is at the center of an octahedron of (OH) radicals. Octahedral coordination of Si by O is rare in minerals formed at ordinary pressures and is ascribed in this instance to the effect of electronegative H atoms attached to the oxygens. This was the first proved example of six-coordination of silicon by hydroxyl. *AC(B) 27:594(1971)*. *25-128*: 9.56$_{10}$ 5.51$_4$ 3.78$_2$ 3.41$_2$ 3.18$_2$ 2.71$_1$ 2.57$_1$ 2.16$_1$. Massive, rarely as small crystals, fibrous crystalline masses consisting of acicular prismatic crystals. Snow white; dull to earthy, or silky, luster. Cleavage in traces, subconchoidal fracture, brittle. H = $3\frac{1}{2}$. G = 1.881. In CT decrepitates and affords much water. Uniaxial (−); N$_O$ = 1.507, N$_E$ = 1.468. West Paterson and Great Notch, Passaic Co., NJ, where it is sometimes pseudomorphous after anhydrite. Beaver Co., UT. Crestmore, Riverside Co., CA, with spurrite, ettringite, or monticellite. *Bjelke mine, near Åreskutan*, and Kalls Parish, Jemtland, *Sweden*, also Långban, Värmland. Sulitelma, Norway. Haslach, Baden, Germany. Halap, Hungary. N'Chwaning mine, Kuruman manganese district, Cape Prov., South Africa(!), in gemmy yellowish crystals to 15 mm. RG *AMG 2(HI-2):137(1957)*, *AM 45:1277(1960)*.

32.4.5.1 Hotsonite Al$_5$(SO$_4$)(PO$_4$)(OH)$_{10}$ · 8H$_2$O

Named in 1984 for the locality. Ettringite group, which see TRIC $P1$ or $P\bar{1}$. $a = 11.23$, $b = 11.66$, $c = 10.55$, $\alpha = 112.53°$, $\beta = 107.53°$, $\gamma = 64.45°$, $Z = 2$, $D = 2.026$. *AM 69:979(1984)*. *38-366*: 10.1$_{10}$ 8.45$_4$ 5.20$_1$ 5.01$_1$ 4.63$_2$ 4.43$_1$ 4.34$_1$ 3.67$_1$. Compact masses, chalklike; under high magnification, lathlike crystals. Colorless to white, dull to silky luster. Earthy fracture. H = $2\frac{1}{2}$. G = 2.06. Biaxial; N$_x$ = 1.519, N$_z$ = 1.521. From the farm *Hotson 42, about 65 km W of Pofadder, Bushmanland, Cape Prov., South Africa*, in a sillimanite quarry, encrusting zaherite. RG *MM 53:385(1989)*, *ZVMO 119:121(1990)*, *AM 76:1734(1991)*.

32.4.6.1 Chessexite Na$_4$Ca$_2$Mg$_3$Al$_8$(SO$_4$)$_{10}$(SiO$_4$)$_2$(OH)$_{10}$ · 40H$_2$O

Named in 1982 for Ronald Chessex (b.1929), Swiss petrologist and professor, University of Geneva. Ettringite group, which see ORTH Space group unknown. $a = 13.70$, $b = 27.96$, $c = 9.99$, $Z = 2$, $D = 2.21$. *AM 69:406(1984)*. *35-692*: 13.9$_{10}$ 4.85$_9$ 3.98$_6$ 3.45$_4$ 3.42$_{10}$ 3.32$_4$ 2.91$_3$ 2.16$_3$. Minute crystals, square or rectangular plates. White, silky luster. Biaxial (+); XYZ= acb; N$_x$ = 1.456, N$_y$ = 1.460, N$_z$ = 1.480; 2V = 47°. *Maine mine, Autun, Saône et Loire, France*, coating fluorite. RG *SMPM 62:337(1982)*.

===== Class 33 =====

Selenates and Tellurates

33.1.1.1 Schmiederite $Cu_2Pb_2(SeO_3)(SeO_4)(OH)_4$

Named in 1962 for Oskar Schmieder (b.1891), German geographer, Universidad Nacional at Cordoba, Argentina. Isostructural with linarite. MON $P2_1/m$. $a = 9.922$, $b = 5.712$, $c = 9.396$, $\beta = 101.97°$, $Z = 2$, $D = 5.63$. *MP 36:3(1987)*. *41-1377*: 4.85_4 4.47_5 3.53_5 3.18_{10} 3.11_8 2.33_3 2.11_3 1.819_3. Fracture fillings and small crystals in vugs, showing prismatic habit; also radial-fibrous tufts of acicular crystals. Light blue, greenish blue. *Condor mine, La Rioja, Argentina*(*). El Dragon mine, Potosi, Bolivia. RG *AM 49:1498(1964)*, *SMPM 67:219(1987)*.

33.1.2.1 Xocomecatlite $Cu_3Te^{6+}O_4(OH)_4$

Named in 1975 for the Nahuatl for *bunches of grapes*, in allusion to the appearance of the green spherules of the mineral. Probably ORTH Space group unknown. $a = 12.140$, $b = 14.318$, $c = 11.662$, $Z = 12$, $D = 4.42$. *29-579*: 4.63_{10} 3.44_4 3.32_3 3.10_4 2.83_3 2.67_6 2.63_3 2.43_4. Minute spherules, to 0.15 mm, of radial-fibrous structures. Emerald-green. Tough. $H = 4$, $G = 4.65$. Nonfluorescent. Biaxial $(-)$; $N_x = 1.775$, $N_y = 1.900$ (rich bluish green), $N_z = 1.920$; $2V = 41°$; low dispersion; absorption $Z > X = Y$. Centennial Eureka mine, Tintic district, Juab Co., UT, as emerald-green crystallized balls. *Bambollita mine, Moctezuma, SON, Mexico*(*), in cracks in altered rhyolite. RG *MM 40:221(1975)*, *AM 61:504(1976)*.

33.1.3.1 Khinite $Cu_3PbTe^{6+}O_4(OH)_6$

Named in 1978 for Ba Saw Khin (b.1931), Burmese–American geologist and petrographer. Dimorphous with parakhinite. ORTH $Fddd$. $a = 5.740$, $b = 9.983$, $c = 23.960$, $Z = 8$, $D = 6.69$. *29-1419*: 4.87_8 3.45_9 3.00_7 2.82_6 2.49_{10} 2.20_7 1.92_6 1.56_6. Minute, bipyramidal crystals. Bottle green. Cleavage $\{001\}$, fair; brittle. $H = 3\frac{1}{2}$. $G = 6.5–7.0$. Nonfluorescent. Biaxial $(+)$; $N_x = 2.110$, $N_y = 2.112$, $N_z = 2.165$; $2V = 20°$; strong dispersion; absorption $Z = Y > X$. X, vivid emerald-green; Y,Z, yellowish green. *Tombstone, AZ*, at the *Old Guard mine*; also at the Empire mine(!) in green micaceous plates, with parakhinite, gold, hessite, and torbernite. RG *AM 63:1016(1978)*.

33.1.4.1 Frankhawthorneite $Cu_2Te^{6+}O_4(OH)_2$

Named in 1995 for Frank C. Hawthorne (b.1946), Canadian mineralogist, University of Manitoba, in recognition of his studies of Cu oxysalts. MON $P2_1/n$. $a = 9.095$, $b = 5.206$, $c = 4.604$, $\beta = 98.69°$, $Z = 2$, $D = 5.43$. PD: 4.50_4 4.34_6 3.84_5 2.89_7 2.60_{10} 1.834_4 1.713_4 1.500_4. Minute crystals to 0.1 mm, prismatic to stubby bladed, elongate [001]. Leaf-green, lighter green streak, vitreous luster. Uneven fracture, brittle. $H = 3-4$. Nonfluorescent. Pale gray in reflected light, weakly bireflectant, nonpleochroic, $N_{mean\;calc} = 2.00$; $R = 10.5$ (589 nm). From the mine dumps at the *Centennial Eureka mine, Tintic district, Juab Co., UT*, associated with drusy quartz, mcalpineite, hematite, chrysocolla, connellite, enargite, and other Cu oxysalts, including other new Te species. RG *CM 33:641(1995), AM 81:516(1996)*.

33.2.1.1 Kuranakhite $PbMn^{4+}Te^{6+}O_6$

Named in 1975 for the locality. ORTH Space group unknown. $a = 5.1$, $b = 8.9$, $c = 5.3$, $Z = 2$, $D = 6.7$. *29-779*: 3.41_{10} 3.04_1 2.56_6 2.32_1 2.05_5 1.851_5 1.668_1 1.596_4. Small grains, maximum 0.012 mm. Brownish to nearly black, brown streak, vitreous luster. Brittle. $H = 4-5$. Soluble in hot concentrated HCl or aqua regia. $N_x = 1.952$, $N_z = 2.002$. In RL, light gray with a bluish tint; weakly bireflectant, distinctly in oil, with rose tints. No internal reflections; distinctly anisotropic, blue to brown. $R = 24.4$ (470 nm), 24.0 (546 nm), 24.5 (589 nm), 23.0 (650 nm). *Kuranakh gold deposit, southern Yakutia, Russia*, in quartz–limonite ores associated with supergene Au, electrum, and hydrous ferric oxides. Moctezuma mine, Moctezuma, SON, Mexico. RG *ZVMO 104:310(1975), AM 61:339(1976)*.

33.2.2.1 Montanite $Bi_2Te^{6+}O_6 \cdot 2H_2O$

Named in 1868 for the locality. An inadequately characterized mineral. *38-417*: 3.54_{10} 3.22_5 2.95_3 2.60_9 2.04_2 1.892_8 1.708_5 1.504_3. Earthy to compact. Yellowish brown, greenish, white; dull to waxy luster. Soft. $G = 6.2$. Biaxial (−); $N = 2.09$; abnormally pleochroic, green; 2V small; bireflectance= 0.01; $r < v$, extreme. *Highland, MT*, as an alteration of tetradymite. Davidson Co., NC. Norongo, near Captain's Flat, NSW, Australia, with tetradymite. RG *DII:636, SMMB 1:12(1985)*.

33.2.3.1 Cuzticite $Fe^{3+}Te^{6+}O_6 \cdot 3H_2O$

Named in 1982 from the Nahuatl for *something yellow*, in allusion to its color. HEX Space group unknown. $a = 5.045$, $c = 14.63$, $Z = 2$, $D = 4.01$. *35-664*: 4.87_4 3.26_{10} 2.52_7 2.24_3 1.99_2 1.75_2 1.56_3 1.46_3. Crystallized crusts, scaly, stalactitic. Yellow to brownish yellow, similar to goethite. $H = 3$. $G = 3.9$. Soluble in cold dilute HCl, insoluble in alkalis, in CT affords water. Uniaxial (−); $N_O = 2.06$, $N_E = 2.05$; nondichroic. *Moctezuma mine, Moctezuma, SON, Mexico*. RG *MM 46:257(1982)*.

33.2.4.1 Parakhinite $Cu_3PbTe^{6+}O_4(OH)_6$

Named in 1978 in allusion to its dimorphous relationship to khinite. HEX $P6_222$ or $P6_422$. $a = 5.753$, $c = 17.958$, $Z = 3$, $D = 6.69$. *29-1420:* 4.80_8 3.34_{10} 2.99_5 2.91_7 2.49_{10} 2.25_6 2.00_5 1.56_8. Thin crystalline films and tabular hexagonal euhedra. Bottle green. $H = 3\frac{1}{2}$. $G = 6.5$–7. Uniaxial $(-)$; $N_O = 2.155$, $N_E = 2.120$; strong dispersion. *Tombstone, AZ*, at the *Emerald mine*; also at the Empire mine(!) with khinite, dugganite, gold, and hessite, the habit and color resembling those of paratacamite. RG *AM 63:1016(1978)*.

33.2.5.1 Mcalpineite $Cu_3Te^{6+}O_6 \cdot H_2O$

Named in 1994 for the locality. ISO Space group unknown. $a = 9.555$, $Z = 8$, $D = 6.65$. *MM 58:417(1994)*. PD: 4.76_3 4.26_4 3.91_3 2.76_{10} 2.38_7 1.873_4 1.689_8 1.440_6. Found as crusts, coatings, and millimeter-size cryptocrystalline nodules consisting of micron-size sheaves of fibrous or prismatic crystals. Emerald-green, olive-green, to green-black; lighter green streak; adamantine luster. Uneven fracture, brittle. H and G could not be determined due to limitations on size of available specimens. In RL, gray, with abundant internal reflections, isotropic. $R = 12.8$ (470 nm), 11.6 (546 nm), 11.2 (589 nm), 11.0 (650 nm). First found on the dumps of the *McAlpine mine, Tuolumne Co., CA*, on drusy quartz with chromian muscovite and several primary sulfides and tellurides; and subsequently on dumps of the Centennial–Eureka mine, Juab Co., UT, with xocomecatlite, jensenite, and other hydrated copper tellurates. RG

33.2.6.1 Jensenite $Cu_3Te^{6+}O_6 \cdot 2H_2O$

Named in 1996 for Martin C. Jensen (b.1959), U.S. mineralogist, scanning electron microprobe specialist, and mineral collector, Reno, NV. Dimorphous with cesbronite. MON $P2_1/n$. $a = 9.204$, $b = 9.170$, $c = 7.584$, $Z = 4$, $D = 4.78$. *CM 34:49(1996)*. PD: 6.43_{10} 4.52_3 3.22_7 2.60_4 2.53_5 2.36_3 2.14_4 1.750_4. Subhedral to euhedral crystals, as nearly equant rhombs, some slightly elongate [101], up to 0.4 mm. Emerald green, somewhat lighter streak, adamantine luster. Cleavage $\{\bar{1}01\}$, fair; uneven fracture; brittle. $H = 3$–4. Nonfluorescent. In RL, gray in air, much darker gray in oil, weakly bireflectant and nonpleochroic, nonanisotropic but strong green internal reflections. $R = 10.7$ (470 nm), 10.4 (546 nm), 9.8 (589 nm), 9.15 (650 nm). Found on the dumps of the *Centennial Eureka mine, Tintic district, Juab Co., UT*, associated with mcalpineite, xocomecatlite, and other tellurium-bearing minerals. RG

33.2.7.1 Cesbronite $Cu_3Te^{6+}O_6 \cdot 2H_2O$

Named in 1974 for Fabien Cesbron (b.1938), French mineralogist, Bureau de Recherches Géologiques et Minière, Orléans. Dimorphous with jensenite. ORTH $Cmc2_1$. $a = 2.937$, $b = 11.864$, $c = 8.621$, $Z = 2$, $D = 4.983$. T.S. Ercit, *pers. comm. 27-194*: 5.93_{10} 4.89_7 3.49_9 2.38_4 2.36_7 2.16_3 1.70_3 1.59_3. Small spherules of radiating crystals, or as minute wedge-shaped crystals elongated on *a*. Bright emerald green, bright luster. Cleavage $\{021\}$, good; $\{010\}$, poor;

brittle. H = 3. G = 4.45. Nonfluorescent. Soluble in dilute HCl and HNO_3, insoluble in water or in KOH solution. Biaxial (+); X = a; $N_x = 1.880$, $N_y = 1.928$, $N_z = 2.029$ (Na); 2V = 72°; r > v; absorption $Z \approx Y > X$. X, pale bluish green; Y, rich yellow-green; Z, deep emerald-green. From the Centennial Eureka mine, Tintic district, Juab Co., UT, with xocomecatlite. *Bambollita ("La Oriental") mine, Moctezuma, SON, Mexico*, on fracture surfaces associated with a pea-green amorphous Cu–Fe–Te mineral, carlfriesite, and teineite. RG *MM 39:744(1974), AM 64:653(1979)*.

33.3.1.1 Schieffelinite $Pb(Te,S)O_4 \cdot H_2O$

Named in 1980 for Ed Schieffelin (1847–1897), stagecoach driver and prospector, who about 1880 discovered the Tombstone district. ORTH *Cmcm*. a = 9.67, b = 19.56, c = 10.47, Z = 16, D = 5.15. *33-771:* 9.78_{10} 3.56_5 3.43_6 3.34_5 3.25_6 3.03_5 2.93_5 2.60_5. Platy or scaly; also as minute, sharp prismatic crystals. Colorless to white, adamantine luster. Cleavage {010}, perfect and easy. H = 2. G = 4.98. Nonfluorescent. Soluble in cold dilute HCl or HNO_3, in CT fuses easily and affords water. Biaxial (−); XYZ = *bca*; $N_x = 1.897$, $N_y = 1.940$, $N_z = 1.942$ (Na); 2V = 24°; r < v, marked. *Joe and Grand Central mines, Tombstone, AZ*, associated with girdite, rodalquilarite, bromargyrite, and gold. RG *MM 43:771(1980), AM 66:219(1981)*.

33.3.2.1 Tlalocite $Cu_{10}Zn_6(Te^{4+}O_3)(Te^{6+}O_4)_2Cl(OH)_{25} \cdot 27H_2O$

Named in 1975 for Tlaloc, the god of rain in the Toltec and Aztec civilizations, in allusion to its high water content. ORTH Space group unknown. a = 16.780, b = 19.985, c = 12.069, D = 4.58. *29-590:* 16.8_{10} 8.39_8 4.20_{10} 3.36_6 2.80_3 2.59_3 2.47_3 1.56_4. Velvety crusts and coatings showing fibrous structure. Sky blue (Capri blue). Sectile. H = 1. G = 4.55. Nonfluorescent. Biaxial (−); Z parallel to crystal length; $N_x = 1.758$, $N_y = 1.796$, $N_z = 1.810$; 2V = 64°; absorption Z > Y > X. X, yellowish green; Y,Z, bluish green. *Bambollita mine, Moctezuma, SON, Mexico*, in fractures, with tenorite, azurite, and malachite. RG *MM 40:221(1975), AM 61:504(1976)*.

33.3.3.1 Girdite $Pb_3H_2(Te^{4+}O_3)(Te^{6+}O_6)$

Named in 1979 for Richard Gird (1836–1910), mining engineer and assayer, who made the first rich silver assays and helped open up the Tombstone district. MON Space group unknown. a = 6.241, b = 5.686, c = 8.719, β = 91.68°, Z = 1, D = 5.49. *34-1428:* 3.12_7 3.05_{10} 2.99_7 2.84_8 2.18_5 2.10_7 1.813_5 1.765_8. Spherules up to 3 mm, also as slender tapered prisms of twinned crystals. White. Tough, brittle. H = 2. G = 5.5. Nonfluorescent. Readily soluble in cold dilute HCl and HNO_3. Biaxial (+); Z = b; $Y \wedge c = 34°$; $N_x = 2.44$, $N_y = 2.47$, $N_z = 2.48$; 2V = 70°; r > v, strong. *Grand Central mine, Tombstone, AZ*, with chlorargyrite, gold, anglesite, rodalquilarite, and several other tellurites. RG *MM 43:453(1979), AM 65:809(1980)*.

33.3.5.1 Cheremnykhite $Pb_3Zn_3TeO_6(VO_4)_2$

Named in 1990 for I. M. Cheremnykh, Russian prospector, one of the discoverers of the Kuranakh deposit. Isostructural with kuksite and dugganite. ORTH *Cmmm*, *C222*, *Cm2m*, or *Cmm2*. $a = 8.58$, $b = 14.86$, $c = 5.18$, $Z = 2$, $D = 6.44$. *ZVMO 119:50(1990)*. PD: 3.30_{10} 3.00_9 2.47_4 1.903_5 1.607_6. Tabular, elongate crystals, with well developed {100} and {010}. Greenish yellow, adamantine luster, transparent. Perfect cleavage, brittle. $VHN_{10} = 673$ kg/cm^2. Nonfluorescent. Biaxial $(-)$; $XYZ = abc$; $N_x = 1.986$; $N_z = 1.997$; $2V = 20°$; positive elongation; $r > v$; nonpleochroic. Found at the *Kuranakh gold deposit, central Aldan, Siberia, Russia*, as a late-stage interstitial mineral in calcite, with kuksite, dugganite, yafsoanite, and descloizite. RG *AM 77:446(1992)*.

33.3.5.2 Kuksite $Pb_3Zn_3TeO_6(PO_4)_2$

Named in 1990 for A. I. Kuks, Russian prospector, one of the discoverers of the Kuranakh gold deposit. Isostructural with cheremnykhite and dugganite. ORTH *Cmmm*. $a = 8.50$, $b = 14.72$, $c = 5.19$, $Z = 2$, $D = 6.21$. *ZVMO 119:50(1990)*. PD: 3.29_{10} 3.00_8 2.59_4 1.903_5 1.606_3. Crystals thin tabular {100}. Gray, white streak, adamantine luster, transparent to turbid. Perfect cleavage, brittle. $VHN_{10} = 325$ kg/cm^2. Weakly fluorescent. Biaxial $(-)$; $Z = c$; $N_x = 1.971$, $N_z = 1.981$; positive elongation; $r > v$; nonpleochroic. Found at the *Kuranakh gold deposit, central Aldan, southern Yakutia, Siberia, Russia*, associated with cheremnykhite, dugganite, yafsoanite, and descloizite. RG *AM 77:446(1992)*.

33.3.5.3 Dugganite $Pb_3Zn_3(TeO_6)(AsO_4)(OH)_3$

Named in 1978 for Marjorie Duggan (b.1927), U.S. chemist, who analyzed the mineral. Isostructural with kuksite and cheremykhite. HEX $P6/mmm$. $a = 8.472$, $c = 5.208$, $Z = 1$, $D = 6.33$. *AM 63:1016(1978)*. 29-1429: 4.23_4 3.28_{10} 3.00_8 2.77_5 2.45_6 1.603_6 1.209_4 1.177_5. Prismatic, microscopic crystals. Colorless to green, white streak, adamantine luster. Cleavage {11$\bar{2}$0}, poor; very brittle. H = 3. G = 6.33. Nonfluorescent. Uniaxial $(-)$; $N_O = 1.977$, $N_E = 1.967$. From the *Emerald and other mines, Tombstone, AZ*, associated with parakhinite, rodalquilarite, and other tellurites–tellurates. Kuranakh gold deposit, central Aldan, Yakutia, Siberia, Russia, associated with kuksite, cheremnykhite, yafsoanite, and descloizite. RG

Class 34

Selenites, Tellurites, and Sulfites

34.1.1.1 Molybdomenite PbSeO$_3$

Named in 1882 for the composition: lead and selenium. It is derived from the Greek *molybdos*, lead, and *mene*, moon. The word selenium is derived from a different Greek word meaning "moon". Hence, by rather tortured reasoning, molybdomenite = lead selenite. MON $P2_1$ or $P2_1/m$. $a = 6.86$, $b = 5.48$, $c = 4.50$, $\beta = 112.75°$, $Z = 2$, $D = 7.12$. *CM 8:149(1964)*. 15-462(syn): 3.39_7 3.30_8 3.14_9 2.74_{10} 2.25_4 2.07_6 1.52_4 1.23_5. Flat, bladed crystals. Colorless to white, pearly to adamantine luster. Cleavage {001}, good; {h0l}, fair; brittle. $G_{syn} = 7.07$. Biaxial (−); $N_x = 2.12$, $N_z = 2.14$; $2V = 80°$; positive elongation. Ranwick uranium mine, Montreal River area, ONT, Canada, as an alteration product of clausthalite. Trogtal, Harz, Germany. *Cerro de Cachueta, Mendoza, Argentina*. In Bolivia at the Pacajake mine near Colquechaca, and at the El Dragon mine, Potosi. RG *DII:640, AM 50:812(1965), TMPM 17:196(1972)*.

34.1.1.2 Scotlandite PbSO$_3$

Named in 1984 for the country of origin. MON $P2/m$. $a = 4.505$, $b = 5.333$, $c = 6.405$, $\beta = 106.23°$, $Z = 2$, $D = 6.456$. *MM 48:283(1984)*. 38-409: 4.04_{10} 3.36_7 3.26_7 3.08_9 2.67_4 2.04_3 2.01_3 1.709_2. Minute prismatic chisel-shaped crystals and fan-shaped aggregates. Colorless to yellowish; adamantine luster, pearly on cleavage. Cleavage {100}, perfect; {010}, good. H = 2. G = 6.37. Confirmed as a sulfite by IR absorption bands: $V_3 = 920.9$, $V_1 = 970$, $V_2 = 620.6$, $V_4 = 488.5$; scotlandite was the first naturally occurring sulfite to be recognized. Biaxial (+); $Y = b$; $N_x = 2.035$, $N_y = 2.040$, $N_z = 2.085$ (Na); $2V = 35.4°$; $r \ll v$, strong. *Leadhills, Scotland*, in cavities in massive barite and anglesite, associated with lanarkite, susannite, and leadhillite. The crystals have a thin coating of anglesite, which probably protected them from further oxidation. Argentolle mine, St. Prix, Saône-et-Loire, France. RG *AM 70:876(1985)*.

34.1.2.1 Fairbankite PbTeO$_3$

Named in 1979 for Nathaniel Kellog Fairbank (1829–1903), an important American entrepreneur in the early development of Tombstone. Dimorphous with plumbotellurite (34.1.4.1). TRIC $P1$ or $P\bar{1}$. $a = 7.81$, $b = 7.11$, $c = 6.96$, $\alpha = 117.2°$, $\beta = 93.78°$, $\gamma = 93.4°$, $Z = 4$, $D = 7.45$. *MM 43:453(1979)*. 34-

1253: 3.27_{10} 3.15_6 3.10_6 3.02_5 2.83_6 2.52_5 2.08_3 1.94_3. Small crystals. Colorless, resinous to adamantine luster. No cleavage, brittle. H = 2. Slowly soluble in cold dilute HNO_3 or HCl, rapidly soluble in hot acids, easily fusible. Biaxial; $N_x = 2.29$, $N_y = 2.31$, $N_z = 2.33$; 2V = 86°; negligible dispersion. *Grand Central mine, Tombstone, AZ*, associated with oboyerite, jarosite, and opal. RG *AM 65:809(1980)*.

34.1.3.1 Balyakinite $CuTeO_3$

Named in 1980 for Tatiana Stepanovna Balyakina (1906–1990), Russian teacher of geology, Moscow State University. ORTH *Pnma* or *Pmcn*. $a = 7.60$, $b = 5.83$, $c = 12.69$, Z = 8, D = 5.64. *AM 66:436(1981)*. *22-1092:* 4.35_6 3.43_6 3.18_6 3.09_8 2.93_6 2.85_{10} 2.84_9 2.28_6. Minute, short prismatic crystals. Gray-green to blue-green. No cleavage. H = 3. G = 5.6. Biaxial (−); XYZ= *abc*; $N_x = 2.11$, $N_y = 2.18$, $N_z = 2.22$; weakly pleochroic; 2V = 80°. *Pionersk deposit, eastern Sayan*, and the *Aginsk deposit, central Kamchatka, Russia*, with tellurite and lead–copper tellurites in the oxidation zone of these deposits. RG *DANS 253:1448(1980)*.

34.1.4.1 Plumbotellurite $PbTeO_3$

Named in 1982 for the composition. Dimorphous with fairbankite (34.1.2.1). ORTH Space group unknown. $a = 8.423$, $b = 13.739$, $c = 9.199$, Z = 12, D = 7.16. *AM 67:1075(1982)*. *35-644:* 3.21_3 3.17_{10} 3.10_4 2.98_3 2.86_3 2.73_2 2.11_1 2.10_1. Replacement rims and pseudomorphs after altaite. Gray, yellowish gray, to brown. No cleavage. H = 2. G = 7.2. Biaxial (+); $N_x = 2.19$, $N_y = 2.23$, $N_z = 2.35$; 2V = 50°. *Moctezuma, SON, Mexico*, as gray-white scales. *Zhana–Tyube deposit, northern Kazakhstan*, associated with altaite. RG *DANS 262:1231(1982)*.

34.1.5.1 Moctezumite $Pb(UO_2)(TeO_3)_2$

Named for the locality, which was named after Moctezuma (1466–1520), last king of the Aztecs. MON $P2_1/c$. $a = 7.819$, $b = 7.070$, $c = 13.836$, $\beta = 93.62°$, Z = 4, D = 7.25. *AM 50:1158(1965)*. *18-707:* 5.34_3 3.49_9 3.39_6 3.16_{10} 3.09_7 3.00_8 2.01_5 1.95_4. Prismatic crystals, bladelike, flattened {201}, and elongated [010]. Bright orange, dull luster. Cleavage {100}, perfect. H = 3. G = 5.73. Readily soluble in dilute HCl and dilute NaOH solution. Biaxial (−); N > 2.11; 2V = 5–10°. Sometimes pseudomorphously alters to schmitterite. *Moctezuma mine, Moctezuma, SON, Mexico*(!) (also known as "La Bambolla"), associated with zemannite, burckhardtite, schmitterite, other tellurites, and pyrite. RG *AM 78:835(1993)*.

34.1.6.1 Schmitterite $UO_2(TeO_3)$

Named in 1971 for Eduardo Schmitter Villada (1904–1982), professor of mineralogy, Instituto de Geologia, Universidad Nacional Autonoma de Mexico. ORTH $Pca2_1$. $a = 10.161$, $b = 5.363$, $c = 7.862$, Z = 4, D = 6.91. *AM 56:411(1971)*. Structure: *AC(B) 29:1251(1973)*. *25-1001:* 5.35_9 4.73_8 3.68_{10}

3.17_8 3.10_9 1.97_7 1.73_5 1.55_6. Minute, bladelike crystals, elongated [100], and flattened {001}. Very pale straw yellow, vitreous to pearly luster, glistening. Flexible but not elastic. H = 1. G = 6.88. Readily soluble in dilute HCl and alkalis. Biaxial (−), XYZ= cba; $N_y = 2.05$, $N_z > 2.11$; 2V = 75°. *Moctezuma mine* ("La Bambolla"), *Moctezuma, SON, Mexico*(!), with emmonsite, moctezumite, and other tellurites; also from the San Miguel mine, Moctezuma. Shinkolobwe, Shaba, Zaire. RG *BM 99:334(1976)*.

34.1.7.1 Magnolite Hg_2TeO_3

Named in 1877 for the locality. ORTH *Pma*2. a = 5.958, b = 10.576, c = 3.749, Z = 2, D = 8.12. *CM 27:133(1989)*. *45-1459:* 5.25_5 3.95_7 3.74_4 3.04_{10} 2.59_5 1.986_4 1.944_3 1.757_5. Acicular crystals, sometimes radiating, elongated [001]. Colorless, adamantine luster. Cleavage {010}, perfect; {001}, good; brittle. Soft. Biaxial (+); XYZ= bac; N > 2.0; 2V = ±45°; no dispersion. *Keystone mine*, and the Mountain Lion mine, *Magnolia District, Boulder Co., CO*, as an alteration product of coloradoite, associated with keystonite, melonite, and tellurium. RG *AJS 27:118(1877)*, *AM 75:437(1990)*.

34.1.8.1 Smirnite Bi_2TeO_5

Named in 1984 for Vladimir Ivanovich Smirnov (1910–1988), Russian economic geologist and department chairman, University of Moscow. ORTH *Cm*2*a*. a = 16.447, b = 5.513, c = 11.579, Z = 8, D = 7.72. *AM 70:876(1985)*. *38-420:* 3.23_{10} 2.89_6 2.75_2 2.74_2 1.996_3 1.992_3 1.727_3 1.668_2. Tabular crystals or crusts. Colorless, light gray, or light yellow. Cleavage {001}, perfect; brittle. H = $3\frac{1}{2}$–4. G = 7.78. Biaxial (+); XYZ= cba; $N_x = 2.35$, $N_y = 2.36$, $N_z = 2.46$ (Na); 2V = 35–40°; darkens on exposure to sunlight. *Zod deposit, Armenia*; also from the northern Aksu and the Zakharpathia deposits, Kazakhstan, associated with tellurobismuthite, tetradymite, volynskite, and galena. RG *DANS 278:199(1984)*.

34.2.1.1 Graemite $CuTeO_3 \cdot H_2O$

Named in 1975 for Richard Graeme (b.1941), American mining geologist. ORTH *Pmma*. a = 6.805, b = 25.613, c = 5.780, Z = 10, D = 4.24. *MR 6:32(1975)*. *26-1118:* 12.8_5 6.40_{10} 5.64_3 3.43_8 2.87_4 2.80_3 2.56_5 2.34_4. Crystals short prismatic [100]. Blue-green, vitreous luster. Brittle. H = 3–$3\frac{1}{2}$. G = 4.13. Nonfluorescent. Insoluble in water, easily soluble in cold dilute acids or 40% KOH. Biaxial (+); XYZ= bca; $N_x = 1.920$, $N_y = 1.960$, $N_z = 2.20$ (Na); 2V = 48.5°; no dispersion; absorption Y > Z = X. X, yellowish green; Y,Z, blue-green. *Cole shaft, Bisbee, AZ*, associated with malachite, cuprite, and teineite. Also in AZ from Tombstone, Cochise Co., and from the Dome Rock Mts., Yuma Co. It was formed by the dehydration of teineite. RG *AM 60:486(1975)*.

34.2.2.1 Chalcomenite CuSeO$_3 \cdot$ 2H$_2$O

Named in 1881 from the Greek for *copper* and *moon*. See molybdomenite (34.1.1.1). ORTH $P2_12_12_1$. $a = 6.666$, $b = 9.169$, $c = 7.373$, $Z = 4$, $D = 3.346$. *NJMM:551(1989). 17-523:* 5.39$_{10}$ 4.94$_9$ 3.77$_7$ 3.35$_8$ 3.04$_7$ 2.88$_6$ 2.53$_7$ 2.16$_6$. Crystals prismatic [001] acicular to equant. Bright blue, vitreous luster. No cleavage. H = 2–2$\frac{1}{2}$. G = 3.35. Soluble in acids, in CT affords water and a sublimate of SeO$_2$, easily fusible. XYZ= *acb*; N$_x$ = 1.712, N$_y$ = 1.732, N$_z$ = 1.732 (Na); r < v, strong (Bolivia); also r > v, strong (Argentina). X, pale blue; Y,Z, darker blue. Eagle shaft, Eldorado mine, Uranium City, SK, Canada. Liauzun-en-Olloix, Puy de Dôme, France, in blue incrustations. In *Argentina* at the *Cerro de Cacheuta, Mendoza*, and at the Sierra de Umango and the Sierra de Famatina, La Rioja. In Bolivia at Pacajake, near Colquechaca, and at the El Dragon mine, Potosi. In fine crystals at the Musonoi mine, Kolwezi, Zaire(!), with guilleminite, derriksite, and other selenites. RG *DII:638*, *AM 49:1481(1964)*, *AC 11:377(1958)*.

34.2.2.2 Teineite CuTeO$_3 \cdot$ 2H$_2$O

Named in 1939 for the locality. ORTH $P2_12_12_1$. $a = 6.634$, $b = 9.597$, $c = 7.428$, $Z = 4$, $D = 3.864$. *TMPM 24:287(1977). 17-733:* 5.87$_6$ 5.45$_9$ 4.96$_6$ 3.45$_{10}$ 3.06$_8$ 2.94$_7$ 2.88$_7$ 2.39$_7$. Crystals prismatic [001]; sometimes also flattened {010}; as crusts. Deep sky blue. Cleavage {010}, good, {001}, {100}, poor; brittle. H = 2$\frac{1}{2}$. G = 3.88. Soluble in HCl and HNO$_3$; in CT, affords water. Biaxial (–); XYZ= *abc*; N$_x$ = 1.767, N$_y$ = 1.782, N$_z$ = 1.791; 2V = 36°; absorption Z > Y > X. X, greenish blue; Y, blue; Z, indigo blue. Cole shaft, Bisbee, AZ, associated with malachite, graemite, and cuprite. Bambollita mine, Moctezuma, SON, Mexico. *Teine mine, Hokkaido, Japan*(!), with tellurite, as an alteration product of tetrahedrite and tellurium. RG *DII:635*, *AM 46:466(1961)*, *AC 15:698,863(1962)*.

34.2.3.1 Clinochalcomenite CuSeO$_3 \cdot$ 2H$_2$O

Named in 1980 for its monoclinic symmetry and dimorphous relationship to chalcomenite. MON $P2_1/c$. $a = 8.177$, $b = 8.611$, $c = 6.290$, $\beta = 97.27°$, $Z = 4$, $D = 3.42$. *AM 66:217(1981). 38-479:* 5.89$_{10}$ 4.06$_7$ 3.66$_9$ 2.93$_8$ 2.69$_7$ 2.57$_6$ 2.48$_7$ 1.82$_7$. Crystals prismatic [001], striated [001]. Bluish green, vitreous luster. Cleavage {110}, perfect; brittle. H = 2. G = 3.28. Biaxial (–); Y = *b*; Z $\wedge$ c = 10°; N$_x$ = 1.675, N$_y$ = 1.723, N$_z$ = 1.765; 2V = 78°; r < v; positive elongation. X,Y, colorless; Z, bluish green. From a *uranium deposit in China*, associated with volborthite, chalcomenite, umangite, and malachite. El Dragon mine, Potosi, Bolivia. RG *KT 25:89(1980)*.

34.2.3.2 Cobaltomenite CoSeO$_3 \cdot$ 2H$_2$O

Named in 1882 for the composition: cobalt and selenium. See molybdomenite (34.1.1.1). MON $P2_1/c$. $a = 7.615$, $b = 8.814$, $c = 6.499$, $\beta = 98.85°$, $Z = 4$, $D = 3.42$. *CM 12:304(1974). 25-125*(syn): 5.72$_{10}$ 4.04$_4$ 3.80$_7$ 3.46$_6$ 3.02$_{10}$ 2.95$_4$ 2.74$_6$ 2.51$_4$. Crystals long prismatic [001], or tabular {$\bar{1}$01}. Red to pink, vitreous luster. Cleavage {110}, {103}; brittle. H = 2–2$\frac{1}{2}$. G = 3.37. Biaxial (−); Z∧c = −12°; N_x = 1.681, N_y = 1.728, N_z = 1.769; no dispersion. X, pale pink; Y, pink; Z, red. A complete solid-solution series exists between cobaltomenite and ahlfeldite. Most natural members of the series are in the range Ni/Co 4:1 to 1:2. In Bolivia from the Pacajake mine, Hiaco, near Colquechaca, associated with chalcomenite and ahlfeldite; and from the El Dragon mine near Potosi associated with other selenites. Musonoi mine, Kolwezi, Zaire(!), associated with derriksite, chalcomenite, and other selenites. *Cerro de Cacheuta, Mendoza, Argentina*, with molybdomenite. RG *DII:639, NW 50:333(1963), AM 48:1183(1963)*.

34.2.3.3 Ahlfeldite $NiSeO_3 \cdot 2H_2O$

Named for Federico Ahlfeld (1892–1982), German–Bolivian geologist and mineralogist, who studied the minerals of Bolivia. MON $P2_1/c$. a = 7.519, b = 8.751, c = 6.448, β = 99°, Z = 4, D = 3.51. *CM 12:304(1974)*. 25-1233(syn): 5.66$_{10}$ 3.77$_5$ 3.42$_7$ 2.99$_7$ 2.72$_5$ 2.35$_4$ 2.19$_5$ 1.71$_4$. Crystals prismatic [001] or flattened {110}. Pale green to brownish green, pinkish; vitreous luster. Cleavage {110}, {103}, fair; brittle. H = 2–2$\frac{1}{2}$. G = 3.37. Biaxial; Z∧c = −12°; N_x = 1.703, N_y = 1.744, N_z = 1.786; 2V = 87°; r < v; absorption Z > Y > X. X, very pale green; Y, pale green; Z, green. Forms a complete solid-solution series with cobaltomenite, $CoSeO_3 \cdot 2H_2O$. In *Bolivia* at the *Pacajake mine, Hiaco*, formed as an alteration product of penroseite and other selenides, associated with cerussite, anglesite, chalcomenite, cobaltomenite. Also at the El Dragon mine near Potosi. RG *DII:635, AM 54:449(1969)*.

34.2.4.1 Choloalite $CuPb(TeO_3)_2 \cdot H_2O$

Named in 1981 from the Nahua *choloa*, evasive, in allusion to the fact that the mineral escaped detection in the mine for many years. ISOM Space group unknown. a = 12.519, Z = 12, D = 6.41. *MM 44:55(1981)*. 34-1248: 7.26$_5$ 4.19$_2$ 3.63$_3$ 3.49$_4$ 3.36$_6$ 3.05$_{10}$ 2.96$_3$ 1.966$_2$. Crystals are simple octahedra. Forest green, adamantine luster. No cleavage, irregular fracture, brittle. H = 3. G = 6.4. Soluble in dilute acids. Isotropic; N = 2.04; anomalous birefringence, up to 0.011. Tombstone, AZ, near the Joe and Grand Central shafts, associated with emmonsite, rodalquilarite, and cerussite. *Bambollita mine, Moctezuma, SON, Mexico*. RG

34.2.5.1 Hannebachite $CaSO_3 \cdot 0.5H_2O$

Named in 1985 for the locality. ORTH *Pbcn* or *Pbna*. a = 6.473, b = 9.782, c = 10.646, Z = 8, D = 2.52. *AM 73:928(1988)*. 39-725(syn): 5.56$_5$ 5.34$_2$ 3.80$_2$ 3.16$_{10}$ 2.82$_2$ 2.63$_6$ 1.851$_3$ 1.807$_2$. Crystals elongated [010], flattened {001}, bladelike. Colorless, vitreous luster. Cleavage {110}, perfect. H = 3$\frac{1}{2}$. G = 2.52. IR spectrum has absorption bands characteristic for $(SO_3)^{2-}$. Biaxial (+); XYZ= *acb*; N_x = 1.596, N_y = 1.600, N_z = 1.634; 2V = 38°. From the Quaternary volcano at *Hannebacher Ley, Eifel, Germany*, in cavities in melilite–nepheline–leucite lava, with calcite, aragonite, gypsum, barite, celestine, zeolites, and whewellite. RG *NJMM:241(1985)*.

34.2.5.2 Gravegliaite $MnSO_3 \cdot 3H_2O$

Named in 1991 for the locality. ORTH *Pnma*. $a = 9.763$, $b = 5.635$, $c = 9.558$, $Z = 4$, $D = 2.39$. *ZK 197:97(1991)*. *29-899*(syn)(calc): 6.81_9 4.77_3 4.34_{10} 4.28_5 3.43_4 3.41_5 2.69_4 2.40_4. See also *44-1445*. Prismatic crystals, elongated [010] and prismatic {h0l} with a pyramidal termination; also as sheaflike or radial aggregates. Colorless, white streak, vitreous luster, transparent. Cleavage [010]. Insoluble in water, soluble in strong acids. Nonfluorescent. Biaxial (+), $N_x = 1.590$, $N_y = 1.596$, $N_z = 1.636$; $2V = 41°$; strong dispersion. Found in cavities along fractures that cut Mn ore at the *Gambatesa mine, Val Graveglia, eastern Liguria, Italy*; associated minerals are chalcocite and rare alabandite. RG *AM 77:672(1972)*.

34.3.1.1 Cliffordite UTe_3O_8

Named in 1969 for Clifford Frondel (b.1907), American mineralogist and professor, Harvard University; connoisseur of uranium minerals. ISO *Pa3*. $a = 11.371$, $Z = 8$, $D = 6.76$. *AM 54:697(1969)*. Structure: *AC(B) 27:608(1971)*. *25-999*: 4.63_6 4.02_6 3.27_{10} 2.84_8 2.76_7 2.01_8 1.80_6 1.71_7. Octahedral crystals; as crusts. Bright yellow, adamantine luster. No cleavage. $H = 4$. $G = 6.57$. Nonfluorescent. Soluble in concentrated HCl, insoluble in dilute HCl. Isotropic; $N = 2.25$. *TMPM 29(1):1(1981)*. *San Miguel mine, Moctezuma, SON, Mexico*, associated with barite and mackayite, and in small amounts at the Moctezuma mine. RG *AM 57:597(1972)*.

34.3.2.1 Zemannite $Mg_{0.5}[ZnFe^{3+}(TeO_3)_3] \cdot 4.5H_2O$

Named in 1969 for Dr. Josef Zemann (b. 1923), Austrian mineralogist, University of Vienna, who has studied the tellurium oxysalts extensively. HEX $P6_3/m$. $a = 9.41$, $c = 7.64$, $Z = 2$, $D = 4.34$. *CM 14:387(1976)*. *31-1310*: 8.15_{10} 4.07_8 2.96_6 2.85_6 2.78_9 2.35_5 2.16_5 1.726_6. Small hexagonal prisms $\{10\bar{1}0\}$ terminated by a pyramid, $\{10\bar{1}1\}$; rarely tabular hexagonal $\{0001\}$; occasionally acicular. Dark to light brown; yellowish brown when acicular, with vitreous to adamantine luster. No cleavage, brittle. $H = 2$. Uniaxial (+); $N_O = 1.85$, $N_E = 1.93$; absorption O > E. O, reddish brown; E, yellowish brown. *Moctezuma mine ("La Bambolla"), Moctezuma, SON, Mexico*(!), associated with tellurite, moctezumite, denningite, burckhardtite, spiroffite, and other tellurium oxysalts. RG *TMPM 12:108(1967)*, *AM 55:1448(1970)*, *EJM 7:509(1995)*.

34.3.2.2 Kinichilite $Mg_{0.5}[Mn^{2+}Fe^{3+}(TeO_3)_3] \cdot 4.5H_2O$

Named in 1981 for Kin-Ichi Sakurai (b.1912), Japanese amateur mineralogist and collector, coauthor of *Minerals of Japan* (1938). Isostructural with zemannite and keystoneite. HEX $P6_3/m$. $a = 9.419$, $c = 7.666$, $Z = 2$, $D = 3.96$. *33-609*: 8.15_9 4.08_{10} 3.82_4 3.47_1 2.97_3 2.86_5 2.79_8 1.735_2. Hexagonal crystals with $\{10\bar{1}0\}$ and $\{0001\}$. Dark brown, brown streak, subadamantine luster. No cleavage. Soft. Uniaxial (+); $N_O > 1.80$, $N_E > 1.80$. O, light brown; E, yellowish brown. Kinichilite is the Mn^{2+}-dominant analog of zemannite and keystoneite. Moctezuma mine, Moctezuma, SON, Mexico, with tellurite and other Te oxysalts, *Kawazu mine, Shimada, Izu Penin., Japan*, in a quartz vein. RG *MJJ 10:333(1981), EJM 7:509(1995), AM 67:623(1982)*.

34.3.2.3 Keystoneite $Mg_{0.5}[NiFe^{3+}(TeO_3)_3] \cdot 4.5H_2O$

Named in 1988 for the locality. Isostructural with zemannite and kinichilite. HEX $P6_3/m$ or $P6_3$. $a = 9.358$, $c = 7.608$, $Z = 2$, $D = 4.27$. *CM (in press)(1997)*. PD: 8.12_9 4.05_8 3.81_4 2.95_5 2.84_5 2.77_{10} 1.720_6 1.498_5. Radiating sprays of acicular microcrystals. Golden yellow to orange yellow, adamantine luster, no cleavage, brittle. $G_{calc} = 4.23$–4.35. Uniaxial (+); $N_O = 1.85$, $N_E = 1.99$ (Na). From the *Keystone mine, Magnolia district, Boulder Co., CO*, associated with melonite, tellurium, pyrite, paratellurite, and magnolite. RG *GAC-MAC abstr. A4a(1988), EJM 7:509(1995)*.

34.3.3.1 Emmonsite $Fe_2^{3+}(TeO_3)_3 \cdot 2H_2O$

Named in 1885 for Samuel Franklin Emmons (1841–1911), economic geologist, U.S. Geological Survey. TRIC $P\bar{1}$. $a = 7.90$, $b = 8.00$, $c = 7.62$, $\alpha = 96.7°$, $\beta = 95°$, $\gamma = 84.5°$, $Z = 2$, $D = 4.719$. Structure: *SR 39A:323(1973)*. 7-404: 5.81_7 3.69_8 3.14_{10} 2.87_9 2.67_8 2.52_9 2.29_7 1.80_8. Crystals rare; usually as filiform and vermiform matted crystal aggregates, velvety. Bright yellow-green, grass green, blackish green; vitreous luster. Cleavage $\{010\}$, perfect; $\{100\}$, $\{001\}$, good. $H = 5$. $G = 4.549$. Nonfluorescent. Easily soluble in strong acids; in CT melts easily to a deep red liquid, loses water, and leaves an anhydrous residue. Biaxial (−); $Y = b$; $N_x = 1.962$, $N_y = 2.09$, $N_z = 2.10$; weakly pleochroic; $2V \pm 20°$; $r > v$, strong; absorption $Z > X$. *Tombstone, AZ*, from which mine, unknown. In good crystals at Goldfield, NV(!), with mackayite. Good Hope mine, Vulcan, Gunnison Co., CO, with sonoraite, rickardite, and weissite. Formerly abundant at the Moctezuma mine, Moctezuma, SON, Mexico(!), associated with schmitterite, cuzticite, sonoraite, and other tellurites, where it was a product of the oxidation of a tellurium–pyrite orebody. El Plomo mine, Ojojona district, Tegucigalpa, Honduras ("durdenite"). RG *DII:640, MR 3:82(1972), TMPM 18:157(1972)*.

34.3.4.1 Mandarinoite $Fe_2^{3+}(SeO_3)_3 \cdot 6H_2O$

Named in 1978 for Joseph A. Mandarino (b.1929), American–Canadian mineralogist, Royal Ontario Museum, Toronto. MON $P2_1/c$. $a = 16.810$, $b = 7.880$, $c = 10.019$, $\beta = 98.27°$, $Z = 4$, $D = 3.037$. *CM 22:475(1984)*. 29-719: 8.25_4 7.10_{10} 3.55_5 3.43_4 2.98_7 2.80_2 1.52_2 1.48_2. Prismatic crystals, with $\{100\}$, $\{110\}$, $\{011\}$, and $\{101\}$; usually twinned on $\{100\}$. Light green, similar to gem peridot; vitreous to greasy luster. No cleavage. $H = 2\frac{1}{2}$. $G = 2.93$. Nonfluorescent. Biaxial (−); X parallel to b; $Z \wedge c = 2°$; $N_x = 1.715$, $N_y = 1.80$, $N_z = 1.87$; $2V = 85°$; no dispersion. DeLamar silver mine, ID, with chlorargyrite. El Plomo mine, Ojojona district, Honduras. Skouriotissa mine, Cyprus, on quartz. In *Bolivia* at the *Pacajake mine, near Hiaco, ENE of Colquechaca*, with other selenites on penroseite; also at the El Dragon mine, Potosi, with chalcomenite and kerstenite on krutaite. RG *CM 16:605(1978), AM 70:440(1985)*.

34.4.1.1 Denningite (Ca,Mn)(Mn,Zn)Te$_4$O$_{10}$

Named in 1963 for Reynolds M. Denning (1916–1967), mineralogist, University of Michigan. TET $P4_2/nbc$. $a = 8.82$, $c = 13.04$, $Z = 8$, $D = 5.07$. *CM 7:443(1963). 15-129:* 6.26_5 4.42_{10} 3.38_8 3.12_7 2.62_7 2.03_7 1.53_6 1.51_5. See also *45-1368*(syn). Crystals tabular to platy {001}; also massive, or scaly. Pale apple green to colorless, brownish, purplish; adamantine luster. Cleavage {001}, perfect; conchoidal fracture. H = 4. G = 5.05. Nonfluorescent. Soluble in cold HCl, giving a yellow solution; insoluble in HNO$_3$; slightly soluble in H$_2$SO$_4$. Uniaxial (+); N$_O$ = 1.89, N$_E$ = 2.00. *Moctezuma mine ("La Bambolla"), Moctezuma, SON, Mexico*(!), in a quartz vein associated with tellurite, zemannite, spiroffite, and other tellurites. RG *AM 47:1484(1962), TMPM 10:241(1965)*.

34.4.2.1 Rajite CuTe$_2$O$_5$

Named in 1979 for the initials of Robert Allen Jenkins (b.1944), mineralogist and geologist, Phelps Dodge Corp., who first recognized the species. MON $P2_1/c$. $a = 6.866$, $b = 9.314$, $c = 7.598$, $\beta = 109.1°$, $Z = 4$, $D = 5.77$. *MM 43:91(1979). 33-494:* 4.65_8 3.79_6 3.35_8 3.11_7 3.06_{10} 2.84_5 2.80_5 2.74_7. Minute crystals, bladed or tabular. Bright green, resinous luster. Cleavage {010}, perfect; brittle. H = 4. G = 5.75. Nonfluorescent. Easily soluble in acids, readily fusible. Biaxial (+); Y = b; Z[100] = 3°; N$_x$ = 2.115, N$_y$ = 2.135, N$_z$ = 2.26 (Na); weakly pleochroic; 2V = 40°; absorption Z > Y > X. From a small prospect near the *Lone Pine Au–Te deposit, Catron Co., NM*, associated with mackayite on quartz breccia. RG *AM 64:1331(1979)*.

34.5.1.1 Spiroffite (Mn,Zn)$_2$Te$_3$O$_8$

Named in 1962 for Kiril Spiroff (1901–1981), Bulgarian–American economic geologist, Michigan College of Mining and Technology. MON Cc or $C2/c$. $a = 13.00$, $b = 5.38$, $c = 12.12$, $\beta = 98°$, $Z = 4$, $D = 4.97$. *CM 7:450(1962). 24-740:* 4.96_8 4.66_3 4.04_6 3.30_{10} 2.99_{10} 2.82_5 2.77_3 1.620_3. No euhedral crystals, forms coxcomblike druses of crystals, massive. Deep pink to purplish pink, subadamantine luster. Cleavage in two directions, conchoidal fracture. H = $3\frac{1}{2}$. G = 5.01. Nonfluorescent. Readily soluble in HCl, giving a yellow solution, insoluble in hot HNO$_3$ or H$_2$SO$_4$, readily fusible. Biaxial (+); N$_x$ = 1.85, N$_y$ = 1.91, N$_z$ > 2.10 (N$_{z\,calc}$ = 2.20); 2V = 55°. *Moctezuma mine ("La Bambolla"), Moctezuma, SON, Mexico*(!), with tellurite, denningite, zemannite, other tellurites, and tellurium, in quartz matrix. RG *AM 47:196(1962), MSASP 1:305(1963), NW 54:199(1967)*.

34.5.2.1 Winstanleyite TiTe$_3$O$_8$

Named in 1979 for Betty Jo Winstanley (b.1934), amateur mineralogist, who found the first specimen. ISO $Ia3$. $a = 10.963$, $Z = 8$, $D = 5.632$. *MM 43:453(1979)*. *34-1485:* 4.47_6 3.17_{10} 2.93_4 2.74_7 2.58_4 1.94_8 1.65_8 1.39_4. Crystallizes in cubes, occasionally modified by {111}. Chinese yellow; also tan or cream-colored, adamantine luster. No cleavage, tough. H = 4. G = 5.57. Very slowly soluble in 1:1 HNO$_3$ or 1:1 HCl, in CT readily fusible. Isotropic; N = 2.34; sometimes anomalously birefringent up to 0.011. *Grand Central mine, Tombstone, AZ*, in a silicified granodiorite dike, associated with jarosite, chlorargyrite, and rodalquilarite. RG *AM 65:810(1980)*.

34.5.3.1 Carlfriesite CaTe$_2^{4+}$Te^{6+}O$_8$

Named in 1975 for Carl Fries, Jr. (1910–1965), American geologist, U.S. Geological Survey and Instituto de Geologia, UNAM, who devoted his career to studying the geology of Mexico. MON $C2/c$. $a = 12.576$, $b = 5.662$, $c = 9.994$, $\beta = 115.57°$, $Z = 4$, $D = 5.699$. *AM 63:847(1978)*. *29-333:* 5.06_6 4.83_3 4.09_2 3.37_4 3.17_{10} 3.08_9 2.83_6 2.70_3. Crystals prismatic to bladelike [001], also as botryoidal crusts. Bright yellow. Cleavage {010}, fair; brittle. H = $3\frac{1}{2}$. G = 6.3. Nonfluorescent. Biaxial (−); Y parallel to b; $N_x = 1.982$, $N_y = 2.095$, $N_z = 2.19$ (Na); weakly pleochroic; 2V = 80°; absorption Z > X = Y. From the *Bambollita mine (also known as "La Oriental"), Moctezuma, SON, Mexico*, on fractures in barite–quartz vein material, associated with an amorphous copper tellurite, dickite, and chlorargyrite. RG *MM 40:127(1975)*.

34.5.4.1 Pingguite Bi$_6$Te$_2^{4+}$O$_{13}$

Named in 1994 for the locality. ORTH Space group unknown. $a = 5.689$, $b = 10.791$, $c = 5.308$, $Z = 1$, $D = 8.64$. PD: 3.15_{10} 2.84_8 2.69_2 2.65_9 1.956_{10} 1.891_8 1.695_2 1.631_1. Rounded granular aggregates of minute crystals (grain size 0.3–0.6 μm). Yellowish green, light yellow-green streak, vitreous to adamantine luster, transparent. H = 5–6, VHN$_{50}$ = 510 kg/mm^2. G = 8.44. Nonfluorescent. Biaxial; N > 2.0. R = 13.25 (589 nm). From the oxidation zone of a small gold deposit at *Yangjiava, Pinggu Co. (near Beijing), China*, associated with "limonite," malachite, pyromorphite, bismutite, gold, and other minerals. RG *AMS 14(4):315(1994)*, *MR 27:119(1996)*.

34.6.1.1 Mackayite Fe^{3+}Te$_2$O$_5$(OH)

Named in 1944 for John William Mackay (1831–1902), mine operator, Comstock Lode, and benefactor of the Mackay School of Mines, University of Nevada. TET $I4_1/a$. $a = 11.704$, $c = 14.984$, $Z = 16$, $D = 5.281$. *NJMM: 145(1977)*. *29-730:* 5.56_4 4.95_7 3.76_5 3.33_{10} 3.17_9 2.79_5 2.73_4 1.622_5. Short prismatic to equant crystals, bipyramidal, showing {010} and {112}, often pseudo-dodecahedral. Bottle green to brownish green, adamantine to vitreous luster. No cleavage, subconchoidal fracture; brittle. H = $4\frac{1}{2}$. G = 4.86. Nonfluorescent. Soluble in hot dilute HCl, in CT fuses. Uniaxial (+); $N_O = 2.19$,

$N_E = 2.21$. O, green; E, yellowish green. *Mohawk mine, Goldfield, NV*, associated with emmonsite and sonoraite. Abundant at the San Miguel mine, Moctezuma, SON, Mexico(!), in druses associated with cliffordite, tellurite, and barite. RG *AM 29:211(1944), 55:1072(1970); DII:642; TMPM 13:213(1969)*.

34.6.2.1 Rodalquilarite $H_3Fe_2(TeO_3)_4Cl$

Named in 1968 for the locality. TRIC $P1$ or $P\bar{1}$. $a = 8.89$, $b = 5.08$, $c = 6.63$, $\alpha = 103.17°$, $\beta = 107.08°$, $\gamma = 77.87°$, $Z = 1$. Structure: *AC(B) 25:1551(1969)*. *20-536:* 4.24_{10} 3.46_2 3.31_6 2.97_4 2.85_5 2.62_8 2.17_6 1.873_3. Minute stout crystals and crusts. Grass-green to emerald-green, greasy luster. One good cleavage, very brittle. $H = 2$–3. $G = 5.10$. Biaxial $(-)$; N between 2.1 and 2.2; nonpleochroic; $2V = 38°$. Grand Central shaft, Tombstone, AZ, associated with jarosite, chlorargyrite, and winstanleyite. *Rodalquilar gold deposit, Almeria, Spain*, with jarosite, gold, and emmonsite. Wendy pit, El Indio gold mine, 4th Region, Chile(!), associated with gold and jarosite. RG *BM 91:28(1968), AM 53:2104(1968)*.

34.6.3.1 Quetzalcoatlite $Cu_4Zn_8(TeO_3)_3(OH)_{18}$

Named in 1973 for Quetzalcoatl, a god of the Toltecs, who was associated with the sea, in allusion to the beautiful sea-blue color of the mineral. HEX $P6_322$. $a = 10.097$, $c = 4.944$, $Z = 1$, $D = 6.12$. *MM 39:261(1973)*. *26-485:* 8.75_{10} 5.05_2 3.53_4 3.27_3 2.75_7 2.52_4 2.25_2 1.77_3. Very minute crystals, simple combinations of $\{0001\}$ and $\{10\bar{1}0\}$; as crusts or sprays of needles. Sea-blue. Cleavage $\{10\bar{1}0\}$, fair; brittle. $H = 3$. $G = 6.05$. Nonfluorescent. Uniaxial $(-)$; $N_O = 1.802$, $N_E = 1.740$ (Na). O, blue-green; E, almost colorless. Empire mine, Tombstone, AZ. *Bambollita mine, Moctezuma, SON, Mexico*, associated with dickite, barite, azurite, hessite, and galena. RG *AM 59:874(1974)*.

34.6.4.1 Chekhovichite $Bi_2Te_4O_{11}$

Named in 1987 for Sergei K. Chekhovich, Kazakh geologist, Polytechnical Institute of Alma Ata. MON $P2_1$. $a = 19.00$, $b = 7.982$, $c = 6.938$, $\beta = 95.67°$, $Z = 4$, $D = 7.005$. *AM 74:1400(1989)*. *42-1323:* 3.29_{10} 3.15_9 3.14_{10} 2.73_5 2.002_4 1.998_5 1.686_3 1.683_3. Aggregates of grains; crusts. Gray, with yellowish, whitish, or greenish tints; adamantine luster. Perfect cleavage parallel to elongation of grains, and a second perfect cleavage at $87°$ to it. $H = 4$. $G = 6.88$. Biaxial $(-)$; $X \wedge c = 21°$, $Z \wedge a = 25°$; $N_x = 2.45$, $N_y = 2.50$, $N_z = 2.65$ (Na); $2V = 65°$; $r < v$. From the mines at *Zod, Armenia*, and at *Zhana Tyube and North Aksu, northern Kazakhstan*, with limonite and as pseudomorphs after tellurobismuthite. RG

34.6.5.1 Sophiite $Zn_2(SeO_3)Cl_2$

Named in 1989 for Sophia Ivanovna Noboko (b.1909), Russian vulcanologist and mineralogist, a leading investigator of the volcanoes of Kamchatka. ORTH *Pccn*. $a = 10.251$, $b = 15.223$, $c = 7.666$, $Z = 8$, $D = 3.64$. *AM 75:1211(1990)*.

45-1463: 7.61_{10} 3.81_2 3.24_1 3.06_1 2.92_1 2.73_1 2.54_1 2.13_1. Thin platy crystals, pseudohexagonal on $\{010\}$, sometimes elongated [001]; swallowtail twins on $\{100\}$. Colorless, becoming sky blue on long exposure to air; vitreous to greasy or silky luster. Cleavage $\{010\}$, perfect; $\{201\}$, fair; brittle. H = 2. Biaxial (+); XYZ= bca; $N_x = 1.709$, $N_y = 1.726$, $N_z = 1.750$; 2V = 81°. Found in fumaroles of the *Great Tolbachik Fissure Eruption, Kamchatka, Russia*, with tenorite, cotunnite, gold, ponomarevite, halite, and sylvite. Specimens were collected at 180–230°. RG *ZVMO 118:65(1989)*.

34.6.6.1 Francisite $Cu_3Bi(SeO_3)_2O_2Cl$

Named in 1990 for Glyn Francis (b.1939), quality control officer at the Iron Monarch iron mine. ORTH $Pmmn$. $a = 6.354$, $b = 9.630$, $c = 7.220$, $Z = 2$, $D = 5.42$. Structure: *AM 75:1421(1990)*. PD: 5.31_6 3.39_{10} 3.18_3 2.87_8 2.65_7 2.49_6 1.695_3 1.588_6. Radiating clusters of bladelike crystals, elongated [010]; contact twinning sometimes present. Bright apple-green, adamantine luster. No cleavage, conchoidal fracture. H = 3–4. Biaxial; N > 1.79; francisite reacts with high-refractive-index liquids. From the *Iron Monarch iron deposit, Middleback Ranges, SA, Australia, about 400 km NW of Adelaide*, in cavities in a barite lens in the iron deposit, associated with muscovite, chlorargyrite, bismuth, naumannite, djurleite, and other minerals. RG

34.7.1.1 Sonoraite $Fe^{3+}TeO_3(OH) \cdot H_2O$

Named in 1968 for the locality. MON $P2_1/c$. $a = 10.984$, $b = 10.268$, $c = 7.917$, $\beta = 108.48°$, $Z = 8$, $D = 4.179$. Structure: *SR 40A:311(1974)*. 21-430: 10.4_{10} 6.09_4 5.18_5 4.66_8 3.66_6 3.29_7 3.11_8 3.04_5. Crystals platy $\{100\}$, striated parallel to c, with $\{110\}$ and $\{011\}$. Yellowish green, vitreous to subadamantine luster. H = 3. G = 3.95. Soluble in dilute acids. Biaxial (−); $N_x = 2.018$, $N_y = 2.023$, $N_z = 2.025$; 2V = 20–25°. Mohawk mine, Goldfield, NV; Cripple Creek, CO; Good Hope mine, Vulcan, Gunnison Co., CO; Joe shaft, Tombstone, AZ(!), with jarosite and rodalquilarite. *Moctezuma mine, Moctezuma, SON, Mexico*(!), with emmonsite, goethite, cuzticite, and anglesite. RG *AM 53:1828(1968)*, *TMPM 14:27(1970)*, *MR 3:82(1972)*.

34.7.2.1 Cesbronite $Cu_5(TeO_3)_2(OH)_6 \cdot 2H_2O$

Named in 1974 for Fabien Cesbron (b.1938), French mineralogist, Bureau de Recherches Geologiques et Minière, Orleans. ORTH $Pbcn$. $a = 8.624$, $b = 11.878$, $c = 5.872$, $Z = 2$, $D = 4.455$. *MM 39:744(1974)*. 27-194: 5.93_{10} 4.89_7 3.49_9 2.38_4 2.36_7 2.16_3 1.70_3 1.59_3. Small spherules of radiating crystals, or as minute wedge-shaped crystals elongated on a. Bright emerald-green, bright luster. Cleavage $\{021\}$, good; $\{010\}$, poor; brittle. H = 3. G = 4.45. Nonfluorescent. Soluble in dilute HCl and HNO_3, insoluble in water or in KOH solution. Biaxial (+); X = a; $N_x = 1.880$, $N_y = 1.928$, $N_z = 2.029$ (Na); 2V = 72°; r > v; absorption ZY > X. X, pale bluish green; Y, rich yellow-green; Z, deep emerald-green. From the Centennial Eureka mine, Tintic district, Juab Co., UT, with xocomecatlite. *Bambollita ("La Oriental") mine,*

Moctezuma, SON, Mexico, on fracture surfaces associated with a pea-green amorphous Cu–Fe–Te mineral, carlfriesite, and teineite. RG *AM 64:653(1979)*.

34.7.3.1 Guilleminite $Ba(UO_2)_3(SeO_3)_2(OH)_4 \cdot 3H_2O$

Named in 1965 for Claude Guillemin (1923–1994), French mineralogist and inspector general, Bureau de Recherches Geologiques et Minière. ORTH *Pmna*. $a = 7.25$, $b = 16.84$, $c = 7.08$, $Z = 2$, $D = 5.08$. *AM 50:2103(1965)*. *18-582*: 8.39_{10} 7.29_{10} 6.68_6 6.51_2 3.55_8 3.17_4 3.04_6 2.80_6. Bladelike crystals flattened {010}, yellow coatings, silky masses. Light canary-yellow. Cleavage {100}, perfect; {010}, good; fragile and brittle. $G = 4.88$. Biaxial $(-)$; XYZ= *cba*; $N_x = 1.805$, $N_y = 1.798$, $N_z = 1.720$; $2V = 35°$; $r > v$, strong. X, bright yellow; Y, yellow; Z, colorless. In microcrystals at Liauzun en Olloix, Puy de Dôme, France. *Musonoi mine, Kolwezi, Shaba, Zaire*(!), in cavities in selenian digenite, associated with malachite, derriksite, marthozite, and other selenites. RG *BM 88:132(1965)*.

34.7.4.1 Marthozite $Cu(UO_2)_3(SeO_3)_3(OH)_2 \cdot 7H_2O$

Named in 1969 for Aimé Marthoz (1894–1962), Belgian civil engineer, formerly director general of the Union Minière du Haut-Katanga. ORTH *Pnma* or *Pn2_1a*. $a = 16.40$, $b = 17.20$, $c = 6.98$, $Z = 4$, $D = 4.7$. *AM 55:533(1970)*. *25-320*: 8.65_7 8.23_{10} 4.44_7 3.50_8 3.22_9 3.09_{10} 3.02_8 2.90_9. Crystals flattened on {100}; {100} faces striated parallel to [001]. Yellowish green to greenish brown. Cleavage {100}, perfect; {010}, imperfect. $G = 4.4$. Soluble in 1:1 HNO_3 and in dilute HCl. Biaxial $(-)$; XY= *ab*; $N_y = 1.780$–1.785, $N_z = 1.780$–1.785 (Na); $r > v$. X, yellow-brown; Y, greenish yellow. Dehydrates at room temperature to a "meta" phase, with $a = 15.80$, $b = 17.20$, $c = 6.98$. From the zone of oxidation of the *Musonoi mine, Kolwezi, Shaba, Zaire*, with demesmaekerite, guilleminite, malachite, and chalcomenite, in cavities in selenian digenite. RG *BM 92:278(1969)*.

34.7.5.1 Derriksite $Cu_4(UO_2)(SeO_3)_2(OH)_6$

Named in 1971 for Joseph Derriks (b.1912), Belgian mining geologist, who studied the Shinkolobwe uranium deposits. ORTH *Pn2_1m*. $a = 5.570$, $b = 19.088$, $c = 5.965$, $Z = 2$, $D = 4.72$. Structure: *AC(C) 39:1605(1983)*. *25-319*: 5.35_5 4.78_{10} 4.35_6 4.20_5 4.07_8 3.75_8 3.43_5 3.22_6. Crystals elongated [001], bladelike, flattened on {100}. Green to bottle-green. Cleavage {010}, very good. Biaxial $(-)$; XYZ= *abc*; $N_x = 1.77$, $N_y = 1.85$, $N_z = 1.89$ (520 nm). *Musonoi mine, Kolwezi, Shaba, Zaire*, in cavities in selenian digenite, associated with chalcomenite and demesmaekerite on malachite. RG *BM 94:534(1971)*, *AM 57:1912(1972)*.

34.7.6.1 Demesmaekerite $Cu_5Pb_2(UO_2)_2(SeO_3)_6(OH)_6 \cdot 2H_2O$

Named in 1965 for Gaston Demesmaeker (b.1911), Belgian mining geologist. TRIC *P$\bar{1}$*. $a = 11.955$, $b = 10.039$, $c = 5.639$, $\alpha = 89.78°$, $\beta = 100.37°$, $\gamma = 91.33°$, $Z = 1$, $D = 5.42$. Structure: *AC(C) 39:824(1983)*. *18-692*: 5.89_6 5.42_8 5.14_5

4.72_5 4.67_5 4.58_5 3.34_6 2.97_{10}. Triclinic crystals, elongated [100] and flattened {100}. Bottle green when fresh, turning somewhat brownish upon dehydration. No cleavage. H = 3–4. G = 5.28. Biaxial (+); $N_x = 1.835$, $N_z = 1.910$. *Musonoi mine, Kolwezi, Shaba, Zaire*, in the lower part of the oxidation zone, associated with cuproskodowskite, kasolite, malachite, guilleminite, and chalcomenite in selenian digenite. RG *BM 88:422(1966)*, *AM 51:1815(1966)*.

34.7.7.1 Haynesite $(UO_2)_3(OH)_2(SeO_3)_2 \cdot 5H_2O$

Named in 1991 for Patrick Haynes, U.S. geologist who first collected the mineral. ORTH *Pnc2* or *Pncm*. $a = 8.025$, $b = 17.43$, $c = 6.935$, Z = 2, D = 4.07. *CM 29:561(1991)*. *45-1349*: 8.01_{10} 5.93_3 4.01_7 3.47_6 3.19_5 3.12_7 2.91_8 2.47_4. Crystallizes as tablets elongate [001], and as acicular prismatic rosettes to 3 mm in diameter. Amber-yellow, vitreous luster, transparent. Cleavage {010}, easy. H = 1½–2. G = 4.1. Fluorescence yellowish green. Biaxial (−); XYZ = abc; $N_x = 1.618$, $N_y = 1.738$, $N_z = 1.765$; 2V = 45°; pleochroic, pale yellow to bright yellow. Found at the *Repete mine near Blanding, San Juan Co., UT*, associated with andersonite, boltwoodite, gypsum, and calcite as crusts on mudstones and sandstones of the Morrison formation. RG *AM 77:447(1992)*.

34.8.1.1 Poughite $Fe_3(TeO_3)_2(SO_4) \cdot 3H_2O$

Named in 1968 for Frederick Harvey Pough (b.1906), American mineralogist, curator, and writer. ORTH $Pna2_1$. $a = 9.66$, $b = 14.20$, $c = 7.86$, Z = 4, D = 3.76. *SR 37A:318(1971)*. *21-435*: 7.10_{10} 5.74_{10} 3.56_6 3.34_6 3.24_7 3.05_6 2.94_5 2.87_5. Crystals tabular, diamond-shaped, flattened {010}; also as botryoidal crusts. Yellowish to brownish and greenish yellow, vitreous luster. Cleavage {010}, perfect; {101}, good. H = 2½. G = 3.75. Nonfluorescent. Easily soluble in HCl, slowly soluble in hot dilute H_2SO_4. Biaxial (−); XYZ= bac; $N_x = 1.72$, $N_y = 1.985$, $N_z = 1.990$; 2V = 15–20°. X, colorless; Y, pale greenish yellow; Z, pale yellow. Lone Pine Au–Te deposit, Catron Co., NM, as brown crusts. *Moctezuma mine ("La Bambolla"), Moctezuma, SON, Mexico*, with goethite, emmonsite, jarosite, and pyrite. In small crystals at the *El Plomo mine, Ojojona district, Tegucigalpa, Honduras*. RG *AM 53:1075(1968)*, *TMPM 15:279(1971)*.

34.8.2.1 Tlapallite $H_6(Ca,Pb)_2Cu_3(TeO_3)_4(TeO_6)(SO_4)$

Named in 1978 from the Nahua *tlapalli*, paint, in allusion to its paintlike coating of rock surfaces. MON $a = 11.97$, $b = 9.11$, $c = 15.66$, $\beta = 90.04°$, Z = 4, D = 5.05. *MM42:183(1978)*. *29-319*: 12.0_{10} 5.95_5 4.73_4 3.73_4 3.54_6 3.28_4 2.99_{10} 2.89_5. Paintlike thin crusts, or somewhat botryoidal when the crusts are thicker. Slightly yellowish green; vitreous luster. No cleavage observed. G = 4.65. Nonfluorescent. Pb may substitute for part of the Ca and the Tombstone material is Pb-rich, with Ca/Pb = 5:4, accounting for the higher density, the higher refractive indices, and the slightly larger unit cell of this variety versus that of the Mexican material whose data is given here. A small amount of zinc substitutes for the copper in both Arizona and Sonora material. Biaxial $(-)$; $N_x = 1.815$, $N_y = 1.960$, $N_z = 1.960$; 2V small. At Tombstone, AZ, probably the Lucky Cuss mine, a plumbian tlapallite was found cementing a calc–silicate tactite. At the *Bambollita mine ("La Oriental")*, *Moctezuma, SON, Mexico*, it was found on fracture surfaces adjacent to veins cutting intensely sericitized rhyolites, sometimes associated with carlfriesite. RG *AM 64:465(1979)*.

34.8.3.1 Oboyerite $H_6Pb_6(TeO_3)_3(TeO_6)_2 \cdot 2H_2O$

Named in 1979 for O. (Oliver) Boyer, one of the men who first staked the Grand Central lode claim in the Tombstone district. TRIC $P1$ or $P\bar{1}$. $a = 12.249$, $b = 15.113$, $c = 6.868$, $\alpha = 116.45°$, $\beta = 98.58°$, $\gamma = 85.82°$, $Z = 2$, $D = 6.66$. *AM 66:220(1981)*. *34-1260*: 9.04_3 6.08_3 3.18_7 3.04_{10} 2.98_5 2.93_5 2.86_5 1.80_4. Minute spherulites with radiating fibrous structure. White. One cleavage, perfect. $H = 1\frac{1}{2}$. $G = \pm 6.4$. Nonfluorescent. Readily soluble in cold dilute HNO_3, slowly soluble in 1:6 HCl, easily fusible. Biaxial; $N_x = 2.24$, $N_z = 2.26$; no dispersion. *Grand Central mine, Tombstone, AZ*, associated with jarosite, fairbankite, and opal in altered, mineralized shale. RG *MM 43:453(1979)*, *AM 65:809(1980)*.

34.8.4.1 Mroseite $CaTeO_2(CO_3)$

Named for Mary Emma Mrose (b.1910), American mineralogist, U.S. Geological Survey. ORTH $Pbca$. $a = 6.988$, $b = 11.201$, $c = 10.566$, $Z = 8$, $D = 4.171$. Structure: *CM 13:383(1975)*. *29-309*: 5.60_6 5.14_9 4.20_8 3.35_7 3.14_{10} 3.02_7 2.39_7 1.97_8. Massive, sometimes with radiating structure, crude crystals elongated [001], sometimes platy. Colorless to white, resembling cerussite or barite; adamantine luster. No cleavage. $H = 4$. $G = 4.35$. Nonfluorescent. Soluble with effervescence in cold dilute HCl. Biaxial $(-)$; XYZ= acb; $N_x = 1.79$, $N_y = 1.85$, $N_z = 1.89$; $2V = 74°$; $r < v$, strong. *Moctezuma mine, Moctezuma, SON, Mexico*, associated with tellurite and spiroffite. RG *CM 13:286(1975)*, *AM 61:339(1976)*.

34.8.5.1 Eztlite $Fe^{3+}_6Pb_2(Te^{4+}O_3)_3(Te^{6+}O_6)(OH)_{10} \cdot 8H_2O$

Named in 1982 from the Nahua *eztli*, blood, in allusion to the color of the mineral. MON Space group unknown. $a = 6.58$, $b = 9.68$, $c = 20.52$, $\beta = 90.25°$, $Z = 2$, $D = 4.61$. *MM 46:257(1982)*. *35-665*: 10.3_5 5.13_6 4.04_9 3.43_{10} 3.29_{10} 3.24_9 2.61_6 2.45_7. Thin sparkling crusts of minute crystals. Brilliant red, orange streak. Cleavage {001}, good; very brittle. $H = 3$. $G = 4.5$. Readily soluble in dilute HCl, soluble with difficulty in warm 1:1 HNO_3. $N_x = 2.14$, $N_z = 2.15$; nonpleochroic. From the oxidation zone of the *Moctezuma mine ("La Bambolla"), Moctezuma, SON, Mexico*, associated with cuzticite and kuranakhite. RG *AM 68:471(1983)*.

34.8.6.1 Yecoraite $Bi_5Fe_3^{3+}O_9(Te^{4+}O_3)(Te^{6+}O_4)_2 \cdot 9H_2O$

Named in 1985 for the locality. TET or HEX No material suitable for single-crystal X-ray work was found, and cell constants could not be determined. PD: 5.45_4 3.72_4 3.21_7 2.96_4 2.74_{10} 1.94_3 1.63_3. Cryptocrystalline to glassy masses. Orange to honey colored; pale yellow to brown, resinous to waxy luster. No cleavage, conchoidal fracture, brittle. H = 3. G = 5.59. Nonfluorescent. Readily soluble in cold dilute HCl or HNO_3, fuses easily to a red glass, affords water in CT. Uniaxial (+); $N_O = 1.812$, $N_E = 1.824$; microfibers show parallel extinction and are length-slow. *San Martin de Porres mine, W of Yecora, Son., Mexico*, associated with and derived from the oxidation of tetradymite and pyrite. RG *SMMB 1:10(1985), AM 71:1547(1986)*.

34.8.7.1 Orschallite $Ca_3(SO_3)_2(SO_4) \cdot 12H_2O$

Named in 1993 for P. Orschall, German mineralogist, who discovered the mineral. HEX-R $R\bar{3}c$. $a = 11.350$, $c = 28.321$, Z = 6, D = 1.87. *MP 48:168(1993)*. 45-1484: 8.11_8 5.73_{10} 4.87_3 4.63_3 3.63_6 3.28_4 2.69_8 2.11_4. Crystals are colorless, transparent pseudocubes to 0.3 mm, vitreous luster. No cleavage, irregular fracture. H ≈ 4. G = 1.90. Uniaxial (+); $N_O = 1.494$, $N_E = 1.496$. IR, TGA. Found in cavities in melilite, nepheline, leucitite at *Hannebacher Ley, Eifel, Germany*, with clinopyroxene, phillipsite, and apatite; other cavities contain hannebachite, $CaSO_3 \cdot 0.5H_2O$. RG *AM 79:572(1994)*.

Class 35

Anhydrous Chromates

35.1.1.1 Tarapacaite K₂CrO₄

Named in 1878 for the locality. ORTH $Pmcn$ or $Pnma$. $a = 7.663$, $b = 10.388$, $c = 5.922$, $Z = 4$, $D = 2.736$. Structure: $AC(B)$ 28:2845(1972). 15-365(syn): 3.08_{10} 2.99_8 2.96_4 2.60_2 2.57_2 2.48_3 2.29_3 2.13_2. Crystals thick tabular {001}; pseudohexagonal twins on {110}, analogous to aragonite. Bright canary yellow, vitreous luster. Cleavage {010}, {001}, distinct. $G = 2.74$. Easily soluble in water. Biaxial $(-)$; XYZ= bac; $N_x = 1.687$, $N_y = 1.722$, $N_z = 1.731$; $2V = 52°$; $r > v$, weak. *Tarapacá province, Chile*, and also in the nitrate deposits of Antofagasta and Atacama provinces. RG DII:644, ZPC 35:109(1962).

35.1.2.1 Phoenicochroite Pb₂(CrO₄)O

Named in 1839 from the Greek for *deep red* and *color*, in allusion to the color. MON $C2/m$. $a = 14.001$, $b = 5.675$, $c = 7.137$, $\beta = 115.22°$, $Z = 4$, $D = 7.073$. AM 55:784(1970). 29-768: 6.46_2 6.34_2 3.39_{10} 2.98_8 $2,88_2$ 2.84_4 2.06_2 1.869_2. Crystals tabular, imperfect; also massive and as thin coatings. Dark red, becomes lemon yellow on exposure; resinous or adamantine luster. Cleavage {$\bar{2}$01}, perfect. $H = 2\frac{1}{2}$. $G = 7.01$. Soluble in HCl with separation of lead chloride. Biaxial $(+)$; $Y = a$; $Z \wedge b = 2°$; $N_x = 2.34$, $N_y = 2.38$, $N_z = 2.65$ (Li); $2V = 58°$; $r > v$, strong. Potter–Cramer property and Pack Rat claim, S of Wickenburg, Maricopa Co., and elsewhere in AZ, with mimetite, willemite, hemihedrite, and vauquelinite. *Beresov, Ural Mts., Russia*, with anglesite, galena, and crocoite. Anarak, Iran. Sierra Gorda, Antofagasta, Chile. RG DII:649, BMNHB 2(8)(1974), BM 103:469(1980).

35.2.1.1 Lopezite K₂Cr₂O₇

Named in 1937 for Emiliano López Saa (1871–1959), Chilean mining engineer and mineral collector from Iquique. Artificial K₂Cr₂O₇, besides the triclinic unit cell, also exists in two monoclinic modifications. It has not been established to which of these possibilities Lopezite belongs. TRIC $P\bar{1}$. $a = 13.419$, $b = 7.391$, $c = 7.468$, $\alpha = 90.87°$, $\beta = 96.23°$, $\gamma = 98.13°$, $Z = 4$, $D = 2.682$. NBSM 25(15):47(1978). 27-380(syn): 4.87_5 3.66_9 3.47_9 3.30_{10} 3.06_3 3.03_3 2.88_3 2.86_3. Small ball-like aggregates of crystals. Orange-red to red. Cleavage {010}, perfect; {100}, {001}, distinct. $H = 2\frac{1}{2}$. $G = 2.69$. Easily soluble in water to an orange solution. Biaxial $(+)$; $N_x = 1.714–1.720$, $N_y = 1.732–$

1.738, $N_z = 1.805$–1.820; $2V \approx 51°$; $r > v$, medium. *Huaram, Iquique Pampa, Oficina Rosario, Tarapacá, Chile*, in vugs in massive nitrate rock with tarapacáite, dietzeite, and ulexite. RG *DII:645*, *DANS 172:1068(1967)*, *CJC 46:933(1968)*.

35.3.1.1 Crocoite PbCrO$_4$

Named in 1832 from the Greek for *saffron*, in allusion to the color. MON $P2_1/n$. $a = 7.127$, $b = 7.438$, $c = 6.799$, $\beta = 102.43°$, $Z = 4$, $D = 6.10$. Structure: *AC 19:287(1965)*. *8-209*(syn): 4.96_2 4.38_2 3.48_5 3.28_{10} 3.03_6 3.00_3 2.25_2 2.09_2. Crystals prismatic [001]; also elongated parallel to [$\bar{1}$01]; sometimes octahedral with {111} and {$\bar{1}$11}; crystals frequently are cavernous or hollow; massive; granular. Orange, red; yellow, orange streak, adamantine to vitreous luster. Cleavage {110}, distinct; conchoidal to uneven fracture; brittle; sectile. $H = 2\frac{1}{2}$–3. $G = 6.0$. Fusible; in CT blackens and decrepitates, but recovers its color on cooling. Biaxial (+); $N_x = 2.29$, $N_y = 2.36$, $N_z = 2.66$ (Li); $2V = 57°$. Crocoite is a secondary mineral found in the oxidized zone of lead deposits, especially those associated with ultrabasic rocks. It forms later than anglesite, often replacing it, but usually earlier than cerussite. With wulfenite at the Darwin mines, Inyo Co., CA. In AZ south of Wickenburg, Maricopa Co., at the Potter–Cramer and Phoenix prospects, and the Rat Tail claim. Callenberg, Saxony, Germany(*). Leadhills, Scotland, in microcrystals. Nontron, Dordogne, France. *Beresov, Ekaterinburg, Ural Mts., Russia*(!), wih phoenicochroite, anglesite, galena, cerussite, and so on; also from Mursinsk and at Nizhne Tagilsk. Several mines in Zimbabwe, including the Old West and Proprietary mines, Penhalonga; Pet mine, Victoria; Howard's Luck mine, Umtali. In Brazil from Goyabeira, Congonhas do Campo, MG(*), in good crystals. From the Dundas district, TAS, Australia(!), from the Adelaide, West Coast, Dundas Extension, and other mines in magnificent crystals and groups, some individuals reaching 15 cm in length, associated with chromian cerussite, dundasite, massicot, and Mn oxides. RG *DII:646*, *BMNHB 2:8(1974)*, *ZK 176:75(1986)*.

35.3.2.1 Chromatite CaCrO$_4$

Named in 1963 for the composition, a chromate. TET $I4_1/amd$. $a = 7.242$, $c = 6.290$, $Z = 4$, $D = 3.142$. *NBSC 539(7):13(1957)*. *8-458*(syn): 3.62_{10} 2.88_2 2.68_5 2.56_1 2.38_2 1.85_4 1.81_2 1.50_1. Small grains and coatings. Citron-yellow. No cleavage, conchoidal fracture. Uniaxial (+); $N_O = 1.81$–1.85, $N_E = 1.84$–1.88; nonpleochroic. From the *Jerusalem–Jericho highway, Jordan*, in road cuts, from cleft surfaces in upper Cretaceous limestones, associated with gypsum. RG *NW 50:612(1963)*, *AM 49:439(1964)*.

35.3.3.1 Hashemite Ba(Cr,S)O$_4$

Named in 1983 for the locality, the Hashemite Kingdom of Jordan. Barite group. See barite (28.3.1.1). ORTH *Pnma*. $a = 9.103$, $b = 5.528$, $c = 7.331$, $Z = 4$, $D = 4.52$. Structure: *AC(C) 43:1467(1987)*. *35-642*: 3.67_6 3.52_{10} 3.40_3

3.17_8 2.90_3 2.77_4 2.18_6 2.15_4. Prismatic crystals, stubby to tabular. Dark to light brown, adamantine luster. Cleavage {001}, perfect; {010}, {100}, good. $H = 3\frac{1}{2}$. $G = 4.54$. Analyzed material from the type locality varied from $Ba(Cr_{0.93}S_{0.07})O_4$ (dark material) to $Ba(Cr_{0.64}S_{0.36})O_4$ (light material). Probably a complete solid-solution series exists between $BaCrO_4$ and $BaSO_4$. Biaxial (+); $N_x = 1.952$, $N_y = 1.960$, $N_z = 1.977$ (Na); $2V = 56$–$57°$ (dark brown); $N_x = 1.810$, $N_y = 1.813$, $N_z = 1.824$ (Na); $2V = 35$–$45°$ (light brown); $r < v$, extreme. *Lisdan–Siwaga fault, Hashem region, 60 km SE of Amman, Jordan*, in specimens from building stone quarries in a phosphatic carbonate rock of Cretaceous age, associated with chrome-bearing ettringite, blue-green apatite, and minor calcite. RG *AM 68:1223(1983), 71:1217(1986).*

35.4.1.1 Santanaite $9PbO \cdot 2PbO_2 \cdot CrO_3$

Named in 1972 for the locality. HEX $P6_322$. $a = 9.03$, $c = 39.84$, $Z = 6$, $D = 9.155$. 25-435: 3.54_{10} 2.95_4 2.85_4 2.61_8 2.24_4 2.10_4 2.08_5 1.70_5. Aggregates of minute platelets. Straw yellow, adamantine luster. Cleavage {0001}, perfect; also {$1\bar{2}10$}, {1010}. $H = 4$. Uniaxial (−); $N_O = 2.32$, $N_E = 2.12$. *Santa Clara mine, Caracoles, Sierra Gorda, Chile*, in corroded galena, associated with phoenicochroite, chromian wulfenite, diaboleite, and quartz. RG *NJMM:455(1972), AM 58:966(1973).*

35.4.2.1 Wattersite $Hg_4^{1+}Hg^{2+}Cr^{6+}O_6$

Named in 1991 for Lu Watters (1911–1989), California mineral collector, chef, musician, botanist, inventor, and environmentalist. MON $C2/c$. $a = 11.250$, $b = 11.630$, $c = 6.595$, $\beta = 98.17°$, $Z = 4$, $D = 8.91$. *MR 22:269(1991).* 45-1381: 8.06_8 5.58_5 3.60_5 3.30_6 3.26_6 2.95_5 2.92_5 2.66_{10}. Minute crystals on fracture surfaces, or shell-like aggregates in vugs. Dark reddish brown to black, dark brick-red streak, submetallic luster. Cleavage {010}, good; {001}, fair; conchoidal fracture; brittle. $H < 5.0$. Nonfluorescent. Soluble in cold dilute HCl giving a precipitate of calomel; in CT, gives off Hg, which condenses on the side of the tube, leaving a residue of green Cr_2O_3. In RL, lilac-gray, with deep to bright red internal reflections. Moderately pleochroic, strongly anisotropic. Purple, dark blue, greenish blue, turquoise. $R = 22.4$–23.5 (470 nm), 19.9–26.1 (546 nm), 18.9–23.3 (589 nm), 18.3–21.2 (650 nm). From a prospect pit near the *Clear Creek mercury mine, New Idria district, San Benito Co., CA*, associated with cinnabar and mercury. RG *AM 77:672(1992).*

35.4.3.1 Deanesmithite $Hg_2^{1+}Hg_3^{2+}Cr^{6+}O_5S_2$

Named in 1993 for Deane K. Smith (b.1930), American mineralogist, Pennsylvania State University, who gave long service to the International Centre for Diffraction Data. TRIC $P\bar{1}$. $a = 8.116$, $b = 9.501$, $c = 6.891$, $\alpha = 100.43°$, $\beta = 10.23°$, $\gamma = 82.8°$, $Z = 2$, $D = 8.14$. *CM 31:787(1993).* PD: 5.72_9 3.37_6 3.01_{10} 2.86_5 2.77_5 2.54_5 2.49_5 2.43_6. Bladed to acicular radiating crystal clusters, flattened {100}, elongated [001]. Orange-red, adamantine luster. Cleavage {$\bar{1}10$}, well developed; {001}, fair; irregular to subconchoidal fracture;

brittle to friable. H < 5. In RL, dark bluish gray, with bright orange internal reflections, weakly birefringent, weakly pleochroic. R = 23.4, 23.5 (470 nm), 20.4, 22.7 (546 nm), 20.0, 21.65 (589 nm), 19.65, 20.95 (650 nm). Found in a quartz–magnesite rock from a prospect pit near *Clear Creek mercury mine, New Idria district, San Benito Co., CA*, with cinnabar, edoylerite, edgarbaileyite, and other Hg minerals. RG *AM 79:1009(1994)*.

35.4.4.1 Edoylerite $Hg_3^{2+}Cr^{6+}O_4S_2$

Named in 1993 for Edward H. Oyler (b.1915), American mineral collector, who discovered the mineral. MON $P2_1/a$. $a = 7.524$, $b = 14.819$, $c = 7.443$, $\beta = 118.72°$, Z = 4, D = 7.13. *MR 24:471(1993)*. 45-1452: 5.94_4 4.88_5 3.21_{10} 3.01_6 2.31_4 2.21_4 2.19_4 1.908_3. Crystals acicular to prismatic [001], usually in stellate groupings. Canary- to orange-yellow, adamantine luster. Photosensitive, slowly turns olive-green in daylight. Subconchoidal fracture, brittle. Insoluble in common acids, but dissolves slowly in aqua regia. Biaxial; Z = c; N ≫ 1.78; strongly birefringent; weakly pleochroic; absorption Z > X = Y. In RL, light gray, weakly bireflectant, weakly pleochroic; no internal reflections. R = 17.6, 19.0 (470 nm), 16.95, 19.7 (546 nm), 16.4, 19.55 (589 nm), 16.0, 19.15 (650 nm). *Clear Creek claim, New Idria mining district, San Benito Co., CA*, with cinnabar and secondary mercury minerals on a quartz–ferroan magnesite matrix. RG *AM 79:1009(1994)*.

= Class 36 =

Compound Chromates

36.1.1.1 Iranite $Pb_{10}Cu(CrO_4)_6(SiO_4)_2(F,OH)_2$

Named in 1963 for the country of origin. TRIC $P1$ or $P\bar{1}$. $a = 9.543$, $b = 11.403$, $c = 10.740$, $\alpha = 120.43°$, $\beta = 92.5°$, $\gamma = 55.55°$, $Z = 1$, $D = 6.67$. BM 103:469(1980). 15-683: 4.84_8 4.42_8 3.60_{10} 3.49_{10} 3.44_8 3.28_{10} 3.18_{10} 3.08_{10}. Minute crystals. Saffron-yellow, vitreous luster. $G = 6.1$. Soluble in warm Na_2CO_3 solution. Biaxial; $N_x = 2.25$–2.30 (brownish orange, parallel elongation), $N_z = 2.40$–2.50 (yellow-orange, perpendicular to elongation); 2V very large. *Sebarz mine, NE of Anarak, Iran*, associated with dioptase and fornacite. RG BM 86:133(1963), AM 48:1417(1963), NJMM:406(1972).

36.1.1.2 Hemihedrite $Pb_{10}Zn(CrO_4)_6(SiO_4)_2F_2$

Named in 1967 in allusion to its morphology. TRIC $P1$. $a = 9.497$, $b = 11.443$, $c = 10.841$, $\alpha = 120.5°$, $\beta = 92.1°$, $\gamma = 55.83°$, $Z = 1$, $D = 6.50$. AM 55:1088(1970). Structure: AM 55:1103(1970). 24-1457: 4.87_9 4.36_8 3.30_{10} 3.16_8 3.10_8 2.92_6 2.85_5 2.18_5. Triclinic crystals, with hemihedral symmetry; four twin laws observed. Bright orange, brown, nearly black; yellow streak. $H = 3$. $G = 6.42$. Decomposed by 1:1 NHO_3 or 1:1 HCl with formation of white residues, slowly decomposed by 20% KOH. Biaxial $(-)$; $N_x = 2.105$, $N_y = 2.32$, $N_z = 2.65$; $2V = 85°$; strong dispersion, unsymmetric. *Florence mine, Pinal Co., and Pack Rat claim, Wickenburg, Maricopa Co., AZ*, associated with cerussite, phoenicochroite, vauquelinite, willemite, and wulfenite. RG CM 9:310(1967), NJMM:328(1972), BM 103:469(1980).

36.1.2.1 Macquartite $Pb_3Cu(CrO_4)(SiO_3)(OH)_4 \cdot 2H_2O$

Named in 1980 for Louis Charles Henri Macquart (1745–1803), French chemist, who brought to France from Russia the samples of crocoite in which the element chromium was discovered. MON Space group uncertain, probably $C2/m$. $a = 20.81$, $b = 5.84$, $c = 9.26$, $\beta = 91.8°$, $Z = 4$, $D = 5.58$. AM 66:638(1981). 33-466: 4.82_9 4.63_9 3.16_{10} 3.09_6 2.93_5 2.77_5 2.32_4 2.24_4. Small euhedral crystals, somewhat resembling mimetite. Orange, pale orange streak. Cleavage $\{100\}$, good. $H = 3\frac{1}{2}$. $G = 5.49$. Decomposed by 40% $HClO_4$ or concentrated HCl. Biaxial $(-)$; $Y = b$; $X \wedge c = 35°$; $N_x = 2.28$, $N_y = 2.31$, $N_z = 2.34$; $2V = 85°$. *Mammoth mine, Tiger, Pinal Co., AZ*, encrusting dioptase, associated with willemite and quartz. RG BM 103:530(1980).

Class 37

Anhydrous Acid Phosphates, Arsenates, and Vanadates

37.1.1.1 Monetite CaHPO$_4$

Named by C. U. Shepard in 1882 for the locality. TRIC $P\bar{1}$. $a = 6.910$, $b = 6.627$, $c = 6.998$, $\alpha = 96.34°$, $\beta = 103.82°$, $\gamma = 88.33°$, Z = 4, D = 2.924. Structure: AC(B) 33:1223(1977). 9-80(syn): 3.37_7 3.35_8 3.13_2 2.96_{10} 2.94_4 2.75_2 2.72_4 1.850_2. Crystals small with rough faces, usually flattened {010} with rhombohedral outline, interpenetrating groups; massive aggregates of minute crystals, crusts, or stalactites. Pale yellowish white, synthetic pure white; vitreous luster; translucent. Indistinct cleavage in three directions, uneven fracture, brittle. H = $3\frac{1}{2}$. G = 2.929. Biaxial (+); $N_x = 1.587$, $N_y \approx 1.615$, $N_z = 1.640$; 2V large; r > v, weak. Finely divided monetite takes up H$_2$O from atmosphere forming brushite. Loses H$_2$O, forming Ca$_2$P$_2$O$_7$ at elevated temperature and pressure. Occurs in phosphate-rock deposits in limestone underlying bed of bird guano on *Moneta Is.* and Mona Is., *64 km S of Mayaguez, PR*, with gypsum. Also with whitlockite on Los Monges Is., off Gulf of Maracaibo, Venezuela; on Ascension Is., South Atlantic Ocean, with apatite and newberyite; cave deposit of phosphates at Gunong Jerneh, Malaysia; Enderbury Is., Phoenix Islands. HCWS DII:660, MRB 5:437(1970).

37.1.1.2 Weilite CaHAsO$_4$

Named by Herpin and Pierrot in 1963 for René Weil (b.1901), mineralogist of Strasbourg, France, known for his study of Alsatian minerals. TRIC $P\bar{1}$. $a = 7.059$, $b = 6.891$, $c = 7.201$, $\alpha = 97.43°$, $\beta = 103.55°$, $\gamma = 87.75°$, Z = 4, D = 3.45 AC(B) 26:403(1970). 16-710(syn): 4.55_6 3.48_5 3.43_{10} 3.22_5 3.05_{10} 2.79_7 2.60_5 2.29_6. Powdery incrustations or pseudomorphs after pharmocolite, haidingerite. White, porcelanous, greasy to pearly luster. G = 3.48. Biaxial (−); X ∧ (001) = 20°, Y ∧ (001) = 27°; Z perpendicular to (001)= 34°; $N_x = 1.688$, $N_z = 1.664$ in (001); 2V = 82°. A secondary mineral in oxidized As-rich veins, usual alteration product of pharmacolite (39.1.1.2) and haidingerite (39.1.5.1), which are hydrated Ca arsenates. Found at Getchell mine, Humboldt Co., NV; St.-Marie-aux-mines, Haut-Rhin, France, and at *Schneeberg, Saxony, Germany.* HCWS AM 49:816(1964).

37.1.2.1 Schultenite PbHAsO$_4$

Named in 1926 by L. J. Spencer for Baron August Benjamin de Schulten (1856–1912) of Helsingfors and Paris, who prepared and described artificial crystals of the compound. MON $P2/a$. $a = 5.827$, $b = 6.743$, $c = 4.847$, $\beta = 95.34°$, $Z = 2$, $D = 6.079$ *MM 49:65(1985)*. *29-772*(syn): 6.74_3 4.41_2 3.38_{10} 3.15_7 2.92_2 2.91_2 2.56_2 2.20_2. Crystals tabular parallel to {010}, resembling gypsum in outline with {001} usually, but may show {111}, {130}, {140}, {011}, {121}, {$\bar{3}$22}. Striations parallel to c on {010} and parallel to a on {001}. White, pale yellow, colorless; white streak; vitreous to adamantine luster; transparent; brilliant. Fishtail twins from North Bend up to 3.5 mm. Cleavage parallel to {010}, good; brittle. $H = 2\frac{1}{2}$. $G = 6.07$. Fluoresces dull yellow in LWUV (North Bend). Biaxial (+); X parallel to b; $Y \wedge c = -24°$, $Z \wedge c = 66°$; $N_{x\,calc} = 1.890$, $N_y = 1.910$, $N_z = 1.977$ (Na); $2V = 58.2°$; strong dispersion. Found at *Tsumeb, Namibia*, associated with anglesite and bayldonite; alters to anglesite. Also at North Bend, King Co., WA, Australia, in vugs in quartz–arsenopyrite–galena vein with galena and mimetite and associated with cuprian adamite, keyite, and tennantite. The common constituent of the "lead arsenate" of commerce, widely used as an insecticide. HCWS *DII:661*, *MM 29:287(1950)*.

37.1.3.1 Phosphammite (NH$_4$)$_2$HPO$_4$

Named by C. U. Shepard in 1870 for the composition. See biphosphammite (37.1.4.1). MON $P2_1/c$. $a = 8.03$, $b = 6.68$, $c = 11.02$, $\beta = 113.6°$, $Z = 4$, $D = 1.61$. *AC 10:709(1957)*. *29-111*(syn): 5.57_7 5.05_{10} 4.94_6 4.13_4 3.78_5 3.22_6 3.14_4 3.06_4. Colorless crystals, translucent. $G = 1.62$. Found with biphosphammite in the guano deposits at *Guañape Is., Peru*, 640 km NE of Chincha Is.; also with urea, ammonian aphthitalite, and weddelite in guano at Toppin Hill (28°42′ S, 123°56′ E), WA, Australia. HCWS *DS 6:807(1892)*, *MM 39:346(1973)*.

37.1.4.1 Biphosphammite (NH$_4$,K)H$_2$PO$_4$

Named by C. U. Shepard in 1870 for the composition. See phosphammite (37.1.3.1) and archerite (37.1.4.2). TET $I\bar{4}2d$. $a = 7.4935$, $c = 7.340$, $Z = 4$, $D \approx 1.98$. Structure: *AC(B) 29:2721(1973)*. *37-1479*: 5.32_{10} 3.75_9 3.08_9 3.07_9 2.66_2 2.65_2 2.02_4 2.01_4. Powdery crusts. White to deep brown, white streak; dull, earthy luster. Very soft. $G = 2.04$. Uniaxial (−); $N_O = 1.525$, $N_E = 1.480$; synthetic identical. Found on phosphammite in guano deposits at *Guañape Is., Peru*; crusts on bat guano, Murra-el-Elevyn cave, Cocklebiddy, WA, Australia. HCWS *DS 6:807(1892)*, *MM 38:965(1972)*, *AM 58:806(1973)*.

37.1.4.2 Archerite KH$_2$PO$_4$

Named by P. J. Bridge in 1977 for Michael Archer (b.1945), professor of biology, University of NSW, Sydney, Australia, who drew attention to the deposit. TET $I\bar{4}2d$. $a = 7.453$, $c = 6.974$, $Z = 4$, $D = 2.33$. *NBS M*

$25:21:101(1985)$ Structure: ZK 74:306(1930). 35-807(syn): 5.09_1 3.72_{10} 3.01_1 2.91_8 2.64_2 2.34_1 1.982_1 1.953_5. Tetragonal crystals up to 2 mm long with second-order prisms and pyramids. White, translucent, no distinct cleavage, relatively soft, water soluble. Uniaxial $(-)$; $N_O = 1.511$, $N_E = 1.43$. Found in stalactites at *Petrogale cave, near Madura Motel ($31°54'$ S, $127°$ E), WA, Australia*, with biphosphammite, other phosphates, and saline minerals such as aphthitalite, syngenite, stercordite, weddelite, whitlockite, newberyite, calcite, and halite. HCWS *MM 41:33(1977), AM 62:1057(1977)*.

37.1.5.1 Nahpoite Na_2HPO_4

Named in 1981 after the composition (mnemonic). MON $P2_1/m$ or $P2_1$. $a = 5.47$, $b = 6.84$, $c = 5.45$, $\beta = 116.33°$, $Z = 2$, $D = 2.58$. *CM 19:373(1981)*. 35-735(syn): 3.98_5 3.84_6 3.42_2 2.88_4 2.81_{10} 2.73_5 2.72_3 2.66_2. Fine grained, earthy, white, soft. Microscopic grains are elongate to 4 μm with irregular edges. $N_{max} = 1.505$, $N_{min} = 1.490$; parallel extinction; length-fast. Found Mt. St.-Hilaire, QUE, Canada; in nodules in phosphatic ironstones in *Big Fish R. in northern Richardson Mts., YT, Canada*, with maricite and satterlyite; also in Khibina and Lovozero massifs, Kola Penin., Russia. HCWS

37.1.6.1 Lotharmeyerite $CaZnMn^{3+}(AsO_3OH)_2(OH)_3$

Named by P. Dunn in 1983 for Julius Lothar Meyer (1830–1895), German chemist who developed concepts for the periodic table. Mn^{3+} end member in Mn^{3+}–Fe^{3+} series with ferrilotharmeyerite (37.1.6.2). MON $C2, Cm, C2/m$. $a = 9.066$, $b = 6.276$, $c = 7.408$, $\beta = 116.16°$, $Z = 2$, $D = 4.29$. *MR 15:223(1984)*. 35-620: 4.94_8 3.41_9 3.18_9 2.91_9 2.82_8 2.71_8 2.56_{10} 1.687_8. Equant twinned crystals. Dark reddish orange, orange streak, vitreous luster. $H = 3$. $G = 4.23$. $N_{av} > 1.80$; strongly pleochroic, dark red-orange to light pink-orange. Occurs as drusy incrustations on adamite and Mn oxides (e.g., cryptomelane). Found at *Ojuela mine, Mapimi, DGO, Mexico*. HCWS *AM 68:849(1983), 68:849(1983); MR 14:35(1983)*.

37.1.6.2 Ferrilotharmeyerite $Ca(Zn,Cu)(Fe^{3+},Zn)(AsO_3OH)_2(OH)_3$

Named in 1992 for its relationship to lotharmeyerite (37.1.6.1). MON $C2, Cmn, C2/m$. $a = 8.997$, $b = 6.236$, $c = 7.390$, $\beta = 115.74°$, $Z = 2$, $D = 4.25$. Structure: *CM 30:225(1992)*. PD: 4.95_7 3.40_{10} 3.18_{10} 2.94_{10} 2.82_7 2.70_7 2.54_{10}. Aggregates (to 3 mm) of subhedral crystallites (0.2 mm long) and tabular to wedge- or lozenge-shaped (0.6 mm long). Brownish yellow, pale yellow streak, adamantine to greasy luster, transparent to translucent. Cleavage {001}, good; uneven fracture. $H = 3$. $G = 4.25$. Nonfluorescent. Biaxial $(+)$; $X \approx a$, $Y = b$; $N_x = 1.811$, $N_y = 1.844$, $N_z = 1.88$; $2V = 85°$; strongly pleochroic; olive-green; Y, pale green; Z, colorless inclined dispersion; $r > v$. X. Occurs with tennantite, scorodite, conichalcite, beudantite, and schneiderhönite at the *Tsumeb mine, Tsumeb, Namibia*. HCWS

Class 38

Anhydrous Normal Phosphates, Arsenates, and Vanadates

38.1.1.1 Triphylite   LiFe^{2+}PO$_4$

Named by J. Fuchs in 1834 from the Greek for *threefold* and *family*, in allusion to the composition containing three cations. In a series (Mn–Fe) with lithiophilite (38.1.1.2). Isostructural with olivines. *CIWY 68:290(1970)*. ORTH *Pbnm*. $a = 4.704$, $b = 10.347$, $c = 6.0189$, $Z = 4$, $D = 3.59$. See also *JPD 4:26(1989)*. Structure: AC(B) 32:2761 (1976). *40-1499:* 5.18_3 4.28_8 3.92_3 3.49_7 3.01_{10} 2.78_3 2.53_8 2.46_2. See also *19-721*(syn). Euhedral crystals rare, usually coarse with uneven surfaces, stout prismatic [100]. Grains up to 5 mm have been found in pegmatites. Commonly massive, cleavable. Bluish gray or greenish gray, colorless to grayish white streak, vitreous to subresinous luster, transparent to translucent. Cleavage {001}, nearly perfect (distinctive from olivines); {010}, imperfect; {011}, interrupted; uneven to subconchoidal fracture. $H = 4\frac{1}{2}$. $G = 3.565$. Oriented growth with graftonite, or armored by blue tourmaline. Biaxial (–) at highest Fe–Mn contents; $N_x = 1.694$, $N_y = 1.695$, $N_z > 1.71$ (Na); $2V \approx 0°$; nonpleochroic; see *DII:666*. Oxidation, leaching of Li increases N, while Mg reduces N *(AM28:90(1943))*. Altered by meteoric and hydrothermal solutions with ease, forming ferrisicklerite with concomitant leaching of Li. The oxidation products are often found as oriented single-crystal pseudomorphs after original crystals and may ultimately alter completely to iron oxyhydroxides. Alteration is usually accompanied by the formation of secondary Fe-phosphates such as strengite, metastrengite, dufrenite, frondelite, rockbridgeite, and vivianite. Occurs as primary mineral in granite pegmatites. Many localities worldwide were summarized in *MR 4:103(1973)*. New England sites prominent in the list are mines and quarries in ME: at Mount Apatite and Pulsifer quarry, near Auburn, at Mt. Mica mine, near Paris, and the Black Mt. and Red Hill mines near Rumford; in NH: at E. E. Smith mine, Newport, Sullivan Co., near Alexandria(!!), the Buzzo quarry and Parker Mt. mine near Center Strafford, at mine 4 at Palermo No. 1 and No. 2(!), North Groton, and at the Ruggles mine(!) near Grafton Center. Also found at Cobalt, and *Branchville, CT*; in the pegmatites of the Keystone district in the Black Hills at the Tin Mt. Lode, Big Chief mine, Dan Patch and Bull Moose pegmatite(!), near Custer, SD; at Pala, San Diego Co., CA; and at Pointe du Bois, MB, Canada. European localities: La Vilate, and Huréaux, France; at Hagendorf Süd(!) and Hühnerkobel, Bavaria, Germany, triphylite is embedded in "bull" quartz, and associated with beryl, tantalite, tourmaline, and minor sulfides–arsenides; at Dolai Bory, Moravia, Czech Republic; at Norrö and Varuträsk, Sweden; at Sukula and Tammela, Finland; Kalbinsk pegmatite, Russia; at Aracuai, Lavra Velha, and Galileia, Lavra do Enio, and at Sapucaia(!), MG, Brazil. Collectors often find good micromounts of many phosphate minerals at these locations. HCWS *DII:665*; *SPD 22:347(1978)*; *BM 99:274(1976), 99:271(1976)*; *MR 4:103(1973)*.

38.1.1.2 Lithiophilite $LiMn^{2+}PO_4$

Named in 1878 for the composition and from the Greek for *friend*. See triphylite (38.1.1.1). ORTH *Pbnm*. $a = 4.749, b = 10.454, c = 6.106, Z = 4, D = 3.44$. Structure: *AC 13:325(1960)*. Isostructural with olivine. *33-803*: 5.24_3 4.31_6 3.96_3 3.52_7 3.05_9 2.81_3 2.55_{10} 2.49_3. See also *33-804*(syn). Crystals rare, coarse with uneven surfaces, stout prismatic, commonly massive, cleavable to compact. Clove-brown, yellowish brown to honey-yellow, salmon, exposed surfaces often brown, dark gray to black due to alteration; colorless to grayish white streak; vitreous luster; transparent. Cleavage {001}, nearly perfect; {010}, imperfect; {011}, interrupted; uneven to subconchoidal fracture. $H = 4\frac{1}{2}$. $G = 3.34$. Biaxial (+); XYZ= *acb*; $N_x = 1.669, N_y = 1.673, N_z = 1.682$; $2V \approx 65°$; $r < v$, strong. Mg substitution markedly lowers N *DII:666*. Small amounts of Na substitute for Li. Soluble in acids. Readily undergoes alteration with meteoric or hydrothermal solutions to other phosphate minerals with Mn predominant, such as the Mn end members sicklerite, triploidite, reddingite, eosphorite, fairfieldite, dickinsonite, fillowite, and rhodochrosite. Further oxidation (Mn^{2+} to Mn^{3+}, associated with Fe^{2+} to Fe^{3+}) yields purpurite–heterosite series minerals. The oxidized forms are often pseudomorphic around unaltered cores of lithiophilite, other $Mn^{3+}-Fe^{3+}$ phosphates, and hydrated phosphate phases, such as hureaulite, may be present. Lithiophilite is a primary mineral in granite pegmatites, commonly associated with the alteration minerals listed above, and spodumene, albite, beryl, amblygonite, and graftonite. Found at Tamminen quarry, Greenwood, Oxford Co.; Pulsifer quarry, Auburn, Androscoggin Co. ME; Strickland quarry, Portland, Middlesex Co.; *Fillow quarry*(!), *Branchville, Redding Twp., Fairfield Co., CT*; Foote mine(!), Kings Mt., NC; Animikie Red Ace pegmatite, Florence Co., WI; Custer Mt. pegmatite, Custer, Custer Co., SD; Harding mine, Taos Co., NM; Outpost mine, White Picacho district, AZ; Stewart mine, Pala, San Diego Co., CA; Tanco mine, Bernic Lake, Manitoba, Canada; Mangualde(!), Portugal; Viitaniemi, Eräjärvi, Finland; Sajany, Siberia, Russia; Picuí, Paraíba, Brazil; Karibib, Namibia; Noumas pegmatite, Steinkopf, Namaqualand, South Africa; Wodgina, WA, Australia. HCWS *MR 4:103(1973)*.

38.1.1.3 Natrophilite $NaMn^{2+}PO_4$

Named in 1890 from the Latin *natrium*, sodium, and from the Greek for *friend*. ORTH *Pbnm*. $a = 4.985, b = 10.517, c = 6.314, Z = 4, D = 3.47$. Olivine structure; Na confined to one cation site. *AM 57:1333(1972)*. *25-846*: 4.50_6 4.05_6 3.66_5 3.15_5 2.86_8 2.60_{10} 2.58_{10} 1.831_7. Rare, indistinct crystals, resembling triphylite, commonly as cleavable masses or granular. Deep wine-yellow; adamantine, resinous luster, pearly on cleavage; transparent to translucent. Cleavage {001}, good; {010}, indistinct; {021}, interrupted; conchoidal fracture. H = 4.7. G = 3.41. Biaxial (+); XYZ= acb; N_x = 1.671, N_y = 1.674, N_z = 1.684; 2V = 75°; $r < v$, strong. Fe^{2+} substitutes for Mn^{2+} with maximum Fe/Mn ≈ 1.13. Soluble in acids. Alters to fine-fibrous yellow mineral with silky luster. Found sparingly with lithiophilite, triploidite, eosphorite, hureaulite, fairfieldite, and dickinsonite in granite pegmatite at *Fillow quarry, Branchville, Fairfield Co., CT.* HCWS *DII:670*.

38.1.2.1 Maricite $NaFe^{2+}PO_4$

Named in 1977 for Luka Maric (b.1899), mineralogist, University of Zagreb, Croatia. ORTH $Pmnb$. a = 6.861, b = 8.987, c = 5.045, Z = 4, D = 3.64. Structure: *CM 15:518(1977)*. Pseudohexagonal; buchwaldite (38.1.2.2) is the Ca analog. *29-1216:* 4.40_2 3.71_4 2.73_9 2.71_8 2.57_{10} 2.53_3 1.881_3 1.853_6. Nodular, elongate on a. Ludlamite, quartz, and vivianite may be inclusions within the maricite. Colorless, gray, pale brown; white streak; vitreous luster; transparent. No cleavage. Nonfluorescent. G = 3.66. Biaxial (−); XYZ= abc; N_x = 1.676, N_y = 1.695, N_z = 1.698; 2V = 43°; r > v, weak, nonpleochroic. Found as concretions in shales in sideritic ironstones, occasionally monomineralic, but often with kulanite, baricite, and penikisite. Associated with lazulite, childrenite, and siderite at *Big Fish R., YT, Canada*. HCWS *CM 15:396(1977)*.

38.1.2.2 Buchwaldite $NaCaPO_4$

Named in 1977 for Vagn Buchwald, Department of Metallurgy, Danmarks Tekniske Hojskole, Lyngby, Denmark. ORTH $Pmn2_1$. a = 5.167, b = 9.259, c = 6.737, Z = 4, D = 3.21. *AM 62:362(1977)*. *29-1194:* 3.79_9 2.72_{10} 2.63_6 2.58_5 2.51_4 2.19_4 1.91_8 1.54_3. Minute inclusions mostly < 10 μm across, polycrystalline interlocking masses of fine needles. White, translucent. H < 3. Biaxial (−); N_x = 1.607; N_y = 1.610, N_z = 1.616; 2V ≈ 65°; parallel extinction; positive elongation. A meteorite mineral found within troilite nodules of the *Agpalilik fragment, Cape York Meteorite, Greenland*, associated with single crystals of chromite and two other phosphate phases, one Cr-rich, the other Mn-rich. HCWS

38.1.3.1 Olgite $Na(Sr,Ba)PO_4$

Named in 1980 for Olga Anisimovne-Vorobiova (1902–1974), Russian mineralogist. HEX-R $P3$. a = 5.565, c = 7.050, D = 3.52. *SPC 29:633(1984)*. *33-1212:* 6.99_3 3.97_4 2.84_{10} 2.76_{10} 2.27_3 1.982_6 1.647_3 1.607_4. Crystals up to 2 mm in diameter exhibit (10$\bar{1}$0), (10$\bar{1}$1) dominant, also (0001), (11$\bar{2}$0), (11$\bar{2}$1). Bright blue, bluish green; vitreous luster. H = $4\frac{1}{2}$. G = 3.94. IR. Uniaxial (−); N_O = 1.623, N_E = 1.619. Insoluble in H_2O, dissolves in cold 1:10 HCl. Occurs in nepheline syenite pegmatite in analcime and natrosilite at *Mt. Karnasurt, Lovozero massif, Kola Penin., Russia.* HCWS *ZVMO:109(1981), AM 66:438(1981)*.

38.1.4.1 Ferrisicklerite ($\square_x$,Li_{1-x})(Fe^{2+}_{1-x},Fe^{3+}_x)PO_4

Named in 1937 for the Sickler family of Pala, CA. End member of the series (Fe^{2+}, Fe^{3+})–Mn^{2+} with sicklerite (38.1.4.2); intermediate in alteration of triphylite (38.1.1.1). ORTH $Pbnm$. $a = 4.796$, $b = 10.026$, $c = 5.916$, $Z = 4$, $D = 3.53$. Structure: $AC(B)$ 32:2761(1976). Related to olivines. 29-808: 5.01_7 4.33_4 3.82_2 3.49_3 2.96_{10} 2.49_3 2.44_1 1.479_1. Massive. Yellowish, dark brown; streak and powder light yellow-brown to brown but not purple; subtranslucent to opaque. Cleavage {100}, good. H = 4. G = 3.41. A secondary mineral formed through alteration of triphylite–lithiophilite, usually as rims around these species in crystallographic continuity. Some (OH) may substitute for O of (PO_4). Found in pegmatites associated with hureaulite, stewartite, jahnsite, and metastrengite at Center Strafford, Rochester, and North Groton, NH; at Peru and Stoneham, ME, and other pegmatite localities in New England; at Williams prospect, Carson Co., AL; in Keystone and Custer districts, Black Hills, SD; with sicklerite at *Pala, CA*; in the pegmatite district in Haute-Vienne, Plateau Central, France; at Hühnerkobel and Hagendorf, Bavaria, Germany; Varuträsk, Sweden; Tammela, Finland; Telirio mine, MG, Brazil; Euriowie Range, NSW, Australia. HCWS *DII:672, MR 4:103(1973), BM 99:271(1976)*.

38.1.4.2 Sicklerite ($\square_x$,Li_{1-x})(Mn^{2+}_{1-x},Mn^{3+}_x)PO_4

Named in 1912 for the Sickler family of Pala, CA. End member of a series Mn^{2+}–(Fe^{2+} + Fe^{3+}) with ferrisicklerite (38.1.4.1); intermediate in alteration of lithiophilite (38.1.1.2). ORTH $Pbnm$. $a = 4.794$, $b = 10.063$, $c = 5.947$, $D = 3.50$. *BM 99:274(1976)*. 33-802: 5.04_3 4.33_3 3.50_4 2.97_{10} 2.50_8 2.45_4 1.608_3 1.487_3 (contains Fe). Massive. Yellow, dark brown; streak and powder light yellow-brown to brown, red-brown, but not purple; subtranslucent to opaque. Cleavage {100}, good. H = 4. G = 3.36. Biaxial (−); X = a; $N_x = 1.715$, $N_y = 1.735$, $N_z = 1.745$; 2V medium large; r > v, very strong. Alters to purpurite. Alteration product of lithiophilite in pegmatites in New England and in the weathering zone at *Naylor–Vanderburg mine, Pala, San Diego Co., CA*, associated with hureaulite, stewartite, jahnsite, and metastrengite; at Varuträsk, Sweden; at Eräjärvi near Oriväsi and Tammela, Finland; at Wodgina, WA, Australia. HCWS *DII:672, MR 4:103(1973)*.

38.1.5.1 Beryllonite $NaBePO_4$

Named in 1888 for the composition. See esperite, a silicate analog containing (Pb,Ca,Zn). MON $P2_1/c$. $a = 8.178$, $b = 7.818$, $c = 14.114$, $\beta = 90°$, $Z = 12$, $D = 2.802$. Structure: *TMPM 20:1(1973)*. Puckered PO_4 and BeO_4 in pseudo-ditrigonal rings that parallel (010). Na in channels. 33-1213(syn): 4.09_2 3.90_{10} 3.62_4 3.60_3 2.82_6 2.25_4 1.954_2 1.948_4. PD: 4.41_4 3.90_7 3.62_8 3.61_4 2.823_{10} 2.816_6 2.35_4 2.25_6. *JPD 8:47(1993)*. Tabular {010} to short prismatic [010], crystals often complex with multiple faces in the zone [010] near {100}. Faces often dull or rough may be in oscillatory combination in the zones [100] and [010]. Twin plane {101}, contact and penetration twins, repeating, and in pseudohexagonal stellate forms, also polysynthetic. Crystals often show columnar structure due to presence of hollow canals and fluid-filled cavities parallel to [010]. Colorless to snow white, pale yellowish; vitreous, brilliant luster, pearly on {010}; transparent to translucent. Cleavage {010}, perfect; {100}, good; {101}, indistinct; conchoidal fracture; brittle. H = $5\frac{1}{2}$–6. G = 2.81. Biaxial (−); XYZ = bac; $N_x = 1.552$, $N_y = 1.558$, $N_z = 1.561$ (Na); 2V = 67.95°; r < v, weak. Found at Emmons quarry, Greenwood; with hydroxylherderite and elbaite at Dunton quarry, Newry; with smoky quartz crystals, muscovite, fluorapatite, triplite, beryl, cassiterite, microcline, and albite in a disintegrated *pegmatite at base of Sugarloaf Mt*(!) (erroneously McKean Mt.), Stoneham, Oxford Co., ME; Rio Tinto mine, Ungave, Labrador; Mt. St-Hilaire, QUE, Canada; Viitaniemi, Eräjärvi, Finland; Altai, Russia; Lavra da Ilha(!), Taquaral; Almerindo mine, Linopólis; Lavra do Énio, Galiléia, MG, Brazil; Paroc(!), Kunar, Laghman Prov., Afghanistan. HCWS *DII:677, AM 37:931(1958), MR 4:103(1973)*.

38.1.6.1 Panethite $(Na,Ca,K)_2(Mg,Fe^{2+},Mn^{2+})_2(PO_4)_2$

Named in 1967 for Friedrich Adolph Paneth (1887–1958), director of the Max-Planck Institute for Chemistry, Mainz, Germany, who studied meteorites. MON $P2_1/n$. $a = 10.18$, $b = 14.90$, $c = 25.87$, $\beta = 91.10°$, Z = 4, D = 2.99. *GCA 31:1711(1967). 20-828:* 5.10_6 4.21_4 3.95_4 3.24_5 3.01_{10} 2.75_5 2.71_7 2.55_4. Grains in meteorite, shows simple twinning. Pale amber, transparent. G = 2.9. Biaxial (−); $N_x = 1.567$, $N_y = 1.576$, $N_z = 1.579$ (Na); 2V = 51°. Microprobe analysis; has been synthesized. A meteorite mineral. Occurs with brianite in the *Dayton meteorite, Montgomery Co., OH*. HCWS *AM 53:509(1968)*.

38.1.7.1 Brianite $Na_2CaMg(PO_4)_2$

Named in 1967 for Brian Mason (b.1917), mineralogist, Smithsonian Institution, Washington, DC. MON $P2_1/a$. $a = 13.36$, $b = 5.23$, $c = 9.13$, $\beta = 91.2°$, Z = 4, D = 3.127. *AM 60:717(1975)*. Isostructural with merwinite. *29-1192:* 3.73_9 3.34_7 3.16_6 2.72_8 2.68_9 2.63_{10} 2.23_7 1.875_9. Grains up to 0.2 mm with fine lamellar structure. Colorless, transparent. H = $4\frac{1}{2}$. G = 3.10. Biaxial (−); $N_x = 1.598$, $N_y = 1.605$, $N_z = 1.608$; 2V ≈ 65°; extinct between lamellae 2–3°. Synthesized, melts between 800 and 100°; insoluble in H_2O. Occurs in small pockets in the metallic phase in the *Dayton meteorite, Montgomery Co., OH*, associated with whitlockite, panethite, albite, and enstatite. HCWS *GCA 31:1711(1967), AM 53:508(1968)*.

38.1.8.1 Vitusite-Ce $Na_3(Ce,La,Nd)(PO_4)_2$

Named in 1979 for Vitus Bering (1681–1741), explorer of northern seas. ORTH $Pcmb$ or $Pc2_1b$. $a = 5.342$, $b = 18.680$, $c = 14.062$, Z = 8, D = 3.86 (Ilimaussaq). *NJMA 137:42(1979)*. Indications of superlattice with $a = 58.75$. Structure may be related to glaserite and apatite. *MM 56:235(1992). 33-1232:* 6.58_9 4.67_9 4.63_9 3.51_9 2.81_{10} 2.80_{10} 2.69_9 2.67_9. Pale pink (Lovozero), white to pale green (Ilimaussaq, I), pseudohexagonal, twinned after (160), (120); vitreous luster. Cleavages {100}, {010}, {001}. Biaxial (−); XYZ= abc; $N_x = 1.604$, $N_y = 1.647$, $N_z = 1.649$; 2V = 28.5°(I). Vitusite-Nd has been synthesized. Readily dissolves in cold 1:10 HCl or H_2SO_4. Occurs in the natrolite zone of an alkalic pegmatite at Yubileynayn, *Mt. Karnasurt, Kola Penin., Russia*, associated with steenstrupine, belovite, neptunite, leucosphenite, and sazhinite, and at the *Ilimaussaq intrusion, southern Greenland*, a melanocratic nepheline syenite with arfvedsonite, albite, microcline, nepheline, sodalite, steenstrupine, and two other phosphate minerals. HCWS *AM 65:812(1980), MM 56:235(1992)*.

38.2.1.1 Berzeliite $(Ca,Na)_3(Mg,Mn)_2(AsO_4)_3$

Named in 1840 for Jöns Jacob Berzelius (1779–1848), Swedish chemist. End member of the series with manganberzeliite (38.2.1.2). ISO $Ia3d$. $a = 12.355$, $Z = 8$, $D = 4.088$. Structure: *AC(B) 32:1581(1976)*. Isostructural with garnet. *19-165*: 5.03_6 3.30_5 3.09_6 2.75_{10} 2.63_6 1.780_5 1.712_7 1.648_7. Usually massive or as rounded grains, rarely as trapezohedrons with small modifying faces. Yellow or honey-yellow to orange-yellow and yellowish red; red tint increases with Mn content; white to orange-yellow streak; resinous luster; transparent to translucent. No cleavage, subconchoidal to uneven fracture, brittle. $H = 4\frac{1}{2}$–5. $G \approx 4.08$. $N = 1.702$ (535 nm); colorless to orange, sometimes anomalous birefringence. Na_2O about 4.5 wt % usual. $Na/Ca \approx 2:5$; also Sb/As up to 1 : 10, usually trace amount. Members of the series have been synthesized. Occurs at *Långban*, and Moss mine, *Värmland*, in limestone skarn associated with hausmannite, manganophyllite, and caryinite, and at Sjö mine, Grythytte parish, Örebro, *Sweden*, in veinlets with rhodonite, tephroite, manganosite, pyrochroite, and barite. HCWS *DII:681, AM 53:316(1968)*.

38.2.1.2 Manganberzeliite $(Ca,Na)_3(Mn^{2+},Mg)_2(AsO_4)_3$

Named in 1894 for the composition and similarity to berzeliite (38.2.1.1). Mn end member of the series. ISO $Ia3d$. $a = 12.52$, $Z = 8$, $D = 4.379$. *AM 53:316(1968)*. Isostructural with garnet. *20-1089*(syn): 5.10_4 3.34_3 3.13_5 2.80_{10} 2.67_3 2.56_9 1.736_4 1.673_8. Similar to berzeliite in habit and other physicochemical properties. Reddish tint indicates Mn-rich; synthetic manganberzeliite is brown, dodecahedra, translucent, and optically yellow. $N = 1.77$. Only trace Fe^{2+} in analyses; no Fe analog has been synthesized. Occurs in *Långban, Värmland, Sweden*. HCWS *DII:681, AM 48:663(1963)*.

38.2.1.3 Palenzonaite $(Ca_2Na)Mn_2(VO_4)_3$

Named in 1987 for A. Palenzona, Italian amateur mineralogist, who discovered the mineral. ISO $Ia3d$. $a = 12.534$, $Z = 8$, $D = 3.78$. Structure: *NJMM:136(1987)*. Isostructural with garnet; analogous to manganberzeliite (38.2.1.2), an arsenate garnet. *41-608*: 5.12_1 3.13_6 2.80_{10} 2.67_2 2.56_6 2.46_1 1.738_2 1.675_4. Dodecahedral crystals and anhedral grains up to 6 mm in diameter. Wine red, streak brownish red, adamantine luster, transparent. No cleavage, subconchoidal fracture. $H = 5$–$5\frac{1}{2}$. $G = 3.63$. Nonfluorescent. $N = 1.965$, wine-red in thick section. Soluble in strong acids. Some As and Si substitution with an excess of Ca compared to idealized formula. Occurs in veinlets about 1 mm thick cutting quartz–braunite (Mn) ore and as crystals in a manganoan calcite vein 2 cm wide in the *Molinello manganese mine, Val Graveglia, eastern Lugiria, Italy.* Veinlets also contain saneroite, ganophillite, axinite, or tinaenite. HCWS *AM 73:930(1988).*

ALLUAUDITE-WYLLIEITE GROUPS

The alluaudite–wyllieite minerals have general formula
$$(AB)_5(TO_4)_3$$
where

$A = Ca, Na, Mn^{2+}$, or $\square$
$B = Al, Ca, Cu, Fe^{2+}, Fe^{3+}, H, Mg, Mn^{2+}, Pb, Zn$
$T = As, P$

and crystallize in a monoclinic, pseudoisometric space group.

The basic structure *MM 56:1955(1971)*, is composed of staggered chains of distorted cation-O octahedra stacked parallel to the {101} plane these layers are linked by TO_4 groups into pleated sheets that parallel y. The alluaudite subgroup members have four distinct cation sites called A2, A1, B2 and B1. The wyllieite subgroup minerals have six (called A2, A1a, A1b, B1, B2a and B2b) and show intrachain ordering. Formulae to express cyrstallographic information are based on 48 oxygens: for the alluaudites: $(A2)_4(A1)_4(B1)_4(B2)_8[(P,As)O_4]_{12}$. For the wyllieites: $(A2)_4(A1a)_2(A1b)_2(B1b)_2(B1a)_2(B2a)_4(B2b)_4(PO_4)_{12}$.

ALLUAUDITE SUBGROUP
$(A2)_4(A1)_4(B1)_4(B2)_8[(P,As)O_4]_{12}$
Space Group $C2/c$

Mineral	A2	A1	B1	B2	T	a	b	c	β	
Caryinite*	(Na,Pb)	Ca	(Mn^{2+},Ca)	(Mn^{2+},Mg)	As	12.42	13.17	6.87	114.1	38.2.2.1
Arseniopleite*	Na	Ca	Mn^{2+}	Mn^{2+}	As	12.36	13.06	6.86	114.0	38.2.2.2
Ferrohagendorfite	Na	(Ca,Na)	Fe^{2+}	(Fe^{3+},Fe^{2+})	P					38.2.3.1
Hagendorfite*	Na	(Ca,Na)	Mn^{2+}	(Fe^{3+},Fe^{2+})	P	11.92	12.59	6.52	114.7	38.2.3.2
Varulite	(Na,$\square$)	(Na,Ca, Mn^{2+})	Mn^{2+}	$(Mn^{2+},Fe^{3+}, Fe^{2+})$	P	11.91	12.54	6.46	114.98	38.2.3.3
Maghagendorfite	Na	(Na,Ca)	Mn^{2+}	(Mg,Fe^{3+}, Fe^{2+})	P					38.2.3.4
Ferroalluaudite*	$\square$	Na	Fe^{2+}	(Fe^{3+},Fe^{2+})	P	11.87	12.54	6.46	114.6	38.2.3.5
Alluaudite	$\square$	Na	Mn^{2+}	Fe^{3+}	P	12.004	12.533	6.404	114.4	38.2.3.6
O'Danielite*	$\square$	Na	Zn	Zn	As	12.113	12.445	6.793	112.87	38.2.3.7
Johillerite	$\square$	Na	(Mg,Zn)	Cu	As	11.870	12.755	6.770	113.42	38.2.3.8
Nickenichite	(Na,$\square$)	(Ca,$\square$)	(Cu,$\square$)	(Mn,Fe^{3+},Al)	As	11.882	12.760	6.647	112.81	38.2.3.9

* *Original space group transformed*
Notes:
1. In the alluaudite subgroup, leaching of alkali ions, or the oxidation of cations (i.e., Fe^{3+} rather than Fe^{2+}) usually results in a site that may be vacant ($\square$) or partially vacant. Cation site placement and occupancy or vacancy are proposed from cation—O bond lengths calculated from single-crystal structure determinations.
2. Assignment of cations (or vacancies) to sites requires not only chemical analyses but single crystal structure determinations. Actual chemical analyses on alluaudite give the formula $\square NaMn^{2+}ZnFe_2^{3+}(PO_4)_3$, or without vacancies $NaMn^{2+}ZnFe_2^{3+}(PO_4)_3$. Similarly the formula for O'Daniellite, $\square NaZnZn_2(AsO_4)(AsO_3OH)_2$ may be presented as $NaZn_3H_2(AsO_4)_3$.

Alluaudite-Wyllieite Groups

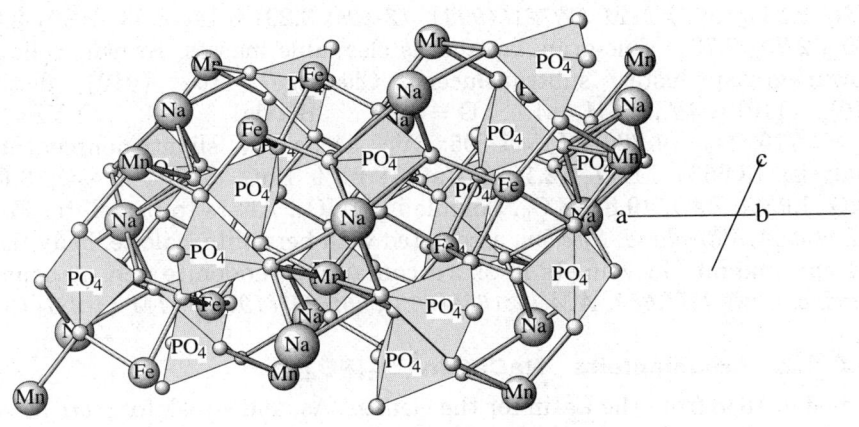

ALLUAUDITE: $\square$NaMn^{2+}Fe$_2^{3+}$(PO$_4$)$_3$

In this a axis projection of alluaudite [presentation depicts Na, and Fe^{3+} ions, and PO$_4$ groups, a simplified version of $\square$Na$_4$Mn$_4$Fe$_8$(PO$_4$)$_{12}$] the staggered chains of distorted (Mn,Fe) octahedra show as layers parallel to (101) linked through the oxygens of PO$_4$ groups. The result is pleated sheets with Na (or other A ions, or $\square$) in the interstices *MM 56:1955 (1971) KB:128 (1989)*.

WYLLIEITE SUBGROUP
Space Group $P2_1/n$, a subgroup of $C2/c$

Mineral	A2	A1a	A1b	B1	B2a	B2b	a	b	c	β	
Ferrowyllieite	(Na,$\square$)	Na	(Fe^{2+},Ca,)	(Fe^{2+},Mg)	(Fe^{2+},Mn)	Al^{3+}	11.834	12.293	6.323	114.20	38.2.4.1
Wyllieite	(Na,$\square$)	(Na,$\square$)	(Mn^{2+}, Ca)	(Mn^{2+}, Fe^{2+})	(Fe^{2+},Mg)	(Al^{3+}, Fe^{3+})	11.868	12.382	6.354	114.52	38.2.4.2
Rosemaryite	$\square$	Na	(Mn^2, Fe^{2+})	(Mn^2,Fe^{2+})	Fe^{3+}	Al	11.977	12.388	6.320	114.45	38.2.4.3
Qingheiite	(Na,$\square$)	Na	Mn^{2+}	(Mn^{2+},Mg)	(Al,Fe^{3+})	(Al, Fe^{3+})	11.856	12.411	6.421	114.27	38.2.4.4
Bobfergusonite*	Na	Na	Mn^{2+}	Mn^{2+}	Fe^{3+}	Al	12.045	12.486	12.773	114.35	38.2.4.5

* *Bobfergusonite original space group $P2_1/n$ transformed to $C2/c$ to show coincidence with other minerals.*

Notes:
1. Al^{3+} is an essential component in the wyllieite subgroup and changes the association of the chains and the space group. Wyllieite has three ordered sites, ferrowyllieite has two ordered sites, and Bobfergusonite is probably the most ordered structure in the subgroup.

38.2.2.1 Caryinite   (Na,Pb)Ca(Mn^{2+},Ca)(Mn^{2+},Mg)$_2$(AsO$_4$)$_3$

Named in 1874 for *in error*, not for the color *nut brown* in Greek. Dana (1892) changed "K" to "c." Alluaudite group and subgroup. See also arseniopleite (38.2.2.2). MON $P2_1/c$. $a = 11.48$, $b = 13.17$, $c = 6.67$, $\beta = 99°$; or by analogy to alluaudites $C2/c$. $a = 12.42$, $b = 13.17$, $c = 6.87$, $\beta = 114.1°$, $Z = 4$, $D = 4.486$. *AMG 2:333(1957) MM 57:721(1993)*. *12-295*: 3.29_2 3.14_2 3.03_3 2.90_2 2.87_{10} 2.85_{10} 2.79_2 2.73_2. Fine-granular and as cleavable masses. Brown, yellowish brown; greasy luster; subtranslucent. Cleavage {110}, {010}, distinct. $(010) \wedge (110) \approx 49°$. $H = 4$. $G = 4.29$. Biaxial (+); XYZ= *cab*; $N_x = 1.776$, $N_y = 1.780$, $N_z = 1.805$; $2V = 41°$; $r > v$, slight; nonpleochroic. Analysis: 18.66% MnO, 12.12% CaO, 9.21% PbO, 5.16% Na$_2$O, 3.09% MgO, 1.03% BaO, 49.8° As$_2$O$_5$. Soluble in HNO$_3$. Alters to berzeliite. Found at *Långban, Värmland, Sweden*, associated with berzeliite, calcite, hedyphane, and hausmannite in veinlets in skarn containing rhodonite and manganoan aegirine. HCWS *DII:683; AM 47:163(1962), 56:1955(1971); MM 51:281(1987)*.

38.2.2.2 Arseniopleite   NaCa(Mn^{2+})$_3$(PO$_4$)$_3$

Named in 1888 from the Latin for the element As, and Greek for *more* because it adds to the number of related minerals already described. Alluaudite group and subgroup. Mn-rich member of solid-solution series with caryinite (38.2.2.1). MON $P2_1/c$. $a = 11.31$, $b = 13.06$, $c = 6.86$, $\beta = 99°$, recalculated to $C2/c$, $a = 12.36$, $b = 13.06$, $c = 6.68$, $\beta = 114.0°$, $Z = 4$. $D = 4.29$ (Chanteloube). *AMG 4:425(1966)*. *20-224*: 6.52_4 4.13_4 3.27_4 2.99_5 2.83_{10} 2.79_4 2.68_6 1.693_5. Massive, granular. Deep brownish red, cherry-red, to apricot; yellowish-brown streak, opaque. Pseudorhombohedral cleavage, conchoidal fracture. $H = 3\frac{1}{2}$. $G = 4.22$. Easily soluble in 1:10 HCl, HNO$_3$. Some Fe^{3+} detected in analyses. Found at Canteloube, France; as irregular seams or fracture fillings of orange crystals with calcite and berzeliite at Långban mine, Filipstad, Värmland, and as fissure material in thin veins and as nodules in marble with rhodonite, tephroite, and hedyphane at *Sjö mine, Grythytte Parish, Örebro, Sweden*. HCWS *DII:844; AMG 2:333(1957), AM 56:1955(1971), 73:666(1988); MM 57:721(1993)*.

38.2.3.1 Ferrohagendorfite   Na(Ca,Na)(Fe^{2+},Fe^{3+})$_3$(PO$_4$)$_3$

Named in 1979 as the iron end member in the series with hagendorfite. Alluaudite group and subgroup. See hagendorfite (38.2.3.2). HCWS *MM 43:227(1979)*.

38.2.3.2 Hagendorfite   Na(Ca,Na)Mn^{2+}(Fe^{2+},Fe^{3+})$_2$(PO$_4$)$_3$

Named in 1954 for the locality. Alluaudite group and subgroup. The Mn^{2+}-containing end member in a series with ferrohagendorphite (38.2.3.1). MON $C2/c$. $a = 11.92$, $b = 12.59$, $c = 6.52$, $\beta = 114.7°$, $Z = 4$, $D = 3.84$. *BGSA 67:1694(1956)*. *29-1191*: 6.11_5 3.42_7 3.08_5 2.85_5 2.69_{10} 2.59_8 2.12_5 2.09_5. Massive. Greenish black. Three cleavages with one good, others indistinct. $H = 3\frac{1}{2}$. $G = 3.71$. Biaxial (−); $N_x = 1.739$, $N_y = 1.742$, $N_z = 1.745$; $2V = 70°$; pleochroic. X=yellowbrown Y=green Z=bluegreen. Found in the

Hagendorf-Süd pegmatite, Waidhaus, Oberpfalz, Bavaria, Germany, associated with triphylite, wolfeite, and hematite; at Norrö, Sweden. HCWS *NJMM:252(1954), MM 43:227(1979) AM 40:1100(1956)*.

38.2.3.3 Varulite NaCaMn^{2+}(Mn^{2+},Fe^{2+},Fe^{3+})$_2$(PO$_4$)$_3$

Named in 1937 for the locality. Alluaudite group and subgroup. The Mn-rich member in series with ferrohagendorfite (38.2.3.1) and with hagendorfite (38.2.3.2). MON $C2/c$. $a = 11.91$, $b = 12.54$, $c = 6.46$, $\beta = 114.98°$, $D = 3.89$. *AMG 2:9(1956)*. 6-487: 6.35$_3$ 5.46$_3$ 3.50$_4$ 3.12$_3$ 2.91$_3$ 2.79$_2$ 2.74$_{10}$ 2.56$_4$. Massive, granular. Dull olive-green, becomes yellowish or brownish on alteration; vitreous luster. Cleavage {100}, good; {010}, distinct. H = 5. G = 3.5. Biaxial (+); $N_x = 1.708$, $N_z = 1.722$; 2V large; r > v; pleochroic. X, yellow-green; Z, grass green. Analysis: 21.06% MnO, 12.01% FeO, 6.44% Fe$_2$O$_3$, 9.72% Na$_2$O, 3.60% CaO, 1.66% H$_2$O. Occurs at *Varuträsk pegmatite, Boliden mine, Skellefteå, Vasterbottea, Sweden*. HCWS *DII:669, AM 35:59(1950), MM 43:227(1979)*.

38.2.3.4 Maghagendorfite NaMn^{2+}(Mg, Fe^{3+}, Fe^{2+})$_2$(PO$_4$)$_3$

Named in 1979 as the Mg-containing member of the series with hagendorfite (38.2.3.2). Alluaudite group and subgroup. Analysis: 20.42% MnO, 9.89% Fe$_2$O$_3$, 8.75% FeO, 6.12% Na$_2$O, 5.06% MgO. Found in the *Dyke lode, Custer, SD*. CS *MM 43:227(1979), AM 65:810(1980)*.

38.2.3.5 Ferroalluaudite □NaFe^{2+}(Fe^{3+},Fe^{2+})$_2$(PO$_4$)$_3$

Named in 1979 as the Fe-rich partially oxidized end member of the series with alluaudite. Alluaudite group and subgroup. Synonym: ferri-alluaudite. MON $I2_1/a$, $a = 11.002$, $b = 12.527$, $c = 6.41$ *AM50 : 713(1965)* or transformed to $C2/c$. $a = 11.87$, $b = 12.54$, $c = 6.46$, $\beta = 114.6°$, Z = 4, D = 3.61. 39-337 6.28$_7$ 5.45$_2$ 4.11$_2$ 3.49$_3$ 3.08$_5$ 2.72$_{10}$ 2.53$_3$ 2.09$_2$. Single and grouped prismatic crystals up to 3 mm across, elongated on {010}, {001} with well-developed prism faces; granular, cleavable masses up to 10 cm across. Greenish black, vitreous luster. H = 5. G = 3.5. Biaxial (+); $N_x = 1.727$, $N_z = 1.738$; emerald green. X, tan; Z, blue-green. Analysis: 28.2% Fe$_2$O$_3$, 9.1% FeO, 7.3% MnO, 2.0% MgO, 1.6% CaO, 6.9% Na$_2$O, 43.1% P$_2$O$_5$. Crystals more or less replaced by ludlamite are found along one side of pod associated with arrojadite embedded in siderite, ludlamite, and vivianite masses which are within quartz–mica–perthite at the core of Palermo No. 1 pegmatite(!), North Groton, NH; at *Pleasant Valley pegmatite, Chester, Custer Co., SD*; at Hühnerkobel, Bavaria, Germany, and Norrö, Ranö, Sweden. HCWS *AM 56:1955(1971), MM 43:227(1979)*.

38.2.3.6 Alluaudite □NaMn^{2+}Fe^{3+}$_2$(PO$_4$)$_3$

Named in 1848 for François Alluaud of Limoges, France, who discovered the Chanteloube material. Alluaudite group and subgroup. The Mn-containing member of a series with ferroalluaudite (38.2.3.5). Synonym: hühnerkobelite. MON $C2/c$. $a = 12.004$, $b = 12.533$, $c = 6.404$, $\beta = 114.4°$,

$Z = 4$, $D = 3.62$. (Buranga) *AM 56:1995(1971)*. *42-581:* 6.27_8 5.47_4 3.50_4 3.07_5 2.87_3 2.73_5 2.72_{10} 2.52_4. Massive, compact granular, radiating fibrous or globular aggregates. Dirty yellow to brownish yellow, greenish black, outwardly brownish black or black due to alteration; dirty yellow streak and powder; subtransparent to opaque. Cleavage {100}, {010}, good. H = 5.2. G = 3.45. Biaxial (+), $N_x = 1.782$, $N_y = 1.802$, $N_z = 1.835$ (Pringle Fe/Mn = 2.29), X = straw yellow, Y,Z = light yellow, 2V = 79° (Na), r > v. Soluble in acids. Alters to heterosite, purpurite with leaching of Na and oxidation of Fe, and Mn. Found as a secondary mineral in pegmatites in the many mines in NH and SD listed under triphylite (38.1.1.1); also at Mangualde, Portugal; *La Vilate quarry, Chanteloube, Haute-Vienne, France*; Hagendorf Süd, Bavaria, Germany; Dolni Bory, western Moravia, Czech Republic; Norrö, Strumpetorp, and Varuträsk, Sweden; Sukula, Finland; Sidi-Bou-Othmane, Morocco; Buranga, Rwanda; Lavra do Enio, Galileia, MG, Brazil. HCWS *DII:674*, *AM 40:1100(1956)*, *MR 4:103(1973)*, *MM 43:227(1979)*.

38.2.3.7 O'Danielite □NaZn$_3$H$_2$(AsO$_4$)(AsO$_3$OH)$_2$

Named in 1980 for Herbert O'Daniel (1903–1977), professor of mineralogy, University of Munich, Germany. Alluaudite group and subgroup. MON $C2/c$. $a = 12.113$, $b = 12.445$, $c = 6.793$, $\beta = 112.87°$, $Z = 4$, $D = 4.49$. Structure: *NJMM 4:155(1981)*. *35-482:* 6.22_{10} 5.59_5 3.56_7 3.26_{10} 2.78_8 2.72_8 1.687_7 1.664_6. Pale violet, vitreous luster. Cleavage {010}, {100}, perfect; also {001}. G = 4.24. Biaxial (+); Z = b; Y ∧ c = 18°; $N_x = 1.745$, $N_y = 1.753$, $N_z = 1.778$; 2V = 60°; no dispersion observed. Analysis: 54.4% As$_2$O$_5$ 33.8% ZnO, 4.7% Na$_2$O, 2.9% MgO, minor CaO, FeO. Found with cuprian adamite, koritnigite, or prosperite at *Tsumeb mine, Tsumeb, Namibia*. HCWS *FM 58:68(1980)*, *AM 66:1276(1981)*.

38.2.3.8 Johillerite □Na(Mg,Zn)$_3$Cu(AsO$_4$)$_3$

Named in 1982 for Johannes Eric Hiller (1911–1972), professor of mineralogy, Stuttgart, Germany. Alluaudite group and subgroup. MON $C2/c$. $a = 11.870$, $b = 12.755$, $c = 6.770$, $\beta = 113.42°$, $Z = 4$, $D = 4.21$. *TMPM 29:169(1982)*. *35-522:* 4.06_5 3.50_4 3.25_8 2.75_{10} 2.64_5 1.952_4 1.682_4 1.660_5. Radial aggregates of thin platelets (010), elongation parallel to c, size 1 mm × 0.3 mm × 0.1 mm. Violet, transparent. Cleavage {010}, perfect; {001}, good. H = 3. G = 4.15. Analysis: 5.4% Na$_2$O, 18.3% MgO, 5.4% ZnO, 15.8% CuO, 55.8% As$_2$O$_5$. Found in cavities with conchalcite and cuprian adamite on specimen of altered tennantite with chalcocite from *Tsumeb, Namibia*. HCWS *AM 67:1075(1982)*.

38.2.3.9 Nickenichite (Na,□)(Ca,□)(Cu,□)(Mg,Fe^{3+},Al)$_3$(AsO$_4$)$_3$

Named in 1993 for the locality. Alluaudite group and subgroup. MON $C2/c$. $a = 11.882$, $b = 12.760$, $c = 6.647$, $\beta = 112.81°$, $Z = 4$. Structure: *TMPM 48:153(1993)*. Two distorted octahedral (Mg, Fe^{3+}, Al)–O$_6$ groups share edges forming zigzag chains in [10$\bar{1}$] which are connected by tetrahedral (AsO$_4$) groups into a three-dimensional framework. Two types of channels parallel to [001] result: one for monovalent, the other for divalent, cations. Multiple opportunities for vacancies: Na site ≈ 0.8, Ca site ≈ 0.4, Cu site ≈ 0.9 occupancy. PD: 4.35_4 4.06_5 3.56_4 3.53_4 3.20_6 3.07_4 2.74_{10} 2.61_4. Acicular crystals < 20 mm long. Bright blue, vitreous luster. Cleavage {010}, perfect; {100}, {10$\bar{1}$}, good. H ≈ 3. Analysis: 9.6% Na_2O, 2.8% CaO, 17.9% MgO, 4.7% CuO, 5.5% Fe_2O_3, 1.1% Al_2O_3, 64.3% As_2O_5. Biaxial (+); $\gamma \wedge c \approx 28°$ on (010); Y parallel to b; $N_x = 1.714$, $N_y = 1.744$, $N_z = 1.783$ (598 nm); $2V = 60°$; $2V_{calc} = 84°$; weak dispersion. Found in a cavity covered by vanadinite in pyroclastic rocks near the village of *Nickenich, in the Eifel, Germany.* HCWS *AM 79:571(1994)*.

38.2.4.1 Ferrowyllieite $Na_2(Fe^{2+},Mn^{2+})_2Al(PO_4)_6$

Named in 1979 for Peter John Wyllie (b.1930), mineralogist and petrologist, California Institute of Technology, Pasadena, as the Fe^{2+}-rich end member of a series with wyllieite. Alluaudite group. Wyllieite subgroup. MON $P2_1/c$. $a = 11.834$, $b = 12.293$, $c = 6.323$, $\beta = 114.20°$, $Z = 4$, $D = 3.68$. Structure: *AM 59:280(1974)*. *39-409:* 6.14_2 3.45_2 3.03_2 2.83_3 2.71_{10} 2.70_6 2.48_3. Crystals rare, usually as masses. Bluish green, grayish green, green; submetallic luster. Cleavage {010}, perfect; {$\bar{1}$01}, poor to good; brittle. H = 4–4.5. G = 3.60. Biaxial (+); $N_x = 1.688$, $N_y = 1.691$, $N_z = 1.696$; $2V \approx 50°$; r < v, strong; pleochroic. X, smoky blue-gray; Y, smoky blue-green; Z, green. DTA. Analysis: 43.8% P_2O_5, 29.2% FeO, 8.0% NaO, 7.9% Al_2O_3, 4.3% MnO, 2.5% CaO, 1.97% MgO. Larger alkali sites in the structure are extensively occupied by Fe^{2+} and Mn^{2+}. Found at *Victor mine, Custer, Custer Co., SD.* HCWS *MM 43:227(1979), AM 65:810(1980)*.

38.2.4.2 Wyllieite $Na_2(Mn^{2+})_2(Al,Fe^{3+})_2(PO_4)_6$

Named in 1973 for Peter John Wyllie (b.1930), mineralogist and petrologist, California Institute of Technology, Pasadena. Alluaudite group. Wyllieite subgroup. Redefined as the Mn end member of the series with ferrowyllieite (38.2.4.1). MON $P2_1/n$. $a = 11.868$, $b = 12.382$, $c = 6.354$, $\beta = 114.52°$, $Z = 4$, $D = 3.68$. Structure: *AM 59:280(1974)*. *26-1378:* 8.12_4 6.15_6 3.34_5 3.05_4 2.85_4 2.69_{10} 2.67_{10} 2.50_5. Euhedral crystals up to 2 cm×10 cm. Bluish green, grayish green, greenish black; dirty olive-green streak and powder; vitreous to submetallic luster. Cleavage {010}, perfect; {$\bar{1}$01}, distinct; brittle. G = 3.601. Analysis: 45.4% P_2O_5, 14.7% FeO, 11.2% MnO, 10.6% Fe_2O_3, 6.5% Al_2O_3, 3.9% NaO, 3.1% CaO, 1.7% MgO. Al is an essential element in this subgroup. Wyllieite has lower Fe^{2+} and higher Fe^{3+} and Mn^{2+} than ferrowyllieite. Found at *Old Mike mine, Custer, Custer Co., SD,* with quartz, muscovite, plagioclase, graftonite, and sarcopide, also arsenopyrite, scorzalite, ludlamite, vivianite, and garnet and at the Kestone Nickel Plate pegmatite mine; at Corrego Frio mine, MG, Brazil. HCWS *MR 4:131(1973), MM 43:227(1979)*.

38.2.4.3 Rosemaryite $\square Na(Mn^{2+},Fe^{2+})Fe^{3+}Al(PO_4)_3$

Named in 1979 for F. Rosemary Wyllie, wife of Peter J. Wyllie, to designate a member in the series with prominent Fe^{3+}. Alluaudite group. Wyllieite subgroup. MON $P2_1/n$. $a = 11.977$, $b = 12.388$, $c = 6.320$, $\beta = 114.45°$. *EJM 7:567(1995)*. PD: 6.19_5 5.45_5 3.49_7 3.02_{10} 2.84_6 2.73_9 2.69_{10} 2.06_6. Bottle green, dark brown to reddish brown; resinous luster. Cleavage observed. Biaxial (−); $N_x = 1.723$, $N_y = 1.742$, $N_z = 1.758$; 2V = 80°; pleochroic. X, brownish yellow; Z, greenish yellow. Analysis: 43.8% P_2O_5, 14.6% Fe_2O_3, 13.6% MnO, 11.6% FeO, 6.9% Al_2O_3, 3.4% Na_2O, 1.8% CuO. Found as an alteration of triphylite at *Rock Ridge pegmatite, Custer, Custer Co., SD*; with trolleite and scorzalite or with montebrasite and alluaudite-Na☐ at *Buranga pegmatite* (also Na-rich), *Gatumba, Gisenyi Prov., Rwanda*. HCWS *MM 43:227(1979)*, *AM 65:810(1980)*.

38.2.4.4 Qingheiite $Na_2(Mn^{2+},Mg,Fe^{2+})_2(Al,Fe^{3+})(PO_4)_3$

Named in 1983 for the locality. Alluaudite group. Wyllieite subgroup. MON $P2_1/n$. $a = 11.856$, $b = 12.411$, $c = 6.421$, $\beta = 114.27°$, $Z = 2$, $D = 3.61$. Structure: *SS 26B:876(1983)*. *41-588*: 8.14_2 6.21_4 3.46_3 3.07_2 2.86_3 2.83_2 2.69_{10} 2.52_3. Irregular, short prismatic grains, tabular up to 4 mm. Jade-green, glassy luster. Cleavage {010}, imperfect; conchoidal fracture; brittle. $H = 5\frac{1}{2}$. $G = 3.718$. Biaxial (+); $N_x = 1.678$, $N_y = 1.684$, $N_z = 1.691$; 2V = 79.6°; strongly pleochroic. X, light yellow-green; Y, jade-green; Z, dark bluish green; $r \gg v$, strong. Analysis: 23.7% MnO, 9.8% MgO, 8.7% Na_2O, 4.5% Al_2O_3, 3.9% FeO, 2.2% Fe_2O_3. Found in muscovite pegmatite at *Oinghe Co., Altay Pref., Uygur Autonomous Region of Xinjiang, China*, associated with muscovite, quartz, perthite, microcline, braunite, and pyrolusite. HCWS *AMS 3:161(1983)*, *AM 69:567(1984)*.

38.2.4.5 Bobfergusonite $Na_2Mn_5Fe^{3+}Al(PO_4)_6$

Named in 1986 for Robert Bury Ferguson (b.1920), Canadian feldspar mineralogist interested in pegmatites. Alluaudite group, Wyllieite subgroup. MON $P2_1/n$, $a = 12.776$, $b = 12.488$, $c = 11.035$, $\beta = 97.21°$ recalculated as $C2/c$. $a = 12.045$, $b = 12.486$, $c = 12.773$, $\beta = 114.35°$, $Z = 4$, $D = 3.57$. Structure: *CM 24:605(1986)*. *40-509*: 3.05_{10} 2.90_4 2.87_7 2.82_4 2.74_4 2.71_5 2.51_5 2.08_7. Anhedral grains. Red to green brown, medium brown; resinous luster. Cleavage {101} perfect, {100} prominent. Biaxial (+); Y parallel to *b*; $X \wedge a = +10°$; $N_x = 1.694$, $N_y = 1.698$, $N_z = 1.715$. X,Y, yellow-orange; Z, orange. Analysis: 45.1% P_2O_5 31.7% MnO, 7.5% Al_2O_3, 6.9% Fe_2O_3, 6.8% Na_2O, 1.2% CaO. Found on an island 5 km NNW of *Cross Lake, MB, Canada*. HCWS.

38.2.5.1 Fillowite $\square_2Na_2Ca(Mn^{2+},Fe^{2+})_7(PO_4)_6$

Named in 1879 for Abijan N. Fillow, the original owner and operator of the Branchville pegmatite mine, CT, who first collected the manganese phosphates for which the locality is known. See also johnsomervilleite (38.2.5.2). HEX-R $R\bar{3}$. $a = 15.282$, $c = 43.507$, $Z = 18$, $D = 3.54$. Structure: *AM 74:918(1989)*. Very complicated structure with nine distinct cation $-O_6$ octahedra, six additional cation polyhedra with five-, seven-, and eight-coordination, and six distinct PO_4 sites. *18-516*: 11.44_3 8.49_2 3.79_4 3.02_7 2.81_{10} 2.55_6

38.3.1.1 Sarcopside

2.16_2 1.888_4. Granular crystalline masses. Waxy yellow, orange to reddish brown, colorless; white streak; subresinous to greasy luster; transparent to translucent. Cleavage {0001}, nearly perfect; uneven fracture; brittle. H = $4\frac{1}{2}$. G = 3.43. Uniaxial (+); $N_O = 1.672$, $N_E = 1.676$; 2V small; r < v; strong. Colorless to yellow. Soluble in acids. In pegmatite at *Branchville, CT*, associated with triploidite, fairfieldite, and reddingite; Kabira, Uganda. HCWS *DII:719; AM 50:1643(1965)*.

38.2.5.2 Johnsomervilleite $\square_2Na_2Ca(Mg,Fe^{2+},Mn^{2+})_7(PO_4)_6$

Named in 1980 for John M. Somerville, who presented the specimen to the Royal Scottish Museum, Edinburgh, Scotland. Mg–Fe analog of fillowite (38.2.5.1). HEX-R. $a = 15.00$, $c = 42.75$, $Z = 18$, $D = 3.41$. *MM 43:833(1980)*. *33-1224:* 11.20_5 4.48_1 3.70_7 3.55_7 2.97_7 2.76_{10} 2.50_4 1.852_2. Anhedral crystals up to 1.5 mm. Dark brown, light brown streak, glassy luster. Cleavage {0001}, perfect; subconchoidal fracture; brittle. H = $4\frac{1}{2}$. G = 3.35. Nonfluorescent. Anomalously biaxial; $N_{av} = 1.655$; low birefringence; nonpleochroic. Found in metamorphic terrane in segregation pods 15–30 cm long in kyanite–silimanite gneiss at *Glenn Cosaidh, Loch Quoich, Scotland*, but not in the pegmatite. Associated with apatite, graftonite, and secondary phosphates such as jahnite, phosphosiderite, rockbridgite, mitridatite, and vivianite; at Sapucaia pegmatite, MG, Brazil. HCWS *AM 66:437(1981)*.

38.2.5.3 Chladniite $(Na,\square)_2CaMg_7(PO_4)_6$

Named in 1994 for Ernst Florens Friedrich Chladni (1756–1827), the "Father of Meteoritics." Mg end member in series with fillowite (38.2.5.1) and johnsomervilleite (38.2.5.2). HEX-R $R3$. $a = 14.967$, $c = 42.595$. Structure: *M 28:394(1993)*. Synthetic and natural material have Na(1) site partially vacant. Polished section shows grain as dark gray, weakly anisotropic. Found in the *Carlton (IIICD) meteorite* as a single large (975 μm × 175 μm) grain within a silicate-bearing inclusion that is mostly chlorapatite; associated with olivine, pyroxene, plagioclase, schreibersite, troilite, and cross cut by FeOOH of terrestrial origin. HCWS *LPS XXV:1337(1994)*.

38.3.1.1 Sarcopside $Fe_3^{2+}(PO_4)_2$

Named in 1868 from the Greek for *flesh* and *view* in allusion to its color. MON $P2_1/a$. $a = 10.437$, $b = 4.768$, $c = 6.026$, $\beta = 90.0°$, $Z = 2$, $D = 3.94$. Structure: *JMC 2:191(1991)*. Isostructural with olivine. *39-341:* 3.91_5 3.52_{10} 3.01_6 2.82_7 2.49_7 2.47_4 1.769_4 1.758_5. Irregular masses with fibrous structure, sometimes distorted hexagonal plates. Flesh-red, to reddish brown, changing on exposure to dark brown, blue, lavender, or green; silky and glistening luster; translucent. Cleavage distinct perpendicular to fibers, another less distinct parallel to fibers. Splintery fracture, fibrous. H = 4. G = 3.73 (Silesia), 3.64 (NH). Biaxial (−); $Z = b$; $X \wedge c = 45°$; $N_x = 1.675$, $N_y = 1.728$, $N_z = 1.730$; 2V small; r > v. Flesh-red in transmitted light. Occurs in pegmatite lens in gneiss on Gingrass farm near Deering, and at Demott mine, East Alstead, G.E. Smith mine, Newport, NH; at the Victory and Bull Moose mines, Custer, SD; in pegmatite *Gory Sowie, near Michelsdorf, Eulengebirge, Silesia, Germany*, intergrown with graftonite, and with hureaulite and vivianite later alteration products; Dolni Bory, Moravia, Czech Republic; Lavra do Enio, Galileia, MG, Brazil. HCWS *DII:858; AM 50:1698(1965), 57:24(1972)*.

38.3.1.2 Farringtonite $Mg_3(PO_4)_2$

Named in 1961 for O. C. Farrington (1864–1935), expert on meteorites, curator of geology, Field Museum, Chicago. See sarcopside (38.3.1.1). MON $P2_1/n$. $a = 7.599$, $b = 8.235$, $c = 5.076$, $\beta = 94.06°$, $Z = 2$, $D = 2.76$. Structure: *ACHS 22:1466(1968). 33-876*(syn): 4.36_3 4.31_2 4.12_3 4.08_3 3.85_9 3.66_3 3.44_{10} 2.41_3. $G = 2.74$. Biaxial (+); $Z \wedge c = 16°$; $N_x = 1.540$, $N_y = 1.544$, $N_z = 1.559$; $2V = 54°$. A meteorite mineral. Some Fe^{2+} in solid solution, but < 500 ppm Si. Found in the Krasnojarsk and *Sringwater pallasite (meteorites)*, where it coexists with olivine, iron metal, and schreibersite and shows alteration to a straw-yellow powder, a hydrous magnesian phosphate. HCWS *GCA 24:198(1961); AM 46:1513(1961), 58:949(1973)*.

38.3.2.1 Xanthiosite $Ni_3(AsO_4)_2$

Named in 1869 by G. J. Adam from the Greek for *yellow* and *sulfur*. MON $P2_1/a$. $a = 10.174$, $b = 9.548$, $c = 5.766$, $\beta = 92.98°$, $Z = 2$, $D = 5.388$. *MM 35:72(1965). 18-874*: 4.32_6 3.90_5 3.67_5 3.46_8 2.75_8 2.69_8 2.62_6 2.53_{10}. Massive. Golden yellow, translucent. $H = 4$. $G = 5.37$. Synthetic investigations show a high-temperature polymorph. Secondary mineral occurring with aerugite in comby quartz vein material in the deeper workings of the *South Terras mine, St. Stephen-in-Brannel, Cornwall, England*, associated with Ni–Co–As ores and decomposing pitchblende; at Johanngeorgenstadt, Saxony, Germany. HCWS *AM 50:2108(1965)*.

38.3.3.1 Graftonite $(Fe^{2+},Mn^{2+},Ca)_3(PO_4)_2$

Named in 1900 for the locality. See also beusite (38.3.3.2), with which it forms an isomorphous Fe–Mn solid solution. MON $P2_1/c$. $a = 8.882$, $b = 11.171$, $c = 6.143$, $\beta = 99.31°$, $Z = 4$, $D = 3.95$. Structure: *AM 67:826(1982)*. Mossbauer, neutron diffraction established occupancies of three distinct cation sites. *27-250*: 3.42_{10} 3.02_5 2.92_5 2.91_4 2.84_8 2.83_7 2.79_3 2.73_6. Stout prismatic and rough composite crystals; massive, cleavable. Often intergrown in laminations with triphylite such that graftonite {010}, [110] is parallel to triphylite {102}, [010]. Salmon pink to reddish brown, dark brown on alteration; white to faintly pink streak; vitreous to slightly resinous luster. Cleavage {010}, good; {100}, fair, but not easily observed; conchoidal fracture. $H = 5$. $G = 3.67–3.79$ varying with Fe/Mn/Ca. Biaxial (+); $X = b$; $Z \wedge c = 36°$; $N_x = 1.70$, $N_y = 1.705$, $N_z = 1.724$; 2V small, faintly pleochroic. $r > v$. X,Y, colorless; Z, pink. Analysis (Grafton, NH): 9.2% CaO, 30.6% FeO, 17.6% MnO. Nickel Plate mine sample with higher Mn, but not >Fe, 6.0% CaO has $N_x = 1.709$, $N_y = 1.7140$, $N_z = 1.736$; $2V = 60°$; $r < v$, strong.

Easily soluble in acids. A primary mineral in complex granite pegmatites often with triphylite in intergrowths in which triphylite may be altered to ferrisicklerite or heterosite. Found on *Melvin Mt., Grafton, Grafton Co., NH*, also at Palermo and Rice mines, N. Groton, Grafton Co., and Center Strafford, NH; with arrojadite in Nickel Plate mine, Pennington Co., SD; at Olgiasca, Lake Como, Italy; Brissago, Tessin, Switzerland; in pegmatites of Kaondakovo district, eastern Siberia, and Altyntau granite, Uzbekistan. HCWS *DII:686, AM 53:72(1968), RZ 6B 58(1985)*.

38.3.3.2 Beusite $(Mn^{2+},Fe^{2+},Ca)_3(PO_4)_2$

Named in 1968 for Alexei A. Beus, Russian mineralogist and geochemist, Moscow Polytechnical Institute. The Mn-rich end member in series with graftonite (38.3.3.1). MON $P2_1/c$. $a = 9.797$, $b = 11.758$, $c = 6.170$, $\beta = 99.31°$, $Z = 4$, $D = 3.715$. Structure: *CM 28:141(1990)*. *36-401*: 3.52_{10} 2.65_4 2.95_2 2.93_3 2.88_6 2.86_6 2.74_2 2.71_6. Rough prismatic crystals up to 30 cm long interlaminated with lithiophyllite. Red-brown, pale pink streak, vitreous luster. Cleavage {010}, good; {100}, fair. $H = 5$. $G = 3.702$. DTA. Biaxial (+); $X = b$; $Z \wedge c = 37°$; $N_x = 1.702$, $N_y = 1.703$, $N_z = 1.722$; $2V = 25°$; $r > v$, strong. Chemical analysis is necessary to distinguish between graftonite and beusite (i.e., Fe/Mn ratio). A Ca-free form has been described from a meteorite, other elements in analyses: Mg, Zn, Ni, Co, Cd. A primary phase in granite pegmatites. Found at a small pegmatite at *Los Aleros*(!), *San Luis Prov., Argentina*, also at San Salvador, Amanda, Cacique Cachueta, Ranquel mine in San Luis Prov.; in euhedral crystals within troilite nodules surrounded by Fe–Ni metal in El Sampal (IIIA) iron meteorite(!) found in 1973 near Nueva Lubuka, Chubut Prov., Argentina. HCWS *AM 53:1799(1968), 76:1985(1991)*.

38.3.4.1 Whitlockite $Ca_9(Mg,Fe^{2+})(PO_3OH)(PO_4)_6$

Named in 1941 for Herbert P. Whitlock (1868–1948), mineralogist and curator of mineralogy, American Museum of Natural History, New York. Forms a series with strontiowhitlockite (38.3.4.2); see also merrillite (38.3.4.3). HEX-R $R3c$. $a = 10.330$, $c = 37.103$, $Z = 3$, $D = 3.102$. Structure: *AM 60:120(1975)*. Analogous to $Ba_3P_2O_8$, and isostructural with cerite-(Ce), whitlockite contains two types of chains, and two cation sites, one of which is partially occupied by H_2. The phase is not equivalent to β-$Ca(PO_4)_2$ [*NAT(DS) 237:30(1972)*] and is distinct from the apatites. For the relationship of whitlockite to merrillite (38.3.4.3), see *EPSL:35(1977)*. *9-169*(syn): 5.21_2 3.45_3 3.21_6 2.88_{10} 2.76_2 2.61_7 1.933_2 1.728_3. Crystals usually rhombohedral, rarely tabular {0001}; coarse-granular to earthy. Colorless to white, gray, or yellowish; vitreous luster; transparent to translucent. No cleavage, uneven to subconchoidal fracture. $H = 5$. $G = 3.12$. DTA. Uniaxial (−); $N_O = 1.629$, $N_E = 1.626$. There is a loss of H near 780°, complete by 1035°. Occurs with siderite, quartz, apatite, ludlamite, fairfieldite, xanthoxenite, and triphylite in pegmatite, *Palermo mine, NH*; Newry, ME; Tip Top mine, Custer, SD; Sabinas, HGO, Mexico; also at the caves at Sebdou, Oran, Algeria, and as a

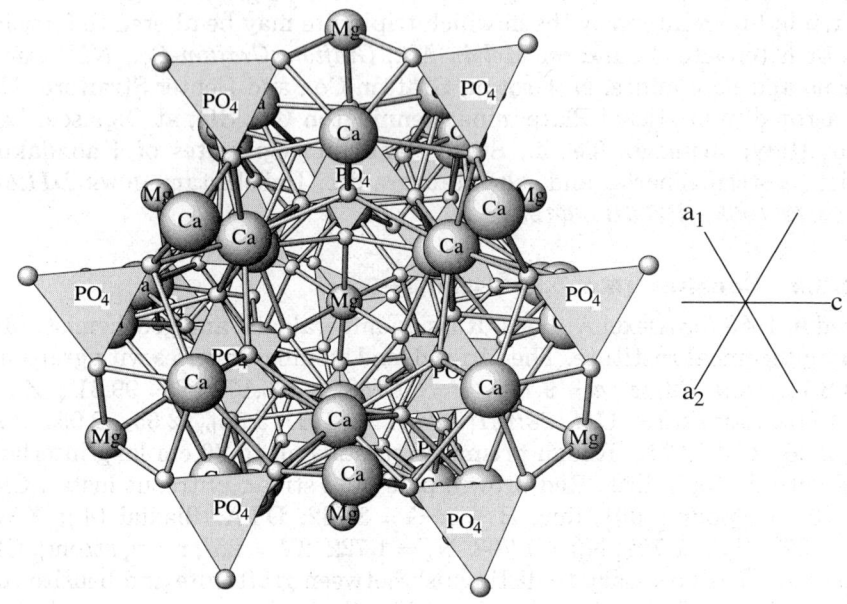

WHITLOCKITE: Ca$_9$(PO$_4$)$_6$Mg(PO$_3$OH)
The structure is composed of two chains running parallel to the c axis which contain PO$_4$, (PO$_3$OH) groups, two different Ca sites and a (Mg,Fe) site. Ca2 on the A chain is coordinated through 6 oxygens from six different PO$_4$ groups on the six B chains that surround it. The A chain also contains randomly distributed octahedrally coordinated (Mg,Fe) ions, a rotated PO$_4$ group and half as many Ca ions (the vacant cation site is assigned as H$_2$) as the B chains. Ca1 on the B chain has eightfold coordination while Ca2 has either eight or nine-fold coordination relative to the rotation of the PO$_3$OH groups. This c axis projection illustrates the hexagonal/rhombohedral character of the structure.

component of calculus, the calcium phosphate deposit in human–animal oral cavities known as dental plaque. HCWS *DII:684, AM 45:645(1960)*.

38.3.4.2 Strontiowhitlockite Sr$_9$Mg(PO$_3$OH)(PO$_4$)$_6$

Named in 1991 after the composition and the similarity to whitlockite (38.3.4.1). HEX-R $R3c$. $a = 10.644$, $c = 39.54$, $Z = 6$, $D = 3.60$. *CM 29:87(1991)*. PD: 5.33_2 3.29_4 3.00_{10} 2.66_8 2.07_3 1.991_3 1.940_3 1.783_4. Rosettes of rounded tabular {0001} crystals up to 8 mm and in pipelike aggregates (2 mm× 20 nm) with the stem enclosing another fibrous phosphate species; flattened on {0001}. White, white streak, dull luster, translucent. Cleavage or parting {0001}. G = 3.64. IR. Uniaxial (−); N$_O$ = 1.601, N$_E$ = 1.598 (589 nm). Analysis shows Ca, Ba, Mg, Mn, and Fe. Found in cavities associated with dolomite, pyrite, strontian, collinsite, and another strontium magnesium phosphate in a carbonatite vein in pyroxenites in *Kovdor deposit, Kola Penin., Russia*. HCWS *AM 76:2024(1991)*.

38.3.4.3 Merrillite-(Ca) $(Ca,\square)_{19}Mg^{2+}{}_2(PO_4)_{14}$
38.3.4.4 Merrillite-(Na) $Ca_{18}Na_2Mg_2^{2+}(PO_4)_{14}$
38.3.4.5 Merrillite-(Y) $Ca_{16}Y_2Mg_2^{2+}(PO_4)_{14}$

The name *merrillite* proposed in 1917 to honor George Perkins Merrill (1854–1929), meteorite specialist at U.S. National Museum, Smithsonian Institution, Washington, DC, was applied in 1977 to the anhydrous end members of calcium phosphates in extraterrestrial specimens to distinguish them from the H-containing, terrestrial whitlockite (38.3.4.1). These minerals not yet approved by IMA. HEX-R $R3c$. $a = 10.362$, $c = 37.106$, $Z = 3$ (Angra dos Reis achondrite). *EPSL 35:347(1977)*. Structure: *JSSC 10:232(1974)*, similar to β-$Ca_3(PO_4)_2$. SCI 137:425(1962): 5.18_6 3.43_6 3.19_6 2.85_{10} 2.58_6 1.92_5 1.71_5 1.54_4. Uniaxial (−); $N_O = 1.623$, $N_E = 1.620$. Lunar merrillite-(Ca) contains Na, REE, and a vacancy instead of H_2; additional compositional variations detected in other extraterrestrial samples. *LS VI:646(1975)*. Reported as very small amounts of anhedral grains in the following meteorites: Kimble Co., Texas, Estacao, Allegan, Bjurbole, Homestead, New Concord, Shergotta, Waconda. Also Earth's moon, lunar sample 12032 and 12036. HCWS *DII:797*, *MR 2:277(1971)*.

38.3.5.1 Stanfieldite $Ca_4(Mg,Fe^{2+})_5(PO_4)_6$

Named in 1967 for Stanley Field (1875–1964), former chairman of the board, Field Museum of Natural History, Chicago. MON Pc or $P2/c$. $a = 17.16$, $b = 10.00$, $c = 22.88$, $\beta = 100.2°$, $Z = 8$, $D = 3.15$. *SCI 158:910(1967)*. *20-223*: 8.31_5 6.01_5 5.01_3 3.85_6 3.75_8 2.82_{10} 2.51_8 1.870_4. See also *11-231*(syn). Irregular grains up to 1 mm in diameter. Reddish to amber, clear, and transparent. No cleavage. $H = 4\frac{1}{2}$. $G = 3.15$. Biaxial (+); $N_x = 1.631$, $N_y = 1.622$, $N_z = 1.619$; $2V = 50°$ (Estherville). A meteorite mineral. Mg/Fe^{2+} varies from 1.5 (Estherville) to 15 in the pallasites. Insoluble in H_2O. Found as grains along walls of cracks in the *Estherville mesosiderite meteorite*; also as thin veinlets, less than 1 mm in width, cutting the olivine in other pallasite meteorites from Santa Rosalia, Albin, Finnmarken, Imilac, Mount Verona, and Newport associated with merrillite, orthopyroxenes, and tridymite. HCWS *AM 53:508(1968)*.

38.3.6.1 Hurlbutite $CaBe_2(PO_4)_2$

Named in 1952 for Cornelius S. Hurlbut, Jr. (b.1906), professor of mineralogy, Harvard University, Cambridge, MA. MON $P2_1/c$. $a = 8.299$, $b = 8.782$, $c = 7.798$, $\beta = 90.5°$, $Z = 4$, $D = 2.899$. Structure: *AM 59:1267(1974)*. Tetrahedral PO_4 groups alternate with tetrahedral BeO_4 groups linked at the vertices in a framework with Ca in seven- and nine-fold coordination in the larger openings created. *6-213*: 3.67_{10} 3.50_7 3.28_6 3.03_9 2.78_9 1.50_7 2.40_6 2.21_9. See also *34-1441*. Short stout prismatic crystals 4–25 mm with striated faces; {001}, {110} predominate, truncated by {201}, {101}. Pale yellow, green, colorless, to gem white; white streak; vitreous to greasy luster, transparent to translucent. No cleavage, conchoidal fracture, brittle. $H = 6$. $G = 2.877$. Nonfluorescent. Biaxial (−); XYZ= bca; $N_x = 1.595$, $N_y = 1.601$, $N_z = 1.604$;

$2V = 70°$; $r > v$, weak. Slowly soluble in acids. Found at Black Mt. pegmatite, Rumford, Oxford Co., ME; as a replacement of beryl at Beuregard pegmatite, Gilsum, Cheshire Co.; G.E. Smith mine, Newport, Sullivan Co., NH; Tip Top pegmatite, Custer, Custer Co., SD; Butte, MT; Otov, Czech Republic; Viitaniemi, Eräjärvi, Finland; Grand Slam, Zimbabwe. HCWS *AM 37:931(1952)*.

38.3.7.1 Stranskiite $CuZn_2(AsO_4)_2$

Named in 1960 for Iwan N. Stranski (b.1897), physicist and chemist from Berlin, Germany. TRIC $P\bar{1}$. $a = 5.073$, $b = 6.669$, $c = 5.267$, $\alpha = 109.85°$, $\beta = 112.14°$, $\gamma = 86.88°$, $Z = 1$, $D = 5.1$. *AM 63:213(1978)*. Structure: *TMPM 26:167(1979)*. Isostructural with mcbirneyite (38.3.10.1). *29-1422:* 6.25_2 4.69_2 4.15_3 3.91_2 3.13_{10} 2.79_8 2.60_3 2.51_6. Radiating aggregates of tablets in subparallel arrays on chalcocite and as anhedral inclusions up to centimeter size in massive tennantite; no twinning. Blue, translucent. Cleavage {010}, perfect; {100}, good; also {001}, {$\bar{1}$01}. $H = 4$. $G = 5.23$. Nonfluorescent. Biaxial (−); $N_x = 1.795$, $N_y = 1.842$, $N_z = 1.874$; $2V = 80°$. Synthesized. The ratio of Zn/Cu in natural crystals can be as low as 1.3. Found in lower oxidation zone, 950 m below the surface at the *Tsumeb mine, Tsumeb, Namibia*, intergrown with schultenite in massive tennantite and associated with adamite, ludlockite, claraite, and tsumcorite. HCWS *NW 47:376(1960); BM: 236(1973), 660(1975)*.

38.3.8.1 Keyite $(Cu,Zn,Cd)_3(AsO_4)_2$

Named in 1977 for Charles Locke Key, mineral dealer, Canton, CT, who first called attention to the mineral. MON $I2$, $I2/m$, or Im. $a = 11.65$, $b = 12.68$, $c = 6.87$, $\beta = 98.95°$, $Z = 6$, $D = 3.495$. *MR 8:87(1977)*. *29-255:* 6.41_7 3.29_9 3.15_6 2.88_9 2.80_{10} 2.32_6 1.700_6 1.644_7. Crystals 0.25 mm × 0.1 mm × 0.04 mm, prismatic [001] to tabular {010} showing {010}, {210}, {011}, {$\bar{2}$01}. Deep sky blue, pale blue streak, translucent. Cleavage {001}, good. $H = 3\frac{1}{2}$–4. Biaxial; $Y = b$; $X \wedge c = 10\frac{1}{2}°$; $N_x = 1.80$, $N_z = 1.87$; strongly pleochroic; absorption $Z > Y > X$. X, pale blue; Y, greenish blue; Z, deep blue. Readily dissolved in concentrated acids. Found in cavities in tennantite ore at *Tsumeb mine, Tsumeb, Namibia*, associated with schultenite and zincian olivenite. HCWS *AM 62:1259(1977)*.

38.3.9.1 Lammerite $Cu_3(AsO_4)_2$

Named in 1981 for Franz Lammer, chemist of Leoben, Austria, who collected the mineral. MON $P2_1$. $a = 5.079$, $b = 11.611$, $c = 5.394$, $\beta = 111.72°$, $Z = 2$, $D = 5.264$. Structure: *AM 71:206(1986)*. *39-385:* 4.06_3 3.80_3 3.07_5 3.00_5 2.89_{10} 2.62_4 2.59_4 2.52_5. See also *35-497*. Tabular {100}, intergrown radial aggregates. Dark green, adamantine luster, transparent. Cleavage {010}, perfect; {100}, good; {001}, trace. $H = 3\frac{1}{2}$–4. $G = 5.18$. Biaxial (+); α parallel to b; $Z \wedge c = 40°$; $N_x = 1.89$, $N_y = 1.90$, $N_z = 1.95$; strongly pleochroic. X, pale blue; Y, sky blue; Z, pale blue-green. Associated with olivenite at *Sica Siva, Laurani, Meta*

Negra, La Paz, Bolivia, also at Tsumeb, Namibia; exhalations of Tolbachinski volcano, Kamchatka, Russia. HCWS *TMPM 28:157(1981)*, *RZ 4B:247(1985)*.

38.3.10.1 Mcbirneyite $Cu_3(VO_4)_2$

Named in 1984 for Alexander Robert McBirney (b.1924), volcanologist, University of Oregon. Isostructural with stranskiite (38.3.7.1). TRIC $P\bar{1}$. $a = 5.342$, $b = 6.510$, $c = 5.180$, $\alpha = 88.61°$, $\beta = 68.11°$, $\gamma = 69.22°$, $Z = 1$, $D = 4.50$. *JVGR 33:183(1984)*. PD: 4.25_6 4.01_8 3.12_{10} 2.82_{10} 2.64_8 2.57_6 2.43_8 2.31_6. See also *16-419*. Anhedral crystals all less than 200 µm. Dark gray, black; metallic luster; opaque. $R = 18.5$ (470 nm), 17.5 (546 nm), 18.7 (589 nm), 20.6 (650 nm). Found as a sublimate from a fumerole in summit crater at *Izalco volcano, El Salvador*, in the outer sulfate zone ($T = 100–200°$) associated with fingerite, thenardite, euchlorine, and other vanadates. HCWS *AM 73:1495(1988)*.

38.4.1.1 Heterosite $Fe^{3+}(PO_4)$

Named in 1826 from the Greek for *another* because it was the second Mn-containing mineral from the same locality. See purpurite (38.4.1.2), the Mn-rich member of the series. ORTH *Pbnm*. $a = 4.786$, $b = 9.823$, $c = 5.824$, $Z = 4$, $D = 3.672$. *34-134*: 4.94_6 4.31_9 3.46_8 2.96_3 2.92_8 2.45_{10} 2.41_4 1.731_3. Massive. Deep rose to reddish purple; paler tints in streak, usually altered appearing dull or earthy in luster, dark brown to brown-black in color; washing in dilute acid restores color, but note solubility; subtranslucent to opaque. Cleavage {100}, good; {010}, imperfect; surfaces often curved or crinkled; uneven fracture; brittle. $H = 4–4\frac{1}{2}$. $G = 3.2–3.4$. Biaxial (+) (Limoge); $X = a$; $N_x = 1.86$, $N_y = 1.89$, $N_z = 1.91$; 2V large; strongly pleochroic; absorption $Z = Y > X$ or $Z > Y \gg X$; strong dispersion may show anomalous green interference colors. X, greenish gray, gray to rose-red; Y,Z, deep blood-red to purplish red. Easily soluble in HCl. Fe/Mn determines identification as heterosite relative to purpurite. Heterosite is the end member in the alteration of a phosphate mineral series beginning with triphylite (38.1.1.1). Alters to pitchy material, a mixture of the oxyhydroxides of Fe^{3+} and Mn^{3+}. Found at Newry, Rumford, and Peru, Oxford Co., ME; Palermo No. 1 pegmatite, Groton, Grafton Co., NH; Huntington (Norwich), MA; Branchville, Fairfield Co., CT; Bull Moose, Pleasant Valley, Tip Top, White Elephant pegmatites, Custer Co.; Gap Lode, Hill City; Dyke lode, Pennington Co., Keystone, SD; *Huréaux, St. Sylvestre, Haute-Vienne, France*; Hagendorf Süd and Hühnerkobel pegmatite, Rabenstein, Bavaria, Germany; Varuträsk, Sweden; Telírio mine, Lavro Do Ênio, Galiléia, MG, Brazil; Angarf Süd, Tazenakht Plain, Morocco; Becota, Alto Ligonha, Mozambique; Okatjimukuju Farm pegmatite, Karibib; Erongo, Namibia; Bakennoje, Kazakhstan; Korgel, Laghman, Afghanistan; Euriowie Range, NSW, Australia. HCWS *DII:675*, *AM 57:45(1972)*, *KT 31:1192(1986)*.

38.4.1.2 Purpurite $Mn^{3+}PO_4$

Named in 1905 from the Latin *purpura*, in allusion to the color. The Mn-rich member of the series with heterosite (38.4.1.1). ORTH *Pbnm*. $a = 4.777$,

$b = 9.766$, $c = 5.824$, $Z = 4$, $D = 3.69$. *37-487*: 4.88_7 4.29_{10} 3.74_3 3.46_7 2.95_4 2.91_6 2.44_7 2.41_3. Physical properties similar to heterosite trending toward the darker and black shades. Biaxial (+); X=a; $N_x = 1.85$, $N_y = 1.86$, $N_z = 1.92$ (Peru, ME); 2V moderate, 38°; strongly pleochroic as heterosite. Soluble in HCl. The phase identified when Mn/Fe > 1. Secondary mineral in sequence with lithiophilite. Noted especially at Faires mine, Kings Mt., Gaston Co., NC; Custer district, Pennington Co., SD; Pala and Rincon, San Diego Co., CA; Feiteira, Portugal; Chanteloube, France; Korgal, Laghman, Afghanistan; Wodgina, WA, Australia. HCWS *DII:675*, *KT 31:1192(1986)*.

38.4.2.1 Berlinite AlPO$_4$

Named in 1868 for N. J. Berlin (b.1812), pharmacologist of the University of Lund, Sweden. HEX-R $P3_121$. $a = 4.941$, $c = 10.960$, $Z = 3$, $D = 2.62$. Structure: *ZK 192:119(1990)*. *10-423*: 4.28_2 3.37_{10} 2.48_1 2.31_1 1.835_2 1.552_1 1.393_1 1.389_1. Artificial crystals similar to quartz with $\{10\bar{1}1\}$, $\{01\bar{1}1\}$, and $\{10\bar{1}0\}$; natural is massive, granular. Pale rose, colorless to grayish; vitreous luster; transparent to translucent. No cleavage, conchoidal fracture. H $\approx 6\frac{1}{2}$. $G_{nat} = 2.64$, $G_{syn} = 2.56$. Uniaxial (+); $N_{O\,syn} = 1.524$, $N_{E\,syn} = 1.530$. Soluble in alkali, insoluble in cold acids. Isostructural with quartz, and AlAsO$_4$, tridymite-like and cristobalite-like polymorphs obtained on heating. Synthetic *MR 3:60(1972)*, *SPC 31:712(1986)*. Found at Inspiration mine, Gila Co., AZ; with augelite, attacolite, and other phosphates at *Vestanå mine, Nästum, Kristianstad, Sweden*. HCWS *DII:696*, *ZK 123:161(1966)*, *NBSC 539(10):3(1960)*, *MR 3:60(1972)*, *SPC 31:712(1986)*.

MONAZITE GROUP

The monazite group consists of phosphates and arsenates corresponding to the general formula

$$ATO_4 \quad \text{or a multiple thereof}$$

where

A = Ce^{3+}, La, Nd, Th^{4+}, Ca, Bi, REE
T = P, As

and crystallizing in the monoclinic space group $P2_1/n$.

The structure is a three dimensional framework with the cations linked through the oxygens of the tetrahedral TO$_4$ groups into a tight compact structure. Compare the tetragonal structures of xenotime (38.4.9.1) and zircon.

The monazite group is but one part of a ternary system REEPO$_4$-CaThPO$_4$-ThSiO$_4$ whose minerals have indentical structures and can be thought of as a series of coupled solid solutions. For example $^{4+}$ ions, such as Th^{4+}, may substitute for M^{3+}, i.e. Ce^{3+}, and the charge will be balanced by anionic substitution, such as SiO$_4{}^{4-}$ for PO$_4{}^{3-}$. Naming the different composition species in the monazite group has followed the suggestions outlined in *AM51:152 (1966)*: the dominant element is appended to an end member

Monazite Group

thereby minimizing the number of mineral names, i.e. monazite-(Ce), monanzite-(Nd). A member of the group without analysis is usually referred to as 'monazite'. Several members of this group often form intergrowths on a submicron scale *MM43:1031 (1980)*.

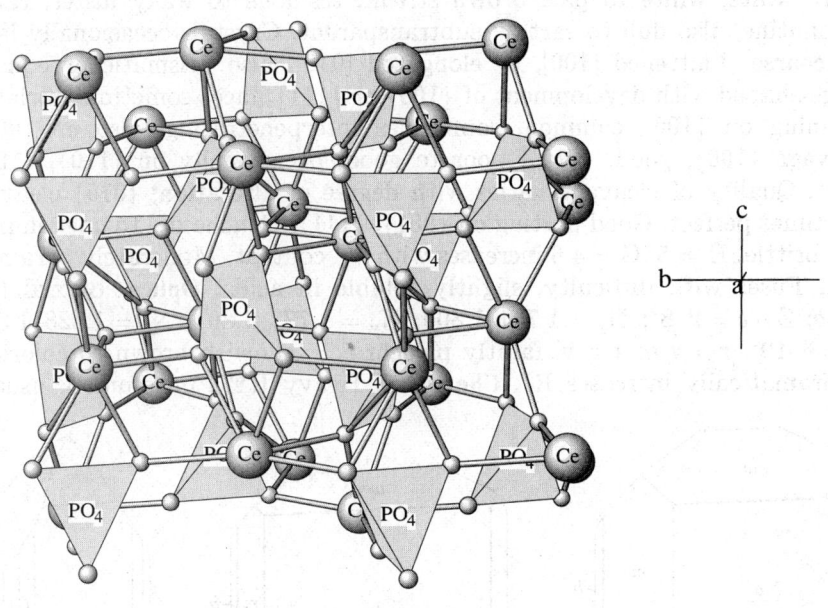

MONAZITE: $CePO_4$

The three dimensional framework structure viewed down the *a* axis of monazite–(Ce) shows that Ce coordinates with nine oxygens CeO_9 from PO_4 groups with no channelways for H_2 or other ions.

MONAZITE GROUP

Mineral	Formula	Space Group	a	b	c	β	
Monazite-(Ce)	$CePO_4$	$P2_1/n$	6.761	6.966	6.478	103.47	38.4.3.1
Monazite-(La)	$LaPO_4$	$P2_1/n$	6.836	7.070	6.503	103.23	38.4.3.2
Cheralite	$(Ca,Th,Ce)PO_4$	$P2_1/n$	6.751	6.962	6.468	103.88	38.4.3.3
Brabantite	$CaTh(PO_4)_2$	$P2_1/n$	6.726	6.933	6.447	103.89	38.4.3.4
Monazite-(Nd)	$NdPO_4$	$P2_1/n$	6.745	6.964	6.435	103.65	38.4.3.5
Gasparite-(Ce)	$CeAsO_4$	$P2_1/n$	6.937	7.137	6.738	104.69	38.4.3.6
Rooseveltite	$BiAsO_4$	$P2_1/n$	6.879	7.159	6.732	104.84	38.4.4.1

Note: Members of this group with identical crystal structures and similar compositions are found together as intergrowths on a submicron scale.

38.4.3.1 Monazite-(Ce) CePO₄

Named by A. Breithaupt in 1829 from the Greek *to be solitary*, in allusion to its rare occurrence in first-known localities. Monazite group. MON $P2_1/n$. $a = 6.761$, $b = 6.966$, $c = 6.45$, $\beta = 103.47°$, $Z = 4$, $D = 5.06$. Structure: *MZ 10:37(1988)*. *29-403:* 3.29_4 3.08_8 2.86_{10} 2.44_3 2.15_4 1.963_5 1.738_4. See also *32-199*(syn). **Physical properties:** Yellowish or reddish brown to brown, greenish, nearly white; white to pale brown streak; resinous to waxy luster, rarely adamantine, also dull to earthy; subtransparent. Crystals occasionally large and coarse. Flattened [100], or elongated [010], also prismatic or equant, wedge-shaped with development of {100} or {111}, faces sometimes striated. Twinning on {100} common, sometimes interpenetrating, also on {001}. Cleavage {100}, good; {010}, poor to good; occasionally on {110}, {101}, {011}. Quality of cleavage varies with degree of alteration; {010} cleavage sometimes perfect. Good parting on {001}, {111}. Conchoidal to uneven fracture, brittle. H = 5. G = 4.6 increases with Th content. Moderately paramagnetic. Fuses with difficulty, slightly soluble in acids. **Optics:** Biaxial (+); X = b; $Z \wedge c = 2$–$8°$; $N_x = 1.774$–1.800, $N_y = 1.777$–1.801, $N_z = 1.828$–1.849, $2V = 6$–$19°$, r > v or r < v, faintly pleochroic, yellowish brown to colorless. Th dramatically increases RI. **Chemistry:** Heavy REE uncommon, usually

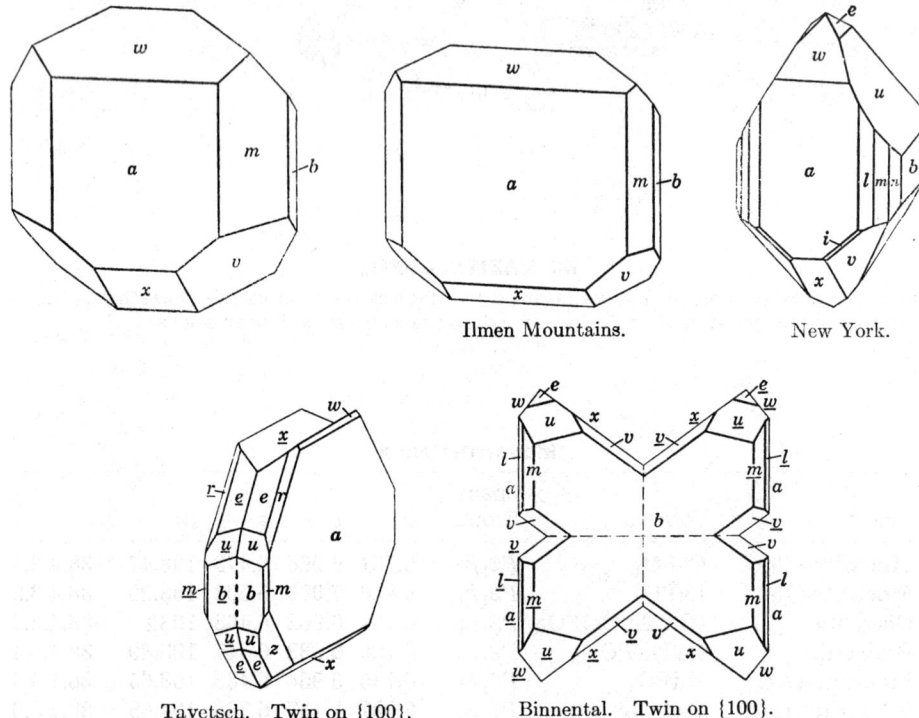

Ilmen Mountains.   New York.

Tavetsch. Twin on {100}.   Binnental. Twin on {100}.

Monazite CePO₄

Forms: a={100}, b={010}, l={210}, m={110}, n={120}, u={021}, w={101}, x={1̄01}, r={111}, v1̄={11}, i={2̄11}, underlined letters refer to second crystal (twin).

38.4.3.1 Monazite-(Ce)

<0.3 wt % to over 2 wt %; ThO_2 may be present to 12% grading to cheralite, Th-free material rare. U, Ca, Fe^{2+}, Fe^{3+}, Mg, and He may also be present. Si is usually low, occasionally to 3 wt %, but a series can extend to huttonite. *MM 43:1031(1980)*. Slowly decomposed in acids. Has been synthesized. Frequently used in radiometric dating. **Occurrence**: Common as an accessory mineral in igneous rocks, especially granite, and gneisses. Monazite may be found in detrital sands in sufficient quantities to be a commercial source of REE. Large crystals in granite pegmatites may be with allanite-(Ce), zircon, xenotime-(Y), gadolinite, samarskite-(Y), fergusonite, magnetite, apatite, columbite, ilmenite, or quartz; also with fluorite and bastnäsite-(Ce). **Localities**: Found at Topsham, Sagadahoc Co., ME; Albany, NH; Yantic Falls, Norwich, CT; Manhattan schist, NYC; with sillimanite near Yorktown Heights, Westchester Co., NY; Boothwyn, Delaware Co., PA; with microlite at Amelia(!), VA; in sands of Brindletown Creek, Burke Co.; Polk Co.; Lincoln Co.; Silver Creek, McDowell Co.; Ellenboro, Rutherford Co.; Hiddenite, Alexander Co.; Double Shoals and Brooks Farm, Shelby, Cleveland Co.; Mars Hill, Madison Co., NC; Cowpens, Spartanburg Co.; Gaffney, Cherokee Co.; Greenville, Greenville Co., SC; beach sands in FL; sands near Centerville, Boise Co., ID; Pikes Peak batholith; pegmatites, S. Platte district, Jefferson Co., CO; Big Horn placer, WY; Elk Mt., San Miguel Co.; Harding mine(!), Taos Co.; Petaca district(!), Rio Arriba Co., NM; Snowdrift claim, Lemhi Co., ID; Southern Pacific quarry, Nuevo, Riverside Co.; Mountain Pass(!), San Bernardino Co.; Himalaya mine(!), San Diego Co., CA; Villeneuve, Ottawa Co., QUE; Parry Sound, ONT, Canada; Nordre Strømfjord, Greenland; Prince of Wales Quarry, Tintagel, Cornwall, England; granitic pegmatites at Arendal, Dillingo, Råde, Østfold, Tvedstrand, Risör in Aust-Agder, Notero on Kristiana Fiord; Hitterö Is.; Dalane No. 2, Rostadheia(!), Eptevan No. 4, Mølland No. 3, and Tveit Farm pegmatites(!), Iveland, Norway; Bjertnes, Krødern; Lilla Holma near Stromstad; Kararfvet near Falun, Kopparberg, Sweden; Sordavala and Impilaks pegmatites, Lake Ladoga, Finland; St. Christophe, near Bourg d'Oisans, Isère, France; in alpine veins with anatase, ilmenite, rutile, or titanite at Piz Blas, Val Nalps; Val Tavetsch and Wälschen Ofen, Binnatal, Valais; Val Cornera, Grisons, Switzerland; in sanidine bombs, Laacher See, Eifel district, Germany; Rauris, Salzburg, Austria; Pisek and Schüttenhofen, Bohemia, Czech Republic; Alakurtti, Chupa, Karelia; Yaurijoki(!), Murmansk Oblast; gold sands, Sanarka River, southern Ural Mts., near Miask(!); *Zlatoust, Ilmenskie Mts., Ural Mts., Russia*; Jaguaraçu pegmatite, Jaguaraçu; Buenópolis (twins!), MG; ES, RJ, BA; Santa Luzia, Paraíba, Brazil; Llallagua, Bolivia; Rosetta, Egypt; placer sands derived from Younger Granites, Nigeria; Van Rhynsdorp, Namaqualand, South Africa; pegmatites at Ampangabé(!); Ambatofotsikely(!); Madiaomby North(!); Mt. Volhambohitra(!), Ankazobe district; black sands, Manajary basin, Madagascar; sands of Embabaan district, Swaziland; Kerala (Travancore)(!), India; west coast of Sri Lanka; tin deposits of Banka and Billiton, Malaysia; Antro Creek, SA; with wolframite at Mt. Bischoff, TAS; Moolyella district, WA, Australia; Ishikawa district, Fukushima Pref.; Yamanoo, Ibar-

agi Pref.; Naegi district, Gifu Pref.; Takehara, Mie Pref.; Kotoge and Daisen, Fukuoka Pref., Japan. HCWS $DII:691$.

38.4.3.2 Monazite-(La) LaPO$_4$

Systematically named by Levinson in 1966. Monazite group, as described in AM $51:152(1966)$. MON $P2_1/n$. $a = 6.836$, $b = 7.070$, $c = 6.503$, $\beta = 103.23°$, $Z = 4$, $D = 5.08$. $35\text{-}731$(syn): 3.53_2 3.31_9 3.12_{10} 3.01_2 2.88_4 2.21_2 2.15_4 1.981_2. Frequently in granular masses, light brown, tan, to orange pink; twinned {100} as well as obliquely, resinous to adamantine luster. $H = 5$–5.5. Biaxial (+); $N_x = 1.785$, $N_y = 1.787$, $N_z = 1.840$; faintly pleochroic. Found at Gaya district, Bihar; Ratnapur, India; with uraninite at Shinkolobwe, Shaba, Zaire; Llallagua, Potosi, Bolivia. HCWS $DII:691$.

38.4.3.3 Cheralite (Ca,Th,Ce)PO$_4$

Named in 1953 for the ancient kingdom Chera (Kerala) in southwestern India. Monazite group. MON $P2_1/n$. $a = 6.751$, $b = 6.962$, $c = 6.468$, $\beta = 103.88°$, $Z = 4$, $D = 5.39$. MM $43:885(1980)$. $33\text{-}1095$: 4.66_2 4.17_3 3.48_3 3.28_6 3.07_{10} 2.86_7 2.18_2 1.731_2. Masses of rough crystals up to 2 in. across. Green, translucent. Cleavage {010}, distinct; {100}, difficult; parting {001}, poor. $H = 5$. $G = 5.28$. Biaxial (+); X=b; $Z \wedge c = 7°$; $N_x = 1.779$, $N_y = 1.789$, $N_z = 1.816$; $2V = 18°$; faintly pleochroic. X,Y, green; Z, yellowish green. Thrich member of the monazite group, also contains some Si. Found in kaolinized pegmatite at *Kultakughi, east of Trivandrum, Travancore, India*, associated with black tourmaline (schorl), chrysoberyl, dark zircon, and smoky quartz. HCWS AM $38:734(1953)$, $39:403(1954)$.

38.4.3.4 Brabantite CaTh(PO$_4$)$_2$

Named in 1980 for the locality. Monazite group. Part of a series with cheralite (38.4.3.3). Synonym: Lingaitukuang. Identical to brabanite, a REE-free cheralite. AM $66:878(1981)$. MON $P2_1/n$. $a = 6.726$, $b = 6.933$, $c = 6.447$, $\beta = 103.89°$, $Z = 2$, $D = 5.27$. $NJMM:247(1980)$. $31\text{-}311$: 4.14_3 3.26_7 3.06_{10} 2.84_{10} 2.42_3 2.16_3 1.945_4 1.853_4. See also $35\text{-}465$. Elongated crystals. Gray-brown, surface alteration to reddish brown. Cleavage {100}, {001}. $H = 5.5$. $G = 5.20$. Biaxial; $N_y = 1.73$, $Z \wedge X = 0.05$. Occurs admixed with brockite in zone with microcrystalline muscovite, thorite, and uraninite in a pegmatite on the *Brabanite farm, Karibib district, Namibia*; Xingjiang, China. HCWS AM $66:878(1981)$.

38.4.3.5 Monazite-(Nd) NdPO$_4$

Named in 1987 in allusion to the chemistry and relationship to monazite. Monazite group. MON $P2_1/n$. $a = 6.745$, $b = 6.964$, $c = 6.435$, $\beta = 103.65°$, $Z = 4$, $D = 5.43$. $SMPM$ $67:103(1987)$. $42\text{-}1363$: 4.15_3 3.28_6 3.08_{10} 2.85_5 2.59_3 2.43_4 2.18_3 1.954_3. Long prismatic crystals, translucent to milky but always bright rose in color. Biaxial (+); $N_x = 1.793$, $N_y = 1.795$, $N_z = 1.860$. Occurs with xenotime, gadolinite, bastnäsite, allanite, and monazite-(Ce) in fissures and druses in aplitic to pegmatitic veins that cut schistose gneiss on moraine of the northern slope of Pta. Glogstafel, Val Formassa, Italy; as fine-grained 1 to 3μm pore fillings and fissure fillings up to 15 μm in brindleyite in *Marmara bauxite deposit, Greece*. HCWS *AM 68:849(1983), 73:1495(1988)*.

38.4.3.6 Gasparite-(Ce) $CeAsO_4$

Named in 1987 for G. Gaspari. Monazite group. MON $P2_1/n$. $a = 6.937$, $b = 7.137$, $c = 6.738$, $\beta = 104.69°$, $Z = 4$, $D = 5.63$. *SMPM 67:103(1987)*. PD: 3.36_8 3.16_{10} 3.04_3 2.97_7 2.71_4 2.52_3 2.00_3 1.811_3. See also *15-772*(syn). Alteration product as rims, or complete pseudomorphs of synchsite crystal aggregates. Light brown-red, white streak. Conchoidal to uneven fracture. $H = 4\frac{1}{2}$. Biaxial (+); $X = b$; $Z \wedge c = 4°$; $N_x = 1.810$, $N_y = 1.825$, $N_z = 1.92$; $2V = 40°$. Occurs associated with rutile, anatase, magnetite, hematite, synchisite-(Ce), chernovite, and cafarsite in fissures in metasedimentary rocks at *Pizzo Cervandone, a mountain on the border of Italy and Switzerland*. HCWS *AM 73:1494(1988)*.

38.4.4.1 Rooseveltite $BiAsO_4$

Named in 1946 for Franklin D. Roosevelt (1882–1945), 32nd president of the United States of America. Monazite group. MON $P2_1/n$. $a = 6.879$, $b = 7.159$, $c = 6.732$, $\beta = 104.84°$, $Z = 4$, $D = 7.21$. *TMPM 17:65(1972)*. Structure: $AC(B)$ *38:1559(1982)*. *25-49*(syn): 4.82_4 4.31_2 3.33_4 3.15_{10} 3.02_2 2.98_4 2.96_7 2.70_3. Thin botryoidal crusts, fine-grained aggregates pseudomorphic after bismuthinite. Gray, adamantine luster. Brittle. $H = 4\frac{1}{2}$. $G = 6.86$. Biaxial; $N_{av} = 2.10$–2.30. Soluble in acids. Has been synthesized. Found at Odsbach bei Oberkirsch, Black Forest, Germany; in cassiterite veinlets in rhyolite and dacite lava flows at *Santiaguillo, Macha, Potosi, Bolivia*; at San Francisco de los Andes, Cerro Negro de la Aguadata, San Juan Prov., Argentina. HCWS *DII:697*.

38.4.5.1 Tetrarooseveltite β-$BiAsO_4$

Named in 1994 as the tetragonal analog of rooseveltite (38.4.4.1). TET $I4_1/a$. $a = 5.085$, $c = 11.69$, $Z = 4$, $D = 7.64$. *NJMM:179(1994)*. Isotypic with scheelite. PD: 4.66_1 3.07_{10} 2.55_1 2.23_1 1.797_1 1.715_1 1.581_1 1.551_2. Fine-grained powdery aggregates with crystals up to 3 mm. White to yellowish white, earthy. $H = 2\frac{1}{2}$. Nonfluorescent both LWUV and SVUV. Uniaxial (+); $N_{mean} = 2.20$ (Na). Synthesized. *AC 1:163(1948)*. Oxidation product found on the fluorite–barite gangue in the *Josef mine, Moldava deposit, Krune Hory Mts., northwestern Bohemia, Czech Republic*, and associated with other secondary minerals, such as bayldonite, malachite, mimetite, presingerite, zavaritskite, bismutite, rooseveltite, mixite, wulfenite, and quartz. HCWS

38.4.6.1 Pucherite $BiVO_4$

Named in 1871 for the locality. One of three polymorphs of $BiVO_4$. See clinobisvanite (38.4.4.2) and dreyerite (38.4.6.1). ORTH *Pnca*. $a = 5.328$, $b = 5.052$, $c = 12.003$, $Z = 4$, $D = 6.69$. *CMP 41:325(1973)*. Structure: *ZK 169:289(1984)*. *12-293*: 4.64_5 3.98_5 3.50_{10} 2.99_4 2.70_{10} 2.13_4 1.992_4 1.934_4. Small crystals, curved faces and marked composite structure, tabular on {001} also acicular; {111} striated parallel to {001}; massive, ocherous, or coatings of minute crystals. Dark brownish red to yellowish brown, yellow streak, vitreous to adamantine luster, opaque to transparent. Cleavage {001}, perfect; subconchoidal fracture; brittle. H = 4. G = 6.69. Biaxial (−); $N_x = 2.41$, $N_y = 2.50$, $N_z = 2.51$ (Li); 2V = 19°; r < v, extreme; yellow brown. Soluble in HCl forming a deep red solution with evolution of Cl, which on dilution becomes green and precipitates yellow basic chloride. Changes to monoclinic form above 500°. A secondary mineral associated with bismutite and beyerite. Found as ocherous alteration products of native bismuth in pegmatite at Stewart and Pala Chief mines near Pala, San Diego Co., CA; in oxidized portions of Bi–Ag–U–Cu vein in the *Pucher shaft of the Wolfgang mine, Schneeberg, Saxony, Germany*, and from the Arme Hilfe at Ullersreuth, Thuringia; in pegmatites in Madagascar with native bismuth at Mt. Bity and with bismuthinite at Ampangabe; at Sao Jose, Brejauba, MG, Brazil. HCWS *DII:1050*.

38.4.7.1 Clinobisvanite BiVO$_4$

Named in 1974 for the composition and the symmetry. Other polymorphs: pucherite (38.4.5.1) and dreyerite (38.4.6.1). MON $I2/a$. $a = 5.186$, $b = 11.708$, $c = 5.100$, $\beta = 90.43°$, Z = 4, D = 6.95. *MM 39:847(1974)*. *24-688*: 4.75_2 4.67_2 3.12_3 3.10_{10} 3.08_9 2.92_2 1.719_2 1.717_2. See also *14-688*(syn). Soft friable crusts, aggregates, reticulated plates to 0.1 mm showing multiple twinning. Yellow powder, orange aggregates, yellow streak, earthy, subvitreous luster. Cleavage {010}, perfect. Soft. $N_{calc} = 2.63$; strong dispersion. Known as synthetic material before mineral discovered and named. Found at Mutala pegmatite, Mozambique, and in several other localities such as Londonderry, Wodgina, Menzies, Westonia, and *Yinnietharp Station, Pyramid Hill, WA, Australia*, where the mineral is often intergrown with bismutite and with polymorph pucherite and on garnet in beryl-bearing spessartine pegmatites which contain bismutite, bimutoferrite, columbite, uvarovite, pyrite, leucophosphate, calcite, zircon, and triphylite. HCWS *NBSM 3:14(1964)*.

38.4.8.1 Dreyerite BiVO$_4$

Named in 1981 for Gerhard Dreyer, professor of mineralogy at Johannes Gutenberg University, Mainz, Germany. Xenotime group. Other polymorphs: clinobisvanite (38.4.4.2) and pucherite (38.4.5.1). TET $I4_1/amd$. $a = 7.303$, $c = 6.458$, Z = 4, D = 6.25. Structure: *NJMM:151(1981)*. Isostructural with zircon. *14-133*: 4.84_4 3.65_{10} 2.91_2 274_6 2.59_2 2.28_2 1.879_5 1.825_2. Plates parallel to (001), 20–50 µm thick to 0.5 mm in diameter. Brownish yellow to orange-yellow, yellow streak, adamantine luster. H = 2.5. Uniaxial (+); N > 2.0. O, bright yellow; E, brownish yellow. Mineral identical with synthetic TET BiVO$_4$. Found in rhyolitic ash tuff associated with quartz, barite, hematite, native silver, bismite, and carnotite at *Hirschorn, Rheinland–Platz, Germany*. HCWS *AM 67:622(1982)*.

38.4.9.1 Ximengite BiPO$_4$

Named in 1990 for the locality. HEX-R $P3_121$. $a = 6.986$, $c = 6.475$, $Z = 3$, $D = 5.534$. Structure: *ZK 117:371(1962)*. *PD:* 6.05$_7$ 4.42$_9$ 3.49$_9$ 3.02$_{10}$ 2.85$_6$. Veinlets and as irregular aggregates of grains < 0.001–0.1 mm. Colorless, white streak, vitreous to greasy luster. Brittle. H = 4. Yellowish fluorescence under electron beam. Nonmagnetic. IR, DTA, TGA. Uniaxial (+); N > 1.78. X-ray data similar to synthetic. Inverts to monoclinic form at ≈ 1000°, then to monazite-type form. Found as an alteration of bismuthinite containing waylandite, monazite with unnamed hydrated Bi phosphates and sulfates; irregularly distributed in cassiterite–tourmaline–quartz and K-feldspar rock in tin mining district of *Ximen, about 420 km SW Kunming, Yunnan Prov., China*. HCWS *MA:4610(1990)*, *AM 76:1436(1991)*.

38.4.10.1 Lithiophosphate Li$_3$PO$_4$

Named in 1957 for the composition. ORTH $Pmn2_1$. $a = 6.116$, $b = 5.234$, $c = 4.845$, $Z = 4$, $D = 2.48$. *NBSM 25(4):21(1966)*, Structure: *SPD 23:287(1978)*. *25-1030*(syn): 5.23$_3$ 3.97$_{10}$ 3.80$_{10}$ 3.55$_5$ 2.64$_7$ 2.62$_4$ 2.42$_5$ 2.32$_2$. Masses up to 5 × 9 cm. White to colorless to light rose. One excellent cleavage, one fair, about 50° apart. H = 4. G = 2.46. Luminesces turquoise-blue in X-ray beam, nonfluorescent in UV. Biaxial (+); $N_x = 1.550$, $N_y = 1.557$, $N_z = 1.567$, 2V = 69°; positive elongation. Slightly soluble in hot water, soluble in strong acids. Hydrothermal alteration product of montebrasite; weathers to mangan-apatite and davisonite. Found in Foote mine, Kings Mt., Cleveland Co., NC; Bernic Lake, MB, Canada; with montebracite in central zone of pegmatites, Kola Penin., Russia, and with microperthite, quartz, spodumene, beryl, tourmaline, pollucite, and epidolite. HCWS *DANS 112:124(1957)*, *AM 42:585(1957)*.

38.4.10.2 Olympite LiNa$_5$(PO$_4$)$_2$

Named to honor the Olympic Games at Moscow in 1980. ORTH $P2_12_12_1$. $a = 10.124$, $b = 14.794$, $c = 10.132$, $Z = 8$, $D = 2.71$. [*CR 39:35(1994)*] for Khibina holotype material; virtually identical to Lovozero sample. *DANS 320:499(1992)*. Structure: *SPC 37:772(1992)*. Replacement of Na by Li is limited by stability constraints; there may be a Li analog of olympite. *33-1272:* 4.18$_9$ 3.70$_5$ 3.58$_6$ 2.58$_{10}$ 2.53$_7$ 2.43$_7$ 2.39$_5$ 1.472$_7$. Oval grains up to 3 mm in diameter. Colorless, vitreous luster, translucent. Conchoidal fracture, brittle. H ≈ 4. G = 2.8. Biaxial (+); $N_x = N_y = 1.510$, $N_z = 1.512$; 2V = 46°. No chemical data presented in the structure references. Readily soluble in cold water, giving strongly alkaline solution. In air, alters rapidly to sodium carbonate and hydrated sodium phosphates. Occurs in alkalic pegmatites at the Lovozero massif and at *Mt. Rasvumchorr, Khibina massif, Kola Penin., Russia*, in association with sidorenkite, villiaumite, aegirine, and another sodium Mn silicate. HCWS *AM 66:438(1981)*, *ZVMO 189:476(1980)*.

38.4.10.3 Nalipoite NaLi$_2$PO$_4$

Named in 1991 for the composition: Na–Li–P–O. ORTH $Pmnb$. $a = 6.884$, $b = 9.976$, $c = 4.927$, $Z = 4$, $D = 2.612$. Structure: *CM 29:565(1991)*, *29:569(1991)*. PD: 4.04$_4$ 4.00$_4$ 3.51$_{10}$ 3.44$_7$ 2.83$_4$ 2.72$_3$ 2.49$_6$ 2.46$_6$. Anhedral to subhedral blocky grains to 2 mm and irregular grains to 0.2 mm. White, pale blue, to pale yellow; white streak; vitreous luster; transparent to translucent. Cleavage {100}, {010}, {001}, {110}, good, and possibly {101}. Uneven fracture, very brittle. $H = 4$. $G = 2.58$. Nonfluorescent. Biaxial $(-)$; XYZ= acb; $N_x = 1.533$, $N_y = 1.540$, $N_z = 1.541$; 2V = 49°. Readily soluble in 1:1 HNO$_3$, less so in 1:1 HCl, slowly in 1:1 H$_2$SO$_4$. X-ray data similar to synthetic. Found in the sodalite–analcime–microcline xenoliths at the *Poudrette quarry in nephaline syenite of Mt. St.-Hilaire, near Montreal, QUE, Canada*. HCWS *AM 77:449(1992)*.

XENOTIME GROUP

The xenotime group are REE-containing minerals conforming to the general formula

$$(REE)TO_4$$

where

REE = rare-earth element = Y, Ce, Bi
T = P, As, V, Si

and crystallize in the tetragonal space group $I4_1/amd$.

The structure is similar to monazite [ATO$_4$] and to zircon: rare-earth-element polyhedra are linked by TO$_4$. The three-dimensional framework structure is composed of two types of chains. One type, parallel to c, consists of alternating REE–O$_8$ and PO$_4$, and the other consists of REE–O$_8$ polyhedra along the a axis. All of the polyhedra are linked by shared edges. Octahedral voids form channels parallel to c.

XENOTIME GROUP

Mineral	Formula	Space group	a	c	
Xenotime-(Y)	YPO$_4$	$I4_1/amd$	6.878	6.036	38.4.11.1
Chernovite-(Y)	YAsO$_4$	$I4_1/amd$	7.039	6.272	38.4.11.2
Wakefieldite-(Y)	YVO$_4$	$I4/amd$	7.105	6.29	38.4.11.3
Wakefieldite-(Ce)	CeVO$_4$	$I4_1/amd$	7.354	6.488	38.4.11.4
Dreyerite	BiVO$_4$	$I4_1/amd$	7.303	6.458	38.4.8.1

*Compare clinobisvanite (38.4.7.1) and pucherite (3.8.4.6.11) other polymorphs of BiVO$_4$.
Note: Dreyerite is a member of the xenotime group. See clinobisvanite (38.4.7.1) and pucherite (38.4.6.1) for other polymorphs of BiVO$_4$.

38.4.9.1 Xenotime-(Y)

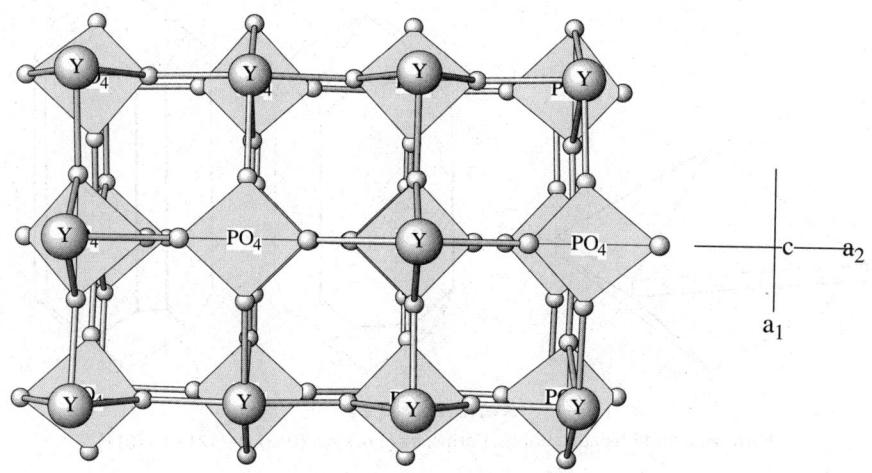

XENOTIME-Y YPO$_4$

An a$_2$ axis projection of the three dimensional framework structure of xenotime. Compare Zircon.

38.4.9.1 Xenotime-(Y)   YPO$_4$

Named in 1832 from the Greek *vain* and *honor*, recalling the fact that Y in it had been mistaken for a new element. Xenotime group. TET $I4_1/amd$. $a = 6.878$, $c = 6.036$, $\dot{Z} = 4$, $D = 4.277$. *MM 39:145(1973)*. Istrostructural with zircon ZrSiO$_4$. 11-254: 4.55$_3$ 3.45$_{10}$ 2.565$_5$ 2.44$_2$ 2.15$_3$ 1.824$_2$ 1.768$_5$ 1.725$_2$. See also *9-377*(syn). Short to long prismatic [001]; equant; pyramidal; radial aggregates of coarse crystals that resemble zircon; rosettes. Often found in parallel growth with zircon. Twinning {111} rare. Yellowish brown to reddish brown, flesh-red, grayish white, wine-yellow, pale yellow, greenish; pale brown, yellowish, or reddish streak; vitreous to resinous luster; transparent to opaque. Cleavage {100}, uneven to splintery fracture, brittle. H = 4–5. G = 4.4–5.1. Paramagnetic. Uniaxial (+); N$_O$ = 1.721, N$_E$ = 1.816 (Na); weakly dichroic. O, pink, yellow, yellow-brown; E, brownish yellow, grayish brown, or greenish. Very lightly attacked by acids. Occurs widely as a minor accessory mineral in both acidic and alkalic igneous rocks and in large crystals in the associated pegmatites; also in mica and quartz-rich gneisses and in the quartz veinlets that form the pegmatitic portions of these rocks. Commonly associated with monazite, zircon, ilmenite, rutile, anatase, magnetite, hematite, sillimanite, feldspars, and occasionally with fergusonite and columbite-tantalites. Found in the Manhattan schist, New York City, NY; as a detrital mineral in placer gold deposits of Polk, McDowell, Burke Co., NC; in similar locations in GA and AL; at Rib Mt., Marathon Co., WI; at Snowbird mine, Mineral Co., MO; with fluorcerite and bastnäsite in Cheyenne Canyon, Pikes Peak district, El Paso Co., CO; at Evans–Lou mine, QUE, Canada. Localities

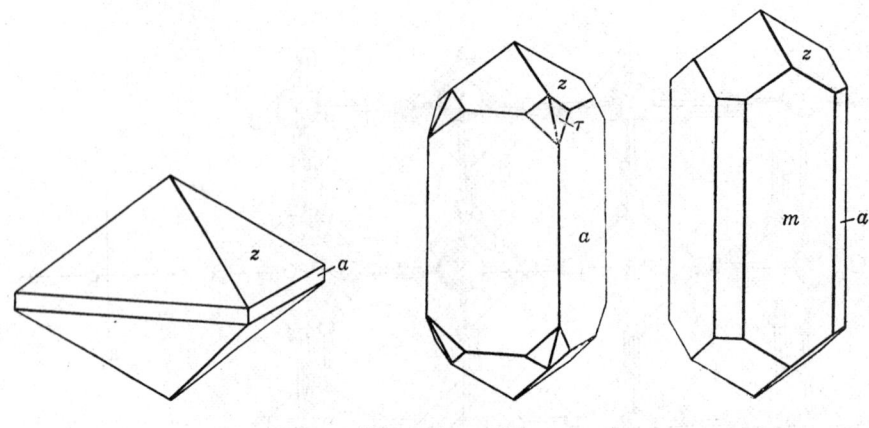

Xenotime -(Y)
Pyramidal and Prismatic habits. Forms: a={110}, m={010}, r={121}, z={011}

in Europe include the granitic pegmatites of southern Norway associated with gadolinite, polycrase–euxenite, allanite, zircon, yttrotantalite, and thorite, such as at Hitterö in West-Agder, Tvedestrand, Garta, and other places near Arendal, Aust-Agder, Raade near Moss, in Osfold, at Kragerö, and Telemark; at Ytterby NE of Stockholm, Sweden; at Koralpe, Packalpe, Salzburg, Kuppergrund, Lohringer q., Rauris, Austria; at Mt. Fibria SW of St. Gotthard, in the Binnenthal, Valais, and in the Tavetschtal, Graubünden, Switzerland; at Königshain near Görlitz and with monazite at Screiberhau, Silesia, Germany; at Bressanone, Le Chiuse, Val Masino, Italy; at Nizyunita Ohari, Marumori, Igu Co., Miyagi Pref., Japan; in the alluvial deposits of Kiracoravo, Madagascar; as an accessory mineral in granites of Cape Prov., South Africa; in Manbhum district, Bihar and Orissa Prov., India; in the river sands North Westland, New Zealand; widely distributed in the alluvial deposits at Dattas, and the Urubu pegmatite, MG, and in the diamond washings of BA, Brazil. HCWS *DII:688*.

38.4.11.2 Chernovite-(Y) YAsO$_4$

Named in 1967 for Aleksandr A. Chernov (1877–1963), Russian geologist and paleontologist, Academy of Science, Komi, Syktyvkar. Xenotime group. TET $I4_1/amd$. $a = 7.039$, $c = 6.272$, $Z = 4$, $D = 4.866$. *ZVMO 96:699(1967)*. *26-999:* 4.64_4 3.50_{10} 2.47_5 2.18_3 1.801_7 1.749_4 1.455_3 1.255_3. See also *13-429*(syn). Bipyramidal crystals up to 1 mm, aggregates elongated on c with inner zone green-yellow, outer bright yellow, also colorless; vitreous luster. Cleavage {100}, indistinct. $N_{av} = 1.753$ (Binnatal), 1.783 (Russian). Contains more La-group elements (Gd, Py, Ho, Er, Yb) than xenotime. Found at *Mjarta-syn-yu R., Telops-iz E, Ural Mts., Russia*, and in the Cervandom massif, and at Binntal, Switzerland. HCWS *AM 53:1777(1968)*.

38.4.11.3 Wakefieldite-(Y) YVO$_4$

Named in 1969 for the locality. Xenotime group. TET $I4_1/amd$. $a = 7.105$, $c = 6.29$, Z = 4, D = 4.25. AM 56:395(1971). 17-341(syn): 4.71_1 3.56_{10} 2.67_5 2.52_1 2.22_1 1.830_5 1.780_1 1.481_1. Pulverulent, tiny, <0.1 µm in length, crystallites have prism faces and are terminated with pyramids and basal faces reminiscent of zircon. Yellow, pale tan, translucent. H ≈ 5. Uniaxial (+); $N_{O\,syn} = 2.00$, $N_{E\,syn} = 2.14$. A secondary mineral filling cavities in a pegmatite along with quartz, hellandite, cenosite, and montmorillonite at the *Evans–Lou mine, St. Pierre near Wakefield, Portland Twp., Papineau, QUE, Canada*, associated with yttrian fluorapatite, thorogummite, or with fergusonite and xenotime. HCWS *NBSM 25:59(1967), NJMM:127(1995)*.

38.4.11.4 Wakefieldite-(Ce) CeVO$_4$

Named "kusuite" in 1977 for the locality, later changed to wakefieldite-(Ce) to coincide with previous mineral described and the dominant rare-earth element. Xenotime group. TET $I4_1/amd$. $a = 7.354$, $c = 6.488$, Z = 4, D = 4.80. BM 109:305(1986). Structure: BM 110:657(1987). 29-398: 4.89_5 3.68_{10} 2.77_9 2.59_4 2.29_3 1.891_7 1.833_2 1.638_4. See also 12-757(syn). Crystals showing {101} to 1 mm in size. Black, rust-brown streak. No cleavage. H = $4\frac{1}{2}$. G = 4.76. N > 2.00, yellow in transmitted light. Equal small amounts of Pb^{2+} and Pb^{4+} present but no other REE than Ce. Found in the oxidation zone of *Kusu Dept., 85 km SW of Kinshasa, Zaire*, associated with mottramite, chervetite, cuprite, dioptase, heyite, malachite, plancheite, and vanadinite. HCWS *BM 100:39(1977), AM 62:1058(1977)*.

38.4.12.1 Kosnarite KZr$_2$(PO$_4$)$_3$

Named in 1993 for Richard A. Kosnar, mineral dealer, of Black Hawk, CO. HEX-R $R\bar{3}c$. $a = 8.687$, $c = 23.877$, Z = 6, D = 3.206. AM 78:653(1993). PD: 6.41_5 4.68_5 4.33_{10} 3.81_9 3.17_4 2.93_9 2.50_5 1.903_5. See also 25-1206(syn), 35-756(syn). Pseudocubic rhombohedral crystals up to 0.9 mm with {102} dominant. Pale blue to blue-green, colorless; white streak; vitreous luster; transparent. Not twinned. Cleavage {102}, perfect; conchoidal fracture; brittle. H = $4\frac{1}{2}$. G = 3.194. Nonfluorescent. Uniaxial (+); $N_O = 1.656$, $N_E = 1.682$; nonpleochroic. Found in vugs with euhedral crystals of siderite, eosphorite, fluorapatite, moraesite, and quartz, part of the late-stage secondary phosphate assemblage from *Mount Mica pegmatite, Paris, and Black Mt. pegmatite, Rumford, Oxford Co., ME*; at the Wycheproof pegmatite, SA, Australia. HCWS *AM 78:653(1993)*.

38.5.1.1 Chursinite Hg^{1+}Hg^{2+}(AsO$_4$)$_2$

Named in 1984 for Ludmilla A. Chursina, Russian actress. MON $P2_1/n$. $a = 8.73$, $b = 5.08$, $c = 15.64$, $\beta = 128.4°$, Z = 4, D = 9.05. Structure: ZVMO 113:341(1984), AC(B) 29:1666(1973). 38-399: 3.05_{10} 2.89_8 2.84_7 2.51_5 2.46_5 2.19_5 2.14_5 1.911_5. Grains to 0.2 mm. Yellow-brown to orange-brown, adamantine luster, transparent. Cleavage parallel to elongation, brittle. Sections become covered with a blue-violet film when exposed to air. H = 3. G = 9.05. N > 2.0; anisotropic, dark brown to blue-green. R = 18.5–22.8 (460 nm), 16.1–20.4 (546 nm), 15.8–19.7 (540 nm), 14.5–18.5 (656 nm). Synthesized. Found in the oxidized portion of Sb–As–Hg ores of *Khaidarken Dept., Kyrgyzstan*, associated with calomel, eglestonite, terlinguaite, corderoite, montroydite, kuzovetsovite, shakhovite, poyarkovite, and native Hg. HCWS *AM 70:871(1985)*.

38.5.2.1 Lyonsite $Cu_3^{2+}Fe_4^{3+}(VO_4)_6$

Named in 1987 for John B. Lyons (b.1916), professor of mineralogy, Darmouth College, Hanover, NH. ORTH *Pmcn*. $a = 10.296$, $b = 17.207$, $c = 4.910$, Z = 2, D = 4.215. Structure: *AM 72:1000(1987)*. *40-1456*: 8.79_2 4.42_4 3.28_{10} 2.78_4 2.72_4 2.53_6 1.591_4 1.550_4. Euhedral, lathlike crystals up to 230 μm long. Black, dark gray streak, metallic luster, creamy white in reflected light. Cleavage {001}, good. On (010) R=17.5–23.1 (481 nm), 16.6–22.3 (547 nm), 14.7–20.7 (591 nm), 14.4–18.8 (644 nm). One of the several minerals sublimating from the Y fumerole of the *Izalco volcano, El Salvador*, associated with thenardite. Other vanadates: stoiberite, fingerite, mcbirneyite, ziesite, blossite, bannermanite, and shcherbinaite. HCWS

38.5.3.1 Howardevansite $NaCu^{2+}Fe_2^{3+}(VO_4)_3$

Named in 1988 for Howard T. Evans (b.1919), mineralogist and crystallographer, U.S. Geological Survey, Reston, VA. TRIC $P\bar{1}$. $a = 8.198$, $b = 9.773$, $c = 6.651$, $\alpha = 103.82°$, $\beta = 101.99°$, $\gamma = 106.74°$, Z = 2, D = 3.814. Structure: *AM 73:181(1988)*. *42-1333*: 3.27_6 3.17_{10} 3.09_{10} 2.68_6 2.60_6 2.05_5 1.659_7 1.433_6. Euhedral, tabular up to 80 μm in diameter, dominant forms (100), (010), (001), (1$\bar{1}$0), (101), ($\bar{1}$01), (0$\bar{1}$1). Black, red-brown streak, metallic luster. Brittle. Reflectance medium gray, moderately anisotropic. R = 15.5–20.1 (481 nm), 15.0–18.4 (547 nm), 14.1–17.2 (591 nm), 13.5–17.1 (644 nm). Found as one of the sublimating vanadates at the fumaroles of *Izalco volcano, El Salvador*; see also lyonsite (38.5.2.1). HCWS

38.5.4.1 Ludlockite $(Fe^{2+},Pb)As_2O_6$

Named in 1970 for F. Ludlow Smith III and C. Locke Key, mineral collectors and dealers in NJ and ME. TRIC $P\bar{1}$ or $P1$. $a = 10.11$, $b = 11.95$, $c = 9.86$, $\alpha = 113.9°$, $\beta = 99.7°$, $\gamma = 82.74°$, Z = 9, D = 4.35. *MR 8:91(1977)*. *29-774*: 10.90_6 8.81_{10} 4.74_6 4.47_6 3.33_7 3.16_7 2.94_9 2.86_7. Crystals elongated [100] flattened {0$\bar{1}$1} with {021}, no terminal faces. Red, light brown streak, subadamantine luster. Sectile, flexible. Cleavage {0$\bar{1}$1}, micaceous, combines with other easy {0kl} cleavages to give fibers. Lamellar twinning on {0$\bar{1}$1}. H = $1\frac{1}{2}$–2. G = 4.40. IR, Raman suggest bonding is similar to the As–O–As bonding in claudetite. Biaxial (+); Z near a; $N_x = 1.96$, $N_y = 2.055$, $N_z > 2.11$; optic axial plane perpendicular to {0$\bar{1}$1}, pleochroic. X, yellow; Y, deep yellow; Z, orange-yellow. Readily dissolves in 1:1 HCl, HNO_3. Found with zincian siderite in a cavity of ore from the "germanite section" at *Tsumeb, Namibia*. HCWS *AM 57:1003(1972)*.

38.5.5.1 Chervetite $Pb_2V_2^{5+}O_7$

Named in 1963 for J. Chervet (1904–1962), French mineralogist. MON $P2_1/a$. $a = 13.37$, $b = 7.16$, $c = 7.11$, $\beta = 106°$, $Z = 4$, $D = 6.38$. Structure: *BM 90:279(1967)*. *18-708:* 4.93_6 4.32_5 3.57_5 3.44_{10} 3.43_{10} 3.21_8 3.16_5 3.08_8. Crystals up to a few centimeters showing (100), (221), (2$\bar{2}$1), (001) always twinned (100), pseudomorphs after francevillite (40.2a.27.1). Cleavage {100}, {010}, doubtful. Colorless, gray, brown, white streak, adamantine luster. H = 3. G = 6.31. Biaxial (−); N = 2.2–2.6; 2V = 65–75°; bireflectance = 0.275; dispersion inclined weak. Found at St. Andreasberg, Harz, Germany; Vrancice, Bohemia, Czech Republic; *Mounana uranium mine, Franceville, Haut Ogooue, Gabon*; Kusu deposit, 85 km SW of Kinshasa, Zaire. HCWS *AM 48:1416(1963)*; *BM 86:117(1963)*, *88:126(1965)*.

38.5.5.2 Ziesite β-$Cu_2V_2^{5+}O_7$

Named in 1973 for Emanuel G. Zies (1884–1981), Geophysical Laboratory, Carnegie Institution, Washington, DC. Polymorphic with blossite (38.5.6.1). MON $C2/c$. $a = 7.689$, $b = 8.029$, $c = 10.106$, $\beta = 110.25°$, $Z = 4$, $D = 3.869$. Structure: *AM 65:1146(1980)*, *NJMM:41(1989)*. *26-569*(syn): 5.34_8 3.60_{10} 3.09_{10} 3.06_{10} 2.30_8 2.11_8 2.09_8 1.606_8. Anhedral equant crystals <200 μm in diameter. Black, reddish-brown streak (like hematite), metallic luster. No cleavage. $G_{syn} = 3.86$. Nonfluorescent. $N_{mean\ calc} = 2.055$. White in reflected light, approximates the reflectivity of acanthite or tetrahedrite. Nonpleochroic, moderately to strongly anisotropic. Grains often composed of several optically discontinuous regions. Found encrusting basaltic breccia fragments 1 m below the surface at the Y fumarole, one of five fumaroles at the *Izalco volcano, El Salvador*. HCWS *CR 177C:1101(1973)*.

38.5.6.1 Blossite α-$Cu_2V_2^{5+}O_7$

Named for F. Donald Bloss (b.1920), professor of mineralogy, Virginia Polytechnic Institute and State University, Blacksburg, VA. Polymorphous with ziesite (38.5.5.2). The natural analog of a previously described synthetic phase and the low-temperature modification of ziesite (38.5.5.2). ORTH *Fddd*. $a = 20.676$, $b = 8.392$, $c = 6.446$, $Z = 8$, $D = 4.051$. Structure: *AM 72:397(1987)*, *AC(B) 29:2737(1973)*. *26-566*(syn): 3.26_{10} 3.22_8 3.08_{10} 2.48_8 2.102_7 2.096_7 1.712_7 1.611_7. Equant, anhedral crystals up to 150 μm. Black, red-brown streak, metallic luster, opaque, white in reflected light. Weakly to moderately anisotropic, gray to creamy brown-gray; moderate bireflectance in shades of creamy white. R = 14.6–15.3 (481 nm), 15.4–16.6 (547 nm), 14.8–16.7 (591 nm), 14.5–15.7 (644 nm). $N_{mean} = 2.05$. Found as a fumarolic sublimate in the Y fumarole of *Izalco volcano, El Salvador*, with shcherbinaite, bannermanite, stoiberite, fingerite, ziesite, and mcbirneyite. HCWS *CR 277C:1101(1973)*.

38.5.7.1 Simferite $Li_{0.5}(Mg_{0.5}Fe^{3+}_{0.03}Mn^{3+}_{0.2})_2(PO_4)_3$

An inadequately described mineral. ORTH $Pbnm$ or $Pbn2_1$. $a = 4.747$, $b = 10.101$, $c = 5.899$, $D = 3.25$. Structure: *DANS 307:1119(1989)*. Aggregates up to 3 mm, individual crystals up to 0.1, commonly twinned. Dark red. $G = 3.22$. Biaxial; $N_x = 1.690$–1.704, $N_y = 1.702$–1.716, $N_z = 1.712$–1.726. Occurs at the contact of REE-bearing pegmatite and phlogopitized tremolitic rock in Russia. HCWS *AM 78:452(1993), SPD 34:669(1989)*.

38.5.9.1 Namibite $CuBi_2(VO_4)O_2$

Named in 1981 for the country in which it was found. MON $C2/m$, Cm, or $C2$. $a = 11.864$, $b = 3.696$, $c = 7.491$, $\beta = 109.70°$, $Z = 2$, $D = 6.76$. *SMPM 61:7(1981). 35-711:* 7.05_3 5.58_7 3.57_7 3.53_4 3.28_3 3.02_{10} 2.67_6 2.61_3. Platy crystals up to 2 mm. Commonly twinned by interpenetration (011), or polysynthetic. Dark green, pistachio-green streak, translucent. Cleavage {100}, good. $H = 4\frac{1}{2}$–5. $G = 6.86$. Nonfluorescent. IR. Biaxial (−); $X \wedge a \approx 12°$; $Z = b$; $N \gg 2.10$; 2V moderate; $Z > Y > X$; $r > v$. X, yellow-green; Y, pistachio-green; Z, dark green. Easily soluble in cold 1:10 acids. Found in cavities in drusy quartz veins at copper mine near *Khorixas (formerly Welwitschia), northwestern Namibia*, associated with beyerite, Bi, bismite, bismutite, and oxidized Cu minerals. HCWS *AM 67:857(1982)*.

38.5.10.1 Aerugite $(\square,Ni)_9(AsO_4)_2(AsO_6)(?)$

Named in 1858 from the Greek for *copper rust*, in allusion to its appearance. See also xanthiosite (38.3.2.1). HEX-R $R\bar{3}m$. $a = 5.951$, $c = 27.568$, $Z = 1$, $D = 5.772$. Structure: *AC(B) 45:201(1989)*. Structure shows layers stacked along [001]. Ni in octahedral coordination, Ni–O_6, site only partially (5/6) occupied to achieve charge balance in layers with octahedral As, As–O_6, and As in tetrahedral coordination. *18-873:* 9.19_7 5.05_8 3.76_9 2.86_8 2.49_8 2.33_8 2.06_{10} 1.485_8. Small isolated crusts. Synthetic crystals tabular on (001). Deep blue-green. $G = 5.85$. Slowly soluble in boiling perchloric acid. Found as the crusts on walls of furnaces and with native Bi, bunsenite, at *South Terras mine, Cornwall, England*, and at Tolgarrick Mill, St.-Stephen-in-Branad, 5 miles W of St. Austell, with xanthosite in comby quartz in a uranium mine with Ni–Co minerals such as rammelsberite and gersdorffite, and also erythrite, annabergite, scodorite, and pharmacosiderite; at Johanngeorgenstadt, Saxony, Germany. HCWS *MM 35:72(1965)*.

Class 39

Hydrated Acid Phosphates, Arsenates, and Vanadates

39.1.1.1 Brushite CaHPO$_4$ · 2H$_2$O

Named in 1865 for George Jarvis Brush (1831–1912), professor of mineralogy, Yale University, New Haven, CT. Gypsum group. See gypsum (29.6.3.1). MON $C2/c$. $a = 6.361$, $b = 15.191$, $c = 5.184$, $\beta = 118.45°$, $Z = 4$, $D = 2.318$. AM 76:1722(1991), Structure: ZK 181:205(1987). 11-293: 7.62$_{10}$ 3.80$_3$ 3.06$_1$ 2.90$_1$ 2.63$_1$ 2.53$_1$ 2.01$_1$ 1.90$_1$. Platy {010}; needlelike or prismatic crystals; cryptocrystalline efflorescences; earthy, powdery, foliated. Colorless to pale yellow; vitreous luster, pearly on cleavage; transparent to translucent. Cleavage {010}, {001}, perfect. H = $2\frac{1}{2}$. G = 2.328. Piezoelectric. DTA. IR. MSLM 4:393,395(1974). Biaxial (+); Z = b; X $\wedge$ c = $-30°$; N$_x$ = 1.540, N$_y$ = 1.546, N$_z$ = 1.552; 2V = 86°; r > v; crossed dispersion notable. Readily soluble in HCl. Loses all water by 100° forming monetite (37.1.1.1). Synthesized. Widespread in small amounts in sedimentary phosphate deposits, agricultural fertilizer runoff, tomb walls, and guano-bearing caves; occurs as incrustations on ancient human and animal bones and human urinary calculi. Found in Pig Hole Cave, Johns Creek Mt., Giles Co., VA; Maury formation, Dekalb Co., TN; Rock Creek, Kankakee Co., IL; Paoha Is., Mono Lake, Mono Co., CA; Rapid Creek, YK, Canada; Mona Island, PR; Sombrero Is., BVI; in guano on *Aves Island, Venezuela*; Wheal Cock, St. Just, Cornwall, England; Begues, Baix Llobregat, Catalonia region, Spain; Quercy near Limoges; on ancient human skeletons, Minerva Cave, Cevennes Mts., Hérault; Grandpré, Ardennes, France; with hydroxylapatite and taranakite in bat guano deposits from Apulian caves, southern Italy; commercially worked in a cave near Oran, Algeria; Moorba Cave, Jurien Bay, WA, Australia; Enderbury Is., Phoenix Islands. HCWS DII:704, AM 41:616(1956), MSLM 4:393,395(1974).

39.1.1.2 Pharmacolite CaHAsO$_4$ · 2H$_2$O

Named in 1800 from the Greek for *poison*, in allusion to the arsenic content. Gypsum group. See gypsum (29.6.3.1). MON $P2/m$. $a = 5.975$, $b = 15.434$, $c = 6.280$, $\beta = 114.83°$, $Z = 4$, $D = 2.731$. Structure: AC(B) 25:1544(1969), but see double cell (Z = 8) comparison with krautite. AM 64:1248(1979). 25-138: 7.7$_9$ 5.1$_7$ 4.30$_{10}$ 3.08$_8$ 3.02$_8$ 2.70$_9$ 2.66$_9$ 2.48$_5$. Delicate silky fibers or acicular clusters; botryoidal and stalactitic; small crystals (rare) flattened on

{010}, or needle-like [001]. Colorless, whitish, grayish; vitreous luster, pearly on cleavage; transparent. Cleavage {010}, perfect; uneven fracture; flexible in thin laminae. H = 2–2$\frac{1}{2}$. G = 2.725. IR. MSLM 4:393(1974). Biaxial (−); Z = b; X ∧ c = −29°; N_x = 1.583, N_y = 1.589, N_z = 1.594 (Na); 2V = 79.4°; r > v. Soluble in acids, insoluble in water. Dehydrates at 60°. Synthesized. Found with other arsenates, especially haidingerite, in oxidized portions of arsenical ores at White Caps mine, Manhattan, NV; with erythrite at O.K. mine, San Gabriel Canyon, Los Angeles Co., CA; at Ste.-Marie-aux-Mines (Markirch), Vosges Mts., Alsace, France; as small crystals at Jachymov, Bohemia, Czech Republic(!), at Andreasberg in the Harz, and at *Grube Anton, Wittichen, Black Forest, Germany*. HCWS *DII:706, BM 87:169(1984)*.

39.1.2.1 Fluckite CaMn^{2+}(HAsO$_4$)$_2$ · 2H$_2$O

Named in 1980 for Pierre Fluck, mineralogist at Louis Pasteur University, Strasbourg, France. TRIC $P\bar{1}$. a = 8.459, b = 7.613, c = 6.968, α = 82.21°, β = 98.25°, γ = 95.86°, Z = 2, D = 3.11 *BM 103:122(1980)*. Structure: *103:129(1980)*. See krautite (39.1.3.1). *33-290*: 8.32$_4$ 7.51$_{10}$ 3.77$_6$ 3.53$_9$ 3.27$_9$ 2.98$_8$ 2.69$_6$ 2.67$_4$. Radiating prismatic crystals with {100}, {110}, {010} dominant faces. Colorless to pale to deep rose. Cleavage {010}, perfect; {100}, easy; {$\bar{1}$01}, difficult. H = 3$\frac{1}{2}$–4. G = 3.05. DTA. Biaxial (+); N_x = 1.618, N_z = 1.642, 2V large. Found at the 60- and 100-m levels of the *Gabe Gottes–St. Jacques vein at Ste.-Marie-aux-Mines, Vosges Mts., Alsace, France*, associated with pharmacolite and picropharmacolite. HCWS *AM 65:1066(1980)*.

39.1.3.1 Krautite Mn^{2+}As^{5+}O$_3$(OH)$_2$ · 2H$_2$O

Named in 1975 for François Kraut, mineralogist at the Musée Nationale d'Histoire Naturelle, Paris. MON $P2_1/n$. a = 8.012, b = 15.956, c = 6.801, β = 96.60°, Z = 8, D = 3.274. Structure: *AM 64:1248(1979)*. Layers of AsO$_4$ and MnO$_6$ polyhedra linked by H$^+$ bonds are the basic structural components, a similar structural motif to that of haidingerite (39.1.5.1) and pharmacolite (39.1.1.2). *29-888*(syn): 7.99$_{10}$ 4.88$_4$ 3.86$_{10}$ 3.30$_{10}$ 3.27$_{10}$ 3.19$_9$ 2.86$_4$ 2.73$_6$. See also *33-895*. Lamellar up to 2 mm in largest dimension and aggregates of the lamellae. Pale rose, translucent. Cleavage {010}, perfect, micaceous; {$\bar{1}$01}, good; {101}, distinct. H = 4. G = 3.30. Biaxial (+); X = b; Z ∧ ($\bar{1}$01) = 16°; N_x = 1.620, N_y = 1.639, N_z = 1.686 (Na); 2V = 65°; positive elongation. Synthesized in 1886. Found at Sacarimb, formerly Nagyag, and Kapnik (Cavic), Transylvania, Romania. HCWS *BM 98:78(1975); AM 61:503(1976), 66:432(1981)*.

39.1.4.1 Koritnigite Zn(AsO$_3$OH) · H$_2$O

Named in 1979 for Sigmund Koritnig (b.1912), professor, Göttingen, Germany. TRIC $P\bar{1}$. a = 7.948, b = 15.829, c = 6.668, α = 90.86°, β = 96.56°, γ = 90.05°, Z = 8, D = 3.56. *TMPM 26:51(1979)*, Structure: *NJMA 138:316(1980)*. *34-1396*: 7.90$_{10}$ 3.95$_4$ 3.83$_7$ 3.16$_9$ 2.93$_4$ 2.68$_4$ 2.46$_6$ 2.19$_5$. Massive. Colorless, transparent. Cleavage {010}, perfect; traces of cleavage parallel to [001] and [100] visible on {010}. H = 2. G = 3.54. IR. Biaxial (+); X = b; Z ∧ c = 22°; $N_x = 1.632$, $N_y = 1.652$, $N_z = 1.693$. Soluble in cold 1:10 HCl, HNO_3. Occurs in cavities in tennantite associated with Cu-adamite, and stranskiite at level 31 of the *Tsumeb mine, Tsumeb, Namibia*. HCWS *AM 65:206(1980)*. *AUF 31:43(1980)*.

39.1.4.2 Cobaltkoritnigite (Co,Zn)(AsO₃OH) · H₂O

Named in 1981 for the similarity to koritnigite and the composition. TRIC $P\bar{1}$. $a = 7.95$, $b = 15.83$, $c = 6.67$, $\alpha = 90.86°$, $\beta = 96.6°$, $\gamma = 90.1°$, $Z = 8$, $D = 3.44$. *NJMM:257(1981)*. *35-499*: 7.94_{10} 7.07_2 3.82_5 3.25_4 3.23_4 3.14_7 2.69_3 2.46_4. Tabular crystals. Purple, vitreous luster. Cleavage {010}, perfect; {100}, good. Biaxial (+); X ∧ b = 3°, Y ∧ a = 12.5°; $N_x = 1.646$, $N_y = 1.668$, $N_z = 1.705$; strongly pleochroic. X, deep violet; Y, reddish violet; Z, bluish violet. Found as an alteration product of cobaltite in the Richelsdorf Mts., and by alteration of glaucodot and associated with sphaerocobaltite, erythrite, pitticite, and quartz at *Erzgeburge, Saxony, Germany*. HCWS *ZAAC 454:134(1979)*, *NJMM:237(1980)*, *AM 67:414(1982)*.

39.1.5.1 Haidingerite Ca(AsO₃OH) · H₂O

Named in 1827 for William Karl von Haidinger (1795–1871), Austrian mineralogist and geologist. ORTH *Pcnb*. $a = 6.904$, $b = 16.161$, $c = 7.935$, $Z = 8$, $D = 2.971$. Structure: *AM 64:1248(1979)*. See also *AC(B) 28:209(1972)*. *18-288*: 8.06_8 5.25_8 5.22_{10} 3.84_8 3.19_8 3.15_8 3.11_7 2.96_9. Also *17-161*. Fine grained botryoidal to fibrous coatings; crystals rare equidimensional to short prismatic [001], [100]. Colorless, white; vitreous luster, pearly on cleavage; transparent. Cleavage {010}, perfect; sectile; flexible thin laminae. H = 2–2½. G = 2.848. Biaxial (+); $N_x = 1.590$, $N_y = 1.602$, $N_z = 1.638$; 2V = 58°; r > v. Loses water in several stages, completely dehydrated to pyroarsenate at 200°. Synthesized; Ba and Sr analogs are known. Found as an oxidation product of arsenical ores in association with pharmacolite at White Caps mine, Manhattan, NV; Schneeberg and Johanngeorgenstadt, Saxony, near Wittichen in the Black Forest, Germany; from *Jachymov, Zapadocesky Kraj, Bohemia, Czech Republic*. HCWS *DII:708*, *BM 89:18(1966)*.

39.1.6.1 Newberyite Mg(PO₃OH) · 3H₂O

Named in 1879 for James Cosmo Newbery (1843–1895) of Melbourne, VIC, Australia. ORTH *Pbca*. $a = 10.203$, $b = 20.678$, $c = 10.015$, $Z = 16$, $D = 2.12$. *AM 51:1755(1966)*, Structure: *TMPM 32:187(1983)*. *35-780*(syn): 5.95_5 4.71_5 4.50_4 4.15_3 3.46_6 3.08_5 3.04_{10} 2.58_3. Crystals equidimensional, short prismatic [001] or tabular {100}; {010} and {111} also seen in artificial crystals. Colorless, vitreous luster, transparent. Cleavage {010}, perfect; {001}, imperfect. H = 3–3½. G = 2.10. IR. *MSLM 4:398(1974)*. Biaxial (+); $N_x = 1.514$, $N_y = 1.517$, $N_z = 1.533$; 2V = 44.78°; r < v, perceptible. Scarcely soluble in cold water, readily soluble in 1:10 HCl. Dehydrates to $MgHPO_4$ at 130°. Synthesized. Found at Paoha Is., Mono Lake, Mono Co., CA; from mammoth tusk with struvite and magnesite, near Dawson, YT, Canada; with collophane (carbonate–apatite) on basalt from roof of cavern on Ascension Is., southern Atlantic Ocean; as small reticulated crystals in cavities of lava tunnel filled with bat guano, Island of Reunion, Indian Ocean; from the *Skipton Caves, near Ballarat, VIC, Australia*(!), in inch-square crystals with hannayite in bat guano; as clear crystals on cracks in guano, Mejillones, Chile. HCWS *DII:709, NJMM:97(1961), BM 94:556(1971)*.

39.1.7.1 Brassite $Mg(AsO_3OH) \cdot 4H_2O$

Named in 1973 for Rejane Brasse, French chemist who synthesized the compound. ORTH *Pbca*. $a = 7.472$, $b = 10.890$, $c = 16.585$, $Z = 8$, $D = 2.326$ *BM 96:365(1973)*. Structure: *AC(B) 32:1460(1976)*. *23-1228*(syn): 5.55_8 5.45_8 4.95_{10} 3.89_8 3.73_8 3.30_8 3.20_8 3.08_{10}. See also *29-860*(calc). Crusts and coatings on museum specimens. White, translucent. Synthetic crystals are tabular with {001} dominant, also {011}, {021}, {102} show cleavage {001}, excellent. $G = 2.28$. DTA. Biaxial (+); $XYZ = bac$; $N_x = 1.531$, $N_y = 1.546$, $N_z = 1.562$; 2V large. Occurs on specimens from *Jachymov, Bohemia, Czech Republic*; Bieber, Saxony, Neurode, Silesia, and Wittichen, Baden, Germany, labeled rösslerite, pharmacolite, haidingite, and "wapplerite" (= rösslerite), and is associated with arsenic, weilite, rauenthalite, and picropharmacolite. HCWS *AM 60:945(1975)*.

39.1.8.1 Dorfmanite $Na_2(PO_3OH) \cdot 2H_2O$

Named in 1980 for Moisei Davidovich Dorfman (b.1908), Russian mineralogist who first reported a sodium phosphate in 1963. ORTH. $a = 10.34$, $b = 16.82$, $c = 6.60$, $Z = 8$, $D = 2.06$. *ZVMO 109:211(1980)*. *10-190*(syn): 8.42_6 5.28_8 4.64_9 3.36_{10} 3.26_8 2.93_4 2.87_7 2.74_5. Powdery aggregates. White. $H = 1-1\frac{1}{2}$. $G = 1.99$. IR, DTA. Biaxial (+); $N_x = 1.454$, $N_y = 1.461$, $N_z = 1.461$; $2V = 65°$; $r > v$, weak. Very soluble in water. Found at Mt. St.-Hilaire, QUE, Canada. Formed by alteration of lomonosovite in drill cores in alkalic pegmatites of *Khibina and Lovozero massifs, Kola Penin., Russia*. HCWS *AM 66:217(1981), IGR 24:111(1982)*.

39.1.9.1 Rösslerite $Mg(AsO_3OH) \cdot 7H_2O$

Named in 1861 for Karl Rössler, of Hanau, Germany. Synonym: wapplerite. MON $C2/c$. $a = 6.692$, $b = 25.744$, $c = 11.538$, $\beta = 95.15°$, $Z = 8$, $D = 1.948$. *ZK 137:194(1973)*. Structure: *AC(B) 29:286(1973)*. *26-1447*(syn): 12.9_6 6.43_9 4.67_8 4.49_9 4.29_8 4.08_{10} 3.31_5 2.86_6. Fine-grained crusts, sometimes hairlike. Small, short, prismatic crystals [001] (rare). Artificial crystals tabular {010}. Colorless, white; vitreous or dull luster; transparent. Cleavage {111}, imperfect. $H = 2-3$. $G_{syn} = 1.943$. Biaxial (+); $Z = b$; $X \wedge c = 14°$; $N_x = 1.525$, $N_y = 1.53$, $N_z = 1.550$; 2V small. Readily soluble in 1:10 HCl, slightly soluble in H_2O. Crystals lose water on exposure to air becoming white, dull. Synthesized under acidic conditions. A secondary mineral with other arsenates in oxidized zones of arsenical ore deposits. Found at *Bieber, near Hanau*, with pharmacolite and erythrite and at Sophie mine near Wittichen, at Wolfgang mine near Alpairsbach, Black Forest, *Germany*; at Jachymov, Bohemia, Czech Republic, in weathered crystals with pharmacolite and haidingerite. HCWS *DII:712*.

39.1.9.2 Phosphorrösslerite Mg(PO$_3$OH)·7H$_2$O

Named in 1939 for its relation to rösslerite, an arsenate, as it is the phosphate analog. MON $C2/c$. $a = 6.574$, $b = 25.36$, $c = 11.32$, $\beta = 95.18°$, $Z = 8$, $D = 1.741$. *ZK 137:246(1973)*. *19-761*(syn): 4.55$_9$ 4.42$_9$ 4.13$_9$ 4.02$_9$ 3.29$_{10}$ 2.82$_{10}$ 2.70$_9$ 2.63$_9$. Crystals equi-dimensional or short prismatic [100], skeletal, crusted. Rarely clear, mostly yellow due to included iron oxides and organic matter; vitreous luster when fresh, dull due to water loss. Conchoidal fracture. $H = 2\frac{1}{2}$. $G = 1.725$. Biaxial $(-)$; $X = b$; $Z \wedge c = 6.5°$; $N_x = 1.477$, $N_y = 1.485$, $N_z = 1.486$; $2V = 38.16°$; $r > v$. Natural crystals more or less dehydrated to pseudomorphs of lower hydrate (3H$_2$O), lose water in dry air. Synthesized. Found in cold (10°), wet muck in abandoned Stüblau mine near *Schellfgaden, Salzburg Prov., Austria*. HCWS *DII:713*.

39.2.1.1 Huréaulite Mn$_5^{2+}$(PO$_3$OH)$_2$(PO$_4$)$_2$·4H$_2$O

Named in 1826 for the locality. MON $C2/c$. $a = 17.587$, $b = 9.127$, $c = 9.497$, $\beta = 96.68°$, $Z = 4$, $D = 3.19$ *BM 99:261(1976)*. Structure: *AC(C) 43:1829(1987)*. *39-146*: 8.75$_4$ 8.09$_7$ 4.55$_4$ 3.20$_3$ 3.15$_{10}$ 2.99$_7$ 2.63$_4$ 2.19$_3$. See also *16-383*. Crystals short prismatic [001], sometimes tabular {100}, or equant; isolated, in groups or fascicled, also massive, compact, scaly, or imperfectly fibrous. Orange, orange-red, brownish orange, reddish to yellowish brown, violet-rose, pale rose, red, amber, also gray to nearly colorless; near-white streak; vitreous to greasy luster; transparent to translucent. Cleavage {100}, good. $H = 3\frac{1}{2}$. $G = 3.19$. Biaxial $(-)$; $X = b$; $Z \wedge c = 75°$; $N_x = 1.647$, $N_y = 1.654$, $N_z = 1.660$; $2V = 75°$; $r < v$, very strong; crossed, strong; pleochroic. X, colorless; Y, yellow to pale rose; Z, reddish yellow to reddish brown. Fe^{2+} substitutes for Mn^{2+} up to at least 33% (Huréaux samples) and affects optical properties. Easily soluble in acids. Synthesized, Cd analog prepared. A secondary mineral found with fairfieldite, dickinsonite, reddingite, and eosphorite as hydrothermal alteration of lithiophilite in Branchville, CT, pegmatite, and at Strickland quarry, Portland, CT; with metastrengite, strengite, and heterosite in weathered triphylite at Palermo No. 1 and Fletcher mines, North Groton, NH; in the Black Hills at Elkhorn, Gap Lode, Hot Shot, John Ross, and Linwood pegmatites, in Custer Co. at Bull Moose Mtn., Rock Ridge, Tip Top, White Elephant mines, and in Pennington Co. at Ferguson Lode, Big Chief pegmatite, and Hesnard mine, SD; with stewartite and metastrengite in altered lithiophilite at Pala and Stewart Lithia mine, San Diego Co., CA; with rockbridgeite, cacoxenite, and vivianite in cavities in heterosite formed by alteration of triphylite in pegmatite at *Huréaux in St. Silvestre* and at *Vilate near Chanteloube, N of Limoges, Haute Vienne, France*; at Hagendorf, Hühnerkobel, Pleystein, Germany; at Vanitrask, Sweden; at Viitanieme near Erajarvi, Finland; at Criminoso mine, Mun. Boa Vista, and at Sapucaia mine, Galileia, MG, Brazil. HCWS *DII:700; AM 40:50(1955), 58:302(1973), 70:395(1985).*

39.2.1.2 Sainfeldite $Ca_5(AsO_3OH)_2(AsO_4)_2 \cdot 4H_2O$

Named in 1964 for Paul Sainfeld (b.1916), Musée des Mineralogie, École des Mines, Paris, who collected the type material in Alsace. MON $C2/c$. $a = 18.781$, $b = 9.820$, $c = 10.191$, $\beta = 97.0°$, $Z = 4$, $D = 3.027$. Structure: *BM 87:169(1964), 95:33(1972). 16-615:* 8.7_8 4.86_4 4.64_6 4.45_4 3.37_{10} 3.18_8 3.06_6 1.744_4. Flattened rosettes or radiating crystals; elongated parallel to c, flattened on (100). Colorless to light pink, transparent. $G = 3.04$. Biaxial; $X = b$; $Y \wedge c \approx 20°$; $N_x = 1.600$, $N_y = 1.610$, $N_z = 1.616$; $2V = 80°$. Found at the 40-m level of the *Gabe Gottes vein, Ste.-Marie-aux-Mines, Vosges Mts., Alsace, France.* HCWS *AM 50:806(1965).*

39.2.1.3 Villyaellenite $Mn_5^{2+}(AsO_3OH)_2(AsO_4)_2 \cdot 4H_2O$

Named in 1984 for Villy Aellen, director of the Natural History Museum of Geneva, Switzerland. MON $C2/c$. $a = 18.015$, $b = 9.261$, $c = 9.770$, $\beta = 96.24°$, $Z = 4$, $D = 3.72$. Structure: *AM 73:1172(1988). 41-1455:* 8.21_5 4.45_3 3.24_4 3.21_{10} 3.13_3 3.06_7 2.98_4 2.34_3. See also *37-344*. Rosettes of prismatic crystals elongated parallel to [001], tabular parallel to {100}, show {100}, {011}, {010}, and rarely {001}. Colorless, light pink to orange-pink; white streak; vitreous luster; transparent. Cleavage {100}, good. $H = 4$. $G = 3.20$. Biaxial (−); $X = b$; $Z \wedge c = 40°$; $N_x = 1.713$, $N_y = 1.723$, $N_z = 1.729$; $2V = 70°$, absorbance $Z \gg X > Y$; pleochroic. X, very pale orange-pink; Y, exceedingly pale orange-pink; Z, pale orange-pink. Soluble in HCl. Found at Sterling Hill, NJ, on calcite-bearing willemite–franklinite ore; as nearly pure end member crystals up to 4 cm length in a vug associated with ogdensburgite, arseniosiderite, adamite, and chalcophanite at Mapimi, DGO, Mexico(!). The original material from *Ste.-Marie-aux-Mines, Vosges Mts., Alsace, France*, had a composition near the midpoint of the series villyaellenite–sainfeldite and was associated with fluckite (39.1.2.1). There is some question as to the degree of solid solution in the Ca–Mn series and whether type material (in Museum, Switzerland) may represent a distinct species compared to the Mexican–NJ mineral. HCWS *SMPM 64:323(1984), AM 71:1547(1971).*

39.2.2.1 Vladimirite $Ca_5(AsO_3OH)_2(AsO_4)_2 \cdot 5H_2O$

Named in 1953 for the locality. MON $P2_1/c$. $a = 5.81$, $b = 10.19$, $c = 22.7$, $\beta = 82.7°$, $Z = 3$, $D = 3.72$ *BM 87:169(1964)*. Structure: *ZK 157:296(1981)* on triclinic modification. *17-162:* 5.09_4 4.80_4 4.15_8 4.00_8 3.20_4 3.04_6 2.87_4 2.79_{10}. Rosettes, radiating needles; acicular. Colorless, pale rose, translucent. $G = 3.14$. Biaxial (−); $Z \wedge c = 37°$; $N_x = 1.661$, $N_y = 1.656$, $N_z = 1.656$; $2V = 70°$; $r > v$, strong. Found at *Vladmiriskoe, Altaie Mts., Russia*, and at Kovuaksi, Tuva Autonomous Republic, Russia; at Irhtem mine, Bou-Azzer, Morocco. HCWS *AM 50:813(1965); ZVMO 82:317(1953), 99:362(1970).*

39.2.2.2 Guerinite $Ca_5(AsO_3OH)_2(AsO_4)_2 \cdot 9H_2O$

Named in 1961 for Henri Guerin (b.1906), French chemist who synthesized the compound. Dimorphic with ferrarisite (39.2.3.1). MON $P2_1/n$. $a = 17.63$, $b = 6.734$, $c = 23.47$, $\beta = 90.6°$, $Z = 5$, $D = 2.74$. BM 87:169(1964), Structure: AC(B) 30:1789(1971). 26-1055: 14.0_{10} 11.7_4 8.7_3 4.84_4 3.89_8 3.49_6 3.01_8 2.90_8. Spherulites and rosettes; feathery. Acicular and wedge-shaped crystals up to 0.3 mm. Colorless or white in aggregates, vitreous luster, translucent to opaque. $H = 1\frac{1}{2}$. $G = 2.76$. Biaxial (−); $N_x = 1.574$, $N_y = 1.582$, $N_z = 1.582$ (Na); $2V = 15°$; $r < v$; positive elongation. Synthesized, many hydrates detected. Found at Sterling Hill Mine, Ogdensburg, Sussex Co., NJ; Gabe Gottes mine, Alsace, France; from *Daniel mine, Schneeberg, Saxony and Richelsdorf, Hesse, Germany*. HCWS AM 47:416(1962), 50:812(1965).

39.2.3.1 Ferrarisite $Ca_5(AsO_3OH)_2(AsO_4)_2 \cdot 9H_2O$

Named in 1980 for Giovanni Ferraris (b.1937) of the Institute of Mineralogy ("G. Spezia"), University of Torino, Italy. Dimorphous with guerinite (39.2.2.2); see also sainfeldite (39.2.1.2). TRIC $P\bar{1}$. $a = 8.294$, $b = 6.722$, $c = 11.198$, $\alpha = 106.16°$, $\beta = 92.94°$, $\gamma = 99.20°$, $Z = 1$, $D = 2.57$ BM 103:533(1980). Structure: BM 103:541(1980). 33-280: 10.81_{10} 6.34_3 5.36_3 4.73_3 4.07_4 3.57_4 3.17_8 2.83_9. Colorless, transparent, becoming white on dehydration. Biaxial (+); $X \wedge c = 17°$; Z perpendicular to (110); $N_x = 1.562$, $N_y = 1.572$, $N_z = 1.585$; $2V = 83°$. Dissolves readily in 1:10 HCl. Synthesized. Occurs at *Gift mine* and Gabes Gottes, *Ste.-Marie-aux-Mines, Vosges Mts., Alsace, France*, associated with picropharmacolite and rauenthalite; at Anton mine, Wittichen, Baden, Germany. HCWS AM 66:637(1981).

39.2.4.1 Picropharmacolite $Ca_4Mg(AsO_3OH)_2(AsO_4)_2 \cdot 11H_2O$

Named in 1819 for the magnesia content and the similarity to pharmacolite. See irhtemite (39.2.5.1), a lower hydrate form of the same composition. TRIC $P\bar{1}$. $a = 13.547$, $b = 13.500$, $c = 6.710$, $\alpha = 99.85°$, $\beta = 96.41°$, $\gamma = 91.60°$, $Z = 2$, $D = 2.60$. Structure: AM 66:385(1981). 14-222: 13.50_{10} 9.20_6 4.84_4 4.44_4 3.78_6 3.18_8 3.066_7 2.92_5. Small, spherical, botryoidal forms with radiating foliated structure; silky fibers or minute acicular crystals; rectangular prisms elongated on [001] with cleavages {100} and {010} perfect. White, slightly pearly luster, opaque. $G = 2.58$. IR, DTA. Formed in the oxidation zone of arsenical ore deposits, where the particular phase is dependent on the pH and Eh values of the solutions. Biaxial (+); $Y = b$; $X \wedge c \approx 37°$; $N_x = 1.631$, $N_y = 1.632$, $N_z = 1.640$; $2V = 40°$; $r < v$, strong. Synthesized. Found at Sterling Hill mine, Ogdensburg, Sussex Co., NJ; as a coating on dolomite in Joplin, MO; at Ste.-Marie-aux-Mines, Vosges Mts., Alsace, France; at Grube Anton, Heubachtal, *Riechelsdorf, Hessen*, and as a recent efflorescence in mine workings at Freiberg, Saxony, *Germany*; at Hovou-Aksi, Russia. HCWS DII:740; BM 84:391(1961), 87:169(1964); AM 59:807(1974), 61:326(1976).

39.2.5.1 Irhtemite $Ca_4Mg(AsO_3OH)_2(AsO_4)_2 \cdot 4H_2O$

Named in 1972 for the locality. MON. $a = 16.73$, $b = 9.48$, $c = 10.84$, $\beta = 97.25°$, $Z = 4$, $D = 3.153$. BM 95:365(1972). 25-158: 9.42_3 6.85_3 5.35_3 5.08_4 3.68_4 3.24_9 2.97_{10} 2.82_9. Spherulites to 1 mm in diameter. White to pale rose; colorless crystals; silky luster. $G = 3.09$. DTA. Biaxial (+); $N_x = 1.634$, $N_z' = 1.642$; negative elongation; extinction angle 25°. Dissolves in 1:10 acids. Synthesized by heating picropharmacolite. Occurs in the *Irhtem ore deposit* and Bou Azzer ore deposit, *Morocco*, associated with sainfeldite and erythrite, probably formed by dehydration of picropharmacolite. HCWS *AM 59:209(1974)*.

39.2.6.1 Chudobaite $Mg_5(AsO_3OH)_2(AsO_4)_2 \cdot 10H_2O$

Named in 1960 for Karl Franz Chudoba (1898–1968), mineralogist at the University of Bonn, Germany. See geigerite (39.6.2.2), Mn analog. Some question as to whether the formula is accurate. Original description showed Na, K, and Zn as well as Mg. TRIC $P\bar{1}$. $a = 7.797$, $b = 11.440$, $c = 6.616$, $\alpha = 115.31°$, $\beta = 95.77°$, $\gamma = 93.87°$, $Z = 1$, $D = 2.90$. NW 63:243(1976). 12-643: 10.2_{10} 3.86_5 3.75_5 3.44_8 3.27_8 2.98_9 2.83_5 2.73_7. Crystals to 0.5 cm showing {100}, {010}, {001} also {110}, {$\bar{1}$20}, {$\bar{1}$80}, {101}. Pink, translucent. Cleavage {010}, perfect; {100}, imperfect. $H = 2\frac{1}{2}$–3. $G = 2.94$. Nonfluorescent in UV. Biaxial (–); $Z \wedge c(010) = -24°$; $N_x = 1.583$, $N_y = 1.608$, $N_z = 1.633$; $2V = 79°$. Loses H_2O at 140°. Soluble in HCl. Found at the 1,000-m level, second oxidation zone in the *Tsumeb mine, Tsumeb, Namibia*, associated with conichalcite, cuproadamite, and zincian olivenite. HCWS *NJMM:1(1960); AM 44:1323(1959), 45:1130(1960), 62:599(1977)*.

39.2.6.2 Geigerite $Mn_5^{2+}(AsO_3OH)_2(AsO_4)_2 \cdot 10H_2O$

Named in 1989 for Thomas Geiger of Weisendangen, Switzerland, who studied the Falotta manganese ores. Mn^{2+} equivalent of chudobaite (39.6.2.1). TRIC $P\bar{1}$. $a = 7.944$, $b = 10.691$, $c = 6.770$, $\alpha = 80.97°$, $\beta = 84.20°$, $\gamma = 81.55°$, $Z = 1$, $D = 3.00$. Structure: *AM 74:676(1989)*. Structure contains equal number of $[AsO_4]$ and $[AsO_3OH]$ groups sharing corners with octahedrally coordinated Mn; no cavities for acid H; one H_2O molecule not linked directly to Mn. PD: 10.45_{10} 7.85_1 4.89_1 3.51_2 3.34_2 3.05_2 3.01_2 2.79_1. Platy, triangular crystals flattened on (010) up to 0.5 mm long. Rose red, white streak, vitreous to pearly luster. Cleavage {010}, perfect. $H = 3$. $G = 3.05$. Biaxial (–); $X \approx b$, $Y \approx a$, $Z \approx c$; on {010} $Z \wedge c = 15°$; $N_x = 1.601$, $N_y = 1.630$, $N_z = 1.660$ (589 nm); 2V large. Weakly pleochroic, colorless to rose-red with $Z > Y \approx X$. Found in abandoned manganese mines near *Falotta, Oberhalberstein, Grisons Canton, Switzerland*, associated with tilasite, sarkinite, manganberzeliite, brandtite, and grischunite. HCWS

39.2.7.1 Lindackerite $Cu_5(AsO_3OH)_2(AsO_4)_2 \cdot 8$–$9H_2O$

Named in 1853 for Joseph Lindacker, Austrian chemist, who made the analysis. MON. $a = 3.95$, $b = 8.02$, $c = 6.28$, $\beta = 100.5°$, $Z = 1$, $D = 3.46$. BM

79:7(1956). 11-166: 10.2_{10} 5.9_4 3.95_6 3.19_8 3.08_6 2.95_4 2.86_4 2.67_4. Minute laths grouped in rosettes or crusts. Green, greenish white, greenish blue; pale green to white streak; vitreous luster; transparent. Cleavage {010}, perfect; {100}, {001}, easy; conchoidal fracture; brittle. H = $2-2\frac{1}{2}$. G = 2.0–2.5. Biaxial (+); $N_x = 1.627$, $N_y = 1.659$, $N_z = 1.729$; 2V = 68°. Dehydrates at 450°. Secondary oxidation zone mineral found at *Eliaas mine, Jachymov, Bohemia, Czech Republic*, associated with erythrite, zippeite, pitticite, annabergite, and other secondary minerals. HCWS *DII:1007.*

39.3.1.1 Stercorite $Na(NH_4)(PO_3OH) \cdot 4H_2O$

Named in 1850 from the Latin for *dung*. Microcosmic salt. TRIC $P\bar{1}$. $a = 10.636$, $b = 6.919$, $c = 6.436$, $\alpha = 90.46°$, $\beta = 98.58°$, $\gamma = 109.20°$, Z = 2, D = 1.578. Structure: *AC(B) 30:504(1974). 24-1048:* 9.93_9 6.53_{10} 4.77_4 4.69_3 4.24_5 3.66_4 2.91_5 2.88_5. Massive and nodular. Artificial crystals simulate monoclinic symmetry due to repeated twinning of triclinic individuals on {010}. Faces even and lustrous without reentrant angles. Crystals colorless, natural material white to yellowish, brownish; vitreous luster; transparent. No cleavage. H = 2. $G_{syn} = 1.574$, G = 1.615. Biaxial (+); $N_x = 1.439$, $N_y = 1.442$, $N_z = 1.468$; optic plane approximately perpendicular to {010}, Z approximately perpendicular to {001}; {010} may show two sets of lamellar twinning at about 90°; 2V = 35.52°; r > v, strong. Soluble in cold H_2O. (NH_4) lost ≈ 200°, H_2O at higher temperatures. Synthesized from 10° to −6° by evaporation from solution. Found in a guano deposit on the *Island of Ichaboe, Namibia*; Guañape Islands, Peru. HCWS *DII:698.*

39.3.2.1 Schertelite $Mg(NH_4)_2(PO_3OH)_2 \cdot 4H_2O$

Named in 1902 for Arnulf Schertel (1841–1902) of the Freiberg Mining Academy, Saxony, Germany. ORTH *Pbca*. $a = 11.49$, $b = 23.66$, $c = 8.62$, Z = 8, D = 1.84. Structure: *AC(B) 28:683(1972). 16-353*(syn): 5.94_{10} 5.21_4 4.31_2 3.91_2 3.46_2 3.02_4 2.97_5 2.80_3. Small indistinct flat crystals. Synthetic stout rods elongated on c showing {110}, {111}; slow growth tabular on {100}. Colorless, transparent. No cleavage. G = 1.82. Biaxial (+); XYZ = *bca*; $N_x = 1.508$, $N_y = 1.515$, $N_z = 1.523$ (Na); optic plane {001}. Dissolves incongruently in H_2O forming struvite. H_2O lost at 120°. Synthesized. Found sparingly at *Skipton Caves, SW of Ballarat, VIC, Australia*. HCWS *DII:699, AM 48:635(1963).*

39.3.3.1 Mundrabillaite $Ca(NH_4)_2(PO_3OH)_2 \cdot H_2O$

Named in 1983 for the locality. Dimorph of swaknoite (39.3.4.1). MON *Pm, P2*, or *P2/m*. $a = 8.643$, $b = 8.184$, $c = 6.411$, $\beta = 98.0°$, Z = 2, D = 2.09. *MM 47:80(1983). 35-728*(syn): 8.58_5 6.32_2 5.01_2 4.79_1 4.29_{10} 3.69_1 3.28_3 3.10_3. Tiny crystals. Colorless, earthy. Very soft. G = 2.05. Biaxial (−); $N_x = 1.522$, $N_y = 1.544$, $N_z = 1.552$; extinction inclined 26°. Synthetic material: Biaxial X = *b*; Y $\wedge$ c = +25°; $N_x = 1.521$, $N_y = 1.542$, $N_z = 1.551$; 2V = 65°. Found in the *Petrogale Cave* (31°54′ S, 127°00 E) *near Mundrabilla*

Station, WA, Australia, in association with other salts. HCWS AM 69:407(1984).

39.3.4.1 Swaknoite Ca(NH$_4$)$_2$ · (HPO$_4$)$_2$ · H$_2$O

Named in 1991 for the acronym for the Suid Wes Africa Karst Navorsing Organisasie (SWAKNO), a speleological association in Namibia. Dimorph of mundrabillaite (39.3.3.1). ORTH. $a = 20.959$, $b = 7.403$, $c = 6.478$, $Z = 4$, $D_{syn} = 1.89$. BSASA 32:72(1991). PD(syn): 10.5$_5$ 6.99$_{10}$ 4.74$_4$ 3.71$_9$ 3.65$_4$ 3.18$_6$. Commonly rosettes composed of needles up to 1 mm long, elongate [001], 10 μm wide, which show {100}, {111}, sometimes {001}. White, vitreous luster. Soft, brittle. $G = 1.91$. Soluble in H$_2$O. Occurs as white coating with mundrabillaite, dittmarite, and arcanite on the dolomite walls of *Arnhem cave*, 150 km E of *Windhoek, Namibia*. HCWS AM 78:1110(1993).

39.3.5.1 Hannayite Mg$_3$(NH$_4$)$_2$(PO$_3$OH)$_4$ · 8H$_2$O

Named in 1878 for James B. Hannay (b.1855), chemist, University of Manchester, England. TRIC $P\bar{1}$. $a = 10.728$, $b = 7.670$, $c = 6.702$, $\alpha = 97.87°$, $\beta = 96.97°$, $\gamma = 104.74°$, $Z = 1$, $D = 2.030$. Structure: AC(B) 32:2842(1976). 16-361(syn): 6.96$_{10}$ 5.15$_3$ 4.64$_2$ 3.75$_2$ 3.46$_7$ 3.29$_2$ 3.13$_1$ 3.00$_2$. Small, slender crystals elongated [001], striated parallel to the same direction, may show {100}, {010}, {110}, {001}, {01$\bar{1}$}, {$\bar{1}$11}; tabular on {100} with prominent {01$\bar{1}$}, modified narrow {110}. Yellowish, translucent. Cleavage {001}, perfect; {110}, {1$\bar{1}$0}, {130}, poor. $H = 1$. $G = 2.03$. Biaxial $(-)$; $N_x = 1.504$, $N_y = 1.522$, $N_z = 1.539$; Y perpendicular to {110}; optic plane {$\bar{2}$21}; r < v, weak. Easily soluble in acids. Loses 21% H$_2$O between 100 and 115°. Found in bat guano at *Skipton Caves, SW of Ballarat, VIC, Australia*, associated with struvite, brushite, newberyite, dittmarite, and schertelite. HCWS DII:699, AM 48:635(1963), MM 39:467(1973).

39.3.5.2 Francoanellite H$_6$(K,Na)$_3$(Al,Fe^{3+})$_5$(PO$_4$)$_8$ · 13H$_2$O

Named in 1976 for Franco Anelli, professor of geography, University of Bari, Italy, who discovered the cave in which the mineral was found. See taranakite (39.3.6.1) for the higher hydrate. HEX-R $R3c$ or $R\bar{3}c$. $a = 8.71$, $c = 82.8$, $Z = 6$, $D = 2.11$. NJMM:49(1976). 29-980: 13.75$_{10}$ 7.44$_3$ 6.86$_4$ 5.57$_3$ 4.29$_3$ 3.41$_5$ 2.81$_5$ 2.78$_4$. Nodular aggregates up to 0.2 mm diameter, pulverulent masses. Yellowish-white. Soft. $G = 2.26$. DTA. Uniaxial $(+)$; $N_O = 1.510$, $N_E = 1.515$. Readily soluble in 1:10 HCl, HNO$_3$. Loses 19% H$_2$O up to 200°. Hydrates to taranakite. Synthesized. Occurs at the contact of "terra rossa" with bat guano in the karst cave *Grotte di Castellana, Puglia, Italy*. HCWS AM 44:138(1959), 61:1054(1976); NJMM:363(1979).

39.3.6.1 Taranakite H$_6$K$_3$Al$_5$(PO$_4$)$_8$ · 18H$_2$O

Named in 1882 for the locality. HEX-R $R3c$ or $R\bar{3}c$. $a = 8.718$, $c = 95.10$, $Z = 6$, $D = 2.11$. AM 76:1722(1991). Structure: AM 61:329(1976). Layers of Al–PO$_4$ sheets perpendicular to c are separated by water molecules, cations. See francoanellite (39.3.5.2), the lower hydrate. *29-981:* 15.89_{10} 7.94_3 7.46_3 4.32_2 3.59_2 3.35_3 3.14_3 2.82_2. Nodular masses, claylike, pulverulent to compact, composed of fine-grained powder or pseudohexagonal thin colorless plates. White, gray, or yellowish white. Very soft, unctuous to the touch. G = 2.15. Uniaxial (−); $N_O = 1.508$, $N_E = 1.502$. Other elements detected: Na, Ca, Fe^{3+}, S, and Rb. Half H_2O lost at 100°; easily soluble in acids. Occurs in caves or on sea coasts as a product of phosphatic solutions derived from bat or bird guano on clays or aluminous rocks. Found at Grotte de Minerve, Aude Valley, Hérault, France; at Monte Alburno, Salerno, and in the caves in the Apulian Platform, Italy; at Misserghin, Oran, Algeria; on the Island of Réunion, Indian Ocean; at Moorba Cave, Jurien Bay, WA, and Jenolan Caves, NSW, Australia; from the bird colonies at *Sugarloaves, Taranaki, New Zealand*; at Islas Leones, Patagonia, Argentina. HCWS *DII:999, MM 28:31(1947), AM 44:138(1959), 60:331(1975), 61:329(1976).*

39.3.7.1   McNearite $NaCa_5H_4(AsO_4)_5 \cdot 4H_2O$

Named in 1981 for Elizabeth McNear, mineralogist and crystallographer, University of Geneva, Switzerland. TRIC $P1$ or $P\bar{1}$. $a = 13.50$, $b = 14.10$, $c = 6.95$, $\alpha = 90.0°$, $\beta = 92°$, $\gamma = 119°$, Z = 2, D = 2.85. *SMPM 61:1(1981).* *35-688:* 12.33_{10} 6.94_5 4.40_3 3.92_6 3.40_2 3.12_6 2.94_2 2.75_4. Radiating fibers up to 2 mm length. White, pearly luster. Cleavage, perfect, parallel elongation. G = 2.60. Biaxial (+); $Y \wedge c = 6\text{--}22°$; $N_x = 1.559$, $N_y = 1.562$, $N_z = 1.572$; $2V = 60°$; positive elongation; strong dispersion. Dissolves in acids. Found in specimens from *Ste.-Marie-aux-Mines, Vosges Mts., Alsace, France,* associated with pharmacolite, guerinite, and haidingerite. HCWS *AM 67:856(1982).*

39.3.8.1   Machatschkiite $(Ca,Na)_6(AsO_4)(AsO_3OH)_3(P,S)O_4 \cdot 15H_2O$

Named in 1977 for Felix Karl Ludwig Machatschki (1895–1976), Austrian professor of mineralogy at Tübingen, Munich, and Vienna. HEX-R $R3c$. $a = 15.127$, $c = 22.471$, Z = 6, D = 2.59. *SMPM 62:177(1982),* Structure: *TMPM 30:145(1982). 29-296:* 8.59_{10} 5.34_8 3.74_7 3.59_8 2.86_8 2.83_8 2.70_8 2.67_8. Crusts showing characteristic rhomb form $\{10\bar{1}2\}$. Colorless, translucent to transparent. No cleavage, conchoidal fracture. H = 2–3. G = 2.50. Uniaxial (−); $N_O = 1.593$, $N_E = 1.585$; weakly biaxial. Dissolves readily in 1:10 HCl, HNO_3. Secondary mineral in crusts on granite with gypsum, pharmacolite, picropharmacolite, and sainfeldite at *Grube Anton, Heubachtal near Schiltach, Black Forest, Germany.* HCWS *AM 62:1260(1977), TMPM 24:125(1977), FM 59:41(1981), AM 67:418(1982).*

39.3.9.1   Kaatialaite $Fe[H_2AsO_4]_3 \cdot 5\text{--}5.5H_2O$

Named in 1984 for the locality. MON $P2_1$ or $P2_1/m$. $a = 15.34$, $b = 19.80$, $c = 4.752$, $\beta = 91.90°$, Z = 4, D = 2.64 *NJMM:337(1986).* Structure(syn): *AC(B) 37:1402(1981). 40-475:* 9.92_8 8.33_{10} 7.67_8 6.07_7 4.47_4 3.76_4 3.41_6 2.81_5. See also *43-681*(syn). Needles to lathlike crystals up to 50 μm long, synthetic material aggregates of tabular, six-sided crystals up to 0.3 mm across, original description powdery coatings. Greenish blue, transparent, fades to pale green, losing transparency, within months, recrystallizes to feathery aggregates of thin laths elongated parallel to c with oblique terminations. $G = 2.64$. TGA, IR. Biaxial (+); $N_x \approx N_y = 1.573$, $N_z = 1.636$; $2V = 13°$. Synthetic: $Z \wedge c \approx 3°$; $N_x = 1.581$, $N_{y\ calc} = 1.582$, $N_z = 1.626$; positive elongation; $2V = 15°$. *BSCF:2575(1972)*. Water content varies with humidity. Found as oxidation coatings with calcite, gypsum, and arsenic minerals, on löllingite in As- and Ag-containing veins in the gabbro of Needer-Beerbach, Odenwald, Germany; at the *Kaatiala granite pegmatite, Kuortane, western Finland*, associated with arsenolite, scorordite, and parasymplesite. HCWS *AM 69:383(1984)*.

Class 40

Hydrated Normal Phosphates, Arsenates, and Vanadates

40.1.1.1 Struvite $Mg(NH_4)(PO_4) \cdot 6H_2O$

Named in 1845 for Heinrich C. G. von Struve (1772–1851), geologist and mineralogist in the Russian diplomatic service. ORTH $Pmn2_1$. $a = 6.955$, $b = 6.142$, $c = 11.218$, $Z = 2$, $D = 1.70$. NBSM *3:4(1964)*, *AC(B) 42:253(1986)*. 15-762(syn): 5.91_4 5.60_6 4.26_{10} 4.14_4 2.92_5 2.80_3 2.69_5 2.66_4. Usually found as distinct crystals up to 3 cm with marked hemimorphic habit, displaying variable shapes, such as equant, wedge-shaped, coffin-shaped, short prismatic, or thick tabular. Figure 40.1.1 (*DII:715*) illustrates the morphology and the peculiar unequal development and distribution of pyramidal and domical faces: {101}, {10$\bar{1}$}, {001}, {010}, {100}; tabular on {100}. Twinned on {001}, composition plane {001}, deep reentrant angles. Colorless, slightly yellowish or brown with impurities, becoming white and chalky on dehydration; vitreous luster; transparent to translucent. Cleavage {001}, good; {100}, poor; subconchoidal to uneven fracture; brittle. Piezo- and pyro-electric. $H = 2$. $G = 1.711$. Biaxial (+); $XYZ = bca$; $N_x = 1.495$, $N_y = 1.496$, $N_z = 1.504$ (Na); $2V = 37.35°$; $r < v$, strong. Easily soluble in acids, very slightly soluble in water. Synthesized as large crystals. Becomes white and pulverulent on standing in dry, warm atmosphere. Formed in deposits of guano or dung, putrescent matter, canned foods, in sediments or soils by bacterial action, and as bladder and urinary concretions in humans. Found

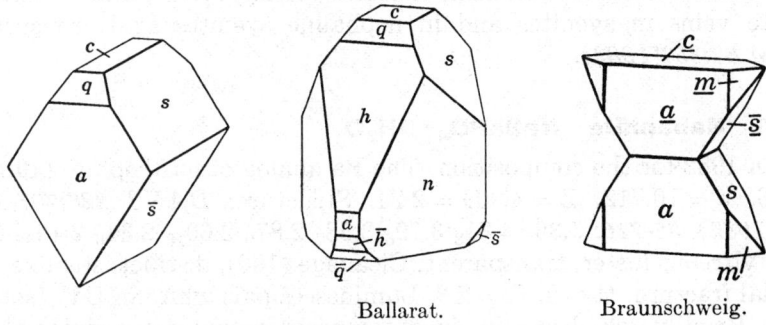

Ballarat.   Braunschweig.

Struvite

Forms: a={100}, c={001}, h={101}, $\bar{q}$={10$\bar{2}$}, $\bar{s}$={01$\bar{1}$}, q={102}. Underlined letters indicate faces of second (twin) crystal.

with newberyite in tooth of mammoth from Quartz Creek near Dawson, YT, Canada; originally from *Hamburg, Germany*, in a bed of peat underlying deposits of organic matter below an old church, and in cattle dung at Homburg v.d. Höhe; in sediment rich in organic remains at Liim Fjord, Denmark; in guano at Saldanha Bay, Cape Prov., South Africa; in bat guano at Skipton Caves SW of Ballarat, Australia; on the Island of Reunion, Indian Ocean. HCWS *DII:715*.

40.1.2.1   Dittmarite $Mg(NH_4)(PO_4)(PO_4) \cdot H_2O$

Named in 1887 for W. Dittmar, chemist of Glasgow, Scotland. ORTH $Pmn2_1$. $a = 5.606$, $b = 8.758$, $c = 4.788$, $Z = 2$, $D = 2.19$. *USGSPP 750A:A115(1971)*. *20-663*: 8.77_{10} 4.72_3 4.20_2 2.92_4 2.80_5 2.50_2 2.28_1 2.12_1. Tiny transparent orthorhombic crystals. Loses water at 100° becoming turbid. Soft. Biaxial; $N_x = 1.549$, $N_y = 1.569$, $N_z = 1.571$; $2V = 40°$. Found with struvite, newberyite, hannayite, and schertelite in bat guano at *Skipton Caves, SW of Ballarat, VIC, Australia*. HCWS *DII:699, AM 57:1316(1972)*.

40.1.2.2   Niahite $(Mn^{2+},Mg,Ca)(NH_4)(PO_4) \cdot H_2O$

Named in 1983 for the locality. ORTH $Pmn2_1$. $a = 5.68$, $b = 8.78$, $c = 4.88$, $Z = 2$, $D = 2.437$. *MM 47:79(1983)*. *35-574*: 8.82_{10} 4.79_5 4.27_6 3.41_4 2.85_8 2.83_9 2.51_3 2.30_4. Radiating subparallel clusters, crystals to 0.5 mm. Pale orange, white streak, translucent. $H = 2$. $G = 2.39$. Biaxial $(-)$; $N_x = 1.582$, $N_y = 1.604$, $N_z = 1.609$; $2V = 54°$. Colorless in thin section. Found at *Niah Great Cave, Sarawak, Malaysia*, in soft newberyite. HCWS

40.1.3.1   Nastrophite $Na(Sr,Ba)(PO_4) \cdot 9H_2O$

Named in 1981 for the composition. ISO $P2_13$. $a = 10.559$, $Z = 4$, $D = 2.12$. *ZVMO 110:604(1981)*. Structure: *SPD 26:1023(1981)*. *25-993*(syn): 5.28_8 4.73_8 3.73_2 3.52_2 2.57_{10} 2.07_3 1.961_3 1.928_3. Crystals mostly 0.2–0.5 but up to 3 mm, and irregular masses to 1 cm in diameter. Colorless, vitreous luster. Conchoidal fracture, brittle. $H = 2$. $G = 2.05$. DTA, IR. Associated with cancrinite, aegirine, vuonnemite, epistolite, mountainite, villiaumite, kogarkoite, and thermonatrite at *Alluaiv Mt., Lovozero massif, Kola Penin., Russia*, in pegmatite veins in syenites and in nephaline syenites at Karnasurt Mt. HCWS *AM 67:857(1982)*.

40.1.3.2   Nabaphite $NaBaPO_4 \cdot 9H_2O$

Named in 1982 for the composition. The Ba analog of nastrophite (40.1.3.1). ISO $P2_13$. $a = 10.712$, $Z = 4$, $D = 2.24$. Structure: *DANS 226:707(1982), 226:624(1982)*. *35-716*: 5.36_8 4.81_8 3.79_6 3.58_7 2.87_9 2.60_{10} 2.34_6 2.10_8. Colorless, dull vitreous luster, transparent. Cleavage (100), distinct; steplike, semiconchoidal fracture. $H \approx 2$. $G = 2.3$. Luminesces pale white in UV. Isotropic; $N=1.504$. Readily dehydrates in air, decomposes in cold water giving alkaline solution; soluble in $1:10$ acids. Occurs in cavities in ijolite–urtite pegmatite, *Yukspor Mt., Khibina massif, Kola Penin., Russia*. HCWS *AM 68:643(1983)*.

40.1.4.1 Pottsite HPbBi(VO$_4$)$_2$ · 2H$_2$O

Named in 1988 for the locality. TET $I4_122$. $a = 11.084$, $c = 12.634$, $Z = 10$, $D = 7.31$. *MM 52:389(1988)*. PD: 4.62_9 4.17_3 3.21_4 3.06_{10} 2.48_4 2.25_3 2.18_3 1.952_3. Crystals up to 1 mm long are dipyramids, stubby prisms terminated by pyramids, {101} dominates also {110}, {101}, {211}. Bright yellow, pale yellow streak, adamantine luster, translucent. Brittle. H = $3\frac{1}{2}$. G = 7.0. Uniaxial (−); N$_O$ = 2.40, N$_E$ = 2.30. Cloudy yellow in thin section. Occurs near *Potts, Lander Co., NV*, in oxide zone of tungsten mine associated with scheelite, clinobisvanite, bismutite, and vanadinite. HCWS *AM 74:503(1989)*.

40.1.5.1 Mahlmoodite Fe^{2+}Zr(PO$_4$)$_2$ · 4H$_2$O

Named in 1993 for Bertha K. Mahlmood, U.S. Geological Survey, Reston, VA. MON $P2_1c$. $a = 9.12$, $b = 5.42$, $c = 19.17$. $\beta = 94.8°$, $Z = 4$, $D = 2.877$. *AM 78:437(1993)*. PD: 9.58_8 4.57_7 4.38_8 4.09_6 3.97_4 3.58_4 3.16_{10} 2.64_7. Orthogonal layer structure forms a series with many AZr(PO$_4$)$_2$ · 2H$_2$O synthetic compounds. *IC 8:431(1969), 23:4679(1984); JCSD:2115(1976)*. Spherules of radiating thin, flat plates (<0.05 mm long), creamy white, H ≈ 3. Uniaxial (−); N$_O$ = 1.652, normal to plate <1.646. Ca, Si, and Ti also detected by microprobe. Spherulites show white core of Si-containing material. Found in vugs as last deposited species on black sodic pyroxene crystals at *Wilson Springs, Garland Co., AR*; Kerriac Cove, Porthtown, Cornwall, England. HCWS

40.2.1.1 Anapaite Ca$_2$Fe^{2+}(PO$_4$)$_2$ · 4H$_2$O

Named in 1910 for the locality. TRIC $P\bar{1}$. $a = 6.447$, $b = 6.816$, $c = 5.898$, $\alpha = 101.63°$, $\beta = 104.24°$, $\gamma = 70.76°$, $Z = 1$, $D = 2.811$. Structure: *BM 102:314(1979)*. 34-148: 6.39_4 4.55_4 3.76_{10} 3.24_4 3.19_5 3.16_7 2.88_8 2.78_4. Tabular {110}; rosettes or crusts of subparallel crystals. Green to greenish white, white streak, vitreous luster, transparent. Cleavage {001}, perfect; {010}, distinct. Biaxial (+); N$_x$ = 1.602, N$_y$ = 1.613, N$_z$ = 1.649; 2V = 54°; r > v, perceptible. Loses H$_2$O above 120°, altering to another hydrous calcium and ferric phosphate. Easily soluble in HCl, HNO$_3$. Found at Palermo No. 1 pegmatite, Groton, Grafton Co., NH; in well cores at Kings Co., CA; in nodular concretions in Miocene clay at Bellevor de Cerdana, Lerida, and Prats-Sampsor, Spain; rims on messelite at Messel, Hesse, Germany; Allori and Castelnuovo mines, near San Giovanni, Valderno, Firenze, Italy; in crevices in oolitic iron ore at *Sheljesny Rog near Anapa, Taman Peninsula (Black Sea), Crimea, Ukraine*. HCWS *DII:731*.

FAIRFIELDITE/ROSELITE GROUP

The general formula

$$AB_2(TO_4)_2 \cdot 2H_2O \text{ or } AB_2(OH,H_2O)_2(TO_4)_2$$

is appropriate for several groups and subgroups of related minerals. In the fairfieldite and roselite subgroups

A = Ca,
B = Mg, Fe^{2+}, Mn^{2+}, Co, Ni, Zn, and
T = P or As.

The structure of the minerals in the two subgroups is dominated by chains of tetrahedra (TO_4) and octahedra [(cation-$O_4(H_2O)_2$] that parallel the c axis.

The roselite subgroup minerals, all arsenates, crystallize in the monoclinic space group $P2_1/c$. The AsO_4 tetrahedra and B–$O_4(H_2O)_2$ octahedra are interlocked into sheets parallel to (010) by Ca–$O_6(H_2O)$ polyhedra, the direction that corresponds to the plane of perfect cleavage. Roselite is isostructural with kröhnkite, $Na_2Cu(SO_4)_2 \cdot 2H_2O$, a hydrated sulfate mineral.

The fairfieldite group, which contains both phosphate and arsenate species, crystallizes in the triclinic space group $P\bar{1}$. The B-octahedra are compressed, resulting in chain disorder, and there is different hydrogen bonding of the H_2O molecules which is expressed in a less distinct cleavage.

As might be anticipated from the structural and chemical similarities, dimorphism between the two groups is common and the following pairs have been identified (the first named is the monoclinic crystalline form followed by the triclinic modification): roselite- β-roselite, $Ca_2Co(AsO_4)_2 \cdot 2H_2O$; brandite-parabrandite $Ca_2Mn(AsO_4)_2 \cdot 2H_2O$; and talmessite-wendwilsonite, $Ca_2Mg(AsO_4)_2 \cdot 2H_2O$. It is essential that both composition and structure be determined to assure accurate identification of a particular species.

In both subgroups elements of like size substitute one for another, for example Fe^{2+} for Mn^{2+}, and though these compositional variations are small they are expressed in different optical as well as unit cell parameters.

	ROSELITE SUBGROUP	SPACE GROUP $P2_1/c$				
Name	Formula	a	b	c	β	
Roselite	$Ca_2Co(AsO_4)_2 \cdot 2H_2O$	5.801	12.989	5.617	107.42	40.2.3.1
Brandite	$Ca_2Mn(AsO_4)_2 \cdot 2H_2O$	5.889	12.968	5.684	108.05	40.2.3.2
Zincroselite	$Ca_2Zn(AsO_4)_2 \cdot 2H_2O$	5.832	12.889	5.644	107.72	40.2.3.3
Wendwilsonite	$Ca_2Mg(AsO_4)_2 \cdot 2H_2O$	5.806	12.912	5.623	107.40	40.2.3.4

40.2.2.1 Fairfieldite

FAIRFIELDITE SUBGROUP   SPACE GROUP $P\bar{1}$

Name	Formula	a	b	c	α	β	γ	
Fairfieldite	$Ca_2Mn(PO_4)_2 \cdot 2H_2O$	5.79	6.57	5.51	102.25°	108.67	90.30°	40.2.2.1
Messelite	$Ca_2Fe(PO_4)_2 \cdot 2H_2O$	5.8	6.6	5.5	102.	109.	90.	40.2.2.2
Collinsite	$Ca_2Mg(PO_4)_2 \cdot 2H_2O$	5.73	6.78	5.44	97.29	108.56	107.28	40.2.2.3
Cassidyite	$Ca_2Ni(PO_4)_2 \cdot 2H_2O$	5.71	6.43	5.41	96.8	107.3	104.6	40.2.2.4
Talmessite	$Ca_2Mg(AsO_4)_2 \cdot 2H_2O$	5.87	6.94	5.54	92.3	108.7	108.1	40.2.2.5
Gaitite	$Ca_2Zn(AsO_4)_2 \cdot 2H_2O$	5.90	7.61	5.57	111.67	70.	119.47	40.2.2.6
β-Roselite	$Ca_2Co(AsO_4)_2 \cdot 2H_2O$	5.89	7.69	5.56	112.6	70.8	119.4	40.2.2.7
Parabrandite	$Ca_2Mn(AsO_4)_2 \cdot 2H_2O$	5.89	7.03	5.64	96.77	109.32	108.47	40.2.2.8

Fairfieldite and roselite should be compared with:
1) the brackebuschite group 40.2.8.1, where A can be Ca, Ba, Pb or Sr and in addition to P or As, T may contain V, Cr, and S;
2) the helmutwinklerite group 40.2.11, all arsenates (T=As) that contain only Pb, Cu, Zn and Fe in the A and B sites.
3) the bearthite group 41.10.3.2 where A is Ca or Sr but B contains Al and T is P.
4) the fornacite group 43.4.3.2 in which Pb and Cu are paired with T=As, P, S, Cr.

Note that the anion and cation associations and structural expression of the tetrahedral phosphates and arsenates of the fairfieldite/roselite subgroups are also expressed in the sulfate and silicate mineral groups and species.

40.2.2.1 Fairfieldite $Ca_2Mn(PO_4)_2 \cdot 2H_2O$

Named in 1879 for the locality. Fairfieldite/Roselite group. TRIC $P\bar{1}$. $a = 5.79$, $b = 6.57$, $c = 5.51$, $\alpha = 102.25°$, $\beta = 108.67°$, $\gamma = 90.30°$, $Z = 1$, $D = 3.095$. *NJMM:78(1957)*, *AC(B) 26:333(1970)*. 10-390: 6.40_9 3.23_{10} 3.20_7 3.03_8 2.86_7 2.66_7 2.63_7 2.46_7. Crystals prismatic to equant, often composite; foliated to lamellar aggregates sometimes resembling gypsum; occasionally curved or fibrous; radiating masses. White, greenish white, pale straw-yellow, salmon-yellow; white streak; transparent. Cleavage {001}, perfect; {010}, good; {1$\bar{1}$0}, distinct; uneven fracture; brittle. $H = 3\frac{1}{2}$. $G = 3.08$. Biaxial (+); $N_x = 1.636$, $N_y = 1.644$, $N_z = 1.654$ (Branchville); 2V very large; $r > v$, moderate. Soluble in acids. Oxidizes to white or greenish products; also found as alteration of dickinsonite, pseudomorphous after rhodochrosite. Found at Nevel quarry(!), Newry; Bennett quarry, Buckfield, Oxford Co.; with rhodochrosite, eosphorite, reddingite, and other manganese phosphates at Berry quarry, Poland, Androscoggin Co., ME; with eosphorite, triploidite, dickinsonite, reddingite, fillowite, or lithiophilite at *A. N. Fillow quarry, Branchville, Redding, Fairfield Co., CT*; Foote pegmatite(!), Kings Mt., Cleveland Co., NC; Tip Top mine, Custer Co., SD; White Picacho district, Yavapai Co., AZ; Mangualde, Portugal; Hühnerkobel, Rabenstein, Bavaria, Germany; Voroni'y Tundry, Kola Penin., Russia. HCWS *DII:720*, *AM 40:828(1955)*.

40.2.2.2 Messelite $Ca_2Fe^{2+}(PO_4)_2 \cdot 2H_2O$

Named in 1889 for the locality. Fairfieldite/Roselite group. TRIC $P\bar{1}$. $a \approx 5.8$, $b \approx 6.6$, $c \approx 5.5$, $\alpha \approx 102°$, $\beta \approx 109°$, $\gamma \approx 90°$, $Z = 1$. *AM 40:828(1955)*. *10-389*: 9.00_5 6.34_{10} 5.07_5 3.17_{10} 3.02_8 2.79_6 2.68_8 2.57_7. Granular, prismatic, or platy fibrous, sheaflike aggregates. Colorless, pale green to white, greenish gray; white streak; vitreous to pearly luster; transparent. Cleavage {001}, perfect; {010}, good; {1$\bar{1}$0}, distinct; uneven fracture; brittle. $H = 3\frac{1}{2}$. $G = 3.16$. Found at Palermo No. 1 mine, North Groton, NH; at Cobalt, CT; at Foote mine, King's Mt., Cleveland Co., NC; with hureaulite at Bull Moose mine, and in altered triphylite at the Tip Top mine, Custer Co., as large crystals with siderite, ludlamite at Big Chief mine(!), and with triphylite at Ingersol, Dan Patch mines, Pennington Co., SD; as late hydrothermal product in granite pegmatites rimmed with anapaite near Prinz von Hesse coal mine, *Messel, Hessen, Germany*; at Lavra do Enio, Galileia, MG, Brazil. HCWS *AM 44:469(1959), RM 54:255(1979)*.

40.2.2.3 Collinsite $Ca_2Mg(PO_4)_2 \cdot 2H_2O$

Named in 1927 for William Henry Collins (1878–1937), director of the Geological Survey of Canada. Fairfieldite/Roselite group. TRIC $P\bar{1}$. $a = 5.734$, $b = 6.780$, $c = 5.441$, $\alpha = 97.29°$, $\beta = 108.56°$, $\gamma = 107.28°$, $Z = 1$, $D = 2.955$. *MM 39:577(1974)*. *26-1063*: 5.00_4 3.24_5 3.14_5 3.04_{10} 3.02_5 2.74_6 2.71_8 2.68_9. See also *27-83* (Reaphook), *35-635*. Radial blades or lathlike fibers elongated [100]. Light brown, translucent, aggregates silky. Cleavage {001}, {010}, fair. $H = 3\frac{1}{2}$. $G = 2.99$. Biaxial (+); $N_x = 1.632$, $N_y = 1.642$, $N_z = 1.657$; $2V = 80°$; pale yellow brown to colorless. Easily soluble in acids. Type material probably contains some Fe^{2+}. Found with dark brown, fibrous carbonatian fluorapatite at *François Lake, BC, Canada*, as concentric alternate crusts with radial fibrous structure on fragments of andesite in a veinlike deposit and associated with asphaltum. Abundant at Lavrada Ilha Taquaral, MG, Brazil, with eosphorite and other phosphates; also at Reaphook Hill, SA, and Mulgun Station, WA, Australia. HCWS *DII:722, AJC 27:653(1974)*.

40.2.2.4 Cassidyite $Ca_2Ni(PO_4)_2 \cdot 2H_2O$

Named in 1967 for William A. Cassidy, who mapped Wolf Creek crater, WA, Australia, where the mineral was found. Fairfieldite/Roselite group. In a series Ni–Mg with collinsite (40.2.2.3). TRIC $P\bar{1}$. $a = 5.71$, $b = 6.43$, $c = 5.41$, $\alpha = 96.8°$, $\beta = 107.3°$, $\gamma = 104.6°$, $Z = 1$, $D = 3.26$. *AM 52:1190(1967)*. *20-228*: 3.49_4 3.23_6 3.13_5 3.03_9 2.70_{10} 2.67_8 2.23_4 1.660_5. Crusts and small spherules. Pale to bright green, translucent. $G \approx 3.1$. Biaxial; $N_{x'} = 1.64$, $N_{z'} = 1.67$; optically colorless, finely fibrous, length-slow. Contains some Mg. Occurs in the cavities and cracks in weathered meteorites at *Wolf Creek, WA, Australia*. HCWS

40.2.2.5 Talmessite $Ca_2Mg(AsO_4)_2 \cdot 2H_2O$

Named in 1960 for the locality. Fairfieldite/Roselite group TRIC $P\bar{1}$. $a = 5.874$, $b = 6.943$, $c = 5.537$, $\alpha = 97.3°$, $\beta = 108.7°$, $\gamma = 108.1°$, $Z = 1$, $D = 3.42$. Structure: *BM 100:230(1977)*. *38-451*: 5.11_5 4.62_3 3.59_5 3.35_4 3.21_9 3.08_{10} 2.82_4 2.78_8. See also *17-164*. Fine crystalline aggregates. Green, colorless, translucent. $H = 5$. $G = 3.57$. TGA. Biaxial; $N_x = 1.672$, $N_y = 1.685$, $N_z = 1.698$; $2V \approx 90°$. Found at Gold Hill, UT; at Cross Creek, near Savana, BC, Canada; with barite at Lucéram, Alpes-Maritimes, France; at the *Talmessi mine, 35 miles W of Anarak, Iran*, associated with dolomite; at Bou-Azzur, Morocco. HCWS *AM 45:1315(1960), 50:813(1965); BM 83:118(1960), 87:169(1964)*.

40.2.2.6 Gaitite $Ca_2Zn(AsO_4)_2 \cdot 2H_2O$

Named in 1980 for Robert Irwin Gait (b.1938), curator of mineralogy, Royal Ontario Museum, Toronto, Canada. Fairfieldite/Roselite group TRIC $P\bar{1}$. $a = 5.90$, $b = 7.61$, $c = 5.57$, $\alpha = 111.67°$, $\beta = 70.83°$, $\gamma = 119.47°$, $Z = 1$, $D = 3.80$. *CM 18:197(1986)*. *33-598*: 5.05_4 3.58_4 3.35_5 3.21_5 3.08_8 2.78_{10} 2.75_7 1.72_6. Coatings and small crystals up to 1 mm showing {010}, {001}, {0$\bar{1}$1}, {001} (small). White to colorless, vitreous luster, translucent. Cleavages {010}, {001}, {0$\bar{1}$1}, good but not easily observed. $H \approx 5$. $G = 3.18$. DTA. Biaxial (+); $N_x = 1.713$, $N_y = 1.730$, $N_z = 1.730$; $2V = 88°$. When Zn/Mg = 1:1: Biaxial (−); $N_x = 1.689$, $N_y = 1.707$, $N_z = 1.727$; $2V = -85°$. Found on prosperite, adamite, and austinite at *Tsumeb mine, Tsumeb, Namibia*. HCWS *AM 66:1274(1981)*.

40.2.2.7 β-Roselite $Ca_2Co(AsO_4)_2 \cdot 2H_2O$

Named in 1955 for its dimorphic relationship to roselite (40.2.3.1). Fairfieldite/Roselite group. TRIC $P1$ or $P\bar{1}$. $a = 5.89$, $b = 7.69$, $c = 5.56$, $\alpha = 112.6°$, $\beta = 70.8°$, $\gamma = 119.4°$, $Z = 1$, $D = 3.77$. *AM 40:828(1955)*. *17-166*: 5.07_3 3.56_6 3.33_4 3.20_8 3.07_8 2.78_{10} 2.16_4 1.714_8. Granular masses. Dark rose-red, vitreous luster, transparent to translucent. Cleavage {010}, perfect. $H = 3\frac{1}{2}-4$. $G = 3.71$. Biaxial (−); $N_x = 1.723$, $N_y = 1.737$, $N_z = 1.756$; $2V = 80-90°$. Found with roselite and calcite in quartz vein at *Schneeberg, Saxony, Germany*; with erythrite at Bou-Azzer, Morocco; at Tsumeb, Namibia. HCWS *AM 45:1315(1960)*.

40.2.2.8 Parabrandtite $Ca_2Mn(AsO_4)_2 \cdot 2H_2O$

Named in 1987 to show its dimorphic relationship with brandtite. Fairfieldite/Roseite group. TRIC $P\bar{1}$ or $P1$. $a = 5.89$, $b = 7.03$, $c = 5.64$. $\alpha = 96.77°$, $\beta = 109.32°$, $\gamma = 108.47°$, $Z = 1$, $D = 3.60$. *NJMM 157:113(1987)*. *40-1457*: 5.11_4 3.61_7 3.37_4 3.23_5 3.09_8 2.81_{10} 2.78_5 2.04_4. Aggregates to 1.5 mm in diameter of crystals in parallel growth. Colorless, white streak, vitreous luster, transparent. Cleavage {010}, {110}, perfect. $H = 3-4$. $G = 3.55$. Nonfluorescent in UV. Biaxial (+); $c = [001]$, b^* perpendicular to (010); $N_x = 1.701$, $N_y = 1.721$, $N_z = 1.751$; $2V = 79.9°$; $r > v$, weak. Found in a cavity with sarkinite in a vein of primary ore containing franklinite, pink calcite, and red willemite at *Sterling Hill mine, Sussex Co., NJ.* HCWS *AM 73:1496(1988)*.

40.2.3.1 Roselite $Ca_2(Co^{2+},Mg)(AsO_4)_2 \cdot 2H_2O$

Named in 1825 for Gustav Rose (1798–1873), professor of mineralogy, University of Berlin, Germany. Fairfieldite/Roselite group MON $P2_1c$. $a = 5.801$, $b = 12.989$, $c = 5.617$, $\beta = 107.42°$, $Z = 2$, $D = 3.617$. Structure: *CM 15:36(1977)*. *29-315*: 6.44_2 5.09_2 3.75_1 3.40_3 3.35_6 3.22_8 2.99_{10} 2.76_5. Druses or spherical aggregates of short prismatic [001], or thick tabular {001} crystals with large {122}, {$\bar{1}22$}. Twinned on {100} with {100} as composition plane, often as fourlings. *DII:724*. Dark rose-red to pink, vitreous luster, transparent to translucent. Cleavage {010}, perfect, easy. $H = 3\frac{1}{2}$. $G = 3.62$. Biaxial $(-)$; $X = b$; $Y \wedge c = 12$–$20°$, $Z \wedge c = -70$–$78°$; $N_x = 1.725$, $N_y = 1.728$, $N_z = 1.735$; $2V = 60°$; $r < v$, weak. X, dark rose; Y, pale rose; Z, paler rose. Indices, and deep color, vary with Co content, which substitutes for Ca and Mg and may produce zonation within crystals. Found at Molviza, Spain; at *Daniel and Rappold mines, Schneeberg, Saxony, Germany*; at Bou Azzer, Morocco(!), with β-roselite and erthyrite; at Tsumeb, Namibia. HCWS *DII:723*, *AM 72:217(1987)*.

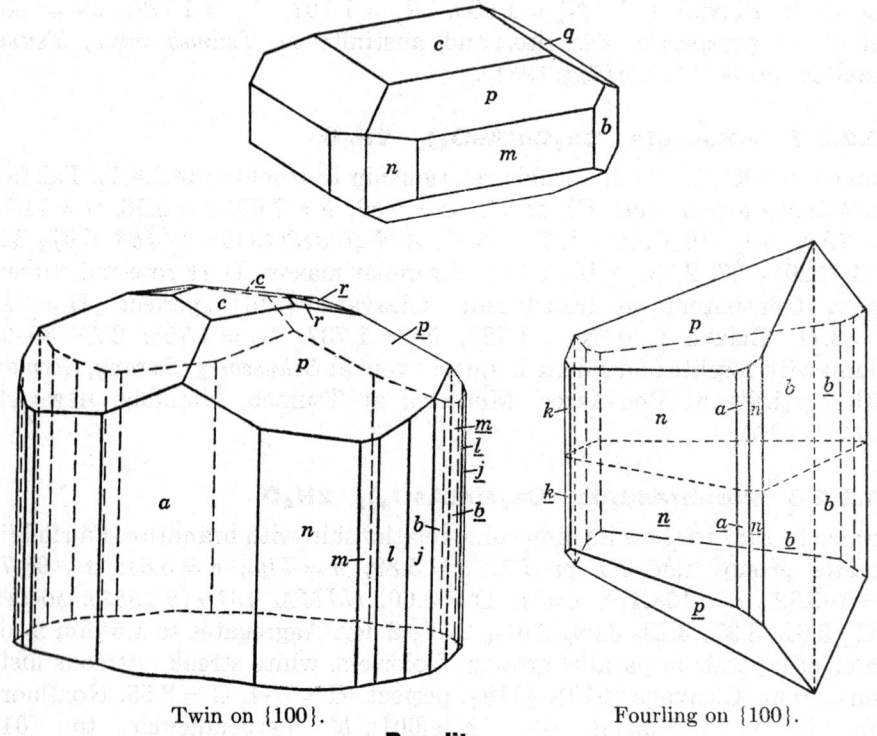

Twin on {100}.   Fourling on {100}.
Roselite

Forms: a={100}, b={010}, c={001}, k={8.10.0}, l={130}, m={110}, n={140}, p={122}, r={T38}. Underlined letters refer to faces of second crystals (twin).

40.2.3.2 Brandtite $Ca_2(Mn^{2+},Mg)(AsO_4)_2 \cdot 2H_2O$

Named in 1888 for Georg Brandt (1694–1768), Swedish chemist. Fairfieldite/Roselite group. MON $P2_1/c$. $a = 5.899$, $b = 12.968$, $c = 5.684$, $\beta = 108.05°$, $Z = 2$, $D = 3.70$. CM 15:36(1977). Structure: SR III:653(1963). 29-348: 6.48_2 5.15_2 3.79_2 3.42_3 3.37_5 3.24_7 3.01_{10} 2.27_1. Stout prismatic [001] with many prism forms visible. {100} striated [001]. Often united in radiating groups; rounded or reniform masses with radial fibrous structure. Twinning on {100} common. Colorless to white, vitreous luster, transparent to translucent. Cleavage {010}, perfect. $H = 3\frac{1}{2}$. $G = 3.67$. Biaxial (+); X= b; $Y \wedge c = 6°$, $Z \wedge c = 84°$; $N_x = 1.709$, $N_y = 1.711$, $N_z = 1.724$; $2V = 23°$; r < v, strong. Soluble in 1 : 10 acids. Found in vug at Sterling Hill, NJ, associated with rhodochrosite, chalcopyrite, franklinite, black willemite, calcite, and sphalerite; at *Harstig mine, Pajsberg near Persberg, Värmland, Sweden*, associated with barite, calcite, sarkinite, flinkite, caryopilite, galena, and native lead; at Rappold and Daniel mines, Schneeberg, Saxony, Germany. HCWS DII:725, AM 25:738(1952).

40.2.3.3 Zincroselite $Ca_2Zn(AsO_4)_2 \cdot 2H_2O$

Named in 1986 for its composition and relationship to roselite. Fairfieldite/Roselite group. MON $P2_1/c$. $a = 5.832$, $b = 12.889$, $c = 5.644$, $\beta = 107.72°$, $Z = 2$, $D = 3.77$. NJMM: 523(1986). 40-498: 6.44_5 5.10_6 3.40_5 3.36_5 3.22_8 3.13_5 3.00_{10} 2.78_8. Aggregates up to 12 mm × 5 mm × 4 mm of intergrown lathlike crystals elongated and strongly striated [001]. Commonly twinned on (100) with {hk0} and {hkl} present. Colorless to white, translucent. Brittle. $H = 3$. $G = 3.75$. Nonfluorescent in UV. Biaxial (+); $Y = b$; $Z \wedge c = 0$–$5°$ in acute β; $N_x = 1.703$, $N_y = 1.710$, $N_z = 1.720$; $2V = 50°$; r < v, strong. Found in a small cavity in oxidized copper ore at *Tsumeb mine, Tsumeb, Namibia*. HCWS AM 73:932(1988).

40.2.3.4 Wendwilsonite $Ca_2Mg(AsO_4)_2 \cdot 2H_2O$

Named in 1987 for Wendell E. Wilson (b.1946), mineralogist, editor, and publisher of the *Mineralogical Record*. Fairfieldite/Roselite group. See also erythrite (40.3.6.3). MON $P2_1/c$. $a = 5.806$, $b = 12.912$, $c = 5.623$, $\beta = 107.40°$, $Z = 2$, $D = 3.57$. AM 72:217(1987). 41-607: 5.09_5 3.40_3 3.36_4 3.23_6 2.99_{10} 2.77_8 2.59_4 2.10_3. Crystals up to 6 mm in size, stout prismatic, elongate on [100] with large {011}, {111}, narrow {110}, {010}. Twinned by reflection on {100} with {100} composition plane; lamellar parallel to {011}, {111}, {010} also possibly the result of twinning. Pale pink to intense red, pale pink streak, vitreous luster, transparent. Cleavage {010}, perfect; uneven fracture; no parting. $H = 3$–4. $G = 3.52$. Nonfluorescent. Biaxial (+); Y parallel to b; $Z \wedge c = 92°$ in obtuse β; $N_x = 1.694$, $N_y = 1.703$, $N_z = 1.717$; $2V = 87°$; r < v; pleochroic. X, violet pink; Y, rose pink; Z, colorless. Found at *Sterling Hill mine, Ogdensburg, NJ*, with calcite crystals on serpentine in franklinite–willemite ore; at Coahuila, COAH, Mexico; associated with talmessite and erythrite on altered ore and calcite gangue at Arhbar mine, Bou Azzer, Morocco. HCWS

40.2.4.1 Prosperite $CaZn_2(AsO_4)_2 \cdot H_2O$

Named in 1979 for Prosper J. Williams (b.1910), South African–Canadian mineral dealer from Toronto, Canada. MON $C2/c$. $a = 19.238$, $b = 7.731$, $c = 9.765$, $\beta = 104.47°$, $Z = 8$, $D = 4.408$. *CM 17:87(1979)*. Structure: *ZK 158:33(1982)*. *33-599*: 3.86_5 3.78_7 3.36_6 3.11_8 2.99_9 2.79_5 2.72_{10} 2.61_6. Radiating sprays of prismatic crystals up to 10 mm long and 1 mm in diameter; sheaflike subparallel aggregates. Crystals elongated parallel to [001] show $\{100\}$, $\{\bar{1}01\}$, $\{110\}$, $\{\bar{1}11\}$, $\{421\}$, $\{540\}$, $\{210\}$, $\{310\}$, $\{301\}$, $\{\bar{1}12\}$, $\{311\}$. White to colorless, white streak, brilliant vitreous to silky luster. $H = 4\frac{1}{2}$. $G = 4.31$. DTA. Nonfluorescent. Biaxial (+); $Y = b$; $Z \wedge c = 27°$; $N_x = 1.746$, $N_y = 1.748$, $N_z = 1.768$; $2V = 34°$; $r \gg v$. Found as a secondary mineral in vugs in partly altered sulfide ore at *Tsumeb mine, Tsumeb, Namibia*, associated with chalcocite, mercurian silver, cuprite, conichalcite, adamite, austinite, and koritnigite. HCWS *AM 65:208(1980)*.

40.2.5.1 Parascholzite $CaZn_2(PO_4)_2 \cdot 2H_2O$

Named in 1981 for its dimorphic relationship to scholzite (40.2.6.1). MON $C2/c$ or Cc. $a = 17.864$, $b = 7.422$, $c = 6.74$, $\beta = 106.45°$, $Z = 4$, $D = 3.10$. *AM 66:843(1981)*. Cell corresponds to subcell of scholzite. *35-495*: 8.55_{10} 4.53_3 4.16_5 3.41_4 3.15_2 2.80_8 2.78_4 2.59_3. Crystals less than 1 mm, tabular, prismatic, show (100), (110), (310), (410), (001), $(\bar{1}01)$, and $(\bar{1}11)$; mostly polysynthetically twinned by reflection on $\{100\}$ giving microscopic lamellae. White to colorless, white streak, vitreous luster. No cleavage, parting along twin composition plane $\{100\}$. $H = 4$. $G = 3.12$. DTA, TGA. Nonfluorescent. Biaxial (+); X parallel to b, $Z \wedge c = 13°$; $N_x = 1.603$, $N_y = 1.588$, $N_z = 1.587$ (Na); $2V = 25°$; $r > v$. Found in association with scholzite on specimens from *Hagendorf Süd and Nord, Bavaria, Germany*; probably present with scholzite everywhere. HCWS

40.2.6.1 Scholzite $CaZn_2(PO_4)_2 \cdot 2H_2O$

Named in 1950 for Adolph Scholz (1894–1950), mineral collector and chemist of Regensburg, Germany. Dimorphous with parascholzite (40.2.5.1). ORTH $Pbcm$ or $Pbc2_1$. $a = 17.149$, $b = 22.236$, $c = 6.667$, $Z = 12$, $D = 3.110$. Structure: *AM 60:1019(1975)*, layered substructure; see also *ZK 198:239(1992)*. *27-95*: 8.65_{10} 4.56_3 4.29_4 3.71_2 3.40_3 3.33_3 2.81_6 2.27_3. See also *13-445* (Hagendorf), *29-1412*. Well-developed crystals up to 1 cm long with 1 mm cross section, elongated parallel to [001] showing $\{100\}$, $\{110\}$, $\{130\}$, $\{231\}$, $\{601\}$. Distinctive habits observed in close proximity; all crystals appear to include lamellae of parascholzite. White to colorless, white streak, vitreous luster. No cleavage, parting along plane of syntaxy $\{100\}$ with parascholzite. $H = 4$. $G = 3.13$. Nonfluorescent. Biaxial (+); XYZ= bac; $N_x = 1.597$, $N_y = 1.586$, $N_z = 1.585$; $2V = 33°$; $r > v$ (Reaphook). Analyses show minor

Mn, Mg, and trace Fe. Synthesized *NJMM:25(1977)*. Found as primary product of crystallization in Zn- and phosphate-bearing pegmatites at Otov, near Domazlice, Bohemia, Czech Republic; as secondary mineral from alteration of phosphates in pegmatites at *Hagendorf, Bavaria, Germany*; at Zabaiykalya, Russia; as a secondary mineral produced during weathering (with enrichment through groundwaters) of sedimentary rocks containing low-grade Zn and phosphate minerals at Richelle, Belgium, and at Reaphook Hill, Flinders Ranges, SA, Australia. HCWS *AM 36:382(1951), 66:843(1981)*.

40.2.7.1 Phosphophyllite $Zn_2(Fe^{2+},Mn^{2+})(PO_4)_2 \cdot 4H_2O$

Named in 1920 in allusion to the phosphate composition and having a perfect cleavage. MON $P2_1/c$. $a = 10.378$, $b = 5.084$, $c = 10.553$, $\beta = 121.14°$, $Z = 2$, $D = 3.058$. *JAC 9:503(1976)*. Structure: *AM 62:812(1977)*. *29-1427:* 8.86_8 4.55_2 4.44_{10} 3.38_6 2.83_5 2.82_4 2.22_3 1.481_3 (Potosi). See also *17-774* (Hagendorf). Thick tabular {100} with a variety of faces showing. *DII:738*. Twinned on {100} sometimes polysynthetic, also rarely on {$\bar{1}$02}. Colorless to pale bluish green; vitreous luster; transparent. Cleavage {100}, perfect; {010}, {$\bar{1}$02}, distinct; brittle. $H = 3–3\frac{1}{2}$. $G = 3.08$. Fluoresces violet in SWUV. Biaxial $(-)$; $Z = b$; $Y \wedge c = 50°$; $N_x = 1.595$, $N_y = 1.614$, $N_z = 1.616$; $2V \approx 45°$; $r > v$. Soluble in acids. Found at Dunton and Bell pegmatites, Newry; Red Hill(!), Rumford, Oxford Co., ME; Palermo No. 1 pegmatite, Groton, Grafton Co., NH; in pegmatite with triplite, triphylite, sphalerite, apatite, vivianite, rockbridgeite, strengite, fairfieldite, or phosphosiderite at *Hagendorf Süd(!), near Pleystein, Oberpfälz, Bavaria, Germany*; in sharp, light green-blue crystals to 14 cm at Unificada mine, Cerro de Potosi; Siglo XX mine, Llallagua, Bolivia; Broken Hill, Zambia; Reaphook Hill, SA, Australia. HCWS.

BRACKEBUSHITE GROUP

The minerals in the brackebushite group conform to the general formula

$$A_2B(H_2O,OH)(TO_4)_2$$

where
A = Ba, Ca, Pb, Sr
B = Al, Cr^{2+}, Fe^{2+}, Fe^{3+}, Mn^{2+}, Mn^{3+}, Zn
T = As, Cr, P, S, V (and possibly Si)
and crystallize in the monoclinic system, space group $P2_1/m$.

The structure is composed of $B-(O,OH)_6$ octahedra, two nonequivalent TO_4 tetrahedra, and two different irregular polyhedra of large cations one in 6- or 8- and the other in 10- or 11-fold coordination with O, OH, or H_2O. Chains formed from the octahedra link through the apex oxygens of $TO_{4(2)}$, while the polyhedra form double chains parellel to [010] through edge-sharing with $TO_{4(1)}$. The result is a tight three-dimenstional structure *TMPM 25:153(1978)*. Infrared investigations on brackebushite and gamagarite indicate strong hydrogen-bonded OH which is probably the dominant species in these minerals. *AM 69:803(1984)*.

BRACKEBUSHITE GROUP

Mineral	Formula	Space Group	a	b	c	β	
Brackebushite	$Pb_2(Mn^{2+}, Fe^{2+})(VO_4)_2$	$P2_1/m$	8.810	6.185	7.681	111.50	40.2.8.1
Arsenobrackebushite	$Pb_2(Fe^{2+}Zn)(AsO_4)_2$	$P2/m$	7.763	6.046	9.022	112.5	40.2.8.2
Gamagarite	$Ba_2(Fe^{3+},Mn^{3+})(OH,H_2O)(VO_4)_2$	$P2_1/m$	9.121	6.142	7.838	112.88	41.10.4.3
Goedkenite	$(Sr,Ca)_2Al(PO_4)_2(OH)$	$P2_2/m$	8.45	5.74	7.26	113.7	41.10.4.1
Bearhite	$Ca_2Al(PO_2)_2(OH)$	$IP2_2/m$	7.231	5.734	8.263	112.57	41.10.4.2
Tsumebite	$Pb_2Cu(PO_4)(SO_4)(OH)$	$P21/m$	8.69	5.78	7.86	111.87	43.4.2.1
Arsentsumebite	$Pb_2Cu(AsO_4)(SO_4)(OH)$	$P2_1/m$	8.85	5.92	7.84	112.60	43.4.2.2
Vanquelinite	$Pb_2Cu(PO_4)(CrO_2)(OH)$	$P2/n$	13.68	5.83	9.53	93.97	43.4.3.1
Fornacite	$Pb_2Cu(AsO_4)(CrO_4)(OH)$	$P2_1/c$	8.022	5.906	17.564	110.4	43.4.3.2
Molybdofornacite	$Pb_2Cu(AsO_4)(MoO_4)(OH)$	$P2_2/c$	8.100	5.946	17.65	109.17	43.4.3.3

Notes: Accumulating species from other classes accentuates the crystallographic similarities between different composition materials.
1. The elongate, acicular, or fibrous morphology with positive optical sign shown by these minerals is indicative of the condensed chainlike motif.
2. See also the adelite–descoloisite group (41.5.1.1).
3. The tornenbohmitie minerals (52.4.5.1, and 52..4.5.2) are related.

40.2.8.1 Brackebuschite $Pb_2(Mn^{2+},Fe^{2+})(VO_4)_2 \cdot H_2O$

Named in 1880 for Luis Brackebusch (1849–1906), professor of mineralogy, University of Cordoba, Argentina. Brackebuschite group. MON $P2_1/m$. $a = 8.810$, $b = 6.185$, $c = 7.681$, $\beta = 111.50°$, $Z = 2$, $D = 6.11$. *MM 39:69(1973)*. *6-284*: 4.95_8 3.25_{10} 3.08_5 2.98_6 2.76_8 2.13_4 2.07_4 1.720_6. Acicular [010] with {100}, {001}, striated [010], and sometimes flattened parallel to elongation; as tufts or groups of crystals; dendritic or botryoidal. Dark brown to black, yellow streak, submetallic luster, translucent to nearly opaque. $G = 6.05$. Biaxial (+); $N_x = 2.28$, $N_y = 2.36$, $N_z = 2.48$, 2V large; r > v, strong; pleochroic. X, colorless; Y, dark reddish brown; Z, clear reddish brown. Found with descloizite, vanadinite in oxidized zone of lead–zinc veins in the western part *Sierra de Cordoba, Cordoba Prov., Argentina*. HCWS

40.2.8.2 Arsenbrackebuschite

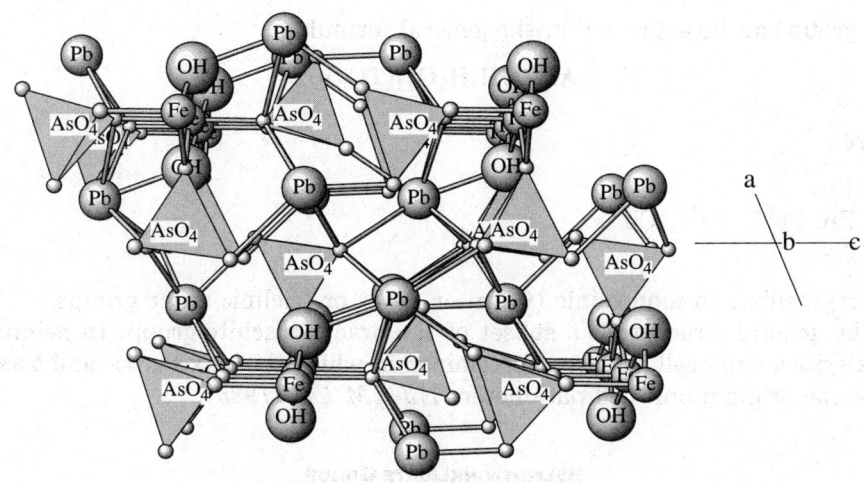

BRACKEBUSCHITE: $Pb_2(Mn^{2+},Fe^{2+})(VO_4)_2 \cdot H_2O$

A 'b' axis projection two irregular A polyhedra, B-(O,OH)$_2$ octahedra and two non-equivalent TO$_4$ groups shows the three dimensional framework structure.

40.2.8.2 Arsenbrackebuschite $Pb_2(Fe^{2+},Zn)(AsO_4)_2 \cdot H_2O$

Named in 1978 in analogy to brackebuschite being the arsenic analog, but note the presence of Fe rather than Mn. Brackebuschite group. MON $P2/m$. $a = 7.763$, $b = 6.046$, $c = 9.022$, $\beta = 112.5°$, $Z = 2$, $D = 6.54$. *NJMM:193(1976)*, Structure: *TMPM 25:153(1978)*. 29-1428: 4.92_2 3.68_2 3.44_2 3.27_9 3.02_{10} 3.00_{10} 2.78_6 2.31_3. Flat laths or platelets to 0.5 mm × 0.2 mm × 0.1 mm. Honey-yellow, light brownish-yellow streak, resinous to adamantine luster. Cleavage {010}, perfect. H = 4–5. Biaxial (−); $N_{x'} < 2.0$, $N_{z'} > 2.04$; pleochroic, honey-yellow to bright yellow. Occurs in the Clara mine, Black Forest, Germany, associated with anglesite, bayldonite, mimetite, and stolzite, intimately intergrown with beudantite; at *Tsumeb mine, Tsumeb, Namibia*. HCWS *AM 63:1282(1978)*.

HELMUTWINKLERITE SUBGROUP

The group can be expressed in the general formula

$$AB_2(OH,H_2O)_2(TO_4)_2$$

where

A = Pb
B = Zn, Fe^{2+}, Fe^{3+}, Cu
T = As,

and crystallizes in monoclinic (C2/m or C2/c) or triclinic space groups.

The general structure is a subset of the brackebuschite group. In helmutwinklerite a supercell has been determined in which the a axis is 4× and b axis is 3× the original unit cell parameters *NJMM 446 (1985)*.

HELMUTWINKLERITE GROUP

Name	Formula	Space group	a	b	c	α	β	γ
Tsumcorite	$Pb_2Fe^{2+}(AsO_4)_2 \cdot 2H_2O$	C2/m	9.124	6.329	7.577		115.3	
Helmutwinklerite	$PbZn_2(OH,H_2O)_2(AsO_4)_2$	P$\bar{1}$, P1	22.40	22.44	7.62	70.2	70.0	69.2
Thometzekite	$PbCu_2(AsO_4)_2 \cdot 2H_2O$							
Mawbyite	$Pb(Fe^{2+}_{1.5},Zn_{0.5})(AsO_4)_2$ $(OH_{1.5},H_2O_{0.5})$	C2/c	9.05	6.277	7.580		114.7°	

40.2.9.1 Tsumcorite $PbZnFe^{2+}(AsO_4)_2 \cdot H_2O$

Named in 1971 for the Tsumeb Corporation. MON $C2/m$. $a = 9.124$, $b = 6.329$, $c = 7.577$, $\beta = 115.3°$, $Z = 2$, $D = 5.39$. *NJMM:446(1985)*, Structure: *AC(B) 29:2789(1973)*, in series with helmutwinklerite (40.2.11.1) and thometzekite (40.2.9.2). *25-399:* 4.66_9 3.24_{10} 3.02_6 2.86_9 2.74_7 2.57_5 2.54_4 2.51_4. Radiating and spherulitic fibrous crusts. Red-brown, yellow streak. $H = 4\frac{1}{2}$. $G = 5.2$. DTA. Biaxial; $N \approx 1.90$; $2V \approx 90°$; dichroic, yellow to yellow-green. Dissolves in HCl. Found at the Richmond–Sitting Bull mine, Galena, Lawrence Co., SD; at the *Tsumeb mine, Tsumeb, Namibia*, in the oxidized zone associated with beudantite, anglesite, mimetite, and other arsenates; on a sample from Thasos, Greece, in the British Museum. HCWS *NJMM:305(1971)*.

40.2.9.2 Helmutwinklerite $PbZn_2(AsO_4)_2 \cdot 2H_2O$

Named in 1980 for Helmut G. F. Winkler (1915–1980), University of Göttingen, Germany. End member in the Zn–Cu series with thometzekite (40.2.9.3); see also tsumcorite (40.2.10.1). TRIC $P1$ or $P\bar{1}$. $a = 22.40$, $b = 22.44$, $c = 7.62$, $\alpha = 70.2°$, $\beta = 70.0°$, $\gamma = 69.2°$, $Z = 1$, $D = 5.29$. *NJMM:118(1980)*, *446(1985)*. Supercell related to tsumcorite and thometzekite. *33-781:* 6.95_4

4.70_8 3.46_4 3.27_7 3.08_7 2.77_5 2.58_{10} 1.735_5. Crystals show {001}, {010}, {1$\bar{1}$0}, {11$\bar{1}$}, pseudomonoclinic. Sky blue, vitreous to resinous luster. No cleavage. H = $4\frac{1}{2}$. G = 5.3. Biaxial (+); $N_x = 1.72$, $N_y = 1.80$, $N_z = 1.98$. Dissolves in cold 10 N HCl. Found in cavities in tennantite, associated with quartz, willemite at *Tsumeb, Namibia*. HCWS *AM 65:1067(1980)*.

40.2.9.3 Thometzekite $PbCu_2(AsO_4)_2 \cdot 2H_2O$

Named in 1985 for W. Thometzek, mining director at Tsumeb from 1912–1922. End member in the Cu–Zn series with helmutwinklerite (40.2.11.1). See also tsumcorite (40.2.10.1). Probably MON. *39-340:* 4.69_8 4.50_5 3.27_{10} 3.16_4 2.96_7 2.87_7 2.58_3 2.52_7. Aggregates of tabular crystals 20 μm long and 1 μm thick. Blue-green to green, earthy. TGA. Biaxial; $N_{av} = 1.855$; light gray; wavy extinction; nonpleochroic. Some Zn and Fe also detected on microprobe analysis. Found on specimen of massive gypsum from *Tsumeb, Namibia*. HCWS *NJMM:446(1985), AM 77:931(1988)*.

40.2.9.4 Mawbyite $Pb(Fe^{2+}Zn)(AsO_4)_2(OH)_3 \cdot H_2O$

Named in 1989 for Maurice Mawby (1904–1977) to honor his contribution to the knowledge of minerals at Broken Hill, NSW, Australia. In series with tsumcorite (40.2.9.1); pure Fe end member may be dimorphous with carminite (41.10.7.1). MON $C2/c$. $a = 9.05$, $b = 6.277$, $c = 7.580$, $\beta = 114.57°$, $Z = 2$, $D = 5.53$. *AM 74:1377(1989)*. PD: 4.65_{10} 4.46_3 3.25_{10} 3.14_3 2.86_4 2.72_7 2.55_5 2.50_3. Drusy crusts of "dogtooth" to prismatic crystals up to 0.2 mm long showing {110}, {$\bar{1}$01}, {001}; rarely tabular; hemispherical, cylindrical, and sheaflike aggregates with platy, spongy appearance. Twins V-shaped, composition plane (100). Orange brown (Zn/Fe = 1:1) to bright reddish brown (pure Fe end member), orange-yellow streak, adamantine luster, transparent to translucent. Cleavage {001}, conchoidal fracture. H = 4. Nonfluorescent. Biaxial; N > 1.94–>2.00 for Fe/Zn = 3:1, for Fe end member N > 2.00; length-fast; pleochroic, brown to reddish brown. Formation probably related to the pH of circulating waters in oxidation zone. Found at *Kintore Opencut, Broken Hill, NSW, Australia*, in the oxidized zone on spessartite, quartz rocks as a reaction halo associated with members of the corkite–beudanite series, adamite, olivenite, duftite, mimetite, balydonite, hidalgoite, pharmacosiderite, and Mn and Fe oxyhydroxides. HCWS

40.2.10.1 Wicksite $NaCa_2(Fe^{2+},Mn^{2+})_4MgFe^{3+}(PO_4)_6 \cdot 2H_2O$

Named in 1981 for Frederick John Wicks (b.1937), curator of mineralogy, Royal Ontario Museum, Toronto, Canada. ORTH *Pbca*. $a = 12.896$, $b = 12.511$, $c = 11.634$, $Z = 4$, $D = 3.58$. *CM 19:377(1981)*. *35-512:* 3.50_2 3.02_8 2.91_8 2.87_3 2.84_3 2.75_{10} 2.57_4 2.12_6. Massive; plates parallel to {010}, striated parallel to [100]. Dark blue almost black, green streak, submetallic luster. Cleavage {010}, good. H = $4\frac{1}{2}$–5. G = 3.54. Nonfluorescent. Biaxial (+); XYZ = *bac*; $N_x = 1.713$, $N_y = 1.718$, $N_z = 1.728$; 2V = 66°; r < v, strong; pleochroic; X = Y > Z. X, blue; Y, greenish blue; Z, pale yellowish brown.

May show anomalous biaxial with extinction to 5°. Found with wolfeiite, atterlyite, marcite, pyrite, and quartz in nodules in *Big Fish area, YT, Canada*. Type locality at 68°28'30" N, 136°29' W. HCWS *AM 67:1077(1982), MR 17:301(1986)*.

40.2.11.1 Grischunite $NaCa_2Mn_5^{2+}Fe^{3+}(AsO_4)_6 \cdot 2H_2O$

Named in 1984 for the locality. ORTH *Pcab*. $a = 12.855$, $b = 13.487$, $c = 12.047$, $Z = 4$, $D = 4.144$. *SMPM 64:1(1984)*, Structure: *AM 72:1225(1987)*. *38-387*: 6.04_3 4.24_3 3.62_7 3.15_9 3.02_8 2.94_6 2.84_{10} 2.68_3. Anhedral grains and lathlike crystals up to 1 mm elongated parallel to b. Dark red-brown, yellow-red to yellow-brown streak, vitreous luster, translucent. Cleavage (010), perfect. $H \approx 5$. $G = 3.8$. IR. Biaxial; $XYZ = bac$; $N_x = 1.784$, $N_y = 1.785$, $N_z = 1.790$; $2V = 40$–$50°$; pleochroic; absorption $Z \approx Y \gg X$, $r \ll v$. X, yellow green; Y, yellow-brown; Z, dark red-brown. Found as an alteration product of sarkinite, with which it is intimately intergrown, in the manganese deposits near *Falotta, Oberhalbstein, Grisons Canton, Switzerland*, with brandtite, sarkinite, Mn-berzeliite, and tilasite. HCWS *AM 71:227(1986)*.

AUTUNITE GROUP

The autunite group minerals contain uranium as expressed in the formula

$$A[(UO_2)_2(TO_4)_2] \cdot nH_2O$$

where

A = Ca, Ce, Ba, K, NH_4, Sr, Pb, Mg, Na, Co, Zn, or H_3O^+
T = P, As, V
n = variable, related to the number of bonded and unbonded water molecules, and the cation(s) species and charge balance

and the mineral crystallize with tetragonal or pseudotetragonal monoclinic symmetry.

The structure is dominated by sheets composed of the XO_4 tetrahedral groups linked to dumbbell-shaped $(UO_2)^{2+}$ ions. The axis of the UO_2 dumbbell is perpendicular to the sheet, with the uranyl ion having an alternate upward or downward displacement from the plane of the X position atoms. The puckered $[(UO_2)(TO_2)]$ sheets are stacked perpendicular to the c axis and are separated by a layer composed of H_2O and cations, usually symmetrically arrayed, so that they form a mirror plane. This site, parallel to {001}, may also contain unbonded H_2O molecules which are easily mobilized on dehydration, and it is a locus of structural weakness, as is evident in the perfect cleavage of these minerals.

The autunite group is divided into two subgroups, autunites and meta-autunites, based on crystal structure differences initially determined on synthetic materials. *RC 57:155(1938)*. Both subgroups contain PO_4 and AsO_4

mineral species and are characterized by solid solutions for similarly charged and coordinated A cations in addition to substitution in the X site. The variable size and valence of the cations in the water layer of the structure influence not only the composition but also the stacking and bonding schemes. Members of the meta-autunite subgroup represent the lower, and often more stable, hydration states of a particular cation. *USGSB: 1036(1956), 1964(1958); AM 43:799(1958); CE 24:254(1965).*

In the fully hydrated species (autunite, uranospinite, torbernite, etc.), the UO_2–PO_4 sheets have maximum separation, and the unit cell has the largest c axis. When subjected to a different temperature and humidity, the minerals dehydrate and the result is a decrease in c-axis length, but the planar character, weakness at water layers, and perfect cleavage remain.

For a specific cation there are usually multiple hydrates, and in most cases a lower hydrate is the stable phase at room temperature and humidities. Related hydrated species retain the name of the fully hydrated version; that is, autunite in lower hydrate(s) is indicated by the prefix "meta-," giving meta-autunite. In cases where there are many hydrates a roman numeral has been used, or most recently, the c-axis spacing is appended. Uranocircite I, for example, has $c = 22.6$ Å, or may be referred to as "uranocircite 22 Å"; while uranocircite II, with $c = 20$ Å, may be called "uranocircite 20 Å." For this specific species even lower hydrates, meta-urancircite I and II, have been designated.

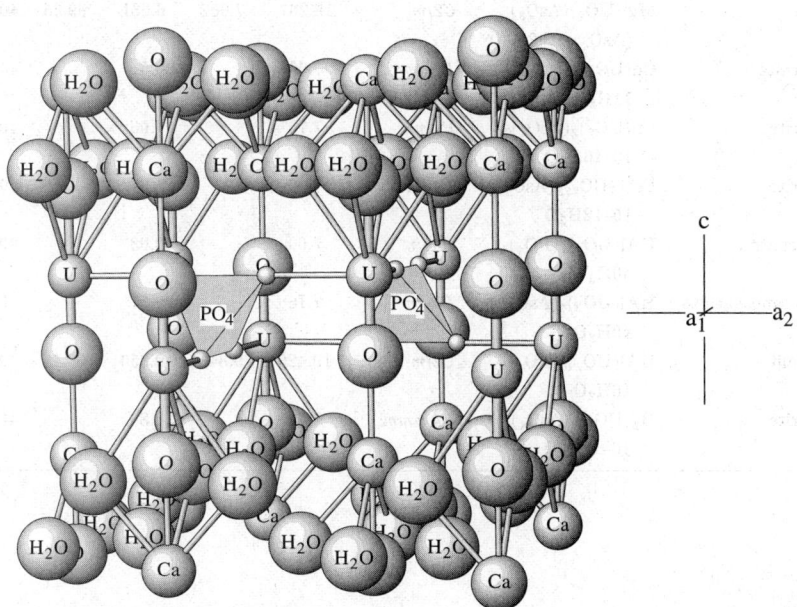

AUTUNITE: Ca[$(UO_2)_2$ $(PO_4)_2$] · 10-12H_2O

The planar character of the [$(UO_2)_2(PO_4)_2$] layers and cation -H_2O layers in this 'a' axis projection.

AUTUNITE GROUP: AUTUNITE SUBGROUP

Mineral	Formula	Space Group	a	b	c	β	
Autunite	$Ca(UO_2)_2(PO_4)_2 \cdot 10\text{–}12H_2O$	$I4/mmm$	7.009		20.736		40.2a.1.1
Uranospinite	$Ca(UO_2)_2(AsO_4)_2 \cdot 10H_2O$	$I4/mmm$	7.17		20.61		40.2a.2.1
Uranocircite	$Ba(UO_2)_2(PO_4)_2 \cdot 10H_2O$	$P4/nnc$	7.01		20.46		40.2a.3.1
Heinrichite	$Ba(UO_2)_2(AsO_4)_2 \cdot 10\text{–}12H_2O$	$P4/nnc$	7.13		20.56		40.2a.4.1

AUTUNITE GROUP: AUTUNITE SUBGROUP (CONT'D)

Mineral	Formula	Space Group	a	b	c	β	
Sodium autunite	$Na_2(UO_2)_2(PO_4)_2 \cdot 8H_2O$	$P4/nmm$	6.97		8.96		40.2a.5.1
Novacekite	$Mg(UO_2)_2(AsO_4)_2 \cdot 10 \text{ or } 12H_2O$	$P4_2/n$	14.30		22.00		40.2a.10.1
Saleeite	$Mg(UO_2)_2(PO_4)_2 \cdot 10H_2O$	$P2/m$	6.951	19.947	9.896	135.17	40.2a.11.1
Seelite	$Mg[(UO_2)(AsO_3)_x(AsO_4)_{1-x}] \cdot 7H_2O$	$C2/m$	18.207	7.062	6.661	99.65	40.2a.12.1
Torbernite	$Cu(UO_2)_2(PO_4)_2 \cdot 11H_2O$	$I4/mmm$	7.06		20.5		40.2a.13.1
Zeunerite	$Cu(UO_2)_2(AsO_4)_2 \cdot 10\text{–}16H_2O$	$P4/nnc$	7.18		21.06		40.2a.14.1
Kahlerite	$Fe^{2+}(UO_2)_2(AsO_4)_2 \cdot 10\text{–}12H_2O$	$P4_2/n$	14.30		21.97		40.2a.15.1
Uranospathite	$HAl(UO_2)_4(PO_4)_4 \cdot 40H_2O$	$P4_2/n$	7.00		30.03		40.2a.22.1
Arsenuranospathite	$HAl(UO_2)_4(AsO_4)_4 \cdot 40H_2O$	$P4_2/n$	7.16		30.37		40.2a.23.1
Sabugalite	$HAl(UO_2)_4(PO_4)_4 \cdot 16H_2O$	$C2/m$	19.426	9.843	9.850	96.16	40.2a.24.1
Trögerite	$H_2(UO_2)_2(AsO_4)_2 \cdot 10\text{–}12H_2O$	$P4/nmm$	7.16		8.80		40.2a.32.1

Autunite Group: Meta-Autunite Subgroup

Mineral	Formula	Space Group	a	b	c	β	
Meta-autunite	$Ca(UO_2)_2(PO_4)_2 \cdot 2\text{--}6H_2O$	$Pmmm$	6.551	7.053	8.164		40.2a.1.2
Meta-uranospinite	$Ca(UO_2)_2(AsO_4)_2 \cdot 6H_2O$	$P4/nmm$	7.14		17.00		40.2a.2.2
Meta-uranocircite	$Ba(UO_2)_2(PO_4)_2 \cdot 8H_2O$	$P4_2/n$	6.953		17.634		40.2a.3.2
Meta-heinrichite	$Ba(UO_2)_2(AsO_4)_2 \cdot 8H_2O$	$P4_2/?$	7.08		17.66		40.2a.4.2
Sodium uranospinite	$(Na_2,Ca)(UO_2)_2(AsO_4)_2 \cdot 5H_2O$	$P4/nmm$	7.12		8.70		40.2a.6.1
Uramphite	$(NH_4)_2(UO_2)_2(PO_4)_2 \cdot 6H_2O$	$P4/nnc$	7.02		18.08		40.2a.7.1
Meta-ankoleite	$K_2(UO_2)_2(PO_4)_2 \cdot 6H_2O$	$P4/nmm$	6.993		8.891		40.2a.8.1
Abernathyite	$K(UO_2)(AsO_4)_2 \cdot 3H_2O$	$P4/ncc$	7.176		18.126		40.2a.9.1

Autunite Group: Meta-Autunite Subgroup (cont'd)

Mineral	Formula	Space Group	a	b	c	β	
Meta-novacekite	$Mg(UO_2)_2(AsO_4)_2 \cdot 8H_2O$	$P4/n$	7.16		8.58		40.2a.10.2
Meta-torbernite	$Cu(UO_2)_2(PO_4)_2 \cdot 8H_2O$	$P4/n$	6.969		17.306		40.2a.13.2
Meta-zeunerite	$Cu(UO_2)_2(AsO_4)_2 \cdot 8H_2O$	$P4_2/n$	7.12		17.45		40.2a.14.2
Meta-kahlerite	$Fe^{2+}(UO_2)_2(AsO_4)_2 \cdot 8H_2O$	$P4/nmm$	7.16		8.62		40.2a.15.2
Bassetite	$Fe^{2+}(UO_2)_2(PO_4)_2 \cdot 8H_2O$	$P2_1/m$	6.98	17.07	7.01	90.5	40.2a.16.1
Lehnerite	$Mn^{2+}(UO_2)_2(PO_4)_2 \cdot 8H_2O$	$P2_1/n$	7.04	17.16	6.95	90.18	40.2a.16.2
Meta-kirscheimerite	$Co(UO_2)_2(AsO_4)_2 \cdot 8H_2O$	$P4/nmm$	7.17		17.36		40.2a.17.1
Meta-lodevite	$Zn(UO_2)_2(AsO_4)_2 \cdot 10H_2O$	$P4_2/n$	7.16		17.20		40.2a.18.1
Chernikovite	$H_3O_2(UO_2)_2(PO_4)_2 \cdot 6H_2O$	$P4/nmm$	7.020		9.043		40.2a.19.1
Trögerite	$H_2(UO_2)_2(A_5O_4)_2 \cdot 8H_2O$	$P4/nmm$	7.16		8.80		40.2a.20

Notes:
1. Dehydration of autunite subgroup minerals to meta-autunite subgroup results in a reduction of the c axis unit cell parameter (interlayer spacing), an increase in level of fluorescence under UV irradiation, and an increase in the index of refraction, often with the appearance of anomalous biaxiality. These changes indicate a departure from the $P4/mmm$ pseudocell which usually can be verified only by detailed crystal structure determinations on rare superior quality crystals.
2. The dehydrated phases often display a grating-like twinning that appears as striae and rectangular domains parallel to (001) and (110) with the crystals resembling checkerboards under crossed polars. The twinning is due to ordering of the interlayer water.
3. Compare these minerals with the members of the carnotite group (40.2a.25–30) which have $[UO_2, V_2O_8]$ layers. The carnotites have different anion geometry.
4. See also uranophane, a hydrated silicate.

40.2a.1.1 Autunite Ca(UO$_2$)$_2$(PO$_4$)$_2$ · 10–12H$_2$O

Named in 1852 for the locality. Autunite group. See meta-autunite (40.2a.1.2), sodium autunite (40.2a.5.1). TET $I4/mmm$. $a = 7.009$, $c = 20.736$, $Z = 2$, $D = 3.10$. RC 57:155(1938), Structure: AM 38:476(1953). 41-1353: 10.4$_9$ 5.19$_7$ 4.93$_4$ 3.57$_{10}$ 3.50$_4$ 3.32$_4$ 2.73$_4$ 2.07$_6$. See also 12-418. Thin tabular {001}, approaching torbernite in form and angles; subparallel growth common; scaly foliated aggregates; thick crusts with serrated surface composed of crystals standing on edge. Twinning on {110}. Oriented growth with torbernite common. Lemon yellow to greenish yellow to pale green, yellowish streak, vitreous luster to pearly on {001}, transparent to translucent. Cleavage {001}, perfect; {100}, indistinct; not brittle. H = 2–2$\frac{1}{2}$. G = 3.15. Strongly fluorescent yellow-green. Anomalously biaxial (−); Z = c; Y parallel to diagonal of plates; N$_x$ = 1.553, N$_y$ = 1.575, N$_z$ = 1.577 (Autun); 2V = 0–53°. X, colorless to pale yellow; Y,Z, yellow to dark yellow. Biaxiality dependent on H$_2$O content. Small amounts of Ba and Mg substitute for Ca with trace amounts of Pb, V, and other elements. Water content decreases with mild heating formation of meta-autunite, the form always found in museum specimens. Synthesized. A secondary mineral in the zone of oxidation derived from alteration of uraninite or other uranium-containing minerals. Found in pegmatites in New England and elsewhere, including the Dunton Gem pit, Newry, ME; the Ruggles mine, Grafton Center, and the Palermo mine, N. Groton, NH; Bedford, Westchester Co., NY; Spruce Pine, Penland, and the Flat Rock mine in Mitchell Co., NC; La Sal mine, Boulder Co., CO; White Signal district, Grant Co., NM; near Keystone, Black Hills, SD; Daybreak and Triple H & J mines, Mt. Spokane, Pend Orielle Co., WA; at Redruth and St. Austell in Cornwall, England; at Sabugal SE of Guarda and at Vizeu, Portugal; at Dom Benito U deposit, Spain; at *L'Ouche de Jau, St. Symphorien, and other localities near Autun, Saône-et-Loire, France*; in the Johanngeorgenstadt district and at Falkenstein, Saxony, and the pegmatites of Hagendorf and Pleystein, Bavaria, Germany; at Ningyo Pass, Okayama/Tottori Pref., Japan; in the Shaba district, Musoni, Zaire; at Kaying-Paksing, western Siang district, Arunachal Pradesh, India; in the Flinders Ranges, SA, Australia; at Malacacheta, MG, Brazil. HCWS *DII:984*; *USGSB 1064:160(1958)*; *AM 45:99(1960)*, *46:812(1961)*.

40.2a.1.2 Meta-Autunite Ca(UO$_2$)$_2$(PO$_4$)$_2$ · 2–6H$_2$O

Named in 1938 for its relation to autunite. Autunite group. Several hydrates have been described. 2H$_2$O (or 8 Å): ORTH $Pmmm$. $a = 6.551$, $b = 7.053$, $c = 8.164$, $Z = 1$, $D = 3.55$. 14-73: 8.17$_{10}$ 4.81$_1$ 4.14$_3$ 4.08$_6$ 3.51$_1$ 2.96$_1$ 2.72$_1$ 2.04$_1$. N$_O$ = 1.630, N$_E$ = 1.61–1.62 (Ningyo Pass, Japan). 3H$_2$O (or 9 Å): TET $P4/mmm$. $a = 6.988$, $c = 8.459$, $Z = 1$, $D = 3.31$. 39-1351: 8.46$_{10}$ 5.39$_2$ 4.23$_2$ 3.62$_6$ 3.50$_2$ 2.62$_3$ 2.12$_4$ 1.601$_3$. N$_O$ = 1.604, N$_E$ = 1.596 (Oudots mine, Saône et Loire, France, heated 2 hours at 52°). 3H$_2$O (9 Å): TET $P4_2$. $a = 19.72$, $c = 16.92$, $Z = 16$, $D = 3.53$. [Strong pseudocell ($a = 6.97$, $c = 8.46$) rotated 45° to true cell noted. *GSAPA 13:588(1981)*.] See also *28-*

40.2a.2.1 Uranospinite

247(syn), *12-432*. Meta-autunites appear naturally, and reversibly, on slight heating or drying, and are virtually always the usual mineral form identified. Properties similar to autunite: light to dark green and yellow, vitreous luster. H = 1. G = 3.44. Uniaxial (−); $N_O = 1.604$, $N_E = 1.596$. Sr-, Co-, Fe-, Li-, and Ni-meta-autunites have been prepared. *AM 44:702(1959), 48:1589(1963); GSAPA 31002:727(1983)*. Meta-autunite occurs at most of the localities listed for autunite and forms irreversibly when autunite is heated above ≈ 80°. Dark green (may contain admixed uraninite) meta-autunite was identified at *Daybreak mine, Mt. Spokane, WA*. HCWS *DII:985, AM 46:812(1961)*.

40.2a.1.3 Pseudoautunite $(H_3O)_4Ca_2(UO_2)_2(PO_4)_4 \cdot 5H_2O$

Named in 1964 for its similarity to autunite and meta-autunite in composition and occurrence. Composition questioned, name unacceptable. *MM 36:1144(1968)*. ORTH. $a = b \approx 6.964$, $c \approx 12.85$, Z = 2, D = 3.29. *AM 50:1505(1965)*. *18-1084:* 6.20_{10} 3.38_8 3.25_{10} 2.95_8 2.19_8 2.06_8 1.915_9 1.294_8. Fine platy, powdery crusts that consist of intergrowths of small, up to 0.1 mm, pseudohexagonal plates with micalike cleavage. Pale yellow to white. G = 3.28. Intensely yellow-green luminescence in SWUV with persistence, weakly luminescent LWUV. DTA. Biaxial (−); $N_x = 1.541$, $N_{y,\ calc} = 1.568$, $N_z = 1.570$; plane of optic axis parallel to (001), Y perpendicular to (001); extinction parallel or 2° to (010); r > v, distinct. Found in fissures and cavities of albite–acmite veins in the contact zone of ultrabasic–alkalic rocks of *northern Karelia, Russia*, associated with oxonium pyrochlore, calcite, limonitized sulfides, and sometimes apatite. HCWS *MG 1:31(1964)*.

40.2a.2.1 Uranospinite $Ca(UO_2)_2(AsO_4)_2 \cdot 10H_2O$

Named in 1873 for uranium and the Greek for *siskin*, in allusion to its composition and color. Autunite group. See meta-uranospinite (40.2a.2.2). TET *I4/mmm*. $a = 7.17$, $c = 20.61$, Z = 2, D = 3.25. *JLCM 44:99(1976), TMPM 9:252(1964)*. *29-390*(syn): 10.0_8 5.21_8 4.97_8 3.62_{10} 3.41_9 2.93_6 2.55_6 2.28_6. Thin rectangular plates {001}; aggregates to 20 μm. Lemon yellow to siskin green, pearly luster on {010}. Cleavage {001}, {100}, distinct. H = 2–3. G = 3.45. LWUV fluoresces yellow-green. DTA. Anomalously biaxial (−); X = c; $N_x = 1.560$, $N_y = 1.582$, $N_z = 1.587$ (Schneeberg); 2V = 0–5°; r > v, moderate; pleochroic. X, colorless; Y,Z, pale yellow for natural material ≈ $10H_2O$. Samples from Schneeberg often show cores of meta-zeunerite, zones of uniaxial or biaxial optics. Water content varies with humidity and temperature. Synthesized. A secondary mineral found at Honeycomb Hills, Fish Springs mining district, Juab Co., and near Pahreah in Kane Co., UT; at *Weisser Hirsch mine, Neustädtl, Schneeberg, Saxony, Germany*, associated with walpurgite, meta-uranocircite, trögerite, meta-zeunerite, and at Clara Stollen Halde, Wittichen; at Radium Hill, Mount Painter, SA, Australia; as an alteration product of uraninite associated with niccolite at Solitaria mine, Argentina. HCWS *DII:990, USGSB 1064:183(1958)*.

40.2a.2.2 Meta-uranospinite $Ca(UO_2)_2(AsO_4)_2 \cdot 6$ or $8H_2O$

Named in 1957 for its relationship to uranospinite. Autunite group. Two hydrates with different unit cell parameters have been described. $6H_2O$ (17 Å): TET $P4/nmm$. $a = 7.14$, $c = 17.00$, $Z = 2$, $D = 3.84$. *TMPM 9:252(1965)*. *18-309:* 8.65_{10} 5.53_8 5.08_4 4.36_4 3.67_4 3.57_{10} 3.31_9 3.00_5. $N_z = 1.618$. $8H_2O$ (9 Å): TET $P4/nmm$. $a = 7.19$, $c = 8.81$, $Z = 1$, $D = 3.65$. *AM 38:1159(1953)*. *8-319*(syn): 8.85_{10} 5.57_6 5.10_7 3.74_5 3.59_9 3.34_8 2.55_5 1.804_5. Biaxial $(-)$; $N_x = 1.591$, $N_y = 1.619$, $N_z = 1.621$; $2V = 0-5°$. Sometimes uniaxial $(-)$; $N_O = 1.586$, $N_E = 1.560$; weakly pleochroic. X, colorless; E, pale yellow. Cleavage {001}, perfect; {100}, distinct. For other physical properties, see uranospinite (40.2a.2.1). Rehydrates easily to uranospinite. Found at Clara Stollen Halde and the *Sophia mine, Wittichen, Black Forest, Germany*. HCWS *AM 45:254(1960)*.

40.2a.3.1 Uranocircite $Ba(UO_2)_2(PO_4)_2 \cdot 12$ or $10H_2O$

Named in 1877 for the composition and from the Greek for *falcon* because it was discovered at Falkenstein, Germany. Autunite group. Two hydrates have been described, both rapidly dehydrate irreversibly to meta-uranocircite. $12H_2O$ (also known as I) has $c = 22.59$ Å. $10H_2O$ (II): TET $P4/nnc$. $a = 7.01$, $c = 20.46$, $Z = 2$, $D = 3.46$ (Mentenschwand, Schneeberg, Germany). *JGLBW 6:113(1963)*. *18-199:* 10.14_6 5.10_{10} 2.16_5 2.04_{10} 1.645_5 1.576_5 1.527_5 1.462_6. Crystals resemble autunite, thin tabular {001}. Yellow-green, pearly luster on {001}, transparent to translucent. Twin lamellae parallel to {100}, {010}. Cleavage {001}, perfect; {100}, {010}, distinct. $H = 2\frac{1}{2}$. $G = 3.53$. Fluoresces green. Anomalously biaxial $(-)$; $X = c$; $N_x = 1.574$, $N_y = 1.583$, $N_z = 1.588$; $2V \approx 70°$; pleochroic. X, colorless; Y, Z, pale canary yellow. Found at Cameron area, Coconino Co., AZ; at Annie Creek Mine, Lawrence Co., SD; at *Bergen, near Falkenstein in Voigtland, Saxony, Germany*, as a secondary mineral in quartz veins, and at Wölsendorf, Bavaria, with parsonsite and other secondary uranium minerals in a fluorite deposit. Reported from Rosmaneira, Spain; Srdnia Gora, Banat, Hungary; at Mounana mine, Gabon. HCWS *DII:987*.

40.2a.3.2 Meta-Uranocircite $Ba(UO_2)_2(PO_4)_2 \cdot 6$ or $8H_2O$

Named in 1982 to show relationship to uranocircite. Autunite group. Two hydrates described, $6H_2O$ or II, has been synthesized. $8H_2O$ (I, 18 Å): TET $P4_2/n$. $a = 6.953$, $c = 17.634$, $Z = 2$, $D = 3.94$. *36-407:* 8.82_{10} 5.46_2 4.41_6 3.72_5 3.23_1 2.71_2 2.21_2 2.10_1. Uniaxial $(-)$; $N_O = 1.626$; $N_E = 1.615$. Sample from Streuber, Bergen, Germany. $6H_2O$ (II, 17 Å): TET $P4/nmm$. $a = 6.94$, $c = 16.83$, $Z = 2$, $D = 4.00$. *17-759:* 8.55_{10} 5.39_7 4.91_5 4.25_6 3.61_9 3.48_5 3.21_7 2.60_4. Uniaxial $(-)$; $N_O = 1.626$, $N_E = 1.611$. Sample from Menzenschevand, Germany. But note the synthetic $6H_2O$ [II, 17Å(syn)]: MON (pseudo-ORTH) $P2_1/a$. $a = 9.855$, $b = 9.756$, $c = 16.84$, $\beta = 90.6°$, $Z = [4]$, $D = 4.0$. *25-1468*(syn): 8.34_9 5.35_3 4.27_4 4.23_4 3.60_{10} 3.47_3 3.21_3 2.61_3. Structure of $6H_2O$ species shows corrugated UO_2PO_4 layers parallel to (001) with monoclinic distortion. Ba is in ninefold coordination, and two of the six H_2O molecules are not bonded to Ba. TMPM *23:183(1982)*. Platy crystals flattened on {001} with rectangular outline, subparallel aggregates, fanlike groups. Synthetic crystals to 0.6 mm, micaceous "books." Lemon-yellow, pearly luster on {001}, transparent to translucent. G = 3.964. Strongly fluorescent green. Easily synthesized from cation exchange starting with schoepite through chernikovite. MM *56:367(1992)*. Found at Danvier mine, Lawrence Co., in uraniferous lignite, Slim Buttes, Harding Co., and in Chandrom formation, SD; at Kerségelac, Brittany, France; in the Wittichen district., Heubachtal, and at Schmiedestollenhalde, Germany; at Mounana, Gabon; in an alluvial deposit near Antsitabe, Madagascar; and other localities listed under uranocircite (40.2a.3.1). HCWS USGSB *1064:211(1958)*.

40.2a.4.1 Heinrichite $Ba(UO_2)_2(AsO_4)_2 \cdot 10-12H_2O$

Named in 1958 for Eberhardt William Heinrich (1918–1991), mineralogist, University of Michigan, Ann Arbor, MI. Autunite group. See meta-heinrichite (40.2a.4.2). TET $P4/nnc$. $a = 7.13$, $c = 20.56$, $Z = 2$, $D = 3.50$. TMPM *9:252(1965)*. *29-210*(syn): 10.0_{10} 5.09_7 3.53_{10} 3.35_8 2.50_4 2.25_6 1.78_5 1.60_7. See also *18-198*. Small scaly crystals and aggregates. Oregon samples are tablets up to 1 mm long, not more than 0.1 mm thick, showing {001}, {100}, {110}. Black Forest crystals elongated, resemble bassetite (40.2a.15.1). Yellow-green; vitreous luster, pearly on cleavage; transparent to translucent. $H = 2\frac{1}{2}$. Fluoresces bright green to greenish yellow. DTA. Uniaxial (−); $N_O = 1.605$, $N_E = 1.573$ (Grube Anton); $2V = 32°$ (max.). O, bright yellow; E, colorless. Readily dehydrates to meta-heinrichite (40.2a.4.2). Synthesized. Found as fracture coatings in silicified rhyolite tuff in the *White King Mine, 22 km NW of Lakeview, Lake Co., OR*; as an alteration product of pitchblende associated with zeunerite, novacekite, arseniosiderite, pitticite, and erythrite at Grube Anton, Schiltach, and Reinerzau in the Black Forest, Wittichen district, Heubachtal, Schmiedestollenhalde, Germany. HCWS *AM43:1134(1958)*.

40.2a.4.2 Meta-Heinrichite $Ba(UO_2)_2(AsO_4)_2 \cdot 8H_2O$

Named in 1958 for its relation to heinrichite. Autunite group. TET Space group incompletely determined, $P4_1/$. $a = 7.08$, $c = 17.66$, $Z = 2$, $D = 4.09$. TMPM *9:252(1965)*. *24-128*: 8.90_{10} 5.54_7 4.99_5 4.42_6 3.75_8 3.55_6 3.28_6 2.98_5. Yellow-green; vitreous, pearly luster. G = 4.04 (Oregon material dried). Uniaxial (−); $N_O = 1.641$, $N_E = 1.608$, anomalously biaxial $2V \approx 28°$ (Black Forest). Found at *White King mine, Lakeview, Lake Co., OR*; at localities in Black Forest region, Germany, where heinrichite occurs. HCWS *AM 43:1134(1958)*.

40.2a.5.1 Sodium Autunite $Na_2(UO_2)_2(PO_4)_2 \cdot 8H_2O$

Named in 1958 for its relation to autunite and the composition. Autunite group. TET $P4/nmm$. $a = 6.97$, $c = 8.96$, $Z = 1$, $D = 3.89$. AM *43:383(1958)*. *29-1283*: 8.57_5 3.67_{10} 3.23_7 2.68_8 2.12_6 1.639_7 1.566_8 1.540_8. See also *29-*

1284(syn). Plates, foliated and radiating masses. Lemon yellow–lettuce yellow, pearly luster on {001}. Cleavage {001}, perfect; also {100}; brittle. H = 2–2$\frac{1}{2}$. G = 3.584. DTA. Strongly luminescent yellow-green. Uniaxial (−); N_O = 1.564, N_E = 1.558; after 2 days at 40°, N_O = 1.585, N_E = 1.578; weakly pleochroic. O, light yellow; E, pale yellow. Soluble in acids. Synthesized (3H$_2$O): a = 7.00, c = 8.71. *JLCM 44:99(1976)*. Found in lignite deposits in Cave Hills and Slim Buttes areas, Harding Co., SD; in granodiorite massif, *Russia*. HCWS *AM 14:7(1929)*.

40.2a.6.1 Sodium Uranospinite (Na$_2$,Ca)(UO$_2$)$_2$(AsO$_4$)$_2$ · 5H$_2$O

Named in 1953 for the relationship to uranospinite and the composition. Autunite group. TET $P4/nmm$. a = 7.12, c = 8.70, Z = 1, D = 3.80. *AM 38:1159(1953)*. *28-1165*(syn): 8.67$_{10}$ 5.56$_7$ 4.34$_6$ 3.72$_{10}$ 3.58$_8$ 3.32$_9$ 3.01$_6$ 2.68$_6$. See also *8-446*(syn). Tablets up to 2 cm, radiating fibrous aggregates, square crystals pseudomorphic after meta-zeunerite. Lemon yellow to straw yellow; vitreous luster, pearly on {001}. Cleavage {001}, perfect; {100}, {010}, distinct. H = 2$\frac{1}{2}$. G = 3.846. DTA. Fluoresces light yellow-green, less bright in SWUV. Uniaxial (−); N_O = 1.617, N_E = 1.586. O, colorless; E, yellowish. May be anomalously biaxial. Synthesized: *AM 36:322(1951)*; 3.5 H$_2$O: *JLCM 44:99(1976)*. Found in pitchblende-S deposits, unspecified locality *Russia*, associated with arsenopyrite, pyrite, galena, and meta-zeunerite. Sodium uranospinite is the most abundant secondary U mineral at this locality and may be replaced in part by uranophane. HCWS *AM 43:383(1958)*.

40.2a.7.1 Uramphite (NH$_4$)$_2$(UO$_2$)$_2$(PO$_4$)$_2$ · 6H$_2$O

Named in 1957 for the composition. Autunite group. TET $P4/ncc$. a = 7.02, c = 18.08, Z = 4, D = 3.26. *NBS 93:557(1988)*. *42-384*(syn): 9.08$_{10}$ 5.56$_2$ 4.35$_1$ 3.80$_5$ 3.51$_2$ 3.28$_2$ 2.78$_3$ 2.16$_2$. Square tablets up to 0.2 mm, rosettes, lichenlike. Bottle to pale green, vitreous luster. Cleavage distinct in two directions. G = 3.7. Fluoresces yellow-green. DTA. Anomalously biaxial (−); N_x = 1.585, $N_{y=z}$ = 1.564. X, colorless; Y,Z, pale green. Soluble in 1:10 cold HCl. Synthesized. Occurs in the oxidized zone of U-coal deposits in fractures 20–30 m below the surface, *Russia*. HCWS *AM 44:464(1959)*.

40.2a.8.1 Meta-Ankoleite K$_2$(UO$_2$)$_2$(PO$_4$)$_2$ · 6H$_2$O

Named in 1966 for the locality. Autunite group. TET $P4/nmm$. a = 6.993, c = 8.891, Z = 1, D = 3.54. *GSGB B25:49(1966)*. *19-1008*: 8.92$_{10}$ 5.47$_4$ 4.93$_5$ 4.32$_4$ 3.73$_7$ 3.49$_5$ 3.25$_6$ 2.95$_3$. See also *29-1061*(syn). Plates up to 1 mm. Yellow, translucent. Cleavage {001}, micaceous, perfect; {100}, distinct. Fluoresces yelllow-green. N_{av} = 1.580. Synthesized (3H$_2$O): a = 6.99, c = 8.89. *JLCM 44:99(1976)*. Found in a beryl-bearing pegmatite *(Mungenyi)*, Ankole district, *Uganda*, associated with muscovite and quartz; at Sebungive district, Zimbabwe, associated with sericitized microcline in Karoo sandstones. HCWS *AM 52:560(1967)*.

40.2a.9.1 Abernathyite $K(UO_2)(AsO_4)_2 \cdot 3H_2O$

Named in 1956 for Jess Abernathy, operator of the mine where the mineral was found. Autunite group. TET $P4/ncc$. $a = 7.176$, $c = 18.126$, $Z = 4$, $D = 3.572$. Structure: *AM 49:1578(1964)*. *16-386:* 9.1_{10} 5.64_7 5.08_6 4.53_6 3.83_9 3.59_8 3.34_8 2.79_7. Thick tablets up to 0.5 mm showing $\{100\}$, $\{110\}$. Yellow, vitreous luster. $H = 2–3$. $G = 3.32$. Uniaxial $(-)$; $N_O = 1.608$, $N_E = 1.570$. Found disseminated in uraniferous lignite in Lower Ludlow member, Fort Union formation, at Cave Hills and Slim Buttes, Harding Co., SD; at *Fuemrole mine No. 2, Temple Mt., Emery Co., UT*, associated with scorodite coating fractures in sandstone. HCWS *AM 41:82(1956)*.

40.2a.10.1 Novacekite $Mg(UO_2)_2(AsO_4)_2 \cdot 10$ or $12H_2O$

Named in 1951 for Radim Nováček (1905–1942), Czech mineralogist. Autunite group. Two hydrates known: both have been synthesized. See metanovacekite (40.2a.10.2). $12H_2O$ [I or 22 Å(syn)]: TET $P4_2/n$. $a = 14.30$, $c = 22.00$, $Z = 8$, $D = 3.13$. Anomalously biaxial $(-)$; $N_x = 1.540$, $N_y = 1.570$, $N_z = 1.575$; $2V = 35–44°$ (Grube Anton). *17-147*(syn): 10.9_{10} 5.49_6 5.03_4 4.76_4 4.07_4 3.54_{10} 3.22_7 2.50_4. $10H_2O$ (II or 20 Å): $P4_2/n$. $a = 7.16$, $c = 20.19$, $Z = 2$, $D = 3.28$. Anomalously biaxial $(-)$; $N_x = 1.620$, $N_y = 1.637$, $N_z = 1.637$; $2V = 0–15°$ (Schneeberg). *8-286:* 10.2_{10} 6.80_4 5.06_8 3.58_9 3.35_5 2.52_3 2.26_3 1.787_3. But note synthetic $10H_2O$ [II or 20 Å (syn)]: TET $P4_2/n$. $a = 7.10$, $c = 20.01$. Biaxial $(-)$; $N_x = 1.563$, $N_y = 1.591$, $N_z = 1.595$; $2V = 0–50°$. *17-148*(syn): 10.0_{10} 5.02_8 3.57_{10} 3.35_4 2.52_3 2.46_3 1.774_4 1.588_5. Thick tablets showing $\{001\}$, rounded $\{h0l\}$, forming lamellar subparallel aggregates often intergrown with zeunerite. Light yellow, lemon yellow, greenish yellow; waxy luster; translucent. Cleavage $\{001\}$, perfect. $H = 2\frac{1}{2}$. $G = 3.25$. Fluoresces dull green. Found at Woodrow area, Laguna Res., Valencia Co., NM; at Twin Mt., Wichita Mts., near Quanah Parker Lake, Comanche Co., OK; at Aldama, CHIH, Mexico; at *Schneeberg, Saxony, Germany*, associated with zeunerite, uranophane, and a phosphatian novacekite ($a = 7.12$, $c = 20.14$, $N_{y=z} = 1.620–1.623$), and at Wittichen district, Heubachtal; crystals up to 2.5 cm at Pedra Preta(!), Brumado district, Bahia, Brazil. HCWS *AM 36:680(1951)*, *USGSB 1064:177(1958)*, *TMPM 9:111(1964)*.

40.2a.10.2 Meta-Novacekite $Mg(UO_2)_2(AsO_4)_2 \cdot 8H_2O$

Named for the relationship to novacekite I and II, being the third and lowest hydrate. Autunite group. TET $P4/n$. $a = 7.16$, $c = 8.58$, $Z = 1$, $D = 3.72$. *TMPM 9:111(1964)*. *17-152:* 8.52_{10} 4.29_5 3.57_9 3.02_4 2.53_5 2.27_4 2.14_6 1.791_5. Rectangular plates flattened on $\{001\}$, aggregates of thin plates, crusts. Pale dull yellow, opaque. Cleavage $\{001\}$, perfect; also $\{010\}$, $\{110\}$. $H = 2\frac{1}{2}$. $G = 3.51$. DTA. Fluoresces dull green. Synthesized. Uniaxial $(-)$; $N_O = 1.632$, $N_E = 1.595$. Found at *Anton mine, Black Forest, Germany*; in Russia, and the localities listed under novacekite (40.2a.10.1). HCWS *JLBW 3:17(1958)*, *ZVMO 103:606(1974)*.

40.2a.11.1 Saleeite $Mg(UO_2)_2(PO_4)_2 \cdot 10H_2O$

Named for Achille Salée (1883–1932), mineralogist, Université Louvain, Belgium. Autunite group. Meta-saleeite ($8H_2O$) has been described, novacekite is the As end member. MON $P2_1/n$. $a = 6.951$, $b = 19.947$, $c = 9.896$, $\beta = 135.17°$, $Z = 2$, $D = 3.12$ BM $103:630(1980)$. Structure: ZK $177:247(1986)$. 8-313: 9.85_{10} 4.95_8 4.51_4 3.49_9 3.23_5 2.95_5 2.45_6 2.19_7. See also 29-874(syn)= $9H_2O$: $P4/nmm$. $a = 7.047$, $c = 19.652$. $G = 3.20$. JLCM $44:99(1976)$. Thin rectangular tablets flattened on $\{001\}$; interleaved coherently to form porous aggregates with subparallel growth; oriented parallel growth with torbernite. Pale canary yellow, lemon- or straw-yellow; transparent, translucent when partially dehydrated. Cleavage $\{001\}$, perfect; $\{010\}$, $\{110\}$, indistinct. H = 2–3. G = 3.27. Fluoresces yellow-green. Biaxial (−); $N_x = 1.559$, $N_y = 1.570$, $N_z = 1.574$ (Katanga); $2V = 61$; $r > v$, strong. O, pale greenish yellow; E, colorless. Uniaxial (−); $N_O = 1.574$, $N_E = 1.559$ (Schneeberg). Crystals show division into two or four sectors with $Z = c$ and Y perpendicular to the edge of the plate in each sector. Meta-saleeite ($8H_2O$): TET $P4/mmm$. $a = 7.22$, $c = 17.73$, $Z = 2$, $D = 3.23$. 41-389: 8.84_{10} 6.32_4 4.55_6 4.43_4 3.78_3 3.57_4 3.35_5 3.28_3. Yellow-green. Uniaxial (−); $N_O = 1.572$, $N_E = 1.562$ (Urucum, MG, Brazil). Found disseminated in carnotite-bearing sandstone Gull Mine, Fall River Co., and in lignite at Cave Hills and Slim Buttes, Harding Co., SD; at Sue Mine, Gila Co., AZ; at Mina da Quinta, Sabugal, Portugal (arsenic free); at Villaviejabe, Spain; found with uranophane and zeunerite at Schneeberg, Germany (contains arsenic); at *Shinkolobwe*(!), *Shaba, Zaire*, with torbernite and dewindite; at Rum Jungle(!), Australia; at Lavra do Enio, Galileio, MG, Brazil. HCWS $DII:988$; USGSB $1036:130(1956)$, $1064:177(1958)$.

40.2a.12.1 Seelite $Mg[UO_2(AsO_3)_x(AsO_4)_{1-x}]_2 \cdot 7H_2O$

Named in 1993 for mineral collectors Paul (1904–1982) and Hilde (1901–1987) Seel of Belgium. MON $C2/m$. $a = 18.207$, $b = 7.062$, $c = 6.661$, $\beta = 99.65°$, $Z = 2$, $D = 3.60$ (Talmessi). MR $24:463(1993)$. Structure: AC(C) $47:2013$ (1991). Used to partition As^{3+}, As^{5+}. AM $78:453(1993)$. PD: 9.05_{10} 4.85_5 4.44_8 3.99_5 3.52_6 1.92_8. (Talmessi). Tufted spherules, rosettes of bright yellow tabular crystals; elongate [010], flattened on (100) showing $\{100\}$, $\{001\}$ prism faces terminated by $\{010\}$ or $\{011\}$, $\{01\bar{1}\}$. Colorless streak, vitreous luster, transparent, irregular fracture. H = 3. G = 3.70. Nonfluorescent. Biaxial (−); $N_x = 1.602$ parallel to a (colorless), $N_y = 1.610$ parallel to b (yellow), $N_z = 1.753$ parallel to c (yellow), $2V_{meas.} = 41°$, OAP(010) Z:c = 5° extreme $r > v$, strongly pleochroic (Talmessi). $N_x = 1.610$, $N_y = 1.730$, $N_z = 1.740$, $2V_{meas.} = 30°$ (Rabejac). Found in the oxidation zones of *Talmessi mine, 35 km W of Amarak, central Iran*, and in the *Rabejac uranium deposit, 7 km SSE of Lodève, Hérault, France*. HCWS AM $79:1012(1994)$.

40.2a.13.1 Torbernite Cu(UO$_2$)$_2$(PO$_4$)$_2$ · 11H$_2$O

Named in 1793 for Torbern Olaf Bergmann (1735–1784), Swedish chemist and mineralogist. Autunite group. See meta-torbernite (40.2a.13.2), the lower hydrate (8H$_2$O). TET $I4/mmm$. $a = 7.06$, $c = 20.5$, $Z = 2$, $D = 3.16$. *USGSB 1036:138(1956)*, *AM 42:905(1957)*. *8-360*(syn): 10.3$_{10}$ 6.61$_4$ 5.18$_3$ 4.94$_9$ 4.48$_4$ 3.67$_4$ 3.58$_9$ 3.51$_8$. Thin to thick tabular {001} usually square in outline, rarely pyramidal, subparallel growth common; foliated, micaceous, or scaly aggregates. Twinning on {100} rare. Parallel growth with autunite, uranospathite, zeunerite, and bassetite. Emerald to grass-green, leek-green, apple-green, or siskin-green; streak paler than color; vitreous to subadamantine luster, pearly on {001}; transparent to translucent. Cleavage {001}, perfect, micaceous; {100}, indistinct; brittle lamellae. H = 2–2$\frac{1}{2}$. G = 3.22. Nonfluorescent. Uniaxial (−); N$_O$ = 1.592, N$_E$ = 1.582. O, green; E, blue. Some As substitutes for P; Pb for Cu; very minor amounts of V, Ba, Mn detected on analysis. H$_2$O content varies with humidity and temperature. Torbernite deposited from aqueous solutions up to ≈ 75°, above that temperature the phase is meta-trobernite. Soluble in acids. Secondary mineral associated with autunite and other uranium minerals as an oxidation product of uraninite in gossans that also contain Cu-sulfide minerals. Found at Haddam Neck, CT, as an alteration product of uraninite in a pegmatite; in other New England pegmatites usually in association with autunite; at Spruce Pine, Mitchell Co., NC; at Hannibal mine, Terry, Lawrence Co., near Keystone, Pennington Co., SD; in the La Sal Mts., San Juan Co., and in the plateau carnotite-bearing sandstones of UT and CO; in NM in the White Signal district, Grant Co., and the San Lorenzo district, Socorro Co., NM; in Sonora, Mexico. Occurs widely at mines at Redruth, St. Austell, St. Just, St. Agnes, Cornwall(!!), England; at Vizeu and Sabugal, Portugal; in Cacéres Prov., Estramadura, Spain; at Cap Garonne and in the Puy de Dôme district, France; at Schlaggenwald and Zinnwald, Johanngeorgenstadt and Schneeberg, Saxony, Germany; with pitchblende in vein deposits at *Joachimsthal, Zapadocesky Kraj, Bohemia, Czech Republic*; at Totimoto, Tamura Co., Hukusima Pref., Japan; at Kasolo, Shinkolobwe, and elsewhere in the Shaba district(!), Zaire; at Mt. Painter, Flinders Ranges, SA, Australia. HCWS *DII:981*, *USGSB 1064:170(1958)*.

40.2a.13.2 Meta-Torbernite Cu(UO$_2$)$_2$(PO$_4$)$_2$ · 8H$_2$O

Named in 1916 for the relationship to torbernite, being the lower and more stable hydrate. Autunite group. TET $P4/n$. $a = 6.969$, $c = 17.306$, $Z = 2$, $D = 3.70$. Structure: *AM 49:1603(1964)*. *36-406*: 8.66$_{10}$ 5.43$_2$ 4.93$_2$ 3.68$_7$ 3.48$_2$ 3.23$_1$ 2.67$_1$ 2.17$_1$. See also *16-404*. Tablets on {001}, rosettes or sheaflike aggregates of irregularly curved and composite crystals. Pale to dark green; vitreous, subadamantine luster, pearly on {001}; transparent to translucent. Cleavage {001}, perfect; brittle. H = 2$\frac{1}{2}$. G = 3.70. Nonfluorescent. Uniaxial (+); N$_O$ = 1.624, N$_E$ = 1.626; occasionally biaxial with small 2V. O, green; E, blue. Stable phase precipitated from solutions above 75°. Soluble in acids. Synthesized. Secondary, and probably also a hydrothermal, mineral at Spruce

Pine, NC; with copper–uranium deposits in Colorado Plateau, and in uraniferous lignite at North Cave Hills, Harding Co., SD; in the San Rafael Swell, UT; in quartzite, in Wilson Creek area, Gila Co., and at Silver Bell mine, AZ; at Nicholson mine, Lake Athabasca, SAS, and at Wilberforce, ONT, Canada; at Luz del Cobra mine, San Antonio de la Huerta, SON, Mexico; at Anyiar, Spain; at Gunnis Lake mine, Calstock(!), and at Wheal Basset, Cornwall, England; at Puy de Dôme and Aveyron Margabel, France; at Krinkelbachtel, Menzenschwand, and Erzgebirge, Saxony, and in the Black Forest, Germany; in Bohemia, Czech Republic; in the Piedmont of Italy at Lurisia; in the Shaba district(!), Musoni mine and Shinkolobwe, Zaire; at Chuquicamata, Antofagasta, Chile. HCWS $DII:991$.

40.2a.14.1 Zeunerite $Cu(UO_2)_2(AsO_4)_2 \cdot 10$–$16H_2O$

Named in 1872 for Gustav Anton Zeuner (1828–1907), German physicist, director, School of Mines, Freiberg, Germany. Autunite group. TET $P4/nnc$. $a = 7.18$, $c = 21.06$, $Z = 2$, $D = 3.40$. $TMPM$ $9:111(1964)$. 17-150: 10.3_{10} 5.20_7 4.98_5 3.59_{10} 2.39_5 2.08_6 1.924_6 1.794_5. See also 4-90(syn). Rectangular tablets $\{001\}$. Pale to emerald green, transparent. $G = 3.39$. Nonfluorescent. Uniaxial $(-)$; $N_O = 1.615$, $N_E = 1.586$. Variable water content. At 65° in air zeunerite transforms to meta-zeunerite with $8H_2O$. Synthesized. Found at Centennial Eureka mine, Tintic district, and Dexter Mine, Calf Mesa, San Raphael Swell, UT; at Grandview mine, Coconino Co., AZ; at Anton mine, Black Forest, and the *Weisser Hirsch mine, Neustädtel, Schneeberg, Saxony, Germany*; at Miyosi mine, Kurasiki Co., Okayavaa Pref., Japan; at Prada Preta(!), Brumado, BA, Brazil, in crystals to 4 cm, and at other localities listed under meta-zeunerite. HCWS AM $42:905(1957)$, $DII:989$.

40.2a.14.2 Meta-Zeunerite $Cu(UO_2)_2(AsO_4)_2 \cdot 8H_2O$

Named in 1937 for the relationship to zeunerite, being the lower hydrate. Autunite group. TET $P4_2/n$. $a = 7.12$, $c = 17.45$, $Z = 2$, $D = 3.85$. Structure: AM $49:1619(1964)$. See also $CJP(B)$ $10:169(1960)$. 17-146: 8.86_{10} 5.57_8 5.10_6 3.73_{10} 3.57_7 3.30_8 2.98_4 1.561_6. Tabular $\{001\}$, resembling torbernite; scaly; acute pyramidal. Forms other than $\{001\}$ are striated, rough. Oriented growth parallel to axes with trogerite, uranospinite; overgrowths with meta-uranocircite. Grass- to emerald-green; vitreous luster, pearly on $\{001\}$. Cleavage $\{001\}$, perfect; $\{100\}$, distinct; uneven fracture; brittle. $H = 2$–$2\frac{1}{2}$. $G = 3.64$. Fluoresces yellow green. Uniaxial $(-)$; $N_O = 1.643$, $N_E = 1.623$ (Schneeberg). O, grass-green; E, pale green. Indices vary with H_2O content. Easily soluble in acids. Does not convert to zeunerite when placed in H_2O. Synthesized. Found disseminated in uraniferous lignite in the Cave Hills and Slim Buttes areas, Harding Co., SD; with other secondary uranium minerals (i.e., torbernite, trögerite, uranospinite, walpurgite) and with erythrite and mimetite in sandstone uranium deposits of Colorado Plateau and especially at the Centennial Eureka mine, Tintic district, UT; at Grandview mine, AZ; at the Myler mine(!), Majuba Hill, Pershing Co., NV; at the Nicholson mines, Lake Athabasca, SAS, Canada; at Wheal Garland, Cornwall, England; with chalcophyllite, olivenite, pharmacosiderite at Cap Garonne, Var, France; at the Anton mine, Black Forest, Baden, and *Weisser Hirsch mine, Schneeberg, Saxony, Germany*, and at Joachimstal and Zinnwald, Bohemia, Czech Republic; at Cala Mästra, Island of Monte Cristo, Italy; at Baly Lo copper mine, Ashburton Downs, WA, Australia; in crystals to 3.5 cm at Pedra Preta pit(!!), Brumado, Bahia, Brazil. HCWS *DII:993, USGSB 1064:215(1958), TMPM 9:111(1964)*.

40.2a.15.1 Kahlerite $Fe^{2+}(UO_2)_2(AsO_4)_2 \cdot 10-12H_2O$

Named in 1953 for Franz Kahler, geologist, with Carinthian Landesmuseum, Klagenfurt, Austria. Autunite group. TET $P4_2/n$. $a = 14.30$, $c = 21.97$, $Z = 8$, $D = 3.22$. 17-145(syn): 11.1_8 5.55_5 3.59_5 3.53_{10} 3.20_2 2.53_2 1.763_3 1.603_4. Tabular to 2 mm showing (111), (021), (012), (011), (010). Lemon yellow, translucent. Basal cleavage. Anomalously biaxial; $N_x = 1.632$, $N_z = 1.634$; 2V = 9–33°. Found at Sophie mine, Wittichen, Germany, and at *Hüttenberg, Carinthia, Kärnten, Austria*, in siderite deposit. HCWS *AM 39:1038(1954), TMPM 9:111(1964)*.

40.2a.15.2 Meta-Kahlerite $Fe^{2+}(UO_2)_2(AsO_4)_2 \cdot 8H_2O$

Named in 1964 for the relationship to kahlerite (40.2a.15.1), being the lower hydrate. Autunite group. TET $P4/nmm$. $a = 7.16$, $c = 8.62$, $Z = 1$, $D = 3.83$. *TMPM 9:111(1964)*. Synthetic: space group uncertain, probably $P4_2$. $a = 20.25$, $c = 17.20$, $Z = 16$, $D = 3.77$. *AM 71:1037(1986)*. 17-151: 8.86_{10} 5.07_4 3.59_{10} 3.03_4 2.54_6 2.28_6 1.788_6 1.608_7. See also 12-576. Scales. Sulfur-yellow, translucent. Cleavage {001}, perfect; {100}, good. G = 3.77. Nonfluorescent. DTA, TGA, IR, Mossbauer. Anomalously biaxial; $N_x = 1.642$, $N_z = 1.608$; 2V up to 22°. X, pale yellow; Z, colorless. Both the Fe^{2+} and the Mn^{2+} end members have been synthesized as has a fully oxidized (Fe^{3+}) form ($P1$ or $P\bar{1}$; $a = 7.44$, $b = 9.94$, $c = 7.37$, $\alpha = 94.7°$, $\beta = 102.7°$, $\gamma = 90.3°$, $Z = 16$, $D = 3.18$; $G = 3.19$). Found in the siderite deposit of Hüttenberg, Kärnten, Austria; at *Sophia mine, Wittichen, Baden, Black Forest, Germany*. HCWS *AM 45:254(1960)*.

40.2a.16.1 Bassetite $Fe^{2+}(UO_2)_2(PO_4)_2 \cdot 8H_2O$

Named in 1915 for the locality. Autunite group. MON $P2_1/m$. $a = 6.98$, $b = 17.07$, $c = 7.01$, $\beta = 90.5°$, $Z = 2$, $D = 3.66$ (Basset). *MM 30:343(1954), AM 69:967(1984)*(syn). 7-288: 8.59_6 4.89_{10} 4.24_3 4.05_3 3.46_{10} 2.96_3 2.85_2 2.20_6. Fanlike or chessboardlike groupings of thin tablets with almost rectangular outline, flattened on {010}. Often twinned with two or more individuals arranged at 90° with {010} in common. {001} of one individual is parallel to {100} of the other with {010} the contact surface in some, and {110} against {011} in others. Yellow, transparent. Cleavage {010}, perfect; {100}, {001}, distinct. G = 3.10. Nonfluorescent. Mossbauer. Biaxial $(-)$; $X = b$, $Z \wedge c = -4°$; natural with low H_2O: $N_x = 1.603$, $N_y = 1.610$, $N_z = 1.617$,

2V = 90°; synthetic: $N_x \approx 1.561$, $N_y = 1.574$, $N_z = 1.580$ (Na); 2V ≈ 62°. X, pale yellow; Y,Z, deep yellow. Loses H_2O at RT with the plates breaking into four sectors with axial planes at right angles; indices of refraction and extinction angles increase to ≈ 20°. Soluble in acids. Synthesized, as has a fully oxidized (Fe^{3+}) bassetite (MON $a = 6.958$, $b = 6.943$, $c = 21.052$, $\beta = 90.92°$, $Z = 2$; $G = 3.00$). Found at Fuemrol mines, Temple Mt., and Denise mine, Green River, Emery Co., UT; in the uranium-bearing quartzites of Dripping Springs, Gila Co., AZ; at the *Wheal Basset mine, Redruth, Cornwall, England*, as oriented growths with uranospathite, torbernite in pyrite–uraninite vein; at LaCrouzille, Puy de Dôme, France. HCWS *DII:994*.

40.2a.16.2   Lehnerite $Mn^{2+}(UO_2)_2(PO_4)_2 \cdot 8H_2O$

Named in 1988 (reviving a name that was used in 1925 for a mineral subsequently named ludlamite) for Ferdinand Lehner (1868–1943), mineral collector of Hagendorf minerals from Pleystein, Bavaria, Germany. Autunite group. Mn^{2+} analog of bassetite (40.2a.16.1); see also fritzschite (40.2a.25.1), a manganese vanadate. MON $P2_1/n$. $a = 7.04$, $b = 17.16$, $c = 6.95$, $\beta = 90.18°$, $Z = 2$, $D = 3.717$. *AUF 39:209(1988)*. PD: 8.56_{10} 4.96_6 3.50_8 2.48_3 2.23_7 2.17_3 1.746_3 1.375_4. Pseudotetragonal prisms up to 1 mm, platy along (010) with layers perpendicular to [010], show in addition {001}, {100}, {101}, {$\bar{1}$01}. Yellow, bronze-yellow; whitish yellow streak; glassy to nacreous luster. Cleavage {010}, perfect; {101}, {$\bar{1}$01}, {100}, less perfect. H = 2–3. G = 3.5. Nonfluorescent. Biaxial (−); $X = b$; $Y \wedge a = 8$–10°, $Z \wedge c = 8$–19°; $N_x = 1.599$, $N_y = 1.607$, $N_z = 1.607$ (589 nm); 2V = 45°; r > v, strong; pleochroic, shades of yellow with Z darker than Y. Some zoning noted with extinction more oblique on outer edges of crystals, intermediate zones may be uniaxial, probably the result of variable H_2O content. Found at *Hagendorf-Süd pegmatite, Oberpfalz, Bavaria, Germany*, associated with decomposed zwieselite, corroded rockbridgeite on the 67-m level. HCWS *AM 75:221(1990), 75:1433(1990)*.

40.2a.17.1   Meta-Kirchheimerite $Co(UO_2)_2(AsO_4)_2 \cdot 8H_2O$

Named in 1958 for Franz Waldemar Kirchheimer (b.1911), director, Geologisches Landesamt für Wurttemberg-Baden, Germany. Autunite group. TET $P4/nmm$. $a = 7.17$, $c = 17.36$, $Z = 2$, $D = 3.80$. *JGLBW 3:17(1958)*. 12-586: 8.78_{10} 5.08_6 4.30_6 3.57_{10} 3.42_5 3.01_6 2.52_5 2.26_4. Crusts and tabular crystals. Pale rose, transparent. Cleavage {001}, perfect, pearly luster. H = 2–2½. G = 3.33. Uniaxial (−); $N_O = 1.644$, $N_E = 1.617$. Anomalously biaxial; Synthesized: $P4_2/m$; $a = 6.979$, $b = 16.93$. *BM 100:272(1977)*. Found at *Sophia mine, Wittichen, Black Forest, Germany*, on pitchblende. HCWS *AM 44:466(1959), TMPM 9:111(1964)*.

40.2a.18.1   Meta-Lodevite $Zn(UO_2)_2(AsO_4)_2 \cdot 10H_2O$

Named in 1972 for the locality. Autunite group. TET $P4_2/n$. $a = 7.16$, $c = 17.20$, $Z = 2$, $D = 4.01$. *BM 95:360(1972)*. 25-1239: 8.66_7 5.09_4 3.59_{10}

3.50_3 2.98_6 2.55_3 2.29_3 2.17_3. Crystals to 0.2 mm flattened on (001), striated parallel to [100]; aggregates. Pale yellow, vitreous luster, transparent. Cleavage {001}. Fluoresces feebly yellow-green. Anomalously biaxial $(-)$; $N_x = 1.638$, $N_y = 1.635$, $N_z = 1.615$; $2V = 27$–$37°$. X,Y, pale yellow; Z, brownish. Shows inclusions of pitchblende. Found at *Lodève, Herault*, and at St. Martin du Bosc, France. HCWS

40.2a.19.1 Chernikovite $(H_3O)_2(UO_2)_2(PO_4)_2 \cdot 6H_2O$

Named in 1988 for A. P. Chernikov (b.1927), Institute of Mineralogy, Geochemistry, and Rare Elements, Moscow. Autunite group. Previously synthesized [*AM 40:917(1955)*] and named hydrogen-autunite. TET $P4/nmm$. $a = 7.020$, $c = 9.043$, $Z = 1$, $D = 3.27$. *8-296*(syn): 9.03_{10} 5.56_5 3.80_9 3.51_7 3.27_8 2.96_6 2.77_7 2.16_5. See also *29-670*. $4H_2O$ form, a material with high degree of proton conductivity. *MRB 12:701(1977)*, Structure: *AC(B) 34:3732(1978)*. Micaceous plates elongated on [010], synthetics have square or octagonal outline. Pale lemon yellow, vitreous luster, transparent. Cleavage {001}, perfect; {100}, distinct. $G_{syn} = 3.40$; Russian samples, $G = 3.259$. Fluoresces intense yellow-green. Uniaxial $(-)$; $N_{O\,syn} = 1.579$, $N_{E\,syn} = 1.568$; slightly pleochroic. Synthesized from schoepite; easily transformed into parsonsite. *NJMM:551(1991)*. Found in two localities in *Russia*; at Perus, SP, Brazil. HCWS *MR 19:249(1988)*.

40.2a.20.1 Trögerite $H_2(UO_2)_2(AsO_4)_2 \cdot 8H_2O$ (?)

Named in 1871 for R. Tröger, mining official at Schneeberg, Germany. Autunite group. TET $P4/nmm$. $a = 7.16$, $c = 8.80$, $Z = 1$, $D = 3.55$ (see below). *8-326*(syn): 8.59_{10} 5.50_7 4.35_7 3.79_9 3.30_8 2.70_7 2.19_7 2.01_7. Thin elongate tabular {001}; nonpinacoidal forms asymmetrically developed (implies monoclinic symmetry). {001} often composite and uneven, other faces small and horizontally striated; subparallel aggregates. Lemon yellow, pearly luster on {001}. Cleavage {001}, perfect, micaceous; {100}, good; also {011}. $H = 2$–3. $G = 3.55$. IR. Natural reported to be nonfluorescent. Zeunerite commonly found in parallel overgrowths with trögerite. Uniaxial $(-)$; $N_{O\,nat} = 1.585$, $N_{E\,nat} = 1.630$; $N_{O\,syn} = 1.580$, $N_{E\,syn} = 1.620$–1.624. *DANS 197:109(1971)*. Anomalously biaxial $(-)$; $X = c$; $Z \wedge a = 12°$; $N_x = 1.585$, $N_y = 1.624$, $N_z = 1.630$; 2V very small. *AM 36:322(1951)*, *38:1159(1953)*. Soluble in acids. Composition close to synthetic hydrogen uranospinite, $H_2(UO_2)_2(AsO_4)_2 \cdot 10H_2O$ [*DII:967*, *DANS 197:109(1971)*], which forms as tabular, square-outline crystals that fluoresce bright lemon-green. Synthetic studies show a reversible, nonquenchable, phase transition $\approx 25°$ (not dehydration) to lower symmetry (pseudotetragonal), resulting in the formation of domains with polysynthetic rectalinear twinning. *AM 59:763(1974)*. Found at Bald Mt. district, Pennington Co., SD; with walpurgite and zeunerite in the *Walpurgis vein of the Weisser Hirsch mine, Neustädtl, Schneeberg, Saxony, Germany*. HCWS *DII:966*.

40.2a.21.1 Przhevalskite $Pb(UO_2)_2(PO_4)_2 \cdot 4H_2O$

Named in 1946 for Nikolai M. Przhevalsky (1839–1888), Russian explorer. ORTH (possibly TET) $a = 7.24$, $c = 18.22$, $Z = 2$, $D = 3.38$. *AM 43:381(1958)*. *29-787*(syn): 9.49_5 9.08_9 3.61_{10} 2.63_5 1.960_5 1.619_6 1.530_6 1.349_5. Foliated aggregates of tabular crystals to 1 mm. Bright yellow with greenish tinge; adamantine luster, pearly on cleavage {001}. Biaxial (−); $N_x = 1.739$, $N_y = 1.749$, $N_z = 1.752$; $2V \approx 30°$; parallel extinction; positive elongation. X, colorless; Y, pale yellow; Z, deep yellow. Soluble in acids. Both Si and Al in analysis. Synthesized. Found in the oxidation zone of a pitchblende–sulfide deposit with torbernite, autunite, uranospinite, and hydrated iron oxides in *Russia*. HCWS *AM 41:816(1956)*.

40.2a.22.1 Uranospathite $HAl(UO_2)_4(PO_4)_4 \cdot 40H_2O$

Named in 1915 for the composition and from the Greek for a *broad blade*, in allusion to the crystal habit. Autunite group. See also arsenuranospathite (40.2a.23.1). TET $P4_2/n$. $a = 7.00$, $c = 30.03$, $Z = 1$, $D = 2.49$. *MM 42:117(1978)*. *31-587*: 15.22_{10} 7.60_{10} 4.93_{10} 4.48_6 4.08_4 3.50_8 3.43_4 2.21_6. Fanlike groups of rectangular plates and laths flattened {001}, striated parallel to elongation {010}; fibrous. Sometimes twinned on {110}, in cruciform groups. Yellow to pale green, translucent. Cleavage {001}, perfect; {100}, good; {010}, doubtful. $H = 2\frac{1}{2}$. $G = 2.50$. Fluoresces yellow-green. Anomalously biaxial (−); $N_x \approx 1.492$, $N_y = 1.511$, $N_z = 1.521$ (Na); $2V = 76°$. Loses water at RT and nonhumid conditions, or in desiccator, converting to a lower hydrate that is identical with sabugalite (40.2a.24.1). Soluble in acids. Forms oriented growths with bassetite. Found at *Basset mine, Redruth, Cornwall, England*; at Menzenschwand, Black Forest, Germany. HCWS *DII:990*, *MM 30:343(1954)*, *USGSB 1064:194(1958)*, *AM 64:465(1979)*.

40.2a.23.1 Arsenuranospathite $HAl(UO_2)_4(AsO_4)_4 \cdot 40H_2O$

Named for its relationship to uranospathite (40.2a.22.1) being the arsenic analog. Autunite group. TET Space group uncertain, probably $P4_2/n$. $a = 7.16$, $c = 30.37$, $Z = 1$, $D = 2.54$. *MM 42:117(1978)*. *31-586*: 14.6_{10} 7.62_{10} 5.03_8 3.59_6 3.49_9 3.24_4 2.48_2 2.25_3. Lathlike or wedge-shaped crystals showing {100}, {010}, {110}, {001}, orthorhombic habit. White to pale yellow, translucent. Cleavage {001}, perfect; {100}, {010}, good. $H = 2$. Fluoresces greenish. Anomalously biaxial (−); $N_{x'} = 1.538$, $N_{z'} = 1.542$; $2V = -52°$; $r > v$; straight extinction; length positive. Converts under normal room-temperature conditions to hydrate with half amount of H_2O; a 32 H_2O phase ($a = 7.15$, $c = 20.5$, $Z = 2$, $D = 3.20$; $N_x = 1.564$, $N_y = 1.594$, $N_z = 1.596$; $2V \approx 28°$; $r > v$; X, colorless; Y,Z, yellow) has been synthesized. Found at *Sophia mine, Menzenschwand, Black Forest*, and at Wittichen, Germany. HCWS *AM 64:465(1979)*.

40.2a.24.1 Sabugalite $HAl(UO_2)_4(PO_4)_4 \cdot 16H_2O$

Named in 1951 for the locality. Autunite group, but see uranospathite (40.2a.22.1). This mineral may be "meta-uranospathite." MON Space group uncertain, probably $C2/m$. $a = 19.426$, $b = 9.843$, $c = 9.850$, $\beta = 96.16°$, $Z = 2$, $D = 3.150$. *PCM 9:23(1983)*. Pseudotetragonal, probably isostructural with fully hydrated autunite. *39-313*(syn): 9.63_{10} 4.86_8 4.55_5 4.40_5 3.47_9 3.36_7 2.45_4 2.18_5. See also *5-107*. Crusts of minute plates to 1 mm square or rectangular in shape. Yellow, translucent. Cleavage {001}. $H = 2\frac{1}{2}$. $G = 3.20$. Fluoresces bright lemon yellow in UV. IR, TGA, zeta potential. Biaxial; $N_x = 1.564$, $N_y = 1.581$, $N_z = 1.582$; $2V = 0°$ to moderate. *AM 36:671(1951)*. Found in sandstone-type uranium deposits at Huskon, Arrow Head claims, near Cameron, AZ; at Happy Jack mine, White canyon, San Juan Co., UT; at Lucky Mc No. 20 mine, and in Blarco group, Freemont Co., and several claims in Johnson Co. and Crook Co., WY; at Margnac II, Haute Vienne, France; associated with meta-autunite, saléeite, phsophuranylite, milky quartz, and feldspars at *Mina de Quarta Feira, Guarda district, Sabugal, Beira Alta Prov., Portugal*, and at Kariz, Minho Parish, and at Mina de Coitos, Bandada Parish; at Pedro Alvaro, Salamanca, Spain; at Corrego Frio mine, MG, Brazil. HCWS *USGSB 1064:196(1958)*, *BGSC 60:285(1985)*.

CARNOTITE GROUP

The carnotite minerals are uranium-vandates that are chemically similar to the autunite group minerals but have a distinctive structure. They conform to the chemical formula:

$$A(UO_2)_2V_2O_8 \cdot nH_2O$$

The structure is layered and consists of sheets of divanadate groups (V_2O_8) composed of two pyramids (V^{5+} surrounded by 5 oxygens) joined at a common edge, and linear UO_2 groups whose axis, as in the autunite group minerals, is perpendicular to the extension of the anion group. These puckered sheets have a composition $[(UO_2)_2V_2O_8]$. Cation and water molecules occur in layers between the $[(UO_2)_2V_2O_8]$ sheets with the possibility that different hydrates are stable *Adv. ucl. End. 2:294 (1954)*. The parallelism with the autunite group is evidenced in the two species tyuyamunite, meta-tyuymunite which have identical cation composition but are different hydrates.

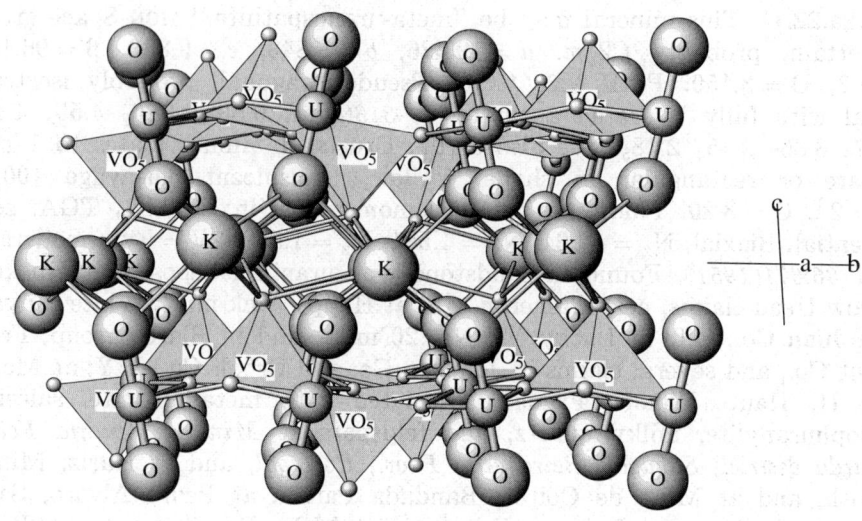

CARNOTITE: $K_2(UO_2)_2(V_2O_8)_2 \cdot 1\text{-}3H_2O$

The puckered uranyl-varadate sheets effectively perpendicular to the c axis are linked through K, H_2O layers. The water content is variable. Compare autunite (40.2a.1.1).

CARNOTITE GROUP

Mineral	Formula	Space Group	a	b	c	β	
Fritzscheite	$Mn^{2+}(UO_2)_2V_2O_8 \cdot 4H_2O$	Pnam	10.59	8.25	15.54		40.2a.25.1
Tyuyamunite	$Ca(UO_2)_2V_2O_8 \cdot 5\text{-}8H_2O$	Pnam	10.63	8.36	20.48		40.2a.26.1
Meta-tyuyamunite	$Ca(UO_2)_2V_2O_8 \cdot 3\text{-}5H_2O$	Pnam	10.63	8.36	16.96		40.2a.26.2
Francevillite	$Ba(UO_2)_2V_2O_8 \cdot 5H_2O$	Pcan	10.419	8.510	16.763		40.2a.27.1
Curienite	$Pb(UO_2)_2V_2O_8 \cdot 5H\_2O$	Pcan	10.42	8.49	16.41		40.2a.27.2
Carnotite	$K_2(UO_2)_2V_2O_8 \cdot 1\text{-}3H_2O$	$P2_1/a$	10.47	8.41	6.91	103.66	40.2a.28.1
Margaritasite	$Cs_2(UO_2)_2V_2O_8 \cdot H_2O$	$P2_1/a$	10.514	8.425	7.25	106.01	40.2a.28.2
Strelkinite	$Na_2(UO_2)_2V_2O_8 \cdot 6H_2O$	Pnnm or Pnm2	10.64	8.36	36.72		40.2a.29.1
Vanuranylite	$(H_3O,Pb,K,Ca)_2 (UO_2)_2V_2O_8 \cdot 4H_2O$	C2/c(?)	10.49	8.37	20.30	90.0	40.2a.30.1

40.2a.25.1 Fritzscheite $Mn^{2+}(UO_2)_2V_2O_8 \cdot 4H_2O$

Named in 1865 for Karl Julius Fritzsche (1808–1871), German chemist. Carnotite group. ORTH $Pnam$. $a = 10.59$, $b = 8.25$, $c = 15.54$, $Z = 4$, $D = 4.39$. BM 93:$320(1970)$. 23-1249(syn): 7.77_{10} 6.52_8 5.30_8 4.13_9 3.88_8 3.65_8 3.25_{10} 3.00_{10}. Rectangular plates with {001} cleavage. Red-brown to hyacinth-red, vitreous to pearly luster, transparent. $H = 2$–3. Found at Autun, France; at Neuhammer, near Neudeck, Bohemia, Czech Republic; and at Steinig and *Georg Wagsfort mine, Johanngeorgenstadt, Saxony, Germany*. HCWS DII:984.

40.2a.26.1 Tyuyamunite $Ca(UO_2)_2V_2O_8 \cdot 5$–$8H_2O$

Named in 1912 for the locality. Carnotite group. ORTH $Pnna$. $a = 10.63$, $b = 8.36$, $c = 20.48$, $Z = 4$, $D = 3.4$. AM 39:$323(1954)$, 41:$187(1956)$. 6-17: 10.2_{10} 6.62_3 5.02_9 4.02_2 3.37_3 3.20_5 3.12_3 2.04_4. Scales and laths flattened {001}, elongated {001}, radial aggregates. Crystal faces dull, often curved. Massive, compact to cryptocrystalline, pulverulent. Canary yellow, lemon yellow to greenish yellow; crystals adamantine pearly on {001}, massive material waxy; transparent to opaque. Cleavage {001}, perfect, micaceous; {010}, {100}, distinct; not brittle; cakes on grinding. $H \approx 2$. $G = 3.6$. DTA. Biaxial (+); XYZ= cba; $N_x = 1.72$, $N_y = 1.868$, $N_z = 1.953$. (Henry Mts., UT); $2V = 48°$; higher indices with dehydration; $r < v$, weak. X, colorless; Y, pale canary yellow; Z, canary yellow. Some K detected in natural samples on analysis. Insoluble in 1:10 acetic acid, soluble in HCl, HNO_3, and H_2SO_4. Synthesized via substitution of Ca for K in carnotite. Found in carnotite-bearing district of the western United States, notably at Paradox Valley, Montrose Co., and at Thompson's, Grand Co., CO; at Hudspeth Co., TX; at Richardson, NW side of La Sal Mts., the Henry Mts., Garfield Co., Red Creek, Browns Park, Uinta Co., UT, often as fine-grained aggregates with fossil wood in sandstone associated with carnotite; also at Bisbee, AZ; at *Tyuya Muyun on Aravan R., 65 km SE of Fedchenko, Ferghana, Uzbekistan*, as veinlets or disseminated in limestone, in the karst caverns associated with malachite, turanite, barite, calcite, and other V minerals. HCWS DII:1045; $USGSB$ 1009-$B(1954)$, 1064:$248(1958)$.

40.2a.26.2 Meta-Tyuyamunite $Ca(UO_2)_2V_2O_8 \cdot 3$–$5H_2O$

Named in 1950 for its relationship as the lower hydrates of tyuyamunite. Carnotite group. ORTH $Pnam$. $a = 10.63$, $b = 8.36$, $c = 16.96$ ($\approx 4.5H_2O$), $Z = 4$, $D = 4.04$. (May Day mine, CO). AM 41:$187(1956)$. 43-1457: 8.41_{10} 4.20_4 3.75_2 3.29_3 3.26_3 3.06_2 3.03_3 2.58_4. See also 8-287. Pulverulent masses; disseminated in sandstone, limestone mixed with carnotite; radial aggregates to 3 mm on gypsum. Canary yellow, greenish yellow; adamantine luster. Cleavage {010}, {100}, distinct; not brittle. $G = 3.8$–4.0. Biaxial (−); $N_x = 1.62$, $N_y = 1.842$, $N_z = 1.899$ (May Day mine, Mesa Co., CO); X perpendicular to plate; $2V = 45°$; $r < v$. X, colorless; Y, very pale canary yellow; Z, pale canary yellow. Found at over 35 localities in the Colorado Plateau area, often with carnotite and tyuyamunite, such as at the *Jo Dandy mine*,

Montrose Co., CO, Mayday or Small Spot mines, Mesa Co.; in the Shiprock and Grants districts, NM, and at other localities worldwide with tyuyamunite (40.2a.26.1). Unaltered samples of uranium minerals are good candidates for estimating the age of rocks: U–Ra analyses were made on meta-tyuyamunite. HCWS *DII:1045, AM 39:1037(1954), USGSB 1064:254(1958)*.

40.2a.27.1 Francevillite $Ba(UO_2)_2V_2O_8 \cdot 5H_2O$

Named in 1975 for the locality. Carnotite group. The Ba end member in the Ba–Pb series with curienite (40.2a.27.2). ORTH *Pcan*. $a = 10.419$, $b = 8.510$, $c = 16.763$, $Z = 4$, $D = 4.46$. *DANS 220:137(1975)*, Structure: *NJMM 552(1986)*. *21-381*(syn): 8.4_{10} 5.20_6 4.26_6 4.20_4 3.30_4 3.21_4 3.00_{10} 2.13_6. Plates exhibiting {001}, {201} and vicinal forms {111}, {221} possibly other {hkl}. Yellow, green, or orange; translucent. $G = 4.42$. Biaxial (−); $N_{x'} = 1.945$, $N_{z'} = 1.975$ (Na); 2V= 48°. Natural material contains Pb up to Ba/Pb 2:1. Synthesized. Found in the Otish Mts., QUE, Canada; at the South Terras mine, St. Stephen in Brannel, Cornwall, England; at the *Mounana mine, Franceville, Haut Ogooué, Gabon*, with curienite. HCWS *BM 91:453(1968)*.

40.2a.27.2 Curienite $Pb(UO_2)_2V_2O_8 \cdot 5H_2O$

Named in 1968 for Hubert Curien (b.1924), crystallographer, University of Paris (Sorbonne), France. Carnotite group. Pb end member of the Pb–Ba series with francevillite (40.2a.27.1). ORTH *Pcan*. $a = 10.42$, $b = 8.49$, $c = 16.41$, $Z = 4$, $D = 4.98$. *BM 91:453(1968)*. Structure: *BM 94:8(1971)*; see also *NJMM:552(1986)*. *22-402*: 8.19_{10} 6.11_3 5.13_6 4.22_6 4.108_8 3.23_4 3.01_{10} 2.12_4. Microcrystals show cleavage {010}. Canary yellow, translucent. $G = 4.88$. DTA. Synthesized. Biaxial; $N > 2.00$; 2V = 66°. Found as a powder on francevillite at *Mounana mine, Franceville, Haut Ogooué, Gabon*, in mineralized sulfide–uranium–vanadium deposits. HCWS *AM 54:1220(1969)*.

40.2a.28.1 Carnotite $K_2(UO_2)_2V_2O_8 \cdot 1\text{--}3H_2O$

Named in 1899 for Marie-Adolphe Carnot (1839–1920), mining engineer and chemist, France. Carnotite group. MON $P2_1/a$. $a = 10.47$, $b = 8.41$, $c = 6.91$, $\beta = 103.66°$, $Z = 2$, $D = 4.91$ *AM 43:799(1948)*. Structure: *AM 50:825(1965)*. Structure similar to meta-torbernite: layers parallel to {001} composed of divanadate groups, two tetrahedra joined at common edge, linked to perpendicular, linear, UO_2 groups have composition $[(UO_2)_2V_2O_8]$. K and H_2O are found between the layers. *8-317*: 6.56_{10} 4.25_3 3.53_5 3.25_3 3.12_7 2.57_2 2.16_3 1.942_2. See *11-338*(syn), anhydrous form with $c = 6.59$. Powdery or loosely coherent microcrystalline aggregates; compact; disseminated; rarely crusts of imperfect platy crystals flattened {001}, also rhomboidal {110}, lathlike [110] with {100}, {110}, or {120}; (110) ∧ ($\bar{1}$10) ≈ 78°. Twinned. Bright yellow to lemon yellow, greenish yellow; dull or earthy luster, pearly and silky when coarsely crystalline. Cleavage {001}, perfect. $G = 4.70$. Biaxial (−); XYZ= *cba*; $N_x = 1.750$, $N_y = 1.925$, $N_z = 1.950$ (Long Park, CO); 2V ≈ 40°; pleochroic; X, colorless; Y,Z, canary yellow. $N > 2.06$ (Olary, SA, Australia).

Indices, increase with dehydration. Small amounts of Ca, Ba, Mg, Fe^{2+}, and Na reported on analysis. Easily soluble in acids. Synthesized. Converted to tyuyamunite by treatment with calcium bicarbonate solution. Common in the plateau region of southwestern CO and in adjoining UT, AZ, and NM, disseminated in cross-bedded sandstones, concentrated around petrified or carbonized tree trunks or other vegetal matter. Formed through the action of meteoric waters on preexisting uranium–vanadium minerals; associated with tyuyamunite, volborthite, calciovolborthite, rossite, hewettite, vanoxite, pintadoite, uvanite, rauvite, and organic materials. Found in the Pottsville conglomerate, Jim Thorpe, Carbon Co., PA; in the Paradox Valley, Roc and La Sal creeks, Montrose Co., along the Rio Dolores SW of Gateway, Mesa Co., near Placerville and Cedar, San Miguel Co., and at Meeker, Rio Blanco Co., CO; on San Rafael Swell W of Green River, Emery Co., in the Henry Mts., Garfield Co., La Sal Mts., near Richardson, and at Thompson's, Grand Co., N of Monticello, San Juan Co., UT; at Anderson mine, AZ; at Blue Bird mine, Imperial Co., CA; in the red sandstone at Shaba, Zaire; at Radium Hill, Olary, SA, Australia. HCWS *DII:1043*.

40.2a.28.2   Margaritasite $Cs_2(UO_2)_2V_2O_8 \cdot H_2O$

Named in 1982 for the locality. Carnotite group. MON $P2_1/a$. $a = 10.514$, $b = 8.425$, $c = 7.25$, $\beta = 106.01°$, $Z = 2$, $D = 5.40$. AM *67:1273(1982)*, Structure: *50:825(1965)*. *35-378*: 6.97_{10} 3.60_7 3.49_6 3.34_7 3.23_5 3.16_5 3.13_4 2.70_2. See also *25-1218* (calc, no water). Aggregates up to 1 mm of fine-grained crystallites as linings surrounding relict phenocrysts. Individual tabular crystallites 1–3 μm. X-ray diffraction; Cs content distinguishes this mineral from carnotite. Yellow. Nonfluorescent. Biaxial; $N_x < 1.83$, $N_y = 2.49$, $N_z = 2.79$; $2V = 45.5°$ (synthetic anhydrous analog). Some K indicated during microprobe analyses, but no evidence of a solid-solution series with carnotite from synthetic studies. Probably a hydrothermal product. Found at *Margaritas No. 2 U deposit, Sierra Peña Blanca, W of Chihuahua City, Mexico*. HCWS

40.2a.29.1   Strelkinite $Na_2(UO_2)_2V_2O_8 \cdot 6H_2O$

Named in 1974 for M. F. Strelkin (1905–1965), mineralogist interested in uranium ores, Russia. Na analog of carnotite (40.2a.28.1); note difference in H_2O content. ORTH *Pnmm* or *Pnm2*. $a = 10.64$, $b = 8.36$, $c = 32.72$, $Z = 8$, $D = 4.22$. ZVMO *103:576(1974)*. *27-822*: 8.18_4 7.68_{10} 408_6 3.95_8 3.55_3 3.20_5 3.10_3 2.01_4. Fine powdery crusts along fractures; plates nearly isometric in form up to 1.5 mm long. Gold-yellow to canary yellow, silky to pearly luster, translucent. Cleavage {001}, perfect. Weakly luminesces light green. H = 2–$2\frac{1}{2}$. G = 4–4.2. DTA. Biaxial (−); XYZ= *cba*; $N_x = 1.770–1.674$, $N_y = 1.907–1.855$, $N_z = 1.915–1.880$; 2V medium; pleochroic Y, yellow; Z, pale yellow. Spectrographic analyses show Ni, Co, Ti, Y, Cr, Mo, Cu, Pb, and Tl. Found in carbonacous–siliceous shales in desert regions in *Russia*, also in effusive rocks in the arid zones associated with iron oxyhydroxides, clay minerals, quartz, and calcite. HCWS *AM 60:488(1975)*.

40.2a.30.1 Vanuranylite $(H_3O,Pb,k,Ca)_2(UO_2)_2V_2O_8 \cdot 4H_2O$

Named in 1965 for the composition. Carnotite group. Incompletely characterized MON Space group uncertain, probably $C2/c$. $a = 10.49$, $b = 8.37$, $c = 20.30$, $\beta = 90.0°$, $Z = 4$, $D = 3.25$. *ZVMO 94:437(1965)*. Name rejected as too close to vanuralite (42.11.13.1). *MM 36:1144(1968)*. *19-1417*: 5.3_4 5.0_{10} 3.43_4 3.23_{10} 3.12_4 2.11_8 2.05_5 1.97_8. Pattern resembles tyuyamunite (40.2a.26.1). Crusts of crystals to 0.03 mm, platy, pseudohexagonal. Intense yellow, translucent. Cleavage {100}, perfect. H = 2. G = 3.644. DTA. Nonfluorescent. Biaxial (−); $N_x < 1.710$, $N_y = 1.92$–1.95, $N_z > 1.95$; 2V = 15°; weakly pleochroic. X, pale yellow; Y,Z, yellow. Contains some Pb, K, and Ca; detailed analysis lacking. Easily soluble in acids. Forms as crusts in fractures with uranophane and soddyite in the oxidation zone in sandstone in *Russia*. HCWS *AM 51:1548(1966)*.

40.2a.31.1 Parsonsite $Pb_2(UO_2)(PO_4)_2 \cdot 2H_2O$

Named in 1923 for Arthur Leonard Parsons (1873–1957), mineralogist, University of Toronto, Canada. TRIC $P\bar{1}$. $a = 6.849$, $b = 10.360$, $c = 6.671$, $\alpha = 101.84°$, $\beta = 98.23°$, $\gamma = 86.36°$, $Z = 2$, $D = 6.34$. *NJMM:551 (1991)*, Structure: *SR 22:422(1958)*. *12-259*: 4.23_7 4.16_3 3.41_7 3.28_{10} 3.25_{10} 3.16_5 2.94_3 2.78_4. Crusts and powdery aggregates of minute lathlike crystals which are elongated [001], flattened {010} with {001}, {101}, {100}; (001) ∧ (100)≈ 81°, (101) ∧ (100) ≈ 132°. Pale citron yellow, subadamantine luster, transparent to translucent. No cleavage. H = $2\frac{1}{2}$–3. G = 5.37. Nonfluorescent. Biaxial (−); Y = b; Z (or X) ∧ c = 6–23°; $N_x \approx 1.870$, $N_z \approx 1.890$ (Ruggles). Variations in optical orientation and indices. Nonpleochroic. Soluble in acids. Synthesized with $0.5H_2O$. G = 6.29 for anhydrous form. Found at Ruggles mine, Grafton Center, NH, with autunite and phosphuranylite; as crusts on quartz in Huachuca Mts., Cochise Co., AZ; at Lachaux, Puy de Dôme, Grury, Saône et Loire, France, with torbernite; at Wölsendorf, Bavaria, Germany, with uranocircite; at Shinkolobwe and *Kasolo, Shaba, Zaire*, associated with torbernite, kasolite, dewindite. HCWS *AM 35:245(1950)*, *DII:913*, *USGSB 1064:235(1958)*.

40.2a.32.1 Hallimondite $Pb_2(UO_2)(AsO_4)_2$

Named in 1965 for Arthur Francis Hallimond (1890–1968), British mineralogist, who contributed to the knowledge of secondary uranium minerals. TRIC $P\bar{1}$. $a = 7.123$, $b = 10.469$, $c = 6.844$, $\alpha = 100.57°$, $\beta = 94.80°$, $\gamma = 91.20°$, $Z = 2$, $D = 6.40$. *AM 50:1143(1965)*. *18-706*(syn): 4.41_5 4.24_4 3.40_{10} 3.01_4 2.84_9 2.14_4 1.736_6 1.696_5. Small, fine-grained coatings in cavities and fractures of quartz. Crystals are up to 0.4 mm long, flattened on {110}, {100} elongated on c; forms observed {100}, {010}, {001}, {110}, {610}, {001}, {018}, {111} with {001} striated. Oriented intergrowths with hügelite on common c axis. Yellow, pale yellow streak, subadamantine luster, transparent to translucent. No cleavage, conchoidal fracture. H = $2\frac{1}{2}$–3. G = 6.39. Nonfluorescent. DTA. Biaxial (+); $N_x \approx 1.882$, $N_z \approx 1.915$; 2V = 80°; r > v; weakly pleochroic. Ba,

Si, Fe, Cu, Ca, Sr but no P on analysis. Synthesized. Found at the *Michael mine (Pb–Zn), Reichenbach, Black Forest, Germany*, associated with hügelite, widenmannite, weilerite, quartz, barite, mimetite, and galena. HCWS

40.2a.33.1 Ulrichite $CaCu(UO_2)(PO_4)_2 \cdot 4H_2O$

Named in 1988 for George Henry Frederick Ulrich (1830–1900), who contributed descriptions of minerals from VIC, Australia. MON $C2/m$. $a = 12.79$, $b = 6.85$, $c = 13.02$, $\beta = 91.03°$, $Z = 4$, $D = 3.71$. *AUSM 3:125(1988)*. PD: 6.39_{10} 5.68_1 5.52_1 4.50_1 3.48_1 3.19_4 2.79_2 2.37_1. Radiating sprays of acicular crystals; flat prisms, some twinned with complex pyramidal terminations. Pale apple-lime green, vitreous luster, translucent to transparent. No cleavage. $H = 3\frac{1}{2}$. Nonfluorescent. Probably biaxial (−); $N_x = 1.622$, $N_{y,z} = 1.634$; length-slow; nonpleochroic. Readily soluble in 1:10 HCl, HNO$_3$. Found at *Lake Boga, VIC, Australia*, with turquoise and chalcosiderite in granite pegmatites associated with fluorapatite, libethenite, cyrilovite, torbernite, saleeite, sampleite, and iron phosphates. HCWS *AM 75:243(1990)*.

40.2a.34.1 Triangulite $Al_3(UO_2)_4(PO_4)_4(OH)_5 \cdot 5H_2O$

Named in 1982 for the triangular habit of the crystals. TRIC $P\bar{1}$ or $P1$. $a = 10.39$, $b = 10.56$, $c = 10.60$, $\alpha = 116.4°$, $\beta = 107.8°$, $\gamma = 113.4°$, $Z = 1$, $D = 3.68$. *BM 105:611(1982)*. 35-577: 7.80_{10} 4.70_3 3.87_8 3.74_2 3.63_2 3.39_2 3.15_7 2.99_5. Aggregates of flat, triangular, or rhombohedral crystals up to 0.2 mm. Twinned about $\{01\bar{1}\}$ and $\{011\}$. Bright yellow. $G = 3.7$. Nonfluorescent. Biaxial (+); $Z \approx [01\bar{1}1]$, $Y \approx [011]$, X perpendicular to [100]; $N_{x\,calc} = 1.639$, $N_y = 1.665$, $N_z = 1.704$; $2V = 80°$; pleochroic Y, pale green-yellow; Z, bright yellow. Found within a quartz–K–feldspar–columbite zone in a complex pegmatite associated with beryl, metamict zircon, meta-autunite, phosphuranylite, and ranunculite at *Kobokobo, Kivu, Zaire*. HCWS *AM 69:212(1984)*.

40.3.1.1 Warikahnite $Zn_3(AsO_4)_2 \cdot 2H_2O$

Named in 1979 for Walter Richard Kahn, German mineralogist, who worked in Namibia and was interested in the crystal structures of secondary minerals. TRIC $P1$ or $P\bar{1}$. $a = 6.710$, $b = 8.989$, $c = 14.533$, $\alpha = 105.59°$, $\beta = 93.44°$, $\gamma = 108.68°$, $Z = 4$, $D = 4.29$ *NJMM:389(1979)*. Structure: *TMPM 27:187(1980)*. A complicated structure in which Zn is in six-, five-, and four-fold coordination and in five different combinations with AsO$_4$ and H$_2$O. 33-1468: 13.80_5 8.12_6 6.21_8 3.45_7 3.04_6 2.87_8 2.82_{10} 2.30_7. Bladed crystals to 3 mm × 0.5 mm × 0.5 mm, elongated on a; radiating masses. Pale yellow to colorless, transparent to translucent. Cleavage {001}, perfect; {100}, distinct. $H = 2$. $G = 4.24$. IR, TGA. Biaxial (+); $Z \wedge a = 47°$; $N_x = 1.747$, $N_y = 1.753$, $N_z = 1.768$; $2V = 75°$; no dispersion. Dissolves in hot HCl, HNO$_3$. Occurs in the second oxidation zone at *Tsumeb mine, Tsumeb, Namibia*, on corroded tennantite associated with adamite, stranskiite, koritnigite, claudetite, tsumcorite, and ludlockite. HCWS *AM 65:408(1980)*.

PHOSPHOFERRITE GROUP

Secondary hydrated phosphates the phosphoferrite group formula can be expressed as

$$A_3(PO_4)_2 \cdot n\ H_2O$$

where
$A = Fe^{2+}, Fe^{3+}, Mg, Mn^{2+}$
n is variable, dependent on the oxidation state of the cations, especially Fe and the temperature of formation, and crystallize in the orthorhombic space group Pbna. The structure consists of kinked edge-sharing octahedrally coordinated cations (A) that forms chains parallel to the c axis which are linked by corner sharing into sheets parallel to {100}. The sheets are heldl together along the c axis by $(PO_4)^{3-}$ tetrahedra.

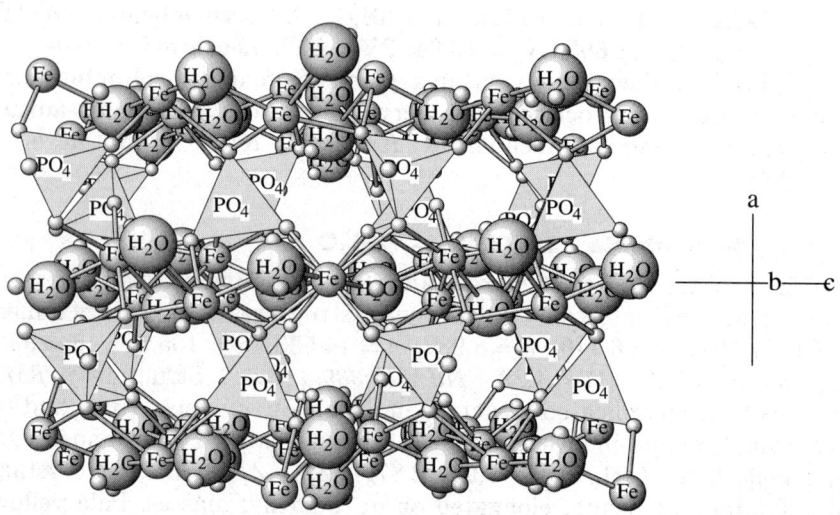

PHOSPHOFERRITE: $(Fe^{2+}, Mn^{2+})_3(PO_4)_2 \cdot 3H_2O$

This b axis projection of phosphoferrite (all Fe^{2+}) shows tetrahedral PO_4 groups joining sheets parallel to {100} composed of chains of octahedrally coordinated cations and water along the c axis.

40.3.2.2 Kryzhanovskite

PHOSPHOFERRITE Group

Mineral	Formula	Space Group	a	b	c	
Phosphoferrite	$(Fe^{2+},Mn^{2+})_3(PO_4)_2 \cdot 3H_2O$	Pbna	9.460	10.024	8.670	40.3.2.1
Kryzhanovskite	$(Mn^{2+},Fe^{3+})_2(PO_4)_2(OH)_2 \cdot 3H_2O$	Pbna	9.460	10.024	8.670	40.3.2.2
Reddingite	$(Mn^{2+},Fe^{2+})_3(PO_4)_2 \cdot 3H_2O$	Pbna	9.49	10.08	8.70	40.3.2.3
Landesite	$(Fe^{3+}Mn^{2+})_2(PO_4)_2(OH) \cdot 2H_2O$	Pbna	9.458	10.185	8.543	40.3.2.4
Garyansellite	$(Mg,Fe^{3+})_3(PO_4)_2(OH)_{1.5} \cdot 1.5H_2O$	Pbna	9.452	9.890	8.198	40.3.2.5

Note: compare with vivianite group (40.3.6.1).

40.3.2.1 Phosphoferrite $(Fe^{2+},Mn^{2+})_3(PO_4)_2 \cdot 3H_2O$

Named in 1926 in allusion to the composition. Phosphoferrite group. Part of the Fe–Mn series with reddingite. ORTH $P2na$ or $Pbna$. $a = 9.460$, $b = 10.024$, $c = 8.670$, $Z = 4$, $D = 3.32$. MM *43:789(1980)*. Structure: IC *15:316(1976)*. *9-479*: 4.25_7 3.18_{10} 2.72_8 2.64_7 2.41_7 2.22_7 1.615_7 1.550_7. Octahedral crystals with large {111}, or tabular on {010} often in parallel groups; massive, granular, coarsely fibrous. Pale green to red-brown from alteration, vitreous to subresinous luster, transparent to translucent. Cleavage {010}, poor; uneven fracture; brittle. H = 3–3½. G = 3.1. Biaxial (+); XYZ= abc; $N_x = 1.672$, $N_y = 1.680$, $N_z = 1.700$; $2V = 68°$; r > v, distinct; pleochroic (Hagendorf Fe/Mn = 2.7:1). Soluble in acids. Found at Palermo No. 1 mine, North Groton, NH; at Bull Moose mine, Custer and Dan Patch pegmatite, Keystone, SD; *Hagendorf Süd, near Pleystein, Oberpfälz, Bavaria, Germany*, with triploidite, ludlamite, and vivianite in pegmatite; at Lavra do Énio, Galileia, MG, Brazil. HCWS *DII:727*, NJMA *87:185(1954)*.

40.3.2.2 Kryzhanovskite $(Mn^{2+},Fe^{3+})_2(PO_4)_2(OH)_2 \cdot H_2O$

Named in 1950 for Vladimir Ilyitch Kryzhanovsky (1881–1947), curator of the Mineralogical Museum, Russian Academy of Sciences, Moscow. Phosphoferrite group. The Fe oxidized equivalent of phosphoferrite (40.3.2.1); in series with garyansellite (40.3.2.5). ORTH $Pbna$. $a = 9.460$, $b = 10.024$, $c = 8.670$, $Z = 4$, $D = 3.464$. Structure: AM *56:1(1971)*, MM *43:789(1980)*. *24-731*: 5.00_7 4.70_5 4.25_5 3.16_{10} 2.72_5 2.53_5 2.21_5 1.503_5. Poorly formed prismatic crystals up to 3 cm in diameter. Brown, green-brown, bronze on cleavage surfaces; yellow-brown streak; vitreous to dull luster. Cleavage {001}, perfect; uneven fracture. H = 3½–4. G = 3.31. Biaxial (+); $N_x = 1.79$, $N_z = 1.82$; $2V = 40$–$45°$; axial plane perpendicular to (001); r ≪ v; absorption X > Y > Z. X, wine-yellow; Y, orange-brown; Z, red-brown. Found with ludlamite at Palermo No. 1 pegmatite, Groton, Grafton Co., NH; Bull Moose and Tip Top pegmatites, Custer, Custer Co.; with diadochite, hureaulite, or siderite at Big Chief pegmatite, Glendale, Pennington Co.; Dan Patch pegmatite, Keystone Co., SD; with ludlamite and arrojadite at Rapid Creek, YT, Canada; Hagendorf Süd, Bavaria, Germany; with altered triphylite in the *Kolbinsk pegmatite, Russia*; Okatjimukuju Farm pegmatite, Karibib, Namibia. HCWS *DANS 72:762(1950); AM 36:382(1951), 59:896(1974).*

40.3.2.3 Reddingite $(Mn^{2+},Fe^{2+})_3(PO_4)_2 \cdot 3H_2O$

Named in 1878 for the locality. Phosphoferrite group. End member of the Mn–Fe series with phosphoferrite (40.3.2.1). ORTH *Pbna*. $a = 9.49$, $b = 10.08$, $c = 8.70$, $Z = 4$, $D = 3.262$. Structure: *MM 43:789(1980), ZK 118:327(1963)*. *9-496*: 4.28_7 3.20_{10} 2.74_8 2.66_7 2.42_7 2.23_7 1.625_7 1.560_7. Octahedral with large {111}, tabular {010}, often parallel groups; massive, granular, coarsely fibrous. Pinkish white or pale rose to yellowish white and colorless; vitreous to subresinous luster; translucent to transparent. Cleavage {010}, poor; uneven fracture. $H = 3-3\frac{1}{2}$. $G = 3.24$ (Mn:Fe 3:1). Biaxial (+); XYZ $= abc$; $N_x = 1.651$, $N_y = 1.656$, $N_z = 1.683$ (Branchville); $2V = 41°$; $r > v$, distinct. X, colorless; Y, pinkish brown; Z, pale yellow. Soluble in acids. Found at Black Mt. mine, Landford, and at the Feldspar, and Paul Bennet quarries, and Nevell Tourmaline mine, Buckfield, ME; at *A.N. Fillow quarry, Branchville, Redding Twp., Fairfield Co., CT*, the alteration product of lithiophilite–triphylite. HCWS *DII:727, NJMA 87:185(1954).*

40.3.2.4 Landesite $(Fe^{3+}Mn^{2+})_2(PO_4)_2(OH) \cdot 2H_2O$

Named in 1930 for Kenneth Knight Landes (b.1899), professor of geology, University of Michigan. Phosphoferrite group. ORTH *Pbna*. $a = 9.458$, $b = 10.185$, $c = 8.543$, $Z = 1$, $D = 3.210$. Structure: *MM 43:789(1980)*. *16-603*: 5.37_6 5.10_8 4.73_5 4.28_8 3.21_{10} 3.16_7 3.09_5 2.63_8. Rough octahedral-like crystals. Dirty yellow-brown, translucent. Cleavage {010}, good; {001}, poor. $H = 3-3\frac{1}{2}$. $G = 3.026$. Biaxial (−); $N_x = 1.720$, $N_y = 1.728$, $N_z = 1.735$; X perpendicular to {001} cleavage, Z perpendicular to {010}; 2V large. Found as an alteration product of reddingite in granite pegmatite at *Black Mt. mine, Berry quarry, Poland, Androscoggin Co., ME*, with rhodochrosite, eosphorite, fairfieldite, lithophilite, and apatite. HCWS *DII:729, AM 49:1122(1964).*

40.3.2.5 Garyansellite $(Mg,Fe^{3+})_3(PO_4)_2(OH)_{1.5} \cdot 1.5H_2O$

Named in 1984 for H. Gary Ansell, associate curator of the National Mineral Collection, Geological Survey of Canada. Phosphoferrite group. Mg analog of kryzhanovskite (40.3.2.2). ORTH *Pbna*. $a = 9.452$, $b = 9.890$, $c = 8.198$, $Z = 4$, $D = 3.154$. *AM 69:207(1984)*. *35-638*: 4.93_5 4.26_3 3.16_{10} 3.10_4 2.71_6 2.54_3 2.21_3 2.19_3. Platy crystals parallel to {010} or elongate on [100] with large {111}, {011} showing. Clove-brown, brown streak, vitreous with bronze luster on cleavage surfaces. Cleavage {001}, good. $H = 4$. $G = 3.16$. DTA/TGA. Biaxial (−); XYZ$= bac$; $N_x = 1.733$, $N_y = 1.757$, $N_z = 1.761$; $2V = 55°$; absorption $Z = X > Y$. X, reddish brown; Y, yellow;

Z, reddish brown. Found with kryzhanovskite in fractures in the *iron formations, northeastern YT, Canada*, with ludlamite, arrojadite, quartz, vivianite, meta-vivianite, and souzalite. HCWS

40.3.3.1 Parahopeite $Zn_3(PO_4)_2 \cdot 4H_2O$

Named in 1908 for its dimorphic relation to hopeite (40.3.4.1). TRIC $P\bar{1}$. $a = 5.768$, $b = 7.550$, $c = 5.296$, $\alpha = 93.42°$, $\beta = 91.18°$, $\gamma = 91.37°$, $Z = 1$, $D = 3.32$. Structure: *ZK 130:261(1969)*. *39-1352*: 7.53_4 4.44_5 2.98_{10} 2.88_7 2.78_6 2.72_5 2.67_5 2.51_4. See also *24-1461* and *34-156*. Crystals elongated [001], tabular {100} often as subparallel aggregates, tufted or fan-shaped. Twinning polysynthetic on {100} common. Colorless; vitreous luster, pearly on cleavage; transparent. Cleavage {010}, perfect. $H = 3\frac{1}{2}$–4. $G = 3.31$. Biaxial (+); X near a; Y $\wedge c$ on {100} $= 30°$; $N_x = 1.614$, $N_y = 1.625$, $N_z = 1.637$; $2V \approx 90°$, $r < v$. Reaphook Hill samples show zoning: center enriched in Fe and Mn, edges Mg-rich. Synthesized. Found at Hudson Bay mine, Salmo, Nelson district, BC, Canada, with hopeite, spencerite, and hemimorphite; at the Broken Hill mine, Zambia; at Reaphook Hill, SA, Australia, associated with scholzite, chalcophanite, gypsum, and hemimorphite. HCWS *DII:733; MM 36:621(1968), 39:684(1974)*.

40.3.4.1 Hopeite $Zn_3(PO_4)_2 \cdot 4H_2O$

Named in 1908 for Thomas Charles Hope (1766–1844), professor of chemistry, University of Edinburgh, Scotland. Dimorph of parahopeite (40.3.3.1) ORTH *Pnma*. $a = 10.594$, $b = 18.333$, $c = 5.029$, $Z = 4$, $D = 2.94$. Structure: *AM 61:987(1976) JSSC 93:9(1991)*. Sheets made up of zigzag chains of corner-sharing ZnO_4 parallel to c and PO_4 tetrahedra parallel to (010). The Zn atoms occupy two nonequivalent sites. Zn_1 is octahedrally coordinated with four O's from the water molecules, and two O's from PO_4 groups and is probably not fully occupied. Zn_2 is tetrahedrally coordinated. *37-465*: 9.12_6 4.85_2 4.57_{10} 4.41_2 3.46_3 3.39_3 2.85_7 1.939_2. See also *33-1474*(syn). Tabular {010} to prismatic {001} crystals that are single, tufted, or divergent aggregates, crusts, reniform masses, compact; faces development irregular; not twinned. Colorless to grayish white, pale yellow; white streak; vitreous luster, pearly on cleavage; transparent to translucent. Cleavage {010}, perfect; {100}, good; {001}, poor; uneven fracture; brittle. $H = 3\frac{1}{2}$. $G = 3.05$. Biaxial (−); XYZ= acb; $N_x = 1.589$, $N_y = 1.598$, $N_z = 1.599$ (Zambia); $2V = 37°$; $r < v$. Optical examination of sections on {100} shows distinct zones. The zones may represent distinct crystallites within a single grain with the same composition but different hydrogen-bonding schemes. *AM 61:987(1976)*. Earlier work had proposed two polymorphs: α and β, based on the distinct stabilities on heating; that is they had different water contents. *MM 15:1(1908)*. Synthesized. Found with spencerite at Hudson Bay mine, Salmo, Nelson district, BC, Canada; associated with hemimorphite in Zn mine at *Altenberg, Moresnet district, Belgium*; at the mines named α, β, and Globe & Phoenix at Broken Hill,

Zambia, in bone breccia in a limestone cave associated with tarbuttite, hemimorphite, smithsonite, and vanadinite. HCWS *DII:734*, *MJ 7:289(1973)*, *NBSM 25(6):85(1979)*.

40.3.5.1 Ludlamite $(Fe^{2+},Mg,Mn)_3(PO_4)_2 \cdot 4H_2O$

Named in 1877 for Henry Ludlam (1824–1880), English mineralogist and collector. MON $P2_1/a$. $a = 10.541$, $b = 4.646$, $c = 9.324$, $\beta = 100.43°$, $Z = 2$, $D = 3.176$. *AC 4:412(1951)*. 35-634: 4.92_{10} 4.59_8 3.96_8 3.75_4 3.06_6 3.00_6 2.77_5 2.56_5. Tabular {001}, sometimes parallel aggregates, massive, granular. Bright green to apple green, pale greenish-white streak; vitreous luster, translucent. Cleavage {001}, perfect; {100}, indistinct. $H = 3\frac{1}{2}$. $G = 3.165$. Biaxial (+); $Y = b$; $Z \wedge c = -67°$; $N_x = 1.653$, $N_y = 1.675$, $N_z = 1.697$ (Cornwall); $2V = 82°$, $r > v$. Soluble in acids. Found at Red Hill, Rumford, Oxford Co., ME; with phosphoferrite, triphylite, carbonate hydroxylapatite at Palermo No. 1 pegmatite, Groton, Grafton Co., NH; Victory, Big Chief, Bull Moose, Tip Top, Hesnard, Dan Patch, Etta, and Ferguson pegmatites, SD; Blackbird mine(!), Lemhi Co., ID, with vivianite; San Pedro mine, Pala, San Diego Co., CA; Rapid Creek, YT, Canada; crystals to 3 cm at San Antonio mine(!), Santa Eulalia district, CHIH, Mexico; *Wheal Jane*(!), *Kea, near Truro, Cornwall, England*; Brunita mine, Murcia, Spain; Stosgen, near Linz, and Hagendorf Süd, Bavaria, Germany; Stari Trg mine(!), Trepča, Kosovo, Serbia; Lavra do Énio, Galiléia, MG, Brazil; Morococala(!), Dalence Prov.; to 4 cm at Huanuni(!), Oruro Dept., Bolivia; Ashio mine, Tochigi Pref., Japan. HCWS *DII:952*.

40.3.5.2 Metaswitzerite $Mn_3^{2+}(PO_4)_2 \cdot 4H_2O$

Named in 1986 as the stable, lower hydrate of switzerite (40.3.5.4). MON $P2_1/n$. $a = 8.496$, $b = 13.173$, $c = 17.214$, $\beta = 96.65°$, $Z = 8$, $D = 2.96$. Structure: *TMPM 26:266(1979)*, *AM 71:1224(1986)*. 20-713: 8.55_{10} 7.13_4 6.78_4 3.17_4 2.93_4 2.84_4 2.76_4 2.59_6. Bladed crystals, micaceous masses, flattened and elongated on {100}, up to 3 mm long, showing {100}, {210}, {410}, {101}, {1̄01}, {111}, {1̄11}. Light to darker brown; adamantine luster, pearly on cleavage. Cleavage {001}, perfect; {010}, fair. $H = 2\frac{1}{2}$. $G = 2.95$. Biaxial (−); $Y = b$; $Z \wedge c = 10.5°$; $N_{x\,calc} = 1.602$, $N_y = 1.628$, $N_z = 1.632$; $2V = 42°$, $r < v$, slight. Dark brown material has $N_y = 1.666$, $N_z = 1.670$. Fe^{2+} substituting for Mn is probably responsible for the brown color and on oxidation the color darkens. Found at *Foote Mineral Co. spodumene mine, Kings Mt., Cleveland Co., NC*; at Viitaniemi, Finland; at Reaphook Hill, SA, Australia. HCWS

40.3.5.3 Sterlinghillite $Mn_3^{2+}(AsO_4)_2 \cdot 4H_2O$

Named in 1981 for the locality. An inadequately described mineral. Only one specimen known. 35-478: 11.1_{10} 6.39_3 5.01_1 3.69_3 3.21_{10} 2.88_4 2.85_4 2.75_6. Hemispherical clusters to 0.1 mm of randomly intergrown tabular crystals. White, light pink; white streak; silky to pearly luster. Cleavage parallel to elongation. H = 3. G = 2.95. Nonfluorescent. Uniaxial (−); $N_O = 1.671$, $N_E = 1.656$. Small amounts of Zn, Mg, and Fe detected in analysis. Formed by decomposition of loellingite. Found on loellingite–franklinite–willemite–calcite ores at the *Sterling Hill mine, Ogdensburg*. HCWS *AM 66:182(1981)*.

40.3.5.4 Switzerite $(Mn^{2+},Fe^{2+})_3(PO_4)_2 \cdot 7H_2O$

Named in 1967 for George Shirley Switzer (b.1915), mineralogist, National Museum of Natural History, Smithsonian Institution, Washington, DC. Unstable hydrate of metaswitzerite (40.3.5.2). MON $P2_1/a$. $a = 8.545$, $b = 13.164$, $c = 11.878$, $\beta = 110.12°$, $Z = 4$, $D = 2.545$. *AM 71:1221(1986)*. *41-598*: 11.12_{10} 8.47_5 3.37_9 2.94_3 2.35_4 2.14_5 1.914_2 1.612_3. Crystals must be completely hydrated, as when exposed to air they may disintegrate explosively to meta-switzerite. Thin tabular on {001}. Pale pink, vitreous to pearly luster, translucent. Cleavage {001}, perfect; brittle. Soft. G = 2.535. Biaxial (−); Z = b; X perpendicular to (001); $N_x = 1.560$, $N_y = 1.574$, $N_z = 1.580$; 2V = 70°. Occurs at the lowest levels of the *Foote Mineral Co. mine, Kings Mt., Cleveland Co., NC*, only in freshly exposed rock. HCWS

VIVIANITE GROUP

The members of the vivianite group have a general formula

$$A_3(XO_4)_2 \cdot 8H_2O$$

where A may be Fe^{2+}, Fe^{3+}, Mn, Mg, Zn, Co, or Ni, X is either P or As,

and monoclinic or triclinic spacegroups.

The vivianite structure has chains of Fe octahedra and PO_4 tetrahedra that form sheets parallel to {100}. The sheets are held together by hydrogen bonding from the H_2O ligands, a weak bonding that accounts for the perfect (010) cleavage found in these minerals.

Two arsenate compositions (Fe and Zn) show polymorphism. Different hydrates are also possible and to distinguish between them the prefix 'meta' is used. In general, the lower H_2O content minerals have a triclinic rather than a monoclinic (C2/m or I2/m) space group *CM 9:993(1968) BM 103:135(1980)*.

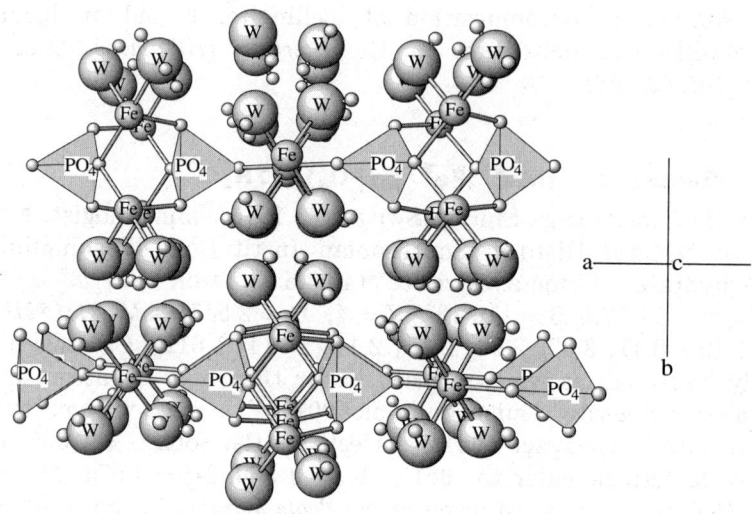

VIVIANITE: $Fe_3(PO_4)_2 \cdot 8H_2O$

In this c axis projection the two types of Fe sites are easily seen. Fe1 is coordinated with 4 H_2O and 2 oxygens of the $(PO_4)^{3-}$ groups. Fe2 coordinates with 2 H_2O and 4 oxygens of the $(PO_4)^{3-}$ groups and forms a double octahedra though sharing a common O–O edge. The Fe-Fe distance in the double Fe2 octahedra is only 2.96Å. Sheets of Fe1 octahedra, PO_4 tetrahedra, and the double Fe2 octahedra parallel {010}, and are held together by hydrogen bonding.

VIVIANITE GROUP

Name	Formula	Space Group	a	b	c	α	β	γ	
Vivianite	$Fe_3^{2+}(PO_4)_2 \cdot 8H_2O$	$C2/m$	10.086	13.441	4.703		104.27°		40.3.6.1
Baricite	$MG_3(PO)_2 \cdot 8H_2O$	$C2/m$	10.075	13.416	4.760		104.87		40.3.6.2
Erythrite	$Co_3(AsO_4)_2 \cdot 8H_2O$	$I2/m$	10.118	13.433	4.762		101.90		40.3.6.3
Annabergite	$Ni_3(AsO_4)_2 \cdot 8H_2O$	$I2/m$	10.054	13.303	4.716		102.10		40.3.6.4
Köttignite	$Zn_3(AsO)_2 \cdot 8H_2O$	$C2/m$	10.24	13.405	4.57		105.21		40.3.6.5
Parasymplesite	$Fe^{2+}(AsO_4)_2 \cdot 8H_2O$	$C2/m$	10.335	13.491	4.777		105.04		40.3.6.6
Hörnesite	$Mg_3(AsO_4)_2 \cdot 8H_2O$	$C2/m$	10.26	13.44	4.645		102.41		40.3.6.7
Arupite	$Ni_3(PO_4)_2 \cdot 8H_2O$	$I2/m$	9.889	13.225	4.645		102.41		40.3.6.8
Bobierrite	$Mg_3(PO_4)_2 \cdot 8H_2O$	$A2/2$	10.067	27.926	4.667		105.01		40.3.7.1
Manganese-Hörnesite	$Mn_3(AsO)_2 \cdot 8H_2O$	$P2_1/c$	10.38	28.09	4.77		105.67		40.3.7.2
Symplesite	$Fe^{2+}(AsO_4)_2 \cdot 8H_2O$	$P1$	7.86	9.35	4.75	93.7	98.1	106.5	40.3.8.1
Metaköttigite	$Zn_3(AsO_4)_2 \cdot 8H_2O$	$P1$	7.96	9.44	4.72	95.6	97.0	107.8	40.3.8.2
Metavivianite	$(Fe_{3-x}^{2+},Fe_x^{3+})_3(PO_4)(OH)_x \cdot 8\text{-}x\,H_2O$	$P\bar{1}, P1$	7.81	9.08	4.65	94.77	97.15	107.37	40.3.9.1

40.3.6.1 Vivianite $Fe_3^{2+}(PO_4)_2 \cdot 8H_2O$

Named in 1817 for J. G. Vivian, English mineralogist, who discovered the mineral. Vivianite group. MON $C2/m$. $a = 10.086$, $b = 13.441$, $c = 4.703$, $\beta = 104.27°$, $Z = 2$, $D = 2.696$. *BM* *103:135(1980)*. Structure: *BM* *105:147(1982)*. *30-662*(syn): 7.93_1 6.73_{10} 4.90_1 4.08_1 3.21_2 2.99_1 2.96_1 2.73_1. Prismatic [001], also flattened {010}, rarely {100}, equant. Crystals often rounded by vicinal development into bladelike or lanceolate shapes, also as stellate groups, reniform, globular, or tabular masses or concretions, incrusting with bladed or fibrous structure, earthy, pulverulent. Twinning lamellar on {010}, corresponds $\approx$ {304}. Colorless and transparent when fresh, becomes pale blue to greenish blue on oxidation and darkens on further exposure in air to bluish black; colorless to bluish white streak, darkening to dark blue or brown; vitreous luster, pearly on {010}, or dull and earthy; transparent to translucent. Cleavage {010}, perfect; traces on {$\bar{1}$06}, {100}; fibrous fracture $\approx$ perpendicular to [001]; flexible when in thin laminae; sectile. Translation gliding T(010), t[001]. $H = 1\frac{1}{2}$–2. $G = 2.68$. Mossbauer. Biaxial (+); $X = b$; $Z \wedge c = +28\frac{1}{2}°$; $N_x = 1.579$, $N_y = 1.602$, $N_z = 1.637$ (Na); $2V = 83\frac{1}{2}°$, $r < v$, weak. X, blue; Y,Z, pale yellowish green. Increasing oxidation (deepening of color) increases indices, decreases birefringence and pleochroism on {010} becomes stronger. Easily soluble in acids. Easily synthesized. Small amounts of Mn^{2+}, Mg, and Ca substitute for Fe^{2+}; amount of Fe^{3+} indicates oxidation level. Mossbauer analyses suggests that the Fe in the paired octahedra is the most resistant to oxidation. The oxidation reaction Fe^{2+} to Fe^{3+} leads to a reduction in H_2O, loss of OH^-, and changes in the hydrogen-bonding scheme. At some juncture during the loss of H bonding there would be collapse of the structure. Vivianite undergoing oxidation turns dark blue and at high Fe^{3+} content may disintegrate to a red-brown powder. *AM* *65:361(1980)*. There has been discussion as to whether the ferri-ferrous compositional distinctions (coupled with H_2O changes) are fully expressed by the less hydrated triclinic species meta-vivianite or whether kerchite, also triclinic, probably a more fully oxidized ferri-ferrous hydroxyphosphate, and a discredited mineral species, should be reinstated. *MM* *50:687(1986)*. A widespread secondary mineral, vivianite occurs in the gossan of metallic ore deposits and in pegmatites, in clays, recent alluvial, and glauconitic sediments, in pebble phosphates such as those composed of bone, in decayed wood, or in other organic sediments, such as in lignite, peat, and bog irons. It is often mistaken for turquoise as the coloring material of fossil teeth or bones. Occurs abundantly in the pegmatites in Newry, ME; at Palermo No. 1 mine, North Groton, NH; at Allentown, Shrewsbury, Monmouth Co., and Mullica Hill, Gloucester Co., NJ; in green sand at Middletown, New Castle Co., DE; at the Ideal Cement quarry, Castle Hayne, NC; at Stewart Co., GA; at Clear Springs mine, Barston, FL; at Ibex Mine, Lake Co., CO; at the Big Chief, Bull Moose, and Tip Top pegmatites, SD; in diatomite in a Tertiary lake bed near Burney, Shasta Co., CA; in bog-iron ore at Côte St. Charles, Vaufreuil Co., QUE, Canada; at the San Antonio mine, Santa Eulalia, CHIH, Mexico, in blue-green gem-quality crystals to 8 cm(!); at Cornwall, and Millersdale, Derbyshire, England; in the coal measures, France; in the limonite ores in Amberg–Auerbach, and the pegmatites of Hagendorf, Bavaria, Germany; at Stari Trg mine, Trepca, Serbia; in the sedimentary iron ores in the Kerch and Taman Penins. on the Black Sea, Russia; as crystals at the Ashio mines(!), Shimotsuke Pref., Japan; at the Wannon R., VIC, Australia; in many localities in Bolivia, such as at Colquechaca, and the Santa Fe mine, Morocoala, at Avicaya, Ichucollo, Poopo, and Porvenienr mines, Ovuro, and in especially large crystals (to 10 cm) at the Siglo XX mine(!!), Llallagua, Potosi; at Aracuai, Lavra Velha, and in Galiléia, Lavra do Enio(!), at the Telido mine, and at Larada Ilha, Taquaral, MG, Brazil. HCWS *DII:742*.

40.3.6.2 Baricite $(Mg^{2+},Fe^{2+})_3(PO_4)_2 \cdot 8H_2O$

Named in 1976 for Ljudevit Baric, mineralogist, University of Zagreb, Croatia. Vivianite Group. MON $C2/m$. $a = 10.075$, $b = 13.416$, $c = 4.760$, $\beta = 104.87°$, $Z = 2$, $D = 2.448$. *AM 61:1053(1976)*. 29-705: 6.71_{10} 3.20_4 2.96_6 2.70_7 2.53_5 2.42_3 2.22_3 1.676_3. Large plates up to 12 cm in maximum dimension and up to 5 cm thick. Colorless to pale blue. Cleavage {010}, perfect. H = $1\frac{1}{2}$–2. G = 2.42. Biaxial (+); X = b; Z $\wedge$ c = +32°; $N_x = 1.554$, $N_y = 1.564$, $N_z = 1.595$; 2V = 59°; r < v, weak. Occurs in fractures in a siderite iron formation in *Big Fish–Blow R. area, northeastern YT, Canada*. HCWS *CM 14:403(1976)*.

40.3.6.3 Erythrite $Co_3(AsO_4)_2 \cdot 8H_2O$

Named in 1832 from the Greek for *red*. The Co end member in the Co–Ni series with annabergite (40.3.6.4). Vivianite group. MON $I2/m$. $a = 10.118$, $b = 13.433$, $c = 4.762$, $\beta = 101.90°$, $Z = 2$, $D = 3.14$. *NBSM 25(19):39(1970), CM 9:493(1968)*. 33-413(syn): 7.96_3 6.72_{10} 4.40_2 3.26_5 3.23_4 2.74_3 2.71_2 2.46_2. Prismatic to acicular [001], flattened {010}. Crystals deeply striated, furrowed on [001] or {010} parallel to {h0l} or {$\bar{h}$0l}; radial or stellate groups, globular or reniform with drusy surface; coarse-fibrous, earthy, pulverulent. Crimson-red and peach-red, lighter colors indicate more Ni content, may show color banding; streak paler than color; weakly adamantine luster, pearly on {010}, dull or earthy; transparent to translucent. Cleavage {010}, perfect; {100}, {$\bar{1}$02}, indistinct; flexible in thin {010} laminae; translation gliding T{010}, t{001}; sectile. H = $1\frac{1}{2}$–$2\frac{1}{2}$. G = 3.06 (Schneeberg). Biaxial (+); X = b; Z $\wedge$ c = +31°; $N_x = 1.629$, $N_y = 1.663$, $N_z = 1.701$ (Schneeberg); 2V very large; r > v [may also be biaxial (–)]. X, pale pinkish; Y, pale violet; Z, red. Substitutions of Ca, Fe, Zn, and Mg have led to the description of several varieties. Synthesized. A secondary mineral usually formed by the oxidation of arsenides of Co (and Ni), erythrite is associated with symplesite, morenosite, retgersite, malachite, adamite, and scorodite. Occurs at Blackbird district, Lemhi Co., ID; at Cottonwood Canyon, Stillwater Range, NV; in the Black Hills, Yavapai Co., AZ; with annabergite in the Black Hawk district, Grant Co., NM, at the Bishop mine, E of Long Lake, Inyo Co., and mines in

San Gabriel Canyon, Los Angeles Co., CA; in the Cobalt area(!), Timiskiming district, ONT, Canada; Sara Alicia mine, Alamos, SON, Mexico; earthy varieties at Chalanches near Allemont, Isère, and at Markirch, Alsace, France; in the Heubachtal, Wittichen district, Black Forest, Baden, and at Schmiedestollenhalde, and Schneeberg, Saxony, Germany; at Joachimsthal, Bohemia, Czech Republic; at the Arhbar mine No. 4 vein, Bou Azzer, Morocco(!). HCWS *DII:747*.

40.3.6.4 Annabergite $Ni_3(AsO_4)_2 \cdot 8H_2O$

Named in 1852 for the locality. The Ni end member of the Ni–Co series with erythrite (40.3.6.3). Vivianite Group. MON $I2/m$. $a = 10.054$, $b = 13.303$, $c = 4.716$, $\beta = 102.10°$, $Z = 2$, $D = 3.22$. Structure: *BM 105:332(1951)*. *34-141*(syn): 7.91_3 6.66_{10} 4.36_2 3.20_4 2.98_5 2.72_3 2.69_2 2.44_2. See also *35-568*, Mg-rich mineral from Laurium, Greece. Similar in habit and physical properties to erythrite, annabergite is usually found as fine-crystalline coatings or earthy crusts. Pale rose, light gray to apple green in the highest Ni varieties. Biaxial (+); $X = b$; $Z \wedge c = +36°$; $N_x = 1.622$, $N_y = 1.658$, $N_z = 1.687$; $2V = 84°$; $r > v$; pleochroic. Calcian variety (Creetown, Kirkudbrightshire, Scotland) is white and pulverulent; Mg-containing varieties were among the earliest described for this mineral. Less common than erythrite, annabergite occurs with retgersite at Lovelock mine, Cottonwood Canyon, Humbolt Co., NV; at Silver Cliff, Custer Co., CO; notably at Gowganda and Keeley mine, Cobalt in the Timinkiming district, ONT, Canada; at Schneeberg and *Annaberg, Saxony, Germany*; at *Leogang, Salzburg, Austria*; from the Arburese mine, Gonnosfanadiga, Sardinia; Mg varieties at Sierra Carera, Almeria Prov., Spain, and at Chalanches, Isère; at the Plaka mine, Laurium, Attica, Greece(!). HCWS *DII:747*.

40.3.6.5 Köttigite $Zn_3(AsO_4)_2 \cdot 8H_2O$

Named in 1850 for Otto Köttig (b.1824), chemist of Schneeberg, Saxony, Germany. Part of a series with erythrite and annabergite. Vivianite group. MON $C2/m$. $a = 10.24$, $b = 13.405$, $c = 4.757$, $\beta = 105.21°$, $Z = 2$, $D = 3.24$. Structure: *AM 64:376(1979)*. Transition metals are randomly distributed over the two octahedral sites. *33-1467*(syn): 7.97_2 6.72_{10} 3.23_5 3.01_3 3.00_4 2.74_3 2.71_3 2.33_2. See also *33-430*(calc). Prismatic [001], flattened {010}, massive, crusts with crystalline or fibrous structure. Carmine or peach blossom-red, reddish-white streak, silky luster on fracture, translucent. Cleavage {010}, perfect. H = $2\frac{1}{2}$-3. G = 3.33. Biaxial (+); $X = b$; $Z \wedge c = 37°$; $N_x = 1.622$, $N_y = 1.638$, $N_z = 1.671$; $2V = 74°$. Soluble in acids. Synthesized. A secondary mineral that occurs at the Sterling Hill mine, Ogdensburg, Sussex Co., NJ; at the Ojuela mine, Mapimi, DGO, Mexico, in sprays of bladed crystals to 6 mm(!); and known from the alteration of smaltite and sphalerite at the *Daniel mine, Schneeberg, Saxony, Germany*. HCWS *DII:751*.

40.3.6.6 Parasymplesite $Fe_3^{2+}(AsO_4)_2 \cdot 8H_2O$

Named in 1837 from the Greek *to bring together*, in allusion to its relations to other minerals. Dimorphous with symplesite (40.3.8.1). Vivianite group. MON $C2/m$. $a = 10.335$, $b = 13.491$, $c = 4.777$, $\beta = 105.04°$, $Z = 2$, $D = 3.043$. *BM* *100:310(1977)*. *35-461:* 7.91_7 6.68_{10} 4.41_4 3.91_3 3.24_5 3.01_9 2.75_5 2.47_4. See also *42-536*, zinc-rich mineral from Ojuela mine, Mapimi, Mexico. Spherical aggregates with coarsely fibrous, radial structure; crystals elongated [001], sometimes flattened $\{1\bar{1}0\}$; twinned on $\{1\bar{1}0\}$. Light green to leek-green, becoming greenish black or deep indigo-blue on oxidation, bluish-white streak, vitreous luster, transparent when fresh. Cleavage $\{1\bar{1}0\}$, perfect, uneven fracture, brittle; translation gliding $T\{1\bar{1}0\}$, $t[001]$. $H \approx 2\frac{1}{2}$. $G = 3.01$. Biaxial $(-)$; X perpendicular to $(1\bar{1}0)$; $Z \wedge c = 31\frac{1}{2}°$; $N_x = 1.635$, $N_y = 1.668$, $N_z = 1.702$; $2V = 86\frac{1}{2}°$, $r > v$. X, deep blue; Y, nearly colorless; Z, colorless. With alteration the extinction angle increases, absorption colors go from blue to green to brown and decrease in intensity. Soluble in acids. Turns red-brown in H_2O_2, KOH solutions. A secondary mineral associated with pharmcosiderite, scorodite, erythrite, annabergite, and iron oxyhydroxides. Found at Sterling Hill mine, Ogdensburg, Sussex Co., NJ; at Ojuela mine, Mapimi, DGO, Mexico, in crystals to 5 cm(!); at Lobenstein, and Saubach in Voigtland, with roselite at Neustädtel near Schneeberg, Saxony, Germany; with quartz in cavities in hornstone at Felsöbanya, Romania; at Pizzo Cipolla, Mandanici, Messina Prov., Italy; at Kiura Ohita, Japan; at the Magnet mine, TAS, Australia. HCWS *DII:752*.

40.3.6.7 Hörnesite $Mg_3(AsO_4)_2 \cdot 8H_2O$

Named in 1860 for Moritz Hoernes (1815–1868), curator of the Hof-Mineralienkabinett, Vienna, Austria. Vivianite group. MON $C2/m$. $a = 10.26$, $b = 13.44$, $c = 4.645$, $\beta = 102.41°$, $Z = 2$, $D = 2.602$. *NJMM:349(1966)*. *35-856*(syn): 7.97_3 6.73_{10} 4.40_5 3.99_3 3.21_5 3.00_6 2.74_4 2.71_3. Prismatic [001] and flattened $\{010\}$, columnar, radially foliated. White, transparent, pearly on cleavage. Cleavage $\{010\}$, perfect; $\{100\}$, poor; flexible. $H = 1$. $G = 2.73$ (Jachymov). Biaxial $(+)$; $X = b$; $Z \wedge c = 31°$; $N_x = 1.563$, $N_y = 1.571$, $N_z = 1.596$; $2V = 60°$. Soluble in acids. Synthesized. Found at Sterling Hill mine, Ogdensburg, Sussex Co., NJ; at Långban, Värmland, Sweden; at *Csiklova in the Banat, Romania*; and with nagyagite from Nagyág; at Jachymov, Bohemia, Czech Republic; at Fiano, near Naples, Italy, with nocerite, fluorite, and hydromagnesite in marble. HCWS *DII:755, AM 52:1588(1967)*.

40.3.6.8 Arupite $Ni_3(PO_4)_2 \cdot 8H_2O$

Named in 1990 for Hans Arup (b.1928), director of the Danish Corrosion Center, Copenhagen, Denmark. Ni analog of vivianite. Vivianite group. MON $I2/m$. $a = 9.889$, $b = 13.225$, $c = 4.645$, $\beta = 102.41°$, $Z = 2$, $D = 2.85$. *NJMM:76(1990)*. *33-951*(syn): 7.77_4 6.62_{10} 4.81_6 4.48_3 3.80_5 2.92_7 2.68_5 2.66_4. Earthy aggregates up to 2 mm in diameter, grains are short prisms up to 5 μm long. Sky blue to turquoise-blue, earthy luster, translucent.

$H = 1\frac{1}{2}$–2. Biaxial; $N_x = 1.632$, $N_z = 1.680$. X, blue; Z, colorless. Synthesized. Occurs in the weathered *Santa Catharina Ni-rich Fe meteorite* found in 1875 on the coast of southern *Brazil*, associated with honessite and reevesite. HCWS *NBSM 25(19):64(1982)*.

40.3.7.1 Bobierrite $Mg_3(PO_4)_2 \cdot 8H_2O$

Named by J. Dana in 1868 for Pierre Adolphe Bobierre (1823–1881), French agricultural chemist who described the mineral. Vivianite group. MON $A2/a$. $a = 10.067$, $b = 27.926$, $c = 4.667$, $\beta = 105.01°$, $Z = 4$, $D = 2.133$. Structure: *AM 71:1229(1986)*. I and II forms observed; I (mineral): unit cell data given; II (synthetic): $C2/m$, has slightly smaller unit cell dimensions. *16-330*(syn): 8.04_2 6.96_{10} 3.02_1 2.94_3 2.81_1 2.57_1 2.41_1 2.13_1. Minute acicular or fibrous crystals elongated [001] flattened {010}, rosettelike flattened aggregates, massive, lamellar. Colorless, white; weakly vitreous luster; transparent. Cleavage {010}, perfect. $H = 2$–$2\frac{1}{2}$. $G = 2.195$. Biaxial (+); $Y = b$; $Z \wedge c = 29°$; $N_x = 1.510$, $N_y = 1.520$, $N_z = 1.543$ (Minnesota); $2V = 71°$; $r < v$. Synthesized. Found on a fossil elephant tusk in Pleistocene gravel, Edgarton, Pipeston Co., MN; with apatite at *Odengärden, Bramle, Norway*; at Wodgina, Pilbara district, WA, Australia; in guano on the island of Mejillones, Chile. HCWS *DII:753, AM 48:635(1963)*.

40.3.7.2 Manganese-Hörnesite $(Mn^{2+},Mg)_3(AsO_4)_2 \cdot 8H_2O$

Named in 1951 for the composition and relationship to hörnesite. Vivianite group. MON $P2_1/c$. $a = 10.38$, $b = 28.09$, $c = 4.77$, $\beta = 105.67°$, $Z = 4$, $D = 2.76$. *AM 39:159(1954)*. *8-141*: 8.19_8 7.01_{10} 3.25_6 3.09_6 3.02_7 2.88_4 2.72_4 2.41_7. Needlelike crystals elongated and flattened on {010}, resembling gypsum. Radiating aggregates, stellate, up to 1 cm in diameter. White to colorless, white streak, silky luster on cleavage, transparent. Cleavage {010}, perfect; {100}, {201}, {$\bar{2}$01}, imperfect. $H = 1$. $G = 2.64$. Biaxial (+); $N_x = 1.579$, $N_y = 1.589$, $N_z = 1.609$; $2V = 65$–$70°$. Found at Sterling Hill mine, Ogdensburg, Sussex Co., NJ; at *Långban, Värmland, Sweden*, with rhodochrosite. HCWS *AMG 1:333(1951)*.

40.3.8.1 Symplesite $Fe_3^{2+}(AsO_4)_2 \cdot 8H_2O$

Named in 1837 from the Greek *to bring together*. Dimorphous with parasymplesite (40.3.6.6); see also ferrisymplesite (42.10.1.1). Vivianite group. TRIC $P\bar{1}$. $a = 7.86$, $b = 9.35$, $c = 4.75$, $\alpha = 93.7°$, $\beta = 98.1°$, $\gamma = 106.5°$, $Z = 1$, $D = 3.01$. Structure: *NJMA 138:94(1980), SR 63:303(1950)*. *8-172*: 8.97_2 7.50_2 6.79_{10} 5.03_1 4.07_1 4.01_1 3.74_1 3.40_1. Similar in physical and chemical properties to parasymplesite. Biaxial (−); $N_x = 1.635$, $N_y = 1.668$, $N_z = 1.702$; $2V = 87°$. Found as sprays to 4 mm Ojuela mine, Mapimi, DUR, Mexico; Voigtland, Germany; at Felsobanya, Rumania, at Bou Azzer, Morocco. HCWS DII: 752.

40.3.8.2 Metaköttigite $(Zn,Fe^{2+},Fe^{3+})_3(AsO_4)_2 \cdot 8H_2O$

Named in 1982 for the relationship to köttigite. The Zn analog of symplesite (40.3.8.1). Vivianite group. TRIC $P\bar{1}$. $a = 7.96$, $b = 9.44$, $c = 4.72$, $\alpha = 95.6°$, $\beta = 97.0°$, $\gamma = 107.8°$, $Z = 1$, $D = 3.03$. *NJMM:506(1982). 35-563:* 8.90_4 6.91_{10} 4.99_2 3.93_3 3.79_2 3.11_2 3.00_3 2.83_2. Minute bluish-gray crystals in oriented intergrowth on common c axis with köttigite. Twinned on $\{1\bar{1}0\}$. Cleavage $\{1\bar{1}0\}$ perfect parallel to (010) cleavage of köttigite. Biaxial; $N_x = 1.648$, $N_y = 1.680$, $N_z = 1.716$; $2V = 93°$; strongly pleochroic. X, dark blue; Y, yellow; Z, pale yellow. Microprobe and Mossbauer analyses showed minor Co and Fe. Found on massive samples of fine-grained goethite and smithsonite at *Ojuela mine, Mapimi, DGO, Mexico*, associated with köttigite and adamite. HCWS *AM 68:1039(1983)*.

40.3.9.1 Metavivianite $(Fe^{2+}_{3-x},Fe^{3+}_x)(PO_4)_2(OH)_x \cdot 8 - xH_2O$

Named in 1974 to reflect the relationship to vivianite (40.3.6.1). Synonyms: kertschenite, oxykertschenite. Vivianite group. TRIC $P1$ or $P\bar{1}$. $a = 7.81$, $b = 9.08$, $c = 4.65$, $\alpha = 94.77°$, $\beta = 97.15°$, $\gamma = 107.37°$, $Z = 1$, $D = 2.69$. *AM 50:896(1974). 29-1137:* 8.59_4 6.71_{10} 4.86_4 4.27_1 3.87_3 3.07_1 2.90_2 2.77_3. Flat prisms $\{110\}$ elongated on c, striated parallel to c. Twinned on $\{110\}$, equivalent to the mirror plane $\{010\}$ in (monoclinic) vivianite structure. Leek-green, opaque to translucent. Cleavage $\{110\}$, perfect. Biaxial $(-)$; X perpendicular to (010); Y parallel to (110); $N_x = 1.579$, $N_y = 1.603$, $N_z = 1.629$; X, blue, blue-green; Y,Z, yellowish green (similar to vivianite). Metavivianite is not a dimorph of vivianite, as the oxidation of Fe is coupled with changes in H_2O content, which influences the OH^- and H^+ available for bonding and the stability of the structure. See discussion under vivianite (40.3.6.1). *MM 50:687(1986)*. Occurs intimately intergrown with dark red kryzhanovskite in solution cavities in massive triphyllite boulders at *Big Chief pegmatite, Glendale, Pennington Co., SD*. HCWS

40.3.10.1 Volborthite $Cu_3V_2O_7(OH)_2 \cdot 2H_2O$

Named in 1838 for Alexander von Volborth (1800–1876), Russian paleontologist who first noted the mineral. MON $C2, Cm$ or $C2/m$. $a = 10.606$, $b = 5.874$, $c = 7.213$, $\beta = 94.9°$, $Z = 2$, $D = 3.52$. *NJMM:385(1988)*. Structure: *JSSC 85:220(1990)*. Pyrovanadate ions connected by Cu octahedra into layers parallel to (001) with H_2O between the layers, similar to thortveitite. *26-1119:* 7.16_{10} 4.10_5 3.09_5 3.00_5 2.89_5 2.64_7 2.57_7 2.39_7. Scaly, spongy, or fibrous crusts, rosettelike aggregates, also reticulated. Scales may have triangular or hexagonal outline. Dark olive-green to yellowish green, vitreous luster, subtranslucent. Perfect cleavage one direction. $H = 3\frac{1}{2}$. $G = 3.42$. Biaxial $(-)$; $N_x = 2.01$, $N_y = 2.04$, $N_z = 2.07$ (Uzbeck); $2V$ large; $r < v$. Soluble in acids. A secondary mineral found in the sandstones at Richardson in the canyon of the Colorado R., Grand Co., UT; at Monument No. 1 mine(!), Monument Valley, Navajo Co., AZ, in a siltstone with carnotite, hewettite, metatyuyamunite, rauvite, and tyuyamunite; at *Syssersk and Nizhne Taglisk, Ural Mts.*,

Russia; at Woskressensk, Perm, and in Ferghana, Uzbekistan; at La Verde mine, Copiapo, Chile. HCWS *DII:818, AM 59:372(1974)*.

40.3.11.1 Rauenthalite $Ca_3(AsO_4)_2 \cdot 10H_2O$

Named in 1964 for the locality. TRIC $P\bar{1}$. $a = 12.564$, $b = 12.169$, $c = 6.195$, $\alpha = 89.09°$, $\beta = 79.69°$, $\gamma = 118.58°$, $Z = 2$, $D = 2.362$. *BM 105:327(1982)*. Structure: *AC(B) 39:4(1983). 35-569*: 10.8_{10} 6.2_6 5.40_4 4.79_4 3.95_4 3.36_8 2.44_7 2.06_6. Rounded irregular polycrystalline spherulites, individual crystals flattened, prismatic. Snow white, colorless, translucent. $G = 2.36$. Biaxial (+); X, Y in plane of flattening; $Y \wedge c$ (elongation)= 5–10°; $N_{x\ calc} = 1.540$, $N_y = 1.552$, $N_z = 1.570$; $2V = 85°$. Found in the *Rauenthal vein system, at the 40-m level of Gabe Gottes vein, Ste.-Marie-aux-Mines, Vosges Mts., Alsace, France*. HCWS *BM 87:169(1964), AM 50:805(1965)*.

40.3.12.1 Phaunouxite $Ca_3(AsO_4)_2 \cdot 11H_2O$

Named in 1982 for the French name for the locality, *Phaunoux*. See rauenthalite (40.3.11.1). TRIC $P\bar{1}$. $a = 12.563$, $b = 12.181$, $c = 6.205$, $\alpha = 88.94°$, $\beta = 91.67°$, $\gamma = 113.44°$, $Z = 2$, $D = 2.294$. *BM 105:327(1982)*, Structure: *AC(B) 39:4(1983). 41-581*: 11.49_{10} 6.23_9 5.42_8 3.28_9 3.10_5 2.96_6 2.44_7 2.03_6. Acicular crystals showing {100}, {010}, fan-shaped aggregates. Colorless, vitreous luster, translucent. $G = 2.28$. Biaxial (+); $N_x = 1.532$, $N_y = 1.542$, $N_z = 1.556$; $2V \approx 80°$. At room temperature dehydrates slowly to rauenthalite. Found in two portions of the *Gabe Gottes vein, Ste.-Marie-aux-Mines, Vosges Mts., Alsace, France*. HCWS *AM 68:850(1983)*.

40.3.13.1 Ferrazite $(Pb,Ba)_3(PO_4)_2 \cdot 8H_2O$ (?)

Named in 1919 for Jorge Palmiro de Araujo Ferraz (1883–1926) of the Geological Survey of Brazil. Defined on chemical analysis of compact, dark yellowish-white disk-shaped pebbles (favas) found in the Brazilian diamond deposits near Diamantina, Brazil, but there have been no additional physical or chemical data. HCWS *AJS (4)48:353(1919)*.

VARISCITE GROUP

The members of the variscite group conform to the general formula

$$ATO_4 \cdot 2H_2O$$

where

$A = Al, Fe^{3+}$, Sc, other trivalent cations
$T = P, As$

and the structure is orthorhombic in space group *Pbca* or is monoclinic in space group $P2_1/n$.

The structure is built of alternate TO_4 tetrahedra and octahedrally coordinated cations that associate into a three-dimensional network. The cations have a distorted octahedral array bonding with O from four PO_4 groups and O from two water molecules. The arrangement presents limited opportunities for cleavage except where H_2O molecules are concentrated. Tilting of the tetrahedra and a somewhat different hydrogen-bonding arrangement distinguish the monoclinic from the orthorhombic subgroups.

There is complete solid solution between Fe^{3+} and Al in the PO_4 members variscite and strengite, and probably between the AsO_4 members mansfieldite and scorodite.

VARISCITE GROUP: ORTHORHOMBIC SUBGROUP

Mineral	Formula	Space Group	a	b	c	
Variscite	$AlPO_4 \cdot 2H_2O$	$Pbca$	9.822	8.561	9.630	40.4.1.1
Strengite	$Fe^{3+}PO_4 \cdot 2H_2O$	$Pcab$	10.07	9.82	8.67	40.4.1.2
Scorodite	$Fe^{3+}AsO_4 \cdot 2H_2O$	$Pcab$	9.996	10.278	8.937	40.4.1.3
Mansfieldite	$AlAsO_4 \cdot 2H_2O$	$Pcab$	10.10	9.80	8.79	40.4.1.4
Yanomanite	$InAsO_4 \cdot 2H_2O$	$Pbca$	10.446	9.068	10.345	40.4.1.5

VARISCITE GROUP: MONOCLINIC SUBGROUP

Mineral	Formula	Space Group	a	b	c	β	
Metavariscite	$AlPO_4 \cdot 2H_2O$	$P2_1/n$	5.178	9.514	8.454	90.35	40.4.3.1
Phosphosiderite	$Fe^{3+}PO_4 \cdot 2H_2O$	$P2_1/n$	5.30	9.77	8.73	90.6	40.4.3.2
Kolbeckite	$ScPO_4 \cdot 2H_2O$	$P2_1/n$	5.418	10.25	8.893	90.7	40.4.3.3

40.4.1.1 Variscite $AlPO_4 \cdot 2H_2O$

Named in 1837 for Variscia, the ancient name of the Voigtland district in Germany where the mineral was discovered. Variscite group. Two orthorhombic modifications, 1O and 2O, and a monoclinic form, metavariscite (40.4.3.1), have been described. See also strengite (40.4.1.2) with which there is a series Al–Fe^{3+}, and metavariscite (40.4.3.1). ORTH $Pbca$. $a = 9.822$, $b = 8.561$, $c = 9.630$, $Z = 8$, $D = 2.59$. Structure: $AC(B)$ 33:263(1977), JSC 32:631 (1991). 33-33: 5.36_7 4.82_3 4.26_7 3.90_3 3.63_2 3.04_{10} 2.91_5 2.87_4 (variscite 1O: $Pbca$, $a = 9.822$, $b = 8.558$, $c = 9.622$). See also 25-18 (variscite 2O: $Pcab$, $a = 9.90$, $b = 9.66$, $c = 17.18$, $Z = 16$). Rare crystals octahedral {111} with {001} (= Lucin-type 1O), predominantly as cryptocrystalline aggregates, fine-grained masses, nodules, veinlets or crusts, chalcedonic to opaline (= Messbach type 2O). Pale to emerald-green, blue-green, to colorless; white streak; vitreous to waxy luster in the dense varieties; transparent to translucent. Cleavage {010}, good; {001}, poor; massive variscite has splintery fracture, conchoidal in glassy types. $H = 4\frac{1}{2}$. $G = 2.57$ (crystals); G lower for microcrystalline varieties. Biaxial (−); XYZ= acb; $N_x = 1.563$, $N_y = 1.588$, $N_z = 1.594$ (Lucin); 2V moderate; r < v, weak. Crystalline Al end members insoluble in HCl, soluble in alkalies. Fe^{3+} substitution in variscite leads to higher indices of refraction. On heating green samples may turn purplish red. High-Al members typically deposited under surface conditions in cavities or breccias as the result of phosphatic meteoric waters acting on aluminous rocks. Associated minerals are wavellite, crandallite, metavariscite, apatite, chalcedony, and iron oxyhydroxides. Occurs at Moore's Mill, Cumberland Co., PA; at Carolina pyrophyllite mine and Winston-Salem, NC: at Cole Shaft, Bisbee, AZ; at the Little Green Monster mine, Fairfield(!), Utah Co., in nodules in brecciated phosphatic sediments associated with other phosphates, at Amatrice Hill, Stockton, and at Lucin, Box Elder Co., UT; at Champion mine, White Mts., Mono Co., CA; at *Messbach Plauen, Voigtland (Variscia), Thuringia, Germany*, and at Langenstriegis, Frankenberg, Saxony; at Mixnitz, and at Brandberg, Austria; at Zeleznik, Trenice, Zajecow near St. Benigna, Bohemia, Czech Republic; at Iron Knob, Iron Monarch mine, SA, Australia; with apatite in insular rock phosphates on Clipperton Atoll in the Pacific Ocean and in the Antilles, Gran Roque Is., Venezuela; at de la Perle Island near Martinique; Malpelo Is., Colombia. HCWS *DII:756*, *AM 57:36(1972)*.

40.4.1.2 Strengite $Fe^{3+}PO_4 \cdot 2H_2O$

Named in 1877 for Johann August Streng (1830–1897), mineralogist, University of Giessen, Germany. Variscite group. See phosphosiderite (40.4.3.2) and AsO_4 species scorodite (40.4.1.3). ORTH *Pcab*. $a = 10.07$, $b = 9.82$, $c = 8.67$, $Z = 8$, $D = 2.90$. *NJMA 98:1(1962)*. *33-667*: 5.50_9 4.38_7 4.00_5 3.11_{10} 3.00_5 2.95_5 2.55_5 2.53_4. See also *15-391*. Spherical and botryoidal aggregates with radial fibrous structure and drusy surface, or as crusts; rarely as fine crystals to 3 mm. Peach-blossom red, carmine, violet, colorless; white streak; transparent to translucent. Cleavage {010}, good; {001}, poor. $H = 3\frac{1}{2}$. $G = 2.87$. DTA. Biaxial (+); XYZ= acb; $N_x = 1.707$, $N_y = 1.719$, $N_z = 1.741$ (Pleystein); 2V small. Optics variable with Al substitution. Soluble in HCl, but not in HNO_3. Complete series extends between Al (variscite) and Fe^{3+} (strengite) although most analyses appear to be close to end member composition. Synthesized. Surface products formed by alteration of iron-containing phosphates such as triphylite, or dufrenite, in pegmatites, also common in gossans of iron ore deposits. Associated minerals are beraunite, cacoxenite, vivianite, metastrengite, frondelite-rockbridgeite, dufrenite, microcrystalline apatite. Occurs as an oxidation product of triphylite at Palermo No. 1 mine, and Fletcher mine, North Groton, NH; at Mullica Hill, NJ; at Moor's Mill, Cumberland Co., and Noblis mine, Bearville, Lancaster Co., PA; with rockbridgeite near Midval, Rockbridge Co., VA; at Jacksonville, Calhoun Co., Indian Mt., Cherokee Co., Rock Run Station, Indian Mt., AL; in Polk Co., AR; at Woods mine, Burnett Gap, Locke Co., TN; in the Bull Moose pegmatite, Custer Co., SD(!); at Champion mine, White Mts., Mono Co., and a Mn-rich variety at the Stewart Lithia mine, Pala, San Diego Co., CA. Al-rich varieties found at Manhattan, Nye Co., NV; at Mangualde, Beira Prov., Portugal. Also found in the *Eleonore iron mine, Giessen, near Waldgirmes, Thüringia*, alteration product at the pegmatites at Pleystein and Hagendorf, Bavaria, *Germany*; in the magnetite ore at Kirunavaara, Sweden; at the Iron Monarch mine, Iron Knob, SA, Australia; at Sapucaia Velha pegmatite, MG, Brazil. HCWS *DII:756*.

40.4.1.3 Scorodite Fe^{3+}AsO$_4$ · 2H$_2$O

Named in 1817 from the Greek for *garlic-like*, in allusion to its odor when heated. Variscite group. See also the Al end member of the series, mansfieldite (40.4.1.4). ORTH *Pcab.* $a = 9.996$, $b = 10.278$, $c = 8.937$, $Z = 8$, $D = 3.39$. *MM 35:776(1966)*, *AC(B) 32:2891(1976)*. *37-468:* 5.61$_8$ 5.02$_4$ 4.47$_{10}$ 4.09$_3$ 3.18$_9$ 3.06$_5$ 3.00$_4$ 2.60$_4$. See also *18-654, 26-778*. Crystals usually pyramidal {111}, tabular {001}, or prismatic [010], often aggregates as irregular groups or crusts, also massive, crystalline or porous and sinterlike, dense to earthy. Pale leek-green or gray-green to liver-brown, colorless, bluish, violet, yellow; earthy material is pale green, grayish, or brownish green; vitreous to subadamantine, subresinous luster; transparent in crystals. Cleavage {201}, imperfect; {001}, {100}, in traces; subconchoidal fracture. DTA. Biaxial (+); XYZ= *acb*; $N_x = 1.784$, $N_y = 1.795$, $N_z = 1.814$ (Durango); 2V ≈ 75°; faintly pleochroic; r > v, strong; Z > X, Y. Biaxial (+); $N_x = 1.888$, $N_y = 1.895$, $N_z = 1.915$; 2V = 62° (Japan, Al-rich). Indices, birefringence, 2V decrease with decreasing Fe/Al. Soluble in acids, in strong alkali mineral decomposes with separation of hydrous iron oxide. Synthesized. There is probably a complete Al–Fe series to mansfieldite, some substitution of PO$_4$ for AsO$_4$ (toward strengite). Found as a primary mineral at hydrothermal deposits such as at Saubach, Voigtland, Germany, but mostly as a secondary phase in gossans from the oxidation of arsenopyrite, or other As-containing minerals. Associated with pharmacosiderite, beudanite, vivianite, gypsum, chalcedonic quartz, iron oxyhydroxides, clays. Occurs at Sterling Hill mine, Ogdensburg, Sussex Co., NJ; at Gold Hill, Toole Co., UT; at Grandview mine, AZ; at the Mina Noche Bueno, Mazapil, ZAC(!), and similarly at the Ojuela mine, Mapimi, DGO(!), Mexico, in sharp, lustrous blue-green crystals to 3 cm; at Grube Clara, Randlachtal and Schwarzenberg, Saxony, Germany; at Kamereza mine, Attica, Greece; at Bhilwara district pegmatites, Rajasthan, India; at Tsumeb mine, Tsumeb, Namibia; at Preamimma mine, Callington, SA, Australia; at Geralfo Candinha mine, MG, Brazil; at Chuquicamata, Antofagasta, Chile. HCWS *DII:763*.

40.4.1.4 Mansfieldite AlAsO$_4$ · 2H$_2$O

Named in 1948 for George Rogers Mansfield (1875–1947), geologist, U.S. Geological Survey. Variscite group. The Al end member of an Al–Fe^{3+} series with scorodite. ORTH *Pcab.* $a = 10.10$, $b = 9.80$, $c = 8.79$, $Z = 8$, $D = 3.08$. *NJMM:79(1963)*. *22-123:* 5.53$_{10}$ 4.92$_3$ 4.41$_{10}$ 4.02$_4$ 3.12$_{10}$ 3.00$_9$ 2.53$_7$ 2.45$_4$.

Spherulitic crusts made up of fibers; crusts up to 20 cm in diameter, porous, cellular masses. Light green to pale gray. H = $3\frac{1}{2}$–4. G = 3.02. Biaxial (−); $N_x = 1.627$, $N_z = 1.657$; 2V = 68° (Oregon). Biaxial (+); $N_x = 1.622$, $N_y = 1.624$, $N_z = 1.642$; 2V = 30°; r > v. *AM 33:122(1948)*. Found at Hobart Butte, Lane Co., OR, with realgar and scorodite and intergrown with quartz and kaolinite; at Neubulach, Black Forest, Germany. HCWS *DII:763, AM 33:122(1948)*.

40.4.1.5 Yanomamite $InAsO_4 \cdot 2H_2O$

Named in 1994 for the Yanomami Indian people of the Amazon basin. Vacasite group. ORTH *Pbca*. $a = 10.446$, $b = 9.086$, $c = 10.345$, Z = 8, D = 3.876. *EJM 6:245(1994)*. PD: 5.70_7 4.53_{10} 4.16_5 3.87_6 3.25_6 3.11_5 2.66_4 2.54_4. Idiomorphic crystals in aggregates, green-yellow, always coated with thin film of In-rich scorodite and intimately intermixed. Biaxial; $N_{mean} = 1.65$. Al and Fe^{3+} present on analysis. Appears to be cogenetic with In-bearing scorodite as an alteration of arsenopyrite and In-rich sphalerite. A secondary mineral found in the Sn deposits of *Mangabeim, Goiáa, Brazil*. HCWS

40.4.2.1 Koninckite $Fe^{3+}PO_4 \cdot 3H_2O$

Named in 1883 for Laurent Guillaume de Koninck (1809–1887), Belgian geologist. TET Space group unknown. $a = 11.95$, $c = 14.52$, Z = 16, D = 2.625. *BM 91:487(1968)*. *41-1489*: 8.47_{10} 6.00_1 4.52_1 3.85_2 3.79_3 3.66_1 3.36_1 2.99_2. See also *22-339*. Small spherical aggregates up to 0.8 mm in diameter of radiating needles about 40 μm long terminated by an oblique plane. Yellow-white, light yellow streak, transparent. Cleavage transverse to elongation. H = $3\frac{1}{2}$–4. G = 2.40. Biaxial; $N_{x'} = 1.645$, $N_{z'} = 1.656$; 2V not very large; elongated on X. May also be isotropic, N = 1.65. Easily soluble in hot HCl, HNO_3. Found at *Richelle, Visé, near Liège, Belgium*. Specimen with some Al from Nagano Pref., Japan. HCWS *DII:763*.

40.4.3.1 Metavariscite $AlPO_4 \cdot 2H_2O$

Named in 1925 for its dimorphic relationship to variscite (40.4.1.1). Variscite group. See also phosphosiderite (40.4.3.2), the Fe analog. MON $P2_1/n$. $a = 5.178$, $b = 9.514$, $c = 8.454$, $\beta = 90.35°$, Z = 4, D = 2.535. *AC(B) 29:2292(1973)*. *33-32*: 6.33_2 4.76_{10} 4.55_7 4.23_6 3.50_6 2.71_9 2.29_2 1.585_2. See also *15-311*. Crystals thin to thick tabular {010}, equant or slightly elongate [001] or [100] producing laths, rarely prismatic [100], also pseudo-orthorhombic, granular masses. Light green, white streak, vitreous luster, transparent, masses translucent. Cleavage {010}. H = $3\frac{1}{2}$. G = 2.54. Biaxial (−); $N_x = 1.551$, $N_y = 1.558$, $N_z = 1.582$ (crystals); 2V = 55°; r < v. Possibly a complete isomorphous substitution to phosphosiderite. Found as crystals associated with variscite crystals in cavities of variscite nodules at *Edison-Bird mine, Utahlite Hill, Lucin, Box Elder Co., UT*; massive with variscite at Candelaria, NV; in phosphatized andesite on Malpelo Island, (Pacific Ocean), Colombia. HCWS *DII:767*.

40.4.3.2 Phosphosiderite $Fe^{3+}PO_4 \cdot 2H_2O$

Named in 1890 for the composition and from the Greek for *iron*. Variscite group. Also known as metastrengite, as it is the dimorph of strengite (40.4.1.2). MON $P2_1/n$. $a = 5.30$, $b = 9.77$, $c = 8.73$, $\beta = 90.6°$, $Z = 4$, $D = 2.76$. AM 37:362(1952), Structure: AM 51:168(1966). A more open structure than strengite. 33-666: 4.91_3 4.69_{10} 4.57_2 4.36_7 3.61_5 2.79_8 2.57_4 2.01_2. Crystals tabular {010} or stout prismatic [001], botryoidal or reniform masses, crusts with a radial fibrous structure. Light violet-pink, peach blossom red, moss green; aggregates bright rose-red to nearly colorless; vitreous to subresinous luster; translucent. $H = 3\frac{1}{2}$–4. $G = 2.74$. Biaxial (−); $X \wedge c \approx 4°$; $Y = b$; $N_x = 1.692$, $N_y = 1.725$, $N_z = 1.738$ (Na); $2V = 62°$; $r > v$, very strong. Al substitutes for Fe^{3+}, a complete series to metavariscite probable. Synthesized. Found at the Fletcher and Palermo No. 1 mines, North Groton, and at Fitzgibbon mine, Alstead, NH; at Mullica Hill, NJ; at Graves Mt., GA; at Williams properties, Coosa Co., AL; at Bull Moose pegmatite, Custer, Highland Lode, Etta mine, Keystone, and White Elephant pegmatite, SD; at Stewart Lithium mine, Pala, San Diego Co., CA; at Chanteloub, France; at *Kaltenborn mine, near Eisenfeld, Siegen, Germany*, in iron ore, also in Bavaria with pharmacosiderite at Kreuzberg, Pleystein, Hagendorf, and the other pegmatites in Bavaria; at Lavra do Enio, Galileia, and Sapucaia Velha pegmatite, Galileia, MG, Brazil; El Criollo pegmatite, Argentina. HCWS *DII:767*.

40.4.3.3 Kolbeckite $ScPO_4 \cdot 2H_2O$

Named in 1926 for Friedrich L. W. Kolbeck (1860–1943), mineralogist, Mining Academy, Freiberg, Germany, and reexamined (1959) to establish the composition. Variscite group. MON $P2_1/n$. $a = 5.418$, $b = 10.25$, $c = 8.893$, $\beta = 90.7°$, $Z = 4$, $D = 2.344$. 38-431: 5.1_3 4.78_{10} 4.66_2 4.44_{10} 3.71_4 2.88_7 2.85_5 2.61_6. See also *2-177*. Crystals short prismatic [001] with {110}, {011}, and {001}; twinned on {100}, appearing orthorhombic. Yellowish apple-green to colorless. Cleavage {010}, distinct; conchoidal fracture; brittle. $H = 3\frac{1}{2}$–4. $G = 2.35$. Biaxial (−); $N_{x'} = 1.570$, $N_{z'} = 1.610$. Synthesized. Found at Potash Sulfur Springs and Magnet Cove, AK; at Fairfield, UT; in a quartz wolframite vein in the Sadisdorf copper mine, *Schmiedeberg, Saxony, Germany*; at Klause quarry, Gleichenberg, Styria, Austria; at Felsobanya, Hungary; at Sakpur, Damnagar Talaq, Boroda, Gujarat, India; in the W–Sn ores, Tigrinoye deposit, Silchote–Alin Range, China. HCWS *DII:1015*, *BGSA 70:1648(1959)*, *MM 46:493(1982)*, *DANS 322:1963(1992)*.

40.4.4.1 Kankite $Fe^{3+}AsO_4 \cdot 3.5H_2O$

Named in 1976 for the locality. MON. $a = 18.803$, $b = 17.490$, $c = 7.633$, $\beta = 92.71°$, $Z = 16$, $D = 2.732$. *MJ 12:6(1984)*. 29-694: 12.8_{10} 7.56_3 7.22_2 4.76_4 4.26_3 3.70_2 2.63_2 2.55_2. Botryoidal coatings and crusts up to 7 mm thick; spherules of very thin tablets less than 10 μm across. Yellowish-green, earthy, dull to vitreous luster. H = 2–3. G = 2.70. DTA, TG, IR. Biaxial; $N_{max} = 1.680$, $N_{min} = 1.664$; length-fast. Some S substitutes for As. Insoluble in H_2O but readily dissolves in 1:10 HCl. Found on the old (thirteenth- to fifteenth-century) dumps at *Kank, Kutno Hora district, Stredocesky Kraj, Bohemia, Czech Republic*; at Suzukura mine, Enzan City, Yamana-shi Pref., Japan. HCWS *NJMM:426(1976); AM 62:594(1974), 70:220(1985)*.

40.4.5.1 Steigerite $AlVO_4 \cdot 3H_2O$

Named in 1935 for George Steiger (1869–1944), chief chemist, U.S. Geological Survey. MON $P2_1/M$, P2. $a = 11.840$, $b = 25.00$, $c = 11.040$, $\beta = 110.16°$, $Z = 24$, $D = 2.58$. *ZVMO 116:100(1987), AM 44:322(1959). 39-406*: 12.52_5 5.67_3 3.09_4 2.96_3 2.93_{10} 2.21_4 2.18_5 1.426_4. See also *29-20*. Pulverulent coatings composed of fibers, gumlike masses, and occasionally flat plates. Canary-yellow to olive-green, waxy luster. H = $2\frac{1}{2}$–3. G = 2.56. Biaxial (+); $N_x = 1.730$, $N_y = 1.734$, $N_z = 1.752$; 2V = 58°. Easily soluble in 1:10 acids to a deep cherry-red solution. Insoluble in H_2O. Small amounts of Cr, Fe, Ca, P, and S detected on analysis. Synthesized. Found with fervanite and gypsum coating fractures in and around nodular concretions of corvusite in sandstone at *Sullivan Brothers claim, Gypsum Valley, San Miguel Co., CO*; at northwestern Karatau Mt. range, Kazakhstan. HCWS *DII:1049*.

40.4.6.1 Churchite-(Y) $YPO_4 \cdot 2H_2O$

Named in 1865 for Arthur Herbert Church (1834–1915), English chemist who described and analyzed the mineral. Gypsum group. Synonym: weinschenkite. MON. $a = 5.61$, $b = 15.14$, $c = 6.19$, $\beta = 115.3°$, $Z = 4$, $D = 3.25$. *MM 30:211(1953), ZK suppl. 3:258(1991). 8-167*: 7.5_9 4.70_6 4.21_{10} 3.74_6 3.02_9 2.82_7 2.62_6 2.17_6. Crusts of subparallel platy aggregates or acicular radiating aggregates. Yellowish-gray, tinged with flesh red; vitreous luster, pearly on cleavage; transparent. Cleavage parallel to {010}, perfect; {100}, {001}, distinct. H = 3. G = 3.27. Biaxial (+); $N_x = 1.623$, $N_y = 1.631$, $N_z = 1.657$; 2V = 15°; Z perpendicular to plates. Analysis showed other rare earths Y: (Er, La, Nd, Ce) = 3:1, also Fe^{3+}, Ca, Nd-rich samples of churchite (26–42% of RE elements relative to 50% Y) were found in the mantle of weathered metamorphic rocks in Kazakhstan. Crystallites from the yellowish or pinkish-white oolitic segregations up to 12 μm long and 2μm wide with rectangular cross section gave slightly different indices of refraction and X-ray diffraction powder pattern. *39-1385*: 6.12_3 3.49_4 3.02_{10} 2.82_7 2.15_7 1.92_3 1.858_9 1.743_3. Associated minerals were apatite, fluorite, amphiboles, and acicular transparent zeolites, possibly natrolite, *DANS 268:139(1983)*. Synthesized. Found at Indian Mt., Cherokee Co., NC; at Vesuvius, Rockbridge, Co., VA; at Sausalito, CA; at Wheal Pendres tin mine, Camborne, with limonite, quartz, and barite, and at *Trefoil mine, near Bodmin, Cornwall, England*; in

Bavaria, Germany; in laterite of Chutukou, Enisei Range, Yenisei, Russia; at Mt. Weld, WA, Australia, in carbonaceous laterites where churchite occurs as void fillings in crandallite group minerals, as fine grains, or intergrown with limonite and associated with goyazite, gorceixite, florencite, crandallite, monazite, apatite, and cerianite. HCWS DII:773, DANS 268:139(1983), MM 51:468(1987).

RHABDOPHANE GROUP

The rhabdophane minerals are phosphate monohydrates corresponding to the general formula

$$APO_4 \cdot H_2O \quad \text{or a multiple thereof}$$

where

$A =$ Ce, La, Nd, Th, Ca, U, Fe^{3+}

and crystallizing in hexagonal or orthorhombic space groups.

Thy hydrated equivalents of minerals in the monazite group the rhabdophane group minerals have a very open structure formed of alternating cations and phosphate groups in columns along the c axis. The linkage of each column to four neighbours results in open channels with a diameter of 5.3Å along the unique hexagonal, or pseudohexagonal, direction. The channels are sites for variable amounts of the "zeolitic" water molecules MM 48:146 (1984).

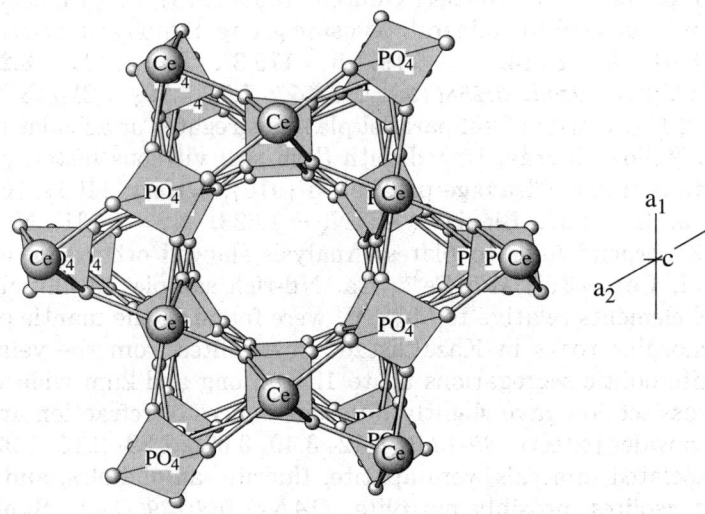

RHABDOPHANE: - (Ce)CePO$_4$ · H$_2$O

This c axis projection shows the open channels and structure of rhabdophane formed from columns of alternating cation and PO$_4$ groups stacked along c. Note no water molecules are shown in this diagram.

40.4.7.3 Rhabdophane-(Nd)

RHABDOPHANE GROUP

Mineral	Formula	Space Group	a	b	c	
Rhabdophane-(Ce)	$CePO_4 \cdot H_2O$	$P6_222$	6.960		6.372	40.4.7.1
Rhabdophane-(La)	$LaPO_4 \cdot H_2O$	$P6_222$	6.98		6.39	40.4.7.2
Rhabdophane-(Nd)	$NdPO_4 \cdot H_2O$	$P6_222$				40.4.7.3
Grayite	$ThPO_4 \cdot H_2O$	$P6_222$	6.957		6.396	40.4.7.4
Brockite	$(Ca,Th,REE)PO_4 \cdot H_2O$	$P6_222$	6.98		6.40	40.4.7.5
Tristramite	$(Ca,U^{4+},Fe^{3+})PO_4 \cdot H_2O$	$P6_222$	6.913		6.422	40.4.7.6
Ningyoite	$(U,Ca,Fe)_2(PO_4)_2 \cdot 1-2H_2O$	$P222$	6.78	12.10	6.38	40.4.8.1

40.4.7.1 Rhabdophane-(Ce) $CePO_4 \cdot H_2O$

Named in 1878 from the Greek *rod* and *to appear*, in allusion to the characteristic bands shown in its spectrum. Rhabdophane group. HEX $P6_222$. $a = 6.960$, $c = 6.372$, $Z = 3$, $D = 4.72$. *MM 48:146(1984)*, *AC 3:339(1950)*. 35-614: 6.03_5 4.38_5 3.48_5 3.01_9 2.82_{10} 2.35_2 2.15_5 1.853_5. Incrustations, botryoidal, globular or stalactitic with radiated-fibrous structure. Brown, pink, yellow-white; greasy luster; translucent. Uneven fracture. $H = 3\frac{1}{2}$. $G = 3.97$. DTA. Uniaxial (+); $N_O = 1.688$, $N_E = 1.744$ (Oberwolfach); positive elongation. Easily soluble in HCl. Analyses show distinctions among the amount and relative proportions of RE elements. Ce-rich forms from Cornwall also show Y, (Al,Fe), and Si, while samples from Oberwolfach show Nd, La, Pr, Sm, Gd, and Eu, higher indices of refraction and larger unit cell parameters. The compositional range of REE phosphates with rhabdophane type structure is probably large. See rhabdophane-(La) (40.4.7.2). Found on the *mine dumps of Fowey Consols, Cornwall, England*; at the Clara mine, Oberwolfach, Middle Black Forest, Germany, in druses on barite, fluorite, chalcopyrite, and tetrahedrite. HCWS *DII:774*, *CE 38:331(1978)*, *AM 65:1065(1980)*, *MM 47:393(1983)*.

40.4.7.2 Rhabdophane-(La) $LaPO_4 \cdot H_2O$

40.4.7.3 Rhabdophane-(Nd) $NdPO_4 \cdot H_2O$

Named in 1959 to distinguish La/Nd-rich forms of rhabdophane-(Ce). Rhabdophane group. HEX $P6_222$. $a = 6.98$, $c = 6.39$, $Z = 3$, $D = 3.99$. *AM 44:633(1959)*. 12-277: 6.07_5 4.40_8 3.49_6 3.02_{10} 2.83_8 2.36_4 2.15_8 1.859_6. See also *MM 47:933(1983)*. Uniaxial (+); $N_O = 1.654$, $N_E = 1.703$. Similar in properties to Ce form. The data above from thin incrustations found at *Salisbury, CT*, which show RE elemental composition mainly of Nd, La, Y, and low Ce. High La and Nd variety also obtained from Monumental Summit, Valley Co., ID. In both localities the mineral is associated with hydrated iron hydroxides, clays, and quartz in highly weathered environments. HCWS *DII:774*, *USGSPP 600:B48(1964)*, *MM 48:146(1984)*, *AM 70:440(1985)*.

40.4.7.4 Grayite $ThPO_4 \cdot H_2O$

Named in 1957 for Anton Gray, mining engineer, advisor to U.K. Atomic Energy Authority. Rhabdophane group. HEX $P6_222$. $a = 6.957$, $c = 6.396$, $Z = 3$, $D = 6.41$. *USGSPP 424C:339(1961)*. *42-1389:* 6.05_2 4.35_5 3.46_1 3.03_{10} 2.82_8 2.36_1 2.15_8 1.856_5. Cryptocrystalline aggregates, powdery. Dark reddish brown, resinous luster. Conchoidal fracture. $N_{av} = 1.66–1.69$; moderately birefringent. Less than 2% RE in analyses; Pb, Ca detected. Found in lithia pegmatite, Gooddays mine, Mtoko, and the Mahaka claim, Wedza, *Zimbabwe*; in Gunnison Co., CO. HCWS *AM 47:419(1962)*.

40.4.7.5 Brockite $(Ca,Th,REE)(PO_4) \cdot H_2O$

Named in 1962 for Maurice Brock of the U.S. Geological Survey, who mapped the area in which the mineral was found. Rhabdophane group. See also grayite (40.4.7.4). HEX $P6_222$. $a = 6.98$, $c = 6.40$, $Z = 3$, $D = 4.28$. *AM 47:1346(1962)*. *15-248:* 6.06_4 4.37_7 3.47_5 3.03_{10} 2.83_7 2.37_3 1.86_5 1.69_4. Fine-grained massive aggregates and nodules up to 35 mm in diameter; radially fibrous with individuals up to 50 μm long, earthy coatings. Yellow or red-brown (from hematite stain), greasy luster, translucent. $G = 3.90$. Uniaxial (+); $N_O = 1.680$, $N_E = 1.695$. High percentage Nd relative to Ce, La, and Y in analysis. Found in inclusions of pyrite in veins of altered granitic rocks at the *Bassick mine, Wet Mts., Custer Co., CO*, with quartz, albite, hematite, and apatite, also at Hardwick and Nightingale mines with thorite and barite. HCWS

40.4.7.6 Tristramite $(Ca,U^{4+},Fe^{3+})PO_4 \cdot 2H_2O$

Named in 1983 for Sir Tristram, the Arthurian mythological character. Rhabdophane group. HEX $P6_222$. $a = 6.913$, $c = 6.422$, $Z = 3$, $D = 4.18$. *MM 47:393(1983)*. *35-613:* 5.99_4 4.37_4 3.46_3 2.99_{10} 2.83_{10} 2.35_2 2.14_5 1.850_5. Acicular to fibrous crystals up to 80 μm long. Pale yellow to greenish yellow. $G = 3.8–4.2$. Uniaxial (+); $N_O = 1.644$, $N_E = 1.664$. Found intergrown with goethite in fine fractures in mines such as *Wheal Trewavas, Breage, Cornwall, England*. HCWS *AM 69:815(1984)*.

40.4.8.1 Ningyoite $(U,Ca,Fe)_2(PO_4)_2 \cdot 1–2H_2O$

Named in 1959 for the locality. Rhabdophane group. ORTH $P222$. $a = 6.78$, $b = 12.10$, $c = 6.38$, $Z = 3$, $D = 4.63$. *AM 44:633(1959)*. *12-273:* 5.99_4 4.33_6 3.38_5 3.02_{10} 2.81_8 2.35_4 2.13_8 1.845_4. Acicular, lozenge-shaped, crystals up to 5 μm long. Brown, green-brown. $N_{av} \approx 1.64$; extinction parallel to elongation. Synthesized. $N_x = 1.69–1.70$, $N_z = 1.70–1.71$. X, pale green; Z, green. Found at *Ningyo-toge mine, Tottori Pref., Honshu, Japan*, where it coats pyrite and other minerals and fills cracks. HCWS

40.4.8.2 Lermontovite $(U^{4+})(PO_4)(OH) \cdot H_2O$

Named in 1955 for Mikhail Yurevich Lermontov (1814–1841), Russian poet. ORTH $P2_12_12_1$. $a = 9.72$, $b = 18.89$, $c = 10.13$, $Z = 12$, $D = 3.94$. *MZ*

5:82(1983). 41-1399: 4.87_8 4.69_7 4.12_8 3.92_{10} 3.83_8 3.58_8 3.29_9 3.15_7. Radiating fibrous aggregates, botryoidal. Gray-green, dull to silky luster on fracture, transparent. Very brittle. Rapidly disintegrates in electron microscope beam. G = 4.0–4.5. Biaxial; $N_x = 1.686$–1.690, $N_y = 1.707$, $N_z = 1.724$–1.726; Z = c =elongation parallel extinction. Analysis shows Tl in addition to U. Found under sharply reducing conditions in the hydrothermal deposits of the *Kola Penin., Russia*, associated with molybdenum sulfates, marcasite, and "thallium ochre." HCWS *AM 43:379(1958), 69:214(1984)*.

40.4.8.3 Vyacheslavite $U^{4+}(PO_4)(OH) \cdot 2.5H_2O$

Named in 1984 for Vyacheslav Gavilovich Melkov (b.1911), Russian mineralogist. ORTH Space group $Cmcm$, $Cmc2_1$, or $C2cm$. $a = 6.96$, $b = 9.10$, $c = 12.38$, Z = 6, D = 5.01. *ZVMO 113:360(1984)*. 38-397: 6.19_{10} 5.03_2 4.56_6 4.13_6 3.68_5 3.04_3 2.71_5 2.69_7. Spherulitic aggregates of tabular crystallites, 8 μm long, 1.5 μm wide. Green, dark green. Cleavage {010}. G = 4.90. Biaxial (−); $N_x = 1.700$, $N_y = 1.728$, $N_z = 1.730$; 2V = 15°; Z perpendicular to elongation; 2V small; positive elongation; weakly pleochroic in greens. Some Ca detected on analysis. Found on quartz crystals associated with octahedral pyrite in *Uzbekistan*. HCWS *AM 69:1195(1984), 70:828(1985)*.

40.5.1.1 Fransoletite $Ca_3Be_2(PO_4)_2(PO_3OH)_2 \cdot 4H_2O$

Named in 1983 for André-Mathieu Fransolet, University of Liège, France. MON $P2_1/a$. $a = 7.354$, $b = 15.07$, $c = 7.055$, $\beta = 96.41°$, Z = 2, D = 2.53. Structure: *BM 106:499(1983), AM 77:848(1992)*. 35-629: 7.52_4 6.55_3 4.57_3 3.47_9 3.32_7 3.04_{10} 2.60_4 2.32_5. Crystals with mosaic texture, aggregates to 3 mm in diameter. Colorless to whitish, white streak, subvitreous luster, transparent in thin sections, otherwise turbid. Cleavage {010}, imperfect, easily produced; irregular fracture; not brittle. H ≈ 3. Nonfluorescent. Biaxial (+); Y = b; $Z \wedge c = 14°$; $N_x = 1.560$, $N_y = 1.566$, $N_z = 1.586$; 2V = 25°; nonpleochroic; no dispersion visible. Found in the *Tip Top pegmatite, Black Hills, near Custer, SD*, associated with beryl and other secondary minerals. See parafransoletite (38.2.9.1). HCWS

40.5.2.1 Parafransoleite $Ca_3Be_2(PO_4)_2(PO_3OH)_2 \cdot 4H_2O$

Named for its relationshp to fransoletite, the dimorph. TRIC $P\bar{1}$. $a = 7.327$, $b = 7.696$, $c = 7.061$, $\alpha = 94.90°$, $\beta = 96.82°$, $\gamma = 101.87°$, Z = 1, D = 2.56. Structure: *AM 77:843(1992), 77:848(1992)*. Dominant structural elements are tetrahedral chains parallel to [100] in which four-membered rings of alternating BeO_4 and PO_4 tetrahedra link through BeO_4 tetrahedra. The chains are linked through Ca and by H bonding. The difference between the dimorphs is the placement of one of two nonequivalent Ca atoms relative to the chains. Structure similar to that of ehrleite (40.5.7.1). *PD*: 7.52_5 3.62_8 3.36_2 3.03_{10} 2.60_6 2.49_3 2.33_4 1.870_3. Crystals are thin spear-shaped blades to 0.4 long, and as sheaf- and bowtielike aggregates, radial sprays to 2 mm in diameter showing subparallel crystal intergrowths. Most crystals flattened on {010}, elongated parallel to [100], with the following forms, showing {010}, {011}, {13$\bar{1}$}, {$\bar{1}$31}, {14$\bar{4}$}; often doubly terminated and showing simple contact twinning on {010}. White, white streak, silky luster. H = $2\frac{1}{2}$. G = 2.54. Nonfluorescent. Biaxial (+); XYZ $\approx$ abc; $N_x = 1.562$, $N_y = 1.564$, $N_z = 1.588$; $2V_{calc} = 33°$. Dissolves readily in cold 1:1 HCl. Conditions for the formation of a particular dimorph are not known. Found at the *Tip Top pegmatite, Black Hills, Custer, SD* with other secondary phosphates covering fracture surfaces in large beryl crystals and associated with mitridatite, whitlockite, montgomeryite, and also with hurlbutite, roscherite, robertsite. HCWS

40.5.3.1 Faheyite $(Mg,Mn^{2+})Fe_2^{3+}Be_2(PO_4)_4 \cdot 6H_2O$

Named in 1953 for Joseph John Fahey (1901–1980), geochemist, U.S. Geological Survey. HEX $P6_322$. $a = 9.43$, $c = 16.00$, $Z = 3$, $D = 2.670$. *AM 49:395(1968)*. 6-109: 7.28_9 5.72_{10} 3.96_5 3.24_6 3.09_6 3.03_6 2.72_3 2.67_3. Botryoidal, flat rosettes and tufts composed of individual crystals 0.08 mm long × 0.01 mm wide and very thin, often terminated by pyramidal face giving the appearance of a pointed end. White, bluish white, brownish white. Cleavage perfect parallel to *c*. H = 3. G = 2.66. Uniaxial (+); $N_O = 1.631$, $N_E = 1.652$. Dissolves slowly in hot 1:10 acids. Found at Noumas mine, Blesberg, Namaqualand, Cape Prov., South Africa(!); at the *Sapucaia Velha pegmatite, Galileia, MG, Brazil*, where it coats muscovite, quartz, variscite, and frondelite and occurs wtih roscherite. HCWS *AM 38:263(1953)*.

40.5.4.1 Gainesite $Na_2Zr_2Be(PO_4)_4 \cdot 1-2H_2O$

Named in 1983 for Richard V. Gaines (b.1917), of Earlysville, VA, economic geologist, mineralogist, and collector of pegmatite, especially Be minerals. See mccrillisite (40.5.4.2), and selwynite (40.5.4.3). TET $I4_1/amd$. $a = 6.567$, $c = 17.119$, $Z = 2$, $D = 2.84$. Structure: *AM 68:1022(1983)*. An open-framework structure [zeolite-like] based on BeP_4O_{16} clusters [similar to zunyite cluster (Si_5O_{16})], coordinated with ZrO_6, and two Na sites, one favored by the larger alkalis (Na, K, Rb, Cs). Cation sites only partially occupied, H_2O interstitial. 35-604: 6.07_{10} 4.24_3 3.27_9 3.02_5 2.88_4 2.17_3 2.01_3 1.673_3. Also 35-724. Crystals simple bipyramids up to 1 mm, {111} dominant. Pale lavender-blue, vitreous luster. Conchoidal fracture. H = 4. G = 2.94. Uniaxial (+); $N_O = 1.618$, $N_E = 1.630$. Found in small crevices in cleavelandite (albite variety) associated with roscherite and eosphorite at *Nevel Quarry (Twin Tunnels) on Pumbago Mt., Newry, Oxford Co., ME*. HCWS *CM 32:839(1994)*.

40.5.4.2 Mccrillisite $NaCs(Be,Li)Zr_2(PO_4)_4 \cdot 1-2H_2O$

Named to honor the McCrillis family, who have mined the Oxford Co., ME, pegmatites for over 90 years. The Cs analog of gainesite (40.5.2.1). See also wycheproofite (40.2.9.1) and selwynite (40.5.2.3). TET $I4_1/amd$. $a = 6.159$, $c = 17.28$, $Z = 2$, $D = 3.30$. *CM 32:839(1994)*. PD: 6.16_9 4.33_8 4.10_4 3.28_8 3.06_{10} 2.90_3 1.849_3. Bipyramidal crystals up to 1.2 mm maximum dimension {111}, dominant; {001}, minor. White to colorless, white streak, vitreous luster, translucent to transparent. No cleavage, conchoidal fracture. H = 4–4$\frac{1}{2}$. G = 3.125. Nonfluorescent. Uniaxial (+); N_O = 1.634, N_E = 1.645; nonpleochroic. Close to gainesite (40.5.2.1) in composition with CsO 15%, HfO_2 2.5%. Occurs as an alteration phase in the pegmatites of *Newry quarries, Pumbago Mt., Oxford Co., ME.* HCWS *CM 30:1178(1992).*

40.5.4.3 Selwynite NaK(Be,Al)Zr$_2$(PO$_4$)$_4$ · 2H$_2$O

Named in 1995 for A. R. C. Selwyn, founding director of the Geological Survey of Victoria, Australia. Related to gainesite (40.5.2.1), the Na,Na member, and mccrillisite (40.5.2.2), the Na,Cs member. $I4_1/amd$. a = 6.570, c = 17.142, Z = 2. *CM 33:55(1995).* PD: 6.16$_{10}$ 4.29$_3$ 3.29$_5$ 3.04$_3$ 2.90$_2$ 2.60$_1$ 2.43$_1$ 2.17$_1$. Irregular crusts of radiating anhedral crystals. Deep purple-blue, may weather to lavender; pale lavender streak; vitreous luster; transparent. No distinct cleavage, conchoidal fracture. H = 4. G = 2.94. Uniaxial (+); N_O = 1.624, N_E = 1.636. O, bluish lavender; E, pale bluish lavender. Small amount of Mn the probable cause of the purple-blue color. Occurs as irregular infillings in <8-mm cavities in a quartz–orthoclase–albite–muscovite–schorl pegmatite vein in Devonian granite at *Wycheproof, northwestern VIC, Australia*, closely associated with wardite and eosphorite. Other minerals present: kosnarite, wycheproofite, rockbridgeite, kidwellitelike mineral, saleeite, and montmorillonite. If weathered, the mineral remains and is associated with limonite and clays. HCWS.

40.5.5.1 Wycheproofite NaAlZr(PO$_4$)$_2$(OH)$_2$ · H$_2$O

Named for the locality, an aboriginal name. $P1$ or $P\bar{1}$. a = 10.926, b = 10.986, c = 12.479, α = 71.37°, β = 77.39°, γ = 87.54°, Z = 6, D = 2.81. *MM 58:635(1994).* PD: 8.87$_4$ 4.13$_8$ 3.71$_7$ 3.47$_6$ 3.24$_4$ 2.60$_{10}$. Compact aggregates of finely fibrous crystals 5–10 μm wide up to several millimeters long. Pale pink to brown-orange, colorless streak, vitreous to pearly luster, transparent. No cleavage, rough fracture. H = 4–5. G = 2.83. N = 1.62 (perpendicular to length), –1.64 (parallel to length); parallel extinction; length-slow; nonpleochroic. Found with a variety of other phosphates, such as wardite, eosphorite, cyrilovite, kidwell-like PO$_4$, rockbridgeite, leucophosphite, and saleeite, in a pegmatite vein in granite at *Wycheproof, northwestern Australia.* HCWS *AM 78:653(1978), 80:847(1995).*

40.5.6.1 Ehrleite Ca$_2$ZnBe(PO$_4$)$_2$(PO$_3$OH) · 4H$_2$O

Named in 1985 for Howard Ehrle of Miles City, MT, who found the holotype specimen. TRIC $P\bar{1}$. a = 7.130, b = 7.430, c = 12.479, α = 94.31°, β = 102.07°, γ = 82.65°, Z = 1, D = 2.62. *CM 23:1507(1985).* Structure: *25:767(1987).* 38-402: 12.41$_7$ 6.95$_5$ 6.58$_3$ 4.31$_4$ 3.16$_{10}$ 3.01$_3$ 2.72$_6$ 2.52$_3$. Thick tabular crystals up to 2 mm × 2 mm × 0.1 mm that show {100}, {010}, {001}. White to colorless, vitreous luster. Uneven to subconchoidal fracture, parting on (001). H = 3$\frac{1}{2}$. G = 2.64. Nonfluorescent. Biaxial (+); X parallel to c; Y $\wedge$ b = 18°, Z $\wedge$ a = 9°; N_x = 1.556, N_y = 1.560, N_z = 1.580; 2V = 62°. Soluble in cold 1 : 5 acids.

Found at *Tip Top mine, Custer Co., SD*, on a matrix of beryl and quartz associated wtih mitridatite, roscherite, hydroxyl–herderite, and goyazite–crandallite. HCWS *AM 71:1544(1986)*.

40.5.7.1 Pahasapaite $(Ca,Li,K,Na,\square)_{11}Li_8Be_{24}(PO_4)_{24} \cdot 38H_2O$

Named in 1987 for the Lakota Sioux *pahasapa*, the Black Hills. ISO *I*23. $a = 13.781$, $Z = 1$, $D = 2.241$. *NJMM:433(1987)*. Structure: *AM 74:1191(1989)*. An ordered (1:1) framework of $(PO_4)^{3-}$ and $(BeO_4)^{6-}$ groups forms a structure similar to the synthetic zeolite "rho" and the faujasit group. The other cations and water are found in, or associated with, the distorted openings (cages) created by the linkages of the tetrahedral groups. *41-1384*: 9.6_{10} 5.61_3 4.35_4 3.68_9 3.25_9 2.94_6 2.70_6 2.24_4. Euhedral crystals up to 1 mm showing {110}, {111}. Colorless, light pink, vitreous luster, transparent. No cleavage. $H = 4\frac{1}{2}$. $G = 2.28$. DTA. Isotropic; $N = 1.523$ (Na). Found at *Tip Top pegmatite mine, Custer Co., SD*, with other beryllophosphates in fractured beryl crystals. Associated minerals: montgomeryite, tiptopite, eosphoite–childrenite, and roscherite. HCWS *AM 73:1496(1988)*.

40.5.8.1 Smolianinovite $(Ci,Ca,Mg)(Fe^{3+},Al)(AsO_4)_4 \cdot 11H_2O$

Named in 1956 for Nikolai A. Smoljaninov (1885–1957), mineralogist, Moscow University, Russia. Possibly the Co analog of the Co–Zn series with faheyite (40.5.1.1). ORTH. $a = 6.40$, $b = 11.72$, $c = 21.9$, $Z = 2$, $D = 2.2$. *MM 41:385(1977)* (Mt. Cobalt). *43-663*(diffuse): 23.08_{10} 12.00_9 9.95_9 7.85_{10} 6.85_5 4.48_5 3.25_8 2.89_7. See also *43-676*. Dense aggregates of fine fibers. Yellow; earthy luster, silky on cleavage. $H = 2$. $G = 2.43$–2.49. DTA. $N_{av} = 1.625$; parallel extinction; positive elongation; birefringence ≈ 0.006. Analysis of Bou Azzer shows high Ni content. Found at Bou Azzer, Morocco; at Tsumeb, Namibia; at Mt. Cobalt, Q, Australia. HCWS *AM 42:307(1956), 59:1141(1974)*; *NJMM:167(1988)*.

40.5.8.2 Fahleite $Zn_5CaFe_2^{3+}(AsO_4)_6 \cdot 14H_2O$

Named in 1988 for Rolfe Fahle of Munich, Germany, a mineral dealer specializing in Tsumeb specimens. May be the Zn analog of smolianinovite (40.5.3.1). ORTH. $a = 6.40$, $b = 11.72$, $c = 21.9$, $Z = 2$, $D = 3.16$. *NJMM:167(1988)*. *42-1417*: 22.0_{10} 11.0_{10} 3.2_8 2.9_5 1.65_2. Minute fibrous aggregates or spherules with crystals greater than 1 cm long but several microns thick, may be twisted or curved; feltlike intergrowths. Gray, straw-yellow, bright green; silky, pearly luster; transparent. Cleavage normal to fibers perfect, very soft, sectile. IR confirms presence of H_2O. All diffraction diffuse. Biaxial (+); $N_x = 1.628$, $N_y = 1.631$, $N_z = 1.656$; $2V = 39°$; Z parallel to fiber length. Analysis shows some Mn. Readily soluble in cold acid. Found at *Tsumeb, Namibia*, with gypsum, smolianinovite, lavendulan, cuproadamite, conichalcite, tsumcorite, quartz, and calcite. HCWS *AM 74:501(1989)*.

40.5.9.1 Walpurgite $(BiO_4)_2(UO_2)(AsO_4)_2 \cdot 2H_2O$

Named in 1871 for the locality. TRIC $P\bar{1}$. $a = 7.135$, $b = 10.462$ (corrected in *42-587*), $c = 5.494$, $\alpha = 101.47°$, $\beta = 110.82°$, $\gamma = 88.20°$, $Z = 1$, $D = 6.69$. Structure: *TMPM 30:1129(1982)*. $(UO_2)O_4$ octahedra link with AsO_4 groups to form $UO_2(AsO_4)$ chains parallel to c. Bi^{3+} in two independent sites, eight- and nine-fold coordination polygons, form $Bi_4O_4(AsO_4)_2 \cdot H_2O$ layers parallel to {010}. *42-587*(calc): 10.24_{10} 5.70_3 3.28_6 3.133_6 3.129_4 3.11_5 3.06_6 2.74_4. See also *8-324*. Yellow, light yellow, colorless; adamantine, greasy luster. $H = 3\frac{1}{2}$. $G = 5.95$. Biaxial $(-)$; $N_x = 1.871$, $N_y = 1.975$, $N_z = 2.005$; $2V = 52°$. Some Mn, but no Co, Ni, Mg, or Si detected on analysis. Found at *Walpurgis vein, Weisse Hirsch mine, Neustädtel, Schneeberg, Saxony, Germany*, and also at Heubachtal, Wittichen district. HCWS *DII:796, USGSB 1036:149(1956)*.

40.5.9.2 Orthowalpurgite $(BiO)_4(UO_2)(AsO_4)_2 \cdot 2H_2O$

Named in 1995 for its polymorphic relation to walpurgite (40.5.4.1). ORTH *Pbcm*. $a = 5.492$, $b = 13.324$, $c = 20.685$, $Z = 4$, $D = 6.51$. Structure: *EJM 7:1313(1995)*. Bi in fourfold (edge) and fivefold (corner) coordination with O forms Bi_8O_8 layers which are connected via $[(UO_2)(AsO_4)_2]$ chains to form a three-dimensional network. PD: 10.354_9 6.659_3 5.610_4 3.277_6 3.208_{10} 3.088_8 2.999_5 2.852_5. Crystals up to 0.3 mm tabular on {010} and elongate parallel to [100]. Yellow, transparent. Indistinct cleavage parallel to {001}, conchoidal fracture. $H = 4\frac{1}{2}$. Biaxial $(+)$; $XYZ = cab$; $N_x = 1.91$, $N_y = 2.100$, $N_{z\,calc} = 2.05$; $2V = 70°$. EPMA (wt %): UO_3, 17.86; Bi_2O_3, 64.21; As_2O_5, 16.11; H_2O_{calc}, 2.43. Occurs on the dumps of a Co–Ni–Ag–Bi–U mine at *Schmiedestollen, Wittichen, Black Forest, Germany*, which was famous in the eighteenth century for native Ag occurrences. Over 150 mineral species have been found in this locality. This mineral is associated with preisingerite spherules, quartz, and tabular anatase. HCWS

40.5.10.1 Walentaite $(Ca,Mn^{2+},Fe^{2+})Fe_3^{3+}(AsO_4)(PO_4)_3(PO_3OH) \cdot 7H_2O$

Named in 1984 for Kurt Walenta (b.1927), mineralogist, University of Stuttgart, Germany, who has studied many phosphate minerals. ORTH *Immm, Ibam, Ibca*, or *Imma*. $a = 26.24$, $b = 10.31$, $c = 7.38$, $Z = 1$, $D = 2.74$. *NJMM:169(1984)*. *43-680*: 12.9_{10} 9.6_1 6.56_2 4.82_2 4.43_3 3.00_5 2.93_2 2.78_2. Rosettelike aggregates of thin-bladed crystals 20 µm × 60 µm × 2 µm elongate [001] flattened perpendicular to b so that {010} dominant; coatings. Bright yellow, yellow streak, vitreous luster, translucent. Cleavage {010}, perfect; brittle. $G = 2.72$. DTA. Nonfluorescent. Biaxial $(+)$; $XYZ = bac$; $N_y = 1.738$, $N_z = 1.779$; $Z < Y$. Y, pale yellow-green; Z, medium yellow-green. Microprobe analysis shows both As and P. Found at *White Elephant mine, Pringle, Custer Co., SD*, where primary phosphates are extensively altered. Occurs with frondelite, rockbridgeite, loellingite, spessartine, quartz, tridymite, and muscovite. HCWS *AM 69:1193(1984)*.

40.5.11.1 Canaphite CaNa$_2$P$_2$O$_7$ · 4H$_2$O

Named in 1988 for the composition. MON Pc. $a = 10.529$, $b = 8.48$, $c = 5.67$, $\beta = 106.13°$, $Z = 2$, $D = 2.27$. Structure: *AM 73:168(1988)*. *41-410:* 8.47$_8$ 5.44$_8$ 5.06$_6$ 4.36$_7$ 3.87$_6$ 3.28$_6$ 3.06$_{10}$ 2.61$_9$. Clusters of prismatic crystals up to 0.5 mm long [001], 100 μm wide; tabular on {010}, exhibit {010}, {001}, {100}. Colorless, white streak, vitreous luster. Cleavage {010}, perfect. H = 2. G = 2.24. Biaxial (−); N$_x$ = 1.496, N$_y$ = 1.502, N$_z$ = 1.506; 2V = 52°. Found as coatings on stilbite at *Haledon, Passaic Co., NJ*, and at Great Notch quarry, Little Falls. HCWS *MR 16:467(1985)*.

40.5.12.1 Geminite Cu$_2$As$_2$O$_7$ · 3H$_2$O

Named in 1990 from the Latin *gemini*, twins, in allusion to the usual character of the crystals. TRIC $P1$ or $P\bar{1}$. $a = 6.639$, $b = 8.110$, $c = 15.732$, $\alpha = 92.01°$, $\beta = 93.87°$, $\gamma = 95.02°$, $Z = 4$, $D = 3.71$. *SMPM 70:309(1990)*. *44-1454:* 7.83$_{10}$ 4.02$_3$ 3.93$_{10}$ 3.26$_7$ 3.11$_4$ 3.07$_7$ 2.82$_4$ 2.61$_5$. Euhedral intergrown crystals up to 0.3 mm × 0.06 mm. Tabular {001}, slightly elongate [100] showing {001}, {100}, {010}, {hk0}, {$h\bar{k}$0}, {$\bar{h}k$0}, {$\bar{h}\bar{k}$0}. Twinned polysynthetically parallel to {001}. Light green to sea-green, pale green streak, vitreous luster, transparent. Cleavage {001}, perfect; irregular fracture. H = 3–3$\frac{1}{2}$. G = 3.70. Nonfluorescent. IR, TGA. Biaxial (+); X′ ∧ (001) ≈ 90°, Y′ ∧ a = 17–18.5°, Z ∧ b = 13°; N$_x$ = 1.656, N$_y$ = 1.692, N$_z$ = 1.770; 2V = 75°; r > v, weak. X, Y, colorless; Z, pale green to gray. Soluble in 1:10 HCl. Synthesized. Found at *Cap Garonne mine, Var, France*, on quartz associated with tennantite, covellite, chalcanthite, lavendulan, antlerite, and brochantite. HCWS *AM 77:670(1992)*.

40.5.13.1 Kolovratite Zn–Ni–hydrous vanadate

Named in 1922 for Lev S. Kolovrat-Chervinsky (1884–1921), a Russian physicist. An inadequately described mineral. *15-102:* 11.6$_{10}$ 5.83$_{10}$ 3.88$_{10}$ 2.62$_6$ 2.57$_6$ 2.39$_6$ 1.52$_6$ 1.47$_4$. Incrustations and botryoidal crusts with hollow centers, fibrous felted aggregates with divergent fibers. Yellow, greenish yellow; light yellow streak; vitreous luster; transparent. Conchoidal fracture, brittle. H = 2–3. N$_{av}$ = 1.577; parallel extinction; low birefringence. Cu, Fe, and Mn also found on analysis. Found at *Kara-Chagyr* and *Uch-Kurgan, Ferghana, Turkestan*, distributed in the quartz schists and carbonaceous slates. HCWS *DII:1048, CM 7:322(1962)*.

Class 41

Anhydrous Phosphates, Arsenates, and Vanadates Containing Hydrogen or Halogen

41.1.1.1 Chlorophoenicite $(Mn^{2+},Mg)_3Zn_2(AsO_4)(OH)_7$

Named in 1924 for the Greek for *green* and *purple-red* in allusion to the color changes of the mineral in natural and artificial light. See magnesium-chlorophoenicite (41.1.1.2), the Mg analog. MON $C2/m$. $a = 22.98$, $b = 3.32$, $c = 7.323$, $\beta = 106.0°$, $Z = 2$, $D = 4.7$. *CM 19:333(1981)*. Structure: *AM 53:1110(1968)*. 25-1159: 6.87_5 3.71_7 3.11_5 2.99_4 2.64_{10} 2.43_1 1.822_2 1.758_3. Long prismatic [010], crystals deeply striated [010], with terminal faces dull, etched, {100} relatively smooth; {h0l} faces uneven or warped. Light gray-green in natural light, pink to purple-red in strong artificial light, vitreous, translucent. Cleavage {100} good, brittle. $H = 3\frac{1}{2}$, $G = 3.46$. Biaxial $(-)$; $N_x = 1.682$, $N_y = 1.690$, $N_z = 1.697$, $2V = 83°$, $r > v$ strong. Soluble in acids. Found in massive franklinite–willemite ore at *Franklin, Sussex Co., NJ*, with calcite, leucophoenicite, tephroite, gageite, pyrochroite, and in veinlets with calcite, barite in the ore at Sterling Hill mine. HCWS *DII:779*.

41.1.1.2 Magnesium-chlorophoenicite $(Mg,Mn^{2+})_3Zn_2(AsO_4)(OH)_7$

Named in 1935 for the composition and relationship to chlorophoenicite. MON $C2/m$. $a = 22.99$, $b = 3.236$, $c = 7.299$, $\beta = 106.5°$, $Z = 2$, $D = 3.18$. *JPD 2:225(1987)*. 38-1438: 6.98_2 6.87_2 3.71_5 3.09_4 2.98_3 2.61_{10} 1.756_2 1.480_1. See also *34-190*. Rosettes and radial fibrous aggregates. Colorless to white. Fibers have perfect lengthwise cleavage. $G = 3.37$. Biaxial $(+)$; $N_x = 1.669$, $N_y = 1.672$, $N_z = 1.677$, $r < v$ strong. Spectrographic analysis shows maximum Mn found = Mg/Mn = 1.4:1.2, end member not identified in nature. Found in veinlets with zincite, willemite, haidingerite, calcite at *Franklin, Sussex Co., NJ*. HCWS *DII:780, CM 19:333(1981)*.

41.1.1.3 Jarosewichite $Mn_3^{2+}Mn^{3+}(AsO_4)(OH)_6$

Named in 1982 for Eugene Jarosewich (b. 1926), chief chemist, Department of Mineral Sciences, National Museum of Natural History, Smithsonian Institution, Washington, DC. ORTH $C_2m^2/m^2/n$,, C222, or Cmm2. $a = 6.56$, $b = 25.20$, $c = 10.00$, $Z = 8$, $D = 3.70$. *AM 67:1043(1982)*. Pronounced substructure with $a' = 3.28$, $b' = 12.60$, $c' = 10.00$. *41-580:* 6.29_2 3.91_6 2.67_{10} 2.50_3 1.788_5 1.567_2 1.558_3 1.501_2. Coatings and aggregates of barrel-shaped crystals with rough irregular surfaces up to 0.5 mm long in radial divergent sprays. Flattened on a, tabular on b, dominant face is {010} with minor occurrence of {021}. Dark red, reddish-orange streak, subvitreous luster. No cleavage. H = 4. G = 3.66. Nonfluorescent. Biaxial (−); XYZ= abc; $N_x = 1.780$, $N_y = 1.795$, $N_z = 1.805$; $2V_{calc} = 78°$; weakly pleochroic; Z > X. X, medium brownish red; Z, dark brownish red. Reacts with index liquids. Microprobe analysis Mn^{3+} content inferred from associated minerals, calculated. Found at *Franklin, Sussex Co., NJ*, associated with andradite, franklinite, flinkite, cahnite, and hausmannite. HCWS

41.1.2.1   Theisite $Cu_5Zn_5[(As,Sb)O_4]_2(OH)_{14}$

Named in 1982 for Nicholas J. Theis, geologist, Bendix Corp., who discovered the original sample. ORTH Pseudo hexagonal. $a = 8.225$, $b = 7.123$, $c = 14.97$, $Z = 2$, $D = 4.45$. *MM 46:49(1982). 35-527:* 14.97_9 7.48_5 4.11_4 3.74_{10} 3.00_4 2.53_9 1.830_5 1.553_5. Aggregates to 2mm across. Pale blue-green, white streak, translucent luster. Cleavage (001), perfect; sectile. H = $1\frac{1}{2}$. G = 4.3. Biaxial (−); $N_x = 1.755$, $N_y = 1.785$, $N_z = 1.785$; $2V = 0°$; nonpleochroic. Chemical analysis shows some Sb. Found as the only mineral in thin seams at a uranium prospect near *Durango, La Plata Co., CO*, with uraninite, galena, malachite, azurite, and the secondary Cu minerals partzite, covellite, chalcosite, anglesite, cerussite, zeunerite, and duftite. HCWS

41.1.3.1   Georgiadesite $Pb_{16}(AsO_4)_4Cl_{14}O_2(OH)_2$

Named in 1907 for Georgiadès, director of the mine at Laurium, Greece. MON $P2_1/c$. $a = 13.803$, $b = 7.910$, $c = 10.8112$, $\beta = 102.68°$, $Z = 1$, $D = 6.44$. *MM 47:219(1983). 35-689:* 6.33_3 5.30_3 4.03_2 3.96_5 3.77_3 3.16_5 3.10_{10} 2.97_3. Stubby pseudohexagonal tablets with platy composite structure parallel to [100]. Grooved on {001}, striated on {010}. White or brownish yellow, resinous luster. No cleavage. H = $3\frac{1}{2}$. G = 6.30. Biaxial (+); $N_x = 2.17$, $N_y = 2.17$, $N_z = 2.18$; 2V very large; r < v, strong. Soluble in HNO_3. Found as coatings on altered slag at *Laurium, Greece*, with laurionite, fiedlerite, matlockite, and nealite. HCWS *DII:791*.

41.1.4.1   Sahlinite $Pb_{14}(AsO_4)_2O_9Cl_4$

Named in 1934 for Carl Andreas Sahlin (1861–1943), chemist and manager of the iron works at Laxa, Sweden. Kombatite (41.1.4.2) is V analog. MON $C2/c$ or Cc. $a = 12.710$, $b = 22.498$, $c = 11.360$, $\beta = 118.99°$, $Z = 4$, $D = 8.096$. *NJMM:127(1985)*. A pronounced pseudotetragonal subcell: $a_1 = 3.89$, $a_2 = 4.04$, $c = 22.50$, $\alpha = \beta = 90°$, $\gamma = 90.5°$ in which the fourfold axis is coincident with the monoclinic twofold axis. *42-600:* 3.01_{10} 2.95_{10} 2.81_9 2.25_7 2.02_4 1.754_8 1.593_4 1.581_4. Aggregates of thin scales up to 2 mm. Pale sulfur-yellow, translucent. Cleavage {010}, perfect. H = 2–3. G = 8.00. Biaxial (−); X = b; $N_x \gg 3.3$; Found at *Långban, Filipstad, Värmland, Sweden*, in a dolomite rock with hausmannite, manganhumite, and forsterite. HCWS $DII:775$, AM $71:231(1986)$.

41.1.4.2 Kombatite $Pb_{14}(VO_4)_2O_9Cl_4$

Named in 1986 for the locality. V analog of sahlinite (41.1.4.1). MON $C2/c$ or Cc. $a = 12.552$, $b = 22.495$, $c = 11.512$, $\beta = 118.99°$, $Z = 4$, $D = 7.979$. $NJMM:519(1986)$. 40-497: 3.01_{10} 2.96_{10} 2.81_{10} 2.25_7 2.01_4 1.989_5 1.756_7 1.590_6. Anhedral grains 0.2 mm in diameter, mostly turbid. Bright yellow, light yellow streak, adamantine luster. Cleavage $\{010\}$, perfect. $H = 2$–3. Nonfluorescent. Biaxial $(-)$; $N > 1.90$; nonpleochroic. Found at *Kombat mine, 78 km S of Tsumeb, Otavi, Namibia*. HCWS AM $73:928(1986)$.

41.2.1.1 Allactite $Mn_7(AsO_4)_2(OH)_8$

Named in 1884 for the Greek *to change*, in allusion to the pleochroism. MON $P2_1/a$. $a = 7.257$, $b = 6.017$, $c = 7.734$, $\beta = 111.88°$, $Z = 2$, $D = 3.94$. Structure: AM $53:733(1968)$. Edge-linked bands of double $Mn^{2+}O_6$ octahedra $[Mn_2^{2+}(O,OH)_{10}]$ held together by vertex-linked $Mn^{2+}O_6$ octahedra form flat sheets parallel to $\{100\}$. The sheets are held together by chains of highly distorted $Mn^{2+}(O,OH)_6$ octahedra and AsO_4 tetrahedra parallel to $\{010\}$. 17-748: 4.95_5 3.71_7 3.39_5 3.28_6 3.23_5 3.06_{10} 2.93_5 2.51_5. Elongated [010] slender prisms or tabular $\{100\}$ crystals; rosette-like aggregates. Dark to light purplish red, brownish red, streak greyish to faint brown, vitreous to greasy on fracture surfaces; translucent. Cleavage $\{001\}$, distinct; uneven fracture; brittle. $H = 4\frac{1}{2}$. $G = 3.83$. Biaxial $(-)$; y=b $X \wedge c \approx 51.3°$, $N_x = 1.755$, $N_y = 1.772$, $N_z = 1.774$, $2V \approx 0°$ (Långban). X=blood-red; Y=pale yellow; Z=sea-green; $r > v$, strong at (Na) and wavelengths >573 mμ. The crystals appear uniaxial at 573 mμ and at <573 mμ the axial plane is perpendicular to $\{010\}$. Easily soluble in HCl. Trace amounts (<0.5%) Ca, Fe detected. Found at Franklin and at Sterling Hill, Sussex Co., NJ with calcite, willemite, franklinite, fluorite in small veinlets in the ore; with fluorite, synadelphite, hematolite, hausmannite, pyrochroite at the *Moss mine, Nordmark, SWEDEN* and at Långban with similar minerals and native lead in veinlets and crevices $DII:785$. HCWS

41.2.2.1 Heyite $Pb_5Fe_2^{2+}(VO_4)_2O_4$

Named in 1973 for Max Hutchinson Hey (1904–1984), chemist and mineralogist, British Museum (Natural History), London, England. MON $P2_1/m$. $a = 8.910$, $b = 6.017$, $c = 7.734$, $\beta = 111.88°$, $Z = 1$, $D = 6.284$. MM $39:65(1973)$. 25-1404: 8.28_3 4.87_5 3.67_3 3.25_{10} 3.01_4 2.97_7 2.77_6 2.31_4. Crystals to 0.4 mm elongated parallel to [010] or tabular parallel to [100] showing $\{001\}$, $\{100\}$, $\{110\}$, $\{\bar{1}01\}$. Twinned on $\{110\}$. Yellow-orange, yellow streak, translucent. $H = 4$. $G = 6.3$. Biaxial $(+)$; $X \wedge [001] = 36°$; $Y = b$; $N_x = 2.185$, $N_y = 2.219$, $N_z = 2.266$; $2V = 89°$; nonpleochroic. Found on or replacing wulfenite at *Betty Jo claim, Ely, White Pine Co., NV*, with galena, chalcopyrite, pyrite, pyromorphite, crysocolla, and cerussite. HCWS AM $59:382(1974)$.

41.3.1.1 Clinoclase $Cu_3(AsO_4)(OH)_3$

Named in 1868 from the Greek *to incline* and *to break*, in allusion to the oblique basal cleavage. MON $P2_1/c$. $a = 7.257$, $b = 6.457$, $c = 12.378$, $\beta = 99.51°$, $Z = 4$, $D = 4.42$. *MJ 1:89(1954)*. Structure: *AC(C) 46:2291(1990)*. *37-447:* 7.17_1 3.58_{10} 3.14_3 3.06_1 2.99_1 2.91_1 2.39_1 2.31_1. See also *42-1357*. Crystals elongated on [010], or on [001] with {100} prominent; tabular [001], rhombohedral in aspect. Isolated crystals or aggregated as rosettes; crusts or coatings with fibrous structure. Dark greenish black to greenish blue, bluish-green streak, vitreous to pearly luster on cleavage, transparent to translucent. Cleavage {001}, perfect; uneven fracture; brittle. $H = 2\frac{1}{2}$–3. $G = 4.38$. Biaxial (−); $Y = b$; Z near a; $N_x = 1.756$, $N_y = 1.874$, $N_z = 1.896$; $2V = 50°$; $r \ll v$. X, pale blue-green; Y, light blue-green; Z, benzol-green. Small substitution of P for As, P/As = 1:11. Synthesized. Found at the Mammoth mine, Tintic district, Juab Co., UT; at Majuba Hill, Pershing Co., NV; associated with olivenite and other secondary Cu minerals in *Wheal Gorland*, Wheal Unity, Ting Tang mines, and St. Day, *Cornwall, England*, and at Bedford United mines, Tavistock, Devonshire; at Markirch, in the Vosges, France; at Dome Rock, SA, Australia; at Tsumeb mine, Namibia. HCWS *DII:787*.

41.3.2.1 Cornetite $Cu_3(PO_4)(OH)_3$

Named in 1917 for Jules Cornet (1865–1929), Belgian geologist. ORTH *Pbca*. $a = 10.865$, $b = 14.071$, $c = 7.098$, $Z = 8$, $D = 4.12$. *37-449:* 5.07_3 4.54_3 4.30_6 3.68_7 3.18_7 3.04_{10} 3.03_4 2.95_6. See also *37-483, 9-495*. Short prismatic [001] to equant crystals, {210} usually rounded; crusts of minute crystals. Twinned on {h0l}. Deep blue or peacock-blue to greenish blue, vitreous luster, transparent to translucent. No cleavage. $H \approx 4\frac{1}{2}$. $G = 4.10$. Biaxial (−); XYZ= *bac*; $N_x = 1.765$, $N_y = 1.81$, $N_z = 1.82$; $2V \approx 33°$; nonpleochroic; $r < v$, strong. Soluble in cold HCl. Minor Fe, Mg, Al, Si, and As. Found at the Blue Jay and Empire–Nevada mines, Yerington, NV; at *L'Étoile du Congo mine, Elizabethville, Shaba Prov., Zaire*, associated with pseudomalachite in fine-grained argillaceous sandstone; at Bwana M'kubwa, northern Zambia. HCWS *DII:789*, *AM 35:365(1950)*.

41.3.3.1 Flinkite $Mn_2^{2+}Mn^{3+}(AsO_4)(OH)_4$

Named in 1889 for Gustav Flink (1849–1931), Swedish mineralogist and collector. ORTH *Pnma*. $a = 9.55$, $b = 13.11$, $c = 5.25$, $Z = 4$, $D = 3.76$. *CM 7:547(1963)*. Structure: *AM 52:1603(1967)*. *12-400:* 4.73_{10} 4.39_{10} 3.82_3 3.18_8 2.71_1 2.66_{10} 2.51_3 1.538_2. Thin tabular {001}, slightly elongated {010}, crystals often rounded in [100], {001} striated parallel to [100]; featherlike aggregates. Greenish brown to dark green, vitreous to greasy luster, transparent. Brittle. $H = 4\frac{1}{2}$. $G = 3.87$. Biaxial (+); XYZ= *bca*; $N_x = 1.783$, $N_y = 1.801$, $N_z = 1.834$; 2V large; $r > v$, weak. X, pale brownish green; Y, yellow-green; Z, orange-brown. Soluble in HCl. Occurs at Franklin, Sussex Co., NJ; in veinlets in magnetite ore associated wtih sarkinite, brandtite, caryopilite, nadorite, native lead, manganoan calcite, and barite at the *Harstig mine at Pajsberg, near Persberg, Värmland, Sweden.* HCWS *DII:793*.

41.3.4.1 Retzian-(Ce) $Mn_2^{2+}Ce(AsO_4)(OH)_4$

Named in 1894 for Anders Jahan Retzius (1742–1821), Swedish naturalist. Redefined in 1982 to accommodate the compositional variations for other rare earth elements. See retzian-(La) 41.3.4.3 and retzian-(Nd). ORTH *Pban*. $a = 5.670$, $b = 12.03$, $c = 4.863$, $Z = 2$, $D = 4.57$. *AM 67:841(1982)*. Structure: *AM 52:1603(1967)*. *20-731:* 6.04_4 4.84_4 3.53_8 2.72_{10} 2.34_3 1.895_3 1.848_5 1.619_4. Prismatic [001], tabular {010}. Dark chocolate-brown to chestnut-brown, light brown streak, vitreous to greasy luster, subtranslucent. Conchoidal to uneven fracture. $H = 4$. $G = 4.15$. Biaxial (+); XYZ= cba; $N_x = 1.777$, $N_y = 1.788$, $N_z = 1.800$; 2V large; $Z > Y > X$; $r < v$, weak. X, colorless to orange-yellow; Y, yellowish brown to dark brown; Z, red-brown to crimson. Soluble in acids. REE analysis confirms Ce dominant over La and Nd, also present: Y, Pr, Sm, Gd, Eu. Found at Sterling Hill mine, Sussex Co., NJ; at *Moss mine, Nordmark, Värmland, Sweden.* HCWS *DII:794*.

41.3.4.2 Retzian-(Nd) $Mn_2^{2+}Nd(AsO_4)(OH)_4$

Named in 1982 for its relationship to retzian-(Ce) 41.3.4.1. ORTH *Pban*. $a = 5.690$, $b = 12.12$, $c = 4.874$, $Z = 2$, $D = 4.49$. *AM 67:841(1982)*. *35-686:* 6.05_2 4.89_3 3.53_6 3.30_1 2.73_{10} 1.857_3 1.625_2 1.463_2. Crystals and aggregates of crystals, twins, trillings, sixlings with a distinct spokelike appearance. Reddish brown to pinkish brown, color zoned with interior darker; light brown streak; vitreous luster on fracture surfaces to dull on crystal faces. No cleavage, parting along parallel growth planes, uneven fracture. $H = 3–4$. $G > 4.2$. DTA, TGA. Biaxial (+); XYZ= cba; $N_x = 1.774$, $N_y = 1.782$, $N_z = 1.798$; $2V = 69°$; $Z > Y \gg X$; $r < v$, weak. X, yellow; Y, reddish brown; Z, brown. Microprobe analyses show Nd > Ce, La, Pr, low Mg, and virtually no substitutions for As. Found at *Sterling Hill mine(!), Ogdensburg, Sussex Co., NJ*, as clove-brown crystals, coating, and embedded in light brown rhodochrosite in the willemite–franklinite ore. HCWS

41.3.4.3 Retzian-(La) $Mn_2^{2+}La(AsO_4)(OH)4$

Named in 1982 to distinguish the La-dominant retzian. ORTH *Pban*. $a = 5.670$, $b = 12.011$, $c = 4.869$, $Z = 2$, $D = 4.49$. *MM 48:533(1988)*. *38-380:* 5.98_3 4.84_3 3.51_8 2.72_{10} 2.06_2 1.848_5 1.615_3 1.456_4. Crystals to 0.5 mm showing {001}, {010}, {110}, {150}, pseudohexagonal. Reddish brown, vitreous luster. No cleavage. $H = 3–4$. $G = 4.49$. Biaxial (+); XYZ= cba; $N_x = 1.766$, $N_y = 1.773$, $N_z = 1.788$; $2V = 82°$; weakly pleochroic; $Z > Y > X$; $r > v$, strong. Microprobe analyses show La > Ce > Nd with Pr, Sm, Y plus some Mg, Zn. Found at *Sterling Hill mine, Ogdensburg, Sussex Co., NJ*, with willemite, calcite, and franklinite. HCWS *AM 67:841(1982)*.

41.3.5.1 Viitaniemiite Na(Ca,Mn^{2+})Al(PO$_4$)(F,OH)$_3$

Named in 1981 for the locality. MON $P2_1$ or $P2_1/m$. $a = 6.832$, $b = 7.143$, $c = 5.447$, $\beta = 109.36°$, $Z = 2$, $D = 3.242$. *AM 66:1102(1981)*. Structure: *AM 69:961(1984)*. *33-1226*: 4.89$_4$ 3.22$_5$ 2.97$_3$ 2.94$_6$ 2.88$_{10}$ 2.57$_4$ 2.16$_4$ 1.915$_3$. Fan-shaped tabular crystals from 0.02 to 0.2 mm wide; inclusions or as rims on other minerals. Gray to white, vitreous luster. Cleavage {10$\bar{1}$}, distinct. $H = 5$. $G = 3.245$. Nonfluorescent. Biaxial (−); $Y = b$; $N_x = 1.557$, $N_y = 1.565$, $N_z = 1.571$; optic axial plane = (010); $2V = 81°$. Readily dissolved by HNO$_3$ and H$_2$SO$_4$. Mn-free crystals [Francon: *CM 21:499(1983), 35-598*] elongated [010], flat (100), twinned (100). Found as an inclusion in esosphorite at *Viitaniemi granite pegmatite, Eräjärvi, Orivesi, Finland*; at Francon quarry, Montreal, QUE, Canada, in cavities of a silicocarbonatite sill. HCWS *ZVMO 120:74(1991)*.

41.3.6.1 Nacaphite Na$_2$Ca(PO$_4$)F

Named in 1980 for the composition. TRIC $P1$. $a = 10.565$, $b = 24.443$, $c = 7.102$, $\alpha = 89.99°$, $\beta = 90.0°$, $\gamma = 90.01°$, $Z = 2$, $D = 2.88$. Structure: *DANP 34:9(1989)*. Earlier investigations suggested ORTH space group. *33-1222*: 4.00$_2$ 3.51$_2$ 3.054$_4$ 3.049$_4$ 2.65$_{10}$ 2.01$_4$ 1.762$_3$ 1.533$_2$. Inclusions about 1 mm in diameter in thermonatrite. Colorless, vitreous luster, transparent. Conchoidal fracture. $H = 3$. $G = 2.85$. IR, DTA. Biaxial (−); $N_x = 1.508$, $N_y = 1.515$, $N_z = 1.520$; $2V = 80°$; $r > v$, weak. Analyses show minor K, Mn, and Sr. Occurs associated with apatite, aegirine, barytolampophyllite in ijolite pegmatite at *Mt. Rasvumchorr, Khibiny massif, Kola Penin., Russia*. HCWS *ZVMO 109:50(1980), AM 66:218(1981)*.

41.3.7.1 Gerdtremmelite ZnAl$_2$(AsO$_4$)(OH)$_5$

Named in 1985 for Gerd Tremmel, who first recognized the mineral. TRIC $P1$ or $P\bar{1}$. $a = 5.169$, $b = 13.038$, $c = 4.931$, $\alpha = 98.78°$, $\beta = 100.80°$, $\gamma = 78.73°$, $Z = 2$, $D = 3.66$. *NJMM:1(1985)*. *38-373*: 12.77$_{10}$ 4.80$_4$ 4.70$_4$ 4.22$_4$ 4.08$_4$ 3.88$_5$ 3.63$_5$ 3.14$_5$. Spherulitic aggregates of tabular crystals up to 3 μm. Yellowish brown to dark brown, white streak, adamantine luster, transparent. No visible cleavage or fracture. TGA. Nonfluorescent. $N_{av} = 1.73$–1.74, highly birefringent. Insoluble in HCl. Found in the deeply oxidized ore zone at *Tsumeb, Namibia*. HCWS *AM 71:845(1986)*.

41.3.8.1 Paulkellerite Bi$_2$Fe^{3+}(PO$_4$)O$_2$(OH)$_2$

Named in 1988 for Paul Keller (b.1940), University of Stuttgart, Germany, for his contributions to the mineralogy of secondary minerals. MON $C2/c$. $a = 11.382$, $b = 6.690$, $c = 9.666$, $\beta = 114.73°$, $Z = 4$, $D = 6.17$. *AM 73:870(1988)*. *41-1468*: 5.13$_2$ 3.91$_2$ 3.34$_2$ 3.12$_{10}$ 2.80$_3$ 2.24$_2$ 2.16$_4$ 2.04$_2$. Structure: Single zig-zag chains of alternating PO$_4$ tetrahedra and Fe^{3+} (O,OH)$_6$ octahedra extend along c and form layers parallel to (100) with Bi bonding to 3 tetrahedral oxygens, 3 octahedral oxygens and one (OH) resulting in Bi (O,OH)$_7$ polyhedra *AM 73:873(1988)*. Compare zairite, BiFe^{3+}(PO$_4$)O$_2$(OH)$_2$

(41.5.12.3), a member of the crandellite group, exhibiting a different structure. *41-1468* 5.13_2 3.91_2 3.34_2 3.12_{10} 2.80_3 2.24_2 2.16_4 2.04_2. Wedge-shaped, euhedral, crystals with slightly curved faces up to 0.8 mm displaying {110}, {011}, and poorly developed {$\bar{1}$01}. Greenish yellow, light yellow streak, vitreous to adamantine luster, transparent. No cleavage. H = 4. Nonfluorescent. Biaxial (+); X ∧ a = 25° in obtuse β; YZ= bc; $N_x = 1.762$, $N_y = 1.767$, $N_z = 1.825$; 2V = 37°. Found at *Neuhilfe mine, Schneeberg, in the Erzgebirge area, Germany*, with native bismuth, quartz, mixed skutterudite group minerals, erythrite, and bismutoferrite. HCWS

41.4.1.1 Arsenoclasite $Mn_5(AsO_4)_2(OH)_4$

Named in 1931 from the Greek for *arsenic* and *cleavage*. ORTH $P2_12_12_1$. a = 18.29, b = 5.75, c = 9.31, Z = 4, D = 4.21. *AMG 4:425(1965)*, Structure: *AM 53:1539(1971)*. *20-704:* 5.57_4 4.92_4 4.55_7 3.06_6 2.93_{10} 2.84_7 2.74_8 1.631_5. Massive, granular. Red, translucent. Cleavage {010}, perfect. H = 5–6. G = 4.16. Biaxial (−); $N_x = 1.787$, $N_y = 1.810$, $N_z = 1.816$; 2V = 54°. Found at *Långban, Filipstad, Värmland, Sweden*, with calcite, sarkinite, and adelite along fissures in dolomite and associated with hausmannite. HCWS *DII:801*.

41.4.1.2 Gatehouseite $Mn_5(PO_4)_2(OH)_4$

Named in 1993 for Brian M. K. C. Gatehouse (b. 1932), crystal chemist, Monash University, Melbourne, Australia. The P analog of arsenoclasite (41.4.1.1.). ORTH $P2_12_12_1$. a = 9.097, b = 55.693, c = 18.002, Z = 4, D = 3.854. *MM 57:309(1993)*. *PD:* 4.48_1 4.03_1 2.90_{10} 2.85_7 2.80_5 2.70_8 2.02_2 1.608_2. Radiating to divergent bladed crystals up to 100 μm long, elongated on [010] showing {102}, {110}, {001}, occasionally twinned on {001}. Pale yellow to pale brownish orange, pale yellow streak, adamantine luster, transparent. Cleavage {010}, distinct; splintery fracture. H = 4. Biaxial; $N_{x'} = 1.74$, $N_{z'} = 1.76$; length-slow; parallel extinction; pleochroic; brown to colorless. Found as overgrowths on arsenoclasite in cavities with hematite, hausmannite, barite, and carbonates at *Iron Monarch iron ore deposit, northern Middleback Ranges, SA, Australia*. HCWS *AM 79:185(1994)*.

41.4.2.1 Cornubite $Cu_5(AsO_4)_2(OH)_4$

Named in 1959 from the Roman name for Cornwall, England. See dimorph cornwallite (41.4.3.1) and pseudomalachite, 41.4.4.1, the P analog which has two polymorphs. TRIC $P\bar{1}$. a = 6.121, b = 6.251, c = 6.790, α = 92.93°, β = 111.30°, γ = 107.47°, Z = 1, D = 4.85. *FM 62:231(1984)*. Structure: *BGSF 57:119(1985)*. Sheets of edge shared $(Cu(O,OH)_6$ octahedra parallel to $(0\bar{1}1)$ connect by AsO_4 groups via common corners (3 oxygens in the octahedral sheet, the fourth in the adjacent sheet) and H bonds. *38-441:* 4.72_{10} 3.49_8 2.87_7 2.69_9 2.56_{10} 2.49_{10} 2.30_7 1.575_7. See also *12-288, 41-1460*(calc). Crystals to 0.3 mm × 0.3 mm × 0.05 mm thick (Reichenbach, Germany), fibrous botryoidal aggregates or porcellaneous crusts. Apple-green to dark green; vitreous, glassy luster; transparent. G = 4.64. Biaxial (−); $N_x = 1.87$,

$N_z = 1.90$ (Cap Garrone). *BM 84:318(1961)*. Found at Centennial mine, Eureka, Tintic district, UT; at Potts Gill, Caldbeck, Cumberland, *Wheal Carpenter, Gwinear, St. Day, Cornwall, England*, in drusy quartz associated with olivenite, malachite, cuprite, cornwallite, and liroconite; at Reichenbach, Odenwald, Hessen, Germany. HCWS *MM 32:1(1959)*.

41.4.2.2 Cornwallite $Cu_5^{2+}(AsO_4)_2(OH)_2$

Named in 1847 for the locality. Dimorph of cornubite (41.4.2.1); pseudomalachite (41.4.3.1) is the P analog. MON $P2_1/a$. $a = 17.33$, $b = 5.82$, $c = 4.60$, $\beta = 92.2°$, $Z = 2$, $D = 4.645$. *MM 32:1(1959)*. *39-1357*: 4.81_4 3.53_5 3.20_5 3.02_4 2.89_5 2.74_5 2.46_5 2.41_{10}. See also *12-287*. Botryoidal crusts with radial-fibrous structure, drusy. Pale green, verdigris-green, emerald-green. Conchoidal fracture, not brittle. $H = 4\frac{1}{2}$. $G = 4.52$. Biaxial (+); $N_x = 1.796$, $N_y = 1.815$, $N_z = 1.850$; 2V small; negative elongation. Decomposed in oils containing As_2S_3, S. Some substitution of P for As, As/P = 7:1. Found at Gold Hill, Tooele Co., UT; with temorite, olivenite in *Wheal Gorland, St. Day, Cornwall, England*; at Vosges, France; at Grube Clara, Rankachtal, Germany. HCWS *DII:925*, *AM 36:484(1951)*.

41.4.3.1 Pseudomalachite $Cu_5(PO_4)_2(OH)_4$

Named in 1813 from the Greek for *false* and *malachite* because of the close resemblance to that mineral. Trimorph with reichenbachite (41.4.3.2), and ludjibaite (41.4.3.3). MON $P2_1/c$. $a = 4.473$, $b = 5.747$, $c = 17.032$, $\beta = 91.04°$, $Z = 2$, $D = 4.367$. *AM 66:176(1981)*. *36-408*: 4.48_{10} 3.47_2 3.12_2 3.06_3 2.98_3 2.44_3 2.42_2 2.39_3. Single crystals rare, prismatic [001] usually uneven, small, or in subparallel aggregates with drusy surface or in hemispherical forms; reniform, botryoidal, massive with radial-fibrous structure and concentric banding, fibers elongated [010]; foliated, microcrystalline or dense; colloform. Twinned on {100}. Dark emerald-green to blackish green, fibrous material lighter in color, bluish green; streak paler than color; vitreous luster; translucent to subtranslucent. Cleavage {010}, distinct, but difficult; splintery fracture. $H = 4\frac{1}{2}$–5. $G = 4.35$; fibers, $G = 4.0$. Biaxial (−); $X \wedge c = 21°$; $Z = b$; $N_x = 1.789$, $N_y = 1.835$, $N_z = 1.845$ (Rheinbreitbach); $2V \approx 50°$; r < v, strong [also biaxial (+), r > v]; weakly pleochroic. X, bluish green; Y, yellowish green; Z, deep bluish green. Soluble in acids. A secondary mineral found with malachite, chrysocolla, tenorite, pyromorphite, chalcedony, and iron oxyhydroxides in the oxidized zones of Cu ore deposits. Occurs at Wheatley mines, Chester Co., and at Ecton, Perkiomen mine, Montgomery Co., PA; at Cabarrus Co., NC; in AZ at Ajo, New Cornelia mine, Lavender pit, Bisbee, Cochise Co., Harquahala mine, La Paz Co., Table Mt. mine, Mammoth Co., Silver Bell mine; at Liskeard and elsewhere in Cornwall, England; at La Coste mine Alban-la-Fraysse, Tarn Prov., France; at Ehl, near Linz, Virneberg, near Rheinbreitbach and Hirschberg, Thuringia, Germany; at Libethen and Rezbanya, Czech Republic; at Moldawa, Romania; at Mindouli, Congo Republic; in Shaba, Zaire; at Rokana mine, Zambia; at Burra, SA,

Rum Jungle, NT, and West Bogan mine, Tottenham, NSW, Australia. HCWS
DII:799, AM 35:365(1950).

41.4.3.2 Reichenbachite $Cu_5(PO_4)_2(OH)_4$

Named in 1984 for the locality. Trimorph with pseudomalachite (41.4.3.1) and ludjibaite (41.4.3.3). MON $P2_1/a$. $a = 9.198$, $b = 10.691$, $c = 3.376$, $\beta = 92.42°$, $Z = 2$, $D = 4.35$. AM 72:404(1987). Structure: 62:115(1977). Synthetic. 40-502: 4.47_{10} 3.13_3 3.01_3 2.95_2 2.81_3 2.41_{10} 2.36_2 2.24_2. Crystals lance-shaped to 0.3 mm × 0.3 mm × 0.6 mm. Dark green, light green streak, vitreous luster, translucent. No cleavage, irregular fracture. $H = 3\frac{1}{2}$. Nonfluorescent. Biaxial (−); $Y = b$, $Z \wedge a = 29°$; $N_x = 1.782$, $N_y = 1.833$, $N_z = 1.867$; $2V = 76°$; weakly pleochroic; $r < v$, weak. X, bright emerald-green; Z, emerald-green. Synthesized. AM 62:115(1977). Found in silicified barite vein at *Reichenbach, Odenwald, Germany*, associated with quartz, bayldonite, mimetite, malachite, duftite, goethite, and pseudomalachite; at Brown's prospect, Rum Jungle, NT, Australia, with pseudomalachite. HCWS NJMM:281(1991), AM 70:881(1985).

41.4.3.3 Ludjibaite $Cu_5(PO_4)_2(OH)_4$

Named in 1988 for the locality. TRIC $P\bar{1}$. $a = 4.446$, $b = 5.871$, $c = 8.680$, $\alpha = 103.9°$, $\beta = 90.3°$, $\gamma = 93.2°$, $Z = 1$, $D = 4.36$. BM 111:167(1988). Structure: AM 66:169(1981). PD: 4.46_{10} 3.02_2 2.46_5 2.41_2 2.35_2 2.22_1 2.02_2 1.572_2. Aggregates of blue-green blades up to 0.3 mm, white to pale blue streak, vitreous luster. Biaxial (+) [or (−)]; $N_{x'} = 1.786$, $N_{z'} = 1.840$; 2V large; weakly pleochroic, blue to pale blue. Found on pseudomalachite in association with libethenite in the *Ludjiba deposit, Zaire*; at Lubietová, Slovakia. HCWS AM 73:1495(1988), NJMM:281(1991).

41.4.4.1 Turanite $Cu_5(VO_4)_2(OH)_4$

Named in 1909 for the locality. An inadequately characterized mineral. 12-522: 7.25_8 4.76_{10} 2.70_4 2.56_6 2.47_3 2.29_3 2.11_3 1.97_3. Reniform crusts, spherical concretions of radial fibrous structure. Olive-green, translucent. $H = 5$. Biaxial (−); $N_x = 2.00$, $N_y = 2.01$, $N_z = 2.02$; 2V medium; positive elongation; $r > v$, strong; pleochroic: X,Y, brown; Z, green. Found with a variety of vanadium and uranium minerals in cavities in limestone at *Tyuya Muyun, Ferghana, Turan Region, Uzbekistan*. HCWS DII:818, BM 79:219(1956).

41.4.5.1 Stoiberite $Cu_5(VO_4)_2O_2$

Named in 1979 for Richard Edwin Stoiber (b.1911), geologist, Dartmouth College, NH, who studied Central American volcanoes and their fumarolic deposits. MON $P2_1/n$. $a = 15.654$, $b = 6.054$, $c = 8.385$, $\beta = 102.29°$, $Z = 4$, $D = 4.96$. AM 64:941(1979). Structure: AC(B) 29:1338(1973). 33-504: 7.92_{10} 4.80_7 2.76_8 2.69_7 2.60_7 2.54_6 2.10_8 1.956_7. Polycrystalline aggregates <100 mm in diameter, platelike {100} with striae on (100) possibly parallel to {hll} cleavage trace. Synthetic material may show needlelike habit. Black, reddish-brown streak, metallic luster, transmits light in thin sections. $G_{syn} = 5.0$. Reflected light: medium, < galena, very light gray, slightly bluish-white pleochroic. Some Cr^{6+} substitutes for V. Found at *the Y fumarole, Izalco volcano, El Salvador*, in the oxide zone coating basaltic breccia fragments. HCWS

41.4.6.1 Nealite $Pb_4Fe^{2+}(AsO_4)_2Cl_4$

Named in 1980 for Leo Neal Yedlin (1908–1977), mineral (micromount) collector of New Haven, CT. TRIC $P\bar{1}$. $a = 6.537$, $b = 10.239$, $c = 5.582$, $\alpha = 96.20°$, $\beta = 89.39°$, $\gamma = 97.74°$, $Z = 1$, $D = 5.88$. *MR 11:299(1980)*. 33-755: 10.09_6 6.48_8 4.24_7 3.92_5 3.54_{10} 3.25_6 2.97_5 2.78_6. Prismatic, bladed clusters tabular on {010}, elongated parallel to [001]. Orange, light orange streak, translucent. No cleavage, brittle. Nonfluorescent. Biaxial; $N > 2.00$; length-slow; nonpleochroic. Some Zn detected on microprobe analysis. Found at Cap Garonne, France and on the slags at *Laurion, Attica, Greece*. HCWS

41.4.7.1 Tancoite $LiNa_2H[Al(PO_4)_2(OH)]$

Named in 1980 for the locality. ORTH *Pbcb*. $a = 6.948$, $b = 14.089$, $c = 14.065$, $Z = 8$, $D = 2.724$. *CM 18:185(1980)*. Structure: *TMPM 31:121(1983)*. 33-602: 4.67_{10} 3.41_9 3.15_{10} 2.48_9 2.03_3 1.760_4 1.438_4 1.429_4. Isolated crystals up to 1 mm long and druses of columnar individuals with (100), (010), (111), dominant; (021), (001), minor. Colorless to pale pink, vitreous luster. Cleavage {010}, {001}, fair; conchoidal fracture. $H = 4$–$4\frac{1}{2}$. $G = 2.752$. IR, DTA. Biaxial (−); XYZ= *abc*; $N_x = 1.541$, $N_y = 1.563$, $N_z = 1.564$; $2V = -23°$, $r < v$, weak. Dissolves in 1:10 HCl, HNO_3. Occurs in the *Tanco spodumene-bearing pegmatite mine at Bernic Lake, MB, Canada*. HCWS *AM 66:1278(1981)*, *69:215(1984)*.

41.4.8.1 Reppiaite $Mn_5(VO_4)_2(OH)_4$

Named in 1992 for a village near the locality. MON $C2/m$. $a = 9.604$, $b = 9.558$, $c = 5.393$, $\beta = 98.45°$, $Z = 2$, $D = 3.91$. Structure: *ZK 201:223(1992)*. PD: 4.76_8 3.37_4 3.00_3 2.68_{10} 2.66_5 2.16_5 1.565_4 1.510_3. Minute tabular crystals flattened on (100) of irregular outline up to 50 μm thick, 300 μm diameter, striated on (100). Orange-red, orange-yellow streak, vitreous luster, transparent. No observable cleavage. $H < 3$, $VHN_{15} = 90$–110 on (100). $G = 3.92$. Nonfluorescent. Biaxial (−); $N_{x'} = 1.893$, $N_{z'} = 1.810$; 2V large; pleochroic on (100) weak yellow-orange to deep orange; medium dispersion. Some As substitutes for V. Found at *Gambatesa mine, near Reppia, Val Graveglia, Italy*, in manganese chert ores with hausmannite, tephroite, Mn-containing carbonates, and silicates. HCWS

ADELITE–DESCLOIZITE GROUP

Members of the adelite–descloizite group fall into two closely related subgroups. The general formula is

$$(AB)(OH)(TO_4)$$

where

A = Ca, Pb
B = Co, Cu, Fe^{2+}, Mg, Mn, Zn, Ni
T = As, V (and minor amounts of P in solid solution)

Both subgroups crystallize with orthorhombic symmetry. The Adelite subgroup in space group $P2_12_12_1$ and the other in space group $Pnam$.

The structure [$NJMM:300(1989)$] is composed of chains created from edge-sharing octahedra, BO_6, and very distorted $A(O,OH)_8$ polyhedra linked through TO_4 into a tight three-dimensional network. The minor structural differences between members of the two subgroups are related to a shift in the position of an oxygen atom of a TO_4 group linked to the framework by a hydrogen bond. See the brackebuschite (40.2.8.1) group, which has a related structure.

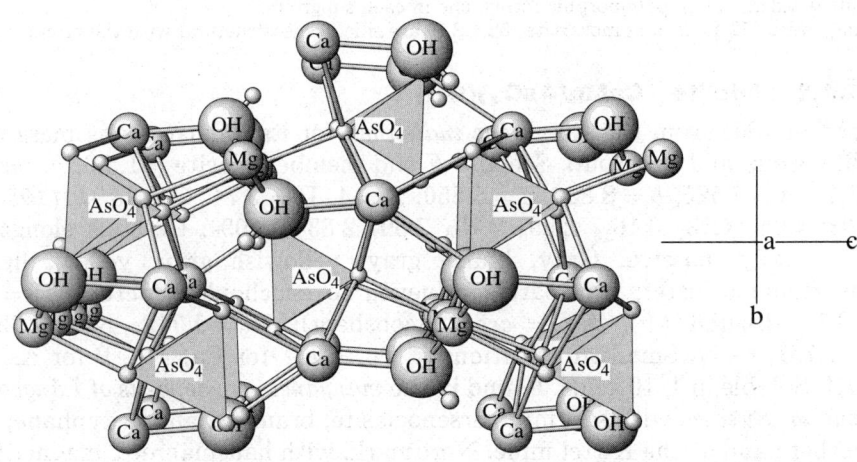

ADELITE: $CaMg(OH)(AsO_4)$

An a axis projection of adelite shows the distorted $Ca(O,OH)_8$ polyhedra, MgO_6 octahedra and A_5O_4 groups in a tight network. Distractions in the network due to OH bonding is the difference between the adelite and descloisite groups.

Adelite Subgroup

Mineral	Formula	Space Group	a	b	c	D	
Adelite	CaMg(OH)(AsO$_4$)	$P2_12_12_1$	7.525	8.895	5.850	3.74	41.5.1.1
Conichalcite	CaCu(OH)(AsO$_4$)	$P2_12_12_1$	7.393	9.220	5.830	4.338	41.5.1.2
Austinite	CaZn(OH)(AsO$_4$)	$P2_12_12_1$	7.505	9.037	5.921	4.323	41.5.1.3
Duftite-β	PbCu(OH)(AsO$_4$)	$P2_12_12_1$	7.49	9.36	5.91	6.12	41.5.1.4
Gabrielsonite	PbFe(OH)(AsO$_4$)	$P2_1m$	7.86	8.62	5.98	6.69	41.5.1.5
Tangeite	CaCu(OH)(VO$_4$)	$P2_12_12_1$	7.44	9.24	5.90	3.82	41.5.1.6
Nickelaustinite	CaNi(OH)(AsO$_4$)	$P2_12_12_1$	7.455	8.955	5.916	4.27	41.5.1.7
Cobaltaustinite	CaCo(OH)(AsO$_4$)	$P2_12_12_1$	7.498	9.006	5.920	4.24	41.5.1.8
Arsendescloizite	PbZn(OH)(AsO$_4$)	$P2_12_12_1$	7.634	9.358	6.075	6.57	41.5.1.9

Descloizite Subgroup

Mineral	Formula	Space Group	a	b	c	D	
Descloizite	Pb(Zn,Cu)(OH)(VO$_4$)	$Pnam$	7.593	9.416	6.057	6.202	41.5.2.1
Mottramite	PbCu(OH)(AsO$_4$)	$Pnam$	7.711	9.249	6.035	6.19	41.5.2.2
Pyrobelonite	PbMn(OH)(VO$_4$)	$Pnam$	7.644	9.508	6.182	5.82	41.5.2.3
Cechite	Pb(Fe^{2+},Mn)(OH)(VO$_4$)	$Pnam$	7.607	9.441	6.096	5.99	41.5.2.4
Duftite-α	PbCu(OH)(VO$_4$)	$Pnam$	7.788	9.223	6.001	6.57	41.5.2.5

Notes:
1. Duftite exists in two polymorphic forms, one in each subgroup.
2. Vuagnatite (52.4.2.2) and mozartite (52.4.2.1) are silicates isostructural with the group.

41.5.1.1 Adelite CaMg(AsO$_4$)(OH)

Named in 1804 from the Greek for *indistinct*, for its occurrence as massive. Adelite group and subgroup. See also F end member tilasite (41.5.6.1). ORTH $P2_12_12_1$. $a = 7.525$, $b = 8.850$, $c = 5.850$, $Z = 4$, $D = 3.74$. *CM 18:191(1980)*. *24-208:* 4.96_6 4.13_7 3.16_{10} 2.98_6 2.78_6 2.59_7 2.33_7 1.609_7. Crystals elongate [100], usually massive. Gray, bluish gray, yellowish gray, yellow, light green; resinous luster; transparent. Uneven to conchoidal fracture. H = 5. G = 3.73. Biaxial (+); XYZ= *acb* (Jacobsberg); $N_x = 1.712$, $N_y = 1.721$, $N_z = 1.731$; r < v. Small substitutions of Pb, Mn^{2+} for Ca, Mg, P for As, F for OH. Soluble in 1:10 acids. Found in the *manganese ore deposits of Långban, Värmland, Sweden*, with sarkinite, arsenoclasite, braunite, and hedyphane; at Jacobsberg and at the Kittel mine, Nordmark, with hausmannite, magnetite, and native copper. HCWS *DII:804*, *AMG 4:499(1968)*.

41.5.1.2 Conichalcite CaCu(AsO$_4$)(OH)

Named in 1849 from the Greek for *powder* and *lime*, in allusion to its appearance. Adelite group and subgroup. The Cu end member in Cu–Zn series with austinite (41.5.1.3). ORTH $P2_12_12_1$. $a = 7.393$, $b = 9.220$, $c = 5.830$, $Z = 4$, $D = 4.338$. *AM 56:1359(1971)*. Structure: *CM 7:561(1963)*. *37-448:* 5.78_3

4.11_3 3.12_8 2.89_3 2.84_{10} 2.60_7 2.59_6 2.55_5. See also *11-306*. Crystals equant to short prismatic [010]; botryoidal or reniform crusts with radiating fibrous structure. Grass-green to yellowish green, pistachio-green, emerald-green; green streak; vitreous luster; subtranslucent. No cleavage, uneven fracture, brittle. H = $4\frac{1}{2}$. G = 4.33. Biaxial (−) [or (+)]; XYZ= *cba*; N_x = 1.800, N_y = 1.831, N_z = 1.846 (Bisbee); r > v, strong. Y, yellow-green; Z, blue-green. Other substitutions noted: Mg for Cu, P and V for As. Fibrous varieties may show excess of H_2O. Soluble in HCl, HNO_3. A secondary mineral found in the oxidized zones of copper deposits. Occurs at Tuckers Tunnel prospect, La Plata Co., CO; at Ajo, New Cornelia mine, Bisbee, Cochise Co., Higgins mine, Shattuck shaft, Twilight claim, White-Tailed Deer mine, AZ; in the Tintic district, Juab Co., and at Gold Hill, Western Utah mine, UT; at Gallinas, Red Cloud Copper mine, and at Copper Hill, Taos Co., NM; at Bristol mine, Lincoln Co., NV, with tenorite and chrysocolla, and elsewhere in Mineral Co., and Lyon Clark Co.; at Zinc Hill mine, CA; at *Hinojosa de Cordoba, Andalusia, Spain*; at Graul near Schwarzenberg, Saxony, Germany; with tennantite at Miedzianka near Kielce, Poland; at Guchab near Otavi, Namibia; at Dome Rock, SA, Australia; at Collahuasi, Tarapacá, Chile. HCWS *DII:806, AM 36:484(1951), CM 25:406(1987)*.

41.5.1.3 Austinite CaZn(AsO$_4$)(OH)

Named in 1935 for Austin Flint Rogers (1877–1957), mineralogist, Stanford University, CA. Adelite group and subgroup. The Zn end member of the Cu–Zn series with conichalcite (41.5.1.2). ORTH $P2_12_12_1$. a = 7.505, b = 9.037, c = 5.921, Z = 4, D = 4.323. *AM 56:1359(1971)*. Structure: *NJMM:159 (1988)*. *37-455:* 5.78_3 4.96_5 4.14_3 3.17_9 2.80_{10} 2.64_8 2.60_4 2.53_4. Minute bladed or acicular crystals elongated [100], sometimes with scepterlike terminations; radially fibrous crusts and nodules. Colorless, white to pale yellowish white; subadamantine luster, to silky in fibrous aggregates. Cleavage {011}, good; brittle. H = 4–$4\frac{1}{2}$. G = 4.13. Biaxial (+); XYZ= *acb*; N_x = 1.759, N_y = 1.763, N_z = 1.783 (Na); 2V ≈ 47°. Found at Sterling Hill mine, Ogdensburg, Sussex Co., NJ; at Tucker Tunnel prospect, La Plata Co., CO; at *Gold Hill, Tooele Co., UT*, with adamite, quartz, and iron oxyhydroxides; at Kanaveza mine, Attica, Greece; with barthite from Guchab, and at Tsumeb, Namibia; as fibrous veinlets with calcite and chalcedony in the Lilli mine, Lomitos, Sica Sica Prov., Bolivia. HCWS *DII:809*.

41.5.1.4 Duftite-β PbCu(AsO$_4$)(OH)

Named in 1920 for G. Duft, director of mines at Tsumeb, Namibia. Adelite group and subgroup. In 1956 duftite was shown to have two forms, α and β. The β form appears to contain some Ca and may form a series with conichalcite (41.5.1.2). See duftite-α (41.5.2.6). ORTH $P2_12_12_1$. a = 7.49, b = 9.36, c = 5.91, Z = 4, D = 6.12. *BM 79:7(1956)*. PD: 4.18_7 3.17_{10} 2.91_8 2.63_9 2.26_5 1.868_5 1.750_5 1.637_7. *MM 29:609(1951)*. Found as small prismatic crystallites up to 0.05 mm; fibrous, stalactitic, mammiliary. Olive-green to yellow-green, vitreous luster. No cleavage, conchoidal fracture. H = $4\frac{1}{2}$. G = 5.86. DTA. N_{av} = 1.97; nonpleochroic. Synthesized. Samples show intergrowth of duftite-α (end member) with calcian duftite-β. Found at Ojuéla mine, Mapimi, DGO, Mexico, with adamite, wulfenite, and mimetite; at Brandy Gill, Caldbeck Falls, Cumberland, England; at Anozel, Vosges, and Cap Garonne, France, with malachite, azurite, olivenite, and mottramite; at *Tsumeb, Namibia*, as pseudomorphs of mimetite with duftite-α, malachite, bayldonite, and azurite. HCWS *MA 29:609(1951), MR 8:17(1977)*.

41.5.1.5 Gabrielsonite PbFe(AsO$_4$)(OH)

Named in 1967 for Olof Erik Gabrielson (b.1912), mineralogist, Swedish Natural History Museum, Stockholm, Sweden. Adelite group and subgroup. ORTH $P2_1m$. a = 7.86, b = 8.62, c = 5.98, Z = 4, D = 6.69. *AMG 4:401(1967)*. *20-583*: 4.91_3 3.93_3 3.19_{10} 3.07_{10} 2.71_4 2.65_4 1.622_4 1.438_4. Aggregates and masses of crude, rounded crystals up to centimeter size; nodules. Black, pale chocolate-brown streak, adamantine luster, transparent. No cleavage, brittle. H = $3\frac{1}{2}$. G = 6.67. N_{av} > 2.00. Soluble in cold 1:1 HCl. Found with romeite, nadorite, calcite, and barite in the *Hindenburg stope at Långban, Värmland, Sweden*. HCWS *AM 53:1063(1968)*.

41.5.1.6 Tangeite CaCu(VO$_4$)(OH)

Named in 1848 for its relationship to volborthite. Adelite group and subgroup. Previously called calciovolborthite. ORTH $P2_12_12_1$. a = 7.44, b = 9.24, c = 5.90, Z = 4, D = 3.82. *BM 37:219(1956)*. Structure: *NJMM:300(1989)*. *31-265*: 5.86_4 4.15_8 3.73_4 3.14_6 2.88_{10} 2.60_{10} 1.73_3 1.61_8. Scaly aggregates, acicular crystals to 0.5 mm long, fibrous. Gray-green, yellow-green to olive-green; vitreous to pearly luster; translucent. Cleavage (010), (001). H = 3–4. G = 3.75. Biaxial (+); N_x = 2.01, N_y = 2.05, N_z = 2.09 (Na); 2V = 83°; nonpleochroic; positive elongation. Soluble in acids. Synthesized. A secondary mineral in sandstones and in oxidized zones of deposits containing vanadium minerals. Found with carnotite and tyuyamunite in the sandstones of southwestern CO at Naturita, Paradox Valley, and Uravan, Montrose Co., and in UT at Grand R.; at Tombstone, Cochise Co., AZ; at Mammoth copper mine, Grindstone Creek, Glenn Co., Camp Signal, San Bernardino Co., CA; at *Friedrichroda, Thuringia, Gemany*, with crednerite and manganese oxides; at Tyuya Muyun, Ferghana district, Uzbekistan. HCWS *DII:817*.

41.5.1.7 Nickelaustinite CaNi(AsO$_4$)(OH)

Named in 1987 for the composition and structural relationship to austinite. Adelite group and subgroup. ORTH $P2_12_12_1$. a = 7.455, b = 8.955, c = 5.916, Z = 4, D = 4.27. Structure: *CM 25:401(1987)*. *41-1425*: 3.15_9 2.77_9 2.63_{10} 2.58_8 2.51_8 2.47_5 2.34_5 2.33_5. Fibrolamellar, flattened on (110), elongate on [001] to 0.2 mm. Yellowish green to grass-green, silky luster, translucent. Cleavage {110}, good. H = 4. Biaxial (−); X = c; N_x = 1.770, $N_{z'}$ = 1.778 in (110); nonpleochroic. Microprobe analysis also showed Cu, Zn, Fe, Mg, and Co.

Found on dolomite with roselite, calcite at *Bou Azzer, Morocco.* HCWS *AM* 73:930(1987).

41.5.1.8 Cobaltaustinite CaCo(AsO_4)(OH)

Named in 1988 for the composition and the relationship to austinite. Adelite group and subgroup. ORTH $P2_12_12_1$. $a = 7.498$, $b = 9.006$, $c = 5.920$, Z = 4, D = 4.24. *AUSM 3:53(1988)*. PD: 4.97_3 4.13_7 3.16_{10} 2.63_8 2.60_8 2.53_4 2.35_3 1.609_{10}. Coatings of crystallites up to 25 μm in diameter. Green, pale green streak, dull luster, translucent. Conchoidal fracture, brittle. H = $4\frac{1}{2}$. Biaxial (+); $N_x = 1.777$, $N_{z'} = 1.802$; 2V large; nonpleochroic. Possible complete solid solution Co–Cu with conichalcite (41.5.1.2). Soluble in HCl, HNO_3. Occurs with erythrite and other arsenates at *Dome Rock copper deposit, 42 km N of Mingary, SA, Australia.* HCWS *AM* 74:501(1989).

41.5.1.9 Arsendescloizite PbZn(AsO_4)(OH)

Named in 1982 to reflect the chemical and close crystallographic relationship to descloizite (41.5.2.1); the As end member of the AsO_4–VO_4 series. Adelite group and subgroup. ORTH $P2_12_12_1$. $a = 7.634$, $b = 9.358$, $c = 6.075$, Z = 4, D = 6.57. *MR 13:155(1982)*. 35-668: 4.67_5 4.23_6 3.23_{10} 2.88_{10} 2.60_8 2.09_6 1.656_6 1.559_8. Crystals tabular on {100} up to 1 mm × 0.4 mm × 0.5 mm with {101}, {110}, {100}, {111}; rosette-like aggregates. Pale yellow, white streak, brilliant subadamantine luster, transparent. No cleavage. H ≈ 4. Biaxial (–); XYZ= *bac*; $N_x = 1.990$, $N_y = 2.030$, $N_z = 2.035$; 2V ≈ 30°; r > v. Found with willemite, chalcocite, mimetite, quartz, and goethite, on a matrix of tennantite and chalcocite at *Tsumeb mine, Namibia.* HCWS *AM* 68:280(1983).

41.5.2.1 Descloizite Pb(Zn,Cu)(VO_4)(OH)

Named in 1854 for Alfred Louis Oliver Legrande Des Cloizeaux (1817–1897), French mineralogist and crystallographer. Adelite group–descloizite subgroup. Zincian end member in a Zn–Cu series with mottramite (41.5.2.2). ORTH *Pnam*. $a = 7.593$, $b = 9.416$, $c = 6.057$, Z = 4, D = 6.202. *MM (1956)31:289*, Structure: *AC(B) 35:717(1979)*. 12-537: 5.12_8 4.25_6 3.23_{10} 2.90_8 2.69_8 2.62_8 2.30_8 1.652_8. See 31-1467, 41-1369(cuprian). Often pyramidal {111}, prismatic [001], rarely tabular {100} or short prismatic [100]. Crystal faces usually uneven and rough, subparallel growth frequent. Also as drusy crusts of intergrown crystals, stalactitic, massive with coarse fibrous structure, or massive granular, compact to friable, mammillary or botryoidal. Brownish red, orange, blackish brown; orange to brownish red streak; greasy luster; transparent to nearly opaque. Often showing zonal growth with varying color, optical properties. No cleavage, conchoidal to uneven fracture, brittle. H = $3-3\frac{1}{2}$. G = 6.2. Biaxial (–); XYZ= *cba*; $N_x = 2.185$, $N_y = 2.265$, $N_z = 2.35$; 2V ≈ 90° (Mammoth, AZ); , r > v, strong; fibers negative elongation. In addition to the complete Zn–Cu, descloizite–mottramite series, Mn and Fe^{2+} substitute (<1%) for (Zn, Cu), As for V. Trace amounts of P and Cl are found on analysis. Easily soluble in acids. May be found as pseudomorphs after vanadinite (41.8.4.3). Secondary minerals; descloizite and mottramite may be found together in the oxidized ore zones of deposits containing vanadinite, pyromorphite, mimetite, and cerussite, and in sandstones as deposits of meteoric waters. Found with mottramite at Mammoth mine, Pinal Co., AZ; in notable deposits in Georgetown district, Grant Co., and in the Caballos Mts., Hillsboro, NM; cuprian descloizite occurs as large reniform masses and crusts in San Luis Potosi, and at the Ahumada mine, Los Lamentos, CHIH, Mexico; at Pim Hill, Shrewsbury, England; at Niederschletten, Lauterthal, Bavaria, and near Freiberg, Baden, Germany; as coatings on lava at Vesuvius, Italy; at Sah Changi mine, Iran; in large deposits (with mottramite) in sandy pockets in the limestone and dolomite at Abenab, Grootfontein(!) (Berg Aukas mine), Olifantsfontein, Guchab(!), Tsumeb, Uitsub, and elsewhere in the Otavi region, Namibia; also at Broken Hill mine, Zimbabwe; in the Katanga district, Zaire; at Itaccoambi mine, MG, Brazil; in the *Sierra de Cordoba, Argentina*. HCWS *DII:811*, *MM 50:137(1986)*, *NJMM:465(1991)*.

41.5.2.2 Mottramite   PbCu(VO$_4$)(OH)

Named in 1876 for the locality. Adelite group descloizite subgroup. The Cu end member in the Zn–Cu series with descloizite (41.5.2.1). ORTH *Pnma*. $a = 7.711$, $b = 9.249$, $c = 6.035$, Z = 4, D = 6.19. *MM 50:137(1986)*. *43-1463:* 5.06_6 4.23_4 3.25_{10} 2.87_9 2.69_4 2.66_5 2.59_4 2.30_4. See also *12-538*. Druses of intergrown (with descloisite) crystals often showing zones of different color and optical properties. Grass-green, olive-green to siskin-green; yellowish streak; greasy luster; transparent to opaque. H = 3–3$\frac{1}{2}$. G = 5.9. Biaxial (−) [or (+)]; $N_x = 2.17$, $N_y = 2.26$, $N_z = 2.32$; 2V ≈ 73° (Bisbee, AZ); , r > or < v, weak to strong; pleochroic: X,Y, canary to greenish yellow; Z, brownish yellow. A secondary mineral associated with descloizite and other vanadates in oxidized zones, especially sandstones. Found in the Silver Star district, MT; at Red Cloud tungsten mine, Gallinas Mt., NM; at Apache mine, and the Shattuck shaft, Bisbee, Cochise Co., the 79 mine, Gila Co., Old Yuma mine, Pima Co., Mammoth–St. Anthony mine, Tiger, and in the Tombstone district, AZ; in the Lake Valley district, Sierra Co., NM; at *Mottram–St. Andrew, Cheshire, England*; at Bena de Podru, Sassaro, Sardinia; at Mouanan, Gabon; at Broken Hill mine, Zambia; with descloizite in many parts of the Tsumeb district, Namibia. HCWS *DII:811*, *MM 31:289(1956)*.

41.5.2.3 Pyrobelonite   PbMn(VO$_4$)(OH)

Named in 1919 from the Greek for *fire* and *needle*, in allusion to the color and habit. Adelite group descloizite subgroup. ORTH *Pnam*. $a = 7.644$, $b = 9.508$, $c = 6.182$, Z = 4, D = 5.82. *CM 10:117(1969)*. Structure: *AM 40:580(1955)*. *20-588:* 5.14_7 4.27_4 3.25_{10} 2.91_5 2.67_6 2.64_7 2.33_7 2.13_5. Acicular [001]. Fire-red, orange-yellow or reddish streak, adamantine luster, transparent. H = 3$\frac{1}{2}$.

$G = 5.58$. Biaxial $(-)$; $N_x = 2.32$, $N_y = 2.36$, $N_z = 2.37$; $2V = 29°$, $r > v$. Reflectance in plane-polarized light, gray with bireflectance; under crossed polars strongly anisotropic in grays with red and orange internal reflections. Mn may substitute for Pb up to Mn/Pb 1:1. Found at Franklin, Sussex Co., NJ; at Tyloch, near South Cornelly, Mid Glamorgan, Wales, with vanadinite, manganocalcite, dolomite, and manganese oxides; at *Långban, Filipstad, Värmland, Sweden*, associated with hausmannite, pyrochroite, barite, and calcite. HCWS *MM 41:85(1977)*.

41.5.2.4 Cechite $Pb(Fe^{2+},Mn)(VO_4)(OH)$

Named in 1981 for Frantisek Cech (b.1944), mineralogist, Charles University, Prague, Czech Republic. Adelite group–Descloizite subgroup. ORTH *Pnam*. $a = 7.607$, $b = 9.441$, $c = 6.096$, $Z = 4$, $D = 5.99$. *NJMM:520(1981)*, Structure: *NJMM:35(1989)*. Isotypic with pyrobelonite (41.5.2.3). *35-530*: 5.12_8 4.72_4 4.25_5 3.35_4 3.23_{10} 2.91_8 2.66_6 2.63_8. Granular masses to 3 cm across, crystals up to 3 mm long show {110}, {010}. Black, black streak, submetallic to resinous luster. Uneven to conchoidal fracture, brittle. $H = 4\frac{1}{2}$–5. $G = 5.88$. IR. Magnetic. In RL, blue light gives gray-white with yellowish tint, yellow light gives brown-yellow to beige. Low birefringence, anisotropic with color shift from light gray to dark gray with brownish tint. $R = 18.5$, 16.0 (460 nm), 17.6, 14.8 (540 nm); 17.7, 14.4 (580 nm); 18.9, 15.1 (660 nm). Some question about the presence and relative amounts of Fe^{3+}. Found at *Vrancice, Pribram, Stredocsky Kraj, Cechy, Bohemia, Czech Republic*. HCWS *AM 67:1074(1982)*.

41.5.2.5 Duftite-α $PbCu(AsO.4)(OH)III$

Named in 1956 for its relationship to the original duftite. See duftite-β 42.5.1.4. Adeite group, descloizite subgroup. ORTH Pnam $a = 7.788$, $b = 9.223$, $c = 6.001$, $Z = 4$, $D = 6.57$. *BM 79:7(1956)*. *41-1444* 5.03_5 4.22_6 3.26_{10} 2.85_7 2.67_8 2.58_7 2.30_5 1.653_6. In spheroidal fibrous aggregates to 0.5 mm; octahedra to 0.2 mm in mammillary crusts. Green-black to light green, water green, translucent to opaque. Conchoidal fracture. $H = 3$, $G = 6.40$. Biaxial $(-)$; $N_x = 2.04$, $N_y = 2.08$, $N_z = 2.10$, $2V =$ large. Synthesized. Found at Cap Garonne, France; at *Tsumeb, Namibia* with olivenite, malachite, mottramite *MR 8:17(1977)*. HCWS

41.5.3.1 Babefphite $BaBe(PO_4)F$

Named in 1966 for the composition. TRIC $P1$. $a = 6.889$, $b = 16.814$, $c = 6.902$, $\alpha = 90.01°$, $\beta = 89.99°$, $\gamma = 90.32°$, $Z = 8$, $D = 4.325$. *DANS 167:93(1966)*. Structure: *SPC 25:28(1980)*. *18-157*: 3.67_6 3.19_{10} 2.76_8 2.44_7 2.76_{10} 2.03_7 1.516_{10} 1.135_7. Equant or flattened tabular grains to 1 mm × 1.5 mm. No faceted crystals seen. White, greasy luster. No cleavage. $H = 3\frac{1}{2}$. $G = 4.31$. Nonluminescent. Appears uniaxial $(+)$; $N_{min} = 1.629$, $N_{max} = 1.632$; negative elongation. Found in the alluvial deposits directly above a rare-metal fluorite deposit in *Siberia*, along with the heavy concentrates which contain zircon, ilmenorutile fluorite, phenakite, and scheelite. HCWS *AM 51:1547(1966)*.

41.5.4.1 Herderite CaBe(PO$_4$)(F)

Named in 1828 for Siegmund August Wolfgang von Herder (1776–1838), mining official in Freiberg, Saxony, Germany. The F end member (may not exist in nature) in the F–OH series with hydroxylherderite (41.5.4.2). MON $P2_1/c$. $a = 8.82$, $b = 7.70$, $c = 9.82$, Z = 4, D = 2.94. *ZK 93:146(1936)*. *6-338*: 3.43_5 3.14_{10} 3.00_6 2.86_8 2.55_6 2.26_5 2.20_7 1.650_5. Biaxial (−); $N_x = 1.556$, $N_y = 1.578$, $N_z = 1.589$. For properties, see hydroxylherderite. Found in the tin mines at *Ehrenfriedersdorf, Saxony, Germany*, and as a gemstone of unknown provenance from MG, Brazil. HCWS *AM 37:938(1952)*.

41.5.4.2 Hydroxylherderite CaBe(PO$_4$)(OH)

Named in 1894 for the composition and relationship to herderite (41.5.4.1). MON $P2_1/a$. $a = 9.789$, $b = 7.661$, $c = 4.804$, $\beta = 90.02°$, Z = 4, D = 3.00. Structure: *AM 59:919(1974)*. *34-147*: 3.76_2 3.43_3 3.14_{10} 3.00_3 2.87_4 2.55_2 2.21_3 1.996_2. Mainly euhedral as stout prismatic [100] or [001], thick tabular {001}, pseudo-orthorhombic, or pseudohexagonal crystals; botryoidal or spheroidal aggregates with radial-fibrous structure. Usually twinned, although not obvious, with twin plane {001} or {100}. Colorless to pale yellow, greenish white; vitreous luster; transparent to translucent. Cleavage {110}, interrupted; subconchoidal fracture. H = 5–5$\frac{1}{2}$. G = 2.95. Biaxial (−); $N_x = 1.59$–1.615, $N_y = 1.61$–1.634, $N_z = 1.62$–1.643; 2V = 70°. Indices decrease with increase in F. *AM 63:913(1978)*. Soluble in acids. Little replacement of Ca by Fe, Mg, or Mn but may have some Si (toward the silicate mineral ellestadite). Found associated with quartz, albite, muscovite, and clays in pegmatite localities in ME such as Beryllium Corp. mine, Buckland, Harvard quarry, Mt. Mica mine, Bell pit, Dunton Gem, Newry mine, where it is an alteration product of beryllonite, at Paul Bennet quarry, Oxford Co., and Berry quarry, Lord Hill, near Poland, Androscoggin Co., and at Stoneham; at Charles Davis, Fletcher, Palermo No. 1 mines, North Groton, NH; at Blue Chihuahua mine, Himalya mine, San Diego Co., CA; at the pegmatites at Epprechstein or Reinersreuth, Bavaria; at Viitaniemi, Finland; in pegmatites at Mursinsk, Ural Mts., Russia; in MG, Brazil, at Lavra do Enio, Galileia, and the Xanda mine(!), Virgen da Lapa, in violet-colored crystals to 15 cm associated with lepidolite, elbaite, topaz, and microlite, and at the Golconda mine(!), Governador Voladares district, in crystals and twins to 7 cm, at Alto Bernado, Picui, and Paraiba, Brazil; also at Mursinsk, Ural Mts., Russia. HCWS *DII:820*.

41.5.4.3 Väyrynenite $Mn^{2+}Be(PO_4)(OH,F)$

Named in 1954 for Heikki Allan Väyrynen (1888–1956), mineralogist and geologist, Technical High School, Helsinki, Finland. The Mn analog of hydroxylherderite (41.5.4.2). MON $P2_1/a$. $a = 5.41$, $b = 14.49$, $c = 4.73$, $\beta = 102.7°$, $Z = 4$, $D = 3.23$. *ZK 112:275(1959)*. Structure: *117:16(1962)*. 12-707: 7.25_9 4.96_3 4.40_6 3.45_{10} 2.89_9 2.85_4 2.66_4 2.64_3. Rose-red to light pink, vitreous luster, transparent. Cleavage {001}, good. H = 5. G = 3.183. Biaxial (−); $N_x = 1.640$, $N_y = 1.662$, $N_z = 1.667$. *AM 39:848(1954)*. Microprobe analyses show Fe^{2+}, Ca, Be, Na, K, and Al. Euclase is the Al–Si analog. Found at the pegmatite in *Viitaniemi, Erajarvi, Finland*, associated with herderite, hurlbutite, beryllonite, microcline, muscovite, and morinite: in the Broghul Pass area(!), northern Chitral Valley, Pakistan–Afghanistan, in crystals to 2.5 cm. HCWS *AM 41:371(1956)*, *ZK 143:309(1976)*.

41.5.4.4 Bergslagite $CaBe(AsO_4)(OH)$

Named in 1984 for the locality. MON $P2_1/c$. $a = 4.882$, $b = 7.809$, $c = 10.127$, $\beta = 90.16°$, $Z = 4$, $D = 3.40$. *NJMM:257(1984)*. Structure: *ZK 166:73(1984)*. Isostructural with hydroxylherderite (41.5.4.2). 35-650: 6.18_5 4.88_7 3.91_7 3.52_8 3.20_{10} 3.05_7 2.92_7 2.41_6 Centimeter-size aggregates of elongated crystals. Colorless, whitish, or grayish; vitreous luster; translucent. No cleavage, uneven fracture. H = 5. G = 3.4. Fluoresces pale green, yellowish brown, or blue under SWUV. Biaxial (−); XYZ= abc; $N_x = 1.659$, $N_y = 1.681$, $N_z = 1.694$ (598 nm); 2V = 70°. Found in the *Bergslagen region, Långban, Värmland, Sweden*, in thin veins in fine-grained hematite ore associated with manganoan pyroxene, manganberzeliite, svabite, barite, and calcite. HCWS *AM 70:436(1985)*.

TILASITE GROUP

The tilasite group minerals are described by the general formula

$$ABZ(TO_4)$$

where

A = Na, Ca
B = Al, Fe^{3+}, Mg
T = As, P
Z = F, OH

and crystallize in monoclinic space groups.

The structure is similar to titanite, $CaTiSiO_5$ (52.4.3.1) with PO_4 or AsO_4 in place of SiO_4. The B(O, OH, F) octahedra are cross-linked by the TO_4 groups, resulting in an open-framework arrangement. Ca or Na in seven-fold coordination are located within the large cavities created.

TILASITE GROUP

Mineral	Formula	Space Group	a	b	c	β	
Lacroixite	$NaAl(PO_4)F$	$C2/c$	6.414	8.207	6.885	115.47	41.5.5.1
Durangite	$NaAl(AsO_4)F$	$C2/c$	6.54	8.48	7.31	119.34	41.5.5.2
Maxwellite	$NaFe^{3+}(AsO_4)F$	Aa or $A2/a$	7.161	8.780	6.687	114.58	41.5.5.3
Tilasite	$CaMg(AsO_4)F$	$C2/c$	6.688	8.944	7.570	121.17	41.5.6.1
Isokite	$CaMg(PO_4)F$	$A2/a$	6.909	8.746	6.518	112.2	41.5.6.2
Panasqueiraite	$CaMg(PO_4)(OH,F)$	$C2/c$	6.535	8.753	7.498	121.40	41.5.6.3

Note: In addition to titanite, $CatiSiO_4O$ (52.4.3.1) related minerals include vuagnatite, $CaAl(SiO_4)(OH)$ (52.4.2.2); and malayite, $CaSnSiO_4O$.

41.5.5.1 Lacroixite $NaAl(PO_4)F$

Named in 1914 for Francois Antoine Alfred Lacroix (1863–1948), French mineralogist. Tilasite group. The phosphate analog of durangite (41.5.5.2). MON $C2/c$. $a = 6.414$, $b = 8.207$, $c = 6.885$, $\beta = 115.47°$, $Z = 4$, $D = 3.29$. *CM 27:211(1989)*. Structure: *AM 70:849(1985)*. 39-325: 4.73_5 4.63_3 3.25_2 3.16_{10} 2.90_{10} 2.48_6 2.17_4 1.578_3. Fragments of crystals. Gray or colorless, vitreous to resinous luster, transparent. Cleavage (111), ($1\bar{1}1$), indistinct. $H = 4\frac{1}{2}$. $G = 3.29$. Biaxial (−); $N_x = 1.546$, $N_y = 1.563$, $N_z = 1.580$; $2V = 89°$; slight dispersion. Easily soluble in HCl, H_2SO_4. Coupled substitution of Li, OH for Na,+ F noted in Buranga specimens. Found at Strickland quarry, Portland, CT; at Montebras, France; at Jihlava, Jeclov, Konigswart, Bohemia, Czech Republic; in druses in lithium-bearing pegmatite in granite intimately mixed with viitaniemiite, its alteration product, and associated with jezekite, apatite, childrenite, roscherite, and tourmaline at Greifenstien near *Ehrenfriedersdorf, Saxony, Germany*; at Buranga pegmatite, Rwanda, in montebrasite host. HCWS *DII:783*.

41.5.5.2 Durangite $NaAl(AsO_4)F$

Named in 1869 for the locality. Tilasite group. MON $C2/c$. $a = 6.54$, $b = 8.48$, $c = 7.31$, $\beta = 119.34°$, $Z = 4$, $D = 3.616$. *CM 23:241(1985)*. Structure (syn): *ZK 99:38(1938)*. 35-424: 4.76_8 3.54_4 3.34_5 3.24_8 2.97_{10} 2.56_6 2.55_6 1.533_4. Pyramidal crystals with $\{\bar{1}11\}$, $\{110\}$ predominating, faces usually dull, rough. Synthetic crystals prismatic [101], or tabular with interpenetration twins on $\{001\}$ common. Orange-red, light to dark, synthetics are green; cream-yellow streak; vitreous luster; translucent. Cleavage $\{110\}$, distinct; uneven fracture. $H = 5$. $G \approx 4.0$. Biaxial (−); $X \wedge c = -25°$; $Z = b$; $N_x = 1.634$, $N_y = 1.673$, $N_z = 1.685$; $2V = 45°$, $r > v$. Soluble in H_2SO_4. Synthesized. Found in northern Black Range, southwestern NM, in veins cutting alkali rhyolites; with amblygonite and cassiterite in pegmatite near Lake Ramsay, New Ross, Luneberg Co., NS, Canada; with cassiterite, hematite, and topaz at the *Bar-*

ranca tin mine, *DUR*, Mexico; on a joint face at the Cheesewring quarry, Linkinhorne, Cornwall, England. HCWS *DII:829*.

41.5.5.3   Maxwellite   NaFe(AsO$_4$)F

Named in 1991 for Charles Henry Maxwell (b.1923), U.S. Geological Survey (Denver). Tilasite group. MON Aa or $A2/a$. $a = 7.161$, $b = 8.780$, $c = 6.687$, $\beta = 114.58°$, $Z = 4$, $D = 3.95$. *NJMM:363(1991)*. PD: 4.84_7 3.64_4 3.29_{10} 3.04_8 2.91_4 2.66_3 2.64_5 2.61_8. Aggregates to 3 mm of short, blocky prisms; subhedral crystals to 1 mm showing {013}, {526}, {5̄26}, {011}. Dark red, medium to pale red-orange streak, vitreous luster, transparent. Cleavage {110}, good; irregular to conchoidal fracture. $H = 5–5\frac{1}{2}$. $G = 3.90$. Nonfluorescent. Biaxial (+); $X \wedge c = 11°$; $Y = b$; $Z \wedge a = 35.5°$; $N_x = 1.748$, $N_y = 1.772$, $N_z = 1.798$; $2V = 86°$; $Z > X > Y$; $r > v$, strong. X,Y, medium yellow-orange; Z, dark orangish red. Slowly soluble in HCl, HNO$_3$, readily soluble in hot H$_2$SO$_4$. Found in the *Black Range tin district, Catron Co., NM*. HCWS *AM 77:449(1992)*.

41.5.6.1   Tilasite   CaMg(AsO$_4$)F

Named in 1895 for Daniel Tilas (1712–1772), Swedish mining engineer. Tilasite group. MON $C2c$. $a = 6.688$, $b = 8.944$, $c = 7.570$, $\beta = 121.17°$, $Z = 4$, $D = 3.722$. Space group revised *NJMM: 289 (1994) MR 9:385(1978)*. Structure: *AM 57:1880(1972)*. 2-485: 3.26_{10} 3.07_{10} 2.86_7 2.69_{10} 2.63_7 2.27_8 1.74_9 1.69_9. Crystals elongated [100], sometimes flattened {010}, occasionally in subparallel groups; massive. Twinning {001} common. Gray to violet-gray (Långban), olive-green to apple-green (India); resinous luster on cleavage surfaces, vitreous; translucent. $H = 5$. $G = 3.77$. India crystals piezoelectric. Biaxial (−); $X \wedge c \approx 30°$; $Z = b$; $N_x = 1.640$, $N_y = 1.660$, $N_z = 1.675$; $2V = 83°$; $r < v$ (India). Readily soluble in HCl, HNO$_3$. Found at Sterling Hill mine, Sussex Co., NJ; at White Tailed Deer mine, Bisbee, Cochise Co., AZ; as grains or veinlets with berzeliite, barite in dolomitic limestone with hausmannite at *Långban, Värmland, Sweden*; at Cherbudung, Falotta, Grisons, Binntal, Switzerland; at Kajlidongri, Jhabua State, India; at Guettera, Algeria. HCWS *DII:827*.

41.5.6.2   Isokite   CaMg(PO$_4$)F

Named in 1955 for the locality. Tilasite group. MON $A2/a$. $a = 6.909$, $b = 8.746$, $c = 6.518$, $\beta = 112.2°$, $Z = 4$, $D = 3.248$. *MM 56:227(1992)*. 7-406: 3.19_{10} 3.02_{10} 2.78_5 2.63_{10} 2.59_5 2.30_8 2.22_5 1.720_8. Patchy spherulitic aggregates of fibrous or platy crystals; short prismatic. White, light brown, pink. Cleavage {010}, good. $H = 5$. $G = 3.29$. Biaxial (−); $N_x = 1.590$, $N_y = 1.595$, $N_z = 1.615$; $2V = 51°$. Dissolves in cold HCl. Found at Benson mines, west-central Adirondack highlands, NY, in magnetite–hematite deposits whose veins, dikes, and fractures are filled with wagnerite, replaced by isokite. Associated minerals: ankerite, dolomite, Sr- and F-containing apatite, pyrite, magnetite, ilmenite, monazite, pyrochlore, and sellaite; in the emerald mines in the

Ural Mts., Russia; in ankeritic dolomite (carbonatite), *Nkumbwa Hill, near Isoka, northern Kumbwa, Zambia.* HCWS *AM 40:776(1955), MM 30:68(1955).*

41.5.6.3 Panasqueiraite CaMg(PO$_4$)(OH,F)

Named in 1981 for the locality. Tilasite group. MON $C2/c$. $a = 6.535$, $b = 8.753$, $c = 7.498$, $\beta = 121.40°$, $Z = 4$, $D = 3.213$. *CM 23:131(1985).* 35-511: 3.20$_7$ 3.02$_9$ 2.78$_3$ 2.63$_{10}$ 2.58$_5$ 2.11$_2$ 1.722$_4$ 1.658$_3$. Aggregates several centimeters in diameter with individual anhedral grains about 1 mm. Pink, white streak, vitreous luster, translucent. Cleavage {010}, poor. H = 5. G = 3.27. Nonfluorescent but cathodoluminescent in electron beam (12 kV). Biaxial (+); $X \wedge c = 22°$; $Z = b$; $N_x = 1.590$, $N_y = 1.596$, $N_z = 1.616$; $2V = 51°$; nonpleochroic. Occurs with thadeuite, fluorapatite, wolfeite, topaz, muscovite, sphalerite, quartz, chalcopyrite, pyrrhotite, siderite, arsenopyrite, chlorite, vivianite, and althausite in vein selvages at *Panasqueira, Portugal.* HCWS *CM 19:389(1981), AM 67:859(1982).*

41.5.7.1 Brazilianite NaAl$_3$(PO$_4$)$_2$(OH)$_4$

Named in 1945 for the country in which it was found. MON $P2_1/n$. $a = 11.233$, $b = 10.142$, $c = 7.097$, $\beta = 97.37°$, $Z = 4$, $D = 2.998$. *SMPM 41:407(1961).* Structure: *AC(B) 30:1311(1974).* 42-1354: 5.78$_2$ 5.07$_{10}$ 3.75$_1$ 3.29$_1$ 2.99$_3$ 2.87$_2$ 2.74$_2$ 2.69$_3$. See also *14-379* and *27-630*(calc). Equant to short prismatic [001] with prism zone striated [001], elongated [100]; many forms showing; globular with radial-fibrous structure. Chartreuse-yellow to pale yellow, colorless streak, vitreous luster, transparent. Cleavage [010], good; conchoidal fracture; brittle. H = 5$\frac{1}{2}$. G = 2.983. Biaxial (+); $X \wedge c = -20°$; $Y = b$; $N_x = 1.602$, $N_y = 1.609$, $N_z = 1.621$ (Na) (Brazil); $2V = 75°$, r < v, faint. Slowly decomposed by HF and by hot H$_2$SO$_4$. Found at the Harvard quarry and Nevel tourmaline mine, Newry, ME, G.E. Smith mine, Newport, and with whitlockite and apatite in the Fletcher mine and Palermo No. 1 mine(!), North Groton, NH; in cavities in a pegmatite near *Corrego Frio*(!) and Lavra de Telirio(!), *Linopolis,* also Mendes Pimental(!), *MG, Brazil,* with muscovite, albite, apatite, and tourmaline. HCWS *DII:841.*

41.5.8.1 Amblygonite LiAl(PO$_4$)F

Named in 1817 from the Greek for *blunt* and *angle,* in allusion to the appearance of crystalline samples. The F-rich member of the F–OH series with montebrasite (41.5.8.2) and the Li end member of the Li–Na series with natromontebrasite (41.5.8.3). TRIC $P\bar{1}$. $a = 6.645$, $b = 7.733$, $c = 6.919$, $\alpha = 90.35°$, $\beta = 117.44°$, $\gamma = 91.20°$, $Z = 4$, $D = 3.08$. *ZVMO 118:47(1989).* Structure: *AM 75:992(1990).* 22-1138: 4.64$_8$ 3.87$_8$ 3.30$_5$ 3.24$_6$ 3.15$_{10}$ 2.96$_8$ 2.38$_5$ 1.935$_6$ 1.728$_5$. Equant to short prismatic [010] crystals, ordinarily rough; cleavable masses; columnar; compact. Twinning common on {$\bar{1}\bar{1}1$} with composition plane {$\bar{1}\bar{1}1$}. Twins usually tabular parallel to {$\bar{1}\bar{1}1$} with individuals about equal in size, also tabular {110} with unequal individuals; twin {111} rare; lamellar. White to milky or creamy white, yellowish, beige, salmon-pink, greenish, bluish, gray, occasionally colorless and water clear; vitreous to greasy luster; transparent to translucent. Cleavage {100}, perfect; {110}, good; {0$\bar{1}$1}, distinct; {001}, imperfect; uneven to subconchoidal fracture; brittle. H = 5$\frac{1}{2}$–6. G = 3.11. Biaxial (−); $N_x = 1.591$, $N_y = 1.605$, $N_z = 1.613$ (Na) (Utö, analysis shows Na and OH); r > v. Indices decrease with increasing F and Na. Soluble with difficulty in acids. Complete solid-solution series for Na–Li and F–OH, and most samples show all these elements on analysis. Usually found in Li-bearing granitic pegmatites, often as very large crystals with variable composition, even from a single site. Associated minerals: spodumene, lithiophilite–triphylite, apatite, lepidolite, petalite, pollucite, and tourmaline. Major occurrences include transparent crystals found at New Pit, Newry(!!); other forms at the many pegmatite localities in ME and NH, usually compositionally montebrasite; giant crystals in the pegmatites of the Custer district(!), SD, and elsewhere in the Black Hills, where masses weighing hundreds of tons have been mined; at Pala, Mesa Grande, and elsewhere in San Diego Co., and San Bernardino Co., CA; at the pegmatites in the Yellowknife–Beaulieu area and in the Oiseaux R. district, MAN, Canada. In Europe the early finds were at Churgsdorf and Arnsdorf near Penig and with cassiterite at Geyer in Saxony, Germany, with several members in the series known from *Montebras, Creuse, France*; on the island of Utö near Stockholm, Sweden; at many localities in MG, Brazil, such as Araçuai, Urubu, Itinga, Lavra de Ilha(!), Taquaral; Mendes Pimentel, Mantena. HCWS *DII:823, MM 37:414(1969)*.

41.5.8.2 Montebrasite LiAl(PO$_4$)(OH)

Named in 1872 for the locality. The (OH) end member of a (OH)–F series with amblygonite (41.5.8.1) and the Li end member in the Li–Na series with natromontebrasite (41.5.8.3). TRIC P$\bar{1}$. $a = 6.713$, $b = 7.708$, $c = 7.019$, $\alpha = 91.31°$, $\beta = 117.93°$, $\gamma = 91.77°$, Z = 4, D = 3.00. *ZVMO 118:47(1989)*. Structure: *AM 75:992(1990)*. 12-448: 4.68$_9$ 3.33$_4$ 3.27$_5$ 3.23$_5$ 3.20$_6$ 3.16$_9$ 2.97$_{10}$ 2.40$_4$. Habit and physical and chemical properties similar to those of amblygonite, often occurring in crystals of gigantic size and associated with lithiophilite–triphylite, apatite, lepidolite, petalite, pollucite, and tourmaline. Composition variations between OH–F and Li–Na often detected in single locality. Biaxial (+); $N_x = 1.594$, $N_y = 1.608$, $N_z = 1.616$ (Na) (Karibib); 2V = 75°; r < v. (analysis shows Li/Na = 9:0.4, OH/F ≈ 1:1). Found at New Pit(!), Newry, Oxford Co., ME; at Branchville and Portland, CT; at Royal Gorge, CO; as giant crystals in the pegmatites of the Custer district and in the Black Hills, Peerless, and Bob Ingersol mines(!!) near Keystone, at Gaint–Volney mine near Tinton, at Beecher Lode and Tin Mt., pegmatites, SD; at Yellowknife–Beaulieu area, NWT; and at New Ross, Lunenberg Co., NS, Canada; at *Montebras, Creuse, France*; in large masses in the pegmatite at Varuträsk, Sweden; in the lepidolite pegmatites at Karibib district, Namibia, with petalite; at Lemonade Springs pegmatite, Greenbushes, and at Ravensthorpe and

Ubini, WA, Australia; at Mogi das Cruzes, São Paulo, Brazil. HCWS *DII:823*, *MM 37:414(1979)*.

41.5.8.3 Natromontebrasite   (Na,Li)Al(PO$_4$)(OH,F)

Named in 1913 for the composition and resemblance to montebrasite (41.5.8.2) and amblygonite (41.5.8.1). $P\bar{1}$. $a = 5.266$ $b = 7.174$ $c = 5.042$, $\alpha = 112.3°$, $\beta = 97.70°$, $\gamma = 67.13$, Z = 2, D = 3.31. *44-1429:* 4.56$_5$ 3.34$_2$ 3.27$_2$ 3.23$_3$ 3.15$_{10}$ 2.96$_{10}$ 2.50$_8$ 2.39$_3$. Biaxial (+); N$_x$ = 1.594, N$_y$ = 1.603, N$_z$ = 1.615 (Freemont, CO); 2V very large (analysis of this sample shows Na/Li = 1.69:1 and OH/F ≈ 1:1). Found at Canyon City, and Eight Mile Park, Fremont Co., CO; at Jeclov, Moravia, Czech Republic; at Presidente Buena, MG, Brazil. HCWS *DII:823*.

41.5.9.1 Tavorite   LiFe^{3+}(PO$_4$)(OH)

Named in 1955 for Elysairio Tavora (b.1911), mineralogist, University of Brazil, Rio de Janeiro, Brazil. The Fe^{3+} analog of montebrasite. TRIC P1 or $P\bar{1}$. $a = 5.138$, $b = 5.307$, $c = 7.442$, $\alpha = 67.48°$, $\beta = 67.72°$, $\gamma = 81.98°$, Z = 2, D = 3.33. *AM 40:952(1953)*, Structure: *DANP 29:27(1984)*. Tetrahedral PO$_4$ groups coordinate with octahedral FeO$_6$ groups, creating channels in which Li occurs. *41-1376:* 4.97$_6$ 4.77$_5$ 4.66$_6$ 3.94$_4$ 3.32$_7$ 3.27$_{10}$ 3.04$_9$ 2.46$_5$. Very fine grained, flakes. Yellow, green-yellow to green, dark green. G = 3.29. N$_{av}$ = 1.807. Found at Fletcher and Palermo No. 1 mines, North Groton, NH; at pegmatite localities in SD, such as Bull Moose, Custer Mt., Elkhorn pegmatite, Tip Top and White Elephant mines; at Hagendorf, Bavaria, Germany; at *Sapucaia pegmatite, Galileia, MG, Brazil*, in veinlets and disseminated in porous triphylite and heterosite, admixed with barbosalite. HCWS *NJMM:78(1957)*.

FLORENCITE/ARSENOFLORENCITE GROUP

The Florencite/Arsenoflorencite minerals are a special set of the Crandallite subgroup with a general formula:

$$AB_3(TO_4)_2(OH)_{5 \text{ or } 6}$$

where

A = REE, Ba, Bi, Ca, Th, Pb, ☐
B = Al, Fe^{3+}
T = As, P
(OH) amount depends on the charge of the A ion.

The minerals crystallize in the hexagonal-rhombohedral space group $R\bar{3}m$. The structure *NJMM:227(1990)* resembles alunite but the tetrahedral TO$_4$ groups coordinated with B(O,OH)$_6$ polyhedra are both irregular. The apical

41.5.10.2 Florencite-(Ce)

T–O bond of the TO_4 group is longer than the other three T–O bonds. The A ions are in 12 fold coordination in the intersticies of this scaffold-like structure.

FLORENCITE/ARSENOFLORENCITE GROUP

Name	Formula	a	c	
Dussertite	$BaFe_3^{3+}(AsO_4)_2(OH)_5$	7.424	17.494	41.5.10.1
Florencite-(Ce)	$CeAl_3(PO_4)_2(OH)_6$	6.927	16.261	41.5.10.2
Florencite-(La)	$LaAl_3(PO_4)_2(OH)_6$	6.987	16.248	41.5.10.3
Florencite-(Nd)	$NdAl_3(PO_4)_2(OH)_6$	6.992	16.454	41.5.10.4
Arsenoflorencite-(Ce)	$CeAl_3(AsO_5)(OH)_6$	7.029	16.517	41.5.11.1
Arsenoflorencite-(Nd)	$NdAl_3(AsO_4)(OH)$			41.5.11.2
Arsenoflorencite-(La)	$LaAl_3(AsO_4)(OH)$			41.5.11.3
Waylandite	$(Bi,Ca)Al_3(PO_4,SiO_4)_2(OH)_6$	6.983	16.175	41.5.12.1
Eylettersite	$(Th,Pb,\square)Al_3(PO_4,SiO_4)_2(OH)_6$	6.99	16.70	41.5.12.2
Zairite	$BiFe_3^{3+}(PO_4)_2(OH)_6$	7.015	16.335	41.5.12.3

41.5.10.1 Dussertite $BaFe_3^{3+}(AsO_4)_2(OH)_5$

Named in 1925 for D. Dussert, French mining engineer. Florencite group, see crandallite subgroup 42.7.3.1. HEX-R $R\bar{3}m$. $a = 7.424$, $c = 17.494$, $Z = 3$, $D = 4.09$. *TMPM 11:1215(1966)*. *35-621*: 6.07_7 3.72_6 3.13_{10} 2.58_3 2.33_5 2.00_6 1.850_6 1.560_3. Minute crystals flattened {001} grouped as rosettes or crusts. Green to pistachio-green, translucent. No cleavage. $H = 3\frac{1}{2}$. $G = 3.75$. Uniaxial (–); $N_O = 1.87$, $N_E = 1.85$ (Algeria) (occasionally biaxial, $2V = 15–20°$). Occurs with arseniosiderite and carminite at Mapimi, DGO, Mexico; on spongy or lamellar quartz at *Djebel Debar, northeastern Hammam Meskhoutine, Constantine, Algeria.* CS *DII:839*.

41.5.10.2 Florencite-(Ce) $CeAl_3(PO_4)_2(OH)_6$

Named in 1899 for William Florence (1864–1942), Brazilian mineralogist, who made a preliminary chemical analysis of the mineral. Florencite group. Compare crandallite group (42.4.3.1); see also arsenoflorencite (41.5.11.1–3), the As analog. HEX-R $R\bar{3}m$. $a = 6.972$, $c = 16.261$, $Z = 3$, $D = 3.73$. Structure: *NJMM:227 (1990)*. *43-673*: 5.65_8 3.48_6 2.93_{10} 2.42_2 2.20_3 2.17_4 1.886_4 1.742_3. Sample from Backbone range, McKenzie Mts., NWT, Canada *CM 19:535(1981)*. See also *8-413* (Mata dos Creoulos, Diamantina, MG, Brazil). Small rhombohedral crystals exhibiting {02$\bar{2}$1} or {10$\bar{1}$1}, pseudocubic clear, pale yellow. Luster greasy to resinous. Transparent splintery fracture. $H = 5–6$ $G = 3.586$ (Brazil). Uniaxial (+) $N_O = 1.695$, $N_E = 1.705$ (Africa). Partly soluble in HCl. Found at Kangankunde Hill, Malawi; with microcline and fluorite in pegmatite at Klein Spitzkopje, Namibia; in mica schists at Morro do Caizambu, near Ouro Preto, and as a rare constituent of the sands of Tripuhy, Ouro Preto, associated with cinnabar, monazite, and xenotime; in rounded grains in diamond-bearing sands from Matta de Creoulos, Rio Jequetinhonha, near Diamantina, MG, Brazil HCWS *DII:838, MM 33:281(1962)*.

41.5.10.3 Florencite-(La) $LaAl_3(PO_4)_2(OH)_6$

Named in 1981 as the La end member of the florencite series. See Florencite-(Ce) 10.2. HEX-R $R\bar{3}m$. $a = 6.987$, $c = 16.248$. *38-347*: 5.67_9 3.49_7 2.93_{10} 2.17_8 1.884_7 1.743_7 1.601_7 1.190_7. Sample from Shituru Cu deposit, Shaba, Zaire, contains La > Ce, trace Nd cs *CM 18:301(1980)*. Physical and chemical properties as Florencite-(Ce). HCWS

41.5.10.4 Florencite-(Nd) $NdAl_3(PO_4)_2(OH)_6$

Named in 1986 as the Nd end member of the florencite series. Florencite-group (Ce) 41.5, 10.2. HEX-R $R\bar{3}m$. $a = 6.992$, $c = 16.454$,. *PD 1:330(1986)*. *39-333*: 5.69_9 3.50_9 2.95_{10} 2.84_2 2.19_5 1.895_5 1.748_6 1.286_3. Sample from Sausalito, Marin Co., CA. Uniaxial (+) $N_O = 1.66$–1.666, $N_E = 1.688$–1.694. *DANS 314: 195(1990)*. HCWS

41.5.11.1 Arsenoflorencite-(Ce) $CeAl_3(AsO_4)_2(OH)_2$

Named in 1987 for the composition and relationship to florencite. The As analog of florencite-(Ce), arsenoflorencite group. Related to the crandallite (42.7.3.1) group, with which there is limited solid solution, also isomorphous with arsenogoyazite (42.7.4.3) and with gorceixite (42.7.3.2). HEX-R $R\bar{3}m$. $a = 7.029$, $c = 16.517$, $Z = 3$, $D = 4.091$. *MM 51:605(1987)*. *42-1390*: 3.51_6 2.96_{10} 2.75_3 2.20_4 1.905_5 1.753_4 1.627_3 1.482_3. Scalenohedral crystals showing {102}, {101}, and fragments to 0.5 mm in diameter. Colorless to light brown, clear to cloudy, translucent. No cleavage, conchoidal fracture, brittle. $H = 3$. $G = 4.096$. Nonfluorescent, no cathodoluminescence. $G = 4.096$. Uniaxial (+); $N_O = 1.739$, $N_E = 1.745$ (589 nm). Microprobe analysis (decomposes in the microprobe beam) shows for florencite-(Ce) Ce > La, Nd with minor amounts of Pr, Gd, Sr, P, and trace amounts of Sm, S. In REE proxying for Sr, the several chemical varieties can be distinguished only by microprobe analysis. Arsenoflorencite-(Ce) is found in the sands examined from the *Kimba mapsheet area, Eyre Penin., SA, Australia*, associated with magnetite, tourmaline, garnet, sillimanite, ilmenite, apatite, zircon, spinel, and xenotime; also in sands at the NT/Q border with cassiterite, rutile, zircon, anatase, and monazite. May have inclusions and/or covering of quartz or alunite. CS

41.5.11.2 Arsenoflorencite-(Nd) $NdAl_3(AsO_4)_2(OH)_6$

Identified and named for its relationship to arsenoflorencite and florencite arsenoflorencite group. Microprobe analysis: Nd dominant with minor amounts of Fe^{3+}, P, S, Ba, La, Ce, Pb, Pr, Ca, and Sm. HCWS

41.5.11.3 Arsenoflorencite-(La) $LaAl_3(AsO_4)_2(OH)_6$

Identified and named for its relationship to arsenoflorencite and florencite. Arsenoflorencite group. Microprobe analysis: La dominant with minor amounts of Fe^{3+}, P, S, Ca, Sr, Ce, Nd. La-rich arsenoflorencite is found as cores of zoned crystals with florencite-(Ce), also as 30-μm irregular aggregates with arsenogoyazite, as outer zones of 5- to 10-μm grains with cores of crandallite, and with arsenoflorencite-(Nd) in the cements of a Cretaceous sandstone in drill cores from *Holicky, Stráz, and Osecná uraniferous deposits, northern Bohemia, Czech Republic*. Coprecipitation of the several chemical varieties of arsenoflorencite and florencite is usual. HCWS *CMG 36:103(1991), AM 78:672(1993)*.

41.5.12.1 Waylandite $(Bi,Ca)Al_3(PO_4,SiO_4)_2(OH)_6$

Named in 1962 for Edward James Wayland, first director of the Uganda Geological Survey. Florencite group. HEX-R $R\bar{3}m$. $a = 6.983$, $c = 16.175$, $Z = 3$, $D = 4.08$ *MM 50:730(1986). 39-367:* 5.67_8 3.49_8 2.94_{10} 2.20_8 2.17_{10} 1.888_9 1.743_8 1.434_6. Massive, compact, powdery, fine-grained; grains show cyclic twinning. White, colorless, bluish; vitreous to dull luster; translucent. Uneven fracture. $G = 3.86$. Uniaxial (+); $N_O = 1.748$, $N_E = 1.774$. Found as crystalline bands, veinlets, and crusts on a matrix of bismutite, or replacing bismutotantalite, at Restormel iron mine, Lostwithiel, and Wheal Coates, St. Agnes, Cornwall, England; at *Wampiro Hill, Busiro Co., Buganda, Uganda*; at Kobokobo pegmatite, Kivu, Kinshasa, Zaire. HCWS *AM 48:216(1963)*.

41.5.12.2 Eylettersite $(Th,Pb\square)_{1-x}Al_3(PO_4,SiO_4)_2(OH)_6$

Named in 1972 for Mme Van Wambeke, the wife of the discoverer, L. Van Wambeke. Florencite group. HEX-R $R\bar{3}m$. $a = 6.99$, $c = 16.70$, $Z = 3$, $D = 3.44$. *AM 56:1366(1971). 26-991:* 5.7_5 3.51_6 2.95_{10} 2.85_2 2.27_2 2.19_4 1.899_3 1.748_2. Pulverulent nodules of creamy white, usually heavily altered. $G = 3.38$. Fluoresces brown-cream in LWUV, greenish yellow in SWUV. Radioactive, first thorium phosphate found in nature. DTA. $N_{av} = 1.635$, range depending on alteration level 1.61–1.66; nearly isotropic. Readily dissolved by H_2SO_4. Analysis shows U, Ca, Ba, Zr, minor Sr, and Fe. Considerable cation deficiency (Th position), possibly balanced by substitution of H_4O_4 for SiO_4. Altered specimens are found at *Kobokobo pegmatite, Kivu, Kinshasa, Zaire*, associated with feldspar, cryolite, columbite, apatite, candallite plumbogummite and phosphuranylite. HCWS *BM 95:98(1972), AM 59:208(1974)*.

41.5.12.3 Zairite $BiFe_3^{3+}(PO_4)_2(OH)_6$

Named in 1975 for the country of origin. The ferric iron analog of waylandite (41.5.12.1). Florencite group. HEX-R $R\bar{3}m$. $a = 7.015$, $c = 16.365$, $Z = 3$, $D = 4.42$. *BM 98:351(1976). 29-226:* 5.71_{10} 4.87_1 3.50_4 2.95_9 2.85_1 2.73_2 2.16_1 1.877_1. Small masses. Greenish, translucent. $H = 4\frac{1}{2}$. Uniaxial (−); $N_{av} = 1.82$–1.83; weakly to moderately birefringent. Found in the zone of weathering at *Eta-Etu, northern Kivu, Zaire*, where there are quartz veins containing wolframite, mica, and other Bi-containing phosphates. HCWS *AM 62:194(1977)*.

41.5.13.1 Vesignieite BaCu$_3$(VO$_4$)$_2$(OH)$_2$

Named in 1955 for Colonel Jean Paul Louis Vésignié (1870–1954), French mineral collector and president of the Mineralogical Society of France. Balydonite (41.5.14.1) is the Pb analog. MON $C2/c$. $a = 10.270$, $b = 5.911$, $c = 7.711$, $\beta = 116.42°$, $Z = 2$, $D = 4.69$. *MM 56:67(1992)*. Structure: *AGS 64:302(1990)*. *31-136*: 3.36$_3$ 3.22$_{10}$ 2.95$_4$ 2.71$_8$ 2.56$_6$ 2.30$_6$ 1.614$_4$ 1.479$_4$. See also *12-519*. Lamellar aggregates, polysynthetically twinned, giving pseudohexagonal outline, up to 0.5 mm in diameter. Yellow-green to dark olive-green, vitreous luster, transparent. Cleavage {001}, good. H = 3–4. G = 4.43. Biaxial (−); N$_x$ = 2.053, N$_y$ = 2.129, N$_z$ = 2.133. Analyses also showed minor Zn, Ni, Ag, As, Mo, Ca, Mg, K, Na, Fe^{3+}, Si, Al. Synthesized. Found as geodes in Mn ore at *Friedrichsroda, Thuringen, Germany*, with calcite and barite; at Perm, Ural Mts., Russia; at Agalik, Uzbekistan; at Shensi Prov., China. HCWS *AM 40:942(1955)*, *BM 79:219(1956)*.

41.5.14.1 Bayldonite Cu$_3$PbO(AsO$_3$OH)$_2$(OH)$_2$

Named in 1865 for John Bayldon of England. Vesignieite (41.5.13.1) the Ba, V analog. MON $A2/a$. $a = 14.083$, $b = 5.893$, $c = 10.152$, $\beta = 106.10°$, $Z = 4$, $D = 5.82$. *MM 39:716(1974)*, *AC(B) 35:819(1979)*. *26-1410*: 4.52$_6$ 3.23$_8$ 3.15$_{10}$ 2.93$_8$ 2.72$_6$ 2.66$_5$ 2.48$_3$ 2.26$_5$. Minute mammillary concretions with fibrous structure, drusy surface; massive, fine-granular to powdery; crusts. Siskin-green to apple-green, yellow-green; resinous luster, subtranslucent. H = 4$\frac{1}{2}$. G = 5.5. IR, DTA. Biaxial (+); X = b, Y ∧ elongation ≈ 25°; N$_x$ = 1.95, N$_y$ = 1.97, N$_z$ = 1.99; 2V large; r < v. Soluble with difficulty in HCl. Found at *St. Day, Cornwall, England*, and at the Penberthy Creft uranium mine, Grampound Rd., St. Hilary; at several sites (i.e., Verriere, Ardillat, Rohe, Rebasse, Ceilhes) in the Hérault Dept., France; at Dzeskazghan, Kazakhstan; at Tsumeb, Namibia. HCWS *DII:929*, *AM 66:148(1981)*.

41.5.15.1 Curetonite Ba$_4$Al$_3$Ti(PO$_4$)$_4$(O,OH)$_6$

Named in 1979 for Forrest Cureton, mineralogist, and Michael Cureton, of Tucson, AZ, who found the mineral. MON $P2_1/m$. $a = 6.957$, $b = 12.55$, $c = 5.22$, $\beta = 102.02°$, $Z = 1$, $D = 4.31$. *MR 10:219(1979)*. *33-134*: 4.29$_5$ 3.29$_8$ 3.23$_{10}$ 2.99$_6$ 2.82$_6$ 2.25$_6$ 2.20$_6$ 1.686$_7$. Crystals up to 3 mm show {100}, {010}, {001}, {011}, {$\bar{2}$01}. Bright yellow-green to nickel-green, white streak, translucent. Polysynthetic twinning common on {100}. Cleavage {011}, good; parting on {010}. H = 3$\frac{1}{2}$. Biaxial (+); X = b, Z ∧ c = 30°; N$_x$ = 1.676, N$_y$ = 1.680, N$_z$ = 1.693; 2V = 60°; splotchy pleochroic in yellow; X > Y = Z; r < v, weak; strong inclined dispersion. Found in a barite mine at *Golconda, NV*, where barite replaces sericitic, phosphatic, and black organic cherts and shales. Massive barite is cut by veins of coarse-grained barite, euhedral adularia, and curetonite. HCWS *AM 65:206(1980)*.

41.5.16.1 Thadeuite $(Ca,Mn^{2+})(Mg,Fe^{2+},Mn^{2+})_3(PO_4)_2(OH,F)_2$

Named in 1979 for Décio Thadeu, Instituto Superior Téchnico, Lisbon, Portugal, who described the geological setting. ORTH $C222_1$. $a = 6.412$, $b = 13.563$, $c = 8.545$, $Z = 4$, $D = 3.21$. *AM 64:359(1979)*. Structure: *AM 69:120(1982)*. *33-284:* 3.61_2 3.38_{10} 3.00_3 2.79_2 2.63_2 2.19_2 1.879_1 1.696_1. Massive, coarse-grained. Yellow-orange, white streak, vitreous luster, translucent. Cleavage {010}, good, plus another at 90°. H = 4. G = 3.25. Nonfluorescent, no cathodoluminescence. Biaxial (−); XYZ= *cba*; $N_{x\,calc} = 1.568$, $N_y = 1.597$, $N_z = 1.600$; 2V = 33°. Found intergrown with fluorapatite, danasqueiraite along the edges of hydrothermal tin–tungsten veins at *Panasqueira mine, Beira Baixa, Portugal*, associated with wolfeite, topaz, quartz, muscovite, sphalerite, chalcopyrite, pyrrhotite, siderite, arsenopyrite, chlorite, althausite, and vivianite. HCWS

41.5.17.1 Leningradite $PbCu_3(VO_4)_2Cl_2$

Named in 1990 for the city where many volcanic sublimates have been studied and described. ORTH *Ibam*. $a = 8.988$, $b = 11.083$, $c = 9.360$, $Z = 4$, $D = 4.97$. *DANS 303:948(1990)*. PD: 5.55_5 3.49_3 3.42_{10} 3.24_6 2.76_9 2.55_7 2.36_7 1.847_5. Rhombic tablets or flakes to 0.3 mm and globules up to 0.6 mm in diameter with radial or tangential orientation of crystals. Crystals show {010}, {101}, {100}, {001} with prism interfacial angle 87.9°. Red, orange-red streak, vitreous luster, transparent. Cleavage {010}, perfect; brittle. H = $4\frac{1}{4}$. G = 4.81. Biaxial; XYZ = *bac*; N_x unknown (all grains are cleavage sections), $N_y = 2.29$, $N_z = 2.35$; 2V large; parallel or symmetric extinction; no dispersion; nonpleochroic. Found as a product of fumarolic activity at the *Great Tolbachik fissure eruption (1975–1976), Kamchatka, Russia*, associated with anglesite, hematite, lammerite, and tolbachite. HCWS *AM 76:1434(1991)*.

41.5.18.1 Arctite $Na_2Ca_4(PO_4)_3F$

Named in 1981 for the arctic region, where it was found. HEX-R $R\bar{3}m$. $a = 7.078$, $c = 41.203$, $Z = 6$, $D = 3.11$. *ZVMO 110:506(1981)*. Structure: *SPD 29:5(1984)*. *33-1228:* 13.8_2 3.54_2 3.43_3 3.06_3 2.80_3 2.75_{10} 2.72_2 1.769_2. Colorless, vitreous luster. Cleavage {001}, perfect. H ≈ 5. G = 3.13. Uniaxial (−); $N_O = 1.578$, $N_E = 1.577$ (anomalously biaxial). Analyses also show Ba, K, Fe, Zr, S, Si. Insoluble in H_2O, soluble in 1:20 HCl. Found in a drill core from the valley of the *Vuonnemi R., Khibina alkalic massif, Kola Penin., Russia*, associated with rasvumite, villiaumite, aegirine, and thenardite. HCWS *AM 67:621(1982)*.

TRIPLITE–TRIPLOIDITE GROUP

The general formula for the triplite–triploidite group minerals is

$$A_2(TO_4)Z$$

where

$A = Fe^{2+}, Mn^{2+}, Mg$
$T = P, As$
$Z = F, OH$

and crystallize in the monoclinic system but appear pseudoisometric.

The structure [*BM 104:677(1981)*] is analogous to that of titanite, and similar to that of alluaudite (38.2.3.6) and wyllieite (38.2.4.2). It is characterized by two types of chains, formed from distorted AO_6 octahedra which are connected into a three-dimensional framework by sharing vertices with TO_4 tetrahedra. One chain is parallel to b, while the chains of $[AO_4(OH)F]$ octahedra extend along the a axis.

The F/OH sites appear to have an average occupancy of 50% and could be disordered. However, OH/F ordering results in a doubling of the b unit cell axis and the different space group ($P2_1/c$) for triploidite compared with triplite ($I2/a$).

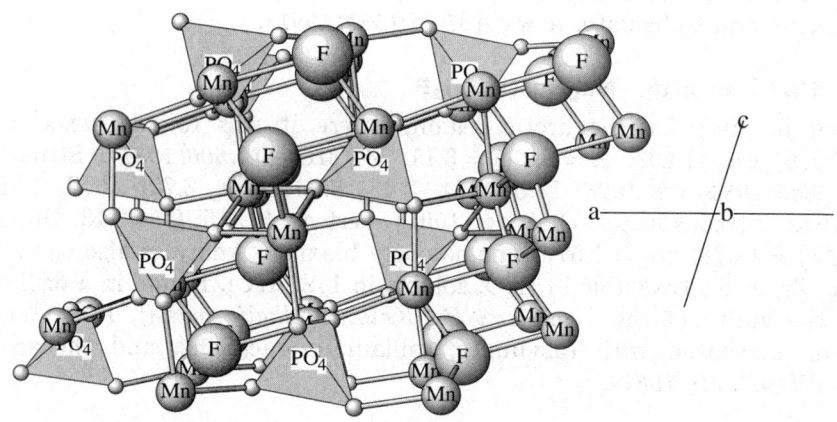

TRIPLITE: $Mn_2(PO_4)F$

A 'b' axis projection illustrates the two types of MnF_6 chains connected through PO_4 groups. One chain is parallel to 'b' the other along 'a' axis forms a tight framework,

41.6.1.2 Triplite

TRIPLITE SUBGROUP

Mineral	Formula	Space Group	a	b	c	β	
Zweiselite	$Fe_2^{2+}(PO_4)F$	$I2/a$	12.046	6.446	9.888	107.42	41.6.1.1
Triplite	$Mn_2^{2+}(PO_4)F$	$I2/a$	12.065	6.545	9.937	107.09	41.6.1.2
Magniotriplite	$(Mg,Fe^{2+},Mn^{2+})_2(PO_4)(F,OH)$	$I2/a$	12.035	6.432	9.799	108.12	41.6.1.3

TRIPLOIDITE SUBGROUP

Mineral	Formula	Space Group	a	b	c	β	
Wagnerite	$Mg_2(PO_4)F$	$P2_1/c$	11.945	12.717	9.70	108.18	41.6.2.1
Wolfeite	$Fe_2^{2+}(PO_4)(OH)$	$P2_1/a$	12.319	13.230	9.840	108.40	41.6.3.1
Triploidite	$Mn_2^{2+}(PO_4)(OH)$	$P2_1/a$	12.366	13.296	9.94	108.23	41.6.3.2
Sarkinite	$Mn_2^{2+}(AsO_4)(OH)$	$P2_1/a$	12.779	13.596	10.208	108.98	41.6.3.3

Note: Three distinct cleavages are found in the triplite subgroup and reflect the chain-composed framework; triploidite subgroup minerals show acicular or fibrous morphology.

41.6.1.1 Zweiselite $Fe_2^{2+}(PO_4)F$

Named in 1841 for the locality. Triplite group and subgroup. The Fe end member of the Fe^{2+}–Mn series with triplite (41.6.1.2). See also the Mg end member magniotriplite (41.6.1.3). MON $I2/a$. $a = 12.046$, $b = 6.446$, $c = 9.888$, $\beta = 107.42°$, $Z = 8$, $D = 4.00$. *MM 31:587(1957)*. 30-654: 3.60_5 3.38_4 3.20_7 3.03_9 2.86_{10} 2.59_3 2.12_4 1.986_4. See also *21-811*. Dark brown. See triplite (41.6.1.2) for physical properties. Biaxial (+); $N_x = 1.680$, $N_y = 1.689$, $N_z = 1.702$; $2V_{calc} = 80°$. Analyses also show Ca, Mn^{2+}, Mg, Ti. Light brown samples have higher Ca, lower indices of refraction. Found at Black Hills, SD; at *Rabenstein, Zweisel, Bavaria, Germany*; at Slavkov, Bohemia, and Cyrilov, western Moravia, Czech Republic. HCWS *DII:849*.

41.6.1.2 Triplite $Mn_2^{2+}(PO_4)F$

Named in 1813 from the Greek for *threefold*, probably in allusion to the three cleavages. Triplite group and subgroup. Forms a Mn–Fe^{2+} series with zweiselite (41.6.1.1) and a Mn–Mg series with wagnerite (41.6.2.1) and magniotriplite (41.6.1.3). MON $I2/a$. $a = 12.065$, $b = 6.545$, $c = 9.937$, $\beta = 107.09°$, $Z = 8$, $D = 3.61$. Structure: *ZK 130:1(1969)*. 25-1080: 3.69_2 3.48_2 3.31_4 3.05_{10} 2.90_3 2.88_5 2.62_1 2.05_6. Rough, undeveloped, massive crystals. Dark brown to chestnut-brown, reddish brown, to salmon-pink when the Mn content is high; becomes brown-black on alteration; white to brown streak; vitreous to resinous luster; subtranslucent to opaque. Cleavage {001}, good; {010}, fair; difficult to see in altered material; uneven to subconchoidal fracture. $H = 5-5\frac{1}{2}$. $G = 3.5-3.9$. Biaxial (+); $Y = b$; $Z \wedge c = -41°$; $N_x = 1.643$, $N_y = 1.647$, $N_z = 1.668$; $2V = 25°$ (Mica Lode, CO); $r > v$, moderate to strong; pleochroic in distinct shades of yellow-brown or reddish brown;

X > Y, Z. Indices decrease with increasing Mg, Ca. Soluble in acids. Fe^{2+} usually present as well as small amounts of Ca, Mg, and (OH) for F. Synthesized. A primary mineral in granite pegmatites which are phosphate-rich, associated with wolfeite, triphylite–lithiophilite, apatite, and tourmaline. Often weathers to vivianite, dufrenitelike minerals or manganese oxyhydroxides. Found in ME at Stoneham, Hebron, Mt. Apatite near Auburn; at Chatham, Branchville, and at Strickland quarry, Portland, CT; at Amelia, VA; at Elk Ridge mine, Custer Co., SD; near Aurum, Reagan district, White Pine Co., NV; at Mica Lode pegmatite, Royal Gorge, Fremont Co., and Dead Man's Gulch, Colorado Springs, El Paso Co., CO; in the Eureka district, Yavapai Co., AZ; at Pala, San Diego Co., CA. Found at Mangualda, Beira Prov., Portugal; at *Limoges near Chanteloube, Haute Vienne, France*; at Hagendorf, Kreuzberg, Pleystein, Germany; at Horrsjöberg in Värmland, Sweden, with lazulite, apatite, rutile, kyanite, and at Skrumpetorp; in Moravia at Cyrillhof, Wien, Moravia, Czech Republic; in the Sn–W deposit of Tigrinoye, Sikjote Alin Range, Ural Mts., Russia; at Korgal, Laghman, Afghanistan; in the Karibib area, Namibia; in the Lomagundi district and elsewhere in Zimbabwe; at Sierra de Zapata, Catamarca Prov., Calamuchite Dept., Cordoba, Argentina. HCWS *DII:849*, *AM 36:256(1951)*, *MM 31:587(1957)*.

41.6.1.3 Magniotriplite $(Mg,Fe^{2+},Mn^{2+})_2(PO_4)(F,OH)$

Named in 1951 for its relationship chemically and crystallographically to triplite. Triplite group and subgroup. In the Mg–Mn series with triplite (41.6.1.2) and wagnerite (41.6.2.1). MON $I2/a$. $a = 12.035$, $b = 6.432$, $c = 9.799$, $\beta = 108.12°$, $Z = 8$, $D = 3.68$. *BM 104:672(1981)*, Structure: *BM 104:677(1981)* (Mg ≈ Fe). F:OH=6:1 42-582: 3.59_5 3.24_5 3.16_7 3.02_8 2.86_{10} 2.79_6 2.73_2 2.098_3. See also *8-140*. Crystalline masses to 0.5 cm × 8 cm. Reddish brown, vitreous luster. Cleavage {001}, uneven fracture. $H = 4$. $G = 3.59$. Biaxial (+); $N_x = 1.648$, $N_y = 1.653$, $N_z = 1.664$; $2V = 40°$. Analyses also show minor Ca and Ti. Found at *Turkestan ridge, Kazakhstan*. HCWS *AM 37:359(1951)*.

41.6.2.1 Wagnerite $Mg_2(PO_4)F$

Named in 1821 for F. M. von Wagner (1768–1851), mining official, Munich, Germany. Triplite group and triploidite subgroup. See magniotriplite (41.6.1.3). MON $P2_1/c$. $a = 11.945$, $b = 12.717$, $c = 9.70$, $\beta = 108.18°$, $Z = 16$, $D = 3.291$. *MM 56:227(1992)*. Structure: *ZK 177:15(1986)*. 42-1330: 3.29_6 3.11_6 2.97_{10} 2.84_9 2.75_3 2.47_2 1.557_2 1.553_2. Prismatic [001] with prism planes vertically striated or vicinally rounded, crystals sometimes large and coarse; rarely tabular; lenticular masses to 4 cm × 8 cm. Yellow, grayish, flesh-red, greenish; vitreous to resinous luster, translucent to opaque in altered material. Cleavage {100}, {120}, imperfect; {001}, in traces; uneven or subconchoidal to splintery fracture. $H = 5-5\frac{1}{2}$. $G = 3.15$. Biaxial (+); $N_x = 1.585$, $N_y = 1.588$, $N_z = 1.601$; $2V = 51°$; $r < v$, weak; pleochroic, colorless to pale yellow. Soluble in acids. May contain Fe^{2+} and Ca in substitution for Mg. Synthesized. Found at Benson mine, west-central Adirondack highlands, NY, with isokite in magnetite–hematite deposits associated with sillimanite–garnet–pyroxene-rich paragneiss; at Quartzsite, AZ; at *Höllgraben, near Werfen, Salzburg, Austria*, associated with ferroan magnesite, lazulite, and chlorite; at Havredal, Porsgrund, Telemark, Norway; in lava of 1872 on Vesuvius, Italy. HCWS *DII:845*; *CM 10:919(1971)*, *12:346(1974)*.

41.6.3.1 Wolfeite $Fe_2^{2+}(PO_4)(OH)$

Named in 1949 for Caleb Wroe Wolfe (1908–1980), crystallographer, Boston University, MA. Triplite group and triploidite subgroup. The Fe end member in the Fe^{2+}–Mn series with triploidite (41.6.3.2). MON $P2_1/a$. $a = 12.319$, $b = 13.230$, $c = 9.840$, $\beta = 108.40°$, $Z = 16$, $D = 3.66$. *JPD 4:34(1989)*, *AM 34:692(1949)*. *41-1411*: 3.65_3 3.38_2 3.19_8 3.10_{10} 2.93_7 2.88_3 2.82_6 2.57_5. See also *5-612*. Massive, granular, or fibrous aggregates, elongated on [001] to 2.5 cm, showing {120}, {100}. Red-brown to clove-brown, rarely green; nearly white streak, vitreous to silky luster (fibers); transparent to translucent. Cleavage {100}, good; {010}, fair; {120}, {110}, poor; uneven fracture. H = $4\frac{1}{2}$–5. G = 3.83 (est.). Biaxial (+); X = b; $N_x = 1.741$, $N_y = 1.742$, $N_z = 1.746$ (Palermo); 2V moderate; faintly pleochroic; Z > X,Y; r > v, strong; marked dispersion of bisectrices. Soluble in acids. Minor amounts of Mn and Ca usually present, also Na, Li, and Zn. Reports of Fe^{3+} indicate oxidation of Fe^{2+} and mineral alteration. Found at *Palermo mine, North Groton, NH*, as the alteration product of triphylite; at Ross lode, Custer Co., SD; with triplite at Skumpetorp, Sweden; at Hagendorf-Süd, Bavaria, Germany; at Cyrillhof, Moravia, Czech Republic; at Lavra do Enio, Galileia, MG, Brazil. HCWS *DII:853*.

41.6.3.2 Triploidite $Mn_2^{2+}(PO_4)(OH)$

Named in 1878 from the Greek for *form* and its resemblance to triplite in physical characters and composition. Triplite group and tripoilite subgroup. The Mn end member of a Mn–Fe series with wolfeite, and the OH end member of the OH–F series with triplite. MON $P2_1/a$. $a = 12.366$, $b = 13.296$, $c = 9.94$, $\beta = 108.23°$, $Z = 16$, $D = 3.80$. *AM 34:692(1949)*, Structure: *ZK 131:1(1970)*. *26-1240*: 3.41_5 3.19_8 3.10_9 2.94_{10} 2.58_5 2.31_5 2.15_5 1.80_6. See also *26-1239*(calc). Habit and physical properties similar to Fe^{2+} end member wolfeite. Pinkish, wine-yellow to yellow-brown for material high in Mn, specific gravity, and indices of refraction lower for Mn end members. G = 3.66 (est.). Biaxial (+); X = b; $Z \wedge c = -4°$; $N_x = 1.725$, $N_y = 1.723$, $N_z = 1.730$ (Branchville, CT, Mn/Fe = 3.3:1); 2V moderate; r $\gg$ v. Both triploidite and wolfeite occur in granite pegmatites associated with triplite, lithiophilite, and triphilite. Found at *Branchville, Fairfield Co., CT*, associated with eosphorite, dickinsonite, and rhodochrosite in lithiophilite; at Ross lode, SD; as an alteration of triplite at Wien, Moravia, Czech Republic. HCWS *DII:853*.

41.6.3.3 Sarkinite $Mn_2^{2+}(AsO_4)(OH)$

Named in 1885 from the Greek for *made of flesh*, in allusion to the blood-red color. Triplite group and triploidite subgroup. The As analog of triploidite (41.6.3.2). MON $P2_1/a$. $a = 12.779$, $b = 13.596$, $c = 10.208$, $\beta = 108.98°$, $Z = 16$, $D = 4.20$. *MM 43:681(1969)*. *14-214:* 6.0_3 3.48_8 3.29_9 3.18_{10} 3.04_{10} 2.90_7 2.65_6 2.37_3. Crystals usually thick tabular {100}, elongate [010]; short prismatic [010] or shortened on b, tabular on {100}, {001}, often uneven with {001} striated [100]; rudely spherical or granular. Flesh-red to dark blood-red, rose-red, reddish yellow to yellow; rose-red to yellow streak; greasy luster. Cleavage {100}, distinct; subconchoidal to uneven fracture. $H = 4$–5. $G \approx 4.10$. Biaxial $(-)$; $X \wedge c = -54°$; $Y = b$; $N_x = 1.793$, $N_y = 1.807$, $N_z = 1.809$; $2V = 83°$; weakly pleochroic; $X > Z > Y$; no dispersion. Easily soluble in dilute acids. Very little P or Sb substitution for As, some Ca, Mg, and Fe^{2+} usually present on analysis. Found in NJ at Franklin, Sussex Co., and at Sterling Hill mine, Ogdensburg; at *Hartigen mine*(!), *Pajsberg, Sweden*, associated with native lead, bementite, brandtite, calcite, and barite; at the Sjö mine, Grythytte parish, Örebro, with magnetite, jacobsite, and tephroite; at Långban with hausmannite and hedyphane. HCWS *DII:855*, *AMG 4:499(1968)*.

41.6.4.1 Satterlyite $Fe_2^{2+}(PO_4)(OH)$

Named in 1978 for Jack Satterly (b.1906), geologist at Ontario Department of Mines, Canada. Dimorph of wolfeite, 41.6.3.1, see also holtedahlite 41.6.4.2. HEX-R $P\bar{3}m$, $P31m$, $P312$. $a = 11.361$, $c = 5.041$. $Z = 6$, $D = 3.60$. *CM 16:411(1978)*. *35-622* 4.49_4 3.77_1 3.52_5 2.99_3 2.84_6 2.47_{10} 1.885_4 1.640_3. Grains up to $1 \times 1 \times 40$ mm elongated parallel to [0001], in radiating aggregates or nodules up to 10 cm diameter. Pale yellow to pale brown, streak pale yellow, vitreous. No cleavage. $H = 4\frac{1}{2}$–5. $G = 3.68$. Non-fluorescent. Uniaxial $(-)$; $N_O = 1.721$, $N_E = 1.719$ dichroic when thick O=pale yellow, E=brownish-yellow, absorption $E > O$. Some Fe^{3+} detected on analysis, as are Na, Mg, Mn, trace of Si. Found in the shales along the *Big Fish R., NE YT, Canada* associated with quartz, pyrite, wolfeite, maricite *AM 64:657(1979)*.

41.6.4.2 Holtedahlite $Mg_2PO_4(OH)$

Named in 1979 for Olaf Holtedahl (1885–1975), geologist, University of Oslo, Norway. Dimorph of althausite (41.6.5.1). HEX-R $P321$, $P3m1$, or $P\bar{3}m1$. $a = 11.186$, $c = 4.977$, $Z = 6$, $D = 2.936$. *LIT 12:283(1979)*. Structure: *MP 40:91(1989)*. *33-879:* 3.72_9 3.48_5 3.23_3 2.95_2 2.80_3 2.44_{10} 2.18_3 1.859_3. Patches up to 0.5 cm $\times$ 1 cm. Colorless, vitreous luster. No cleavage, uneven fracture. $H = 4\frac{1}{2}$–5. $G = 2.94$. Uniaxial $(-)$; $N_O = 1.599$, $N_E = 1.597$. Analyses also show CO_2, minor Na, Mn, and F. Occurs in serpentine–magnesite deposits of Tingelstadtjen at *Modum, Norway*, associated with apatite, althausite, serpentine, and talc, cut by veins of szaibelyite. CS *AM 65:809(1979)*.

41.6.5.1 Althausite $Mg_2(PO_4)(OH,O,F)$

Named in 1975 for Egon Althaus (b.1933), mineralogist, Karlsruhe University, Germany. ORTH $Pnma$. $a = 9.258$, $b = 6.054$, $c = 14.383$, $Z = 8$, $D = 2.91$. LIT 8:215(1975). Structure: AM 65:488(1980). 29-869: 3.59_{10} 3.42_4 3.32_9 3.02_8 2.89_4 2.79_6 2.70_3 2.64_6. Crystals elongated on c, flattened on b show {010}, {110}, {131}. Light gray, translucent. Cleavage {001}, perfect; {101}, distinct. H = $3\frac{1}{2}$ in c direction, 4 in a. G = 2.97. IR, TGA. Synthesis studies suggest complete series to the F end member. Biaxial (+); XYZ= cba; $N_x = 1.588$, $N_y = 1.592$, $N_z = 1.598$; 2V = 70°. Found in the serpentine–magnesite deposits of *Tingelstadtjen, at Modum, Norway*, and at Overtjern associated with apatite, talc, magnesite, and enstatite. HCWS AM 61:502(1976).

41.6.6.1 Olivenite $Cu_2(AsO_4)(OH)$

Named in 1824 in allusion to the color. ORTH $Pnnm$. $a = 8.615$, $b = 8.240$, $c = 5.953$, $Z = 4$, $D = 4.45$. Structure: AC(B) 34:715(1978), 33:2628(1977). 42-1353: 5.94_6 4.89_8 4.82_6 4.20_6 2.98_{10} 2.66_6 2.46_5 2.41_4. See also 41-1354. Variable habit: elongated [100], short prismatic to acicular [001], tabular on {011}, {100}, {001}; globular, reniform with fibrous structure, fibers straight and divergent, rarely irrregular, curved lamellar; massive, granular to earthy. Shades of olive-green, brownish green, straw-yellow, or grayish white; olive-green to brown streak; adamantine luster; translucent to opaque. Cleavage {011}, {110}, indistinct; conchoidal to irregular fracture. H = 3. G = 3.9–4.5. Biaxial (+) [or (−)]; $N_x = 1.772$, $N_y = 1.810$, $N_z = 1.863$ (Tintic); 2V ≈ 90°; (+): r < v, strong; (−): r > v, strong. Soluble in acids and ammonia. Zn and Pb substitute for Cu. There is probably a complete solid-solution series (AsO_4–PO_4) to libethenite (41.6.6.2). A secondary mineral found in the oxidized zone of many ore deposits. Occurs in the Blackbird district, Lemhi Co., ID; in the Tintic district at American Eagle, Mammoth mines, UT, associated with clinoclase, tyrolite, and conichalcite; at New Cornelia mine, Ajo, 79 mine, Gila Co., Grand Reef mine, Graham Co., Tiger mine, Mammoth–St. Anthony, AZ; at Majuba Hill, Pershing Co., NV, with cornetite; at Yaquino Head, Newport, OR; at *Cartharrack mine, Gwennap, Cornwall*, and many other mines as crystals or fibrous aggregates, also at Alston Moor, Cumberland, and Tavistock, Devonshire, *England*; Grube Clara, Rankachtal, Germany; at Kamarezu mine, Attica, Greece; at Tsumeb, Namibia, where Zn and Fe varieties were described; at Callington, Preamimma mine, SA, and Balilo Cu mine, Ashburton Downs, WA, Australia; at Chuquicamata, Antofagasta, Chile. HCWS DII:859, AM 36:484(1951).

41.6.6.2 Libethenite $Cu_2(PO_4)(OH)$

Named in 1823 for the locality. ORTH $Pnnm$. $a = 8.062$, $b = 8.384$, $c = 5.881$, $Z = 4$, $D = 3.972$. NJMA 134:147(1979). Structure: CM 16:153(1978). 36-404: 5.81_8 4.82_{10} 4.76_7 3.72_5 2.91_8 2.64_4 2.63_6 2.41_3. Short prismatic [001], slightly elongated [100], equant; crystals usually have composite structure. {110} vertically grooved or striated, {011} striated parallel to edge with

{111}. Light to dark olive-green, deep green to black-green; vitreous luster on crystal faces, greasy on fracture surfaces; translucent. Cleavage {100}, {010}, indistinct; conchoidal, uneven fracture. H = 4. G = 3.97. Biaxial (−); $N_x = 1.701$, $N_y = 1.743$, $N_z = 1.787$ (Libethen); 2V ≈ 90°; r > v, strong; faintly pleochroic. X, pale yellowish blue; Z, pale greenish blue. Easily soluble in acids and ammonia. Synthesized. Secondary mineral in oxidized zone of Cu ore deposits with malachite, azurite, and pyromorphite. Found at Perkiomen mine, Montgomery Co., PA; at Yerrington, Lyon Co., NV; in AZ at the Clifton–Morenci district, Greenlee Co., and Silver Hill mine, Copper Creek, Pima Co., Silver Bell mine, Ray mine, Pinal Co.; at Wheal Phoenix, Cornwall, England; at Montebras, Plateau Centrale, France; at *Libethen, near Neusohl, Stredoslovensky Kraj, Czech Republic*; at Nizhne Taglisk and Wissokaya Gora, Ural Mts., Russia; at Broken Hill mine, Rokana mine, Kabua, Zambia; at West Bogen mine, Tottenham, NSW, Australia; at Chuiquicamata, and the Mercedes mine, Coquimbo, Chile. HCWS *DII:862, BM 79:219(1956)*.

41.6.6.3 Adamite $Zn_2(AsO_4)(OH)$

Named in 1866 for Gilbert Joseph Adam (1795–1881), French mineralogist, who supplied the first specimens. ORTH *Pnnm*. $a = 8.306$, $b = 8.524$, $c = 6.043$, $Z = 4$, $D = 4.45$. *AC(B) 34:715(1979)*. Structure: *AM 61:979(1976)*. *39-1354*: 4.89_7 4.25_4 2.98_{10} 2.70_7 2.47_7 2.46_6 2.45_6 2.42_7. See also *6-536*. Radiating crusts, fan-shaped rosettes, crystals elongated parallel to [010] to 0.8 mm, showing {101}, {120}, {110}, {210}, {010}; [010] striated parallel to [101]. Light to honey- and brownish yellow, pale green, rarely colorless, white; Cu-rich bright green; Co-rich violet to rose; vitreous luster; transparent to translucent. Cleavage {101}, good; {010}, poor; brittle. H = $3\frac{1}{2}$. G = 4.43. May fluoresce yellow-green. Biaxial (+); XYZ= *acb*; $N_x = 1.722$, $N_y = 1.742$, $N_z = 1.763$ (Mapimi); 2V = 88°; r < v, strong. Indices and color vary widely with substitutions of Cu and Co for Zn. Easily soluble in 1:10 acids. Synthesized. Secondary mineral in oxidized portions of Zn- and As-rich primary deposits. Found at Franklin, Sussex Co., NJ; at Grandview mine, AZ; at western Utah mine, Gold Hill, Clarke Mts., Tooele Co., UT; at Zinc mine, Mohawk, CA: at Ojuela mine(!), Mapimi, DUR, Mexico, associated with scorodite, mimetite, hemimorphite, calcite and "limonite"; at Cap Garonne, near Hyères, Var, France, where cobaltian and cuproan varieties form crusts in sandstones; at Reichenbach near Lahr, Black Forest, Germany; at Mt. Valerio near Campiglia Marittima, Tuscany, Italy; at the Kamareza mine, Attica, and at Larium, Greece, filling drusy cavities in cellular smithsonite; on the Island of Thasis, Turkey; at Tsumeb(!), Namibia; at *Chanarcillo, Chile*. HCWS *DII:864*.

41.6.6.4 Eveite $Mn_2(AsO_4)(OH)$

Named in 1967 for its structural and morphological similarities to adamite. Dimorph of sarkinite (41.6.3.3). ORTH *Pnnm*. $a = 8.57$, $b = 8.77$, $c = 6.27$, $Z = 4$, $D = 3.73$. *AMG 4:473(1968)*. Structure: *AM 53:1841(1968)*. *22-1166*:

6.10_7 5.09_8 4.39_{10} 3.14_4 3.06_9 2.80_5 2.55_4 2.53_6. Tabular, sheaflike crystals, crusts. Apple-green, white streak, translucent. Cleavage {101}, fair. H = 4. G = 3.76. Biaxial (+); $N_x = 1.700$, $N_y = 1.715$, $N_z = 1.732$; 2V = 65°; r > v, medium. Soluble in cold 1:1 HCl. The low-pressure form of $Mn_2(AsO_4)(OH)$. A secondary mineral found encrusting cavities and fractures in deposits containing Fe–Mn oxides and carbonates; associated minerals hausmannite, dolomite, and calcite. Found at Buckwheat open pit, Franklin, Sussex Co., NJ; at *Långban, Värmland, Sweden*. HCWS *AM 55:319(1970)*.

41.6.7.1 Tarbuttite $Zn_2(PO_4)(OH)$

Named in 1907 for Percy Coventry Tarbutt, director of the Broken Hill Exploration Co., Zambia, who collected the first specimens. TRIC $P\bar{1}$. $a = 5.552$, $b = 5.700$, $c = 6.471$, $\alpha = 102.7°$, $\beta = 102.8°$, $\gamma = 86.9°$, Z = 2, D = 4.11. *MM 39:684(1974)*. Structure: *DANS 282:314(1985)*. *36-410*: 6.17_{10} 3.70_2 3.09_1 2.98_1 2.97_1 2.88_2 2.78_3 2.06_3. Equant to short prismatic [001] crystals, rounded, deeply striated; sheaflike or compact aggregates up to 5 mm; crusts. Pale shades of yellow, brown, red, green, colorless; white streak; vitreous luster, pearly on cleavage; translucent. H = $3\frac{1}{2}$–4. G = 4.12. Biaxial (−); $N_x = 1.660$, $N_y = 1.705$, $N_z = 1.713$; 2V = 50°; strong dispersion of bisectrices. Found at *Kabwe, Broken Hill mines, northwestern Zambia*, with pyromorphite, hopeite, descloizite, vanadinite, hemimorphite, cerussite, smithsonite, and "limonite"; at Reaphook Hill, SA, Australia. HCWS *DII:869, ZK 123:321(1966)*.

41.6.7.2 Paradamite $Zn_2(AsO_4)(OH)$

Named in 1956 as the dimorph to adamite; the P analog of tarbuttite (41.6.7.1). TRIC $P\bar{1}$. $a = 5.638$, $b = 5.827$, $c = 6.692$, $\alpha = 103.25°$, $\beta = 104.37°$, $\gamma = 87.92°$, Z = 2, D = 4.595. *SCI 123:1039(1956)*. Structure: *AC(B) 35:720(1979)*. *33-1469*: 6.32_7 5.46_5 3.70_{10} 2.99_{10} 2.84_{10} 2.57_5 2.51_5 2.48_7. See also *12-223*. Sheaflike aggregates, round striated equant crystals up to 5 mm. Pale yellow, vitreous luster, transparent. Cleavage {010}, perfect. G = 4.55. Biaxial (−); $N_x = 1.726$, $N_y = 1.771$, $N_z = 1.780$; 2V = 50°. Found with legrandite, plattnerite, and murdochite at *Mapimi, DGO, Mexico*. HCWS *AM 41:958(1957), 65:353(1980)*.

41.6.8.1 Augelite $Al_2(PO_4)(OH)_3$

Named in 1868 from the Greek for *luster*. MON $C2/m$. $a = 13.124$, $b = 7.988$, $c = 5.088$, $\beta = 112.25°$, Z = 4, D = 2.702. *SMPM 41:407(1961)*. Structure: *AM 53:1096(1968)*. Al occurs in two different octahedral sites: $Al(OH)_6$ linked through sharing OH edges, and AlO_6 that link to PO_4 through sharing of corner O atoms. *34-151*: 4.71_3 4.17_3 4.01_3 3.52_6 3.49_3 3.34_{10} 3.16_1 2.49_1. Thick tabular {001}; prismatic to acicular [001]; thin triangular plates flattened {110}; massive. Dominant forms: {110} striated [001], {001}, {$\bar{2}$10}. Colorless to white, yellowish, pale rose; white streak; vitreous luster, pearly on {110} cleavage; transparent. Cleavage {110}, perfect; {$\bar{2}$10}, good; {001},

{$\bar{1}01$}, imperfect; uneven fracture; brittle. H = $4\frac{1}{2}$–5. G = 2.696. Biaxial (+); $N_x = 1.574$, $N_y = 1.576$, $N_z = 1.588$; 2V = 50.9°; no dispersion. Slowly soluble in hot 10 M HCl. Found at the pegmatites in Newry, ME; G.E. Smith mine, Newport, Palermo No. 1 mine, North Groton, NH; Hugo pegmatite, Keystone, SD; at Champion mine, White Mts., Mono Co., CA; at Big Fish R., YT, Canada; at *Westanå iron mine, Näsum, Kristianstad, Sweden*; from a quartz vein at Chernaya Mt. and at Starubba Mt., polar Ural Mts., Russia; at Buranga pegmatite, Rwanda; at Mbale, Uganda, with amblygonite; at Socavon mine, Oruro, with arsenopyrite, stannite, pyrite, and quartz, and at Tatasi and Portugalete, Potosi Prov., Bolivia. HCWS *DII:871*.

41.6.9.1 Arsenobismite $Bi_2(AsO_4)(OH)_3$

Named in 1916 for the composition. An inadequately described mineral. ORTH 7-358: 6.06_9 3.72_6 3.11_{10} 3.00_6 2.86_5 2.28_6 1.997_7 1.843_8. Cryptocrystalline aggregates, pulverulent, friable, ocherous masses. Yellowish green, yellowish brown. H $\approx$ 3. G $\approx$ 5.7. N > 2.04. Found in the oxidized ore at the 200-m level in *Mammoth mine, Tintic district, UT*; with pucherite at Schneeberg, Saxony, Germany; with bindheimite and bismutite at Tazna, Bolivia. HCWS *DII:907, AM 28:536(1943)*.

41.6.10.1 Angelellite $Fe_4^{3+}(AsO_4)_2O_3$

Named in 1959 for Victorio Angelelli (b.1908), mining geologist and director of the Geological Survey of Argentina. TRIC $P\bar{1}$. $a = 6.461$, $b = 6.594$, $c = 5.036$, $\alpha = 106.21°$, $\beta = 98.35°$, $\gamma = 108.82°$, Z = 1, D = 4.762. *NJMM:152(1959)*, Structure: NJMM 132:91(1978). 13-121: 5.29_3 3.15_{10} 3.00_7 2.96_4 2.955_4 2.86_5 2.49_5 2.07_5. Globular aggregates and crystals usually tabular with {001} dominant, other forms present {100} {02$\bar{1}$}, {10$\bar{1}$}, {03$\bar{1}$}, {$\bar{1}\bar{1}1$}, {101}. Blackish brown, red-brown streak, adamantine to submetallic luster, translucent. Cleavage {001}, good; conchoidal fracture; brittle. H = $5\frac{1}{2}$. G = 4.95. Biaxial (+); $N_x = 2.13$, $N_y = 2.2$, $N_z = 2.40$; Z > X; strongly pleochroic, blood-red to red-brown; strongly anisotropic in reflected light. Some Sb on analysis. Found intergrown with hematite at *Cerro Pululus tin mine, Jujuy, Argentina*, possibly precipitated from fumarolic vapors. HCWS *AM 44:1322(1959)*.

41.6.11.1 Spodiosite $Ca_2(PO_4)F$

Named in 1872 from the Greek for *ash gray*, in allusion to the color. Inadequately described mineral. ORTH. Crystals flattened {010}, elongated [001]. Ash gray to brown, white streak, dull porcelainlike luster, vitreous. Cleavage. {010}, distinct; [001], indistinct. Biaxial (+); $N_x = 1.663$, $N_y = 1.674$, $N_z = 1.699$; $2V_{calc} = 69°$; r > v, strong. Soluble in acids. Found at the *Nyttsta Kran mine, Filipstad, Värmland, Sweden*, and at Nordmark with serpentine, chondrodite, magnetite, and calcite. HCWS *DII:848*.

41.7.1.1 Palermoite (Sr,Ca)(Li,Na)$_2$Al$_4$(PO$_4$)$_4$(OH)$_4$

Named in 1952 for the locality. ORTH *Imcb*. $a = 11.556$, $b = 15.84$, $c = 7.315$, $Z = 4$, $D = 3.26$. Structure: *AM 60:460(1975)*. Octahedral AlO$_6$ dimers form chains parallel to the a axis and connect into slabs through tetrahedral PO$_4$ to form the group [Al$_4^{3+}$(OH)$_4$(PO$_4$)$_4$] parallel to {010}. The other cations occupy openings in the structure: Li in fourfold and Sr in eightfold coordination. *18-950:* 4.36$_6$ 3.32$_5$ 3.13$_6$ 3.09$_{10}$ 2.91$_5$ 2.60$_5$ 2.44$_5$ 1.659$_5$. Fibrous aggregates; occasional small crystals show {001}, {010}, {011}, {130}, {110}, vertically striated prisms. Colorless to white; white streak; vitreous, subadamantine luster. Fibrous to subconchoidal fracture, brittle. $H = 5$. $G = 3.22$. Fluoresces white in X-ray beam. Biaxial (–); XYZ $= cab$; $c =$ elongation; $N_x = 1.627$, $N_y = 1.642$, $N_z = 1.644$; $2V \approx 20°$; $r < v$, moderate. Found as small crystals in cavities at the *Palermo No. 1 mine, North Groton, NH*. HCWS *AM 38:354(1953), 50:777(1965)*.

41.7.1.2 Bertossaite Li$_2$CaAl$_4$(PO$_4$)$_4$(OH)$_4$

Named in 1966 for Antonio Bertossa, director of the Geological Survey of Rwanda. The Ca analog of palermoite (41.7.1.1). ORTH *Imaa* or *Iaa2*. $a = 11.48$, $b = 15.73$, $c = 7.23$, $Z = 4$, $D = 3.11$. *CM 8:668(1966). 41-1450:* 4.63$_6$ 3.30$_6$ 3.10$_8$ 3.06$_{10}$ 2.88$_6$ 2.58$_4$ 2.41$_6$ 1.592$_3$. Light pink; vitreous, glassy luster. Cleavage {100}, good; uneven to subconchoidal fracture. $H = 6$. $G = 3.10$. Biaxial (–); X $= a$, Y $= c$; $N_x = 1.624$, $N_y = 1.636$, $N_z = 1.642$; $2V_{calc} = 53°$; $r < v$, moderate. Dissolves slowly in HNO$_3$. Found at the *Buranga lithium pegmatite, Gatumba, Rwanda*. HCWS

41.7.2.1 Arrojadite KNa$_4$CaFe$_{14}^{2+}$Al(PO$_4$)$_{12}$(OH,F)$_2$

Named in 1925 for Miguel Arrojado Ribeiro Lisbôa (1872–1932), Brazilian geologist. The Fe^{2+} end member in the Fe^{2+}–Mn^{2+} series with dickinsonite (41.7.2.2). MON $C2/m$. $a = 24.730$, $b = 10.057$, $c = 16.526$, $\beta = 105.78°$, $Z = 4$, $D = 3.538$. Structure: A complex structure of multiple cation coordination polyhedra, some partially occupied or disordered and containing Na, Ca, K with six distinct PO$_4$ group sites. *AM 66:1034(1981). 34-149:* 5.93$_2$ 5.55$_2$ 5.03$_2$ 3.22$_5$ 3.04$_{10}$ 2.77$_3$ 2.71$_9$ 2.55$_3$. Large cleavable masses. Dark green, vitreous luster, translucent. Cleavage {001}, good; {201}, poor; uneven to subconchoidal fracture. $H = 5$. $G = 3.55$. DTA. Biaxial (–); X $= b$; Y $\wedge c = 21.5°$; $N_x = 1.664$, $N_y = 1.670$, $N_z = 1.675$; $2V = 80°$; $r < v$; pleochroic. X, colorless; Y, pale green; Z, pale yellow-green. The most chemically complex of primary pegmatitic phosphates, K, Ca, and Ba, may substitute for Na, although there may also be vacancies and disorder at these sites; Mn^{2+}, Mg, Al for Fe^{2+}, some Fe^{3+}, and H$_2$O are usually reported. Found in high-temperature ($\approx 800°$) granitic pegmatites at Palermo No. 1 and No. 2 mines, Nancy No. 2, Rice mine, NH; at Victory pegmatite, Custer Co., and as large masses in the pegmatite at *Nickel Plate mine, Keystone district, Pennington Co., SD*, associated with graftonite, cassiterite, spodumene, beryl, and muscovite; at Hess

R., YT, Canada; at Serra Branca, Picui, Paraiba, Brazil. HCWS *DII:679*, *AM 35:59(1950)*, *MM 43:227(1979)*.

41.7.2.2 Dickinsonite $KNa_4AlMn^{2+}_{14}(PO_4)_{12}(OH,F)_2$

Named in 1878 for William Dickinson of Redding, CT, who collected the rare minerals in this locality. The Mn end member of the $Mn^{2+}-Fe^{2+}$ series with arrojadite (41.7.2.1). MON $C2/m$. $a = 24.940$, $b = 10.131$, $c = 16.722$, $\beta = 105.60°$, $Z = 4$, $D = 3.426$. Structure: *AM 66:1034(1981)*. PD: 7.73_2 5.98_3 5.58_2 5.06_3 3.43_3 3.23_5 3.06_{10} 2.72_7. Tabular {001}, often pseudorhombohedral with triangular striations on {001}; micaceous; curved lamellar, radiated stellate, disseminated scales. Olive-green, grass-green, yellowish to brownish green; nearly white streak; transparent to translucent. Cleavage {001}, perfect, micaceous; uneven fracture; very brittle. $H = 3\frac{1}{2}-4$. $G = 3.41$. DTA shows inversion to alluaudite $\approx 500°$. Biaxial (+); $X = b$; $Y \wedge c = 15°$; $N_x = 1.658$, $N_y = 1.662$, $N_z = 1.671$ (Branchville); 2V moderate; $r > v$, strong; pleochroic. X, pale olive-green; Y, paler olive-green; Z, very pale yellowish green. Soluble in acids. Substitutions, especially Fe^{2+} for Mn^{2+}, common. Found at Nevel tourmaline mine, Newry, and Berry mine, Poland, ME; at *Branchville, Fairfield Co., CT*, associated with eosphorite, triploidite, lithiophilite, rhodochrosite, reddinggite, and fairfieldite. HCWS *DII:717*, *AM 50:1647(1965)*.

APATITE GROUP

The apatite minerals conform to the general formula

$$A_5(TO_4)_3Z$$

where

A = Ca, Sr, Pb, Ba

T = P, As, or V in the traditional designation, but also Si, S, and possibly CO_3

Z = OH, F, Cl, and possibly CO_3

and crystallize with hexagonal or pseudohexagonal monoclinic symmetry.

The apatite minerals are very common, and the structure, known since 1930 [*ZK(A) 75:323(1930)*, *75:387(1930)*; *AM 74:870(1989)*, *76:1857(1991)*], accommodates an extraordinarily wide range of both cations and anions. In fact, half the elements in the periodic table have been shown to occur in the apatites.

The dominant characteristic of the structure is the hexagonal disposition of tetrahedral groups (AsO_4, VO_4, or PO_4) that create the 6_3 screw axis, and the channels parallel to the unique c-axis direction in the crystals where the Z atoms are located.

Apatite Group

There are two distinct cation sites, one connected to the tetrahedral backbone oxygens and the channels and the other related to a trigonal axis in the center of the cell. The ratio of cations in these two sites is A_3/A_2, or more appropriately for the unit cell contents and $P6_3/m$ space group, A_6/A_4. The cation sites can be ordered or disordered; that is, a single element can fill both sites, or both sites can be randomly occupied by a variety of cationic species, or be partially occupied, or vacant. The variations of possible coordination number and size of the two sites means that not only Ca, but many other elements, such as Pb^{2+}, Na, REE, U, and Th, may be accommodated by a general expansion of the unit cell. The distribution and occupancy of specific elements in the two distinct cation sites allows classification of the apatites into a number of subgroups.

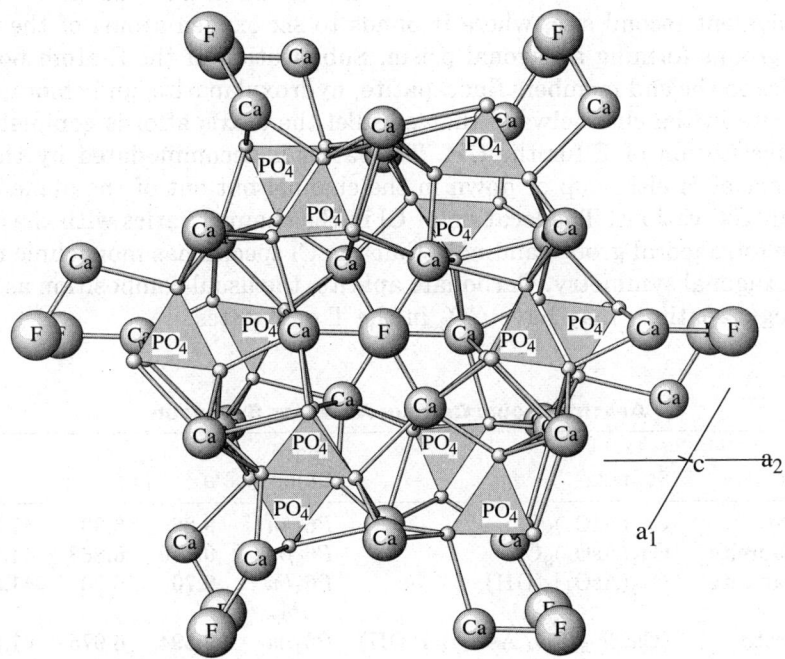

FLUORAPATITE: $Ca_5(PO_4)_3(OH)_3$

This c axis projection shows the 6_3 screw axis of fluorapatite and the channels parallel to the unique 'c' direction where the F atoms are located. There are two distinct cation sites, one connected to five oxygens from (PO_4) groups and the F in the channels, the other in octahedral coordinationn with tetrahedral backbone oxygens away from the channels (at the center of the unit cell). The variety displayed by group members show that the cation sites can be ordered or disordered, that, a single element can fill both sites, or both sites can be randomly occupied with a variety of cations, be partially occupied or vacant.

APATITE GROUP: CALCIUM PHOSPHATE SUBGROUP

Mineral	Formula	Space Group	a	c	
Fluorapatite	$Ca_5(PO_4)_3F$	$P6_3/m$	9.367	6.884	41.8.1.1
Chlorapatite	$Ca_5(PO_4)_3Cl$	$P6_3/m$	9.598	6.776	41.8.1.2
Hydroxylapatite	$Ca_5(PO_4)_3(OH)$	$P6_3/m$	9.418	6.875	41.8.1.3
Carbonate-fluorapatite	$Ca_5(PO_4,CO_3)_3F$	$P6_3/m$,	9.343	6.887	41.8.1.4
Carbonate-Hydroxylapatite*	$Ca_5(PO_4,CO_3)_3(OH)$	$P6_3/m$	9.309	6.927	41.8.1.5

Note:*Synthetic.

The calcium phosphate apatites have Ca atoms in a trigonal array with each Ca in sixfold coordination: with five oxygens from PO_4 groups and one with the F, OH, or Cl located in the channels. Ca is also the cation in the nonequivalent second site, where it bonds to six oxygen atoms of the tetrahedral groups forming a trigonal prism. Substitution of the Z-atom position gives rise to the end members fluorapatite, hydroxylapatite, and chlorapatite. The Z site in the channelways that parallel the c axis affords complete and easy substitution of F for the OH. The latter is accommodated by the displacement of H either up or down in the channel but out of the plane of the trigonal (Ca) cations. The location of Cl in the channel varies with the cation and the tetrahedral group, and note that the Cl species has monoclinic rather than hexagonal symmetry. Carbonate apatite, the usual composition ascribed to biologic apatites, may have CO_3 in the T or Z sistes.

APATITE GROUP: CALCIUM ARSENATE SUBGROUP

Mineral	Formula	Space Group	a	c	
Svabite	$Ca_5(AsO_4)_3F$	$P6_3/m$	9.80	6.90	41.8.3.1
Turneaureite	$Ca_5(AsO_4)_3Cl$	$P6_3/m$	9.810	6.868	41.8.3.2
Johnbaumite	$Ca_5(AsO_4)_3(OH)$	$P6_3/m$, $P6_3$	9.70	6.93	41.8.3.3
Fermorite	$(Ca,Sr)_5(PO_4,AsO_4)_3(F,OH)$	$P6_3/m$	9.594	6.975	41.8.3.4

APATITE GROUP: STRONTIUM CONTAINING SUBGROUP

Mineral	Formula	Space Group	a	c	
Belovite	$(Sr,Ce,Na,Ca)_5(PO_4)_3(OH)$	$P6_3/m$	9.692	7.201	41.8.1.6
Strontium-apatite	$(Sr,Ca)_5(PO_4)_3(OH)$	$P6_3/m$	9.565	7.111	41.8.1.7

Apatite Group: Lead-Containing Subgroup

Mineral	Formula	Space Group	a	c	
Hedyphane	$Ca_2Pb_3(AsO_4)_3Cl$	$P6_3/m$	10.140	7.185	41.8.2.1
Pyromorphite	$Pb_5(PO_4)_3Cl$	$P6_3/m$	9.976	7.351	41.8.4.1
Mimetite	$Pb_5(AsO_4)_3Cl$	$P6_3/m$	10.212	7.419	41.8.4.2
Vanadinite	$Pb_5(VO_4)_3Cl$	$P6_3/m$	10.317	7.338	41.8.4.3
Clinomimetite†	$Pb_5(AsO_4)_3Cl$	$P2_1/b$	10.180	7.46	41.8.6.1
Finnemanite*	$Pb_5(As^{3+}O_3,\square)(PO_4)_3Cl$	$P6_3/m$			

†Monoclinic polymorph of mimetite, but strongly pseudohexagonal.
*Listed chemically as an arsenite-phosphate, treated under class 43; related structurally and chemically to the apatite group.

The subset pyromorphite, mimetite, and vanadinite are respectively the phosphate, arsenate, and vanadate end members that contain A = Pb and Z = Cl. They are found in the oxidized portions of lead deposits, and exhibit little solid solution toward the calcium phosphate apatites. That is, calcium is not found to any great extent substituting in pyromorphite, nor is much lead accommodated in the calcium phosphate apatites despite the similarities in composition, identical space group, and crystal habit. Hedyphane is an intermediate, so to speak, in the Ca–Pb cation solid-solution range. It has been shown to be an ordered structure with one cation site preferentially occupied by Ca, with Pb in the other, but there is a miscibility gap at the high-Pb end of the Pb–Ca series.

Apatite Series: Barium-Containing Subgroup

Mineral	Formula	Space Group	a	c	
Morelandite	$(Ba,Ca,Pb)_5(AsO_4, PO_4)_3Cl$	$P6_3/m$	10.169	7.315	41.8.5.1
Alforsite	$Ba_5(PO_4)_3Cl$	$P6_3/m$	10.75	7.64	41.8.5.2

Phase Relations. Several of the chemical systems mentioned above have been investigated for their interesting and useful physical and chemical properties [J. R. van Wazer, *Phosphorus and Its Compounds*, John Wiley & Sons, New York, 1966; *AJS 273:545(1973)*]. Many exotic end members of the apatite group minerals have been synthesized and the ranges, and limits for cationic- and anionic-site solid solutions are now well known. The composition range of the natural apatites, though large, has some intriguing limits, and not only because of the geochemical availability of the elements. The apatite structure forms with Ca and PO_4 and AsO_4, but a naturally occurring $Ca_5(VO_4)_3(Z)$ apatite has not been described. Svabite, $Ca_5(AsO_4)_3Cl$, the arsenate analog of chlorapatite, contains very low PO_4, and the inverse is also true: most chlorapatites contain virtually no AsO_4.

The Pb-containing apatite subgroup members appear to prefer Cl in the Z position, with only trace amounts of F or OH. Fluorapatite and hydroxylapatite end members of the calcium phosphate apatites are very common, but chlorapatite is rare and found only in metamorphic locations, despite the availability of Cl in seawater or human serum, where precipitation of apatite dominates.

Other Structurally Related Minerals. Cesanite, a sulfate with the apatite structure, is described with the sulfates (30.3.3.1). Minerals with composite anionic groups (i.e., containing PO_4 and SO_4 or SiO_4) are described under the dominant anion group: for example, under silicates britholite-(Ce) is described. *DII:878;* W.R. Deer, R. Howie, and J. Zussman, *Rock Forming Minerals*, Vol. V, *Nonsilicates*, John Wiley & Sons, New York, 1966, pp. 323–338; *MM 57:709(1993);* J. C. Elliott, *Structure and Chemistry of the Apatites and Other Calcium Phosphates*, Elsevier, New York, 1994.

41.8.1.1 Fluorapatite Ca$_5$(PO$_4$)$_3$F

Apatite was first named in 1788 from the Greek for *to deceive*, as members of this subset of the apatite group resemble many other common minerals (aquamarine, amethyst, olivine, fluorite, etc.) in color and habit. Fluorapatite, named in 1860 to distinguish the composition, is the F-containing end member, which forms an isomorphous series with hydroxylapatite. Apatite group. Chlorapatite, the Cl end member, does not form a continuous series. See the discussion under chlorapatite (41.8.1.2). See also carbonate-fluorapatite

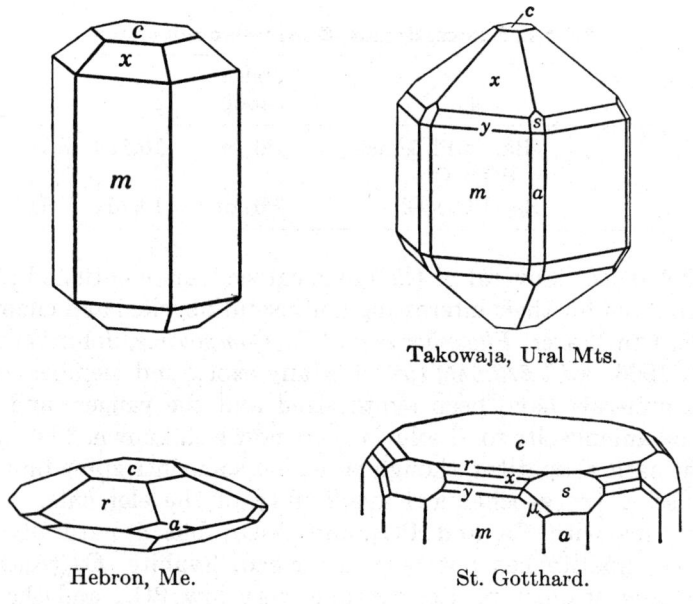

Takowaja, Ural Mts.

Hebron, Me.   St. Gotthard.

Forms a $\{11\bar{2}0\}$ c $\{0001\}$ m $\{10\bar{1}0\}$ r $\{10\bar{1}2\}$ s $\{11\bar{2}1\}$ x $\{10\bar{1}1\}$ y$\{20\bar{2}1\}$ μ$\{21\bar{3}1\}$

41.8.1.1 Fluorapatite

(41.8.1.4) and carbonate-hydroxylapatite (41.8.1.5). HEX $P6_3/m$. $a = 9.367$, $c = 6.884$, $Z = 2$, $D = 3.18$. *AM 75:295(1990)*. *15-876:* 3.44_4 2.80_{10} 2.77_6 2.70_6 2.62_3 2.25_2 1.937_3 1.837_3. See also *35-496, 34-11*. **Habit:** Crystals short to long prismatic [0001] with $\{10\bar{1}0\}$, $\{10\bar{1}1\}$ dominant; thick tabular $\{0001\}$ with $\{10\bar{1}0\}$, large $\{0001\}$, and other low pyramidal forms produce quite complex crystals (Figure 41.8.1.1*B*); massive, coarse granular to compact, sometimes globular to reniform with subfibrous, scaly, or columnar structure; stalactitic; earthy; nodular concretions. Rarely twinned, contact with twin plane $\{11\bar{2}1\}$, other twin planes $\{10\bar{1}0\}$, $\{11\bar{2}3\}$, $\{10\bar{1}3\}$. Commonly found with inclusions, such as needles of rutile where the *c* axes of both phases are parallel, and also monazite. Crystals often zoned from F-rich to OH-rich apatite, which may show color differences. **Physical properties:** Usually sea-green to other shades of green, blue, violet-blue, amethyst, colorless, white, yellowish, gray, brown, flesh, or rose-red. Manganoan varieties are dark green and blue-green. Violet color is lost on heating. White streak, may show bluish opalescence; vitreous luster, transparent. Cleavage of variable ease and quality; $\{0001\}$, indistinct; $\{10\bar{1}0\}$, in traces; conchoidal to uneven fracture; brittle. $H = 5$. $G = 3.1$–3.2. Usually fluorescent under UV, cathode, or X-radiation; phosphorescent, may be strongly thermoluminescent, not piezoelectric. Uniaxial $(-)$; $N_O = 1.6325$, $N_E = 1.630$ (Na) (synthetic) *DII: 879*; $N_O = 1.646$, $N_E = 1.642$ (588 nm) (Zillertal). Indices highly sensitive to substitution especially OH for F (McConnell. D 1973 *Apatite*, Springer Verlag, N.Y. Colored crystals show strong to weak dichroism with absorption $E > O$: higher indices with Mn, or with OH-containing fluorapatite crystals, lower in CO_3-containing apatites. **Chemistry:** REE, especially Ce, are common substituents for Ca, as are Mn and Sr [see belovite (41.8.1.6), strontium-apatite (41.8.1.7)]. An extremely wide variety of cations including both two or three positive valence states (Me^{2+} and Me^{3+} cations) have been detected on chemical analyses and many end members have been produced synthetically. However, natural crystals containing more than trace amounts of Mg and Fe^{2+} are rare, and alkalis usually occur as separate isomorphous phases with SiO_4 or SO_4 partially substituting for PO_4. Arsenic is usually absent on analysis probably due to geochemical nonavailability, as AsO_4 forms a complete series with PO_4 (to svabite, fermorite). Synthetic studies suggest Al, Cr^{3+}, and Cr^{6+} easily fit into the anionic site in the apatite structure. **Occurrence:** Fluor- and hydroxylapatites are very common, occurring in all types of igneous rocks as accessory minerals usually in microscopic euhedral crystals; in both acidic and basic pegmatites; in magnetite deposits; in hydrothermal veins; in both regionally and contact metamorphosed rocks, especially recrystallized limestones associated with titanite, zircon, pyroxene, amphiboles, spinel, vesuvianite, and phlogopite; in talc and chlorite schists; as extensive deposits of marine origin with fine-grained massive form; as replacement beds of limestone or coral via solutions derived from guano, or as nodules disseminated in nearshore sediments, and in coal measures. **Localities:** Notable crystals of fluor- and hydroxylapatite were found in the United States at Newry(!!), and Mt. Apatite(!!), Auburn, ME; in the Adirondack Mts., NY, along with magnetite; in crystalline limestones in NJ (at Franklin); at Tip Top mine, Custer Co., SD; at Clay Canyon, Fairfield, UT. Apatite has been described from Pulsifer quarry, Auburn, ME; Palermo No. 1 mine, North Groton, NH; Keystone quarry, Cornog, PA; Foote mine, Kings Mt., NC; Jeffrey quarry, Little Rock, AK; Bald Mt., Jefferson Co., MT; Hugo pegmatite, Keystone, SD; Climax mine, Lake Co., Urad mine, Clear Creek Co., and at the Pikes Peak batholith, CO; Warren mining district, Bisbee, Cochise Co., Mammoth–St. Anthony mine, Tiger, Twin Buttes mine, AZ; Harding mine, Dixon, Taos Co., NM; Champion mine, White Mts., Mono Co., Gem mine, San Benito Co., Clark, and Himalya mines, San Diego Co., CA. In Canada, fluorapatite was commonly found in the Grenville marble as irregular veins and crystals, even as large crystals(!!!; up to 200 kg) in Ottawa Co., Lanark Co., Renfrew Co., Frontenac Co., QUE. More recent finds are on the Lievre R., at Mt. St-Hilaire(!), at Yates mine, Pontiac Co., QUE; at Ottertail Range, Ice River complex, BC; at several localities in Greenland, such as Iglunguak, Narsarsuk, and Kangerdluarsuk. Fine yellow crystals were found in the magnetite deposit of Cerro Mercado(!!) DGO, Mexico. In Europe F- and OH-apatites were found in andesite–tuff at Jumilla, Murcia, Spain; and more recently in Portugal at Panasqueira, at Mt. Redondo quarry. The tin veins at Tavistock, Devonshire, and Wheal Franco, Cornwall, England, and Villeder, France, were early-noted occurrences, as were the pegmatites at Zillertal, Tyrol, Austria, and at Epprechstein and Waldstein, Fichtel Mts., the tin mines at Ehrenfriedersdorf, Saxony, Germany, and most recently at Mendig, Laacher See district; at Schwartzenstein, Salzburg, Untersulzbachtal, Austria, and at several localities in the Piedmont of Italy; at Gletch on the Rhône glacier, Switzerland; at Schlaggenwald, Czech Republic. Enormous deposits are known from the alkalic igneous rocks in the Khibina tundra, Kola peninsula(!), and from the mica schists east of Ekaterinburg(!), Ural Mts., where apatite occurs with emerald and chrysoberyl. A new deposit has been found at Slyudyanka, Siberia, Russia. The Onganja mine(!), Namibia, also has produced some fine specimens. Fluorapatite occurs in Australia at Iron Monarch mine, Iron Knob, SA; at the San Jose mine, Oruro, and the Sligo XX mine, Llallagua, Bolivia; at several mines in MG, such as Lavra do Enio, Pamaro, Santo do Encoberto, Lavra da Ilha, Virgem da Lapa, and at Paraiba and in the Brumado district, BA, Brazil. *DII:879, AM 76:1857(1991)*. Uses: The largest-volume use of apatite materials is to make the apatitic halophosphate phosphors ($\approx$ 10 tons/day) for fluorescent lighting. The luminescence depends on the small amount (<2 wt %) of Sb^{3+} incorporated into the apatite lattice ("doping"). As of this writing the actual site(s) within the structure where Sb resides are not known. Phosphatic deposits are also mined for agricultural fertilizer and feed for animals. HCWS

41.8.1.2 Chlorapatite $Ca_5(PO_4)_3Cl$

Named in 1860 for the relationshp to apatite and the composition. Apatite group. HEX $P6_3/m$. $a = 9.598$, $c = 6.776$, Z = 2, D = 3.199. Structure: *AM*

74:870(1989). The Cl atoms may occur above or below the mirror plane that passes through the Ca(2) triangle (see Figure 41.8.1.2). This disposition changes the Ca(2) bond lengths and effectively prohibits OH and F from occupying the centered positions in the same column. Such shifts suggest that the end member F-, OH-, and Cl-apatites are immiscible (i.e., do not have mutual solid solutions). It is possible that the Z atoms may be segregated in individual columns (producing monoclinic structure) and if the column distribution is disordered throughout the crystals, the general structure will remain hexagonal. 33-271(syn): 8.33_1 2.86_9 2.78_9 2.77_{10} 2.32_2 2.31_2 1.963_3 1.836_3. Crystals pale yellow, vitreous luster, transparent; appear pinkish white with talc inclusions. H = 5. G = 3.181. Uniaxial (−); $N_O = 1.6684$, $N_E = 1.6675$ (synthetic). Many apatite analyses show very small amounts of Cl (i.e., Cl-containing fluorapatite). However, the totally Cl end member, chlorapatite, is extremely rare but has been synthesized and may show monoclinic symmetry: $P2_1/a$ with $a = 19.210$, $b = 6.785$, $c = 9.605$, $\beta = 120°$; biaxial (−); $N_x = 1.667$, $N_z = 1.665$; $2V \approx 10°$. Found in veins in calc–silicate rocks and marbles at *Bob's Lake, Öso Twp., Frontenac Co., Canada*, with actinolite, diopside, calcite, quartz, and talc. Other localities: Odegarden, Kragerö, Norway, where the mineral is associated with rutile, ilmenite, sphene, magnetite, pyrrhotite, hornblende, wernerite, and also at Kurokura, Kanagawa Pref., Japan, have considerably lower Cl contents. Chlorapatite has also been found in meteorites. HCWS *DII:879*, *CM 10:252(1970)*.

41.8.1.3 Hydroxylapatite $Ca_5(PO_4)_3(OH)$

Named in 1856 to distinguish the (OH) end member of the minerals known generally as apatite. Apatite group. See fluorapatite (41.8.1.1) for general outline of habit and physical and chemical properties; see carbonate F/OH apatite for biological mineral. HEX $P6_3/m$. $a = 9.418$, $c = 6.875$, $Z = 2$, D = 3.16 (Holly Springs). *AM 74:870(1989)*. 9-432(syn): 3.44_4 2.82_{10} 2.79_6 2.72_6 2.63_3 2.26_2 1.943_2 1.841_3. See also 24-33(calc), 34-10, 25-166. Uniaxial (−); $N_O = 1.651$, $N_E = 1.644$ (Na). High-temperatures (300–600°) and high-pressure (2 kbar) studies in the system $CaO-P_2O_5-H_2O$ show that hydroxylapatite is stable over the pH range 4–12. The synthetic hydroxylapatite crystals are stable to temperatures in excess of 1,200° in humid environments, but lose H_2O in vacuo by 850°. *AJS 273:545(1973)*, *JPC 79:2017(1975)*. Low-temperature (<100°) precipitation of hydroxylapatite does not provide well-crystallized materials, and the assumption is that CO_2 in the environment inhibits crystal growth. The addition of F to the solution increases crystallinity and is the rational for fluoridation, adding a few parts per million to drinking water, which appears to have the salutory effect of reducing caries in the general populace. Especially fine, clear, light yellow-green euhedral hydroxylapatite crystals are found at Holly Springs, Cherokee Co., GA.; at Eagle, CO; in the talc schists of the alpine region in Switzerland; in the Kop Dag area, Kop Daglari, Turkey; at Milgun Station, WA, Australia. HCWS *DII:879*, *GT 4:208(1988)*.

41.8.1.4 Carbonate-Fluorapatite $Ca_5(PO_4,CO_3)_3F$
41.8.1.5 Carbonate-Hydroxylapatite $Ca_5(PO_4,CO_3)_3(OH)$

Named in 1906 to accommodate the intimate association of carbonate with apatite in materials that were ultrafine-grained, massive, earthy, pulverulent, or opaline, and formed as crusts, spherules, and nodules, in sedimentary horizons, as bulk phosphatic rock, or as the mineral portion of bones and teeth, all vertebrates, animal and human, both fossil and modern. *Collophane* was the original name for such a species. Francolite, named in 1859 for the locality, Wheal Franco, Devonshire, England, has been used for high-F-containing carbonate-apatites, while dahllite, named for T. and J. Dahl, Norwegian geologists and brothers [*ZK 17:426 (1890)*] is applied to carbonate-hydroxylapatite. Apatite group. *31-267*: 8.04_2 4.04_2 3.43_2 3.05_4 2.79_6 2.69_{10} 2.24_5 1.783_3. See also *19-272*(syn). Soluble in HCl, HNO_3 with slight effervescence. Despite research that shows up to 6 wt % CO_2, in some carbonate-containing apatite mineral materials the actual disposition of the CO_3 group at specific site(s) in the apatite lattice is not yet known. It has been proposed that collophane and bone mineral: (1) are mixtures, the coprecipitation of $CaCO_3$ plus apatite; or (2) CO_3^{2-} or CO_2 may be adsorbed on the surfaces of very fine grains of F- or OH-apatite; or (3) a $[CO_3-F]$ group substitutes for PO_4 in the structure. The last hypothesis has been supported by chemical analyses of carbonate-containing reasonably well-crystallized apatites which show greater amounts of (F,OH,Cl) than would be expected from the stoichiometric formula. *CMP 101:394(1989)*. Uniaxial $(-)$; $N_O = 1.628$, $N_E = 1.619$ (Na); very variable; higher birefringence than that of F- and OH-apatites. Francolites were found at *Wheal Franco, Devonshire*, at Callington–Gunnislake, Cornwall, at Caldbeck Fells, Cumbria, associated with Skiddaw granites; as concretions in a light brown groundmass in argillaceous sediments with quartz, illite, pyrite, and associated with magnesian siderite at Hepworth Iron mine, Westphalia, Yorkshire, *England*. Dahllite is encountered in the skarns at Magnet Cove, AK; as fibers in Ödergarden, Bamle, Norway; at the carbonatites in Palabora, northeastern Transvaal, South Africa. In addition to the fossilized animal and normal human bone samples in sedimentary horizons, there are the massive phosphorite deposits, such as the Phosphoria formation in the western United States or Florida and in Morocco and Russia. Much of the commercial phosphate material comes from these horizons and is used in the production of phosphoric acid and fertilizers, washing detergents, and animal feed. HCWS *DII:879*, *SCI 155:1409(1964)*, *AM 53:445(1968)*, *MM 41:M4(1977)*.

41.8.1.6 Belovite $(Sr,Ce,Na,Ca)_5(PO_4)_3(OH)$

Named in 1955 for Nikolai Vassilievich Belov (1891–1982), mineralogist and crystallographer, Institute of Crystallography, Moscow, Russia. Apatite group. One of two species that are Sr analogs of apatite. Strontium-apatite (41.8.1.7) contains > 50 at % Sr, while belovite contains < 50 at %. HEX $P6_3/m$. $a = 9.692$, $c = 7.201$, $Z = 2$, $D = 4.131$. *MZ 9:45(1987)*. *31-1350*: 3.17_7 2.89_{10} 2.79_7 2.15_5 2.01_7 1.957_4 1.913_7 1.470_6. Crystals prismatic up to

2 cm showing $\{10\bar{1}0\}$, $\{0001\}$. Honey-yellow to light green, vitreous to greasy luster, transparent. Imperfect cleavage; irregular fracture; brittle. H = 5. G = 4.19. Uniaxial $(-)$; $N_O = 1.660$, $N_E = 1.664$. Soluble in 1:10 HCl, HNO_3. Contains < 50 at % Sr, with REE, Ca, and Na substitution. Occurs in central portions of the pegmatites in nepheline syenites at *Mt. Karnasurt, Lovozero massif, Kola Penin., Russia*, in ussingite with nepheline, aegirine, natrolite, and in association with minor sodalite, microcline, schizolite, erikite, murmanite, neptunite, and steenstrupine. HCWS *AM 40:367(1955)*, *DANS-ESS 142:113(1962)*.

41.8.1.7 Strontium-Apatite $(Sr,Ca)_5(PO_4)_3(F,OH)$

Named in 1962 for the composition and resemblance to the apatites. Apatite group. The name given to species with > 50 at % Sr see belovite (41.8.1.6). HEX $P6_3/m$. $a = 9.565$, $c = 7.115$, $Z = 2$, $D = 3.95$. Structure: *SPC 32:524(1989)*. Large ions (Sr, Ba)/(Ca, Mg, Na, Fe, Th, REE) usually have the ratio 6:4, conforming to the two possible cationic sites in the structure. *33-1348*(syn): 3.18_2 2.93_{10} 2.92_{10} 2.82_4 2.03_3 1.932_3 1.819_2 1.469_2. Prismatic crystals to 4 cm length, although mostly smaller 0.2 mm × 1 mm, or minute, poorly formed, oval in shape with corroded $\{10\bar{1}0\}$, $\{10\bar{1}1\}$; sugary. Pale green, pale yellow-green, colorless; vitreous to greasy luster; transparent. Imperfect cleavage parallel to prism faces, uneven fracture. H = 5. G = 3.84. Uniaxial $(-)$; $N_O = 1.651$, $N_E = 1.637$. Soluble in acids. Microprobe analysis shows Ba, La, Ce, Pr, Nd, Sm, Eu, Gd, Tb, Dy, Er, and Yb. Synthetic studies had originally suggested a complete solid-solution Ca–Sr, but documentation of the ordering of Ca into one cation site in arsenate-containing apatites, and more recent studies of high-Ca-containing portions of the series (e.g., Ca_9Sr to Ca_6Sr_4) suggest that these compositions are usually not formed, nor have they been found in nature. The results for the Ca–Sr solid solution are similar to the range found for the Ca–Pb series, hedyphane–turneaureite. *AM 69:920(1984)*. Occurs in interstices of albite–eckermanite–aegirine veins in alkali pegmatites that cut the dunite core of the *Inaglinsky massif, southern Yakutia, Russia*, intergrown with bathysite, ramseyite, and eudialyte, and in association with microcline, batisite, innelite, and lorenzenite. HCWS *DANS 142:113(1962)*, *AM 47:808(1962)*.

41.8.2.1 Hedyphane $Ca_2Pb_3(AsO_4)_3Cl$

Named in 1830 from the Greek for *beautifully bright*, in allusion to the high luster. Apatite group. HEX $P6_3/m$. $a = 10.140$, $c = 7.185$, $Z = 2$, $D = 5.99$. Structure: *AM 69:920(1984)*. Structure determination confirms that Ca and Pb are regularly disposed in the two different cation structure sites. Therefore, hedyphane is an ordered intermediate species between $Ca_5(AsO_4)_3Cl$ and $Pb_5(AsO_4)Cl$, mimetite (41.8.4.2). *36-396*: 8.77_4 5.07_4 4.14_5 3.59_7 3.32_4 3.01_{10} 2.93_8 1.943_8. Prismatic [0001], pyramidal to thick tabular $\{0001\}$; massive. White, yellow-white, bluish; white streak; bright, greasy to resinous luster; translucent. Cleavage $\{10\bar{1}1\}$, barely discernable; even fracture; brittle.

$H = 4\frac{1}{2}$. $G = 5.82$. Weak light yellow fluorescence in UV. Uniaxial (+); $N_O = 1.958$, $N_E = 1.948$ (Na); or uniaxial (−); $N_O = 2.03$. $N_E = 2.01$, both Franklin, NJ, samples. Soluble in HNO_3. Sr, Ba, and P detected on analysis, with Sr the dominant substituent at Franklin and at Ba, Långban. A solid-solution series has been demonstrated synthetically between hedyphane–mimetite, the Ca_2Pb_3–Pb_5 portion of the Ca–Pb system, but there is a miscibility gap in natural materials for the portion represented by hedyphane–turneaureite: Ca_2Pb_3–Ca_5. Found at Franklin, NJ, as the most abundant nonsilicate lead mineral, occurring occasionally as fine dipyramidal or prismatic crystals(!) and associated with calcite, willemite, native copper, native lead, hancockite, rhodonite, manganaxinite, and apatite in veinlets in massive ore; at *Långban, Värmland, Sweden*, with barite, baylite, and rhodonite, as crystals coating calcite crystals in veinlets and associated with allactite and hausmannite, also at Harstig mine, Pajsberg, with tephroite and calcite in magnetite; at Tsumeb, Namibia; at Puttappi mine, Flinders Ranges, SA, Australia. HCWS *DII:901*.

41.8.3.1   Svabite $Ca_5(AsO_4)_3F$

Named in 1891 for Anton Svab (1703–1768), Swedish mining official. Apatite group. HEX $P6_3/m$. $a = 9.80$, $c = 6.90$, $Z = 2$, $D = 3.708$. *DANS (1966) 166:134. 19-215*: 3.99_7 3.43_6 2.91_{10} 2.84_9 2.82_9 2.67_7 2.23_5 1.999_6. Short prismatic [0001] notable hexagonal outline; massive. Colorless, yellowish white, gray, grayish green, pale lilac; vitreous to subresinous luster; transparent to translucent. Cleavage $\{10\bar{1}0\}$, indistinct; brittle. $H = 4$–5. $G = 3.5$–3.8 with Pb-rich samples the highest value. Fluorescent LWUV red-orange, in cathodoluminescence pale pink. DTA. Uniaxial (−); $N_O = 1.706$, $N_E = 1.698$ (Harstig). Soluble in 1:10 acids. Pb, Mg, Mn^{2+} substitute for Ca; Cl and (OH) for F, P for As with the potential for series toward fluorapatite, mimetite, or hedyphane. Found as rough hexagonal prisms with rounded terminations embedded in franklinite ore, Franklin, NJ; at *Harstig mine, Pajsberg, Värmland, Sweden*, with manganoan acmite, garnet, brandtite, sarkinite, and massive at Jakobsberg, Nordmark district, and at Långban with hematite, mica, and garnet; in contact metasomatic deposits in Siberia, and in the Ural Mts. with vesuvianite in pyroxene–garnet assemblages and in zones with spodumene, which alters to szaibelyite and brucite, Russia. HCWS *DII:899*.

41.8.3.2   Turneaureite $Ca_5(AsO_4)_3Cl$

Named in 1985 for Frederick Steward Turneaure (1899–1986), University of Michigan, Ann Arbor. Apatite group. The Cl analog of svabite and johnbaumite, As analog of chlorapatite, and Ca analog of morelandite and mimetite. HEX $P6_3/m$. $a = 9.810$, $c = 6.868$, $Z = 2$, $D = 3.63$. *CM 23:251(1985). 38-383*: 3.98_5 3.43_6 2.91_{10} 2.83_9 2.67_5 1.995_3 1.864_4 1.718_3. Prismatic crystals to 1.5 mm long, showing $\{10\bar{1}0\}$, $\{0001\}$; massive. Colorless, white streak, vitreous to greasy luster. No cleavage, uneven fracture. $H = 5$. $G = 3.60$. Fluorescent bright orange in SWUV. Weak phosphorescence in Franklin samples. Uniaxial

(−); $N_O = 1.708$, $N_E = 1.700$ (Na). Microprobe analysis shows Mn, P, F. Found in high-grade marble terranes (e.g., the manganese-rich siliceous marble at Balmat, St. Lawrence Co., NY) with donpeacorite, tirodite, ferrian braunite, dravite, anhydrite, and manganoan dolomite; as gray-white anhedral masses at Franklin, Sussex Co., NJ; at *Långban mine, Värmland, Sweden*(!), with calcite and andradite, in massive andradite–magnetite ore. HCWS *AM 71:1280(1980)*.

41.8.3.3 Johnbaumite $Ca_5(AsO_4)_3(OH)$

Named in 1980 for John L. Baum (b.1916), resident geologist of the New Jersey Zinc Co. and curator of the Franklin Mineral Museum, Franklin, NJ. Apatite group. As analog of hydroxylapatite and (OH) analog of svabite. HEX $P6_3/m$. or $P6_3$. $a = 9.70$, $c = 6.93$, $Z = 2$, $D = 3.73$. *AM 65:1143(1980)*. *33-265:* 3.98_5 3.47_5 2.90_{10} 2.82_7 2.80_7 2.68_5 2.03_4 1.879_4. Massive, anhedral grains to 8 mm. Colorless, white streak, vitreous to greasy luster on fracture surfaces, transparent. Cleavage $\{10\bar{1}0\}$, distinct. $H = 4\frac{1}{2}$. $G = 3.68$. DTA/TGA. Fluorescent pink-orange in SWUV only. Uniaxial (−); $N_O = 1.687$, $N_E = 1.684$. Microprobe analysis shows minor P, Fe^{2+}, Mg, F, Cl; also H_2O. Found at *Palmer Shaft, Franklin mine, Franklin, Sussex Co., NJ*, with yeatmanite, diopside, andradite, franklinite, native copper, and romeite. HCWS

41.8.3.4 Fermorite $(Ca,Sr)_5(PO_4,AsO_4)_3(F,OH)$

Named in 1911 for Lewis Leigh Fermor (1880–1954), director of the Geological Survey of India. Apatite group. MON $P2_1/m$. $a = 9.594$, $b = 6.975$, $c = 9.597$, $\beta = 119.97°$, $Z = 1$, $D = 3.608$. Structure: *327(1991) NJMM:*. Strongly pseudohexagonal. *14-215:* 3.95_3 3.49_5 3.12_3 2.86_{10} 2.75_6 2.67_3 1.971_4 1.867_4. Massive, granular. Pale pinkish white, white streak, greasy luster, translucent. Even fracture. $H = 5$. $G = 3.518$. Uniaxial (−); $N_{av} = 1.660$; weakly birefringent. Soluble in acids. Analysis shows significant substitution of As for P, maximum $As/P = 1:1$, which probably influences crystallization in monoclinic space group, minor Sr, trace H_2O. Found at *Sitapar, Chindwara district, Central Provinces, India*. HCWS *DII:904*.

41.8.4.1 Pyromorphite $Pb_5(PO_4)_3Cl$

Named in 1813 from the Greek for *fire* and *form* because after melting the mineral sample to a globule, a crystalline shape forms on cooling. Apatite group. Pyromorphite forms a complete series, P–As, with mimetite, and is the end-member name used when P > As. HEX $P6_3/m$. $a = 9.976$, $c = 7.351$, $Z = 2$, $D = 7.109$. Structure: *CM 27:189(1989)*. *19-701*(syn): 4.13_5 3.38_3 3.27_4 2.99_{10} 2.96_{10} 2.89_6 2.064_3 1.861_3. Crystals prismatic [0001], generally simple, showing $\{10\bar{1}0\}$, $\{0001\}$, $\{10\bar{1}1\}$; equant, barrel shaped; spindle shaped, or with cavernous terminations; rarely tabular $\{0001\}$ or pyramidal; branching groups of prismatic crystals in parallel position, tapering to a point; globular, reniform, wartlike with subcolumnar structure; granular. Crystals often show concentric growth zones, probably due to As/P variations. Dark green to green-yellow, shades of brown, waxy yellow, orange-yellow, white, colorless; white streak; resinous luster; subtransparent to translucent. Cleavage $\{10\bar{1}1\}$ in traces, uneven to subconchoidal fracture, brittle. $H = 3\frac{1}{2}$. $G = 7.04$ (lower with substitutions). Crystals from Ems (and mimetite from Tsumeb) are piezoelectric and optically biaxial. Uniaxial $(-)$; $N_O = 2.058$, $N_E = 2.048$ (589 nm); faintly pleochroic; biaxial when As content is high. Ca frequently substitutes for Pb, but in natural minerals a complete series to chlorapatite does not exist [see hedyphane (41.8.2.1)], minor Fe^{2+}, Cr, REE, Ba, Mn substitution reported; trace amounts of F and OH substitute for Cl. Easily synthesized below 100°. Synthetic studies concluded that a complete P–As–V solid solution was possible [*AM 51:1712(1966)*], but crystal structure investigations suggest that the larger VO_4 group will probably be of limited extent, especially in natural systems. Pseudomorphs of pyromorphite after galena and cerussite are common; galena may be found pseudomorphic after pyromorphite. Occurs in oxidized portions of lead deposits with mimetite, cerussite, iron oxyhydroxides, smithsonite, hemimorphite, anglesite, malachite, leadhillite, caledonite, vanadinite, wulfenite, descloizite, mottramite, and bayldonite. Found at Loudville Mine, Hampshire Co., MA; at Wheatly mines, Phoenixville, Chester Co., PA(!); in the Leadville area, CO; at the many mines in the Coeur d'Alene district(!), Shoshone Co., ID; at Bisbee, Cochise Co., and elsewhere in AZ, UT, and NM; in Canada at Mt. St.-Hilaire, QUE, and in the Moyie mines, BC; at Chihuahua, Los Lamentos, CHIH, Mexico; at mines in Cumberland, Derbyshire, Somerset, and at Bwich-Glas mine, Dyfed, Wales; at Leadhills, Wanlockhead district, Scotland; at El Horcajo, Castilla Province, Spain; at Les Farges mine, Ussel, Correze, France; in Germany at Friedrichssegen mine, Ems, Grube Phillipseck, Munster; at lead mining areas in Russia; at Broken Hill, NSW, and in the mines at Dundas, TAS, and elsewhere in NT and SA, Australia; at Touissit mine, Oujda, Morocco; at Broken Hill mine, Zambia; at Boquira mine, BA, Brazil. HCWS *DII:889*, *NJMA 99:113(1963)*.

41.8.4.2 Mimetite $Pb_5(AsO_4)_3Cl$

Named in 1835 from the Greek for *imitator* because of its resemblance to pyromorphite. Apatite group. Dimorphous with clinomimetite (41.8.6.1). HEX $P6_3/m$. $a = 10.212$, $c = 7.419$, $Z = 2$, $D = 7.80$. Structure: *CM 29:369(1991)*. Not analogous to chlorapatite, where Ca(2) bonds to one Cl as in mimetite Pb(2) bonds to two Cl's in the Z position. 19-683(syn): 4.44_2 3.36_4 3.06_{10} 3.01_9 2.96_7 2.11_3 1.994_2 1.905_2. See also *13-124* (P-containing species) and 19-683 (syn). Crystals simple barrel shapes showing $\{10\bar{1}0\}$, $\{0001\}$; botryoidal, globular, reniform, granular, rarely tabular, acicular. Pale yellow to yellowish brown, orange-yellow, orange-red, white, colorless; white streak; resinous luster; subtransparent. Zoned growth As alternates with P. Cleavage $\{10\bar{1}1\}$ in traces, subconchoidal fracture, brittle. $H = 3\frac{1}{2}$–4. $G = 7.28$. IR. Uniaxial/biaxial $(-)$; $X = c$; $N_{O'} = 2.147$, $N_{E'} = 2.128$ (589 nm); 2V = 41°; r < v; 2V decreases with increasing P. Anomalous optics may denote twinning, or sector growth. Soluble in HNO_3, KOH. Synthesized. In natural minerals there is a complete series As–P, pyromorphite–mimetite. The latter name is used when there is > 50 mol % As. Vanadinite forms a series up to V/As = 1:1. Ca substitutes for Pb, but see hedyphane (41.8.2.1). Less common than pyromorphite, mimetite occurs in oxidized zones of lead deposits and other formations where lead and arsenic occur. Found in association with pyromorphite at many of the lead mines in AZ, such as at Bisbee and Tombstone, and in Gila, Maricopa, Pima, and Yuma Cos., at Mammoth–St. Anthony, Tiger, and at Red Cloud Copper mine, Gallinas Mts., NM; at Santa Eulalia, CHIH, Mexico; at Millers Dale, Derbyshire, Merehead quarry, Somerset, England, with curved barrel-shaped crystals originally from Drygill, Cumberland(!), but found elsewhere; at Les Farges mine, France; at Johanngeorgenstadt, Germany; at Kamareza mine, Attica, Greece; at Seh Changi mine, Iran; at Tsumeb, Namibia; at Broken Hill, NSW, Australia. HCWS *DII:889, AM 51:1712(1966), ZK 191:125(1990)*.

41.8.4.3 Vanadinite $Pb_5(VO_4)_3Cl$

Named in 1839 for the composition. Apatite group. HEX $P6_3/m$. $a = 10.317$, $c = 7.338$, $Z = 2$, $D = 6.953$. Structure: *CM 27:189(1989)*. *43-1461*: 4.22_3 3.40_3 3.38_4 3.07_9 2.99_{10} 2.98_8 2.11_3 1.976_3. See also *19-684*(syn): *13-585*. Crystals short to long prismatic [0001] with smooth faces, sharp edges; acicular to hairlike; occasionally, hollow prism crystals with rounded domes in parallel grouping, such as pyromorphite. Orange-red, ruby-red, brownish red, brown-yellow to pale straw-yellow; whitish-yellow streak; subresinous luster; subtransparent to nearly opaque. May be concentrically zoned with composition variations. Uneven fracture, brittle. $H = 2\frac{1}{2}$–3. $G = 6.88$, decreasing with Ca substitution. Uniaxial (−); $N_O = 2.416$, $N_E = 2.350$ (598 nm); pleochroic with E < O. Ca, P substitution will reduce indices. Easily soluble in HNO_3, giving a yellow solution. Soluble in HCl, giving a green solution, deposition of lead chloride. Small amounts of Ca, Zn, Cu substitute for Pb; substantial amounts of P and As substitute for V. Synthesized. A secondary mineral found in the oxidized zones of lead deposits. Occurs in many of the mines in AZ: Apache mine, Globe, C & B mine, Gila Co., Old Yuma mine(!), Pima Co., Grey House mine, Pinal Co., Red Cloud mine(!), Rowley mine, Mammoth–St. Anthony mine, Tiger, Hamburg mine, Yuma Co.; at the Chalk Mt. mine, Churchill Co., NV; in the Hillsboro(!) and Lake Valley district(!), and Caballos Mt.(!), Sierra Co., at Red Cloud Copper mine, Gallinas Mts., NM; at Camp Signal, San Bernardino Co., and El Dorado mine, Riverside Co., CA; at Ahumada mine, Los Lamentos, CHIH, Mexico; at Leadhills, Scotland; at M'Fis and Acif mines, Morocco; at Broken Hill, Zambia; at Tsumeb, Namibia; at Proprietary mine, Broken Hill, NSW, Australia; at Itacarambi mine, MG, Brazil. HCWS *DII:895, CM 7:301(1962)*.

41.8.5.1 Morelandite (Ba,Ca,Pb)$_5$(AsO$_4$,PO$_4$)$_3$Cl

Named in 1978 for Groven C. Moreland (1912–1978), supervisor of sample preparation laboratory, National Museum of Natural History, Smithsonian Institution, Washington, DC. Apatite group. See mimetite (41.8.4.2). HEX $P6_3$ or $P6_3/m$. $a = 10.169$, $c = 7.315$, $Z = 2$, $D = 5.30$. *CM 16:601(1978)*. *31-132*: 8.86$_2$ 5.09$_3$ 3.66$_5$ 3.18$_3$ 3.03$_{10}$ 2.97$_7$ 2.94$_6$ 1.965$_3$. Irregular masses. Light yellow to almost pure gray, white streak, greasy to vitreous luster, translucent. Weak cleavage parallel to {0001}. H = $4\frac{1}{2}$. G = 5.33. Nonfluorescent. Uniaxial (+); $N_O = 1.880$, $N_E = 1.884$. Soluble in cold 1:1 HCl, turning opaque, chalky white, on exposure to HCl. Microprobe showed some Ca, P, minor Ca, Mn, Fe, only a trace of H$_2$O detected. Found at *Jakobsberg mine, near Nordmark, Sweden*. HCWS

41.8.5.2 Alforsite Ba$_5$(PO$_4$)$_3$Cl

Named in 1981 for John T. Alfors (b.1930), geologist, California Division of Mines and Geology. Apatite group. HEX $P6_3/m$. $a = 10.75$, $c = 7.64$, $Z = 2$, $D = 4.83$. *AM 66:1050(1981)*. *35-691*(syn): 3.37$_2$ 3.06$_{10}$ 2.95$_3$ 2.13$_4$ 2.03$_3$ 1.928$_3$ 1.566$_3$ 1.336$_3$. Minute subhedral grains up to 0.2 mm but usually < 0.05 mm in diameter. Colorless, distinguished by low birefringence, high relief in thin sections. Intense violet fluorescence in 10- to 15-kV electron beam. Not fluorescent in UV. Uniaxial (−); $N_O = 1.696$, $N_E = 1.694$ (Synthetic). An accessory mineral in metamorphic rocks of *Incline, Mariposa Co., and Big Creek, Fresno Co., CA*, associated with pink gillespite, sanbornite, witherite, celsian, quartz, and fluorapatite. HCWS

41.8.6.1 Clinomimetite Pb$_5$(AsO$_4$)$_3$Cl

Named in 1991 for its dimorphic relationship to mimetite. Apatite group. See also finnemanite Pb(A$_5$O$_3$)$_3$Cl, and arsenite. MON $P2_1/b$. $a = 10.180$, $b = 7.46$, $c = 20.372$, $\beta = 119.88°$, $Z = 4$, $D = 7.32$. Structure: *CM 29:369(1991)*. Strongly pseudohexagonal. Small changes in Pb bonding with O of PO$_4$ group as a result of the Cl (Z atom) position in the column are probably responsible for the monoclinicity 47-1808: 3.34$_5$ 3.05$_{10}$ 3.01$_7$ 2.95$_7$ 2.11$_5$ 2.00$_4$ 1.961$_5$ 1.903$_5$. Physical and chemical properties similar to mimetite (41.8.4.2). Spontaneous and reversible transition between $P6_3/m$ and $P2_1/b$ occurs ≈ 100° and has been demonstrated in both natural and synthetic materials. Synthetic studies show continuous series P-As Pyromorphite-mimetite but not to V, therefore there is no series to vanadinite. H = $3\frac{1}{2}$. G = 7.36. Found at *Johanngeorgenstadt, Erzgebirge, Saxony, Germany*. HCWS *NJMM:359(1968), 64(1969); NJMA 113:219(1970)*.

BJAREBYITE GROUP

Bjarebyite minerals are phosphates with the general formula

$$AB_2C_2(PO_4)_3(OH)_3$$

where

A = Ba, Sr
B = Fe^{2+}, Mn^{2+}, Mg
C = Al, Fe^{3+}

and crystallize with monoclinic, or triclinic pseudomonoclinic, symmetry.

The structure [AM 59:567(1974)] is composed of pairs of edge-sharing octahedra $[C_2O_6(OH)_3]$ joined at the corners forming chains parallel to the b axis. The chains share edges with distorted pairs of BO_6 octahedra, and are bridged by PO_4 tetrahedra. The A atoms occupy holes in this framework and have 11-fold coordination.

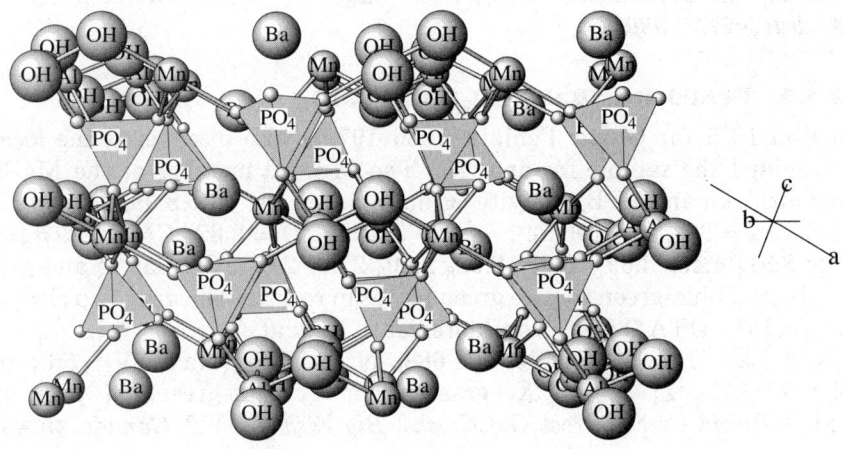

BJAREBYITE: $BaMn_2Al_2(PO_4)_3(OH)_3$

This projection of the bjarebyite structure shows the large A ions in the intersticies of a framework composed of Mn and $Al(Op,OH)_6$. Octahedra connected through PO_4 groups.

BJAREBYITE GROUP

Mineral	Formula	Space Group	a	b	c	β	
Kulanite*	$BaFe_2^{2+}Al_2(PO_4)_3(OH)_3$	$P\bar{1}$	9.032	12.119	4.936	100.38	41.9.1.1
Penikisite*	$BaMg_2Al_2(PO_4)_3(OH)_3$	$P\bar{1}$	8.999	12.069	4.921	100.52	41.9.1.2
Bjarebyite	$(Ba,Sr)(Mn^{2+},Fe^{2+},Mg)_2$ $Al_2(PO_4)_3(OH)_3$	$P2_1/m$	8.930	12.073	4.917	100.15	41.9.1.3
Perloffite	$Ba(Mn^{2+},Fe^{2+})_2Fe_2^{3+}$ $(PO_4)_3(OH)_3$	$P2_1/m$	9.223	12.422	4.995	100.39	41.9.1.4

*Triclinic but α and γ are 90°.

41.9.1.1 Kulanite $BaFe_2^{2+}Al_2(PO_4)_3(OH)_3$

Named in 1976 for Alan Kulan (1921–1977), who collected the first specimen. The Fe^{2+} end member of a Fe^{2+}Mg series with penikisite. Bjarebyite group. TRIC $P\bar{1}$. $a = 9.032$, $b = 12.119$, $c = 4.936$, $\alpha = 90°$, $\beta = 100.38°$, $\gamma = 90°$, $Z = 2$, $D = 3.92$. CM 14:127(1976). 29-170: 8.84_6 4.51_5 3.58_5 3.11_{10} 3.04_7 2.93_8 2.69_6 2.66_7. Thin tablets flattened on $\{\bar{1}01\}$, forming rosettes up to 3 mm × 3 mm × 0.5 mm. Blue-green, vitreous luster, translucent to transparent. H = 4. G = 3.91. DTA/TGA. Nonfluorescent. Biaxial (+); $Y \approx b$; $Z \wedge c = -8°$; $N_x = 1.703$, $N_y = 1.705$, $N_z = 1.723$; $2V = 32°$; $r \gg v$, very strong; pleochroic. X, brown-green; Y, green; Z, pale brown. Found with quartz, siderite, and apatite at *Rapid Creek (Cross Cut Creek), Big Fish River, YT, Canada*, associated with lazulite, wardite, vivianite, childrenite, ludlamite, metavivianite, arrojadite, augelite, and brazilianite. HCWS *IMA(abstr.):274(1986)*.

41.9.1.2 Penikisite $BaMg_2Al_2(PO_4)_3(OH)_3$

Named in 1977 for Gunar Penikis (1936–1979), who discovered the locality and provided the sample for analysis. The Mg end member of the Mg-Fe^{2+} series with kulanite. Bjarebyite group. TRIC $P\bar{1}$. $a = 8.999$, $b = 12.069$, $c = 4.921$, $\alpha = 90°$, $\beta = 100.52°$, $\gamma = 90°$, $Z = 2$, $D = 3.82$. CM 15:393(1977). 29-169: 8.81_6 4.49_6 3.09_{10} 3.03_6 2.92_8 2.69_6 2.68_6 2.65_7. Similar to and grades into kulanite, blue-green to pale green plates in rosettes showing two cleavages $\{010\}$, $\{110\}$. DTA/TGA. Nonfluorescent. Biaxial (+); $Y = b$, $Z \wedge c = -6°$ (range 0–19°); $N_x = 1.684$, $N_y = 1.688$, $N_z = 1.705$ (Na); $2V = 56°$; pleochroic; $X \approx Y > Z$; $r \gg v$. X, grass-green; Y, blue-green; Z, pale pink. Found at *Rapid Creek (Cross Cut Creek), Big Fish R., YT, Canada*. HCWS

41.9.1.3 Bjarebyite $(Ba,Sr)(Mn^{2+},Fe^{2+},Mg)_2Al_2(PO_4)_3(OH)_3$

Named in 1973 for Gunnar Bjareby (1899–1967), Swedish–American artist, naturalist, and amateur mineralogist who collected the sample. Bjarebyite group. The name is applied to the Al end member, in which Mn > Fe, and Ba > Sr of the (Al–Fe^{3+}) series with perloffite (41.9.1.4). MON $P2_1/m$. $a = 8.930$, $b = 12.073$, $c = 4.917$, $\beta = 100.15°$, $Z = 2$, $D = 4.02$ for Fe = Mn.

MR 4:$282(1973)$. Structure: AM 59:$567(1974)$. 29-171: 8.95_{10} 3.12_8 3.06_4 2.94_4 2.90_5 2.71_5 2.68_4 1.790_4. See also 28-415(calc). Crystals complex spear-shaped up to 3 mm long, showing many forms: {001}, {100}, {010}, {110}, {120}, {111}, {121}, {131}, {141}, {411}, {211}, {021}, {011}. Forms usually pitted and etched and all negative forms missing. Emerald-green with bluish tinge, white streak, subadamantine luster, translucent. Cleavage {010}, {100}, perfect. H = 4+. G = 3.95. Biaxial (+); $N_x = 1.692$, $N_y = 1.695$, $N_z = 1.710$; 2V ≈ 35°; r > v, strong; pleochroic, grayish tan to pale yellow-green. Microprobe analysis shows minor Ca. Found at the *Palermo No. 1 pegmatite, North Groton, NH*, as crystals in cavities between amblygonite–scorzalite and Fe–Mn oxides. Additional associated minerals: augelite, childrenite, siderite, quartz, and palermoite. HCWS AM 59:$873(1974)$.

41.9.1.4 Perloffite $Ba(Mn^{2+},Fe^{2+})_2Fe_2^{3+}(PO_4)_3(OH)_3$

Named in 1977 for Louis Perloff (b.1907), amateur mineralogist and micromount collector, Tryon, NC. Bjarebyite group. The Fe^{3+} analog of bjarebyite (41.9.1.3). MON $P2_1/m$. $a = 9.223$, $b = 12.422$, $c = 4.995$, $\beta = 100.39°$, Z = 2, D = 3.996. MR 8:$112(1977)$. 29-184: 5.12_3 4.69_3 3.17_{10} 3.11_4 2.98_5 2.94_4 2.73_8 2.07_7. Spear-shaped crystals showing {001}, {$\bar{1}$01}, {021}, {$\bar{1}$31} to 1 mm long. Dark brown to greenish brown to black, greenish-yellow streak, vitreous to subadamantine luster. Cleavage {100}, perfect. H ≈ 5. Biaxial (−); X = b; Y ∧ c ≈ 42°; $N_x = 1.793$, $N_y = 1.803$, $N_z = 1.808$; 2V = 70–80°; r < v, strong; pleochroic. X,Z, dark greenish brown; Y, light greenish brown. Slowly dissolved by cold 1:1 HCl. Microprobe analyses also show minor Ca, Mg, and Al. Found at *Big Chief pegmatite, Glendale, SD*, as crystals on ludlamite, hureaulite, siderite, and in vugs in ludlamite formed from the hydration of triphylite. HCWS AM 62:$1059(1977)$.

41.9.2.1 Rockbridgeite $Fe^{2+}Fe_4^{3+}(PO_4)_3(OH)_5$

Named in 1949 for the locality. The Fe^{2+} end member of the Fe–Mn^{2+} series with frondelite (41.9.2.2). See also dufrenite (42.9.1.2). ORTH $Bbmm$. $a = 13.846$, $b = 16.782$, $c = 5.185$, Z = 4, D = 3.60. Structure: AM 55:$135(1970)$. Clusters of triple face-sharing octahedra of FeO_6 join through shared edges to form chains parallel to the b axis. These chains are linked to corner-sharing octahedral doublets whose composition can be expressed as $Fe_2(OH)_5O_6 + O$ from PO_4, which also secure the cluster triplet. 34-150: 3.59_4 3.44_4 3.38_4 3.19_{10} 3.02_2 2.42_2 2.406_2 1.592_2. See also 22-356. Subparallel plates {010} beveled by {100}, {110}, {001}, with a rounded dome {011}, often form sheafs or botryoidal crusts with fibrous structure. Aggregates often show color banding. Dark to olive-green; if oxidized, turns red-brown; vitreous to dull luster; subtranslucent. Cleavage {100}, excellent; {010}, good; {001}, fair; uneven fracture; brittle. H = $4\frac{1}{2}$. G = 3.3–3.49 depending on Fe/Mn ratio.. Biaxial (+); X = c; $N_x = 1.873$, $N_y = 1.880$, $N_z = 1.895$ (Rockbridge); 2V moderate; r < v; Z > Y > X; pleochroic. X, pale yellow-brown; Y, bluish green; Z, dark bluish green. Absorption less marked at high Fe^{3+};

indices increase with oxidation (Fe^{2+} to Fe^{3+}) and with increase in Fe/Mn ratio. Complete series $Fe^{2+}-Mn^{2+}$: rockbridgeite–frondelite. Small substitution of Ca, Mg for (Fe, Mn), Al for Fe^{3+}, As for P. Soluble in HCl but not in HNO_3 or H_2SO_4. A secondary mineral earlier confused with dufrenite and found as an alteration product of triphylite or other manganese phosphates in pegmatites. Associated minerals: vivianite, strengite, metastrengite, heterosite, and goethite. Found at Palermo No. 1 and Fletcher mines, North Groton, Grafton Co., NH; with limonite at Mullica Hill, NJ; at *South Mt., near Midvale, Rockbridge Co., VA*; at Polk Co., AR; at Jacksonville, Calhoun Co., Indian Mt., Cherokee Co., AL; at Big Chief and Tip Top pegmatites, SD; at Chanteloube, France; at Kreuzberg, Pleystein, and Hagendorf, Bavaria, Germany; at Reaphook Hill, SA, Australia; at Sapucaia Velha(!), and at Lavro do Enio, Galileia, Lavra do Ilha, Taquaral, MG, Brazil. HCWS *AM 34:541(1949)*, *DII:867*.

41.9.2.2 Frondelite $Mn^{2+}Fe_4^{3+}(PO_4)_3(OH)_5$

Named in 1949 for Clifford Frondel (b.1907), mineralogist, Harvard University, Cambridge, MA, and author of Vol. II (1956) and Vol. III (1962), 7th Edition, *Dana's System of Mineralogy*. The Mn analog of rockbridgeite. ORTH Bbmm. $a = 13.810$, $b = 16.968$, $c = 5.182$, $Z = 4$, $D = 3.55$. 35-625: 3.62_4 3.45_3 3.38_9 3.19_{10} 3.05_2 2.42_2 1.979_3 1.598_2. Similar to rockbridgeite in physical and chemical properties. Biaxial (−); $X = c$; $N_x = 1.869$, $N_y = 1.880$, $N_z = 1.893$ (Sapucaia); 2V moderate; $r > v$; pleochroic. X, pale yellow-brown; Y,Z, orange-brown. Analysis shows Al, minor Ca, Mg. Occurs at Fletcher mine, North Groton, NH, with metastrengite; at Sapucaia Velha, Galileia, MG, Brazil, and localities described for dufrenite. HCWS *DII:867*, *SMPM 41:407(1961)*.

41.9.3.1 Griphite $Mn_4Ca_6(Mn^{2+},Fe^{2+},Mg)Li_2Al_8(PO_4)_{24}(F,OH)_8$

Named in 1891 from the Greek for *enigma*, in allusion to the chemical composition. ISO $Pa3$. $a = 12.205$, $Z = 4$, $D = 3.65$. Structure: *BM 101:543(1978)*. Structure resembles garnet: two interpenetrating networks one of PO_4 tetrahedra and AlO_6 octahedra that are expressed as the PO_4 tetrahedra alternating with AlO_6 octahedra and CaO_6F_2 cubes. 42-1351: 7.05_1 3.39_1 3.05_3 2.96_2 2.73_{10} 2.49_3 2.27_2 1.693_1. See also 41-582. Massive, reniform. Yellow, brown, and black; resinous to vitreous luster; translucent; slightly metamict. No cleavage, conchoidal fracture, brittle. $H = 5\frac{1}{2}$. $G = 3.64$. $N_{var} = 1.63-1.66$. Readily soluble in acids. Analyses show Na, Li, Al, Mg, Fe^{3+}, H_2O. Found in masses up to 125 kg in granite pegmatite at *Everly mine, Riverton Lode, Harney City, Pennington Co., SD*; at Alberes massif, East Pyrenees, France; at Mt. Ida, NT, Australia. HCWS *AM 27:452(1942)*, *BM 101:536(1978)*.

LAZULITE GROUP

The lazulite group are phosphate mineral with the general formula

$$AB_2(PO_4)_2(OH)_2$$

where

$A = Cu, Fe^{2+}, Mg$
$B = Al, Fe^{3+}$

and the minerals crystallize in the monoclinic system.

The structure [*NJMM:410(1983)*] is made up of polyatomic triplet clusters composed of one AO_6 octahedron sandwiched by two corner-sharing BO_6 octahedra disposed in the c–a plane. The clusters are connected through the hydroxyls of the BO_6 polyhedra and the oxygens of the tetrahedral PO_4 groups into a compact network.

LAZULITE GROUP

Mineral	Formula	Space Group	a	b	c	β	
Lazulite	$MgAl_2(PO_4)_2(OH)_2$	$P2_1/c$	7.228	7.278	7.144	120.51	41.10.1.1
Scorzalite	$Fe^{2+}Al_2(PO_4)_2(OH)_2$	$P2_1/n$	7.250	7.301	7.155	120.58	41.10.1.2
Hentschelite	$CuFe_2^{3+}(PO_4)_2(OH)_2$	$P2_1/n$	7.266	7.786	6.984	117.68	41.10.1.3
Barbosalite	$Fe^{2+}Fe_2^{3+}(PO_4)_2(OH)_2$	$P2_1/n$	7.397	7.489	7.313	118.52	41.10.1.4

41.10.1.1 Lazulite $MgAl_2(PO_4)_2(OH)_2$

Named in 1795 from the old German *lazurstein*, blue stone. Lazulite group. The Mg end member in a Mg–Fe^{2+} series with scorzalite. MON $P2_1/c$. $a = 7.228$, $b = 7.278$, $c = 7.144$, $β = 120.51°$, $Z = 2$, $D = 3.09$. Structure: *NJMM:410(1983)*. *34-136*: 3.24_{10} 3.21_9 3.15_6 3.08_7 2.55_4 1.985_2 1.572_4 1.549_3. Crystals acute pyramidal with large $\{111\}$, $\{\bar{1}11\}$, small $\{101\}$, tabular on $\{\bar{1}11\}$ or $\{101\}$; massive compact or granular. Twinning on $\{100\}$ common, sometimes polysynthetic or lamellar, composition plane $\{001\}$ or $\{100\}$ usually shows reentrant angle, rarely on $\{223\}$ and other planes. Azure-blue, sky-blue, bluish white to bluish green; white streak; vitreous luster; subtranslucent to opaque, occasionally transparent and of gem quality. $H = 5\frac{1}{2}$–6. $G = 3.08$–3.38. Biaxial (−); $X \wedge c = 9\frac{1}{2}°$; $Y = b$; $N_x = 1.604$, $N_y = 1.626$, $N_z = 1.637$; $2V = 69.7°$; $r < v$. Increase in Fe^{2+} increases indices, and birefringence, 2V decreases slightly. Slowly soluble in hot acids. Some Mn, Ca, Fe^{2+}, Fe^{3+} (substituting for Al) reported in analyses. Occurs in aluminous high-grade metamorphic rocks with quartz as disseminated grains, in quartz veins or dikes in such rocks, in granite pegmatites. Found at Scott mine, ME; at G.E. Smith mine, Newport; Palermo No. 1 mine, North Groton, NH; with corundum at Chubbs Mt. and Chowders Mt., Gaston Co., NC; with kyanite and rutile in quartzite at Graves Mt., Lincoln Co., GA; with andalusite, rutile in garnetiferous quartzite, Markleeville, Alpine Co., and

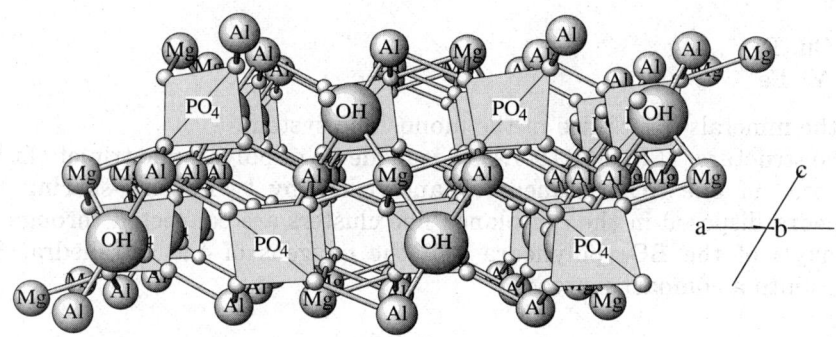

LAZULITE: MgAl$_2$(PO$_4$)$_2$(OH)$_2$

This *b* axis projection of lazulite shows the intimate connections through the oxygens from the (PO$_4$) group and octahedrally coordinated cations. There are two B–(O,OH)$_6$ and one A–(O,OH)$_6$ octahedra in this compact structure. Compare this structure with bjarebyite (41.9.1.3).

with andalusite at Champion mine, White Mts., Mono Co.(!), CA; at Blow R. near Mt. Fitton, YT, Canada; with svanbergite at Horrsjöberg near Ny(!!), with berlinite at Westanå, Sweden; with wagnerite in quartz carbonate veins in clay slates near Werfen, Salzburg(!), Austria, and at Vorau near Graz, in the Müztal, Styria; near Zermatt, Valais, Switzerland; in quartz veins in lower Triassic quartzite of Tribec, Slovakia, Czech Republic; in quartz veins on Mt. Bity and at Ranomainty SE of Betafo, and Jamalezo, Madagascar; at the Barunda pegmatite, Rwanda; as pseudomorphs with sericite after orthoclase at Real Socavon, Potosi, Bolivia; at Fazeuda Modelo, and in quartz veins with itacolumite at Dattas, MG, Brazil. HCWS *DII:908; AM 47:157(1962), 67:610(1982)*.

41.10.1.2 Scorzalite Fe^{2+}Al$_2$(PO$_4$)$_2$(OH)$_2$

Named in 1947 for Evaristo Pena Scorza (b.1899), mineralogist, Mineral Survey of Brazil. Lazulite group. The Fe^{2+} end member of the Fe^{2+}–Mg series with lazulite. MON $P2_1/n$. $a = 7.250$, $b = 7.301$, $c = 7.155$, $\beta = 120.58°$, $Z = 2$, $D = 3.32$. Structure: *AC 12:695(1959)*. *35-632*: 6.17$_2$ 4.75$_2$ 3.25$_8$ 3.21$_{10}$ 3.15$_5$ 3.08$_4$ 2.55$_2$ 2.22$_2$. Physical properties similar to lazulite. Biaxial (−); $N_x = 1.639$, $N_y = 1.670$, $N_z = 1.680$; $2V = 58.3°$ (Fe/Mg = 4:1); $r < v$; Z>Y>X. Strongly pleochroic. X, colorless; Y, blue; Z, darker blue. High-Fe^{2+}-content samples are rare. Occurrences the same as lazulite and especially at Victory pegmatite, Custer Co., SD; *Corrego Frio pegmatite, MG, Brazil.* HCWS *DII:908; AM 34:83(1949), 35:1(1950), 67:619(1982).*

41.10.1.3 Hentschelite $CuFe_2^{3+}(PO_4)_2(OH)_2$

Named in 1987 for Gerhard Hentschel (b.1930), Hessisches Landesamt für Bodenforschung, Wiesbaden, Germany. Lazulite group. MON $P2_1/n$. $a = 7.266$, $b = 7.786$, $c = 6.984$, $\beta = 117.68°$, $Z = 2$, $D = 3.79$. *AM 72:404(1987).* Structure: *AC(C) 43:1855(1987). 40-501:* 4.81_5 3.33_{10} 3.27_5 3.06_2 2.40_2 2.11_2 1.669_2 1.648_2. Disseminated, clusters and drusy crystals to 1 mm, showing {100}, {101}, {011}, {11$\bar{1}$}; twinned, composition plane {102}. Dark green, light green streak, vitreous luster, translucent. No cleavage, irregular fracture. $H = 3\frac{1}{2}$. Biaxial (+); $Y = b$; $X \approx [101]$; $N_x = 1.843$, $N_y = 1.848$, $N_z = 1.945$ (598 nm); $2V = 15°$; $r \gg v$. Microprobe analysis shows some As, Al, minor Si. Found as replacement mineral in silicified barite veins at *Reichenbach, near Benshien, Hessen, Germany*, with mimetite, beudantite, goethite, and quartz. HCWS *MM 52:408(1986).*

41.10.1.4 Barbosalite $Fe^{2+}Fe_2^{3+}(PO_4)_2(OH)_2$

Named in 1955 for Aliuzio Licinio de Mirande Barbosa (b.1896), geologist, Escola de Minas, MG, Brazil. Lazulite group. Dimorph of lipscombite. MON $P2_1/n$. $a = 7.397$, $b = 7.489$, $c = 7.313$, $\beta = 118.52°$, $Z = 2$, $D_{(recalc)} = 3.71$. Structure: *AC 12:695(1959). 33-668:* 4.84_6 3.36_{10} 3.31_8 3.24_5 3.16_6 2.61_3 2.33_4 1.309_4. Massive, very fine-grained crystalline aggregates. Dark blue-green to black, opaque, appears dark green in thin section. $H = 6$. $G = 3.60$. Biaxial (+); $N_x = 1.77$, $N_y = 1.79$, $N_z = 1.835$; pleochroic. X,Y, dark blue-green; Z, dark olive-green. Slightly soluble in hot HCl, not soluble in 1:10 HNO_3, H_2SO_4; attacked by Clerici solution. Microprobe analysis shows Mn, Mg and very minor Na, Ti. Found at Palermo No. 1 mine, North Groton, NH; at Williams property, AL; at Bull Moose pegmatite, SD; at Thackaringa pegmatite, 40 km SW of Broken Hill, NSW, Australia, where barbosalite replaces manganese oxides and is associated with wolfeite, arsenopyrite, scorodite, and quartz; at *Sapucaia pegmatite, Galileia, MG, Brazil*, with heterosite, triphylite, hureaulite, and tavorite. HCWS *AM 40:952(1955), MM 43:505(1979).*

41.10.2.1 Lipscombite $(Fe^{2+},\square)Fe_2^{3+}(PO_4)_2(OH)_2$

Named in 1953 for William Nunn Lipscomb (b.1909), mineralogist, University of Minnesota, MN. Dimorph of barbosalite (41.10.1.4). TET $P4_32_12$. $a = 7.310$, $c = 13.212$, $Z = 4$, $D_{syn} = 3.68$. Structure: *AM 74:456(1989)* (synthetic). Two disordered partially filled Fe^{2+} positions, and one Fe^{3+} position, all octahedrally coordinated $[Fe(O, OH)_6]$ with O from PO_4 tetrahedra and (OH) groups. Octahedra form chains parallel to [110]. *PD*(syn): 4.82_5 3.35_{10} 3.30_5 3.23_5 3.15_4 2.61_2 2.32_3 2.28_2. See also *14-569*(syn), *14-310* (manganoan, Mn/Fe = 2:1: $P4_12_12$, $a = 7.40$, $c = 12.81$, $G = 3.66$). Massive. Olive-green to black, splendent. $G = 3.8$. DTA. $N = 1.67$–1.80. Manganoan (Mn^{2+}) sample also contains Cu, Pb, Mg, Zn, Be, and Ni. Synthesized. Tetragonal–monoclinic phase change between 110 and 290°. Found at Williams property, AL; manganoan at *Sepucaia pegmatite, Galileia, MG, Brazil*, in vugs with frondelite, leucophosphite, and metastrengite. HCWS *AM 38:612(1953), 47:353(1962)*.

41.10.3.1 Jagowerite $BaAl_2(PO_4)_2(OH)_2$

Named in 1973 for John Arthur Gower (1921–1972), mineralogist, University of British Columbia, Canada. TRIC $P\bar{1}$ or $P1$. $a = 6.049$, $b = 6.964$, $c = 4.971$, $\alpha = 116.51°$, $\beta = 86.06°$, $\gamma = 112.59°$, $Z = 1$, $D = 4.05$. Structure: *AM 59:291(1973)*. A network of PO_4 tetrahedra that share common corners with pairs of $Al(O, OH)_6$ octahedra. *35-636:* 5.53_4 3.26_6 3.14_5 3.00_{10} 2.94_3 2.91_3 2.17_3 2.02_2. See also *26-136, 28-123*(calc). Crystalline masses up to 2.5 cm. Light green, vitreous luster, translucent. Cleavage {100}, {0$\bar{1}$1}, good; {0$\bar{2}$1}, fair. $H = 4\frac{1}{2}$. $G = 4.01$. Fluoresces greenish white in LWUV. IR. Biaxial (+); $N_x = 1.672$, $N_y = 1.6923$, $N_z = 1.710$; $2V \approx 80°$. Insoluble in HCl. Analyses show Fe^{3+}, S, and < 1000 ppm of Ca, Cr, Ti, V, Ta, Mn, Cu, Be, Sr, Si. Found *16 miles N of Hess River, west-central YT, Canada*. HCWS *CM 12:135(1973), AM 60:945(1975)*.

41.10.4.1 Goedkenite $(Sr,Ca)_2Al(PO_4)_2(OH)$

Named in 1975 for Virgil Linus Goedken (b.1940), chemist, University of Chicago, IL. Brackebuschite group. See (40.2.8.1). Close to the Sr end member of a Sr–Ca series with bearthite (41.10.4.2). MON $P2_1/m$. $a = 8.45$, $b = 5.74$, $c = 7.26$, $\beta = 113.7°$, $Z = 2$, $D = 3.83$. *AM 60:957(1975)*. *29-383:* 7.76_5 4.33_4 3.42_4 3.06_{10} 2.84_6 2.81_4 2.59_7 1.609_5. Lozenge-shaped to spear-shaped crystals up to 1 mm occurring in parallel growth with {001} bent to feathery tips, forms: {001}, {011}, {012}, {$\bar{1}$01}, {111}; tabular parallel to {001}, elongated parallel to [100]. Colorless to pale yellow, subadamantine luster, translucent. $H = 5$. Biaxial (+); X parallel to b; $N_x = 1.669$, $N_y = 1.673$, $N_z = 1.692$; $2V = 45–50°$. Microprobe analysis shows Sr > Ca, minor Mg, Mn, Fe, Na. Secondary mineral found at *southern wall of Palermo No. 1 pegmatite, North Groton, NH*, replacement of goyazite, palermoite, and associated with apatite, whitlockite, quartz, siderite, bjarebyite, and childrenite. HCWS

41.10.4.2 Bearthite $Ca_2Al(PO_4)_2(OH)$

Named in 1993 for P. Bearth (1902–1989), Swiss petrologist. Brackebuschite group. See (40.2.8.1). MON $P2_1/m$. $a = 7.231$, $b = 5.734$, $c = 8.263$, $\beta = 112.57°$, $Z = 2$, $D = 3.25$. Structure: *SMPM 73:1(1993)*. PD: 4.58_2 3.05_{10} 2.87_6 2.75_3 2.63_2 2.57_4 2.44_2. Aggregates up to 1 mm in size, consisting of flat prismatic crystals and smaller anhedral grains. Yellowish, white streak, transparent. Poor prismatic cleavage, uneven fracture. $H < 5$. Biaxial (+); X parallel to b; $N_x = 1.662$, $N_y = 1.671$, $N_z = 1.696$; $2V = 65°$; $r < v$. Microprobe analysis: Sr, Mg, Fe^{2+}, Ce, La, Si, S, F, Cl. Soluble in HCl. Synthesized.

Found in the high-pressure terranes of alpine metamorphism corroded by lazulite in concordant quartz segregations of the gneiss and schist at *Monte Rosa massif, Zermatt Valley, Switzerland*; in pyrope–phengite quartzite and in coesite-bearing metapelite in Dora Maira massif, western Alps, Italy. HCWS *AM 78:1314(1993)*.

41.10.4.3 Gamagarite $Ba_2(Fe^{3+},Mn^{3+})(VO_4)_2(OH)$

Named in 1943 for a ridge of hills adjacent the locality. Isostructural with brackebushite (40.2.8.1.) but contains oxidized cations: Fe^{3+} and Mn^{3+}. See also goedkenite, 41.10.4.1, bearthite, 41.10.4.2, and tsumebite, 43.4.2.1. MON $P2_1/m$. $a = 9.121$ $b = 6.142$ $c = 7.838$ $\beta = 112.88°$ $Z = 2$ $D = 4.71$ (Italy). Structure: *NJMM 295 (1987)*. Chains of $(Fe,Mn)O_{6 \text{ or } 8}$ polyhedra parallel to [010] are connected by two non-equivalent tetrahedral groups $VO_4(2)$ and $VO_4(1)$ into a framework that includes two distinct and irregular polyhedra: $Ba(O,OH)_8$ and $Ba(O,OH)_{11}$. *41-1413* 3.71_2 3.30_{10} 3.07_3 3.04_5 2.80_5 2.34_3 2.16_2 1.651_2. Crystals prismatic, needles up to a cm long, flattened {001}. Dark brown, nearly black, streak red-brown, adamantine. Cleavages {001}. {100} distinct, {$\bar{1}$01} indistinct, brittle. Magnetic. $H = 4\frac{1}{2} - 5$ $G = 4.62$. Biaxial (+); $N_x 2.106$, $N_y 2.040$, $N_z 2.130$; $x \wedge c = 43°$; $y = b$, elongation $2V = 46 - 62°$, $r > v$ strong. Pleochroc x = red brown, y = very deep red-brown, z = light salmon buff. Absorbtion $y > x > z$. Minute opaque inclusions of hematite, sitaparite (S. Africa). Analysis: 48.3% BaO, 3.60% SrO, 7.80% Fe_2O_3, 6.11% Mn_2O_3, 29.05% V_2O_5, 3.47% As_2O_5, H_2O by difference 1.58% (Italy). IR suggests a strong hydrogen-bonded OH, which is probably the dominant hydrogen species in this mineral, not an acid vanadate *AM 60:803 (1984)*. Found associated with sitaparite, diaspore, at *Gloucester farm, Postmasburg dist., near Gamagara Ridge, Cape Prov., South Africa*; at Molinello, Val Graveglia, N. Appennines, E. Liguria, Italy *AM 28:329 (1943)*. HCWS

41.10.5.1 Melonjosephite $CaFe^{2+}Fe^{3+}(PO_4)_2(OH)$

Named in 1973 for Joseph Mélon, mineralogist, University of Liège, Belgium. ORTH *Pbam*. $a = 9.542$, $b = 10.834$, $c = 6.374$, $Z = 4$, $D = 3.615$. Structure: *AM 62:60(1977)*. *25-1454*: 5.42_9 3.19_3 3.05_{10} 2.91_4 2.71_9 2.62_6 2.38_3 2.19_4. Fibrous masses. Dark green to black, brilliant to resinous luster, translucent. Splintery cleavage along fibers, transverse imperfect; brittle. $H \approx 5$. IR. Biaxial (−); XYZ= cab; $N_x = 1.720$, $N_y = 1.770$, $N_z = 1.800$; $2V = 80–85°$; strong dispersion; pleochroic. X, deep brown; Y, greenish brown; Z, yellow to greenish yellow. Readily dissolved by HCl. Analyses show Al, Mn, Mg, Na, Li, and traces of Pb, Mo, Ni. Found in the *Angarf–South pegmatite, Morocco*, in the zone surrounding triphylite. HCWS *BM 96:135(1973)*, *AM 60:946(1975)*.

41.10.6.1 Carminite $PbFe_2^{3+}(AsO_4)_2(OH)_2$

Named in 1850 for the color. ORTH *Amaa*. $a = 16.595$, $b = 7.580$, $c = 12.295$, $Z = 8$, $D = 5.46$. *AM 44:663(1959)*. Structure: *AM 48:1(1963)*. *39-1355*: 4.58_4 3.52_5 3.23_{10} 3.04_4 2.95_5 2.72_4 2.55_{10} 1.805_3. See also *12-278*. Lathlike crystals, flattened on {010}, elongated [001]; tufted or satiny aggregates of fine needles, massive with radiating structure; radially fibrous aggregates. Carmine-red to tile-red, reddish brown; reddish-yellow streak; vitreous luster; translucent. Cleavage {110}, distinct; brittle. H = $3\frac{1}{2}$. G = 4.10 (Mapimi). Biaxial (+); XYZ= cab; N_x = 2.070, N_y = 2.070, N_z = 2.080; 2V medium; r < v, strong; pleochroic: X, pale yellowish red; Y,Z, dark carmine-red. Slowly soluble in HCl with separation of $PbCl_2$, completely soluble in HNO_3. Found in the United States at Buckwheat quarry, Franklin, Sussex Co., NJ; at Eureka, NV, and Tintic district, UT; at Hingston Down Consols mine, Calstock, Cornwall, England, with scorodite, mimetite, and pharmacosiderite; at Mapimi, DUR, Mexico, with scorodite, dussertite, arsenosiderite, anglesite, and cerussite; at *Luise iron mine, Horhausen, Rhine Prov., Germany*, with beudantite and at Ems, Hesse-Nassau; at Wyloo, Ashburton district, WA, Australia; at Tsumeb, Namibia. HCWS *DII:912*.

41.10.7.1 Mounanaite $PbFe_2^{3+}(VO_4)_2(OH)_2$

Named in 1969 for the locality. TRIC $P\bar{1}$. a = 5.55, b = 7.66, c = 5.56, α = 111.02°, β = 112.12°, γ = 94.15°, Z = 1, D = 4.89. *BM 92:196(1969)*. 22-657: 4.66_{10} 4.56_9 3.26_8 2.82_8 2.77_8 2.60_5 2.32_5 1.711_6. Crystals elongated on c, tabular on [010], common forms {010}, {100}, {110}, {011}, {11$\bar{1}$}, {12$\bar{1}$}, {02$\bar{1}$}, {01$\bar{1}$}, {021}. Most crystals twinned on [001], or {1$\bar{1}$1}. Reddish brown, translucent. G = 4.85. DTA. Biaxial; N > 2.09; extinction on {010} = 38° to c; pleochroic, brown-red to brown-yellow at 90° in this orientation. Synthesized. Found at *Mounana mine, Franceville, Haute Ogooué, Gabon*, associated with goethite, francevillite, and curienite. HCWS *AM 54:1738(1969)*.

41.10.8.1 Samuelsonite $(Ca,Ba)Fe_2^{2+}Mn_2^{2+}Ca_8Al_2(PO_4)_{10}(OH)_2$

Named in 1975 for Peter B. Samuelson (b.1941), prospector, Rumney, NH. MON $C2/m$. a = 18.495, b = 6.805, c = 14.000, β = 112.75°, Z = 2, D = 3.355. *AM 60:957(1975)*, Structure: *AM 62:229(1977)*. The unit $[Ca_4(PO_4)_{12}]$ forms double columns (similar units are found in the apatite structure) which are linked with chains of $[Al(OH)_2O_4]$ and cation-O_6 octahedra. 29-154: 3.39_5 3.25_4 3.06_{10} 3.03_6 2.73_5 2.66_7 2.56_4 1.706_6. Crystals lamellar, bladed, lath-shaped or striated parallel to [010], prisms up to 4 cm long and 1 cm in width showing {001}, {100}, {101}, {$\bar{1}$01}, {$\bar{2}$11}, {$\bar{1}$12} with etched terminations. Colorless to pale pink stained yellow and yellow-brown, subadamantine luster. H = 5+. G = 2.353. Biaxial (+); Z ∧ c = 22°; N_x = 1.645, N_y = 1.650, N_z = 1.655; 2V large. Trace quantities of Cu, Ti, Ga detected on spectrographic analysis. Found at *Palermo No. 1 mine, North Groton, NH*, embedded in black ferric and manganic oxides and in whitlockite–carbonate apatite rock; in the Buranga pegmatite, Rwanda. HCWS

41.10.9.1 Preisingerite $Bi_3O(OH)(AsO_4)_2$

Named in 1981 for Anton Preisinger (b.1925), mineralogist, Department of Crystallography and Structural Chemistry, University of Technology,

Vienna. TRIC $P\bar{1}$. $a = 9.993$, $b = 7.404$, $c = 6.937$, $\alpha = 87.82°$, $\beta = 115.01°$, $\gamma = 111.01°$, $Z = 2$, $D = 7.24$. Structure: *AM 67:833(1982)*. $Bi_6O_2(OH)_2$ groups link into a framework with AsO_4 tetrahedra. There are three different Bi sites. *35-543:* 4.52_4 3.28_{10} 3.26_{10} 3.19_9 3.09_8 3.02_4 2.61_2 1.967_3. Porous aggregates of tiny (up to 0.02 mm) tabular crystals, pulverulent masses. White to grayish white, white streak, subadamantine luster, translucent. Indistinct cleavage, conchoidal fracture, brittle. $H = 3-4$. $G \approx 6$. TGA (decomposes $\approx 400°$). Biaxial (+); $N_x = 2.130$, $N_y = 2.16$ (calc), $N_z = 2.195$ (589 nm); $2V \approx 90°$. Some substitutions of Pb and P. Found in weathering zones in the Mammoth mine, Tintic district, UT; Tazna, Bolivia; and at *San Francisco de Los Andes and Cerro Negro de la Aguadita, Calingasta, San Juan Prov., Argentina*, with rooseveltite. HCWS *FM 59:15(1981); AM 28:536(1943), 67:416(1982)*.

41.10.9.2 Schumacherite $Bi_3O(OH)[(V,P,As)O_4]_2$

Named in 1983 for Friedrich Schumacher, professor, University of Freiberg and Bonn, Germany. See also preisingerite (41.10.9.1). TRIC $P\bar{1}$. $a = 10.05$, $b = 7.46$, $c = 6.90$, $\alpha = 87.7°$, $\beta = 115.3°$, $\gamma = 111.5°$, $Z = 2$, $D = 6.90$. *TMPM 31:165(1983)*. *35-587:* 6.21_4 4.57_6 4.13_4 3.28_{10} 3.19_8 3.09_8 3.02_4 1.976_5. Crystals to 0.1 mm, tabular {010} showing {110}, {101}; crusts. Yellow to yellow-brown, adamantine luster, translucent. No cleavage, conchoidal fracture. $H \approx 3$. Biaxial (+); $X' \wedge c \approx 22°$, $Z' \wedge a \approx 15°$; $N_x > 2.20$, $N_z < 2.30$–2.42; $2V \approx 90°$; optic plane approximately parallel to $\{11\bar{1}\}$; extinction on (010). Microprobe analyses show As, P with V/As/P highly variable, probably part of a series with preisingerite. Crystal structure change $\approx 800°$, similar to preisingerite. Found as crusts on quartz-rich vein specimens from *Schneeberg, Germany*. HCWS *AM 70:438(1985)*.

41.10.10.1 Petitjohnite $Bi_3O(OH)(PO_4)_2$

Named in 1993 for K. Petitjohn, a mineral collector, Odenwald, Germany, who found the mineral. The P analog of preisingerite (As) and schumacherite (V). TRIC $P1$. $a = 9.789$, $b = 7.250$, $c = 6.866$, $\alpha = 88.28°$, $\beta = 115.27°$, $\gamma = 110.70°$, $Z = 2$, $D = 6.99$. *NJMM:487(1993)*. PD: 4.44_5 3.25_9 3.19_{10} 3.14_{10} 3.03_8 2.95_5. Crusts and small spherical aggregates with dominant forms {010}, {100}, {001}, and {101}; rotation twins on [010] with preisingerite. White to dark brown (V-bearing samples are yellow), vitreous to adamantine luster. $H = 4\frac{1}{2}$. Biaxial (+); $N_{x'} = 2.06$, $N_{z'} = 2.13$; $2V = 75°$; $r > v$. Forms complete solid solution with preisingerite and schumacherite. Found in association with bismuthite, mixite, reichenbachite, pyromorphite, and malachite in silicified barite veins at *Gaderheim* and *Reichenbach, near Bensheim, Odenwald, Hesse, Germany*, and on the mine dumps on the Pucher-Schacht, Saxony. HCWS

41.10.11.1 Drugmanite $Pb_2(Fe^{3+},Al)H(PO_4)_2(OH)_2$

Named in 1979 for Julien Drugman (1875–1950), Belgian mineralogist. MON $P2_1/a$. $a = 11.111$, $b = 7.986$, $c = 4.643$, $\beta = 90.41°$, $Z = 2$, $D = 5.55$. BM 111:431(1988). Structurally related to the silicates: gadolinite group. 33-732: 4.63_9 3.75_{10} 3.35_8 3.25_8 2.91_9 2.71_7 2.39_6 2.13_6. Platy crystals to 0.2 mm in aggregates, showing {001}, {110}. Colorless, transparent. H < 6. G = 4.10. $N_{mean} = 1.87$; 2V = 33°; X approximately perpendicular to {001}; optic axis plane = (010); r < v, strong. Occurs in vugs in silicified limestone at *Richelle, Belgium*. HCWS AM 65:809(1980).

41.11.1.1 Trolleite $Al_4(PO_4)_3(OH)_3$

Named in 1868 for Hans Gabriel Trolle-Wachtmeister (1782–1871), Swedish chemist. MON $I2/a$. $a = 18.894$, $b = 7.161$, $c = 7.162$, $\beta = 99.99°$, $Z = 4$, $D = 3.08$. Structure: AM 59:974(1974). Structurally similar to lazulite (41.10.1.1). 26-1009: 6.67_4 3.34_4 3.21_{10} 3.10_9 3.08_5 2.52_5 1.983_4 1.547_3. Polycrystalline nodules, lenticular masses and blebs, massive, lamellar. Pale green, vitreous luster, translucent. Indistinct cleavage in two directions at an angle of 110.9°, even to conchoidal fracture. H = $5\frac{1}{2}$–6. G = 3.10. Biaxial with optic plane parallel to best of cleavages; 2V ≈ 80°; r > v, weak. Synthesized. High-pressure studies show trolleite relationship to berlinite (38.4.2.1), variscite (40.4.1.1), and metavariscite (40.4.3.1). Found at White Mts., Mono Co., CA; in polycrystalline nodules in kyanite-bearing specularite–muscovite–quartzite with berlinite, attacolite, pyrophyllite, augelite, and lazulite, at *Westanå, Kristianstad, Sweden*. HCWS DII:911, AM 64:1175(1979).

41.11.2.1 Jamesite $Pb_2Zn_2Fe_5^{3+}(AsO_4)_5O_4$

Named in 1981 for Christopher James, mining engineer, Tsumeb, Namibia. TRIC $P\bar{1}$. $a = 5.622$, $b = 9.593$, $c = 10.279$, $\alpha = 109.80°$, $\beta = 90.54°$, $\gamma = 97.69°$, $Z = 1$, $D = 5.10$. CE 40:105(1981). 35-509: 9.67_8 4.70_8 3.40_{10} 3.26_8 3.04_9 2.92_6 2.68_5 2.04_6. Massive, disseminated. Reddish brown, subadamantine luster. H ≈ 3. Biaxial (−); Y ∧ a on (001) ≈ 5°; Y ∧ a on (010) = 15°; $N_x = 1.960$, $N_y = 1.995$, $N_z = 2.020$; 2V ≈ 75°; r > v; strongly pleochroic: X,Y, pale brown; Z, deep reddish brown. Found in the oxidized zone at *Tsumeb, Namibia*. HCWS AM 66:1275(1981).

41.11.3.1 Fingerite $Cu_{11}(VO_4)_6O_2$

Named in 1985 for Larry W. Finger (b.1940), mineralogist, Geophysical Laboratory, Carnegie Institute of Washington, DC. TRIC $P\bar{1}$. $a = 8.158$, $b = 8.269$, $c = 8.044$, $\alpha = 107.14°$, $\beta = 91.39°$, $\gamma = 106.44°$, $Z = 1$, $D = 4.776$. AM 70:193(1985), Structure: AM 70:197(1985). CuO_6 octahedrally coordinated forms a sheet with holes cross-linked by VO_4 tetrahedra and CuO_5 bipyramids. Cu sites show Jahn–Teller effect. 38-378: 6.48_{10} 4.74_5 3.89_6 3.62_6 3.29_6 3.04_8 2.82_7 2.03_7. See also 36-431. Crystals up to 150 μm. Black, dark reddish-brown streak, metallic luster, opaque. No cleavage. Nonfluorescent. $N_{mean\ calc} = 2.124$. In RL in air, gray, nonpleochroic, two extinctions on complete rotation of stage; in blue-filtered white light $= 21°$. Found as a fumarolic deposit at *Izalco volcano, El Salvador*, with thenardite and euchlorine. HCWS

41.11.4.1 Nefedovite $Na_5Ca_4(PO_4)_4F$

Named in 1983 for E. I. Nefedov (1910–1976), Russian scientist. TRIC $P1$ or $P\bar{1}$. $a = 5.401$, $b = 11.647$, $c = 16.484$, $\alpha = 134.90°$, $\beta = 90.04°$, $\gamma = 89.96°$, $Z = 2$, $D = 3.05$. *ZVMO 112:479(1983)*. Structure: *SPD 29:700(1984)*. Pseudotetragonal: $a = 11.65$, $c = 5.40$. *37-422:* 5.83_4 3.73_8 2.77_{10} 2.70_7 2.51_8 2.29_8 2.14_4 1.877_6. Irregular grains 0.01- to 0.5-mm form aggregates to 5 mm. Colorless, vitreous luster, transparent. Conchoidal fracture. $H \approx 4\frac{1}{2}$. $G = 3.01$. Uniaxial (+); $N_O = 1.571$, $N_E = 1.590$. Insoluble in H_2O, dissolves at room temperature in 1:10 HCl with no residue. Microprobe analysis shows minor K. Found with nacaphite, adularia, alkali amphibole, and titanite in pegmatitic segregations in agpaitic nepheline syenites of the *Khibinski massif, Kola Penin., Russia*. HCWS *AM 69:812(1984)*.

41.11.5.1 Atelestite $Bi_8(AsO_4)_3(OH)_5O_5$

Named in 1832 from the Greek *incomplete*, presumably because its composition was unknown when first described. MON $P2_1/a$. $a = 10.88$, $b = 7.42$, $c = 6.98$, $\beta = 107.13°$, $Z = 1$, $D = 6.05$. *CM 7:547(1963)*. *15-735:* 6.62_2 4.23_2 3.24_{10} 3.12_4 2.73_3 2.53_2 2.20_2 2.12_2. Minute (to 1-mm) crystals, tabular on $\{100\}$, showing $\{100\}$, $\{\bar{1}01\}$, $\{101\}$, $\{313\}$; the last rounded, other faces smooth and brilliant; spherical or mammillary crystalline aggregates smooth on the surface. Sulfur-yellow to yellowish green or wax-yellow, resinous to adamantine luster, transparent to translucent. $H = 4\frac{1}{2}$–5. $G = 6.82$. Biaxial (+); $N_x = 2.14$, $N_y = 2.15$, $N_z = 2.18$ (Na) (Neuhilfe); $2V = 44°$; $r < v$, strong. Easily soluble in HCl, with difficulty in HNO_3. Found with bismutite, eulytite, and quartz at *Neuhilfe mine, Schneeberg, Germany*, and at Neustädtl, where associated with walpurgite, torbernite, and bismutite. HCWS *DII:792*.

Class 42

Hydrated Phosphates, Arsenates, and Vanadates Containing Hydroxyl or Halogen

42.1.1.1 Liskeardite $(Al,Fe^{3+})_3(AsO_4)(OH)_6 \cdot 5H_2O$

Named in 1878 for the locality. Probably ORTH. *11–146:* 17.6_{10} 12.2_6 8.65_8 7.85_6 7.44_6 4.25_4 3.33_{10} 2.37_6. Massive, thin crusts with radial fibrous structure. White to greenish, bluish, brownish white. Cleavage parallel to elongation. Soft. G = 3.10. Biaxial (+); $N_x = 1.661$, $N_y = 1.675$, $N_z = 1.689$; $2V \approx 90°$. Found coating scorodite, arsenopyrite, chalcopyrite, pyrite, and quartz at *Liskeard, Cornwall, England*. Similar(?) mineral from Cap Garrone, Hyères, Var, France, HCWS *DII:924, BM 75:71(1952).*

42.1.2.1 Evansite $Al_3(PO_4)(OH)_6 \cdot 6H_2O$

Named in 1864 for Brooke Evans (1797–1862), mining geologist, Birmingham, England, who brought the specimens from Romania in 1855. Massive, opaline, reniform coatings, sometimes showing concentric, colloform structure; stalactitic. Colorless to milky white, tinged with blue, green-yellow; white streak; vitreous to resinous, waxy luster; transparent to translucent. H = 3–4. G = 1.8–2.2. $N_{av} = 1.485$ (Sirk). Easily soluble in acids. Possibly amorphous. A secondary mineral found with "limonite" and allophane. Occurs at Columbiana, and Coalville, Coosa coal field, AL; at Goldberg, Custer Co., ID; in fissures in shale at Yoredale Rocks, Macclesfield, East Cheshire, England; at Teis near Vigo, Spain; at Epernay, Marne, France; in graphite deposit at Gross-Tresny, Moravia, at Litosice and Nizna-Slava, Bohemia, Czech Republic; with variscite at Vashegy, Also-Sajo(!), Verespatak, and Offenbach, Hungary; at *Mt. Zeleznik, Sirk, Com. Gömör, Slovakia*; in graphitic gneiss in the Vatoinandry district, eastern Madagascar; at Mt. Zeehan, TAS, Australia, HCWS *DII:923.*

42.2.1.1 Liroconite $Cu_2^{2+}Al(AsO_4)(OH)_4 \cdot 4H_2O$

Named in 1825 from the Greek for *pale* and *powder*, in allusion to the color of the streak. MON *I2/a*. a = 12.664, b = 7.563, c = 9.914, β = 91.32°, Z = 4, D = 3.04. Structure: *AC(C) 49:916(1991)*. *12-526:* 6.52_{10} 6.03_8 3.95_1 3.91_2 3.01_5 2.79_2 2.70_3 2.21_2. Crystals thin or lenticular [001] with flat octahedral appearance showing {001}, {010}, {100}, {110}, {011} with striations parallel to prism face edges. Sky-blue to green, vitreous luster, transparent to translucent. Cleavage {110}, {011}, indistinct; uneven to conchoidal fracture. H = 2–2$\frac{1}{2}$. G = 2.9. Biaxial (−); $Y = b$; $Z \wedge a = 25°$; $N_x = 1.612$, $N_y = 1.652$, $N_z = 1.675$; 2V = 72°; r < v; nonpleochroic. Soluble in acids. P substitutes for As up to 1:4. Found in oxidized zones of copper deposits: at Cerro Gordo mine, Inyo Co., CA, with linarite and caldonite; at *Wheal Gorland* with cornubite and cornwallite, at other Redruth and St. Day areas, *Cornwall*, and in Devonshire, *England*. HCWS *DII:921*, *AM 36:484(1951)*.

42.2.2.1 Ranunculite HAl(UO$_2$)(PO$_4$)(OH)$_3$ · 4H$_2$O

Named in 1979 for the color (*ranunculus*, buttercup). MON (pseudo-ORTH) Space group unknown. $a = 11.1$, $b = 17.7$, $c = 18.0$, $\beta \approx 90°$, Z = 14, D = 3.39. *MM 43:321(1979)*. 33-972: 9.00$_{10}$ 4.70$_5$ 4.45$_3$ 3.53$_2$ 3.34$_2$ 3.13$_8$ 2.98$_4$ 1.850$_4$. Nodules up to 1 mm in diameter on flexible plates flattened on [010], often having rounded outline; botryoidal crusts. Gold-yellow to canary-yellow. Cleavage parallel to [001], [100] irregular. H = 3. G = 3.4. DTA. Nonfluorescent. Biaxial (−); $N_x = 1.643$, $N_y = 1.664$, $N_z = 1.670$; 2V$_{calc}$ = 56°; extinction 20° to elongation of plates; strongly pleochroic. X, Z, pale green-yellow; Y, yellow. Slowly soluble in 1:1 HNO$_3$, not in HCl. Found in the U-rich portions of the pegmatite at *Kobokobo, Kivu, Zaire*, with meta-autunite, phosphuranylite, and other aluminum-uranyl phosphates, HCWS *AM 65:407(1980)*, *BSBG 86:183(1977)*.

42.2.3.1 Veszelyite (Cu^{2+},Zn)$_3$(PO$_4$)(OH)$_3$ · 2H$_2$O

Named in 1874 for A. Veszelyi (1820–1888), Hungarian mining engineer, who discovered the mineral. MON $P2_1/c$. $a = 9.828$, $b = 10.224$, $c = 7.532$, $\beta = 102.18°$, Z = 4, D = 3.42. *BM 79:219(1956)*, Structure: *AM 59:573(1974)*. 12-525: 7.29$_3$ 6.96$_4$ 4.49$_2$ 3.64$_{10}$ 3.48$_3$ 2.96$_3$ 2.77$_3$ 2.48$_3$. Short prismatic [001] and thick tabular {100}; equant or octahedral with large {100}, {011}; granular aggregates of indistinct crystals. Greenish blue to dark blue, vitreous luster, translucent. Cleavage {001}, {110}. G = 3.4. Biaxial (+); $Y = b$; $Z \wedge c = -35°$; $N_x = 1.618$, $N_y = 1.622$, $N_z = 1.658$; 2V = 38.5°; r < v, weak to strong; pleochroic: X, greenish blue; Z, blue. Soluble in acids. Found with the secondary copper minerals. Occurs at Black Pine mine(!), Philipsburg, MT; at *Moravicza (Vaskö) in the Banat, Romania* and at Ocna de Fier, Caras Severia; at Arakawa mine, Ugo Prov., Japan; at Broken Hill Mine, Zambia; at Kipushi, Shaba, Zaire. HCWS *DII:916*.

42.2.4.1 Kipushite (Cu^{2+},Zn)$_5$Zn(PO$_4$)$_2$(OH)$_6$ · H$_2$O

Named in 1985 for the locality. The P end member in the P–As series with philipsburgite (42.2.4.2). MON $P2_1/c$. $a = 12.197$, $b = 9.156$, $c = 10.667$, $\beta = 96.77°$, Z = 4, D = 3.904. Structure: *CM 23:35(1985)*. 38-375: 12.2$_5$ 6.06$_4$ 4.03$_{10}$ 3.39$_5$ 2.97$_6$ 2.65$_4$ 2.55$_9$ 1.531$_6$. Prismatic crystals, elongated aggregates to 3 mm showing {111}, {102}, {01$\bar{2}$}, {11$\bar{1}$}. Emerald-green, vitreous luster, transparent to translucent. No distinct cleavages, uneven fracture. H ≈ 4. G = 3.8. TGA. Biaxial (−); $N_x = 1.693$, $N_y = 1.738$, $N_z = 1.740$; $2V_{calc} = 23°$; pleochroic: X, colorless; Y, blue; Z, bright blue. Cu/Zn varies between 2.6:1 and 1.1:1 on different grains. Found in the oxidation zone of the *Kipushi deposit, Shaba, Zaire*, with pseudomalachite, malachite, hemimorphite, pyromorphite, veszelyite, vauquelinite, libethenite, quartz, iron oxides. HCWS *AM 71:228(1986)*.

42.2.4.2   Philipsburgite $(Cu^{2+}, Zn)_6(AsO_4,PO_4)_2(OH)_6 \cdot H_2O$

Named in 1985 for the locality. The As analog of kipushite (42.2.4.1). MON $P2_1/c$. $a = 12.33$, $b = 9.20$, $c = 10.69$, $\beta = 96.92°$, $Z = 4$, $D = 4.04$. *CM 23:255(1985)*. *38-384*: 12.2_8 6.21_4 4.29_4 4.05_9 3.41_5 2.67_6 2.56_{10} 1.534_6. Crystals are elongate on {010} curved with chisel-like terminations, composite faces {100} dominant; microscopic spherules with crystal terminations providing rough exterior; as druses, coatings to 0.5 mm thick. Bright emerald-green, light green streak, vitreous luster, transparent to opaque. No discernible cleavage or parting, not brittle. H = 3–4. G = 4.07. TGA. Nonfluorescent. Biaxial (−); $Z = b$, $Y \wedge c = 7°$; $N_x = 1.729$, $N_y = 1.774$, $N_z = 1.775$; $2V = 16°$; pleochroic. X, pale green; Y, Z, medium green; absorbtion Y = Z > X; r >> v. Anomalous blue interference color when (100) in plane of microscope stage. Found in the zone of secondary minerals at the *Black Pine mine, Flink Creek Valley, John Long Mts., 14.5 km W of Philipsburg, MT*, associated with arthurite, duftite, bayldonite, tsumebite, and veszelyite and with carbonates, arsenates, sulfates, and antimonates. HCWS *AM 71:1279(1986)*.

42.3.1.1   Rusakovite $(Fe^{3+}, Al)_5(VO_4,PO_4)_2(OH)_9 \cdot 3H_2O$

Named in 1960 for Mikhail Petrovich Rusakov (1892–1963), geologist, Kazakhstan, Russia. *14-60*: 5.17_4 4.20_5 3.21_{10} 2.95_9 2.44_8 2.14_7 1.569_6 1.366_3. *ZVMO 89:440(1960)*. Reniform concretions, crusts, veinlets. Yellow-orange to reddish-yellow, ochre streak, dull luster. $H = 1\frac{1}{2}$. G = 2.80. DTA. $N_{av} = 1.833$. Weakly polarizing. Soluble in 1:10 acids. Found in the oxidized layers of carbonaceous shales with apatite, colophane, mica, Pb-, Zn-, and Cu-sulfides, steigerite, fervanite, satpaevite, and alvanite at *Balasauskandyk, Kara Tau, Kazakhstan*. HCWS *AM 45:1316(1960)*.

42.4.1.1   Akrochordite $Mn_4^{2+}Mg(AsO_4)_2(OH)_4 \cdot 4H_2O$

Named in 1922 from the Greek for *wart*, in allusion to the form of aggregation. MON $P2_1/a$. $a = 6.832$, $b = 17.627$, $c = 5.682$, $\beta = 99.49°$, $Z = 2$, $D = 3.26$. *AMG 4:425(1968)*, Structure: *AM 74:256 (1989)*. *20-715*: 8.79_8 5.31_4 4.40_{10} 3.62_4 3.11_4 3.06_4 2.75_5 2.36_4. Wartlike or spherical aggregates of minute crystals. Red-brown with yellow tint, translucent. Cleavage {010}, perfect. $H = 3\frac{1}{2}$. G = 3.194. Biaxial (+); $N_x = 1.672$, $N_y = 1.676$, $N_z = 1.683$; $2V = 50°$; r < v, strong. Easily soluble in 1:10 H_2SO_4, giving a purple solution. Found at Sterling Hill mine, Ogdensburg, NJ; at "Japan workings," *Långban*,

Värmland, Sweden, with pyrochroite and barite in hausmannite ore. HCWS *DII:927*, *AM 53:1779(1968)*.

42.4.2.1 Morinite $NaCa_2Al_2(PO_4)_2(F,OH)_5 \cdot 2H_2O$

Named in 1891 for Mr. Morineau, director of the tin mine at Montebras, France. (The Na end member known as jezekite, is not an accepted mineral.) MON $P2_1/m$. $a = 9.454$, $b = 10.692$, $c = 5.444$, $\beta = 105.46°$, $Z = 2$, $D = 2.981$. Structure: *CM 17:93(1979)*. 33-1219(calc): 9.11_5 4.70_7 3.75_6, 3.47_{10}, 3.00_7 2.95_9 2.89_8 2.63_6. See also *11-666*. Massive, prismatic crystals up to 1 cm with striated faces. Colorless, pink; white streak; vitreous luster; translucent. $H = 4\frac{1}{2}$. $G = 2.962$. Biaxial ($-$); $N_x = 1.551$, $N_y = 1.563$, $N_z = 1.565$; $2V = 43°$. Found at Hugo mine, Keystone, SD, associated with montebraesite, apatite, augelite, wardite, crandallite, quartz, and clays; at *Montebras, Soumans, Creuse, France*; at Viitaniemi, Finland; with viitaniemiite in the stanniferous greisens of Odinokoe, Siberia, Russia; at Broken Hill, NSW, Australia. HCWS *DII:783; AM 43:585(1958), 47:398(1962); BM 108:533(1985)*.

42.4.3.1 Tyrolite $CaCu_5(AsO_4)_2(CO_3)(OH)_4 \cdot 6H_2O$

Named in 1845 for the locality. ORTH $Pmma$. $a = 10.50$, $b = 54.71$, $c = 5.59$, $Z = 8$, $D = 3.606$. *BM 59:7(1956)*. 11-348: 28.0_{10} 14.1_8 5.60_3 5.27_4 4.85_7 4.44_5 2.98_8 2.70_8. Fan-shaped and closely foliated aggregates, lath-shaped crystals or scales flattened on [010] and elongated [100], [001]; crusts, reniform masses with radiating fibrous structures, drusy, divergent fibers. Pale apple-green to sky-blue, paler streak, vitreous luster, translucent. Cleavage {001}, perfect; thin laminae are flexible; sectile. $H \approx 2$. $G = 3.0$–3.25. Biaxial ($-$); $XYZ = bca$; $N_x = 1.694$, $N_y = 1.726$, $N_z = 1.730$; $2V = 36°$; $r > v$, strong. Soluble in acids and ammonia. A secondary mineral in oxidation zone of copper deposits associated with erythrite, malachite, azurite, chrysocolla, aurichalcite, cuprite, iron oxyhydroxides. Found in the Tintic district, UT, with chalcopyllite, conichalcite; at Matlock, Derbyshire, England; at Linares in the Sierra Morena, and Pennamellera, Spain; at Cap Garrone, France; at Saalfeld, Thuringia, and Bieber and Riechelsdorf, Hesse, and Schneeberg, Saxony, Germany; at *Tyrol, Austria*, in several localities such as at Kogel near Brixlegg, on the Falkenstein near Scwaz, and at Rattenberg; at Herrengrund, Hungary; at Libethen, at Pojnik near Meusohl, Czech Republic.; at Nerchinski, Transbaikalia, Russia. HCWS *DII:925*.

42.4.4.1 Clinotyrolite $Ca_2Cu_9^{2+}[(As,S)O_4]_4(O,OH)_{10} \cdot 10H_2O$

Named in 1980 for its relationship to tyrolite (42.4.3.1). MON P/a or $P2/a$. $a = 27.61$, $b = 5.56$, $c = 10.513$, $\beta = 94.0°$, $Z = 2$, $D = 3.214$. *AGS 54:134(1980)*. 33-273: 13.7_6 4.81_9 4.40_8 3.67_7 3.50_7 2.96_{10} 2.67_{10} 1.748_8. Radiating aggregates of flaky crystals flattened parallel to (001), elongated [100]. Emerald green, silky. $G = 3.22$. IR and DTA. Biaxial ($-$); $Y = b$, $Z \wedge c = 7°$; $N_x = 1.666$, $N_y = 1.686$, $N_z = 1.695$; $2V = 65°$. Chemical analysis

46.04 CuO, 28.09 As_2O_3, 7.24 CaO, 1.28 SO_3, 17.63 H_2O. Found at *Dongchuan copper mine, Yunnan Prov., China*, with tyrolite. HCWS *MA:80-4909*.

42.4.5.1 Dumontite $Pb_2(UO_2)_3O_2(PO_4)_2 \cdot 5H_2O$

Named in 1924 for André Hubert Dumont (1809–1857), Belgian geologist. See also hügelite (42.4.5.2). MON $P2_1/m$. $a = 8.118$, $b = 16.819$, $c = 6.983$, $\beta = 109.03°$, $Z = 2$, $D = 5.66$. Structure: *BM 111:439(1988)*. Resembles Renardite (42.4.9.2) *12-158:* 8.38_4 6.14_6 4.27_{10} 3.74_6 3.48_7 3.31_6 3.00_9 2.95_9. Small crystals elongated [001], flattened {010}. Ochre-yellow, ochre-yellow streak. $G = 5.65$. Fluoresces greenish. Biaxial (+); XYZ = *acb*; $N_x = 1.88$, $N_z = 1.89$; 2V large; r < v; pleochroic. X, pale yellow; Z, deep yellow. Soluble in acids, H_2O entirely lost by 300°. Found with torbernite at *Shinkolobwe, Shaba, Zaire*. *DII:928, BM 81:63 (1958)*. HCWS

42.4.5.2 Hügelite $Pb_2(UO_2)_3(AsO_4)_2(OH)_4 \cdot 5H_2O$

Named in 1913 for F. Hügel. The As analog of dumontite (42.4.5.1). MON $P2_1/m$. $a = 8.13$, $b = 17.27$, $c = 7.01$, $\beta = 109.0°$, $Z = 2$, $D = 5.80$. *TMPM 26:11(1979)*. *34-1476:* 4.33_6 3.73_{10} 3.06_9 3.00_7 2.89_7 2.70_6 2.16_6 1.833_7. Crystals prismatic [001] to 3 mm long, showing {110}, {001}, {011}. Orange-yellow to yellow-brown, pale yellow streak, greasy to adamantine luster, transparent and translucent. $H = 2\frac{1}{2}$. $G \approx 5.1$. Biaxial (+); Y = elongation (for blue); $N_x = 1.898$, $N_z = 1.915$; 2V = 25° (at 592 μm), 25° (for red); extreme dispersion; r <<< v; anomalous interference colors; incomplete extinction; pleochroic. X, yellow; Y, yellow-orange; Z, colorless to pale yellow. Found at Michael mine, Weiler, and at *Reichenbach, near Lahr, Baden–Württemberg, Germany*, filling cavities with hallimondite, widenmanite, zeunerite, mimetite, and cerussite. HCWS *DII:815; AM 47:418(1962)*.

42.4.5.3 Bergenite $(Ba,Ca)_2(UO_2)_3(PO_4)_2(OH)_4 \cdot 5.5H_2O$

Named in 1959 for the locality. MON $P2_1/c$. $a = 23.32$, $b = 17.19$, $c = 20.63$, $\beta = 93.0°$, $Z = 18$, $D = 4.09$. *BM 104:16(1981)*. Similar to synthetic "barium phosphuranylite." *AM 41:915(1956)*. *43-668:* 8.54_4 7.73_{10} 3.84_8 3.74_4 3.42_4 3.05_6 2.87_5 2.83_5. See also *20-154*. Thin tabular plates. Yellow, translucent. $G = 4.10$. Fluoresces pale green in SWUV and LWUV. Biaxial (−); $N_x = 1.660$, $N_y \approx 1.690$, $N_z \approx 1.698$; 2V = 45°. Found at *Streuberg, Bergen on the Trieb, Vogtland, Saxony, Germany*, associated with uranocircite, torbernite, renardite, autunite, and barium-containing uranophane. HCWS *NJMM:2326(1959); AM 45:909(1960), 66:1102(1981)*.

42.4.6.1 Phuralumite $Al_2(UO_2)_3(PO_4)_2(OH)_6 \cdot 10H_2O$

Named in 1979 for the composition. MON $P2_1/a$. $a = 13.836$, $b = 20.918$, $c = 9.428$, $\beta = 112.44°$, $Z = 4$, $D = 3.52$. Structure: *AC(B) 35:1880(1979)*. *33-38:* 10.4_{10} 8.0_2 5.17_7 3.47_4 3.40_5 3.08_8 2.95_2 2.66_2. Prismatic crystals to 0.5 mm. Lemon-yellow, translucent. $H = 3$. $G = 3.5$. Nonfluorescent. Biaxial (−); X = *b*, Y ≈ elongation; $N_x = 1.559$, $N_y = 1.616$, $N_z = 1.624$; 2V = 40°;

pleochroic, colorless to pale yellow on Y and Z. Found at *Kobokobo, Kivu, Zaire*, in beryl–columbite pegmatites associated with meta-autunite, phosphuranylite, threadgoldite, and other Al- and U-containing phosphates. HCWS *BM 102:333(1979), AM 65:208(1980)*.

42.4.7.1 Phurcalite $Ca_2(UO_2)_3(O_2PO_4)_2 \cdot 7H_2O$

Named in 1978 for the composition. ORTH *Pbca*. $a = 17.415$, $b = 16.035$, $c = 13.598$, $Z = 8$, $D = 4.220$. Structure: *CM 29:95(1991)*. *30-285*: 8.02_{10} 7.63_1 4.24_3 3.99_2 3.83_2 3.39_3 3.10_4 2.88_3. See *30-384*(calc). Tablets flattened on [010] elongated [001] up to 1 mm long. Yellow, vitreous to adamantine luster, translucent. Cleavage {001}, {010}, perfect; also {100}. $H = 3$. $G = 4.22$. Nonfluorescent. Biaxial (−); $N_{x\,calc} = 1.690$, $N_y = 1.730$, $N_z = 1.749$; positive elongation; strongly pleochroic. X, bright yellow; Y, very pale yellow. Found at *Streuberg, Bergen, Vogtland, Saxony, Germany*. HCWS *BM 101:356(1978), AM 64:243(1979)*.

42.4.8.1 Phosphuranylite

Named in 1879 for the composition. ORTH *Bmmb*. $a = 15.835$, $b = 17.324$, $c = 13.724$, $Z = 4$, $D = 4.58$. *EJM 3:69(1991)*, Structure: *AC(B) 47:439(1991)*. *19-898*: 10.4_5 7.96_8 2.86_8 4.43_8 3.87_8 3.16_{10} 3.09_{10} 2.88_{10}. Scaly coatings or crusts, tiny rectangular laths. Deep yellow to golden yellow, translucent. Cleavage {001}, perfect. $H \approx 2\frac{1}{2}$. $G = 4.10$. Biaxial (−); $N_{x\,or\,E} = 1.690$, $N_y = 1.718$, $N_{z\,or\,O} = 1.718$ (Flat Rock); Z perpendicular to flattening; $2V = 5$–$20°$; $r > v$, strong; pleochroic. X, colorless to pale yellow; Y, Z, golden yellow. Easily soluble in acids. A secondary mineral associated with autunite, uranophane, beta-uranophane, opal in weathered zones of granitic pegmatites that contain uraninite. Found at Newry, ME; at Ruggles pegmatite, Grafton Center, and Palermo pegmatite, North Groton, NH; at Branchville pegmatite, Fairfield Co., CT; at Bedford, Westchester Co., NY; at *Flat Rock pegmatite, Mitchell Co., NC*; at Rosmaneira and Carrasca, Sabugal and at Urgeirica, Cannas de Senhorim, Portugal; at Wölsendorf, Bavaria, Germany; at Pamplonita, Santander do Norte, Colombia; at Saddle Ridge mine, s. Alligator R., NT, Australia; at Lavra do Enio, Galileia, at Cruzeiro mine, Sao Jose da Safira, and Lavra do Telirio, MG, Brazil. HCWS *DII:876, AM 39:444 (1954), 76:1734(1991), AMG 4:267(1966)*.

42.4.9.1 Arsenuranylite $Ca(UO_2)_4(AsO_4)_2(OH)_4 \cdot 6H_2O$

Named in 1958 for its relationship to phosphuranylite (42.4.8.1) as the end member. ORTH *Bmmb*. $a = 15.40$, $b = 17.40$, $c = 13.77$, $Z = 6$, $D = 4.25$. *ZVMO 87:589(1958)*. *14-268*: 8.41_8 7.72_{10} 3.85_{10} 3.42_7 3.13_8 1.778_7 1.729_7 1.612_7. Scales, lichenlike deposits. Yellow, orange, translucent. Biaxial (−); $N_x = 1.738$, $N_y = 1.761$, $N_z = 1.774$; $2V_{calc} = 74°$. Found in the oxidation zone of As-containing sulfide deposits associated with metazeunerite, uranospinite, and novacekite, replaced by schoepite and paraschoepite, Russia. HCWS *AM 44:208(1959)*.

42.4.9.2 Renardite Pb(UO$_2$)$_4$(PO$_4$)$_2$(OH)$_4$ · 7H$_2$O

Named in 1928 for Alphonse François Renard (1842–1903), mineralogist, University of Ghent, Belgium. ORTH $Bmmb$. $a = 16.01$, $b = 17.5$, $c = 13.7$, $Z = 6$, $D = 4.34$. AM $39:448(1954)$. PD: 10.25_5 7.95_{10} 5.83_6 4.43_6 3.96_7 3.40_4 3.11_9, 2.88_8. Drusy crusts of minute crystals flattened on {100}, slightly elongated {001}, showing {010}, {001}, {101}. Yellow, transparent. Cleavage {100}, perfect. G > 4. Biaxial (−); XYZ = acb; N$_x$ = 1.715, N$_y$ = 1.736, N$_z$ = 1.739; 2V = 41°; r > v; pleochroic. X,Y, yellow; Z, colorless. Soluble in acids. Occurs with dewindite, dumontite, and torbernite, *Kaslo mine, Shinkolobwe, Shaba, Katanga, Zaire.* HCWS *DII:928*.

42.4.9.3 Kivuite (Th,Ca,Pb)H$_2$(UO$_2$)$_4$(PO$_4$)$_2$(OH)$_8$ · 7H$_2$O

Named in 1958 for the locality. Incompletely described mineral. Possibly a thorian vanmeersscheite. ORTH. $a = 15.88$, $b = 17.24$, $c = 13.76$, $Z = 6$. AM $44:1326(1959)$. 13-419: 10.3_{10} 7.96_9 5.88_6 4.43_6 3.94_6 3.86_6 3.08_8 2.87_8. Earthy masses, orthorhombic plates. Yellow. Nonfluorescent. Biaxial (−); N$_x$ = 1.618, N$_y$ = 1.654, N$_z$ = 1.655; 2V = 0–5°. X, colorless; Y,Z, greenish yellow. Decomposed in HNO$_3$. Found at the *Kobokobo pegmatite, Kivu, Zaire*, associated with phosphuranylite, renardite, zircon, columbotantalite, and apatite. HCWS

42.4.10.1 Glucine CaBe$_4$(PO$_4$)$_2$(OH)$_4$ · 0.5H$_2$O

Named in 1963 for the alternative name for beryllium: glucinum. 15-781: 10.8_{10} 2.67_6 2.41_{10} 1.948_8 1.529_8 1.390_{10} 1.229_7 1.199_7. $ZVMO$ $92:691(1963)$. Massive, encrusting, micron-size needles in concretions. H = 5. G = 2.23–2.40. DTA. N$_{av}$ = 1.555, needles N$_{x'}$ = 1.547, N$_{z'}$ = 1.571; positive elongation parallel to extinction. Dissolved in 1:20 HCl. Spectrographic analysis also shows Sr, Ba, Pb, Cu, Na, Y. Associated with hydromuscovite, "limonite," quartz, moraesite, fluorite, and rutile in mica–fluorite veins, *Ural Mts., Russia.* HCWS AM $49:1152(1964)$.

42.4.11.1 Asselbornite (Pb,Ba)(BiO)$_4$(UO$_2$)$_6$(AsO$_4$)$_2$(OH)$_{12}$ · 3H$_2$O

Named in 1983 for Eric Asselborn (b.1954), mineral collector and surgeon, Montrevel-en-Bresse, France. ISO $Im3m$ or $I432$. $a = 15.66$, $Z = 5$, $D = 5.7$. $NJMM:417(1983)$. 35-605: 6.39_3 5.54_4 4.52_7 4.19_{10} 3.69_6 3.50_4 3.20_8 2.61_4. Cubic crystals up to 0.3 mm. Brown to lemon yellow, greasy to adamantine luster, translucent. TGA. N ≈ 1.9, light yellow in thin section. Slowly soluble in 1:10 HCl. Ba increases as Pb decreases toward exterior of grains. Found in association with metauranospinite, uranophane, and uranosphaerite in quartz gangue from *Walpurgis vein, Weisser Hirsch, Neustadtel, Schneeberg, Saxony, Germany.* HCWS AM $69:565(1984)$.

42.4.12.1 Wallkilldellite $Ca_4Mn_6^{2+}(AsO_4)_4(OH)_8 \cdot 18H_2O$

Named in 1983 for the dell of the Wallkill R., NJ, where the Sterling Hill mine is located. The As and MN^{2+} analog of kittatinnyite (silicate). HEX. $a = 6.506$, $c = 23.49$, $Z = 4$, $D = 2.90$. *AM 68:1029(1983)*. *35-608:* 11.5_{10} 5.61_9 4.56_4 2.84_6 2.75_5 2.55_5 2.35_4. Flattened radial clusters of platy crystals to 0.5 mm in diameter. Dark red; light orange streak; vitreous luster, resinous on fracture; translucent. Cleavage {0001}, perfect. H ≈ 3. G = 2.85. Nonfluorescent. Uniaxial (−); $N_O = 1.728$; N_E ND because of thinness of plates; O > E; pleochroic. O, reddish orange; E, light pink-orange. Found only rarely at *Sterling Hill mine, Ogdensburg, Sussex Co., NJ*, in specimens of the typical massive franklinite–willemite ore with carbonates and adamite group minerals. HCWS

42.4.13.1 Althupite $ThAl(UO_2)_7(PO_4)_4O_2(OH)_5 \cdot 15H_2O$

Named in 1987 for the composition. TRIC $P\bar{1}$. $a = 13.504$, $b = 18.567$, $c = 10.953$, $\alpha = 95.79°$, $\beta = 111.80°$, $\gamma = 72.64°$, $Z = 2$, $D = 3.98$. Structure: *BM 110:65(1987)*. *40-503:* 10.2_{10} 6.67_4 5.80_5 5.09_7 4.91_4 4.41_5 4.07_4 2.90_5. Thin tablets flattened (100), elongated {001} to 0.1 mm. Yellow, transparent. G = 3.9. Biaxial (−); X ≈ [100]; Z ∧ $c = 15°$; $N_x = 1.620$, $N_y = 1.661$, $N_z = 1.665$; 2V = 31°; r < v, strong; pleochroic: X, pale yellow; Y,Z, darker yellow. Found at *Kobokobo, Kivu, Zaire*, in a pegmatite with beryl and columbite. HCWS *AM 73:189(1988)*.

42.4.14.1 Françoisite-(Nd) $Nd[(UO_2)_3O(OH)(PO_4)_2] \cdot 6H_2O$

Named in 1988 for Armand François (b.1922), geologist and director of Gécamine, Zairian mining company. MON $P2_1/n$. $a = 13.668$, $b = 15.605$, $c = 9.298$, $\beta = 112.77°$, $Z = 4$, $D = 4.63$. *BM 111:443(1988)*. *42-1314:* 7.79_{10} 5.76_5 4.44_4 4.33_4 3.88_5 3.13_5 2.87_4 2.84_4. Aggregates of tabular (010) crystals, elongate [001], up to 0.3 mm long showing {100}, {$\bar{1}$02}; twinned on (100). Yellow, white streak, vitreous luster, translucent. Cleavage {010}, easy; uneven fracture. H ≈ 3. G > 4.06. Nonfluorescent. Biaxial (−); X = b; Y ∧ 3 = 14.5°, Z ∧ a = 8.3°; $N_{x\ calc} = 1.65$, $N_y = 1.74$, $N_z = 1.75$; 2V = 35°. Microprobe analyses show Y, La, Ce, Pr, Sm, Dy; gas chromatography for H_2O. Found in the *Kamoto-Est copper–cobalt deposit, 6 km W of Kolwezi, Shaba, Zaire*, in association with uraninite, schoepite, uranophane, curite, schuilingite, and kamotoite-(Y). HCWS

42.4.15.1 Mrazekite $Bi_2Cu_3(OH)_2O_2(PO_4)_2 \cdot 2H_2O$

Named in 1992 for Zdenek Mrázek (1952–1984), who suggested that this was a new mineral on samples from the ancient copper mine at Lubietová, Czech Republic. MON $P2/n$. $a = 9.065$, $b = 6.340$, $c = 21.239$, $\beta = 101.57°$, $Z = 4$, $D = 5.00$. Structure: *CM 32:365(1994)*. Two independent irregular $Cu-O_6$ octahedra and tetrahedral PO_4 groups form sheets parallel to {$20\bar{1}$} which are connected by Bi and H bonds. Note: a pseudocell has been reported in *CM 30:215(1992)*. *45-1391:* 7.63_8 5.41_4 5.20_5 5.15_5 3.17_4 3.04_{10} 3.01_6 2.92_8.

Needles or plates elongated on b, parallel to $\{20\bar{1}\}$, in sheaflike aggregates up to 2 mm. Bright, cerulean blue, similar to azurite; vitreous luster; translucent. Cleavage $\{20\bar{1}\}$, good. H = 2–3. G = 4.90 (low due to quartz impurity). IR. Biaxial $(-)$; X $\wedge$ $c = 27°$, Y $\wedge$ $a = 27°$, Z parallel to c; N =1.86–1.879. Found in the secondary assemblage in a quartz matrix at the *Reinera mine, Lubietová, western Slovenské Rudorhorie Mts., near Banská Bystrica, Slovakia, Czech Republic*, associated with many other arsenate and phosphate minerals: specifically, libenthenite, pseudomalachite, and euchroite. HCWS

42.5.1.1 Mixite $BiCu_6^{2+}(AsO_4)_3(OH)_6 \cdot 3H_2O$

Named in 1880 for A. Mixa, mining official at Jachymov, Bohemia, Czech Republic. HEX $P6_3/m$. $a = 13.646$, $c = 5.920$, $Z = 2$, $D = 4.045$. *NJMM:223(1960), SR 54A:250(1987)*. *13-414*: 12.0_{10} 4.47_5 4.18_5 3.57_8 2.95_7 2.86_6 2.70_6 2.46_9. See also *13-413*, non-Bi-containing sample, called Chlorotile. Hairlike crystals with hexagonal outline elongated [0001], deeply striated vertically; tufted groups, compact reniform spherical masses with concentric fibrous structure; cross-fiber veinlets. Emerald-green to blue-green to whitish, streak lighter in color, brilliant luster on crystals to dull on aggregates. H = 3–4. G = 3.79. Uniaxial (+); $N_O = 1.730$, $N_E = 1.810$ (Mammoth, Tintic); pleochroic. O, colorless; E, bright green. Analysis also shows Zn, Fe^{2+}, Ca, P. About 0.5 H_2O lost by heating to 175°. Found at Mammoth mine, Tintic dist., UT; at Mammoth-St Anthony mine, Tiger, AZ; with erythrite in barite at St. Anton mine, and Humbachtal, Wittichen dist.; and at Grube Clara, Rankachtal; Freundenstadt and Bulach, in Baden; and at Schneeberg, Graul, and Neustädtl, Saxony, Germany; at *Jachymov, Bohemia, Czech Republic*, with bismutite, smaltite, native bismuth; at Kamarezu mine, Attica, Greece; at Sprint Creek, SA, Australia; at Tsumeb, Namibia. HCWS *DII:943, BM 92:420(1969), NJMM:487(1991)*.

42.5.1.2 Agardite-(Y) $(Y,Ca)Cu_6^{2+}(AO_4)_3(OH)_6 \cdot 3H_2O$

Named in 1969 for Jules Agard, geologist, Bureau de Recherches Géologiques et Minières, Orléans, France. Agardite-(La) and agardite-(Ce) described, but no quantitative data presently available. *AM 70:871(1985)*. HEX $P6_3/m$. $a = 13.55$, $c = 5.87$, $Z = 2$, $D = 3.66$. *BM 92:420(1969)*, Structure: *AC(C) 41:161(1985)*. *25-183*: 11.7_{10} 4.43_7 3.54_7 3.25_5 2.94_8 2.69_6 2.56_6 2.45_8. Rosettes, spherical aggregates, and acicular crystals to 5 mm long. Blue-green, translucent. G = 3.72. DTA. Uniaxial (+); $N_O = 1.701$, $N_E = 1.782$. Dissolved by HCl. REE predominate in these members of the mixite group, with names reflecting the dominant species. Found in Red Cloud fluorite mine, Gallinas Mts., NM; at *Bou Skour mine, Jebel Sarhro, Ouazazate, Morocco*, in the oxidation zone associated with azurite, malachite, cuprite, native Cu, and quartz; Laurium, Attica, Greece, associated with smithsonite, aurichalcite, hydrozincite, and cuproadamite. HCWS *BM 92:420(1969), AM 55:1447(1970), NJMM:487(1991)*.

42.5.1.3 Goudeyite (Al,Y)Cu$_6^{2+}$(AsO$_4$)$_3$(OH)$_6$ · 3H$_2$O

Named in 1978 for Hatfield Goudey (1906–1985), mining geologist and mineral collector of San Mateo, CA. The Al analog of agardite. HEX. $a = 13.472$, $c = 5.902$, Z = 2, D = 3.58. *AM 63:704(1978)*. *29-526:* 11.63$_{10}$ 4.41$_5$ 3.53$_2$ 3.37$_2$ 3.24$_5$ 2.92$_5$ 2.68$_4$ 2.55$_5$. Hairlike hexagonal crystals 0.5 mm long (c axis), 2 μm in diameter in tufted groups, encrusting, and in cross-fiber veins. Yellow-green, translucent. H = 3–4. G = 3.50. Uniaxial (+); N$_O$ = 1.704, N$_E$ = 1.765. O, pale yellow-green; E, green; E > O. Found at *Majuba Hill mine, Pershing Co., NV.* HCWS

42.5.2.1 Petersite-(Y) (Y,Ce,Nd,Ca)Cu$_6^{2+}$(PO$_4$)$_3$(OH)$_6$ · 3H$_2$O

Named in 1982 for Thomas A. Peters (b.1947) and Joseph Peters (b.1951), curators of minerals at the Paterson, NJ, museum and the American Museum of Natural History, NY, respectively. HEX $P6_3/m$. $a = 13.288$, $c = 5.877$, Z = 2, D = 3.40. *NJMM:487(1991)*. See also *44-1434:* 11.6$_{10}$ 4.36$_5$ 3.49$_4$ 2.88$_4$ 2.51$_3$ 2.43$_6$ 1.961$_2$ 1.742$_1$. Prismatic crystals showing {10$\bar{1}$0}, {0001}, as radiating clusters and sprays up to 0.1 mm in diameter. Yellow-green; vitreous luster on prism faces, dull on pinocoids. Brittle. G = 3.41. Nonfluorescent. Uniaxial (+); N$_O$ = 1.666, N$_E$ = 1.747; strongly dichroic; E >> O. E, green; O, light yellowish green. Soluble in 1:1 HCl solution, yellow-green. Microprobe analysis also showed Ca, Fe, Ce, Nd, Sm, La. Found in minute cavities between brecciated hornfels in a traprock quarry *Laurel (Big Snake) Hill, Secaucus, Hudson Co., NJ*, on altered chalcopyrite with diadochite, chrysocolla, and hematite, and encrusting malachite and opal. HCWS *AM 67:1039(1982)*.

42.5.3.1 Zapatalite Cu$_3^{2+}$Al$_4$(PO$_4$)$_3$(OH)$_9$ · 4H$_2$O

Named in 1972 for Emiliano Zapata (1879–1919), hero of the Mexican Revolution. TET. $a = 15.22$, $c = 11.52$, Z = 6, D = 3.017. *MM 38:541(1972)*. *25-261:* 11.6$_{10}$ 7.62$_{10}$ 6.82$_2$ 5.75$_8$ 4.58$_5$ 3.04$_5$ 2.95$_4$ 2.53$_5$. Massive. Pale blue, light blue streak, translucent. Cleavage basal, good; sectile. H = 1$\frac{1}{2}$. G = 3.016. Uniaxial (−); N$_O$ = 1.646, N$_E$ = 1.635; weakly dichroic; E > O; occasionally biaxial. Readily dissolves in 1:10 HNO$_3$, HCl, decomposed in 20% KOH. Found at *Cerro Morita Naco, 27 km SW of Agua Prieta, SON, Mexico*, as a replacement for libethenite and pseudomalachite and may alter to chrysocolla; associated with chenevixite, alunite, and beaverite; Villa Vicosa, Estremoz, Portugal. HCWS *AM 57:1911(1972)*.

42.6.1.1 Moraesite Be$_2$(PO$_4$)(OH) · 4H$_2$O

Named in 1953 for Luciano Jacques de Moraes (1896–1968), Brazilian geologist and mineral collector. The P end member of a P–As series with bearsite (42.6.1.2). MON Cc or $C2/c$. $a = 8.553$, $b = 12.319$, $c = 7.155$, $\beta = 97.93°$, Z = 4, D = 1.80. Structure: *ZK 201:253(1992)*. *6-58:* 7.0$_{10}$ 6.15$_4$ 4.24$_6$ 3.28$_6$ 3.02$_6$ 2.82$_6$ 2.33$_4$ 2.05$_4$. Spherulites and drusy crusts of distinct acicular or fibrous crystals to 1 mm. Needles elongated on [001] show {110}, {111}; tabular

{100}, striated parallel to c. White, white streak, translucent. Cleavage {010}, {001}. G = 1.805. Biaxial (−); Z = b; Y $\wedge$ c = 11°; N_x = 1.462, N_y = 1.482, N_z = 1.490; 2V = 65°. Soluble in 1:10 acids. Decrepitates when heated. Found at Dunton Gem mine, Newry, ME; at Palermo No. 1, North Groton, NH; at Feldspar quarry, Londonderry, WA, Australia; at the Humanita pegmatite, Itinga, in the oxidation zone with beryl, albite, quartz, muscovite, and frondelite, and at the *Sapucaia pegmatite, Galileia, MG, Brazil*. HCWS *AM* 38:1126(1953), CM 25:419(1987).

42.6.1.2 Bearsite $Be_2(AsO_4)(OH) \cdot 4H_2O$

Named in 1962 for the composition, the first known Be arsenate. The As analog of moraesite (42.6.1.1). MON $C2/c$ or Cc. a = 8.55, b = 36.90, c = 7.13, β = 97.82°, Z = 12, D = 2.199. *ZVMO 91:442(1962)*. 15-378: 6.95_{10} 4.23_6 3.31_8 3.02_6 2.88_5 2.15_5 1.956_5 1.889_5. Fine encrustations, tangled fibrous aggregates to 0.1 mm. Prismatic, longitudinally striated. White, translucent. G > 1.8. Biaxial (−); Z $\wedge$ c = 8–10°; N_x = 1.490, N_y = N_z = 1.502; $2V_{calc}$ = 0°. Microprobe analysis also shows Al, Fe, Ca, Mg, and Si. Found in the zone of oxidation of a polymetallic deposit that contains arsenopyrite, molybdenite, galena, pyrite, sphalerite, realgar, orpiment, and pitchblende in felsic porphyry with beryl in *Kazakhstan*, coating pharmacosiderite and arsenosiderite, and with scorodite–mansfieldite, conichalcite, and tyrolite. HCWS *AM* 48:210(1963).

42.6.2.1 Isoclasite $Ca_2(PO_4)(OH) \cdot 2H_2O$

Named in 1870 from the Greek for *equal* and *fracture*, in allusion to the cleavage. An inadequately described species. MON (inferred from optical properties). Dull, minute crystals, prismatic [001] and as cotton-like fibers. Colorless to snow white, vitreous to pearly luster on cleavage surfaces. Cleavage {010}, distinct. H = $1\frac{1}{2}$. G = 2.92. Biaxial (+); X = b; Z $\wedge$ c, small; N_x = 1.565, N_y = 1.568, N_z = 1.580; 2V medium. Easily soluble in acids. Found at *Jachymov, Zapadocesky Kraj, Bohemia, Czech Republic*, with chalcedony and dolomite. HCWS *DII:933*.

42.6.3.1 Euchroite $Cu_2^{2+}(AsO_4)(OH) \cdot 3H_2O$

Named in 1823 from the Greek for *beautiful color*. ORTH $P2_12_12_1$. a = 10.056, b = 10.506, c = 6.103, Z = 4, D = 3.47. *AM 36:484(1951)*, Structure: *AC(C) 45:1479(1989)*. 39-1358: 7.29_{10} 5.26_7 5.23_{10} 5.04_7 3.71_7 2.83_9 2.65_5 2.63_5. See *4-222*. Short prismatic [010] showing {110}, {302}, {201} striated parallel to [010], also {010}, {100}, {011}, {021},{101}; equant or thick tabular {100}. Bright emerald-green or leek-green, vitreous luster, transparent to translucent. Cleavage {101}, {110}, slight; uneven to subconchoidal fracture; brittle. H = $3\frac{1}{2}$–4. G = 3.44. Biaxial (+); $XYZ = cab$; N_x = 1.695, N_y = 1.698, N_z = 1.733; 2V = 29°; r > v. Weakly pleochroic. Soluble in acids. Found at Cramer Creek, Missoula Co., MT; as crystals lining crevices in mica schist at *L'ubietova, E. of Banka Bystrica, Slovakia*, with olivenite. HCWS *DII:934*.

42.6.4.1 Legrandite $Zn_2(AsO_4)(OH) \cdot H_2O$

Named in 1932 for Mr. Legrand, from Belgium, the mine manager who collected the original specimen. MON $P2_1/c$. $a = 12.805$, $b = 7.933$, $c = 10.215$, $\beta = 104.39°$, $Z = 8$, $D = 4.01$. AM 56:1147(1971), Structure: SPC 17:747(1973). 42-1356: 6.67_6 5.92_8 4.20_6 4.13_5 4.07_{10} 3.09_7 3.02_5 2.98_6. See also 16-607. Radiating aggregates of prismatic [001] crystals, up to 28 cm × 7.5 mm (Ojuela), showing {001}, {100}, {110}, {$\bar{1}$11}. Colorless to canary yellow, vitreous luster, transparent. Cleavage {100}, fair. $H \approx 4\frac{1}{2}$. $G = 3.975$. TGA. Biaxial (+); $X = b$; $Z \wedge c = 36$–$40°$; $N_x = 1.702$, $N_y = 1.709$, $N_z = 1.740$; $2V = 50°$ (Ojuela); $r < v$; pleochroic. X,Y, colorless to yellow; Z, yellow. Contains minor Fe and Mn. Found at Ljuela mine, Mapimi, DGO., the *Flor de Pena mine, Lampazos, NL*, with siderite, mimetite, pyrite on massive sphalerite, and at Potosi mine, Sta. Eulalia, CHIH, *Mexico*; at Tsumeb, Namibia. HCWS DII:958, AM 48:1258(1963).

42.6.4.2 Spencerite $Zn_4(PO_4)_2(OH)_2 \cdot 3H_2O$

Named in 1916 for Leonard James Spencer (1870–1959), British mineralogist, editor, and curator of minerals at the British Museum of Natural History, London, England. MON $P2_1/a$. $a = 11.191$, $b = 5.276$, $c = 10.439$, $\beta = 116.81°$, $Z = 2$, $D = 3.242$. Structure: MM 38:687(1972). Zn in octahedral and tetrahedral coordination forms complex sheets with tetrahedral PO_4 groups. These layers connected by H_2O molecules. Structure explains prominent twinning. 35-631: 9.34_{10} 4.59_2 3.86_1 3.49_5 3.42_1 3.11_1 3.04_1 2.33_2. See also 13-195. Crystals tabular {100}, elongated [001], often lancelike showing {$\bar{1}$02}, {001}, {$\bar{1}$01}, {$\bar{1}$21}, {$\bar{2}$21}; {h0l} striated [010] due to polysynthetic twinning with composition surface {100}. Also massive, stalactitic with columnar or platy structure. White; pearly luster on cleavages, to vitreous luster. Cleavage {100}, perfect; {010}, good; {001}, distinct. $H = 3$. $G = 3.14$. Biaxial (−); $X \approx a$; $Z = b$; $N_x = 1.586$, $N_y = 1.602$, $N_z = 1.606$; $2V = 49°$; $r > v$. Polysynthetic twin lamellae show maximum extinction on {100} of $\approx 6°$. Readily soluble in acids. Found with hemimorphite, hopeite, salmonite, and clay in cave in oxidized zinc ore at *Hudson Bay mine, Salmo, Nelson mining district, BC, Canada*. HCWS DII:931.

42.6.5.1 Strashimirite $Cu_8(AsO_4)_4(OH)_4 \cdot 5H_2O$

Named in 1968 for Strashimir Dimitrov, Bulgarian petrographer. MON $P2/m$, $P2$, or Pm. $a = 9.71$, $b = 18.85$, $c = 8.94$, $\beta = 97.20°$, $Z = 3$, $D = 3.76$. ZVMO 97:470(1968). 21-289: 18.7_{10} 9.46_8 8.97_9 4.79_8 4.21_8 3.35_8 3.13_9 2.86_{10}. Fine platy to fibrous aggregate in spherulites to 0.5 mm. White, pale green, pearly to greasy luster. Biaxial (−); $Z \wedge$ elongation $\approx 5°$; $N_x = 1.726$, $N_y \sim 1.740$, $N_z = 1.747$; $2V = 70°$; positive elongation; weakly pleochroic. Y, pale yellow-green; Z, yellow-green. Microprobe analysis also shows Sb, Mn, Cd, Pb, Tl, Ba, Ca, Mg, Si, Al but no P,S,V. Found at Majuba Hill, Pershing Co.; Burrus mine, Washoe Co., NV; Gold Hill, Tooele Co., UT; Wheal Gorland and Wheal Unity, Gwennap, Cornwall, England; Horcajo mine, Ciudad Real, Spain;

Kamsdorf, Thüringia; Clara mine, Oberwolfach, Schwarzwald; Schweina, Bad Liebenstein, Germany; Novoveská Huta, Slovakia; with azurite, olivenite, malachite, or euchroite and replaces tyrolite and cornwallite at *Zapachitsa copper mine, western Stara-Planina, Bulgaria*. HCWS *AM 54:1221(1969)*.

42.6.5.2   Arhbarite $Cu_2^{2+}(AsO_4)(OH) \cdot 6H_2O$

Named in 1982 for the locality. MON Space group unknown. $a = 16.988$, $b = 15.901$, $c = 6.100$, $\beta = 97.5°$. *NJMM:529(1982)*. *35-562:* 4.57_{10} 4.51_9 3.72_6 3.25_4 2.63_4 2.60_5 2.47_5. Fibrous to spear-shaped crystallites 10 μm long and 2 μm wide in spherulitic aggregates to 0.3 mm. Blue, vitreous luster. IR. Biaxial; $N_{x'} = 1.720$, $N_{z'} = 1.740$; $\beta = 90°$ (parallel and perpendicular to fiber axis), extinction 45° to length; pleochroic. X', turquoise-blue; Z', deep turquoise-blue. Found on massive dolomite associated with hematite, löllingite, pharmacolite, erythrite, talc, and mcguinnessite at *Arhbar mine, Bou-Azzer, Morocco*. HCWS *AM 68:1038(1983)*.

42.6.6.1   Delvauxite $CaFe_4^{3+}(PO_4,SO_4)_2(OH)_8 \cdot 4-6H_2O$
42.6.6.2   Bořickite $CaFe_4^{3+}(PO_4,SO_4)_2(OH)_8 \cdot 3.5H_2O$

Delvauxite named in 1838 for J. S. P. J. Delvaux de Feuffe (1782–1863), Belgian chemist who analyzed the mineral, and borickite for Emanuel Boricky (1840–1881), Czech petrographer. These two minerals are probably identical to one another and to picite and possibly forcherite. See *DII:935,915*. Amorphous. Concretionary or botryoidal masses, and crusts. Chestnut to red brown, yellow streak, waxy luster. Conchoidal fracture. $H = 2\frac{1}{2}$. $G \approx 2-2.5$. DTA/TCA, IR. Isotropics, $N = 1.726$ (Na) (Litosice); picite samples, $N = 1.677$. Easily soluble in HCl. Found and described as hydrous ferric phosphate at *Berneau, near Visé, Liege, Belgium*; at Leoben, Styria, Austria; at Litosice, and Zeleznik, in gossan at *Nenacovic, S of Kladno*, Bohemia, Czech Republic; in the oolitic iron ores of Kerch Penin., Russia. HCWS *TMPM 26:79(1978)*, *AM 65:813(1980)*.

42.6.7.1   Senegalite $Al_2(PO_4)(OH)_3 \cdot H_2O$

Named in 1976 for the country in which it was found. Bolivarite (42.6.8.1) is the higher hydrate, and bulachite (42.6.7.2) is the As end member of the P–As series. ORTH $Pna2_1$. $a = 9.693$, $b = 7.596$, $c = 7.660$, $Z = 4$, $D = 2.57$. Structure: *AM 64:1243(1979)*. Two types of Al-O polyhedra form infinite chains that run parallel to [101] and [1̄01]. (PO_4) tetrahedra associate with three chains forming an open framework structure with (OH) groups at the points of chain linkage. H_2O is linked to Al, therefore not zeolitic. *36-409:* 5.40_7 4.09_6 3.83_{10} 3.61_6 3.54_3 3.21_4 2.99_4 2.35_3. Crystals show (100), (001), (010), (110), (210), (011), (502), (111). Colorless to pale yellow, vitreous luster. Cleavage {100}, imperfect. $H = 5\frac{1}{2}$. $G = 2.552$. IR, DTA. Nonfluorescent. Biaxial (+); $XYZ = bca$; $N_x = 1.562$, $N_y = 1.566$, $N_z = 1.587$ (Na); $2V = 53°$; $r > v$, weak. Analyses also show minor Fe, trace Cu. Found in the zone of oxidation at *Kourondicko iron ore (magnetite) deposit, Falémé R. basin, eastern Senegal*, associated with turquoise, augelite, wavellite, and crandallite. HCWS *LIT 9:165(1976), AM 62:595(1977)*.

42.6.7.2 Bulachite $Al_2(AsO_4)(OH)_3 \cdot 3H_2O$

Named in 1985 for the locality. ORTH *Pmnm, $P2_122_1$, Pma2,* or *$Pmn2_1$*. $a = 15.53$, $b = 17.78$, $c = 7.03$, $Z = 10$, $D = 2.55$. *AUF 34:445(1983). 37-482:* 8.97_7 7.78_{10} 7.14_5 6.55_7 5.92_7 3.75_7 3.49_8 2.73_6. Very fine needles in radial aggregates. White, silky luster, translucent. No discernible cleavage. $H \approx 2$. $G = 2.60$. Biaxial $(-)$; $N_x = 1.540$, $N_{y\,calc} = 1.546$, $N_z = 1.548$; $2V \approx 66°$; $r > v$. Found at *Neubulach, northern Black Forest, Germany,* associated with arsenocrandallite, malachite, azurite, Ba-pharmacosiderite, and goethite. HCWS *AM 70:214(1985)*.

42.6.8.1 Bolivarite $Al_2PO_4(OH)_3 \cdot 4-5H_2O$

Named in 1921 for Ignacio Bolivar y Urrutia (b.1850), Spanish entomologist. Amorphous. Cryptocrystalline crusts. Pale yellow-green, vitreous luster. Conchoidal fracture, brittle. $H = 2\frac{1}{2}$. $G = 2.05$. Strongly fluorescent bright green. $N = 1.506$. Feebly birefringent. $Al_2O_3/P_2O_5/H_2O = 2:1:12.5$. Found in crevices in granites near *Pontevedra, Spain*; at Kobokobo, Kivu, Zaire. HCWS *DII:872; MM 32:419(1960), 38:418(1971)*.

42.6.9.1 Fluellite $Al_2(PO_4)(OH)F_2 \cdot 7H_2O$

Named in 1824 in allusion to the composition as the discoverer overlooked the presence of phosphate. ORTH *Fddd*. $a = 8.546$, $b = 11.22$, $c = 21.158$, $Z = 8$, $D = 2.16$. *MR 8:392(1977)*, Structure: *AM 51:1579(1966). 37-450:* 6.51_{10} 4.98_1 3.25_4 3.09_2 2.75_1 2.67_1 2.45_1 2.16_1. See also *19-38*. Crystals dipyramidal {111}, modified by {010}. Colorless to white, vitreous luster, transparent. Cleavage {010}, {111}, indistinct. $H = 3$. $G = 2.14-2.17$. Biaxial $(+)$; $N_x = 1.473$, $N_y = 1.490$, $N_z = 1.511$ (Stenna Gwyn); 2V very large; $r > v$, strong. Found at Pyrophyllite mine, Carolina Co., NC; at *Stenna Gwyn, near St. Austell, Cornwall, England,* as minute crystals on quartz with fluorite, arsenopyrite, torbernite, tavistockite, and wavellite; at the pegmatites of Kreuzberg, Pleystein, in Oberpfalz, and at Hagendorf, with phosphosiderite and strengite, as an alteration of triplite at Konigswart near Marienbad, Bavaria, Germany; at Port Chester, Wolfdene, Q, and at the following quarries in SA: Moculta in Angaston, St. Johns in Kapunda, and Tom's, Australia. HCWS *DII:124*.

42.6.10.1 Sengierite $Cu_2(UO_2)_2V_2O_8(OH)_2 \cdot 6H_2O$

Named in 1949 for Edgar Sengier (1879–1963), executive director, Société Generale de Belgique, Managing Director, Union Miniere du Haut-Katanga, Zaire. MON $P2_1/a$. $a = 10.599$, $b = 8.903$, $c = 10.085$, $\beta = 103.42°$, $Z = 2$, $D = 4.10$. *BM 79:219(1956)*, Structure: *BM 103:176(1980). 34-172:* 9.82_{10} 5.75_4 4.91_8 3.74_6 3.20_6 3.18_6 3.14_6 3.09_6. See also *8-398*. Crystals thin plates flattened {001} with six-sided outline show {111}, {010}, {110}, {011}.

Yellowish green, light green streak, vitreous to adamantine luster, transparent. Cleavage {001}, perfect. H = $2\frac{1}{2}$. G = 4.05. TGA. Radioactive. Biaxial (−); XYZ = cba N_x = 1.77, N_y = 1.94, N_z = 1.94; 2V ≈ 37°; r < v, strong; pleochroic. X = blue-green; Y = olive-green; Z = yellow-green. Soluble in acids. Found at Cole shaft, Bisbee, Cochise Co., AZ; at *Luiswishi, Elizabethville-Jadotville area, Shaba, Zaire*, with volborthite, vandenbrandeite, malachite, and black oxides of Co. HCWS *DII:1047; AM 34:109(1949), 66:220(1981)*.

42.6.11.1   Tiptopite $K_2(Na,Ca)_2Li_3Be_6(PO_4)_6(OH)_2 \cdot H_2O$

Named in 1985 for the locality. HEX $P6_3$. a = 11.655, c = 4.692, Z = 1, D = 2.52. *CM 23:43(1985)*, Structure: *AM 72:816(1987)*. Isostructural with cancrinite group minerals. *38-374*: 3.81_4 3.36_5 2.97_{10} 2.53_9 2.35_7 2.22_6 2.08_5 1.459_6. Prismatic needles, up to 2 mm long and 0.1 mm wide, occasionally as cavernous tubes, in radial and divergent spray fasicles. Colorless, vitreous luster, transparent. No apparent cleavage, uneven fracture, brittle. TGA. Nonfluorescent. Uniaxial (+); N_O = 1.551, N_E = 1.559 (589 nm). Microprobe and AA analyses showed Na, Ca, Mn, Al. Found in two different assemblages: (1) with whitlockite, red and yellow montgomeryite, englishite, hurlbutite, red and orange roscherite, whiteite, robertsite, fairfieldite, and fransoleite, and (2) with mitidatite, light brown to colorless montgomeryite, eosphorite-childrenite, dark olive-green roscherite, in the outer intermediate zone of the *Tip Top pegmatite, Custer, Black Hills, SD*. HCWS *AM 71:230(1986)*.

42.6.12.1   Yingjiangite $(K,Ca)(UO_2)_3(PO_4)_2(OH) \cdot 4H_2O$

Named in 1990 for the locality. ORTH $C222_1$. a = 13.73, b = 15.99, c = 17.33, Z = 8, D = 4.17. *AMS 10:102(1990)*. PD: 8.03_{10} 5.90_4 3.99_9 3.88_4 3.45_4 3.17_7 3.10_7 2.89_6. Compact microcrystalline aggregates with grain size to 0.15 mm. Golden yellow to yellow, subadamantine to resinous luster, transparent to translucent. H = 3–4. G = 4.15. Fluoresces weak yellow-green. IR, DTA. Biaxial (−); N_x = 1.669, N_y = 1.692, N_z = 1.710; 2V = 83°; length slow; pleochroic. X, colorless; Y, light yellow; Z, yellow. Found as a secondary product in the oxidized zone containing uraninite and uranothorite in *Tongbiguan village, Yingjiang Co., Yunnan Prov., China*. HCWS *AM 76:1731(1991)*.

42.7.1.1   Childrenite $Fe^{2+}Al(PO_4)(OH)_2 \cdot H_2O$

Named in 1823 for John George Children (1777–1852), English chemist and mineralogist. Forms a Fe^{2+}–Mn^{2+} series with eosphorite (42.7.1.2). ORTH *Bba*2. a = 10.395, b = 13.394, c = 6.918, Z = 8, D = 3.15. *AM 35:793(1950)*, Structure: *NJMM:263(1984)*. *11-621*(manganoan childrenite): 5.27_4 4.40_3 3.54_3 3.39_3 2.81_{10} 2.42_4 2.39_3 1.522_4; *42-1420*(strontian childrenite); *16-632*(possibly a triclinic modification). Equant, pyramidal to short prismatic [001], thick tabular {010} or platy {100}; radial groups of distinct crystals grading into botryoidal masses or crusts with fibrous structure, also occasionally massive. Prism zone striated [001]. Brown to yellowish brown, white streak, vitreous luster, transparent to translucent. Cleavage {100}, poor; subconchoidal to uneven fracture. H = 5. G = 3.25. Biaxial (−); XYZ = bac; $N_x = 1.649$, $N_y = 1.683$, $N_z = 1.691$; r < v, strong; pleochroic. X, yellow; Y, pink; Z, pale pink to colorless. Soluble in acids. A complete solid solution Fe–Mn probably exists but analyses report mostly end member; composition Ca, Sr, and occasionally Fe^{3+} are also reported. Found in mines at Black Mt., Cole, Mt. Mica, Tourmaline, near Newry, and at Paul Bennett and Berry quarries, Poland, at Ryerson Hill, and Tamminen mine, ME; at G.E. Smith, Newport, and Palermo No. 1 mine, North Groton, NH; at Branchville, CT; at Foote mine, Kings Mt., Cleveland Co., NC; at Helen Beryl mine, Custer Co., Hugo mine, Keystone Co., Pennington, SD; at *Tavistock, Devonshire, England*, and at Wheal Crebor, and the George and Charlotte mines nearby associated with siderite, quartz, pyrite, and apatite in hydrothermal veins; also at St. Austell, Cornwall, with zinnwaldite, tourmaline, and apatite; at the granitic pegmatite in Greifenstein, Ehrenfriedersdorf, Saxony, Germany, and Hagendorf near Pleystein, Bavaria; at Beryl mine, Sapucaia, at Larva do Enio, Galileia, and Corrego Frio, MG, Brazil. HCWS *DII:936, NJMM:78(1957)*.

42.7.1.2 Eosphorite $Mn^{2+}Al(PO_4)(OH)_2 \cdot H_2O$

Named in 1878 from the Greek for *dawn-bearing*, in allusion to the pink color. The Mn end member in the Mn–Fe^{2+} series with childrenite (42.7.1.1). ORTH Bbam. $a = 10.436$, $b = 13.495$, $c = 6.923$, Z = 8, D = 3.12. Structure: *AC 13:384(1960). 36-402:* 6.75_3 5.22_6 4.87_3 4.38_4 4.13_4 3.41_3 2.82_{10} 2.43_3. Short to long prismatic {001} with prism zone striated [001], twinning; radial groups, massive with coarse fibrous structure. Pink, rose-red, yellowish brown; white streak; vitreous to resinous luster; transparent to translucent. Cleavage {100}, poor; uneven fracture. H = 5. G = 3.06. Biaxial (−); XYZ = bac; $Z \wedge C = 5\text{-}6°$, reported suggesting monoclinic rather than orthorhoubie symmetry. $N_x = 1.628$, $N_y = 1.648$, $N_z = 1.657$; $2V \approx 50°$; r < v, strong; pleochroic. X, yellow; Y, pink; Z, pale pink. Microprobe analyses show minor Ca and trace Cu. Found at many of the localities listed for childrenite especially at Newry, Poland, Buckfield, Mt. Mica, Black Mt., Hebron, ME, and at Palermo mine, North Groton, NH. Originally from *Branchville, Fairfield Co., CT*, associated with rhodochrosite, lithiophyllite, triploidite, dickinsonite in granitic pegmatite; at Hagendorf Süd, Oberpfalz, Germany; at Mawi Laghman, Afghanistan; at Lavra da Ilha, Taquaral, and at Mendes Pimentel, MG, Brazil. HCWS *DII:936, AM 35:1793(1950), NJMM 78:(1957)*.

42.7.1.3. Ernstite $(Mn^{2+}_{1-x}Fe^{3+}_x)Al(PO_4)(OH)_{2-x}O_x$

Named in 1970 for Theodor K. A. Ernst (b.1904), mineralogist, University of Erlangen, Germany. MON $C2/c$ or Cc. $a = 13.32$, $b = 10.497$, $c = 6.969$, $\beta = 90.37°$, Z = 8, D = 3.086. *NJMM:289(1970). 24-730:* 5.24_3 4.36_4 3.52_4 2.84_8 2.83_{10} 2.44_5 2.42_4 2.00_4. Radiating aggregates to 15 mm long, crystals untwinned. Yellow-brown, translucent. Cleavage {010}, {100}, good. H = 3–$3\frac{1}{2}$. G = 3.07. Biaxial (−); Z = b; $Y \wedge c = -4°$; $N_x = 1.678$, $N_y = 1.706$, $N_z = 1.721$ (Na); r > v; pleochroic: X, yellow-brown; Y, red-brown; Z, pale yellow. Found as an oxidation product of eosphorite in granitic pegmatite at *Daib-East farm, Karibib, Namibia*. HCWS *AM 56:637(1971)*.

42.7.1.4 Sinkankasite $H_2Mn^{2+}Al(PO_4)_2(OH) \cdot 6H_2O$

Named in 1984 for John Sinkankas (b.1915), Captain, U.S. Navy, and author of books on minerals and gems. TRIC $P\bar{1}$ or $P1$. $a = 9.58$, $b = 9.79$, $c = 6.88$, $\alpha = 108.1°$, $\beta = 99.6°$, $\gamma = 98.7°$, $Z = 2$, $D = 2.25$. *AM 69:380(1984)*. *42-597*: 9.2_{10} 5.41_5 5.06_6 4.13_2 2.94_3 2.83_4 2.70_4 2.43_3. Crystals flattened on {100}, elongated on [001], multiply twinned on {100}. Colorless, vitreous luster, transparent. Cleavage {100}, perfect but difficult; parting {100} along twin boundaries good; very brittle. $H = 4$. $G = 2.27$. Nonfluorescent. Biaxial $(-)$; $Y \wedge c \approx 11°$; $N_x = 1.511$, $N_y = 1.529$, $N_z = 1.544$; $2V_{(calc)} = 84°$; $r < v$. Analyses also showed Fe, Mg, Ca, minor amounts of Zn, Na, trace Si, Pb, Ti, Be, Li. Found at Palermo pegmatite, North Groton, NH; at *Barker pegmatite, SW of Keystone, Pennington Co., SD*, where it occurs as pseudomorphs after triphylite and as cavity fillings from dissolution of triphylite in quartz, microcline, albite, and muscovite rock, and associated with vivianite, hureaulite, carbonate–apatite, strengite, barbosalite, and fluellite. HCWS

42.7.2.1 Foggite $CaAl(PO_4)(OH)_2 \cdot H_2O$

Named in 1975 for Forrest F. Fogg (b.1920), wildlife manager and micromount mineral collector, Penacook, NH. ORTH $A2_122$. $a = 9.270$, $b = 21.324$, $c = 5.190$, $Z = 8$, $D = 2.771$. *AM 60:957(1975)*, Structure: *60:965(1975)*. *29-282*: 10.7_4 6.96_5 4.24_{10} 4.20_5 3.11_7 2.69_8 2.51_5 2.29_4. Spherulites, radial aggregates, and as platy (parallel to {010}, elongated parallel to [001]) crystals up to 0.2 mm, in bunches like a partly opened book. Snow white to colorless, translucent. $H \approx 4$. $G = 2.78$. Biaxial $(+)$; $XYZ = cba$; $N_x = 1.610$, $N_y = 1.610$, $N_z = 1.611$; $2V = 40–45°$. Found at *Palermo No. 1 mine, North Groton, Grafton Co., NH*, on quartz, childrenite, and siderite; at Milgun Station, WA, Australia. HCWS

CRANDALITE GROUP

Minerals are a subset of the ALUNITE GROUP (DANA #) with the general formula

$$AB_3(TO_4)_2(OH)_5 \cdot H_2O$$

where

A=Ca, Sr, Ba, PB
B=Al, Fe^{3+}
T=As, P

and crystallize in hexagonal/rhombohedral (R3m) or monoclinic pseudo-hexagonal (Cm) space groups. See also florencite/arsenoflorencite subgroups 41.10.5.1.

42.7.3.1 Crandallite

CRANDALLITE GROUP

Mineral	Formula	a	b	c	β
Crandallite	$CaAl_3(PO_4)_2(OH)_5 \cdot H_2O$	7.005	16.192		
Gorceixite	$BaAl_3(PO_3OH)(OH)_6$	12.217	7.056	7.061	125.21
Goyazite	$SrAl_3(PO_4)(OH)_5 \cdot H_2O$	7.021	16.505		
Lusungite	$(Sr,Pb)Fe^{3+}(PO_4)_2 \cdot H_2O$	7.04	16.80		
Plumbogummite	$PbAl_3(PO_4)_2(OH)_5 \cdot H_2O$	7.033	16.789		
Arsenocrandallite	$(Ca,Cr)Al_3[(As,P)O_4]_2$ $(OH)_5 \cdot H_2O$	7.08	17.27		
Philipsbornite	$PbAl_3(AsO_4)_2(OH)_5 \cdot H_2O$	7.11	17.05		
Arsenogoyazite	$(Sr,Ca,Ba)Al_3[(As,P)O_4]_2$ $(OH)_5 \cdot H_2O$	7.10	17.16		
Segnitite	$PbFe^{3+}H(AsO_4)_2(OH)_6$	7.359	17.113		

42.7.3.1 Crandallite $CaAl_3(PO_4)_2(OH)_5 \cdot H_2O$

Named in 1917 for Milan L. Crandall, Jr., engineer, Knight Syndicate, Provo, UT. HEX-R $R\bar{3}m$. $a = 7.005$, $c = 16.192$, $Z = 3$, $D = 2.997$. Structure: *AM* 59:41(1974). Structure resembles alunite. Sheets of corner-shared distorted Al octahedra $[Al(O,OH)_6]$ form six-membered trigonal rings that link with PO_4. The sheets are stacked parallel to (0001). Ca atoms bonded to 12 (O,OH) atoms are in large cavities between the sheets. The perfect {0001} cleavage is an expression of the structure. *33-257:* 4.87_2 3.51_3 2.98_4 2.94_{10} 2.16_6 1.897_3 1.752_4 1.429_3. See also *42-1420*, strontian crandallite; *16–632*, possibly a triclinic modification. Minute trigonal prisms terminated by {0001}, rosettes of fibers elongated perpendicular to [0001], usually massive, nodular aggregates with concentric layers composed of subparallel or matted fibers; spherules with radial fibrous structure. Yellow, yellowish white, white, gray; vitreous to dull or chalky luster; transparent to opaque. Cleavage {0001}, perfect. H = 5. G = 2.92. Uniaxial (+); $N_O = 1.618$, $N_E = 1.623$ (Fairfield, fibrous). Sr, Ba, and REE may substitute for Ca and Fe^{3+} for Al. May be found as a pseudomorph after gordonite (42.11.14.4). Crandallite is a secondary mineral that occurs in lateritic phosphate deposits and as an alteration product in pegmatites. Found at Fletcher and Palermo mines, North Groton, NH; in the phosphate deposits of the Bone Valley formation, central FL, with clays, millisite, and wavellite, and in the phosphorite unit of the Hawthorne formation; at Glass Sand quarry, Gore, VA; at Tip Top mine, Custer Co., Everly mine, Pennington Co., Hugo mine, Reyson, SD; in UT at the *Brooklyn mine, near Silver City, Tintic district, Juab Co., UT,* with quartz and barite; and at Fairfield, Utah Co., associated with deltaite, millisite, wardite, variscite, gordonite, englishite, montgomeryite, overite, sterrettite, and apatite; at Esmeralda Co., NV; at Bajo de Santa Fe, Guatemala; in alluvial sediments in Barandin, Czech Republic; at the Buranga pegmatite, Rwanda; at Iron Knob, Monarch quarry, SA, Australia; at Alto Benedito pegmatite, Picui, Paraiba, Brazil. HCWS *DII:835; AM 57:473(1972), 65:953(1980).*

42.7.3.2 Gorceixite $BaAl_3(PO_4)(PO_3OH)(OH)_6$

Named in 1906 for Henrique Gorceix (1842-1919), first director of the School of Mines, Ouro Preto, Brazil. Crandallite subgroup forms a Ba–Sr series with goyazite (42.7.3.3). MON Cm. $a = 12.217$, $b = 7.056$, $c = 7.061$, $\beta = 125.21°$, $Z = 2$, $D = 3.41$. *NJMM:113(1990)*, Structure: *NJMM 446(1982)*. *41-1459*: 5.77_9 3.53_6 3.01_{10} 2.29_4 2.23_2 1.920_3 1.763_2 1.506_2. See also *33-130, 19-535*. Microcrystalline grains. Brown to salmon, mottled, vitreous to dull luster. Commonly found as fine-grained rolled pebbles locally called *favas*. Porcelaneous fracture. H = 6. G = 3.04–3.19. Uniaxial (+); $N_O = 1.618$, $N_E = 1.625$; length-fast, undulose extinction. Ca, Ce, and Sr substitute for Ba. Found at similar localities as crandallite in NH, SD, and UT and as late-stage mineralization with fluorite, apatite, and barite in tin–tungsten deposits in the Ozarks at the Silvermine area, Madison Co., MO; at Big Fish and Hess R., YT, Canada; at Lengenbach quarry, Binn Valais, Switzerland; at Felsobanya, Baia Sprie, Romania; in the graphite deposit of Petrovskoye, Ukraine; at Dompim, Bonsa R., Gold Coast; at Oiyi district, Sierra Leone; and in Triassic gravels at Somabula, Zimbabwe; at Issineru, Mazaruni district, Guiana; in the diamond sands of *Rio Abaeto, MG, Brazil*, and at Serra de Congonhas in the Diamantina. HCWS *DII:833*; *AM 43:688(1958)*, *69:985(1984)*; *PD 4:227(1989)*.

42.7.3.3 Goyazite $SrAl_3(PO_4)_2(OH)_5 \cdot H_2O$

Named in 1884 for the state of Goias, Brazil. Crandallite group. HEX-R $R3m$. $a = 7.021$, $c = 16.505$, $Z = 3$, $D = 3.265$. *MM 47:221(1983)*. *34-152*: 5.70_5 3.51_6 2.97_{10} 2.77_2 2.21_5 1.903_2 1.756_1 1.457_1. Small rhombohedral crystals, pseudocubic $\{10\bar{1}2\}$ or tabular $\{0001\}$ with rhomb faces striated horizontally; pebbles or rounded grains. Colorless, pink, honey-yellow; greasy to resinous luster on $\{001\}$ pearly; transparent. Cleavage $\{001\}$, perfect. H = $4\frac{1}{2}$–5. G = 3.26. Uniaxial (+); $N_o = 1.629$, $N_E = 1.639$ (Na) (Brazil). May show biaxiality with zonal growth, thick sections pleochroic. Slowly soluble in acids. Found at Black Mt., Harvard quarry, Lord Hill, Mt. Mica mine, Mt. Rubellite mine, Newry, Bell Mt., Paul Bennett, ME; Palermo mine, North Grafton, NH; Home Sweet Home mine, Park Co., CO; at Big Fish R., Rapid Creek, YT, Canada; on anhydrite in the Simplon tunnel, Switzerland; with barite, pyrite, clayey inclusions, and quartz in the limestones of Belaya Kalitva region, Donetz basin, and in the breccias of Romny and Issachki salt domes, Ukraine; at Alto Bernardino, Picui, Paraiba, in the diamond-bearing sands of *Goias, and at Serra de Congonhas, near Diamantina, MG, Brazil*. HCWS *DII:834*, *MJ 13:390(1987)*, *NJMM:113(1990)*.

42.7.3.4 Lusungite $(Sr,Pb)Fe_3^{3+}(PO_4)_2(OH)_5 \cdot H_2O$

Named in 1958 for the Lusungu River, Zaire. Crandallite group. HEX-R $R3m$. $a = 7.04$, $c = 16.80$, $Z = 3$, $D = 4.12$. *AM 44:906(1959)*. *14-58*: 5.77_9 3.53_6 2.98_{10} 2.48_4 2.26_2 2.22_2 2.20_1 1.913_2. Crusts. Yellow-brown, translucent. Found at *Kobokobo, Kivu, Zaire*, with iron oxyhydroxides, "limonite." HCWS

42.7.3.5 Plumbogummite $PbAl_3(PO_4)_2(OH)_5 \cdot H_2O$

Named in 1832 for the Greek for *lead* for the composition and *gum* and the appearance. Crandallite subgroup. HEX-R R3m $a = 7.033$ $c = 16.789$ Z = 3 D = 4.025 *NJMM 113(1990)*. *35-623* 5.7_9 5.57_2 3.50_6 3.44_2 2.97_{10} 2.22_3 1.908_3 1.751_2. Botryodial, reniform, stalctitic, globular, crusts and masses with concentric structure; compact massive resembling 'gum'; radially fibrous or spherulitic; rare crystals have hexagonal outline. Greyish-white, yellowish grey to yellow to reddish brown, greenish, bluish, streak white, dull to resinous, translucent. Fracture uneven to subconchoidal, brittle. H = $4\frac{1}{2}$-5 G = 4.01. Uniax $+N_o 1.653$ $N_e 1.675$ may be isotropic. Soluble in hot acids, may give off organic odor. A secondary mineral often associated, and may be pseudomorpic after, pyromorphite or barite in lead deposits. Found with marcasite at Canton mine, GA; at Mine La Motte, MO; at Roughten Gill, Gry Gill, near Caldbeck, Cumberland, England with pyromorphite and mimetite; at Huelgoat, Finistère, Brittany, Nussière near Beaujeu with pyromorphite, cerussite, anglesite, wulfenite, France; as rolled fragments (favas) with fibrous structure in diamond-bearing alluvium near Diamantina, Brazil *DII;831, MM 36:550(1967)*.

42.7.3.6 Kintoreite $PbFe_3(PO_4)_2(OH,H_2O)_6$

Named in 1995 for the locality. HEX-R $R\bar{3}$ $a = 7.325$ $c = 16.900$ Z = 3 D = 4.34 *MM 59:143(1995)*. PD 5.96_9 3.67_6 3.07_{10} 2.97_4 2.82_4 2.54_5 2.26_5 1.979_5. Clusters and coatings of crystals showing rhomb {112} to several mm. also globular, hemispheric, Cream, yellowish-green and brownish-yellow. Crystals translucent, vitreous to adamantine; globules transparent to translucent with waxy luster. Streak pale yellow green. Cleavage {001} good, brittle, rough fracture. H ≈ 4 G > 4.2. Uniaxial (−) N between 1.935 and 1.955; pleochroic yellow green to yellow. Some Zn, Cu, Ba, As, S, on microprobe analysis, and H_2O and CO_2 analyses to 98.9 total wt %. Occurs with segnitite, goethite, pyromorphite, mimetite, hinsdalite, libethinite and apatite in the *Kintore open pit* oxidized portion, *Broken Hill, NSW, Australia* and at the Clara mine near Oberwolfach and Igelschlatt mine near Grafenhausen, Germany. *AM 80:1073(1995)*. HCWS

42.7.4.1 Arsenocrandallite $(Ca,Sr)Al_3[(As,P)O_4]_2(OH)_5 \cdot H_2O$

Named in 1981 for the composition and relation to crandallite. Crandallite group. HEX-R $R3m$. $a = 7.08$, $c = 17.27$, Z = 3, D = 3.30. *SMPM 61:23(1981)*. *35-647*: 5.84_8 5.02_3 3.55_9 2.99_{10} 2.23_4 1.919_5 1.769_6 1.302_3. Reniform crusts, spherulitic aggregates up to 0.1 mm. Blue, bluish green; vitreous luster; translucent. No cleavage, conchoidal fracture. H ≈ $5\frac{1}{2}$. G = 3.25. Uniaxial; $N_{av} = 1.625$ (range 1.600–1.650). May show triangular sectors. Partially dissolved by cold 1:1 HCl or HNO_3, and dissolves slowly in hot 1:1 HCl. Microprobe analysis shows Sr, Ba, Bi, Cu, Al, Fe^{3+}, Si. Spectrographic analysis shows Na, K, Cl. Found on dumps in the *Neubulach mining district, Black Forest, Germany*, with broachantite, chalcophyllite, parnauite, arsenosiderite, mansfieldite, and corroded tennantite. HCWS *AM 67:854(1982)*, *AUF 30:241(1979)*.

42.7.4.2 Philipsbornite PbAl$_3$(AsO$_4$)$_2$(OH)$_5$ · H$_2$O

Named in 1982 for Helmut von Philipsborn (1892–1983), mineralogist, University of Bonn, Germany. Crandallite group. HEX-R $R\bar{3}m$ or $R3m$. $a = 7.11$, $c = 17.05$, $Z = 3$, $D = 4.33$. *NJMM:1(1982). 35-540:* 5.85$_6$ 5.73$_2$ 3.59$_5$ 3.54$_2$ 3.04$_{10}$ 2.87$_2$ 2.28$_3$ 1.944$_3$. Massive to earthy crusts of fine-grained aggregates. Grayish green. No cleavage, conchoidal fracture. H ≈ 4$\frac{1}{2}$. G > 4.1. Isotropic; N$_{av}$ = 1.790. Microprobe showed Fe^{3+}, Zn, Cu, Mn, Cr, S. Found at Tsumeb, Namibia; associated with crocoite at *Dundas district, TAS, Australia*. HCWS *AM 67:859(1982)*.

42.7.4.3 Arsenogoyazite (Sr,Ca,Ba)Al$_3$(As,P)O$_4$)$_2$(OH,F)$_5$ · H$_2$O

Named in 1984 for the chemical composition and relationship with goyazite. Crandallite group. HEX-R $R\bar{3}m$ or $R3m$. $a = 7.10$, $c = 17.16$, $Z = 3$, $D = 3.33$. *SMPM 64:11(1984). 38-386:* 5.84$_7$ 5.03$_2$ 3.56$_8$ 3.03$_{10}$ 2.31$_4$ 2.27$_4$ 1.933$_5$ 1.777$_4$. Reniform crusts with rhombohedronlike crystal faces; small tabular crystals show {0001}. White, yellowish, pale green to gray-green; vitreous luster; translucent. No cleavage, conchoidal fracture. H ≈ 4. G = 3.35. Isotropic; N$_{av}$ = 1.64; weakly birefringent. Found at *Clara mine, Oberwolfach, Black Forest, Germany*, on quartz and barite, in association with malachite, brochantite, olivenite, Ba-pharmacosiderite, and SO$_4$-free weilerite. HCWS *AM 71:845(1986)*.

42.7.4.4 Segnitite PbFe^{3+}H(AsO$_4$)$_2$(OH)$_6$

Named in 1992 for E. Ralph Segnit (b.1923), mineralogist and petrologist, Adelaide University, Adelaide, SA, Australia. Crandallite group. Fe^{3+} analog of philipsbornite; related to beudantite (43.4.1.1). HEX-R $R\bar{3}m$. $a = 7.359$, $c = 17.113$, $Z = 3$, $D = 4.77$. *AM 77:656(1992). 45-1392:* 5.97$_5$ 3.68$_4$ 3.09$_{10}$ 2.85$_2$ 2.28$_3$ 2.25$_2$ 1.992$_3$ 1.840$_3$. Pseudo-octahedral crystals up to 1 mm across, as aggregates, crusts, and clusters of rhombs up to 5 mm high showing {112},{001}. Greenish brown to yellowish brown, pale yellow streak, adamantine luster, translucent to transparent. Cleavage good on {001}, rough fracture. H = 4. G > 4.2. Found at Kintore Pit, *Broken Hill, NSW, Australia*, in the oxidized zone of the lead–zinc sulfide orebodies, where it overgrows beudantite on a matrix of goethite encrusting bluish-gray quartz and small spessartine crystals. It is also associated with mimetite, carminite, baylonite, agardite-(Y), and mawbyite. HCWS

42.7.5.1 Nissonite Cu$_2$Mg$_2$(PO$_4$)$_2$(OH)$_2$ · 5H$_2$O

Named in 1966 for William H. Nisson (1912–1965), amateur mineralogist, Petaluma, CA. MON $C2/c$. $a = 22.523$, $b = 5.015$, $c = 10.506$, $\beta = 99.62°$, $Z = 4$, $D = 2.782$. Structure: *AM 75:1170(1990). 40-478:* 10.8$_{10}$ 4.36$_7$ 3.72$_6$ 2.77$_{10}$ 2.68$_6$ 2.53$_6$ 2.22$_7$ 1.859$_5$. Crusts or minute diamond-shaped, crystals tabular

{100}, elongated [001] showing {001}, {100}, {$\bar{1}$11}. Bluish green, translucent. Cleavage (100), fair. H = $2\frac{1}{2}$. G = 2.73. Biaxial (−); Z = b; Y ∧ c = 6°, X ∧ a = 15°; N_x = 1.584, N_y = 1.620, N_z = 1.621; $2V_{calc}$ = 19°; r > v, very strong; pleochroic. X, colorless; Y,Z, turquoise-blue. Easily soluble in 1:10 acids. Analysis shows minor V and Fe. Found in small copper prospect in the *Franciscan formation, Panoche Valley, San Benito Co., CA*, in association with malachite, azurite, libethenite, turquoise, chrysocolla, cuprite, carite, calcite, gypsum, riebeckite, and crossite. HCWS *AM 52:927(1967), PD 1:33(1986)*.

42.7.6.1 Uralolite $Ca_2Be_4(PO_4)_3(OH)_2 \cdot 5H_2O$

Named in 1964 for the region (Ural Mts.) in which it was discovered. MON $P2_1/n$. a = 6.550, b = 16.005, c = 15.969, β = 101.64°, Z = 4, D = 2.197. Structure: *EJM 6:887(1994)*. Corregated layers of [$Be_4(PO_4)_3(OH)_3$] parallel (010). The layers consist of 4 corner-sharing Be(O,OH)$_4$ groups linked with PO$_4$ into rings with 3, 4, and 8 members (first example of a Be-Be-Be-P ring). Ca polyhedra [CaO$_5$(H$_2$O)] link the layers. There is one free H$_2$O. *PD:* 7.82$_6$ 7.12$_{10}$ 5.59$_7$ 3.56$_7$ 3.20$_4$ 3.17$_4$ 3.03$_6$. See also *16-718*. Concretions, radial fibrous spherulites with diameter to 3 mm, or sheaf-like growths. Colorless to white, sometimes stained brown with iron oxides; silky needles have vitreous luster. Concretions, radial-fibrous spherulites or rare sheaflike growths. Cleavage {100}, {010}, poor. H = 2. G = 2.1. Biaxial (−); X parallel to b; Z ∧ a = 9°; N_x = 1.510–1.512, N_y = 1.525–1.526, N_z = 1.533–1.536; birefringence = 0.021–0.026; 2V = 66–80°; positive elongation; extinction [001] ∧ 20°. Found in alteration sequence of beryllonite to hydroxylherderite to (brown) ferroan roscherite to white uralolite, *Dunton quarry, Newry, Oxford Co., ME*; *Taquaral and Galiléia, MG, Brazil*; in late fractures with other Be-phosphates at *Weinebene, Koralpe, Carinthia, Austria*; in kaolin–muscovite rocks containing fluorite, beryl, apatite, crandallite, moraesite, and glucine at *Boyevka, central Ural Mts., Russia*. VK *ZVMO 93:156(1964), AM 49:1776 (1964), MR 9:99(1978)*.

42.7.6.2 Weinebenite $CaBe_3(PO_4)_2(OH)_2 \cdot 4H_2O$

Named in 1990 for the locality. See also uralolite (42.7.6.1). MON Cc. a = 11.897, b = 9.707, c = 9.633, β = 95.76°, Z = 4, D = 2.17. Structure: *EJM 4:1275(1992)*. A framework beryllophosphate (the first reported with three-membered rings) with Ca and H$_2$O in the cavities. *PD:* 5.92$_6$ 5.74$_4$ 4.85$_4$ 4.33$_5$ 3.42$_7$ 2.96$_6$ 2.95$_5$ 2.51$_{10}$. Platy crystals to 0.1 mm × 0.3 mm × 0.5 mm tabular on {001}, elongated on [100], show {001}, {00$\bar{1}$}, {010}, also {100}, {$\bar{1}$00}, {111}, {$\bar{1}$11}, {113}, {$\bar{1}$13}; xenomorphic rosettes up to 20 mm in diameter. No twinning. Colorless, white streak, vitreous luster, transparent to translucent. No cleavage, splintery fracture, brittle. H = 3–4. G = 2.15. Nonfluorescent. Biaxial (+); Z ∧ c = 42° in acute β; N_x = 1.520, N_y = 1.520, N_z = 1.530; 2V < 10°; nonpleochroic. A secondary mineral found in small fractures in a spodumene-bearing pegmatite at *Weinebebe, Koralpe, Carinthia, Austria*, associated with fairfieldite, roscherite, and uralolite. HCWS *AM 78:847(1993)*.

42.7.7.1 Roscherite-A $Ca(Mn^{2+},Mg,Fe^{2+},Fe^{3+},Mn^{3+})_2Be_3(PO_4)_3$ $(OH)_3 \cdot 2H_2O$

42.7.7.2 Roscherite-M

Named in 1914 for Walter Roscher, mineral collector of Ehrenfriedsdorf, Germany. Triclinic (A) and monoclinic (M) modifications have been described, but compositional end members not yet clearly defined. It appears that the structure responds to the amount and site occupancy of trivalent cations (e.g., Fe^{3+}), which in turn affects the OH/H_2O ratio. Structure consists of $Fe-O_6$ octahedral chains cross-linked with PO_4 tetrahedra with oxidation of the Fe affecting chain continuance. *TMPM 24:169(1977)*. See zanazziite (42.7.7.3). Roscherite-A: TRIC $C\bar{1}$. $a = 15.921$, $b = 11.965$, $c = 6.741$, $\alpha = 91.07°$, $\beta = 94.35°$, $\gamma = 89.99°$, $Z = 4$, $D = 2.89$ (high-Mn, low-Mg sample from Foote, CA). *30-173*: 9.54_{10} 5.98_9 4.84_4 3.35_2 3.18_9 3.01_4 2.80_5 2.65_4. Roscherite-M: MON $C2/c$. $a = 15.844$, $b = 11.854$, $c = 6.605$, $\beta = 95.34°$, $Z = 4$, $D = 2.78$. Structure: *TMPM 22:266(1975)* (high-Mn, high-Mg sample from Lavra da Ilha). *30-172*: 9.48_{10} 5.93_8 3.16_5 3.14_5 3.05_5 2.94_4 2.77_6 2.63_4 (Lavra da Ilha). See also *11-355* (Fe^{3+}-containing sample from Sapucaia). Short prismatic [001], with eight- or six-sided cross section; thin tabular {001}, elongated [010] with rectangular outline due to large {100}, {010}, minor {110}; vermicular aggregates of thin plates. Dark brown to olive-green, translucent. $H = 4\frac{1}{2}$. $G = 2.77$ (Lavra) to 2.93 (Sapucaia). Biaxial $(-)$; $X = b$; $Y \wedge c = -15°$; $N_x = 1.624$, $N_y = 1.639$, $N_z = 1.643$; 2V large; crossed dispersion; r >> v; abnormal interference colors; pleochroic. X, yellow to olive-green; Y, yellow-brown, greenish; Z, chestnut-brown. Soluble in acids. Mn^{2+}, Fe^{2+} form series with analyses showing $Ca/Mn/Fe = 10:10:7$; Fe^{3+}, Al also detected on analysis. Found at Newry and Black Mt. pegmatites, ME; at Foote mine, NC; with jezekite, lacroixite, childrenite, apatite, and tourmaline in drusy cavities in a granite at *Greifenstein(!), near Ehrenfriedersdorf, Saxony, Germany*; at Sapucaia pegmatite, and Lavra da Ilha, Taquaral, MG, Brazil. HCWS *DII:968*; *AM 43:824(1958)*, *63:427(1978)*; *NBSM 25:100(1979)*; *MR 17:237(1986)*, *21:413(1990)*.

42.7.7.3 Zanazziite $Ca_2(Mg,Fe^{2+})(Mg,Fe^{2+},Al)_4Be_4$ $(PO_4)_6(OH)_4 \cdot 6H_2O$

Named in 1990 for Pier F. Zanazzi (b. 1939), crystallographer, Universita delgi Studi di Perugia, Italy. See roscherite (42.7.7.1). MON $C2/c$. $a = 15.874$, $b = 11.854$, $c = 6.605$, $\beta = 95.35°$, $Z = 2$, $D = 2.77$. *MR 21:413(1990)*. PD: 9.50_9 5.91_{10} 3.16_7 3.05_5 2.77_5 2.68_4 2.20_4 1.642_5. Barrel-shaped crystals showing {100}, {110} with {001} irregular and rounded; bladed; crystal rosettes to 4 mm. Pale to dark olive-green, zoned, white streak, vitreous to slightly pearly luster on cleavage surfaces. Cleavage {100}, good; {010}, distinct. $H = 5$. $G = 2.76$. Nonfluorescent. Biaxial(+); $X = b$; $Z \wedge [100] = 3°$ in obtuse β; $N_x = 1.606$, $N_y = 1.610$, $N_z = 1.620$; $2V = 72°$. Microprobe and other analyses show minor Mn, Fe^{3+}. Found at *Lavra da*

Ilha, Tequaral, MG, Brazil, associated with eosphorite, wardite, roscherite, and rose quartz in pegmatite. HCWS *AM 76:1732(1991)*.

42.7.8.1 Cyrilovite NaFe$_3^{3+}$(PO$_4$)$_2$(OH)$_4 \cdot$ 2H$_2$O

Named in 1953 for the locality. See also wardite (42.7.8.2) the Al end member. TET $P4_12_12$ or $P4_32_12$. $a = 7.313$, $c = 19.315$, $Z = 4$, $D = 3.115$. Structure: *MP 37:1(1987)*. *31-1308*: 5.85$_3$ 4.85$_{10}$ 3.60$_5$ 3.19$_8$ 3.10$_7$ 2.66$_8$ 2.02$_4$ 1.553$_5$. See *43-672*, an Al-rich sample. Orange, orange-brown, yellow, translucent. G = 3.088. Uniaxial $(-)$; N$_O$ = 1.803, N$_E$ = 1.769. TGA. Analyses show K, Ca, and Al, but there is not a complete Fe^{3+}–Al solid solution to wardite (42.7.8.2). Found at Palermo No. 1 mine, North Groton, NH; at Hagendorf Süd, Bavaria, Germany; Rochefort-en-Terre, Morbihan, Bretagne, France; *Cyrilov, Velke Mezirici, western Moravia, Czech Republic*; at Iron Monarch mine, Iron Knob, SA, Australia; at Sapucaia pegmatite, Ze Berto mine, Galileia, MG, Brazil. HCWS *AM 42:204(1954)*, *BM 104:785(1981)*.

42.7.8.2 Wardite NaAl$_3$(PO$_4$)$_2$(OH)$_4 \cdot$ 2H$_2$O

Named in 1896 for Henry Augustus Ward (1834–1906), mineral dealer and collector, Rochester, NY. See Cyrilovite 42.7.8.1 the other Fe^{3+} end member. TET $P4_12_12$. $a = 7.03$, $c = 19.04$, $Z = 4$, $D = 2.805$. Structure: *MM 37:598(1970)*. *33-1202*: 4.77$_5$ 4.73$_{10}$ 3.12$_6$ 3.09$_8$ 3.00$_8$ 2.83$_4$ 2.59$_8$ 2.11$_4$. See also *11-330*. Pyramidal {102} or {114} with {001} usually present, {102} striated horizontally, all forms except {001} are usually uneven; granular aggregates; crusts; parallel aggregates of coarse fibers; radially fibrous and concentrically banded spherulites. Blue-green to pale green to colorless, vitreous luster, transparent. Cleavage {001}, perfect. H = 5. G = 2.81–2.87. Uniaxial (+); N$_O$ = 1.590, N$_E$ = 1.599 (Fairfield). Difficultly but completely soluble in acids. Found at Newry, ME; in Beryl Mt. pegmatite(!), West Andover, NH; at Hugo pegmatite, SD; in variscite nodules at *Clay Canyon, Fairfield, Utah Co.*, with millisite and crandallite, at Amatrice Hill, and Lucin, UT; at Rapid Creek(!), Big Fish R., YT, Canada, in crystals to 4 cm; as an alteration product of amblygonite at Montebras, Soumans, Dept. Creuse, France; at Iron Monarch mine, Iron Knob, SA, and Milgun Station, WA, Australia; at Lavra da Ilha(!), Taquaral, MG, Brazil. HCWS *DII:940*; *AM 37:849(1952)*, *42:208(1957)*.

42.7.9.1 Millisite (Na,K)CaAl$_6$(PO$_4$)$_4$(OH)$_9 \cdot$ 3H$_2$O

Named in 1930 for F. T. Millis of Lehi, UT, who found the first specimens. TET $P4_12_12$. $a = 7.00$, $c = 19.07$, $Z = 2$, $D = 2.88$. *AM 45:547(1960)*. *13-371*: 6.6$_4$ 4.84$_{10}$ 3.93$_3$ 3.48$_3$ 3.40$_3$ 3.09$_6$ 2.98$_8$ 2.81$_8$. See also *13-370*. Chalcedonic crusts or spherules with fine-fibrous structure. White to light gray, translucent. H = 5$\frac{1}{2}$. G = 2.83. Uniaxial (+); N$_O$ = 1.602, N$_E$ = 1.584; or biaxial; negative elongation. Found with apatite and Na/K = 8:1 as microcrystalline plates growing in the cement of an altered phosphate deposit in Bone Valley, west-central FL; with dehrnite, lewistonite, and green wardite in variscite nodules at Little Green Monster mine, *Clay Canyon, Fairfield, Utah Co., UT*; at Iron Knob, SA, Australia; at Thies, Senegal, with Na/K = 30:1. HCWS *DII:941*.

42.7.10.1 Chenevixite $Cu_2Fe_2^{3+}(AsO_4)_2(OH)_4 \cdot H_2O$

Named in 1866 for Richard Chenevix (1774–1830), French chemist, who in 1801 analyzed an arsenate of copper and iron from Cornwall, England. See luetheite (42.7.10.2). MON $P2_1/m$. $a = 15.006$, $b = 5.189$, $c = 5.724$, $\beta = 102.25°$, $Z = 2$, $D = 4.594$. *MM 41:27(1977)*. *29-553*: 7.32_4 4.22_5 3.56_{10} 2.99_5 2.59_3 2.55_7 2.50_4 2.45_5. Massive, compact earthy to opaline; acicular to lathlike crystals. Greenish yellow to olive-green and dark green, yellow-green streak. Subconchoidal fracture, brittle. H = $3\frac{1}{2}$–$4\frac{1}{2}$. G = 3.93. Easily soluble in acids. Found at American Eagle, other mines in Tintic district, UT, with olivenite; at Alum Gulch, Patagonia, AZ; at Las Animas, SON, Mexico; at *Wheal Gorland, St. Day, Cornwall, England*; at Tsumeb, Namibia; at Bali Lo copper mine, Ashburton Downs, WA, Australia; at Antifagasto, Chuquicamata, Chile. HCWS *DII:840*.

42.7.10.2 Luetheite $Cu_2Al_2(AsO_4)_2(OH)_4 \cdot H_2O$

Named in 1977 for R. D. Luethe (b.1944), geologist at Phelps Dodge Corp., who found the mineral. The Al analog of chenevixite (42.7.10.1). MON $P2_1/m$ or $P2_1/c$. $a = 14.743$, $b = 5.093$, $c = 5.598$, $\beta = 101.82°$, $Z = 2$, $D = 4.40$. *MM 41:27(1977)*. *29-527*: 7.21_7 3.50_{10} 2.55_3 2.51_5 2.45_3 1.803_5 1.471_3 1.270_4. Tabular crystals on {100} up to 0.2 mm. Indian-blue to greenish, translucent. Cleavage {100}, distinct; brittle. H = 3. G = 4.28. Biaxial; X parallel to [010]; Y ∧ [001]; = 10° in obtuse β; $N_x = 1.752$, $N_y = 1.775$, $N_z = 1.796$; $2V_{calc} = 88°$; pleochroic, feebly pale blue. Found as minute crystals in silicified rhyolite porphyry with chenevixite and hematite at the Patagonia mining district, *Santa Cruz Co., AZ*. HCWS *AM 62:1058(1977)*.

42.7.11.1 Olmsteadite $K_2Fe_2^{2+}(Nb,Ta)(PO_4)_2O_2 \cdot 2H_2O$

Named in 1976 for Milo Olmstead, micromount mineral collector, Rapid City, SD. See johnwalkite (42.7.11.2). ORTH $Pb2_1m$. $a = 7.512$, $b = 10.000$, $c = 6.492$, $Z = 2$, $D = 3.41$. Structure: *AM 61:5(1976)*. Octahedrally coordinated $Nb^{5+}O$ and $Fe^{2+}O$ form chains parallel to c and link with PO_4 tetrahedra into an open-framework structure. K resides in a large pocket and has distorted cubic coordination with O atoms. *35-633*: 7.5_2 6.54_2 6.00_{10} 4.17_3 3.05_7 3.00_9 2.87_4 2.57_3. Thick prismatic crystals up to 1 mm, thin tabular to 0.3 mm elongated parallel to [010]; showing {100}, {010}, {111}, {1$\bar{1}$1}, {110}, {101}. Deep brown, red-brown, to black with bronzy surfaces; olive-green streak; subadamantine luster; translucent. Cleavage {100}, {001}, good. H = 4. G = 3.36. Biaxial (+); XYZ = *cab*; $N_x = 1.725$, $N_y = 1.755$, $N_z = 1.815$ (Big Chief); 2V ≈ 60°; pleochroic. X, blue-green; Y, yellow; Z, brown X >> Z > Y. Difficultly soluble in warm 1:1 HCl coloring solution, pale green, and producing a white flocculent residue. Mn^{2+} substitutes for Fe^{2+}. Found as a product of hydrothermal leaching of triphylite–lithiophilite, columbite–tantalite at *Big Chief pegmatite, near Glendale, Pennington Co., SD*, and at Hesnard pegmatite, probably as an oxidized form containing Fe^{3+} near Custer, Custer Co., with rockbridgeite. HCWS

42.7.11.2 Johnwalkite $K(Mn^{2+},Fe^{2+},Fe^{3+})_2(Nb,Ta)(PO_4)_2O_2(H_2O,OH)_2$

Named in 1986 for R. Johnson (b.1936) and F. Walkup (1943–1993), mineral preparators, Natural Museum of Natural History, Smithsonian Institution, Washington, DC. See olmsteadite (42.7.11.1). ORTH $Pb2_1m$. $a = 7.516$, $b = 10.023$, $c = 6.502$, $Z = 2$, $D = 3.44$. *NJMM:115(1986)*. *39-343:* 7.54_3 6.51_4 6.01_{10} 3.26_3 3.05_7 3.01_8 2.86_7 2.56_4. Prismatic crystals elongated on b to 6 mm in radiating aggregates. Dark reddish brown, vitreous luster, translucent. Cleavages {001}, {100}, good. $H \approx 4$. $G = 3.40$. Biaxial (+); $XYZ = cab$; $N_x = 1.748$, $N_y = 1.763$, $N_z = 1.84$; $2V = 53°$; pleochroic. X, blue-green; Y, yellow to pale brown; Z, brown $X \gg Z \gg Y$. Found at *Champion mine, in the Expectation pegmatite, SE of Keystone, Pennington Co., SD*. HCWS *AM 72:223(1987)*.

42.7.12.1 Upalite $Al(UO_2)_3(PO_4)_2O(OH) \cdot 7H_2O$

Named in 1979 for the composition. MON $P2_1/a$. $a = 13.704$, $b = 16.82$, $c = 9.332$, $\beta = 111.5°$, $Z = 4$, $D = 3.94$. *BM 103:333(1979)*, Structure: *BM 106:383(1983)*. *35-619:* 8.4_{10} 6.03_5 4.24_6 4.18_8 3.43_8 3.17_7 3.08_7 2.90_8. Needles up to 0.33 mm. Amber-yellow. $G = 3.5$. Nonfluorescent. Biaxial (−); $XYZ = bac$; $N_x = 1.649$, $N_y = 1.666$, $N_z = 1.676$; $2V_{calc} = 74°$; pleochroic. X, colorless; Y,Z, canary-yellow. Found with phuralumite in *beryl–columbite pegmatite at Kobokobo, Kivu, Zaire*, associated with meta-autunite, phosphuranylite, and threadgoldite. HCWS *AM 65:208(1980)*.

42.7.13.1 Mundite $Al(UO_2)_3(PO_4)_2(OH)_3 \cdot 5.5H_2O$

Named in 1981 for Walter Mund (1892–1956), radiochemist, University of Louvain, Belgium. ORTH $P2_1cn$ or $Pmcn$. $a = 17.08$, $b = 30.98$, $c = 13.76$, $Z = 16$, $D = 4.295$. *BM 104:669(1981)*. Structurally related to phosphuranylite. *35-535:* 7.80_{10} 5.74_4 4.23_3 3.87_6 3.37_7 3.06_5 3.00_4 2.13_3. Rectangular plates flattened on {010} elongated [001]. Pale yellow, translucent. Cleavage {010}, {100}, {001}, perfect. Biaxial (−); $XYZ = bac$; $N_{x\ calc} = 1.62$, $N_y = 1.682$, $N_z = 1.688$; $2V = 33°$; weakly pleochroic; negative elongation. Found at *Kobokobo, Kivu, Zaire*, with phuralumite, upalite, and ranunculite. HCWS *AM 67:624(1982)*.

42.7.14.1 Vanmeersscheite $U^{6+}(UO_2)_3(PO_4)_2(OH)_6 \cdot 4H_2O$

42.7.14.2 Metavanmeerscheite $U^{6+}(UO_2)_3(PO_4)_2(OH)_6 \cdot 2H_2O$

Named in 1982 for Maurice Van Meerssche (1923–1990), crystallographer, University of Louvain, Belgium. Vanmeersscheite: ORTH $P2_1mn$. $a = 17.06$, $b = 16.76$, $c = 7.023$, $Z = 4$, $D = 4.49$. *35-547:* 8.39_{10} 5.96_8 5.30_5 4.18_7 3.91_5

3.51_5 3.07_7 2.89_7. Biaxial (−); $N_{x\,calc} = 1.704$, $N_y = 1.715$, $N_z = 1.718$; 2V = −56°. Metavanmeersscheite: ORTH $Fddd$. $a = 34.18$, $b = 33.88$, $c = 14.074$, Z = 32, D = 4.49. 35-548: 8.49_{10} 6.01_9 5.38_5 4.44_4 4.23_5 3.52_5 3.07_7 2.89_6. Biaxial (−); $N_x \approx 1.67$, $N_y \approx 1.68$, $N_z \approx 1.69$; 2V = −83°. Both minerals occur as plates elongated on [001], showing {010}, {100}, {101}, {$\bar{1}$01}. Yellow, translucent. Cleavage {010}, good; {100}, less good. Both strongly fluorescent green. Found at Dunton quarry, Newry, Oxford Co., ME; *Kobokobo pegmatite, Kivu, Zaire*, associated with studtite. HCWS *BM 105:125(1982)*.

42.7.15.1 Richelsdorfite $Ca_2Cu_5^{2+}Sb^{5+}(AsO_4)_4(OH)_6Cl \cdot 6H_2O$

Named in 1983 for the locality. MON $C2/m$. $a = 14.079$, $b = 14.203$, $c = 13.470$, $\beta = 101.05°$, Z = 4, D = 3.30. *NJMM:145(1983)*, Structure: *ZK 179:323(1987)*. 35-585: 6.80_3 4.91_7 4.39_6 3.05_{10} 2.67_5 1.753_6 0.834_3 0.795_5. Spheroidal aggregates up to 0.2 mm in diameter; tabular crystals to 0.5 mm. Turquoise-blue to sky-blue, vitreous luster, translucent. Cleavage {001}, perfect. H = 2. G = 3.2. Biaxial (−); X ∧ c; Y ≈ parallel to a, Z parallel to b; $N_x = 1.689$, $N_y = 1.765$, $N_z = 1.799$; $2V_{calc} = 69°$; r > v. X, pale blue; Y, greenish blue; Z, light greenish blue. Readily soluble in 1:10 HCl. Microprobe analyses also show Zn, Fe. Found in Permian sandstone in *Richelsdorf Mts., Hesse, Germany*, with calcite, duftite, tirolite, and tetrahedrite, also in cavities in barite, as incrustations on blocks of Cu-rich schist, in cavities in quartz with calcite, tetrahedrite, galena, brochantite, and devillite at St. Andreasberg, Harz. HCWS *AM 69:211(1984)*.

42.7.16.1 Camgasite $CaMg(AsO_4)(OH) \cdot 5H_2O$

Named in 1989 for the composition. MON $P2_1/m$. $a = 9.18$, $b = 7.63$, $c = 16.27$, $\beta = 128.0°$, Z = 4, D = 2.345. *AUF 40:369(1989)*. PD: 7.28_{10} 6.42_{10} 4.00_8 3.28_8 3.21_8 3.02_8 2.96_8 2.54_8. Crusts, crystals prismatic up to 0.15 mm, elongate [010], showing {100}, {010}, {001}. Colorless, white streak, vitreous luster, transparent to translucent. Polysynthetic twinning observed in section parallel to {010}. Cleavage {100}, {001}, good; conchoidal fracture. H ≈ 2. G = 2.40. Nonfluorescent. Biaxial (+); X = b; Z approximately parallel to crystal edge; $N_x = 1.540$, $N_{y\,calc} = 1.548$, $N_z = 1.563$ (589 nm); 2V = 74°. Easily soluble in 1:10 HCl, HNO_3. Found at lower adit of the *Johann mine, Wittichen, central Black Forest, Germany*, as an intimate intergrowth with monohydrocalcite and associated with calcite, gypsum, rapidcreekite, novackite, erythrite, and hörnesite on hydrothermally altered granite with U and Co mineralization. HCWS *AM 76:2021(1991)*.

42.8.1.1 Pharmacosiderite $KFe_4^{3+}(AsO_4)_3(OH)_4 \cdot 6\text{-}7H_2O$

Named in 1813 from the Greek for *poison*, in allusion to the As content, and for iron, to distinguish the composition. See also alumopharmacosiderite (42.8.1.2), barium pharmacosiderite (42.8.1.3), and sodium pharmacosiderite (42.8.1.5). ISO $P\bar{4}3m$. $a = 7.982$, Z = 1, D = 2.60. *NJMM: 183(1984)*,

Structure: ZK 125:192(1967). Zeolite-like framework of AsO_4 tetrahedra and edge-sharing $Fe(OH)_3O_3$ octahedra with other cations and water molecules in channels whose smallest opening is 1.35 Å. Cations and H_2O molecules exchangeable 34-155: 8.0_9 4.61_{10} 3.26_7 2.82_7 2.41_8 2.30_6 1.784_7 1.596_6. Crystals are cubes with faces diagonally striated or replaced by vicinal trapezohedra; granular or earthy. Olive-green, honey-yellow, yellow-brown, dark brown, hyacinth-red, brown-red, grass-green, emerald-green; adamantine to greasy luster; transparent to translucent. Cleavage {001}, imperfect to good; uneven fracture; sectile. $H = 2\frac{1}{2}$. $G = 2.797$. TGA. $N = 1.693$ (Na) (Cornwall), varies with color. Anomalously biaxial (+), $r \ll v$; or biaxial (−), $r \gg v$; 2V large, weakly birefringent. May show six biaxial sectors on {001} with optical orientation of sectors varying in different specimens. Lamellar twinning parallel to sides of border sectors. Soluble in HCl. In NH_4OH green crystals turn red (take up NH_4) and back to green on reimmersion in 1:1 HCl. Cation substitution shows P for As to 1:11, Al for Fe^{3+}, Na or Ba for K. Exchange studies with Na, K, Rb, Cs, NH_4, and HCl showed ease and parallel color changes. Zeolitic water varies from 2.4 to 7.3%, assuming four Fe in the formula unit. A secondary mineral occurring as hydrothermal deposit or oxidation product of arsenic-rich minerals, may alter to "limonite." Found at Sterling Hill mine, Ogdensburg, Sussex Co., NJ; at mines in the Tintic district, UT, with scodorite and copper arsenates derived from enargite; at Lavender pit, Bisbee, Cochise Co., AZ; at the Carharract and Tincroft mines, Cornwall, England, and at St. Day, Liskeard, Redruth, Calstock; at Vaulry, Puy-les-Vignes near St. Leonard, Haute-Vienne, France; at Bösenbrunn, Pöhla, and Schwarzenberg, Saxony, and Grube Clara, Rankachtel, Oberwolfach, and with scordorite in Voigtland, Germany; at Kamareza mine, Attica, Greece; at Capo Bianco, Elba, Italy; at Mouzaia, Algiers, Algeria; at Bou-Azzer, Morocco; at Tsumeb, Namibia; at Geralfo Caudinho mine, MG, Brazil; at Chuquicamata, Antofagasta, Chile. HCWS DII:995.

42.8.1.2 Alumopharmacosiderite $KAl_4(AsO_4)_3(OH)_4 \cdot 6.5H_2O$

Named in 1981 for the relationship to pharmacosiderite and the composition. The Al analog of pharmacosiderite. ISO $P\bar{4}3m$. $a = 7.745$, $Z = 1$, $D = 2.676$. NJMM:97(1981). 35-670: 7.77_{10} 4.48_5 3.87_4 3.16_5 2.74_6 2.45_3 2.34_4 2.24_2. Polycrystalline, incrustation. White, translucent. $N = 1.565$. Transforms to $AlAsO_4$ at 800°. Found with blue ceruleite, green schlossmacherite, olivenite, and white mansfieldite at Guanaco, northeastern Chile. HCWS AM 66:1099(1981).

42.8.1.3 Barium Pharmacosiderite $BaFe_8^{3+}(AsO_4)_6(OH)_8 \cdot 14H_2O$

Named in 1966 for the relationship to pharmacosiderite and the composition probably P_43N, I_43M, P_42M, or I_42M. TET. Suggestion of BA ordering to give superstructures NJMM:183(1984). $a = 7.97$, $c = 8.10$, $Z = 1$, $D = 3.09$. TMPM 11:121(1966). 34-154: 8.08_8 7.95_{10} 3.26_7 2.83_{10} 2.81_6 2.55_5 2.52_6 2.31_5. See also 34-153. Cubic crystals to 1 mm. Yellow-brown, translucent.

Cleavage {100}, good. H = 2–3. G = 3.00. TGA, Mossbauer. Anomalously biaxial (−); $N_x = 1.718$, $N_z = 1.728$; 2V = 0–39°; nonpleochroic. Minor Zn, trace of Al, Ca, Cu, K, Mg, P, Si detected. Readily soluble in warm 1:1 HCl. Found at Sterling Hill mine, NJ; Tintic district, UT; Cornwall, England; on limonite and barite at *Clara mine, Black Forest, Germany*, and at Scholkrippen, Spessart Mts.; Robinson's Reef mine, Port Philip, VIC, Australia. HCWS *AM 52:1585(1967)*.

42.8.1.4 Barium-alumopharmacosiderite $BaAl_4(AsO_4)_3(OH)_5 \cdot 5 H_2O$

Named as the aluminium end member with barium pharmcosiderite, but see *MM 38:103(1971)*. ISO $P\bar{4}3m$ $a = 7.7134$ Z = 1 D = 3.03 *TMPM 11:21(1966)*. *19-94* 7.8_{10} 4.46_6 3.87_7 3.16_7 2.73_8 2.44_4 2.32_5 2.33_4. Space group in doubt will depend on whether Ba site is ordered leading to superstructure.

42.8.1.5 Sodium Pharmacosiderite $(Na,K)_2Fe_4^{3+}(AsO_4)_3(OH)_5 \cdot 7H_2O$

Named in 1985 for the relationship to pharmacosiderite and the composition. ISO $P\bar{4}3m$. $a = 8.012$, Z = 1, D = 2.90. *MR 16:121(1985)*. *38-388*: 7.99_{10} 4.61_5 4.00_4 3.27_8 2.83_6 2.67_3 2.53_5 2.42_6. Pale green, vitreous luster, translucent. Cleavage {001}, imperfect; uneven fracture; brittle. H ≈ 3. N = 1.705. Found with yellow pharmacosiderite, scodorite, arseniosiderite, a yellow member of the jarosite group in vugs formed in dissolving at *Marda, WA, Australia*. HCWS *AM 71:230(1986)*

42.8.2.1 Kidwellite $NaFe_9^{3+}(PO_4)_6(OH)_{10} \cdot 5H_2O$

Named in 1978 for Albert Lewis Kidwell (b.1919), Houston, Texas. MON A2/M, Am or A_2. $a = 10.61$, $b = 5.15$, $c = 13.75$, $\beta = 112.64°$, Z = 2, D = 3.34. *MM 42:137(1978)*. *30-1202*: 9.41_{10} 6.43_3 4.02_4 3.81_4 3.41_6 3.19_4 3.17_4 3.08_2. Fine feathery crystals elongated parallel to [010] in mats or wads. Pale chartreuse-green to greenish yellow, greenish white, bright yellow; silky luster; translucent. Cleavage {100}, perfect. H = 3. Biaxial (−); X parallel to *b*; $N_x = 1.787$, $N_y = 1.800$, $N_z = 1.805$; 2V large; nonpleochroic. Analyses also showed Si, Al, Mg, Mn. Found at Irish Creek, Rockbridge Co., VA; at Indian Mt., AL; at *Coon Creek mine, Shady, Polk Co., AR*, associated with rockbridgeite, dufrenite, beraunite, and strengite in novaculite deposits; at Rotläufchen mine, Waldirme, Germany. HCWS *AM 64:242(1979)*.

42.8.3.1 Schoonerite $Fe_2^{2+}ZnMn^{2+}Fe^{3+}(PO_4)_3(OH)_2 \cdot 9H_2O$

Named in 1977 for Richard Schooner (b.1925), mineral collector, Woodstock, CT. ORTH *Pmab*. $a = 11.119$, $b = 25.546$, $c = 6.437$, Z = 4, D = 2.79. *62:246(1977)*, Structure: *AM 62:250(1977)*. *29-709*: 12.8_{10} 8.35_7 6.43_4 5.52_3 3.76_4 3.18_4 2.77_9 1.600_4. Rosettes, mats, scales, and laths up to 2 mm long; thin tabular parallel to {010}, elongated parallel to [100], showing {001},

{010}, {100}, although most crystals are curved and crinkly. Brown to reddish brown to coppery on oxidized material, pale brown streak, vitreous luster, transparent to translucent. Cleavage {010}, micaceous; {001}, good. H ≈ 4. G = 2.87–2.92. Found as part of the alteration assemblage in association with mitridatite, laueite, strunzite, whitmoreite, and Fe–Mn oxyhydroxides at *Palermo No. 1 mine, North Groton, NH*, as part of the rock alteration of pods containing triphylite, pyrite, pyrrhotite, sphalerite, galena, ludlamite, and vivianite; at Hagendorf Süd, Bavaria, Germany. HCWS

42.8.4.1 Mitridatite $Ca_2Fe_3^{3+}(PO_4)_3O_2 \cdot 3H_2O$

Named in 1914 for the locality. Forms a Fe^{3+}–Mn^{3+} series with robertsite (42.8.4.2), and is the P end member of the P–As series with arseniosiderite (42.8.4.3). MON $A2/a$. $a = 17.52$, $b = 19.35$, $c = 11.25$, $\beta = 95.92°$, $Z = 12$, $D = 3.249$. *AM 59:48(1974)*, Structure: *IC 16:1096(1977)*. Pronounced pseudorhombohedral cell. 26-1057: 8.64_{10} 5.55_6 3.20_4 2.88_4 2.72_7 2.56_4 2.17_4 1.612_4. Massive, nodules, gumlike masses; plates up to 0.2 mm show {100}, {001}, {$\bar{4}$23}. Greenish brown, deep-red to bronze-red; olive-green streak; dull to resinous luster. H = $2\frac{1}{2}$. G = 3.24. Biaxial (−); X approximately perpendicular to {100}; $N_x = 1.785$, $N_y \approx N_z = 1.85$; 2V = 5–10°; extremely pleochroic. X, pale greenish yellow; Y,Z, deep greenish brown Y, Z >> X. Soluble in hot acids. Found as the dull green stains, and occasionally as crystals, formed after weathering of primary phosphate-bearing pegmatites at Palermo No. 1 mine, North Groton, NH; at Big Chief, Tip Top, Linwood, and White Elephant(!) pegmatites, SD, associated with leucophosphite, hureaulite, collinsite, hydroxylapatite, bermanite, laueite, jahnsite, segelerite, and robertsite; in sedimentary Oolitic iron ores at *Mithridat Hill, Kerch, Crimea*, and Taman peninsulas, *Russia*, with vivianite. HCWS *DII:955*.

42.8.4.2 Robertsite $Ca_2Mn_3^{3+}(PO_4)_3O_2 \cdot 3H_2O$

Named in 1974 for Willard Lincoln Roberts (1923–1987), mineralogist, South Dakota School of Mines. Dimorph of pararobertsite (42.8.5.2), in series with arseniosiderite, mitridatite. MON $A2/a$. $a = 17.36$, $b = 19.53$, $c = 11.30$, $\beta = 96.0°$, $Z = 12$, $D = 3.05$. *AM 59:48(1974)*, Structure: *IC 16:1096(1977)*. Pronounced pseudohexagonal cell. 26-1067: 8.63_{10} 5.61_5 3.27_4 2.88_3 2.75_6 2.59_4 2.16_3 1.623_5. Fine feathery aggregates, botryoidal with fibrous structure, plates {100} up to 5 mm in clusters to 7 cm, wedge-shaped crystals which may show {100}, {001},{031}. Fibers resemble goethite, plates lepidocrocite. Crystals twinned by rotation perpendicular to {100} in multiples of $\pi/3$. Deep red to bronze, black; chocolate-brown streak; shiny luster; translucent. Cleavage {100}, good. H = $3\frac{1}{2}$. G = 3.13. Biaxial (−); X approximately perpendicular to {100}; $N_x = 1.775$, $N_y = N_z = 1.82$; 2V ≈ 8°; extremely pleochroic. X, pale reddish pink; Y,Z, deep reddish brown Y,Z >> X. Found in tryphilite-containing pegmatites with rockbridgeite, such as at *Tip Top(!)*, Linwood, White Elephant, Gap Lode, *near Custer, SD*; at Hagendorf Süd, Germany. HCWS

42.8.4.3 Arseniosiderite $Ca_2Fe_3^{3+}(AsO_4)_3O_2 \cdot 3H_2O$

Named in 1842 for the composition and from the Greek for *iron*. See also mitridatite (42.8.4.1) and robertsite (42.8.4.2). MON $A2/a$. $a = 17.76$, $b = 19.53$, $c = 11.30$, $\beta = 96.0°$, $Z = 12$, $D = 3.60$. *AM 59:48(1974)*, Structure: *IC 16:1096(1977)*. *26-1002*: 8.84_{10} 5.62_5 3.28_4 3.22_4 2.95_5 2.77_8 2.21_4 1.643_4. Massive radial-fibrous aggregates with fibers flattened {001}, easily separable; granular; pseudomorphs after siderite, scodorite. Golden-yellow, yellow-brown, reddish brown, brownish black, black; ochre-yellow streak; submetallic to silky luster; opaque except in small grains. Cleavage {001}. $H = 4\frac{1}{2}$, $1\frac{1}{2}$ in fibers. $G = 3.60$. Biaxial $(-)$; X perpendicular to {100}; $N_x = 1.80$, $N_y = N_z = 1.88$; $2V \approx 0°$; Y, extremely pleochroic. X, colorless; Y,Z, dark reddish brown Z $>>$ X. Easily soluble in hot acids. A low-temperature oxidation product of löllingite, arsenopyrite, and replaces scorodite. Found at Sterling Hill mine, Ogdensburg, and at Franklin, Sussex Co., NJ; in the Tintic district, UT; as concretions at Tagish Lake, YT, Canada; at Jesus Maria mine, Mazapil district, ZAC, Mexico, with pharmacolite, chrysocolla, and calcite and as pseudomorphs after scoodrite; also at Ojuela mine, Mapimi, DUR; in the manganese deposit at *Romanêche-Thorins, Macon, Saône et Loire, France*, with quartz, goethite, romanechite, and psilomelane; at Schneeberg, Saxony, and Wittichen, Bavaria, Germany, with erythrite and roselite; as felted masses with scorodite, symplesite, pitticite, and pharmacolite on löllingite at Hüttenberg, Carinthia, Austria; at Kamare mine, Attica, Greece; at Tsumeb, Namibia. HCWS *DII:853*.

42.8.5.1 Kolfanite $Ca_2Fe_3^{3+}O_2(AsO_4)_3 \cdot 2H_2O$

Named in 1982 for the abbreviation of Kola Filial Akad. Nauk. MON $a = 17.86$, $b = 19.66$, $c = 11.11$, $\beta = 96°$, $Z = 12$, $D = 3.75$. *MZ 4:90(1982)*. See arseniosiderite (42.8.4.3). *35-662*: 8.90_{10} 5.64_5 3.29_9 2.95_9 2.72_{10} 2.22_8 1.646_8 1.503_5. Massive or thin plates. Red, orange to yellow; adamantine luster. One cleavage, brittle. $H = 2\frac{1}{2}$. $G = 3.3$. Biaxial $(-)$; $N_x = 1.810$, $N_y = 1.923$, $N_z = 1.933$, $2V = 5-7°$; pleochroic. X, pale yellow; Y,Z, dark orange. Microprobe analysis shows minor P, Sb, Al, and Si. Found in hydrothermally altered granite pegmatites, *Kola Penin., Russia*, formed by alteration of holtite and associated with mitridatite, arseniosiderite, and laueite. HCWS *AM 68:280(1983)*.

42.8.5.2 Pararobertsite $Ca_2Mn_3^{3+}(PO_4)_3O_2 \cdot 3H_2O$

Named in 1989 for the demorphic relationship to robertsite. MON $P2_1/c$. $a = 8.825$, $b = 13.258$, $c = 11.087$, $\beta = 101.19°$, (unconditional setting to compare with robertsite) $Z = 4$, $D = 3.21$. *CM 27:451(1989)*. Pronounced subcell $b' = \frac{1}{2}b$. PD: 8.69_{10} 5.66_6 5.44_5 3.18_5 2.88_6 2.83_5 2.61_6 2.16_6. Plates and clusters of plates up to 0.2 mm and < 0.02 mm thick. Tabular {100}. Red, brownish-red streak, vitreous luster, translucent. Cleavage {100}, perfect; brittle. Soft. $G = 3.22$. Nonfluorescent. Biaxial $(-)$; Y $= b$; Z $\wedge$ c $= 4°$; $N_x = 1.79$, $N_y = 1.81$, $N_z = 1.83$; Y, pleochroic. X, yellowish brown; Y,Z, reddish brown Z > X. Found on whitlockite encrusting CO_3-apatite on quartz in secondary seams at *Tip Top pegmatite, Custer, SD*. HCWS *AM 75:1211(1990)*.

42.8.6.1 Yukonite $Ca_2Fe_3^{3+}(AsO_4)_4(OH) \cdot 12H_2O$

Named in 1913 for the locality. An inadequately characterized mineral. Gel-like waxy aggregates, brownish-black color, and as dark-red pseudomorphs after köttingite–parasymplessite. *35-553:* 5.60_8 3.25_{10} 2.79_8 2.52_3 2.33_3 2.23_3 1.63_4 1.51_2. Pattern resembles arseniosiderite. Found associated with ogdensburgite at Sterling Hill mine, Ogdensburg, Sussex Co., NJ; at *Tagish Lake, YT, Canada*. HCWS *CM 8:667(1966), MM 46:261(1982), AM 68:474(1983)*.

42.8.7.1 Mapimite $Zn_2Fe_3^{3+}(AsO_4)_3(OH)_4 \cdot 10H_2O$

Named in 1981 for the mining district where it was found. MON *Cm*. $a = 11.415$, $b = 11.259$, $c = 8.661$, $\beta = 107.71°$, $Z = 2$, $D = 3.02$. *BM 104:582(1981)*, Structure: *AC(B) 37:1040(1981)*. *35-515:* 8.24_{10} 7.83_9 4.66_6 3.88_5 3.45_4 3.33_4 3.22_4 2.83_4. Rectangular crystals to 3 mm, flattened {001}, bounded {110}. Polysynthetic twinning (001) with [104] pseudoaxis. Blue, blue-green, green; vitreous luster; translucent. Cleavage {001}, {010}. $H = 3$. $G = 2.95$. DTA. Biaxial (+); $Y = b$; $Z \wedge a = 13°$; $N_x = 1.672$, $N_y = 1.678$, $N_z = 1.712$; $2V = 50°$; $r < v$; strongly pleochroic: X, pale yellow; Y, greenish yellow; Z, deep prussian-blue. Readily soluble in 1:10 acids, fuses to a brown scoria. Some Fe^{2+} detected on analysis. Found at the *Ojuela mine, Mapimi, DUR, Mexico*, associated with scorodite, adamite, and smithsonite on "limonite" matrix. HCWS *AM 67:623(1982)*.

42.8.8.1 Shubnikovite $Cu_8^{2+}Ca_2(AsO_4)_6(OH)Cl \cdot 7H_2O$

Named in 1953 for Aleksei V. Shubnikov (1887–1971), crystallographer, Institute of Crystallography, Moscow, Russia. An inadequately described mineral. Platy, light blue, translucent. $H = 2$. Biaxial (−); $N_x = 1.640$, $N_y = N_z = 1.690$; 2V small; $r < v$; pleochroic: X, light blue; Y,Z, greenish blue. Found in an oxidation zone in *Russia*. HCWS *ZVMO 82:311(1953), AM 40:552(1955)*.

42.9.1.1 Burangaite $(Na,Ca)_2(Fe^{2+},Mg)_2Al_{10}(PO_4)_8(OH,O)_{12} \cdot 4H_2O$

Named in 1977 for the locality. See also dufrenite (42.9.1.2) and natrodufrenite (42.9.1.3). MON $C2/c$. $a = 25.09$, $b = 5.048$, $c = 13.45$, $\beta = 110.91°$, $Z = 2$, $D = 3.31$. *BGSF 49:33(1977)*. *29-1190:* 11.7_{10} 4.86_4 4.03_4 3.32_4 3.12_7 3.08_9 2.04_3 1.675_3. Long prismatic crystals showing {301}, {102}, {104}, {$\bar{1}$01}, {$\bar{2}$23}, {331}, {$\bar{2}$23}, {$\bar{3}$1$\bar{1}$}, {40$\bar{1}$} and hourglass growth. Blue core, colorless margin. Bluish, bluish green, slightly bluish streak. Cleavage {100}, perfect. $H = 5$. $G = 3.05$. Biaxial (−); $X \wedge c = 11°$ in acute β; $Z = b$; $N_x = 1.611$, $N_y = 1.635$, $N_z = 1.643$; $2V = 58°$; $r > v$; ; pleochroic. X, light blue; Y, dark blue; Z, colorless Y > X > Z. Found in the *Buranga pegmatite, Rwanda*, associated with bertossaite, trolleite, scorzalite (which it replaces), apatite, bjarebyite, and wardite. HCWS *AM 63:793(1978)*.

42.9.1.2 Dufrenite (Ca,☐) Fe^{2+}Fe$_5^{3+}$(PO$_4$)$_4$(OH)$_6$ · 2H$_2$O

Named in 1833 for P. A. Dufrénoy (1792–1857), French mineralogist. The Fe end member in the Fe^{2+}–Ca–Na series with burangaite (42.9.1.1) and natrodufrenite (42.9.1.3). MON $C2/c$. $a = 25.84$, $b = 5.126$, $c = 13.78$, $\beta = 111.20°$, $Z = 4$, $D = 3.407$. *MM 54:419(1990)*, Structure: *AM 55:135(1970)*. (Fe^{2+},Fe^{3+})–(O,OH)$_6$ octahedra are joined by face-sharing with PO$_4$ into triplets with formula Fe$_3$(OH)$_4$(O$_{P-group}$)$_8$ and via corner sharing into a second triplet Fe$_3$(OH)$_6$(H$_2$O)$_2$(O$_{PO_4-group}$)$_8$ to form a slab parallel to [100]. The remainder of the structure has cavities that may be occupied, or partially occupied, by other ions, such as Ca. The PO$_4$ groups plus triplets along [100] account for the perfect cleavage direction and the fiber axis. *8-155:* 5.05$_9$ 4.15$_4$ 3.42$_9$ 3.24$_8$ 3.17$_{10}$ 2.88$_5$ 2.44$_5$ 2.11$_6$. See also *22-1143*, a Ca-containing dufrenite. Botryoidal masses and crusts with radial-fibrous structure whose surface may show drusy round-ended crystals or exfoliated terminations; rounded crystals in subparallel or sheaflike aggregates. Dark green, olive-green, to greenish black becoming reddish with oxidation; vitreous to silky luster; translucent to nearly opaque. Cleavage perfect parallel to [100], and fiber length, brittle. H = $3\frac{1}{2}$–$4\frac{1}{2}$. G = 3.1–3.4. Biaxial (+); N$_x$ = 1.837, N$_y$ = 1.845, N$_z$ = 1.895 (Wheal Phoenix); 2V small; r > v, extreme; pleochroic. X, pale yellow-brown; Y, pale brown; Z, red-brown Z > Y > X. Indices increase with Fe^{3+} content. Soluble in 1:10 acids. Oxidation analogous to the triphyllite series. A secondary mineral found in gossans and in iron ore deposits. Found at Palermo No. 1, North Groton, NH; at Indian and Wolden Mts., Rock Run, Cherokee Co., AL; at Wheal Phoenix(!), Cornwall, England; at Hureaux, St. Silvester, France; at Hirschberg, Loebenstein in Thuringia, and Ullersreuth and Hauptmannsgrün near Reichenbach, Saxony, and Seigen, Westphalia, Germany; at Kangas farm, Aggeneys district, Bushmanland, South Africa. HCWS *DII:873, AM 34:513(1949)*.

42.9.1.3 Natrodufrenite Na(Fe^{3+},Fe^{2+})(Fe^{3+},Al)$_5$(PO$_4$)$_4$(OH)$_6$ · 2H$_2$O

Named in 1982 for the relationship to dufrenite and the composition. The Na-rich end member of the Na–Fe^{2+} series with dufrenite (42.9.1.2). MON $C2/c$. $a = 25.83$, $b = 5.150$, $c = 13.772$, $\beta = 111.53°$, $Z = 4$, $D = 3.20$. *BM 105:321(1982)*. *35-570:* 12.04$_8$ 5.04$_6$ 4.12$_5$ 3.40$_8$ 3.20$_8$ 3.15$_{10}$ 2.99$_5$ 2.86$_5$. Spheres of compact radiating fibers up to 0.5 mm in diameter. Pale blue-green, translucent. G = 3.23. DTA. Biaxial; N$_x$ = 1.756, N$_z$ = 1.775; parallel or subparallel extinction; positive elongation; pleochroic: X, pale yellow; Z, dark green. Found at *Rochefort-en-Terre, France*, where it occurs in a limonitic bed with cyrilovite and goethite. HCWS *AM 68:1039(1983)*.

42.9.2.1 Souzalite (Mg,Fe^{2+})$_3$(Al,Fe^{3+})$_4$(PO$_4$)$_4$(OH)$_6$ · 2H$_2$O

Named in 1949 for J. A. de Souza (1896–1961), director of Departamento Nacional da Producao Mineral, Brazil. Forms a Mg–Fe^{2+} series with gormanite (42.9.2.2). TRIC $P1$ or $P\bar{1}$. $a = 11.74$, $b = 5.11$, $c = 13.58$, $\alpha = 90.83$,

$\beta = 99.08°$, $\gamma = 90.33°$, $Z = 2$, $D = 3.07$. *CM 19:381(1981)*. *33-863:* 6.72_3 4.76_6 3.39_{10} 3.34_3 3.15_6 3.06_4 2.92_8 2.55_9. Coarse fibrous masses. Green, dark to bluish green; vitreous luster; translucent. Polysynthetic twinning, composition plane parallel to good cleavage. Cleavage in two directions at right angles: one good, the other fair. $H = 5\frac{1}{2}-6$. $G = 3.087$. Biaxial $(-)$; X perpendicular to cleavage; Z || elongation; $N_x = 1.618$, $N_y = 1.642$, $N_z = 1.652$; $2V = 68°$; $r \gg v$; pleochroic. X, green; Y, blue; Z, yellow. Found at Bell pit, Newry, Oxford Co., ME; Rapid Creek, Big Fish R., Blow R. area, YT, Canada, with gormanite; as hydrothermal alteration product of scorzalite at *Córrego Frio pegmatite, Linopólis district, N of Divino das Laranjeiras, MG, Brazil*. HCWS DII:911; *AM 34:83(1949)*, *AM 55:135(1970)*.

42.9.2.2 Gormanite $Fe_3^{2+}Al_4(PO_4)_4(OH)_6 \cdot 2H_2O$

Named in 1981 for Donald Herbert Gorman (b.1922), University of Toronto, Canada. The Fe^{2+} analog of souzalite (42.9.2.1). TRIC $P\bar{1}$ or $P1$. $a = 11.79$, $b = 5.11$, $c = 13.57$, $\alpha = 90.83°$, $\beta = 99.25°$, $\gamma = 90.08°$, $Z = 2$, $D = 3.12$. *CM 19:381(1981)*. *36-403:* 6.70_{10} 4.79_6 3.39_{10} 3.34_7 3.16_7 3.13_8 3.10_6 2.55_7. See also *33-638*. Radiating aggregates, and blades elongated parallel to [010] up to 3 mm × 0.5 mm × 0.1 mm, with {001} dominant, and {$\bar{1}$00}, {102}, {$\bar{1}$02}, {010}. Twinned, composition plane {001}, axis [010]. Blue-green, pale green streak, vitreous luster. Cleavage {001}, poor. $H = 4-5$. $G = 3.13$. DTA/TGA. Nonfluorescent. Biaxial $(-)$; X approximately perpendicular to {001}; $Z \wedge b$ (elongation) $= 14°$; $N_x = 1.619$, $N_y = 1.653$, $N_z = 1.660$; $2V = 53°$; pleochroic. X, colorless; Y, blue; Z, colorless $X = Z < Y$. Analyses also show Mg, Ca, Mn, Fe^{3+}. Strong chemical zoning toward souzalite. Found in outcrops of phosphate ironstone in *Rapid Creek and Big Fish areas, YT, Canada*, with quartz, siderite, ludlamite, arrojadite, kryzhanovskite, vivianite (oxidized), souzalite. HCWS *AM 67:622(1982)*.

TURQUOISE GROUP

The name *turquoise* has been in use since the sixteenth century for the blue-green mineral materials brought through Turkey to Europe from Iran (Persia). But several minerals of the group have only recently been more fully described and the distinctive end members elucidated. The general formula is

$$A_{0-1}B_6(PO_4)_{4-x}(PO_3OH)_x(OH)_8 \cdot 4H_2O$$

where

$A = \square$, Cu, Fe^{2+}, Zn
$B = Al, Fe^{3+}, Cr$
$x = 0$ to 2

and the minerals are usually massive botryoidal aggregates crystallizing in the triclinic space group $P\bar{1}$.

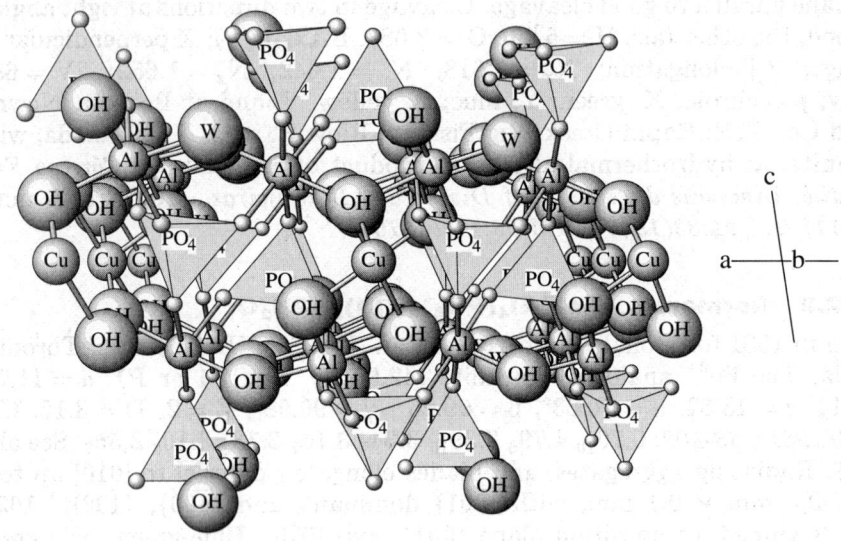

TURQUOISE: $CuAl_6(PO_4)_4(OH)_8 \cdot 4H_2O$

This b axis projection of turquoise shows chains (paralleling b) of Al (can also be Fe^{3+})-)O,OH)$_6$ octahedra which connect through shared vertices of two PO_4 tetrahedra. Divalent cations, such as Cu^{2+} join two or four chains, or these sites may be vacant.

The basic units of the structure [*NJMM:227(1989)*] are chains along b built up of alternate pairs of octahedra (either AlO_6 or $Fe^{3+}O_6$) joined by an edge and connected through shared vertices by two PO_4 tetrahedra. Cations in octahedral, or tetragonal dipyramidal, coordination with oxygens join two or four chains. In several of the end members, the A sites are vacant.

There is probably a continuous solid-solution series between Al and Fe^{3+}, but some Fe^{2+} is usually present. Most "turquoise" is deficient in Cu, and only a few of the materials presently called turquoise contain Cu as a principal constituent of the A site. In the (Cu,Fe^{3+}) end member, chalcosiderite, Cu atoms alternate with a vacancy. Electrostatic equilibrium of the structure is maintained by very small differences in the coordination of the OH around the "holes." From structure determinations it appears that the Cu atoms preferentially bond with groups that have hydroxylated oxygens PO_3OH and with H_2O molecules. Planerite, whose composition shows virtually no Cu and high concentrations of acid phosphate (PO_3OH) or hydronium ions (H_3O^+), suggests a structure where acid groups balance the charge, and the bond distribution varies to accommodate the lack of A-site atoms. *ZK 121:87(1965), NJMM:227(1989).*

42.9.3.1 Turquoise

TURQUOISE GROUP

Mineral	Formula	Space Group	a	b	c	α	β	γ	
Turquoise	$CuAl_6(PO_4)_4(OH)_8 \cdot 4H_2O$	$P\bar{1}$	7.690	9.950	7.490	110.57	115.40	68.40	42.9.3.1
Faustite	$(Zn,Cu)Al_6(PO_4)_4(OH)_8 \cdot 4H_2O$	$P\bar{1}$	7.670	9.890	7.440	110.35	115.60	72.80	42.9.3.3
Chalcosiderite	$CuFe_6^{3+}(PO_4)_4(OH)_8 \cdot 4H_2O$	$P\bar{1}$	7.885	10.199	7.672	110.84	115.12	67.51	42.9.3.4
Ayhelite	$(Fe^{2+},Zn)Al_6(PO_4)_4(OH)_8 \cdot 4H_2O$	$P\bar{1}$ or $P1$	7.408	9.891	7.627	110.94	115.05	69.89	42.9.3.5
Planerite	$\square Al_6(PO_4)_2(PO_3OH)_2(OH)_8 \cdot 4H_2O$	$P\bar{1}$	7.704	10.109	7.390	110.84	115.07	70.76	42.9.3.6

42.9.3.1 Turquoise $CuAl_6(PO_4)_4(OH)_8 \cdot 4H_2O$

Named in antiquity for the "source" of the mineral materials: Turkey. Turquoise group. In Al–Fe^{3+} series with chalcosiderite (42.9.3.4). TRIC $P\bar{1}$. $a = 7.690$, $b = 9.950$, $c = 7.490$, $\alpha = 110.57°$; $\beta = 115.40°$, $\gamma = 68.40°$, $Z = 1$, $D = 2.88$. Structure: *ZK 121:87(1965)*. *6-214*: 9.09_5 6.17_7 4.80_6 3.68_{10} 3.44_7 3.28_7 2.91_8 2.02_6. *25-260* (Ferrian-turquoise): 6.22_4 4.83_5 3.70_{10} 3.46_7 3.31_7 2.93_{10} 2.91_7 2.54_4. Crystals are rare, short prismatic [001] with large {001}, {010}, {1$\bar{1}$0}; usually massive, dense, cryptocrystalline to fine-granular; crusts, veinlets, stalactitic, concretionary. Massive turquoise is sky-blue, bluish green to apple-green, green-gray; waxy luster; subtranslucent to opaque. Crystals bright blue; white, greenish to pale green streak; vitreous luster; transparent. Cleavage {001}, perfect; fracture (of massive) conchoidal to smooth. $H = 5$–6. $G = 2.6$–2.8 (massive), 2.84 (crystals). DTA, IR, Mossbauer, ESR. *AM 64:449(1979)*, *NJMM: 469(1992)*. Biaxial (+); $N_x = 1.61$, $N_y = 1.62$, $N_z = 1.65$ (Na); $2V = 40°$; $r < v$, strong; weakly pleochroic in thick grains. X, colorless; Z, pale blue or pale green. Analyses always show some Fe^{3+}. Other elements found in traces (most are probably impurities): Si, Mg, Ca, Na, Ti, Mn, Ag, B, Ba, Be, Co, Cr, Ni, Pb, Sr, Zn, Zr. Soluble with difficulty in HCl, readily after low heat, which gives a blue solution when treated with ammonia. Pseudomorphs after apatite, bone, teeth, and orthoclase have been described, although fossil bone coated with a thin coating of vivianite may be mistaken for true turquoise. Synthesized from malachite. Imitation turquoise obtained through melting precipitated aluminum phosphate plus copper phosphate (Gilson turquoise), or by staining chalcedony, calcite, or glass. Natural turquoise may be treated artificially to enhance color. See *G&G 15:226(1977)*, *ZDGG 29:152(1980)*, and *JG 17:386(1981)* for studies of synthetic and imitation materials. A secondary mineral found with chalcedony, kaolin, iron oxides, and oxyhydroxides; in arid regions, turquoise forms on aluminous igneous or sedimentary rocks, probably derived from the weathering of apatite and copper sulfides. Found in fine 1- to 2-mm crystals at Bishop mine, Lynch station, Campbell Co., VA; in crusts at Wood Copper mine, Micaville and Erin, AL; at Turquoise Chief mine, St. Kevin district, Leadville, Lake Co., the Hall mine, Villa Grove, Saguache Co., CO; at Mt. Chalchuitl, Los Cerruos, NM (originally worked by Aztecs); at New Cornelia mine, Ajo, the Cole shaft and Lavender pit, Cochise Co., on the Fort Apache Indian Res., Mineral Park mine, Mohave Co., in the Helvetia district, Santa Rita Mts., AZ; at Pilot Mt. mine, Lincoln Co., NV; at Baker, Fresno Co., CA. In Europe, crystals at Bunny mine and Stowe's mine, Wheal Phoenix, Linkinhorne, and at Hensbarrow, and at Wheal Remfry china clay pits, St. Austell Moor, Cornwall, England; as 1- to 2-mm crystals in manganiferous quartz veins at Ottré near Vielsalm, Ardennes, Belgium; crusts with wavellite at Montebras, Creuse, France; at Jordansmühl, Silesia, Poland; at Poniklá, near Jilemnice, northern Bohemia, Czech Republic; at Ölsnitz and Messbach, Vogtland, Saxony, Germany. Also from the Sinai Peninsula at Wadi Maghara, Egypt, in seams with limonite in porphyry; at two classical localities, Nishapur and Damghan, northeastern Iran, and on the southern slopes of the Ali-Mirsa-Kuh Mts. NW of Maden, Khorasan, where it is in narrow seams and irregular patches in brecciated trachyte; at Sar Chesmeh Cu–Mo porphyry in south-central Iran; in the Vathi area, Macedonia; in the Madneuli deposit, Georgia, Russia; a more greenish-blue variety from Karkaralinsk, Semipalatinsk, Siberia; in the Kara-Tube Mts. S of Samarkand, Turkestan; in the Burkantau deposit, central Kyzulkum, and the Kalmakyr deposit, Almalyk ore field, Uzbekistan; in Husei Prov., China; at Got, near Angolola, Ethiopia; crystalline materials at Grant's Lookout, Narooma, NSW, Australia; in the banded iron formation at Itatiaiuçu, and in Consecheiro Mata region, Diamantina, MG, Brazil; at Chuquicamata, Antofagasta, Chile. Used for jewelry since ancient times, a common item of barter. HCWS *DII:946, AM 38:9634(1953), MM 40:640(1976), NJMM:469(1992)*.

42.9.3.2 Coeruleolactite $(Ca,Cu)Al_6(PO_4)_4(OH)_8 \cdot 4–5H_2O$

Named in 1871 from the Latin *coeruleus*, blue, and *lac, lactis*, milk, in allusion to its color. Questionable mineral. Extensive search for turquoiselike material containing Ca has not been successful. Since the ionic radius of Ca^{2+} (0.99) is substantially larger than that of Cu, Zn, or Fe^{2+}, the formation of a species with such a composition is unlikely (E. E. Foord, personal communication). Possibly a mixture of a turquoise group mineral and calcite. TRIC $P\bar{1}$. $a = 7.652$, $b = 10.153$, $c = 7.748$, $\alpha = 111.9°$, $\beta = 115.9°$, $\gamma = 67.6°$, $Z = 1$, $D = 2.99$. *MP 6:182(1958), AM 43:1224(1958)*. 12-166: 6.11_4 4.77_3 3.70_9 3.48_5 3.34_5 2.96_{10} 2.52_3 1.907_4. Cryptocrystalline, microcrystalline, fibrous, veinlets, crusts, botryoidal aggregates. Milk-white to light blue, white streak, translucent. Uneven to conchoidal fracture. H = 5. G = 2.57. Biaxial (+) (almost uniaxial); $N_x = 1.580$, $N_z = 1.588$ (PA). Soluble in acids and alkalis. Found at General Trimble mine, East Whiteland Twp., Chester Co., PA, with variscite-L, wavellite, mataleite, gibbsite, cacoxenite, and goethite in plates to

10 μm; at Wheal Phoenix mine, Linkinhorne, Cornwall, England; in "limonite" at *Rindsberg mine, Katzenellnbogen, Nassau, Germany*; at Cruzeiro mine, MG, Brazil. HCWS *DII:961, IMA:102(1986)*.

42.9.3.3   Faustite $(Zn,Cu)Al_6(PO_4)_4(OH)_8 \cdot 4H_2O$

Named in 1953 for George Tobias Faust (1908–1985), mineralogist, U.S. Geological Survey. Turquoise group. The Zn analog of turquoise. TRIC $P\bar{1}$. $a = 7.670$, $b = 9.890$, $c = 7.440$, $\alpha = 110.35°$, $\beta = 115.60°$, $\gamma = 72.80°$, $Z = 2$, $D = 2.99$. *6-216*: 6.70_7 6.14_7 3.68_{10} 3.44_6 3.28_6 2.92_6 2.89_8 2.05_7. Very fine grained, compact, nodular, veinlets. Apple-green to white when pure, white to pale yellow streak, waxy, dull luster. Conchoidal fracture, brittle. H = $4\frac{1}{2}$–$5\frac{1}{2}$. G = 2.92. Nonfluorescent. $N_{av} = 1.613$ (Na). Soluble in cold 1:1 HCl. Analyses show Fe^{3+}, and minor Ca, Si, V, Ca, Mg, Ba, with traces of Cr, Ti, Co, Ni, Sc, Zr, Sr, Mn, Bi. Only Ca and Fe were considered important to the chemistry of the mineral; the remainder were suggested to be impurities. *AM 38:964(1953)*. Found at *Copper King mine, Maggie Creek, Eureka Co., NV*, as apple-green veinlets in cherts and shales associated with montmorillonitic clays, quartz, kaolinite, and alunite; at Wheal Phoenix mine, Linkinhorne, Cornwall, England; in quartz veins in altered latite–andesite, Spahievo ore field, eastern Rhodope Mts., Bulgaria; cuprian faustite at Heyshapur, Iran. HCWS *GMP 16:55(1982)*.

42.9.3.4   Chalcosiderite $CuFe_6^{3+}(PO_4)_4(OH)_8 \cdot 4H_2O$

Named in 1814 by J. C. Ullman from the Greek *chalcos*, copper, and *sideros*, iron. Turquoise group. The Fe^{3+} end member of turquoise. TRIC $P\bar{1}$. $a = 7.885$, $b = 10.199$, $c = 7.672$, $\alpha = 110.84°$, $\beta = 115.12°$, $\gamma = 67.51°$, $Z = 1$, $D = 3.26$. Structure: *NJMM:227(1989)*. Pseudomonoclinic, I-centered. *37-446*: 6.39_3 4.93_2 3.76_{10} 3.54_4 3.38_4 3.01_4 2.95_3 2.52_3. See also *8-127*. Crusts or sheaflike groups of distinct crystals, short prismatic [001] with large {001}, {010}, {1$\bar{1}$0}, {$\bar{1}\bar{1}$1}. Light siskin-green to dark green, vitreous luster, transparent. Cleavage {001}, perfect; {010}, good. H = $4\frac{1}{2}$. G = 3.22. IR. Biaxial (−); $N_x = 1.775$, $N_y = 1.840$, $N_z = 1.844$; 2V = 22°; r > v, strong, crossed; weakly pleochroic. X, colorless; Z, pale green. Some Al substitutes for Fe^{3+}, minor As, Zn detected on analysis. *ZVMO 108:701(1979)*. Occurs as a secondary mineral in the oxidized zone of copper ore deposits. Found at Cole and Shattuck shafts, Bisbee, Cochise Co., AZ; with dufrenite, andrewsite, and goethite in gossan of Wheal Phoenix mine, Cornwall, England; with pharmacosiderite at Schneckenstein, Saxony, and with dufrenite at *Siegen, Westphalia, Germany*. HCWS *DII:947, MR 12:349(1981)*.

42.9.3.5   Aheylite $(Fe^{2+},Zn)Al_6(PO_4)_4(OH)_8 \cdot 4H_2O$

Named in 1986 for Allen V. Heyl (b.1918), economic geologist, U.S. Geological Survey. Turquoise group. The Fe^{2+} analog of turquoise. TRIC $P\bar{1}$ or $P1$. $a = 7.408$, $b = 9.891$, $c = 7.627$, $\alpha = 110.94°$, $\beta = 115.05°$, $\gamma = 69.89°$, $Z = 1$, $D = 2.89$. *IMMA:102(1986)*. PD: 6.15_4 3.67_{10} 3.44_4 3.40_3 3.27_4 2.91_3 2.89_7

2.02_3. Botryoidal, spheres to 2 mm of radiating concentric aggregates of crystals to 3 μm. Pale blue to blue-green, white streak, porcelaneous–subvitreous luster, translucent. Hackly and splintery fracture. Nonfluorescent. H = $5-5\frac{1}{2}$. G = 2.85. Biaxial(+); N_{mean} = 1.63. Found at *Mira Flores mine, Huanuni district, Oruro, Bolivia*, associated with cassiterite, pyrite, quartz, sphalerite, variscite, wavellite, and vivianite. HCWS

42.9.3.6 Planerite $\square Al_6(PO_4)_2(PO_3OH)_2(OH)_8 \cdot 4H_2O$

Named in 1862 for D. I. Planer, director of mines, Gumeshevsk, Ural Mts., Russia. Turquoise group. TRIC $P\bar{1}$. $a = 7.704$, $b = 10.109$, $c = 7.390$, $\alpha = 110.84°$, $\beta = 115.07°$, $\gamma = 70.76°$, Z = 1, D = 2.71. Structural data indicate vacancy for A site and some Fe^{3+} (Mossbauer) substituting for Al (B site). *IMMA:102(1986)*. *42-1318*: 6.82_9 6.20_2 4.73_2 3.75_{10} 3.70_3 3.29_1 3.09_2 2.55_1. Subcrystalline, botryoidal crusts. White to light green, dull luster, translucent. H = 5. G = 2.68. TGA. N_{av} = 1.517. Soluble in boiling NaOH. Found at General Trimble's mine, East Whiteland Twp., Chester, PA; Kelly Bank mine, Vesuvius, VA; Mauldin Mt. quarry, Mt. Ida, Montgomery Co., AR; Camelback Mt., AZ; in the *copper mines at Gumeshevsk, Ural Mts., Russia*; at Toyoda, near Kochi City, Shikoku Is., Japan, with wavellite, quartz, vermiculite, and iron hydroxides. HCWS *DII:762*, *NJBA 96:1(1961)*, *JMPEG 83:141(1988)*.

42.9.4.1 Sampleite $NaCaCu_5(PO_4)_4Cl \cdot 5H_2O$

Named in 1942 by C. Hurlbut for Mat Sample, superintendent at the mine, Chuquicamata, Chile. See lavendulan (42.9.4.2), the As analog. ORTH (dipyramidal, 2/m 2/m 2/m, pseudotetragonal with C_{+et}=borth). $a = 9.72$, $b = 38.48$, $c = 9.67$, Z = 8, D = 3.26. *MM 42:369(1978)*. *11-349*: 9.60_{10} 6.85_7 4.30_8 3.89_7 3.04_{10} 1.79_7 1.71_8 1.45_7. Lathlike crystals flattened {010}, elongated {001}; wedge-shaped aggregates subparallel to [010] to 0.2 mm × 0.2 mm × 0.02 mm. Light blue to blue-green (turquoise), pearly luster on {010}, transparent. Cleavage {010}, perfect; {100}, {001}, good. H ≈ 4. G = 3.20. Biaxial (−); XYZ = *bac*; N_x = 1.625, $N_{y,z}$ = 1.674; 2V = 5–10°; r > v; analyses also show Mg O,K_2O pleochroic. X, turquoise-green; Y,Z, benzol-green X < Y,Z. Easily soluble in acids. Found at Jinqemia cave, WA, Australia, formed as a reaction of CuS with guano and associated with atacamite, weddelite, gypsum, goethite, Mn oxides, birnessite, dolomite, halite, apatite, and malachite; in highly oxidized ore at Mantos Blancos with atacamite, calcite, halite, and chrysocola, and at *Chuquicamata, Chile*, with gypsum, atacamite, jarosite, and iron oxyhydroxides; as a corrosion product on old bronzes. HCWS *DII:945*, *BM 79:7(1956)*.

42.9.4.2 Lavendulan $NaCaCu_5(AsO_4)_4Cl \cdot 5H_2O$

Named in 1837 for the color. Sampleite (42.9.4.1) is the P analog. ORTH. $a = 9.73$, $b = 41.0$, $c = 9.85$, Z = 8, D = 3.64. *BM 79:7(1956)*. *31-1280*: 9.77_{10} 7.01_4 4.88_5 4.41_4 3.11_7 2.76_2 2.48_2 1.97_2. Thin botryoidal crusts of minute radiating fibers or plates. Lavender blue, vitreous to waxy luster, translucent. H = $2\frac{1}{2}$-3. G = 3.84. DTA Biaxial $(-)$; $N_x = 1.645$, $N_y = 1.715$, $N_z = 1.725$; Z inclined to fiber length. Analyses show some K, Co, Ni. Found at *Jachimov, Bohemia, Czech Republic*; at Laurium, Greece; at Bou Azzer, Morocco; at Alice Mary Cu mine, Rundip, High Ranges, Bali Low Cu mine, Capricorn Ranges, Francis Furnace mine, Marvel Loch, WA, and at Preamimma mine, Callington, SA, Australia; at Tsumeb, Namibia; at San Juan, Chile. HCWS *DII:750*, *TRSSA 97:135(1973)*.

42.9.4.3 Zdenekite $NaPbCu_5(AsO_4)_4Cl \cdot 5H_2O$

Named in 1995 after Zdenek Johan (b.1935), mineralogist and director of Scientific Affairs, Bureau de Recherches Géologiques, Minières (BRGM), France. TET $P4_122$ or $P4_322$. $a = 10.066$, $c = 39.39$, $Z = 8$, $D = 4.08$. *EJM 7:553(1995)*. PD: 9.83_{10} 7.11_3 4.93_6 4.48_5 3.13_9 2.77_4 2.52_5 1.78_4. Individual tabular tetragonal crystals up to 0.1 mm, flattened on {001}, showing {001}, {100}, and {110}; as spherules and aggregates to 0.3 mm. Intense blue color, pale blue streak, vitreous luster, translucent. Fragile, soft. Cleavage (001), perfect; irregular fracture. G = 4.1. Nonfluorescent in UV. Occurs at the *Cap Garonne mine, Var, France*, associated with anglesite, tsumcorite-like mineral, and olivenite in quartz conglomerate that has been mineralized with tennantite and covellite. HCWS

42.9.5.1 Duhamelite $Cu_4Pb_2Bi(VO_4)_4(OH)_3 \cdot 8H_2O$

Named in 1981 for J. E. DuHamel, geologist, Phelps Dodge Corp., who found the mineral. ORTH *P*. $a = 7.49$, $b = 9.66$, $c = 5.87$, $Z = 1$, $D = 5.99$. *MM 44:151(1981)*. 35-500: 5.01_4 4.83_3 4.18_2 3.49_5 3.16_7 2.95_{10} 2.64_9 1.755_3. Prismatic crystals to 0.2 mm $\times$ 0.4 mm and in barrel-shaped bundles. Green, yellow-green streak. Brittle. H = 3. G = 5.80. Biaxial; $N_x = 2.08$, $N_y = 2.11$ (Na); weakly pleochroic in yellows; $X = Y < Z$. Soluble in cold 1:10 HCl, HNO_3, not in H_2O. Found with chrysocolla, malachite, fornacite, wulfenite, and bismutite in Au-bearing quartz veins cutting Precambrian greenstones 5 km SW of *Payson, at Lousy Gulch, Gila Co., AZ*. HCWS *AM 67:414(1982)*.

42.9.6.1 Santafeite $NaMn_3(Ca,Sr)(V,As)_3O_{13} \cdot 4H_2O$

Named in 1958 for the Santa Fe Railroad Co. ORTH $B22_12$. $a = 9.239$, $b = 29.978$, $c = 6.314$, $Z = 4$, $D = 3.34$. 11-169: 14.88_{10} 7.47_5 3.66_3 3.39_3 2.91_3 2.79_4 2.70_5 1.667_4. Small rosettes (to 4 mm in diameter) of acicular crystals radially oriented. Black, brown streak, subadamantine luster, translucent. Cleavage {010}, perfect; {110}, distinct; very brittle. G = 3.379. $N_{av} = 2.01$; $X = c$; pleochroic: red to yellow-brown; distinct dispersion. Small amounts of K, Fe^{3+}, Ni, Cu, U, and CO_2 as well as traces of REE in analyses. The oxidation state(s) of Mn not fully defined but Mn^{4+} appears dominant. Found on Todilto limestone in the uranium district at *Grants dist., McKinley Co., NM*. HCWS *AM 43:677(1958), MM 50:299(1986)*.

42.9.7.1 Ogdensburgite $Ca_2(Zn,Mn^{2+})Fe_4^{3+}(AsO_4)_4(OH)_6 \cdot 6H_2O$

Named in 1981 for the locality. ORTH $Bmmm$, $B222^*$. $a = 11.351$, $b = 14.837$, $c = 6.555$, $Z = 2$, $D = 3.39$. AM $72:409(1987)$. Significant site vacancies are suggested as calculated and measured densities are vastly different. 35-518: 14.8_{10} 7.39_2 5.30_2 4.51_3 3.28_3 2.79_2 2.73_2 2.65_4. See also 38-482. Rectangular blades up to 1 mm long; encrustations of thin (0.1-mm) plates (micaceous). Dark brownish red, resinous luster on cleavage surfaces, translucent. Cleavage {001}, perfect; {010}, fair; {100}, poor. $H = 2$. $G = 3.11$. Nonfluorescent. Biaxial $(-)$; $XYZ = cba$; Y and Z parallel to {001} cleavage edges; $N_x = 1.715$, $N_y = 1.783$, $N_z = 1.785$; $2V = 5$–$10°$; $Z > Y >>> X$; strongly pleochroic. X, yellow; Y,Z, red-brown. Found at *Sterling Hill mine, Ogdensburg, NJ*, with para-symplesite, köttigite, and other iron arsenates; at Ojuela mine, Mapimi, DUR, Mexico, with villyaellenite crystals(!) in vugs coated with arseniosiderite and chalcophanite in limonitic gossan that contains adamite and manganese oxides. HCWS AM $67:858(1982)$.

42.9.8.1 Dewindtite $Pb_3(UO_2)_6H_2(PO_4)_4O_4 \cdot 12H_2O$

Named in 1922 for Jean Dewind, Belgian geologist. ORTH $Bmmb$. $a = 16.031$, $b = 17.264$, $c = 13.605$, $Z = 4$, $D = 5.12$. Structure: EJM $2:399(1990)$. Structure contains $[H(UO_2)_3O_2(PO_4)_2]$ layers parallel to (100) connected through Pb in eightfold coordination. 39-1350: 7.92_{10} 5.85_3 3.96_{10} 3.36_3 3.13_6 3.09_5 2.88_5 2.09_3. Microscopic rectangular tablets flattened {100} terminated by {001}, striated [001]; pulverulent, compact. Canary-yellow, translucent. Cleavage {100}. $G = 5.03$. Fluoresces green. Biaxial $(-)$; $XYZ = bca$; $N_x = 1.716$, $N_y = 1.736$, $N_z = 1.740$; $2V = 40°$; $r < v$; nonpleochroic. Soluble in acids. Found at Grury, Saône-et-Loire, France; at Wölsendorf, Bavaria, Germany; associated with metatorbernite and other secondary U minerals at *Kasolo, Shaba, Zaire*, and at Shinkalobwe. HCWS $DII:875$, AM $39:444(1954)$.

42.10.1.1 Ferrisymplesite $Fe_3^{3+}(AsO_4)_2(OH)_3 \cdot 5H_2O$

Named in 1924 as the oxidized equivalent of symplesite (40.3.8.1). An inadequately described mineral. 1-119(syn): 9.0_2 6.7_{10} 3.99_4 3.70_4 3.36_2 3.18_4 2.95_4 2.82_4. Small irregular masses with fibrous structure. Amber-brown, resinous luster. Found intimately admixed with erythrite and anabergite at *Hudson Bay mine, Cobalt, ONT, Canada*. HCWS $DII:753$, AUF $30:213(1979)$.

42.10.2.1 Wavellite $Al_3(PO_4)_2(OH,F)_3 \cdot 5H_2O$

Named in 1805 for William Wavell (d.1829), physician, Horwood Parish, Devonshire, England, who discovered the mineral. ORTH $Pcmn$. $a = 9.621$, $b = 17.363$, $c = 6.994$, $Z = 4$, $D = 2.325$. Structure: ZK $127:21(1968)$. Octahedrally coordinated Al (AlO_6) forms two distinct (Al1 and Al2) zigzag chains parallel to the c axis that are linked by phosphate tetrahedra into layers parallel to (010). Water molecules occupy cavities created. This structural configuration is responsible for the fibrous habit and the perfect cleavages. *25-20:* 8.67_{10} 8.42_{10} 5.65_5 4.81_3 3.42_4 3.22_6 2.57_3 2.11_2. Crystals rare, stout to long prismatic [001], with {110} striated [001]; hemispherical or globular aggregates with radial-fibrous or stellate structure, as crusts or stalactitic, rarely chalcedonic, opaline masses. Greenish white, green to yellow, yellowish brown, brown, brownish black, blue, white, colorless; white streak; vitreous to pearly and resinous luster; translucent. Cleavages {110}, {101}, perfect; {010}, distinct; subconchoidal to uneven fracture: brittle. $H = 3\frac{1}{2}$–4. $G = 2.36$. Biaxial (+); $XYZ = bac$; $N_x = 1.535$, $N_y = 1.543$, $N_z = 1.561$(Cernovice); $2V \approx 71°$; r > v, weak; X > Z; weakly pleochroic. X, greenish; Z, yellowish. Indices increase with increasing Fe^{3+} for Al and F for OH. Easily soluble in acids. A secondary mineral common in aluminous low-grade metamorphic rocks, in limonitic or phosphate rock deposits, and as a late-formed hydrothermal mineral in veins. Found at Amity, Orange Co., NY; at General Trimble's mine, Chester Co., Moore's Mill, Cumberland Co., Hellertown, Northampton Co., PA; at Kelly Bank mine, Vesuvius, Virginia Glass Sand Co. quarry, VA; at Carolina pyrophyllite mine, NC; at Brewer mine, Cedartown, GA; in phosphate rocks at Dunellen, Marion Co., FL; at Dug Hill, and Montgomery Co., Hot Springs, Garland Co., Magnet Cove, Hot Spring Co., in the novaculite beds, and in Silver City region, Montgomery Co., AR; near Coal City, St. Clair Co., Jacksonville, Calhoun Co., Erin, Indian Mt., Wolfden Mt., AL; at Mineral Park mine, Mohave Co., AZ; at Clonmel, County Tipperary, Tracton and Kinsale, County Cork, Ireland; in a clay slate at *High Down quarry, Barnstaple, Devonshire, England*, and at Stenna Gwyn mine, St. Austell, Cornwall; with amblygonite in tin veins at Montebras, Creuse, France; at Langenstriegis, Saxony, Waldgirmes, and the phosphate deposits of the Lahn–Dill region, Hesse, Nassau, Germany; at Cernovice, Zbirow, and neighboring localities in Bohemia, Czech Republic; at Kapnik, Romania; in Shaba, Zaire; at Iron Monarch quarry, Iron Knob, SA, Australia; abundant in the tin veins of Silgo XX mine, Llallagua, Bolivia, with apatite, vauxite, and paravauxite; also at Itos, Porvenir, San José mines, Oruro, Potosi. HCWS *DII:962, USGSPP 800G:53(1972)*.

42.10.3.1 Kingite $Al_3(PO_4)_2(OH,F)_3 \cdot 9H_2O$

Named in 1957 for D. King (1926–1990), geologist, Department of Mines, SA, Australia. TRIC $P\bar{1}$ or $P1$. $a = 9.15$, $b = 10.00$, $c = 7.24$, $\alpha = 98.6°$, $\beta = 93.6°$, $\gamma = 93.2°$, $Z = 2$, $D = 2.465$. *AM 55:515(1970)*. *24-10:* 9.1_{10} 5.43_3 5.28_5 3.48_7 3.45_8 3.38_3 3.17_4 3.11_3. Platy or irregular crystals to 1.0 µm in nodules to 5 cm in diameter. White, translucent. $G = 2.2$. DTA. $N_{av} = 1.514$. Very minor Na, K, F detected on analysis. Dehydrates under electron beam to meta-kingite. Found coated with a brown film of goethite and talc at the phosphate deposits of *Fairview mine, Robertstown, Burra Co., SA, Australia*, associated with vivianite, fluorapatite, leucophosphite, also occurs at Clinton 100 km west-southwest of this locality. HCWS *MM 31:351(1957)*.

42.10.4.1 Gutsevichite $Al_3(PO_4)_2(OH)_3 \cdot 8H_2O$

Named in 1959 for V. P. Gutsevich, geologist in Kazakhstan. ISO. *14-139*: 4.08_{10} 3.51_5 2.51_9 1.82_8 1.45_7. Irregular crusts and concretions. Yellow-olive to dark brown, waxy, dull luster. $H = 2\frac{1}{2}$. $G = 1.90$–2.0. DTA. N varies with color: 1.560 (yellow) to 1.60 (dark brown). Easily soluble in cold 1:10 acids. Analyses showed Mg, Ca, Ba, S, Si. Found in *northwestern Kazakhstan* as cavity fillings in zone of oxidation of V-containing shales. HCWS *AM 46:1200(1961)*.

OVERITE GROUP

The overite group minerals are hydrated phosphates with the general formula

$$ABC(PO_4)_2(OH) \cdot 2\text{–}4H_2O$$

where

A = Ca, Mn, Zn
B = Mg, Fe^{2+}, Mn^{2+}
C = Al, Fe^{3+}

and a layered structure that crystallizes with orthorhombic symmetry.

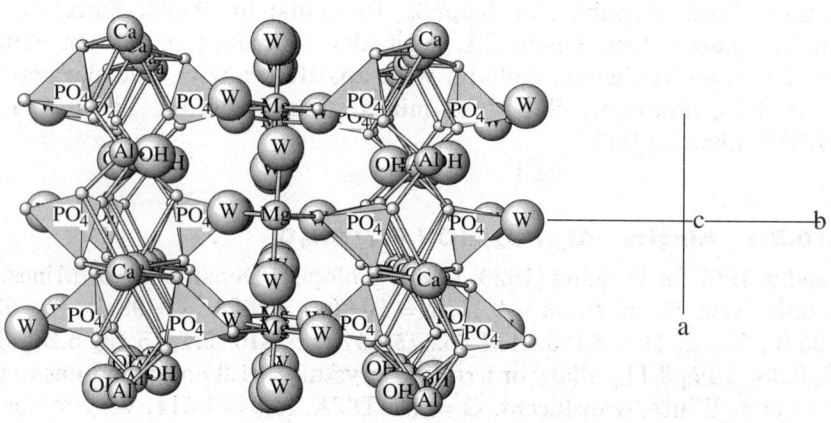

OVERITE: $CaMgAl(PO_4)_2(OH) \cdot 4H_2O$

The *c* axis projection of overite shows slabs consisting of $Ca(Op)_6$ alternating with $Al(Op)_2(OH)_4$ on {010} and parallel layers of $Mg(Op)_2(H_2O)_4$.

42.11.1.2 Segelerite

The structure is closely related to the jahnsite–whiteite group (42.11.2.1) with octahedral corner-sharing chains of $C(O,OH)_6+Ca(Op)_6$ forming slabs with PO_4 tetrahedra which are connected through $B(OH,H_2O)_6$ octahedra and other cations. In addition to distinctions between the several species depending on Al or Fe^{3+} content, there are variations in the general structure related to a twisting of the octahedral chains and distortion of the B-octahedra. The identification of a particular species in the overite group (and in the jahnsite–whiteite group) usually requires single-crystal structure examination.

OVERITE GROUP

Mineral	Formula	Space Group	a	b	c	
Overite	$CaMgAl(PO_4)_2(OH)$ $\cdot 4H_2O$	$Pbca$	14.723	18.746	7.107	42.11.1.1
Segelerite	$CaMgFe^{3+}(PO_4)_2$ $(OH)\cdot 4H_2O$	$Pbca$	14.826	18.75	7.307	42.11.1.2
Mangano-segelerite	$CaMn^{2+}Fe^{3+}(PO_4)_2$ $(OH)\cdot 4H_2O$	$Pcba$	14.89	18.79	4.405	42.11.1.3
Lun'okite	(Mn^{2+},Ca) (Mg,Fe^{2+},Mn^{2+}) $Al(PO_4)_2(OH)$ $\cdot 4H_2O$	$Pbca$	14.95	18.71	6.96	42.11.1.4
Wilhelmvierlingite	$CaMn^{2+}Fe^{3+}(PO_4)_2$ $(OH)\cdot 4H_2O$	$Pbca$	14.80	18.70	7.31	42.11.1.5
Kaluginite	$(Mn,Ca)MgFe^{3+}$ $(PO_4)_2(OH)\cdot 4H_2O$	$Ccca$	15.05	37.37	7.18	42.11.1.6

42.11.1.1 Overite $CaMgAl(PO_4)_2(OH) \cdot 4H_2O$

Named in 1940 for Edwin Over (1905-1963), mineral collector at Fairfield, UT. Overite group. The Al analog of segelerite (42.11.1.2). ORTH $Pbca$. $a = 14.723$, $b = 18.746$, $c = 7.107$, $Z = 8$, $D = 2.51$. Structure: AM $62:692(1977)$. 16-157: 9.4_8 5.29_6 4.98_2 3.70_5 3.42_4 2.89_6 2.83_{10} 1.975_5. Crystals flattened on $\{010\}$ elongated $[001]$ with platy to lathlike shape; subparallel growths on $\{010\}$; massive aggregates of coarse plates. Light apple-green to colorless, vitreous luster, transparent. Cleavage $\{010\}$, perfect. $H = 3\frac{1}{2}$–4. $G = 2.53$. Biaxial $(-)$; $XYZ = cab$; $N_x = 1.568$, $N_y = 1.574$, $N_z = 1.580$; $2V = 75°$; $r > v$, weak. Easily soluble in hot HNO_3. Found in cavities in variscite nodules at *Clay Canyon, Fairfield, Utah Co., UT*, associated with crandallite, apatite. HCWS $DII:979$, AM $59:48$ (1974).

42.11.1.2 Segelerite $CaMgFe^{3+}(PO_4)_2(OH) \cdot 4H_2O$

Named in 1974 for Curt G. Segeler (1901-1989), amateur mineralogist, New York City. Overite group. See overite (42.11.1.1), the Al analog. ORTH $Pbca$.

$a = 14.826$, $b = 18.75$, $c = 7.307$, $Z = 8$, $D = 2.61$. *AM 59:48(1974)*, Structure: *AM 62:692(1977)*. *AM 26-1061*: 9.31_9 5.34_6 4.97_4 4.65_4 3.73_5 3.42_5 2.87_{10} 2.58_4. Crystals up to 0.5 mm striated parallel to prism length [001] with square outline showing {100}, {010}, {110}, {121}. Pale yellow-green (chartreuse) to colorless, vitreous luster, transparent. Cleavage {010}, perfect. $H = 4$. $G = 2.67$. Biaxial (−); $XYZ = bac$; $N_x = 1.618$, $N_y = 1.635$, $N_z = 1.650$; 2V large; pleochroic. X,Y, colorless; Z, yellow $Z > X$, Y. Found at *Tip Top pegmatite, near Custer, Custer Co., SD*; at Milgun Station, WA, Australia. HCWS

42.11.1.3 Manganosegelerite $CaMn^{2+}Fe^{3+}(PO_4)_2(OH) \cdot 4H_2O$

Named in 1992 for the composition and relationship to segelerite (42.11.1.2). Overite group. ORTH *Pcba*. $a = 14.89$, $b = 18.79$, $c = 4.405$, $Z = 8$, $D = 2.74$. *ZVMO 121:95(1992)*. PD: 9.39_{10} 4.70_5 2.97_4 2.86_9 2.60_4 2.02_4 1.966_5 1.880_5. Aggregates to 2 mm across of prismatic individuals up to 0.05 mm across. Yellow to yellow-green, yellow streak, vitreous luster, transparent in thin grains. Cleavage {001}, imperfect. $H = 3$–4. $G = 2.76$. IR shows both OH and H_2O. Biaxial (−); $YXZ = abc$; $N_x = 1.657$, $N_y = 1.668$, $N_z = 1.691$; $2V_{calc} = 70°$; marked dispersion; $r < v$; pleochroic. X, yellow; Z, light yellow. Found on mitradatite in late fractures in granitic pegmatites of the *Kola Penin., Russia*, also locally developed from lunjokite, replacing it as pseudomorphs and associated with eosphorite, kingsmountite, and Mn-gordonite. HCWS *AM 79:185(1994)*.

42.11.1.4 Lun'okite $(Mn^{2+},Ca)(Mg,Fe^{2+},Mn^{2+})Al(PO_4)_2(OH) \cdot 4H_2O$

Named in 1983 for the locality. Overite group. ORTH *Pbca*. $a = 14.95$, $b = 18.71$, $c = 6.96$, $Z = 8$, $D = 2.69$. *ZVMO 112:232 (1983)*. *37-456*: 9.39_9 5.15_5 3.48_6 2.92_7 2.81_{10} 1.877_5 1.568_5 1.555_5. Radiating aggregates to 1 mm; crusts. Colorless to white, yellowish, transparent. Cleavage {010}, perfect; {001}, imperfect. $H = 3$–4. $G = 2.66$. DTA, IR. Biaxial (+); $XYZ = cab$; $N_x = 1.603$, $N_y = 1.608$, $N_z = 1.616$; $2V = 70°$. Found as crusts on nodules of mitridatite and in fractures in granitic pegmatites at *Lun'ok R., Kola Penin., Russia*, associated with eosphorite, laueite, and kingsmountite. HCWS *AM 69:210(1984)*.

42.11.1.5 Wilhelmvierlingite $CaMn^{2+}Fe^{3+}(PO_4)_2(OH) \cdot 2H_2O$

Named in 1983 for Wilhelm Vierling (1901–1995), Weiden, Bavaria, Germany. Overite group. Mn analog of segelerite (42.11.1.2). ORTH *Pbca*. $a = 14.80$, $b = 18.70$, $c = 7.31$, $Z = 8$, $D = 2.60$. *AM 69:568(1984)*. *38-429*: 9.34_7 5.00_6 4.67_4 3.44_3 2.86_{10} 2.58_4 1.98_5 1.96_4. Fine-grained radial-fibrous aggregates. Pale yellow to brownish, translucent. Cleavage {010}, perfect. $H = 4$. $G = 2.58$. IR. Biaxial (−); $XYZ = bac$; $N_x = 1.637$, $N_y = 1.664$, $N_z = 1.692$; $2V_{calc} = 45°$; pleochroic. X, Y, light yellow; Z, dark yellow. Substitution of Zn for Ca detected by microprobe analysis. Found with rockbridgeite, zwieselite, and other secondary phosphates in the *Hagendorf pegmatite, Bavaria, Germany.* HCWS *DA 34:267(1983)*.

42.11.1.6 Kaluginite (Mn,Ca)MgFe^{3+}(PO$_4$)$_2$(OH) · 4H$_2$O

Named in 1991 for Aleksandr V. Kalugin (1857–1933), mineralogist, Ural Mts., Russia. Overite group. ORTH *Ccca*. $a = 15.05$, $b = 37.37$, $c = 7.18$, $Z = 16$, $D = 2.70$. *ZVMO 120:108(1991)*. 44-1403: 9.38$_{10}$ 4.59$_5$ 5.00$_7$ 3.52$_7$ 2.97$_5$ 2.84$_9$ 2.60$_8$ 1.88$_7$. Platy crystals to 1 mm, flattened {100}, elongate [001], showing {001}, {100}, {010}, {111}, {211}. Light yellowish green to greenish yellow, white streak, vitreous to greasy luster, transparent to translucent. Cleavage {010}, imperfect; uneven fracture; brittle. H = $3\frac{1}{2}$. G = 2.69. Biaxial (−); XYZ = *bac*; N$_x$ = 1.627, N$_y$ = 1.642, N$_z$ = 1.658; 2V = 90°; r > v; pleochroic: X,Y, colorless; Z, greenish yellow. Found in a granite pegmatite vein in *Ilmenskij National Park, Ural Mts., Russia*, associated with products of decomposition of triplite, francolite, ushkovite, matveevite, and Fe and Mn oxides. HCWS *AM 78:450(1993)*.

JAHNSITE–WHITEITE GROUP

The jahnsite–whiteite minerals are phosphate minerals whose general formula can be represented by

$$XABC(OH)_2(H_2O)_8[TO_4]_4 \cdot 8H_2O$$

where

X = Ca, Mn^{2+}
A = Fe^{2+}, MN^{2+}, Mg
B = Mg, Mn^{2+}
C = Fe^{3+}, Al

and crystallize with layered structures in the monoclinic, space group *P2/a*, although the crystals are usually twinned and appear pseudo-orthorhombic. See Overite group (42.11.1.1).

The structure [*AM 59:964(1974), 59:1292(1974)*] is dominated by corner-linked chains with a 7-Å repeat formed through octahedral coordination of six different and distinct cation sites. These chains form slabs parallel to {001} when connected through TO$_4$ tetrahedra. See also the laueite group (42.11.10.1).

The two subgroups, jahnsite and whiteite, are distinguished by the dominant element in the C site: Fe^{3+} or Al. End members are often named by adding a suffix indicating X, A, and B cations, that is, jahnsite-(CaMnMg). The familiar structural motif which produces virtually identical cell constants for the several cations in solid solution means that accurate identification of specific species requires precise crystal chemical analysis.

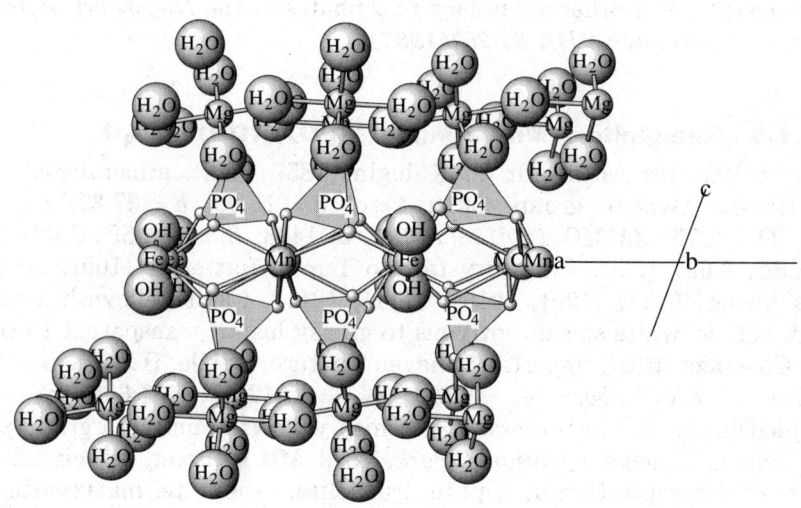

JAHNSITE: $CaMnMgFe_2(PO_4)(OH)_2 \cdot 8H_2O$

This c axis projection of jahnsite-(CaMnMg) shows slabs paralleling {001} formed of chains with Ca in eightfold coordination, $Fe^{3+} - (OH)_4(O_p)_2$ and $Mn^{2+}(O_p)_6$ connected through the tetrahedral PO_4 and with an octahedrally coordinated $Mg(H_2O, O_p)_6$ layer. The whiteite group replaces Fe^{3+} with Al. See also overite 42.11.1.1, montgomeryite 42.11.8.1 and laueite 42.11.10.1.

Jahnsite Subgroup, $C=Fe^{3+}$

Mineral	Formula	Space Group	a	b	c	β	
Jahnsite-(CaMnMg)	$CaMnMg_2Fe_2^{3+}(PO_4)_4(OH)_2 \cdot 8H_2O$	$P2/a$	14.94	7.14	9.93	110.16	42.11.2.1
Jahnsite-(CaMnFe)	$CaMnFe_2^{2+}Fe_2^{3+}(PO_4)_4(OH)_2 \cdot 8H_2O$	$P2/a$	15.01	7.15	9.87	111.23	42.11.2.2
Jahnsite-(CaMnMn)	$CaMnMn_2Fe_2^{3+}(PO_4)_4(OH)_2 \cdot 8H_2O$	$P2/a$	14.887	7.152	9.966	109.77	42.11.2.3
Jahnsite-(MnMnMn)	$MnMnMn_2Fe_2^{3+}(PO_4)_4(OH)_2 \cdot 8H_2O$	$P2/a$	15.02	7.23	18.75	100.42	42.11.2.4
Keckite	$Ca(Mn^{2+},Zn)_2Fe_3^{3+}(PO_4)_4(OH)_3 \cdot 2H_2O$	$P2/a$	15.02	7.19	19.74	110.30	42.11.4.1

42.11.2.4 Jahnsite-(MnMnMn)

WHITEITE SUBGROUP, C = Al

Mineral	Formula	Space Group	a	b	c	β	
Whiteite-(CaFeMg)	Ca(Fe^{2+},Mn^{2+})Mg$_2$Al$_2$(PO$_4$)$_4$(OH)$_2$·8H$_2$O	P2/a	14.90	6.98	10.13	113.12	42.11.3.1
Whiteite-(MnFeMg)	(Mn^{2+},Ca)(Fe^{2+},Mn^{2+})Mg$_2$Al$_2$(PO$_4$)$_4$(OH)$_2$·8H$_2$O	P2/a	14.99	6.96	10.14	113.32	42.11.3.2
Whiteite-(CaMnMg)	CaMn^{2+}Mg$_2$Al$_2$(PO$_4$)$_4$(OH)$_2$·8H$_2$O	P2/a	14.842	6.996	10.109	112.59	42.11.3.3

WHITEITE SUBGROUP, C = Al (cont'd)

Mineral	Formula	Space Group	a	b	c	β	
Rittmannite	(Mn^{2+},Ca)Mn^{2+}(Fe^{2+},Mn^{2+})$_2$(Al,Fe^{3+})$_2$(PO$_4$)$_4$(OH)$_2$·8H$_2$O	P2/a	15.01	6.89	10.16	112.82	42.11.3.4

Notes:
1. The overite group (42.11.1.1) is a related group with the linkages of the slabs all *trans*-[Mg(O$_p$)$_2$(H$_2$O)$_4$], while in jahnsite they are alternately *cis* and *trans*. AM 62:692(1977).
2. These minerals are secondary phosphates (alteration products) found in granitic pegmatites.
3. Jahnsite (FE^{3+}-containing species) have higher indices of refraction (1.64-1.71 and are beaxial – whereas whiteites (Al -containing species are Biaxial($^+$) with indices 1.580-1.595 except Rittmanite (some FE^{3+}) 1.62-1.65)

42.11.2.1 Jahnsite-(CaMnMg) CaMnMg$_2$Fe$_2^{3+}$(PO$_4$)$_4$(OH)$_2$·8H$_2$O

42.11.2.2 Jahnsite-(CaMnFe) CaMnFe$_2^{2+}$Fe$_2^{3+}$(PO$_4$)$_4$(OH)$_2$·8H$_2$O

42.11.2.3 Jahnsite-(CaMnMn) CaMnMn$_2$Fe$_2^{3+}$(PO$_4$)$_4$(OH)$_2$·8H$_2$O

42.11.2.4 Jahnsite-(MnMnMn) MnMnMn$_2$Fe$_2^{3+}$(PO$_4$)$_4$(OH)$_2$·8H$_2$O

Named in 1974 for Richard H. Jahns (1915–1983), mineralogist and pegmatite expert, Stanford University, Palo Alto, CA. The several minerals above are the Fe^{3+} end members in the Fe^{3+}–Al series with the whiteite (42.11.3.1-3) group. Jahnsite-whiteite group. (CaMnMg): Original description of jahnsite. MON P2/a. $a = 14.94$, $b = 7.14$, $c = 9.93$, $\beta = 110.16°$, Z = 2, D = 2.715. Structure: AM 59:48(1974), 59:964(1974). 26-1062: 9.27$_{10}$ 4.91$_6$ 4.63$_4$ 3.52$_5$ 2.95$_5$ 2.83$_8$ 2.58$_4$ 1.945$_4$. See also 30-265(calc). Biaxial (−); Z parallel to b; X ∧ [001] = 18°; N$_x$ = 1.640, N$_y$ = 1.658, N$_z$ = 1.670; 2V large; Y >> Z > X; pleochroic. X, pale purple; Y, deep purplish brown; Z, greenish yellow. (CaMnFe): Fe^{2+} end member in CaMn (Fe^{2+}–Mg) series. MON P2/a. $a = 15.01$, $b = 7.15$,

$c = 9.87$, $\beta = 111.23°$, $Z = 2$, $D = 2.88$. *MM 42:309(1978)*. *30-266:* 9.15_{10} 4.96_2 4.84_1 4.60_4 3.48_3 3.45_3 2.81_5 1.874_2. **(CaMnMn):** Mn^{2+} end member in the CaMn (Mn^{2+}–Mg) series. MON $P2/a$. $a = 14.887$, $b = 7.152$, $c = 9.966$, $\beta = 109.77°$, $Z = 2$, $D = 2.798$. *AM 75:401(1990)*. PD: 9.40_{10} 5.74_2 5.02_2 4.70_3 3.53_2 3.46_2 2.99_2 2.87_8. **(MnMnMn):** Mn^{2+} end member of all jahnsite species. MON $P2/a$. $a = 15.02$, $b = 7.23$, $c = 18.75$, $\beta = 100.42°$, $Z = 2$, $D = 2.90$. *GSAPA:76A(1955)*, *MM 42:309(1978)*. Biaxial (−); $N_x = 1.682$, $N_y = 1.695$, $N_z = 1.707$; $r > v$, strong; pleochroic. X, pale yellow-brown; Y, yellow-brown; Z, dark yellow-brown. Long to short prismatic [010] crystals with striations parallel to [010] up to 5 mm long. Often as parallel aggregates twinned by reflection on {001}, pseudo-orthorhombic appearance. Forms showing {001}, {100}, {201}, {$\bar{2}$01}, {$\bar{1}$01}, {011}, {110}, {$\bar{1}$11} with striae on {$\bar{2}$01}, {100}. Granular masses up to 5 cm across. Nut-brown, purplish brown, yellow, yellow-orange, greenish yellow. Cleavage {001}, good. H = 4. G = 2.706–2.90. The original description of jahnsite is from (CaMnMg) material found at *Tip Top pegmatite, Custer, SD*. The (CaMnFe) Fe^{2+} end member was found at *Fletcher mine, North Groton, NH*, with laueite and strunzite; the (CaMnMn) Mn^{2+} end member was found at *Mangualde pegmatite, Beira, Portugal*, and the (MnMnMn) Mn^{2+} end member of all jahnsite series minerals was found at *Stewart mine, Pala, San Diego Co., CA*. Other localities are the Palermo No. 1 mines, North Groton, NH; the White Elephant, Bull Moose, Big Chief, and Linwood pegmatites, Custer Co., SD, where they are associated with triphylite, ferrisicklerite–sicklerite series, heterosite–purpurite series, and rockbridgeite–frondelite series in vugs, which invariably contain leucophosphite, hureaulite, and may be accompanied by collinsite, hydroxylapatite, whitlockite, bermanite, laueite, mitridatite, segelerite, and robertsite; also found at Hagendorf Süd, Bavaria, Germany; at Corrego Frio mine, MG, Brazil. HCWS *AM 62:692(1977)*.

42.11.3.1 Whiteite-(CaFeMg) $Ca(Fe^{2+},Mn^{2+})Mg_2Al_2(PO_4)_4(OH)_2 \cdot 8H_2O$

42.11.3.2 Whiteite-(MnFeMg) $(Mn^{2+},Ca)(Fe^{2+},Mn^{2+})Mg_2Al_2(PO_4)_4(OH)_2 \cdot 8H_2O$

42.11.3.3 Whiteite-(CaMnMg) $CaMn^{2+}Mg_2Al_2(PO_4)_4(OH)_2 \cdot 8H_2O$

Named in 1978 for John S. White, Jr. (b.1933), editor of the *Mineralogical Record* and associate curator of minerals and gems, U.S. National Museum of Natural History, Washington, DC. The minerals above are the Al end members of the whiteite series within the whiteite group. Bunched aggregates of tabular crystals (to 5 mm) to thick tabular canoe-shaped (warped) crystals (to 2 cm) showing {001}, {$\bar{1}$11} and invariably twinned by reflection on {001} producing a pseudo-orthorhombic appearance. H = 3–4. **(CaFeMg):** Original description, Ca,Fe-containing member of the series. MON $P2/a$. $a = 14.90$, $b = 6.98$, $c = 10.13$, $\beta = 113.12°$, $Z = 2$, $D = 2.51$. *MM 42:309(1978)*. *30-254:* 9.27_{10} 4.82_7 4.64_7 3.48_4 2.94_7 2.78_8 1.943_5 1.552_5. Flattened on {001} with large {001} to 5 mm in length. Tan. G = 2.58.

(MnFeMg): Low Ca member of the series. MON $P2/a$. $a = 14.99$, $b = 6.96$, $c = 10.14$, $\beta = 113.32°$, $Z = 2$, $D = 2.62$. MM 42:309(1978). 43-1467: 9.32_{10} 5.60_3 4.82_5 4.64_4 3.52_4 3.25_4 2.95_5 2.78_9. Canoe-shaped outline with {001} and {$\bar{1}$11} nearly balanced (to 1.5 cm in length). The curved aspect of {$\bar{1}$11} facets is most pronounced in Ca-poor species and gives almond-shaped cross section. Chocolate-brown. $G = 2.67$. (CaMnMg): Low-Fe^{2+}, high-Ca member of the series. MON $P2/a$. $a = 14.842$, $b = 6.996$, $c = 10.109$, $\beta = 112.59°$, $Z = 2$, $D = 2.64$. CM 27:699(1989). Tabular pseudohexagonal subparallel aggregates in nodules to 0.2 mm × 0.3 mm × 0.04 mm. Pale yellow. $G = 2.81$. Biaxial (+); $N_x = 1.580$, $N_y = 1.584$, $N_z = 1.59$; $2V = 81°$. Whiteite can be found at the Fletcher and Palermo mines, North Groton, NH. Whiteite-(CaMnMg) was found at the *Tip Top pegmatite, Custer, Custer Co., SD*; the original composition, whiteite-(CaFeMg), and also whiteite-(MnFeMg) were found at *Ilha de Taquaral, MG, Brazil*, associated with large childrenite–eosphorite crystals, wardite, green phase related to roscherite, upon large rose quartz crystals, and along joints and fractures in quartz and albite; at Blow R., YT, Canada. HCWS

42.11.3.4 Rittmannite $(Mn^{2+},Ca)Mn^{2+}(Fe^{2+},Mn^{2+})_2(Al,Fe^{3+})_2$ $(PO_4)_4(OH)_2 \cdot 8H_2O$

Named in 1989 for Alfred Rittman (1893–1980), German vulcanologist. Jahnsite–whiteite group whiteite subgroup. First whiteite structure type mineral with Fe^{2+} dominant in B site. MON $P2/a$. $a = 15.01$, $b = 6.89$, $c = 10.16$, $\beta = 112.82°$, $Z = 2$, $D = 2.83$. CM 27:447(1989). PD: 9.38_{10} 5.66_5 4.93_5 4.85_5 4.69_7 3.53_5 3.46_5 2.80_{10}. Tabular crystals in subparallel aggregates to 0.3 mm × 0.3 mm × 0.04 mm. Pseudohexagonal appearance. Pale yellow, vitreous luster, translucent. $H = 3\frac{1}{2}$. $G = 2.81$. Biaxial (+); X parallel to b; $Z \wedge c = 7°$; $N_x = 1.622$, $N_y = 1.626$, $N_z = 1.654$; nonpleochroic. Analyses may also show Mg. Found in altered phosphatic nodules on kryzhanovskite, frondelite, huraeulite, and adularia in two mica pegmatites at *Mansualde, Viseu district, Beria Alta, Portugal*. HCWS

42.11.4.1 Keckite $Ca(Mn^{2+},Zn)_2Fe_3^{3+}(PO_4)_4(OH)_3 \cdot 2H_2O$

Named in 1979 for Erich Keck, collector of Hagendorf minerals. Jahnsite–whiteite group Jahnsite subgroup. MON $P2/a$. $a = 15.02$, $b = 7.19$, $c = 19.74$, $\beta = 110.30°$, $Z = 2$, $D = 2.682$. NJMA 134:183(1979). 33-289: 9.3_8 4.98_5 4.63_3 3.51_5 2.86_{10} 2.59_4 2.34_3 1.879_4. Aggregates of crystal plates up to 2 mm long. Brown, yellow-brown, dirty grayish brown; dull luster. Cleavage {001}, {100}. $H = 4$. $G = 2.6$. IR. Biaxial (−); $X \wedge c = 15-22°$; $Z = b$; $N_y = 1.692$, $N_z = 1.699$; undulatory extinction; pleochroic: X, reddish brown; Y, yellow; Z, brighter yellow X >> Y > Z. Found in the oxidation and weathering zone of phosphophyllite or rockbridgeite in *Hagendorf pegmatite, Oberfalz, Bavaria, Germany*. HCWS AM 64:1330(1979).

42.11.5.1 Minyulite $KAl_2(PO_4)_2(OH,F) \cdot 4H_2O$

Named in 1933 for the locality. ORTH $Pba2$. $a = 9.337$, $b = 9.740$, $c = 5.522$, $Z = 2$, $D = 2.472$. *BM 96:67(1973)*, Structure: *AM 62:256(1977)*. Two AlO_6 octahedra share one F-vertex and two PO_4 tetrahedra, forming a dimeric cluster that links into sheets parallel to {001}. Cavities in the sheets are occupied by K^+. Hydrogen bonds from H_2O molecules (coordinated to Al) form the only link between sheets. *37-467:* 6.75_4 5.52_{10} 3.40_5 3.35_6 3.07_4 2.68_3 2.61_2 2.14_3. Radial aggregates of needles; short prismatic crystals in subparallel growth. Colorless to white, silky luster, transparent. Cleavage probably parallel to elongation, brittle. $H = 3\frac{1}{2}$. $G = 2.45$. Biaxial (+); $N_x = 1.531$, $N_y = 1.534$, $N_z = 1.538$. Traces of Fe^{3+} on analysis. Soluble in hot 1:1 acids and warm 1:10 NaOH. Synthesized below 95°. Found near *Minyulo well, Dandaragan, WA*, with dufrenite in altered outcrops of glauconite-containing phosphate beds, at Wait's quarry, Noarlunga, Moculta quarry, Angaston, SA, and with fluellite at Wolffdene quarry, Q, *Australia*. HCWS *DII:970*.

42.11.6.1 Leucophosphite $KFe_2^{3+}(PO_4)_2(OH) \cdot 2H_2O$

Named in 1932 from the Greek *leuco*, white, in allusion to the color and for the composition. The Fe^{3+} end member in the $K-NH_4$ series with spheniscidite (42.11.6.3): Tinsleyite (42.11.6.2) is the Al analog. MON $P2_1/n$. $a = 9.782$, $b = 9.658$, $c = 9.751$, $\beta = 102.24°$, $Z = 4$, $D = 2.911$. Structure: *AM 57:397(1972)*. Tetramers of FeO_6 octahedra, clusters of composition $[Fe_4^{3+}(OH)_2(H_2O)_2(O)_{16}]$, are joined through the PO_4 groups to produce a structure with channelways where K^+ resides. *37-466:* 6.77_7 6.12_4 5.97_{10} 4.27_4 3.06_7 3.05_5 2.91_6 2.82_6. See also *9-446*. Massive amorphous, tiny pseudo-orthorhombic crystallites; columnar aggregates of prismatic crystals to 1 mm. Buff, white, greenish white (purplish gray in incandescent light); chalky, translucent. Excellent cleavage parallel to (100), friable. $G = 2.948$. Biaxial (+); $X = b$, $Z \wedge c = 26°$; $N_x = 1.707$, $N_y = 1.721$, $N_z = 1.739$; $2V = 62°$; $r < v$, very strong. Soluble in 1:1 HCl, H_2SO_4. Synthesized below 145°, pH 2.5–6.0, and $[PO_4]$ 1.0–3.5 M. Experiments show limited Na substitution in leucophosphite as separate Na phases form. Found at Palermo No. 1, North Groton, NH; at Tip Top mine, Custer, Custer Co., SD, with rockbridgeite; at Bomi Hill, and Bambuta, western Liberia, as a result of solutions from bat dung; at *Weelhamby Lake, Ninghanboun Hill, WA, Australia*, with variscite, chalcedony, and opal in veinlets in serpentine, presumably via solutions of bird guano on serpentine; at Sapucaia pegmatite, MG, Brazil, with cyrilovite, metastrengite, and frondelite. HCWS *DII:936*, *AM 42:214(1957)*.

42.11.6.2 Tinsleyite $KAl_2(PO_4)_2(OH_2) \cdot 2H_2O$

Named in 1984 for Frank C. Tinsley (b.1916), mineral collector, Black Hills specialist, of Rapid City, SD. Al analog of leucophosphite (42.11.6.1). MON Pn. $a = 9.602$, $b = 9.532$, $c = 5.543$, $\beta = 103.16°$, $Z = 4$, $D = 2.62$. *AM 69:374(1984)*. Note space group difference from leucophosphite. *38-351:*

6.68_{10} 5.91_8 4.16_5 3.72_5 3.01_7 2.84_5 2.77_5 2.62_6. Tabular on $\{\bar{1}01\}$ prismatic crystals show $\{\bar{1}01\}$, $\{\bar{1}11\}$, $\{011\}$, $\{110\}$. Composite crystals with colorless Fe-rich leucophosphite interior and tinsleyite as the outermost zone (epitaxial). Dark violet-red, pink streak, vitreous luster, translucent. Brittle. H ≈ 5. G = 2.69. Nonfluorescent. Biaxial (+); X parallel to b; $N_x = 1.591$, $N_y = 1.597$, $N_z = 1.604$; 2V = 86°; strongly pleochroic. X, pale orange-brown; Y, light purple; Z, dark purplish red Z > Y > X. Microprobe analyses show Fe^{3+}, and Mn^{3+} or $^{4+}$ oxidation state suggested because of mineral association. Found in the highly altered triphylite pods within the intermediate zone of the *Tip Top pegmatite, 8.8 km SE of Custer, Custer Co., SD*, in association with leucophosphite, robertsite, rockbridgeite–frondelite, tavorite, perthite, quartz, muscovite, fluorapatite, albite, beryl, elbaite, columbite–tantalite, carbonate-apatite, and laueite; at *Sapucaia pegmatite, MG, Brazil*. hcws

42.11.6.3 Spheniscidite $(NH_4,K)(Fe^{3+},Al)_2(PO_4)_2(OH) \cdot 2H_2O$

Named in 1986 from the Latin *sphenisciformes*, to designate of the order for the animal group containing penguins. The NH_4 member in the K–NH_4 series with leucophosphite (42.11.6.1). mon $P2_1/n$. $a = 9.75$, $b = 9.63$, $c = 9.70$, $\beta = 102.57°$, Z = 4, D = 2.71. *MM 50:291(1986)*. *41-593*: 7.62_4 6.79_{10} 6.09_3 5.99_9 4.75_4 4.26_4 3.36_4 3.05_5. Fine grained (< 200 µm) micaceous aggregates. Brown, earthy, greasy luster, translucent. Soft. Cleavage $\{100\}$. DTA, IR. Biaxial (+); X = b; OAP perpendicular to (010); $N_{x\,syn} = 1.706$, $N_{y\,syn} = 1.724$, $N_{z\,syn} = 1.741$; N ≈ 1.7 in mineral with moderate birefringence. Soluble in acids, insoluble in H_2O. Analysis also shows Ca, Mg. Synthesized with $NH_4 > K$, $Fe^{3+} > Al$. Found as the rim material on phyllitic rock in ornithogenic soils (guano-containing) associated with struvite, hydroxylapatite, taranakite, and minyulite at *Elephant Island, South Shetland Is., British Antarctic Territory*. hcws

42.11.7.1 Giniite $Fe^{2+}Fe_4^{3+}(PO_4)_4(OH)_2 \cdot 2H_2O$

Named in 1980 by Paul Keller for his wife, Gini Keller. mon $P2/a$. $a = 14.253$, $b = 5.152$, $c = 10.353$, $\beta = 111.30°$, Z = 2, D = 3.42. *NJMM:561(1980)*. *33-669*: 4.07_4 3.36_{10} 3.20_7 2.80_5 2.28_6 2.04_7 1.679_6 1.604_7. Crystals to 0.5 mm × 0.2 mm × 0.05 mm, showing $\{100\}$, $\{201\}$, $\{210\}$, $\{103\}$, $\{010\}$. Twinned with plane $\{\bar{1}12\}$. Blackish green to blackish brown, olive streak, vitreous to greasy luster, translucent. No cleavage, conchoidal fracture. H = 3–4. G = 3.41. Biaxial (−); X = c; $N_x = 1.775$, $N_y = 1.803$, $N_z = 1.812$; 2V ≈ 55°; pleochroic. X, light brown; Y, dark brown; Z, dark blue-green. Fe^{2+} and Fe^{3+} influence color and pleochroism. Found in the pegmatite at *Sandamab, near Usakos, Namibia*, associated with hureauite, tavorite, leucophosphite, and another phosphate formed by the alteration of triphylite. hcws *NJMM:49(1980), AM 65:1066(1980)*.

MONTGOMERYITE GROUP

The montgomeryite group minerals have the general formula

$$Ca_4AB_4(PO_4)_6(OH)_4 \cdot 12H_2O$$

where

$A = Fe^{2+}, Mg, Mn^{2+}$
$B = Al, Fe^{3+}$

and crystallize with monoclinic or orthorhombic space groups.

The structure is related to that of other Fe^{3+}–Al containing phosphates, and it is composed of Fe^{3+}–Al octahedra corner-linked with PO_4 tetrahedra that form chains with a 7-Å repeat (see the jahnsite–whiteite group). In montgomeryite the A site is partially occupied but ordered with the Mg cations bonded to two O(4) atoms and to four O's from H_2O molecules. The result is that there are no shared edges with other MgO_6 groups to form Mg octahedral sites. The Ca cation is located between the chains.

The space group for montgomeryite was determined to be $C2$, a subset of $C2/c$, which is probably appropriate for the other members of the group as

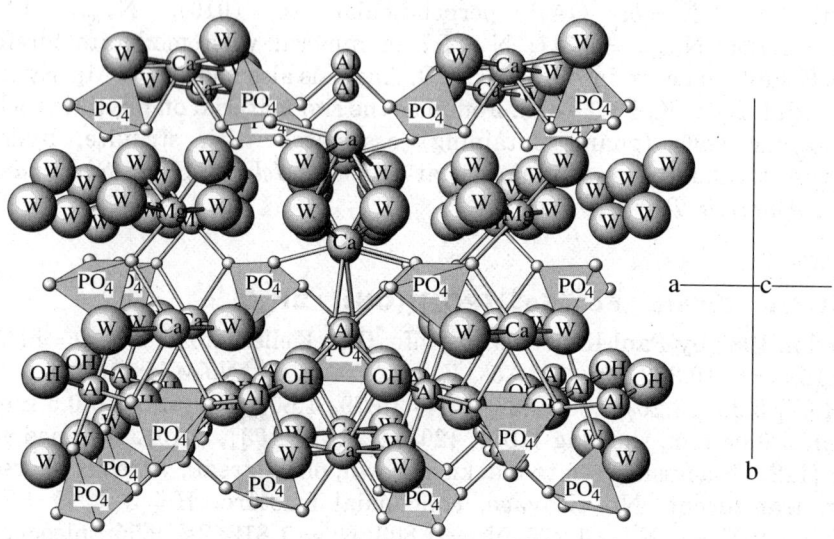

MONTGOMERYITE: $Ca_4MgAl_4(PO_4)_6(OH)_4 \cdot 12H_2O$

A c axis projection that shows slabs composed of Ca polyhedra, $Al(OH,O_p)_6$ and tetrahedral PO_4 groups. A second $Ca(O_p,H_2O)$ polyhedra and $Mg(H_2O)_6$ groups link the slabs. Compare with other Class 42 Type 11 structures.

well, although further detailed crystallographic work needs to be done. *AM 59:843(1974), 61:12(1976), 77:19(1992); DHZ 14:231(1963); FF 444(1984).*

Montgomeryite Group

Mineral	Formula	Space Group	a	b	c	β	
Montgomeryite	$Ca_4MgAl_4(PO_4)_6(OH)_4 \cdot 12H_2O$	$C2$	10.004	24.083	6.235	91.55	42.11.8.1
Kingsmountite	$(Ca,Mn^{2+})_4(Fe^{2+},Mn^{2+})Al_4(PO_4)_6(OH)_4 \cdot 12H_2O$	$C2$	10.029	24.46	6.31	91.16	42.11.8.2
Calcioferrite	$Ca_4Mg(Fe^{3+}Al)_4(PO_4)_6(OH)_4 \cdot 12H_2O$	$C2$	10.34	24.20	6.31	91.5	42.11.8.3
Zodacite	$Ca_4MnFe_4^{3+}(PO_4)_6(OH)_4 \cdot 12H_2O$	$C2/c$ or Cc	10.152	24.14	6.308	91.14	42.11.8.4

42.11.8.1 Montgomeryite $Ca_4MgAl_4(PO_4)_6(OH)_4 \cdot 12H_2O$

Named in 1940 for Arthur Montgomery (b.1909), mineralogist, teacher, collector, and founder of Friends of Mineralogy. He and Edwin Over mined the Fairfield, UT, phosphate deposit and recognized the mineral as new. Montgomeryite group. See kingsmountite (42.11.8.2), the Fe^{2+} end member in the Fe^{2+}-Mg series of the Al-containing members of the montgomeryite group. MON $C2$. $a = 10.004$, $b = 24.083$, $c = 6.235$, $\beta = 91.55°$, $Z = 2$, $D = 2.523$. *AM 61:12(1976)*, Structure: *AM 59:843(1974)*. *35-624*: 12.0_{10} 6.03_2 5.11_7 3.13_2 2.90_5 2.87_3 2.62_3 2.58_3. See also *28-225*(calc). Dark green, pale green, bright red, colorless; vitreous luster; translucent. H = 4. G = 2.530. Biaxial (−); $N_x = 1.572$, $N_y = 1.579$, $N_z = 1.582$; 2V = 75°. *MR 14:195(1983)*. Microprobe analysis shows minor Mn. Found at the Tip Top pegmatite, Custer, Custer Co., SD, also at the Etta pegmatite, near Keystone, SD, replacing mitridatite; at *Little Green Monster Mine, Clay Canyon, Fairfield, Utah Co., UT*, with wardite, englishite, and gordonite replacing variscite, and crandallite and apatite; at Milgun Station, WA, Iron Monarch quarry, Iron Knob, SA, Australia; at Lavra da Ilha, Taquaral, Brazil. HCWS

42.11.8.2 Kingsmountite $(Ca,Mn^{2+})_4(Fe^{2+},Mn^{2+})Al_4(PO_4)_6(OH)_4 \cdot 12H_2O$

Named in 1979 for the locality. Montgomeryite group. The Fe^{2+} analog of montgomeryite (42.11.8.1). MON $C2$. $a = 10.029$, $b = 24.46$, $c = 6.31$, $\beta = 91.16°$, $Z = 2$, $D = 2.58$. *CM 17:579(1979)*. *35-694*: 12.23_5 9.30_2 6.33_3 5.15_{10} 3.31_3 2.95_4 2.92_3 2.62_6. Radiating plates, fibers in subparallel aggregates, spherules to 2.0 mm in diameter. White, light brown; pearly luster;

translucent. H = $2\frac{1}{2}$. G = 2.51. Nonfluorescent. TGA. Biaxial (−); lath has X parallel to one edge; Y perpendicular to lath; Z $\wedge$ elongation 35°; $N_x = 1.575$, $N_y = 1.581$, $N_z = 1.583$, $2V = 62°$. Microprobe analysis shows Mn > Fe, minor Mg. Found at *Foote Mineral Co. spodumene mine, Kings Mt., Cleveland Co., NC*, with mitridatite, birnessite; at Hagendorf pegmatite, Bavaria, Germany. HCWS

42.11.8.3. Calcioferrite $Ca_4Mg(Fe^{3+},Al)_4(PO_4)_6(OH)_4 \cdot 12H_2O$

Named in 1858 for the composition. The Fe^{3+} analog of montgomeryite (42.11.8.1). MON C2. $a = 10.34$, $b = 24.20$, $c = 6.31$, $\beta = 91.5°$, $Z = 2$, $D = 2.65$. *MR 16:477(1985)*. 39-380: 12.1_{10} 6.05_6 5.17_6 3.15_7 2.90_9 2.65_9 2.59_9 1.578_9. Foliated or reniform masses. Sulfur-yellow, greenish yellow, siskin-green, yellowish white; sulfur-yellow to white streak, pearly luster on cleavage surfaces; translucent. Cleavage parallel to foliation perfect, second at right angles, third oblique; brittle. H = $2\frac{1}{2}$. G = 2.53. Biaxial (−); $N_x = 1.572$, $N_y = 1.579$, $N_z = 1.583$; $2V = 75°$. Analysis shows Cu, Al, Si, Cr. Found at *Battenberg, Bavaria, Germany*, in a bed of Tertiary clay; at Karatay, Kazakhstan. HCWS

42.11.8.4 Zodacite $Ca_4MnFe_4^{3+}(PO_4)_6(OH)_4 \cdot 12H_2O$

Named in 1988 for Peter Zodac (1894–1967), founder and editor of *Rocks and Minerals*. Montgomeryite group. The Mn and Fe^{3+}- containing end member. MON $C2/c$ or Cc. $a = 10.152$, $b = 24.14$, $c = 6.308$, $\beta = 91.14°$, $Z = 2$, $D = 2.65$. *AM 73:1179(1988)*. 41-1454: 12.0_6 9.38_2 6.31_3 5.59_2 5.18_{10} 3.15_4 2.91_5 2.66_4. Isolated crystals up to 0.2 mm, flattened, radial arrays. Light to medium yellow, tiny crystals may be colorless; light yellow to white streak; vitreous luster. Cleavage assumed parallel to {010}. H ≈ 4. G = 2.68. Nonfluorescent. Biaxial (−); X = b; Y $\wedge$ $c = 24°$; Z $\wedge$ $a = 23°$; $N_x = 1.598$, $N_y = 1.601$, $N_z = 1.602$ (Na); weakly pleochroic. Y, very pale green; Z, pale green Y < Z. Found at the *Mangualde pegmatite, Beira, Portugal*, in association with jahnsite-(Mn,Mn,Mn), hureaulite, and phosphosiderite on altered varulite. HCWS

LAUEITE–PARAVAUXITE GROUP

The laueite–paravauxite group of phosphate minerals contains two subgroups, plus three allied species, conforming to the formula

$$AB_2(PO_4)_2(OH)_2 \cdot nH_2O$$

where

A = Mg, Fe^{2+}, Mn, Fe^{3+}
B = Al, Fe^{3+}, Cr^{3+}
n = 5 to 8, but is usually 8

and usually crystallize is usually with triclinic symmetry in space group $P\bar{1}$ or $P1$.

Laueite-Paravauxite Group

The structure [lauetite: *MP 38:201(1988)*; paravauxite: *NJMM:430(1969)*] comprises chains of BO_6 octahedra connected by PO_4 tetrahedra into heteropolyhedral sheets with composition $[(Fe^{3+}/Al)(PO_4)_2(OH)_2(H_2O)_2]^{2+}$. The sheets are cross-linked of the sheets is through octahedrally coordinated cations $A-O_6$. Additionally, at least two H_2O molecules are incorporated as part of the H-bonded network.

Variations in the bonding and orientation of the groups that make up the infinite chains and the associated cations give rise to different linkages and hydration states, resulting in different species, and specifically polymorphism. For example, in laueite there are two different Fe^{3+}-O octahedral linkages, resulting in two types of chains. In stewartite there are three nonequivalent $Fe^{3+}-O_6$ octahedra, while pseudolaueite has all $Fe^{3+}-O_6$ octahedra equivalent, with the result chains have consistent components and configuration.

A strunzite subgroup has lower hydration states, while the paravauxite subgroup has $Al-O_6$ octahedra and linkages. *AM 59:1272(1974), 53:1025(1968), 54:1312(1969).*

LAUEITE SUBGROUP, B = Fe^{3+}

Mineral	Formula	Space Group	a	b	c	α	β	γ	
Laueite	$Mn^{2+}Fe_2^{3+}(PO_4)_2$ $(OH)_2 \cdot 8H_2O$	$P\bar{1}$	5.28	10.66	7.14	107.92	114.98	71.12	42.11.10.1
Stewartite	$Mn^{2+}Fe_2^{3+}(PO_4)_2$ $(OH)_2 \cdot 8H_2O$	$P\bar{1}$	10.398	10.672	7.223	90.10	109.10	71.83	42.11.10.2
Pseudolaueite	$Mn^{2+}Fe_2^{3+}(PO_4)_2$ $(OH)_2 \cdot 7\text{-}8H_2O$	$P2_1/a$	9.647	7.428	10.194		104.63		42.11.10.3
Ushkovite	$MgFe_2^{3+}(PO_4)_2$ $(OH)_2 \cdot 8H_2O$	$P\bar{1}$	5.20	10.70	7.14	108.6	106.9	72.7	42.11.10.4

PARAVAUXITE SUBGROUP, B = Al

Mineral	Formula	Space Group	a	b	c	α	β	γ	
Metavauxite	$Fe^{2+}Al_2(PO_4)_2$ $(OH)_2 \cdot 8H_2O$	$P2_1/c$	10.23	9.59	6.94		98.03		42.11.11.1
Vauxite	$Fe^{2+}Al_2(PO_4)_2$ $(OH)_2 \cdot 6H_2O$	$P\bar{1}$	9.13	11.59	6.14	98.3	92.0	108.4	42.11.14.1
Paravauxite	$Fe^{2+}Al_2(PO_4)_2(OH)_2$ $\cdot 8H_2O$	$P\bar{1}$	5.233	10.541	6.96	106.9	110.8	72.1	42.11.14.2
Sigloite	$(Fe^{2+}, Fe^{3+})Al_2(PO_4)_2$ $(OH)_3 \cdot 8H_2O$	$P\bar{1}$	5.190	10.419	7.033	105.00	111.31	70.87	42.11.14.3
Gordonite	$MgAl_2(PO_4)_2(OH)_2$ $\cdot 8H_2O$	$P\bar{1}$	5.246	10.532	6.975	107.51	111.03	72.21	42.11.14.4
Mangangordonite	$(Mn, Fe, Mg)Al_2$ $(PO_4)_2(OH)_2$ $\cdot 8H_2O$	$P\bar{1}$	5.257	10.363	7.040	105.44	113.07	78.69	42.11.14.5

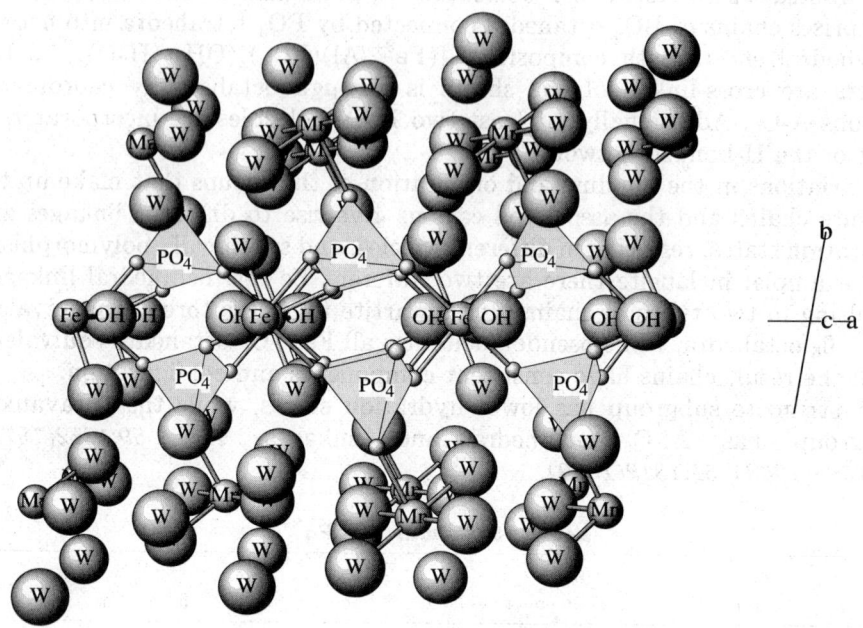

LAUEITE: MnFe$_2$(PO$_4$)$_2$(OH)$_2$ · 8H$_2$O

This c axis projection allows the similarity between laueite and jahnsite to be appreciated. In laueite the Fe octahedra are connected by PO$_4$ groups and linked by Mn(O$_p$)$_2$(H$_2$O) octahedra.

STRUNZITE SUBGROUP

Mineral	Formula	Space Group	a	b	c	α	β	γ	
Strunzite	Mn^{2+}Fe$_2^{3+}$(PO$_4$)$_2$(OH)$_2$ · 6H$_2$O	$P\bar{1}$	10.228	9.837	7.284	90.17	98.44	117.44	42.11.9.1
Ferrostrunzite	Fe^{2+}Fe$_2^{3+}$(PO$_4$)$_2$(OH)$_2$ · 6H$_2$O	$P\bar{1}$ or P1	10.23	9.77	7.37	89.65	98.28	117.26	42.11.9.2
Ferristrunzite	Fe^{3+}Fe$_2^{3+}$(PO$_4$)$_2$(OH)$_3$ · 5H$_2$O	$P\bar{1}$ or P1	10.01	9.73	7.334	90.5	97.0	116.4	42.11.9.3

1. Note increase in index of refraction with Fe^{3+} and optic sign change between Ferroit ferri-strunzite.

42.11.9.1 Strunzite Mn^{2+}Fe$_2^{3+}$(PO$_4$)$_2$(OH)$_2$ · 6H$_2$O

Named in 1958 by C. Frondel for Hugo Strunz, mineralogist (b.1910), Berlin and Regensburg, Germany. Laueite–paravauxite group strunzite subgroup. See also ferrostrunzite (42.11.9.2), ferristrunzite (42.11.9.3), and laueite (42.11.10.1). TRIC $P\bar{1}$. $a = 10.228$, $b = 9.837$, $c = 7.284$, $\alpha = 90.17°$, $\beta = 98.44°$, $\gamma = 117.44°$, $Z = 2$, $D = 2.581$. *NJMM:222(1957)*, Structure: *TMPM (ser. 3)25:77(1978)*. 11-133: 9.02$_{10}$ 5.32$_8$ 4.50$_5$ 4.35$_6$ 3.35$_4$ 3.29$_6$

3.23_6. Tufts of hairlike or lathlike crystals flattened {010}, twinned {100}. Straw-yellow to brown-yellow, translucent. G = 2.52. Biaxial (−); Z ∧ c = 10°; $N_x = 1.619$, $N_y = 1.670$, $N_z = 1.720$; 2V moderate; faintly pleochroic, yellow with Z > X, Y. Found as surface weathering product of triphylite at Nevel tourmaline mine, Newry, BB mine, Norway, Red Hill mine, Rumford, ME; Fitzgibbon mine, Alstead, Buzzo quarry, Center Strafford, G.E. Smith mine, Newport, Fletcher and Palermo No. 1 mines, North Groton, NH; at Cobalt, CT; at Mullica Hill, NJ; at Williams Property, AL; at Big Chief mine, Custer Co., Hesnard mine, Etta mine, Keystone Co., SD; as films in weathered outcrops of PO_4 rock in the Phosphoria formation, Rasmussen Valley, ID; at *Hagendorf pegmatite, Oberfalz, Bavaria, Germany*. HCWS *AM 43:793(1958)*.

42.11.9.2 Ferrostrunzite $Fe^{2+}Fe_2^{3+}(PO_4)_2(OH)_2 \cdot 6H_2O$

Named in 1983 for the composition and relationship to strunzite (42.11.9.1). Laueite–paravauxite group strunzite subgroup. TRIC $P\bar{1}$ or P1. a = 10.23, b = 9.77, c = 7.37, α = 89.65°, β = 98.28°, γ = 117.26°, Z = 2, D = 2.57. *NJMM: 524(1983)*, Structure: *NJMM 207(1992)*. There two distinct structural sites for Fe, but Mossbauaer suggests some overlap of Fe^{2+} Fe^{3+} in sites$^+$ as Fe^{2+}/Fe^{3+}=0.34. Rather than 0.50 if strictly segregated. *42-595:* 8.94_8 5.29_{10} 4.47_3 4.33_2 4.24_2 3.45_3 3.28_4 3.21_3. Radiating sprays to 0.5 mm of fibrous prismatic crystals to 150 μm: flattened {100}, elongated [001]; needles to 1 mm with inclined top face. Colorless, faint straw-yellow, light brown color and streak; vitreous luster; translucent. Cleavages: parallel to length, perpendicular to optic normal, and perpendicular to B×a; very brittle. H ≈ 4. G = 2.50. Mossbauer. Biaxial (−); Z ∧ c = 3–4°; $N_x = 1.628$, $N_y = 1.682$, $N_z = 1.723$; 2V = 80°; pleochroic: X, pale yellow-green; Z, pale red-yellow. Moderate and asymmetric dispersion, minimum absorption parallel to Z. No Al detected on analysis. Found at *Mullica Hill, NJ*, where the sprays encrust rockbridgeite that has replaced fossil belemnites in Cretaceous sediments; at Bethel Church, IN; at Perranporth, Cornwall, England; at Arnsberg, Sauerland, Germany. HCWS *AM 69:811(1984)*.

42.11.9.3 Ferristrunzite $Fe^{3+}Fe_2^{3+}(PO_4)_2(OH)_3 \cdot 5H_2O$

Named in 1987 for the composition and relationship to strunzite, the completely oxidized version. Laueite–paravauxite group strunzite subgroup. TRIC $P\bar{1}$ or P1. a = 10.01, b = 9.73, c = 7.334, α = 90.5°, β = 97.0°, γ = 116.4°, Z = 2, D = 2.55. *NJMM:453(1987)*, Structure: *NJMM:176(1990)*. *41-1383:* 8.87_8 5.34_{10} 4.48_2 4.37_2 4.20_3 3.44_3 3.39_3 3.27_4. Matted aggregates of acicular [001] crystals (to 30 μm) up to 0.5 mm long. Radiating sprays up to 2.6 mm in diameter; twinned {110}. Light brownish yellow, light yellow streak, translucent. Cleavage parallel to X–Z optic plane, good. G = 2.5. Biaxial (+); Z ∧ c ≈ 17° with Z approximately parallel to elongation; $N_x = 1.664$, $N_y = 1.698$, $N_z = 1.757$; 2V = 77°; strongly pleochroic: X, greenish yellow; Z, brownish yellow. Found associated with crandallite, diadochite, allophane-evansite, minyulite, strengite, phosphosiderite, cacoxenite, beraunite in argillaceous and clastic sediments exposed in a canal near *Blaton, Belgium.* HCWS *AM 74:502(1989)*.

42.11.10.1 Laueite $Mn^{2+}Fe_2^{3+}(PO_4)_2(OH)_2 \cdot 8H_2O$

Named in 1954 for Max Felix Theodor von Laue (1879–1960), crystallographer, Berlin, Germany. Laue was the first to use crystals as diffraction gratings (in 1912), thus initiating a technique to study the atomic structure of crystals. Laueite–paravauxite group. See dimorph stewartite (42.11.10.2) and the closely related minerals pseudolaueite (42.11.10.3), strunzite (42.11.9.1), gordonite (42.11.14.4), and paravauxite (42.11.14.2). TRIC $P\bar{1}$. $a = 5.28$, $b = 10.66$, $c = 7.14$, $\alpha = 107.92°$, $\beta = 114.98°$, $\gamma = 71.12°$, $Z = 1$, $D = 2.56$. *AM 50:1884(1965)*, Structure: *AM 54:1312(1969)*. Two types of chains, $[Fe(1)O_4(OH)]^{6-}$ and $[Fe(2)O_2(OH)(H_2O)_2]^{2-}$, are linked by PO_4 groups into sheets parallel to (010). The sheets are connected through Mn^{2+} atoms in octahedral coordination. The latter site may be partially occupied, or a portion of the Mn may be Mn^{3+}. The two remaining H_2O molecules are not coordinated to cations. *14-246*: 9.91_{10} 6.57_7 4.95_8 4.02_5 3.93_5 3.28_9 3.12_5 2.88_6. Crystals to 2 mm showing {100}, {010}, {110}, {1$\bar{1}$0}, {011}. Honey-brown. H = 3. G = 2.46. Biaxial (−); $N_x = 1.612$, $N_y = 1.658$, $N_z = 1.682$; 2V = 50°. Analyses also show Ca, Al, Fe^{2+} and Mg, some Mn^{3+} inferred from crystal structure determination. Found at Nevel tourmaline mine, Paul Bennet quarry, Newry, ME; at Fitzgibbon, Palermo Nos. 1 and 2, Fletcher mines, North Groton, NH; at Big Chief pegmatite, Custer Co., Hesnard, High Climb, Etta, Linwood, and White Elephant pegmatites, Keystone Co., SD; at *Hagendorf Süd pegmatite, Bavaria, Germany*; at Lavra do Enio, Galileia, MG, Brazil. HCWS *NJMA 123:148(1975)*.

42.11.10.2 Stewartite $Mn^{2+}Fe_2^{3+}(PO_4)_2(OH)_2 \cdot 8H_2O$

Named in 1912 for the locality. Laueite–paravauxite group. Polymorph of laueite (42.11.10.1). TRIC $P\bar{1}$. $a = 10.398$, $b = 10.672$, $c = 7.223$, $\alpha = 90.10°$, $\beta = 109.10°$, $\gamma = 71.83°$, $Z = 2$, $D = 2.48$. *AM 43:1148(1958)*, Structure: *AM 59:1272(1974)*. Three nonequivalent Fe^{3+} octahedra linked through (OH) corner sharing form chains parallel to {100}. The chains are bridged by PO_4 tetrahedral groups to form sheets parallel to {010}. Mn^{2+} connects sheets. Two H_2O molecules are not directly connected to cations. *5-110*: 10.0_{10} 6.71_7 5.86_3 5.01_5 3.93_5 2.99_4 2.65_{24} 2.49_2. Minute crystals and tufts of fibers. Brownish yellow, translucent. Cleavage {010}, perfect. G = 2.94. Biaxial (−); $N_x = 1.63$, $N_y = 1.658$, $N_z = 1.66$; 2V = 60°; r < v, strong; pleochroic: X, colorless; Y, pale yellow; Z, yellow. Found at Newry, ME; at Fletcher and Palermo no. 1 mines, North Groton, NH; at Williams Property, AL; at Tip Top pegmatite, Custer Co., and Hesnard, Linwood and White Elephant pegmatites, Keyston, SD; at *Stewart lithia mine, Pala, San Diego Co., CA*, with palaite, hureaulite as an alteration product of lithiophilite; at Chanteloube,

France; at Hagendorf Süd, Bavaria, Germany. HCWS *DII:730, NJMA 123:148(1975)*.

42.11.10.3 Pseudolaueite $Mn^{2+}Fe_2^{3+}(PO_4)_2(OH)_2 \cdot 8H_2O$

Named in 1956 for the polymorphic relationship to laueite (42.11.10.1). Laueite–paravauxite group. MON $P2_1/a$. $a = 9.647$, $b = 7.428$, $c = 10.194$, $\beta = 104.63°$, $Z = 2$, $D = 2.51$. *NATW 43:128(1956)*, Structure: *AM 54:1312(1969)*. Similar to laueite but all $Fe^{3+}O,OH)_6$ octahedra are equivalent and form similar chains parallel to {010} bridged by PO_4 tetrahedral groups into heteropolygonal sheets parallel to {001}. Mn^{2+} connects sheets into a unique three-dimensional structure. *12-294:* 9.93_{10} 5.87_7 4.68_3 3.91_3 3.47_4 3.19_3 3.07_3 2.55_3. Prismatic, or thick, tabular, crystals showing {100}, {001}, {110}, {011}, {$\bar{2}$01}. Orange-yellow. $G = 2.46$. Biaxial (+); $X \wedge c = 2°$; $Y \wedge a = 12°$; $Z = b$; $N_x = 1.626$, $N_y = 1.650$, $N_z = 1.868$; $2V = 80°$; pleochroic: X,Y, pale yellow; Z, yellow. Found at *Hagendorf, Bavaria, Germany*. Crystals may show core of stewartite and are associated with Mn and Fe oxides. HCWS *AM 41:815(1956)*.

42.11.10.4 Ushkovite $MgFe_2^{3+}(PO_4)_2(OH)_2 \cdot 8H_2O$

Named in 1983 for S. I. Ushkov (1880–1951), naturalist who studied the Il'men National Forest in the Ural Mts., Russia. Laueite–paravauxite group. TRIC $P\bar{1}$. $a = 5.20$, $b = 10.70$, $c = 7.14$, $\alpha = 108.6°$. $\beta = 106.9°$, $\gamma = 72.7°$, $Z = 1$, $D = 2.47$. *ZVMO 112:42(1983)*. *35-685:* 9.86_{10} 6.57_8 4.95_5 4.85_5 4.01_4 3.28_6 3.20_8 3.12_4. Short prismatic crystals (to 2 mm) showing (001), (010), (1$\bar{1}$0) as principal faces; {100}, {110} minor faces. Pale yellow to orange-yellow, light brown; vitreous luster to pearly on cleavage, greasy on fracture; translucent. Cleavage (010), perfect; brittle. $H = 3\frac{1}{2}$. $G = 2.38$. IR, DTA. Biaxial (−); $N_x = 1.584$, $N_y = 1.637$, $N_z = 1.670$; $Y \wedge c = 26°$; $2V = -50°$; $r > v$, strong; pleochroic: X, yellow; Y, orange-yellow; Z, light brown. Dissolved by 1:10 acids. Found as the weathering product of triplite in granite pegmatites in the *Il'men Mts., Ural Mts., Russia*, associated with Mn hydroxides, francolite, mitridatite, and beraunite. HCWS *AM 69:212(1984)*.

42.11.11.1 Metavauxite $Fe^{3+}Al_2(PO_4)_2(OH)_2 \cdot 8H_2O$

Named in 1922 for the chemical relationship to vauxite (42.11.14.1). Laueite–paravauxite group. Dimorph of paravauxite (42.11.14.2). MON $P2_1/c$. $a = 10.23$, $b = 9.59$, $c = 6.94$, $\beta = 98.03°$, $Z = 2$, $D = 2.36$. Structure: *NJMA 123:148(1975)*. *33-639:* 10.12_7 5.10_5 4.69_{10} 4.34_8 3.99_5 2.76_7 2.75_3 2.27_3. Prismatic to acicular [001], subparallel radial aggregates. Colorless, white, pale green; vitreous luster to silky in fibers; transparent to translucent. Brittle. $H = 3$. $G = 2.345$. Biaxial (+); $X = b$; $Z \wedge c \approx 17°$; $N_x = 1.550$, $N_y = 1.561$, $N_z = 1.577$ (Na). Ca and Mg may substitute for Fe. Found with vauxite, paravauxite, and wavellite in oxidation zone at *Siglo XX mine, Llallagua, Bolivia*. HCWS *DII:971*.

42.11.12.1 Gatumbaite CaAl$_2$(PO$_4$)$_2$(OH)$_2$ · H$_2$O

Named in 1977 for the locality. MON $P2_1/m$, $P2$, or Pm. $a = 6.907$, $b = 5.095$, $c = 10.764$, $\beta = 91.05°$, $Z = 2$, $D = 2.95$. *NJMM:561(1979)*. *29-283:* 6.9$_5$ 4.21$_{10}$ 3.21$_5$ 3.16$_5$ 2.77$_7$ 2.30$_7$ 2.24$_{10}$ 1.726$_8$. Sheaves and rosettes with radial fibrous structure up to 10 mm in diameter. Pure white, pearly luster, translucent. Asbestiform splinters, longitudinal cleavage, cross fractures, brittle. H < 5. G = 2.92. Nonfluorescent. IR. Biaxial (−); Z = b; OAP perpendicular to (010); N$_x$ = 1.610, N$_y$ = 1.63$_{(est)}$, N$_z$ = 1.639. Found in the *Buranga pegmatite, near Gatumba, Gisenyi Prov., Rwanda*, associated with trolleite, scorzalite, apatite, bjarebyite, and a Be-phosphate. HCWS *AM 63:793(1978)*, *59:1140(1974)*.

42.11.12.2 Kleemanite ZnAl$_2$(PO$_4$)$_2$(OH)$_2$ · 3H$_2$O

Named in 1979 for Alfred William Kleeman, petrologist, University of Adelaide, SA, Australia. MON $P2,P2_1$, P_2/M or P_{21}/M. $a = 7.29$, $b = 7.194$, $c = 9.762$, $\beta = 110.20°$, $Z = 2$, $D = 2.76$. *MM 43:93(1979)*. *33-1465:* 9.09$_6$ 5.66$_5$ 4.76$_{10}$ 3.88$_5$ 3.64$_5$ 3.33$_5$ 3.30$_6$ 3.09$_8$. Very small crystallites and fibers. Ochre-colored. TGA. Biaxial; N$_x$ = 1.598, N$_z$ = 1.614; colorless; inclined extinction to 40° of fiber axis. Coatings or thin (1 to 2 mm thick) veins on manganiferous iron ore at *Iron Knob, SA, Australia*. HCWS *AUSM 4:123(1989)*, *AM 64:1331(1979)*.

42.11.13.1 Vanuralite Al(UO$_2$)$_2$(VO$_4$)$_2$(OH) · 11H$_2$O

Named in 1963 for the composition: vanadium and uranium. See also metavanuralite (42.11.13.2). MON $A2/a$. $a = 10.55$, $b = 8.44$, $c = 24.52$, $\beta = 103°$, $Z = 4$, $D = 3.16$. *BM 93:242(1970)*. *23-769:* 12.0$_{10}$ 6.53$_4$ 5.98$_9$ 5.14$_7$ 3.98$_8$ 3.27$_5$ 3.23$_8$ 3.18$_8$. Microcrystalline. Greenish yellow, translucent. H = 2. G = 3.62. Biaxial (−); N$_x$ = 1.65, N$_y$ = 1.85, N$_z$ = 1.90; 2V = 44°. Found at *Mounana mine, Franceville, Haute Ogooué, Gabon*. HCWS *AM 48:1415(1963)*, *56:639(1971)*.

42.11.13.2 Metavanuralite Al(UO$_2$)$_2$(VO$_4$)$_2$(OH) · 8H$_2$O

Named in 1970 as the lower hydrate of vanuralite (42.11.13.1). TRIC $P\bar{1}$ or $P1$. $a = 10.46$, $b = 8.44$, $c = 10.43$, $\alpha = 75.88°$, $\beta = 102.83°$, $\gamma = 90°$, $Z = 2$, $D = 3.66$. *BM 93:242(1970)*. *23-770:* 9.92$_{10}$ 5.10$_7$ 4.17$_9$ 4.09$_8$ 3.24$_8$ 3.20$_7$ 3.16$_9$ 3.07$_8$. Greenish yellow, translucent. Stable at 20° in the relative humidity range 28–47%. Found at the *Mounana mine, Franceville, Haut Ogooué, Gabon*. HCWS *AM 56:637(1971)*.

42.11.13.3 Threadgoldite Al(UO$_2$)$_2$(PO$_4$)$_2$(OH) · 8H$_2$O

Named in 1979 for Ian M. Threadgold (b.1929), University of Sydney, NSW, Australia. The P analog of metavanuralite (42.11.13.2). MON $C2/c$. $a = 20.168$, $b = 9.847$, $c = 19.719$, $\beta = 110.71°$, $Z = 8$, $D = 3.33$. Structure: *TMPM 30:111(1982)*. *33-39:* 9.43$_{10}$ 5.35$_5$ 4.93$_4$ 4.10$_5$ 3.73$_4$ 3.47$_8$ 3.37$_6$ 2.20$_6$. Micaceous tabular crystals elongated on b to 1 mm, showing {100}, {001}, {010}, {012}. Greenish yellow, translucent. G = 3.4. DTA. Fluorescent green in LWUV, pale green in SWUV. Biaxial (−); Y = b; Z $\wedge$ c = 4°; N_x = 1.573, N_y = 1.583, N_z = 1.588; negative elongation; 2V = 70°; pleochroic, pale yellow to colorless. Found at *Kobokobo, Kivu, Zaire*, with phuralumite and upalite. HCWS *AC(B) 35:3017(1979), AM 65:209(1980)*.

42.11.14.1 Vauxite $Fe^{2+}Al_2(PO_4)_2(OH)_2 \cdot 6H_2O$

Named in 1922 for George Vaux, Jr. (1863–1927), mineral collector, Bryn Mawr, PA. Laueite–paravauxite group, paravauxite subgroup. TRIC $P\bar{1}$. a = 9.13, b = 11.59, c = 6.14, α = 98.3°, β = 92.0°, γ = 108.4°, Z = 2, D = 2.41. Structure: *AM 53:1025(1968)*. See also paravauxite (42.11.14.2) and metavauxite (42.11.11.1). *33-640:* 10.85_{10} 6.07_1 5.91_1 5.46_2 3.05_1 2.96_1 2.88_1 2.72_1. Minute tabular {010} crystals, elongated [001] or [101], showing {010}, {1$\bar{1}$0}, {$\bar{1}$01}, {130}, {$\bar{2}$11}, {$\bar{1}$32}; subparallel to radial aggregates and nodules. Sky-blue to venetian-blue, becoming greenish on exposure, white streak, vitreous luster, transparent. No cleavage, brittle. H = $3\frac{1}{2}$. G = 2.39. Biaxial (+); Z approximately parallel to {010}; N_x = 1.551, N_y = 1.555, N_z = 1.562; 2V = 32°; r > v, strong; pleochroic. X,Z, colorless; Y, blue. Ca and Mg substitute for Fe^{2+}. Found associated with wavellite and paravauxite in the Siglo XX tin mines of *Llallagua, Potosi*, and at the Miraflores mine, Huanumi, *Bolivia*. HCWS *DII:974*.

42.11.14.2 Paravauxite $Fe^{2+}Al_2(PO_4)_2(OH)_2 \cdot 8H_2O$

Named in 1922 for the chemical similarity to vauxite. Polymorph of metavauxite (42.11.11.1) and vauxite (42.11.14.1). Laueite–paravauxite group, paravauxite subgroup. TRIC $P\bar{1}$. a = 5.233, b = 10.541, c = 6.96, α = 106.9°, β = 110.8°, γ = 72.1°, Z = 1, D = 2.479. Structure: *NJMM:430(1969)*. *29-1424:* 9.89_{10} 6.40_7 4.93_3 4.79_3 3.19_3 3.09_2 2.85_2 2.58_2. Short prismatic {001} to thick tabular {010}, subparallel or radial aggregates, showing {010}, {100}, {110}, {001}, {0$\bar{1}$1}, {0$\bar{5}$4}, {$\bar{2}$11}. Colorless to pale greenish white; white streak; vitreous luster, pearly on cleavages; transparent to translucent. Cleavage {010}, conchoidal fracture, brittle. H = 3. G = 2.36. DTA. Biaxial (+); N_x = 1.554, N_y = 1.558, N_z = 1.573 (Na); 2V medium. Minor Mg, Na, Si detected on analysis. Found with vauxite, metavauxite, and wavellite as a later secondary phosphate in the tin mine at *Llallagua, Potosi, Bolivia*, and at Oruro. HCWS *DII:972, AM 47:1(1962)*.

42.11.14.3 Sigloite $Fe^{3+}Al_2(PO_4)_2(OH)_3 \cdot 8H_2O$

Named in 1962 for the locality. Laueite–paravauxite group, paravauxite subgroup. Oxidized equivalent of paravauxite TRIC $P\bar{1}$. a = 5.190, b = 10.419, c = 7.033, α = 105.00°, β = 111.31°, γ = 70.87°, Z = 1, D = 2.500. Structure: *MP 38:201(1988)*. *14-171:* 9.69_{10} 6.46_9 4.86_4 4.68_4 3.91_4 3.23_7 2.82_6 2.56_5. Crystals pseudomorphic after paravauxite. Straw-yellow, light brown, translucent. Cleavage {100}, perfect; {001}. H = 3. G = 2.35. Biaxial (+);

$N_x = 1.563$, $N_y = 1.86$, $N_z = 1.619$; 2V = 76°; r < v. *AM 47:1(1962)*. Found at *Siglo XX mine, Llallagua, Potosi, Bolivia*, on wavellite that overgrows quartz as secondary minerals in cassiterite deposit that contains metavauxite, crandallite, and childrenite. HCWS

42.11.14.4 Gordonite $MgAl_2(PO_4)_2(OH)_2 \cdot 8H_2O$

Named in 1930 for Samuel G. Gordon (1897–1952), mineralogist, Academy of Natural Sciences, Philadelphia, PA. Laueite–paravauxite group, paravauxite subgroup. See mangangordonite (42.11.14.5). TRIC $P\bar{1}$. $a = 5.246$, $b = 10.532$, $c = 6.975$, $\alpha = 107.51°$, $\beta = 111.03°$, $\gamma = 72.21°$, Z = 1, D = 2.319. Structure: *NJMM:265(1988)*. Structurally similar to paravauxite but not a series with other group members despite compositional similarities. *14-313:* 9.78_{10} 6.32_5 4.76_5 3.17_8 3.07_5 2.83_7 2.56_6 2.39_4. Bundles of sheaflike aggregates with all individuals in one group similarly terminated. Crystals are prismatic [001] to platy {010}, strongly striated [001], less marked [100]. Forms {010}, {100}, {1$\bar{1}$0}, {0$\bar{1}$1}, {001}, rarely doubly terminated. Smoky white, colorless, occasionally tipped in green or pink; white streak; vitreous luster, pearly on {010}; transparent. Cleavage {010}, perfect; {100}, fair; {001}, poor; conchoidal fracture; brittle. H = $3\frac{1}{2}$. G = 2.319. Biaxial (+); $N_x = 1.534$, $N_y = 1.543$, $N_z = 1.558$; 2V = 73°; r > v. Soluble in acids. Found with variscite in phosphate nodules at *Clay Canyon, Fairfield, Utah Co., UT*; at Milgun Station, WA, Australia. HCWS *DII:975, AM 47:1(1962)*.

42.11.14.5 Mangangordonite $(Mn,Fe,Mg)Al_2(PO_4)_2(OH)_2 \cdot 8H_2O$

Named in 1991 for the composition and relationship to gordonite (42.11.14.4). Laueite–paravauxite group, paravauxite subgroup. TRIC $P\bar{1}$. $a = 5.257$, $b = 10.363$, $c = 7.040$, $\alpha = 105.44°$, $\beta = 113.07°$, $\gamma = 78.69°$, Z = 1, D = 2.32. *NJMM:169(1991)*, Structure: *NJMM:265(1988)*. PD: 9.96_6 6.39_8 4.77_{10} 3.90_3 3.18_7 2.86_5 2.59_4 2.12_3. Equant to bladed crystals to 2 mm; aggregates of bladed crystals. Tabular on {010}, elongate [001] forms {010}, {100}, {1$\bar{1}$0}, {0$\bar{1}$1}. Colorless to faintly yellow, white streak, vitreous luster, translucent. Cleavage {010}, perfect. H = 3. G = 2.36. Biaxial(+); Z ≈ c; (010) ∧ OAP = −45°; $N_x = 1.556$, $N_y = 1.561$, $N_z = 1.571$; 2V = 70°; r < v, distinct. Reacts with microprobe beam. Found at Dunton Gem mine, Newry, ME; at *Foote Mineral Co., spodumene mine, Kings Mt., NC*, in vugs in microcline–spodumene–quartz pegmatite. HCWS *AM 76:2022(1991)*.

42.11.15.1 Xanthoxenite $Ca_4Fe_2^{3+}(PO_4)_4(OH)_2 \cdot 3H_2O$

Named in 1920 in allusion to the color and possible resemblance to cacoxenite. TRIC $P\bar{1}$ or $P1$. $a = 6.70$, $b = 8.85$, $c = 6.54$, $\alpha = 92.1°$, $\beta = 110.2°$, $\gamma = 93.2°$, Z = 1, D = 3.38. *MM 42:309(1978)*. *30-258:* 6.27_8 3.49_9 3.34_4 3.24_7 3.14_7 3.06_{10} 2.97_4 2.74_5. Crusts and masses of distinct platy or lathlike crystals. Pale yellow to brownish yellow, dull to waxy luster, translucent. Cleavage {010}, perfect. H ≈ $2\frac{1}{2}$. G = 2.8-2.97. Biaxial (−); $N_x = 1.704$, $N_y = 1.715$, $N_z = 1.724$; 2V large; r < v, strong; weakly pleochroic, pale yellow to pale lemon yellow. Found at Newry, ME; at Fletcher and Palermo No. 1 mines, North Groton, NH; at Lynwood, Tip Top, Custer Co., SD; at Hühnerkobel pegmatite, Rabenstein, Bavaria, Germany (may be stewartite); at Pribyslavice, Otov, Czech Republic; at Lavra da Ilha, Taquaral, MG, Brazil. HCWS DII:977, AM 34:692(1949).

42.11.16.1 Beraunite $Fe^{2+}Fe_5^{3+}(PO_4)_4(OH)_5 \cdot 4H_2O$

Named in 1841 for the locality. MON $C2/c$. $a = 20.953$, $b = 5.171$, $c = 19.266$, $\beta = 90.34°$, $Z = 4$, $D = 2.89$. CM 27:441(1989), Structure: ZK 201:263(1992). 22-631: 10.34_{10} 9.58_5 7.23_5 4.83_6 4.42_5 3.75_3 3.19_4 3.08_6. Rare crystals: tabular [100] elongated [010] striated [010]; radiated, foliated globules and crusts; radial aggregates or discoidal concretions with coarse fibrous structure. Twinned on [100], occasionally penetration twins. Reddish brown to hyacinth-red, blood-red, rarely greenish brown or dull green; yellow to olive-drab streak; vitreous to pearly luster on cleavage surfaces; translucent. Cleavage {100}, good. $H = 3\frac{1}{2}-4$. $G = 3.00$. Biaxial (+); $Y \wedge c = 1\frac{1}{2}-5°$; $Z = b$; $N_x = 1.775$, $N_y = 1.786$, $N_z = 1.815$ (Giessen); 2V medium large; $r > v$; pleochroic. X,Y, pale flesh; Z, carnelian red. Easily soluble in HCl. One of several phosphate minerals in secondary iron ore deposits and also an alteration product of triphylite in pegmatites. Found at Fletcher and Palermo No. 1 mines, North Groton, NH; at Mullica Hill, and Middletown, NJ; at Bachman mine, Hellertown, Northampton Co., Moore's Mills, Cumberland Co., PA; at Jacksonville, Calhoun Co., and Rock Run Station, Indian Mt., Cherokee Co., AL; at White Elephant, Custer Co., Big Chief, Hesnard, Pennington Co., SD; *Hrbek mine, Beraun, Svatá Dobrotivá, Czech Republic*; at Hagendorf Süd, Huhnerkubel, Kreuzberg, Germany; at Corrego Frio, MG, Brazil. HCWS DII:959, AM 55:135(1970).

42.11.17.1 Bermanite $Mn^{2+}Mn_2^{3+}(PO_4)_2(OH)_2 \cdot 4H_2O$

Named in 1936 for Harry Berman (1902–1944), mineralogist, Harvard University, Cambridge, MA. MON $P2_1$. $a = 5.446$, $b = 19.25$, $c = 5.428$, $\beta = 110.29°$, $Z = 2$, $D = 2.867$. Structure: AM 61:1241(1976). 20-712: 9.63_{10} 5.10_2 4.82_3 3.67_2 3.26_3 2.91_4 2.11_1 1.828_1. Tabular {001} to 0.5 mm; fan-shaped or subparallel rosettes of crystals; lamellar masses. Pseudo-orthorhombic, polysynthetic and other twinning. AM 53:416(1968). Pale red to reddish brown, darkening on exposure; vitreous to resinous luster; translucent. Cleavage {001}, perfect; {110}, imperfect. $H = 3\frac{1}{2}$. $G = 2.85$. DTA, heated products magnetic. Biaxial (−); $X = b$; $Y \wedge c = -36\frac{1}{2}°$; $Z \wedge a = 2°$; $N_x = 1.690$, $N_y = 1.729$, $N_z = 1.750$; 2V = 72°; $r < v$; pleochroic. X, light red; Y, pale yellow; Z, deep red. Soluble in HNO_3. Small amounts of Fe^{3+} substitute for Mn^{3+}; Mg, Ca, Na for Mn^{2+}. Occurs as a secondary mineral in pegmatites. Found at Fletcher and Palermo No. 1 mines, North Groton, NH; at Williams property, Coose Co., AL; at the Bull Moose, Tip Top, and White Elephant pegmatites, Custer Co., Hesnard, Pennington Co., SD; in druzy cavities in triplite in pegmatite on *7U7 Ranch near Bagdad copper mine, 40 km W of*

Hillside, Yavapai Co., AZ, with other phosphates; at Stewart Lithia mine, San Diego Co., CA; at Mangualde, Portugal; at Hagendorf Süd, Bavaria, Germany; at the Flinders Ranges, SA, Australia; at Sapucaia, MG, Brazil; as the alteration product of triplite embedded in a black Mn deposit at El Criollo, Tanti, Cordoba, Argentina. HCWS *DII:967*.

WHITMOREITE GROUP

The whitmoreite group minerals can be expressed by the general formula

$$A(H_2O)_4Fe_2^{3+}(OH)_2(TO_4)_2$$

where
A = Cu, Fe^{2+}, Mn, Zn,□, and possibly some Fe^{3+}
T = As, P, S

The structure [*AM 59:900(1974)*] is composed of octahedrally coordinated Fe sheets into slabs parallel to {100} with composition $Fe_2^{3+}(OH)_2(O_p)_6$, (where O_p = oxygen from a phosphate group). The slabs are linked through

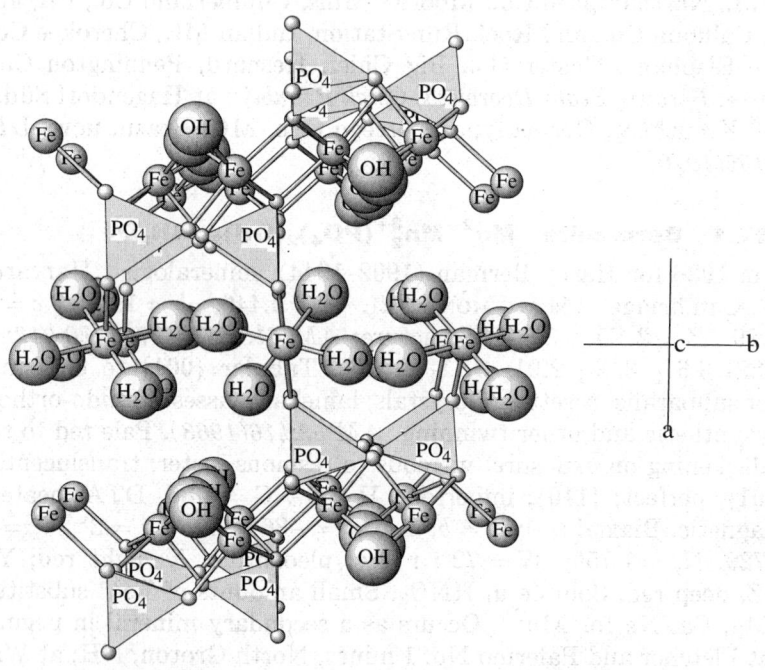

WHITEMOREITE: $Fe^{2+}(H_2O)_4[Fe^{3+}(OH)_2(PO_4)_2]$

A c axis projection that illustrates the structure of whitmoreite: slabs of $Fe_2^{3+}(OH)_2(O_p)_6$ and tetrahedral PO_4 groups in staggerred array connected by water rich layers that contain Fe^{2+} or other elements.

the remaining PO$_4$ apical oxygens by [A(O$_p$)$_2$(H$_2$O)] octahedra. The A site may be partially occupied or contain several different ions.

WHITMOREITE GROUP

Mineral	Formula	Space Group	a	b	c	β	
Whitmoreite	Fe^{2+}(H$_2$O)$_4$[Fe$_2^{3+}$(OH)$_2$(PO$_4$)$_2$]	$P2_1/c$	10.00	9.73	5.471	93.8	42.11.18.1
Arthurite	Cu(H$_2$O)$_4$[Fe$_2^{3+}$(OH)$_2$(AsO$_4$)$_2$]	$P2_1/c$	10.189	9.649	5.598	92.27	42.11.18.2
Ojuelite	Zn(H$_2$O)$_4$[Fe$_2^{3+}$(OH)$_2$(AsO$_4$)$_2$]	$P2_1/c$	10.247	9.665	5.569	94.37	42.11.18.3
Earlshannonite	Mn^{2+}(H$_2$O)$_4$[Fe$_2^{3+}$(OH)$_2$(PO$_4$)$_2$]	$P2_1/c$	9.910	9.669	5.455	93.95	42.11.18.4

42.11.18.1 Whitmoreite Fe^{2+}(H$_2$O)$_4$[Fe$_2^{3+}$(OH)$_2$(PO$_4$)$_2$]

Named in 1974 for Robert W. Whitmore (b.1936), Weare, NH, micromount mineral collector and owner of the Palermo mine, NH. Whitmoreite group. MON $P2_1/c$. $a = 10.00$, $b = 9.73$, $c = 5.471$, β $= 93.8°$, Z $= 2$, D $= 2.85$. Structure: *AM 59:900(1974)*. *26-1138:* 10.1$_{10}$ 7.01$_7$ 4.98$_7$ 4.42$_6$ 4.21$_7$ 3.48$_5$ 2.88$_5$ 2.80$_7$. Acicular crystals parallel to [001] $\approx$ 10 × longer than wide, to 2 mm long, show rhomboidal cross section with {100}, {110}, {011}, {021}, {112}, usually twinned by reflection parallel to {100} imparting an orthorhombic appearance; fans or sprays, radial aggregates. Pale tan, dark brown, green-brown; vitreous to subadamantine luster; translucent. Cleavage {100}, fair. H $= 3$. G $= 2.87$. Biaxial (−); Y parallel to b, Z parallel to c; N$_x$ = 1.676, N$_y$ = 1.725, N$_z$ = 1.745; 2V = 60–65°; pleochroic. X,Y, light green-brown; Z, dark greenish brown. Microprobe analyses show Mn. Found at Newry, ME; as hydrothermal alteration product of triphylite in cavities in quartz-bearing pegmatite at *Palermo No. 1 mine, North Groton, NH*, perched on siderite rhombs with ludlamite, strunzite, laueite, beraunite, mitridatite, and ferric oxyhydroxides, also at the Fitzgibbon pegmatite, East Alstead; at Big Chief pegmatite, near Glendale, SD; at Hagendorf Süd, Bavaria, Germany. HCWS

42.11.18.2 Arthurite Cu(H$_2$O)$_4$[Fe$_2^{3+}$(OH)$_2$(AsO$_4$)$_2$]

Named in 1964 for Arthur Edward Ian Russell (1878–1964) and W. G. Kingsbury (1906–1968), mineralogists and mineral collectors, England. Whitmoreite group. The As and Cu end member. MON $P2_1/c$. $a = 10.189$, $b = 9.649$, $c = 5.598$, β $= 92.27°$, Z $= 2$, D $= 3.376$. *MM 33:937(1964)*, Structure: *NJMA 133:291(1978)*. *36-400:* 10.14$_8$ 6.98$_{10}$ 4.81$_6$ 4.30$_5$ 2.93$_4$ 2.81$_5$ 2.77$_3$ 2.71$_3$. See also *16-397*. Thin crusts. Pale olive-green, translucent. G $= 3.2$. IR. Biaxial (+); N$_x$ = 1.746, N$_y$ = 1.774, N$_z$ = 1.806; 2V$_{calc}$ = 88°. Analyses also show P and S. Found at Mylar mine, and Copper Slope, Majuba Hill, Pershing Co., NV; at *Hingston Down Consols mine, Calstock, Cornwall, England*, mixed with pharmacosiderite, scorodite, other phosphates, and quartz. HCWS *MM 37:570(1969)*.

42.11.18.3 Ojuelaite $Zn(H_2O)_4[Fe_2^{3+}(OH)_2(AsO_4)_2]$

Named in 1981 for the locality. Whitmoreite group. The As and Zn end member. See also mapimite (42.8.7.1). MON $P2_1/c$. $a = 10.247$, $b = 9.665$, $c = 5.569$, $\beta = 94.37°$, $Z = 2$, $D = 3.39$. *BM 104:582(1981)*. *35-516:* 10.2_7 7.03_8 4.83_8 4.52_5 4.25_{10} 2.90_6 2.87_9 2.63_6. Fibers with elongation on c axis. Greenish yellow, translucent. $H = 3$. $G = 3.39$. DTA. Biaxial (+); $N_x = 1.696$, $N_y = 1.730$, $N_z = 1.798$; $2V_{calc} = 73°$; extinction; fiber length parallels to pleochroic: yellow $Z = X > Y$. Soluble in 1:10 acids. Analyses show traces of Ca. Found in the oxidation zone at *Ojuela mine, Mapimi, DUR, Mexico*, associated with scorodite, adamite, smithsonite, and "limonite"; at Tsumeb mine, Namibia. HCWS *AM 67:623(1982)*.

42.11.18.4 Earlshannonite $Mn^{2+}(H_2O)_4[Fe_2^{3+}(OH)_2(PO_4)_2]$

Named in 1984 for Earl V. Shannon (1895–1981), mineralogist and chemist at U.S. National Museum, Smithsonian Institution, Washington, DC. Whitmoreite group. The Fe^{3+} analog of whitmoreite (42.11.18.1). MON $P2_1/c$. $a = 9.910$, $b = 9.669$, $c = 5.455$, $\beta = 93.95°$, $Z = 2$, $G = 2.92$. *CM 22:471(1984)*. *38-364:* 9.8_{10} 6.9_8 4.95_4 4.38_4 4.18_6 3.45_6 2.86_6 2.79_7. Hemispheric aggregates of radial prismatic crystals to 0.5 mm length [001] showing {110}, {100}, {011}. Twinned on {100} showing parallel growth with {001} in common. Red-brown, vitreous luster, transparent. Two cleavages {100}, good; even fracture; brittle. $H = 3–4$. $G = 2.90$. Nonfluorescent. Biaxial (−); $N_x = 1.696$, $N_y = 1.745$, $N_z = 1.765$; $2V = 64°$; pleochroic. X,Y, light yellow-brown; Z, yellow-brown. Microprobe analysis shows small amounts of Fe^{2+}, Al, Mg. Found on the dumps at *Foote spodumene mine, Kings Mt., Cleveland Co., NC*, associated with mitridatite, laueite, rockbridgeite, jahnsite, quartz, and Mn oxides; at Hagendorf pegmatite (bright yellow), Waidhaus, Oberpfalz, Bavaria, Germany. HCWS *AM 70:871(1985)*.

42.11.19.1 Sincosite $Ca(V^{4+}O)_2(PO_4)_2 \cdot 5H_2O$

Named in 1924 for the locality. TET $Pmmm$ or $P4/nmm$. $a = 8.895$, $c = 12.727$, $Z = 2$, $D = 2.970$. *AM 70:409(1985)*. Isostructural with meta-autunite (40.2a.1.2). *39-318:* 6.35_{10} 3.84_3 3.51_6 3.19_8 2.91_4 2.68_5 2.10_4 1.264_2. Rectangular scales, or thin plates showing {001} (striated parallel to [100], {100}, possibly {110}; aggregates of stacked plates in nearly parallel position, forming scaly irregular masses; radial aggregates and nodules admixed with gypsum and carbonaceous matter. Occasionally, twinned on {110}. Leek-green, blue-green, yellow-green; green streak; vitreous to submetallic luster in altered crystals; transparent in thin crystals. Cleavage {001}, good; {100}, {110}, poor; brittle. Soft. $G \approx 2.84$. Uniaxial (−); $N_O \approx 1.680$, $N_E \approx 1.655$; may be biaxial (−) if dehydrated with $N_x = 1.675$, $N_{y\,calc} = 1.690$, $N_z = 1.693$; pleochroic: O, or Z, gray-green; E, or X, colorless to pale yellow. May alter to lower hydrate on exposure with partial oxidation of V^{4+} to V^{5+} (submetallic material). Readily soluble in dilute acids to a blue solution, insoluble in H_2O. Found in a black carbonaceous shale near *Sincos, Dept. Junin, Peru*, about 160 km E of Lima as small veinlets, usually parallel to bedding planes but also as irregular gypsum-bearing nodular masses in particular shale beds. HCWS *DII:1057*.

42.11.20.1 Furongite $Al_2(UO_2)(PO_4)_2(OH)_2 \cdot 8H_2O$

Named in 1976 for the poetic description of the Hunan Province, China, where the mineral was found. TRIC $P1$ or $P\bar{1}$. $a = 19.271$, $b = 14.173$, $c = 12.316$, $\alpha = 67.62°$, $\beta = 115.45°$, $\gamma = 94.58°$, $Z = 7$, $D = 2.90$. *SMPM 65:1(1985)*, Structure: *AC(A) 37C:186(1981)*. 29-98: 10.2_{10} 8.62_8 5.55_3 5.10_3 4.31_5 3.64_4 3.52_3 2.87_4. See also *44-1463*. Tabular crystals; cryptocrystalline aggregates. Bright yellow to lemon yellow, vitreous luster, translucent. One perfect cleavage, two others distinct; brittle. $G = 2.82$–2.90. Radioactive. Fluoresces strong pale yellowish green. DTA, TGA, DTG. Biaxial $(-)$; $N_x = 1.543$–1.549, $N_y = 1.564$–1.567, $N_z = 1.570$–1.575; $2V = 65°$; pleochroic in yellow; oblique extinction, small angle. Found at Kobokobo, Kivu, Zaire; in the oxidation zone of an alluvial-type U deposit in Lower Cambrian carbonaceous shale, *western Hunan Prov., China*, associated with variscite, evansite, opal, halloysite, limonite, and autunite. HCWS *AM 63:425(1978), 73:198(1988)*.

42.11.21.1 Paulkerrite $K(Mn^{2+},Mg)_2Fe_2^{3+}Ti(PO_4)_4(OH)_3 \cdot 15H_2O$

Named in 1984 for Paul Francis Kerr (1897–1981), mineralogist, Columbia University, New York City. Mantiennéite (42.11.21.2) is the Al analog. ORTH *Pbca*. $a = 10.49$, $b = 20.75$, $c = 12.44$, $Z = 4$, $D = 2.36$. *MR 15:303(1984)*. 38-390: 10.3_9 7.46_8 6.20_{10} 5.22_2 3.95_3 3.75_4 3.13_7 2.87_4. Yellow-brown, vitreous luster, translucent. $H = 3$. $G = 2.36$. Biaxial $(-)$; $N_x = 1.598$, $N_y = 1.624$, $N_z = 1.643$; $2V = 80°$. Microprobe analysis shows Al, Mn > Mg, F. Found at *Bagdad copper mine, Hillside, Yavapai Co., AZ*. HCWS *AM 70:875(1985)*.

42.11.21.2 Mantiennéite $K(Mg,Fe^{2+})_2(Al,Fe^{3+})_2Ti(PO_4)_4(OH)_3 \cdot 15H_2O$

Named in 1984 for Joseph Mantienné (b.1929), mineralogist and collections specialist of the Bureau de Recherches Géologiques et Minières, Orléans, France. The Al end member in the Al–Fe^{3+} series with paulkerrite (42.11.21.1). ORTH *Pbca*. $a = 10.409$, $b = 20.330$, $c = 12.312$, $Z = 4$, $D = 2.25$. *BM 107:737(1984)*. 37-438: 10.18_4 7.41_3 6.16_{10} 3.10_3 3.08_6 2.95_3 2.84_3 2.05_2. Spheres of radiating fibers up to 1 mm diameter. Yellow-brown, bright, translucent. Cleavage {001}, perfect; {010}, poor; brittle. $G = 2.31$. DTA, TGA. Biaxial $(-)$; $XYZ = bca$; $N_x = 1.564$, $N_z = 1.598$; $2V \approx 55°$; $r > v$, distinct; pleochroic. X, colorless; Z, very pale yellow. Contains some Mn, Ca, Na. Found in the vivianite deposit at *Anloua, Cameron*, associated with quartz, siderite, and kaolinite as the matrix of sandy layers intercalated in black shales of lacustrine origin. HCWS *AM 70:1330(1985)*.

42.11.21.3 Matveevite $KTiMn_2Fe_2^{3+}(PO_4)_4(OH)_3 \cdot 15H_2O$

Named in 1991 for K. K. Matveev (1875–1954). The Mn end member in the series with paulkerrite (42.11.21.1). ORTH $Pnam$ or $Pna2_1$. $a = 12.42$, $b = 20.52$, $c = 10.52$, $D = 2.40$. $ZVMO\ 120:108(1991)$. 44-1404: $10.4_9\ 7.50_8\ 6.23_{10}\ 3.15_9\ 2.86_4$. Isometric or thick platy crystals to 0.2 mm showing $\{010\}$, $\{100\}$, $\{111\}$. White to colorless, yellowish; white streak; vitreous luster, pearly on $\{010\}$; transparent to translucent. Cleavage $\{010\}$, perfect; uneven fracture; brittle. $H = 2\frac{1}{2}$. $G = 2.32$. Biaxial (+); $XYZ = bac$; $N_x = 1.574$, $N_y = 1.580$, $N_z = 1.618$; $2V = 44°$. Found in a granitic pegmatite vein in *Ilmenskij National Park, Ural Mts., Russia*, associated with decomposition products of triplite, francolite, ushkovite, kaluninite, and Fe and Mn hydroxides. HCWS $AM\ 78:451(1993)$.

42.11.21.4 Benyacarite $(H_2O,K)Ti(Mn^{2+},Fe^{2+})_2(Fe^{3+},Ti)_2(PO_4)_4(O,F)_2 \cdot 14H_2O$

Named in 1993. An inadequately described mineral. Compare matveevite (42.11.21.3). The Mn, H_2O analog of paulkerrite (42.11.21.1). ORTH $Pbca$. $a = 10.561$, $b = 20.585$, $c = 12.51$, $Z = 4$, $D = 2.37$. Structure: $ZK\ 208:57(1993)$. Occurs with other phosphates and with F-rich minerals such as fluellite and pachnoite in a granitic pegmatite at *Cerro Blanco, near Tanti, Cordoba, Argentina*. HCWS $AM\ 79:763(1994)$.

42.11.22.1 Mcauslanite $HFe_3^{2+}Al_2(PO_4)_4F \cdot 18H_2O$

Named in 1988 for David A. McAuslan, exploration manager, Shell Canada Resources Ltd., who discovered the locality. TRIC $P1$. $a = 10.055$, $b = 11.568$, $c = 6.888$, $\alpha = 105.84°$, $\beta = 93.66°$, $\gamma = 106.47°$, $Z = 1$, $D = 2.17$. $CM\ 26:917(1988)$. PD: $10.6_9\ 9.53_8\ 6.55_7\ 5.41_2\ 4.96_{10}\ 4.77_2\ 3.44_3\ 2.81_5$. Clusters up to 4 mm across of radiating bladed or fibrous crystals up to 1 mm long, 0.2 mm wide; elongate [001], showing $\{010\}$, $\{100\}$, $\{001\}$. Yellow-white, white streak, vitreous luster, translucent. Good cleavage parallel to [001], brittle. $H = 3\frac{1}{2}$. $G = 2.22$. TG/EGA. Biaxial (−); $X \wedge b = 17°$ in β obtuse, $Y \wedge c = 18°$ in α obtuse, $Z \wedge a = 14°$ in γ obtuse; $N_x = 1.522$, $N_y = 1.531$, $N_z = 1.534$. Found as part of a complex supergene enrichment area that includes vivianite, phosphophyllite, and childrenite–eosphorite filling joints and shear zones near the surface of *East Kemptville tin mine, Yarmouth Co., NS, Canada*. HCWS $AM\ 75:707(1990)$.

42.11.23.1 Vochtenite $(Fe^{2+},Mg)Fe^{3+}[(UO_2)(PO_4)]_4(OH) \cdot 12$–$13H_2O$

Named in 1989 for Renard F. C. Vochten (b.1933), mineralogist, State University of Antwerp, Antwerp, Belgium. MON $a = 12.606$, $b = 19.990$, $c = 9.990$, $\beta = 102.52°$, $Z = 3$, $D = 3.663$. $MM\ 53:473(1989)$. 42-1358: $10.0_{10}\ 4.89_5\ 3.48_7\ 3.33_5\ 3.09_4\ 2.21_4\ 2.15_5\ 2.11_5$. Aggregates of subparallel crystals each 1 mm and pseudoquadratic in outline due to intersection of $\{001\}$ and $\{100\}$. Brown, brown streak, bronzy, translucent but opaque in thick fragments. Cleavage

{010}, distinct. H = 3. G = 3.650. TGA. Massbauer Shaws $Fe^{2+} \approx Fe^{3+}$. Nonfluorescent. Biaxial (−); X = b; Z ∧ c = small; $N_x = 1.575$, $N_y = 1.589$, $N_z = 1.603$ (589 nm); 2V = 89°. Soluble in 1:1 HCl. Zeolitic H_2O is implied by the formula. Found at *Bassett mine, Redruth, Cornwall, England*. HCWS *AM 75:1212(1990)*.

42.12.1.1 Vantasselite $Al_4(PO_4)_3(OH)_3 \cdot 9H_2O$

Named in 1987 for René Van Tassel (b.1916), mineralogist of phosphate and sulfate minerals, Brussels and Louvain, Belgium. ORTH *Pman, Pma2,* or $P2_1am$. a = 10.528, b = 16.541, c = 20.373, Z = 8, D = 2.312. *BM 110:647(1987)*. Sheet structure similar to that of vashegyite (42.12.2.1) and matulaite (42.13.3.1). *42-1382:* 10.2_{10} 4.87_3 3.72_3 3.40_4 3.21_4 2.89_5 2.74_3 2.39_4. Rosettes up to 8 mm of thin blades elongate [100], flattened {001}, terminated by {120}. White, pearly luster, translucent. Cleavage {001}, perfect. H = 2–$2\frac{1}{2}$. G = 2.30. TGA. Biaxial (−); $N_{x\,calc} = 1.511$, $N_y = 1.560$, $N_z = 1.579$; 2V = 61°. Wet chemical and microprobe analyses show >5%(wt), Fe_2O_3<0.4% S. Found at *Stavelot massif, Bihain, Belgium*, on the dumps of a quartzite quarry, where it occurs in the schistosity planes or as lamellae in quartz veinlets within metapelites associated with wavellite, variscite, cacoxenite, turquoise, clinochlore, muscovite, and lithiophoite. HCWS *AM 73:931(1988)*.

42.12.2.1 Vashegyite $Al_{11}(PO_4)_9(OH)_6 \cdot 38H_2O$ or $Al_6(PO_4)_5(OH)_3 \cdot 23H_2O$

Named in 1909 for the locality. ORTH Pna2, $Pna2_1$ a = 10.773 b = 14.991 c = 20.626 Z=2 D=1.934 for Al_{11} (type locality). *Pnam, Pna2,* a = 10.754 b = 14.971 c = 22.675 Z=4 D=2.005 for Al_6 (Chvaletice) *CM 21:489(1983)*. *35-704* 10.3_{10} 7.49_4 7.04_9 4.90_4 3.46_5 2.92_9 2.42_5 2.13_5 (type). See also *29-68*. *35-597* 11.3_{10} 7.50_8 6.26_7 5.39_4 3.75_4 3.30_6 2.91_9 2.83_4. (Chvaletice). Massive, compact, porous, microcystalline with radial fibrous styructure in hemispheres. Diamond-shaped platy crystals showing {110}, {010}, (001) (Chvaletice) with irregular stacking. White, pale green to yellow, brownish, dull, opaque to subtranslucent. Cleavage {001} perfect. H=2–3 G=1.93–1.99. DTA IR. Biaxial (−) $N_x1.470$ $N_y1.477$ $N_z1.482$ (Chvaletice); *xyz=cba*. Minute fibers + elongation, birefring ≈ 0.02, $N_{av}1.48$. Easily soluble in acids. Variable water between the sheets in the structure probably accounts for the differences in physical properties. Found in altered slate with variscite near Manhattan, Nye Co., NV; at *Zeleznik (Vashegy) mine, Sirk, R., Slovakia* with variscite, evansite; at Chvaletice, Bohemia, Czech Rep. intimately mixed with apatite, iron hydroxides; at Lavra do Enio, Galileia, MG, Brazil *DII:999, MM 39:802(1974)*. HCWS

42.12.3.1 Moreauite $Al_3(UO_2)(PO_4)_3(OH)_2 \cdot 13H_2O$

Named in 1985 for Jules Moreau (b.1931), mineralogist, Catholic University, Louvain, Belgium. MON $P2_1/c$. a = 23.41, b = 21.44, c = 18.34, β = 92.0°,

$Z = 16$, $D = 2.61$. *BM 108:9(1985). 39-317:* 14.0_6 11.7_8 10.8_{10} 9.13_7 5.43_4 4.51_4 3.04_6 2.93_7. Nodules, books of plates flattened on {100}, sometimes elongated parallel to b to 0.2 mm. Greenish yellow, vitreous luster, translucent. Cleavage {100}, good. G = 2.64. TGA. Biaxial $(-)$; $X \approx a$, $Y = b$, $Z \approx c$; $N_x = 1.540$, $N_y = 1.552$, $N_z = 1.558$; $2V_{calc} = 70°$; negative elongation; parallel extinction. Found at *Kobokobo, Kivu, Zaire*, associated with furongite, renunculite, and phosphosiderite. HCWS *AM 70:1330(1985)*.

42.12.4.1 Tinticite $Fe_4^{3+}(PO_4)_3(OH)_3 \cdot 5H_2O$

Named in 1946 for the locality. MON *P2, Pm*, or *P2/m*. $a = 13.65$, $b = 6.542$, $c = 12.31$, $\beta = 91.2°$, $Z = 3$, $D = 2.97$. *NJMM:446(1988). 8-151:* 6.7_6 6.07_6 5.67_8 4.56_6 3.91_{10} 3.28_{10} 3.01_9 2.96_5. See also *44-1448*. Claylike earthy nodules of small (≈ 1.5 μm) platy crystals; compact porcelaneous. Cream white to slightly yellow-green, dull luster. $H \approx 2\frac{1}{2}$. G = 2.94. $N_{av} = 1.745$; birefringence $= 0.0005$. Insoluble in H_2O, H_2SO_4, HNO_3, slowly soluble in HCl. Found at *Tintic Standard mine, Tintic district, UT*, with jarosite, "limonite," on limestone cave wall, possibly the result of guano leaching and availability of Fe; at Gavà close to Banco and in Barcelona, Spain, where nodules up to 5 cm in diameter are covered with montgomeryite. HCWS *AM 31:395(1946), MJJ 15:261(1991)*.

42.12.5.1 Tooeleite $Fe_{8-2x}^{3+}[(As_{1-x}^{5+},S_x^{2+})O_4]_6 \cdot 5H_2O$

Named in 1992 for the county where the mineral was found. ORTH $Pbcm$ or $Pbc2_1$. $a = 6.416$, $b = 19.45$, $c = 8.941$, $Z = 2$, $D = 4.15$. *MM 56:71(1992). 44-1468:* 9.75_{10} 4.48_4 3.66_3 3.46_3 $3{,}24_3$ 3.21_9 3.05_5 2.68_4. Powdery crusts up to 10 mm thick, earthy. H = 3. G = 4.23. Biaxial $(-)$; $N_x = 1.94$, $N_y = 2.05$, $N_z = 2.06$; 2V small. Microprobe analysis shows minor Mn, Al, and Mg, possible trace of Mo. Occurs in voids in massive scorodite that also contain massive jarosite in the oxidized waste dump at the *U.S. mine, Tooele Co., UT*. Tooeleite encrusts both jarosite and scorodite; other minerals present: kaatialaite, galena, sphalerite, and goethite. HCWS

42.13.1.1 Aldermanite $Mg_5Al_{12}(PO_4)_8(OH)_{22} \cdot 32H_2O$

Named in 1981 for Arthur Richard Alderman (1901–1980), mineralogist and petrologist, University of Adelaide, SA, Australia. ORTH $a = 15.000$, $b = 8.330$, $c = 26.60$, $Z = 2$, $D = 2.15$. *MM 44:59(1981). 35-676:* 13.4_{10} 7.98_8 5.70_3 5.55_6 4.96_3 3.47_2 2.84_5 2.66_3. Talclike flakes only 0.1 μm thick × 0.1 mm across. Colorless, pearly luster, translucent. $H \approx 2$. $N_{av} = 1.500$. Soluble in mineral acids. Found with fluellite in the Moculta P deposit at *Angaston, 60 km NE of Adelaide, SA, Australia*, with apatite, beraunite, cacoxenite, childrenite, crandallite, cyrilovite, dufrenite, gorceixite, leucophosphite, minyulite, rockbridgeite, variscite, and wavellite. HCWS *AM 66:1099(1981)*.

42.13.2.1 Richellite $Ca_3Fe^{3+}_{10}(PO_4)_8(OH)_{12} \cdot nH_2O$

Named in 1884 for the locality. Data for heated material, as original sample amorphous: TET $P4$. $a = 5.18$, $c = 12.61$, $Z = 1$, $D = 1.87$. *AM 48:300(1963)*. *15-632:* 5.99_6 4.35_6 4.14_6 3.58_7 3.24_{10} 3.15_7 2.27_6 1.590_8. Massive, compact, or foliated; radially fibrous globules. Reddish to yellowish brown, similar streak, greasy to hornlike. $H = 2–3$. $G \approx 2$. Readily soluble in acids. Analysis shows $\approx$ 1% HfO. Found at *Richelle, near Visé in Liége, Belgium*, associated with halloysite, allophane, and koninckite. HCWS *DII:956*.

42.13.3.1 Matulaite $Ca[Al_9(H_2O)_{10}(OH)_{10}(PO_4)_6]_2 \cdot 24H_2O$

Named in 1980 for Margaret Matula (b. 1925) of Allentown, PA, who found the sample. MON $P2_1/c$. $a = 20.4$, $b = 16.7$, $c = 10.6$, $\beta = 98.2°$, $Z = 2$, $D = 2.33$. *AUF 31:55(1980)*. *33-258:* 9.96_{10} 6.37_4 4.83_3 4.42_4 3.79_3 3.66_3 3.07_3 2.40_4. Small rosettes of thin tabular crystals, or botryoidal. Colorless to white, pearly luster, translucent. Cleavage {100}, perfect. $H = 1$. $G = 2.330$. Biaxial $(-)$; $Y = b$; $Z \wedge 8°$; $N_x = 1.576$, $N_z = 1.582$; $2V \approx 60°$; $r < v$, strong. Analysis shows in Si, Fe^{3+}. Found at the *Bachman mine, Hellertown, PA*, as an incrustation on chert with rockbridgeite, dufrenite, cacoxenite, strengite, and wavellite; at LCA pegmatite, Gaston Co., NC; at Rotläufchen iron mine, Waldgrimes, Germany. HCWS *AM 65:1067(1980)*.

42.13.4.1 Jungite $Ca_2Zn_4Fe^{3+}_8(OH)_9(PO_4)_9 \cdot 16H_2O$

Named in 1980 for Gerhard Jung, German mineral collector who found the mineral. ORTH $Pcmm$, $Pcm2_1$, or $Pc2m$. $a = 11.98$, $b = 20.37$, $c = 9.95$, $Z = 2$, $D = 2.849$. *AUF 31:55(1980)*. *35-655:* 5.09_{10} 4.91_6 3.79_8 3.37_{10} 3.30_{10} 3.08_6 2.94_6 1.443_8. Rosettes up to 1 cm in diameter of thin tabular, bent crystals. Dark green, yellow streak, silky to vitreous luster, translucent. Cleavage {010}, perfect. $H = 1$. $G = 2.843$. Biaxial $[-(?)]$; $Y = a$ $Z = C$; $N_x = 1.658$, $N_z = 1.664$; $2V \approx 60°$; $r < v$, strong. Found at the *Hagendorf Süd pegmatite, Bavaria, Germany*, associated with mitridatite and Mn oxides. HCWS *AM 65:1067(1980)*.

42.13.5.1 Cacoxenite $(Fe^{3+},Al)_{25}O_6(PO_4)_{17}(OH)_{12} \cdot \approx 75H_2O$

Named in 1825 from the Greek for *bad guest* because the P content interferes with the quality of iron made when the mineral occurs in limonite ore. HEX $P6_3/m$. $a = 27.559$, $c = 10.550$, $Z = 2$, $D = 2.217$. Structure: *NAT 306:356(1983)*. A framework of $[(Al,Fe^{3+})_4O_6(OH)_{12}(PO_4)_{17}$ $(H_2O)_{24}]$ with zeolitic water in large (14 Å) channels. *PD:* 22.0_8 11.94_{10} 9.83_2 9.06_2 6.92_3 4.90_4 3.35_5 2.79_2. See also *14-331*. Tiny acicular crystals [0001], sometimes with hexagonal cross section and indistinct pyramidal faces; tufted or radial aggregates; fibrous coatings; spherulitic. Yellow to brownish yellow, golden yellow, reddish or greenish yellow; silky luster; translucent. No cleavage. $H = 3–4$. $G = 2.3$. Uniaxial $(+)$; $N_O = 1.575$, $N_E = 1.635$(Hellertown); dichroic: O, pale yellow; E, canary-yellow to orange-yellow. Easily soluble in acids. Found at Palermo No. 1, North Groton, NH; on earthy hematite at Antwerp, NY; at Mullica Hill, NJ; at Hellertown, Northampton Co., Moore's Mill, Cumberland Co., Noble's mine, Lancaster Co., PA; at Snow Camp, Alaman Co., NC; at Brewer mine, Cedartown, GA; at Jacksonville, Calhoun Co., Indian Mt. Iron mine, Rock Run Station, Cherokee Co., AL; at White Elephant mine, Custer Co., SD; at Tonopah, NV; at Rochefort-en-Terre, Morbihan, France; at Eleonore mine near Giessen, Rothlaufchen mine, Waldirmes, Rhine, and at the Amberg–Auerbach limonite mine, Hagendorf, Rabenstein, Bavaria, Germany; at *Hrbek mine, Svatá Dobrotivá (St. Benigna), Bohemia, Czech Republic*; in the magnetite ore, Kirunavaara, Sweden. HCWS *DII:997; AM 35:132(1950), 51:1811(1966)*.

42.13.6.1 Ceruleite $Cu_2Al_7(AsO_4)_4(OH)_{13} \cdot 12H_2O$

Named in 1900 for its color. TRIC $P\bar{1}$ or $P1$. $a = 14.359$, $b = 14.687$, $c = 7.440$, $\alpha = 96.06°$, $\beta = 93.19°$, $\gamma = 91.63°$, $Z = 2$, $D = 2.734$. *NJMM:418(1976)*. *29-525*: 7.30_8 5.93_7 5.65_{10} 4.88_5 4.76_7 3.55_6 3.24_5 2.65_6. Massive, compact, claylike aggregates to 10 cm diameter of rodlike crystals up to 5 μm long. Turquoise blue. $H = 5-6$. $G = 2.70$. DTA. Soluble in HCl, HNO_3, KOH, but not in H_2O. Found at the *Emma Luisa gold mine, Huanaco, Taltal Prov., Chile*; and in southern Bolivia, associated with quartz, barite, goethite, and mansfieldite. HCWS *DII:927, AM 62:598(1977)*.

42.13.7.1 Rosiérésite Hydrous phosphate of Al,Pb,Cu

Named in 1910 for the locality. An inadequately described mineral. Amorphous. Opaline stalactitic masses with concentric structure. Greenish yellow, yellow, light brown, soluble in HNO_3. $N \approx 1.50$. Found as a recent deposit in abandoned workings of Cu mine at *Rosiérès, Carmaux, Tarn, France*. HCWS *DII:924*.

42.13.8.1 Englishite $Na_2K_3Ca_{10}Al_{15}(PO_4)_{21}(OH)_7 \cdot 26H_2O$

Named in 1930 for George L. English (1864–1944), American mineral dealer and collector. MON $A2$ or Aa. $a = 38.43$, $b = 11.86$, $c = 20.67$, $\beta = 111.27°$, $Z = 8$, $D = 2.69$. Structure: *MM 40:863(1976)*. *29-1037*: 17.7_9 8.94_{10} 5.57_6 3.85_4 2.96_6 2.84_8 2.68_6 2.34_4. Aggregates and layers of subparallel plates; friable puffy masses. Colorless; vitreous luster, pearly on micaceous cleavage. Cleavage {001}, perfect. $H \approx 3$. $G = 2.68$. Biaxial $(-)$; $X = c$, $Y \approx a$, $Z \approx b$; $N_x = 1.570$, $N_z = 1.572$; 2V small. Found at Tip Top pegmatite, Custer Co., SD, associated with crandallite and other phosphates; at *Clay Canyon, Fairfield, Utah Co., UT*, with montgomeryite, wardite, millisite, and crandallite in variscite nodules. HCWS *DII:957, CM 22:469(1982)*.

42.13.9.1 Natrophosphate $Na_7(PO_4)_2F \cdot 19H_2O$

Named in 1972 for the composition. ISO $Fd3c$. $a = 27.755$, $Z = 32$, $D = 1.77$. Structure: *AC(B) 30:2218(1974)*. *25-831*: 8.12_8 4.94_6 4.03_6 3.07_6 2.90_6 2.68_{10} 2.43_9 1.950_5. Aggregates up to 5 cm × 3 cm. Colorless to white, vitreous to greasy luster, translucent. Cleavage {111}, perfect; conchoidal fracture.

$H = 2\frac{1}{2}$. $G = 1.71$–2. Fluoresces orange (weak). Alters on contact with air, becoming covered with film of villiaumite and secondary phosphates. $N = 1.460$–1.462. Spectrographic analyses showed traces of Mn, Ba, Fe, Ca, Sr. Synthesized. Found in the central cavernous zone of pegmatite in ijolite-urtite at *Yukspor Mt., Khibina massif, Kola Penin., Russia*. Nodules may be surrounded by a rim of villiaumite. HCWS *ZVMO 101:80(1972)*, *AM 58:139(1973)*.

42.13.10.1 Phosphofibrite $KCuFe^{3+}_{15}(PO_4)_{12}(OH)_{12} \cdot 12H_2O$

Named in 1984 for the composition and crystal habit. ORTH $Pbmn$ or $Pnmn$. $a = 14.40$, $b = 18.76$, $c = 10.40$, $Z = 2$, $D = 2.94$. *CE 43:11(1984)*. Structurally related to kidwellite (42.8.2.1). *38-353:* 9.50_{10} 4.35_5 3.96_4 3.48_4 3.23_6 3.13_6 2.99_5 2.10_5. Radial aggregates of fibrous crystals to 0.5 mm long. Yellow, yellow-green; vitreous luster; translucent. Pinacoidal cleavage, good. $H \approx 4$. $G = 2.90$. Biaxial $(-)$; $N_x = 1.755$, $N_z = 1.790$; $2V = 40°$. R<<V Soluble in cold HCl, insoluble in warm 1:10 HNO_3. Analyses show some Al. Found at *Clara mine, Wolfach, Black Forest, Baden–Württenberg, Germany*. HCWS *AM 69:1192(1984)*.

42.13.11.1 Kamitugaite $PbAl(UO_2)_5(PO_4)_2(OH)_9 \cdot 9.5H_2O$

Named in 1984 for the mining center at the locality. TRIC $P\bar{1}$ or $P1$. $a = 10.98$, $b = 15.96$, $c = 9.068$, $\alpha = 95.1°$, $\beta = 96.1°$, $\gamma = 89.0°$, $Z = 2$, $D = 4.47$. *BM 107:15(1984)*. *37-425:* 15.9_3 7.95_{10} 4.31_3 3.97_8 3.49_3 3.27_4 3.18_4 2.88_2. Thin plates flattened (010) to 0.5 mm, showing $\{100\}$, $\{\bar{1}01\}$, $\{001\}$. Yellow, transparent. Cleavage $\{010\}$, $\{001\}$, easy. $G = 4.03$. Nonfluorescent. Biaxial $(-)$; $X \approx b^*$; $Y \wedge a \approx 2°$; Z approximately perpendicular to a and b^*; $N_{x\,calc} = 1.709$, $N_y = 1.735$, $N_z = 1.744$; $2V = 60°$; pleochroic. X,Y, light yellow; Z, yellow-green. Microprobe analyses also show As. Found in a quartz–albite–muscovite–beryl–columbite pegmatite associated with minor triangulite, threadgoldite, dumontite, and studtite at *Kobokobo, Kivu, Zaire*. HCWS *AM 70:437(1985)*.

42.13.12.1 Sieleckiite $Cu_3Al_4(PO_4)_2(OH)_{12} \cdot 2H_2O$

Named in 1988 for Robert Sielecki (b.1958), Australian geologist. TRIC $P\bar{1}$ or $P1$. $a = 9.41$, $b = 7.56$, $c = 5.95$, $\alpha = 90.25°$, $\beta = 91.27°$, $\gamma = 104.02°$, $Z = 1$, $D = 2.94$. *MM 52:515(1988)*. PD: 9.12_5 5.04_{10} 4.76_2 3.85_{10} 3.28_3 2.83_5 2.46_5 2.57_2. Spheres up to 0.5 mm in diameter showing fine radiating fibrous structure with crystals to 100 μm long × 2 μm wide. Deep sky-blue to royal-blue, pale blue streak, pearly luster, translucent. No cleavage, subconchoidal fracture. Not twinned. $H \approx 3$. $G = 3.02$. Nonfluorescent. Biaxial; $N = 63$–1.66; length-slow; weakly pleochroic, colorless to pale blue. Found at the *Mt. Oxide Cu mine, 150 km N of Mt. Isa, Q, Australia*, in oxidation zone of massive pyrite body that shows enrichment in chalcosite and secondary Cu minerals. Mineral in fractures in varicite boulder with turquoise, libethenite, and pseudomalachite. HCWS *AM 74:1401(1989)*.

42.13.13.1 Alvanite $(Zn,Ni)Al_4(VO_3)_2(OH)_{12} \cdot 2H_2O$

Named in 1959 for the composition. MON $P2_1/n$. $a = 17.808$, $b = 5.132$, $c = 8.881$, $\beta = 92.11°$, $Z = 2$, $D = 2.492$. *MM 54:609(1990)*, Structure: *NJMM:385(1990)*. *13-386:* 4.8_5 4.48_{10} 2.97_4 1.982_8 1.911_6 1.686_5 1.542_4 1.484_9. Micaceous flakes with hexagonal form showing {001}, {010}, {100}, {101}. Light blue-green, blue-black; white streak; vitreous to pearly luster on cleavage; translucent. Cleavage {010}, perfect; also polysynthetic twinning parallel to (010). $H = 3-3\frac{1}{2}$. $G = 2.49$. Biaxial; X ∧ cleavage = 14°; $N_x = 1.658$, $N_z = 1.714$; $2V = 80-85°$; $r < v$; both positive and negative elongation observed. Found at *Kurumsak ore field, Kara-Tua, Kazakhstan*. HCWS *ZVMO:157(1959), AM 44:1325(1959)*.

Class 43

Compound Phosphates, Arsenates, and Vanadates

43.1.1.1 Ardealite Ca$_2$(HPO$_4$)(SO$_4$) · 4H$_2$O
Named in 1932 for Ardeal, the old Romanian name for Transylvania. MON Cc. $a = 5.726$, $b = 30.95$, $c = 6.265$, $\beta = 117.29°$, $Z = 4$, $D = 2.32$. *41-585:* 7.73$_{10}$ 3.93$_2$ 3.87$_4$ 3.15$_2$ 3.08$_6$ 2.85$_2$ 2.81$_3$ 1.81$_3$. Powdery masses and crusts. Yellow to brown, pale yellow streak, earthy luster. Soft. G = 2.30. Biaxial (−); N$_x$ = 1.531, N$_y$ = 1.539, N$_z$ = 1.546; 2V = 86°. Occurs in limestone caves. La Guangola, near Altamura, Italy; *Cioclovina, Transylvania, Romania*; Onino-Iwaya, Hiroshima Pref., Japan; Moorba, Jurien Bay, WA, Australia. BM *DII:1010*, AM *63:520(1978)*, BM *105:621(1982)*, NJMM:*461(1984)*.

BRADLEYITE GROUP

The bradleyite group minerals are carbonate–phosphate minerals of the general formula

$$Na_3M^{2+}(PO_4)(CO_3)$$

where

M = Mg, Mn, Fe^{2+}, Sr

and where the structure is in a monoclinic space group.

BRADLEYITE GROUP

Mineral	Formula	Space Group	a	b	c	β	
Bradleyite	Na$_3$Mg(PO$_4$)(CO$_3$)	$P2_1/m$	8.85	6.63	5.16	90.42	43.2.1.1
Sidorenkite	Na$_3$Mn(PO$_4$)(CO$_3$)	$P2_1/m$	8.979	6.729	5.150	90.10	43.2.1.2
Bonshtedtite	Na$_3$Fe^{2+}(PO$_4$)(CO$_3$)	$P2_1/m$	8.921	6.631	5.151	90.42	43.2.1.3
Crawfordite	Na$_3$Sr(PO$_4$)(CO$_3$)	$P2_1$	9.187	6.707	5.279	89.98	43.2.1.4

43.2.1.1 Bradleyite Na$_3$Mg(PO$_4$)(CO$_3$)
Named in 1941 for Wilmot H. Bradley (1899–1979), U.S. geologist, who investigated the Wyoming trona deposits. Bradleyite group. MON $P2_1/m$. $a = 8.85$,

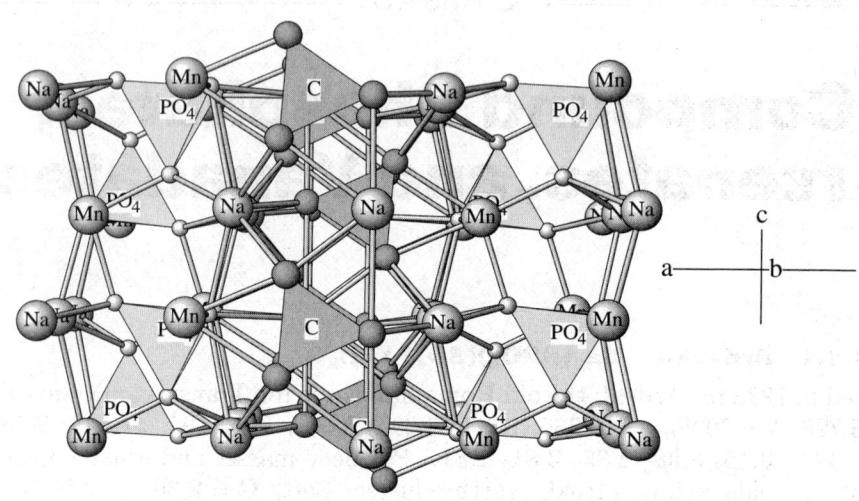

SIDORENKITE: $Na_3Mn(PO_4)(CO_3)$ (Bradleyite group)

This structure can be considered as alternating layers of Na carbonate and Mn phosphate on (100), although Mn and Na are each bonded to both carbonate and phosphate groups.

$b = 6.63$, $c = 5.16$, $\beta = 90.42°$, $Z = 2$, $D = 2.72$. *22-478:* 8.85_5 4.47_2 3.69_3 3.31_7 2.66_{10} 2.58_4 1.84_3 1.66_3. Colorless white to gray masses. $H = 3\frac{1}{2}$. $G = 2.72$. Biaxial $(-)$; $X = b$; $Y \wedge c = 7°$; $N_x = 1.487$, $N_y = 1.546$, $N_z = 1.560$; $2V = 49°$; $r < v$. *Westvaco mine, Sweetwater Co., WY*, in trona deposit. BM *DII:295*.

43.2.1.2 Sidorenkite $Na_3Mn(PO_4)(CO_3)$

Named in 1979 for Alexander V. Sidorenko (1917–1982), Russian mineralogist. Bradleyite group. MON $P2_1/m$. $a = 8.979$, $b = 6.729$, $c = 5.150$, $\beta = 90.10°$, $Z = 2$, $D = 2.98$. *33-1266:* 8.97_2 3.36_{10} 2.99_1 2.69_2 2.58_1 1.87_1 1.68_2 1.50_1. Prismatic crystals to 2 cm and irregular grains. Pale rose, vitreous to pearly luster. Cleavage {100}, {010}, perfect; brittle. $H = 2$. $G = 2.90$. Biaxial $(-)$; $N_x = 1.521$, $N_y = 1.563$, $N_z = 1.585$; $2V = 68°$. *Mt. Alluaiv, Lovozero massif, Kola Penin., Russia*, in syenite pegmatite. BM *ZVMO 108:56(1979)*, *AM 64:1332(1979)*.

43.2.1.3 Bonshtedtite $Na_3Fe(PO_4)(CO_3)$

Named in 1982 for Elza M. Bonshtedt-Kupletskaya (1897–1974), Russian mineralogist. Bradleyite group. MON $P2_1/m$. $a = 8.921$, $b = 6.631$, $c = 5.151$, $\beta = 90.42°$, $Z = 2$, $D = 3.05$. *35-678:* 8.92_2, 4.47_1 3.32_{10} 2.66_3 2.58_2 2.15_2 1.85_2 1.66_2. Crystals up to 5 mm and fine-grained aggregates. Colorless with rose, yellow, or green tint; vitreous to pearly luster. Cleavage {100}, {010}, perfect;

brittle. H = 4. G = 2.95. Biaxial (−); XYZ = bca; N_x = 1.520–1.525, N_y = 1.568–1.570, N_z = 1.591–1.597; 2V = 68°. *Khibina and Kovdor massifs, Kola Penin., Russia.* BM *ZVMO 111:486(1982), AM 68:1038(1983)*.

43.2.1.4 Crawfordite $Na_3Sr(PO_4)(CO_3)$

Named in 1994 for Adair Crawford (1748–1795), Scottish chemist, who showed in 1790 that strontianite contained a new element. Bradleyite group. MON $P2_1$. a = 9.187, b = 6.707, c = 5.279, β = 89.98°, Z = 2, D = 3.08. *ZVMO:123:* 3.35_5 2.71_{10} 2.65_9 2.17_{10} 1.89_8 1.42_7 1.13_6 1.11_6. Colorless grains to 1 mm, vitreous luster. Conchoidal fracture. H = 3. G = 3.05. Biaxial (−); N_x = 1.520, N_y = 1.564, N_z = 1.565; 2V = 20°. Fluoresces bright yellow-green in UV. Occurs in pegmatites in *Khibiny massif, Kola Penin., Russia.* BM *ZVMO 123(3):107(1994), AM 80:1328(1995)*.

43.2.2.1 Gartrellite $Pb(Cu,Fe)_2(AsO_4,SO_4)_2(CO_3,H_2O)_{0.7}$

Named in 1989 for Blair Gartrell (1950–1994), Australian collector, who discovered the mineral. TRIC Probably $P1$. a = 5.454, b = 7.664, c = 5.685, α = 98.0°, β = 110.0°, γ = 111.1°, Z = 1, D = 5.38. *AM 75:* 4.61_7 3.34_7 3.20_{10} 2.96_{10} 2.92_7 2.52_6. Fine-grained coatings and cavity fillings. Yellow to greenish yellow, yellow streak, chalky luster. Soft and friable. Biaxial; N = 1.94–2.00; strongly birefringent. Centennial Eureka mine, Juab Co., UT; *Ashburton Downs, WA*, and Kintore pit, Broken Hill, NSW, *Australia*. BM *AM 75:932(1990)*.

43.3.1.1 Schoderite $Al_2(PO_4)(VO_4) \cdot 8H_2O$

Named in 1962 for William P. Schoder (1900–1977), U.S. chemist. MON $P2_1$ or $P2_1/m$. a = 16.26, b = 30.59, c = 12.56, β = 91.83°, Z = 18, D = 1.95. *34-1263:* 16.3_{10} 15.3_7 11.2_2 7.64_4 5.69_3 5.41_3 2.89_4 2.84_3. Microscopic platy crystals. Yellow to yellow-orange. H = 2. G = 1.92. Biaxial (−); X = b; Y ∧ c = 26°; N_x = 1.560, N_y = 1.563, N_z = 1.565; 2V = 42°; pleochroic. X, pale yellow; Y, deep yellow; Z, yellow. *Fish Creek Range, near Eureka, NV;* Wilson Springs vanadium deposit, Garland Co., AR. BM *AM 47:637(1962), 64:713(1979)*.

43.3.1.2 Metaschoderite $Al_2(PO_4)(VO_4) \cdot 6H_2O$

Named in 1962 for its relationship to schoderite. MON $P2/m$. a = 11.4, b = 14.9, c = 9.2, β = 101.0°, Z = 4, D = 1.61. *14-217:* 14.9_6 11.1_4 9.6_2 8.6_1 7.5_{10} 5.68_1 4.92_1 3.02_2. Microscopic platy crystals. Yellow-orange. H = 2. G = 1.88. Biaxial (+); Z = c; Y ∧ b = 20°; N_x = 1.598, N_y = 1.604, N_z = 1.626; 2V = 56°. *Fish Creek Range, near Eureka, NV.* BM *AM 47:637(1962), 64:713(1979)*.

43.3.2.1 Embreyite $Pb_5(PO_4)_2(CrO_4)_2 \cdot H_2O$

Named in 1972 for Peter G. Embrey (b.1929), English mineralogist. MON $P2_1/m$. a = 9.755, b = 8.018, c = 7.135, β = 103.08°, Z = 1, D = 6.41. *25-436:*

4.75_6 3.56_3 3.48_3 3.17_{10} 2.82_6 2.21_3 2.19_3 1.92_5. Microscopic tabular crystals. Dull orange, yellow streak, resinous or dull luster. No cleavage, irregular fracture, brittle. H = $3\frac{1}{2}$. G = 6.45. Biaxial (−); Y = b; N_x = 2.20, N_y, N_z = 2.36; 2V = 0–11°; pleochroic. X, honey-yellow; Y,Z, amber. *Berezov, Ural Mts., Russia*, Argent mine, Bushveld Complex, Transvaal, South Africa. BM *MM 38:790(1972)*.

43.3.2.2 Cassedanneite $Pb_5(VO_4)_2(CrO_4)_2 \cdot H_2O$

Named in 1988 for Jacques P. Cassedanne (b.1928), Brazilian mineralogist. MON $A2/m$ or $A2$. a = 7.693, b = 5.763, c = 9.795, β = 115.93°, Z = 1, D = 6.45. *42-1370:* 4.83_6 3.45_2 3.22_6 3.15_{10} 2.87_5 2.83_3 2.12_3 1.91_3. Small platy crystals. Red-orange, resinous luster. H = $3\frac{1}{2}$. *Beresov, Ural Mts., Russia*, in association with embreyite. BM *CR 306:125(1988)*, *AM 73:1493(1988)*.

BEUDANTITE GROUP

The beudantite minerals are isostructural sulfate-arsenates and sulfate-phosphates with the general formula

$$AB_3(XO_4)_2(OH)_6$$

where

A = Ba, Ca, Pb, Sr
B = Al, Fe
X = S, As, or P

and the space group $R\bar{3}m$.

The minerals of this group are isostructural not only within themselves but also with the crandallite and the alunite groups. See the alunite group for a description of the structure.

BEUDANTITE GROUP

Mineral	Formula	Space Group	a	c	D	
Beudantite	$PbFe_3(AsO_4)(SO_4)(OH)_6$	$R\bar{3}m$	7.320	17.020	4.49	43.4.1.1
Corkite	$PbFe_3(PO_4)(SO_4)(OH)_6$	$R\bar{3}m$	7.220	16.660	4.42	43.4.1.2
Hidalgoite	$PbAl_3(AsO_4)(SO_4)(OH)_6$	$R\bar{3}m$	7.071	16.975	4.24	43.4.1.3
Orpheite	$Pb_3Al_9(PO_4)_4(SO_4)_2(OH)_{18}$	$R\bar{3}m$	7.016	16.730	4.02	43.4.1.4
Hinsdalite	$(Pb,Sr)Al_3(PO_4)(SO_4)(OH)_6$	$R\bar{3}m$	6.990	16.800	4.07	43.4.1.5
Svanbergite	$SrAl_3(PO_4)(SO_4)(OH)_6$	$R\bar{3}m$	6.975	16.597	3.29	43.4.1.6
Kemmlitzite	$SrAl_3(AsO_4)(SO_4)(OH)_6$	$R\bar{3}m$	7.027	16.510	3.57	43.4.1.7
Woodhouseite	$CaAl_3(PO_4)(SO_4)(OH)_6$	$R\bar{3}m$	6.979	16.214	3.02	43.4.1.8
Weilerite	$BaAl_3(AsO_4)(SO_4)(OH)_6$	$R\bar{3}m$	7.110	17.380	3.93	43.4.1.9

43.4.1.1 Beudantite $PbFe_3(AsO_4)(SO_4)(OH)_6$

Named in 1826 for Francois S. Beudant (1787–1850), French mineralogist. Beudantite group. HEX-R $R\bar{3}m$. $a = 7.320$, $c = 17.020$, $Z = 3$, $D = 4.49$. *19-689*; 5.99_8 3.67_7 3.08_{10} 2.84_5 2.54_5 2.27_6 1.98_6 1.83_6. Rhombohedral crystals, often pseudocubic. Dark green, brown, or black; green to gray streak; vitreous to resinous luster. Cleavage {0001}, good. H = $3\frac{1}{2}$–$4\frac{1}{2}$. G = 4.48. Uniaxial (−), sometimes anomalously biaxial; $N_O = 1.957$, $N_E = 1.943$. Bisbee, Cochise Co., Tiger, Pinal Co., and Pine Grove district, Yavapai Co., AZ; Wheal Gorland, St. Day, Cornwall, England; Mine de la Garonne, Var, France; *Horhausen, Rheinland-Pfalz*, and Dernbach, Nassau, Germany; Kamareza mine, Laurion, Greece; Kipushi, Shaba, Zaire; Tsumeb, Namibia. BM *DII:1001*, *CM 26:923(1988)*, *NJMM:27(1989)*.

43.4.1.2 Corkite $PbFe_3(PO_4)(SO_4)(OH)_6$

Named in 1862 for the locality. Beudantite group. HEX-R $R\bar{3}m$. $a = 7.220$, $c = 16.660$, $Z = 3$, $D = 4.42$. *17-471*: 5.86_7 3.61_4 3.03_{10} 2.79_4 2.51_3 2.24_6 1.96_3 1.48_4. Small rhombohedral crystals, usually pseudocubic, and massive. Yellow to dark green, yellow to green streak, vitreous to resinous luster. Cleavage {0001}, perfect. H = $3\frac{1}{2}$–$4\frac{1}{2}$. G = 4.30. Uniaxial (−); $N_O = 1.93$; weakly birefringent. Silver Queen mine, Galena, SD; Black Pine mine, Philipsburg, MT; Frisco, Beaver Co., UT; San Felix mine, Caborca, SON, Mexico; *Glendore iron mine, Cork, Eire*; Dernbach, Nassau, Germany; St. Antoniodi Gennemari, Sardinia, Italy; Mt. Isa, Q, Australia; Tsumeb, Namibia. BM *DII:1002*, *AM 72:178(1987)*, *NJMM:71(1987)*.

43.4.1.3 Hidalgoite $PbAl_3(AsO_4)(SO_4)(OH)_6$

Named in 1953 for the locality. Beudantite group. HEX-R $R\bar{3}m$. $a = 7.071$, $c = 16.975$, $Z = 3$, $D = 4.24$. *39-1359*: 5.76_4 3.54_5 3.00_{10} 2.48_2 2.26_4 2.23_2 1.92_4 1.77_3. Massive. White, white streak, dull luster. H = $4\frac{1}{2}$. G = 3.96. Uniaxial (+); $N_O = 1.730$, $N_E = 1.735$. Gold Hill, Tooele Co., UT; *San Pasquale mine, Zimapan mining district, HGO, Mexico*; Cap Garonne, Var, France; Kintore cut, Broken Hill, NSW, Australia; Tsumeb, Namibia. BM *AM 38:1218(1953)*, *MR 2:212(1971)*.

43.4.1.4 Orpheite $Pb_3Al_9(PO_4)_4(SO_4)_2(OH)_{18}$

Named in 1971 for Orpheus, the mythical singer of the Rhodope Mts. Beudantite group. Probably synonymous with hinsdalite. HEX-R $R\bar{3}m$. $a = 7.016$, $c = 16.730$, $Z = 1$, $D = 4.02$. *29-756*: 5.66_{10} 4.91_2 3.50_7 3.45_2 2.97_{10} 2.21_2 1.90_4 1.75_3. Colorless, yellow-green, pale blue, or gray; vitreous luster. Cleavage {0001}, indistinct. H = $3\frac{1}{2}$. G = 3.75. Uniaxial (−); $N_O = 1.682$–1.704, $N_E = 1.670$–1.691. *Madjarova deposit, Rhodope Mts., Bulgaria*. BM *AM 61:176(1976)*.

43.4.1.5 Hinsdalite $(Pb,Sr)Al_3(PO_4)(SO_4)(OH)_6$

Named in 1911 for the locality. Beudantite group. HEX-R $R\bar{3}m$. $a = 6.990$, $c = 16.800$, $Z = 3$, $D = 4.07$. *16-711:* 5.7_5 5.6_7 3.50_4 3.45_2 2.96_8 2.78_{10} 2.22_5 1.90_3. Pseudocubic crystals, also massive, granular. Colorless, vitreous to greasy luster. Cleavage {0001}, perfect. $H = 4\frac{1}{2}$. $G = 3.65$. Uniaxial (+); $N_O = 1.671$, $N_E = 1.689$; also anomalously biaxial. *Golden Fleece mine, Hinsdale Co., CO;* Mineral Park mine, Mohave Co., AZ; Sylvester mine, Zeehan, and Comet mine, Dundas, TAS, Australia. BM *DII:1004; MR 8:391(1977), 11:243(1980).*

43.4.1.6 Svanbergite $SrAl_3(PO_4)(SO_4)(OH)_6$

Named in 1854 for Lars F. Svanberg (1805–1878), Swedish chemist. Beudantite group. HEX-R $R\bar{3}m$. $a = 6.975$, $c = 16.597$, $Z = 3$, $D = 3.29$. *39-1361:* 5.68_3, 3.49_2 2.95_{10} 2.77_3 2.21_9 1.89_5 1.74_3 1.45_4. Rhombohedral, often pseudocubic crystals, also massive. Colorless to yellow, rose, or red-brown; vitreous to adamantine luster. Cleavage {0001}, distinct. $H = 5$. $G = 3.22$. Uniaxial (+); $N_O = 1.631$–1.635, $N_E = 1.646$–1.649. Dover mine, near Hawthorne, Mineral Co., NV; Champion mine, White Mts., Mono Co., and Blue Bird Hill, Imperial Co., CA; Chizeuil, near Chalmoux, Saône et Loire, France; *Horrsjöberg, Värmland,* and Westanå, Skåne, *Sweden;* Rio São Jose, BA, Brazil. BM *DII:1005; MR 2:225(1971), 8:480(1977).*

43.4.1.7 Kemmlitzite $SrAl_3(AsO_4)(SO_4)(OH)_6$

Named in 1969 for the locality. Beudantite group. HEX-R $R\bar{3}m$. $a = 7.027$, $c = 16.510$, $Z = 3$, $D = 3.57$. *22-1248:* 5.71_7 3.51_9 2.96_{10} 2.75_5 2.20_8 1.90_9 1.76_8 1.29_6. Microscopic pseudocubic crystals. Colorless to pale brown. Cleavage {0001}, indistinct. $H = 5\frac{1}{2}$. $G = 3.63$. Uniaxial (+); $N_O = 1.701$, $N_E = 1.707$. *Kemmlitz, near Oschatz, Saxony, Germany,* in kaolin deposit. BM *NJMM:201(1969), AM 55:320(1970).*

43.4.1.8 Woodhouseite $CaAl_3(PO_4)(SO_4)(OH)_6$

Named in 1937 for Charles D. Woodhouse (1888–1975), U.S. mineral collector. Beudantite group. HEX-R $R\bar{3}m$. $a = 6.979$, $c = 16.214$, $Z = 3$, $D = 3.02$, *37-469:* 5.66_2 4.85_4 3.49_3 2.97_4 2.93_{10} 2.70_3 2.16_5 1.89_2. Pseudocubic and tabular crystals. Colorless to pale orange, vitreous to pearly luster. Cleavage {0001}, good. $H = 4\frac{1}{2}$. $G = 3.01$. Uniaxial (+); $N_O = 1.636$, $N_E = 1.647$. *Champion mine, White Mts., Mono Co., CA,* in vugs in quartz veins; Richelle, Belgium; Iron Knob, SA, Australia. BM *DII:1006; MR 8:477(1977), 12:106(1981).*

43.4.1.9 Weilerite $BaAl_3(AsO_4)(SO_4)(OH)_6$

Named in 1961 for the locality. Beudantite group. HEX-R $R\bar{3}m$. $a = 7.110$, $c = 17.380$, $Z = 3$, $D = 3.93$. *35-648:* 5.84_8 3.55_8 3.02_{10} 2.90_4 2.30_6 1.93_6 1.77_5 1.52_5. Spherulitic aggregates and earthy crusts. White to yellow or green. Uniaxial (+); $N_O = 1.688$, $N_E = 1.698$. *Weiler* and Neubulach, *Black*

Forest, Germany; Magnet mine, Dundas, TAS, Australia. BM *AM 47:415(1962), 52:1588(1967); SMPM 61:23(1981)*.

43.4.2.1 Tsumebite $Pb_2Cu(PO_4)(SO_4)(OH)$

Named in 1912 for the locality. Brackebuschite group. See (40.2.8.1). MON $P2_1/m$. $a = 8.69$, $b = 5.78$, $c = 7.86$, $\beta = 111.87°$, $Z = 2$, $D = 6.22$. *29-568:* 4.70_3 3.92_2 3.24_{10} 2.94_4 2.90_7 2.72_4 2.62_3 2.27_4. Crusts of intergrown crystals. Emerald-green, vitreous luster. Uneven fracture, brittle. H = $3\frac{1}{2}$. G = 6.13. Biaxial (+); $N_x = 1.885$, $N_y = 1.920$, $N_z = 1.956$; 2V near 90°; r < v, strong; faintly pleochroic. Morenci, Greenlee Co., and Mammoth mine, Pinal Co., AZ; Carrock Fell, Cumberland, England; Kipushi, Shaba, Zaire; *Tsumeb, Namibia*. BM *DII:918, AM 51:258(1966)*.

43.4.2.2 Arsentsumebite $Pb_2Cu(AsO_4)(SO_4)(OH)$

Named in 1935 for its relationship to tsumebite. Brackebuschite group. See (40.2.8.1). MON $P2_1/m$. $a = 8.85$, $b = 5.92$, $c = 7.84$, $\beta = 112.60°$, $Z = 2$, $D = 6.39$. *25-456:* 4.80_7 3.64_5 3.25_{10} 3.01_3 2.96_4 2.76_6 2.27_3 2.09_3. Crystalline crusts. Emerald-green, vitreous to dull luster. G = 6.46. Biaxial (−); $N_x = 1.970$, $N_y = 1.992$, $N_z = 2.011$; 2V = 88°; pleochroic. X, pale pistachio-green; Y,Z, bottle-green. *Tsumeb, Namibia*. BM *DII:918, AM 51:258(1966)*.

43.4.3.1 Vauquelinite $Pb_2Cu(PO_4)(CrO_4)(OH)$

Named in 1818 for Louis N. Vauquelin (1763–1829), French chemist and discoverer of chromium. Brackebushite group. See (40.2.8.1). MON $P2_1/n$. $a = 13.68$, $b = 5.83$, $c = 9.53$, $\beta = 93.97°$, $Z = 4$, $D = 6.18$. *13-302:* 4.73_7 4.00_3 3.31_{10} 2.96_3 2.89_6 2.77_3 2.31_5 1.90_5. Small wedge-shaped crystals, also granular and compact. Green to brown color and streak, resinous to adamantine luster. Uneven fracture, brittle. H = $2\frac{1}{2}$–3. G = 6.02. Biaxial (−); $N_x = 2.11$, $N_y = 2.22$, $N_z = 2.22$; 2V near zero; pleochroic. X, pale green; Y,Z, pale brown. Near Benson, Cochise Co., near Wickenberg, Maricopa Co., Mammoth mine, Pinal Co., AZ; Killie mine, Spruce Mt., Elko Co., NV; Leadhills and Wanlockhead, Scotland; Pontgibaud, Puy-de-Dôme, France; *Beresov, Ural Mts., Russia*; Dundas, TAS, Australia; Congonhas do Campo, MG, Brazil. BM *DII:650, ZK 126:443(1968), BM 103:469(1980)*.

43.4.3.2 Fornacite $Pb_2Cu(AsO_4)(CrO_4)(OH)$

Named in 1915 for Lucien L. Forneau (1867–1930), French colonial governor. Brackebushite group. See (40.2.8.1). MON $P2_1/c$. $a = 8.022$, $b = 5.906$, $c = 17.564$, $\beta = 110.4°$, $Z = 4$, $D = 6.39$. *15-200:* 8.22_5 4.80_9 4.00_5 3.31_{10} 2.98_{10} 2.88_{10} 2.80_{10} 2.71_9. Microscopic prismatic crystals. Olive-green, yellow streak. G = 6.27. Biaxial (+); $N_x = 2.14$, $N_z = 2.24$. Bisbee, Cochise Co., Tonopah–Belmont mine, Maricopa Co., Old Yuma mine, Pima Co., Mammoth mine, Pinal Co., and New Water Mts., Yuma Co., AZ; Cnah Khuni and Seh Cnangi mines, Iran; Smolianinova, Kirghizstan; *Djoue, Congo*. BM *DII:652, BM 85:309(1962), ZK 124:385(1967), MR 11:296(1980)*.

43.4.3.3 Molybdofornacite $Pb_2Cu(AsO_4)(MoO_4)(OH)$

Named in 1983 for its relationship to fornacite. Brackebushite group. See (40.2.8.1). MON $P2_1/c$. $a = 8.100$, $b = 6.946$, $c = 17.65$, $\beta = 109.17°$, $Z = 4$, $D = 6.57$. *35-693:* 4.84_5 3.41_4 3.32_{10} 2.98_6 2.94_4 2.85_6 2.76_6 2.36_5. Minute lath-like crystals. Pale green, yellow streak, adamantine luster. No cleavage, conchoidal fracture. H = 2–3. Biaxial; $N_x = 2.05$, $N_z = 2.15$; pleochroic. X,Y, pale yellow; Z, yellow-green. Tres Hermanas Mts., Luna Co., NM; Tsumeb, Namibia. BM *NJMM:289(1983), AM 69:567(1984).*

43.4.4.1 Cahnite $Ca_2[B(OH)_4](AsO_4)$

Named in 1927 for Lazard Cahn (1865–1940), U.S. mineral collector and dealer. TET $I\bar{4}$. $a = 7.110$, $c = 6.200$, $Z = 2$, $D = 3.16$. *13-158:* 5.02_2 4.67_2 3.56_{10} 2.83_2 2.64_5 2.51_2 2.34_2 1.82_6. Pseudotetrahedral crystals and cruciform twins. Colorless or white, vitreous luster. Cleavage {110}, perfect; brittle. H = 3. G = 3.156. Uniaxial (+); $N_O = 1.662$, $N_E = 1.663$; strong dispersion; shows anomalous interference colors. Franklin, NJ; Klodeborg mine, Arendal, Norway; Capo di Bove, near Rome, Italy; Solongo, Transbaikalia, Russia. BM *DII:386, MM 32:666(1960), AM 46:1077(1961).*

43.4.5.1 Seamanite $Mn_3[B(OH)_4](PO_4)(OH)_2$

Named in 1930 for Arthur E. Seaman (1858–1937), US mineralogist. ORTH *Pbnm*. $a = 7.81$, $b = 15.11$, $c = 6.69$, $Z = 4$, $D = 3.13$. *25-536:* 7.55_6 6.92_{10} 4.20_6 3.78_7 3.47_6 2.84_8 2.50_7 2.39_6. Acicular crystals. Pale yellow to wine-yellow. Cleavage {001}, distinct; brittle. H = 4. G = 3.08. Biaxial (+); $N_x = 1.640$, $N_y = 1.663$, $N_z = 1.665$; 2V = 40°; r < v. Chicagon mine, Iron Co., MI. BM *DII:388, AM 56:1527(1971).*

43.4.6.1 Hematolite $(Mn,Mg,Al)_{15}(AsO_3)(AsO_4)_2(OH)_{23}$

Named in 1884 from the Greek for *blood*, in reference to the color. HEX-R $R3$. $a = 8.275$, $c = 36.60$, $Z = 3$, $D = 3.50$. *33-868:* 12.2_4 5.64_3 4.22_2 4.14_{10} 3.42_2 2.96_2 2.41_2 2.39_2. Flattened rhombohedral crystals. Blood-red, red-brown streak, vitreous to submetallic luster. Cleavage {0001}, perfect; uneven fracture; brittle. H = $3\frac{1}{2}$. G = 3.49. Uniaxial (−); $N_O = 1.733$, $N_E = 1.714$. Moss mine and Långban, Värmland, Sweden. BM *DII:777, AM 63:150(1978).*

43.4.7.1 Holdenite $(Mn,Mg)_6Zn_3(AsO_4)_2(SiO_4)(OH)_8$

Named in 1927 for Albert F. Holden (1866–1913), U.S. mining engineer and mineral collector. ORTH *Abma*. $a = 11.99$, $b = 31.46$, $c = 8.697$, $Z = 8$, $D = 4.18$. *29-903:* 5.74_5 3.58_8 3.41_6 2.84_{10} 2.79_5 2.58_8 2.46_7 1.53_8. Tabular crystals and fibrous veinlets. Pink to deep red, vitreous luster. Cleavage {010}, poor; subconchoidal fracture. H = 4. G = 4.11. Biaxial (+); XYZ = *cba*; $N_x = 1.769$, $N_y = 1.770$, $N_z = 1.785$; 2V = 30°; r > v. Franklin and Sterling Hill, NJ. BM *DII:775, AM 62:513(1977), MR 12:373(1981).*

43.4.8.1 Kolicite $Mn_7Zn_4(AsO_4)_2(SiO_4)_2(OH)_8$

Named in 1979 for John Kolic (b.1943), Franklin miner, who discovered the mineral. ORTH *Cmca*. $a = 18.59$, $b = 8.789$, $c = 12.04$, $Z = 4$, $D = 4.20$. *33-918:* 4.12_1 3.58_6 2.97_{10} 2.82_4 2.61_5 2.48_4 2.34_4 1.54_7. Platy aggregates. Orange, pale orange streak, vitreous luster. No cleavage, brittle. $H = 4\frac{1}{2}$. $G = 4.17$. Biaxial $(-)$; $XYZ = bca$; $N_x = 1.779$, $N_y = 1.786$, $N_z = 1.790$; $2V = 78°$; $r < v$, strong; strongly pleochroic. X, colorless or pale yellow; Y, yellow-orange; Z, pale yellow. Franklin and *Sterling Hill, NJ*. BM *AM 64:708(1979), 65:483(1980)*.

43.4.9.1 McGovernite $Mn_9Mg_4Zn_2As_2Si_2O_{17}(OH)_{14}$

Named in 1927 for J. J. McGovern (d.1915), Franklin miner and mineral collector. HEX-R $R\bar{3}c$. $a = 8.22$, $c = 205.5$, $Z = 18$, $D = 3.65$. *25-531:* 4.06_3 3.43_3 3.24_3 2.80_5 2.66_6 2.37_5 2.35_6 1.54_{10}. Micaceous veinlets. Red-brown, pale brown streak, vitreous to pearly luster. Cleavage {0001}, perfect. $H = 3$. $G = 3.72$. Uniaxial $(+)$; $N_O = 1.757$; very weakly birefringent. *Sterling Hill, NJ*. BM *AM 12:373(1927), 45:937(1960), 65:957(1980), 73:1182(1988)*.

43.4.10.1 Kraisslite $Mn_{44}Zn_6Mg_4Fe_2(AsO_3)_4(AsO_4)_6(SiO_4)_{12}(OH)_{36}$

Named in 1978 for Frederick Kraissl (1899–1986) and Alice Kraissl (1905–1986), U.S. mineral collectors. HEX $P6_322$. $a = 8.22$, $c = 43.88$, $Z = 1$, $D = 3.92$. *29-1432:* 10.9_1 4.39_5 3.65_4 3.13_2 2.74_{10} 2.44_6 2.19_6 1.22_1. Micaceous veinlets and fracture fillings. Deep copper-brown, golden brown streak, submetallic luster. Cleavage {0001}, perfect. $H = 3–4$. $G = 3.876$. Uniaxial $(+)$; $N_O = 1.805$; weakly birefringent. *Sterling Hill, NJ*. BM *AM 63:938(1978), 65:957(1980)*.

43.4.11.1 Heneuite $Mg_5Ca(CO_3)(PO_4)_3(OH)$

Named in 1986 for Henrich Neumann (1914–1983), Norwegian mineralogist. TRIC $P\bar{1}$. $a = 6.307$, $b = 10.839$, $c = 8.674$, $\alpha = 95.01°$, $\beta = 93.41°$, $\gamma = 101.04°$, $Z = 2$, $D = 3.01$. *39-362:* 3.67_4 3.19_3 2.88_9 2.85_4 2.84_4 2.79_8 2.70_{10} 2.60_4. Nodular masses. Pale blue-green, white streak, vitreous luster. Cleavage {001}, good. $H = 5\frac{1}{2}$. $G = 3.016$. Biaxial $(-)$; $N_x = 1.586$, $N_y = 1.620$, $N_z = 1.630$; $2V = 50°$. *Tingelstadtjern quarry, Modum, Norway*, in a serpentine–magnesite deposit. BM *NJMM:343(1986), AM 73:440(1988)*.

43.4.12.1 Cervandonite-(Ce) $(Ce,Nd,La)(Fe,Ti,Al)_3(Si,As)_3O_{13}$

Named in 1988 for the locality. MON *Cm* (?). $a = 11.3$, $b = 19.5$, $c = 7.2$, $\beta = 121°$, $Z = 6$, $D = 4.9$. *SMPM 68:* 5.39_8 3.58_4 3.25_9 3.08_8 2.88_{10} 2.79_6 2.70_5 1.63_4. Rosettelike aggregates. Black, brown-black streak, adamantine luster. Cleavage {001}, poor; conchoidal fracture; brittle. $H = 5$. Biaxial; $N \sim 2.0$; pleochroic, yellow to brown. *Pizzo Cervandone, Alpe Devero, Italy*, in fissure veins in gneiss. BM *SMPM 68:125(1988)*.

43.4.13.1 Attakolite (Ca,Sr)Mn(Al,Fe)$_4$(PO$_4$)$_3$(HSiO$_4$)(OH)$_4$

Named in 1868 from the Greek for *salmon*, in allusion to the color. MON $C2/m$. $a = 17.188$, $b = 11.477$, $c = 7.322$, $\beta = 113.83°$, $Z = 4$, $D = 3.23$. *18-146:* 6.57$_2$ 4.80$_1$ 4.33$_3$ 3.12$_3$ 3.08$_{10}$ 2.91$_2$ 2.87$_2$ 2.40$_4$. Massive, indistinctly crystalline. White to pale red, white streak. G = 3.09–3.23. Biaxial (−); Y = b; X ∧ $a = 24°$; N$_x$ = 1.650, N$_y$ = 1.654, N$_z$ = 1.661; 2V = 75°; r > v. Found with berlinite and lazulite in iron ore at *Västanå, Nästum, Kristianstad, Sweden.* BM *DII:845, AMG 3:537(1965), CM 8:668(1966), AM 77:1285(1992).*

43.5.1.1 Sarmientite Fe$_2$(AsO$_4$)(SO$_4$)(OH) · 5H$_2$O

Named in 1941 for Domingo F. Sarmiento (1811–1888), Argentinian statesman. MON $P2_1/c$. $a = 6.55$, $b = 18.55$, $c = 9.70$, $\beta = 97.65°$, $Z = 4$, $D = 2.58$. *22-342:* 9.29$_{10}$ 6.13$_3$ 4.87$_4$ 4.64$_9$ 4.26$_8$ 3.43$_4$ 3.06$_7$ 2.60$_5$. Microscopic prismatic crystals and nodular masses. Pale yellow-orange, dull luster. G = 2.58. Biaxial (+); XY = ab; Z ∧ $c = 12°$; N$_x$ = 1.628, N$_y$ = 1.635, N$_z$ = 1.698; 2V = 38°. Winnemucca, Humboldt Co., NV; *Santa Elena mine, Barreal Dept., Argentina.* BM *DII:1013, AM 53:2077(1968).*

43.5.1.2 Bukovskyite Fe$_2$(AsO$_4$)(SO$_4$)(OH) · 7H$_2$O

Named in 1967 for Antonin Bukovsky (1865–1950), Czech mineralogist. TRIC $P1$ or $P\bar{1}$. $a = 10.722$, $b = 14.079$, $c = 10.284$, $\alpha = 93.50°$, $\beta = 115.96°$, $\gamma = 90.27°$, $Z = 4$, $D = 2.34$. Microcrystalline nodular masses. Pale yellow-green to gray-green, dull luster. G = 2.33. Biaxial (+); N$_y$ = 1.570–1.582, N$_z$ = 1.626–1.631. *Kank, near Kutna Hora, Czech Republic,* an alteration product of arsenopyrite. BM *AM 54:576,991(1969), NJMM:445(1986).*

43.5.1.3 Sanjuanite Al$_2$(PO$_4$)(SO$_4$)(OH) · 9H$_2$O

Named in 1968 for the locality. TRIC $P1$ or $P\bar{1}$. $a = 11.314$, $b = 9.018$, $c = 7.376$, $\alpha = 93.07°$, $\beta = 95.78°$, $\gamma = 105.32°$, $Z = 2$, $D = 1.96$. *20-47:* 10.8$_{10}$ 8.66$_3$ 5.28$_4$ 4.43$_3$ 4.32$_4$ 4.27$_3$ 4.13$_6$ 3.45$_4$. Compact masses. White, white streak, silky to dull luster. H = 3. G = 1.94. Biaxial; N$_x$ = 1.484, N$_z$ = 1.499. *Sierra Chica de Zonda, San Juan Prov., Argentina.* BM *AM 53:1(1968), MM 53:385(1989).*

43.5.2.1 Diadochite Fe$_2$(PO$_4$)(SO$_4$)(OH) · 5H$_2$O

Named in 1837 from the Greek for *successor*, presumably in allusion to its secondary origin. Synonym: destinezite, TRIC $P1$ or $P\bar{1}$. $a = 9.584$, $b = 9.748$, $c = 7.338$, $\alpha = 98.78°$, $\beta = 108.00°$, $\gamma = 63.87°$, $Z = 2$, $D = 2.33$. *42-1364:* 8.77$_9$ 8.27$_8$ 4.38$_{10}$ 4.08$_7$ 3.93$_{10}$ 3.21$_4$ 2.94$_8$ 2.92$_4$. Nodular and colloform crusts and masses. Yellow to yellow-green and brown, earthy or dull luster. H = 3–4. G = 2.2. Biaxial (+); N$_x$ = 1.615, N$_y$ = 1.627, N$_z$ = 1.670; 2V = 55°; r < v, strong. New Idria mine, San Benito Co., CA: Visé, Argenteau, Belgium; *Ansbach, Thuringia, Germany;* Litošice, Chvaletice, Nucic, and other localities in Bohemia, Czech Republic; Leoben, Steiermark, Austria; Eisenbach, Hungary; Blyava, Ural Mts., Russia; Kara-tan, Kazakhstan; La Mure mine, Zin, Pakistan. BM *DII:1011*.

43.5.3.1 Pitticite Fe,AsO$_4$,SO$_4$,H$_2$O

Named in 1813 from Greek for *pitch*, for its resinous appearance. Amorphous. Massive and as opaline and earthy crusts. Red-brown, vitreous or resinous luster. H = 2–3. G = 2.37–2.72. N = 1.634–1.765. Tintic, UT; Manhattan, NV; Redruth, Cornwall, England; *Freiberg, Saxony, Germany*; Jachymov and Pribram, Bohemia, Czech Republic; Djebel Debar, Constantine, Algeria. BM *DII:1014, MM 46:261(1982)*.

43.5.3.2 Zykaite Fe$_4$(AsO$_4$)$_3$(SO$_4$)(OH) · 15H$_2$O

Named in 1978 for Vaclav Zyka (b.1926), Czech geochemist. ORTH $P1$ or $P\bar{1}$. $a = 20.85$, $b = 7.033$, $c = 36.99$, $Z = 8$, $D = 2.51$. *29-695:* 10.6$_7$ 10.4$_{10}$ 6.92$_4$ 5.61$_4$ 3.81$_4$ 3.52$_3$ 3.25$_3$ 2.83$_4$. Nodules and cavity fillings. Gray-white, dull luster. Uneven fracture. H = 2. G = 2.50. Biaxial (−); N$_x$ = 1.632, N$_z$ = 1.646; 2V = 60°. *Kank, near Kutna Hora, Czech Republic*. BM *NJMM:134(1978), AM 63:1284(1978)*.

43.5.4.1 Sasaite (Al,Fe)$_{14}$(PO$_4$)$_{11}$(SO$_4$)(OH)$_7$ · 84H$_2$O

Named in 1978 for the South African Speleological Society. ORTH $P222$ or $Pmm2$ or $Pmmm$. $a = 10.81$, $b = 15.031$, $c = 23.064$, $Z = 1.25$, $D = 1.75$. *31-20:* 11.5$_{10}$ 7.51$_2$ 7.13$_2$ 6.99$_3$ 6.30$_2$ 4.21$_2$ 3.26$_2$ 2.90$_4$. Chalky nodules to 2 cm. White, white streak. G = 1.75. Biaxial (−); XYZ = cba; N$_x$ = 1.465, N$_y$ = 1.473, N$_z$ = 1.477; 2V = 15°. *West Dreifontein cave, near Carlstonville, Transvaal, South Africa*. BM *MM 42:401(1978), CM 21:489(1983)*.

43.5.5.1 Coconinoite Fe$_2$Al$_2$(UO$_2$)$_2$(PO$_4$)$_4$(SO$_4$)(OH)$_2$ · 18H$_2$O

Named in 1966 for the locality. MON $C2/c$ or Cc. $a = 12.50$, $b = 12.97$, $c = 23.00$, $\beta = 106.6°$, $Z = 4$, $D = 2.85$. *25-16:* 12.3$_1$ 11.1$_{10}$ 5.64$_2$ 5.56$_4$ 4.59$_1$ 4.31$_1$ 3.71$_1$ 3.30$_2$. Aggregates of microscopic platy to lathlike grains. Pale cream-yellow. Soft. G = 2.70. Biaxial (−); N$_x$ = 1.550, N$_y$ = 1.588, N$_z$ = 1.590; 2V = 40°; pleochroic. X, colorless; Y,Z, pale yellow. *Sun Valley mine, Coconino Co., and Blackwater No. 4 mine, Apache Co., AZ*; White Canyon area, San Juan Co., UT, and other localities on the Colorado Plateau; Kyzylkumi, Uzbekistan. BM *AM 51:651(1966), DANS 329:792 (1993)*.

43.5.6.1 Xiangjiangite (Fe,Al)(UO$_2$)$_4$(PO$_4$)$_2$(SO$_4$)$_2$(OH) · 22H$_2$O

Named in 1978 for the locality. MON $C2/c$ or Cc. $a = 12.54$, $b = 12.98$, $c = 23.08$, $\beta = 108.6°$, $Z = 4$, $D = 2.87$. *AM 64:* 11.1$_{10}$ 5.58$_5$ 4.62$_6$ 3.74$_8$ 2.94$_7$ 2.18$_5$ 2.07$_5$. Microcrystalline platy aggregates. Bright yellow, pale yellow streak, silky luster. H = 1–2. G = 2.9–3.1. Biaxial (−); N$_x$ = 1.558, N$_y$ = 1.576, N$_z$ = 1.593; 2V = 87°; weakly pleochroic, yellow. *Xiangjiang R., Hunan, China*. BM *AM 64:466(1979), DANS 329:792(1993)*.

43.5.7.1 Kribergite $Al_5(PO_4)_3(SO_4)(OH)_4 \cdot 4H_2O$

Named in 1945 for the locality. TRIC $P1$ or $P\bar{1}$. $a = 18.126$, $b = 13.519$, $c = 7.500$, $\alpha = 70.50°$, $\beta = 117.87°$, $\gamma = 136.23°$, $Z = 2$, $D = 1.94$. *20-48:* 11.6_{10} 6.62_2 5.85_1 5.72_1 5.37_1 5.02_3 2.93_1 2.86_1. Compact veinlet. White, white streak, dull luster. $G = 1.92$. Biaxial; $N_y = 1.484$; birefringence 0.002. *Kristineberg mine, Västerbotten, Sweden.* BM *DII:1011, MM 53:385(1989).*

43.5.8.1 Kovdorskite $Mg_5(PO_4)_2(CO_3)(OH)_2 \cdot 4.5H_2O$

Named in 1980 for the locality. MON $P2_1/c$. $a = 4.74$, $b = 12.90$, $c = 10.35$, $\beta = 102.0°$, $Z = 2$, $D = 2.30$. *33-861:* 7.96_{10} 6.45_3 5.44_5 4.32_4 2.82_6 2.66_3 2.53_3 2.26_7. Pale rose, in part transparent. Conchoidal or uneven fracture. $H = 4$. $G = 2.60$. Biaxial $(-)$; $Z \wedge c = 1–3°$; $N_x = 1.527$, $N_y = 1.542$, $N_z = 1.549$; $2V = 80°$; $r > v$, very weak. *Kovdor massif, Kola Penin., Russia,* in magnetite–forsterite rocks. BM *ZVMO 109:341(1980), AM 66:437(1981).*

43.5.9.1 Viséite $CaAl_3(PO_3OH)(SiO_3OH)(OH)_6$

Named in 1942 for the locality. HEX-R $R\bar{3}m$. $a = 6.89$, $c = 18.065$, $Z = 3$, $D = 2.17$. *5-616:* 5.68_4 3.46_5 2.92_{10} 2.20_2 1.89_3 1.74_6 1.20_2 1.16_2. Botryoidal clusters and massive. Chalky white or blue. $H = 3–4$. $G = 2.2$. Isotropic; $N = 1.530$. *Champion mine, Mono Co., CA; Visé, Belgium.* BM *MA 9:88(1945), MM 41:437(1977).*

43.5.10.1 Perhamite $Ca_3Al_7(SiO_4)_3(PO_4)_4(OH)_3 \cdot 16.5H_2O$

Named in 1977 for Frank C. Perham (b.1934), U.S. geologist and pegmatite miner. HEX $P6/mmm$. $a = 7.022$, $c = 20.182$, $Z = 1$, $D = 2.53$. *29-284:* 6.71_4 6.08_5 5.80_7 3.51_5 3.12_5 3.01_4 2.88_{10} 1.76_5. Spherulites of platy crystals. White to pale brown, white streak, vitreous luster. Cleavage $\{0001\}$, perfect. $H = 5$. $G = 2.64$. Uniaxial $(+)$; $N_O = 1.564$, $N_E = 1.577$; $r > v$, moderate. *Bell Pit, Newry Hill, Newry, ME,* in a pegmatite vug; *Tom's quarry, Kapunda, SA, Australia.* BM *MM 41:437(1977).*

43.5.11.1 Lünebergite $Mg_3B_2(OH)_6(PO_4)_2 \cdot 6H_2O$

Named in 1870 for the locality. TRIC $P\bar{1}$. $a = 6.348$, $b = 9.803$, $c = 6.298$, $\alpha = 84.46°$, $\beta = 106.40°$, $\gamma = 96.40°$, $Z = 1$, $D = 2.204$. *25-1155:* 4.98_{10} 4.85_7 3.23_4 3.03_3 2.96_6 2.82_5 2.51_4 2.01_3. Flattened masses and nodules. White, white streak, vitreous to dull luster. Cleavage $\{010\}$, fair. $H = 2$. $G = 2.07$. Biaxial $(-)$; $N_x = 1.522$, $N_y = 1.541$, $N_z = 1.549$; $2V = 63°$. Occurs in marine evaporites. Permian salt basin, New Mexico and west Texas; *Lüneberg, Hannover, Germany;* Bela Stena, Serbia; Kerch Penin., Ukraine; Mejillones, Tarapaca, Chile. BM *DII:385, AM 76:1400(1991).*

43.5.12.1 Synadelphite $Mn_9(AsO_3)(AsO_4)_2(OH)_9 \cdot 2H_2O$

Named in 1884 from the Greek for *with* and *brother*, referring to its occurrence with several related minerals. ORTH $Pnma$. $a = 10.754$, $b = 18.865$, $c = 9.884$,

$Z = 4$, $D = 3.59$. *24-725:* 9.36_7 8.72_{10} 5.27_5 3.84_4 3.08_5 2.64_9 2.09_7 1.60_7. Short prismatic crystals, reniform crusts, and massive. Red-brown to black, vitreous to adamantine luster. Cleavage {010}, imperfect; conchoidal to uneven fracture; brittle. $H = 4\frac{1}{2}$. $G = 3.57$. Biaxial (+); $N_x = 1.750$, $N_y = 1.751$, $N_z = 1.761$; $2V = 37°$; $r > v$; pleochroic, X,Y, colorless; Z, pale brown. Sterling Hill, NJ; *Moss mine* and Långban, *Värmland, Sweden*. BM *DII:780*, *AM 55:2023(1970)*.

43.5.13.1 Parnauite $Cu_9(AsO_4)_2(SO_4)(OH)_{10} \cdot 7H_2O$

Named in 1978 for John L. Parnau (1906–1990), U.S. mineral collector, who discovered the mineral. ORTH $P2_122$. $a = 14.98$, $b = 14.223$, $c = 6.018$, $Z = 2$, $D = 3.22$. *29-533:* 14.3_{10} 10.4_3 7.14_1 6.42_1 4.52_6 4.00_2 2.85_2 2.80_1. Rosettes of lathlike crystals, and scales and crusts. Pale blue to green, pale green streak, vitreous to dull luster. $H = 2$. $G = 3.09$. Biaxial (−); $XYZ = bac$; $N_x = 1.680$, $N_y = 1.704$, $N_z = 1.712$; $2V = 60°$; pleochroic, X, pale green; Y, yellow-green; Z, blue-green. *Majuba Hill mine, Pershing Co., NV;* Grandview mine, Coconino Co., AZ; La Garonne, Var, France. BM *AM 63:704(1978)*.

43.5.14.1 Chalcophyllite $Cu_{18}Al_2(AsO_4)_3(SO_4)_3(OH)_{27} \cdot 33H_2O$

Named in 1847 from the Greek for *copper* and *leaf*, referring to its composition and micaceous structure. HEX-R $R\bar{3}$. $a = 10.77$, $c = 57.5$, $Z = 3$, $D = 2.69$. *39-1356:* 9.51_{10} 4.76_6 2.73_2 2.70_3 2.59_5 2.34_3 2.05_2 1.53_2. Tabular hexagonal crystals and foliated masses. Blue-green to emerald-green, pale green streak, vitreous to pearly luster. Cleavage {0001}, perfect. $H = 2$. $G = 2.67$. Uniaxial (−); $N_O = 1.618$, $N_E = 1.552$; pleochroic, O, blue-green; E, almost colorless. Widely distributed as a secondary copper mineral. Tintic, UT; Bisbee, AZ; Majuba Hill, NV; Redruth and other localities in Cornwall, England; Cap Garonne, Var, France; Schmiederberg, Germany; Schwaz, Austria; Nizhni Tagilsk, Ural Mts., Russia; Tsumeb, Namibia; Teniente mine, near Rancagua, Chile. BM *DII:1008, ZK 151:129(1980)*.

43.5.15.1 Peisleyite $Na_3Al_{16}(SO_4)_2(PO_4)_{10}(OH)_{17} \cdot 20H_2O$

Named in 1982 for Vincent Peisley (b.1941), Australian mineral collector, who discovered the mineral. MON. $a = 13.310$, $b = 12.620$, $c = 23.15$, $\beta = 110.00°$, $Z = 2$, $D = 2.08$. *35-674:* 12.63_{10} 10.85_2 7.82_4 7.59_3 6.43_2 5.41_4 4.35_2 4.11_2. Compact masses. White, white streak, dull luster. $H = 3$. $G = 2.12$. Biaxial; $N_{av} = 1.510$; weakly birefringent. *Tom's quarry, near Kapunda, SA, Australia.* BM *MM 46:449(1982)*.

43.5.16.1 Voggite $Na_2Zr(PO_4)(CO_3)(OH) \cdot 2H_2O$

Named in 1990 for Adolf Vogg (b.1931), Canadian mineral collector, who discovered the mineral. MON $I2/m$. $a = 12.251$, $b = 6.557$, $c = 11.755$, $\beta = 116.12°$, $Z = 4$, $D = 2.704$. *CM 28:* 10.2_{10} 5.58_8 5.29_3 4.06_6 3.89_7 2.55_4 2.04_4 1.94_3. Small acicular crystals. Colorless to white, white streak, vitreous luster. Cleavage {010}, poor; brittle. $G = 2.70$. Biaxial (+); $X = b$,

$Z \wedge a = 22°$; $N_x = 1.569$, $N_y = 1.594$, $N_z = 1.622$; $2V = 81°$; $r < v$, strong. *Francon quarry, Montreal, QUE, Canada.* BM *CM 28:155(1990)*.

43.5.17.1 Girvasite $NaCa_2Mg_3(PO_4)_2[PO_2(OH)_2](CO_3)(OH)_2 \cdot 4H_2O$

Named in 1990 for the locality. MON $P2_1/c$. $a = 6.507$, $b = 12.267$, $c = 21.403$, $\beta = 90.37°$, $Z = 4$, $D = 2.529$. *42-1397:* 10.72_{10} 3.57_8 3.42_3 3.08_3 2.82_3 2.11_3 2.02_4. Spherulites up to 1.5 mm. White, white streak, silky to vitreous luster. Cleavage [001], perfect; very brittle. $H = 3\frac{1}{2}$. $G = 2.46$. Biaxial $(-)$; $Y = b$; $Z \wedge a = 31°$; $N_x = 1.541$, $N_y = 1.557$, $N_z = 1.565$; $2V = 60°$. *Lake Girvas, Kovdor, Kola Penin., Russia*, in dolomitic carbonitite. BM *MZ 12(3):79(1990)*, *AM 77:207(1992)*.

43.5.18.1 Hotsonite $Al_{11}(SO_4)_3(PO_4)_2(OH)_{21} \cdot 16H_2O$

Named in 1984 for the locality. TRIC $P1$ or $P\bar{1}$. $a = 11.23$, $b = 11.66$, $c = 10.55$, $\alpha = 112.54°$, $\beta = 107.53°$, $\gamma = 64.45°$, $Z = 1$, $D = 2.08$. *38-366:* 10.05_{10} 8.45_4 5.20_1 5.01_1 4.63_2 4.43_1 3.91_1 3.67_1. White compact masses. Silky to dull luster. $H = 2\frac{1}{2}$. $G = 2.06$. Biaxial; $N_x = 1.519$, $N_z = 1.521$. *Hotson farm, Pofadder, South Africa.* BM *AM 69:979(1984)*.

43.19.1 Quadruphite $Na_{14}CaMgTi_4(Si_2O_7)_2(PO_4)O_4F_2$

Named in 1992 for the Latin *quadruplex* and the element P, the quantity of PO_4 groups in the formula. TRIC $P1$. $a = 5.415$, $b = 7.081$, $c = 20.34$, $\alpha = 86.85°$, $\beta = 94.40°$, $\gamma = 89.94°$, $Z = 1$, $D = 3.11$. *ZVMO 121:105(1992)*. PD: 2.88_{10} 2.70_8 2.64_7 2.05_5 1.622_4 1.600_5. Irregular plates on {001} 1–2 mm thick and up to 3 mm across, some in epitaxial intergrowths with plates of lomonosovite, sobolevite. Light brown; white streak; vitreous luster, pearly on basal cleavage; translucent, transparent in thin sections. Brittle, step-like fracture; perfect {001}, less perfect {110}, {100} cleavages. $H = 5$. $G = 3.12$. Nonfluorescent. Biaxial $(-)$; $N_x = 1.630$ (colorless), $N_y = 1.678$ (yellow), $N_z = 1.697$ (yellow); $2V = 62°$; $r \ll v$; pronounced pleochroism. Found with K-feldspar, nepheline, sodalite, arfvedsonite, aegirine, cancrinite, analcite, albite, ussingite, makatite, grumantite, lomonosovite, vuonnemite, sobolevite, and other rare minerals at *Mount Alluaiv in the NW part of the Lovozero alkaline massif, Kola Penin., Russia.* HCWS *AM 78:1316(1993)*.

43.20.1 Polyphite $Na_{17}Ca_3Mg(Ti,Mn)_4(Si_2O_7)(PO_4)_6O_2F_2$

Named in 1992 from the Greek for *many* and the element P, indicating the quantity of PO_4 groups in the formula unit. See quadruphite (42.13.14.1), a very similar mineral and in intimate association. TRIC $P1$. $a = 5.412$, $b = 7.079$, $c = 26.56$, $\alpha = 95.21°$, $\beta = 93.51°$, $\gamma = 90.10°$, $Z = 1$, $D = 3.00$, *ZVMO 121:105(1992)*. PD: 2.94_{10} 2.70_9 2.66_8 2.05_8 1.771_5 1.730_5. Biaxial $(-)$; $N_x = 1.600$, $N_y = 1.658$, $N_z = 1.676$. $2V = 56°$. X, colorless; Y,Z, yellowish. $G = 3.07$. Found with quadruphite and other minerals in alkali pegmatites of the *Lovozero massif, Kola Penin., Russia.* HCWS *AM 78:1316(1993)*.

Class 44

Antimonates

44.1.1.1 Stibiconite $Sb^{3+}Sb_2^{5+}O_6(OH)$

Named in 1832 for the composition, stibium (antimony), and from the Greek for *powder*, since it often occurs in this form. ISO $Fd3m$. $a = 10.27$, $Z = 8$, $D = 5.86$. *10-388:* 5.93_9 3.09_7 2.96_{10} 2.57_4 1.81_8 1.55_6 1.18_4 1.15_4. Pseudomorphs after stibnite or massive. White to yellow or brown, white streak, earthy luster. H = 3–5. G = 3.3–5.6. Isotropic; N = 1.64–2.05. Variable composition; the cation positions may be only partially filled, and the Sb^{3+} may be partially replaced by Ca (up to 14% CaO, var. hydroromeite). Soluble in concentrated HCl + KI. A secondary mineral formed by the oxidation of stibnite and other antimony minerals; occurs in many antimony deposits throughout the world. Important localities include Manhattan, Nye Co., NV; Antimony Peak, Kern Co., CA; many localities in Mexico: El Antimonio, SON, Catorce* and Charcas*, SLP, Paletos and Mapimi, DGO, Tejocotes, OAX; Allemont, France; Moros, Zaragoza, Spain; Poggio Fuoco, Manciano, Italy; Baia Sprie, Romania; Derekdy*, Gonyuh, Turkey; Hunan*, China; Ain Kerma, Algeria; Cochabaniba*, Bolivia. BM *DI:597, AM 37:982(1952)*.

44.1.1.2 Bindheimite $Pb_2Sb_2O_6(OH)$

Named in 1868 for Johann J. Bindheim (1750–1825), German chemist, who made the first analysis on material from Nerchinsk, Siberia. ISO $Fd3m$. $a = 10.47$, $Z = 8$, $D = 7.50$. *18-687:* 3.03_{10} 2.62_6 1.85_8 1.58_8 1.20_4 1.17_4 1.07_4 0.884_5. Massive. Yellow to brown, yellow streak, earthy luster. H = 4–5. G = 4.6–5.6. Variable composition; calcian, cuprian, and argentian varieties exist. Isotropic; N = 1.84–1.87. Widespread as an oxidation product of lead–antimony sulfides. Wamsley mine, Mineral Co., and Lovelock, Pershing Co., NV; Cerro Gordo mine, Inyo Co., CA; Mapimi, DGO, Mexico; Endellion, Cornwall, England; La Sanguinede, Herault, France; *Nerchinsk, Siberia, Russia*; Broken Hill and Thackaringa, NSW, Australia; Ouro Preto, MG, Brazil; Machacamarca, Bolivia. BM *DII:1018, MM 30:102(1953)*.

44.1.1.3 Roméite $Ca_2Sb_2O_6(O,OH,F)$

Named in 1841 for Romé de Lisle (1736–1790), French crystallographer. ISO $Fd3m$. $a = 10.26$–10.31, $Z = 8$, $D = 4.85$. *27-89:* 6.0_8 3.09_7 2.95_{10} 2.57_6 1.83_8 1.55_7 1.18_5 1.15_4. Small octahedral crystals, also massive. Yellow to brown, yellow streak, vitreous to adamantine luster. Cleavage {111}, imperfect. H = $5\frac{1}{2}$–$6\frac{1}{2}$. G = 4.7–5.4. Variable composition; the following varieties have been recognized: atopite (sodian), lewisite (titanian), mauzeliite (plumbian), schneebergite (ferroan), and weslienite (sodian). Isotropic; N = 1.82–1.87. An accessory mineral, usually in metamorphosed manganese ores. Franklin, NJ; Långban and Jakobsberg, Värmland, Sweden; *San Marcel mine, Piedmont*, and Schneeberg, near Vipiteno, *Italy*; Ngan Hui, Kwangsi, China; Miguel Burnier, MG, Brazil. BM *DII:1020*.

44.1.1.4 Monimolite $(Pb,Ca)_2Sb_2O_7$

Named in 1865 from the Greek for *stable*, because decomposed chemically with great difficulty. ISO *Fd3m*. $a = 10.47$, $Z = 8$, $D = 7.50$. X-ray powder pattern identical with bindheimite (44.1.1.2). Small cubic and octahedral crystals. Yellow to brown, yellow streak, greasy to adamantine luster. Cleavage {111}, indistinct; conchoidal or splintery fracture. H = 5–6. G = 5.9–7.3. Isotropic; N > 2.06. *Harstig mine, Värmland, Sweden*. BM *DII:1023, MM 30:102(1953)*.

44.1.1.5 Stetefeldite $Ag_2Sb_2O_6(O,OH)$

Named in 1867 for Carl A. Stetefeld (1838–1896), German–American mining engineer. ISO *Fd3m*. $a = 10.46$, $D = 5.38$. *8-12*: 3.02_{10} 2.61_7 1.85_7 1.58_7 1.51_3 1.20_5 1.17_4 0.884_5. Massive. Yellow, blackening on exposure; yellow streak, earthy luster. H = $3\frac{1}{2}$–$4\frac{1}{2}$. G = 4.6. Isotropic; N = 1.95. *Belmont, Nye Co.*, and *Silver Peak, Esmeralda Co., NV*; Doigs, near Mt. Monger, WA, Australia; Huancavelica, Peru. BM *DI:598, MM 30:105(1953)*.

44.1.1.6 Bismutostibiconite $Bi_2(Sb,Fe)_2O_7$

Named in 1983 for the composition. ISO *Fd3m*. $a = 10.38$, $D = 6.97$. *42-591*: 3.01_{10} 2.60_7 1.83_7 1.57_7 1.50_4 1.30_3 1.19_5 1.16_5. Massive. Yellow to brown, yellow streak, earthy luster. H = 4–5. G = 7.38. Isotropic; N = 2.09. *Clara mine, Black Forest, Germany*. BM *CE 42:77(1983), AM 69:1190(1984)*.

44.1.1.7 Partzite $Cu_2Sb_2O_6(OH)$

Named in 1867 for A. F. W. Partz, who discovered the mineral. ISO *Fd3m*. $a = 10.25$, $Z = 8$, $D = 5.95$. *7-303*: 5.91_9 3.08_7 2.95_{10} 2.56_5 1.81_8 1.73_4 1.54_7 1.33_5. Massive. Olive-green with black tarnish, pale green streak, earthy luster. H = 3–4. G = 3.0–4.0. Isotropic; N = 1.61–1.83. *Blind Spring mining district, Mono Co.*, and Benton, Inyo Co., *CA*; Ashburton Downs, WA, Australia. BM *DI:599, MM 30:105(1953), MR 24:215(1993)*.

44.1.2.1 Ingersonite $Ca_3MnSb_4O_{14}$

Named in 1988 for F. Earl Ingerson (1906–1993), U.S. geochemist. HEX-R Probably *P3m*. $a = 7.287$, $c = 17.679$, $Z = 3$, $D = 5.42$. *AM 73*: 5.89_2 3.10_2 2.97_{10} 2.57_4 1.82_5 1.81_5 1.55_6 1.54_4. Subhedral crystals to 3 mm. Yellow-brown color and streak, vitreous luster. Cleavage {0001}, perfect. H = $6\frac{1}{2}$. Uniaxial (−); $N_O = 1.93$, $N_E = 1.92$. *Långban, Värmland, Sweden*, a rare accessory in manganese ores. BM *AM 73:405(1988)*.

44.2.1.1 Byströmite MgSb$_2$O$_6$

Named in 1952 for Anders Byström (1916–1956), Swedish crystal chemist. TET $P4_2/mnm$. $a = 4.68$, $c = 9.21$, $Z = 2$, $D = 5.99$. *15-684:* 4.63_4 4.19_7 3.32_{10} 2.57_9 2.34_5 1.73_9 1.65_4 1.48_4. Massive. Blue-gray color and streak, earthy luster. H = 7. G = 5.7. Uniaxial (+); $N_O = 1.908$, $N_E = 1.915$. *El Antimonio, SON, Mexico*. BM *AM 37:53(1952)*.

44.2.1.2 Ordoñezite ZnSb$_2$O$_6$

Named in 1955 for Ezequiel Ordoñez (1867–1950), Mexican geologist. TET $P4_2/mnm$. $a = 4.67$, $c = 9.24$, $Z = 2$, $D = 6.73$. *11-214:* 3.26_9 2.55_8 2.31_4 1.72_{10} 1.64_5 1.39_4 1.38_6 1.19_6. Tetragonal crystals to 2 mm; pale to dark brown, yellow streak, adamantine luster. No cleavage, conchoidal fracture. H = $6\frac{1}{2}$. G = 6.635. Uniaxial (+); N > 1.95. *Santín mine, GTO*, and *Serrita Morita, SON, Mexico*. BM *AM 40:64(1955)*.

44.2.1.3 Tripuhyite FeSb$_2$O$_6$

Named in 1897 for the locality. TET $P4_2/mnm$. $a = 4.63$, $c = 9.14$, $Z = 3$, $D = 6.08$. *7-349:* 3.28_{10} 2.56_9 2.32_4 1.72_9 1.64_5 1.52_3 1.47_4 1.38_3. Detrital grains or massive. Yellow to brown, yellow streak, vitreous to earthy luster. H = 7. G = 5.6–5.8. Anomalously biaxial (+); $N_x = 2.19$, $N_y = 2.20$, $N_z = 2.33$; 2V small; r < v. *Mopung Hills, Churchill Co., NV; El Antimonio, SON, Mexico; Clara mine, Black Forest, Germany; Faletta, Graubunden, Switzerland; Cetine, near Siena, Italy; Hammam N'bail, near Constantine, Algeria; Tripuhy, MG, Brazil; Doncellas, Jujuy, Argentina; Ashburton Downs, WA, Australia*. BM *DII:1024, MM 30:107(1953), MR 24:216(1993)*.

44.3.1.1 Swedenborgite NaBe$_4$SbO$_7$

Named in 1924 for Emanuel Swedenborg (1688–1772), Swedish philosopher. HEX $P6_3mc$. $a = 5.42$, $c = 8.80$, $Z = 2$, $D = 4.28$. *23-656:* 4.7_7 4.4_7 4.2_{10} 3.22_8 2.72_9 2.51_9 2.32_9 2.28_7. Hexagonal crystals to 8 mm. Colorless to yellow, white streak, vitreous luster. Cleavage {0001}, distinct; subconchoidal fracture. H = 8. G = 4.285. Fluorescent intense blue under SWUV, deep red under LWUV. Uniaxial (−); $N_O = 1.772$, $N_E = 1.770$. BM *Långban, Värmland, Sweden*. *DII:1027*.

44.3.2.1 Manganostibite Mn$_7$Sb^{5+}As^{5+}O$_{12}$

Named in 1884 for the composition. ORTH *Ibmm*. $a = 8.727$, $b = 18.847$, $c = 6.062$, $Z = 4$, $D = 5.00$. *23-1236:* 9.31_4 4.97_6 4.69_4 4.38_6 2.82_4 2.65_{10} 2.62_4 2.18_5. Small rounded prismatic crystals. Black, brown streak, greasy luster. G = 4.949. Biaxial (−); $N_x = 1.92$, $N_y = 1.95$, $N_z = 1.96$; 2V small. *Moss mine, Värmland, Sweden.* BM *DII:1027*, *AM 55:1489(1970)*.

44.3.3.1 Parwelite (Mn,Mg)$_5$Sb(As,Si)$_2$O$_{12}$

Named in 1970 for Alexander Parwel (1906–1976), Swedish chemist. MON *Aa*. $a = 9.76$, $b = 19.32$, $c = 10.06$, $\beta = 95.9°$, $Z = 8$, $D = 4.69$. *29-346:* 9.65_3 4.83_5 3.42_5 2.92_{10} 2.73_{10} 2.42_4 2.31_3 1.74_3. Stubby prismatic crystals to 10 mm. Yellow-brown, yellow streak, greasy luster. Cleavage {010}, fair; subconchoidal fracture. $H = 5\frac{1}{2}$. $G = 4.62$. Biaxial (+); $N_x, N_y = 1.85$, $N_z = 1.88$; $2V = 27°$; $r > v$. *Långban, Värmland, Sweden.* BM *AMG 4:467(1969)*, *AM 55:323(1970)*.

44.3.4.1 Långbanite Mn$_4^{2+}$Mn$_9^{3+}$SbSi$_2$O$_{24}$

Named in 1887 for the locality. HEX-R *P*31*m*. $a = 11.563$, $c = 11.100$, $Z = 3$, $D = 4.946$. *14-195:* 2.80_8 2.75_9 2.55_{10} 1.99_4 1.67_7 1.54_8 1.43_5 1.03_4. Hexagonal prismatic crystals to 1 cm. Black, brown streak, metallic luster. Cleavage {0001}, good; conchoidal fracture; brittle. $H = 6\frac{1}{2}$. $G = 4.9$. Uniaxial (−); $N_O = 2.36$, $N_E = 2.31$; weakly pleochroic, red-brown. *Långban, Värmland, Sweden*, in manganophyllite skarn; Gozaisho mine, Fukushima Pref., Japan. BM *SPC 18:320(1973)*, *AM 76:1408(1991)*, *NJMM:193(1991)*.

44.3.5.1 Katoptrite Mn$_{13}$Al$_4$Sb$_2$Si$_2$O$_{28}$

Named in 1917 from the Greek for *mirror*, in allusion to the luster on cleavages. Synonym: catoptrite. MON *C*2/*m*. $a = 5.617$, $b = 23.02$, $c = 7.079$, $\beta = 101.38°$, $Z = 2$, $D = 4.53$. *19-274:* 8.88_7 4.43_5 2.96_{10} 2.81_4 2.60_5 2.49_5 2.15_3 1.94_4. Platy crystals. Black, red-brown streak, submetallic luster. Cleavage {100}, perfect, micaceous; brittle. $H = 5\frac{1}{2}$. $G = 4.56$. Biaxial (−); $N_x = 1.92$, $N_y, N_z = 1.95$; 2V small, $r > v$; strongly pleochroic, yellow-red to red-brown. *Brattfors, Moss, and Sjö mines, Värmland, Sweden.* BM *DII:1029*, *AM 51:1494(1966)*, *NJMA 127:47(1976)*.

44.3.6.1 Yeatmanite Mn$_9$Zn$_6$Sb$_2$Si$_4$O$_{28}$

Named in 1938 for Pope Yeatman (1861–1953), U.S. mining engineer. TRIC *P*$\bar{1}$. $a = 5.604$, $b = 11.602$, $c = 9.058$, $\alpha = 92.17°$, $\beta = 100.90°$, $\gamma = 77.30°$, $Z = 1$, $D = 5.04$. Platy anhedral crystals to 13 mm. Clove-brown, pale brown streak, vitreous luster. Cleavage {100}, good. Brittle. $H = 4$. $G = 5.02$. Biaxial (−); $N_x = 1.864$, $N_y = 1.895$, $N_z = 1.905$; $2V = 52°$; $r > v$. *Franklin and Sterling Hill, Sussex Co., NJ.* BM *AM 65:196(1980)*, *MJJ 13:53(1986)*.

44.3.7.1 Bahianite Al$_5$Sb$_3$O$_{14}$(OH)$_2$

Named in 1976 for the locality. MON *C*2/*m*. $a = 9.406$, $b = 11.541$, $c = 4.410$, $\beta = 99.94°$, $Z = 2$, $D = 5.26$. *29-2:* 4.71_7 3.24_{10} 3.19_{10} 2.46_7 2.41_7 2.16_8 1.65_7 1.64_7. Water-worn pebbles (favas). Cream to tan, white streak, adamantine luster. Cleavage {100}, perfect. $H = 9$. $G = 4.89$–5.46. Biaxial (−); $N_x = 1.81$,

$N_y = 1.87$, $N_z = 1.92$; 2V large; r > v. *Paramirim region, BA, Brazil.* BM *NJMA 126:113(1976), MM 42:179(1978).*

44.3.8.1 Shakhovite $Hg_4SbO_3(OH)_3$

Named in 1980 for F. N. Shakhov (1894–1971), Russian geologist. MON *Im* $a = 4.871$, $b = 15.098$, $c = 5.433$, $\beta = 98.86°$, $Z = 2$, $D = 8.35$. *37-460:* 3.88_{10} 3.33_8 2.69_6 2.63_5 2.55_5 2.03_3 1.93_4 1.89_3. Platy crystals to 1 mm. Yellow-green to olive-green, yellow streak, adamantine luster. $H = 3-3\frac{1}{2}$. $G = 8.38$. Probably biaxial; $N > 2.03$. *Kelyan deposit, Buryat ASSR, Russia;* Khaidarkan deposit, Kyrgyzstan; Moschallandsberg, Palatinate, Germany. BM *MM 46:525(1982); TMPM 30:227(1982); AM 68:1041(1983), 73:1499(1988).*

44.3.9.1 Cyanophyllite $Cu_5Al_2(SbO_4)_3(OH)_2 \cdot 12H_2O$

Named in 1981 for the blue color and platy form. ORTH *Pmmb.* $a = 11.82$, $b = 10.80$, $c = 9.64$, $Z = 2$, $D = 3.12$. *35-507:* 9.67_6 4.84_{10} 2.59_6 2.44_5 1.92_4 1.56_3 1.48_3 1.21_2. Spherulitic platy aggregates. Blue-green, blue streak, pearly to silky luster. Cleavage {001}, perfect. $H = 2$. $G = 3.10$. Biaxial $(-)$; $X = c$; $N_x = 1.640$, $N_y = 1.664$, $N_z = 1.675$; 2V = 67°. *Clara mine, Black Forest, Germany.* BM *CE 40:195(1981), AM 66:1274(1981).*

44.3.10.1 Cualstibite $Cu_6Al_3(SbO_4)_3(OH)_{12} \cdot 10H_2O$

Named in 1984 for the composition. HEX-R Space group uncertain, probably *P3.* $a = 9.20$, $c = 9.73$, $Z = 1$, $D = 3.25$. *38-362:* 4.89_{10} 4.63_4 4.17_8 3.35_5 2.65_7 2.33_9 1.79_8 1.53_4. Microscopic trigonal crystals and massive. Blue-green, pale blue streak, vitreous luster. No cleavage, conchoidal fracture. $H = 2$. $G = 3.18$. Uniaxial $(-)$; $N_O = 1.672$, $N_E = 1.644$; pleochroic, colorless to pale blue-green. *Clara mine, Black Forest, Germany.* BM *CE 43:255(1984), AM 70:1329(1985).*

44.3.11.1 Camerolaite $Cu_4Al_2(HSbO_4,SO_4)(OH)_{10}(CO_3) \cdot 2H_2O$

Named in 1991 for Michel Camerola, French mineral collector. MON Space group uncertain, probably $P2_1$. $a = 10.765$, $b = 2.903$, $c = 12.527$, $\beta = 95.61°$, $Z = 1$, $D = 3.09$. *44-1441:* 5.62_5 5.16_9 4.28_{10} 3.57_4 2.38_4 2.33_4 2.14_3 1.92_3. Radiating fibrous aggregates to 2 mm. Blue-green, pale green streak, silky luster. Cleavage {100}, {001}, good; fibrous fracture; brittle. $G = 3.1$. Biaxial $(+)$; $N_x = 1.626$, $N_y = 1.646$, $N_z = 1.682$; 2V = 77°; r < v, strong; pleochroic, colorless to blue-green. *Cap Garonne, Var, France.* BM *NJMM:481(1991), AM 77:1116(1992).*

44.3.12.1 Brizziite $NaSbO_3$

Named in 1994 for Giancarlo Brizzi (1936–1992), Italian mineral collector. HEX-R $R\bar{3}$. $a = 5.301$, $c = 15.932$, $Z = 6$, $D = 4.95$. *9-23*(syn): 5.31_{10} 4.40_4 3.97_4 3.01_4 2.65_6 2.37_4 1.88_4 1.53_2. Microscopic aggregates of platy crystals. Yellow or pale pink, pearly luster. Cleavage {0001}, perfect; flexible. $VHN_{15} = 57$.

$G = 4.82$. Uniaxial $(-)$; $N_O = 1.84$, $N_E = 1.631$. *Cetine mine, Tuscany, Italy,* an alteration product of stibnite, together with stibiconite and mopungite. BM *EM 6:667(1994)*.

Class 45

Acid and Normal Antimonites and Arsenites

45.1.1.1 Reinerite $Zn_3(AsO_3)_2$

Named in 1959 for Willy Reiner (b.1904), chemist, Tsumeb Corporation, who analyzed the mineral. ORTH $Pbam$. $a = 6.092$, $b = 14.407$, $c = 7.811$, $Z = 4$, $D = 4.28$. *11-158:* 4.00_{10} 3.43_7 3.40_7 3.20_8 2.64_8 2.60_7 2.40_7 1.95_7. Prismatic crystals to 5 cm, also massive. Blue to yellow-green, vitreous to adamantine luster. Cleavage {110}, good; {011}, {111}, fair. $H = 5-5\frac{1}{2}$. $G = 4.27$. Biaxial (−); $N_x = 1.74$, $N_y = 1.79$, $N_z = 1.82$; $2V = 80°$. *Tsumeb, Namibia.* BM *NJMM:160(1958); AM 44:207(1959), 62:1129(1977).*

45.1.2.1 Trigonite $Pb_3Mn(AsO_3)_2(AsO_2OH)$

Named in 1920 for the shape of the crystals. MON Pn. $a = 7.258$, $b = 6.822$, $c = 11.088$, $\beta = 94.45°$, $Z = 2$, $D = 6.34$. *23-330:* 4.9_6 4.5_4 3.65_4 3.24_7 3.08_9 2.99_{10} 2.76_4 2.71_4. Thick tabular crystals. Yellow to brown, yellow streak, vitreous to adamantine luster. Cleavage {010}, perfect; {101}, good; uneven fracture. $H = 2-3$. $G = 6.1$. Biaxial (−); $N_x = 2.07$, $N_y = 2.10$, $N_z = 2.12$; 2V large, r < v. *Långban, Värmland, Sweden.* BM *DII:1032, TMPM 25:95(1978), MA 40:1661(1989).*

45.1.2.2 Rouseite $Pb_3Mn(AsO_3)_2 \cdot 2H_2O$

Named in 1986 for Roland Rouse (b.1943), U.S. mineralogist. TRIC $P1$ or $P\bar{1}$. $a = 6.36$, $b = 7.29$, $c = 5.54$, $\alpha = 97.3°$, $\beta = 114.2°$, $\gamma = 106.0°$, $Z = 1$, $D = 5.70$. *40-472:* 6.81_{10} 3.38_6 3.25_6 3.06_7 2.98_8 2.85_9 2.75_5 2.02_6. Microscopic crystals. Yellow-orange, pale yellow streak, vitreous luster. Cleavage {010}, {100}, perfect. $H = 3$. $G > 4.2$. Biaxial (−); $N = 1.8-1.9$; $2V = 46°$, r > v. *Långban, Värmland, Sweden.* BM *AM 71:1034(1986).*

45.1.3.1 Asbecasite $Ca_3Be_2Si_2(Ti,Sn)(AsO_3)_6O_2$

Named in 1966 for the composition. HEX-R $P\bar{3}c1$. $a = 8.364$, $c = 15.304$, $Z = 2$, $D = 3.71$. *19-87:* 4.04_5 3.84_5 3.23_{10} 2.41_6 1.75_6 1.57_7 1.32_6 1.15_7. Rhombohedral crystals to 5 mm. Lemon yellow, pale yellow streak, adamantine luster. Perfect rhombohedral cleavage. $H = 6\frac{1}{2}-7$. $G = 3.70$. Uniaxial (−); $N_O = 1.86$, $N_E = 1.83$. *Binntal, Switzerland,* in a fissure in gneiss; *Pizzo Cervandone, Italy.* BM *SMPM 46:367(1966); AM 52:1583(1967), 66:819(1981).*

45.1.4.1 Cafarsite $(Ca,Mn)_8Fe_3Ti_3(AsO_3)_{12} \cdot 4H_2O$

Named in 1966 for the composition. ISO $Pn3$. $a = 15.984$, $Z = 4$, $D = 3.95$. *19-197:* 9.51_4 3.69_5 3.15_7 2.83_{10} 2.75_8 1.72_5 1.63_6 0.979_3. Cubo-octahedral crystals to 3 cm. Dark brown, yellow-brown streak, adamantine luster. No cleavage, conchoidal fracture, brittle. H = $5\frac{1}{2}$–6. G = 3.90. Isotropic; N $\geqslant$ 2.0. *Binntal, Switzerland;* Pizzo Cervandone, Italy. BM *SMPM 46:367(1966), 57:1(1977); AM 52:1584(1967).*

45.1.5.1 Trippkeite $CuAs_2O_4$

Named in 1880 for the Polish mineralogist Paul Trippke (1851–1880), who discovered the mineral. TET $P4/mbc$. $a = 8.59$, $c = 5.57$, $Z = 4$, $D = 4.49$. *31-451:* 6.07_6 3.16_{10} 3.04_5 2.72_4 2.53_2 1.95_4 1.64_3 1.39_2. Short prismatic crystals. Blue-green, pale blue streak, adamantine luster. Cleavage {100}, perfect; {110}, good. G = 4.8. Uniaxial (+); $N_O = 1.90$, $N_E = 2.12$. Near *Copiapó, Chile.* BM *DII:1034, TMPM 22:211(1975).*

45.1.6.1 Schafarzikite $FeSb_2O_4$

Named in 1921 for Ferenc Schafarzik (1854–1927), Hungarian mineralogist. TET $P4/mbc$. $a = 8.59$, $c = 5.91$, $Z = 4$, $D = 5.53$. *25-1406:* 3.22_4 2.03_4 1.67_8 1.31_5 1.24_4 1.16_5 1.05_{10} 1.01_{10}. Prismatic crystals. Red to red-brown, brown streak, metallic luster. Cleavage {110}, perfect; {100}, good. H = $3\frac{1}{2}$. G = 4.3. Uniaxial (+); N ~ 2.1; weakly birefringent; pleochroic, pale yellow to yellow-brown. Lac Nicolet mine, South Ham, QUE, Canada; *Pernek, Slovakia.* BM *DII:1035, TMPM 22:236(1975).*

45.1.7.1 Leiteite $ZnAs_2O_4$

Named for Luis Texeira-Leite (b.1942), Portuguese mineral dealer, who discovered the mineral. MON $P2_1/c$. $a = 4.542$, $b = 5.022$, $c = 17.597$, $\beta = 90.81°$, $Z = 2$, $D = 4.62$. *29-740:* 4.83_3 3.32_4 3.30_3 3.16_8 3.13_{10} 2.94_2 2.18_2 1.69_5. Cleavable masses to 7 cm. Colorless to brown, white streak, pearly luster. Cleavage {100}, perfect; cleavage lamellae flexible, inelastic. H = $1\frac{1}{2}$–2. G = 4.31. Biaxial (+); Z $\wedge$ $c = 10°$; $N_x = 1.87$, $N_y = 1.880$, $N_z = 1.98$; 2V = 27°; r < v, very strong. *Tsumeb, Namibia.* BM *MR 8(3):97(1977); AM 62:1259(1977), 72:629(1987).*

45.1.8.1 Paulmooreite $Pb_2As_2O_5$

Named in 1979 for Paul B. Moore (b.1940), U.S. mineralogist, investigator of Långban minerals. MON $P2_1/a$. $a = 13.584$, $b = 5.650$, $c = 8.551$, $\beta = 108.78°$, $Z = 4$, $D = 6.86$. *33-736:* 4.26_5 3.42_5 3.30_{10} 3.02_7 2.91_8 2.02_6 1.77_7 1.73_6. Small platy crystals. Colorless to pale orange, white streak, adamantine luster. Cleavage {001}, perfect; very brittle. H = 3. G = 6.95. Biaxial (+); N > 1.9; birefringence = 0.110; 2V = 65°; r > v, very strong. *Långban, Värmland, Sweden.* BM *AM 64:352(1979), 65:340(1980).*

45.1.9.1 Apuanite $Fe^{2+}Fe_4^{3+}Sb_4O_{12}S$

Named in 1979 for the locality. TET $P4_2/mbc$. $a = 8.367$, $c = 17.959$, $Z = 4$, $D = 5.23$. 33-642: 5.90_2 4.18_2 3.17_{10} 2.98_4 2.65_2 2.43_2 1.92_5 1.65_3. Platy crystals and massive aggregates. Black, black streak, metallic luster. Cleavage {110}, perfect. $H = 4$. $G = 5.33$. *Buca della Vena mine, Apuan Alps, Italy.* BM *AM 64:1230(1979)*.

45.1.10.1 Versiliaite $Fe_4^{2+}Fe_8^{3+}Sb_{12}O_{32}S_2$

Named in 1979 for the locality. ORTH Space group probably $Pbam$. $a = 8.499$, $b = 8.326$, $c = 11.935$, $Z = 1$, $D = 5.20$. 33-698: 5.94_3 3.20_{10} 3.17_{10} 2.97_8 2.68_4 2.64_2 1.95_4 1.92_3. Platy crystals. Black, black streak, metallic luster. Cleavage {110}, perfect. $H = 4\frac{1}{2}$. $G = 5.12$. *Buca della Vena mine, Versilia valley, Apuan Alps, Italy.* BM *AM 64:1230(1979)*.

45.1.11.1 Stibivanite-2M $VOSb_2O_4$

Named in 1980 for the composition. MON $C2/c$. $a = 17.989$, $b = 4.792$, $c = 5.500$, $\beta = 95.15°$, $Z = 4$, $D = 5.26$. 33-122: 4.64_6 3.50_7 3.17_7 3.00_{10} 2.11_4 2.03_3 1.87_4 1.71_3. Radiating fibrous crystals to 2 mm. Yellow-green, white streak, adamantine luster. $H = 4$. Biaxial $(-)$; $N_x > 1.87$, $N_z < 1.89$; $2V = 85°$; $r > v$, strong; strongly pleochroic, emerald-green to olive-green. *Lake George antimony mine, NB, Canada.* BM *CM 18:329(1980)*.

45.1.11.2 Stibivanite-2O $VOSb_2O_4$

Named in 1989 for the composition. ORTH $Pmcn$. $a = 17.916$, $b = 4.790$, $c = 5.509$, $Z = 4$, $D = 5.26$. *CM 27*: 9.00_5 4.62_5 3.63_2 3.10_{10} 2.99_7 2.51_2 1.87_5 1.75_3. Acicular and fibrous crystals to 2 mm. Emerald green, white streak, adamantine luster. Parting parallel to c. $H = 4$. Biaxial $(+)$; $N > 1.87$; 2V very small; pleochroic, emerald-green to olive-green. *Buca della Vena mine, Apuan Alps, Italy.* BM *CM 27:625(1989)*.

45.1.12.1 Schneiderhöhnite $Fe^{2+}Fe_3^{3+}As_5O_{13}$

Named in 1973 for Hans Schneiderhöhn (1887–1962), German mineralogist. TRIC $P\bar{1}$. $a = 8.924$, $b = 10.016$, $c = 9.103$, $\alpha = 59.91°$, $\beta = 112.41°$, $\gamma = 81.69°$, $Z = 2$, $D = 4.33$. 35-462: 7.27_{10} 4.14_7 3.93_6 3.64_8 3.35_8 2.97_{10} 2.91_{10}. Platy crystals to 7 mm. Dark brown to black, brown streak, adamantine to metallic luster. Cleavage {100}, perfect. $H = 3$. $G = 4.3$. Biaxial $(+)$; $N_x > 2.11$, $N_z < 2.13$; pleochroic, yellow to red-brown. *Tsumeb, Namibia.* BM *NJMM:517(1973), AM 59:1139(1974), CM 23:675(1985)*.

45.1.13.1 Fetiasite $(Fe^{2+},Fe^{3+},Ti)_3O_2(As_2O_5)$

Named in 1994 for the composition. MON $P2_1/m$. $a = 10.614$, $b = 3.252$, $c = 8.945$, $\beta = 108.95°$, $Z = 2$, $D = 4.74$. *AM 79*: 3.53_3 2.99_7 2.81_9 2.75_{10} 2.39_8 1.78_5 1.75_3 1.71_4. Microscopic tabular crystals and globular aggregates. Brown to black, metallic to submetallic luster. Cleavage {100}, perfect;

uneven to conchoidal fracture. H ≈ 5, $VHN_{50} = 440$–490. G = 4.6. R = 15.3–15.8 (546 nm). *Pizzo Cervandone, Italy,* and *Binntal, Switzerland,* in Alpine fissure veins. BM *AM 79:996(1994).*

45.1.14.1 Ludlockite $PbFe_4^{3+}As_{10}O_{22}$

Named in 1977 for mineral dealers Ludlow Smith and Locke Key, who discovered the mineral. TRIC $P\bar{1}$. $a = 10.426$, $b = 12.074$, $c = 18.349$, $\alpha = 101.84°$, $\beta = 100.21°$, $\gamma = 90.60°$, Z = 4, D = 4.37. *29-774:* 10.9_6 8.81_{10} 4.74_6 4.47_6 3.33_7 3.16_7 2.94_9 2.86_7. Needles to 4 mm. Red-brown, pale brown streak, subadamantine luster. Cleavage $\{0\bar{1}\bar{1}\}$, $\{021\}$, perfect. H = $1\frac{1}{2}$–2. G = 4.40. Biaxial (+); $N_x = 1.96$, $N_y = 2.055$, $N_z = 2.11$; pleochroic, X, yellow; Y, deep yellow; Z, orange-yellow. *Tsumeb, Namibia,* in a cavity in sulfide ore. BM *MR 8:91(1977), CM 34:79(1996).*

===== Class 46 =====

Basic or Halogen-Containing Antimonites and Arsenites

46.1.1.1 Finnemanite $Pb_5(AsO_3)_3Cl$

Named in 1923 for Karl J. Finneman (1880–1953), mine worker at Långban, who recognized many new minerals. HEX $P6_3/m$. $a = 10.322$, $c = 7.055$, $Z = 2$, $D = 7.30$. Isostructural with apatite. *14-187*: 4.42_4 3.35_6 3.03_{10} 2.88_9 2.07_6 2.02_4 1.96_5 1.92_5. Small prismatic crystals. Gray to black, gray streak, subadamantine luster. Cleavage $\{10\bar{1}1\}$, distinct. $H = 2\frac{1}{2}$. $G = 7.265$. Uniaxial $(-)$; $N_O = 2.295$, $N_E = 2.285$. *Långban, Värmland, Sweden;* Beltana mine, near Copley, SA, Australia. BM *DII:1038, TMPM 26:95(1979)*.

46.1.2.1 Nanlingite $CaMg_4(AsO_3)_2F_4$

Named in 1976 for the locality. HEX-R $R3m$ or $R\bar{3}m$. $a = 10.24$, $c = 25.76$, $Z = 15$, $D = 3.91$. *30-264*: 8.35_9 4.45_5 3.29_5 2.86_6 2.78_{10} 2.43_7 1.78_8 1.70_8. Tabular crystals. Red-brown, pale yellow streak, vitreous luster. $H = 2$. $G = 3.927$. Uniaxial $(-)$; $N_O = 1.82$, $N_E = 1.78$. *Nanling area, southern China*, in a contact zone between granite and dolomitic limestone. BM *AM 62:1058(1977)*.

46.1.3.1 Magnussonite $Mn_{10}(AsO_3)_6(OH,Cl)_2$

Named in 1956 for Nils H. Magnusson (1890–1976), Swedish geologist, who monographed the Långban deposit. ISO $Ia3d$. $a = 19.70$, $Z = 16$, $D = 4.61$. *10-407*: 8.01_1 4.02_1 3.13_3 2.85_{10} 2.47_3 1.74_2 1.48_1 1.42_1. Fine-grained incrustations. Grass-green to emerald-green, white streak, vitreous luster. $H = 3\frac{1}{2}-4$. $G > 4.4$. Isotropic; $N = 1.983$. *Sterling Hill, Sussex Co., NJ; Brattfors mine* (a tetragonal variety, $a = 19.58$, $c = 19.72$) and *Långban, Värmland, Sweden*. BM *AMG 2:133(1956); AM 64:390(1979), 69:800(1984)*.

46.1.4.1 Stenhuggarite $CaFe^{3+}SbAs_2O_7$

Named in 1970 from the Swedish *stenhuggar,* (stone)mason, referring to Brian Mason (b.1917), New Zealand–U.S. mineralogist, who worked on Långban minerals. TET $I4_1/a$. $a = 16.144$, $c = 10.706$, $Z = 16$, $D = 4.56$. *24-199*: 8.0_2 6.0_2 4.8_2 3.66_3 2.98_{10} 2.54_4 1.84_5 1.54_4. Equant crystals to 3 mm. Orange, yellow streak, vitreous luster. No cleavage. $H = 4$. $G = 4.63$. Uniaxial; $N > 2.00$. *Långban, Värmland, Sweden*. BM *AMG 5:55(1970), AM 56:636(1971), AC(B) 33:1807(1977)*.

46.1.5.1 Freedite $Pb_8Cu(AsO_3)_2O_3Cl_5$

Named in 1985 for Robert L. Freed (b.1950), U.S. mineralogist. MON $C2/m$. $a = 13.578$, $b = 20.099$, $c = 7.465$, $\beta = 105.73°$, $Z = 4$, $D = 7.43$. 39-326: 6.51_4 3.83_5 3.58_4 2.95_{10} 2.83_8 2.73_8 2.10_5 1.70_4. Radiating aggregates. Yellow-green, yellow streak, vitreous luster. Cleavage {100}, perfect. H = 3. G = 7.0. Biaxial; N > 1.90. Långban, Värmland, Sweden. BM AM 70:845(1985), TMPM 36:85(1987).

46.1.6.1 Nealite $Pb_4Fe(AsO_3)_2Cl_4 \cdot 2H_2O$

Named in 1980 for Neal Yedlin (1908–1977), U.S. mineralogist, who discovered the mineral. TRIC $P\bar{1}$. $a = 6.548$, $b = 10.243$, $c = 5.587$, $\alpha = 96.2°$, $\beta = 89.6°$, $\gamma = 97.7°$, $Z = 1$, $D = 5.89$. 33-755: 10.09_6 6.48_8 4.24_7 3.92_5 3.54_{10} 3.25_6 2.97_5 2.78_6. Microscopic prismatic or bladed crystals. Bright orange, pale orange streak, adamantine luster. No cleavage, very brittle. Biaxial; N > 2.00; nonpleochroic. Laurion, Greece, in weathered lead slag. BM MR 11:299(1980), TMPM 48:193(1993), NJMM:278(1993).

46.2.1.1 Ecdemite $Pb_6As_2O_7Cl_4$

Named in 1877 from the Greek for unusual, in reference to the composition. Possibly dimorphous with heliophyllite. TET Space group unknown. $a = 10.80$, $c = 25.62$, $Z = 8$, $D = 7.32$. 23-343: 3.66_8 2.85_{10} 2.72_8 2.07_7 1.92_6 1.67_4 1.65_5 1.59_7. Small tabular crystals or foliated masses. Yellow, yellow streak, vitreous to greasy luster. Cleavage {100}, distinct. H = $2\frac{1}{2}$–3. G = 7.14. Uniaxial (−); $N_O = 2.32$, $N_E = 2.25$. Mammoth mine, Pinal Co., AZ; Långban and Harstig mines, Värmland, Sweden. BM DII:1036.

46.2.2.1 Heliophyllite $Pb_6As_2O_7Cl_4$

Named in 1888 from the Greek for sun and leaf, in reference to its color and structure. Possibly dimorphous with ecdemite. ORTH Space group unknown. $a = 10.823$, $b = 10.783$, $c = 25.580$, $Z = 8$, $D = 7.33$. 20-471: 3.66_4 3.19_2 2.84_{10} 2.06_3 1.91_2 1.65_3 1.59_3 1.25_1. Pyramidal crystals and massive, foliated or granular. Yellow, yellow streak, vitreous to greasy luster. Cleavage {011}, good. H = 2. G = 6.89. Biaxial (−); low birefringence; 2V large; r < v, strong. Långban, Harstig, and Jakobsberg mines, Värmland, Sweden. BM DII:1037.

46.2.3.1 Tomichite $(V,Fe)_4Ti_3AsO_{13}(OH)$

Named in 1979 for Stephan A. Tomich (b.1914), consulting geologist, Perth, WA, Australia, who discovered the mineral. MON $A2/m$. $a = 7.119$, $b = 14.176$, $c = 4.992$, $\beta = 105.05°$, $Z = 2$, $D = 4.42$. 34-1401: 4.94_2 3.15_2 3.09_4 2.83_9 2.66_{10} 2.49_2 2.02_3 1.71_3. Tabular grains to 3 mm. Black, black streak, metallic luster. H = 6. G = 4.16. R = 16.3 (546 nm); anisotropic, yellow-gray to dark gray. Hemlo gold deposit, near Marathon, ONT, Canada; Kalgoorlie, WA, Australia. BM MM 43:469(1979), AM 72:201(1987).

46.2.3.2 Derbylite $Fe_4Ti_3SbO_{13}(OH)$

Named in 1897 for Orville A. Derby (1851–1915), director of the Geological Survey of Brazil. MON $P2_1/m$. $a = 7.160$, $b = 14.347$, $c = 4.970$, $\beta = 104.61°$, $Z = 2$, $D = 4.76$. 35-599: 4.00_2 3.83_2 3.19_5 3.12_4 2.85_{10} 2.67_8 2.48_3 2.39_3. Prismatic crystals to 2 mm, and detrital grains. Dark brown to black, brown streak, resinous to metallic luster. No cleavage, conchoidal fracture, very brittle. $H = 5$. $G = 4.5$–4.6. Biaxial (+); $N_x = 2.45$, $N_y = 2.45$, $N_z = 2.51$; 2V small. Buca della Vena mine, Apuan Alps, Italy; Tripuhy, MG, Brazil. BM DII:1025, NJMA 126:292(1976), CM 21:513(1983).

46.2.4.1 Hemloite $(Ti,V,Fe)_{12}(As,Sb)_2O_{23}(OH)$

Named in 1989 for the locality. Contains up to 1.4% Al_2O_3. TRIC $P\bar{1}$. $a = 7.158$, $b = 7.552$, $c = 16.014$, $\alpha = 89.06°$, $\beta = 104.32°$, $\gamma = 84.97°$, $Z = 2$, $D = 4.61$. CM 27: 3.05_7 2.92_{10} 2.80_8 2.72_9 2.67_9 2.50_7 1.77_6 1.62_6. Microscopic subhedral to anhedral grains. Black, black streak, submetallic to metallic luster. No cleavage, irregular fracture. $H = 6\frac{1}{2}$–7. Hemlo gold deposit, near Marathon, ONT, Canada. BM CM 27:427(1989).

46.2.5.1 Gebhardite $Pb_8OCl_6(As_2O_5)_2$

Named in 1983 for Georg Gebhard, German chemist and mineral collector. MON $P2_1/c$. $a = 6.724$, $b = 11.20$, $c = 34.19$, $\beta = 85.2°$, $Z = 2$, $D = 6.07$. 35-611: 4.55_{10} 3.34_7 3.15_8 2.28_8 2.15_6 2.07_4 1.93_8 1.74_6. Fibrous crystal groups to 5 mm. Brown, white streak, adamantine luster. Cleavage {001}, perfect; {010}, good. $H = 3$. Biaxial (−); $N_x = 2.08$, $N_z = 2.12$. Tsumeb, Namibia. BM NJMM:445(1983), AM 70:215(1985).

46.2.6.1 Manganarsite $Mn_8As_2O_4(OH)_4$

Named in 1986 for the composition. HEX-R Space group uncertain, probably $P312$. $a = 11.451$, $c = 7.252$, $Z = 4$, $D = 3.60$. 40-484: 5.74_3 4.49_3 3.62_5 2.66_{10} 2.24_3 1.85_7 1.65_6 1.53_4. Hexagonal crystals to 1 mm, also massive, fine-grained. Pale pink-brown, pale pink streak, vitreous luster. Cleavage {001}, perfect. $H = 3$. $G = 3.64$. Biaxial (−); $N_x = 1.78$, N_y, $N_z = 1.81$; $2V = 28$–$43°$; $r > v$. BM Långban, Värmland, Sweden. AM 71:1517(1986).

46.2.7.1 Armangite $Mn_{26}As_{18}O_{50}(CO_3)(OH)_4$

Named in 1920 for the composition. HEX-R $P\bar{3}$. $a = 13.491$, $c = 8.855$, $Z = 1$, $D = 4.41$. 33-888: 4.88_1 3.95_2 2.95_6 2.92_4 2.77_{10} 2.44_4 1.76_3 1.75_2. Short prismatic crystals. Brown to black, brown streak, adamantine luster. Cleavage {001}, fair. $H = 4$. $G = 4.43$. Uniaxial (−); $N_O = 2.01$, $N_E = 1.99$. Långban, Värmland, Sweden. BM DII:1031, AM 64:748(1979).

46.2.8.1 Dixenite $Mn_{14}CuFe(AsO_4)(AsO_3)_5(SiO_4)_2(OH)_6$

Named in 1920 from the Greek for *two* and *stranger*, for the unique association of silicate and arsenite. HEX-R $R3$. $a = 8.233$, $c = 37.499$, $Z = 3$, $D = 4.38$. 35-

520: 4.10_9 3.90_5 2.96_8 2.92_{10} 2.83_5 2.40_6 2.37_8 1.55_6. Micaceous plates. Red-brown, pale brown streak, resinous to submetallic luster. Cleavage {0001}, perfect. H = 3–4. G = 4.36. Uniaxial (−); N = 1.96; weakly birefringent. *Långban, Värmland, Sweden.* BM *AM 66:1263(1981)*.

46.2.9.1 Seelite $Mg(UO_2)(AsO_3)_x(AsO_4)_{1-x} \cdot 7H_2O$ □ (x = 0.7)

Named in 1993 for Paul Seel (1904–1982) and Hilde Seel (1901–1987), U.S. mineralogists. MON $C2/m$. $a = 18.207$, $b = 7.062$, $c = 6.661$, $\beta = 99.65°$, Z = 2, D = 3.60. MR *24:* 9.05_{10} 6.48_3 5.72_3 4.85_5 4.44_8 3.99_5 3.52_6 1.92_8. Tufted spherules of tabular crystals to 1 mm. Bright yellow, white streak, vitreous luster. Irregular fracture. H = 3. G = 3.70. Biaxial (−); Z ∧ c = 5°; $N_x = 1.602$, $N_y = 1.737$, $N_z = 1.753$; 2V = 41°; extreme dispersion, r > v; strongly pleochroic. X, colorless; Y,Z, yellow. Occurs in oxidation zones of uranium deposits at Rabejc, Herault, France, and *Talmessi, central Iran.* BM *MR 24:463(1993), AM 79:1012(1994)*.

Class 47

Vanadium Oxysalts

47.1.1.1   Rossite $Ca(VO_3)_2 \cdot 4H_2O$

Named in 1927 for Clarence S. Ross (1880–1975), mineralogist, U.S. Geological Survey. TRIC $P\bar{1}$. $a = 8.552$, $b = 8.576$, $c = 7.028$, $\alpha = 101.50°$, $\beta = 114.96°$, $\gamma = 103.39°$, $Z = 2$, $D = 2.40$. 36-422: 7.26_9 6.64_{10} 6.02_5 3.93_3 3.86_{10} 3.43_6 3.03_6 3.00_5. Glassy coatings on sandstone. Yellow, yellow streak, vitreous to pearly luster. Brittle. $H = 2-3$. $G = 2.45$. Slowly soluble in water. Biaxial $(-)$; $Y \wedge b = 45°$; $N_x = 1.720$, $N_y = 1.783$, $N_z = 1.842$; $2V = 85°$. Occurs in sedimentary U-V deposits of the Colorado Plateau. *Bull Pen Canyon* and Burro mine, *San Miguel Co.*, Club mine, Montrose Co., Small Spot mine, Mesa Co., CO; Yellow Cat district, Grand Co., UT. BM *DII:1053*, *CM 7:713(1963)*, *MM 49:140(1985)*.

47.1.1.2   Metarossite $Ca(VO_3)_2 \cdot 2H_2O$

Named in 1927 as the dehydration product of rossite. TRIC $P\bar{1}$. $a = 6.215$, $b = 7.065$, $c = 7.769$, $\alpha = 92.58°$, $\beta = 96.39°$, $\gamma = 105.47°$, $Z = 2$, $D = 2.80$. 29-392: 7.69_4 6.77_6 5.93_{10} 5.03_6 3.84_2 3.10_2 3.04_5 2.52_3. Powdery coatings and veinlets. Pale yellow color and streak, pearly to dull luster. Biaxial $(+)$; $N_x = 1.840$, N_y, $N_z > 1.85$; 2V large. Occurrence and localities as for rossite. BM *DII:1054*, *CM 6:448(1960)*.

47.1.2.1   Delrioite $CaSr(VO_3)_2(OH)_2 \cdot 3H_2O$

Named in 1959 for Andres M. del Rio (1764–1849), Mexican mineralogist, who discovered vanadium in 1801. MON I/a or $I2/a$. $a = 17.170$, $b = 7.081$, $c = 14.644$, $\beta = 102.48°$, $Z = 8$, $D = 3.16$. 22-528: 6.53_{10} 4.39_6 4.19_3 3.54_8 3.26_6 2.79_6 2.57_3 2.17_5. Radial aggregates of acicular crystals. Pale yellow-green color and streak, vitreous to pearly luster. $H = 2$. $G = 3.1$. Biaxial $(-)$; $N_x = 1.783$, $N_y = 1.834$, $N_z = 1.866$; 2V medium to large; pleochroic, colorless to yellow. As coatings on sandstone, *Jo Dandy mine*, Montrose Co., CO. BM *AM 44:261(1959)*, *55:185(1970)*.

47.1.2.2   Metadelrioite $CaSr(VO_3)_2(OH)_2 \cdot H_2O$

Named in 1970 as the dehydration product of delrioite. TRIC Space group probably $P\bar{1}$. $a = 7.343$, $b = 8.382$, $c = 5.117$, $\alpha = 119.65°$, $\beta = 90.27°$, $\gamma = 102.82°$, $Z = 2$, $D = 4.52$. 22-600: 4.94_{10} 4.73_6 3.46_{10} 2.68_6 2.55_5 2.52_5

1.89_5 1.86_5. Fibrous crystals intergrown with delrioite. Pale yellow-green color and streak. H = 2. G = 4.3. *Jo Dandy mine, Montrose Co., CO.* BM *AM 55:185(1970)*.

47.1.3.1 Munirite $NaVO_3 \cdot 2H_2O$

Named in 1983 for Munir A. Khan, chairman, Pakistan Atomic Energy Commission. MON $P2_1/a$. $a = 16.72$, $b = 3.636$, $c = 8.015$, $\beta = 111.0°$, Z = 4, D = 2.31. *35-615:* 7.85_{10} 6.74_5 4.13_7 3.30_9 2.98_7 2.85_3 2.68_8 1.82_5. Radiating fibrous aggregates. Pearly white with a tinge of apple-green. G = 2.43. Biaxial (−); XYZ = cab; $N_x = 1.692$, $N_y = 1.756$, $N_z = 1.770$; 2V = 75°. *Bhimber area, Azad Kashmir, Pakistan,* on sandstone. BM *MM 47:391(1983), 52:716(1988)*.

47.1.3.2 Metamunirite $\beta\text{-}NaVO_3$

Named in 1991 for its relationship to munirite. ORTH *Pnma*. $a = 14.134$, $b = 3.648$, $c = 5.357$, Z = 4, D = 2.92. *32-1198*(syn): 7.09_1 5.01_{10} 3.53_3 3.25_2 3.02_1 2.95_3 2.68_1 2.54_1. Microscopic colorless needles. Soft, friable. $G_{syn} = 2.877$. Biaxial (+); XYZ = acb; $N_x = 1.780$, $N_y = 1.800$, $N_z = 1.800$; 2V = 30–40°. *Burro mine, near Slick Rock, San Miguel Co., CO,* in sandstone. BM *MM 55:509(1991)*.

47.2.1.1 Pascoite $Ca_3(V_{10}O_{28}) \cdot 17H_2O$

Named in 1914 for the locality. MON C2. $a = 16.834$, $b = 10.156$, $c = 10.921$, $\beta = 93.13°$, Z = 2, D = 2.46. *21-171:* 8.8_6 7.3_7 5.5_{10} 5.1_8 4.67_{10} 4.45_6 3.34_5 3.01_7. Granular crusts. Orange, yellow streak, vitreous to subadamantine luster. Cleavage {010}, distinct; conchoidal fracture. H = $2\frac{1}{2}$. G = 1.87. Biaxial (−); X = b; $N_x = 1.775$, $N_y = 1.815$, $N_z = 1.825$; 2V = 50°; r > v, very strong; pleochroic, X, pale yellow; Y, yellow; Z, orange. Soluble in water. Widely distributed in sedimentary U–V deposits of the Colorado Plateau. Paradox Valley, San Miguel Co., Rifle, Garfield Co., Uravan, Montrose Co., CO; Grants, Valencia Co., NM; Temple Rock, Emery Co., Mi Verde mine, San Juan Co., UT; *Minasragra, Pasco Prov., Peru.* BM *DII:1055, AC(A) 70:21(1966)*.

47.2.2.1 Hummerite $K_2Mg_2(V_{10}O_{28}) \cdot 16H_2O$

Named in 1951 for the locality. TRIC $P\bar{1}$. $a = 10.81$, $b = 11.01$, $c = 8.85$, $\alpha = 106.1°$, $\beta = 107.8°$, $\gamma = 65.67°$, Z = 2, D = 2.53. *14-155:* 9.3_3 8.2_{10} 7.4_7 7.0_5 3.31_4 3.13_4 2.73_6 2.11_4. Platy crusts on sandstone. Orange, yellow streak, pearly to dull luster. H = 2. Biaxial (−); $N_y = 1.81$; strong dispersion. Soluble in water. *Hummer mine, Montrose Co., CO;* Gold Quarry mine, Eureka Co., NV. BM *AM 36:326(1951), 40:315(1955)*.

47.2.3.1 Huemulite $Na_4Mg(V_{10}O_{28}) \cdot 24H_2O$

Named in 1966 for the locality. TRIC $P\bar{1}$. $a = 11.770$, $b = 11.838$, $c = 9.018$, $\alpha = 107.22°$, $\beta = 112.17°$, $\gamma = 101.5°$, $Z = 2$, $D = 2.40$. *18-1225:* 10.6_9 10.2_5 9.11_5 8.20_4 7.61_{10} 3.79_2 3.06_3 2.88_4. Granular coatings on sandstone. Yellow to orange, yellow streak, dull luster. $H = 2\frac{1}{2}$–3. $G = 2.39$. Biaxial $(-)$; $N_x = 1.679$, $N_y = 1.734$, $N_z = 1.742$; $2V = 20$–$25°$; $r > v$, strong; pleochroic, X, pale yellow; Y, golden yellow; Z, yellow-orange. Soluble in water. *Huemul mine, Mendoza Prov., Argentina.* BM *AM 51:1(1966)*.

47.2.4.1 Sherwoodite $Ca_9Al_2V_{28}O_{80} \cdot 56H_2O$

Named in 1958 for Alexander M. Sherwood (b.1888), chemist, U.S. Geological Survey. TET $I4_1/amd$. $a = 28.06$, $c = 13.56$, $Z = 4$, $D = 2.58$. *11-191:* 12.3_{10} 10.0_{10} 9.3_8 7.8_4 7.1_4 4.65_5 2.61_7 2.10_7. Equant crystals and polycrystalline aggregates. Dark blue-black, pale blue streak, vitreous to earthy luster. Subconchoidal to uneven fracture. $H = 2$. $G = 2.8$. Uniaxial $(-)$; $N_O = 1.765$, $N_E = 1.735$; pleochroic. O, green; E, blue. Occurs in sandstone in many U–V mines in the Colorado Plateau. *Peanut mine, Montrose Co.*, Matchless mine, Mesa Co., Fall Creek mine, San Miguel Co., Shadyside mine, San Juan Co., CO; Wilson Mesa, Grand Co., UT; Cerro de Pasco, Peru. BM *AM 43:749(1958), 63:863(1978)*.

47.3.1.1 Hewettite $CaV_6O_{16} \cdot 9H_2O$

Named in 1914 for D. Foster Hewett (1881–1971), geologist, U.S. Geological Survey. MON $P2_1/m$. $a = 12.290$, $b = 3.590$, $c = 11.174$, $\beta = 97.24°$, $Z = 1$, $D = 2.67$. *35-564:* 11.03_{10} 5.63_1 5.53_4 3.66_1 3.35_1 3.09_2 2.57_1 2.16_3. Microscopic needles and earthy aggregates. Deep red color and streak. Silky or dull luster. Soft. $G = 2.62$. Biaxial $(-)$; $N_x = 1.77$, $N_y = 2.18$, $N_z = 2.35$; $2V = 55°$; pleochroic, orange-yellow to dark red. Slightly soluble in water. Widely distributed in U–V deposits of the Colorado Plateau. Hummer and Jo Dandy mines, Montrose Co., Laverick Mesa, Mesa Co., CO; Grants, Valencia Co., NM; Temple Rock, Emery Co., Polar Mesa, Red Co., Cactus Rat mine, Grand Co., UT; Monument Valley, AZ; Bloomington, Bear Lake Co., ID; Gold Quarry mine, Eureka, NV; *Minasragra, Peru.* BM *DII:1060, CM 27:181(1989), AM 75:508(1990)*.

47.3.1.2 Metahewettite $CaV_6O_{16} \cdot 3H_2O$

Named in 1914 for its relationship to hewettite. MON $P2_1/m$. $a = 12.15$, $b = 3.590$, $c = 18.44$, $\beta = 118.03°$, $Z = 2$, $D = 3.05$. *33-317:* 8.19_{10} 5.36_1 3.58_3 3.06_2 2.81_3 2.30_2 2.21_2 2.03_1. Powdery masses or coatings. Deep red color and streak, dull luster. Soft. $G = 2.94$. Biaxial $(-)$; $Z = b$; $N_x = 1.70$, $N_y = 2.10$, $N_z = 2.23$; $2V = 52°$; pleochroic, orange-yellow to dark red. Occurrence and localities as for hewettite. BM *DII:1061, AM 75:513(1990)*.

47.3.1.3 Barnesite $Na_2V_6O_{16} \cdot 3H_2O$

Named in 1963 for William H. Barnes (b.1903), Canadian crystal chemist. MON $P2_1/m$. $a = 12.17$, $b = 3.602$, $c = 7.78$, $\beta = 95.03°$, $Z = 1$, $D = 3.23$. *16-601:* 12.2_3 7.9_{10} 3.70_1 3.45_4 3.12_7 2.92_2 2.27_3 1.80_3. Microscopic fibrous to bladed crystals. Dark red color and streak, adamantine luster. Soft. $G = 3.15$. Biaxial $(-)$; $Z = b$; $X \wedge c = 5°$; $N_x = 1.797$, N_y, $N_z > 2.0$; pleochroic, X, yellow; Y, orange-yellow; Z, red. *Cactus Rat mine, Grand Co., UT.* BM *AM 48:1187(1963), 75:514(1990).*

47.3.1.4 Hendersonite $Ca_{1.3}V_6O_{16} \cdot 6H_2O$

Named in 1962 for Edward P. Henderson (1898–1992), chemist, U.S. National Museum, who described several new vanadium minerals. ORTH *Pnam*. $a = 12.40$, $b = 3.59$, $c = 18.92$, $Z = 1$, $D = 2.80$. *15-277:* 9.43_{10} 6.20_1 5.18_1 4.70_2 3.79_1 3.11_4 2.31_1 2.26_2. Microscopic bladed to fibrous crystals and veinlets in sandstone. Greenish black to black, pearly to subadamantine luster. $H = 2\frac{1}{2}$. $G = 2.78$. Biaxial $(-)$; $N_x < 2.0$, N_y, $N_z > 2.10$; 2V medium; $r > v$, weak; moderately pleochroic, X, yellow-green; Y, green; Z, brown. *J. J. mine, Montrose Co., CO*; Eastside mine, San Juan Co., NM. BM *AM 47:1252(1962), 75:514(1990).*

47.3.1.5 Grantsite $(Na,Ca)_{2.2}V_6O_{16} \cdot 4H_2O$

Named in 1964 for the locality. MON $C2/m$. $a = 12.429$, $b = 3.604$, $c = 17.542$, $\beta = 95.33°$, $Z = 2$, $D = 2.95$. *16-408:* 12.4_2 8.74_{10} 4.37_1 3.61_3 3.01_1 2.87_1 2.73_1 2.28_1. Microscopic fibrous to bladed crystals. Dark olive-green to green-black, silky to subadamantine luster. Soft. $G = 2.94$. Biaxial $(-)$; $N_x = 1.82$, N_y, $N_z > 2.0$; pleochroic, X, green; Y, green-brown; Z, brown. *F-33 mine, near Grants, NM*; Golden Cycle and La Salle mines, Montrose Co., CO; Parco 25 mine, Grand Co., UT. BM *AM 49:1511(1964), 75:514(1990).*

47.3.2.1 Straczekite $(Ca,K,Ba)V_8O_{20} \cdot 3H_2O$

Named in 1984 for John A. Straczek (b.1914), U.S. geologist. MON $C2/m$. $a = 11.679$, $b = 3.661$, $c = 10.636$, $\beta = 100.53°$, $Z = 1$, $D = 3.21$. *42-598:* 10.45_5 5.74_1 3.49_{10} 3.26_1 2.49_1 1.94_3 1.83_5 1.80_1. Microscopic fibrous crystals forming foliated masses. Greenish-black color and streak, greasy luster. Cleavage {100}, perfect. Very soft. $G = 3.19$. Biaxial $(-)$; $N \sim 2.0$; 2V moderate to large; slightly pleochroic, green. *Union Carbide mine, Wilson Springs, AR*; Monument No. 2 mine, Apache Co., AZ. BM *MM 48:289(1984), AM 75:515(1990).*

47.3.2.2 Corvusite $(Na,Ca,K)V_8O_{20} \cdot 4H_2O$

Named in 1933 from the Latin *corvus*, raven, in allusion to the color. MON $C2/m$. $a = 11.706$, $b = 3.644$, $c = 11.10$, $\beta = 103.46°$, $Z = 1$, $D = 3.27$. *AM 75:* 10.82_{10} 3.59_1 3.46_2 3.26_1 2.90_1 2.60_1 1.93_1 1.82_1. Fibrous veinlets and masses. Blue-black color and streak; silky to dull luster. $H = 2\frac{1}{2}$–3.

G = 2.82. Widely distributed in U–V deposits of the Colorado Plateau. Ponto No. 3 claim, Gypsum Valley, San Miguel Co., and Hummer and Club mines, Montrose Co., CO; *Jack claim, Grand Co.*, and Temple Rock, Emery Co, *UT*; Monument No. 2 mine, Apache Co., AZ. BM *DI:603; AM 44:331(1959), 75:516(1990); CM 32:339(1994)*.

47.3.2.3 Fernandinite (Ca,Na,K)$V_8O_{20} \cdot 4H_2O$

Named in 1915 for Eulagio E. Fernandini, former owner of the Minasragra deposit. MON $C2/m$. $a = 11.680$, $b = 3.654$, $c = 11.023$, $\beta = 105.00°$, $Z = 1$, $D = 3.07$. *AM 75:* 10.68_{10} 5.65_1 3.55_1 3.48_3 2.83_1 2.55_1 1.95_1 1.84_8. Massive with flaky crystalline surfaces. Dark green color and streak, submetallic luster. Soft. G = 2.78. Biaxial; N ~ 2.05. *Minasragra, Peru.* BM *DII:1062; AM 75:516(1990), CM 32:339(1994)*.

47.3.2.4 Bariandite $Al_{0.6}V_8O_{20} \cdot 9H_2O$

Named in 1971 for Pierre Bariand (b.1933), French mineralogist. MON $C2/c$. $a = 11.70$, $b = 3.63$, $c = 29.06$, $\beta = 101.50°$, $Z = 2$, $D = 2.50$. *25-1006:* 14.2_{10} 7.08_5 5.72_7 3.48_8 3.43_8 2.85_7 1.94_7 1.83_7. Small fibrous crystals. Black, black streak. Cleavage {001}, perfect. G = 2.7. N > 1.85; strongly pleochroic, brown-gray. *Mounana, Gabon*, in U–V deposit in sandstone. *BM 94:49(1971), MR 6:239(1975), AM 75:517(1990)*.

47.3.2.5 Bokite $(Al,Fe)_{1.3}(V,Fe)_8O_{20} \cdot 4.7H_2O$

Named in 1963 for Ivan I. Bok (b.1898), Kazakh mineralogist. MON $C2/m$. $a = 11.85$, $b = 3.650$, $c = 11.11$, $\beta = 110.6°$, $Z = 1$. *AM 75:* 10.47_{10} 3.45_3 3.18_1 2.91_1 2.76_1 2.59_1 1.97_1 1.82_1. Reniform crusts with radiating fibrous structure. Black, black streak, submetallic to dull luster. One perfect cleavage. H ~ 3. G = 2.97–3.10. Biaxial; $N_x = 2.01$, $N_z = 2.06$; strongly pleochroic, dirty olive to deep red-brown. Union Carbide mine, Wilson Springs, Garland Co., AR; Monument No. 2 mine, Apache Co., AZ; *Kurumsak area, Kazakhstan.* BM *ZVMO 92:51(1963); AM 48:1180(1963), 75:517(1990)*.

47.3.2.6 Kazakhstanite $Fe_5V_{15}O_{39} \cdot 8.5H_2O$

Named in 1989 for the locality. MON $C2/m$. $a = 11.84$, $b = 3.650$, $c = 21.27$, $\beta = 100.0°$, $Z = 1$, $D = 3.52$. *ZVMO 118:* 10.51_{10} 3.48_6 3.18_2 2.92_3 2.76_3 2.61_4 2.10_2 1.97_2. Microscopic grains in veinlets and aggregates. Black, black streak, adamantine to dull luster. Cleavage {001}, perfect. H = $2\frac{1}{2}$. G = 3.4–3.6. Gold Quarry mine, Eureka Co., NV; *Northwestern Karatau, Kazakhstan*, in black shale. BM *ZVMO 118(5):95(1989), AM 76:667(1991)*.

47.3.3.1 Schubnelite $Fe_2V_2O_4(OH)_4$

Named in 1970 for Henri J. Schubnel (b.1935), French mineralogist. TRIC $P\bar{1}$. $a = 6.59$, $b = 5.43$, $c = 6.62$, $\alpha = 125°$, $\beta = 104°$, $\gamma = 84.72°$, $Z = 1$, $D = 3.34$. *24-542:* 6.41_8 5.15_9 4.47_{10} 3.21_{10} 3.19_9 3.07_8 2.73_5 2.68_6. Small crystalline aggregates. Black, brown streak, adamantine luster. Soft. G = 3.28. Gold Quarry mine, Eureka Co., NV; *Mounana, Gabon*, on sandstone. BM *BM 93:470(1970), AM 75:518(1990)*.

47.3.4.1 Fervanite $Fe_4V_4O_{16} \cdot 5H_2O$

Named in 1931 for its composition. MON Space group unknown. $a = 9.02$, $c = 6.65$, $\beta = 103.33°$. *27-257*: 8.79_{10} 6.45_9 3.29_2 3.22_8 3.15_7 2.99_5 2.93_5 2.91_2. Fibrous aggregates. Golden brown, adamantine luster. Biaxial (−); $N_x = 2.186$, $N_y = 2.222$, $N_z = 2.224$; 2V very small. Widely distributed in the U–V deposits of the Colorado Plateau. Wilson Springs, Garland Co., AR; Gypsum Valley, San Miguel Co., and Hummer mine, Montrose Co., CO; *Polar Mesa, Grand Co., UT*; Gold Quarry mine, Eureka Co., NV; *Mounana, Gabon*. BM *DII:1049; AM 44:336(1959), 75:518(1990)*.

47.3.5.1 Bannermanite $Na_{0.7}V_6O_{15}$

Named in 1983 for Harold M. Bannerman (1897–1976), U.S. economic geologist. MON $C2/m$. $a = 15.413$, $b = 3.615$, $c = 10.066$, $\beta = 109.29°$, $Z = 2$, $D = 3.55$. *24-1155*(syn): 9.49_8 7.25_{10} 4.75_6 3.38_9 3.20_7 3.07_7 2.92_6 1.81_7. Microscopic lathlike crystals. Black, gray-black streak, submetallic luster. Cleavage {100}, perfect; {010}, fair; brittle. G = 3.5. Biaxial; $N_{mean} = 2.2$. *Izalco volcano, El Salvador*, in fumaroles. BM *AM 68:634(1983), 75:519(1990)*.

47.3.6.1 Melanovanadite $CaV_4O_{10} \cdot 5H_2O$

Named in 1921 for its color and composition. TRIC $P\bar{1}$. $a = 6.360$, $b = 18.090$, $c = 6.276$, $\alpha = 110.18°$, $\beta = 101.62°$, $\gamma = 82.86°$, $Z = 2$, $D = 2.53$. *39-321*: 8.4_{10} 4.21_6 4.12_4 3.18_4 3.11_5 3.01_4 2.97_5 2.47_5. Prismatic crystals 1–2 mm. Black, dark red-brown streak; submetallic luster. Cleavage {010}, perfect; brittle. $H = 2\frac{1}{2}$. G = 2.55. Biaxial (−); $Z = b$; $Y \wedge c = 15°$; $N_x = 1.73$, $N_y = 1.96$, $N_z = 1.98$; 2V medium; pleochroic, red-brown. Peanut mine, Montrose Co., CO; *Minasragra, Peru*. BM *DII:1058, AM 72:637(1987)*.

47.4.1.1 Satpaevite $Al_{12}V_8O_{37} \cdot 30H_2O$

Named in 1959 for Kanysh I. Satpaev (1899–1964), Kazakh geologist. Probably ORTH Space group unknown. *13-476*: 5.87_6 3.91_7 2.50_4 2.37_9 1.93_{10} 1.55_6 1.47_8 1.27_4. Foliated masses and granular aggregates. Yellow, yellow streak, pearly to dull luster. Perfect pinacoidal cleavage. $H = 1\frac{1}{2}$. G = 2.4. Biaxial (+); $N_x = 1.676$, $N_y = 1.681$, $N_z = 1.690$; 2V near 70°; weakly pleochroic. *Kurumsak, Kara-Tau, Kazakhstan*. BM *ZVMO 88(2):157(1959), AM 44:1325(1959)*.

47.4.2.1 Vanalite $NaAl_8V_{10}O_{38} \cdot 30H_2O$

Named in 1962 for the composition. MON $P2/m$. $a = 12.616$, $b = 10.673$, $c = 10.923$, $\beta = 95.22°$, $Z = 1$, $D = 2.15$. *25-782*: 10.7_{10} 8.52_4 7.90_4 6.33_3 5.34_3 3.31_3 3.12_2 2.25_3. Microscopic wedge-shaped crystals. Yellow, yellow streak, waxy to vitreous luster. G = 2.3–2.4. Biaxial; $N_x = 1.710$, $N_z = 1.735$; weakly pleochroic, yellow. *Kara-Tau, Kazakhstan*. BM *ZVMO 91:307(1962); AM 48:1180(1963), 57:597(1972)*.

47.4.3.1 Pintadoite $Ca_2V_2O_7 \cdot 9H_2O$

Named in 1914 for the locality. Space group unknown. Thin coatings on sandstone. Yellow-green color and streak. Weakly pleochroic, yellow-green; moderately to highly birefringent. *Canyon Pintado, San Juan Co.*, and *Temple Rock, Emery Co., UT*; *Bull Canyon, San Miguel Co., CO*. BM *DII:1053*.

47.4.4.1 Rauvite $Ca(UO_2)_2(V_{12}O_{32}) \cdot 16H_2O$

Named in 1922 for its constituent elements: Ra, U, and V. Space group unknown. *8-288:* 10.7_{10} 5.83_1 3.87_2 3.49_4 2.95_5 2.62_3 2.22_1 2.10_1. Dense masses and botryoidal crusts. Purple-black to blue-black, olive to yellow-brown streak. Biaxial (−); $N_{mean} = 1.88$. *Temple Rock, Emery Co., UT*. BM *DII:1058*.

47.4.5.1 Uvanite $U_2V_6O_{21} \cdot 15H_2O$

Named in 1914 for the composition. Space group unknown. *8-322:* 5.9_6 4.6_4 2.94_{10} 2.24_8 1.71_9 1.64_4 1.48_7 1.12_4. Minutely crystalline masses and coatings. Yellow-brown. Two pinacoidal cleavages. Biaxial (+); $N_x = 1.817$, $N_y = 1.879$, $N_z = 2.057$; $2V = 52°$; pleochroic. X, light brown; Y, dark brown; Z, yellow-green. *Temple Rock, Emery Co., UT*. BM *DII:1056*.

47.4.6.1 Vanoxite $V_6O_{13} \cdot 8H_2O$

Named in 1925 for the composition. No X-ray data. Fine-grained in sandstone. Black. *Jo Dandy claim, Montrose Co.*, and *Small Spot mine, Mesa Co., CO*. BM *DI:601*.

47.4.7.1 Doloresite $H_8V_6O_{16}$

Named in 1957 for the locality. MON $C2/m$. $a = 19.07$, $b = 2.99$, $c = 4.83$, $\beta = 90.3°$, $Z = 1$, $D = 3.44$. *11-368:* 4.7_{10} 3.83_5 3.16_4 2.98_2 2.45_5 1.93_3 1.80_2 1.73_1. Radiating fibrous crystals. Black, green-black streak, submetallic luster. Lamellar twinning, twin plane {100}. G = 3.27–3.33. Biaxial; $N_{mean} = 1.90$; pleochroic, yellow-red to red-brown. Occurs in U–V deposits of the Colorado Plateau. *Dolores R., Montrose Co.*, and *Black Mama* and other mines in *Mesa Co., CO*; *Section 33 mine, Grants, Valencia Co., NM*; *Mi Vida mine, San Juan Co., UT*; *Monument No. 2 mine, Apache Co., AZ*. BM *AM 42:587(1957), 45:1144(1960); ZK 116:482(1961)*.

47.4.8.1 Simplotite $CaV_4O_9 \cdot 5H_2O$

Named in 1958 for Jack R. Simplot (b.1909) of the Simplot Mining Co., Boise, Idaho. MON $A2/m$. $a = 8.39$, $b = 17.02$, $c = 8.37$, $\beta = 90.42°$, $Z = 4$, $D = 2.66$. *11-267:* 8.51_{10} 4.26_1 3.42_1 3.14_2 2.84_1 2.62_3 2.52_1 1.88_1. Micaceous plates and hemispherical aggregates. Dark green, brown-black streak, vitreous luster.

Cleavage {010}, perfect. H = 1. G = 2.64. Biaxial (−); X = b; Z ∧ c = 58°; $N_x = 1.705$, $N_y = 1.767$, $N_z = 1.769$; 2V = 25°; r > v, weak; pleochroic, X, yellow; Y,Z, green. *Sundown Claim, San Miguel Co.*, and Peanut and J. J. mines, Montrose Co., *CO*; Vanadium Queen mine, San Juan Co., UT. BM *AM* *43*:16(*1958*).

Class 48

Anhydrous Molybdates and Tungstates

WOLFRAMITE GROUP

The wolframite group consists of isostructural minerals of the general formula

$$AWO_4$$

where

A = Mn, Fe, Zn

and the structure is in $P2/c$.

For example, the structure of ferberite, the principal member of the group, is based on W and Fe octahedra forming infinite zig-zag chains in the c axis direction. Each chain contains just one type of cation, and each octahedron is joined to the next by a common edge. Each W-octahedral chain is attached by common corners to 4 Fe chains, and each Fe chain is also surrounded by 4 W chains.

A complete solid-solution series exists between ferberite and hübnerite. Wolframite was originally considered a species name and referred to intermediate members of that series. Wolframite has been dropped as a species name but is still widely used in commerce to refer to tungsten ores and concentrates containing principally ferberite–hübnerite mineral, normally Fe rich. The nearly pure Mn end member is comparatively uncommon. Zincian members are rare, and sanmartinite has been found at only one location. Mo does not substitute for W in the structure. *ZK 127:61(1968), 144:238(1976)*.

WOLFRAMITE GROUP

Mineral	Formula	Space Group	a	b	c	β	D	
Hübnerite	$MnWO_4$	$P2/c$	4.829	5.759	4.998		7.234	48.1.1.1
Ferberite	$FeWO_4$	$P2/c$	4.72	5.70	4.96	90	7.55	48.1.1.2
Sanmartinite	$(Zn,Fe)WO_4$	$P2/c$	4.691	5.720	4.925		7.87	48.1.2.1

48.1.1.1 Hübnerite $MnWO_4$

Named in 1865 for Adolph Hübner, German mining engineer of Freiberg, Saxony, who first analyzed the mineral. Wolframite group. MON $P2/c$.

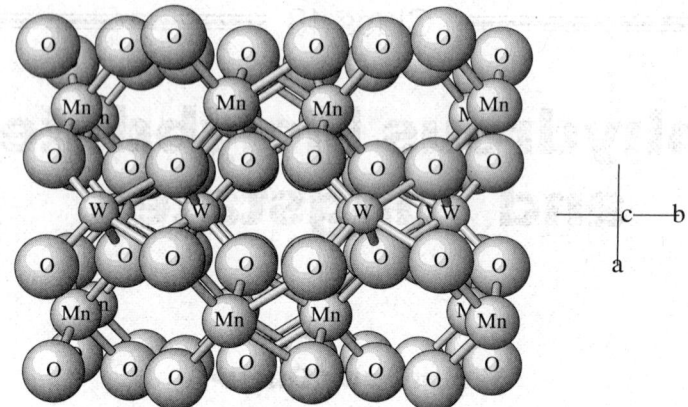

HEUBNERITE: MnWO$_4$

Huebnerite contains hexagonal closest-packed layers of oxygen stacked on (100), with all cations in octahedral interstices. Double chains of W and Mn edge-sharing octahedra are parallel to the c axis.

$a = 4.829$, $b = 5.759$, $c = 4.998$, $Z = 2$, $D = 7.234$. *NBSM 25(2):24(1963)*. *13-434*(syn): 4.84_6 3.78_6 3.70_5 3.00_{10} 2.95_9 2.88_3 2.50_5 1.73_3. Commonly prismatic to long prismatic [001], and flattened or tabular {100}. Usually striated [001]. As groups of parallel or subparallel crystals, or as radiating groups. Twins common on {100} as contact twins. Reddish brown, yellow to reddish-brown streak, submetallic–adamantine to resinous luster. Cleavage {010}, perfect; irregular fracture. $H = 4$–$4\frac{1}{2}$. $G = 7.12$. Near end member hübnerite is nonmagnetic. Biaxial (+); $X = b$; $Z \wedge c = 17$–$21°$; $N_x = 2.17$, $N_y = 2.22$, $N_z = 2.32$. $2V_{calc} = 73°$. X, bright yellow to orange-red; Y, greenish yellow to orange-red; Z, olive-green to brick-red. In RL, gray-white, weakly anisotropic, weakly pleochroic. Strong internal reflections, red-brown to yellow. R = 17.9, 14.85 (470 nm); 16.6, 13.9 (546 nm); 16.25, 13.65 (589 nm); 15.95, 13.45 (650 nm). *P&J:394*. Hübnerite is most commonly found in mesothermal veins, occasionally in epithermal ones, associated with pyrite, other sulfides, rhodochrosite, fluorite, and quartz. *Erie and Enterprise veins, Ellsworth, Mammoth District, NV*. Adams mine, San Juan Co., CO(!), in good crystals. Blue Wing district, Lehmi Co., ID. Little Dragoon Mts., Cochise Co., AZ, with scheelite and fluorite. In spectacular crystals and groups with quartz crystals from the Pasto Bueno mine(!), Ancash Dept., Peru. An ore of tungsten. RG *DII:1064*; *ZK 125:120(1967)*, *144:238(1976)*; *NJMA 113:13(1970)*; *NJMM:477(1976)*.

48.1.1.2 Ferberite FeWO$_4$

Named in 1863 for Moritz Rudolph Ferber (1805–1875), German factory owner, amateur mineralogist, and benefactor, in Gera. Wolframite group. MON $P2/c$. $a = 4.72$, $b = 5.70$, $c = 4.96$, $\beta = 90°$, $Z = 2$, $D = 7.55$. *ZK 127:61(1978)*. *21-436*: 4.69_8 3.75_6 3.65_5 2.94_{10} 2.47_6 2.470_6 2.195_5 2.189_5.

48.1.1.2 Ferberite

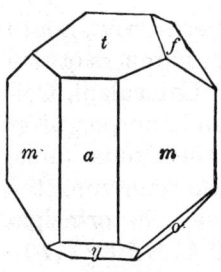
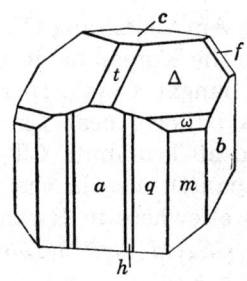
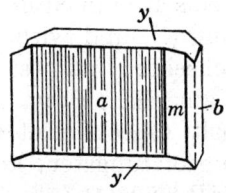

Wolframite. Bolivia.   Wolframite. Bolivia. Twin on {100}.

Ferberite

Forms: a{100}, b{010}, c{001}, r{120}, m{110}, l{210}, f{011}, y{$\bar{1}$02}, ω{111}, Δ{112}, t{102}, q{830},

Habit: Crystals commonly elongated [010] and flattened {100}. Usually striated [001], with {001} often striated [010]. Crystals often have a wedge-shaped appearance; groups of bladed crystals. Also forms excellent pseudomorphs after scheelite ("reinite"). Twins on {100} common, usually as simple contact twins; also as interpenetration twins. Twin plane {023} also common. **Physical properties:** Black, black to brownish-black streak, submetallic/adamantine luster. Cleavage {010}, perfect; uneven fracture; brittle. H = 4–4$\frac{1}{2}$. G = 7.51. Weakly magnetic. **Tests:** Easily fusible. Slowly decomposed by hot concentrated H_2SO_4 or HCl; decomposed by aqua regia with separation of tungstic oxide. **Chemistry:** A complete solid-solution series exists between ferberite and hübnerite; in most commercial deposits, the mineral is an intermediate member of the series, which would formerly have been called wolframite. On the other hand, zincian members are rare and sanmartinite has been found at only one location. Mo does not substitute for W in the structure. **Optics:** N ≈ 2.40; X = b; Z ∧ c = 17–21°; pleochroic. X, red-brown; Y, red-brown; Z, red brown-black. In RL, gray-white, similar to sphalerite, with brownish-red internal reflections and weak pleochroism. R = 18.6, 16.05 (470 nm); 18.7, 16.0 (546 nm); 18.6, 15.8 (589 nm); 18.5, 15.55 (650 nm). P&J:393. **Occurrence:** Ferberite–hübnerite deposits are found in greisen or quartz-rich veins and pegmatites immediately associated with granitic intrusive rocks; in high-temperature hydrothermal ("hypothermal") veins, associated with cassiterite, arsenopyrite, apatite, tourmaline, topaz, fluorite, specular hematite, molybdenite, and bismuth; in mesothermal veins, with cassiterite and sulfides, scheelite, bismuthinite, and siderite. **Localities:** Ferberite was formerly mined in the Nederland district, Boulder Co., CO(*), where it was found in finely crystallized druses in mesothermal veins. Cave Creek district, Maricopa Co., AZ. Elizabethtown district, Colfax Co., NM. At several mines in the Black Hills, SD. It was first observed in quartz veins near *Aquilas, Sierra Almagrera, Spain*. Large deposits, often beautifully crystallized, are mined at Panasqueira, Beira Baixa, Portugal(!), associated with arsenopyrite, apatite, schorl, and quartz. In Germany at Schneeberg and Zinnwald, Saxony. Horni Slavkov, Bohemia, Czech Republic. Cornwall, England, with tourmaline in cassiterite veins. Baia Sprie, Romania(*), in good crystals. Tae Hwa mine, Neung Am Ri, Chung Chong northern Prov., South Korea(!), and other localities on the Korean peninsula. Large numbers of rich veins are mined in Hunan and Jiangxi, China. In Bolivia at Calacalani, Colquiri(!), and with cassiterite at Viloco, near La Paz. Pseudomorphs after scheelite ("reinite") were found at Trumbull, CT, and at the Otome mine, Yamanashi Prov., Honshu, Japan(!); also in the Central African tungsten belt, especially at Gifurve and elsewhere in Rwanda(!). Use: The principal ore of tungsten. RG *DII:1064*, *GSAM 85:223(1962)*, *NJMA 113:13(1970)*, *MR 8:393(1977)*.

48.1.1.3 Sanmartinite $(Zn,Fe)WO_4$

Named in 1948 for the locality. Wolframite group. MON $P2/c$. $a = 4.691$, $b = 5.720$, $c = 4.925$, $Z = 2$, $D = 7.87$. *NSBM 25(2):40(1963)*. 15-774(syn): 4.69_4 3.73_4 3.62_4 2.93_{10} 2.91_9 2.86_3 2.472_4 2.464_4. Fine-grained masses; microscopic crystals, tabular {001}. Dark brown; brownish black, resinous luster. Cleavage {010}, perfect. G = 6.70. FeO and MnO enter the composition to the observed extent of about 8% and 1%, respectively. Found only at *Los Cerillos, SW of San Martin, San Luis, Argentina*, in a quartz vein, as an alteration of scheelite, associated with willemite. RG *DII:1072*, *AM 33:653(1948)*, *MM 42:281(1978)*.

48.1.2.1 Scheelite $CaWO_4$

Named in 1821 for Karl Wilhelm Scheele (1742–1786), Swedish chemist and co-discoverer (with Priestley) of oxygen, who proved that scheelite contains tungstic oxide. TET $I4_1/a$. $a = 5.242$, $c = 11.372$, $Z = 4$, $D = 6.120$. *NBSC 539(6):23(1956)*. Structure: In scheelite, Ca and W form a tetragonal framework. This framework can be said to be constructed of two cubes resting one on the other. The base of the bottom cube consists of four Ca atoms at the corners and one W atom at the center; at the centers of two opposite side faces there are Ca atoms, and of the other two side faces, W atoms. The top of the bottom cube consists of four W atoms at the corners and one Ca atom at the center; this also constitutes the base of the upper cube. The upper cube is a

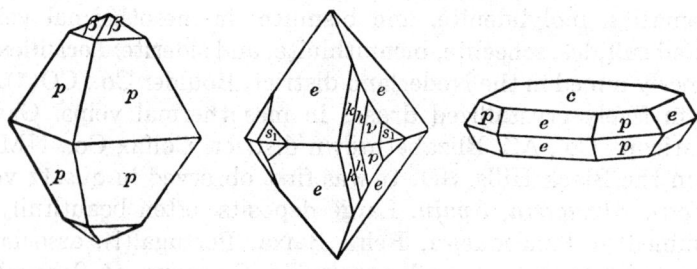

Scheelite
Forms: o{102}, e{101}, β{113}, p{111}, k{515}, h{313}, s{131}

48.1.2.1 Scheelite

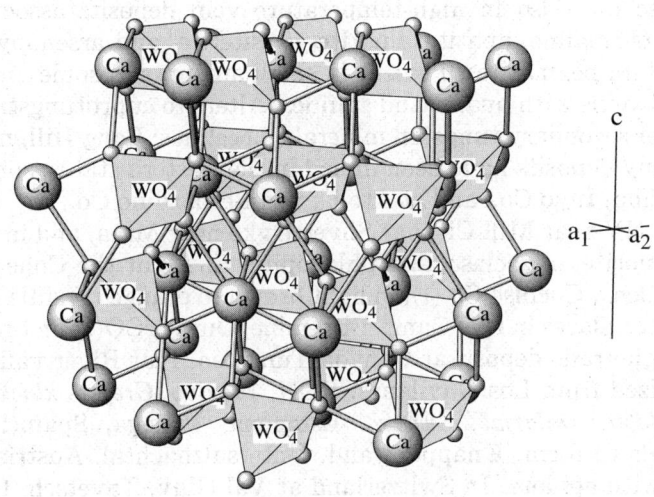

SCHEELITE: CaWO$_4$

Ca atoms and WO$_4$ tetrahedra strictly alternate in the [110] directions, but form interrupted chains in the c direction.

mirror image of the bottom cube, except rotated 90° with respect to the side face atoms. The W atoms are in tetrahedral coordination with oxygen atoms, the tetrahedra being considerably flattened in the direction of c. *DJ(1959), SR 29:331(1964). 41-1431*(syn): 4.77$_8$ 3.11$_{10}$ 3.07$_3$ 2.84$_4$ 2.62$_2$ 2.30$_2$ 1.928$_4$ 1.688$_2$. **Habit**: Crystals usually bipyramidal {011} or {112}. As discrete grains; massive. Twinning on {110} common, as penetration, or, less often, contact twins; composition plane {110} or {001}. **Physical properties**: White, colorless, yellow, pale green, orange, and so on; white streak; vitreous/adamantine luster. Cleavage {101}, distinct; {112}, interrupted; uneven to subconchoidal fracture. H = $4\frac{1}{2}$–5. G = 6.10. Fluoresces bright blue-white under SWUV when nearly pure CaWO$_4$; the presence of even small amounts of Mo substituting for W causes the fluorescence to become yellowish, up to about 2% when the fluorescence is pure yellowish white. Above 2% Mo there is no further change in color. Thermoluminescent. **Tests**: Decomposed by HCl or HNO$_3$ leaving a yellow powder of hydrous tungstic oxide which is soluble in ammonia. Fusible with difficulty. **Chemistry**: A partial solid-solution series exists between scheelite and powellite, but natural material with over 4 mol % MoO$_3$ is not known; some older analyses show higher MoO$_3$ but it is not certain that only one phase was present. Nevertheless, synthetic scheelites and powellites have been prepared having a broad range of substitution, and it is probable that a complete series exists synthetically. **Optics**: Uniaxial (+); N$_O$ = 1.918, N$_E$ = 1.935. **Occurrence**: Scheelite is typically found in contact metamorphic deposits, especially where granitic rocks have been intruded into limestones with the formation of skarn or tactite associated with grossular, andradite, wollastonite, diopside, tremolite, vesuvianite, chalcopyrite, epidote, quartz, pyrite, and so on. Also in high-temperature vein deposits associated with wolframite, tourmaline, apatite, fluorite, cassiterite, and arsenopyrite. Also, rarely, in certain pegmatites. Present in small amounts in some mesothermal to epithermal veins with quartz and sulfides. Alters to cuprotungstite, anthoinite, and other secondary tungsten minerals. **Localities**: Long Hill, near Trumbull, CT. Many deposits have been mined in the western states, some of large size: near Bishop, Inyo Co., and at Atolia, San Bernardino Co., CA, in epithermal veins. In NV near Mill City; at Silver Dyke near Mina, and in pegmatite with beryl, fluorite, oligoclase, and phlogopite. In AZ at the Cohen tungsten mine, E of Wilcox, Cochise Co.(!), in light brown to orange crystals up to 7 kg, and many other places in AZ. Camp Bird mine, Ouray, CO(!), in fine crystals. In a large, high-grade deposit at Canada Tungsten, Flat River valley, NWT. Well crystallized from Los Gavilanes, BCN, Mexico. *Gregers klack, Bispberg iron mine, Säter, Dalarme, Sweden.* Estepona, Malaga, Spain(!), gemmy orange crystals to 5 cm. Knappenwand, Untersulzbachtal, Austria(!), colorless crystals with epidote. In Switzerland at Val Giuv, Tavetsch, Grisons(*), and at Kammegg, Guttannen, Uri(!), crystals to 10 cm. In Germany in fine specimens from Zinnwald, Saxony(!), and similarly from Cinovec, Bohemia, Czech Republic. Felbertal, Austria. In Italy in large fine crystals in serpentine from Traversella, Piedmont(!). At Tae Wha, Korea, in magnificent sharp brown crystals(!) up to 1 kg, associated with quartz, wolframite, and bertrandite. From Sichuan, China(!), in rich orange crystals to 10 cm with "goshenite" (colorless beryl) crystals, muscovite, and quartz. Gifurve, Rwanda, and elsewhere in the Central African tungsten belt, in large crystals often pseudomorphously altered to ferberite ("reinite"). In Brazil in large metasomatic deposits at Currais Novos, RGN, and in a hypothermal gold vein, the Morro Velho mine, Nova Lima, MG(!), in sharp orange-red crystals associated with pink apatite, pyrrhotite, siderite, dolomite, and quartz. Use: An important ore of tungsten. RG *DII:1074, AM 37:719(1952), JCP 40:504(1964)*.

48.1.2.2 Powellite CaMoO$_4$

Named in 1891 for John Wesley Powell (1834–1902), American explorer and geologist. TET $I4_1/a$. $a = 5.226$, $c = 11.430$, $Z = 4$, $D = 4.255$. *NBSC 539(6):22(1956)*. 29-351(syn): 4.76_3 3.10_{10} 2.86_1 2.61_2 1.929_3 1.848_1 1.694_1 1.588_2. Crystals pyramidal $\{111\}$; more rarely tabular $\{001\}$; also massive, with foliated structure pseudomorphous after molybdenite. White, yellow, brown, gray; subadamantine to greasy luster. Indistinct cleavage, uneven fracture. $H = 3\frac{1}{2}$–4. $G = 4.27$. G increases markedly with increase in W content. Fluoresces creamy yellow in both SWUV and LWUV. Decomposed by HCl and HNO$_3$. W may substitute for Mo up to Mo/W = 9:1, but a complete series to CaWO$_4$ does not exist in natural material. Uniaxial (+); N$_O$ = 1.967, N$_E$ = 1.978. Powellite is usually a secondary mineral, often formed by the alteration of molybdenite. Also sometimes formed in low-temperature hydrothermal environments, as with zeolites in vugs in basalt or in some low-temperature copper or mercury mines. More rarely formed in skarns and contact metamorphic environments with scheelite. Peacock lode, Seven Devils district, Adams Co., ID. In vugs in altered rhyolite at the Tonopah Divide mine, Nye Co., NV. South Hecla and Isle Royale mines, Houghton Co., MI, with native copper and epidote. Pine Creek tungsten mine, Inyo Co., CA. Huahuaxtla mercury mine, GRO, Mexico. In metasomatic deposits with scheelite and molybdenite in the Nuratinsky range, Uzbekistan. In India in vugs in traprock quarried on Pandulena Hill near Nasik, Maharashtra(!), with apophyllite and zeolites, forming magnificent clear sharp yellow-orange crystals up to 6 cm. In Chile at Tierra Amarilla, Atacama, in green crusts resembling smithsonite, with szenicsite. RG $DII:1079$, SPC $13:414(1968)$.

48.1.3.1 Wulfenite $PbMoO_4$

Named in 1841 for Franz Xavier Wulfen (1728–1805), Austro-Hungarian Jesuit and mineralogist, who wrote a monograph on the lead ores of Carinthia. TET $I4_1/a$. $a = 5.435$, $c = 12.11$, $Z = 4$, $D = 6.815$. NBSC $539(7):23(1957)$. Structure: ZK $121:158(1965)$. 44-1486(syn): 4.96_1 3.25_{10} 3.03_2 2.72_2 2.02_2 1.921_1 1.787_1 1.653_2. Crystals commonly square tabular $\{001\}$, often very thin, or with a flat vicinal pyramid replacing $\{001\}$; less frequently pseudo-octahedral; rarely cuboidal or short prismatic pyramidal. Sometimes shows hemimorphic habit, possibly due to twinning. Twinning on $[001]$, common but rarely evident through the usual morphology. AM $40:857(1955)$. Also massive, granular. Yellow, orange, red, gray, brown, blue, and so on, with subadamantine to greasy luster. Cleavage $\{011\}$, distinct; $\{001\}$, $\{013\}$, indistinct; subconchoidal to uneven fracture; not very brittle. $H = 2\frac{1}{2}$–3. $G = 6.7$–7.0. W substitutes for Mo, with W/Mo up to at least 1:1, indicating at least a partial solid-solution series with stolzite; however, most wulfenites have only traces of W. Ca substitutes for Pb, and a partial series exists with the isostructural mineral powellite. Small amounts of V, Cr, and As may also (rarely) substitute for Mo. Uniaxial $(-)$; $N_O = 2.405$, $N_E = 2.283$. Wulfenite is a secondary mineral found in the oxide zone of deposits of Pb- and Mo-containing minerals. When mined to the primary zone, some wulfenite deposits have shown no obvious primary Mo-containing mineral, and in these cases the Mo is thought to have been introduced by laterally moving supergene solutions that have leached traces of Mo out of the surrounding formations. Common associated minerals are mimetite, cerussite, pyromorphite, vanadinite, goethite, and manganese oxides. Found in many places in the United States: Loudville mines, MA; Wheatley mine, Phoenixville, PA; Sheman tunnel, Leadville, CO; in NM at the Stephenson–Bennett mine, Dona Ana Co.(!), and in the Hillsboro district, Sierra Co.(*); especially in AZ, where in Santa Cruz Co. it is found in spectacular large thin plates, some crystals measuring > 10 cm on edge, at the Glove mine(!), Tyndall district, and in the Palmetto district, Patagonia Mts.; in Pinal Co. abundant at the Mammoth–St Anthony mine(!), and at the Florence mine; in Yuma Co. at the Red Cloud mine(!), in large brilliant orange-red crystals considered by many to be the world's finest, and at the Melissa mine(*); in Pima Co. at the Old Yuma mine(!) and the Total Wreck mine; in Cochise Co. in the Turquoise district, at the Silver Bill/mystery tunnel(*), Gleeson(!), in the Tombstone district(!), and at the Hilltop mine(!); in Gila Co. at the 79 mine(!) and in the Castle Dome district(*); and in Maricopa Co. at the Rowley mine(!). In UT in the Lucin district, Box Elder Co. Many localities in Mexico, especially in CHIH at Los Lamentos(!), in large toffee-colored thick square plates, at the San Carlos mine(!), at San Pedro Corralitos(!), and at the La Aurora mine, Cuchillo Parado; in SON at Cerro Prieto (= San Francisco mine), Magdalena(!), in large yellow crystals with orange-red mimetite, and at Cucurpe; and in DGO at the Ojuela mine(!), Mapimi. *Bleiberg*(!) and *Schwarzenbach*(!), *Carinthia, Austria*. In Slovenia at Crna(!), often twinned; and at Mežica(!)(= Mies). Příbram, Bohemia, Czech Republic. In Iran at the Tchah Mille, Tchah-Kharboze (near Anarak)(*) in fine red-orange crystals, Nakhlak, and Seh-Changi mines. Gansu Province, China, in thick dark red crystals. In Morocco at Les Dalles mine(!), Mibladen, and the Zelidja mine(!), Touissit. In Zaire at M'Fouati(!), and at Musonoi mine, Shaba. Broken Hill mine, Zambia. Tsumeb mine(!), Namibia, sometimes highly tungstenian and in a wide variety of habits and colors. In Australia at the Junction mine, Broken Hill, NSW, and a tungstenian variety ("chillagite") at the Christmas Gift mine. Use: A minor ore of lead, of decreasing importance, as it is found only in the now depleted zones of oxidation. RG *DII:1081, AM 51:1212(1966)*.

48.1.3.2 Stolzite PbWO$_4$

Named in 1845 for Joseph Alexis Stolz (1803–1896) of Teplice, Bohemia, who first drew attention to the species. Dimorphous with raspite. TET $I4_1/a$. $a = 5.462$, $c = 12.049$, $Z = 4$, $D = 8.408$. *NBSM 25(5):34(1967). 19-708:* 3.25_{10} 3.01_2 2.73_3 2.02_4 1.931_2 1.782_2 1.661_4 1.626_2. Usually pyramidal with {111} or {101}; also thick tabular {001}; infrequently prismatic. Reddish brown, yellow, red, brown, green; uncolored streak; resinous to subadamantine luster. Cleavage {001}, imperfect; {011), indistinct; conchoidal to uneven fracture; brittle. H $= 2\frac{1}{2}$–3. G $= 8.34$. Soluble in HCl with separation of yellow tungstic acid. Natural stolzites usually show little substitution of Mo for W. There is a partial solid-solution series from about W/Mo 1:1 to pure wulfenite, and stolzite from the Lost Gulch area, Globe District, Gila Co., AZ, contains about 9% MoO$_3$, indicating that a complete series may exist. *AM 43:156(1958)*. Uniaxial ($-$); N$_O$ = 2.269, N$_E$ = 2.182 (Na). Manhan lead mine, Southampton, MA. Wheatley mine, Phoenixville, PA. Boerich claims, Dragoon Mts., and at the Reef mine, Huachuca Mts., Cochise Co., AZ. Cariboo district, BC, Canada, with scheelite and ferritungstite. *Cínovec (Zinnwald), Krušné Hory (Erzgebirge), Čechy (Bohemia), Czech Republic*. Brandy Gill lead mine, Carrock Fell, Cumberland, England. Mine de Sainte Lucie, Saint Léger de Peyres, Lozère, France(!), thin to thick tabular to 8 cm and octahedra to 10 mm. Tsumeb, Namibia(!), in crystals to 20 mm. Formerly abundant and in large crystals (to 25 mm) at the Proprietary mine, Broken Hill, NSW, Australia(!), in a manganese gossan. Sumidoro mine, near Mariana, MG, Brazil, with raspite. RG *DII:1087*, *SPC 15:928(1971)*, *BSE 14:23(1991)*.

48.1.4.1 Raspite PbWO$_4$

Named in 1897 for Charles Rasp (1846–1907), German–Australian prospector, discoverer of the Broken Hill ore deposit. Dimorphous with stolzite. MON $P2_1/n$. $a = 13.555$, $b = 4.976$, $c = 5.561$, $\beta = 107.63°$, $Z = 4$, $D = 8.45$. Structure: *AC(B) 33:162(1977)*. *29-784:* 3.63_{10} 3.60_2 3.49_4 3.23_4 2.92_3 2.76_2 2.71_6 1.85_2. Crystals usually tabular {100}, often elongated [100] or thin tabular {$\bar{1}$01}; {100} striated parallel to [010]. Twinning on {100} common; also on {$\bar{1}$02}. Yellowish brown, light yellow, gray; adamantine luster. Cleavage {100}, perfect. $H = 2\frac{1}{2}$–3. $G = 8.46$. Decomposed by HCl with separation of yellow tungstic acid; fusible. Biaxial (+); $Y = b$; $Z \wedge c \approx 30°$; $N_x = 2.27$, $N_y = 2.27$, $N_z = 2.30$; $2V \approx 0°$. Broken Hill, NSW, Australia, with stolzite in manganiferous gossan. Sumidoro mine, Municipo de Mariana, MG, Brazil, with stolzite. RG *DII:1089*.

48.2.1.1 Russellite (BiO)$_2$WO$_4$

Named in 1938 for Arthur Russell (1878–1964), British mineralogist. TET $I\bar{4}2d$ or $I4/amd$. $a = 5.48$, $c = 11.5$, $Z = 2$, $D = 7.40$. *MM 37:705(1970)*. *26-1044:* 3.12_{10} 2.71_8 1.92_8 1.64_9 1.58_3 1.25_4 1.22_3 1.11_3. Fine-grained, compact masses. Pale yellow, yellowish green. $H = 3\frac{1}{2}$. $G = 7.35$. Castle an Dinas mine, St. Columb Major, Cornwall, England, associated with wolframite, topaz, Li-mica, tourmaline, cassiterite, bismuth, and bismuthinite. Poona, Murchison division, WA, Australia, associated with bismite, koechlinite, and bismutite in pegmatite. RG *DII:604*.

48.2.2.1 Koechlinite (BiO)$_2$MoO$_4$

Named in 1914 for Rudolf Koechlin (1862–1939), Austrian mineralogist. ORTH $Pca2_1$. $a = 5.482$, $b = 16.199$, $c = 5.509$, $Z = 4$, $D = 8.28$. Structure: *AC(C) 40:2001(1984)*. *21-102:* 3.15_{10} 2.75_2 2.741_2 2.701_2 1.942_2 1.925_2 1.652_2 1.634_2. Thin rectangular plates and laths flattened {010} and striated [010]; massive, earthy. Twin plane {101}, both contact and penetration. Greenish yellow. Cleavage {010}, perfect; very brittle. Easily soluble in HCl, with difficulty in HNO$_3$. Biaxial (−); $XYZ = cab$; $N_x = 2.52$, $N_y = 2.61$, $N_z = 2.67$ (Li); weakly pleochroic; 2V large; $r < v$, rather strong. Daniel mine, Schneeberg, Saxony, Germany, in cavities in vein material composed of quartz, bismuth, and smaltite. Dunallan mine, Coolgardie, WA, Australia, as an alteration product of tetradymite, and at Bygoo, NSW, with bismoclite. RG *AM 28:536(1943)*, *DII:1092*, *NJMM:1(1951)*, *FM 60:162(1982)*.

48.2.3.1 Tungstibite $(SbO)_2WO_4$

Named by Walenta in 1995 for the composition. ORTH $P22_12_1$. $a = 8.59$, $b = 9.58$, $c = 6.12$, $Z = 4$. PD 3.32_{10} 3.06_{10} 2.98_4 2.73_6 2.46_5 1.919_4 1.829_3 1.718_3. Forms globular aggregates to 1mm diam., composed of thin tabular crystals. Flattened (001) and elongated along one axis, sometimes spear-shaped. Green to dark green, streak greenish, luster dull to pearly, translucent to opaque. Cleavage (001) perfect, irregular fracture. H = 2. G = 6.90. Biaxial (+); $N_x = 2.285$ (dark green to brownish green), $N_y = 2.40$ (dark green), $N_z = 2.58$ (pale green to yellowish); $2V_{calc} = 82°$; dispersion r < v, strong, Z = c. From the *Clara mine, near Oberwolfach, central Black Forest, Germany*. Associated minerals are quartz, barite, fluorite, and occasionally with tetrahedrite, chalcopyrite, or cervantite. *CE 55:217(1995)*.

48.3.1.1 Lindgrenite $Cu_3(MoO_4)_2(OH)_2$

Named in 1935 for Waldemar Lindgren (1860–1939), American mining geologist and teacher. MON $P2_1/n$. $a = 5.394$, $b = 14.023$, $c = 5.608$, $\beta = 98.5°$, $Z = 4$, $D = 4.310$. *CM 6:31(1957)*. Structure: *NJMM:234(1985)*. *36-405:* 4.16_5 3.58_3 3.52_{10} 3.48_7 2.78_2 2.72_3 2.68_4 2.67_2. Crystals platy {010}; rarely acicular [101]. {010} striated parallel to [001]. Pistachio green, dark yellow-green; vitreous luster. Cleavage {010}, perfect; {101}, {100}, poor. $H = 4\frac{1}{2}$. G = 4.26. Easily soluble in HCl and HNO_3. Biaxial (−); X ∧ c = 9°; Z = b; $N_x = 1.930$, $N_y = 2.002$, $N_z = 2.020$; 2V = 71°; r > v. Inspiration copper mine, Gila Co., AZ. Helena mine, NE of Cuprum, Seven Devils district, Adams Co., ID, with chrysocolla, brochantite, and limonite. Brandy Gill, Carrock Fell, Cumberland, England, associated with wulfenite and stolzite. *Chuquicamata, Antofagasta, Chile*, with antlerite, limonite, and hematite; also with szenicsite at Tierra Amarilla, Chile. RG *DII:1094*, *MM 30:723(1955)*.

48.3.2.1 Cuprotungstite $Cu_3(WO_4)_2(OH)_2$

Named in 1869 for the composition. TET Space group unknown. $a = 8.93$, $c = 14.48$, $Z = 12$, $D = 7.06$. *MM 43:448(1979)*. *34-1297:* 7.76_5 4.03_8 3.80_8 2.53_{10} 2.45_4 2.35_4 1.73_5 1.46_7. Microcrystalline masses and crusts. Pistachio-green to olive-green; vitreous, waxy, or earthy luster. Easily soluble in acids. Uniaxial (−); $N_O = 2.18$, $N_E = 2.10$. Cuprotungstite is a common mineral in the oxidized zone of scheelite deposits containing primary copper sulfides. Green monster mine, near White River, Kern Co., CA. In *Mexico* from near *La Paz, BCS*, and abundant at the Guadalupana mine, Potrero de Bojorquez, CHIH. Villa Salto, Sardinia, Italy. Sorpresa, Montoro, Spain. Mitake mine, Kai, and at Ofuku, Nagato, Japan. At several mines in Georgiana Co., NSW, Australia. Llamuco copper mines, near Santiago, Chile. RG *DII:1091*, *BM 92:497(1969)*, *AUM 3:61(1988)*, *AM 75:713(1990)*.

48.3.3.1 Anthoinite $AlWO_3(OH)_3$

Named in 1947 for Raymond Anthoine (1888–1971), Belgian mining engineer, who wrote on prospecting of alluvial deposits. TRIC $P1$ or $P\bar{1}$. $a = 9.21$,

$b = 11.36$, $c = 8.26$, $\alpha = 94.75°$, $\beta = 90°$, $\gamma = 92.58°$, $D = 4.84$. 25-1489: 5.63_{10} 4.33_3 4.19_8 4.12_3 3.97_7 3.65_2 3.06_7 2.46_6. Chalky masses resembling kaolin. White, powdery luster. $H = 1$. $G = 4.8$. Slowly attacked by HCl and HNO_3. Easily soluble in strong KOH solution. *Mt. Misobo tungsten mine, Kalima district, Maniema, Zaire*, in eluvium; also quite common as an oxidation–leaching product of scheelite throughout the Central African tungsten belt. Kara mine, SSW of Burnie, TAS, Australia, as an alteration product of scheelite, associated with mpororoite. RG *DII:1097, BGSF 43:95(1970), MR 12:81(1981), MM 48:397(1984)*.

48.3.4.1 Deloryite $Cu_4(UO_2)(MoO_4)_2(OH)_6$

Named in 1992 for Jean Claude Delory, French mineral collector, who found the mineral. MON *C2, Cm*, or *C2/m*. $a = 19.83$, $b = 6.112$, $c = 5.529$, $\beta = 103.9°$, $Z = 2$, $D = 4.84$. *NJMM:58(1992)*. PD: 4.82_8 4.43_4 4.28_4 4.10_{10} 3.73_9 3.25_4 2.63_4 2.48_6. Crystals tabular {010}, elongate [001], to max. 3 mm, in rosettes. Dark green to black, green streak, vitreous to greasy luster, transparent to nearly opaque. Cleavage {010}, {100}, perfect; {001}, good; conchoidal fracture. $H = 4$. $G = 4.9$. Nonfluorescent. Soluble in HCl. Biaxial (+); $X \wedge c = 21.9°$; $Y \wedge a = 36.2°$; $Z = b$; $N_x = 1.90$, $N_y = 1.93$, $N_z = 1.96$, $r > v$, strong. Found at the *Cap Garonne mine near Pradet, Var, France*, on a dendritic quartz gangue, associated with metazeunerite, atacamite, paratacamite, malachite, tourmaline, and barite. Crystallographically related to derriksite. RG *AM 77:1305(1992)*.

48.3.5.1 Szenicsite $Cu_3MoO_4(OH)_4$

Named in 1994 for Terry Szenics (b.1947) and Marissa Szenics (b.1955), U.S. miners and mineral collectors, who found the mineral. ORTH *Pnnm*. $a = 8.449$, $b = 12.527$, $c = 6.067$, $Z = 4$, $D = 4.30$. *MR 25:76(1994)*. PD: 5.06_5 3.76_{10} 2.77_6 2.59_7 2.13_3. Bladed crystals are flattened {100}, elongate [001] up to 3 cm × 1 cm, usually in radial aggregates. Dark green, adamantine luster. $H = 3\frac{1}{2}-4$. $G = 4.26$. Nonfluorescent. Biaxial (+); XYZ = *bac*; $N_x = 1.886$, $N_y = 1.892$, $N_z = 1.903$; $2V = 74°$, nonpleochroic, $r > v$, strong. From the *Jardinera I mine near Inca do Oro, 100 km S of Tierra Amarilla, Atacama, Chile*, where it was derived from the oxidation of bornite and molybdenite and associated with abundant green powellite, chrysocolla, brochantite, hematite, and quartz. RG *AM 79:1210(1994)*.

48.4.1.1 Sedovite $U(MoO_4)_2$

Named in 1965 for Georgii Yakolevich Sedov (1877–1914), Russian polar investigator. ORTH Space group unknown. $a = 3.36$, $b = 11.08$, $c = 6.42$, $Z = 8$, $D = 5.11$. 18-1425: 11.0_9 5.53_8 3.70_8 3.37_9 3.19_{10} 3.06_9 2.78_6 2.56_6. Powdery, or radial-fibrous bundles of acicular crystals. Brown, reddish brown. Cleavage parallel to elongation. $H = 3$. $G = 4.2$. Soluble in concentrated HCl, HNO_3, and H_2SO_4 only with difficulty, on boiling. Biaxial; $Z \wedge$ elongation, $38°$; $N > 1.79$; weakly pleochroic; positive elongation. In

RL, gray with a brownish tint; strong red internal reflections; slightly anisotropic; R = 11.8 (589 nm). From *Russia* (?), in the supergene zone of a U–Mo deposit, associated with pitchblende, molybdenite, gypsum, wulfenite, and mourite. RG *ZVMO 94:548(1965), AM 51:530(1966)*.

48.4.2.1 Jixianite $Pb(W,Fe^{3+})_2(O,OH)_7$

Named in 1979 for the locality. ISO $Fd3m$. $a = 10.359$, $Z = 8$, $D = 7.89$. *AM 64:1330(1979)*. 33-760: 5.97_5 2.98_{10} 2.59_6 1.83_8 1.56_7 1.19_5 1.16_5 0.997_5. Crusts or honeycomblike microcrystalline aggregates; rarely in octahedral crystals. Red; brownish red, yellow streak, resinous luster. No cleavage. H = 3. G = 6.04. Weakly magnetic. Soluble in hot concentrated H_3PO_4, but not in other mineral acids. Isotropic; N = 2.26–2.31. *Jixian, Ji county, Hebei, China*, in the oxidized zone of a quartz vein containing primary wolframite, cassiterite, pyrite, chalcopyrite, and scheelite. Associated minerals are wulfenite, silver, copper, goethite, and malachite. RG

Class 49

Basic and Hydrated Molybdates and Tungstates

49.1.1.1 Hydrotungstite $H_2WO_4 \cdot H_2O$

Named in 1944 for the composition. MON $P2/m$. $a = 7.379$, $b = 6.901$, $c = 3.748$, $\beta = 90.37°$, $Z = 2$, $D = 4.655$. 18-1420: 6.96_{10} 3.77_3 3.70_3 3.47_3 3.31_4 3.26_4 2.65_1 2.63_1. Platy microcrystals. Dark green; yellowish green, vitreous to dull luster. Cleavage {010}, imperfect. $H = 2$. $G = 4.60$. Insoluble in acids, readily soluble in NH_4OH. In CT, yields abundant water, turns yellow, then bleaches. Biaxial (+); $N_x = 1.70$, $N_y = 1.95$, $N_z = 2.04$; $2V = 52°$; $r < v$; $Z > Y > X$. X, colorless; Y, yellow-green; Z, dark green. Forms from the oxidation of ferberite under acid conditions. *Calacalani mine, Oruro, Bolivia,* in crystals. Also reported from the Central African tungsten belt. RG *AM 29:192(1944), 48:935(1963); BGSF 43:89(1971); MR 12:81(1981).*

49.1.2.1 Moluranite $H_4U^{4+}(UO_2)_3(MoO_4)_7 \cdot 18H_2O$

Named in 1959 for the composition. Amorphous. 29-1371(heated): 4.17_{10} 3.44_5 2.58_5 2.12_5 2.03_5 1.97_5 1.31_6 1.30_6. Colloform accumulations in fractures. Black; translucent brown in thin splinters; resinous luster. Brittle. $H = 3$–4. $G = 4$. Soluble with difficulty in acids, more easily when hot. When heated changes to iriginite. Isotropic; $N = 1.97$–1.98. In RL, light gray, similar to pitchblende. From *Russia (?),* in thin fissures in granulated albitite, associated with molybdenite, chalcopyrite, galena, brannerite, and other U–Mo compounds. RG *AM 43:380(1958), 45:258(1960); ZVMO 88:564(1959).*

49.2.1.1 Ferrimolybdite $Fe_2^{3+}(MoO_4)_3 \cdot 8H_2O$

Named in 1914 for the composition. ORTH Space group unknown. 41-1388: 9.81_9 8.24_{10} 6.66_4 4.16_2 3.35_3 3.17_2 3.04_2 2.78_2. Fibrous crusts; tufted aggregates; powdery, earthy; as radiating rosettes of acicular microcrystals. Canary-yellow; adamantine, or silky to earthy luster. $H = 1$. $G = 4.46$. The water content is readily lost in very dry atmospheres or above $50°$ and is lost in several stages. The formula may be written $Fe_2(MoO_4)_3 \cdot nH_2O$. Variation of the H_2O content also causes small variations in the physical properties and optics. Biaxial (+); $N_x = 1.79$–1.81, $N_y = 1.81$–1.83, $N_z = 1.99$–2.01; $2V = 28°$; $r < v$, marked; $Z = $ elongation. X,Y, colorless; Z, gray to canary-yellow. A secondary mineral, ferrimolybdite is widespread in the oxidation zone of molybdenum-bearing deposits. Abundant at the Climax molybdenum mine, near Leadville, CO; Westmoreland, NH; Granite Hill, Cuttingsville, VT; Pioneer Mills, Cabarrus Co., NC; Red River District, Taos Co., NM; Kingman, Mojave Co., AZ; Red Mt., Ritter Range, Mojave Co., CA. In *Siberia, Russia*; Bivongi, Calabria, Italy; Pitkaranta, Lake Ladoga, Finland. In Australia at the Kingsgate, Bald Knob, and other molybdenite mines in NSW, and from Mulgine, WA. RG *DII:1095; AM 45:1111(1960), 48:14(1963); BSBG 80:159(1971)*.

49.2.2.1 Umohoite $(UO_2)(MoO_4) \cdot 4H_2O$

Named in 1953 for the composition. MON $P2_1c$. $a = 6.38$, $b = 7.50$, $c = 14.30$, $\beta = 99.1°$, $Z = 4$, $D = 4.93$. Also reported as ORTH; cell dimensions are somewhat variable, depending on the state of hydration and especially on the method of preparation of the sample for X-ray study. *AM 42:657(1957).* 11-375: 14.1_2 7.1_{10} 4.74_2 3.22_5 3.18_2 3.03_2 2.84_2 2.04_2. Delicate rosettes of tabular plates. Blue-black; dark green, splendent luster. $G = 4.55$. Biaxial (−); $X \wedge a = 9°$; $YZ = bc$; $N_x = 1.66$, $N_y = 1.831$, $N_z = 1.915$ (Na); $2V = 65°$. X, dark blue; Y, light blue; Z, olive-green. *Freedom No. 2 mine, Marysvale, Piute Co., UT.* Lucky Mc mine, Gas Hills, Fremont Co., WY, with gypsum, uranium oxides, and iron sulfides in sandstone; Alyce Tolino mine, Cameron, AZ. Also found in Russia (locality unspecified). RG *AM 44:920, 1248(1959); IANS, Ser. Geol.:131(1985)*.

49.2.3.1. Iriginite $(UO_2)(Mo_2^{6+}O_7) \cdot 3H_2O$

Named in 1956. MON Space group unknown. $a = 8.58$, $b = 12.87$, $c = 7.48$, $\beta = 107.67°$, $Z = 3$, $D = 4.00$. *AM 49:409(1964)*. 18:1426: 6.35_{10} 5.25_4 4.28_5 3.33_4 3.21_{10} 3.11_4 2.62_5 1.84_4. Very fine grained, dense aggregates. Canary yellow, vitreous to dull luster. Conchoidal to uneven fracture. $H = 1–2$. $G = 3.84$. Nonfluorescent. Soluble in hot acids and in strong bases. Biaxial (+); $YZ = bc$; $N_x = 1.73$, $N_y = 1.82$, $N_z = 1.93$; $2V = 60°$; parallel extinction. White River badlands, Pennington Co., SD, in sandstone. *Russia (?)*, in albitite, associated with brannerite and with other U–Mo minerals. RG *BM: 81:157(1958); AM 43:379(1958), 45:257(1960); ZVMO 88:564(1959). AM 79:574(1994).*

49.2.4.1 Phyllotungstite $CaFe_3^{3+}H(WO_4)_6 \cdot 10H_2O$

Named in 1984 for the crystal habit and composition. ORTH $P222$, $Pmm2$, or $Pmmm$. $a = 7.29$, $b = 12.59$, $c = 19.55$, $Z = 3$, $D = 5.26$. *AM 71:846(1986).* 38-381: 6.33_7 6.01_6 4.91_6 3.25_8 3.13_{10} 2.92_7 2.45_7 1.82_8. Crusts of tabular or scaly orthorhombic crystals, flattened {001}. Yellow, pearly luster. Cleavage {001}, perfect; irregular fracture. $H = 2$. Biaxial (−); $Z = c$; $N_x = 2.10$, $N_z = 2.185$; $2V = 18°$; sometimes uniaxial. *Clara Mine, Oberwolfach, Schwarzwald, Baden–Württemberg, Germany*, as crusts on quartz associated with ferritungstite, scheelite, and pyrite. RG *NJMM:529(1984)*.

49.3.1.1 Calcurmolite $Ca(UO_2)_3(MoO_4)_3(OH)_2 \cdot 11H_2O$

Named in 1962 for the composition. An inadequately characterized mineral. *AM 49:1152(1964)*. *16-145:* 8.34_5 7.85_{10} 4.29_3 3.89_6 3.56_3 3.21_8 1.99_5 1.86_4. Prismatic crystals or massive platy aggregates. Honey-yellow. Fluoresces bright yellow-green under UV light. Biaxial $(-)$; $N_x = 1.770$, $N_y = 1.816$, $N_z = 1.856$. X, colorless; Y, pale yellow; Z, yellow. In radial-fibrous, yellow spherulites to 3 mm at Has d'Alary and Rabèjac, Lodève, Hérault, France. From *Russia*. It forms pseudomorphs after uraninite; associated with uranophane, uranospinite, halloysite, betpakdalite, jarosite, and ferrimolybdite in the oxidation zone of a U–Mo ore deposit. RG *BM 82:564(1959)*.

49.3.1.2 Tengchongite $CaU_6^{6+}Mo_2^{6+}O_{25} \cdot 12H_2O$

Named in 1986 for the locality. ORTH $A2_122$ or $C222_1$. $a = 15.616$, $b = 13.043$, $c = 17.716$, $Z = 4$, $D = 4.24$. *AM 73:195(1988)*. *41-1491:* 8.84_{10} 7.66_3 5.37_5 4.27_5 3.65_4 3.38_7 3.17_8 2.04_4. Irregular grains, tabular {001}. Yellow, vitreous luster. Cleavage {001}, perfect. $H = 2-2\frac{1}{2}$. $G = 4.25$. Nonfluorescent. Soluble in H_2SO_4 and HNO_3, slowly soluble in 1% HCl, insoluble in KOH solution. Biaxial $(-)$; $N_x = 1.663$, $N_y = 1.760$, $N_z = 1.762$; $2V = 16°$. *Tengchong County, Yunnan, China*, in the zone of oxidation of a uranium deposit, associated with studtite, kivuite, and calcurmolite. RG *KT 31:396(1986)*.

49.3.2.1 Cousinite $Mg(UO_2)_2(MoO_4)_2(OH)_2 \cdot 5H_2O$ (?)

Named for Jules Cousin (1884–1965), Belgian, president of the Union Minière du Haut Katanga. An inadequately characterized mineral. Crystals are thin blades. Black, vitreous luster. *Shinkolobwe, Shaba, Zaire*, as an alteration product of ores containing uraninite and molybdenite. RG *AM 44:910(1959)*.

49.3.3.1 Cerotungstite-(Ce) $(Ce,Y,Nd)(W_2O_6)(OH)_3$

Named in 1970 for the composition. MON $P2_1/m$. $a = 7.070$, $b = 8.700$, $c = 5.874$, $\beta = 105.45°$, $Z = 2$, $D = 6.244$. *AM 57:1558(1972)*. PD: 6.83_4 3.41_{10} 3.34_2 2.63_2 2.27_5 1.767_2. Bladed, radiating crystals; twinning on {001}. Orange-yellow. Cleavage {100}, perfect. $H \approx 1$. Biaxial; $X = a$; Y perpendicular to {001}; $Z = b$; $N_x = 1.89$, $N_y = 1.95$, $N_z = 2.02$. *Kirwa and Nyamulilo mines, Kigezi district, Uganda*, associated with ferberite. RG *MR 12:81(1981)*.

49.3.3.2 Yttrotungstite-(Y) $Y(W_2O_6)(OH)_3$

Named in 1950 for the composition. MON $P2_1/m$. $a = 6.95$, $b = 8.64$, $c = 5.77$, $\beta = 104.93°$, $Z = 2$. *MM 38:261(1971)*. *26-1396:* 6.73_7 4.98_8 4.69_{10} 3.36_7 3.26_{10} 2.87_7 2.79_7 2.03_9. Earthy; as monoclinic laths, elongated [001], and flattened {100}. Always twinned on {100} to pseudo-orthorhombic symmetry. Orange yellow, earthy luster. Cleavage {010}, good; {10$\bar{1}$}, fair. $H \approx 1$. $G = 5.96$. Biaxial $(-)$; $X \approx c$; $Z = b$; $N_x = 1.89$, $N_y = 1.98$, $N_z = 2.02$; $2V = 68°$. X, pale yellow; Y,Z, dark yellow. *Kramat Pulai mine, Kinta district, Perak,*

Malaysia, associated with quartz, alumotungstite, stolzite, and raspite. RG
BGSF 42:223(1970), MR 12:81(1981).

49.3.4.1 Mpororoite $AlWO_3(OH)_3 \cdot 2H_2O$

Named in 1972 after the locality. TRIC $P1$ or $P\bar{1}$. $a = 9.40$, $b = 11.46$, $c = 8.20$, $\alpha = 94.33°$, $\beta = 89.75°$, $\gamma = 95.17°$. *MR 12:81(1981). 25-1496:* 8.21_{10} 6.18_2 5.69_2 4.73_1 4.20_2 4.09_2 3.08_9 2.36_1. Powdery, very fine grained; under SEM crystals are platy. Greenish yellow, powdery luster. *Mpororo tungsten deposit, Kigezi, Uganda*, as an alteration product of scheelite, associated with ferberite and ferritungstite. *Kara mine, 40 km SSW of Burnie, TAS, Australia*, in a similar environment but with additional associated anthoinite. RG *BGSF 44:107(1972), AM 58:1112(1973), MM 48:397(1984)*.

49.3.5.1 Uranotungstite $(Fe,Ba,Pb)(UO_2)_2(WO_4)(OH)_4 \cdot 12H_2O$

Named in 1985 for the composition. ORTH $P222_1$ or $Pmm2$. $a = 9.22$, $b = 13.81$, $c = 7.17$, $Z = 2$, $D = 4.27$. *AM 71:1547(1986). 41-587:* 7.67_2 6.91_{10} 4.60_5 3.46_3 3.20_8 3.04_7 2.80_3 2.04_1. Minute, lathlike crystals, in spherulitic aggregates. Yellow, orange, brown; dull luster. Cleavage {010}, perfect; irregular fracture. $H = 2$. $G = >4.03$. Nonfluorescent. Soluble in cold 1:1 HCl, decolorized but not dissolved by NHO_3 or H_2SO_4. Biaxial $(-)$; $N_x = 1.682$, $N_y = 1.845$, $N_z = 1.855$; $2V = 42°$. X, colorless; Y,Z, yellow; *Menzenschwand*, and *Clara Mine, Wolfach, Schwarzwald, Baden-Württemberg, Germany*, as crusts on quartz, associated with meta-uranocircite and meta-heinrichite, bergenite, schoepite, meta-zeunerite, and meta-torbernite. RG *TMPM 34:25(1985)*.

49.3.6.1 Rankachite $CaFe^{2+}V_4^{5+}W_8^{6+}O_{36} \cdot 12H_2O$

Named in 1984 for the locality. ORTH $Pmmn$. $a = 8.17$, $b = 42.02$, $c = 5.45$, $Z = 2$, $D = 4.5$. *AM 70:876(1985). 38-367:* 10.6_{10} 5.44_7 4.57_6 4.08_6 3.51_6 3.26_6 3.04_6 2.63_6. Crusts and rosettes of acicular crystals. Dark brown, yellow, resinous to subadamantine luster. Cleavage {100} or {010}, nearly perfect; uneven fracture. $H \approx 2\frac{1}{2}$. $G > 4.03$. Biaxial $(-)$; Y perpendicular to {100} or {010}; $Z = c$; $N_x = 1.770$, $N_y = 1.925$, $N_z = 1.970$; $2V = 58°$; $r \ll v$. X,Y, yellow-brown; Z, deep reddish brown. *Clara Mine, Wolfach, Rankach Valley, Schwarzwald, Baden-Württemberg, Germany*, encrusting quartz, associated with marcasite, pyrite, and scheelite. RG *NJMM:289(1984)*.

49.4.1.1 Betpakdalite $(H,K)_6Ca_4Fe_6^{3+}As_4^{5+}Mo_{16}^{6+}O_{74} \cdot 28H_2O$

Named in 1961 for the locality. MON $C2/m$. $a = 19.44$, $b = 11.10$, $c = 15.25$, $\beta = 131.28°$, $Z = 1$, $D = 2.70$. *AM 70:1333(1985). ·37-434:* 11.5_4 9.69_3 8.91_{10} 7.30_5 3.65_4 3.10_3 3.03_3 2.95_3. Powdery finely crystalline aggregates; also as tiny octahedral crystals with {001}, {20$\bar{1}$}, {011}. Bright lemon-yellow, greenish yellow; waxy–vitreous luster. Cleavage {001}, very good. $H = 3$. $G = 3.0$. Biaxial $(+)$; $X \wedge c = 12°$; $Y = b$; $N_x = 1.809$, $N_y = 1.821$, $N_z = 1.857$; $2V = 60°$; extreme, inclined dispersion; $Z > Y > X$ (Kazakhstan). X, pale yellow; Y, greenish yellow; Z, green. Alternatively, N_x, N_y, $N_z = 1.782$, 1.797, 1.850 (Tsumeb). *Bet-Pak-Dal desert, Kazakhstan*, in quartz–hübnerite–pyrite–arsenopyrite veins, in the oxidation zone, associated with jarosite, ferrimolybdite, opal, gypsum, and limonite. Tsumeb, Namibia, in the lower oxidation zone, with scorodite, chalcocite, powellite, adamite, and other minerals. RG *ZVMO 90:425(1961), AM 47:172(1962), IC 16:1096(1977), NJMM:393(1984)*.

49.4.2.1 Obradovicite $H_4(K,Na)Cu^{2+}Fe_2^{3+}(AsO_4)(MoO_4)_5 \cdot 12H_2O$

Named in 1986 for Martin T. Obradovic, Chilean mineral collector from whose collection the type material came. ORTH *Pcnm* or *Pnma*. $a = 15.046$, $b = 14.848$, $c = 11.056$, $Z = 4$, $D = 3.68$. *AM 72:1026(1987). 40-468:* 10.6_8 8.91_{10} 7.42_8 5.73_5 3.69_4 2.97_6 2.90_5 2.76_5. Crystals are platy {100}, in clusters. Pea green. No cleavage. $H = 2\frac{1}{2}$. $G = 3.55$. Nonfluorescent. Soluble in 1:1 HCl or in hot 1:1 HNO_3, insoluble in cold HNO_3, turns dull orange in 40% KOH. Biaxial (+); $XYZ = bca$; $N_x = 1.790$, $N_y = 1.798$, $N_z = 1.811$ (Na); $2V = 81°$; extreme dispersion; $Z > X = Y$. *Chuquicamata, Antofagasta, Chile*, associated with quartz, jarosite and minor wulfenite. RG *MM 50:283(1986)*.

49.4.3.1 Melkovite $H_6CaFe^{3+}(MoO_4)_4(PO_4) \cdot 6H_2O$

Named in 1969 for Vyacheslav Gavrilovich Melkov (b. 1911), Russian mineralogist. MON Space group unknown. $a = 17.46$, $b = 18.48$, $c = 10.93$, $\beta = 94.5°$, $Z = 4$, $D = 2.512$. *AM 55:320(1970). 23-384:* 8.42_7 3.54_9 3.04_7 2.92_{10} 2.42_5 1.99_6 1.79_8 1.54_5. Powdery films and platy microcrystals. Lemon yellow; brownish yellow, dull to waxy luster. One perfect cleavage, brittle. $H = 3$. $G = 2.971$. Readily soluble in dilute acids. $N \approx 1.838$; weakly pleochroic, colorless to pale green. *Shunak Mts., Kazakhstan*, in a molybdenite–fluorite deposit, associated with powellite, iriginite, ferrimolybdite, jarosite, autunite, and limonite. RG *ZVMO 98:207(1969), IGR 12:1411(1970), IC 16:1096(1977)*.

49.4.4.1 Sodium Betpakdalite $(Na_2Ca)Fe_2^{3+}(As_2Mo_6O_{28}) \cdot 15H_2O$

Named in 1971 for its chemistry and similarity to betpakdalite. MON Space group unknown. $a = 11.28$, $b = 19.30$, $c = 17.67$, $\beta = 94.5°$, $Z = 4$, $D = 3.32$. *AM 57:1312(1972). 39-398:* 9.64_6 8.73_{10} 3.63_9 3.24_7 2.94_7 2.75_7 2.07_7 1.84_8. Microcrystalline or powdery films and crusts. Lemon yellow, dull luster. $G = 2.92$. Soluble in dilute mineral acids, decomposed by KOH and Na_2CO_3 solutions. Biaxial; $Z \wedge$ elongation $= 38°$; $N_x = 1.792$, $N_z = 1.810$. X, pale yellow; Z, yellow. *Kazakhstan* (?), in the zone of oxidation of a molybdenum deposit, associated with goethite, natrojarosite, gypsum, opal, and ferrimolybdite. RG *ZVMO 100:477(1971), IGR 14:473(1972)*.

49.4.5.1 Mendozavilite Na(Ca,Mg)$_2$Fe$_6^{3+}$(PO$_4$)$_2$(P^{5+}Mo$_{11}^{6+}$O$_{39}$)(OH,Cl)$_{10}$ · 33H$_2$O

Named in 1986 for H. Mendoza Avila, Mexican geologist, who found the first specimen. MON or TRIC Space group unknown. *41-1437:* 11.6$_3$ 9.46$_8$ 8.77$_{10}$ 5.44$_3$ 3.68$_5$ 3.12$_4$ 1.820$_5$ 1.552$_4$. Fine-grained masses. Yellow, orange; vitreous luster. H = 1$\frac{1}{2}$. G = 3.85. Nonfluorescent. Readily soluble in dilute mineral acids. Biaxial (+); N$_x$ = 1.762, N$_y$ = 1.763, N$_z$ = 1.766; 2V = 5–15°; r > v, very strong; Z > Y > X. X, pale yellow. *Cumobabi molybdenum deposit, SW of Cumpas, SON, Mexico*, on quartz in a Mo-rich pegmatite, associated with paramendozavilite and other secondary Mo minerals. RG *AM 73:193(1988)*.

49.4.6.1 Paramendozavilite NaAl$_4$Fe$_7^{3+}$(PO$_4$)$_5$(P^{5+}Mo$_{12}^{6+}$O$_{40}$)(OH)$_{16}$ · 56H$_2$O

Named in 1986 for the similarity to mendozavilite. MON or TRIC Space group unknown. *41-1438:* 14.4$_{10}$ 12.9$_4$ 10.2$_6$ 9.48$_{10}$ 7.98$_5$ 7.38$_7$ 7.16$_4$ 6.56$_5$. Fine-grained powdery masses. Twinning polysynthetic, coincident with cleavage. Pale yellow, vitreous luster. One perfect cleavage. H = 1. G = 3.35. Readily soluble in dilute mineral acids. Biaxial (–); N$_x$ = 1.686, N$_y$ = 1.710, N$_z$ = 1.720; 2V = 60°; extinction oblique to cleavage; Z > Y > X. *Cumobabi, SW of Cumpas, SON, Mexico*, associated with biotite in a biotite-rich molybdenum-bearing pegmatite, with mendozavilite and other secondary molybdenum minerals. RG *BMM 2:13(1986), AM 73:194(1988)*.

49.4.7.1 Paraniite-(Y) Ca$_2$Y(AsO$_4$)(WO$_4$)$_2$

Named in 1992 for Fausto Parani, Italian mineral collector, who discovered the mineral. TET $I4_1/a$. a = 5.135, c = 33.882, Z = 4, D = 5.95. *SMPM 74:155(1994)*. Structurally related to scheelite, but with c tripled in paranite-(Y), and consists of layers of composition YAsO$_4$ and CaWO$_4$ stacked in an orderly manner along [001] in a molar ratio 1:2. PD: 4.67$_2$ 3.05$_{10}$ 2.57$_2$ 1.899$_3$ 1.816$_2$ 1.671$_2$ 1.560$_3$. Crystals dipyramidal, elongate, to 3 mm. Creamy yellow, vitreous luster. Cleavage {001}, distinct; uneven to subconchoidal fracture. Fluoresces orange-yellow in SWUV, nonfluorescent in LWUV. Uniaxial (+); N$_O$ = 1.87, N$_E$ = 1.92. Besides the content of Y$_2$O$_3$, other REEs detected are Dy, Er, Gd, and Yb. Found in a fissure in gneiss from *Pizzo Cervandone*, from the *Alpe Devero-Ossola valley area*, on the *Italian side of the Swiss–Italian border*. RG *AM 78:452(1993), 80:631(1995)*.

Class 50

Salts of Organic Acids

50.1.1.1 Whewellite $CaC_2O_4 \cdot H_2O$

Named in 1852 for William H. Whewell (1794–1866), English natural scientist and philosopher. MON $P2_1/n$. $a = 6.290$, $b = 14.583$, $c = 10.116$, $\beta = 109.47°$, $Z = 8$, $D = 2.218$. Structure: *AM 65:327(1980)*. *20-231*(syn): 5.93_{10} 5.79_3 3.65_7 2.97_5 2.49_2 2.36_3 2.35_1 2.08_1. Known only as crystals, which are usually equant or short prismatic [001]. Twinning very common; twin plane $\{\bar{1}01\}$, either heart-shaped or prismatic pseudo-orthorhombic. Colorless, yellowish, brownish; vitreous luster, pearly on {010}. Cleavage $\{\bar{1}01\}$, good; {010}, imperfect; conchoidal fracture; brittle. $H = 2\frac{1}{2}$–3. $G = 2.23$. Insoluble in water, soluble in acids. Biaxial (+); $N_x = 1.491$, $N_y = 1.554$, $N_z = 1.650$ (Na); $2V = 84°$. Whewellite is ordinarily of organic origin but has also been found as a primary hydrothermal deposit in ore veins. Type locality not known. In a septarian nodule in the bed of the Huron River, Erie Co., OH; similarly, near Havre, MT, in a septarian limestone concretion, on calcite crystals. Burgk, near Dresden, Saxony, Germany, in large crystals in a coal seam. Also with calcite and silver in a vein at Freiberg, Saxony. In lignite coal at Kladno-Pchery, and elsewhere in the Czech Republic. With tetrahedrite, carbonates, and quartz in veins of the Saint-Sylvestre mine, Urbeis, Alsace, France. In calcite veins in shale on the Yareg River, southern Timan, Russia. A common constituent of human urinary calculi. RG *DII:1099*; *AM 39:208(1954), 53:455(1968)*.

50.1.2.1 Weddellite $CaC_2O_4 \cdot 2H_2O$

Named in 1942 for the locality. TET $I4/m$. $a = 12.371$, $c = 7.357$, $Z = 8$, $D = 1.936$. Structure: *AM 65:327(1980)*. *17-541*(syn): 6.18_{10} 4.42_3 3.68_1 2.82_1 2.78_7 2.41_2 2.24_3 1.90_2. Crystals pyramidal {101}, sometimes aggregated into groups. Colorless, white; also yellowish, brown due to included organic matter. No cleavage, subconchoidal fracture. $H = 4$. $G = 1.94$. Insoluble in water. Decomposes to $CaCO_3$ at about 270°. Uniaxial (+); $N_O = 1.523$, $N_E = 1.544$. *Weddell Sea, Antarctica*, as small crystals embedded in the bottom muds. Weddellite is a principal ingredient in human urinary calculi. RG *DII:1101*, *MM 34:256(1965)*, *AC 18:917(1965)*.

50.1.3.1 Humboldtine $Fe^{2+}C_2O_4 \cdot 2H_2O$

Named in 1821 for Alexander von Humboldt (1769–1859), German explorer and naturalist. MON $C2/c$. $a = 9.707$, $b = 5.556$, $c = 9.921$, $\beta = 104.5°$. *NBSM*

$25(10):24(1972)$. Structure: SR $21:505(1957)$. 23-293(syn): 4.8_{10} 4.7_6 3.88_2 3.60_2 3.00_5 2.65_3 2.62_2 1.82_2. Crystals prismatic {001}, or platy. Usually capillary or botryoidal forms; granular, earthy. Amber, yellow; dull to resinous luster. Cleavage {110}, perfect; {100}, {010}, imperfect. H = $1\frac{1}{2}$–2. G = 2.28. Soluble in acids; in CT, affords water. Biaxial (+); XYZ = abc; $N_x = 1.494$, $N_y = 1.561$, $N_z = 1.692$; 2V large. X, pale yellowish green; Y, pale greenish yellow; Z, intense yellow. Kettle point, Bosanquet Twp., ONT, Canada. *Korozluky, SW of Bilina, Bohemia, Czech Republic*, with gypsum in brown coal; also at Cermniky. In Germany at Grossalmerode, Hesse-Nassau, and in coal at Potschappel near Dresden. RG *DII:1102*, *PM 2:269(1957)*, *BM 82:50(1959)*.

50.1.3.2 Glushinskite $MgC_2O_4 \cdot 2H_2O$

Named in 1960 for P. I. Glushinskii, Russian coal petrographer. MON $C2/c$. $a = 12.675$, $b = 5.406$, $c = 9.984$, $\beta = 129.45°$, Z = 4, D = 1.865. *MM 43:837(1980)*. 28-625(syn): 4.89_{10} 3.17_8 2.55_4 2.38_6 2.09_4 2.04_8 1.863_6 1.529_4. Minute crystals showing distorted pyramidal habit. Creamy white, colorless. H = 2. G = 1.85. Soluble in water, easily dissolved by HCl. Biaxial (−); $N_x = 1.365$, $N_y = 1.530$, $N_z = 1.595$. Mill of Johnston, near Insch, NE Scotland, on a lichen–rock interface on serpentinite. Island of Rhum, Inner Hebrides. *Russian Arctic*, in chalky clays and in Cretaceous coals. RG *ZVMO 91:204(1962)*; *AM 47:1482(1962)*, *66:439(1981)*; *MM 51:327(1987)*.

50.1.4.1 Minguzzite $K_3Fe^{3+}(C_2O_4)_3 \cdot 3H_2O$

Named in 1955 for Carlo Minguzzi (1910–1953), Italian mineralogist and professor, University of Pavia. MON $P2_1/c$. $a = 7.66$, $b = 19.87$, $c = 10.27$, $\beta = 105.1°$, Z = 4, D = 2.156. Structure: *BM 81:245(1958)*. 14-720(syn): 10.2_5 6.9_{10} 4.96_6 3.61_7 3.16_5 3.03_4 2.67_6 2.18_7. Tabular, minute crystals. Green, yellow-green; vitreous luster. Cleavage {010}, perfect. $G_{nat} = 2.142$, $G_{syn} = 2.092$. Easily soluble in water. Biaxial (−); $N_x = 1.498$, $N_y = 1.554$, $N_z = 1.594$; 2V = 78°. X, yellow-green; Y,Z, emerald-green. *Cape Calamita, Elba, Tuscany, Italy*, in limonite, associated with humboldtine. *ACH 10:488(1938)*, *ANLR 18:392(1955)*, *AM 41:370(1956)*.

50.1.5.1 Oxammite $(NH_4)_2C_2O_4 \cdot H_2O$

Named in 1870 for the composition: oxalate and ammonium. ORTH $P2_12_12$. $a = 8.035$, $b = 10.309$, $c = 3.801$, Z = 2, D = 1.498. *NBSC 539(7):5(1957)*. Structure: *AC(B) 28:3340(1972)*. 14-801(syn): 6.32_{10} 3.80_7 3.26_6 3.06_{10} 2.86_6 2.67_{10} 2.61_5 2.59_5. Lamellar masses, pulverulent. Crystals rare; artificial crystals are sphenoidal. Colorless, yellowish white; powdery luster. Cleavage {001}, distinct. H = $2\frac{1}{2}$. G = 1.5. Biaxial (−); XYZ = cab; $N_x = 1.438$, $N_y = 1.547$, $N_z = 1.595$ (Na); 2V = 62°; r < v, distinct. *Guañape Islands, Peru*, in the guano deposits, associated with mascagnite. RG *DII:1103*, *AM 36:590(1951)*.

50.1.6.1 Moolooite $CuC_2O_4 \cdot nH_2O$ ($n < 1$)

Named in 1986 for the locality. ORTH Space group unknown. $a = 5.35$, $b = 5.63$, $c = 2.56$, $Z = 1$, $D = 3.43$. *21-297:* 3.88_{10} 2.48_2 2.43_1 2.317_1 2.304_1 1.937_1 1.774_2 1.756_2. As microconcretions. Blue, green; dull/waxy luster. Soluble in warm dilute HCl. $N = 1.57$ (perpendicular to elongation), $N = 1.95$ (parallel to elongation). Ste.-Marie-aux-Mines, Vosges, France. *Bunbury Well, Mooloo Downs, WA, Australia*, in quartz, associated with sampleite and liebethenite, on the outcrop of a vein bearing copper sulfides. RG *MM 50:295(1986), AM 72:1025(1987).*

50.1.7.1 Stepanovite $NaMgFe^{3+}(C_2O_4)_3 \cdot 8\text{--}9H_2O$

Named in 1953 for Pavel Ivanovich Stepanov (1880–1947), Russian geologist. HEX-R Space group unknown. $a = 9.28$, $c = 36.68$, $Z = 6$, $D = 1.69$. Granular aggregates, xenomorphic grains. Greenish, vitreous luster. No cleavage, irregular fracture. $H = 2$. $G = 1.69$. Soluble in water, fuses with difficulty. Uniaxial $(-)$; $N_O = 1.515$, $N_E = 1.417$. O, yellowish green; E, colorless. *Tyllakh coal deposit, Lena River, Siberia, Russia.* RG *ZVMO 82:311(1953); AM 40:551(1955), 49:442(1964).*

50.1.7.2 Zhemchuzhnikovite $NaMg(Al,Fe)(C_2O_4)_3 \cdot 8H_2O$

Named in 1963 for Yurii Apollonovich Zhemchuzhnikov (1885–1957), Russian geologist. HEX-R Space group unknown. $a = 16.67$, $c = 12.51$, $Z = 6$, $D = 1.66$. Crystals acicular to fibrous, [0001]. Smoky green to violet; synthetic crystals are green in daylight, amethyst violet under artificial light; vitreous luster. Cleavage {0001}, good. $H = 2$. $G = 1.69$. Soluble in water. Uniaxial $(-)$; $N_O = 1.479$, $N_E = 1.408$. O, greenish yellow; E, reddish violet. *Chaitumusuk deposits, Lena River, Siberia, Russia.* RG *AM 47:1483(1962), 49:442(1964).*

50.1.8.1 Wheatleyite $Na_2Cu(C_2O_4)_2 \cdot 2H_2O$

Named in 1986 for the locality. TRIC $P\bar{1}$. $a = 7.559$, $b = 9.665$, $c = 3.589$, $\alpha = 76.65°$, $\beta = 103.67°$, $\gamma = 109.1°$, $Z = 1$, $D = 2.250$. *AM 71:1240(1986). 39-1500:* 7.03_6 6.52_{10} 3.65_5 3.47_9 3.16_8 2.80_4 2.50_3 2.34_3. Aggregates of acicular crystals. Bright blue, vitreous luster. Cleavage {100}, perfect; brittle. $H = 1\text{--}2$. $G = 2.27$. Nonfluorescent. Soluble in water. Biaxial $(+)$; $N_x = 1.400$, $N_y = 1.499$, $N_z = 1.667$; $2V = 83°$; $r < v$; $Z > Y > X$. X, colorless; Y, pale blue; Z, dark blue. *Wheatley mine, Phoenixville, Chester Co., PA. Nishimomaki mine, Gumma, Japan.* RG

50.2.1.1 Mellite $Al_2[C_6(COO)_6] \cdot 16H_2O$

Named in 1793 from the Greek for *honey*, in allusion to the color. TET $I4_1/acd$. $a = 15.549$, $c = 23.209$, $Z = 8$, $D = 1.605$. Structure: *AC(B) 29:26(1973). 42-1501:* 7.98_{10} 5.80_8 5.17_6 4.24_{10} 3.46_4 3.39_6 2.99_4 2.59_5. Massive, nodular, as coatings. Crystals rare, prismatic [001] or pyramidal {011}. Honey-yellow, reddish, brownish, white; resinous/vitreous luster. Cleavage {011}, indistinct;

conchoidal fracture; somewhat sectile. H = 2–$2\frac{1}{2}$. G = 1.64. Fluoresces blue in SWUV; pyroelectric. Insoluble in water and alcohol; soluble in HNO_3; in CT, affords water. Uniaxial (−); $N_O = 1.539$, $N_E = 1.511$ (Na). *Arten, Thuringia, Germany*, as a secondary mineral in crevices in brown coal and lignite. Luschitz, near Bilin, Bohemia, Czech Republic. Malowka, Bogoroditsk, Tula, Russia. With gypsum in lignitic clay, Paris basin, Seine, France. RG *DII:1104*, *MM 35:542(1965)*, *PD 4:172(1989)*.

50.2.2.1 Earlandite $Ca_3(C_6H_5O_7)_2 \cdot 4H_2O$

Named in 1936 for Arthur Earland, English oceanographer. Synonym: calcium citrate. MON Space group unknown. $a = 30.94$, $b = 5.93$, $c = 10.56$, $\beta = 93.73°$, $Z = 4$, $D = 1.96$. *28-2003*(syn): 15.5_{10} 8.5_1 7.7_5 6.4_1 5.2_3 4.74_1 4.50_1 3.09_1. Warty, fine-grained nodules. White, pale yellow. $G = 1.95$. $N \approx 1.56$. *Weddell Sea, Antarctica*, in oceanic bottom samples at a depth of 2,580 m. RG *DII:1105*.

50.2.3.1 Julienite $Na_2Co(CNS)_4 \cdot 8H_2O$

Named in 1928 for Henri Julien (d. 1920), Belgian scientist. Synonym: sodium cobalt thiocyanate. TET $P4_2/n$. $a = 9.24$, $c = 5.57$, $Z = 1$, $D = 1.68$. Also MON $P2_1/c$. $a = 18.941$, $b = 19.209$, $c = 5.460$, $\beta = 91.63°$, $Z = 4$, $D = 1.610$. *ZK 91-229(1935)*. *2-372*: 6.5_6 4.6_6 3.55_8 3.23_8 1.375_8 1.220_8 0.935_{10} 0.903_{10}. Thin crusts of minute needles elongated [001]. Blue. Cleavage [001], good. $G = 1.648$. Soluble in water or alcohol, yielding a rose-colored solution. Uniaxial (+); $N_O = 1.556$, $N_E = 1.645$. *Chamitumba, near Kambove, Shaba, Zaire*, with cobaltian wad in a white talc schist. RG *DII:1106*, *TMPM 3:376(1953)*.

50.2.4.1 Calclacite $CaCl_2 \cdot Ca(C_2H_3O_2)_2 \cdot 10H_2O$

Named in 1945 for the composition: Ca, Cl, acetate. MON $P2_1/c$. $a = 11.51$, $b = 13.72$, $c = 6.82$, $\beta = 116.7°$, $Z = 2$, $D = 1.55$. Structure: *AC 11:745(1958)*. *12-869*(syn): 8.27_{10} 6.87_8 6.15_5 4.16_8 3.24_{10} 2.43_{10} 2.30_8 2.04_8. Hairlike efflorescences. White. $G = 1.5$. Biaxial (+); Z = elongation; $N_x = 1.468$, $N_y = 1.484$, $N_z = 1.515$; $2V = 80°$. Found in oak museum cases, encrusting Ca-rich objects. The acetic acid was derived from the oak wood. RG *DII:1107*.

50.2.5.1 Kafehydrocyanite $K_4Fe^{2+}(CN)_6 \cdot 3H_2O$

Named in 1973 for the composition. Natural origin in doubt. TET $I4_1/a$. $a = 9.394$, $c = 33.72$, $Z = 8$, $D = 1.89$. *25-1354*: 4.80_8 3.62_7 3.12_5 2.93_{10} 2.83_5 2.79_5 2.21_8 2.10_6. Aggregates of crystallites with square and rectangular plates. Lemon yellow. Cleavage {001}, good. H = 2–$2\frac{1}{2}$. $G = 1.98$. Appreciably soluble in water. Uniaxial (+); $N_O = 1.577$, $N_E = 1.584$. *Medvezhii Log gold mine, Olkhovsk ore field, Sayan, Russia*, and in several other deposits, as stalactites in mine workings, associated with melanterite, gypsum, and various unidentified cyanides and oxalates. RG *AM 59:209(1974)*.

50.3.1.1 Kratochvilite $C_{13}H_{10}$ (Fluorene)

Named in 1937 for Joseph Kratochvil (1878–1958), Czech, professor of petrography, Charles University, Prague. ORTH $Pnma$. $a = 8.49$, $b = 5.721$, $c = 18.97$, $Z = 4$, $D = 1.197$. Structure: SR $19:583(1955)$. $28\text{-}2011$(syn): 9.39_7 4.68_{10} 4.21_7 3.79_5 3.38_9 2.54_6 2.45_5 2.37_5. Colorless. $G = 1.206$. Biaxial; $N_x = 1.557$, $N_y = 1.725$. From burning coal heaps in Kladno, Rozpravy, Czech Republic. RG AM $23:667(1938)$.

50.3.2.1 Ravatite $C_{14}H_{10}$

Named in 1993 for the locality. Synonym: phenanthrene. MON $P2_1$. $a = 8.392$, $b = 6.181$, $c = 9.558$, $\beta = 98.8°$, $Z = 2$, $D = 1.207$. EJM $5:699(1993)$. PD: 9.43_{10} 4.94_1 4.72_1 4.03_1 3.37_1. Thin platelike individuals of irregular shape; aggregates form crusts generally < 1 mm thick. Colorless to white to pale gray, translucent, vitreous to waxy luster, waxlike tenacity. Cleavage {001}, perfect. $G = 1.11$. Biaxial, $N = 1.75$–1.95. From *Ravat, about 100 km N of Dushanbe, Northwestern Tadzhikistan*, in a Middle Jurassic brown coal seam. Weathering of finely dispersed pyrite and the bacterial decomposition of the coal have caused lignite seams to ignite spontaneously, reaching temperatures above 800°. Ravatite forms as a sublimation product in the outer, cool zones, with liquid bitumen and rarely, native Se. RG AM $79:389(1994)$.

50.3.3.1 Simonellite $C_{19}H_{24}$

Named in 1919 for Vittorio Simonelli (1860-1929), Italian geologist and paleontologist, University of Bologna, who first observed the mineral. Synonym: 1,1-dimethyl-7-isopropyl-1,2,3,4-tetrahydrophenanthrene. ORTH $Pnaa$. $a = 9.231$, $b = 9.134$, $c = 36.01$, $Z = 8$, $D = 1.104$. $30\text{-}2001$: 9.00_3 8.21_3 6.45_{10} 6.11_2 4.82_2 4.57_2 4.07_1 4.05_1. *Fognano, Tuscany, Italy*. RG $ANLR$ $47:41(1969)$, AM $55:1818(1970)$.

50.3.4.1 Fichtelite $C_{19}H_{34}$

Named in 1841 for the locality. MON $P2_1$. $a = 10.75$, $b = 7.51$, $c = 13.08$, $\beta = 126.9°$, $Z = 2$, $D = 1.032$. Shining scales, flat crystals, thin layers. White, greasy luster. Brittle. $H = 1$. $G = 1.03$. Easily soluble in ether, less so in alcohol. M.P. 46°. Soluble in concentrated H_2SO_4 and in HNO_3. Reported from Alabama in recent pine logs. *Fichtel Mts., Redwitz, Bavaria, Germany*, in altered pine wood from a peat bog; also elsewhere in Germany in a similar environment. In England at Handforth in Cheshire in pine logs in moss. RG $D6:1000$, NW $49:9(1962)$.

50.3.5.1 Dinite $C_{20}H_{36}$

Named in 1852 for Olinto Dini (1802–1866), Italian teacher and professor of physics, University of Pisa. ORTH $P2_12_12_1$. $a = 12.356$, $b = 12.762$, $c = 11.427$, $Z = 1$, $D = 1.02$. AM $77:674(1992)$. PD: 8.92_7 8.32_7 7.00_8 5.53_{10} 5.06_{10} 4.02_7 3.87_5 3.79_5. Massive, crystalline. Colorless to yellowish. No cleavage, fragile.

$G = 1.01$. Soluble in carbon tetrachloride. Found with lignite at *Castelnuovo Garfagnana, northern Tuscany, Italy.* RG *D6:1001, EM 3:855(1991)*.

50.3.6.1 Evenkite $C_{24}H_{50}$

Synonym: n-tetracosane. Named in 1953 for the locality. MON $P2_1/a$. $a = 7.52$, $b = 4.98$, $c = 32.50$, $\beta = 94°$, $Z = 2$, $D = 0.926$. *AM 50:2109(1965)*. *28-2004*: 4.18_{10} 3.74_9 3.02_5 2.52_7 2.25_8 2.12_6 1.91_5 1.75_6. Pseudohexagonal tabular crystals, flattened perpendicular to c, {001}. Polysynthetic twinning. Colorless; yellowish, waxy luster. Perfect cleavage, micalike. $H = 1$. $G = 9.20$. Soluble in ether. Melting point $50.7°$. Biaxial (+); $N_x = 1.504$, $N_y = 1.504$, $N_z = 1.553$. *Evenki region, Lower Tunguska River, Siberia, Russia,* in vugs in a quartz vein cutting vesicular lava, associated with chalcedony, calcite, and various sulfides. RG *DANS 88:717(1953), AM 40:368(1955), NJMM:19(1965)*.

50.3.7.1 Karpatite $C_{24}H_{12}$

Named in 1955 for the locality, the Carpathian mountains. Synonyms: coronene; "pendletonite." Pendletonite has been shown to be identical to karpatite, and the name *pendletonite* has been discontinued. MON $P2_1/c$ or $P2/c$. $a = 10.035$, $b = 4.695$, $c = 16.014$, $\beta = 69°$, $Z = 2$, $D = 1.416$. *AM 54:329(1969)*. *28-2007*(syn): 9.44_{10} 7.43_{10} 5.09_1 4.84_1 3.96_4 3.49_6 3.03_4 1.989_1. Acicular crystals and fibrous, radiating aggregates. Pale yellow, red-violet; vitreous luster. Cleavage {100}, {001}, {$\bar{2}$01}, perfect; flexible, almost plastic. $H < 1$. $G = 1.35$–1.40. Melts about $430°$. Soluble in aniline, from which it can be recrystallized, insoluble in acids, sublimes on heating, flammable. Biaxial (−); $X = b$; $Z \wedge c = 21°$; $N_x = 1.780$, $N_y = 1.977$–1.982, $N_z = 2.05$–2.15; 2V large; v > r, extreme; axial plane parallel to X. *Picachos mercury mine, New Idria, San Benito Co., CA,* associated with quartz and cinnabar. *Transcarpathia, Ukraine,* associated with idrialite and organic material in cavities at the contact of diorite porphyry with argillites. RG *AM 42:120(1957), 52:611(1967)*.

50.3.8.1 Idrialite $C_{22}H_{14}$

Named in 1832 for the locality. Synonyms: picene, "curtisite." Curtisite has been shown to be identical with idrialite, and the name *curtisite* is discontinued. ORTH Space group unknown. $a = 8.07$, $b = 6.42$, $c = 27.75$, $Z = 4$, $D = 1.29$. *AM 50:2109(1965)*. *28-2006*: 7.08_2 4.94_{10} 4.43_2 4.04_6 3.40_8 2.48_3 2.06_2 1.92_1. Tabular crystals. Light brown, pistachio-green; colorless when recrystallized and pure. Fluoresces blue-white in UV light. Soluble in chloroform, soluble in concentrated H_2SO_4 producing an intense blue-green color. M.P. $319°$. *Skaggs Springs, Sonoma Co., CA. Idria, Slovenia,* in a mercury mine, mixed with cinnabar and clay. RG *NJMM:19(1965), MP 40:111(1989)*.

50.4.1.1 Refikite $C_{20}H_{32}O_2$

Named in 1852 for Refik Bey (d.1865), Turkish journalist, for his interest in the sciences (he was a founder of the New Ottoman Society). Synonym: Δ13-dihydro-*d*-pimaric acid. ORTH $P2_12_12$. $a = 10.43$, $b = 22.35$, $c = 7.98$, $Z = 4$, $D = 1.09$. *AM 50:2110(1965)*. *28-2009*: 11.2_1 7.97_3 6.09_9 5.50_{10} 5.20_7 4.82_1 4.28_2. Acicular crystals, small scales. White. Fragile, very soft. With concentrated H_2SO_4 + concentrated acetic acid produces an orange to brown solution. Soluble in ether and alcohol. M.P. 182°. *Montorio, near Feramo, Abruzze Mts., Italy*, in lignite. In a swamp in southern Bavaria, Germany, as crystals in the fossil roots of a spruce. RG *D6:1006, NJMM:19(1965)*.

50.4.2.1 Hoelite $C_{14}H_8O_2$

Named in 1922 for Adolf Hoel (b.1879), geologist, leader of a Norwegian expedition to Spitzbergen. Synonym: anthraquinone. MON $P2_1/c$. $a = 15.81$, $b = 3.967$, $c = 7.876$, $\beta = 102.67°$, $Z = 2$, $D = 1.43$. *28-2002*(syn): 7.69_9 6.16_5 3.84_3 3.79_1 3.52_{10} 3.36_8 3.14_1 3.05_1. Orthorhombic needles. Yellow. G = 1.42. *Spitzbergen, Norway*, from a burning coal seam. RG

50.4.3.1 Flagstaffite $C_{10}H_{22}O_3$

Named in 1920 for the locality. Synonym: *cis*-terpin hydrate. ORTH $Fdd2$. $a = 18.60$, $b = 23.00$, $c = 10.86$, $Z = 16$, $D = 1.088$. *AM 50:2109(1965)*. *28-2014*: 8.72_8 7.16_8 5.89_5 5.65_5 4.88_{10} 4.73_5 4.36_6 3.53_5. Well-formed prismatic crystals, with {110} and {111}; also {010} and {011}. Yellowish, resinous luster. Cleavage {110}, imperfect. G = 1.09. M.P. 116°. Gives a strong orange-yellow solution with concentrated H_2SO_4. Biaxial (+); $N_x = 1.505$, $N_y = 1.512$, $N_z = 1.524$ (Na); $2V = 77°$; r > v; optic axial plane {010}. From fossil logs near *San Francisco Peaks, N of Flagstaff, AZ*, as crystals in radial cracks in the logs. RG *NJMM:19(1965)*.

50.4.4.1 Uricite $C_5H_4N_4O_3$

Named in 1807 for the composition. Synonym: uric acid. MON $P2_1/c$. $a = 14.44$, $b = 7.403$, $c = 6.208$, $\beta = 65.1°$, $D = 1.851$. Structure: *AC 19:286(1965)*. *28-2016*(syn): 6.55_4 5.63_2 4.91_5 3.86_4 3.18_5 3.093_{10} 3.087_7 2.87_2. White. G = 1.851. From guano deposits, *North Chincha Island, Peru*. Dingo Donga cave, (31°51′ S, 126°44′ E), WA, Australia, in guano associated with biphosphammite, brushite, and syngenite. RG *NBSM 8:154(1970), MM 39:889(1974)*.

50.4.5.1 Guanine $C_5H_3(NH_2)N_4O$

Named in 1844 for its derivation from and association with guano. MON $P2_1/c$. $a = 16.510$, $b = 11.277$, $c = 3.645$, $\beta = 96.8°$, $Z = 4$, $D = 1.489$. *28-2012*(syn): 7.9_4 6.32_7 5.01_4 3.53_5 3.22_{10} 2.63_5 2.00_5 1.80_5. White. Soft. *North Chincha Island, Peru*, in guano derived from seabirds. *Murra-el-elevyn cave, Cockabiddy Nullabor Plain, WA, Australia*, formed as a reaction to bat guano and urine with limestone in the cave, associated with biphosphammite, hannayite, aphthitalite, syngenite, monetite, whitlockite, gypsum, and organic materials. RG *MM 39:467(1974), 39:889(1974)*.

50.4.6.1 Urea $CO(NH_2)_2$

Named in 1973 for the composition. TET $P\bar{4}2_1m$. $a = 5.645$, $c = 4.704$, $Z = 2$, $D = 1.330$. *NBSC 539(7):61(1957)*. Structure: *AC 17:544(1964)*. PD 4.70_1 3.98_{10} 3.61_2 3.04_3 2.82_1 2.52_1 2.17_1 1.837_1. Elongated pyramidal crystals. Pale yellow to pale brown. $G = 1.33$. Uniaxial (+); $N_O = 1.484$, $N_E = 1.603$. Toppin Hill, WA, Australia, in a cave, on guano, associated with phosphammite, weddellite, and ammonian aphthitalite. RG *MM 39:346(1973), AM 59:874(1974)*.

50.4.7.1 Acetamide CH_3CONH_2

Named in 1975 for its identity with synthetic acetamide. Synonym: ethanamide. HEX-R $R3c$. $a = 11.44$, $c = 13.50$, $Z = 18$, $D = 1.15$. *39-1851*: 5.74_{10} 3.63_1 3.56_2 3.32_1 3.29_2 2.88_3 2.174_2 2.166_1. Granular aggregates of crystals with a hexagonal outline, with prism $\{11\bar{2}0\}$ prominent. Colorless, gray. Fine conchoidal fracture. $H = 1$–$1\frac{1}{2}$. $G = 1.17$. Volatilizes in a few hours in sunlight. Uniaxial (−); $N_O = 1.495$, $N_E = 1.460$. Lvov-Volynskii Basin, Ukraine, in waste piles from a coal shaft, in areas enriched in ammonia, shut off from oxygen and daylight. Seasonal, appearing only in dry weather. RG *ZVMO 104:326(1975), AM 61:338(1976)*.

50.4.8.1 Kladnoite $C_6H_4(CO)_2NH$

Synonym: phthalimide. Named in 1942 for the locality. MON $P2_1/c$. $a = 22.83$, $b = 7.651$, $c = 3.810$, $\beta = 91.37°$, $Z = 4$, $D = 1.469$. Structure: *AC(B) 28:415(1971)*. *28-2013*(syn): 11.4_2 6.35_9 5.70_{10} 5.39_1 3.74_1 3.38_1 3.28_3 3.13_1. Bladelike crystals, flattened $\{100\}$. $G = 1.47$. M.P. 234°. Biaxial (+); $Y \wedge c = 16°$; $Z = b$; $N_x = 1.501$, $N_y = 1.519$, $N_z = 1.755$. Libusin, Kladno coal basin, near Prague, Czech Republic, from burning coal heaps. RG *AM 31:605(1946)*.

50.4.9.1 Abelsonite $NiC_{31}H_{32}N_4$

Synonym: nickel porphyrin. Named in 1975 for Philip H. Abelson, president, the Carnegie Institution, Washington, DC, a pioneer in organic geochemistry. TRIC $P1$ or $P\bar{1}$. $a = 8.44$, $b = 11.12$, $c = 7.28$, $\alpha = 90.88°$, $\beta = 113.75°$, $\gamma = 79.57°$, $Z = 1$, $D = 1.45$. *AM 63:930(1978)*. *30-1840*: 10.9_{10} 7.63_5 6.63_3 5.79_4 5.51_3 4.90_1 3.77_8 3.14_4. Aggregates of platy crystals. Pink–purple, reddish brown; submetallic/adamantine luster. Cleavage $\{1\bar{1}1\}$; fragile. $H < 3$, very soft. $G = 1.45$. Nonfluorescent. Insoluble in H_2O and in dilute HCl and HNO_3, soluble in benzene and acetone. Absorption very strong, reddish brown to reddish black. The mineral is rapidly attacked by index of refraction liquids. *Green River formation, eastern Uintah Co., UT*, from a drill core at a number of locations, in fractures, associated with albite, orthoclase, pyrite, quartz, dolomite, and analcime. RG *AM 61:502(1976), SCI 223:1075(1984)*.

Class 51
Insular SiO₄ Groups Only

PHENAKITE GROUP

Phenakite minerals are tectosilicates characterized by the general formula

$$R_2SiO_4$$

where

R = Be, Zn, LiAl

and where the structure is based on a close packing of oxygens with all cations in tetrahedral coordination.

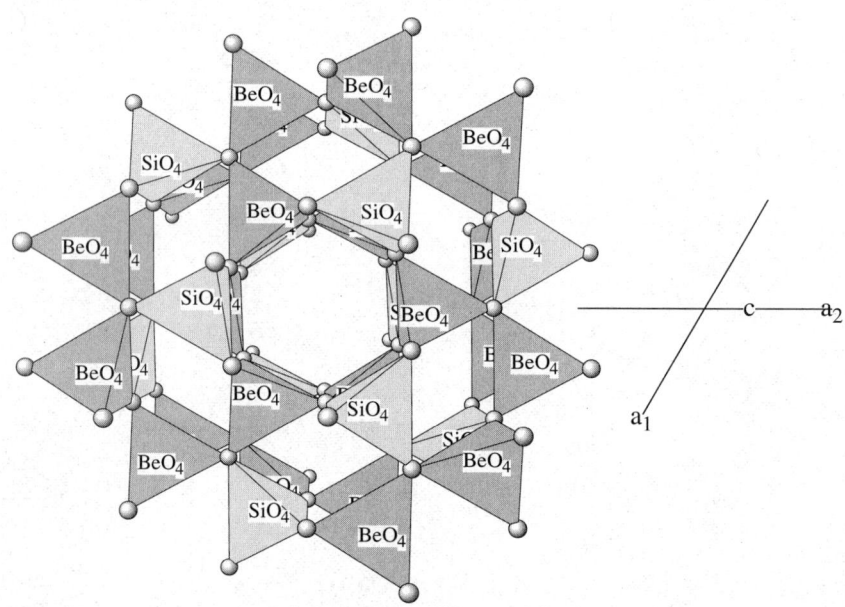

PHENAKITE: Be_2SiO_4
Be and Si tetrahedra form a triply linked framework—each oxygen is shared among two Be and one Si tetrahedra. This view shows the larger hexagonal rings and smaller rhombic rings.

51.1.1.1 Phenakite

PHENAKITE GROUP

Mineral	Formula	Space Group	a	c	
Phenakite	Be$_2$SiO$_4$	$R\bar{3}$	12.438	8.231	51.1.1.1
Willemite	Zn$_2$SiO$_4$	$R\bar{3}$	13.93	9.31	51.1.1.2
Eucryptite*	LiAlSiO$_4$	$R\bar{3}$	13.471	8.998	51.1.1.3

*In eucryptite, Li occupies the Si sites of phenakite, and Al and Si are disordered in the R sites. The structure is thus a tektosilicate but is retained here because of its derivation from a simpler phenakite structure.

Note: Liberite, Li$_2$BeSiO$_4$ (5.1.2.1), although stoichiometrically distinct, has a structure based on approximate close packing of oxygens with all cations in tetrahedral coordination sites.

51.1.1.1 Phenakite Be$_2$SiO$_4$

Named in 1833 by Nordenskiöld from the Greek for *deceiver*, in allusion to its being mistaken for quartz. Phenakite group. HEX-R $R\bar{3}$. $a = 12.438$, $c = 8.231$, $Z = 18$, $D = 2.96$. *9-431:* 6.24$_4$ 3.66$_8$ 3.60$_3$ 3.12$_{10}$ 2.52$_7$ 2.36$_7$ 2.19$_6$ 2.08$_5$. Structure: *JPCM 14:426(1987)*. Flattened rhombohedra, often highly modified; dominant forms $\{11\bar{2}0\}$, $\{10\bar{1}0\}$, $\{10\bar{1}1\}$, $\{12\bar{3}2\}$, $\{11\bar{2}3\}$, $\{2\bar{1}\bar{1}3\}$, $\{01\bar{1}2\}$; less commonly prismatic. Penetration twins on $\{\bar{1}010\}$, twin axis [0001]. Colorless, white, yellow, pale rose; vitreous luster. Cleavage $\{11\bar{2}0\}$, distinct; $\{10\bar{1}1\}$, imperfect; conchoidal fracture. H = $7\frac{1}{2}$–8. G = 2.98. Uniaxial (+); N$_O$ = 1.654, N$_E$ = 1.670. Fusibility 7; insoluble in acids. Near the ideal composition, with little substitution. Found in granite pegmatites with microcline, topaz, quartz, and so on; also in schists; as a product of beryl alteration. At Lords Hill, Oxford Co., ME; Bald Face Mt., Carroll Co., NH; Morefield mine, Amelia Co., VA; at Rib Mt., Marathon Co., WI; at Mt. Antero, Chaffee Co.(!), as equant crystals and rare twins, and in pegmatites of the Pikes Peak granite as at Crystal Peak, near Florissant, and Harris Park, Teller Co., and Crystal Park, St. Peter's Dome, El Paso Co., CO. At Kragerö, Norway; Framont, Vosges, France; at Reckingen, Valais, Switzerland; at Tschirnitz, Silesia, and Strzegom, Wroclaw, Poland; at Rauris, Salzburg, Austria; at Ober Neusattel, Bohemia, Czech Republic. With emerald and chrysoberyl as large crystals(!) in schist at *Takovaya, about 90 km E of Yekaterinburg, Sverdlovsk, Ural Mts., Russia*, and in the Ilmen Mts., near Miask. Found in

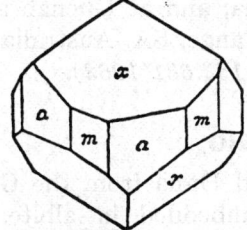

PHENAKITE, Mt. Antero, CO.
Forms: m, $\{10\bar{1}0\}$; a, $\{11\bar{2}0\}$; x, $\{12\bar{3}2\}$.

the Naegi district, Gifu Pref., Japan. At Klein Spizkopje, Swakopmund, Namibia, and in the Usagara district, Tanzania; as prismatic crystals(!) at Anjanabonoina, Madagascar. At San Miguel de Piracicaba(!) as sharp, flat rhombohedra to 12 cm, and as equant transparent rhombs(!) at Marambaia, MG, Brazil. An incidental ore of beryllium and a collectors' gemstone. AR *SPC* 15:1021(1970), *MR* 9:184(1978), *PCM* 13:69(1986).

51.1.1.2 Willemite Zn$_2$SiO$_4$

Named in 1830 by Levy in honor of William I (Willem), King (1813–1840) of the Netherlands. The variety troostite was named in 1833 for Gerard Troost (1776–1850), Dutch-born American mineralogist and state geologist of Tennessee. Phenakite group. HEX-R $R\bar{3}$. $a = 13.93$, $c = 9.31$, $Z = 18$, $D = 4.17$. *37-1485*(syn): 4.10_2 4.02_3 3.49_7 2.84_{10} 2.63_9 2.32_5 1.860_4 1.552_2. Structure: *SPC* 15:387(1970). Massive, granular, less commonly short to long prismatic to fibrous [0001]. Light green, yellow-green, yellow-brown; red-brown, colorless, blue-violet (cobaltoan); vitreous to resinous luster; translucent to transparent. Cleavage {11$\bar{2}$0}, good to poor; {0001}, poor; conchoidal to uneven fracture; brittle. $H = 5-5\frac{1}{2}$. $G = 4.05-4.20$. Uniaxial (+); $N_O = 1.691-1.714$, $N_E = 1.719-1.732$. Fluoresces bright green in LWUV and SWUV, often phosphorescent; weaker or nonfluorescent in high-purity material or with high transition-element substitution. Fusibility 3–4; gelatinizes with acids. Mn^{2+} and Fe^{2+} replace zinc (troostite); Co^{2+} occurs rarely; refractive indices increase with these substitutions. Willemite with up to 30 mol % phenakite has been prepared in the laboratory. Found in oxidized zones of zinc deposits and in metamorphosed environments; rarely in syenites. Found at Sterling Hill and *Franklin, Sussex Co., NJ*(!), where it was found six years prior to being named, both massive and in fine crystals and as large, brownish, rough crystals (troostite); in microcrystals at Mammoth–St. Anthony mine, Pinal Co, AZ, and in CO, NM, and UT; at the Cerro Gordo and Ygnacio mines, Inyo Co., CA. At the Poudrette quarry(!), Mt. St.-Hilaire, QUE, Canada. At Los Lamentos, CHIH, Mexico. At Ilimaussaq, Greenland. At *Altenberg, Moresenet, Belgium*; Långban, Värmland, Sweden; Kamareza mine, Attica, Greece. Beryllian (0.5 wt % BeO) willemite is reported from an undisclosed locality in Russia (?). At the Seh Chengi mine, eastern Iran; containing up to 25% CoO at Xingsao, Hunan, China; in Huang Hae Prov., North Korea. Near Satif, northeastern Algeria; at Mindouli, Congo; at Mumbwa and Kabuse (Broken Hill), Zambia; and at Guchab and Tsumeb, Namibia. At the Puttapa mine, Flinders Range, SA, Australia. A locally important zinc ore. AR *DT* 601(1932), *DANS* 153:681(1963).

51.1.1.3 Eucryptite LiAlSiO$_4$

Named in 1880 by Brush and Dana from the Greek for *well concealed*, in allusion to its occurrence embedded in albite. Phenakite group. HEX-R $R\bar{3}$. $a = 13.471$, $c = 8.998$, $Z = 18$, $D = 2.66$. *14-667*: 6.74_6 4.20_3 3.96_{10} 3.89_4 3.37_9 2.74_8 2.55_6 2.38_3. Structure: *ZK* 172:147(1985). Usually massive, less commonly in equant euhedral crystals with $\{10\bar{1}0\}$, $\{0001\}$, $\{11\bar{2}0\}$. Colorless, white, tan; vitreous luster; translucent to transparent. Cleavage $\{10\bar{1}1\}$, indistinct; conchoidal fracture. H = $5\frac{1}{2}$–$6\frac{1}{2}$. G = 2.66. Uniaxial (+); N_O = 1.572, N_E = 1.586. Fluoresces pink in UV; gelatinizes in acids. In pegmatites, often as graphic intergrowths with albite derived from alteration of spodumene. From *Branchville, CT*; Center Stafford, NH; Foote mine, Cleveland Co., NC; Harding mine, Taos Co., NM. From the Tanco pegmatite, Bernic Lake, MB, Canada. From the Nanling Range, Hunan–Guangdong Prov., China. From Bikita, Zimbabwe(*), where it is mined for lithium. AR *D6 426(1892)*, *AM 47:557(1962)*.

51.1.2.1 Liberite Li₂BeSiO₄

Named in 1964 by Chao for the component elements: lithium and beryllium. Equivalent to β-Li₂BeSiO₄. MON *Pn*. a = 4.68, b = 4.95, c = 6.13, β = 90.3°, Z = 2, D = 2.69; the original description has the a and c axes reversed. *29-799*(syn): 3.82_{10} 3.71_{10} 3.40_7 2.59_{10} 2.47_9 2.35_9 2.27_9 2.19_7. The structure is based on close-packed oxygen with all cations in tetrahedral coordination. *SR 31A:230(1966)*. Minute aggregates occasionally display pinacoidal faces. Pale yellow to brown; vitreous luster on cleavage, somewhat greasy on fractures. Cleavage $\{010\}$, perfect; $\{100\}$, $\{001\}$, distinct; brittle. H = 7. G = 2.69. Biaxial (−); $2V_{calc}$ = 66.3°; N_x = 1.622, N_y = 1.633, N_z = 1.638; $Z \wedge c(a$ in structure cell) = 41°. Found with eucryptite, lepidolite, natrolite, cassiterite, and scheelite in veins cutting lepidolite–fluorite–magnetite "ribbon rock" in the *Nanling Range, Hunan–Guangdong Prov., China*. AR *AM 50:519(1965)*, *BM 98:6(1975)*.

51.2.1.1 Trimerite CaMn₂Be₃[SiO₄]₃

Named in 1890 by Flink from the Greek for *three parts*, in allusion to the optically observable radial segments. MON $P2_1/c$. a = 8.09, b = 7.61, c = 14.06, β = 90.0°, Z = 4, D = 3.47. *17-477*: 3.56_4 2.76_{10} 2.33_3 2.23_3 2.05_2 1.897_2 1.784_2 1.420_3. Structure: *ZK 147:46(1977)*. Crystals short prismatic to tabular, pseudohexagonal; twinned on $\{100\}$, $\{101\}$, and $\{10\bar{1}\}$, cyclic. Colorless, pink, orange-red; vitreous luster; transparent. Cleavage $\{001\}$, distinct; conchoidal fracture. H = 6–7. G = 3.474. Biaxial (−); 2V = 83°; N_x = 1.715, N_y = 1.720, N_z = 1.725; X perpendicular to $\{001\}$; basal sections display three radial segments due to twinning. Gelatinizes with HCl. Occurs in crystals up to 10 mm as intergrowths with calcite in inclusions in a magnetite–pyroxene–garnet skarn at the *Harstigen mine, Pajsberg, Värmland, Sweden*, and similarly at Jacobsberg; also with calcite in fine-grained hematite at Långban. AR *DT:600(1932)*, *VLA 2:117(1966)*.

51.2.1.2 Larsenite PbZnSiO₄

Named in 1928 by Palache et al. for Esper Signius Larsen, Jr. (1879–1961), petrologist and Professor of Geology, Harvard University. ORTH *Pnam*. a = 8.244, b = 18.963, c = 5.06, Z = 8, D = 6.12. *20-607*(syn):

4.88_8 4.19_6 4.11_5 4.03_5 3.78_3 3.19_{10} 3.04_8 2.85_9. Structure: *ZK 124:115(1967)*. Slender prismatic [001](?), sometimes tabular {010}. White; greasy, adamantine luster; transparent. Cleavage {120}, good; brittle. H = 3. G = 5.90. Biaxial (−); 2V = 80°(?); $N_x = 1.92$, $N_y = 1.95$, $N_z = 1.96$; X = a. Found in veinlets with willemite, clinohedrite, esperite, hardystonite, hodgkinsonite, and so on, in ores of *Franklin, Sussex Co., NJ*. At Tsumeb, Namibia, and at the Beltana lead–zinc deposit, Puttapa, SA, Australia. AR *AM 13:142(1928)*, *52:1077(1967)*; *MR 13:142(1982)*.

51.2.1.3 Esperite $Ca_3PbZn_4[SiO_4]_4$

Named in 1965 by Moore and Ribbe for Esper Signius Larsen, Jr. (1879–1961), petrologist and professor of geology, Harvard University. Earlier described as, and synonymous with, calcium-larsenite. MON $B2_1/m$. $a = 17.628$, $b = 8.270$, $c = 30.52$, $\beta = 90°$, Z = 12, D = 4.25. *16-373:* 7.62_4 3.36_2 3.02_{10} 2.96_2 2.88_3 2.54_7 2.37_4 1.944_4. Structure: *AM 50:1170 (1965)*. Massive to coarse granular. White, vitreous to greasy luster. Cleavage {010} and {100}, distinct; brittle. H = 5. G = 4.28. Biaxial (−); 2V = 40°; $N_x = 1.762$, $N_y = 1.770$, $N_z = 1.774$; also 2V = 5–40°: $N_x = 1.760$, $N_y = 1.769$, $N_z = 1.769$. Fluoresces bright yellow in SWUV. Occurs with willemite, franklinite, and so on, in the zinc ores of *Franklin, Sussex Co., NJ*. AR *USGSPP 180:81(1935)*.

51.2.2.1 Sverigeite $NaMn^{2+}MgSn^{4+}Be_2(OH)[SiO_4]_3$

Named in 1984 by Dunn et al. for the country of origin. ORTH *Imma*. $a = 10.815$, $b = 13.273$, $c = 6.818$, Z = 4, D = 3.61. *38-392:* 6.63_5 5.77_7 4.35_7 2.98_6 2.88_{10} 2.83_9 2.71_5 2.64_6. Isolated SiO_4 tetrahedra are linked to Be tetrahedra to form three- and four-membered rings, $[Be_2SiO_8(OH)]$ and $[Be_2Si_2O_{11}(OH)]$, respectively, that are linked in turn to form crenulated chains parallel to [100]. Sodium is in an usual planar four-coordination. The original description interchanges a and c. Structure: *AM 74:1343(1989)*. Yellow, translucent. Cleavage {010}. H = $6\frac{1}{2}$. G = 3.60. Biaxial (+); 2V = 67°; $N_x = 1.678$, $N_y = 1.684$, $N_z = 1.699$; Z = b; r > v, strong; pleochroic, X > Y ≈ Z; yellow to pale yellow. Found in a mine dump specimen as platy aggregates with mimetite, jacobsite, and an amphibole at the *Långban mine, Värmland, Sweden*. AR *AM 70:1332(1985)*.

OLIVINE GROUP

The olivine group minerals are isostructural silicates conforming to

$$A_2(SiO_4)$$

where A cations are distributed between two sites:

M1 = at a center of symmetry = Zn, Ni, Mg, Fe^{2+}, Mn, Ca
M2 = on a mirror plane = Zn, Ni, Mg, Fe^{2+}, Mn, Ca

and a structure based on a hexagonal close packing of oxygen atoms.

51.3.1.1 Fayalite

The olivines consist of hexagonal close-packed oxygen atoms with one-half of the octahedrally coordinated sites occupied by A atoms and one-eighth of the tetrahedrally coordinated sites occupied by silicon. The A atoms do not occupy a single set of equivalent lattice positions: half are located at centers of symmetry and half are on mirror planes. The olivines contain serrated, or zigzag, chains of two size populations of edge-sharing distorted occupied octahedra (M1 < M2) parallel to the c axis. They are bridged by isolated SiO_4 tetrahedra. An equivalent chain of octahedra is displaced by $b/2$.

The olivines may be divided into two subgroups: one subgroup is essentially disordered with respect to A-site occupancy, while the other is ordered. There is a large miscibility gap between ordered and disordered olivines, intermediate compositions being virtually unknown. *AM 60:292(1975), 68:1174(1983), 69:1110(1984); JG 59:193(1951); NJBM:205(1977); RIM 5:275(1980); DHZ 1A:1(1982).*

OLIVINE GROUP

Mineral	Formula	Space Group	a	b	c	β	
Fayalite	Fe_2SiO_4	*Pbnm*	4.756–4.817	10.195–10.477	5.981–6.105		51.3.1.1
Forsterite	Mg_2SiO_4	*Pbnm*	4.756–4.817	10.195–10.477	5.981–6.105		51.3.1.2
Liebenbergite	Ni_2SiO_4	*Pbnm*	4.727	10.191	5.955		51.3.1.3
Tephroite	Mn_2SiO_4	*Pbnm*	4.86–4.908	10.59–10.772	6.22–6.302		51.3.1.4
Laihunite	$(Fe^{3+}_{0.5}\square_{0.5})Fe^{3+}SiO_4$	$P2_1/b$	4.805–4.83	10.189–10.27	5.801–5.90	90.3–91.3	51.3.1.5
Monticellite	$CaMgSiO_4$	*Pbnm*	4.822–4.825	11.108–11.111	6.382–6.383		51.3.2.1
Kirschsteinite	$Ca(Fe,Mg,Mn)SiO_4$	*Pbnm*	4.859–4.875	11.132–11.159	6.420–6.441		51.3.2.2
Glaucochroite	$MnCaSiO_4$	*Pbnm*	4.913–4.93	11.14–11.19	6.488–6.60		51.3.2.3
Calcio-olivine	Ca_2SiO_4	*Pbnm*	5.06	11.28	6.78		

Notes:
1. Ringwoodite and wadsleyite are both cubic polymorphs of forsterite.
2. Bredigite and larnite are polymorphs of calcio-olivine.
3. Merwinite (Ca_3MgSiO_4) is related to monticellite.
4. The humite group minerals are structurally related.

51.3.1.1 Fayalite $Fe_2(SiO_4)$

Named in 1840 by Gmelin for the type locality. Olivine group. See forsterite (51.3.1.2). Very manganoan fayalite has been called knebelite and can grade into tephroite (51.3.1.4). ORTH *Pbnm*. $a = 4.756–4.817$, $b = 10.195–10.477$, $c = 5.981–6.105$, $Z = 4$; manganoan (knebelite): $a = 4.826$, $b = 10.500$, $c = 6.102$. *31-633*(magnesian fayalite): 5.21_2 3.95_2 3.54_3 2.81_{10} 2.55_6 2.49_7 2.29_3 1.77_4; *12-220*(manganoan): 5.31_3 4.02_2 3.58_3 3.08_2 2.86_{10} 2.59_3 2.53_8 1.79_3; *34-178*(syn): 3.56_6 2.83_9 2.63_3 2.565_5 2.50_{10} 1.78_8 1.77_7 1.515_3. Forms a

complete solid-solution series with forsterite (51.3.1.2). ZnO has been observed in fayalite (10.96 wt %; Ogdensburg, NJ). *USGSPP 180:(1937)*. Fe_2O_3 generally less than 2 wt %; higher amounts are due to oxidation, alteration, or inclusions of iron oxides. A complete series extends from fayalite to tephroite (51.3.1.4), with intermediate iron-rich compositions known as knebelite. Ca occupancy of M1 is unfavorable, and CaO is generally less than 0.5 wt %. Ni is preferentially partitioned to forsterite rather than fayalite and is less than 0.4 wt % in fayalite. Pale yellow to olive-green, grading to olive-brown to black, color intensifying with increasing iron contents; colorless to white streak grading to gray; oily or vitreous luster [forsterite (Fo)] grading to dull or submetallic (fayalite). Cleavage {010}, moderate; {100}, weak. $H = 6\frac{1}{2}–7$. $G_{syn} = 4.39$(fayalite). Fluorescence (white–tan, SWUV) noted in low-iron forsterites from contact-metamorphosed deposits (e.g., Crestmore, CA; Gabbs, NV; Balmat, NY). IR spectra: *PCM 5:327(1980)* USGS OFR 87-263(1987). Biaxial; $Fo_{100}–Fo_{88}$ (+, $2V = 87–90°$, r < v), $Fo_{87}–Fo_0$ (−, $2V = 90–74°$, r > v); parallel extinction. Fe-rich fayalite is usually pleochroic: X, pale yellow; Y, orange-yellow, reddish brown; Z, pale yellow. Forsterite: $N_x = 1.635–1.730$, $N_y = 1.651–1.758$, $N_z = 1.670–1.772$, $2V_z = 82°(+)–74°(−)$; Fayalite: $N_x = 1.730–1.827$, $N_y = 1.758–1.869$, $N_z = 1.772–1.879$, $2V_z = 74°–46°(−)$. Forsterite–fayalite can be an accessory in many low-silica igneous rocks. Occasionally, iron-rich composition crystals found in granites, also late-stage cavities in extrusive rocks or in xenoliths; in some rhyolites and ash-flow tuffs. Found in many worldwide localities; only noteworthy localities are mentioned here: Hurricane Mt., Conway, Carroll Co., NH; near-end-member composition from Rockport, Essex Co., MA; Ogdensburg(!), Sussex Co., NJ; Pigeon Point, MN; Cheyenne Mt., CO; Obsidian Cliff(!), Lake of the Woods, Yellowstone Park, WY; Coso Hot Springs(!), Inyo Co., CA; Sisters Mt., McKinley Pass, ID; Salt Lake, Oahu Is., HI; Kiglapait and Nain intrusions, Labrador; Forsythe mine, Hull, QUE, Canada; Cerro de las Navajas, Mexico; Skaergaard intrusion, Kangerdlugssuaq, Greenland; Rauthaskritha, Iceland; Slavcarrach, Skye Is., and Mull and Rhum Is., Scotland; Mourne Mts., Northern Ireland; Tullisenlampi, Lemi, Finland; *Fayal (Faial) Is., Azores*; Capucin, Mt. Dore, France; Forgia Vecchia, Lipari Is., Cuddia Mida, Pantelleria Is., Italy; Künstlich, Schlakken, Germany; Dannemora, Sweden; Younger Granites, Nigeria; Bushveld complex, South Africa; Billiton, Indonesia; Broken Hill, NSW, Australia; Lake Taupo, North Island, New Zealand; Kajiya, Kanagawa-Ken, Honshu, Japan.
VK, EF

51.3.1.2 Forsterite $Mg_2(SiO_4)$

Named in 1824 by M. Lévy for Adolarius Jacob Forster (1739–1806), noted English mineral collector. Olivine group. The original name of any olivine mineral that is definitely recognizable as an olivine from its description is chrysolite, named in 1747 by Wallerius and the name of priority use for Mg-rich compositions as early as by Dana in 1892. *Peridot ordinaire* was

51.3.1.2 Forsterite

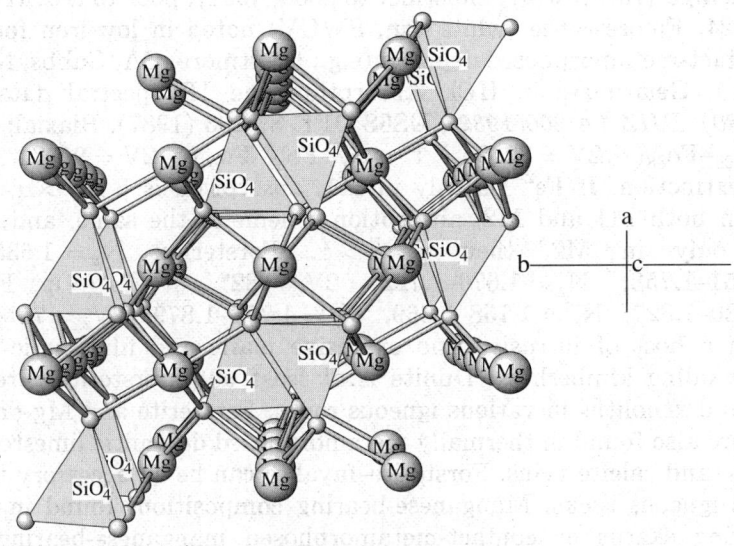

FORSTERITE: Mg₂SiO₄

This view shows the edge-sharing chains of M2 octahedra, flanked by M1 octahedra. The octahedra are linked by Si tetrahedra which face alternately up and down in the a direction.

named by d'Argenville in 1755, and *peridot* has survived as the gem name for olivine, usually forsterite. The forsterite–fayalite series has been given names for its intermediate compositions: forsterite (90–100% Fo), chrysolite (70–90%Fo), hyalosiderite (50-70% Fo), hortonolite (30-50% Fo), ferrohortonolite (10-30% Fo), fayalite (0-10% Fo). ORTH $Pbnm$. $a = 4.756$–4.817, $b = 10.195$–10.477, $c = 5.981$–6.105, $Z = 4$, $D = 3.22$. *31-795*(ferroan forsterite: $Mg_{0.64}Fe_{0.36}$): 3.92_4 3.52_3 2.79_{10} 2.53_6 2.48_6 2.29_3 2.27_3 1.76_5; *34-189*(syn): 3.88_8 2.765_7 2.51_8 2.46_{10} 2.27_6 2.25_4 1.75_7 1.48_3. A complete solid-solution series exists between forsterite (Fo) and fayalite (Fa). Forsterites of composition $Fo_{89\pm05}$ are very common in basaltic rocks for petrogenetic reasons. Al_2O_3 contents as high as 1.68 wt % have been found in forsterite. Generally, only tenths of 1 wt % Fe_2O_3 are present in fresh olivine; larger amounts are usually due to oxidation or to the presence of minute Fe_2O_3 inclusions. Alkalis are rarely present, both 0.X wt.% or lower. Ca occupancy of M1 is unfavorable, but the content of CaO may reach several wt %. MnO content in forsterite and Fe-bearing forsterite is < 1 wt %. Ni is preferentially partitioned to forsterite relative to fayalite and reaches as much as 0.6 wt % NiO. Crystals are generally short prismatic, elongated on c and flattened on b. Depending on growth velocities and other factors, crystal habit may be dendritic, skeletal, and so on. Dominant forms are: {110}, {010}, {021}, {101}, {001}, {111}, {120}. Twinning on {011}, {012}, {031}. Pale yellow to olive-green grading to olive-brown to black, color intensifying with increasing iron content; synthetic material is colorless; colorless to white streak grading to gray; oily or vitreous luster (forsterite) grading to dull or submetallic (fayalite). Cleavage {100}, {010}, indistinct to good; {001}, poor to fair. H = $6\frac{1}{2}$–7. G_{syn} = 3.24. Fluorescence (white–tan, SWUV) noted in low-iron forsterites from contact-metamorphosed deposits (e.g., Crestmore, CA; Gabbs, NV; Balmat, NY). Gelatinizes in HCl. Absorption and IR spectral data: *PCM 5:327(1980)*, *DHZ 1A:200(1982)*, US6S OFR 87-263 (1987). Biaxial; XYZ = *bca*; Fo_{100}–Fo_{88}(+, 2V = 82–90°, r < v), Fo_{87}–Fo_0 (−, 2V = 90–74°, r > v); parallel extinction. If Fe^{2+} is only in M1, absorption is β > γ > α; if Fe is present in both M1 and M2, absorption scheme is the same, and if Fe is present only in M2, then γ > β > α. Forsterite: N_x = 1.635–1.730, N_y = 1.651–1.758, N_z = 1.670–1.722; $2V_z$ = 82° (+)–74°(−); Fayalite: N_x = 1.730–1.827, N_y = 1.758–1.869, N_z = 1.772–1.879; $2V_z$ = 74°–46°(−). Found in a host of intrusive and extrusive mafic and ultramafic igneous rocks, including kimberlites. Dunite is at least 90% forsterite. Present as nodules and xenoliths in various igneous rocks. Forsterite and Mg-dominant olivines are also found in thermally metamorphosed dolomitic limestones and dolomites, and calcite veins. Forsterite–fayalite can be an accessory in many low-silica igneous rocks. Manganese-bearing compositions found in manganese-bearing skarns or contact-metamorphosed manganese-bearing rocks. Found in some meteorites (e.g., pallasites = stony irons). Ubiquitous, only noteworthy localities are mentioned here: Bolton, Worcester Co., MA; Balmat, St. Lawrence Co., NY; Palisades Sill, Bergen Co., NJ; Balsam Gap(!) and Webster (dunite), Jackson Co., NC; Wichita Complex, OK; Kilbourne Hole, 20 miles SW of Las Cruces, NM; San Carlos Reservation(!), Gila Co., AZ; Gabbs, Nye Co., NV; Crestmore(!), Riverside Co., CA; Snake River plain, ID; Jumbo Mt.(!), Snohomish Co., WA; Mauna Loa, Hawaii Is., HI; crystals to 6 cm × 20 cm × 30 cm at the Parker mine, Notre Dame du Laus, and Oka complex, QUE, Canada; Skaergaard intrusion, Kangerdlugssuaq, Greenland; Lanzarote, Canary islands; Coire a Gretaich, Cullins and Bornaskitaig, Skye, Scotland; Puy de la Denise(!), Puy de Dôme district, France; Traversella(!), Piedmont, *Vesuvius(!) and Monte Somma(!)*, Campania Naples, Baccano, Albani Hills, Mt. Etna, Sicily, Italy; Findelen moraine, Zermatt, Switzerland; Bellerberg, Laacher See, Dreiser Weiher, and Porstberg, Eifel district, Unkel(!), Rhine, Germany; Triblitz, Bohemia, Czech Republic; Pfunders, Tirol, Austria; Vourinos complex, Greece; Møre(!) and Snarum(!), Norway; Alnö island, Sweden; gem peridot occurs on Zabargad Is.(!) (St. John's Is.), Egypt; crystals to >10 cm from the Kovdor massif, Kola Penin.; Nikolai–Maximillian mine, Achmatovsk, and Kyshtym, Zlatoust district, Ural Mts., Russia; very fine, large crystals to 14 cm of "peridot" have recently been found near Sumput (Suppatt), between Kamila and Naran, near Nanga Parbat, Mansehra district, NWFP, Pakistan. Crystals as heavy as >2 kg and of variable color ranging from superb green and of transparent gem quality (> 100 ct) to yellowish green and grayish green; Jacupiranga(!), São Paulo, Brazil; Nyiragongo volcano, Zaire; Ethiopia; Klaasvoogds, Cape Prov., South Africa; Salem(!), Tami Nadu, India; Sri Lanka; Mogok district. Crystallized and gem peridot occur in Myanmar (Burma); Vaitunamea, Tahiti, Society Is.; Kalka and Ewarara intrusions, Australia; Dun Mt.

(dunite), New Zealand; Vietnam; Nishinotake, Japan. Used as a minor gemstone (tumbled and faceted gems). VK, EF *G&G 28:16(1992)*.

51.3.1.3 Liebenbergite Ni$_2$SiO$_4$

Named in 1973 for William Roland Liebenberg (1919–1988), South African metallurgist (ore dressing), mining official, and mineralogist. Olivine group. ORTH *Pbnm*. (Nat): $a = 4.727$, $b = 10.191$, $c = 5.955$ [*AM 58:733(1973)*]; (nat); $a = 4.731$, $b = 10.180$, $c = 5.941$ [*AM 66:770(1981)*]; (syn): $a = 4.725$–4.728, $b = 10.117$–10.118, $c = 5.908$–5.9125 [*NJBM 30 (1963); AC(B) 39:692(1983)*]; $Z = 4$, $D = 4.585$–4.60(nat), $D = 4.92$(syn). 15-0388 (nat): 5.09_3 3.47_6 2.76_9 2.50_8 2.44_{10} 2.25_3 1.74_9, 1.473_3; (syn): $4.28_{2.5}$ 3.84_2 3.47_9 2.74_9 2.55_4 2.49_8 2.43_{10} 1.73_6. Ni is strongly partitioned from coexisting fluids by olivine [*NAT:1147(1966)*] and is strongly ordered into the M1 site [*AM 60:292(1975)*], potentially leading to ordered NiM2SiO$_4$ intermediate compositions [*AM 66:770(1981)*]. Natural liebenbergite is fully ordered with M1 occupied only by Ni, whereas synthetic material is only partially ordered (because of rates and temperatures of equilibration). Structure of synthetic: *AC(B) 39:692(1983)*. Granular (most <150 μm), yellowish green, or bright green grains to 1 mm. Cleavage {010}, good to poor; {100}, poor. H = 6–6$\frac{1}{2}$. IR spectra: *PCM 5:327(1980)*. Ni is a frequent minor element in forsterite in the range 0.10–0.30 wt %. *JG 78:304(1970)*. End member liebenbergite not yet known. Analysis of type material (wt %): SiO$_2$, 29.39; NiO, 56.32; MgO, 6.50; FeO, 4.37; CoO, 1.80. Biaxial (−); XYZ = *bca*; N$_x$ = 1.820, N$_y$ = 1.854, N$_z$ = 1.888; 2V = 80–88°; OAP = (001); r > v; parallel extinction; slightly pleochroic. X,Y, colorless to pale green; Z, greenish yellow. Is variably replaced by Ni-serpentine (nepouite). Occurs in a serpentinized ultramafite containing abundant trevorite. From two assemblages variously containing bonaccordite, bunsenite, gaspéite, goethite, millerite, nepouite, nimite, reevesite, trevorite, violarite, and willemeite. Found at the *Scotia Talc mine, Bon Accord, Barberton Mountain Land, Transvaal, South Africa*. VK, EF

51.3.1.4 Tephroite Mn$_2$SiO$_4$

Named in 1823 by Breithaupt in allusion to the ash-colored nature of the original material. Olivine group. ORTH *Pbnm*. (Nat): $a = 4.86$–4.908, $b = 10.59$–10.772, $c = 6.22$–6.302; (syn): $a = 4.903$, $b = 10.604$, $c = 6.258$; $Z = 4$, $D = 3.97$(nat), $D = 4.12$(syn). 31-0823(magnesian): 3.57_8 3.09_2 2.84_9 2.58_{10} 2.53_9 1.79_8 1.55_5 1.53_5; 35-0748(syn): 3.63_5 2.87_9 2.69_3 2.67_7 2.56_{10} 1.81_7 1.80_3 1.56_3; 45-1396(calcian): $4.10_{1.5}$ 3.64_5 $2.89_{8.5}$ 2.72_2 $2.63_{8.5}$ 2.57_{10} 1.82_5 1.815_5. Twins on {110}, also polysynthetic twinning and rarely trillings. Elongated on *c*, sometimes on *a*. Dominant forms: {100}, {010}, {110}, {101}, {021}, {111}. Crystals are scarce at most deposits, tephroite being either individual grains enclosed in matrix only rarely grading to euhedral (to 2.5 cm) (Ogdensburg) or compact and massive. Elongated crystals of complex form found lining fractures. The color of the end member tephroite is variable: bluish to greenish gray, and grades to gray (daylight); with increasing elemental substitutions grades to smoky gray to black, frequently brown to red or purple (incandescent light); gray streak; vitreous to subviteous luster grading to dull; high iron compositions can have a nearly submetallic luster. Cleavage: {010}, good; {001}, moderate; {011}, poor; conchoidal fracture. H = $5\frac{1}{2}$–6. G = 3.78–4.1. Gelatinizes in mineral acids. Optical absorption spectra: *IMA Papers and Proc., 5th Gen. Meet., Cambridge: 22–35(1968)*. IR data: *AM 49:1388(1964)*. Raman spectra: *JCS(A):1372(1969)*. Grades into fayalite with intermediate compositions called knebelite, while compositions grading to forsterite are scarce and have been termed picrotephroite. Compositions between Fo–Fa–Te have been rationalized by plots of N_y vs. d_{130}. *CM 14:479(1976)*. ZnO substitutes to 18.90 wt % (Ogdensburg), while CaO has been observed to 3.8 wt % (Japan). Both Al_2O_3 (to 0.97 wt %, Japan) and Fe_2O_3 (to 4.57 wt %) substitute in tephroite, but may be due to alteration. Ferroan tephroites tend to have a wide variety of additional substituents. Sometimes intergrown with alleghanyite. Biaxial (−); XYZ = *bca*; N_x = 1.760–1.795, N_y = 1.790–1.810, N_z = 1.807–1.830; 2V = 55–70°. X < Z < Y; OAP = (001). Sometimes pleochroic; X, brownish; Y, reddish; Z, greenish blue. Also colorless to pale yellow and brownish. Z, bluish; X, pale yellow. RI values increase with increasing Fe^{2+} substitution but are greatly lowered by Mg substitution (e.g., N_x = 1.750, N_y = 1.766, N_z = 1.779). Mg substitution increases 2V; r > v. Sometimes formed as a metasomatic reaction of rhodonite with rhodochrosite or manganoan calcite. Also suggested as a reaction of a manganoan carbonate with quartz, producing tephroite and amphibole. Found in iron- and manganese-rich ores and skarns, sometimes in late-stage veins. Sometimes found in contact-metamorphosed manganese-bearing sediments. Found frequently with manganoan calcite and with other manganese silicates, such as rhodonite, spessartine, bustamite, rarely alleghanyite, friedelite, or glaucochroite. Found with willemite and zincite ores at Franklin and Ogdensburg. Numerous worldwide localities; only localities having noteworthy material are mentioned here. Occurs at the Betts mine, Plainfield, MA; Sterling Mine(!), Ogdensburg, and Mine Hill(!), Franklin, Sussex Co., NJ; Bald Knob mine, Sparta, Alleghany Co., NC: Alamo Crossing, Mohave Co., AZ; Caldwell mine, Caldwell, Mariposa Co., CA; Benallt mine, Rhiw, Carnarvonshire, Wales; Meldon, Okehampton, Devonshire, Treburland mine, Altarnun, Cornwall, England; Bonneval-sur-Arc, Haute Maurienne, France; Bjelkes, Kolegii, Loka, Nya, and Stor mines, Långban, Brattfors, Eastern Moss, and Jakobsberg mines, Nordmark, Harstigen and Pajsberg mines, Pajsberg, Värmland, Sjö mine, Grythyttan, Ørebro, and Dannemora, Sweden; near Davos and St. Moritz, Switzerland; northern Appennines, Italy; near end-member composition from Möserboden, near Salzburg, Austria; Laacher See, Germany; Chvaletice, Bohemia, Czech Republic; Sebes-Cibin Mts., Romania; cuprian (to 29.6%) from Bulgaria; Primorye, Russia; Lake Balkhash, Kazakhstan; Dzhumart, Kamys, and Karadzhal deposits, central Kazakhstan; La Loma area, Santa Barbara, Antioquia Dept., Bolivia; Moanda, Gabon; Wessels mine(!), Black Rock, Kalahari Field, South Africa; Xihucum, Chanping, Beijing, China; Ikenotsura mine,

Kumamoto Pref., Kaso and Kyurazawa mines, Tochigi Pref., Hijikuzu, Tamagawa (magnesian, 6.6 wt % MgO), and Toyoguchi mines, Himegamori, Iwate Pref., Japan; calcian (3.8 wt % CaO) from Kanoiri mine, Kanuma Station, Tochigi Pref., Japan; Consolidated mine, Broken Hill and Danglemah, Tamworth district, NSW, Australia; Clark Peninsula, Wilkes Land, Antarctica; Serra do Navio, Amapa, Brazil; Andra Pradesh, India; Fotadrevo, Madagascar. VK, EF *MIN 3(1):199(1972)*, *LIT 29:57(1992)*, *DHZ 1A:337(1982)*, *NJFM 159:101(1988)*, *BNSM 17:119(1991)*.

51.3.1.5 Laihunite ($Fe^{2+}_{0.5}\square_{0.5}$)$Fe^{3+}SiO_4$

Named in 1976 for the type locality. Olivine group. May also be formulated as ($\square_x Fe^{2+}_{2-3x} Fe^{3+}_{2x} SiO_4$). MONO $P2_1/b$. (1M): $a = 4.805$–4.83, $b = 10.189$–10.27, $c = 5.801$–5.90, $\alpha = 90.9$–$91.3°$ (to conform with *Pbnm* setting of olivine); (syn 2M): $a = 4.81$, $b = 10.43$, $c = 5.93$, $\beta = 91.0°$; (syn 3M): $a = 4.81$, $b = 10.44$, $c = 5.99$, $\beta = 90.3°$; $Z = 2$, $D = 4.10$–4.24. *AM 70:737(1985)*. Superlattice reflections of $2c$ and $3c$ exist, and are called 2M and 3M, respectively (Ramsdell notation). 30–664: 3.78_6 3.47_9 2.78_8 2.52_{10} 2.405_6 2.26_5 1.75_7 1.44_4; the powder data differ from fayalite in intensity, extinction rules, and smaller d values. *AM 62:1058(1977)*. A nonstoichiometric, distorted fayalite-type structure. Excess charge caused by the substitution of Fe^{3+} for Fe^{2+} is compensated at M1 by vacancies (octahedral sites). This accounts for Fe deficiency in the $M^{2+}_2 SiO_4$ olivine structure. All M2 positions are occupied by Fe^{3+}, whereas M1 positions are alternately vacant and occupied by Fe^{2+}. Crystals < 1 mm thick, tabular to short, prismatic. Black, pale brown streak, metallic to submetallic luster, opaque. Weakly magnetic. Cleavage {010}, perfect; {001}, good; {100}, perfect. Microscopically twinned, and with development of antiphase domains. Intergrowths of laihunite with magnetite are frequent but not universal; also intergrown with hematite. $H = 5\frac{1}{2}$–$6\frac{1}{2}$. $G = 3.92$–3.97. Mössbauer spectra and magnetic and electrical properties: *AM 70:729(1985)*, *70:576(1985)* *SS 30(4):373(1995)*. Thermal data: *AM 62:1058(1977)*, *63:424(1978)*. Essentially a ferric analog of fayalite with compensating vacancies in M_1. Analysis (wt %): SiO_2, 31.31; Fe_2O_3, 44.29; FeO, 23.89; MgO, 0.67; CaO, 0.34. Substitutions suggest compositions grading to magnesian or manganoan varieties [e.g., ($Fe^{2+}0.5_{0.5}$)(Fe^{3+}, Mg, Mn)SiO_4]. Observed substitutions (wt %) include: Al_2O_3, 0.065; MgO, 5.91; MnO, 4.31; CaO, 0.47. Opaque in standard thin sections, brownish red when a few microns thick. Calculated refractive indices range between 2.01 and 2.04. In RL, gray to grayish black, parallel extinction, weakly anisotropic, slightly heterogeneous extinction. $R = 2.39$ (650 nm), 12.39 (589 nm), 13.43 (green); internal pleochroic reflections gray to grayish black. Laihunite plus magnetite intergrowths can form from the oxidation of fayalite, with which they can be in contact as microscopic rods or as discrete patches. Originally found in a quartz–hypersthene granulite associated with a metamorphosed banded iron deposit; laihunite was found with quartz, enstatite, and magnetite. Also reported as migmatite and/or eulysite (China). Found as inclusions in black fayalite with magnetite from syenite, granite, and granite pegmatites (e.g., CO). Found as individual crystals in lithophysae with tridymite, fayalite, and hematite in dacite (Japan). Found at Stove Mountain, St. Peters Dome, Crystal Park, and Cheyenne Canyon, El Paso Co., CO, Lake Albert, OR; Mourne Mts., Northern Ireland; Pantelleria Is., Sicily, Italy; *Lai-He village, Qianan Co., Liaoning Prov.*, Jixi Co., Anhui Prov., and Hubei Prov., *China*; Kamitaga, Atami, Shizuoka Pref., Yugawara-cho, Kanagawa Pref., and Naegi, Japan. Possibly remotely sensed on Mars and Venus. VK, EF *G 1:115(1982); MJJ 11:382(1983), 12:376(1985); AM 69:154(1984)*.

51.3.2.1 Monticellite CaMgSiO$_4$

Named in 1831 by Brooke for Teodoro Monticelli (1759–1845), noted Italian mineralogist. Olivine group. ORTH *Pbnm*. (Syn): $a = 4.815$, $b = 11.108$, $c = 6.37$; (nat): 2.676_{10} 3.64_8 2.60_6 3.19_4 2.94_4 4.19_3 2.40_3 1.819_3; 35-0590(syn): 3.63_5 3.18_3 2.93_4 2.66_{10} 2.58_6 2.40_3 1.82_6 1.59_3. Of the two available octahedral sites, Ca atoms occupy sites on the mirror planes (M2) and Mg atoms are at the centers of symmetry (M1). Monticellite is frequently near-end-member composition, but grades toward kirchsteinite ($M_{75}K_{25}$) in rare cases (Muck). Minor substitutions observed (wt %) include Na$_2$O, 0.14; K$_2$O, 0.25; SrO, 0.08; BaO, 0.18; TiO$_2$, 0.13; Al$_2$O$_3$, 0.75; Fe$_2$O$_3$, 1.70; MnO, 1.17. Some Crestmore material contains as much as 4.6 wt % SO$_3$ *BCGF 157:7(1952)*. Crystals are scarce and then usually rounded, short prismatic on *c*. Dominant forms: {010}, {120}, {110}, {111}, {121}, others are rare. Twinning observed on {031}. Colorless to white to gray, also greenish gray, greenish yellow, or amber yellow; white streak; vitreous to oily luster. Cleavage {010}, poor; conchoidal fracture. H = $5\frac{1}{2}$. G = 3.04–3.31. IR data: *AM 49:1388(1964)*. Biaxial (−); XYZ = *bca*; N$_x$ = (1.641 syn) 1.641–1.654 (1.674 near kirchsteinite), N$_y$ = (1.646 syn) 1.649–1.664 (1.694 near kirchsteinite), N$_z$ = (1.652 syn) 1.655–1.674 (1.706 near K); 2V = (85–90° syn), 70°–90° (decreasing with increasing Fe^{2+} or Mn substitution); colorless; OAP parallel to (001); r > v, parallel extinction, sometimes as optically continuous overgrowths on forsterite. Sometimes thermoluminescent. Observed to be replaced by serpentine and/or fassaitic pyroxene. Formed in skarns and marbles during high-temperature metasomatism or metamorphism. Found in some ultrabasic rocks, mellilites, peridotites, nepheline basalt, alnöite, lamprophyres, or carbonatites. Found in association with apatite, biotite, calcite, diopside, dolomite, forsterite, gehlenite, magnetite, merwinite, perovskite, scapolites, spinel, spurrite, vesuvianite, or wollastonite, rarely with clinohumite or cuspidine. Notable localities include Cascade slide(!), Essex Co., NY; Magnet Cove(!), Hot Spring Co., AR; Omaha oil field, Gallatin Co., IL; Haystack Butte, Highwood Mts., MT; Wind River Range, WY; Tres Hermanas Mts., Luna Co., NM; Crestmore(!), Riverside Co., Dewey mine, Valley Wells, Clark Mt. district, San Bernardino Co., CA; York River skarn, Bancroft, ONT, Montreal mine and Lake of Two Mountains, Oka, and Isle Cadieux, QUE, Mineral Mt. and Copper Butte, BC, Elwin Bay diatreme, NWT, Canada; Camas

Malag, Broadford, Skye Is., and Camas Mòr, Muck Is., Inverness, Scotland; Carlingford district, County South, Ireland; Schälbischen Alb, Württemburg, Künstlich, Schlakken, and Dillenburg, Nassau, Germany; Monte Somma and Mt. Vesuvius, Naples, Campania, Italy; Toal dei Rizzoni, Mt. Monzoni, and Pesmeda Alp, Tirol, Austria; Severnyi mine, Enski district, and Kovdor massif, Murmansk Oblast, Kola Penin., Malo-Murun massif, Yakutia, Tazheran massif, Lake Baikal, and Anakit district, Lower Tunguska R., Siberia, and Zagorny massif, Kuznetsk Alatau, Akhmatovsk mine, Southern Ural Mts., Lespromkhoznoye skarn, Patynsk, Gornaya Schoriya, and Taknakh, Russia; Hatrurim Formation, Israel; Salpetre Kop, South Africa; Borra, Visakhapatnam district, and Nanjangud, Mysore, India; Xujiachong picrite body, Jingshan Co., Hubei Prov., China; Hoki-zawa, Kanagawa Pref., Kushiro, Hiroshima Pref., Japan; Hobart, TAS, Australia. VK, EF *MIN 3(1):212(1972)*, *DHZ 1A:352(1982)*.

51.3.2.2 Kirschsteinite Ca(Fe,Mg,Mn)SiO$_4$

Named in 1957 by Sahama and Hytönen for Egon Kirschstein, German geologist and pioneer in geological work in North Kivu, Zaire. Olivine group. ORTH Pbnm. (Syn): $a = 4.875$, $b = 11.159$, $c = 6.441$, $Z = 4$, $D = 3.56$. 34-0098(nat): 2.949_{10} $2.68_{8.5}$ 2.604_8 3.658_7 1.839_6; (syn): 5.58_3 3.67_4 2.96_7 2.79_3 2.68_7 $2.61_{10}1.835_9$ 1.61_5. Can use d_{130} X-ray spacing to distinguish members of the monticellite–kirschsteinite series from Ca-free or Ca-poor olivine. *AM 43:862(1958)*. Type material is anhedral, but sharp translucent crystals to 1 mm are found at Mt. Nyiragongo. Yellowish to greenish gray; streak not reported, probably colorless; vitreous to oily luster presumed. Cleavage not reported, but monticellite has a poor cleavage. H = 5.5. G = 3.31–3.43. Soluble in acids. Mossbauer spectra *SSC 5:267(1967)* shows that Fe^{2+} is in a single site (M2). Extremely magnesian compositions to near monticellite occur. A subcalcic variety of kirschsteinite has been analyzed. Observed substitutions (wt %) include: TiO$_2$, 0.28; Al$_3$O$_e$, 0.71; Fe$_2$O$_3$, 1.87; MnO, 1.65; Na$_2$O, 0.38; K$_2$O, 0.36. As much as 2.55 wt % Al$_2$O$_3$ in monticellite–kirschsteinite solid-solution series. Biaxial (−); XYZ = bca; N$_x$ = 1.674–1.696 (Syn), N$_y$ = 1.694–1.734 (Syn), N$_z$ = 1.706–1.743 (Syn); 2V = 51–65°; parallel extinction. Found at the type locality in a melilite nephelinite associated with clinopyroxene, apatite, biotite, combeite, gehlenite, götzenite, hornblende, kalsilite, nepheline, perovskite, and sodalite. Also with clinopyroxene, leucite, magnetite, melilite, nepheline, and pyrrhotite (Mt. Nyiragongo). Occurs in calcareous skarn xenoliths in alkali syenite (Tazheran). Observed in peridotites, meteorites, and lunar samples. Formed during underground nuclear bomb detonation. *AM 51:1192(1966)*. Found in the Minnesota River Valley, MN; magnesian from Tazheran massif, Lake Baikal region, Russia; top of Mt. Nyiragongo and *northeastern Mt. Shaheru, Nyiragongo area, Kivu, Zaire*; manganoan (Kr$_{55}$–Gl$_{45}$) from Wessels mine, Kuruman, Kalahari district, South Africa; magnesian in Angra dos Reis, Brazil, achondrite; Lewis Cliff meteorite. VK, EF *MIN 3(1):218(1972)*, *DHZ 1A:352(1982)*.

51.3.2.3 Glauchchroite MnCaSiO$_4$

Named in 1899 by Penfield and Warren from the Greek for *blue* and *color*, in allusion to the blue color of some specimens. Olivine group. ORTH *Pbnm*. $a = 4.913$–4.93, $b = 11.14$–11.19, $c = 6.488$–6.60, $Z = 4$, $D = 3.49$. *14-0376:* 4.23_4 3.69_6 2.96_6 2.69_8 2.63_8 1.85_{10} 1.13_5 1.11_5; Wessels mine: 5.53_4, 4.26_2 3.02_5 2.70_{10} 2.58_4 2.46_4 1.845_5 1.61_{10}. Ca is in M2 and Mn is in M1. *AM 63:365(1978)*. Crystals are elongated, to 1 cm prismatic on *c*. Forms include {100}, {010}, {110}, {120}, {103}, {021}, and others. Gray-lilac, bluish green through brown to brownish rose and red-orange, rare crystals pale pinkish lilac (Franklin); white streak; vitreous to dull (fine-grained) luster. Cleavage {100}, poor; conchoidal fracture; brittle. Twinning observed on {011} (Franklin). $H = 6$. $G = 3.41$–3.48. Moderately fusible; gelatinizes in acids, particularly HCl. Optical absorption spectra: *IMA Papers and Proc. 5th Gen. Meet., Cambridge, 1966; MSL:22-35(1968)*. IR spectra: *AM 49:1388(1964)*. Generally near end member with minor MgO (to 3.17 wt %, Franklin). Substitutions observed (wt %) include: Al_2O_3, 0.5–1.85 (Wessels); PbO, 1.74 (Franklin); ZnO, 2.35; minor alkalis (Franklin). Biaxial $(-)$; $XYZ = bca$; $N_x = 1.682(1.685$ syn$)$–1.695, $N_y = 1.710$–$1.723(\text{syn})$, $N_z = 1.725$ – $1.736(\text{syn})$; $2V = 61°$; parallel extinction; $r > v$; weak or no pleochroism. Color in thin section: orange-red (Wessels) or colorless; X,Y, light yellow-orange; Z, red-orange (Wessels). Partially derived from metasomatism of manganese silicates. *AM 72:423(1987)*. Weathers to a black sooty material. Found in a manganese-rich hydrothermally worked, skarnlike assemblage (Franklin) associated with manganoan andradite, barite, bustamite, clinohedrite, diopside, franklinite, hardystonite, leucophoenicite, nasonite, willemite, or zincite. Also with calcite, andradite, diopside–hedenbergite in a dolerite–limestone skarn (Russia). With diopside, wollastonite, and other minerals in calc–silicate rocks with manganese minerals in the Wessels mine, Kalahari manganese field, South Africa. Occurs at the *Parker shaft, Franklin, Sussex Co., NJ*; Anakit skarns, Nizhnaya, Lower Tunguska R., Siberia, Russia; Wessels mine, Kuruman, Kalahari Field, South Africa; Guning Bijih (Ertsberg) skarn, Irian Jaya, Indonesia. VK, EF *MIN 3(1):220(1972)*.

51.3.3.1 Ringwoodite Mg$_2$SiO$_4$

Named in 1969 by Binns et al. for Alfred Edward Ringwood (b.1930), petrologist, University of Canberra, Australia. ISO *Fd3m*. $a = 8.075$(syn Mg$_2$SiO$_4$), 8.113(ferroan)–8.122(nat), 8.234(syn Fe$_2$SiO$_4$); $Z = 8$, $D = 3.90$. *21-1258*(ferroan): 2.87_2 2.45_{10} 2.03_4 1.56_2 1.43_6 1.06_1 1.01_1 0.828_1; *13-0230*: 4.74_4 2.91_3 2.48_{10} 2.06_1 1.94_1 1.89_1 1.68_5 1.58_5. Ringwoodite is a polymorph of forsterite and wadsleyite and has the normal spinel structure. See also *CJG 9:99(1990)*. The original material is a ferroan ringwoodite with 23.2–23.4 wt % FeO with no other substituents. Catherwood varies in FeO by 22.5–24.1 wt %. The name *ringwoodite* is currently applied to the entire series (Fe$_2$SiO$_4$–Mg$_2$SiO$_4$), but if iron-dominant material is found, nomenclature changes will be necessary. Anhedral, purple, bluish gray, or smoky gray 100-μm grains. G = 3.90. Isotropic; N = 1.768, colorless to pale purple. Found as anhedral grains replacing ferroan forsterite or as dark veinlets in the Tenham chondrite meteorite. Presumed to exist in Earth's mantle. Also found in extraterrestrial microspherules of melt glass in a granite in western Suzhou, China. *CSB 38:1651(1993)*. Found in the Catherwood, SAS, Canada, L6 olivine–hypersthene chondrite; Peace River, AL, Canada, L6 chondrite; Coorara, WA, chondrite, and *Tenham, Australia, chondrite;* Pampa de Infierno chondrite, Argentina. VK, EF *N 221:943(1969)*, *MIN 3(1):13(1972)*, *GCA 46:1903(1982)*.

51.3.4.1 Wadsleyite -(Mg, Fe^{2+})$_2$SiO$_4$

Named in 1983 by Price et al. for A. D. Wadsley, crystallographer. ORTH *Imma*. $a = 5.70$, $b = 11.51$, $c = 8.24$, $Z = 8$, $D = 3.84$. *37-0415*: 2.89_5 2.69_4 2.64_3 2.45_{10} 2.04_8 1.57_3 1.55_3 1.44_8. Polymorph of forsterite and ringwoodite. *PEPI 3:166(1970)*. Light grayish brown, translucent. Type wadsleyite is very ferroan (Mg$_{1.5}$Fe$_{0.5}$SiO$_4$). Analysis (wt %): Mg, 23.05; Si, 18.08; Fe varies from 16.73 to 17.78. Minor substitutions (wt %) include: Ca, 0.11; Cr, 0.07; Mn, 0.31–0.36; Ni, 0.07–0.13; Zn, 0.19. Anisotropic, fine-grained aggregates show RI = 1.76, low first-order interference colors. Found as microcrystalline aggregates pseudomorphing previously existing forsterite grains in veinlets in a meteorite believed to have experienced high-pressure shock. *Peace River, Alberta, Canada, L6 chondrite*. EF *CM 21:29(1983)*.

51.4.1.1 Bredigite Ca$_7$Mg(SiO$_4$)$_4$

Named in 1948 by Tilley and Vincent for Max Albrecht Bredig (b.1902), German-born American chemist, who studied polymorphism of calcium silicates. ORTH *P2nn*. $a = 10.909$, $b = 18.34$, $c = 6.739$, $Z = 4$, $D = 3.412$. *35-260*(syn): 2.74_{10} 2.73_{10} 2.67_{10} 2.23_3 2.22_2 1.924_3 1.899_2 1.562_2. The structure allegedly corresponds to that of α'-Ca$_2$SiO$_4$; the Ca sites, however, are not all equivalent, and the formula may be written as (Ca,Ba)Ca$_{13}$Mg$_2$(SiO$_4$)$_8$. *AM 61:74(1976)*. Found as rounded grains, 50–100 μm; twinned on {110}. Colorless to gray, vitreous luster. Cleavage {130}, distinct. G = 3.40. Biaxial (+); 2V variable; N$_x$ = 1.712, N$_y$ = 1.716, N$_z$ = 1.725; r < v, weak; for Israel material N$_x$ = 1.695, N$_z$ = 1.705. Barium may replace up to 29% of the calcium in the larger Ca site, but overall BaO is < 0.3 wt %. Gelatinizes with acids. Found in thermally metamorphosed aureoles in intruded limestones, and in near-surface thermal metamorphism of carbonaceous limestones. With larnite at Marble Canyon, Culberson Co., TX. At *Scawt Hill, County Antrim, Northern Ireland;* Island of Muck, Inverness-shire, Scotland. In surface exposures of the Hatrurim formation at several localities in Israel. AR *MM 28:255(1948)*, *GSIB 70:30(1977)*.

51.4.2.1 Merwinite $Ca_3Mg(SiO_4)_2$

Named in 1921 by Larsen and Foshag for Herbert Eugene Merwin (1878–1963), an American mineralogist. MONO $P2_1/a$. $a = 13.254$–13.26, $b = 5.293$–5.301, $c = 9.328$–9.336, $\beta = 91.9$–$92.8°$; (syn): $a = 13.298$, $b = 5.305$, $c = 9.352$, $\beta = 92.09°$; $Z = 4$, $D = 3.31$–3.33. PD(nat): 2.674_{10} 2.452_8 4.70_5 2.652_5 2.211_4 2.752_4 1.909_3 2.031_3; 35-591(syn): 2.76_3 2.69_{10} $2.67_{6.5}$ 2.65_5 2.21_2 2.03_1 $1.91_{3.5}$ 1.88_2. Merwinite has a distorted glaserite-type structure. Isolated [MgO_6] octahedra are linked at all apices with [SiO_4] tetrahedra, showing pseudohexagonal arrangement. Ca resides within three different sites in layers slightly above and below the [MgO_6] octahedral layer, one Ca site is 12-coordinated and two are ideally 10-coordinated but are frequently distorted to eight- to nine-coordinated. Noteworthy features include layers of Ca sites which are coplanar oxygen positions. *AM* 57:1355(1972). Euhedral crystals rare (to 3 mm), usually granular; may be prismatic along [011] or [01$\bar{1}$] or tabular parallel to (100); prominent forms are: {21$\bar{1}$}, {011}, {001}, {101}; polysynthetic twinning on {100}, twin axis [013], common. Colorless, white to light green; vitreous to oily luster; translucent. Cleavage {100}. H = 6. G = 3.15–3.32. IR data: *MM* 31:187(1956). Readily soluble in HCl. Essentially end-member compositions observed, with some substitution of iron to several weight percent. Biaxial (+); $N_x = 1.706$–1.714, $N_y = 1.711$–1.720, $N_z = 1.718$–1.729; $2V = 65$–$74°$; $r > v$. Found in high-temperature, low-pressure contact zones of eruptive rocks and carbonate rocks: metamorphosed marbles, persistent to high pressures, and is associated with calcite, clintonite, gehlenite, grossular, larnite, monticellite, spinel, spurrite, vesuvianite, wollastonite, or xonotlite. The mnemonic for the order of appearance of reaction products of progressive contact metamorphism of marbles—tremolite, forsterite, diopside, periclase, wollastonite, monticellite, åkermanite, spurrite, merwinite, larnite—is "Tremble, for dire peril walks, monstrous acrimony's spurning merwinite's laws." *JG* 48:225(1940). A component of portland cement, also found in silicate slags. Found at Marble Canyon, Culberson Co., and in the Christmas Mountains, Brewster Co., TX; *Wet Weather quarry*, Jensen, and other quarries, *Crestmore, Riverside Co., CA*; Niehart, Little Belt Mts., MT; Guardarraya, Santa Maria, and Terneras mines, Velardeña district, DUR, and Cerro Mazahua, Susupuato district, MICH, Mexico; Camas Mòr, Muck Is., Inverness-shire, and near Kilchoan, Ardnamuvchan, Argyllshire, Scotland; Scawt Hill, Larne, County Antrim, Northern Ireland; Killala Bay, County Sligo, Ireland; Iglika deposit; Elkhovsko dist., Bulgaria; Anakit district, Lower Tunguska R., Kochumdek R. and Kuzmovka areas, Stony Tunguska R., Siberia, and Gavasai deposit, Kirghizia, Central Asia, Russia; Guneyce–Ikizdere area, Turkey; Nanjangud, Mysore, India; Hatrium Formation, Israel. VK, EF

GARNET GROUP

The garnet group of 15 species consists of minerals conforming to

$$A_3B_2(TO_4)_3$$

where

A = Ca, Fe^{2+}, Mg, Mn^{2+}, with minor or trace amounts of Zn, Y, and Na
B = Al, Fe^{3+}, Cr^{3+}, V^{3+}, Ti, Zr, with minor or trace amounts of Mn^{3+}, Ti^{3+}, Zr, and Si
T = Si, but also occasionally, Fe^{3+}, Al, Ti, P, and OH

and where the mineral crystallizes close to the ideal cubic structure in space group $Ia3d$.

The structure consists of a continuous network of alternating corner-sharing octahedra and tetrahedra. The oxygen framework of this arrangement also

GARNET GROUP

Mineral	Formula	Space Group	a	D	
Pyrope	$Mg_3Al_2(SiO_4)_3$	$Ia3d$	11.455	3.56	51.4.3a.1
Almandine	$Fe_3Al_2(SiO_4)_3$	$Ia3d$	11.526	4.31	51.4.3a.2
Spessartine	$Mn_3Al_2(SiO_4)_3$	$Ia3d$	11.63	4.18	51.4.3a.3
Knorringite	$Mg_3Cr_2(SiO_4)_3$	$Ia3d$	11.622	3.86	51.4.3a.4
Majorite	$Mg_3(Fe, Al, Si)_2Si_4O_3$	$Ia3d$	11.54	4.00	51.4.3a.5
Calderite	$(Mn^{2+}, Ca)_3(Fe^{3+}, Al_2)(SiO_4)_3$	$Ia3d$	11.82	4.45	51.4.3a.6
Andradite	$Ca_3Fe_2^{3+}(SiO_4)_3$	$Ia3d$	12.059	3.855	51.4.3b.1
Grossular	$Ca_3Al_2(SiO_4)_3$	$Ia3d$	11.851	3.59	51.4.3b.2
Uvarovite	$Ca_3Cr_2(SiO_4)_3$	$Ia3d$	11.999	3.85	51.4.3b.3
Goldmanite	$Ca_3(V, Al, Cr, Fe^{3+})_2(SiO_4)_3$	$Ia3d$	12.070	3.77	51.4.3b.4
Schorlomite	$Ca_3(Ti^{4+}, Fe^{3+}, Al)_2[(Si, Fe^{3+}, Fe^{2+})O_4]_3$	$Ia3d$	12.17	3.84	51.4.3c.1
Kimzeyite	$Ca_3(Zr, Ti)_2[(Si, Al, Fe^{3+})_4O]_3$	$Ia3d$	12.456	3.96	51.4.3c.2
Morimotoite	$Ca_3(Ti, Fe^{2+}, Fe^{3+})_2(Si, Fe^{3+})_3O_{12}$	$Ia3d$	12.162	3.80	51.4.3c.3
Hibschite	$Ca_3Al_2(Si, OH)_3O_{12}$*	$Ia3d$	≈ 11.97	≈ 3.26	51.4.3d.1
Katoite	$Ca_3Al_2(Si, OH)_3O_{12}$**	$Ia3d$	12.57	2.53	51.4.3d.2

* Also rendered as $Ca_3Al_2(SiO_4)_{3-x}(OH)_{2x}$, $0.2 < x < 1.5$.

** Also rendered as $Ca_3Al_2(SiO_4)_{3-x}(OH)_{2x}$, $1.5 < x < 3.0$.

Note: Optical data suggest that the calcium garnets are not truly cubic, due to ordering of the B cations and, perhaps, noncubic ordering of the hydroxyl groups. These garnets are orthorhombic, triclinic, or tetragonal. Ordering of Al and Fe^{3+} on the B sites and ordering of Ca and Fe^{2+} on the A sites are other causes of deviation from cubic symmetry.

encloses large voids shaped like triangular dodecahedra which contain the A sites. *EJM 1:363(1989), NJMA 136:146(1979), AM 69:1116(1984)*.

51.4.3a.1 Pyrope Mg$_3$Al$_2$(SiO$_4$)$_3$

Named in 1803 by Werner from the Greek for *fire* and *to appear*, in allusion to its red color. Garnet group. Rhodolite is a garnet intermediate between pyrope and almandine, typically about two-thirds pyrope and one-third almandine. ISO $Ia3d$. $a = 11.455$(syn), $Z = 8$, $D = 3.56$(syn). *15-742*(syn): 2.87_6 2.56_{10} 2.44_4 2.34_2 2.25_2 1.86_2 1.59_3 1.53_5. End-member synthetic material is colorless, but all natural material contains some Fe^{2+}, which is responsible for the characteristic red color. Colors vary depending on composition, principally substitution of Fe, Mn, Ti, Cr, and other elements. Dark red, violet-red, red, rose-red (rhodolite), and reddish orange. Some chromium-rich pyrope (generally more than 3–4 wt % Cr_2O_3) shows color change (i.e., alexandrite effect): bluish green or green in daylight or fluorescent light, reddish purple in incandescent light. Such pyrope is colored by both Cr^{3+} and Fe^{2+} and shows a change of color from purple to green when heated to about 200°. The "Alexandrite effect" is caused by a total of less than 1 wt % Cr^{3+} and V^{3+} in octahedral coordination, in conjunction with Mn^{2+} in intermediate pyrope–spessartine garnets (malaia garnets). *G&G:200(1984)*. Inert to LWUV and SWUV light. $H = 7-7\frac{1}{2}$. $G = 3.65-3.82$. IR reflectance and Raman spectra: *PCM 17:503(1991)*. Dielectric constants: *AM 77:94(1992)*. Pyrope from eclogites, peridotites, and kimberlites contains variable amounts of Cr. Minor amounts (usually less than several tenths of 1%) of Na and P are found in

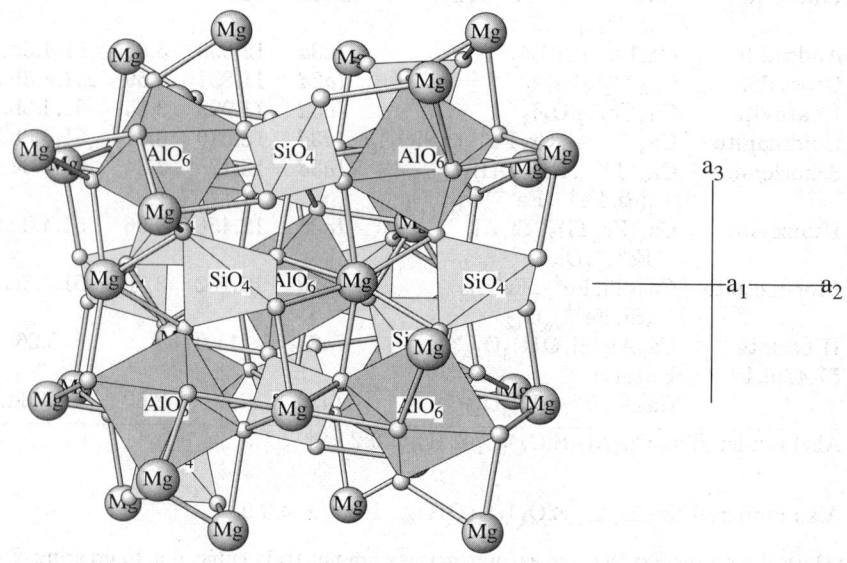

PYROPE: Mg$_3$Al$_2$Si$_3$O$_{12}$

Pyrope, a structure typical of the garnet group, in perspective view approximately along an a-axis.

pyropes and other high-pressure garnets. Malaia garnet is of intermediate composition between pyrope and spessartine. Isotropic; N = 1.714 (syn), most natural pyropes are between 1.74 and 1.77. Colorless to shades of pink, pinkish red, and red in thin section. Pyrope is a high-pressure mineral and occurs in schists and eclogite, magmatic ultrabasic rocks, kimberlites, and peridotites. An "indicator" mineral for diamondiferous kimberlites or peridotites. Also found in alluvial deposits, and as a detrital mineral in sedimentary rocks. Widespread throughout the world; noteworthy localities include: pyrope and almandine at the Barton mine, Gore Mountain, Warren Co., NY; fine rhodolite at Mason Mt., Macon Co., NC; also on Mason's Branch as alluvial material. Sugarloaf Mtn., Jackson CO., NC. In alkalic rocks at Magnet Cove, AR. Cr-bearing (color-change) pyrope occurs in the sloan Kimberlite diatremes, Larimer CO., In alluvial and colluvial deposits associated with kimberlite breccia pipes in northern Arizona and New Mexico (e.g., Garnet Ridge field, Apache Co., AZ; Buell Park area, Fort Defiance, McKinley Co., NM). Also in the Moses Rock and Mexican Hat fields, San Juan Co., UT. Upper Canada mine, Gauthier, ONT, Canada, Elie, Fyfeshire, Scotland, UK. Classic localities in Bohemia, Czech Republic (Bohemian garnet) in the Ceske Stredohori Mts.: at Trebenice (Trebnitz); in gravels derived from weathering of serpentine derived from peridotite at Merunice (Meronitz); also at Krems near Budweis and Podsedice. The largest Bohemian pyropes are the size of pigeon or hens' eggs (Rouse Garnet, 1986). Cut stones over 10 ct are rare. Bohemian pyropes from Linhorka show the "Alexandrite effect." At Zoblitz, Saxony, and Fichtelgebirge, Bavaria, Germany. Some gem Cr-bearing, color-change pyrope on the island of Otterøy near Modle, western Norway, and at Sandvik, Gurskoy Is., western Norway. Gem material at Kimberley, South Africa; Harts Range, NT, Lowood, Q, and Bingara, NSW, Australia. Gem color-change pyrope–spessartine (malaia) garnet in the Umba Valley region and Makonde, Tanzania; Vaal R., Transvaal, and Kimberley Cape Prov., RSA; Orissa, and in the Sittampundi complex, Salem dist., India. Color-change pyrope–almandine garnet at Ratnapura, Sri Lanka. Gem pyrope also in Burma, Brazil, and Russia. Some star (asteriated) rhodolite is known from Tanzania and India. Principal uses are as an abrasive for cutting and grinding tools and as a semiprecious gemstone. Commercial rhodolite is produced from Tanzania, Sri Lanka, Zimbabwe, and India. EF *ROU (1986), DHZ 1A:468(1982)*.

51.4.3a.2 Almandine $Fe_3Al_2(SiO_4)_3$

Known since antiquity; described by Agricola in 1546 and by Karsten in 1800. The name is derived from the locality at Alabanda, Turkey, where garnets were cut and polished in antiquity. Garnet group. ISO $Ia3d$. $a = 11.526$(syn), sometimes noncubic (e.g., tetragonal, $I4_1/acd$, $a = 11.6207$, $c = 11.6230$), $Z = 8$, $D = 4.318$. *9-427*(nat): 4.04_3 2.87_4 2.57_{10} 1.87_3 1.66_3 1.60_4 1.54_5 1.26_3. Color is generally red, orange-red, or violet-red; also brown or rarely black. $H = 7\frac{1}{2}$. $G = 3.85$–4.20 for most natural samples. Soluble with difficulty in HF; IR reflectance and Raman spectra data: *PCM 17:503(1991)*. Solid solution is common toward pyrope (51.4.3a.1) or spessartine (51.4.3a.3) (hence "pyralspite" garnets), with less common substitution to grossular (51.4.3b.2) or andradite (51.4.3b.1). Noncubic almandines are due to slight ordering between (Fe,Mn,Mg) and grandite (Ca) components. High pressure favors substantial solid solution, and relatively low temperature prevents complete disorder. *AM 77:399(1992)*. Sectored birefringent (about 0.005) almandine–grossular garnet occurs in the Gagnon Terrane, Labrador, Canada. *CM 32:105(1994)*. This garnet also shows a split in the (400) reflection indicating a slight departive from true cubic symmetry. Isotropic; $N_{syn} = 1.830$, usually shades of red or red-orange in thin section. Report N for 75 mol.% or greater almandine garnet is between 1.813–1.785; those with high pyrope and/or grossular components have N as low as 1.766. Almandine is the typical garnet of metamorphic rocks, particularly garnet–mica schists resulting from the regional metamorphism of argillaceous sediments also found in blueschists (glaucophane-bearing). Stable over a wide range of temperature and pressure; thus found in both contact and regional metamorphic rocks. Also occurs in acid plutonic and volcanic igneous rocks as xenocrysts and as a primary magmatic constituent in lithophysal rhyolites, aplites, granites, and pegmatites. Magmatic garnets are usually almandine–spessartine. Widespread as a detrital mineral in sedimentary rocks. Occurs in Precambrian iron formations. Occurs rarely as inclusions in diamonds (almandine–pyrope–grossular) and as a constituent of eclogites (almadine-pyrope) and granulites. Also found in one lunar sample. Abundant worldwide, and only classic, well-known, and distinctive localities are mentioned here. Excellent crystals to 25 cm from pegmatic at Topsham, ME. Fine material from Russell, MA; from near Roxbury, CT, and Leiperville, PA. Occurs as large crystals to several feet across at the Barton and other mines, North Creek and Gore Mountain, Warren Co., NY; a famous 4.4-kg near-perfect trapezohedral almandine was discovered at 35th Street and Broadway in New York City and is now on display at the American Museum of Natural History. Substantial amounts of abrasive-grade almandine were produced from High Peak, Burke (Laurel Creek, Morganton), Caldwell, and Catawba Cos., NC. Huge, euhedral crystals to 30 cm found at Stony Creek, Thermal City, Rutherford Co., NC and to 10 cm from the Little Pine Mine, Redmon, Madison Co., NC. Occurs in large (to 14 lb) crystals in skarn from the Calumet mine, near Salida, Chaffee Co., CO. Asteriated "star" almandine crystals to 5 cm occurs on Emerald Creek, Fernwood, Benewah and Latah Counties., ID, and other nearby localities. Stars may be four- or six-rayed. At Orofino, Clearwater Co., ID. Occurs in lithophysal rhyolite from Ruby Mountain, Nathrop, Chaffee Co., CO, and the Thomas Range, Juab Co., UT. Also occurs as euhedral crystals ($Alm_{73}Spess_{27}$) to several cm in lithophysal rhyolite at Garnet Hill, near Ruth, White Pine Co., NV. A classic locality for almandine is near Fort Wrangell on the Stikine River, 1.75 miles SE of Sergief Island, AK. Crystals from this locality occur in a mica schist and are as much as 2 inches across. Albert Harbor, Baffin Island, NWT, Canada; southern Greenland in the region east of Godthaab (e.g. Ameralik Fjord-Kangersunek Fjord) and on the west coast from Disko Bay South. Garnet mases to 25 kg have been found in the Godthaab district. Crinkle Crags, Cumbria, England; large crystals occur in pegmatite at Irchenrieth, Bavaria, Germany; also in Eppenreuth, Bavaria; in rhyolite from the Saar–Pfalz region, Germany. Noted from Obergurgl im Ötzal, and Rossrucken, Zillertal, Tyrol, Austria. Alpe Siviglia, Val Codera, Italy. Large dodecahedral crystals occur at Falun, Sweden. From Bodö, Nordland, Norway; from Přibyslavice, Bohemia, Czech Rep. Optically anomalous material from Mt. Saroba, Kazahstan; *Alabanda, Turkey*. From granitic pematites (Dusso, and Shingus, Haramosh Mountains) in the Gilgit Div., NWFP, Pakistan; from Mtoko, Zimbabwe; also Ampandramaika and Ampanihy, Madagascar. Gem material is noted from Sri Lanka; Nambunju mine, Namaputa in Lindi Province, Tanzania; also from Namapura, Masasi dist., Tanzania; Zimbabwe; lower Umba Valley, Kenya; Trincomalee, Sri Lanka; a source of almandine since antiquity is Rajputana, northern India. Other Indian localities include; Meja Udaipur, Sarwak, Kishengurh; Kakoria, Jaipur; Kondapali, Godavari dist., and Gharibpeth, Hyderabad; Bhadrachalam, central provinces; Mahanadibott, Orissa; Pallavaram, Madras; star almandine occurs in India as well as Idaho. Maude, Florence, and Hale Rivers, No. Territory; Fishery Bay, Eyre Pen., South Australia, Australia; the states of Minas Gerais (e.g. Minas Novas and Poaia, Lavra Presidente), Bahia, and Rio de Janeiro, Brazil. Principal uses are as an abrasive for cutting, grinding, and drilling tools, sandpaper, and as a semiprecious gemstone. The Barton mine supplies about 90% of all crushed garnet worldwide. *MSA RIM 5:25(1980), DHZ 1A:468(1982), Rouse Garnet (1986), AM 73:568(1988).*

51.4.3a.3 Spessartine $Mn_3Al_2(SiO_4)_3$

Named in 1832 by Beudant for the type locality. Garnet group. ISO $Ia3d$. $a = 11.63$(syn), 11.628(F-bearing), 11.650(andradite molecule present), $Z = 8$, $D = 4.18$. *10-354*: 2.91_2 2.60_{10} 2.37_2 2.13_2 1.89_2 1.68_2 1.61_3 1.56_4. F substitutes for O in Si-absent tetrahedra similar to OH substitution in hydrogrossular. *AM 75:314(1990)*. Si is 10% deficient (for material with 5.6 wt % F). Usually in solid solution principally with almandine, but spessartines with appreciable amounts of grossular also occur (to 33% of the grossular molecule). More rarely the andradite molecule may also be found. A calcian spessartine contained 2.6 wt % H_2O. *AM 58:508(1973)*. Spessartines sometimes have elevated REE contents; one from Suishoyama, Japan, has 3.2 wt % total REE and one from Pyörönmaa, Finland, has 3.35 wt % total RE_2O_3. REEs substitute according to the following schemes: $Mn^{2+} + Si^{4+} \leftrightarrow Y^{3+} + Al^{3+}$ and $2Mn^{2+} \leftrightarrow Na^+ + Y^{3+}$. Vanadium (to 1.4 wt % V_2O_3) also occurs in spessartine (Fuji mine, Fukui Pref., Japan). As much as 0.36% ZnO occurs in spessartine from Franklin, NJ. Rare, color-change (alexandrite-effect) spessartine–grossular garnet exists. *G&G:100(1982)*. Very pale yellow, orange, reddish orange, orange-red, rose, hyacinth-red, and reddish brown; the orange color is due to divalent manganese in a site with distorted cubic coordination. Sometimes confused with "hessonite," an orange grossular. Crystals may be as much as 5 in. or more across. H = $7–7\frac{1}{4}$. G = 3.90–4.20. IR data: *CM 31:381(1993)*, *PCM 17:503(1990)*. Soluble with difficulty in HF. Isotropic; N = 1.79–1.81. Sometimes (water-bearing) weakly birefringent with a birefringence of up to 0.008. The anisotropy observed in some spessartines is believed to be due to the presence of a significant amount of the andradite molecule or ordering of Fe^{3+} in octahedral positions. *CMG 29:279(1984)*. Found in Mn-rich metamorphic rocks and in igneous rocks: most commonly in granites, aplites, and pegmatites. Pure spessartine occurs in fractionated granitic pegmatites and some metamorphic rocks. Also occurs as a lithophysal mineral in topaz-bearing rhyolites. Often occurs as etched masses in pegmatites. Occurs at many worldwide localities, only some of the best are noted here. Fine spessartine crystals to 7 cm occur in pegmatites in Leiper's quarry, Springfield Township, and the upper and lower quarries at Avondale, Delaware Co., PA. In pegmatite at the Rutherford mine, Amelia, Amelia Co., VA; the "Rutherford Lady" weighs 2829 ct and another specimen of rough spessartine from the Rutherford No. 2 mine weighs 1.34 kg (6720 ct). Also occurs at the Moorefield mine at Amelia. Occurs in the Bald Knob manganese deposit, NC. Henderson Mo porphyry deposit, Clear Creek Co., CO (F-bearing); in rhyolite at Ruby Mt., Chaffee Co., CO; in pegmatites of the Sawtooth batholith, e.g. Imogene Lake, Custer Co., ID; Victoria dist., Luna Co., NM; in rhyolite at Ash Creek, Hayden, AZ; in rhyolite from the Thomas Range, Juab Co., UT; fine orange specimen and gem quality (stones to greater than 10 ct) material occurs in several pegmatite dikes at the Little Three mine, Ramona, San Diego Co., CA; zoned almandine–spessartine and nearly pure (99.5%) spessartine occur in the Himalaya pegmatite–aplite dike system, Mesa Grande district, San Diego Co., CA; Quebec, Canada; Glen Skiag, Scotland, UK; Salm Chateau, Belgium; in alpine at Byneset, Trondheim, Norway; *at Aschaffenburg, Spessart, northwestern Bavaria, Germany*; also from Ilfeld, Germany; Tyrol, and Pfitsch, Austria; St. Marcel, Piedmont, Italy; in pegmatite at San Piero in Campo, Elba, Italy; Y-bearing at Pyörönmaa pegmatite, Kangasala, southwestern Finland; in pegmatite at Kimito, Finland; in pegmatite at Gola, Poland; in pegmatite at Budislav, Bohemia, optically anomalous spessartine occurs in hornfelses east of Chvaletice, east Bohemia, also in pegmatites at Drahonín and Puklice, Moravia, Czech Republic; Ilmen Mtns. and Bagaryak, Urals, Russia; to 10 cm across from pegmatites in the Shingus area, Gilgit area, and Shigar Valley, NWF, Pakistan; fine gem-quality material occurs in metamorphic schist at Marienfluss, Namibia; clear orange material occurs in pegmatites at Tsilaizina, near Antsirabe, Madagascar; also at Maevetanana; gem-quality material (malaia garnet, ranging from ferroan spessartine to manganoan pyrope) occurs in the Umba River area, Tanzania; orange spessartine occurs in the Taita Hills, Kenya; gem-quality spessartine–pyrope occurs near Ratnapura, Sri Lanka; also in northern Burma; in pegmatite at Qibeiling mine, Altai, Xinjiang Uygur AR; China; Y-bearing in pegmatite at Suishoyama, Kawamata-machi, Fukushima Pref., Japan; Broken Hill, NSW, Australia; in gem gravels of the Santa Maria and Abaete rivers as well as at Arassuahy, Registo and Ceará, also Rio Grande do Norte, MG, Brazil. Cubic (habit) spessartine crystals (to 2 mm) occur in pegmatite gem pockets at Linopolis, MG, Brazil. Sometimes alters to black manganese oxides and hydroxides. Principal use is as a gemstone. Malaia (also spelled Malaya) garnet is a garnet of intermediate spessartine–pyrope composition. Some malaia garnets contain trace amounts of V and Cr and are color-change garnets. A polished 708-ct red spessartine is reported from Brazil. EF *DHZ 1A:590(1982), ROU 1986*.

51.4.3a.4 Knorringite $Mg_3Cr_2(SiO_4)_3$

Named in 1968 by Nixon and Hornung for Oleg von Knorring (1915–1994), mineralogist, Leeds University, England. Expert on pegmatites and on southern Africa. Garnet group. ISO $Ia3d$. $a = 11.596$(nat), 11.622(calc. end member), $Z = 8$, $D = 3.86$ (3.835 calc. end member). 35-536(syn): 2.90_8 2.59_{10} 2.47_4 2.37_4 2.27_2 1.83_2 1.61_4 1.55_6. Subhedral, rounded grains. Blue-green to lilac(nat) or dark green(syn). $H = 7$. $G = 3.76$. Insoluble in common acids. Isotropic. $N = 1.790$–1.803 (1.875 for calc. end member). Forms a series with pyrope (51.4.3a.1). Analysis (wt.%):SiO_2 39.9, TiO_2 0.11, Al_2O_3 9.7, Cr_2O_3 17.5, Fe_2O_3 1.2, FeO 6.5, MnO 0.6, MgO 17.0, CaO 8.1. Occurs in kimberlite pipes. *Kao kimberlite, Lesotho;* as an inclusion in a type IIa diamond from a kimberlite in the Mirny area, Yakutia, Russia. EF *AM 53:1833(1968)*.

51.4.3a.5 Majorite $Mg_3(Fe,Al,Si)_2(SiO_4)_3$

Named in 1970 by Smith and Mason for Alan Major, who participated in studies on the synthesis of garnet from pyroxenes. Garnet group. ISO. $Ia3d$. $a = 11.524$(at high temperature and pressure), 11.54(nat); transforms to TET $I4_1/a$ at low temperature and pressure; $Z = 8$, $D = 4.00$. 25-843(tetragonal): 2.88_7 2.58_{10} 2.45_5 2.35_3 2.26_4 2.04_3 1.60_4 1.54_6. End member Mg-majorite (syn) is cubic at $> 2350°$ and 22 GPa but is tetragonal below this temperature, probably through ordering of octahedral Mg and octahedral Si atoms. One Si atom is in sixfold coordination. $(Mg,Fe)SiO_3$ pyroxenes transform to the garnet structure at high pressures. Natural majorite with $[Fe_{tot}/(Mg + Fe)_{tot} \geq 0.2]$ found in meteorites is cubic. Purple-violet, colorless, or yellowish brown. No cleavage. $H = 7$–$7\frac{1}{2}$. Occurs as minute grains. Synthesis:EPSL12:411(1971). Composition is near that of hypersthene–bronzite. SiO_2 52.0, Al_2O_3 2.6, Cr_2O_3 0.7, FeO 16.9, MgO 27.5, Na_2O 0.7. Al_2O_3 to 4.6 and CaO to 1.6. Found in the Earth's mantle and in meteorites. A product of shock transformation of olivine and pyroxene. *Coorara meteorite, Rawlina, WA, Australia;* Tenham chondrite meteorite; Catherwood chondrite meteorite; Pampa del Infierno chondrite meteorite; Peace R. meteorite, ALB, Canada. EF *DHZ 1A:512(1982), CIWY 81:279(1982), GCA 46:1903(1982), JGR(B) 86:6171 (1981), CM 21:29(1983), AM 78:1165(1993)*.

51.4.3a.6 Calderite $(Mn^{2+},Ca)_3(Fe^{3+},Al)_2(SiO_4)_3$

Described in 1850 by Piddington as a massive silicon–iron–manganese rock from India; named for J. Calder. Garnet group. In 1909, Fermor used the name *calderite* for a mineral of the garnet group having the ideal composition $Mn_3Fe_2(SiO_4)_3$. ISO $Ia3d$. $a = 11.81–11.82$, $Z = 8$, $D = 4.42–4.45$ (for pure calderite) and $0 = 4.07$ for NAT. *10-367*: 2.92_9 2.62_{10} 2.39_9 1.91_8 1.63_9 1.57_{10} 1.47_7 0.97_9. Dark yellow to yellow-brown or reddish brown. $G = 4.05–4.08$. Isotropic; $N = 1.872–1.890$(nat), and $N = 1.97$(calc. for pure calderite). Ideal end member calderite is stable only at high pressure (>22 GPa) and low temperature. Complete solid solution exists among calderite, andradite, and spessartine. Compositions approaching ideal calderite are unknown. Some Mn is trivalent (Mn_2O_3 to 1.5 wt %); CaO to 2.7 wt %. Calderitic andradite is found in high-pressure metamorphic Fe–Mn-rich quartzites and cherts. Also in metamorphosed Mn-silicate–oxide and silicate–carbonate rocks. In areas of subduction-zone metamorphism. Found at Panoche Pass, and at the Laytonville quarry, Laytonville, CA; occurs in the Wabush iron formation, LAB, Canada; *Sausar Group at Netra, Balaghat dist. Madhya Pradesh, India*; Otjosondu, Namibia; calderite and/or calderitic spessartine occurs at the Kodaira iron deposit, Fukushima Pref., Japan. Calderitic andradite occurs on Andros Island, Greece; calderitic spessartine in the Sausar Group at Chikla, Maharashtra, Katkamsandi, western India, and Tirodi, Madhya Pradesh, India; calderitic spessartine also occurs at Brezovica, Yugoslavia. EF *DHZ 1A:591(1982), CMP 84:199(1983), MM 51:577(1987), ZK 189:43(1989)*.

51.4.3b.1 Andradite $Ca_3Fe_2^{3+}(SiO_4)_3$

Named in 1868 by Dana for J. B. d'Andrada e Silva (1763–1838), a Brazilian mineralogist, who examined and described a variety of this mineral. Garnet group. Forms two series, one with schorlomite (51.4.3c.1) and one with grossular (51.4.3b.2). Sometimes in solid solution with spessartine (51.4.3a.3) (to 11.4 wt % MnO from Pajsberg, Sweden) and uvarovite (51.4.3b.3) (to more than 3.0 wt % Cr_2O_3. May contain H_2O (as OH) as "hydroandradite" component. Some andradite from Wessels mine, South Africa, has >2 wt % H_2O. May also contain significant Mn_2O_3 (henritermierite component, $CaMn^{3+}$). Varieties of andradite include topazolite, which is a topaz-yellow color, and demantoid, which is yellow-green to emerald-green in color. Melanite is a titanian andradite with less than about 15 wt % TiO_2 (<1 atom of Ti) and grades into schorlomite with increasing TiO_2 content. ISO $Ia3d$. $a = 12.051–12.059$(syn), $Z = 8$, $D = 3.855$(syn). Hydrous andradite has $a = 12.340$. Usually cubic but sometimes orthorhombic, *Fddd* or triclinic, IT, because of partial ordering of octahedral and dodecahedral cations, and of noncubic arrangement of OH groups. Other factors, such as twinning or strains at compositional boundaries, also occur. Tweedlike structure, interpreted to be an exsolution texture formed after crystal growth, is also observed in grandite garnets. *AM 71:1210(1986)*. Distinct compositional differences exist between the Fe-richer lamellae and Fe-poorer host. Violation of $Ia3d$ symmetry is

51.4.3b.1 Andradite

clearly evident on X-ray precession films. *10-288*(syn): 3.01_6 2.70_{10} 2.46_4 1.96_2 1.67_2 1.61_2 1.32_2 1.20_2; *30-253*(hydroxyl-bearing andradite, 5.29 wt % H_2O^+). Hydrous andradite (EJM7:1221(1995) is composed of disordered microdomains containing SiO_4 and (H_4O_4) tetrahedral units. Many different colors: brownish red, brownish yellow, brown, yellow to yellow-green (topazolite), grayish green, medium to deep green (demantoid), black (melanite). Pale yellowish green is due to Fe^{3+} in octahedral coordination; yellow to orange to black is due to charge-transfer processes involving titanium and iron. The green color of demantoid is due to low concentrations of Cr^{3+} in octahedral coordination. Vitreous to adamantine luster. G = 3.70–4.10; hydroxyl-bearing, G = 3.45. IR reflectance and Raman spectral data: US6SOF.87-263(1987) *PCM 17:503(1991)*. Mössbauer spectral data for Fe–Ti-bearing andradite: *AM 65:142(1980), PCM 13:198(1986)*. Melanite contains 0.03–0.16 $Ti^{3+}/\Sigma Ti$. XPS studies confirm that Ti^{3+} is in octahedral coordination and Ti^{4+} is in octahedral and tetrahedral sites. *EJM 7:847(1995)*. Predominant substitution modes in melanite are: $^{VI}Ti^{4+} + {}^{VI}M^{2+} \leftrightarrow 2{}^{VI}M^{3+}$ and $^{VI}Ti^{4+} + {}^{VI}Fe^{3+} \leftrightarrow {}^{IV}Si^{4+} + {}^{VI}M^{3+}$, where $M^{3+} = Fe^{3+}$ and Al^{3+}. As much as 4 to 5.8 wt % SnO_2 has been reported in andradite–grossular garnets, and the replacement $Sn^{4+} + Fe^{2+} \leftrightarrow 2Fe^{3+}$ has been proposed. *MSA RIM 5:33(1980)*. Y_2O_3 to 3.4 wt %. Colorless to pale yellow birefringent overgrowths on melanite and schorlomite from Magnet Cove, AR, are fluorohydrograndites (F to about 1.0 wt % in grossular-rich andradite). A titanian F-bearing (1.2 wt % F) and water-bearing (2.3 wt % H_2O) andradite occurs in altered basalt, Land's End, England. *BM 109:613(1986)*. Zoning between andradite and grossular may be extreme [e.g., And_{100} (host) and And_{50}-$Gros_{50}$ (lamellae) from Crested Butte, CO]. *AM 58:840(1973)*. Isotropic; N = 1.887(syn), 1.88–1.94(nat). Sometimes anisotropic and biaxial. Usually, garnets near both end members (andradite and grossular, respectively) are isotropic. Anisotropy is due to formation of order–disorder growth sectors. Various degrees of Fe^{3+}/Al ordering in their octahedral cation sites are produced on the side faces of surface steps during growth. AM *69:328(1984)*. The dispersion of demantoid is higher than that of diamond or titanite (0.057 vs. 0.044 and 0.051, respectively). Pale yellow to red-brown, rarely shades of green in thin section. demantoid is often characterized by "horsetail" inclusions of "byssolite" (tremolite–actinolite). Some grandite garnets are iridescent, and andradite from the Adelaide mining district, near Golconda, NV, shows compositional lamellar zoning (And_{87} and And_{78}, respectively) and a moiré-like type of postcrystallization exsolution texture lamellae. *AM 67:1242(1982), 71:123(1986)*. A hydrous andradite, $Ca_3Fe_{1.54}Mn_{0.20}Al_{0.26}$ $(SiO_4)_{1.65}(O_4N_4)_{1.35}$, occurs at the Wessels Mine, RSA. Cornwall, Lancaster Co., PA. Mostly found in calcium and iron-rich contact-metamorphic rocks and iron-rich skarns and less commonly in alkalinic igneous rocks.. Also in ultrabasic rocks (Demantoid) and Ti-rich alkalic rocks (melanite). May be associated with low-temperature zeolites in cavities in basalt. In alluvial deposits. Some may be of low-temperature hydrothermal origin, such as in deep-sea sediments. Occurs in many localities worldwide and only well-known and/or distinctive localities are listed here. In dodecahedral crystals to 13 cm at Franklin, Sussex Co., NJ. Melanite occurs at Magnet Cove, AR. Strongly compositionally zoned andradite occurs at Crested Butte, Gunnison Co., CO. At Stanley Butte, and Sierra Ancha, Graham Co., AZ. Triclinic andradite in skarn deposits in the Sonoma Range, Nevada; topazolite occurs at the green fire mine and melanite occurs near the Dallas Gem mine, San Benito Co., CA. Zoned crystals with epidote at Garnet Hill, Calaveras Co., CA. Crystals to 7 cm occur near the Jumbo Mine, Prince of Wales Is., AK. Melanite occurs in the Ice R. alkalic complex, QUE, Canada. Also from Marmora, ONT. At Pic d'Espada and Erslios, Barèges, Hautes Pyrénées, France. *Dramnen and Feiringen, Norway.* Large dodecahedral crystals at Arendal, Augst-Agder, and also Stoko, Langsundfjord Norway. At Långban, Värmland, Sweden. High-MgO-containing andradite at Sala, Sweden. At Pitkaranta, Lake Lagoda, Finland. Fine topazolite at the Mussa Alp, Ala Valley, Piedmont, Italy; also at Zermatt, Valais, Switzerland. An old classic locality is that of Knappenward, Untersuizbachtal, Salzburg, Austria. Also from Pfitschtal, Tyrol, Austria. Topazolite at Schwarzenberg, Saxony, and Wurlitz, Germany. Demantoid also in Saxony. Andradite at Beitenbrunn and Schwarzenberg. Melanite at Rieden am Laacher See and Kaiserstuhl, Germany. Zr-bearing melanite–schorlomite in the Polino carbonatite, Terni, Umbria, Italy. At Pfitschtal, Trentino, Italy. Melanite at Monte Somma (Vesuvius), and in the Albanhius, Rome, Italy. Demantoid at Val Malenco, (Sondrio) Frascati, Italy; also in serpentinites at Dobsiná, Slovakia. At Moravicza and Dognascska, Romania. Mega Xhorid, Serifas, Greece. Demantoid (to >1 cm) in alluvial placers (Bobrovka R.) at Elizavetinskoye and in situ at Nizhni Tagil, central Ural Mts., Russia. Demantoid also in the Syssert district, near Poldnevaya, Chutosky Mtns., Chutosky Pen., Russia. Exceptionally bright black melanite crystals to 7 cm from Mina Ojos Españoles, Lazaro Cardenas, CHIH, Mexico. Olive-green crystals to 20 cm from Trapichillos, COL, Mexico. Fine melanite from Fanshan complex, Hebei Prov., China. Demantoid in Zaire. Abundant at Black Rock, RSA. Mn-bearing garnets intermediate between calderite and andradite, but closer to andradite, in metamorphosed manganese deposits at Otjosondu, Namibia, and Andros Is., Greece. Primary use is as a semiprecious gemstone, particularly topazolite and demantoid. Cut stones of demantoid >4 ct are very rare. One exceptionally fine specimen of demantoid from Val Malenco, Italy, would, if cut, yield a stone of approximately 15 ct. EF *MM 36:775(1968); AM 56:1983(1971), 62:475(1977), 62:646(1977), 74:840(1989), 74:1307(1989); DHZ 1A 617(1982); ZK 158:53(1982); G&G 19:202(1983); ROU (1986); NJBM:76(1991); EJM 5:59(1993).*

51.4.3b.2 Grossular $Ca_3Al_2(SiO_4)_3$

Named in 1811 by Werner in allusion to the greenish color of the gooseberry (*Ribes grossularia*). Garnet group. Generally forms in solid solution with

51.4.3b.2 Grossular

andradite and if Cr is present, with uvarovite. However, solid solution to varying extent occurs with other garnets. ISO $Ia3d$, but many grossular garnets exhibit lowered symmetry on long range order due to order–disorder growth mechanisms. Depending on orientation, the symmetry is ORTH, MON, or TRIC. An ordered arrangement of Al and Fe^{3+} (octahedral), Ca and Fe^{2+} (dodecahedral), and/or a noncubic (OH) distribution is responsible. Sectoral twins are common: sectors contain differently ordered arrangements of Al and Fe^{3+}. Changes in the ordering scheme or degree of ordering take place during growth. Strain in the crystal results from lattice mismatch, thereby reducing the symmetry from cubic. Single-crystal IR spectroscopy showed five tringent grossant andradite garnets to be cubic measurement of individual unit cells (AM 78.957 (1993). $a = 11.851$(syn), a grossular from asbestos, QUE is both TRIC (IT) (before heating): a 11.845, b 11.858, c 11.846, $\alpha 89.98°$, $\beta 89.96°$, $\gamma 89.97$, and monoclinic (after heating): a11.837, b 11.840, c 11.836, $\alpha 90.01°$, $\beta 89.98°$, $\gamma 90.00$. $Z = 8$, $D = 3.59$(syn). *39-368:* 2.96_4 2.65_{10} 2.42_2 2.32_2 2.16_1 1.92_2 1.64_2 1.58_3. Many colors, most commonly colorless, white, pink, cream, green, yellow, orange, red, cinnamon, honey, brown, and rarely black. Hessonite (or essonite) is a brownish-orange variety and tsavorite, discovered in the late 1960s (= tsavolite), is a pale to dark emerald-green variety colored by V^{3+} in octahedral coordination. $H = 7$, $6 = 3.42$–3.80 (depending on composition), Hessonite is inert to LWUV and SWUV light; colorless to light green stones (including tsavorite) may fluoresce a weak to moderate orange in LW and weak yellow-orange in SW; yellow stones may fluoresce a weak orange to both LW and SW. IR and Raman spectra data: USGS of 87–263(1987) *PCM 17:503(1991), AM 76:1153(1991)*. Isotropic; $N = 1.734$(syn). Most hessonite has $N = 1.733$–1.760, while grossular spans $N = 1.731$–1.754. *G&G 18:204(1982)*. Anisotropic ugrandite garnets have a birefringence generally less than 0.005. Occasionally iridescent, with the iridescence being caused by submicroscopic twinning parallel to crystal surfaces with a periodicity of about 1000 Å. With an increase in iron content, hessonite goes from yellow to orange to red-brown. The color of tsavorite varies from pale yellow green to dark emerald green with increasing V (and some Cr) content. May contain F (e.g., 3.5 wt % in a hydrogrossular on melanite from the Diamond Jo Quarry, Magnet Cove, AR). Grossular is especially characteristic of both thermally and regionally metamorphosed impure calcareous rocks, and also occurs in rocks that have undergone calcium metasomatism. *DHZ 1A:614(1982)*. Also occurs in association with serpentinite and rodingites. Rarely occurs in meteorites. Ubiquitous, and only localities producing fine-quality or unusual material are mentioned here. Fine crystals to 12 cm occur at the Pitts-Tenney quarry, Minot, and red crystals to 5 cm from the Sanford deposit, York Co., ME; world-class orange crystals to 1 cm or more across at Belvidere Mtn., Lowell (Eden Mills), Lamoille Co., VT; Hunting Hill Q., Rockville, MD; Italian Mountain, Gunnison Co., CO; 2.5 mi. E of Wah Wah Summit, Beaver Co., UT; Ramona, San Diego Co., CA; hydroxyl-bearing grossular occurs at Crestmore, Riverside Co., CA; exceptional cubo-dodecahedral crystals (to 3 cm) occur at Vesper Peak, Snohomish

Co., WA; nearly pure colorless or light orange brown grossular, as well as green chromian grossular occurs at the Jeffrey mine, Asbestos, QUE in fine crystals to 3 cm; chromian grossular also occurs at the B.C. asbestos mine at Black Lake; colorless and other colors occur at the Beaver mine and other mines at Thetford, ONT; Orford nickel mine, Orford Twp., QUE; to 8 cm from the McBride property, Wakefield Twp., Hull Co., QUE; York River skarn near Bancroft, ONT, Canada; fine quality pink to red and dark cinnamon-red (due to Mn^{3+} in octahedral coordination)–pale-grey green to colorless, dodecahedral crystals to 5.5 cm occur associated with vesuvianite and wollastonite in the Las Cruces Mtns. at Sierra de la Cruz, and at Sierra Mojada at Rancho San Juan, Xalostoc, Morelos, COAH., Mexico (also occurs similarly at Lake Jaco, Chihuahua); fine material from Concepcion del Oro, Zacatecas, Mexico; Isle of Mull Scotland; Beiaren, Nordland, and Telemark, Norway; Malsjö, Vårmland, Sweden; Mittagshorn, Saasthal and Zermatt, Switzerland; ambered-colored crystals from Ala Valley, Piedmont, Italy; honey-yellow octahedra (to several mm) from Elba and Monte Somma (Vesuvius), Italy; also from Monzoni, Val di Fassa, Trentino, Italy; Bellecombe, Val d'Aosta, Italy; Auerbach, Germany; Jordanow, Gleinitz, lower Silesia, Poland; Oravicza, Csiklova, Vasko and Dognaska, Romania; Kimito, Finland; *Vilui River area, Yakutia, Russia* (with vesuvianite); Achmatovsk, near Kussink, Zlatoust district, Urals, Russia; gem green tsavorite occurs in graphite-bearing gneisses and marbles associated with scapolite and epidote at the Lualenyi and Scorpion mines, near Voi, Taita Hills, Kenya (near Tsavo National Park on the Kenya-Tanzania border) and also at Kide Hill, Taita Hills District. Crystals contain as much as 1.6 wt % V_2O_3, and rarely exceed 3 carats; the largest gem being about 20 cts. Gem tsavoritie in well-formed crystals ococurs at the Karo tanzanite pit in the Merelani Hills, south of Arusha and Moshi, Tanzania. Colorless, yellow and orange grosslular also occurs in abundance in East Africa, e.g. Lelatima Mtns. Gem brownish-green (exterior) and yellow-green (interior) grossular (with about 20% andradite component) has been found near Diakon and Dionboko, Mali; stones over 1 carat are rare. Large yellow-green to brown grossular-andradite crystals to 14 cm in diameter occur near Sandare, Nioro du Sahel, mali. The most important source of gem-quality hessonite is grom gem-gravels in Sri Lanka. Fine material at the Cocoktau mine, Altai, Xinjian Uygur AR, China. *DHZ RFM 1A 468(1982), AM 69:328(1984), AM 71:779(1986), Rouse (Garnet) 1986, AM 73:568(1988), AM 74:151(1989), AM 74:859(1989)*

51.4.3b.3 Uvarovite $Ca_3Cr_2(SiO_4)_3$

Named in 1832 by Hess for Count Sergei Semeonovich Uvarov (1786–1855), a Russian nobleman, Imperial Academy of St. Petersburg. Garnet group. ISO $Ia3d$. $a = 11.999$(syn), $Z = 8$, $D = 3.85$. *11-696*(syn): 3.00_7 2.68_{10} 2.56_2 2.45_6 2.35_3 1.95_2 1.66_3 1.60_6. Complete solid solution exists with grossular. Some minor solid solution with andradite also occurs. Generally deep rich green in color but may be a "rusty" green. Very dark green in large crystals. The characteristic green is due to Cr^{3+} in octahedral coordination. H = $7\frac{1}{2}$. G = 3.40–3.83, depending on Cr content. Isotropic; N = 1.865(syn), 1.80–1.87(nat). May be slightly anisotropic. Uvarovite is associated with chromite and chromite deposits, serpentinized olivine, ultramafic rocks, and other chromium-bearing rocks. May also occur in xenoliths of calc–silicate rocks in kimberlites and in contact-metamorphic limestones and skarns. Only selected and well-documented occurrences of uvarovite are given below. Occurs in serpentinite at Wood's chrome mine, TX; Lancaster Co., PA; in serpentinite at Riddle, OR; New Idria, San Benito Co., and at least 20 other localities in CA; Kiglapait, LAB, and the Gun prospect, ITSI Mtns., Macmillan Pass, Yukon, in nephrite at Cassiar, BC, Canada; in serpentinite near Röros, Norway; at several localities in Karelia, Finland: famous from Outokumpu, where crystals to more than 2 cm across occur in skarn; some is very high in the uvarovite molecule. Also found at Kuusjarvi and Luikonlahti. Occurs at Jordanow, Silesia, Poland; from the chromium deposits of the central Ural Mts., Russia: Syssert, Nizhni Tagil, and Mt. Saranovsk; in chromium ore near *Bissersk, near Kyshtym, NE of Zlatoust*. Associated with kammererite at the Kop Krom mine, Kop Daglari, Turkey. Occurs in the Bushveld complex, Transvaal, South Africa. Principal use is as a gemstone. Stones are usually less than several carats. EF *CM 16:205(1978), 27:565(1989); DHZ 1A:642(1982); ROU (1986)*.

51.4.3b.4   Goldmanite $Ca_3(V,Al,Cr,Fe^{3+})_2(SiO_4)_3$

Named in 1964 by Moench and Meyrowitz for Marcus Isaac Goldman (1881–1965), American sedimentary petrologist, U.S. Geological Survey. Garnet group. The vanadium analog of grossular and andradite. "Yamatoite" is the synthetic Mn analog of goldmanite but has not been discovered in nature. ISO $Ia3d$. a = 12.070(syn), 12.01–12.06(nat), 11.947(chromian), 11.97(manganoan), Z = 8, D = 3.77. *16-714:* 3.01_7 2.69_{10} 2.45_4 2.36_2 2.19_2 1.95_2 1.67_2 1.61_5. Dark green to brownish green or grass green (highest V content), colorless streak, vitreous luster. No cleavage, brittle. H = 7. G = 3.74–3.91. IR data: *APM 8:265(1989)*. Contains as much as 28.7 wt % V_2O_3 and 96.3 mol %. Complete solid solution exists between grossular and goldmanite. Some have some F to 1.3 wt %, Cr_2O_3 to 10.1 wt %, MnO to 15.9 wt % ("yamatoite"). Contains some FeO and Fe_2O_3 as well. Isotropic; N = 1.834(syn), range for natural material is 1.792–1.834 with manganoan material having N = 1.855. Rarely slightly anisotropic. Occurs in vanadium-bearing sedimentary rocks or contact-metamorphosed carbonaceous and calcareous shales and sandstones; also in magnetite-bearing skarn deposits. Mn-bearing goldmanite occurs in Mn ore. In metapelite and skarn and in veins in muscovite-bearing marble. *Laguna, Grants district, NM;* Hemlo, ONT, Canada; detrital material (with maximum amount of V_2O_3 of 28.7 wt %) in the U.K. sector of the North Sea; Coat-an-Noz, Belle Isle-en-Terre (Cotes du Nord), Brittany, France; Pezinok–Pernek complex, Lesser Carpathian Mts., Slovakia; 1- to 2-mm crystals occur at Tetetice (St Runadlo), Klatovy, Czech Republic; crystals to 2.7 mm at the Tashelginskiye magnetite deposits, Gornaya Shoriya, western Siberia, and the Mramornoye deposit, Kuznetsk–Alatau, Russia; Ishimskaya Luka, northern Kazakhstan; chromian goldmanite from the Tongshan Cu–Mo deposit, Jurong Co., Jiangsu Prov., China; manganoan from Yamato mine, Kagoshima Pref., Japan; kelyphitic rims on garnet and blue zoisite from the Lualenyi mine, Kenya, are grossular–goldmanite. EF *AM 49:644(1964)*, *DHZ 1A:628,641(1982)*, *MM 56:117(1992)*, *CM 32:319(1994)*.

51.4.3c.1 Schorlomite $Ca_3(Ti^{4+},Fe^{3+},Al)_2[(Si,Fe^{3+},Fe^{2+})O_4]_3$

Named in 1846 by C. U. Shepard for the mineral schorl and from the Greek *homos*, the same, in allusion to its resemblance to schorl. Garnet group. Closely related to melanite (titanian andradite), andradite, and morimotoite. ISO $Ia3d$. $a = 12.128(13$ wt % $TiO_2)$, $12.17(17$ wt % $TiO_2)$, $Z = 8$, $D = 3.84$. *33-285*(actually a melanite, with 13 wt % TiO_2): 4.29_3 3.04_8 2.71_{10} 2.48_6 2.38_2 1.97_2 1.68_2 1.62_5. Fe^{3+} is in octahedral [Y] and tetrahedral [Z] sites. Fe^{2+} is in eightfold triangular dodecahedral [X] and octahedral [Y] sites. Also, Fe^{2+} [X] ↔ Fe^{3+} [Z] electron delocalization (charge transfer). Some tetrahedral Fe^{2+} replacement of Si occurs. *AM 80:27(1995)*. No Ti^{3+} is present according to XANES studies. Oxygen is disordered over two positions as the result of occupancy of the Z site by cations of very different radii (Fe and Si). *CM 33:627(1995)*. Black to very dark brown, dark brown streak, vitreous to metallic luster. $H = 7-7\frac{1}{2}$. $G = 3.77-3.93$. Isotropic; $N = 1.94-1.98$. Mossbauer study of natural Fe–Ti garnets: *AM 65:142(1980)*. Synthetic material with composition $Ca_3Ti_2Fe_2^{3+}SiO_{12}$ is end-member composition. Highest TiO_2 content in natural schorlomite is about 27.4 wt %. Minor amounts of Al_2O_3 (to about 4 wt %) are present. Occurs principally in alkaline igneous rocks, such as syenites, ijolite, melteigite, and fenites. Found at *Magnet Cove, Hot Springs Co., AR*; Rainy Creek alkaline complex, Libby, MT; Oka, QUE, and crystals to more than 8 cm diameter from the Ice River alkaline complex, Yoho National Park, BC, Canada; Camphouse, Ardnamurchan, Scotland; Horningsholm, and Pottäng, Alnö Is., Sweden; Iivaara and Kuusamo, Finland; Zr-bearing (to 15.8 wt % ZrO_2) from Polino carbonatite, Terni, Umbria, Italy; Podzamek, Poland; Africanda massif, Kola Penin., and Morotu, Sakhalin Is., Russia (27.4 wt % TiO_2). EF *AM 52:773(1967)*, *62:475(1977)*, *62:646(1977)*; *MM 36:775(1968)*; *DHZ 1A: 468(1982)*.

51.4.3c.2 Kimzeyite $Ca_3(Zr,Ti)_2[(Si,Al,Fe^{3+})O_4]_3$

Named in 1961 by Milton et al. for the Kimzey family, long known in connection with the mineralogy of Magnet Cove, AR. The mineral was discovered by Joe Kimzey. Garnet group. ISO $Ia3d$. $a = 12.365-12.46$(nat), 12.456(syn Ca–Al kimzeyite), 12.598(syn Ca–Fe kimzeyite), $Z = 8$, $D = 3.96$. *13-130*:

4.42_4 3.12_6 2.79_8 2.54_9 1.73_4 1.67_{10} 1.395_4 1.137_4. Dark brown, light brown streak, vitreous luster. No cleavage. H = 7. G = 3.94–4.00. Dominant forms are {211} and {110}. The end member "ferric kimzeyite" is $Ca_3Zr_2(Fe_2Si)O_{12}$ and the end member kimzeyite is $Ca_3Zr_2(Al_2Si)O_{12}$. Two garnets exsolve from intermediate Ca–Fe and Ca–Al compositions; below 650°, line splitting is present. *MJJ 16:371(1993)*. Isotropic, brown; N = 1.94. Both Al- and Fe-dominant kimzeyites occur in nature. Occurs in carbonatite at the Kimzey farm, *Magnet Cove, AR*, as crystals to 5 mm associated with calcite, apatite, monticellite, magnetite, perovskite, phlogopite, pyrite, and other minerals. In shoshonitic basalt at Stromboli, Aeolian Is., Italy. In lamprophyre dikes at the Marathon dikes, McKellar Harbor, northwestern ONT, Canada. EF *AM 46:533(1961), 52:773(1967), 56:1983(1971), 64:546(1979), 65:188(1980); DHZ 1A:617(1982); MM 56:581(1992)*.

51.4.3c.3 Morimotoite $Ca_3(Ti,Fe^{2+},Fe^{3+})_2(Si,Fe^{3+})_3O_{12}$

Described in 1995 by Henmi et al. and named for Nobuo Morimoto (b.1925), emeritus professor of mineralogy and crystallography, Osaka University, Japan. Garnet group. ISO $Ia3d$. $a = 12.162$, Z = 8, D = 3.80. *MM 59:115(1995)*: 3.04_7 2.72_{10} 2.59_2 2.48_5 $2.385_{2.5}$ 1.97_2 1.69_3 1.63_6. Detailed structure not reported. Black euhedral or subhedral grains to 15 mm with or without rims of grandite garnet; adamantine luster. No cleavage. H = $7\frac{1}{2}$. G = 3.75. IR spectra are very similar to those for schorlomite. Analysis (wt %): SiO_2, 26.9; TiO_2, 18.5; ZrO_2, 1.48; Al_2O_3, 0.97; Fe_2O_3, 11.4; FeO, 7.78; MnO, 0.23; MgO, 0.87; CaO, 31.35. The mineral is derived from andradite by the substitution $Ti + Fe^{2+} = 2Fe^{3+}$. However, as defined, morimotoite is indistinguishable from schorlomite! Isotropic; Na(D): N = 1.955. Occurs in endoskarn at *Fuka, Bitchu-Cho, Okayama Pref., Japan*. Associated with calcite, vesuvianite, grossular, wollastonite, hematite, prehnite, fluorapatite, perovskite, zircon, baddeleyite, and calzirtite. EF

51.4.3d.1 Hibschite $Ca_3Al_2(SiO_4)_{3-x}(OH)_{4x}$ (x = 0.2–1.5)

Named in 1905 by Cornu for J. E. Hibsch. Garnet group. Formerly known as hydrogrossular. Synonym: Plazolite. ISO $Ia3d$. $a = 11.90$–12.04 (most samples), but as high as 12.287; Z = 8, D = 3.24–3.28 most samples, but as low as 2.92. *31-250*: 3.00_8 2.68_{10} 2.46_5 2.19_5 1.95_6 1.73_5 1.66_5 1.61_8. *45-1447*(nat): 5.01_4 3.07_5 2.75_{10} 2.41_3 2.24_6 1.99_7 1.70_4 1.64_4. See also *42-570*(calc. pattern) and *3-801*(Crestmore, CA). Structure: NJBM 404(1983)EJM1:363(1989). Colorless, white, gray, green to bluish green, pink. H = 7. G = 3.06–3.25. N = 1.67–1.750, TGA data. *BM 107:605(1984)*. Generally nonfluorescent; however, a grossular or "hydrogrossular" from Mosjoen, Nordland, Norway, shows a clear orange fluorescence in SWUV. Some S may substitute for Si. Solid solution exists with andradite (and "hydroandradite"), pyrope, and traces of the spessartine molecule may be present. Soluble in HF and with difficulty in HCl or HNO_3. Occurs principally in rodingites, contact-metamorphosed marls, and skarn deposits. Often occurs as mantles on andradite or grossular. Nearly pure "plazolite" occurs at Crestmore, Riverside Co., CA; Ayrshire, England; Aubenas, Vivarais, France; Mariods Weisach, Northern Bavaria, Germany; S-bearing (3.1 wt % SO_3) hibschite rims occur on Ti-bearing "hydroandradite" at *Mariánská, near Ustí nad Labem, near Bohemia, Czech Republic*; Nikortzminda, Caucasus, and Lopan, Georgia, and Bug R., Russia; "hydroandradite" in the Hatrurim formation, Israel; pink and green (Cr-bearing) "hydrogrossular" (= Transvaal jade or South African jade), Buffelsfontein, Tuiffontein, and Wolhuterskop farms, Rustenburg, Bushveld complex, Transvaal, South Africa; Dun Mt., Champion Creek, Waimea district, and Wairere, North Is., New Zealand; MgO-bearing (3.86 wt %) from Kushiro, Hiroshima Pref., Japan. Some material (e.g., Transvaal jade from about 40 miles W of Pretoria) is used as a gemstone; massive white material for carving comes from an unspecified locality in China. EF *AM 76:1153(1991), DHZ 1A:649(1982), NJBM:251(1983), EJM 1:639(1989).*

51.4.3d.2 Katoite $Ca_3Al_2(SiO_4)_{3-x}(OH)_{4x}$

Named in 1984 for Akira Kato (b.1931) of the Natural Science Museum, Tokyo, former chairman of the CNMMN IMA. Garnet group. Hydrogrossular is now a group name for the hydrous garnet series hibschite–katoite. Katoite forms a series with grossular and hibschite. ISO $Ia3d$. $a = 12.57$(syn), 12.358(nat), $Z = 8$, $D = 2.76$(nat), 2.53(syn). *38-368*(nat): 5.05_4 3.30_3 3.09_5 2.76_{10} 2.26_6 2.00_6 1.71_3 1.65_4. *24-217*(syn): 5.13_9 4.44_4 3.36_6 3.14_5 2.81_8 2.30_{10} 2.04_{10} 1.68_5. In katoite, the Si position (of grossular) is vacant, and charge balance is achieved by H atoms located slightly outside the faces of an empty O_4 tetrahedron. $4H^+$ for Si^{4+}. Thin crusts, rounded microcrystals, and larger (>0.3 mm) octahedral crystals. Colorless or milky white. IR and TGA data: *BM 107:605(1984)*. Isotropic, sometimes very faintly birefringent; $N = 1.632$(nat), 1.605(syn). Some substitution of S for Si in tetrahedral coordination. Occurs in vugs within a phonolite emplaced in argillaceous marls at *Campomorto quarry, Pietramassa, Montalto di Castro, Viterbo, Italy*. Associated with other hydrated Ca-silicates and aluminates. Also reported from basalt xenoliths at Maroldsweisach, Northern Bavaria, Germany. EF *BM 108:1(1985), AM 72:756(1987), CM 28:87(1990).*

51.4.4.1 Henritermierite $Ca_3(Mn,Fe,Al)_2(SiO_4)_2(OH)_4$

Named in 1969 for Henri F. E. Termier (b.1897), French geologist and professor at the Sorbonne, Paris. TET $I4_1/acd$. $a = 12.39$, $c = 11.91$, $Z = 8$, $D = 3.40$. *22-150*: 6.21_3 4.92_3 4.37_6 3.09_6 2.98_6 2.75_{10} 2.52_8 1.61_5. Each oxygen of the SiO_4 tetrahedron is shared with an $MnO_4(OH)$ octahedron, four of the six oxygens of this octahedron being shared with SiO_4 tetrahedra while the other two are OH groups. Ca atoms lie in distorted cubic cavities. *BM*

92:126(1969). Henritermierite is tetragonal rather than cubic because of Jahn–Teller distortion along c caused by Mn^{3+}, which disrupts the cubic symmetry. Brown, clove, or apricot-brown, dark orange to red; pale brown streak; vitreous luster. No cleavage, conchoidal fracture. G = 3.34. Slowly dissolved by cold HCl, easily in warm HCl. HNO_3 attacks it superficially, leaving a coating of black MnO_2. DTA and TGA show that decomposition begins at about 500°. Uniaxial (+); $N_O = 1.800$, $N_E = 1.765$. O, very pale yellow; E, lemon yellow. Sometimes anomalously biaxial (+) with a small 2V. Commonly twinned on {101}, giving four sectors. Sometimes has andradite rims. Occurs in manganese deposits. Rounded trapezohedral crystals to 3 mm in calcite-rich skarn from Långban, Värmland, Sweden; *Tachgagalt, Anti-Atlas, Morocco*; 1- to 3-mm sharp amber crystals associated with sturmanite, calcite, and andradite from the Wessels and N'Chwaning II mines, Kuruman, Kalahari manganese area, Cape Prov., South Africa. EF *BM 92:185(1969)*, *GSSA:93(1989)*, *MR 22:279(1991)*, *DHZ 1A:655(1982)*.

51.4.5.1 Wadalite $Ca_6Al_5Si_2O_{16}Cl_3$

Structure published in 1993, but origin of name not specified. ISO $I\bar{4}3d$. $a = 12.001$, Z = 4, D = 3.06. Powder pattern not given but is stated to be indistinguishable from grossular. G = 3.01. Other physical properties not given. Occurs in a skarn xenolith in two-pyroxene andesite at *Tadano, Koriyama City, Fukushima Pref., Japan*, and in skarn at the La Negra mine, Querétaro, Mexico. EF *AC(C) 49:205(1993)*.

51.5.1.1 Larnite Ca_2SiO_4

Named in 1929 by Tilley for the locality. Equivalent to β-Ca_2SiO_4. Bredigite, which usually contains significant Mg, is its high-temperature polymorph. MON $P2_1/n$. $a = 5.48$, $b = 6.76$, $c = 9.28$, $\beta = 94.55°$, Z = 4, D = 3.33. *33-302*(syn): 2.79_{10} 2.78_{10} 2.75_8 2.72_3 2.61_4 2.28_2 2.19_5 1.982_2. Structure: *AC 5:307(1952)*. Crystals tabular, subhedral, also granular and massive. Polysynthetic twinning on {100}. White, colorless; vitreous luster. Cleavage {100}, distinct. H ≈ 6. G = 3.28. Biaxial (+); 2V = 60–63°; $N_x = 1.700$–1.715, $N_y = 1.715$–1.723, $N_z = 1.725$–1.740 (higher values are for synthetic material); $Z = b$, $X \wedge c = 14°$; r > v, weak. Gelatinized by dilute HCl. Occurs in contact zone of syenite–monzonite intrusion at Marble Canyon, Culberson Co., TX; similarly in the Tres Hermanas Mts., Luna Co., NM. At *Scawt Hill, near Larne, County Antrim, Northern Ireland*; at Ardnamurchan, and in metamorphosed calcitic amygdules in a plug on Mull, Scotland. In the Hatrurim at various localities in Israel. In a contact zone at Tokatoka, North Auckland, New Zealand. AR *MM 22:77(1929)*, *26:229(1942)*, *28:90(1947)*; *AM 51:1766(1966)*.

ZIRCON GROUP

The zircon group consists of five orthosilicate members: zircon, hafnon, thorite, thorogummite and coffinite. Behierite, $(Ta, Nb)BO_4$ and xenotime, YPO_4 also have the zircon structure. Zircon is a ubiquitous mineral, and xenotime is common, but the others are rare, with behierite being known from only one locality.

Name	Formula	Space Group	a(Å)	c(Å)	Z
Zircon	$ZrSiO_4$	$I4_1/amd$	6.604	5.979	4
Hafnon	$HfSiO_4$	$I4_1/amd$	6.573	5.964	4
Thorite	$ThSiO_4$	$I4_1/amd$	7.133	6.319	4
Thorogummite	$(Th,U)(SiO_4)_{1-x}(OH)_{4x}$	$I4_1/amd$	7.068	6.260	4
Coffinite	$USiO_4$	$I4_1/amd$	6.994	6.263	4
Behierite	$(Ta,Nb)BO_4$	$I4_1/amd$	6.213	5.468	4
Xenotime-(Y)	YPO_4	$I4_1/amd$	6.885	5.982	4

Good references for the zircon group minerals include: MSA RIM 5:67 and 113(1980), DHZ RFM 1A:418(1982).
The principal structural unit in zircon is a chain of alternating edge-sharing $[SiO_4]$ tetrahedra and $[ZrO_8]$ triangular dodecahedra extending parallel to c. The chains are joined laterally by edge-sharing dodecahedra. Octahedral voids are present but contain no cations as in the case of garnets (MSA RIM 5:67(1980)).

51.5.2.1 Zircon $ZrSiO_4$

Named from the Arabic *zargun*, which was derived from the Persian *zar*, gold, and *gun*, color. Known in antiquity but first modern description is that of Werner (1783). Zircon group. Synonyms: cyrtolite (high U content and metamict), hyacinth (gem variety: red-brown, etc.), malacon (high Th content and metamict). See hafnon, $HfSiO_4$ (51.5.2.2); thorite, $ThSiO_4$ (51.5.2.3); coffinite (51.5.2.4); and thorogummite (51.5.2.5). Other minerals with the zircon structure are xenotime and behierite. TET $I4_1/amd$. (Syn): $a = 6.604$, $c = 5.979$. *JACS 50:549(1967)*. (Nat): $a = 6.6042$, $c = 5.9796$. *AM 64:196(1979)*. $Z = 4$ $D = 4.714$. 6-266(nat): 4.43_4 3.30_{10} 2.52_4 2.07_2 1.91_1 1.75_1 1.71_4 1.65_1. **Habit:** The principal structural unit in zircon is a chain of alternating edge-sharing (SiO_4) tetrahedra and (ZrO_8) triangular dodecahedra extending parallel to c. The chains are joined laterally by edge-sharing dodecahedra. *AM 56:782(1971), 64:196(1979); CIWY 73:544(1974)*. Water (as OH) in zircon is in at least three sites. *AM 76:1533(1991)*. Metamictization: *AM 76:74(1991)*. **Physical properties:** Occurs usually in varying shades of brown, to white, yellow, orange, green, or blue. Characteristic reddish-brown color is believed due to Nb^{4+} ions produced by irradiation-induced reduction of Nb^{5+} substituting for Zr. *AM 55:428(1970)*. Metamict varieties usually leaf green to olive or brownish green; sometimes orange. Prismatic to dipyramidal, with {100}, {110} prisms and {101} and {112} dipyramids most prominent. Other forms: *KL (1980), MIN 3(1):98(1972), DF (1932)*. Crystals to 730 cm; Twinning on {101} produces geniculate (elbow) growth twins. Twins in zircon with {011} as the composition plane and either a{100} or m{110} as the dominant prism plane. *MM 39:587(1974)*. White streak, adamantine to oily luster. Cleavage

51.5.2.1 Zircon

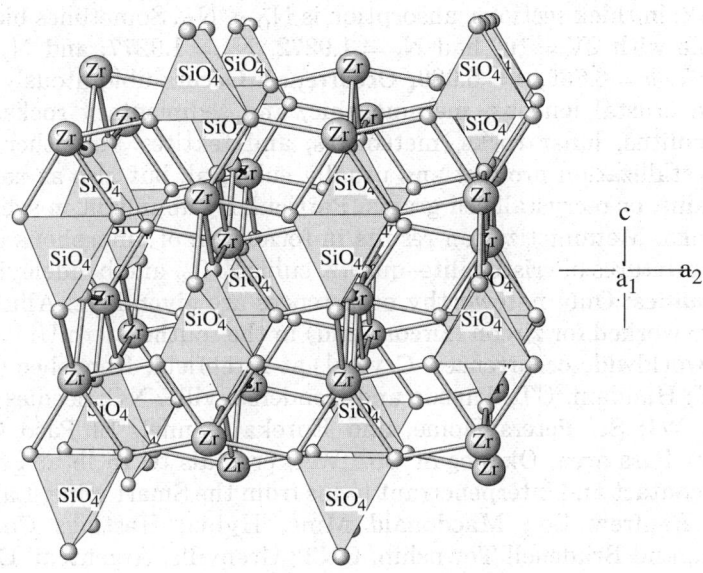

ZIRCON: ZrSiO$_4$

The Zr eight-cornered polyhedra share edges with Si tetrahedra to form ribbons running both parallel and perpendicular to the c axis.

{110}, imperfect; {111}, poor; conchoidal fracture; brittle. H = $7\frac{1}{2}$(fresh), 7(intermediate), 6–7(metamict). G = 4.6–4.71(normal), 4.2–4.6(intermediate), 3.9–4.2(metamict). **Chemistry:** Slowly attacked by hot concentrated H$_2$SO$_4$. IR: *AM 76:1533(1991)*. Raman: *JCS(A):1372(1970)*. Degree of metamictization: *EJM 7:471(1995)*. Zircon is usually luminescent and fluorescent. Fluorescence colors are orange-yellow, yellow, orange, and sometimes green under SWUV and LWUV. Always contains some Hf; normally about 1 wt %. Hf becomes progressively enriched with continued differentiation and fractionation. A complete solid-solution series exists between zircon and hafnon but not between zircon and thorite or zircon and coffinite. Water in zircon is chiefly present as (OH) and not molecular water. *AM 76:1533(1991), 70:1224(1985)*. The maximum amount of (OH) observed in metamict material was about 0.1 wt %. Molecular water is present in fluid inclusions. Some P substitutes for Si. REE's substitute for Zr and are dominated by the HREE's (Y-group). Other elements, such as Fe, Sn, Nb, Na, and others, are probably due chiefly to inclusions of other minerals. U and Th are often present and result in metamictization of the zircon. Total REE as much as 10 wt %, U to 7 wt %, and Th to 10 wt %. Compositional zoning is often present: several generations of zircon may be present. May be epitaxial growth with zenotime(y). **Optics:** Usually uniaxial (+) with very strong dispersion; N$_O$ = 1.960, N$_E$ = 2.01 (Na, syn). Normal zircon has N$_O$ = 1.924–1.934, N$_E$ = 1.970–1.977; intermediate zircon has N$_O$ = 1.903–1.927, N$_E$ = 1.921–1.970; metamict zircon has N$_O$ = 1.782–1.864, N$_E$ = 1.827–1.872. Pleochroism is very weak; in thick sections, absorption is $N_O < N_E$. Sometimes biaxial (+); one example with $2V = 10°$ had $N_x = 1.9272$, $N_y = 1.9277$, and $N_z = 1.982$, and $a = 6.67$, $b = 6.730$, $c = 6.080$. **Occurrence:** Occurs ubiquitously in small amounts in crustal igneous, metamorphic, and sedimentary rocks. Also in mantle xenoliths, lunar rocks, meteorites, and tektites. As either primary igneous crystallization product and usually euhedral, but also as xenocrysts, detrital grains, or recrystallized grains. Particularly abundant in syenitic and granitic rocks. Metamictization results in formation of amorphous glass and sometimes mixtures of cristobalite–quartz, cubic ZrO_2, and baddeleyite (*Mon.* ZrO_2). **Localities:** Only noteworthy occurrences are given here. Alluvial sand deposits are worked for zircon (zircon sand) in the southeastern United States and other worldwide occurrences. Crystals at Litchfield, Kennebec Co., ME; Rossie, NY; Haddam, CT; Tuxedo and Hendersonville, NC; needles to 20 cm at Mellen, WI; St. Peters Dome, and Eureka Tunnel, El Paso Co., CO; Washington Pass area, Okanogan Co., WA; crystals to 15 lb at Sebastopol Twp., fine contact and interpenetrant twins from the Smart mine, Lake Clear, Eganville, Renfrew Co.; Macdonald Mine, Hybia; Hastings Co., Perth, Lanark Co., and Brudenell Township, ONT; Grenville, Argenteuil Co., Templeton, Hull Co., and Mt. St.-Hilaire, QUE, Canada; in fine crystals and twins from Mina La Panchita, Ayoquesco, OAX, Mexico; in southwestern Norway (e.g., Iveland and Langesundfjord) in zircon syenites and pegmatites; also from Seiland Is., Alta Fjord, northern Norway, in fine orange-brown crystals to 2.5 in. or more in length; Riou Pezzouliou, Expailly, Puy de Dome, France; Laacher See, Eifel, Germany; Miask, and Ilmen Mts. (to 10 lb) in the Ural Mts., and fine crystals to 5 cm at Slyudanka, Irkutsk oblast, Russia; Jemaal Kaduna, Nigeria; Mt. Ampanobe, Fianarantsoa, and Itrongay, Betroka, Madagascar; Sri Lanka, in alluvial gem gravels; Chiang Mai dist., Thailand; Tamilnad, India; multicolored gem-quality crystals to 5.5 lb from the Mud Tank area, Harts Range, NT, Australia; Pocos da Caldas, MG, and Quruipil, Goias, Brazil. Used in foundry sands, refractories, and ceramics because of its thermal properties. Lesser amounts are used for abrasives, gemstones, and the production of Zr and Hf; also for radiometric and thermoluminescence age dating. Heat treatments are used to produce blue zircons for gems. EF *EPSL 44 390(1970)*, *MIN 3(1):98(1972)*, *MM 39:587(1974)*, *MSASC 5: 67(1980)*, *DHZ 1A:418(1982)*.

51.5.2.2 Hafnon $HfSiO_4$

Named in 1974 by Correia-Neves et al. for the composition, a Hf mineral; in a series with zircon, thus the similar ending. Zircon group. TET $I4_1/amd$. $a = 6.5725$(syn), $c = 5.9632$(syn), $Z = 4$, $D = 6.97$. *20-467*(syn): 4.43_6 3.29_{10} $2.64_{2.5}$ 2.51_7 2.32_2 2.06_2 1.74_2 $1.70_{5.5}$. At high pressure and temperature, transforms to the scheelite structure. Material ranges from 50:50 to nearly pure $HfSiO_4$ (1.2% ZrO_2). Uniaxial; $N = 1.946$(calc). Occurs in tantalum-bearing pegmatites, associated with cookeite and cleavelandite. Occurs in the Himalaya pegmatite–aplite dike system, Mesa Grande district, San Diego Co., CA, as small millimeter-size amber to buff-colored crystals; at the *Morro Conca and Muiane mines, Morrua area, Zambesia district, Mozambique*. AR *CMPM 48:73(1974)*, AM *67:804(1982)*.

51.5.2.3 Thorite (Th,U)SiO$_4$

Named in 1829 by Berzelius for the composition: contains thorium. Zircon group. See coffinite (51.5.2.4), the U dominant analog. TET $I4_1/amd$. $a = 7.1328$, $c = 6.3188$, $Z = 4$, $D = 6.67$. $11\text{-}419$(syn): $4.72_{8.5}$ 3.55_{10} $2.84_{4.5}$ $2.68_{7.5}$ 2.52_3 2.22_3 1.89_3 $1.83_{6.5}$. Voids in the structure connected parallel to c permit incorporation of molecular H$_2$O + metamictization. Most is partly or completely metamict. *AC(B) 34:1074(1978)*. Yellowish gray, light olive to olive-gray, light brown, brownish yellow, pale yellow, green, orange-brown, brownish black, reddish brown, and so on; white, brown, yellow-brown streak; vitreous luster, resinous to greasy when fresh, dull when altered. To >8 cm in diameter, usually short prismatic with {110} + {111}, {100} + {101} dominant, also pyramidal {101} with prism faces small or absent, also massive, reniform masses, anhedral grains and masses. Cleavage {100}, distinct in non-metamict crystals; conchoidal fracture in metamict crystals; conchoidal to splintery fracture; brittle. H $\approx 4\frac{1}{2}$–5, fresh, less when altered. G = 4.1(with H$_2$O)–6.7(no H$_2$O) and 6.63(syn). IR data: *MSASC 5:113(1980)*. Raman: *MSASC 5:113(1980)*. Paramagnetic. Fresh material is insoluble in common acids, and partly metamict or completely metamict material is attacked by hot acids. Usually contains appreciable U. Other elements, such as Pb, Ca, REE (e.g. >8.1 wt % Y$_2$O$_3$), Fe and H$_2$O substitute. Nearly pure material occurs in the Thomas Range, Utah, as well as uranoan (about 25 wt % UO$_2$) thorite. Phosphatian varieties known: 13.6% P$_2$O$_5$ from Massif Central, France. Total REE's to 20 wt %. Fe in many cases is due to contamination. P$_2$O$_5$ substitutes for Si. Sulfur and arsenic also substitute: 2.17 wt % As$_2$O$_5$, 4.27 wt % S. Uniaxial (+) and isotropic; N = 1.664–1.87 when isotropic and metamict; N$_O$ = 1.78–1.827(syn), N$_E$ = 1.79–1.885(syn); N$_E$ > N$_O$; sometimes biaxial with 2V = 5–6°. O, pale green; E, yellow-green. Occurs in detrital grains in Witwatersrand. Uranoan and regular variety in lithophysae of rhyolites. Widespread primary mineral. In granitic pegmatites, metasomatized zones in impure limestone, hydrothermal veins, and detrital deposits. Accessory mineral in granites and alkaline rocks. Associated with monazite, zircon, allanite, and others. In gneisenized and albitized rocks. Occurs in many worldwide localities and only a few are mentioned here: a PO$_4$-rich thorite occurs at the Freeman mine on Green River, Henderson Co., NC; Barringer Hill, Llano Co., TX; at St. Peters Dome, El Paso Co., and the Powderhorn Dist., Gunnison Co., CO., yellow to deep green crystals in rhyolite from the Thomas Range, UT; metamict material from the Harding Mine, Taos Co., NM; uranoan thorite occurs at Bancroft, large crystals in pegmatite at the Macdonald mine, Hybla, and in pyroxene scarn at Kemp uranium mines, Wilberforce, ONT, Canada; numerous pegmatites in Norway and Sweden: *Lomo, Brevig, Langesundfjord, Norway*; Barkevik, Iveland, Kragerö, Landsbo, and

Arendal (to >6 cm), Norway; Bjorko, Sweden; Vishneveye Gor, Ilmen Mts., Ural Mts., Batum region, Caucasus, Polar Urals, and Tongul and Krakhol massifs, western Tuva, Russia; Dara-i-Pioz, Tadzhikistan; Naegi, Gifu, Japan; fine crystals from Ambatofotsy massif, Ardrotsato, and Befarita, Madagascar; Wodgina, Western Australia. An ore of Th. EF *MIN 3(1):127(1972)*, *CM 27:643(1989)*.

51.5.2.4 Coffinite $U[SiO_4]_{1-x}(OH)_{4x}$

Named in 1956 by Stieff et al. for Reuben Clare Coffin (1886–1972), pioneer in the study of uranium deposits of the Colorado Plateau. Zircon group. TET $I4_1/amd$. $a = 6.995$, $c = 6.263$, $Z = 4$, $D = 6.9$ (for $x = 0.6$). PD: 4.66_{10} 3.47_{10} 2.78_2 2.64_5 2.18_2 1.841_2 1.801_5 1.737_2. Also *11-420*(syn) and *17-460*(yttrian). Natural material is usually metamict. Structure: *MZ 3:81(1989)*. Most commonly as aggregates of 5- to 50-μm grains and masses; rarely as prismatic crystals with pyramidal termination. Black, yellow to brown in thin section; brownish-black streak; dull to adamantine luster. H = 5–6. G = 3.5–5.1. Isotropic (metamict); N = 1.64–1.92. Uniaxial (+/−); $N_O \approx N_E \approx 1.73$–$1.75$. Tetrahedral $(OH)_4$ clusters replace the insular $[SiO_4]$ groups of the zircon structure, but the presence of water rather than hydroxyl ion clusters, with the formula $U[SiO_4] \cdot nH_2O$ has also been suggested. The latter view may well be valid; it is supported by chemical and IR data, but correlation of those observations with degree of crystallinity and cell size is not presented. In any given sample, both chemical interpretations may be playing a role, thus accounting for the great variation in refractive indices and the discrepancy between observed and calculated density. Thorium and RE may replace uranium; Pb present as a decay product; possible complete series with thorogummite. Other cations and anions have been observed but may represent mechanical impurities. Formed from uraninite in a supergene, reducing, alkaline environment, as found in sedimentary environments and fossil placers, by absorption of silica and hydration. Usually as aggregates of 5- to 50-μm grains and masses; rare crystals are prismatic or prismatic–pyramidal. Found at the *La Sal No. 2 mine, Lumsden Canyon, Gateway district, Mesa Co., CO*, and throughout the uranium deposits of the Colorado plateau (AZ, CO, UT); also at the Woodrow, Crownpoint, and Jackpile mines, Valencia Co., NM, and in TX, WY, OK, and SD. Found also at Schneeberg and Marienberg, Saxony, and Wölsendorf and Pfalz, Bavaria, Germany; in Carinthia, Austria; Kovar, Poland; and in Spain, Yugoslavia, and Russia. In the fossil placers of the Dominion Reef, Witwatersrand, and other sites in South Africa; Australia. A significant ore mineral and important component of sedimentary-environment uranium deposits. AR *AM 41:675(1956)*, *USGSB 1064(1958)*, *CM 27:643(1989)*.

51.5.2.5 Thorogummite $(Th,U,Ce)(SiO_4)_{1-x}(OH)_{4x}$

Named in 1889 by Hidden and Mackintosh for the composition and appearance and from the Latin *gummi*, gum. Zircon group. TET $I4_1/amd$. $a = 7.029$–

7.068, $c = 6.232$–6.260, $Z = 4$, $D = 5.75$; *8-440*(nat uranoan): 4.7_9 3.53_{10} 2.82_4 2.65_6 2.20_4 2.00_4 1.82_6 1.17_4. Black (contains U^{4+}), yellowish brown, greenish brown tan, white, greenish gray; usually fine-grained, rarely as acicular radiating crystals to 3 mm, nonmetamict to partly metamict; earthy, subvitreous to resinous luster. Conchoidal to splintery fracture. $G = 3.2$–5.4. Readily soluble in HNO_3. Analysis (wt %), Barringer Hill: UO_2, 22.4; ThO_2, 41.4; Ln_2O_3, 6.69; CaO, 0.41; SiO_2, 13.1; PbO, 2.16; Al_2O_3, 0.97; Fe_2O_3, 0.85; P_2O_5, 1.18; H_2O^+, 7.88; H_2O^-, 1.23. Some may contain appreciable $P_2O_5 \approx 10\%$. May contain many minor and trace elements: REE, Mn, Mg, Ca, Nb, Ta, Zr, CO_2. Some have major Fe, to $>12\%$ Fe_2O_3. $\bar{N} = 1.54$–1.64. Also given as 1.74–1.77 when metamict, and uniaxial (+) with $N_O = 1.665$, $N_E = 1.685$ to $N_O = 1.692$, $N_E = 1.710$. *USGSB: 1627(1984)*. Occurs in pegmatites. A product of alteration of primary Th minerals. Always fine-grained. Most often forms from thorite, sometimes from yttrialite, also from uraninite. Alteration: replacing thorite. Found in a large number of worldwide occurrences, only some of which are given here: Barringer Hill pegmatite, Blufton, Llano Co., TX; from thorianite at Easton, PA; to 1-cm masses at Mt. St.-Hilaire, QUE, and Hybla, ONT, Canada; REE-bearing (9.15 wt % RE_2O_3) from Transbaikal, Russia; Vishnevye Mt., Ilmen Mts., Ural Mts., and Lovozero masif, Kola Penin., Russia; Haicheng Pref., southern Manchuria; Wodgina, WA, Australia; from yttrialite at Iisaka-mura, Date-gun, Fukushima Pref., Tateiwa mine, Syobu, Hozyo town, Ehime Pref., and Hayamadake, Tokiwa town, Fukushima Pref., Japan; several localities in Madagascar. EF *AM 38:1007(1953), CM 27:643(1989)*.

51.5.3.1 Huttonite $ThSiO_4$

Named in 1950 for Colin Osborne Hutton (1910–1971), New Zealand–American mineralogist, Stanford University. See monazite. A dimorph of thorite. Always nonmetamict. MON $P2_1/n$. $a = 6.7759$–6.80, $b = 6.948$–6.974, $c = 6.4982$–6.54, $\beta = 104.92$–104.99°, $Z = 4$, $D = 7.18$(syn)–7.27. *34-188*(syn): 4.66_4 4.20_7 4.04_3 3.49_3 3.28_6 3.07_{10} 2.88_3 2.86_6. Isostructural with monazite: structure of synthetic material consists of SiO_4 monomers and a compact arrangement of edge-sharing ThO_9 polyhedra. *AC(B) 34:1074(1978)*. Colorless, white. Grains with poor development of faces, to 0.2 mm (New Zealand), {100} dominant. Imperfect cleavage on {100} and {001}, and may be parting; conchoidal fracture. $G = 7.10$–7.20. In SWUV fluoresces a dull white with a pink tinge. Unaffected by hot concentrated HCl, very finely powdered material is slowly dissolved in concentrated H_2SO_4. Analysis (wt %): ThO_2, 76.6; SiO_2, 19.7; Fe_2O_3, 1.2; Ce_2O_3 etc., 2.6. Solid solution with monazite takes place if $Th/Si + F_4 + (OH)_4$ and $(REE + Ca)/P = 1$. Substitution of Ca, F + (OH) takes place. *MM 43:1031(1980)*. Series is continuous to at least 27 wt % of Th at the monazite end and to at least 20% REE at the huttonite end. Biaxial (+); Y parallel to b, Z near c; $N_x = 1.898$–1.900(syn), $N_y = 1.900$, $N_z = 1.922$–1.930(syn); colorless; dispersion, r < v, moderate. Occurs in beach sands, thought to be derived from low-grade schists or less likely from sparsely distributed pegmatite veins. Holiday Mine, Hawthorne, Mineral CO., NV; from the val d'Aosta region, Italy; Silesia, Poland; *Gillespie's Beach, South Westland, New Zealand;* also in sands from five other beaches to the north and south of the Waikukupa R., New Zealand. EF

51.5.3.2 Tombarthite-(Y) $Y_4(Si,H_4)_4O_{12-x}(OH)_{4+2x}$

Named in 1968 by Neumann and Nilsson for Thomas Fredrik Weiby Barth (1899–1971), Norwegian geochemist, Oslo University. Partial substitution of 4H atoms for a Si atom. One of the ideal end members of tombarthite-(Y), $YSiO_3(OH)$ can be regarded as a Y,Si analog of monazite-Ce (Z = 1). MON $P2_1/n$. $a = 7.12$, $b = 7.29$, $c = 6.71$, $\beta = 102.8°$, $Z = 1$, $D = 3.64$–3.68. *21-1314*(nat): 7.32_3 6.55_{10} 3.54_3 3.42_8 3.23_7 2.97_6 2.89_5 2.40_4. Structure not determined. Brownish black, partly to completely metamict, masses to 1 cm wide, dull luster. H = 5–6. G = 3.51–3.65. DTA: *LIT 1:113(1968)*. Contains some MnO (to 10%), FeO (to 5.5%), CaO (to 4.9%), and ThO_2 (to 6%). $\bar{N} = 1.639$; isotropic, partly metamict. Occurs in contact with or intergrown with thalenite in a granite pegmatite at *Høgetveit, Evje, Norway*. In pegmatite at Vest-Agder, Reiarsdal, Norway. A very late replacement (after britholite) mineral. EF *MIN 3(1):529(1972), NGT 59:265(1979)*.

51.5.4.1 Eulytite $Bi_4[SiO_4]_3$

Named in 1827 by Breithaupt from the Greek *eulytos*, easily dissolved or fused. Synonym: eulytine. ISO $I\bar{4}3d$. $a = 10.3$, $Z = 4$, $D = 6.76$. *33-215*(syn): 4.20_{10} 3.26_9 2.75_8 2.57_1 2.10_4 2.02_4 1.765_1 1.670_3. Structure: *ZK 123:73(1966)*. Small, complex tetrahedral crystals with dominant {211} and {2$\bar{1}$1} and less commonly {100}, {111}, {511}, {110}. Penetration twins by reflection on {100}. Also in spherical forms. White, yellow, gray, brown, colorless; adamantine luster; translucent to opaque, rarely transparent. Conchoidal fracture, brittle. Cleavage {110}, imperfect. $H = 4\frac{1}{2}$. $G = 6.1$, 6.6. Isotropic; $N = 2.05$; also anomalously uniaxial (−) with OA perpendicular to {111}; also biaxial (−) with birefringence < 0.002. Fusibility 2, yielding yellow, then red mass, and finally metallic bismuth on charcoal with Na_2CO_3; fused mass acid soluble. Found with clinobisvanite and namibite in pegmatite at the Elizabeth R mine, Pala district, San Diego Co., CA; at the Evans–Lou mine, Hull, QUE, Canada. Complex tetrahedra (0.56 mm) and spherical aggregates (2 mm) at Buckbarrow Beck, Cumbria, England, and reported from Southwick Cliffs, near Dalbeattie, Scotland; with quartz at Johanngeorgenstadt, and with bismuth at the *Neu Glück mine, Schneeberg, Saxony, Germany*; and at Dognacska, Hungary. AR *ZK 78:136(1931), 106:34(1945); DT:591(1932); AM 28:536(1943); MM 57:754(1993)*.

Class 52

Insular SiO$_4$ Groups and O, OH, F, and H$_2$O

52.1.1.1 Beryllite Be$_2$SiO$_4$(OH)$_2$ · H$_2$O

Named in 1954 by Kuzmenko in allusion to the beryllium content. ORTH *13-411*: 4.01$_{10}$ 3.64$_9$ 3.39$_7$ 3.19$_7$ 2.90$_5$ 2.34$_{10}$ 2.12$_7$ 1.351$_8$. Also *20-166*. Fibrous in spherulites and crusts. White, silky luster. H = 1. G = 2.196. Biaxial (−); 2V < 45°, 2V$_{calc}$ = 74°; N$_x$ = 1.541, N$_y$ = 1.553, N$_z$ = 1.560; length-slow and found at Ilimaussaq, Greenland. In alkaline pegmatites with albite and epididymite in the *Lovozero massif, Kola Penin., Russia*. AR *AM 40:787(1955), ZMGP 1:191(1959), VMRE 2:96(1966)*.

52.2.1.1 Euclase BeAl[SiO$_4$](OH)

Named in 1792 by Haüy from the Greek *eu* and *klasis*, good fracture, in allusion to its perfect cleavage. MON $P2_1/c$. a = 4.632, b = 14.334, c = 4.781, β = 100.33°, Z = 4, D = 3.085. Structure: *SMPM 72:159(1992)*. *14-65*: 7.15$_{10}$ 3.84$_3$ 3.22$_5$ 2.77$_3$ 2.54$_2$ 2.44$_3$ 1.991$_2$ 1.880$_2$. Ordered Be and Si tetrahedra form composite corner-sharing chains parallel to [001]. Isostructural with väyrynenite (41.5.4.3), BeMn^{2+}[PO$_4$](OH, F). The axial setting for the quoted cell is in agreement with the original description; an earlier structure paper reverses a and c. Habit commonly prismatic and striated parallel to [001] with {120}, {110}, {100}, {010} dominant; terminations complex with dominant {111}, {1̄31}, {021}. Less commonly flattened {010}. Colorless, white, pale green, pale to deep blue, the latter often patchy; vitreous luster, sometimes pearly on cleavage; transparent to translucent. Cleavage {010}, perfect; conchoidal fracture; brittle. H = 7$\frac{1}{2}$. G = 3.05–3.10. Biaxial (+); 2V = 48–52°; N$_x$ = 1.650–1.652, N$_y$ = 1.655–1.656, N$_z$ = 1.671–1.672; Y = b, Z∧c = 42°; r > v; distinctly pleochroic in darker blue material. Insoluble in acids; fusibility 5.5. Minor iron and fluorine may be present. A low-temperature hydrothermal mineral in granite pegmatites and Alpine veins; also found in chlorite schist and phyllite, and as a detrital mineral. Found at the Boomer mine, Park Co., CO. At Pizzo Giubine, near St. Gothard Pass, Ticino, Switzerland; in the Austrian Alps in the Glosslockner region, and near Hochnarr, also at Sonnblick Mt., near Salzburg; at Epprechstein, Fichtelgebirge, Bavaria, Germany; at Dobschütz, near Görlitz, Silesia; Norway; in the auriferous sands of the Orenburg district (Sanarka R.), southern Ural Mts. Russia(!). Deep blue crystals(!) at the Last Hope mine, Karai, Miami district, Zimbabwe. First brought to Europe from South America in 1785; *Villa Rica, Ouro Preto district, MG*, may be type locality; Santana do Ecoberto, Mun. São Sebastão Maranhão(!) in colorless crystals to 10 cm, and other pegmatite deposits in MG; at the Santino, Mamoes, and Amancio pegmatites (blue), near Ecuador, and at the Jacú mine, RGN, Brazil. In placer deposits in Guyana. AR *D6:508(1892), ZK 117:16(1972), MR 2:12(1971)*.

52.2.1.2 Clinohedrite CaZnSiO$_4$ · H$_2$O

Named in 1898 by Penfield and Foote from the Greek for *inclined plane*, in allusion to the morphology of the point group symmetry. MON Cc. $a = 5.131$, $b = 15.928$, $c = 5.422$, $\beta = 103.39°$, Z = 4, D = 3.29. *17-214*: 7.81$_5$ 3.97$_5$ 3.23$_7$ 2.76$_{10}$ 2.50$_6$ 2.47$_4$ 2.36$_5$ 1.990$_4$. Structure: *SP 22:614(1978)*. The classic and oft-illustrated example of point group symmetry m. Consequently, the original morphological indexing is retained; the transformation matrix to the structural cell is 002/020/100. Short prismatic [001] with {110}, {$\bar{1}$10}, {$\bar{5}$51}, {$\bar{1}$3$\bar{1}$} dominant; also wedge-shaped with {$\bar{1}$10}, {$\bar{7}$71}, {111}, {11$\bar{1}$} dominant; massive, granular. Colorless to white to pale amethystine, vitreous luster. Cleavage {010}, perfect. H = 5$\frac{1}{2}$. G = 3.33. Gelatinized in HCl; fusibility 4. Biaxial (−); 2V$_{calc}$ = 64°; N$_x$ = 1.662, N$_y$ = 1.667, N$_z$ = 1.669; Y = b; Z∧c = 61°; r > v. Found with willemite, axinite, hancockite, and other minerals at the *Trotter mine, Franklin, Sussex Co., NJ*. AR *USGSPP 180:106(1935), ZK 144:377(1976)*.

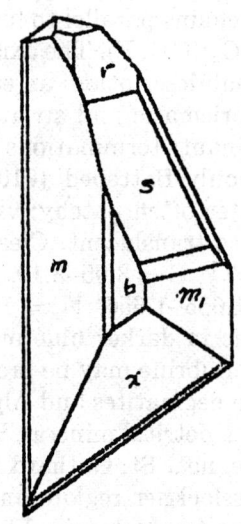

CLINOHEDRITE

Clinohedrite, New Jersey. View along b-axis. dominant forms: b, {010}; m, {110},; m$_1$, {110}; s, {$\bar{5}$51}; r, {$\bar{3}$31}; x, {$\bar{1}$31}

52.2.1.3. Hodgkinsonite MnZn$_2$(SiO$_4$)(OH)$_2$

Named in 1913 (Palache and Schaller) for H. H. Hodgkinson, assistant underground supervisor of Franklin mine, who discovered it in 1913. MON $P2_1/a$. $a = 8.171$, $b = 5.316$, $c = 11.761$, $\beta = 95.25°$, $Z = 4$, $D = 4.07$. *35-554:* 11.8_1 3.94_3 3.75_4 3.64_4 3.24_4 3.02_3 2.97_8 2.87_{10}. Structure: *ZK 119:117(1963)*. Prismatic [001] crystals with, in the structural cell, {110}, {443}, {553}, and {023} dominant. The structural c axis is $1\frac{1}{2}$ times the morphological c axis; the transformation matrix, morphology to structure, is 200/020/003. Pink to brownish pink, vitreous luster, translucent to transparent. Cleavage {001}, perfect. H = $4\frac{1}{2}$–5. G = 4.06–4.08. Biaxial (−); 2V = 52°; N$_x$ = 1.724, N$_y$ = 1.742, N$_z$ = 1.746; Y = b; Z∧c = 38°; r > v, strong; weakly pleochroic X = Z, lavender; Y, pale green. Gelatinizes with acids, easily fusible. Weak pink fluorescence in LWUV. Found in rich ores at the Sterling Hill mine, Ogdensburg, and at *Franklin*(!), *Sussex Co., NJ.* AR *USGSPP 180:108(1935), MR 13:229(1982)*.

52.2.1.4 Gerstmannite (Mg,Mn^{2+})$_2$Zn(SiO$_4$)(OH)$_2$

Named in 1977 by Moore and Araki for Ewald Hans Gerstmann (b.1918), prominent German–American collector, of Franklin, NJ, who drew attention to the original material. ORTH *Bbcm.* $a = 8.185$, $b = 18.65$, $c = 6.256$, $Z = 8$, $D = 3.66$. *29-867:* 9.33_9 4.81_5 3.42_8 2.98_6 2.76_8 2.60_{10} 2.47_5 2.33_8. Structure: *AM 62:51(1977)*. The structure consists of [(Mg,Mn)O$_3$(OH)$_2$] octahedral sheets and [ZnSiO$_4$] tetrahedral sheets, and may be considered a "zincosilicate." Prismatic crystals, unterminated, with {100}, {010}, {110}, forming mats and sprays up to 10 mm in length. White to pale pink, vitreous luster, transparent to opaque. Cleavage {010}, good. H = $4\frac{1}{2}$. G = 3.68. Biaxial (−); 2V = 50–60°; N$_x$ = 1.665, N$_y$ = 1.675, N$_z$ = 1.678; XZ = bc. Occurs with calcite, manganpyrosmalite, and sphalerite in veins cutting franklinite–willemite ore at the *Sterling Hill mine, Ogdensburg, Sussex Co., NJ.* AR

ANDALUSITE GROUP

The members of the andalusite group all share the same chemical formula,

$$Al_2O(SiO_4)$$

or a minor variation of it. Although not isostructural, they are related by a common structural feature of chains of edge-sharing AlO$_6$ octahedra, which results in a common c-axis length of approximately 5.6 Å. The cross-linking of the chains varies in each of the species, resulting in four-, five-, and six-coordinated aluminum and a variety of structures. *AJS 271:97(1991), AM 76:677(1991)*.

ANDALUSITE GROUP

Mineral	Formula	Space Group	a	b	c	α	β	γ	
Sillimanite	$Al_2O(SiO_4)$	$Pnma$	7.484	7.672	5.77				52.2.2a.1
Mullite*	$Al_{4+2x}Si_{2-2x}O_{10-x}$	$Pbam$	7.578	7.682	2.886				52.2.2a.2
Andalusite	$Al_2O(SiO_4)$	$Pnnm$	7.798	7.903	5.556				52.2.2b.1
Kanonaite	$(Mn^{3+},Al)(Al,Mn^{+3})O(SiO_4)$	$Pnnm$	7.959	8.047	5.616				52.2.2b.2
Yoderite*	$(Mg,Al,Fe)_8O_2(SiO_4)_4(OH)_2$	$P2_1$	8.10	5.78	7.28		106		52.2.2b.3
Kyanite	$Al_2O(SiO_4)$	$P\bar{1}$	7.126	7.852	5.572	89.99	101.11	106.03	52.2.2c.1
Staurolite*	$(Fe^{2+},Mg,Fe^{3+})_4Al_{17}O_{13}[(Si,Al)O_4]_8(OH)_3$	$C2/m$	7.870	16.623	5.661		90.12		52.2.3.1

*Although not a member of the group in a strict sense, these minerals are related.

Notes:
1. The triple point for the three $Al_2O(SiO_4)$ polymorphs is uncertain but close to 520° and 4 kbar.
2. All the andalusite minerals are characteristic of metamorphic environments.

52.2.2a.1 Sillimanite $Al_2O(SiO_4)$

Named in 1824 by Bowen for Benjamin Silliman (1779–1864), American chemist and geologist at Yale University. Andalusite group. Fibrolite refers to the columnar to fibrous form, but is used interchangeably with sillimanite in the gem trade. Polymorphs: andalusite, kyanite. See also mullite (52.2.2a.2). ORTH $Pnma$. $a = 7.484$, $b = 7.672$, $c = 5.77$, $Z = 4$, $D = 3.239$. *38-471:* 3.84_1 3.42_{10} 3.37_4 2.68_2 2.54_2 2.42_2 2.20_3 1.520_1. Structure: *AM 77:374(1992)*. The six-coordinated aluminum–oxygen chains parallel [001] and are cross-linked by four-coordinated aluminum and (SiO_4) tetrahedra. Prismatic to acicular, frequently with square cross section, {110}, striated parallel to [001] and with poor terminations; columnar to fibrous aggregates, sometimes radiating. Colorless to white; pale yellow to brown; pale blue, green, and violet; vitreous to silky luster; transparent to translucent. Cleavage {010}, perfect; uneven fracture. H = 7. G = 3.23–3.27. Biaxial (+); 2V = 21–30°; $N_x = 1.654$–1.661, $N_y = 1.658$–1.662, $N_z = 1.673$–1.683; XYZ = abc; r > v, moderate. Pleochroic in thick sections. X, pale yellow; Y, brown or light green; Z, dark brown or blue. Infusible and insoluble in acids. Usually close to the ideal composition, with Fe_2O_3 the only significant substituent (up to 1.5 wt %). Characteristic of high-temperature regional and thermal metamorphism of pelitic rocks; amphibolite to granulite facies. Formed with K-feldspar by breakdown of micas, and by transformation of andalusite or kyanite at high temperature and moderate pressure. Sometimes in parallel growth with andalusite; frequently associated with cordierite and corundum. Common worldwide in the appropriate metamorphic terrains (i.e., the metamorphic belts from New England to Georgia). At Chester, Middlesex Co., CT; in distinct crystals at Yorktown, Westchester Co., NY; at Chester and Mineral Hill, Delaware Co., PA; at Brandywine Springs, DE; at Culsagee mine, Macon Co., NC(!); 3 km N of Seneca, Oconee Co., SC; in Custer and Pennington Cos., SD; and the Silver Hill tin mine, Spokane Co., WA. At Romaine, QUE, Canada. At Ariège and Pontgibaud, France; in metamorphosed Ross of Mull granite, Scotland; at Lisens Alp,

52.2.2a.2 Mullite

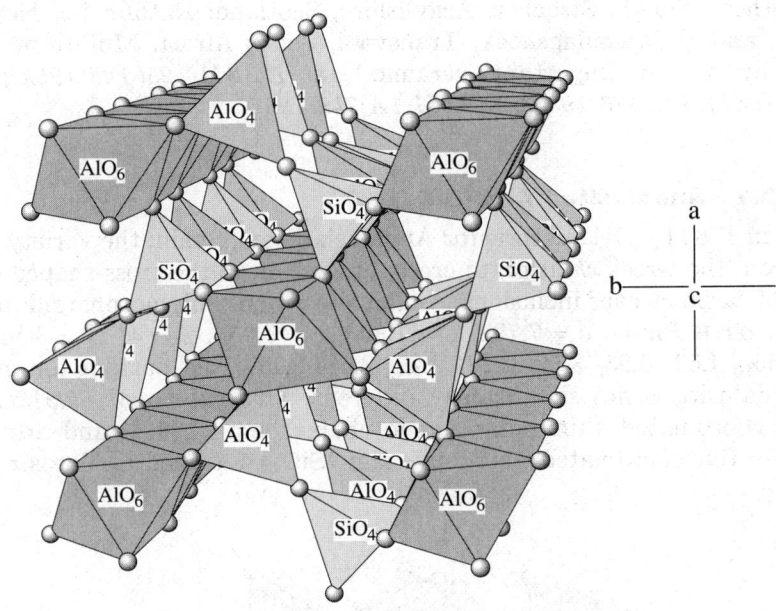

SILLIMANITE: Al$_2$SiO$_5$

In sillimanite, the cross-linking Al atoms are tetrahedral, which leads to the formation of double chains running parallel to c. These chains consist of four-membered rings with strictly alternating Si and Al, unlike the double chains in amphibole which consist of six-membered rings.

Tyrol, Austria; at *Vltava, Susice, Czech Republic*. At Relli, Vizagapatam district, and in the Khasi Hills, Assam, India. Pale blue to violet, gemmy material from Sri Lanka, and from the Mogok district, Myanmar. In the diamond gravels of Diamantina, MG, Brazil. Used in the manufacture of mullite refractory ceramics; rarely as faceted gems and chatoyant cabochons for the gem collector. AR *NJMA 111:74(1969), DHZII 1A:719(1982)*.

52.2.2a.2 Mullite Al$_{4+2x}$Si$_{2-2x}$O$_{10-x}$

Named in 1924 by Bowen et al. for the locality. Andalusite group. ORTH $Pbam$. $a = 7.578$, $b = 7.682$, $c = 2.886$, $Z = 1$, $D = 3.02$. 15-776(syn): 5.39_5 3.43_9 3.39_{10} 2.69_4 2.54_5 2.21_6 2.12_2 1.524_3. Structure: *AM 76:332(1991)*. The structure is closely related to that of sillimanite, with aluminum partially replacing silicon. Fine-grained prismatic [001]; massive (synthetic material). White, gray, pale pink; vitreous luster. Cleavage {010}, distinct. H = 6–7. G = 3.11–3.26. Biaxial (+); $2V = 20$–$60°$; $N_x = 1.638$–1.653, $N_y = 1.642$–1.657, $N_z = 1.653$–1.671; XYZ = abc. Formula x has observed value of 0.17–0.59; x = 0.25 for commonly given Al$_6$Si$_2$O$_{13}$ (3:2 mullite, 3Al$_2$O$_3 \cdot$ 2SiO$_2$); natural material lies between 3:2 and 4:1. In high-temperature thermally metamorphosed argillaceous rocks. Formed by heating any of the Al$_2$O(SiO$_4$) polymorphs. In argillaceous inclusion in eruptive rock on *Isle of Mull, Scotland*;

from Sithean Sluigh, Stracher, Argyllshire, Scotland; Rathlin Is., Northern Ireland; and at Maandagshoek, Transvaal, South Africa. Mullite produced artificially is an important ceramic. AR *NBSM 25(3):3(1964); AM 66:142(1981), 71:1476(1986); DHZII 1A:742(1982).*

52.2.2b.1 Andalusite Al$_2$O(SiO$_4$)

Named in 1789 by Delamethrie for Andalusia Prov., Spain; the variety chiastolite from the Greek *chiastos*, to cross, in allusion to the cross-shaped distribution of carbonaceous inclusions. Andalusite group. Polymorphs: sillimanite, kyanite. ORTH *Pnnm*. $a = 7.798$, $b = 7.903$, $c = 5.556$, Z = 4, D = 3.144. *39-376*: 5.55_{10} 4.53_8 3.93_3 3.52_3 2.77_7 2.27_3 2.18_3 2.17_4; the pattern for manganian andalusite is not significantly different. Structure: *AM 69:513(1984).* The six-coordinated aluminum–oxygen chains parallel [001] and are cross-linked by five-coordinated aluminum and (SiO$_4$) tetrahedra. The structure

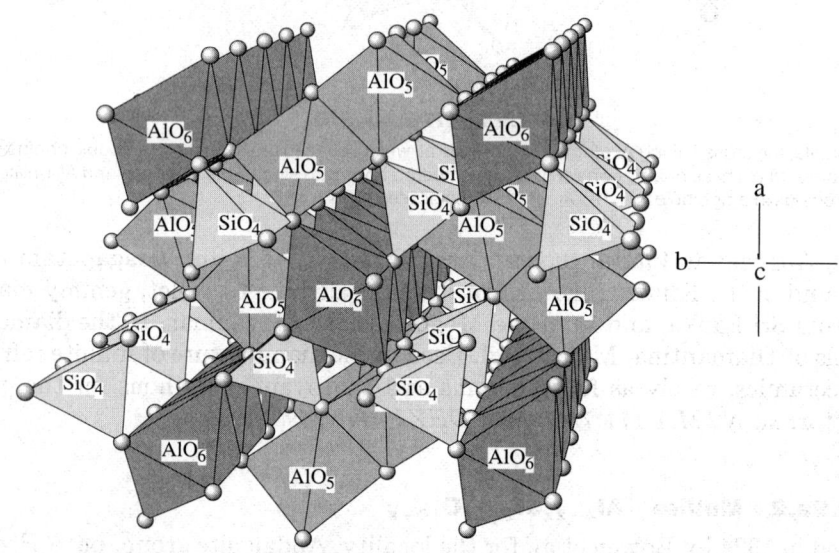

ANDALUSITE: Al$_2$SiO$_5$

In andalusite, the cross-linking Al atoms are 5-coordinated, so andalusite is structurally intermediate between kyanite and sillimanite. All three Al$_2$SiO$_5$ polymorph have chains of AlO$_6$ octahedra running parallel to c along the cell edges and at or near the center of the cell, which will be called here the "A chains." The A chains are very similar in sillimanite and andalusite, and in both the center and edge chains are linked with isolated silicate tetrahedra but otherwise are very different. In sillimanite the linking tetrahedra are in double chains consisting of four-membered rings with alternating Al and Si tetrahedra. These are the structural units in terms of strong bonds and explain why sillimanite is often fibrous. In andalusite the tetrahedral chains are replaced by pairs of AlO$_5$ polyhedra, and the SiO$_4$ tetrahedra are in quite another place. Si tetrahedra can also be considered to link the AlO$_5$ polyhedra into a three-dimensional framework. In kyanite the octahedral A chains are quite different from those in sillimanite and andalusite and the central chain is pushed to one side by a B layer of AlO$_6$ octahedra. The "central" A chain in kyanite is related to the corner ones by a center of symmetry (in the B layer) rather than by glide planes as in sillimanite and andalusite.

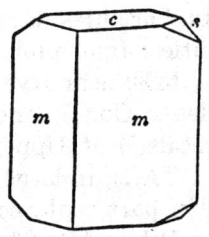

ANDALUSITE

Andalusite. Forms: c, {001}; m, {110}; s, {011}

is analogous to that of adamite, $Zn_2(OH)(AsO_4)$, and related species. **Habit:** Most commonly as prismatic [001] crystals of approximately square cross section with {110} and {001} dominant: chiastolite crystals often subhedral, elongated, and tapered; massive. Twinning on {101} uncommon. **Physical properties:** Pale to deep pink or reddish brown, white, gray, violet, yellow, green, blue; vitreous luster; transparent to opaque. Cleavage {110}, good to perfect; {100}, poor; conchoidal fracture, $H = 6\frac{1}{2}$–$7\frac{1}{2}$. $G = 3.13$–3.21. Usually biaxial $(-)$; $2V = 73$–$86°$; $N_x = 1.627$–1.637, $N_y = 1.635$–1.644, $N_z = 1.639$–1.650; $XYZ = cba$; $r < v$; weakly to strongly pleochroic. X, pale pink to red-brown; $Y \approx Z$, yellow to olive-green. Mn- and Fe-rich varieties (viridine) are light to dark green with a gray-green streak; biaxial $(+)$; $2V = 52$–$77°$; N_x reaching 1.679, N_z reaching 1.727; $XYZ = abc$; strongly pleochroic. $Y > Z > X$, in shades of green and yellow. Infusible and insoluble in acids. IR Spectra: USGS OF 87-263(1987). **Chemistry:** Mn_2O_3 and Fe_2O_3 are usually present in small amounts and may reach 4% and 20%, respectively, influencing color and optical properties. A complete series may exist between andalusite and its manganese–aluminum analog, kanonaite. A blue andalusite from Belgium contains P^{5+} (up to 0.31%) in substitution for Si^{4+} coupled with Fe^{2+} for aluminum. May contain up to 2% Cr_2O_3. Alters readily to sericite, particularly in the chiastolite variety; may alter in part to kaolinite and/or pyrophyllite, and to kyanite paramorphs; alumina may be leached to leave a porous residue, largely quartz. **Occurrence:** Characteristic of low-grade metamorphism of pelitic sediments, primarily in contact zones, but also in regional metamorphism, and is associated with cordierite in both environments. Less commonly in granites and pegmatites, possibly through contamination by argillaceous sediments. Frequently as a detrital

Andalusite (var. chiastolite); Successive sections showing carbonaceous inclusions.

mineral in arenaceous sediments. **Localities:** Common worldwide in appropriate terrains. Noteworthy localities: fine, pink crystals(!) at Standish, and chiastolite at southern Berwick, ME; near Rye (chiastolite) and at Charleston, NH; at Westford, Middlesex Co.(!), and at Lancaster and Sterling (chiastolite), MA; in large crystals(!) at Upper Providence, Delaware Co., PA; at Altavista, Campbell Co., VA(!), in large, square crystals up to 30 cm, associated with corundum and in part replaced by sillimanite and kyanite; abundant(*) at Champion Mine, White Mts., Inyo Co., CA. Found in *Andalusia Prov., Spain*; near Var, Provence, France; blue phosphatic andalusite at Ottré, Vann-Stavelot massif, Belgium; viridine at Ultevis, Jokkmokk, Sweden; as colorless and pink crystals at Steinbach, Bavaria, and at Darmstadt, Germany; in large crystals with kyanite at Lisens Alp, Tyrol, Austria; at Mursinsk, Ural Mts., and chiastolite at Nankova, Nerchinsk district, Transbaikalia, Russia. Nearly black viridine at Manzabar, Purulia district, India; as fine gem pebbles(!) in the Ratnapura district, Sri Lanka. At Juhuagou, near Beijing, and chiastolite in Henan Prov., China; chiastolite(!) at Mt. Howden, near Bimbowrie, SA, Australia. Chromian material from corundum–fuchsite rocks in Zimbabwe. Gem-grade stream pebbles and crystals(!) at Santa Tereza, ES, and gemmy crystals at Santa Maria, Cruzeiro Novo, MG; (viridine) with bahianite and cassiterite at Serra das Almas, Paramirim, Bahia, Brazil. **Uses:** In the manufacture of mullite ceramics; as a minor faceted gemstone notable for its strong pleochroism in red and green tones, and rarely (Brazil) as cat's-eye gems. AR *DHZII 1A:759(1982), BM 107(5):587(1984), MM 54:75(1990)*.

52.2.2b.2 Kanonaite $(Mn^{3+},Al)(Al_2,Mn^{3+})O(SiO_4)$

Named in 1978 by Vran et al. for the type locality. Andalusite group. ORTH *Pnnm*. $a = 7.959$, $b = 8.047$, $c = 5.616$, $Z = 4$, $D = 3.39$. *42-575:* 5.66_{10} 4.59_6 3.99_2 2.83_4 2.52_4 2.30_2 1.513_2. Structure: *AM 66:561(1981)*. Most of the manganese preferentially replaces six-coordinated aluminum. Greenish black, gray-green streak, vitreous luster. Cleavage {110}, poor. $H = 6\frac{1}{2}$. Biaxial (+); $2V = 53°$; $N_x = 1.702$, $N_y = 1.730$, $N_z = 1.823$; $XYZ = abc$. Pleochroic; X, yellow-green; Y, blue-green; Z, deep golden yellow. The type material contains 32.2 wt % Mn_2O_3; the hypothetical end member would contain 44.7%. As rare porphyroblasts in gahnite–chlorite–coronadite–quartz schist at *Kanona, Serenje, Zambia*. AR *CMP 66:325(1978)*.

52.2.2b.3 Yoderite $(Mg,Al,Fe)_8O_2(SiO_4)_4(OH)_2$

Named in 1959 by McKie and Radford for Hatton S. Yoder (b.1921), petrologist and director of the Geophysical Laboratory, Washington, DC. Andalusite group. MON $P2_1$. $a = 8.10$, $b = 5.78$, $c = 7.28$, $\beta = 106°$, $Z = 1$, $D = 3.36$. *12-625:* 3.50_{10} 3.23_7 3.03_9 2.91_7 2.61_8 2.58_7 2.00_8. Structure: *AM 67:76(1982)*. The high-temperature structure has the characteristic chain of edge-sharing octahedra with additional cations occupying five-fold trigonal–bipyramidal sites, plus silicon in tetrahedral sites. Found as anhedral laths up $\frac{1}{2}$ in., elongated on [010]. Dark blue with pale blue streak, the color being attributed to charge transfer between Fe^{2+} and Fe^{3+}; also emerald green; vitreous luster. Parting {001}, good; {100}, poor. H = 6. G = 3.39. Biaxial (+); 2V = 25°; $N_x = 1.689$, $N_y = 1.691$, $N_z = 1.715$; Y = b; $Z \wedge c \approx 7°$; pleochroic, Y > X > Z; prussian blue to light olive-green. Iron is present in both valence states with Fe^{2+} predominant. Occurs intergrown with kyanite in quartz–kyanite–talc schist at *Mautia Hill, Kongwa, Tanzania*. AR *MM 32:282(1959)*, *NAT 210:1148 (1966)*, *AM 76:1052(1991)*.

52.2.2c.1 Kyanite Al$_2$O(SiO$_4$)

Named in 1789 by Werner from the Greek *kuanos*, dark blue. *Cyanite* is a former spelling. Andalusite group. Synonyms: Now obsolete, *disthene*, from the Greek *dis*, two, and *sthenos*, strength, in allusion to the differing hardness as a function of orientation. Polymorphs: andalusite, sillimanite. TRIC $P\bar{1}$. $a = 7.126$, $b = 7.852$, $c = 5.572$, $\alpha = 89.99°$, $\beta = 101.11°$, $\gamma = 106.03°$, Z = 4, D = 3.666. *11-46*: 3.35_{10} 3.18_2 2.70_4 2.51_5 2.35_4 2.16_2 1.960_5 1.929_8. Structure: *AM 64:573(1979)*. The six-coordinated aluminum–oxygen chains parallel [001] and are cross-linked by six-coordinated aluminum and (SiO$_4$) tetrahedra. **Habit**: Commonly bladed, elongated [001], and flattened {100};

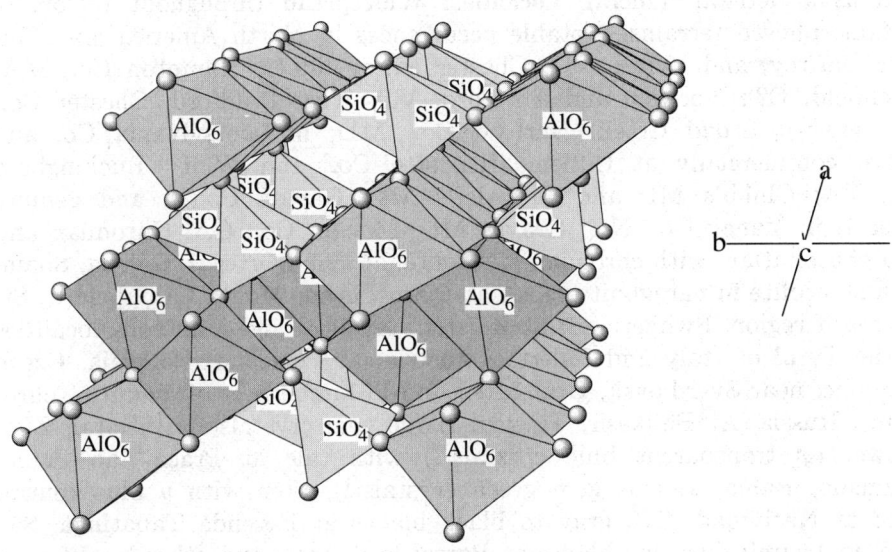

KYANITE: Al$_2$SiO$_5$

Perspective view from left and below. All aluminum is in six-coordination (octahedra).

dominant forms {100} and {010}, rarely terminated, with {001} most frequent; columnar aggregates, often radiating. Twinning polysynthetic on {100} with twin axis perpendicular thereto or parallel to [010] or [001]; also multiple twins on {001}. Parallel growth with its polymorphs. **Physical properties:** Blue, green, colorless, gray due to inclusions; color often darker toward center of laths; vitreous luster; translucent to transparent. Cleavage {100}, perfect; {010}, good; parting {001}; splintery fracture. H variable, 4–5 on {100} parallel to [001], 6–7 on {100} parallel to [010], 7–$7\frac{1}{2}$ on {010}, $5\frac{1}{2}$–$6\frac{1}{2}$ on {001}. G = 3.56–3.67. Biaxial (−); 2V = 82–83°; X approximately perpendicular {100}; $Z' \wedge c \approx 30°$ on {100} and $\approx 7°$ on {010}; $N_x = 1.712$–1.718, $N_y = 1.721$–1.723, $N_z = 1.727$–1.734. Weakly pleochroic (thick sections), Z > Y > X. IR Spectra: USGS of 87-263 (1987). **Chemistry:** Small amounts of Fe^{2+}, Fe^{3+}, and Ti may be present, the blue color being attributed to charge transfer involving these ions. Cr_2O_3 may substitute for Al_2O_3 up to 12.9 wt %, contributing green color. P^{5+} may replace silicon to a small degree. Converted to mullite and glass at $\approx 1300°$. May alter to pyrophyllite, muscovite, and sericite, and may invert to andalusite or sillimanite with a change in metamorphic environment. **Occurrence:** Typically in regional metamorphism of pelitic, less commonly psammitic, rocks, and commonly used as a zonal mineral in pelitic assemblages. May be derived from pyrophyllite and paragonite, and by transformation of anadalusite when thermally metamorphosed rocks are followed by regional metamorphism. May also be found in some eclogites and kyanite amphibolites. Not uncommon as a detrital mineral. **Localities:** Widespread throughout regionally metamorphosed terrains. Notable occurrences in North America are Hanover, Jaffrey, and Lyme, NH; Chester emery mines, Hampton Co., MA; Litchfield, CT; Norwich and Thetford, VT; West Bradford, Chester Co., PA (green); Broad Creek, Harford Co., MD; in Spotsylvania Co. and mined commercially at Cullen, Charlotte Co., and Willis, Buckingham Co., VA; Clubb's Mt. and Crowder's Mt., Gaston Co.(!), and gemmy green(!) in Yancy Co., NC; Graves Mt., Lincoln Co., GA. Chromian and also phosphatian, with chromian staurolite, at Cabo Ortegal, Galicia, Spain; with staurolite in paragonite schist at Pizzo Forno, Monte Camplone(!), St. Gotthard region, Switzerland; at *Zillertal, Austria*, and numerous localities in the Tyrol of Italy and Zillertal, Austria; at Petschau, Bohemia, Czech Republic; near Sverdlovsk, Ural Mts., and in kimberlites of Yakutia (chromian), Russia. At Balikesir, Turkey. Large gem crystals(!) at Karai, Zimbabwe. As transparent blue crystals(!) with talc at Praça São Paulo, Brumado, Bahia, and as gem green crystals(!) often with a blue central band at Natividad, GO; gray to black blades at Fazenda Tabatinga, São José do Jacuni; fine blue blades at Barra do Salinas; and other localities in MG, Brazil. **Uses:** In the manufacture of mullite ceramics and as a collector's gemstone. AR *NJMM 111:74(1969)*; *MR 5:229(1974)*, *9:23(1978)*; *DHZII 1A:780(1982)*.

52.2.3.1 Staurolite (Fe^{2+}, Mg, Fe^{3+})$_4$Al$_{17}$O$_{13}$[(Si,Al)O$_4$]$_8$(OH)$_3$

Although known earlier, named in 1792 by Delamethrie from the Greek *stauros*, a cross, in allusion to the common cruciform twins. Andalusite group. MON (pseudo-ORTH) $C2/m$. $a = 7.870$, $b = 16.623$, $c = 5.661$, $\beta = 90.12°$, $Z = 1$, $D = 3.79$. *41-1484*: 4.15$_2$ 3.01$_3$ 2.77$_8$ 2.69$_4$ 2.40$_5$ 2.37$_4$ 1.982$_5$ 1.968$_{10}$. Structure: *AC(B)* 46:292(1990). An idealized structure may be thought of as layers lying parallel to {010}, of composition alternating between Al$_2$OSiO$_4$ (kyanite) and HFe^{2+}AlO$_4$, thus yielding a hypothetical unit cell composition of H$_2$Fe$_4^{2+}$Al$_{18}$Si$_8$O$_{48}$. Aluminum, with minor iron and magnesium, is situated in octahedral sites (M1–M3) corresponding to those of kyanite; iron is situated in additional octahedral sites (M4) and tetrahedral sites (T2) with vacancies existing in these sites and in M3; silicon and aluminum alone occupy the remaining tetrahedral sites (T1). **Habit:** Short to long prismatic [001] with {110}, {010}, {001}, and equally developed {201} and {$\bar{2}$01}, resulting in pseudo-orthorhombic appearance. Twinning common on {031}, yielding cruciform penetration twins at close to 90°, and on {231} yielding penetration twins at approximately 60°, sometimes repeated to form trillings; combined twins also known. Earlier morphological descriptions

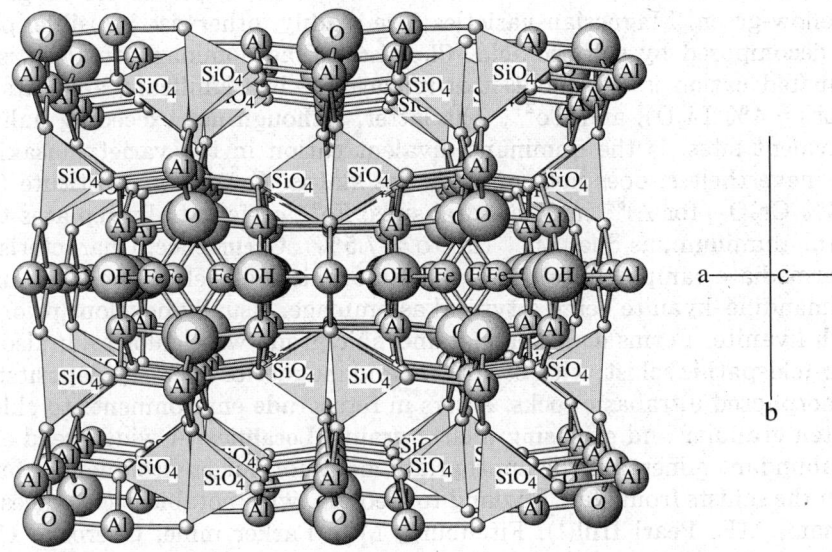

STAUROLITE: Fe$_2$Al$_9$Si$_4$O$_{22}$(OH)$_2$

Staurolite can be described broadly as slabs of Al-oxide/hydroxide parallel to (010), held together with isolated silicate tetrahedra. There are several types of Al octahedra and several types of octahedral chains running parallel to c. Ferrous iron atoms occupy large tetrahedral sites in the middle of the slabs.

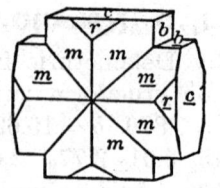 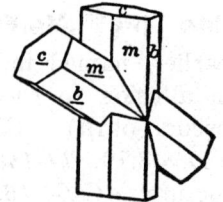

STAUROLITE

Staurolite twins: cruciform on {031} (left); oblique on {231}. Forms: b, {010}; c, {001} {; m, {110}; r, {201} (pseudo-orthorhombic indexing).

are based on orthorhombic symmetry and doubled c axis; the transformation matrix, old to new, is 200/020/001. Staurolite–kyanite parallel growth with $c_s = c_k$ and $\{010\}_s$ parallel to $\{100\}_k$ is known. **Physical properties:** Medium to dark brown, reddish brown; colorless, gray streak; vitreous luster; transparent to opaque. Cleavage {010}, fair; conchoidal fracture. $H = 7-7\frac{1}{2}$. $G = 3.74-3.83$. Biaxial (+); $2V = 80-90°$; $N_x = 1.736-1.747$, $N_y = 1.742-1.753$, $N_z = 1.748-1.761$; $XYZ = bac$; $r > v$, weak. Strongly pleochroic; X, colorless; Y, pale yellow; Z, yellow to orange. Indices increase and 2V decreases with increase in iron content. Cobaltoan staurolite (lusakite) is blue-black with $N_x = 1.739$, $N_y = 1.746$, $N_z = 1.753$. Strongly pleochroic; X, cobalt blue; Y, violet-blue; Z, violet. Chromian staurolite is pleochroic in green and yellow-green. Magnesian varieties fuse readily, otherwise infusible; partially decomposed by sulfuric acid. **Chemistry:** The dominant divalent, six-coordinated cation is Fe^{2+}, but it is replaceable by significant amounts of Mg, Li ($\approx 4\%$ Li_2O), and Co^{2+}. The latter, although not exceeding half of the divalent sites, is the dominant divalent cation in the variety lusakite, which, nevertheless, does not enjoy species status. Cr^{3+} may substitute (up to $\approx 6\%$ Cr_2O_3) for Al^{3+} in octahedral sites; Fe^{3+} preferentially replaces tetrahedral aluminum, as does Zn^{2+} (up to $\approx 7.5\%$). **Occurrence:** Characteristic of intermediate (amphibolite)-grade metamorphism of pelitic rocks, staurolite–almandine–kyanite being a typical assemblage. Usually develops prior to or with kyanite. Forms at lower metamorphic grade with chloritoid; also in quartz–feldspathic schist; with corundum and magnetite in emery deposits; in metamorphosed ultrabasic rocks. Alters in retrograde environments to chlorite, often granular and enclosing quartz grains. **Localities:** A widespread and often abundant mineral in the appropriate metamorphic environment. Abundant in the schists from New England to Georgia, with notable occurrences at Windham, ME; Pearl Hill(!), Fitchburg, MA; Parker mine, Cherokee Co., NC; Fannin Co., GA, commonly twinned, loose in the soil but often weathered (locally called "fairy stones"), and including zincian material. In Rio Arriba and Taos Cos., and in the Truchas Mts. (Fe- and Li-bearing), NM; at various localities in SD, AZ, and WA. Found at numerous localities in Great Britain; as large twins in the schists of Brittany, France; at Cabo Ortega, northwestern Spain (chromian and magnesian); at Mt. Campione, Switzerland(!), in schist as long prisms with kyanite; at Mt. Greiner, Zillerthal, and elsewhere in the Austrian and Italian Tyrol. In twinned crystals at Keivy, Kola Penin., Russia. As pseudomorphs after quartz, corundum, and kyanite in the Dunghai district, China. At Broken Hill, NSW, Australia (zincian); at Fiordland, New Zealand (chromian). Cobaltoan staurolite (lusakite) at a site about 120 km E of Lusaka, Zambia; in Uganda (zincian), and at Manantsahala, Madagascar (chromian). At Rubelita, MG, Brazil, as large, sharp, occasionally gemmy crystals(!), often twinned on both laws. Uses: An index mineral in the interpretation of metamorphic grade. Cruciform twins used ornamentally and as amulets. As a collector's gemstone (rarely). AR *AM 20:316(1935), 67:292(1982), 71:1461(1986), 76:1910(1991)*.

52.3.1.1 Topaz $Al_2SiO_4(F,OH)_2$

Known since antiquity. The history of the use of the name *topaz* is both curious and confusing. According to Pliny, the Greek work *Topazos* or *Topazion* referred to an island in the Red Sea now known as Geziret Zabarget and formerly as St. John's Island, just off the south coast of Egypt. This island was and is a source of gem periodot (olivine), sometimes called chrysolite. It is probable that the gem we call peridot today was formerly called topaz ["topasius vulgaris = chrysolithos veterum," de Boot, Gemm. (1636)]. The first use of the name *topaz* for the species we now know by that name was by

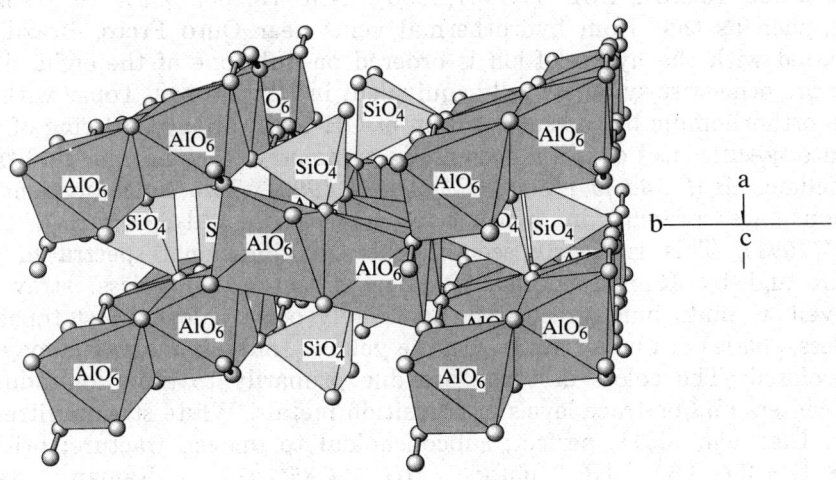

TOPAZ: $Al_2SiO_4(F,OH)_2$

Topaz is similar in some respects to the Al_2SiO_5 minerals. Like kyanite, it has multiple chaisn of octahedra (parallel to a) cross-linked by Si tetrahedra. Its density is almost as great as that of kyanite.

Henckel in 1737, when he described the deposits of Saxony. The term *topaz* has commonly been applied to any yellowish gemstone, including citrine ("false topaz," "quartz topaz," "Madeira topaz"); yellow sapphire ("oriental topaz"); yellow andradite ("topazolite"); even smoky quartz ("smoky topaz"), and others. In the gem trade true topaz of a reddish orange-yellow to sherry color is often referred to as "Imperial topaz." ORTH *Pbnm* or TRIC *P*1. For F-rich topaz, Thomas Range, UT (0.1 wt % H_2O^+), $a = 4.6512(4)$, $b = 8.795(8)$, $c = 8.3993(7)$. For OH-bearing topaz, Ouro Preto, Brazil (2.6 wt % H_2O^+), $a = 4.6673(5)$, $b = 8.8415(7)$, $c = 8.3947(7)$, $Z = 4$, $D = 3.56$(OH-free topaz), 3.37 (OH-rich topaz, 25 mol % OH for F). *GAC-MAC Ann. Meet., 16:A38(1991)*. Syn $Al_2SiO_4(OH)_2$ has $a = 4.724$, $b = 8.947$, $c = 8.390$. *AM 78:285(1993)*. 12-765(nat): 3.69_6 3.195_6 3.04_3 2.94_{10} 2.38_3 2.36_4 2.105_4 1.67_3. **Habit**: Frequently in crystals prismatic to equant (with crystallographic setting $a < c < b$), with {110} or {120} predominating. Other forms present, in decreasing order of abundance, are {112}, {011}, {021}, {001}, {111}, {113}, {101}, {010}. In pegmatitic topaz {021}, {011}, {001} often prominent. Twins rare, "contact" on (010). The key structural units are crankshaft-like chains of edge-sharing $[AlO_6]$ octahedra parallel to c cross-linked by corner-sharing $\{SiO_4\}$ tetrahedra. Structurally related to andalusite and danburite. Orthorhombic topaz contains eight symmetry-equivalent positions for F and OH. H is disordered over these sites. Three of the eight sites (38%) are Raman active and permit substitution of OH for F. Natural topaz with more than about 30 mol % substitution of OH for F has not been found. Topaz with complete replacement of F by OH has been synthesized at high temperature (to 1000°) and pressure (55–100kb). *EOS 71:1707(1990), AM 78:285(1993)*. In triclinic topaz, such as that from hydrothermal veins near Ouro Preto, Brazil, H associated with the hydroxyl ion is ordered on only one of the eight sites, which are otherwise symmetrically equivalent in *Pbnm* topaz. Topaz with no OH is orthorhombic but as substitution of OH for F occurs, ordering of the OH into specific sites causes a decrease in symmetry to monoclinic and then to triclinic. *MM 43:943(1980), CM 28:827(1990)*. Two nonequivalent H positions are present in synthetic high-pressure $Al_2SiO_4(OH)_2$. *AM 79:401(1994)*. This is clearly seen in the OH-vibrational spectra in the mid-IR and by Raman spectra. **Physical properties**: Colorless, straw to wine-yellow, pink, blue, green, red, and rarely other colors. Most topaz is colorless, shades of blue or straw to wine yellow. Individual crystals may be multicolored. The colors of topaz are due primarily to radiation-induced color centers and/or trace levels of transition metals. White streak, vitreous luster. Cleavage {001}, perfect; subconchoidal to uneven fracture; brittle. $H = 8$. $G = 3.4$–3.6. IR data: *MG 4:85(1973)*. Raman data: *JCS(A):1327(1969), PCM 15:148(1987)*. Rarely fluorescent yellow, white, orange, greenish yellow under both SWUV and LWUV. Pyroelectric and piezoelectric because of ordered OH for F substitution. Color is sometimes changed by heating or exposure to light due to presence of radiation-induced color centers. *NJMA 130:288(1977)*. The variation of density, unit cell

52.3.1.1 Topaz

parameters (particularly b), indices of refraction, and 2V for topaz are a consequence of OH for F substitution. *AM 52:1890(1967), 56:1812(1971); MM 37:717(1970); GAC-MAC Ann. Meet., 16:A38(1991)*. Partially attacked by H_2SO_4. **Chemistry:** Topaz usually does not contain many trace or minor elements. As much as about 700 ppm Ge has been reported. More than 100 ppm each of B, Li, and Na has been reported from rhyolite-hosted topaz. Some rhyolitic topaz is pink and is colored by >600 ppm Mn. As much as several thousand ppm Cr has been reported from pink-red topaz from hydrothermal veins, and Mn^{3+} as well as Fe^{3+} may also play a role in topaz coloration. Topaz from topaz rhyolites contains less than 0.1 wt % H_2O^+. That from pegmatites and greisens contains from 0.2 to 0.9 wt % H_2O^+ and topaz from hydrothermal veins contains from about 1.5 to 2.7 wt %. F ranges from about 20 wt % in volcanic topaz down to 15 wt % in hydrothermal vein topaz. **Optics:** Biaxial (+); XYZ = abc; N_x = 1.606–1.634, N_y = 1.609–1.637, N_z = 1.616–1.644; $2V_{z\,meas}$ 48–68°. For orthorhombic topaz, OAP = (010), BXA = [001]. Optics for triclinic (= optically anomalous) topaz: *MM 43:237(1979)*. **Occurrence:** Topaz is a product of crystallization from F-rich vapors in rhyolites. It also crystallizes as a liquidus phase from rhyolites and certain silicic magmas known as ongonites. It is found chiefly in veins and druses in peraluminous granites, greisens, tin veins, and pegmatites. Topaz also occurs in hydrothermal veins and rarely in high-temperature sillimanite-bearing metamorphic rocks. Cr-bearing topaz is found in hydrothermal veins that transect rocks containing elevated levels of Cr or are closely associated with such rocks. **Localities:** Only noteworthy worldwide occurrences of topaz are given here. Occurs at Lord Hill, Stoneham, ME, in fine crystals; Moat Mt., and Conway region (e.g., Baldface Mt.), NH, in miarolytic cavities in granite. Unusual fine-grained, massive, low-F-bearing topaz occurs at the Brewer gold mine, Jefferson, Chesterfield Co., SC. Blue topaz occurs at the Morefield and Rutherford mines, Amelia district, Amelia Co., VA. Blue topaz(!) is found in pegmatites in the Llano uplift, Mason Co., TX. In CO in pegmatites in the Pikes Peak region (including the Lake George intrusive center, Tarryall region, and Devils Head area), and in lithophysae in rhyolite at Ruby Mt., Nathrop. In lithophysae in rhyolite from the Thomas Range, Juab Co., UT, as fine sherry-colored crystals(!!) which become colorless upon exposure to sunlight, associated with red beryl, bixbyite, pseudobrookite, and hematite. Zoned pink crystals, colored by Mn^{3+}, occur at the northern end of the Thomas Range. In pegmatites at the Little Three mine, Ramona(!), and at the Blue Lady mine, Aguanga Mt., San Diego Co., CA(!). In pegmatites of the Sawtooth Batholith, Salmon, ID. In Mexico, as orange crystals in rhyolites near Cerritos, SLP, and at Tepetate, ZAC, and in DGO. Coarse, fibrous topaz (pycnite) from Altenberg, Schneckenstein, Saxony, Germany. In Ukraine from the Volyn region(!!). In Russia in pegmatites at Miask, Ilmen Mts.(!). In superb nearly equant crystals of sky-blue color from the Alabaschka–Mursinka region, Ural Mts.(!!); from the Adun-Tschilon Mts., Nertschinsk, and Urulga R. area, Transbaikal region, Siberia. Cr-bearing, pink to purplish-red crystals are found in placer and gem gravels derived from quartz veins as well as in the quartz veins and vugs in limestone, near Sanarka, Orenburg district, Ural Mts., Russia. Cr-bearing topaz(!!) occurs at Katlang, Mardan district, Pakistan, in deep-pink gem crystals with quartz and calcite in a hydrothermal vein. The mineral occurs in fine colorless, amber-yellow, and sherry wine–colored crystals(!!) in pegmatites near Shingus and Bulechi in the Gilgit area and at Niyit Bruk and Gone, near Dassu in the Skardu district. In Japan at Tanokamiyama, Omi Pref.(!). In Nigeria at Jos in veins with cassiterite and columbite. In Zimbabwe in fine blue crystals(!) at St. Ann's mine, Karoi district. In Namibia at Klein Spitzkopje, Usakos. In Brazil in fine yellow, orange, and pink to violet gem crystals(!!) ("Imperial topaz") in hydrothermal veins in the Ouro Preto district, MG, associated with euclase, hematite, and rutile; also in MG abundant in the Marambaia district in fine colorless crystals(!); at Virgem da Lapa, in pegmatites in fine blue crystals(!!), one of which weighed 28 kg, associated with lepidolite and herderite; and at the Fazenda do Funil, Sta. Maria da Itabira, in magnificent large crystals(!!) associated with gem beryl and cassiterite. One of these crystals weighed over 270 kg and several exceeded 100 kg. Three of these crystals are on exhibit at Harvard, AMNH, and the USNM. A large crystal of Topaz, about 350 tons, was described in 1944 from the São Domingos Mine, Mugui, Espirito Santo, Brazil. Pseudomorphs are sometimes formed by hydrothermal action, causing alteration to clays or muscovite and quartz; also to fluorite, margarite, or kaolinite. Use: Principal use is as a gemstone. Irradiation (with gamma rays or with neutrons or electrons in linear accelerator facilities); treatment after cutting is commonly employed to produce "electric blue" or "London blue" stones from colorless material. Some especially large gems are: Champagne Topaz (36,853 ct), American Golden (22,875 ct), and the Brazilian Princess (21,005 ct). EF *MIN 3(1):274(1972), MSASC 5:215(1980), NJMM:465–470(1982), DHZ 1A:801(1982), AM 71:1186(1986), TOPAZ 207 pp (1992), MR 26:5(1995).*

HUMITE GROUP

The humite group consists of neosilicate minerals conforming to the general formula

$$A_n(SiO_4)_m(F,OH)_2$$

where

A = Mg, Ca, Mn, Fe^{2+}, or Zn; n=3,5,7,9; m=1,2,3,4

Humite Group

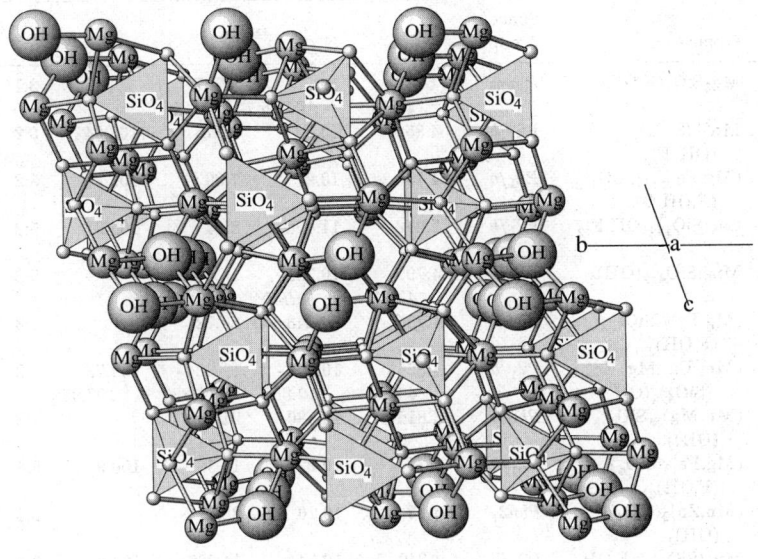

CHONDRODITE: $Mg_5Si_2O_8(OH,F)_2$
Perspective view along *a*, showing the olivine and brucite layers typical of the humite group.

and where the structure is monoclinic or orthorhombic and based on hexagonal close-packed anions.

The humite group minerals (and the olivine group minerals) have structures based on hexagonal close-packed anion (O, OH, F) arrays in which one-half of the octahedral sites are filled by divalent cations. The key structural unit is a serrated chain of edge-sharing M-octahedra parallel to the *c* axis. Members with straight chains are orthorhombic, and those with diagonal chains are monoclinic. (The angle is α rather than β because of the cell orientation. The orthorhombic minerals are generally described in a nonconventional orthorhombic setting, *Pbnm*, to show the structural relationship with the olivine group.) Adjacent chains in the (001) plane are related by the translation vector *b* and are cross-linked by sharing corners with up- and down-pointing (+ and −) tetrahedra. Each (+) tetrahedron shares its three basal edges [i.e., those approximately parallel to (100)] with octahedra in the layer below, each (−) tetrahedron with octahedra in the layer above. Octahedral chains in adjacent (001) layers are related by the *b* glide operation.

HUMITE GROUP

Mineral	Formula	Space Group	a	b	c	α	M^{2+}/Si	
Norbergite	$Mg_3SiO_4(F,OH)_2$	$Pbnm$	4.706–4.710	10.271–10.275	8.727–8.820		3:1	52.3.2a.1
Alleghanyite	$Mn_5^{2+}(SiO_4)_2(OH,F)_2$	$P2_1/b$	4.85	10.72	8.275	108.64	5:2	52.3.2b.1
Chondrodite*	$(Mg,Fe,Ti)_5(SiO_4)_2(F,OH,O)_2$	$P2_1/b$	4.75	10.27	7.80	109.2	5:2	52.3.2b.2
Reinhardbraunsite	$Ca_5(SiO_4)_2(OH,F)_2$	$P2_1/b$	5.052	11.458	8.84	108.91	5:2	52.3.2b.3
Ribbeite	$Mn_5(SiO_4)_2(OH)_2$	$Pbnm$	4.799–4.811	10.732–10.740	15.672–15.70		5:2	52.3.2b.4
Humite**	$(Mg,Fe)_7(SiO_4)_3(F,OH)_2$	$Pbnm$	4.735	10.243	20.72		7:3	52.3.2c.1
Leucophoenicite	$(Mn, Ca, Mg, Zn)(SiO_4)_3(OH)_2$	$P2_1/b$	4.826–4.828	10.842–10.85	11.324–11.380	103.73–103.93	7:3	52.3.2c.2
Manganhumite	$(Mn,Mg)_7(SiO_4)_3(OH)_2$	$Pbnm$	4.815	10.580	21.448		7:3	52.3.2c.3
Clinohumite	$(Mg,Fe)_9(SiO_4)_4(F,OH)_2$	$P2_1/b$	4.74	10.226	13.58	100.9	9:4	52.3.2d.1
Jerrygibbsite	$(Mn,Zn)_9(SiO_4)_4(OH)_2$	$Pbn2_1$	4.85	10.70	28.17		9:4	52.3.2d.2
Sonolite	$Mn_9(SiO_4)_4(F,OH)_2$	$P2_1/b$	4.849	10.544	14.002	100.3–100.97	9:4	52.3.2d.3

* Data for synthetic F end member.
** Data for synthetic OH end member.

COMPOSITION OF HUMITE GROUP MINERALS BASED ON THE M^{2+}/Si RATIO

M^{2+}/Si	Ca	Mg (Humite Subgroup)	Mn (Mn-Humite Subgroup)	Mn (Leucophoenicite Subgroup)
3:1		Norbergite	(unknown)	(unknown)
5:2	Reinhardbraunsite	Chondrodite	Alleghanyite	Ribbeite
7:3		Humite	Manganhumite	Leucophoenicite
9:4		Clinohumite	Sonolite	Jerrygibbsite

Notes:
1. All the members of the humite group have a very limited paragenesis, and their occurrence is, with rare exceptions, restricted to metamorphosed and metasomatized limestones and dolomites and to skarns associated with ore deposits at contacts with acid and, less frequently, plutonic rocks. *DHZ 1A:337(1982)*.
2. Excellent summaries of the humite group are given in *MSASC 5:231(1980)* and *DHZII 1A:377(1982)*.

52.3.2a.1 Norbergite $Mg_3SiO_4(F,OH)_2$

Named in 1926 by Geijer for the locality. Humite group. ORTH $Pbnm$. $a = 4.706–4.710$, $b = 10.271–10.275$, $c = 8.727–8.802$, $Z = 4$, $D = 3.14–3.19$; *11-686*(syn F end member): 4.37_3 3.06_{10} 2.64_7 2.41_3 2.34_3 2.26_7 2.23_8 1.72_5. Structure: *AM 54:376(1969)*, *MSASC 5:231(1980)*. Crystals rare, flattened on a, typically granular no twinning observed. Light yellowish brown, white, rose, yellow; vitreous luster, uneven to subconchoidal fracture, brittle. $H = 6–6\frac{1}{2}$. $G = 3.13–3.20$. IR data: *MIN 3(1):300(1972)*; material from many localities (e.g. Balmat, NY; Franklin and Sparta, NJ; Pargas, Finland) Balmat, NY, fluoresces bright yellow only under SWUV. Soluble in HCl, producing a silica gel. Nearly pure from Balmat, NY. Fe substitutes for Mg and OH substitutes for F. Small amounts of Fe, Mn, Al, and Ti are present in natural material. Biaxial (+); XYZ = acb; N_x = 1.558–1.567, N_y = 1.563–1.579, N_z = 1.581–1.593; 2V= 44–50°; OAP = (001); X > Y > Z; r > v. X, colorless or weakly colored; Y, pale yellow, weak pale yellow; Z, colorless. Occurs in contact metasomatic rocks, Marbles Mg-rich skarns, and as a contact mineral between calcite–dolomite marble and granite. Rarely in Fe and Fe–Zn ores and Ca-rich skarn zones. Sometimes with other members of the humite group and other Mg-silicates (e.g., forsterite, diopside, phlogopite, brucite, etc). Mg–Zn skarns: Franklin, NJ. Occurs at Balmat, St, Lawrence and around Edenville, Orange Co., NY; Franklin, Sussex Co., NJ: *Östanmossa mine, Norberg, Västmanland, Sweden*; Parainen and Pargas, Finland; Monte Sommarand, Mt. Vesuvius, Naples, Italy; Pitkyaranta, Karelia, Russia; several places, including Lyangar in Uzbekistan; Bihar, India. EF *AM 64:1027(1979), 64:1156(1979), 67:538(1982), 67:545(1982); DHZ 1B:378 (1982)*.

52.3.2b.1 Alleghanyite $(Mn^{2+})_5(SiO_4)_2(OH,F)_2$

Named in 1932 by Ross and Kerr for the locality Humite group. See chondrodite (52.3.2b.2) (Mg-dominant analog). Polymorph: ribbeite. MON $P2_1/b$. (Nat Mg-rich): a = 4.827, b = 10.613, c = 8.116, α = 108.65°; (nat): a = 4.850, b = 10.720, c = 8.275, α = 108.64°, Z = 2, D = 3.96; for magnesian, D = 3.75. *43-683*(nat): 3.10_{10} 2.84_7 2.75_4 2.70_4 2.59_4 2.33_5 1.79_6. Calc. 25-1187; Calc magnesian 39-1348. Structure (for magnesian): *AM 70:182(1985)*. Crystals are slender, platelike, and deeply striated by twinning, usually 0.3–1 mm; also unstriated stout platy crystals, rounded, anhedral grains; twinning may be present, polysynthetic on {001}, also rarely on 015 and 035. Strawberry-pink, grayish red, clove-brown; pale pink streak; vitreous to resinous luster. Cleavage (001), good to fair; conchoidal fracture; brittle. H – 5–5$\frac{1}{2}$. G(nat) = 3.93–4.02, G(magnesian) = 3.75. Soluble in 1:1 HCl, leaving a silica gel. A high-Mg variety exists with up to 15.0 wt % MgO. The Mg-rich alleghanyite has $(Mn_{0.6}Mg_{0.4})$. Some ZnO (to $\approx$ 3.6 wt %) is also present. F to 3.8% for material from Sterling Hill, NJ. Near-end-member material is known from Bald Knob, NC. Biaxial (–); $X \wedge a$ = –35°, Z = b, $Y \wedge c$ = –35°; in the nonstandard olivine-type setting: $X \wedge b$ = –35°, Z = a, $Y \wedge c$ = –35°; N_x = 1.756–1.762, N_y = 1.780–1.782, N_z = 1.792–1.793; 2V = 70–90°; OAP perpendicular to (100); r > v, weak. X, brown; Y, light brown; Z, colorless. Occurs in Mn-rich skarns with other Mn-silicates and carbonates, and others. Also in Mn veins with tephroite and spessartine. Occurs at *Bald Knob, Alleghany Co., NC*; a high-Mg variety occurs at Sterling Hill, Ogdensburg, and Franklin, Sussex Co., NJ; Great Gossan, VA; Sunnyside mine, and Eureka Gulch, Silverton district, San Juan Co., CO; near San Jose, Santa Clara Co., and Germois prospect, Fiddletown, Amador Co., CA; Brown Mule claim, Mason Co., WA; Benallt mine, Rhiw, Carnarvonshire, Wales; Langtan and the Brattfors mine, Nordmark, Värmland, Sweden; northern end of the Inylchek Range, southeastern Kirgizia; several localities in Japan: Mukayama and Sono in Kyoto Pref.; Kasugi, Yamaguti Pref.; Tamagawa, Iwate. Pref., and others in Aichi, Nagono, and Tochigi Prefectures. EF *MIN 3(1):319(1972); MR 15:299(1984), 12:167(1981); AM 70:379(1985), 68:951(1983)*.

52.3.2b.2 Chondrodite $(Mg,Fe,Ti)_5(SiO_4)_2(F,OH,O)_2$

Named in 1817 by d'Ohsson from the Greek for *granule*, in allusion to its occurrence as isolated grains. Humite group. MON $P2_1/b$. (Nat): $a = 4.730$, $b = 10.255$, $c = 7.867$, $\alpha = 109.0°$; (nat titanian): $a = 4.727$, $b = 10.318$, $c = 7.905$, $\alpha = 109.33°$; (syn F end member): $a = 4.75$, $b = 10.27$, $c = 7.80$, $\alpha = 109.2°$; (syn OH end member): $a = 4.752$, $b = 10.350$, $c = 7.914$, $\alpha = 108.71°$; $Z = 2$; $D = 3.38$(titanian) and 3.18(calc F end member). *12-527*(nat): 4.84_3 3.56_3 3.02_4 2.76_3 2.51_4 2.29_3 2.26_{10} 1.74_7; *33-865*(titanian): 3.49_6 3.03_6 2.77_5 2.68_5 2.62_7 2.52_7 2.26_8 1.74_{10}; *14-10*(syn F end member): 4.85_4 3.55_4 3.01_5 2.67_5 2.61_6 2.27_{10} 2.25_{10} 1.74_9. Structure: *AM 55:1182 (1970), 63:535(1978)*. Ti is ordered into M(3). Crystals to 2 cm or more, granular, equant, well-formed crystals, many are modified; polysynthetic twinning on {001}; titanian material appears to show more twinning, also simple, lamellar twinning on {001}, rare twinning on {031} and {0$\bar{3}$5}. Yellow, orange, brownish red, reddish or greenish brown, rarely colorless, often translucent; white-gray, yellow streak; vitreous to greasy luster. Cleavage (001), poor to good; conchoidal to uneven fracture; brittle. $H = 6-6\frac{1}{2}$. $G = 3.1-3.23$ (Tilley Foster) and 3.16 (F syn). IR data: *JRNBS 65A:415(1961)*. Sometimes luminesces yellow or brown-orange, sometimes fluorescent in LWUV a yellow-white to yellow. Soluble in HCl and H_2SO_4. May contain appreciable Ti (to $\approx 9-9.6\%$ TiO_2) (e.g., Buell Park, AZ); Franklin and Sterling Hill chondrodite contains ZnO (to 11.5% ZnO) and MnO (to 36% MnO), and grades to alleghanyite. B_2O_3 to 0.71 wt %, M + 2 (OH,F) $\leftrightarrow$ Ti + 2(O); $(Si_{1.88}B_{0.2})_2$ is maximum B content. Usually has F/OH $\approx 1:1$ or $> 1:1$. FeO to 10.5%. Biaxial (+); OAP perpendicular to (100); $Z = a$; $X \wedge b = 22-31°$; regular range: $N_x = 1.592-1.617$, $N_y = 1.602-1.635$, $N_z = 1.621-1.646$; syn F end member: $N_x = 1.582$, $N_y = 1.594$, $N_z = 1.612$; Syn OH end members $N_x 1.630$, Ny 1.642, $N_z 1.658$; Tilley Foster mine: $N_x = 1.615$, $N_y = 1.627$, $N_z = 1.646$; titanian: $N_x = 1.714$, $N_y = 1.717$, $N_z = 1.750$; $X > Y \geq Z$; $r > v$ weak. X, medium yellow; Y, colorless; Z, colorless. Occurs in ultramafic rocks, serpentinite, kimberlite, ultramafic contact-metamorphic rocks, marbles. In contact silicate and carbonate rocks. Magnesian skarns. May be found with other members of the chondrodite–humite group. Occurs in many localities worldwide but only noteworthy localities are given here: Tilley Foster mine! Brewster, Putnam, and Amity, Orange Cos., NY; Franklin and Sterling Hill, and Sparta, Sussex Co., NJ; Ti-rich material from Buell Park kimberlites, AZ; Crestmore, Riverside Co., CA; Cardiff, Ont, Canada; Broadford, Skye, Scotland; Aker, Kareltorp, and Nordmarken, Sweden; Sorfinnset and Kristiansund, Norway; *Pargas*, Hangelby, and Sibbo, *Finland*; West Tauern, Austria; Monte Somma, Vesuvius, Naples, Italy; Wunsiedel and Passau, Germany; Zlotystok, Poland; Slyudanka, Irkutsk Oblast, Russia; southern Yakutia Fe deposits; Uzbekistan; Aldan shield, Siberia; Assam, India: gemmy red crystals to 10 mm occur in the Phalaborwa Complex, South Africa; nearly pure (57 wt % MgO) material from Kamioka, Japan. Occasionally used as a gemstone (e.g., material from the Tilley Foster mine). EF *MIN 3(1):304(1972); AM 64:1027(1979), 70:379(1985), 73:547(1988); DHZ 1A:377(1982)*.

52.3.2b.3 Reinhardbraunsite $Ca_5(SiO_4)_2(OH,F)_2$

Named in 1983 by Hamm and Hentschel for Reinhard Brauns (1861–1937), former professor of mineralogy, University of Bonn. Humite group. Known earlier as synthetic "calciochondrodite." MON $P2_1/b$. (Nat fluorian): $a = 5.052$, $b = 11.458$, $c = 8.840$, $\alpha = 108.91°$, $Z = 2$, $D = 2.885$; (syn OH end member): $a = 5.076$–5.077, $b = 11.448$–11.465, $c = 8.915$–8.921, $\alpha = 108.32$–$108.66°$, $Z = 2$, $D = 2.82$; (syn F end member): $a = 5.046$, $b = 11.492$, $c = 8.777$, $\alpha = 109.09°$, $D = 2.92$. *37-414*(nat fluorian): 3.81_4 3.32_5 3.04_8 2.94_5 2.90_8 1.90_{10} 1.89_5 1.66_5. *29-380*(syn OH): 3.83_7 3.35_9 3.04_{10} 2.95_6 2.92_9 2.91_8 2.78_9 1.91_7. Structure: *TMPM 31:137(1983) NJBM:119(1983)*. Anhedral to subhedral grains to 3 mm across, twinning parallel to (001). Light pink to light brown, translucent to transparent, also light blue; white streak; vitreous luster. $H = 5$–6. $G = 2.85$. Biaxial (–); $Z = a$, $X \wedge c = 18°$; $N_x = 1.606$, $N_y = 1.617$, $N_z = 1.620$; $2V = 44$–$50°$; $r > v$, distinct. DTA and IR data: *NJBM:119(1983)*. Soluble in HNO_3, gelatinizes with HCl. Occurs from scoriae of a volcano. Occurs in the *Ettringer Bellerberg volcano, Mayen, Laacher See, Eifel, Germany*; in a Ca-rich xenolith at Üdersdorf, Emmelberg, West Eifel. EF

52.3.2b.4 Ribbeite $Mn_5(SiO_4)_2(OH)_2$

Named in 1987 by Peacor et al. for Paul Hubert Ribbe (b.1935), professor of mineralogy, Virginia Polytechnic Institute, Blacksburg, VA. Humite group. Dimorph: alleghanyite. ORTH *Pbnm*. $a = 4.799$–4.811, $b = 10.732$–10.740, $c = 15.672$–15.70, $Z = 4$, $D = 3.84$–4.05. *AM 72:213(1987)*. *40-491*(nat): 2.92_7 2.87_8 2.82_7 2.69_6 2.55_8 2.51_4 2.36_4 1.80_{10}. Structurally related to leucophoenicite in that it contains serrated, edge-sharing chains of Mn octahedra and two distinct types of Si tetrahedral sites: (1) a single, isolated, fully occupied Si tetrahedron, and (2) a pair of edge-sharing half-occupied Si tetrahedra, which are related by an inversion center. Ribbeite is the unit cell twinned dimorph of alleghanyite; the disordered edge-sharing tetrahedra in ribbeite are a direct consequence of mixed unit cell glide operations in alleghanyite: statistically, one-half of the glides are alternate, half are en echelon. *AM 78:190(1993)*. 0.5 mm in diameter, anhedral grains. Pink, light pink streak, vitreous luster. Cleavage not observed. $H = 5$. $G = 3.90$. Analysis (wt %): $SiO_2, 24.3$; $FeO, 0.3$; $MgO, 5.2$; $MnO, 65.1$; $CaO, 0.2$; $H_2O, 4.9$. Biaxial

(+); XYZ = bac; $N_x = 1.780$, $N_y = 1.792$, $N_z = 1.808$; 2V = 81°, Na(D); Z > X = Y; r > v, moderate. X,Y, colorless; Z, light pink. Occurs at the *Kombat mine, 49 km south of Tsumet, Namibia*; from a metamorphosed Mn deposit; Cu and Pb sulfides were then deposited in essentially unmetamorphosed dolomite. Lenses of Fe and Mn oxides are present, including ribbeite. Lenses of ribbeite are up to 5 cm thick. Alleghanyite is also present. EF

52.3.2c.1 Humite $(Mg,Fe)_7(SiO_4)_3(F,OH)_2$

Named in 1813 by Bournon for Abraham Hume (1749-1838), English connoisseur and collector of works of art, precious stones, and minerals, and president of the Geological Society of London. Humite group. ORTH $Pbnm$. (Syn OH end member): $a = 4.735$, $b = 10.243$, $c = 20.72$; (ferroan): $a = 4.741$, $b = 10.258$, $c = 20.853$, also reported as: $a = 4.78$, $b = 10.22$, $c = 21.02$ (*AM*80 : 638(1995) Z = 4, D = 3.20(syn). *12-755*(syn F-member): 3.64_5 2.69_5 2.57_4 2.44_7 2.26_{10} 2.11_4 1.74_6 1.48_7. Structure: Ordering of Fe and Mg. Humite group minerals often intergrown in parallel growth. *AM 56:1155(1971)*. Granular, anhedral, well-formed crystals rare, crystals 2-10 mm, >20 forms observed, dominant: {001}, {010}, {100}, {110}, {120}; twinning present but rare. Yellow to dark orange, reddish orange; yellow to orange streak; vitreous luster. Cleavage {001}, poor. H = $6-6\frac{1}{2}$. G = 3.24(syn), 3.20-3.32(nat). IR data: *JRNBS 65A(5):415(1961)*. EPR and Mossbauer data: *CA 97(10):75846(1982)*. Soluble in HCl and H_2SO_4 with production of silica gel; gives up F in concentrated H_2SO_4. Manganoan humite from Franklin and Sterling Hill, NJ. Solid-solution exists between manganhumite and humite. Analysis: FeO to 7.8%, Fe_2O_3 to 2.7%, Al_2O_3 to 2.9%. Biaxial (+); XYZ = cab; OAP = {010}, for olivine setting; 2V = 65-84°; $N_x = 1.607-1.643$, $N_y = 1.619-1.655$, $N_z = 1.639-1.675$, F end member: 2V = 59°; syn F end member: $N_x = 1.598$, $N_y = 1.606$, $N_z = 1.630$; for syn OH end member: $N_x 1.635$, Ny 1,640, $N_z 1.661$, $2V_2 = 56.1°$; absorption X > Z > Y; r > v, weak. X,Z, pale yellow-golden yellow; Y, colorless-pale yellow. Occurs in metamorphosed and metasomatized impure carbonate sediments adjacent to acid and less frequently alkaline plutonic rocks. Occurs at a number of worldwide localities but only selected and noteworthy localities are given here: Tilley Foster Iron Mine, Brewster, Putnam Co, NY; Franklin and Sterling Hill, Sussex Co., NJ; Skye, Scotland; South Harris, Outer Hebrides, Scotland; *Monte Somma, Mt. Vesuvius, Naples, Italy*; Sillböle, and Hermala and Lohja, Finland; Nordmarken, Norberg area, and The Ladu Mine, Persberg, Värmland, Sweden; Sjoffin-Syor Finset, Glomfiord, Norway; Wallis, Switzerland; Yeno-Kovdor deposit, Kola Penin., and Lesnoi-Barak pit, Kyaranta, Karelia, Russia, in marble with spinel at Anzahamarono, Malagasy Republic (Madagascar). EF *MIN 3(1):309(1972); AM 64:1027(1979), 70:379(1985); MSASC 5:231(1980); DHZ 1A:376(1982); PCM 8:55(1982).*

52.3.2c.2 Leucophoenicite $(Mn, Ca, Mg, Zn)_7(SiO4)_3(OH)_2$

Named in 1899 by Penfield and Warren from the Greek for *pale* and *purple-red*, in allusion to its color. Humite group. Dimorph of manganhumite. MON $P2_1/b$. $a = 4.826\text{-}4.828$, $b = 10.842\text{-}10.85$, $c = 11.324\text{-}11.380$, 10.842-10.85, 11.324-11.380 $\alpha = 103.73\text{-}103.93°$ $Z = 2$, $D = 4.01$. *AM 69:546(1984)*: 3.64_4 3.28_3 2.89_{10} 2.72_5 2.69_9 2.63_3 2.45_5 1.81_{10}; *22-1168*: 4.36_5 3.61_5 2.88_9 2.68_8 2.62_4 2.44_4 2.36_5 1.81_{10}. Structure: closely related to risseire. see risseire for details. *AM 55:1146(1970)*. Granular crystalline masses, crystals rare, slender, prismatic, striated, elongated on a, also lath-shaped and flattened short crystals; twinning = polysynthetic on {001}. Brown, violet-red, light pink to deep pink, red-brown; white, pale pink streak; vitreous luster. Cleavage {001}, poor. H = $5\frac{1}{2}$-6. G = 3.85. Gelatinizes with acids. Analysis (wt %): FeO to 0.3%; MgO, 2.7-5.5; CaO, 2.4-2.8; ZnO, 3%; ZnO to 5.4%. *USGSPP:180(1935)*. Biaxial (−); X perpendicular to cleavage; $N_x = 1.751\text{-}1.760$, $N_y = 1.771\text{-}1.778$, $N_z = 1.782\text{-}1.790$; $2V = 74°$; colorless or light to deep pink; r > v, weak. Occurs in Mn deposits (metamorphosed). In hydrothermal veins at Franklin, NJ. Occurs at the *Franklin mine, Franklin, Sussex Co., NJ*, with willemite and and vesuvianite; Påjsberg, Värmland, Sweden; Val Sesia-Val Tournanche area, Italian Alps, Italy; Avontuur and Leinster basins, and Wessels mine, Kalahari Mn field, Cape Prov., South Africa; Kombat mine, Namibia. EF *AC(B) 39:10(1983)*; *AM 68:1009(1983)*, *69:546(1984)*, *70:379(1985)*, *71:985(1986)*.

52.3.2c.3 Manganhumite $(Mn,Mg)_7(SiO_4)_3(OH)_2$

Named in 1978 by Moore for the composition: like humite, but with predominant Mn. ORTH *Pbnm*. $a = 4.815\text{-}4.822$, $b = 10.54\text{-}10.580$, $c = 21.448\text{-}21.45$, $Z = 4$, $D = 3.84$. *33-870*, *AM 63:874(1978)*; *29-866, MM 42:133(1978)*; *33-870*(nat): 3.74_4 3.39_9 2.82_6 2.77_5 2.64_7 2.51_8 1.78_{10}. Structure: magnesium-rich material was ordered with respect to distribution of Mg and Mn. *AM 63:874(1978)*. Anhedral, to 1 mm; Franklin material is massive, occurring in > 1-cm pieces; pale to deep brownish orange, medium brown; yellow-brown streak; subadamantine luster. Cleavage {010}, perfect. H = 4. G = 3.83. Dissolves easily in warm diluted HCl, leaving a silica gel. F absent in Swedish material, 1.2% F in Franklin material. Solid solution between Mg and Mn. Traces of CaO and FeO. Material from Bald Knob, NC, is nearly pure. Magnesian manganhumite with 14.2 wt % MgO: $(Mn_{0.68}Mg_{0.30})$. Biaxial (+); XYZ = *cba*; $N_x = 1.761$, $N_y = 1.772$, $N_z = 1.781$; magnesian manganhumite: $N_x = 1.707$, $N_y = 1.712$, $N_z = 1.732$; $2V = 84°$, 2V for magnesian manganhumite = 37°; nonpleochroic, very pale orange; OAP = (010), weak absorption, pale orange; r > v, perceptible. Occurs as a skarn mineral with katoptrite, manganostibite, magnussonite, galaxite, etc. At Franklin and Sterling Hill, is associated with franklinite, willemite, and so on. Occurs at Franklin and Sterling Hill, Sussex Co., NJ; near-end-member composition material occurs at Bald Knob, Sparta, Alleghany Co., NC; *Brattfors mine, Nordmark, and*

Långban, Värmland, Sweden. EF *MSASC 5:231(1980); MR 12:167(1981); PCM 8:167(1982); AM 68:951(1983), 70:379(1985).*

52.3.2d.1 Clinohumite $(Mg,Fe)_9(SiO_4)_4(F,OH)_2$

Named in 1876 by Des Cloizeaux for the crystal system, monoclinic, and its position in the humite group. Humite group. MON $P2_1/b$. $a = 4.740$–4.745, $b = 10.226$–10.283, $c = 13.582$–13.699, $\alpha = 100.9$–$101.00°$, $Z = 2$, $D = 3.38$ (Ti- and Fe-bearing) and 3.28 (normal). *33-867*(titanian clinohumite calculated pattern) and *14-9*(syn $Mg_9F_2SiO_4)_4$); (titanian nat): $3.71_{6.5}$ 2.77_9 $2.76_{4.5}$ 2.55_9 2.52_9 2.26_8 1.75_{10} 1.74_5; (syn F end member): 3.87_3 $3.70_{4.5}$ 3.49_3 2.77_6 2.54_6 2.51_6 2.26_{10} 1.74_8. Structure: Ti substitutes into M3 site. Ordered distribution of Ti. See *CM 12:39(1973)* and *AM 63:535(1978)*. Equant, sometimes elongated on a, usually highly modified, to 1 cm; twinning on $\{001\}$, often on $\{0\bar{1}1\}$, sometimes polysynthetically twinned on $\{0\bar{1}1\}$, rarely on $\{021\}$, $\{0\bar{1}3\}$ as well. Yellow, brown, white, orange; streak varies from colorless to yellow, yellow-orange, honey-brown, to reddish brown; vitreous luster. Cleavage (001), poor; uneven to subconchoidal fracture; brittle. $H = 6$. $G = 3.17$–3.35. IR data: *MIN 3(1):313(1972)*.Kukh-i-Lal material is very brightly fluorescent yellow orange in SW UV, and material from Crestmore, CA and Lake Albano, Rome, Italy is moderately fluorescent yellow orange to orange. Clinohumite from alkalic and ultramafic rocks may have as much as 6% TiO_2. There is an inverse relationship between F and Ti content. Exchange operation $TiO_2 \cdot Mg^- F^{2-}$ occurs. FeO to 11.6% in titanian clinohumite. May have minor Al, Ca, Cr, Ni, and Mn. May have BeO to 1.7%. Ti to 0.5 at % content. Total of FeO + Fe_2O_3 to 14%, MnO 1–15% max. May contain minor B: B_2O_3 to 0.8%. Biaxial (+); OAP perpendicular to (100); $X \wedge b = 7$–$15°$ in obtuse β; $Z = a$; for Ti-rich; $X \wedge b = 20°$; (Ti- and Fe-bearing): $N_x = 1.632$–1.707, $N_y = 1.643$–1.714, $N_z = 1.664$–1.734; (syn F end member): $N_x = 1.608$, $N_y = 1.618$, $N_z = 1.636$; (Syn OH and member): N_x 1.638, N_y 1.641, N_z 1.669; $2V = 73$–$88°$; for Ti-rich material, $2V = 52$–$62°$; for syn F end member, $2V = 59°$ (calc. $2V = 79°$); $N_x > N_y > N_z$; $r < v$, weak. X, yellow–golden yellow; Y, pale yellow; Z, colorless–pale yellow. Occurs in kimberlite, carbonatite; peridotites, metadunite, and serpentinites in orogenic belts such as the Alps, Apennines, and Ruby Range MT (Montana). Usually in metamorphosed and metasomatized limestones and dolomites and in skarns associated with ore deposits at contacts with acid and, less frequently, alkaline plutonic rocks. Also occurs in late-magmatic veins in Tertiary alkaline intrusions (e.g., Gardiner complex, Greenland). Widely distributed worldwide but only selected localities are given here: Tilley Foster iron mine, Brewster, Putnam Co., NY; titanian clinohumite in the Buell Park diatreme, AZ, and in the Moses Rock kimberlite, UT; Dillon, MT; Cotopaxi, CO; Twin Lakes, Fresno, Monterey, and Crestmore, Riverside Cos., CA; Cargill Lake, Ont., Canada; Ti-rich clinohumite from the Gardiner Plateau Complex, East Greenland; Broadford, and Camas Malag, Skye, Scotland; *Monte Somma, Mt. Vesuvius, Naples, Italy*; Nordmark and Talgruvan, Sweden; Hameenkyla, and Pargas,

Finland; Llanos de Juanar, Malaga, Spain; Ala Valley, Piedmont, Italy; Val Malenco, and Lake Albano, Rome, Italy; Laperwitzbach, Tyrol, Austria; Vuorijarvi, Karelia, Yeno–Kovdor deposits, Kola Penin., Akhmatovsk and Shishimsk mines, Chelyabinsk Oblast, southern Ural Mts., in the emerald mines of Yekaterinburg Oblast, central Ural Mts., Aldan Shield, Siberia, central Transbaikalia, Siberia, Russia; kimberlites in southern Yakutia (Ti-bearing); fine orange-brown crystals to 2 cm from Kukh-i-Lal, Pamirs, Tadjikistan; Mt. Tanzawa, Kanagawa Pref., Japan; titanian clinohumite rims on forsterite inclusions from Morro da Mina, Jacupiranga carbonatite, São Paulo, Brazil. EF *AM 64:1027(1979)*, *MSASC 5:231(1980)*, *PCM 8:55(1982)*, *DHZ 1A:377(1982)*.

52.3.2d.2 Jerrygibbsite $(Mn,Zn)_9(SiO_4)_4(OH)_2$

Named in 1984 by Dunn et al. for Gerald V. Gibbs (b.1929), Virginia Polytechnic Institute, Blacksburg, VA. Humite group. Dimorphous with sonolite. ORTH $Pbn2_1$. $a = 4.85$–4.875, $b = 10.70$–10.709, $c = 28.17$–28.18, $Z = 4$, $D = 4.07$. *38-352*, *AM 69:546(1984)*; (nat): 2.87_8 2.75_5 2.70_5 2.66_3 2.56_{10} 2.36_4 1.81_{10} 1.55_3. The structure is unit cell twinned sonolite by a $b/4$ glide plane. The structure has serrated chains of edge-sharing octahedra. *NJMM 9:410(1989)*. Massive, interlocking anhedral crystals to 0.5 mm × 2.0 mm. Medium violet-pink, light pink streak, vitreous luster on cleavage and fracture surfaces, cleavage parallel to {001}, imperfect. $H = 5\frac{1}{2}$. $G = 4.00$. No fluorescence or cathodoluminescence. Is F-free. Analysis (wt%): $SiO_2 \approx 27.0$; $FeO \approx 0.3$; $MgO = 1.1$–3.0; $CaO = 0.4$; $ZnO = 3.2$–5.3; $MnO = 61.8$–64.1; $N_2O \approx 2$. Biaxial $(-)$; ZXY = abc; $N_x = 1.772$, $N_y = 1.783$, $N_z = 1.789$; $2V = 72°$; nonpleochroic, light pink; $r > v$, moderate to strong. Occurs at the *Franklin mine, Sussex Co., NJ*, with zincite, willemite, tephroite, franklinite, and other minerals. Intimately intergrown with leucophoenicite and also sonolite also found at the Wessels Mine, Kalahari Manganese Field, South Africa. EF *AM 70:379(1985)*.

52.3.2d.3 Sonolite $Mn_9(SiO_4)_4(OH,F)_2$

Named in 1963 by Yoshinaga for the locality. Humite group. Mn analog of clinohumite. Dimorphous with jerrygibbsite. MON $P2_1/b$. $a = 4.849$–4.905, $b = 10.544$–10.680, $c = 14.022$–14.333, $\alpha = 100.3$–$100.97°$, $Z = 2$, $D = 4.09$; high Zn (17.6 wt % ZnO): $a = 4.79$, $b = 10.51$, $c = 13.94$, $\alpha = 100.83°$. Malocidelnikovskoye: 3.63_4 2.87_{10} 2.85_7 2.71_4 2.66_5 2.47_5 1.81_6 1.76_5 *22-728* (zincian): 2.82_7 2.80_5 2.59_3 2.56_4 2.45_4 2.40_4 2.30_5 1.77_{10}; *22-725*(nat): 3.62_4 2.87_{10} 2.70_5 2.65_7 2.61_7 2.50_4 2.46_5 1.74_{10}. Structure: Majority of Mg is concentrated in M3, the smallest octahedral site. Possibly ordered distribution of Mn, Mg, and Zn. *AM 70:379(1985)*, *NJMM 9:410(1989)*, *MM 58:333(1994)*. Euhedral aggregates, anhedral masses to 2 cm at Bald Knob, tablets or equant, masses to 5–6 cm in Kirgizia; twinning, simple to polysynthetic on {001} and {101}. Dark brown, pink, brownish pink, brownish red, brownish violet; pale reddish orange streak; striated parallel to length = [001], vitreous luster. No cleavage. H = $5\frac{1}{2}$. G = 3.77 (17.6% ZnO) and G = 3.87 (regular). IR: *ZVMO* *119:98(1990)*. Gelatinizes with acids. Three compositional groups from Franklin, NJ: (1) close to end member with minor Zn, Mg, Ca; ZnO to 4% (about 1% F is present); (2) ZnO ≈ 3%, MgO = 5–8%, Fe = 1%; and (3) MgO = 16–19%, ZnO ≈ 10–11%, MnO = 36–41%. F to 2%, FeO ≈ 1%. ZnO to 17.6% has been found at Sterling Hill. F not found in Japanese material. F increases with increasing Mg in NJ samples. Bald Knob material is nearly pure with only traces of Fe (0.83% FeO) and Mg (0.45% MgO) present. Japanese material: to 3.6% Al_2O_3, 4.97% MgO, 1.0% CaO, 2% H_2O, 0.3% F. Discontinuous solid solution with clinohumite. *AM 70:379(1985)*. Hokkejino mine sample has appreciable Ca: $(Mn_{7.6} \; M_{10.9} \; Ca_{0.4}Fe_{0.1})_{9.0} \; (SiO_4)_4(OH)_2$. Biaxial (−); X ∧ c = 9°; Z = a; N_x = 1.695 (zincian)–1.766, N_y = 1.716(zincian)–1.783, N_z = 1.725 (zincian)–1.794; 2V = 70–82°; nonpleochroic, colorless; r > v, weak. Occurs in Mn-bearing skarn deposits, associated with other Mn minerals, such as rhodonite, spessartine, rhodochrosite, tephroite, braunite, jacobsite, and so on. Also in Mn-cherts, (e.g., Alps). Occurs at Sterling Hill, Ogdensburg, and at Franklin, Sussex Co., NJ (some is high in Mg and Zn); Bald Knob, Sparta, Alleghany Co., NC; Pennsylvania mine, San Antonio Valley, Santa Clara Co, CA; Långban and Brattfors, and Harsrigen in the Påjsbeerg ore field, Värmland, Sweden; Haute-Maurienne, Savoie, France; with rhodochrosite, rhodonite, and other Mn minerals from the Malocidelnikovskoye deposit, Yekaterinburg region, Ural Mts., Russia; northern end of the Inylchek Range, SE Kirgizia; Suao, Taiwan; at least 11 different localities in *Japan*: type locality is the *Sono mine, Kyoto Pref.*, Hokkejino mine, Kyoto Pref.; also Hanawa, Iwate Pref.; Takonomiya, Ibaraki Pref.; Kuratomi Mine, Oita Pref., Kaso, Totigi Pref.; Vagi, Kusugi, and Takamori in Yamaguti Pref. EF *MSASC 5:231(1980), PCM 8:167(1982), DHZ 1A:379(1982), ZVMO 122:78(1993)*.

CHLORITOID GROUP

The chloritoid group minerals are layer-structured alumino-orthosilicates having the general formula

$$R^{2+}R_2^{3+}O(TO_4)(OH)_2$$

where

R^{2+} = Fe, Mg, Zn
R^{3+} = Al, Fe, Ga
T = Si, Ge

The structure consists of a brucite-type layer with two nonequivalent octahedral sites (occupied by R^{2+} and Al plus vacancies) alternating with a corundum-like layer whose octahedral sites are occupied only by R^{3+}. The two octahedral layers are joined by T^{4+}, usually silicon in isolated tetrahedral groups, and by hydrogen bonds. The distribution of R^{3+} in the layer types may best be illustrated by writing the formula as $[(R^{2+})_2R^{3+}(OH)_4][Al_3O_2][TO_4]_2$.

Chloritoid Group

Triclinic and monoclinic polytypes are known for several of the species, and a centered cell is chosen for the triclinic type to maintain correspondence with the centered monoclinic cell. The monoclinic scheme can be thought of as equivalent to the triclinic cell, being twinned on {001} by a twofold screw axis parallel to [010].

Substitution of Fe^{3+} for aluminum is ordered preferentially in the larger of the octahedral sites. Complete solid solution between chloritoid and carboirite is likely. The two polytypes commonly coexist as mixed-layer phases, and they cannot generally be distinguished by optical properties because of similar orientation and frequent twinning.

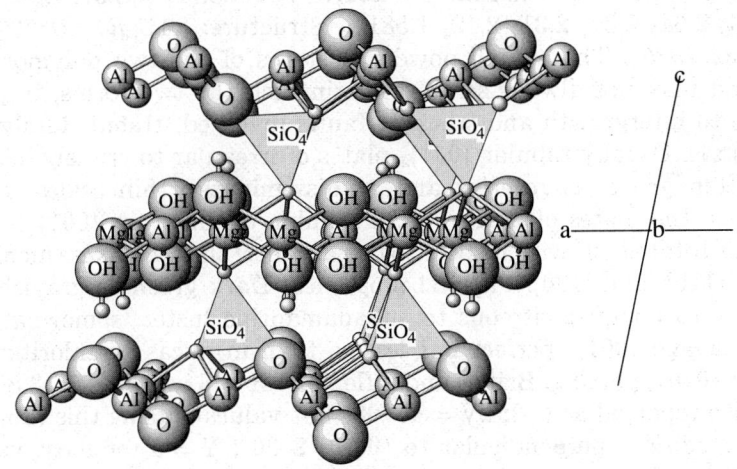

MAGNESIOCHLORITOID-M: $(Fe, Mg)_2Al_4Si_2O_{10}(OH)_4$

Perspective view along b, and parallel to the alternating, actahedral brucite and aluminum oxide layers joined by SiO_4 tetrahedra.

CHLORITOID GROUP

Mineral	Formula	Space Group	a	b	c	α	β	γ	
Chloritoid-A	$(Fe^{2+},Mg,Mn)(Al,Fe^{3+})_2 O[SiO_4](OH)_2$	$C\bar{1}$	9.46	5.50	9.15	97.05	101.56	90.10	52.3.3.1
Chloritoid-M	$(Fe^{2+},Mg,Mn)(Al,Fe^{3+})_2 O[SiO_4](OH)_2$	$C2/c$	9.482	5.484	18.182		101.74		52.3.3.1
Magnesiochloritoid-A	$(Mg,Fe^{2+},Mn)(Al,Fe^{3+})_2 O[SiO_4](OH)_2$	$C\bar{1}$	9.43	5.44	9.13	96.4	101.1	90.0	52.3.3.2
Magnesiochloritoid-M	$(Mg,Fe^{2+},Mn)(Al,Fe^{3+})_2 O[SiO_4](OH)_2$	$C2/c$	9.460	5.471	18.182		101.4		52.3.3.2
Ottrelite-M*	$(Mn^{2+},Fe^{2+},Mg)(Al,Fe^{3+})_2 O[SiO_4](OH)_2$	Cc or $C2/c$	9.505	5.484	18.214		101.77		52.3.3.3
Carboirite	$Fe^{2+}(Al,Ga)_2[GeO_4](OH)_2$	$C\bar{1}$	9.513	5.569	9.296	96.08	101.52	89.45	52.3.3.4

*Triclinic and monoclinic polymorphs are commonly intergrown. Data are given for monoclinic or intergrowths of the two.

52.3.3.1 Chloritoid-A,M $(Fe^{2+},Mg,Mn)(Al,Fe^{3+})_2O[SiO_4](OH)_2$

Named in 1837 by Rose in allusion to its superficial resemblance to minerals of the chlorite group. Chloritoid group. *Ottrelite*, a term now reserved for manganese-dominant chloritoids, was formerly applied to any manganoan chloritoid. The variety *sismondine* is a magnesian chloritoid. Polymorphs: The triclinic and monoclinic forms may not be polytypes in the strictest sense, there being an apparent shift in atomic positions in the successive layers; hypothetical 3T and 6H polytypes have been proposed. TRIC $C\bar{1}$. $a = 9.46$, $b = 5.50$, $c = 9.15$, $\alpha = 97.05°$, $\beta = 101.56°$, $\gamma = 90.10°$, $Z = 4$, $D = 3.61$. *14-344*: 4.45_{10} 3.25_6 2.97_8 2.77_4 2.70_7 2.46_9 2.40_5 2.14_6. MON $C2/c$. $a = 9.482$, $b = 5.484$, $c = 18.182$, $\beta = 101.74°$, $Z = 8$, $D = 3.57$. *14-62*: 4.47_{10} 3.08_4 2.96_9 2.64_5 2.37_7 2.31_7 2.12_4 1.581_8. Structure: *AC(B) 31:780(1975)*, *AM 65:534(1980)*. The overall powder patterns of the two polymorphs are similar and may not always serve to distinguish the two forms, in part, at least, due to intergrowth and stacking faults involved. **Habit:** Rarely in distinct crystals. Usually tabular {001}, plates of irregular to crudely hexagonal outline often group as rosettes and as disseminated thin scales. Foliated masses with the plates often curved. Twinning common on {001}, with the individuals rotated 60° with respect to one another; the twin axes (monoclinic) are [100], [110], and [130]. **Physical properties:** Dark green to grayish green, translucent to opaque; vitreous to subadamantine luster, somewhat pearly {001}. Cleavage {001}, perfect but less so than in micas or chlorite; {110}, moderate; {010}, parting. Brittle and inflexible. $H = 6\frac{1}{2}$. $G = 3.51-3.80$. Biaxial (+), also reported as (−); $2V = 45-68°$ but values outside this range have been reported; $Z \wedge$ perpendicular to $(001) = 2-30°$; $Y = b$ for most varieties, but sometimes $X = b$, and the orientation may vary within a given sample. $N_x = 1.705-1.730$, $N_y = 1.708-1.734$, $N_z = 1.712-1.740$; $r > v$, strong; anomalous interference colors. Strongly pleochroic with X, pale gray-green to green, Y, gray-blue to indigo, Z, colorless to pale yellow. The indices decrease significantly with substitution of Mg for Fe^{2+} (see magnesiochloritoid). Infusible, becoming magnetic; decomposed by sulfuric acid. **Chemistry:** Mn^{2+} readily replaces Fe^{2+} to form a solid solution with ottrelite; similarly substitution of Mg^{2+} yields a solid solution with magnesiochloritoid. Minor replacement of aluminum by Fe^{3+} in the brucite-like layers is common. Mn^{2+} has a distinct preference for the chloritoid structure over other associated phases. A solid solution probably exists with the isostructural, monoclinic carboirite in which Si is replaced by Ge, and Ga substitutes for part of the Al, but its existence in nature has not been established due to the low crustal abundance of Ge and the rarity of its minerals. **Occurrence:** Common in low to medium grade, regionally metamorphosed pelitic sediments, particularly those rich in Al and Fe^{3+}, where it is rarely free of inclusions of quartz and garnet, and is usually associated with muscovite, chlorite, garnet, and staurolite. It usually begins to form in the biotite zone, reaching its maximum development in the garnet zone and beginning to diminish in the staurolite zone. It is known from unsheared, metasomatized rocks. Occurs in nonstress environments such as pyrophyllite lenses in tuffs, rhyolites and dacites, quartz and quartz carbonate veins, hydrothermally altered lavas, and so on. The triclinic form appears to be the common one in hydrothermal, nonstress environments, and both are common in metamorphic environments. Increased stress and temperature appear to favor the monoclinic form. **Localities:** Widespread worldwide in the relatively common appropriate terrain. In the United States at Natick, Kent Co., RI; at the emery mines at Chester, Hampton Co., MA; Lancaster Co., PA; Deep River area, NC; Marquette Co., MI. In large embedded crystals at Vanlup, Shetland Islands, Scotland; Ottré, Ardennes, and at Salm Chateau (salmite), Belgium; at Morbihon and Ile de Croix, France; at St. Marcel, Piedmont, Italy (sismondine). At Güme Dagh, east of Ephesus, Turkey. Originally (both forms) at *Kosoibrod, near Yekateriburg (Sverdlovsk), Ural Mts., Russia*. Near Hualien, Taiwan. **Uses:** Of no commercial value; used in the interpretation of metamorphic petrology. AR *JP 2:49(1961)*, *DHZII 1A:867(1982)*.

52.3.3.2 Magnesiochloritoid-A,M $(Mg,Fe^{2+},Mn)(Al,Fe^{3+})_2$ $O[SiO_4](OH)_2$

Named in 1983 by Chopin et al. in allusion to the composition and its relation to chloritoid. Chloritoid group. The term *sismondine*, named for Professor Sismonda of Turin, was used for magnesian chloritoid. Triclinic and monoclinic polymorphs are commonly intergrown; data are for monoclinic, magnesiochloritoid-M, or intergrowths of the two. TRIC $C\bar{1}$. (Syn): $a = 9.43$, $b = 5.44$, $c = 9.13$, $\alpha = 96.4°$, $\beta = 101.1°$, $\gamma = 90.0°$; increasingly dominant and more stable with increasing temperature. MON $C2/c$. $a = 9.460$, $b = 5.471$, $c = 18.182$, $\beta = 101.4°$, $Z = 8$, $D = 3.332$. PD: 4.46_{10} 2.96_2 2.46_1 2.36_1 2.31_1 1.574_1 1.482_1 1.362_1. The two forms related as in chloritoid. Structure: *PCM 18:483(1992)*. Dark blue, vitreous luster, translucent. Cleavage {001}, perfect. Brittle and inflexible. $H = 6$–7. $G = 3.25$. Biaxial (+); $2V = 40$–$50°$; $N_x = 1.687$, $N_y = 1.690$, $N_z = 1.702$. Pleochroic, $X > Y > Z$; light blue to pale yellow. Found in rocks of the *Monte Rosa massif, Ghiacciaio di Verra, Mezzalama refuge, Italy*, and in the Val d'Ayas and the Palaborna mine, Val D'Aosta. From Zermatt, Valais, Switzerland. AR *BM 106:715(1983)*, *AM 73:358(1988)*, *EJM 4:67(1992)*.

52.3.3.3 Ottrelite-A,M $(Mn^{2+},Fe^{2+},Mg)(Al,Fe^{3+})_2O[SiO_4](OH)_2$

Named in 1819 by Dethier for the type locality. Chloritoid group. The original material was probably a manganoan chloritoid; subsequent to the later naming of chloritoid, the term was used for manganoan chloritoid, and is now reserved for the Mn > Fe species. Triclinic and monoclinic polymorphs are commonly intergrown; data are for monoclinic, ottrelite-M, or intergrowths of the two. MON Cc or $C2/c$. $a = 9.505$, $b = 5.484$, $c = 18.214$, $\beta = 101.77°$, $Z = 8$, $D = 3.49$.   *39-397*: 8.91_1 4.46_{10} 2.97_8 2.69_4 2.46_2 2.44_4 2.38_2 1.585_4. The two polymorphs are undoubtedly related as in chloritoid. Structure: *BM 101:548(1978)*. Green to yellow-green, vitreous to subadamantine luster, translucent to opaque. Cleavage {001}, perfect; brittle and inflexible. H = 6–7. G = 3.56. Biaxial (+); 2V = 60–70°; $N_x = 1.709$, $N_y = 1.712$, $N_z = 1.716$, but these values clearly must be for a high-Mg phase; X = b; Z perpendicular to (001) = 10°; r > v, strong; anomalous interference colors; pleochroic, X > Y > Z; olive-yellow to colorless. Infusible; soluble in sulfuric acid. Occurrence is as for chloritoid—in low- to medium-grade regional metamorphism. Found as scales and plates commonly displaying polysynthetic twinning on {001}, in metamorphic rocks of the Vann-Stavelot massif at *Ottré, Ardennes, Belgium*. The majority of other reported occurrences are chronologically early and probably represent manganoan chloritoid. AR *D6 642(1892)*, MA *30:285(1979)*.

52.3.3.4 Carboirite $Fe^{2+}(Al,Ga)_2O[(Ge,Si)O_4](OH)_2$

Named in 1983 by Johan et al. for the type locality. Chloritoid group. TRIC $C\bar{1}$ a = 9.513, b = 5.569, c = 9.296, α = 96.08°, β = 101.52°, γ = 89.45°, Z = 4, D = 3.95 (for Ge/Si = 3:1). *35-586*: 4.79_5 4.53_{10} 3.02_8 2.73_9 2.48_9 1.854_6 1.608_7 1.384_6. Pseudohexagonal plates {001} with {110} and {010}. Green; vitreous luster. Cleavage {001}, perfect. H ≈ 6. Biaxial (+); 2V = 60–75°; $N_x = 1.731$, $N_y = 1.735$, $N_z = 1.740$; Y perpendicular to (010), Z ∧ perpendicular to (001) = 7°; r > v, strong. Strongly pleochroic. X, blue-green, Y, light blue, Z, colorless-yellow. Contains up to 30% GeO_2 and forms solid solution with chloritoid; significant gallium and zinc may substitute for Al and Fe, respectively. May be replaced, in part, by a germanium analog of lepidomelane. Found in sphalerite associated with Ge-bearing quartz in the zinc deposits at *Carboire, Ariège, central Pyrenees, France*. AR *TMPM 31:97(1983)*, MA *35:320(1984)*.

52.4.1.1 Örebroite $Mn_6(Fe^{3+},Sb^{+5},As)_2Si_2(O,OH)_{14}$

Named by Dunn et al. in 1986 for the locality. See welinite (52.4.1.2) and franciscanite (52.4.1.4). HEX-R *P3*. a = 8.183, c = 4.756, Z = 1, D = 4.77; *40-486*(nat): 4.08_3 3.94_2 3.10_{10} 2.84_5 2.68_4 2.33_7 1.78_9 1.55_5. Dark brown, reddish-brown streak, vitreous luster. No cleavage, irregular fracture. H ≈ 4. Analysis (wt %): SiO_2, 15.2; MnO, 54.6; MgO, 0.8; Fe_2O_3, 9.2; Al_2O_3, 0.3; Sb_2O_5, 15.8; As_2O_5, 1.6; H_2O, 2.8. Uniaxial (+); N_O = 1.85–1.857, N_E = 1.87–1.875; nonpleochroic. O,E, nearly opaque. Occurs in association with calcite in Precambrian metamorphic rocks at the *Sjö mine, Grythyttan, Örebro, Sweden*. EF *AM 71:1522(1986)*.

52.4.1.2 Welinite $Mn_6^{2+}(W^{6+},Mg)_2SiO_2(O,OH)_{14}$

Named by Moore in 1967 for Eric Welin (b.1923), mineralogist and geochronologist. HEX-R *P3*. a = 8.155, c = 4.785, Z = 2, D = 4.41. *20-1389*(nat): 7.0_5 4.07_6 3.95_3 3.10_8 2.84_4 2.33_8 1.78_{10} 1.54_6. Structure: *AM 53:1064(1968)*; AKMG *4:407(1967)*, *4:459(1968)*. Up to 2-cm crystal sections. Reddish black to deep red-brown, red-brown streak, resinous luster. Poor to distinct {00.1} cleavage, brittle. H = 4. G = 4.47. Analysis (wt %): MnO,

55.0; SiO_2, 15.7; MgO, 2.5; Fe_2O_3, 0.8; WO_3, 21.7; Sb_2O_5, 1.6; H_2O, 2.9; $\Sigma = 100.7$. Uniaxial (+); $N_O = 1.864$, $N_E = 1.880$; nonpleochroic; some grains show a small 2V while others are almost isotropic. Occurs in manganese ore associated with calcite, barite, sarkinite, and adelite. In fissures cutting hausmannite ore. Occurs at *Långban, Värmland,* and the Sjö mine, Grythyttan, Sweden. EF *MIN 3(1):95(1972)*.

52.4.1.3   Kittatinnyite $Ca_2Mn_2^{3+}Mn^{2+}Si_2O_8(OH)_4 \cdot 9H_2O$

Named by Dunn *et al.* in 1983 from the Algonquin for *endless hills*, in allusion to the topography of the Franklin, NJ, area. See: wallkilldellite, the As analog (???). HEX $P6_3/mmc$, $P6_3mc$, or $P\bar{6}2c$. $a = 6.498$, $c = 22.78$, $Z = 2$, $D = 2.62$. *35-607* (nat): 11.2_{10} 5.61_6 5.03_3 3.80_3 2.82_5 2.73_6 2.52_4 2.28_4. Thin plates; dominant forms are $\{0001\}$ and $\{10\bar{1}\gamma\}$, 0.2 mm in diameter, crystals 2.0 µm thick. Bright yellow, light yellow streak, vitreous luster. Cleavage $\{0001\}$, perfect; brittle. $H \approx 4$. $G = 2.61$. Nonluminescent. Uniaxial $(-)$; $N_O = 1.727$, not able to measure N_E; $E \geq O$. O, medium yellow; E, moderate yellow. Associated with bostwickite and carbonates at the Franklin mine, Franklin, NJ. EF *AM 68:1029(1983)*.

52.4.1.4   Franciscanite $Mn_6^{2+}(V^{5+},\square)_2Si_2(O,OH)_7$

Named in 1986 by Dunn *et al.* for the Franciscan Complex of California where it was found. Compare welinite and örebroite. HEX $P3$, $a = 8.148$, $c = 4.804$, $Z = 1$, $D = 3.93$. PD 4.07_2 3.97_3 3.11_9 2.84_9 2.67_4 2.33_{10} 1.785_7 1.538_5. Cherry-red when fresh, darkening with time to a brownish red; streak brownish-red; transparent to translucent; luster vitreous; fracture uneven; $H \approx 4$, $G = 4.1$. Uniaxial (+); $N_O = 1.856$, $N_E = 1.882$, but fresh material has nominally lower indices of $N_O = 1.859(sic)$ and $N_E = 1.876$ suggesting that Mn^{2+} oxidizes on exposure to air; pleochroism strong with O wine red and E dark red to nearly black. Found as irregular, glassy grains up to 1.0 mm in diameter in sheared sonolite-hausmannite-gageite assemblage in chert of the Franciscan Complex, on the dump of the *Pennsylvania mine* (type material), *on the southwest side of the San Antonio Valley, Santa Clara Co., CA.* EF, RG *AM 71:1522(1986)*.

52.4.2.1   Mozartite $CaMn(SiO_4)(OH)$

Named in 1993 by Basso et al. for Wolfgang Amadeus Mozart, the mineral having been first observed in 1991, the 200th anniversary of the composer's death, and in recognition of the association of the mineralogical sciences with his music, especially the opera "The Magic Flute." Compare vuagnatite (52.4.2.2). ORTH $P2_12_12_1$. $a = 5.838$, $b = 7.224$, $c = 8.690$, $Z = 4$, $D = 3.68$. PD: 5.56_9 3.61_4 3.07_7 2.58_{10} 2.51_6 2.44_6 2.20_3 1.668_3 1.564_5. Isostructural with vuagnatite (52.4.2.2) and the arsenates and vanadates of the adelite group. Structure: *CM 31:331(1993)*. Reddish brown, deep red; red streak; vitreous luster; transparent. $G = 3.63$. Biaxial (+); $2V = 50°(52.8_{calc})$; $N_x = 1.840$, $N_y = 1.855$, $N_z = 1.920$; $Z = a$; length-slow. Strongly pleochroic; X, yellow-brown; Y, yellow; Z, orange-red. Water was determined by difference; minor aluminum is present and a trace of magnesium. Found as aggregates of minute anhedral crystals intergrown with pectolite and minor calcite, quartz, and hausmannite in veins cross-cutting massive beraunite at the *Cerchiara mine, near Faggiona, Val di Vara, La Spezia, eastern Liguria, Italy.* AR *TN 3:9(1991)*.

52.4.2.2 Vuagnatite $CaAl(SiO_4)(OH)$

Named in 1976 by Sarp et al. for Marc Bernard Vuagnat (b.1922) of the Department of Mineralogy, University of Geneva, who studied ophiolitic rocks. See mozartite (52.4.2.1). ORTH $P2_12_12_1$. $a = 7.055$, $b = 8.542$, $c = 5.683$, Z = 4, D = 3.42. *29-289:* 3.94_4 2.99_{10} 2.64_7 2.52_6. Isostructural with mozartite and the arsenates and vanadates of the adelite group. Structure: *AM 61:831(1976)*. Anhedral grains and, less commonly, prismatic [001] crystals displaying {110}, {001}, {100}, and {010} (Turkey); also dispenoidal forms, {111}, {$\bar{1}$11} with {011} and {021} (California). Colorless, white, pale blue, to pale brown; vitreous luster; transparent to translucent. Conchoidal fracture, brittle. H = 6. G = 3.20–3.42. Biaxial (−); 2V = 44–48°; N_x = 1.700–1.702, N_y = 1.724–1.726, N_z = 1.727–1.730; XYZ = *cba*; r < v, strong. Insoluble or only slightly soluble in hot aqua regia. Found in veinlets in sheared metagabbro, mainly hydrogrossular and chlorite with minor copper and chalcocite, in the Red Mountain area, Lake and Mendocino Cos., and in a cobble from the Russian R., Mendocino Co., CA. Guatemala. Found with chantalite, prehnite, hydrogrossular, chlorite, and calcite in rodingitic dikes of the ophiolite zone at *Doğanbaba, Burdur, Taurus Mts., in southwestern Turkey*. As pale blue lenses in pectolite veinlets in serpentinite at Shiraki, Toba, Mie Pref., Japan. AR *AM 61:825(1976), SMPM 57:149(1977), MA 78:3469(1978)*.

52.4.3.1 Titanite $(Ca,REE)(Ti,Al,Fe)SiO_4(O,OH,F)$

Named in 1795 by Klaproth in allusion to the dark brown or black color of the original specimen, and from the composition. Synonym: sphene. See vuagnatite, $CaAlSiO_4(OH)$ (52.4.2.2); malayaite, $CaSnSiO_5$ (52.4.3.2), and the tilasite group. MON $A2/a$ and $P2_1/a$. The vast majority of samples are $A2/a$. $A2/a$(syn): $a = 7.066$, $b = 8.705$, $c = 6.561$, β = 113.93°; $P2_1/a$(syn): $a = 7.069$, $b = 8.772$, $c = 6.566$, β = 113.86°. *AM 61:238(1976)*. Range: $a = 7.039$–7.088, $b = 8.643$–8.740, $c = 6.527$–6.584, β = 113.74–114.15°. *MSASC 5:137(1980)*. Z = 4, D = 3.52 (for pure $CaTiSiO_5$). *25-177*(syn $A2/a$): 4.95_1 3.24_{10} 3.00_3 2.61_3 2.59_3 2.27_1 2.06_1 1.745_1; *31-295*(natural yttrian titanite): 3.60_3 3.36_3 3.27_9 3.02_9 2.92_3 2.66_{10} 2.07_2 1.66_5. A reversible, displacive (second order) paraelectric–antiferroelectric phase transformation takes place upon heating from $P2_1/a$ to $A2/a$ at about 220°. *AM 61:435(1976), PCM 17:591(1991)*. High-Al titanites are all $A2/a$ (high = Al/ (Ti + Al + Fe) ≥ 0.25). At the unit cell level, titanites are $P2_1/a$, but formation of domains due to chemical substitution in the octahedral chains causes the "average" structure to be disordered to $A2/a$. Substitution of REE's for Ca stabilize the $A2/a$ dimouph as lower substitution levels than

52.4.3.1 Titanite

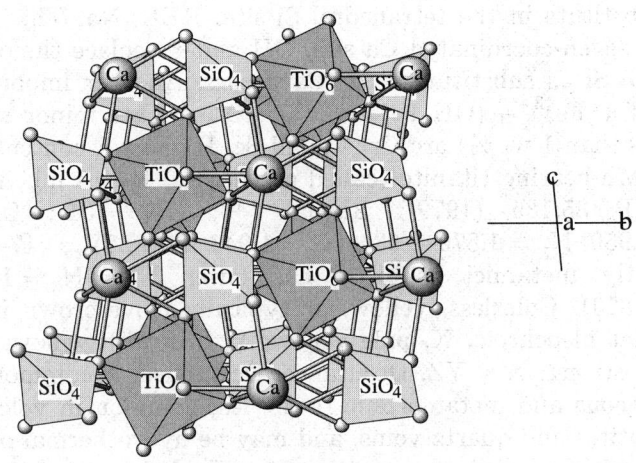

TITANITE(sphene): CaTiSiO$_5$

Like garnet, this is a framework of corner-sharing tetrahedra and octahedra (Ti) with Ca in the interstices. However, unlike garnet, some of the octahedra share corners with each other, forming chains in the a-axis direction. Ca and Si alternate in between these chains.

required for octahedral substitutions. Sometimes metamict. A natural $P2_1/a$ titanite with $Al_2O_3 = 0.07$ wt % has been reported from the Tyrol. *EJM 3:777(1991)*. Titanite may be variably metamict. *AM 76:370(1991)*. Generally wedge-shaped, flattened parallel to {001} or prismatic by elongation on {110}, sometimes fibrous. Summary of forms found given in *DF (1932)* and *MIN 3(1):336(1972)*. Simple contact and penetration twins have TP parallel to {001}, occasionally lamellar twinning on {122}. Physical Properties: Colorless, yellow, yellow-brown, green, brown, black, pink, red, blue; white to pale brown streak; vitreous to greasy luster. Cleavage {011}, imperfect to conchoidal fracture H = 5–5$\frac{1}{2}$. G = 3.48–3.60. IR data: *AM 76:370(1991)*, *DHZ 1A:443(1982)*. Raman data: *JCS(A):1372(1969)*. Thermal data: *AM 76:370(1991)*. Heating restores XRD patterns but not IR patterns. Slowly attacked by HF, warm concentrated H_2SO_4, or aqua regia. Chemistry: Can incorporate many elements other than Ca, Ti, Si, and O. An Sn content of more than 10 wt % has been reported (malayaite is the tin analog of titanite). An Sb-rich Titanite (with 12.6 wt % Sb_2O_5), equivalent to 0.165 Sb^{5+} per 1.00 Si^{4+} occurs with greenschist facies minerals in Italy (MM 59:717(1995)). High Al-containing titanites (formerly called grothite) occur in rocks without anorthite-rich plagioclase. Al substitution is favored by high P environments, but pressure is not the only factor. Al_2O_3 may be > 9.6 wt %, Ta_2O_5 to 16 wt %, and Nb_2O_5 to 6.5 wt %. Ferrous iron may substitute for Ca or Ti. May be enriched in both the LREE's and HREE's. *EJM 4:307(1992)*. REE-bearing titanite originally called keilhauite. Substitution mechanism is: $Ca^{2+} + Ti^{4+} \leftrightarrow REE^{3+} + (Fe^{3+}, Al)$. Small amounts of

Ti and Al substitute in the tetrahedral Si site. REE, Na, Th, U, etc. substitute in the seven-coordinated Ca site. OH and F replace the one O anion not bonded to Si as substitution for Ca occurs. Another important subst. scheme is $(Al + Fe)^{3+} + (OH, F)^- \leftrightarrow Ti^{4+} + O^{2-}$. Other minor substituents (generally less than 1 wt %) are Cr, V, and Sc. Highest F content is about 3 wt %. Pink, Mn-bearing titanite (called greenovite) has as much as 1 wt % MnO RSIMP 35:135 (1979). Biaxial (+); $Z \wedge c = 51°$; OAP (010); $N_x = 1.843–1.950$, $N_y = 1.870–2.034$, $N_z = 1.943–2.110$; $2V_z = 17–40°$. Metamict or partly metamict material has lower N's ($N_x = 1.795–1.800$, $N_z = 1.835–1.850$). Colorless, yellow, or typically clove-brown in thin section. Somewhat pleochroic. X, pale yellow; Y, brownish yellow; Z, orange-brown. r > v, strong; X < YZ. Occurance: Occurence: A common accessory mineral in igneous and metamorphic rocks; less common in volcanic rocks. Also in pegmatites and quartz veins, and may be hydrothermal or a detrital mineral. May alter to anatase or rutile + quartz. Other alteration products known. Localities: Noteworthy U.S. occurrences include honey-yellow crystals from Franklin, Sussex Co., NJ; as fine transparent, greenish crystals at the Tilley Foster mine, Brewster, Putnam Co., NY; as gemmy yellow-green to brown crystals(!) up to 2.5 in. long at Bridgewater Station, and upper Chichester Twp., Delaware Co., PA; black, high-Sn titanite from Kings Mt., NC; Magnet Cove, Garland Co., AR; Crestmore, Riverside Co., CA; large titanites to 15 cm × 15 cm (usually with apatite) from several localities in ONT, Canada (e.g., Wilberforce, haliburton Co., Eganville, Renfrew Co.; Otter Lake, Turner's Island in Lake Clear; Bear Lake Road, Elliott Mine, Tory Hill, Monmouth Twp.). Also in fine black twinned crystals at Litchfield, Pontiac Co., QUE; much gem material (to 4 × 6 × 1.5 in in pegmatite from Pino Solo; also from El Rodeo and La Huerta, BC Norte, Mexico; Bourg d'Oisans and Isère at Maronnne, France; in Norway in Aust-Agder at Arendal, and Risör, Kragerö. Nördmark, Värmland, Sweden, Laacher See, Paluenscher Grund, and Eifel, Germany; Grisons, Ticino in the St. Gotthard region at Val Sella, Val Maggia, in Valais at Zermatt, Kreuzlital in Val Tavetsch, and Binnental, Switzerland; at Schwarzenstein and Rotenkopf, Zillertal, Tyrol, and Sulzbachtal, Austria; in Italy at Pfunders and in the Pfitschtal, Trentino; in Piedmont in the Ala Valley; at St. Marcel (greenovite); Val Maggia and Passodi Vizze, and Monte Somma, Vesuvius. At Achmatovsk, and Saran Paul, Polarurals, Russia also from the Khibiny and Lovozero massits, Kola Pen. Gem-quality material from Sri Lanka; Campo do Boa, Capelinha, MG, Brazil. Other noteworthy localities are in Greenland; Finland; Dassu, Pakistan; Midongy, Ambalavaokely near Betroka and Naevatanana in Madagascar (Malagasy Republic); New Zealand. Uses: Has minor use as a gemstone. EF *DAN-ESS 186:125(1969)*, *MIN 3(1):336(1972)*, *AM 61:878(1976)*, *DHZ RFM 1A 443 (1982)*.

52.4.3.2 Malayaite CaSnSiO$_5$

Named by Alexander and Flinterin 1965 for the country of origin. MON $A2/a$. (Nat): $a = 7.149$, $b = 8.906$, $c = 6.667$, $\beta = 113.3°$. *AM 62:801 (1977)*. (Syn): $a = 7.173$, $b = 8.876$, $c = 6.688$, $\beta = 113.71°$, $Z = 4$, $D = 4.49$–4.56. 25-176(syn): 5.05_5 3.28_{10} 3.06_3 2.665_3 2.64_4 2.41_2 2.10_2 1.76_1. Has the structure of the high-temperature form of titanite because of substitution of Sn for Ti in the [MO$_6$] octahedra. *AM 62:801(1977)*. Wedge-shaped crystals to 3 cm. Also massive coatings. Colorless, pale yellow to orange-yellow, or tan-cream; white streak; vitreous luster. H = $3\frac{1}{2}$–4. G = 4.3–4.55. Both natural and synthetic material fluoresce a bright yellowish green in SWUV. Soluble in HCl and HF. Substitution of Ti (as much as 30 mol % titanite, about 10 wt % TiO$_2$) for Sn. Biaxial ($-$); $N_x = 1.764$–1.767, $N_y = 1.783$–1.785, $N_z = 1.798$–1.802; $2V_x = 85°$. Colorless to pale yellow. Associated with hydrothermal alteration of a cassiterite–quartz assemblage and associated with varlamoffite in wollastonite–diopside and other Sn-bearing skarns. Also occurs in granitic pegmatites. Found in the Himalaya and San Diego mines, Mesa Grande district, San Diego Co., CA; from the JC claims, near Dorsey Lake, Cassiar Mts., and Ti-bearing crystals to 3 cm from the Ash Mountains, Tuya Range, near McDame BC, Canada [*EG 73:269(1978)*]; essentially pure material occurs 4 km north of Ash Mountain itself; Redaven Valley, Meldon Alpe Rosso, Devonshire, England; Val Vigezzo, Novara, Piedmont, Italy; Kolyma R., and Kanonskoye deposit, Magadan oblast, Russia; *Sunsei Lah, Chenderiang, Perak, Malaysia*, and Kinta Valley, Malaysia; Pinyak mine, Thailand; Gumble, and Bourke, NSW, Australia; Toroku and Sampo mine Mirate mines, Miyazaki Pref., Okayama Pref., and several other localities, Japan [*DHZ 1A:443(1982)*]. EF *MM 35:622(1965)*, *MIN 3(1):351(1972)*, *DANS-ESS 212:126(1973)*, *JACES 59:455(1976)*, *MSASC 5:137(1980)*.

52.4.3.3 Vanadomalayaite CaVOSiO$_4$

Described and named in 1994 by Basso et al. for the composition. The mineral is also the V analog of titanite. MON $C2/c$. $a = 6.526$–6.532, $b = 8.691$–8.692, $c = 7.032$–7.039, $\beta = 113.88°$, $Z = 4$, $D = 3.61$. PD: 4.90_2 3.22_{10} 2.97_7 2.61_6 2.59_4 2.27_2 2.06_2 1.64_2. The structure is analogous to malayaite and titanite. It consists of zigzag chains of corner-sharing VO$_6$ octahedra running parallel to the [001] axis, cross-linked by isolated SiO$_4$ tetrahedra. The large cavities of the resulting framework are occupied by Ca atoms in irregular sevenfold coordination. *NJBM:489(1994)*. Subhedral crystals, prismatic, with good {110} cleavage, not exceeding 0.4 mm in size. No twinning observed. Deep red, red streak, vitreous luster. G = 3.60. For Na(D), $N_z = 2.105$ and $N_x \approx 1.95$. Dispersion is strong and the pleochroism varies from Z, deep greenish blue, to X $\approx$ Y, brownish red-orange. The mean chemical composition is Ca$_{1.00}$(V$_{0.71}$Ti$_{0.26}$Fe$_{0.01}$Al$_{0.01}$)$_{\Sigma 0.99}$OSi$_{1.01}$O$_4$. The mineral occurs with calcite, quartz, and haradaite in a small vein cutting ophiolitic metacherts at the *Gambatesa mine, Reppia, eastern Liguria, Italy*. EF *16th IMA Gen. Meet. Abst.: 31–32(1994)*.

52.4.4.1 Natisite $Na_2TiOSiO_4$

Named in 1975 by Men'shikov et al. for the components: natrium, titanium, and silicon. Dimorphous with paranatisite. TET $P4/mmm$. $a = 6.50$, $c = 5.07$, $Z = 2$, $D = 3.15$. *29-1279:* 5.05_8 3.96_5 2.71_{10} 2.52_5 2.35_6 1.689_7 1.611_5 1.100_6. Yellow-green to greenish gray, vitreous to adamantine luster. Cleavage {001}, perfect; {100}, less perfect. $H = 3$–4. Uniaxial $(-)$; $N_O = 1.756$, $N_E = 1.680$. Insoluble in dilute acids, readily fusible. Found as single grains and rosettes with chkalovite, aegirine, and vuonnemite in natrolite–ussingite veins cutting alkalic rocks at *Mt. Karnarsut, Lovozero massif, Kola Penin., Russia*. AR *ZVMO 104:314(1975)*, *AM 61:339(1976)*.

52.4.4.2 Paranatisite Na_2TiSiO_4

Named in 1992 by Khomyakov et al. for its relationship to natisite and the composition. ORTH $Pmma$. $a = 9.827$, $b = 9.167$, $c = 4.799$, $Z = 4$, $D = 3.07$. PD: 2.74_{10} 2.257_3 1.720_3 1.680_3 1.660_2 1.475_3 1.443_4. Yellow to orange-brown, vitreous to subadamantine luster, transparent to translucent. $H = 5$. $G = 3.12$. Biaxial $(+)$; $2V = 23°$; $N_x = 1.740$, $N_y = 1.741$, $N_z = 1.765$; $XYZ = bac$; $r > v$, weak; Pleochroic, $X = Y > Z$; brown to yellow. Found as interstitial grains (1 mm) and aggregates (5 mm) in pegmatites on *Mt. Yukspor, Khibiny massif, Kola Penin., Russia*. AR *ZVMO 121(6):133(1992)*, *AM 79:764(1994)*.

52.4.4.3 Yftisite-(Y) $(Y,Dy,Er)_4(Ti,Sn)O(SiO_4)_2(F,OH)_6$

Named in 1965 by Shipovalov and also used in 1977 by Balko and Bakakin for the chemical composition (Y,F,Ti,Si). Formula is not certain because of incomplete chemical analysis, and the mineral was discredited by the CNMMN IMA in 1987. This is probably a new mineral species, but a complete chemical analysis must be done. ORTH $Cmcm$. $a = 14.949$, $b = 10.626$, $c = 7.043$, $Z = 4$. *33-1462*(nat): 3.73_7 3.10_{10} 2.67_{10} 1.82_9 1.77_{10} 1.71_{10} 1.56_8 1.44_8. Infinite columns of Ti octahedra and Si tetrahedra parallel to c. Columns are linked by two seven- and eight-coordinated independent Y polyhedra. *ZSK 16:837(1975)*. 6–7 mm long. Yellowish, vitreous to greasy luster. No cleavage. $H = 3\frac{1}{2}$–4. $G = 3.96$. Biaxial $(-)$; $N_x = 1.690$, $N_y = 1.705$, $N_z = 1.711$; birefringence = 0.021, 2V large; $r > v$, distinct; anomalous blue interference colors. Occurs in silicified alkalic granite, near a contact with gabbro-anorthosite, *Kola Penin., Russia*; also occurs at an unspecified locality in *Kazakhstan*. EF

52.4.5.1 Törnebohmite-(Ce) $(Ce,La,Nd)_2Al(SiO_4)_2(OH)$

52.4.5.2 Törnebohmite-(La) $(La,Ce)_2Al(SiO_4)_2(OH)$

Named in 1921 by Geijer for Alfred Ellis Törnebohm (1839–1911), former director of the Geological Survey of Sweden. MON $P2_1/c$. $a = 7.383$, $b = 5.673$, $c = 16.937$, $\beta = 112.04°$, $Z = 4$, $D = 5.12$. 2.82_9 2.62_4 2.18_5 2.01_6

52.4.6.1 Cerite-(Ce)

1.783_5; 34-160(calc for Ce): 4.60_7 3.48_4 3.09_{10} 2.84_4 2.82_7 2.63_6 2.18_4 2.00_3; also 37-62[La(?)]. Structure: AM 67:$1021(1982)$. The structure determination is allegedly for Swedish type material in which lanthanum earth slightly exceed cerium earths. Isostructural with fornacite and related to the brackebuschite and pyrobelonite groups. 14-257(Sweden): 4.57_4 3.50_4 3.08_{10}. Massive or fine-grained. Bright olive-green, probably adamantine luster. Biaxial (+); 2V = 26°; $N_x = 1.808$, $N_y = 1.812$, $N_z = 1.838$; also $N_x = 1.845$, $N_y = 1.852$, $N_z = 1.878$; r < v, strong; pleochroic. X, Y > X = Z; pink or greenish yellow; Y, bluish green; Z, pink. Slowly dissolved by hot acids. Both Ce and La dominant phases occur intergrown at the type locality, and correlation of unit cell and optical data with composition is not clear. Törnebohmite (Ce) is found at the Jamestown district, Boulder Co., and at the Black Cloud pogmatite, Teuer Co., CO; at the *Bastnäs mine, Riddarhyttan, Västmanland, Sweden*; and in the Ural Mts., Russia. AR *DT 634(1932), AM 51:152(1966), USGSB 167:192,195(1984).*

52.4.6.1 Cerite-(Ce) $(Ce)_9(\Box,Ca)(Fe^{3+},Mg)(SiO_4)_6(SiO_3OH)(OH,F)_3$

Named in 1803 by Hisinger and Berzelius for the asteroid Ceres; discovered in 1801, two years before the mineral and the element cerium were named. The element Ce was discovered in this mineral. HEX-R $R3c$. $a = 10.75$–10.81, $c = 37.71$–38.29, Z = 6, D = 4.86. 11-126(nat): 3.47_4 $3.31_{3.5}$ 3.11_3 2.95_{10} 2.83_4 $2.80_{2.5}$ 2.69_4 1.95_5. Nearly isostructural with whitlockite, $Ca_9Mg(PO_4)_6$ (PO_3OH) $(\Box)_3$. AM 68:$996(1983)$. Usually fine-grained, pseudo-octahedral with $\{0001\} + \{10\bar{1}2\}$, plates, tabular on $\{0001\}$; 0.5- to 1-mm crystals at Mt. St.-Hilaire, 2–7 mm at Mt. Pass. Usually dark rose-red, also pale pink, pink, brown, yellow, green, colorless; streak greyish white; vitreous-adamantine (resinous) luster, dull when fine grained. Cleavage not apparent, uneven fracture, brittle. H = $5\frac{1}{2}$. G = 4.78–4.9. Nonfluorescent. Gelatinizes with acids. Analysis: CaO, 4%; REE_2O_3, 70%; Al_2O_3, < 1%; SiO_2, 20%; H_2O, 2–5%. Uniaxial (+) and anomalously biaxial (+); $N_O = 1.806$–1.814, $N_E = 1.808$–1.820; $N_x = 1.815$–1.817, $N_y = 1.815$–1.818, $N_z = 1.821$–1.820; 2V small, 11–25°; r < v, very strong. O, colorless; E, pale brownish red. Occurs in pegmatite at Mt. St. Hilaire, QUE, Canada. Bands in silicate zones (skarn) of quartz-banded hematite. Originally reported in silicate skarns at Jamestown, but is actually britholite-(Ce). Fenitized pegmatites related to a Ne-syenite (Tuva). May alter to bastnaesite, lanthanite, and others. Jamestown, Boulder Co., CO; in rare-earth-bearing hydrothermal quartz-bearing carbonate veins in shonkinite at the Mountain Pass REE deposit, San Bernardino Co., CA, particularly at the Birthday claims and in vein No. 6; Mt. St.-Hilaire, and Villeneuve mine, Papineau Co., QUE, Canada; *Bastnäs, Ridarhyttan, Västmanland,* and Östanmossa mine, Norberg, *Sweden;* in natrolite veins in the Lovozero massif, Kola Penin., Mochalin Log, Kushtym, Ural Mts., and Ulan-Erge massif, Sangilen, southwestern Tuva, Russia. EF *MIN 3(1):492(1972), ZVMO 118:47(1989), AM 43:460(1958).*

52.4.7.1 Afwillite $Ca_3(SiO_3)_2(OH)_2 \cdot 2H_2O$

Named by Parry and Wright in 1925 for Alpheus Fuller Williams (1874–1953) of Kimberley, South Africa, manager of DeBeers Consolidated Mines and an expert on diamonds, who found the mineral. MON Cc, $a = 16.278$, $b = 5.632$, $c = 13.236$, $\beta = 134.90°$, $Z = 4$, $D = 2.65$. *29-330:* 6.52_5, 5.06_3, 3.17_{10}, 2.83_{10}, 2.82_6 2.73_7, 2.15_3 1.94_3. Structure: *AC(B) 32:475(1976)*. Massive, tabular, flattened, vertically striated; crystals to 3–4 cm with masses to 11 cm, also in radial fibrous spherulites; no twinning observed; {001}, {100}, {110}, {101}, and so on. Using $a/b/c = 2.071:1:2.351$, $\beta = 98.83°$, crystals are elongated parallel to {010}, tabular parallel to {101} may be prismatic, elongated on b. White to colorless, vitreous luster, transparent to translucent. Cleavage {001}, perfect; {100}, good. $H = 4-4\frac{1}{2}$. $G = 2.62-2.63$. Pyroelectric and piezoelectric; gelatinizes in HCl or H_2SO_4. Biaxial (+); $Y = b$, $X \wedge c = 30-31°$ for $a/b/c$ of 2.071:1:2.351 and $\beta = 98.83°$; $N_x = 1.617-1.619$, $N_y = 1.620$, $N_z = 1.634$, colorless; $2V = 53-56°$; $r < v$; positive or negative elongation. Occurs in the contact zone of eruptive rocks with sedimentary rocks and in xenoliths of sedimentary rocks in volcanic rocks. In carbonate hornfels; also in stratabound Mn deposits; may also form from tilleyite. Occurs at Crestmore in fractures associated with quartz, thaumasite, merwinite, gehlenite, and calcite. At Viterbo, Italy, associated with tobermorite, calcite, and ettringite. A hydrothermal mineral. In a dolerite inclusion at Dutoitspan. May form from spurrite. Commercial quarry, Crestmore, San Bernardino Co., CA; Scawt Hill, County Antrim, Northern Ireland; Schellkopf, near Brenk, Eifel Dist., and as Zeilberg, Maroldsweisach, Bavaria, Germany, Camponorro, Montaldo di Castro (Viterbo), Italy; Anakit, Lower Tungvskar region, Siberia, Russia; *Dutoitspan diamond mine, Kimberley, South Africa*; clear, white, prismatic crystals to 4.5 cm were found in 1989 in the Wessels mine, near Kuruman, Kalahari manganese field; Cape Prov., South Africa; Nahal Ayalon, Israel; Fuka, Bitchu, Okayama Pref., Japan. Used as an additive to portland cement. EF *MIN 3(1):233(1972)*, *RI 4:97(1982)*, *MJJ 14:279(1989)*, *MR 22:279(1991)*.

52.4.7.2 Bultfonteinite $Ca_2[SiO_3(OH)]F \cdot H_2O$

Named in 1932 by Parry et al. for the type locality. TRIC $P1$. $a = 10.99$, $b = 8.18$, $c = 5.67$, $\alpha = 93.95°$, $\beta = 91.32°$, $\gamma = 89.85°$, $Z = 4$, $D = 2.74$. *8-223:* 8.16_7 3.51_5 3.46_5 2.92_{10} 2.89_{10} 2.83_4 2.78_4 1.930_8. The structural study suggests that the tetrahedral unit has the composition $[SiO_2(OH_{0.5})_2]^{3-}$. *AC 16:551(1963)*. Radiated acicular crystals; twinning on {010} and {100} common. Colorless to pink, vitreous luster, transparent. Cleavage {010}, {100}, distinct. $H = 4\frac{1}{2}$. $G = 2.73$. Biaxial (+); $2V = 70°$ and on (100) $= 47°$; $N_x = 1.587$, $N_y = 1.590$; $Z \wedge c$ on (010) $= 28°$; $r > v$. Gelatinized by acids. Found at the Crestmore quarry, Riverside Co., CA; in the H̲atrurim formation, Israel; at Fuka, near Bicchu, and the Mihara mine, Okayama Pref., Japan; with calcite, apophyllite, and natrolite at the *Bultfontein mine* and the Dutoitspan and Jagersfontein mines, *Kimberley, South Africa*; also in the

N'Chwaning and Wessels Mines, near Kuruman, Cape Prov. AR MM $23:145(1932)$, AM $40:900(1955)$.

52.4.7.3 Hatrurite Ca$_3$O[SiO$_4$]

Named in 1977 by Gross for the locality. Corresponds to the "alite" phase of Törnebohm found in portland cement. HEX-R (or pseudo-HEX) $R3m$. $a = 7.150$, $c = 25.56$, $Z = 9$, $D = 3.02$. $16\text{-}406$(syn): 3.07_9 3.01_3 2.84_7 2.79_{10} 2.65_{10} 2.22_5 1.789_5 1.513_2. Colorless, vitreous luster. Uniaxial $(-)$; $N_O = 1.723$, $N_E = 1.718$; length-fast; also biaxial $(-)$; 2V small; $N_x = 1.716$, $N_y = 1.723$, $N_z = 1.723$. Decomposed by H$_2$O. Found as very small; (50 μm) acicular [0001] crystals in larnite–brownmillerite–mayenite rock in the H̲atrurim formation, exposed in the H̲atrurim syncline, west of the Dead Sea, Israel, and elsewhere where that formation is exposed. AR $GSIB$ $70:35(1977)$, AM $63:425(1978)$.

52.4.7.4 Jasmundite Ca$_{11}$O$_2$(SiO$_4$)$_4$S

Named by Dent Glasser and Lee in 1981 for Karl Jasmund (1913–1991), director of the Mineralogical-Petrographical Institute, University of Cologne. TET $I\bar{4}m2$ $[AC(B)$ $37:803(1981)]$ or $I4/mmm$ $[SPC$ $34:40(1989)]$. $a = 10.452\text{–}10.461$, $c = 8.698\text{–}8.813$, $Z = 2$, $D = 3.01$. $33\text{-}296$(nat): 3.24_4 2.83_{10} 2.76_2 2.61_4 1.92_2 1.85_4 1.70_2 1.55_2. Two structure solutions are reported: SPC $34(1):40(1989)$, $AC(B)$ $37:803(1981)$. Mostly irregular grains to several millimeters; development of {101} gives a rhombohedral appearance; pseudoisometric habit, with dominant forms {110} and {101}; {100} and {001} are subordinate. Dark brown, greenish brown, brownish green; white streak; resinous luster. No cleavage, conchoidal fracture. H = 5. G = 3.03. Soluble in dilute HCl, with slight evolution of gas. Uniaxial $(-)$; $N_O = 1.750$, $N_E = 1.728$ [from $33\text{-}296$]; uniaxial $(+)$; Contains small amounts of AL, Fe and Mg. $N_O = 1.715$, $N_E = 1.728$ (from $NJMM:337(1983)$). Optics need to be redetermined. Occurs as a constituent of metamorphosed limestone inclusions in basalt. Associated with mayenite, brownmillerite, larnite, portlandite, and ettringite. Ettringer Feld, a lava flow of the Bellerberg volcano, Mayen, Eifel, Rheinland–Pfalz, Germany. EF

52.4.8.1 Trimounsite-(Y) Y$_2$Ti$_2$SiO$_9$

Named in 1990 by Piret et al. for the locality. MON $P2_1/c$. $a = 12.299$, $b = 11.120$, $c = 4.858$, $\beta = 95.62°$, $Z = 4$ $[EJM$ $2:725(1990)]$, $D = 4.85$. PD (nat): 3.44_9 2.82_{10} 2.78_4 2.36_4 2.12_5 1.96_6 1.64_6 1.63_5. A REE silico-titanate, a nesosilicate, Y$_2$Ti$_2$(O)$_5$SiO$_4$; structure not close to those of any other minerals. Y surrounded by seven oxygens, Ti-octahedra form (Ti$_2$O$_6$)$_n$ double chains. Prismatic to 5 mm long, elongate on [001] with {110}, {130}, {010}, {211}, {121}, {011}, striated on {110} and {130}; no twinning. Light brown, white streak, adamantine luster, translucent or slightly smoky and transparent. No cleavage, brittle. H = 7. G = 5.0. Nonfluorescent, strong bluish cathodoluminescence in electron beam. Insoluble in acids. Biaxial $(-)$; 2V not measurable, Na(D): all refractive indices > 2.10; r > v, strong. Occurs with allanite, bastnaesite, and other REE minerals in cavities with calcite and dolomite crystals in white dolomitic rock in a talc deposit at *Trimouns, 6 km N of Luzenac, Ariége, France*. EF

Britholite subgroup of the apatite group:

Recent structural studies *(ZK 191:249(1990), AMS 12:131(1992), ZK 206:233(1993))* of britholite and related minerals indicate that there are both hexagonal and monoclinic polymorphs of both britholite-(Ce) and britholite-(Y). In addition, F-dominant analogues of both monoclinic polymorphs have been identified. Uniaxial and biaxial optics reported for britholites can be explained by the existence of the two dimorphs. A systematic study on the nomenclatures and redefinitions of "hydroxylbritholite-(Ce)", "hydroxylbritholite-(Y)", "fluorbritholite-(Ce)", "fluorbritholite-(Y)", their monoclinic dimorphs, "lessingite" and "abukumalite" is necessary.

52.4.9.1 Britholite-(Ce) $(Ce,REE,Ca,Na)_5[(Si,P)O_4]_3(OH,F)$

Named in 1901 by Winther from the Greek for *weight*, in allusion to its relatively high specific gravity. See the apatite group. Synonyms: beckelite, lessingite, pravdite. Melanoccrite, $(Ce,Ca)_5(Si,B)_3O_{12}(ON, F) \cdot hH_2O$, may belong to the Aparite group. HEX $P6_3/m$ or $P6_3$. $a = 9.597$–9.638 (Sr-bearing), $c = 7.04$–7.08 (Sr-bearing), $Z = 2$, $D = 4.02$–4.72; "lessingite" (the monoclinic dimorph) is $P2_1$ and has $a = 9.580$–9.628, $b = 9.590$–9.630, $c = 6.980$–7.050, $\beta = 120.02$–$120.08°$. 17-724(heated at $900°$): 4.12_6 3.93_6 3.48_8 3.21_6 3.11_6 2.84_{10} 2.81_8 1.86_8. In hexagonal britholite, there is an ordered distribution of Ca and REE's in the A1 and A2 sites: (two Ca and two La) in A1 and six La in A2. A structural formula for ordered britholite-(Ce) is $(Ca,Ce)_2(Ce,Ca,Th,U,Sr)_3(SiO_4)_3$ (O,OH,F). In natural Monte Somma material *[EJM 1:723(1989)]*, Ca^{2+} is in A1 (nine-coordinated, all Ca) and A2 (seven-coordinated) is filled by 50% REE, 25% Th, and 25% Ca. The decrease in symmetry from $P6_3/m$ to $P6_3$ is accompanied by the introduction of additional atomic sites. The Ca1 position in $P6_3/m$ corresponds to two REE positions (REE1) and (REE2), both of which have site symmetry 3. The oxygen atoms are also not related by a mirror plane. *ZK 191:249(1990)*. The change to $P2_1$ results from the shifts of all the atoms from the pseudohexagonal equivalent positions. Synthetic $Na_{1.25}La_{8.55}Si_6O_{24}(F_{0.9}O_{1.1})$ *[NJBM:311(1992)]* is $P2_1$. *CSS 9:422(1981), ZK 206:233(1993)*. Pseudohexagonal or hexagonal prisms to 1 cm, tabular to anhedral grains, acicular needles. Elongated along c, and also tabular on {001}. Tan to dark brown, dark red, pale blue, yellow, greenish gray or brownish green; brown to very pale tan streak. Adamantine to vitreous luster. No cleavage, conchoidal to uneven fracture, brittle. H $\approx$ 5. Usually G is > 3.9 (metamict or partially metamict) to 4.8. Luminesces emerald green in Hg-vapor lamp, characteristic of REE content. Occasionally, fluorescent pale blue in SWUV. Gelatinizes in acids. IR and DTA data: *MS 26:306(1972)*. The chemistry is complex, and many substitutions occur. Often chemically zoned. "Alumo-britholite" has as much as 14.9 wt % Al_2O_3, ThO_2 to 20.7 wt %, SrO to 6.2 wt %, P_2O_5 to 13.5 wt % or more, CaO 4.68 to 19.9 wt %, Na_2O to $\approx$ 2 wt %. H_2O (total) to > 10 wt % (metamict varieties). Principal substitutions are $2Ca^{2+} \leftrightarrow Na^+ + REE^{3+}$ and $Ca^{2+} + P^{5+} \leftrightarrow REE^{3+} + Si^{4+}$. Uniaxial (−) and biaxial (−); XY = ca; N_O = 1.752–1.823, N_E = 1.748–1.817; 2V to 44°, Sr-containing britholite (Khibiny) is (+), 2V $\approx$ 0°; sometimes pleochroic. O, brown; E, colorless. Sr-containing britholite is very pale pink to colorless. Isotropic to anisotropic. May also be isotropic (i.e., metamict), with N as low as 1.72. Occurs in alkalic rocks and their cogenic pegmatites. Rarely in carbonatite complexes and lamproites. In sanidine-rich ejecta block (Monte Somma). Shonkin Sag, Montana; Rusty Gold prospect, Jamestown, Boulder Co., CO; Iron Hill, Powderhorn district, Gunnison Co., CO; Mt. St.-Hilaire (crystals to 1 cm) and Kipawa complex, and intermediate between apatite and britholite (metamict) from Oka, QUE, Canada; *Naujakasik, Ilimaussag, Greenland*; Tredaien, Norway; San Vito quarry, Monte Somma, Vesuvius, Italy; Cancarix, Albacete Prov., Spain. Kychtym, Tuva ("lessingite"), and Vishnevye Gor, Ural Mts., Khibiny massif, Kola Penin.; Kalchik R. and Priazov, Ukraine; northern Predgori Ganzurinsk Range, western Transbaikalia; Kisky massif, western Yenesei region; Kuznetsk Alatau, Berikulsk region, Russia. Borsuk, Kazakhstan; Pibnesburg complex, Rustenburg, Transvaal, South Africa; Nam Se, northern Vietnam; Madagascar. EF *VLA:297(1966), JSSC 26:383(1978), ZK 191:249(1990), SPC 36:19(1991), MM 57:203(1993)*.

52.4.9.2 Britholite-(Y) $Ca_2Y_3Si_3O_{12}(OH,F)$

Named in 1966 by Levinson as the Y-dominant analog of britholite-(Ce). Former name was abukumalite. See apatite group. HEX $P6_3$ or $P2_1$. *AMS 12:131(1992), ZK 206:233(1993)*. Hexagonal material: a = 9.39–9.44, c = 6.79–6.81, Z = 2, D = 4.05–4.54. Monoclinic material: a = 9.362 (F-dominant analog)–9.504, b = 9.364(F-dominant analog)–9.414, c = 6.731 (F-dominant analog)–6.922, γ = 119.71–120.03°. *31-315*(nat): $4.08_{2.5}$ 3.39_3 3.13_5 3.09_5 2.81_{10} 2.75_9 2.73_8 1.885_3. The split of the Ca1 site in apatite into two REE_{1a} and REE_{1b} sites with different occupancies of rare-earth atoms in the monoclinic britholite structure causes the lowering of symmetry from hexagonal to monoclinic. Rounded masses to 2.5 cm; also flattened pseudohexagonal crystals; often metamict or partly so. Black, dark reddish brown, light tan to chocolate-brown; pale brown streak; greasy to adamantine luster. Imperfect cleavage on {0001} and {10$\bar{1}$0}, uneven to splintery or conchoidal fracture, brittle. H = 5–6. G = 4.05–4.57. Soluble in cold dilute acids. Y-group REE predominate over the Ce group. Uniaxial (−) and biaxial (−); VLA:297(1966) N_O = 1.732–1.780, N_E = 1.728–1.783; 2V to 40°; light brown or yellowish to reddish brown. For an F-dominant sample from Fusamata: uniaxial (+); N_O = 1.773, N_E = 1.777. O, colorless–light brown; E, light brown. US6SB1627(1984) gives uniax (+), N_O1.728–1.780, N_C1.730–1.783. Found in alkalic rocks and their cogenic pegmatites. Chiefly in pegmatites with other

Y-bearing minerals and REE-bearing minerals, such as yttrialite, thorogummite, tengerite, or allanite. May alter to bastnaesite and churchite. Found in the Wet Mountains, Custer and Fremont Cos., CO; Powderhorn district, Gunnison Co., CO; Reiarsdal, Vest-Agder, southern Norway (crystals to 5–20 cm); also possibly at Tvedalen, Larvik, Norway; eastern Baltic Shield, Ukraine, and Kola Penin., Russia; *Suishoyama, Fusamata, Iisaka, Kawamata-machi, Fukushima Pref., Japan.* EF *MRE II:297(1996), NJBM:311(1992).*

52.4.9.3 Fluorellestadite $Ca_5[(Si,P,S)O_4]_3(F,OH,Cl)$

Named in 1982 by Rouse and Dunn for Reuben B. Ellestad (1900–1993), American analytical chemist of Minneapolis, MN. F-dominant. Name also used by Chesnokov et al. [*ZVMO 116:743(1987)*]; found in nature by them. See hydroxyellestadite (52.4.9.4) and chlorellestadite (52.4.9.5) see Apatite group. HEX $P6_3/m$. (Nat): $a = 9.53$–9.561, $c = 6.91$–6.920, $Z = 2$, $D = 3.05$–3.38; (syn): $a = 9.423$–9.442, $c = 6.895$–6.939, $D = 3.11$–3.14. *3-708*(nat): 3.46_3 2.86_{10} 2.80_4 2.76_6 2.66_3 2.30_3 1.97_6 1.86_6; *43-575, 45-0009*(syn): 3.45_3 2.81_{10} 2.78_4 2.71_6 2.63_3 2.26_3 1.95_3 1.85_5. 5–8 cm long and 2–3 mm wide. Pink, bright blue to medium blue-green; white to pale blue streak. $G = 3.07$. Sometimes fluorescent; some material from Crestmore is weakly fluorescent white to blue-white or yellow-white in SWUV and shows medium white–yellow–brown fluorescence in LWUV. *ZVMO 116:743(1987)*: SO_3, 20.75; SiO_2, 15.30; P_2O_3, 1.31; CO_2, 0.66; CaO, 55.0; MnO, 0.18; MgO, 1.38; Al_2O_3, 1.84; Fe_2O_3, 0.11; Na_2O, 0.33; K_2O, 0.1; H_2O^+, 0.30; F, 3.60. $\Sigma 100.86$-1.52=9934. Uniaxial ($-$); $N_O = 1.655$, $N_E = 1.650$. A skarn mineral: occurs associated with diopside, wollastonite, idocrase, monticellite, okenite, calcite, and other minerals at *Crestmore, Riverside Co., CA*; also occurs at Kopeysk in the *Chelyabinsk coal basin, southern Ural Mts., Russia* in burned fragments of petrified wood in coal dumps.. EF *AM 67:90(1982)*.

52.4.9.4 Hydroxylellestadite $Ca_5(SiO_4,SO_4)_3(OH,F,Cl)$

Named by Harada et al. in 1971 for the composition. See fluorellestadite (52.4.9.3) and chlorellestadite (52.4.9.5) see apatite group. HEX $P6_3/m$. $a = 9.491$, $c = 6.921$, $Z = 2$. Also MON $P2_1/m$. $a = 9.476$, $b = 9.508$, $c = 6.919$, $\gamma = 119.53°$, $D = 3.08$. *25-173*(nat): 3.46_4 2.84_{10} 2.80_5 2.74_6 2.66_5 1.96_2 1.85_5 1.48_2. From the study of one crystal, it is not known if all hydroxylellestadite is monoclinic or if both hexagonal and monoclinic structures exist. *AC(B) 36:1636(1980)*. Anhedral masses to 100 kg, cleavable. Pale purple, violet; vitreous luster; translucent. $H \approx 4\frac{1}{2}$. $G = 3.02$. IR, DTA, and TGA data: *AM 56:1507(1971)*. Rapidly soluble in cold, dilute HCl (1 N). Analysis (wt %), original ellestadite, Crestmore, CA: CaO, 55.2; MgO, 0.47; MnO, 0.01; CO_2, 0.61; P_2O_5, 3.06; SiO_2, 17.31; SO_3, 20.69; Al_2O_3, 0.13; Fe_2O_3, 0.22; F, 0.57; Cl, 1.64; H_2O^+, 0.53; H_2O^-, 0.10; $\Sigma = 100.52 - 0.61 = 99.91$. *AM 22:977(1937)*. Analysis (wt %): CaO, 54.51; CO_2, 1.65; SiO_2, 17.30; SO_3, 21.56; H_2O^+, 2.04; H_2O^-, 0.72; minor Mn, Sr, Na, K, P, Fe, F, and Cl present; MnO, 0.04; SrO, 0.28; Na$_2$O, 0.3; K$_2$O, 0.07; CO$_2$, 1.65; P$_2$O$_5$, 0.66; Fe$_2$O$_3$, 0.21; F, 0.28; Cl, 0.91. Uniaxial $(-)$; Na(D): Na; N$_O$ = 1.654, N$_E$ = 1.650. Occurs at the Crestmore Quarry, Riverside Co., CA. Occurs in pre-ore skarn, associated with diopside, wollastonite, xanthophyllite, vesuvianite, and calcite at the *Chichibu mine, Saitama Pref., Japan*. An unspecified ellestradite occurs in basalt xenoliths at Marords-Welsach, Northern Bavaria, Germany, and may be hydroxylellestadite based on the composition of assoc. minerals. EF *AM 56:1507(1971), DA 46:69(1995)*.

52.4.9.5 Chlorellestadite Ca$_5$(SiO$_4$,SO$_4$,PO$_4$)$_3$(Cl,OH,F)

Named in 1982 for the composition. The composition can also be rendered as Ca$_{10}$(SiO$_4$)$_{3-x}$(SO$_4$)$_{3-x}$(PO$_4$)$_{2x}$(Cl,OH,F)$_2$ (if $x = 0$: ellestadite, and if $x = 3$: apatite) *AM 67:90(1982)*. Compare: fluorellestadite (52.4.9.3) and hydroxylellestadite (52.4.9.4). See Apatite group. HEX $P6_3/m$. (Nat): $a = 9.510$, $c = 6.897$, and (syn): $a = 9.669$–9.688, $c = 6.849$–6.853, Z = 2, D = 3.20. *25-167*(nat): 3.45$_4$ 2.84$_{9.5}$ 2.80$_4$ 2.75$_{10}$ 2.65$_3$ 2.28$_4$ 1.96$_3$ 1.85$_3$ and *NJBM 7:295(1992)*(syn). Ca$_5$(SO$_4$)$_{1.5}$(SiO$_4$)$_{1.5}$Cl 3.43$_2$ 2.87$_{10}$ 2.80$_8$ 2.79$_8$ 2.32$_3$ 1.98$_3$ 1.85$_3$ 1.13$_2$. Structure: *AC(B) 36:1636(1980)* analogous to that of hydroxylellestadite. Finely granular, compact aggregates < 1 mm. Pink, pale rose or orange. G = 3.07. IR, DTA, and TGA data: *AM 56:1507(1971)*. Rapidly soluble in cold, dilute HCl (1N). Analysis (wt %): CaO 52.69, P$_2$O$_5$ 15.9, SiO$_2$ 11.99, SO$_3$ 13.47, Cl 1.44, H$_2$O$^+$ 0.36, H$_2$O$^-$ 0.29, minor Sr, Fe, F, SrO 0.86, Na$_2$O 0.06, K$_2$O 0.06, CO$_2$ 2.18, Fe$_2$O$_3$ 0.29, F 0.43, Σ = 99.57. Uniaxial $(-)$. N$_O$ = 1.652-1.655, N$_E$ = 1.649-1.650. Occurs in skarn associated with Vesuvianite, diopside, wollastonite, idocrase and pale blue calcite at the Chickibu Mine, *Takiue, Saitama Pref., Japan*. EF *AM 56:1507(1971), DA 46:69(1995)*.

52.4.9.6 Mattheddleite Pb$_5$(SiO$_4$, SO$_4$)$_3$Cl

Named by Livingstone *et al.* in 1987 for Matthew Foster Heddle (1828–1897), Scottish mineralogist. Apatite group. HEX $P6_3/m$. $a = 9.948$–10.01, $c = 7.464$–7.504, Z = 2, D = 6.86–6.96. *41-610*(nat): 4.31$_4$ 4.16$_4$ 3.44$_2$ 3.26$_3$ 3.00$_{10}$ 2.88$_4$ 1.88$_2$. Hexagonal prisms, to 3 mm, rosettiform aggregates. White (creamy), white streak, adamantine luster. *SJG 23:1(1987)*. Fluoresces dull yellow in SWUV. Analysis (wt %): PbO, 83.6; SiO$_2$, 7.65; SO$_3$, 6.00; Cl, 2.4. Uniaxial $(-)$; (Na light): N$_E$ = 1.999, N$_O$ = 2.017; colorless, length-fast. Occurs with quartz and secondary minerals, including caledonite, cerussite, pyromorphite, leadhillite, susannite, and macphersonite, in an area of hydrothermal Pb–Zn mineralization. *Leadhills Dod, Strathclyde region, Lanarkshire, Scotland*. Also do the Brae Fells, Red gill, and Roughton Gill Mines, Cardbeck Fells, Cumbria, and from a number of localities in Wales. EF *U.K. jour. of Mines and Minerals, 5:21(1988)*.

52.4.9.7 Karnasurtite-(Ce) (Ce,La,Th)(Ti,Nb)(Al,Fe)$(Si,P)_2O_7(OH)_4 \cdot 3H_2O$

Named in 1959 by Kuzmenko and Kozhanov for the locality. This species is considered a mineral by some sources but not by others. The mineral may be britholite-(Ce). The composition is very close to that of "hydrocerite" of Vlasov et al. Amorphous (metamict), possibly HEX $a = 10.6$, $c = 7.3$. PD (nat): no X-ray diffraction data; when heated to 900°, it gives a pattern similar to that of huttonite—$ThSiO_4$; (heated): 3.10_7 2.88_7 3.29_6 3.49_5 1.72_5. Grains to 1 cm diameter or aggregates of platy crystals to 10 cm × 6 cm × 2 cm. Honey yellow (fresh) to pale yellow (altered), yellowish streak, greasy luster. One good cleavage, brittle. H = 2. G = 2.89–2.95. Dehydration curve shows one-half of H_2O lost to 150°, the rest gradually to 650°. Four chemical analyses given in *AM 45:1133(1960)*; P_2O_5 to 6.8%, MgO to 3.4%, CaO to 3.2%. Uniaxial (−); $N_O = 1.617$, $N_E = 1.595$; nonpleochroic, pale yellow; parallel extinction; sometimes anomalously biaxial. Occurs in a zoned pegmatitic stock, in intermediate replacement zone of microcline. Associated with other alkalic minerals. Occurs in *Vein No. 2, Mt. Karnasurt, Lovozero massif, Kola Penin., Russia*. Also on Mt. Punkaruaiv, Lovozero massif; Red Wine alkalic complex, central LAB, Canada. EF *MIN 3(1):847(1972)*, *GSCB 294:1(1981)*, *AM 45:1133(1960)*.

52.4.9.8 Fluorbritholite-(Ce) $(REE,Ca)_5(Si,P)_3O_{12}F$ Britholite subgroup of the apatite group

Named by Gu et al. in 1994 for the composition. HEX. $P6_3/ma$ 9.537Å c 7.008Å z=2, D=4.66; 2.84_{10} 2.82_3 2.75_3 1.87_3 (Jour. of Wuhan Univ. of Technology, 9(3), 9-14; abstracted in *AM 81:1013(1996)*). Pale yellow, tan, or reddish brown aggregates and patches consisting of radiating to subparallel groups; crystals are prismatic (to 0.5 mm) to equant; colorless to pale brown streak, adamantine luster, opaque to translucent, brittle, even to conchoidal fracture, perfect {0001} cleavage, H=5, G=4.66. Uniax. (−), N_o 1.792 N_e 1.786; pale yellow, nonpleochroic. Three grains analyzed, one of which has Ce>Ca (atomic): Ce_2O_3 33.14, La_2O_3, 19.78, Pr_2O_3 1.20, Nd_2O_3 8.78, FeO 0.10, CaO 1.92, SrO 3.74, MnO 1.48, Na_2O 3.04, ThO_2 1.09, U_3O_8 0.31, SiO 16.59, Al_2O_3 0.08, P_2O_5 5.12, F 2.70, O for F 1.14, Σ99.20. Occurs in vugs in nepheline syenite, marble xenoliths, sodalite syenite xenoliths, and pegmatite dikes at Mont Saint-Hilaire, Quebec.

52.4.10.1 Ellenbergerite $(Mg,Ti,\square)_2Mg_6Al_6Si_6Si_2O_{38}H_{10}$ [*16th IMA Mtg.:74(1994)*]

Named by Chopin et al. in 1986 for François Ellenberger, Paris, for his geological work in the western Alps. A phosphorus-rich variety (16% P_2O_5) also was found. HEX $P6_3m$. $a = 12.25$, $c = 4.93$, Z = 1, D = 3.10–3.17. *40-507*(nat): 6.15_3 3.61_7 3.54_{10} 3.06_7 2.65_9 2.34_4 2.19_7 1.78_4. The structure consists of single chains of face-sharing octahedra partially occupied by Mg + Ti or Zr, and double chains consisting of couples (pairs) of face-sharing octahedra occupied by Mg + Al linked by edges along the c axis. Isolated SiO_4 tetrahedra interconnect the octahedral chains. EJM 5:819(1993). Anhedral, rarely as prisms to 10 mm long, with hexagonal cross section. Purple to lilac $Fe^{2+}-Ti^{4+}$ change transfer is responsible for color; white streak; vitreous luster; transparent. Easy fracture, brittle. $H = 6\frac{1}{2}$. $G = 3.15$–3.22. Nonfluorescent in UV, faint bluish Cl in electron beam. Analysis (wt%): SiO_2, 39.1; P_2O_5, 0.45; Al_2O_3, 25.1; TiO_2, 4.0; MgO, 22.2; FeO, 0.2; H_2O, 8.0. Original paper has range SiO_2, 32.5–38.6; P_2O_5, 0–8.3; Al_2O_3, 20.4–24.9; TiO_2, 0.5–4.1; ZrO_2, 0–3.1; MgO, 21.8–25.8; FeO, 0.15–0.43. P is incorporated by SiAl ↔ PMg. Formula also given as: [Mg, (Ti,Zr),□]$_2$ Mg$_6$(AL,Mg)$_6$(Si,P)$_8$ O$_{28}$(OH)$_{10}$. Uniaxial (−); at 589 nm: $N_O = 1.6533$–1.6789, $N_E = 1.6538$–1.6697; vividly pleochroic. O, colorless; E, colorless to deep lilac. Highest O values are for samples rich in Ti, and the color zoning is related to Ti for Zr substitution. Variations in E are more dependent on the P content. Occurs in low-temperature, high-pressure conditions in a phengite–quartzite layer of a massif in the western Alps. Occurs as inclusions in pyrope. Coesite associated. 25–30 kbar and 700–800°. *Dora-Maira massif, Parigi, near Martiniana Po, Italy.* EF *CMP 92:316(1986), BM 111:17(1988).*

52.4.11.1 Sitinakite Na$_2$KTi$_4$Si$_2$O$_{13}$(OH)·4H$_2$O

Named in 1992 by Menshikov *et al.* for the composition (Si-Ti-Na-K). TETR. $P4_2/mcm$ a 7.819-7.821Å c 12.021-12.099Å, Z=2 D=2.88; PD 7.84$_{10}$ 6.02$_{10}$ 5.04$_4$ 3.36$_5$ 3.25$_8$ 2.80$_5$ 2.61$_6$ 2.00$_7$. Structural formula: Na$_2$(H$_2$O)$_2$[Ti$_4$O$_5$(OH)(SIO$_4$)$_2$]·K(H$_2$O)$_2$. DAN SSSR 307:114(1989). Crystals with anom. optics may be indexed on Pcc2, and symmetry reduction may be related to high concentrations of Nb, or to filling of zeolite-like channels by K and H$_2$O. Colorless to pale rose or dirty white. Primary hydrothermal crystals are equant to short prismatic, 1 × 1 × 2 mm with some to 3-4 mm. pseudormorphs are as much as 0.6 × 3 × 4 cm. Euhedral, pseudo-cubic. Luster vitreous, streak white. No fourescence. Excellent {100} and fair {001} cleavage. H $4\frac{1}{2}$, G 2.86. IR, DTA and TGA data given in *ZVMO 121:94(1992)*.. No dissolution observed in water, HC1 or HNO$_3$ at 25°C. Alkali sites contain Na, K, Sr, Ba and Ca and the Ti site contains Ti, Nb and Fe. Nb$_2$O$_5$ 5.1wt. %. Uniax. (+) with N_o 1.780 N_o 1.988. Anom. Biax., 2V=10–11°. Occurs on *Mount Kukisvumchorr, Khibiny massif, Kola Peninsula, Russia*; in a hydrothermal vein and in two pegmatite veins in meltigite-urtite. Assoc. with natrolite, vinogradovite, aegirine and others. Forms with vinogradovite as pseudomorphs after an unknown mineral.

Class 53

Insular SiO₄ Groups and Other Anions or Complex Cations

53.1.1.1 Spurrite Ca₅(SiO₄)₂CO₃
Named by Wright in 1908 for Josiah Edward Spurr (1870–1950), American geologist. Synonyms: β-spurrite; α-spurrite is the high-temperature, orthorhombic form (>1200°). See the dimorph, paraspurrite (53.1.2.1). MON $P2_1/a$. $a = 10.49$, $b = 6.705$, $c = 14.15$, $\beta = 101.3°$, $Z = 4$, $D = 3.03$. 13-496(nat): 3.81_3 3.02_7 2.70_{10} 2.66_5 2.64_7 2.61_3 2.17_4 1.90_3. Structure: *AC* 13:434(1960). Granular masses, sphenidiscoidal masses to 15 cm; polysynthetic twinning on {001}, {101}, and {205}. White, pale blue, or yellow; white streak; vitreous luster. Cleavage {001}, distinct; {100}, poor. H = 5. G = 3.00–3.02. Thermal data: *DHZI:1B(1986)*. Green luminescence reported. *DHZI:1B(1986)*. Effervesces with dilute HCl, yielding a gelatinous silica residue. Quite pure. Traces of Al, Fe, Mg, Na, P. Biaxial (−); $X = b$, $Y \wedge a = 45°$, $Z \wedge c = 33°$; $N_x = 1.637–1.641$, $N_y = 1.672–1.676$, $N_z = 1.676–1.681$; colorless; 2V = 35–41°; r > v, weak, crossed. Occurs with hillebrandite in a contact-metamorphic limestone in Mexico, also with minerals such as larnite. In skarns. High-temperature contact metamorphism. Also in areas of spontaneous combustion of organic matter in bituminous siliceous limestones. Alteration: to scawtite. Occurs in the Little Belt Mountains, MT, and the Christmas Mountains, TX; Tres Hermanas Mountains, Luna Co., NM; Crestmore quarry, Riverside Co., CA; Scawt Hill, and Carneal, County Antrim, and Carlingford, Co. Louth, Northern Ireland; Camas Mòr, Isle of Muck, and at Camphouse, Ardamur-chan, Argyllshire, Scotland; Cornet Hill, Romania; Anakit area, lower Tunguska R. and Kochumdek R. area, Siberia, Russia; Güneyce–Ikizdere area, Trabzon Prov., Turkey; Hatrurim area, Israel; Magarin area, Jordan; Kushiro, and Hirata, Hiroshima Pref., Fuka, Bicchu, Okayama Pref., Japan; Terneras mine, *Velardeña, DUR*, Cerro Mazahua, Zitacuaro, MICH, and Maconi, QUER, *Mexico*; Golden Gully, Tokatoka dist., New Zealand. EF *MIN 3(1):358(1972), AC 13:434(1986)*.

53.1.2.1 Paraspurrite Ca₅(SiO₄)₂CO₃
Named in 1977 by Colville and Colville from the Greek *para*, besides, and for spurrite, a dimorphous pair. See the polymorph, spurrite (53.1.1.1). MON

$P2_1/a$. $a = 10.473$, $b = 6.706$, $c = 27.78$, $\beta = 90.58°$, $Z = 8$, $D = 3.03$. *29-307* (nat): 6.92_8 6.03_4 4.62_4 3.47_{10} 2.72_7 2.70_5 2.65_4 1.98_7. Structure is very close to that of spurrite (53.1.1.1). a and b remain essentially the same, but $c = 27.78$ and $\beta = 90.58°$. In effect, paraspurrite has $2 \times d_{001}$ of spurrite (53.1.1.1). Crystals average 0.5 cm across, maximum size 2 cm; polysynthetic twinning on (001). White, white streak. Poor cleavage parallel to (001). $G = 3.00$. Quite pure: traces of Al, Fe, Mg, P. Biaxial (−); $X = b$, $Y \wedge a = 30°$, $Z \wedge c = 30°$; $N_x = 1.650$, $N_y = 1.672$, $N_z = 1.677$; colorless; $2V = 47°$. Occurs in high-temperature contact-metamorphic rocks. Limestones. In roof pendant of siliceous carbonate rock in syenite. With gehlenite, vesuvianite, apatite, and larnite. *Darwin, Inyo Co., CA;* Cerro Mazahua, Zitacuaro, MICH, Mexico. EF *AM 62:1003(1977)*.

53.1.3.1 Iimoriite-(Y) $Y_2(SiO_4)(CO_3)$

Named in 1970 by Kato and Nagashima for Satoyasu Iimorii (1885–1982), Japanese analytical chemist and REE specialist, and his son Takeo Iimorii (died in World War II), who described many REE minerals. TRIC $P\bar{1}$. $a = 6.524$–6.573, $b = 6.537$–6.651, $c = 6.417$–6.454, $\alpha = 116.36$–$116.94°$, $\beta = 92.34$–$92.75°$, $\gamma = 95.51$–$95.92°$, $Z = 2$, $D = 4.91$. *35-640*(nat): $4.54_{2.5}$ $3.30_{2.5}$ 3.02_4 2.95_8 2.88_{10} $2.84_{3.5}$ 2.78_4 2.70_3. All oxygen atoms in iimoriite-(Y) are bonded in anionic complexes, either SiO_4 tetrahedra or CO_3 carbonate groups. The atomic arrangement of iimoriite consists of alternating $(01\bar{1})$ slabs of orthosilicate groups and carbonate groups, with no sharing of oxygen atoms between adjacent slabs. Subhedral to anhedral grains to 0.5 mm. Buff-tan, white–pale cream, yellow; white streak; vitreous luster. One good cleavage on {011}. $H = 5\frac{1}{2}$–6. $G = 4.47$; IR data: *AM 69:196(1984)*. Slightly soluble in cold HCl. Kola Peninsula sample is highest in Y_2O_3. Biaxial (−); $N_x = 1.753$–1.786, $N_y = 1.820$–1.827, $N_z = 1.827$–1.830; colorless; $2V = 31$–$37°$ (Alaska and Kola), $2V = 5$–$15°$ (Japan); OAP parallel to a cleavage. Occurs with fluorite, thalenite-(Y), keivyite-(Y), and vyuntspakhite-(Y) in amazonite–zinnwaldite pegmatite on the Kola Peninsula. In a thorite–uraninite-bearing quartz–albite vein, Bokan Mt. In quartz–microcline pegmatite, Fukushima Pref., Japan, and the nearby Suishoyama pegmatite, as an alteration product of thalenite-(Y). Occurs at Bokan Mt., Prince of Wales Island, AK; in the Ploskogorskaya amazonite pegmatite, Keivy area, Kola Penin., Russia; *Fusamata and Suishoyama, Kawatamachi, Fukushima Pref., Honshu, and Soraku, Kyoto Pref., Japan.* EF *AM 58:140(1973)*.

53.2.1.1 Saryarkite-Y $Ca(Y,Th)Al_5(SiO_4)_2(PO_4,SO_4)_2(OH)_7 \cdot 6H_2O$

Named in 1964 by Krol et al., but etymology not given. Synonym: saryarkite. TET $P42_12$ or $P4_22_12$. $a = 8.213$, $c = 6.55$, $Z = 4$, $D = 3.35$. *16-712:* 3.45_9 3.01_{10} 2.83_{10} 2.56_5 2.36_5 2.14_9 1.854_{10} 1.312_8. White, dull to greasy luster. $H = 3\frac{1}{2}$–4. $G = 3.07$–3.15. Uniaxial (+); $N_O = 1.606$, $N_E = 1.620$. Occurs as massive material at an unspecified locality (possibly Sary-Arka) in *Kazakhstan (?)*, associated with barite, molybdenite, pyrite, hematite, and other iron oxides in propylitized acid effusive Devonian rocks and altered granitic rocks. AR *ZVMO 93:147(1964)*, *AM 49:1775(1964)*.

53.2.2.1 Nagelschmidtite $Ca_7(SiO_4)_2(PO_4)_2$

Named in 1977 by Gross for G. Nagelschmidt, who first described the synthetic compound in 1937. Probably HEX Space group unknown. (Syn): $a = 5.38$, $c = 7.10$, $Z = 1$, $D = 3.49$. *3-706* (syn): 3.93_3 3.52_3 2.86_{10} 2.70_{10} 1.960_{10} 1.761_5 1.485_5 1.346_5. Apparently consists, in part, of a mixture of two polymorphs. The clear α-phase is colorless and transparent; uniaxial (+); $N_O = 1.680$, $N_E = 1.698$; presumably close to the ideal composition. The more common, cloudy β-phase is slightly yellow to brown and translucent; uniaxial (+) to biaxial (+) with small 2V; $N_O = 1.638$, $N_E = 1.652$. In thin section, displays a complex set of lamellae intersecting at close to 60°. The composition is deficient in phosphorus, the Si/P ratio being near 4:1 rather than 1:1; traces of Na and K are present. Occurs as anhedral grains, up to 150 μm in diameter in the lowermost part of the *Hatrurim formation, Israel*, where it may constitute up to 30% of the rock and is associated with gehlenite, rankinite, perovskite, titanian andradite, and magnetite. AR *GSIB 70:31(1977)*, *AM 63:425(1978)*.

URANOPHANE GROUP

The uranophane minerals are hydrated uranyl silicates with the general formula

$$R_p[UO_2SiO_4]_q \cdot nH_2O$$

where

$$R_p = Pb^{2+}, Ca^{2+}, K, Na, Cu^{2+}, Mg^{2+}, Co^{2+}, U^{6+}$$

and with a structure based on parallel chains of composition $[UO_2SiO_4]_n^{2n-}$ built of edge-sharing pentagonal bipyramidal uranium polyhedra and SiO_4 tetrahedra with a fundamental repeat distance of approximately 7 Å. In the structure, the chains are cross-linked (Si–O–U) to form sheets that are bonded to the remaining cations.

The species are not isostructural, as their symmetry and structural details are influenced by the number and nature of the associated cations as well as the orientation of the silicate tetrahedra within the chain. The chain conformation is such that in uranophane the apical oxygens of silicate tetrahedra point up along one side of the chain and down on the other; whereas in β-uranophane, the apical oxygens point alternately up and down on each side of the chain.

Except for the case of kasolite, the cations do not provide the required formal charge balance, necessitating the treatment of part of the water present as hydrogen, hydronium, or hydroxyl, the last as a substitution for a

Uranophane Group

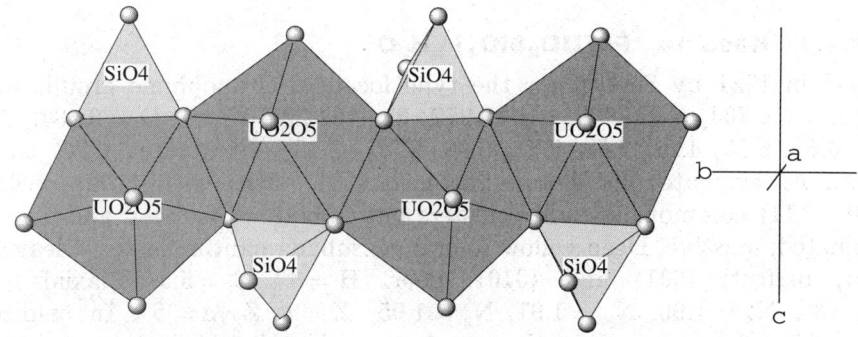

URANOPHANE: $Ca(U(OH)_2)_2(SiO_3OH)_2 \cdot 4H_2O$

The uranyl silicate chain in kasolite, parallel to the ≈ 7Å axis, that is common to most members of the uranophane group. In β-uranophane, silicate tetrahedra point alternately up and down on each side of the chain.

silicate apical oxygen. Structural studies suggest a variety of charge-balancing mechanisms, but the presence of either $SiO_3(OH)$ tetrahedra or (H_3O) is likely. Variability in water content in analyses of these species suggests that at least part of the water may be zeolitic.

URANOPHANE GROUP

Mineral	Formula	Space Group	Axis	Chain Length (Å)	
Kasolite	$Pb[UO_2SiO_4] \cdot H_2O$	$P2_1/a$	b	6.932	52.3.1.1
Uranophane	$Ca[(UO_2)_2(SiO_3OH)]_2 \cdot H_2O$	$P2_1$	b	7.002	53.3.1.2
Sklodowskite	$Mg(H_3O)_2[UO_2SiO_4]_2 \cdot 4H_2O$	$C2/m$	b	7.047	53.3.1.3
Cuprosklodowskite	$Cu[(UO_2)(SiO_3OH)]_2 \cdot 6H_2O$	$P\bar{1}$	a	7.052	53.3.1.4
Boltwoodite	$(H_3O)K[UO_2SiO_4]$	$P2_1$	b	7.064	53.3.1.5
Sodiumboltwoodite	$(H_3O)(Na, K)[UO_2SiO_4] \cdot H_2O$	$P2_12_12_1$	b	7.02	53.3.1.6
Oursinite*	$(Co,Mg)(H_3O)_2[UO_2SiO_4]_2 \cdot 3H_2O$	Aba2 or Abam	c	7.050	53.3.1.7
Swamboite*	$H_6U^{6+}[UO_2SiO_4]_6 \cdot 30H_2O$	$P2_1/a$	b/3	7.004	53.3.1.8
β-Uranophane**	$Ca[(UO_2)(SiO_3OH)_2] \cdot H_2O$	$P2_1/a$	a/2	6.983	53.3.1.9

* Chain conformation not yet determined. All others, except β-uranophane, have the uranophane conformation.
** Dimorph of uranophane.

Notes:
1. The same fundamental chain structure, with additional independent SiO_4 tetrahedra, appears in weeksite, haiweeite, and metahaiweeite. In weeksite the additional silicate tetrahedra are described as forming "walls" (layers?); nevertheless, these species are classified here because of the uranophane-like chains. Similar chains exist in soddylite (53.3.3.1) but with one set at 90° and cross-linked to form a three-dimensional framework.
2. See autunite for other $[UO_2TO_4]$ minerals.

53.3.1.1 Kasolite Pb[UO$_2$SiO$_4$] · H$_2$O

Named in 1921 by Schoep for the type locality. Uranophane group. MON $P2_1/a$. $a = 6.704$, $b = 6.932$, $c = 13.252$, $\beta = 104.22°$, $Z = 4$, $D = 6.534$. 29-788: 6.51_2 6.15_4 4.21_6 3.52_6 3.24_6 3.06_8 2.92_{10} 2.18_2. Structure: *CSC 6:617 (1977)*. As elongated [010] and flattened (001) laths with {100}, {001}, {110}, {111} common, in radial clusters but sometimes isolated and doubly terminated; massive. Deep yellow to orange, subadamantine luster. Cleavage {100}, perfect; {001} and {010}, poor. H = $4\frac{1}{2}$. G = 5.0. Biaxial (+); $2V = 43°$; $N_x = 1.90$, $N_y = 1.91$, $N_z = 1.95$; $X = b$, $Z \wedge a = 5°$. In oxidized zones of uranium ores with other secondary minerals of uranium; a constituent of the uraninite alteration product "gummite." At the Ruggles mine, Grafton Center, NH; East Walker R. area, Lyon Co., and Goodsprings, Clarke Co., NV, and at various localities in NY, PA, UT, and AZ; Nicholson mine, Lake Athabaska, SK, and Great Bear Lake, NWT, Canada; Kersegalec, Ligud, Marbihon, and Grury, Saône-et-Loire, and Puy de Dôme, France; Norway; Italy; Franceville, Gabon; in fine but small crystals at *Kasolo, Shaba, Zaire*, and at Menda, Swambo, Luiswishi, Kasmbove, Kalongwe and Musonoi. An incidental ore of uranium. AR *CR 173:1476(1921), MSUZ:105(1981)*.

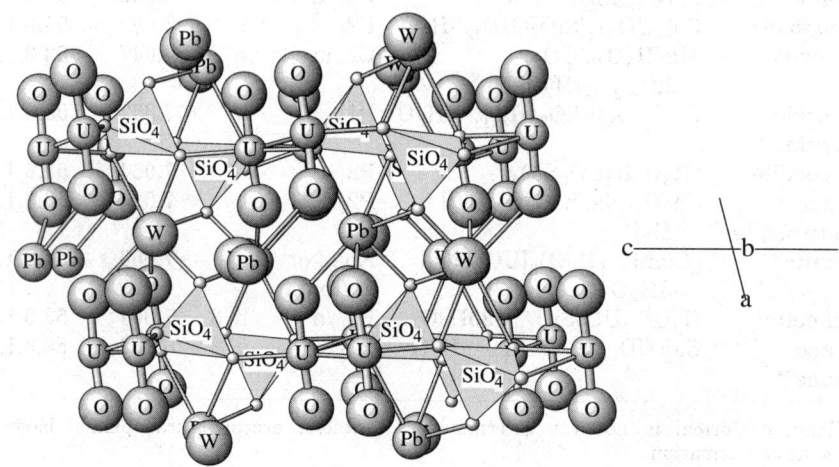

KASOLITE: Pb(UO$_2$)SiO$_4$ · H$_2$O

The uranyl silicate minreals have sheets of uranyl ions (UO$_2$) and isolated Si tetrahedra, with Pb and water in between. U atoms are five-coordinated, and this polyhedron shares one edge with an Si tetrahedron.

53.3.1.2 Uranophane Ca[(UO$_2$)(SiO$_3$OH)]$_2 \cdot$ 5H$_2$O

Named in 1853 by Websky for the composition and from the Greek *to appear*. Synonyms: uranotile, α-uranophane, and uranophane-α. Dimorphous with β-uranophane. MON $P2_1$. $a = 15.909$, $b = 7.002$, $c = 6.665$, $\beta = 97.27°$, $Z = 2$, $D_{calc} = 3.78$. *39-1360:* 7.90$_{10}$ 3.95$_9$ 3.60$_2$ 2.63$_4$ 2.10$_2$ 1.975$_5$ 1.961$_4$ 1.771$_2$. Structure: *AC(C) 44:421(1988)*. As acicular crystals [010] and rosettes, frequently with tapered terminations and as fine-grained, earthy masses. Pale to canary yellow, vitreous to silky luster. Cleavage {100}, perfect; conchoidal fracture. H = 2½–3. G = 3.83–3.90. Biaxial (−); 2V = 37°; $N_x = 1.642$–1.648, $N_y = 1.661$–1.667, $N_z = 1.667$–1.675; X ≈ a, Y $\wedge c$ ≈ 10°, Z = b; r < v, strong; anomalous interference colors. Variability in refractive indices may reflect loss of zeolitic water. Forms as an alteration product of uraninite, constituting much of the yellow fine-grained material adjacent to the orange-to-brown "gummite" alteration of pegmatitic uraninite. As well-crystallized needles and rosettes in oxidized uranium deposits. In pegmatites at Grafton, NH; Avondale, Delaware Co., PA; Spruce Pine district, Mitchell Co., NC; and the Black Hills, SD, as alteration haloes around uraninite and "gummite." At Lusk and Pumpkin Buttes, WY; as superb acicular crystals to 20 mm in vugs and as radial aggregate fracture coating in the Todilto limestone at Haystack Butte and adjoining areas in McKinley and Valencia Cos., NM. As silky, fibrous, radial clusters to several centimeters at the Faraday mine, Bancroft, ONT, Canada; at La Virgen Mine, Puerto del Aire, CHIH, Mexico. Occurs abundantly (uranotile) at Wolsendorf, and at Schneeberg, Saxony, Germany; at Oberpfalz; and *Miedzianca (Kupferberg), Silesia, Poland*. At Mt. Painter, Flinders Range, SA, Australia. At most uranium occurrences in Shaba, Zaire, as pseudomorphs after uraninite (Shinkolobwe) and as superb crystals of several centimeters and as rosettes in association with hydrated uranium oxides, kasolite, rutherfordine, and less commonly with soddyite (Kolwezi and Swambo). AR *AM 24:324(1939), 40:634(1955), 66:610(1981), 71:1489(1986)*.

53.3.1.3 Sklodowskite Mg(H$_3$O)$_2$[UO$_2$SiO$_4$]$_2 \cdot$ 4H$_2$O

Named in 1924 by Schoep for Maria Sklodowska Curie (1867–1934), Polish-born French chemist, Nobelist in physics, 1904, and chemistry, 1911. Uranophane group. MON $C2/m$. $a = 17.382$, $b = 7.047$, $c = 6.610$, $\beta = 105.9°$, $Z = 2$, $D = 3.638$. *29-875:* 8.42$_{10}$ 5.91$_5$ 4.19$_8$ 4.00$_5$ 3.52$_6$ 3.27$_7$ 3.00$_6$ 2.87$_5$. Structure: *CSC 6:611(1977)*. As acicular to fibrous [010] crystals with {100}, {001}, {301}, {011} but commonly with tapering termination; as rosettes and spherulites; massive. Pale to lemon yellow to greenish yellow, almost white in fibrous samples; vitreous luster; transparent. Cleavage {100}, perfect; conchoidal fracture. H = 2–3. G = 3.64. Biaxial (−); 2V large; $N_x = 1.613$, $N_y = 1.635$, $N_z = 1.657$; Y = b (elongation); r > v, very strong; strongly pleochroic, yellow to colorless Z > Y > X. In oxidation zones of uranium ores, and often replacing uraninite; commonly associated with other secondary uranium minerals. At the New Haven mine, Crook Co., WY; at Oyler Tunnel claim, Wayne Co., and Honeycomb Hills, Juab Co., UT; at Grants, NM; at Naica, CHIH, Mexico; at *Shinkolobwe, Shaba, Zaire*, in fine, acicular crystals and as pseudomorphs after uraninite, and at Luiswishi, Musonoi, and Swambo; associated with novacekite and zeunerite at the Pedra Preta magnesite mine, Brumado, BA, Brazil(!). AR *CR 179:413(1924), AM 66:610(1981)*.

53.3.1.4 Cuprosklodowskite $Cu[(UO_2)(SiO_3OH)]_2 \cdot 6H_2O$

Named in 1933 by Buttgenbach in allusion to the composition and similarity to sklodowskite. Uranophane group. Synonyms: jachymovite (jachimovite). TRIC $P\bar{1}$. $a = 7.052$, $b = 9.267$, $c = 6.655$, $\alpha = 109.23°$, $\beta = 89.83°$, $\gamma = 110.02°$, $Z = 1$, $D = 3.815$. *19-413:* 8.16_{10} 6.06_7 4.82_9 4.12_3 3.53_7 3.29_4 2.96_4 2.92_4. Structure: *AM 60:448(1975)*. As acicular [100] crystals with {010}, {011}, {001}, {110}; as rosettes and crystalline masses. Twins with [100] axis and (010) composition common; also with composition plane ($2\bar{1}0$). Yellow-green to emerald-green, vitreous luster. Cleavage {010}, perfect; conchoidal fracture. $H = 3\frac{1}{2}$. $G = 3.83$. Biaxial $(-)$; 2V small; $N_x = 1.655$, $N_y \approx N_z = 1.667$ $X \wedge a$ small; distinctly pleochroic. In oxidation zones of uranium ores containing copper sulfides, and in association with various secondary uranium and copper minerals. In Red Canyon and White Canyon, San Juan Co., and at Seven Mile Canyon, near Moab, Grand Co., UT; in limestone at Grants, NM; at the Nicholson mine, Lake Athabaska, SK, Canada; at Johanngeorgstadt, Saxony; at Joachimsthal (Jachymov), Czech Republic; at Amelal, Argana-Biguadine, Morocco; at *Kalongwe, Shaba, Zaire*, and as excellent acicular crystals(!) up to several centimeters long at Musonoi, and at Shinkolobwe and other sites in Shaba, Zaire, with vandenbrandeite, kasolite, sklodowskite, and malachite. AR *SGBA 56:B331(1933), BM 90:259(1967), AM 66:610(1981)*.

53.3.1.5 Boltwoodite $(H_3O)K[UO_2SiO_4] \cdot H_2O$

Named in 1956 by Frondel and Ito for Bertram B. Boltwood (1870–1927), professor of radiochemistry, Yale University. Uranophane group. Synonym: nenadkevite. MON $P2_1$. $a = 7.073$, $b = 7.064$, $c = 6.638$, $\beta = 105.75°$, $Z = 2$, $D_{calc} = 4.372$. *35-490:* 6.81_{10} 6.40_5 3.54_7 3.40_9 2.95_8 2.91_7 1.900_6 1.764_6. Structure: *AM 66:610(1981)*. Acicular [010]; fine, matted fibers; massive. Twinned. Light yellow, vitreous to earthy luster. Cleavage {010}, perfect; {001}, imperfect. $H = 3\frac{1}{2}$–4. $G = 3.60$. Biaxial $(-)$; 2V large; $N_x = 1.668$–1.670, $N_y = 1.695$–1.696, $N_z = 1.698$–1.703; $Z = b$; $r < v$. Pleochroic; X, colorless, Y,Z, pale yellow. Anomalous interference colors. May have often been misidentified as uranophane. Interstitially in quartz sandstones; as an alteration product of uraninite. At Williams quarry, Easton, PA, as an alteration of thorian uraninite; Lookout No. 22 mine, Gunnison Co., CO; New Method mine and *Pick's Delta mine, Emery Co., UT*; Arandis mine, Namibia, in radial aggregates up to 2 cm long; Myponga, SA, Australia; Alto Boqueirao, Parelhas, RGN, Brazil; in the outer zones of "gummite" alteration at La Chiquita mine, Quebrado, Cordoba, Argentina. AR *SCI 124:931(1956), AM 46:12(1961)*.

53.3.1.6 Sodium-Boltwoodite $(H_3O)(Na,K)[UO_2SiO_4] \cdot H_2O$

Named in 1975 by Chernikov et al. in allusion to composition and similarity to boltwoodite. Uranophane group. ORTH $P2_12_12_1$. $a = 27.40$, $b = 7.02$, $c = 6.65$, $Z = 8$, $D_{calc} = 4.4$. *29-1044:* 6.92_7 6.71_{10} 4.70_8 3.49_8 3.37_8 3.10_8 2.92_{10} 1.920_7. Powdery aggregates. Pale yellow, dull luster. Perfect cleavage, reported as {010} but probably parallel [010]. $H = 3\frac{1}{2}$–4. $G = 4.1$. Biaxial; 2V large; $N_x = 1.613$, $N_y = 1.63$, $N_z = 1.645$, also reported 1.645, 1.66, and 1.72, respectively. Pleochroic. X, colorless; Z, pale yellow. The described material contains sodium and potassium, Na/K $\approx 7:3$. An oxidation-zone mineral. Found in unspecified *arid regions of the former USSR.* AR *DANES 221:144 (1975); AM 61:1054(1976), 66:610(1981).*

53.3.1.7 Oursinite $(Co,Mg)(H_3O)_2[UO_2SiO_4]_2 \cdot 3H_2O$

Named in 1983 by Deliens and Piret from the French *oursin*, sea urchin, in allusion to the radiating habit of the acicular crystals. Uranophane group. ORTH *Aba2* or *Abam.* $a = 12.74$, $b = 17.55$, $c = 7.050$, $Z = 4$, $D = 3.67$. *35-600:* 8.73_{10} 7.20_7 5.16_5 4.55_3 4.14_7 3.53_4 3.27_3 2.85_9. Structure: *BM 106:305 (1983).* As radiating aggregates of acicular [001] crystals of 1 mm maximum length. Pale yellow, vitreous luster, transparent to translucent. Cleavage parallel to elongation. Biaxial (−); $2V_{calc} = 76°$; $N_x = 1.624$, $N_y = 1.640$, $N_z = 1.650$; Y = *c*(elongation). In oxidation zones of uranium ores at *Shinkolobwe, Shaba, Zaire,* in association with soddyite, sklodowskite, kasolite, lepersonnite, bijvoetite, and other secondary uranium minerals. AR *AM 69:567(1984).*

53.3.1.8 Swamboite $H_6U^{6+}[UO_2SiO_4]_6 \cdot 30H_2O$

Named in 1981 by Deliens and Piret for the type locality. Uranophane group. MON $P2_1/a$. $a = 17.64$, $b = 21.00$, $c = 20.12$, $\beta = 103.4°$, $Z = 6$, $D = 4.064$. *35-660:* 8.67_{10} 5.85_5 4.76_8 4.49_5 4.32_8 3.51_6 3.26_5 2.98_6. Structure: *CM 19:553 (1981).* Acicular [010] with beveled terminations. Pale yellow. $G = 4.03$. Biaxial (−); $2V = 30°$ $N_x = 1.640$, $N_y = 1.661$, $N_z = 1.663$; X $\approx a$, Y $= b$, Z $\wedge c = 13°$. Pleochroic; X, colorless; Y, pale yellow. As an alteration product of other uranium minerals at the Jo Mac mine, San Juan Co., UT, and at *Swambo, Shaba, Zaire.* AR

53.3.1.9 Beta-uranophane $Ca[(UO_2)(SiO_3OH)]_2 \cdot H_2O$

Named in 1935 by Novacek to correspond with its polymorph, uranophane. Uranophane group. Synonyms: β-uranotile, uranophane-β. Dimorphous with uranophane. MON $P2_1/a$. $a = 13.966$, $b = 15.443$, $c = 6.632$, $\beta = 91.38°$, $Z = 4$, $D = 4.117$. *8-442:* 7.49_{10} 5.04_8 4.53_8 3.83_9 3.51_8 3.17_7 3.02_8 2.80_9. Structure: *AM 71:1489(1986).* The structure is similar to that of uranhophane except for the sequence of orientation of the SiO_4 tetrahedra along the chain. Acicular [100], laths; fine-grained, fibrous; massive. Twinning on {010} common. Yellow, pale yellow streak, vitreous to dull luster. Cleavage {010}, perfect; {100}, poor. $H = 2\frac{1}{2}$–3. $G = 3.9$. Variable composition due to zeolitic water may account for the variability in optical properties. Biaxial $(-)$; $X = b$, $2V = 45-71°$; $N_x = 1.661-1.671$, $N_y = 1.682-1.694$, $N_z = 1.689-1.702$; $Z \wedge c = 32-57°$; $r > v$, strong; crossed extinction, often incomplete. Pleochroic: X, colorless; Y,Z, deep yellow. An oxidation product of uraninite found in oxidized ores and pegmatites; pseudomorphs after uraninite. May be more common than reported, due to similarity to uranophane and similarly colored uranium minerals. Newry, ME; Ruggles and Palermo mines, Grafton Center, NH; Bedford, NY; Black Hills, SD; Faraday mine, Bancroft, ONT, Canada; France; Portugal; Wölsendorf, Bavaria, Germany; *Joachimsthal (Jachymov)*, Czech Republic; Japan; Argentina. An incidental ore of uranium. AR *AM 24:324(1939), 40:634(1955), 66:610(1981); GSAM 135:281(1972).*

53.3.2.1 Weeksite $(K,Na)_2(UO_2)_2(Si_5O_{13}) \cdot 3H_2O$

Named in 1960 by Outerbridge et al. for Alice Dowse Weeks (1909–1991), professor of geology, Temple University. ORTH *Cmmm*. $a = 7.092$, $b = 17.888$, $c = 7.113$, $Z = 2$, $D = 3.71$. Also *Pnnb*. $a = 14.26$, $b = 35.88$, $c = 14.20$, $Z = 16$ for $K_2(UO_2)_2(Si_2O_5)_3 \cdot 4H_2O$. *12-462:* 8.98_8 7.11_{10} 5.57_9 3.55_7 3.30_7 3.20_5 2.37_5 2.91_6. Structure: *SPD 30:435(1985)*. The structure consists of walls of silicate tetrahedra joined by uranophane-like chains. Small spherulites of radiating lath. Yellow, waxy to silky luster. $G \approx 4.1$. Biaxial $(-)$; $2V \approx 60°$; $N_x = 1.596$, $N_y = 1.603$, $N_z = 1.606$; $XYZ = bca$; weakly pleochroic. X; colorless; Y, pale yellow-green; Z, pale green. In opal veinlets in rhyolite; with carbonates and gypsum. Mammoth mine, near Presidio, TX; Jackpile mine, Laguna Co., NM; at Goodwill claim and *Autunite No. 8 claim, Thomas Range, Juab Co., UT*; Muggins Mountains, Yuma Co., AZ; Haiwee reservoir, Coso Mountains, Inyo Co., CA. At the Margaritas, Sierra Peña, CHIH, Mexico. From Le Bois–Noirs, Loire, France, and Russia Afghanistan, Japan and Yinnietharra, WA, Australia. AR *AM 45:39(1960), 66:610(1981); ZVMO 106:553(1977).*

53.3.2.2 Haiweeite $Ca(UO_2)_2Si_6O_{15} \cdot 5H_2O$

Named in 1959 by McBurney for the type locality. Synonym: ranquilite. MON $P2/c$. $a = 15.4$, $b = 7.05$, $c = 7.10$, $Z = 2$, $D = 4.88$. *12-721:* 9.26_{10} 7.97_2 7.09_3 4.53_8 4.41_5 3.54_3 3.30_3 3.18_4. Structure: *AM 66:610(1981)*. Acicular, very small; spherulitic aggregates; flakes. Yellow, silky to dull luster. Cleavage $\{001\}$, good. $H = 3\frac{1}{2}$. $G = 3.35$. Biaxial $(-)$; $2V \approx 15°$ $N_x = 1.571$, $N_y = 1.575$, $N_z = 1.578$ (Haiwee res.); $N_x = 1.560, N_y = 1.580$, $N_z = 1.582$ (Bad Gastein); $r > v$, strong; pleochroic; $X > Z$; anomalous interference colors. A secondary, oxidation product. At *Haiwee reservoir, Coso Mountains, Inyo Co., CA*, in granite and voids in lake-bed sediments; in gneiss at Bad Gastein, Salzburg, Austria; Hingyotoge deposit, Tottori Pref., Japan; at Nishapur, Kuh-e-Madan, Iran: in tourmaline granite at Perus, 25 km N of São Paulo, Brazil. AR *AM 44:839(1959), 54:966(1969).*

53.3.2.3 Metahaiweeite Ca(UO$_2$)$_2$Si$_6$O$_{15}$ · nH$_2$O $n < 5$

Named in 1959 by McBurney from the parent phase, haiweeite. MON $P2/c$(?). *12-722:* 8.81$_5$ 6.98$_{10}$ 5.85$_5$ 4.54$_4$ 3.52$_5$ 3.28$_4$ 3.16$_5$ 2.90$_5$, for material heated to redness. Flakes and spherical clusters. Pale yellow to greenish yellow, pearly luster. Cleavage {100}, distinct. H = $3\frac{1}{2}$. G = 3.35. Dull green in UV. Reported biaxial (+), 2V$_{calc}$ = 21°; N$_x$ = 1.611, N$_y$ = 1.620, N$_z$ = 1.645. A dehydration product of haiweeite; data are derived largely from laboratory-heated specimens of haiweeite. *Haiwee reservoir, Coso Mountains, Inyo Co., CA.* AR *AM 44:839(1959)*.

53.3.3.1 Soddyite (UO$_2$)$_2$SiO$_4$ · H$_2$O

Named in 1922 by Schoep for Frederick Soddy (1877–1956), British radiochemist and physicist. ORTH *Fddd. a* = 8.32, *b* = 11.21, *c* = 18.71, Z = 8, D = 4.75. *35-491:* 6.30$_{10}$ 4.81$_5$ 4.56$_{10}$ 3.35$_{10}$ 3.26$_2$ 2.99$_5$ 2.81$_4$ 2.72$_{10}$. Structure: *AM 66:610(1981)*. Thick tabular {001} to pyramidal with {111}, {110}, {001} dominant; fibrous; massive. Cleavage {001}, perfect; {111}, good. Lemon yellow (opaque) to amber-yellow (transparent); vitreous to resinous luster, earthy. H = $3\frac{1}{2}$. G = 4.70. Biaxial (−); 2V large. X; N$_x$ = 1.650–1.654, N$_y$ = 1.685, N$_z$ = 1.699–1.715 XYZ = *cba*, colorless; Y, pale yellow; Z, pale yellow-green. In oxidized uranium ores. As pseudomorphs after uraninite at *Ruggles mine, Grafton Center, NH; Lucky Mc mine, Fremont Co., WY; Honeycomb Hills, UT; Jackpile mine, Laguna Co., NM; Steel City mine, Yavapai Co., AZ;* in fine crystals up to 3 mm at *Shinkolobwe, Shaba, Zaire,* and at *Musonoi* and *Kolongwe;* in pegmatite at *Norabees, Namaqualand, South Africa.* AR *CR 174:1066(1922), AM 37:386(1952), SPD 24:315(1980)*.

53.3.3.2 Uranosilite (UO$_2$)Si$_7$O$_{15}$

Named in 1983 (Walenta) in allusion to the composition. ORTH *P*22$_1$2$_1$, *Pmmb*, or *Pmcb. a* = 11.60, *b* = 14.68, *c* = 12.83, Z = 6, D$_{calc}$ = 3.25. *37-417:* 11.6$_9$ 7.30$_{10}$ 6.19$_9$ 3.50$_8$ 3.42$_6$ 3.20$_6$ 3.08$_6$ 2.77$_7$. Minute acicular [001] crystals. Yellowish white, vitreous luster. Biaxial (−); N$_x$ = 1.570, N$_z$ = 1.584; Y or Z = *c*; slightly anomalous interference colors. Intimately intergrown with uranophane and studtite at *Menzeschwand, Schwarzwald, Germany.* AR *NJMM:259(1983)*.

Borosilicates and Some Beryllosilicates

54.1.1.1 Grandidierite (Mg,Fe)Al$_3$BSiO$_9$

Named in 1902 by Lacroix for Alfred Grandidier (1836–1912), French explorer, who described the geography and natural history of Madagascar. ORTH *Pbnm.* $a = 10.335$–10.36, $b = 10.978$–10.94, $c = 5.75$–5.760, $Z = 4$, $D = 3.10$. *18-581*(nat): 5.48_8 5.17_{10} 5.04_{10} 3.71_5 2.96_4 2.74_8 2.58_5 2.17_6. The structure is related to andalusite. The only silicate in which significant amounts of Fe^{2+} are in five-fold coordination. There is a distorted trigonal bipyramid polyhedron around the (Mg,Fe) site. *AC(B) 24:1518(1968)*. Prismatic crystals or grains; ≈ 0.1 mm or less, a few $= 0$–0.4 mm long, original Madagascar specimens to 8 cm long. Greenish blue or bluish green, pale blue-green streak, vitreous luster on cleavage. Bluish green, greenish blue, vitreous luster. Cleavage parallel to (100), good; parallel to (010), poor; uneven fracture; brittle. $H = 7\frac{1}{2}$. $G = 2.93$–2.98. IR data: *BGSF 41:71(1969)*, *BM 93:224(1970)*. Mössbauer data: *AM 62:547(1977)*. Infusible, unattacked by acids. Analysis (wt %): MgO, 2.4–14.0; FeO, 1.1–18.5; traces of Mn, Ca; Fe$_2$O$_3$ is always very low (< 0.9 wt %), Fe^2/Fe^3 $\approx$ 1:1 to 3:1, the only substitutions are Fe^{2+} $\leftrightarrow$ Mg and Al $\leftrightarrow$ Fe^{3+}; X$_{Mg}$ range is 0.98–0.19. Thus there are two end members: grandidierite-(Fe) and grandidierite-(Mg). Biaxial ($-$); XYZ parallel to *acb*; N values vary with Fe/Mg ratio: $N_x = 1.587$–1.602, $N_y = 1.618$–1.636, $N_z = 1.622$–1.639; $2V = 27$–$32°$; $X > Z > Y$ or $Z > X > Y$; OAP parallel to (001); $r > v$, moderate to strong, with some $r < v$. X, pale blue to blue; Y, colorless to yellow-brown; Z, very pale blue to blue-green or deep blue; iron-rich grandidierite is deeper colored. Most grandidierite occurs in ancient metamorphic terranes, in pelitic gneisses and quartzite. Also occurs in aluminous metasedimentary xenoliths in acid volcanics as well as in pegmatites. In hornfels close to intrusions: low pressure, high temperature ($>600°$). Stable, coexisting grandidierite–kornerupine assemblage requires high pressure and high temperature. Occurs at somewhat more than 25 worldwide localities: it occurs replacing serendibite at Russell, NY [*MM 54:133(1990)*]; with kornerupine at Opinicon Lake, southeastern ONT, and at Morton and Brewer's Mills, ONT, Canada; Haddo House complex, northeastern Scotland; iron-rich material associated with Ti-rich dumortierite, and tourmaline at Almjotheia, 12 km N of Moi, Rogaland, and Vestpolltind, southwestern Norway; Nickenicher Sattel, Eifel, Germany;

Mount Amiata, Tuscany, and Mount Cimino, Latium, Italy; a locality in Hungary; Transbaikalia, Russia; Tizi-Ouchen, Bejaia, Algeria; Zimbabwe; Lukusuzi R. area, eastern province, Zambia; Mchinji, Malawi; with werdingite and kornerupine at Bok Se Puts, Namaqualand, and at Port Shepstone, Natal, South Africa; in cordierite and garnet in gneisses at Ihosy, southern Madagascar; also at *Fort Dauphin area, Andrahomana*; Sakatelo, Ampamatoa, Vohiboly, Sahakondro, and other localities, Madagascar; at Paderu, Vishakhapatnam district, Andhra Pradesh, India; Blanket Bay and Landing Bay, Cuvier Island, New Zealand; Maratakka, Surinam; with kornerupine at McCarthy Point and Seal Cove, Larsemann Hills, Prydz Bay, eastern Antarctica. EF *MM 43:651(1980), 54:131(1990), 59:327(1995); NJBA 143:249(1982); CMP 98:502(1988); AM 75:415(1990)*.

54.1.2.1 Dumortierite $(Al,Fe^{3+})_7(BO_3)(SiO_4)_3O_3$ (anhydrous) and $(Al_{6.75}\square_{0.25})BSi_3[O_{17.25}(OH)_{0.75}]$ (hydrous)

Named in 1881 by Gonnard and named for Vincent Eugene Dumortier (1802–1873), French paleontologist, Lyons, France. See the closely related mineral holtite (54.1.3.1). ORTH *Pmcn*. $a = 11.77$–11.82, $b = 20.18$–20.25, $c = 4.70$–4.71, $Z = 4$, $D = 3.39$. 7-71: 5.89_9 5.84_9 5.09_9 $3.45_{7.5}$ $3.23_{8.5}$ 2.93_9 2.89_7 2.55_{10}. Structure: *NJBA 132:231(1978)*. Generally occurs as fibrous to fan-shaped radiating aggregates to >5 cm long. Acicular on c, prismatic, usually massive columnar, fibrous. Dominant forms are: {100}, {010}, {110}, {130}. May show lamellar twinning with (001) as the twin plane; also twinned on (110). Usually pink-red or blue-purple violet, but sometimes brown or greenish; white or bluish-white streak; vitreous to dull luster, sometimes silky. Cleavage {100}, distinct; {110}, imperfect; brittle. $H = 7$–8. $G = 3.25$–3.41 (high Fe). IR data and synthetic material: *CMP 105:11(1990)*. Insoluble in all acids except HF. Sometimes luminescent bright blue under electron microprobe beam (e.g., Ti-bearing material from Zambia). Sometimes fluorescent blue in SWUV light. MgO to 8 wt % in dumortierite from the western Alps [see magnesiodumortierite (54.1.2.2)]. P_2O_5 to 0.5 wt %, Fe_2O_3 to 1.5 wt %, CaO to 1.0 wt %, TiO_2 to >4.6 wt %; As_2O_5 to 1.67 wt %, from Sri Lanka. The color variation (pink to blue) is possibly due to the relative concentrations of Ti and Fe and/or charge-transfer processes. Optics are variable. Biaxial (−); XYZ = cba; $N_x = 1.659$–1.686, $N_y = 1.684$–1.722, $N_z = 1.683$–1.723; $2V = 13$–$52°$. OAP generally parallel to (010); however, in material from Kutna Hora, Czech Republic, Riverside, CA, and Madya Pradesh, India, the OAP is (100), not (010). The pleochroic scheme is quite variable and is not due to just the variation in TiO_2 content. Examples are: X, deep blue; Y, yellow; Z, colorless; X, pink, Y, colorless, Z, colorless; X, deep purple; Y, light tan; Z, colorless. Negative elongation. Dispersion is $r < v$, medium to $r > v$, weak. A minor to major constituent in pegmatitic, aplitic, and granitic rocks. Also occurs in pneumatolytic–hydrothermal deposits and regional metamorphosed quartzites, gneisses, and granulites; in Al-rich metamorphic rocks; and in high-pressure metamorphic rocks (e.g., Alps). May alter to muscovite–sericite, kaolinite; also may form pseudomorphs after andalusite. Only selected major localities are given here: Topsham and Woolwich, Sagadahoc Co., ME; Fremont Co., CO; Petaca, Rio Arrita, Co., NM., Clip, near Yuma and Quartzsite, La Paz Co., AZ; Ruby Range, east of Dillon, MT; in large quantities in Gypsy Queen Canyon, Oreana, Humboldt Co., NV; Lincoln Hill, Rochester district, Pershing Co., NV; Woodstock, WA; Corona, Riverside Co., CA; Dehesa, San Diego Co., CA; also from near Ogilby, in Imperial and Kern Cos., CA; Ashby Twp., Addington and Lennox Cos., ONT, Canada; Land's End, Cornwall, England; Aberdeen, Scotland; Mg-bearing from Bamble sector, southern Norway; with Fe-rich grandidierite at Almjotheia, Moi, Rogaland, southwestern Norway; Tvedestrand, Augst-Agder, Norway; *Beaunan, Chaponost, near Lyon, Rhône, France*; Malsburg, Schwartzwald, Germany; in pegmatite to 5 cm long from Kaňk, near Kutna Hora, Czech Republic; Wolfschau, Silesia, Poland; Yelshits, Bulgaria; Mursinka, Ural Mts., Aldan and Taezhno regions, Yakutia, and Mt. Zhanet in Kounrad region of north-western Pribaikal, Russia; Lyailyok R., Uzbekistan; Erong Mtns., Namibia, blue lapidary-quality titanian dumortierite occurs at the farm N'Rougas Suid, near Kenhardt, northern Cape Prov., South Africa; Soavina and Riamfotsi near Tananarive, and Saharina Madagascar; Nabikura mine, Yamaguchi Pref., Japan; Jaipur, Rajasthan, and Mogra Bhandara, India; Nacozari, SON, Mexico; Serra da Verada, Makoba, and Bokira, Bahia, Brazil (to 10 cm). Used as insulators and ceramics and sometimes for gemstones (cabochons). EF *MIN 3(1):329(1972)*, *NJBM:22(1979)*, *NJBA 143:249(1982)*, *AM 71:786(1986)*, *MM 55:563(1991)*, *CM 30:137(1993)*.

54.1.2.2 Magnesiodumortierite [(Mg,Ti,☐)Al$_4$(Al,Mg)$_2$][Si(Si)$_2$O$_{15}$(OH)$_3$]B

Named in 1995 by Ferraris et al. for the compositional relationship to dumortierite. ORTH *Pmcn*. $a = 11.91$, $b = 20.40$, $c = 4.730$, $Z = 4$, $D = 3.22$. PD(nat): 5.91$_9$ 3.47$_7$ 3.25$_{10}$ 3.09$_5$ 2.915$_6$ 2.105$_9$ 1.69$_6$ 1.345$_6$. *EJM 7:525(1995)*. $R = 3.1\%$. Mg is the dominant cation in the single chain of face-sharing M1 octahedra, where Ti and vacant sites are also present. With respect to dumortierite, some substitution of Al also takes place in the second face-sharing octahedron M4, but not in M3 and M2, which only share corners and edges. Occurs as mostly anhedral grains a few tens to 300 μm across. Pink to red, vitreous luster, transparent. Streak, hardness, and density could not be measured. H should be between 7 and 8. Analysis (wt %): SiO$_2$, 31.7; P$_2$O$_5$, 0.07; Al$_2$O$_3$, 47.3; TiO$_2$, 4.3; ZrO$_2$, 0.06; FeO, 0.09; MgO, 7.8; B$_2$O$_3$, 5.1; H$_2$O, 3.5. The mineral varies in composition from the Ti-free end member (Mg$_{1-x}$,☐$_x$)Al$_2$Al$_4$Si$_3$O$_{17-2x}$(OH)$_{1+2x}$B ($x \approx 0.33$) toward (Mg$_{0.33}$, Ti$_{0.33}$,☐$_{0.33}$) (Al$_{0.5}$Mg$_{0.5}$)$_2$ Al$_4$Si$_3$O$_{16}$(OH)$_2$B. Biaxial ($-$); $X = c$; probably with $YZ = ba$; Na(D): $N_x = 1.678$, $N_y = 1.700$, $N_z = 1.701$; $2V = 38.5°$; $X \gg Y = Z$. X, pale pink to red; Y,Z, colorless. Color is due to Fe^{2+} ↔ Ti^{4+} charge transfer. Occurs as rare inclusions in pyrope megablasts with other magnesian minerals, such as talc, clinochlore, ellenbergerite, and kyanite in a coesite-bearing quartz–phengite-rich "white schist" within a high-pressure terrane, the *Dora-Maira massif in the western Alps, Italy*; specifically at San Giacomo near Brossasco in Val Varaita and Case Romello near Parigi (Martiniana Po) in Val Po. EF *EJM 7:167(1995)*.

54.1.3.1 Holtite $(Si,Sb,As)_3B[Al_6(Al,Ta,\square)O_{15}(O,OH)_{2.25}]$

Formula *MM 53:457(1989)* and $(Ta,Nb,Sb,Al,Fe,Mn,Mg,Ca)\ Al_{6-n}[\{Al,As\}_k Si_{1-k}\}\ O_4]_3\ [B_{1-m}O_3][O,OH]_{2-3}$, where $m+n < 1$, $k = (m+n)/3$ [*ZVMO 106:337 (1977)*] Named by Pryce in 1971 for Harold Edward Holt (1908–1967), prime minister of Australia. See dumortierite (54.1.2.1), a very closely related borosilicate. ORTH *Pnma*. $a = 4.691$, $b = 11.896$, $c = 20.383$, $Z = 4$. *MM 53:457(1989)*. 25-1209(nat) [*MM 38:21(1971)*]: 10.3_{10} 5.93_3 5.89_3 5.12_2 5.08_3 2.94_4 2.36_2 2.34_2; *ZVMO 106(3):337(1977)*: 10.26_{10} 5.13_9 4.27_5 3.90_7 3.46_6 3.34_5, 3.23_8 2.09_7. The structure is closely related to dumortierite but with extensive substitutions of Si by Sb in holtite. *MM 53:457(1989)*. Weak spots appear to indicate a supercell with $2a$, $2b$, c. *MM 38:21(1971)*. Fibrous, acicular prisms, pseudohexagonal; crystals to 3 cm × 0.8 cm; multiple twinning on {110}. Pale brownish white to brown, greenish gray, buff; vitreous to resinous or dull luster. Cleavage {010}, distinct. $H = 8\frac{1}{2}$. $G = 3.90$ (Australia) and 3.60–3.67 (Kola Peninsula). IR, DTA, TGA data: *ZVMO 106(3):337(1977)*. Fluoresces dull orange under SWUV and bright yellow in LWUV. X-ray diffraction data unchanged after heating to 1000°. Two types of holtite have been distinguished from the Kola Peninsula: holtite-I: $Al_2O_3 \approx 48\%$, $Sb_2O_5 \approx 7\%$, $Ta_2O_5 \approx 14\%$, $Nb_2O_5 \approx 0.2\%$, $SiO_2 \approx 22\%$, $As_2O_5 \approx 3.5\%$; holtite-II: $Al_2O_3 = 47\%$, $Sb_2O_5 = 20\%$, $Ta_2O_5 \approx 11\%$, $Nb_2O_5 \approx 0.2\%$, $SiO_2 = 17.5\%$, $As_2O_5 \approx 2.5\%$. Australian material is chemically similar but with $B_2O_3 = 4$–5%. Biaxial $(-)$; $X = c$, OAP = {010}; $N_x = 1.705$–1.709 (Kola), 1.743–1.746 (Australia), $N_y = 1.728$–1.729 (Kola), 1.756–1.759 (Australia), $N_z = 1.730$–1.737 (Kola), 1.758–1.761 (Australia); $2V = 49$–55° (Australia) and $2V = 20$–30° (Kola); $r < v$. (Australia): X, shades of yellow; Y, colorless; Z, colorless. Kola holtite is nonpleochroic. Occurs in granitic pegmatites and as an alluvial mineral from pegmatites. In pollucite-bearing complex granitic pegmatites associated with albite, lepidolite, zircon, rubellite, and amblygonite. At Greenbushes as coatings on stibiotantalite and replacing tantalite. Voronyi Tundra area and Ploskogorskaya (Flat Mountain) pegmatite, Keivy, Kola Penin., Russia; *Bunbury Gully, Greenbushes, West Australia*. EF

DATOLITE–GADOLINITE GROUP

The datolite–gadolinite group consists of nine members plus additional species which are related but only partially accepted as members of the group. The general formula is

$$A_2BC_2(TO_4)_2X_2$$

where

A = REE, Ca, Bi
B = Fe^{2+}, Fe^{3+}, Mn, Ca, $\square$
C = Be, B
T = Si, P, As
X = O, OH, F

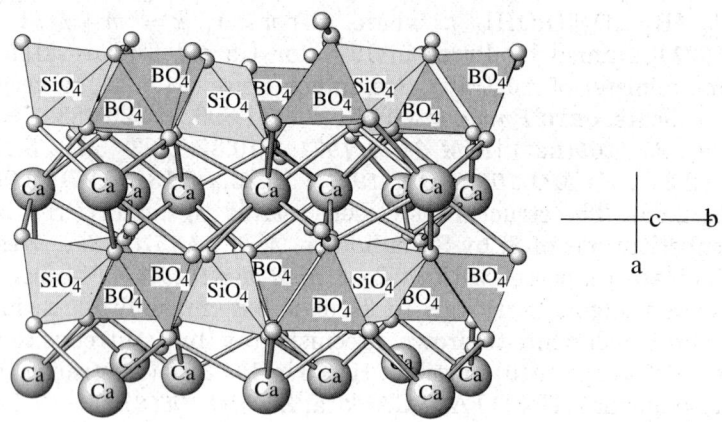

DATOLITE: $CaBSiO_4OH$

Datolite has sheets of alternating Si and B tetrahedra, with Ca in between layers. The sheets are composed of four- and eight-membered rings. The OH atoms are at the apices of the B tetrahedra.

DATOLITE–GADOLINITE GROUP

Mineral	Formula	Space Group	a	b	c	β	
Datolite	$Ca_2(\square)B_2Si_2O_8(OH)_2$	$P2/_1a$	9.63	7.61	4.83	90.2	54.2.1a.1
Hingganite-(Ce)	$(Ce,Y)_2(\square,Fe^{2+})Be_2Si_2O_8(OH,O)_2$	$P2/_1a$	9.996	7.705	4.792	90.06	54.2.1a.2
Hingganite-(Y)	$(Y,Ce)_2(\square,Fe)_2Be_2Si_2O_3(OH)_2$	$P2/_1a$	9.86	7.605	4.720	89.65	54.2.1a.3
Hingganite-(Yb)	$(Yb,Y)_2(\square,Fe^{2+})Be_2Si_2O_8(OH)_2$	$P2/_1a$	9.888	7.607	4.740	90.45	54.2.1a.4
Bakerite	$Ca_4B_5Si_3O_{15}(OH)_5$	$P2/_1a$	9.53	7.57	4.79	90.4	54.2.1b.1
Gadolinite-(Ce)	$(Ce,La,Nd,Y)_2Fe^{2+}Be_2Si_2O_{10}$	$P2/_1a$	10.01	7.58	4.82	90.78	54.2.1b.2
Gadolinite-(Y)	$Y_2Fe^{2+}Be_2Si_2O_{10}$	$P2/_1a$	10.00	7.565	4.768	90.31	54.2.1b.3
Calciogadolinik	$CaRee(Fe^{3+})Be_2$	$P2/_1a$	9.988	7.566	4.696	9.000	54.2.1b.4
Homilite	$Ca_2(Fe,Mg,Mn)B_2Si_2O_{10}$	$P2/_1a$	9.786	7.621	4.776	90.61	54.2.1b.5
Minasgeraisite-(Y)	$Y_2CaBe_2Si_2O_{10}$	$P2/_1a$	9.833	7.562	4.702	90.46	54.2.1b.6

54.2.1a.1 Datolite

The structure of gadolinite and minerals isostructural with gadolinite can be described in terms of a sheet structure built up from alternating SiO_4 and BeO_4 (or BO_4) tetrahedra. The sheet extends parallel to the (001) plane at Z=0.5. The rare earth and iron atoms are located at Z=0, and connect the sheets. The silicon atom exhibits the usual tetrahedral coordination by four oxygen atoms, O(1), O(2), O(3), and O(4). Similarly, the Be or B atom is tetrahedrally coordinated by four oxygen atoms, O(2), O(3), O(4), and O(5). These tetrahedra make corner-sharing linkages at O(2), O(3), and O(4) sites to form the sheet structure. The structure of datolite can be derived from that of gadolinite simply by replacing Be with B, Y with Ca, and O with OH. The iron site at (0,0,0) is empty in datolite. The structure of herderite is isotypic with that of datolite *(AM 69:948(1984))*.

DATOLITE–GADOLINITE RELATED SPECIES

Mineral	Formula	Space Group	a	b	c	β	
Hellandite-(Y)	$(Ca,Y,\square)_4(Y,\square)_2,(Al,Fe)B_4Si_4O_{13}(OH,O)_4(OH)_2$	$P2/a$	18.99	4.71	10.30	111.4	54.2.2.1
Tadzhikite-(Y)	$(Ca,Ce)_{3-4}(Y,Ce)_2(Ti,Al,Fe^{3+})B_4Si_4O_{22}(O,OH)_2$	$P2/a$	18.946	4.741	10.35	111.7	54.2.2.2
Cappelenite-(Y)	$Ba(Y,REE)_6[Si_3B_6O_{24}]F_2$	$P3$	10.67		4.680		54.2.3.1
Stillwellite-(Ce)	$(Ce,La,Th,Ca)BSiO_5$	$P3_1$	6.84		6.70		54.2.3.2
Okanoganite-(Y)	$(Na,Ca)_3(Y,Ce)_{12}Si_6B_2O_{27}F_{14}$	Unknown	10.72		27.05		54.2.4.3
Garrelsite	$Ba_3NaSi_2B_7O_{16}(OH)_4$	$C2/c$	14.65	8.48	13.46	114.30	54.2.4.1
Howlite	$Ca_2B_5SiO_9(OH)_5$	$P2/a$	8.61	9.35	12.81	104.83	52.2.4.2

54.2.1a.1 Datolite $CaBSiO_4(OH)$ or $Ca_2(\square)B_2Si_2O_8(OH)_2$

Named in 1806 by Esmark from the Greek *to divide*, in allusion to the granular character of some of its varieties. Datolite–gadolinite group. MON $P2_1/a$. $a = 9.620–9.636$, $b = 7.600–7.610$, $c = 4.820–4.835$, $\beta = 90.08–90.40°$, $Z = 2$ (for 10 oxygen atoms), $D = 3.00$. *36-429*(nat): 3.755_5 3.40_7 3.12_{10} 2.86_7 2.52_5 2.25_3 2.19_3 1.88_5. Structure: *AM 58:909(1973)*. Generally platy to short, prismatic; crystals to 14cm in length, also equant, may occur as spherical aggregates or as massive, granular to compact and cryptocrystalline masses; dominant forms are {011} and {110}; sometimes tabular parallel to (102). Colorless, white, gray, yellowish, green, pale pink, bluish white, greenish white; white streak; vitreous to greasy luster. Cleavage (001), imperfect; uneven to conchoidal fracture; brittle. $H = 5-5\frac{1}{2}$. $G = 2.90–3.0$. Sometimes shows blue or white luminescence in SW UV. Partly soluble in HCl, HNO_3, and H_2SO_4. May have some MgO (to 1.3 wt %), FeO (to 0.8%), Al_2O_3 (to 0.5%), Na_2O (to 0.2%). Biaxial (–); $Y = b$, $X \wedge c = 1°$, $Z \wedge a = 1-3°$; $N_x = 1.623–1.626$, $N_y = 1.650–1.654$, $N_z = 1.667–1.670$; $2V = 74°$; $r > v$, weak. Occurs as a hydrothermal, postvolcanic mineral in veins and cavities in basic rocks. Often associated with calcite, prehnite, and zeolites. Also associated with danburite; in gneiss, diorite, serpentine, and other rock types. Sometimes in beds of iron ores. Occurs associated with diabase of Connecticut River valley, as at Westfield, Hampden Co., MA; in CT in Hartford Co., in large crystals at Tariffville and from near Hartford; at Meriden, New Haven Co. In New Jersey in the amygdaloidal basalts of Bergen Hill, Great Notch, Essex Co., Paterson, and Prospect Park, Passaic Co., and elsewhere, in splendid crystals and associated with various zeolites. At the Goose Creek Quarry, near Icesburg, Loudon Co., VA. Occurs in both crystalline and compact opaque varieties associated with the Keweenaw Basalts of the Michigan copper district, Keweenaw Co., MI. Clear Creek, San Benito Co., and along the Russian River, Mendocino Or., CA. Occurs at Fifeshire, Scotland; found at Andreasberg, Harz Mts., Germany, in diabase and in veins of silver ore, and at Haslach, Black Forest. From the Alpe di Siusi, Trentino; on the Seiser Alpe (Mont da Sous) and at Tiso (Theiss) northeast of Chiusa (Klausen); in geodes in amygdaloidal basalt; in granite at Baveno, Lake Maggiore; at Toggiano near Modena, Italy, in serpentine and from Serra del Zanchetti in the Apennines near Bologna in highly complex crystals. A datolite hornfels occurs at Repiste near Paskov, and veins of datolite occur at Zermanice Dam, northern Moravia, Czech Republic. Occurs at Arendal and Utö in Aust-Agder, Norway, in a bed of magnetite; fine pale blue to green crystals from Dalnegorsk, Primorsky Krai, Russia; Wessels mine, Kalahari Mn field, Cape Prov., South Africa. Fine crystals from Charcas, San Luis Potosi, Mexico; Rosebery, Montagu Co., Tasmania; Iwato Mine, Miyazaki Pref., Japan. Sometimes used as precious stone. EF

54.2.1a.2 Hingganite-(Ce) $(Ce,Y)_2(\square,Fe^{2+})Be_2Si_2O_8(OH,O)_2$

Named in 1987 by Miyawaki *et al.* Datolite–gadolinite group. See hingganite-(Y) (54.2.1a.3) and hingganite-(Yb) (54.2.1a.4). MON $P2_1/a$. $a = 9.996$, $b = 7.705$, $c = 4.792$, $\beta = 90.06°$, $Z = 2$, $D = 4.66$. *40-1452*(nat): 6.09_5 $4.80_{7.5}$ 3.16_{10} $2.87_{7.5}$ $2.58_{7.5}$ $2.565_{7.5}$ 1.99_8. Structure is like that of gadolinite. However, the octahedral Fe site is empty or nearly so. One of the corners of the BeO_4 tetrahedra is mainly occupied by (OH) in hingganite. Idiomorphic crystals, 1–5 mm long. Light red brown. Analysis (wt %): FeO, 5.65; Y_2O_3, 10.9; La_2O_3, Ce_2O_3, 16.77; Pr_2O_3, 3.5; Nd_2O_3, 9.79; Sm_2O_3, 4.70; Gd_2O_3, 4.18; Tb_2O_3, 0.5; Dy_2O_3, 3.82; Ho_2O_3, 1.08; Er_2O_3, 1.84; Yb_2O_3, 1.00; Lu_2O_3, 0.3; CaO, 0.39; BeO, 9.27; SiO_2, 22.27; H_2O, 1.9. Occurs in pegmatite (drusy) associated with quartz, feldspar, mica, SnO_2, stokesite CaF_2, and so on. Occurs in the *Iwaguro Sekizai quarry, Tahara area, Hirukawa, Ena, Gifu Pref., Japan.* EF *KZ 18:17(1987), AM 75:432(1990).*

54.2.1a.3 Hingganite-(Y) $Y_2(\square)Be_2Si_2O_8(OH)_2$ or $(Y,Ce)_2(\square,Fe)_2Be_2Si_2O_8(OH)_2$

Named in 1983 by Xiaoshi et al. for the locality. Datolite–gadolinite group. Synonyms: yttroceberysite [*GRC 27:459(1981)*], and yberisilite *CCCM(1972)*. MON $P2_1/a$. *AMS 5:289(1985)*.(Nat): $a = 9.895$–9.930, $b = 7.606$–7.676, $c = 4.733$–4.768, $\beta = 90.17$–$90.35°$, $Z = 2$; *26-812*(syn): $a = 9.861$, $b = 7.605$, $c = 4.720$, $\beta = 89.65°$, $D = 4.53$–4.72. *26-812*(syn): 3.12_8 3.11_8 2.84_{10} 2.55_5 2.54_5 2.20_3 2.19_3 1.97_3. Structure is like that of gadolinite. *AMS 5:289(1985)*, *CM 30:127(1993)*. Vacancy exists for the divalent cation site. Subhedral grains 0.1–0.2 mm, crystals to 1 mm long and fan to sheaflike aggregates to 2 mm from Zomba, Malawi; a few are stout prisms; forms developed: {011}, {111}, minor {100}. Milky white, colorless, light yellow or light green, pale blue and pale rose; white streak; vitreous to greasy luster. Uneven fracture. $H = 5$–$5\frac{1}{2}$, 4.08–4.57 (China). Biaxial (+); $Z \wedge c = 6$–$13°$ in acute β, $Y = b$, $X \wedge a = 14°$; $N_x = 1.744$–1.748, $N_y = 1.752$–1.765, $N_z = 1.765$–1.783; colorless; $2V = 80°$; $r < v$ strong. Generally in agpaitic granitic pegmatites. Occurs in druse pegmatite with quartz, feldspar, mica, cassiterite, stokesite, fluorite, chlorite, and titanite in Japan; in a Be- and REE-bearing granophyre in Manchuria, China; in amazonite pegmatite on Kola Penin., Russia, in granite pegmatite in albitized part with muscovite, spodumene, and pyrochlore. Occurs at Trimouns, Ariège, France; Glogstafelberg, Val Formazza, Switzerland; Saga, Tvedalen, and Høydalen, Tørdal, Langesundfjord, Norway; Tuva, and Ploskogorskaya, Kola Penin., Russia; *Heilongjiang, and greater Khingan area, Manchuria, China*; Tahara area, Hirukawa-mura, Enagun, Gifu Pref., Japan; Zomba–Malosa complex, southern Malawi. EF *SPC 8:539(1964)*, *AM 59:700(1974)*, *DAN SSSR 270:1188(1983)*, *KZ 18:17(1987)*, *NJBM:185(1994)*.

54.2.1a.4 Hingganite-(Yb) $(Yb,Y)_2(\square)Be_2Si_2O_8(OH)_2$

Named in 1983 by Voloshin et al. for the relation to hingganite. Datolite–gadolinite group. MON $P2_1/a$. $a = 9.888$, $b = 7.607$, $c = 4.740$, $\beta = 90.45°$, $Z = 2$, $D = 4.83$. *35-705*(nat): 6.06_7, 4.76_6, 3.74_6, 3.45_6, 3.13_{10} 2.85_{10} 2.57_8 2.54_8. Spherical aggregates of fine acicular crystals to 2 mm long, 0.1–0.2 mm wide. Colorless, vitreous luster, transparent. No cleavage. $H = 6$–7. IR data: *DAN SSSR 270:1188(1983)*. Analysis (wt %): Y_2O_3, 8.56; Yb_2O_3, 34.07; Gr_2O_3, 8.22; Dy_2O_3, 2.47; Lu_2O_3, 4.50; Tm_2O_3, 3.10; Ho_2O_3, 1.03; CaO, 1.14; SiO_2 22.11; BeO, 10.90; H_2O, 3.74. Biaxial (+); $Z \wedge c = 23°$, $X \wedge c = 20°$; $N_x = 1.725$, $N_y = 1.738$, $N_z = 1.760$; $2V = 65°$. OAP = (010). Occurs in an amazonite pegmatite on plumbomicrolite and violet fluorite, as a product of very late stage replacement reactions. *Ploskogorskaya, Keivy district, Kola Penin., Russia*. EF *K 28:457(1983)*.

54.2.1b.1 Bakerite $Ca_4B_5Si_3O_{15}(OH)_5$

Named in 1903 by Giles for Richard C. Baker (1858–1937), Nutfield, Surrey, one of the managing directors of Borax Consolidated Ltd. Baker provided

Giles with the material. Datolite–gadolinite group. MON $P2_1/a$. $a = 9.533$–9.610, $b = 7.569$–7.596, $c = 4.794$–4.814, $\beta = 90.00$–90.41°, $Z = 1$, $D = 2.94$–2.99. $36\text{-}428$(nat): 3.72_5 3.10_{10} 3.08_6 2.97_5 2.84_{10} 2.49_5 2.23_7 2.17_5; MJJ $17\text{:}111(1994)$: 3.75_5 3.40_3 3.11_{10} 2.98_5 2.85_8 2.52_4 2.24_4 2.18_4. The latter data while reported for bakerite are most likely from datalite. The structure has not been determined in detail. Usually massive but crystals to 0.2 mm occur as short rhombic prisms with oblique terminations or as thin diamond-shaped tablets; dominant forms are $\{11\bar{1}\}$, $\{100\}$, $\{210\}$, and less commonly $\{011\}$. Gray to white, colorless, sometimes greenish white, may contain clay inclusions; white streak; vitreous to porcelaneous or dull luster. No cleavage. $H = 4\frac{1}{2}$. $G = 2.88$. DTA, TG, and IR: MJJ $17\text{:}111(1994)$. Readily soluble in hot dilute HCl. In bakerite $SiO_2/B_2O_3 = 1.0$. Formula also given as: $Ca_4B_4(BO_4)(SiO_4)_3(OH)_3 \cdot H_2O$. Biaxial $(-)$; $Y = b$, $2V_{OBS}$ and $2V_{calc}$, $Z \wedge c = 44°$; $N_x = 1.623$–1.626, $N_y = 1.635$–1.648, $N_z = 1.654$–1.658; $2V = 67°$; various authors report higher $2V$'s and also optically $(-)$, however, calculated signs is always $(+)$! birefringence $= 0.03$; negative elongation. A low-temperature hydrothermal mineral. Occurs as irregular veins in altered volcanic rocks (Inyo and Los Angeles Cos., CA); in thin veins in diabase spilite (Turkey); in slag heaps of old Fe and Cu mines (Italy); also in skarns as a vein-forming mineral (Japan). Ocurs with datolite at the Sterling borax mine, Tick Canyon, Los Angeles Co., Corkscrew Canyon, Black Mts. W of *Furnace Creek District, Death Valley, Inyo Co., CA*; either bakerite or datolite occurs at Tory Hill, Bancroft, ON, Canada; Charcas, SLP, Mexico; Sivas, Golcuk Plateau, Turkey; with fluorapophyllite in veins in limestone at Fuka, Okayama Pref., Japan. EF $DII\text{:}363(1951)$; AM $41\text{:}689(1956)$, $47\text{:}919(1962)$, $56\text{:}1109(1971)$; MIN $3(1)\text{:}395(1972)$; RI $(2)\text{:}67(1985)$.

54.2.1b.2   Gadolinite-(Ce) $(Ce,La,Nd,Y)_2Fe^{2+}Be_2Si_2O_{10}$

Named in 1978 for Johan Gadolin (1760–1852), Finnish chemist and discoverer of yttrium. Datolite–gadolinite group. The Ce-dominant analog of gadolinite-(Y). MON $P2_1/a$. $a = 10.01$, $b = 7.58$, $c = 4.82$, $\beta = 90.78°$, $Z = 2$, $D = 4.90$. $29\text{-}1408$, $29\text{-}1409$(heated)(nat): 4.81_9 3.59_6 3.18_7 3.17_6 3.00_7 2.88_{10} 2.60_8 2.59_8. Some is metamict. Irregular masses to 20 mm diameter (Norway). Black, vitreous luster. Conchoidal fracture. $G = 4.20$–4.22 (metamict, Norway). Norwegian material is chemically zoned, with higher LREE contents on the rims. Biaxial; $N_y = 1.78$. Norwegian material is olive-green. Occurs in alkalic pegmatites and alkalic granites. Pale blue nonmetamict material occurs at Liberty Bell Mountain, WA; near *Skien, Bjørkedalen valley, SW Oslo region*, Bakken quarry, Tvedalen, and Fyresdal, Telemarken, *Norway*. EF AM $63\text{:}188(1978)$.

54.2.1b.3   Gadolinite-(Y) $Y_2Fe^{2+}Be_2Si_2O_{10}$

Named in 1800 by Klaproth for J. Gadolin (1760–1852), Finnish chemist, who first separated yttrium in 1794. Datolite–gadolinite group. MON $P2_1/a$. (Nat): $a = 10.000$, $b = 7.565$, $c = 4.768$, $\beta = 90.31°$, $Z = 2$ [AM $69\text{:}948(1984)$] and

54.2.1b.3 Gadolinite-(Y)

$a = 9.900$–9.970, $b = 7.510$–7.598, $c = 4.737$–4.759, $\beta = 90.41$–$90.79°$ (for 12 Alpine samples); (syn): $a = 9.920$, $b = 7.484$, $c = 4.747$, $\beta = 90.40°$ (*26-1134*), D = 4.41. See also *22-990*, *22-991*. *AM* 69:948(1984)(nat): 4.78_6 3.14_7 3.06_3 2.97_5 2.85_9 2.84_{10} 2.58_4 2.56_5; *26-1134*(syn): 4.74_8 3.12_5 3.11_5 2.93_7 2.82_{10} 2.81_{10} 2.56_6 2.54_6. Some vacancy exists at the Fe^{2+} site. Have OH for O to compensate for the lack of (+) charge at the Fe site. $RE_{2.00}Fe^{2+}Be_{2.00}Si_{2.00}O_{9.72}(OH)_{0.28}$. Sheets parallel to (001) plane, which is formed by corner-sharing linkages of SiO_4 and BeO_4 tetrahedra. The two REE atoms are situated between the eight-membered rings of the tetrahedra to form eight-coordinated square antiprisms. The Fe atoms are at the origins of the unit cells. *AM* 69:948(1984), *CM* 30:127(1993). Most common forms are {100}, {001},{110}, {120}, {011}, prismatic or rough crystals to >20 cm, commonly in compact masses; often metamict or partly so; radioactive, contains U and Th. Pale green, green, blue-green when fresh, unaltered, and nonmetamict; black or red-brown when partly altered and metamict; greenish-gray streak; vitreous to greasy luster. No or very poor cleavage on {100}, conchoidal to splintery fracture, brittle. H = $6\frac{1}{2}$–7. G = 4.21–4.77, 4.33(syn), 3.42(metamict), 4.77(fresh). IR data: *MIN* 3(1):403(1972). Gelatinizes in acids, especially if metamict; dissolves with difficulty upon heating in HCl. Magnetic properties: *MZ* 6:64(1984). B or Be substitution for Si is very small or nonexistent. In fresh material, all Fe is Fe^{2+}, and FeO >> Fe_2O_3. Ca substitutes for Fe^{2+} usually 1–2 wt % CaO, as much as 4% CaO. B substitutes for Be (as much as 4.2% and possibly 5.0% B_2O_3). Also (OH) substitutes for O. ThO_2 to >3%. One gadolinite from Kragerø (Lind Vikskollen), Norway, has about 7% CaO and 5% FeO. Metamict varieties may have several wt % each of UO_2 and ThO_2. *MIN* 3(1):403(1972). Biaxial (+); N_y parallel to b; X $\wedge$ $c = 3$–$12°$, Z $\wedge$ $a = 3$–$12°$; $N_x = 1.665$–1.82, $N_y = 1.774$–1.812, $N_z = 1.787$–1.824; 2V = 80–85°; rarely biaxial (−), 2V = 85°; r < v; OAP = (010). X, olive; Y,Z, green in nonmetamict material, also brownish green to greenish brown. Occurs most often in alkalic granitic pegmatites or granites. Also occurs in alpine-type veins. Associated with other REE minerals and F-bearing minerals. Often partly or completely metamict. Occurs at many worldwide localities; only noteworthy localities are mentioned here. Masses to 95 kg at the Baringer Hill and Rhode Ranch pegmatites, Llano Co., TX; White Cloud mine, South Platte district, Jefferson Co., and Devils Head, Douglas Co., CO; near Kingman, Mohave Co., AZ; Golden Horn Batholith, WA; Loughborough Twp., Frontenac Co., ONT, Canada; in at least 15 locations in *Sweden: Ytterby*, and at Kopparberg, Brodbø, Finbo, and Korarfvet near Falun. Segregations to 500 kg from Frikstad, Iveland; nonmetamict material from Hundholmen, Tysfjord; at Hitterö and Sätersdal, Vest Agder; and Bidjovagge, Finnmark, Norway. Lovböle, Kimito, Finland. Fresh and unaltered material at several localities in the Swiss Alps: Val Vigezzo (Ossola), Rauris Valley, and Markogen pegmatite near Villach; Piz Blas, Val Nalps, Graubunden, Switzerland; Baveno, Italy; Radautal, Harz, Germany; Shklvarska-Porema, Poland; western Keivy and Kanozero, Kola Penin., Russia. Occurs in at least 10 localities in Japan (e.g., nonmetamict material from Miyazuma, Yokkaichi; Takehinata, Kohu City, Yamanachi Pref.; Fusamata, Aichi Pref.; Hazumachi; Shinden, and Kurokawa, Gifu Pref.). Alto Ligonha, Mozambique. EF *AM 59:700(1974), MIN 3(1):403(1972).*

54.2.1b.4 Calciogadolinite CaREE(Fe^{3+})Be$_2$Si$_2$O$_{10}$

The name is not approved by the CNMN IMA. Datolite–gadolinite group. MON $P2_1/a$. $a = 9.988$, $b = 7.566$, $c = 4.696$, $\beta = 90.00°$ for synthetic material. *AM 59:700(1967)*; $a=9.97$, $b=7.56$, $c=4.69$, $\beta=90.0°$ *(CCCM1:1972)* Z=2. General formula: (REE$_{2-x}$Ca$_x$) (Fe$^{2+}_{1-x}$Fe$^{3+}_x$)Be$_2$Si$_2$O$_{10}$ *AM 63:188(1978)*. Tadati (Tadachi) village, Nagano Pref., Japan [*CSJB 13:591(1938)*;] possible calciogadolinite with B site (Fe site) partially vacant, from Lindvikskollen, Kragerø, Norway. EF *NGT 52:197(1972)*.

54.2.1b.5 Homilite Ca$_2$(Fe,Mg,Mn)B$_2$Si$_2$O$_{10}$

Named in 1876 by Paikjull from the Greek for *to occur together*, in allusion to its association with meliphanite. Datolite–gadolinite group. See other members of the group. MON $P2_1/a$. *AC(C) 41:13(1985)*: $a = 9.786$, $b = 7.621$, $c = 4.776$, $\beta = 90.61°$, $Z = 2$, $D = 3.45$; *36-430*: $a = 9.787$, $b = 7.614$, $c = 4.780$, $\beta = 90.56°$. *36-430*(nat): 3.13_3 3.11_{10} 2.98_7 2.86_6 2.85_5 2.55_9 2.53_4 2.19_3. The structure may be derived from datolite by placing the Fe atom in the octahedral vacancy and by charge-compensation replacement of OH$^-$ with O^{2-}. Common forms are {001}, {100}, {110}, {210}, and {011}; often tabular parallel to c (001) or octahedral; largest crystal was ≈ 2 in. long, another weighed 50 g; twinning on (001), also on (100) and (032). Pale green, black, blackish brown; gray-white streak; resinous–vitreous luster. Distinct cleavage not developed, poor to conchoidal fracture, brittle. H = 5. G = 3.28–3.38. IR data: *AM 59:700(1974)*. Readily soluble in common acids. Analysis (wt %): Na$_2$O, ≈ 1; MgO, to 0.5; Al$_2$O$_3$, to 2.7; REE$_2$O$_3$, to 2.5; and so on. Biaxial (+); $Z = b$, Y nearly parallel to c ($Z \wedge c = 1–4°$, Brögger); N$_x$ = 1.710–1.715, N$_y$ = 1.717–1.725, N$_z$ = 1.728–1.738; 2V = 76–80°; Y > X > Z; r > v, distinct; OAP perpendicular to (010); strong horizontal dispersion of the bisectrix. X, bluish green; Y, medium brownish red to brownish gray; Z, smoky gray to brownish yellow. Occurs in veins in augite Ne syenite with meliphanite and also zircon, titanite, löllingite, and others. Alteration: altered material may be isotropic or anisotropic with variable optical properties. Occurs on the *island of Stokö and on neighboring islands (e.g., Arø) near Brevik, Langesundfjord, Telemark Co., Norway.* EF *GFSF 3:229(1876)*, *DS 6:505(1920)*, *MIN 3(1):399(1972)*.

54.2.1b.6 Minasgeraisite-(Y) (Y,Bi)$_2$CaBe$_2$Si$_2$O$_{10}$

Named in 1986 by Foord *et al.* for the locality. Datolite–gadolinite group. MON $P2_1/a$. $a = 9.833$, $b = 7.562$, $c = 4.702$, $\beta = 90.46°$, $Z = 2$ [*AM 71:603(1986)*], $D = 4.90$(end member) and 4.29(observed composition). *39-344*(nat): 5.99_3 3.71_3 3.41_3 3.11_{10} 2.83_{10} 2.54_9 2.25_3 2.19_3. The structure is of the gadolinite-type. Single and multiple rosettes, curved sheaves; individual crystals are 3–5 µm across, 0.2–1.0 mm across as single and multiple rosettes. Pale to medium purple, pale purple streak, earthy to subvitreous luster. One excellent cleavage, probably {100}; a second good cleavage, {001}. H = 6–7. G > 4.25. Nonfluorescent under either SWUV or LWUV. Soluble in acids, gelatinizes in hot H_2SO_4 or HCl. Is zoned into Bi-rich and lesser-Bi portions (14.7–28.5% Bi_2O_3). MnO, 2.8–3.5%; P_2O_5, 1.2%; B_2O_3, 1.5%. For the Bi-richest substitution, $(REE_{0.86}Bi_{0.65}Ca_{0.49})$. Rims are Bi and Mn rich. Biaxial (+); $N_x = 1.740(4)$, $N_y = 1.754(4)$, $N_z = 1.786(4)$; 2V = 50–72°, av. 68°; Z > Y > X; r > v, very weak. X, colorless; Y, pale grayish yellow, 50–72°; Z, lavender purple. Occurs as a sparse, accessory, late-stage mineral in small druses in a zoned, complex granitic pegmatite. Associated minerals are Y-bearing milarite, muscovite, quartz, and albite. Occurs at the *Lavra de Sr. Jose Pinto, Jaguaracú, Municipio de Timoteo, MG, Brazil.* EF

54.2.2.1 Hellandite-(Y) $(Ca,Y,\square)_6(Al,Fe)B_4Si_4O_{18}(OH,O)_4(OH)_2$ or $(CaY,\square)_4(Y,\square)_2 \cdots$

Named in 1903 by Brögger for Amund Theodor Helland (1841–1918), Norwegian geologist. Datolite–gadolinite group. Canadian material has Y > Ca. Tadzhikite-(Y), Ca_3Y_2 (Ti, Al, Fe) $B_4Si_4O_{22}$, is probably the Ti analog of hellandite-(Y). MON $P2/a$. $a = 18.824$–18.99, $b = 4.684$–4.715, $c = 10.248$–10.30, $\beta = 111.4$–111.7°, Z = 2, D = 3.645. *25-184*(nat): 4.69_8 3.44_7 3.20_7 3.07_6 2.88_6 2.81_{10} 2.635_8 2.60_8. *CM 11:760(1972)*. The main structural features are infinite silicoborate chains $[Si_4B_4O_{20}(OH)_2]^{-14}$. Two such chains are parallel to c in the (010) plane, with Si and B atoms at Y ≈ $\frac{1}{2}$, and are related to each other by the glide plane. Fe and Al are in octahedral coordination and Ca, Y, and REE are in square antiprismatic coordination. All are located at Y ≈ 0 and connect the silicoborate chains. Closely related to Tadzhikite. *AM 62:89(1977)*. Prismatic to tabular. Twinning on (100). Brownish red to black, brown, sometimes colorless; vitreous luster. Cleavage (100), distinct. H between $4\frac{1}{2}$ and $5\frac{1}{2}$. G = 3.35–3.63. Gelatinizes with acids (metamict material). Japan: $(Ca_{5.75}Y_{3.70}Nd_{0.24}Dy_{0.21}\cdots)\Sigma 10.65$ $(Al_{1.45}Mn_{0.45}Ti_{0.21}Fe_{0.11})_{2.22}$ $(OH)_4$ $[Si_8 B_8 O_{40.22}(OH)_{3.78}$, Italy: $(Ca_{5.10}Y_{3.01}\Sigma REE_{4.56}))_{9.66}(Al_{1.12}Fe_{0.89}^{3+}Mn_{0.18})\cdots$, Norway: $(Y_{3.46}Ca_{3.7}\Sigma REE_{4.92})_{8.62}(Al_{0.91}Fe_{0.89}^{3+}Mn_{0.18})\cdots$, Canada: $(Y_{4.62}Ca_{3.26}\Sigma REE_{6.60})_{9.86}(Al_{1.77}Fe_{0.56}^{+3}Mn_{0.15})$; all on Si = 8.0 atoms. B_2O_3 is low in Norwegian and Canadian material: 5.44 atoms and 4.79 atoms instead of 8.0. H_2O goes up in this same material. Possible Mn analog at the Belo Horizonte pegmatite, Riverside Co., CA. Biaxial (−); X = b; $N_x = 1.656$, $N_y = 1.662$–1.749, $N_z = 1.668$; maybe metamict, with $\bar{N} = 1.75$; 2V = 87–90°. Occurs in druse pegmatite with hingganite, quartz, fluorite, chlorite, and zeolites. Colorless crystals cover hingganite (Japan). In granite pegmatites at Kragerø, Norway, associated with tourmaline, thorite, allanite, apatite, phenakite, and zircon. In granite at Predazzo, Italy. Belo Horizonte pegmatite, Anza, and Jensen quarry, Riverside Co., CA; Evans–Lou mine, Wakefield Lake, southern QUE, Canada; Predazzo, Trentino-alto Adige, and Lago Vico, Viterbo, Italy; *Lindvikskollen, Kragerø, Norway*; Trimouns Tak deposit, near Luzenac

Ariège, France; in ores of the Zaaiplaats tin mine, northern Transvaal, South Africa; Tahara area, Gifu Pref., Japan. EF *NGT 44:35(1964), TMPM 10:125(1965), MIN 3(1):845(1972), CM 11:760(1972), KZ 18:17(1987)*.

54.2.2.2 Tadzhikite-(Y) $(Ca,Ce)_{3-4}(Y,Ce)_2(Ti,Al,Fe^{3+})B_4Si_4O_{22}(O,OH)_2$ or $Ca_7Y_5(Al,Ti)_2(OH)_4Si_8B_8O_{44}$ or $Ca_7Y_5(Al,Ti)_2O_2Si_8B_8O_{44}$

Named in 1970 by Efimov et al. for the locality. Datolite–gadolinite group. Hellandite is hydrated. See hellandite (54.2.2.1), the Al analog. MON $P2/a$. $a = 18.946$–18.97, $b = 4.71$–4.714, $c = 10.302$–10.39, $\beta = 111.6$–$111.8°$, $Z = 2$, $D = 3.73$–3.77. *24-137*(nat): 4.97_3 3.48_2 2.94_3 $2.86_{2.5}$ 2.65_{10} $2.19_{2.5}$ $1.91_{5.5}$ 1.66_3. Also TRIC. Two morphological types found in tadzhikite: (1) spherulites of bent flakes not more than 1.5 cm in diameter, and (2) flattened prismatic crystals with rhombic prism and {010} pinacoid, also botryoidal, flaky, radiating lamellar, foliated masses; polysynthetic twinning present in both types, small deviations of α and γ from 90° indicate an initially monoclinic mineral that has become triclinic with the development of polysynthetic twinning; probably all monoclinic, but ordering of Ti would explain the change to triclinic symmetry. Yellowish brown to tan, beige, very pale pink and white, grayish brown to dark brown, colorless to brown; dull to pearly luster, sometimes vitreous. No cleavage is present on {100}; parting is parallel to (010). $H = 6$. $G = 3.86$ (type 1) and 3.73 (type 2). IR data: *DANS 264:342(1982)*. DTA data: *DANS-ESS 195:136(1970)*. Fuses at 1000°. Analysis (wt %), type 1: SiO_2, 24.7; TiO_2, 6.53; CaO, 18.3; ΣREE, 32.4; type 2: SiO_2, 23.4; TiO_2, 3.72; Al_2O_3, 2.30; CaO, 18.3; ΣREE, 34.1; Y dominant. Biaxial; $Z \wedge a = 7°$, $Z \wedge b = 4°$, $Z \wedge c = 25°$; type 1: $N_x = 1.750$–1.760, $N_z = 1.763$–1.772; type 2: $N_x = 1.761$, $N_z = 1.772$; $2V = -88$ to $+80°$ for type 1; very slightly pleochroic, nearly colorless to brown; wavy extinction. Occurs in alkalic pegmatites (former USSR and Norway) and as an accessory mineral in marble xenolith cavities (Mt. St.-Hilaire) and in igneous breccia cavities. Type 1 is in an albitized outer zone of pegmatite in the form of spherulites associated with ekanite, titanite, and eudialyte. Type 2 is in polylithionite–quartz replacement zones in an inner blocky zone of a pegmatite with arfvedsonite–riebekite, and accessory pyrochlore and tienshanite, and stillwellite. Mt. St.-Hilaire, QUE, Canada; Tvedalen, Langesundfjord, Norway; Cimini Hills, near Rome, Lazio, Italy; *Alaisk Ridge, Dara-i-Pioz massif, Tien Shan, Tadzhikistan.* EF *AM 62:89(1977), DAN SSSR 264:342(1982), CSREM 11:72(1987)*.

54.2.3.1 Cappelenite-(Y) $Ba(Y,REE)_6[Si_3B_6O_{24}]F_2$

Named in 1885 by Brögger for D. Cappelen (1856-?) of Holden, Norway. Datolite–gadolinite group. HEX-R (trigonal) $P3$. $a = 10.67$, $c = 4.680$, $Z = 1$, $D = 4.40$. *27-47*(nat): 4.7_8 3.48_9 2.80_{10} 1.94_9 1.84_6 1.77_7 1.71_7 1.66_8. See also *39-1349*(calc) pattern. Structure: *AM 69:190(1984)*. B_6O_{15} rings are linked by isolated SiO_4 tetrahedra to form a sheet perpendicular to {001}. Topologically, the mineral is $P6mm$. Pure crystals of end-member composition may

54.2.4.1 Garrelsite

satisfy this space group. Prismatic, two principal forms $\{10\bar{1}0\}$ hexagonal prism and pyramid $\{10\bar{1}1\}$ or $\{10\bar{1}3\}$. Often doubly terminated; two crystals from Norway; one is a 2 cm × 1.5 cm free crystal. Metamict. Reconstituted at 725° for two weeks. White, yellow-white, greenish brown; white streak. No cleavage, conchoidal fracture, vitreous luster. H = 6. G = 4.20–4.40. Readily soluble in acids. Uniaxial (−); $N_O = 1.751$, $N_E = 1.748$ (Na light); pale brownish yellow; not dichroic. Contains 0.87% ThO_2—cause for metamict state. Substitution of Ca for Ba. Occurs in nepheline–syenite pegmatite veins associated with other alkalic minerals: wöhlerite, rosenbushite, catapleiite, lävenite, and so on. *Lille Arø island, Langesundfjord, Telemark, Norway*; Kazakhstan. EF *MRE 2:248(1966)*.

54.2.3.2 Stillwellite-(Ce)   (Ce,La,Th,Ca)BSiO₅

Named by McAndrew and Scott in 1955 for Frank Leslie Stillwell (1888–1963), an Australian mineralogist. Datolite–gadolinite group. HEX-R $P3_1$. $a = 6.832$–6.841, $c = 6.653$–6.702, Z = 3, D = 4.76. *25-1447*(nat): 4.44_7 3.43_{10} 2.96_{10} 2.71_6 2.24_6 2.23_6 2.13_8 1.86_7; *26-349*(syn): 4.44_9 3.42_9 2.97_8 2.91_{10} 2.20_5 2.13_8 2.09_6 1.86_{10}. Structure: *NJBM:49(1992)*, *CM 31:147(1993)*. Structure is probably $P3_121$ at high temperature and undergoes a ferroelectric phase transition to $P3_1$ upon cooling. Hexagonal tabular with prism $\{11\bar{2}0\}$ and rhombohedron $\{10\bar{1}1\}$; habit is very similar to quartz. Twinned on [100]. Crystals from Tadzhikistan are as large as 5 cm × 1 cm, and fist-size pieces exist of the Australian material. Sometimes partly metamict (e.g., Mary Kathleen mine, Mt. Isa district, Q, Australia). Streak white, dark brown, light brown, pink, maroon red, brownish yellow, and waxy white; vitreous to greasy luster (metamict material is resinous). No cleavage, subconchoidal to conchoidal fracture. H = $5\frac{1}{2}$–$6\frac{1}{2}$. G = 4.60–4.70. No reaction in acids. IR data: *MS 24:1(1970)*. Sometimes shows an alexandrite effect: pinkish tan to greenish brown. Uniaxial (+); $N_O = 1.76$–1.778, $N_E = 1.78$–1.787; negative elongation. Sometimes anomalously biaxial with $2V_2 = 0$–$6°$. Contains as much as 5.4 wt % ThO_2 and variable amounts of CaO, minor Al_2O_3, and sometimes minor UO_2. Occurs with uraninite and allanite in garnet–diopside skarns formed between granodiorite and granulite. Also in magnetite ore, in pegmatites and hydrothermal veins associated with alkalic rocks, and in volcanic ejecta (syenite and sanidinite). Mineville, Essex Co., NY; Mt. St.-Hilaire, QUE, and the Desmont Mine, Bancroft, ONT, Canada; Langesundfjord area, Norway; Vico Volcano, Vetralla, Rome, Italy; Bassano Romano, Viterbo (Latium), Italy; Dara-i-Pioz, Alai Range, Tadjikistan(!); Inagli massif, southern Yakutia; Synnysh Massif, Siberia, Russia; *Mary Kathleen mine, Mt. Isa district, northwestern Q, Australia*. EF *MIN 3(1):272(1972)*.

54.2.4.1   Garrelsite   Ba₃NaSi₂B₇O₁₆(OH)₄

Named by Milton et al. in 1955 for Robert Minard Garrels (1916-1988), American geochemist and educator. Related to the gadolinite group. MON C2/c a 14.639-14.655 b 8.466-8.480 c 13.460-14.438 β 114.21-114.30°, Z=4,

D=3.89. 26-1369 6.13_3 4.23_3 3.94_3 3.64_8 3.05_{10} $2.87_{4.5}$ 2.03_6. The structure is a three-dimensional framework composed of silicoborate sheets and Ba-O polyhedral layers both running parallel to (001). A silicoborate with the pentaborate $[B_5O_{12}]^{9-}$ polyanion ACB 32:824(1976). Steep, bipyramidal, prismatic, commonly striated (direction) crystals to 3 mm. Dominant forms are {110}, {1$\bar{1}$0}, {21$\bar{1}$}, {2$\bar{1}\bar{1}$}, etc. Colorless to white, streak white, perfect {001} cleavage, vitreous luster, H about 6, G 3.68 (on impure material). Biaxial (−), 2V 55-72°, N_x 1.620, N_y 1.633, N_z 1.640, Y=b Z∧[101] (the edge of the {$\bar{1}$11} form) = 33°; one optic axis is nearly normal to this edge. An authigenic mineral, assoc. with nahcolite, shortite, searlesite, etc. in dolomitic shale of the Green River Formation. From several drill holes in *Uintah Co., Utah*. Also occurs in the Kramer borate deposit, Kern Co., and at Searles Lake, San Bernardino Co., CA. USGSJR 2:213(1974), SMPM 53:199(1973).

54.2.4.3 Okanoganite-(Y) $(Na,Ca)_3(Y,Ce)_{12}Si_6B_2O_{27}F_{14}$

Named by Boggs in 1980 for the locality. Related to the datolite group. Compare an unnamed REE-fluorosilicate from Siberia. ZVMO 95:339(1966). Has Al_2O_3 and P_2O_5 and no B_2O_3. It may be the P and Al analog of okanoganite or the same species if B, Al, and P do not occupy a separate site but only substitute for Si. HEX-R Space group unknown. $a = 10.72$, $c = 27.05$, $Z = 3$, $D = 4.37$. 35-483(nat): 4.35_4 3.11_5 2.97_{10} $2.94_{9.5}$ 2.93_5 2.68_3 $1.98_{3.5}$ $1.78_{4.5}$. Structure unknown. Pseudotetrahedral crystals to 4 mm across, dominant form is {0001}, twinning: fourlings with TP {01$\bar{1}$4}, the c face of each individual forming the faces of the pseudotetrahedra, also rare interpenetration twins with the composition plane = {10$\bar{1}$2}. Rotation of 180° about a twin axis perpendicular to {01$\bar{1}$4}. Opaque white to reddish orange, light pink, yellow, tan; white streak; dull to brilliant luster. H = 4. G = 4.35. Nonluminescent. Substitutions: Ca + Th ↔ REE and Ca ↔ Na; TiO_2, 0.7%, ThO_2, to 3.2%, UO_2, to 0.3%, FeO, to 2.4%, MnO, 0.5%, and so on. HREE > LREE. Uniaxial (−); wavelength used: white light; $N_O = 1.753$, $N_E = 1.740$; colorless. Occurs in isolated unaltered miarolitic cavities of an alkaline granite. Occurs with arfvedsonite, astrophyllite, bastnäesite, polylithionite, zektzerite, zircon, and other minerals. Several hundred specimens found in ≈ 2 ft³ of rock. Occurs in the *Golden Horn batholith, near Washington Pass, Okanogan Co., WA*; occurs as inclusions in yttrian fluorite at Tysfjord, Norway. EF AM 65:1138(1980), R. C. Boggs (1984), Ph.D. thesis, UCSB, 187 pp.

54.2.5.1 Tritomite-(Ce) $(Ce,La,Ca,Y,Th)_5(Si,B)_3(O,OH,F)_{13}$

Named in 1849 by Weibye and Berlin from the Greek for *three* and *to cut*, in allusion to the trihedral cavities left in the host rock by the crystals. Synonymous with the original tritomite. See also tritomite-(Y) (54.2.5.2). HEX $P6_3/m$. $a = 9.35$, $c = 6.88$, $Z = 2$, $D = 4.94$. 14-174: 3.44_4 3.08_3 2.81_{10} 2.70_3 1.94_3 1.84_4 1.74_3 1.25_4. The morphology of the type material crystals is distinctly 3m, in the form of pseudotetrahedrons or steep trigonal pyramids with

54.2.5.3 Melanocerite-(Ce)

a basal pedion. The morphological cell has six times the axial ratio of the X-ray cell that is based on a heated metamict sample. The structure is thought to be related to the apatite structure or to that of datolite. Dark brown, amber, black vitreous to resinous luster. Conchoidal fracture, brittle. $H = 5\frac{1}{2}$. $G = 4.15$–4.25 as low as 3.7. Uniaxial, but usually metamict and isotropic with $N = 1.73$–1.76 N as low as 1.69. Gelatinizes in HCl. In nepheline syenite on *Låven Island, Langesundfjord, near Brevik, Norway*, and also on Stokø and Arø, near Barkevik. AR *AM 47:9(1962), 51:152(1966)*.

54.2.5.2 Tritomite-(Y) $(Y,Ca,La,Fe^{2+},Ca)_5(Si,B,Al)_3(O,OH,F)_{13}$

Named as Spenaite in 1962 by Jaffe and Molinski Tritomite is from the Greek for *three* and *to cut*. Synonym: spenceite, named in honor of Hugh S. Spence, who collected the material in 1934. Name changed to Tritomite-(Y) in 1966. See also tritomite-(Ce) (54.2.5.1). HEX $P6_3/m$. $a = 9.32$, $c = 6.84$, $Z = 2$. *27-1063*(heated): 3.16_5 2.83_{10} 2.73_5 1.96_4 1.93_4 1.85_7 1.76_5 1.11_5. Irregular grains and masses, to 2.5 cm. The pattern of heated material is very similar to that of fluorapatite. Dark greenish black, reddish brown, brownish black; vitreous to resinous luster. Conchoidal fracture, brittle. $H = 5$-$6\frac{1}{2}$(NJ), $3\frac{1}{2}$(ONT). $G = 3.05$(ONT), 3.40(NJ). Isotropic (metamict); $N = 1.63$–1.685. Found as subhedral grains with magnetite, zircon, and apatite in granite pegmatite at *Cranberry Lake, Sussex Co., NJ*; also massive in pegmatite in vuggy pyroxenite with calcite, apatite, diopside, wernerite, and fluorite in Cardiff Twp., and altered at the Faraday mine, Bancroft, Haliburton Co., ONT, Canada. AR *AM 47:9(1962), 51:152(1966); CM 12:66(1973)*.

54.2.5.3 Melanocerite-(Ce) $(Ce,TH,Ca)_5(Si,B)_3O_{12} \cdot nH_2O$
comparte tritomite-(Ce) & tritomite-(Y)

Named in 1887 (Brøgger) from the Greek meaning black and the composition. May be equivalent to tritomite-(Ce). HEX-R $a=9.35$ $c=6.88$ $Z=2$ $D=?$. *38-457* 3.44_4 2.81_{10} 1.84_4 1.24_4 corresponding to the strongest lines of tritomite-(Ce) but commonly metamict and giving only diffuse pattern. Tabular {0001} crystals. Brown, dark brown, reddish brown, black; streak light brown; opaque to somewhat translucent; luster greasy, vireous or resinous; fracture conchoidal to uneven; $H=5$-6 $G=4.1$-4.7. Radioactive, containing up to 13.6 wt% ThO_2; B_2O_3 up to 4.6 wt%; in addition to the elements shown in the formula Al, Fe, Mg, Mn, Pb, Ti, Ta, and Zr may be present in significant amounts. Soluble in 1:1 HCl leaving a silica residue; also soluble in concentrated H_2SO_4. Uniax $(-)N_o \approx 1.73$ $N_g \approx 1.72$; sometimes anomalously Biax, 2V=5-8°; metamict material isotropic with $N \approx 1.71$-1.77. Found in alkalic rocks and their cogenetic pegmatites; also in metamorphosed alkalic rocks. Sometimes altered to bastnaesite. Found at Wilberforce, ON, Canada; at *Aro Island and Kjeo Island, Barkevik, Langesundfjord, Norway*; in Russia in the Afrikanda massif, Kola Peninsula, in the Tatrsk massif, Krasnoyarsk region, and the Burpala massif, Northern Pribaikal. *ZVMO 91:573(1962), Vlasov 2:301(1966), Minerali 3-1:510(1972)*.

54.3.1.1. Vicanite-(Ce)
$(Ca,REE,Th)_{15}As^{5+}(As^{3+}{}_{0.5}Na_{0.5})Fe^{3+}Si_6B_4O_{40}F_7$

Described in 1995 by Maras *et al.* and named for the locality, the Vico volcanic district, Latium, Italy. Comments: cell parameters are very close to those of okanoganite-(Y). HEX.-R R3m a 10.795Å c 27.336Å Z=3 D=4.73; PD (no ICDD data) (nat.): 7.70_5 4.42_5 2.99_{10} 2.95_7 2.70_5 1.84_5 1.80_5 1.69_5. Structure: the main feature is the presence of a new polyanion $(Si_3B_3O_{18})^{-15}$. It is made up of a ring of three BO_4 tetrahedra, linked via an apex to three SiO tetrahedra. These polyanions are linked to each other by octahedrally coordinated Fe^{3+}, which also links iolated SiO_4 tetrahedra to form a complex layer of polyhedra perpendicular to the c direction. Both anionic and cationic interlayer positions may be vacant due to local charge balance *(EJM 7:439(1995))*. Transparent, yellowish green, anhedral grains or rare rice-grain crystals consisting of a vertical prism terminated by a pyramid, up to 0.3 mm in size, streak white, vitreous, conchoidal fracture, no cleavage, sometimes twinned giving a pseudocubic habit; non-fluorescent; H≈5-6, D<4.2; IR data given. SiO_2 13.8 wt.%, Al_2O_3 0.16, TiO_2 0.14, P_2O_5 0.38, AS_2O_3 4.49, Fe_2O_3 1.66, CaO 17.07, Na_2O 0.14, ThO_2 18.24, UO_2 1.96, RE_2O_3 30.98, B_2O_3 3.1, F 7.50, Σ101.81, O=F 3.16, Σ98.65. Uniax. (−), Na(D) N_o1.757 N_e1.722, non-pleochroic. Occurs in small miarolytic cavitgies associated with other rare minerals such as thorium-rich hellandite, asbecasite, and stillwellite-(Ce), inside volcanic ejecta from a pyroclastic formation of the *Vico volcanic complex at Tre Croci, Vetralla, Viterbo Prov., northern Latium, Italy. RI 3:207(1994)*.

================ Class 55 ================

Si$_2$O$_7$ Groups, Generally with No Additional Anions

55.1.1.1 Barylite BaBe$_2$Si$_2$O$_7$
Named in 1876 by Blomstrand from the Greek for *heavy*, in allusion to its high specific gravity. ORTH $Pn2_1a$. $a = 9.82$–9.835, $b = 11.65$–11.67, $c = 4.673$–4.695, $Z = 4$, $D = 3.99$. *20-119*(syn): 5.81$_5$ 4.52$_4$ 3.39$_9$ 3.05$_9$ 2.99$_{10}$ 2.92$_{10}$ 2.34$_4$ 2.21$_5$. Structure: *AM 62:167(1977)*. Colorless to white or pale blue, with {010}, {001}, {110}, and {100} dominant, tabular to bladed, elongated parallel to [001], flattened on {010}; piezoelectric; white streak; vitreous to waxy luster, sometimes dull. Cleavage (for most localities): {001}, {100}, perfect; and brittle. $H = 7$. $G = 4.02$–4.07. Some crystals (e.g., from Franklin, NJ, and Långban, Sweden) are fluorescent bluish but most are not fluorescent. Insoluble in acids. Traces of MgO, CaO to 1.8 wt %. Biaxial (−) or (+); for Långban and Seal Lake, Labrador, XYZ = bac; for Franklin, and Norwegian material, XYZ = bca; (Långban): $N_x = 1.691$, $N_y = 1.696$, $N_z = 1.703$, $2V_z = 81°$; (Franklin): $N_x = 1.695$, $N_y = 1.702$, $N_z = 1.708$, $2V_x = 70°$; OAP parallel to (001); r > v, weak to moderate or not observed. Occurs in alkalic pegmatite and miarolitic cavities at Mt. St.-Hilaire. Also in alkaline granite pegmatite at Lake George, CO. In alkaline pegmatite at Dara-i-Pioz. Contact aureoles of granites and in pegmatites. Also in nepheline syenite pegmatites and alkaline intrusions. Occurs at Franklin, NJ; in fine crystals (to > 10 cm long) with microlline, barite and fluorite in the Lake George ring-dike structure, Lake George, Park Co., CO; at Mt. St-Hilaire (very rare), QUE; in the Deadhorse Creek diatreme complex, Marathon, ONT; with eudialyte at Seal Lake, and in the Red Wine alkaline complex, central LAB, Canada; at Narsarssuk, Greenland; Bratthagen farm, Lågen Valley, and Brevig, on Arø Island, Langesundfjord area, Norway; with hedyphane at *Långban, Vårmland, Sweden*; above the Hackman valley on the southwestern slope of Mt. Yukspor, Khibiny massif, Kola Penin., Vishnevye Gor (Cherry Mountain), Ural Mts., Russia; Dara-i-Pioz, Tadjikistan; Pokrovo-Kirovsk, Kazakhstan. EF *NGT 46:335(1966)*, *MIN 3(1):560(1972)*, *MZ 2:84(1980)*, *TMPM 27:35(1980)*.

THORTVEITITE GROUP

The thortveitite group consists of four isostructural or nearly isostructural minerals conforming to the formula

ABSi$_2$O$_7$

A = Ca, Zr, Y, Yt, Sc
B = A

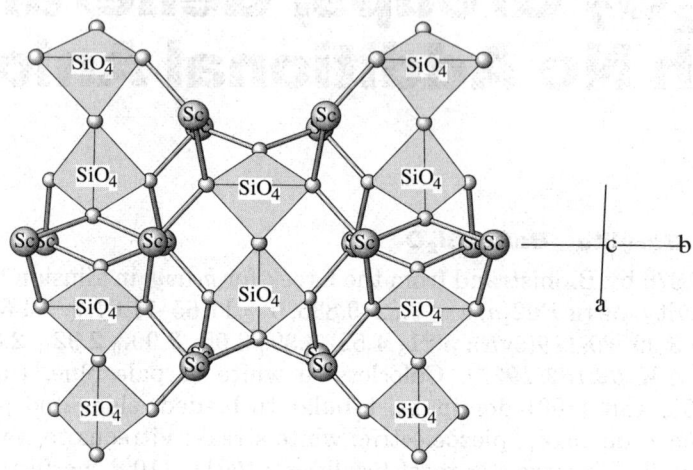

THORTVEITITE: (Sc,Y)$_2$Si$_2$O$_7$
Thortveitite has Si$_2$O$_7$ groups, interconnected with Sc. It is a rare example of linear Si–O–Si arrangement.

THORTVEITITE GROUP

Mineral	Formula	Space Group	a	b	c	β	
Gittinsite	CaZrSi$_2$O$_7$	C2	6.852	8.659	4.686	101.69	55.2.1.1
Keivyite-(Y)	(Y,Yb)$_2$Si$_2$O$_7$	C2/m	6.875	8.970	4.721	101.74	55.2.1.2
Keivyite-(Yb)	Yb$_2$Si$_2$O$_7$	C2/m	6.840	8.916	4.745	102.11	55.2.1.3
Thortveitite	(Sc,Y)$_2$Si$_2$O$_7$	C2/m	6.530	8.521	4.681	102.63	55.2.1.4

Note: Related structures include thalenite-(Y) (55.2.1.5), rosenhahnite (55.2.1.6), rowlandite-(Y) (55.2.1.7), and yttrialite-(Y) (55.2.1.8).

55.2.1a.1 Gittinsite CaZrSi$_2$O$_7$

Named in 1980 by Ansell et al. for John Gittins (1932–?), professor, Department of Geology, University of Toronto. Thortveitite group. Compare other members of the thortveitite group. MON C2. $a = 6.852$–6.878, $b = 8.659$–8.674, $c = 4.686$–4.697, $\beta = 101.69$–101.74°, a sample from Mongolia has: $a = 6.895$, $b = 8.691$, $c = 4.709$, $\beta = 100.81°$, Z = 2, D = 3.62. *33-322*(nat): 5.32$_6$ 4.62$_4$ 3.23$_8$ 3.16$_{10}$ 3.03$_8$ 2.66$_8$ 2.23$_5$ 1.68$_7$. Structure is a modification of the thortveitite structure, which involves reduction of symmetry to C2 from C2/m of thortveitite. *CM 27:703(1989)*. Generally fine-grained, also rarely as radiating sheaves of prismatic crystals 0.1–0.3 mm long. White or gray-white; white streak; chalky, gray-white fine-grained intergrowths with apophyllite, or vlasovite. H = $3\frac{1}{2}$–4 for intergrowth with apophyllite, 6–$6\frac{1}{2}$ for mongolian material. G=3.56. Orange cathodoluminescence. IR spectrum is similar to that of Thortveitite. May have 0.5% Fe, to 1.3% Sn, 1.2% Pb, 0.9–1.2% MnO, ≈ 40% SiO_2, ≈ 35.5% ZrO_2, 1.2% HfO_2, 17% CaO. Biaxial (−); nearly parallel extinction to length of fibers, X ∧ c = 5–10°, Y = b, X ∧ a = 3–8°, elongation = c, Z and Y nearly perpendicular to it; N_x = 1.720, N_y = 1.736, N_z = 1.738; Mongolian material has N_x = 1.716, N_y = 1.738, N_z = 1.760; 2V = 30°; colorless. Occurs in pegmatitic lenses and gneisses of agpaitic peralkaline rocks. Associated with other exotic minerals: eudialyte, elpidite, armstrongite, gadolinite-(Y), kainosite-(Y), and unnamed minerals. A rock-forming mineral at Strange Lake. May replace earlier Zr-bearing minerals similar to armstrongite. May occur with elpidite also. May be nonpseudomorphic, with quartz and allanite at Strange Lake. Alteration: Gittinsite + aegirine + Fe-oxides isomorphously replace elpidite at Strange Lake. Occurs as a rare alteration product of eudialyte in alkalic rocks at Parjarito Mountain, Mescalero Apache Indian Reservation, Otero Co., NM; an alteration product of vlasovite in nepheline-syenite pegmatites at *Kipawa River, Villedieu Twp., Temiscaming Co., Quebec, Canada*. Also in the Strange Lake complex, southeast of Lac Brisson, Quebec-Labrador, Canada; as pseudomorphs to several cm after elpidite, along with quartz, fluorite, hematite and sometimes aegirine, Khaldzan-Buregteg intrusions, Mongolian Altai, Mongolia. *CM 18:201(1980), CM 30:191(1992), DRAN 331:82(1993)*.

55.2.1a.2 Keiviite-(Y) $(Y,Yb)_2Si_2O_7$

Named in 1985 by Voloshin et al. for the area where it was found. Thortveitite group. MON $C2/m$. (Syn): a = 6.875, b = 8.970, c = 4.721, β = 101.74°, Z = 2; (nat): a = 6.845, b = 8.960, c = 4.734, β = 101.65°, Z = 2, D = 4.48. 38-440(nat): 4.65_9 3.23_{10} 3.04_8 2.28_7 ⋯ (three weak lines have hkl values that do not fit the assumed space group); (syn): 5.38_7 4.62_9 3.25_7 3.22_{10} 3.03_9 2.73_7 2.27_9 2.14_9 *TJBCS 85:17(1986)*. Prismatic, 0.05–1 mm long and < 0.15 mm thick. Colorless to white, white streak, vitreous luster. No cleavage, uneven fracture. H= 4–5, VHN= 1300. G = 4.45. No fluorescence, but has yellow-green cathodoluminescence. Biaxial (−); X ∧ c = 4°, Z = b, Y ∧ a = 7°; Na(D); N_x = 1.713, N_y = 1.748, N_z = 1.758; 2V = 56° (calc. 55°); r < v. Occurs in unspecified amazonite-bearing granitic pegmatites of the *Kola Penin., Russia*, in fissures in quartz and fluorite associated with thalenite, xenotime, bastnaesite, and kuliokite-(Y). Also on keiviite-Yb. Unnamed nonmetamict $REE_2Si_2O_7$ minerals having the thortveitite structure were described from Norway [*NGT 51:1(1971)*] and should be either keiviite-(Y) or keiviite-(Yb): Åskagen, Värmland; Högetveit, Evje, and Ivedal, Iveland, southern Norway; Ånneröd, Vaaler, Norway. EF *MZ 7:79(1985)*.

55.2.1a.3 Keiviite-(Yb) $Yb_2Si_2O_7$

Named in 1983 by Voloshin et al. for the locality. Thortveitite group. See Keiviite–(Y) (55.2.1.2). MON $C2/m$. $a = 6.840$, $b = 8.916$, $c = 4.745$, $\beta = 102.11°$, $Z = 2$ [MZ 5:94(1983)], $D = 6.04$. 37-458(nat): 4.64_8 3.24_{10} 3.20_{10} 3.03_9 2.72_7 2.67_7 2.26_7 2.17_6. First generation is elongated platy and prismatic crystals, and second generation is small spherulites; polysynthetic twins are common. Colorless, white streak, vitreous luster, transparent. Cleavage {110}, perfect; {001}, imperfect. $H = 5$–6. $G = 5.95$. IR data are similar to synthetic $Yb_2Si_2O_7$. Faint green cathodoluminescence. Dissolves in cold HCl. Some Y substitutes for Yb. Other HREE's are present. $Yb > Er \approx Lu > Tm > Dy$; variable Y_2O_3, 1.0–15.4 wt %. Biaxial (–); $X = b$, $Y \wedge a = 7$–$8°$, extinction angle of polysynthetic twins $= 33°$, $X \wedge c = 3$–$5°$; $N_x = 1.723$, $N_y = 1.758$, $N_z = 1.768$; $2V = 58°$; $r < v$, strong. Occurs near *Keivy, Kola Penin., Russia*, in amazonite pegmatite, within violet fluorite. Two generations of the mineral are present. Associated with fluorite, bastnaesite, and hingganite. EF

55.2.1a.4 Thortveitite $(Sc,Y)_2Si_2O_7$

Named in 1911 by Schetelig for Olaus Thortveit (1872–1917), Norwegian farmer and miner, who found the mineral. Thortveitite group. MON $C2/m$. (Nat): $a = 6.530$, $b = 8.521$, $c = 4.681$, $\beta = 102.63°$, $Z = 2$, $D = 3.50$; (syn): $a = 6.508$, $b = 8.506$, $c = 4.677$, $\beta = 102.7°$. 19-1125(nat): 5.18_6 4.58_2 $3.18_{4.5}$ 3.14_{10} $2.97_{6.5}$ 2.63_3 2.60_5 $2.20_{2.5}$; 20-1037(syn): 5.09_2 3.13_{10} 3.11_{10} $2.93_{4.5}$ 2.58_2 2.17_2 2.04_2 1.64_2. The structure consists of isolated Si_2O_7 groups. *AM* 73:601(1988), *CM* 31:337(1993). Crystals may occur singly or in radial aggregates, also flattened plates to elongate prismatic, crystals may be more than 35 cm long and 450 g in weight; {110} dominant; simple rotational twinning on (110) is common; rare twins occur on (150), and polysynthetic twinning is sometimes seen. Color is variable: white-gray, brown, gray-black, dirty green, grayish green; white-gray streak; vitreous–adamantine luster. Cleavage (110), good; uneven to conchoidal fracture; brittle. $H = 6\frac{1}{2}$. $G = 3.50$–3.58. Slightly soluble in boiling HCl and also HF. May contain > 17.7 wt % Y_2O_3, and 10 wt % ZrO_2. Minor substitution of Fe, Mn, Mg, Al, Hf, and Y-group REE occurs. Biaxial (–); $Y = b$, $X \wedge c = -5°$; $N_x = 1.750$–1.756, $N_y = 1.789$–1.793, $N_z = 1.802$–1.809; colorless to green-brown; birefringence $= 0.05$; $2V = 60$–$70°$; $r > v$, distinct. Most often occurs in granitic pegmatites, but also occurs in hydrothermal veins. Sometimes in greisens and in alaskitic granites. Occurs in commercial quantities at Crystal Mountain, Darby, Ravalli Co., MT, as millimeter-size grains included in fluorite; occurs in the Deadhorse Creek diatreme Complex, Marathon, ONT, Canada; several localities in *Norway:* Y-bearing from *Iveland*, Sätersdalen, Tuftan, Frikstad; Evje, Landsverk, and Rampeitroll sinken pegmatites; Eptevann, Fen Complex, southern Norway; Bidjovagge, Finnmark, northern Norway; Kola Penin., and Tagilo-Baranchinsk massif, Ural Mts.,

Russia; central Kazakhstan; Zr-bearing from Befanamo, Madagascar; Isango mine, Kobe, Kyoto Pref., Japan. Used as an ore of scandium which is used in the metals industry and the aircraft industry. EF *MIN 3(1):576(1972)*, *DAN SSSR 318:972(1991)*.

55.2.1b.1 Rowlandite-(Y) $(Y,TR)_4Fe(Si_2O_7)F_2$

Named in 1891 by Hidden for Henry Augustus Rowland (1848–1901), American physicist, Johns Hopkins University, Baltimore, MD. TRIC $P1$ or $P\bar{1}$. $a = 6.59$, $b = 8.65$, $c = 5.53$, $\alpha = 99.03°$, $\beta = 104.13°$, $\gamma = 91.47°$, $Z = 1$. *13-565*(nat heated): 4.87_6 3.59_5 $3.51_{5.5}$ 3.06_{10} $2.61_{3.5}$ $2.52_{3.5}$ $2.44_{3.5}$ $2.08_{4.5}$. The pattern is very similar, if not identical, to those for melanocerite and tritomite heated to 1000°. Structure is metamict. Heating produces multiple phases. Y_2O_3 and cristobalite and yttrialite (high-temperature form). At 950 to 1050°, both metamict and crystalline rowlandite are converted to another phase, perhaps thalenite. Is stated to be triclinic, $P1$ or $P\bar{1}$. Occurs as irregular masses. Grayish green, red, bluish green, greenish brown, grayish white; vitreous luster. No cleavage, conchoidal fracture, brittle. $H = 5\frac{1}{2}$. $G = 4.39$ (unheated), 4.55 (heated), and ≈ 4.57 (Russian). Three analyses (wt %): Fe is present as Fe^{2+}, Y_2O_3 plus other REE, 61.9; FeO, 4.69; CaO, 0.19; UO_3, 0.40; SiO_2, 25.98; LOI, 2.01; $F \approx 5\%$. Baringer Hill: $Y^{3+} + O^{2-} \leftrightarrow Fe^{2+} + F^-$, $Y^{3+} + F^- \leftrightarrow Fe^{2+}$. Uniaxial (+) to biaxial (+); most has N = 1.72–1.726, 1.704(isotropic), 1.76(heated). Occurs with gadolinite and yttrialite at Baringer Hill pegmatite. Also in postmagmatic veins, related to an alkaline aegirine–arfvedsonite granite. *Baringer Hill, Llano Co., TX*; Kola Penin., Russia; Kazakhstan. EF *CM 6:576(1961)*, *ZVMO 95:339(1966)*, *TMMAN 21:119(1972)*, *MA 27:67(1976)*, *MIN 3(1):590(1972)*.

55.2.1b.2 Thalénite-(Y) $Y_3Si_3O_{10}(OH)$

Named in 1898 by Benedicks for Tobins Robert Thalén (1827–1905), Swedish physicist. Compare with members of the thortveitite group. Thalenite may include rowlandite and yttrialite. *AM 71:188(1986)*. MON $P2_1/n$. $a = 10.34$–10.38, $b = 11.09$–11.22, $c = 7.29$–7.33, $\beta = 96.9$–97.3°, $Z = 4$, $D = 4.25$. *38-439*(nat): 5.50_2 $3.79_{2.5}$ 3.27_2 3.16_2 3.10_{10} 2.81_4 2.75_3 2.24_3. Structure is often metamict. Cannot be reconstituted by heating. Heating dehydrates it and converts it to thortveitite. *DAN SSSR 202:1324(1972)*. A fluorine-dominant thalenite was described [*MZ 7:79(1985)*] and the structure determined. *SPC 33:356(1988)*. Tabular or prismatic crystals, flattened on b, and sometimes only slightly elongated on c with {010}, {120}, {121}, and {$\bar{1}11$} dominant, also compact, massive, crystals to 2 cm across, resembles quartz. Usually pink or yellowish pink, also colorless, yellowish white, cream, grayish brown to brownish black (related to Fe^{3+} content); white streak; luster is resinous on fresh fracture surface, also greasy or vitreous. No cleavage, uneven to splintery fracture, and sometimes conchoidal, brittle. $H = 6\frac{1}{2}$. $G = 4.16$–4.41. Metamict varieties have G as low as 3.6. DTA and TGA: *VLA 2:243(1966)*. IR:*MIN 3(1):583(1972)*. Some is luminescent yellow under X-rays. Electromagnetic. Thalenite from Siberia and Sweden is dielectric. Soluble in HCl and H_2SO_4. Chemistry is summarized in *AM 71:188(1986)*, fresh material has little H_2O (0.75 wt %), fairly pure material has SiO_2, 30–32%; Fe_2O_3, 0.2–0.4; REE_2O_3, 66–69%; $Al_2O_3 \approx 0.3$. Other thalenites have variable low levels of CaO, MnO, Na_2O, ThO_2, CO_2 and other elements. Biaxial (−); OAP perpendicular to (010), Y nearly parallel to c, [*AM 71:188(1986)*]. Biaxial (−) parallel to b, OAP perpendicular to c; $N_x = 1.709$–1.731, $N_y = 1.716$–1.739, $N_z = 1.723$–1.748; colorless; 2V = 65–70°, Na(D). Most often occurs in granitic pegmatites, both alkalic and subalkalic. Occurs as replacement veinlets in yttrofluorite, associated with allanite at the White Cloud pegmatite, Jefferson Co., CO, and at the Snowflake pegmatite, Teller Co., CO; Guy Hazen claims, Boulder Springs, Mojave Co., 4.5 miles S of Kingman, AZ; *Österby pegmatite, Dalarne, Sweden*; large crystals from Åskagen, also Torsbacken, and Forskeelbak, Värmland, Sweden. Hundholmen, Tysfjord, northern Norway; Reiarsdal, Vest-Agder; Høgetveit, Evje; and Iveland, Rosså Setesdalen, Norway. Pyörönmaa pegmatite, Kangasala, Finland; Keivy area, Kola Penin., and unspecified location in eastern Siberia, Russia; Suishoyama, Kawamatamachi, Fukushima Pref., and Nosen village, Yamanashi Pref., Japan. EF *NGT 59:265(1979)*.

55.2.1b.3 Yttrialite-(Y) $(Y,Th)_2Si_2O_7$

Named in 1889 by Hidden and Mackintosh [*AJS 38:474(1989)*] for the composition, the yttrium group REE's being the chief cations. Usually metamict; heating produces several phases. Heating for 1 hour at 1000° produces low-temperature γ-phase yttrialite. MON $P2_1/m$ and $P2_1/n$ (heated). (Syn): $a = 7.34$–7.50, $b = 8.06$, $c = 5.02$, $\beta = 108.50$–$112.00°$, Z = 2; (heated 1000°): $a = 7.37$, $b = 8.10$, $c = 5.07$, $\beta = 108.30°$. 24-1427(nat): $4.35_{2.5}$ $3.31_{3.5}$ 3.01_{10} $2.91_{4.5}$ $2.86_{2.5}$ 2.81_3 2.72_2 1.81_2; 24-1428(syn): $4.67_{2.5}$ $4.03_{2.5}$ $3.47_{2.5}$ $3.45_{2.5}$ 3.05_{10} 2.62_4 $2.07_{3.5}$ 1.74_4. Structure: *K 16:905(1971)*. R factor between 8 and 16% on artificial material. Anhedral to subhedral masses, and equant crystals to 5 cm. Orange-yellow (when altered), olive-green, greenish and brownish black, blackish or reddish brown; greenish white or white streak; resinous and vitreous to waxy-powdery luster. No cleavage, fracture uneven to conchoidal. Brittle. H = $5\frac{1}{2}$–$6\frac{1}{2}$. G = 4.31–4.56. G sometimes as low as 4.02. Strongly electromagnetic, forms a gel in hot HCl. IR and DTA: *AM 58:545(1973)*. Analysis (wt %): Baringer Hill: SiO_2, 30.2; Al_2O_3, 1.0; Fe_2O_3, 16.8; Y_2O_3, 32.7; Ln_2O_3, 15.8; ThO_2, 13.6; CaO, 3.1; MgO, 0.5; Na_2O, 0.3; P_2O_5, 0.3; Nb_2O_5, 0.2. Material from Iveland, Norway, has ThO_2 6.97. Th and U substitute for REE, their higher valences being balanced by Ca, Fe^{2+}, and Mn. Biaxial (−); N = 1.760 (isotropic) down to 1.660 (variable hydration); $N_x = 1.731$, $N_y = 1.738$, $N_z = 1.744$; 2V = 68°. Occurs in granite pegmatites, associated with gadolinite-(Y) and tengerite-(Y). Alteration: to Y-carbonates, tengerite, and others. Occurs at the Baringer Hill pegmatite, Blufton, Llano Co., TX, in masses to 10 lb; also at the nearby Rhode Ranch pegmatite; Ivedal, Iveland, southern Norway; Högetveit, Evje, Norway; and Åskagen, Värmland, Sweden; Siberia; Komenono, Tateiwa Hozyo, Ehime Pref., and Yashima, Kagawa Pref.; also Fusamata and Suishoyama, Iisaka, Kawamata-machi, Fukushima Pref., Japan. EF *MRE II:243(1966)*, *NGT 51:1(1971)*, *AM 58:545(1973)*.

55.2.2.1 Keldyshite (Na,H$_3$O$^+$)$_2$ZrSi$_2$O$_7$, or Na$_{2-x}$H$_x$ZrSi$_2$O$_7$ · nH$_2$O

Named in 1962 by Gerasimovskii for Mstislav Vsevolodovich Keldysh (1911–1978), Russian mathematician and former president of the USSR Academy of Sciences. Polymorph: parakeldyshite. TRIC $P\bar{1}$. $a = 9.01$, $b = 5.34$, $c = 6.96$, $\alpha = 92°$, $\beta = 116°$, $\gamma = 88°$, $Z = 2$; $D = 3.26$; also $a = 9.31$, $b = 5.42$, $c = 6.66$, $\alpha = 94.3°$, $\beta = 115.3°$, $\gamma = 89.6°$. *DAN SSSR 238:573(1977)*. *24-1097*(nat): 4.16_7 3.95_6 2.91_{10} 2.70_7 2.20_6 1.93_5 1.70_5 1.60_5. Note: According to *PD 2:176(1987)*, these powder data are just a poorly measured variant of *29-1293* (parakeldyshite). Grains to 4 mm, 6-mm aggregates; fine polysynthetic twinning. White, white streak, vitreous–greasy luster. Two cleavages at 90°, irregular fracture, very brittle. H = 4. G = 3.30. Decomposed by HCl, HNO$_3$ + H$_2$SO$_4$. Analysis (wt %): H$_2$O$^+$, 0.95; H$_2$O$^-$, 0.35; K$_2$O, 0.94; Na$_2$O, 16.03; FeO + Fe$_2$O$_3$, 0.31; TiO$_2$, 0.60; ZrO$_2$, 40.35; SiO$_2$, 39.39 (original analysis). Biaxial (− and +); N$_x$ = 1.670, N$_z$ = 1.710; 2V (−): 60° red light, 78° for yellow, and 112° for blue, so the mineral is (+) in blue light; strong dispersion. Occurs at Lågendalen, near Larvik, Norway; in the *Tavaiok and Angoundasiok rivers, Mt. Alluaiv, Lovozero massif, Kola Penin., Russia*, with parakeldyshite, in foyaites; a primary mineral, associated with lorenzeite. EF *DANSSSR 189:166(1969)*, *MIN 3(1):595(1972)*.

55.2.2.2 Khibinskite K$_2$ZrSi$_2$O$_7$

Named in 1974 by Khomyakov et al. for the locality. See keldyshite (55.2.2.1). MON (pseudo-TRI) $B2/m$. (Nat): $a = 19.22$, $b = 14.10$, $c = 11.10$, $\beta = 116.5°$, $Z = 16$, $D = 3.33$; (syn): $a = 19.188$, $b = 14.072$, $c = 11.075$, $\beta = 117.07°$. *26-0928*(nat): 4.27_4 2.95_7 2.76_{10} 2.13_5 1.63_7 1.595_4 1.38_5 1.25_5; *24-0710*(syn): 6.38_5 4.31_6 4.27_3 3.40_5 3.00_{10} 2.80_5 2.77_8 2.16_3. The structure has three nonequivalent sites for Zr and six each for Si and K. All Zr atoms are octahedrally coordinated. The K coordinations are varied, two six-coordinated, three seven-coordinated, and one eight-coordinated. *SPD 21:696 (1977)*. In 2- to 3-cm ovoids, crystals are in irregular grains to 3 mm. White (syn), pale yellowish cream or colorless (nat); vitreous, dull, or greasy luster; translucent. Cleavage (001), (100), ($\bar{1}$11), excellent; planar, steplike fracture. H = $4\frac{1}{2}$–$5\frac{1}{2}$. G = 3.30–3.40. Analysis (wt %): SiO$_2$, 33.8; ZrO$_2$, 37.8; TiO$_2$, 0.6; trace CaO; K$_2$O, 27.0. Biaxial (−); OAP perpendicular to (010), Z = b, X and Y lie in (010); N$_x$ = 1.665, N$_y$ = 1.715, N$_z$ = 1.715; 2V = 6–16°. Occurs in the *Gakman Valley, Khibiny massif, Kola Penin., Russia*, as ovoids 2–3 cm in diameter in aegirine-rich metasomatic rocks. Outer zone is eudialyte, inner zone is zircon and khibinskite. EF *ZVMO 103:110(1974)*, *DANSSSR 231:1351(1976)*.

55.2.2.3 Parakeldyshite $Na_2ZrSi_2O_7$

First described in 1972 and named in 1977 by Khomyakov for a dimorph-polymorph of keldyshite. Para = beside in Greek – associated together. A high temperature polymorph. If keldyshite is heated at 500–600°C, one obtains parakeldyshite. TRIC. $P\bar{1}$. $a = 6.629$Å, $b = 8.814$, $c = 5.434$, $\alpha = 92.70°$, $\beta = 94.25°$, $\gamma = 71.46°$, $Z = 2$, $D = 3.40$ (synth.) and $D = 3.38$–3.44; PD 39-209 (synth.): 6.02_3 4.23_5 4.18_9 3.98_{10} 2.91_{10} 2.71_6 2.68_4 2.65_9. See also PD 2:176(1987). Structure: (Jour. Struct. Chem. (China) 11:866(1970). Colorless, streak white; 1–6 mm grains and masses at Mont Saint-Hilaire; vitreous; cleavage = 3 excellent to very good: {001}, {110}, {1$\bar{1}$0}; H $5\frac{1}{2}$–6. G 3.39; fluoresces orange-white in SW UV, strong cream fluorescence in SW. Biax. (–), $2V = 84°$, N_α 1.670, N_β 1.692–1.697, N_γ 1.713–1.718; colorless; $X \approx b$, $Y \approx c$. Occurs in alkalic pegmatites in Kola Peninsula. Occurs with keldyshite inbedded in sodalite in sodalite xenoliths at Mont Saint-Hilaire. Occurs at Mont Saint-Hilaire, Quebec, Canada; Lagedalen, southern Norway; *Umbozero region, Lovozero Massif, Kola Peninsula, Russia*; Khibiny Massif, Kola Peninsula, Russia. CM 15:102(1977), DAN SSSR 237:703(1977).

55.2.3.1 Barysilite $Pb_8Mn[Si_2O_7]_3$

Named in 1888 by Sjögren and Lundström from the Greek for *heavy* and for the composition (Si). HEX-R $R\bar{3}c$. $a = 9.821$, $c = 38.38$, $Z = 6$, $D = 6.84$. 23-404(syn): 4.57_5 3.88_6 3.22_5 3.21_5 3.16_6 2.96_{10} 2.77_8 2.68_8. Structure: AC 20:357(1966). The [Si_2O_7] dimers are nonlinear; the cell size for synthetic material is somewhat larger, yielding a volume greater by $\approx 7\%$. White to pinkish white, darkens on exposure to light; adamantine luster, pearly on cleavage. Cleavage {0001}, distinct; uneven fracture; brittle. H = 3. G = 6.55–6.72. Uniaxial (–); $N_O = 2.033$–2.07, $N_E = 1.015$–2.05. Gelatinizes in HCl with separation of $PbCl_2$; easily fusible. Minor substitution for Mn by Ca, Mg, and Zn. Found in thin films and veinlets in zinc ore from the Parker shaft, Franklin, Sussex Co., NJ. Found as tabular, rhombohedral crystals at Långban, and as embedded, curved lamellar masses in iron ore at the *Harstig mine, Pajsberg, Värmland, Sweden*. AR DT 584(1932), AM 54:510(1969), MR 4:62(1973).

55.3.1.1 Rankinite $Ca_3[Si_2O_7]$

Named in 1942 by Tilley for George Atwater Rankin (b.1884), American physical chemist, Geophysical Laboratory, Washington, DC, who first synthesized the compound. Dimorphous with kilchoanite (58.3.2.1). MON $P2_1/a$. $a = 10.557$, $b = 8.885$, $c = 7.858$, $\beta = 119.59°$, $Z = 4$, $D = 2.99$. 22:539: 4.48_7 3.84_7 3.20_5 3.18_8 3.03_6 2.90_5 2.72_{10} 1.819_6. Structure: MJJ 8:240(1976). Colorless to white, vitreous luster, transparent to translucent. H = $5\frac{1}{2}$. G = 2.96–3.0. Biaxial (+); $2V = 60$–$69°$; $N_x = 1.640$–1.643, $N_y = 1.644$–1.647, $N_z = 1.650$–1.652; $Y = b$, $Z \wedge c = 15°$; $r > v$. Alters to its dimorph, kilchoanite. Gelatinizes with acids; infusible. Found as rounded or irregular grains, and massive in high-temperature, contact-metamorphosed limestones. Found in the Christmas Mts., Brewster Co., TX; in northern

55.3.2.1 Jaffeite

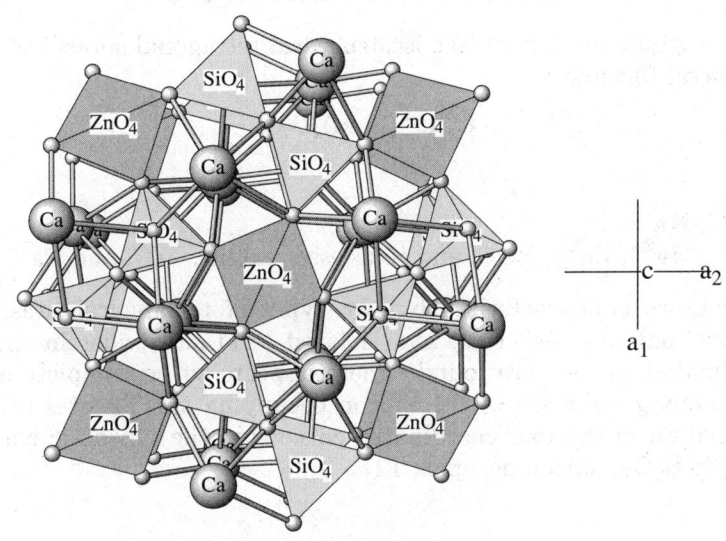

HARDYSTONITE: $Ca_2ZnSi_2O_7$
Perspective view along c. The structure is analogous to that of other tetragonal members of the group except for the cation occupancy of the larger tetrahedral site.

COAH, Mexico; at *Scawt Hill, Country Antrim, Northern Ireland*; Camas Mòr, Muck, Inner Hebrides, and from Ardnamurchan, Scotland. Found in the Hatrurim formation, at Hatrurim and Ma'aleh Adumim, Israel; in the Anakitsk massif, Lower Tunguska, Siberia, Russia; at Fuka, Bicchu-cho, Okayama Pref., Japan; at Takatoka, New Zealand. AR *MM 26:190(1942), 37:517(1969)*.

55.3.2.1 Jaffeite $Ca_6Si_2O_7(OH)_6$

Named in 1989 by Sarp and Peacor for Howard Jaffe (b.1919), University of Massachusetts. Natural analog of tricalcium silicate hydrate, a component of cement. HEX $P\bar{3}$. $a = 10.026$, $c = 7.482$, $Z = 2$, $D = 2.58$. 29-375(syn): 8.66_{10} 3.28_5 3.00_9 2.89_7 2.83_9 2.47_3 2.08_3 1.76_3. Structure: *DAN SSSR 219:340(1974)* on synthetic material. Elongated parallel to [001], with hexagonal cross section, {0001} and {10$\bar{1}$0} only; euhedral to subhedral crystals, average length 0.4 mm, diameter 0.25 mm. Colorless, white streak, vitreous luster, transparent. Cleavage {10$\bar{1}$0}, imperfect; conchoidal fracture; brittle. $G = 2.65$. Dissolved by HCl. Analysis (wt %): CaO, 65.0; SiO_2, 24.0; H_2O, 11.0. Uniaxial (+); $N_O = 1.596$, $N_E = 1.604$; colorless; birefringence = 0.008. Occurs in a manganese deposit with defernite and hausmannite and minor brucite, and so on, in low-grade metamorphic rocks. *Kombat mine, 49 km south of Tsumeb, Namibia*; Cerro Mazahua, Zitacuaro, MICH, Mexico. EF *AM 74:1203(1989)*.

MELILITE-FRESNOITE GROUPS

The melilite group consists of five isostructural tetragonal sorosilicates, having the general formula

$$R_2^{[8]}R^{[4]}(Si,Al)_2O_7$$

where

$R^{[8]} = Ca^{2+}, Na^+$
$R^{[4]} = Al^{3+}, Mg^{2+}, Zn^{2+}, Be^{2+}$, and to a lesser extent, Fe^{2+} and Fe^{3+}

The structure is characterized by two types of tetrahedral sites, double tetrahedron "paired sites" (T2) and isolated (T1) sites wherein the larger four-coordinated cations are found. There appears to be complete ordering of cations among these sites, with silicon occupying the T2 sites as long as it is the smallest of the four coordinated cations; in the synthetic boron-gehlenite, $Ca_2Si(B_2O_7)$ silicon occupies T1.

MELILITE-FRESNOITE GROUP: MELILITE SUBGROUP

Mineral	Formula	Space Group	a	c	D	
Åkermanite	$Ca_2Mg(Si_2O_7)$	$P\bar{4}2_1m$	7.835	5.010	2.944	55.4.1.1
Gehlenite	$Ca_2Al(SiAlO_7)$	$P\bar{4}2_1m$	7.677	5.059	3.054	55.4.1.2
Melilite*	$(Ca,Na)_2(Mg,Al,Fe)$ $[(Si,Al)_2O_7]$	$P\bar{4}2_1m$	7.789	5.018	3.09	55.4.1.3

*The sodium-bearing intermediate members are referred to as melilite, which does not enjoy species status. Soda-melilite (Sm), $CaNaAl(Si_2O_7)$, has been synthesized, and it is this component that substitutes in the intermediate phases. Consequently, melilite is treated here as a species.

MELILITE-FRESNOITE GROUP: FRESNOITE SUBGROUP

Mineral	Formula	Space Group	a	c	D	
Fresnoite	$Ba_2TiO(Si_2O_7)$	$P4bm$	8.518	5.211	4.520	55.4.2.1
Hardystonite	$Ca_2Zn(Si_2O_7)$	$P\bar{4}2_1m$	7.827	5.014	3.42	55.4.2.2
Jeffreyite	$(Ca,Na)_2(Be,Al)[Si_2(O,OH)_7]$	$C222_1$	14.90*	40.41	2.98	55.4.2.3
Leucophanite	$CaNaBe(Si_2O_6F)$	$P2_12_12_1$ or $P1$	**		2.948	55.4.2.4
Meliphanite	$(Ca,Na)_2Be[(Si,Al)_2O_6(F,OH)]$	$I4$	10.516	9.887	3.024	55.4.2.5
Gugiaite	$Ca_2Be(Si_2O_7)$	$P\bar{4}2_1m$	7.419	4.988	3.113	55.4.2.6

*Orthorhombic, but pseudotetragonal $a = b = 14.90$.
**Orthorhombic and triclinic cells reported with essentially identical cell constants: $a = 7.401$, $b = 7.412$, $c = 9.990$, $\alpha = \beta = \gamma = 90°$.

55.4.1.1 Åkermanite $Ca_2Mg(Si_2O_7)$

Named in 1884 by Vogt for Anders Richard Åkerman (1837–1922), Swedish metallurgist. Melilite–fresnoite group. TET $P\bar{4}2_1m$. $a = 7.835$, $c = 5.010$, $Z = 2$, $D = 2.944$. 35-592(syn): 5.54_1 3.72_1 3.09_2 2.87_{10} 2.48_1 2.034_1 1.776_1 1.761_2. Structure: *NJMM: 1(1981)*. Anhedral, equant grains; massive. Twinned on {100} and {001}. Colorless, gray, green, brown; vitreous luster; translucent to transparent. Cleavage {001}, distinct; conchoidal fracture. H = 5–6. G = 2.90–2.97. Uniaxial (+), sometimes anomalously biaxial with small 2V; $N_E = 1.638$, $N_O = 1.631$, with indices increasing and converging to 1.652 at composition $Ak_{52}Ge_{48}$; indices may increase with iron substitution, decrease and converge with substitution of sodium to 1.625 at composition $Ak_{65}Sm_{35}$; anomalous interference colors. Near-end-member åkermanite is very rare, possibly unknown in nature. Substitution of aluminum for magnesium and silicon is common, as is substitution of sodium for calcium. Ferrous and ferric iron may also replace magnesium. See the discussion of melilite (55.4.1.3). Gelatinizes with HCl. No verified natural occurrences of end-member phases; commonly found in slags and as a synthetic product. Aluminian varieties found in contact zones and meteorites. Found in the Cascade Slide xenolith, Adirondack Mts., Essex Co., NY; in the Tres Hermanas Mts., Luna Co., NM. Found in the Allende, Mexico, meteorite and other carbonaceous chondrites. From Kilchoan, Ardnamurchan, Argyllshire, Scotland, and from Scawt Hill, County Antrim, Northern Ireland. Has been reported from Vesuvius and Monte Somma, Campania, Italy; from the Wessels mine, near Kuroman, Cape Prov., South Africa. The initial description by Vogt was from a slag. AR *D6:476(1892)*, *DHZ 1B:285(1986)*.

55.4.1.2 Gehlenite $Ca_2Al(SiAlO_7)$

Named in 1815 by Fuchs for Adolph Ferdinand Gehlen (1775–1815), German chemist. Melilite–fresnoite group. TET $P\bar{4}2_1m$. $a = 7.677$, $c = 5.059$, $Z = 2$, $D = 3.054$. 35-755(syn): 3.71_2 3.06_2 2.85_{10} 2.43_2 2.41_1 2.39_1 1.754_3 1.750_2. Structure: *NJMA 44:254(1982)*. Short prismatic to equant crystals with {100}, {001} dominant; usually as anhedral grains; massive. Twinning on {001} and {100}. Colorless, gray, straw-yellow, brown; vitreous luster. Cleavage {001}, distinct; {110}, poor; conchoidal fracture. H = 5–6. G = 3.04. Uniaxial (−), sometimes anomalously biaxial with small 2V; $N_O = 1.667$, $N_E = 1.657$ for pure gehlenite, and decreasing and converging to 1.652 at composition $Ak_{52}Ge_{48}$. Indices decrease with substitution of sodium and increase with substitution of iron. Complete solid solution can form with åkermanite and partial solid solution with soda melilite. Ferric and ferrous iron may also substitute as well as minor cations. Naturally occurring low-Mg gehlenites are usually low in sodium and iron content. Gelatinizes in HCl. Occurs primarily as a component of skarns formed at the contact of siliceous limestones and monzonitic to dioritic intrusives, associated with melilite, wollastonite, larnite, spinel, magnetite, and others. In chondritic meteorites. At Iron Hill, Gunnison Co., Co; in Luna Co., NM, light brown massive; at Crestmore, Riverside Co., CA; in chondrules in the Allende meteorite, Mexico (ca. $Ak_{30}Ge_{70}$), and other carbonaceous chondrites; at Carneal and Chalk, Scawt Hill, County Antrim, Northern Ireland; at Monte Monzoni, Val di Fassa, Trentino–Alte Adigo (Tyrol), Italy; at Mayen, Eifel, Germany; Oravicza, Romania; Santorini volcano, Greece; Kushiro, Tojo-cho, Hiroshima Pref., Japan. AR *D6:476(1892)*, CM *10:822(1971)*, DHZII *1B:285(1986)*.

55.4.1.3 Melilite   (Ca,Na)(Mg,Al,Fe)[(Si,Al)$_2$O$_7$]

Named in 1796 by Delamétherie from the Greek *meli*, honey, in allusion to the color. Melilite–fresnoite group. Synonyms: humboldtilite, velardeñite. TET $P\bar{4}2_1m$. $a = 7.789$, $c = 5.018$, $Z = 2$, $D = 3.09$. The lattice constants vary from 7.68 and 5.06 for pure gehlenite to 7.84 and 5.01 for pure åkermanite, and decrease with substitution of sodium or iron. For $Ak_{33}Ge_{33}Sm_{34}$ the constants are 7.72 and 5.04. **Habit:** Equant to short prismatic crystals with {100} and {001} dominant; laths; more commonly as anhedral grains and massive. Twinning on {001} and {100}, also on {110} and {1$\bar{1}$0} on a fine scale. **Physical properties:** Colorless, pale yellow, light brown; vitreous luster; translucent. Cleavage {001}, distinct {110}; conchoidal fracture. $H = 5-6$. $G = 2.9-3.1$. Uniaxial $(-)$; isotropic, uniaxial $(+)$, sometimes anomalously biaxial with small 2V; anomalous interference colors serve to distinguish from apatite; $N_O = 1.631-1.667$, $N_E = 1.638-1.657$ for sodium- and iron-free melilites, converging to 1.652 at $Ak_{52}Ge_{48}$; for gehlenite-free melilite $N_O = 1.631$, $N_E = 1.638$ decreasing and converging to 1.626 at $Ak_{64}Ge_{36}$; for low-magnesium melilite the indices decrease to $N_O = 1.639$, $N_E = 1.633$ at $Ge_{65}Sm_{35}$; iron substitution may increase the highest of these indices to about 1.675. As a result of the complex substitutions, optical properties are not diagnostic for melilite composition. **Chemistry:** Soda-melilite is present up to 40% in natural melilites. Igneous rock melilite tends to compositions near $Ak_{70}Sm_{30}$ and decrease in sodium is accompanied by decrease in magnesium; metamorphic melilites have a broader range of composition, but significant sodium is generally present. Iron is commonly present in either oxidation state producing the yellow and brown coloration. Barium, strontium, manganese, nickel, zinc, and rare earths may also be present. Gelatinizes in cold, dilute HCl. **Occurrence:** In ultrabasic igneous rocks (uncompahgrites) melilite is a dominant constituent in association with pyroxene, magnetite, perovskite, and apatite; in potassic rocks with leucite, kalsilite, and olivine; in melilitites as phenocrysts and in the groundmass; in volcanic rocks. In skarns at the contact of impure limestones and igneous intrusives, primarily monzonites and diorites, but granites as well, with typical minerals of this environment. Åkermanite is considered indicative of the sanidinite metamorphic facies. Sometimes replaced by clinopyroxene; alters to calcite and diopsidic pyroxene and to bicchulite. **Localities:** Adirondack Mts., NY; Magnet Cove, AR; in Uvalde Co., TX; Iron Hill, Gunnison Co., CO; Crestmore, Riverside Co.,

CA; Oahu, HI; Oka, QUE, Canada; Velardeña, DGO, Mexico; Gardiner intrusion, Kangerlussuak, eastern Greenland; Scawt Hill and Carneal, County Antrim, Northern Ireland; Massif Centrale, France; Monte Somma, Vesuvius, and *Capo di Bove, near Rome, Italy*; Santorino volcano, Greece; Kola and Turiy Penins., and Yakutia, Russia; Lake Baga-Dzoz-Nur, Mongolia; at Ḥatrurim and other localities in Israel; at Kushiro, Hiroshima-ken, and other localities in Japan; Merapi, Java; Etinde, Cameroons; Nyiragongo volcano, Zaire; Mt. Elgon, Uganda; Saltpetre Kop, South Africa. See also gehlenite (55.4.1.2). AR *D6:474(1892)*, *DHZII 1B:285(1986)*, *JISI 174:121(1953)*, *AM 38:643(1953)*, *CIWY: 359(1973–74)*.

55.4.2.1 Fresnoite $Ba_2TiO(Si_2O_7)$

Named in 1965 by Alfors et al. for the locality. Melilite–fresnoite group. TET $P4bm$. $a = 8.518$, $c = 5.211$, $Z = 2$, $D = 4.520$. *22-513*(syn): 3.82_3 3.30_5 3.08_{10} 2.70_3 2.61_2 2.15_2 2.07_2 1.874_2. Structure: *ZK 130:438(1969)*. The structure contains titanium in an unusual tetragonal–pyramidal five-coordination. Lemon- to canary-yellow, vitreous luster. Fluoresces yellow (SWUV). H =3–4. G = 4.43. Uniaxial (−); $N_O = 1.775$, $N_E = 1.765$. E, yellow; O, colorless. Found as subhedral to euhedral grains (0.1–3 mm) in sanbornite–quartz rock and sanbornite-poor quartzite with celsian, taramellite, diopside, and pyrrhotite near contact with granodiorite at *Big Creek, Rush Creek area, Fresno Co., CA*, and from the Victor mine, Clear Creek, and the Gem mine, San Benito Co.; from the Gunn claim, near Macmillan Pass, YT, Canada; Eifel district, Germany. AR *AM 50:314(1965)*.

55.4.2.2 Hardystonite $Ca_2Zn(Si_2O_7)$

Named in 1899 by Wolff for the locality. Melilite–fresnoite group. TET $P\bar{4}2_1m$. $a = 7.827$, $c = 5.014$, $Z = 2$, $D = 3.42$. *35-745*(syn): 5.02_1 3.72_4 3.09_6 2.87_{10} 2.48_4 2.04_1 1.762_4 1.750_2. Structure: *ZK 130:427(1969)*. The larger zinc ions are in T1. Massive, granular, and disseminated grains. White to pink or light brown with substitution of Mn or Fe, vitreous luster. Cleavage {001}, distinct; {100}, {011}, poor. H =3–4. G = 3.4. Uniaxial (−); $N_O = 1.669$, $N_E = 1.657$. Occurs in fine-grained banded ore with willemite, rhodonite, and franklinite at *North Hill mine, Hardyston Twp., Sussex Co., NJ*. AR *USGSPP 180:93(1935)*.

55.4.2.3 Jeffreyite $(Ca,Na)_2(Be,Al)[Si_2(O,OH)_7]$

Named in 1984 by Grice and Robinson for the locality. Melilite–fresnoite group. See leucophanite (55.4.2.4), meliphanite (54.4.2.5), and gugiaite (55.4.2.6). Possibly dimorphous with gugiaite and meliphanite. ORTH (pseudo-TET) $C222_1$. $a = 14.90$, $b = 14.90$, $c = 40.41$, $Z = 64$, $D = 2.98$. *38-365*: 5.00_4 2.99_9 2.77_{10} 2.54_6 2.36_4 2.23_4 1.866_4 1.755_5. Colorless, vitreous luster. Cleavage {001}, {110}, perfect; brittle. H = 5. G = 2.99. Biaxial (−); 2V = 40°. $N_x = 1.625$, $N_y = 1.641$, $N_z = 1.643$. Found as thin, micaceous, pseudotetragonal plates at the *Jeffrey pit, Asbestos, Richmond Co., QUE, Canada*, in a cavity in a rodingitized dike. AR *CM 22:443(1984)*.

55.4.2.4 Leucophanite CaNaBe(Si$_2$O$_6$F)

Named in 1840 by Esmark from the Greek for *white* and *to appear*. Melilite–fresnoite group. See jeffreyite (55.4.2.3), meliphanite (55.4.2.5), and gugiaite (55.4.2.6), to which it may have a polymorphous relationship. ORTH $P2_12_12_1$. $a = 7.401$, $b = 7.412$, $c = 9.990$, Z = 4, D = 2.948 or, as given in a later study, TRIC $P1$ with essentially identical cell constants. *18-711:* 5.94_2 3.60_4 2.97_3 2.76_{10} 2.32_3 1.988_2 1.705_2. Structure: *ZK 202:71(1992)*. Sodium and calcium are ordered in contrast to melilite; half the Si occupies the T1 site, the balance being ordered with beryllium in the T2 site as in meliphanite. Short prismatic to tabular, pseudotetragonal, spherulitic. Twinned on {001} and {110}, penetration. Colorless, pale yellow, light green; vitreous luster; transparent to translucent. Cleavage {001}, perfect; {100}, {010}, distinct. H = 3–4. G = 2.96. Biaxial (−); 2V ≈ 40°; N_x = 1.570–1.571, N_y = 1.594–1.595, N_z = 1.596–1.598; XYZ = *cab*; r > v, weak. Found in nepheline–syenite pegmatites. As excellent, transparent crystal up to 4 cm at Mt. St.-Hilaire, QUE, Canada; Narssarssuk, Greenland; at *Låven, Langesundfjord, Norway*, and on Arø and Stokkø, and in Telemark and Tvedalen; Lovozero massif, Kola Penin., Russia. AR *D6:417(1892)*, *MA 19:267(1968)*, *CM 27:193(1989)*.

55.4.2.5 Meliphanite (Ca,Na)$_2$Be[(Si,Al)$_2$O$_6$(F,OH)]

Named in 1852 by Scheerer from the Greek *honey* and *to appear*. Melilite–fresnoite group. See jeffreyite (55.4.2.3), leucophanite (55.4.2.4), and gugiaite (55.4.2.6) that may have a polymorphic relationship. TET $I\bar{4}$. $a = 10.516$, $c = 9.887$, Z = 8, D_{calc} = 3.024. *31-304:* 3.59_4 2.96_5 2.75_{10} 2.35_4 2.34_4 2.32_4 1.978_4 1.702_5. Structure: *AC 23:260(1967)*. Half the silicon is in the T1 site, the balance being ordered with beryllium in the T2 sites of the double tetrahedral group. Thin tabular {001} or short, square prisms; as tabular or lamellar aggregates. Yellow in various shades, less commonly red, also colorless; vitreous luster; transparent to translucent. Cleavage {010}, perfect; {001}, distinct; uneven and brittle fracture. H = 5–5$\frac{1}{2}$. G = 3.01–3.03. Uniaxial (+/−); N_O = 1.611–1.664, N_E = 1.592–1.672; distinctly pleochroic in dark-colored examples. Reported both as gelatinized and as insoluble in HCl. Found in nepheline–syenite pegmatite at Mt. St-Hilaire, QUE, Canada, and at Julianehaab, Greenland, but recent lists of minerals from these localities do not include it. Found at *Frederiksvärn, Norway*, and on several islands of the Langesundfjord; in the Sakhariok massif, Kola Penin., Russia; reported from

Gugia, China, but this may be gugiaite. AR *ZK 16:279(1890), D6:418(1892), MR 24:60(1993).*

55.4.2.6 Gugiaite Ca$_2$Be(Si$_2$O$_7$)

Named in 1962 by Peng et al. for the locality. Melilite–fresnoite group. See jeffreyite (55.4.2.3), leucophanite (55.4.2.4), and meliphanite (55.4.2.5). TET $P\bar{4}2_1m$. $a = 7.419$, $c = 4.988$, $Z = 2$, $D_{calc} = 3.113$. *15-199:* 5.25_4 2.97_4 2.77_{10} 2.36_4 2.32_4 2.21_4 1.709_7 1.485_7. Structure: *NJMA 143:210(1982).* Cation distribution is ordered with Be in T1 and Si in T2. Tabular crystals with {001}, {011}, {111}, {110}. Colorless, vitreous luster, transparent. Cleavage {100}, perfect; {001}, distinct. $G = 3.03$. Uniaxial (+); $N_O = 1.664$, $N_E = 1.672$. At *Gugia, China,* an inadequately defined locality, in skarn with orthoclase, vesuvianite, aegerine, sphene, apatite, and prehnite. AR *SS 11:977(1962), AM 48:211(1963), VLA 2:129(1966).*

55.5.1.1 Edgarbaileyite Hg$_6^{1+}$Si$_2$O$_7$ [*AM 75:1192(1990)*]

Named in 1990 by Roberts et al. for Edgar H. Bailey (1914–1983), geologist and mercury commodity specialist, U.S. Geological Survey. MON $C2/m$, Cm, or $C2$. $a = 11.755$, $b = 7.678$, $c = 5.991$, $\beta = 111.73°$, $Z = 2$ [*AM 75:1192(1990)*], $D = 9.13$. *45-1438*(nat): 6.28_2 3.16_{10} 3.03_3 2.95_3 2.77_2 2.72_6 2.32_2 1.87_4. *MR 21:215(1990).* Structure determination shows that all Hg is present as (Hg$_2$)$^{2+}$ dimers. Rounded to mammillary; platy, flattened on {100}; cryptocrystalline and anhedral to 0.2-mm platy crystal aggregates with {100} dominant; polysynthetic twinning with twofold rotation about c, the composition plane is {100}, polysynthetic and coherent twinning. Photosensitive; fresh—lemon yellow to orangish yellow, exposed—dark olive-green to lighter yellowish green to dark green-brown; light green streak with yellow tinge; crystals are vitreous, masses are resinous; translucent to opaque. Well-developed {100} cleavage, irregular to subconchoidal fracture. $H = 4$. $G = 9.4$. IR data for region 250–4000 cm^{-1}. Heating to low red heat destroys the structure, native Hg is liberated, and the mineral turns white. X-ray diffraction = opal-CT. Heating to 1000–1100° for 14 hours produces tridymite and cristobalite. Soluble in cold 1:10 HCl and 1:1 HNO$_3$. Biaxial; all $N > 2.0$ (2.10–2.58 calculated); weakly pleochroic, lemon yellow, calculated RI values yield a minimum of 2.10 at 590 nm, and a maximum of 2.58; strong absorption; VHN = VHN$_{100}$ = 192; color in plane-polarized light is gray to slightly lighter gray; internal reflections pale lemon yellow; weakly to strongly bireflectant; nonpleochroic. $R_{air} = 16.1$–22.8 (470 nm), 14.5–20.9 (546 nm), 14.1–20.3 (589 nm), 13.8–19.75 (650 nm); $R_{oil} = 4.69$–9.12 (470 nm), 3.90–7.95 (546 nm), 3.65–7.50 (589 nm), 3.48–7.16 (650 nm). Occurs at the California localities as thin crusts on fracture surfaces, as disseminated rounded to mammillary masses in small cavities and as hollow mammillary nodules. At Terlingua, Brewster Co., TX, occurs as ill-formed tabular aggregates and sheaves of crystals in cavities and fractures associated with HgO, terlinguaite, eglestonite, and Hg. At the *Socrates mine, Sonoma Co., CA*, is associated with quartz, chalcedony, native Hg, HgS, and montroydite. At the Clear Creek claim, San Benito Co., CA, associated with HgO and Hg. Also at the San Luis Mine, Huahuaxtla, Guerro, Mexico. Secondary product formed most probably by reaction of native Hg and quartz. EF

Class 56

Si_2O_7 with Additional O, OH, F, and H_2O

56.1.1.1 Bertrandite $Be_4[Si_2O_7](OH)_2$

Named in 1883 by Damour for its discoverer, the French mineralogist, Émile Bertrand (1844–1909) designer of the Bertrand lens (petrographic microscope). ORTH $Cmc2_1$. $a = 8.716$, $b = 15.255$, $c = 4.565$, $Z = 4$, $D = 2.61$. *12-452:* (nat.) 4.385_6 4.35_4 3.91_2 3.81_{10} 3.16_5 2.54_8 2.52_4 2.28_2; also *17-515* (replaced by *12-542*): 4.38_{10} 3.94_4 3.19_9 2.54_8 2.28_6 1.555_3 1.305_4. The structure contains dimeric (Si_2O_7) and (Be_2O_6OH) groups that are linked into a three-dimensional framework through shared oxygens (AM 72:979(1987) and NJBM 13(1992)). Sphaerobertrandite and gelbertrandite are two poorly characterized varieties. Habit is predominantly tabular {001} with {001} dominant, but may be platy or prismatic. Dominant forms are {001} and {00$\bar{1}$}, and {101}, {010}, {110}, with {100} less so. May also be flattened on {100} or {010}. The earlier, morphological setting gives a c axis twice that of the X-ray cell. The hemimorphic character is not generally obvious. Twinned (polysynthetic and heart or V-shaped) on {021} and {011}; other twins occur as well. Granular aggregates, also granular, massive and earthy. Rarely in sheaves or sprays with crystals to several cm. Colorless to white, pale yellow, pale pink, light brown; streak white, transparent to translucent; luster vitreous, somewhat pearly on {001}. Cleavage {001}, perfect; {010}, {110}, good; fracture conchoidal, brittle. $H = 6\frac{1}{2}$–7. $G = 2.57$–2.63. Pyroelectric. Soluble in HF, H_2SO_4, and NaOH. Partly soluble in HCl. Usually contains trace minor amounts of Fe, Al, and Ca. Biaxial (−); $2V = 73$–81; $N_x = 1.584$–1.591, $N_y = 1.598$–1.605, $N_z = 1.611$–1.614; $XYZ = abc$; $r < v$ weak. Widespread but sparse in granitic pegmatites-apiaites, pneumatolitic and hydrothermal veins as a late-stage product of beryl resorption and alteration, and may form pseudomorphs after beryl. Also in W and Sn-bearing veins and greisens. A fine-grained component along with fluorite in rhyolite and tuffs, and derived sediments, e.g., Spor Mountain, Utah. Found in a wide number of world-wide localities, only noteworthy and unusual localities are mentioned here. Occurs in a number of pegmatites at Stoneham, and world-class twins are from the Hayes Ledge quarry! and Waisanen quarry, Greenwood, Oxford Co. and Auburn, Androscoggin Co., ME; Strickland quarry, Portland, Middlesex Co., CT; Amelia, Amelia Co., VA; the Dan Patch pegmatite, near Keystone, and the Helen beryl mine, near Custer, SD; crystals to 1 cm or more, assoc. with beryl and phenakite, from Mount Antero, Chaffee Co., CO; in greisen at the Boomer mine, Park Co., CO; crystals to 3-4 cm from amazonite pegmatites in the Lake George Ring, Lake George, Park and Teller Counties, CO; an ore mineral at Spor Mountain, Juab Co., UT; Mt.Wheeler tungsten deposit, White Pine Co., NV; in pegmatites of the Pala district, San Diego Co., CA; Sierra de Aguachile, Acuña, COAH., MEX; Cornwall, England; *Barbin quarries, Petit-Port, near Nantes, Loire-Atlantique, France*; St. Gotthard, Switzerland; Pizzo Marcio, Val Vigezzo, Piedmont, and from San Pietro di Campo, Elba, Italy; Henneberg, Germany; Vest-Agder, Iveland, Norway; Eräjärvi, Finland; Písek, Bohemia, Czech. Rep.; Maršíkov, Ctidružice and Rožná, Moravia, Czech Rep.: in the Altai, Urals (emerald mines) and Transbaikal regions, Russia; good crystals to 2.5 cm from Kara-Oba, Akchatau and Kounrad, Kazakhstan; as flat plates to 3 cm(!) associated with beryl, scheelite, wolframite, and dolomite in vein deposits at the Tae Hwa mine, Neung Am Ri, Chung Cheong North Prov., South Korea; Klein Spitzkopje, Namibia; as sheaves (to 5 cm) of thick, tabular crystals(!) from the Golconda mine, 30 km from Governador Valadares, and as crystals to >4 cm from Conselheiro Peno, MG, and as blocky crystals to 1.5 cm(!) at the Ermo mine, Parelahs, RGN, Brazil. Mt. Isa, Queensland, Australia. Currently, the major world source of beryllium (Spor Mountain, Juab Co., UT.) EF and AR *D6:545(1892), AM 45:1300(1960), MIN 3(1):627(1972), MAC SC 8:135(1982), PCM 13:69(1986)*.

56.1.2.1 Hemimorphite $Zn_4[Si_2O_7](OH)_2 \cdot H_2O$

Named in 1853 by Kenngott for the crystal morphology, but the mineral has been known from antiquity. Synonym: calamine, which may be a corruption of Cadmia or possibly from Latin *calamus*, reed, in allusion to a stalactitic habit. ORTH $Imm2$. $a = 8.367$, $b = 10.730$, $c = 5.115$, $Z = 2$, $D = 3.484$. *5-555*: 6.60_9 5.36_5 4.62_4 3.30_7 3.29_7 3.10_{10} 2.56_5 2.40_5. $[Si_2O_7]$ dimers lie approximately parallel to [010]. ZnO_3OH tetrahedra form sheets parallel to {010} that are cross-linked by the silicate dimers and water molecules. **Habit:** As isolated or radiating groups of distinctly hemimorphic crystals, prismatic [001] and flattened to greater or lesser degree parallel to {010}. The analogous end is usually highly modified, while the antilogous end is less complex but less frequently observed due to preferential implanting on [00$\bar{1}$]. Dominant forms are {010}, {110}, {001}, {12$\bar{1}$}, {301}. Twinning on (00$\bar{1}$), thus joining the antilogous ends, and not optically detectable. Also found as thick, botryoidal encrustations of thick to almost fibrous crystals; granular and massive as well. **Physical properties:** Colorless to white, less commonly, pale yellow, pale green, sky-blue, brown; vitreous luster, pearly on {001}; transparent to translucent. Cleavage {110}, perfect; {101}, good; {001}, poor; uneven fracture; brittle. $H = 4\frac{1}{2}$-5. $G = 3.4$-3.5. Biaxial (+); $2V = 40$-$46°$ $N_x = 1.614$, $N_y = 1.617$, $N_z = 1.636$ for Na$_D$, the indices being fairly constant; $XYZ = bac$. Strongly pyroelectric. **Chemistry:** Dehydration experiments indicate that half of the total water content is lost continuously and below 500°

56.1.2.1 Hemimorphite

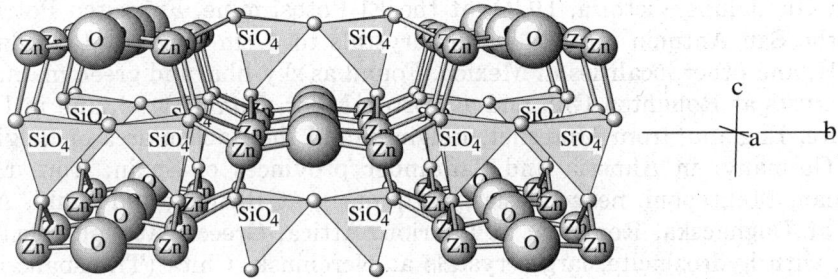

HEMIMORPHITE: $Zn_4Si_2O_7(OH)_2$
Rows of disilicate groups running down the a axis are linked by Zn. OH atoms also are bonded to Zn. The lack of centrosymmetry is due to the silicate groups, which are all in the same orientation.

without destruction of the crystals; the balance is lost only at significantly higher temperatures. Composition is close to the ideal, but often impure, due to admixture of smithsonite, clays, and iron oxides, the last, as hematite, often staining transparent crystal to a red hue. Gelatinizes in acids. **Occurrence:** A widespread, almost ubiquitous phase in the oxidation zone of zinc deposits. **Localities:** A type locality cannot be defined. Found as thick crusts of thick crystals with notched multiple terminations at Sterling Hill, Sussex Co., NJ; at Friedensville, Lehigh Co., PA, where it was mined extensively in the past; formerly abundant and mined at Austin's mines, Wythe Co., VA; with zinc ores of the Tri-State district, especially at Granby, Newton Co., MO; in fine crystals at Elkhorn, Jefferson Co., MT; at Leadville, Lake Co., CO; at the

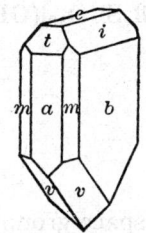

HEMIMORPHITE
Hemimorphite. Forms: a, {100}; b, {010}; c, {001}, m, {110}; t, {301}; i, {031}; v, {12$\bar{1}$}.

Bunker Hill mine, Kellogg, Shoshone Co., ID; at the 79 mine, Hayden, Gila Co., AZ; from the Organ Mts., Dona Ana Co., NM; in greenish-blue mammillary forms at the Emma mine, Salt Lake Co., UT. At Mapimí, DGO, as fine white to colorless crystals, and as sky-blue botryoidal crusts at the Santo Niño mine, Guadalupe Victoria, DGO; at the El Potosí mine, Francisco Potrillo, and the San Antonio mine(!) in fine crystals to 4 cm long, Santa Eulalia, CHIH, and other localities in Mexico. Found as sky-blue and green mammillary crusts at Roughten Gill, and at Alston Moor, Cumberland, and in Derbyshire, England; from Moresnet, Belgium, and the Aachen area of Belgium and Germany; in Almeria and Santander provinces of Spain; from Mte. Agruian, Monteponi, near Iglesias, Sardinia(!); at Bleiberg and Raibl, Austria; at Dognacska, Romania; at Lavrion, Attica, Greece. At Yadz, central Iran, with hydrozincite; large crystals at Nerchinsk, Chita (Transbaikalia), Russia. At Broken Hill, NSW, Australia. In fine crystal(!) at Djebel Guergour, NW of Setif, Algeria; at Tsumeb, Namibia. Use: A minor but significant ore of zinc. AR $D6:539(1892)$, $DT\ 632(1932)$, $ZK\ 146:241(1977)$.

56.2.1.1 Junitoite $CaZn_2[Si_2O_7] \cdot H_2O$

Named in 1976 by Williams for Jun Ito (1926–1978), American mineral chemist. ORTH $Bbm2$. $a = 6.318$, $b = 12.510$, $c = 8.561$ (original setting), $Z = 4$, $D = 3.516$. $29\text{-}394$: 6.25_4 4.70_5 3.53_{10} 2.82_{10} 2.54_{10} 2.52_5 2.35_7 1.540_6. Structure: $MM\ 49:91(1985)$. Structurally related to hemimorphite. Distinctly hemimorphic crystals, tabular $\{010\}$ and elongated $[001]$, 4–5 mm long and 0.5 mm thick with dominant forms $\{010\}$, $\{001\}$, $\{13\bar{1}\}$, $\{111\}$, $\{101\}$. Colorless to milky white. Cleavage $\{010\}$, good; $\{100\}$, $\{101\}$, poor. $H = 4\frac{1}{2}$. $G = 3.5$. Biaxial (+); $2V \approx 90°$; $N_x = 1.656$, $N_y = 1.664$, $N_z = 1.672$; $r < v$, weak. Strongly pyroelectric. Found with kinoite, apophyllite, and xonotlite in a retrogressively altered tactite zone at Christmas mine, Gila Co., AZ. AR $AM\ 61:1255(1976)$.

AXINITE GROUP

Axinite group minerals have the chemical formula

$$A_4M_2C_4[B_2Si_8O_{30}](OH)_2$$

where

A = Ca, Mn
M = Fe^{2+}, Mg, Mn
C = Al_2

and the structure is in the triclinic space group $P\bar{1}$.

The structure consists of Si_2O_7 groups linked in offset finite chains by BO_4 tetrahedra. The result is an Si_2O_7 group with an "apex-down" corner-shared BO_4 group. The remaining corners of this BO_4 group are shared with two

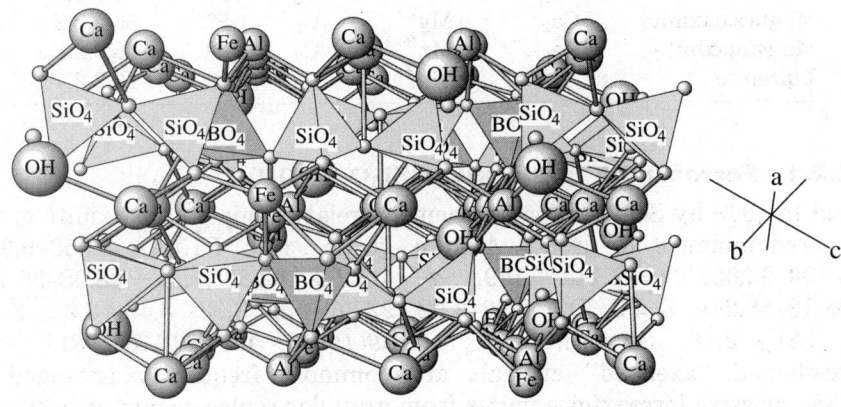

FERROAXINITE: $(Fe, Mg)_2Ca_4Al_4(OH)_2(B_2Si_8)O_{30}$

Axinite is not a true shee silicate, but contains $B_2Si_8O_{30}$ units—a ring of B_2Si_4 plus pairs of Si tetrahedra branching off. These are arranged in sheets with all the cations in between. The rings are compressed laterally, and unlike standard sheet silicates, the OH atoms are not at the center of the 6-rings. As in pyroxenes and amphiboles, the Ca atoms are coordinated partly by the bases (bridging oxygens) of tetrahedra, while smaller cations are coordinated by apices (nonbridging oxygens).

additional Si_2O_7 groups. The chain narrows, but the Si_2O_4 groups share corner oxygens with a second "apex-up" borate tetrahedron, and continues with a fourth Si_2O_4 group rotated 180° with respect to the first group. The prominent motif of the structure is a $B_2Si_8O_{30}$ polyanion with a midlength six-membered ring organized in sheets. Two kinds of octahedra, also organized in sheets, separate the polyanionic $B_2Si_8O_{30}$ sheets. One type consists of relatively small edge-sharing but symmetrical octahedra with hydroxyl substituting for oxygen [AlO_5OH]. A larger distorted octahedron [$M^{2+}O_6$] containing Ca^{2+} shares an edge with an [AlO_5OH] octahedron. The two octahedra form a finite chain of four small [AlO_5OH]octahedra terminated on both ends by large [$M^{2+}O_6$] octahedra containing Fe^{2+}, Mn^{2+}, and Mg^{2+}. Two types of large six-membered antiprisms roughly parallel the short, essentially colinear octahedral chains. One antiprism is "isolated" and one is edge sharing.

All of the axinite minerals are [all] triclinic, $P\bar{1}$, with dimensions approximately $a = 8.97$, $b = 9.17$, $c = 7.15$, $\alpha = 102.8°$, $\beta = 98.2°$, $\gamma = 88.1°$. *AM* 64:635(1979), 66:428(1981); *ZK* 140:389(1974), 140:289(1974); *DHZ* 1B:603(1986); *MIN* 3(22):213(1981).

Axinite Group

Mineral	Formula A	M	C	Space Group	
Ferroaxinite	Ca_2	Fe^{2+}	Al_2	$P\bar{1}$	56.2.2.1
Magnesioaxinite	Ca_2	Mg^{2+}	Al_2	$P\bar{1}$	56.2.2.2
Manganaxinite	Ca_2	Mn^{2+}	Al_2	$P\bar{1}$	56.2.2.3
Tinzenite	CaMn	Mn^{2+}	Al_2	$P\bar{1}$	56.2.2.4

56.2.2.1 Ferroaxinite $Ca_2FeAl_2[BSi_4O_{15}](OH)$

Named in 1909 by Schaller for its chemical relationship to the axinite group (i.e., predominance of iron). Axinite group. TRIC $P\bar{1}$. $a = 8.953$–9.962, $b = 9.190$–9.203, $c = 7.144$–7.159, $\alpha = 102.63$–$102.80°$, $\beta = 98.08$–$98.19°$, $\gamma = 88.18$–$88.24°$, $Z = 2$, $D = 3.29$–3.32. 27-76: 6.3_7 3.68_6 3.46_8 3.28_6 3.16_9 3.00_6 2.81_{10} 2.16_7. Structure: *ZK 140:289(1974)*, *AM 66:428(1981)*. Thin wedge-shaped "axehead" crystals are common, frequently arranged in rosettes; massive ferroaxinite varies from granular replacements or in-fillings to lamellar groups of closely intergrown crystals. Usually gray to bluish gray, brown, honey-brown or violet, occasionally green; uncolored streak or extremely faintly similar to body color; vitreous luster. Cleavage {100}, good; {001}, {110}, {011}, poor; uneven to conchoidal fracture brittle. $H = 6\frac{1}{2}$–7. $G = 3.23$–3.32. Generally nonfluorescent. Intermediate compositions approaching manganaxinite are common, although a complete series could extend to magnesioaxinite. Compositions with as much as 2.4 wt % MgO are rare, but even 1.2 wt % MgO greatly affects optical properties (e.g., decreasing refractive indices). Ferroaxinite from Obira is exceptional due to a small amount of Al substituting for Si. Si- and β-deficient ferroaxinite with an excess of trivalent cations (Al,Fe) was reported from Strzegom *MPOL 25:63(1994)*. Ti^{4+} to 0.19 wt % (Meldon, UK), while Na_2O to 0.33 wt % (Jokioinen, Finland) and K_2O to 0.21 wt % (Meldon, UK) have also been observed. Biaxial $(-)$; $N_x = 1.674$–1.683, $N_y = 1.682$–1.691, $N_z = 1.685$–1.694; $2V = 67$–$73°$. Due to the frequent substitution of Mg in iron-bearing axinites, the refractive indices overlap with that of Mn-rich axinites, which are usually Mg-poorer. Found in quartz–calcite veins, alpine-type veins, shear-zone mineralization, especially in high-grade metamorphic zones; hydrothermal metasomatism of granite or aplite, rarely in granitic pegmatites; at igneous contacts with sedimentary rocks. Can be associated with epidote–clinozoisite, grossular, vesuvianite, ilvaite, hedenbergite, hastingsite, datolite, danburite, prehnite, titanite, plagioclase, microcline, arsenopyrite, pyrite, zeolites, and others. Rarely replaced by clinochlore–chamosite and/or smectite. Occurs in many worldwide localities; notable localities include: Oxford and Franklin, and pink crystals at Bridgeville, NJ; Kibblehouse quarry, Perkiomenville, PA; Luck quarry, Goose Creek, VA; Foote quarry, Kings Mt., NC. Moosa canyon, Bonsall, San Diego Co.; Coarse Gold, Madera Co.; clear purple-brown crystals to 10 cm at the New Melones Dam(!), Calaveras Co.; at

Riverside, Stinson Beach, and Yreka, CA. Elkhorn, Jefferson Co., MT; Hope, BC, and Moneta mine, Timmins, ONT, Canada; Los Galvalontos(!) and Trinidad, Baja California; San Sebastian mine, Charcas, SLP, Mexico; Botallack mine and Roscommon Cliff, St. Just (Tremore), Bodmin, and Liskeard, Cornwall, Meldon, Okehampton, Devon, England; Jokioinen and Petsamo, Finland; Kongsberg, Buskerud, Norway; St. Cristophe(!), Bourg d'Oisans and La Balme D'Auris(!), Isère (Dauphiné), France; Alpe Scopi(!), Luckmanier Pass, Berg Sroyi, and Piz Vallatcha, Grisons, Sankt Jakob, Uri, Switzerland; Freseberg and Pferdekopf, Harz Mts., Ehrenfriedersdorf and Thum, Saxony, Raudautal, Germany; Davle and Zbraslav, and Libodrice, west of Kolín, Czech Republic; Knappenwand, Tirol, Austria; Monzoni, Trentino-Alto Adige, Val Tremola, Baveno, Novara, Italy; 5- to 7-cm crystals at Val D'Ossola near Premia, Italy; Strzegom, Wroclaw, Poland; extremely fine reddish-purple brown, sharp crystals to 15 cm from the Puiva deposit, Saranpaul, Tyumen district, polar Ural Mts., and Berkutskaya Sara Kreis, Zlatoust, Ural Mts., Russia; Fazenda Baixa da Brauna(!), Santa Rosa, and Victoria da Conquista, BAH, Brazil; Colebrook Hill(!), Tasmania, and Bowling Alley Point, western New England, NSW, Australia; Obira(!), Bungo, Oita Pref., Iwatamura, Nishius, Oki Gun, Hicoga Pref., Hajikami and Yamura, Hyugo, and Toroku mine, Iwate, Miyazaki Pref., Japan. VK, EF *AM 64:635(1979)*, *MIN 3(2):213(1981)*, *DHZ 1B:603(1986)*.

56.2.2.2 Magnesioaxinite $Ca_2MgAl_2[BSi_4O_{15}](OH)$

Named in 1975 by Jobbins et al. for its chemical relationship to the axinite group (i.e., dominance of Mg). Axinite group. TRIC $P\bar{1}$. $a = 8.933$, $b = 9.155$, $c = 7.121$, $\alpha = 102.59°$, $\beta = 98.28°$, $\gamma = 88.09°$, Z = 2, D = 3.18. Cell constants of magnesioaxinite from the type locality are anomalous for the axinite group. *AM 64:643(1979)*. 29-344: 6.29_3 3.44_7 3.27_2 3.14_7 2.80_{10} 2.56_3 2.18_3 2.15_3. Magnesioaxinite from the type locality is nearly end member, while analyses of the mineral from Luning, Nevada, grade from manganoan ferroan magnesioaxinite into manganoan ferroaxinite. *MR 11:13(1980)*. Crystals of magnesioaxinite are thin wedge-shaped "axeheads"; striations are common. Violet-blue to light gray-brown to pink, white streak to an extremely pale shade of its body color; vitreous luster. Cleavage {100}, good; {011}, {110}, {001}, poor; brittle. $H = 6\frac{1}{2}$. $G = 3.18$–3.26. As much as 0.13 wt % V_2O_3 is present in material from Tanzania and is responsible for the color and pleochroism. Generally fluorescent red to orange-red in both SWUV and LWUV. Biaxial (− or +); $N_x = 1.667$–1.669, $N_y = 1.673$–1.675, $N_z = 1.678$–1.680; $2V \approx 82°(−)$–$109°$ (+); $r > v$, strong; pleochroic, shades of pale blue to pale violet to pale gray. Originally identified from a faceted gemstone. First found in alluvium, later found in contact-metasomatic skarns. Associated with epidote, calcite, tremolite–actinolite, vesuvianite, and prehnite. Occurs at Luning, Mineral Co., Nevada; Arusha dist., *Tanzania*; London Bridge, near Queanbeyan, NSW, Australia. VK, EF *JG 14:368(1975)*, *AM 61:503(1976)*, *MIN 3(2):213(1981)*, *DHZ 1B:603(1986)*.

56.2.2.3 Manganaxinite Ca$_2$MnAl$_2$[BSi$_4$O$_{15}$](OH)

Named in 1909 by Fromme for its chemical relationship to the axinite group (i.e., predominance of manganese). Axinite group. TRIC $P\bar{1}$. $a = 7.157$–7.162, $b = 9.13$–9.213, $c = 8.938$–8.978, $\alpha = 91.80$–$91.87°$, $\beta = 98.07$–$98.80°$, $\gamma = 76.77$–$77.30°$, $Z = 2$, $D = 3.31$. 27–84: 6.31_2 3.69_1 3.47_7 3.44_2 3.16_{10} 2.89_2 2.81_5 2.19_2. Thin wedge-shaped "axehead" crystals are common, frequently arranged in rosettes; massive manganaxinite varies from granular replacements or in-fillings to lamellar groups of congested crystals. Usually yellow to golden brown, also gray-brown, occasionally green; uncolored streak or extremely pale, similar to body color; vitreous luster. Cleavage {100}, good; {001}, {110}, {011}, poor; brittle. H = $6\frac{1}{2}$–7. G = 3.30–3.36. Frequently fluorescent red in both SWUV and LWUV. Frequently low in MgO (<0.5 wt %). Franklin, NJ, manganaxinite can have Fe$_2$O$_3$ (to 0.83 wt %), CuO (0.12 wt %), PbO (0.09 wt %), and ZnO (4.49 wt %). Biaxial $(-)$; N$_x$ = 1.672–1.684, N$_y$ = 1.678–1.692, N$_z$ = 1.687–1.696; 2V = 60–$77°$; r < v, strong; pleochroic, yellow–colorless to amber–pale yellow. Found in quartz–calcite veins; igneous contact zones especially with mafic rocks; occurs in some granitic pegmatites. Associated minerals include quartz, plagioclase, microcline, chamosite, spessartine, andradite, johannsenite, rhodonite, and other manganese silicates, tourmaline, datolite, also with many rare minerals. Many worldwide localities are known and only notable or unusual localities are given here: occurs at the Franklin Mine(!), Franklin, and Sterling Hill mine, Ogdensburg, NJ; Avondale, Delaware Co., Cornog, Chester Co., Easton, Northampton Co., and Camels Hump, Bethlehem, PA; McKinney mine, Mitchell Co., NC; Mitchell mine, Babbitt, MN; Iron Cap mine, Landsman Camp, Graham Co., and Huachuca Mts., Cochise Co., AZ; Consummes mine, Amador Co., Greystone mine, Genesee Valley, Plumas Co., and Little Three mine, Ramona, San Diego Co., CA; Marmora, ONT, Canada; Guadalcazar, Mexico; Aarvold Valley, Norway, Barèges, Pyrennes, France; Radautal, Germany; Monte Pù, Ligurian Apennines, Italy; Takanosu mine, Kami Kasuo, Awano-machi, Kamitsuga-gun, Tachigi Pref., Obiva mine, Bungo, Oita Pref., Anawai mine, Kochi-ken, Shikoku, Japan; Julcani mine, Julcani district, Peru. VK, EF. *MIN 3(2):213(1981)*, *DHZ 1B:603(1986)*.

56.2.2.4 Tinzenite (CaMn)MnAl$_2$[BSi$_4$O$_{15}$](OH)

Named in 1923 by Jakob for the type locality. Axinite group. Formerly called severginite (along with manganaxinite) in Russia. TRIC $P\bar{1}$. $a = 7.153$–7.162, $b = 9.103$–9.184, $c = 8.945$–8.958, $\alpha = 91.86$–$91.97°$, $\beta = 98.31$–$98.68°$, $\gamma = 76.97$–$77.29°$, $Z = 2$, $D = 3.45$. 6–444: 6.30_7 3.46_8 3.28_6 3.14_7 2.98_7 2.81_{10} 2.17_7 2.01_7. Structure: *PM 42:369(1973)*. Mn occupies one of the two Ca positions. Thin wedge-shaped "axehead" crystals are common, frequently encased in quartz or as tightly clustered vein linings of terminated crystals. Usually light to dark yellow or orange to brownish orange, but may be orange-red or rose; colorless streak; vitreous luster. Cleavage {100}, good; {001}, {110}, {011}, poor; brittle. H = $6\frac{1}{2}$–7. G = 3.29–3.36. Nonfluorescent. Near-end-member compositions reported, minor MgO (to 0.20 wt %), FeO to 2.48 wt %. Biaxial (−); $N_x = 1.690$–1.694, $N_y = 1.697$–1.701, $N_z = 1.698$–1.706; $2V = 63$–$88°$; $r > v$, strong; pleochroic. X, yellow-green or colorless; Y, yellow; Z, colorless to pale yellow. Found in quartz veins and in metamorphosed Mn-bearing ophiolites. Occurs at the Cassagna and Gambatesa mines, near Chiavari, Liguria, Italy; *Alpe Parsettens, near Tinzen, Val d'Err, Grisons*, and at Tinzen, Graubunden, *Switzerland*; Camis Fe–Mn deposit, central Kazakhstan, Kazakh Republic; Akatore, New Zealand. Sometimes altered to black manganese oxides. VK, EF *AM 84:635(1979)*.

56.2.3.1 Lawsonite CaAl$_2$[Si$_2$O$_7$](OH)$_2$ · H$_2$O

Named in 1895 by Ransom for Andrew Cowper Lawson (1861–1952), American geologist, University of California. Dimorphous with partheite. Isostructural with hennomartinite (56.2.3.2). ORTH *Ccmm*. $a = 8.795$, $b = 5.847$, $c = 13.142$, $Z = 4$, $D = 3.088$. *15-533:* 4.84_6 3.66_6 2.73_7 2.68_5 2.62_{10} 2.43_6 2.13_6 1.550_8. Structure: *AM 63:311(1978)*. Calcium is in a distorted octahedral coordination with two additional, more distant oxygens; aluminum–oxygen octahedra form edge-sharing chains parallel to [010] that are bridged by the [Si$_2$O$_7$] dimers. The cell is dimensionally similar to that of ilvaite and may correspond to its high-temperature orthorhombic form. Crystals tabular {010} with {010} and {101} dominant, or prismatic [010] with {101}, {011}, and {010} dominant; lamellar twinning on {101}. Data for transformation to other morphological and optical settings are inconsistent. Also as subhedral tablets, massive, and granular. Colorless, white, gray, pale blue, pinkish; vitreous to greasy luster, translucent. Cleavage {100}, {010}, perfect; {101}, imperfect. H = 6. G = 3.05–3.12. Biaxial (+); $2V = 76$–$87°$ $XYZ = cab$; but also reported as $XYZ = abc$; $N_x = 1.663$–1.665, $N_y = 1.672$–1.675, $N_z = 1.682$–1.686; $r > v$, strong. Pleochroic in thick sections; X, light blue; Y, yellowish; Z, colorless. Insoluble in acids. Stoichiometrically close to the ideal formula with negligible substitution of Al for Si, and only minor substitution of Fe^{3+} for Al; an occurrence in the French Alps is reported with Cr^{3+}/Al $\approx$ 9:2, which should properly be a distinct species. Characteristic of low-temperature metamorphism and a common component of the blueschist facies; commonly in glaucophane schists with chlorite, epidote, and pumpellyite. Found at the *Tiburon Peninsula, Marin Co., CA*, and in euhedra to 5 cm(!) near Covelo in Mendocino Co., and other localities in that state; near Darrington, Snohomish Co., WA. In a xenolith in serpentinite at Santa Clara, Cuba. At Anglesey, Wales; in recrystallized gabbro in the Roche Noire massif in the French Alps (chromian), and in other Alpine regions of Europe; in metamorphosed pillow lavas of northern Calabria, Italy; Corsica; Crete. In the Tavsanli region of northwestern Turkey; the Penzha Range, Kamchatka, eastern Russia. At Kamuikotan, Hokkaido, at Ogose, Saitama Pref., and other sites in Japan; in New Caledonia; in the metagreywackes and slates of Maitai Valley, Nelson, New Zealand; South Shetland Is., Antarctica. AR *AJS 258:689(1960)*, *DHZII 1B:180(1986)*.

56.2.3.2 Hennomartinite $SrMn_2^{3+}[Si_2O_7](OH)_2 \cdot H_2O$

Named in 1993 by Armbruster et al. for Henno Martin, German geologist who escaped to South West Africa during the Hitler regime; he later worked for the Geological Survey of South Africa, and subsequently became director of the Geology–Paleontology Institute, University of Göttingen. Compare lawsonite (56.2.3.1). ORTH $Cmcm$. $a = 6.245$, $b = 9.031$, $c = 13.404$, $Z = 4$, $D = 3.68$. PD: 4.80_2 4.29_6 3.37_7 2.83_{10} 2.81_8 2.72_6 2.70_{10} 2.40_7. Structure: EJM 4:17(1992). Isostructural with lawsonite. Yellow-brown, vitreous luster, translucent. $H \approx 4$. Biaxial; $2V \approx 63°$; $N > 1.82$; strongly pleochroic, yellowish brown to dark red-brown. Barium and ferric iron are very minor substituents for strontium and manganese, respectively. Found, with kornite, as poikiloblasts overgrowing sugilite in a felted, serandite–pectolite rock at the Wessels mine, Kalahari manganese field, South Africa. AR SMPM 73:349(1993).

56.2.3.3 Ilvaite $CaFe^{2+}Fe_2^{3+}O[Si_2O_7](OH)$

Named in 1811 by Steffens for the locality (Latin Ilva, Elba). Synonym: lievrite. MON (below 70°) $P2_1/a$. $a = 13.010$, $b = 8.804$, $c = 5.852$, $\beta = 90.21°$, $Z = 4$, $D = 4.05$. 12-149: 7.31_7 3.26_6 2.87_7 2.84_{10} 2.72_7 2.68_{10} 2.18_6 2.17_6. ORTH (above 70°) $Pnam$. $a = 13.005$, $b = 8.818$, $c = 5.853$, $Z = 4$ (transposed to agree with monoclinic cell). Structure: ZK 144:161(1976), 179:415(1987). Ferrous and ferric iron lie near the centers of two (A and B) tetragonally distorted octahedral sites that form, by edge sharing, double chains parallel to [001]. Calcium is seven-coordinated, and Si is in $[Si_2O_7]$ dimers. The displacive transformation from high-temperature orthorhombic to low-temperature monoclinic appears to be the result of electron ordering in the A sites. The morphological and physical property descriptions are in terms of the orthorhombic symmetry, with a and b reversed from the values above (i.e., in the traditional setting for this species). Crystals are of short to long prismatic [001] habit and frequently striated in that zone. Dominant forms are {110}, {120}, {010}, {111}, {101}, sometimes with many additional terminating forms. Coarsely crystalline, massive, and granular. Black to dark grayish black, greenish to brownish black streak, nearly opaque. Cleavage {010},

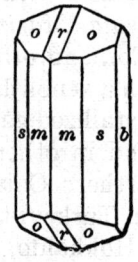

ILVAITE

Ilvaite. forms (orthorhombic indexing): b, {010}; m, {110}; s, {120}; r, {101}, o, {111}.

56.2.3.3 Ilvaite

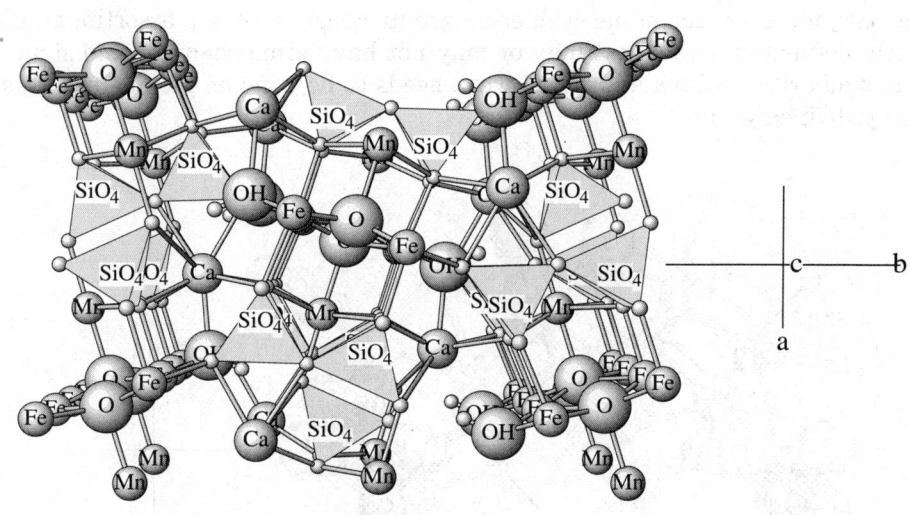

ILVAITE: $CaFe_3OSi_2O_7OH$

Fe^{2+} and Fe^{3+} octahedra form chains parallel to c, cross-linked by Ca and Si_2O_7 groups.

{001}, distinct; {100}, indistinct; brittle. H = $5\frac{1}{2}$–6. G = 4.0. Biaxial (−); 2V = 20–30°; $N_x = 1.727$, $N_y = 1.870$, $N_z = 1.853$; XYZ = cba; r < v, strong; pleochroic, X > Y > Z dark green to yellow brown; but also $N_x = 1.830$, $N_y = 1.92$, $N_z = 1.925$ with the same orientation, and also $N_y = 1.89$, $N_z = 1.92$–1.93 with XYZ = bac. Gelatinizes with HCl, fusible at 2.5 to a magnetic bead. Up to ≈ 8.5 wt % MnO has been reported. Occurs in contact-metasomatic iron, zinc, and copper ores; in sodalite–syenites; and associated with zeolites. Found at Balmat, NY; North Clear Creek, Gilpin Co., CO; at the Laxey mine(!), South Mountain, Owyhee Co., ID; in the Fierro–Hanover district, Grant Co., NM; in the Dragoon Mts., Cochise Co., AZ; in Fresno, Shasta, and Sonoma Cos., CA. Found in sodalite–syenites of the Ilimaussaq complex, Julianehaab district, Greenland; at Thyrill, Iceland; at Fossum, near Skeen, Norway. At Andreasberg, Harz; near Herborn (manganoan), Hessen–Nassau; and other sites in Germany. In Italy at Mte. Mulatto near Predazzo, Trentino, in the Italian Tyrol, and at the Temperino mine, Tuscany; in large crystals(!) and aggregates with pyroxene in dolomite at Capo Calamita and *Rio Marina, Elba*; and on Seriphos, Cyclades, Greece. In lustrous black crystals(!) at the Sovietskiy mine No. 1, Dal'negorsk, Primor'ye, eastern Russia. Found at the Chichibu mine, Saitama Pref., the Hideya mine, Niigata Pref., Honshu, and other localities in Japan. At Cap Bougaroun, Constantine, Algeria. AR. *D6:541(1892), SMPM 52:57(1972), PCM 14:151(1987)*.

Cuspidine–Wöhlerite Group

The cuspidine or cuspidine–wöhlerite group consists of an assortment of poorly defined minerals that may or may not have some chemical and structural similarities. Much additional work needs to be done on members of this poorly defined group.

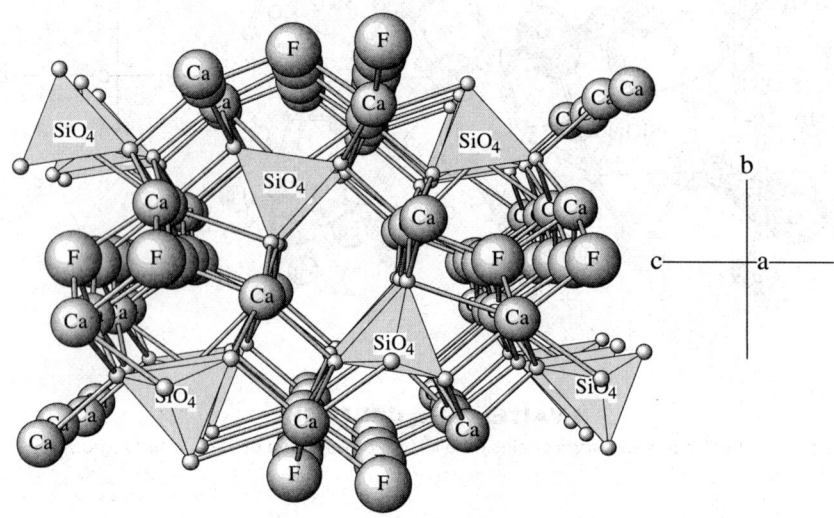

CUSPIDINE
Perspective view down a.

Mineral	
Baghdadite	56.2.4.1
Burpalite	56.2.4.2
Cuspidine	56.2.4.3
Låvenite	56.2.4.4
Wöhlerite	56.2.4.5
Niocalite	56.2.4.6
Hiortdahlite (I and II)	56.2.4.7
Rosenbuschite	56.2.4.8
Hainite	56.2.4.9
Janhaugite	56.2.4.10
Jennite	56.2.4.11
Komarovite	56.2.4.12
Na-komarovite	56.2.4.13
Suolunite	56.2.4.14
Mongolite	56.2.4.15
Killalaite	56.2.9.2
Foshallasite	56.2.9.3

56.2.4.1 Baghdadite $Ca_3Zr(Si_2O_7)O_2$

Named in 1986 by Al-Hermezi et al. for the city of Baghdad, Iraq. Cuspidine–wöhlerite group. See other members of the group. MON $P2_1/a$. $a = 10.36$–10.42, $b = 10.15$–10.16, $c = 7.36$–7.38, $\beta = 90.7$–$91.1°$, $Z = 4$, $D = 3.48$. 16-0155(nat): $7.30_{4.5}$ 3.23_8 $3.04_{7.5}$ $2.98_{8.5}$ 2.88_7 2.84_{10} $2.43_{3.5}$ 1.70_4; 39-195(syn): 7.26_3 3.22_{10} 3.02_5 2.98_5 2.98_5 2.86_5 2.84_8 1.99_3 1.97_5. Stumpy, prismatic; to 250 μm across, very rare; twinning = contact with b as the twin axis. Colorless, vitreous luster. No cleavage. H ≈ 6. G = 3.46(syn $Ca_3ZrSi_2O_9$). Cathodoluminescence color is dull gray with a greenish tint; slowly soluble in concentrated HCl and insoluble in concentrated HNO_3 and concentrated H_2SO_4. Analysis (wt %): SiO_2, 29.3; ZrO_2, 27.0; TiO_2, 2.1; Fe_2O_3, 0.11; Al_2O_3, 0.03; MgO, 0.05; CaO, 41.4; Na_2O, 0.02; $\Sigma = 100.02 + \approx 0.16\%$ HfO_2. Biaxial (+); $XZ = ca$, Y parallel to b; $N_x = 1.652$, $N_y = 1.658$, $N_z = 1.670$; 2V = 72°; birefringence =0.018; no dispersion. Occurs at *Dupezeh Mountain, Qala–Dizeh region, Iraq*; in melilite skarn in contact with a banded diorite. Associated with calcite, perovskite, wollastonite, and melilite. EF *DAN SSSR 142:639(1962), MM 50:119(1986)*.

56.2.4.2 Burpalite $Na_2CaZrSi_2O_7F_2$

Named in 1990 by Merlino et al. for the locality. Cuspidine–wöhlerite group. Polytypes: "orthorhombic låvenite" (*DAN SSSR ESS 146:128(1966)*) and *TMMAN SSSR 24:203(1975)*. Probably a polytype of burpalite. Same locality as burpalite, same formula. $b = 10.05$, $a = 11.11$, $c = 7.23$, $\beta = 109.0°$, b of burpalite = $(a \sin \beta = 10.30)$. MON $P2_1/a$. $a = 10.1173$, $b = 10.4446$, $c = 7.2555$, $\beta = 90.039°$, $Z = 4$ [*EJM 2:177(1990)*], $D = 3.27$. 42-1322(nat): $5.23_{2.5}$ $4.56_{1.5}$ 3.30_4 3.20_2 2.95_{10} 2.87_6 $2.84_{2.5}$ 1.78_2; *EJM 2:177(1990)*: 3.31_m 3.03_m 2.96_{vs} 1.89_{ms} 1.79_s 1.68_m 1.56_{ms}. One of 10 possible structure types derived for the cuspidine–wöhlerite family. R = 0.067. Order–disorder character present. Domains with låvenite-type structure are often present in burpalite crystals. *EJM 2:177(1990)*. Platy on (010) elongated parallel to [001]; crystals to 1 mm × 3 mm × 5 mm, forms: {110}, {001}, {011}, {101}, and {111}. Colorless, white; vitreous luster. No cleavage, no parting, conchoidal fracture, brittle, strong tenacity. H = 5–6. G = 3.33. Glows weak yellow orange in X-rays. Decomposed by 10% HCl at 25°. Analysis (wt %): Na_2O, 13.9; CaO, 14.5; MnO, 0.6; FeO, 0.43; SiO_2, 31.8; TiO_2, 1.06; ZrO_2, 31.11; Nb_2O_5, 0.22; Y_2O_3, 0.32; H_2O, 1.23; F, 8.1. Biaxial (−); $XYZ = bca$; Na(D): $N_x = 1.627$, $N_y = 1.634$, $N_z = 1.639$; colorless, 2V = 82°; r < v, weak. Occurs in fenitized sandstone in the contact zone of the *Burpalinsky alkaline massif, Maygunda R., northern Transbaikal, Russia*, 120 km NE of the northern extremity of Lake Baikal. Associated minerals are albite, nepheline, aegirine, alkali amphibole, biotite, catapleiite, astrophyllite, fluorite, and assessory loparite. EF

56.2.4.3 Cuspidine $Ca_4(Si_2O_7)F_2$

Named in 1876 by Scacchi from the Latin *cuspis*, spear, in allusion to the characteristic habit of the crystals. Cuspidine–wöhlerite group. MON $P2_1/a$. $a = 10.906$-10.909, $b = 10.521$-10.547, $c = 7.518$-7.539, $\beta = 109.3$-$109.6°$, $Z = 4$

[MJJ 8:286(1977)], D = 2.98. 41-1474(nat): 3.26_2 3.06_{10} 2.95_2 2.93_2 2.90_3 2.87_5 2.02_1 1.88_2. A sorosilicate in which Si_2O_7 groups are elongated parallel to the a axis. The Ca atoms that connect these Si_2O_7 groups have six to eight nearest neighbors of oxygen atoms. $ZVMO$ 84:159(1955), MJJ 8:286(1977). Earthy claylike masses, fine crystals to 1 cm; twinning present on {102} and sometimes polysynthetic twinning is present. Colorless, white, pale rose, rose-red, green; white streak; vitreous luster. Cleavage {100}, good; {$\bar{1}22$}, rare. H = 5–6. G = 2.97–2.99 (Franklin). Franklin material fluoresces strong yellow to light violet in LWUV; fluorescence in SWUV is similar but weaker; Crestmore material fluoresces a weak grey-tan in SWUV and weak gray in LWUV. Soluble in HNO_3, also soluble in HCl with production of a silica gel. Analysis (wt %): Ca, 43.8; Si, 15.25; F, 9.91; O, 30.37. Biaxial (+); Y = b, X $\wedge$ c = 5–6° in acute angle β; N_x = 1.586–1.594, N_y = 1.589–1.596, N_z = 1.598–1.606(?); 2V = 59–71°, colorless; OAP = (010); r > v, moderate. Occurs in a number of environments: Precambrian zinc ore; at contact of eruptive rocks with limestone; contact-metasomatic deposits. Associated with wollastonite, garnet, diopside, monticellite, and others. Occurs on Cascade Mtn., Adirondack Mts., Esses Co., NY; at Franklin, Sussex Co., NJ; Alder dist., Custer Co., Idaho; Crestmore quarry, Riverside Co., CA; Carlingford, Ireland; Camas Malag, Skye, and Camas Mór, Muck, Scotland; Beuerberg Volcano, Mayen, Eifel, Germany; *Monte Somma, Vesuvius, Campania, Italy*; Belogorsk deposit, southern Primorye, Russia; in a kimberlite pipe at Novinka, Yakutia, Russia; Anakit massif, Lower Tunguska region, Yakutia, Russia; Fuka, near Bitchu, and in the Sanpo and Mihara mines, Kawakami, Okayama Pref., Japan; San Luis Potosi, and Cerro Mazahua, Zitacuaro, MICH, Mexico. EF MIN 3(1):803(1972), CMP 74:143(1980), AM 72:423(1987).

56.2.4.4 Låvenite $(Na,Ca)_2(Mn,Fe^{2+},Ca,Ti)(Zr,Ti,Nb)Si_2O_7(O,OH,F)_2$

Named in 1885 by Brøgger for the locality. Cuspidine–wöhlerite group. See baghdadite (56.2.4.1), burpalite (56.2.4.2). Ti-rich låvenite has been called titan-låvenite. MON $P2_1/a$. a = 10.54–10.95, b = 9.86–10.01, c = 7.14–7.23, β = 108.1–110.3°, Z = 4, D = 3.84 (for *14-0586*), also 3.50. *14-0586*(nat): 6.79_3 3.99_3 3.22_5 2.89_{10} 2.81_9 2.41_3 2.01_3 1.79_3. Structure: $TMPM$ 28:99(1981). Elongated on c, oblique prismatic, tabular parallel to (100), to 1–2 cm; to 1–2 mm at Mt. St.-Hilaire; twinning = {100} simple, lamellar on c or perpendicular to (100). Colorless to brownish yellow to brown, brown-red, Ti-rich låvenite is orange-brown; yellow-white streak; vitreous luster. Cleavage {100}, good; uneven fracture; brittle. H = $5\frac{1}{2}$–6. G = 3.4–3.55. Slowly attacked by HCl. 3–5.2% Nb_2O_5 + Ta_2O_5. Major feature is the range in content of Zr and Ti. ZrO_2 30.0–35.0 wt % and TiO_2 2.0–19.0 wt %, MnO to 13.0 wt %. Biaxial (−); Z $\wedge$ a = 40–41°, Y = b, X $\wedge$ c = 20°; N_x = 1.645–1.728, N_y = 1.652–1.746, N_z = 1.656–1.760; 2V = 68–86°; OAP parallel to (010); r > v, v > r, weak; pleochroic. X, colorless, pale yellow; Y, colorless, pale greenish yellow; Z, very pale yellow, golden, orange-yellow, or brownish yellow. Absorption: Z > Y > X. Ti-rich låvenite has $N_x = 1.741$ and $N_z = 1.807$; X, orange-yellow; Z, dark brown. Occurs in a biotite-rich xenolith and in the contact of eudialyte-rich altered pegmatites at Mt. St.-Hilaire. In alkalic rocks and their pegmatites. In xenoliths in phonolitic pyroclastics. In nepheline syenite. Magmatic, early. Mt. St.-Hilaire, and St.-Amable sill, Varennes, and in the Montreal mine, Oka, QUE, Canada; Kangerdlugssuaq, eastern Greenland; Klein Arö, Langesundfjord, Norway; *Isle of Låven, Langesundfjord, Norway*; Mendig, Eifel region, Germany; Ischia Island, Naples, Campania, Italy; Ti-rich låvenite from Koklukhtiuai and Mt. Morua, Lovozero massif, Kovdor massif and Khibiny massif, Kola Penin., Burpala massif, northern Baikal, Siberia, Dugdinsky massif, eastern Tuva, and Ti-rich låvenite from Zaangarya, Yenesei Range, Siberia, Russia; Ile de Rouma, French Guyana; São Miguel, Azores; Ti-rich låvenite from Tenerife, Canary Islands; Serro do Tingua, MG, Brazil; 20 km N of Zomba, Chilwa Province, Malawi. EF *MIN 3(1):797(1972), IGR 19:217(1977)*.

56.2.4.5 Wöhlerite NaCa$_2$(Zr,Nb)Si$_2$O$_7$(O,OH,F)$_2$

Named in 1843 by Scheerer for Friedrich Wöhler (1800–1882), German chemist and professor, Göttingen. Cuspidine–wöhlerite group. See baghdadite (56.2.4.1), burpalite (56.2.4.2), cuspidine (56.2.4.3), niocalite (56.2.4.6), and other members of the group. MON $P2_1$. $a = 10.823$–10.869, $b = 10.244$–10.26, $c = 7.296$–7.920, $\beta = 109.00°$, $Z = 4$, $D = 3.56$. *10-0462*(nat): 7.26$_3$ 5.02$_3$ 3.25$_6$ 3.00$_7$ 2.97$_5$ 2.84$_{10}$ 2.01$_4$ 1.69$_4$. Structure: *DAN ESS 146:128(1962)*. To 3 cm, tabular or prismatic, flattened parallel to (100); 1- to 5-mm crystals from Mt. St.-Hilaire; twinning common, parallel to (100); sometimes polysynthetic. Pale yellow–orange yellow, brown, gray; greasy to vitreous luster; transparent–translucent. Cleavage {010}, distinct; {100}, {110}, weak; conchoidal to splintery fracture; brittle. $H = 5\frac{1}{2}$–6.0. $G = 3.41$–3.44. Luminescence: some material is CL blue-green. CL is dependent on Fe content, no CL if >1.7% FeO. Soluble in HCl. Analysis (wt %) in addition to major elements: TiO$_2$, 0–2; MgO, 0–0.3; MnO, 0.2–1.5; FeO, 0.15–2.1; Ce$_2$O$_3$, 0.0–0.7; SrO, 0–0.4; Ta$_2$O$_5$, 0–0.45; HfO$_2$, 0–0.6; FO, −3.9. Biaxial (−); $c \wedge a = 45°$, $Z = b$; $N_x = 1.700$–1.705, $N_y = 1.716$–1.720, $N_z = 1.726$–1.728; OAP perpendicular to (010) and almost parallel to (101); Z > Y > X; $2V = 71$–79°; r < v, weak. X, colorless; Y, light yellow; Z, dark wine-yellow. Occurs in syenites, agpaitic and miaskitic Ne-syenite, and their pegmatite derivatives. Also in fenites associated with alkaline massifs and in ijolite–pyroxenite, carbonatite, and silicocarbonatite. Reported from basalt, trachyte, and phonolite. Commonly found with other Nb-bearing minerals. Red Hill complex, Moultonborough, Carroll Co., NH; Magnet Cove, Hot Springs Co., AR; Prairie Lake, ONT, Mont Saint-Hilaire, and St. Lawrence pit, Oka, QUE, Canada; *Brevik, Langesundfjord, Norway*; Sangilene massif, Tuva, Khibiny massif, Kola Penin., eastern Sayan region, Yenesei region, Yakutia, and Burpala massif, northern Baikal, Siberia, Russia; Tchivira Mtns., Quilengues, Angola; Muri Mts., Guyana; from carbonatites at In Imanal and Wadi Anezrouf, Mali; Los

Islands, East Africa. EF *TMPM 26:109(1979)*, *ZVMO 109:594(1980)*, *CM 27:709(1989)*.

56.2.4.6 Niocalite $Ca_{14}Nb_2(Si_2O_7)_4O_6F_2$

Named in 1956 by Nickel from the composition *nio*bium and *cal*cium. Cuspidine-wöhlerite group. MON Pa, previously given as $P2_1/a$: $a = 10.830$, $b = 10.420$, $c = 7.380$, $\beta = 109.67°$; (Pa): $a = 10.863$, $b = 10.431$, $c = 7.370$, $\beta = 110.1°$, $Z = 1$, $D = 3.30$. *11-0622* (nat): 7.31_3 3.24_5 3.01_{10} 2.89_6 2.85_6 2.58_3 2.43_3 1.84_4. Structure: $P2_1/a$ SPD *11:197(1966)* and Pa *TMPM 30:249(1982)*. Niocalite is isostructural with cuspidine and låvenite, but topologically distinct from the related species wöhlerite. Crystals to 10 mm, prismatic with square cross sections; closely spaced (100–1000Å) polysynthetic twinning with (100) as the composition plane. Pale yellow, vitreous luster. Conchoidal fracture, brittle. H = 6. G = 3.32. Cathodoluminescence spectrum in *CM 27:709(1989)*. Analysis (wt %): Na_2O, 0.78; K_2O, 0.02; MgO, 0.28; CaO, 47.50; SrO, $\approx 1\%$; MnO, 1.28, $Al_2O_3 + RE_2O_3 + ZrO_2$, 1.31; Fe_2O_3, 0.54; SiO_2, 29.7; TiO_2, 0.22; P_2O_5, 0.60; Nb_2O_5, 16.56; H_2O, 0.16; H_2O^+, none; F, 1.70. Biaxial (−); $X = b$, $Z \wedge a = 8°$ in acute $\angle\beta$, $Y \wedge a = 32°$; $N_x = 1.700$–1.701, $N_y = 1.714$–1.721, $N_z = 1.720$–1.730; $2V = 34$–$56°$. Occurs at *Oka, QUE, Canada*, in carbonatites associated with apatite, diopside, biotite, pyrochlore, perovskite, and calcite. Early magmatic. EF *AM 41:785(1956)*, *CM 6:264(1958)*, *MIN 3(1):802(1972)*.

56.2.4.7 Hiortdahlite $(Ca,Na)_3(Zr,Ti,Nb)Si_2O_7(O,OH,F)_2$

Named in 1889 by Brøgger for Thorstein Hallager Hiortdahl (1839–1925), Norwegian mineralogist of Christiania. Cuspidine–wöhlerite group. Merlino and Perchiazzi [*TMPM 34:297(1985)*, *MP 37:25(1987)*] have distinguished two crystallographic forms of hiortdahlite, called I and II, respectively. TRIC Both $P\bar{1}$. (I): $a = 11.012$, $b = 10.940$, $c = 7.354$, $\alpha = 109.35°$, $\beta = 109.88°$, $\gamma = 83.43°$, $Z = 4$, $D = 3.22$; (II): $a = 11.012$, $b = 10.342$, $c = 7.359$, $\alpha = 89.91°$, $\beta = 109.21°$, $\gamma = 90.06°$, $Z = 4$, $D = 3.287$. *CM 12:241(1974)*. *27-668*(type I)(nat): $3.28_{4.5}$ 3.00_9 2.87_{10} $2.51_{2.5}$ $2.03_{2.5}$ 1.84_3 $1.80_{2.5}$ 1.70_4. Type II has Si_2O_7 diorthogroups and two symmetry-independent octahedral walls; a wall built up by alternating Ca1-, Zr-, NaCa1-octahedra and Ca2-polyhedron parallel to (210) and another wall built up by Ca3-Ca4-Y and NaCa2-octahedra parallel to (210). To 3–4 cm, tabular parallel to (100), most common forms are {101}, {1$\bar{1}$1} and {010}, {100}; polysynthetic {100} twins, twin axis coincides with vertical axis c. Light brown, honey-yellow, yellow-brown, light yellow; yellow streak; vitreous to greasy luster. Cleavage {110}, {1$\bar{1}$0}, distinct; {100}, {010}, weak; brittle. H = $5\frac{1}{2}$–6. G = 3.256(II), = 3.26(I), 3.31 (New South Wales material). Cathodoluminescence spectrum: *CM 27:709(1989)*. Gelatinizes with acids. It is very difficult to distinguish minerals of the group from each other by means of powder patterns because of the similar assembly of the common structural units in all the minerals of this

56.2.4.8 Rosenbuschite

family. Biaxial (+)(I), (−)(II); (I): $N_x = 1.639$, $N_y = 1.643$, $N_z = 1.646$; (II): $N_x = 1.652$, $N_y = 1.658$, $N_z = 1.665$; $2V = 80°$(I), $2V = 70°$(II); $Z > Y > X$; negative elongation; $r < v$, strong, distinctly pleochroic. X, colorless; Y, yellowish; Z, wine-yellow. Occurs in alkalic rocks and pegmatites associated with feldspar, nepheline, aegirine, meliphane, bismuth, astrophyllite, and titanite. Also in veins and in miarolitic cavities in sanidinites. Mainly in nepheline syenites and sodalite sanidinites. Kipawa R., Villedieu Township, Temiscaming Co., QUE, Canada (I and II); Kangerd-luggssvag Fjord, Greenland; *Middle Arø, Lille Arø, and Stokkø Islands, Langesundfjord, Norway*; Monte Somma, and Vesuvius, Campania, Italy; Laacher See volcanics, Eifel dist., Germany; Korgeredab massif, Sangilen, Tuva, Russia; Los Islands, East Africa; Jingera, NSW, Australia (type I). EF *MIN 3(1):815(1972)*, *CM 12:241(1974)*, *JGSA 26:81(1979)*.

56.2.4.8 Rosenbuschite $(Ca,Na)_3(Zr,Ti,Nb)Si_2O_8F$

Named in 1887 by Brögger for Karl Harry F. Rosenbusch (1836–1914), German geologist and mineralogist, Heidelberg. Cuspidine–wöhlerite group. TRIC $P\bar{1}$. *SPC 8:406(1964)*. $a = 10.12$–10.15, $b = 11.29$–11.39, $c = 7.27$–7.358, $\alpha = 91.33$–$92.2°$, $\beta = 99.7$–$101.15°$, $\gamma = 111.6$–$112.02°$, $Z = 4$, $D = 3.27$. 14-447(nat): 3.96_4 3.06_8 2.94_{10} 2.63_4 2.48_4 2.20_3 1.89_6 1.82_4. Structure: *SPC 8:406(1964)*. This structure solution places rosenbuschite in the rinkite–seidozerite group rather than the cuspidine–wöhlerite group. *SPC 16:1026(1972)* considers götzenite and rosenbuschite isostructural. A redetermination of the structure is necessary. Acicular to prismatic, probably parallel to [010]; radial fibrous, sometimes massive, 1- to 10-mm crystals; pale yellow to beige, gray, orange; white streak; vitreous luster. Cleavage {010}, perfect, but not readily apparent; also {100}; uneven fracture; brittle. $H = 5$–6. $G = 3.30$. Gelatinizes with acids. To 0.7% Nb_2O_5, 1.2% FeO, 1.4% MnO. Biaxial (+); Z ∧ perpendicular to cleavage (100) $= 28\frac{1}{2}°$, X ∧ $c = 0$–$9°$, parallel to length of needles; $N_x = 1.657$–1.678, $N_y = 1.664$–1.687, $N_z = 1.675$–1.705; $2V = 78$–$80°$; weakly pleochroic. X,Y, colorless; Z, yellow. $Z > Y > X$. Occurs in miarolitic cavities in nepheline syenite at Mt. St.-Hilaire, QUE, Canada. On analcime crystals, as divergent sprays. Associated with natrolite, paranatrolite, calcite, wöhlerite, zircon, and aegirine. In nepheline syenites and early pegmatites in Norway and NH. Red Hill, Moultenboro, Carroll Co., NH; Ilimaussaq, Greenland; *Lille and Middle Arø Islands, and Skudesundskaar, near Brevik, Langesundfjord, Norway* in fluorite-bearing syenite; Gränna!, Norra Kärr dist., Småland, Sweden, with eudialyte, fluorite and arfvedsonite; Lovozero and Khibiny massifs, Kola Penin.; Ti-rich rosenbuschite from Transangara, Yenisei Range, Tuva, Russia; 20 km N of Zomba, Chilwa Prov., Malawi. EF *NGT 42:179(1962)*, *MIN 3(1):838(1972)*.

56.2.4.9 Hainite $Na_2Ca_4[(Ti,Zr,Mn,Fe,Nb,Ta)_{1.5}\square_{0.5}](Si_2O_7)_2F_4$

Named in 1893 by Blumrich for the locality. Cuspidine–wöhlerite group. See rosenbuschite (56.2.4.8) and götzenite (56.2.5.4). TRIC $P\bar{1}$. $a = 5.676$, $b = 7.259$, $c = 9.586$, $\alpha = 101.08°$, $\beta = 101.14°$, $\gamma = 90.27°$, Z = 1, D = 3.157. 45-1453(nat): 3.96_5 3.07_9 2.96_{10} 2.63_5 2.49_5 1.90_7 1.82_5 1.68_4. Isostructural with götzenite; the formula of hainite is derived from that of götzenite by the substitution $2Ca^{2+} \rightarrow (Ti,Zr)^{4+} + \square$. Needles and plates, elongate [001] with {010} + {100} modified by {$\bar{1}$10}; crystals < 1 mm long; twinning on (100), space group by analogy to götzenite and rosenbuschite. Wine-yellow, honey-yellow, colorless; vitreous to adamantine luster. Cleavage {010} good, {100} poor; brittle. H = 5. G = 3.148. Readily soluble in diluted HCl, HF, and H_2SO_4. Analysis (wt %): Na_2O, 7.5; CaO, 32.6; MnO, 2.2; TiO_2, 8.3; ZrO_2, 6.8; Ce_2O_3, 1.10; Al_2O_3, 0.04; Fe_2O_3, 1.37; La_2O_3, 0.6; Nb_2O_5, 0.8; Ta_2O_5, 0.3; SiO_2, 32.1; F, 10.8; O = F, 4.56. Biaxial (+); $N_y \approx 1.7$; birefringence = 0.012; OAP and B×a perpendicular to (010) and oblique to (100) or Z perpendicular to (010); r > v, strong; weakly pleochroic. X, colorless; Z, wine-yellow. Z > Y > X. Occurs in cavities and groundmass of phonolite and tinguates, with aegirine. *Vrchni Hajn (formerly "Hohe Hain") near Frydlant, northern Bohemia; Krásné, Hradiště, and Kamenice, Bohemia, Czech Republic.* EF CR 308(II):1237(1989), AM 75:936(1990).

56.2.4.10 Janhaugite $(Na,Ca)_3(Mn,Fe^{2+})_3(Ti,Zr,Nb)_2(Si_2O_7)_2(O,OH,F)_4$

Named in 1983 by Raade and Mladek for Jan Haug, Norway, an amateur mineralogist, who discovered the mineral. Cuspidine–wöhlerite group. MON $P2_1/n$. $a = 10.668$, $b = 9.787$, $c = 13.931$, $\beta = 107.82°$, Z = 4, D = 3.71. 35-641(nat): 6.65_3 3.92_4 3.20_6 2.84_{10} 2.83_9 2.78_9 2.74_4 2.72_4. Structure: *NJBM: 7-18(1985)*. Prismatic, lamellar aggregates and sprays of prisms elongated on [001], flattened on {010} with prisms {110}, {120}, {130}, rough {001}; to 4 mm long, crystals may be curved and bent, vertically striated, sprays to 1.5 cm; twinning present, twin plane is probably {100}. Reddish brown, light brown streak, vitreous luster. Cleavage {010}, distinct; very brittle. H = 5. G = 3.60. IR data: *AM 68:1216(1983)*. Nonfluorescent. 5.00wt% Nb_2O_5, 0.26wt% Ta_2O_5, substitute for Ti. Biaxial (+); X ∧ c = +15°, Z = b; in white light: $N_x = 1.770$, $N_y = 1.828$, $N_z = 1.910$(calc); 2V = 80°; X < Y, weakly pleochroic. X, nearly colorless; Y, beige; negative elongation. Occurs as an early to late magmatic mineral in pegmatites in alkaline granite. Also in miarolitic cavities, associated with pyrophanite, elpidite, monazite, and also dalyite. Fairly widespread and in places is a rock-forming constituent at *Gjerdingen, Nordmarka, northern Oslo region, Norway.* EF

56.2.4.11 Jennite $Ca_9H_2Si_6O_{18}(OH)_8 \cdot 6H_2O$

Named in 1966 by Carpenter et al. for Clarence Marvin Jenni (1896–1973), American mineral collector and director of the Geological Museum, University of Missouri. Cuspidine–wöhlerite group. TRIC $P1$ or $P\bar{1}$. $a = 10.593$,

$b = 7.284$, $c = 10.839$, $\alpha = 99.67°$, $\beta = 97.65°$, $\gamma = 110.11°$, $Z = 1$ [*AM 62:365(1977)*], $D = 2.36$. *18-1206*(nat): 10.5_{10} 6.46_5 3.47_5 3.29_5 3.04_6 2.92_8 2.83_6 2.66_6. Blade-shaped crystals elongated on b or platy, to 6 mm long. White. Cleavage {001}, vitreous luster. $G = 2.32$–2.33. Material from Crestmore fluoresces a weak white in SWUV and weak to medium white in LWUV. IR and DTA data: *AM 51:56(1966)*. Loses 7 wt % water on moderate 290° heating to give meta-jennite, $9CaO \cdot 6SiO_2 \cdot 7H_2O$, $a = 10.590$, $b = 7.278$, $c = 9.511$, $\alpha = 101.03°$, $\beta = 105.74°$, $\gamma = 110.10°$; not considered a species. Biaxial ($-$); $X \wedge c = 90°$, $Y \wedge b = 35$–$40°$; $N_x = 1.548$–1.552, $N_y = 1.562$–1.564, $N_z = 1.570$–1.571; $2V = 74°$, positive elongation. A late-stage openspace filling mineral in contact rocks at Crestmore. Associated with 14-Å tobermorite and scawtite and calcite. *Crestmore quarry, Riverside Co., CA*; in xenoliths in basalt from Maroldsweisach, northern Bavaria, Germany; Bellerberg volcano, Mayen, Eifel dist., Bavaria; Campomorto, Montalto di Castro, Viterbo, Italy; Wessels mine, Kalahari manganese field, Cape Prov., South Africa; Fuka, Bitchu, Okayama Pref., Japan; hatrurim formation in Israel; Mina La Negra, Zimapan, Velardeña, DUR, and Cerro Mazahua, Zitacuaro, MICH, Mexico. EF *MIN 3(1): 814(1972)*, *AM 62:365(1977)*, *MJ 14:279(1989)*, *DA 46:69(1995)*.

56.2.4.12 Komarovite (H,Ca)$_2$(Nb,Ti)$_2$Si$_2$O$_{10}$(OH,F)$_2$ · H$_2$O

Named in 1971 by Portnov et al. for Vladimir M. Komarov (1927–1967), Russian cosmonaut, who was killed during his return flight on April 23, 1967. Cuspidine–wöhlerite group. See Na-komarovite (56.2.4.13), labuntsovite (60.1.3.2) and nenadkevichite (60.1.3.1). ORTH $a = 14.87$, $b = 12.29$, $c = 10.38$ [*DANS ESS 24:248(1979)*], $Z = 8$, $D = 3.04$. *25-163*[*ZVMO 100:599(1971)*](nat): 12.3_7 6.3_3 5.43_3 3.17_{10} 3.13_4 $2.75_{2.5}$ $2.72_{2.5}$ $1.78_{4.5}$. Silicified pyrochlore. Unit cells of pyrochlore and komarovite are closely related. *DANS ESS 24:248(1979)*. Foliated, platy, filling fractures in natrolite. Pale rose, white streak, dull luster. One fair cleavage on {001}. $H = 1\frac{1}{2}$–2. $G = 3.0$. IR data: *GZ 36:54(1976)*. Thermal DTA: *ZVMO 100:599(1971)*. Wet analysis (wt %): Nb$_2$O$_5$, 47.0; TiO$_2$, 2.5; SiO$_2$, 23.5; Fe$_2$O$_3$, 1.5; Al$_2$O$_3$, 1.0; MnO, 5.0; CaO, 4.7; Na$_2$O, 0.85; K$_2$O, 0.3; H$_2$O, 12.0; F, 1.21. Probe analysis (wt %): Nb$_2$O$_5$, 50.0; TiO$_2$, 2.5; SiO$_2$, 28.0; CaO, 6.0; Na$_2$O, 1.0; H$_2$O, 12.4; F, 1.34. Biaxial (+); $2V = 48°$ XYZ $= acb$; $N_x = 1.730$, $N_y = 1.766$, $N_z = 1.850$. Occurs on *Mt. Karnasurt, Lovozero massif, Kola Penin., Russia*; in alkalic syenites associated with late albite and natrolite. EF

56.2.4.13 Na-Komarovite (Na,Ca,H)$_2$(Nb,Ti)$_2$Si$_2$O$_{10}$(OH,F)$_2$ · H$_2$O

Named in 1979 by Krivokoneva et al. for V. M. Komarov. Described in 1969; an incompletely described mineral. Cuspidine-wöhlerite group; the Na analog of komarovite. See komarovite (56.2.4.12), the Ca-dominant member. Probably ORTH $a = 15.00$, $b = 12.31$, $c = 10.48$, $Z = 8$, $D = 3.26$. *33-608*(nat): 12.4_5 6.39_3 $5.49_{3.5}$ 3.18_{10} 3.15_5 2.75_4 2.74_3 1.79_4. Silicified pyrochlore. The unit cells of Na-komarovite and pyrochlore can be related. *DANS ESS 248:127(1981)*. G = 3.3. Wet analysis (wt %): Nb_2O_5, 50.2; TiO_2, 5.3; SiO_2, 16.5; RE_2O_3, 0.5; Fe_2O_3, 0.5; Al_2O_3, 1.4; CaO, 5.3; Na_2O, 13.7; H_2O, 4.6; F, 2.5. Probe analysis (wt %): Nb_2O_5, 52.0; TiO_2, 3.0; SiO_2, 24.0; CaO, 6.0; Na_2O, 8.0; H_2O, 4.6; F, 2.5. *Ilimaussaq massif, Greenland*. EF

56.2.4.14 Suolunite $Ca_2Si_2O_5(OH)_2 \cdot H_2O$

Named in 1965 by Huang for the locality. Cuspidine–wöhlerite group. ORTH *Fdd2*. $a = 11.02$–11.15, $b = 19.67$–19.82, $c = 6.00$–6.08, Z = 8, D = 2.69. *26-307*(nat): 4.13_{10} 3.17_8 $2.85_{6.5}$ 2.68_5 2.64_3 2.55_4 2.50_4 2.24_4; *MM 48:143(1984)*: 4.13_{10} 3.17_8 2.85_6 $2.69_{8.5}$ 2.65_6 1.99_5 1.85_4 1.71_7. Structure described in two Chinese papers (1966 and 1974) listed in *MM 48:143(1984)*. Stellate concretionary aggregates, fine-grained granular masses, crystals 0.1 to 2–3 mm, bladed. Colorless, white yellow; white streak; vitreous to resinous luster. No cleavage. H = $3\frac{1}{2}$. G = 2.63–2.68. Distinctly piezoelectric. DTA data show a strong endotherm at 440°. *SS 23:7(1965)*. Dissolves in HCl (dilute) and leaves gelatinous silica. Analysis: SiO_2 43.35, CaO 42.22, H_2O^+ 14.05 fluoresces white in SWUV. Biaxial (–); XYZ = *acb*; $N_x = 1.610$–1.612, $N_y = 1.620$, $N_z = 1.623$; 2V = 30–40°. Occurs in a 2- to 3-cm vein cutting harzburgite rocks in an ultrabasic massif. Associated with faults in arid or semiarid ultramafic environments. Alteration: To calcite. Spa Kulasi, near Doboj, Bosnia (Yugoslavia); Wadi Semali, Al Khawd, near Muscat, Oman; *Suolun, Mongolia, China*. EF *MIN 3(1):844(1972)*.

56.2.4.15 Mongolite $Ca_4Nb_6Si_5O_{24}(OH)_{10} \cdot 5H_2O$

Named in 1985 by Vladykin et al. for the locality. Cuspidine–wöhlerite group. TET Space group unknown. $a = 7.00$, $c = 29.01$, Z = 2, D = 3.51. *39-330* [*ZVMO 114:374(1985)*](nat): $9.67_{4.5}$ 6.78_3 $5.82_{4.5}$ 3.16_{10} $3.09_{6.5}$ 2.97_7 $2.90_{3.5}$ 2.66_4. Microflaky, mica-like aggregates, 40-μm crystals. Violet to grayish violet, silky luster. One micaceous cleavage. H = 2. G = 3.15. DTA and TGA data: *ZVMO 114:374(1985)*. Not soluble in common acids. Average of two analyses (wt %): SiO_2, 22.2; Al_2O_3, 0.83; Nb_2O_5, 52.45; MnO, 1.01; CaO, 12.88; BaO, 0.72; ZnO, 0.66; SrO, 1.97; H_2O, 7.72; minor TiO_2, MgO, Na_2O, K_2O. Uniaxial (–); $N_E = 1.74$, $N_O = 1.80$. Occurs in the *Khan–Bogdinsky massif, Gobi desert, Mongolia, China*; agpaitic pegmatites in alkalic granite. Associated with quartz, arfvedsonite, aegerine. EF

56.2.5.1 Mosandrite $Na(Na,Ca)_2(Ca,Ce)_4(Ti,Nb,Zr)(Si_2O_7)_2(O,F)_2F_2$

Named in 1841 by Erdmann for Carl Gustav Mosander (1797–1858), Swedish chemist and mineralogist, who researched and discovered several REE elements. Formerly considered part of the rinkite group, but has been shown to be equivalent to rinkite itself. Synonyms: rinkite (now discontinued) = rinkolite = johnstrupite. *CSREM 108:249(1993)*. MON $P2_1$. $a = 5.664$–5.679,

$b = 7.412\text{--}7.437$, $c = 18.835\text{--}18.843$, $\alpha = 101.26\text{--}101.4°$, $Z = 2$, $D = 3.36$. $31\text{--}313(\text{nat})$: 3.58_2 3.07_{10} 2.95_4 2.80_4 2.70_7 2.58_2 2.00_2 1.68_2. Material fromTuva is stated to be TRIC with $a = 9.66$, $b = 5.75$, $c = 7.34$, $\alpha = 90°$, $\beta = 102°$, $\gamma = 101°$. *ZVMO 109:594(1980)*. Also given as TRIC (*MT:1989*): $a = 18.5$, $b = 7.44$, $c = 5.65$, $\alpha = 90.4°$, $\beta = 91.3°$, $\gamma = 102°$. Structure given in CSREM 5th supplement, p. 12. Some cation ordering is present. *AC(B) 27:1277(1971)*, *SPC 36:349(1991)*. Prismatic (pseudohexagonal), acicular and thin tabular radiating groups, acicular parallel [100]; to 3.5 cm at Mt. St.-Hilaire. Colorless to lemon yellow, pale orange-yellow, mauve, beige, yellow-brown, brown; white streak; vitreous to resinous luster. Cleavage {001}, distinct; {010}, poor; uneven to conchoidal fracture; brittle. $H = 5$. $G = 3.29\text{--}3.46$. Nonluminescent. Gelatinizes with acids. Analysis (wt%): Ce_2O_3 8.1, Y_2O_3 0.72, La_2O_3 9.92, U_3O_8 0.33, ThO_2 1.18, Nb_2O_5 0.45, TiO_2 11.25, SiO_2 31.73, Al_2O_3 2.55, FeO 0.28, CaO 21.91, SrO 0.89, K_2O 0.84, Na_2O 4.20, F 4.72, H_2O^+ 2.71, H_2O^- 1.34. Biaxial (+); $Y = b$, $X \wedge a = 3°$; $N_x = 1.643\text{--}1.665$, $N_y = 1.645\text{--}1.668$, $N_z = 1.651\text{--}1.681$; $2V = 43\text{--}80°$; $Z > Y > X$; $r > v$, strong. Occurs in eudialyte-rich pegmatites, sodalite xenoliths, and nepheline syenite at Mt. St.-Hilaire. In sodalite syenites in Greenland. Occurs at Mt. St.-Hilaire, and St.-Amable sill (phonolite), Varennes, QUE, Canada; Kangerdluarssuk area, Ilimaussaq, Julianehaab district, Greenland; also at Naujakasik and Narsarsuk, Greenland; *Låven, and Stokkø, near Barkevik, Langesundfjord, Norway*; Khibiny and Lovozero massifs, Kola Penin., Sangilen massif, Tuva, southwest of Lake Baikal, Siberia, and central Aldan, Yakutia, Russia; Saima, Liaoning Prov., China; Pilanesberg complex, Saulspoort, South Africa. EF *DAN SSSR 263:435(1982)*, *MR 21:284(1990)*.

56.2.5.2 Nacareniobsite-(Ce) $Na_3Ca_3(Ce,La,Nd)Nb(Si_2O_7)_2OF_3$

Named in 1989 by Petersen et al. for the major constituents: Na, Ca, RE, Nb, Si. Related mineral is rinkite (mosandrite): $2Ca^{2+} \leftrightarrow REE^{3+} + Na^+$. The Nb analog of $(Ca,Na,REE)_7TiSi_4O_{14}OF_3$ (rinkite). MON $P2_1/a$. $a = 18.901$, $b = 5.683$, $c = 7.462$, $\beta = 101.29°$, $Z = 2$, $D = 3.43$. $PD(\text{nat})$: 3.09_5 3.08_{10} 2.96_5 2.81_3 2.03_5 1.87_4 1.69_4 1.55_5. Structure not yet solved, but probably isostructural with mosandrite. Rectangular, thin bladed crystals to 3 mm × 0.4 mm × 0.04 mm; elongated || [001]; prominent {100}, {410} and {310}; colorless, vitreous luster, transparent. Cleavage {011}, good; uneven fracture. $H = 5$. $G = 3.45$. Nonluminescent. Analysis (wt %): SiO_2, 29.6; TiO_2, 2.79; Nb_2O_5, 11.61; Ta_2O_5, 0.34; Na_2O, 10.01; CaO, 19.92; SrO, 0.27; V_2O_3, 0.78; La_2O_3, 4.09; Ce_2O_3, 10.32; Pr_2O_3, 1.42; Nd_2O_3, 4.19; Sm_2O_3, 0.05; F, 6.87. Biaxial (+); $XY = bc$, $Z \wedge a = 11°$ in obtuse β; $N_x = 1.6618$, $N_y = 1.6706$, $N_z = 1.6924$; $2V = 66°$. Occurs in dissolution cavities in arfvedsonite-repheline syenite (lujavrite), *Kvanefjeld tunnel, Ilímaussaq complex, southern Greenland*. Associated with sodalite, eudialyite, and villiaumite. EF *NJBM:84(1989)*.

56.2.5.3 Fersmanite $(Ca,Na)_4(Ti,Nb)_2Si_2O_{11}(F,OH)_2$

Named in 1929 by Labuntsov for Alexander Yevgenievich Fersman (1883–1945), Russian mineralogist and geochemist. TRIC $P\bar{1}$ or $P1$. $a = 7.210$, $b = 7.213$, $c = 20.451$, $\alpha = 95.15°$, $\beta = 95.60°$, $\gamma = 89.04°$, $Z = 4$, $D = 3.44$. 29-1446(nat): 3.06_{10} 2.82_6 $2.53_{3.5}$ $1.91_{2.5}$ 1.80_5 $1.69_{4.5}$ 1.55_4 $1.52_{5.5}$. Structure: SPC 29:31(1984). This determination indicated a monoclinic cell with $a = 10.212$, $b = 20.450$, $c = 10.198$, $\beta = 97.22°$. Space group Bb or B2/b. It was found to be a diorthosilicate, $(Ca,Na)_4(Ti,Nb)_2(Si_2O_7)O_4(F,OH)_2$. Pseudotetragonal, distorted crystals; euhedral crystals not common; crystals to 2–3 cm across; twinning present possibly on {001}. Dark brown, much resembling titanite, light to cinnamon brown; golden yellow; transparent to translucent, pale tan streak, vitreous luster. No cleavage, uneven fracture. $H = 5$–$5\frac{1}{2}$. $G = 3.44$–3.46. Soluble in acids. Main substitution scheme is CaTi ↔ NaNb, but some OH may replace O. Analysis: K_2O, to $\approx 1\%$; MgO, to $\approx 1\%$; SrO, to 0.5%; FeO, to 1.4%; Fe_2O_3, to 3.5%; Ta_2O_5, to 0.8%; F, $\approx 4\%$. Biaxial ($-$); $N_x = 1.873$–1.886, $N_y = 1.930$, $N_z = 1.914$–1.939; golden yellow. $2V_{calc} = 48°$, also given as 0–7°. Occurs in aegirine-rich pegmatite veins in nepheline syenite assoc. with other alkalic minerals. Hydrothermal in origin. Sometimes alters to lamprophyllite. Occurs just above the *Vuonnemijok River, on the southwestern slope of Mt. Eveslogchorr, Khibiny massif, Kola Penin., Russia*, and at a second locality about 2 km to the east; also occurs on Mt. Yukspor and in the apatite deposit on Mt. Kukisvumchorr, as well as Mt. Kuel'por and in the Vorkeuai R. valley. EF MRE 2:564(1966), MIN 3(1):353(1972), CM 15:87(1977).

56.2.5.4 Götzenite $(Ca,Na)_3(Ti,Al)Si_2O_7(F,OH)_2$ or $Na_2Ca_5Ti(Si_2O_7)_2F_4$

Named in 1957 by Sahama and Hytönen for Gustav Adolf Von Götzen (1866–1910), German traveler and first European to climb Mt. Shaheru, North Kivu, Zaire, the type locality. Synonym: calcium rinkite. See rosenbuschite (56.2.4.8) and hainite (56.2.4.9). TRIC $P\bar{1}$. $a = 9.589$–9.667, $b = 7.23$–7.334, $c = 5.731$–5.758, $\alpha = 90.00$–$92.2°$, $\beta = 100.2$–$101.3°$, $\gamma = 99.4$–$101.1°$, $Z = 2$, $D = 3.04$. 12-536(nat): 3.99_2 3.10_{10} 2.99_{10} 2.65_4 2.51_3 2.26_2 1.91_5 1.69_3. The main feature of the crystal structure of götzenite is the presence of "walls" of octahedra and eight-cornered polyhedra parallel to (100) which are connected by "ribbons" of octahedra parallel to [010] and by Si_2O_7 groups. SPC 16(6):1026(1972). Tablets, and prismatic, acicular parallel to [001] (probably); to 1 cm long; most show lamellar polysynthetic twinning with twin axis = b and the composition plane perpendicular to (001). MM 31:503(1957). Very pale yellow, tan or off-white, colorless, white; white streak; dull, vitreous to greasy luster. Cleavage {100}, perfect; {001}, good; brittle. $H = 5\frac{1}{2}$–6. $G = 3.14$–3.20. IR data: MIN 3(1):831(1972). Easily soluble in hot dilute HCl. Analysis (wt %): major Na_2O, CaO, SiO_2, TiO_2 and F; K_2O, 0.1–0.7; MgO, trace to 0.6; SrO, to 3.8; MnO, to 2.1; FeO, to 0.5; Al_2O_3, to 4.3; Fe_2O_3, to 0.9; TR_2O_3, to 6.4; ThO_2, to 0.4; ZrO_2, to 2.9; Nb_2O_5 +

Ta_2O_5, to 12.5; H_2O, to 2%. Biaxial (+); $X \approx c$; $Z' \wedge c \approx 58°$; $N_x = 1.651$–1.662, $N_y = 1.653$–1.665, $N_z = 1.659$–1.672; $2V = 38$–$74°$; $r > v$, strong; length-fast or length-slow; pleochroic, from colorless to pale yellow to pale rose. Occurs in nephelinite lava (Zaire), pegmatites (Norway), and in marble xenolith cavities (Mt. St.-Hilaire) associated with pectolite and other minerals. In nepheline syenites and cogentic pegmatites, associated with other alkaline minerals. In olivine meltigeite porphyry. In sanidine–anorthoclase rocks. Occurs at Mt. St.-Hilaire, QUE, Canada; near Kirchberg, Saxony, Üdersdorf, Daun, West Eifel, and Kaiserstuhl, Baden-Württemberg, Germany; to 10 cm long from near Barkevik, Langesundfjord, Norway; Sangilen massif, Tuva, southern Baikal, Azov area, Ukraine; Mt. Yukspor and Mt. Eveslogchorr, Khibiny massif, and Gornoozerna massif, Cape Turi, Kola Penin., Russia; *Nyiragongo crater, Mount Shaheru, Kivu Prov., Zaire.* EF *MIN 3(1):831(1972), ZVMO 109:594(1980), DHZ RFM 113:343(1986), MR 21:284(1990).*

56.2.6a.1 Seidozerite $(Na,Ca)_4MnTi(Zr_{1.5}Ti_{0.5})(Si_2O_7)_2O_2(F,OH)$

Named in 1958 by Semenov et al. for the locality. Calcian seidozerite has $Na_2(Na,Ca)_2(Ca,Mn)TiZr_2O_2[Si_2O_7]_2(F,OH)_2$. MON $P2/c$. $a = 5.53$–5.54, $b = 7.10$, $c = 18.30$–18.36, $\beta = 102.68$–$102.7°$, $Z = 2$, $D = 3.53$–3.60. *13-576*(nat): 2.97_{10} 2.87_7 2.58_4 2.43_3 2.25_3 1.83_7 1.76_3 1.63_4. Structure: CR *DAN SSSR 122:473(1958); SPC 8:406(1964), 10:505(1966), 16:1026(1972).* Elongate on b, striated parallel to b, prismatic to fan-shaped intergrowths, equant flattened crystals to 5 cm, also spherulites, dominant forms are {001}, {100}, {$\bar{2}$03}. Brownish red, dark red, yellow to brown; strong vitreous luster. Cleavage (001), perfect; brittle. H= $4\frac{1}{2}$. G = 3.42–3.47. IR data: *IM: 208(1975), K 6:126(1961).* No piezoelectric effect, electromagnetic. Soluble with difficulty in HCl. Analysis (wt %): major Na_2O, MnO, SiO_2, TiO_2, ZrO_2, F; CaO, 2.8 to 9.7; FeO + Fe_2O_3, to 4.17; Al_2O_3, to 1.4; Nb_2O_5, to 0.6. Biaxial (+); $2V = 62$–$68°$; $Z = a$, $Y \wedge c = 13°$; $N_x = 1.718$(calcian)–1.725, $N_y = 1.758$, $N_z = 1.772$(calcian)–1.830; OAP = (001); $X > Y > Z$; $r > v$, strong. X, dark red to reddish brown; Y, red; Z, pale yellow to yellow. Occurs in poikilitic nepheline syenite pegmatites at Lovozero. In albitized pegmatites and stringer zones and in aegirine–microcline pegmatites in the Burpala massif. Alteration to catapleite. Occurs in the *Lake Seidozero region of the Lovozero massif, Kola Penin., Russia*; Ca-rich (10% CaO) variety from the Burpala massif, northern Baikal region, and Tuva region, southern Baikal region, Siberia, Russia. EF *MS 19:131(1965), MIN 3(1):835(1972).*

56.2.6b.1 Bafertisite $Ba(Fe,Mn)_2TiO(Si_2O_7)(OH,F)_2$

Described in 1959 by Semenov and P'ei-Shan Chang and named for the composition: Ba, Fe (ferrum), Ti, Si. Hejtmanite is the Mn analog. MON Cm. $a = 10.60$, $b = 13.64$, $c = 12.47$, $\beta = 119.5°$. *DANS 149:1416(1963).* These data match the cell of *SPC 36:186(1991)*, however, the true cell may be $P2_1/m$ as for hejtmanite. *EJM 4:35(1992).* $Z = 8$. Another proposed cell

[*CA 60:75(1964)*] is: $a = 10.98$, $b = 6.80$, $c = 5.36$, $\beta = 94°$, $P2_1/m$. Originally stated to be orthorhombic. *AM 45:754,1317(1960)*. *14-541:* 2.65_{10} 2.52_3 2.23_3 2.11_4 2.07_3 1.75_3 1.72_4 1.57_3. Structure is related to micas. Bright red, yellowish red to light brown, or orange. Cleavage {001}, perfect; {010}, poor; brittle. $H = 5$. $G = 3.96$–4.27. IR data: *IM:208(1975)*. Mössbauer spectra: *APMA 1:23(1982)*)Contains minor Nb, Al, Fe^{3+}, Mg, Ca, Na, and K. Material from Burpala massif is Mn-rich (12.8 wt % MnO). Biaxial $(-)$; $X \wedge c = 24°$ in obtuse β, $Z \wedge a = 5.5°$; $N_x = 1.786$–1.808, $N_y = 1.813$–1.835, $N_z = 1.852$–1.862; $2V = 54$–$86°$; OAP = (010); $X > Z$. X, yellow-red or brownish red; Y, yellow; Z, pale yellow or greenish yellow. Occurs in alkaline igneous rocks or in rocks associated with alkaline rocks (i.e., alkali granite and alkali syenite), hydrothermal veins associated with aegirine, fluorite, barite, and bastnaesite (Bayan Obo); in a dike of microclinized granite–aplite (Burpala massif). Fountain Quarry, Pitt County, NC, with an unidentified Ba–Ca–Mn–Fe–Ti silicate; Mn-rich variety from the Burpala massif, northern Baikal area, Russia; Tarbagata, eastern Kazakhstan; *Bayan Obo REE deposit, Inner Mongolia, Jiangsu Prov., China.* EF *TMMAN SSSR 16:293(1965)*, *MIN 3(1):635(1972)*, *AM 57:1005(1972)*.

56.2.6b.2 Hejtmanite $Ba(Mn,Fe^{2+})_2TiO(Si_2O_7)(OH,F)_2$

Described in 1992 by Vrána et al. and named for Bohuslav Hejtman (b.1911), emeritus professor of petrology, Charles University, Prague. First found in Kirgizia but not named. The mineral is the Mn analog of bafertisite. MON $P2_1/m$. $a = 11.748$, $b = 13.768$, $c = 10.698$, $\beta = 112.27°$, $Z = 8$, $D = 4.29$. (*EJM 4:35(1992)*). Erroneously stated to be triclinic. *ZVMO 118:81(1989)*. Alternative monoclinic cells are: $a = 5.361$, $b = 6.906$, $c = 12.556$, $\beta = 119.8°$, $P2_1/m$ and $a = 10.723$, $b = 13.812$, $c = 12.563$, $\beta = 119.9°$, Cm. *SPC 36:186(1991)*. *EJM 4:35(1992)* and *44-1470:* 5.47_3 3.65_1 3.45_2 3.22_2 2.73_{10} 2.53_1 2.18_2 2.07_1. Crystals are lath-shaped to platy, or prismatic to 1 mm long. {100} prominent, elongated parallel to [001]. Brownish yellow to golden yellow, brownish-yellow streak, vitreous luster, transparent. Cleavage {100}, perfect; {0kl}, poor, parting; irregular fracture; brittle. $G = 4.02$. Biaxial $(-)$; $X = b$, $Y \wedge c = 38°$ in obtuse β, $Z \wedge a = 16°$. $N_x = 1.814$, $N_y = 1.846$, $N_z = 1.867$; $2V = 76°$; $Y > Z = X$. X, light green-yellow; Y, dark golden yellow; Z, light yellow. Optical orientation is different from bafertisite. Occurs in manganoan arfvedsonite crystals that are 10–30 cm across, along with albite and microcline in pegmatoid veins. Also in a rhodonite–tephroite–spessartine–quartz assemblage. Northern end of Inylchek Range, southeastern Kirgizia; *Mbolwe Hill, Mkushi River area, Central Prov., Zambia.* EF

LAMPROPHYLLITE GROUP

The lamprophyllite group consists of four dimensionally equivalent, and probably isostructural, species and a displacively transformed polymorph. A simplified general formula for the group is

$$A_2B_3C_3[Si_2O_7]_2(O,OH,F)_4$$

where

A = Sr, Ba, K (12-coordinated)
B = Na, Ca, Mn, Ti^{4+}, Fe^{3+} (6-coordinated)
C = Ti^{4+}, Fe^{3+} (5-coordinated)

The structure has not been elucidated for all species. The [Si_2O_7] groups are linked to tetragonal–pyramidal five-coordinated ions to form sheets, two of which are linked to either side of distorted, close-packed octahedral sheets forming triple layers similar to 2:1 phyllosilicates. The sheets are parallel to (100) with intervening 12-coordinated cations, thus accounting for the perfect {100} cleavage. Structure: *MJJ 10:107(1980); MZ 12:25(1990)*.

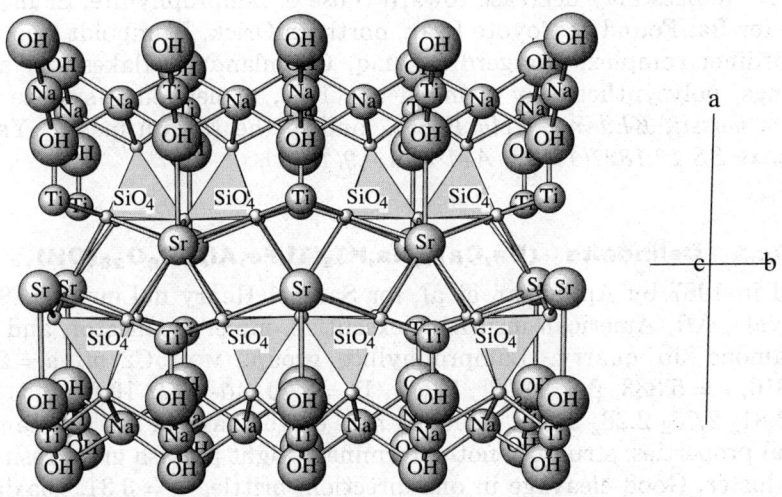

LAMPROPHYLLITE: $Sr_2Na_3Ti_3(Si_2O_7)_2(OH)_4$

Si_2O_7 groups combined with Ti octahedra can be considered to form layers parallel to (100). Ti and Na form an octahedral layer joining apices of the tetrahedra, and Sr joins the back side (bases) of the tetrahedra.

LAMPROPHYLLITE GROUP

Mineral	Formula	Space Group	a	b	c	β	
Barytolamprophyllite	$Ba_2Na_3(Fe^{3+},Ti^{4+})[Si_2O_7]_2(O,OH,F)_4$	C2/m	19.833	7.089	5.393	96.66	56.2.6c.1
Delindeite	$(Ba,Ca)_4(Na,K)_3(Ti,Fe,Al)_6Si_8O_{26}(OH)_{14}$	C2/m	21.617	6.816	5.383	94.03	56.2.6c.2
Ericssonite	$BaMn_2^{2+}Fe^{3+}O[Si_2O_7](OH)$	C2/c	20.46	7.03	5.34	95.5	56.2.6c.3
Lamprophyllite	$Sr_2Na_3Ti_3[Si_2O_7]_2(O,OH,F)_4$	C2/m	19.431	7.086	5.386	96.75	56.2.6c.4
Orthoericssonite	$BaMn_2^{2+}Fe^{3+}O[Si_2O_7](OH)$	Pnmn or Pn2n	20.230	6.979	5.392		56.2.6c.5

56.2.6c.1 Barytolamprophyllite $Ba_2Na_3(Fe^{3+},Ti^{4+})_3(Si_2O_7)_2(O,OH,F)_4$

Named in 1965 by Peng and Chang for the composition and similarity to lamprophyllite). Lamprophyllite group. Isomorphous with lamprophyllite; the existence of an orthorhombic dimorph has been suggested. MON $C2/m$. $a = 19.833$, $b = 7.089$, $c = 5.393$, $\beta = 96.66°$, $Z = 2$, $D = 3.70$. 34-313: 3.45_7 3.29_5 2.80_{10} 2.69_5 2.15_9 1.790_7 1.601_8 1.482_9. An orthorhombic dimorph with $a = 19.32$, $b = 7.05$, $c = 5.43$ is alleged to exist at the (100) twin boundaries of the monoclinic structure. Structure: KT 29:$237(1984)$. Golden brown, light brown streak, vitreous to submetallic luster, translucent. Cleavage {100}, perfect; {011}, distinct; {010}, poor. $H = 2$–3. $G = 3.62$–3.66. Biaxial (+); $2V = 30°$; $N_x = 1.742$–1.743, $N_y = 1.754$, $N_z = 1.776$–1.778 for Na_D; $X = b$, $Z \wedge c = 6$–$7°$ r > v; strongly pleochroic, $Z > Y > X$ brown to light yellow; refractive indices may decrease toward those of lamprophyllite. Sr may substitute for Ba. Found at Coyote Peak, northern Orick, Humboldt Co., CA; in the Gardiner complex, Kangerdlugssuaq, Greenland; as flakes and stellate groupings, polysynthetically twinned on (100), in nepheline–syenite of the *Lovozero massif, Khibiny, Kola Penin.*, and in the Murun massif, Yakutia, Russia. AR SS 12:$1827(1965)$, AM 51:$1549(1966)$.

56.2.6c.2 Delindeite $(Ba,Ca)_4(Na,K)_3(Ti,Fe,Al)_6Si_8O_{26}(OH)_{14}$

Named in 1987 by Appleman et al. for Samuel Henry deLinde (b.1923) of Mabelvale, AR, American insurance executive, mineral collector, and owner of Diamond Jo quarry. Lamprophyllite group. MON $C2/m$. $a = 21.617$, $b = 6.816$, $c = 5.383$, $\beta = 94.03°$, $Z = 1$, $D = 3.70$. 40-1454: 10.8_{10} 3.54_2 3.08_3 2.89_3 2.81_2 2.75_2 2.26_2 2.21_1. Included here on the basis of cell constants and physical properties; structure not determined. Light pinkish gray, resinous to pearly luster. Good cleavage in one direction, brittle. $G = 3.31$. Biaxial (+); $2V_{calc} = 54°$; $N_x = 1.790$, $N_y = 1.825$, $N_z = 1.982$. Found as micron-size grains and laths in spherulitic aggregates to 1 mm in diameter at the *Diamond Jo quarry, Magnet Cove, near Hot Springs, AR*. AR MM 51:$417(1987)$.

56.2.6c.3 Ericssonite $BaMn_2^{2+}Fe^{3+}O[Si_2O_7](OH)$

Named in 1971 by Moore for John E. Ericsson (1803–1883), Swedish-born American engineer and inventor, designer of the iron-clad ship *Monitor*. Lamprophyllite group. Dimorphous with orthoericssonite. MON $C2/c$. $a = 20.46$, $b = 7.03$. $c = 5.34$, $\beta = 95.5°$, $Z = 4$, $D = 4.38$. *29-186:* 10.1_6 5.08_5 3.51_{10} 3.40_5 2.78_6 2.03_5 1.752_6 1.597_6. Reddish black, brown streak, submetallic luster. Cleavage {100}, perfect; {011}, distinct; very brittle. $H = 4\frac{1}{2}$. $G = 4.21$. Biaxial (+); optical properties not distinguished from those of orthoericssonite. Weakly magnetic. Massive and as embedded plates, intimately intergrown with the more abundant orthoericssonite in schefferite–rhodonite–tephroite skarn at the *Långban mine, near Filipstad, Värmland, Sweden*. AR *LIT 4:137(1971)*, *AM 56:2157(1971)*.

56.2.6c.4 Lamprophyllite $Sr_2Na_3Ti_3[Si_2O_7]_2(O,OH,F)_4$

Named in 1894 by Ramsey and Hackmann from the Greek for *shining leaf*, in allusion to the luster and cleavage. Lamprophyllite group. Isomorphous with barytolamprophyllite. MON $C2/m$. $a = 19.431$, $b = 7.086$, $c = 5.386$, $\beta = 96.75°$, $Z = 2$, $D = 3.54$. *17-751:* 3.73_4 3.43_6 3.27_4 2.87_4 2.77_{10} 2.66_4 2.13_5 1.479_5. Structure: *SPD 28:206(1983)*. Crystals tabular {100} and elongated [001], as isolated flakes and as stellate aggregates. Polysynthetic twinning on {100} common. Yellow-brown, light brown streak, vitreous to submetallic luster, translucent. Cleavage {100}, perfect; {010}, poor. H = 2–3. $G = 3.45$–3.54. Biaxial (+); $2V = 23$–$41°$; $N_x = 1.733$, $N_y = 1.740$, $N_z = 1.769$ for Na_D; $X = b$, $Z \wedge c = 5$–$6°$; $r > v$, distinctly pleochroic, $Z > Y > X$; yellow-orange to pale yellow, pleochroism serves to distinguish from astrophyllite, for which $Y > Z$. The refractive indices increase toward those of barytolamprophyllite. Ba and K may substitute for Sr, and Ca and Mn^{2+} for Na. Found in nepheline–syenites in the Bearpaw Mts., Hill Co., MT; at Mt. St.-Hilaire, Canada; Langesundfjord, Norway; *Lujavur-Urt, Kola Penin., Russia*; and Pilansberg, Transvaal, South Africa. AR *AM 27:397(1942)*, *SS 14:1829(1966)*.

56.2.6c.5 Orthoericssonite $BaMn_2^{2+}Fe^{3+}O[Si_2O_7](OH)$

Named in 1971 by Moore from the symmetry and relationship to ericssonite. Lamprophyllite group. Dimorphous with ericssonite. ORTH $Pnmn$ or $Pn2n$. $a = 20.230$, $b = 6.979$, $c = 5.392$, $Z = 4$, $D = 4.26$. *29-185:* 5.08_6 3.61_2 3.38_{10} 2.94_1 2.82_1 2.68_3 2.54_4 2.03_1. Reddish black, brown streak, submetallic luster. Cleavage {100}, perfect; {011}, distinct; very brittle. $H = 4\frac{1}{2}$. $G = 4.21$. Biaxial (+); $2V = 30°(67°_{calc})$; $N_x = 1.807$, $N_y = 1.833$, $N_z = 1.891$ (Sweden); $XYZ = bca$; $2V \approx 50°(83°_{calc})$: $N_x = 1.802$, $N_y = 1.840$, $N_z = 1.888$ (Japan). Strongly pleochroic; X, pale greenish tan or yellow-brown; Y, red-brown; Z, deep brown. Weakly magnetic. Massive and as embedded plates, intimately intergrown with the less abundant ericssonite in schefferite–rhodonite–tephroite skarn at the *Långban mine, near Filipstad, Värmland, Sweden*; at the Hijikuzu mine, Iwate Pref., Honshu, Japan. AR *LIT 4:137(1971)*; *AM 56:2157(1971)*; *MJJ 7:513(1975), 10:107(1980)*.

56.2.6c.6 Andremeyerite $BaFe(Fe,Mn,Mg)Si_2O_7$

Named in 1973 by Sahama for André Marie Meyer (1890–1965), Belgian mining engineer, honorary director general of mining, who collected the original specimens in Zaire. MON (pseudo-ORTH) $P2_1/c$. $a = 7.488$, $b = 13.785$, $c = 7.085$, $\beta = 118.23°$, $Z = 4$, $D = 4.31$ [4.14 using *BGSF 45:1(1973)*]. *26-1031*: 4.63_4 3.29_6 3.20_2 3.12_8 3.06_{10} 3.02_2 2.81_4 2.47_6. Structure: *AM 73:608(1988)*. Related to lamprophyllite group. Prismatic, <0.2 mm long and 0.1 mm in diameter, {100} and {010} dominant, {011} and {120} also present. Always shows multiple {100} twinning. Pale emerald green. Cleavage {100}, {010}, perfect. $H = 5\frac{1}{2}$. $G = 4.15$. Contains minor amounts of K, Na, Al, Ce, N, Mg, Mn. Biaxial (+); $Z = b$, $X \wedge c = 2°$ at 670 nm to 61° at 470 nm; $N_x = 1.740$, $N_y = 1.740$, $N_z = 1.760$; 2V near 0° below 490 nm, increasing to 40° at 540 nm, and decreasing again to near 0° at 585 nm; $X > Y \approx Z$; very strong dispersion. X, slight bluish green; Y,Z, colorless with faint brownish tint. Occurs in vesicles of a melilite–leucite–nephelinite block in nephelinite lava. Associated with a green Fe-rich glass and with kirschsteinite and melilite. *Nyiragongo volcano, Zaire.* EF

56.2.7.1 Epistolite $Na_2(Nb,Ti)_2Si_2O_9 \cdot nH_2O$

Found in 1897 by Boggild and named in 1900; name originated from the Greek for *letter*, in allusion to the flat rectangular form and white color. Ideal formula is $Na_2Ti_2Nb_2(Si_2O_7)_2O_2(OH)_4 \cdot 2H_2O$. Related to murmanite, considered by some sources to be the Ti-dominant analog. See murmanite (56.2.7.2). TRIC $P1$. $a = 5.41$–5.570, $b = 7.08$–7.16, $c = 11.821$–12.10, $\alpha = 103.05$–$104.39°$, $\beta = 96.00$–$96.31°$, $\gamma = 88.52$–$88.85°$; alternate cell: $a = 5.570$, $b = 7.044$, $c = 11.821$, $\alpha = 96.53°$, $\beta = 100.04°$, $\gamma = 89.34°$ [*MT:76(1989)*]; $Z = 2$, $D = 2.88$–2.91. *39-407*(nat): 12.0_8 5.90_7 4.32_{10} 2.96_9 2.87_9 2.71_5 2.15_6 1.79_7. The structure is related to that of murmanite. [*NJBA 155:289(1986)*]. Sharp, platy cleavage masses to 6 cm (Mt. St.-Hilaire); lamellar to micaceous, flattened on (001), {001} dominant; to 5 cm in length, to 10 cm in diameter and 5 mm thick in Greenland. Tan, silvery on {001}, pinkish beige, yellow-gray, white, cream-colored, brown; white streak; pearly to silky luster, vitreous to adamantine at Mt. St.-Hilaire; transparent. Cleavage {001}, perfect; one or two other good cleavages as well; *MT:76(1989)* says {100} cleavage is perfect; uneven fracture; brittle; sometimes readily scratched with a finger. $H = 2$–3; *MIN 3(1):673(1972)* says 1–$1\frac{1}{2}$; $G = 2.89$ (Greenland) and 2.65 (Lovozero). Thermal data: *MIN 3(1):673(1972), NJBA 155:289(1986)*. Nonfluorescent. Some Ca substitutes for Na. Biaxial (−), $2V = 80°$, rarely (+) with $2V = 84°$; $N_x = 1.610$, $N_y = 1.650$, $N_z = 1.682$; material from Lovozero has $N_x = 1.72$ and $N_z = 1.77$ with 2V (−) of 60°, colorless; $Y = b$, $Z \wedge c = 7°$; $r > v$ or $r < v$, perceptible. Occurs in pegmatites and hydrothermal veins and disseminated in lujavrite + naujaite. A rare mineral of nepheline–syenite massifs. Associated with rare alkaline species, rinkite, and so on. Ussingite pegmatites in Lovozero. In pegmatites in nepheline–syenite in Greenland. At Mt. St.-Hilaire, some epistolite may be an alteration product of vuonnemite. Alters to leucoxene, nenadkevichite, gerasimovskite, and other minerals. Found at Mt. St.-Hilaire, QUE, Canada; *Tugtupagtakorfia, Kvanefjeld, and Taseq slope, Ilimaussaq intrusion, Julianehaab area, southern Greenland*; also at Kangerdluarssuk and Mt. Nakalik. Punkaruaiev area, Lovozero massif, Kola Penin., Russia. EF *KSM 10:96(1976)*.

56.2.7.2 Murmanite $Na_3(Ti,Nb)_4(Si_2O_7)_2O_4 \cdot 4H_2O$

Named in 1923 by Fersman, apparently for the locality near Murmansk. Is a supergene, an alteration product of lomonosovite, and also is primary. TRIC $P\bar{1}$ or $P1$ (probably $P1$). $a = 8.700$, $b = 8.728$, $c = 11.688$, $\alpha = 94.31°$, $\beta = 98.62°$, $\gamma = 105.62°$, $Z = 2$ [*K 31:82(1986)*], D = 2.90–3.00. *40-479*(nat): 11.56_9 5.81_9 4.22_{10} 3.76_6 2.87_{10} 2.64_4 2.48_4 1.77_4. Structure: *SPC 31:44(1986)*. R = 9.1% (anisotropic). Scaly or tabular on (001), crystals and plates to 7 cm; aggregates to 10 cm across. Pink, deep violet; cherry-red streak; pearly to submetallic luster. Three cleavages present; one is perfect (001) and micaceous parallel to tabular development of crystals. H = 2–3. G = 3.34, sometimes lower. IR, DTA, TGA data: *NJBA 155:67(1986)*. Decomposed by HCl and H_2SO_4, leaving a silica residue. "Pink murmanite" has lower Na_2O than that of regular murmanite. It has higher MnO, TiO_2, and Nb_2O_5. Analysis: ZrO_2, to >4%; Fe_2O_3, to 73.5%; MnO, to ≈3%; CaO, to ≈3%; MgO, 0.75%; Na_2O, major; $TiO_2 + SiO_2$, major; K_2O, to ≈1%; Nb_2O_5, to >7%; total H_2O= 10%. Biaxial (−); N_x = 1.682–1.735(calc), N_y = 1.765–1.770, N_z = 1.807–1.839; 2V = 57–64°; OAP perpendicular to (001), X perpendicular to (001); Z > Y > X; r < v or r > v, weak; strongly pleochroic. X, pale rose to rose; Y, light brown or light yellow-brown; Z, dark brown or brown with violet tint. Occurs in alkalic syenites and their derivative pegmatites; also in altered and sheared lava and gabbro intruded by lujavrite. A supergene alteration product of lomonosovite. Alteration: readily susceptible to alteration; sometimes to labuntsovite, also to belyankinite and other products. Occurs at Mont Saint-Hilaire, QUE, Canada; Kvanefjeld, Steenstrupsfjeld, and Skovfjord, near Illimaussaq, Greenland; Mt. Punkaruaiev, Mt. Suoluaiev, Ninchurt, Mt. Mannepakh, Mt. Alluaiv, Chingulusuai R., Sengischorr cirque, and other localities in the *Lovozero massif, near Murmansk, Kola Penin., Russia*; also in the Khibiny massif at Mt. Rasvumchorr, Mt. Rischorr, Mt. Karnasurt, Mt. Kukisvumchorr, Mt. Yukspor, and others. EF *MIN 3(1):668(1972)*, *KSM 10:96(1976)*.

56.2.8.1 Chevkinite-(Ce) $(Ce,Ca,Th)_4(Fe,Mg)(Ti,Mg,Fe)_4Si_4O_{22}$

Named in 1839 by Rose for Konstantin Vladimirovich Tschevkin (1802–1875), chief of Russian mining corps. Compare perrierite (56.2.8.3) and strontiochevkinite (56.2.8.2). Not strictly polymorphous with perrierite. Synthetics are polymorphous for certain compositions. MON $C2/m$ or $P2_1/a$ [synthetic material is $P2_1/a$: AM *59:1277(1974)*]. Natural material from Mianxi, Sichuan Prov., China, was studied by TEM techniques and SAED and CBED showed space group $C2/m$ rather than $P2_1/a$. JCUG *2:75(1991)*. This discrepancy may be due to compositional differences between synthetic and natural materials used. Heated metamict chevkinite from Zambia was refined in $P2_1/a$. AUCG:*181(1983)*. $a = 13.26$–13.60, $b = 5.67$–5.82, $c = 11.04$–11.35, $\beta = 100.53$–$100.87°$, $Z = 2$, $D = 4.88$–5.10; *42-1394*(heated nat): 3.48_4 3.47_4 3.19_9 3.15_6 3.00_4 2.87_5 2.76_6 2.72_{10}; (unheated nonmetamict): 4.61_3 3.19_8 3.15_8 2.88_7 2.76_5 2.72_{10} $2.175_{3.5}$ 1.98_3. NJBM:*358(1994)*. Is often metamict or partly so. Structure: AM *59:1277(1974)*. Anhedral to subhedral masses to 12 cm or more, or short prismatic crystals with {100}, {001}, and {201} dominant, {110} minor; elongated on b, sometimes twinned on {001}. Black, dark brown or dark reddish brown, gray-brown streak, vitreous–adamantine luster. No or very poor cleavage on (001), conchoidal fracture, brittle. $H = 5$–6. $G = 4.4$–4.9. Soluble in HCl, HNO_3, HF, and AR, but not in bases. Metamict material is more readily soluble. Analysis: ThO_2 to 24.2% (Ilmen Mts., Urals); Nb_2O_5 to 7.4% (Ilmen Mts., Urals); UO_2 to 2.1%; may contain elevated MnO. See perrierite (56.2.8.3) for the general formula and possible chemical substitutions. Biaxial (−); $Z = b$, $X \wedge c = 11.5$–$25.75°$ in acute β; $2V = 70°$, N_x as high as 2.00 when fresh, N_z as high as 2.05 when fresh. Metamict material has N as low as 1.88. $Z > Y > X$; $r > v$. Opaque to deep reddish brown; sometimes $X =$ very pale brown-yellow; $Y =$ pale red-brown; $Z =$ dark red-brown. R at four standard wavelengths is about 11%. Occurs as a minor accessory mineral in volcanic ash beds and rhyolites, but is more commonly present as an accessory mineral in alkalic or peralkaline granites, granite pegmatites, nepheline–syenites, syenites, and syenitic pegmatites. Rarely in fenites, and also carbonatites. May alter to various secondary Ti- and REE-bearing minerals. Occurs widely throughout the world, but only selected localities are given here: Stark, NH; Martin's store, Nelson Co., VA; in volcanic ash beds throughout the western United States; accessory mineral in Pikes Peak Granite, Front Range, CO; Questa, NM; Aquarius Mountains, Mohave Co., AZ; Panamint Range, Inyo Co., CA; Golden Horn Batholith, Cascades, WA; St. Kilda, Scotland; Stokkøya, Sandefjord, and Bjørkedalen in the Oslo region, Norway; Novopoltavsk, Ukraine; Khibiny and Lovozero massifs, and west Keivy region, Kola Penin., Russia; *Kuchtym and Vishnevye Gor (Cherry Mt.), Ilmen Mts., Ural Mts., Russia*; Mianxi, Sichuan Prov. and Fujian Prov., China; Sabaragamuwa Prov., Ceylon (Sri Lanka); Mbolwe Hill, Central Prov., Zambia; Zomba and Mulanje, Chilwa Alkaline Prov., southern Malawi; Itorendrika, Betroka, and other localities in Madagascar; Kanjamali Hill, Salem district, Madras, and Orissa; Jokalandi, India; Kogendo region, Korea; Cape Ashizuri, Shikoku, Japan; southeastern Q, Australia. EF *MRE 3:309(1966)*; *AM 52:1094(1967), 56:307(1971)*; *MIN 3(1):784(1972)*.

56.2.8.2 Strontiochevkinite
$Sr_2(La,Ce,Ca)_2(Fe^{2+},Fe^{3+})(Ti,Zr)_2Ti_2Si_4O_{22}$

Named in 1983 by Haggerty and Mariano as the Sr-dominant analog of chevkinite. See chevkinite (56.2.8.1) perrierite (56.2.8.3). MON $P2_1/a$. $a = 13.56$, $b = 5.70$, $c = 11.10$, $\beta = 100.32°$, $Z = 2$, $D = 5.44$. 38-443(nat): 3.21_1 3.01_{10} 2.85_3 2.73_3 2.51_4 2.19_7 1.97_8 1.67_1. Structure not yet determined but is of the chevkinite type. Occurs as black rounded grains to 1.5 mm in diameter. Thick parallel lamellar or interpenetration twinning present. Submetallic luster. H $\approx$ 5. G = 5.4. Analysis (wt %): TiO_2, 23.2; SiO_2, 20.45; ZrO_2, 10.3; Al_2O_3, 0.11; FeO, 6.02; MnO, 0.08; CaO, 2.05; SrO, 19.6; BaO, 0.38; La_2O_3, 9.18; Ce_2O_3, 9.35. The mineral is opaque in standard thin sections. In RL, dark gray, strongly anisotropic in shades of gray with flesh-pink coloration. Deep red internal reflections. Reflectivities are 10% in air and 2% in oil. Occurs in rheomorphic fenite dikes in carbonatite, associated with lamprophyllite and strontian loparite. *Sarambi, Paraguay.* EF *CMP 84:365(1983)*.

56.2.8.3 Perrierite-(Ce)
$(Ca,Ce,Th)_4(Fe,Mg,Mn)(Ti,Mg,Fe,Al)_2Ti_2Si_4O_{22}$

Named in 1950 by Bonatti and Gottardi for Carlo Perrier (1886–1948), an Italian mineralogist. See chevkinite (56.2.8.1) and strontiochevkinite (56.2.8.2). Strictly speaking, chevkinite and perrierite are not polymorphs. The phase change of chevkinite to perrierite, and vice versa, is controlled primarily by composition. The two minerals do not coexist. Dimorphism occurs but only within fairly narrow chemical limits. Perrierite transforms to chevkinite with increasing temperature. *CMP 84:365(1983)*. MON $C2/m$. $a = 13.52$–13.95, $b = 5.62$–5.73, $c = 11.63$–11.73, $\beta = 113.3$–$114.5°$, $Z = 2$, $D = 4.86$; 19-302(nat): 4.10_6 3.56_8 2.99_{10} 2.96_{10} 2.84_7 2.69_6 2.18_8 1.95_8. May be metamict or partly so. Structure: *AM 59:1277(1974)*. Anhedral masses to subhedral masses, also euhedral crystals resembling epidote with dominant forms: {001}, {100}, and {111}; twinning frequent on (001); crystal size to more than 1 cm. Brownish black or black to brownish red, brownish-gray streak, resinous to vitreous luster. No cleavage or one indistinct cleavage, uneven to conchoidal fracture, brittle. H = $5\frac{1}{2}$–6. G = 4.06–4.55. Gelatinizes in HCl and dissolves in hot H_2SO_4. General formula is $A_4^{3+}B^{2+}C_2^{3+}Ti_2Si_4O_{22}$, where A = large cations: REE, Sr, Ca, Th, Na, K; B = Fe^{2+}, Mn, Mg, and sometimes Ca and C = Ti, Mg, Mn, Fe^{3+}, Fe^{2+}, Al. May contain appreciable Sc_2O_3 (to 4.17 wt %); also appreciable SrO (to 7.7 wt %); ThO_2 to >4 wt %; ZrO_2 to 5.6 wt %; may also be high in MgO + Al_2O_3 (Σ10 wt %). Biaxial ($-$); Z = b, X $\wedge$ a = 24° in obtuse β; N_x = 1.90–1.95, $N_{y\ calc}$ = 2.01, N_z = 2.02–2.06; 2V = 60° or less; Z $\geq$ Y $>>$ X, very strong; r > v, strong. X, yellow to yellow-brown; Y, opaque to violet-red or medium red; Z, opaque to deep brown or red-brown. A minor accessory mineral (e.g., chevkinite) in volcanic ash beds, but is more commonly found in alkalic or peralkaline granites, granite pegmatites, syenites, and syenitic pegmatites. Also occurs in tuffaceous sands, as at Nettuno, Italy. Rarely in granodiorites (e.g., VA) or anorthosite and gabbro. Alters to various secondary REE- and Ti-bearing minerals. Burley farm, NW of Amherst, and Wares Gap, Amherst Co., Chamblissburg, Bedford Co., Roseland, Nelson Co., and Hat Creek, Massie's Mills, Nelson Co., VA; many locations in the western United States in volcanic ash; Bancroft, ONT, Canada; *Nettuno, Latium, Italy*; Zr-rich from La Mola, Sabatini complex, also Monte Amiata, Tuscany, Italy; Oslo region, Norway [e.g., Storgangen (Sogndal anorthosite) and Bjørkedalen]; Mänty Harju, Finland; Burpala massif, northern Baikal area (Sr-bearing), Ural Mts., and Chatkal-Kuramin Mts., Central Asia, Russia; Orlovsk Massif, Rudnyi Altai, and Tarbagata, Kentsk, Kazakhstan (scandian variety); found with apatite ores in the apatitovoye deposit Mushgay ore field, southern Mongolia; Tete, Mozambique; Salem and Bangalore, India; Shiroisi, Kobe, Ohmiya town, Kyoto Pref., Japan; Vestfold Hills, eastern Antarctica. EF *AM 52:1094(1967), 53:1558(1968), 56:307(1971); MIN 3(1):779(1972); MM 58:607(1994).*

56.2.8.4 Orthochevkinite (not approved by CNMMN IMA)

A possible orthorhombic form of chevkinite was reported by Ungemach [*BSFM 39:5-37(1916)*] and by Lacroix [*CRASP 171:594(1920)*]. Also mentioned by Lacroix [*MMAD 1: 585(1922), 3:313(1923)*]. Masses to nearly 1 kg were recovered and crystals to 10 cm. They are partly or completely metamict. Dominant forms are: {001}, {100}, {110}, {210}. Found in alkaline syenite pegmatites in the region of Ifazina–Itorendrika, with aegirine, alkali amphibole, and bastnaesite, and in granitic pegmatites of Betroka and Itrongay. Studied by Bonatti and Gottardi [*RSMI 9:242(1953)*]. Summarized in *MIN 3(1):791(1972)*. Cell dimensions are: $a = $ ca. 13.6, $b = 5.8$, $c = 22.0$. Strongest X-ray diffraction lines of unheated material [*AM 44:115(1959)*]: 4.15_{10} 3.78_7 3.54_7 2.88_4 2.70_7 2.53_4 1.69_5 1.49_4. Heating metamict material at 700° and 1000° in nitrogen produces chevkinite. At 1000° in air, heating produces cubic CeO_2 and perrierite. This material needs to be reexamined carefully by modern methods. EF

56.2.9.1 Tilleyite $Ca_5Si_2O_7(CO_3)_2$

Named in 1933 by Larsen and Dunham for Cecil Edgar Tilley (1894–1973), British petrologist, Cambridge University. MON $P2_1a$. $a = 15.108$–15.111, $b = 10.241$–10.242, $c = 7.577$–7.579, $\beta = 105.15$–105.17° [*ZK 132:288(1970)*], $Z = 4$, $D = 2.87$. *24-184:* $3.09_{8.5}$ 3.00_{10} $2.97_{5.5}$ 2.96_4 $2.50_{2.5}$ $2.10_{4.5}$ 1.91_3 $1.90_{6.5}$. See also *25-159*. Structure: *AC 6:9(1953), ZK 132:288(1970)*. Usually fine-grained, to 15 cm long at Red Cap Creek; twinning = simple {100}, often lamellar X $\wedge$ twin plane $\approx 24°$; not all tilleyite shows twinning. Deep gray, white, light-bluish gray, colorless; white streak; dull luster. Cleavage {$20\bar{1}$}, perfect; {100}, {010}, poor. H = 5. G = 2.84. Effervesces in HCl and forms a silica gel. Usually pure. Traces of Ti for Si and Mg for Ca, traces of Al and Fe present as well. Biaxial (+); X $\wedge$ $c = 24°$, $Y = b$, Z $\wedge$ {$20\bar{1}$} cleavage trace $\approx 12°$; $N_x = 1.605$–1.617, $N_y = 1.626$–1.635, $N_z = 1.651$–1.654; 2V = 85–90°; r < v. Occurs in magmatic skarns, 800°. Marble–igneous rocks. Skarns formed at the igneous rock–marble contact. As opposed to vein skarns–fractures in marble. In carbonate-bearing rocks of sanidinite-facies environments. Sometimes replaced by wollastonite–vesuvianite–melanite, calcite. Also foshagite and calcite. To calcite and hillebrandite, also cuspidine. North End Peak, Iron Mountain, Sierra Co., NM; Helena, MT; *Crestmore quarry, Riverside Co., CA*; Carlingford, Ireland; Camas Mor, Island of Muck, Inverness-shire, and Kilchean, Ardnamurchan, Isles of Skye and Rhum, Scotland; Cornet Hill, Romania; Anabarsky intrusive, Anakit district, Lower Tunguska region, and Kuzmovka, Stony Tunguska R., Yakutia, Russia; Guneyce–Ikizdere area, Turkey; Kushiro, Hiroshima Pref., Japan; to 25 cm from the Akagene mine, Iwate Pref., Japan; coarse-grained, 2- to 15-cm prisms from Redcap Creek, northern Q, Australia. EF *MIN 3(1):808(1972), DHZ 1B 278(1986)*.

56.2.9.2 Killalaite $Ca_3Si_2O_7 \cdot H_2O$ and $2Ca_3Si_2O_7 \cdot H_2O$

Named in 1974 by Nawaz for the locality. Cuspidine–wöhlerite group. MON $P2_1/m$. $a = 6.807$, $b = 15.459$, $c = 6.811$, $\beta = 97.76°$ [*MM 41:363(1977)*], Z = 2 or 4; (pseudocell): $a = 6.80$, $b = 15.47$, $c = 6.82$, $\beta = 98.3°$ [*CA 98:182738(1983)*] (additional very weak reflections indicating a B-centered monoclinic true cell with $2a$ and $2c$), D = 2.88–2.94. *29-332* (calc pattern): 3.86_2 3.37_3 3.09_5 2.82_{10} 2.71_4 2.57_4 2.54_3 2.22_3. See also *26-1070*(nat). Structure: H positions not located. $Ca_{3.2}(H_{0.6}Si_2O_7)$ (OH). *MM 41:363(1977)*. Euhedral crystals to 2 mm long; "bowtie" twinning present, penetrative. Colorless. Cleavage {100}, perfect; {010}, {001}, good. G = 2.9. Analysis (wt %): CaO, 57.9; SiO_2, 39.8; H_2O, 3.2 (diff). Biaxial (−); Y = b, Z ∧ [001] ≈ 16°, in obtuse β, Z ∧ c = 16–19°; $N_x = 1.634$–1.635, $N_y = 1.646$, $N_z = 1.642$–1.648; 2V = 26°; elongation [010]; OAP (010). Occurs in a thermally metamorphosed limestone as a secondary mineral. In cavities and veins with calcite or afwillite in spurrite–wollastonite rocks. Basalt–dolerite dikes caused the metamorphism: 350–550° in a CO_2-deficient environment. In skarns in Turkey with hillebrandite, tobermorite, and vesuvianite. Alteration to xonotlite. *Killala Bay, near Inish Crone, County Sligo, Ireland*; Guneyce–Ikizdere region, eastern Pontids, Turkey. EF *MM 39:544(1974)*.

56.2.9.3 Foshallasite $Ca_3Si_2O_7 \cdot 3H_2O$

Named in 1936 by Chirvinsky for William Frederick Foshag (1894–1956), American mineralogist, and for the mineral centr*allasite*. Cuspidine–wöhlerite group. Gyrolite has replaced the name *centrallasite*. An incompletely described mineral. Probably MON No X-ray diffraction data available. Scaly, sometimes as spherical aggregates, tabular parallel to {100}, elongated and striated parallel to c. White, white streak, pearly luster. Perfect tabular cleavage {100}. H = $2\frac{1}{2}$–3. G = 2.5. Analysis (wt %): SiO_2, 32.65; R_2O_3, 1.89; CaO, 45.45; Na_2O, 0.40; H_2O, 0.16; ignition loss 16.66. Biaxial (−); X perpendicular to cleavage; $N_x = 1.535$, $N_y = 1.542$, $N_z = 1.549$. 2V = 12–18°. Occurs on *Mt. Yukspor, Khibiny massif, Kola Penin., Russia*; in veinlets associated with calcite and mesolite in lovchorrite, an alkaline nepheline syenite. EF

56.2.10.1 Melanotekite $Pb_2(Fe^{3+},Mn^{3+})_2O_2(Si_2O_7)$

Named in 1880 by Lindström from the Greek *melanos* and *tektos*, black glass, from the appearance of the mineral. See kentrolite (56.2.10.2), the Mn analog. ORTH *Pbcn*. (Nat): $a = 6.992$, $b = 11.022$, $c = 10.054$; (syn): $a = 6.93$, $b = 10.98$, $c = 10.06$, $Z = 4$ [*AM 76:1389(1991)*], $D = 6.22$(nat manganoan), 6.30 (pure Fe end member). *15-633*(nat manganoan): 5.52_8 5.03_6 3.72_6 3.49_8 3.25_8 2.91_{10} 2.86_6 2.73_{10}; *20-585*(syn Fe): 5.49_3 5.04_3 3.70_3 3.47_3 3.24_5 2.91_{10} 2.86_6 2.73_8. Structure: ferrian kentrolite done [*AM 76:1389(1991)*]. Prismatic, also massive. Blackish gray, black, synthetic pure Fe is dark brown, also reddish brown; metallic to greasy luster. Cleavable; two unequal cleavages. $H = 6\frac{1}{2}$. $G = 5.73–6.19$. Mössbauer data: *GFF 107:221(1985)*. Decomposed by HNO_3. Some contain Mn as a solid solution to kentrolite, as well as minor amounts of TiO_2 and Al_2O_3. Biaxial (+); for manganoan material: $N_x = 2.12$, $N_y = 2.17$, $N_z = 2.31$; $2V = 67°$; $r > v$, strong. X, nearly colorless; Y, pale red-brown; Z, deep reddish brown. Occurs in peralkaline pegmatite at Dara-i-Pioz. In Precambrian metamorphosed Mn deposits at Långban and elsewhere in Sweden. Fine, sharp crystals of nearly end-member composition from Hillsboro, Sierra Co., NM; Cerro Gordo, Inyo Co., CA; *Långban, Värmland*, also at the Harstig mine, Påjsberg, *Sweden*; just about end-member composition from Jakobsberg mine, Filipstad district, Värmland, Sweden; Dara-i-Pioz, Tadzhikistan. EF *MM 58:172(1994)*.

56.2.10.2 Kentrolite $Pb_2Mn_2^{3+}O_2[Si_2O_7]$

Named in 1880 by Damour and Vom Rath from the Greek *kentron*, thorn or spike, in allusion to the crystal habit. Polymorph: melanotekite, the Fe analog. ORTH *Pbcn*. (Nat): $a = 6.961$, $b = 11.018$, $c = 9.964$; (syn.): $a = 6.98$, $b = 11.04$, $c = 9.96$, $D = 6.27$; (syn): $Z = 4$, $D = 6.26$. *20-586*(nat): 5.56_6 3.71_8 3.51_8 3.24_8 2.96_5 2.90_{10} 2.86_{10} 2.74_{10}; *20-587*(syn): 5.52_5 4.98_5 3.49_5 3.26_7 2.89_{10} 2.86_8 2.84_6 2.73_{10}. Structure is now refined to $R = 0.047$. An $[(Mn,Fe)_2O_2Si_2O_7]$ fraction is the simple part, based on infinite $[M^{3+}O_4]$ edge-sharing octahedral chains parallel to [001]. The $6S^2Pb^{2+}$ lone-pair cations are split. *AM 76:1389(1991)*. Prismatic, also massive, sheaflike. Dark reddish brown (nat) black on surface, black (syn), also deep red; yellowish brown streak; greasy luster. Cleavage {110}, distinct; brittle. $H = 5$. $G = 6.19–6.2$. Soluble in concentrated HCl. No reaction to KOH, NaOH, H_2O_2, or HNO_3. Analysis (wt %): PbO, 59.6; ZnO, 0.05; MgO, 0.05; Fe_2O_3, 6.62; Mn_2O_3, 13.55; Al_2O_3, 0.30; TiO_2, 0.38; SiO_2, 16.45. Biaxial (+); $X = a$. $N_x = 2.10$, $N_y = 2.20$, $N_z = 2.31$; $2V = 88°$; $r < v$, strong; OAP parallel to (010). Occurs in metamorphosed Mn–Zn deposits. Near-end-member kentrolite occurs at Franklin, Sussex Co., NJ; type locality is an unspecified locality in southern *Chile* and/or *Långban, Värmland, Sweden*; also occurs at Jakobsberg, Filipstad, Nordmark district, and at Klintgruvan, near Krylbo, central Sweden; Påjsberg and Harstigen, Sweden; Bena de Padru near Ozieri, Sassari, Sardinia; Ushkatin mines, Atasuisk region, central Kazakhstan; very rare at Wessels mine, associated with sturmanite, henritermierite, calcite, and so on, Kalahari manganese field, Cape Prov., South Africa. EF

56.2.11.1 Nasonite $Pb_6Ca_4Si_6O_{21}Cl_2$

Named in 1899 by Penfield and Warren for Frank Lewis Nason (1856–1928), American geologist, Geological Survey of New Jersey. HEX $P6_3/m$ or $P6_3$. $a = 9.94$, $c = 13.08$, $Z = 2$, $D = 5.63$. Crystals minute and rare, prismatic [0001], with $\{10\bar{1}0\}$, $\{11\bar{2}0\}$, $\{10\bar{1}1\}$, and $\{0001\}$; usually massive, granular. White, greasy to adamantine luster. Cleavage $\{0001\}$, good, prismatic distinct. $H = 4$. $G = 5.55$. Uniaxial (+); $N_O = 1.971$, $N_E = 1.945$. Occurs in association with barysilite, datolite, prehnite, axinite, hancockite, clinohedrite, and manganophyllite below the 800 level in pilar 910, south of the Palmer shaft, and associated with glaucochroite, garnet, axinite, and barite at the Parker shaft, *Franklin, Sussex Co., NJ*; found in cacite-filled fissures with schefferite, lead, apophyllite, margarosanite, and thaumasite at Långban, Värmland, Sweden. AR *USGSPP 180:92(1935)*; *AM 36:534(1951)*, *56:1174(1971)*; *AC(B) 43:171(1987)*.

56.3.1.1 Danburite $CaB_2Si_2O_7(O)$

Named in 1839 by C. U. Shepard for the locality. ORTH *Pnam*. $a = 8.038$–8.049, $b = 8.752$–8.765, $c = 7.730$–7.737, $Z = 4$, $D = 3.00$. *29-304*(nat): $3.65_{3.5}$ 3.57_{10} $3.44_{3.5}$ 2.96_8 2.74_7 2.73_4 2.66_6 1.44_4. The structure consists of a tetrahedral framework of ordered B_2O_7 and Si_2O_7 groups which accommodates Ca in irregular coordination. Structurally similar to feldspars. *AM 59:79(1974)*. Habit is granular to prismatic, resembling topaz, $\{110\}$, $\{100\}$, $\{010\}$ dominant, elongated on c. Colorless, amber, yellow, pink, gray, white, yellow-brown; white streak; glassy–vitreous–greasy luster; transparent–translucent. No distinct cleavage, one very indistinct on (001); subconchoidal to uneven fracture. $H = 7$–$7\frac{1}{2}$. $G = 2.97$–3.03. IR data: *MS 23:359(1969)*, *PCM 14:441(1987)*. Sometimes fluorescent blue to blue-green in LWUV, with red thermoluminescence. Soluble in acids. Analysis (wt %): CaO, 22.8; B_2O_3, 28.4; SiO_2, 48.8. Biaxial (−); $XZ = ba$; $N_x = 1.630$, $N_y = 1.633$, $N_z = 1.635$; colorless; OAP = (001); $2V = 80$–$88°$; $r < v$, strong. A contact-metamorphic mineral, occurring in moderate- to low-temperature and low-Fe-bearing conditions. Associated with calcite, datolite, and nifontovite at Charcas. Associated with axinite in Japan. In metamorphosed limestone at Danbury, CT. Also in pegmatites and greisens. In evaporites and salt domes in the Texas Gulf. In a gypsum bed in Bolivia. Occurs in dolomite marble at the extinct White's Factory excavation, *Danbury, Fairfield Co., CT*; crystals to 4 in. long at Russell, St. Lawrence Co., NY; in pegmatite at DeKalb, NY; Little 3 mine, Ramona, San Diego Co., CA; Silver Harbor, ONT, Canada; with smoky quartz in gneiss at Piz Valatscha, Val Medels, Uri, Graubunden, Switzerland; Kragerø district, Norway; with anhydrite at Zechstein, Germany; Ligurian Brianconnais, Maritime Alps, Italy; Maglovec, near Presov, Czech Republic; large crystals to >30 cm long and 10 cm wide from Dalnegorsk, Primorsky

Krai, Russia; gem-quality pebbles from marble near Mogok, Myanmar; at Maharitra on Mt. Bity, and at Imalo and Sahasonjo, Madagascar; at Obira, Bungo, Oita Pref., Japan; with axinite at the Toroku mine, Miyazake Pref., Kyushu, Japan; abundant and fine-quality prismatic crystals from lead–zinc mines at Charcas, SLP, Mexico; also from the La Sirena mine, Zimapan, Mexico; to 15 cm long in pegmatites in northern Baja California, Mexico; in evaporites and volcanic ejecta at Cochabamba, Bolivia. Minor use as a gemstone. A 138-ct pale yellow stone is in British Museum. Best faceted stones are from Burma (yellowish to brownish). EF *AUCG 1–2:111(1973)*, *NJBM: 289(1987)*.

56.3.2.1 Werdingite $(Mg,Fe)_2Al_{12}(Al,Fe)_2Si_4(B,Al)_4O_{37}$

Named in 1990 by Moore et al. for Günter Werding (b.1929), Ruhr-University, Bochum, Germany. Related to sillimanite, mullite, and grandidierite. TRIC $P\bar{1}$. $a = 7.995$, $b = 8.152$, $c = 11.406$, $\alpha = 110.45°$, $\beta = 110.85°$, $\gamma = 84.66°$, $Z = 1$, $D = 3.07$. PD(nat): 5.43_8 5.23_{10} 4.98_8 3.44_3 3.39_5 2.71_6 2.19_5 1.53_4; *44-1467*(syn): 5.44_8 5.25_{10} 5.03_7 3.40_9 $3.31_{5.5}$ 2.72_9 2.53_5 $2.20_{7.5}$. Chains of AlO_6 octahedra parallel to c, cross-linked by Si_2O_7 groups, Al and Mg in fivefold coordination, Al and Fe tetrahedra, and B triangles. *AM 76:246(1991)*. Maximum size ≈ 3 mm; simple twins common, lamellar twins of three or four individuals also occur. Translucent—brownish yellow (honey colored); buff-white streak; vitreous luster. Conchoidal fracture. H = 7. G = 3.04. Does not fluoresce in UV. Analysis (wt %): MgO, 4.46; FeO, 5.06; Al_2O_3, 59.49; TiO_2, 0.05; SiO_2, 19.83; B_2O_3, 10.19. Biaxial (−); X = c; in white light: $N_x = 1.614$, $N_y = 1.646$, $N_z = 1.651$; 2V = 33°; prism elongation and the CP of simple twins are parallel to c; r > v, moderately strong; horizontal dispersion present; pleochroic. X, colorless; Y, yellow; Z, colorless. Occurs in a supracrustal granulite facies gneissic sequence at *Bok Se Puts, Namaqualand, South Africa*; with sillimanite–hercynite, kornerupine, grandidierite, and Fe–Ti oxides. Alteration: grandidierite and hercynite replace werdingite. EF

56.4.1.1 Lomonosovite $Na_5Ti_2O_2(Si_2O_7)(PO_4)$

Named by Gerasimovskii in 1950 for Mikhail V. Lomonosov (1711–1765), Russian mineralogist and naturalist. TRIC $P\bar{1}$. $a = 5.393$–5.49, $b = 7.11$–7.121, $c = 14.420$–14.50, $\alpha = 100.02$–$101°$, $\beta = 96$–$96.44°$, $\gamma = 90$–$90.44°$, $Z = 2$, $D = 3.04$. *NJBA 148:83(1983)*(nat): 4.14_6 2.90_8 2.83_{10} 2.73_{10} 2.68_6 2.65_8 1.77_{10} 1.75_6; data also given on *15-155* and *17-542* for natural material. Structure: *DAN SSSR 197:81(1971)*, *235:1064(1977)*; *SPD 22:422(1977)*. Formula may also be written as $Na_2Ti_2Si_2O_9 \cdot Na_3PO_4$. Crystals to 15 cm × 10 cm, a axis lies in the tabular direction, also lamellar and flaky aggregates; tabular crystals bound by {001}, {013}, {012}, {01$\bar{2}$}, {01$\bar{4}$}, {201}, {20$\bar{1}$}, {110}; simple contact to finely polysynthetic twinning present with (001) as the composition plane and the a axis as the twin axis. Dark brown, orange-brown, black; medium red-brown streak; vitreous–adamantine luster on cleavages, greasy on fractures.

56.4.1.2 Polyphite

Cleavage {001}, perfect; uneven fracture; brittle. $H = 4\frac{1}{2}$, decreasing with alteration. $G = 3.12$–3.15, lower when altered and hydrated, e.g., to 2.88. DTA, TGA, IR, Mössbauer data: *NJBA 148:83(1983)*. Contains Fe^{3+} in octahedral coordination and Fe^{2+} in two different octahedral sites. Easily soluble in HCl or H_2SO_4, leaving a silica residue. Composition of Ilimaussaq material (avg. wt %): Na, 20.1; Ca, 0.92; Ti, 16.4; Nb, 1.73; Fe, 1.43; Mn, 0.18; Si, 10.94; P, 6.04; O, 39.7. Cinnamon-colored material from Lovozero: P_2O_5, 12.8; $Nb_2O_5 + Ta_2O_5$, 3.0; TiO_2, 24.4; SiO_2, 24.1; ZrO_2, 2.1; Fe_2O_3, 2.4; MnO, 3.2; MgO, 0.6; CaO, 0.8; Na_2O, 26.1; H_2O, 0.3. Biaxial (−); $N_x = 1.654$–1.708, $N_y = 1.736$–1.759, $N_z = 1.764$–1.778; $2V = 56$–$69°$; $Z > Y > X$; $r > v$, strongly pleochroic. X, nearly colorless; Y, dark brown to lilac; Z, yellowish brown. The color intensity increases with increasing content of Ti and decreasing content of Nb. For dark brown material: X, yellow to pale yellow; Y, orange-yellow or brownish yellow; Z, dark brown to violet. $P_{(001)}$ $N_x = 38°$, $N_y = 70°$, $N_z = 59°$. One of the optical axes lies in the *a–c* crystallographic plane and this axis forms an angle of 74° with the *a* axis. The optic normal is perpendicular to the face pairs (201) and ($\bar{2}$01). Occurs in pegmatites and in several nepheline syenite varieties in the Lovozero massif; in Greenland, the mineral occurs in fine-grained lavas and in fine- to coarse-grained gabbros near the contact with intrusive nepheline syenites, composed of arfvedsonite, aegirine, microcline, and/or albite, nepheline, eudialyte, and a number of accessory minerals. Kvanefjeld area, Ilímaussaq intrusion, southern Greenland; in the Jubilee pegmatite on Mt. Karnasurt, and the Chinglusuai R., Lovozero massif, Kola Penin., Russia; also in the Khibiny massif at Mt. Yukspor, Mt. Rasvumchorr, Mt. Koashva, and Mt. Kukisvumchorr. Alters to murmanite. Beta-lomonosovite is an indefinite intermediate stage in the alteration of lomonosovite to murmanite by leaching out of PO_4 and hydration. Originally described in *DAN SSSR 142:670(1962)*. Lomonosovite may also be replaced by narsarsukite, mangan–neptunite, and mountainite; sometimes replaced by bornemanite. EF *DAN SSSR 70:83(1950)*, *MIN 3(1):661(1972)*.

56.4.1.2 Polyphite $Na_{17}Ca_3Mg(Ti,Mn)_4[Si_2O_7]_2(PO_4)_6O_2F_6$

Named in 1992 by Khomyakov et al. from the Greek for *many*, in allusion to the many elements with multiple phosphate groups in the formula. Related to quadruphite, lomonosovite, and sobolevite. TRIC $P1$. $a = 5.412$, $b = 7.079$, $c = 26.56$, $\alpha = 95.21°$, $\beta = 93.51°$, $\gamma = 90.10°$, $Z = 1$, $D = 3.00$. PD: 4.37_4 4.13_4 2.94_{10} 2.70_9 2.66_8 2.05_8 1.77_5 1.73_5. Structure: *DAN SSSR 294(2):357(1987)*, *ZVMO 121:105(1992)*. {001} plates, 1–2 mm and sometimes to 3 mm, epitaxially on lomonosovite (with or without sobolevite intergrowths). Light brown; tan streak; vitreous luster, pearly on cleavage to specular–metallic. Cleavage {001}, excellent; less so on {110} and {100}; brittle. $H = 5$. $G = 3.07$. Nonfluorescent. Soluble in cold 10% HCl. Analysis (wt %): Na_2O, 28.0; SrO, 0.3; BaO, 0.9; CaO, 8.3; MgO, 1.3; MnO, 5.6; FeO, 0.2; SiO_2, 13.2; ZrO, 1.6; TiO_2, 12.0; Nb_2O_5, 2.7; P_2O_5, 23.3; F, 5.1; O for F, 2.1. Biaxial (−); $N_x = 1.600$, $N_y = 1.658$, $N_z = 1.676$; $P_{(001)}$: $N_x = 30°$, $N_y = 64°$,

$N_z = 76°$; $2V_{obs} = 56°$, $2V_{calc} = 57°$; $Z \geq Y > X$; $r < v$, strong. X, colorless; Y,Z, yellowish. Found in ultra-agpaitic pegmatitic rocks, associated with many other alkalic minerals. See also quadruphite (56.4.1.3). Found at Mt. Alluaiev, in the northern part of the Lovozero massif, Kola Penin., Russia. EF *AM 78:1316(1993), MA 45:240(1994)*.

56.4.1.3 Quadruphite $Na_{14}CaMgTi_4(Si_2O_7)_2(PO_4)O_4F_2$

Named in 1992 by Khomyakov et al. for the Latin *quadruplex* and the element P, the quantity of PO_4 groups in the formula. TRIC $P1$. $a = 5.415$, $b = 7.081$, $c = 20.34$, $\alpha = 86.85°$, $\beta = 94.40°$, $\gamma = 89.94°$ (for all acute cell), $Z = 1$, $D = 3.11$. *ZVMO 121:105(1992)*. PD: 2.88_{10} 2.70_8 2.64_7 2.05_5 1.77_2 1.71_3 1.66_4 1.600_5. Structure: *MZ 9(3):28(1987)* and *ZVMO 121(1):105(1992)*. Irregular plates flattened on {001} 1–2 mm thick and up to 3 mm across, some in epitaxial intergrowths with plates of lomonosovite, and sobolevite. Light brown; white to tan streak; vitreous luster, pearly on basal cleavage to specular-metallic; translucent, transparent in thin sections. Cleavage {001}, perfect; {110}, {100}, less perfect; steplike fracture; brittle. $H = 5$. $G = 3.12$. Nonfluorescent. Soluble in cold 10% HCl. Analysis (wt%): Na_2O 28.1, SrO 0.4, BaO 1.1, CaO 5.3, MgO 1.1, MnO 4.3, FeO 0.3, SiO_2 16.5, ZrO_2 4.1, TiO_2 13.7, Nb_2O_5 3.8, P_2O_5 19.2, F 3.4, O=F 1.4. Biaxial $(-)$; $N_x = 1.630$, $N_y = 1.678$, $N_z = 1.697$; $2V_{obs} = 62°$, $2V_{calc} = 63°$; $P_{(001)}$: $N_x = 28°$, $N_y = 71°$, $N_z = 70°$; $Z \geq Y > X$; $r < v$; pronouncedly pleochroic. X, colorless; Y,Z, yellow. Found in ultra-agpaitic pegmatitic rocks with K-feldspar, nepheline, sodalite, arfvedsonite, aegirine, cancrisilite, analcite, albite, ussingite, makatite, grumantite, lomonosovite, vuonnemite, sobolevite, and other rare minerals at *Mt. Alluaiv in the northwestern part of the Lovozero alkaline massif, Kola Penin., Russia*. EF *AM 78:1316(1993), MA 45:240(1994)*.

56.4.1.4 Sobolevite $Na_{11}(Na,Ca)_4(Mg,Mn)Ti_4[Si_2O_7]_2[PO_4]_4O_3F_3$

Named in 1983 by Khomyakov et al. for Vladimir Stepanovich Sobolev (1908–1982), Russian petrologist and former president of the IMA. TRIC $P1$. $a = 7.078$, $b = 5.4115$, $c = 40.618$, $\alpha = 90.01°$, $\beta = 93.19°$, $\gamma = 90.00°$, $Z = 2$, $D = 2.96$–3.03. *36-439*(nat): 4.29_3 3.18_4 2.91_5 2.90_{10} 2.69_7 1.77_5 $1.72_{3.5}$ 1.67_5. Originally described as monoclinic, but is actually triclinic. *DAN SSSR 302(5):1112(1988)*, *ZVMO 112:456(1983)*. Lomonosovite group. Related to polyphite, quadruphite, and lomonosovite. Brown; tan streak; metallic luster, pearly on (001), resinous. Cleavage {001}, perfect; {110}, distinct. $H = 4\frac{1}{2}$–5. $G = 3.03$. Nonfluorescent. Readily decomposed by cold 10% HCl. IR data: *ZVMO 112:456(1983)*. Analysis (wt %): SiO_2, 17.1; Na_2O, 29.7; CaO, 6.4; MgO, 0.6; TiO_2, 15.2; Nb_2O_5, 4.4; MnO, 4.0; Fe_2O_3, 0.6; P_2O_5, 19.9; F, 0.7; O for F, 0.3. Biaxial $(-)$; $N_x = 1.627$, $N_y = 1.686$, $N_z = 1.690$; $P_{(001)}$: $N_x = 35°$, $N_y = 90°$, $N_z = 55°$; $2V_{obs} = 29°$(Na light), $2V = 25°$(red), $33°$(blue); $Z = Y > X$; $r < v$, strong. X, nearly colorless; Y,Z, yellowish brown. Found in ultra-agpaitic pegmatitic rocks associated with other alkalic mineral and rare species. Pegmatite cutting sodalite–cancrinite syenite. Associated with lomonosovite and lamprophyllite. Found on *Mt. Alluaiev, in the northwestern part of the Lovozero massif, Kola Penin., Russia.* EF *AM 76:305(1991)*.

56.4.2.1 Bornemanite BaNa$_4$Ti$_2$NbSi$_4$O$_{17}$(F,OH) · Na$_3$PO$_4$

Named in 1975 by Men'shikov et al. for Irina Dmitrievna Borneman-Starynkevich (b.1890), Russian mineralogist, Institute of Ore Deposits, Moscow, and student of the mineralogy of the Khibiny and Lovozero massifs. ORTH *Ibmm* or *Ibm2*. $a = 5.48$, $b = 7.10$, $c = 48.2$, $Z = 4$, $D = 3.49$. 29-1176: 24.1$_{10}$ 8.04$_{10}$ 3.44$_{10}$ 3.02$_{10}$ 2.68$_8$ 1.81$_7$ 1.610$_8$. Pale yellow, pearly luster, translucent to transparent. Cleavage {001}, perfect. H = $3\frac{1}{2}$–4. G = 3.47–3.50. Biaxial (+); $2V = 40°$, $2V_{calc} = 66°$; $N_x = 1.682$–1.683, $N_y = 1.687$–1.695, $N_z = 1.718$–1.720; XYZ = *cba*; Pleochroic; X,Y, colorless; Z, pale yellow-brown. Found as a late hydrothermal product occurring as platy aggregates up to 10 mm × 8 mm × 0.2 mm on the surface of or on cleavage cracks in lomonosovite and in natrolite in the *Yubileinaya (Jubilee) alkalic pegmatite, Mt. Karnarsut, Lovozero massif, Kola Penin., Russia.* AR *ZVMO 104:322(1975), AM 61:338(1976)*.

56.4.3.1 Vuonnemite Na$_5$Nb$_2$Ti^{3+}(Si$_2$O$_7$)$_2$O$_2$F$_2$·2Na$_3$PO$_4$

Named by Bussen et al. in 1973 for the locality. Related to murmanite, epistolite, and lomonosovite. TRIC. P$\bar{1}$ $a = 5.498$–5.50Å $b = 7.161$–7.162Å $c = 14.440$–14.450Å $\alpha = 92.60$–92.63° $\beta = 95.30$–95.33° $\gamma = 90.57$–90.60° $Z = 1$ $D = 3.17$; PD 26-792 (nat. Kola Pen.): 7.14$_6$ 5.16$_5$ 4.77$_5$ 4.24$_5$ 3.58$_6$ 2.87$_{10}$ 2.39$_5$ 1.79$_6$; (NJBM 451(1983) (nat. Greenland): 14.35$_6$ 7.15$_7$ 4.25$_{10}$ 2.87$_{10}$ 2.77$_{10}$ 2.74$_{10}$ 2.66$_6$ 1.79$_8$. Pale yellow to lemon yellow or greenish yellow, streak white, vitreous to greasy luster, plates to 2 cm across and bladed crystals to more than 10 cm, perfect {001} cleavage and two other excellent cleavages; brittle, H 2-3 G3.11-3.13. IR and DTA data *(ZVMO 102:423(1973))*. Some crystals fluoresce intense greenish yellow in SW UV or whitish with a light green hue and pale yellow or no fluorescence under LW UV. Faint bluish green phosphorescence for a short period. Decomposed by H$_2$O. SiO$_2$ 22.3-22.6 wt. %, TiO$_2$ 7.5-8.1, Al$_2$O$_3$ 0.3-0.5, Nb$_2$O$_5$ 23.2-27.9, FeO 0.1-0.3, MnO 0.1-0.5, MgO 0.1, CaO 0.3, Na$_2$O 30.2-32.0, K$_2$O 0.1, P$_2$O$_5$ 13.1-13.5, LOI 1.0, Σ99.7-100.1. Biax. (+), 2V=53-54°, also to 86°; N$_\alpha$ 1.636-1.639, N$_\beta$ 1.651-1.656, N$_\gamma$ 1.680-1.683; OAP={001}, Y ∧ c 4°. Orientation of Ilimaussaq material: P ∧ α 57°, P ∧ β 33°, and P ∧ γ 89°, where P is the normal to the {001} cleavage plane. Ocuurs in albitized alkalic rocks of both the Khibiny and Lovozero massifs, assoc. with albite, microcline, nepheline, aegirine, lorenzenite, cancrinite, etc. In a transition zone between two nepheline syenites (lujavrite) in Ilimaussaq. Mont Saint-Hilaire, Quebec; Kvanefjeld plateau, Ilimaussaq intrusion, SW Greenland; *Vuonnemi River, Khibiny massif, Kola Peninsula, Russia*; in the Jubilee pegmatite on Mount Karnasurt, and in the Umbozero mine on Mount Alluaiev, Lovozero massif, Kola Peninsula, Russia. Zorite may replace and pseudomorph vuonnemite. EF *MR 21:284(1990)*, Ercit and Hawthorne 1987 GAC-MAC v. 12, p.41; Ercit and Hawthorne (1990)

Class 57

Insular Si_3O_{10} and Larger Noncyclic Groups

57.1.1.1 Aminoffite $Ca_3Be_2Si_3O_{10}(OH)_2$

Named in 1937 for Gregori Aminoff (1883–1947), Swedish mineralogist and crystallographer, Riksmuseum, Stockholm. TET $P4_2/n$ or $I4$. $a = 9.865$, $c = 9.930$, $Z = 4$, $D = 2.90$. 23-80(nat): 6.97_7 4.40_7 4.02_8 3.48_7 3.11_7 2.84_9 2.61_{10} 2.14_8. Dipyramidal and flattened tablets, with {001}, {111}, {100} common; twinning present; piezoelectric; crystals to 0.5–1.0 mm. Pale yellow or colorless, white streak, vitreous luster. Cleavage {001}, weak; conchoidal fracture. $H = 5\frac{1}{2}$. $G = 2.94$–3.00. IR data: *GZ 36:76(1976)*. Not soluble in common acids. May contain some minor Na, and F. Uniaxial $(-)$; $N_O = 1.647$–1.654, $N_E = 1.628$–1.637; colorless; sometimes has anomalous 2V with $2V = 0$–$15°$. Occurs in cavities in magnetite–limonite skarn; also in calcite veins with barite and fluorite. An accessory mineral in fluorite veins. In a contact zone between granite and marble (at Tuva), with vesuvianite, analcite, and others. *Långban, Värmland, Sweden*; in fluorite veins in the Dugdinsk Massif, Tuva, Siberia, Russia; Bayan Kol dike field. EF *MIN 3(2):8(1981); AM 49:212(1964); AC 23:255(1967), 23:260(1967); DANS 209:260(1973); GZ 36:76(1976)*.

57.1.1.2 Harstigite $Ca_6Be_4(Mn,Mg)[SiO_4]_2[Si_2O_7]_2(OH)_2$

Named in 1886 by Flink for the locality. ORTH *Pcmn*. $a = 13.830$, $b = 13.636$, $c = 9.793$ (original setting), $Z = 4$, $D = 3.19$. 20-200: 4.35_4 3.22_4 2.82_5 2.79_5 2.70_{10} 2.27_5 1.790_4 1.331_4. Structure: The structure has been related to that of aminoffite as well as that of epidote. Colorless, vitreous luster, transparent. $H = 5\frac{1}{2}$. $G = 3.16$. Biaxial (+); $2V = 52°$; $N_x = 1.678$, $N_y = 1.68$, $N_z = 1.683$; $XYZ = cba$; $r < v$, weak. Prismatic [001] crystals at the *Harstigen mine, Pajsberg, Värmland, Sweden*. AR *D6:532(1892), AM 53:309(1968), ZK 177:143(1986)*.

57.1.2.1 Kinoite $Ca_2Cu_2(Si_3O_{10}) \cdot 2H_2O$

Named in 1970 by Anthony and Laughon for Eusebio Francisco Kino (1645–1711), Jesuit explorer of the southwestern United States. MON $P2_1/m$. $a = 6.990$, $b = 12.890$, $c = 5.654$, $\beta = 96.08°$, $Z = 2$, $D = 3.19$. Structure: *AM 56:193(1971)*. 23-946: 6.44_7 4.72_{10} 3.95_3 3.14_3 3.05_8 2.32_3 2.12_4 2.07_2. The $[Si_3O_{10}]^{8-}$ unit consists of three tetrahedra forming a roughly U-shaped group. As small (1 mm) tabular {100} crystals elongated [001] with [001] zone striated and curved. Azure blue, light blue streak, vitreous luster, translucent to transparent. H ≈ 5. G = 3.16. Biaxial (−); 2V = 68°; $N_x = 1.638$, $N_y = 1.665$, $N_z = 1.676$; X = b, Z ≈ c; r > v. Pleochroic; X, pale green; Y, blue; Z, dark blue. Decomposed by HCl. Found enclosed in quartz and calcite at the Laurium mine and Osceola shaft, Kearsarge lode, Calumet, MI; associated with apophyllite at *Helvetia, Santa Rita Mts., Pima Co., AZ*, and with apophyllite and ruizite at the Christmas mine, Hayden, Gila Co.; at the Bawana mine, Beaver Co., UT. AR *AM 55:709(1970)*.

57.1.3.1 Trabzonite $Ca_4Si_3O_{10} \cdot 2H_2O$

Named in 1986 by Sarp and Burri for the locality. MON $P2_1$ or $P2_1/m$. a = 6.895, b = 20.64, c = 6.920, β = 98.0°, Z = 4, D = 3.08. *40-513*: 5.71_3 3.44_6 3.06_{10} 2.91_3 2.85_5 2.64_5 2.59_9 2.06_2. Colorless, vitreous luster. G = 2.9. Biaxial (+); 2V = 55°; $N_x = 1.632$, $N_y = 1.634$, $N_z = 1.640$; Y = b, X ∧ c = 8°; r > v, weak to moderate. A retrograde mineral found as grains with rustumite, tobermorite, killalite, garnet, and molybdenite *near Ikizdere, Varda Yaylasi, near Trabzon, Turkey*. AR *AM 73:1497(1988)*.

57.1.4.1 Rosenhahnite $Ca_3Si_3O_8[(OH)_{2-4x}O_x(CO_3)_x]$

Named in 1967 by Pabst et al. for Leo Rosenhahn (1904–1991), American amateur mineralogist, who found the mineral. TRIC $P\bar{1}$. a = 6.946–6.962, b = 9.474–9.492, c = 6.802–6.812, α = 108.82–108.64(2)°, β = 94.72–94.85°, γ = 95.72–95.89(2)°, Z = 2, D = 2.895–2.095. *29-378*(nat): 3.43_3 3.36_3 $3.20_{7.5}$ $3.04_{6.5}$ 2.97_{10} 2.77_4 2.66_3 2.285_3. First studied experimentally. *AJS 261:79(1943)*. High-P polymorph of xonotlite occurring below 450° and above 20 kbar. Structure consists of (Ca1)O_9, (Ca2)O_6 + (Ca3)O_7 polyhedra and insular [$Si_3O_8(OH)_2$] groups. *NAT 241:42(1973), AM 62:503(1977)*. Flat, tabular to lathlike with {010} by far most predominant, elongated parallel to c, flattened parallel to b, and breaks on {001} cleavage; maximum size of 5 mm × 2 mm × 1 mm, to +5 mm in Japan. Colorless to buff-white, light yellow, clear and etched; white streak. Cleavage {001}, less common on {100} and {010}. H = $4\frac{1}{2}$–5. G = 2.89. TGA: 0–300° no loss, most water lost at 400–600°, not much more after 600°. Russian R. material fluoresces yellowish green in SWUV, Japanese shows no fluorescence, NC material fluoresces light orangish pink in LWUV, none in SWUV. Only slightly soluble in concentrated acids. Upon heating, single crystals are converted to single crystals of wollastonite in perfect topotactic relation. Japanese has 1.6 wt % CO_2 and 0.4% FeO. CA material has traces of MgO, FeO, Na_2O, BaO (0.77%), and 1.3% CO_2. Biaxial (−); Na(D): $N_x = 1.624$–1.625, $N_y = 1.640$–1.641, $N_z = 1.646$–1.647; colorless; 2V = 62–64°. Detailed orientation in *AM 52:336(1967)*. Occurs in Franciscan greywacke, in veins and cavities cutting brecciated metasediments, associated with diopside, hydrogrossular, tremolite, titanite, and veins of pectolite and xonolite. Irregular segregations intergrown with prehnite, gyrolite, apophyllite, and okenite. Also in metasomatized gabbro and rodingite in serpentinite, and in limestone-contaminated granitic rock. Alteration: to opal (pseudomorphs) and to calcite and other phases. In traprock at the Durham quarry, Durham, Wake Co., NC; boulders in the Russian R. streambed, Cloverdale, Mendocino Co., CA; locality in Bohemia, Germany; Ozersk massif, West Baikal region, Russia; Engyoji, Kochi City, and Kushiro, Hiroshima Pref., Japan; Wairere, New Zealand. EF *AM 52:336(1967)*, *MIN 3(2):525(1981)*, *BNSM(C) 10:1(1984)*.

57.2.1.1 Tiragalloite $Mn_4AsSi_3O_{12}(OH)$

Named in 1980 for Paolo Tiragallo (b.1905). MON $P2_1/n$. $a = 6.66$, $b = 19.92$, $c = 7.67$, $\beta = 95.7°$, $Z = 4$, $D = 3.86$. *33-889*(nat): 3.26_{10} 3.15_8 3.03_7 2.74_6 2.61_7 2.50_5 2.49_6. Structure contains nonisolated AsO_4 groups. *AM 65:947(1980)*. Grains to 1.5 mm long, but usually (0.2–0.4 mm); symmetric twinning, twin plane coincident with cleavage plane. Orange, subadamantine luster. Cleavage {100}, good; distinct parting perpendicular to elongation. $G = 3.84$. Insoluble in HCl or H_2SO_4, slightly soluble in HNO_3. Analysis (wt %): SiO_2, 32.4; MnO, 48.3; CaO, 0.75; As_2O_3, 16.1; V_2O_5, 1.7. Biaxial (+); $XYZ = abc$; in white light: $N_x = 1.745$, $N_y = 1.751$, $N_z = 1.760$; $2V = 38–46°$; angle between α and the cleavage pole is 5–6°; positive elongation; inclined axial dispersion; nonpleochroic, orange to yellow. Occurs in veinlets cutting massive black aggregates of braunite and quartz. Occurs at an abandoned manganese mine at *Molinello, Chiavari, Val Graveglia, Liguria, Italy*. EF *AC(B) 35:2287(1979)*.

57.2.2.1 Ruizite $Ca_2Mn_2^{3+}[Si_4O_{11}](OH)_4 \cdot 2H_2O$
[*AM 70:171(1985)*] and $Ca_2Mn_2^{3+}[Si_4O_{11}(OH_2)](OH)_2(H_2O)_2$ [*TMPM 33:135(1984)*]

Named for Joe Ana Ruiz (b.1924), Mammoth, AZ, pharmacist and American amateur mineralogist, who discovered the mineral. MON $C2/m$. *AM 70:171(1985)*. $a = 9.064$, $b = 6.171$, $c = 11.976$, $\gamma = 91.38°$. Also $A2$. *TMPM 33:135(1984)*. $a = 11.984$, $b = 6.175$, $c = 9.052$, $\beta = 91.34°$, $Z = 2$, $D = 2.89$. *29-350*(nat): 11.95_{10} 5.09_5 4.19_7 3.64_4 3.12_6 2.95_4 2.59_4 2.13_4; *36-435*(calc pattern): 11.97_{10} 5.10_2 4.20_2 3.12_2 2.96_1 $2.74_{1.5}$ $2.55_{1.5}$ 2.335_1. A sorosilicate with an $[Si_4O_{11}(OH)_2]$ linear cluster of corner-sharing tetrahedra. A heteropolyhedral framework of Mn octahedra and Si tetrahedra with Ca occupying the [7] coordinate interstices between the octahedral–tetrahedral sheets. Crystals show dominant forms: {001}, {201}, {$\bar{2}01$}, {111}, {110}; generally occur as spherules of radially arranged acicular crystals; 0.5- to 1.0-mm spheres, clusters to 4 mm; simple twinning common on {001}. Orange, but varies to red-brown, brown, and white; pale apricot streak. $H \approx 5$. $G = 2.89–2.90$. Solubility in cold 1:1 HNO_3, 1:1 HCl, and 40% KOH is negligible; when heated, it readily dissolves in all three. Analysis (wt %): CaO, 20.6; Mn_2O_3, 23.4; SiO_2, 39.1; H_2O, 16.0. Biaxial (−); $Z \wedge [100] = 44°$, in obtuse β; $N_x = 1.663$, $N_y = 1.715$, $N_z = 1.734$; $2V = 60°$; β parallel [010]; length-fast or length-slow; 1 O.A. is nearly perpendicular to {001}, crystals lying on this face do not extinguish, also because of twinning; $Y > Z > X$; $r > v$, inclined, strong.

X, pale Mars orange; Y, Mars orange; Z, Dresden yellow. Occurs at the *Christmas mine, Gila Co., AZ*, in a calc–silicate assemblage. In veinlets and on fracture surfaces in meta-limestones that have been converted to calc–silicates during contact metamorphism by a porphyry copper stock. Associated with wollastonite, grossular, diopside, and vesuvianite. Also occurs at the N'Chwaning II and Wessels mines, Kuruman, Cape Prov., South Africa, associated with inesite, orientite, quartz, datolite, and others. EF *MM 41:429(1977), MR 9:137(1978)*.

57.2.3.1 Akatoreite $Mn_9^{2+}Al_2Si_8O_{24}(OH)_8$

Described in 1971 by Read and Reay and named for the locality. TRIC $P\bar{1}$. $a = 8.337$–8.344, $b = 10.358$–10.367, $c = 7.627$–7.629, $\alpha = 104.46$–$104.48°$, $\beta = 93.63$–$93.81°$, $\gamma = 103.95$–$104.18°$; $Z = 1$, $D = 3.47$–3.51. *25-533*(nat): 9.68_6 4.67_{10} 3.47_5 3.31_9 3.06_5 2.87_5 2.84_4 2.21_8. Structure: R = 2.9%, a sorosilicate with a very complex structure. *CM 31:321(1993)*. Masses or columnar aggregates; radial aggregates, or massive; crystals are prismatic, striated along [100], with {0kl} predominating; lamellar twinning with {0$\bar{2}$1} twin plane. Orange-brown to orange yellow, vitreous luster. Cleavage {010}, excellent to very good; {0$\bar{1}$2}, imperfect. H = 6. G = 3.48. Traces of Fe, Ca, and Mg substitute for Mn. Biaxial (+); X ∧ {010} = 58°, Y ∧ {010} = 30°, Z ∧ {P010} = 13°. $N_x = 1.698$, $N_y = 1.704$, $N_z = 1.720$; 2V = 65°. X, colorless; Y, pale yellow; Z, canary yellow. Associated with pyroxmangite and rhodochrosite in metamorphosed limestone and chert horizons of the Haast schist group. Also associated with rhodonite, spessartine, quartz, huebnerite, alabandite, tinzenite, and others. *Akatore Creek, Eastern Otago, South Island, New Zealand*; Haeste field, Norberg ore district, central Sweden. EF *NJBA 157:225(1987)*.

57.3.1.1 Zunyite $Al_{13}O_4(Si_5O_{16})(OH,F)_{18}Cl$

Named in 1884 by Hillebrand for the type locality. ISO $F\bar{4}3m$. $a = 13.880$, $Z = 4$, $D = 2.89$. *14-698:* 8.07_{10} 4.21_{10} 4.02_6 3.20_3 2.84_5 2.68_9 2.01_5 1.640_9. Structure: *AC(B) 38:390(1982)*. The fundamental structural element is a pentamer involving four silicon–oxygen tetrahedra sharing an oxygen each with a fifth, central one. Tetrahedra, {111} modified by {11$\bar{1}$} and {100}; sometimes pseudo-octahedral. Twinning common. Colorless, white, light grey, pinkish tan; vitreous luster; translucent to transparent. Cleavage {111}, distinct. H = 7. G = 2.88. Isotropic; N = 1.563–1.600, also weakly anisotropic. Infusible; insoluble in acids. Found embedded in "guitermanite" [baumhauerite (?)] as crystals up to 5 mm(!) at *Zuni mine, Anvil Mt., Silverton, San Juan Co., CO*, and in altered feldspathic rocks at Charter Oak mine, Ouray Co., and the Bonanza district, Saguache Co.; at the San Manuel mine, Pinal Co., and Big Bertha Extension mine, Yuma Co., and in the Dome Rock Mts., near Quartzsite, AZ. At the Emberton quarry, near Cockermouth, Cumberland, England. At Beni-Embarek, Algeria, and in highly aluminous shales at Postmasburg, South Africa. As transparent crystals at Kuni, Gumma Pref., Japan at Brumado, Bahia, Brazil. AR *D6:436(1892)*; *BM 85:407(1962), 97:271(1974)*.

57.4.1.1 Medaite $(Mn,Ca)_6(V,As)Si_5O_{18}(OH)$

Named in 1982 by Gramaccioli et al. for Francesco Meda (1926–1977), Turin, Italy, amateur mineralogist. MON $P2_1/n$. $a = 6.712$, $b = 28.948$, $c = 7.578$, $\beta = 95.40°$, $Z = 4$, $D = 3.75$. *36-433*(calc): 3.26_{10} 3.16_{10} 3.09_7 3.01_7 2.96_7 2.94_7 2.79_6 2.61_{10}. No natural powder diffraction pattern exists because of intimate intergrowth with tiragalloite. Contains a vanadopentasilicate ion $[VSi_5O_{18}(OH)]^{12-}$, comprising six tetrahedra linked together to form a chain fragment. Anhedral grains to 1.5 mm long, most are 0.2–0.4 mm; twin plane is coincident with the cleavage plane. Orange-yellow to brown, dark orange, brownish red; light orange streak; subadamantine luster. Cleavage {100}, good; parting perpendicular to elongation. $G = 3.70$. Insoluble in H_2O, HCl, and HNO_3. Contains Fe, 0.3%; CaO, 1.3%; As_2O_5, 2.1%. Biaxial (+); $XYZ \approx abc$, $X \wedge$ cleavage normal $= 3°$; in white light: $N_x = 1.77$, $N_y = 1.78$, $N_z = 1.80$; $2V_{calc} = 71°$, 2V not able to be measured because of very strong absorption; positive elongation; faintly pleochroic, dark to clear orange. Occurs at *Molinello, Val Graveglia, near Chiavari, eastern Liguria, Italy*, in a manganese mine intimately associated with tiragalloite. Occurs in veinless cutting massive aggregates of quartz and braunite. Also near Lavagna, Genoa, Italy. EF *RSIMP 36:159(1980), AC(B) 37:1972(1981), AM 67:85(1982)*.

Class 58

Insular, Mixed, Single, and Larger Tetrahedral Groups

58.1.1.1 Kornerupine ($\square$,Mg,Fe)(Mg,Fe,Al)$_9$(Si,Al,B)$_5$(O,OH,F)$_{22}$
MM56 : 247(1992)

Named in 1884 by J. Lorenzen for Andreas Nikolaus Kornerup (1857–1881), Danish geologist. Space group unknown. $a = 15.999$–16.117, $b = 13.707$–13.746, $c = 6.704$–6.749; as B is substituted into kornerupine, the volume and $a + c$ parameters decrease; $Z = 4$, $D = 3.25$. *29-852*(nat): 10.4_4 7.99_3 6.84_4 3.35_5 3.01_{10} 2.62_{10} 2.10_6 2.08_5. Structure: *AM 76:1824(1991)*. 0–0.69 B per formula unit of 22 O atoms. 0–69% occupancy of one site by B. One eightfold: five octahedral and three tetrahedral sites. *AM 76:1824(1991)*. Crystals to >5 cm, rodlike to fibrous, long prisms. Dark green, sea-green, cream, white, colorless, blue, gray, pink, greenish yellow, brown, black; streak usually white; striated, with {110}, {100}, {021}; vitreous luster. Cleavage {110}, distinct prismatic; {001}, poor. $H = 6\frac{1}{2}$–7.0. $G = 3.27$–3.45. Moderately magnetic. Insoluble in common acids but soluble in HF. Formula *MM 56:247(1992)*. A boron-poor variety exists, but the mineral usually has some B; up to 5.2 wt % B_2O_3, to ≈ 15% FeO, 13.4% MgO, and 0.6% F; type material has 0.5–1.44% B_2O_3. Most localities have B-bearing material. Some Fe_2O_3 may also be present. Three localities for β-free material are known. Fe-richest material (about 15 wt % FeO) is from Port Shepstone and Homagama; Norwegian material has 20% MgO, 5.0% Fe_2O_3. As much as 0.2 wt % BeO and 0.3 wt % Sc_2O_3 may be present. Biaxial (−); $Z = b$; $N_x = 1.660$–1.682, $N_y = 1.674$–1.696, $N_z = 1.675$–1.699; $2V = 3$–$48°$; $r < v$, also $r > v$, weak; negative elongation; pleochroism variable. X, colorless to yellow to dark green; Y, colorless to pale brown to reddish blue; Z, light amber to light blue to dark green. Occurs in Si-poor, Mg–Al-rich gneisses. Often Precambrian in age. In aluminous upper amphibolite and granulite-facies rocks. Also in magnesian metapelites and in pegmatite. Some is alluvial. In high-grade contact- and regional metamorphic rocks. Formed with or instead of tourmaline if sufficient B is present. Associated cordierite, sapphirine, corundum, sillimanite, and others. Alteration: kornerupine may alter to a mixture of sapphirine and cordierite, gedrite ± orthopyroxene. May replace or be replaced by grandidierite. There are 50–60 localities worldwide. Opinicon Lake and two other areas in southeastern ONT, and Lac Sainte Marie, Gatineau Co., QUE, Canada; to more than 23 cm long from *Fiskenaesset region, western Greenland*, some is B-poor, and to 50 cm long elsewhere in Greenland; Hullvannet, Bamble sector, southern Norway; Waldheim, Saxony, Germany; Kola Penin., Russia; Sar-e-Sang, Afghanistan; with werdingite and grandiderite at Bok se Puts, Namaqualand, Port Shepstone, KwaZulu-Natal, and from the Cavan farm, Soutpansberg district, northern Transvaal, South Africa; B-poor from Sinyoni claims, Beit Bridge, Zimbabwe; Kwale district, southern Kenya; Mautia Hill, Tanzania; Labwor Hills, Uganda; Lukusuzi, Eastern Prov., Zambia; clear, sea-green, large gem-quality crystals from Itrongay and at Betroka, Madagascar; gem gravels at Talatu Oya, eastern Khandy, Sri Lanka (Ceylon); Fe-rich material from Homagama, Sri Lanka; B-poor material from the Reynolds Range, NT, and in the Harts and Anmatjiras Ranges, Australia; with grandidierite at McCarthy Point and Seal Cove, Larsemann Hills, Prydz Bay, eastern Antarctica. Some Greenland material is of gem quality, 5.88 to 0.09 ct. Also Itrongay, Madagascar, and Sri Lanka. The largest Sri Lankan stone is 9.2 ct. EF *MIN 3(1):290(1972)*; *MM 51:695(1987), 55:563(1991), 56:247(1992), 58:27(1994), 59:163(1995), 59:327(1995)*; *EJM 7:623(1995)*.

58.1.2.1 Davreuxite $MnAl_6Si_4O_{17}(OH)_2$

Named in 1878 by De Koninck for Charles Joseph Davreux (1800–1863), Belgian pharmacist and natural scientist. MON $P2_1/m$. $a = 9.518$, $b = 5.753$, $c = 12.05$, $\beta = 108.0°$, $Z = 2$. *37-431*: 8.51_3 5.72_4 4.29_4 3.82_3 3.51_{10} 3.10_5 2.87_6 2.84_4. The structure consists of isolated and double, $Si_2O_6(OH)$, tetrahedral silicate units. Aluminum is in six- and four-coordination, the latter forming with silicate tetrahedra, single and double chains. Aggregates of slender laminae and bundles of fine fibers [010], asbestiform. White with yellow to reddish tinge, pearly to dull luster. Cleavage {100}, good. $H = 2$–3. $G = 3.30$–3.38. Biaxial $(-)$; $2V = 48°$; $N_x = 1.660$, $N_y = 1.684$, $N_z = 1.692$; X approximately perpendicular to {100}, $Z = b$; . Magnesium and, to a lesser degree, iron may replace manganese. With quartz and pyrophyllite in low-grade metamorphosed Ordovician shales. Alters to kaolinite. At *Ottré, Stavelot massif, Ardennes, Belgium*, also at Rect, Regne, Vielsam, and Sart-Close. AR *CR(D) 283:295(1976)*, *AM 69:777(1984)*.

58.1.3.1 Queitite $Zn_2Pb_4[SiO_4][Si_2O_7](SO_4)$

Named by Keller et al. in honor of Clive S. Queit (b.1946), Tsumeb, Namibia. MON $P2_1$. $a = 11.362$, $b = 5.266$, $c = 12.655$, $\beta = 108.16°$, $Z = 2$, $D = 6.07$. *33-782*: 3.77_5 3.59_5 3.18_{10} 2.99_5 2.82_5 1.635_8 1.490_6 1.486_6. Structure: *ZK 151:287(1980)*. Elongated [010], tabular {001} crystals up to 10 mm × 3 mm × 0.5 mm; {001} dominant with {112}, {110}, and {101} prominent. Twinned on {100} and {001}. Colorless to pale yellow, greasy luster. Cleavage {010}, {001}, traces. $H = 4$. Biaxial $(+)$; $2V \approx 90°$; $N_x = 1.899$, $N_y = 1.901$, $N_z = 1.903$; $r > v$, strong; anomalous interference colors. Found on corroded galena, sphalerite, and tennantite with willemite, alamosite, melanotekite, and leadhillite at *Tsumeb, Namibia*. AR *AM 65:407(1979)*.

EPIDOTE GROUP

The epidote group consists of 13 isostructural silicates conforming to the formula

$$A_1A_2M_1M_2M_3(SiO_4)(O,OH)(OH,F)$$

where

A1 = Ca
A2 = Ca, REE, Pb, Sr
M1 = Al, Fe, Mn, Mg
M2 = Al
M3 = Al, Fe^{3+}, Mn^{3+}, Cr^{3+}

and the minerals crystallize with monoclinic or orthorhombic symmetry.

The structure consists of chains of edge-sharing octahedra of two types: a single chain of M2 octahedra and a multiple or zigzag chain composed of central M1 and peripheral M3 octahedra. The chains are cross-linked by SiO_4 and Si_2O_7 groups. Finally, there remain between the chains and cross-links relatively large cavities which house the A1 and A2 cations.

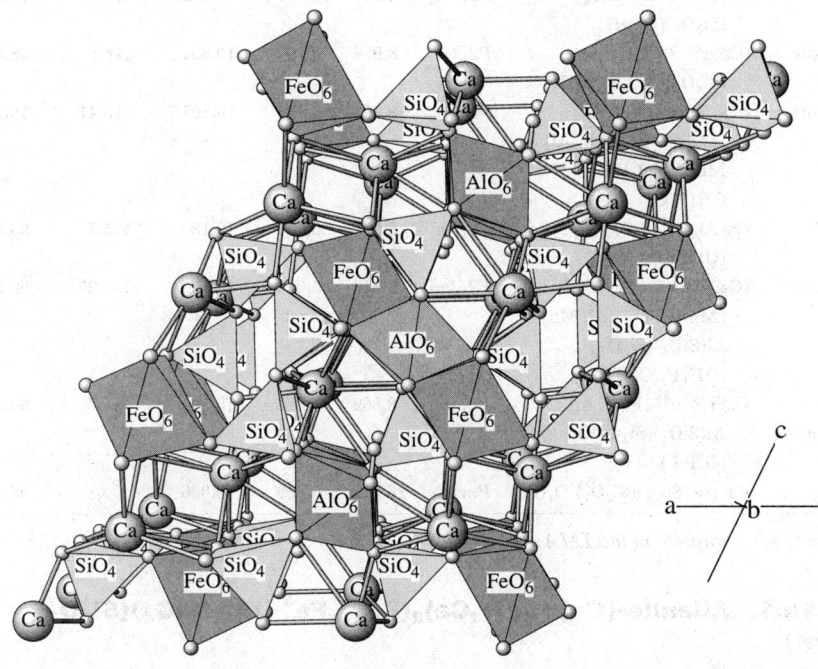

EPIDOTE: $Ca_2(Al, Fe^{3+})_3O(SiO_4)(Si_2O_7)(OH)$

Many piemontites and epidotes as named in the literature are designated incorrectly: for a discussion, see *NJBM 8:357–366(1989)*. A detailed study of 23 members of the epidote group has further elucidated the effects of Al ↔ Fe^{3+} ↔ Fe^{2+} substitution and the entry of REE^{3+}. *MM 53:133(1995)*. In the structure of allanite and REE-bearing epidotes, the entry of trivalent REE in the A sites is balanced by the presence of divalent cations, mainly Fe^{2+} and Mg. Divalent iron in allanites occurs mainly in M3.

Mineral	Formula	Space Group	a	b	c	β	
Allanite-(Ce)	$(Ce,Y,Ca)_2(Fe^{2+},Fe^{3+})Al_2$ $(SiO_4)(Si_2O_7)(O)(OH)$	$P2_1/m$	8.927	5.761	10.150	114.77	58.2.1a.1
Allanite-(La)*	$(La,Y,Ca)_2(Fe^{2+},Fe^{3+})$ $Al_2(SiO_4)(Si_2O_7)(O)(OH)$	$P2_1/m$	8.905	5.699	10.131	114.9	58.2.1a.2
Allanite-(Y)*	$(Y,Ce,Ca)_2(Fe^{2+},Fe^{3+})$ $Al_2(SiO_4)(Si_2O_7)(O)(OH)$	$P2_1/m$					58.2.1a.3
Clinozoisite	$Ca_2(Al,Fe^{3+})Al_2(SiO_4)$ $(Si_2O_7)(O)(OH)$	$P2_1/m$	8.879	5.583	10.149	115.43	58.2.1a.4
Dissakisite-(Ce)	$Ca(Ce,La)MgAl_2$ $(SiO_4)(Si_2O_7)(O)(OH)$	$P2_1/m$	8.905– 8.916	5.684– 5.700	10.112	114.62– 114.72	58.2.1a.5
Dollaseite-(Ce)	$Ca(Ce,REE)(Mg)(Mg,Al)$ $(SiO_4)(Si_2O_7)O(F)$	$P2_1/m$	8.934	5.721	10.176	114.31	58.2.1a.6
Epidote	$Ca_2(Fe^{3+},Al)_3(SiO_4)$ $(Si_2O_7)(O,OH)_2$	$P2_1/m$	8.914	5.640	10.162	115.4	58.2.1a.7
Hancockite	$(Ca,Pb,Sr)_2(Al,Fe)_3$ $(Si_2O_7)(SiO_4)(O,OH)_2$	$P2_1/m$	8.958	5.665	10.304	114.4	58.2.1a.8
Khristovite-(Ce)	$(Ca,REE)(Ce,REE)$ (Mg,Fe,Cr,Ti,V,Al) $MnAl(SiO_4)(Si_2O_7)$ $OH(F,O)$	$P2_1/m$	8.903	5.748	10.107	113.41	58.2.1a.9
Mukhinite	$Ca_2Al_2V^{3+}(SiO_4)(Si_2O_7)$ $(O)(OH)$	$P2_1/m$	8.90	5.61	10.15	115.5	58.2.1a.10
Piemontite	$(Ca,Mn)(Ca,Sr,REE)$ $(Mn^{3+},Fe^{3+})(Al,Mn)$ $Al(SiO_4)(Si_2O_7)$ $(OH,F)(O)$	$P2_1/m$	8.843	5.664	10.150	115.25	58.2.1a.11
Strontio-piemontite	$CaSr(Mn^{3+},Fe^{3+},Al)$ $Al(SiO_4)(Si_2O_7)O$ (OH,F)	$P2_1/m$	8.862	5.682	10.191	114.7	58.2.1a.12
Zoisite	$Ca_2Al_3(SiO_4)(Si_2O_7)O(OH)$	$Pnma$	16.212	5.559	10.036		58.2.1b.1

*Name not yet approved by the IMA.

58.2.1a.1 Allanite-(Ce) $(Ce,Y,Ca)_2(Fe^{2+},Fe^{3+})Al_2(SiO_4)(Si_2O_7)$ (O)(OH)

Named in 1810 by Thomas Thomson for Thomas Allan (1777–1833), Scottish mineralogist, who first observed the mineral. Epidote group. Synonym: orthite. Ce is the dominant REE present. MON $P2_1/m$. $a = 8.927$, $b = 5.761$,

$c = 10.150$, $\beta = 114.77°$, $Z = 2$, $D = 4.11$. *25-169:* 3.53_4 2.92_{10} 2.89_3 2.71_6 2.63_4 2.56_2 2.18_3 2.16_2. Also *9-474*(heated metamict allanite). REE's in larger A2 site, Ca in A1 site. *AM 56:447(1971)*. Light brown to black, light brown streak, resinous or pitchy submetallic luster. Cleavage {001}, imperfect; {100}, {110}, poor; conchoidal to uneven fracture; brittle. $H = 5-6\frac{1}{2}$. $G = 3.4-4.2$. Often metamict with even lower density. Gelatinizes with HCl. May be halogen-bearing (0.95 wt % Cl and 0.76 wt % F; e.g., Hemlo deposit). Coupled substitution $Ca + (Al + Fe^{3+}) = REE + M^{2+}$ is dominant; however, as $REE + M^{2+}$ increases, $3Ca^{2+} = 2REE^{3+} + \square$ operates. Vanadian allanite-(Ce) with about 8% V_2O_3 exists in the Hemlo gold deposit. Chromium may also substitute for Al and Fe^{3+} in allanite. MnO to 6.7 wt %, BeO to 3.8 wt %, P_2O_5 to 6.5 wt %, ThO_2 commonly to about 2 wt % and rarely to about 5%, SrO to 5 wt %, F to >1.1 wt %. One rare occurrence of Ge-bearing allanite with 10.6 wt % GeO_2 and 2.5 wt % Ge_2O_3 from a Zn deposit in the Pyrenees. Biaxial $(-)$ or $(+)$, majority are $(-)$; $X \wedge c = 1-42°$, $Y = b = $ OAP, sometimes perpendicular to (010) = parallel to (100), $Z \wedge a = 26-67°$. $N_x = 1.690-1.813$, $N_y = 1.700-1.857$, $N_z = 1.706-1.891$; $2V_x = 40-123°$; $Y > Z > X$ or $Z > Y > X$; $r > v$, moderate to strong; pleochroic colors are quite variable in brown, reddish brown, yellow, green, and yellow-green. Metamict allanites are isotropic with N as low as 1.61. A common accessory mineral in many igneous rocks, particularly those that are felsic, and metamorphosed igneous rocks. Also in pegmatites, skarns, and tactites. Not common in volcanic rocks. A detrital mineral. Of the many localities, only noteworthy sites are mentioned. Franklin, NJ; Amelia, VA; Moriah, Essex Co., NY; Monroe, Orange Co., NY; large crystals to 45 cm long occur in the Pacoima pegmatite, Pacoima Canyon, Los Angeles Co., CA; Macdonald mine near Hybla, Renfrew, and Tory Hill, and vanadian varieties from Hemlo, ONT, Canada; *Avigeit, eastern Greenland*; Trimouns mine, Luzenac, France; Ytterby, Riddarhyttan, and Finbo, and Mg-rich variety from Koberg mine, Bergslagen, Sweden; Hittero, Hundholmen, Iveland, and Ardendal, Norway; Vaarala, Finland; Ural Mts., and Vishnevye Mts., Russia; Sama, Madagascar. EF *AM 49:1159(1964)*, *MIN 3(1):750(1972)*, *DHZ 1B:151(1986)*, *CM 28:67(1990)*, *MM 55:497(1991)*.

58.2.1a.2 Allanite-(La) $(La,Y,Ca)_2(Fe^{2+}, Fe^{3+})Al_2(SiO_4)(Si_2O_7)(O)(OH)$

Epidote group. *Note*: Not formally approved by CNMMN IMA. MON $P2_1/m$. $a = 8.905-8.934$, $b = 5.699-5.760$, $c = 10.131-10.236$, $\beta = 114.9-114.97°$, $Z = 2$, $D = 3.84-4.06$. *CM 30:159(1993)*: 4.74_2 3.54_3 3.24_5 3.19_2 2.94_{10} 2.90_3 2.73_5 2.64_4; *45-1352:* $3.51_{4.5}$ 2.93_3 2.91_{10} 2.85_5 2.70_6 2.69_4 2.61_6 2.175_2. Black, brown streak. Cleavage {100}, poor; conchoidal fracture. $H = 5-5\frac{1}{2}$. $G = 3.69-4.02$. La_2O_3, 12.1 wt %; Ce_2O_3, 10.9 wt %. Biaxial $(+)$; $N_x = 1.730$, $N_y = 1.740$, $N_z = 1.770$; $2V_x = 129°$. Occurs at Waterford, CT; Mitchell Twp., QUE, Canada; in skarn at Serifos Island, Cyclades, Greece; Karelia, Russia. EF *AM 51:152(1966)*.

58.2.1a.3 Allanite-(Y) (Y,Ce,Ca)$_2$(Fe^{2+},Fe^{3+})Al$_2$(SiO$_4$)(Si$_2$O$_7$)(O)(OH)

Originally described in 1949 as lombaardite from the Zaaiplaats tin mine, Potgietersrust district, Transvaal, South Africa. Epidote group. Unit cell given with 2a, b, 2c of epidote on the basis of oscillation and Weissenberg photographs. Epidote-type pseudocell noted. Detailed structure solution not done. May be $P2$ or $P2/m$. Dark brown needles. Biaxial (+); $Y = b$; $N_x = 1.756$, $N_y = 1.761$, $N_z = 1.777$; $2V_{z\,meas} = 60°$, $2V_{z\,calc} = 58°$, $2V_z = 70$–$80°$ for Swedish material; $r > v$, strong; $Y > Z > X$; OAP parallel to (010). X, nearly colorless; Y, amber brown; Z, tawny olive. $D_m = 3.85$, $H = 6$. In an explosion breccia in a cassiterite-bearing pipe in granite, in hydrothermally altered arkoses, and in granitic pegmatite. Reported from Åsklagen, Värmland, Sweden. The mineral also occurs at Isle of Skye, Scotland. Possible allanite-(Y) occurs at Sinyaya Pala, northern Karelia, Russia. Also present at the Gole quarry, Murchison Twp., ONT, Canada. A Be- and Y-bearing (6.1 wt % Y$_2$O$_3$) allanite was reported from Skuleboda, Uddevalla, Sweden, but it contains elevated contents of water and CO$_2$ and is altered. EF *AM 36:381(1951), 48:1420(1963); EPSL 48:97(1980); MIN 3(1):750(1972); DHZ 1B:151(1986)*.

58.2.1a.4 Clinozoisite Ca$_2$(Al,Fe^{3+})Al$_2$(SiO$_4$)(Si$_2$O$_7$)(O)(OH)

Described in 1896 by Weinschenk; name is from the crystal system, the monoclinic dimorph of zoisite, and from its resemblance to zoisite. Epidote group. MON $P2_1/m$. $a = 8.879$–8.882, $b = 5.583$–5.604, $c = 10.149$–10.155, $\beta = 115.43$–$115.50°$, $Z = 2$, $D = 3.31$–3.36. *44-1400*(nat): 2.89_{10} 2.80_5 2.68_2 2.67_2 2.53_2 2.51_2 2.39_3 2.16_3. Only Al is present in M1 and M2, with Al, Fe, Mn, and others in M3. *AM 53:1882(1968)*. Prismatic, elongated on b, often striated, may be granular, massive, or fibrous, lamellar twinning on {100} not common. Both orthorhombic and monoclinic varieties of zoisite are considered thulite. *NJBM:325(1989)*. Colorless, pale yellow-gray, yellow-green, rose to red (rare), due to Mn^{3+} (thulite); white streak; vitreous luster. Cleavage {001}, {100}, perfect; uneven fracture. $H = 6\frac{1}{2}$. $D_m = 3.21$–3.38. Spectral data: *DHZ 1B:44(1986)*. Insoluble in HCl. Replacement of Ca by Fe^{2+}, Mn, and Ca is comparatively small. May contain many elements other than Ca, Al, Si, and O (e.g., SrO to 4.7 wt %). As much as 15.4 wt % Cr$_2$O$_3$ is present in clinozoisite from Outoukumpu, Finland. *CM 27:565(1989)*. With increasing substitution of Fe for Al, the mineral grades to epidote. Biaxial (+); $X \wedge c = 0$–$7°$, $Z \wedge a = 18$–$25°$, $Y = b$; $N_x = 1.670$–1.718, $N_y = 1.670$–1.725, $N_z = 1.690$–1.734; $2V_z = 14$–$90°$, 2V basically correlates with Fe content; $Z \geq Y > X$; $r < v$, strong; OAP (010); usually, nonpleochroic, Mn-containing material is pleochroic in colorless–light pink. Most common in regional metamorphic rocks, particularly in greenschists–epidote–amphibolite facies. Also in pegmatites and veins. In acid igneous rocks contaminated with calc–silicate material. As an alteration product of plagioclase. Contact zones. Also in rodingites (with usually very low iron contents). Found in many localities. Typical occurrences are at Eden Mills, VT; Allens Park, Boulder Co., CO; in the Nightingale district, Pershing Co., NV; Los Angeles Co. and elsewhere in CA; Timmins, ONT, Canada; Pinos Altos, BC, Gavilanes, BCN, Arroyo Puerto Nuevo, BCS, and Alamos, SON, Mexico; *Goslerwand, Prägratten, Tyrol, Austria*; other localities in the Tyrol and Piedmont in Switzerland, Austria, Italy, and Malagasy Republic (Madagascar). EF *DHZ 1B:44(1986); MIN 3(1):724(1972).*

58.2.1a.5 Dissakisite-(Ce) $Ca(Ce,La)MgAl_2(SiO_4)(Si_2O_7)O(OH)$

Described in 1991 and named from the Greek for *twice over*, for an Mg analog of allanite, having been discovered twice. MON $P2_1/m$. $a = 8.91$, $b = 5.69$, $c = 10.112$, $\beta = 114.7°$, $Z = 2$, $D = 3.97$. *AM 76:1990(1991)*: 9.1_4 3.50_5 2.91_9 2.84_5 2.70_{10} 2.62_6 2.18_4 2.14_4. Ca is ordered in A1 and lanthanides are ordered in A2. Mg and minor Fe^{2+} occupy M3, M2 is occupied only by Al. *CM 30:153 (1993)*. Dark brown, pale yellow-brown in thin section; vitreous luster. No cleavage observed. $G = 3.75$. No cathodoluminescence. Chromian variety from Outokumpu, Finland. Östanmossa subaluminous ferroan dissakisite has 0.9 wt % F. Biaxial (+); Y parallel to b, $Z \wedge a = 24°$ in obtuse angle β. $N_x = 1.735$, $N_y = 1.741$, $N_z = 1.758$; $2V_{z\ obs} = 64°$, $2V_{z\ calc} = 62°$; $X < Y = Z$; $r < v$, medium. X, pale brown; Y,Z, light yellow-brown. Occurs in Mg-rich metamorphic rocks such as micaceous schists, garnet–corundum gneisses, and in marble associated with calcite, phlogopite, spinel, and so on. Also in tremolite and magnesian skarn and magnetite ore. Trimoun talc deposit, Ariège, France; Östanmossa mine, Norberg district, Sweden; Outokumpu, Finland; Fedorovskoye and Emeldzhak deposits, southern Yakutia, Russia; Donghai district, Jiangsu Prov., China; *Balchen Mountain, eastern Antarctica.* EF *CM 25:413(1987), AM 73:48(1988).*

58.2.1a.6 Dollaseite-(Ce) $Ca(Ce,REE)(Mg)(MgAl)(SiO_4)(Si_2O_7)O(F)$

Named in 1988 for Wayne A. Dollase (b. 1938), American mineralogist, University of California at Los Angeles. Epidote group. Called magnesium orthite by P. Geijer [*SGU 20(4):1(1927)*]. MON $P2_1/m$. $a = 8.934$, $b = 5.721$, $c = 10.176$, $\beta = 114.31°$, $Z = 2$, $D = 3.87$. *41-1469*: 9.29_2 8.25_1 5.13_1 3.52_2 2.915_{10} 2.85_3 2.71_7 2.15_2. Mg is ordered over two equipoints, Ca and the REE's are ordered over two different structural sites, and F and OH are each ordered over two different sites. *AM 73:838(1988)*. Brown. $G = 3.9$. Contains 3.0–3.3 wt % F. Substitution $Mg + F$ for $M^{3+} + O$ in allanite suggests that such substitutions may occur in other members of the group. Two paired substitutions in dollaseite-(Ce) are $REE + Mg \leftrightarrow Ca + Al$ and $Mg + F \leftrightarrow Al + O$. Biaxial (+); $N_x = 1.715$, $N_y = 1.718$, $N_z = 1.733$, $2V_{z\ calc} = 49°$. Occurs in metamorphic rocks associated with tremolite, norbergite, and calcite. *Östanmossa mine, Norberg district, Sweden.* EF

58.2.1a.7 Epidote $Ca_2(Fe^{3+},Al)_3(SiO_4)(Si_2O_7)(O,OH)_2$

Described in 1801 by Haüy and named from the Greek for *increase*, because the base of the prism has one side longer than the other. Epidote group. MON $P2_1/m$. $a = 8.914$, $b = 5.640$, $c = 10.162$, $\beta = 115.4°$ (for 0.81 atom of Fe^{3+}); range: $a = 8.884–8.916$, $b = 5.621–5.662$, $c = 10.154–10.175$, $\beta = 115.2–115.5°$, $Z = 2$, $D = 3.45$. *17-514*(nat): 4.02_5 3.40_4 2.90_{10} 2.82_4 2.69_7 2.68_{10} 2.60_5 2.46_5; *45-1446*: $2.90_{8.5}$ 2.69_6 $2.40_{6.5}$ 1.64_{10} 1.63_6 1.46_7 1.41_6 1.40_6. The epidote structure is summarized in *AM 56:447(1971)*. Crystals are short to long prismatic, often striated; also thick tabular or acicular. More than 200 different forms are known. Commonly massive, coarse to fine granular or fibrous. Twinning {100} lamellar, relatively common and sometimes on {001}. Green, yellow, gray, yellow-green, brownish green, greenish black, black, and bright Cr green (variety *tawmawite*); the characteristic green color of epidote is associated with the presence of Fe^{3+} in M3 and is determined by $O_2 \leftrightarrow Fe^{3+}$ transfer; white-gray streak; vitreous to somewhat pearly or resinous luster. Cleavage {001}, perfect; {100}, good; conchoidal, uneven, or splintery fracture. $H = 6\frac{1}{2}$. $G = 3.38–3.49$. IR and other spectral data: *DHZ 1B:44(1986)*, *AM 76:602(1991)*. Sometimes fluorescent weak red in SWUV. Partly soluble in hot concentrated HCl, soluble in HF. May contain many elements other than Ca, Fe, Al, Si, and O. Most common are REE's, grading to allanite. Additional elemental substitutions are Mg, Mn, Sn, Sr (to 8.5 wt % SrO), and others. The variety *tawmawite*, first described in 1907, is characterized by elevated Cr content. The highest reported Cr content, in epidote from Hemlo, ONT, is 11.8 Cr_2O_3 *[CM 27:565(1989)]*. A new Cr end member in the epidote group would be $Ca_2Cr(Al,Fe)Si_3O_{12}(OH)$. *Tawmawite* should be redefined as that epidote family mineral with $X^{oct}_{Cr} = 0.33$ and $X^{oct}_{Al} = 0.67$ *[MM 51:593(1987)]*. Within epidote, there are strong site preferences for substitution of Cr, Fe, and Mn for Al. Cr has a strong preference for M1, and with substitution of V grades to mukhinite. Biaxial (−); $X \wedge c = 0–8°$, $Y = b$, $Z \wedge a = 25–33°$; $N_x = 1.708–1.751$, $N_y = 1.714–1.784$, $N_z = 1.718–1.797$; $2V_x = 64–90°$; $r > v$, moderate to strong; usually $Y \geq Z > X$, sometimes $Z > Y > X$; pleochroism increases with Fe content. X, colorless, pale yellow, pale green; Y, yellowish green, brownish green, greenish yellow; Z, yellowish green. Other pleochroic colors as well, depending on chemical composition (e.g., Cr, V, Mn, etc.). Boundary between optically (+) clinozoisite and optically (−) epidote is between 15.0 and 15.9 mol % $Ca_2Fe_3^{3+}Si_3O_{12}(OH)$; about 7.7 wt % Fe_2O_3. OAP parallel to b. For *tawmawite* (Cr-epidote), $N_x = 1.695$, $N_y = 1.705$, $N_z = 1.715$; $2V_x = 40°$ (note that $2V_{calc} = 90°$). Epidote is a ubiquitous mineral in regional and contact-metamorphic rocks but is particularly characteristic of greenschist and epidote–amphibolite facies rocks. Also occurs in lower-grade rocks with zeolites, prehnite, pumpellyite, lawsonite, and stilpnomelane. In blueschists and eclogites as well. In metasomatic skarns, calcsilicate rocks, and epithermal ore deposits. In low-temperature veins, amygdules, fillings, and some hot-spring systems. A hydrothermal alteration product of plagioclase. Occurs as a primary magmatic mineral in some basic to granitic rocks and pegmatites. Cr-bearing epidote occurs in contact-metamorphic rocks. Notable occurrences include: Splendid crystals from Haddam, CT. Good crystals from Calumet mine, Chaffee Co., CO, also from Garnet Hill, Calaveras County, CA. Fine, splendant crystals from the Green Monster mine, Prince of Wales Island, Sulzer, AK(!). Hemlo, ONT, Canada; Sn-bearing epidote from Land's End, Cornwall, England; beautiful crystallizations from Bourg d'Oisans, Isère, France(!); gem-quality and complex crystals from Knappenwand, Untersulzbachtal, Salzburg, and Zillertal, Tyrol, Austria(!). Traversella and Ala Valley, Piemonte, Italy; St. Gottard, Ticino, and Zermatt, Switzerland; Arendal, Norway; Zöptau, Moravia, Czech Republic; Cr-bearing material from the Outokumpu area, Finland; Achmatovsk, Kussinsk, Ural Mts., Russia; gem crystals from Gilgit, Tormiq, and Dusso, northwestern Pakistan(!); Tawmaw, Myanmar (Burma); Namibia(!). Sometimes used as a semiprecious gemstone and for lapidary purposes (unakite-epidotized granitic rocks). EF *MIN 3(1):728(1972), KLOK:696(1980), AM 66:974(1981), DHZ IB:44(1986).*

58.2.1a.8 Hancockite $(Ca,Pb,Sr)_2(Al,Fe)_3(Si_2O_7)(SiO_4)(O,OH)_2$

Named in 1899 for Elwood P. Hancock (1836–1916), Burlington, NJ, a mineral collector. Epidote group. MON $P2_1/m$. $a = 8.958$, $b = 5.665$, $c = 10.304$, $\beta = 114.4°$, $Z = 2$, $D = 4.03$. 17-212: 4.62_3 3.49_5 3.27_3 2.91_{10} 2.81_4 2.71_4 2.18_4 1.90_4. Pb (along with Sr and Ba) is in A2. Crystals are usually about 0.5 mm long and 0.15 mm wide, but may be as large as 2 cm × 0.2 cm. Lathlike crystals show dominant forms {001}, {100}, {101}, {$\bar{1}$01}, {$\bar{1}$11}. Striated parallel to [010]; usually massive and masses are brownish red or maroon. Crystals are red to red-brown but may be yellowish brown to yellow-green. Vitreous luster, translucent. Cleavage {001}, perfect; uneven fracture; brittle. $H = 6\frac{1}{2}$–7. $G = 4.03$. Analysis (wt %): PbO, 18.5; SrO, 3.9 (Franklin, NJ); PbO, 32.4 (Jakobsberg, Sweden). Biaxial (−); $Y = b$; $N_x = 1.788$, $N_y = 1.810$–1.825, $N_z = 1.830$; $2V_x = 38$–50°; OAP = (010); $Z > X$; $r > v$; strongly pleochroic. X, colorless, pale rose, greenish yellow; Y, pale brownish yellow to yellow; Z, pale rose, greenish yellow or green. Occurs in fractures in willemite–franklinite ore and as masses in the contact-metasomatic Zn deposit at *Franklin, Sussex Co., NJ*. Associated with garnet, hendricksite, and as inclusions in manganaxinite, datolite, willemite, and hyalophane. Also occurs in a metamorphosed manganese–iron orebody in skarns enclosed in dolomitic marble at Jakobsberg, Filipstad, Värmland, Sweden [*MM 58:172(1994)*], associated with melanotekite, garnet, and hematite. EF *AM 56:447(1971), MIN 3(1):749(1972), MM 49:721(1985), DHZ 1B:44(1986).*

58.2.1a.9 Khristovite-(Ce) (Ca,REE)(Ce,REE)(Mg,Fe,Cr,Ti,V,Al)MnAl$(SiO_4)(Si_2O_7)(OH)(F,O)$

Described in 1991 [*SPC 36(2):172(1991)*] but not named until 1993 [*ZVMO 122:103(1993)*]. Named for Evgenia Vladimirovicha Khristova (b.1933), Russian geologist and a specialist in Tien-shan geology. Epidote group. MON $P2_1/m$. $a = 8.903$, $b = 5.748$, $c = 10.107$, $\beta = 113.41°$, $Z = 2$, $D = 4.11$. PD: 9.32_2 5.23_2 4.67_2 3.52_4 2.91_{10} 2.73_7 2.63_8 1.437_2. The Mn analog of dollaseite-(Ce), but with some differences and with Mn^{2+} predominant in M3. $R = 3.5\%$. Prismatic crystals up to 1.5 mm long; dominant forms are {001}, {100}, and subordinate {101} and {102}. Brown to dark brown or dark grayish brown, light brown streak, vitreous luster, transparent. No cleavage observed. $H = 5$. $G = 4.05$. IR. Analysis (wt %): SiO_2, 29.9; Al_2O_3, 9.5; MgO, 2.7; CaO, 5.6; TiO_2, 1.6; V_2O_3, 1.1; Cr_2O_3, 1.5; FeO, 1.8; MnO, 11.8; Ce_2O_3, 13.6; La_2O_3, 8.7; Nd_2O_3, 4.2; Sm_2O_3, 0.6; Dy_2O_3, 3.1; Pr_2O_3, 1.4; F, 2.0; $(H_2O)_{calc}$, 1.5; O for F, 0.8. MnO to 14.5 wt %. Biaxial (−); in sections perpendicular to B×a, Y ∧ elongation = 1.5–3°, in sections perpendicular to B×o, extinction is parallel; $Y = b$; OAP is perpendicular to the elongation direction; $N_x = 1.773$, $N_y = 1.790$, $N_z = 1.803$; $2V_x = 83°$; $Y > Z \gg X$; $r < v$, medium. Occurs in rhodonite in a garnet–biotite–quartz-rich exocontact of the subalkaline granites of the Inyl'chek massif, on the *northern slope of the Inyl'chek Range, southwestern Tien-shan, Kirgiziya, Russia*. Occurs with rhodonite, tephroite, rhodochrosite, hyalophane, barite, hejtmanite, hübnerite, and an unidentified Cl-bearing Mn-silicate. EF

58.2.1a.10 Mukhinite $Ca_2Al_2V^{3+}(SiO_4)(Si_2O_7)(O)(OH)$

Named in 1969 for Alekseia Stepanovna Mukhin (1910–1973), Russian geologist, Western Siberia Geological Administration. Epidote group. MON $P2_1/m$. $a = 8.90$, $b = 5.61$, $c = 10.15$, $\beta = 115.5°$, $Z = 2$, $D = 3.47$. *22-1066:* 2.89_{10} 2.68_8 2.60_8 2.40_8 2.33_7 1.67_4 1.405_8 1.39_8. A vanadian clinozoisite in which one Al is replaced by V. Brown-black, brownish-gray streak, vitreous luster. Cleavage {001}, perfect; {100}, very good; brittle. $H = 8$. $G = 3.3$. Difficultly soluble in acids. Contains 11.3 wt % V_2O_3. Biaxial (+); $Y = b$, $Z \wedge a = 32°$; $N_x = 1.723$, $N_y = 1.733$, $N_z = 1.755$; $2V_z = 88°$. X, pale olive-green; Y, pale reddish brown; Z, reddish brown. Occurs with goldmanite, muscovite, and common sulfides in marble. Occurs near Lewiston, ID. *Tashelginsk iron deposit, Gornaya Shoriya, western Siberia, Russia*. EF *DAN SSSR 185(6):1342(1969)*.

58.2.1a.11 Piemontite (Ca,Mn)(Ca,Sr, REE)(Mn^{3+},Fe^{3+})(Al,Mn) Al(SiO_4)(Si_2O_7)(OH,F)(O)

Named in 1853 by Kenngott for the locality. Epidote group. MON $P2_1/m$. $a = 8.843$, $b = 5.664$, $c = 10.150$, $\beta = 115.25°$, $Z = 2$. *29-288:* 5.02_4 4.00_5 3.50_4 2.90_{10} 2.83_4 2.69_4 2.60_5 2.41_5. Also *19-897*. The b and c dimensions increase with increasing substitution of $Mn^{3+} + Fe^{3+}$. *MM 38:64(1971)*. M2 octahedral site filled completely and uniquely by Al. M1 and M3 contain Al, Fe, Mn, and Fe and Mn preferentially occupy M3. Crystals are prismatic, elongate on b. Twinning lamellar on {001}, not common. Reddish brown, black, or red; cherry-red streak; vitreous luster. Cleavage {001}, perfect. $H = 6$. $G = 3.38$–3.61. Slowly attacked by HF only. Mn_2O_3 to 22 wt %. Compositional field is large. Substitution between $Ca_2Fe^{3+}Al_2 \ldots -Ca_2Mn^{3+}Al_2 \ldots$ and $Ca_2Mn_2^{3+}Al \ldots$. Mg may substitute as well. Some material from the type locality is strontiopiemontite because of the high SrO (12 wt %) content. Overall, on a worldwide basis, Sr is low to absent. REE-bearing (with as much as about 8 wt % total REE_2O_3) piemontites exist confirming solid solution toward manganian allanites. *EJM 4:23(1992)*. REE substitution in A2 balanced by (Fe^{2+},Mg, Mn^{2+}) substitution in the octahedral framework. Biaxial (−) and (+); $X \wedge c = 2$–$9°$, $Y = b$, $Z \wedge a = 27$–$32°$; $N_x = 1.730$–1.794, $N_y = 1.740$–1.807, $N_z = 1.762$–1.829; $2V_z = 64$–$106°$; OAP parallel to (010); $r < v$, weak, to $r > v$, strong; $Z > X > Y$. X, yellow, lemon yellow, yellow-orange; Y, pale pink, amethyst, violet-pink; Z, crimson red, magenta. In low-grade regional metamorphic rocks, altered (oxidized) volcanic rocks, and as hydrothermal product in certain Mn deposits. From greenschist and blueschist facies rocks to amphibolite and occasionally, higher facies. Also a late crystallization phase in acid and intermediate volcanics. High fO_2 (partial pressure of oxygen) is necessary for formation. Also in pegmatites and quartz veins. Noteworthy occurrences include: Pilar, NM; San Gorgonio Pass, Riverside Co., CA; Glen Coe, Scotland; Haute Maurienne, France; Sorharas, Långban, and Ultevis, Sweden; *Praborna Mn mine, St. Marcel, Val d'Aosta, Piemonte*, and Monte Brugiana, Alpi Apuane, *Italy*; Treppes, Euboea Island, Greece; Atta, Alai range, Russia: Swat, Pakistan; Hidaka Mts., Hokkaido, and Abukuma plateau, Japan; Haast R. (Westland) and Lake Wakatipu, Otago, South Island, New Zealand; Djebel Dekhan, Egypt. Piemontite schist in New Zealand is used as an ornamental stone. EF *AM 54:710(1969)*, *MIN 3(1):744(1972)*, *DHZ 1B:135(1986)*, *PIMA 13:635(1986)*.

58.2.1a.12 Strontiopiemontite $CaSr(Mn^{3+},Fe^{3+},Al)Al(SiO_4)(Si_2O_7)O(OH,F)$

Named in 1990 as the Sr analog of piemontite. Epidote group. MON $P2_1/m$. $a = 8.862$, $b = 5.682$, $c = 10.191$, $\beta = 114.7°$, $Z = 2$, $D = 3.73$. PD: 3.49_5 2.92_{10} 2.84_5 2.68_4 2.65_3 2.60_5 2.12_4 1.59_5. Structure: *EJM 2:519(1990)*. A(2) occupied by Sr. Prismatic, elongate on b. To 0.5 mm long. Deep red, purple-brown streak, vitreous luster. Cleavage {001}, perfect; often twinned on {100} with (010) as the TP. H = 6. G = 3.65–3.73. Biaxial (+); $Y = b$; $\bar{N} = 1.763$, strongly pleochroic. X, yellow-orange; Y, violet; Z, reddish violet. In an Mn ore deposit in veinlets 3–4 mm thick cutting braunite ore. Associated with calcite, rhodochrosite, rhodonite, and ganophyllite. *Val Graveglia, Liguria, Italy*; also from the Praborna Mn mine, St. Marcel, Val d'Aosta, Piemonte, Italy; large masses of deep red-brown strontiopiemontite and small crystals to 1 mm occur in the Wessels mine, Kalahara manganese field, Cape Prov., South Africa, associated with pectolite. EF *EJM 2:519(1990)*.

58.2.1b.1 Zoisite $Ca_2Al_3(SiO_4)(Si_2O_7)O(OH)$

Named in 1805 by Werner for Siegmund Zois, Baron von Edelstein (1747–1819), Austrian natural scientist. Epidote group. ORTH *Pnma*. $a = 16.212$,

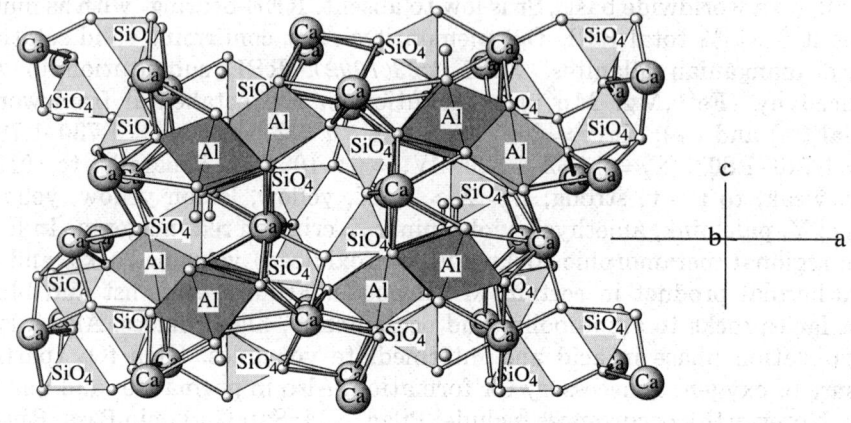

ZOISITE: $Ca_2Al_3Si_3O_{12}OH$
Perspective view along b-axis

$b = 5.559$, $c = 10.036$, $Z = 4$. $13\text{-}562$(syn): 8.09_4 5.01_3 4.03_5 2.87_6 2.79_3 2.69_{10} 2.02_3 1.60_3. Consists of $Al(O,OH)_6$ chains parallel to b, which are linked by SiO_4 and Si_2O_7 groups and Ca polyhedra. *AM 53:1883(1968)*. Prismatic, vertically striated, twinning not observed. Yellowish green, green, brown, blue (tanzanite), colorless, purple, gray, pink (thulite); white streak; vitreous luster, sometimes pearly on cleavage. Cleavage {100}, perfect; {010}, difficult; conchoidal to uneven fracture; brittle. H = 6–7. G = 3.355(tanzanite), G = 3.15(syn), range(nat) = 3.15–3.36. Insoluble in common acids but soluble in HF. Heat-treated tanzanite crystals become sapphire blue in color in two optical directions. Replacement of Al by Fe is very limited; rarely more than 10% of (Al,Fe) octahedra. α-Zoisite has as much as 4.0 wt % Fe_2O_3. No Al–Fe replacement series as in epidote. There is no orthorhombic equivalent of piemontite. Thulite is a pink to rose-colored variety of zoisite and clinozoisite and has 0.05–0.9 wt % MnO. *MM 44:45(1981)*. SrO contents to 7.4 wt %. *TMPM 34:87(1985)*. Zoisite may contain some Cr (0.3–0.4 wt % Cr_2O_3). May contain some F (to 1.67 wt %). Tanzanite contains 0.2 wt % V_2O_3. *AM 54:702(1969)*. Two varieties exist: regular (β-zoisite) and ferrian (α-zoisite). Both forms may be intergrown. Biaxial (+); $N_x = 1.685\text{–}1.705$, $N_y = 1.688\text{–}1.710$, $N_z = 1.697\text{–}1.725$; $2V_z = 0\text{–}80°$ [rarely higher, to 90–120°(?)]. For tanzanite, $N_x = 1.6925$ (red-violet), $N_y = 1.6943$ (deep blue), $N_z = 1.7015$ (yellow-green), $2V_z = 53°$. After heating, X, violet-red; Y,Z, deep blue. Thulite shows X, pale to medium pink; Y, colorless to pink; Z, pale yellow to yellow.

Rhombic dispersion, r > v (zoisite); strong rhombic dispersion, r ≪ v (Fe-zoisite). For β-zoisite, XYZ = abc; ferrian zoisite, XYZ = bac. In β-zoisite, OAP is perpendicular to (100) cleavage and parallel to (010), and in ferrian zoisite, OAP is parallel to (100) cleavage. May show anomalous blue interference colors. Optically (−) zoisite with dispersion r ≫ v has been reported. *MM 49:97(1985)*. Chiefly in regionally and, to a lesser extent, in contact-metamorphosed rocks. Also in quartz veins, pegmatites, and some eclogites. α-Zoisite is generally considered to be the high-temperature zoisite, occurring as a primary mineral in eclogitic rocks, whereas β-zoisite forms during retrograde metamorphism. Occurs at Pilar, NM (thulite); thulite occurs in at least 15 states; Ducktown, TN; thulite also occurs at Remigny, QUE, Canada; Souland and Lexviken, Norway (thulite); Vipiteno and Passiria, Italy; Gefrees, Germany; Zermatt, Switzerland; Rauris and Saualpe, Austria; Galicia, northwestern Spain; Tytyri, Lohja, Finland; Borzovka, Ural Mts., Russia (thulite); Matabatu Mts., Tanzania; gem-quality vanadian zoisite (*tanzanite*) occurs in the Merelani Hills, Tanzania; green, sharp, gemmy crystals to 6 cm from Alchuri village, between Shigar and Dassu, Baltistan, Pakistan. Also occurs (α-zoisite) near Turmik as colorless to gray crystals to several centimeters in length. Nagatoro, Saitawa Pref., Japan. Thulite is used as an ornamental stone, and tanzanite is used as a gemstone. EF *DHZ 1B:4(1986), MIN 3(1):709(1972), AJS 264:364(1966)*.

PUMPELLYITE GROUP

The general formula for the pumpellyite minerals can be given as

$$W_8X_4Y_8Z_{12}O_{56-n}(OH)_n$$

where

W = Ca, K, Na
X = Mg, Mn^{2+}, Fe^{2+}, Al, Cr
Y = Al, Fe^{3+}, Cr, Mn^{3+}, Ti
Z = Si

and the structure is monoclinic in space group $A2/m$.

The structure [*AM 70:1011(1985)*] is related to that of the epidote group in that it contains chains of edge-sharing octahedra running parallel to the [010] direction, and these are linked laterally by independent SiO_4 tetrahedra and Si_2O_6 double tetrahedra. The octahedral chains are of two kinds not related by symmetry, one containing Al only and the other containing 50% Al, 35% Mg, and 15% Fe. There are four of the former and two of the latter in the unit cell. There are two independent Ca atoms, each in sevenfold coordination. *DHZ 1B:1996, CM 12:219(1973), LIT 4:93(1971), MM:113(1953)*.

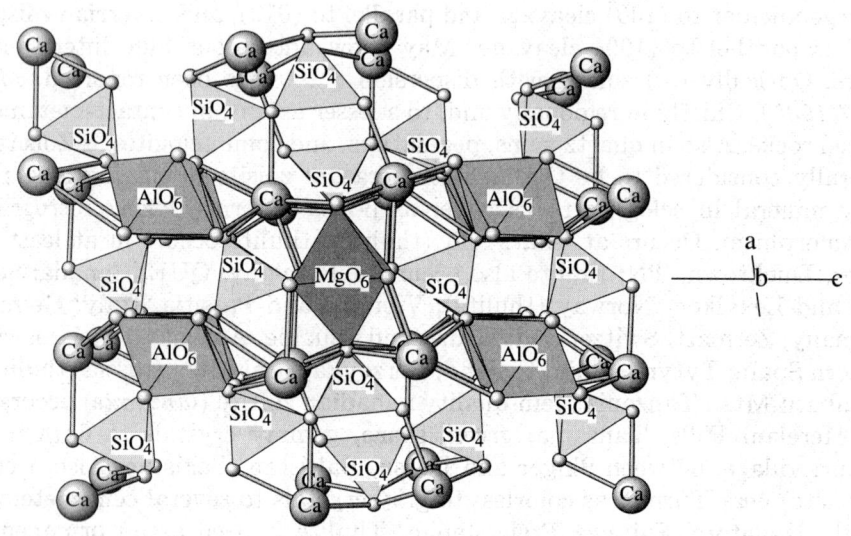

PUMPELLYITE-(Mg)

PUMPELLYITE GROUP

Mineral	Formula	Space Group	a	b	c	β	
Julgoldite-(Fe^{2+})	$Ca_2Fe^{2+}(Fe^{3+},Al)_2(SiO_4)$ $(Si_2O_7)(OH)_2 \cdot H_2O$	$A2/m$	8.949	6.047	19.426	97.58	58.2.2.1
Julgoldite-(Fe^{3+})*	$Ca_2Fe^{3+}(Fe,Al)_2^{3+}(SiO_4)$ $(Si_2O_7)(O,OH)_2 \cdot H_2O$	$A2/m$	8.92	6.047	19.37	97.38	58.2.2.2
Okhotskite-(Mg)*	$Ca_2(Mg,Mn^{2+})$ $(Mn^{3+},Al,Fe^{3+})_2(SiO_4)$ $(Si_2O_7)(OH)_2 \cdot H_2O$	$A2/m$	8.887	6.000	19.19	97.08	58.2.2.3
Okhotskite-(Mn^{2+})*	$Ca_2(Mn^{2+},Mg)$ $(Mn^{3+},Al,Fe^{3+})_2(SiO_4)$ $(Si_2O_7)(OH)_2 \cdot H_2O$	$A2/m$	8.887	6.000	19.19	97.08	58.2.2.4
Pumpellyite-(Fe^{2+})	$Ca_2Fe^{2+}(Al,Fe^{3+})_2(SiO_4)$ $(Si_2O_7)(OH)_2 \cdot H_2O$	$A2/m$	8.49	5.99	19.40	100.4	58.2.2.5
Pumpellyite-(Fe^{3+})	$Ca_2Fe^{3+}Al_2(SiO_4)(Si_2O_7)$ $(OH,O)_2 \cdot H_2O$	$A2/m$	8.825	5.945	19.131	97.45	58.2.2.6
Pumpellyite-(Mg)	$Ca_2MgAl_2(SiO_4)(Si_2O_7)$ $(OH)_2 \cdot H_2O$	$A2/m$	8.812	5.895	19.116	97.41	58.2.2.7

58.2.2.1 Julgoldite-(Fe^{2+})

PUMPELLYITE GROUP (CONT'D)

Mineral	Formula	Space Group	a	b	c	β	
Pumpellyite-(Mn^{2+})	$Ca_2(Mn^{2+}, Mg)$ $(Al, Mn^{3+}, Fe)_2(SiO_4)$ $(Si_2O_7)(OH)_2 \cdot H_2O$	$A2/m$	8.923	5.995	19.156	97.13	58.2.2.8
Shuiskite-(Mg)	$Ca_2(Mg, Al)(Cr, Al)_2(SiO_4)$ $(Si_2O_7)(OH)_2 \cdot 2H_2O$	$A2/m$	8.897	5.843	19.41	98	58.2.2.9

*Name not CNMMN IMA approved.
Notes:
1. The nomenclature for the group has been defined in two ways: see Moore (1971) and Passaglia and Gottardi (1973) in references above. The latter system is adopted here.
2. See also the related minerals sursassite (58.2.3.1) and macfallite (58.2.3.2).

58.2.2.1 Julgoldite-(Fe^{2+}) $Ca_2Fe^{2+}(Fe^{3+},Al)_2(SiO_4)(Si_2O_7)(OH)_2 \cdot H_2O$

Named in 1971 by Moore for Julian Royce Goldsmith (b. 1918), American geochemist, University of Chicago. Pumpellyite group. Forms a series with pumpellyite-(Fe^{2+}) and pumpellyite-(Mg). See other members of the group. The material from Långban is actually julgoldite-(Fe^{3+}) if the nomenclature of Passaglia and Gottardi (1973) is used. A structure determination [*MM* 39:271(1973)] showed the mineral to be julgoldite-(Fe^{3+}). MON $C2/m$. (Nat): $a = 19.426$, $b = 6.047$, $c = 8.949$, $\beta = 97.58°$, $Z = 4$ [*NGT* 64:251(1984)]; 23-117(calc pattern): $a = 8.922$, $b = 6.081$, $c = 19.432$, $\beta = 97.60°$, $Z = 4$, $D = 3.55$. 24-198(nat): 4.80_7 3.84_8 2.95_{10} 2.78_6 2.57_7 2.38_5 1.62_6 1.52_6. 23-117 and 24-198 both actually give data for julgoldite-(Fe^{3+}). $R = 4.3\%$, pumpellyite type. *MM* 39:271(1973). Flat, prismatic to bladed, up to more than 10 mm long; elongated on b or c, and flattened parallel to {100}; twinning on {001} common, repeated; dominant forms are {001}, {100}, {111}. Black, greenish-olive streak with bluish tinge, almost submetallic luster. Cleavage {100}, {001}, good to perfect. $H = 4\frac{1}{2}$. $G = 3.58$–3.60. IR data: *MM* 40:761(1976). Insoluble in cold 1:1 HCl but rapidly soluble in hot 1:1 HCl, leaving a gel and a crystalline precipitate. Analysis (wt %): SiO_2, 33.1; TiO_2, 0.0; Al_2O_3, 5.9; Fe_2O_3, 21.5; FeO, 12.6; MnO, 0.2; MgO, 0.5; CaO, 20.7; Na_2O, 0.25; H_2O^+, n.d. *NGT* 64:251(1984). Optical properties for pumpellyite-(Fe^{2+}) are apparently not available. Occurs in altered or metamorphosed basalts and diabases associated with prehnite and laumontite. Occurs in fissures in Precambrian granitic gneiss (Norway). Also occurs in quartz dolerite (Scotland). Occurs in the Reydarfjordur drill hole, eastern Iceland; the Ratho and Auchinstarry quarries, near Edinburgh, Scotland; Siljan area, central Sweden; Tafjord, Sunmøre district, western Norway; Khondivili quarry, Bombay, India. EF *NJBM* 8:367(1989), *EJM* 2:875(1990).

58.2.2.2 Julgoldite-(Fe^{3+}) $Ca_2Fe^{3+}(Fe,Al)_2^{3+}(SiO_4)(Si_2O_7)$ $(O,OH)_2 \cdot H_2O$

Named in 1971 by Moore as oxyjulgoldite and redefined in 1973 by Passaglia and Gottardi as julgoldite-(Fe^{3+}). Pumpellyite group. This name is not currently approved by the CNMMN IMA. See pumpellyite-(Mg) (58.2.2.7) for details about the name. MON $A2/m$. $a = 8.92$–8.949, $b = 6.047$–6.09, $c = 19.37$–19.426, $\beta = 97.38$–$97.5°$, $Z = 4$, $D = 3.56$–3.60. MM 39:271(1973): 4.80_7 3.84_8 2.95_{10} 2.78_6 2.57_7 2.385_5 $1.62_{5.5}$ $1.52_{5.5}$. Structure: MM 39:271(1973). Flat, prismatic, bladed crystals with dominant forms {100}, {111}, also {001}, {$\bar{1}$02}, {$\bar{1}$01}, and {$\bar{1}$11}, commonly striated parallel to b, elongated parallel to b, flattened parallel to a; usually twinned on c, {001} = TP. Deep lustrous black, greenish olive with bluish tinge; almost submetallic luster. Cleavage {100}, good. H = 4.5. G = 3.60. Insoluble in cold 1:1 HCl, rapidly dissolved by hot 1:1 HCl, producing a gel. Analysis (wt %): SiO_2, 34.0; Al_2O_3, 1.3; Fe_2O_3, 29.6; TiO_2, 0.1; CaO, 22.0; MgO, 0.2; FeO, 8.7; MnO, 0.2; H_2O, 4.69. LIT 4:93(1971). Also SiO_2, 32.4; Al_2O_3, 1.0; total Fe as Fe_2O_3, 40.6; CaO, 19.3. MM 39:271(1973). This material may actually be julgoldite-(Fe^{2+}), depending on the exact ferrous/ferric iron ratio, but structure solution has ferric iron predominant over ferrous iron. Biaxial (–); $N_x = 1.776$, $N_y = 1.814$; $2V = 50$–$70°$, $2V_{calc} = 73°$; $Z \gg Y > X$; strongly pleochroic. X, pale brown; Y, pale brownish green; Z, deep emerald-green. Occurs as cavity fillings and embedded in plates of apophyllite, in a fracture filling in hematite–magnetite ore from the Amerika stope at *Långban, Värmland, Sweden*. EF

58.2.2.3 Okhotskite-(Mg) $Ca_2(Mg,Mn^{2+})(Mn^{3+},Al,Fe^{3+})_2(SiO_4)$ $(Si_2O_7)(OH)_2 \cdot H_2O$

58.2.2.4 Okhotskite-(Mn^{2+}) $Ca_2(Mn^{2+},Mg)(Mn^{3+},Al,Fe^{3+})_2$ $(SiO_4)(Si_2O_7)(OH)_2 \cdot H_2O$

Named in 1987 by Togari and Akasaka for the Sea of Okhotsk, along which the mine is located. Pumpellyite group. The names *okhotskite-(Mg)* and *okhotskite-(Mn)* are not yet approved by the CNMMN IMA. Compare other members of the group. A complete solid-solution series exists between pumpellyite-(Mn) and okhotskite-(Mn). MON $A2/m$. $a = 8.887$–8.98, $b = 6.000$, $c = 19.19$–19.53, $\beta = 97.08$–$98.1°$, $Z = 4$, $D = 3.40$ and 3.58. 42-1387 [MM 51:611(1987)](nat): 4.76_6 3.87_7 2.96_{10} 2.72_7 $2.66_{4.5}$ $2.55_{4.5}$ $2.38_{4.5}$ $1.81_{4.5}$. Prismatic; to 0.2 mm long. Deep orange, pale orange streak, vitreous luster, transparent. One set of cleavages. H = 6. Mossbauer spectra: MM 51:611(1987). Nonfluorescent. Biaxial (–); Y = b, Z∧c = 9–14° in acute β; $N_x = 1.782$, $N_y = 1.8201$, $N_z = 1.827$; $2V = 46°$; $X > Y > Z$, $X < Y < Z$ (Japan); parallel extinction; no dispersion. Distinctly pleochroic: X, light yellow–yellow; Y, light pink–deep orange; Z, rose pink–deep orange. Occurs in MnFe-oxide ores (metamorphosed). In veinlets cutting hematite ore in metamorphic rocks. Associated with pumpellyite and piemontite. Occurs at Gowari Wadhona, Madhya Pradesh, India; *Kokuriki mine, Hiyoshi, Tokoro district, eastern*

Hokkaido, Japan, associated with other Mn minerals; possibly at Ochiai mine, Japan. EF *BM 104:396(1981)*, *AM 76:241(1991)*.

58.2.2.5 Pumpellyite-(Fe^{2+})
Ca$_2$Fe^{2+}(Al, Fe^{3+})$_2$(SiO$_4$)(Si$_2$O$_7$)(OH)$_2$ · H$_2$O

Named by Moore in 1971 as ferropumpellyite and later redefined in 1973 by Passaglia and Gottardi. Pumpellyite group. Note that what was described as pumpellyite by Moore in 1971 is actually julgoldite-(Fe^{3+}) according to the nomenclature scheme of Passaglia and Gottardi. MON $A2/m$. $a = 8.49$, $b = 5.99$, $c = 19.40$, $\beta = 100.4°$, $Z = 4$, $D = 3.27$. *36-437*(nat): 4.76$_3$ 4.43$_3$ 3.79$_6$ 2.91$_{10}$ 2.75$_6$ 2.66$_3$ 2.47$_3$ 2.22$_4$; *10-477(Pump-Mg-Fe^{2+})*: 4.38$_2$ 3.79$_5$ 2.90$_{10}$ 2.74$_5$ 2.64$_3$ 2.45$_4$ 2.21$_3$ 1.60$_3$. Twinning sometimes on {001}, CP = TP, X ∧ TP(001) = 12°; other physical properties not given but by analogy should be similar to other members of the group. $G = 3.31$. Analysis (wt %): SiO$_2$, 34.83; TiO$_2$, 0.1; Al$_2$O$_3$, 10.10; Fe$_2$O$_3$, 18.05; FeO, 9.09; CaO, 20.5; MgO, 0.94; MnO, 0.02; Na$_2$O, 0.18; H$_2$O$^+$, 5.62 (Norilsk, Russia). This analysis actually calculates to a julgoldite-(Fe)$^{2+}$. Biaxial (+); N$_x$ = 1.682, N$_y$ = 1.685, N$_z$ = 1.694; 2V = 56°, normally r < v, but in some brownish varieties and those with high RI it is r > v; the optical properties of members of the pumpellyite group as well as pumpellyite *sensu strictu* vary systematically with Fe^{2+} + Fe^{3+} + Mn + Cr content. They are both (+) and (−), because two orientations exist [*MM 30:113(1953)*]: (1) OAP parallel to (010); β = 1.71–1.73; 2V$_x$ = 80–20°; r > v, strong; X, very pale bluish green; Y, pale to deep golden yellow. (2) OAP perpendicular to (010), Z parallel to fiber length; Z to 1.75; 2V$_x$ = 0–60°; strong dispersion; strongly pleochroic, same as above. In very iron-rich samples, the OAP may be perpendicular to (010) or parallel to (010); 2V$_x$ = 80–30°. Occurs in low-grade metamorphic rocks, granites, and volcanic rocks (spilites and keratophyres). Occurs at Calumet, Houghton, MI; Hicks Ranch, Sonoma Co., CA; Siljan area, central Sweden; Långban, near Filipstad, Värmland, Sweden (actually julgoldite); Ivakin Creek, Norilsk district, Siberia, Russia; Coloured Series, NE of Neyriz, Iran; Khondivilli Quarry, Bombay, India; East Taiwan ophiolite, Taiwan. EF *CM 12:219(1973)*; *AM 61:176(1976), 68:1250(1983)*; *NJBM:367(1989)*; *EJM 2:875(1990)*.

58.2.2.6 Pumpellyite-(Fe^{3+})
Ca$_2$Fe^{3+}Al$_2$(SiO$_4$)(Si$_2$O$_7$)(OH,O)$_2$ · H$_2$O

Named in 1973 by Passaglia and Gottardi. Pumpellyite group. MON $A2/m$. $a = 8.825$, $b = 5.945$, $c = 19.131$, $\beta = 97.45°$, $Z = 4$, $D = 3.35$. *39-1368*(nat): 4.68$_5$ 4.38$_7$ 3.79$_9$ 2.91$_{10}$ 2.90$_{10}$ 2.74$_7$ 2.52$_5$ 2.19$_7$. Physical properties not given but are presumably analogous to other members of the pumpellyite group. Analysis (wt %): SiO$_2$, 35.75; TiO$_2$, 0.12; Al$_2$O$_3$, 18.70; Fe$_2$O$_3$, 11.10; FeO, 2.80; MnO, 0.05; MgO, 3.12; CaO, 20.72; Na$_2$O, 0.44; H$_2$O$^+$, 6.63. Biaxial; N$_y$ = 1.713; for optics for Fe-rich varieties, see pumpellyite-(Fe^{2+})

(58.2.2.5). Occurs in metamorphic rocks. Occurs at *Långban, near Filipstad, Värmland, Sweden* (actually julgoldite); Bulla Stream, Val Gardena, Bolzano, Italy. EF *MM 30:113(1953), PM 41:273(1972), CM 12:219(1973)*.

58.2.2.7 Pumpellyite-(Mg) $Ca_2MgAl_2(SiO_4)(Si_2O_7)(OH)_2 \cdot H_2O$

Named in 1925 by Palache and Vassar for Raphael Pumpelly (1837–1923), a pioneer student of the Cu deposits of the Keweenaw Peninsula, Michigan, the locality from which it was first described. Pumpellyite group. MON $A2/m$. *AM 70:1011(1985)*: $a = 8.812$, $b = 5.895$, $c = 19.116$, $\beta = 97.41°$; *10-447*: $a = 8.81$, $b = 5.94$, $c = 19.14$, $\beta = 97.6°$; *25-156*: $a = 8.820$, $b = 5.904$, $c = 19.12$, $\beta = 97.4°$; $Z = 4$, $D = 3.17$–3.25. *10-447*[nat pumpellyite-(Mg-Fe^{2+})]: 4.38_2 3.79_5 2.90_{10} 2.74_5 2.64_3 2.45_4 2.21_3 1.60_3; *25-156*[pumpellyite-(Mg)]: $4.66_{2.5}$ 4.37_3 3.79_3 $2.92_{5.5}$ 2.90_{10} 2.73_3 2.51_3 $2.45_{2.5}$. Structure: *AC(B) 27:1871(1971)*, *AM 70:1011(1985)*. Needles to plates, fibrous, flattened on {001}; twinning polysynthetic, with {100} as the twin plane, also on {001} common. Blue-green, olive green, brown, colorless; streak usually gray-green, green; vitreous luster. Cleavage (001), perfect; {100}, less so; subconchoidal fracture. $H = 5\frac{1}{2}$–6. $G = 3.18$–3.2. DTA and TGA data: *DHZ 1B(1986)*. Insoluble in HCl. Analysis (wt %): Hicks Ranch, CA: Ca, 7.60; Mg, 1.60; Fe^{2+}, 0.77; Fe^{3+}, 0.18; Al, 10.80; Si, 11.20; H, 14.7; Calumet, MI: SiO$_2$, 37.18; Al$_2$O$_3$, 23.50; Fe$_2$O$_3$, 5.29; FeO, 2.09; MnO, 0.13; MgO, 3.18; CaO, 23.08; Na$_2$O, 0.19; H$_2$O$^+$, 6.28; H$_2$O$^-$, 0.06. Cr- and Ti-bearing pumpellyites occur in a metagabbro pebble (Miocene) on Chita Peninsula, Aichi Pref., Japan: Cr$_2$O$_3$ to 8 wt % and TiO$_2$ to 5.6 wt %. Inferred to be a pseudomorph after Cr-rich spinel; occurs with chlorite, non-Cr-bearing pumpellyite, albite, and rutile. Vanadium-bearing pumpellyite-(Mg) with 1.7 to 14 wt % V$_2$O$_3$ occurs in the Hemlo gold deposit, ONT, Canada. As much as 4.8 wt % As$_2$O$_5$ and 1.0% F are present. As substitutes for Si. V goes into pumpellyite-group minerals by Al $\leftrightarrow$ V, with minor contribution from $2Al^{3+} \leftrightarrow V^{4+} + (Mg, Fe)^{2+}$ or $3Al^{3+} \leftrightarrow V^{5+} + 2(Mg, Fe)^{2+}$. Biaxial (+); $N_x = 1.670$–1.698, $N_y = 1.672$–1.706, $N_z = 1.684$–1.720; 2V usually ≈ 10–$50°$ (Calumet has $2V = 80°$, and Hicks Ranch has $2V = 41°$); strongly pleochoric for vanadoan (1.7–14% V$_2$O$_3$). X, colorless to weak red; Y, brown or deep brown with a violet tint; Z, pale brown. *CM 30:153(1992)*. Occurs in metamorphic rocks, of hydrothermal origin. Occurs at *Calumet, Houghton Co., Keweenaw Penin., MI*; Hicks Ranch, Sonoma Co., CA; V-rich in mica schist from the main ore zone of the Hemlo gold deposit, ONT, Canada; Tiso, Bolzano, Italy; Königskopf, Germany; Jordanov, Poland; Lotru, Romania; Blagodat, Ural Mts., Russia; Luohnsha ophiolite suite, Yalu Tsampo River, southern Tibet; Witwatersrand, South Africa; Toba, Sinsen-mura, Japan; Sada mine, Takasino, Saitama Pref., Japan; Watiba, Tukikawa, Saitama Pref., Japan; Cr- and Ti-bearing from Toyoura area, Chita Penin., Minamichita-cho, Japan; Lake Wakatipu, New Zealand. EF *CM 12:219(1973), MJJ 13:90(1986)*.

58.2.2.8 Pumpellyite-(Mn^{2+})   Ca$_2$(Mn^{2+},Mg)(Al,Mn^{3+},Fe)$_2$(SiO$_4$)(Si$_2$O$_7$)(OH)$_2 \cdot$ H$_2$O

Named in 1981 by Kato et al., see pumpellyite-(Mg) (58.2.2.7) for origin of name. Pumpellyite group. See other members of the group. MON $A2/m$. $a = 8.923$, $b = 5.995$, $c = 19.156$, $\beta = 97.13°$, $Z = 4$, $D = 3.34$. $35\text{-}0314$(nat): 4.75$_{6.5}$ 4.43$_{3.5}$ 3.84$_{6.5}$ 2.93$_{10}$ 2.725$_9$ 2.65$_{5.5}$ 2.53$_5$ 2.20$_{4.5}$. Platelets elongated on b, tabular to {001} or granular, crystal size to 0.1 mm. Light grayish, brownish pink; light grayish-pink streak; vitreous luster. Cleavage {001}, perfect. H = 5. Analysis (wt %): SiO$_2$, 35.66; TiO$_2$, 0.02; Al$_2$O$_3$, 13.40; Fe$_2$O, 2.43; Mn$_2$O$_3$, 7.74; MnO, 13.41; MgO, 0.89; CaO, 20.69; H$_2$O$^+$, 5.75 (by difference). Biaxial (−); N$_x$ = 1.752, N$_y$ = 1.795, N$_z$ = 1.800; 2V = 40–50°; parallel extinction; positive elongation; B×a nearly perpendicular to {001}; distinctly pleochroic. X, pale pink; Y,Z, brownish pink. Occurs in Mn-oxide ores of Precambrian pelitic and dolomitic rocks (Sausar Group, India). In braunite ore, with caryopilite, quartz, rhodochrosite, and late calcite in Japan. Occurs in eastern Liguria, Italy; *Gowari Wadhona, Madhya Pradesh, India; Ochiai mine, Barazawa, Kohsai-cho, Yamanashi Pref., Japan.*
EF *BM 104:396(1981), NJBM 12:563(1983), AM 76:241(1991).*

58.2.2.9 Shuiskite-(Mg)   Ca$_2$(Mg,Al)(Cr,Al)$_2$(SiO$_4$)(Si$_2$O$_7$)(OH)$_2 \cdot$ 2H$_2$O

Named in 1981 by Ivanov et al. for V. P. Shuisk, lithologist at the Ural Scientific Center, Yekaterinburg, Russia. Pumpellyite group. MON $A2/m$. $a = 8.897$, $b = 5.843$, $c = 19.41$, $\beta = 98°$, $Z = 4$; (H$_2$O poor): $a = 8.780$, $b = 5.963$, $c = 19.01$, $\beta = 98.07°$, $Z = 4$, $D = 3.63$. $39\text{-}1369$(nat, H$_2$O poor): 4.69$_6$ 2.88$_{10}$ 2.83$_{10}$ 2.37$_6$ 2.34$_7$ 2.09$_8$ 1.92$_6$ 0.96$_6$; (with H$_2$O): 2.90$_9$ 2.73$_7$ 2.64$_5$ 2.52$_5$ 2.46$_5$ 2.22$_4$ 1.59$_{10}$ 1.49$_8$. Prismatic radiating fibrous aggregates; at Saranov, 7 mm × 0.5 mm × 0.1 mm; twinning on (100). Dark brown with a violet tint, dark violet; light greenish-brown or grayish-green streak; vitreous luster. Cleavage {001}, perfect, or not observed; irregular and conchoidal fracture. H = 6. G = 3.24–3.32. IR data: *ZVMO 114:49(1985)*. DTA, TGA data: *ZVMO 110:508(1981), 114:49(1985)*. 26.5% Cr$_2$O$_3$ in Roche Noir sample, 28.6% Cr$_2$O$_3$ in Saranov sample. Some samples are H$_2$O poor (7.03 to 3.77 wt %). Some vacancy exists in the W site. Biaxial (−); Z ∧ c = 0° in zone of maximum birefringence, Z ∧ c = 3.5–6° in zone perpendicular to X; (H$_2$O poor): N$_x$ = 1.755–1.772, N$_y$ = 1.775–1.795, N$_z$ = 1.781–1.810; (H$_2$O-bearing): 2V = 40–45°, N$_x$ = 1.725–1.733, N$_y$ = 1.762–1.772, N$_z$ = 1.769–1.775; positive elongation; r < v, strong; strongly pleochroic. X, violet-blue; Y, yellowish green; Z, dark violet to dark violet-rose. Occurs on fractures in chromitite. Associated with uvarovite, chromian–chlorite, and chromian–titanite. In high-pressure metamorphosed gabbro in France. At Saranovsk, in calcite veins. *Bisersk deposit, Laki Station, Gorozavod region, Perm Prov.,* and Saranov chromite deposit, Ural Mts. (lower water content), *Russia;* Roche Noir Massif, French Alps, France [could be called shuiskite-(Cr)]. EF

58.2.3.1 Sursassite $Mn_2Al_3(SiO_4)(Si_2O_7)(OH)_3$

Named in 1926 by Jakob from *Sursass*, the name for Oberhalbstein, Switzerland. Possibly of the ardennite group. MON $P2_1/m$. $a = 8.704$–8.885, $b = 5.795$–5.827, $c = 9.787$–9.827, $\beta = 108.65$–$108.97°$, $Z = 2$, $D = 3.19$–3.57. *18-1286*(nat): 4.56_5 3.72_5 2.88_7 2.82_{10} 2.66_6 2.57_7 2.44_5 2.14_6. Also *37-479*. HRTEM studies and structure refinement show the presence of stacking disorder and intergrowth with pumpellyite. *PCM 10:99(1984)*. Crystals are fibrous and occur in masses, elongated on b. Reddish brown to copper-colored, cinnamon-brown streak, silky to somewhat dull luster. Cleavage {101}, distinct. $G = 3.26$–3.44. Contains minor amounts of Ca, Mg, Fe. Biaxial $(-)$; $Y = b$, $X \wedge c = 55°$; $N_x = 1.720$–1.736, $N_y = 1.737$–1.755, $N_z = 1.745$–1.766; $2V = 65°$; $r < v$, strong; pleochroic. X,Z, colorless to pale yellow; Y, golden to reddish brown. May show abnormal interference colors. Occurs in a glaucophane schist in George F. Canyon in the northwestern Palos Verdes Hills, CA; occurs in veinlets cutting Fe–Mn ores at the Strategic manganese mine, Plymouth, Woodstock, NB, Canada, and in cracks in radiolarian cherts of the Mn deposits of the *Val d'Err, Grisons, Graubunden, Switzerland*, where it occurs in Mn ores with tinzenite, parsettensite, rhodonite, quartz, albite, barite, and manganocalcite. Also occurs in the northern Apennines, Italy. EF *MIN 3(1):768(1972), RSIMP 35:151(1979), FM 62:3(1984)*.

58.2.3.2 Macfallite $Ca_2(Mn^{3+},Al)_3(SiO_4)(Si_2O_7)(OH)_3$

Named in 1979 for Russell P. MacFall, a dedicated amateur mineralogist. See pumpellyite (58.2.2.5 to 58.2.2.8) and sursassite (58.2.3.1). MON $P2_1/m$. $a = 10.235$, $b = 6.086$, $c = 8.970$, $\beta = 110.75°$, $Z = 2$, $D = 3.51$. *35-471*(nat): 4.76_4 4.45_3 3.89_7 3.40_{10} 2.96_5 2.71_5 2.44_3 2.19_3; *42-601*(calc pattern): 4.79_8 3.92_9 2.98_{10} 2.71_5 2.68_3 2.57_7 2.45_3 2.40_5. Structure solution gave $R = 18\%$ because structural disorder (domains, intergrowths) and/or solid solution probably affect the structure of this mineral as well as orientite and ruizite, and true single crystals of these and related compounds are encountered very infrequently. *AM 70:171(1985)*. Acicular to elongated crystals, often in aggregates; twinning by reflection on {100} is universal. Brown–reddish brown, maroon; brown streak with reddish tint; silky to subadamantine luster. Cleavage (001), perfect. $H > 5$. $G = 3.43$. Analysis (wt %) (Manganese Lake material): CuO, 1.2–1.9; Al_2O_3, 3.95–8.45; V_2O_5, 0.0–2.2; SO_3, 0.4–0.6. Substitutions: Cu^{2+} for Mn^{3+}, Al for Mn^{3+}, V^{5+} for Si^{4+}, S for Si^{4+}. MgO, 0.73 wt %, Al_2O_3, 1.03 wt % in Italian material. Biaxial $(-)$; $Z = b$; $N_x = 1.773$–1.775, $N_y = 1.795$, $N_z = 1.810$–1.815; $2V = 86°$, birefringence $= 0.035$–0.042. X, colorless to yellow; Y, light brown; Z, brownish red to dark brown. Occurs at *Manganese Lake, Copper Harbor, Keweenaw Co., MI*, in Keweenaw basalt, associated with manganite, braunite, orientite, and pyrolusite. The assemblage replaces calcite in fissures and lenses in the basalt. Occurs at the Cerchiara mine, Faggiona, eastern Liguria, Italy, in metacherts of an ophiolite sequence, interbedded layers of braunite and Mn-silicates, intergrown with orientite. EF *MM 43:325(1979), NJBM:455(1989)*.

58.2.4.1 Vesuvianite $(Ca,Na,REE,Pb,Sb^{3+})_{19}(Al,Mg,Fe^{3+},Fe^{2+},Ti,Mn,Cu,Zn)_{13}(SiO_4)_{10}(Si_2O_7)_4(B,\square)_{0.5}O_{68}(OH,F,O)_{10}$

Described in 1795 by Werner and named for the locality. Synonym: idocrase. TET (High vesuvianite): $P4/nnc$. $a = 15.51$–15.80, $c = 11.70$–11.92, $Z = 2$, $D = 3.35$. REE-bearing (17 wt % REE_2O_3) vesuvianite from San Benito Co., CA, has the largest cell dimensions reported for natural vesuvianites. High-precision cell dimensions for low vesuvianite from Lowell (Eden Mills), VT, are: $a = 15.5498$, $b = 15.5594$, $c = 11.8276$. TET $P4/n$, $P4nc$, $P\bar{4}$, and possibly as low as $P2/n$ (monoclinic). *GSAPA 27(6):A-363(1995)*. For some vesuvianites, the sum of Y cations is >13.0 (mean sum = 13.57), $P4/nnc$, an additional site (T1) is tetrahedrally coordinated by four oxygen atoms. This site is occupied by (B,Al,Fe), which replace two H atoms at adjacent H1 positions. The substitution is accompanied by the incorporation of a vacancy at an adjacent X3 site. For B-bearing samples, $a = 15.70$, $c = 11.75$. F-bearing samples are $P4/nnc$. *38-473*(Asbestos, QUE): 3.04_1 2.95_4 2.76_1 2.75_{10} 2.60_4 2.59_6 2.45_4 1.76_1; *38-474*(ferrian): 3.49_1 3.01_1 2.96_4 2.75_{10} 2.61_4 2.60_6 2.46_4 2.13_1; also *25-145*(calc pattern). High-symmetry vesuvianite has a long-range disordered structure with $P4/nnc$ average symmetry. Low symmetry vesuvianite has $P2/n$ or Pn space group symmetry, based on optical and X-ray diffraction data. Vesuvianite probably undergoes a ferroelastic phase transition at high temperature (>500°) from a $P4/nnc$ structure to a $P2/n$ or Pn structure. Crystals containing equal volumes of oppositely ordered domains exhibit $P4/nnc$ pseudosymmetry (long-range order). SHA (second harmonic analysis) shows high vesuvianite to be centric and low vesuvianite to be acentric. Some vesuvianites exhibit weak to intense reflections, violating the various glide-plane extinctions. In addition, some vesuvianites are biaxial, indicating that they are not even tetragonal. Violations exist because of ordering of partially occupied (Y1 and X4) sites. Because of diffuseness of the violating reflections, there is domain structure. TEM studies show that there is a twin domain structure in at least some low-symmetry vesuvianites. SAED patterns are consistent with $P4/n$. Domain structure is formed by transformation twinning resulting from ordering of the disordered high-temperature structure with space group $P4/nnc$. Domains are 10–50 nm wide; some exceed 1 μm. Domain structure is often lamellar parallel to {001}. Thus the true symmetry is not tetragonal; it is probably $P2/n$ or Pn. Ordering of channel cations over nonequivalent pairs of Y1 and X4 sites cannot be the order parameter that drives the $P4/nnc \rightarrow P2/n$ transition, as it does not transform as the active irreducible representation of this transition. 010 and 011 are occupied by monovalent anions (i.e., OH, F, Cl). B-free vesuvianite has $10(H + F)$ with $F \geq 1.0$ atoms per formula unit. For B-bearing vesuvianite, use 19 X cations to normalize to, and for B-free vesuvianite, use 50 total cations per formula unit. T1 and T2 sites (occupied by B) are empty. Occupancy of T1 and T2 sites precludes occupancy of two H1 and H2 positions. Fe and Al can also occur at the T positions; therefore B-bearing vesuvianites usually have "excess" Y-group cation sums. EPR studies indicate that Fe^{3+}

substitutes on the fivefold-coordinated square-pyramidal Y1 site. *NJBM:264(1995)*. **Habit:** Prismatic along [001]. Forms {001}, {010}, {110}, {111}, and others. **Physical properties:** Usually green or chartreuse, but may be white, pink, yellow, yellow-brown, yellow-green, brown, red, black, purple, or blue ("cyprine"); white to pale greenish brown streak; vitreous to resinous luster. Cleavage {110}, poor; {100}, {001}, very poor; subconchoidal to uneven fracture; brittle. H = 5–6$\frac{1}{2}$. G = 3.32–3.43. **Chemistry:** Insoluble in common acids or partly soluble in HCl; soluble in HF. IR spectroscopy in the OH region. *CM 33:609(1995)*. Tetrahedral Z site has only Si in it, no Al substitution. X site is eight- or nine-coordinated, dominantly Ca. For B-bearing vesuvianite, B + Mg ↔ 2H + Al. B-bearing vesuvianites have much less OH (2–5 atoms per formula unit) than B-free (8.5–10 atoms per formula unit). Fe to as much as 7.4 wt % FeO, MgO to 4.75 wt %, TiO_2 to 6.7 wt %, B_2O_3 to about 4.0 wt %, BeO to 1.1 wt %, F to 3.15 wt %. Unusual Cu-bearing (to 2.2 wt % CuO) and Pb-bearing (to 2.1 wt % PbO) vesuvianite from Franklin, NJ (cyprine). Unusual bright red vesuvianite from Långban, Sweden, had 2.5 wt % Bi_2O_3 and as much as 3.8 wt % MnO. May have as much as 0.6 wt % Cr_2O_3. ZnO to 4.5 wt % at Franklin, NJ. May also contain S (to 1.0 wt % SO_3) from Lake Jaco, Mexico. Cl to 1.2 wt %. B-bearing vesuvianite occurs at Templeton Twp., QUE, Canada; Ariccia, Italy; Lake Jaco, CHIH, Mexico; Tulare Co., CA; Achtaragda R., Vilui River area, Yakutia, Russia. ThO_2 to 1.5 wt %. About 17 wt % REE_2O_3 in vesuvianite from San Benito Co., CA. An unusual Sb-bearing vesuvianite (15.5 wt % Sb_2O_3) has been reported. Low-symmetry vesuvianite has very low F and Cl contents, and high-symmetry vesuvianites have considerable F and Cl. Uniaxial (−) and (+), also anomalously biaxial (−). All B-bearing vesuvianites are (+) and all B-free samples are (−). N_O = 1.703–1.752, N_E = 1.700–1.746. An Fe- (10.9 wt % FeO + Fe_2O_3) and high-REE-containing vesuvianite (16.7 wt % REE_2O_3) has N_O = 1.762 and N_E = 1.750. The unusual Sb-bearing (15.7 wt % Sb_2O_3) vesuvianite from Sarawak, Malaysia has N_O = 1.795 and N_E = 1.775. **Optics:** Weakly dichroic, usually brownish yellow to yellowish brown, but may be green, brown, or blue. Strong dispersion. Metamict material may have N as low as 1.66. Characteristically shows abnormal interference colors in blue and brown. Three groups of vesuvianite are recognized on the basis of optical properties: (1) normal crystals showing uniform extinction and small (0–5°) 2V; (2) blocky crystals showing irregularly shaped areas of variable birefringence in a (001) section with 2V values in the range 5–35°; and (3) sector-zoned crystals showing {001}, {101}, and {100} sectors with low (about 5°), intermediate (20–35°), and high (40–60°) values of 2V, respectively. The different optical types of vesuvianite arise from different relationships between the temperature range of crystallization and the temperature of the ferroelastic phase transition. **Occurrence:** A rock-forming or accessory mineral in skarns, rodingites, and altered alkali syenites. Low vesuvianite occurs in metamorphic rocks formed at < 300° compared to those containing high vesuvianite (400–800°). Ordering to low vesuvianite presumably takes place during crystal growth rather than by ordering transformation on cooling. High vesuvianite is in skarns and calc-silicate hornfels and calc-schists. Low vesuvianite is found in veins, rodingites, and altered alkali syenites, and metamorphosed limestones. May be metamict or partly so. **Localities:** Localities for metamict material include Seward Peninsula, AK, to 1.5 cm, in nepheline syenite. Also in a skarn near Lake George, Park Co., CO; to 2-cm crystals. An alkaline pegmatite in the Tuva district, Russia, also contains metamict material. A reasonably abundant mineral that occurs worldwide; only notable localities are mentioned here. Crystals to 7 cm occur at Sanford(!), York Co., ME. Also at Auburn, Androscoggin Co., ME. Olmsteadville, NY(!); at Lowell (Eden Mills)(!), VT, associated with grossular. Be-bearing vesuvianite and "cyprine" at the Parker shaft, Franklin, Sussex Co., NJ. Rockville, Montgomery Co., MD; Magnet Cove, Garland Co., AR; Italian Mountain, Gunnison Co., CO; Crestmore quarry, Riverside Co., CA; San Benito Co., CA (REE-bearing); massive material called "californite" from localities in Siskiyou, Butte, and Fresno Cos., CA. The best "californite" is from Pulga, Butte Co. Fine vesuvianite (some of lilac-purple color) at the Jeffrey mine, Asbestos; gem material at Laurel, Argenteuil Co.; large, brownish-yellow crystals at Calumet Falls, Litchfield, Pontiac Co., and from Templeton, Ottawa Co.,; as masses and minute crystals of a bright pink color at the Montreal chrome pit, Black Lake, Colerain Twp., Megantic Co., QUE, Canada. Blue, Cu-bearing (1.1% CuO) vesuvianite at Sauland, Telemark, and Eger(!), Norway; *Monte Somma, Vesuvius, Italy.* Sharp green-brown "toenails," blocky and striated crystals to 3 cm at Mt. Banchetti, Bellecombe, Aosta Valley, Italy. Transparent green or brown brilliant crystals at Mussa Alp, Ala Valley, Piedmont, Italy. Vilui River region (viluite), Yakutia, Russia; Hindubagh and Muslimbagh, Pakistan. **Uses:** Vesuvianite has minor use as a lapidary material and as a gemstone. EF *MIN 3(1):597(1972); DHZ 1A:699(1982); AM 76:397(1991); MP 46:163(1992); CM 30:1(1992), 30:19(1992), 30:1065(1992), 31:357(1993), 31:617(1993), 32:497(1994), 32:505(1994); MJ 17:1(1994).*

58.2.5.1 Rustumite $Ca_{10}[Si_2O_7]_2[SiO_4](OH)_2Cl_2$

Named in 1965 by Agrell for Rustum Roy (b.1925), materials chemist, Pennsylvania State University. MON $C2/c$. $a = 7.62$, $b = 18.55$, $c = 15.51$, $\beta = 104.35°$, $Z = 4$, $D = 2.835$. *18-305:* 3.19_8 3.03_{10} 2.89_9 2.75_5 2.63_5 1.752_7 1.661_5 1.613_5. Crude tabular crystals to 2 mm in length; polysynthetic twinning on {100}. Colorless to white, transparent to translucent. Cleavage {100}, poor; {010}, {001}, very poor. $G = 2.86$. Biaxial $(-)$; $2V = 80°$; $N_x = 1.640$, $N_{y\,calc} = 1.645$, $N_z = 1.651$; $X \wedge c = 4$–$6°$, $Y = b$. Found at the La Negra mine, Maconi, QRO, Mexico; in the skarns of *Kilchoan, Ardnamurchan, Scotland*, with åkermanite, merwinite, larnite, spurrite, rankinite, and kilchoanite; Carlingford, Co. Louth, Ireland; in skarn near Güneyce–Ikizdere, Turkey. AR *MM 34:1(1965), AM 64:659(1979).*

58.2.6.1 Yoshimuraite $(Ba,Sr)_2Mn_2^{2+}(Ti,Fe^{3+})O(Si_2O_7)[(P,S)O_4]$ (OH)

Named by Watanabe in 1961 for Toyofumi Yoshimura (1905–1988), Japanese mineralogist and economic geologist, Kyushu University. TRIC $P\bar{1}$. $a = 5.386$, $b = 6.999$, $c = 14.748$, $\alpha = 95.50°$, $\beta = 93.62°$, $\gamma = 89.98°$, $Z = 2$, $D = 4.16$–4.21. 36-411(nat): 4.89_6 3.39_3 3.32_2 $3.23_{2.5}$ 3.12_3 2.93_{10} $2.77_{3.5}$ $2.68_{2.5}$. Isostructural with innelite. The structure is strongly layered parallel to {001}, consisting of a brucite layer of MnO_6 octahedra, a layer of TiO_5 tetragonal pyramids cross-linked by Si_2O_7 groups, and a layer of edge- and face-sharing PO_4 and BaO_{11} polyhedra. Linking of all three layers along {001} results in a mica-type structure. Si and P are well ordered and there is no sharing of tetrahedral elements between SiO_4 and PO_4. Blades or micalike forms to 5 cm long. Polysynthetic twinning on {010} common. Brown, orange-red, or dark brown; transparent to translucent. Cleavage {010}, perfect; {10$\bar{1}$}, {101}, distinct; brittle. $H = 4\frac{1}{2}$. $G = 4.13$–4.19. Analysis (wt %): SiO_2, 17.2–18.25; TiO_2, 7.47–10.0; Al_2O_3, 0.21; Fe_2O_3, 1.32–3.48; FeO, 1.47–3.16; MnO, 15.83–17.64; MgO, 0.31–0.56; CaO, 1.45; ZnO, 0.50; Na_2O, 0.1–0.16; K_2O, 0.01–0.03; BaO, 33.51–38.11; SrO, 3.03–4.62; P_2O_5, 3.98–4.62; SO_3, 3.84–5.40; Cl, 0.41; H_2O, 1.06–2.34. Biaxial (+); $N_x = 1.763$–1.768, $N_y = 1.777$, $N_z = 1.785$–1.794; $2V = 80$–$90°$; $X < Y \leq Z$; $r > v$; x' is nearly parallel to a; z' nearly parallel to c. X, bright yellow; Y, orange-brown; Z, brown. Occurs in coarse-grained pegmatite along the boundary between massive manganese ore and hornfels. Associated with barian K-feldspar, quartz, richterite and riebeckite, aegirine, and rhodonite. Exclusively from bedded manganese deposits (metamorphosed). Occurs exclusively at several localities in *Japan*: Taguchi mine, Shidara, Kitashidara-gun, Aichi Pref., Honshu; at the *Noda-Tamagawa mine, Tamagawamura, Iwate Pref.*; at the Hijikuzu and Tanohata mines, Iwate Pref. EF *MJJ 2:408(1959)*, *3:156(1961)*; *JMSJ 6:230(1963)*; *MSJSP 1:74(1971)*; *GAC-MACP 19:A73(1994)*.

58.2.6.2 Innelite $Na_2Ba_4CaTi_3O_4[Si_2O_7]_2(SO_4)$

Named in 1961 by Kravchenko et al. from the Yakut *Inneli*, the Inagli River. TRIC $P\bar{1}$. $a = 14.76$, $b = 7.14$, $c = 5.38$, $\alpha = 90°$, $\beta = 95°$, $\gamma = 99°$, $Z = 1$, $D = 4.08$. 15-71: 6.31_5 5.36_5 3.92_{10} 3.04_6 2.95_6 1.964_6 1.845_6 1.735_6. Structure: *SPC 16:65(1971)*. Tabular crystals up to several centimeters. Polysynthetic twinning. Pale yellow to brown, vitreous to oily luster. Cleavage {010}, {110}, {10$\bar{1}$}, perfect; {001}, less so; brittle. $H = 5$. $G = 3.96$. Biaxial (+); $2V = 82°$; $N_x = 1.726$, $N_y = 1.737$, $N_z = 1.766$; $Z = a$; $r > v$, strong. Pleochroic, $Z > Y = X$; pale yellow-brown to yellow. Anomalous blue interference colors. Nearly insoluble in acids. Found in pulaskites and aegirine–eckermannite–microcline pegmatites of the *Inagli massif, Aldon region*, and in shonkinite of the Yakutsk massif, *southern Yakutia, Russia*. AR *AM 47:805(1962)*.

58.2.7.1 Samfowlerite $Ca_{14}Mn_3Zn_2(Zn,Be)_2Be_6[SiO_4]_6[Si_2O_7]_4$ $(OH,F)_6$

Named in 1994 by Dunn et al. for Sam Fowler (1779–1844), a physician and pre-eminent figure in the history of the Franklin mining district. MON $P2_1/c$ $a = 9.068$, $b = 17.992$, $c = 14.586$, $\beta = 104.86°$, $Z = 2$, $D = 3.31$. PD 2.86_{10} 2.77_4 2.65_5 2.39_5 2.27_3 1.860_2 1.832_3 1.803_2. Structure: CM 32:43(1994). The SiO_4 and Si_2O_7 groups are isolated from one another but, together with tetrahedra occupied by beryllium and zinc, form layers roughly perpendicular to [100]. Colorless, white in aggregate, streak white, vitreous luster; cleavage not detected, but sample too small for verification of any cleavage. $H \leq 3$. $G = 3.28$. Biaxial (−); $2V = 29°$; $N_x = 1.674$, $N_y = 1.680$, $N_z = 1.681$, $Y = b$, $X \wedge a = 44°$, $Z \wedge c = 29°$. Fluoresces very weak red in both LWUV and SWUV. Substitution of Be for Zn in a tetrahedral site is in a ratio varying from 5/8 to 3/8, suggesting that the substitution may be ordered and that the true symmetry is lower. Found as minute (0.05 mm) crystals in an andradite lined vug in franklinite-willemite ore at *Franklin, Sussex Co., NJ*. AR CM 32:43(1994).

58.3.1.1 Ardennite $(Mn^{2+},Ca,Mg)_4[(Al,Fe^{3+})_5Mg](AsO_4)_2$ $(Si_3O_{10})(OH)_6$

Named in 1872 by Lasaulx and Bettendorf for the locality. Ardennite group. Polytypic family includes ardennite, pumpellyite, sursassite, and one hypothetical structure type: four maximum-degree-of-order polytypes. At least two polytypes: "normal" 18.5-Å c (e.g., Alps—medium-grade metamorphic rocks), and new 37-Å c (e.g., Greece—low-grade metamorphic rocks); has both 18- and 37-Å polytypes with stacking faults present. ORTH Pnmm. $a = 8.692$–8.730, $b = 5.811$–5.828, $c = 18.521$–18.560, $Z = 2$, $D = 3.67$–3.72. 18-141(nat): 4.21_6 3.76_5 3.15_6 2.91_7 2.87_6 2.57_{10} 1.45_6 2.245_5; 40-477(high-As): $4.28_{1.5}$ 2.94_{10} $2.61_{7.5}$ $2.32_{2.5}$ $2.26_{1.5}$ $2.12_{1.5}$ 2.02_2 1.54_3; 41-1391: 3.18_3 2.95_7 2.91_7 2.78_4 2.60_{10} 2.03_4 1.61_3 1.58_4 1.46_5. The structure consists of chains of edge-sharing (Al,Mg, Fe^{3+})-octahedra connected by orthosilicate and trisilicate groups. A slightly larger tetrahedron is occupied by $As^{5+}(V^{5+})$ cations. Mn^{2+}, Ca^{2+}, and Mg^{2+} occur in the cavities of the structure in six- and sevenfold coordination. A mixed-anion silicate. Order-disorder built up from two different kinds of layers occurring in regular alternation. Linear groups of three Si-O tetrahedra (Si_3O_{10}) rather than double tetrahedra (like epidote). As and V are in an independent $(As,V)O_4$ tetrahedron. AC(B) 24:845(1968), 27:1871(1971). Prismatic blades and fibers, elongate along b, {101} dominant, crystalline aggregates and small rosettes of tapered acicular radiating crystals; to 7 mm long (New Zealand). Yellow to yellowish brown, black, yellow-green, pale golden yellow, orange; subadamantine luster. Cleavage {010}, perfect; {110}, distinct; {100}, good; very brittle. $H = 6$–7. $G = 3.60$–3.63. Nearly insoluble in acids. As content is variable. In Greek material, most of the As is replaced by Si. Greek material has Mn^{3+} and Mn^{2+}. Some V replaces As. One sample from Salm Chateau had 4.84 wt %

V_2O_5 and 4.14 wt % As_2O_5. NZ material also has Mn^{2+} and Mn^{3+}; most is Mn^{2+}. Traces of B_2O_3 (0.23 wt %) in material from Otago. Sanbagawa: CaO to 4.6 wt %, V_2O_5 to 7.3 wt %. Highest As_2O_5 from Salm-Chateau is 13.25% MgO, to 5.26 wt % from Kajlidongri, India. Biaxial (+); Z = a, X or Y = b; $N_x = 1.734$–1.753, $N_y = 1.735$–1.753, $N_z = 1.751$–1.770; 2V usually small but ranges from 0 to 70°; r > v or r < v, strong; weakly to strongly pleochroic. X, brownish yellow; Y, golden yellow; Z, pale yellow (weak to strong). *MP 48:295(1993)* gives X parallel to b, Y parallel to a, and Z parallel to c. Vanadium-rich material from Sanbagawa, Japan has: $N_x = 1.798$ and $N_z = 1.822$. Occurs in low-pressure, low-temperature rocks; in oxidized Mn–Al-rich greenschist facies metasediments, shales, and quartz schists. Metamorphic pressure and temperature conditions span 300–400° and 2–14 kbar to 500–630° and 14–28 kbar. Associated minerals include quartz, spessartine, hematite, rutile, clinochlore, piemontite, and others. Also occurs in quartz veins. In Todilto limestone on a joint surface, with calcite and cuprosklowskite. Occurrences include: in Todilto limestone 20 miles NW of Grants, NM; Shepton Mallet, Somerset, England; best and most abundant material found at *Salm-Chateau*, also at Bihain, *Ardennes, Belgium*; Zermatt, Switzerland; Haute-Maurienne, French Alps, and Bonneval-sur-Arc, Savoie, France; Val d'Aosta, Ala Valley, Piedmont, Italy; Andros and Evvia Is., Greece; Kajlidongri mine, Jhabua district, Madhya Pradesh, India; Asemi-gawa area, Motoyama, Kochi Pref., and Ohnara, Sanbagawa, and Kamugawa, Gumma Pref., Shikoku, Japan; Arrow Junction, western Otago, New Zealand. EF *MIN 3(1):716(1972)*, *EJM 3:819(1991)*, *BNSM(C) 13:1(1987)*, *NJBA 158:89(1987)*.

58.3.1.2 Orientite $Ca_8Mn^{3+}_{10}[(SiO_4)_3(Si_3O_{10})_3(OH)_{10}] \cdot 4H_2O$

Named in 1921 by Hewett and Shannon for the locality. Ardennite group. ORTH *Bbmm* or *P2mm* [*AM 71:176(1986)*]. For *Bbmm*: $a = 9.074$, $b = 19.130$, $c = 6.121$, $Z = 4$, $D = 3.48$; for *P2mm*: $a = 9.044$, $b = 6.091$, $c = 19.031$. *NJBM:455(1989)*: $a = 6.094$, $b = 19.042$, $c = 9.026$. *41-589*(calc pattern): 9.56_4 4.78_4 4.10_9 3.97_{10} 2.68_4 2.61_5 2.58_8 2.45_4; *35-472*(nat): 9.45_{10} 4.74_{10} $4.51_{2.5}$ $4.39_{1.5}$ $4.07_{3.5}$ 3.27_3 $3.03_{4.5}$ $2.70_{1.5}$. See also *18-941*. Two structure refinements exist: *AM 70:171(1985)* and *AM 71:176(1986)*. Chains of edge-sharing Mn^{3+} octahedra run along [010] and are connected by (SiO_4) orthosilicate and (Si_3O_{10}) trisilicate groups. The mineral shows varying degrees of order. Some show $2c = 38$ Å, thus a new polytype. Thin to thick tabular parallel to {001} prisms and tabular grains. Dominant forms: {110}, {001}, {010}. Crystals to 1 mm in Italy; polysynthetic twinning. Deep red to brown, maroon; crystals are transparent reddish brown; brown streak; resinous–subvitreous luster. Cleavage or parting {001}, imperfect. $H = 4\frac{1}{2}$–5. $G = 3.33$. Soluble in hot HCl, with evolution of Cl and separation of residual insoluble silica; insoluble in HNO_3. Manganese Lake material has 0.4–1.5 wt % V_2O_5 and Al_2O_3 is absent. Some Al_2O_3 and Fe_2O_3 in Oriente Prov. material. CuO is as high as 2.6 wt %, SO_3 is as high as 1.2 wt %. V^{5+} substitutes for Si^{3+} in tetrahedral coordination; Cu^{2+} substitutes for Mn^{3+}, and S substitutes for Si^{4+}. Italian material has 0.89 wt % MgO, 0.4 wt % Al_2O_3. Biaxial $(-)$; XYZ = bca; N_x = 1.756–1.765, N_y = 1.777–1.791, N_z = 1.794–1.810; 2V = 68–87°; birefringence = 0.038–0.045; OAP parallel to (001) and transverse to elongation, B × A perpendicular to (100); r < v, very strong; pleochroic. X, colorless to pale yellow; Y, reddish brown or yellow-brown; Z, deep red-brown. *NJBM:455(1989)*. *AM 46:226(1961)* gives X, red-brown; Y, yellow; Z, yellow-brown. The mineral is of low-temperature origin and low-alumina conditions, occurring with jasper, psilomelane, manganite, and barite. Occurs with manganite, braunite, macfallite, and pyrolusite, which replace calcite in fissures and lenses in Keweenaw Basalt at Lake Manganese, Copper Harbor, Keweenaw Co., MI; within volcanic breccias, tuffs, andesite, and latite at *Costa Manuel and Vicente claims, 6 miles S of Bueycito and at the Santa Rosa prospect, near Banes, Oriente Prov., Cuba*. In Mn-bearing metacherts of an ophiolite sequence at the Cerchiara mine, eastern Liguria, Italy, where the orientite is intergrown with macfallite or as lamellar intergrowths. EF *AM 46:226(1961)*.

58.3.2.1 Kilchoanite $Ca_3[SiO_4][Si_3O_{10}]$

Named in 1961 by Agrell and Gay for the locality. Dimorphous with rankinite (55.3.1.1). ORTH $I2cm$. a = 11.42, b = 5.09, c = 21.95, Z = 8, D = 3.002. 29-370(syn): 3.56_7 3.07_8 2.89_{10} 2.68_9 2.55_5 2.49_5 2.36_7 1.964_7. Structure: *MM 38:26(1971)*. The structure contains insular $[SiO_4]$ and insular $[Si_3O_{10}]$ groups, in contrast with rankinite, which has only $[Si_2O_7]$ dimers. Colorless to white, vitreous luster. G = 2.99. Biaxial $(-)$; 2V = 60°; N_x = 1.647, $N_{y\,calc}$ = 1.649, N_z = 1.650; r > v. Anomalous interference colors. Found as a retrograde replacement of rankinite at *Kilchoan, Ardnamurchan, Scotland*; at Carlingford, County Louth, Ireland; Ḥatrurim, Israel; from a xenolith in the Ozereskii gabbro, Priol'khonye, western Baikal, Russia; at Fuka, Bicchu-cho, Okayama Pref., Japan; and at Takatoka, New Zealand. AR *NAT 189:743(1961)*, *DHZII 1B:272(1986)*.

Class 59

Three-Membered Rings

BENITOITE GROUP

Benitoite and Catapleiite structures conform to the formula

$$ABSi_3O_9 \cdot nH_2O \text{ or } A_2BSi_3O_9 \cdot nH_2O$$

where

A = Ba, K_2, Na_2, K, Na, Ca
B = Ti, Zr, Sn
n = 0, 2, or 3

and the structure is similar or identical to the hexagonal bazirite structure. The structure of the benitoite group minerals consists of octahedrally coordinated quadrivalent cations that link (Si_3O_9) rings into an octahedral–tetrahedral framework with tubular one-dimensional cavities that contain the A cations. *NJBM:16(1987), ZK 129:222(1969)*.

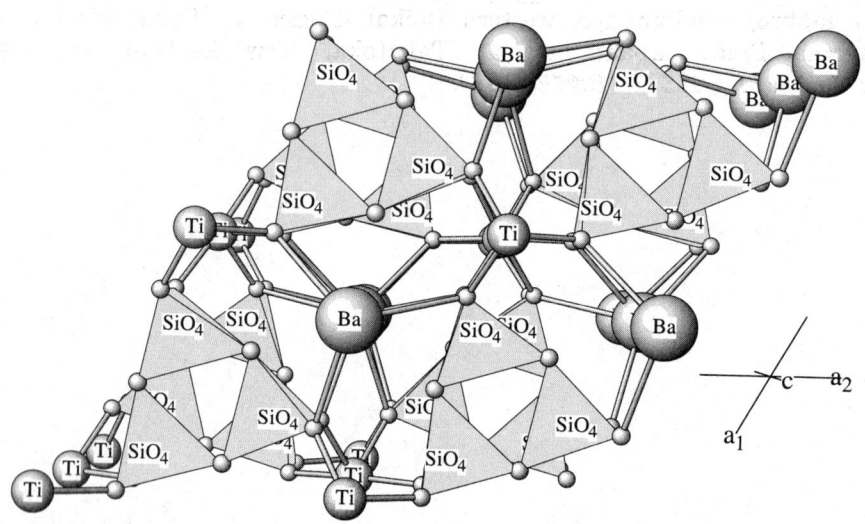

BENITOITE: $BaTiSi_3O_9$
Perspective view along c.

Benitoite Group

BENITOITE GROUP SENSU STRICTU: HEXAGONAL, $n = 0$

Mineral	Formula	Space Group	a	c	
Bazirite	$BaZrSi_3O_9$	$P\bar{6}c2$	6.75	9.98	59.1.1.1.
Benitoite	$BaTiSi_3O_9$	$P\bar{6}c2$	6.6409	9.759	59.1.1.2
Pabstite	$BaSnSi_3O_9$	$P\bar{6}c2$	6.703	9.824	59.1.1.3
Wadeite	$K_2ZrSi_3O_9$	$P\bar{6}$	6.93	10.18	59.1.1.4

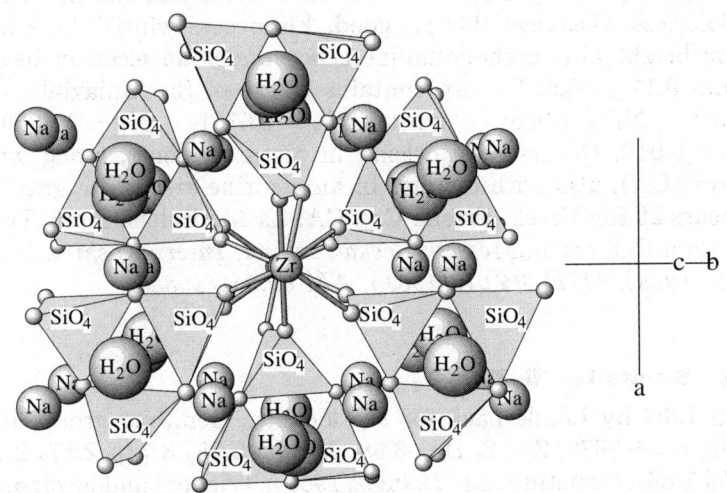

CATAPLEIITE: $Na_2ZrSi_3O_9 \cdot 2H_2O$
Triangular Si_3O_9 rings form sheets parallel to (001). These are interleaved with sandwiches of Zr/Na–H_2O.

CATAPLEIITE SUBGROUP:, $n > 0$

Mineral	Formula	Space Group	a	b	c	β	
Catapleiite	$(Na, Ca, \square)_2ZrSi_3O_9 \cdot 2H_2O$	$B2/b$	23.917	20.148	7.432	147.46	59.2.2.1
Calcium catapleiite	$CaZrSi_3O_9 \cdot 2H_2O$	unknown	7.32		10.15		59.2.2.2
Gaidonnayite	$(Na, K)_2ZrSi_3O_9 \cdot 2H_2O$	$P1_2nb$	11.78	12.82	6.92		59.2.2.3
Georgechaoite	$KNaZrSi_3O_9 \cdot 2H_2O$	$P2_1nb$	11.836	12.940	6.735		59.2.2.4
Hilairite	$Na_2ZrSi_3O_9 \cdot 3H_2O$	$R32$	10.556		15.855		59.2.3.1
Calciohilairite	$CaZrSi_3O_9 \cdot 3H_2O$	$R32$	20.87		16.00		59.2.3.2

59.1.1.1 Bazirite BaZrSi$_3$O$_9$

Named in 1978 by Young for the composition. Benitoite group. HEX $P\bar{6}c2$.
$a = 6.737$–6.769, $c = 9.997$–10.020, $Z = 2$, $D = 3.82$; (syn): $a = 6.755$, $c = 9.980$, $Z = 2$, $D = 3.85$. 29-214(syn): 5.82_4 3.79_{10} 3.38_4 $2.92_{3.5}$ 2.80_{10} $2.21_{2.5}$ $2.01_{3.5}$ $1.60_{2.5}$. The structure is the same as that of benitoite. Octahedrally coordinated quadrivalent cations link (Si$_3$O$_9$) rings into an octahedral–tetrahedral framework with tubular one-dimensional cavities that contain the Ba cations. *NJBM:16(1987)*. {0001} + {10$\bar{1}$0}, crystals generally anhedral to subhedral, basal plates, prismatic; to 0.25 mm across and 0.65 mm long at Big Creek. Colorless. Cleavage {0001}, good. Fluoresces whitish blue in SWUV and shows bright blue cathodoluminescence under an electron beam. TiO$_2$ varies from 0.17 to 0.51%, also contains traces of Sn. Uniaxial (+); (nat): $N_O = 1.679$, $N_E = 1.688$; (syn): $N_O = 1.6751$, $N_E = 1.6850$; birefringence = 0.010. Occurs with celsian in metasediments, along with other Ba-silicates (CA); also with elpidite in an aegirine–riebeckite granite (Scotland). Occurs at Big Creek, Fresno Co., CA; La Madrelena mine, Tres Pozos, northern Baja California, Mexico; *Rockall Island, Inverness-shire, Scotland*. EF *MM 42:35(1978)*, *MIN 3(2):18(1981)*, *AM 69:358(1984)*.

59.1.1.2 Benitoite BaTiSi$_3$O$_9$

Named in 1907 by Louderback for the locality. Benitoite group. HEX $P\bar{6}c2$.
$a = 6.6409$, $c = 9.7579$, $Z = 2$, $D = 3.68$. *38-464*: 5.75_2 3.72_{10} 2.87_2 2.74_8 2.17_2 1.98_2 1.97_2 1.86_2. Structure: *ZK 129:222(1969)*. Trigonal and hexagonal plates and platelets: {0001}, {0$\bar{1}$11}, {1011}, {10$\bar{1}$0}, {1012}, {10.1.9.10}. Rotation twins rare on [0001]. Blue to dark blue, also colorless and white (largest well-formed crystal is 5.6 cm across); rarely pink (to 2-mm crystals) or purplish blue; white streak; vitreous luster. Cleavage {10$\bar{1}$1}, imperfect; conchoidal to uneven fracture. $H = 6\frac{1}{2}$. $G = 3.64$. IR data: *TMPM 21:47(1974)*. Raman data: *JCS:1372(1969)*. Bright blue-white fluorescence under SWUV; cores fluoresce brighter than blue edges of crystals, colorless cores fluoresce dull red in LWUV and rims are nonfluorescent. Insoluble in HCl, but easily attacked by HF. Stannian benitoite with up to 4.1 wt % SnO$_2$ has been reported. White cores and blue rims are the same, but 0.05 wt % Fe present in deep blue material may be the chromophore. Uniaxial (+); $N_O = 1.756$–1.757, $N_E = 1.802$–1.804; strongly dichroic. O, colorless; E, deep blue to purplish blue. Occurs in natrolite with neptunite and joaquinite in hydrothermally altered serpentine (CA). Crossite inclusions are sometimes present. Occurs in pectolite in gas cavities in pseudoleucite syenite (AR). Occurs at the Diamond Jo quarry, Magnet Cove, AK; *Dallas gem mine (Victor claim), San Benito Co., CA*; Numero Uno mine (pink crystals) 6 miles W of Dallas gem mine; Rush Creek area, Fresno Co., CA (stannian benitoite); Esneux and Streupas, L'Ourthe Valley, Belgium (subsequently shown to be misidentified). Also found near Ohmi, Niigata Pref., Japan, Korea, and Texas. Used as a gemstone. Largest cut gem is 7.6 ct (USNM). Also used as a fluor-

escent standard for electron microprobes. EF *AM 57:85(1972)*, *MR 6:442(1977)*, *MIN 3(2):15(1981)*, *LJ 44:46, 61(1990)*.

59.1.1.3 Pabstite BaSnSi$_3$O$_9$

Named in 1965 for Adolph Pabst (1899–1990), American mineralogist, University of California, Berkeley. Benitoite group. HEX $P\bar{6}c2$. (Nat): $a = 6.703$, $c = 9.824$, $Z = 2$, $D = 4.07$; (syn): $a = 6.7329$, $c = 9.8514$, $Z = 2$, $D = 4.16$–4.17. *18-196*(syn): $5.85_{3.5}$ 4.93_2 3.77_{10} 3.37_5 2.92_2 2.78_9 2.46_2 1.99_3; *43-633*(syn): 5.83_4 3.76_{10} 3.37_3 2.92_2 2.78_{10} 1.99_2 1.94_1 1.88_1. Structure: *NJBM:16(1987)*. Anhedral grains usually <2 mm, aggregates do not exceed 10 mm. Colorless to white, white streak, vitreous luster. H = 6. G = 4.03. Fluoresces bright blue in SWUV. Kalkar quarry sample: Ba(Sn$_{0.77}$Ti$_{0.23}$)Si$_3$O$_9$. Uniaxial (−); N$_O$ = 1.685, N$_E$ = 1.674; nondichroic; birefringence = 0.011. Occurs in recrystallized siliceous limestone, associated with tremolite, witherite, Sn-bearing sulfosalts, phlogopite, diopside, and forsterite. Fracture filling and disseminated grains. *Kalkar quarry, Santa Cruz Co., CA*. EF *AM 50:1164(1965)*, *69:358(1984)*.

59.1.1.4 Wadeite K$_2$ZrSi$_3$O$_9$

Named in 1939 by Prider for Arthur Wade (1878–1951), Australian geologist, who collected the mineral. Benitoite group. HEX $P\bar{6}$ *SPC 22:31(1977)*. $a = 6.926$–6.931, $c = 10.177$–10.197, $Z = 2$, $D = 3.11$. *35-31*(syn): 5.99_8 5.16_6 3.88_8 3.46_6 2.96_8 $2.87_{10}$$2.55_6$; *43-231*: 6.00_4 5.17_1 3.89_5 3.47_3 3.28_2 2.96_4 2.87_{10} 1.86_2. The structure is similar to that of benitoite, but in wadeite the [Si$_3$O$_9$] rings do not superimpose in projection but alternately occupy different positions between the [ZrO$_6$] octahedra to leave finite cavities for the remaining large cations. Hexagonal prismatic, tabular crystals with development of prism, basal pinacoid, and some pyramids; crystals to 3 cm in diameter at Mt. Yukspor. Colorless, light pink, lilac, white; white streak; vitreous luster. No cleavage, conchoidal fracture, brittle. H = $5\frac{1}{2}$–6. G = 3.10. IR data: *GZ 36:54(1976)*. Shows bright blue cathodoluminescence. Nonmagnetic. Insoluble in hot acids. Contains minor TiO$_2$. Uniaxial (+); N$_O$ = 1.624–1.627, N$_E$ = 1.653–1.658; birefringence = 0.030. Occurs in leucite-bearing volcanic rocks at West Kimberly. In carbonatite veins in Kovdor massif; in albitized fenites. In nepheline syenite pegmatites on Mt. Yukspor. Partly replaced by calcite along pyramidal cleavage. Replaced by lepidomelane plus zircon or catapleiite at Mt. Yukspor. Also alters to ancyclite and zeolites. Occurs at Middle Table Mt., Leucite Hills, WY; Mt. St.-Hilaire, QUE, Canada; Mts. Eveslogchorr, Marchenko, and Yukspor, Khibiny massif, and Kovdor massif, Kola Penin., and Murun massif, Aldan Shield, Yakutia, Russia; *Wolgidee Hills, West Kimberly area, WA, Australia*. EF *CM 30:1153(1992)*, *CMP 72:191(1980)*, *SPC 22:31(1977)*.

59.1.2.1 Walstromite $Ca_2BaSi_3O_9$

Named in 1965 for Robert E. Walstrom (b.1920), American mineral collector, who discovered the mineral. See the Pb analog, margarosanite (59.1.2.3). TRIC $P\bar{1}$. *ZK 130:185(1969)* (syn, reduced cell): $a = 6.723$, $b = 9.616$, $c = 6.733$, $\alpha = 83.01°$, $\beta = 77.67°$, $\gamma = 69.62°$, $Z = 2$, $D = 3.73$; *AM 50:314(1965)* (nat): $a = 6.743$, $b = 9.607$, $c = 6.687$, $\alpha = 69.8°$, $\beta = 102.2°$, $\gamma = 97.1°$; $Z = 2$. *18-162*(nat): 6.58_2 4.40_2 3.35_2 3.20_2 3.06_2 2.99_{10} 2.78_2 2.70_2. A ring structure with Si_3O_9 rings. *AM 53:9(1968)*. Grains are from <1 to 15 mm, 3 mm average, equigranular crystals to 1.5 cm, best forms are {010}, {100}, {011}. White or colorless; white to colorless streak; subvitreous luster, pearly on cleavage surfaces. Cleavage {011}, {010}, perfect; {100}, less perfect. $H = 3\frac{1}{2}$. $G = 3.67$. IR data: *MZ 1:3(1979)*. SWUV, faint pink; LWUV, distinct pink or pinkish orange, also white and cream reported for SWUV and LWUV. Easily soluble in cold dilute acid with production of a silica gel, no reaction with bases. Contains traces of Mn, Sr, and Al. Biaxial $(-)$; $N_x = 1.668$, $N_y = 1.684$, $N_z = 1.685$; $2V = 30°$; colorless; weak dispersion. Occurs in metamorphosed metasediments. Associated with sanbornite, wollastonite, celsian, taramellite, pyrrhotite, pyrite, fresnoite, and many other Ba-silicates. In metasomatic rocks. A major constituent in some of the rocks of *Big Creek and Rush Creek areas, Fresno Co., CA.* EF *MIN 3(2):33(1981)*, *AM 69:358(1984)*.

59.1.2.2 Margarosanite $Pb(Ca,Mn)_2Si_3O_9$

Named in 1916 by Ford and Bradley from the Greek for *pearl* and *tablet*, in allusion to its luster and habit. See the Ba analog, walstromite (59.1.2.2). TRIC $P\bar{1}$. $a = 6.768$, $b = 9.575$, $c = 6.718$, $\alpha = 110.36°$, $\beta = 102.98°$, $\gamma = 83.02°$, $Z = 2$, $D = 4.31$. *ZK 128:213(1969)*. Also, $a = 6.77$, $b = 9.64$, $c = 6.75$, $\alpha = 110.6°$, $\beta = 102°$, $\gamma = 88.5°$, $D = 4.30$. *AM 48:698(1963)*. *27-79*(calc pattern): 6.59_8 5.10_8 4.38_6 3.22_5 3.18_6 3.05_8 3.03_9 2.99_{10}; *16-356*(nat): 5.08_3 3.20_3 3.16_3 3.03_5 2.98_{10} 2.69_3 2.67_4 2.61_3. Structure: *ZK 128:213(1969)*. Aggregates of flat plates or rarely in stalklike to spikelike crystals to 2 mm thick and 2 cm in diameter. Colorless, white streak, pearly luster on faces of cleavage flakes. Cleavage {010}, perfect; {100}, good; {001}, fair. $H = 2\frac{1}{2}$–3. $G = 4.33$. IR data: *MZ 1:3(1979)*. Mn-bearing material from Franklin fluoresces pink to red or orange in SWUV; usually vivid blue-white SWUV fluorescence according to *MR 22:273(1991)*. Decomposes in HNO_3 with separation of silica. MnO to 2.2 wt %. Biaxial $(-)$; $N_x = 1.727$–1.729, $N_y = 1.771$–1.773, $N_z = 1.798$–1.807; colorless; $2V = 78$–$83°$; $r < v$, strong. Occurs in metamorphosed Precambrian Mn deposits at *Franklin, Sussex Co., NJ*, particularly the Parker shaft and on North Mine Hill; also at Långban, and Jakoberg, Värmland, Sweden. EF *AM 49:781(1964)*, *MIN 3(2):35(1981)*, *MM 49:721(1985)*.

59.2.1.1 Umbite $K_2ZrSi_3O_9 \cdot H_2O$

Named in 1983 by Khomyakov et al. for Lake Umba, 20 km from the locality. Compare dimorph kostylevite (59.2.1.3). ORTH $P2_12_12_1$. $a = 10.208$,

$b = 13.241$, $c = 7.174$, $Z = 4$, $D = 2.79$. $35\text{-}709$(nat): 6.56_6 5.91_9 3.31_7 3.02_{10} 2.87_8 2.16_4 1.80_8 1.55_4. Wollastonitelike infinite chains of ($Si_{2+1}O_9$) with four chains bonded to one Zr-octahedron. *DAN SSSR 257:622(1981)*. Platy, flattened on b with dominant forms: {010}, {101}, {110}, {011}. Colorless, yellow; vitreous luster. Cleavage {010}, micaceous; {100}, less perfect. $H = 4\frac{1}{2}$. $G = 2.79$. IR and thermal data: *ZVMO 112:461(1983)*. In UV, shows a weak yellow-white photoluminescence. It is bright whitish green at the temperature of liquid nitrogen; readily decomposed by cold HCl (10%), leaving a silica skeleton. Analysis (wt %): SiO_2, 42.36; K_2O, 22.41; ZrO_2, 22.28; HfO_2, 0.43; TiO_2, 3.33; Fe_2O_3, 0.12; Na_2O, 0.16; CaO, none; H_2O, 5.03; F, 2.00. Biaxial $(-)$; $XYZ = cba$; $N_x = 1.596$, $N_{y\,calc} = 1.610$, $N_z = 1.619$; $2V = 80°$. Occurs in the *Vuonnemiok River Valley, Khibiny massif, Kola Penin., Russia*; intergrown with kostylevite and wadeite. In alkalic pegmatite. EF

59.2.1.2 Paraumbite $K_3Zr_2HSi_6O_{18} \cdot nH_2O$

Named in 1983 by Khomyakov et al. for the close structural similarity to umbite, with a and b nearly identical and c twice that of umbite. See umbite (59.2.1.1). ORTH Space group unknown. $a = 10.34$, $b = 13.29$, $c = 14.55$, $Z = 4$, $D = 2.63$. $P2/cm$ $35\text{-}710$(nat): 6.46_8 5.95_{10} 4.06_5 3.34_7 3.01_9 2.90_7 2.56_6 1.83_5. Plates (pseudohexagonal); 1–2 mm in sodalite xenoliths at Mt. St.-Hilaire. Colorless to very pale yellow, white to tan; white streak; vitreous to pearly luster on cleavage planes. Cleavage {010}, micaceous; {100}, less perfect. $H \approx 4\frac{1}{2}$. $G = 2.59$. Readily decomposed by cold 10% HCl. Analysis (wt %): SiO_2, 39.6; K_2O, 15.4; Na_2O, 0.1; CaO, none; ZrO_2, 27.9; HfO_2, 0.3; TiO_2, 0.9; Fe_2O_3, 0.1; H_2O, 15.73 (diff.); $n = 7$, but the calculated density corresponds to $n = 3$. Biaxial $(-)$; $XYZ = cba$; $N_x = 1.588$, $N_y = 1.601$, $N_z = 1.610$; $2V = 82°$. Occurs in alkalic pegmatites along with K-feldspar and eudialyte. Eudialyte is partly replaced by wadeite, gaidonnayite, and paraumbite. Alteration: some possibly are pseudomorphs after lovozerite. Mt. St. Hilaire, QUE, Canada; *Mt. Eveslogchorr, eastern Khibiny massif, Kola Penin., Russia*. EF *ZVMO 112:461(1983)*.

59.2.1.3 Kostylevite $K_2ZrSi_3O_9 \cdot H_2O$

Named in 1983 by Khomyakov et al. for Yekaterina Eutikhievna Kostyleva-Labuntsova (1894–1974), Russian mineralogist. Compare the dimorph umbite. MON. $P2_1/a$. $a = 13.171$, $b = 11.727$, $c = 6.565$, $\beta = 105.26°$, $Z = 4$ (*ZVMO 112:469(1983)*) D=2.79–2.82; PD $37\text{-}459$ (nat.): 6.42_5 5.86_3 5.60_6 5.24_3 3.34_5 3.09_{10} 2.86_3 2.80_5. Structure: *DAN SSSR 256:860(1981)*, *AM 69:812(1984)*. Colorless, transparent; streak white, elongated on c, dominant forms: {001}, {010}, {100}, {011} + {110}; twinning on (100); vitreous; cleavage = (110) perfect; $H \approx 5$, G 2.74; readily decomposed by cold 10% HCl. 2.06% TiO_2, SiO_2 42.01%, ZrO_2 23.90%, HfO_2 0.61%, Fe_2O_3 0.02%, K_2O 22.14%, Σ90.74, Na + Ca absent. Biax. (+), $2V = 48°$, $N_\alpha = 1.595$, $N_\beta = 1.598$, $N_\gamma = 1.610$; $X = b$, $Y \wedge c = 45°$; dispersion = r < v weak. Occurs in the *Vuonnemiok River Valley, eastern Khibiny Massif, Kola Peninsula, Russia*, in alkalic pegmatite, intergrown with umbite and wadeite; assoc. with eudialyte, aegirine, and others.

59.2.2.1   Catapleiite $(Na,Ca,\square)_2ZrSi_3O_9 \cdot 2H_2O$

Named in 1850 by Weibye and Sjogren from the Greek for *wholly* and *full*, because it is always accompanied by a number of rare minerals. Benitoite group. See the dimorph, gaidonnayite (59.2.2.3). Forms a solid solution with calcium catapleiite. MON (pseudo-HEX, pseudo-ORTH) $B2/b$. $a = 23.917$, $b = 20.148$, $c = 7.432$, $\gamma = 147.46°$, $Z = 8$, $D = 2.79$. *SPC 26:808(1981)*. Also given as $I2/c$. $a = 12.779$, $b = 7.419$, $c = 20.157$, $\beta = 90.41°$; (syn, stable above 140°): $a = 7.388$, $c = 10.068$. *14-297*(nat): 6.35_6 5.37_5 3.05_{10} 2.96_{10} 2.69_9 1.97_6 1.85_4. The structure is identical to that of wadeite: the $[ZrSi_3O_9]$ framework. In the hexagonal wadeite structure, the cavity occupied by Na is much too large to contain Na at special positions on the threefold axis, so the symmetry decreases to monoclinic. *SPC 26:808(1981)*. Forms present: {001} {100}, {103} {102}, {101}, and so on, tabular to platy crystals, generally lamellar, tabular parallel to {001}, pseudohexagonal edges are bound by a prism and one or more pyramids; from millimeter size to rosettes 15 cm in diameter; polysynthetic twinning, six twin laws reported, most common have the twin plane (10$\bar{1}$0) or {0001}. Tan to beige to pale gray, gray-green, brown to colorless; small crystals are pale yellow, pale orange, pink, and very pale blue-white; vitreous luster, adamantine on base. Cleavage {10$\bar{1}$0}, perfect; {10$\bar{1}$1}, {10$\bar{1}$2}, imperfect; conchoidal fracture; brittle. $H = 5$–6. $G = 2.60$–2.79. Thermal and spectral data: *DHZ 1B:364(1986)*. Dissolves in all strong acids without gelatinizing. May have >6% CaO. Trace to minor K_2O, minor FeO; $2Na \leftrightarrow Ca$; REE, and Nb_2O_5 may be present. Uniaxial (+) or biaxial (+) and (−); $N_x = 1.582$–1.603 (calcium catapleiite), $N_y = 1.582$–1.618, $N_z = 1.600$–1.639 (calcium catapleiite); birefringence 0.018–0.036; $2V_z = 0$–$60°$; most catapleiites are (+); $Z \wedge c = 4°$ in obtuse β, $Y = b$, in biaxial crystals, OAP = (010); for (−) crystals, $X = c$. For (+) crystals, Z is almost perpendicular to (001). When heated, monoclinic lamellae disappear, but reappear when cooled; $r < v$, weak. Occurs in alkaline igneous rocks and is a relatively common minor constituent in nepheline syenite pegmatites and veins. It is rarely a primary mineral and usually occurs as an alteration product of eudialyte. Occurs as pseudomorphs after dalyite and vlasovite at Strange Lake. Magnet Cove complex, Hot Springs Co., AR; Rocky Boy Stock, Bearpaw Mts., MT; Wind Mountain, Otero Co., NM; the best and largest (to >2 cm across and 15 cm long) crystals in the world are found at Mt. St.-Hilaire, QUE (this mineral and serandite are the most sought after); Strange Lake complex, QUE–LAB, Canada; Kangerdlugssuaq, eastern Greenland, and Narssarssuk, Julianehaab district, Greenland; *Låven Island, Langesundfjord area, and Bratthagen, Norway*; primary crystals from Norra Kärr complex, Sweden; Kukisvumchorr, Khibiny massif, and Lovozero massif, Kola Penin., Vishnevye Gor (Cherry Mountain), Ural Mts., and Tuva and Aldan shields, Yakutia, Russia; 20 km N of Zomba, Chilwa Prov., Malawi; Madagascar. Minor use as a gemstone. EF *MIN 3(2):24(1981)*, *MR 21:284(1990)*, *CM 30:191(1992)*.

59.2.2.2 Calcium Catapleiite (Ca,☐)ZrSi$_3$O$_9$ · 2H$_2$O

Named in 1964 by Portnov for the composition and relationship to catapleiite. Benitoite group. HEX Space group unknown, possibly P6$_3$ mmc. $a = 7.32$, $c = 10.15$, $Z = 2$, $D = 2.79$. *16-371*(nat): 6.45$_7$ 5.40$_6$ 3.96$_8$ 3.06$_8$ 2.96$_{10}$ 1.98$_8$ 1.84$_8$ 1.74$_7$. The structure involves replacement of four Na atoms by two Ca atoms and two vacancies. Light yellow, yellowish white, pale yellow–cream, white; white streak; vitreous to dull luster; opaque. Cleavage present. $H = 4\frac{1}{2}$–5. $G = 2.77$. IR and thermal data: *DAN SSSR 154:607(1964)*. Analysis (wt %), Strange Lake: SiO$_2$, 44.5; ZrO$_2$, 30; HfO$_2$, 0.9; Al$_2$O$_3$, 0.5; CaO, 11.0; FeO, 0.2; PbO, 0.9; Na$_2$O, 0.7; Burpala massif: SiO$_2$, 44.5; ZrO$_2$, 31.0; TiO$_2$, 0.06; Al$_2$O$_3$, 0.60; Fe$_2$O$_3$, 0.36; CaO, 13.82; REE$_2$O$_3$, 0.28; Na$_2$O, 0.32; K$_2$O, 0.10; H$_2$O$^+$, 9.15; H$_2$O$^-$, 0.18. A Pb-bearing (to 11.4 wt % PbO) catapleiite occurs as cores to rims of calcium catapleiite at Golden Horn. Uniaxial (+); $N_O = 1.603$, $N_E = 1.639$. Occurs at Strange Lake as a replacement of earlier-formed Zr minerals, especially vlasovite. Also occurs as euhedral grains rimmed by gittinsite. In evolved granite. In syenite pegmatites at Burpala, in cavities between microcline. In miarolitic cavity in arfvedsonite granite at Golden Horn. Occurs in the Washington Pass area, Golden Horn batholith, Okanogan Co., WA; Strange Lake complex, QUE-LAB, and with edingtonite from Ice River, BC, Canada; as Ca-rich rims on catapleiite from Norra Karr, Sweden; and *Burpala massif, northern Baikal area, Siberia, Russia*. EF *DHZ 1B:364(1986)*, *CM 30:191(1992)*.

59.2.2.3 Gaidonnayite (Na,K)$_2$ZrSi$_3$O$_9$ · 2H$_2$O

Named in 1974 by Chao and Watkinson for Gabrielle Hamburger Donnay (1920–1987), crystallographer and mineralogist, McGill University, Montreal. Benitoite group. See the dimorph, catapleiite (59.2.2.1). Georgechaoite is the K analog. ORTH $P2_1nb$. $a = 11.740$–11.778 (potassian), $b = 12.820$–12.842 (potassian), $c = 6.691$–6.693 (potassian), $Z = 4$, $D = 2.70$. *39-353*(potassian): 6.45$_4$ 5.90$_8$ 5.64$_6$ 3.12$_{10}$ 2.95$_7$ 2.83$_4$ 2.50$_4$ 1.90$_4$; *26-1387*(regular): 6.42$_3$ 5.93$_8$ 5.84$_8$ 5.63$_5$ 4.31$_3$ 3.12$_{10}$ 3.09$_8$ 2.93$_4$. Powder data for K-Ca gaidonnayite: *DAN SSSR 320:1220(1991)*. Cell parameters vary linearly with K content between gaidonnayite and georgechaoite. *NGT 71:303(1991)*. Structure: *CM 23:11(1985)*. Crystals equant, blocky, tabular, and wedge-shaped, rhombic, pyramidal to 4 mm across, often striated parallel to [001], dominant forms: {010}, {120}, {011}, {100}, {001}; usually in intergrown, stacked spherical groups and druses; twinning common, [012]$_{180°}$, one individual appears to be rotated 90° about [100] relative to the other. White, colorless, pale yellow, to very pale brown, pale gray, gray-green; white streak; adamantine to vitreous luster. No cleavage observed, conchoidal fracture, brittle. $H = 5$. $G = 2.67$. Thermal data: *CM 12:316(1974)*. Fluoresces bright green under SWUV and LWUV (Mt. St.-Hilaire and Norway); some nonfluorescent material has also been found. No cathodoluminescence. Gelatinizes in acids. Analysis (wt %), Potassian: Na_2O, 9.4; K_2O, 6.1; CaO, 2.1; SiO_2, 42.6; ZrO_2, 29.7; H_2O, 10.1 $(Na_{1.23}K_{0.53}Ca_{0.15})_{1.91}Zr_{0.98}Si_{2.87}O_{8.73} \cdot 2.26H_2O$; minor TiO_2 substitutes for Zr, to 1.8% TiO_2; Norway: TiO_2, 1.36; Nb_2O_5, 0.68; ZnO, 0.10; Narsarssuk: CaO, 0.4; TiO_2, 0.6; Nb_2O_5, 0.9; Mt. St.-Hilaire: Na_2O, 13.1; K_2O, 2.2; SiO_2, 42.5; ZrO_2, 30.2; TiO_2, 0.42; Nb_2O_5, 3.00; H_2O, 9.25. Biaxial (−); XYZ = abc; $N_x = 1.570$–1.575, $N_y = 1.590$–1.593, $N_z = 1.596$–1.605; 2V = 53–59°. Occurs at *Mt. St.-Hilaire, QUE, Canada*, in altered pegmatites and rarely in miarolitic cavities in nepheline syenite and breccia cavities; also in an igneous sill in limestone. In eudialyte-rich pegmatite lenses in the Kipawa syenite complex, Villedieu Twp., Temiscamingue Co., QUE, as a late-stage alteration product of vlasovite. In a vein of aegirine syenite at Narssarssuk, southern Greenland; in nepheline syenite pegmatite at Brønnebukta, Siktesøya, Langesundfjord, Oslo region, Norway; Vuorijarvi, Karelia, Mt. Alluaiev and Mt. Karnasurt, Lovozero massif, Mt. Yukspor and Mt. Evslogchorr, Khibiny massif, Kola Penin., and K-Ca-gaidonnayite at Mt. Mannepakh, northwestern Khibiny massif, Russia; 5- to 6-mm crystals of gaidonnayite at Pocos da Caldas, MG, Brazil. EF *MIN 3(2):556(1981), GSCP 83-1A:480(1983), DANS ESS 321A:111(1994)*.

59.2.2.4 Georgechaoite $KNaZrSi_3O_9 \cdot 2H_2O$

Named in 1985 by Boggs and Ghose for George Y. Chao (b.1930), Canadian mineralogist. Benitoite group. See the Na analog, gaidonnayite (59.2.2.3). ORTH $P2_1nb$. a = 11.836, b = 12.940, c = 6.735, Z = 4, D = 2.69. *39-315*(nat): 6.46_7 5.95_7 5.83_3 5.67_5 3.12_{10} 2.89_2 2.83_2 2.20_2. Isostructural with gaidonnayite, $Na_2ZrSi_3O_9 \cdot 2H_2O$. Georgechaoite shows Na/K ordering of 1:1. *CM 23:5(1985)*. Crystals to 1 mm long, forms: {010}, {011}, {100}, {$\bar{1}$00}, {101}, {$\bar{1}$01}, {120}; twinning $[023]_{180}$ observed, interpenetration twins, with a half-turn about [023], a axes of the two individuals are parallel, b axis of one individual is 14° to the c axis of the second. Colorless to white, white streak, vitreous luster. No cleavage, chonchoidal fracture. H = 5. G = 2.70. No UV response. Biaxial (−); XYZ = abc; $N_x = 1.578$, $N_y = 1.597$, $N_z = 1.606$; 2V = 67°. Associated with microcline, nepheline, analcime, aegirine, chlorite, catapleiite, and monazite in miarolitic cavities in Ne-syenite at *Wind Mt., Otero Co., NM*. What has been reported [*DAN SSSR 320:1220(1991)*] as K-Ca gaidonnayite occurs as a rim of brown fine-grained aggregate around eudialyte and sometimes forms complete pseudomorphs of eudialyte; in a pegmatite vein on Mt. Mannepakh, northwestern Khibiny massif, Kola Penin., Russia. The material is fine-grained and turbid; colors are patchy from light to dark reddish brown. H = 2. Further work needs to be done on better material to confirm the identity of the mineral. EF *CM 23:1(1985)*.

59.2.2.5 Loudounite NaCa$_5$Zr$_4$Si$_{16}$(OH)$_{11}$ · 8H$_2$O

Named in 1983 by Duan and Newbury for the locality. Not indexed, no cell data given. 35-573(nat): 7.37$_6$ 5.81$_5$ 4.56$_3$ 4.06$_7$ 3.53$_4$ 2.93$_{10}$ 2.69$_4$. No structure determination or single-crystal work possible. Spherules of radiating, fibrous crystals; average spherule diameter is 0.05 mm. Have overgrowths of ancylite. Light green, white, translucent; colorless. H = 5. G = 2.48. Luminescence: no response. Analysis (wt %): SiO$_2$, 45.8; Al$_2$O$_3$, 0.8; FeO, 2.0; MgO, 0.3; CaO, 12.1; ZrO$_2$, 25.7; Na$_2$, 0.2; H$_2$O, difference from 100.0. Biaxial with wavy extinction; N$_x$ = 1.536, N$_z$ = 1.550; nonpleochroic; optical orientation scheme not determined; length-slow. Associated with actinolite, chlorite, epidote, stilbite, and ancylite in diabase. Loudounite crystallized last. Occurs at the *Goose Creek quarry, Loudoun Co.*, and at the Fairfax quarry, Centerville, Fairfax Co., *VA*. EF *CM 21:37(1983)*.

59.2.3.1 Hilairite Na$_2$ZrSi$_3$O$_9$ · 3H$_2$O

Named in 1974 by Chao et al. for the type locality. Benitoite group. HEX-R R32. a = 10.556, c = 15.855, Z = 6, D = 2.73. 26-975(nat): 6.0$_6$ 5.28$_{10}$ 3.17$_5$ 3.05$_4$ 2.99$_3$ 2.64$_3$ 2.03$_3$ 1.76$_4$. The structure consists of spiral [Si$_3$O$_9$] chains rather than [Si$_3$O$_9$] rings parallel to c and polyhedra of cations connecting tetrahedra in the chains. Cavities contain Na$^+$ and H$_2$O. The Na sites are partially vacant in hilairite but are fully occupied in the REE analog. *DAN SSSR 260:1118(1981), SPC 37:845(1992)*. Usually short, to elongated prisms {11$\bar{2}$0} terminated by the {01$\bar{1}$2} rhombohedron; 0.5- to 4-mm crystals, many doubly terminated; twinning frequent on two laws: [2$\bar{2}$01]$_{180°}$ and [0001]$_{180°}$; may be multiply twinned according to both laws. Usually pale to dark brown, also pink or rose; white streak; vitreous to waxy luster; translucent, rarely transparent. No cleavage observed, conchoidal fracture, brittle. H > 4. G = 2.72. Nonfluorescent. Slightly attacked by 1:1 HCl and HNO$_3$ but not by 1:1 H$_2$SO$_4$. At Strange Lake, some limited solid solution occurs with calciohilairite. At Mt. St.-Hilaire, minor K and Ca substitute for Na. Uniaxial (−); N$_O$ = 1.609, N$_E$ = 1.596. Occurs in the *Demix and Poudrette quarries, Mt. St.-Hilaire, QUE, Canada*, as an accessory mineral in miarolitic cavities and in some altered pegmatite veins, usually associated with gaidonnayite, elpidite, natrolite, microcline, aegirine, pyrite, rutile, and zircon; and as interstitial anhedral grains in granite of the Strange Lake complex, QUE-LAB. In hydrothermally altered pegmatites at Langesundfjord, Norway; in the Lovozero massif, Kola Penin., Russia. EF *CM 12:237(1974), 30:191(1992)*; *MZ 2:95(1980)*; *MIN 3(2):32(1981)*.

59.2.3.2 Calciohilairite CaZrSi$_3$O$_9$ · 3H$_2$O

Named in 1988 by Boggs as the Ca analog of hilairite. Benitoite group. HEX-R R32. a = 20.870–20.90, c = 16.002–16.05, Z = 24, D = 2.71–2.74. *AM 73:1191(1988)*. 41-1456(nat): 5.99$_1$ 5.23$_{10}$ 3.20$_1$ 3.14$_2$ 3.01$_3$ 2.90$_1$ 2.61$_1$ 1.98$_1$. Structure not yet determined. Dominant forms: {11$\bar{2}$0}, {2$\bar{1}\bar{1}$0}, {10$\bar{1}$2}, {01$\bar{1}$2}, blocky, trigonal prisms; to 2 mm long; no twinning observed. Pale blue to chalky white, white streak, vitreous luster. No cleavage, conchoidal fracture. H = 4. G = 2.68. Nonluminescent. Analysis (wt %): Na_2O, 0.2; CaO, 11.1; CuO, 0.65; FeO, 0.19; SiO_2, 40.0; ZrO_2, 32.6; TiO_2, 0.06; Al_2O_3, 1.3; H_2O, 13.8. Uniaxial (−); $N_O = 1.622$, $N_E = 1.619$. Occurs in miarolitic cavities in peralkaline granite near Liberty Bell Peak, Washington Pass area, in the *Golden Horn Batholith, northern Cascades, WA*. Associated with bastnaesite-(Ce) and malachite, quartz, and feldspar. EF

59.2.3.3 Sazykinaite-(Y) $Na_5YZrSi_6O_{18} \cdot 6H_2O$

Described in 1993 by Khomyakov et al. and named for Lyudmila Borisnova Sazykina (b.1934), Russian mineralogist and artist (using semi-precious and precious stones). The mineral is the Y-dominant analog of hilairite and calciohilairite. It is also related to komkovite. HEX-R $R32$. $a = 10.825$, $c = 15.809$, Z = 3, D = 2.74. *ZVMO 122:76(1993)*: 6.03_3 5.40_6 3.24_8 3.13_9 3.03_{10} 2.71_2 2.02_2 1.805_2. Structure done (R = 3.4%). *DAN SSSR 37(6):158(1992)* and *SPC 37:845(1993)*. Two alkali positions [i.e., $Na_2(Na,K)_3$]. Rhombohedral crystals from 0.5 to 2.0 mm across, and as complex grains showing $\{01\bar{1}2\}$. Light green to yellow, white streak, vitreous luster. Cleavage $\{01\bar{1}2\}$, imperfect; steplike fracture, fragile. H = 5. G = 2.67. IR, and thermal data given. Fluoresces green in UV light. Stable in water but readily soluble in 10% HCl and HNO_3. Analysis (wt %): Na_2O, 15.2; K_2O, 3.05; Y_2O_3, 8.7; Ce_2O_3, 0.2; Nd_2O_3, 0.25; Sm_2O_3, 0.4; Eu_2O_3, 0.2; Gd_2O_3, 1.0; Tb_2O_3, 0.2; Dy_2O_3, 1.3; Er_2O_3, 0.8; Tm_2O_3, 0.2; Yb_2O_3, 0.6; ThO_2, 0.7; ZrO_2, 10.2; TiO_2, 1.4; SiO_2, 40.5; Nb_2O_5, 1.3; H_2O, 12.6. Solid solution of $Y^{3+} + Na \leftrightarrow Zr^{4+}$ is assumed. Uniaxial (−); $N_E = 1.578$, $N_O = 1.585$. Occurs in pegmatites of *Koashva Mt., Khibiny alkaline massif, Kola Penin., Russia*, intergrown with labuntsovite and associated with aegirine, natrolite, alkali feldspars, pectolite, alkali amphibole, astrophyllite, lomonosovite, and sphalerite. EF

59.2.4.1 Komkovite $BaZrSi_3O_9 \cdot 3H_2O$

Named in 1990 by Voloshin et al. for A. I. Komkov (1926–1987), Russian mineralogist and crystallographer. HEX-R $R32$ [*DAN SSSR 320:1384(1991)*], $R3$ [*ZVMO 122:76(1993)*]. $a = 10.526$, $c = 15.736$, Z = 6, D = 3.31. *42-1398*(nat): 5.23_{10} 3.59_8 3.02_8 2.96_9 2.57_6 2.10_6 1.700_5 1.55_5. Structure: *DAN SSSR 320:1384(1991)*. Equant 1–5 mm across with trigonal pyramidal faces. Brown, light brown streak, vitreous luster. No cleavage, parting on (001). H = 3–4. G = 3.31. IR and DTA data: *MZ 12:69(1990)*. Nonfluorescent, light blue cathodoluminescence. Analysis (wt %): BaO, 29; CaO, 0.04; MnO, 0.1; FeO, ≈ 0.2; K_2O, 0.1; ZrO_2, 25; HfO_2, ≈ 0.2; SiO_2, 34; H_2O, ≈ 11. Uniaxial (−); $N_O = 1.671$, $N_E = 1.644$. Occurs in carbonatites of *Vuorijarvi, Kola Penin., Russia*, in dolomite veinlets that cut pyroxenites, associated with dolomite, strontianite, phlogopite, barite, georgechaoite, and pyrite; apparently formed by alteration of catapleiite. EF

Class 60

Four-Membered Rings

JOAQUINITE GROUP

The joaquinite group minerals are titanosilicates conforming to the formula

$$R_6[Ti(Si_4O_{12})]_2(O,OH,F)_2 \cdot H_2O$$

where

R = Na, Fe, Ba, Ce, Sr, Mn

The structure is based on a fundamental element consisting of rings of four silicon–oxygen tetrahedra linked to octahedrally coordinated titanium to form sheets of $[Ti(Si_4O_{12})]^{4-}$ parallel to {001}. The group may be divided into two polymorphic types, monoclinic and orthorhombic. *AM 60:872(1975)*.

JOAQUINITE GROUP—JOAQUINITE SUBGROUP: MONOCLINIC

Mineral	Formula	Space Group	a	b	c	β	
Joaquinite-(Ce)	$NaFe^{2+}Ba_2Ce_2$ $(Ti,Nb)_2[Si_4O_{12}]_2$ $O_2(OH,F) \cdot H_2O$	C2	10.516	9.686	11.833	109.67	60.1.1a.1
Strontio-joaquinite	$(Na,Fe^{2+})_2Ba_2Sr_2Ti_2$ $[Si_4O_{12}]_2(O,OH)_2 \cdot H_2O$	C2	10.516	9.764	11.87	109.28	60.1.1a.2

JOAQUINITE GROUP—ORTHOJOAQUINITE SUBGROUP: ORTHORHOMBIC*

Mineral	Formula	Space Group	a	b	c	
Orthojoaquinite-(Ce)	$NaFe^{2+}Ba_2Ce_2Ti_2[Si_4O_{12}]_2O_2$ $(OH) \cdot H_2O$	Ccmm	10.48	9.66	22.26	60.1.1b.1
Strontio-orthojoaquinite	$(Na,Fe^{2+})_2Ba_2Sr_2Ti_2[Si_4O_{12}]_2$ $(O,OH)_2 \cdot H_2O$	Pbcm	10.517	9.777	22.392	60.1.1b.2
Bario-orthojoaquinite	$Fe_2^{2+}(Ba,Sr)Ti_2[Si_4O_{12}]_2O_2$ $\cdot H_2O$	Ccmm	10.477	9.599	22.59	60.1.1b.3
Byelorussite-(Ce)	$NaMn^{2+}Ba_2Ce_2Ti_2[Si_4O_{12}]_2O_2$ $(F,OH) \cdot H_2O$	Ccmm	10.492	9.640	22.266	60.1.1b.4

Note: The orthorhombic subgroup is related to the monoclinic by a transformation of cell coordinates such that $c_o \approx 2c_m \sin\beta$. The *a* and *b* axes have also been interchanged in some species, to maintain correspondence with the monoclinic subgroup.

60.1.1a.1 Joaquinite-(Ce)
$NaFe^{2+}Ba_2Ce_2(Ti,Nb)_2[Si_4O_{12}]_2O_2(OH,F) \cdot H_2O$

Named in 1909 by Louderback for the type locality. Joaquinite group. Dimorphous with orthojoaquinite-(Ce). MON $C2$. $a = 10.516$, $b = 9.686$, $c = 11.833$, $\beta = 109.67°$, $Z = 2$, $D = 4.11$. 36-385: 4.44_7 3.57_5 3.30_5 3.11_8 2.95_9 2.89_8 2.78_{10} 2.42_5. Structure: AM 60:872(1975). Honey-yellow, orange, brown; vitreous luster. Brittle. $H = 5-5\frac{1}{2}$. $G = 3.62-3.98$. Biaxial (+); $2V \approx 50°$; $N_x = 1.754$, $N_y = 1.767$, $N_z = 1.823$; $XZ = ac$; $r < v$. Pleochroic, $Z > X = Y$; light yellow to colorless. Indices may be lower due to isomorphous substitution. Found at the Diamond Jo quarry, Magnet Cove, AR; as equant crystals up to 3 mm with benitoite, neptunite, and massive natrolite at *Joaquin Ridge, Diablo Range, San Benito Co., CA*, and at the Dallas Gem and Numero Uno mines in the same county; at Mt. St.-Hilaire, PQ, and with eudidymite and barylite at Seal Lake, Labrador, NF, Canada; near Kvanefjeld, Ilimaussaq, Greenland. AR AM 52:1762(1967), 57:85(1972).

60.1.1a.2 Strontiojoaquinite $(Na,Fe^{2+})_2Ba_2Sr_2Ti_2[Si_4O_{12}]_2$ $(O,OH)_2 \cdot H_2O$

Named in 1982 by Wise for the composition and relationship to joaquinite-(Ce). Joaquinite group. Dimorphous with strontio-orthojoaquinite. MON $C2$ (also reported as Primitive). $a = 10.516$, $b = 9.764$, $c = 11.87$, $\beta = 109.28°$, $Z = 2$, $D = 3.68$. 36-384: 4.47_4 4.30_3 3.01_5 2.97_7 2.92_4 2.80_{10} 2.61_4 2.43_4. Green to yellow-green, brown; vitreous luster. Cleavage {001}, good. $H = 5\frac{1}{2}$. $G = 3.68$. Biaxial (+); $2V = 35-45°$; $N_x = 1.710$, $N_y = 1.718$, $N_z = 1.780$; $r > v$, strong. As small pseudo-orthorhombic crystals with {110}, {11$\bar{1}$}, and {001} at the *Numero Uno mine, San Benito Co., CA*. AR AM 67:809(1982).

60.1.1b.1 Orthojoaquinite-(Ce) $NaFe^{2+}Ba_2Ce_2Ti_2[Si_4O_{12}]_2O_2(OH)$ $\cdot H_2O$

Named in 1982 by Wise from symmetry and relationship to its dimorph, joaquinite-(Ce). Joaquinite group. ORTH $Ccmm$ or $Ccm2_1$. $a = 10.48$, $b = 9.66$, $c = 22.26$, $Z = 4$, $D = 4.14$. 26-1034: 4.43_9 3.29_6 3.05_4 2.98_4 2.94_{10} 2.89_8 2.61_6 1.866_5. Light brown, vitreous luster. Cleavage {001}, good. $H = 5\frac{1}{2}$. Biaxial (+); optical properties like those of joaquinite-(Ce). Found intergrown with joaquinite-(Ce) at the *Dallas Gem mine, San Benito Co., CA*, and other nearby localities. AR AM 57:85(1972), 67:809(1982).

60.1.1b.2 Strontio-orthojoaquinite $(Na,Fe^{2+})_2Ba_2Sr_2Ti_2[Si_4O_{12}]_2$ $(O,OH)_2 \cdot H_2O$

Named in 1982 by Wise in keeping with its dimorph, strontiojoaquinite; earlier described but not named in 1974 by Chihara et al. Joaquinite group. ORTH $Pbcm$ or $Pbc2$ (Pca^* in original setting). $a = 10.517$, $b = 9.777$, $c = 22.392$, $Z = 4$, $D = 3.87$. PD: 5.60_3 4.47_3 2.97_4 2.80_{10} 2.61_4 2.44_3 2.43_2 2.24_3. Light yellow, translucent. Cleavage {001}, good. $H = 5\frac{1}{2}$. $G = 3.62$. Optical properties not given; presumably similar to those of strontiojoaquinite. Occurs associated with other members of this group in *San Benito Co., CA*. Earlier described but not named from *Ohmi, Niigita Pref., Japan*. AR *MJJ* 7:395(1974) AM 67:809(1982).

60.1.1b.3 Bario-orthojoaquinite $Fe_2^{2+}(Ba,Sr)_2Ti_2[Si_4O_{12}]_2O_2 \cdot H_2O$

Named in 1982 by Wise for the composition and relationship to joaquinite-(Ce). Joaquinite group. ORTH $Ccmm$ or $Ccm2$. $a = 10.477$, $b = 9.599$, $c = 22.59$, $Z = 4$, $D = 4.14$. *36-386*: 5.64_7 4.30_6 3.20_5 3.00_{10} 2.95_9 2.94_7 2.82_9 2.60_5. Bipyramidal, equant crystals to 8 mm with {111} and {001} prominent. Yellow-brown, vitreous luster. Cleavage {001}, good. $H = 5–5\frac{1}{2}$. $G = 3.96$. Biaxial (+); $2V = 10–15°$; $N_x = 1.735$, $N_y = 1.757$, $N_z = 1.80$; $XYZ = abc$; pleochroic, $Z > Y > X$; very pale yellow to yellow. Found in association with joaquinite-(Ce), benitoite, and natrolite at the *Dallas Gem mine, San Benito Co., CA*. AR *AM 67:809(1982)*.

60.1.1b.4 Byelorussite-(Ce) $NaMn^{2+}Ba_2Ce_2Ti_2[Si_4O_{12}]_2O_2(F,OH) \cdot H_2O$

Named in 1989 by Shpanov et al. for the type locality. Joaquinite group. ORTH $Ccmm$ or $Ccm2$. $a = 10.492$, $b = 9.640$, $c = 22.266$ (a and b transposed from structure description), $Z = 4$, $D = 4.15$. Originally reported as $P2_12_12_1$. $a = 10.57$, $b = 9.69$, $c = 22.38$, $Z = 4$, $D = 4.09$. PD: 4.42_6 4.29_4 3.30_4 3.00_7 2.95_6 2.91_5 2.78_{10} 2.61_5. Structure: *ZK supp. 4:321(1991)*. Pale yellow to pale brown, vitreous luster. Cleavage {001}, perfect; {010}, imperfect; {100}, very imperfect. $H = 5\frac{1}{2}–6$. $G = 3.92$. Biaxial (+); $2V = 58–62°$; $N_x = 1.743$, $N_y = 1.760$, $N_z = 1.820$; $XYZ = abc$. Very weakly pleochroic, $Z > Y \approx X$; colorless to very pale yellow. In metasomatized granosyenite of the *Zhitkovitscheskii horst, Homel district, Belarus (Byelorussia)*. AR *ZVMO 118(5):100(1989), AM 76:665(1991)*.

60.1.2.1 Baotite $Ba_4Ti_4(Ti^{4+},Nb^{5+},W^{6+})_4Si_4O_{28}Cl$

Named in 1959 by Peng as Pao-T'ou-K'uang (baotite by Semenov in 1961) for the locality. TET $I4_1/a$. $a = 19.92–20.02$, $c = 5.908–6.02$, $Z = 4$, $D = 4.33$. *39-390*: 3.55_8 3.34_5 3.17_9 2.89_7 2.50_7 2.24_{10} 1.77_6 1.33_6. There are two Ti sites, nonequivalent: $Ba_{4-x}Ti_4(Ti,Ni,W)_{4-x}Si_4O_{28}Cl$. Vacancies exist in the Ba site and one Ti site. $2Ti^{4+} \leftrightarrow Nb^{5+} + M^{3+}$, $3Ti^{4+} \leftrightarrow W^{6+} + 2M^{3+}$, also $Ti^{4+} + Ba^{2+} \leftrightarrow W^{6+} + \square$. $M^{3+} = Fe^{3+}$, Al, Cr. Al can replace tetrahedral Si. Structure consists of $[Si_4O_{12}]$ rings and (Ti,Nb) octahedral chains. *K 14:602(1969)*. Crystals are short, equant with {110} and {011} dominant; 0.8–4 mm (MT); masses to 10 cm in China. Light brown to black, vitreous luster. Two cleavages at (110), two directions at 90°; hackly fracture; brittle. $H = 6$. $G = 4.71–4.72$. IR data: *MZ 1:3(1979)*. As much as 6 wt % WO_3 (Pierrefitte, France); Nb is very low in French and Czech material. Nb_2O_5 to 11.5 wt % (Mongolia), Ta_2O_5 to 2.0 wt %; Czech: SiO_2, 15.7; TiO_2, 41.5; trace to minor amounts of Na, K, Mg, Ca, Al, Fe. Uniaxial (+);

$N_O = 1.94–1.944$, $N_E = 2.16$ to > 2.20; $E > O$ (Mongolia); strongly dichroic. O, brown, greenish yellow, or colorless; E, dark brown to nearly black, greenish brown, dark brownish to greenish yellow. Diamagnetic. Occurs in a base metal orebody as an accessory mineral included in alstonite and celsian, associated with common sulfides. In quartz veins with albite, aegirine, bastnaesite, and sulfides related to a monzonsyenite intrusion at Bayan-Obo. In carbonatite (MT) with other Nb-bearing species. In metasomatites in Russia, in alkaline dikes. Occurs at Sheep Creek, Ravalli Co., MT; Dallas Gem mine, San Benito Co., CA; W-bearing material at the Garaoulère mine, Pierrefitte, Pyrenees, France; Sebkovice, western Moravia, Czech Republic and at Karlstein and Jarolden, Waloviertel, Austria, in the Bohemian massif, Czech Republic; at two unspecified localities in the western Ural Mts. and in Primorye, Russia; high-Nb bearing at *Bayan-obo REE deposit, Baotou, inner Mongolia, China.* EF *MIN 3(2):44(1981), NDM 30:106(1982), NJBM:31(1987), MP 45:19(1991).*

60.1.3.1 Nenadkevichite (Na,K,Ca,Ba,Mn)(Nb,Ti)(Si$_2$O$_6$)(O,OH) · 2H$_2$O and K-dominant analogue

Named in 1955 by Kuzmenko and Kazakova for Konstantin Antonomovich Nenadkevich (1880–1963), Russian mineralogist and geochemist. Labuntsovite (Ti)–nenadkevichite series (Nb). In some samples, Ti > Nb; such material has been called "titanonenadkevichite." Na is always > K stoichiometrically. ORTH *Pbam*. $a = 7.328–7.408$, $b = 14.123–14.198$, $c = 7.115–7.148$, $Z = 4$, $D = 2.70–2.81$. *37-484*(nat): 7.11_{10} 6.50_4 5.01_4 3.26_9 3.18_5 3.12_3 2.55_4 1.73_4. Columns formed by (Nb,Ti) octahedra with shared transvertices. The octahedral columns are linked by tetrahedral four-membered rings [Si$_4$O$_{12}$]. The statistically distributed K, Na ions and water molecules are located in the channels of the mixed framework. *AC(B) 29:1432(1973); EJM 6:503(1994).* Prismatic and tabular {100}, {010}, {011}; elongated on *c*; single crystals to 3 cm long and 4 mm thick. Colorless, white, pale pink, rose-pink, yellow, orange, and brownish red; pale pink, white streak; vitreous to pearly or dull luster. Cleavage {001}, fair; uneven fracture. Brittle. H ≈ 5. $G = 2.73–2.97$. IR and thermal data: *MIN 3(2):64(1981)*. Readily soluble in H$_2$SO$_4$, difficultly soluble in HCl + HNO$_3$. Biaxial (+); XYZ = *abc*, $N_x = 1.633–1.665$, $N_y = 1.642–1.686$, $N_z = 1.738–1.785$; $2V = 19–46°$; OAP = (010); pleochroic, pale rose to colorless, also brownish. X, colorless; Y, pale yellow; Z, pale rose. Occurs in altered and unaltered pegmatites and very rarely in cavities in marble xenoliths and igneous breccia at Mt. St.-Hilaire. An alteration product of epistolite. In alkaline pegmatites in Kola Penin., Russia. At Mt. St.-Hilaire, and the St.-Amable sill, Varennes, QUE, Canada; Ilimaussaq, southern Greenland; Gjerdingen, Norway; several locations in the Khibiny massif (e.g., Mt. Mannepakh), and at *Vein No. 1, Mt. Karnasurt, Lovozero massif, Kola Penin., Russia;* also in the Jubilee pegmatite on Mt. Karnasurt and Mt. Flor, Lovozero massif. EF *NJBM:49(1986), MT(1989).*

POTASSIUM ANALOGUE OF NENADKEVICHITE:
$(K,NA,CA)_2(NB,TI)_2SI_4O_{13}(OH)_2 \cdot 4H_2O$.

ORTH. $Fmmm$. $a = 14.72$Å, $b = 14.20$Å, $c = 27.87$Å, $Z = 4$. Systematic extinctions show that the mineral can be considered as a superposition of two monoclinic diffraction patterns with a monoclinic axis parallel to axis 'b' in an orthorhombically twinned lattice and with C-centering of each monoclinic cell *(NJBM 103(1996)* and *EJM 6:503(1994))*. Colorless, transparent, mm-sized aggregates of platy/bladed crystals parallel to {001}; crystals are ≤0.05 mm, sometimes epitaxially overgrown on elpidite; {010}, {001}, {101} common, {012} rare, morphologically orthorhombic. Na_2O 3.75, CaO 1.04, K_2O 6.53, BaO 1.28, Ce_2O_3 0.41, Nb_2O_5 35.8, TiO_2 2.54, SiO_2 35.5, Al_2O_3 0.23, H_2O 13.4 (calc. by stoichiometry). Biax. (+), $2V_z = 55$–$65°$, $2V_z$(calc.): $58°$; r < v, (Na light): $N_\alpha = 1.6386$, $N_\beta = 1.6474$, $N\gamma = 1.7680$; $X = c$, $Y = b$, $Z = a$. The optical orientation is different from what is generally accepted for nenadkevichite: $X = a$, and $Z = c$. Occurs at Mont Saint-Hilaire, Quebec, Canada; associated with K-feldspar, aegirine, eudialyte, quartz, neptunite, elpididite, ancylite, epididymite, polylithionite, and pyrochlore at Narssârssuk pegmatite, westernmost margin of the Igaliko complex, South Greenland; Gjerdingen, Norway; in carbonatites at Vuorijarvi, Kola Peninsula, Russia.

60.1.3.2 Labuntsovite $(K,Na,Ba)_8(Ti,Nb)_9(SiO_3)_{16}(O,OH)_{10} \cdot xH_2O$ $(x \approx 7)$

Named in 1926 by Labuntsov and described as titanoelpidite; renamed in 1955 by Semenov and Burova in honor of Aleksandr Nikolaievich Labuntsov (1884–1963) and his wife, Ekaterina Eutikhieva Labuntsova-Kostyleva, Russian mineralogists. Synonym: titanoelpidite. See nenadkevichite (60.1.3.1), the Nb end member (much more abundant at Mt. St.-Hilaire than labuntsovite). MON $C2/m$; possibly ORTH $I2/m$. $a = 15.33$–15.57, $b = 13.70$–13.93, $c = 14.14$–14.33, $\beta = 116.95$–$117.12°$, $Z = 2$, $D = 2.87$ or 2.68 (using ideal formula and unit cell dimensions). 9-498(nat): 3.15_{10} 3.09_8 2.56_9 2.47_7 1.78_7 1.68_8 1.54_9 1.41_9. A question still exists about exact symmetry. Structure needs to be redetermined. *SPC 18:596(1974)*. May be orthorhombic. *MZ 3:49(1981)*. Exact formula is also still uncertain. Prismatic to acicular on [001], sometimes flattened; crystals to 12 mm × 3 mm × 2 mm, dominant forms: {001}, {110}, {011}, {010}, and {111}; spherulites to 3 cm; sometimes twinned. Pale to deep orange, rose to brownish yellow, red; white to pale orange streak; adamantine to vitreous luster. Cleavage {$\bar{1}02$}, perfect; subconchoidal fracture; brittle. H = 6. G = 2.8–3.02. IR data show presence of H_2O, H_3O^+, and OH^-. DTA data: *MZ 3:49(1981)*. Dissolves with difficulty in HCl, HNO_3, or H_2SO_4, readily soluble in HF. Analysis (wt %): SiO_2, 39.6; Al_2O_3, 1.3; Fe_2O_3, 1.56; Nb_2O_5, 1.45; TiO_2, 25.5; MgO, 0.42; MnO, 2.34; CaO, 1.19; BaO, 8.61; K_2O, 7.23; Na_2O, 3.18; H_2O, 7.91 (Lovozero). Na may be greater than K stoichiometrically, and in some samples, K may be > Na: potash labuntsovite. *MZ 3:49(1981)*, *MIN 3(2):68(1981)*. Nb_2O_5 to 7.27 wt

%. Biaxial (+); $Z \wedge c = 63°$, $Y = b$, $X \wedge a = 27°$ in the obtuse angle, $Z = c$; $N_x = 1.684–1.688$, $N_y = 1.693–1.698$, $N_z = 1.804–1.839$; $2V = 20–41°$; $r > v$, distinct; OAP = (010), $Z > Y > X$; pleochroic. X, colorless or yellowish; Y, orange-yellow; Z, brownish yellow, lemon yellow, or colorless. Occurs in cavities in igneous breccia zones and in nepheline syenite and hornfels at Mt. St.-Hilaire; occasionally as epitactic overgrowth on elpidite. In druses in Khibiny (e.g., Mts. Nyorkpakh and Koashva; S. M. Kirov mine) and Lovozero massifs, associated with aegirine, nepheline, eudialyte, lorenzenite, and others. A hydrothermal alteration product of murmanite. Also in carbonatites at Kovdor; at Cape Turiy, Kola Penin. Sometimes alters to anatase. Associated with barite, gypsum, aegirine, and microcline in syenite at the Diamond Jo quarry, Magnet Cove, AR; in Green River formation at Trona, WY; Mt. St.-Hilaire, QUE, Canada; several locations in the *Khibiny massif: Mt. Manepakh, Mt. Kukisvumchorr, and others*, also in the Nyorkpakhk quarry on Suoluaiv Mt. and on Mt. Yukspor (above the Hackman Valley); K-rich material on Mt. Evslogchorr, *Kola Penin., Russia*; Lovozero massif: Mt. Kuftnion, Mt. Karnasurt, Mt. Sengischorr, and several others; Kovdor massif carbonatites; Turi Cape, Kola Penin., Vuorijarvi, Karelia, Murun massif, Yakutia, and central Aldan Shield, Yakutia, Siberia, Russia. EF

60.1.4.1 Papagoite $CaCu^{+2}AlSi_2O_6(OH)_3$

Named in 1960 by Hutton and Vslidis for the Papago Indian tribe that once inhabited the region in which the mining center of Ajo, AZ is situated. MON. $C2/m$. $a = 12.94$Å, $b = 11.52$Å, $c = 4.68$Å, $\beta = 100.5°$, $Z = 4$, $D = 3.24$; PD 13-372 (nat.): 6.33_7 4.61_7 4.29_9 3.85_7 3.44_8 2.87_{10} 2.80_8 2.20_9. Structure: $R = 22\%$! *(BM 88:119(1965))*. Needs to be redetermined. Cerulean blue; very pale blue streak, equidimensional crystals < 1 mm long, slightly flattened || to {001} with faces in the zone [010] well developed. Elongated || b, all faces in the zone [010] are striated || to the zone axis; cleavage = distinct, || to (100); H 5–5½; G 3.25; TGA data *(AM 45:599(1960))*; partly soluble in boiling conc. HCl. Traces of Fe, Mn, etc. Biax. (−), $2V = 78°$, $N_\alpha = 1.607$, $N_\beta = 1.641$, $N_\gamma = 1.672$; pleochroism strong: X = colorless to very pale greenish blue, Y = blue, Z = deep greenish blue; $X \wedge c = 44°$, $x \wedge a = 35°$, both in acute β, $Y \wedge c = 46°$, Y to the perpendicular to (001) is $\approx 36°$, $Z = b$; absorption, $Z > Y > X$. Dispersion = very faint $R > V$. Occurs at the *New Cornelia mine, Ajo, Pima Co., AZ*; with ajoite in metasomatically altered quartz-albite rock; dark blue inclusions in quartz crystals from the Messina mines, notably the no. 5 shaft, RSA. EF *Minerali 3(2):74(1981)*.

60.2.1.1 Kainosite-(Y) $Ca_2(Y,Ce)_2Si_4O_{12}(CO_3) \cdot H_2O$

Named in 1886 by Nordenskiöld from the Greek for *unusual*, in allusion to the composition: a silicate with CO_3. Synonym: cenosite. ORTH $Pmnb$. $a = 13.011$, $b = 14.310$, $c = 6.757$, $Z = 4$, $D = 3.54$. *NJBM 4:153(1989)*. *14-332*(nat): 4.83_5 3.45_7 3.29_8 3.19_8 2.76_{10} 2.17_7 2.11_5 1.93_6. The Ca and (Y,REE) polyhedra form large, alternating sheets along b. The connection, by shared edges and

60.2.1.1 Kainosite-(Y)

corners of the polyhedra, forms large "hexagonal" holes in which the Si_4O_{12} rings and CO_3 groups are accommodated in the (001) plane across the (010) sheets. Y and Ca independently occupy two crystallographic sites. Prismatic on [001] to 8 mm, rarely to 2 cm long. {001}, {100}, {010} pinacoids and {110} prism. Elongated on c, also flattened on {100}; 3–5 mm long at Mt. St.-Hilaire, to several centimeters from Norway. Pale yellowish brown-brown, sometimes colorless or white, also rose-pink; white, pale rose-pink, or pale brown streak; vitreous–greasy luster, pearly on some faces. Cleavage at 90°, (100) and (010); sometimes no cleavage is developed; uneven to subconchoidal fracture; brittle. H = 5–5$\frac{1}{2}$. G = 3.52. IR data: *MIN 3(2):54(1981)*, *DAN SSSR 294:948(1987)*. Nonfluorescent. Soluble in acids. In HCl, gives off CO_2. Often contains minor Na. Biaxial $(-)$; XYZ = cba; $N_x = 1.658$–1.667, $N_y = 1.681$–1.689, $N_z = 1.683$–1.692; 2V = 40°; r < v, weak to moderate; nonpleochroic. Most often occurs in granitic pegmatites or alpine veins; also in skarns. In cavities in breccia at Mt. St.-Hilaire, and in granite at Baveno, Italy, and Golden Horn batholith, WA. Associated with quartz–feldspar-zinnwaldite; with diopside, clinochlore, magnetite, and apatite at Nordmark in an iron mine. In thorium–quartz veins at Porthill, ID. May form from yttrialite, also as pseudomorphs after yttrialite. Occurs in granite with kamphaugite-(Y) at Mount Desert Is., ME; in dikes on SR30 N of Long Lake village, NY; with epidote, fluorite, calcite and quartz; also found with calcite in galena mines at Rossie, NY; in the Cotopaxi skarn, Fremont Co., CO; at Porthill, Boundary Co., ID; the Golden Horn batholith, Okanogan Co., WA; very rare at Mt. St.-Hilaire, QUE, North Burgess Twp., Lanark Co., and Bicroft and Grayhawk mines, Bancroft, ONT, and in the Strange Lake Complex, Quebel-Labrador, Canada; *Igeltjern, Hitterö Island, Norway*; Ko mine, Nordmark, Värmland, Sweden; Mt. Blanc massif, Chamonix, Haute-Savoie, France; Baveno, Novara, Italy; Grimsel region, Aar massif, Switzerland; also in the Gotthard massif; crystals to 2 cm in the Curnera Valley, Switzerland; Itkinsk massif, western Tuva, Russia. EF *AM 49:1736(1964)*, *CM 8:1(1964)*, *DAN SSSR 294:948(1987)*.

= Class 61 =

Six-Membered Rings

BERYL GROUP

The beryl group consists of beryllium, magnesium, and iron aluminosilicate minerals. The general formula is

$$A_{2\text{-}3}B_2[Si_5(Si,Al)]O_{18}$$

where

A = Be, Mg, Fe
B = Al, Sc, Fe

and where the mineral crystallizes in a hexagonal or in a closely related orthorhombic space group.

The principal features of the beryl structure are hexagonal rings made up of six Si-O tetrahedra. The rings have their centers on sixfold axes and are stacked above one another. Within the rings, two of the oxygen atoms in each SiO_4 are shared by the SiO_4 groups on either side, thus giving the meta-silicate ratio SiO = 1:3. The unit cell has reflection planes parallel to the base at heights of 0, $c/2$, and c. The hexagonal rings lie with the silicon atoms and shared oxygen atoms on these planes: between them lie Al and Be atoms, each Al coordinated with an octahedral group of six oxygen atoms on a distorted tetrahedron. In these positions they link the oxygen atoms of the Si_6O_{18} rings, and the whole structure is linked laterally and vertically. The structure is like a honeycomb with open channels parallel to the c axis; no atomic center is nearer than 2.55 Å to the centers of these channels. Large alkali cations and water molecules, if present, are contained within the channels. *DHZ 1B:372(1986)*.

BERYL-CORDIERITE GROUP

Mineral	Formula	Space Group	a	b	c	
Beryl	$Be_3Al_2Si_6O_{18}$	$P6/mcc$	9.208		9.175	61.1.1.1
Bazzite	$Be_3(Sc,Fe)_2Si_6O_{18}$	$P6/mcc$	9.552		9.165	61.1.1.2
Indialite	$Mg_2Al_4Si_5O_{18}$	$P6/mcc$	9.80		9.35	61.1.1.3
Cordierite	$Mg_2Al_4Si_5O_{18}$	$Cccm$	17.062	9.721	9.339	61.2.1.1
Sekaninaite	$(Fe^{2+},Mg)_2Al_4Si_5O_{18}$	$Cccm$	17.234	9.824	9.298	61.2.1.2

61.1.1.1 Beryl

Note: Indialite, cordierite, and sekaninaite are also referred to as α-cordierite, cordierite-(Mg), and cordierite-(Fe), respectively.

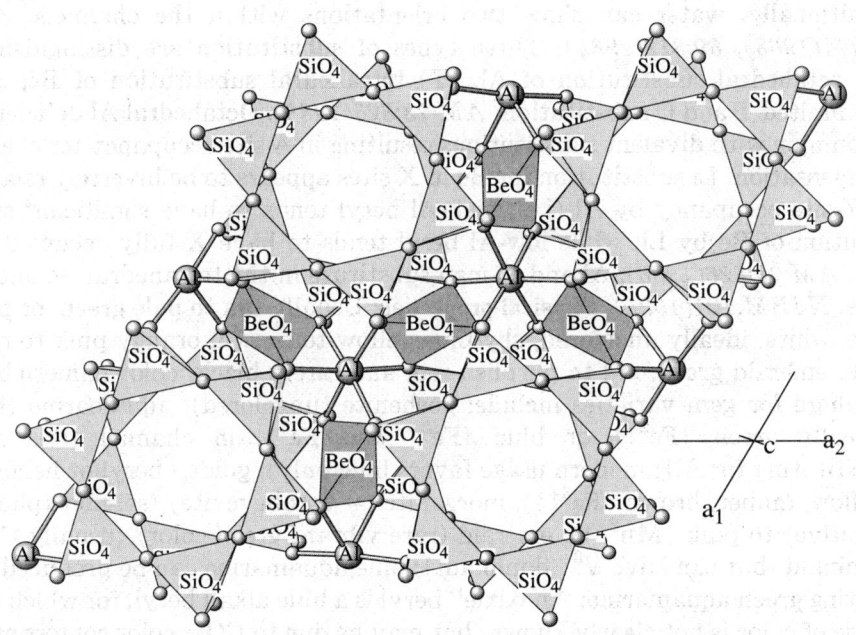

BERYL: Al$_2$Be$_3$Si$_6$O$_{18}$

61.1.1.1 Beryl Be$_3$Al$_2$Si$_6$O$_{18}$

Named from the Greek *beryllos*, an ancient name that was applied to many green minerals; its original significance is unknown. Beryl group. HEX $P6/mcc$. $Z = 2$, D(syn) = 2.64, for ideal synthetic: $a = 9.208$, $c = 9.175$; natural range: $a = 9.17 - 9.28$, $c = 9.14 - 9.254$. *9-430*(nat): 7.98$_9$ 4.60$_5$ 3.99$_5$ 3.25$_{10}$ 3.01$_4$ 2.87$_{10}$ 2.52$_3$ 1.99$_2$. High-iron (6.2 wt % Fe$_2$O$_3$3) beryl has $a = 9.28$, $c = 9.18$, with a significantly different powder pattern. *NJBM:193(1993)*. For: [A] X$_3$Y$_2$T$_6$O$_{18}$) A =☐, H$_2$O, OH, Na, Ca, Fe^{2+}, K, Cs, Rb, Li(?); X = B3, Al, Li, Si; Y = Al, Fe^{2+}, Mg, Fe^{3+}, Ca, Mn^{2+}, Cr^{3+}, V^{5+}, Te^{4+}, Sc; T = Si. Recent structure determinations include *CM 29:271(1991)* and *AM 78:762(1993)*. The silicate rings have a minimal opening of ~2.8 Å. The void between two stacked rings is maximally ~5.1 Å × 5.1 Å in (0001) and ~4.59 Å along c. The channels (A sites) within the silicate rings are ideally vacant, but alkalis, water, CO_2, and so on, can enter the site. Water and large cations such as Cs, Rb, or K occupy the central volume between adjacent rings while smaller substituents such as Na or OH occupy the minimum aperture of single rings. Additionally, water can show two orientations within the channels. *AM 53:777(1968), 69:319(1984)*. Three types of substitution are distinguished: (O) octahedral substitution of Al, (T) tetrahedral substitution of Be, and (N) limited T and O substitution. *AM 73:826(1988)*. Octahedral Al deficiency is common with divalent substitutions resulting in A site occupancy for charge compensation. Li substitution for Be in X sites appears to be inversely related to Y-site occupancy by Al (i.e., high-Al beryl tends to have significant substitution of Be by Li, while low-Al beryl tends to have X fully occupied by Be). *AM 73:826(1988)*. Al and Si may substitute in the tetrahedral Be and Si sites. *NJBM:121(1994)*. **Physical properties:** Usually tan to pale green, or pale blue, white, ideally uncolored, sky blue yellow to brown, orangy pink to rose pink, emerald green, red to purplish red, and rarely black, color names/chromophore for gem varieties include: goshenite (uncolored); aquamarine (historically green $\{Fe^{2+}\}$ or blue $\{Fe^{2+}$ and Fe^{3+}, in channel sites and substituting for Al$\}$; modern usage favors blue only), golden beryl or heliodor (yellow, amber, brown $\{Fe^{3+}\}$), morganite (= vorobeyevite) (salmon {photosensitive} to pink $\{Mn^{2+}\}$); emerald (rare vibrant green color) {usually Cr^{3+} dominant, but can have V^{3+} dominant}; blue aquamarine can be produced by heating green aquamarine; "maxixie" beryl is a blue alkali beryl, for which the cause of color is not clearly known, but may be due to $CO_3{-}$ color centers and/or gamma irradiation of Cs in the beryl. Color zoning frequently observed. Cleavage (0001), poor; twinning $\{31\bar{4}1\}$, $\{11\bar{2}0\}$, rare; $\{40\bar{4}1\}$, doubtful; brittle. H = $7\frac{1}{2}$–8, rarely as low as ~ 7. G = 2.63–2.78 rarely ranging to 2.92 [high alkalis, *DHZIB(1986)*]. Insoluble in common acids, not very soluble in HF. Soluble in NaOH or KOH. Soluble in NaOH or KOH.Spectral data (IR, NMR, Raman, and others): *DHZIB:372(1986), CM 32:55(1994)*. May fluoresce white in SWUV and/or LWUV light, and the fluorescence appears to be related to primary crystal growth. Beryl crystals are ordinarily columnar hexagonal prisms with [$10\bar{1}0$] dominant in combination with a basal pinacoid $\{0001\}$. Dominant forms include $\{0001\}$, $\{10\bar{1}0\}$, $\{10\bar{1}1\}$, $\{11\bar{2}1\}$, with the second order prism $\{11\bar{2}0\}$ being common only for some emeralds. Pyramidal faces are generally small. Crystals from gem pockets in evolved pegmatites can be complexly terminated and modified by a variety of pyramidal faces, commonly $\{11\bar{2}1\}$ and $\{10\bar{1}1\}$; also by second order and dihexagonal prisms. Such crystals (particularly those with a high alkali content) often have a tabular habit, being greatly flattened on c. Etch pits, growth features on faces, dissolution tubes, and so on, are common. Pegmatite crystals frequently reach 1 m, with one crystal of 18 m × 3.5 m (Malakialina). Organized clusters of beryl crystals are rare: in one locality in MG, Brazil, terminated crystals (to

5 cm × 1 cm) forming partly spherical clusters of well-terminated individuals. Three-dimensionally dendritic emeralds ("trapiche") from Mina Peña Blanca, Muzo (not Chivor), Colombia, were described. *AM 55:416,1808(1970)*. Chemistry: The BeO content of ideal beryl is 13.96 wt %, but contents between 12.5 and 13.2 wt % are common, with a low amount of 9.6 wt % due to alkali (Cs_2O to 6.7 wt %, Na_2O to 2.5 wt %, K_2O to 2.9 wt %, Li_2O to 1.3 wt %) and other substitutions. Beryls, especially from granite pegmatite miarolitic cavities, frequently have rims richer in alkalis with a consequent increase in the indices of refraction. Color zoning sometimes observed due to trace chromophores. Alkali substitution in beryl is limited to one per six Si or one divalent ion over the two octahedral sites, while Be ↔ Li is limited to 0.333 Li. *AM 73:826(1988)*. Chemical substitutions in beryl are generally either octahedral or tetrahedral; any resulting charge imbalances of one site preclude valence change in the other. Water average about 2 wt % and a maximum of about 4 wt %) can be present up to one molecule per six Si: pegmatite beryl generally having 0.3–0.6 per six Si with a maximum observed in a beryl (Altai Mts.) of 0.9 H_2O. Only the red beryls from Utah rhyolite are virtually anhydrous. Observed total alkalis can amount to over 11 wt %. *CM 11:714(1973)*. Trace elements include Co, Ni, Cu; other components (wt %): P_2O_5, to 3.60; B_2O_3, to 0.39; WO_3, to 0.39; ZrO_2, 0.80; Nb_2O_5, 1.75; Ta_2O_5, to 0.72; TiO_2, 0.05; MnO, to 0.75; Sn, 1; also CO_2, He, Ar: emerald (0.14–>1.2 wt % Cr_2O_3, generally 0.2–0.3 wt % [*EP(1989)*]; some emeralds colored by as much as 0.9 wt % V_2O_3); aquamarine (0.1–0.3 wt % FeO + Fe_2O_3). FeO can be as high as 2.2 wt %, but generally under 0.5; Fe_2O_3 has been observed to 6.2 wt % but is generally under 1 wt %. MgO may be as much as 3.2 wt %. Red beryl contains elevated amounts of Fe, Mn, Ti, Cs, and other elements. *GG 20:208(1984)*. Sc can be present in sufficient quantity, resulting in the species bazzite, but Sc is generally a trace element (less than 0.1 wt %) and intermediate beryl-bazzite specimens are almost unknown. Intrachannel cation exchangeability is known. *CM 27:663(1989)*. Many studies of beryl synthesis indicate beryl stability in saturated H_2O conditions ranging from ~ 230–680° at 1 atm to ~ 500–735° at 10 kbar; also phase relations of beryl and related Be silicates have been studied. *AM 71:277(1986)*. Emerald crystals for jewelry applications have been synthesized by flux melt techniques (Chatham, San Francisco; Lennix and Gilson, France; Kyocera, Japan) and by hydrothermal methods (Linde-Regency, Biron, etc.). **Optics:** Uniaxial (−). General refractive indices of beryl have been well recorded due to the interest of the gem industry in identifying gem materials. *MR 7:211(1976), EOB(1981)*. Some extreme values have also been recorded outside these values. (Low alkali, uncolored beryl): N_O = 1.568. N_E = 1.564, birefringence = 0.004; (emerald): N_O = 1.576–1.593, N_E = 1.569–1.586, birefringence = 0.005–0.007; (aquamarine): N_O = 1.575–1.585, N_E = 1.570–1.580, birefringence = 0.006; (red beryl): N_O = 1.576, N_E = 1.570, birefringence = 0.006; (high-alkali beryl): N_O = 1.585–1.598, N_E = 1.578–1.1589; (high-alkali–Fe–Mg beryl):

$N_O = 1.608$, $N_E = 1.599$; birefringence = 0.006–0.009. Length-fast with parallel extinction. Anomalous biaxial optics (2V to 17°) rarely observed, due to strain. Emerald may be slightly pleochroic. Maxixie beryl has $N_O = 1.592$, cobalt blue and $N_E = 1.584$, colorless, the reverse of the normal absorption scheme. The color fades to yellowish on exposure to light or heat. **Occurrence:** Occurs principally in granite pegmatites, less common in skarns and greisens, schists and gneisses, rhyolite (only source of red beryl), hydrothermal quartz veins (especially with Sn–W–mo minerals such as cassiterite, scheelite, wolframite, and molybdenite), calcite veins (emerald—Muzo), rarely in metasomatized peridotites and nepheline syenites. Emerald occurrences are restricted to sources of available Cr and/or V, in hydrothermal veins, pegmatites cutting ultramafic rocks, and schists. Beryl can be replaced by kaolinite or muscovite, rarely chlorite or margarite; sometimes by Be silicates, principally bertrandite, rarely bavenite, and so on; Be phosphates (hydroxyl-herderite, moraesite, hurlbutite, beryllonite, etc.), usually derived from dissolution of beryl, but sometimes occurring within replacements. **Localities:** Industrial beryl: *large crystals*: Bumpus quarry, Albany, ME; Beryl Mt. mine, Acworth, McInnes mine, Rumney, NH; Bob Ingersoll No. 1 mine, Keystone, SD; A-4 mine (379.5 metric tons), Malakialina district, Malagasy; Jackalswater district, Namaqualand, South Africa. For the following varieties of beryl, only noteworthy localities are listed: *Aquamarine*: Melrose Farm, Stoneham, ME; Rice mine, Groton, NH; Beryl Hill mine, Royalston, MA; Centreville district, ID; Mt. Antero and Mt. White, Chaffee Co., CO; Chimney Rock locality, Mourne Mts., Co. Down, Northern Ireland; Nerchinsk district(!) and Adun Chilon(!) and Sherlova Gora(!), Transbaikalia, Russia; Mimoso do Sul mine, Mimoso do Sul, Salinas, and at Ponte do Reis, Fazenda do Funil, Santa Maria da Itabira, MG, Brazil; Dassu (Dusso)(!) and Shingus(!), Baltistan and Chumar Bakhoor(!), Nagar, NWF, Pakistan, and other areas around Haramosh. Jos(!), Nigeria; Kleine Spitzkopje, Namibia; Ambatoharanana, Marijao, Tongafeno locality, Tsaramanga, Malagasy; *Green beryl*: Monetnaya Dacha mine, Tshaitansk and Miass district(!), Ural Mts., Russia; Ariranho, Rio Salinas placer(!), Barra de Salinas, Papamel placer(!), Marambaia, Mimoso do Sul mine, Mimoso do Sul, Pedroso placer, Padre Paraiso, Maxixe placer(!), Piaui valley, Resplendor(!), Rio Jequitinhonha placers (Araçuai, Salinas, Forteleza, Jequitinhonha), Ilha Alegre, São Pedro de Jequitinhonha, Teófilo Otoni(!), MG, Brazil; Muiâne, Alto Ligonha, Mozambique; Fefena, Malagasy. *High alkali and/or Fe–Mg beryl*: (6.7 wt % Cs_2O) in pegmatite at Bountiful Beryl prospect (Bluebird prospect), 15 miles S of Peach Springs, Mohave Co., AZ; Lassur mine, Ariège, France; Otov and Pisek, Bohemia; Skály, near Rýmařov, northern Moravia, Czech Republic. *Emerald*: Emerald and Hiddenite mine and Ellis mine, Hiddenite, Crabtree mine, Little Switzerland, and Turner mine, Shelby, Alexander Co., NC and mining has proceeded sporadically since that time, with much tally hoo but little commercial success; in pegmatites at Lake Mjosa, Byrud, north of Eidsvoll, Norway (vanadian, 0.9 wt % V_2O_3); Leckbach Ravine, Habachtal, Salzburg, Austria; Marinsky, Stretensky, and Troitsky mines, Takovaya district, Ural Mts., Russia; Panjsher Valley, Kapisa district, Afghanistan; Mingora, Charbagh, Makhad, and Gujarkili in the Swat district, Khaltaro in the Gilgit district, Barang and Gandao (vanadian beryls with 0.66 wt % V_2O_3), Mohmand district, NWFP, Pakistan; Jebel Sikait localities, Egypt; Taita district, Kenya (vanadian); Zeus mine, Sandawana, Mayfield, and others, Zimbabwe; Lake Manyara mine, Manyara district, Tanzania; Kafubu, Kitwe, and Miku, Zambia; Cobra, BVD, Gravelotte, and Somerset Hill mines, Gravelotte district, Transvaal, South Africa; Leysdorp, Tranvsvaal, South Africa; Novello claims, Victoria district, Zimbabwe (with Alexandria), Ajmer-Merwara field, India; Ankadilalana, and near Mananjary, Madagascar; Morrua area, Marropino district, Mozambique; Poona mine, Murchison Gold district, and Menzies, WA, Australia. Bode and Grota Funda mines, Carnaiba, Brumado and Salininha (vanadian), Bahia and Santana dos Ferros, Belmont, Itabira, and Capoeirana, near Nova Era, MG, and Santa Terezinha, Goias, Brazil; Chivor mine(!), Chivor, Muzo mines(!) and trapiche crystals from Mina Peña Blanca, Muzo, Buenavista mine, Ubalá, Gachalá mine(!), Vega de San Juan, and Burbar, Colombia; *Goshenite*: Tshaitansk(!) and Mursinka(!), Ural Mts., Russia; Resplendor(!), MG, Brazil; Ampangabe and Sahanivotry, Malagasy; Dassu(!), Baltistan, Pakistan. *Heliodor or golden beryl*: Trenton mine(!), Topsham, Maine; Slocum mine, East Hampton, Roebling mine, Upper Merryall, CT; NC; PA. Adun Chilon(!), Urgachen Mt., and Sherlova Gora, Transbaikalia, Russia and Wolodarsk-Wolynsk(!), Volyn, Ukraine, Zelatoya Voda, near Rangkul, Tadjikistan; Kokto-Kai, Xinjiangar, China; Magli, Nigeria; Rössing(!), Namibia; Kunar, Afghanistan. Golden beryl occurs in Brazil at Serra da Mesa, Minacu, GO; Sapucaia do Norte, M6. *Morganite and cesian beryl*: Bennett quarry(!), Buckfield, Tubbs Ledge prospect, Norway, ME; Gillette mine, Haddam, CT; Himalaya and San Diego mines, Mesa Grande and Stewart(!) and White Queen mines(!), Pala, CA; Facciatoia, Fonte del Prete, Grottia d'Oggi, Speranza, and Stabbiali quarries, San Piero di Campo, Elba, Italy I; Province, Afghanistan; Drot-Balachi, near Shengus, Gilgit-Skardu Road, NWF, Pakistan; Golconda mine, Governador Valadares, Corrego do Urucum(!), Galileia, Urubú mine(!), Teófilo Otoni, and large crystals to 25 kg from Resplendor, MG, Brazil; Muiâne(!), Mozambique; Ampangabe, Anjanabonoina, Marihitra district, Tsaravovonana(!), Tsilaizina, Vohidahy, Malagasy. *Red beryl*: Wildhorse Spring, Starvation Canyon, Topaz Valley prospects, and Pismire Wash, Thomas Range, Juab Co. and Violet claims(!), Wah Wah Mts., Beaver Co., UT; Black Mesa, Grant Co., and Paramount Canyon, Sierra Co., NM; San Luis Potosi, Mexico. *Rubidian beryl*: Mt. Mica mine, Paris, Oxford Co., ME. Uses: as an ore of Be [for beryllium copper (2% Be) and Be metal]. Also as a collectible mineral and art object, and as gemstones. The finest quality cut red beryl is >10,000/ct for stones >1ct. The separation of natural from synthetic emerald is usually made on the basis of refractive index and specific gravity. Both are generally lower for flux synthetics than natural stones. Fluorescence and examination of inclusions under magnification are also used for this distinction. Chemical distinction [e.g., *GG 20:141(1984)*, *JG 18:530(1983)*] may also be used but involves destructive techniques. "Maxixie-type" beryl is produced by exposing pale blue beryl to gamma rays. VK, EF, RVG *AM 47:672(1962)*, *MIN 3(2):83(1981)*, *BM 105:615(1982)*, *DHZ IB:372(1986)*.

61.1.1.2 Bazzite $Be_3Sc_2Si_6O_{18}$

Described in 1915 by Artini and named for Alessandro Eugenio Bazzi (1862–1929), discoverer of the mineral. Beryl group. HEX $P6/mcc$. Z = 2; D = 2.82; (nat): a = 9.501–9.549, c = 9.152–9.210; (syn): a = 9.552, c = 9.165. *AM 53:943(1968)*. *20-165*(syn): 9.15_1 8.27_9 4.58_3 4.01_5 3.31_{10} 2.96_8 2.58_2 1.76_2. Structure is of the beryl type, but with Y filled by Sc and variable amounts of Fe^{3+} and sometimes Mg. *MP 52:113(1995)*. The octahedra are strongly compressed parallel to the c axis. This results in a larger a cell dimension than for beryl. A complete series to beryl is possible, but intermediate bazzite–beryl compositions are rare. Minor substitutions of Fe^{2+}, Fe^{3+}, Al^{3+}, Na, Cs, Mn. Analysis (wt %): Cs_2O, to about 3; Fe_2O_3, to >5.6. Crystals frequently occur as hexagonal prisms with {0001}, to several centimeters in length. Usually deep blue or shades of blue to blue-green, white streak, vitreous luster. Cleavage (0001), poor; brittle. H = 6.5–7. G = 2.77–2.8. IR spectral data: Raman spectral data: *PCM 17:395(1990)*. Insoluble in ordinary acids. Deep blue bazzite is electromagnetic. Uniaxial (−); N_O = 1.623–1.628, N_E = 1.602–1.607; dichroic; birefringence ≈ 0.02; E > O. E, azure to medium blue; O, pale greenish yellow or colorless. Cesian and ferrian bazzite has N_O = 1.637 and N_E = 1.622. Found in miarolitic granite cavities; granite pegmatite, or alpine veins frequently associated with albite, microcline, clinochlore, muscovite, bertrandite, beryl, fluorite, and rarely euclase. Occurs at the Government Pit, Albany, NY; Sugarloaf Mountain, Bethlehem, NH; Pea Ridge mine, AR; Mount Antero, Chaffee Co., CO; Hefrejern, Tørdal, Telemark, Norway; Monte Cervandonne, Val Strem, Grigioni, and Pizzo Cornera, Alpe Deverno, *Baveno, Lake Maggiore, Piedmont, Italy*; Val Strem (Graubunden), Tavetsch, Oberaarkraftwerkstollen, and Grosseidelhorn in Grimsel Pass, Bern, La Fibbia, and Monte Prosa in St. Gotthard Pass, Tessin, Furkabasistunnel, Uri–Wallis, Witen Alpstock, and Stremhörner, near Sedrun, Uri, Stollen Oberaar, Grimsel Pass, Gurtnellen, Reusstal, Uri, Aushaub–Naters, Brig and Mittel Hohtenn Tunnel, Wallis, Switzerland; Grandiorbruchsteinerleinbach, Röhrenbach, Germany; Heiligenblut, Carinthia, Austria; Kent massif, central Kazakhstan; Turé Penin., Russia. EF, VK *AM 53:943(1968)*, *SMPM 50:445(1970)*, *CA 93:171112(1980)*, *MP 43:131(1990)*, *MIN 3(2):100(1981)*.

61.1.1.3 Indialite $Mg_2Al_4Si_5O_{18}$

Described in 1954 by Miyashiro and Iiyama and named for the country of origin. Beryl group. See the Fe-dominant analog, sekaninaite (61.2.1.2); also cordierite (61.2.1.1), the low-temperature form. Synonym: α-cordierite. HEX $P6/mcc$ (Na and Cs) or $P\bar{6}$ [for K dominant, *AM 77:407(1992)*]. $a = 9.770$–9.812, $c = 9.345$–9.365, $Z = 2$, D(syn) = 2.51. *13-293*(syn): 8.48_{10} 4.89_3 4.09_5 $3.38_{5.5}$ $3.14_{6.5}$ $3.03_{8.5}$ 1.875_2 1.69_3. There is a more or less random distribution of Al and Si between the ring and linking tetrahedra. Ordering of Al and Si in all tetrahedra produces cordierite (ORTH *Cccm*). Pure Mg member is stable at 1 atm between 1455 and 1465°. Disordered high cordierite also crystallizes metastably below 1455° and can be retained by quenching. Short prismatic crystals. Colorless, vitreous luster. H = 7. G = 2.6. Natural material from India has the composition $(Mg_{1.40}Fe_{0.60})Al_{4.00}(Si_{4.89}Al_{0.11})O_{18}$. Uniaxial (–); $N_O = 1.526$, $N_E = 1.522$; also $N_O = 1.538$, $N_E = 1.535$, colorless or blue. Occurs in a sedimentary rock fused by the heat of a burning coal seam (India). Associated with sillimanite, spinel, and anorthite. in a polymetamorphosed pelitic rock. Intergrown cordierite and indialite occur in Japan. Bellerberg volcano, Mayen, Eifel region, Germany; *Bokaro coal field, Hazaribagh, Bihar, India*; Daimonji Yama, Kyoto Shi, Kyoto Pref., and Unazuki area, Toyama Pref., Japan. EF *CM 15:43(1977), CMP 80:110(1982), DHZ 1B:410(1986)*.

LOVOZERITE GROUP

The lovozerite group minerals consist of several isotypic species with the general formula

$$A_3B_3C_2M[Si_6O_{18}]$$

where

A = Na, Ca, H
B = Na, H, Ca
C = H, Mn, Ca, Fe, Zr
M = Zr, Ti, Ca, Fe

and a structure the same as, or similar to, that of lovozerite-R.

The structures of the lovozerite group involve a fundamental pseudocubic block approximately 7.5 Å on edge and having $\bar{3}m$ symmetry. The block contains all the components described by the general formula, with an isolated ring of six silica tetrahedra lying at the center of, and normal to, the trigonal axis of the block. The remaining cation sites, usually not fully occupied, lie at edge centers (A), face centers (B), on the triad (C), and at the corners (M) of the block.

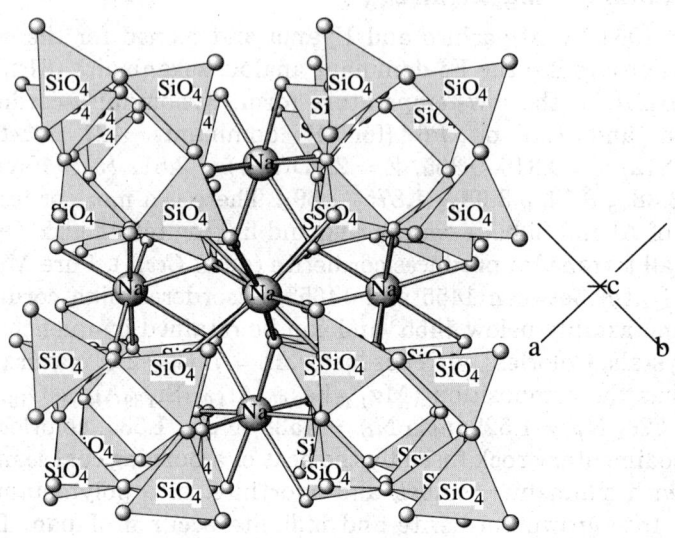

LOVOZERITE: $(Na, Ca)_2(Zr, Ti)Si_6O_{18} \cdot 3H_2O$

Highly contorted six-membered silicate rings (they may look four-membered in projection) form layers parallel to (001). Layers are linked by Zr, and Na occurs both between rings in the layers and in interlayer positions.

LOVOZERITE GROUP

Mineral	Formula	Space Group	a	c	
Lovozerite	$H_4Na_2Ca(Zr, Ti)[Si_6O_{18}]$	$R\bar{3}m$	10.174	13.053	61.1.2a.1
Kazakovite	$Na_6(Mn, H_2)Ti[Si_6O_{18}]$	$R\bar{3}m$	10.162	13.056	61.1.2a.2
Tisinalite	$H_3Na_3(Mn^{2+}, Ca, Fe^{2+})Ti[Si_6O_{18}] \cdot 2H_2O$	$R\bar{3}m$	10.14	13.08	61.1.2a.3
Zirsinalite	$Na_3Na_3CaZr[Si_6O_{18}]$	$R\bar{3}m$	10.29	13.11*	61.1.2a.4
Combeite	$Na_4Ca_4[Si_6O_{18}]$	$R\bar{3}m$	10.429	13.146	61.1.2a.5

*Also reported as $c = 26.31$.

The same pseudocubic blocks, with slight shifting of cation sites and consequent reduction in symmetry to $2/m$ or 2, may be stacked in several different sequences to yield the following orthorhombic and monoclinic species.

Lovozerite Group: Related Structures

Mineral	Formula	Space Group	a	b	c	β	
Imandrite	$Na_{12}Ca_3Fe_2^{3+}[Si_6O_{18}]_2$	$Pmnn$	10.331	10.546	7.426		61.1.2b.1
Koashvite	$Na_6(Ca,Mn)(Ti,Fe)[Si_6O_{18}] \cdot H_2O$	$Pmnb$	10.169	20.899	7.335		61.1.2b.2
Petarasite	$Na_5Zr_2[Si_6O_{18}](Cl,OH) \cdot 2H_2O$	$C/2m$	10.796	14.493	6.623	113.21	61.1.2b.3

Notes:
1. Structure determinations are lacking or only inferred for many of the species, and physical property data are lacking or incomplete, particularly for the dimorphic combeite and lovozerite.
2. Additional species, not included here, may be based on the same or similar structural elements.

SPC 35:227(1990).

61.1.2a.1 Lovozerite $H_4Na_2Ca(Zr,Ti)[Si_6O_{18}]$

Named in 1939 by Gerasimovskii for the locality. Lovozerite group. Rhombohedral and monoclinic dimorphs have been reported, but their descriptions are poor and they may not be distinct phases. The names lovozerite-R (lovozerite-T) and lovozerite-M have been used for the possible dimorphs. HEX-R $R\bar{3}m$. $a = 10.174$, $c = 13.053$, $Z = 3$. 28-1201: 7.33_5 5.27_8 3.64_7 3.32_7 3.21_{10} 2.97_5 2.65_5 1.826_5. Also reported as MON $C2$. $a = 10.48$, $b = 10.20$, $c = 7.33$, $\beta = 92.5°$, $Z = 2$, $D = 2.63$. PD: 3.31_4 3.24_5 2.64_9 2.57_8 1.842_{10} 1.543_4 1.498_6 1.369_4. Structure: DANES 131:379(1960). Occurs as small equant crystals with pseudocubic, pseudododecahedral, or rhombohedral habit. Simple and polysynthetic twinning is common. Massive and irregular grains, and as pseudomorphs after eudialyte. Brown to black; vitreous to resinous luster; opaque, less often translucent. Cleavage undefined in two directions, brittle. $H = 5$. $G = 2.64$. Uniaxial $(-)$; $N_O = 1.560$, $N_E = 1.549$; also biaxial $(-)$; 2V small; $N_x = 1.549$, $N_y = 1.560$, $N_z = 1.561$; weakly pleochroic. Altered material usually lighter in color and friable. Found in sodalite xenoliths at Mt. St.-Hilaire, PQ, Canada; at Ilimaussaq, Greenland; at several localities in the Khibiny massif, and the *Lovozero massif, Kola Penin., Russia*. AR ZVMO 103:551(1974), AM 59:633(1974), MR 21:319(1990).

61.1.2a.2 Kazakovite $Na_6(Mn,H_2)Ti[Si_6O_{18}]$

Named in 1974 by Khomyakov et al. for Maria Efremovna Kazakova (b. 1913), Russian chemist who analyzed the material. Lovozerite group. HEX-R $R\bar{3}m$. $a = 10.162$, $c = 13.056$, $Z = 3$, $D = 2.97$. 26-1385: 3.60_7 3.28_6 3.17_6 2.60_{10} 2.52_8 1.816_8 1.529_6 1.480_7. Structure: SPD 24:132(1979). Rhombohedral crystals, 0.01–2.0 mm, with $\{11\bar{2}1\}$ and $\{11\bar{2}4\}$. Simple and polysynthetic twinning with twin axis perpendicular to $\{11\bar{2}4\}$. Pale yellow, vitreous to greasy luster. Uneven to subconchoidal fracture. $H = 4$. $G = 2.84$. Uniaxial $(-)$; $N_O = 1.648$, $N_E = 1.625$. Disseminated in ussingite, and associated with nordite, belovite, and vuonnemite in sodalite syenite at *Mt. Karnasurt, Lovozero massif, Kola Penin., Russia.* AR *ZVMO 103:342(1974), AM 60:161(1975).*

61.1.2a.3 Tisinalite $H_3Na_3(Mn^{2+},Ca,Fe^{2+})Ti[Si_6O_{18}]\cdot 2H_2O$

Named in 1980 by Kapustin et al. for the composition: titanium, silicon, and natrium. Lovozerite group. HEX-R $R\bar{3}m$. $a = 10.14$, $c = 13.08$, $Z = 3$, $D = 2.682$. 33-607: 5.19_7 3.60_{10} 3.26_6 3.18_8 2.92_3 2.59_6 2.51_5 1.802_5. Yellow-orange, vitreous luster. Uneven to conchoidal fracture. H = 5. Uniaxial (−); $N_O = 1.624$, $N_E = 1.590$–1.592. Found as crude crystals up to 1 mm in diameter and 1 cm aggregates formed by alteration of lomonosovite and barian lamprophyllite, and associated with koashvite and shcherbakovite, in drill core from *Mt. Koashva, Khibiny massif, Kola Penin., Russia.* AR *ZVMO 109:223(1980), AM 66:219(1981).*

61.1.2a.4 Zirsinalite $Na_6(Ca,Mn^{2+},Fe^{2+})Zr[Si_6O_{18}]$

Named in 1974 by Kapustin et al. for the composition: zirconium, silicon, and natrium. Lovozerite group. HEX-R $R\bar{3}m$. $a = 10.29$, $c = 13.11$, $Z = 3$, $D = 3.08$. 27-670: 4.22_4 3.68_4 3.33_4 3.26_6 2.64_9 2.57_8 1.842_{10} 1.539_4. Colorless to pale yellowish gray, vitreous luster. Conchoidal fracture. H = $5\frac{1}{2}$. G = 2.90. Uniaxial (−); $N_O = 1.610$, $N_E = 1.605$. Associated with aegirine, barian lamprophyllite, and lomonosovite in drill core from *Mt. Koashva, Khibiny massif, Kola Penin., Russia.* AR *ZVMO 103:551(1974), AM 60:489(1975).*

61.1.2a.5 Combeite $Na_4Ca_4[Si_6O_{18}]$

Named in 1957 by Sahama and Hytonen for Arthur Delmar Combe, Geological Survey of Uganda. Lovozerite group. The designations combeite-I and combeite-II are later (1990, Tamazyan and Malinkovskii). Phases I and II correspond to the high- and low-temperature polymorphs of the synthetic compound. Phase II was described in 1983 (Fischer and Tillmanns). **Combeite-I:** HEX-R $R\bar{3}m$. $a = 10.429$, $c = 13.146$, $Z = 3$, $D = 2.826(?)$. 25-800: 4.38_3 3.72_5 3.35_4 3.30_7 2.66_{10} 2.61_8 1.861_4 1.544_2; this powder pattern is given for the original description of combeite from Zaire. Structure: *NJMM:49(1983), AC(C) 43:1852(1987).* **Combeite II** (not formally named): HEX $P3_1$. $a = 10.464$, $c = 13.176$, $Z = 3$, $D = 2.79$ for German material; this later modified to $P3_121$. No powder data are given. The difference in the two structures is that the rings are highly symmetrical in phase I but not in phase II. The morphology of crystals from both sources is trigonal or hexagonal prismatic, markedly different from the rhombohedral habit of other $\bar{3}m$ lovozerites, suggesting that phase II may be present at both localities. Colorless. No cleavage. G = 2.84. Uniaxial (+/−) or nearly isotropic; $N_O = 1.598$, $N_E = 1.598$. Gelatinized by hot HCl. Found at the *Meiener field, Eifel, Germany* (combeite-II), and in melilite–nephelinites from *Mt. Shaheru, Rutshuru territory, Kivu, Zaire* (combeite-I and possibly, combeite-II); at the Oldoinyo Lengai volcano, Tanzania. AR *MM 31:503(1957), JG 97:367(1989).*

61.1.2b.1 Imandrite $Na_{12}Ca_3Fe_2^{3+}[Si_6O_{18}]_2$

Named in 1979 by Khomyakov et al. for the locality. Lovozerite group. ORTH $Pmnn$. $a = 10.331$, $b = 10.546$, $c = 7.426$, $Z = 1$, $D = 2.92$. PD: 3.97_3 3.73_5 3.33_6 3.29_2 2.63_{10} 2.59_2 1.853_7 1.520_5. Honey yellow, vitreous luster. H = 4. Biaxial (+); 2V = 75°; $N_x = 1.605$, $N_y = 1.608$, $N_z = 1.612$. Found as 1- to 3-mm anhedral, isolated grains and rims around and within fractures in eudialyte in alkalic pegmatite of the *Khibiny massif, near Lake Imandra, Kola Penin., Russia*. AR *MZ 1(1):89(1979), AM 65:810(1980), SPD 25:337(1980)*.

61.1.2b.2 Koashvite $Na_6(Ca,Mn)(Ti,Fe)[Si_6O_{18}] \cdot H_2O$

Named in 1974 by Kapustin et al. for the locality. Lovozerite group. ORTH $Pmnb$. $a = 10.169$, $b = 20.899$, $c = 7.335$, $Z = 4$, $D = 2.99$. 27-669: 3.66_5 3.57_2 3.28_5 2.62_4 2.58_{10} 1.820_7 1.755_3 1.294_4. Structure: *MZ 2(5):40(1980)*. Pale yellow, vitreous luster, transparent. Conchoidal fracture. H = 6. G = 2.98–3.02. Biaxial (−); 2V = 82–84°; $N_x = 1.637$, $N_y = 1.643$, $N_z = 1.648$; r > v, weak. Some references do not include water in the formula. In alkalic pegmatite as veinlets replacing lomonosovite and associated with pectolite, villiaumite, and tisinalite at *Mt. Koashva, Khibiny massif, Kola Penin., Russia*. AR *ZVMO 103:559(1974), AM 60:487(1975)*.

61.1.2b.3 Petarasite $Na_5Zr_2[Si_6O_{18}](ClOH) \cdot 2H_2O$

Named in 1980 by Chao et al. for Peter Tarasoff (b.1934), assistant director, Noranda Research, and amateur mineralogist, Dollard-des-Ormeaux, Quebec, Canada. Lovozerite group. MON $C2/m$. $a = 10.796$, $b = 14.493$, $c = 6.623$, $\beta = 113.21°$, $Z = 2$, $D = 2.86$. 33-1310: 7.25_7 6.09_4 4.96_1 4.10_{10} 3.22_3 3.04_3 2.92_{10} 1.729_2. Structure: *CM 18:503(1980)*. As short prismatic [001] crystals, 0.5 mm to 6.5 cm long. Forms {010}, {001}, {110}, {20$\bar{1}$}, and {$\bar{1}$11} dominant. Orange, orange-yellow, greenish yellow, brown, rarely rose to violet; opaque to translucent, rarely transparent. Cleavage {110}, perfect; {010}, {001}, good. H = 5–5$\frac{1}{2}$. G = 2.88. Biaxial (+); 2V = 29°; $N_x = 1.596$, $N_y = 1.598$, $N_z = 1.632$ (all for N_O); X = b, Z ∧ c = 41.5°; r < v, weak, pleochroic, Z = Y > X; pale greenish yellow to colorless. Decomposes to parakeldyshite at 1100°. Found in interstices of biotite crystals with apatite, catapleiite, and zircon in a xenolith in nepheline syenite at the *Demix quarry, Mt. St.-Hilaire, PQ, Canada*. AR *CM 18:497(1980), MR 21:329(1990)*.

61.1.3.1 Dioptase $Cu_6(Si_6O_{18}) \cdot 6H_2O$

Named in 1797 by Haüy from the Greek for *through* and *to see*, in allusion to the visibility of internal cleavage planes. HEX-R $R\bar{3}$. $a = 14.566$, $c = 7.778$, $Z = 3$, $D = 3.296$. 33-0487: 7.29_8 4.90_4 4.06_4 2.60_{10} 2.44_4 2.11_5 1.506_4 1.420_5. Structure: *AM 62:807(1977)*. The formula is chosen to show the six-membered ring structure. Long to short prismatic [0001], dominant forms {11$\bar{2}$0} and {02$\bar{2}$1}, {13$\bar{4}$1}; crystalline aggregates; massive. Twinning on {10$\bar{1}$1} uncommon. Bright emerald green, vitreous to greasy luster. Cleavage {10$\bar{1}$1}, perfect. Uneven to conchoidal fracture. H = 5. G = 3.28–3.35. Uniaxial (+);

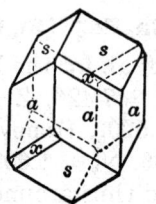

DIOPTASE
Forms: a, {11$\bar{1}$0}; s, {02$\bar{2}$1}; x, {13$\bar{4}$1}.

$N_O = 1.652$–1.658, $N_E = 1.697$–1.709; weakly pleochroic, O > E; rarely biaxial in sectors with OAP parallel to {11$\bar{2}$0}. Gelatinized in HCl. A frequent secondary mineral in many oxidized zones of copper deposits. Found at the Mammoth–St. Anthony mine, Tiger, and other localities in Pinal Co., at the Christmas mine, Gila Co., and in Cochise, Greenlee, and Yuma Cos., AZ; Soda Lake Mts., San Bernardino Co., CA; Nishapur, Iran; *Altyn Tube, Khirghiz Steppes, Kazakhstan*(!). At Reneville(!), the Sanda mine at Mindouli(!), 'Mbumba, and Pimbi, southern Congo; at Mavoyo, Angola; Guchab, Kaokoveld, and Tsumeb, Namibia, at the last in exceptional, brilliant crystals up to 5 cm long. At the Malpaso mine, Cordoba, Argentina, and at Ihca de Oro, Copiapó, Chile. An incidental ore of copper; crystal clusters used ornamentally; faceted as a collector gem. AR *D6:463(1892)*, *DANES 155:125(1964)*, *ZK 187:15(1989)*.

61.1.4.1 Baratovite $KLi_3Ca_7(Ti,Zr)_2[Si_6O_{18}]_2F_2$

Named in 1975 by Dusmatov et al. for Rauf Baratovich Baratov (b.1921), Academician and Soviet petrographer, Tadjik Academy of Sciences, Dushanbe, Tadjikistan. Compare katayamalite (61.1.4.2). MON $C2/c$. $a = 16.941$, $b = 9.746$, $c = 20.907$, $\beta = 112.50°$, $Z = 4$, $D = 2.912$. *29-821*: 4.81_1 3.54_1 3.22_{10} 3.02_1 2.41_2 1.92_2 1.60_1 1.49_1. Structure: *AM 64:383(1979)*. The structure consists of paired layers of six-tetrahedron rings linked by a layer of six-coordinated calcium; the resultant slabs are linked by the remaining cations. White; vitreous luster, pearly on {001}. Cleavage {001}, perfect. $H = 3\frac{1}{2}$. $G = 2.92$. Biaxial (+); $2V \approx 60°$; $N_x = 1.672$, $N_y = 1.672$, $N_z = 1.673$; $X \wedge c = 50°$; birefringence very low; r > v, strong. Found as platy aggregates with muscovite, ekanite, and sphene in quartz–albite–aegirine veinlets in syenite at the *Dara–Pioz alkaline massif, Alai Range, Tien Shan, Tadjikistan*. AR *ZVMO 104:580(1975)*, *AM 61:1053(1976)*.

61.1.4.2 Katayamalite $(K,Na)Li_3Ca_7Ti_2[Si_6O_{18}]_2(OH,F)_2$

Named in 1983 by Murakami et al. for Nobuo Katayama (b.1910), prominent Japanese mineralogist, University of Tokyo. See baratovite (61.1.4.1). MON $C2/c$. $a = 16.94$, $b = 9.72$, $c = 20.91$, $\beta = 112.4°$, $Z = 4$, $D = 2.90$. *37-470*: 3.23_{10} 3.06_3 2.94_3 2.90_3 2.42_3 1.933_4 1.841_2 1.382_2. Structure: *EJM*

4:839(1992). Isostructural with baratovite and possibly a variety thereof; originally described as TRIC $C\bar{1}$. $a = 9.721$, $b = 16.932$, $c = 19.942$, $\alpha = 91.43°$, $\beta = 104.15°$, $\gamma = 89.94°$. As tabular {001} crystals and frequent twins on {001}. White; vitreous luster, pearly on {001}. Cleavage {001}, perfect. $H = 3\frac{1}{2}$–4. $G = 2.91$. Biaxial (+); $2V \approx 32°$; $N_x = 1.670$, $N_y = 1.671$, $N_z = 1.677$; $Y \approx b$, $Z \wedge$ cleavage $\approx 54°$; $r > v$, strong. Bright blue-white fluorescence in SWUV. May be dimorphous with baratovite, but differs from the latter in the lack of zirconium and the predominance of hydroxyl ion over fluorine. Found in aegirine syenite with albite, aegirine, and pectolite on Iwagi Islet, Ehime Pref., Japan. AR *MJJ 11:261(1983)*, *12:206(1985)*; *AM 69:811(1984)*.

61.1.5.1 Odintsovite $K_2Na_4Ca_3Ti_2Be_4Si_{12}O_{38}$

Named by Konev et al. in 1995 for Prof. M. M. Odintsov (1911–1979), well-known Siberian scientist and a founder of the Institute of the Earth's Crust, Irkutsk, Russia. ORTH $Fddd$ $a = 14.243$Å $b = 13.045$Å $c = 33.484$Å $Z = 8$ $D = 2.91$; PD 9.23_9 4.15_{10} 3.30_{10} 3.16_{10} 2.53_{10} 2.42_{10} 1.67_8 1.63_8. Based on a crystal structure determination (K 40:253(1995); Eng. Trans. as Crystallography Reports 40:228(1995)), the structural formula is: $K_2(Na_{3.75}Li_{0.25})Ca_3(Ti_2O_2[Be_4(Si_6O_{18})_2]$. Six-membered Si–O rings, discrete Be-tetrahedra, and edge-sharing Ti-octahedra form a mixed framework with the channels extended along the two orthogonal directions. The K atoms fill the channels, with the Ca and Na atoms located between them. Transparent, pink with a brownish tint, dark red, seldom white; anhedral 0.1–1 mm grains, and aggregates of grains (0.5–4 mm) up to 7 mm. Vitreous, brittle, no cleavage observed, uneven fracture, $G = 2.94$–2.98, $H = 5$–$5\frac{1}{2}$. Non-fluorescent, no thermal effects to 1000°C, and neither OH or H_2O were detected by IR. Biaxial (+), $N_\alpha = 1.630$, $N_\beta = 1.644$, $N_\gamma = 1.675$ in white light, $2V_{meas}$ 70°, $XYZ = abc$, $X =$ colorless, $Y = Z =$ pink. Mean analytical results: Na_2O 8.57 (wt. %), K_2O 6.86, BeO 6.75, CaO 11.51, MnO 0.05, SrO 1.21, FeO 0.42, Al_2O_3 0.11, TiO_2 11.42, SiO_2 52.56, $\Sigma 99.46$. Occurs in a vein in alkaline syenite pegmatite and in kalsilite–syenite in the Malomurun massif (220 km south of Olekminsk), Aldan Shield, Russia. Associated minerals are: aegirine, K–feldspar, titanite, barytolamprophyllite, strontianite, wadeite, galena, and an unidentified mineral. EF ZVMO 124(5):92(1995).

61.2.1.1 Cordierite $(Mg,Fe^{2+})_2Al_3[AlSi_5O_{18}] \cdot H_2O$

Described in 1813 by Lukas and named for Pierre Louis A. Cordier (1777–1861), French mining engineer and geologist. Beryl group. Synonyms: iolite and dichroite. Dimorphous with indialite (completely disordered high-temperature form), which is truly hexagonal and is isostructural with beryl. ORTH *Cccm*. (Nat): $a = 17.088$, $b = 9.734$, $c = 9.359$ [*AM 64:337(1979)*]; (syn Mg-cordierite): $a = 17.062$, $b = 9.721$, $c = 9.339$, $Z = 4$, $D = 2.51$. *13-294*(syn Mg-cordierite): 8.52_9 8.45_{10} 4.09_5 3.38_5 3.13_6 3.04_6 3.01_5. A framework silicate with structural channels formed by stacking of six-membered Si,Al-ordered rings of corner-sharing tetrahedra ($Si_4Al_2O_{18}$) along the c axis. In natural specimens, these channels may contain molecules such as H_2O, CO_2, N_2, He, and Ar. Structurally closely related to beryl. A so-called distortion index, $\delta = 2\theta_{131} - (2\theta_{511} + 2\theta_{421})/2$, is a measure of the degree of departure from hexagonality. Maximum value for δ is about 0.31. Undergoes a first-order phase transition to indialite. *AM 68:60(1983)*. An intermediate short-range ordered domain structure exists (modulated structure). *AM 64:337(1979)*. **Habit:** Crystals are short prismatic, granular, or massive aggregates. {110} and {310} simple, lamellar and cyclic twinning may be present. Crystals may be several centimeters or more in size. **Physical properties:** Blue, blue-green, gray to violet; blue color is believed due to Fe^{2+}–Fe^{3+} charge transfer; white streak; vitreous–greasy luster. Cleavage {100}, moderate; {001}, {010}, poor; conchoidal to uneven fracture; brittle. H = 7. G = 2.53–2.65. Spectral and thermal data: *DHZ IB:410(1986)*. **Chemistry:** Partially decomposed by acids. Mossbauer spectral studies of cordierites show that alkali-free cordierites have Fe^{2+} present in the octahedral position. Those with the largest concentrations of alkali cations display in addition a second Fe^{2+} position (i.e., channels or tetrahedral sites). *GSAAP 27(6):A-364(1995)*. IR (LIT 32:95(1994)) shows both type I and II water and CO_2 as fluid constituents in the channel sites. Substitutions of the type $Be^{2+} + Na^+ \leftrightarrow Al^{3+}$ or $Na + Li \leftrightarrow Mg$ occur. Most cordierites are rich in Mg and specimens with Fe/(Fe + Mg) > 0.5 are rare (sekaninaite). Fe-rich cordierites are mostly in pegmatites. $FeO \gg Fe_2O_3$; BeO to 1.77 wt %; may contain trace to minor amounts of Na and K also, CO_2 to 2.2 wt %, H_2O to 0.5 atom. Complete formula is Na_{x+2} $(Mg,Fe,Mn)_{2-y}$ $(Al_{4-x}Be_xSi_5)$ $O_{18} \cdot n(H_2O, CO_2)$. **Optics:** Biaxial (+) or (−), most are (−); $2V_x = 35$–$106°$; (syn Mg-cordierite): $N_x = 1.522$, $N_z = 1.527$; (nat): $N_x = 1.527$–1.560, $N_y = 1.532$–1.574, $N_z = 1.537$–1.578. Mg-rich: X, pale yellow or green; Y, violet or violet-blue; Z, pale blue; Y > Z > X; v > r, weak; OAP = (010), XYZ = cba. **Occurrence:** Occurs in thermally metamorphosed rocks, particularly those derived from argillaceous sediments. Also a common product of high-grade regional metamorphism and may occur in schists, gneisses, and granulites. It may also be found in some basic igneous rocks and in some granites. In these rocks, the formation of cordierite is generally considered to be a result of assimilation of Al-rich sediments by acid or basic magmas. The mineral occurs more rarely in pegmatites and quartz veins. The most common alteration products of cordierite are chlorite–sericite aggregates (pinnite). **Localities:** Many worldwide localities. Several localities (Haddam, New London, Guilford, Plymouth) in CT; Garnet Island, NWT, Canada(!), in flawless crystals to 5 cm; notable localities include: Bodenmais and Wechselburg, Bavaria, Germany; beryllian cordierite occurs in pegmatites at Biskupice, Dolní Bory, and Věžná, Czech Republic; Orijarvi, and Pielavesi, Finland; Kragerö and Arendal, as well as the Bamble sector, Norway; Naversberg, Sweden; Cabo de Gato, Spain; gem-grade material occurs in the Sasyksky area, Muzkolsk metamorphic complex in the eastern Pamirs, SW of Rankul, Tadzikhistan; Mt. Bity, Madagascar (gem material); in gem gravels in Sri Lanka; fine, massive, gem-quality material from the Blue Dragon mine, Orange R., Namaqualand, Cape Prov., and Guadon Farm, 70 km east of Steinkopf, North Cape, South Africa(!); White Well, WA, Australia; gem material occurs at Umburang, Paraiba, Brazil; beryllian cordierite occurs at Soto, Argentina. A nearly high cordierite is present in the Allende meteorite (chondrite), Parral, CHIH, Mexico. Uses: Cordierite (iolite) is used as a gemstone; however, most is used in ceramics applications because of its very low thermal expansion coefficient and resistance to thermal shock. Insulator for spark plugs, and refractory coating on metals for internal combustion and turbine engines. Gem material shows fine pleochroism from colorless to yellow to deep blue (when viewed perpendicular to c). The principal gem source is Madras, India; however it is produced in Sri Lanka from gem gravels; Myanmar (Burma); Tanzania, and Mr. Biry, MAlagasy Republic. EF *Armbruster, 1986, Crystal Chem. of Mins.; Proc. 13th Gen. Meet. IMA:485(1986), MIN 3(2):110(1981)*; ZVMO 124:76(1995)

61.2.1.2 Sekaninaite $(Fe^{2+},Mg)_2Al_4Si_5O_{18}$

Named in 1975 by Stanek and Miskovsky for Josef Sekanina (b.1901), Czech mineralogist, who discovered the mineral in 1928. Beryl group. See cordierite (61.2.1.1), the Mg analogue. Much of the literature still refers to Fe-cordierite (sekaninaite) and Mg-cordierite. Cordierites with $Fe^{2+}/(Fe^{2+}+ Mg) > 0.5$ are not that rare. ORTH $Cccm$. (Nat): $a = 17.186$–17.230, $b = 9.817$–9.835, $c = 9.298$–9.313; (syn): $a = 17.234$, $b = 9.824$, $c = 9.298$; BMP 11:297(1965); (syn Fe-cordierite): $a = 17.251$, $b = 9.834$, $c = 9.281$ [*AM 62:395(1977)*]; $Z = 4$, $D = 2.78$. *31-616*(nat): 8.58_{10} 4.08_6 3.39_{10} 3.38_{10} 3.16_4 3.14_6 3.07_7 3.04_5; *31-615*(syn): 8.63_7 8.55_7 4.09_7 3.39_{10} 3.38_6 3.14_4 3.07_5 3.06_5. For structural details, see cordierite (61.2.1.1). Sekaninaite has differing degrees of Al–Si order. *AM 64:337(1979)*. Mössbauer data indicate that divalent iron is strongly predominant over trivalent iron and is in the octahedral site only. The color and pleochroism of Fe-bearing cordierite and sekaninaite have been interpreted to arise from charge transfer between octahedral Fe^{2+} and channel (?), tetrahedral or octahedral Fe^{3+}. Short, prismatic, pseudohexagonal; twinning present on {110} and {310}; crystals may be 60–70 cm in length. Light blue to blue violet, white streak, vitreous luster. Cleavage {100}, good; indistinct parting on {001} and {010}; brittle. H = 7–7$\frac{1}{2}$. G = 2.76–2.77. IR, DTA, TGA data: *AM 62:395(1977)*. Other spectral data: *DHZ IB:410(1986)*. Mössbauer spectral data (CM 35:167(1997)). See *DHZ IB:410(1986)* and CM 35:167(1997) for chemical analyses. May contain some Li (0.1–0.54 wt % Li_2O), Na_2O (0.13–1.57 wt %), Ca 0.0x wt % and water (about 3%) in addition to major constituents. Li is incorporated by the substitution $Li^{VI}Na^{ch}(Mg, Fe)^-$. Fe_2O_3 is probably about 0.1 wt %. Biaxial (−); XYZ = cba; $N_x = 1.561$, $N_y = 1.572$, $N_z = 1.576$; $2V_x = 66$–$68°$; Y > Z > X; v > r, weak. X, colorless; Y, blue; Z, pale blue. Usually occurs in pegmatites emplaced in granulitic gneisses, but also in peraluminous granites and metamorphic xenoliths. Alters to septechamosite, andalusite, chloritoid, ferrogedrite, and siderophyllite. *GG 34:141(1993)*. Occurs at Brockley, Rathlin Island, County Antrim, Northern Ireland; in simple beryl–columbite subtype pegmatites at *Dolni Bory, western Moravia, Czech Republic*; at Gåsborn, Bergslagen, Sweden; Kemiö, Finland; in Japan, from Ide, Kyoto Pref.; Ishii, Yamaguchi Pref.; Kitasugama, Fukushima Pref.; Sasago and Doshi, Yamanashi Pref.; and from the Hanoka mine, Kagoshima Pref. EF *AM 65:522(1980)*.

61.2.2.1 Jonesite (Na,K)$_2$Ba$_4$Ti$_4$[AlSi$_5$O$_{18}$]$_2$ · 6H$_2$O

Named in 1977 by Wise et al. for Francis Tucker Jones (1905–1993), Berkeley, California, mineralogist and chemical microscopist, U.S. Department of Agriculture, who first observed crystals of the species. ORTH $B22_12$. $a = 13.730$, $b = 25.904$, $c = 10.608$, $Z = 4$, $D = 3.24$. *29-983:* 13.0$_{10}$ 3.03$_4$ 3.01$_2$ 2.65$_3$ 2.60$_2$ 2.23$_2$ 2.16$_2$ 2.07$_2$. Laths elongated [001] and flattened {010} with {010}, {310}, {210}, and {101} dominant. Colorless, vitreous luster. Cleavage {010}. H = 3–4. G = 3.25. Biaxial (+); 2V = 76–78°; N$_x$ = 1.641, N$_y$ = 1.660, N$_z$ = 1.682; XYZ = *bac*. Fluoresces dull orange under SWUV. Found associated with neptunite and joaquinite, but not benitoite, in natrolite veins at the *Benitoite Gem mine, San Benito Co., CA*. AR *MR 8:453(1977)*.

61.2.3.1 Lourenswalsite (K,Ba)$_2$Ti$_4$(Si,Al)$_6$O$_{14}$(OH)$_{12}$

Named in 1987 by Appleman et al. for Lourens Wals (b.1939), a leading Belgian mineral collector. Pseudo-HEX Space group unknown. $a = 5.244$, $c = 20.49$, $Z = 1$, $D = 3.199$. *40-1453:* 10.22$_2$ 4.08$_1$ 3.93$_2$ 3.43$_1$ 2.61$_{10}$ 2.25$_2$ 1.515$_8$ 1.311$_2$. Silver-gray to light brownish gray, pearly to dull luster. Friable; cleavage {001}, good. G = 3.17. Biaxial (−); 2V ≈ 0°; N$_x$ = 1.815, N$_y$ = 1.840, N$_z$ = 1.840; nonpleochroic. Formed under oxidizing conditions as rosettes of pseudo–hexagonal plates at the *Diamond Jo quarry, Magnet Cove, near Hot Springs, AR*. AR *MM 51:417(1987), AM 73:1493(1988)*.

TOURMALINE GROUP

The tourmaline group is an important group of minerals corresponding to the general formula

$$XY_3Z_6(BO_3)_3Si_6O_{18}(OH,O)_3(OH,F)$$

where

X = Na, Ca, □, sometimes Mg, Fe^{2+}, Mn^{2+}, rarely K, possibly H$_3$O$^+$
Y = Fe^{2+}, Mg, Al, Li, Fe^{3+}, Mn^{2+}, Mn^{3+}, Cr^{3+}, Ti^{4+}, Cu, Zn, trace REE
Z = Al, Fe^{3+}, Fe^{2+}, Cr^{3+}, V^{3+}, Mn^{3+}, Ti^{3+}, Ti^{4+}
O$_1$ = OH, O, rarely Cl
O$_2$ = OH, F, rarely Cl
T = Si, very minor Al and/or B, rarely very minor Ti^{4+}, P

Tourmaline Group

The space group is $R3m$, and Si and O atoms are organized into columns of six-membered hexagonal silicate rings $[Si_6O_{18}]$. T sites have three oxygens per tetrahedron coplanar with three oxygens of every other tetrahedron of the ring, with the tetrahedral apices toward $[000\bar{1}]$. The hexagonal rings overlay three mutually edge-sharing octahedra (Y sites), both centered on the threefold axis. Distorted triangular BO_3 groups infill indentations of the Y octahedral clusters. Large cations (Na or Ca) or vacancies occur within the hexagonal silicate rings (X site). Relatively small octahedral sites (Z sites, usually containing Al and/or Fe^{3+} but also Li) bridge the columns of silicate and borate rings and the relatively larger Y octahedral clusters.

Two halogen- or oxygen-bearing sites are significant. The O1 site occurs within the silicate ring and is centered on the threefold axis: it contains monovalent F or OH only. The O3 site contains oxygen or hydroxyl, depending on the local charge-balance requirements of the contents of the Y site.

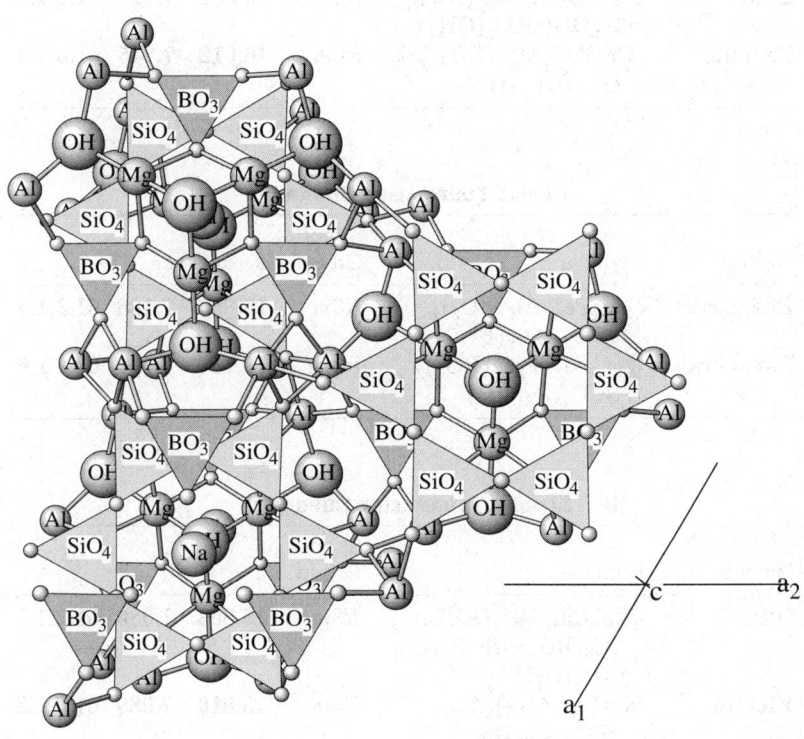

DRAUITE: $NaMg_3Al_6(BO_3)_3Si_6O_{18}(OH)_3(OH)$
Drauite, a typical tourmaline, in perspective view approximately along c.

Alkali-Deficient Tourmaline Subgroup

Mineral	Formula	Space Group	a	c	
Foitite	$\square[Fe_2^{2+}(Al, Fe^{3+})]Al_6(BO_3)_3Si_6O_{18}(OH)_4$	$R3m$	15.967	7.126	61.3.1.1
(unnamed)	$\square(Mg_2(Al,Fe^{3+})]Al_6(BO_3)_3Si_6O_{18}(OH)_4$	$R3m$			
(unnamed)	$\square(Al,Li)_3Al_6(BO_3)_3Si_6O_{18}(OH)_4$	$R3m$			

Calcic Tourmaline Subgroup

Mineral	Formula	Space Group	a	c	
Liddicoatite	$Ca(Li_{1.5}Al_{1.5})_3Al_6(BO_3)_3Si_6O_{18}(OH_{0.66}, O_{0.33})_3(F)$	$R3m$	15.867	7.135	61.3.1.2
Uvite	$CaMg_3(Al_5Mg)(BO_3)_3Si_6O_{18}(OH)_3(OH)$	$R3m$	15.881	7.17	61.3.1.3
Feruvite	$(Ca)Fe_3^{2+}Al_6^{3+}(BO_3)_3Si_6O_{18}(OH)_3(OH)$	$R3m$	16.012	7.245	61.3.1.4

Ferric Tourmaline Subgroup

Mineral	Formula	Space Group	a	c	
Buergerite	$(Na)Fe_3^{3+}Al_6(BO_3)_3Si_6O_{18}(O)_3(F)$	$R3m$	15.869	7.188	61.3.1.5
Povondraite	$(Na)Fe_3^{3+}Fe_6^{3+}(BO_3)_3Si_6O_{18}O_3(OH)$	$R3m$	16.186	7.444	61.3.1.6

Lithian Tourmaline Subgroup

Mineral	Formula	Space Group	a	c	
Olenite	$(Na, Ca, \square)_{1-x}(Al, Li)_3Al_6(BO_3)_3(Si, B)_6O_{18}(O)_3(OH)$	$R3m$	15.803	7.086	61.3.1.7
Elbaite	$Na(Li_{0.5}Al_{0.5})_3Al_6(BO_3)_3Si_6O_{18}(OH)_3(F)$	$R3m$	15.810	7.085	61.3.1.8

61.3.1.2 Liddicoatite

Sodic Tourmaline Subgroup

Mineral	Formula	Space Group	a	c	
Dravite	$NaMg_3Al_6(BO_3)_3Si_6O_{18}(OH)_3(OH)$	$R3m$	15.910	7.210	61.3.1.9
Schorl	$NaFe_3^{2+}Al_6(BO_3)_3Si_6O_{18}(OH)_3(OH)$	$R3m$	15.992	7.172	61.3.1.10
Chromdravite	$NaMg_3(Cr_5^{3+}Fe^{3+})(BO_3)_3Si_6O_{18}(OH)_3(OH)$	$R3m$	16.11	7.27	61.3.1.11

The tourmaline group has been well studied and there are many good references. Structures: *CM 33:849(1995), 27:199(1989); AC(B) 25:1524(1969); AM 56:101(1971), 78:433(1993), 78:1299(1993), 78:265(1993)*. Infrared: *K 14:370(1969), GCA 48:1331(1984), CIWYB 62:166(1993)*. Composition and substitution: *DAN SSSR 307:1461(1989); ZVMO 117:70(1988); CMP 20:235(1969), 62:109(1977)*. Synthetic tourmalines: *AM 32:680(1947), 64:180(1979); GCA 48:1331(1984). DTTG 1985; DHZ IB 559(1986)*.

61.3.1.1 Foitite $\square[Fe_2^{2+}(Al,Fe^{3+})]Al_6(BO_3)_3Si_6O_{18}(OH)_3(OH)$

Named in 1993 by MacDonald et al. for Franklin F. Foit, Jr. (b.1942), mineralogist, Washington State University. HEX-R $R3m$. $a = 15.967$, $c = 7.126$, $Z = 3$, $D = 3.14$. *AM 78:1299(1993)*: 6.34_8 4.21_5 3.99_4 3.45_9 2.94_7 2.57_{10} 2.04_3 1.27_3. Structure is typical of that of the tourmaline group but with X-site vacancy. Prismatic crystals, generally elongated and striated along [0001], terminated, crystals to several centimeters from pegmatite, crystals less than 1 mm in veins or fillings to 1 cm from tuffs. Dark indigo with purple tints grading to black, grayish white streak, vitreous to oily luster, translucent. $H = 7$. $G = 3.17$. No cleavage, brittle. Nonfluorescent. Alkali-deficient, iron-bearing tourmaline; minor substituents include Mn, Mg, Ca, Na, Li. Foitite and foitite-related compositions have been synthesized. *AM 64:180(1979), GCA 48:1331(1984)*. Alkali deficiency arises from coupled substitutions: $\square + R^{3+} \leftrightarrow R^+ + R^{2+}$. Uniaxial (−); (Na-light): $N_E = 1.642$, $N_O = 1.664$; strongly dichroic; O < E. O, pale lavender; E, dark blue. From miarolitic cavities in granite pegmatite (White Queen mine, Dobrá Voda); in miarolitic cavities in granite (Elba); found in hydrothermally altered dumortierite-bearing tuffs. *AM 74:1317(1989)*. White Queen mine, Pala, San Diego Co., CA; Jack Creek deposit, near Basin, Jefferson County, MT; San Piero in Campo, Filone della Speranza, Elba, Italy; Dobrá Voda, Western Moravia, Czech Republic, Ben Lomond deposit, Hervey Range, Q, Australia. VK, EF

61.3.1.2 Liddicoatite $Ca(Li_{1.5}Al_{1.5})_3Al_6(BO_3)_3Si_6O_{18}(OH_{0.66}, O_{0.33})_3(F)$

Named in 1977 by Dunn et al. for Richard T. Liddicoat (b.1918), gemologist and president of the Gemological Institute of America. Tourmaline group.

HEX-R $R3m$. $a = 15.867$, $c = 7.135$, $Z = 3$, $D = 3.05$. *30-748:* 4.94_3 4.20_5 3.96_6 3.45_5 2.93_{10} 2.56_9 2.025_4 1.905_4. Liddicoatite is isostructural with all members of the tourmaline group and indistinguishable from elbaite by XRD. Ca substitutes for Na in the X site. Charge balance is maintained by a number of substitutions: Ca by Na, Al by Li, and OH by O. End-member liddicoatite has not yet been found. Prismatic crystals, as large as $0.25\,\text{m} \times 0.25\,\text{m} \times 0.75\,\text{m}$, generally elongated and striated along [0001], sometimes acicular on large crystals due to etching. Hemimorphic, usually with pedion opposite one or two pyramids, crystals often to many centimeters; usually smoky brown, but also transparent pink or green. Color zoning is abundant at the type locality with little change in Ca/Na ratio between color zones. *AM 62:1121(1977)*. The terminations of liddicoatite are oscillatory color zoned, sometimes with adjacent pyramidal forms selectively colored. When the pyramidal end of the crystal is sliced perpendicular to the c axis, the selective internal color zones vary from triangles to triskelions. Pale brown, light pink or green, nearly white; streak lighter than mass color. $H = 7\frac{1}{2}$. $G = 3.02$. Cleavage $\{10\bar{1}1\}$, poor or absent; brittle. Nonfluorescent. Liddicoatite is the Ca analog of elbaite. Minor amounts of metals (generally less than 1 wt %) substitute in the Y site (e.g., Mn, Fe^{2+}, Ti^{4+}, Mg). Forms a series to elbaite and probably to olenite. Uniaxial $(-)$; $N_O = 1.637$, $N_E = 1.621$; dichroic; $O > E$. O, dark brown, pink; E, light brown, pale pink. Color-zoned crystals reported from the Blizna graphite mine, Blizna, Bohemia, Czech Republic. Some sodian liddicoatite occurs in the Malkhan (Krasny Chikoy)pegmatite field, Transbaikalia, Russia; found in fresh pegmatite and in decomposed pegmatite-derived elluvium and alluvium, *Anjanabonoina, Tsilaizina, Antsirabé, Madagascar (Malagasy Republic)*; individual crystals from an in situ source in an unspecified granite pegmatite in Goias, Brazil. VK, EF

61.3.1.3 Uvite $CaMg_3(Al_5Mg)(BO_3)_3Si_6O_{18}(OH)_3(OH)$

Named in 1929 by Kunitz for the locality. Tourmaline group. HEX-R $R3m$. $a = 15.88$, $c = 7.17$, $Z = 3$, $D = 3.01$. *29-342:* 4.23_7 3.99_7 3.49_6 2.97_{10} 2.58_9 2.04_8 1.92_6 1.46_6. Crystals frequently equant, with one or two pyramids opposite a pedion sometimes modified by pyramid faces, to several centimeters. Green or brown to black, grayish green or grayish brown to gray streak, vitreous to oily luster, translucent. Nonfluorescent, except when low-Fe containing and then fluoresces yellow in SW UV. $H = 7\frac{1}{2}$. $G = 2.96\text{--}3.06$. No cleavage, brittle. An Mg- and Ca-bearing tourmaline with complete compositional variability to dravite (Fe^{2+}); minor substituents include Fe^{3+}, Mn^{2+}, Ti^{4+}, Na, Cr^{3+}, V^{3+}, trace Zn. Several analyses show considerable F, while most analyses have not reported on its presence. Three analyses, all of which contain enough F to dominate a site have been noted. *MR 8:100(1977)*. Uniaxial $(-)$; $N_O = 1.634$(low iron)-1.668, $N_E = 1.619$(low iron)-1.639; strongly dichroic; $O > E$. O, dark green, dark brown, yellow; E, olive-green, light brown, nearly colorless. Type material is from placer gravels; frequently found in marble associated with diopside, tremolite, phlogopite, or meionite;

sometimes with quartz. Localities include: East Hampton, Middlesex Co., CT; Canton, DeKalb, Gouverneur(!), Horicon, Macomb; Powers locality(!), Pierrepont; Actinolite locality(!), West Pierrepont, St. Lawrence and Warren Cos., NY; Franklin marble(!), Franklin, Hamburg, and McAfee, Sussex Co., NJ; Renfrew and Madoc, ONT, Canada; Santa Cruz, SON, Mexico; Hnúšta, Slovakia; *Uva province, Sri Lanka*; Dnieper Basin, Russia; Uzbekistan; chromian uvite from Landani, Tanzania(!); chromian uvite from Sagaing district, Myanmar (Burma); Brumado(!), BA, Brazil, in equant to tabular crystals to 5 cm, some being transparent and shades of green to orange-red; New Zealand. VK, EF *CE 4:208(1929)*.

61.3.1.4 Feruvite (Ca)Fe$_3^{2+}$Al$_6^{3+}$(BO$_3$)$_3$Si$_6$O$_{18}$(OH)$_3$(OH)

Named in 1989 by Grice and Robinson for the compositional relationship to uvite. Tourmaline group. HEX-R $R3m$. $a = 16.012$, $c = 7.245$, $Z = 3$, $D = 3.21$. *CM 27:199(1989)*: 6.43$_4$ 4.24$_6$ 4.00$_6$ 3.50$_6$ 2.98$_8$ 2.59$_{10}$ 2.05$_5$ 1.93$_4$. Subhedral crystals, generally equant, individual crystals to 2 mm. Black, gray streak, vitreous to dull luster. $H = 7$. $G = 3.21$. No cleavage, brittle. Nonfluorescent. Iron- and calcium-bearing tourmaline; major substituents include Fe^{3+}, Mg, Al, Ti^{4+}; trace amounts of Mn, K. Type material is alkali-deficient and also strongly compositionally zoned with dravite and schorl intergrowths. Charge balance maintained by alkali deficiency and low-valence Y and Z cations. Uniaxial ($-$); N$_O$ = 1.687, N$_E$ = 1.669; strongly dichroic; O > E. O, very dark brown, E, light brown. In the Sullivan Pb-Ag-Zn deposit, BC; zoned crystals with feruvite cores and schor-dravite rims occur with biotite and muscovite from rare element pegmatites at Red Cross Lake, Manitoba, Canada; also in pegmatitic hydrothermally altered and tourmalinized quartz veins associated with microcline, chlorapatite, and pyrite at *Cuvier Island, New Zealand*. VK, EF

61.3.1.5 Buergerite (Na)Fe$_3^{3+}$Al$_6$(BO$_3$)$_3$Si$_6$O$_{18}$(O)$_3$(F)

Named in 1966 by Donnay et al. for Martin J. Buerger (1903–1986), institute professor of mineralogy, MIT, pioneer of crystal structure analysis and inventor of X-ray precession camera permitting undistorted photography of the reciprocal lattice. Tourmaline group. HEX-R $R3m$. $a = 15.869$, $c = 7.188$, $Z = 3$, $D = 3.29$. *25-703*: 6.33$_5$ 4.20$_4$ 3.96$_5$ 3.47$_5$ 2.95$_6$ 2.56$_{10}$ 2.03$_4$ 1.91$_3$. Structure *AC(B) 25:1524(1969)*, *AM 56:101(1971)*. Crystals prismatic, stout to slightly elongated, with minor striations along [001]; terminated crystals with pyramids opposite a pedion, flattened subparallel to divergent crystal clusters to several centimeters, individual crystals to 10 mm. Bronzy brown, brown streak, vitreous to oily luster. $H = 7$. $G = 3.30$. Cleavage with yellow-brown schiller, (11$\bar{2}$0), good; brittle. Nonfluorescent. A ferric iron-bearing tourmaline; minor substituents include Fe^{2+}, Ca, Mg, K, Al, Ti^{4+}, minor B for Si; aluminum analog of povondraite. Uniaxial ($-$); N$_O$ = 1.735, N$_E$ = 1.655; strongly dichroic; O > E; O, yellow-brown; E, very pale yellow.

Found in hydrothermally altered tuffs in clay seams in rhyolite at *Mexquitic, SLP, Mexico*. VK, EF *AM 51:198(1966)*.

61.3.1.6 Povondraite $(Na)Fe_3^{3+}Fe_6^{3+}(BO_3)_3Si_6O_{18}(O)_3(OH)$

Named in 1993 by Grice et al. for Pavel Povondra (b.1924), mineralogist, Charles University, Prague, Czech Republic. Tourmaline group. HEX-R $R3m$. $a = 16.186$, $c = 7.444$, $Z = 3$, $D = 3.33$. *AM 78:433(1993)*: 6.63_9 5.13_7 4.71_7 4.32_7 4.05_9 3.61_8 3.05_9 2.63_{10}. Structure: *AM 78:433(1993)*. Prismatic crystals, generally equant and striated along [0001], terminated crystals with pyramid opposite a pedion, poorly crystallized, flattened crystal clusters to 1 cm, individual crystals to 3 mm. Black, brown streak, vitreous to oily luster. H = 7. G = 3.26. No cleavage, brittle. Nonfluorescent. Aluminum-deficient, iron-bearing tourmaline; minor substituents include Fe^{2+}, Mg (≈ 6 wt % MgO), K, Al; trace amounts of Cu, Pb, Sn. Ferric analog of buergerite. Originally thought to be the ferric analog of dravite [i.e., ferridravite, *AM 64:945(1979)*], but Mg is not dominant in either the Y or Z site. Uniaxial (–); $N_O = 1.800$ (to 1.82 with highest Fe content), $N_E = 1.743$–1.751; strongly dichroic; O > E. O, brown to opaque; E, light brown. Found in hydrothermally altered basaltic clasts (to 10 cm) entrained within a metamorphosed evaporite [*RASPA 21:41(1994)*] associated with microcline, talc, dolomite, danburite, and rare boracite at *Cristalmuya gorge, 200 m downhill from the San Francisco asbestos mine, 50 km from Villa Tunari, Chapare Prov., Cochabamba Dept., Bolivia*. VK, EF

61.3.1.7 Olenite $(Na,Ca,\square)_{1-x}(Al,Li)_3Al_6(BO_3)_3(Si,B)_6O_{18}(O)_3(OH)$

Named in 1986 by Sokolov et al. for the locality. Tourmaline group. HEX-R $R3m$. $a = 15.775$–15.803, $c = 7.079$–7.086, $Z = 3$, $D = 3.12$. *39-336*: 6.33_3 4.18_4 3.95_7 3.43_8 3.39_7 2.92_3 2.55_{10} 2.02_3. Structure solution shows that B substitutes for tetrahedral Si (0.50 atoms). Acicular crystals, elongated and striated along [0001], also crystals to several centimeters from pegmatite. Pale pink, white streak, vitreous to oily luster, translucent. H = 7. G = 3.01. No cleavage, brittle. Nonfluorescent. An aluminum-rich tourmaline, can be slightly alkali-deficient; major substituents include Li for Al in Y. Minor substituents include Fe^{3+}, Mn^{2+}, Mg, Ca, K, F. Alkali deficiency in X can be compensated by Y-site low-valence substituents (e.g., Li) and by hydroxylation. *GCA 48:1331(1984)*. Many analyses (without citing specific localities) show essentially continuous solid solution with elbaite. ZnO to 3.8 wt %. *ZVMO 117:70(1988)*, *AM 73:441(1988)*. Uniaxial (–); $N_O = 1.654$, $N_E = 1.635$; dichroic; O > E. O, bright pink; E, pinkish yellow. Occurs in highly fractionated granitic pegmatites; type material is from a zone within an "elbaite" crystal. Also found within zoned crystals from Elba. *Black Mountain quarry, Rumford, Oxford Co., ME*; *Belo Horizonte No. 1 pegmatite, Anza, Riverside Co., CA*; *Decembrist pegmatite field, Olenëk River Basin, Russia*; *San Piero di Campo, Elba, Italy*. VK, EF *ZVMO 115:119(1986)*

61.3.1.8 Elbaite Na(Li$_{0.5}$Al$_{0.5}$)$_3$Al$_6$(BO$_3$)$_3$Si$_6$O$_{18}$(OH)$_3$(F)

Named in 1908 for the locality. Tourmaline group. HEX-R $R3m$.
$a = 15.810$(ideal)–16.23, $c = 7.085$(ideal)–7.26, Z = 3, D = 3.07. *26-964*(nat): 6.32$_{2.5}$ 4.20$_6$ 3.96$_8$ 3.45$_7$ 2.93$_9$ 2.56$_{10}$ 2.03$_5$. **Habit**: Prismatic crystals generally elongated and striated along [0001], frequently acicular, commonly with etched terminal faces; hemimorphic, usually with pedion opposite one or two pyramids, crystals often to many centimeters; twinning rare. **Physical properties**: Usually green, frequently pink, red, or blue grading to black; sometimes brown or colorless, rarely yellow or orange, often multicolored longitudinally and/or concentrically; streak is white to very pale shade of the massive mineral; vitreous to oily luster; usually transparent to translucent. Cleavage {11$\bar{2}$0}, {10$\bar{1}$1}, poor; conchoidal fracture; brittle. H = 7$\frac{1}{2}$. G = 3.05–3.10 (low iron compositions). Insoluble in virtually all acids except HF; piezoelectric, pyroelectric; rarely fluorescent weak blue-white SWUV. IR data: *K 14:370(1969)*. **Chemistry**: Numerous simple trace element chromophoric octahedral substitutions (Fe^{2+}[green], Fe^{3+}[yellow], Fe$^{2+}_{0.5}$ ↔ Fe$^{3+}_{0.5}$[blue], Mn^{2+}[pink], Mn^{3+}[red], rarely Cu^{2+}[blue]); if present, iron in various valences generally dominates chromophoric effects, other substitutions F ↔ OH, Ca$_{0.5}$ ↔ Na, coupled substitution (Li$_{0.5}$Al$_{0.5}$)(OH) for Al(O); concentric and longitudinal color zoning common; enormously variable repeated concentric zoning along prisms with red triskelions near terminations (Antsirabé and Anjanabonoina), sometimes slightly alkali deficient with increasing Al or Fe^{3+} substitution, extensive substitutions grading into schorl with increasing iron or to olenite with decreasing lithium. Analyses commonly show Li ≈ 1.3 or fewer atoms per formula unit. *CMP 62:109(1967)*. In the elbaite–schorl series, MnO has been found to substitute into the composition, and a manganoan schorl–elbaite from Tsilaisina, Madagascar, which contained about 6 wt % MnO (as much as 8.2 wt % MnO was found) was called "tsilaisite." *CE 4:208(1929)*. Recently, a yellowish-green elbaite with 8.8 wt % MnO was described. *AM 71:1214(1986)*. However, even this quantity of MnO falls short of the approximate 10.8 wt % MnO required to constitute a separate species with a majority of the Mn occupying the Y position, so the name *tsilaisite* should not be used. *AM 71:1214(1986)*. The role of radioactivity and the intensity of color was demonstrated in pink-red elbaites. *AM 73:822(1988)*. **Optics**: Uniaxial (−); N$_O$ = 1.619–1.620, N$_E$ = 1.634–1.641 (for colorless, pink, light green varieties, *G&G 15:19(1975)*); N$_O$ = 1.631–1.634, N$_E$ = 1.653–1.656 (for dark blue, dark green varieties); dichroic; O > E. **Uses**: The use of tourmaline as a gem dates from the sixteenth century with the discovery of green gem tourmaline, which at first was thought to be emerald, in Brazil. Any transparent tourmaline can be used as a gem, but most species are black or too dark or of unattractive colors; practically speaking, only elbaite and to a minor extent liddicoatite are used. The varietal names rubellite (for pink or red), indicolite (for blue), verdelite (for green), and achroite (for colorless) have been used in gemology, but only *rubellite* and to a minor extent *indicolite* are so used today; the terms *elbaite*,

verdelite, and *achroite* are not now used to designate gems. Some chromian uvites and some dravites make attractive gems when transparent, but they are rare. Bicolored elbaite, with pink centers and green rinds, is called "watermelon tourmaline." Generally, in late-stage cavities or cavity margins associated with lepidolite and albite in granite pegmatites; rarely from aplite, schists, or dolomites. Sometimes replaced by muscovite, lepidolite, or cookeite. **Localities:** Localities for fine quality and noteworthy material include: Dunton mine(!), Newry and Mt. Mica mine(!), Paris, Oxford Co., ME; Gillette mine, Haddam, Middlesex Co., CT; Pala Chief(!), Stewart(!), and Tourmaline Queen mines(!), Pala; Little Three mine(!), Ramona and Himalaya mine(!!), Mesa Grande, San Diego Co., CA; *Grotto d'Oggi, San Piero di Campo, Elba, Italy*; Lengenbach quarry, Binn, Switzerland; Penig, Saxony, Germany; Utö and Varuträsk, Sweden; Mursinka, Pervoural'sk (Shaitanka), Lipovaya, and Alabashka, Ural Mts., and Hoppevskaya Gora, Sherlovaya Gora, and Soktuj Gora prospects, Adun-Chalon Mt. district and Dorogoy-Utess prospects, Borshchovochnoi Mt. district, both near Nerchinsk, Siberia, Russia; Malkhan district, Krasny Chikoi, Transbaikalia, and Chita region, Siberia; Mawi, Korgay and Paprock, Laghman(!), and Kunar(!), Nuristan Valley, Afghanistan; Sri Lanka; Chainpur and Sankhuwa Sabha district, Nepal(!); Stak Nala(!), NWF, Pakistan; Jos(!), Nigeria; Alto Ligonha(!), Mozambique; Otjua mine(!), Karibib, and Usakos, Namibia; Anjanabonoina(!) and Shatany Valley(!), Antandrokomby, Mt. Bity district, Madagascar; in MG, Brazil, at Coronel Murta(!) and at Urubu mine, Itinga (fluorescent white); Barra do Salinas(!), Salinas; Jonas mine(!), Itatiaia, Conselheiro Pena (gem rubellite crystals to 1.07 m long); Corrego Frio mine, Divino das Laranjeiras, Galileia; Golconda mine(!), Coroaci; Cruzeiro and Aricanga mines(!), São Jose da Safira; Santa Rosa mine(!), Itambacuri; Viadinho mine(!), Marilac; Xanda and Limoero mines(!), Virgem da Lapa, and in Paraiba at São José da Batalha (brilliant sky-blue gems colored by Cu; CuO as much as 2.4 wt %); native Cu and tenorite inclusions have been found in the cuprian elbaites [*GG 30:178(1994)*]; Capoeira, RGN; Gregorio pegmatite, Serra dos Quintos, RGN; Lavra da Ilha, Itinga; Novo Cruzeiro, Rodrigo Silva mine, Resplendor, Rio Doce, Santana do Encoberto. VK, EF

61.3.1.9 Dravite NaMg$_3$Al$_6$(BO$_3$)$_3$Si$_6$O$_{18}$(OH)$_3$(OH)

Named in 1883 by Tschermak for the locality. Tourmaline group. HEX-R $R3m$. $a = 15.910$, $c = 7.210$, $Z = 3$, $D = 3.02$–3.07. *14-76:* 6.38_3 4.22_7 3.99_9 3.48_6 2.96_9 2.58_{10} 2.04_5 1.92_4. Vanadium-rich (4.9 wt % V$_2$O$_3$) dravite has $a = 15.960$, $c = 7.189$. *PD 7:236(1992), 44-1457*. Chromian dravite has $a = 15.92$, $c = 7.15$. *25-1307, 45-1340*. 45-1340: 4.60_1 4.22_1 3.98_{10} 3.46_1 3.12_1 2.95_1 2.57_3 1.59_1. Structure: *AM 78:265(1993)*. Partial Z site occupancy by Mg. Small crystals and grains, usually elongated, crystals usually have one pyramid with hexagonal prism with a pedion or another pyramid; Yinnietharra crystals show apparent, higher, nonhemimorphic, symmetry. Crystals from micrometers to many centimeters. Brown to black, pale brown to gray streak, vitreous to oily luster, translucent. H = 7–7$\frac{1}{2}$. G = 2.9–3.3. No cleavage, brittle. Low iron compositions may be fluorescent SWUV yellow to orange. Magnesium- and sodium-bearing tourmaline; minor substituents include Fe^{2+}, Mn^{2+}, Ti^{4+}, Ca, Fe^{3+}, Cr^{3+}, V^{3+}. Complete compositional variability from dravite through schorl or uvite is common and that to chromdravite exists. Uniaxial (−); N_O = 1.635(low iron)–1.661, N_E = 1.610(low iron)–1.632; strongly dichroic; O > E. O, brown to yellow; E, yellow to colorless. V-rich material has N_E = 1.646, N_O = 1.676; E, light yellow green; O, medium brown. Commonly found in magnesian marbles or schists, sometimes slates, sometimes authigenic. Many worldwide localities and only localities for fine material are given here: Dunton mine(!), Newry and Black Hawk mine, Blue Hill, ME; Franconia mine(!), Franconia, Cornerstone Mountain mine, Orford, Warren, NH; Cod Fish Hill mine, Bethel, CT; McLear mine, De Kalb, St Lawrence Co., Gomer Jones locality(!), Gouverneur, Macomb, Powers locality(!), Pierrepont, Reese locality(!), Richville, 2$\frac{1}{2}$ mine, Talcville, NY; Franklin marble, Franklin, Lambertville, and Limecrest quarry, Sparta, NJ; Adams mine(!), Hiddenite and Crabtree mine, Spruce Pine, NC; Colfax, and Crestmore, Riverside Co., CA; V-rich (4.9 wt % V_2O_3) at Silver Knob, Mariposa Co., CA; Desmont mine, Monmouth, ONT, Canada; Santa Cruz(!), SON, Mexico; Bjordam mine, Bamble, Fredriksvärn, Kjenlid and Snarum, Norway; Simplon Tunnel and Feldbach, Binntal, Switzerland; Cyrilov, Moravia, Czech Republic; Prevali and Windisch Kappel, Carinthia and *Unterdrauburg, Dobrowa, Drava R., Carinthia, Austria*; chromian dravite (to 13.9 wt % (V_2O_3) occurs in chromite ores from the Vardos area, northern Greece; green, prismatic crystals to 4.5 cm from Mikhailovskoye, Chita Oblast, Transbaikal, Russia; chromian dravite (3.70 wt % Cr_2O_3) from Sarykulboldy chrysoprase deposit, central Kazakhstan; Osarara, Narok district, Kenya; Gujarkot, Bheri zone, Nepal; chromian dravite (to 23 wt %) (V_2O_3) occurs at Nabsahi, Orissa, India; crystals to 15 cm from the prismatic crystals from Brumado, BA, Brazil. Soklich mine(!), Yinnietharra Station, WA, Australia. VK, EF MM 41:408(1977) *AM 70:1(1985), DANS-ESS 315:225(1990).*

61.3.1.10   Schorl $NaFe_3^{2+}Al_6(BO_3)_3Si_6O_{18}(OH)_3(OH)$

Early (1524) name, perhaps indicating *worthless*; by 1747, used specifically for a tourmaline. Tourmaline group. HEX-R $R3m$. a = 15.992, c = 7.172, Z = 3, D = 3.30. *43-1464:* 6.37$_5$ 4.23$_5$ 4.00$_7$ 3.47$_5$ 2.96$_5$ 2.58$_{10}$ 2.17$_2$ 2.04$_4$. Structure: *AM 74:422(1989)*. Prismatic crystals, generally elongated and striated along [0001], sometimes acicular and in radial groups, terminated crystals usually with one or two pyramids opposite a pedion, crystals to many centimeters from pegmatite. Black, sometimes with very dark brown, green, or blue internal reflections, occasionally blue translucent; grayish white to bluish white streak; vitreous to oily luster. H = 7$\frac{1}{2}$. G = 3.13. No cleavage, brittle. Nonfluorescent. Iron- and sodium-bearing tourmaline, usually slightly alkali-deficient; minor substituents include Fe^{3+}, Mn^{2+}, Mg, Ca, Li, Cr^{3+}, Ti^{4+}. F can dominate or fill O3 site. Alkali deficiency can arise from coupled substitutions:

vacancy + $R^{3+} \leftrightarrow R^+ + R^{2+}$. *AM 74:1317(1989)*. Complete series observed with dravite/uvite or elbaite. Uniaxial (−); $N_{Eav} = 1.638$, $N_{Oav} = 1.663$; $N_O = 1.640$–1.698, $N_E = 1.622$–1.675; blue and green schorls have lower refractive indices than black schorls; strongly dichroic; O > E. O, gray, dark brown, blue, or green; E, light gray, brown, blue, or green. Type material believed to be from a German tin and/or tungsten deposit. Frequently found in granite pegmatite and granite. Often found in quartz veins; also greisens, schists, gneisses, phyllites; can be authigenic. Sometimes replaced by muscovite; rarely chamosite. Occurs in many worldwide localities; only well-known and notable localities are given here: Noyes and Uncle Tom Mts., Greenwood; Mount Mica and Whispering Pines(!) quarries, Paris; Hinckley prospect, Pownal; Barbour prospect, Stoneham, ME; Rice mine(!), Groton, Grafton Co., NH; Rollstone Hill quarry, Fitchburg, MA; York mine, Overlook, NY; Southern Pacific mine, Nuevo, Riverside Co.; Blue Chihuahua mine(!), Chihuahua Valley; Fano Simmons(!) and Mile Down(!) mines, Little Cahuilla Mt.; Hercules(!) and Little Three mines(!), Ramona, San Diego Co., CA; Tregarden mine, Luxulyan; Goonbarrow mine, St. Austell, and numerous other localities in Cornwall, England; Písek, Bohemia and Dolni Bory, Moravia, Horni Bory, and Cyrilov, Moravia, Czech Republic; Vitosha Mountain, Sofia, Bulgaria; Kaatiala, Erajarvi, Finland; Andreasberg, Saxony, Germany; Arendal, Aust-Agder, and Kragerø, Telemark, Norway; Alabashka, Mursinka district, Yekaterinburg Oblast, Russia; Dassu(!), Baltistan and Stak Nala(!), Pakistan; Paprock, Nuristan, Afghanistan; Barra do Salinas(!), Conseheiro Pena; Salinas; Divino das Larenjeiras, Galileia; Cruzeiro mine(!), São Jose da Safira, MG, Brazil. VK, EF

61.3.1.11 Chromdravite $NaMg_3(Cr_5^{3+}Fe^{3+})(BO_3)_3Si_6O_{18}(OH)_3(OH)$

Named in 1983 by Rumantseva for compositional relationship to dravite. Tourmaline group. HEX-R $R3m$. $a = 16.11$, $c = 7.27$, $Z = 3$, $D = 3.38$. *35-717:* 6.57_5 5.10_4 4.31_4 4.05_5 3.58_8 3.04_8 2.62_{10} 2.08_5. Small crystals and grains, nearly equant, crystals to several millimeters. Dark green to black, grayish green streak, vitreous to oily luster, translucent. $H = 7$. $G = 3.39$–3.40. No cleavage, brittle. Nonfluorescent. IR data: *ZVMO 112:222(1983)*. Magnesium- and chromium-bearing tourmaline; minor substituents include Mn, Ti^{4+}, Na, V^{3+} (to 13.3 wt %). *ZVMO 114:55(1985)*. Uniaxial (−); $N_O = 1.778$–1.780, $N_E = 1.722$–1.724; strongly dichroic; O > E. O, dark green; E, yellow-green. Type material occurs in micaceous metasomatites associated with chromian phengite, taeniolite, vanadian muscovite, quartz, and dolomite. Found in the *Onezhkii depression, central Karelia, Russia*. EF

61.4.1.1 Abenakiite-(Ce) $Na_{26}REE_6(SiO_3)_6(PO_4)_6(CO_3)_6(S^{4+}O_2)O$

Named in 1994 by McDonald et al. for the Abenaki Indian tribe, which inhabited the area around the type locality. HEX-R $R\bar{3}$. $a = 16.018$, $c = 19.761$, $Z = 3$, $D = 3.27$. *PD:* $11.4_{7.5}$ $8.04_{8.5}$ $6.55_{8.5}$ $4.65_{7.5}$ 3.77_9 3.59_8 3.15_7 2.67_{10}. Structure: *CM 32:843(1994)*. Occurs as a single 2 mm × 1 mm pale brown ellipsoidal grain in association with aegirine, eudialyte, manganneptunite, polylithionite, serandite, and steenstrupine-(Ce). Also pinkish grains. White streak, vitreous luster. Cleavage {0001}, poor; conchoidal fracture; brittle. H > 4. G = 3.21. Weakly effervescent in 1:1 HCl. Nonfluorescent. Uniaxial (−); $N_O = 1.589$, $N_E = 1.586$; nonpleochroic. Occurs in a xenolith of sodalite syenite in nepheline syenite at the *Poudrette quarry, Mt. St.-Hilaire, Rouville Co., QUE, Canada.* EF

61.4.2.1 Steenstrupine-(Ce) $Na_{14}Ce_6Mn^{2+}Mn^{3+}Fe_2^{3+}(Zr,Th)(OH)_2(PO_4)_7(Si_6O_{18})_2 \cdot 3H_2O$

Named in 1881 by Lorenzen for Knud Johannes Vogelius Steenstrup (1842–1913), geologist, Copenhagen, Denmark. HEX-R $R\bar{3}m$. $a = 10.456$–10.460, $c = 44.985$–45.479, Z = 3, D = 3.63–3.67. 41-1419(nat): 15.1_6 3.27_{10} 3.17_8 3.04_5 2.88_5 2.74_8 2.60_9 2.11_4. Generally partly to completely metamict. Structure: *TMPM 31:47(1983)*. Steenstrupine can be described as a rod structure consisting of two rods having different sequences. In rod 1, T is a disordered (PO_4) group and the sequence Na1–M1–Na1 is a face-sharing octahedral trimer and M3–M2 a face-sharing octahedral dimer. A (Si_6O_{18}) six-membered ring occurs, with approximate point symmetry {$C3i$}. Steenstrupine has one of the largest unit cell volumes known for a mineral species. Crystals to > 4 cm. Black to dark brown, brownish red; white, rose, or tan streak; vitreous to metallic luster or greasy and sometimes dull. Cleavage {0001}, poor; uneven to conchoidal fracture; brittle. H = 4–5. G = 3.25–3.50. IR, DTG, DTA data: *NJBA 140:300(1981)*. Not soluble in acids. A complex REE-Th silicophosphate. Analysis (wt %): Na_2O, 0.5–8.3; K_2O, 0.0–1.2; BeO, to 1.9; MgO, 0.0–1.6; CaO, 1.85–8.4; MnO, 0.0–29.0; PbO, 0.0–1.0; Al_2O_3, 0.0–2.4; Mn_2O_3, 0.0–6.8; Fe_2O_3, 0.65–9.7; REE_2O_3, to 33; SiO_2, 7–32.4; ThO_2, 2.1–12.1; $Ta_2O_5 + Nb_2O_5$, 0.2–4.4; P_2O_5, 0.0–10.7; H_2O, 3.4–18.6; F, to 3.7. Uniaxial (−) or isotropic; $N_O = 1.665$–1.668, $N_E = 1.660$–1.663; colorless to pale yellow, brownish red, and yellowish brown. Pleochroism is weak in browns and yellows. Yellow-brown parallel to c and brown or dark brown perpendicular to c. A characteristic mineral of ultra-agpaitic pegmatites of nepheline syenites and sodalite syenites. Found in pegmatoid veins, and late hydrothermal veins associated with natrolite, albite, aegirine, sodalite, epistolite, and others. Often contains inclusions of other minerals. Found at Mt. St.-Hilaire, QUE, Canada; at *Tunugdliarfik, Tugtup Agtakorfia*, and nearby Igdlunguak, and Tupersuatsiak in the Ilimaussaq intrusion, southern Greenland; also in the Kangerdluarssuk area, all in the Julianehaab district; Mt. Punkaruaiev, Jubilee pegmatite on Mt. Karnasurt (Mn-rich), Mt. Nepka, Mt. Alluaiev, Mt. Sengischorr, and Kitkuai R. valley, Lovozero massif, Kola Penin., Russia. EF
M0G 167:1(1962) *MIN 3(1):497(1972)*.

61.4.2.2 Thorosteenstrupine $(Ca,Th,Mn)_3Si_4O_{11}F \cdot 6H_2O$

Named in 1963 by Kuprianova et al. from the composition: like steenstrupine, but with predominant Th. May be a mixture of ekanite $ThCa_2Si_8O_{20}$ and cheralite $(Ce,Th)PO_4$ as all X-ray lines could be indexed for these two minerals. *CM 20:65(1982)*. See steenstrupine (61.4.2.1). Amorphous, recrystallizes upon heating, no ICDD data. PD(heated, nat): 4.08_{10} 3.31_6 3.25_{10} 3.06_9 2.84_8 2.73_6 2.61_{10} 2.13_6 1.94_7 1.91_7 1.79_8. Fine platy crystals 2–5 mm, sometimes to 1 cm long. Dark brown, nearly black; greasy–vitreous luster. Conchoidal fracture, brittle. H ≈ 4. G = 3.02. IR and thermal data: *ZVMO 91:325(1963)*. Analysis (wt %): Al_2O_3, 0.31; Fe_2O_3, 0.65; Σ Ce_2O_3 and Y_2O_3, 1.12; SiO_2, 31.87; TiO_2, none; ThO_2, 35.70; H_2O, 13.77; F, 2.43; P_2O_5, none; MnO, 7.75; CaO, 8.38. $\bar{N}$ = 1.63–1.66; isotropic. Occurs in metasomatic veins, associated with microcline, albite, aegirine–augite, quartz, fluorite, thorite, and REE-bearing miserite. *Unspecified locality in eastern Siberia, Russia.* EF *MIN 3(1):501(1972)*.

Class 62

Eight-Membered Rings

62.1.1.1 Muirite $Ba_{10}Ca_2Mn^{2+}TiSi_{10}O_{30}(OH,Cl,F)_{10}$

Named in 1965 by Alfors et al. for John Muir (1838–1914), American geologist and explorer, who made early observations in geology in the Sierra Nevada Mountains, California. TET $P4/mmm$. $a = 14.000$–14.022, $c = 5.625$–5.627, $Z = 1$, $D = 3.88$. *18-161*(nat): 4.42_8 3.73_6 3.60_5 3.51_6 3.31_6 2.91_{10} 2.81_4 2.74_4. Structure: *SCI 173:3961(1971), DAN SSSR 221:343(1975)*. Grains to 1 mm, aggregates to 3 mm. Generally anhedral grains, but subhedral and euhedral grains to 3 mm occur. {001} + {100} dominant with lesser {110} and {h0l} forms. Orange, pale orange streak, subvitreous luster. Cleavage (001), (100), indistinct. H ≈ $2\frac{1}{2}$. G = 3.86. Decomposed by dilute HCl and HNO_3, grains lose their color and disintegrate to a white or colorless residue; no effect from H_2SO_4, acetic acid, or NaOH. Analysis (wt %): SiO_2, 22.15; TiO_2, 4.17; Al_2O_3, 0.53; FeO, 0.4; MnO, 2.04; MgO, 0.11; CaO, 4.67; SrO, 0.13; BaO, 59.6; K_2O, 0.1; F, 1.4; Cl, 4.5; H_2O, 1.8. Uniaxial (+); $N_O = 1.697$, $N_E = 1.704$. O, orange; E, colorless. Anomalous interference colors. Occurs at *Rush Creek and Big Creek, Fresno Co., CA*, in metasediments surrounded by quartz diorite; in sanbornite–quartz rock. Large brown grains (to several centimeters) in sanbornite with hedenbergite, fresnoite, uvarovite, and others at the Gunn Claims, Itsy Mts., Macmillan Pass, YT, Canada. EF *MIN 3(2):43(1981), AM 69:358(1984)*.

62.1.2.1 Megacyclite $Na_8KSi_9O_{18}(OH)_9 \cdot 19H_2O$

Named in 1993 by Khomyakov et al. from the Greek *mega*, large, and *cyclos*, rings, in allusion to the size of the structure of insular circular radicals made of 18 SiO_4 tetrahedra. MON $P2_1/c$. $a = 24.91$, $b = 11.94$, $c = 14.92$, $\beta = 94.47°$, $Z = 4$, $D = 1.87$. PD(nat): 7.42_2 6.92_2 4.26_6 3.08_{10} 2.94_7 2.65_6 2.40_4 2.29_4. Structure: *K 37:334(1992)*. Anhedral grains 1–3 mm across and aggregates to 3–5 mm across. Colorless, dull vitreous luster. Cleavage {100}, excellent, {001}, less so; gradual (steplike) fracture. H = 2. G = 1.82. IR data: *ZVMO 122:125(1993)*. Relatively easily soluble in water at 25° giving an alkaline fluid. Turns white in air after prolonged exposure and with production of poorly known secondary alteration products. Analysis (wt %): H_2O, 33.2; Na_2O, 19.75; K_2O, 3.62; SiO_2, 43.4. Biaxial (−); $Y = b$, $X \wedge c = 30°$; $N_x = 1.460$, $N_y = 1.478$, $N_z = 1.481$; $2V = 43°$; $r > v$, strong. Occurs on Mt.

Rasvumchorr, Khibiny massif, Kola Penin., Russia; in veinlike bodies of pegmatitic rock. A hydrothermal mineral formed at very low temperature during final stages of crystallization. Occurs as inclusions 1–3 mm across intergrown with revdite in fenaksite–delhayelite veins. EF

= Class 63 =

Condensed Rings

63.1.1.1 Steacyite $K_{0.6-0.8}(Ca,Na)_2ThSi_8O_{20}$
Named in 1982 by Perrault and Szymanski for Harold R. Steacy (b.1923), Geological Survey of Canada, in recognition of his work on radioactive minerals of Canada. Originally described in 1970 as desourdyite, changed in 1971 to ekanite, and redefined (1982) as steacyite. Compare iraqite-La (63.1.1.2). TET $P4/mcc$. $a = 7.583$, $c = 14.763$, $Z = 2$, $D = 3.32$. $35\text{-}531$(Mt. St.-Hilaire): 5.37_1 5.30_4 3.38_{10} 3.32_5 2.64_4 2.16_2 2.00_3 1.821_2; $39\text{-}408$(Guinea): 7.57_3 7.38_3 5.29_{10} 3.37_{10} 3.32_5 2.64_8 2.53_3 1.99_5. Structure: $AC(B)\ 28{:}1794(1972)$. Called ekanite in the published structure. The silicate unit is an isolated, double four-membered ring of composition $[Si_8O_{20}]^{8-}$ lying perpendicular to c. The unit cell is dimensionally equivalent to that of ekanite (72.1.1.1), whose structure can also be interpreted as having condensed four-tetrahedron rings. Isostructural with iraqite-La. Tabular {001} to prismatic [001], 1- to 3-mm crystals displaying {001}, {100}, {010}, {101}; sometimes in spherical aggregates. Cruciform twins. Dark gray, gray-brown, brown, sometimes color-zoned with darker cores; greasy–vitreous luster; transparent. $H \approx 5$. $G = 2.95$. Uniaxial $(-)$; $N_O = 1.573$, $N_E = 1.572$. Found as an accessory mineral in pegmatite veins, marble xenoliths, hornfels, and in cavities in igneous breccia at *Mt. St.-Hilaire, QUE, Canada*; from peralkaline pegmatite of the Dara-i-Pioz massif, Tien Shan, Tajikistan; on Rouma Isle, Los Islands., Guinea. Some material designated as ekanite may actually be steacyite. AR $CM\ 20{:}59(1982)$, $NJMM{:}233(1987)$, $EJM\ 5{:}971(1993)$.

63.1.1.2 Iraqite-La $K(Ca,Na)_4(La,Ce,Th)_2(Si,Al)_{16}O_{40}$
Named in 1976 by Livingstone et al. for the locality. Compare ekanite (72.1.1.1) and steacyite (63.1.1.1). TET $P4/mcc$. $a = 7.61$, $c = 14.72$, $Z = 1$, $D = 3.28$. $29\text{-}995$: 7.62_3 7.36_8 5.28_{10} 3.40_6 3.38_8 3.31_{10} 2.64_{10} 2.17_4. Isostructural with steacyite. Pale greenish yellow, dull to pearly luster, translucent. $H = 4\tfrac{1}{2}$. $G = 3.27$. Uniaxial $(-)$; $N_O = 1.590$, $N_E = 1.585$. Found in granite at dolomitic marble contact at *Shakhi-Rash Mt., Hero, Qala-Diz, northern Iraq*. AR $MM\ 40{:}441(1976)$.

MILARITE GROUP

The milarite group, also known as the osumilite group, consists of tectosilicates having the general formula

$$AM_2(T2)_3(T1)_{12}O_{30} \cdot nH_2O \text{ or a related formula}$$

$$^{[6]}A_2{}^{[9]}B_2{}^{[12]}C^{[18]}D^{[4]}T2_3{}^{[4]}T1_{12}O_{30}$$

where

A = K, Na, Mg, H_2O, H_3O^+, □, Ca, Ba
M = Mg, Ca, Fe^{3+}, Ti, Sn, Fe^{2+}, Al, Zr, Na, Y, REE
T2 = tetrahedral site = Al, Be, Mg, Fe, Li, B, Mn^{2+}, Si, Zn
T1 = tetrahedral site = Si, Al

and the structure is a framework structure, usually hexagonal.

The structure consists of double six-membered rings of linked $(Si,Al)O_4$ tetrahedra (T1 tetrahedra). The double rings are linked to each other both laterally and vertically, not only by $(Mg,Fe^{2+},Mn)O_4$ octahedra (M sites) but also by $(Al,Fe)O_4$ tetrahedra (T2 tetrahedra), to produce a three-dimensionally linked framework structure. Si occurs only in ring-forming and not in linking tetrahedra. Each of the T1 tetrahedra shares three of its corners with adjacent T1 tetrahedra to form the double rings, which have approximate composition $Si_{10}Al_2O_{30}$. The fourth corner is shared with a T2 tetrahedron and an M octahedron. Molecules.

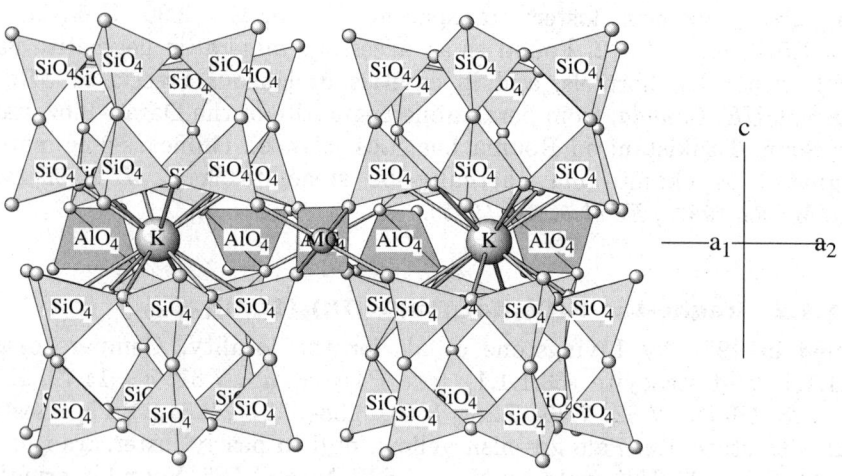

OSUMILITE: $KMg_2Al_3Si_{10}Al_2O_{30}$

Osumilite is similar to beryl and cordierite, but the six-membered rings are stacked in pairs along the c axis direction. There are also K atoms between silicate layers, centered on the rings.

63.2.1a.1 Brannockite

The structure contains channels parallel to z which can be occupied by large cations and possibly by water molecules.

Review papers on the mineralogy of the group include: *MIN 3(2):139(1981)*, *DHZ 1B:541(1986)* and *AM 76:1836(1991)*.

MILARITE GROUP

Mineral	Formula	Space Group	a	c	
Brannockite	$KSn_2Li_3Si_{12}O_{30}$	$P6/mcc$	10.01	14.25	63.2.1a.1
Chayesite	$K(Mg, Fe^{2+})_4 Fe^{3+} [Si_{12}O_{30}]$	$P6/mcc$	10.15	14.39	63.2.1a.2
Darapiosite	$KNa_2Zr[Li(Mn, Zr)_2 Si_{12}O_{30}]$	$P6/mcc$	10.32	14.39	63.2.1a.3
Eifelite	$KNa_3Mg[Mg_3Si_{12}O_{30}]$	$P6/mcc$	10.14	14.22	63.2.1a.4
Merrihueite	$(K, Na)_2(Fe, Mg)_2 [(Fe, Mg)_3Si_{12}O_{30}]$	$P6/mcc$	10.16	14.32	63.2.1a.5
Osumilite	$(K, Na)(Fe, Mg)_2[(Al, Fe)_2 (Si, Al)_{12}O_{30}] \cdot nH_2O$	$P6/mcc$	10.15	14.29	63.2.1a.6
Osumilite-(Mg)	$(K, Na)(Mg, Fe)_2[(Al, Fe)_2 (Si, Al)_{12}O_{30}] \cdot nH_2O$	$P6/mcc$	10.18	14.32	63.2.1a.7
Poudretteite	$KNa_2B_3Si_{12}O_{30}$	$P6/mcc$	10.24	13.49	63.2.1a.8
Sugilite	$KNa_2(Fe^{3+}, Fe^{2+}, Mn^{2+}, Al)_2 Li_3Si_{12}O_{30} \cdot nH_2O$	$P6/mcc$	10.00	14.00	63.2.1a.9
Yagiite	$(Na, K)_{1.5}Mg_2(Al, Mg)_3 (Si, Al)_{12}O_{30}$	$P6/mcc$	10.09	14.29	63.2.1a.10
Dusmatovite	$K(K,Na,\Box)(Mn,Zr,Y)_2 (Zn,Li)_3Si_{12}O_{30}$	$P6/mcc$	10.20	14.28	63.2.1a.11
Milarite	$K_2Ca_4Al_2Be_4Si_{24}O_{60} \cdot H_2O$	$P6/mcc$	10.41	13.82	63.2.1a.11
Sogdianite	$(K, Na)_2(Zr, Fe, Ti) [Li_2(Li, Al)Si_{12}O_{30}]$	$P6/mcc$	10.09	14.24	63.2.1a.12
Roedderite	$(Na, K)_2(Mg, Fe)_2 [(Mg, Fe)_3Si_{12}O_{30}]$	$P\bar{6}2c$	10.14	14.30	63.2.1a.13
Armenite	$BaCa_2Al_6Si_9O_{30} \cdot 2H_2O$	$Pnna^*$	13.87	10.70	63.2.1b.1

*$b = 18.661$ Å.

63.2.1a.1 Brannockite $KSn_2Li_3Si_{12}O_{30}$

Named in 1973 by White et al. for Kent Combs Brannock (1923–1973), chemist and mineralogist, Tennessee Eastman Company, Kingsport, TN. Milarite group. HEX $P6/mcc$. $a = 10.0167$–10.002, $c = 14.2452$–14.263, $Z = 2$ [*AM 73:595(1988)*], $D = 3.08$. 26-0853(nat): 8.69_6 7.14_8 5.50_7 4.34_8 4.11_{10} 2.905_9 2.68_6 2.50_6. A double-ring silicate. The Na site of sugilite is vacant in brannockite. *AM 73:594(1988)*. Thin hexagonal plates to 1 mm: {100}, {110}, {114}, and {001} dominant; twinning on [210] twin axis, perpendicular to (001) plane, twin plane is (210) or (140). Colorless, white streak, vitreous luster, brittle. No cleavage. $G = 2.980$. Fluoresces bright blue-white under SWUV. Contains minor Na for K. Uniaxial $(-)$; $N_O = 1.567$, $N_E = 1.566$. Occurs in vugs and open fissures in pegmatite with bavenite, tetrawickmannite, pyrite, titanite, albite, and quartz at the *Foote mine, Kings Mt., Bessemer*

City, NC. Formed in a low-temperature hydrothermal environment. EF MR 4:73(1973).

63.2.1a.2 Chayesite K(Mg,Fe^{2+})$_4$Fe^{3+}[Si$_{12}$O$_{30}$]

Named in 1989 by Velde et al. for Felix Chayes (1916–1993), Geophysical Laboratory, Washington, DC. Milarite group. See roedderite (63.2.1a.14), related by the substitution Fe^{3+} + □ = Fe^{3+} = Fe^{2+} + (Na, K). A solid-solution series exists between the two. HEX Probably $P6/mcc$. $a = 10.120$–10.153, $c = 14.305$–14.388, $Z = 4$, $D = 2.68$. 45-1472(nat): 7.14$_8$ 5.08$_{10}$ 4.14$_5$ 3.75$_{10}$ 3.35$_5$ 3.24$_{10}$ 2.93$_5$ 2.78$_8$. Tabular with {0001}, {10$\bar{1}$0}, {11$\bar{2}$0}, {10$\bar{1}$2}; tabular crystals to 100 μm. Deep blue, white streak, vitreous luster, transparent. No cleavage. Analysis (wt %), Utah: SiO$_2$, 69.2; Al$_2$O$_3$, 0.20; FeO + Fe$_2$O$_3$, 11.5; MgO, 12.70; Na$_2$O, 0.31; K$_2$O, 5.2; MnO, 0.29; TiO$_2$O, 0.25; Cancarix: SiO$_2$, 70.3; FeO (total Fe), 12.6; MnO, 0.2; MgO, 11.4; Na$_2$O, 0.04; K$_2$O, 4.48. Uniaxial (+); Na(D): N$_O$ = 1.575, N$_E$ = 1.578. O, sky-blue; E, colorless. Occurs as a rock-forming silicate in lamproite (both Utah and Spain). Associated with potassian richterite diopside and K-feldspar. Found at *Moon Canyon, E of Francis, Summit Co., Utah*; Cancarix, Albacete Prov., Spain. EF *AM 74:1368(1989)*, *MM 58:655(1994)*.

63.2.1a.3 Darapiosite KNa$_2$Zr[Li(Mn,Zr)$_2$Si$_{12}$O$_{30}$]

Named in 1975 by Semenov et al. for the locality. Milarite group. HEX $P6/mcc$. $a = 10.32$, $c = 14.39$, $Z = 2$, $D = 2.80$. 29-825(O assigned because of poor agreement between d_{obs} and d_{calc})(nat): 7.09$_6$ 4.43$_4$ 4.13$_5$ 3.75$_4$ 3.26$_{10}$ 2.93$_{6.5}$ 2.76$_{4.5}$ 2.56$_{5.5}$. Isometric appearing crystals to 0.5 cm across. Colorless, white, rarely brownish or bluish; white streak; oily, greasy luster. H = 5. G = 2.92. IR data: *ZVMO 104:583(1975)*. 0.90% Nb$_2$O$_5$ substitutes for Zr. Uniaxial (−); N$_O$ = 1.580, N$_E$ = 1.575; usually not pleochroic, a blue variety has O, violet, E, blue. Occurs at *Dara-i-Pioz, Alai Range, Tien Shan, northern Tadzhikistan*, in peralkaline rocks (massif) and their cogenetic pegmatites. Associated with aegirine, quartz, sogdianite, eudialyte, schizolite, and polylithionite. EF *AM 61:1053(1976)*.

63.2.1a.4 Eifelite KNa$_3$Mg[Mg$_3$Si$_{12}$O$_{30}$]

Described in 1980 and named in 1983 by Abraham et al. for the Eifel region of Germany. Milarite group. Very similar to roedderite; only chemical analyses can positively distinguish the two minerals. A complete solid-solution series exists between eifelite and roedderite, with 2Na ↔ Mg. Roedderite is KNaMg$_5$Si$_{12}$O$_{30}$. See the milarite group minerals. HEX $P6/mcc$. $a = 10.137$–10.150, $c = 14.223$, $Z = 2$, $D = 2.67$. 35-588(nat): 7.07$_5$ 5.11$_5$ 4.43$_6$ 4.14$_5$ 3.75$_9$ 3.26$_{10}$ 2.91$_5$ 2.76$_5$. Eifelite has octahedral Na (six-coordinated) with eight cations per formula unit outside of the hexagonal rings; it has the highest cation occupancy of any milarite group mineral. *CMP 82:252(1983)*. Hexagonal, platy to prismatic; dominant forms: {10$\bar{1}$0}, {0001}, subordinate {11$\bar{2}$0} and {10$\bar{1}$2}; euhedral crystals to 1 mm in diameter. Colorless to very light yellow or green, white streak, vitreous luster. No cleavage observed. Insoluble in HCl + H_2SO_4. Analysis (wt %): SiO_2, 71.06, 69.06; TiO_2, 0.06, 0.03; Al_2O_3, 0.79, 0.57; Cr_2O_3, 0.06, 0.00; FeO, 0.48, 0.19; MnO, 0.46, 0.53; CuO, 0.08, 0.16; ZnO, 0.34, 0.43; MgO, 16.25, 15.47; K_2O, 4.18, 4.16; Na_2O, 6.48, 8.07. Uniaxial (+); $N_O = 1.5430-1.5445$, $N_E = 1.5443-1.5458$; colorless; birefringence = 0.001 compared to 0.003–0.007 for roedderite. Occurs at *Bellerberg volcano, Mayen, eastern Eifel, Germany*, in peralkaline rocks. Melt-coated cavities in gneissic xenoliths ejected in leucite–tephrite lavas. Associated with pyroxene, amphibole, tridymite, hematite, quartz, sanidine, and pseudobrookite. EF *FM 58:3(1980)*, *DA 44:89(1993)*.

63.2.1a.5 Merrihueite $(K,Na)_2(Fe,Mg)_2[(Fe,Mg)_3Si_{12}O_{30}]$

Named in 1965 by Dodd et al. for Craig M. Merrihue (1936–1965), Smithsonian astrophysical laboratory. Milarite group. See roedderite (63.2.1a.14), the Na and Mg analog of merrihueite. HEX $P6/mcc$. (Nat): $a = 10.16$, $c = 14.32$, $Z = 2$ [*SCI 149:972(1965)*]; (syn): $a = 10.222$, $c = 14.152$, $Z = 2$, $D = 2.87$. *21-1270*(nat): 7.13_8 5.50_{10} 5.03_8 3.73_{10} 3.23_9 2.92_6 2.77_{10} 2.01_2; PD(syn): *AC(B) 28:267(1971)*. Structure solution for synthetic material: *AC(B) 28:267(1971)*. Greenish blue. Other physical properties were not determined. Analysis (wt %): SiO_2, 61.8; Al_2O_3, 0.3; FeO, 23.7 (total Fe); MgO, 4.4; MnO, 0.5; CaO, 0.3; K_2O, 3.8; Na_2O, 2.0. Uniaxial or biaxial with $2V = 5–10°$; $N_y = 1.559-1.592$; colorless to greenish blue, N values variable with Fe content; anomalous blues to purples in normal thin sections suggesting a low to moderate birefringence. Occurs as inclusions, to 150 μm in diameter, in clinopyroxenes in chondrite in the Mezö–Medaras chondrite, *Mezö Medaras, Transylvania, Romania*; Iwo Jima volcano, Kyushu, Japan. EF

63.2.1a.6 Osumilite $(K,Na)(Fe,Mg)_2[(Al,Fe)_{2-3}(Si,Al)_{12}O_{30}] \cdot nH_2O$

63.2.1a.7 Osumilite-(Mg) $(K,Na)(Mg,Fe)_2[(Al,Fe)_{2-3}(Si,Al)_{12}O_{30}] \cdot nH_2O$, and crystal chemical formula $CA_2(T_2)_3(T_1)_{12}O_{30}$, where C = K,Na, A = Fe,Mg, T_2 = Al,Fe,Mg, and T_1 = Si,Al

Named in 1953 by Miyashiro for the old province of Osumi in Sakkabira, Kyushu, Japan; osumilite-(Mg) was named by Schreyer et al. in 1983 on the basis of its composition. Milarite group. HEX $P6/mcc$. $a = 10.08-10.17$, $c = 14.26-14.35$, $Z = 2$; (osumilite): $a = 10.150$, $c = 14.289$, $Z = 2$; [osumilite-(Mg)]: $a = 10.078$, $c = 14.319$, $Z = 2$, $D = 2.63-2.74$. *25-0658*(nat osumilite): 7.13_7 5.53_6 5.09_7 4.13_8 3.75_8 3.22_{10} 2.92_9 2.77_9; *30-942*[osumilite-(Mg)]: 7.1_4 5.01_6 4.11_6 3.72_6 3.22_{10} 2.92_6 2.77_6 2.53_2; *29-1016*[syn osumilite-(Mg)]: 7.18_5 5.04_6 4.12_4 $3.73_{4.5}$ 3.58_5 3.21_{10} 2.92_6 $2.77_{6.5}$. Structure is similar to cordierite and beryl. Channels parallel to c formed by a superposition of double hexagonal rings. Tetrahedral rings are linked by octahedra and additional tetrahedra. A framework rather than a ring silicate. Channels contain large cations and possibly H_2O. A tectosilicate. Possible variable Si–Al order in the double hexagonal rings of the Nain, Labrador, osumilite. *AM 73:585(1988)*, *DHZ*

IB:541(1986). **Habit:** Prismatic to tabular, with dominant {0001}, {10$\bar{1}$0}, {11$\bar{2}$0}, {10$\bar{1}$2}; to 10 mm across in Eifel region; twinning, sometimes on (21$\bar{3}$0) Eifel. **Physical properties:** Black to deep blue—usual color, pink from Nain and Norway, some dark brown or gray or colorless. Generally no cleavage; sometimes a prismatic cleavage is present. G = 2.64 (osumilite), 2.63 [osumilite-(Mg)], 2.60 [osumilite-(Mg) syn]; range 2.58–2.68. IR and Mössbauer data: *AM 63:490(1978)*. **Chemistry:** Osumilite has $Fe^{2+} + Fe^{3+}$. IR work sometimes shows no H_2O. Blue color due to intervalence charge transfer between octahedral Fe^{2+} and channel Fe^{3+}. Some Fe is in the channels. Alkali position may be partly vacant when octahedral Al is present: $^{VI}Al^{3+} + {}^{XII}\square \leftrightarrow {}^{VI}(Mg,Fe^{2+}) + {}^{XII}(K,Na)$. $Si + R^{2+} \leftrightarrow 2R^{3+}$. Mainly $(Fe^{2+} + Mg^{2+}) + Si \leftrightarrow 2Al$. As much as ≈ 11 wt % FeO in Eifel material. Typical analyses (wt %): SiO_2, $\approx$ 60–62; Al_2O_3, 19.5–24; FeO, 0.45–10.95; MgO, 1.3–9.3; Na_2O, 0.10–1.0; K_2O, 2.4–4.3; TiO_2, 0.02–0.80; MnO, 0.2–1.5. **Optics:** Uniaxial (+) and (−); (range): N_O = 1.536–1.565, N_E = 1.544–1.563; (osumilite): N_O = 1.546, N_E = 1.550; (osumilite with 12.8 wt % FeO): N_O − 1.565, N_E = 1.563; [osumilite-(Mg)]: N_E = 1.541, N_O = 1.547. O, light blue, bluish purple, or light pink; E, colorless to brown. For (+) material, sometimes metamorphic osumilites are biaxial with 2V = 13–40° (Nain osumilite) due to Fe in channels. One sample from the Eifel region has been reported as (−), with E blue, and O brown to olive-brown, N_E = 1.563, N_O = 1.565; osumilite-(Mg) is all (+), birefringence increases with increasing Fe in the A site; N_O increases faster than N_E, reducing the birefringence. Osumilite with >90% (Fe,Mn,Ti) in A will be (−); optical orientation scheme: E parallel to c; absorption = O > E; low Fe-colorless; medium Fe-pink; high Fe-blue; heating pink Nain material produced blue color. Some Fe^{2+} oxidized to Fe^{3+}, more likely charge transfer between ferrous and ferric iron, Fe^{3+} in $T_2 + Fe^{2+}$ in A. **Occurrence:** Occurs in acid volcanic rocks (e.g., rhyolite, dacite), as anhedral phenocrysts (i.e., magmatic) in the groundmass and as cavity-lining crystals associated with tridymite and quartz. Also in contact-metamorphic xenoliths in volcanic rocks (e.g., Eifel region); in fused hornfelses (buchites). In high-grade contact-metamorphic aureoles. Thus conditions of crystallization range from near-solidus volcanic to granulite-facies metamorphic. Osumilite is not stable in hydrous, low-temperature conditions. Stable in anhydrous high-temperature and high-pressure conditions, and in hydrous magmas at >1 kbar. Also in the Colomera Fe-meteorite. Present in regional metamorphic rocks. Found in a trachyte flow containing sanidinite xenoliths from Italy. Rarely in volcanic ash deposits. Alteration: will break down retrogressively to cordierite, mica, and quartz. As of 1988, about 24 natural occurrences were known. **Localities:** Localities for osumilite include: Obsidian Cliffs and North Sister Mt., Lane Co., OR; Fe-richest material from Bellerberg volcano, Mayen, Eifel, Germany; Nickenicher Sattel volcano, Eifel region, Germany; Funtanafigu quarry, Monte Arci, Cagliari, Sardinia, Italy; in Japan at *Sakkabira, Kagoshima Pref., Kyushu*; Hayato-cho, Kagoshima Pref.; in volcanic ash at Izumo-zaki and Oguni, Niigata Pref., Haneyama, Oita Pref.; Asama volcano, Yamanashi Pref. Those for osumilite-(Mg) include: Ikkinikulluit Brook, Nain complex, LAB, Canada; *Tieveragh, County Antrim, Northern Ireland*; Vikeså, Rogaland, southwestern Norway; Fosso Ricomero, Viterbo, Latium, Italy; Malyi Nepiskol volcano, Caucasus Mts., Russia; in ultrapotassic lamprophyre dike at Chili Palca Hacienda, Puno Dept., southern Peru; in Japan at Shimizu and Iriki, Kagoshima Pref.; Haneyama, Oita Pref; Akagi volcano, Gumma Pref.; Rishiri Is., Hokkaido, Japan; about six areas, including Reference Peak and Mt. Riiser-Larsen, in Enderby Land, Antarctica. Localities for which the Fe/Mg ratio is not specified include: various localities, such as Lac Ste.-Marie, QUE, in the Grenville Prov., Canada; Kuppe buchite, Blauen, Eschwege, Germany; Ruaperhu and White Is., New Zealand. EF *AM 55:875(1970), 73:585(1988); TMPM 31:215(1983); DA 44:89(1993); EJM 6:657(1994)*.

63.2.1a.8 Poudretteite KNa$_2$B$_3$Si$_{12}$O$_{30}$

Named in 1987 by Grice et al. for the locality. Milarite group. HEX $P6/mcc$. $a = 10.239–10.253$, $c = 13.485–13.503$, $Z = 2$, $D = 2.53$. *42-1376*(nat): 6.74$_3$ 5.13$_{10}$ 4.07$_3$ 3.25$_{10}$ 2.96$_4$ 2.81$_6$ 2.69$_5$ 1.82$_4$. Structure: *CM 25:763(1987)*. General formula: ^{9}B^{12}C^{6}A$_2^4$(T$_2$)$_3^4$(T$_1$)$_{12}$O$_{30}$; the B site is empty in poudretteite. K in a [12]-coordinated C site, Na in an octahedrally coordinated A site, B in a tetrahedrally coordinated T2 site, Si in a tetrahedrally coordinated T1 site. Prismatic; short parallel to [001]; equant, deeply barrel-shaped crystals, to 5 mm long, that look like etched quartz or fluorapophyllite; no twinning observed. Colorless to very pale pink, streak white; vitreous luster. No cleavage, splintery to conchoidal fracture. H = 5. G = 2.51. No response to UV light or cathodoluminescence. Analysis (wt %): SiO$_2$, 77.7; B$_2$O$_3$, 11.4; BeO, 0.0; Li$_2$O, 0.0; K$_2$O, 5.2; Na$_2$O, 6.2. Uniaxial (+); (Na D): N$_O$ = 1.516, N$_E$ = 1.532; colorless. Occurs in cavities in marble xenoliths in nepheline–syenite at the *Poudrette quarry, Mt. St.-Hilaire, QUE, Canada*. Associated with pectolite, apophyllite, and aegirine. EF *MR 21:284(1990)*.

63.2.1a.9 Sugilite KNa$_2$(Fe^{3+},Fe^{2+},Mn^{2+},Al)$_2$Li$_3$Si$_{12}$O$_{30}$·nH$_2$O

Named in 1976 by Murakami et al. for Ken-ichi Sugi (1901–1948), Japanese petrologist; found in 1944 by Sugi and Kitsuna, but not identified. Milarite group. See the other members of the milarite group. HEX $P6/mcc$. $a = 9.986–10.040$, $c = 13.936–14.062$, $Z = 2$, $D = 2.80$. *29-824*(nat)[*MJJ 8:110(1976)*]: 6.98$_1$ 4.32$_{10}$ 4.06$_6$ 3.50$_2$ 3.19$_8$ 2.88$_5$ 2.725$_1$ 2.50$_2$; Wessels mine: 4.34$_{10}$ 4.08$_5$ 3.50$_3$ 3.28$_3$ 3.25$_3$ 3.19$_{10}$ 2.87$_8$ 2.51$_2$. A double-ring silicate. Na shows disordered ninefold coordination. *AM 73:595(1988)*. Often in large compact, pure masses, and granular to prismatic; to 2 cm. Pale pink, light brownish yellow, purple: manganoan variety, reddish purple (magenta), pink, deep pink; white streak; vitreous luster. No cleavage or {0001}, poor. H = $5\frac{1}{2}$–$6\frac{1}{2}$. G = 2.73–2.79. IR data: *G&G 23:78(1987)*. DTA data: *NJBM 10:443(1990)*. Melts at ≈ 895° to a pale brown glass. Pink to purple color due to the presence of trivalent manganese. *MM 58:681(1994)*. Pink material is Al-rich and purple material is Fe-rich. Analysis (wt %), Wessels mine material: FeO, 0.72; Fe$_2$O$_3$,

10.30; Li_2O, 3.72; Indian material: Mn_2O_3 to 3.6; some H_2O^+ is present (to about 0.9 wt %), as well as H_2O^-; SiO_2, 68.2–73.3; Al_2O_3, 0.48–9.7; Fe_2O_3, 3.2–13.9; Na_2O, 5.5–6.2; K_2O, 3.8–4.6; MnO, 0.1–3.0; Li_2O, 3.1–4.1. Uniaxial (−); $N_O = 1.579$–1.611 (Mn-bearing), $N_E = 1.577$–1.607; generally colorless but Mn-rich material may be pale pink and pleochroic; birefringence = 0.003–0.006. Refractive indices tend to increase from Al-rich, Fe-poor to Al-poor, Fe-rich material. Occurs in stratabound metamorphosed manganese deposits (e.g., India, South Africa, and Australia) and in peralkaline igneous rocks. Also in marble xenoliths, as at Mt. St.-Hilaire, QUE, Canada. At Iwagi Islet, Japan, occurs in aegirine syenite, with albite, aegirine, pectolite, and a Ca–K–Ti silicate. In the Kalahari manganese field, Cape Prov., South Africa, with braunite, hausmannite, manganite, and others as bedded seams to 3 cm thick. Occurs at Mt. St.-Hilaire, where it is associated with poudretteite, quartz, polylithionite, and pectolite; Al-bearing (to 9.7 wt % Al_2O_3) from Cerchiara mine, near Faggiona, northern Apennines, Italy; fine crystals to 6 mm long and some crystals to 2 cm across at the Wessels and N'Chwaning mines, near Kuruman, Kalahari manganese field, Cape Prov., South Africa, often intergrown with variable amounts of chalcedony; at the Hoskins mine, Grenfell, and Woods mine, 30 km NNE of Tamworth, NSW, Australia; Madhya Pradesh, India; Mn-free from *Iwagi Islet*, and Mn-bearing from the Furumiya mine, *Ehime Pref., Japan*. Massive sugilite from the Wessels mine, called "lavulite," is used as a popular gemstone. May be partly replaced by aegirine (retrogressive reaction) and other minerals. EF *NJBM 10:443(1990); MM 58:671(1994), 58:679(1994)*.

63.2.1a.10 Yagiite $(Na,K)_{1.5}Mg_2(Al,Mg)_3(Si,Al)_{12}O_{30}$ or $(Na,K)_3Mg_4(Al,Mg)_6(Si,Al)_{24}O_{60}$

Named in 1969 by Bunch and Fuchs for Kenzo Yagi (b.1914), professor of geology, Hokkaido University, Sapporo, Japan. Milarite group. HEX $P6/mcc$. $a = 10.09$, $c = 14.29$, Z = 2, or 1 [*AM 54:14(1969)*], D = 2.70. *21-1365*(nat): 7.12_3 $5.06_{6.5}$ 3.73_5 $3.57_{2.5}$ 3.23_{10} 2.91_4 2.765_5 2.00_3. Grains < 0.8 mm. Colorless, white; vitreous luster. Analysis (wt %): SiO_2O, 61.7; Al_2O_3, 19.1; TiO_2, 0.8; FeO, 2.4; MgO, 10.5; Na_2O, 3.7; K_2O, 1.4; CaO, 0.1; Cr_2O_3, 0.1; MnO, 0.2. Uniaxial (+); $N_O = 1.536$, $N_E = 1.544$; birefringence = 0.008. O, very light blue, E, colorless. Occurs in a small 0.8-mm silicate inclusion in the *Colomera iron meteorite*. Associated with diopside, whitlockite, tridymite, and plagioclase. EF

63.2.1a.11 Dusmatovite $K(K,Na,\square)(Mn^{2+},Y,Zr)_2(Zn,Li)_3Si_{12}O_{30}$

Described in 1995 and 1996 by Sokolova and Pautov, and Pautov and others and named for Vyacheslav Djuraevitch Dusmatov (1936–), a Russian mineralogist and geologist who has done much work at the type locality. A member of the milarite group. HEX. $P6/mcc$. $a = 10.196$, $c = 14.284$, Z = 2, D = 2.98. PD: 7.13_3 $4.15_{4.5}$ 3.75_5 3.25_{10} 2.92_4 2.78_3 2.55_5. Structure: *DAN 344(5):607(1995)*. Dark blue, dirty blue, violet brown, streak light blue, translucent, no forms observed, no twinning observed, no cleavage, brittle, uneven fracture, non-fluorescent, $H = 4\frac{1}{2}$, $G = 2.96$. Uniax. $(-)$, $N_O = 1.590$, $N_E = 1.586$; pleochroism strong: O = light violet, E = light blue. Analysis (wt. %): Li_2O 1.10, Na_2O 0.61, K_2O 6.16, MnO 8.78, FeO 0.45, ZnO 15.51, Mn_2O_3 1.13, Yb_2O_3 0.54, Y_2O_3 1.51, SiO_2 64.40, ZrO_2 1.55. Occurs in a pegmatite boulder in a glacial moraine in the *Dara-i-Pioz alkaline massif, Tien Shan, Tajikistan*. Associated minerals are: quartz, microcline, aegirine, tadzhikite-(Y), cesium-kupletskite, hyalotekite, betafite, and polylithionite. EF Vestnik Moscow University Series 4 Geology(2):54(1996).

63.2.1a.12 Milarite $K_2Ca_4Al_2Be_4Si_{24}O_{60} \cdot H_2O$ (Z = 1) (end member), and $(K,Na)Ca_2[(Be_2Al)Si_{12}O_{30}] \cdot nH_2O$ ($n \approx 3/4$) [AM 76:1836(1991)]

Named by Kenngott in 1870 for the locality, Val Milar, Switzerland, although the original specimen came from Val Giuf. Milarite group. HEX $P6/mcc$. $a = 10.342$–10.42, $c = 13.76$–13.845, Z = 2, D = 2.56–2.61. *12-450*(nat)[*CM 18:41(1980)*]: 5.22_4 4.16_5 3.78_2 3.46_2 3.32_{10} 3.01_3 2.88_7 2.745_5; *35-459*(nat): 5.20_2 4.51_4 4.14_3 3.42_2 3.31_{10} 3.01_6 2.86_7 2.73_3; [nat Y and HREE-rich, *CM 29:533(1991)*]: 6.90_4 5.17_3 4.48_3 4.13_8 3.29_{10} 2.985_4 2.87_7 2.73_3. The structure of milarite has been well determined. There are 15 minerals with a simple four-connected three-dimensional net. Similar to that of bazzite and beryl. The difference is the number of Si_6O_{18} rings. Bazzite has one ring while milarite has double Si_6O_{18} rings. *AM 76:1836(1991)*, *EJM 1:353(1989)*. Prismatic parallel to [001], hexagonal prisms of varying length with forms: {11$\bar{2}$0}, {0001}, {10$\bar{1}$1}, {0001}; at Mt. St.-Hilaire 2–3 mm, Strzegom to 4 mm, Brazil to +5 cm. Also radial fibrous aggregates. White to colorless, pale green; white streak; vitreous–silky luster (because of striations). Cleavage (0001), indistinct; conchoidal fracture. H = 5–6. G = 2.51–2.56. IR data: *AM 76:1836(1991)*, *MM 50:271(1986)*. Blue-white fluorescence in SWUV (Mt. St.-Hilaire) and bright green cathodoluminescence (Strzegom); generally insoluble in common acids, some is slightly soluble in HCl, soluble in HF. Range of Be/Al proportions: $(Be_{1.91}Al_{1.1})$ to $(Be_{2.55}Al_{0.45})$ resulting from the substitution $Be^{2+} + (Na, K) \leftrightarrow Al^{3+} + \square$; yttrian milarite is very heterogeneous. Y + REE substitute for Ca. Y + REE increase with Be/Be + Al. $Ca^{2+} + Al^{3+} \leftrightarrow (Y, REE)^{3+} + Be^{2+}$. Y_2O_3 contents to ≈ 7.9 wt %. Water is present as H_2O. A simplified general formula is: $KNa_{1-x}Ca_2(Be_{3-x}Al_x)Si_{12}O_{30}[H_2O]$. Biaxial $(-)$; $N_E = 1.529$–1.552, $N_O = 1.532$–1.555; $2V = 34°$ in {0001} to $64°$ in {10$\bar{1}$1}; usually optically anomalous, showing biaxial character, and not truly uniaxial, showing wavy or patchy extinction, sector-zoning or moiré texture. West Věžná material shows true *uniaxial* optics. OAP azimuths related to the external hexagonal symmetry of the crystals. Anomalous optical properties are characteristic of milarite. Good summary in *MM 50:271(1986)*. Possibly due to variation in channel H_2O contents. Positional disorder. Synthetic Mn-milarite is not $P6/mcc$. Twelve reflections violate the space group. *SPC 22:453(1977)*. Occurs at Mt. St.-Hilaire on pectolite crystals in marble xenoliths. At Strange Lake, occurs in extensively mineralized peralkaline granite. In granite pegmatite at Jaguaraçú, MG, Brazil (Y-bearing). In vugs in plutonic rocks, pegmatites, hydrothermal ore deposits, and alpine veins. A low-temperature hydrothermal mineral crystallizing in range 200–250° and at low pressures. In pegmatites at Strzegom. May form from breakdown of beryl and cordierite. Occurs in pegmatites in the Conway granite at the government pit and Moat Mt., near Conway, Carroll Co., NH; at the Foote mine, Kings Mt., Cleveland Co., NC; Mt. St.-Hilaire, QUE, and Strange Lake, QUE–LAB border, Canada; Tanno pegmatite, Chiavenna, Sondrio, Italy; Tittling, Bavaria, and Henneberg, Thuringia, Germany; Säulkopf, west Tyrol, and Habachtal, Gasteiner, Austria; Val Giuf, Grissons, Graubünden, and about 25 other localities in Switzerland, including Val Mila, Grigioni; Tysfjord, Drag, Nedre Lapplagret, Norway; Strzegom, Poland; West and East Věžná, Radkovice, and Maršíkov, Czech Republic; Yermakorsk, eastern Siberia (pink); Kola Penin., and Central Asia, Russia; Kent, central Kazakhstan; fine crystals from Rössing, Swakopmund, Namibia (gem-quality); La Valencia mine, Guanajuato, Mexico; zoned Y-bearing and Y-poor fine crystals from the José Pinto pegmatite, Jaguaraçú, near Coronel Fabriciano, MG, Brazil. A rare gemstone. EF *SMPM 50:445(1970)*.

63.2.1a.13 Sogdianite $(K,Na)_2(Zr,Fe,Ti)[Li_2(Li,Al)Si_{12}O_{30}]$

Named in 1968 by Dusmatov et al. for the ancient state of Middle Asia, Sogdiana. Milarite group. HEX $P6/mcc$. $a = 10.079$–10.09, $c = 13.98$–14.338, $Z = 2$, $D = 2.82$; 21-501(nat): 4.51_6 4.09_9 3.52_5 3.20_{10} 2.90_{10} 1.84_8 1.52_7 1.33_8. For structure, see osumilite (63.2.1a.6,7). Flaky grains, platy deposits to 10 cm × 7 cm × 4 cm in a vein. Violet, purple-pink, violet-pink; vitreous luster; transparent. Cleavage {0001}, perfect. $H = 6$–7. $G = 2.80$–2.90; DTA: exotherm at 950°, endotherm at 1050°. *DAN SSSR 182:137(1968)*. Analysis (wt %), Alai Range: SiO_2O, 68.83; ZrO_2, 9.78; Li_2O, 3.73; TiO_2, 2.88; Al_2O_3, 1.04; Fe_2O_3, 4.61; FeO, 1.22; MgO and MnO, traces; CaO, none; BaO, 0.32; K_2O, 4.84; Na_2O, 2.81; H_2O, none; F, none. Golden Horn has more ZrO_2 (≈ 15–18%) and less $Na_2O \approx 1.5$. Uniaxial $(-)$; $N_O = 1.608$, $N_E = 1.606$; colorless; birefringence $= 0.002$. Occurs in pegmatite associated with zektzerite at Washington Pass in the Golden Horn batholith, Okanogan Co., WA; in peralkaline rocks, usually in pegmatites. Associated with quartz, aegerine, REE minerals, thorite and stillwellite at *Dara-i-Pioz massif, Alai Range, Tien Shan, Tadzhikistan*. EF *GSAPA 19(7):594(1987)*, *AM 76:1836(1991)*.

63.2.1a.14 Roedderite $(Na,K)_2(Mg,Fe)_2[(Mg,Fe)_3Si_{12}O_{30}]$

Named in 1966 by Fuchs et al. for Edwin Woods Roedder (b.1919), U.S. Geological Survey, who synthesized $K_2O \cdot 5MgO \cdot 12SiO_2$. Milarite group. Forms series with eifelite and chayesite. See the other members of the group. HEX $P\bar{6}2c$. $a = 10.139$–10.142, $c = 14.269$–14.275, $Z = 2$, $D = 2.65$. 23-76(nat)[*AM 52:1519(1967)*]: 5.06_8 4.38_9 4.13_8 3.73_9 3.57_8 3.23_{10} 2.92_9 2.76_9. Structure done at 100 and 300 K. An ordered arrangement of Na atoms reduces the space group from the normal $P6/mcc$ to $P\bar{6}2c$. EJM $1{:}715(1989)$. Hexagonal tablets with dominant $\{0001\}$, $\{10\bar{1}2\}$, $\{10\bar{1}0\}$; $\{11\bar{2}0\}$ is rare. Also prismatic with same forms, and $\{10\bar{1}0\}$ dominant; to ≈ 6 mm across in Eifel region. Colorless, medium dark blue, dark green, orange, yellow. No cleavage. $G = 2.60\text{--}2.63$. Analysis (wt %), Indarch meteorite: SiO_2, 71.0; Al_2O_3, 0.4; MgO, 19.5; FeO, 2.0; K_2O, 3.3; Na_2O, 4.0; Eifel–Bellerberg: SiO_2, 71; Al_2O_3, 0.4–0.8; FeO, 0.5–2.5; MgO, 19–17.6; Na_2O, 3.5–4.0; K_2O, ≈ 4.5; TiO_2, 0.03–0.12; MnO, 0.15–0.4; Cancarix, Spain: FeO, 5.8–13.9; MgO, 10.4–16.0; K_2O, 4.2–4.6; Na_2O, 0.5–2.6, and so on. K or Na may dominate in the alkali position (C site). Main substitution between chayesite and roedderite is: $(K,Na) + Fe^{2+} \leftrightarrow \square + Fe^{3+}$. Uniaxial and biaxial $(+)$; $N_O = 1.537$, $N_E = 1.542$; $N_x = 1.536$, $N_y \approx 1.536$, $N_z = 1.543$. O,E, colorless. Sometimes biaxial with $2V = 5\text{--}8°$ $(+)$. Occurs in peralkaline volcanic rocks and meteorites (enstatite chondrites and octahedrites). In altered gneisses in leucite tephrite lava. In xenoliths in Eifel volcanics, associated with clinopyroxene, quartz, enstatite, tridymite, hematite, amphibole, and so on. In lamproite in Spain. Occurs in the Wichita Co., KA, octahedrite meteorite; in silicate inclusions in the Canyon Diablo meteorite, AZ; from Cancarix, Albacete Prov., Spain; in the Bellerberg volcano, Mayen, east Eifel region, Germany, and at about 10 localities (e.g., Emmelberg) in the East and West Eifel region; in the *Indarch enstatite chondrite (fell April 7, 1891) in Transcaucasia, Russia*; possibly at the Wessels mine, Kuruman district, Kalahari manganese field, Cape Prov., South Africa; Qingzhen enstatite chondrite, China. EF *CMP 82:252(1983), DA 44:89(1993), MM 58:655(1994), METEO 29:707(1994)*.

63.2.1b.1 Armenite $BaCa_2Al_6Si_9O_{30} \cdot 2H_2O$

Named in 1939 by Neumann for the locality. Milarite group. Synonym: calciocelsian (now discredited). A double-ring analog to cordierite (single six-membered ring). If heated for 4 days at 900–1000°, H_2O is expelled and optics become uniaxial with decreased refractive index values. The twinning pattern also disappears. Superstructure reflections disappear and the crystals become metrically hexagonal. However, based on order–disorder relationships in cordierite of (Si,Al), there probably is still not complete disorder of (Si,Al). So, crystals are probably still orthorhombic $(Amaa)$. ORTH $Pnna$. (Nat): $a = 13.874$, $b = 18.661$, $c = 10.697$, $Z = 4$; (at 1000°): $a = 14.024$, $b = 18.383$, $c = 10.608$, $Z = 4$, $D = 2.75$ and 2.79; the hexagonal parameters are approximately: $a = 10.69\text{--}10.74$, $c = 13.87\text{--}13.909$. *37-432* [see *CM 22:453(1984)*]: could not resolve any splitting, also as hexagonal; *20-112*(nat): $9.31_{5.5}$ 6.94_8 $4.25_{6.5}$ 3.86_{10} 3.41_9 3.10_4 2.91_9 2.78_8. *37-432* is very similar; *SMPM 71:301(1991)*, best data: 9.28_3 6.93_{10} 4.26_5 3.87_3 3.86_6 3.41_3 2.91_6 2.78_5; also *45-1481*. A double-ring silicate. Deviation from true hexagonal symmetry to orthorhombic symmetry is due to (1) ordering of H_2O and (2) partial Si–Al ordering. *SMPM 71:301(1991), AM 77:422(1992)*. Prismatic, crystals to 1.5 cm, aggregates to several centimeters across, pseudohexagonal morphology formed by fine lamellar penetration twins. Gray-green to colorless, white streak, vitreous luster. Three cleavages: {010}, {011}, distinct. H = 7–8. G = 2.74–2.77. TGA data: *NGT 21:19(1941)*. IR data: *CM 22:453(1984)*. Usually fairly pure. Traces of Na_2O and K_2O. Biaxial (−); $N_x = 1.5505$–1.551, $N_y = 1.557$–1.559, $N_z = 1.559$–1.562; colorless; 2V = 42–66°; birefringence = 0.011; also uniaxial (−): $N_O = 1.556$, $N_E = 1.550$. Occurs in Ba-rich environments. Hydrothermal veins and hydrothermally altered rocks. In sulfidic calc–silicate-rich quartz rocks (schists) and in granulitic gneisses as porphyroblasts. Low- and high-temperature environments. Occurs near Rémigny, QUE, Canada; Coire Loch Kander, Scotland; *Armen mine, near Kongsberg, Norway*; Zufuru mine, southwestern Sardinia, Italy; Simplon area, Wasenalp, Vallais, Switzerland; Chvaletice, 20 km W of Pardubice, Czech Republic; Northern Dneiper region, Ukraine; Broken Hill and Purna Moota, 30 km N of Broken Hill, NSW, Australia. EF *MZ 9:83(1987); MM 51:317(1987), 55:135(1991); NJBM:49(1989)*.

Class 64

Rings with Other Anions and Insular Silicate Groups

64.1.1.1 Eudialyte $Ca_6Zr_3(\square,Fe)_3Si_{24}(Si,\square_3)(Ti,\square)(Nb,Al,\square)$-$(Fe,Mn,\square)_3(Na,K,Sr,REE,Mg)_3(\square,Sr,Mg)_3(Na,\square)_{36}O_{69}/(O,\square)_6(OH,\square)_6(Cl,\square)_2$ [*SPC 33:207(1988)*]; simplified formula: $Na_{15}Ca_7Fe_3Zr_3Si(Si_3O_9)_2(Si_9O_{27})_2(OH)_2Cl_2$ [*PD 5:89(1990)*]

Described in 1801 by F. Stromeyer and named from the Greek for *readily decomposable*, in allusion to its easy reaction with acids. The names *eucolite* (from T. Scheerer in 1847) (high content of heavy multivalent cations such as Fe, Mn, REE, Nb, Ca, etc., and optically negative) and *barsanovite* (a piezoelectric variety of eucolite) are synonymous with eudialyte. HEX-R $R\bar{3}m$ or $R3m$. $a = 13.95$–14.39, $c = 29.91$–30.26, $Z = 3$ (for simplified formula), $D = 2.67$ (REE-free). *41-1465*(nat): 7.12_4 $5.71_{4.5}$ $4.31_{5.5}$ 3.40_4 3.22_5 $3.165_{3.5}$ 2.98_8 2.86_{10}. A monoclinic variety of eudialyte was reported from Pocos da Caldas, MG, Brazil: $a = 24.641$, $b = 14.188$, $c = 21.761$, $\beta = 112.44°$. *AABC 52:243(1980)*. A cyclosilicate, with both threefold (Si_3O_9) and ninefold [$Si_9O_{24}(OH)_3$] rings of SiO_4 tetrahedra. *TMPM 16:105(1971)*. The structure is related to that of zeolites, in that the framework of eudialyte is the same in the different varieties, but the atoms in the zeolitic cavities are individual and influence the properties in a definite way. Differences in occupancies by heavy cations produce accentricity. *ZVMO 119:65(1990)*. **Habit:** Crystals may be as much as 8–10 cm across from the Khibiny and Lovozero massifs and also from Greenland. Crystals may be pseudo-octahedral, equant; also rhombohedral, short prismatic. Dominant forms are: {0001}, {11$\bar{2}$0}, {10$\bar{1}$0}, {10$\bar{1}$1}. **Physical properties:** Color is variable: carmine-red, orange-red, orange, pink, cherry-red, brownish red, yellowish brown, brown, yellow, violet, or green. The color appears to be controlled by the amounts of Fe^{3+} and Mn^{3+} present. White to pale pink streak, vitreous to greasy luster. Cleavage {0001}, distinct; {11$\bar{2}$0}, imperfect; {10$\bar{1}$0}, {10$\bar{1}$4}, lesser; uneven to conchoidal fracture; brittle. $H = 5$–6. $G = 2.70$–3.10. IR data (Moenke), optical absorption spectra, and Mössbauer spectra: *PCM 18:117(1991)*. Sometimes luminescent orange-red. Weakly electromagnetic, gelatinizes with acids. **Chemistry:** Fe^{2+} is responsible for the red color in eudialyte (optically +). In optically (−) eucolites, Fe^{2+} ions are in a tetragonal pyramid based on the planar quadrangle. Eucolites have Fe^{3+} as well. Eucolite has planar four- and fivefold (pyramid) Fe^{2+}. REE-free eudialyte exists. The change in optical properties may also result from complex coupled substitutions involving Ca and REE. The chemistry varies widely. A eucolite from Tanzania had no REE, U, and Th, while a eucolite on Ascension Island had 10–12 wt % REE. Either Ce or Y group REE's may be dominant. A eudialyte from Malawi has 3.6 wt % Nb_2O_5, and 6.50 wt % MnO. A K_2O-rich (to 6.4 wt %) and H_3O^+-bearing eudialyte was recently described from the Khibiny massif. *DAN-ESS 318:154(1993)*. **Optics:** Optical properties are variable: uniaxial (+) or (−), also isotropic; N = 1.602 (isotropic), N_O = 1.588(hydronium-bearing)–1.623, N_E = 1.594(hydronium-bearing)–1.633(optically positive eudialyte) and N_O = 1.606–1.630, N_E = 1.602–1.627(optically negative eudialyte). Birefringence = 0.00–0.02. Colorless or pleochroic with N_O pale yellow, yellowish pink, and N_E pale yellow or pink. Crystals may have (−), (+), and isotropic zones. Reversals in optic sign are induced by heating: Na-rich samples (eudialyte) become (+) and Ca-rich samples (eucolite) become (−). Eudialyte is often anomalously biaxial, but most eucolites are uniaxial. "Barsanovite" has N_x = 1.624–1.633, N_y = 1.628–1.633, N_z = 1.628–1.639, with $2V_x$ = 12–17°. **Occurrence:** Occurs in alkaline and peralkaline igneous rocks (e.g., Ne-syenites) and their cogenetic pegmatites. It is a primary magmatic mineral, crystallizing at >720°. Also occurs rarely in silica-saturated rocks. May alter to catapleiite, zircon, feldspar, acmite, analcime, clays, and other minerals. **Localities:** Localities include: Magnet Cove, AR; Pajarito Mt., Otero Co., NM; Cornudas Mts., NM; Bearpaw Mts., MT; in alkali granite 32 miles SSW of Farewell, AK; Mt. St.-Hilaire, and Kipawa Lake, Temiscamingue Co., QUE; Seal Lake area and Red Wine alkaline complex, LAB, Canada; *Kangerdluarssuk (Ilimaussaq)*, and to 10 cm from Naujakasik, Julianehaab district, Greenland; Langesundfjord, southern Norway; many localities in the Lovozero (e.g., Mts. Karnasurt and Vavnbed) and Khibiny (e.g., Mts. Nyorkpakhk, Kukisvumchorr, and Takhtarvumchorr) massifs (barsanovite was described from near Petrelius Mt.), the largest crystal being approx. 8 cm × 3.5 cm; Cape Turi, and crystals of an unusual green color are found in the "Mica" quarry, Kovdor massif, Kola Penin., and Tuva, and Burpala massif, Siberia, Russia; Ampasibitika, Madagascar; Oldoingo Lengai, Tanzania; southeastern Q, Australia; Pocos da Caldas, MG, Brazil. EF *DAN SSSR 153:1164(1963); MIN 3(2):227(1981); MM 46:421(1982), 56:428(1992); TMPM 32:153(1983); DHZ IB(1986)*.

64.1.1.2 Alluaivite $Na_{19}(Ca,Mn^{2+})_6(Ti,Nb)_3Si_{26}O_{74}Cl \cdot 2H_2O$

Named in 1990 by Khomyakov et al. for the location. Eudialyte group. HEX-R $R\bar{3}m$. a = 14.046, c = 60.60, Z = 6, D = 2.78. *PD:* 7.14_8 4.30_7 3.36_5 2.96_{10} 2.82_{10} 2.14_7 1.76_8 1.36_7. Groups of $[Si_3O_9]$ and two types of $[Si_{10}O_{28}]$ groups are in the structure. Irregular grains to 1 mm. Colorless to pale pink with a brownish tint, white streak, vitreous luster, transparent. Conchoidal fracture, brittle. H = 5–6. G = 2.76. IR data are similar to eudialyte. Vivid red-orange fluorescence in UV light. Mineral slowly leaches in 10% HCl, lowering the refractive index values. Analysis (wt %): Na_2O, 18.6; K_2O, 0.2; CaO, 8.6; SrO,

1.0; BaO, 0.6; MnO, 3.6; La_2O_3, 0.2; Ce_2O_3, 0.8; SiO_2, 53.3; TiO_2, 6.0; ZrO_2, 0.2; Nb_2O_5, 3.9; Cl, 0.8; H_2O, 1.7. Uniaxial (+); $N_O = 1.618$, $N_E = 1.626$; colorless. Occurs in ultra-agpaitic pegmatites, associated with nepheline, sodalite, potassium feldspar, arfvedsonite, aegirine, a cancrinite-like mineral, and others. Closely intergrown with eudialyte. *Alluaiv Mt., Lovozero massif, Kola Penin., Russia.* EF *ZVMO* 119:117(1990), *DAN SSSR* 312:1379(1990), *AM* 76:1728(1991).

64.2.1.1 Scawtite $Ca_7(Si_3O_9)_2(CO_3) \cdot 2H_2O$

Named in 1929 by C. E. Tilley for the locality. MON Space group unknown, probably $I2/m$. $a = 10.118$–10.121, $b = 15.180$–15.187, $c = 6.623$–6.626, $\beta = 100.55$–$100.66°$, $Z = 2$, $D = 2.765$. Redetermined structure [*CSB* 37(11):930(1992)] indicates that the space group is Cm with cell dimensions: $a = 10.039$, $b = 15.194$, $c = 6.634$, $\beta = 115.64°$. 31-261(nat): 3.20_6 3.03_5 3.02_{10} 2.99_8 2.96_4 2.49_4 2.23_4 1.89_4. Two structure determinations [*AC(B)* 29:73(1973), *ZC* 13:195(1973)] state that the CO_3 group is disordered. However, the R factor was only 11.7%. A 1992 Chinese determination, with R = 2.2%, indicates that the CO_3 group is ordered in the structure. Platy crystals to 1.5 mm in subparallel and slightly diverging groups, also granular aggregates, seven main forms: {010}, {100}, {130}, {120}, {110}, {011}, {101}. Colorless or white, gray-white, pale flesh, shows piezoelectric effect; vitreous luster. Cleavage {010}, {001}, perfect; uneven or conchoidal fracture. $H = 4\frac{1}{2}$–5. $G = 2.71$–2.73. Some material from Crestmore fluoresces weak white in SWUV and LWUV. IR data for synthetic material: *SILI* 22(2):151(1978), *MSLM* 4:445(1974). DTA data: *MIN* 3(2):194(1981). Sometimes fluoresces bluish white in SW. Soluble in cold dilute HCl with formation of gelatinous residue. Analyses from four localities show variable amounts of SiO_2 and CO_2. Biaxial (+); $N_x = 1.595$–1.603, $N_y = 1.605$–1.609, $N_z = 1.618$–1.622; $2V = 75°$; colorless, OAP parallel to (010); $Z \wedge c = 30°$ [in setting of Murdoch, *AM* 40:505(1955)]. Forms as pseudomorphs after spurrite. Occurs in hybrid rocks formed by assimilation of limestone by dolerite. In Northern Ireland, associated with melilite, bultfonteinite, calcite, and zeolites; in North America and New Zealand, with larnite, spurrite, tobermorite, gehlenite, garnet, vesuvianite, calcite, wollastonite, and others; in Japan, with grossular, vesuvianite, and calcite. In regions of metamorphosed siliceous limestones. Alteration: to tobermorite and tacharanite. Occurs in the Little Belt Mts., MT; Crestmore quarry, Riverside Co., CA; *Scawt Hill, Co. Antrim, N. Ireland*; Ballycraigy, Co. Larne, N. Ireland; Kearny, Aberdeen, Scotland; Maroldsweisach, northern Bavaria, Germany; Kansay deposit, southwestern Karamazar, Russia; Yamato and Mihara mines, and at Fuka, Bicchu, Okayama Pref., Japan; Kushiro, Hiroshima Pref., Japan; Fushan mine, Hebei, northern China; Java Sea; near Rehia, Tokatoka, New Zealand; Cerro Mazahua, Zitacuaro, Michoacan, Mexico. EF

64.2.2.1 Roeblingite $Pb_2Ca_6(SO_4)_2(OH)_2(H_2O)_4[Mn(Si_3O_9)_2]$ [AM 69:1173(1984)]

Named in 1897 by Penfield and Foote for Washington Augustus Roebling (1837–1926), civil engineer and appreciative mineral collector, Trenton, NJ, a founder of the Mineralogical Society of America. MON $C2/m$. $a = 13.208$, $b = 8.287$, $c = 13.089$, $\beta = 106.65°$, $Z = 2$, $D = 3.44$. 18-292(nat): 3.2_7 3.06_8 3.00_9 2.94_{10} 2.88_7 2.53_7 2.20_7 2.11_7. Structure: AM 69:1173(1984). Usually fine-grained, also platy, lathlike, micaceous or fibrous; dense, white nodular aggregates; Franklin—nodular masses, > 10 cm in diameter, composed of fine-grained lathlike crystals, and Långban—prismatic crystals in parallel growth. Usually white, but may be very pale pink, due to inclusions or alteration; white streak; porcelaneous. Cleavage {001}, perfect; {100}, imperfect. H = 3. G = 3.43–3.50. Gelatinizes with acids. IR and thermal data: MM 49:756(1985). At least 10 chemical analyses given in the literature. Contains some SrO as well (0.7–2.8% SrO), S is as SO_3 (i.e., sulfate). Biaxial (+); $N_x = 1.64$, $N_y = 1.64$, $N_z = 1.66$; $N_x = 1.654$, $N_y = 1.660$, $N_z = 1.678$; colorless; optics need to be redetermined; plane of the cleavage contains X and Y. 2V small to 61°; r < v, slight. Occurs in Zn–Mn ores. Precambrian metamorphosed sediments. Franklin, Sussex Co., NJ; Långban, Värmland, Sweden. EF MIN 3(1):370(1972), MM 46:341(1982), MM 49:721(1985).

64.3.1.1 Taramellite $Ba_4(Fe^{3+},Ti^{4+},Fe^{2+},Mg,V^{3+})_4(B_2Si_8O_{27})O_2Cl_x$ ($x = 0–1$)

Named in 1908 by Tacconi for Torquato Taramelli (1845–1922), Italian geologist. See titantaramellite (64.3.1.2) and nagashimalite (64.3.1.3), the V analog. ORTH $Pmmn$. $a = 12.125–12.217$, $b = 13.929–13.981$, $c = 7.130–7.136$, $Z = 2$. 17-479(nat): 3.83_5 3.68_3 3.30_4 3.16_4 3.01_{10} 2.78_3 2.58_6 2.48_5. Main structural feature is a borosilicate radical $(B_2Si_8O_{27})^{-16}$, which is formed by the rings of four Si tetrahedra connected by a B_2O_7 group sharing two oxygen atoms with each ring. Cl atoms occur, with incomplete occupancy, between adjacent borosilicate groups. In the Rush Creek, CA, sample, the principal octahedral Me cation (titantaramellite) is Ti^{4+}; in Candoglia, Italy, it is Fe^{3+}. AM 65:123(1980). {100} dominant, tabular to equidimensional crystals, elongated parallel to c, rarely elongated on a. Bronze-purple, reddish brown, brownish violet; brown streak; vitreous or greasy luster. Cleavage {010}, perfect. H = $5\frac{1}{2}$. G = 3.92. IR data: MIN 3(2):52(1981). Insoluble in acids. V-rich taramellite from the Mogurazawa mine (Kiryu, Japan), became nagashimalite. See AM 69:358(1984) for details on chemistry. Biaxial (+); $N_x = 1.770$, $N_y = 1.774$, $N_z = 1.83$ [low Ti ($\approx$ 7% TiO_2)]; X = Y = pale pinkish or nearly colorless; Z = nearly opaque (because of Fe–Ti charge-transfer processes); X is normal to the plates; XYZ parallel to cba using $a = 12$ Å, $b = 13.9$ Å, and $c = 7.1$ Å. 2V = 77°; r > v, distinct. Occurs in limestone at contact with gneiss at Candoglia. A contact-metamorphic mineral. Associated with calcite, magnetite, chalcopyrite, pyrite, pyroxene, actinolite, and celsian. With other Ba-silicates and Ba–Ti-silicates in metamorphic rocks at or close to contacts with large granitic bodies. Sanbornite is the dominant associated mineral. Clear Creek area, San Benito Co., Tulare Co., and Trumbull Peak, Incline, Mariposa Co., CA; Ross River, and the Gun Prospect, Macmillan Pass, Itsi Mountains, YT, Canada; La Madrelena mine, Rancho Tres Pozos, La Rumarosa, Baja Norte California, Mexico; *Candoglia, Valle del Toce, Piedmont, Italy.* EF *TMPM 25:245(1978)*.

64.3.1.2 Titantaramellite $Ba_4(Ti,Fe^{3+},Mg,Fe^{2+})_4(B_2Si_8O_{27})O_2Cl_x$ $x = 0–1$, with Ti > Fe)

Named in 1984 by Alfors and Pabst for taramellite and dominance of Ti as the octahedral cation. See Fe-dominant taramellite (64.3.1.1) and nagashimalite (64.3.1.3), the V analog. ORTH *Pmmn*. $a = 12.053–12.220$, $b = 13.904–14.005$, $c = 7.12–7.141$, $Z = 2$. No ICDD data. See X-ray diffraction data for taramellite (64.3.1.1). Structure solution exists for titanian taramellite: *AM 65:123(1980)*. Tabular to equidimensional, tabular parallel to {100}, ≈ 30 forms known, {100} is dominant. Physical properties are as for taramellite (64.3.1.1). $H = 6$. $G = 4.00–4.05$. Heating produces no changes to 920°, changes to fresnoite at >960°. Wet analysis (wt %): SiO_2, 33.0; Fe_2O_3, 6.79; FeO, 3.60; MgO, 0.55; TiO_2, 10.63; MnO, 0.09; B_2O_3, 3.64; Cl, 1.83; BaO, 40.45; H_2O, nil; O for Cl, 0.41; probe analysis: Al_2O_3, 0.26; V_2O_3, 0.37. In other samples: V_2O_5, to 3.2%; MgO, 2.1–4.6; FeO, 1.45–5.25; eight analyses give TiO_2, 9.10–12.53. Biaxial (+); XYZ parallel to *cba*; Big Creek: $N_x = 1.747–1.751$, $N_y = 1.756–1.761$, $N_z = 1.780$; Rush Creek: $N_x = 1.753$, $N_y = 1.757$, $N_z = 1.782$; $2V = 45–59°$; pleochroic. X,Y, yellow-brown; Z, opaque. Occurs in metamorphosed limestone associated with sulfides and sulfosalts (Kalkar quarry). Victor claim is in ultramafic body in Franciscan enclaves. Others are at or close to contacts of large granitic bodies where zones or lenses within metamorphic rocks are largely Ba-silicates. Associated with sanbornite (dominant) and rare pellyite. *Rush Creek, Fresno Co.*, Big Creek, Fresno Co., Chickencoop Canyon, Tulare Co., Kalkar quarry, Santa Cruz Co., and Victor Claim, San Benito Co., CA. EF

64.3.1.3 Nagashimalite $Ba_4(V^{3+},Ti)_4Si_8B_2O_{27}Cl(O,OH)_2$

Named in 1980 by Matsubara and Kato for Otokichi Nagashima (1890–1969), called the "pioneer" of Japanese amateur mineralogists. See taramellite (64.3.1.1) and titantaramellite (64.3.1.2). ORTH *Pmmn*. $a = 13.937$, $b = 12.122$, $c = 7.116$, $Z = 2$, $D = 4.14$. *33-188*(nat): 3.85_3 3.32_2 3.17_2 3.03_6 3.02_{10} 2.79_2 2.59_3 1.78_2. Structure: analogous to taramellite (64.3.1.1). *MJJ 10:131(1980)*. Tabular {001}, elongated on *b*, subparallel aggregates, to 1.5 cm long. Greenish black, green streak, submetallic to vitreous luster. No cleavage. $H \approx 6$. $G = 4.08$. Analysis (wt %): BaO, 41.4; V_2O_5, 16.7; TiO_2, 2.75; MnO, 0.48; SiO_2, 32.4; B_2O_3, 4.0; Cl, 1.73; H_2O, 0.77; −0.39 O for Cl. Biaxial (+); XYZ = *acb*; $N_x = 1.750$, $N_y = 1.753$, $N_z = 1.780$; $2V = 30°$; Z > Y > X; r > v, strong; strongly pleochroic. X, greenish yellow; Y, green; Z, greenish brown. Occurs at the *Mogurazawa mine, Kiryu City, Gumma Pref., Japan*; in bedded

Mn ore (rhodonite, massive). Associated with rhodochrosite, barite, barian roscoelite, alabandite, and suzukiite. EF *MJJ 10:122(1980)*.

64.3.2.1 Strakhovite $NaBa_3(Mn^{2+},Mn^{3+})_4Si_6O_{19}(OH)_3$

Described in 1994 by Kalinin et al. and named for Academician N.M. Strakhov (1900–1978), well-known for his work on the genesis of manganese ore deposits. ORTH *Pnma*. $a = 23.41$Å, $b = 12.266$Å, $c = 7.181$Å, $Z = 4$, $D = 3.84$; PD 7.00_3 4.58_5 3.30_9 3.00_{10} 2.715_5 2.655_{10} 2.16_4 1.65_5 *(ZVMO 123:94(1994))*. Structure solved to R = 6.5%. Contains $[Si_4O_{12}]$ and $[Si_2O_7]$ groups *(AM 78:675(1993))*. The structure determined formula is: $NaBa_3Mn^{2+}_{2.3}Mn^{3+}_{1.7}[Si_4O_{10}(OH)_2](Si_2O_7)O_2(F,OH)\cdot H_2O$. Black with a greenish tint (0.2–0.4 and rarely to 0.7 mm), equant to slightly elongated grains, in thin pieces is dark olive green, sometimes occurs as well-formed crystals in druses, vitreous to greasy luster, no cleavage, brittle, H = 5–6, G = 3.86. Analysis (wt. %): SiO_2 30.03, Al_2O_3 0.23, Fe_2O_3 0.03, Mn_2O_3 11.65, MnO 14.16, CaO 0.03, Na_2O 2.85, BaO 39.2, H_2O^+ 2.40, H_2O^- 0.30, Σ100.88. Biax. (−), 2V 60–65°, $N_x = 1.767$, N_y(calc.) = 1.793, $N_z = 1.871$ at 578 nm; biref. = 0.104, dispersion r > v, strong; X = c, Y = b, Z = a. Strong pleochroism: X = light green, Y = green, Z = dark olive green to brown. Occurs in the *Ir-Nimi manganese deposit, on a tributary of the Udi River, on the SW side of the Taikansk Range, Russian Far East, Russia*. It occurs at the contact between braunite ore and alkali basalt. The mineral occurs in two out of 10 parts of the deposit: Djavodi and Zaoblachno. Deposit occurs in veins in early to middle Cambrian volcanic rocks. Assoc. minerals include: braunite, taikanite, namansilite, pectolite, manganiferous amphiboles, and others. EF

64.4.1.1 Tienshanite $Na_2BaMn^{2+}TiB_2Si_6O_{20}$

Named in 1967 by Dusmatov et al. for the locality. HEX $P6/m$. $a = 16.772$, $c = 10.434$, $Z = 6$, $D = 3.23$. *20-1291*: 10.5_3 4.19_{10} 3.89_3 3.47_8 3.33_2 3.18_9 2.80_5 2.42_5. Structure: *SPD 22:544(1977)*. The structure contains silicate six-membered rings and layers of corner-sharing $[Si_2O_7]$ and $[B_2O_7]$ dimers. Pistachio green, vitreous luster. Cleavage {0001}, distinct; brittle. H = 6–6½. G = 3.29. Uniaxial (−); $N_O = 1.666$, $N_E = 1.653$. The formula above is based on the original description, whereas the formula based on the brief structure description is $KNa_9Ba_6Ca_2Mn^{2+}_6Ti_6B_{12}Si_{36}O_{120}(OH)_2$. Found with pyrochlore, astrophyllite, bafertisite, stillwellite, galena, sphene, and calcite, and in part replaced by datolite, in narrow veins of quartz–microcline–aegirine pegmatite cutting alkalic syenite in the *Dara-i-Pioz massif, Tien Shan, southern Tajikistan*. AR *AM 53:1426(1968)*, *EJM 5:971(1993)*.

64.5.1.1 Traskite $Ba_{24}Ca(Fe,Mn,Ti)_{16}[Si_{12}O_{36}][Si_2O_7]_6(O,OH)_{30}Cl_6 \cdot 14H_2O$

Named in 1965 by Alfors et al. in honor of John Boardman Trask (1824–1879), first state geologist of California. HEX $P\bar{6}m2$. $a = 17.89$, $c = 12.33$, $Z = 1$, $D = 3.52$. *18-171*: 15.4_5 3.51_4 3.27_2 3.21_2 3.08_3 2.96_{10} 2.90_2 2.53_2. Struc-

64.5.1.1 Traskite

ture: *SPD 21:426(1976)*. The structure involves a 12-tetrahedron silicate ring, and six [Si_2O_7] dimers. Brownish red, pale red-brown streak, vitreous luster. Conchoidal fracture. H ≈ 5. G = 3.71. Uniaxial (−); $N_O = 1.714$, $N_E = 1.702$; Pleochroic; O, brownish red; E, colorless to pale straw. Insoluble in acids. Fusibility 3.5. Found as euhedral crystals, 0.1–3 mm, with {0001} and {10$\bar{1}$0} prominent, in sanbornite-bearing metamorphic rocks along a granodiorite contact at *Big Creek and Rush Creek, Fresno Co., CA.* AR *AM 50:314(1965), 50:1500(1965)*.

Class 65

Single-Width, Unbranched Chains, $W = 1$

PYROXENE GROUP

The pyroxene group minerals are silicates with the general formula

$$M2M1(Si_2O_6)$$

where

M1 = site occupied preferentially by smaller cations in nearly ideal six-coordination

M2 = site occupied preferentially by larger cations in distorted six- to eight-coordination

and an orthorhombic or monoclinic structure in which the silicate chain has a periodicity of 2, with N equal to 2, and a width of 1.

The structure consists of $(Si_2O_6)^{4-}$ chains that may be extended (nearly linear) or may be shortened by rotation of tetrahedra. The chains may be of one or two types (designated A and B), and may consist of equivalent or nonequivalent tetrahedra. For the fully extended chains, the angle O_3–O_3–O_3 made by the shared oxygens is 180°. Tetrahedra rotation may be in opposite senses, designated O– and S–.

In all species the chains are parallel to the c axis, and their conformation is such that the unshared, apical oxygens are on one side of the chain opposite a nearly planar arrangement of the triangular tetrahedra bases. Parallel to {010}, the chains may be joined via the M1 sites, apex to apex, to form *I-beams*. Alternatively, the chains may also be joined base to base via the M2 sites. Successive layers of I-beams parallel to [010] are shifted with respect to one another by half an I-beam height in the (010) plane. The resultant staggering of the I-beams leads to the prismatic, {110} or {210}, cleavage at approximately 93° and 87°.

Structural details vary with the cations in the M1 and M2 sites, leading to different space group symmetries. The current pyroxene classification is crystallochemical in nature. In this presentation, however, the space group symmetry is the primary factor in establishing subgroups. Note that in the subgroup tables that follow, the M2 cation is given first if it differs from M1.

Pyroxene Group

For the clinopyroxene subgroup, chain shortening by rotation occurs when both M1 and M2 sites are occupied by smaller, six-coordinated divalent cations. The chains are not equivalent.

PYROXENE GROUP: $P2_1/C$ CLINOPYROXENE SUBGROUP

Mineral	Formula	Space Group	a	b	c	β	
Clinoenstatite	$Mg_2[Si_2O_6]$	$P2_1/c$	9.620	8.825	5.188	108.33	65.1.1.1
Clinoferrosilite	$Fe_2(Si_2O_6)$	$P2_1/c$	9.71	9.087	5.228	108.43	65.1.1.2
Kanoite	$MnMg[Si_2O_6]$	$P2_1/c$	9.739	8.939	5.260	108.56	65.1.1.3
Pigeonite*	$(Mg, Fe^{2+}, Ca)_2[Si_2O_6]$	$P2_1/c$	9.706	8.950	5.246	108.59	65.1.1.4

*Species status is not universally accepted.

Orthopyroxenes are dimorphous with the clinopyroxenes. In the orthopyroxenes chains are nonequivalent and rotated, and the tetrahedra are nonequivalent. The relationship of the cell constants is $a_{ortho} = 2a_{clino} \sin \beta_{clino}$, and the cell contents of the orthorhombic cell are double those of the monoclinic.

PYROXENE GROUP: ORTHOPYROXENE SUBGROUP

Mineral	Formula	Space Group	a	b	c	
Enstatite	$Mg_2[Si_2O_6]$	$Pbca$	18.235	8.818	5.179	65.1.2.1
Ferrosilite	$(Fe, Mg)_2[Si_2O_6]$	$Pbca$	18.418	9.078	5.237	65.1.2.2
Donpeacorite	$(Mn, Mg)Mg[Si_2O_6]$	$Pbca$	18.384	8.878	5.226	65.1.2.3

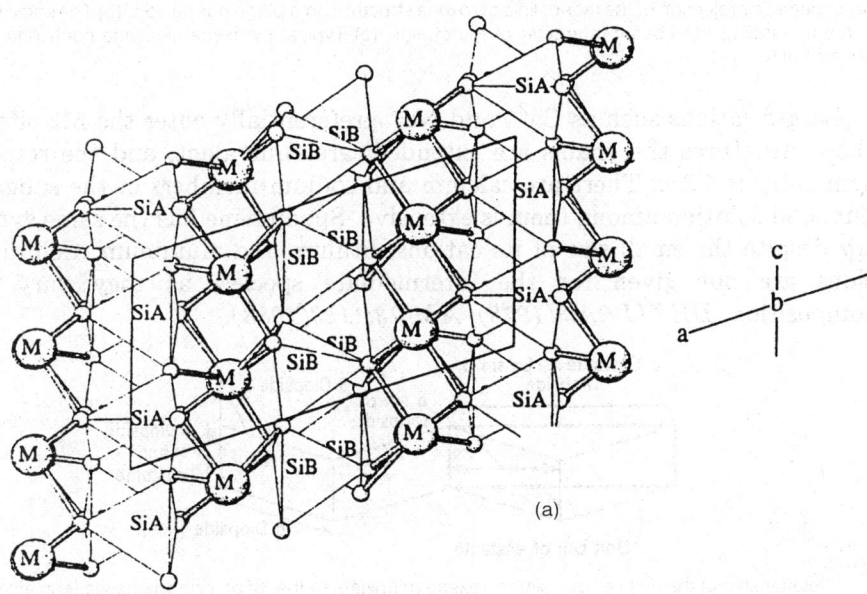

(a)

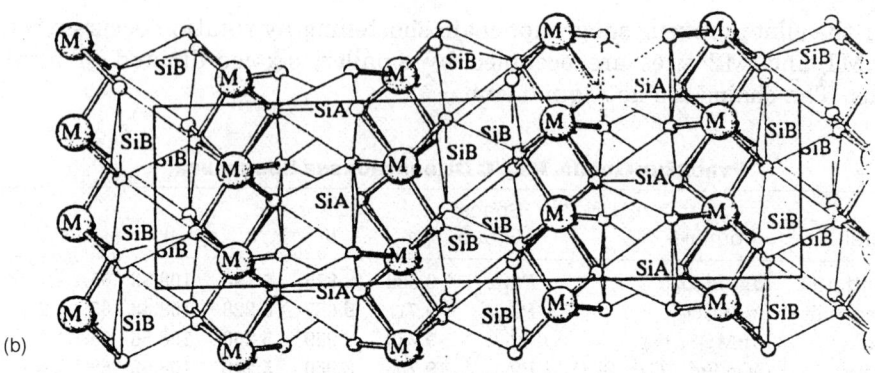

Pyroxene Group

The structures of clinoenstatite (a), and enstatite (b) viewed along the b-axis. In both structures there are two types of silicate chains, SiA and SiB. In C2/c pyroxenes such as diopside the silicate chains are alike, SiA.

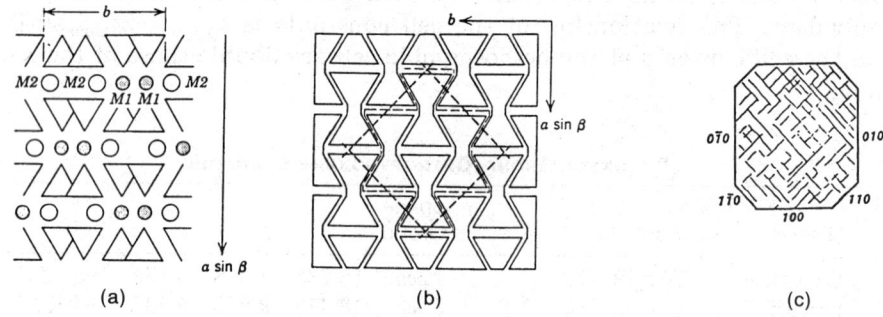

Pyroxene Group

(a) Schematic projection of the monoclinic pyroxene structure on a plane normal to c. (b) The same projection emphasizing the I-beam character of the chains. (c) Typical pyroxene cleavage controlled by the I-beams of (b).

Larger cations such as Ca^{2+} and Na^+ preferentially enter the M2 sites. In these structures the chains are extended, are equivalent, and the resultant symmetry is $C2/c$. There are calcium and sodium members of the subgroup, but solid solution among them is extensive. Spodumene has the same symmetry despite the small size of its cations, lithium and aluminum. Cell dimensions are not given for the intermediate species, as they vary with composition. *DHZII 2A:3(1978), AM 73:1123(1988).*

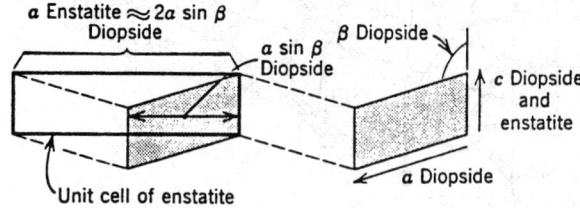

Relationship of the unit cell of a clinopyroxene (diopside) to that of an orthopyroxene (enstatite).

65.1.1.1 Clinoenstatite

PYROXENE GROUP: C2/c CLINOPYROXENE SUBGROUP

Mineral	Formula	Space Group	a	b	c	β	
Diopside	CaMg[Si$_2$O$_6$]	C2/c	9.739	8.913	5.253	106.02	65.1.3a.1
Hedenbergite	CaFe^{2+}[Si$_2$O$_6$]	C2/c	9.852	9.031	5.242	104.84	65.1.3a.2
Augite*	(Ca, Mg, Fe^{2+}, Fe^{3+}, Al, Ti)$_2$[(Al, Si)$_2$O$_6$]	C2/c					65.1.3a.3
Johannsenite	CaMn[Si$_2$O$_6$]	C2/c	9.978	9.156	5.293	105.48	65.1.3a.4
Petedunnite	CaZn[Si$_2$O$_6$]	C2/c	9.82	9.00	5.27	105.6	65.1.3a.5
Esseneite	CaFe^{3+}[AlSiO$_6$]	C2/c	9.79	8.822	5.37	105.81	65.1.3a.6
Omphacite*	(Ca, Na)(R^{2+}, Al)[Si$_2$O$_6$]	C2/c					65.1.3b.1
Aegirine-augite*	(Ca, Na)(R^{2+}, Fe^{3+})[Si$_2$O$_6$]	C2/c					65.1.3b.2
Jadeite	NaAl[Si$_2$O$_6$]	C2/c	9.418	8.562	5.219	107.58	65.1.3c.1
Aegirine	NaFe^{3+}[Si$_2$O$_6$]	C2/c	9.658	8.795	5.295	107.42	65.1.3c.2
Namansilite	NaMn^{3+}[Si$_2$O$_6$]	C2/c	9.513	8.615	5.356	105.12	65.1.3c.3
Kosmochlor	NaCr^{3+}[Si$_2$O$_6$]	C2/c	9.550	8.712	5.273	107.44	65.1.3c.4
Natalyite	Na(V^{3+}, Cr^{3+})[Si$_2$O$_6$]	C2/c	9.58	8.72	5.27	107.16	65.1.3c.5
Jervisite	NaSc^{3+}[Si$_2$O$_6$]	C2/c	9.853	9.042	5.312	106.62	65.1.3c.6
Spodumene	LiAl[Si$_2$O$_6$]	C2/c or C2	9.45	8.39	5.215	110.10	65.1.4.1

*Species status is not universally accepted for these intermediate pyroxenes.

Notes:
1. The name *pyroxene* was coined by Haüy in 1799 from the Greek *pyros*, fire, and *xenos*, stranger, on the basis of his belief that these crystals in lavas were accidental inclusions of a mineral foreign to that environment. In fact, the pyroxene minerals are second only to feldspars in their role in igneous rocks. Despite his misreading of their role, Haüy must be credited for recognizing the relationship among what were previously regarded as unrelated species.
2. For the major species, composition may be described in terms of the dominant end members in the format: En (enstatite), Fs (ferrosilite), Wo (wollastonite), Di (diopside), Hd (heidenbergite), and Ae (aegirine).

65.1.1.1 Clinoenstatite Mg$_2$[Si$_2$O$_6$]

Named in 1906 by Wahl in allusion to its symmetry and chemical identity with enstatite. Pyroxene group. Synonym: clinohypersthene for intermediate members of the clinoenstatite–clinoferrosilite series. Polymorphs: enstatite (orthorhombic), protoenstatite (orthorhombic). MON $P2_1/c$. $a = 9.620$, $b = 8.825$, $c = 5.188$, $\beta = 108.33°$, $Z = 4$, $D = 3.19$. *35-610*(syn): 3.28$_3$ 3.17$_5$ 2.98$_7$ 2.87$_{10}$ 2.54$_2$ 2.52$_2$ 2.46$_3$ 2.12$_3$. Structure: *PCM 10:217(1984)*. Chains are of types A and B with rotations S and O, respectively. Sites M1 and M2 are both occupied by magnesium in six-coordination. Usually as rock-bound, euhedra and subhedra. Commonly displays lamellar twinning on {100}, probably as the result of inversion from protoenstatite. Colorless, yellowish to greenish brown; vitreous luster; translucent to transparent. Cleavage {110}, good to perfect. H = 5–6. Biaxial (+); 2V = 53–35°; $N_x = 1.651$–1.705, $N_y = 1.653$–1.707, $N_z = 1.660$–1.727; X = b, Z ∧ c = (−)22–41°; 2V decreases, indices and extinction angle increase for the composition range Es$_{100}$Fs$_0$ to Es$_{50}$Fs$_{50}$. Substitution of calcium results in a decrease in 2V and slight increase in refractive indices. Complete solid solution with clinoferrosilite exists; solid solution with the wollastonite molecule is limited to a few mol %; additional Ca substitution yields pigeonite, and then $C2/c$ pyroxene; minor substitution of other divalent cations exists. An uncommon component of volcanic rocks and in chondritic and achondritic meteorites. Found in the Mt. Stuart batholith, Cascade Mts., WA; as phenocrysts at Cape Vogel, Papua; Bonin Is., Japan; and the Giles complex, WA, Australia. AR *ZK 114:120(1960), DHZII 2A:30(1978)*.

65.1.1.2 Clinoferrosilite Fe$_2$[Si$_2$O$_6$]

Named in 1935 by Bowen in allusion to its symmetry and chemical identity with ferrosilite. Pyroxene group. Synonym: clinohypersthene for intermediate members of the clinoenstatite–clinoferrosilite series. Polymorph: ferrosilite (orthorhombic). MON $P2_1/c$. $a = 9.71$, $b = 9.087$, $c = 5.228$, $\beta = 108.43°$, $Z = 4$, $D = 4.00$. *17-548*(syn,calc): 6.47_4 4.61_6 3.35_8 3.23_8 3.03_{10} 2.91_6 2.60_3 2.59_2. Colorless to tan, vitreous luster, translucent to transparent. Cleavage {110}, perfect. $G = 3.96$. $H = 5$–6. Biaxial (+); $2V = 34$–$21°$; $N_x = 1.705$–1.727, $N_y = 1.707$–1.753, $N_z = 1.727$–1.790; $X = b$, $Z \wedge c = (-)41$–$30°$; 2V and extinction angle decreasing, and indices increasing for the composition range Es$_{50}$Fs$_{50}$ to Es$_0$Fs$_{100}$; slight calcium substitution leads to similar changes in optical properties. The iron end member probably does not exist in nature; Mg substitutes for iron to form a complete solid solution with clinoenstatite. Occurs as acicular crystals in lithophysae in obsidian at *Obsidian Cliff, Yellowstone National Park, WY*; in the Coso Mts., Inyo Co., CA; at Hrafntiun-uhriggur, Iceland; Lake Naivasha, Kenya. AR *IMA 334(1966), DHZII 2A:30(1978)*.

65.1.1.3 Kanoite MnMg[Si$_2$O$_6$]

Named in 1977 by Kobayashi for Hiroshi Kano, petrologist, Akita University. Pyroxene group. Polymorphous with the hypothetical end member for donpeacorite (65.1.2.3). MON $P2_1/c$. $a = 9.739$, $b = 8.939$, $c = 5.260$, $\beta = 108.56°$, $Z = 4$, $D = 3.60$. *29-865*: 3.21_{10} 3.02_9 2.92_8 2.91_9 2.57_3 2.49_4 2.14_3 1.627_4. Light pinkish brown, vitreous luster. Cleavage {110}, perfect. $H = 6$. $G = 3.66$. Biaxial (+); $2V = 40$–$42°$; $N_x = 1.715$, $N_y = 1.717$, $N_z = 1.728$; $Y = b$, $Z \wedge c = 48°$. The composition is close to the ideal formula with Mn$^{2+}_{1.02}$Mg$_{0.88}$Fe$^{2+}_{0.10}$. Found as grains in a vein in a cummingtonite–pyroxmangite rock at the *Tatehira mine, Oshima Pref., Hokkaido, Japan*. AR *JGSJ 88:537(1977), AM 63:598(1978)*.

65.1.1.4 Pigeonite (Mg,Fe^{2+},Ca)$_2$[Si$_2$O$_6$]

Named in 1900 by Winchell for the locality. Pyroxene group. The name *ferropigeonite* has been applied to iron-rich varieties. Polymorphs: some subcalcic augite approximates pigeonite composition. High-pigeonite ($C2/c$) and protopigeonite ($P2_1/c$) are high-temperature polymorphs that may exist prior to cooling. MON $P2_1/c$. $a = 9.706$, $b = 8.950$, $c = 5.246$, $\beta = 108.59°$, $Z = 4$,

65.1.1.4 Pigeonite

$D = 3.44$ for pigeonite with $En_{39}Fs_{52}Wo_9$. *13-421:* 3.21_8 3.02_{10} 2.91_8 2.90_{10} 2.58_6 1.626_6 1.493_6 1.391_6. Structure: *AM 55:1195(1970)*. The nonequivalent A and B chains have shared-oxygen angles of 170° and 149° and S and O rotation, respectively. M2 is a sevenfold site with occupancy (Fe,Mg,Ca); M1 is close to regular octahedral coordination with occupancy (Mg,Fe). Domain structures may develop on cooling through inversion from the equivalent-chain $C2/c$ structure of high pigeonite. At decreased temperature, high-Ca pigeonite may exsolve augite.

Habit: Crystals generally rock-bound as anhedral to subhedral grains. Twinning common; simple or lamellar, on {100} or {001}. Exsolution lamella of augite parallel {001}.

Physical properties: Brown, greenish brown to black; colorless to pale green or brown in thin section; vitreous luster. Cleavage {110}, perfect; parting {100}, {010}, and {001}; uneven to conchoidal fracture; brittle. $H = 6$. $G = 3.17-3.46$.

Optics: Biaxial (+); $2V = 0-30°$; $N_x = 1.682-1.732$, $N_y = 1.684-1.732$, $N_z = 1.705-1.751$; $X = b$ with OAP perpendicular to (010) but rarely $Y = b$ with OAP parallel to (010), $Z \wedge c = 32-44°$; $r > v$ or $r < v$, moderate; pleochroism absent to weak, $Y > X > Z$; brown or greenish brown to colorless. Indices increase with increasing iron content; 2V decreases slightly with increasing iron, and at first decreases to 0° with increasing calcium and then increasing as OAP orientation changes. Distinguished from other clinopyroxenes by its small optic angle, and from orthopyroxenes by its higher birefringence and oblique extinction.

Chemistry: Calcium is present in the range Wo_{2-16}, magnesium in the range En_{25-65}, and iron in the range Fs_{65-25}. Complex concentric and sector zoning is common, and pigeonite mantled by augite is known. Increased calcium may be present at elevated temperatures but is exsolved on cooling. The total of aluminum, ferric iron, titanium, and alkalis rarely exceeds 0.2 ion per six oxygens of the formula unit. Pigeonite with 13 wt % MnO is known.

Occurrence: The dominant pyroxene in many rapidly cooled lavas and minor intrusives, particularly in andesites, dacites, and diabases, forming in the groundmass as well as phenocrysts. Relict pigeonite is often found inverted to orthopyroxene with exsolved augite lamellae. Iron-rich pigeonite is uncommon in metamorphic rocks. Also found in meteorites and Lunar materials.

Localities: Widespread throughout the world, but often inverted to orthopyroxene. A few noteworthy localities are: in diabase at Lambertsville, NJ; in a diabase pegmatite at Goose Creek, VA; in the Moore Co. meteorite, NC; in the Beaver Bay diabase at *Pigeon Point, Lake Co., MN*, and Mn-rich pigeonite as a product of prograde metamorphism in the Biwabik iron formation. In adamellite and monzonite of the Nain anorthosite massif, Labrador, NF, Canada; in the Allende, Mexico, meteorite; at Skaergaard, Kangerdlugssuaq, Greenland; in andesite from Mull, Scotland; in andesite from Weiselberg, Germany. In andesites from Hakone, and in basalt at Mihara-yama, Oshima,

Japan; Mt. Wellington, Tasmania; in numerous intrusive bodies of South Africa. In Lunar Mare basalts (Apollo 11, 12, 16).

Use: Of no economic importance, but an important phase in the study of igneous rocks. AR *DHZII 2A:162(1978)*.

65.1.2.1 Enstatite $Mg_2[Si_2O_6]$

Named in 1855 by Kenngott from the Greek for *opponent*, in allusion to its infusibility before the blowpipe. Pyroxene group. Synonym: orthoenstatite. Varietal names within the enstatite–ferrosilite series are based primarily on composition; enstatite (Fs_{0-10}); *bronzite* (Fs_{10-30}), in allusion to the color and luster of some examples; *hypersthene* (Fs_{30-50}), from the Greek meaning *over* and *strength*, in allusion to its greater hardness than hornblende, with which it was originally confused; *ferrohypersthene* (Fs_{50-70}) in *sensu strictu* a variety of ferrosilite. Polymorphs: Clinoenstatite (low temperature, monoclinic), protoenstatite (high temperature, orthorhombic). ORTH *Pbca*. $a = 18.235$, $b = 8.818$, $c = 5.179$, $Z = 8$, $D = 3.204$ (for $En_{100}Fs_0$). *19-768*(syn, En_{100}): 4.41_2 3.17_7 3.15_5 2.87_{10} 2.53_2 1.518_1 1.484_2 1.411_2. Structure: *ZK 176:159(1986)*. The unit cell parameters show approximate linear increase with Fe^{2+} substitution for Mg. The chains are of two types, both having O-rotation. The shared oxygen angles for chains A and B at room temperature are 167° and 145°, respectively. The M1 site is close to ideally octahedral; the M2 is a distorted octahedral site $(4+2)$, which becomes a seven-coordinated $(4+3)$ site with increasing iron substitution. Some observed lamellar structures may be due to stacking disorder developed during phase transformation rather than to twinning or exsolution. Lower-symmetry ($P2_1ca$) domains have been observed.

Habit: Rarely well crystallized as short prisms with {100} and {210} dominant with complex terminations. Commonly rock-bound in anhedral to euhedral grains. Twinning rare; lamellar with composition plane {014}; stellate with {201} as composition plane. Exsolution lamellae of diopside parallel {100} with {100} and [010] of the two phases parallel. Sometimes with kinked and displaced domains, and "hourglass" exsolution sectors.

Physical properties: Colorless, pale yellow, pale green darkening with increasing iron content to greenish brown, dark brown, to nearly black; uncolored, gray, or pale brown streak; vitreous luster, sometimes submetallic (bronzite, hypersthene) due to Schiller effect. Cleavage {210}, good to perfect; parting, {100} and {010}. $H = 5-6$. $G = 3.21-3.60$ for $En_{100}Fs_0-En_{50}Fs_{50}$.

Optics: Biaxial $(+)$ for Fs_0-Fs_{12} with $2V = 50-90°$; biaxial $(-)$ for $Fs_{12}-Fs_{50}$ with $2V = 90-55°$; $N_x = 1.649-1.710$, $N_y = 1.651-1.723$, $N_z = 1.657-1.726$ over the range 0–50 mol % Fs; dispersion weak to moderate; $r < v$ to Fs_{15}, $r > v$ to Fs_{50}; $XYZ = bac$. Pleochroism absent to moderate and probably related to Ti content; X, pink to reddish browns; Z, pale to gray-green, light blue. Chatoyant, asteriated, and labradorescent examples are known, particularly in the bronzite range.

Chemistry: Complete solid solution exists with ferrosilite. Zoned crystals with Mg-richer cores are known. Other cations rarely exceed to 10 mol %.

MnO may be present up to about 2.0 wt %, higher in ferrohypersthene. NiO and Cr_2O_3 may be present up to about 1%, particularly in meteorites and peridotites. A few percent Fe_2O_3 may be present, and Al_2O_3 may reach nearly 10% in enstatites of metamorphic rocks. CaO is rarely greater than 1%, but exsolution lamellae of diopside or augite are often present and may lead to erroneously high values. TiO_2 rarely exceeds 0.5%. Alteration to serpentine or talc is common.

Occurrence: Orthopyroxenes commonly occur in basic and ultrabasic plutonic rocks, intermediate and basic volcanics, and in high-grade metamorphics of both regional and thermal origin formed from both igneous and sedimentary antecedents. Magnesium-rich orthopyroxene is probably an important component of the upper mantle, where it is associated with forsterite, diopside, spinel, and pyrope. Known also in meteorites.

Localities: Widespread throughout the world; only a few noteworthy localities are given. Enstatite: in crystals at Tilly Foster mine, Brewster, Putnam Co., NY; in large crystals(!) often altered to talc at Bamle, Telemark, Norway; at Dawros, Connemara, Ireland; mostly altered to serpentine at Baste, Harz, Germany (Schiller spar); in crystals at Aloysthal, Moravia, Czech Republic; colorless(!) from Embilipitiya, and green to brown from Ratnapura, Sri Lanka; as centimeter-size crystals and platy groups(!) from Pirajá, Praça São Paulo, Cabeceiras, and Cordeiros, Brumado district, BA, Brazil; in the Shallowater, TX, meteorite. Bronzite: in eucritic norite in Delaware Co., PA; from the Stillwater Complex, MT; from Mysore, India (four-ray asterism); metamorphic bronzites are found at Eilean Carrach, Ardnamurchan. Scotland, and at Pahaoja, Sotajokli, Lapland, Finland; in the Kesen chondrite, Japan. Hypersthene: in norites from Delaware Co., PA; in hypersthene–granulite at Mt. Tremblant Park, PQ, Canada; as phenocrysts in pitch stone at Holmanes, Reydarfjordor, Iceland; fine crystals at Brugard, Norway (!); as phenocrysts in andesites of several volcanoes in Japan, and in amphibolite at Hokizawa, Kanagawa Pref., Japan; in the Mt. Padbury, WA, Australia, meteorite.

Uses: Although of great petrologic significance, its only economic role is as a lapidary material. Abundant Schiller material is found in Canada; asteriated, four-ray gems are produced from Nammakal, India (black star of India); colorless to light green or brown faceted stones have been produced from Arizona, Burma, and Sri Lanka. AR *DHZII 2A:20(1978)*, *PCM 10:217(1984)*.

65.1.2.2 Ferrosilite $(Fe,Mg)_2[Si_2O_6]$

Named in 1935 by Bowen for the composition: ferrous silicate. Pyroxene group. Synonym: orthoferrosilite. Intermediate compositions are given the names *ferrohypersthene* (Fs_{50-70}) and *eulite* (Fs_{70-90}), a corruption of the rock name *eulysite*, from the Greek for *easily broken* or *dissolved*. Dimorphous with clinoferrosilite. ORTH *Pbca*. $a = 18.418$, $b = 9.078$, $c = 5.237$, $Z = 8$, $D = 4.00$. *21-721*(syn): 6.46_3 4.61_9 3.23_7 3.00_8 2.91_{10} 2.59_4 2.52_3 1.781_2; *31-635*(magnesian): 3.18_{10} 2.96_4 2.88_5 2.84_2 2.72_3 2.56_4 2.51_3 2.48_3. Structure: *AM 61:38(1976)*. Isostructural with enstatite, with cell constants diminishing with substitution of Mg. Usually as rock-bound grains or phenocrysts. Possibly as prismatic crystals (ferrohypersthene). Brown, greenish brown to nearly black; gray to pale brown streak; vitreous luster, less commonly pearly or submetallic on cleavage. Cleavage {210}, good; parting, {100}, {010}. H = 5–6. G = 3.60–4.00. Biaxial (−) for Fs_{50}–Fs_{88} with $2V = 50$–$90°$; biaxial (+) for Fs_{88}–Fs_{100} with $2V = 90$–$55°$; $N_x = 1.710$–1.767, $N_y = 1.723$–1.770, $N_z = 1.726$–1.788; r < v; XYZ = *bac*. Pleochroism absent to weak; X, pinkish brown; Z, greenish brown. Forms a complete isomorphous series with enstatite, but composition with $F_5 > 90\%$ values not known in nature. Ionic substitution is similar to that found in the enstatite half of the series; Al_2O_3 may exceed 10 wt % in some metamorphic occurrences; a few percent Fe_2O_3 may be present, usually somewhat higher than in enstatite–hypersthene and particularly in metamorphic material. Found in basic and ultrabasic plutonic rocks and in acidic to basic volcanics; also in iron-rich, high-grade metamorphics. Widespread throughout the world; only a few well-documented occurrences are given. For ferrohypersthene: in gabbro of the Guadalupe complex, CA; in rhyolite pumice at Taupo, New Zealand; in hornfels at Aavold quarry, Oslo district, Norway; Al-rich in granulite at Satnur, Mysore, India. Eulite: in granite at Rubideaux Mt., Riverside, CA; in garnet–orthopyroxene gneiss at Bear Mt., Popolopan Lake Quadrangle, NY; at an unspecified locality in the former *USSR*; in granulite from Broken Hill, NSW, Australia; in eulysite from Madial, Sudan. AR *DHZII 2A:20(1978)*, *MM 52:535(1988)*.

65.1.2.3 Donpeacorite (Mn,Mg)Mg[Si_2O_6]

Named in 1984 (Petersen et al.) for Donald Ralph Peacor (b.1937), Department of Geological Sciences, University of Michigan. Pyroxene group. Polymorph: clinopyroxene kanoite (65.1.1.3). ORTH *Pbca*. $a = 18.384$, $b = 8.878$, $c = 5.226$, Z = 8, D = 3.403. *38-358:* 4.03_1 3.18_{10} 3.09_1 2.96_1 2.90_6 2.51_1 1.495_1 1.479_1. Manganese is ordered in the M2 site and exceeds 50% of that site occupancy. As interlocking anhedral grains, 1–3 mm in diameter, and displaying no twin or exsolution lamellae. Pale buff to yellow-orange, vitreous luster. Cleavage {210}, perfect. H = 5–6. G = 3.36. Biaxial (−); 2V = 88°; $N_x = 1.677$, $N_y = 1.684$, $N_z = 1.692$; XYZ = *bac*; pleochroism not described. Mn^{2+} makes up 56–63% of the M2 site occupancy; minor calcium, aluminum, and Fe^{2+} are present. Found with tirodite, tourmaline, ferrian beraunite, manganoan dolomite, hedyphane–like apatite, and anhydrite in a manganiferous pod of marble at the 2,500-ft level of the *Balmat No. 4 mine*, Balmat, St. Lawrence Co., NY. AR *AM 69:472(1984)*.

65.1.3a.1 Diopside CaMg[Si_2O_6]

Named in 1800 by d'Andrada from the Greek for *double* and *appearance*, in allusion to its variable appearance. Pyroxene group. Numerous varietal names

65.1.3a.1 Diopside

exist, of which only those utilized petrographically need be preserved: *salite* for the range $Di_{90}Hd_{10}$–$Di_{50}Hd_{50}$ is named for Sala, Sweden; *endiopside* with excess Mg; *diallage* from Greek for *difference*, in allusion to its parting. MON $C2/c$. $a = 9.739$, $b = 8.913$, $c = 5.253$, $\beta = 106.02°$, $Z = 4$, $D = 3.278$. *41-1370:* $3.22_5\ 2.99_{10}\ 2.94_6\ 2.89_3\ 2.56_2\ 2.54_2\ 2.51_3\ 2.13_2$. Structure: *AM 76:1141(1991)*. As with all $C2/c$ pyroxenes the chains are equivalent, with O-rotation and a shared oxygen angle of $\approx 166.5°$. Site M1 (Mg) is octahedral; site M2 (Ca) is eight-coordinated. The cell edges a and b increase toward hedenbergite (Fe^{2+} substitution) but c remains nearly the same; all cell edges decrease toward jadeite (coupled Na–Al substitution).

Habit: Crystals equant to prismatic with {010}, {100}, {111}, {001} dominant or with {110} better developed than {100} and {010}; less commonly tabular with {001} dominant. Commonly as rock-bound, rounded grains; anhedral to euhedral. Less commonly in fibrous masses, or masses of large crystals with prominent tabular habit. Also as exsolution lamellae in pigeonite and other high-Mg pyroxenes. Simple or lamellar twinning on {100} common.

Physical properties: White, pale to dark green, rarely violet; white to pale green streak; vitreous luster. Cleavage {110}, good to perfect; parting sometimes prominent, {100} and {010}. $H = 5\frac{1}{2}$–$6\frac{1}{2}$. $G = 3.22$–3.45. Optics: Biaxial (+); $2V = 49$–$64°$, $N_x = 1.664$–1.697, $N_y = 1.672$–1.706, $N_z = 1.694$–1.728, increasing with iron content; $Y = b$, $Z \wedge c = +35$–$48°$; increasing with iron content; $r > v$, weak to moderate; pleochroism none to weak in iron-rich examples, $Z > Y > X$; pale brownish green to pale green. Chatoyancy, and four-ray asterism in darker varieties, hedenbergite in part.

Chemistry: A complete solid-solution series exists between diopside and hedenbergite. MnO may be present to a few percent, with ferroan diopside and hedenbergite being richer in that component, and a small amount of ZnO may also be present; at least a partial solid solution with johannssenite and petedunnite exists. Coupled substitutions NaAl and $NaFe^{3+}$ for CaMg involve the end-member pyroxenes jadeite and aegirine, respectively. Chrome diopside involves the coupled substitution of NaCr for CaMg and CrAl for MgSi. Endiopside contains an excess of magnesium over the ideal 1:1 Ca/Mg

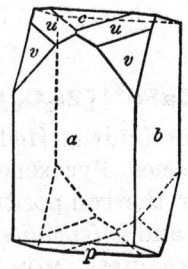

Diopside
New York. Forms: $a\{100\}$, $b\{010\}$, $c\{001\}$, $u\{111\}$, $v\{221\}$, $p\{\bar{1}01\}$,

ratio. Substitution of Al for Si in the chains is usually minor and may accompany minor Ti^{4+} in the M1 site. Extensive substitution outside the diopside hedenbergite series leads to augite, omphacite, and aegirine–augite (65.1.3b.2).

Occurrence: Chrome diopside and endiopside are common pyroxenes in nodules and xenocrysts of kimberlites and alkali olivine basalts. Diopside and salite are typical of hypabyssal crystallization from alkali olivine basalt. Relatively pure diopside is found in thermally metamorphosed siliceous dolomite and in skarns; ferroan diopside occurs in similar environments. **Localities:** Diopside is of worldwide occurrence, but only a few noteworthy or well-documented localities are cited. Excellent, light green, transparent crystals(!) at De Kalb, St. Lawrence Co., and as green granules with andradite in wollastonite at Willsboro, Essex Co., and diallage and salite at Amity, Orange Co., NY; zincian salite at Franklin, Sussex Co., NJ; chromian material at Sloan II diatreme, Larimer Co., and as crystals to 3 cm with prominent {661} at Italian Mt. S of Aspen, CO; large, opaque crystals at Crestmore, CA. Found at the Jeffrey mine, Asbestos, PQ; at Birds Creek, near Bancroft, and at Dog Lake, Storrington Twp., ON, Canada. Crystals to 30 cm in the Gardiner complex, near Kangerdlugssuak Fjord, Greenland. Manganoan material at Långban, Värmland, Sweden; deep green chromian diopside from Outokumpo, Finland; as fine light green crystals(!) with grossular and clinochlore at Ala Tal and Aosta Tal, at St. Marcel (violet), and other sites in Piemonte, Italy; at Katzenbuckel, Odenwald, Germany (ferrian); and at Zillerthal, Tyrol, Austria. Found in Yakutia and at Slyudyanka, Lake Baikal, Russia [green and emerald green(!) chromian material]; dark green chatoyant and asteriated from Nammakal, Rajasthan, India; transparent yellow and chatoyant from Burma; fine, light green crystals(!) resembling those of DeKalb, NY, from the Kunlun Mts.(?), Xinjiang Uygur, China. Dark green, transparent from Taolanaro, Madagascar, and chromian from Kenya. At Setuba, and near Malacacheta, MG, Brazil (chromian). **Uses:** When transparent, the low-iron, light green (NY, China, Madagascar, Italy) and chromian, emerald green (Russia, Brazil) varieties are faceted as minor or collector's faceted gems; the dark green (hedenbergite in part) chatoyant or asteriated varieties are cut en cabochon. AR *DHZII 2A:198(1978); MR 15:82(1984), 21:535(1990); CE 44:299(1985).*

65.1.3a.2 Hedenbergite $CaFe^{2+}[Si_2O_6]$

Named in 1819 by Berzelius for Ludwig Hedenberg, Swedish chemist, who analyzed and described the species. Pyroxene group. Petrographically used terms are *ferrosalite*, in use for the composition range $Di_{50}Hd_{50}$–$Di_{10}Hd_{90}$, and *ferrohedenbergite; schefferite* and *jeffersonite* have been used for manganoan and manganoan–zincian ferrosalites. MON $C2/c$. $a = 9.852$, $b = 9.031$, $c = 5.242$, $\beta = 104.84°$, $Z = 4$, $D = 3.65$. *41-1372:* 6.57_3 4.77_2 3.28_3 3.01_{10} 3.00_9 2.59_3 2.56_5 2.41_3. Structure: *ZVMO 107:113(1978)*. The chains are equivalent, with O-rotation and a shared oxygen angle of $\approx 164.5°$. Site M1

(Fe^{2+}) is octahedral; site M2 (Ca) is eight-coordinated. The cell edges a and b decrease toward diopside (Mg substitution), but c remains nearly the same; a and b decrease toward aegirine (coupled Na/Fe^{3+} substitution) while c increases slightly. **Habit:** Crystals equant to prismatic and as rock-bound grains; essentially the same as for diopside. Simple or lamellar twinning on {100} common. **Physical properties:** Pale to dark green, brownish green, to brownish or greenish black; pale green to tan streak; vitreous to resinous luster. Cleavage {110}, good, occasionally with {100} parting. H = 6. G = 3.45–3.56. Biaxial (+); 2V = 52–64°; N_x = 1.697–1.726, N_y = 1.706–1.730, N_z = 1.728–1.751; Y = b, Z $\wedge$ c = 47–48°, diminishing with increasing iron; r > v, strong; weakly pleochroic, Z > Y > X, green or yellow-green to pale green or bluish green. Chatoyancy and four-ray asterism is known (in part diopside). **Chemistry:** Complete solid solution exists with diopside (65.1.3a.1). Manganese may substitute to a greater degree than in diopside, as is the case for zinc. Excess Fe^{2+} may be present in the M2 site (ferrohedenbergite). Solid solution with sodic pyroxenes, and Al substitution in the chain tetrahedral sites yields, as with diopside, augite, and aegirine–augite. **Occurrence:** Hedenbergite and manganoan hedenbergite are found in limestone–dolomite skarns, thermally metamorphosed iron-rich sediments, and in metasomatically emplaced ores. Ferrohedenbergite and hedenbergite are found in quartz syenites, granophyres, and ferrodiorites. **Localities:** Worldwide distribution; some notable localities are: jeffersonite and schefferite in the zinc ores of Sussex Co., NJ; in marble enclosed in anorthosite from the Ausable quadrangle, Adirondack Mts., NY; Iron Hill, Gunnison Co., CO; in the Hanover district, Grant Co., NM; fine crystal(!) at the Laxey mine, South Mt., ID. At Mt. St.-Hilaire, PQ, and ferrosalite in the Lower Wabush iron formation, Labrador, NF, Canada. Ferrohedenbergite in granophyres of the Skaergaard complex, eastern Greenland; at *Tunaberg, Sweden*, and from Långban (schefferite), the Harstig mine, Pajsberg, and at Nordmark(!); manganoan at the Treburland manganese mine, Altarnun, Cornwall, England; in skarn at Tignitoio iron deposit, Elba, Italy; at Mega Xhorio, Seriphos, Greece, as prismatic crystals and as inclusions in quartz. Also in the Kola Penin., and at Dal'negorsk, Primorye, Russia; fine crystals from Skardu area, Pakistan; Andhra Pradesh, India; the Kamioka mine, Gifu Pref., Japan; and at Broken Hill, NSW, Australia. **Use:** Ferroan diopside and hedenbergite are cut en cabochon when they display chatoyancy or asterism. AR *DHZII 2A:198(1978), MJJ 11:84(1982)*.

65.1.3a.3 Augite (Ca,Mg,Fe^{2+},Fe^{3+},Al,Ti)$_2$[(Al,Si)$_2$O$_6$]

Named in 1792 by Werner from the Greek for *luster*, in allusion to the appearance of the cleavage surface. Pyroxene group. MON $C2/c$. a = 9.707, b = 8.858, c = 5.274, β = 106.52°, Z = 4, D = 3.31 for an augite of composition $\approx$ $Na_{0.06}$ $Ca_{0.75}Mg_{0.79}Fe^{2+}_{0.18}Mn_{0.01}Fe^{3+}_{0.09}Ti_{0.04}Al_{0.38}Si_{1.73}O_6$; the cell constants vary considerably with composition. *41-1483*(aluminian augite): 3.22_7 2.99_{10} 2.94_6 2.90_3 2.56_4 2.53_3 2.51_5 2.13_3. Structure: *MSASP 2:31(1969)*. The structure is essentially like that of diopside with a shared oxygen angle in the chain of $\approx 166°$. Ca is constrained to the M2 site and Al to the M1 site, but Mg and Fe^{2+} may be in either and their distribution is less ordered than in coexisting pigeonite and orthopyroxene.

Habit: Short prismatic [001] crystals somewhat flattened {100}, with {110}, {100}, and {$\bar{1}11$} dominant. Simple contact twins on {100} common, but also lamellar; {001} lamellar. Rock-bound anhedral to euhedral and interstitial crystals common, and as exsolution lamellae; also as large cleavable masses.

Physical properties: Greenish black to black, dark green, light to dark brown; pale brown to greenish gray streak. Cleavage {110}, good; parting {100}, {001}. $H = 5\frac{1}{2}$–6. $G = 3.19$–3.56. Biaxial (+); $2V = 25$–$61°$ ($88°$ reported for sodian varieties); $N_x = 1.671$–1.735, $N_y = 1.672$–1.741, $N_z = 1.703$–1.774, the values generally increasing with increased iron; $Y = b$, $Z \wedge c = 35$–$48°$ ($62°$ reported for sodian augite); $r > v$, weak to moderate but strong in titanian augite. Weakly to moderately pleochroic, increasing with depth of color and Ti content; X, light green, yellow-green, reddish; Y, reddish, red, violet; Z, greenish yellow, reddish, violet, the reddish or violet tones being typical of titanian augite. In thin section commonly showing compositional zoning and hourglass structure.

Chemistry: Augite represents a complex solid solution involving the end members diopside, hedenbergite, jadeite, aegirine, and the Mg–Fe orthopyroxenes, as well as lesser amounts of the less common end members, but presence of high aegirine leads to aegirine–augite (65.1.3b.2), which is not universally accepted as a species. Al commonly substitutes for Si in the tetrehedral sites, and there is usually sufficient Al + Si to satisfy the tetrahedral sight occupancy. Fe^{3+} probably occupies tetrahedral sites in the few instances when the Al + Si sum is deficient. Na, Al, and Fe^{3+}, although almost always present, each rarely exceed 10% of the M1 + M2 cations, and titanium may be present to the same extent (titanaugite). Significant Cr_2O_3 may be present in Mg-rich augites of basic rocks. Exsolution lamellae of calcium-poor orthopyroxene or pigeonite commonly form in initially subcalcic augites of basic plutonic rocks; these are most commonly oriented parallel to {100}. Metamorphic

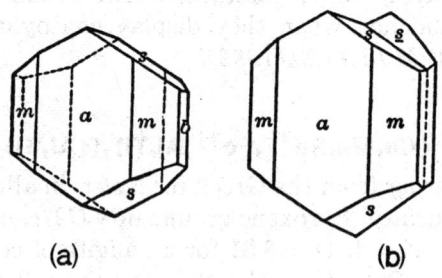

Augite
(a) Forms: a,{100}; b, {010}; m,{110}; s,{$\bar{1}11$}. (b) Contact twin on {100}.

augites may have exsolution lamellae of orthopyroxene parallel to {100} and pigeonite lamellae on irrational planes close to {100} or/and {001}.

Occurrence: Occurs in a wide variety of igneous rocks; a characteristic constituent of gabbros, dolerites, and basalts; less commonly found in ultrabasic rocks; intergrown with ilmenite in kimberlites. Titanian augite is characteristic of basic alkaline rocks. In metamorphic rocks, particularly of the granulite facies, and in metamorphosed impure and phosphatic limestones.

Localities: A ubiquitous mineral in the above-described environments. Some noteworthy and well-documented observations are: with garnet and oligoclase in granulite at Elizabethtown, NY; in crystals at Twin Peaks, Millard Co., UT. Found as crumbly crystals with scapolite in marble at La Panchita mine, near Ayoquezco, OAX, Mexico. Found as crystals and as large cleavable masses, sometimes associated with scapolite and fluorapatite, at Turner's Is., Lake Clear; the Smart mine, near Eganville; near Bancroft; and on the Cardiff property near Wilberforce, all in ON, Canada. Ferroan augite is found in the Skaergaard complex of Greenland. At Arendal, Norway, and at Ersby, near Pargas, Finland. In the Albani Mts., south of Rome, and in the lavas of Vesuvius (sometimes ferrian), Campania, Italy. In good crystals in basalt tuff SE of Teplice, Czech Republic, and in the Laacher See district of Germany. Ferrian augite is found at Khibiny, Kola Penin., and sodian augite in monzonite of Tyaki, Morotu district, Sakhalin, Russia. Common in many volcanic rocks of Japan. AR *DHZII 2A:294(1978), MSARM:7(1980)*.

65.1.3a.4 Johannsenite CaMn[Si$_2$O$_6$]

Named in 1932 by Schaller for Albert Johannsen (1871–1962), geologist and petrologist, University of Chicago. Pyroxene group. MON $C2/c$. $a = 9.978$, $b = 9.156$, $c = 5.293$, $\beta = 105.48°$, $Z = 4$, $D = 3.522$. *38-413:* 3.28_2 3.02_{10} 2.92_2 2.60_3 2.55_7 2.16_2 2.13_2 2.03_2. Reported cell constants vary only slightly for c, but show a range for $a = 9.83$–9.98 and $b = 9.04$–9.16. Structure: *AM 52:709(1967)*. The silicate chains are equivalent and they are slightly more kinked than in diopside with a shared oxygen angle of 163.8°; Mn^{2+} is in site M1 with almost ideal octahedral coordination, and Ca is in M2 with an eightfold $(4+4)$ coordination. Usually massive or columnar consisting of matted, radiating, or spherulitic aggregates of prisms or fibers. Rarely in prismatic crystals. Twinning, simple or lamellar, on {100} common. Brown, gray, green, and, less commonly, colorless to bluish; often black on surface due to Mn oxides; usually colorless in thin section; translucent to opaque. Cleavage {110}, good to perfect; parting on {100} and {001}; conchoidal to splintery fracture; brittle. H = 6. G = 3.37–3.54. Biaxial (+); 2V = 58–72°; $N_x = 1.699$–1.710, $N_y = 1.710$–1.719, $N_z = 1.725$–1.738; $X = b$, $Z \wedge c = 46$–$55°$; r > v or r < v, weak; ferroan material weakly pleochroic with $Z > Y > X$; the indices increase and 2V decreases with Fe^{2+} substitution. The chief substituents for Mn^{2+} in the M1 site are Fe^{2+}, Mg, and minor Al; those for Ca in M2 are Na and Mn^{2+}; only minor substitution of Al for Si is reported. Characteristic of metasomatized limestones adjacent to acid and intermediate intrusions, where it is often associated with rhodonite and bustamite, and in skarns; less commonly with calcite veins in rhyolite. Noteworthy occurrences are at Franklin, Sussex Co., NJ, with rhodonite in interstices of andradite–manganophyllite breccia; at the Star mine, Vanadium, NM, and at the Black Hole prospect, Graham Co., AZ; in the *Bohemia district, Lane Co., OR*; as veins in rhodonite at the Brown Mule claim, Mason Co., WA. Found with xonotlite and calcite at Tetela de Ocampo, PUE, Mexico; in limestone at Monte Civilliana, Recoaro, and the *Schio–Vincenti mine, Venezia, Italy*; in the Madan district of Bulgaria. Often highly altered to Mn oxides and opaline silica, it is the major component of Mn ores at Teragochi, Okayama Pref., Japan. Ferroan material with manganpyrosmalite and bustamite at Broken Hill, NSW, Australia. AR *DHZII 2A:415(1978)*.

65.1.3a.5   Petedunnite $CaZn[Si_2O_6]$

Named in 1987 by Essene and Peacor for Pete Dunn (b.1942), mineralogist, Smithsonian Institution, Washington, DC. Pyroxene group. MON $C2/c$. $a = 9.82$, $b = 9.00$, $c = 5.27$, $\beta = 105.6°$, $Z = 4$, $D = 3.68$. PD: 6.49_1 3.02_{10} 2.96_4 2.59_3 2.54_8 2.23_1 2.14_1 2.02_3. Dark green, vitreous luster, translucent. Cleavage {110}. H ≈ 6. Biaxial (+); $2V = 80°$; $N_x = 1.68$, $N_y = 1.69$, $N_z = 1.70$; $Y = b$, $Z \wedge c = 40°$; $r > v$, strong. Pleochroic; X,Y, light yellow; Z, light green. Found in a single specimen as a mosaic of 10- to 100-mµ grains that make up pseudo–single crystals, associated with calcite at *Franklin, Sussex Co., NJ*. AR *AM 72:157(1987)*.

65.1.3a.6   Esseneite $CaFe^{3+}[AlSiO_6]$

Named in 1987 by Cosca and Peacor for Eric J. Essene (b.1939), metamorphic petrologist, University of Michigan. Pyroxene group. MON $C2/c$. $a = 9.79$, $b = 8.822$, $c = 5.37$, $\beta = 105.81°$, $Z = 4$, $D = 3.54$. PD: 3.21_2 3.00_{10} 2.96_6 2.58_3 2.55_4 2.53_7 1.545_3 1.430_2. Reddish brown, vitreous luster, translucent. Cleavage {110}, perfect. H ≈ 6. Biaxial (−); $2V_{calc} = 70.5°$; $N_x = 1.795$, $N_y = 1.815$, $N_z = 1.825$; $Y = b$, $Z \wedge c = 9°$; strong inclined dispersion; pleochroic, $Z > Y > Z$, lemon yellow to apple-green. Resembles aegirine but distinguished from it by its extinction angle. Natural material has Si/Al ≈ 1.2:0.8. Found as 2- to 8-mm grains with mullite, anorthite, and magnetite–hercynite in association with naturally combusted coal at *Durham Ranch, S of Gillette and NE of Reno Junction, WY*. AR *AM 72:148(1987)*.

65.1.3b.1   Omphacite $(Ca,Na)(R^{2+},Al)[Si_2O_6]$

Named in 1815 by Werner from the Greek for *unripe grape*, in allusion to its usual light green color. Pyroxene group. The variety chloromelanite, named for its dark green color, lies, in part, within or close to the omphacite field. MON $C2/c$. $a = 9.561$, $b = 8.730$, $c = 5.249$, $\beta = 107.00°$, $Z = 4$, $D = 3.39$. *42-568*: 4.34_5 3.14_3 2.95_{10} 2.86_{10} 2.52_3 2.51_3 2.45_4 2.09_4. Structure: *AM 76:1141(1991)*. In addition to the $C2/c$ structure, several omphacite structures have been refined in space groups $P2$ and $P2/n$, the latter probably the valid assignment. The $C2/c$ structure is essentially the same as that of other pyroxenes assigned to this space group. In the $P2/n$ structure the chains are equivalent but have two different Si sites and two distinct M1 and M2 sites; Al and Mg are ordered among the two M1 sites, Ca and Na are partially ordered among the two M2 sites. The ordering thus requires a composition close to $Di_{50}Jd_{50}$ for the $P2/n$ structure, but many omphacites ($C2/c$) are well outside this 1:1 ratio. Green to dark green, colorless in thin section; vitreous luster; translucent. Cleavage {110}, good; parting {100}. H = 5–6. G = 3.16–3.43. Biaxial (+); 2V = 56–84°; N_x = 1.662–1.701, N_y = 1.670–1.712, N_z = 1.685–1.723; Y = b, Z ∧ c = 34–48°; r > v, moderate; weakly pleochroic, Z > Y > X, colorless to pale green, more pronounced in iron-rich varieties. The compositional field lies within the limits $(Di + Hd)_{75}Jd_{25}$, $(Di + Hd)_{60}Jd_{20}Ae_{20}$, $(Di + Hd)_{20}Jd_{60}Ae_{20}$, and $(Di + Hd)_{25}Jd_{75}$. Minor substitution of titanium and chromium occurs as well as some aluminum for silicon; only trace amounts of manganese are reported. Chloromelanite usually contains additional acmite. Omphacite is characteristic of high-pressure igneous or metamorphic environments; it is the common pyroxene of eclogites, but also occurs without garnet in glaucophane schists and amphibolites. Notable occurrences are: in the glaucophane schists of the Franciscan Formation, CA, and from kimberlite at Garnet Ridge, AZ; in eclogite of the Motagua fault, Guatemala. Found in eclogite at Naustdal and Vanelvsdalen, Norway; at Ile de Groix, France (chloromelanite); in kyanite eclogite at *Kupperbrunn, Saualpe, Austria*; in corundum and kyanite eclogites at several localities in Yakutia, Russia. Chloromelanite is found at Suberi-dani, Tokushima Prefecture, Shikoku, Japan, and in New Caledonia. Found in eclogite nodules of several kimberlite pipes of South Africa; from eclogite at Puerto Cabello, Venezuela, and Guajira Penin., Colombia. Chloromelanite finds limited use as an ornamental carving material in the Chinese tradition. AR *DHZII 2A:424(1978)*.

65.1.3b.2 Aegirine-augite $(Ca,Na)(R^{2+},Fe^{3+})[Si_2O_6]$

Species status is not universally assigned to this composition range; see also augite (65.1.3a.3) and aegirine (65.1.3c.2). The name is derived from its intermediate pyroxene composition. Pyroxene group. The pyroxene variety chloromelanite, named for its dark green color, lies in part within or close to the aegirine–augite field. MON $C2/c$. a = 9.68–9.74, b = 8.79–8.94, c = 5.26–5.36, β = 105.0–106.85°. The powder X-ray diffraction pattern is variable and intermediate between that of ferroan augite and aegirine. Cell edges a and b decrease with higher aegirine content and c increases slightly, as does the β angle, resulting in diminished cell volume. Crystals usually rock-bound, short prismatic to acicular. Twinning on {100}, lamellar or simple, common. Dark green to black, green, yellow-green; pale yellow-green or gray-green streak; vitreous luster; translucent to opaque. Cleavage {110}, good to perfect; parting {100}; conchoidal to splintery fracture. H = 6. G = 3.40–3.60. Biaxial (−)/(+); $2V_x$ = 70–110°; N_x = 1.700–1.760, N_y = 1.710–1.800, N_z = 1.730–1.813;

$Y = b$, $X \wedge c = -(0\text{--}20°)$; $r > v$, moderate to strong. Weakly to moderately pleochroic; X, bright to deep green; Y, yellow-green; Z, yellow to greenish brown. The boundary between aegirine and aegirine–augite is placed at about Ae_{80}, but alternately at the change in optic sign occurring at about Ae_{50}. Intermediate in composition between augite and aegirine; generally low in Ti, but significant Zr and minor V and Sc may be present. Chloromelanite lies, in part, within the aegirine–augite field, but most examples contain over 20 mol % each of aegirine, augite, and jadeite. A characteristic mineral of alkaline rocks, particularly syenites; also in iron-rich metasediments, and in metasomatic environments. Worldwide occurrence. Noteworthy, well-documented localities are: at Iron Hill, CO, and in shonkinite in the Bearpaw Mts., MT; in paragneiss at Seal Lake, Labrador, NF, Canada; in syenites of Greenland as at Kangerdlugssuaq. Found in aegirine granulite at Glen Lui, Aberdeenshire, Scotland; at Alnö, Sweden; in alkaline complexes of the Kola Penin., and the Ilmen Mts., Russia. In India at Harohalli, Mysore, and at Khamman, Andhra Pradesh; at the Bessi mine, Hodono, Ehime Pref., Japan. Found in iolite pegmatite, Homa Bay area, Kenya. AR *DHZII 2A:482(1978)*.

65.1.3c.1 Jadeite $NaAl[Si_2O_6]$

Named in 1863 by Damour by derivation from jade, of which it is but one type; from the Spanish *piedra de yjada*, stone of the side, in reference to its alleged curative effect over kidney ailments. Pyroxene group. MON $C2/c$. $a = 9.418$, $b = 8.562$, $c = 5.219$, $\beta = 107.58°$, $Z = 4$, $D = 3.345$. *22-1338*: 4.29_4 3.10_3 2.92_7 2.83_{10} 2.49_2 2.42_2 2.07_3 1.572_2. Structure: *CMP 83:257(1983)*. The equivalent silicate chains are the least kinked of the clinopyroxenes with a shared oxygen angle of $174.7°$. Al is in nearly ideal octahedral coordination in site M1, and the eight-coordinated M2 site has nearly equal Na–O distances. Crystals are rare in vugs in massive material as short prismatic crystals with the pinacoids dominant. Usually in compact masses of granular to short fibrous, anhedral crystals. White, colorless, pale to deep emerald-green, yellow, gray, bluish gray, lavender to violet; vitreous luster; translucent to opaque. Cleavage {110}, good; conchoidal fracture. $H = 6\text{--}7$. $G = 3.24\text{--}3.43$. Biaxial (+); $2V = 60\text{--}90°$; $N_x = 1.640\text{--}1.681$, $N_y = 1.645\text{--}1.684$, $N_z = 1.652\text{--}1.692$; $Y = b$, $Z \wedge c = +35\text{--}55°$; $r > v$, moderate to strong, sometimes yielding anomalous interference colors. Chromium, manganese, and ferric and ferrous iron are commonly present in small amounts and yield the brightly colored jadeite (Cr = emerald green, Mn = violet and lavender); in massive, multicolored material the chromophore-substituted material occurs in random, irregular patches. Compositionally zoned jadeite with iron-richer cores is known. When considerable diopside is present, the term *diopside-jadeite* is often used although the composition is usually within the omphacite field. Jadeite is generally restricted to metamorphic rocks of high-pressure origin, but has also been observed in lower-pressure metamorphics. Almost invariably associated with albite, but also quartz, glaucophane, lawsonite, actinolite, and others; found in metagreywackes where the albite content is low, it having been mostly converted to jadeite and quartz. Reported from pillow lavas and as nodules in serpentinite. Widespread in metamorphic rocks of tectonically active, subductive plate boundaries. Notable localities: found with glaucophane, lawsonite, and other minerals in the Franciscan formation, as at Clear Springs in San Benito Co., Eel R. in Trinity Co., Valley Ford in Sonoma Co., CA, both massive and in crystals; along the Motagua fault, Guatemala. Reported from Switzerland and possibly other Alpine terrains of central Europe. A violet jadeite-bearing rock is reported from Turkey, and a gray material from Pakistan. In varicolored, massive material(!) from around Tawmaw, Kachin Hills, northern Burma, in an area about 100 km NW of Moguang and lying between the Uru and Chindwin rivers; reported, but rare, with tremolite–actinolite (nephrite) from the Kunlun Mts., Xinjiang Uygur, China. Found at Kotaki, Niigata Pref., and elsewhere in Japan; at the Koesek R., Celebes. Widely used as an ornamental material to produce carved sculptures and jewelry stones. The majority of such materials derive from the Burmese occurrence; all colors, including black due to impurities, are worked, but the lavender and emerald-green materials are most highly prized for jewelry, the latter, when uniform and translucent, commanding very high prices, and commonly but not exclusively referred to as *imperial jade*; the same is also applied to any jade, jadeite or nephrite, whose provenance is attributed to the Imperial Palace, China. AR *DHZII 2A:461(1978)*, *JADE:266(1991)*.

65.1.3c.2 Aegirine $NaFe^{3+}[Si_2O_6]$

Named in 1835 by Berzelius for Aegir, the Scandinavian god of the sea, having first been observed in Norway. Pyroxene group. Synonym: acmite, named from the Greek for *point*, in allusion to its crystal habit. MON $C2/c$. $a = 9.658$, $b = 8.795$, $c = 5.295$, $\beta = 107.42°$, $Z = 4$, $D = 3.577$. *31-1309*: 6.38_{10} 4.41_2 2.98_3 2.91_6 2.55_1 2.48_1 2.12_1 1.730_1. Structure: *MSASP 2:31(1969)*. The silicate chains are of one type and relatively unkinked, with the shared oxygen angle being 174°. Fe^{3+} is in the almost regular octahedral M1 site, and Na is in eight-coordination $(6 + 2)$ in the M2 site. Substitution of Ca and (Mg,Fe^{2+}) for Na and Fe^{3+}, respectively, leads to a greater kinking of the chains.

Habit: Crystals, whether free-standing or rock-bound, tend to be prismatic to decidedly acicular [001]. Crystals often reach several tens of centimeters in length. The dominant forms are {110}, {100}, and {010} terminated by {111} and often with prominent {221} and {661}, the latter resulting in the pointed habit to which the term *acmite* was originally applied; *aegirine* was originally applied to crystals with blunt terminations. The prism zone is often striated but lustrous, whereas termination faces are often etched and dull. Twinning on {100} common.

Physical properties: Dark green, reddish brown, black; the brown or yellow tones diminish with substitution of hedenbergite or diopside components; pale tan to yellow-green to pale green streak; vitreous luster. Cleavage {110}, good to perfect; parting {100}; conchoidal fracture. H = 6. G = 3.50–3.60. Biaxial (−); 2V = 60–70°; N_x = 1.750–1.776, N_y = 1.780–1.820, N_z = 1.795–1.836 decreasing with Ca pyroxene substitution; Y = b, X ∧ c = 0–10°; r > v, moderate to strong; negative elongation. Moderately pleochroic; X, emerald-green, deep green; Y, yellow to yellow-green; Z, yellow, yellow-brown, brownish green, green; the pleochroism intensifies with Ca pyroxene substitution. Distinguished from most other pyroxenes by its negative elongation and small extinction angle.

Chemistry: Complete solid solution exists with hedenbergite and diopside; aluminum content is usually low and substitution of Al for Si is very limited. Ti content is generally low, but it may occupy as much as one-fourth of the M1 site; zirconium is often present and may reach 0.65 wt % ZrO_2. The boundary between aegirine and aegirine–augite is placed at about Ae_{80}, but alternatively, at the change in optic sign, at about Ae_{50}.

Occurrence: Aegirine, as well as aegirine–augite, are pyroxenes characteristic of alkaline igneous rocks, particularly syenites and syenite pegmatites. It is also found in alkali granites and related rocks; in metamorphosed iron-rich sediments, schists, and metasomatic environments.

Localities: Occurs worldwide in alkaline rock complexes; noteworthy and well-documented localities are: zirconian and cerian at Quincy, MA;(!) at Magnet Cove, Garland Co., AR; in the iron formation of the Cuyuna Range, MN; authigenic aegirine in the Green River formation in CO, UT, and WY. In all the rocks of Mt. St.-Hilaire, PQ, Canada, and as exceptional(!) crystals in the pegmatitic phases. In the syenite pegmatites and alkaline rocks of Greenland, particularly the Narssarssuk pegmatite(!) of the Julianehaab district. Found at *Rundemyr, Eger, near Kongsberg, Norway*. A common constituent of the alkaline complexes of the Kola Penin., Russia; at Gout Creek, Westland, New Zealand; titanian aegirine is found in NSW, Australia. Exceptional(!), lustrous, prismatic crystals often twinned and in subparallel growth from Mt. Maloosa, Zamba district, Malawi. Found in cancrinite–albite syenite at Itapurapua, SP, Brazil. AR *DHZII 2A:482(1978), MR 25:29(1994)*.

65.1.3c.3 Namansilite NaMn^{3+}[Si$_2$O$_6$]

Named in 1992 by Kalinin et al. for the composition: natrium manganese silicate. Pyroxene group. MON $C2/c$. (Syn): a = 9.513, b = 8.615, c = 5.356, β = 105.12°, Z = 4, D = 3.61. PD: 6.30_6 4.31_4 2.92_{10} 2.89_9 2.59_3 2.50_3 2.07_2 1.694_2. Structure: $AC(C)$ *43:605(1987)*. Red, vitreous luster, translucent. Cleavage {110}, perfect. H = 6–7. For Italian material, biaxial (−); 2V = 85°; N_x = 1.750, N_y = 1.795, N_z = 1.835; Y = b, Z ∧ c = 19°; strong dispersion. Pleochroic; X, pale yellow-orange; Y,Z, red-violet. With braunite, pectolite, and K-spar as euhedral, prismatic crystals at the Cerciara mine, Val diVera, northern Apennines, Italy, and at *Irimni braunite deposit, Taikan Mts., Khabarovsk, Russia*. Publication of the Italian occurrence, without assignment of a name, appeared prior to publication of the Russian occurrence. AR *NJMM:59(1989), ZVMO 121(1):89(1992)*.

65.1.3c.4 Kosmochlor NaCr^{3+}[Si$_2$O$_6$]

Named in 1897 by Laspeyres from the Greek for *cosmos* and *green*, in allusion to its color and its being found in a meteorite. Pyroxene group. Synonyms: cosmochlore; ureyite, for Harold Clayton Urey (1893–1981), American chemist and Nobel laureate. MON $C2/c$. $a = 9.550$, $b = 8.712$, $c = 5.273$, $\beta = 107.44°$, $Z = 4$, $D = 3.603$. *26-1484*(syn): 6.3_3 4.35_3 2.96_{10} 2.87_7 2.52_3 2.51_4 2.45_4 2.10_2. Structure: *MSASP 2:31(1969)*. Dark emerald green, vitreous luster, translucent. Cleavage {110}, good; parting {001}. H ≈ 6. Biaxial (−); 2V = 53°; $N_x = 1.740$–1.766, $N_y = 1.756$–1.778, $N_z = 1.762$–1.781; $Y = b$, $X \wedge c = 8$–$22°$. Pleochroic; Z > Y > X; yellow-green to emerald-green. In meteorite occurrences Cr^{3+} occupies 66–98% of the M1 octahedral site. Experimental work indicates that complete solid solution is possible with jadeite and with diopside. In small amounts in solid solution with jadeite and diopside, it accounts for the emerald-green color of jadeite. Found at Williams Creek, Mendocino Co., CA. Occurs rarely in the *Toluca meteorite*, *Xiquipilco, MEX, Mexico*, and in the Coahuila and Hex R. meteorites, Mexico. Reported, but not well documented, in jadeite–albite–natrolite rocks at Maw-sit-sit, upper Burma, but is unquestionably a component of emerald-green jadeites from the region. AR *DHZII 2A:520(1978)*, *AM 72:126(1987)*, *MM 52:535(1988)*.

65.1.3c.5 Natalyite Na(V^{3+},Cr^{3+})[Si$_2$O$_6$]

Named in 1985 by Reznitskii et al. for Nataliya Vasil'evna Frovola (1907–1960), Russian geologist. Pyroxene group. MON $C2/c$. $a = 9.58$, $b = 8.72$, $c = 5.27$, $\beta = 107.16°$, $Z = 4$, $D = 3.55$. *31-653*: 6.24_6, 4.36_6 2.96_{10} 2.87_8 2.52_{10} 2.46_7 2.19_6 1.391_8. Bright yellowish green, vitreous to silky luster, translucent. Cleavage [110}, good; fracture (parting ?) {001}. H = 7. Biaxial (−); 2V = 8–12°; $N_x = 1.741$, $N_y \approx N_z = 1.762$. Strongly pleochroic; X, greenish yellow; Y,Z, emerald-green. Karelian material is near the end-member vanadium phase. Found as 1 mm × 0.3 mm grains associated with eskolaite–karelianite, uvarovite–goldmanite, Cr–V tourmaline, pyrite, and apatite, in Cr- and V-rich rocks of the Slyudyanka Precambrian metamorphic complex, *Lake Baikal region*, *Transbaikal, Russia*, and from the Onezhkii (Zaonezhki) Penin., Karelia. AR *ZVMO 114:630(1985)*, *123:55(1994)*; *AM 72:223(1987)*.

65.1.3c.6 Jervisite NaSc[Si$_2$O$_6$]

Named in 1982 by Mellini et al. for William P. Jervis (1831–1906), English-Italian scientist, curator of the Musèo Industriale Italiano, Turin. Pyroxene group. MON $C2/c$. (Syn): $a = 9.853$, $b = 9.042$, $c = 5.312$, $\beta = 106.62°$, $Z = 4$, $D = 3.29$. *35-542*: 6.51_2, 4.51_2, 3.39_2 3.26_2 3.04_{10} 2.98_5 2.54_5 1.647_5. Structure: *AC(B) 29:2615(1973)*. Light green, vitreous luster, translucent. Cleavage {110}, perfect. Biaxial (−); 2V = large (56° calc). $N_x = 1.683$, $N_y = 1.715$, $N_z = 1.724$. Natural material contains 18.48% Sc$_2$O$_3$. Found as elongated (0.2 mm) plates with cascandite, albite, orthoclase, and quartz in a vug in granite at *Cava Diverio, Baveno, Italy*. AR *AM 67:599(1982)*.

65.1.4.1 Spodumene LiAl[Si$_2$O$_6$]

Named in 1800 by d'Andrada from the Greek *spodumenos*, reduced to ashes, in allusion to its usual grayish-white (ash) color. Pyroxene group. Synonyms: α-spodumene and triphane, from the Greek for *three appearances*, in allusion to the cleavage and parting surfaces, the latter presently used for colorless or yellow, transparent material. Varieties: kunzite, named for American mineralogist and gemmologist George F. Kunz (1856–1932), and hiddenite, named for American mineralogist William E. Hidden (1853–1918), the latter defined as the chromian variety but often incorrectly used for pale green, chromium-free spodumene. Polymorphs: β-Spodumene and γ-spodumene exist at elevated temperatures and have structures related to keatite and high-quartz, respectively. MON $C2/c$ or $C2$. $a = 9.45$, $b = 8.39$, $c = 5.215$, $\beta = 110.10°$, $Z = 4$, $D = 3.184$. *33-876:* 6.12_4, 4.36_4 4.21_8 3.44_4 3.19_4 2.92_{10} 2.79_9 2.45_3. Structure: *AM 60:919(1975)*. The chains are equivalent with rotation opposite to that of the other $C2/c$ pyroxenes, and a shared oxygen angle of $\approx 170.5°$; in synthetic LiFe^{3+}Si$_2$O$_6$, the shared oxygen angle is $\approx 180°$. The M1 site (Al) is octahedral; the M2 site is distorted octahedral.

Habit: Prismatic [001], often flattened (100); dominant forms {100}, {110}, {010} terminated by {001}, {021}, {$\bar{1}$11}, and more complex forms. Twinning on {100} common, often lamellar. Crystals sometimes attaining enormous size (in excess of 10 m). Often deeply etched and corroded, especially in transparent varieties, and displaying prominent etch patterns and growth hillocks. Less commonly in bladed, cleavable masses.

Physical properties: Gray, white, tan, colorless, yellow, pale green, bright green (hiddenite), pink to lilac to violet (kunzite), the latter rarely blue-gray or purple; vitreous luster. Cleavage {110}, perfect; parting {100}, common; conchoidal to splintery fracture. H = $6\frac{1}{2}$–7. G = 3.03–3.23.

Optics: Biaxial (+); 2V = 58–68°; N$_x$ = 1.648–1.663, N$_y$ = 1.655–1.669, N$_z$ = 1.662–1.679; Y = b, Z $\wedge$ c = 20–26°; r < v. Pleochroic in thick sections of colored varieties with X $\approx$ Y > Z, purple, pink, lilac, green, to colorless.

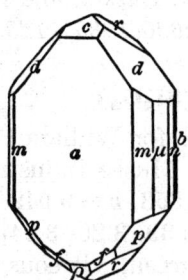

Huntington, MA

Spodumene

Forms: a{100}, b{010}, c{001}, m{110}, u{120}, n{130}, d{021}, p{111}, r{$\bar{2}$21}, t{211}

Color may be fugitive in some light green and bluish gray to violet varieties; color may be intensified by irradiation but is usually fugitive.

Chemistry: The composition is usually close to the ideal, with only minor substitution in any of the cation sites. Minor iron may account for yellow, manganese for pink and violet, and chromium for deeper green varieties, but light green may be due to charge transfer between Fe^{2+} and Mn^{4+}. Some sodium may substitute for lithium despite its larger size; most other cations are present only in trace amounts.

Occurrence: Common in lithium pegmatite and associated granite and gneiss. Usually in the intermediate zone of such pegmatite as embedded prisms and laths, but also in later pneumatolitic stages, often as transparent and etched crystals. Commonly associated with lepidolite, muscovite, albite, etc. Alteration to kaolin, eucryptite, lithium mica, sericite, and others is common.

Localities: Widespread throughout the world; some notable localities are: at Branchville(!), Fairfield Co., CT, and in many pegmatites of New England; at Kings Mt., Gaston Co., and hiddenite(!) with beryl (emerald) at Stony Point, Alexander Co., NC; in crystals in excess of 10 m in length(!) at the Etta mine, Pennington Co., SD; in laths to 2 m long at the Harding mine, Dixon, Taos Co., NM; kunzite in San Diego Co., CA, particularly the Himalaya and Pala Chief mines(!). At the Tanco pegmatite, Bernic Lake, MB, Canada; in granite at Killiney Bay, near Dublin, Ireland; at Varuträsk and Utö Is., Sweden. Light green hiddenite(?) at Kabbur, Hassan district, Mysore, India; kunzite(!) abundant in pegmatites of Kunar, Lagham, and Nuristan, Afghanistan; Burma; in Xinjiang and Szechuan Provs., China. Found at Katumba, Congo, and Bikita, Zimbabwe; occurs, including kunzite(!), in several pegmatites of the Malagasy Republic, particularly at Anjanabonoina, Mt. Bity, and Tilapa. Fine kunzite(!) occurs in many pegmatites of MG, Brazil, particularly at Urupuca, at Lavra do Ze Mario, Corrego do Urucum, Galileia, and also as pale pink and pale green thick crystals at Resplendor. Uses: An important source of lithium. Transparent varieties, particularly pink and lavender (kunzite) and green (hiddenite), but yellow and colorless as well, are faceted as gemstones, sometimes of considerable size (>1000 ct). AR CR 278D:1541(1974); DHZII 2A:521(1978); MR 17:313(1986), 20:197(1989), 21:97(1990).

CARPHOLITE GROUP

The carpholite group minerals are isostructural and conform to the formula

$$R^{2+}Al_2(Si_2O_6)(OH)_4$$

where

R = Mn, Fe, Mg, Li, Ba

and crystallize in space group *Ccca*. The structure is characterized by pairs of single-chain four-coordinated silicon with periodicity 2 (approximately 5.1 Å) lying parallel to [001] that are joined by six-coordinated aluminum into I-beams similar to those found in pyroxenes but linked through additional six-coordinated cations in a manner completely different from that of the pyroxenes, to form a structure with several large cavities which may be partially occupied by larger cations. *NJMM 8:337(1992)*.

CARPHOLITE GROUP

Mineral	Formula	Space Group	a	b	c	D	
Carpholite	$MnAl_2(Si_2O_6)(OH)_4$	$Ccca$	13.718	20.216	5.132	3.071	65.1.5.1
Ferrocarpholite	$(Fe, Mg)Al_2(Si_2O_6)(OH)_4$	$Ccca$	13.797	20.20	5.116	3.05	65.1.5.2
Magnesiocarpholite	$(Mg, Fe)(Al, Fe)_2(Si_2O_6)(OH)_4$	$Ccca$	13.694	20.204	5.108	2.82	65.1.5.3
Balipholite	$LiBaMg_2Al_3(Si_2O_6)_2(OH)_4F_4$	$Ccca$	13.587	20.164	5.144	3.403	65.1.5.4

Note: Although the unit cells of the carpholite minerals are dimensionally orthorhombic and the structures have been successfully refined in space group $Ccca$, optical properties and spectroscopic studies suggest that the symmetry may be only monoclinic.

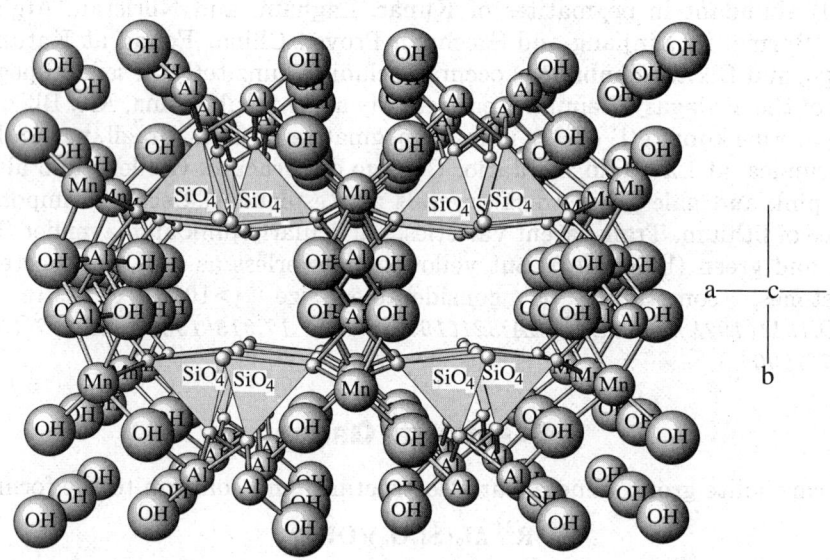

CARPHOLITE: $MnAl_2(Si_2O_6)(OH)_4$
Carpholite. Perspective view along c, the chain direction.

65.1.5.1 Carpholite $MnAl_2(Si_2O_6)(OH)_4$

Named in 1817 by Werner from the Greek *karphos*, straw, in allusion to its color. Carpholite group. ORTH *Ccca*. $a = 13.718$, $b = 20.216$, $c = 5.132$, $Z = 8$, $D = 3.071$. PD: 5.73_9 5.13_2 3.81_3 3.40_4 3.04_5 2.62_{10} 2.37_4 2.05_8. Structure: *SR 45A:369(1979)*, *AM 74:1084(1989)*. Acicular to fibrous [001] radiating tufts. Straw-yellow, silky luster. Cleavage {010}, perfect; {110}, distinct. $H = 5$–$5\frac{1}{2}$. $G = 2.9$–3.04. Biaxial $(-)$; $2V \approx 50°$; $N_x = 1.575$–1.613, $N_y = 1.590$–1.630, $N_z = 1.594$–1.635; $XYZ = bac$; $r > v$, strong. Pleochroic, $X = Y > Z$, straw-yellow to colorless. Insoluble in acids. Mg, Fe^{2+}, K, and Li may substitute for manganese, and fluorine for hydroxyl, the substitutions accounting for the variable refractive indices. Occurs in metamorphosed shales of low grade. Found at Caldbeck Fells, Cumbria, and at several mines in Cornwall, England; at Meuville, Ardennes, Belgium; Wippura, Harz Mts., Germany; originally found with fluorite and quartz in the tin deposits of *Slavkov (Schlaggenwald), Cechy (Bohemia), Czech Republic*; at Radobilny, near Prilep, Macedonia, Yugoslavia; at Fukuzumi mine, Kyoto Pref., Honshu, Japan. AR *D6 549(1892)*, *DANES 195:152(1970)*, *MA 18:200(1967)*.

65.1.5.2 Ferrocarpholite $(Fe,Mg)Al_2(Si_2O_6)(OH)_4$

Named in 1951 by de Roever for the composition and relationship to isostructural carpholite. Carpholite group. ORTH *Ccca*. $a = 13.797$, $b = 20.20$, $c = 5.116$, $Z = 8$, $D = 3.05$. *33-655*: 5.69_7 5.04_{10} 3.44_1 3.36_3 3.02_2 2.60_2 1.855_1 1.680_1. Prismatic to acicular [001] crystals in aggregates to 1 cm long. Observed forms {010}, {100}, {110}. Gray-green to dark green, pale green streak, silky luster. Cleavage {010}, perfect; {110}, indistinct. $H = 5\frac{1}{2}$. $G = 3.04$. Biaxial $(-)$; $2V = 56°$; $N_x = 1.614$–1.621, $N_y = 1.630$–1.636, $N_z = 1.635$–1.640; $XYZ = bac$; $r > v$, strong. Pleochroic; X,Y, yellow-green; Z, blue-green. Found in blueschists with quartz, glaucophane, lawsonite, jadeite, and other minerals at Haute-Ubaye, French Alps, France; at Ruwi, Oman; *west of Tomata, east central Celebes, Indonesia*; at Diahote, New Caledonia. AR *AM 36:736(1951)*, *SMPM 57:157(1977)*, *NJMM 8:337(1992)*.

65.1.5.3 Magnesiocarpholite $(Mg,Fe)(Al,Fe)_2(Si_2O_6)(OH)_4$

Named in 1979 by Goffe et al. for the composition and isostructural carpholite. Carpholite group. ORTH *Ccca*. $a = 13.694$, $b = 20.204$, $c = 5.108$, $Z = 8$, $D = 2.82$. *35-489*(syn): 5.66_{10} 5.02_2 3.79_3 2.73_3 2.59_6 2.35_2 2.04_2 1.848_2. Gray to green, silky luster. Cleavage {010}, perfect; {110}, indistinct. $G = 3.04$. Biaxial $(+)(?)$; $N_x = 1.59$, $N_y = 1.60$. A poorly defined species. From *Vanois, French Alps, France*. AR *AM 65:406(1980)*, *CMP 76:260(1981)*.

65.1.5.4 Balipholite $LiBaMg_2Al_3(Si_2O_6)_2(OH)_4F_4$

Named in 1974 anonymously for barium, lithium carpholite. Carpholite group. ORTH *Ccca*. $a = 13.587$, $b = 20.164$, $c = 5.144$, $Z = 4$, $D = 3.403$. *33-787*: 10.1_{10} 6.79_1 4.06_{10} 3.37_9 2.60_2 2.52_1 2.39_3 2.21_1. Structure: *SS(B) 30:779(1987)*. Barium is in partial occupancy of the large cavities of the carpholite structure type. Yellowish white, silky luster. Cleavage {010}, perfect; {100}, {110}, distinct. G = 3.4. Biaxial (−); 2V = 70°; $N_x = 1.581$, $N_y = 1.596$, $N_z = 1.601$; XYZ = bca. Found as fibrous [001] tufts with Li-mica in miarolitic cavities in quartz at *Hsianghua area, Linwu, Hunan, China*. AR *AM 76:338(1976)*.

65.1.6.1 Lorenzenite $Na_2Ti_2O_3(Si_2O_6)$

Named in 1897 by Flink for Johannes Theodor Lorenzen (1855–1884), Danish mineralogist and student of Greenland mineralogy. Synonymous with and has precedence over ramsayite. Lorenzenite group. ORTH *Pbcn*. $a = 8.713$, $b = 5.233$, $c = 14.487$, Z = 4, D = 3.44. *33-1298*(syn): 5.57_3 3.35_9 3.25_1 3.03_2 2.75_{10} 2.58_1 2.50_1 2.46_2; also *18-1262*. Structure: *AM 72:173(1987)*. The silicate structural element is a pyroxene–like chain (W = 1, P = 2) parallel to [010]. The commonly given morphological orientation differs from that of the structure cell with a, b, c(struc) = b, c, a(morph). As prismatic [001], flattened {100} crystals with {100}, {210}, {111} (morphological setting) prominent; also acicular and as radiating and fibrous groups. White, colorless, light blue, green, pink, pale violet, yellow-brown to dark brown; white to pale brown streak; vitreous luster, submetallic in dark brown varieties; transparent to nearly opaque. Cleavage {100}, perfect; {110}, good. H = 6. G = 3.45. Biaxial (−); 2V = 25–42°; $N_x = 1.92$–1.95, $N_y = 2.01$–2.04, $N_z = 2.02$–2.06; XYZ = abc; r > v, pleochroic, X > Y ≈ Z; orange-yellow to pale yellow. Pale cream fluorescence in SWUV. Morphologically similar to amphiboles and vinogradovite; fibrous varieties resemble ashcroftine. Found in alkalic syenites and pegmatites. Noted from Point of Rocks near Springer, Colfax Co., NM; common at Mt. St.-Hilaire, PQ, Canada, as pink, violet, and colorless acicular crystals; and Tenerife, Canary Islands. *Narssarsuag, Julianehaab, Greenland*; Lake Gjerdingen, Nordmark, and in the Bratthagen nepheline syenite pegmatites, Lågendalen, Oslo region, Norway; as brown, submetallic crystals(!) from Khibiny massif, Kola Penin., and the Inagli, Konder, and Murun massifs, Aldan, Yakutia, Russia. AR *DT:692(1932), AM 32:59(1947), MR 21:319(1990)*.

65.1.6.2 Kukisvumite $Na_6ZnTi_4Si_8O_{24} \cdot 4H_2O$ (?)

Named in 1991 by Yakovenchuk et al. for the locality. Lorenzenite group. ORTH *Pccn*. $a = 28.889$, $b = 8.604$, $c = 5.215$, Z = 2, D = 2.95. *PD*: 14.5_9 6.42_6 4.82_8 4.30_5 3.72_6 3.61_3 3.01_{10} 1.604_3. Structurally and chemically related to lorenzenite, $a_k \approx 2c_l$, $b_k \approx a_l$, $c_k \approx b_l$. Colorless to silvery, transparent. Splintery fracture, breaking into fibers parallel to elongation. H = $5\frac{1}{2}$–6. G = 2.90. Biaxial (−); $2V_{calc} = 77°$; for Na(D): $N_x = 1.676$, $N_y = 1.746$, $N_z = 1.795$. As fan-shaped intergrowths of prismatic [001], flattened {100}, crystals up to 7 mm in an arfvedsonite–microcline pegmatite as partial to complete pseudomorphs after lamprophyllite at the *Kukisvumchorr apatite deposit, Khibiny massif, Kola Penin., Russia*. AR *MZ 13-2:63(1991), AM 77:1116(1992)*.

65.1.6.3 Lintisite Na$_3$LiTi$_2$O$_2$[Si$_2$O$_6$]$_2$ · 2H$_2$O

Named in 1990 by Khomyakov et al. from symbols of the component cations. Lorenzenite group. MON $C2/c$. $a = 28.583$, $b = 8.600$, $c = 5.219$, $\beta = 91.03°$, $Z = 4$, $D = 2.824$. PD: 14.3$_{10}$ 6.39$_5$ 4.77$_5$ 3.69$_5$ 3.00$_{10}$ 2.74$_5$ 2.71$_5$ 1.650$_5$. Alternately-facing, pyroxene–like chains lie parallel to [001] in the manner of pyroxene (similar b and c axes), but with different cation coordination. The Li analog of kukisvumite, with 2Li$^+$ replacing Zn^{2+},□. Occurs as fibrous to columnar aggregates up to 1.5 mm long, of acicular, [001], crystals twinned on {100}. Pale yellow; vitreous luster, pearly on cleavage; transparent to translucent. Cleavage {100}, {010}, perfect; splintery fracture. H = 5.5. G = 2.77. Biaxial (−); 2V = 85°; N$_x$ = 1.672, N$_y$ = 1.739, N$_z$ = 1.802; Z ∧ c = 2°, Y = b; r < v, strong; positive elongation. Weak yellow fluorescence in UV. Dehydrates in steps to yield lorenzenite pattern at 600°. Soluble with silica residue in dilute HCl. Found in nepheline–syenite pegmatites at Mt. St.-Hilaire, PQ, Canada, and *Alluaiv Mt., Lovozero massif, Kola Penin., Russia*. AR *ZVMO 119:76(1990), ZK 193:137(1990), AM 76:1730(1991)*.

65.1.7.1 Shattuckite Cu$_5$(SiO$_3$)$_4$(OH)$_2$

Named in 1915 by Schaller for the type locality mine. ORTH $Pcab$. $a = 9.885$, $b = 19.832$, $c = 5.383$, $Z = 4$, $D = 4.128$. 20-356: 4.97$_6$ 4.43$_{10}$ 3.64$_4$ 3.50$_7$ 3.31$_5$ 2.78$_5$ 2.74$_4$ 2.37$_7$. Structure: *AM 62:491(1977)*. The structure consists of pyroxene-type chains joined to the surface of brucite-like (CuO$_2$)$_n$ sheets; adjacent complex sheets are cross-linked by copper in square, four-coordination. Acicular [001], 1- to 2-mm crystals with {100}, {010}, {110} dominant; radiating, spherulitic aggregates, granular, massive. Medium to dark blue, vitreous to silky luster, translucent. Cleavage {010}, {100}, perfect. H = 3½. G = 3.9–4.1. Biaxial (+); 2V = 88°; N$_x$ = 1.753, N$_y$ = 1.782, N$_z$ = 1.815; XYZ = cba (also reported Z ≈ c, X = b). Pleochroic, Z > Y = X; dark to light blue. Found as an alteration product of other secondary copper minerals at the Eagle Eye mine and the *Shattuck mine, Bisbee, Cochise Co., AZ*, and in the Clifton–Morenci district, Greenlee Co., the Moon–Archer mine and Potter–Cromer property, south of Wickenburg, Maricopa Co., with ajoite at the New Cornelia mine, Ajo, Pima Co., the San Manuel and Roadside mines, Pinal Co., and the New Water Mts., Yuma Co., AZ. At Tsumeb, Namibia, and reported from M'sesa and Tantara, Shaba, Zaire. AR *AM 49:1234(1964)*.

65.1.8.1 Nchwaningite Mn$_2^{2+}$SiO$_3$(OH)$_2$ · H$_2$O

Described in 1995 by Nyfeler et al., and named for the type locality, the N'chwaning mine, Kalahari manganese field, South Africa. ORTH $Pca2_1$ $a = 12.672$Å, $b = 7.217$Å, $c = 5.341$Å, $Z = 4$, $D = 3.202$; PD (*AM 80:377(1995)*) (nat.): 7.22$_6$ 4.08$_6$ 3.01$_{10}$ 2.55$_8$ 2.50$_6$ 2.46$_8$ 2.44$_8$ 1.55$_8$. Structure: $R = 2.1\%$. The structure consists of double layers of laterally linked so-called truncated pyroxene-building units formed by a double chain of octahedra, topped with a Zweier single chain of Si tetrahedra. Symmetry-equivalent units are linked laterally but turned upside down. This yields a double-layer structure with H bridges linking the layers. A striking feature of the structure is that one MnO_6 corner is formed by a water molecule. Light brown, transparent, radiating balls, to 5 mm, of needles which average $1.0 \times 0.1 \times 0.05$ mm in size; platy $\parallel$ (010), streak white, vitreous luster, two perfect cleavages parallel to the long axis (c axis) of the crystals, i.e. parallel to (100) and (010); no fluorescence, G = ?, H = ?. SiO_2 28.0 wt. %, Al_2O_3 0.27, FeO 0.27, MnO 60.02, MgO 1.93, CaO 0.19, H_2O not determined. Biaxial $(-)$, $2V = 54°$; $N_x = 1.681$, $N_y = 1.688$, $N_z = 1.690$, colorless, $Y = a$, $X = b$, $Z = c$. Occurs in one vug in the *N'chwaning mine, Kalahari manganese field, northern Cape Province, South Africa*; also in another vug in the Wessels mine. Associated with calcite, bultfonteinite and chlorite.

WOLLASTONITE GROUP

The wollastonite group minerals conform to the general formula

$$R_3[Si_3O_{8-9}(OH)_{1-0}]$$

where in the simplest cases, R is a divalent cation and no OH is present, but R may also represent monovalent or trivalent cations plus vacancies. The structure is characterized by chains with $P = 3$ (≈ 7.3 Å) lying parallel to b. The conformation of the chains may be envisioned as consisting of alternating double and single tetrahedra. The species in the group are not strictly isostructural: polytypes exist, cation-site occupancies vary, and chains may shift relative to one another. *AC 12:177(1959), AC(A) 37:273(1984)*.

WOLLASTONITE GROUP

Mineral	Formula	Space Group	a	b	c	α	β	γ	
Wollastonite-1A	$CaSiO_3$	$P\bar{1}$	7.926	7.320	7.0675	90.05	95.22	103.43	65.2.1.1a
Wollastonite-2M	$CaSiO_3$	$P2_1$ (?)	15.409	7.322	7.063		95.30		65.2.1.1b
Wollastonite-3A	$CaSiO_3$	$P\bar{1}$	23.2	7.30	7.06	90.0	95.5	94.6	65.2.1.1c
Wollastonite-4A	$CaSiO_3$	$P\bar{1}$	31.2	7.30	7.06	90.0	95.5	96.8	65.2.1.1c
Wollastonite-5A	$CaSiO_3$	$P\bar{1}$	38.6	7.30	7.06	90.0	95.5	98.0	65.2.1.1c
Wollastonite-7A	$CaSiO_3$	P_1	54.3	7.30	7.08	90.0	95.5	92.1	65.2.1.1c
Bustamite	$Ca(Mn, Fe^{2+})(SiO_3)_2$	$A\bar{1}$	7.736	7.157	13.824	90.52	95.58	103.87	65.2.1.2
Ferrobustamite	$Ca(Fe, Ca, Mn)(SiO_3)_2$	$A\bar{1}$	7.862	7.253	13.967	89.73	95.47	103.48	65.2.1.3
Pectolite-1A*	$NaCa_2[Si_3O_8(OH)]$	$P\bar{1}$	7.980	7.023	7.018	90.54	95.14	102.55	65.2.1.4a,b
Serandite	$Na(Mn, Ca)_2[Si_3O_8(OH)]$	$P\bar{1}$	7.683	6.889	6.747	90.53	94.12	102.75	65.2.1.5
Cascandite	$Ca(Sc, Fe^{3+})[Si_3O_8(OH)]$	$P\bar{1}$	7.529	7.051	6.755	92.12	93.67	104.65	65.2.1.6
Denisovite	$(K, Na)Ca_2[Si_3O_8(F, OH)]$		30.92	7.20	18.27		95		65.2.1.7

*A monoclinic polytype has also been reported; however, it may be the result of microscale twinning. Denisovite may also be related to a possible polytype of pectolite.

65.2.1.1a Wollastonite-1A

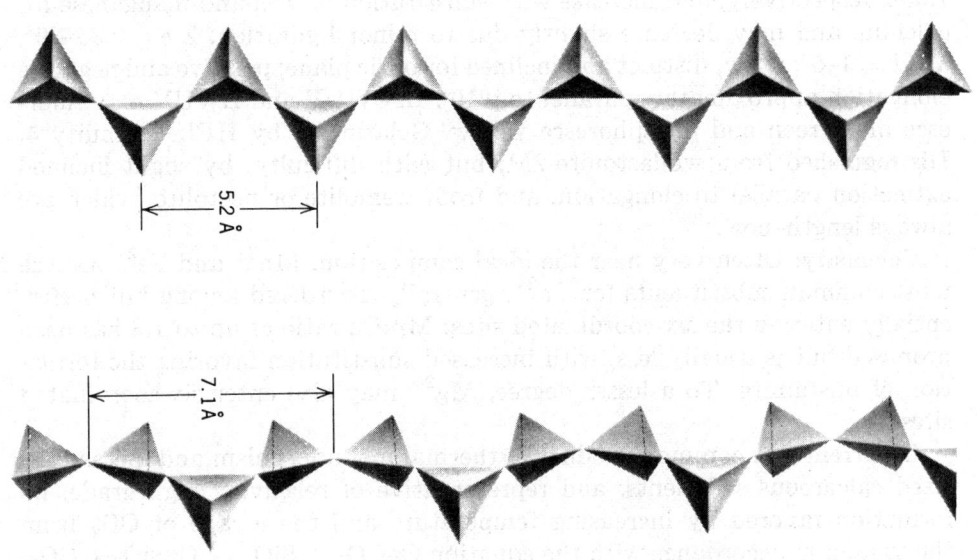

WOLLASTONITE: CaSiO₃
Idealized silicate chain periodicity in pyroxenes (top), and in the wollastonite group (bottom).

65.2.1.1a   Wollastonite-1A   CaSiO$_3$

Named in 1818 by Léman for William Hyde Wollaston (1766–1828), noted English mineralogist and chemist. Wollastonite group. Together with wollastonite-2M, it is considered the equivalent to α-CaSiO$_3$, and corresponds to the originally described wollastonite. The "1A" notation refers to the polytype (single layer, triclinic) that was formerly rendered "1T." Other polytypes are 2M, 3A, 4A, 5A, and 7A. β-CaSiO$_3$, pseudowollastonite, is a structurally distinct, high-temperature polymorph. TRIC $P\bar{1}$. $a = 7.926$, $b = 7.320$, $c = 7.0675$, $\alpha = 90.05°$, $\beta = 95.22°$, $\gamma = 103.43°$, Z = 6, D = 2.915. *29-372*(syn): 3.84_6 3.51_7 3.31_8 3.08_7 2.98_{10} 2.47_6 2.18_7 1.826_7. The diffraction pattern does not allow easy distinction among polytypes, but see wollastonite-2M (65.2.1.1b). Structure: *JPCM 10:217(1984)*. The calcium ions are in two six-coordinated sites (M1 and M2) and one seven-coordinated site (M3).

Habit: Usually massive, often in coarse granular to bladed and foliated forms; fibrous. Rarely well crystallized, elongated [010] and flattened with {001} or {100} prominent, and lesser {$\bar{1}$02}, {540}, and {102}.

Physical properties: White, gray, to pale green; vitreous luster, somewhat pearly on cleavage, silky when fibrous; translucent to transparent. Cleavage {100}, perfect; {001}, {$\bar{1}$02}, good; uneven to splintery fracture. H = $4\frac{1}{2}$–5. G = 2.86–3.09.

Optics: Biaxial $(-)$; $2V = 36$–$60°$; $N_x = 1.616$–1.631, $N_y = 1.628$–1.644, $N_z = 1.631$–1.646, the indices for the pure compound being 1.618, 1.630, and

1.632, respectively, and increase with substitution of iron and manganese for calcium, and may decrease slightly due to minor hydration; $Z \wedge c \approx 36\text{--}39°$, $Y \wedge b = 1\text{--}5°$; $r > v$, distinct and inclined for optic plane; positive and negative elongation approximately parallel to [010]. In SWUV and LWUV may fluoresce blue-green and phosphoresce yellow. Gelatinized by HCl. Fusibility 4. Distinguished from wollastonite-2M, but with difficulty, by slight inclined extinction parallel to elongation, and from tremolite or pectolite, which are always length-slow.

Chemistry: Often very near the ideal composition. Mn^{2+} and Fe^{2+} are the most common substituents for Ca^{2+}, generally disordered among but preferentially entering the six-coordinated sites; Mn:Ca ratio of up to 1:3 has been proposed but is usually less, with increased substitution favoring the formation of bustamite. To a lesser degree, Mg^{2+} may also enter six-coordinated sites.

Occurrence: A common product of thermal metamorphism and metasomatized calcareous sediments, and representative of relatively high grade, its formation favored by increasing temperature and the escape of CO_2 from the system in accordance with the equation $CaCO_3 + SiO_2 \rightleftharpoons CaSiO_3 + CO_2$. Formed in igneous rocks primarily as a result of contamination from calcareous inclusions. In amygdules of thermally metamorphosed basalts, and in the amphibolite and granulite facies of regionally metamorphosed rocks. Rarely as a constituent of alkaline igneous rocks and carbonatites.

Localities: The commonly occurring environment of formation leads to a worldwide distribution; some notable localities are: in coarse, granular masses with andradite and diopside at Willsboro, Essex Co., and in crystals(!) at Diana, Lewis Co., NY; minor, fluorescent material at Franklin, Sussex Co., NJ; at Isle Royal, Keweenaw Co., MI; at the Christmas mine, Gila Co., and the Holbrook shaft, Bisbee, Cochise Co., AZ; at Crestmore, Riverside Co., and in Inyo and Kern Cos., CA. Found at Hermosillo, ZAC, Jalostoc, Ayala, MOR, Concepcion del Oro, ZAC, Pichucalco, CHIH, and elsewhere in Mexico. Found as massive and fibrous material at the Jeffrey mine, Asbestos, and at Grenville, Mt. St.-Hilaire, Wakefield, and the Oka carbonatite complex, PQ; in the Ice River complex, Ottertail Range, BC, Canada. In metabasalts on Mull, Scotland; at Corneal, County Antrim, Northern Ireland; at Smedsgarden Farm, Alnö, Sweden, and at Pargas and Remonmaki, Finland. Good crystals(!) at Mte. Somma, Vesuvius, and Monti Peloritani, Sicily, Italy; also well-crystallized(!) at Santorini volcano, Greece; at *Dognecea, SE of Tinisoara, Banat, Romania*. Manganoan material is found at Hijikuzu mine in Japan; found in Kenya and Tanzania, and the North and South mines, Broken Hill, NSW, Australia.

Uses: Wollastonite is used extensively in the ceramics industry because of its low fusibility, absence of volatiles, and fluxing action. Fibrous material has found increasing use as a replacement for asbestos in insulation applications.
AR *MJJ 7:180(1973), DHZII 2A:547(1978)*.

65.2.1.1b Wollastonite-2M CaSiO₃

Named in 1935 by Peacock, the derivation being as for wollastonite-1A. Wollastonite group. Synonymous with parawollastonite, a discredited name, and included with wollastonite-1A as α-CaSiO$_3$. Other polytypes are 3A, 4A, 5A, and 7A. Polymorphic with pseudowollastonite. MON Space group unknown, possibly $P2_1$(?). $a = 15.409$, $b = 7.322$, $c = 7.063$, $\beta = 95.30°$, $Z = 12$, $D = 2.92$. 27-88(calc): 3.84_2 3.52_2 3.32_3 3.09_2 2.98_{10} 2.48_2 2.18_2 1.839_2. The diffraction pattern does not allow easy distinction among polytypes, but the weak line adjacent to the strong 3.84 Å is alleged to be 4.07 Å in 1A but 4.37 Å in 2M. Structure: *ZK 174:309(1986)*. The structure is built of two units of the 1A polytype repeated along the a axis. Similar to wollastonite-1A in habit, the original description having been based on monoclinic symmetry; usually massive, granular, or fibrous. White, gray, to pale yellow; vitreous luster, somewhat pearly on cleavage, silky when fibrous; translucent to transparent. Cleavage {100}, perfect; {001}, {$\bar{1}$02}, good; uneven to splintery fracture. $H = 4\frac{1}{2}$-5. $G \approx 3.0$. Biaxial (−); $2V = 35$–$43°$; $N_x = 1.614$–1.621, $N_y = 1.629$–1.633, $N_z = 1.631$–1.635; $Z \wedge c \approx 32°$, $Y \wedge b = 0°$; $r > v$, weak and distinctly inclined for optic plane; positive and negative elongation approximately parallel to [010]. Gelatinized by HCl, fusibility 4; distinguished from wollastonite-1A by extinction parallel to elongation, and from tremolite or pectolite, which are always length-slow. Found in the same environments as wollastonite-1A (i.e., primarily in thermally metamorphosed siliceous limestones). Found at Crestmore, Riverside Co., and several localities in the Sierra Nevada Range, CA; at Mte. Somma, Vesuvius, Italy; and at Cziklova, Banat, Romania. Its uses are essentially as those of the much more common 1A polytype, with which it is often admixed and from which it is, therefore, not separated. AR *DHZII 2A:547(1978)*.

65.2.1.1c Wollastonite-3A,4A,5A,7A CaSiO₃

Named in 1983 (3A, 4A, 5A) and 1978 (7A) by Henmi et al., the derivation being as for wollastonite-1A, the polytype notations referring to the number of 1A units repeated along the a axis. Wollastonite group. TRIC 3A: $P\bar{1}$, $a = 23.2$, $b = 7.30$, $c = 7.06$, $\alpha = 90.0°$, $\beta = 95.5°$, $\gamma = 94.6°$, $Z = 18$, $D = 2.93$; 4A: $P\bar{1}$, $a = 31.2$, $b = 7.30$, $c = 7.06$, $\alpha = 90.0°$, $\beta = 95.5°$, $\gamma = 96.8°$, $Z = 24$, $D = 2.91$; 5A: $P\bar{1}$, $a = 38.6$, $b = 7.30$, $c = 7.06$, $\alpha = 90.0°$, $\beta = 95.5°$, $\gamma = 98.0°$, $Z = 30$, $D = 2.95$; 7A: $P1$, $a = 54.3$, $b = 7.30$, $c = 7.08$, $\alpha = 90.0°$, $\beta = 95.5°$, $\gamma = 92.1°$, $Z = 42$, $D = 2.90$. Structure: *MJJ 9:169(1978)*, *AM 68:156(1983)*. In physical appearance and optical properties, essentially indistinguishable from the more common 1A polytype. Polytypes 3A, 4A, and 5A are known only from *Kushiro, Hiroshima Pref., Japan*. Polytype 7A is known from a single sample from a skarn at *Fuka, Okayama Pref., Japan*. AR *AM 64:658(1979)*.

65.2.1.2 Bustamite $Ca(Mn,Fe^{2+})(SiO_3)_2$

Named in 1826 by Brongniart for Anastasio Bustamante (1780–1853) of Mexico. Wollastonite group. TRIC $A\bar{1}$. $a = 7.736$, $b = 7.157$, $c = 13.824$, $\alpha = 90.52°$, $\beta = 95.58°$, $\gamma = 103.87°$, $Z = 6$, $D = 3.326$. Alternative settings, only in part based on polytypes, may be found. *27-86*(syn): 3.45_5 3.24_5 3.04_5 2.99_{10} 2.73_5 2.59_5 2.42_5 1.791_8; *26-1066*(ferroan): 3.72_4 3.43_3 3.23_{10} 3.01_3 2.64_3 2.48_3 2.24_5 1.72_2. Structure: *AM 48:588(1963)*. Four cation sites exist; Mn^{2+} and Fe^{2+} occupy the six-coordinated M3 site; additional Mn^{2+} next enters site M1, and then M2, both sites having 6+ coordination; site M4 is eight-coordinated and is occupied by calcium alone. Habit and physical properties relate to old morphological setting. Crystals usually tabular {001} or equant to prismatic, frequently with rounded edges and corners. Commonly massive, often compact and fibrous. Twinning on {110} rare. Light pink to brownish red, vitreous luster, translucent to transparent. Cleavage {100}, perfect; {110}, {1$\bar{1}$0}, good; {010}, poor. $H = 5\frac{1}{2}$–$6\frac{1}{2}$. $G = 3.32$–3.43. Biaxial $(-)$; $2V = 34$–$60°$; $N_x = 1.640$–1.695, $N_y = 1.651$–1.708, $N_z = 1.653$–1.710, the indices being lower in high Ca varieties; OAP and X approximately perpendicular to {100}; $r < v$, weak to strong with distinct crossed dispersion. Weakly pleochroic; X, Z, orange; Y, rose. The composition range approximately $Ca:(Mn,Fe) = 5:1$ to $1:2$. Minor zinc may also enter the octahedral site. Typically results by metamorphism and/or metasomatism of Mn-bearing ore bodies; typical assemblages are bustamite–rhodonite–tephroite, bustamite–tephroite–calcite, and at higher grade, glaucochroite–tephroite–bustamite. Found at Franklin and Sterling Hill, Sussex Co., NJ (zincian); at *Tetelha de Xonotla, PUE, Mexico*, but this was later shown to be a mixture of johannsenite and rhodonite; at Långban, Sweden; coarsely fibrous material at Treburland mine, Altarnum, Cornwall, and at Meldon Railway quarry, Okehampton, Devon, England; as rosettes of slender needles at Culmage, Harz Mts., Germany. Found at the Nodatamagawa mine, Iwate Pref., and other sites in Japan; with the lead ores of Broken Hill, NSW, Australia; from the N'Chwaning mine, Kuroman, Cape Prov., South Africa. AR *AM 63:274(1978), DHZII 2A:575(1978), MM 49:37(1985)*.

65.2.1.3 Ferrobustamite $Ca(Fe,Ca,Mn)(SiO_3)_2$

Named in 1937 by Tilley for the composition and relationship to bustamite. Wollastonite group. Original name was iron-wollastonite and later ferrowollastonite. TRIC $A\bar{1}$. $a = 7.862$, $b = 7.253$, $c = 13.967$, $\alpha = 89.73°$, $\beta = 95.47°$, $\gamma = 103.48°$, $Z = 6$, $D = 3.07$ (for the approximate composition $Wo_{82}Fs_{18}$). *29-336*: 7.67_2 3.84_5 3.47_6 3.27_{10} 3.05_8 2.95_2 2.70_3 2.28_6. Structure: *AM 62:1216(1977)*. Site occupancy is as for bustamite but with $Fe^{2+} > Mn^{2+}$. Massive to coarsely fibrous. White, greenish, light brown; vitreous luster. Cleavage is as for bustamite (65.2.1.2). $H = 5$–6. $G = 3.09$–3.39. Biaxial $(-)$; $2V \approx 60°$; $N_x = 1.640$, $N_z = 1.653$ for composition $Ca:(Fe,Mn) \approx 5:1$; the indices increase with increasing iron content and overlap those of bustamite. The iron end member is stable only above $800°$ and may not be found in nature. Found in contact-metamorphic and metasomatic environments. Found around chert nodules in contact-metamorphosed dolomite at *Camas Malag, Skye, Scotland*; from Shimane, Yamaguchi, and Yoyama Prefs., Japan; and the Wessels mine, near Kuroman, Cape Prov., South Africa. AR *MM* 24:569(1937).

65.2.1.4a Pectolite-1A NaCa$_2$[Si$_3$O$_8$(OH)]

65.2.1.4b Pectolite-M2abc

Named in 1828 by Kobell from the Greek for *congealed*, in allusion to its translucent and compact appearance. Wollastonite group. The variety schizolite is manganese-rich and intermediate with serandite (65.2.1.5). Monoclinic (?) polytypes synonymous with parapectolite. The several M polytypes may arise from microscale twinning, but no data of merit provided. TRIC $P\bar{1}$. $a = 7.980$, $b = 7.023$, $c = 7.018$, $\alpha = 90.54°$, $\beta = 95.14°$, $\gamma = 102.55°$, $Z = 2$, $D = 2.906$. *33-1223*: 3.31$_2$ 3.27$_2$ 3.15$_1$ 3.08$_4$ 3.06$_2$ 2.90$_{10}$ 2.73$_1$ 2.16$_2$; *33-1227*(manganoan): 3.25$_3$ 3.23$_3$ 3.05$_4$ 3.01$_2$ 2.87$_{10}$ 2.67$_1$ 2.55$_1$ 2.25$_2$. Structure: *ZK 146:281(1977)*. The structure is similar to that of wollastonite-1A but with a considerably shortened b axis due to hydrogen bonding within the chains parallel to b. Calcium is in distorted octahedral sites and sodium is in $3+5$ coordination in a distorted square antiprism configuration. Crystals elongated [010] and flattened {100} with {100} and {001} dominant; also described as prismatic [001] (Mt. St.-Hilaire); more commonly acicular in sprays or compact, radially fibrous masses. Twinning rare with [010] twin axis and composition near (100). White, pale tan, pinkish, pale blue; vitreous to silky luster. Cleavage {100}, {001}, perfect; uneven, brittle fracture. H = $4\frac{1}{2}$–5. G = 2.84–2.90. Biaxial (+); 2V = 50–63°; X $\wedge$ c = 10–19°, Y $\wedge$ a = 10–16°; N$_x$ = 1.592–1.610, N$_y$ = 1.603–1.615, N$_z$ = 1.630–1.645; Z $\wedge$ b = 2°; r < v, weak to strong; indices increase beyond this range with increasing Mn content; OAP near (001). Complete isomorphous substitution on Mn for Ca forms the pectolite–serandite series; minor iron and magnesium may be present. Rare earth substitution is not uncommon in Mn-rich varieties. Partly decomposed and gelatinized by HCl. Fibrous varieties yield dangerously sharp, brittle splinters. Fusibility 2. Sometimes triboluminescent. A hydrothermal mineral typical of cavities and fracture surfaces of basic and syenitic igneous rocks; it is common in basalt cavities in association with zeolites, calcite, and quartz; also in serpentinites and mica peridotites. Found with zeolites in the Triassic Watchung basalts of northern NJ, particularly at Paterson and Bergen Hill, and also at Franklin and Sterling Hill; manganoan material in syenite at Magnet Cove, Garland Co., AR; in mica peridotite in Woodson Co., Kansas. In Canada at Thetford mines, at the Jeffrey quarry(!), Asbestos, and at Mt. St.-Hilaire(!), PQ, and in the Keweenawan traps at Lake Nipigon, ONT. Blue, compact, fibrous masses are found west of Baoruco, Dominican Republic. Occurs at Kangerdluarssuk, Julienhaab, Greenland; at Ratho quarry and near Castle Rock, near Edinburgh, Scotland; at *Mt. Baldo, Verona*, and *Mt. Monzoni, Trento, Italy*. In syenitic rocks at Yuksporlak, Kola

Penin., and veins cutting ultrabasics of the Voikaro–Syninsk massif, Polar Ural Mts., Russia; in serpentinite at Horokanai, Sorachi Pref., Japan. Compact blue material from the Dominican Republic is used as an ornamental and for cabochons and is marketed under the name *larimar*. AR *D6 373(1892)*, *ZK 144:401(1976)*, *DHZII 2A:564(1978)*, *AM 63:401(1978)*.

65.2.1.5 Serandite Na(Mn,Ca)$_2$[Si$_3$O$_8$(OH)]

Named in 1931 by Lacroix for J. M. Sérand, West African mineral collector, who assisted in the collection of the type material. Wollastonite group. TRIC $P\bar{1}$. $a = 7.683$, $b = 6.889$, $c = 6.747$, $\alpha = 90.53°$, $\beta = 94.12°$, $\gamma = 102.75°$, $Z = 2$, $D = 3.46$ (for a composition with Mn:Ca $\approx$ 11:1). *25-723*: 7.51_3 6.72_3 3.16_9 2.98_{10} 2.84_7 2.60_4 2.50_5 2.19_6. Structure: *AM 61:229(1976)*. Serandite is isostructural with pectolite, and there is a direct relationship between increasing Mn content and diminishing cell constants over the pectolite–serandite composition range. Crystals are commonly flattened {001} prisms [010] to equant, often in slightly radiating, subparallel groups. Twinning with [010] twin axis and (100) composition, both contact and penetration; also by reflection (110). Orange to pinkish orange to rose; vitreous luster, pearly on cleavage surface; transparent to translucent. Cleavage {100}, {001}, perfect. H = $4\frac{1}{2}$–$5\frac{1}{2}$. G = 3.0–3.4. Biaxial (+); 2V = 33°; $N_x = 1.680$, $N_y = 1.682$, $N_z = 1.705$ by extrapolation; natural material usually has decreasing indices and increasing 2V as calcium content increases. Complete isomorphous substitution to form the pectolite–serandite series. High-Mn members of the series commonly contain significant rare earths, with yttrium predominant for the highest-Mn members. Alters to dark brown to black birnessite; pseudomorphs of elpidite, nenadkevichite, birnessite, and rhodochrosite are known. Occurs in alkaline rocks and syenite pegmatites. Found at Point of Rocks, near Springer, Colfax Co., NM; as superb(!) salmon-pink to orange crystals to 20 cm long and in groups at Mt. St.-Hilaire, PQ, Canada. Found on *Rouma, Los Is., Guinea*; at the Ezuri mine, Akita Pref., and with banalsite at the Shiromaru mine, Tokyo Pref., Japan; in a phonolite dike at Mt. Goonerineggeringgi, Q, Australia. Mt. St.-Hilaire material has been fashioned into small, faceted gems for collectors. AR *CR 192:187(1931)*, *BNSM(C) 13:107(1987)*, *MR 21:335(1990)*.

65.2.1.6 Cascandite Ca(Sc,Fe^{3+})[Si$_3$O$_8$(OH)]

Named in 1982 by Mellini et al. for the composition: calcium and scandium. Wollastonite group. TRIC $P\bar{1}$. $a = 7.529$, $b = 7.051$, $c = 6.755$, $\alpha = 92.12°$, $\beta = 93.67°$, $\gamma = 104.65°$, $Z = 2$, $D = 3.10$. *35-541*: 7.22_2 6.75_2 4.46_2 3.84_2 3.62_5 3.10_5 2.97_5 2.82_{10}. Structure: *AM 67:604(1982)*. The referenced structure uses a $C\bar{1}$ cell with $a = 9.791$, $b = 10.420$, $c = 7.076$, $\alpha = 98.91°$, $\beta = 102.63°$, $\gamma = 84.17°$, $Z = 4$, but the cell analogous to pectolite–serandite is retained here. Calcium occupies the larger of the octahedral sites, and Sc and Fe the smaller; the site occupied by Na in pectolite–serandite is generally vacant. Crystals with maximum dimension of 0.2 mm, elongated [100] and flattened {100}; dominant forms are {001}, {100}, {120}, {5$\bar{3}$0}. Pale pink, vitreous luster. Cleavage {100}, {001}, good. Biaxial; minimum and maximum indices measured on an {001} plate are 1.663 and 1.684. Found in cavities in granite with quartz, orthoclase, albite, and jervisite at *Cava Diverio, near Baveno, Italy.* AR *AM 67:599(1982).*

65.2.1.7 Denisovite (K,Na)Ca$_2$[Si$_3$O$_8$(F,OH)]

Named in 1984 by Men'shikov for Aleksandr Petrovich Denisov (1918–1972), mineralogist at Geological Institute, Kola Branch, Academy of Sciences USSR, at Apatity. Wollastonite group. MON $a = 30.92$, $b = 7.20$, $c = 18.27$, $\beta = 95°$, $Z = 20$, $D = 2.81$. *37-440:* 3.32_{10} 3.24_9 3.08_8 3.03_9 2.79_8 2.78_8 2.75_{10} 2.69_8. May be related to a possible polytype of pectolite. Parallel columnar aggregates, 10–15 cm across, consisting of acicular crystals. Greenish gray, colorless to faintly yellow; pearly luster. Cleavage or parting perpendicular to elongation. H = 4–5. G = 2.76. Biaxial (+); 2V = 20–30°; N$_x$ = 1.567, N$_y$ = 1.568, N$_z$ = 1.576; parallel extinction; positive elongation. Found in rischorrite rocks of *Mt. Eveslogchorr and Mt. Yukspor, Khibiny massif, Kola Penin., Russia*; and in the Malyy Murun block of the Murun pluton. AR *ZVMO 113:718(1984), AM 70:1329(1985), DANS 293:196(1987).*

65.2.2.1 Foshagite Ca$_4$[Si$_3$O$_9$](OH)$_2$

Named in 1925 by Eakle for William Fredrick Foshag (1894–1956), mineralogist and former curator at U.S. National Museum, who studied the mineralogy at the type locality. TRIC $P\bar{1}$. $a = 10.32$, $b = 7.36$, $c = 7.94$, $\alpha = 117.6°$, $\beta = 104.5°$, $\gamma = 90.0°$, $Z = 2$, $D = 2.74$. *29-377*(syn): 6.68_2 4.96_3 3.36_3 2.95_{10} 2.31_3 2.16_3 1.829_3 1.746_3. Structure: *AC 13:785(1960).* The structure is related to that of wollastonite with chains parallel to *b*, but staggered differently. White, silky luster. H = 3(?). G = 2.36. Biaxial (+); N$_x$ = 1.594, N$_y$ = 1.595, N$_z$ = 1.598; positive elongation. Infusible. Gelatinizes with acids. Occurs as fibers and fiber masses in contact and/or thermally metamorphosed limestones. Found at the Bingham open pit, Salt Lake Co., UT; as veins with fibers to several inches long with idocrase, thaumasite, and blue calcite at *Crestmore, Riverside Co., CA*; from Mull and Morven, Scotland; Hatrurim, Israel; Kushiro, Japan. AR *AM 43:1(1958), GSIB 40:45(1977).*

65.2.3.1 Hillebrandite Ca$_2$(SiO$_3$)(OH)$_2$

Named in 1908 for William Francis Hillebrand (1853–1925), American geochemist, U.S. Geological Survey. ORTH $Cmc2_1$. $a = 3.6389$–3.6427Å, $b = 16.311$–16.319Å, $c = 11.829$–11.840Å, $Z = 6$, $D = 2.69$. *42-538* (nat.) 4.81_3 3.36_2 3.03_3 2.93_{10} 2.84_2 2.79_3 2.25_2 1.82_4. Structure: *(DAN SSSR 123:741(1958)), (MA 14:179(1959))* and *AM 80:841(1995).* The structure consists of a three-dimensional network of Ca–O polyhedra that accommodates wollastonite-type Si–O tetrahedral chains. It is a natural analogue of calcium silicate hydrate phases in Portland cement. Colorless, white, faint green; streak white; radiating, fibrous crystals to 0.6 mm, elongate on *b*, {010} cleavage, crystal data and indices relate to a pseudo-cell, true cell is probably monoclinic, $\beta = 90°$ with doubled b, and $Z = 12$; $H = 5\frac{1}{2}$; $G = 2.66$–2.69, sometimes fluorescent: weak white in SW UV and medium yellow-white in LW UV; IR data = (ZNK 5:2727(1960)); DTA (ZVMO 90:90(1961)); in HCl, separates some silica, but otherwise dissolves. Generally pure, but may contain trace to minor amounts of Fe, Al, and Mg. Biax. $(-)$; $2V = 51°$, $N_x = 1.598$–1.607, $N_y = 1.603$–1.61, $N_z = 1.603$–1.614; $Z = c$, $(+)$ elongation; disper. $r < v$ strong. Shows ultra blue interference colors. Occurs in altered gabbroic rocks in Russia. In a contact zone between limestone and diorite in Mexico assoc. with wollastonite, garnet, gehlenite and spurrite. Occurs assoc. with foshagite at the Crestmore quarry, Crestmore, Riverside Co., CA; *Terneras mine. Velardeña, Durango, Mexico*; Carlingford, Co. Louth, Ireland; Ozersk massif, west Baikal area, Russia; Mt. Chapchachi, West Azgir, Caspian Sea region, Kazakhstan; in the Hatrurim Formation, Israel; Güneyce-Ikizdere area, Trabzon Prov., Turkey; Nugrah, Saudi Arabia; Yamato and Mihara mines, Kushiro, Hiroshima Pref., and in the Mihari mine and at Fuka near Bicchu, Okayama Pref., Japan; N'Chwaning mine, Kalahari Mn Field, northern Cape Prov., RSA. EF *MM 30:150(1953)*.

65.2.4.1 Sorensenite $Na_4SnBe_2[Si_3O_9]_2 \cdot 2H_2O$

Named in 1965 by Semenov et al. for Henning Sørensen (b.1926), geologist, University of Copenhagen. MON $C2/c$. $a = 20.698$, $b = 7.442$, $c = 12.037$, $\beta = 117.28°$, $Z = 4$, $D = 2.92$. *19-1171*: 6.31_6 3.41_8 3.06_7 2.96_8 2.92_{10} 2.84_5 2.68_5 2.12_4. Structure: *AC(B) 32:2553(1976)*. The silicate unit is a single chain $P = 3$, parallel to [010], similar to that of wollastonite. Colorless to pinkish altering to milky white; luster vitreous, silky. Cleavage in two directions at 63°; brittle. $H = 5\frac{1}{2}$. $G = 2.9$. Biaxial $(-)$; $2V$ small to 75°; $N_x = 1.576$, $N_y = 1.581$, $N_z = 1.584$; anomalous yellow-brown and blue interference colors. Found as prismatic crystals up to 10 cm × 1 cm × 1 cm at *Nakalaq in the Ilimaussaq intrusion complex, Greenland*, with analcime, microcline, sodalite, neptunite, and aegirine in veins cross-cutting nepheline syenite; also at Kvanefjeld with nepheline, analcime, and arfvedsonite. AR *AM 51:1547(1966)*.

65.3.1.1 Haradaite $SrVO(SiO_3)_2$

Named in 1974 by Watanabe et al. for Zunpei Harada (1898–1992), emeritus professor, Hakkaido University, Sapporo. See also suzukiite (65.3.1.2). ORTH $Ama2$. $a = 7.06$, $b = 14.64$, $c = 5.33$, $Z = 4$, $D = 3.80$. *18-1284*(syn): 7.3_3 3.65_4 3.20_{10} 2.88_9 2.65_4 2.55_3 2.12_4 2.05_4. Structure: *MJ 5:98(1967)*. The structure is characterized by an infinite kinked chain of periodicity 4 and a repeat length of 7.06 Å lying parallel to [100]. V^{4+} in square-pyramidal five-coordination bridges adjacent chains to form sheets of composition $[(VO)(SiO_3)_2]^{2-}$. Anhedral, tabular {010}, mostly 2–3 mm in diameter; massive. Emerald green, bluish green; very pale green streak; vitreous luster; translucent. Cleavage {010}, perfect; {100}, {001}, distinct; conchoidal fracture. $H = 4\frac{1}{2}$. $G = 3.80$. Biaxial $(-)$; $2V = 75°$; $N_x = 1.713$, $N_y = 1.721$, $N_z = 1.734$; $XYZ = abc$; $r < v$, strong. Strongly pleochroic; X, colorless to pale green; Y, colorless to yellow-green; Z, bluish green. Type material contains 4.90% BaO; a complete solid solution with suzukiite is possible. Found in veinlets cutting rhodonite–manganoan goldmanite ores at *Yamato mine, Kagoshima Pref., Kyushu*, and at Noda Tamagawa mine, Iwate Pref., Honshu, *Japan*. AR *MJJ 4:299(1965), IJM:78(1970), AM 56:1123(1971)*.

65.3.1.2 Suzukiite $BaVO(SiO_3)_2$

Named in 1982 by Matsubara et al. for Jan Suzuki (1896–1970), mineralogist and petrologist at Hokkaido University, Sapporo. See also haradaite (65.3.1.1). ORTH $Ama2$. $a = 7.089$, $b = 15.261$, $c = 5.364$, $Z = 4$, $D = 4.03$. *MJ 11:15(1982)*. *35-681:* 7.63_{10} 3.82_6 3.35_6 3.28_3 2.65_2 2.39_4 2.02_2 1.526_2. The isostructural barium analog of haradaite with which a solid solution is possible. Small flakes {010} and aggregates. Bright green, pale green streak, vitreous luster, translucent. Cleavage {010}, perfect; {100}, {001}, distinct. $H = 4\frac{1}{2}$. $G = 4.0$. Biaxial (−); $2V \approx 90°$; $N_x = 1.730$, $N_y = 1.739$, $N_z = 1.748$; strongly pleochroic, $Z > Y > Z$; pale green to blue-green. Contains $\approx 3.2\%$ SrO. In massive rhodonite–rhodochrosite ore with barite, alabandite, and nagashimalite at the *Mogurazawa mine, Kiry City, Gumma Pref., Honshu, Japan*. AR *AM 68:282(1983)*.

65.3.2.1 Balangeroite $(Mg,Fe^{2+})_{21}O_3(OH)_{20}[Si_4O_{12}]_2$

Named in 1983 by Compagnoni et al. for the type-locality serpentinite. See gageite (65.3.2.2a,b). MON $B2$, Bm or $B2/m$. (First setting): $a = 19.40$, $b = 19.40$, $c = 9.65$, $\gamma = 88.9°$, $Z = 2$, $D = 3.089$. *35-567:* 9.59_4 6.77_8 3.38_4 3.28_4 3.20_3 2.71_{10} 2.67_7 2.52_4. Structure: *AM 72:382(1987)*. An orthorhombic cell, P^{***}, with $a = 13.85$, $b = 13.58$, $c = 9.65$, has also been proposed and may represent a polytype. The first setting for monoclinic crystals is used to maintain correspondence with earlier orthorhombic cell of gageite. The silicate chain with $W = 1$ and $P = 4$ is kinked and lies parallel to [001], the fiber elongation. Isostructural with gageite-2M. Brown, vitreous to silky luster. $G = 2.98$. Biaxial; $N = 1.680$ perpendicular to [001]. Pleochroic; dark brown parallel to [001], yellow-brown perpendicular to [001]. Solid solution with gageite is likely. Found as fibrous to asbestiform masses associated with chrysotile, magnetite, iron–nickel, olivine, and antigorite in the *Balangero serpentinite, at the San Vittore mine, Lanzo Valley, Piemonte, Italy*. AR *AM 68:214(1983), EJM 3:559(1991)*.

65.3.2.2a Gageite-2M $(Mn,Mg)_{21}O_3(OH)_{20}[Si_4O_{12}]_2$

65.3.2.2b Gageite-1A $(Mn,Mg)_{21}O_3(OH)_{20}[Si_4O_{12}]_2$

Named in 1911 by Phillips for Robert Burns Gage (1875–1946), Trenton, NJ, who analyzed the first sample. See balangeroite (65.3.2.1). MON $B2$, Bm or $B2/m$. (First setting): $a = 19.42$, $b = 19.42$, $c = 9.84$, $\gamma = 89.5°$, $Z = 2$, $D = 3.599$. TRIC $P1$. $a = 14.17$, $b = 14.07$, $c = 9.84$, $\alpha = 76.5°$, $\beta = 76.6°$, $\gamma = 86.9°$, $Z = 1$, $D = 3.61$. *20-723:* 6.87_{10} 3.44_2 3.25_3 2.56_8 2.71_8 2.61_5 2.56_6

1.674_6 (for undifferentiated gageite). Structure: *AM 72:382(1987)*. Unit cell originally described as ORTH, *Pnnm*, which may constitute yet another polytype. The monoclinic polytype is isostructural with balangeroite. Colorless to pink, vitreous to silky luster. H = 3(?). G = 3.58. Biaxial (−); $N_x = 1.723$, $N_y = 1.734$, $N_z = 1.736$; Z = c; r < v, moderate; positive elongation. Infusible. Decomposed by HCl. Probable isomorphous series with balangeroite. Found as acicular crystals with a maximum 0.1 mm thickness and as fibrous aggregates in open cavities associated with late-stage, low-temperature hydrothermal activity at *Franklin, Sussex Co., NJ*. AR *AM 53:309(1968), 54:1005(1969)*.

65.3.3.1 Taikanite $BaSr_2Mn_2^{3+}O_2[Si_4O_{12}]$

Named in 1985 by Kalinin et al. for the locality. MON C2. $a = 14.600$, $b = 7.759$, $c = 5.142$, $\beta = 93.25°$, Z = 2, D = 4.81 (South Africa). *39-335*: 3.44_4 3.27_7 3.16_5 2.91_{10} 2.83_9 2.57_8 1.946_6 1.718_4. Structure: *AM 78:1088(1993)*. The structure consists of silicate chains of periodicity 4 lying parallel to [010], linking chains of edge-sharing MnO_6 octahedra that parallel [001]; Ba^{2+} and Sr^{2+} occupy the resulting channels. The abstract for Russian type material gives $C2/m$ and reverses the values of a and b. Equant, to slightly elongated grains up to 1.5 mm, and grain aggregates. Dark green, vitreous to greasy luster, translucent. Cleavage {001}, perfect; brittle, conchoidal fracture. G = 4.72 (Russia). For Russian material, biaxial (+); 2V = 74–80°; $N_x = 1.775$, $N_y = 1.792$, $N_z = 1.814$; XY = ab, $Z \wedge c = 44°$; r > v, strong. Strongly pleochroic; X, violet to black; Y,Z, emerald-green. Occurs with braunite and manganese amphibole in the *Taikan Mts., Khabarovsk, Russia*. In a sugilite-cemented breccia of braunite and hausmannite at the Wessels mine, Kalahari manganese field, NE of Kuroman Hill, Cape Prov., South Africa. AR *ZVMO 114:635(1985), AM 72:226(1987)*.

65.3.4.1 Batisite $(Ba)(K)(Na)_3Ti_2Si_4O_{14}$

Named in 1960 by Kravchenko et al. for the composition: containing Ba, Ti, and Si. See shcherbakovkite (65.3.4.2), the K analog. ORTH *Ima*2 (acentric) or *Imam* (centric). $a = 10.499$, $b = 13.913$, $c = 8.087$, Z = 4, D = 3.49. *14-636*(nat): 3.4_5 3.21_3 2.92_{10} 2.62_3 2.16_5 2.09_4 1.68_5 1.56_3. Structure solution obtained same residual (R) in acentric and centric space groups. *NJBM:107(1987)*. Crystals are elongated on c and flattened on b. Most abundant forms are: {001}, {010}, {100}, with lesser development of {150}, {021}, {031}, {011}, {310}. Dark brown, yellow-brown; rosy brown streak; vitreous luster. Cleavage {100}, fair; {010}, {001}, poor. H = 6. G = 3.43. IR data: *ZVMO 93:641(1964)*. Insoluble in common acids. Contains minor Al_2O_3, Nb_2O_5, Ta_2O_5, ZrO_2, Fe_2O_3, and CaO. Biaxial (+); XYZ = abc; $N_x = 1.728–1.730$, $N_y = 1.733–1.735$, $N_z = 1.790–1.791$; 2V = 7°; r < v, strong, to r > v, weak. X, colorless to yellow-brown; Y, yellow-brown; Z, red-brown. Occurs associated with nepheline (zeolitized), aegirine, arfvedsonite, uranothorite, lorenzenite, eudialyte, apatite, and orthoclase in nepheline–syenite pegmatite in the *Inagalinsk massif, central Aldan region*,

Russia; also occurs in fissures in two nepheline–leucite lavas at Üdersdorf and Altburg near Daun and at Liley in the west Eifel region, Germany. EF *MIN 3(1):569(1972), LAP 14:14(1989)*.

65.3.4.2 Shcherbakovite $(K,Na,Ba)_3(Ti,Nb)_2(Si_2O_7)_2$

Named in 1954 by Eskova and Kazakova for Dmitrii Ivanovich Shcherbakov (1893–1966), Russian mineralogist and geochemist. *DANS 99:837(1954)*. See batisite (65.3.4.1), the Na-dominant analog. ORTH $Ima2$. $a = 10.55$–10.57, $b = 13.92$–13.95, $c = 8.10$–8.12, $Z = 4$, $D = 3.59$. *AM 70:455(1985), ZVMO 93:641(1964)*. 31-1324(nat): 3.40_7 3.20_8 2.91_{10} 2.63_7 2.19_6 2.10_6 1.68_6 1.57_6. Structure: *AM 70:455(1985)*. Crystals as large as 5.0 cm × 1.5 cm × 1.0 cm, elongated and prismatic. Dark brown to reddish brown, rose or red-brown streak, vitreous luster. One or two good cleavages, brittle. $H = 6\frac{1}{2}$. $G = 3.34$. IR data: *IM:208(1975), ZVMO 93:641(1964)*. Insoluble in HNO_3 and HCl, partly soluble in H_2SO_4 on heating. Khibiny material has minor Al, Zr, Fe, Mg, Ca. Biaxial $(-)$; $XYZ = bac$; $N_x = 1.706$–1.714, $N_y = 1.745$–1.746, $N_z = 1.765$–1.772; $2V_x = 64$–$90°$; $r < v$, distinct; positive elongation; distinctly pleochroic. $Z > Y \geq X$; X, colorless–pale yellow; Y, yellow; Z, brownish yellow or golden yellow. In pegmatite veins in alkalic rocks, also in lamproites, and kimberlites. Emmons Mesa, Zirkel Mesa, and Black Butte, Leucite Hills, WY; *Mt. Rasvumchorr, Khibiny massif, Kola Penin., Russia*; also in the dumps of the Koashva mine; Wolgidee Hills, West Kimberly region, WA, Australia. EF *MM 54:645(1990)*.

65.3.4.3 Ohmilite $Sr_3(Ti,Fe^{3+})(Si_2O_6)_2(O,OH) \cdot 2$–$3H_2O$

First described in 1973 by Komatsu et al.; named in 1983 by Mizota et al. for the locality. MON $P2_1/m$. $a = 10.979$, $b = 7.799$, $c = 7.818$, $\beta = 100.90°$, $Z = 2$, $D = 3.39$. 26-1388(nat): 10.8_7 5.39_8 4.62_{10} 3.83_9 3.26_9 3.04_8 2.73_6 2.60_9. Has batisite-type Si_4O_{12} chains, also found in haradaite and batisite. *AM 68:811(1983)*. Aggregates of fibers, elongate on b; spherulites composed of radially arrayed fine needles or fibers. Fibers are <10 μm in length. Light pink, white streak. Cleavage {100}, perfect. $H = 3\frac{1}{2}$. $G = 3.38$. IR and thermal data: *AM 68:811(1983)*. Analysis (wt %): SiO_2, 34.79; TiO_2, 10.27; Fe_2O_3, 0.20; SrO, 43.37; H_2O, 6.68. Biaxial (sign not determined); $YZ = ab$; $c \wedge Y$ not det., $N_x = 1.649$, $N_z = 1.715$; 2V and sign not determined because of small grain size and elongated nature parallel to [010]; weakly pleochroic, colorless to light pink. Occurs at *Ohmi, Niigata Pref., Japan*; in riebeckite–albite units in serpentinite. EF *MJJ 7:298(1973)*.

RHODONITE GROUP

The rhodonite group minerals, classically part of the pyroxenoids, consist of seven essentially isostructural species conforming to

$$R_5[Si_5O_{14}(O,OH)]$$

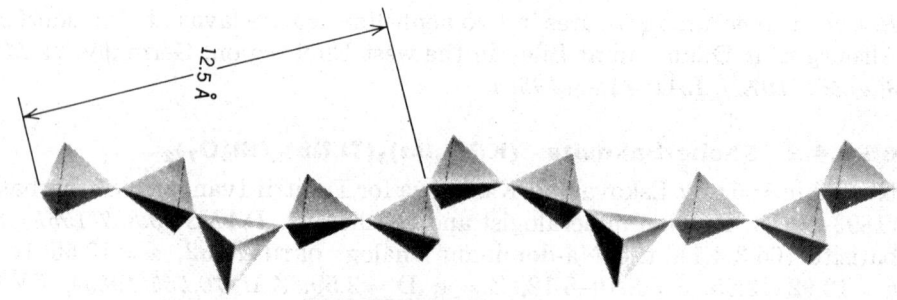

RHODONITE: MnSiO$_3$
Idealized silicate chain periodicity in the rhodonite group.

where R$_5$ = five cation sites, three of which (M1,M2,M3) are octahedral, and two of which (M4,M5) are octahedral to eight-coordinated.

The structure is characterized by an approximately linear chain of five silica tetrahedra lying parallel to one another. All species are triclinic, and thus have no symmetry constraint to choice of unit cell. Consequently, there have been diverse cell assignments, the most common of which have been as follows:

Space Group	$a \approx$	$b \approx$	$c \approx$	$\alpha \approx$	$\beta \approx$	$\gamma \approx$	Chain Parallel to:
$C\bar{1}$	9.7	10.5	12.2	109	103	86	c
$P\bar{1}$	7.6	12.2	6.7	86	94	111	b
$P\bar{1}$	7.6	11.8	6.7	92	94	106	[110]

The first cell bears the closest relationship to the long-established morphological setting of rhodonite and babingtonite. The transformation matrix, morphology to c cell, is 010/100/002. In the species entries, the reported cell constants are those of the latest structure reference found. However, to facilitate comparison, the following table gives the constants for the last of the three settings. *DHZ II 2:586(1987)*.

RHODONITE GROUP

Mineral	Formula	Space Group	a	b	c	α	β	γ	
Rhodonite	(Mn, Ca)SiO$_3$	$P\bar{1}$	7.545	11.782	6.663	92.69	94.32	105.71	65.4.1.1
Babingtonite	Ca$_2$(Fe^{2+}, Mn)Fe^{3+} [Si$_5$O$_{14}$(OH)]	$P\bar{1}$	7.497	12.225	6.710	86.18	93.90	112.27	65.4.1.2
Mn-Babingtonite	Ca$_2$(Mn, Fe^{2+})Fe^{3+} [Si$_5$O$_{14}$(OH)]	$P\bar{1}$	7.492	11.745	6.701	92.54	93.90	103.97	65.4.1.3
Nambulite	(Li, Na)Mn$_4^{2+}$[Si$_5$O$_{14}$(OH)]	$P\bar{1}$	7.621	11.761	6.731	92.77	95.08	106.87	65.4.1.4

65.4.1.1 Rhodonite

RHODONITE GROUP (CONT'D)

Mineral	Formula	Space Group	a	b	c	α	β	γ	
Natronambulite	$(Na,Li)Mn_4^{2+}[Si_5O_{14}(OH)]$	$P\bar{1}$	7.620	11.762	6.737	92.81	95.55	106.87	65.4.1.5
Marsturite	$NaCaMn_3[Si_5O_{14}(OH)]$	$P1$ or $P\bar{1}$	7.70	11.73	6.78	92.2	94.10	108.8	65.4.1.6
Lithiomarsturite	$LiCa_2Mn_2[Si_5O_{14}(OH)]$	$P1$ or $P\bar{1}$	≈7.7	≈11.8	≈6.8	≈92	≈94	≈109	64.4.1.7

65.4.1.1 Rhodonite $(Mn,Ca)_5(Si_5O_{15})$

Named in 1819 by Jasche from the Greek for *rose*, in allusion to its characteristic pink color. The variety fowlerite is named for Samuel Fowler. Rhodonite group. Dimorphous with hypothetical iron-free pyroxmangite (65.6.1.1). TRIC $P\bar{1}$. $a = 7.545$, $b = 11.782$, $c = 6.663$, $\alpha = 92.69°$, $\beta = 94.32°$, $\gamma = 105.71°$, $Z = 2$, $D = 3.55$. *13-138*: 3.34_3 3.14_3 3.10_3 2.98_7 2.92_7 2.77_{10} 2.22_2 2.18_3. Structure: *AM 73:798(1988)*. For the given cell the $[Si_5O_{15}]$ chains lie parallel to [110]; pairs of chains with apical oxygens pointing in opposite directions form strips parallel to $(1\bar{1}1)$, which in turn are joined to strips of cation polyhedra. Of the five cation sites, M1, M2, and M3 are near octahedral, M4 is highly distorted octahedral, and M5 is seven-coordinated. The last is preferentially occupied by Ca and limits the extent of its substitution. **Habit:** In the old morphological setting, crystals are commonly tabular {001} to equant with {001}, {1$\bar{1}$0}, {110} dominant, but also highly modified. Crystal edges often rounded; faces often with rough or pitted surfaces. Twinning polysynthetic on (010). Granular, massive. **Physical properties:** Pink to rose-red to brownish red, often veined or coated black due to alteration; vitreous luster, somewhat pearly on cleavages. Cleavage (morph. setting) {110}, {1$\bar{1}$0}, perfect at 92.5°; {001}, good; conchoidal to uneven fracture. $H = 5\frac{1}{2}$–$6\frac{1}{2}$. $G = 3.55$–3.76. **Optics:** Biaxial (+); $2V = 61$–$87°$; $N_x = 1.711$–1.738, $N_y = 1.716$–1.741, $N_z = 1.724$–1.751; $X \wedge a \approx 5°$, $Y \wedge b \approx 20°$, $Z \wedge c \approx 25°$; $r < v$, crossed; inclined extinction in all sections. Weakly pleochroic; X, yellowish red; Y, pinkish red; Z, pale yellowish red. Refractive indices decrease with both Ca and Mg substitution; Fe substitution has little effect on indices. **Chemistry:** Some calcium is always found in substitution for manganese, but there is a discontinuity between bustamite and rhodonite, and magnesium

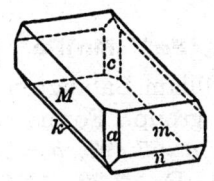

RHODONITE

Rhodonite, New Jersey. Forms (old morphological setting): a, {100}; c, {001}; m, {110}; M, {1$\bar{1}$0}; k, {$\bar{2}$21}; n, {$\bar{2}\bar{2}$1}

may also substitute in small amounts as denoted by the formula $(Mn,Mg)_{1-x}Ca_x(SiO_3)$, with $0 < x < 0.2$, with reported values having $x \approx 0.05$–0.18. Up to 43 mol % $MgSiO_3$ and 26 mol % $FeSiO_3$ have been reported, and the variety fowlerite has been reported with up to 10 mol % $ZnSiO_3$. Minor Fe^{3+} has been reported, and Al levels are low. Blackens and fuses at 2.5. Alters through weathering directly, or via rhodochrosite, to pyrolusite and/or coronadite. **Occurrence:** Widespread in manganese orebodies, and is usually associated with metasomatic activity. Commonly results from contact metamorphism of rhodochrosite by metasomatic introduction of silica, or possibly by simple load-metamorphism of rhodochrosite + quartz in buried sediments. May result from the introduction of manganese into argillaceous sediments. Rarely in pegmatites, where it has probably formed by assimilation of Mn-rich country rock; in ore pockets associated with submarine extrusion of spilites and pillow lavas. As a primary gangue mineral in Pb–Zn deposits. Commonly associated with other manganese silicates, such as tephroite, johannsenite, bustamite, and pyroxmangite. **Localities:** A widespread mineral; notable localities are: in the zinc ores of Franklin, Sussex Co., NJ, in large, blocky to tabular crystals(!), often with rounded edges embedded in calcite, and as small transparent crystals in fractures with manganaxinite and willemite; also in ME, MA, MT, and CA. Magnesian rhodonite is found in regionally metamorphosed Wabush iron formation, western Labrador, NF, Canada. As brilliant crystals at the Harstig mine, Pajsberg, Sweden; brownish red material in quartzite at Simsiö, Lapua, Finland; massive at the B.R. quarry, Meldon, Okehampton, Devonshire, England; at Elbingerode, Harz, Germany; with tetrahedrite at Kapnik and Rezbanya, Hungary; extensive deposits of massive material in the Ykaterinburg district, Ural Mts., Russia. Found at several localities in Japan. Particularly noteworthy is the occurrence in the ores of Broken Hill, NSW, Australia, where gemmy crystals(!) of blocky to prismatic habit up to nearly 10 cm in length have been found. In the Kuroman manganese district of South Africa, and at Mufulira, Zambia. Gemmy crystals have been found at Morro da Mina, Conselheiro Lafaiete, MG, Brazil; in compact groups of crystals(!) to 6 cm long at Chiurocu mine, near Huanzala, Peru. **Uses:** A carving material for ornamental and decorative purposes; occasionally cut en cabochon or, less commonly, into faceted collector gems. AR *D6:378(1892), AC 12:182(1959), DHZII 2A:586(1978)*.

65.4.1.2   Babingtonite $Ca_2(Fe^{2+},Mn)Fe^{3+}[Si_5O_{14}(OH)]$

Named in 1824 by Lévy for William Babington (1757–1833), Irish physician and mineralogist. Rhodonite group. Forms an isomorphous series with manganbabingtonite. TRIC $P\bar{1}$. $a = 7.497$, $b = 12.225$, $c = 6.710$, $\alpha = 86.18°$, $\beta = 93.90°$, $\gamma = 112.27°$, $Z = 2$, $D = 3.39$. *14-321:* 6.69_4 4.07_4 3.12_7 2.95_6 2.87_8 2.75_{10} 2.47_5 2.17_6. Structure: *MJJ 15:8(1990)*. The structure is essentially analogous to that of rhodonite, but chain periodicity faults occur such that $P = 7$ and $P = 3$ sequences interrupt the normal $P = 5$ sequence.

Approximately 20% vacancies exist among the cation sites. Crystal habit (original morphological setting that is equivalent to original rhodonite) is as equant to short prismatic crystals with {001}, {110}, {1$\bar{1}$0} dominant and modified by {100}, {010}, {221}, {2$\bar{2}$1}, or tabular {001}. Dark greenish black; greenish to brownish gray streak; vitreous, splendent luster; nearly opaque. Cleavage {1$\bar{1}$0}, perfect; {110}, {001}, less so; uneven fracture. H = $5\frac{1}{2}$–6. G = 3.34–3.48. Biaxial (+); 2V = 76–88°; N_x = 1.715–1.719, N_y = 1.727–1.734, N_z = 1.749–1.753; r > v, strong; OAP nearly normal to (001). Pleochroic; X, deep green; Y, lilac-brown; Z, pale to deep brown. Insoluble in acids. Fusibility 3. Manganese substitutes for Fe^{2+}. Occurs in cavities in granitic rocks and with zeolites in "trap rocks." Found at Blueberry Mt., Middlesex Co., at Athol, Worcester Co., and in excellent crystals(!) at the Lane Trap Rock quarry, Westfield, Hampden Co., MA; at Wallingford, CT; at the Upper New Street quarry, Paterson, Passaic Co., NJ, and several other quarries in the region; the Goose Creek quarry, Leesburg, Loudon Co., VA. Found as distinct crystals at the *Brastad mine, near Arendal, Norway*; at Herborn, near Dillenburg, Germany; and in granite from Baveno, Novara, Italy; at the Khandivali quarry, near Bombay, Maharashtra, India. Designated the state mineral of Massachusetts. AR *D6:381(1992)*, *CM 19:269(1981)*, *AC(A) 37:617(1981)*.

65.4.1.3 Manganbabingtonite $Ca_2(Mn,Fe^{2+})Fe^{3+}[Si_5O_{14}(OH)]$

Named in 1967 by Vinogradova and Plyusinna for the composition and relationship to babingtonite. Rhodonite group. Forms isomorphous series with babingtonite. TRIC $P\bar{1}$. a = 7.492, b = 11.745, c = 6.701, α = 92.54°, β = 93.90°, γ = 103.97°, Z = 2, D = 3.33. *42-1400*(Japan): 6.93_6 6.67_7 3.47_9 3.34_4 3.13_8 3.02_8 2.96_{10} 2.45_4; *12-313*(Russia): 3.44_8 3.10_{10} 3.00_{10} 2.94_{10} 2.16_{10} 1.654_7 1.427_6. No structure determination, but presumably isostructural with babingtonite. Reddish brown to greenish black, vitreous luster, essentially opaque. Cleavage {001}, perfect; {1$\bar{1}$0}, imperfect. H = 6. G = 3.452. Biaxial (+); 2V = 86°; N_x = 1.716, N_y = 1.730, N_z = 1.746; r > v, strong. Pleochroic; X, green; Y, pale rose; Z, rose-brown. Properties are for material with 7.9 wt % MnO and 4.5 wt % FeO. Fusibility 3. Found with epidote, calcite, and quartz at the *Rudnyi Kaskad deposit, eastern Sayan, Siberia, Russia*, and at Mitani, Kochi City, Shikoku, Japan. AR *AM 53:1064(1968)*, *BNSM(C) 15:81(1989)*.

65.4.1.4 Nambulite $(Li,Na)Mn_4^{2+}[Si_5O_{14}(OH)]$

Named in 1972 by Yoshi et al. for Matsuo Nambu (b.1917), mineralogist, Tohoko University, Japan. Rhodonite group. Forms an isomorphous series with natronambulite. TRIC $P\bar{1}$. a = 7.621, b = 11.761, c = 6.731, α = 92.77°, β = 95.08°, γ = 106.87°, Z = 2, D = 3.55. *29-833*(sodian): 3.34_4 3.17_7 3.14_5 3.09_6 3.07_6 2.97_8 2.96_{10} 2.91_7; *27-1253*(syn, sodium-free): 3.17_6 3.05_5 2.95_7 2.92_{10} 2.70_9 2.62_4 2.22_5 2.20_4. Structure: *AC(B) 31:2422(1975)*. Distorted octahedral sites M1, M2, and M3 are occupied by Mn as is the seven-coordinated site M4; Li and Na occupy eight-coordinated site M5. Red-brown with orange tint, pale yellow streak, vitreous to subadamantine luster, transparent to translucent. Biaxial (+); 2V = 30°; $N_x = 1.707$, $N_y = 1.710$, $N_z = 1.730$; r > v, weak. Found as coarse prismatic crystals, up to 8 mm, associated with albite, neotocite, and rhodochrosite in veinlets up to 5 cm thick at the *Funakozawa mine, Kitakami Mts., Iwate Pref., Japan*, and at the Gozaisho mine, Iwaki-shi, Fukushima Pref. Found as transparent, flawless crystals(!) at the Kombat mine, 37 km E of Otavi, Namibia, but these are, at least in part, natronambulite. Significant sodium is found in the natural materials. A few crystals have been faceted into collector gems. AR *MR 8:391(1977), 22:423(1991); AM 75:413(1990).*

65.4.1.5 Natronambulite (Na,Li)Mn$_4^{2+}$[Si$_5$O$_{14}$(OH)]

Named in 1985 by Matsubara et al. for the composition and its relationship to nambulite, with which it forms an isomorphous series. Rhodonite group. TRIC $P\bar{1}$. $a = 7.620$, $b = 11.762$, $c = 6.737$, $\alpha = 92.81°$, $\beta = 95.55°$, $\gamma = 106.87°$, Z = 2, D = 3.50. *39-332:* 7.13$_5$ 6.70$_4$ 3.56$_{10}$ 3.35$_4$ 3.08$_5$ 2.97$_3$ 2.51$_4$ 2.20$_2$. Sodium occupies the larger M5 site. Pinkish orange, very pale orange streak, vitreous luster, translucent. Cleavage {100}, {001}, perfect. H = $5\frac{1}{2}$–6. G = 3.51. Biaxial (+); 2V = 45°; $N_x = 1.706$, $N_y = 1.710$, $N_z = 1.730$. Found as coarse-grained aggregates up to 7 mm in diameter in the dump of the *Matsumaezawa orebody, Tanohata mine, Iwate Pref., Japan*, associated with manganoan aegirine, manganoan arfvedsonite, rhodonite, and serandite; as transparent, flawless crystals(!) at the Kombat mine, 37 km E of Otavi, Namibia, but these are, at least in part, nambulite. AR *MJJ 12:332(1985); AM 72:224(1987), 75:413(1990).*

65.4.1.6 Marsturite NaCaMn$_3$[Si$_5$O$_{14}$(OH)]

Named in 1978 by Peacor, Dunn, and Sturman for Marion Stuart of Bellvue, ID, who has been a staunch supporter of the preservation of natural history specimens, particularly those of geological interest. Rhodonite group. TRIC $P1$ or $P\bar{1}$. $a = 7.70$, $b = 12.03$, $c = 6.78$, $\alpha = 85.26°$, $\beta = 94.10°$, $\gamma = 111.04°$, Z = 2, D = 3.465. *31-585:* 3.18$_4$ 3.11$_4$ 3.00$_8$ 2.92$_9$ 2.73$_{10}$ 2.22$_3$ 1.695$_3$ 1.441$_4$. Chain-periodicity faults are probably present; probable site occupancies are: M1 and M3 = Mn; M2 = Mn,Ca; M4 = Ca; M5 = Na. Crystals are 0.5-mm laths elongated [010] and flattened on {001} with dominant {100} and {001} and crudely terminated by {010}; also as epitaxial growths on rhodonite with coincidence of all axes and the contact surface being (010) in the marsturite setting. Light pink to white, vitreous luster, transparent to translucent. Cleavage {100}, {001}, imperfect (equivalent to {110} and {1$\bar{1}$0} of rhodonite). H ≈ 6. G = 3.46. Biaxial (+); 2V = 60°; $N_x = 1.686$, $N_y = 1.691$, $N_z = 1.708$ for Na$_D$; XZ ≈ cb; r > v, weak. Minor Fe^{2+}, Mg, and Zn are present. Known from only a few specimens from *Franklin, Sussex Co., NJ*, where it is found associated with rhodonite, willemite, and manganaxinite; at the Tsurumaki mine, 25 km NW of Ogaki, Gifu Pref., and the Shiromaru mine,

Hatonosu, Tokyo Pref., Japan. AR *AM 63:1178(1978), 75:413(1990); MR 17:123(1986).*

65.4.1.7 Lithiomarsturite LiCa$_2$Mn$_2$[Si$_5$O$_{14}$(OH)]

Named in 1990 by Peacor et al. for the composition and relationship to marsturite. Rhodonite group. TRIC $P1$ or $P\bar{1}$. $a = 7.652$, $b = 12.119$, $c = 6.805$, $\alpha = 85.41°$, $\beta = 94.42°$, $\gamma = 111.51°$, $Z = 2$, $D = 3.27$. PD: 6.79_2 3.19_9 3.08_5 3.01_{10} 2.91_9 2.81_2 2.77_6 2.21_5. Structure determination incomplete, due to common chain periodicity faults. Site occupancies are: M1 and M3 = Mn; M2 = Mn,Ca; M4 = Ca; M5 = Li. Euhedral crystals, 1–3 mm in diameter, with {100}, {010}, {001}. Light pinkish brown to light yellow, vitreous luster. Cleavage {100}, {001}, good. H = 6. G = 3.32. Biaxial (–); 2V = 60°; $N_x = 1.645$, $N_y = 1.660$, $N_z = 1.666$; $X \approx b$, $Y \approx [100]$, $Z \approx c$; r > v, moderate. Fe^{2+} substitutes for Mn^{2+}, as does lesser Mg. Found in vugs in pegmatite at the *Foote mine, Kings Mt., NC,* with albite, fluorapatite, bavenite, brannockite, parsettensite, and tetrawickmanite. AR *AM 75:409(1990).*

65.4.2.1 Santaclaraite CaMn$_4^{2+}$[Si$_5$O$_{14}$(OH)](OH)·H$_2$O

Named in 1984 by Erd and Ohashi for the locality. TRIC $B\bar{1}$. $a = 15.633$, $b = 7.603$, $c = 12.003$, $\alpha = 109.71°$, $\beta = 88.61°$, $\gamma = 99.95°$, $Z = 4$, $D = 3.398$. *35-369:* 7.69_6 7.04_{10} 4.80_4 3.85_6 3.52_4 3.15_8 3.00_8 2.20_4. Structure: *AM 66:154(1981).* Structurally related to rhodonite, but the $P = 5$ chains are shifted relative to one another; the five cation sites are six-coordinated except for M5, occupied by Ca, which is seven-coordinated as in rhodonite. One (OH) is bound to Si as in the (OH)-bearing members of the rhodonite group. Radiating lamellar aggregates of thin prismatic to tabular crystals and as individual prismatic crystals. Twinning on {100} common. Pale pink to reddish orange, vitreous luster, transparent to translucent. Cleavage {100}, {010}, good. H = $6\frac{1}{2}$. G = 3.31. Biaxial (–); 2V = 83°; $N_x = 1.681$, $N_y = 1.696$, $N_z = 1.708$ for Na$_D$; r > v, moderate; on {100} sections $Z \wedge c = 2°$ and $X \wedge b = -21°$; on {010} sections $Z \wedge c = 16°$ and $Y \wedge a = -14.5°$; pleochroic with Y > Z > X in pale red tones. Chemically akin to a hydrated rhodonite, but on ignition yields bustamite. Found in masses and in vugs in pink to tan veins in Franciscan chert of the *Diablo Range, Santa Clara and Stanislaus Cos., CA.* AR *AM 69:200(1984).*

65.5.1.1 Stokesite CaSn(Si$_3$O$_9$)·2H$_2$O

Named in 1899 by Hutchinson for Sir George Gabriel Stokes (1819–1903), British mathematician and physicist, Cambridge University. ORTH *Pnna.* $a = 14.465$–14.485, $b = 11.606$–11.630, $c = 5.224$–5.250, $Z = 4$, $D = 3.18$. *13-109:* 7.25_8 5.82_6 4.54_6 3.99_{10} 3.55_6 3.43_6 2.89_{10} 1.556_7. Structure: *MM 33:615(1963).* The silicate structural element is a six-tetrahedron chain. Pyramidal to tabular, {100}, with {100}, {211}, {423} dominant, in the structure setting. The transformation matrix, original morphology to X-ray, is 010/001/100 aggregates to >3 cm across. Colorless, white, pale cream to pale blue;

vitreous luster; transparent to translucent. Cleavage {101}, perfect; {100}, imperfect. H = 6. G = 3.19–3.23. Biaxial (+); 2V = 67–70°; $N_x = 1.609$–1.617, $N_y = 1.612$–1.620, $N_z = 1.619$–1.627; XYZ = cab; r < v. Minor Mn and Fe may substitute for Sn. Found in complex granitic pegmatites and desilicated pegmatites, often in association with cassiterite in the Himalaya mine, and possibly derived from it. Occurs with cassiterite at Mesa Grande, San Diego Co., CA; at the Halvasso quarry, and at *Roscommon Cliff, St. Just, Cornwall, England* with axinite; at Ctidružice and Vĕžná, Moravia, Czech Republic; Pitkaranta, Karelia, Russia; at Kerokubo, Ena, and the Iwaguro Sekizai quarry, Tuhara, Gifu Pref., Japan; in tin-rich garnet skarn at the El Hamman fluorite deposit, central Morocco; and in fine spheroidal aggregates at Córrego do Urucum, Galiléia, MG, Brazil. AR, EEF *MM 32:433(1960), MM 41:411(1977), MR 8:294(1977).*

65.5.2.1 Chkalovite $Na_2Be(Si_2O_6)$

Named in 1939 by Gerasimovskii for Valery Pavlovich Chkalov (1904–1938), Russian aviator who made first nonstop, polar-route flight from Moscow to the United States. ORTH $F2dd$. $a = 21.188$, $b = 21.129$, $c = 6.881$, Z = 24, D = 2.68. *42-572*: 5.29_1 4.04_{10} 3.34_3 3.27_3 2.78_3 2.49_6 2.46_3 1.801_2. Structure: *DANS 248:727(1979).* The silicate structural element is a six-tetrahedron chain linked Be-O tetrahedra to form a cristobalitelike framework. White; vitreous luster, somewhat greasy; transparent to translucent. Cleavage {001}, good; {111}, poor; conchoidal fracture. H = 6. G = 2.662. Biaxial (+); 2V = 78–81°; $N_x = 1.540$–1.543, $N_y = 1.543$–1.550, $N_z = 1.549$–1.553; Y = c; r > v. Rare at Mt. St.-Hilaire, PQ, Canada, as millimeter-size, equant, rounded crystals with villauminite, ussingite, lovozerite, and other minerals; in similar associations at Ilimaussaq, Julianehaab district, Greenland; at *Punkaruyav Mt.*, and elsewhere, in the *Lovozero massif, Kola Penin., Russia.* AR *AM 25:380(1940), MR 21:303(1990).*

65.6.1.1 Pyroxmangite $(Mn,Fe^{2+})_7[Si_7O_{21}]$

Named in 1913 by Ford and Bradley for the composition and relationship to other pyroxenoids. Polymorphic with low-Ca rhodonite. Complete isomorphous series with pyroxferroite (65.6.1.2). TRIC $P\bar{1}$. $a = 6.721$, $b = 7.563$, $c = 17.380$, $\alpha = 113.17°$, $\beta = 82.27°$, $\gamma = 94.13°$, Z = 2, D = 3.77. *29-895*(syn): 4.73_4 3.47_3 3.04_3 3.02_3 2.97_{10} 2.68_4 2.61_3 2.19_4; *25-147*(ferroan): 3.32_2 3.29_2 3.11_5 2.95_{10} 2.67_6 1.168_4 1.708_3 1.666_2. Structure: *AM 73:798(1988).* The structure is characterized by chains of $W = 1$ and $P = 7$ lying parallel to c. There are seven polyhedral sites: M1, M2, and M3 are octahedral and occupied by Mn; sites M4 to M7 are distorted and contain the majority of substituting cations. A $C\bar{1}$ cell can be chosen so as to better correspond to the morphology. Crystals tabular to prismatic; as embedded grains; cleavable masses and compact. Pale pink to rose, often red-brown due to alteration; vitreous to pearly luster; transparent to translucent. Cleavage (morphological setting) {110}, {1$\bar{1}$0}, perfect; {010}, {001}, poor. H = $5\frac{1}{2}$–6.

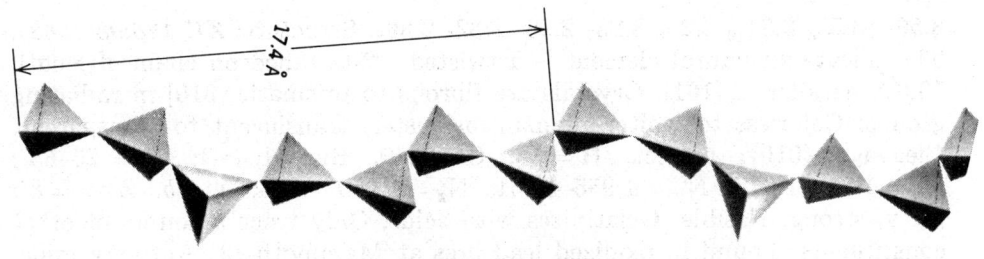

PYROXMANGITE
SiO_3 chain in pyroxmangite.

$G = 3.61$–3.80. Biaxial (+); $2V = 35$–$45°$; $N_x = 1.728$–1.748, $N_y = 1.732$–1.742, $N_z = 1.746$–1.764; $r > v$, moderate; $Z \wedge c \approx 45°$ on (100) sections. Insoluble in acids; easily fusible to black bead. Complete solid solution exists with pyroxferroite. Limited substitution of Ca and Mg for Mn is common. Typical in metamorphic and metasomatic, Mn-rich assemblages. First described from *Iva, Anderson Co., SC*, but this occurrence is actually a Mn-rich pyroxferroite; Silverton district, San Juan Co., and the Galena district, Hinsdale Co., CO; near Boise, ID; and near Randsburg, Kern Co., CA. Found at Simsiö, Lapua, Finland; at Glen Beag, Glenelg, Scotland; and the Bernina area, Rhaetic Alps, Switzerland; from the Ajiro mine, Gifu Pref., the Taguchi mine, Aichi Pref.(!), and from pegmatite at Iwazumi-machi, Iwate Pref., and other sites in Japan. In large crystals(!) from Broken Hill, NSW, Australia. AR *MJJ 8:329(1977)*, *DHZII 2A:600(1978)*.

65.6.1.2 Pyroxferroite $(Fe^{2+},Mn)_7[Si_7O_{21}]$

Named in 1970 by Chao et al. for the composition and analogy with pyroxmangite (65.6.1.1), with which it forms a complete isomorphous series. TRIC $P\bar{1}$. $a = 6.632$, $b = 7.563$, $c = 17.380$, $\alpha = 114.31°$, $\beta = 82.75°$, $\gamma = 94.58°$, $Z = 2$, $D = 3.85$. *20-1:* 6.55_3 4.68_4 3.09_5 3.01_3 2.93_{10} 2.67_6 2.62_3 2.16_4. Structure: *SCI 168:364(1970)*. Yellow, light orange to orange-pink; vitreous luster; transparent to translucent. Cleavage (morphological setting) {110}, {1$\bar{1}$0}, perfect; {010}, {001}, poor. $H = 5$–6. $G = 3.68$–3.76. Biaxial (+); $2V = 30$–$40°$; $N_x = 1.746$–1.756, $N_y = 1.750$–1.758, $N_z = 1.764$–1.768. Insoluble in acids; easily fusible to a magnetic bead. Complete solid solution with pyroxmangite; Ca and Mg commonly present. Manganoan material found at Iva, Anderson Co., SC, was originally described as pyroxmangite; at Vester Silfberg, Dalecarlia, Sweden; the Isanago mine, Kyoto Pref., Japan. Discovered in microgabbros of the *Sea of Tranquility, Moon*, and in the Oceanus Procellarum. AR *DHZII 2A:600(1978)*.

65.7.1.1 Alamosite $PbSiO_3$

Named in 1909 by Palache and Merwin for the locality. MON $P2/n$. $a = 12.247$, $b = 7.059$, $c = 11.236$, $\beta = 113.12°$, $Z = 12$, $D = 6.32$. *29-782:*

3.56_{10} 3.53_8 3.34_{10} 3.25_6 3.23_7 2.98_7 2.82_6 2.30_8. Structure: *ZK 126:98(1968)*. The silicate structural element is a twisted, 12-tetrahedron chain of length 19.6 Å parallel to [10$\bar{1}$]. Crystals are fibrous to prismatic [010] in radiating groups. Colorless to white, adamantine luster, translucent to transparent. Cleavage {010}, perfect. H = $4\frac{1}{2}$. G = 6.49. Biaxial (−); 2V = 25–65°; $N_x = 1.945$–1.947, $N_y = 1.955$–1.961, $N_z = 1.959$–1.968; Y = b, Z∧c = 8°; r < v, strong. Fusible. Gelatinizes with acids. Only trace amounts of other constituents. Found in oxidized lead ores at Mammoth–St. Anthony mine, Pinal Co., and at the Rawhide mine, Maricopa Co., AZ; at *Alamos, SON, Mexico*; at Tsumeb, Namibia(!). AR *DT:567(1932), NJMA 123:138(1974)*.

Class 66

Double-Width, Unbranched Chains, W = 2

AMPHIBOLE GROUP

The amphibole group consists of some 70 naturally occurring inosilicate species and hypothetical end members, sharing a complex formula of

$$A_{0-1}B_2C_5T_8O_{22}(OH,F,O)_2$$

where

A = Na, K
B = Ca, Na, Mn
C = Mg, Fe^{2+}, Mn, Fe^{3+}, Al, Ti, Li, etc.
T = Si, Al, Be

and sharing a common structure. The name amphibole, from the Greek *amphibolos* meaning ambiguous, was coined by Haüy in 1801, and has proved to be a good choice because the complexity of the group's chemistry results in a diversity of appearances and yet a number of common characteristics.

The fundamental structural unit consists of infinite ribbons, two tetrahedrons wide, that can be envisioned as a condensation of a pair of pyroxene-type chains. The ribbon, of $W = 2$ and $P = 2$ and of composition $(Si_4O_{11})^{6-}$ lies in the *b–c* plane and parallel to the approximately 5.3-Å *c* axis common to all amphiboles. The ribbon may be viewed as a series of linked six-tetrahedron rings of ditrigonal symmetry. Of the 11 oxygens in the repeat unit, five are bridging oxygens that link the tetrahedra together and six are nonbridging, two lateral and four apical in their positions. In all the amphiboles, pairs of ribbons face one another, apex to apex, with their apical oxygens octahedrally coordinated to C cations to form I-beam units broader than, but akin to, those of the pyroxenes. The I-beams lie base to base along [100] so that the ditrigonal rings are approximately aligned from unit to unit; furthermore, they are staggered, apex plane to basal plane, along the *b* axis. The resultant lattice dimensions for all the monoclinic amphiboles are approximately $a \approx 9.9$ Å, $b \approx 18$ Å, $c \approx 5.3$ Å, $\beta \approx 104°$.

The octahedral coordination sites (position C) between the apical, nonbridging oxygens are of three types: the two M1 and one M3 sites are each coordinated to four apical oxygens and two (OH), whereas the two M2 sites are coordinated to four apical oxygens and two lateral oxygens of an adjacent I-beam. The two cations of position B, the M4 site, are surrounded by, but not necessarily coordinated to, two apical oxygens of one I-beam plus two lateral and four bridging oxygens of an adjacent I-beam. The M2 and M4 cations thus form the linkage between the I-beam along [010]. Finally, the A position, not always occupied, lies between the facing six-tetrahedron rings and is thus structurally surrounded by 12 bridging oxygens. This very elegant though complex structural scheme allows for the accommodation of the great diversity of the common six- and eight-coordinated cations, as well as those of higher coordination (K^+) and four-coordination (Al^{3+}, Be^{2+}).

Most of the monoclinic amphiboles crystallize in space group C_2/m, in which all the tetrahedral ribbons are alike. A few polymorphs and synthetic phases crystallize in $P2_1/m$ with a very similar structure but whose six-tetrahedron rings have near-hexagonal symmetry and the ribbons are of two types. A $P2/a$ structure with equivalent ribbons and ditrigonal rings has but one known example, joesmithite.

The orthorhombic amphiboles, $Pnma$, have cell constants related to the monoclinic amphiboles as follows: $a_o = 2a_m \sin \beta$, $b_o = b_m$, $c_o = c_m$. Dimensionally, this corresponds to a twinning on (010) on a unit cell scale, but the

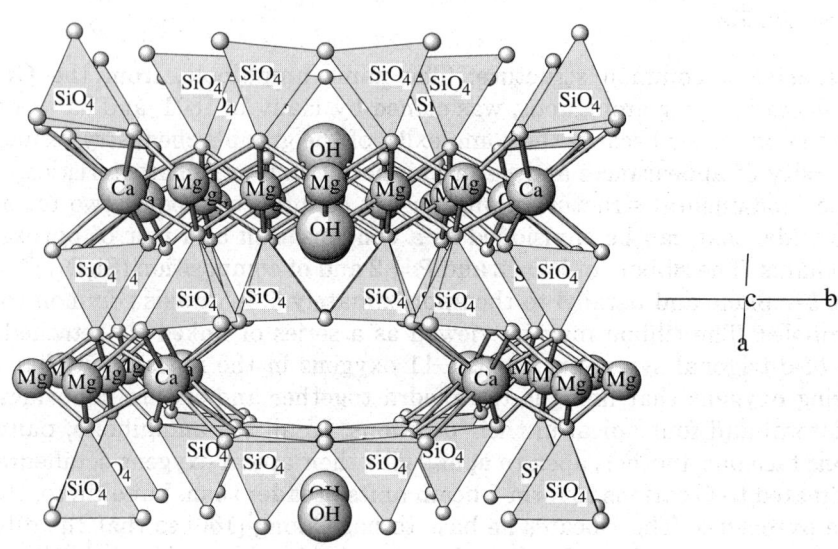

TREMOLITE: $Ca_2Mg_5Si_8O_{22}(OH)_2$
Tremolite, a typical amphibole; perspective view along the double width silicate chains (parallel to c).

structure has two different ribbon coordinations. The M4 site in this structure does not readily accommodate the larger cations (i.e., Ca). Orthorhombic amphiboles crystallizing in space group $Pnmn$ (protoanthophyllite) are not known in natural systems.

AMPHIBOLE GROUP: CUMMINGTONITE SUBGROUP—$(Mg,Fe^{2+},Mn)_7[Si_8O_{22}](OH)_2$

Mineral	Formula			Compositional Limits*	
	A	B + C	T		
Magnesiocummingtonite		$(Mg, Fe^{2+})_7$	Si_8	$Mg \geq 0.7$	66.1.1.1
Cummingtonite		$(Mg, Fe^{2+})_7$	Si_8	$0.7 > Mg > 0.3$	66.1.1.2
Grunerite		$(Fe^{2+}, Mg)_7$	Si_8	$Fe \geq 0.7$	66.1.1.3
Tirodite		$(Mn, Mg, Fe^{2+})_7$	Si_8	$Mn \geq 0.1, Mg \geq Fe$	66.1.1.4
Dannemorite		$(Mn, Fe^{2+}, Mg)_7$	Si_8	$Mn \geq 0.1, Mg < Fe$	66.1.1.5
Magnesioclinoholmquistite		Li_2Mg_3, Al_2	Si_8	$Li \geq 1, Mg \geq 0.9$	66.1.1.6
Clinoholmquistite		$Li_2(Mg, Fe^{2+})_3Al_2$	Si_8	$Li \geq 1,$ $0.9 < Mg > 0.1$	66.1.1.7
Ferroclinoholmquistite**		$Li_2Fe_3^{2+}Al_2$	Si_8	$Li \geq 1, Fe \geq 0.9$	66.1.1.8

*$Mg = Mg/(Mg + Fe^{2+})$, $Fe^{2+} = Fe^{2+}/(Mg + Fe^{2+})$, $Mn = Mn/(Mn + Mg + Fe^{2+})$, for Li the value is the number of ions.
**End member whose natural occurrence is unknown or doubtful.

AMPHIBOLE GROUP: ANTHOPHYLLITE SUBGROUP—$Na_x(Mg,Fe^{2+})_{7-y}$ $[Al_{x+y}Si_{8-x-y}O_{22}](OH)_2$

Mineral	A	B + C	T	Compositional Limits*	
Magnesioanthophyllite	Na_x	$(Mg, Fe^{2+})_{7-y}Al_y$	$Al_{x+y}Si_{8-x-y}$	$x + y < 1, Na < 0.5,$ $Mg \geq 0.9$	66.1.2.1
Anthophyllite	Na_x	$(Mg, Fe^{2+})_{7-y}Al_y$	$Al_{x+y}Si_{8-x-y}$	$0.9 > Mg > 0.1$	66.1.2.2
Ferroanthophyllite**	Na_x	$(Mg, Fe^{2+})_{7-y}Al_y$	$Al_{x+y}Si_{8-x-y}$	$Fe \geq 0.9$	66.1.2.3
Sodium-anthophyllite**	Na_x	$(Mg, Fe^{2+})_{7-y}Al_y$	$Al_{x+y}Si_{8-x-y}$	$x \geq 0.5, Al_T > 0.99$	66.1.2.4
Magnesiogedrite	Na_x	$(Mg, Fe^{2+})_{7-y}Al_y$	$Al_{x+y}Si_{8-x-y}$	$x + y > 1, Al_T > 0.99,$ $Mg \geq 0.9$	66.1.2.5
Gedrite	Na_x	$(Mg, Fe^{2+})_{7-y}Al_y$	$Al_{x+y}Si_{8-x-y}$	$0.9 > Mg > 0.1$	66.1.2.6
Ferrogedrite	Na_x	$(Mg, Fe^{2+})_{7-y}Al_y$	$Al_{x+y}Si_{8-x-y}$	$Fe \geq 0.9$	66.1.2.7
Sodium-gedrite**	Na_x	$(Mg, Fe^{2+})_{7-y}Al_y$	$Al_{x+y}Si_{8-x-y}$	$x \geq 0.75, Al_T > 0.99$	66.1.2.8
Magnesioholmquistite**		$Li_2Mg_3Al_2$	Si_8	$Li \geq 0.1, x = y = 0,$ $Mg \geq 0.9$	66.1.2.9
Holmquistite		$Li_2(Mg, Fe^{2+})_3Al_2$	Si_8	$Li \geq 0.1, x = y = 0,$ $0.9 > Mg > 0.1$	66.1.2.10
Ferroholmquistite**		$Li_2Fe_3^{2+}Al_2$	Si_8	$Li \geq 0.1, x = y = 0,$ $Fe \geq 0.9$	66.1.2.11

*$Mg = Mg/(Mg + Fe^{2+})$, $Fe^{2+} = Fe^{2+}/(Mg + Fe^{2+})$, $Mn = Mn/(Mn + Mg + Fe^{2+})$; for Li the value is the number of ions.
**End member whose natural occurrence is unknown or doubtful.

AMPHIBOLE GROUP: TREMOLITE SUBGROUP—CALCIC CLINOAMPHIBOLES WITH $(CA + NA)_B \geq 1.34$ AND $(NA + K) < 0.67$

Mineral	A	B	C	T	Compositional Limits**	
Tremolite		Ca_2	Mg_5	Si_8	$Mg \geq 0.9$	66.1.3a.1
Actinolite		Ca_2	$(Mg, Fe^{2+})_5$	Si_8	$0.5 > Fe > 0.1$	66.1.3a.2
Ferroactinolite		Ca_2	$(Fe^{2+}, Mg)_5$	Si_8	$Fe^{2+} \geq 0.5$	66.1.3a.3
(Alumino)magnesiohornblende		Ca_2	Mg_4Al	$AlSi_7$	$Mg > 0.5$	66.1.3a.4
(Alumino)ferrohornblende		Ca_2	$Fe_4^{2+}Al$	$AlSi_7$	$Fe^{2+} \geq 0.5$	66.1.3a.5
(Alumino)tschermakite		Ca_2	Mg_3Al_2	Al_2Si_6	$Mg > 0.5$	66.1.3a.6
Ferro(alumino)tschermakite		Ca_2	$Fe_3^{2+}Al_2$	Al_2Si_6	$Fe^{2+} \geq 0.5$	66.1.3a.7
Ferritschermakite**		Ca_2	$Mg_3Fe_2^{3+}$	Al_2Si_6	$Mg > 0.5$	66.1.3a.8
Ferroferritschermakite		Ca_2	$Fe_3^{2+}Fe_2^{3+}$	Al_2Si_6	$Fe^{2+} \geq 0.5$	66.1.3a.9
Edenite	Na	Ca_2	Mg_5	$AlSi_7$	$Mg > 0.5$	66.1.3a.10
Ferroedenite	Na	Ca_2	Fe_5^{2+}	$AlSi_7$	$Fe^{2+} \geq 0.5$	66.1.3a.11
Paragasite	Na	Ca_2	Mg_4Al	Al_2Si_6	$Mg > 0.5$	66.1.3a.12
Ferroparagasite	Na	Ca_2	$Fe_4^{2+}Al$	Al_2Si_6	$Fe^{2+} \geq 0.5$	66.1.3a.13
Magnesiohastingsite	Na	Ca_2	Mg_4Fe^{3+}	Al_2Si_6	$Mg > 0.5$	66.1.3a.14
Hastingsite	Na	Ca_2	$Fe_4^{2+}Fe^{3+}$	Al_2Si_6	$Fe^{2+} \geq 0.5$	66.1.3a.15
Magnesiosadanagaite	K,Na	Ca_2	$(Mg, Fe^{2+}, Al, Ti)_5$	$(Si, Al)_8$	$Mg > 0.5$	66.1.3a.16
Sadanagaite	K,Na	Ca_2	$(Fe^{2+}, Mg, Al, Ti)_5$	$(Si, Al)_8$	$Fe^{2+} \geq 0.5$	66.1.3a.17
Kaersutite	Na	Ca_2	Mg_4Ti	Al_2Si_6	$Mg > 0.5$	66.1.3a.18
Ferrokaersutite	Na	Ca_2	$Fe_4^{2+}Ti$	Al_2Si_6	$Fe^{2+} \geq 0.5$	66.1.3a.19

*$Mg = Mg/(Mg + Fe^{2+})$, $Fe^{2+} = Fe^{2+}/(Mg + Fe^{2+})$, $Mn = Mn/(Mn + Mg + Fe^{2+})$; for all other ions (Ca, Na, etc.) the value is the number of ions.
**End member whose natural occurrence is unknown or doubtful.

AMPHIBOLE GROUP: RICHTERITE SUBGROUP—CA-NA CLINOAMPHIBOLES WITH $(CA + NA)_B \geq 1.34$ AND $1.34 > NA_B > 0.67$

Mineral	A	B	C	T	Compositional Limits*	
(Alumino)winchite		CaNa	Mg_4Al	Si_8	$(Na, K)_A < 0.5, Mg > 0.5$	66.1.3b.1
Ferro(alumino)winchite**		CaNa	$Fe_4^{2+}Al$	Si_8	$(Na, K)_A < 0.5, Fe^{2+} > 0.5$	66.1.3b.2
Ferriwinchite		CaNa	Mg_4Fe^{3+}	Si_8	$(Na, K)_A < 0.5, Mg > 0.5$	66.1.3b.3
Ferroferriwinchite**		CaNa	$Fe_4^{2+}Fe^{3+}$	Si_8	$(Na, K)_A < 0.5, Fe^{2+} > 0.5$	66.1.3b.4
(Alumino)barroisite**		CaNa	Mg_3Al_2	$AlSi_7$	$(Na, K)_A < 0.5, Mg > 0.5$	66.1.3b.5
Ferro(alumino)barroisite**		CaNa	$Fe_3^{2+}Al_2$	$AlSi_7$	$(Na, K)_A < 0.5, Fe^{2+} > 0.5$	66.1.3b.6
Ferribarroisite**		CaNa	$Mg_3Fe_2^{3+}$	$AlSi_7$	$(Na, K)_A < 0.5, Mg > 0.5$	66.1.3b.7
Ferroferribarroisite**		CaNa	$Fe_3^{2+}Fe_2^{3+}$	$AlSi_7$	$(Na, K)_A < 0.5, Fe^{2+} > 0.5$	66.1.3b.8
Richterite	Na	CaNa	Mg_5	Si_8	$Mg > 0.5$	66.1.3b.9
Ferrorichterite	Na	CaNa	Fe_5^{2+}	Si_8	$Fe^{2+} \geq 0.5$	66.1.3b.10
Magnesioaluminokatophorite**	Na	CaNa	Mg_4Al	$AlSi_7$	$(Na, K)_A > 0.5, Mg > 0.5$	66.1.3b.11
Aluminokatophorite**	Na	CaNa	$Fe_4^{2+}Al$	$AlSi_7$	$(Na, K)_A > 0.5, Fe^{2+} > 0.5$	66.1.3b.12
Magnesioferrikatophorite	Na	CaNa	Mg_4Fe^{3+}	$AlSi_7$	$(Na,K)_A > 0.5, Mg^{2+} > 0.5$	66.1.3b.13
Ferrikatophorite	Na	CaNa	$Fe_4^{2+}Fe^{3+}$	$AlSi_7$	$(Na, K)_A > 0.5, Fe^{2+} > 0.5$	66.1.3b.14
Magnesioaluminotaramite**	Na	CaNa	Mg_3Al_2	Al_2Si_6	$(Na, K)_A > 0.5, Mg > 0.5$	66.1.3b.15
(Alumino)taramite**	Na	CaNa	$Fe_3^{2+}Al_2$	Al_2Si_6	$(Na, K)_A > 0.5, Fe^{2+} > 0.5$	66.1.3b.16
Magnesioferritaramite**	Na	CaNa	$Mg_3Fe_2^{3+}$	Al_2Si_6	$(Na, K)_A > 0.5, Mg > 0.5$	66.1.3b.17
Ferritaramite	Na	CaNa	$Fe_3^{2+}Fe^{3+}$	Al_2Si_6	$(Na, K)_A > 0.5, Fe^{2+} > 0.5$	66.1.3b.18

*$Mg = Mg/(Mg + Fe^{2+})$, $Fe^{2+} = Fe^{2+}/(Mg + Fe^{2+})$, $Mn = Mn/(Mn + Mg + Fe^{2+})$; for all other ions (Ca, Na, etc.) the value is the number of ions.
**End member whose natural occurrence is unknown or doubtful.

66.1.1.3 Grunerite

AMPHIBOLE GROUP: GLAUCOPHANE SUBGROUP—ALKALI CLINOAMPHIBOLES WITH $Na_B \geq 1.34$

Mineral	A	B	C	T	Compositional Limits*	
Glacuophane		Na_2	Mg_3Al_2	Si_8	$(Na, K)_A < 0.5$, $Mg > 0.5$	66.1.3c.1
Ferroglaucophane		Na_2	$Fe_3^{2+}Al_2$	Si_8	$(Na, K)_A < 0.5$, $Fe^{2+} > 0.5$	66.1.3c.2
Crossite		Na_2	$(Mg, Fe^{2+})_3(Al, Fe^{3+})_2$	Si_8	$0.3 < Fe^{3+} < 0.7$	66.1.3c.3
Magnesioriebeckite		Na_2	Mg_3Fe^{3+}	Si_8	$0.3 < Fe^{3+} < 0.7$, $Mg > 0.5$	66.1.3c.4
Riebeckite		Na_2	$Fe_3^{2+}Fe_2^{3+}$	Si_8	$0.3 < Fe^{3+} < 0.7$, $Fe^{2+} \geq 0.5$	66.1.3c.5
Eckermannite	Na	Na_2	Mg_4Al	Si_8	$(Na, K)_A < 0.5$, $Mg > 0.5$	66.1.3c.6
Ferroeckermannite**	Na	Na_2	$Fe_4^{2+}Al$	Si_8	$(Na, K)_A < 0.5$, $Fe^{2+} > 0.5$	66.1.3c.7
Magnesioarfvedsonite	Na	Na_2	Mg_4Fe^{3+}	Si_8	$(Na, K)_A < 0.5$, $Mg > 0.5$	66.1.3c.8
Arfvedsonite	Na	Na_2	$Fe_4^{2+}Fe^{3+}$	Si_8	$(Na, K)_A < 0.5$, $Fe^{2+} > 0.5$	66.1.3c.9
Kozulite	Na	Na_2	$Mn_4(Fe^{3+}, Al)$	Si_8	$(Na, K)_A < 0.5$	66.1.3c.10
Nyboite	Na	Na_2	Mg_3Al_2	$AlSi_7$	$(Na, K)_A < 0.5$	66.1.3c.11
Leakeite	(Na,K)	Na_2	$(Mg, Fe^{3+}, Li, Mn^{3+})$	$(Si, Al)_8$	$(Na, K)_A > 0.5$	66.1.3c.12
Kornite	(Na,K)	Na,Li	$(Mg, Mn^{3+}, Fe^{3+}, Li)$	Si_8		66.1.3c.13
Ungarettite	Na	Na_2	$Mn_2^{2+}Mn_3^{3+}$	Si_8		66.1.3c.14
Fluroferroleakeite	Na	Na_2	$(Fe_2^{2+}Fe_2^{3+}Li)$	Si_8		66.1.3c.15

*$Mg = Mg/(Mg + Fe^{2+})$, $Fe^{2+} = Fe^{2+}/(Mg + Fe^{2+})$, $Mn = Mn/(Mn + Mg + Fe^{2+})$; for all other ions (Ca, Na, etc.) the value is the number of ions.
**End member whose natural occurrence is unknown or doubtful.

The presence of a variety of cation sites that can accommodate almost the full range of available cation sizes leads to extensive isomorphism. The certain establishment of immiscibility gaps for natural systems is difficult. The assignment of a specific amphibole, particularly the more varied clinoamphiboles, can be done with certainty only with a full chemical analysis. Extensive solid solution allows, in most cases, only tentative assignment on the basis of optical properties and/or X-ray diffraction. The extent of solid solution is illustrated by Figures 1a;b, which show the compositional spread for some members of the tremolite subgroup (calcic amphiboles).

The arrangement of the amphiboles into the subgroups and series reflected in the tables above is based on structural and crystallochemical considerations, but is in some respects also an artifact of convenience: considerable solid solution exists between one series and another, and even between one subgroup and another. The compositional limits in the tables are as defined by the IMA subcommittee on the amphibole group. *AM 63:1023(1978)*, *MSAMR 9A:1(1981)*.

66.1.1.1 Magnesiocummingtonite $Mg_7[Si_8O_{22}](OH)_2$

66.1.1.2 Cummingtonite $(Mg,Fe)_7[Si_8O_{22}](OH)_2$

66.1.1.3 Grunerite $Fe_7[Si_8O_{22}](OH)_2$

Cummingtonite named in 1824 by Dewey for the locality. Grunerite named in 1853 by Kenngott for Louis Emmanuel Gruner (1809–1883), Swiss–French

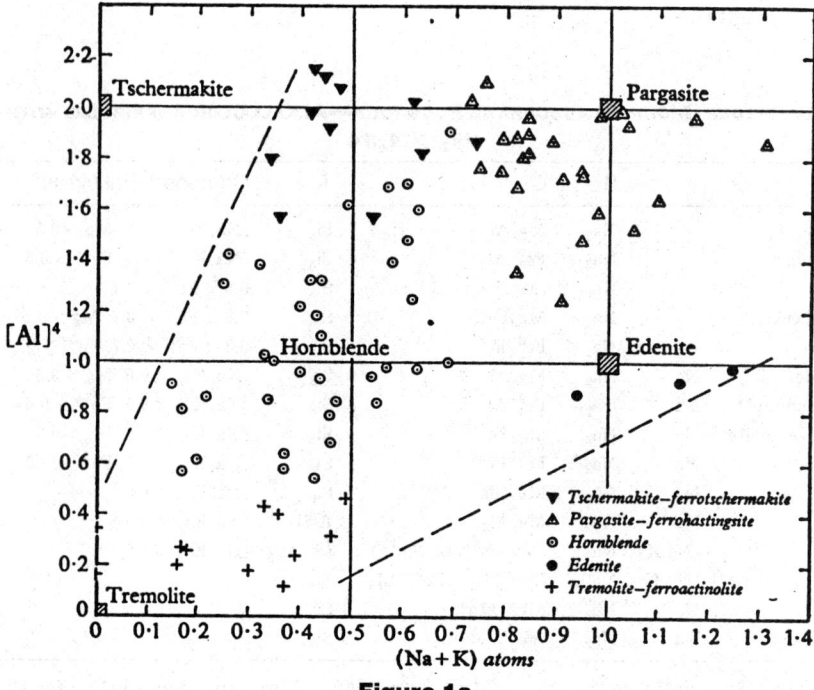

Figure 1a

The chemical variation of the calcium rich amphiboles expressed as the numbers of (Na, K) and $[Al]^4$ atoms per formula unit.

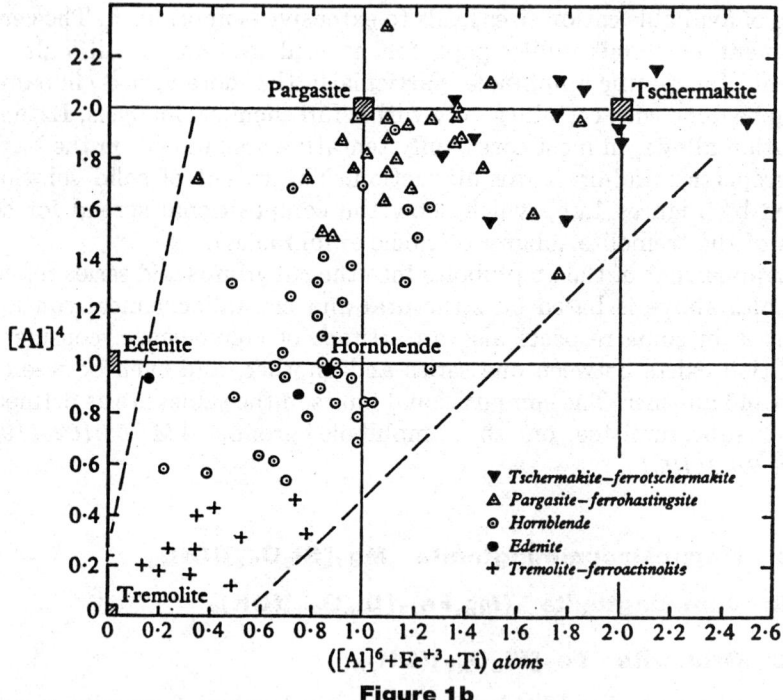

Figure 1b

The chemical variation of the calcium rich amphiboles expressed as the numbers of ($[Al]^6 + Fe^{+3} + Ti$) and $[Al]^4$ atoms per formula unit

66.1.1.3 Grunerite

chemist, who first analyzed the mineral. Magnesiocummingtonite named in 1969 by Kisch. Synonym: amosite, an acronym for "asbestos mines of South Africa," is used for asbestiform members of this series and the gedrite series. Amphibole group, cummingtonite subgroup. Polymorphous with the magnesioanthophyllite–ferroanthophyllite series, $P2_1/m$ cummingtonite, and synthetic, orthorhombic protoanthophyllite. MON $C2/m$. (Magnesiocummingtonite): $a = 9.583$, $b = 18.091$, $c = 5.315$, $\beta = 102.6°$, $Z = 2$, $D = 3.18$; (cummingtonite): $a = 9.604$, $b = 18.124$, $c = 5.325$, $\beta = 102.6°$, $Z = 2$, $D = 3.195$; (grunerite): $a = 9.564$, $b = 18.393$, $c = 5.339$, $\beta = 101.9°$, $Z = 2$, $D = 3.54$. 17-727(Mg-cumm.): 9.03_8 8.30_9 3.43_6 3.25_8 3.07_{10} 2.73_7 1.649_6 1.406_8; 31-636(cumm.): 9.12_5 8.30_{10} 4.55_4 4.14_4 3.26_8 3.06_9 2.75_7 2.62_5; 31-631(grun.): 9.21_5 8.33_{10} 3.88_5 3.47_5 3.07_8 2.77_9 2.64_7 2.51_6. Structure: AC 14:$622(1961)$. Fe^{2+} is strongly ordered in M4, the B site, as is any Ca or Na. **Habit:** Most commonly bladed, fibrous, to asbestiform; rarely displaying distinct {110} faces. Simple and lamellar twinning on (100) common. **Physical properties:** Brown to dark green, vitreous to silky luster, transparent to nearly opaque. Cleavage {110}, good to perfect at $\approx 55°$. H = 5–6. G = 3.10–3.60. **Optics:** *Magnesiocummingtonite*: Biaxial (−), (+); $2V_x = 66$–$97°$; $N_x = 1.621$, $N_y = 1.632$, $N_z = 1.643$; $r < v$ weak; $Y = b$. Pleochroic, X = Y colorless and Z pale yellow green. *Cummingtonite*: Biaxial (+), (−); $2V_z = 79$–$92°$; $N_x = 1.632$–1.662, $N_y = 1.640$–1.678, $N_z = 1.658$–1.698; $Y = b$, $Z \wedge c = 15$–$21°$ decreasing with Fe increase; $r > v$, weak. Pleochroism absent to weak; X,Y, colorless; Z, pale green. *Grunerite*: Biaxial (−); $2V = 82$–$88°$; $N_x = 1.662$–1.687, $N_y = 1.678$–1.709, $N_z = 1.698$–1.729; $Y = b$; $Z \wedge c = 10$–$15°$ decreasing with Fe increase, $r < v$, weak. Distinctly pleochroic; X,Y, very pale yellow to brown; Z, pale greenish brown. Petrographic distinction is often made at $2V = 90°$. In asbestiform varieties elongation is always positive, $Z \wedge c = 0$–$18°$, indices may increase to over 1.80, color intensified and pleochroism more distinct for samples that have been subjected to heat. Insoluble in HCl. **Chemistry:** Compositional limits are given in Amphibole group description. Magnesiocummingtonite is likely only as a hypothetical end member; in natural occurrences, iron content is probably limited to the low-iron end of the cummingtonite range. Aluminum substitution for Si is generally less than one-eighth of the tetrahedral sites; Al and Fe^{3+} substitute to a limited degree in the C position; minor Ti may be present; Mn^{2+} may be present in significant amounts and complete series with dannemorite and tirodite is possible. Ca and Na occupy less than half the B sites. **Occurrence:** Found chiefly in contact or regionally metamorphosed basic rocks, commonly associated with hornblende and plagioclase. Cummingtonite and Ca-rich plagioclase may result from the breakdown of aluminous amphiboles. It has been reported as a primary mineral in hornblende and hypersthene dacites. Grunerite is characteristic of iron-rich siliceous sediments over a wide range of metamorphic conditions. **Localities:** Members of the series are found at numerous localities worldwide, only a few of which are noted here. *Magnesiocummingtonite*: from Cima di Gagnone, Ticino, Switzerland, and *Cooma*,

NSW, Australia; occurrences with other series members from the Wabush iron formation, Labrador, NF, Canada; from *Buskerud, Kongsberg, Norway*, but appears to be outside the end-member range. *Cummingtonite*: the type material from *Cummington, Hampshire Co., MA*, is actually grunerite; abundant in the host rocks of the Homestake gold deposit, Lead, Lawrence Co., SD; Fiskenaesset, southwestern Greenland; from the Sutherland area, Scotland; from Isopää and Teisko, Finland; from Krivoi Rog, Ukraine; from Miyamori, Tochigi Pref., and the Hitachi mine, Ibaragi Pref., Japan; at the Mikouni R., Westland, and at Ringaringa, Stewart Is., New Zealand. *Grunerite*: found at Rockport, MA; with cummingtonite in the Homestake formation, Lawrence Co., and in the Rochford formation, Pennington Co., SD; and elsewhere in the United States; at *La Mallière, Collobrières, France* (type ?); at Nurmo and Vittinki, Finland; asbestiform material(*) in banded ironstones at Penge and Weltevreden districts, Transvaal, South Africa. Uses: Asbestiform varieties of this series have found extensive use as an insulating material to which, together with asbestiform gedrite, the terms *amosite* and *brown asbestos* have been applied. This application has been sharply curtailed in recent years due to the carcinogenic character of the fibers. AR *DHZI 2:234(1963), AM 49:963(1964)*.

66.1.1.4 Tirodite $Mn_2(Mg,Fe^{2+})_5[Si_8O_{22}](OH)_2$

66.1.1.5 Dannemorite $Mn_2(Fe^{2+},Mg)_5[Si_8O_{22}](OH)_2$

Tirodite named in 1938 by Dunn and Roy and dannemorite in 1855 by Kenngott for their respective type localities. Amphibole group, cummingtonite subgroup. Some materials described as manganoan cummingtonite may belong here. MON $C2/m$. $a = 9.606$–9.64, $b = 17.953$–18.34, $c = 5.287$–5.35, $\beta = 104.30$–$102.45°$, $Z = 2$, $D_{tir} = 3.25$, $D_{dan} = 3.64$; tirodite with $P2_1/m$ symmetry has been reported. PD(calc tir.): 9.00_9 8.28_9 3.41_6 3.24_6 3.06_8 2.96_5 2.72_{10} 2.51_6; *23-603*(tir.): 8.42_{10} 4.50_2 3.40_1 3.27_3 3.13_9 2.96_1 2.81_3 2.71_3; *38-465*(dan.): 8.36_{10} 3.45_2 3.28_4 3.09_9 2.76_4 2.63_2 2.52_2 2.20_2; cell dimensions and powder pattern spacings may be strongly affected by calcium substitution. Structure: *CM 15:309(1977)*. Crystals prismatic or acicular [001]; usually in radiating or compact fibrous aggregates; twinned on {100}. Pale green, golden yellow, to reddish brown, the color being darker for Fe-rich examples; vitreous luster, sometimes silky; transparent to translucent. Cleavage {110}, perfect; parting on {001}. $H = 6$–$6\frac{1}{2}$. $G_{tir} = 3.07$–3.13, $G_{dan} = 3.34$. Tirodite: biaxial (+); $2V = 73$–$90°$; $N_x = 1.630$–1.638, $N_y = 1.644$–1.651, $N_z = 1.652$–1.665; $X = b$, $Z \wedge c = 18$–$20°$; dannemorite: biaxial (+,−); $2V_z = 89$–$92°$; $Z \wedge c = 15$–$16°$. A complete solid solution may exist, and apparent discontinuities in properties are probably a consequence of the small number of occurrences. Manganese, which must satisfy the condition $Mn/(Mn + Mg + Fe) \geq 0.1$, is strongly ordered in the M4 site, as is any substituting calcium. Found in manganese deposits, particularly those subject to metamorphism. Tirodite has been found at Belmont, ME; Talcville, NY; Bald Knob, Alleghany Co., NC; in the Wabush iron formation, Labrador, NF,

Canada; at Bonne-val-sur-Arc, France; Chikla, Rejasthau, and at Negpur, Maharashtra, and at *Tirodi, Madhya Pradesh, India*. Calcian material from granite pegmatite at Kamiotomo, Iwaizumi, Iwate Pref., and in contact-metamorphosed Mn ore at the Kusugi mine, Yamaguchi Pref., Japan. Dannemorite is found at several localities, including *Uttersvik*, in *Söndermanland* near *Gnythyttan, Örebro*, and at *Dannemora, Uppsala, Sweden*, and at Broken Hill, NSW, Australia; from Guarultes São Paulo, Brazil. AR *DT:573(1932)*, *GSAM 122:397(1969)*, *BNSM 11:97(1985)*.

66.1.1.6 Magnesioclinoholmquistite $Li_2Mg_3Al_2[Si_8O_{22}](OH)_2$

66.1.1.7 Clinoholmquistite intermediate

66.1.1.8 Ferroclinoholmquistite $Li_2Fe_3^{2+}Al_2[Si_8O_{22}](OH)_2$

Clinoholmquistite named in 1965 by Ginsburg for its relationship to holmquistite. End members named in 1978 by Leake et al. Amphibole group, cummingtonite subgroup. MON $C2/m$. $a = 9.80$, $b = 17.83$, $c = 5.30$, $\beta = 109.1°$, $Z = 2$, $D = 3.02$. The original paper reports space group $P2/m$ with the identical constants. *25-498*(clinoholm.): 7.93_{10} 4.40_9 3.47_5 2.99_{10} 2.70_{10} 1.61_8 1.57_7 1.37_9. Li$^+$ in the B position (M4 site). Light blue, violet; translucent. Cleavage {110}, perfect. H = 5–6. G = 3.02. Biaxial (−); Y = b; 2V = 55–61°;$N_x = 1.610$, $N_y = 1.627$, $N_z = 1.655$; Z $\approx$ c, X $\wedge$ a = 15–16°. Found as prismatic crystals in pegmatite at an *unspecified locality in Siberia, Russia*. AR *AM 52:1585(1967)*.

66.1.2.1 Magnesioanthophyllite $Mg_7[Si_8O_{22}](OH)_2$

66.1.2.2 Anthophyllite $(Mg,Fe^{2+})_7[Si_8O_{22}](OH)_2$

66.1.2.3 Ferroanthophyllite $Fe^{2+}[Si_8O_{22}](OH)_2$

66.1.2.4 Sodium-anthophyllite $Na(Mg,Fe^{2+})_7[AlSi_7O_{22}](OH)_2$

Anthophyllite named in 1801 by Schumacher from the neo-Latin *anthophyllum*, clove, in allusion to its frequent clove-brown color; magnesio- and ferroanthophyllite were named in 1921 by Shannon, the latter being a hypothetical end member in current use. Amphibole group, anthophyllite subgroup. Dimorphous with corresponding members of the cummingtonite series. Protoanthophyllite is a synthetic phase with symmetry $Pnmn$. ORTH $Pnma$. (Magnesioanthophyllite): $a = 18.61$, $b = 18.01$, $c = 5.24$, $Z = 4$, $D = 2.95$; (anthophyllite): $a = 18.554$, $b = 18.026$, $c = 5.282$, $Z = 4$, $D = 3.09$. *16-401*(syn Mg-anth.): 8.33_7 4.67_2 4.49_3 3.66_1 3.23_5 3.06_{10} 2.83_2 1.500_2; *9-455*(anth.): 8.9_3 8.26_5 3.65_3 3.24_6 3.05_{10} 2.84_4 2.69_3 2.54_4; *42-455*(asbestiform): 9.37_9 9.01_5 8.25_6 4.59_8 4.50_{10} 3.23_7 2.59_3 2.54_5. Cell constants and powder pattern may be markedly affected by limited Na, Ca, and Al substitution. Iron preferentially enters sites M1 and M3; the structure appears to be destabilized when additional iron enters the M2 and M4 sites, resulting in cummingtonite. Structure: ZK 188:237(1988). **Habit:** Prismatic [001] crystals rare and usually unterminated; columnar to fibrous, compact to radiating aggregates; asbestiform. **Physical properties:** Clove-brown to dark brown, pale green, gray, white; vitreous to silky luster; transparent to nearly opaque. Cleavage {210}, perfect at 54.5°, slightly smaller than most clinoamphiboles; {010}, {100}, imperfect. H = $5\frac{1}{2}$–6 but often deceptively softer due to intergrowth with talc. G = 2.85–3.57 increasing with iron content. **Optics:** For Mg/(Mg + Fe^{2+}) = 1.0–0.60, biaxial (−,+); 2V$_z$ = 111–78°; N$_x$ = 1.587–1.654, N$_y$ = 1.602–1.660, N$_z$ = 1.613–1.667; XYZ = abc; r > v or r < v, weak; weakly pleochroic; Z = Y > X or Z > Y = X. X, colorless to yellow-brown; Z, gray-green to brown to lilac. Change from (−) to (+) at Mg/(Mg + Fe^{2+}) ≈ 0.78; 2V = 59° reported in a sample with Mg/(Mg + Fe^{2+}) = 0.62. By extrapolation N ≈ 1.73 for ferroanthophyllite. Sodium and calcium substitution appear to increase indices slightly, as does aluminum substituting tetrahedral sites. **Chemistry:** Compositional limits are given in the group description. Sodium and calcium may substitute to a limited degree in position B (M4 site). In low-aluminum varieties iron is limited to 43 mol % ferroanthophyllite, corresponding to the ratio of (M1 + M3)/(M1 + M2 + M3 + M4). Aluminum may enter both tetrahedral and octahedral sites to form solid solution with members of the gedrite series. Minor titanium and manganese may be present. **Occurrence:** Restricted to metamorphic and metasomatic environments. Formed by Mg–Fe metasomatism or by metamorphic segregation, as evidenced by concentration in crests and troughs of minor folds. Frequent associations are with cordierite–plagioclase, with cordierite–corundum–sillimanite, with hastingsite in metagabbro, and with talc in metamorphosed ultrabasic rocks. **Localities:** Worldwide distribution. *Magnesioanthophyllite* is found at Edwards, St. Lawrence Co., NY, and at Vormlitjern, Søndeled, Norway. *Anthophyllite* (including undifferentiated material) is found in Macon and Yancey Cos., NC; asbestiform at Sall Mt., GA; at Cherry Creek, Madison Co., MT; as mass-fiber asbestos at Kamiah, Idaho Co., ID; as cross-fiber asbestos in serpentinite at Coffee Creek, Carville, CA; and elsewhere in the United States; with hastingite at Hastings, ON; as long prismatic crystals at the Evans–Lou pegmatite, 20 miles N of Hull, PQ, Canada. Found at the Fiskenaesset and Sukkertoppen districts, Greenland; Kjernerud and Trondheim, Norway; Orijärvi, and at Paakkilla(*) (asbestiform), Finland; in the Outer Hebrides and at Glen Urquhart, Scotland; at Lizard, Cornwall, England; at Vedretta, Ultimo, Italy; at Miask, Ural Mts., Russia. Found with corundum and sillimanite in the anorthosite belt of the Sittampundi complex, India; at the Miyamori district, Iwate Pref., and elsewhere in Japan. **Uses:** Asbestiform varieties used as an insulating material. The asbestos industry does not generally differentiate asbestiform anthophyllites and gedrites. AR *DHZI 2:211(1963), AM 63:1023(1978), CM 21:173(1983).*

66.1.2.9 Magnesioholmquistite $Li_2Mg_3Al_2[Si_8O_{22}](OH)_2$

66.1.2.10 Holmquistite $Li_2(Mg,Fe^{2+})_3Al_2[Si_8O_{22}](OH)_2$

66.1.2.11 Ferroholmquistite $Li_2Fe_3^{2+}Al_2[Si_8O_{22}](OH)_2$

Holmquistite named in 1913 by Osann for Per Johan Holmquist (1866–1946), petrologist, Stockholm, Sweden; the end members named in 1978 by Leake. Amphibole group, anthophyllite subgroup. Polymorphous with the corresponding members of the clinoholmquistite series. ORTH $Pnma$. $a = 18.254$, $b = 17.636$, $c = 5.270$, $Z = 4$, $D = 3.06$. 13-401: 8.11_{10} 4.43_7 3.61_5 3.34_6 2.80_4 2.54_5 2.13_4 1.572_4. Structure: ZK $188{:}95(1989)$. Li^+ is ordered in the B position (M4 site). Sky-blue to deep violet, light bluish gray streak, vitreous luster, transparent to translucent. Cleavage {210}, perfect at $\approx 54°$; partings {001}, {112}, {113}. $H = 5\frac{1}{2}$. $G = 3.06$–3.13. Biaxial $(-)$; $2V = 44$–$56°$; $N_x = 1.613$–1.625, $N_y = 1.634$–1.645, $N_z = 1.646$–1.666; $XYZ = abc$; pleochroic; $Z = Y > X$ or $Z > Y = X$. X, colorless to yellow-green; Z, blue to violet. Documented analyses lie between 40 and 70 mol % ferroholmquistite end member. Al in tetrahedral sites is minor. Found as prismatic to acicular [001] crystals, as sheaflike aggregates, and fibrous masses in wall rock and contact zones of lithian pegmatites with basic or ultrabasic country rock. Found in Alexander and Yancey Cos., NC; at the Harding mine, Taos Co., NM; at Barrante, PQ, Canada. Found at Utö Is., Söndermanland, Sweden; the Kola Penin., and eastern Sayan Mts., Russia; as crystals and radiating laths(!) to 15 cm long S of Greenbushes, WA, Australia; at the Benson mine, Mtoko, Zimbabwe. AR $DHZI$ $2{:}230(1963)$, MR $9{:}83(1978)$, AM $63{:}1023(1978)$.

66.1.3a.1 Tremolite $Ca_2Mg_5[Si_8O_{22}](OH)_2$

66.1.3a.2 Actinolite $Ca_2(Mg,Fe^{2+})_5[Si_8O_{22}](OH)_2$

66.1.3a.3 Ferroactinolite $Ca_2(Fe^{2+},Mg)_5[Si_8O_{22}](OH)_2$

Tremolite named in 1790 by Hopfner for the locality; actinolite named in 1794 by Kirwan from the Greek for *ray* in allusion to the radiating prismatic habit; ferroactinolite named in 1946 by Sundius in reference to the composition. Amphibole group, tremolite subgroup. Varietal names: byssolite from the Greek for *flax*, in allusion to the fibrous habit; hexagonite in allusion to the mistaken notion of hexagonal symmetry; nephrite from the Latin for *kidney*, in reference to its alleged beneficent effect over kidney ailments; uralite for the Ural Mountains. Synonym: Ferrotremolite is equivalent to ferroactinolite. MON $C2/m$. $a = 9.84$–9.92, $b = 18.02$–18.24, $c = 5.27$–5.31, $\beta \approx 104.7°$, increasing with iron content, $Z = 4$, $D = 3.01$–3.49. 13-437(tremolite): 8.38_{10} 3.38_4 3.27_8 3.12_{10} 2.94_4 2.81_5 2.71_9 2.53_4; 41-1366(actin.): 9.04_2 8.42_8 3.39_3 3.28_5 3.12_{10} 2.94_4 2.71_6 2.53_3; 23-118(Fe-actin.): 8.58_{10} 4.57_2 3.41_1 3.16_3 2.73_4 2.61_2 2.55_2 2.03_3. Structure: CM $14{:}334(1976)$. Asbestiform varieties and some nephrites display chain-width disordering. **Habit:** Short to long prismatic [001] when in euhedral crystals with {110}, {011}, {010}, {100}, {101} {$\bar{1}$01}

66.1.2.5 Magnesiogedrite $Mg_5Al_2[Al_2Si_6O_{22}](OH)_2$

66.1.2.6 Gedrite $(Mg,Fe^{2+})_5Al_2[Al_2Si_6O_{22}](OH)_2$

66.1.2.7 Ferrogedrite $Fe_5^{2+}Al_2[Al_2Si_6O_{22}](OH)_2$

66.1.2.8 Sodium-gedrite $Na(Mg,Fe^{2+})_6Al[Al_2Si_6O_{22}](OH)_2$

Gedrite named in 1836 by Dufrenoy for the locality; ferrogedrite named in 1939 by Tilley; magnesiogedrite and sodium-gedrite named in 1978 by Leake. Amphibole group, anthophyllite subgroup. ORTH $Pnma$. (Magnesiogedrite): $a = 18.531$, $b = 17.741$, $c = 5.249$, $Z = 4$, $D = 3.18$; (gedrite): $a = 18.601$, $b = 17.839$, $c = 5.284$, $Z = 4$, $D = 3.20$; (ferrogedrite): $a = 18.51$, $b = 17.94$, $c = 5.315$, $Z = 4$, $D = 3.56$. $13\text{-}506$(gedrite): 8.97_5 8.27_8 4.48_4 3.65_4 3.35_4 3.23_7 3.06_{10} 2.50_4; $33\text{-}617$(ferroged.): 8.95_1 8.23_{10} 4.63_1 3.22_2 3.04_7 2.68_1 2.58_1 2.51_1. Structure: AM $55:1945(1970)$. Aluminum plays a significant role in the silicate chains; the resultant distortion allows for a greater iron replacement than in the anthophyllite series. Occurs rarely as prismatic [001] crystals; as lamellar to fibrous aggregates; asbestiform. Pale greenish gray to brown, vitreous to silky luster, transparent to translucent. Cleavage {210}, perfect at slightly smaller angle than most clinoamphiboles; {100}, {001} imperfect. $H = 5\frac{1}{2}\text{-}6$. $G = 3.18\text{-}3.57$. Biaxial $(+,-)$; $2V_z = 83\text{-}105°$; $N_x = 1.642\text{-}1.694$, $N_y = 1.649\text{-}1.710$, $N_z = 1.661\text{-}1.722$ for composition from 21 to 98 mol % of iron end member; $N_x = 1.613$, $N_y = 1.625$, $N_z = 1.634$ reported for magnesiogedrite; $XYZ = abc$; pleochroism absent to weak. Magnesiogedrite is rare or of doubtful occurrence in nature. Reported analyses of gedrites range from 21 mol % to nearly pure ferrogedrite. One to two aluminum atoms occupy tetrahedral sites, and Al is also present in octahedral coordination. Sodium is almost always present as $0.0 < Na < 0.5$ in the B position (site M4); calcium substitution is low; Mn and Fe^{3+} occupy less than one of the five C position sites. The term *ferrogedrite* has been used for gedrites with 50 mol % of the iron end member. Sodium-gedrite ($Na \geq 0.75$) is not known to occur in nature. Occurs in metamorphic and metasomatic environments similar to those for anthophyllite, but unlike it, is often associated with kyanite and garnet, an assemblage suggesting a higher-pressure field than for the anthophyllite–cordierite assemblage. Gedrite is of worldwide occurrence, the following being typical: Franklin, NC; Ruby Dam area, Madison Co., MT; Orofino, Clearwater Co., ID; Isopää, Finland; Sutherland, Scotland; gedrite at Héas, Gèdres, France; at Schueretsky, Karelia, Russia; at Tottiyaminttam, Madras, India; at the Wakamatsu mine, Tottori Pref., and Iratsuyama, Ehime Pref., and ferrogedrite at Yakushiyama, Iwate Pref., Japan. Asbestiform material at Penge, Transvaal, South Africa. Asbestiform gedrites are used as an insulating material. In industry, gedrite asbestos is not always differentiated from anthophyllite or from "amosite" (usually cummingtonite–grunerite). AR *DHZI 2:211(1963), AM 63:1023(1978)*.

dominant; more commonly in unterminated bladed crystals or in aggregates with {110} dominant, the blades often lying near a common plane, but also as radiating groups. Magnesium-rich members may be asbestiform and the flexible fibers may exceed 25 cm in length; asbestiform actinolite fibers often isolated and brittle (byssolite); somewhat porous, matted aggregates of short asbestiform material (mountain leather, mountain cork, etc.). Both tremolite and actinolite occur in compact, tough aggregates of short, tightly interlocking fibers with a subparallel alignment to random orientation (nephrite); with pronounced parallelism and increased fiber length may display chatoyancy (cat's-eye actinolite). Twinning, simple or lamellar, on (100) common; lamellar on (001) rare. Rarely epitactically enclosing diopsidic pyroxene with common b and c axes. **Physical properties:** Colorless, white, gray, gray-green (tremolite), green to dark green and nearly black in mass (actinolite to ferroactinolite); rarely bright green due to Cr^{3+}, pink to violet due to Mn^{2+}; vitreous luster, silky in asbestiform varieties, rarely chatoyant, dull to waxy in nephrite; transparent to translucent. Cleavage {110}, perfect; brittle to splintery fracture, fibers brittle to flexible. H = 5–6, perhaps $6\frac{1}{2}$ in nephrite which has a high toughness. G = 2.89–3.44 but mountain leather may trap air so as to float in water. **Optics:** Biaxial (−); 2V = 65–90°; N_x = 1.598–1.684, $N_y \approx$ 1.610–1.699, N_z = 1.622–1.702, the upper limits interpolated; Y = b, $Z \wedge c$ = 10–21° but asbestiform fibers may show parallel extinction; r < v, weak. Darker members pleochroic; X, pale yellow to yellow-green; Y, pale yellow-green to green; Z, pale to deep greenish blue. With increasing Fe^{2+} content the indices increase approximately linearly, while 2V, $Z \wedge c$, and birefringence decrease. For synthetic fluortremolite the indices are N_x = 1.581, N_y = 1.593, N_z = 1.602. **Chemistry:** Complete isomorphism exists within the series, but there is inconsistency in the limiting compositions for the three species. The IMA Commission boundaries (1978) are given in the group description. Isomorphism between these species and other members of the tremolite subgroup occurs at least to a limited degree, but coexistence of actinolite and "hornblende" suggests a possible immiscibility gap. Aluminum is generally constrained to no more than $\frac{1}{2}$ an atom per formula unit, as is the case for Fe^{3+}; only minor sodium is present in either the B or A position. Hydroxyl ion may be replaced by F^-, and a pure fluorine end member has been synthesized. Minor Cr^{3+}, Mn^{2+}, and Ti may be present. Ferroactinolite is relatively rare and but a few examples have been reported with >70 mol % of the iron end member. Structural water is lost between 930° and 988°. **Occurrence:** A typical product of thermal and regional metamorphism. Tremolite is an early product of thermal metamorphism of siliceous dolomites, the excess calcium appearing as calcite with the liberation of CO_2; at higher metamorphic grade the tremolite is unstable and diopside and, if silica is depleted, forsterite are formed. Actinolite may form in similar but ferruginous environments. Actinolite is characteristic of regionally metamorphosed (greenschist facies) ultrabasic rocks, where it is commonly associated with talc and serpentine. Actinolite may be intimately associated with hornblendes in metamorphic rocks, the compositions of the coexisting phases having corresponding changes in refractive index. Actinolite is also a component of some glaucophanitic regionally metamorphosed rocks. Actinolite (uralite) is also a common alteration of pyroxene in many basic rocks. Asbestiform varieties are associated with late-stage crystallization of diabase intrusives, with metamorphosed serpentinites, and in Alpine clefts. Nephrite (tremolite or actinolite jade) is commonly formed in metamorphosed ultrabasics associated with talc and serpentine, and in regionally metamorphosed terrains where dolomites have been intruded by intermediate and basic rocks. **Localities:** Tremolite and actinolite are of widespread occurrence and only a few more noteworthy localities are listed. *Tremolite* is found in dolomite at Lee, Bircher Co., MA; as white crystals and radiating aggregates at Canaan, Litchfield Co., CT. Found in NY as large, white crystals at Diana, Lewis Co.; as radiating aggregates and asbestiform at Phillipstown, Putnam Co.; at Hastings and Yonkers, Westchester Co., NY; at Rossie, at DeKalb with diopside, and manganoan tremolite (hexagonite) as pink to light violet crystalline masses at Edwards, St. Lawrence Co., NY; at Chestnut Hill, Philadelphia, and as very long, fiber masses stiffened by impregnating calcite at the Williams quarry, Easton, PA; at Sacramento Hill, Bisbee, Cochise Co., AZ; and numerous other localities in the United States. Abundant in the metamorphosed limestones of the Grenville province, eastern Canada, as at Irondale(!), Snowdon Twp., and Keller Farm, Hardwood Lake, ON, and at Calumet Falls, Litchfield Co., QUE; fluortremolite at Gatineau Park, PQ. Asbestiform and mountain leather types are found extensively in the Alps of Europe as at Bourg d'Oisans, Isère, France, at Zermatt, Valais, Switzerland, and in the Zillerthal and elsewhere in the Tyrol, Austria. Type tremolite is from *Val Tremola, St. Gotthard, Switzerland*. Large (10 cm), transparent to translucent, poorly terminated prisms(!) at Kantiwa, Noriatak, Afghanistan. Occurs as bladed crystals partly altered to talc near Chung Ju, Chung Cheon N. Prov., South Korea. Chromian material found at Dilma, eastern Sierra Leone. Found as curved crystals(!) to 20 cm in length at the Pedra Preta pit, Brumado, BA, Brazil. *Actinolite* occurs as radiating masses in the soapstone quarries at Windham, Reedsboro, and New Fane, NH; at Carlisle, Pelham, Windsor, and Lee, MA; at Mineral Hill, Delaware Co., and at Kennett, Chester Co., PA. Byssolite is abundant at the French Creek mines, St. Peters, Chester Co., PA. Found in serpentine at Bare Hills, Baltimore Co., MD; at Willis's Mt., Buckingham Co., and at the Fairfax diabase quarry (byssolite), Centreville, VA. Occurs at the New Cornelia mine, Ajo, and at Bisbee, AZ; found as 1 cm very pale green fibers (byssolite) covering areas up to 60 cm × 30 cm associated with albite, and also as parallel aggregates of fibers up to 15 cm long at New Melones Lake, Calaveras Co., CA; fine crystals(!) in talc near Wenatchee Lake, Chelan Co., WA; at Prince of Wales Island, AK. Found in the Grenville Prov. of Canada at many of the same localities as for tremolite; asbestiform at the Jeffrey quarry, Asbestos, PQ; in the Ice River Complex, Ottertail Range, BC, Canada. Occurs as dark green acicular crystals up to 15 cm long and as byssolite in dolerite at Monte Redondo, 155 km N of Lisbon, Portugal. Found in the schists of the central and eastern Alps, particularly at Greiner, in the Zillerthal, at Pfitsch, and in fibrous masses (mountain leather) with epidote, in the Knappenwald, Tyrol; at Traversella, Italy. Asbestiform actinolite is found with nephrite at Fengtien, Hualien, Taiwan; occurs in glaucophane rocks in the Kanto Mts., and with coexisting hornblende in the Nakoso district, Abukuma Plateau, Japan. *Ferroactinolite* is found in the Biwabik iron formation, Mesabi district, MN; at the Tamarack mine, ID; a phase with $(Fe^{2+} + Mn) > Mg$, but $Mg > Fe^{2+}$ occurs at the Kaso mine, Tochigi Pref., Japan. *Nephrite*, both tremolitic and actinolitic, is found worldwide in regional metamorphic terrains. Many reported occurrences of nephrite, particularly of dark green or black material, are neither texturally nor mineralogically true nephrites, despite the dominance of actinolite in their makeup. Bright green to nearly black nephrite occurs in Bull Canyon, Crooks Mt., the Granite and Rattlesnake Mts., in the area between Lander and Rawlins, WY, where it is associated with feldspars, zoisite, and quartz, the last often replaced by nephritic actinolite; locally, special names have been given to "jades" containing large amounts of the associated minerals. Light to dark green pebbles and boulders are common in Mariposa (also in situ with talc), Humboldt, Trinity, and Siskiyou Cos., CA, and light green, botryoidal surface pebbles and boulders are found in Monterey and Morro Bays, in Monterey Co.; green nephrite is found along the Kobuk R., ca. 200 km from its mouth at Kotzbue Sound, AK. In Canada nephrite has been found at Milan Arm at the northern tip of NF; extensive deposits are found in BC in the Omineca Field (Mt. Ogden), the Cassiar Field (Dease Lake), and to a lesser extent at the Lillooet Field; these nephrites are in part similar to those of the Irkutsk and Taiwan, and vary from medium to deep green, sometimes with very uniform color and texture and often enclose bright green (grossular ?) and black (magnetite or chromite) spots. Found in place at Reichenstein and Jordansmühl, Silesia, and as float at Sannthal and Murthal, Styria, and at Schwemmsal, near Leipzig, Poland. Light green nephrite(!), with abundant magnetite inclusions, is found near Irkutsk, Lake Baikal region of Siberia, and white material with brown markings has recently been reported from the same region. White, pale yellow, to pale green nephrite occurs abundantly as stream boulders and in place on the north slopes and drainage of the Kun Lun Mts., between Yarkand and Khotan, Xinjiang Prov. (Chinese Turkestan), China, where it has been mined since antiquity. Medium to dark green nephrite of variable quality is found at Fengtien, Hualien, Taiwan, where it was extensively mined in the 1960s. White to light gray-green material, often with brown markings, is found in place in metamorphosed dolomite near contact with basic or intermediate metaintrusives at several localities near Chun Cheon, about 100 km N of Seoul, South Korea, the deposits probably extending through North Korea to Liaoning Prov., China; much of this has a relict granular structure characteristic of a marble, a trait that is also seen in some of the Xinjiang nephrite. Light to medium green, streaked and mottled nephrite is found near Cowell, SA, Australia; in often deeply weathered boulders at several localities, particularly the Westland Field and near Lake Wakatipu, Otago, in New Zealand, this material having been used by the Maori peoples, and forming the basis for a distinctive, modern lapidary industry. Uses: Tremolite and, to a lesser extent, actinolite of asbestiform habit have been used as thermal insulating materials, and also as chemical filtering agents; these applications are now much curtailed due to the carcinogenic nature of the fibers. Crushed tremolite is used as a filler and whitener in construction materials. Nephrite, jade in part, has a long history of use as an ornamental material. Artifacts have been found in prehistoric sites in Switzerland and elsewhere in Europe; the carving of light-colored nephrite from Xinjiang was developed into a fine art in China and India, where objects of great delicacy or boldness have been produced, internal and surface color variation being incorporated into the design. Green nephrite, perhaps from Siberia, came late on the Chinese scene, but the abundance of material from Canada, Taiwan, New Zealand, and Wyoming has dominated present-day carving and gemstone activity; light-colored material from South Korea and Siberia is comparable in appearance to that of Xinjiang. Cabochons and small carvings have not enjoyed the popularity of the brightly colored jadeite jades, but inclusion-free, apple-green gems from Siberia or Wyoming are highly desirable. Chatoyant actinolite (nephrite) has been produced in Taiwan and to a lesser degree in British Columbia. AR *SGU 40:1(1946)*, *DHZI 2:249(1963)*, *MSARM 9A:237(1981)*, *MR 13:297(1982)*, *JADE:208,220,296(1991)*.

66.1.3a.4 (Alumino)magnesiohornblende $Ca_2(Mg,Fe^{2+})_4Al[AlSi_7O_{22}](OH)_2$

66.1.3a.5 (Alumino)ferrohornblende $Ca_2(Fe^{2+},Mg)_4Al[AlSi_7O_{22}](OH)_2$

Hornblende is from an old German term, meaning *horn* and *blind* or *to deceive*, applied to a prismatic mineral occurring in ore deposits but yielding no metal. There being essentially continuous solid solution between these species and the members of the tschermakite, edenite, and pargasite series, the name has commonly been used to encompass a broader range of composition than that of hornblende *sensu strictu*. *Oxyhornblende* refers to hornblende that is (OH)-deficient and with Fe^{3+} predominating over Fe^{2+}, but also applies to most kaersutites. Amphibole group, tremolite subgroup. MON $C2/m$. $a \approx 9.9$, $b \approx 18.1$, $c \approx 5.31$, $\beta \approx 105°$, $Z = 2$, $D = 3.02$–3.23, the cell constants increasing with iron content. *20-481*(magnesiohornblende): 8.40_{10} 4.50_1 3.39_1 3.26_2 3.10_7 2.94_1 2.79_1 2.70_2; *21-149*(ferroan Mg-hornb.): 8.51_6 3.29_3 3.14_{10} 2.82_2 2.72_3 2.61_1 2.17_2 1.656_2. Structure: *AM 74:1097(1989)*. Habit: Short to long prismatic [001] when in euhedral crystals with {110}, {011}, {010}, {100}, {101}, {$\bar{1}$01} dominant, and often displaying a pseudohexagonal cross section; more commonly in unterminated bladed crystals, usually subparallel in gneiss or schist, or in aggregates with {110} dominant; also in radiating groups, coarse-grained cleavable masses; fibrous. Simple and lamellar twinning is common on (100). Physical properties: Green, dark green, brownish green to black; pale gray-green streak; vitreous luster; translucent to opaque. Cleavage {110}, perfect; brittle. H = 5–6. G = 3.02–3.45. Optics: Biaxial $(-,+)$; $2V_x = 27$–$95°$, usually negative and large; $N_x = 1.615$–1.705, $N_y = 1.618$–1.714, $N_z = 1.632$–1.730, the indices increasing with iron content; $Y = b$, $Z \wedge c = 13$–$34°$ generally decreasing with iron increase; $r > v$ or $r < v$. Distinctly to strongly pleochroic; $Z \geq Y > X$ or $Y > Z > X$; pale yellow-green to green to blue-green or greenish brown. *Oxyhornblende* is Biaxial $(-)$; $2V = 60$–$82°$, $N_y = 1.662$–1.690, $N_y = 1.672$–1.730, $N_y = 1.680$–1.760, $r < v$, $Z \wedge c = 0$–$18°$; pleochroism strong with $Z > Y > X$, yellow to dark brown or red-brown. **Chemistry:** The relationship of hornblende composition to that of other members of the tremolite subgroup is clearly illustrated in Figures (1a,1b); there is a near continuity between tremolite, hornblende, and pargasite series, with a lesser, but significant number of examples approaching the tschermakite and edenite series species. This diversity and continuity in composition reinforces the use of *hornblende* in its broader sense because distinctions are difficult on the basis of optical properties, requiring chemical analysis for absolute species assignment. High Ca, often in excess of two atoms per formula unit, is characteristic of hornblende in the broader sense of the term; sodium is almost invariably present, occupying the B position together with calcium and part of the A position; potassium is often present in small amount. Aluminum is almost invariably present in both four- and six-coordination. Small amounts of titanium and manganese commonly occupy part of the C position. In oxyhornblende Fe^{3+} exceeds Fe^{2+} and Al in four-coordination may be near 2, placing it near the ferritschermakite range; furthermore, (OH) is well below 2 per formula unit. A slight deficiency in (OH) is common and fluorine may replace hydroxyl. There are, however, instances of excess hydrogen; these have been variously attributed to H_2O or $(H_3O)^+$ occupying part of the A position or (OH) replacing O in the chains. Among trace elements in metamorphic hornblende, Cr and Ni decrease and Ti increases with increased metamorphic grade. **Occurrence:** Hornblende may be formed over a wide range of temperature–pressure conditions. It is a common component of intermediate members of calc–alkalic igneous rocks and is common in many hybrid rocks; it generally precedes biotite in the Bowen discontinuous reaction series. In basic igneous rocks it may appear in coronas between olivine and plagioclase. Oxyhornblende occurs in a wide variety of volcanic rocks, particularly andesite, latite, basanite, and tephrite. Hornblende is common in metamorphic rocks of the greenschist to lower granulite facies, and often constitutes the bulk of hornblende schists, hornblende gneisses, and amphibolites. With increasing metamorphism the sequence tends to be actinolite–hornblende–pargasitic hornblende–pargasite, often with the morphological sequence fibrous–acicular–prismatic. Pyroxenes of igneous rocks may undergo late-stage alteration to amphiboles, in part hornblende, that are collectively called uralite. Found also in carbonatites and marble. **Localities:** Hornblende is a common, worldwide mineral, but many reported occurrences are not supported by adequate chemical data. Localities mentioned here are but a few, chosen for their representation of modes of occurrence or exceptional size of crystals. *Magnesiohornblende* is common in the hornblende gneisses and schists throughout the Piedmont metamorphic

province of eastern PA and DE, and in metagabbros of that area; in anorthositic gneiss at South Russell, Adirondack Mts., NY; in tonalite associated with the Idaho Batholith, ID; in diorite pegmatite in San Bernardino Co., CA; in the Green River Formation, UT. Large euhedral crystals(!) and groups and coarse, cleavable masses in carbonatites at the Silver Crater mine, between Bancroft and Wilberforce, and in the Lake Clear, Kuel Lake area, Renfrew Co., ONT, Canada; in recrystallized amphibolite at Lillfjalsgruven, Sweden; in fine-grained gabbro at Garabal Hill, Argyllshire, and in injected syenite hybrid at Glenelg–Ratagain, Inverness-shire, Scotland; chromian in chromatite at Karungalpatti, Madras, India; associated with corundum (ruby)-bearing rocks in the Harts Range, NT, Australia. *Ferrohornblende* occurs as a major component of amphibolite in the Yates member of the Poorman formation, Black Hills, SD; at Barkevik and in pegmatite in the Oslo region, Norway; in chlorite-rich schist at Kyykkä, Kontiolahti, Finland; in garnet–epidote amphibolites at Achahoish, Scotland; in clinopyroxene amphibolite at Yamatama, Nokoso, Fukushima Pref., and in axinite-bearing rocks at Kannonyama, Iwate Pref., Japan; in quartz monzonite–syenite at Sokuchankoge, Kogen-do, Korea. *Oxyhornblende* in prismatic crystals to several centimeters has been found at Korretsberg, near Andarnach, Lacher See volcanic field, Germany. Uses: Hornblende-rich metabasalts containing significant magnetite and plagioclase and having a fine, granular texture have been used for ornamental purposes and sometimes marketed, incorrectly, as nephrite jade.
AR *DHZI 2:263(1963)*, *MSARM 9A:85,325(1981)*, *MR 13:201,213(1982)*.

66.1.3a.6 (Alumino)tschermakite $Ca_2(Mg,Fe^{2+})_3Al_2[Al_2Si_6O_{22}](OH)_2$

66.1.3a.7 Ferro(alumino)tschermakite $Ca_2(Fe^{2+},Mg)_3Al_2[Al_2Si_6O_{22}](OH)_2$

66.1.3a.8 Ferritschermakite $Ca_2(Mg,Fe^{2+})_3Fe_2^{3+}[Al_2Si_6O_{22}](OH)_2$

66.1.3a.9 Ferroferritschermakite $Ca_2(Fe^{2+},Mg)_3Fe_2^{3+}[Al_2Si_6O_{22}](OH)_2$

Named in 1945 by Winchell for Gustav Tschermak von Sessenegg (1836–1927) Austrian mineralogist. Amphibole group, tremolite subgroup. MON $C2/m$. Aluminotschermakite: $a = 9.813$, $b = 18.055$, $c = 5.321$, $\beta = 104.97°$, $Z = 2$; ferrotschermakite: $a = 9.818$, $b = 18.106$, $c = 5.331$, $\beta = 105.0°$, $Z = 2$, $D = 3.302$. *43-665*(ferrotschermakite): 8.40_{10} 3.39_2 3.11_3 2.94_1 2.72_5 2.60_2 2.56_3 2.34_2. Structure: *MM 39:36(1973)*. Commonly occurs as rock-bound laths or prisms with {110} dominant when euhedral; massive. Twinning on {100} common. Green, dark green, brownish green to black; pale gray-green streak; vitreous luster; translucent to opaque. Cleavage {110}, perfect; parting may develop on {001} or {100}; subconchoidal fracture; brittle. $H = 5–6$. $G = 3.13–3.35$. Biaxial $(-)$; $2V = 60–90°$; $N_x = 1.642–1.671$, $N_y = 1.648–1.688$, $N_z = 1.660–1.694$; $Y = b$, $Z \wedge c = 13–28°$; indices tend to increase with iron content while 2V and extinction angle decrease; pleochroic, Z > Y > X, pale yellow to green and brownish green similar to hornblende. In composition most members of this group lie between the tschermakite–hornblende–endenite limits. Sodium is invariably present to some degree, and the aluminum content is generally less than in the aluminotschermakite end member. Low (OH) content is common. Tschermakites and tschermakitic hornblendes occur in regionally metamorphosed rocks commonly of somewhat higher grade than that where common hornblende is present; they are prevalent in aluminous rocks and may be associated with kyanite. Mg-rich tschermakites are the more common, and known ferrotschermakites are close to the Fe^{3+}–$Al^{[6]}$ boundary. Widespread in occurrence but not so much so as hornblende. Some examples, supported by chemical analyses, are: in eclogite at Hurry Inlet, eastern Greenland; in kyanite–garnet–amphibolite at Glenelg, Scotland; in "hornblende" schist at Airolo, Ticino, Switzerland; Na-rich from inclusions in basalt at Saigertshausen, northern Hesse, Germany. An Al-poor ferrotschermakite from gabbrolike amphibolite at Perniö, Finland. Ferrotschermakite in banded amphibolite at Kayal, Kohistan, Pakistan. Near midpoint composition ferrotschermakite occurs in amphibolite schist from the Amanaous R., northwestern Caucasus, Russia, and in medium-grained, gneissose quartz diorite at Furudono, Gosaisyo-Takanuki district, southern Abukuma Plateau, Japan; at Lake Grave, Broken Hill, NSW, Australia. AR *AM 30:27(1945), 34:220(1949)*; *DHZI 2:263(1963)*.

66.1.3a.10   Edenite $NaCa_2(Mg,Fe^{2+})_5[AlSi_7O_{22}](OH)_2$

66.1.3a.11   Ferroedenite $NaCa_2(Fe^{2+},Mg)_5[AlSi_7O_{22}](OH)_2$

Edenite named in 1839 by Glocker for the locality; ferroedenite named in 1946 by Sundius for the composition and relationship to edenite. Amphibole group, tremolite subgroup. MON $C2/m$. $a = 9.80$–10.00, $b = 17.95$–18.22, $c = 5.28$–5.31, $\beta = 104.4$–$105.5°$, $Z = 2$, $D = 3.06$–3.53, cell constants and density increasing with iron content. *23-1405:* 9.01_1 8.43_8 4.50_1 3.38_4 3.27_4 3.12_{10} 2.80_2 2.70_2; *31-1282*(syn): 8.47_4 3.38_4 3.28_5 3.15_{10} 2.95_4 2.82_4 2.70_6 2.59_3. Short prismatic [001] crystals with prominent {110}; massive, granular, fibrous. White, gray, pale to dark green, also brown and pale pinkish brown; vitreous luster; translucent. Cleavage {110}, perfect; parting {100} and {001}; uneven to subconchoidal fracture; brittle. H = 5–6. G = 3.0–3.28. Biaxial (−); 2V = 65–75°; $N_x = 1.634$–1.666, $N_y = 1.645$–1.678, $N_z = 1.658$–1.684; $Y = b$, $Z \wedge c = 18$–$32°$; nonpleochroic to distinctly so with X, yellow; Y, pale green Z dark green. Mn-rich has $Z \wedge c = 33°$; $N_x = 1.620$, $N_y = 1.632$, $N_z = 1.642$; $2V_{calc} = 84°$. Pleochroic with X, colorless; Y, pale orange-brown; Z, pale pinkish brown. Lighter varieties easily confused with tremolite or actinolite. Minor potassium may be present in the A site. Manganese may replace magnesium, but may also occupy as much as half of the B sites. Naturally occurring, true ferroedenites are not documented by chemical analyses. Edenite occurs in metamorphic terrains and particularly in marbles, as in the Grenville series of New York and Ontario. Noteworthy localities: *edtenite* is found as brown crystals(!) to 8 cm long at Amity, and at *Edenville, Orange Co., NY*; Mn-rich with braunite in Grenville marble at the Arnold open pit, Fowler, St. Lawrence Co., NY; fibrous material with tremolite and kaolinite at the Lowell shaft, Bisbee, Cochise Co., AZ; at Spanish Peaks, Plumas Co., CA. Found in Canada at Eganville and Wilberforce, ON, and fluorine-rich edenite at Mt. St.-Hilaire, PQ, Canada. Edenitic hornblende is associated with jadeite at Kotaki, Niigata Pref., Japan. *Ferroedenite* is found at *La Tabatière, PQ, Canada*, and in the Tibchi ring complex Nigeria. AR *DHZI 2:263(1963)*, *CM 21:81(1983)*, *23:447(1988)*, *AM 76:1431(1991)*.

66.1.3a.12   Pargasite $NaCa_2(Mg,Fe^{2+})_4Al[Al_2Si_6O_{22}](OH)_2$

66.1.3a.13   Ferropargasite $NaCa_2(Fe^{2+},Mg)_4Al[Al_2Si_6O_{22}](OH)_2$

66.1.3a.14   Magnesiohastingsite $NaCa_2(Mg,Fe^{2+})_4Fe^{3+}[Al_2Si_6O_{22}]$ $(OH)_2$

66.1.3a.15   Hastingsite $NaCa_2(Fe^{2+},Mg)_4Fe^{3+}[Al_2Si_6O_{22}](OH)_2$

Pargasite named in 1814 by Steinheil, hastingsite named in 1896 by Adams and Harrington, both for their localities; ferropargasite named in 1961 by Wilkinson, magnesiohastingsite named in 1928 by Billings. Amphibole group, tremolite subgroup. Currently defined nomenclature does not always agree with the original compositional descriptions; thus ferrohastingsite is equivalent to current hastingsite, and the name *hastingsite* has been used for intermediate compositions between the current pargasite and hastingsite. MON $C2/m$. $a = 9.805$–9.967, $b = 17.96$–18.269, $c = 5.302$–5.347, $\beta = 104.93$–$105.27°$, $Z = 2$, $D = 3.18$–3.54; the lattice constants and density increase with both ferrous and ferric iron increase. $41\text{-}1430$(syn pargasite): 8.42_5 3.36_7 3.27_7 3.13_{10} 2.92_7 2.69_7 2.54_5 2.34_5; $39\text{-}373$(potassian parg.): 8.47_6 3.38_1 3.28_3 3.14_{10} 2.95_1 2.82_4 2.39_2 2.35_1; $26\text{-}1327$(syn Fe-parg.): 8.51_0 3.40_3 3.30_2 3.15_8 2.72_6 2.61_4 2.57_4 2.36_3; $20\text{-}469$(hastingsite): 8.43_{10} 3.39_5 3.28_5 3.13_7 2.95_4 2.71_6 2.59_5 2.56_5; $20\text{-}378$(K-,Cl-hast.): 8.52_{10} 3.32_3 3.16_6 2.98_3 2.75_3 2.59_4 2.37_3 2.20_4. Structure: *EJM 3:485(1991)*. **Habit:** Usually as massive, bladed to fibrous cleavable masses. When well crystallized, the dominant forms are $\{100\}$, $\{110\}$, $\{011\}$, $\{010\}$. **Physical properties:** Light brown, brown, greenish brown to dark green and black; pale gray-green streak to brownish green in darker examples; vitreous luster. Cleavage $\{110\}$, perfect; parting on $\{100\}$, $\{001\}$; subconchoidal to splintery fracture; brittle. H = 5–6. G = 3.07–3.50. **Optics:** Biaxial $(+,-)$; $2V_x = 120$–$0°$; $N_x = 1.613$–1.705, $N_y = 1.618$–1.731, $N_z = 1.635$–1.732; $Y = b$, $Z \wedge c = 11$–$32°$. Pleochroic. X, colorless to yellow-green; Y, light brown to green to deep greenish blue or olive-green; Z, light brown to bluish green to olive-green or smoky blue-green. Optical properties for series members, supported by selected chemical analyses, are: pargasite: biaxial $(+,-)$; $2V_x = 120$–$75°$; $N_x = 1.613$–1.662, $N_y = 1.618$–1.673, $N_z = 1.635$–1.678; $Z \wedge c = 13$–$26°$; $r > v$, weak; ferroparga-

66.1.3a.15 Hastingsite

site: biaxial $(-)$; $2V = 60$–$50°$; $N_x = 1.669$–1.697, $N_y = 1.686$–1.713, $N_z = 1.691$–1.723; $Z \wedge c = 14$–$25°$; magnesiohastingsite: biaxial $(-)$; $2V = 78$–$68°$; $N_x = 1.652$–1.675, $N_y = 1.664$–1.688, $N_z = 1.675$–1.694; $Z \wedge \approx 30°$; hastingsite: biaxial $(-)$; $2V = 55$–$0°$; $N_x = 1.690$–1.702, $N_y = 1.702$–1.713, $N_z = 1.705$–1.730; $Z \wedge c = 11$–$15°$. Most pargasites are distinguishable from most other amphiboles, except cummingtonite and some ortho-amphiboles, in that they are biaxial $(+)$, but Fe-rich pargasites are biaxial $(-)$. **Chemistry:** Naturally occurring examples cover almost the entire compositional spectrum of the series. Mn rarely exceeds 0.5 atom per formula unit; and Ti rarely over 0.2 atom per formula unit; Pb may be present in significant amounts in the B position; K is almost always present in the A position, and K-dominant pargasite has been reported but not submitted to the IMA names committee; F is almost always present, and in type pargasite exceeds the (OH) content. **Occurrence:** Members of this series are found worldwide in a diversity of metamorphic environments and igneous rocks. **Localities:** Localities are given only for well-crystallized specimens or chemically documented examples of the varied types of occurrence. *Pargasite* has been found in hornblende–gabbro at Burlington, PA; sharp, well-terminated, brown crystals to 8 cm(!) in calcite in massive pargasite at the Jensen quarry, Riverside Co., CA; lustrous black crystals have been found near Wolfe Lake, Bedford Twp., ON, Canada; in metamorphosed limestone in the *Pargas valley, Finland*; in amphibole-rich inclusion in the Tiree marble, Tiree, Argyllshire, Scotland; in eclogite from Sau Alpe, Carinthia, Austria; with tridymite and cristobalite in druses in augite–hypersthene dacite at Ishikani-yama, Kamamoto Pref., Japan; and in an amphibole-rich layer in peridotite from Tinaquillo, Venezuela. Potassium-dominant pargasite has been reported from Einstödingen, Lützow–Holm Bay, Antarctica. *Ferropargasite* occurs in diorite porphyry in the Mt. Ellen complex, Henry Mts., Garfield Co., UT; in sodalite-syenite from Square Butte, Highwood Mts., MT; in granite from Rubideaux Mt., near Riverside, CA; and in biotite–epidote–hornblende schist at Mituisi(?), Japan. *Magnesiohastingsite* has been found in a xenolith in diorite porphyry in the Mt. Hilliers complex, Henry Mts., Garfield Co., UT; both plumbian and normal in skarns at Långban, Sweden. *Hastingsite* has been noted from hornblende granite gneiss in the Stark complex, Degrasse, Adirondack Mts., NY; in a vug in monzonite at Beaver Creek, MT (near magnesiohastingsite); in schist from Bidwell Bar; Clipper Hills quadrangle, CA; in metagabbro in *Dungannon Twp., Hastings Co., ON, Canada*, but some analyzed material from the area falls in the pargasite field; in granite at Tysfjord, Norway; in nepheline syenite from Almunge, Sweden; in rapakivi granite at Uuksonjoki, Salmi, Finland (near end member); in granite gneiss from the Goubensky massif, southern Ural Mts., Russia; and with hedenbergite, andradite, and vesuvianite at the Obira mine, Oita Pref., Kyushu, Japan. AR *DHZI 2:264(1963); AJS 264:698(1966); MM 41:43(1977), 49:703(1985); MR 13:73(1982), 15:279(1984); AM 74:1097(1989).*

66.1.3a.16 Magnesiosadanagaite $(K,Na)Ca_2(Mg,Fe^{2+},Al,Ti)_5$ $[(Si,Al)_8O_{22}](OH)_2$

66.1.3a.17 Sadanagaite $(K,Na)Ca_2(Fe^{2+},Mg,Al,Ti)_5 [(Si,Al)_8 O_{22}](OH)_2$

Named (both) in 1984 by Shimazaki et al. for Ryoichi Sadanaga (b.1920), professor emeritus, University of Tokyo. Amphibole group, tremolite subgroup. MON $C2/m$ ($C2$, Cm). $a = 9.922–10.00$, $b = 18.01–18.06$, $c = 5.352–5.355$, $\beta = 105.30–105.55°$, $Z = 2$, $D = 3.27–3.30$. *31-359*(Mg-sad.): 7.48_{10} 3.39_4 3.28_{10} 3.15_7 2.95_5 2.77_4 2.71_6 2.16_5. Dark brown to black, very light brown streak, vitreous luster, transparent in thin section. Cleavage {110}, perfect. H ≈ 6. Biaxial (+); $2V_{obs} = 80–90°$; $N_x = 1.673–1.674$, $N_y = 1.684–1.686$, $N_z = 1.697–1.699$; $Y = b$, $Z \wedge c = 26–28°$. Pleochroic. X, colorless to pale brown; Z, brownish yellow to greenish brown. (It is noted that the higher reported indices are stated to be for magnesiosadanagaite, the Mg-richer sample.) Markedly silica deficient; the original five published analyses are all close to the midpoint of the series with $Mg/Fe^{2+} = 0.76–1.34$. *Magnesiosadanagaite* in lenses with titanian diopside (fassaite), and a similar mineral suite at *Myojin Island, Ehime Pref., Japan*. Sadanagaite found as grains in banded skarns and lenses in recrystallized limestones, associated with vesuvianite, spinel-hercynite, sphene, ilmenite, titanian diopside, perovskite, and apatite at *Yuge Island*, and in the Nōgō Hakusan area, Fukui Pref., Japan. AR *AM 69:465(1984)*, *MM 53:99(1989)*.

66.1.3a.18 Kaersutite $NaCa_2(Mg,Fe^{2+})_4Ti[Al_2Si_6O_{22}](OH)_2$

66.1.3a.19 Ferrokaersutite $NaCa_2(Fe^{2+},Mg)_4Ti[Al_2Si_6O_{22}](OH)_2$

Named in 1884 by Lorenzen for the locality. Amphibole group, tremolite subgroup. Kaersutites are usually (OH) deficient and so may be regarded as "oxyhornblendes." MON $C2/m$. $a = 9.89$, $b = 19.06$, $c = 5.315$, $\beta = 105.4°$, $Z = 2$, $D = 2.94$. *17-478*: 8.38_6 3.36_6 3.26_5 3.11_8 2.93_5 2.69_{10} 2.59_5 2.55_7. Structure: *NJMM:137(1989)*. Crystals are usually short prismatic with {110} dominant; usually as embedded euhedral to anhedral grains. Twinning on (100) common. Dark brown to black, pale brownish gray streak, vitreous to slightly resinous luster, translucent to nearly opaque. Cleavage {110}, perfect; parting on {100} and {001}; uneven to subconchoidal fracture. H = 5–6. G = 3.2–3.3. Biaxial (−); 2V = 66–82°; $N_x = 1.670–1.689$, $N_y = 1.690–1.741$, $N_z = 1.700–1.772$; $Y = b$, $Z \wedge c = 0–19°$; $r > v$; pleochroic, $Z \geq Y > X$; pale yellow to dark reddish brown. The refractive indices and pleochroism are similar to low Ti, (OH)-deficient hornblendes. Natural occurrences of ferrokaersutites have not been documented. Titanium is an essential component, usually 0.4–1.1 atoms per formula unit; Fe^{3+} is usually low and rarely exceeds 1.0 atom per formula unit; potassium may be present in the A position; fluorine may replace hydroxyl, which is often less than 1 per formula unit. Occurs predominantly in volcanic rocks and is an abundant component of camptonites. *Kaerutite* is found as phenocrysts to 10 cm in diameter in camptonite dikes near Boulder Dam, and in alkali–basalt on the San Carlos Reservation, AZ; in the Ice River Complex, Ottertail Range, BC, Canada; at *Qaersut, Umanq district, northern Greenland*; in hornblende monchiquite at Wart Holm, Copinshay, Orkney Is.; as megacrysts in rocks of the Shama volcanic field, northeastern Jordan: in monzonite at Tyaki, Morotu, Sakhalin, Russia; in a basaltic dike at Tikashi, Dōgo, Oki Is., and numerous other volcanics of Japan; as crystals at Yohodo (?), Korea; in trachybasalt at Dunedin district, New Zealand. *Ferrokaersutite* is reported from Koraput, Orisse, India. AR *DHZI 2:321(1963), MM 39:390(1973)*.

66.1.3b.1 (Alumino)winchite $CaNa(Mg,Fe^{2+})_4Al[Si_8O_{22}](OH)_2$

66.1.3b.2 Ferro(alumino)winchite $CaNa(Fe^{2+},Mg)_4Al[Si_8O_{22}](OH)_2$

66.1.3b.3 Ferriwinchite $CaNa(Mg,Fe^{2+})_4Fe^{3+}[Si_8O_{22}](OH)_2$

66.1.3b.4 Ferroferriwinchite $CaNa(Fe^{2+},Mg)_4Fe^{3+}[Si_8O_{22}](OH)_2$

Named in 1906 by Fermor for Howard J. Winch, Geological Survey of India, its discoverer. Amphibole group, richterite subgroup. MON $C2/m$. $a = 9.834$, $b = 18.062$, $c = 5.300$, $\beta = 104.4°$, $Z = 2$, $D = 2.96$. *20-1390*: 8.40_9 4.48_7 3.40_7 3.12_5 2.98_4 2.70_{10} 2.58_3 2.53_9; *35-460*(potassian): 8.46_4 3.28_8 3.16_{10} 2.96_4 2.34_4 1.916_5 1.660_4 1.448_5. Cobalt blue to blue-violet, pale blue-gray streak, vitreous luster, translucent. Cleavage {110}, perfect. $H = 5\frac{1}{2}$. $G = 2.97$. Biaxial (+); $2V = 82°$; $N_x = 1.636$, $N_y = 1.646$, $N_z = 1.658$ (ferriwinchite); also biaxial (−); $2V = 77°$; $N_x = 1.612$, $N_y = 1.623$, $N_z = 1.627$ (richterite/ferriwinchite); $Y = b$, $Z \wedge c = 18-20°$; pleochroism varied with $Y > Z > X$ or, less commonly, $Z > Y > X$. Considerably lower indices, $N_x = 1.576-1.596$ and $N_z = 1.590-1.600$, have been reported for fluorine- and potassium-rich asbestiform material from Texas. Published analyses show $Fe^{3+} > Al^{[6]}$, suggesting that ferririchterite may be the more common member of the series. Sodium is often relatively low within the defined range of the richterite subgroup. Manganese is commonly present up to 0.5 atom per formula unit. Later analyses of material from the type locality yield divers results, in part ferriwinchite and in part magnesioarfvedsonite; possibly both coexist and thus suggest a miscibility gap. Occurs as prismatic crystals, blades, radiating aggregates, fibrous, and asbestiform. Asbestos from the Diablo prospect, Allammoore talc district, TX, lies in the K-richterite to K-winchite range; at ward creek, Cazadero, Sonoma Co., CA. Light blue, low-sodium richterite from Långban, Sweden, is similar to winchite–ferriwinchite. Originally found in manganese deposits at *Kajlidongri, Jhabua, Madhya Pradesh, India*, where amphiboles in the ferriwinchite–magnesioarfvedsonite–magnesioriebeckite appear to coexist; in granite gneiss and pegmatite (near ferribarroisite) at Chakli, Bhandara, Maharashtra, India, probably formed through incorporation of Mn from local ore deposits. A "soda tremolite" found in altered pyroxenite at Libby, MT, is intermediate between richterite and ferriwinchite. AR *MM*

14:413(1907), 40:395(1975), 50:173(1986); DHZI 2:352(1963); CM 18:101(1980).

66.1.3b.5 (Alumino)barroisite   CaNa(Mg,Fe^{2+})$_3$Al$_2$[AlSi$_7$O$_{22}$](OH)$_2$

66.1.3b.6 Ferro(alumino)barroisite   CaNa(Fe^{2+},Mg)$_3$Al$_2$[AlSi$_7$O$_{22}$](OH)$_2$

66.1.3b.7 Ferribarroisite   CaNa(Mg,Fe^{2+})$_3$Fe$_2^{3+}$[AlSi$_6$O$_{22}$](OH)$_2$

66.1.3b.8 Ferroferribarroisite   CaNa(Fe^{2+},Mg)$_3$Fe$_2^{3+}$[AlSi$_7$O$_{22}$](OH)$_2$

Named in 1922 by Murgoci with no mention of etymology, and described as intermediate between hornblende and glaucophane. Amphibole group, richterite subgroup. Allegedly synonymous with carinthine, from the locality, a term introduced by Werner in 1817. MON $C2/m$. $a = 9.90$, $b = 18.04$, $c = 5.33$, $\beta = 106.7°$, $Z = 2$, $D = 3.16$. Structure: *TMPM 6:215(1957)*. Blue-green, green; translucent. $G = 3.21$. Biaxial $(-)$; 2V variable but small; also uniaxial $(-)$; $Z \wedge c = 12\text{--}15°$. Pleochroic; c, blue-green; b, blue to violet; a, gray-green (*Sic*). Natural occurrences falling within the current boundaries of this series are not supported by adequate chemical or physical data. *Barroisite* reported from Riffel Alp, Zermatt, Switzerland; uniaxial, from the Taunus Mts., Germany; carinthine from Sau Alpe, Carinthia, Austria. *Ferribarroisite* found at Condrey Mt., Klamath Mtns, Del Norte Co., CA; at Cauro-Bastelica, Corsica; and in the Iskou ring Complex, Aïr, Nigeria, close to ferribarroisite. AR CR *175:373,426(1922)*.

66.1.3b.9 Richterite   Na(CaNa)(Mg,Fe^{2+})$_5$[Si$_8$O$_{22}$](OH)$_2$

66.1.3b.10 Ferrorichterite   Na(CaNa)(Fe^{2+},Mg)$_5$[Si$_8$O$_{22}$](OH)$_2$

Richterite named in 1865 by Breithaupt for Theodor Richter (1824–1898), German mineralogist. The name *ferrorichterite* was coined in 1946 (Sundius). Amphibole group, richterite subgroup. MON $C2/m$. Richterite: $a = 9.902$, $b = 17.980$, $c = 5.269$, $\beta = 104.22°$, $Z = 2$, $D = 2.87$; *25-808*(syn, calc): 8.47$_5$ 3.39$_7$ 3.28$_6$ 3.15$_6$ 2.96$_5$ 2.71$_{10}$ 2.58$_4$ 2.53$_7$; *31-1082*(potassian): 8.56$_8$ 3.39$_5$ 3.30$_8$ 3.18$_{10}$ 2.85$_7$ 2.59$_4$ 1.927$_5$ 1.663$_3$; ferrorichterite: $a = 9.982$, $b = 18.223$, $c = 5.298$, $\beta = 103.73°$, $Z = 2$, $D = 3.46$; *26-1373*(syn): 8.58$_{10}$ 3.43$_4$ 3.32$_3$ 3.18$_7$ 3.00$_3$ 2.74$_7$ 2.62$_4$ 2.54$_5$. Structure: *EJM 4:425(1992)*. In ferrian richterites there are apparent vacancies in the C position, total cations being as low as 4 rather than 5. Prismatic [001], often flattened {100} with dominant {110}, {100}, when terminated {011} and {$\bar{1}$01} are prominent; commonly in cleavable prismatic to fibrous aggregates, or as rock-bound prisms. Twinning on (100). Brown, yellow, brownish red, rose-red, pale to dark green, gray-brown; vitreous luster; transparent to translucent. Cleavage {110}, perfect; parting on {100} and {001}; uneven fracture; brittle. H = 5–6. G = 2.97–3.45. Biaxial $(-)$; 2V = 66–87°; for Mg-rich richterites $N_x = 1.605\text{--}1.624$,

$N_y = 1.615$–1.631, $N_z = 1.622$–1.641; $Y = b$, $Z \wedge c = 15$–$40°$; $r < v$; pleochroic; variable absorption, usually $Z > Y > X$. X, colorless, yellow to pale green or pinkish lilac; Y, pale yellow, orange, reddish, blue, pale lilac or green; Z, colorless, yellow-green, blue-green, to violet-brown. The deeper colors appear with increasing iron; for ferrian richterite indices reach $N_x = 1.685$, $N_y = 1.700$, $N_z = 1.712$, and absorption $Z > Y > X$. Magnesium-rich richterites are common, but natural occurrences of ferrorichterite are not documented, although it has been synthesized. Manganese is often present, up to 1 atom per formula unit. Fluorine commonly replaces (OH), and the total of these is often significantly less than 2. In ferrian richterites the substitution of $Fe_2^{3+}\square$ for Mg_3 appears to take place. Many richterites are intermediate toward ferriwinchite, ferribarroisite, or magnesioarfvedsonite. Magnesium-rich richterites have a paragenesis similar to that of tremolite and pargasite, and occur in thermally metamorphosed limestones; also found in skarns associated with iron-manganese deposits; in carbonate veins in pyroxenites; potassium-rich richterites are found in leucite rocks, often displaying a poikilitic texture enclosing diopside and leucite. *Richterite*, intermediate with magnesioarfvedsonite, is a typical amphibole of the alkaline complex at Iron Hill, Gunnison Co., CO; in ordenite of the Leucite Hills, WY; in altered pyroxenite at Libby, MT. Fine fluorine-rich, gray-brown to nearly black crystals(!) in calcite at the Earle Farm, Wilberforce, and at Tory Hill, ON, and as small, transparent, light green crystals in breccia vugs at Mt. St.-Hilaire, PQ, Canada. In the manganese ores of *Långban and Pajsberg, Värmland, Sweden*; potassian at Murcia, Spain; in jadeite rock at Sanka, Myanmar; ferrian richterite occurs in pegmatite at Chakli, Bhandara, Maharashtra, India; potassian richterite in leucite lamproite in the West Kimberley area, WA, Australia; at Imeria, Madagascar. *Ferrorichterite* found at La Tabatière, PQ, Canada. AR *DHZI 2:352(1963); AM 59:518(1974); MR 13:73(1982), 21:383(1990)*.

66.1.3b.11 Magnesio(alumino)katophorite Na(CaNa)(Mg,Fe^{2+})$_4$Al[AlSi$_7$O$_{22}$](OH)$_2$

66.1.3b.12 (Alumino)katophorite Na(CaNa)(Fe^{2+},Mg)$_4$Al[AlSi$_7$O$_{22}$](OH)$_2$

66.1.3b.13 Magnesioferrikatophorite Na(CaNa)Mg,Fe^{2+})$_4$Fe^{3+}[AlSi$_7$O$_{22}$](OH)$_2$

66.1.3b.14 Ferrikatophorite Na(CaNa)(Fe^{2+},Mg)$_4$Fe^{3+}[AlSi$_7$O$_{22}$](OH)$_2$

Named in 1894 by Brøgger from the Greek for *carrying down*, in allusion to its volcanic origin. Amphibole group, richterite subgroup. Originally cataphorite, and defined as intermediate between ferrohornblende and arfvedsonite. MON $C2/m$. $Z = 2$. Cell constants and powder X-ray diffraction data not available for chemically verified members of the series. Rose-red, dark red-brown, to bluish back; pale tan to gray streak; translucent. Cleavage {110}, perfect; {010}, parting. H = 5. G = 3.2–3.5. Biaxial (−); 2V = 0–50°; $N_x = 1.640$–1.681, $N_y = 1.658$–1.688, $N_z = 1.660$–1.692; $X \wedge c = 30$–$70°$; r < v strong for magnesioferrikatophorite with Y = b, r > v strong for ferrikatophorite with Z = b, but the exact composition at which the orientation changes is uncertain. Strongly pleochroic; Z < Y > X, X, yellow to light brown; Y, greenish brown to deep brown; Z, orange to greenish brown for magnesioferrikatophorite; pleochroism strong with Z > Y > X in the range pale yellow through blue-green to nearly black for ferrikatophorite. The optical properties are similar to those of the eckermannite–arfvedsonite series, but extinction angles are generally larger; pleochroism in brown tones may be confused with ferrohornblende, oxyhornblende, and kaersutite. Natural occurrences of aluminokatophorites have not been observed. Katophorites are found in more basic alkaline rocks; in metamorphic environments arfvedsonite–riekbeckite are favored over katophorites. *Magnesiokatophorite* found in theralite in the Shields River basin, MT where it forms rims on aegirine and is the last ferromagnesian mineral to crystallize; at Khibiny, Kola Peninsula, Russia; at Imeria, Madagascar. *Katophorite* is found at La Tabatiere, PQ, Canada; in the *Christiania region, Norway*; the Rallier-du-Baty Peninsula, Kergualen Is., Indian Ocean; and reported from Tawmaw, Kachin, Myanmar. *Ferrikatophorite* (actually nearer ferritaramite in composition) at Mariupol, Ukraine; in many phonolites and some trachytes of the Rungwe volcanics, Lake Nyasa, Tanzania; it is the most abundant ferromagnesian mineral in katophorite trachyte of Lake Naivasha, Kenya. AR *D6 Appl*:14(1914), *DHZI* 2:359(1963).

66.1.3b.15 Magnesio(alumino)taramite $Na(CaNa)(Mg,Fe^{2+})_3Al_2[Al_2Si_6O_{22}](OH)_2$

66.1.3b.16 (Alumino)taramite $Na(CaNa)(Fe^{2+},Mg)_4Al_2[Al_2Si_6O_{22}](OH)_2$

66.1.3b.17 Magnesioferritaramite $Na(CaNa)(Mg,Fe^{2+})_3Fe_2^{3+}[Al_2Si_6O_{22}](OH)_2$

66.1.3b.18 Ferritaramite $Na(CaNa)(Fe^{2+},Mg)_3Fe_2^{3+}[Al_2Si_6O_{22}](OH)_2$

Named in 1923 by Morozewicz for the locality. Amphibole group, richterite subgroup. MON $C2/m$. $a = 9.923$, $b = 18.134$, $c = 5.352$, $\beta = 104.8°$, Z = 2, D = 3.47 (ferritaramite). *20-734*(potassian): 8.53_7 3.29_3 3.15_{10} 2.73_5 2.61_3 2.35_3 2.18_3 1.447_3. Structure: *CM* 16:53(1978). Blue-green, translucent. Cleavage {110}, perfect; subconchoidal fracture. H = 5–6. Biaxial (−); $Y = b$, $Z \wedge c = 11°$; $N_x = 1.684$–1.705, $N_y = 1.700$–1.713, $N_z = 1.703$–1.715; 2V = 46–54°; pleochroic. X, yellow to greenish yellow; Y, violet-gray to blue-green; Z, blue. Indices are for potassian taramite; type material is actually ferritaramite and close to ferrikatophorite; small amounts of Mn may be present; fluorine replacement of (OH) may be significant. Occurs as rockbound, subhedral to euhedral crystals in nepheline syenite. *Magnesiotaramite* has been found in the Nybö eclogite pod at Nordfjord, Norway. *Taramite* (near ferritaramite) in nepheline syenite at *Wali-tarama, Mariupol, Ukraine*; at the Darkainle nepheline-syenite complex, Borama district, Somalia; *potassian ferritaramite* in foyaite dike cutting nepheline-bearing gneiss at Mbozii, Tanzania, and the York Ruver area, Bancroft, ON, Canada. AR *MM 33:1057(1964)*.

66.1.3c.1 Glaucophane $Na_2Mg_3Al_2[Si_8O_{22}](OH)_2$

66.1.3c.2 Ferroglaucophane $Na_2Fe_3^{2+}Al_2[Si_8O_{22}](OH)_2$

66.1.3c.3 Crossite $Na_2(Mg,Fe^{2+})_3(Al,Fe^{3+})_2[Si_8O_{22}](OH)_2$

66.1.3c.4 Magnesioriebeckite $Na_2Mg_3Fe_2^{3+}[Si_8O_{22}](OH)_2$

66.1.3c.5 Riebeckite $Na_2Fe_3^{2+}Fe_2^{3+}[Si_8O_{22}](OH)_2$

Glaucophane named in 1845 by Hausmann from the Greek for *blue* and *to appear*, in allusion to its color; ferroglaucophane and magnesioriebeckite introduced in 1957 by Miyashiro; crossite named in 1894 by Palache for Charles Whitman Cross (1854–1949), geologist, U.S. Geological Survey; riebeckite named in 1888 by Sauer for Emil Riebeck (1853–1885), German explorer. Amphibole group, glaucophane subgroup. Possible polymorphism of glaucophane has been suggested. MON $C2/m$. $Z = 2$. Glaucophane: $a = 9.531$, $b = 17.759$, $c = 5.303$, $\beta = 103.59°$, $D = 3.13$; *20-453*: 8.26_{10} 4.45_3 3.38_3 3.22_2 3.06_7 2.94_3 2.69_6 2.52_3; ferroglaucophane: $a = 9.587$, $b = 17.832$, $c = 5.315$, $\beta = 103.47°$, $D = 3.224$; *31-1307*: 8.31_{10} 4.47_2 3.23_2 3.06_8 2.95_1 2.76_3 2.70_3 2.53_1; crossite: $a = 9.647$, $b = 17.905$, $c = 5.316$, $\beta = 103.6°$, $D = 3.22$; *20-376*: 8.95_1 8.31_{10} 4.87_2 4.48_2 3.40_2 3.08_7 2.71_4 2.53_2; magnesioriebeckite: $a = 9.700$, $b = 17.944$, $c = 5.273$, $\beta = 103.53°$, $D = 3.13$; *20-656*: 8.45_{10} 4.50_2 3.27_3 3.12_9 2.89_6 2.71_3 2.54_1 1.655_1; riebeckite: $a = 9.811$, $b = 18.013$, $c = 5.326$, $\beta = 103.68°$, $D = 3.37$; *19-1061*: 8.40_{10} 4.51_2 3.42_1 3.27_1 3.12_6 2.80_2 2.73_4 2.18_2. Structure: *EJM 3:485(1991)*. **Habit:** Crystals prismatic [001] to lathlike, often striated parallel to elongation; columnar, fibrous, granular, or massive. Simple and lamellar twinning on (100) common. Magnesioriebeckite and riebeckite may be asbestiform (crocidolite). **Physical properties:** Gray to lavender-blue (glaucophanes); light blue to blue-black (riebeckites); pale gray to bluish gray streak; vitreous luster, silky in asbestiform varieties; translucent to opaque. Cleavage {110}, good to perfect. H = 5–6 riebeckite possibly being the softer. G = 3.02–3.42. **Optics:** Glaucophane and crossite are Biaxial (−); 2V = 0–50° decreasing with iron content; $N_x = 1.606$–1.661, $N_y = 1.622$–1.667, $N_z = 1.627$–1.670, birefringence = 0.008–0.022; $Y = b$, $Z \wedge c = 4$–$14°$ (OAP parallel to (010)); r < v but also r > v in some crossite. Pleochroic; Z > Y > X, colorless to lavender-blue to blue, distinct and becoming more so with increasing iron content. Magnesioriebeckite and riebeckite are Biaxial (−); 2V = 40–90° increasing with iron content, less commonly Biaxial (+) with 2V large; $N_x = 1.654–1.701$, $N_y = 1.662–1.711$, $N_z = 1.668–1.717$; birefringence = 0.006–0.016; Z = b, X ∧ c = 3–21° (OAP perpendicular to (010)); r > v or r < v strong with marked dispersion of the bisectrices leading frequently to anomalous interference colors. Pleochroism strong; Z < Y ≤ X, X prussian blue to indigo and Z blue-grey to yellowish green. 1.606Z ∧ c = 4–14° [OAP parallel to (010)]; $N_x = 1.606–1.661$, $N_y = 1.622–1.667$, $N_z = 1.627–1.670$; 2V = 0–50° decreasing with iron content; birefringence = 0.008–0.022; r < v but also r > v in some crossite; Z > Y > X; pleochroic from colorless to lavender-blue to blue, distinct and becoming more so with increasing iron content. Magnesioriebeckite and riebeckite are biaxial (−); 2V = 40–90°, increasing with iron content, less commonly biaxial (+) with 2V large, Z = b, X ∧ c = 3–21° [OAP perpendicular to (010)]; $N_x = 1.654–1.701$, $N_y = 1.662–1.711$, $N_z = 1.668–1.717$, birefringence = 0.006–0.016, r > v or r < v, strong with marked dispersion of the bisectrices leading frequently to anomalous interference colors; Z < Y ≤ X; strongly pleochroic. X, prussian-blue to indigo; Z, blue-gray to yellowish green. The accompanying figure gives the refractive indices for this series. Asbestiform varieties usually have parallel extinction and are length-fast, distinct from other asbestiform amphiboles. **Chemistry:** Near-end-member glaucophanes do not appear to occur naturally, Fe^{2+} always being present in significant amounts, and Fe^{3+} substitution for Al generally paralleling Fe^{2+} substitution for Mg; ferroglaucophanes are all compositionally close to the glaucophane boundary. Increasing $Fe^{2+} + Fe^{3+}$ substitution leads to crossites, but few pass the midpoint between glaucophane and riebeckite. The entire range of magnesioriebeckite–riebeckite compositions is represented in

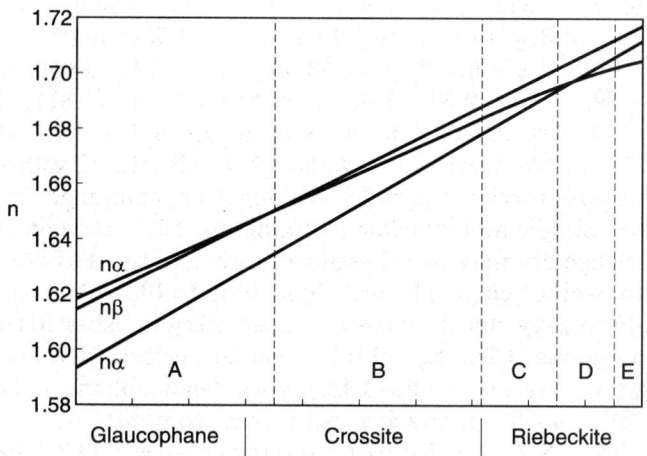

Figure A-4

Approximate refractive indices and optical orientation for the glaucophane-riebeckite series; after CPM 15:67(1967).

nature. Published analyses and the seemingly abrupt change in optical properties suggest a possible discontinuity between crossite and the riebeckite minerals, but this is not borne out in all occurrences. In some riebeckites the sum of B and C site occupancies is greater than 7; in others it is significantly less. Titanium is usually present as no more than 0.2 atom per formula unit, as is the case for manganese. Fluorine substitution is variable and the total of F and (OH) may be significantly less than 2 per formula unit. Asbestiform riebeckite, in part oxidized, may be partially or completely replaced by quartz with retention of the parallel fiber structure as bands of iron-rich impurities. **Occurrence:** Glaucophane and crossite occur only under the conditions of the glaucophane facies, the schists in which they occur presumably having been formed by metamorphism of sodium-rich rocks or by metasomatic addition of sodium. Commonly associated minerals in glaucophane assemblages are lawsonite, epidote, chlorite, stilpnomelane, jadeite, and almandine. In the Franciscan series of California, glaucophanic rocks are formed by retrograde metamorphism of eclogites as well as by recrystallization of sediments. Riebeckite is often of igneous origin, appearing in granites (riebeckite granites of northern Nigeria), syenites, and to a lesser extent in nepheline syenites; also in acid volcanics, particularly Na-rich rhyolites, where it may be associated with arfvedsonite and aegirine or aegerine–augite. Magnesioriebeckite forms as an authigenic mineral in the Green River formation of the western United States, where it occurs around detrital grains of brown hornblende and is associated with aegirine and shortite; and also in a low-temperature environment with opal in Permian clays of Kazakhstan. Asbestiform riebeckite (crocidolite) forms from ironstone formations with little or no addition of material under conditions of moderate temperature and pressure. **Localities:** Widespread throughout the world; only a sampling of well-documented localities follows. *Glaucophane* has been found at Iron Hill, Gunnison Co., CO; is abundant in the schists of the Coast Ranges of CA, particularly as long prisms at Laytonville; the North Cascades, WA; at Zermatt, Switzerland; at *Syra Island, Cyclades, Greece*. Found at Akuwara, Kanto Mts., Honshu; in the Kanto district, Tokushima Pref.; Horokanai, Ishikari Pref., Hokkaido; and elsewhere in Japan. Crossite occurs at *Berkeley, Alameda Co., CA*, and other areas of the Franciscan formation; also at the Dallas Gem Mine, San Benito Co., CA; at The Monument, Anglesey, Wales, and at Knockormal, Ayrshire, Scotland; in Japan at various localities as for glaucophane. *Magnesioriebeckite* occurs in the Green River formation of WY, CO, and UT; in granulite at Glen Lui, Aberdeenshire, Scotland; and at Huppu, Kanto Mts., Honshu, Japan; in *Japan*, but the exact area not specified. *Riebeckite* occurs near Quincy, Norfolk Co., MA; at Cumberland Hill, Providence, RI; St. Peters Dome, El Paso Co., CO; in crystals(!) in miarolitic cavities of the Golden Horn batholith, Okanogan Co., WA; at Narssarssuk, Greenland; in riebeckite felsite at Shetland, Scotland; and in riebeckite–aegirine granite in the Kigon Hills, Nigeria; on *Socotra Is. (Yemen), Indian Ocean*. Crocidolite (asbestiform riebeckite) occurs in Henan and other provinces of China; in dolerite and dolomite as long fibers (magnesioriebeckite) at Robertstown, SA, and in the Hammersley Range,

WA, Australia; in the Cape Asbestos Field, Griqualand West, South Africa(*), where it has been produced commercially on a large scale, and where occur also its silica replacements, cat's eye (blue) and tiger eye (golden brown); and at Cochabamba, Bolivia. Uses: Asbestiform varieties, principally riebeckite, marketed under the trade name Blue Asbestos, have been used extensively as a thermal insulation, although the application is diminishing due to the highly carcinogenic character of the fibers. Silica-replaced materials from South Africa have been popular lapidary materials, their natural colors grayish blue and golden brown; dyed or bleached material in straw-yellow, green, red, and other colors is readily available. AR *DHZI 2:333(1963)*, *CMP 15:67(1967)*, *CM 16:187(1978)*, *MSARM 9A:283(1981)*.

66.1.3c.6   Eckermannite $Na_3Mg_4Al[Si_8O_{22}](OH)_2$

66.1.3c.7   Ferroeckermannite $Na_3Fe_4^{2+}Al[Si_8O_{22}](OH)_2$

66.1.3c.8   Magnesioarfvedsonite $Na_3Mg_4Fe^{3+}[Si_8O_{22}](OH)_2$

66.1.3c.9   Arfvedsonite $Na_3Fe_4^{2+}Fe^{3+}[Si_8O_{22}](OH)_2$

Eckermannite named in 1942 by Adamson for Claes Walther Harry von Eckermann (1886–1969), Swedish petrologist; ferroeckermannite named in 1964 by Phillips and Layton; arfevedsonite named in 1923 by Brooke for Johan A. Arfvedson (1792–1841), Swedish chemist; magnesioarfvedsonite coined in 1957 by Miyashiro. Amphibole group, glaucophane subgroup. MON $C2/m$. $Z = 2$. Eckermannite: $a = 9.762$, $b = 17.892$, $c = 5.284$, $\beta = 103.17°$, $D = 2.97$; *20-386*(syn): 8.35_4 3.40_7 3.25_7 3.10_{10} 2.97_5 2.71_8 2.50_6 2.16_5; magnesioarfvedsonite: $a = 9.796$, $b = 17.932$, $c = 5.289$, $\beta = 104.99°$, $D = 3.17$; *42-1369*: 8.49_6 3.93_1 3.39_1 3.28_3 3.15_{10} 2.82_2 2.71_2 1.909_1; arfvedsonite: $a = 10.007$, $b = 18.077$, $c = 5.332$, $\beta = 104.10°$, $D = 3.40$; *14-633*: 8.51_7 3.42_5 3.16_{10} 2.73_8 2.60_4 2.55_3 2.35_3 2.19_4. Structure: *CM 14:346(1976)*. **Habit:** Long prismatic [001] crystals, often tabular {010} or {100}, with poor to good terminations, {001} and {$\bar{1}$01}; prismatic aggregates. Crystals commonly deeply etched and striated [001]; bladed aggregates; sometimes fibrous. Twinning on (100) common. **Physical properties:** Blue-green (eckermannite) to greenish black, blue-black and black (arfvedsonite); pale to dark blue-gray streak; dull to vitreous luster, occasionally silky; translucent to opaque. Cleavage {110}, perfect; parting {010}; uneven fracture. H = 5–6. G = 3.0–3.5. **Optics:** Eckermannite is biaxial (−); $2V = 80-15°$; $N_x = 1.612–1.638$, $N_y = 1.625–1.652$, $N_z = 1.630–1.654$; $Y = b$, $X \wedge c = 53-18°$ [OAP parallel to (010)]; r > v, very strong; indices increasing while 2V and extinction decrease with increasing iron; $Z < Y < X$; pleochroic. X, bluish green; Z, pale yellow-green to nearly colorless. Arfvedsonite is biaxial (−); $2V = 0-50°$; $N_x = 1.674–1.700$, $N_y = 1.679–1.709$, $N_z = 1.686–1.710$; $Z = b$, $X \wedge c = 0-30°$ [OAP perpendicular to (010)]; r < v, very strong; pronounced anomalous extinction; pleochroic. X, variable absorption, usually $Z < Y < X$; blue-green to indigo; Y, gray-violet, blue to orange-yellow; Z, light yellow-green to blue-gray. The relationships of optical properties of eckermannite–arfvedsonite to one another are similar to those of glaucophane–riebeckite, and the distinction between arfvedsonites and riebeckites may sometimes be difficult. The optical properties of magnesioarfvedsonite are intermediate to those given above, but OAP is usually perpendicular to (010). **Chemistry:** Ferroeckermannite has not been reported from natural systems, the substitution $Fe^{2+} \rightarrow Mg$ generally being accompanied by $Fe^{3+} \rightarrow Al^{[6]}$. The sum of Ca, Na, and K (normally, A and B site ions) frequently exceeds 3.0, and the total of C site ions is often significantly less than 5.0. Mn may reach ≈ 1.0 per formula unit. Fluorine substitution for (OH) is common, as is (OH) deficiency. Published analyses of members of this series are frequently intermediate with ferritaramite and members of the glaucophane–riebeckite series. **Occurrence:** Eckermannites are relatively rare, occurring in alkaline plutonic rocks and in soda-rich metamorphic environments (with jadeite). Arfevedsonites are characteristic components of alkaline plutonic rocks and their associated pegmatites. They occur in both quartz-bearing syenites and nepheline–syenites, generally forming late in their crystallization. Often associated with aegirine or aegirine–augite, the pyroxene and amphibole sometimes intergrown with their c axes common. They also occur as overgrowths on earlier formed hastingsites. **Localities:** *Eckermannite* occurs in a shear zone between granite and monzonite (asbestiform, soda asbestos) at Camp Albion, Boulder Co., CO; at *Norra Kärr, Gränna, Sweden* (lithian); in jadeite rock at Tawmaw, Myanmar. *Magnesioarfvedsonite* occurs in the Green River formation of CO, WY, and UT; at Mt. St.-Hilaire, PQ, Canada; in shonkinite at Katzenbuckel, Odenwald, Germany; in syenite pegmatite at Mariupol, Ukraine (fluor-taramite); in albite-rich pegmatite at Lovozero, Kola Penin., Russia; and in Japan. *Arfvedsonite* occurs at Red Hill, near Moultonboro, NH; fluorine-rich arrvelsonite in pegmatite at Hurricane Mt., Carroll Co., NH; St. Peter's Dome, El Paso Co., CO; abundant as crystals(!) to 20 cm long in pegmatite dikes at Mt. St.-Hilaire, PQ, Canada; in pegmatites and related rocks at *Ilimaussaq* in crystals exceeding 20 cm, at Kangerdluarssuk, at Narssarssuk in coarse intergrown crystals in part altered to asbestiform riebeckite, and at *Tupersuatsiaat, Julianehaab district, Greenland*; in nepheline syenites and related rocks of the Langesundfjord district and Oslo region, Norway; in nepheline–syenite gneiss at Kiihtelysvaara, Finland; in nepheline–syenite at Urma-Waraka, Kola Peninsula, and in syenite at the Morotu River, Sakhalin, Russia; manganoan arfvedsonite in the manganese deposits of Chikla, Bhandara, Maharashtra, India. AR *DHZI 2:364(1963); MR 5:117(1974), 21:296(1990), CM 34:1011 and 1015(1996).*

66.1.3c.10   Kozulite $NaNa_2Mn_4(Fe^{3+},Al)[Si_8O_{22}](OH)_2$

Named in 1969 by Nambu for Shukusuke Kozu (1880–1955), Japanese mineralogist, Tohoku University. Amphibole group, glaucophane subgroup. MON $C2/m$. $a = 9.914$, $b = 18.111$, $c = 5.308$, $\beta = 104.5°$, $Z = 2$, $D = 3.36$. *25-850:* 8.51_{10} 4.53_1 3.40_1 3.30_2 3.15_6 2.83_3 2.72_1 2.17_1. Structure: $AC(A)$

28:S71(1972). Reddish black, vitreous luster, transparent to translucent. Cleavage {110}, perfect. H = 5. G = 3.30. Biaxial (−); 2V = 35°; $N_x = 1.685$, $N_y = 1.717$, $N_z = 1.720$; $Y = b$, $Z \wedge c = 65°$; r > v, weak. Pleochroic; X, yellow-brown; Y, reddish brown; Z, dark brown. Found as short prismatic to bladed crystals displaying {110}, {010}, and {011} in bedded manganese deposits in metamorphosed Cretaceous quartzite intruded by granodiorite at the *Tanahota mine, Iwate Pref., Japan*; from Woodsmine, 30 km north-northeast of Tamsworth, NSW, Australia. AR *AM 55:1815(1970)*.

66.1.3c.11 Nyböite $NaNa_2Mg_3Al_2[AlSi_7O_{22}](OH)_2$

Named in 1981 by Ungaretti et al. for the locality. Amphibole group, glaucophane subgroup. MON $C2/m$. $a = 9.665$, $b = 17.752$, $c = 5.303$, $\beta = 104.11°$, $Z = 2$, $D = 3.031$. Bluish, vitreous luster, translucent. Cleavage {110}, perfect; brittle. H = 5–6. Biaxial, indices not reported. Pleochroic; X, pale violet; Z, colorless. Found in eclogite environments, sometimes as porphyroblasts. Originally from the *Nybö pod, Nordfjord, Norway*; in eclogite from the Donghai area, Jiangsu Prov., China. AR *BM 104:400(1981)*, *EJM 4:171(1992)*.

66.1.3c.12 Leakeite $(Na,K)Na_2(Mg,Fe^{3+},Li,Mn^{3+})_5$ $[(Si,Al)_8O_{22}](OH)_2$

Named in 1992 by Hawthorne et al. for Bernard Elgey Leake (b.1936), geologist, University of Glasgow, in recognition of his work on the chemistry of the amphiboles. Amphibole group, glaucophane subgroup. MON $C2/m$. $a = 9.822$, $b = 17.836$, $c = 5.286$, $\beta = 104.37°$, $Z = 2$, $D = 3.107$. PD: 8.40_6 4.46_1 3.38_2 3.25_2 3.12_{10} 2.80_5 2.70_1 1.431_1. Structure: *AM 77:1112(1992)*. Deep red, vitreous luster, translucent to transparent. Cleavage {110}, perfect. H ≈ 6. G = 3.11. Biaxial (−); 2V = 59–71°; but reported indices compatible with biaxial (+); Na(D): $N_x = 1.667$, $N_y = 1.675$, $N_z = 1.691$; $Z = b$, $X \wedge c = +10°$; r < v, strong; strongly pleochroic. X ≈ Y, dark mauve to red; Z, light pink-red. Li substitution in C position is approximately 0.8; Fe^{3+} predominates over Mn^{3+} distinguishing it from kornite; some F substitutes for (OH). Occurs as prismatic grains associated with albite, braunite, and bixbyite in a dark red rock of semimetallic appearance in the metasediments of the *Kajlidongri manganese mine, Jhabua district, Madhya Pradesh, India*. AR *AM 78:733(1993)*.

66.1.3c.13 Kornite $(Na,K)(Na,Li)_2(Mg,Mn^{3+},Fe^{3+},Li)_5$ $[Si_8O_{22}](OH)_2$

Named in 1993 by Armbruster et al. for Hermann Korn (d.1946), German geologist who lived part of his life in South West Africa. Amphibole group, glaucophane subgroup. MON $P2_1/m$ or $P2/a$. $a = 9.94$, $b = 17.80$, $c = 5.302$, $\beta = 105.5°$, $Z = 2$. PD: 8.89_M 8.43_M 5.08_M 4.44_M 3.26_S 3.13_S 2.81_S 2.55_S. The primitive lattice is based on weak reflections of the type $(0k1)$ with k odd. Brownish lilac to dark red, vitreous luster, transparent. Biaxial (−,+); $2V_x = 88–92°$; $N_x = 1.654$, $N_y = 1.675$, $N_z = 1.696$; $Y = b$, $Z \wedge c = 60–65°$.

Pleochroic; X, pink; Y, dark red; Z, orange-red. Mn^{3+} is dominant over Fe^{3+}, distinguishing it, apart from the primitive lattice, from leakeite. Occurs as fine fibers with maximum length of 200 μm with hennomartinite, sugilite, and serandite–pectolite at the *Wessels mine, Kalahari manganese field, South Africa*. AR *SMPM 73:349(1993)*.

66.1.3c.14 Ungarettiite NaNa$_2$(Mn$_2^{2+}$Mn$_3^{3+}$)[Si$_8$O$_{22}$]O$_2$
glaucophane subgroup

Named in 1995 by Hawthorne et al. in honor of Luciano Ungaretti, Professor of Mineralogy, University of Pavia, Italy in recognition of his insights into the crystal chemistry of rock-forming minerals. MON $C2/m$. $a = 9.89$, $b = 18.04$, $c = 5.29$, $\beta = 104.6$, $Z = 2$, $D = 3.45$. PD 8.40_5 4.48_5 3.26_6 2.99_4 2.71_{10} 2.59_4 2.54_9 2.17_3. Found as rock bound grains; extracted fragments show typical amphibole cleavage prism with irregular fracture terminations. Cherry red to very dark red, streak presumably red, vitreous luster; cleavage {110} perfect. H ≈ 6. G = 3.52. Biaxial (−); 2V = 51°$_{obs}$, 57°$_{calc}$; $N_x = 1.717$, $N_y = 1.780$, $N_z = 1.800$, r < v distinct but difficult to observe due to color; X ∧ a = −2°, Y = b, Z ∧ c = 17° in obtuse β. Pleochroic with X orange red, and Y ≈ Z very dark red. Near the end member composition, it is unusual in that the mid-size C cations are almost entirely manganese, and that it is an anhydrous amphibole. Found in massive silicate/Mn oxide rock associated with other Mn-bearing amphiboles (leakeite, manganoan arfvedsonite, manganoan katophorite), braunite, clinopyroxene, norrishite, and sometimes quartz at the *Hoskins mine, an abandoned Mn deposit, 3 km west of Grenfell, NSW, Australia*. AM *80:165(1995)*.

66.1.3c.15 Fluor-ferro-leakeite NaNa$_2$(Fe$_2^{2+}$Fe$_2^{3+}$Li)[Si$_8$O$_{22}$]F$_2$
glaucophane subgroup

Named in 1996 (Hawthorne et al.) for the composition and relationship to leakeite. MON $C2/m$. $a = 9.792$, $b = 17.938$, $c = 5.313$, $\beta = 103.87$, $Z = 2$, $D = 3.34$. PD 8.43_5 4.48_5 3.40_6 3.12_3 2.99_4 2.71_{10} 2.59_4 2.54_9. Bluish-black to black; vitreous luster; cleavage {110} perfect, brittle; H ≈ 6. G = 3.37. Biaxial (+); 2V = 87°$_{obs}$, 81°$_{calc}$ for Na$_D$ light; $N_x = 1.675$, $N_y = 1.683$, $N_z = 1.694$; X ∧ c = 10°$_{in\ obtuse\ \beta}$, Y = b, Z ∧ a = 4°$_{in\ obtuse\ \beta}$. Pleochroic with X very dark indigo blue, Y grey blue, and Z yellow green. X > Y > Z Potassium replaces sodium in the A position and significant Mg is present in the C positions, and (OH) replaces 30 at. % of the fluorine. Occurs as small (0.05–0.2 mm) irregular patches in the quenched groundmass as well as sparse elongate grains in peralkaline porphyry in the *Cañada Pinabete pluton, Questa, Taos Co., NM*; AM *81:226(1996)*.

66.1.4.1 Joesmithite (Ca,Pb)Ca$_2$(Mg,Fe^{2+},Fe^{3+})$_5$[Be$_2$Si$_6$O$_{22}$](OH)$_2$

Named in 1968 by Moore for Joseph Victor Smith (b.1928), professor of mineralogy, University of Chicago. MON $P2/a$. $a = 9.885$, $b = 17.875$, $c = 5.227$, $\beta = 105.67°$, $Z = 2$, $D = 4.02$. 22-531: 5.03_4 3.70_5 3.33_{10} 2.90_5

2.74_5 2.68_5 2.56_6 2.53_6. Structure: *MSASP 2:111(1969)*. The structure is distinct from other amphiboles in that Be is ordered into one of the four tetrahedral sites of the ribbon silicate unit, and that the A site is occupied entirely by divalent ions. Black to dark brown, pale brown streak, translucent. Cleavage {110}, perfect. H = $5\frac{1}{2}$. G = 3.83. Biaxial (+); 2V large; $N_x = 1.747$, $N_y = 1.765$, $N_z = 1.78$. Pleochroic, Y > X = Z; olive-brown to brown. Insoluble in cold acids. Occurs as well-formed prismatic, [001], crystals displaying {010}, {100}, {110} terminated by {011}, {$\bar{1}$12}, {$\bar{1}$13}, lining cavities in or enclosed by later calcite in a hematite–magnetite–schefferite skarn at the *Långban mine, near Filipstad, Värmland, Sweden*. AR *AM 54:577(1969)*.

66.1.5.1 Ershovite $Na_4K_3(Fe,Mn,Ti)_2Si_8O_{20}(OH)_4 \cdot 4H_2O$

Named in 1993 by Khomyakov et al. for Vadim Viktorovich Ershov (1939–1989), Professor and founder of the mineralogical museum at the Moscow Mining Institute, and a specialist in the area of applied geology and mining. TRIC. $P\bar{1}$. $a = 10.244$, $b = 11.924$, $c = 5.276$, $\alpha = 103.49°$, $\beta = 96.96°$, $\gamma = 91.945°$, Z = 1, D = 2.73; (nat.) 11.57_{10} 3.39_2 3.01_2 2.99_3 2.72_2 2.60_3 2.46_1 2.16_1. Structure is related to amphibole-type (K 36:892(1991)). Olive green with a cinnamon, brownish and yellowish shading, transparent; elongate on c, 0.3–0.5 mm, aggregates to 1–3 cm, 0.1– 1 × 2–3 to 3–5 × 10 mm; vitreous; excellent {100} and {010} cleavage; fibrous; H = 2–3; G = 2.75; IR data in ZVMO 122:116(1993); no UV response; readily soluble at 25°C in weak solutions of HCl, HNO_3, and H_2SO with preservation of the silicate skeleton, no change to 300°C, at 500°C is isotropic, and X-ray amorphous. SiO_2 47.1, Na_2O 12.4, K_2O 13.6, CaO 0.1, MgO 0.5, MnO 4.7, FeO 6.1, TiO_2 3.0, H_2O 12.5, = 100.0. Biaxial (+); 2V = 58°, $2V_{calc} = 59°$; $N_x = 1.569$, $N_y = 1.574$, $N_z = 1.590$; strong pleochroism with X and Y = light green or yellow, and Z = dark olive green; angle between c and X = 86°, Y $\wedge$ c = 73°, N_z = 17°, fibers have (+) elongation; absorption = Z > Y > X; disper. = moderate r > v. Occurs in pegmatite and pegmatitic rocks in Khibiny alkaline massif. Associated with rare alkaline minerals. Pegmatitic-hydrothermal origin. EF *Mt. Rasvumchorr and Mt. Koashva, Khibiny massif, Kola Peninsula, Russia*.

66.2.1.1 Plancheite $Cu_8(Si_4O_{11})_2(OH)_4 \cdot xH_2O$

Named in 1908 by Lacroix for a Mr. Planché, who furnished the African material for study. ORTH *Pcnb*. $a = 19.043$, $b = 20.129$, $c = 5.269$, Z = 4, D = 3.815. *29-576*: 10.1_{10} 9.56_4 6.94_7 4.87_5 4.06_8 3.95_4 3.57_3 3.31_4. Structure: *AM 62:491(1977)*. The structure is similar to that of shattuckite (65.1.7.1) except that the chains joined to the six-coordinated copper sheets are amphibole-like. The water appears to be zeolitic with x < 0.48 for the structure-study material. Commonly fibrous [001]; compact radial aggregates. Pale to dark blue, silky luster. H = 5. G = 3.4. Biaxial (+); 2V = 57°; $N_x = 1.645$, $N_y = 1.660$, $N_z = 1.715$; Z = c. Pleochroic; Z, dark blue; X, pale blue. Indices of up to $N_x = 1.697$ and $N_z = 1.741$ have been reported, and probably reflect the variable water content. Found as an oxidation-zone mineral, as at the

Algomah mine, Ontanagon Co., MI; the Shattuck mine, Bisbee, Cochise Co.; at the Table Mt. mine with compact conichalcite, and elsewhere in Pinal Co.; and at the New Cornelia mine, Pima Co., AZ; in the Beaver Lake district, Beaver Co., UT; at Caldbeck Fells, Cumbria, and St. Austell, Cornwall, England; at the Kisanmori mine, Akita Pref., Japan. Found at *Mindouli, Renneville, Congo Republic*; at Tantara and Mindigi, near Kambove, Shaba, Zaire; at Tsumeb, Guchab, and Okatumba, Namibia. AR *BM 84:267(1961)*.

66.2.2.1 Vlasovite Na$_2$Zr[Si$_4$O$_{11}$]

Named in 1961 by Tikhonenkova and Kazakova for Kuzma Aleksevich Vlasov (1905–1964), Russian mineralogist and geochemist who studied the Lovozero massif. TRIC $P\bar{1}$ (<29°C). $a = 10.96$, $b = 10.01$, $c = 8.53$, $\alpha = 89.70°$, $\beta = 100.40°$, $\gamma = 90.48°$, $Z = 4$, $D = 3.07$. MON $C2/c$ (>29°C). $a = 10.98$, $b = 10.00$, $c = 8.52$, $\beta = 100.40°$. *39-208*(syn, monoclinic): 5.44_4 5.30_6 5.06_7 3.88_4 3.67_5 3.40_6 3.24_{10} 2.96_9. Structure: *DANES 141:1294(1961)*. The silicate element is a ribbon composed of four-tetrahedron rings linked by sharing an apical oxygen to form a ribbon of composition $[Si_4O_{11}]^{6-}$, distinct from the isostoichiometric amphibole ribbon. Colorless; greasy luster, vitreous to pearly on cleavage; transparent. Cleavage {010}, distinct; irregular fracture; brittle. $H = 6$. $G = 2.97$. Biaxial (−); $2V = 50$–$56°$; $N_x = 1.607$, $N_y = 1.623$, $N_z = 1.628$; OAP parallel to (010); $r > v$, distinct. Nonfluorescent in UV, but altered material fluoresces bright orange. Up to 1.7 wt % HfO_2 is reported. Found in ejecta of peralkaline granite on Ascension Is. Large crystals (15 cm) in the Scheffield Lake complex, Villedieu Twp., PQ, Canada. Associated with feldspars, amphiboles, pyroxenes, eudialyte, apatite, and fluorite at contact of syenite pegmatite and fenite, at *Mt. Vavnbed, Lovozero massif, Kola Penin., Russia*, and in the Burpala massif, 120 km N of Lake Baikal. AR *AM 46:1202(1961)*, *MM 36:233(1967)*, *CM 12:211(1973)*.

66.3.1.1 Xonotlite Ca$_6$[Si$_6$O$_{17}$](OH)$_2$

Named in 1866 by Rammelsberg for the type locality. Synonym: jurupaite. MON $P2/a$. $a = 17.03$, $b = 7.356$, $c = 7.007$, $\beta = 90.2°$, $Z = 2$, $D = 2.705$. *29-379*(calc): 7.04_6 4.13_4 3.56_6 3.24_7 3.05_{10} 2.80_6 2.68_5 1.911_4. Structure: *ZK 154:271(1981)*. The silicate structural element is a ribbon of $W = 2$ and $P = 3$, formed by the condensation of two wollastonite chains sharing every third tetrahedron. Polytypes may exist. Habit is prismatic [010] with {100}, {001}, {011}, {01$\bar{1}$}, {010} prominent, but usually in compact fibrous masses. Twinned on {001}. Chalky white, colorless, pale pink, gray; dull, vitreous, or pearly luster; tough; transparent to translucent. Cleavage {001}, perfect. $H = 5$–6. $G = 2.71$. Biaxial (+); 2V small; $N_x = 1.583$, $N_y = 1.583$, $N_z = 1.594$; $Z = b$, OAP $\approx$ (100); $r < v$, weak. Decomposed by acids with separation of silica. Principal occurrence is as veinlets in serpentine or in contact zones. Found at Franklin, Sussex Co., NJ; near Leesburg, Loudoun Co., VA; near Durham, NC; Isle Royale, MI; at the Christmas mine, Gila Co., AZ; and in Mendocino, Monterey, Riverside, San Francisco, and Santa

Barbara Cos., CA. Found at the Jeffrey mine, Asbestos, PQ, Canada; at Tetela de Ocampo, PUE, and *Tetela de Xonotla, HGO, Mexico*; near Yauco, Puerto Rico. At Broadford, Skye, Scotland, and at Scawt Hill, County Antrim, Northern Ireland. With kilchoanite in a xenolith in the Ozerskii gabbro, Priol'khonye, West Baikal region, Russia. As well-formed, 5- to 50-μm crystals at Heguri, Chiba Pref., and at the Mihara mine, Okayama Pref., Japan. AR *AM 10:12(1925), 38:860(1953); BNSM 1:61(1975)*.

66.3.1.2 Zorite $Na_6Ti_5[Si_{12}O_{34}](O,OH)_5 \cdot 11H_2O$

Named in 1973 by Merkov et al. from the Russian for *rosy tint of the dawn sky*. ORTH *Cmmm*. $a = 23.241$, $b = 7.238$, $c = 6.955$, $Z = 2$, $D = 2.23$. *25-1298:* 11.6_8 6.90_{10} 3.38_8 3.07_8 2.98_8 2.59_8 1.752_8 1.701_6. Structure: *SPC 24:686(1979)*. Original description gives C^*c^* with c doubled. The silicate unit is a double chain formed by condensation of two wollastonite-like chains that are cross-linked by six-coordinated Ti to form a zeolite-like framework. The pair of wollastonitelike chains are related by a diad, as opposed to those in xonotlite, which are mirror related. Rose, vitreous luster. Cleavage {010}, {001}, perfect; {110}, good. H = 3–4. G = 2.36. Presumably biaxial. Pleochroic; X, rose; Y, colorless; Z, bluish. Found with aegirine, mountainite, and natrolite in nepheline-filled fractures and cavities in alkalic pegmatite, the same one in which ilmajokite and raite were found, in the *Jubilee pegmatite, Lovozero massif, Kola Penin., Russia*. AR *ZVMO 102:54(1973), AM 58:1113(1973)*.

66.3.1.3 Eudidymite $NaBeSi_3O_7(OH)$

Named in 1887 by Brögger from the Greek for *good* and *twin*, in allusion to its occurrence in twinned crystals. A dimorph of epididymite. MON $C2/c$. $a = 12.568–12.69$, $b = 7.37–7.375$, $c = 13.976–13.99$, $\beta = 103.7–103.78°$, $Z = 8$, $D = 2.56$. *14-201*(nat): 6.35_6 3.69_5 3.40_8 3.16_{10} 3.07_8 3.00_6 2.85_6 2.01_5. Structure: *AM 32:442(1947)*. Thin, tabular, flattened on {010}, with {010}, {100}, {001}, and others; aggregates of crystals to 5 cm from Mt. Malosa, Zomba district, Malawi; V-shaped, stellate twins that mimic maple leaves; twinning by reflection on (100). Colorless to white, pale pink; white streak; vitreous to silky luster, faint bluish iridescence, pearly on cleavage. Cleavage {001}, perfect; {100}, distinct; uneven fracture; brittle. H = 6–7. G = 2.55–2.58. Nonfluorescent. Insoluble in common acids. Be may substitute for Si and OH for O. Biaxial (+); $Y = b$, $Z \wedge c = 54–55°$; $N_x = 1.544–1.545$, $N_y = 1.545–1.546$, $N_z = 1.549$; $2V = 23–30°$; $r > v$, weak to distinct. Occurs in alkaline syenites and their derivative pegmatites. Associated with other alkaline minerals. Often in Na-rich parageneses. Occurs in pegmatite veins and rarely in breccia cavities at Mt. St.-Hilaire, QUE, and in the Red Wine alkaline complex, Seal Lake, central LAB, Canada; in alkalic pegmatites in the Narsarsuk area, southern Greenland; *Övre Arö, Larvik, Langesundfjord area, Vestfold Co., Norway*; in several alkaline massifs, Kola Penin., Russia; Burpala massif, eastern Siberia, Russia; Tallask Mts., Kirghizia; Mt. Malosa, Zomba district, Malawi. EF *MR 25:29(1994)*.

66.3.1.4 Epididymite NaBeSi$_3$O$_7$(OH)

Named in 1893 by Flink from the Greek for *near* and *didymite*, in allusion to the fact that it is dimorphous with eudidymite. See eudidymite (66.3.1.3), the monoclinic dimorph. ORTH *Pnam.* $a = 12.49$–12.74, $b = 7.33$–7.36, $c = 13.48$–13.65, $Z = 8$, $D = 2.58$–2.60. *14-64*(nat): 6.3_7 3.65_7 3.40_{10} 3.09_{10} 2.99_{10} 2.49_7 1.80_8 1.28_8. Structure: *AM 55:1541(1970)*. Generally fibrous, acicular, bladed, and prismatic, but may also be porcelaneous (in masses to 15 cm diameter); sometimes blocky, equant, or pseudohexagonal. Tabular on {001}, flattened on {001}. Dominant forms are: {001}, {110}, {010}, {100}, and others. Elongated on [010]. Frequently, polysynthetically twinned and deeply striated. Reticulate, interpenetrant, and trilling twins common, by rotation of 60° about [001]. Trillings as much as 1 cm across from Mt. St.-Hilaire. {110} and {1$\bar{1}$0} are the twin planes. Crystals to 2 cm from Narsârsuk, and to 6 cm from Zomba. Colorless, white, bluish, violet or greenish; white streak; vitreous to silky luster. Cleavage {001}, perfect; {100}, micaceous and distinct; uneven to splintery fracture; brittle. $H = 6$–7. $G = 2.55$–2.58. IR data: *ZSK 8:233(1967)*. Insoluble in common acids, soluble in Hf. Analysis (wt %): K_2O, to 0.8; CaO, to 1.15; Al_2O_3, to 1.9; traces of Fe. Biaxial $(-)$; $X = a$, $Z = b$, OAP = (001)for $(-)$; $N_x = 1.539$, $N_y = 1.543$, $N_z = 1.544$; $2V = 0$–$25°$. Also biaxial $(+)$; $N_x = 1.543$, $N_y = 1.544$, $N_z = 1.546$; $2V = 22$–$30°$. Dispersion $r > v$, weak. One of five minerals to show crossed axial plane dispersion: red light: OAP = (001), $2V = 26°$, violet light: $2V = 0°$, dark violet light: OAP = (010), $2V = 16°$. Occurs in alkaline igneous rocks and their cogenetic pegmatites as a late hydrothermal mineral. May form from pre-existing Be minerals, and also may alter to bertrandite and beryllite. Localities include: near Quincy, Norfolk Co., MA; Mt. St.-Hilaire, Quebec; *Narssârssuk (Narsârsuk), Greenland*; Ilímaussaq, Greenland as pseudomorphs after chkalovite; Lille Arö, Langesundfjord, Norway; Věžná, Czech Republic; as pseudomorphs after chkalovite from Mt. Alluaiv; also from Mt. Karnasurt and several other localities in the Lovozero massif, Kola Penin., Russia; Mt. Kukisvumchorr, Khibiny massif, Kola Penin., Russia; Talas Mts., Siberia, Russia; slender, prismatic to ruler-shaped crystals to 6 cm in pegmatites from Mt. Malosa, Zomba district, Malawi. EF *MM 33:450(1963)*, *CM 9:706(1969)*, *MIN 3(3):311(1981)*, *MZ 6:84(1984)*, *MR 25:29(1994)*.

66.3.1.5 Yuksporite (K,Ba)(Na,Sr)Ca$_2$(Si,Ti)$_4$O$_{11}$(F,OH) · nH$_2$O

Named in 1922 by Fersman for the locality. Not well characterized. Note difference between D and G. ORTH Space group unknown. $a = 24.869$, $b = 16.756$, $c = 7.057$, $Z = 9(?)$, $D = 2.79$. *41-1400*(has $Z = 10$, $D = 2.64$)(nat): $6.22_{5.5}$ $4.18_{3.5}$ $3.19_{3.5}$ 3.10_9 3.00_9 2.915_7 2.80_{10} $2.74_{3.5}$. Platy or fibrous masses to 10 cm ; forms veinlets, concretions, and nodules in feldspar. Pink to red or yellowish rose. H = 5, G = 3.05. IR data show that it may be related to astrophyllite group and bafertisite. Analysis (wt %): SiO_2, 38.4; TiO_2, 11.0; Al_2O_3, 0.07; Fe_2O_3, 0.75; MnO, 0.29; CaO, 18.9; SrO, 5.87; BaO, 8.60; Na_2O, 3.84; K_2O, 6.15; H_2O^+, 2.20; F, 3.05; Cl, 0.80. Incomplete optical properties: $N_x = 1.644$, $N_z = 1.660$. X = pale rose-yellow, Y = Z = rose-yellow. Occurs in alkalic syenite massifs. First found on *Yukspor Mt., Khibiny massif, Kola Penin., Russia*; also on Mt. Eveslogchorr; Murun massif, Aldan Shield, Siberia, Russia. EF *MZ 7(4):74(1985)*.

66.3.3.1 Inesite $Ca_2Mn_7^{2+}[Si_{10}O_{28}](OH)_2 \cdot 5H_2O$

Named in 1887 by Schneider from the Greek for *flesh fibers*, in allusion to the color and fibrous structure. TRIC $P\bar{1}$. $a = 8.889$, $b = 9.247$, $c = 11.975$, $\alpha = 88.15°$, $\beta = 132.07°$, $\gamma = 96.64°$, $Z = 1$, $D = 3.03$. *21-151*: 9.16_{10} 6.54_4 4.59_5 4.01_5 2.92_8 2.84_8 2.73_7 2.19_6. Structure: *AM 63:563(1978)*. The chains are similar to those found in the $P = 5$ rhodonite group, but are condensed to form double-wide ribbons with unit composition $[Si_{10}O_{28}]$ that contain alternate six- and eightfold rings. These ribbons are joined to bands of edge-sharing polyhedra consisting of seven Mn octahedra and two Ca pentagonal bipyramids. Hydroxyl ion and water are not bonded to silicon. The structure is thus decidedly different from that of the $P = 5$ pyroxenoids. The habit is acicular to long prismatic [001] with {100} and {010} dominant and {1$\bar{1}$0} less so, and terminated by combinations of {001}, {201}, etc., sometimes leading to acute, chisel-like crystals; as acicular, radiating aggregates. Rose to flesh-red, sometimes darkening to brown on exposure; vitreous luster; transparent to translucent. Cleavage {010}, perfect; {100}, good; brittle. H = 6. G = 3.03. Biaxial (−); $2V = 75°$; $N_x = 1.618$, $N_y = 1.639$, $N_z = 1.652$; $X \wedge c = 74°$, $Y \wedge c = 32°$, $Z \wedge c = 62°$; $r > v$, weak; nonpleochroic or only feebly so. Soluble in acids. Yields rhodonite on ignition to 800°. Inesite occurs sparsely in numerous manganese-rich environments. Found near Creede, Mineral Co., CO; with rhodochrosite and bementite in fine crystal aggregates(!) at Hale Creek, Trinity Co., CA; with bementite and hausmannite at the Crescent mine, Clallum Co., WA; at Villa Carona, DGO, Mexico. Occurs at Jacobsberg, Långban, and at the Harstig mine, Pajsberg, Sweden; embedded in calcite at *Nanzenbach, near Dillenburg, Germany*; as radial aggregates of crystals at the Kawazu, Seikoshi, and Yugashima mines, Shizuoka Pref., and the Todoroki mine, Hokkaido, Japan; as fine hemispherical masses(!) and single crystals on level 19, New Broken Hill Consolidated mine, NSW, Australia; in the Kalahari manganese field(!) in rose-red botryoidal masses, radial-fibrous spheres, and crystal groups at the Wessels and Mama Twan mines, Cape Prov., South Africa. AR *D6:564(1892)*, *AM 53:1614(1968)*.

TUHUALITE GROUP

The tuhualite group minerals are isostructural, and they have the general formula

$$ABCD[Si_6O_{15}]$$

where

A = ☐, Na
B = Na, K
C = Fe^{2+}, Li
D = Fe^{3+}, Zr^{4+}

and crystallize in the orthorhombic space group *Cmca*.

The silicate structural element, double that of the empirical formula, consists of a crenulated double-width chain ($W = 2$) having a periodicity of six tetrahedra ($P = 6$). Sites A, B, C, and D are nine-, ten-, four-, and six-coordinated, respectively, and not all may be occupied. In addition to the three species known for this group, there are a number of isostructural synthetic compounds: additional synthetic and natural compounds can be predicted. *AM 63:304(1978)*.

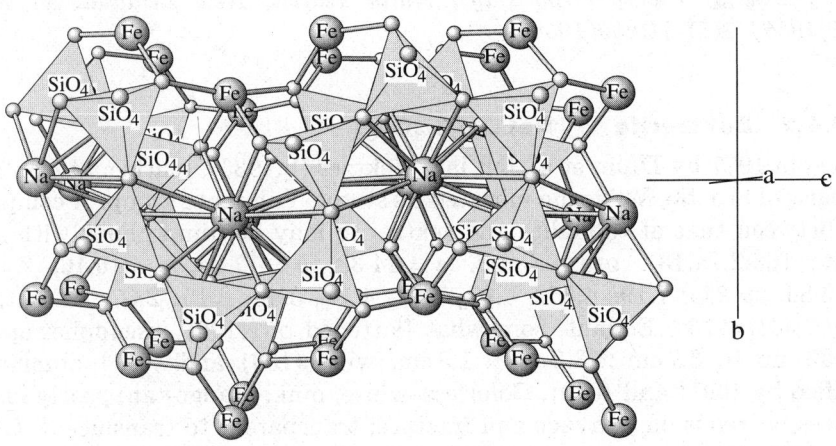

TUHUALITE: $NaFe_2(Si_6O_{15})$

This structure has extremely serpentine double chains parallel to c, which in turn form sheets parallel to (100). Na and Fe are both between the sheets, although the large cavities in the bends in the rings may be occupied in related minerals.

TUHUALITE GROUP

Mineral	Formula	Space Group	a	b	c	
Tuhualite	$\square(Na,K)^+Fe^{2+}Fe^{3+}[Si_6O_{15}]$	$Cmca$	14.31	17.28	10.11	66.3.4.1
Zektzerite	$\square NaLiZr[Si_6O_{15}]$	$Cmca$	14.33	17.35	10.16	66.3.4.2
Emeleusite	$Na_2LiFe^{3+}[Si_6O_{15}]$	$Cmca$	14.004	17.337	10.072	66.3.4.3

Note: Minerals of the milarite group are chemically analogous and essentially polymorphs of the tuhualite group, although no 1:1 correspondences in composition exist between the groups.

66.3.4.1 Tuhualite $\square(Na,K)Fe^{2+}Fe^{3+}[Si_6O_{15}]$

Named in 1932 by Marshall for the locality. Tuhualite group. A composition between that of tuhualite and zektzerite may be dimorphous with sogdianite (63.2.1a.13); merrihueite (63.2.1a.5) is close in composition and may be regarded as a dimorph. ORTH $Cmca$. $a = 14.31$, $b = 17.28$, $c = 10.11$, $Z = 8$, $D = 3.03$. *10-440*: 8.62_8 7.16_{10} 5.52_8 4.85_8 4.35_8 4.32_8 3.71_8 3.47_8. Structure: *SCI 166:1399(1969)*. The presence of both ferric and ferrous ion is evidenced by the intense absorption. Dark blue to violet-black, translucent. Cleavage {100}, {010}, {001}, good; very brittle. $H = 3\frac{1}{2}$. $G = 2.89$. Biaxial $(+)$; $2V = 60$–$70°$; $N_x = 1.608$, $N_y = 1.612$, $N_z = 1.621$; $r > v$, positive elongation. Pleochroic; $Z > Y > X$; deep purple, violet, and pale pink. Also given as biaxial $(-)$; $2V = 66°$; $N_x = 1.601$, $N_y = 1.605$, $N_z = 1.607$. Insoluble in acids; alters to a greenish-yellow material. In volcanic rock of *Mayor (Tuhua) Island, Bay of Plenty (Opo Bay), North Island, New Zealand*. AR *MM 31:96(1956)*, *AM 41:959(1956)*.

66.3.4.2 Zektzerite $\square NaLiZr[Si_6O_{15}]$

Named in 1977 by Dunn et al. for Jack Zektzer (b.1936), mathematician and geologist of Seattle, WA, who initiated its study. Tuhualite group. A composition between that of tuhualite and zektzerite may be dimorphous with sogdianite (63.2.1a.13). ORTH $Cmca$. $a = 14.33$, $b = 17.35$, $c = 10.16$, $Z = 8$, $D = 2.80$. *29-835*: 7.19_8 4.85_{10} 4.37_5 4.34_8 4.14_5 3.48_7 3.15_8 2.67_6. Structure: *AM 63:304(1978)*. Equant, somewhat flattened on {100}, pseudohexagonal crystals up to $3.7 \text{cm} \times 3.5 \text{cm} \times 1.5 \text{cm}$, with {100} and {011} dominant, modified by {001} and {010}. Colorless, white, pink, orange-tan; pearly luster on faces, vitreous on cleavage and fracture; transparent to translucent. Cleavage {100}, {010}, perfect. $H = 6$. $G = 2.79$. Biaxial $(-)$; $2V \approx 0°$; $N_x = 1.582$, $N_y = N_z = 1.584$; $XYZ = abc$; $r > v$, weak. Light yellow fluorescence in SWUV. In miarolitic cavities and as isolated grains in alkaline syenite of the Golden Horn batholith *near Washington Pass, Okanogan Co., WA*, with microcline, quartz, riebeckite, zircon, astrophyllite, elpidite, and aegirine; in Dara-i-Pioz massif, Alai Range, Tien Shan, Tajikistan. AR *AM 62:416(1977)*.

66.3.4.3 Emeleusite $Na_2LiFe^{3+}[Si_6O_{15}]$

Named in 1978 by Upton et al. for Charles Henry Emeleus (b.1930), British petrologist, University of Durham. Tuhualite group. ORTH (pseudo-HEX) Cmca. $a = 14.004$, $b = 17.337$, $c = 10.072$, $Z = 8$, $D = 2.81$ (a and c have been reversed from the original, and structure descriptions to maintain correspondence with the rest of the group; subsequent data are for the new orientation). *29-832*: 7.00_5 4.35_{10} 4.09_6 3.50_7 3.21_8 3.19_6 3.09_7 2.88_6. Structure: *ZK 147:297(1978)*. Tabular {010} with {010}, {100}, {011}, {001}. Penetration trillings on {011}. Colorless to creamy pink, vitreous luster, transparent. H = 5–6. G = 2.76. Biaxial (−); 2V small; NaD: $N_x = 1.596$, $N_y = 1.597$, $N_z = 1.597$; XYZ = cab; r > v, strong. Found in peralkaline rocks at *Igdlutalik, Julianehab, Greenland*. AR *MM 42:31(1978)*.

Class 67

Unbranched Chains with $W > 2$

67.1.1.1 Jimthompsonite $(Mg,Fe)_5[Si_6O_{16}](OH)_2$

Named in 1978 by Veblen and Burnham for James Burleigh Thompson, Jr. (b.1921), American petrologist and geochemist, Harvard University, who predicted the existence of such a phase. Dimorphous with clinojimthompsonite. ORTH *Pbca*. $a = 18.626$, $b = 27.23$, $c = 5.297$, $Z = 8$, $D = 3.05$. *31-638*(calc): 13.6_6 8.81_{10} 3.81_3 3.25_4 3.09_5 2.74_3 2.60_4 2.55_3. Structure: *AM 63:1053(1978)*. The structure derives from that of clinojimthompsonite by reflection across (001) such that $a_j = a_{cj} \sin \beta_{cj}$ in a relationship akin to that between clino- and orth-amphiboles. Colorless to light pinkish brown. Cleavage prismatic {210} at 38°. Biaxial (−); $2V = 62°$; $N_x = 1.605$, $N_y = 1.626$, $N_z = 1.633$; $XYZ = abc$; $r > v$, weak. Found as radiating sprays of prismatic [001] crystals up to 5 cm long, and as parallel intergrowth on {010} with anthophyllite and cummingtonite in an ultramafic body at the *Carelton talc quarry, Chester, Windham Co., VT*. AR *AM 63:1000(1978)*.

67.1.1.2 Clinojimthompsonite $(Mg,Fe)_5[Si_6O_{16}](OH)_2$

Named in 1978 by Veblen and Burnham in reference to its symmetry and its dimorph, jimthompsonite. MON $C2/c$. $a = 9.874$, $b = 27.24$, $c = 5.316$, $\beta = 109.47°$, $Z = 4$, $D = 3.03$. *31-639*(calc): 13.6_7 8.81_{10} 4.70_3 4.04_3 3.25_3 3.08_6 2.64_6 2.51_3. Structure: *AM 63:1053(1978)*. The silicate unit is a three-tetrahedron wide ribbon of periodicity 2 (≈ 5.3 Å) lying in the *b–c* plane and parallel to [001]. The apical oxygens of a pair of ribbons are joined by six-coordinated cations to form wide I-beam units in the manner of pyroxenes and amphiboles. Colorless to light pinkish brown. Cleavage probably at {110}. Presumably biaxial; indices not reported. Found as parallel intergrowth on {010} with anthophyllite and cummingtonite in an ultramafic body at the *Carelton talc quarry, Chester, Windham Co., VT*. AR *AM 63:1000(1978)*.

67.1.2.1 Carlosturanite $(Mg,Fe^{2+},Ti,Mn)_{21}(Si,Al)_{12}O_{28}(OH)_{34} \cdot H_2O$

Described in 1985 by Compagnoni et al. for Carlo Sturani (1938–1976), professor of geology, University of Turin, killed doing fieldwork. Related to serpentine group minerals. MON Cm. $a = 36.55$–36.70, $b = 9.31$–9.41, $c = 7.27$–7.291, $\beta = 101.05$–$101.1°$, $Z = 1$, $D = 2.54$–2.61. *38-404*(nat): $18.0_{2.5}$

7.17_{10} 3.64_2 $3.60_{4.5}$ $3.40_{5.5}$ 2.56_4 2.54_2 $2.28_{3.5}$. X-ray data are similar to antigorite (71.1.2a.1). Structure is somewhat disordered. A chain silicate, characterized by the presence of triple chains. *AM 70:773(1985)*. Fibrous, splintery, elongate on [010], to several cm, further crushing produces laths. Green, light brown; nearly white streak; vitreous–pearly luster. Cleavage {001}, excellent; parting on {010}; flexible. G = 2.63–2.68. IR, DTA, TGA data: *AM 70:767(1985)*. Complete chemical analyses exist for carlosturanite from Sampeyre and Taberg. Biaxial (+); $N_z \approx 1.54$ (Taberg) and 1.60 (Sampeyre); 2V = 35°; weakly pleochroic, orange-brown and pale orange-brown parallel and perpendicular to [010] respectively; positive elongation; parallel extinction. Occurs in low-grade metamorphic rocks in a network of veins that crosscut antigorite–serpentinite (Sampeyre). Occurs with chrysotile, diopside, and opaques. *Sampeyre, Val Varaita, Piedmont, Italy*; various localities in the western Alps including Crissolo, Po Valley; Monte Nebin, Maira Valley; and around Viù, Viù Valley, Taberg, Sweden. EF *EJM 3:559(1991), MIN 4(1):205(1992)*.

67.2.1.1 Tinaksite $K_2Na(Ca,Mn)_2(Ti,Fe)O[Si_7O_{18}(OH)]$

Named in 1965 by Rogov et al. from the symbols of some of its component elements: Ti, Na, K, and Si. See tokkoite (67.2.1.2). TRIC $P\bar{1}$. $a = 10.361$, $b = 12.153$, $c = 7.044$, $\alpha = 90.79°$, $\beta = 99.22°$, $\gamma = 92.83°$, Z = 2, D = 2.84. *18-1382*: 3.25_8 3.09_5 3.03_{10} 2.95_5 2.87_5 2.67_4 2.33_5 2.00_4. Structure: *MZ 13:3(1991)*. The silicate element is a crimped band lying parallel to [001], and consisting of a pair of wollastonite-like chains joined by sharing of an oxygen, plus an additional SiO_4 tetrahedron joined to one of the chains by sharing of two oxygens. Thus W = 5 and P = 3. May also be regarded as a xonotlite chain with an appended tetrahedron sharing two oxygens. Isostructural with tokkoite. Pale yellow to pink, vitreous luster, transparent to translucent. Cleavage {010}, perfect; {110}, imperfect. H = 6. G = 2.82. Biaxial (+); 2V = 76°; $N_x = 1.593$, $N_y = 1.621$, $N_z = 1.666$; strong dispersion. Pleochroic; X,Y, colorless; Z, pale orange-yellow. Insoluble in acids. May form a solid solution with tokkoite; Mn^{2+} may substitute for Ca. Found as prismatic crystals and radiating fibrous aggregates as an accessory in K-feldspar metasomatites at the contact of limestone and the *Murun alkalic massif, near Olekminsk, Yakutia, Russia* and at Mt. Rasvumchorr, Kibiny Massif, Kola Peninsula. AR *AM 50:2098(1965), ZK 189:195(1989)*.

67.2.1.2 Tokkoite $K_2Ca_4[Si_7O_{18}(OH)]F$

Named in 1986 by Lazebnick et al. for the locality. See tinaksite (67.2.1.1). TRIC $P\bar{1}$. $a = 10.438$, $b = 12.511$, $c = 7.112$, $\alpha = 89.92°$, $\beta = 99.75°$, $\gamma = 92.89°$, Z = 2, D = 2.77. *40-517*: 10.3_4 3.34_5 3.32_8 3.26_5 3.15_{10} 3.13_8 3.08_6 3.04_9. Structure: *ZK 189:195(1989)*. Isostructural with tinaksite. Pale yellow to light brown, vitreous luster, transparent to translucent. Cleavage {010}, perfect; {110}, nearly so; splintery fracture. H = 4–5. G = 2.76. Biaxial; 2V = 38°; $N_x = 1.570$, $N_z = 1.577$; $Z \wedge c = 0$–15°; r < v, weak. Found as radiating or columnar aggregates of prismatic crystals intimately intergrown with charoite, tinaksite, and miserite in segregations with aegirine and K-feldspar in the southern part of the *Murun massif, between the Charo and Tokko rivers, Yakutia, Russia.* AR *MZ* 8:85(1986), *AM* 73:196(1988).

Class 68

Structures with Chains of More Than One Width

68.1.1.1 Chesterite (Mg,Fe)$_{17}$[Si$_4$O$_{11}$]$_2$[Si$_6$O$_{16}$]$_2$(OH)$_6$
Named in 1978 by Veblen and Burnham for the type locality. A proposed but unnamed monoclinic dimorph is MON $A^*/^*$. $a = 9.867$, $b = 45.31$, $c = 5.292$, $\beta = 109.7°$, $Z = 2$. ORTH $A2_1ma$. $a = 18.614$, $b = 45.306$, $c = 5.297$, $Z = 4$, $D = 3.09$. *31-637*(calc): 11.3$_4$ 8.61$_{10}$ 3.75$_3$ 3.25$_5$ 3.08$_7$ 2.78$_3$ 2.64$_3$ 2.55$_4$. Structure: *AM 63:1053(1978)*. The structure is made up of three-wide I-beams as in jimthompsonite, alternating with amphibole-like, two-wide I-beams; these lie parallel to [001] and alternate along [010]. Colorless to light pinkish brown. Cleavage prismatic, {210} at 45°. Biaxial (−); 2V = 71°; N$_x$ = 1.617, N$_y$ = 1.632, N$_z$ = 1.640; XYZ = *abc*; r > v, weak. Found as radiating sprays of prismatic [001] crystals up to 5 cm long, and as parallel intergrowth on {010} with anthophyllite and cummingtonite in an ultramafic body at the *Carelton talc quarry, Chester, Windham Co., VT*. AR *AM 63:1000(1978)*.

68.1.2.1 Vinogradovite (Na,Ca)$_4$(Ti,Nb)$_4$O$_4$[Si$_2$O$_6$]$_2$[(Si,Al)$_4$O$_{10}$] (H$_2$O,K)$_3$
Named in 1956 by Semenov et al. for Aleksander Pavlovich Vinogradov (1895–1975), Russian geochemist. MON $C2/c$. $a = 24.50$, $b = 8.662$, $c = 5.211$, $\beta = 100.13°$, $Z = 2$, $D = 3.15$. *PD*(type): 3.21$_{10}$ 3.07$_{10}$ 2.72$_7$ 2.48$_6$ 1.614$_8$ 1.558$_7$ 1.494$_7$ 1.434$_7$; *40-480*(Ilimaussak): 12.2$_{10}$ 4.49$_6$ 3.30$_{10}$ 3.22$_8$ 3.07$_9$ 3.04$_5$ 2.34$_9$ 1.476$_6$. Structure: *ZK 200:237(1992)*. The silicate structural elements are a pair of pyroxene-like chains, [Si$_2$O$_6$], and a vlasovite-like ribbon of four-tetrahedron rings linked to one another via two apical oxygens, [AlSi$_3$O$_{10}$]; both parallel to [001]. Spherulites and fibrous aggregates; less commonly as prismatic, [001], and/or tabular, {100}, crystals, with {310} common and terminated by {221} (these indices being different from those of the original description). Colorless, white, mauve, pink; vitreous luster, in part silky to pearly; transparent to translucent. Cleavage {010}, perfect; brittle. H = 4. G = 2.88. Biaxial (−); 2V = 41°; N$_x$ = 1.734, N$_y$ = 1.770, N$_z$ = 1.774; Z $\wedge$ c = 7°, X = *b*; r > v. Beryllian Ilimaussak material: 2V = 81°; N$_x$ = 1.740, N$_y$ = 1.769, N$_z$ = 1.792; X $\wedge$ *a* = 10°, Y = *b*. Beryllium may replace Si with concomitant charge balancing substitution for other cations. Found in the Green River formation, WY; at Mt. St.-Hilaire, PQ, Canada; at Ilimaussaq, Greenland; with natrolite and analcime or replacing lorenzenite

and lamprophyllite at 12 pegmatites, including those at *Takhtarvumchorr, Khibiny massif, and Nepakh, Lovozero massif, Kola Penin., Russia.* AR *AM* *42:308(1957); MR 8:368(1977), 10:99(1979); NJMH 11:481(1990).*

68.1.3.1   Aërinite $Ca_4(Ca,Al,Mg)_{10}(Si_{12}O_{35})(OH)_{12}(CO_3)\cdot 12H_2O$

Named in 1876 by von Lasaulx from the Greek root *aer-*, alluding to atmosphere or sky and hence the color, sky blue. Revalidated in 1988. MON $a = 14.69$, $b = 16.87$, $c = 5.17$, $\beta = 94.75°$, $Z = 1$, $D = 2.47$. PD: 14.7_{10} 4.05_8 3.80_3 3.66_3 3.65_3 3.28_3 2.81_5 2.72_7. Blue, blue-green; translucent. $H = 3$. $G = 2.48$. Biaxial $(-)$; $2V_{calc} = 63°$; $N_x = 1.510$, $N_y = 1.560$, $N_z = 1.580$. Pleochroic; X, bright blue; $Y \approx Z$, pale beige. A hydrothermal mineral of the zeolite facies. Reported from Gunsight Mts., Pima Co., AZ. Originally described from *Casseras, Huesca, Spain*, and also Juseu and Estopiñan, and from Tartaren, Lerida; as fibers with scolecite and prehnite in tholeitic dolerite at *St.-Pandelon, Landes, France* (revalidating material); from Ourika, Morocco. AR *D6:1025(1892), BM 111:39(1988), AM 73:1498(1988).*

Class 69

Chains with Side Branches or Loops

ASTROPHYLLITE GROUP

Members of the astrophyllite group conform to the formula

$$A_3B_{6-7}C_2(Si_8O_{24})(O,OH)_7$$

where

A = Na, K, Cs, Ca, H_3O
B = Fe, Mn, Mg
C = Ti, Nb, Zr

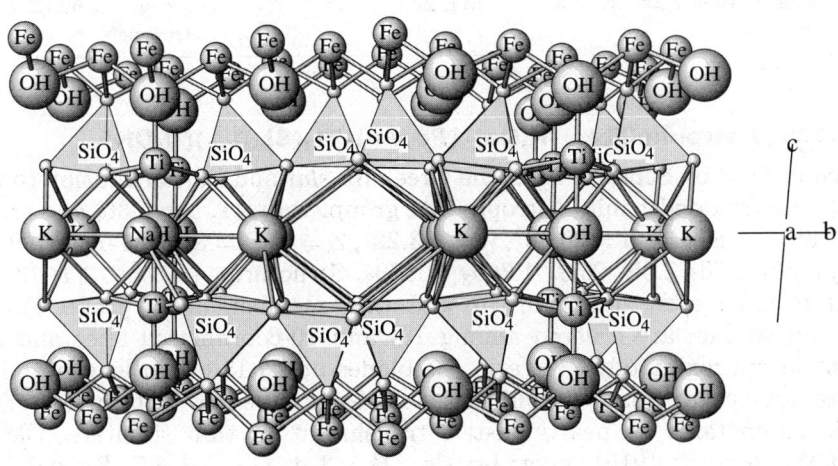

ASTROPHYLLITE
Perspective of the structure typical of the group viewed along the a-axis silicate chain.

The structure is based on infinite $(Si_4O_{12})^{10-}$ chains that are cross-linked via octahedrally coordinated cations $(Ti, Zr, Nb)^{4+}$ to form sheets that lie parallel to {001} and thus account for the micaceous habit and cleavage of the crystals. The periodicity along the chain is approximately 5.4 Å, similar to that in the pyroxenes. The chain conformation is that of a pyroxene but with additional tetrahedra joined to the outward-pointing oxygens of the core, thus yielding a branched chain about three tetrahedrons wide. The outer tetrahedra are cross-linked via the (Ti, Zr, Nb) octahedra to form sheets, pairs of which are joined by octahedral sheets of the B atoms to form slabs. The large A atoms lie between these slabs.

Valence changes involved in some of the substitutions cause slight deviation from the stoichiometry of the general formula.

In the following species descriptions, the unit cells of the original or structural descriptions have, wherever possible, been transposed to reflect the isotypic character. *AC 22:673(1967), MM 39:97(1973)*.

ASTROPHYLLITE GROUP

Mineral	Formula			Space Group	
	A	B	C		
Astrophyllite	Na, K	Fe, Mn	Ti	$A\bar{1}$	69.1.1.1
Kupletskite	K, Na	Mn, Fe	Ti, Nb	unknown	69.1.1.2
Cesium kuplestskite	Cs	Mn, Fe	Ti, Nb	$P1$ or $P\bar{1}$	69.1.1.3
Niobophyllite	K, Na	Fe, Mn	Nb, Ti	$P1$ or $P\bar{1}$	69.1.1.4
Zircophyllite	K, Na	Mn, Fe	Zr	unknown	69.1.1.5
Hydroastrophyllite	H_3O, K, Ca	Fe, Mn	Ti	$P1$ or $P\bar{1}$	69.1.1.6
Magnesium astrophyllite	K, Na	Mg, Fe	Ti	$A2/m$ or triclinic	69.1.1.7

69.1.1.1 Astrophyllite $(K,Na)_3(Fe,Mn)_7Ti_2(Si_8O_{24})(O,OH)_7$

Named in 1854 by Scheerer from the Greek for *star* and *leaf*, in allusion to the stellate and foliated habit. Astrophyllite group. TRIC $A\bar{1}$. $a = 5.36$, $b = 11.76$, $c = 21.08$, $\alpha = 85.13°$, $\beta = 90.00°$, $\gamma = 103.22°$, $Z = 2$, $D = 3.41$. *41-1490*: 10.5_{10} 3.53_{10} 3.20_4 2.78_7 2.65_7 2.58_6 1.765_3 1.579_3. Structure: *AC 22:673(1967)*. A primitive cell is often stated for this and other members of the group. Potassium and sodium are ordered among 13- and 10-coordinated sites and are present in approximately 2:1 ratio. As blades up to 15 cm in length and in stellate aggregates; foliated {001} masses. Golden yellow to bronze, yellow streak, submetallic to pearly luster, translucent in thin splinters. Cleavage {001}, perfect; {010}, poor; brittle. $H = 3-4$. $G = 3.3-3.7$. Biaxial (+), less commonly (−); $2V = 70-85°$; $N_x = 1.663-1.740$, $N_y = 1.691-1.746$, $N_z = 1.716-1.765$; $X \approx b$, $Z \wedge c = 3-19°$; $r > v$, strong. Pleochroic, $X > Y > Z$; yellow; higher indices are concomitant with substitution of heavier atoms in B and C. Substitution of Nb^{5+} for Ti^{4+} in the C position involves vacancies in other sites. Forms a series with kupletskite. Found in nepheline syenites, syenite pegmatites, and alkali granites and gneisses associated with albite, aegirine, arfvedsonite, nepheline, and others. At Magnet Cove, AR; St. Peter's Dome, El Paso Co., CO: Golden Horn batholith, Okanogan Co., WA. In the Red Wine complex, NF, and at Mt. St. Hilaire, PQ, Canada. At Narsarsuk, Julianehaab, and elsewhere in eastern Greenland. *Låven Is., near Brevik, Langesundfjord, Norway*; Pontevedra, Spain; October massif, eastern Azov area, and from Mt. Yukspor and Mt. Eveslogchorr, Khibiny massif, Kola Penin., Russia; at Dara-i-Pioz, Tien Shan, southern Tajikistan. At Mt. Charib, Egypt; at Iles de Los, Guinea; and Pilansberg, Transvaal, South Africa. AR *D6:719(1892), MZ 9(6):77(1987), MM 56:17(1992)*.

69.1.1.2 Kupletskite $(K,Na)_3(Mn,Fe)_7(Ti,Nb)_2[Si_8O_{24}](O,OH)_7$

Named in 1956 by Semenov for Russian geologists Boris Mikhailovich Kupletski (1894–1964) and Elsa Maximilianova Bohnshtedt Kupletskaya (1897–1974). Astrophyllite group. Probably TRIC Space group unknown. *25-6*: 3.51_{10} 3.25_1 3.00_1 2.76_1 2.64_{10} 2.57_5 2.10_4 1.732_4. Tabular–acicular crystals; lamellar masses up to 2 cm consisting of platy crystals. Dark brown to black, brown streak. Cleavage {001}, perfect; also reported as {100}. H $\approx$ 4. G = 3.20–3.23. Biaxial (–); 2V = 79°; N_x = 1.656, N_y = 1.699, N_z = 1.731. Forms a series with astrophyllite. Found in nepheline syenites and syenite pegmatites. Found at 3M quarry, Little Rock, Pulaski Co., AR, with aegirine, analcime, gonnardite, and eggletonite. At Mt. St.-Hilaire, PQ, Canada; in the Ilimaussaq and Werner Berg complexes, Greenland; and in the October massif, eastern Azov area, and the *Lovozero massif, Kola Penin., Russia*, associated with schizolite, neptunite, eudyalite, microcline, and nepheline. AR *DANS 108:933(1956), AM 42:118(1957), CA 65:495(1966), MR 21:284(1990)*.

69.1.1.3 Cesium Kupletskite $Cs_3(Mn,Fe)_7(Ti,Nb)_2[Si_8O_{24}]$ $(O,OH,F)_7$

Named in 1971 by Effimov et al. for the composition and relationship to kupletskite. Astrophyllite group. TRIC $P1$ or $P\bar{1}$. a = 5.41, b = 11.74, c = 21.16, α = 89°, β = 90°, γ = 102.4°, Z = 2, D = 3.62. *25-221*: 10.4_{10} 4.09_3 3.76_3 3.54_8 2.79_8 2.66_8 2.58_6 1.772_4. Dull golden brown; submetallic, dull luster. Cleavage {001}, perfect. H $\approx$ 4. G = 3.68. Biaxial (+); 2V = 75°; N_y = 1.726, N_z = 1.758; Z = a, Y $\wedge$ b $\approx$ 9°. Pleochroic, X < Y < Z; yellow-green to brown. Found as rosettes of curved, platy crystals in a border-zone aegirine-bearing pegmatite in the *Alai alkalic province (Dar-i-Pioz?), Tajikistan*. AR *DANS 197:1394(1971), AM 57:328(1972)*.

69.1.1.4 Niobophyllite $(K,Na)_3(Fe,Mn)_6(Nb,Ti)_2[Si_8O_{24}](O,OH)_7$

Named in 1964 by Nickel et al. in allusion to the composition and foliated character, as in astrophyllite. Astrophyllite group. TRIC $P1$ or $P\bar{1}$. a = 5.391, b = 11.88, c = 21.16, α = 95.2°, β = 87.7°, γ = 103.2°, Z = 2, D = 3.406. *17-742*: 10.5_9 3.74_4 3.51_{10} 3.26_5 3.02_6 2.78_8 2.57_7 2.48_4. Brown. Cleavage {001}, perfect. G = 3.42. Biaxial (–); 2V = 60°; N_x = 1.724, N_y = 1.760, N_z = 1.772;

$X = b$; $Z \wedge c = 12°$. Found as aggregates of {001} flakes in an albite–arfvedsonite paragneiss near *Ten Mile Lake, Seal Lake area, Labrador, NF, Canada*, associated with aegirine, barylite, eudidymite, neptunite, Mn-pectolite, pyrochlore, joaquinite, sphalerite, and galena. AR *CM 8:40(1964)*.

69.1.1.5 Zircophyllite $(K,Na)_3(Mn,Fe)_7Zr_2[Si_8O_{24}](O,OH,F)_7$

Named in 1972 by Kapustin for the composition and foliation, as in astrophyllite. Astrophyllite group. TRIC Space group unknown. *25-856:* 9.80_4 3.75_3 3.50_{10} 3.26_3 2.80_7 2.66_5 2.10_5 1.781_3. Dark brown to black, brown streak, vitreous to adamantine luster. Cleavage {001}, perfect; brittle. $H = 4-4\frac{1}{2}$. $G = 3.34$. Biaxial $(-)$; $2V = 62°$; $N_x = 1.708$, $N_y = 1.738$, $N_z = 1.747$; $r > v$, strong. Pleochroic, $X = Y < Z$; dark yellow to brown. Found as {001} platy aggregates in the natrolite zone of alkalic pegmatite at *Korgereda-binsk, Tuva, Russia*. AR *ZVMO 101:459(1972), AM 58:967(1973)*.

69.1.1.6 Hydroastrophyllite $(H_3O,K,Ca)_3(Fe,Mn)_{5-6}Ti_2[Si_8O_{24}](O,OH)_7$

Named in 1974 for its composition and relationship to astrophyllite. Astrophyllite group. TRIC $P1$ or $P\bar{1}$. $a = 11.86$, $b = 11.98$, $c = 5.42$, $\alpha = 103.42°$, $\beta = 95.15°$, $\gamma = 112.20°$, $Z = 1$, $D = 2.79$. *29-991:* 10.55_9 3.51_{10} 2.87_4 2.78_4 2.64_8 2.58_5 2.12_6 1.768_6. Dark brown. $G = 3.15$. Biaxial $(-)$; $2V = 40°$; $N_x = 1.660$, $N_y = 1.720$, $N_z = 1.728$. Pleochroic; X, yellow; Y, orange-yellow; Z, dull yellow. Found as a weathering product in alkalic pegmatite in *Szechuan Prov., China*. AR *AM 60:736(1975)*.

69.1.1.7 Magnesium Astrophyllite $K_2Na_2Mg_2Fe_5Ti_2[Si_8O_{24}](O,OH,F)_7$

Named in 1963 by Peng and Ma for its composition and relationship to astrophyllite. Astrophyllite group. MON $A2/m$ (also reported as TRIC). $a = 10.56$, $b = 23.00$, $c = 5.35$, $\beta = 102°$, $Z = 2$, $D = 3.32$. *29-1042:* 10.1_8 3.80_6 3.38_{10} 3.08_5 2.76_5 2.55_9 1.818_5 1.463_7. Straw-yellow, vitreous luster. Cleavage {100}, {010}, perfect. $H = 3$. Biaxial $(-)$; $2V \approx 82°$; $N_x = 1.658$, $N_{y\,calc} = 1.687$, $N_z = 1.710$; $Y = b$, $Z \wedge c \approx 5°$; pleochroic in yellows. Found at an unspecified locality near *Khibiny, Kola Penin., Russia*. AR *SGS 1:18(1974), AM 60:737(1975)*.

AENIGMATITE GROUP

The aenigmatite minerals, with the general formula

$$A_2B_6T_6O_{20} \quad \text{or} \quad A_2[M_6O_2][T_6O_{18}]$$

where

A = Ca, Na
B = Mg, Fe^{2+}, Fe^{3+}, Ti, Mn, Sn, Cr, Sb
T = Si, Al, Be

crystallize in space group $P\bar{1}$. The crystal structure of aenigmatite group minerals is based on complex aluminosilicate chains, consisting of pyroxene-like chains with extra tetrahedra attached alternately on opposite sides, having the composition $[T_6O_{18}]_n$. The structure contains six tetrahedrally coordinated sites, seven octahedrally coordinated sites, and two 7- or 8-coordinated sites. Details of the aenigmatite structure are given in AM 56:427(1971) and CM 32:439 (1994).

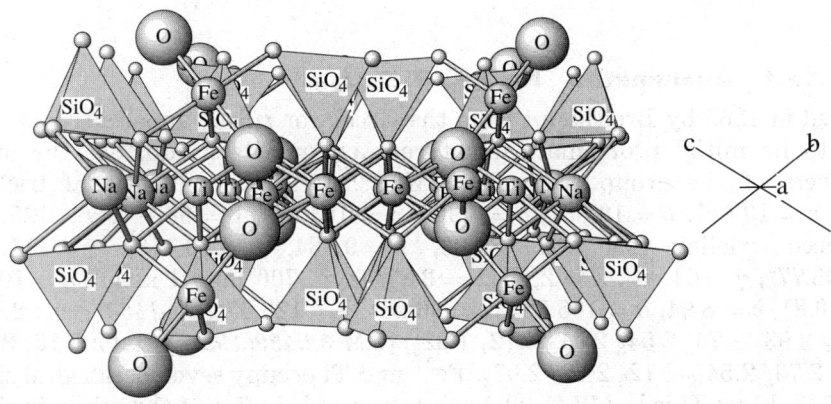

AENIGMATITE: $Na_2Fe_5TiSi_6O_{20}$

This view shows the similarity to pyroxenes and amphiboles. Compare diopside. There is a chain with alternating tetrahedra like pyroxene parallel to a, but there are some single tetrahedra branching off from this chain; thus the structure is in some respects intermediate between pyroxene and amphibole. A unique aspect of aenigmatite is the Fe atoms in the plane of the chains, which weakly link them into sheets, and which in turn introduce nontetrahedral oxygens. There are no OH/F atoms in the center of hexagonal rings of tetrahedra as in amphiboles; the chains are not wide enough.

AENIGMATITE GROUP

Mineral	Formula	a	b	c	α	β	γ	
Aenigmatite	$Na_2Fe_5^{2+}TiSi_6O_{20}$	10.417	10.837	8.929	104.86	96.77	122.54	69.2.1a.1
Dorrite	$Ca_2(Mg_2Fe_4^{3+})(Al,Fe^{3+})_4 Si_2O_{20}$	10.505	10.897	9.019	106.26	95.16	124.75	69.2.1a.2
Høgtuvaite	$Ca_2Fe_4^{2+}Fe_2^{3+}Si_5BeO_{20}$	10.317	10.724	8.855	105.77	96.21	124.77	69.2.1a.3
Krinovite	$NaMg_2CrSi_3O_{10}$	10.238	10.642	8.780	105.15	96.50	125.15	69.2.1a.4
Rhönite	$Ca_2(Mg,Fe^{2+},Fe^{3+},Ti)_6 (Si,Al)_6O_{20}$	10.367	10.756	8.895	105.98	96.04	124.72	69.2.1a.5
Serendibite	$(Ca,Na)_2(Al,Mg,Fe^{2+},Fe^{3+})_6 (Si,B,Al)_6O_{20}$	10.010	10.393	8.631	106.37	96.10	124.38	69.2.1a.6
Welshite	$Ca_2Sb^{5+}Mg_4Fe^{3+}Si_4Be_2O_{20}$	10.167	10.65	8.787	106.5	96.1	124.0	69.2.1a.7
Wilkinsonite	$Na_2Fe_2^{2+}Si_6O_{20}$	10.355	10.812	8.906	105.05	96.63	125.20	69.2.1a.8

Note: Magbasite (69.2.1.3) may be an additional member of the group. Sapphirine (69.2.1b.1) and Surinamite (69.2.1b.2) are related to the Aenigmatite Group.

69.2.1a.1 Aenigmatite $Na_2Fe_5^{2+}TiSi_6O_{20}$

Named in 1865 by Breithaupt from the Greek for *riddle*, apparently in allusion to its initial problematical nature. Aenigmatite group. See the other members of the group. TRIC $P\bar{1}$. *CM 32:439(1994)*(conventional triclinic cell): $a = 10.417$, $b = 10.837$, $c = 8.929$, $\alpha = 104.86°$, $\beta = 96.77°$, $\gamma = 125.54°$; (reduced triclinic cell): $a = 8.930$, $b = 9.734$, $c = 10.416$, $\alpha = 114.31°$, $\beta = 96.77°$, $\gamma = 64.98°$; $Z = 2$, $D_{syn} = 3.87$; (syn, 700° and 1 kbar): $a = 10.35$, $b = 10.81$, $c = 8.94$, $\alpha = 105.28°$, $\beta = 96.25°$, $\gamma = 125.57°$. *22-1453*(syn): 8.09_{10} 3.15_{10} 2.93_7 2.70_8 2.54_8 2.41_6 2.12_7 1.62_6; [*CM 32:439(1994)*] (nat): 8.12_6 3.70_3 2.71_8 2.70_8 2.54_{10} 2.12_3 2.12_4 2.07_3. Fe^{2+} and Ti occupy seven octahedral sites, M1–M7. Most Ti is in M7 (1.99 Å) and the rest is in five of the other six sites. *AM 56:427(1971)*. There are six tetrahedral sites and two seven- or eight-coordinated sites as well. Long prismatic, to > 8 cm; also as irregular segregations; twinning on $\{1\bar{1}0\}$ common, usually lamellar. Black, brown; reddish-brown streak; vitreous to greasy luster. Cleavage {010}, {100}, perfect; uneven fracture; brittle. $H = 5\frac{1}{2}$–6. $G = 3.74$–3.86. Insoluble in acids. Analysis (wt %): SiO_2, 37.3–41.8; Al_2O_3, 0.1–2.7; Fe_2O_3, 1.3–10.0; FeO, 20.3–38.8; TiO_2, 6.8–10.2; MnO, 0.4–4.4; MgO, 0.0–2.2; CaO, 0.0–3.7; Na_2O, 5.6–7.5; K_2O, 0.0–0.6. Ti-free aenigmatite has $(Na_2Fe_4^{2+}Fe_2^{3+}Si_6O_{20})$. Biaxial (+); $Z \wedge c = 40°$; $N_x = 1.79$–1.81, $N_y = 1.805$–1.826, $N_z = 1.87$–1.90; $2V = 27$–55°; r < v. X, yellowish brown to reddish brown; Y, reddish brown to dark brown; Z, reddish brown to brownish black. Occurs in alkalic rocks, particularly pegmatites, also in Na-rich alkalic lavas; peralkaline to transitional lavas produced by advanced differentiation from basalts in oceanic islands and continental rift zones. Also in plutonic rocks, due to differentiation of gabbros.

Occurs worldwide and only selected localities are given here: Point of Rocks, Colfax Co., NM; Picture Gorge Basalt, Spray, OR; Sonoma Co., CA; Mt. St.-Hilaire, QUE, Canada; nearly end-member composition from the Red Wine alkalic complex, central Labrador, Canada; Ilimaussaq and Narssârssuk, Julianehaab district, Greenland; also at Tungliarfik and Naujakasik, Greenland; Vesteroya, Langesundfjord, Norway; Pantelleria Is., Italy; Lovozero, Khibiny, Kovdor, and other alkalic massifs, Kola Penin., Russia; eastern Sayan Mts., Siberia, Russia; Wonchi volcano, Ethiopia; Liruei complex, Nigeria; Mulanje complex, Malawi; Nandewar Mts., NSW, Australia. EF *DHZ IIA:640(1978), MM 55:529(1991)*.

69.2.1a.2 Dorrite $Ca_2(Mg_2Fe_4^{3+})(Al,Fe^{3+})_4Si_2O_{20}$

Named in 1988 by Cosca et al. for John A. Dorr (1922–1986), University of Michigan. Aenigmatite group. Chemically most similar to rhönite. TRIC $P1$ or $P\bar{1}$. $a = 10.505$, $b = 10.897$, $c = 9.019$, $\alpha = 106.26°$, $\beta = 95.16°$, $\gamma = 124.75°$, $Z = 2$, $D = 3.959$. 45-1408(nat): 8.1_2 7.5_2 2.97_{10} 2.56_8 2.52_8 2.12_6 1.51_3 1.48_3. *AM 73:1440(1988)*. Anhedral to prismatic grains to 0.1 mm long; twinning present producing pseudomonoclinic symmetry. Red-brown to brown, gray streak, submetallic luster. Cleavage is parallel to {010} and {001}, irregular fracture, brittle. H ≈ 5. Analysis (wt %): SiO_2, 10.4–12.0; Al_2O_3, 8.6–24.85; Fe_2O_3, 41.65–59.06; FeO, 2.8–3.4; TiO_2, 0.4–1.04; MnO, 0.2–0.7; MgO, 4.3–5.6; CaO, 12.1–13.6; Na_2O, 0.0–0.2; K_2O, 0.0. In some samples Fe^{3+} is $> Al^{3+}$ in the tetrahedral site (to 75%). Biaxial (−); $N_x = 1.82$, $N_y = 1.84$, $N_z = 1.86$; 2V large, near 90°. In ultrathin section: X, red-orange brown; Y, yellowish brown; Z, greenish brown. Occurs in a paralava (pyrometamorphic melt-rock) formed from burning coal beds. Associated with esseneite or titanian andradite. Also from a basalt–limestone contact on Reunion Is. Occurs at the *Durham ranch, 25 km S of Gillette, WY (Powder River Basin)*; St.-Leu, Reunion Is. EF *BM 105:364(1982)*.

69.2.1a.3 Høgtuvaite $(Ca,Na)_2(Fe^{2+},Fe^{3+},Ti,Mg,Mn,Sn)_6(Si,Be,Al)_6O_{20}$; simplified formula: $Ca_2Fe_4^{2+}Fe_2^{3+}Si_5BeO_{20}$

Named in 1994 by Grauch et al. for the locality. Aenigmatite group. See other members of the group. This mineral was also described under the name *makarochkinite* [*SPC 35:818(1990)*], but this name was not approved by the CNMMN IMA. TRIC (pseudo-MON) $P\bar{1}$. $a = 10.317$–10.385, $b = 10.724$–10.751, $c = 8.855$–8.959, $\alpha = 104.76$–$105.77°$, $\beta = 96.16$–$97.03°$, $\gamma = 124.77$–$125.47°$, $Z = 2$, $D = 3.92$–3.98. PD(nat): 8.05_9 3.12_5 2.92_6 2.68_5 2.53_{10} 2.10_6 2.07_5 1.62_3. The structure is that of the aenigmatite group. *CM 32:439(1994)*. Bladed; to 4 cm long, 6 mm circumference maximum, striated parallel to axis of elongation; polysynthetic twinning, with the twin axis perpendicular to (010) (of monoclinic cell). Black; dark green streak; vitreous, nonmetallic luster. Two good cleavage directions at ≈ 55°; pronounced parting present approximately perpendicular to direction of elongation; uneven fracture; brittle. H = $5\frac{1}{2}$. G = 3.85–3.88. Mössbauer spectral data indicate that Fe^{3+}/Fe^{2+} = 0.6; TGA shows 0.25% weight loss at 100° and a minor weight gain (oxidation of Fe^{2+}) at 800°. Contains 2.65 wt % BeO, low Al. Inert to common acids and bases. Soluble in HF. Probably biaxial (−); $N_{x'}$ = 1.78, $N_{z'}$ = 1.82; 2V large; strongly pleochroic. X, green; Z, bronze. Occurs in peraluminous granitic gneiss. Also in a small metamorphic granitic pegmatite and granite pegmatite. Associated minerals include: quartz, albite, microcline, phenakite, biotite, zircon, fluorite, magnetite, and others. Formed in a second metamorphic event. Occurs as poikiloblasts. Found at Høgtuva, Nordland Co., Norway; mine 400, Ilmensky Natural Forest, Russia. EF

69.2.1a.4 Krinovite $NaMg_2CrSi_3O_{10}$

Named in 1968 by Olsen and Fuchs for Evgeny Leonidovich Krinov (1906–1984), Russian student of meteorites. Aenigmatite group. TRIC $P\bar{1}$. a = 10.238, b = 10.642, c = 8.780, α = 105.15°, β = 96.50°, γ = 125.15°, Z = 4, D= 3.46. ZK 187:133(1989). 20-1123(nat): 7.92_6 3.64_6 3.10_6 2.89_8 2.65_9 2.50_{10} 2.08_7 2.05_6. Largest grain is 200 μm. Emerald-green. H = 6–7. G = 3.38. IR data: GZ 36:76(1976). Analysis (wt %): SiO_2, 48.1; Al_2O_3, 0.6; Cr_2O_3, 19.1; FeO, 1.8; TiO_2, 0.5; MnO, 0.1; MgO, 19.7; CaO, 0.1; Na_2O, 9.1; K_2O, 0.0. Biaxial (+); OAP parallel to b (monoclinic cell); Na light: N_x = 1.702, N_y = 1.725, N_z = 1.760; 2V = 61°. X, yellow-green; Y, blue-green; Z, greenish black (anomalous reddish brown). Occurs in meteorite as subhedral grains in graphite nodules in octahedrites: Canyon Diablo, AZ; Wichita Co., KA; and Youndegin, Australia; in association with olivine, roedderite, high albite, richterite, and other minerals. EF ZK 136:81(1972).

69.2.1a.5 Rhönite $Ca_2(Mg,Fe^{2+},Fe^{3+},Ti)_6(Si,Al)_6O_{20}$ and $Ca_4(Fe^{2+}_{3.6}Fe^{3+}_{1.9}Mg_{4.28}Ti_{2.84})_{12}(Si,Al)_{12}O_{40}$

Named in 1907 by Soellner for the locality. Aenigmatite group. For Fe-dominant material: $\sum$ Fe > Mg. Need two names, one for Mg-dominant and one for Fe-dominant material. Several polytypes: polysomes—pyroxene + spinel slabs. TRIC $P\bar{1}$. (Allende meteorite): a = 10.367, b = 10.756, c = 8.895, α = 105.98°, β = 96.04°, γ = 124.72°; (volcanic breccia): a = 10.428, b = 10.807, c = 8.925, α = 105.91°, β = 96.13°, γ = 124.80°; Z = 2 [EJM 2:203(1990)], D = 3.72–3.79. 23-607(nat): $8.14_{2.5}$ 2.95_{10} 2.78_3 $2.69_{6.5}$ 2.55_7 $2.09_{5.5}$ $1.99_{2.5}$ 1.50_4. The structure consists of "octahedral" walls connected through tetrahedral chains with appendices and additional insular octahedra. OD (order–disorder) schemes present. EJM 2:203(1990). As prismatic or tabular crystals to several centimetres or more; also as skeletal to anhedral grains. Forms include {0$\bar{1}$1}, {100}, {010}, {21$\bar{1}$} + others; polysynthetic twinning, with the twin axis perpendicular to (010) plane of the pseudomonoclinic cell. Brown, black; good {010} and {001} cleavage vitreous luster. H = 5–6, G = 3.4–3.75. Ti^{3+} and Ti^{4+} are present in the mineral from the Allende meteorite. Rhönite is of relatively wide compositional range. Aenigmatite is not. Thus a solubility gap exists. Ca + Al ↔ Na + Si. Analysis (wt %): SiO_2,

19.1–30.7; Al_2O_3, 12.3–28.9; Fe_4O_3, 16.0–24.5; Fe_2O_3, 6.9–11.7; FeO, 11.4–17.6; TiO_2, 5.1–16.8; MnO, 0.1–0.4; CaO, 0.4–13.0; Na_2O, 0.7–3.1; K_2O, 0–0.6. Biaxial (+); Na light: $N_x = 1.79$–1.810, $N_y = 1.80$–1.825, $N_z = 1.83$–1.845; $2V = 47$–$90°$; $X < Y < Z$; $r < v$, extreme. X, greenish brown to red-brown; Y, greenish brown–reddish brown or brownish yellow; Z, dark red, black, reddish brown. Occurs in volcanic breccia, the Allende meteorite, a mafic alkaline sill in Texas, in a nepheline basanite in Rhön, Germany, in nepheline–dolerites in Massif Central, France. Also in many other mafic rocks of the Rhön region. In an inclusion of pyroxenite in basalt in Haute-Loire, France. Principally found in sillica-poor, mafic to intermediate rocks. Occurs with titanaugite and feldspathoids. Often an alteration product of amphiboles. Occurs at Big Bend National Park, Brewster Co., TX; north of Kalaheo, Kaui, Hawaii; Hobbs Land, Greenland; Puy Forestier, Haute-Loire, Puy de St.-Sandoux (Auvergne), and Puy de Barneire, Massif Central, France; *Scharnhausen, Platz, Bruckenau, Rhön District, Hessen*, Löbauer Berg, Saxony, Fe-dominant from Sattel volcano, Eich, East Eifel, *Germany*; Snababerg, Skåne, southern Sweden; St.-Leu, Reunion Is.; Hut Point Penin., Ross Is., Antarctica, and several other Antarctic localities; Fe-dominant from Pulling Point, Otago Harbor, Dunedin volcano, Morotu district, New Zealand; Dogo, Oki Islands, Japan, and Sakhalin Is., Russia; high Ti-bearing in the Allende meteorite. EF *TMPM 18:17(1972)*, *AM 70:1211(1985)*, *DA 40:391(1989)*.

69.2.1a.6 Serendibite $(Ca,Na)_2(Al,Mg,Fe^{2+},Fe^{3+})_6(Si,B,Al)_6O_{20}$

Named in 1903 by Prior and Coomáraswámy, for the locality, Serendib, an old Arabic name for Ceylon. Aenigmatite group. See the other members of the group. TRIC $P\bar{1}$. (Fe-poor, 2 wt % FeO): $a = 10.010$, $b = 10.393$, $c = 8.631$, $\alpha = 106.37°$, $\beta = 96.10°$, $\gamma = 124.38°$ [*AM 78:195(1993)*]; (intermediate Fe, 5.8% FeO): $a = 10.043$, $b = 10.428$, $c = 8.662$, $\alpha = 106.37°$, $\beta = 96.02°$, $\gamma = 124.40°$; (high Fe, 13% FeO): $a = 10.094$, $b = 10.478$, $c = 8.694$, $\alpha = 106.36°$, $\beta = 96.00°$, $\gamma = 124.40°$; $Z = 2$, $D = 3.47$. *29-343*(nat): 3.32_4 2.85_{10} 2.60_8 2.46_8 2.04_7 2.03_7 2.02_5. Tetrahedral substitutions are constrained to B for Si and Si for Al. One tetrahedral site is nearly fully occupied by B, whereas the other has a B/Si ratio near 1.5. These two tetrahedra are the only ones in serendibite to share three vertices with other tetrahedra, resulting in B–O–B and B–O–Al bridges, as well as B–O–Si bridges. There is little or no Fe in the tetrahedral site. Fe^{3+} is in octahedral sites along with Fe^{2+}. No evidence for ordering of Fe was found. See *AM 78:195(1993)* and *PNAS 71:4348(1974)* for fundamental details. Oblique tablets or as anhedral grains to 2 cm; fine, polysynthetic twinning on $\{0\bar{1}1\}$ often present. Grayish blue-green, deep blue, sky-blue, dark green, to nearly black and also pale yellow; vitreous luster. Cleavage $\{010\}$, $\{001\}$, good. $H = 6\frac{1}{2}$–7. $G = 3.42$–3.515. Infusible, only slightly attacked by acids. Analysis (wt %): SiO_2, 17.1–26.3; Al_2O_3, 30.7–43.0; B_2O_3, 4.2–8.4; Fe_2O_3, 2.3–5.05; FeO, 2.8–6.1; TiO_2, 0.1–0.3; MnO, 0.2; MgO, 8.45–15.4; CaO, 13.3–17.1; Na_2O, 0.0–0.5; K_2O, 0.2. B_2O_3 may be as high as ≈ 9 wt %. Material from Johnsburg, NY, has 0.9–1.37% Na_2O. Usually biaxial (+); $Z \wedge c = 26$–$40°$; $N_x = 1.700$–1.738, $N_y = 1.703$–1.741, $N_z = 1.706$–1.743; birefringence $= 0.004$–0.006; $2V = 78$–$90°$, also large (−) $2V$; $r < v$ for optically (+) material; dispersion sometimes so strong as to produce anomalous interference colors. X, greenish blue–yellowish green; Y, pale blue, blue-green, pale yellow; Z, blue, indigo-blue, greenish yellow. Occurs typically in contact-metamorphosed calcsilicate rocks, particularly gneisses. Typical associated minerals include diopside, apatite, calcite, clinopyroxene, spinel, pargasite, phlogopite, and plagioclase, but quartz is conspicuously absent. Tourmaline, grandidierite, sinhalite, and other minerals may or may not be directly associated. Typically occurs in zoned metasomatic skarns between marble and feldspathic rock. Serendibite is rare because of the combination of Si-undersaturated skarns, metamorphosed at fairly high temperatures (1–9.5 kbar and 600–825°) during influx of B-bearing fluids. Associated with sinhalite at Johnsburg, Warren Co., Amity, Orange Co., NY; Russell, St. Lawrence Co. (northwestern Adirondacks), NY; New City quarry, Riverside, CA; Melville Peninsula, NWT, Canada; Bug River area, Ukraine; associated with uvite at the Tayozhnoye iron deposit, Aldan Shield, Yakutia, Russia; high-iron-containing (13.0 wt % FeO) from the Ozersk gabbroic massif, Priolkhon (western Pribaikal), Russia; 10 km SE of Ihosy, southern Madagascar (colorless to light green); Ianapera, southwestern Madagascar; Handeni district, Tanzania; in a granulite–limestone contact zone at *Gangapitiya, Sri Lanka (Ceylon)*. EF *MIN 3(1):532(1972); NJBM:435(1974); CM 15:108(1977); MM 54:131(1990), 54:133(1990); AM 76:1061(1991)*.

69.2.1a.7 Welshite $Ca_2Mg_4Fe^{3+}Sb^{5+}O_2[Si_4Be_2O_{18}]$ or $Ca_2Sb^{5+}Mg_4Fe^{3+}Si_4Be_2O_{20}$

Named by Moore in 1978 for Wilfred R. Welsh (b.1915), amateur mineralogist and teacher of the natural sciences, Upper Saddle River, NJ. Aenigmatite group. See the other members of the group. TRIC $P\bar{1}$. $a = 10.167$, $b = 10.65$, $c = 8.787$, $\alpha = 106.5°$, $\beta = 96.1°$, $\gamma = 124.0°$, $Z = 2$, $D = 3.67$. *29-1407*(nat): 7.32_7 4.73_6 2.91_7 2.67_6 2.53_{10} 2.10_6 1.49_7 1.46_7. Aenigmatite-type structure. Thick, prismatic crystals to 3 mm in maximum diameter; polysynthetic twinning on {010} (monoclinic setting). Reddish black to reddish brown, pale brown streak, subadamantine luster. No cleavage, conchoidal fracture. $H = 6$. $G = 3.77$. Insoluble in dilute to strong HCl. Analysis (wt %): SiO_2, 19.6; Al_2O_3, 2.1; BeO, 4.0; As_2O_5, 3.6; Sb_2O_5, 24.8; Fe_2O_3, 9.1; MnO, 1.0; MgO, 15.3; CaO, 14.2; Na_2O, 0.0; K_2O, 0.0. $Sb^{5+} \leftrightarrow Ti^{4+}$ and $Be^{2+} \leftrightarrow Al^{3+}$ in rhönite. Biaxial; $N_x \approx 1.81$, $N_y = 1.81(?)$, $N_z \approx 1.83$; $2E = 45°$; nonpleochroic, deep reddish brown to reddish black. Optical character uncertain. Occurs in dolomite, in fractures in hematite ore. Associated with roméite, adelite, and others. *Långban, Värmland, Sweden*. EF *MM 42:129(1978), CM 32:439(1994)*.

69.2.1a.8 Wilkinsonite $Na_2Fe_2^{2+}Si_6O_{20}$

Named by Duggan in 1990 for John Frederick George Wilkinson (1927–1985), emeritus professor of geology, University of New England, NSW, Australia. Aenigmatite group. See the other members of the group. TRIC $P\bar{1}$. $a = 10.355$, $b = 10.812$, $c = 8.906$, $\alpha = 105.05°$, $\beta = 96.63°$, $\gamma = 125.20°$, $Z = 2$ [*AM* 75:694(1990)], D = 3.89. *PD*(nat): 8.10_{10} 4.82_3 4.18_3 3.702_4 3.149_{10} 2.935_6 2.750_3 2.696_8. Synthesized by Ernst. *JG* 70:689(1962). Small, anhedral grains, < 50 µm in diameter. Black, brown streak, vitreous luster. Conchoidal fracture. H = 5. Forms a solid-solution series with aenigmatite and other members of the group. Substitution of MgO (5.5%), MnO (2.1%), and CaO (0.7%) for Fe^{2+}. Also substitution of minor K_2O (0.75%) for Na_2O. Contains as much as nearly 4% Nb_2O_5. Analysis (wt %): SiO_2, 39.1–41.5; Al_2O_3, 0.03–0.85; Fe_2O_3, 44-48; TiO_2, 0–1.9; ZrO_2, 0.0–0.5; MnO, 1.0–1.4; MgO, 0.0–0.2; CaO, 0.1–0.2; Na_2O, 7.1–7.4; K_2O, 0.0–0.1. Biaxial (+); $N_x = 1.79(1)$, $N_y = 1.79(1)$, $N_z = 1.90(1)$; $2V \approx 10°$; $Z > Y > X$. X, olive-green; Y, gray-brown; Z, dark brown. With increasing contents of TiO_2, the reddish-brown colors of aenigmatite are then observed. Occurs as anhedral groundmass grains associated with anorthoclase, clinopyroxene, sodalite, analcime, and traces of eudialyte and arfvedsonite in a fine-grained silica-undersaturated trachyte. In silica-undersaturated, mildly peralkaline hosts, with very low TiO_2 contents. In syenite ejecta at Wonchi volcano. Occurs as relics surrounded by riebeckite and biotite. Occurs in the Sierra Prieta nepheline–analcime syenite, Diablo Plateau, TX; Wonchi volcano, Ethiopia; *Warrumbungle volcano, NSW, Australia.* EF *CIWY* 72:578(1973), *GSAB* 88:1428(1977), *MM* 55:529(1991).

69.2.1b.1 Sapphirine $[Mg_{4-x}Al_{4+x}][Al_{4+x}Si_{2-x}]O_{20}$

Named in 1819 by Giesecke from *sapphire*, in allusion to its common sapphire-blue color. Most sapphirine is monoclinic (2M polytype), followed by triclinic (1A polytype) and other rarer polytypes: 2A, 3A, 4M, and 5A. *AM* 75:937(1990). A monoclinic 2C superstructure has $a = 11.320$, $b = 14.367$, $c = 19.794$, $\beta = 125.49°$. *AM* 73:1134(1988). MON $P2_1/n$. $a = 11.281$, $b = 14.425$, $c = 9.940$, $\beta = 125.49°$, $Z = 4$, D = 3.48; TRIC $P1$ or $P\bar{1}$. $a = 9.97$–10.04, $b = 10.34$–10.38, $c = 8.62$–8.65, $\alpha = 107.4$–107.55°, $\beta = 95.12$–95.2°, $\gamma = 123.8$–123.92°, $Z = 2$, D = 3.57. *19-750*(nat, 2M polytype): $3.00_{6.5}$ $2.84_{6.5}$ 2.45_{10} 2.35_6 2.02_9 1.44_8 1.42_7 1.41_6; *44-1430*(nat, 1A polytype): 3.01_5 2.85_4 2.78_2 2.58_3 2.46_{10} 2.36_3 2.04_4 2.02_7. An OD (order-disorder structure; "normal" 2M sapphirine has a disordered arrangement of domains. There is a considerable scope for cation order–disorder. There are eight or nine octahedral sites and six tetrahedral sites in the average structures of the two most common polytypes. Sapphirine is stabilized as a phase through a combination of short-range order in T_2 and T_3 and disorder on the remaining tetrahedral sites. It is still unknown if 1A and 2M polytypes differ only in the succession of the structural layers or if they differ in some degree in the structure layers themselves. **Habit**: Usually granular, also occurs as tabular plates, flattened on (010) (for monoclinic); uncommon repeated twinning is present on {100} (monoclinic); in the triclinic case, the composition plane is (010). **Physical properties**: Light to medium blue, blue-green, gray, pale red or maroon (high Cr-containing: 7.5 wt % Cr_2O_3); white to pale blue streak; vitreous luster. Cleavage (monoclinic) {010} moderate; {001}, {100} poor; uneven fracture. $H = 7\frac{1}{2}$. $G = 3.40$–3.60. MAS-NMR studies on ^{27}Al and ^{29}Si and FTIR studies in *AM 77:8(1992)*. Insoluble in acids; decomposed by fusion in Na_2CO_3 or $KHSO_4$. **Chemistry**: There is substitution of Fe^{2+} for Mg and of Fe^{3+} or Cr for Al. Substitution of Fe^{2+} appears to be necessary for stability of the 2M structure. Tschermak substitution in the form $R^{2+}Si \leftrightarrow 2R^{3+}$ is common. This accounts for the deviation from the ideal 2:2:1 formula ratios. May contain as much as 2.2–2.5 wt % BeO; Cr_2O_3 to 7.5 wt % (Greenland); TiO_2 to 1.5 wt %; may also contain as much as 0.75 wt % B_2O_3. **Optics**: Usually biaxial (−), but some are (+); $Z \wedge c = 6$–$12°$, $Y = b$, OAP = (010); $N_x = 1.701$–1.731, $N_y = 1.703$–1.741, $N_z = 1.705$–1.745; $2V_x = 47$–$114°$, most are 50–80°; 2V decreases as Fe content increases, and (+) sapphirine is very low in total Fe; $r < v$, strong. Birefringence = 0.005–0.009. X, colorless, pale red or pink, yellowish green; Y, sky-blue, pale blue, lavender; Z, blue, dark blue, blue-green. **Occurrence**: Occurs in Mg- and Al-rich high-grade metamorphic rocks (e.g., granulites). Also occurs in upper mantle rocks (breccia pipes). Occurs at the contact zone of alpine peridotites, in metagabbros, and in kimberlites. Fine, large crystals occur in calcite–phlogopite rock at Itrongay. Occurs rarely in pegmatites. Characteristic of high-pressure, high-temperature environments. **Localities**: Salt Hill, Cortlandt complex, Peekskill, NY; from Stockdale kimberlite pipe, Riley Co., KA; magmatic in anorthosite from St. Urbain, Charlevoix Co., QUE, Canada; 1A sapphirine from Wilson Lake, LAB, Canada; Cr-bearing from Bjørnesund area, Fiskenaesset region, and regular sapphirine from the *Fiskenaesset Harbor, southwestern Greenland*; Snarum and Vikesa, Norway; Lerc, France; Val Codera, and Sondrio, Chiavenna, Italy; Waldheim, Saxony, Germany; Naxos, Greece; Aldan Shield, Yakutia, Russia; colorless sapphirine with gem spinel from Kuk-i-Lal, southwestern Pamirs, Kazakhstan; Al-rich from Sar-i-sang, Afghanistan; yellow sapphirine from Mautia Hill, Tanganyika; Fe-richest material from the Labwor Hills, Uganda; large crystals from the farms Blinkklip and Randjesfontein, near Messina, Transvaal, and near Okiep, Namaqualand, South Africa; Kakanuru area, eastern Ghats, India; Sitampundi complex, Gangurvapatti, Madura, India; Betroka, Itrongay, and Zahosy, Madagascar, and several other localities in Madagascar (e.g., low-Fe, well-developed sapphirine crystals from the Bekily area); Strangway and Reynolds Ranges, NT, and Dangin, WA, Australia; MacRobertson Land, Be-rich material from Casey Bay, Enderby Land, and Ti-bearing sapphirine (1A) from the Vestfold Hills, Antarctica [*EJM 7:637(1995)*]; Salvador, Bahia, Brazil. EF *CMP 41:23(1973), 68:349,357(1979); ZK 151:91(1980); MIN 3(2):576(1981); MM 46:323(1982); CMP 103:211(1989).*

69.2.1b.2 Surinamite $(Mg,Fe,Mn)_3(Al,Fe)_4BeSi_3O_{16}$

Named in 1976 by deRoever et al. for the locality. MON $P2/n$. (Syn): $a = 9.881$, $b = 11.311$, $c = 9.593$, $\beta = 109.52°$, $Z = 4$, $D = 3.48$; (nat): $a = 9.916$, $b = 11.384$, $c = 9.631$, $\beta = 109.30°$, $Z = 4$, $D = 3.48$. *38-433, 29-702*(syn): 2.91_9 2.89_8 2.65_8 2.43_7 2.42_{10} 1.993_{10} 1.988_9 1.42_8; (nat): 3.08_3 2.91_8 2.67_4 2.635_4 2.42_{10} 2.00_8 1.43_6 1.395_6. *AM 66:1022(1981)*. $R = 4.7\%$. Nine octahedral cations and five tetrahedral cations incorporated in a CCP array of 16 independent oxygen atoms. *AM 68:804(1983)*. Colorless to white (syn) and blue (nat). Platy parallel to (010) with one well-developed cleavage perpendicular to (010); < 0.2-mm crystals in Suriname; one well-developed cleavage perpendicular to (010), also (010). $G > 3.30$. Analyses (wt %): SiO_2, 30.9–32.1; Al_2O_3, 33.7–37.3; FeO, 9.1–13.4; MnO, 0.01–0.2; MgO, 16.4–18.4; Be, 0.5–1; H_2O, 0.66; end member: MgO, 22.8; Al_2O_3, 38.47; BeO, 4.72; SiO_2, 34.00. Biaxial $(-)$; $Y = b$; (syn): $N_x = 1.7015$, $N_y = 1.7035$, $N_z = 1.7055$; (nat): $N_x - 1.738$, $N_y = 1.743$, $N_z = 1.746$; $2V = 68°$. Y, violet for vibrations in the OAP(010); bright blue-green if parallel to the cleavage, but very light colored if perpendicular to it. X, pale yellow to pale yellow-green; Y, bluish, purple; Z, bright green-blue. Optically similar to sapphirine. Very strong dispersion; in (010) the extinction angle between Z and the trace of a cleavage perpendicular to (010) is 31° for violet and 44° for yellow light. Anomalous interference colors. Occurs in Suriname in a mesoperthite gneiss associated with charnokitic granulites. Granulite facies rocks in Australia; same in Antarctica. Forms as pseudomorphs after cordierite in Zambia. In most cases it is formed by alteration of cordierite (Be-bearing). In pegmatites in granulite-facies rocks—Be-rich in Casey Bay, associated with sillimanite. Occurs in the *Bakhuis Mts., western Suriname*; also found in the Chimwala area, near the Malawi border, Eastern Prov., Zambia; eastern Strangways Range, Central Australia; Christmas Point, Casey Bay, Enderby Land, Antarctica. EF *MIN 3(2):584(1981)*, *AM 70:710(1981)*, *CMP 92:113(1986)*.

69.2.1c.1 Magbasite $KBa(Mg,Fe)_6(Al,Sc)Si_6O_{20}F_2$

Named in 1965 by Semenov et al., for the composition: magnesium, barium, silicate. Possibly aenigmatite group. Probably ORTH P lattice. $a = 22.56(4)$, $b = 18.96(4)$, $c = 5.29(2)$, $Z = 6$, $D = 3.99$. *42-571*(nat): 3.57_8 3.13_5 2.98_7 2.84_6 2.71_4 2.56_{10} 1.99_5 1.61_4. Fine fibers. Colorless, light red-violet. $G = 3.41$. Analysis (wt %): SiO_2, 39.7; Al_2O_3, 4.0; Sc_2O_3, 2.1; FeO, 8.9; MgO, 21.4; CaO, 1.7; CaO, 14.8; K_2O, 4.9; F, 5.5. Biaxial $(-)$; $N_x = 1.597$, $N_y = 1.609$, $N_z = 1.615$; $2V = 70°$. Occurs in hydrothermal deposits. Occurs at an unspecified locality in the Asian part of Russia. EF *DAN SSSR 163:718(1965)*, *CA 93:171152(1980)*.

69.2.2.1 Saneroite $Na_2(Mn^{2+},Mn^{3+})_{10}[VSi_{11}(OH)_2](OH)_2$

Named in 1979 by Cortesogno et al. for Edoardo Sanero (b.1901), emeritus professor of mineralogy, University of Genova. TRIC $P\bar{1}$. $a = 9.741$, $b = 9.974$,

$c = 9.108$, $\alpha = 92.70°$, $\beta = 117.11°$, $\gamma = 105.30°$, $Z = 1$, $D = 3.49$. $35\text{-}484$: 3.06_{10} 3.01_6 2.98_6 2.83_{10} 2.70_{10} 2.62_6 2.20_6 1.682_8. Structure: $NJMA$ $138:333(1980)$. The linear component, parallel to [110], of the chain has five tetrahedra; a sixth tetrahedron, statistically occupied by Si and V in a 1:1 ratio, is appended to the chain by one shared corner. Bright orange, resinous to greasy luster. Two perfect cleavages. $G = 3.47$. Biaxial $(-)$; $2V = 40\text{--}48°$; $N_x = 1.720$, $N_y = 1.740\text{--}1.745$, $N_z = 1.745\text{--}1.750$; extinction oblique (15° max). Pleochroic, $X > Z > Y$; deep orange to lemon yellow. Found with barite, cariopilite, and ganophyllite in veins in manganese ores at the *Gambatesa and Molinello mines, Chiavari, Val Greveglia, Liguria Apennines, Italy*. AR $NJMM:161(1981)$, AM $66:1277(1981)$.

69.2.3.1 Howieite $Na(Fe,Mg,Al)_{12}(Si_6O_{17})_2(O,OH)_{10}$

Named in 1964 by Agrell et al. for Robert Andrew Howie (b.1923), British mineralogist and petrologist, King's College, London University. See taneyamalite (69.2.3.2). TRIC $P1$. $a = 10.170$, $b = 9.774$, $c = 9.589$, $\alpha = 91.22°$, $\beta = 70.76°$, $\gamma = 108.09°$, $Z = 1$, $D = 3.35$. $19\text{-}571$: 9.02_{10} 7.79_{10} 3.21_8 3.02_5 2.76_5 2.72_5 2.63_8 2.16_5. Structure: AM $59:896(1974)$. The fundamental structural unit may be regarded as a hybrid single/double chain of tetrahedra, having the conformation of six-membered rings joined by oxygens of opposing pairs of tetrahedra. The resulting ribbons are parallel to one another and to $\{1\bar{2}0\}$. Dark green, pale green streak, vitreous luster, translucent to opaque. Cleavage $\{010\}$, good; $\{100\}$, fair; conchoidal fracture. $G = 3.38$. Biaxial $(-)$; $2V = 65°$; $N_x = 1.701$, $N_y = 1.720$, $N_z = 1.734$; $r < v$, strong. Pleochroic; X, pale golden yellow; Y, dark lilac-gray; Z, dull green. Occurs as bladed crystals in blueschist metamorphic facies, metamorphosed ironstones, and mafic igneous rocks. In the Franciscan formation in the *Laytonville district, Mendocino Co., CA*, associated with glaucophane, lawsonite, deerite, zussmanite, and also at Ward Creek, Sonoma Co., and Panoche Pass, San Benito Co. At Pinchi Lake, BC, Canada; at Brezovica, Yugoslavia; at Maki, Kochi Pref., Shikoku, Japan. AR AM $50:278(1965)$, MM $43:363(1979)$, CM $28:855(1990)$.

69.2.3.2 Taneyamalite $(Na,Ca)(Mn,Mg)_{12}[Si,Al)_6O_{17}]_2(O,OH)_{10}$

Named in 1981 by Matsubara for the Taneyama mine. See howieite (69.2.3.1). TRIC $P1$. $a = 10.198$, $b = 9.820$, $c = 9.484$, $\alpha = 90.50°$, $\beta = 70.53°$, $\gamma = 108.57°$, $Z = 1$, $D = 3.30$. $35\text{-}480$: 9.29_8 7.99_3 4.62_8 3.65_4 3.27_{10} 3.08_5 2.79_3 2.22_3. Isostructural with howieite. Greenish gray-yellow, pale yellow streak, vitreous luster. Cleavage $\{010\}$, perfect. $H = 5$. Biaxial $(-)$; $2V = 70°$; $N_x = 1.646$, $N_y = 1.664$, $N_z = 1.676$; positive elongation. Pleochroic; X,Y, nearly colorless; Z, pale yellow. Found associated with bannisterite as small seams in caryopilite at the *Iwaizama mine, Saitama Pref., Honshu*, and earlier observed but not named at the Taneyama mine, Kumamoto Pref., Kyushu, *Japan*. AR MM $44:51(1981)$, MJ $10:385(1981)$.

69.2.3.3 Deerite (Fe,Mn)$_6$(Fe,Al)$_3$O$_3$(Si$_6$O$_{17}$)(OH)$_5$

Named in 1964 by Agrell et al. for William Alexander Deer (b.1910), British mineralogist and petrologist, Cambridge University. MON $P2_1/a$. $a = 10.786$, $b = 18.88$, $c = 9.564$, $\beta = 107.45°$, $Z = 4$, $D = 3.82$. *19-421:* 9.45$_2$ 9.03$_{10}$ 3.22$_2$ 3.01$_7$ 2.64$_5$ 2.54$_2$ 2.37$_2$ 2.26$_2$. Structure: *AM 62:990(1977)*. The silicate structural unit is a hybrid single/double chain, as in howieite, but the chains are not parallel to one another. Crystals acicular [001] with amphibole-like cross section. Black, transparent on thin edges. Cleavage {110}, good. G = 3.84. Biaxial (+ or −); $N_x = 1.81$–1.84, $N_z = 1.84$–1.87; Z = c. Found in metamorphosed shales, ironstones, and limestones of the Franciscan formation in the *Laytonville district, Mendocino Co., CA*, associated with howieite, zussmanite, and other blueschist facies minerals, and also at Ward Creek, Sonoma Co., and Panoche Pass, San Benito Co.; at Wild Horse Lookout, Curry Co. and the Powers quarry, Coos Co., Oregon. Found at Termignon, Haute-Savoie, France; Eskisehir Prov., Turkey; and Ouega Koumec, New Caledonia. AR *AM 50:278(1965), BM 93:263(1970), MM 43:251(1979)*.

69.2.4.1 Liebauite Ca$_3$Cu$_5$[Si$_9$O$_{26}$]

Named in 1992 by Zöller and Tillmanns for Friedrich Karl Liebau (b.1926), Mineralogisch-Petrologisches Institut, Kiel, in recognition of work on silicate crystal chemistry. MON $C2/c$. $a = 10.160$, $b = 10.001$, $c = 19.973$, $\beta = 91.56°$, $Z = 4$, $D = 3.62$. *PD:* 7.13$_6$ 6.70$_7$ 3.58$_4$ 3.12$_9$ 3.00$_{10}$ 2.45$_6$ 2.41$_7$ 1.78$_5$. Structure: *ZK 200:115(1992)*. The silicate element is a loop chain with $P = 14$ and $N = 18$ consisting of six-membered rings linked by three tetrahedra. Bluish green, vitreous luster, transparent. H = $5\frac{1}{2}$. Biaxial (+); 2V = 35°; $N_x = 1.722$, $N_y = 1.723$, $N_z = 1.734$ for Na(D). Found as minute (<0.03 mm) crystals with several copper minerals in an argillaceous xenolith in scoria of the *Sattelberg cone, near Kruft, Eifel district, Germany*; also in other scoria cones at Osteifel and Westeifel. AR

69.2.5.1 Johninnesite Na$_2$Mg$_9^{2+}$(Mg,Mn)$_7$(OH)$_8$[AsO$_4$]$_2$[Si$_6$O$_{17}$]$_2$

Named in 1986 by Dunn et al. for John Innes, senior mineralogist, Tsumeb Corp. TRIC $P\bar{1}$. $a = 10.485$, $b = 11.065$, $c = 9.654$, $\alpha = 107.11°$, $\beta = 81.17°$, $\gamma = 111.86°$, $Z = 1$, $D = 3.51$. *39-402*(nat): 9.8$_6$ 5.99$_4$ 4.06$_3$ 3.38$_4$ 3.23$_6$ 2.68$_{10}$ 2.48$_4$ 1.54$_4$. A layered structure with a tetrahedral layer consisting of independent AsO$_4$ and Si$_6$O$_{17}$ chains, an octahedral layer of edge-sharing (Mn, Mg)O$_6$ octahedra with one-sixth of the octahedral sites vacant, and a layer with single MnO$_6$ octahedral chains separated by NaO$_6$ distorted octahedra. Very friable aggregates; fibrous, elongate on [001]; flattened on {100}; to 2 cm long on type sample and to 4.5 cm long on others. Light yellow-brown, light brownish yellow streak, vitreous luster. Cleavage {100}, good; {010}, poor. G = 3.48. Analysis (wt %): SiO$_2$, 35.5; FeO, 0.1; MgO, 8.2; MnO, 40.7; As$_2$O$_5$, 10.6; Na$_2$O, 3.1; H$_2$O, 2.6. Biaxial (−); Na(D): $N_x = 1.6742$,

$N_y = 1.6968$, $N_z = 1.6999$; $2V = 42°$; $r > v$, distinct. Occurs at the *Kombat mine, Otavi Valley, Namibia*, in late-stage, low-temperature hydrothermal veins cutting Mn ores. Associated with kentrolite, rhodonite, and richterite. In upper Proterozoic dolostones underlying slates and marls. EF *AM 79:991(1994)*.

===== Class 70 =====

Column or Tube Structures

70.1.1.1 Litidionite KNaCu[Si$_4$O$_{10}$]
Named in 1880 by Scacci from the Greek *lithidion*, pebble, in allusion to the lapilli on which the mineral was found. Synonym: lithidionite. See fenaksite and manaksite. TRIC $P\bar{1}$. $a = 9.80$, $b = 8.01$, $c = 6.97$, $\alpha = 114.12°$, $\beta = 99.52°$, $\gamma = 105.59°$, Z = 2, D = 2.85. *29-1041*: 6.75$_3$ 4.05$_2$ 3.65$_2$ 3.37$_{10}$ 3.22$_7$ 2.84$_2$ 2.68$_4$ 2.41$_8$. Structure: *AM 60:471(1975)*. The silicate structural element consists of a pair of vlasovite-like double chains condensed to form a rod. Essentially isostructural with fenaksite (70.1.1.2). Blue, vitreous luster. H = 5–6. G = 2.75. Biaxial (−); 2V = 56°; N$_x$ = 1.548, N$_y$ = 1.574, N$_z$ = 1.582. Found as crusts with tridymite on lapilli highly modified by fumarolic activity in the *Vesuvius crater, Campania, Italy*. AR *D6:1041(1892)*, BM *104:387(1981)*.

70.1.1.2 Fenaksite (K,Na,Ca)$_4$ (Fe^{2+},Fe^{+3},Mn^{+2})$_2$ Si$_8$O$_{20}$(OH, F), simplified formula: KNaFe^{2+}[Si$_4$O$_{10}$]
Named in 1959 by Dorfman et al. from its component cations: Fe, Na, K, and Si. Compare manaksite, the Mn-analogue. Litidionite group. TRIC $P\bar{1}$. $a = 6.79$–6.98, $b = 8.18$–8.24, $c = 9.97$–9.98, $\alpha = 105.07°$, $\beta = 98.96°$, $\gamma = 114.66°$, Z = 1 *(ZVMO 121:112(1992))*, D = 2.78; PD *13-520* (nat.): 3.50$_7$ 3.03$_{10}$ 2.88$_6$ 2.71$_6$ 2.46$_7$ 1.875$_6$ 1.835$_6$ 1.752$_6$ (note 0 assigned because delta 2 theta = 0.092) and *(ZVMO 121:112(1992))*: 6.86$_5$ 4.67$_5$ 4.48$_5$ 3.55$_7$ 3.03$_{10}$ 2.88$_6$ 2.71$_6$ 2.46$_7$. Structure: *(DAN SSSR 194:818(1970))* and *SPD 15:902(1970))*. Essentially isostructural with litidionite, although the unit cell is chosen differently. Light rose, rose, sometimes with brownish tint, transparent and translucent; white streak; 2–4 cm granular aggregates, some prismatic crystals found: 0.4 × 0.8 × 0.4 cm; luster pearly; two good cleavages at an angle of 122° : {001} + {010}; when abraded, it gives tangled and matted fibrous masses; splintery fracture; brittle; H = 5–5$\frac{1}{2}$. G = 2.73–2.744. IR data = *(MIN 3(3):361(1981))*; Mössbauer spectral data for Fe in *TMMAN 26:198(1978)*; gelatinizes in acids, easily fusible. MgO to 1%, CaO to 0.7%, MnO to 2.5%, traces of Al$_2$O$_3$. Biax. (+) but (−) from indices; 2V = 77°–84, 2V$_{calc}$ = 62°; N$_x$ = 1.541, N$_y$ = 1.560, N$_z$ = 1.567; colorless, Z = b, Y ∧ c = 49°, (+) elongation. Occurs in central zones of ijolite and urtite pegmatites with canasite and lamprophyllite. In rischorrite on the Rasvumchorr Plateau. Associated with alkalic minerals: canasite, delhayelite, pecto-

lite, villaumite. Occurs on *Yukspor mountain, Mt. Rasvumchorr and Kukisvumchorr, Khibiny massif, Kola Peninsula, Russia*. Also in the dumps of the Central mine. AR and EF.

70.1.1.3 Manaksite NaKMnSi$_4$O$_{10}$

Named in 1992 by Khomyakov et al. for the composition: Mn–Na–K–Si. See litidionite and fenaksite, the Fe analog. TRIC $P\bar{1}$. $a = 6.993$, $b = 8.219$, $c = 10.007$, $\alpha = 105.11°$, $\beta = 100.76°$, $\gamma = 114.79°$, Z = 2, D = 2.71. PD(nat): 6.89$_7$ 4.07$_5$ 3.45$_{10}$ 3.26$_9$ 3.05$_8$ 2.88$_7$ 2.71$_7$ 2.46$_7$. Structure not done, but probably like fenaksite (70.1.1.2). *DAN SSSR 194:818(1970)*. Disseminated, irregular grains, 1–3 mm, aggregates to 5 mm. Colorless, creamy, rosy; white streak; vitreous luster. Cleavage (001), perfect; (010), good; steplike and hackly fracture; brittle. H = 5. G = 2.73. IR data available; heating (> 600°) produces serandite, readily soluble in cold 10% HCl, with production of a creamy residue. Analysis (wt %): Na$_2$O, 8.9; K$_2$O, 10.8; SrO, 0.2; CaO, 0.2; MgO, 0.3; MnO, 17.2; FeO, 0.8; SiO$_2$, 62.0. (Na$_{1.11}$K$_{0.89}$Ca$_{0.01}$)$_2$ ... Biaxial (−); white light: N$_x$ = 1.540, N$_y$ = 1.551, N$_z$ = 1.557; 2V = 73°; r > v, moderate; detailed optical orientation given. Occurs on *Mt. Alluaiev, Lovozero massif, Kola Penin., Russia*, in ultra-agpaitic pegmatite, associated with K-feldspar, nepheline, sodalite, cancrinite, and rare accessory minerals. EF *ZVMO 121:112(1992)*.

70.1.1.4 Agrellite NaCa$_2$[Si$_4$O$_{10}$]F

Named in 1976 by Gittins et al. for Stuart O. Agrell, British mineralogist, Cambridge University. TRIC $P\bar{1}$. $a = 7.759$, $b = 18.946$, $c = 6.986$, $\alpha = 89.88°$, $\beta = 116.65°$, $\gamma = 94.32°$, Z = 4, D = 2.887. *33-1230*: 3.83$_3$ 3.44$_8$ 3.33$_8$ 3.19$_{10}$ 3.14$_{10}$ 2.58$_8$ 2.31$_7$ 2.04$_5$. Structure: *AM 64:563(1979)*. The silicate structural element consists of condensed double chains that form tubular or rodlike units parallel to [001]. White, grayish or greenish white; vitreous luster, pearly on cleavage; translucent. Cleavage {110}, {1$\bar{1}$0}, excellent; {010}, poor. H = 5$\frac{1}{2}$. G = 2.90. Biaxial (−); 2V$_{calc}$ = 47°; N$_x$ = 1.567, N$_y$ = 1.579, N$_z$ = 1.581; REE$_2$O$_3$ (lanthanum dominant) may be present to ≈ 2.6 wt %. Found as laths up to 10 cm elongated [001] and flattened {010} or {110} in lenses in alkalic igneous rocks of the Sheffield Lake complex, *Kipawa R., Villdieu Twp., Témiscaming Co., QUE, Canada*; also from the Murun massif, Yakutia, Russia, and the Dara-i-Pioz massif, Alai Range, Tajikistan. AR *CM 14:120(1976), MZ 9:73(1987)*.

70.1.2.1 Narsarsukite Na$_2$(Ti,Fe,Zr)[Si$_4$O$_{10}$](OH,F)

Named in 1990 by Flink for the locality. TET $I4/m$. $a = 10.726$, $c = 7.947$, Z = 4, D = 2.77. *11-478*: 7.61$_3$ 5.37$_{10}$ 3.98$_5$ 3.79$_3$ 3.39$_8$ 3.26$_8$ 2.58$_6$ 2.52$_6$. Structure: *AM 47:539(1962)*. The silicate element is a strip of tetrahedra forming four-membered rings successively turned approximately 90° to one another along the length. Crystals tabular {001} with {001}, {101}, {110} dominant, to prismatic [001] with {100} and {001} dominant. Yellow to brownish gray, greenish yellow to dark green; vitreous luster; translucent. Cleavage {100}, {110}, perfect. H = 6–7. G = 2.78. Uniaxial (+); $N_O = 1.604–1.612$, $N_E = 1.625–1.660$. Weakly pleochroic: O > E; yellow to colorless. As small crystals with aegirine and quartz at Sage Creek and Halfbreed Creek, Sweetgrass Hills, MT; as tabular and prismatic crystals(!) at Mt. St.-Hilaire, PQ, Canada; in fine crystals(!) at Igdlutalik, Ilimaussaq, and *Narsarsuk, Greenland*; at Gjerdingen, Nordmark, Norway; Mt. Flora, Lovozero massif, Kola Penin., Russia; and at Guré, Sudan. AR *AM 59:269(1959), MR 21:324(1990), EJM 3:574(1991), ZVMO 123:58(1994).*

70.1.2.2 Caysichite-(Y) $Y_2(Ca,Gd)_2Si_4O_{10}(CO_3)_3(H_2O,O,OH) \cdot 3H_2O$ Narsarsukite group.

Named in 1974 by Hogarth et al. for the composition, contains Ca, Y, Si, C, H. ORTH. $Ccm2_1$. $a = 13.27–13.283$, $b = 13.91–13.925$, $c = 9.724–9.73$, Z = 4, D = 3.03 (*CM 12:293(1974) 26-1394* (nat.) 6.93_{10} 4.87_4 4.38_6 4.22_6 3.48_6 3.32_9 3.20_3 2.32_4. See also *MA 27:82(1972)*. Structure determined: *CM 16:81(1978)*. Crystals are prismatic, elongated on c and terminated by {001}. Colorless to white, rarely pale yellow or pale green; white streak; vitreous luster; good {010} cleavage, $H = 4\frac{1}{2}$; G = 2.9–3.03; IR data = (*MA 27:82(1972)*); decomposed by cold HCl with evolution of CO_2 + formation of silica residue. Non-fluorescent, but has a faint green cathodoluminescence under the electron beam. Biaxial (−), for colorless material: 2V = 53°, $2V_{calc} = 54.5°$, $N_x = 1.586$, $N_y = 1.614$, $N_z = 1.621$; for yellowish material: 2V = 73°, $2V_{calc} = 62°$, $N_x = 1.589$, $N_y = 1.616$, $N_z = 1.626$. XYZ = *bac*. Occurs in granite pegmatite lining cavities as a dull, white pulverulent coating or as a stain. Also as incrustations, reniform masses and stalagtites to 1 cm long. Occurs in granite as spherulitic crusts and overgrowths assoc. with tengerite-(Y), opal and synchysite-(Y). *Evans-Lou feldspar mine, Portland West Township, Quebec, Canada*; Kazakhstan.

70.1.2.3 Charoite $K_5Ca_8(Si_6O_{15})_2(Si_2O_7)Si_4O_9(OH) \cdot 3H_2O$ and $(K,Na)_5)(Ca,Ba,Sr)_8(Si_{12}O_{30})Si_6O_{16})(OH,F) \cdot nH_2O$ (KSTM, 100(1985) and GSAPA 26:A481(1994) find K 3.5–4.0, Na 1.3–1.7, Ca 6.8–7.1, Ba 0.3–0.4, Sr 0.05–0.1, Mn^{2+} 0.04–0.09, Si 18.01–18.28, F 0.17–0.59, Al, Ti, Fe, Mg, Zr, & Th are <0.01 wt %. Another composition given is: $(K_{2.88}Na_{1.03})(Ca_{5.58}Ba_{0.23}Sr_{0.11}Mn_{0.04})(Si_{14}Al_{0.02})O_{35}(OH)_2 \cdot 3H_2O$ (GSAAP 27(6):A440(1995).

Named by Rogova et al. in 1978 for the impression that it gives: "chary" in Russian means "charms" or "magic" and not for the Chara River, which is 70 km away, Murun Massif, NW Aldan, Yakutia, USSR. MON P*/4. $a = 19.61$Å, $b = 32.12$Å, $c = 7.20$Å; $\beta = 93.76°$; Z = 4, D = 2.68. *42-1402* (nat.) 32.0_6 12.4_{10} 9.39_2 3.90_2 3.35_6 $3.13_{4.5}$ 2.98_2 2.80_2. Structure: not done. Electron difraction studies suggest that the mineral has an orthorhombic-shaped unit cell of dimensions: a 19.6Å, $b = 32$Å, $c = 7.2$Å, a second polymorph, intimately intergrown with the first (monoclinic), has a b-axis repeat of 64Å (GSAAP 27(6):A-

440(1995). The c-axis dimension is consistent with a structure based on linkage of wollastonite-type chains, forming sheets whose b-axis length is defined by two pyroxenoid and one pyroxene chain repeat. Violet-lilac; white; fibrous, elongate on c; vitreous to silky in aggregates; 3 cleavages present, good; H ≈ 5–6 VHN = 412 Kg/mm^2 (50 g load); G = 2.54–2.58; IR and thermal data = ($ZVMO$ $107:94(1978)$); luminescence = very weakly fluorescent in SW & LWUV, bright orangish yellow cathodoluminescence; insoluble in common acids. Charoite does show cation exchange properties (GSAAP 27(6): A440(1995). Analysis (wt.%): SiO_2 57.06, CaO 21.76, BaO 2.68, SrO 0.58, K_2O 9.23, Na_2O 2.00, MnO 0.16, H_2O 5.39, F 0.54; color is due to Mn^{3+}. Analyses given in (Min 3(3):378(1981)). Biaxial (+); 2V = 29°, N_x = 1.550, N_y = 1.553, N_z = 1.559; colorless in section, (+) elongation, $X = b$, $Z \wedge c = 5°$. Occurs in at least 25 areas within a syenite massif, in metasomatic rocks, formed between 200–250°C. Enriched in K at the contact of the massif, with limestone. Associated with canasite, tinaksite, aegirine and other minerals. *Murun massif, Chara River area, Aldan Shield, SW Yakutia (Sakha Republic)*. Alters to 'charoite asbestos', a white to cream-colored felt-like mass, which differs in OH-content from charoite. Extensively used as a semi-precious jewelry stone and for lapidary work; cabochons, vases, boxes, etc. The mineral occurs in a number of different textures (World of Stones, no. 7, 3(1995)).

70.2.1.1 Miserite $K(Ca,Ce)_5[Si_8O_{22}](OH,F)_2$

Named in 1950 by Schaller for Hugh Dinsmore Miser (1884–1969), geologist, U.S. Geological Survey. TRIC $P\bar{1}$. $a = 10.100$, $b = 16.014$, $c = 7.377$, $\alpha = 96.42°$, $\beta = 111.15°$, $\gamma = 76.57°$, $Z = 2$, $D = 2.646$. 22-806: 15.6_6 3.47_5 3.36_5 3.14_{10} 2.83_6 2.38_7 1.845_6 1.641_8. Structure: CM $14:515(1976)$. The structure is based on a tubular unit of four condensed wollastonite-type chains and isolated [Si_2O_7] dimers, but the formula given with the structure determination gives yttrium in place of cerium. Pink to lavender, reddish brown; vitreous or pearly luster; translucent. Cleavage {100}, perfect; {010}, imperfect. H = $5\frac{1}{2}$–6 (but also given as $2\frac{1}{2}$). G = 2.84–2.93. Biaxial (+); 2V = 52–78°; N_x = 1.580–1.586, N_y = 1.583–1.593, N_z = 1.590–1.600; XZ = ca; negative elongation. Insoluble in acids. As fibers and bladed masses with wollastonite in metamorphosed shale at *Wilson Mineral Springs, AR*; at the Kipawa Lake area, Témiscamingue Co., in carbonatite, and at Mt. St.-Hilaire, QUE, Canada; in the Dara–Pioz alkalic massif, Alai Range, Tajikistan. AR AM $35:911(1950)$, CM $11:569(1972)$, MR $21:32(1990)$.

70.2.2.1 Penkvilksite $Na_4Ti_2[Si_8O_{22}] \cdot 5H_2O$

Named in 1974 by Bussen et al. from the Lapp *penk* and *vilkis*, white and curly, in allusion to its appearance. ORTH or MON Space group unknown. $a = 7.48$, $b = 8.77$, $\gamma \approx 90°$ (derived from powder data). 26-1386: 8.20_{10} 3.42_6 3.37_9 3.32_7 3.10_7 3.07_7 2.84_8 1.713_7. White; porcelaneous; dull luster, pearly or silky on fracture. Pefect cleavage in one direction. H = 5. G = 2.58. Biaxial

(+); 2V = 42°; $N_x = 1.637$, $N_y = 1.640$, $N_z = 1.662$. Gelatinizes with acids. Rarely as clusters and rosettes of tabular crystals at Mt. St.-Hilaire, QUE, Canada; as "clotted" masses with mountainite, aegirine, and raite in the *Jubilee pegmatite deposit, Mt. Karnarsut, Lovozero massif, Kola Penin., Russia*. AR *AM 60:340(1975), MR 21:329(1990)*.

70.3.1.1 Ashcroftine-(Y) $K_{10}Na_{10}(Y,Ca)_{24}(Si_{56}O_{140})(CO_3)_{16}(OH)_4 \cdot 16H_2O$

Named in 1933 by Hey and Bannister for Frederick Noel Ashcroft (1878–1949), a prominent mineral collector of London. Originally described as kalithompsonite in 1924 by Gordon. TET $I4/mmm$. $a = 23.994$, $c = 17.512$, $Z = 2$, $D = 2.60$. *22-508:* 17.0_{10} 12.0_9 7.62_6 7.10_2 6.01_3 5.38_3 3.11_5 2.69_5. Structure: *AM 72:1176(1987)*. An unusual structure consisting of ball-like clusters of composition $(Si_{48}O_{128})$ joined to one another by four pairs of (SiO_4) tetrahedra to form columns parallel to [001]. Capillary to fibrous [001] up to 20 mm long with length/width $\approx 100:1$; often bent; powdery. Pale violet, pink, white; silky luster. Cleavage {100}, perfect; {001}, good. H = 5(?), also stated to be soft. G = 2.61. Uniaxial (+); $N_O = 1.536$, $N_E = 1.545$; also biaxial (+); 2V small; $N_x = 1.535$, $N_y = 1.537$, $N_z = 1.545$. Distinguished from lorenzenite by lack of fluorescence in UV. In breccia zones and miarolitic cavities in nepheline syenite at Mt. St.-Hilaire, QUE, Canada, often intergrown with lorenzenite, with which it is easily confused. Found in vugs in aegirine pegmatite in augite–syenite at *Narssarssuk, Greenland*. AR *MM 23:305(1933), 37:515(1969); MR 21:296(1990)*.

70.4.1.1 Neptunite $KNa_2Li(Fe^{2+},Mn)_2Ti_2O_2[Si_8O_{22}]$

Named in 1893 by Flink for *Neptune*, the Roman god of the sea, because it was found with aegirine, which was named for the Scandinavian god of the sea. See manganneptunite (70.4.1.2). MON Cc. $a = 16.46$, $b = 12.50$, $c = 10.01$, $\beta = 115.43°$, $Z = 4$, $D = 3.14$. *14-134:* 9.6_6 3.52_4 3.31_3 3.19_{10} 2.94_3 2.90_3 2.84_3 2.48_3. Structure: *AC 21:200(1966), PCM 18:199(1991)*. The structure consists of two interlaced cages built of linked chains parallel to [110] and [$\bar{1}$10] such that each cage has two 18-tetrahedron rings and four 14-tetrahedron rings, resulting in an $[Si_{16}O_{44}]^{24-}$ element. Crystals prismatic [001], with dominant {110}, {100}, and terminated by {111}, {$\bar{1}$11}, and {$\bar{3}$11}. Black with reddish-brown internal reflection, reddish-brown streak, vitreous luster, nearly opaque. Cleavage {110}, perfect; conchoidal fracture. H = 5–6. G = 3.19–3.23. Biaxial (+); 2V = 45–65°; $N_x = 1.697$–1.711, $N_y = 1.710$–1.721, $N_z = 1.735$–1.744; Y = b, Z $\wedge$ c = 16–20°; r < v, strong. Strongly pleochroic, Z > Y > X; yellow-orange to deep red. Insoluble in acids. Complete solid solution with manganneptunite; Mg may substitute for iron. Alters to brown crossite. Most commonly associated with syenitic rocks. Found at Point of Rocks, near Springer, Colfax Co., NM; with benitoite in natrolite at the Dallas Gem mine, San Benito Co., CA(!); at Mt. St.-Hilaire, QUE, Canada. Found with eudialyte in nepheline syenite at *Narsarsuk, Julianehaab* district, *Greenland*; from the Khibiny and Lovozero massifs, Kola Penin., Russia; the Dara-i-Pioz massif, Alai Range, Tajikistain; at Weal Guk Ri, Dang Myun, Gang Weon Prov., Korea; in the Wood's Reef serpentinite, NSW, Australia. AR *CM 7:679(1963)*, *AM 57:85(1972)*.

70.4.1.2 Mangannetunite $KNa_2Li(Mn,Fe^{2+})_2Ti_2O_2[Si_8O_{22}]$

Named in 1923 by Kurbatov for the composition and relationship to neptunite. See neptunite (70.4.1.1). MON *Cc*. $a = 16.38$, $b = 12.48$, $c = 10.01$, $\beta = 115.4°$, $Z = 4$, $D = 3.26$. *29-823*: 2.95_5 2.90_5 2.84_6 2.49_{10} 2.17_8 1.924_7 1.506_{10} 1.483_9. Isostructural with neptunite. Reddish black, red-brown streak, vitreous luster, translucent to opaque. Cleavage {110}, perfect; conchoidal fracture. $H = 5-6$. $G = 3.23$. Biaxial (+); $2V = 35-45°$; $N_x = 1.691-1.697$, $N_y = 1.700-1.710$, $N_z = 1.725-1.735$; $Y = b$, $Z \wedge c = 16-20°$; $r < v$, strong. Strongly pleochroic; $Z > Y > X$; yellow-orange to deep red. Fusibility 2.5. Insoluble in acids. Complete solid solution with neptunite. Found at Point of Rocks, Colfax Co., NM in syenites and syenite pegmatite at Mt. St.-Hilaire, QUE, Canada, and in the *Lovozero massif, Kola Penin., Russia*. AR *AM 12:96(1927)*, *ZVMO 94:204(1965)*.

70.5.1.1 Chiavennite $CaMn(BeOH)_2Si_5O_{13} \cdot 2H_2O$

Named in 1983 by Bondi et al. and Raade et al. for the Italian locality. Related to zeolites and zeolite-like minerals. Compare tvedalite. ORTH *Pnab* (standard setting), $P2_1ab$ (for $c < a < b$). (Italy): $a = 8.729$, $b = 31.326$, $c = 4.903$; (Norway): $a = 8.866$, $b = 31.34$, $c = 4.787$, (Sweden): $a = 8.905$, $b = 31.229$, $c = 4.775$, $Z = 4$, $D = 2.63-2.66$; PD *35-602* (nat.): 15.7_{10} 7.80_2 4.15_3 3.93_3 3.82_3 3.28_7 2.90_{10} 1.94_3. Structure consists of an interrupted framework of four-connected [SiO$_4$] and three-connected [BeO$_4$] tetrahedra, which consists of zig-zag chains to form a zeolite-like framework. Ca and Mn are in eight- and six-fold coordination respectively. The framework contains channels along [001] formed by 9-membered rings of O atoms, with free channel diameters of 3.9×4.3Å: *(EJM 7:1339(1995)*, *RSIMP 37:994(1981))*. The framework density classifies the mineral as a zeolite. Hemimorphic, spear-shaped crystals flattened on {010} with {161} pyramid, elongated on [100]. Also flat, pseudo-hexagonal, with perf. "basal" cleavage. Spherulites to 2 mm across, with individual crystals 10μ thick. Crystals may be 1 mm long. Pale orange-yellow to reddish orange, or bright orange-orange red, translucent, white streak, vitreous, good to perf. cleavage on {100}, {010}, and {001}, $H = 3$, $G = 2.56-2.64$. *IR* and thermal data *AM 68:623(1983))* and *AM 68:628(1983))*. Insol. in H_2O, HCl, HNO_3 and H_2SO_4. Norwegian material contains Fe(1.5 wt % FeO) which is believed to be the chromophore. Al_2O_3 3.6-6.0 wt % (substitutes for Be and Si), minor to trace Na, Mg, CO_2 and F. Utö material contains 1.0 wt. % B_2O_3 and 2.4 wt. % Fe_2O_3. Italian material is quite pure. There is some variation in T_{Si}(063-068). Biax.(+), $2V_z$(calc) $= 50°$, (Italy) $N_x = 1.581$, $N_z = 1.600$, (Norway) $N_x = 1.596$, $N_y = 1.600$, $N_z = 1.618$. Faintly pleochroic, X = colorless to yellow, Z = yellow orange. Norwegian material is pale yellowish brown, X < Z. X = a, Y = b, Z = c, OAP = (010), disper. r ≫ v. In vugs in syenite pegmatite (Norway), fracture-fillings in massive microcline and analcime (Norway) and coating beryl in granitic pegmatite (Italy). A late-stage, low-temperature hydrothermal mineral. *Tanno, Valle di San Giacomo, Chiavenna, Lombardia, Italy*; Utö, Stockholm, Sweden; Vevja and Heia quarries, Tvedalen, 10 km W of Larvik, and twelve other localities in the Larvik-Oslo region including Langangen, Langesundsfjord, and Bakkane, Brunlanes, Vestfold County, Norway. AR, EF *AM 77:438(1992)* 16th IMA Gen. Meeting Abst., p. 232(1994).

70.5.2.1 Tvedalite $(Ca,Mn)_4Be_3Si_6O_{17}(OH)_4 \cdot 3H_2O$

Named in 1992 by Larsen et al. for the type locality. Compare chiavennite – close in chemistry and cell dimensions. ORTH C face centered lattice. $a = 8.724$Å, $b = 23.14$, $c = 4.923$, $Z = 2$, $D = 2.55$. *PD 44-1465* (nat.): 11.64_9 5.86_7 4.20_5 3.87_8 3.16_7 2.89_8 2.84_{10} 2.49_6. Structure: no single-crystal work was possible. Cream white, pale gray to pale beige, streak white; spherulites, radial platelets, always overgrown by a crust of chiavennite; vitreous; cleavage = {010} perfect. $H = 4\frac{1}{2}$. $G = 2.54$. IR, TGA and DTA data *(AM 77:438(1992))*; non-fluorescent in UV; slightly soluble in hot, concentrated HCl. Zoned: $Ca_{50}Mn_{46}Fe_4 - Ca_{80}Mn_{18}Fe_2$ (in % of divalent atoms). SiO_2 44.2–45.8, Al_2O_3 0.6–1.1, FeO 0.8–1.8, MnO 6.9–15.8, CaO 13.4–24.1, BeO 10.7, H_2O 11.8. Biax, $N_{av} = 1.604$, close to the average value for chiavennite. Occurs in vugs in a nepheline–syenite pegmatite. Occurs as minute tabular crystals in spherulitic aggregates up to 3 mm diameter in nepheline-syenite pegmatite at the *Vevja larvikite quarry, Tvedalen, Brunlanes, Vestfold Co., Norway*. AR, EF.

70.5.3.1 Bavenite $Ca_4Be_2Al_2Si_9O_{26}(OH)_2$ to $Ca_4Be_3Al[Si_9O_{25}(OH)_3]$

Named in 1901 by Artini for the locality. ORTH. *Cmcm (MM 47.87(1983)*: originally given as: $a = 9.72$Å, $b = 11.65$Å, $c = 4.94$Å $Z = 16$; all other sources have a and b doubled: a 19.38–19.480Å b 23.19–23.22Å c 4.94–5.00Å, $Z = 4$, $D = 2.77$; (nat.) *(MM 47:87(1983))*; ICDD 13-535: 4.86_8 3.73_8 3.46_5 3.32_{10} 3.24_{10} 3.12_8 3.05_6 1.94_9. True symmetry is probably monoclinic or triclinic based on inclined and irregular extinction (optical) as being the result of twin formation, along with distinct extinctional dispersion *(NJBM 321(1995))*. Structure determined: *(AC 20:301(1966), ZSK 12:87(1971))*. Structural formula can be given as: $H_nCa_4Be_{2+n}Al_{2-n}(SiO_3)_9 \cdot xH_2O$ where n is between 0.1 and 0.8. White, colorless, pale green, pale rose or pink; crystals are fibrous, prismatic, as well as bladed habit; may also be chalk-like and massive; elongated on c, flattened on a, and in radial lamellar masses. Radiating groups to 6 cm; dominant forms (using $a = 23.19$Å, $b = 5.005$Å, $c = 19.39$Å are: {001}, {100}, {010}, {101}, {201}, {310}, {510} and {111} with {100} dominating. {100} twins known. Streak white, earthy to vitreous or silky; cleavage: {001} excellent, {010} fair to good; brittle, $H = 5\frac{1}{2}$–6, $G = 2.68$–2.76; weakly piezoelectric; IR data: Minerali 3(1) 564:1972. Magnetic properties: MZ 6:64(1984). Not soluble in common acids. DTA, TGA: Minerali 3(1) 564(1972) and RSMI 18:53(1962). Traces of Na_2O, MgO, and Fe_2O_3 are usually present. Bavenite from Ilimaussaq has SiO_2 58.5 wt. %, Al_2O_3 5.60, CaO 24.25, BeO 9.5, H_2O 1.8, $\Sigma = 99.70$, and is the Be-richest and Al-poorest bavenite yet reported. Biaxial (+); $2V = 26-60°$, $N_x = 1.578-1.586$, $N_y = 1.579-1.589$, $N_z = 1.583-1.593$ (USGSB 1627(1984)); for Ilimaussaq bavenite which is Be-richest and Al-poorest (NJBM 321(1995)): $N_x = 1.5845$, $N_y = 1.5845$, $N_z = 1.5957$ with a $2V_z = 18°$ and dispersion r > v to r >> v; colorless; optical orientation: positive elongation, OAP || to (100) and $XYZ = bac$. Occurs in granitic pegmatites and miarolitic cavities in such pegmatites; often pseudomorphous after beryl. Also found replacing emerald – Urals. In skarns, e.g. Baty-stau, Ilimaussaq and Thailand. In some greisens and zones of pneumatolytic alteration. Occurs in minor amounts in many localities. Only noteworthy or unusual occurrences are given here. In pegmatite at Waterville Valley, NH! and at Haddam, CT; at the Rutherford mine!, Amelia, VA; Foote quarry!, Kings Mountain, NC; coating on minerals in miarolitic cavities and as pseudomorphs after beryl in the Himalaya dike system, Mesa Grande district, San Diego Co., CA; Ilimaussaq complex, Kangerdluarssuk Fjord, Greenland; Val D'Ossola and *Baveno, Lago Maggiore, Piedmonte, Italy*; various locations in Switzerland (SMPM 50:445(1970); Striegau (Strzegom)!, Silesia, Poland; Marsíkov, Drahonin and Jeclava, Moravia (after beryl), Czech Rep.; Akoudertia, Siberia; various emerald mines in the Malyshevo district!, Urals, Russia: Batystau, central Kazakhstan; massive bavenite with epidote from Kalisay, Kyrgizia; crystals to 5 cm long from Doi Mok mine, Wiang Pa Pao, north Thailand; Londonderry, W. Australia. *MAC SC 8:135(1982) NJBM 321(1995)*. EF

============ Class 71 ============

Sheets of Six-Membered Rings

KAOLINITE–SERPENTINE GROUP

The kaolinite–serpentine group of minerals are phyllosilicates with the general chemical formula:

$$A_{2-3}Si_2O_5(OH)_4$$

where

A=Al, Fe^{3+}, Mg, Ti, Mn.

The structure is based on a t-o (tetrahedral-octahedral) layer (also called a 1 : 1 layer). The t-o layer itself consists of an infinite sheet of Si–Al tetrahedra made up of linked hexagonal rings of composition Si_2O_5. The apical oxygens of the sheet all point in the same direction and provide the base for a set of octahedral sites. Hydroxyl groups complete the octahedral sites.

The t-o layers are electrically neutral. The entire exterior octahedral side of the t-o layer consists of hydroxyl groups, available for hydrogen bonding to the oxygen-covered exterior of the next "t" layer.

The dimensions of ideal trioctahedral sheets are slightly mismatched to the dimensions of ideal sheets of tetrahedra, and the dimensional mismatch [predicted by Linus Pauling, *PNAS 16:578(1930)*] is relieved by one of four mechanisms that form the basis for the division of the serpentine (trioctahedral) subgroup.

Polymorphism is extensive in the kaolinite–serpentine group. Successive layers may be translated $-(a/3)$, $\pm(b/3)$ and may be rotated $\pm 60°$, $\pm 120°$, or $180°$. References to the polytypes are found in *Electron-Diffraction Analysis of Clay Mineral Structures, Zvyagin(1967), CCM 17:355(1969), AM 53:14(1968), CM 14:314(1976)*.

KAOLINITE SUBGROUP, DIOCTAHEDRAL

Mineral	Formula	
Dickite*	$Al_2Si_2O_5(OH)_4$	71.1.1.1
Kaolinite	$Al_2Si_2O_5(OH)_4$	71.1.1.2
Nacrite*	$Al_2Si_2O_5(OH)_4$	71.1.1.3
Halloysite	$Al_2Si_2O_2(OH)_4 \cdot 2H_2O$	71.1.1.4
Odinite	$(Fe^{3+},Mg,Al,Fe^{2+},Ti,Mn)_{2.42}(Si,Al)_2O_5(OH)_4$	71.1.1.5

*Named polytypes.

SERPENTINE–ANTIGORITE-RELATED SUBGROUP, TRIOCTAHEDRAL

Mineral	Formula	
Antigorite	$Mg_3Si_2O_5(OH)_4$	71.1.2a.1

Strain relieved by an alternating wave of curved t-o sheets.

SERPENTINE–LIZARDITE-RELATED SUBGROUP, TRIOCTAHEDRAL

Mineral	Formula	
Caryopilite	$(Mn,Mg,Zn,Fe)_3(Si,As),O_5(OH,Cl)_4$	71.1.2b.1
Lizardite	$Mg_3(Si,Al)_2O_5(OH)_4$	71.1.2b.2
Nepouite	$Ni_3Si_2O_5(OH)_4$	71.1.2b.3
Greenalite	$Fe_3^{2+}Si_2O_5(OH)_4$	71.1.2b.4

Strain relieved by minor Al and/or Fe^{3+} substitution with the R^{3+} occupying a smaller polyhedron.

SERPENTINE–AMESITE-RELATED SUBGROUP, TRIOCTAHEDRAL

Mineral	Formula	
Amesite	$(Mg_2Al)[SiAl]O_5(OH)_4$	71.1.2c.1
Berthierine	$Fe_2^{2+}Al(SiAl)O_5(OH)_4$	71.1.2c.2
Brindleyite	$Ni_2Al(SiAl)O_5(OH)_4$	71.1.2c.3
Fraipontite	$(Zn,Ca,Al)_3(SiAl)_2O_5(OH)_4$	71.1.2c.4
Kellyite	$Mn_2^{2+}Al[SiAl]O_5(OH)_4$	71.1.2c.5
Manandonite	$Li_2Al_4[(Si_2AlB)O_{10}](OH)_8$	71.1.2c.6
Cronstedtite	$Fe_2^{2+}Fe^{3+}(Si,Fe^{3+})O_5(OH)_4$	71.1.2c.7

Strain relieved by extensive Al substitution in both the t and o layers.

SERPENTINE–CHRYSOTILE-RELATED SUBGROUP, TRIOCTAHEDRAL

Mineral	Formula	
Clinochrysotile	$Mg_3Si_2O_5(OH)_4$	71.1.2d.1
Orthochrysotile	$Mg_3Si_2O_5(OH)_4$	71.1.2d.2
Parachrysotile	$Mg_3Si_2O_5(OH)_4$	71.1.2d.3
Pecoraite	$Ni_3Si_2O_5(OH)_4$	71.1.2d.4

Strain relieved by developing a cylindrically rolled structure.

Reference: *MIN 41:122(1992)*.

Kaolinites show only moderate chemical variations. There has been much discussion of the compositional differences of the Mg serpentines and the extent to which they are polymorphs or are compositionally different. *CM 13:227(1975)*, *AM 55:1025(1970)*. Excess silica is present in many analyses and is attributed to interstratified 2 : 1-layer phyllosilicate impurities or to Mg and OH vacancies. *CM 17:785(1979)*.

71.1.1.1   Dickite $Al_2Si_2O_5(OH)_4$

71.1.1.2   Kaolinite $Al_2Si_2O_5(OH)_4$

71.1.1.3   Nacrite $Al_2Si_2O_5(OH)_4$

Kaolinite is named for the ancient Chinese locality Kauling, *high ridge*. Dickite was named by Ross and Kerr in 1930 for A. B. Dick, British mineralogist, and nacrite was named by Brongniart in 1807 in allusion to its pearly luster. Kaolinite–serpentine group. Polytypes known as kaolinite, dickite, nacrite, and halloysite. Anauxite = kaolinite + silica. **Naorite:** TRIC $C1$ (also $P1$). $a = 5.13\text{–}5.16$, $b = 8.89\text{–}8.95$, $c = 7.25\text{–}7.40$, $\alpha = 90\text{–}91.8°$, $\beta = 104.5\text{–}105°$, $\gamma = 89.8\text{–}90°$, $Z = 1$; dickite: MON Cc. $a = 5.14\text{–}5.15$, $b = 8.94$, $c = 14.40\text{–}14.74$, $\beta = 96.7°$, $103.58°$, $Z = 2$; nacrite: MON Cc. $a = 8.909\text{–}8.96$, $b = 5.146\text{–}5.15$, $c = 14.35\text{–}15.70$, $\beta = (100°)$, $113.70°$, $Z = 2$, $D = 2.63$. *14-164*(kaolinite-1A, nat): 7.17_{10} 4.37_6 $4.19_{4.5}$ 3.58_8 $2.495_{4.5}$ 1.62_7 1.59_6 1.49_9; *29-1488*(kaolinite-1Md, nat): 7.1_{10} 4.41_6 3.56_{10} $2.55_{2.5}$ 2.49_3 2.375_2 2.33_4 1.49_3; *10-446*(dickite-$2M_1$, nat): 7.15_{10} 4.12_7 3.80_6 3.58_{10} 2.51_5 2.33_9 1.975_5 1.49_5; *16-606*(nacrite-$2M_2$, nat): 7.18_{10} 4.41_3 4.36_8 4.13_7 3.59_8 2.43_6 2.40_4 1.49_4. Kaolinite is a 7-Å phyllosilicate which consists of a sheet of silicate tetrahedra polymerized to form nearly hexagonal, ditrigonal openings, due to vacancies in the tetrahedral sheet. The tetrahedral sheet has an essentially coplanar surface of oxygen atoms with the unpolymerized tetrahedral apices directed toward, and forming part of, coordination polyhedra, which, together with additional oxygens and hydroxyls, constitute a layer of octahedral sites two-thirds occupied by Al. Hydroxyls are located below the center of the hexagonal openings of the tetrahedral sheet on the surface of the octahedral sheet and on the entire surface of the octahedral sheet opposite the tetrahedral sheet. Polyhedral distortions and rotations are present to minimize bond energies [*AM 58:471(1973)*] and the layer contains minor corrugations. Kaolinite is dioctahedral with $d_{060} \sim 1.49\text{–}1.50$ Å and has no layer charge. The structure is united by angled hydrogen bonds extending from hydroxyls of the octahedral sheet of one layer to oxygens of the adjacent tetrahedral sheet of the next layer. *CCM 29:316(1981)*, *31:352(1983)*, *32:483(1984)*. Structure of triclinic material: *CCM 36:225(1988)*. The stacking of kaolinite layers is frequently turbostratic, which can result in the loss of definition of XRD peaks in the range 20–24° 2θ, as well as loss of many otherwise permitted hkl reflections. This disordered material, which can be pseudomonoclinic, then gives a broad asymmetrical band-like peak, consisting of three coalescing peaks, which is steep at $\sim 20°$ 2θ and which declines in intensity toward the base of d_{002} at

~ $29°$ 2θ. Additional defects include $\pm^{nb}/3$ translations and $\pm^{2p}/3$ rotations of adjacent layers. Defects also increase with $^{IV}Fe^{3+}$. *BM 105:457(1982)*. The Hinckley Index [*CCM 11:229(1963)*] involves the ratio of X-ray diffraction peaks in the range 20–24° 2θ and is a measure of stacking order, but for low-defect kaolinites, HI > 0.43, the value represents the abundance of the low-defect kaolinite in the entire sample. *CCM 37:203(1989)*. In well-ordered kaolinite, these three peaks are more intense and are well resolved. In practice, however, relatively few peaks are observable, especially in mixtures. Kaolinite derived from plagioclase can have a higher crystallinity index than that derived from alkali feldspar. *DS 27:609(1979)*. Specimen grinding can induce additional disorder. Kaolinite has three polytypes which have historically been given separate names: kaolinite, dickite, and nacrite. **Habit:** Kaolinite is characterized by a stacking sequence of $-a/3$ translations and superimposed layer vacancies. Dickite is a polytype of kaolinite where each layer is shifted by $-a/3$, as in kaolinite proper, but with alternating superpositions of vacancies equivalent to $\pm 120°$ relative rotation of adjacent layers. *CCM 37:297(1989)*. Nacrite was originally thought to be a six-layer polytype, but a two-layer stacking has been derived. The pattern of vacancies in nacrite shifts the symmetry and requires the interchange of a and b axis labels relative to kaolinite and dickite [*AM 48:1196(1963)*], but thus with $-a/3$ stacking sequence and $\pm 60°$ relative rotation of adjacent layers. *CCM 37:297(1989)*. Kaolinite synthesis has been reviewed with temperatures ranging from surface conditions to 400°. *KZ 17:181(1986)*. Kaolinite polytypes are not yet systematically useful indicators of their mode of formation, and all three polytypes have been found in a single deposit. *CCM 29:451(1981)*. NMR studies have been made. *DAN SSSR 261:1169(1981)*, *JCSF I 84:117(1988)*. IR studies are numerous. Kaolinite has been intercalated with K-acetate, hydrazine, amines, urea, dimethyl sulfoxide, and others. Kaolinite polytypes are decomposed and become X-ray amorphous on heating to 550°. **Physical properties:** Usually white, also pale yellow to cream, pale green to pale blue, can be lightly to heavily stained yellow or red to brown. Usually earthy, also waxy to pearly; unctuous, can be plastic when wet and varies from freely slaking to flinty, nonslaking and has even been found as stream cobbles. Crystals pseudohexagonal platy, but usually 2–5 µm visible crystals rare but may be more than 1 mm across. SEM photographs of kaolinite usually show hexagonal platelets or curved stacks of plates, but fibers and spheres have been observed. Mica-law pseudotwinning has been observed. *AM 57:411(1972)*. Cleavage {001}, perfect; flexible but inelastic. H = $2-2\frac{1}{2}$, hardness may be affected by amorphous to crystalline silica impurities. G below 2.0 when porous, but can be up to 2.60–2.68. Infusible. Insoluble in most acids, but can be attacked by HCl. Dickite can show distinct cathode luminescence. *CCM 29:451(1981)*. **Optics:** Kaolinite: biaxial $(-)$; OAP perpendicular to (010), Z = b, X $\wedge$ c = 1–4°, also X $\wedge$ c = -10 to $-13°$, and Y $\wedge$ a = 1–4°; N_x = 1.553–1.565, N_y = 1.559–1.569, N_z = 1.560–1.570; birefringence = 0.005–0.010, usually $\approx$ 0.006; 2V = 24–50°; r $\geq$ v, weak; rarely pleochroic, colorless to pale yellow. Dickite: biaxial $(+)$; Z = b; Y $\wedge$ a = 14–20°; N_x = 1.560–1.564, N_y = 1.561–1.566, N_z = 1.566–

1.570; birefringence = 0.006; 2V = 55–80°; r ≫ v. Nacrite: biaxial (−,+); Z = b, X ∧ c = 10–12°; N_x = 1.557–1.560, N_y = 1.562–1.563, N_z = 1.563–1.566; birefringence = 0.006; 2V = 40–90°; r > v. **Chemistry:** Kaolinite does not have a large chemical variation, most of which can be attributed to a few elements; very minor quantities of other clay impurities, iron oxides, and others. Fine-grained detrital minerals such as quartz, feldspar, rutile, or anatase can therefore be significant chemical contributors. Very minor ^{IV}Al substitutes for ^{IV}Si, usually no more than 0.10 and normally no more than 0.02. Kaolinite has been thought rarely to have additional phyllosilicate interstratifications, such as micalike layers and sometimes much less than 1% of the total layers. *CCM 23:125(1975)*. Ferric iron, Mn^{2+}, and $(VO)^{2+}$ have been recognized in kaolinite using electron spin resonance. *CM 10:313(1975)*. Iron has been observed to be more common in disordered kaolinite than in well-ordered material. Dithionite treatment does not remove all iron oxides. *CCM 29:23(1981)*. Observed analyses (wt %): SiO_2, 37.45–46.79 (higher values indicate free silica in the sample); Al_2O_3, 33.09–43.50; TiO_2, 0.09 (St. Austell) (higher values might be due to detrital Ti minerals); Fe_2O_3, 4.69 (GA); MnO, 1.61 (Jundiai); SO_3, 2.00 (GA); P_2O_5, 1.33 (GA); Na_2O, 0.85 (GA); K_2O, 0.46 (GA); MgO, 0.35 (GA); CaO, 0.23 (Keokuk); H_2O^+, 13.3–13.9. Lithian kaolinite has been synthesized (*CA 114:188488*). CEC is generally negligible to 10 meq per 100 g. Kaolinite has long been known for its small anion-exchange capacity with P and its importance in agriculture; also retains ^{137}Cs. Chromian kaolinite has been observed from three different localities: Urakawa Ni deposit, Shizuoka Pref., Japan; Sonoma region, CA; Takovo, Serbia. As much as 1.2 wt % Cr_2O_3 has been reported. *MIN 4(1):51(1992)*, *CCM 28:295(1980)*. **Occurrence:** Frequently, the result of alteration of plagioclase or alkali feldspars, also micas and smectites. Can be found replacing a variety of minerals, usually feldspars, but also leucite, quartz, sillimanite, tourmaline, and others. Can be associated with a variety of other clay minerals, including halloysite, micas, quartz and silica, feldspar, gypsum, pyrophyllite, rutile, or anatase, rarely alunite, boehmite, gibbsite, diaspore, and a host of detrital minerals. Found in a wide variety of freshwater and marine sedimentary environments, can be an important constituent in coal and, by extension, in coal and fly ash. Frequently, large residual deposits result from deep weathering of granites or feldspathic rocks, but also occur as nearly monomineralic sedimentary layers of economic significance. Abundant in some sandstones. Also may occur in vugs in hornfels. **Uses:** Kaolinite and its ore, kaolin, are consumed in millions of metric tons annually. Used in numerous industrial, agricultural, and other applications when raw, refined, or chemically treated: widely used as a filler and/or coater for paper, also as a pressure-sensitive paper coating for microencapsulated dyes for carbonless copies, used as an absorbant as in dye tracers, also as a pore-forming enhancer for dye uptake, ink retention enhancer, as a simple absorbant for liquid spills (e.g., oils), and as a stain preventive coating on lithographic plates; used in reinforced plastics or rubber, as a filler in some food (e.g., chocolate), as a filler in some talcum powders and paints; used in fireproofing, as a major component of putty and some grouts, cements, glazes, and refractory mortars, in gas barriers for carbonated beverage containers, in a wide variety of gaskets, as a binder in zeolite absorbants and in pencil leads, as a foundry die coating, a nonsticking dispersant for agricultural granules, seeds, and so on, and flinty varieties used as an all-weather coating on primitive roads. Also as catalysts for oxidation or polymerization reactions, can be resin coated for electrolyte fixing, aids extrusion processes, smooths and provides antiabrasion properties for magnetic tape; used as a coagulant in water purification, also for benzene purification, widely used in soaps, cosmetics, and disinfectants, useful in desulfurizing and dedusting agents, as a filler in polyester fibers to improve surface appearance as well as resistance to piling and fibrilation, initially used in the manufacture of porcelain and fine china, also tiles, bricks, and terracotta, and used as a starting material in alumina ceramics manufacture and as a starting run material in the manufacture of some zeolites. **Kaolinite localities:** Brandon, Rutland Co., VT; Blandford, Hampden Co., MA; Chaudiere Falls, PA; Mt. Savage(!), MD; Richmond, Hanover Co., VA; Spruce Pine, Mitchell Co., and near Webster, Jackson Co., NC; hundreds of active and inactive deposits are located around Byron, Peach Co.; Butler, Taylor Co.; Gordon, Wilkinson Co.; Kathleen, Houston Co.; Dry Branch, Twiggs Co.; Griswoldville, Jones Co.; Carrs Station, Hancock Co.; Gibson, Glascock Co,; Hephzibah, Richmond Co.; Buffalo Creek, Washington Co.; and Andersonville, Macon Co., all in Georgia; many locations in SC, including the Dixie Clay Co. mine and the Lamar pit, near Bath, Aikin Co.; Longley and Edgefield, Edgefield Co., SC; Hackleburg, AL; Panola Co., MS; near Murfreesboro, Pike Co., and at Greenwood, Sebastian Co., AR; Blaxland, MO; in quartz geodes at Keokuk(!), Lee Co., IA; Silverton, San Juan Co., CO; Mesa Alta, Rio Arriba Co., NM; Rock Springs formation(!), WY; Lewistown, Fergus Co., MT; Ione, Amador Co., CA; Huberdeau, PQ, and near Walton, NS, Canada; Jacal, Villa Victoria, MICH; Noche Buena and Sombrerte, ZAC; Etzatlan, JAL; Huayacocotla, SLP, Mexico; Almwch(!), Anglesey Is., Wales; exceptional deposits at St. Austell and elsewhere, Cornwall, also in Devon and Stourbridge, England; Ayrshire, Scotland; Yriex and the Limoges district, Haute-Vienne; St. Gilles, La Chartreuse, and Bagatelle, France; Guadalajara to Valencia, Spain; Locchera, Sardinia, Italy; Brand, Dresden, and Kemmlitz, Saxony; Wetro deposit, Bautzen; Tirschenreuth, Diendorf, and elsewhere, Bavaria; also Caminau and Wiesa deposits, Lusatian massif, Germany; Karlovy Vary, Schlan, and elsewhere, Bohemia, Czech Republic; Horna Prievrana, Slovakia; Oldrzychow and Osiecznica, Lower Silesia, Poland; Prosyanovskoe, Donets Basin, and elsewhere, Ukraine; Elena deposit, Ural Mts., Russia; Jari River district, Amapa; Jundiai and Varinhas, SP, Brazil; Barker, Buenos Aires Prov.; Chubut Valley, Chubut Prov., and Santa Cruz Prov., Argentina; Wadi Kalabsha, also Nile delta, Egypt; Kuban delta; Pugu, Tanzania; Grahamstown and Fish Hoek district, Cape Prov.; Inanda, Natal; Koster and elsewhere, Transvaal, South Africa; Mehsana, Gujarat; Malti deposit, Purulia district, western Bengal, India; Guoshan deposit, Fujian Prov., Dapo deposit, Huazhou, Guangdong; Suzhou, Jingsu Prov.; *Kauling*

(Jauchau), Kiangsi Prov., China; Belitung and Banka Is., Indonesia; Moengo area, Suriname; Kanpaku, Tochigi Pref.; Itaya, Okayama Pref.; Amakusa Sarayama deposit, Kyushu, Japan; Weipa, Queensland; Gabbin, WA; Mt. Hope, Imbitcha, and Crawford deposits, SA; Gulgong, NSW; Egerton and Pittong, VIC, Australia; Maungaparerua, North Island, New Zealand. **Dickite localities**: St. Clair(!), Schuylkill Co., PA; Profitt, Albermarle Co.; Willis Mt., Buckingham Co.; Herbb No. 2 pegmatite, Powhatan Co.; Tiberville, Rockingham Co., VA; Rock Springs, WI; Silverton, San Juan Co., and Alma, Park Co.; Red Mt., Ouray, CO; St. George, Washington Co., UT; Bisbee, Cochise Co., AZ; Chon, Nayaret; San Juanito and Cusihuiriachio, Mexico; *Almwch, Anglesey, Wales*; Mas d'Aulry(!), Lodeve, and crystals to 4 mm from the Raberjac mine, Lodev, Herault, France; Crete Nere formation, Sicily; Scabiazza sandstone, Oltrepo Pavese area, Italy; Nowa Ruda, Lower Silesia, Poland; Kara-Chekv massif, Karkaralinsk, Kazakhstan; San Pablo mine, La Paz Dept., Bolivia; Hawksbury sandstone, Sydney Basin, NSW; Williamstown, SA, Australia; Makurasaki, Kagoshima Pref.; Itaya, Okayama Pref.; Syokozan mine, Shokozan; Amakusa Sarayama deposit, Kyushu, Japan. **Nacrite localities**: Middleville, Herkimer Co., NY; Lone Jack quarry, Glasgow, Rockbridge Co., VA; Tracy mine, Marquette Co., MI; Eureka tunnel, St. Peter's Dome, El Paso Co., CO: St. Eustache, QUE; Dresser mine(!), Walton, NS, Canada; Cerro de la Bufa and San Jose de Ranchos, Sombrerte, ZAC; and Xilosintla, OAX, Mexico; Groby, Leicestershire, England; Ottoschwanden, Schwartzwald, Bavaria; Huhnersedel Mt. Baden; Brand and *Freiberg, Saxony, Germany*; Harvivaara, Finland; Eski-Orda, Simferopol, Crimea, Russia; Otege mine, Yamagata Pref.; Itaya, Okayama Pref., Japan. VK, EF *MIN 4(1):25(1992)*.

71.1.1.4 Halloysite $Al_2Si_2O_2(OH)_4 \cdot 2H_2O$

Named in 1826 by Berthier for Belgian Geologist Baron Omalius d'Halloy (1783–1875), the mineral's discoverer. Kaolinite–serpentine group. Halloysite-7 Å (metahalloysite), halloysite-10 Å (hydrated halloysite, endellite; European literature). MON Cc (also TRIC). Cell sometimes given on hexagonal coordinates. $a = 5.14–5.15$, $b = 8.90–8.95$, $c = 7.214–7.37$, $10.1–10.2$, $\beta = 96°$, $99.7–101.9°$, $Z = 2$; (hexagonal coordinates): $a = 5.118–5.133$, $c = 7.16–7.30, 10.03$; $Z = 1$, $D = 2.58–2.62$ (7 Å), $D = 2.14–2.15$ (10 Å). *9-453*(halloysite-7 Å, nat): 7.5_9 4.42_{10} 3.63_9 2.56_8 1.68_8 1.48_9 1.28_7 1.23_7; *29-1487*(halloysite-7 Å, nat): $7.3_{6.5}$ 4.42_{10} 3.62_6 $2.56_{2.5}$ $2.37_{<1}$ 1.68_2 1.48_3; *9-451*(halloysite-10 Å): 10.1_9 4.42_{10} 3.34_9 2.56_8 1.68_8 1.48_9 1.28_7 1.23_7; *29-1489*(halloysite-10 Å): 10.0_{10} 4.36_7 3.35_4 $2.54_{3.5}$ 1.67_1 1.48_3 1.28_1. *PD 4:19(1989)*. Halloysite is a phyllosilicate that has a tubular morphology (generally less than 5 μm), with an average diameter of 0.04 μm, but apparently with the structural topology of kaolinite. Halloysite has a hydrated 10-Å as well as a dehydrated 7-Å modification. The dehydrated 7-Å halloysite is generally regarded as tubular halloysite that has unrolled to form laths or plates. The tube elongation is b, but a has been observed occasionally as the axis coincident with elongation.

CMB 2:133(1954), *AC 7:511(1954)*. The transition from 10 Å to 7 Å was early recognized to occur at about 105° [*ZK 86:340(1934)*], but the dehydration can occur at 70° [B&B 153(1980)], and the gentle heat treatment of halloysite-10 Å is therefore diagnostic. Rehydration is usually not possible, but intermediate hydrates might be restored to full hydration. Halloysite irreversibly becomes X-ray amorphous when heated to 550°. The additional disorder of halloysite compared with disordered kaolinite is useful in distinguishing the two, and tubular morphology has been suggested to be a primary feature of the mineral. *CCM 26:25(1978)*. SEM study generally reveals round tubular morphology for halloysite, but spheres and geometrical platelets have been observed. In Japan, halloysite morphology was related to origin: hydrothermal = tubular and altered pumice = rounded grains [*Proc. Int. Clay Conf.* 257(1975)], but the relationship may have been regional. Platy halloysite is often high in Fe^{3+}. *CM 21:401(1986)*. Polygonal tubes (triangular, pentagonal, hexagonal, and octagonal) have sometimes been interpreted as precursors to round tubes, and spiral layering has been observed. *CCM 39:113(1991)*. The first-order peak of halloysite is generally less intense than the band of peaks observed at $\approx 20°\ 2\theta$ (CuKα). Oriented X-ray diffraction shows considerable disorder for halloysite, with a coalescing of or broadening of peaks. Halloysite expands with ethylene glycol, dimethyl sulfoxide [*CCM 17:57(1978)*], formamide [*CCM 32:241(1984)*], and similar compounds. HWG (hydrazine, water, glycerol) intercalation sequences have been used to distinguish kaolinite, which is usually unaffected, from halloysite, which expands to 11 Å [*Proc. Int. Clay Conf., Tokyo v. 1:3(1969)*], but results can be ambiguous. *CCM 38:591(1990)*. Halloysites have been synthesized only at near-surface conditions. Halloysite is easily damaged by electron beams. **Habit:** Found as an earthy to waxy, rarely pearly, mineral; fine-grained (2–5 µm size common), clay-size particles are typically tubular but sometimes show spherical clustering with thin peel-like exfoliations; indistinct particle morphology may make visual identification (SEM) impossible, rarely rounded polyhedral. Many allophanes could be spherical halloysites. **Physical properties:** White to tan, also bluish and greenish gray, reddish to chocolate brown; pearly to waxy or dull luster. Cleavage {001} probable, conchoidal fracture. H = 1–2. G = 2.0–2.65. Can become translucent when immersed in water. **Optics:** Biaxial (–); $N_x = 1.559$, $N_y = 1.564$, $N_z = 1.565$ (for 7 Å), $N_y = 1.490$ (for 10 Å); 2V = 37°. **Chemistry:** Halloysite does not have a widely variable chemistry and some of the variation might be ascribed to detrital minerals such as anatase and rutile or to additional clays and micas. Analysis (wt %): SiO_2, 42.20–44.46; Al_2O_3, 36.58–38.46; Fe_2O_3, generally less than 0.36, but observed to 12.8; FeO, 0.07; MgO, 0.18; CaO, 0.32; Na_2O, 0.14; K_2O, 0.51; H_2O^+, 13.38–14.59; H_2O^-, 2.58–4.05; TiO_2, 0.15; P_2O_3, 0.18. More effective than kaolinite in fixing soil Cu. Halloysites may have low SiO_2 values if $[OH]_4$ substitutes for $[SiO_4]$ [*Proc. Int. Clay Conf., Jerusalem 1:11(1965)*], and $0.16^{IV}Al$ has been observed. CEC = 20–60 meq per 100 g, but smectite could be present in higher-CEC samples. **Occurrence:** Frequently found as an alteration of basaltic rocks or in hydrothermally altered fissures in mon-

71.1.1.4 Halloysite

zonite, for example; also derived from feldspars, chlorite, micas, rhyolite, granite, volcanic ash, tuffs, and pumice; rarely found in granite pegmatites, also in bauxite or laterite, and in some marine, glacial, or K-T boundary clays. Easily weathers to kaolinite, but can also be derived from it. Sometimes associated with allophane and may be derived from it. In many cases, halloysite is probably the result of low-temperature weathering below the water table but is also the result of hydrothermal alteration of feldspathic igneous rocks. Ongoing formation observed at hot springs. Associated with kaolinite, illite, smectite, allophane, volcanic glass, detrital minerals, and others. Uses are similar to those of kaolinite, especially in high-quality ceramics and high-strength concrete; also as a carrier for catalysts. **Localities:** Only selected occurrences can be listed here: Paris, Oxford Co., ME; Piney R. and (Bunker Hill pegmatite) Buena Vista, Amherst Co., VA; Gusher Knob, Spruce Pine, Alexander, and Corundum Hill mine, Macon Co., NC; Panola Mt. and Taylors Ridge, GA; Stanford and elsewhere, Lincoln Co., also Horse Cove, Hart Co., KY; Gardner mine, Bedford, Lawrence Co., IN; Lawrence Co., MO; Pruden mine, Saline Co., AR; Clayton and Dodge, and Gonzales, Gonzales Co., TX; Wagon Wheel Gap, Mineral Co., CO; San Lorenzo; Gila Hot Springs, Grant Co., NM; Bisbee, Cochise Co., Ray, Pinal Co., and other localities in AZ; Blue Star mine, Eureka Co.; Bullion, NV; New Park, Dragon mine, Eureka, Tintic district, Juab Co., and Wendover, UT; Brownie Butte, Garfield Co., MT; Clearwater Valley–Bitteroot Mts., ID; Cerro Gordo mine, Inyo Co.; Pala, San Diego Co.; Hackberry Mt., San Bernardino Co., CA; Koolau Mts., Oahu Is., also on Lanai and Molokai Is., HI; Vancouver Is., BC, Canada; Los Azufres, MICH, Mexico; Mt. Pelee, Martinique; Guadaloupe; Rossland, Scotland; Bristol district, England; Vuotso-Tankavaara, Finland; with iron, zinc, and lead minerals in limestone at *Angleure, near Liège, Belgium* (now inaccessible); Dordogne; Armorican massif, Brittany, France; Haro, La Rioja, Spain; Valenca, Portugal; Vilco volcano, Rome; Scatola trough, Viterbo, Italy; Altenberg, Stolberg, and Bergnersreuth, near Wunsiedel, Bavaria, Germany; Stanzertal, Styria, Austria; Karlovy Vary, Bohemia, Czech Republic; Michalovce and Cervenica–Dubnik opal deposits, Slovakia; Kovagoors, Hungary; Kubno and Dunino, Lower Silesia; Tarnowitz, Silesia, Poland; Mt. Peres, Israel; Kara Oba and Tsager region, Caucasus Mts., Kazakhstan; Capivarita, RGS, and Chapada do Araripe, Pernambuco, Brazil; Barker, Buenos Aires Prov., Argentina; Copiapo, Chile; Adamwa Plateau and Cameroon Mt., Cameroon; Mt. Kenya, Kenya; Kivu Prov., Zaire; Andhra Pradesh; Panhala, Maharashtra; Colva, Goa, India; Ranong, Thailand; Hong Kong; coastal Guangdong Prov.; Guoshan, Fujian Prov., China; Nanju and Sancheong, South Korea; Talakag, Philippines; Kuantan, Malaysia; Beleitung and Banka Is., Indonesia; Darling Range bauxite, WA; Andamooka opal field and Moculta and other phosphate deposits, SA; Ardlethan mine, NSW; Trial Hill mine, Q, Australia; Maturi Bay, Te Puke, and Maungaparerua, North Island, New Zealand; Ambrym, Vanuatu; New Georgia Sound, Solomon Is.; Inage, Chiba Pref.; Iriki and Oguchi, Kagoshima Pref.; Kusatsu and Joshin, Gunma Pref.; Koganei and

Ina, Nagano Pref.; Yawata, Gifu Pref.; Gosammae, Iwate Pref.; Iki Is., Nagasaki Pref.; Imaichi and Mashiko, Tochigi Pref.; Shigaraki Pref., Japan; Kerguelen Is. VK, EF *ACP II 2:332(1826), AC 31:2851(1975)*.

71.1.1.5   Odinite $(Fe^{3+}, Mg, Al, Fe^{2+}, Ti, Mn)_{2.42}(Si, Al)_2O_5(OH)_4$

Named in 1988 by Bailey for Gilles-Serge Odin, geologist, Paris, France. Kaolinite–serpentine group. Also called phyllite V; originally thought to be berthierine. Two polytypes exist: 1M and 1T, usually mixed in the same sample, with 1M being more abundant. MON and HEX-R Space groups unknown but probably Cm (1M) and $P31m$ (1T). (1M): $a = 5.373$, $b = 9.326$, $c = 7.363$, $\beta = 104.0°$; and (1T, orthogonal axes): $a = 5.366$, $b = 9.334$, $c = 7.161$, $\beta = 90.0°$. *MSASC 19:169(1988)* gives (1M): $a = 5.369$, $b = 9.307$, $c = 7.17$, $\beta = 103.9°$; (1T): $a = 5.371$, $b = 9.316$, $c = 7.361$, $\beta = 90.0°$. X-ray diffraction lines are broad and frequently show a 14-Å peak due to partial conversion to chlorite. A 1 : 1 serpentine-type layer structure, intermediate between dioctahedral and trioctahedral. Ideally dioctahedral and therefore conceptually related to, and the Fe^{3+} analog of, kaolinite. End-member odinite should ideally be $(Fe^{3+})_2Si_2O_5(OH)_4$, but when $\Sigma^{VI}R^{2+} > \Sigma^{VI}R^{3+}$, dioctahedral odinite grades into trioctahedral greenalite. Green, like glauconite; gray-green streak; silky to earthy luster. Mossbauer, IR, DTA data: *CM 23:237(1988)*. Odinite does not expand with ethylene glycol or similar compounds and becomes amorphous on heating to 490°. An Fe^{3+}-rich 1 : 1 clay mineral with 0–10% tetrahedral substitution of Al for Si. Structural formula of "purest" material yielded $Fe^{3+}_{0.784}Mg_{0.772}Al_{0.556}Fe^{2+}_{0.279}Ti_{0.016}Mn_{0.15}\square_{0.443})(Si_{1.788}Al_{0.212})O_5(OH)_4$ (Koukoure R. mouth); also $(Fe^{3+}_{0.72}Al_{0.70}Mg_{0.63}Fe^{2+}_{0.25}\square_{0.70})(Si_2)O_5(OH)_4$ (Guiana). ^{IV}Fe was not observed. The chemistry of odinite is difficult to interpret, due to its admixture with so much extraneous matter. Typical chemical ranges of odinite with Fe^{3+} greater than any other octahedral cation include (wt %): SiO_2, 36.0–39.1*; Al_2O_3, 10.8–12.2; Fe_2O_3, 17.9–19.5**; FeO, 5.0–6.5*; TiO_2, 0.4–0.5; CaO, 0.13–0.7***; MgO, 6.5–11.0*; MnO, 0.0–0.33; P_2O_5, 0.0–0.17; N_2O, trace–0.3; K_2O, 0.35–1.40; H_2O, 2.0–4.10; $H_2O^+(+CO_2)$, 9.4–10.96. *Lower values reported but where $\Sigma R^{2+} > \Sigma R^{3+}$; **higher values reported but where $\Sigma R^{2+} < \Sigma^{3+}$; ***possible impurity. Widespread as infillings or replacements of microtests, bioclasts, fecal pellets, or mineral debris on shallow (15–60 m) marine shelves and reef lagoonal areas in tropical latitudes: e.g. New Caledonia; Ogooue R. prodelta, Congo Republic; the continental shelf between the Amazon and Orinoco rivers, Brazil; Niger R. prodelta, Nigeria; off Martinique Is.; and the *Los Is., in the mouth of the Koukoure R., Guinea*. Authigenic. Associated with detrital quartz, carbonate bioclasts, clays (kaolinite, smectite, illite, etc.), chlorite, pyrite, goethite, and others. Not recognized in rocks older than Recent–Quaternary. Not yet found in oolites in contrast to berthierine. In shallow depths (< 15 m), odinite is usually oxidized yellow to red-brown and probably closer to the end member. Easily alters to chlorite. Informally reported from a wide variety of localities, but chemical data indicate that some samples are ferrian greenalite rather than odinite. VK, EF *CR 301:105(1985)*, *MIN 4(1):176(1992)*.

71.1.2a.1   Antigorite $Mg_3Si_2O_5(OH)_4$

Named in 1840 by E. Schweizer for the locality. Kaolinite–serpentine group. Antigorite is frequently called serpentine and thus confounded with other members of the group; also called williamsite, bowenite, jenkinsite, picrosmine, and picrolite in part. MON *Pm*. $a = 35.4$–47.2 (commonly 43.4 Å, but 33–50 Å and 109 Å as extremes [*AM 42:133(1957)*], usually with a subcell 5.42–5.46 Å, but with a wide range, designated A), $b = 9.2$–9.28, $c = 7.24$–7.28, $\beta = 91.1$–$91.7°$, $Z = 7$–$10, 16$, $D = 2.52$. *21-963*: 7.29_{10} 6.43_4 5.82_4 5.15_4 4.64_4 3.61_8 2.53_{10} 2.46_6. Antigorite deviates from other serpentines in that its sheet-like structure consists of alternating curved layers or waves. The Mg-filled octahedral polyhedra are larger than the normally Si-filled tetrahedral polyhedra, and a curving of the layer relieves the bond-length mismatch with the apical coordinating oxygens, which constitute one side of the octahedral sheet. A plane of hydroxyls forms the remainder of the coordinating ions of the octahedral sheet. The structure is built of alternating up–down orientations of the serpentine layers, and defect structures are common. Antigorite has been observed as a distorted 1T polytype. The radius of curvature of most antigorite is 50–72 Å with a 61 Å average [*ZK 110:282(1958)*], but the tetrahedral–octahedral mismatch of 0.14 Å predicts a curvature with 190 Å radius. *CM 13:227(1975)*. TEM studies confirm the alternating wave structure of antigorite. *CM 17:679(1979)*, *CMP 97:147(1987)*, *AM 78:75(1993)*. Unlike chrysotiles and lizardites, antigorites can show a deficiency of 0.16–0.19 octahedral cations. *USGSPP:384-A(1962)*. The a dimension in antigorite is a large multiple (usually 7–10, ranging to 16 and rarely higher) of 5–5.7 Å, and various layer multiples have been modeled with differing chemical composition. *AC(B) 45:129(1989)*. Subcell X-ray difraction reflections are usually weak. Due to the wavelike pattern of antigorite layers, the dimensional misfit can result in three fewer ^{VI}R for every eight tetrahedral ring units, and a specific formula for antigorite could be $Mg_{3m-3}Si_{2m}O_{5m}(OH)_{4m-6}$, where m is the number of multiples of a. *CMP 97:147(1987)*. Large multiples of a are more stable at low temperatures. *FM 61:283(1983)*. A typical observed structural formula is: $(Mg_{2.87}Al_{0.04}Fe^{3+}_{0.03}Fe^{2+}_{0.02})(Si_{1.99}Al_{0.01})O_5(OH)_4$. IR studies are numerous [e.g. *MJJ 12:299(1985)*]. Antigorite has been synthesized. *AM 58:915(1973)*. A plot of $Fe^{2+}/Mg + Fe^{2+}$ vs. cell edge shows a positive correlation with increasing a subcell and b and a negative correlation with increasing c. *MJJ 12:299(1985)*. Crystal structures have been solved [*MM 30:498(1954)*, *FM 39:206(1961)*] which confirmed suggestions that the antigorite structure was due to regularly occurring distortions. *AM 21:463(1936)*. High ^{IV}Si values (> 4.0) are commonly calculated and are due to neglecting vacancies of Mg and OH. *CM 17:785(1979)*. **Physical properties**: Dark green ranging to many shades of green, including light shades and blue and yellow-greens; also brown to black, and bright yellow to nearly white. Cleavage {001}, perfect; {010},

{100}, good to poor; pseudocubic parting rare, can show some platyness, also splintery to fibrous as in picrolite. Smooth, but not unctuous. H = 2.5–4, rarely 5.5–6 (polishing of antigorite reveals noticeably greater hardness than lizardite and very noticeably greater than chrysotile). *CM 17:785(1979)*. G = 2.4–2.79, slowly decomposed by acids or not at all (chromian varieties unaffected). Fusible with difficulty. **Optics:** Biaxial (−); X = c; N_x = 1.557–1.595, N_y = 1.563–1.603, N_z = 1.564–1.604; birefringence = 0.004–0.014, 2V = 48–61°, decreasing with iron substitution; r > v; parallel extinction. RI increasing with iron substitution. Also chromian (Cr_2O_3, 4.4 wt %): X = c; N_x = 1.567, N_y = 1.577, N_z = 1.578; birefringence = 0.011; 2V = 58°; r ≪ v and abnormal blue interference colors; positive elongation, platy to lathlike appearance common (0.1–4 mm), but not exclusive to antigorite, undulatory to patchy extinction common. N_z = b, N_x ≈ a. Generally nonpleochroic. Fe^{2+} increases N_x and N_y faster than N_z. *MJJ 12:299(1985)*. OAP parallel to (100). **Chemistry:** Ranges of reported antigorite analyses include (wt %): SiO_2, 36.19–45.69; Al_2O_3, 0–5.00 (usually < 0.50); Fe_2O_3, 0–5.47; Cr_2O_3, 0–0.91; TiO_2, 0–0.02; FeO, 0.07–22.73; NiO, 0–0.42; MgO, 21.08–43.45; MnO, 0–2.53; ZnO, 0.14 (Lowell); CaO, 0.02–1.15; Na_2O, 0–0.05; K_2O, 0–0.24; H_2O, 0–0.92; H_2O^+, 10.48–13.30; Cl, 0–0.03; F, 0–0.31. Chemical heterogeneity within a single purified specimen has been noted. *MJJ 12:299(1985)*. **Occurrence:** Antigorite can be found with other serpentines, talc, clinochlore, amphiboles, pyroxenes, brucite, dolomite, magnesite, olivine, magnetite–magnesiochromite, and others. The reactions antigorite → olivine + talc + H_2O [*MM 55:367(1991)*] and antigorite + brucite → forsterite [*GSAB 88:1497(1977)*] have been proposed. Antigorite is stable in the association tremolite, forsterite, and diopside. *CM 17:729(1979)*. Antigorite is stable at 250–550° [*CMP 97:147(1987)*] and may coexist with lizardite over most of that range. Many antigorites form under reducing prograde conditions contrary to many lizardites. *CM 17:785(1979)*. A major component of serpentinites and important in ophiolite sequences. Observed as a vein mineral. Also found with rodingite. The original antigorite specimen may have been an erratic boulder or was from nearby small valleys: Val Devero or Val Bognanco. *GSLQJ 64:152(1908)*. Extraordinary variety of pseudomorphs from Tilly Foster mine(!). Commonly associated with other serpentines, magnetite, talc, magnesite, and/or dolomite. Clinochrysotile may alter to antigorite. **Localities:** Selected localities include: Lowell, Lamoine Co., and elsewhere, VT; Conklin quarry(!), Smithfield, Providence Co., RI; Monroe, Orange Co.; Tilley Foster mine(!), Brewster, Putnam Co.; Staten Is., Richmond Co., NY; Turkey Mt., Montville, Morris Co., NJ; Lows and Cedar Hill mines, Texas, Lancaster Co., PA; Harford Co., Baltimore and Bare Hills, Baltimore Co., MD; Buck Creek, Clay Co., NC; Bisbee, Cochise Co., AZ; Shipton, QUE; Knee Lake, MAN; Tadamagouche Creek, YT, Canada; Tasiussarssuaq fjord, Greenland; Milton and elsewhere, Glen Urquhart, Inverness-shire, Scotland; Mornhausen–Amolose and Bottenhorn, Hesse-Nassau; Waldheim and Zoblitz, Saxony, Germany; Taberg, Sweden; Zermatt, Switzerland; Sterzing, Bolzano Prov.; Malenco Serpentinite, Bergell Aureole, Italy; Valsavaranche,

Aosta; Passo della Rossa and Domodossola, *Antigorio Valley, Novara Prov., Piedmont, Italy*; Steinbach, Burgenland, Austria; Kutna Hora, Bohemia, Czech Republic; Breznicka and elsewhere, Slovakia; Peloponnesus and Kythira Is., Greece; Caracas region, Venezuela; Cana Brava, Goias; Santa Rosa, Bahia, Brazil; Cyclops Mts., Hollandia, Suriname; Upsallata, Mendoza, Argentina; Zeber Kuh Mts., Iran; Karya magnesite deposit, Mysore, Karnataka; Shapur district, Punjab, India; Safed Koh, Kabul R. valley, Afghanistan; Shigar, Kashmir, Pakistan; Fengtien ophiolite, Taiwan; Khotan, Chinese Turkestan; Hsiu-Yen Hsien, Liaoning Prov., Manchuria; Heiluyu, Xichuan Co., Henan Prov.; Xiuyan district, China; Ryumon peridotite, Kii Penin.; Nishisonogi area, Nagasaki Pref.; Jin-Gatou and elsewhere; Sasaguri area, Fukuoka Pref., Japan; Kyongsangpuk Do, South Korea; Griffin Range, Hokitika, Westland; Cross River, Mikonui, New Zealand., VK, EF *MIN 4(1):126(1992)*.

71.1.2b.1 Caryopilite $(Mn,Mg,Zn,Fe)_3(Si,As)_2O_5(OH,Cl)_4$

Described in 1889 by Hamberg and named from the Greek for *nut* and *felt*, in allusion to its brown color and felted structure. Kaolinite–serpentine group. See greenalite (71.1.2b.4), the Fe analog. Two polymorphs exist: 1M and 1T. MON and TET Cm or $C2/m$. (1M): $a = 5.668$–5.692, $b = 9.811$–9.858, $c = 7.518$–7.527, $\beta = 104.52$–$104.6°$, $Z = 2$, $D = 3.00$; (1T): $a = 5.692$, $c = 7.275$, $Z = 1$, $D = 3.00$. *38-422*(nat): 7.3_8 3.64_5 2.82_6 2.52_{10} $2.10_{3.5}$ $1.73_{2.5}$ $1.64_{3.5}$ $1.60_{2.5}$. Pattern for 1T (*41-1446*) is the same as for 1M, along with some other faint lines. Structure not solved in detail. *MJJ 14:165(1980)*. Massive, reniform, stalagtitic aggregates or rosettes of flattened crystals to 3–4 mm, pseudohexagonal plates. Reddish brown, yellow, greenish brown, orange-brown. Cleavage (001), good. H = 3–4. G = 2.80–2.91. IR data: *DAN SSSR 249:513(1979)*. Readily soluble in concentrated acids. Sterling Hill, NJ, material has as much as 6.9 wt % As_2O_3 and 3.4 wt % ZnO. MgO to 7.2 wt %, CaO to 3.6 wt %, FeO to 22.3 wt %. Very Fe-rich material from Primorye (22.3 wt % FeO). Al_2O_3 to 2.4 wt %. Biaxial (–); Z = b, X perpendicular to (001); (1T): $N_x = 1.603$, $N_y = N_z = 1.632$; (1M): $N_x = 1.606$, $N_y = N_z = 1.633$; range given as $N_x = 1.606$–1.620, $N_y = 1.632$–1.65. $2V \approx 0°$; weak dispersion. Forms as an alteration product of primary manganese silicates such as rhodonite. In metamorphosed manganese deposits. Also occurs as granules in sediments. Occurs at Franklin, and Sterling Hill, Sussex Co., NJ; as 0.5-cm-diameter rosettes of radiating yellow-brown plates at Bald Knob, Sparta, Alleghany Co., NC; both 1M and 1T occur at the Hurricane claim, Olympic Penin., WA; Långban, Norbotten, Sweden, and the *Harstig mine, Påjsberg, Värmland, Sweden*; Fallota, Grisons, Switzerland; unspecified locality in Primorye (Russian Far East), Russia; Wessels and N'chwaning mines, Kalahari Mn field, Cape Prov., South Africa; the most Mn-rich known occurs at Tokusawa, Iwate Pref., Japan; Ichinomato mine, Kumamoto Pref., Japan; also Okutama mine, Tokyo; Sagadani mine, Ehime Pref., Niro mine, Shikoku, Japan. VK, EF *CM 20:1(1982), MIN 4(1):199(1992)*.

71.1.2b.2 Lizardite Mg$_3$(Si,Al)$_2$O$_5$(OH)$_4$

Named in 1956 by Whittaker and Zussman for the locality. Kaolinite–serpentine group. Frequently called serpentine and thus confounded with other members of the group; also baumite, bastite, marmolite, and serophite in part. A component of garnierite, picrolite, and retinalite in part. Synonym: orthoantigorite. ORTH Space group unknown. $a = 5.31$, $b = 9.20$, $c = 7.31$, $\beta = 90°$, $Z = 2$. HEX $P31m$. (1T): $a = 5.325$–5.338, $c = 7.257$–7.317, $D = 2.56$; (syn): $a = 5.336$, $c = 7.239$, $Z = 1$, $D = 2.58$; (6T): $a = 5.31$–5.33, $c = 42.62$–43.78, $Z = 6$, $D = 2.57$–2.67. 18-779 (HEX 1T, nat): 7.4$_{10}$ 4.6$_8$ 3.9$_5$ 3.67$_8$ 2.51$_{10}$ 2.16$_8$ 1.538$_8$ 1.310$_{6.5}$, 11-386 (HEX 1T, aluminian): 7.25$_6$ 4.61$_2$ 3.62$_6$ 2.50$_{10}$ 2.15$_4$ 1.778$_2$ 1.541$_3$ 1.507$_2$; 9-444 (Unstable type, nat): 7.33$_{10}$ 4.60$_6$ 3.66$_{10}$ 2.62$_3$ 2.50$_{10}$ 2.34$_7$ 1.963$_7$ 1.535$_8$. Lesser quality data are listed on 13-4 for aluminian 6(2)T lizardite. 31-782 [6(3)T]: 7.12$_{10}$ 4.59$_3$ 4.48$_3$ 3.56$_8$ 2.63$_4$ 2.38$_9$ 2.00$_4$ 1.532$_{5.5}$. $P6_3cm$. (2H$_1$): $a = 5.318$, $c = 14.541$. Lizardite, a trioctahedral 1 : 1 layer silicate, is perhaps the most common member of the serpentine group. Unlike antigorite, it has a "flat" layer structure. The minor substitution of Al might relieve some of the dimensional mismatch of the tetrahedral and octahedral sheets and result in more favorable bond lengths, leading to flat rather than curved layers. Various structural calculations predict that lizardite should have maximum stability at ^{IV}Al = 0.75 or 0.6. *AM 48:368(1963), 60:200(1975)*. A mechanism in one-layer polytypes which can relieve tetrahedral sheet dimensional mismatch with the octahedral sheet involves tetrahedral rotation, but which destroys the coplanar nature of the basal oxygens of that sheet, as well as introducing a minor wave-like modulation of the octahedral sheet *(NJMM: 193(1995))*. Two-layer lizardites have essentially planar layers. Structural formulas include (Mg$_{2.79}$Fe$_{0.12}$Al$_{0.07}$)(Si$_{1.83}$Al$_{0.17}$)O$_5$(OH)$_4$ (Val Sissone) and (Mg$_{2.74}$Fe$_{0.16}$Al$_{0.09}$)(Si$_{1.93}$Al$_{0.07}$)O$_5$(OH)$_4$ (Elba). Additional structural formulas include (Mg$_{2.76}$Fe$^{2+}_{0.04}$Fe$^{3+}_{0.10}$Al$_{0.07}$)(Si$_{1.83}$Al$_{0.17}$)O$_5$(OH)$_4$, while a low-Al lizardite with a particularly good X-ray pattern is (Mg$_{2.825}$Ni$_{0.10}$Fe$^{3+}_{0.07}$Al$_{0.03}$Ti$_{0.001}$)Si$_{2.025}$O$_5$(OH)$_4$. ^{IV}Fe^{3+} can be present in place of Al in lizardite: for example, (Mg$_{2.756}$Fe$^{3+}_{0.172}$Ni$_{0.017}$Ti$_{0.02}$Mn$_{0.004}$Cr$^{3+}_{0.003}$)(Si$_{1.906}$Fe$^{3+}_{0.053}$Al$_{0.041}$)O$_5$(OH)$_4$ (Soroako). Lizardites with ^{IV}Al < 0.15 are probably metastable with respect to antigorite above 400° at 2 kbar [*CM 17:757(1979)*], but nonaluminous lizardite has been synthesized. *AM 60:200(1975)*. Lizardite can show translational and/or rotational disorder. *AC 19:381(1965), CM 13:227(1975)*. Additional crystal structure determinations have been made. *CM 14:314(1976); AM 72:943(1987), 79:194(1994)*. TGA, DTA data: *IGR 17:1035(1975)*. One- and two-layer structures are common and have been synthesized [e.g., *CM 17:757(1979)*]. TEM studies verify the flat layering but show numerous defects [*CM 24:775(1986)*] as well as topotactic replacement orientations. *AM 75:813(1990)*. Unst-type lizardite (6T$_1$), a six-layer polytype, also has been synthesized. *MJJ 6:464(1972)*. Some polygonal serpentine, with or without chrysotile as a core filling in the texture, has the flat-layered lizardite structure [*CM 14:307(1976)*] and formerly has been called Povlen-type chrysotile. IR studies are few. *CM 17:719(1979)*. **Physical properties:** Green to yellow-green, bluish green, also nearly black, rarely yellow or white, can be platy [rarely with visible crystals (0.1–2 mm, Coli, Val Sissone, and Elba)], sometimes unctuous to greasy. H = 2–3 (by polishing, noticeably softer than antigorite and harder than chrysotile). G = 2.55–2.61. Cleavage {001}, perfect. More easily decomposed by 1 N HCl than antigorite, sometimes unaffected by acids. **Optics:** Uniaxial (−); $N_E = 1.545$, $N_O = 1.555$; biaxial (−); $N_x = 1.545$–1.563, $N_y = 1.555$–1.568, $N_z = 1.555$–1.570; birefringence = 0.006–0.010; 2V small or usually 37–61°; parallel extinction; can be nearly isotropic. Nickeloan lizardite can have RI = 1.575–1.580. *AM 48:1227(1963)*. Rarely biaxial (+); $N_x = 1.568$, $N_y = 1.569$, $N_z = 1.581$; birefringence = 0.013; 2V small. Optical data are few. Frequently shows hourglass, mesh, and other textures. Rosette or radial lizardite textures reported. Can be weakly pleochroic light green–light yellow. Bastitic replacements sometimes in antigorite groundmass. **Chemistry:** Observed compositional ranges include (wt %): SiO_2, 31.3–47.8; Al_2O_3, 0–4.5 (13.9–15.9, Stillwater); Fe_2O_3, 0.21–5.68; Cr_2O_3, 0–1.05; FeO, 0–1.81; MgO, 37.75–42.67 (18.5–30.8 when nickeloan; also 32.7, Stillwater); MnO, 0–0.22; NiO, 0–26.2; TiO_2, 0–0.10; CaO, 0–0.24; Na_2O, 0–0.28; K_2O, 0–0.41; Cl, 0.32 (Lizard); H_2O^-, 0.36–1.26; H_2O^+, 12.92–15.03. **Occurrence:** Field evidence suggests that lizardite serpentinization could occur at 75°, which can convert to antigorite at greenschist facies conditions (350°, 3–4 kbar) [*MM 47:301(1983)*], but antigorite and lizardite have been observed to coexist over a wide range of conditions. *CMP 97:147(1987)*. Serpentinization of dunite, peridotite, and ultramafic rocks is common. Can be found replacing olivine in basalt and so on, also as a microscopic replacement in amphibole and uralite. *SCI 206:1398(1979)*. Pseudomorphs of lizardite after enstatite, diopside/augite, or tremolite have been called bastite, but thermal effects can convert lizardite to antigorite or talc + clinochlore. *CM 17:729(1979)*. The reaction lizardite → talc + forsterite + clinochlore + water has been investigated. *CM 17:757(1979)*. Contrary to many antigorites, many lizardites form under oxidizing retrograde conditions. Can be found with other serpentines, talc, chlorite, magnetite–magnesiochromite, dolomite, magnesite, brucite, and so on. **Localities:** Selected localities include: Jim Pond, Franklin Co., ME; Newburyport, Essex Co., MA; Bellows Falls, Windham Co., VT; Aboutville; Talcville, St. Lawrence Co.; Staten Is., Richmond Co., NY; Turkey Mt., Montville, Morris Co.; Franklin, Sussex Co.; Castle Point, Hoboken, Hudson Co., NJ; Mineral Hill, Media, Delaware Co., PA; Lone Mt., Norris, Union Co., TN; Webster, Jackson Co., NC; Burke Mt. complex, Columbia Co., GA; southern Organ Mts., NM; Stillwater complex, MT; New Almaden mine, Santa Clara Co., CA; Riddle, Douglas Co., OR; Sultan area, Cascade Mts., WA; Juan de Fuca submarine ridge; Asbestos, Coleraine, and Thompson Lake, PQ; Setting Lake, MAN; Somerset Is., NWT, Canada; Havana, Havana Prov.; Nicaro, Oriente Prov., Cuba; Loma Pagnerra, Dominican Republic; Mayaguez, Puerto Rico; Unst, Shetland Is., and Polmaily, Glen Urquhart, Inverness-shire, Scotland; *Kennack Cove, The Lizard, Cornwall, England*; Snarum,

Modum Canton, Norway; Kuhmo greenstone belt, Finland; Pernes Massif, Pyrenees Mts., France; Totalp, Davos, Switzerland; Todtmoos, Wehratal, Schwartzwald, Baden; Baste, Harz, Germany; Monte dei Abati(!), Mte. Fico, Elaba(!); Coli(!) and Val Sissone morraine(!), Italy; Tyrol, Austria; Kapaonik and Radusa mines, Korab Mt., Povlen Mts., Serbia; Dognacska and Ocna de Fier, Romania; Ijim, western Sayan, Russia; Morro do Niquel and Morro do Coriso, MG; Cana Brava complex, Goias, Brazil; Sipilou and Moyango massifs, Ivory Coast; Frank Smith kimberlite, South Africa; Valojoro, Madagascar; Hindubagh, Baluchistan; Soroako and Pomalaa, Indonesia; Kwanson, Taitung, Taiwan; Sasaguri, Fukuoka Pref., Japan; Nullagine, WA, Australia. Unst-type lizardite found at the Tracy mine, Negaunee, Marquette Co., MI; Thompson Lake, LAB, Canada; Unst, Scotland; Piz Lunghin, Maloja, Switzerland; Lord Brassy mine, Heazelwood R., Tasmania; Silversheen mine, WA, Australia. VK, EF *MM 31:107(1956)*, *MIN 4(1):126(1992)*.

71.1.2b.3 Nepouite $Ni_3Si_2O_5(OH)_4$

Named in 1906 by Glasser for the locality. Kaolinite–serpentine group. Originally called genthite, also a component of garnierite; has been equated erroneously with nickel-bearing chlorite. ORTH $Ccm2_1$. $a = 5.31$, $b = 9.19$, $c = 14.50$, $Z = 2$, $D = 3.20$; $a = 5.27$–5.31, $b = 9.14$–9.20, $c = 7.24$–$7.27, 14.50$. 25-524(nat): 7.31_{10} 4.55_5 3.63_9 2.89_6 2.50_7 2.31_4 2.20_4 1.53_6. Found in polytypes: 1T [*CM 12:381(1974)*]; 2Or, 2M2, 6H [*MM 44:153(1981)*]. IR data: *IGR 17:1035(1975)*, *AM 48:1227(1963)*. DTA data: *ZVMO 102:143(1972)*. TGA data: *IGR 17:1035(1975)*, *AM 60:863(1975)*. Nickeloan lizardites tend to be better crystallized than nepouites. Nepouites have been synthesized. *AM 46:32(1961)*, *CMP 34:84,346(1972)*. Bright green to slightly yellowish or brownish green (color intensity varying with Ni content), can be earthy to waxy < also pearly, also fibrous, pseudohexagonal crystals (to 0.5 μm). H = 2–2.5. G = 3.18–3.24. Cleavage {001}, perfect with possible secondary cleavages. Biaxial (–); $N_x = 1.60$–1.63, $N_z = 1.635$–1.65; birefringence = 0.036–0.038, 0.05; $N_{av} = 1.62$–1.63; pleochroic. X, green to yellow-green; Z, yellow-green. Optical variation diagram: *IGR 17:1035(1975)*. A range of composition from nepouite to nickeloan lizardite is probably found at a single locality and observed chemical ranges include (wt %): SiO_2, 32.36–47.8; Al_2O_3, 0–10.74; Fe_2O_3, 0.10–2.42; Cr_2O_3, 0.79 (Ba); FeO, 0–5.29; MgO, 1.11–11.80; MnO, 0.02 (Ba); NiO, 27.18–50.70; CaO, 0.05–1.07; TiO_2, 0–0.01; Na_2O, 0.10–1.58; K_2O, 0.02–0.08; CO_2, 1.24 (Ba); H_2O^+, 9.7–11.9. A series to brindleyite is probable. Found in nickel-bearing laterites as a fracture filling in Ni-bearing serpentine or as a contact alteration in limestone. A component in some high-Ni garnierite, but much garnierite consists of nickeloan lizardite. Can be found with willemseite–talc, falcondoite–sepiolite, pimelite–nontronite, goethite, quartz, calcite, halloysite, or others. Might be admixed with poorly crystalline to X-ray amorphous materials. Found at

Woods mine and elsewhere, Lancaster Co., PA; Piliken Chrome mine, El Dorado Co.; Pine Tree and Josephine mines, Mariposa Co.; Kalkar quarry, Santa Cruz Co.; Coyote Mt. Ni mine, Imperial Co.; Venice Hills, Tulare Co., CA; Hanna mine, Riddle, Douglas Co., OR; Rottis, Voigtland, Germany; Frankenstein, Silesia, Poland; Kremze, Bohemia, and Letovice, Czech Republic; Ba, Serbia; Pavlos, Béotie, Greece; Tyulenevskoye and Ackerman deposits, Ural Mts., Russia; Nuggihalli schist belt, Mysore, India; Soroako, Indonesia; Kouaoua, Noumea, Nakety, Thio, and *Reis II mine, Népoui, New Caledonia*. VK, EF *CR 143:1173(1906)*, *MIN 4(1):186(1992)*.

71.1.2b.4 Greenalite $Fe_3^{2+}Si_2O_5(OH)_4$

Named in 1903 by Leith for its color. Kaolinite–serpentine group. Frequently erroneously called glauconite; tosalite. MON *Cm*. $a = 5.54$–5.60, $b = 9.55$–9.70, $c = 7.38$–7.44, $\beta = 104.2$–$104.33°$, $Z = 2$, $D = 3.15$; HEX *P31m*. $a = 5.598$ (syn: 5.490 Å), $c = 7.213$ (syn: 7.140 Å), $Z = 1$, $D = 3.15$. *39-337* (1T): 7.2_{10} 3.61_5 2.78_3 2.61_6 2.49_3 2.21_4 1.62_6 1.58_5; *39-348*(1M): 7.2_{10} 3.61_5 2.78_3 2.61_6 2.49_3 2.21_4 1.62_6 1.58_5 (identical patterns, but not all reflections are accepted by one space group); *45-1353*(1T, syn): 7.16_{10} 4.76_1 $3.57_{6.5}$ 2.56_7 2.44_1 $2.18_{2.5}$ 1.80_2 1.585_2. Greenalite appears to be dominantly trigonal in some samples, but can be mixed with a minor monoclinic form. Found in polytypes: 1M and 1T. Forms a series with lizardite [*AM 47:783(1962)*], but once thought to be related to chamosite. *NAT 181:45(1958)*. Numerous defects occur in greenalite and the structure is organized into 21- to 23-Å domains and a superlattice with $a = 279$ $(5a_0)$, $b = 45.2$ $(4.67b_0)$ has been proposed. CM 20:1(1982). Structural formulas include: $Fe^{2+}1.64Al_{0.75}Mg_{0.23}\square_{0.36})(Si_{1.96}Al_{0.04})O_5(OH)_4$ [*CM 23:447(1988)*] and the nearest to end-member: $(Fe_{2.69}^{2+}\square_{0.275}Mg_{0.02}Mn_{0.01}(Si_{2.0})O_5(OH)_4$. Frequently as rounded granules with a fibrous texture even when indurated. Olive-green to dark green. H = 3. G = 2.7–3.15. Weakly magnetic. Isotropic; N = 1.655–1.675; anisotropic in strong light, yellow-green to blue-green. *AM 20:405(1935)*. Analysis (wt %): SiO_2, 30.08–38.00; Al_2O_3, 0–5.65; Fe_2O_3, 8.40–34.85; FeO, 25.72–46.56; MgO, 0–4.16; MnO, 0–8.52; CaO, 0–0.22; Na_2O, 0–0.22; K_2O, 0–0.10; H_2O^+, 7.04–9.35. Found in slate adjacent to banded iron formations, also as a deep ocean ooze. Ferrian greenalite is currently forming in shallow (15–60 m) tropical sediments. Alters to iron oxides. Can be associated with siderite, chert, iron oxides, and quartz. *Biwabik Iron M (Section 2, T.58N, R.16W), Biwabik*, also Gilbert and Virginia, *Mesabi Range, MN*; Knob Lake, LAB; Gunflint Lake and Belcher Is., Hudson Bay, ONT; Blue Bell mine, Riondell, BC, Canada; Martinique Is. shelf; Arenig, Glenluce, Wigtownshire, Scotland; Mina Brunita, La Union, Murcia, Spain; Hjulsjoe, Bergslagen, Sweden; Amazon and Orinoco R. continental shelves; Congo R. mouth, Ivory Coast; Ogooue R. prodelta; Niger R. prodelta; Casamance estuary sediments and Cap Vert continental shelf, Senegal; east of Mayotte Is. lagoon, Comoro Is.; Toa and Matsua mines, Kochi

Pref., Japan; North Sarawak continental shelf; Weld Range and Dales Gorge member, Hamersley Group, WA, Australia. Found on the moon. VK, EF *USGSM:43(1903), MIN 4(1):169(1992)*.

71.1.2c.1 Amesite $(Mg_2Al)[SiAl]O_5(OH)_4$

Named in 1876 by C. U. Shepard for James Ames, owner of the Chester Emery mine. Kaolinite–serpentine group. TRIC $C1$. $a = 5.31$–5.32 (5.386^*, Fe^{2+}-rich), $b = 9.20$–9.21 (9.291^*), $c = 14.06$–14.07 (14.124^*), $\alpha = 90.01$–$90.09°$, $\beta = 90.25$–$90.27°$, $\gamma = 89.96$–$89.97°$; also $a = 5.31$, $b = 9.21$, $c = 14.40$, $\alpha = 102.11°$, $\beta = 90.2°$, $\gamma = 90.1°$; $Z = 4$. *37-429*: 7.06_{10} 4.63_2 3.53_9 2.64_6 2.51_7 2.33_3 1.938_3 1.549_2; *34-163*: 7.03_{10} 4.37_5 3.52_6 2.61_2 2.48_6 1.934_2 1.930_2 1.535_2. HEX $P6_3$. $a = 5.31$, $c = 14.01$, $Z = 4$, $D = 2.83$. *9-493*: 7.06_{10} 3.52_{10} 2.48_6 2.32_5 1.925_7 1.749_5 1.596_6 1.531_6. Amesite is a trioctahedral 1 : 1 layer hydrous silicate and belongs to the serpentine subgroup of the serpentine–kaolinite group. It shows nearly complete $^{IV\&VI}R$ cation ordering, but the style of ordering can differ with the same kind of polytype [*AM 66:185(1981)*], and IR can falsely suggest extensive disorder. *AM 52:296(1977)*. Amesite is frequently triclinic but can be hexagonal. Occasional disorder has been observed. Mössbauer, IR, DTA, TGA data: *TA 64:83(1983)*. Some detailed studies have erroneously equated amesite with aluminian clinochlore or sudoite [e.g., *EJM 6:279(1994)*]. Found in polytypes: $2H_2$ (most common), 9T [*MM 44:153(1981)*], $6R_2$, $6R_1$, and $2H_1$ (rare). Several polytypes can be present at a single locality. *CCM 27:241(1976), AM 66:185(1981)*. Several crystal structure determinations have been made. *AC 15:510(1962); AM 66:185(1981), 76:651(1991)*. Proton positions have been calculated. *CCM 28:816(1980), 30:409(1982)*. At least one amesite has low ^{IV}Al (New Caledonia), but the distinction between amesite and lizardite lies in $^{IV}Al > 0.5$. Due to charge balance considerations, $^{IV}Al \approx ^{VI}Al$. Light green, pale bluish green, apple-green, also mauve to purple or violet (chromian), foliated, rarely in pseudohexagonal columnar crystals (to several millimeters) elongated parallel to [001]. Cleavage {001}, perfect, with pearly luster. $H = 2\frac{1}{2}$–3. $G = 2.71$–2.80. Slowly decomposed by HCl, infusible. Electronic absorption spectra of chromian amesite: *EJM 7:961(1995)*. Uniaxial (+); $N_O = 1.597$, $N_E = 1.612$; birefringence $= 0.015$; negative elongation; also biaxial (+); $Z \wedge c = 0$–$20°$; $N_x = 1.597$–1.612, $N_y = 1.599$–1.612, $N_z = 1.615$–1.630, with N_x very close to N_y; RI increases with Fe substitution; birefringence $= 0.018$; $2V = 18$–$20°$; $r < v$; negative and positive elongation. Can show sector twinning on {001} or polysynthetic lamellae on {010} [*AM 66:185(1981)*] or hourglass texture. Optical properties of amesite can be confused with chlorites. *MM 46:469(1982)*. Chemistry is moderately variable with some minor element substitution (wt %): SiO_2, 20.11–24.00 (33.51, New Caledonia); Al_2O_3, 31.14–35.21 (21.30, New Caledonia); Fe_2O_3, 0.63 (Saranovsk); Cr_2O_3, 0–3.90; FeO, 0.60–22.48; MgO, 14.59–30.70; NiO, 0–0.31 (3.12, New Caledonia); MnO, 0–0.41; CaO, 0–0.58; Na_2O, 0.38 (Saranovsk); TiO_2, 0.26 (Saranovsk);

H_2O^+, 10.54–13.02. Typical structural formulas include: $(Mg_{1.936}Fe^{2+}_{0.025}Cr_{0.074}\square_{0.022})$ $Al_{0.943}(Si_{1.027}$ $Al_{0.973})O_5(OH)_4$ (Ural Mts.), which can range to $(Mg_{1.132}Fe^{2+}0.979$ $Mn_{0.94}Ca_{0.025})Al_{0.960}$ $(Si_{1.047}Al_{0.953}O_5$ $(OH)_{3.76}$ (Black Lake). Rarely as an alteration of cancrinite and other minerals. At Black Lake, formed during calcium metasomatism of granite, particularly biotite. Found with clinochlore, magnetite, sometimes grossular, diopside, calcite, and clinozoisite; occasionally margarite. *Chester Emery mine, Chester, Hampden Co., MA*; Tracy mine, Negaunee, Marquette Co., MI; Tiger, AZ; Black Lake, PQ, Canada; Cr-bearing from Mt. Sobotka, Lower Silesia, Poland; St. Johns Is., Egypt; Saranovsk chromite deposits, Ural Mts., Russia; Col. de Mouirange, New Caledonia; Dufek massif, Pensacola Mts., Antarctica. VK, EF

71.1.2c.2 Berthierine $Fe_2^{2+}Al(SiAl)O_5(OH)_4$

Named in 1832 by Beudant for Pierre Berthier (1782–1861), redefined by Orcel, Henin, and Caillere (1949). Kaolinite–serpentine group. At one time, erroneously believed to be chamosite and still easily confused with it. MON Cm. $a = 5.25$–5.41, $b = 9.10$–9.33, $c = 7.06$–7.28, $\beta = 104.5°$, $Z = 2$. ORTH $a = 5.42$, $b = 9.38$, $c = 7.11$. HEX $a = 5.42$, $c = 7.11$, $Z = 1$, $D = 3.04$. *31-618:* 7.12_{10} 4.68_5 3.55_{10} 2.53_{10} 2.15_7 1.779_6 1.563_7 1.481_6; *7-315:* 7.05_{10} 3.52_{10} 2.68_4 2.52_9 2.40_4 2.14_6 1.769_4 1.555_7. Found in polytypes 1H and 1M, which might exist together in some samples. Structural formulas include: $(Fe^{2+}_{1.84}Mg_{0.23}Al_{0.84}$ vacancy$_{0.09}Fe^{3+}_{0.01})$ $(Si_{1.33}Al_{0.67})$ $O_5(OH)_4$ (Corby), $(Fe^{2+}_{1.33}Mg_{0.57}Al_{0.82})$ $(Si_{1.28}Al_{0.72})O_5(OH)_4$ (Kursk) [*CCM 30:153(1982)*], and $(Fe^{2+}_{1.14}Zn_{0.83}Al_{0.78}Mg_{0.10}Ca_{0.02})(Si_{1.49}Al_{0.51})O_5(OH)_4$ (Air Mts.). Fe^{3+} in berthierines is usually originally absent and then Fe^{2+} is oxidized in place [*AC 74(1984)*] and ferric berthierines have reduced cell parameters. Heated berthierine (300°) shows collapsed d_{060} 1.548 Å→ 1.513 Å and all reflections disappear at 600°. IR studies have been done and show similarities to iron-bearing chlorites. *CM 19:279(1981)*. Usually earthy, rarely foliated or in columnar pseudohexagonal crystals (to several millimeters at Mt. St.-Hilaire). Dark green, gray, brown, to black; white to light green streak. H = 2–2.5. Cleavage {001}, perfect. Decomposed by hot HCl. Probably biaxial; $N_{av} = 1.62$–1.65; pleochroic, yellow green–dark green. Observed chemical variation includes (wt %): SiO_2, 18.50–26.9; Al_2O_3, 10.99–28.02; Fe_2O_3, 0–6.62, but oxidized specimens to 26.25; Cr_2O_3, 0–0.05; TiO_2, 0–3.63; FeO, 20.20–39.79; MgO, 0.97–8.15; MnO, 0–2.84; ZnO, 19.56 (Air Mts.); Nb_2O_5, 0.78 (Mt. St.-Hilaire); CaO, 0–0.93; Na_2O, 0–0.21; K_2O, 0–0.21; H_2O_{tot}, 9.58–17.4. A "berthierine" with $^{VI}Ti \gg ^{VI}Al(TiO_2 = 20.48$ wt %) (Picton) has been analyzed. *CM 23:213(1985)*. TGA, DTA, IR, Mössbauer data: *MM 30:57(1953), TA 74:291(1984)*. Found in unconsolidated estuarine and marine sediments in iron reduction zones and in arctic soils. Can be found in phosphatic or sideritic sedimentary rocks, also laterites or coal beds, rarely found in granite pegmatite. In banded iron formations with andradite, diopside–hedenbergite, babingtonite, magnetite, stilpnomelane, epidote, and other minerals, also in fault breccias and in fluorite–pyrrhotite veins. Can alter to chamosite. Can be found with kaolinite, chamosite, micas, pyrophyllite, siderite, quartz, or miscellaneous detrital minerals. Selected localities include: Rose Hill Formation, VA; Gunflint Formation, Ely, MN; Wabana, NF; Ruth shale, Howells R. area, LAB; Wakefield and Mt. St.-Hilaire(!), PQ; Gunflint formation; Kidd Creek mine, Timmins and Picton, ON; Clearwater Formation, Cold Lake, ALB; Ellef Ringnes and Ellsmere Is., NWT, Canada; Carnlough and Knowehead, County Antrim, Northern Ireland; Cambokeels mine and Redburn; North Wear tunnel, Weardale, Durham; Horsham, Sussex; Finedon, Stanion Lane mines, Northamptonshire; Corby; Frodingham Ironstone formation, Scunthorpe, England; Ayrshire, Scotland; Kongsberg, Norway; Visingsoe Group, Sweden; Caradocian rocks, Crozon Penin.; Champagne, Bourgogne, and Lorraine; *Hayanges, Moselle, France*; Serra del Cadi, Pyrenees Mts., Spain; Windgalle, Switzerland; Hagendorf Süd, Bavaria, and Pegnitz, Germany; Kank mine, Kutna Hora, Czech Republic; Trojarova deposit, Male Karpaty Mts., Slovakia; Aladag, Turkey; Belgorod district, Kursk and Onega areas; Chara uplift, West Aldan Shield, Siberia, Russia; Orinoco R. delta; Del Corral Hill, Argentina; Niger R. delta and Taghouaji alkaline ring complex, Air Mts., Niger; Sarawak continental shelf; Chashan, Guangxi, China; Hirumatsu, Utatsu, Miyagi Pref.; Ohmine coal field, Yamaguchi Pref.; Kado, Iwate Pref.; Ishikari coal field, Hokkaido, Japan. VK, EF *ACP 35:258(1827)*, BTEM, *MIN 4(1):175(1992)*.

71.1.2c.3 Brindleyite $Ni_2Al(SiAl)O_5(OH)_4$

Described in 1978 by Maksimovic and Bish and named for George William Brindley (1905–1983), English–American clay mineralogist, Pennsylvania State University. Kaolinite–serpentine group. A trioctahedral clay mineral. See berthierine (71.1.2c.2), the Fe–Al analog. Originally called nimesite. *BSCASA 17:224(1972)*. MON $C2/m$ and $P31m$ for 1M and 1T polymorphs. (1M): $a = 5.286$, $b = 9.133$, $c = 7.31$, $\beta = 104.15°$, $Z = 2$, $D = 3.16$. HEX (1T): $a = 5.277$, $c = 7.09$, $Z = 1$, $D = 3.38$. *31-892*(1M): 7.07_{10} 4.54_1 3.54_8 $2.65_{0.5}$ 2.62_2 2.47_2 2.37_2 1.524_2; *42-1311*(1T): ($P3_1m$) 7.07_{10} 4.54_1 3.54_8 2.62_2 2.47_2 2.37_2 2.01_7 1.524_2. Structure unknown. Found in polytypes: 1M-3T, 1T assumed. *AM 63:484(1978)*. Disordered stacking observed. Dark yellowish green (color due to octahedrally coordinated Ni^{2+}), green coatings a few millimeters thick with sporadic thickening to 1.2 cm, on limestone. Has a clayey appearance, with schistosity and lustrous surface, earthy to waxy, microscopically fibrous to platy (generally $< 1\mu m$). Cleavage not observed. $H = 2\frac{1}{2}$–3. $G = 3.01$–3.17. IR, DTA, TGA data: *AM 63:484(1978)*. Biaxial; $N_{av} = 1.635$. Few brindleyite analyses appear in the literature (wt %): SiO_2, 23.61–27.45; Al_2O_3, 23.39–32.18; Fe_2O_3, 0–0.45; Cr_2O_3, 0.05–0.26; REE_2O_3, 0–0.35; TiO_2, 0.99–1.04; FeO, 1.15–3.40; MgO, 2.81–3.54; NiO, 30.18–22.02; MnO, 0–0.18; ZnO, 0.24–0.25; CoO, 0–0.18; CuO, 0–0.72; CaO, 0.07–0.39; Na_2O, 0–0.10; K_2O, 0–0.20; H_2O^+, 11.36–12.59; H_2O^-, 0.70–1.29; CO_2, 0–0.58. CEC = 36 meq per 100 g. Brindleyite can be cation deficient: $(Ni_{1.75}Al_1\square_{0.25})(Si_{1.5}Al_{0.5})$ $O_5(OH)_4$ or $(Ni_{1.31}Mg_{0.26}Fe^{2+}_{0.05}Ti_{0.04}Al_{1.03}\square_{0.31})(Si_{1.49}Al_{0.51})O_5(OH)_4$, but only one locality has published analyses. Found as veinlets in red kaolin and in limestone with bastnaesite-(Ce), malachite, bayerite, and calcite. Also in altered serpentine. Found at Victorio, Grant Co., NM; *Marmara Bauxite mine, Megara, Greece.* EF *AM 58:1112(1973) MSASC 19:169(1988); MIN 4(1):197(1992).*

71.1.2c.4 Fraipontite $(Zn,Cu,Al)_3(Si,Al)_2O_5(OH)_4$; ideal: $[Zn,Cu]_{3-x}Al_x(Si_{2-x}Al_x)O_5(OH)_4$

Described in 1927 by Cesaro and named for Julien Jean Fraipont (1857–1910) and Charles Fraipont, Belgian geologists and paleontologists. Kaolinite–serpentine group. Polytypes include 1H and $2M_1$. Zn analog of berthierine. *MSASC 19:169(1988)* and *MIN 4(1):202(1992)* state that zinalsite is equivalent to fraipontite. MON $C1$ or $C\bar{1}$ (Moresnet): $a = 5.372$, $b = 9.246$, $c = 7.273$, $\beta = 103.55°$; (Defiance): $a = 5.331$, $b = 9.23$, $c = 9.275$, $\beta = 104.15°$, $Z = 2$, $D = 3.44$; another cell for Moresnet material: $a = 5.34$, $b = 9.21$, $c = 14.12$, $\beta = 93.2°$, $Z = 4$; supposedly, orthorhombic material from Laurion has $a = 5.325$, $b = 9.238$, $c = 7.075$, $\beta = 90.0°$. *14-366*(monoclinic): 7.0_{10} 3.52_7 2.63_3 2.48_2 2.36_2 2.25_1 2.12_1 1.53_2; *34-782*(orthorhombic): 7.07_{10} 4.65_1 3.54_6 2.66_1 2.49_2 2.37_1 2.12_1 1.54_1. A hydrous phyllosilicate. Described as Zn-berthierine [*BM 98:235(1975)*] and as a septechlorite [*NJBM 4:155(1977)*] because of much higher Al content. Following Bailey (1980), the term *septechlorite* should be dropped and the Laurion fraipontite could be a high-Al serpentine. Usually earthy and massive, rarely occurs as pseudohexagonal platy crystals to 1 mm. White, yellowish white, light blue-green, bluish gray; white to pale blue or blue-green streak; earthy to silky luster, sometimes waxy. Cleavage {001}, perfect. H = $3\frac{1}{2}$–4 (of aggregate). G = 3.08–3.54. Becomes amorphous on heating to 450°. IR, DTA, TGA data: *BM 98:235(1975), MZ 14:89(1992).* Soluble in HNO_3, gelatinizes upon evaporation; dissolves completely in 10% HCl and 30% solution of NaOH. About eight good analyses of fraipontite exist (wt %): CuO, 0.2–16.8; ZnO, 14.5–51.9; traces of MgO, CaO; Al_2O_3, 16.2–21.7; SiO_2, 19.5–32.8 and H_2O is about 14. Some fraipontites are Zn-bearing and some are cuprian. Solid solution between berthierine and fraipontite exists. *AM 75:909(1990).* Biaxial (−); 2V = 15–20°; indices of refraction not well measured, but the mean ranges from 1.61 to 1.62; positive elongation; OAP parallel to the length of the fibers. A low-temperature clay mineral found with secondary zinc and copper minerals. In slags at Laurion, Greece. On smithsonite at Moresnet. In hydrothermal Cu–Zn base metal deposits. Often associated with sauconite. Forms as pseudomorphs after azurite, cuprite, and other minerals. Coating fluorite, associated with gypsum, jarosite, and quartz in Turkmenia. Found at Sterling Hill, Ogdensburg, Sussex Co., NJ; the Silver Bill and Defiance mines, Gleeson district, Cochise Co., AZ; Bingham(!), Socorro Co., NM; also at the West adit, Mohawk mine, San Bernardino Co., CA; Wensleydale area, Yorkshire,

England; Vielle Montagne, Moresnet, Belgium; Altenberg, near Aachen, Germany; Laurion, Greece; Kugitang-Tau, Kugitang Mts., Turkmenistan; several deposits in Kazakhstan (Akdzhal, Achisai, Bazar-Tube, Maidanshak, Gulshad); Almalik deposit in Kirgizia; Taghouaji complex, Air Mts., Niger; Tsumeb, Namibia. VK, EF *MR 13:137(1982), 14:131(1983)*; *MIN 4(1):202(1992)*.

71.1.2c.5 Kellyite $(Mn,Mg,Al)_3(Si,Al)_2O_5(OH)_4$

Described in 1974 by Peacor et al. for William Crowley Kelley (b. 1929), American geologist, University of Michigan. Kaolinite–serpentine group, Mn analog of amesite. Two polytypes are recognized: $2H_2$ and $6R_2$. HEX $P6_3$. (2H polytype): $a = 5.348$, $c = 14.04$; (6R polytype): $a = 5.44$, $c = 42.13$; *MSASC 19:169(1988)*. *29-885*(2H polytype, nat): 7.0_9 3.51_{10} 2.67_4 2.53_6 2.34_4 2.15_1 1.95_3 1.57_3. See *MSASC 19:169(1988)* for X-ray data for the monoclinic cell. Structure unknown. Micaceous grains to 1 mm, laths and tablets. Golden yellow to lemon yellow; oily to vitreous luster, sometimes pearly. Cleavage {001}, perfect. G = 3.07. Analysis (wt %): SiO_2, 17.6; Al_2O_3, 28.55; Fe_2O_3, 2.18; MnO, 38.84; MgO, 2.97; H_2O (by difference), 10.82. A structural formula is: $(Mn_{1.8}Mg_{0.25}Fe^{3+}_{0.1})\Sigma_{2.15}Al_{0.85}(Si,Al)O_5(OH)_4$. Biaxial (−); X perpendicular to (001); $N_x = 1.639$, $N_y = 1.646$, $N_z = 1.646$; 2V = 16–30°, sometimes as low as 5–6°; r > v, moderately pleochroic. X, colorless to greenish yellow; Y,Z, pale yellow to reddish brown. Irregular to undulatory extinction, possible sector twinning. Occurs in a metamorphic manganese deposit along with Mn-silicates, including alleghanyite, tephroite, sonolite, galaxite–jacobsite, and Mn-carbonates, including kutnahorite. Bald Knob, Sparta, Alleghany Co., NC; Savoie Dept., French Alps. VK, EF *AM 59:1153(1974)*, *BM 101:514(1978)*, *MR 12:167(1981)*, *MIN 4(1):198(1992)*.

71.1.2c.6 Manandonite $Li_2Al_4[(Si_2AlB)O_{10}](OH)_8$

Named in 1912 by Lacroix for the Manandona River in Madagascar. Serpentine–kaolinite group. An Al-, Li-, B-rich analog of amesite. Originally thought to be a chlorite related to cookeite, which it can physically resemble. TRIC C1. $a = 5.070$, $b = 8.776$, $c = 13.778$, $\alpha = 90.09°$, $\beta = 90.12°$, $\gamma = 89.97°$, Z = 2, D = 2.79. *45-1464*(nat): 6.92_{10} $4.36_{2.5}$ 4.16_1 3.45_8 $2.49_{1.5}$ $2.38_{3.5}$ 2.295_1 $1.86_{1.5}$. *EJM 1:633(1989)*. A 1:1 layer silicate (7-Å serpentine–kaolinite group). A two-layer polytype. Refined in space group C1 to R = 5.7%. Tetrahedral Si, Al, and B are partly ordered. B has a greater tendency to order than do Si and Al. There is also partial ordering of the octahedral cations that decreases the ideal $P6_3$ symmetry to P1 (or C1, as used for refinement). Occurs as a $2H_2$ polytype. Dull yellow globules or white isolated platy pseudohexagonal crystals, fanlike aggregates and globules most common. Frequently twinned on (001) as sixfold sectors and warped; white streak; pearly luster, otherwise greasy. Cleavage {001}, excellent. H = $2\frac{1}{2}$. G = 2.78–2.89. IR, TGA, DTA data: *EJM 1:633(1989)*. Analysis (wt %): SiO_2, 23.0; B_2O_3, 7.7; Al_2O_3, 47.85; Fe_2O_3, 0.36; MnO, 0.02; MgO, 0.03; CaO, 0.07; Na_2O, 0.13;

Li$_2$O, 5.38; H$_2$O, 15.40. Biaxial (+); Z perpendicular to (001); N$_x \approx$ N$_y = 1.604$; 2V = 10–15°; no dispersion. Sometimes pseudo-uniaxial (+). Frequently sector-twinned with undulatory extinction. Optics confirm true triclinic symmetry. Found in an Li- and B-rich granite pegmatite with spodumene, red elbaite, albite, quartz, beryl, spessartine, rhodizite, and danburite. Occurs in the *Antandrokomby pegmatite, Manandona R., Sahatany Valley, near Antsirabe, Madagascar*, as crusts in small vugs of albite, lined with elbaite and quartz. VK, EF *CR 155:446(1912), MIN 4(2):172(1992)*.

71.1.2c.7 Cronstedtite Fe$_2^{2+}$Fe^{3+}(Si,Fe^{3+})O$_5$(OH)$_4$

Named in 1821 by J. Steinmann for Axel Cronstedt, Swedish pioneer mineralogist. Kaolinite–serpentine group. MON *CC* (three-layer): $a = 5.49$, $b = 9.51$, $c = 14.29$, $\beta = 82.63°$; *Cm* (three-layer): $a = 5.49$, $b = 9.51$, $c = 7.32$, $\beta = 104.5°$, Z = 2, D = 3.59. HEX: *R*3(six-layer): $a = 5.49$, $c = 42.5$; *P*6$_3$: $a = 5.49$, $c = 14.7$; *P*31*m*(one-layer): $a = 5.49$, $c = 7.085$; *P*6$_3$*cm*(two-layer): $a = 5.49$; $c = 14.7$; *P*31(three-layer): $a = 5.49$, $c = 21.21$; *14-470*(1M): 7.09$_{10}$ 3.54$_9$ 2.72$_5$ 2.44$_4$ 2.31$_2$ 2.04$_2$ 1.680$_2$ 1.586$_4$. Found in polytypes: 1M, 2M$_1$, 2T, 2H$_1$, 2H$_2$, 3T$_1$, 6R$_3$, 6H$_2$. *AC 16:1(1963), 17:404(1964)*. A 9R polytype with large a (9.56 Å) has also been reported. *AM 47:781(1962)*. Twinning is common in some polytypes and frequently occur intergrown in a single specimen and polytype style cannot be predicted by external morphology. *AC 16:1(1963)*. A structural formula is: Fe$_{2.38}^{2+}$Fe$_{0.71}^{3+}$)(Si$_{1.11}$Fe$_{0.81}^{3+}$Al$_{0.07}$)O$_5$(OH)$_4$ (Herja), while a general formula could be (R$_{3-x}^{2+}$Fe$_x^{3+}$)(Si$_{2-x}$Fe$_x^{3+}$)O$_5$(OH)$_4$ with $x \approx 0.5$–1. *AM 60:175(1975)*. Ordering of ^{IV}R^{3+} is present as in amesite, but not ^{VI}R^{3+}. Mössbauer data: *CCM 16:381(1968)*. Frequently in divergent groups of trigonal or hexagonal, striated, tapering crystals (rarely to 1 cm) forming hemispheres, with basal terminations forming the surface, sometimes showing interpenetrating twins, also smooth botryoidal, cylindrical, or massive. Black to brownish black, rarely green or brownish yellow in thin foliae; dark olive-green streak; vitreous. Cleavage {0001}, perfect. Foliae elastic. H = 3½. G = 3.34–3.45. Fusibility 4. Gelatinizes with acid. Uniaxial (−); N$_O$ = 1.80; birefringence large; pleochroic. E, dark reddish brown; O, almost opaque. Also biaxial; N$_y$ = 1.80; 2V = 0°–small. Pleochroic, X, brown; Y,Z, black. Chemical analyses are few (wt %): SiO$_2$, 16.42; Al$_2$O$_3$, 0.90; Fe$_2$O$_3$, 29.72; FeO, 41.86; CaO, 1.32; H$_2$O$^+$, 10.17. Found associated in sulfide veins with limonite and calcite. Long Hill, Trumbull, Fairfield Co., CT; Cornucopia mine, Nye Co., NV; San Antonio and Potosi mines, Santa Eulalia, CHIH, Mexico; Wheal Jane, Kea (almost always three-layer, *P*31, also *Cc* and *Cm*) and Wheal Maudlin, Lanlivery (one-, two-, and six-layer polytypes, including *P*6$_3$ and *P*31*c*), Cornwall, England; Mina Brunita, La Union, Murcia, Spain; Eisleben, Thuringia, Germany; (Kuttenberg), Kutná Hora, Chyňava, and *Příbram, Bohemia, Czech Republic*; Anbruch, Reusengang, Hüttenberg, Carinthia, Austria; Herja(!) (Kisbánya), Transylvania, Romania; Llallagua and Huanuni(!), Bolivia; Conghonas do Campo, MG, Brazil;

Cold Bokkeved, Renazzo, and Cochabamba meteorites. VK, EF *JCP 32:69(1821)*, *MIN 4(1):183(1992)*.

71.1.2d.1 Clinochrysotile $Mg_3Si_2O_5(OH)_4$
71.1.2d.2 Orthochrysotile $Mg_3Si_2O_5(OH)_4$
71.1.2d.3 Parachrysotile $Mg_3Si_2O_5(OH)_4$

Chrysotile was named in 1834 by F. von Kobell from the Greek, in allusion to its golden fibers. Clinochrysotile and orthochrysotile were named in 1951 by Whittaker in allusion to their crystallographic relationship to chrysotile. Parachrysotile was named, but not described, by Whittaker in 1956 in allusion to its crystallographic relationship to chrysotile. Originally called amianthus; also the common asbestos; constitutes in part, aquacreptite, baltimorite, deweylite, jenkinsite, picrolite, retinalite, and other less common materials. Also variously important in schweizerite, metaxite, bastite, marmolite, and rarely, in mountain leather. Clinochrysotile: MON Cc, $C2/m$. $a = 5.30$–5.34, $b = 9.17$–9.25, $c = 14.31$–14.65, $\beta = 93.12$–$93.34°$, $97.15°$(syn), $106.4°(1M_{cl})$; β can vary by up to $1°$ within a specimen. *CM 17:679(1979)*. Orthochrysotile: ORTH $a = 5.32$–5.34, $b = 9.2$, $c = 14.6$–14.63. Parachrysotile: ORTH $a = 5.3$, $b = 9.2$, $c = 14.6$. *MM 31:107(1956)*. Clinochrysotile (syn): Cc. $a = 5.32$, $b = 9.25$, $c = 14.31$, $\beta = 97.15°$, $Z = 4$, $D = 2.63$. *22-1163*: 7.1_9 4.60_2 4.40_2 4.07_1 3.55_6 2.65_3 2.39_{10} 2.001_3; *21-1262*, $C2/m$: 7.28_6 4.54_{10} 3.64_4 2.62_4 2.53_4 2.45_4 1.531_5 1.501_5; *31-808*, $C2/m(2M_{cl})$: $a = 5.31$, $b = 9.12$, $c = 14.64$, $\beta = 93.17°$, $Z = 4$, $D = 2.60$. PD: 7.31_{10} 4.57_5 3.65_7 2.27_3 2.21_3 2.09_3 1.827_3 1.744_3. Orthochrysotile($2Or_{cl}$): $a = 5.34$, $b = 9.25$, $c = 14.20$, $Z = 4$, $D = 2.62$. *22-1162*: 7.1_{10} 4.63_2 4.40_4 3.88_3 3.55_7 2.50_5 2.33_8 1.615_3; *25-645*($2Or_{cl}$): $a = 5.32$, $b = 9.2$, $c = 14.64$, $Z = 4$, $D = 2.57$. PD: 7.36_{10} 4.56_5 3.66_8 2.60_4 2.50_5 2.45_7 2.09_4 1.531_7. Chrysotiles are based on a cylindrical development of curved sheets. Continuous curvature does not result entirely in optimum hydrogen bonds between layers. Chemical composition determines the degree of dimensional mismatch of a and b, and it has been suggested that 88 Å is the ideal radius of a chrysotile cylinder. *AC 10:149(1957)*. The minimum radius of a chrysotile cylinder should be 35–40 Å, while the expected maximum radius would be ≈ 140 Å [*AC 23:704(1971)*], but 400-Å radius has been observed [*MM 50:301(1986)*]. In clinochrysotile and orthochrysotile, a is the fiber axis and c is the interlayer axis, while in parachrysotile b is the fiber axis. Ortho- and clinochrysotile have been oberved in alternating overgrowth in a single fiber. *CCM 32:429(1984)*. All the chrysotiles have been found in polygonal Povlen-type textures, sometimes with lizardite-2H overgrowths, but normal cylindrical morphology might also be found in parachrysotile. *CM 17:693,699(1979), 26:991(1988)*. In some polygonal serpentines, the structure is flat-layered and thus is appropriately called lizardite [*CM 14:308(1976)*], but cores in Povlen-type texture may contain chrysotile. Small chrysotile fibers can transform to polygonal textures. *MM 50:301(1986)*. All forms of chrysotile (clino-, ortho-, and para-) have been observed in a single specimen. Parachrysotile is a growth variety of orthochrysotile, not a polytype variation of it, and powder XRD does not serve to distinguish it from orthochrysotile. Found in one- and two-layer polytypes and which are in the opposite sense of standard serpentine stacking. *CM 13:227(1975)*. Three forms of chrysotile have been given separate names (the subscript c was introduced to indicate cylindrical development): orthochrysotile, clinochrysotile, and parachrysotile. Clinochrysotile is the most abundant polytype and sometimes has orthochrysotile intergrown with it. The polytypes of chrysotile are visually indistinguishable, but ratios of polytypes can be calculated from XRD. *MM 31:107(1956)*. Two kinds of clinochrysotile have been distinguished: normal and Globe-type. *USGSPP:384-A(1962)*. Normal clinochrysotile has a strong d = 2.456 Å line with a diffuse band extending to 2.594 Å. Globe-type has two lines at d = 2.456 Å and 2.604 Å with a weak diffraction band in between, and there is no diffraction line at 2.500 Å. Globe-type might be a mixture of clinochrysotile and orthochrysotile polytypes. Polygonal cross-sectional morphology exists (Povlen-type) and appears to yield sharper X-ray diffraction results than those for fibrous cylindrical material. Disordered chrysotile has been given the symbol D_c. Mössbauer studies of chrysotile indicate $^{IV\&VI}Fe^{3+}$, but $^{VI}Fe^{3+}/^{IV}Fe^{3+} \approx 3:1$. *CM 17:713(1979)*. Chrysotile has been synthesized [*AM 39:957(1954)*, *CM 20:19(1982)*], and increasing Al was found to decrease chrysotile products. **Physical properties:** Green to pale green to yellowish green, also golden yellow to dark brown (WY), rarely ash gray; usually fibrous and weavable (pure orthochrysotile can have a "badly crimped appearance"), also compact or splintery (metaxite), silky to oily luster, but some chrysotile is not visibly fibrous, especially when mixed with other serpentines; sometimes unctuous. H = 2–3 (by polishing, noticeably softer than lizardite and very much softer than antigorite). G = 2.53–2.55. Infusible, insoluble in acids. **Optics:** Biaxial (−); $N_x = 1.545$–1.569, $N_y = 1.546$–1.569, $N_z = 1.553$–1.570; birefringence = 0.001–0.008; 2V = 50°. Fibrous, RI value increases with iron substitution. The refractive indices of the various chrysotiles overlap and are therefore indistinguishable by this method of identification. Can be weakly pleochroic. **Chemistry:** Reported ranges of chrysotile analyses (wt %) include: SiO_2, 40.6–46.20; Al_2O_3, 0–0.96; Fe_2O_3, 0–1.34; Cr_2O_3, 0–0.27; TiO_2, 0–0.07; FeO, 0–1.57; MgO, 35.25–43.6; MnO, 0–0.15; NiO usually $\approx$ 0, but a series can extend to nepouite; CaO, 0–0.85; Na_2O, 0–0.14; K_2O, 0–0.11; H_2O^-, 0.70–4.82; H_2O^+, 12.53–14.04; F, 0–0.31. Aluminian chrysotiles ($^{IV}Al > 0.10$) are probably unstable. *CM 17:757(1979)*. **Occurrence:** Clinochrysotile and orthochrysotile are frequently intergrown in a sample, usually with a ratio of the former to the latter of 30–50 : 1, but some orthochrysotile contents range from 7% (Thetford) to 35% (Shabani) to 81% (Chuddapah) to the essentially pure orthochrysotile that occurs at an unknown locality in Silesia. Methods of establishing the ratio of clinochrysotile to orthochrysotile have been published. *MM 31:107(1956)*. Chrysotiles can be found with other serpentines, talc, chlorite, dolomite, magnesite, brucite, magnetite/magnesiochromite, and so on. Orthochrysotile and parachrysotile can be present mixed with the fibrous brucite called nemalite, and orthochrysotile can occur as cross fibers in dolomite. Clinochrysotile can be found as the major constituent in some mountain leather. Found intergrown with uralite [*SCI 206:1398(1979)*] and rarely with carlosturanite or balangeroite. Parachrysotile can be formed during the incipient serpentinization of forsterite. Can recrystallize to antigorite or weather to smectite. Observed as an authigenic mineral. *CCM 35:43(1987)*. Oxygen isotopes have been used to model chrysotile formation at about 200°. *GI 16(1):170(1979)*. **Localities:** Selected clinochrysotile localities include: Lowell, Lamoine Co., and Eden, Orange Co., VT; Talcville and Balmat, St. Lawrence Co., NY; Castle Point, Hoboken, Hudson Co.; Turkey Mt., Montville, Morris Co., NJ; Woods mine, Cedar Hill mine, and elsewhere, Lancaster Co.; Sylmar West Nottingham Twp., Chester Co., PA; Forest Hill, Bare Hills, Baltimore Co.; Harford Co., MD; Webster, Jackson Co., NC; Prairie Creek, Murfreesboro, Pike Co., AR; Grand View, Grand Canyon, Coconino Co.; Globe, Gila Co., AZ; Current Creek deposit, White Pine Co., NV; Casper, WY; New Almaden mine, Santa Clara Co.; New Idria and Clear Creek Canyon, San Benito Co., CA; Sultan area, Cascade Mts., WA; Baie Verte, NF; Thetford, Black Lake, Asbestos, and elsewhere, QUE; Timmins, ONT; Pipe Lake, MAN; Clinton Creek, YT; Cassiar, BC, Canada; Nicaro and Moa, Oriente Prov., Cuba; The Lizard, Cornwall, England; Glen Urquhart, Scotland; Findeten Glacier, Zermatt, Switzerland; Balangero serpentinite, Lanzo massif; Franscia, Chiesa, and elsewhere, Malenco serpentinite, Bergell Aureole, Italy; *Reichenstein, Silesia, Poland*; Povlen Mts., Serbia; Ocna de Fier, Romania; Aktovrak deposit, Tuva; Bazhenovo, Ural Mts.; Sayan deposit, West Sayan, Russia; Krantzkop, Natal; Premier mine, Kimberley, Transvaal, South Africa; Port Macquarie, WA, Australia; Yamabe mine, Hokkaido, Japan. Orthochrysotile localities studied include: Balmat, St. Lawrence Co., NY; Turkey Mt., Montville, Morris Co., NJ; New Almaden mine, Santa Clara Co., CA; Cuicateco, OAX, Mexico; Unst, Shetland Is., and Milton, Glen Urquhart, Scotland; *Silesia, Poland*; Chuddapah, Andhra Pradesh, India; Shabani mine, Zimbabwe; Havelock mine, Swaziland; Kranskop, Natal; Premier mine, Kimberley and elsewhere, Transvaal, South Africa; Jilin, China; Nunyerry, WA, Australia. Parachrysotile localities studied include: East Broughton, Thetford, and Asbestos, QUE, Canada; Glen Urquhart, Inverness-shire, Scotland; Shabani mine, Zimbabwe; Havelock mine, Swaziland. VK, EF *JPC 2:297(1834), MIN 4(1):126(1992)*.

71.1.2d.4 Pecoraite $Ni_3Si_2O_5(OH)_4$

Named in 1969 by Faust et al. for William Thomas Pecora (1913–1972), former director of the U.S. Geological Survey and student of nickel silicate deposits. Kaolinite–serpentine group. Synonym: Ni-chrysotile. Dimorphous with nepouite. MON Space group unknown. $a = 5.26$, $b = 9.16$, $c = 14.7$, $\beta = 92°$, $Z = 4$, $D = 3.57$. 22-754(nat): 7.43_{10} 4.50_6 3.66_8 2.62_6 2.45_5 1.53_8. Generally poorly crystalline. Found as polytype: $2M_{c1}$. *MM 44:153(1981)*. Synthesis: *AM 46:32(1961)*. Curved plates (to 70 Å), tubes and spirals to 0.4 µm; granular, to 5 mm; also massive. Bright green, blue green to yellow green, earthy to waxy or vitreous luster; transparent to translucent. $G = 3.08$

(with adsorbed water). $N_{av} = 1.565$–1.603 (meteorite), 1.616; also $N_{av} = 1.65$ (dried). Analysis (wt %), Wolf Creek material: SiO_2, 31.0; Al_2O_3, 1.4; FeO, 0.7; MgO, 0.5; NiO, 51.5; CaO, 0.4; H_2O^+, 9.7; H_2O^-, 4.1; Otway material: Ni, 44.0; Mg, 0.6; Fe^{2+}, 0.1; Si, 12.5; O, 20.7. Formed by weathering of Fe–Ni meteorite fragments in a desert environment (Wolf Creek meteorite) in association with goethite, maghemite, cassidyite, reevesite, and quartz. Also found in hydrothermally altered shear zones in serpentinite with millerite, polydymite, gaspéite, magnesite, and rare nickel minerals; also as a replacement of millerite in quartz geodes. Woods mine, Lancaster Co., PA; route 44 roadcut, St. Louis, MO; Hanna nickel mine, Riddle, Douglas Co., OR; Tscheremschanskoye, Russia; Rockys Reward mine, Agnew and Otway prospects, near Spinnaway, Nullagine district, WA; *Wolf Creek meteorite, WA, Australia*. VK, EF *SCI 165:59(1969); AM 54:1740(1969); MM 39:113(1973), USGSPP:384-C(1974); GSAJ 26:61; CM 20:367(1982), 21:341(1983); NJBM:513(1983); AUSM 4:141(1989); MIN 4(1):191(1992)*.

71.1.3.1 Bismutoferrite $BiFe_2^{3+}[SiO_4]_2(OH)$

Named in 1871 by Frenzel in allusion to the composition. See chapmanite (71.1.3.2). MON *Cm.* $a = 5.21$, $b = 9.02$, $c = 7.74$, $\beta = 100.67°$, $Z = 2$, $D = 5.092$. *11-714:* 7.63_{10} 4.18_2 3.87_{10} 3.58_3 3.18_5 2.90_7 2.59_3 2.53_2. Structure: *SPC 22:419(1977)*. The structure consists of t-o layers similar to those found in kaolinite and having the composition $[(Fe_2^{3+}(OH)O_3(Si_2O_5)]^{3-}$, with the iron in the octahedral sites and bismuth occupying the interlayer sites. Isostructural with its antimony analog, chapmanite. Compact, massive; sometimes powdery. Bright to dull yellow-green, waxy to adamantine luster, translucent to opaque. $H = 6$. $G = 4.47$. Biaxial $(-)$; 2V large; $N_x = 1.93$, $N_y = 1.97$, $N_z = 2.01$; . Sb may substitute for Bi to a limited degree; Fe^{2+} may replace Fe^{3+}. Found at Ludi No. 4 Claim Plumas Co., CA; at the South Terras mine, St. Stephan-in-Brannel, Cornwall, and at Buckbarrow Beck, Carney Fell, Cumbria, England; with quartz at Ullersreuth and Gersdorff, and with quartz, bismuth and cobaltite in veins in shale at *Schneeberg, Saxony, Germany*; at Yinnietharra, WA, Australia. AR *AM 43:656(1958)*.

71.1.3.2 Chapmanite $SbFe_2^{3+}[SiO_4]_2(OH)$

Named in 1924 by Walker for Edward John Chapman (1821–1904), Canadian geologist and mineralogist, University of Toronto. See bismutoferrite (71.1.3.1). MON *Cm.* $a = 5.19$, $b = 8.99$, $c = 7.70$, $\beta = 100.67°$, $Z = 2$, $D = 4.29$. *11-135:* 7.63_9 4.17_3 3.88_8 3.58_{10} 3.19_9 2.90_7 2.59_7 2.54_3. Structure: *SPC 22:419(1977)*. Isostructural with its Bi analog, bismutoferrite. Fine-grained aggregates; sometimes granular and microcrystalline to megascopic plates and prisms. Olive-green, yellow-green, yellow; greenish yellow to yellow streak; earthy to waxy, adamantine luster; translucent to opaque. Fracture conchoidal to irregular in masses. $H = 2\frac{1}{2}$. $G = 3.58$–3.57. Biaxial $(-)$; 2V small; $N_x = 1.85$, $N_{y\,calc} = 1.95$, $N_z = 1.96$; X perpendicular to laths and Z parallel to elongation. Limited substitution of Bi for Sb, Fe^{3+} for Sb, and Al for Fe^{3+}. Slightly soluble in HCl and H_2SO_4. On heating yields hematite and and tripuhyite. Occurs as a hydrothermal or secondary mineral in Sb–As deposits and in gneisses with other Sb and As minerals. Found at the McDermitt mercury mine, Humboldt Co., NV; at Venesela Mt., AK; at the *Keeley mine, Timiskaming district, South Lorraine Twp., ON, Canada*; Velardeña, DGO, Mexico; La Bassade (Haute Loire) Vosges, France; at Freiberg, Braunsdorf, and Schneeberg, Saxony, Germany; Tafone, Tuscany, Italy; at Smilkov, Votice, and Borenov, Bohemia, Czech Republic; in the Kadamdzhai deposit, Kirgizia; Suzuyama mine, Kagoshima Pref., Japan. EF *AM 43:656(1958), 49:1499(1964); MIN 3(1):503(1972).*

ALLOPHANE GROUP

The allophanes are poorly crystallized or noncrystalline minerals with some ordering when viewed by electron microscopy. *CLM 18:21(1983), SRF(1988), MIN 29(1994).* Members of the group include the following:

ALLOPHANE GROUP

Mineral	
Allophane	71.1.5.1
Hisingerite	71.1.5.2
Imogolite	71.1.5.3
Neotocite	71.1.5.4

71.1.5.1 Allophane $Al_2Si_{1-2}O_{5-7} \cdot 2-3H_2O$

Named in 1816 by Stromeyer from the Greek *to appear [as] another*, in allusion to its change when blowpiped. Allophane group. Synonym: protoallophane. Amorphous or poorly crystalline. Space group unknown. $b = 9.0$, $c = 7.0$. 38-449(nat): 3.3_{10} 2.25_2 $1.86_{<1}$ 1.40_1 $1.23_{<1}$. Allophane is generally amorphous to X-rays, and is a mineral that has a variable chemical composition. As many as four wide diffraction bands are occasionally observed [*AM 38:634(1953)*] which can disappear on heating to 400° [*CCM 29:124(1981)*] and ^{IV}Al and ^{VI}Al have been determined [*CS 2:1(1964)*], both suggesting some ordering. Numerous TEM studies show that it frequently consists of hollow spheres (~ 35–55 Å with walls 7–10 Å thick and an external water monolayer [*AM 56:465(1971)*], but which can quickly dehydrate and collapse in electron beams. The spheres might be composed of a phyllosilicate structure rolled to reduce bond distortions. *PICC 2:29(1969).* Spherical halloysite can be mistaken for allophane by TEM. *CCM 11:169(1963)* and others. A defect kaolinite structure has been proposed [*DS 27:537(1979)*], but MAS-NMR studies suggest an octahedral sheet similar to that of 2:1 phyllosilicates, but different from that of kaolinite or gibbsite [*CCM 42:276(1994)*]. Dry grinding can affect allophane morphology [*CM 18:101(1983)*], as can electron beams [*AM 61:379(1976)*]. IR studies have been numerous [e.g., *CCM 28:295,328(1980); 29:124(1981)*] and are similar to those of imogolite. *CCM 28:285(1980)*. **Habit:** Generally occurs as hyaline crusts, masses, and coatings; stalactites and rarely as flowstone.**Physical properties:** White to tan, also bright blue, green, or yellow; frequently stained brown; uncolored streak unless stained; waxy to earthy luster, can be unctuous; sometimes translucent. Conchoidal fracture and then showing a shining luster; brittle; sometimes stalactitic. H = 3. G = 1.8–2.78. Infusible. Gelatinizes in HCl. **Chemistry:** Chemistry is variable, in part due to impurities, and has been assigned a composition $Al_2Si_2O_7 \cdot 3H_2O$ to $Al_2SiO_5 \cdot 2H_2O$ [*AM 52:690(1967)*]: $SiO_2/R_2O_3 \approx$ to 2.0 but frequently near 1:1. Analysis (wt %): SiO_2, 19.71–33.22 (34.80–41.29, Japan); Al_2O_3, 28.24–50.5; Fe_2O_3, generally 0.2–1.0 but to 5–6; MgO, 0–0.18; CaO, 0–2.86; P_2O_5, 0–10.57 (Indiana); SO_3, 0–0.44; H_2O^-, 19.26–22.52; H_2O^+, 9.80–25.0. CEC 69–74 meq per 100 g, also 100 meq per 100 g and varies according to cation. *PSSSA 23:210(1959)*. **Optics:** Isotropic; N = 1.47–1.52; RI increases with Fe_2O_3. **Occurrence:** Principally a weathering product of volcanic ash; also a hydrothermal alteration product of feldspars. Allophane may be a precursor to halloysite, but halloysite has been observed to alter from allophane. *CCM 9:315(1962)*. Citric and humic acids can inhibit the formation of allophane. *PICC 221(1985)*. Frequently found with admixed halloysite, imogolite, limonite, opal, gibbsite, cristobalite, occasionally chrysocolla or evansite, but nearly pure material can occur in sulfide veins or as a coating. Found in sedimentary (sediments, laterites, weathered basalt, marl, chalk, limestone, coal beds, etc.) and hydrothermal (sulfide veins, replacement zones, etc.) environments. Variously abundant or insignificant in soils. **Localities:** Selected localities include: Richmond, MA; Bristol, CT; Morgantown, Berks Co.; Friedensville and Allentown, Lehigh Co.; Rohrs Cave, Lancaster Co., Cornwall, Lebanon Co., PA; Polk Co., TN; Lawrence Co., IN; Alabama Street mine and elsewhere, Saline Co., AR; Kelly mine, Socorro Co., NM; Bisbee, and the Maid of Sunshine mine, Gleeson, Cochise Co., AZ; Cerro Gordo, Inyo Co., CA; Trail Bridge and Crescent Lake, OR; Maui Is., HI; Derbyshire; Wheal Hamblyn, Devon; New Charleton, Kent; Hawkswood mine, North Hill, Cornwall, England; Chessy copper mine, Lyons, Rhône, France; Rosas mine, Sulsis, Sardinia; Calabona mine, Alghero; Cerro del Tasca, Mt. Amiata, Italy; Visé, Belgium; Gräfenthal, near Saalfeld, Thuringia; Schneeberg, and Schwarzenberg, Saxony; Dehrn; in marl at *Gräfenthal, Thuringia, Germany*; Jachymov and Chotina, Bohemia; Petrov, Moravia, Czech Republic; Boleslaw mine, Olkusz, Poland; Vyshkovo region, Transcarpathia, Ukraine; Central Aldan and Podolsk district; Cis-Baikal, Yakutia, Russia; Mbobo Mkulu Cave, Transvaal, South Africa; Alotenango, Guatemala; Guoshan deposit, Fujian Prov., China; northwestern Taiwan; Shishigahana, Mt. Chokai; Fukazawa and Iijima, Nagano Pref.; Kakino pumice, Choyo, and Imogo soil, Hitoyoshi, Kumamoto Pref.; Kanuma and Hangadai, Tochigi Pref.; Oze, Gunma Pref.; Bihoro, Hokkaido Pref.; Kitikami, Iwate

Pref.; Kurayoshi, Tottori Pref.; Ando soil, Okamoto, Japan; Bandung, Java; Sulawesi, Celebes, Indonesia; Aoba, Vanuatu; Mt. Schank, South Australia, Australia; Bealey Spur, Silica Springs, and elsewhere, New Zealand. VK, EF *AM 59:1094(1974), CMB 12:289(1977)*.

71.1.5.2 Hisingerite $(Fe,Mn)SiO_3,Fe_2^{3+}Si_2O_7 \cdot 2H_2O$

Named in 1828 by Berzelius for the eminent chemist Wilhelm Hisinger (1766–1852). Allophane group. Synonyms: canbyite, scotiolite, sturtite. Amorphous or MON $a = 5.40$, $b = 9.0$–9.30, $c = 14.99$, $\beta = 98.32°$, $Z = 4$, $D = 3.23$. *26-1140*(nat): 4.23_{10} 2.71_8 2.46_{10} 2.25_8 2.20_8 1.72_{10} 1.57_8 1.55_8. Hisingerite is generally an amorphous substance which frequently yields a small set of wide or diffuse X-ray diffraction peaks. TEM studies [*CCM 35:29(1987)*] reveal that hisingerite consists of hollow spherical particles, generally 100–200 µm but up to 1000 µm in diameter. As sphere size increases, so does smectite content. *NJBM: 321(1982)*. The weak X-ray diffraction peaks observed have been variously attributed to those of a smectite [*CUSSGM 19:1(1974)*], mica, or generalized phyllosilicate, but the longest basal spacings of smectite, micas, and so on, have been observed only rarely (Romania). Does not expand with ethylene glycol. Mössbauer and other more recent studies indicate that hisingerite does not have a layer structure. *CM 18:21(1983)*. IR, DTA, and TGA did not suggest the presence of hydroxyl. *CCM 32:272(1984)*. Hisingerite can be a mixture of amorphous and crystalline materials. *AM 9:1(1924)*. Hisingerite can recrystallize to a mixture of ferrian saponite or nontronite + hematite when heated to 180°. **Habit:** Commonly massive and compact. May be minutely spherical. **Physical properties:** Black to brownish black, amber-brown in thinnest slivers and a streak lighter than the massive material; rarely dark red or dark green and photosensitive; resinous to vitreous luster, rarely greasy. Conchoidal fracture but can have a platy parting. [Trace cleavage observed in canbyite (Brandywine).] H = 3–$3\frac{1}{2}$. G = 2.3–3.0. Decomposed by HCl. Infusible. Can be confused with ferrihydrite. Summary of IR, EPR, Mössbauer data: *MIN 4(2):620(1992)*. **Optics:** Isometric; N = 1.44–1.73; biaxial; $N_x = 1.562$ (maximum range 1.552–1.595), $N_y = 1.580$, $N_z = 1.582$; 2V(–) very small [also $N_x = 1.715$, $N_z = 1.730$, *PASP 44:279(1944)*]; birefringence = 0.020 (0.015); r > v, parallel extinction; B×O perpendicular to cleavage; orange to golden brown, generally nonpleochroic. **Chemistry:** Hisingerite is generally regarded to have $SiO_2/Fe_2O_3 = 2:1$ with minor substitutions [*CMB 2:294(1955)*], while attempts have been made at rationalizing the alkalis into exchangeable positions [*CCM 32:272(1984)*], frequently with the calculating of a smectite-like formula. Chemical analyses are variable [*DS 36:97(1981)*], but given some high values, such as MgO and/or Al_2O_3, suggest that some materials, perhaps unrecognized as new to the group were included in the survey. Inability to distinguish X-ray amorphous mixtures (opal, etc.) and minor crystalline components could contribute to the uncertainty of some extreme analyses (wt %): SiO_2, most commonly 40.35–49.90 ranging to 27.99–56.91; Fe_2O_3, commonly 14.34–29.00, ranging to 4.80–

40.70; Al_2O_3, ranging 0–22.65; TiO_2, 0–1.88; FeO, 0–6.75 (24.64); MgO, 0–25.95; MnO, 0–3.72; CaO, 0–3.80; Na_2O, 0–2.90; K_2O, 0–1.44; H_2O^+, 3.46–12.10; H_2O^-, 5.53–17.92. $CEC_{meas} = 20$–74.5 meq per 100 g. **Occurrence:** Originally occurring in large mammillary masses (to many centimeters) and frequently observed in obvious veins, coatings, or crusts. Increasingly observed as microscopic alterations of iron-bearing rocks, especially igneous rocks and ash or glass. Sometimes observed as a thin amber film or as amber globules in amygdules. Frequently veining fayalite, enstatite, and/or amphiboles, sometimes siderite or wollastonite; also alters from pyrrhotite and possibly from chalcopyrite. A component of chlorophaeite and probably in some iddingsite. Observed in recent sediments. Probably formed by meteoric water, also from hot springs. Alters to nontronite; large hisingerite spheres show nontronite alteration rims. *NJBM: 321(1982)*. Also occurs as pseudomorphs after hedenbergite. **Localities:** Found in a wide variety of occurrences frequently inconspicuous in quantity and due to its nearly amorphous character difficult to recognize. Selected occurrences include: Tilley Foster mine, Brewster, Putnam Co., NY; Gap nickel mine, Lancaster Co., PA; Brandywine quarry, Wilmington, DE; Alexander Co., NC; Montreal mine, Iron Co., WI; Hibbing, Beaver Bay, and Silver Bay, MN; Castle Dome mine, Gila Co., and on the Mildren and Steppe claims, Cababi district, Pima Co., AZ; Bellvue, Blaine Co., ID; Cardinal mine, Stevens Co., WA; Tetrault mine, Montauban-les-Mines and Hull, QUE; Wilcox mine, Parry Sound, ONT; Nicholson mine, Goldfields, SAS, Canada; Lostwithiel and Wheal Jane, Kea, Cornwall, England; Solberg mine, Elvestorp; Brunjo(!), Västmanland; Sjöström mine, Hofors, Gästrikland; Tunaberg, Långban, and Vestra Silfberg, Värmland; *Riddarhyttan, Västmanland, Sweden*; Helsingfors, and Orijärvi; Degerö mine, Helsinki, Finland; Limburg Sasbach, Kaiserstühl, Germany; Gallego and Aragon rivers, Spain; Fagul Cetatii, Balan, and Masca, Iara Valley, Apuseni Mts., Romania; Atlantis II Deep, Red Sea, Israel; Mt. Karnasurt, Lovozero massif, and Mt. Rasvumchorr, Khibiny massif, Kola Penin.; Ilmen Mts., Urals; Terny Astrobleme, Krivoy Rog, Russia; Zaval'ye, Bug region, Ukraine; Dalnegorsk, Primorsky Krai; Talnakh, Kazakhstan; Gran Canaria, Canary Is.; Llallagua, Bolivia; Suzuyama mine, Kagoshima Pref.; Kawayama mine, Yamaguchi Pref.; Sano mine, Wakayama Pref., Japan; Geelong, VIC; Cobar and Broken Hill, NSW, Australia; Aoba, Vanuatu. Remotely sensed on Mars. VK, EF *SUBBSC 19:9(1974), MIN 4(2):620(1992)*.

71.1.5.3 Imogolite $Al_2SiO_3(OH)_4$

Named in 1962 by Yoshinaga and Aomine for the locality. Allophane group. Space group unknown. *38-447*(nat): 21.0_3 11.5_{10} 7.9_8 5.6_2 4.4_1 $3.7_{1.5}$ 3.3_6 2.25_3; *MM 51:327(1987)*: 16_{10} 7.9_7 $5.6_{3.5}$ 4.4_1 4.1_1 3.7_2 $3.3_{6.5}$ $2.25_{2.5}$. Imogolite gives relatively few X-ray diffraction peaks, has a tubular structure (17–21 Å outer diameter and 7–10 Å inner diameter) and can appear as partial webs composed of filaments, usually in bundles, in TEM. Imogolite is sometimes characterized by a sharp 19.7 Å peak and broad peaks at 13.3, 7.6, 5.5–5.7, 3.7, 3.3–3.45,

2.25, 2.1, and 1.40 Å. The main peak might be related to a close packing of tubes and the broad peaks relate to a scattering of single tubes. *DS 27:547(1979)*. MAS-NMR studies suggest that imogolite has an octahedral sheet similar to that of 2:1 phyllosilicates but different from that of kaolinite or gibbsite. *CCM 42:276(1994)*. Dry grinding can affect imogolite crystallinity. *CM 16:139(1981)*. Imogolite has been synthesized [*DS 27:547(1979)*] and might form naturally from allophane. The IR of imogolite is similar to allophane [*CCM 28:285(1980)*] and might represent the crystalline equivalent of allophane. *AM 61:379(1976)*. Earthy, composed of microscopic threadlike grains and bundles of fine tubes, each about 20 Å in diameter. Light brownish yellow, also tan to white, blue, green, brown. Conchoidal fracture, brittle. H = 2–3. G = 2.7. Isometric; N = 1.47–1.51 and similar to allophane. RI increases with Fe_2O_3. Principally in soils derived from volcanic ash. Can be found with allophane, halloysite, vermiculite, goethite, gibbsite, quartz, and detrital minerals. Observed in cracks in weathered plagioclase. Citric and humic acids can inhibit the formation of imogolite. *PICC 221(1985)*. Selected localities include: Adirondack Mts., NY; Hawaii Is., HI; Plastic Lake, ONT, Canada; Roudado basalt, Aurillac, Cantal, France; Luochan loess, Shensi Prov.; Guoshan deposit, Fujian Prov., China; Kitakami, Iwate Pref.; Kanto loam, Ibaroki Pref.; (Fukuwa) Kanumatsuchi ash bed, Kanuma and Hangadai, Tochigi Pref.; Kurayoshi, Tottori Pref.; Mitsutsuchi ash bed, Iijima, Nagano Pref.; Uemura, Choyo, and *Imogo soil, Hitoyoshi, Kumamoto Pref., Japan*; New Hebrides; Papua New Guinea; Tirau, Rangitaua, and Wharepaina, New Zealand. vk, ef *SSPN 8:6(1962)*; *AM 54:50(1969)*; *CM 8:87(1969), 12:289(1977)*; *PCM 12:342(1985)*.

71.1.5.4 Neotocite $Mn^{3+}SiO_3 \cdot H_2O$

Named in 1848 by Nordenskiöld from the Greek for *new origin*, in allusion to its paragenetic position. Allophane group. Synonyms: stratopeite, penwithite. Isostructural with hisingerite. Amorphous or mon Space group unknown. $b = 9.0$–9.30. 14-172(nat): 4.36_{10} 3.59_{10} $2.59_{<1}$ 1.54_{10}. Neotocite is generally an amorphous substance but has been observed to yield at least five X-ray diffraction reflections (6.4, 4.36–4.4, 3.5–3.59, 2.59–2.6, 1.54 Å). Neotocite forms 70- to 100-Å microspheroids as for hisingerite and allophane. *CM 18:21(1983)*. When heated to 1000° for 5 minutes, usually yields braunite and minor jacobsite–magnetite and rarely hausmannite; or pyroxmangite (Påjsberg). Brown, brownish black, rarely dark red, amber brown in thinnest slivers, with a brown to dark brown streak. Generally massive and compact; resinous to greasy luster, rarely vitreous to adamantine. Can be photosensitive, changing from red to brown shades on exposure. Conchoidal fracture, brittle. Rarely shows a platy parting (Klapperud). H = 4. G = 2.04–2.8. Decomposed by HCl. Sensitive to relative humidity. DTA, IRA data: *MM 42:279,M26(1978), MIN 4(2):633(1992)*. Isometric; N = 1.475–1.654, usually pale yellow to reddish brown; sometimes birefringent. 2V to 20°. The composition of neotocite is variable [*MM 42:279,M26(1978)*], especially with respect to iron, and a series might extend to hisingerite. Analysis (wt %): SiO_2, 30.0–43.8; Fe_2O_3, 0.01–18.7; Al_2O_3, 0.10–2.9; Mn_2O_3, 0–33.2; FeO, usually 0, also 9.6 (Gastrikland); MnO, 10.4–37.2; TiO_2, 0.01; MgO, 0.5–9.6; CaO, 0.1–2.3; Na_2O, 0.02–0.3; K_2O, 0.01–0.3; CO_2, 2.2–7.4; H_2O^+, 8.2–11.8; H_2O^-, 7.04 (WI). See also *MIN 4(2):633(1992)*. Associated wtih rhodonite, rhodochrosite, quartz, manganese phosphates, spessartine, bementite, tephroite, and others. Can have admixed birnessite. Frequently observed in obvious veins, coatings, or crusts. Observed as microscopic alterations of manganese-bearing rocks or minerals, also manganese-bearing ophiolite. Frequently veining weathered manganese-bearing pyroxenes, especially rhodonite, and spessartine. Found lining fractures in granite pegmatite and in association with manganese-bearing minerals or rocks, but due to its nearly amorphous character, difficult to recognize. Selected occurrences include: Plainfield, MA; Bald Knob, Alleghany Co.; Foote mine, Kings Mt., Cleveland Co., NC; Montreal mine, Gogebic Range, Iron Co., WI; Batesville, Independence Co., AR; Bromide district, OK; Aravaipa district, Graham Co. and Ajo, AZ; in several locations in CA, including Johe Ranch mine, San Luis Obispo Co., Lake Elsinore area, Riverside Co., and in the Charles Mt. deposit, Humboldt Co.; Anacortes and elsewhere, Olympic Penin., WA; El Fino mine, Pinar del Rio Prov.; Polaris mine, Candelaria district and elsewhere, Bueycito, Oriente Prov., Cuba; Bambolita mine, Moctezuma, SON, Mexico; Llanfaerhys, Rhiw, Wales; Wheal Owles, Penwith, and Geevor mine, St. Just, Cornwall, England; Brevik, West Gothland, Norway; Klapperud, Dalsland; Delecarlia; Gillinge mine, Svärta, Södermanland; at Långban, Jakobsberg, and the Harstig mine, Påjsberg, and Filipstad, Vårmland; *Erik-Ers mine, Torsåker, Gästrikland, Sweden*; also in the Brunsjö mine, near Grythyttan, Örebro; Wittingi, Storkyro, Finland; Herborn, Dillinberg, Germany; Chiavari, Liguria, and Val Graveglia, Genoa, Italy; Litosice, Iron Mts., Czech Republic; Yakobeni, Romania; Malo-Sedelnikovsk, Middle Ural Mts., Russia; Tien Shan, Kirghizstan; Lafaiete district, MG, Brazil; Broken Hill, NSW, Australia; Shidara, Aichi Pref.; Kawazu mine, Shizuoka Pref., Noda–Tamagawa mine, Iwate Pref., Japan. VK, EF *AM 46:1412(1961)*, *MIN 4(2):633(1992)*.

PYROPHYLLITE TALC GROUP

The pyrophyllite talc group minerals are phyllosilicates corresponding to the general formula

$$R_2^{3+}Si_4O_{10}(OH)_2 \quad \text{for the dioctahedral pyrophyllite subgroup}$$

or

$$R_3^{2+}Si_4O_{10}(OH)_2 \quad \text{for the trioctahedral talc subgroup}$$

where

R^{3+} = Al, Fe^{3+}
R^{2+} = Mg, Fe^{2+}, Ni^{2+}

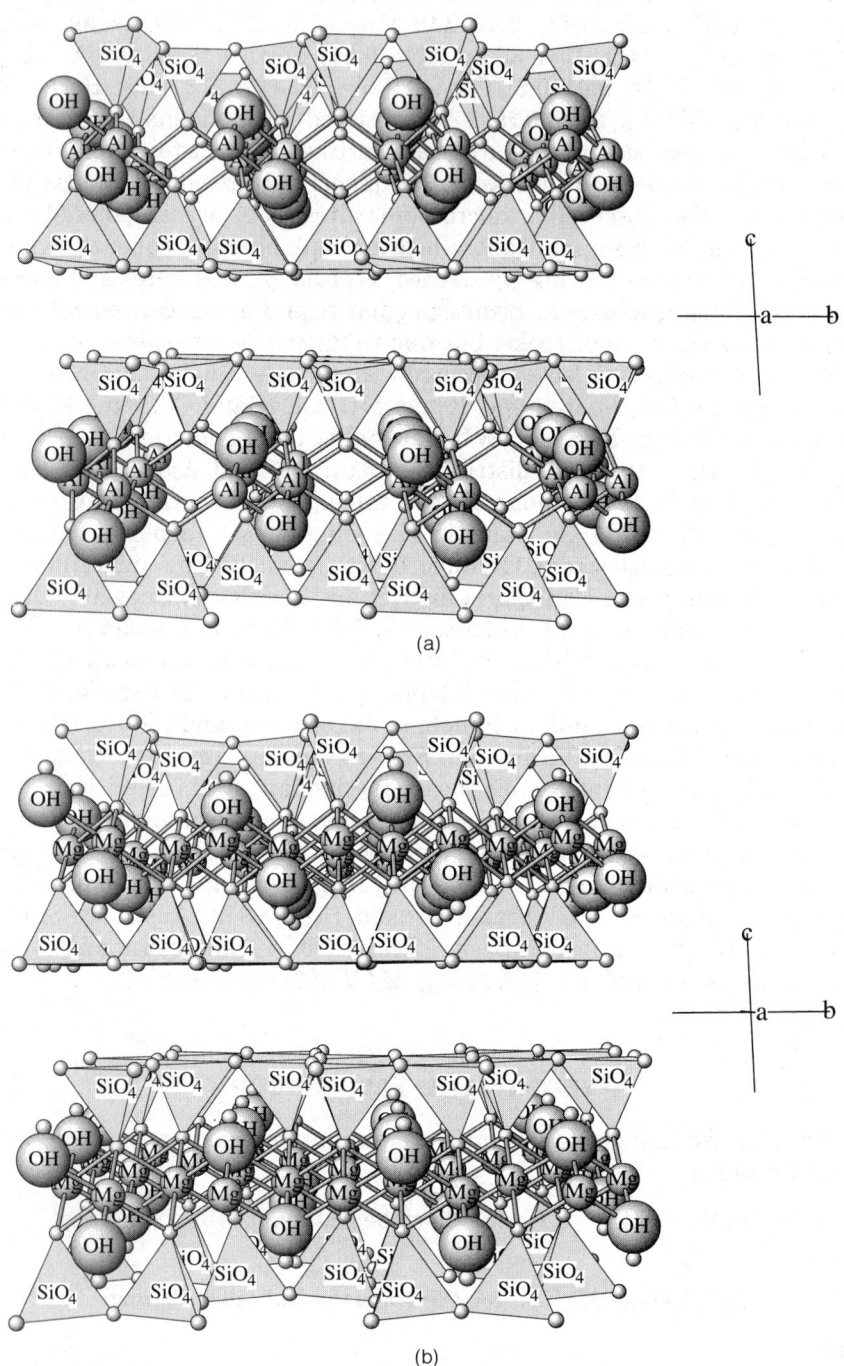

The structure of pyrophyllite (a) and talc (b) in perspective view along a, illustrating the difference between dioctahedral and trioctahedral phyllosilicates.

71.2.1.1 Pyrophyllite-1A, 2M

The structure is based on t-o-t (tetrahedral–octahedral–tetrahedral) layers (also called 2:1 layers). These layers are basically of the same type as in the mica group structures, several of the clay minerals, and in the chlorites.

The t-o-t layer is itself made up of two infinite sheets of six-membered rings of SiO_4 tetrahedra. The apical oxygens of two such layers, oppositely facing, together with (OH), make up the coordination sphere of the octahedral layer. The octahedral layer has one M1 site and two M2 sites per formula unit; layers may have trivalent (R^{3+}) cations in M2 and M1 vacant (dioctahedral), or all sites occupied by divalent (R^{3+}) cations (trioctahedral). These 2:1 composite layers form the basis for the species of this group, the mica group, several of the clay minerals, and are present in the chlorite group.

The t-o-t layers are ideally electrostatically neutral, and thus the bonding between successive t-o-t layers is due essentially to van der Waals forces. This is reflected in the low hardness of the species in the group. The stacking of successive t-o-t layers is shifted by approximately $0.3a$ along one of the pseudohexagonal axes, so that hexagonal rings in successive layers are not aligned, the shift being dictated by Si–Si repulsion. The structural details of the group show considerable variation, due in part to the several possible polytypes and the difficulty of obtaining undeformed crystals of these materials. *MSARM 19:225 (1988)*.

TALC GROUP

Mineral	Formula	Space Group	a	b	c	α	β	γ	
Pyrophyllite-1A	$Al_2Si_4O_{10}(OH)_2$	$C\bar{1}$	5.160	8.966	9.347	91.18	100.46	89.64	71.2.1.1
Pyrophyllite-2M	$Al_2Si_4O_{10}(OH)_2$	$C2/c$	5.138	8.910	18.60		100.02		71.2.1.1
Ferripyrophyllite	$Fe_2^{3+}Si_4O_{10}(OH)_2$	$C2/m$	5.26	9.10	19.1		95.5		71.2.1.2
Talc-1A	$Mg_3Si_4O_{10}(OH)_2$	$C\bar{1}$	5.290	9.173	9.460	90.46	98.68	90.09	71.2.1.3
Willemseite	$(Ni,Mg)_3Si_4O_{10}(OH)_2$	$C2/c$ or Cc	5.316	9.149	18.994		99.96		71.2.1.4
Minnesotaite*	$(Fe^{2+},Mg)_3Si_4O_{10}(OH)_2$	$P\bar{1}$	5.623	9.419	9.624	85.2	95.6	90.0	71.2.1.5

*Minnesotaite is included here but may be placed more appropriately with modulated layer structures.

71.2.1.1 Pyrophyllite-1A, 2M $Al_2Si_4O_{10}(OH)_2$

The name was first used in 1829 by Hermann and derives from the Greek for *fire* and *leaf*, in allusion to its tendency to exfoliate on heating. Talc group. Some agalmatolite, from Greek, *agalma*, image, and *lithos*, stone, is compact, very fine-grained pyrophyllite, but other examples may be talc or muscovite of similar texture. *1A-polytype:* TRIC $C\bar{1}$. $a = 5.160$, $b = 8.966$, $c = 9.347$, $\alpha = 91.18°$, $\beta = 100.46°$, $\gamma = 89.64°$, $Z = 2$, $D = 2.814$. *25-22:* 9.20_9 4.60_3 4.42_{10} 4.23_8 4.06_5 3.07_8 2.53_4 2.42_8. *2M-polytype:* MON $C2/c$. $a = 5.138$, $b = 8.910$, $c = 18.60$, $\beta = 100.02°$, $Z = 4$, $D = 2.85$. *PD:* 9.21_{10} 4.61_4 4.42_6 4.18_9 3.07_{10} 2.53_2 2.42_4 1.647_2. Structure: *AM 66:350(1981)*. The six-membered rings of the tetrahedral layers have a distinctly trigonal conformation. The octahedral sites (M2) occupied by Al are distorted and considerably smaller than the more symmetrical and larger M1 site, which is vacant. Two polytypes are commonly encountered in nature; type 1A is triclinic with a single 2:1 layer per unit cell, and type 2M is monoclinic with two 2:1 layers per unit cell. Twenty-two hypothetical polytypes have been proposed for the idealized pyrophyllite structure, but only 14 of these would yield distinguishable X-ray diffraction patterns. Stacking disorder and mixed phases often make for difficulty in distinguishing the two polytypes described here. **Habit:** Most commonly as granular masses of lamellae or laths flattened on {001}, but often so fine-grained as to be compact, textureless, and commonly incorporating other phases. Less commonly in very fine to coarse laths in radiating aggregates. **Physical properties:** Colorless, white, cream, light blue, gray, brownish green, but also nonuniform and colored by other phases in compact varieties; luster pearly on cleavage to dull (massive); transparent to translucent. Cleavage {001}, perfect, yielding flexible but inelastic scales. H = 1–2. G = 2.7–2.9. **Optics:** Biaxial (−); 2V = 53–62°; N_x = 1.552–1.556, N_y = 1.586–1.589, N_z = 1.596–1.601; pleochroism in colored examples is weak with X < Y = Z. Infusible; only slightly attacked by acids; closely resembles talc but distinguishable therefrom by spot test for Al, by abrasion pH (6 for pyrophyllite, 9 for talc), and by diffraction peak for (060) at ≈ 1.49 Å. Thermal decomposition endotherms occur at ≈ 500° and 675°. **Chemistry:** Published analyses for reasonably uncontaminated examples are very close to the ideal composition. Only traces of Fe^{2+} and Fe^{3+} are found in octahedral sites, and the existence of a series with ferripyrophyllite is doubtful. Large cations are present only in trace amounts. Some analyses suggest that some Al substitution for Si may take place, but this has not been substantiated by structure studies, and these may be due to impure material. **Occurrence:** Forms by low-grade metamorphism of aluminum-rich sediments such as aluminous pelite, bauxites, and aluminous quartzites, and rocks enriched in aluminum by base metal leaching during hydrothermal alteration. It is rare in sedimentary rocks. In prograde metamorphism the sequence is usually kaolinite–pyrophyllite–kyanite, and it normally is unstable beyond the greenschist facies. Also forms in retrograde metamorphism at the expense of other Al-rich phases, such as diaspore and kyanite. Pyrophyllite is stable at temperatures below 130° and pressures of < 10 kbar but may occasionally form at temperatures of 250–300°. **Localities:** Pyrophyllite may be of more widespread occurrence than is realized, it being easily overlooked because of its resemblance to talc and fine-grained muscovite. Some noteworthy occurrences are: in seams in coal at Mahoney City, Schuylkill Co., PA; at Crowder's Mt., Gaston Co., at Cottonstone Mt., Montgomery Co., and compact (agalmatolite) at Deep River, Guilford Co., NC; with lazulite and kyanite in Chesterfield Co., SC; *2M* in masses of coarse, radial clusters(!) with kyanite and rutile at Graves Mt., Lincoln Co., GA; at Bisbee, Cochise Co., and the Dome Rock Mts., near Quartzsite, AZ; and as yellow, radiating needles at Tres Cerritos, Mariposa Co., at Blue Bird mine, Imperial Co., and Champion mine, Mono Co., CA. Found at Horrsjöberg, Värmland, and the Westnå mine, Kristianstad, Sweden; at Liége, Belgium; in the Visp–St. Niklaus–Zermatt area, Valais, Switzerland; between Pyschminsk and Beresovsk, Yekaterinburg district, Ural Mts., Russia. Agalmatolite is found at numerous localities in China, as at Shaoshan, Minhou Co., Fujian Prov., and Quingtian, Zhejiang Prov. Light blue 2M found at Hanami mine, Nagama Pref., Japan; 1A near Coromandel, North Island, New Zealand; an impure, compact variety called "wonderstone" is found at Gestoptefontein, Transvaal, South Africa; found as 5- to 12-cm-long laths and similar-size radial sprays(!) of 1A polytype at Ibitiara, BA, Brazil. Uses: Used as a filler in paints, rubber, and so on, and in cosmetics and other dusting powders in much the manner of talc, and some materials marketed as "talc" may in fact be pyrophyllite. It also finds application as a refractory material and as a thermal and electrical insulator. Compact varieties (agalmatolite) are used as ornamental carving materials, particularly in China. AR DHZI 3:115(1962); MR 16:451(1985), 20:465(1989).

71.2.1.2 Ferripyrophyllite $Fe_2^{3+}Si_4O_{10}(OH)_2$

Named in 1979 by Chukhrov et al. for the composition and relationship to pyrophyllite. Talc group. Inadequate data. MON $C2/m$. $a = 5.26$, $b = 9.10$, $c = 19.1$, $\beta = 95.5°$, $Z = 4$, $D = 3.05$. 45-569: 9.6_8 4.54_{10} 3.17_7 2.62_4 2.47_4 1.725_3 1.665_3 1.518_8. The Fe^{3+} analog of pyrophyllite. Yellow-brown, transparent to translucent. Cleavage {001}, probably perfect. $H = 1\frac{1}{2}$-2. $G = 2.99$. Biaxial (−); $2V = 76°$; $N_x = 1.660$, $N_y = 1.676$, $N_z = 1.686$. German material contains 1.96 Fe^{3+} (octahedral), 0.11 Mg, 0.13 Al, 0.07 Fe^{3+} (tetrahedral), and 0.05 Ca per formula unit. Found in low-temperature hydrothermal deposits at *Strattenschacht, Germany*, and *Mt. Tologay, central Kazakhstan*. AR CE 38:324(1979), MA 31:355(1980), MSARM 19:226(1988).

71.2.1.3 Talc-1A $Mg_3Si_4O_{10}(OH)_2$

The name has been in use since antiquity and is derived from the Arabic. The name *kerolite*, applied to a hydrated and layer-disordered variety, is from the Greek *keros*, wax, and *lithos*, stone, in allusion to its luster. Talc group. *Soapstone* and *steatite* are synonymous and used for massive material, often with admixed serpentine minerals and calcite, the former name in reference to the "soapy" feel, the latter from the Greek through the Latin *steatitis*, precious stone. TRIC $C\bar{1}$. $a = 5.290$, $b = 9.173$, $c = 9.460$, $\alpha = 90.46°$, $\beta = 98.68°$, $\gamma = 90.09°$, $Z = 2$, $D = 2.798$. 24-1493: 9.31_{10} 4.67_2 4.55_6 3.12_9 2.59_2 2.48_3 2.23_1 1.524_3. Structure: ZK 156:177(1981). Alternate space groups have been reported, and a 2M polytype has been reported but not substantiated as occurring in nature. The six-membered rings of the tetrahedral layer have nearly hexagonal symmetry. The M1 and M2 sites are occupied by Mg in nearly ideal octahedral coordination, the average Mg-O bond length in M1 being nearly identical to that in M2. A total of 10 hypothetical polytypes have been proposed, but only seven would be distinguishable using powder diffraction. Habit: Rarely as pseudohexagonal tablets {001} or as free-standing, leafy crystals of irregular outline several square centimeters in area; more commonly in foliated or fibrous masses or stellate aggregates of laths; massive, compact, fine-grained (agalmatolite in part). In extensive masses with serpentine, calcite, and so on (e.g., soapstone, steatite). Often pseudomorphous after forsterite, enstatite, and other Mg-rich silicates. **Physical properties:** Colorless, white, pale to dark green, yellowish to brown; pearly luster, waxy for massive material (kerolite); transparent to translucent. Cleavage {001}, perfect, yielding flexible but inelastic scales; uneven to subconchoidal (massive) fracture. H = 1 resulting in "soapy" feel. G = 2.58–2.83. **Optics:** Biaxial (−); 2V = 0–30°; N_x = 1.539–1.550, N_y = 1.584–1.594, N_z = 1.588–1.602; X approximately perpendicular to (001), Y = b, as expected for a one-layer polytype, but also reported as Z = b, possibly as a result of difficulty in identifying b; r > v, weak; pleochroism evident in darker examples only, with X < Y = Z; refractive indices increase with iron content. Infusible and inert to acids. Often confused with pyrophyllite, but distinguishable from it by tests described under that species. **Chemistry:** Most talc is very close to the ideal composition, the only significant substitutions being Fe^{2+} and Fe^{3+} for Mg, and concomitant Al for Si to maintain charge balance. FeO may be present to ≈ 15 wt %, but reported higher values suggest a series with minnesotaite, formerly regarded as the ferrous analog of talc. Fluorine has been reported to ≈ 0.3 wt %. Dehydration of normal talc takes place as a single endothermic event at ≈ 900°. **Occurrence:** Most commonly found in Mg-rich rocks such as ultramafics and siliceous dolomites, although in the presence of high Al or K content, chlorite and phlogopite may be more likely, and with high Ca as in dolomite, the formation of tremolite may precede the formation of talc. Ferroan talc is not uncommon in meta-ultrabasics, but is less so in talcs derived from siliceous dolomite. Talc is unusual in sedimentary rocks, but Ni- and Fe^{3+}-rich varieties are found in laterites developed on serpentinized ultramafics. Talc is stable at higher temperatures than is pyrophyllite. **Localities:** Talc is of very widespread occurrence, and only a few noteworthy localities are listed. Found at numerous localities in New England, as at Athens (steatite), and Bridgewater(!), Windsor Co., VT; at Chester, Hampton Co., MA(!); at Smithfield, RI. Fine fibrous material found at Edwards, St. Lawrence Co., and at Amity(!), Orange Co., and steatite on Staten Island, NY; formerly extensively quarried (including steatite) at Chestnut Hill, Philadelphia, PA; at Cooptown, MD; steatite quarried in Cherokee Co., NC; and at Bisbee, Cochise Co., AZ. Found at Mt. St.-Hilaire and in Brome Co., PQ, and Hastings Co., ON, Canada. Also found at Malangen, Norway; with serpentine at Lizard Head, Cornwall, England; at St. Gothard, Tecino, and Fiesch, Valais, Switzerland; at Parma, in the Apennines, Italy; steatite at Gröpfersgrün, Bavaria, Germany; at Mt. Grainer, Zillertal, Tyrol, Austria; and at Shabrov, Ural Mts., Russia. Steatite is abundant in Hyderabad, India, and at numerous localities in China; fine-grained, massive, translucent talc, sometimes partially replacing tremolite and rarely with embedded pale green beryl, is mined at Chungju, Chung Cheong N. Prov., South Korea; at Mt. Fittom, SA, Australia. Found in the Barberton district, Transvaal, and the Hotazel, Mamatwon, and Smartt mines, Cape Prov., South Africa. In excellent, free-standing, colorless, leafy crystals(!) of several square centimeters area, and as small, green, pseudohexagonal crystals(!) in magnesite at several mines in the Brumado district, BA, Brazil. Uses: Steatite is used for its thermal and electrical insulating properties; as tabletop and tub material, but now largely replaced by synthetic products. Talc is extensively used as a filler in paints and rubber, in soaps and cosmetics and as a lubricating dusting powder. Fine-grained, compact material (agalmatolite in part) is used as an ornamental stone for carvings, particularly in China, and translucent, light green talc carvings are common in the crafts market, often lacquered to enhance hardness and improve luster. AR *DHZI* 3:121(1962), *AM* 53:751(1968), *MR* 9:196(1978).

71.2.1.4 Willemseite $(Ni,Mg)_3Si_4O_{10}(OH)_2$

Named in 1970 by De Waal for Johannes Willemse (b.1909), geologist, University of Pretoria, South Africa. Talc group. The name *pimelite*, from the Greek *pimele*, fatness, in allusion to the greasy luster and feel, has been applied to Ni-rich talc or the Ni equivalent of kerolite. MON $C2/c$ or Cc. $a = 5.316$, $b = 9.149$, $c = 18.994$, $\beta = 99.96°$, $Z = 4$, $D = 3.348$. 22-711: 9.4_{10} 4.57_2 3.12_3 2.50_2 2.25_1 1.524_1. Light green, greasy to waxy luster, translucent. Cleavage {001}, perfect. $H = 2$. $G = 3.31$. Biaxial $(-)$; $2V = 27°$; $N_x = 1.600$, $N_y = 1.652$, $N_z = 1.655$. Pleochroic, X, yellow-green; Y,Z, blue-green. Mg is present in substitution for Ni, and excess water may be present (pimelite). When Mg is dominant, it is regarded as nickelian talc. Found as massive, fine-grained material at the *Scotia talc mine, Bon Accord, Barberton district, Transvaal, South Africa*. Pimelite is reported from Silesia, Poland. AR *AM* 55:31(1970).

71.2.1.5 Minnesotaite $(Fe^{2+},Mg)_3Si_4O_{10}(OH)_2$

Named in 1944 by Gruner for the state of Minnesota. Talc group. TRIC $P\bar{1}$. $a = 5.623$, $b = 9.419$, $c = 9.624$, $\alpha = 85.2°$, $\beta = 95.6°$, $\gamma = 90.0°$, $Z = 2$, $D = 2.99$. 41-594: 9.54_{10} 4.78_2 3.18_5 2.76_3 2.66_4 2.54_3 2.53_5 2.21_2, the lines with spacing 2.54 and 2.53 commonly unresolved. Structure: *CM* 24:479(1986). The structure is reportedly a modulated layer of distinctive configuration. Alternate unit cells have been reported: (1) P with $a = 28$ Å ($\approx 5 \times 5.6$) for Mg-rich material, and (2) C with $a = 50.6$ Å ($\approx 9 \times 5.6$). Greenish gray to olive-green, dull to greasy luster, translucent. Cleavage {001}, perfect. $H < 3$. $G = 3.01$. Biaxial; 2V small; $N_x = 1.580$–1.592, $N_z = 1.615$–1.632; pleochroic with X > Z, pale green to colorless. Magnesium commonly present and may predominate over Fe^{2+}; if the structure is indeed different from that of talc, Mg-dominant minnesotaite would constitute a distinct species. An alternative formulation is $(Fe,Mg)_{27}Si_{36}O_{86}(OH)_{26}$. Found in low-grade metamorphosed banded iron formation deposits in association with quartz, stilpnomelane, siderite, etc. Found in the *Cuyuna and Mesabi Ranges, St. Luis Co., MN*; in the Sokoman iron formation, Labrador, NF, and at the Blue Bell mine, Riondel, BC, Canada; at Tynagh, Co. Galway, Ireland, and in Cornwall, England. In the Pb–Zn deposits at Sierra de Cartagena, Murcia,

Spain; at Hagendorf, Bavaria, Germany; in the Mara Mamba iron formation, Hamersley basin, Weld Range, WA, Australia; and at the Mamatwan mine, Kalahari manganese field, Cape Prov., South Africa. AR *AM 29:363(1994)*, *MM 47:229(1983)*.

MICA GROUP

The mica minerals comprise some 30 species of phyllosilicates with the general formula

$$AR_{2-3}T_4O_{10}X_2$$

where

A = a large monovalent or divalent cation, Na^+, K^+, $(NH_4)^+$, Cs^+, Ca^{2+}, Ba^{2+}, or a partial vacancy

R = an octahedrally coordinated cation, Li^+, Mg^{2+}, Fe^{2+}, Mn^{2+}, Zn^{2+}, Al^{3+}, Fe^{3+}, Mn^{3+}, V^{3+}

T = a tetrahedrally coordinated cation (dominantly Si^{4+}, with Al^{3+} commonly present, and also B^{3+})

X = an anion, most commonly $(OH)^-$, but also F^- or O^{2-} and rarely S^{2-}

The structure is based on t-o-t layers (also called 2:1 layers) which carry a net negative charge. Most commonly, the net negative charge is the result of substituting Al^{3+} for one-fourth of the tetrahedrally coordinated silicons, but may also be accomplished by positive charge deficiency in the octahedral R sites, or by the substitution of oxygen or sulfur for hydroxyl ion.

The t-o-t layers are of the same type found in the chlorite structures, and (except for the negative charge) in the pyrophyllite and talc structures. In the micas, the net negative charge is compensated by an interlayer of A cations. The octahedral R sites in the t-o-t layers may be either fully occupied, as in talc, to give the trioctahedral micas; or, they may be only two-thirds occupied, as in pyrophyllite, to give the dioctahedral mica series. Micas with R-site occupancy intermediate between 2 and 3 are rarely found: montdorite, whose species status is uncertain, would be a representative.

The t-o-t layers lie parallel to {001} in all the micas. The large interlayer A cations occupy 12-coordinated sites betwen the centers of six-membered rings of tetrahedra in adjacent t-o-t layers. The weak bonds of these cations and the strong intralayer bonding of the t-o-t layers account for the prominent perfect {001} cleavage of these species.

The octahedral t-o-t sheets contain two sites that are similar in size in the trioctahedral micas, but distinctly different in the dioctahedral species. In the latter, two sites (M2) are smaller and occupied by the trivalent cations, and one site (M1) is considerably larger and unoccupied.

The periodicity normal to the stacked sheets is approximately 10 Å or a multiple thereof. The stacking of t-o-t layers may occur in several ways, leading to a number of stacking sequences or polytypes. There is a certain

71.2.1.5 Minnesotaite

degree of composition dependence associated with the various stacking sequences so that they are not *sensu strictu* polytypes but, rather, intermediate between polytypes and polymorphs. There are six standard polytypes, only four of which occur commonly in nature, and several other sequences can be postulated. The stacking sequence may be described in terms of a shift of the layers by a shift of magnitude $\pm x$ accompanied by a rotation.

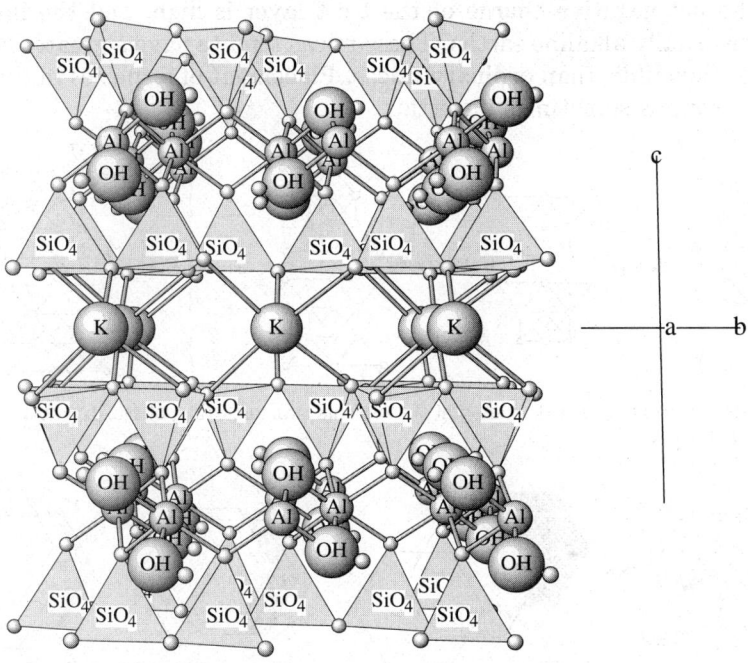

The structure of muscovite, a typical dioctahedral mica, viewed in perspective along *a*, and showing the t-o-t sequence parallel to {001}.

Standard Polytypes of Micas

Type	Shift*	≈ a	≈ b	≈ c	≈ β	Space Group
1M	0°	5.3	9.2	10	100	$C2/m$ or Cm
2O	180°	5.3	9.2	20	90	$Ccm2$
2M$_1$	±120°	5.3	9.2	20	95	$C2/c$
2M$_2$	±60°	9.2	5.3	20	98	$C2/c$
3T	−120°/+120°	5.3		30	90	$P3_122$ or $P3_222$
6H	−60°/+60°	5.3		60	90	$P6_122$ or $P6_522$

*Rotation angle between successive shifts.

Polytype $2M_1$ is dominant in muscovite and tends to be so in most aluminium-rich dioctahedral micas. Polytype 1M is also common in dioctahedral micas, particularly in those with considerable substitution of the octahedral aluminium. The 1M polytype is the most common in the trioctahedral micas. The micas are here divided into three subgroups: (1) the muscovite subgroup of dioctahedral micas, including the intermediate montdorite, and (2) the biotite subgroup of trioctahedral micas, and the brittle mica subgroup (3), which includes both dioctahedral and trioctahedral micas. In the brittle micas, the net negative charge of the t-o-t layer is high, and the interlayer A ions are usually alkaline earths. These micas tend to have a greater hardness and lesser flexibility than ordinary micas. Placement of a species in the brittle mica subgroup is sometimes uncertain.

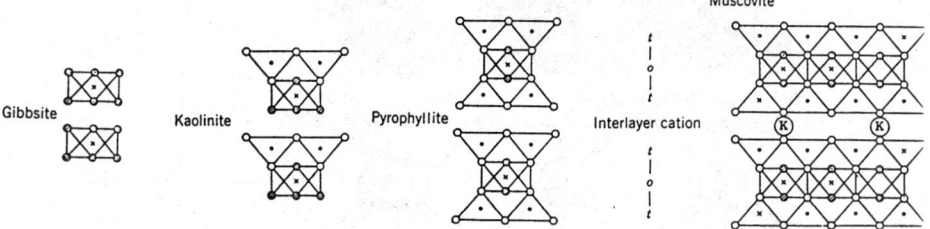

Schematic development of some dioctahedral phyllosilicate structures; (a) gibbsite, (b) kaolinite, (c) pyrophyllite, (d) muscovite

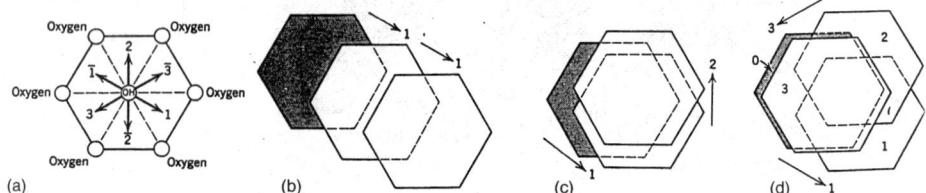

Schematic illustration of some stacking sequences (polytypes) in micas. (a) Three vectors for possible location of (OH) in the $Si_2O_5(OH)$ sheet stacked above or below the hexagonal ring shown. (b) The 1M polytype. (c) The $2M_1$ polytype. (d) The 3T polytype.

MICA GROUP: MUSCOVITE SUBGROUP, DIOCTAHEDRAL MICAS, $A^+R_2[T_4O_{10}](OH,F)_2$

Mineral	Formula	Known Polytypes	
Muscovite	$KAl_2[AlSi_3O_{10}](OH,F)_2$	$2M_1$, 1M, 3T, $2M_2$	71.2.2a.1
Paragonite	$NaAl_2[AlSi_3O_{10}](OH)_2$	$2M_1$, 3T, 1M	71.2.2a.2
Chernykhite	$(Ba,Na)(V^{3+},Al)_2[(Si,Al)_4O_{10}](OH)_2$	$2M_1$	71.2.2a.3
Roscoelite	$K(V^{3+},Al,Mg)[(Si,Al)_4O_{10}](OH)_2$	1M	71.2.2a.4
Glauconite	$(K,Na)(Mg,Fe^{2+})_{0.33}(Fe^{3+},Al)_{1.67}[(Si,Al)_4O_{10}](OH,F)_2$	1M	71.2.2a.5
Celadonite	$K(Fe^{3+},Al,Fe^{2+},Mg)_2[Si_4O_{10}](OH)_2$	1M	71.2.2a.6
Tobelite	$(NH_4,K)Al_2[AlSi_3O_{10}](OH)_2$	1M	71.2.2a.7
Nanpingite	$Cs(Al,Mg,Fe^{2+},Li)_2[AlSi_3O_{10}](OH,F)_2$	$2M_1$	71.2.2a.8
Boromuscovite	$KAl_2[BSi_3O_{10}](OH,F)_2$	1M, $2M_1$	71.2.2a.9
Montdorite	$(K,Na)(Fe^{2+},Mn^{2+},Mg)_{2.5}[Si_4O_{10}](F,OH)_2$	1M, 3T	71.2.2a.10

71.2.1.5 Minnesotaite

Mica Group: Biotite Subgroup, Trioctahedral Micas, $A^+R_3[T_4O_{10}](OH,F,O)_2$

Mineral	Formula	Known Polytypes	
Phlogopite	$KMg_3[AlSi_3O_{10}](F,OH)_2$	1M, 3T, 2M$_1$	71.2.2b.1
Biotite	$K(Fe^{2+},Mg)_3[AlSi_3O_{10}](OH,F)_2$	1M, 3T, 2M$_1$	71.2.2b.2
Annite	$KFe_3^{2+}[AlSi_3O_{10}](OH,F)_2$	1M	71.2.2b.3
Ferriannite	$K(Fe^{2+},Mg)[Fe^{3+}Si_3O_{10}](OH)_2$	1M	71.2.2b.4
Siderophyllite	$KFe_2^{2+}Al[Al_2Si_2O_{10}](OH)_2$	1M	71.2.2b.5
Hendricksite	$K(Zn,Mg,Mn^{2+})_3[AlSi_3O_{10}](OH)_2$	1M	71.2.2b.6
Lepidolite	$K(Li,Al)_3[AlSi_3O_{10}](F,OH)_2$	1M, 2M$_2$, 3T, 3M$_2$, 2M$_1$	71.2.2b.7
Polylithionite	$KLi_2Al[Si_4O_{10}](F,OH)_2$	1M	71.2.2b.8
Taeniolite	$KLiMg_2[Si_4O_{10}]F_2$	1M, 2M$_1$	71.2.2b.9
Zinnwaldite	$KLiFe^{2+}Al[AlSi_3O_{10}](F,OH)_2$	1M, 2M$_1$, 3T	71.2.2b.10
Norrishite	$KLiMn_2^{3+}[Si_4O_{10}]O_2$	1M	71.2.2b.11
Masutomilite	$K(Li,Al,Mn^{2+})_3[(Si,Al)_4O_{10}](F,OH)_2$	1M, 2M$_1$	71.2.2b.12
Sodium phlogopite	$NaMg_3[AlSi_3O_{10}](OH)_2$		71.2.2b.13
Wonesite	$(Na,K)_{0.5}\square_{0.5}(Mg,Fe^{2+})_{2.5}Al_{0.5}[AlSi_3O_{10}](OH,F)_2$	1Md	71.2.2b.14
Preiswerkite	$NaMg_2Al[Al_2Si_2O_{10}](OH)_2$	1Md, 2M$_1$	71.2.2b.15
Ephesite	$NaLiAl_2[Al_2Si_2O_{10}](OH)_2$	2M$_1$, 1M	71.2.2b.16

Mica Group: Margarite (Brittle Mica) Subgroup, Di- and Tri-octahedral Micas, $A^{2+}R_{2-3}[T_4O_{10}](OH,F,O,S)_2$

Mineral	Formula	Known Polytypes	
Margarite	$CaAl_2[Al_2Si_2O_{10}](OH)_2$	2M$_1$	71.2.2c.1
Clintonite	$CaMg_2Al[Al_3SiO_{10}](OH)_2$	1M, 2M$_1$	71.2.2c.2
Bityite	$CaLiAl_2[BeAlSi_2O_{10}](OH)_2$	2M$_1$	71.2.2c.3
Anandite	$(Ba,K)(Fe^{2+},Mg,Fe^{3+})_3[(Si,Fe^{3+})_4O_{10}](OH,S)_2$	2M$_1$, 2O	71.2.2c.4
Kinoshitalite	$(Ba,K)(Mg,Mn^{2+},Al)_3[Al_2Si_2O_{10}](OH)_2$	1M	71.2.2c.5

Morphology: Although characteristically tabular in habit, some common micas display short-to-long [001] prismatic crystals, the prism zone often poorly developed, tapering, or of variable diameter. Early morphological descriptions were based on an axial ratio $a/b/c \approx 0.58:1:3.33$ with $\beta \approx 90°$ that corresponds roughly to the 3T polytype dimensions. The transformation matrix for old setting is $300/030/\bar{1}01$ to the 1M cell, and $200/020/\bar{1}01$ to the 2M$_1$ cell.

Notes:
1. Twinning according to the mica law is common, the twin axis being [310] and the composition plane {001}.
2. Optics: all the micas are optically negative, usually biaxial with small to medium 2V. The acute bisectrix (X) is approximately perpendicular to the {001} cleavage, with $Z = b$ in 2M$_1$ polytypes, and $Y = b$ in 1M polytypes. Extinction is parallel or nearly so relative to the cleavage trace, which is always the slow direction. *D6 611(1892), DHZI 3:1(1962), MSARM 13:1(1984).*

71.2.2a.1   Muscovite-2M$_1$, 1M, 3T, 2M$_2$ KAl$_2$[AlSi$_3$O$_{10}$](OH,F)$_2$

The name muscovite was first used in 1850 by Dana without regard to polytype, and is derived from the term in common use at the time, *Muscovy glass*, from Muscovy Province, Russia. The 2M$_2$ polytype was first defined in 1973 by Zhoukilistov et al. Mica group, muscovite subgroup. Varietal names in common use are: *phengite*, from the Greek *phengites*, a term used by the ancients for transparent and translucent stones used as windows; *fuchsite* (chromian muscovite) is named in honor of Johann Nepomuk von Fuchs (1774–1856), German chemist and mineralogist; *sericite* is from the Greek for *silky*, in allusion to the luster. The 2M$_1$ polytype is the most frequently encountered; 1M and 3T are less common, and disordered 1Md also exists. Phengite is most commonly of the 1M, 1Md, or 3T polytype, with 2M$_1$ less common and 2M$_2$ rarer still. Mixed polytypes exist. For many, probably a majority, of occurrences, the polytype is unreported. 2M$_1$: MON $C2/c$. $a = 5.199$, $b = 9.027$, $c = 20.106$, $\beta = 95.78°$, $Z = 4$, $D = 2.818$. *PD:* 10.01$_{10}$ 5.02$_6$ 4.48$_6$ 4.46$_7$ 3.35$_{10}$ 3.21$_5$ 2.59$_5$ 2.56$_9$; also *6-263:* 9.95$_{10}$ 4.97$_3$ 3.34$_3$ 3.32$_{10}$ 3.19$_3$ 2.99$_4$ 2.57$_6$ 1.993$_5$; also *19-814*(vanadian): similar. 1M: MON $C2/m$. $a = 5.208$, $b = 8.995$, $c = 10.275$, $\beta = 101.6°$, $Z = 2$, $D = 2.805$. *7-25*(syn): 10.1$_{10}$ 5.04$_4$ 4.49$_9$ 3.66$_6$ 3.36$_{10}$ 2.58$_5$ 2.57$_9$ 1.499$_3$; also *21-993*(magnesian): 9.91$_9$ 4.50$_{10}$ 3.62$_7$ 3.33$_6$ 3.29$_6$ 3.06$_2$ 2.59$_5$ 2.56$_8$. 3T: HEX $P3_121$ or $P3_231$. $a = 5.196$, $c = 29.97$, $Z = 3$, $D = 2.831$. *7-42:* 9.97$_{10}$ 4.99$_5$ 4.49$_2$ 4.46$_2$ 3.31$_{10}$ 2.88$_2$ 2.56$_3$ 1.999$_5$; the pattern is essentially like that of the 1M polytype. 2M$_2$: MON $C2/c$. $a = 9.017$, $b = 5.210$, $c = 20.437$, $\beta = 100.4°$, $Z = 4$, $D = 2.80$. *34-175*(calc): 9.98$_{10}$ 4.44$_7$ 3.66$_7$ 3.49$_8$ 3.18$_8$ 3.04$_7$ 2.57$_9$ 2.55$_9$; also *26-649*(calcian), *10-490*(barian). *D6:611(1892), DHZI 3:1(1962), MSARM 13:1(1984)*. Structure: 2M$_1$, *EJM 4:283(1992)*; 1M, *SR 19: 468(1955)*; 3T, *ZK 125:163(1957)*; 2M$_2$, *CCM 21:465(1973)*. In muscovite, as in all dioctahedral micas, the vacant M1 site lies on the mirror plane and is larger than the occupied and distorted M2 sites. **Habit:** Tabular {001} crystals with rhombic to hexagonal outline, bounded most frequently by {221}, {$\bar{1}$11}, and {010}

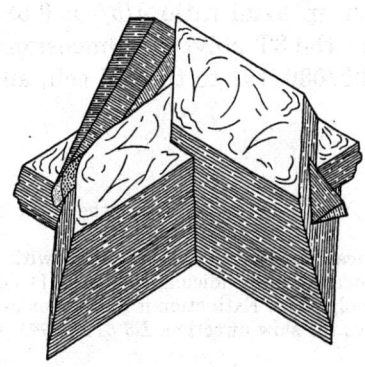

Muscovite displaying rhombic cross-section (Harvard Mineralogical Museum).

(old setting); sometimes in large sheets or "books" of irregular outline; less commonly as pronounced prismatic [001] crystals of crude hexagonal cross section; stellate, fivelings composed of crystals with rhombic cross section. Concentric, curved aggregates; flakes, lamellar masses, plumose aggregates, and fine-grained aggregates. Twinning common according to the mica law; less commonly with composition perpendicular to {001}. **Physical properties:** Colorless, silvery white, pale yellow to light brown, light to medium green, pink to light raspberry-red; vitreous luster, pearly on cleavage, aggregates silky, glistening or dull; transparent to translucent. Cleavage {001}, perfect, yielding flexible elastic sheets; parting on {110} and {010} often well developed by percussion normal to sheet; brittle when torn across cleavage. H = $2\frac{1}{2}$, parallel to cleavage, to $3\frac{1}{2}$, perpendicular to cleavage. G = 2.77–2.88. **Optics:** Common muscovite is biaxial (−); 2V = 28–47°; N_x = 1.552–1.576, N_y = 1.582–1.615, N_z = 1.587–1.618; Z = b, Y ∧ a = 1–3°, OAP perpendicular to {010}; r > v; nonpleochroic to moderately so; Z = Y > X. 3T muscovites are uniaxial (−) with similar indices; chromian muscovites (fuchsites) indices lie within the foregoing range, with distinct pleochroism in light blue-green to green; lithian muscovites have indices as low as N_x = 1.530, N_y = 1.552, N_z = 1.556; vanadian muscovites have higher indices, increasing toward those of roscoelite and display a more pronounced pleochroism; phengites have 2V = 0–30°; N_x = 1.547–1.580, N_y = 1.584–1.620, N_z = 1.587–1.623; rose muscovites have indices lying in the normal range but often display a reverse pleochroism with X, brown-red; Y,Z, light pinkish orange. Refractive indices generally increase with increasing iron or manganese and decreasing tetrahedral aluminum. Insoluble in acids. Fusibility $5\frac{1}{2}$. **Chemistry:** Muscovites approaching the ideal composition are not uncommon, and anhydrous fluormuscovite has been synthesized. Extensive substitution occurs in tetrahedral, octahedral, and interlayer sites leading to partial or complete solid solution with other dioctahedral micas, and to some extent with trioctahedral ones. Sodium is usually low, rarely exceeding 12% but sometimes reaching near 35% of the interlayer cation site; rubidium and cesium may amount to 5% and 1%, respectively, of the interlayer sites; calcium and barium may also substitute here with nearly 10 wt % BaO (oellacherite); interlayer water and potassium deficiency, accompanied by other substitutions, leads to hydromuscovite. Tetrahedral sites ideally have Si : Al = 3 : 1, but excess Si is not uncommon, accompanied by substitution of Mg and Fe^{2+} (phengite) or Li (Li muscovite) in octahedral sites. Substitution of Fe^{2+} and Fe^{3+} in octahedral sites leads to light green and brown muscovites; up to about 3.5 wt % Li_2O may be present with retention of the $2M_1$ polytype; MnO is usually < 1 wt % but may be double that resulting in purple-blue color. Rose muscovites are close to the ideal composition with significant Li and Rb content; the color, however, appears to be a function of a small amount of Mn^{3+}, Fe^{3+} in the tetrahedral sites, or Fe^{3+} and Mn^{2+}, which may result in an election transfer couple. Chromium substitution in octahedral sites, up to 5 wt % Cr_2O_3, is not uncommon in fuchsite, but a Cr-dominant mica with 24.7 wt % Cr_2O_3 has been reported from Outokumpu, Finland. There is evidence of a miscibility gap in the Cr–Al substitution, and a distinct species designation may be appropriate. Substitution of V^{3+} for Al results in olive-green tones, but when significant the 1M polytype predominates, and solid solution with roscoelite exists. Large "books" of muscovite often enclose in their cores, epitactic, reticulated films of red-brown hematite and nearly black magnetite parallel to {001}; these are probably the result of exsolution. **Occurrence:** Muscovite, although generally less common than biotite, is an important constituent of granite and granodiorite, wherein it may have crystallized directly from the fluid at pressures in excess of 3.5 kbar or in the solid state over a wide range of pressure. Muscovite granites tend to be high in quartz. Less commonly it is found in rhyolites, where it is often fluorine-rich. Abundant in granite pegmatites, both simple and complex, where it is associated with biotite and lepidolite. Late-stage replacement of potash feldspars, spodumene, tourmaline, beryl, topaz, etc, by fine-grained muscovite (sericite) is common. Rose muscovites are confined to pegmatites, usually in Na–Li replacement zones, where it is associated with spodumene, topaz, beryl, and tantalum minerals. Muscovite is a common constituent of every progressive zone of regionally metamorphosed sediments, forming initially by recrystallization of the sericitic micas and illite of the original sediments to yield albite–chlorite–sericite schists. With progressive metamorphism, muscovite tends to become coarser-grained and segregated into bands; it may coexist with almandine, but the latter, with K-feldspar, may also be a product of the reaction of muscovite, biotite, and quartz. Fuchsites are found in hydrothermal carbonate replacement zones associated with sulfide-gold deposits, and also in corundum schists and staurolite-zone quartz conglomerates and quartzites. Muscovite is also characteristic of fluorine metasomatism (greisenization), and associated feldspars are often replaced by muscovite. Muscovite in *sensu strictu* is less common in sedimentary rocks than was once believed. Sericite, illite, and various mixed-layer phyllosilicates are the common micaceous minerals of sediments. Those micas that approach true muscovites in such environments are usually of the 1M or 1Md polytypes. **Localities:** As a rock-forming mineral, muscovite is of worldwide, common occurrence, and the same is true for well-crystallized material from pegmatites and metasomatic deposits. Only a sampling of localities for documented analyzed and polytype identified samples, commercially important, or well-crystallized occurrences are given. Unless otherwise noted, the polytype is in all probability $2M_1$. The type locality for undifferentiated muscovite is lost in antiquity. Noteworthy U.S. localities are: Auburn, Androscoggin Co., and Buckfield and Hebron, Oxford Co., ME; fuchsite in staurolite-zone metamorphics at Acworth, Sullivan Co., NH; pink and rose-red at Chesterfield and Goshen, Hampshire Co., MA; large sheets and curved, concentric aggregates at Branchville, Fairfield Co., CT; pink, barian muscovite (oellacherite), 1M and $2M_2$, at Franklin, Sussex Co., NJ; fuchsite at the Line pit, Lancaster Co., PA; Amelia C.T., Amelia Co., VA. In NC, transparent to translucent, green crystals with rutile at Boiling Spring, Cleveland Co.; in fine crystals at Henry, Lincoln Co.; mined commercially in Mitchell, Yancey, Jackson, and Macon Cos. Commercially mined in pegmatites of the Black Hills, SD, the Etta mine, near Keystone, Pennington Co., having yielded large masses of coarse-grained, light yellow, lithian muscovite. Found at Mt. Antero, Chaffee Co., CO; rose muscovite, $2M_1$, is abundant at the Harding pegmatite, near Dixon, Taos Co., NM, where it forms light raspberry-red disseminated scales and crude prismatic crystals to 10 cm long and a few millimeters in diameter embedded in albite, and also replaces K-feldspar in part. In pegmatites of San Diego Co., CA, particularly the Himalaya mine, where it is often epitactically encrusted with lepidolite; phengite, $2M_1$, at Tiburon Peninsula, CA; a 3T polytype is found in the *Sunrise Cu prospect, Sultan Basin, Snohomish Co., WA*. In Canada, a nearly ideal end-member muscovite, $2M_1$, in Methuen Twp., ON, and at Mt. St.-Hilaire, PQ; fuchsite at Pointe du Boise, MB. In Mexico, as well-crystallized material in pegmatites of Baja California. Rose muscovite at Varuträsk, Sweden; fuchsite and Cr-dominant dioctahedral mica at Outokumpu, Finland; at the Penryn granite quarry and Wheal Lock Zawn, Cornwall, England; $2M_2$ calcian muscovite from Albertsweiler, Germany; at numerous localities in the Swiss, Italian, and Austrian Alps, often with adularia, and sodian, $2M_1$ muscovite at Alpe Sponda, Switzerland; magnesian 1M at Borza, Poland; in Russia, in pegmatites of the Ural Mts., particularly at Alabashka, near Mursinka; 1M phengite is reported from Transbaikalia, and $2M_2$ phengite from northern Armenia. Important commercial deposits in India at Hazaribagh, Bihar, and the Inikurti mine, Madras(!), where crystals weighing as much as 85 tons have been found. Fuchsite, $2M_1$, is found in the Harts Range, NT, Australia. $2M_1$ polytype at Archer's Post, Kenya, and $2M_1$ phengite at Bulautad, western (Spanish) Sahara. Abundant at numerous pegmatites of MG, Brazil. Particularly noteworthy are the commercial deposits at Lavra de Enio, Boa Vista; twinned crystals at Laranjeira, between Araçuai and Itaobim; at the Cruzeiro mine, Santa Maria Suaçui, in part lithian; yellow, lozenge-shaped crystals and "star mica" twins at the Riacho Genipapo pegmatite, and as "fishtail" twins at the Urubu pegmatite, near Araçuai; with brazilianite at Corrego Frio, between Linópolis and Mendes Pimentel; as superb reddish-brown tabular crystals(!), sometimes forming "roses," and associated with dark greenish-gray fluorapatite, at the Zé Pinto prospect (Fazenda Santa Elisa), east of Governador Valadares. Uses: The combined properties of perfect cleavage, elasticity, high dielectric constant, and low thermal conductivity make cleavage sheets ideal for many electrical, electronic, and thermal insulation applications. An early application, occasionally still encountered, is as windows, particularly of small size in furnaces. Ground mica (muscovite in large part) is used as a filler and dusting compound. Fine and coarsely ground muscovite finds use in the arts to produce special surface effects in prints and paintings. Scattered fuchsite flakes in quartzite result in one type of aventurine that is used ornamentally when cut into cabochons or beads. AR *DHZI 3:11(1962), MSARM 13:274(1984), MM 51:594(1987), CM 29:481(1991)*.

71.2.2a.2 Paragonite-2M$_1$, 3T, 1M   NaAl$_2$[AlSi$_3$O$_{10}$](OH)$_2$

Named in 1843 by Schafhautl from the Greek for *to mislead*, in allusion to its originally having been mistaken for talc. The 3T polytype described in 1956 by Dietrich and 1973 by Zvyagin et al. Mica group, muscovite subgroup. 2M$_1$: MON $C2/c$. $a = 5.128$, $b = 8.898$, $c = 19.278$, $\beta = 94.35°$, $Z = 4$, $D = 2.893$. *12-165:* 9.7$_8$ 4.9$_6$ 4.44$_{10}$ 4.06$_6$ 2.92$_7$ 2.83$_7$ 2.54$_9$ 2.48$_8$; also *42-602:* 9.63$_3$ 4.44$_4$ 4.39$_4$ 3.21$_{10}$ 2.56$_2$ 2.53$_5$ 1.925$_3$ 1.481$_3$. The substitution of Na for K leads to a distinctly shorter c axis than that found in muscovite. 1M: MON $C2/m$. $a = 5.139$, $b = 8.885$, $c = 9.750$, $\beta = 98.87°$, $Z = 2$, $D = 2.885$. *24-1047*(syn): 9.67$_8$ 4.82$_5$ 4.44$_4$ 4.24$_2$ 3.21$_{10}$ 3.06$_3$ 2.52$_2$ 1.928$_3$. Usually found as massive, fine-grained to scaly aggregates. Colorless to pale yellow, pearly to dull luster, transparent to translucent. Cleavage {001}, perfect. H = $2\frac{1}{2}$. G = 2.78–2.90. Biaxial (−); 2V = 0–40°; N$_x$ = 1.564–1.580, N$_y$ = 1.594–1.609, N$_z$ = 1.600–1.609; X $\wedge$ c ≈ 5°, Y ≈ a, Z = b; r > v. Polytype 3T is uniaxial (−), but 2M$_1$ displays the foregoing range of 2V. Contains ≈ 6 wt % Na$_2$O compared to muscovite, which is usually limited to < 2 wt %; thus complete solid solution is unlikely. Minor Mg and Fe^{3+} substitute for octahedral aluminum; K and Ca may replace Na to a limited extent. Occurs in a broad range of conditions in metamorphic regions in phyllite, schist, and muscovite–biotite gneiss, often associated with staurolite and kyanite. Also in quartz veins and in fine-grained sediments. In feldspar–mica assemblages, sodium has a tendency to enter the feldspar structure preferentially and paragonite is less likely to form than muscovite. Many "sericite" rocks contain significant paragonite. May be more abundant than commonly assumed, due to its association with and misidentification as muscovite. Probably of very widespread occurrence, the following being well documented: in schists of Glebe Mt., VT; in Campbell and Franklin Cos., VA; in the Leadville district, Lake Co., CO; with staurolite and kyanite at *Mte. Campione, Tessin, Switzerland*; at Borgofranco, Piemonte, Italy; and in the southern Ural Mts., Russia. AR *DHZI 3:31(1962)*, *SPC 22:557(1977)*, *AM 69:122(1984)*, *EJM 4:283(1992)*.

71.2.2a.3 Chernykhite-2M$_1$ (Ba,Na)(V^{3+},Al)$_2$[(Si,Al)$_4$O$_{10}$](OH)$_2$

Named in 1972 by Ankinovich et al. for V. V. Chernykh, professor, Leningrad Mining Institute. Mica group, muscovite subgroup. This species may also be considered a brittle mica. MON $C2/c$. $a = 5.29$, $b = 9.182$, $c = 20.023$, $\beta = 95.68°$, $Z = 4$, $D = 3.07$. *25-76:* 3.33$_{10}$ 3.01$_5$ 2.89$_4$ 2.80$_4$ 2.61$_7$ 1.996$_6$ 1.660$_6$ 1.530$_5$. Structure: *SPC 19:70(1974)*. Olive-green, pearly luster, translucent. Cleavage {001}, perfect. H = $3-3\frac{1}{2}$. G = 3.14. Biaxial (−); 2V = 11–12°; N$_x$ = 1.640–1.643, N$_y$ = 1.686–1.691, N$_z$ = 1.702–1.704; Z = b. Contains, by weight, 9.5% BaO, 18.6% V$_2$O$_3$, 5.3% V$_2$O$_4$; the dominance of divalent cations in the interlayer sites makes this species transitional with the brittle mica subgroup. Occurs as veins in carbonate rocks in *Kara Tau, Zhambil, Kazakhstan*. AR *ZVMO 101:451(1972)*, *AM 58:966(1973)*.

71.2.2a.4 Roscoelite-1M $K(V^{3+},Al,Mg)[(Si,Al)_4O_{10}](OH)_2$

Named in 1876 by Blake in honor of Henry Enfield Roscoe (1833–1915), English chemist of Manchester, who first prepared pure vanadium. Mica group, muscovite subgroup. MON $C2/m$. $a = 5.26$, $b = 9.09$, $c = 10.25$, $\beta = 101.0°$, $Z = 2$, $D = 3.08$. *10-496:* 10.0_{10} 4.54_8 3.66_5 3.35_8 3.11_5 2.70_3 2.60_8 1.520_6; also *19-933*(syn). Minute scales, often in small stellate or spheroidal aggregates; occasionally in fine-grained masses. Clove-brown to greenish brown, dark green; pearly to dull, earthy luster; translucent to opaque in mass. Cleavage {001}, perfect. H = 3. G = 2.97. Biaxial (−); 2V = 24–40°; $N_x = 1.59$, $N_y \approx 1.63$, $N_z \approx 1.64$, but also reported with 2V = 10–15°, $N_x = 1.610$, $N_y = 1.685$, $N_z = 1.704$; $X \approx c$, $Y = b$; r < v, strong. Pleochroic; X, olive-green; Y, greenish brown. Interference colors display distinct green tones. Slowly decomposed by HCl. End-member roscoelite would contain ≈ 33.5 wt % V_2O_3, whereas original type-material analyses report ≈ 28.5 wt %. Occurs in quartz veins with gold and as interstitial material in uranium-bearing sandstones. Found extensively in Jurassic sedimentary rocks of the Colorado Plateau, such as at Uravan, Montrose Co., CO, and similar occurrences in UT, AZ, and NM; associated with native gold at Red River, near Sutter's Mill, and at the *Stuckslacker (Sam Sims) mine, on Granite Creek, near Coloma, El Dorado Co., CA*, where several hundred pounds are said to have been found and discarded. Found in seams in porphyry at Kalgoorlie, WA, Australia; and in the uranium deposits at Mounana, Haut-Ogoué, Gabon. A minor but noteworthy ore of vanadium; worked extensively in Colorado Plateau deposits, often as a by-product of uranium recovery. AR *D6:635(1892)*, *AJS 39:253(1955)*, *MJJ 4:299(1965)*, *DHZI 3:11(1962)*, *MR 6:239(1975)*.

71.2.2a.5 Glauconite-1M $(K,Na)(Mg,Fe^{2+})_{0.33}(Fe^{3+},Al)_{1.67}[(Si,Al)_4O_{10}](OH)_2$

Named in 1828 by Keferstein from the Greek *glaukos*, blue-green, in allusion to its color, but the term *glauconie* was used in 1823 by Brongniart. Mica group, muscovite subgroup. The term *glauconite* has also been applied to green pellets in marine sediments that, although typical of the species, commonly incorporate other mineral species or may be entirely lacking in glauconite. Originally thought to be a $2M_1$ mica, the 1M polytype is now firmly established. Disordered (1Md) material and mixed-layer glauconite also exist. A 3T polytype may exist, but cannot be distinguished by X-ray powder diffraction. MON $C2/m$. $a = 5.234$, $b = 9.066$, $c = 10.16$, $\beta = 100.5°$, $Z = 2$, $D = 2.90$. *9-439:* 10.1_{10} 4.53_8 3.63_4 3.33_6 3.09_4 2.59_{10} 2.40_6 1.511_6. The powder pattern often resembles that of trioctahedral micas, but the species is truly dioctahedral. Structure: *AM 20:699(1935)*. **Habit:** Rounded aggregates and pellets of fine-grained, scaly particles; as casts of rhizopod shells; rarely as minute laths or platelets. **Physical properties:** Yellow-green, green, blue-green, rarely colorless; dull, earthy luster; translucent to opaque in mass. Cleavage {001}, perfect. H = 2. G = 2.4–2.95. Biaxial (−); 2V = 0–20°; $N_x = 1.585–1.610$, $N_y = 1.600–1.634$, $N_z = 1.610–1.641$, the indices increasing with increasing total iron content; $X \wedge c \approx 10°$; $Y = b$, $Z \approx a$; $r < v$, distinct; pleochroic, $X < Y = Z$, yellow-green to blue-green. The indices for aluminum-dominant ($Al_2O_3 \approx 18$ wt %) variety are $N_x = 1.559$, $N_y = N_z = 1.586$. Readily decomposed by HCl. **Chemistry:** The interlayer, large cation A sites, are usually not fully occupied or may contain water in place of potassium; in ordered 1M material the A-site occupancy by K^+ is 70–90%, decreasing to 40–70% in disordered, 1Md, glauconite, and further to 10–50% in mixed-layer glauconite that is similar to mixed-layer 2 : 1 clays. Minor Na_2O (< 0.3 wt %) and CaO (< 1.0 wt %) are present. In octahedral sites the Fe^{3+}/Al ratio is usually $> 3:1$; 1–5 wt % FeO is usually present, and 2–5 wt % MgO is usually present; Ni-bearing material has been reported; Mn is usually a minor constituent; Ti is present only in minor amounts. Silica content is usually high compared to muscovites. Water is usually present in amounts significantly in excess of that needed as $(OH)^-$ in the octahedral layer, the amount increasing to as much as 13–14 wt %, being higher in disordered or mixed-layer types. The water may be adsorbed or in part occupy A sites as H_2O or H_3O^+, this excess water being completely lost at 200°. **Occurrence:** The relatively narrow range of composition is consistent with the restricted conditions of formation in shallow-water marine environments under mildly reducing conditions during periods of low or no sedimentation. Association with iron sulfides suggests that the reducing conditions may be those associated with sulfate-reducing bacteria. In marine limestones, siltstones, and sandstones; greenstones are so named because of their color, due to a high glauconite content. The mechanism of formation is not wholly understood. In one hypothesis the proposed conditions are that a precursor 2:1 layer silicate be present, high iron and potassium be available, and a reducing condition exist. Glauconite, however, may form where a layer silicate precursor cannot be identified, so direct precipitation has also been proposed. **Localities:** Glauconite and glauconitic clays are of very widespread occurrence. Formation is presently taking place along many continental margins, particularly along the margins of eastern and western United States, northern Spain, southern Japan, southwestern Australia, eastern and southwestern Africa, and the southern areas of South America. A few well-documented occurrences are: as crystalline grains in dolomite at Bonneterre, St. Francis Co., MO; at Woodburn, County Antrim, Northern Ireland; near Hrodna, Byelarus; aluminum-dominant material from Skole in the eastern Carpathians, Romania; Ni-bearing material is reported from Widgiemooltha(?), Australia; analyzed samples from Table Hill, Kakako Creek, Whare Flat, and the Otepopo tunnel, Otego, New Zealand; from the Vidono formation, Anzoategui, Venezuela. **Uses:** Glauconite has ion-exchange properties which have found it application as a water softener and decolorizer. AR *DHZI 3:35(1962), MSARM 13:545(1984)*.

71.2.2a.6   Celadonite-1M $K(Fe^{3+},Al,Fe^{2+},Mg)_2[Si_4O_{10}](OH)_2$

Named in 1847 by Glocker from the French *celadon*, sea-green, in allusion to its common color. Mica group, muscovite subgroup. Formerly included under glauconite. MON $C2/m$. $a = 5.223$, $b = 9.047$, $c = 10.197$, $\beta = 100.43°$, $Z = 2$, $D = 3.03$. *17-521:* 4.53_9 3.64_8 3.32_7 3.09_8 2.68_8 2.60_7 2.59_{10} 2.40_8. Structure: *MZ 8:32(1986)*. Disordered and mixed-layer varieties are common. Found as earthy aggregates and coatings, or as minute scales. Gray-green to blue-green; dull luster; translucent, opaque in mass. Cleavage {001}, perfect. $H = 2$. $G = 2.95$–3.05. Biaxial (−); 2V small; $N_x = 1.606$–1.644, $N_y = 1.630$–1.662, $N_z = 1.630$–1.663; $X \approx c$, $Y = b$; $r > v$; also uniaxial (−); $N_O = 1.643$–1.663, $N_E = 1.612$–1.644. Pleochroic; X, yellow-green; Y,Z, green to emerald-green. The hypothetical end member is nonaluminous and only Fe^{3+} occupies octahedral sites, but natural material contains ≈ 12–27 wt % Fe_2O_3, ≈ 1–7 wt % Al_2O_3, ≈ 2–10 wt % FeO, and ≈ 0–6 wt % MgO. Decomposed by HCl. Except for its distinctly different mode of occurrence, it is difficult to distinguish from glauconite. Widespread as vesicle lining and coatings in altered volcanics of intermediate to basaltic composition, where it is associated with montmorillonite, quartz, chalcedonic silica, and zeolites. Some noteworthy localities are: at Bisbee, Cochise Co., Horseshoe Dam, Maricopa Co., and elsewhere in AZ; in the Porcupine Hills, Yellowstone National Park, WY; in the John Day Formation, OR; Wind R. area, WA; at Berufjordur, Iceland; Sandoy, Faroe Is., Denmark; at *Mt. Baldo, near Verona, Italy*; and numerous worldwide localities. AR *D6:683(1892)*, *MM 42:373(1978)*, *MSARM 13:545(1984)*.

71.2.2a.7 Tobelite-1M $(NH_4,K)Al_2[AlSi_3O_{10}](OH)_2$

Named in 1982 by Higashi for the locality. Mica group, muscovite subgroup. MON $C2/m$. $a = 5.219$, $b = 8.986$, $c = 10.447$, $\beta = 101.31°$, $Z = 2$, $D = 2.617$. *18-119:* 10.24_{10} 5.12_7 4.49_7 4.36_3 3.69_3 3.41_6 3.10_3 2.57_4; also *42-1321*(syn). Yellowish green, dull luster, translucent. Cleavage {001}, perfect. $G = 2.58$. Biaxial (−); 2V = 28°; $N_x = 1.555$, $N_y = 1.572$, $N_z = 1.581$; $Z = b$. The interlayer site occupancy for Tobe material is $[(NH_4)_{0.53}, K_{0.19}, Na_{0.01}, \square_{0.27}]$. Found as clayey material consisting of minute flakes and crystals up to 0.1 mm in diameter, associated with quartz alone at the *Ohgidani deposit, Tobe, Ehime Pref., Shikoku,* and at the *Horo deposit, Toyosaka, Hiroshima Pref., Honshu, Japan.* AR *MJJ 11:138(1982)*, *AM 68:850(1983)*.

71.2.2a.8 Nanpingite-2M$_1$ $Cs(Al,Mg,Fe^{2+},Li)_2[AlSi_3O_{10}](OH,F)_2$

Named in 1990 by Yang et al. for the locality. Mica group, muscovite subgroup. MON $C2/c$. $a = 5.362$, $b = 8.86$, $c = 21.41$, $\beta = 97.77°$, $Z = 4$, $D = 3.19$. PD: 3.62_5 3.33_6 2.99_5 2.92_5 2.66_{10} 2.65_1 2.13_9 2.12_2. Colorless, white; vitreous luster, pearly on cleavage; transparent to translucent. Cleavage {001}, perfect; more brittle and less elastic than muscovite. $H = 2$–3. $G = 3.11$. Biaxial (−); 2V = 46°; $N_x = 1.551$, $N_y = 1.584$, $N_z = 1.588$; $Z = b$; $r > v$, weak. A cesium analog of muscovite. Occurs as scales and plates up to 10 mm in diameter in radiating aggregates and less commonly as pseudohexagonal crystals, associated with montebrasite, quartz, and apatite in veinlets in the pollucite-bearing middle zone of a muscovite–albite–spodumene pegmatite in the *Nanping area, Fujian Prov., China.* AR *AM 75:708(1990)*.

71.2.2a.9 Boromuscovite-1M, 2M$_1$ KAl$_2$[BSi$_3$O$_{10}$)(OH,F)$_2$
Named in 1991 by Foord et al. from the composition and relationship to muscovite. Mica group, muscovite subgroup. 1M: $C2/m$. $a = 5.077$, $b = 8.775$, $c = 10.061$, $\beta = 101.31°$, $Z = 2$, $D = 2.90$. *PD:* 9.86$_6$ 4.24$_4$ 3.57$_{10}$ 3.29$_4$ 3.01$_8$ 2.51$_8$ 2.49$_4$ 2.20$_4$. 2M$_1$: $C2/c$. $a = 5.075$, $b = 8.775$, $c = 19.815$, $\beta = 95.59°$, $Z = 4$, $D = 2.89$. *PD:* 9.86$_8$ 4.39$_{10}$ 4.01$_5$ 3.29$_5$ 2.52$_5$ 2.34$_4$ 2.10$_4$ 1.46$_5$. For the type material (mixed polytypes), *PD:* 9.86$_6$ 4.39$_8$ 4.24$_4$ 3.57$_{10}$ 3.29$_4$ 3.01$_8$ 2.49$_4$ 2.20$_4$. Structurally analogous to muscovite, the substitution of boron for aluminum in the tetrahedral sites leading to the smaller unit cell. White to cream; dull, earthy luster; translucent. Cleavage {001}, perfect; parting {010}, poor; subconchoidal fracture for masses. H = 2½–3. G = 2.81. Biaxial (−); 2V$_{obs}$ = 44°, 2V$_{calc}$ = 47.5°; N$_x$ = 1.557, N$_y$ = 1.587, N$_z$ = 1.593; Na(D): Z = b; X ∧ c = −1°; r > v, weak. Analysis (wt %): B$_2$O$_3$, 7.0; MgO, 0.15; CaO, 0.10; Rb$_2$O, 0.52. Formed in the temperature range 350–400° as porcelaneous coatings up to 1+ cm thick consisting of minute scales and pseudohexagonal crystals of mixed 1M and 2M$_1$ polytypes in the *New Spaulding Pocket, Main Little Three dike, Little Three mine, Ramona, San Diego Co., CA*, where fragmented topaz, elbaite, and albite are included in the coatings. Reported from Řečice, Czech Republic. AR *AM 76:1998(1991)*.

71.2.2a.10 Montdorite-1M, 3T   (K,Na)(Fe^{2+},Mn^{2+},Mg)$_{2.5}$[Si$_4$O$_{10}$](F,OH)$_2$
Named in 1979 by Robert and Maury for the locality. Mica group, muscovite subgroup. Species status uncertain. MON $C2/c$. $a = 5.310$, $b = 9.20$, $c = 10.18$, $\beta = 99.9°$, $Z = 2$, $D = 3.159$. *33-1016:* 9.97$_4$ 3.41$_4$ 3.34$_{10}$ 3.14$_3$ 2.71$_3$ 2.52$_4$ 2.17$_3$ 1.669$_3$. No data are given for the 3T polytype. Structurally intermediate between di- and trioctahedral, 1M micas. Green to brownish green, translucent. G = 3.15. Biaxial (−); 2V < 1°. N$_x$ = 1.580, N$_y$ ≈ N$_z$ = 1.605; K/Na ≈ 0.82, Fe^{2+} is dominant in the 2.5 octahedral sites occupied, and F/(OH) ≈ 1.65. Occurs as 5- to 25-µm grains in peralkaline rhyolite of the *Mt. Dore stratovolcano, near La Bourboule, France*. AR *CMP 68:117(1979)*.

71.2.2b.1 Phlogopite-1M, 3T, 2M$_1$ KMg$_3$[AlSi$_3$O$_{10}$](F,OH)$_2$
Named in 1841 by Breithaupt from the Greek *phlogos*, firelike, in allusion to the reddish-brown color or internal reflections exhibited by some varieties. Mica group, biotite subgroup. The name *eastonite* applied in part to aluminous phlogopite is from the locality, and may be encountered in petrographic literature. *Manganophyllite*, named from composition and foliated habit, is applied to Mn-rich phlogopite and biotite. The 1M and 1Md polytypes are most common, followed in abundance by 3T and then 2M$_1$. Manganoan phlogopite of polytype 3T has not been reported. 1M: $C2/m$. $a = 5.308$, $b = 9.190$, $c = 10.155$, $\beta = 100.08°$, $Z = 2$, $D = 2.686$. *10-495:* 9.97$_{10}$ 3.39$_2$ 3.35$_{10}$ 2.61$_3$ 2.52$_2$ 2.43$_2$ 2.17$_2$ 2.01$_3$. 3T: $P3_112/P3_212$. $a = 5.317$, $c = 30.618$, $Z = 3$, $D = 2.629$. *10-492:* 10.1$_{10}$ 5.02$_3$ 3.41$_5$ 2.62$_3$ 2.51$_5$ 2.01$_{10}$ 1.673$_4$; taken *in toto* the 1M and 3T powder patterns are almost indistinguishable. 2M$_1$: $C2/c$.

71.2.2b.1 Phlogopite-1M, 3T, 2M₁

$a = 5.347$, $b = 9.227$, $c = 20.252$, $\beta = 95.02°$, $Z = 4$, $D = 2.670$. *10-493:* 10.1_{10} 3.54_4 3.36_1 3.28_4 2.62_{10} 2.18_5 2.02_7 1.677_5 1.538_5. Structure: *AM 58:889(1973)*.
Habit: Pseudohexagonal or rhombic, tabular {001}, short-to-long prismatic [001]; equant crystals bounded by {221}, {$\bar{1}11$}, and {010} (old setting); long prismatic crystals commonly tapering or of nonuniform cross section. Twinning common on {001} according to the mica law. **Physical properties:** Colorless, pale yellow, yellow-brown, reddish to dark brown, green; white to brown-tinted white streak; pearly to submetallic luster, on cleavage surface; transparent to translucent. Cleavage {001}, perfect, yielding flexible and commonly elastic sheets, may develop hexagonal percussion figures on cleavage plates. H = $2-2\frac{1}{2}$. G = $2.76-2.98$. Asterism by transmitted light perpendicular to cleavage is often displayed, and is probably due to inclusions of oxide minerals. **Optics:** Biaxial (−); 2V = 0–20°; $N_x = 1.530–1.590$, $N_y = 1.557–1.637$, $N_z = 1.558–1.637$, the values overlapping those for biotite; for 1M polytype, $X \approx c$, $Y = b$, $Z \wedge a = 0–5°$; r < v. Pleochroism none to moderate; X, pale yellow; Y,Z, yellow to red-brown or green. Indices depth of color increases with increasing iron content. For Mn-rich material, $N_x = 1.575$, $N_y = 1.617$, $N_z = 1.621$; 2V = 30°. Synthetic, end-member fluor-phlogopite has 2V = 14°; $N_x = 1.522$, $N_y = 1.548$, $N_z = 1.549$. Fluorine-rich phlogopites have lower indices than otherwise corresponding OH-rich compositions. **Chemistry:** Phlogopite forms a complete isomorphous series with biotite, annite, and siderophyllite, but solid solution with muscovite appears to be very limited. Members of the series with Mg:Fe ratios of > 2:1 are generally regarded as phlogopites, but the boundary is arbitrary largely because substitutions other than Fe for Mg may also occur. Aluminum may substitute in octahedral sites with concomitant additional Al in tetrahedral sites yielding the variety eastonite. Fe_2O_3 may be present up to ≈ 6 wt %. Significant MnO may be present, and a variety with 7.5 wt % MnO and 5.5 wt % CuO has been reported (Japan). Minor Li and Ti may also be present in octahedral sites. The interlayer cation is predominantly potassium, but significant sodium may be present as well as minor calcium and barium. Examples from metamorphic environments are often F-dominant or F-rich, but those from ultrabasic rocks may contain almost no fluorine. Phlogopite from basic volcanic rocks may be OH and F deficient. **Occurrence:** Commonly occurs in regionally and thermally metamorphosed impure limestones, particularly dolomite marble, the potassium being derived from either detrital potash feldspar or muscovite; excess alumina from the latter may form associated spinel. Fluorine metasomatism at contact zones results in F-rich phlogopite. Phlogopite is also characteristic of ultrabasic rocks and is common in kimberlites, where it may rim pyroxenes and other minerals, and in leucite-rich rocks. **Localities:** Phologopite is of widespread occurrence and only selected well-documented localities are given: found in many of the Grenville metamorphics in NY, such as at Port Henry, Essex Co., and at Gouverneur and as crystals to 60 cm long(!) at Clarke's Hill, St. Lawrence Co.; in marble at Franklin, Sussex Co., NJ; eastonite at Williams quarry, Easton, PA; in pyroxenite at Iron Hill, Gunnison Co., CO. In Canada it is found constituting 20–25% of kimberlite at Bachelor

Lake, PQ; with uraninite and molybdenite in cream-colored calcite at Gatineau Park, PQ; as transparent brown and nearly black crystals of 7 cm average length at Yates mine, Pontiac Co., PQ; as large, F-rich, near-end-member crystals(!) at Burgess, ON. In crystals up to 50 cm across in the Gardiner Complex, Kangerdlugssuak Fjord, Greenland. The $2M_1$ polytype occurs, although rarely, at Lake Gjerdenken, north of Oslo, Norway; formed in chondorodite–spinel marble by F-metasomatism at Manjsö Mt., Sweden; manganophyllite in manganese ores at Långban, Sweden; with monticellite, apatite, vesuvianite, and cuspidine in marble at Carlingford, Co. Louth, Ireland; disseminated and in lenses and bands in marble at contact with Quérigut granite in the Pyrenees Mts., France; 1M polytype at Mte. Braccio, Val Malenco, Italy. Important economic deposits(*)(!), mined since the eighteenth century, occur at Slyudyanka, at south west end of Lake Baikal, Irkutsk region, Russia, where some crystals reach 1 m in length; manganophyllite in a pegmatitic ore band at Chikla, Bhandara district, India; cuprian and manganoan phlogopite occurs at Kamogawa, Boso Peninsula, Honshu, Japan; a constituent of host rock of large dravite crystals at Yinnietharra, WA, and titanian material (≈ 9 wt % TiO_2) at Howes Hill, WA, Australia; at Anxiety Point, Nancy Sound, New Zealand; the $2M_1$ polytype occurs at Saharakara, and undifferentiated, presumably 1M phlogopite, is noted from many localities in Madagascar; (*) in kimberlite at Frank Smith mine, Cape Prov., South Africa; with dravite, magnesite, and talc at the Pedra Preta and Pirajá deposits, Brumado, BA, Brazil. Uses: Low- or iron-free phlogopite has dielectric and thermal conductivity properties akin to those of muscovite, and has application as an electrical and thermal insulating material and as a filler. AR *DHZI 3:42(1962); MR 1:58(1970), 16:492(1985); AM 77:1099(1992).*

71.2.2b.2 Biotite-1M, 3T, $2M_1$ $K(Fe^{2+},Mg)_3[AlSi_3O_{10}](OH,F)_2$

Named in 1847 by Hausmann for Jean Baptiste Biot (1774–1862), French physicist who studied the optical properties of the micas. Mica group, biotite subgroup. The term *lepidomelane*, from Greek for *scale* and *black*, is in common but superfluous use for black biotites, particularly those rich in Fe^{3+}, and may encompass annite and ferriannite. Siderophyllite, now given species status, was formerly included as a variety of biotite. The 1M polytype is by far the most common, followed by 3T and then $2M_1$. A polytype described as $4M_3$ has been reported, but the documentation is poor. 1M: MON $C2/m$. $a = 5.343$, $b = 9.258$, $c = 10.227$, $\beta = 100.26°$, $Z = 2$, $D = 2.81$. *24-867*(syn): 10.2_{10} 3.39_{10} 2.64_3 2.62_9 2.44_5 2.18_4 2.03_4 1.534_4; *42-1437*: 10.05_{10} 3.35_4 2.63_3 2.45_2 2.18_1 2.01_6 1.678_1 1.542_2. 3T: HEX $P3_121/P3_221$. $a \approx 5.3$, $c \approx 30$, $Z = 6$. The powder pattern is essentially equivalent to that of 1M. $2M_1$: MON $C2/c$. $a = 5.315$, $b = 9.22$, $c = 19.95$, $\beta = 95.1°$, $Z = 4$. *42-1339*: 10.01_{10} 3.53_1 3.34_3 3.27_2 3.02_1 2.62_4 2.44_2 2.43_2 2.17_2. $4M_3$: MON Space group unknown. $a = 5.535$, $b = 9.24$, $c = 40.03$, $\beta = 92.54°$, $Z = 8$ (by electron diffraction). Structure: 1M, *AM 75:305(1990)*; $2M_1$, *AM 60:1030(1975)*. **Habit:** Tabular {001} to short prismatic [001], with pseudohexagonal or rhombic cross section; dominant forms

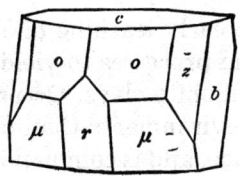

BIOTITE

Biotite. Forms: b, {010}; c, {001}; r, {$\bar{1}$01}; o, {112}; μ, {$\bar{1}$11}; z, {132}.

in old morphological setting are {001}, {221}, {010}, {$\bar{1}$11}, {112}, and {132}; prism faces often in repeated oscillatory combinations; crystals with well-developed {10$\bar{1}$}, {132}, and {1$\bar{3}$2} are pseudorhombohedral. Disseminated grains, scales, and large "books" of irregular outline. Twinning according to the mica law, commonly with {001} composition or, less frequently, perpendicular to {001}. **Physical properties:** Brown, green grading to black, also pale yellow to tan, bronze; white to pale streak; vitreous, splendent luster and sometimes pearly to submetallic on cleavage surface. Cleavage {001}, perfect yielding flexible, commonly elastic but somewhat brittle sheets; six-ray percussion figures may develop on cleavage plates; pressure may develop fractures or pseudofaces on the glide planes {$\bar{2}$05} and {135} resulting in pseudotrigonal–pyramidal forms. $H = 2\frac{1}{2}$–3. $G = 2.7$–3.4. **Optics:** Biaxial $(-)$; $2V = 0$–$25°$; or uniaxial $(-)$; $Z \wedge a = 0$–$9°$; $Y = b(1M)$, $Z = b(2M_1)$; $N_x = 1.565$–1.625, $N_y \approx N_z = 1.605$–1.696, the indices overlapping the ranges of phlogopite and of annite and ferriannite; $r > v$ or $r < v$ (Mg-rich), $r < v$ (Fe-rich), weak. Strongly pleochroic; X, light yellow, brown, or green; Y,Z, dark brown to dark green. Pleochroic halos with more pronounced absorption surround included radioactive phases, particularly zircon. Interference colors strongly influenced by absorption. Color is strongly influenced by the ratio of Ti to Fe^{3+}/(total Fe) such that the lower the ratio, the greener the color. Readily decomposed by strong acids. **Chemistry:** Biotite forms a complete isomorphous series with phlogopite and annite, and at least in part with siderophyllite and ferriannite; solid solution with muscovite is limited. The biotite composition field is arbitrarily defined as that having Mg:Fe ratios between 2 : 1 and 1 : 4, but the division is obscured by substitution of other cations in the octahedral as well as tetrahedral sites. Al^{3+}, Fe^{3+}, and Tl^{4+} may account for up to a third of the octahedral site occupancy, with concomitant excess Al or Fe^{3+} over their usual limit of a fourth of the tetrahedral sites; the charge excess may also be balanced by lithium in octahedral substitution. Up to approximately 5 wt % TiO_2 and approximately 1 wt % Li_2O may be present; Mn^{2+} substitution is usually limited, although some manganophyllite is close to the phlogopite–biotite boundary. Interlayer potassium may be replaced by sodium (20%) or calcium (10%) as well as minor rubidium and cesium; small amounts of barium may also be present, but larger amounts lead to anandite. Fluorine may replace up to half the hydroxyl ions, but is rarely that abundant, in contrast to many phlogopites. **Alteration:** Alteration by hydrothermal fluids tends to leach iron and magnesium to yield sericite pseudomorphs; other products are green biotite, due to early leaching of Ti, chlorites, illite, and kaolinite. Alteration by weathering processes to yield hydrated biotite, vermiculite, and ultimately, montmorillonite clays. **Occurrence:** Biotite is found in a greater variety of geological environments than other micas. It forms over a wide pressure–temperature range, and is common in igneous and metamorphic rocks of diverse composition, but only sparsely in sedimentary environments. It forms early in the thermal metamorphism of argillaceous sediments, and survives into successively higher-grade hornfelses. With increased metamorphism it is gradually replaced by assemblages such as K-feldspar/pyroxene/sillimanite or K-feldspar/cordierite. In regional metamorphism it is the indicator for the relatively low-grade "biotite zone," then giving way to somewhat more magnesian biotite in the "garnet zone," and persisting into still higher metamorphic grades. It may form from iron-rich amphiboles during potassium metasomatism. It is common over a wide range of plutonic rocks, from granites and granite pegmatites through intermediate composition and peralkalic rocks, particularly those which are leucite-rich. Less common in extrusive rocks, it is often partially or completely resorbed, especially in those of low silica content. The biotite of volcanics is commonly poorer in Fe^{2+} and richer in Fe^{3+} and TiO_2 than corresponding plutonics. **Localities:** Biotite is as widespread as the great diversity of rock types in which it is found; only a selection of localities of analyzed samples that exemplify its mode of occurrence is presented: common in the granites, pegmatites, schists, and gneisses of the metamorphic belts of the eastern United States from New England to the Carolinas; black biotite at West Balmat, and reddish-black biotite at Colton, northwestern Adirondack Mts., NY; Ba-rich at Franklin, Sussex Co., NJ; a Rb–Li biotite occurs at Kings Mt., NC, at the contact with spodumene pegmatite; at Rib Mt., Marathon Co., WI; in the Clear Creek pegmatite, Burnet and Llano Cos., TX; biotite and aluminum-rich biotite in granites and pegmatites of the Pikes Peak batholith, Crystal Park, CO; $2M_1$, OH and F-deficient (oxybiotite) at Ruiz Peak, NM; biotite containing 0.15 per formula unit of chlorine in schist from Lemhi Co., ID; in pegmatites, granodiorites, tonalite, and norite of the southern California batholith, CA. Large books of lepidomelane at Davis Hill, Silver Crater, and other localities in the Bancroft area, and both 1M and $2M_1$ polytypes in Burgess Twp., ON; at Pontiac, Litchfield Twp., and at Mt. St.-Hilaire, PQ, Canada. Biotite is found in nepheline syenite at Låven and in nepheline–syenite pegmatite at Brevik, Norway; in kyanite gneiss at Ross of Mull, Scotland; in volcanics of the Laacher See district, Germany. Polytype $4M_3$ is reported from the Khlebodarov quarry, east of the Sea of Azov, Ukraine. Found in biotite schist at Ogawamati, Nakosa, Fukushima Pref., and elsewhere in Japan; in a pegmatite–like lens in gneiss at Charles Sound, New Zealand; lithian lepidomelane in biotite granite and lithian siderophyllite(?) at Liruei, northern Nigeria. AR *DHZI 3:55(1962), AM 76:302(1991).*

71.2.2b.3 Annite-1M $KFe_3^{2+}[AlSi_3O_{10}](OH,F)_2$

Named in 1868 by Dana for the locality. Mica group, biotite subgroup. MON $C2/m$. $a = 5.386$, $b = 9.324$, $c = 10.268$, $\beta = 100.63°$, $Z = 2$, $D = 2.77$, for natural, magnesian material. 42-1413: 10.07_{10} 3.36_4 3.15_2 2.63_4 2.44_3 2.18_2 1.680_2 1.542_2; 33-1016(manganoan): 9.97_5 3.41_4 3.34_{10} 3.14_3 2.91_4 2.52_4 2.17_3 1.669_3. Structure: AM $58:889(1973)$. Black, brownish streak, vitreous to subadamantine luster, translucent to opaque. Cleavage {001}, perfect. $H = 2\frac{1}{2}$-3. $G = 3.17$. Biaxial (−); $2V = 0$–$5°$; $N_x = 1.624$, $N_y = 1.672$, $N_z = 1.672$; $Y = b$; $r < v$. Pleochroic; $X < Y = Z$; brown to dark brown. For synthetic end-member composition $N_x = 1.631$, $N_z = 1.697$, $D = 3.334$. Forms solid solution with biotite and phlogopite and with siderophyllite and ferriannite; in the phlogopite–annite series, annite compositions have a Fe/Mg ratio of 4 : 1 or greater. Natural material contains 32 wt % FeO, 3.1 wt % Fe_2O_3, 3.6 wt % TiO_2. Attacked by strong acids. Found in granite at *Cape Ann, MA*; in the Homestake formation, Lead, SD; and at La Bourboule, France (manganoan); at Sludorudnik, Ural Mts. and at Katugin, Siberia, Russia; at the Kawai mine, Ena, Gifu Pref., Japan. AR JP $3:82(1962)$, AM $77:34(1992)$.

71.2.2b.4 Ferriannite-1M $K(Fe^{2+},Mg)[Fe^{3+}Si_3O_{10}](OH)_2$

Named in 1963 by Wones for the composition and relationship to annite. Mica group, biotite subgroup. MON $C2/m$. $a = 5.402$, $b = 9.237$, $c = 10.306$, $\beta = 99.27°$, $Z = 2$, $D = 3.54$. PD: 10.16_{10} 3.68_5 3.38_5 2.91_1 2.67_1 1.696_5 1.660_3 1.539_1; also 16-169(syn): 10.2_{10} 4.66_1 3.46_1 3.39_7 2.67_5 2.48_2 2.21_1 1.567_2. Structure: AC $17:1369(1964)$. Brown to black, off-white to pale brown streak, vitreous to subadamantine luster; translucent to nearly opaque. Cleavage {001}, perfect. $H = 2\frac{1}{2}$-3. Biaxial (−); $2V = 40°$; $N_x = 1.653$, $N_y = N_z = 1.691$; also $N_x = 1.677$, $N_y = N_z = 1.721$, $2V = 0$–$10°$; also $N_x = 1.660$, $N_y = 1.720$, $N_z = 1.728$. Pleochroic; $X > Y = Z$; X, reddish brown; Y,Z, pale yellow-green to pale greenish brown. The first of the optically described material having FeO, Fe_2O_3, Al_2O_3 equal to 24.5, 8.0, 5.5 wt %, respectively, and the last having those oxides equal to 29.9, 13.2, 1.4 wt %, respectively. Occurs as scales and massive aggregates of tabular grains in association with riebeckite, stilpnomelane, hematite, and magnetite in the banded iron formation of the *Dales George member, Hammersley group, Wittenoom, WA, Australia*. AR AM $67:1179(1982)$.

71.2.2b.5 Siderophyllite-1M $KFe_2^{2+}Al[Al_2Si_2O_{10}](OH)_2$

Named in 1880 by Lewis from the Greek for *iron* and *leaf*, in allusion to the composition and foliated (micaceous) character. Mica group, biotite subgroup. MON $C2/m$. $a = 5.369$, $b = 9.297$, $c = 10.268$, $\beta = 100.06°$, $Z = 2$, $D = 3.27$. 25-1355: 9.99_{10} 3.36_9 3.27_9 2.62_{10} 2.43_8 2.16_7 1.979_7 1.542_9; also 26-909. Tabular crystals and sheets, scales, and flakes. Twinning common according to the mica law. Dark brown to black, also blue-green to gray-green; light tan streak in dark examples; vitreous luster, pearly on cleavage;

transparent to nearly opaque. Cleavage {001}, perfect. H = $2\frac{1}{2}$. G = 3.0. Biaxial (−); 2V small; N_x = 1.582–1.616, $N_y \approx N_z$ = 1.625–1.670; $\hat{Y}$ = b; dark varieties markedly pleochroic with X < Y = Z, brown to dark brown. Some materials designated as siderophyllite lie in or close to the biotite field. End-member siderophyllite as formulated above has not been reported in nature, but Al commonly exceeds one-fourth of the tetrahedral site occupancy and is also present as up to a third of the octahedral sites. Lithium and vacancies may constitute up to a sixth of the octahedral positions. Occurs in pegmatite zones and greisens. Dark brown to black examples found at *Pikes Peak, Park Co., CO.*; green-gray, lithian [biotite (?)] in pegmatite in marble in the Broos Range, AK. Lithium-free, blue-green [biotite (?)] is found with topaz in greisen at Newcastle, County Down, Northern Ireland; at Novara, Baveno, Italy; and at Altenburg in the Erzgebirge, Saxony, Germany. AR *DHZI 3:55(1962), MR 14:165(1983), MSARM 13:581(1984).*

71.2.2b.6 Hendricksite-1M $K(Zn,Mg,Mn^{2+})_3[AlSi_3O_{10}](OH)_2$

Named in 1966 by Frondel and Ito for Sterling Brown Hendricks (1902–1981), American chemist and crystallographer and student of the micas. Mica group, biotite subgroup. MON $C2/m$. a = 5.340, b = 9.524, c = 10.25, β = 100.07°, Z = 2, D = 3.28. *19-544:* 10.20_{10} 5.09_4 3.40_6 2.65_1 2.55_3 2.46_1 1.696_2 1.554_8. Structure: *TMPM 34:1(1985).* Copper-brown, reddish brown, to reddish black; vitreous luster. Cleavage {001}, perfect. H = $2\frac{1}{2}$–3. G = 3.4. Biaxial (−); 2V = 2–8°; N_x = 1.598–1.624; N_y = 1.658–1.686, N_z = 1.660–1.686 Y = b; . For the optical properties reported the compositions span 19.8–21.4 wt % ZnO, 12.5–13.9 wt % MnO, and ≈ 5 wt % Fe_2O_3. Occurs as rough tabular or short prismatic crystals and as anhedral plates to 30 cm in diameter at *Franklin, Sussex Co., NJ.* AR *AM 51:1107(1966).*

71.2.2b.7 Lepidolite-1M, $2M_2$, 3T, $3M_2$, $2M_1$(?)
$K(Li,Al)_3[AlSi_3O_{10}](F,OH)_2$

Named in 1792 by Klaproth from the Greek *lepidos*, scale, and *lithos*, stone, in allusion to its common habit of aggregates of platy crystals. Mica group, biotite subgroup. Although accorded species status, lepidolite actually represents a complex solid solution involving other Li–Al micas, either naturally occurring (polylithionite) or hypothetical (trilithionite), as well as magnesian, ferroan, and manganoan lithium micas. The 1M and $2M_2$ polytypes are most common, followed by 3T and then $2M_1$ and $3M_2$. Other polytypes have been proposed. The $2M_1$ polytype is usually a lithian muscovite rather than a trioctahedral mica. Mixed sequences may occur, and the formation of a particular polytype may be composition dependent. 1M: MON $C2/m$. a = 5.209, b = 9.011, c = 10.149, β = 100.77°, Z = 2, D = 2.83. *38-425:* 9.97_4 4.50_5 3.62_7 3.33_7 3.09_7 2.59_8 2.53_{10} 1.501_5; also *14-565*(ferroan). $2M_2$: MON $C2/c$. a = 9.023, b = 5.197, c = 20.171, β = 99.48°, Z = 4, D = 2.84. *PD:* 10.01_7 5.00_4 4.64_7 3.64_4 3.50_4 3.34_6 3.21_4 2.58_{10b}; also *14-11.* 3T: HEX $P3_121/P3_221$. a = 5.200, c = 29.76, Z = 3, D = 2.85. *42-612:* 9.93_3 4.45_5 3.85_3 3.59_3 3.32_{10b}

New England, as at Auburn(!), Androscoggin Co., ME (1M); $2M_2$ at *South Portland, Cumberland Co., ME*; ferroan lepidolite at Rockport, MA; as blue-gray to lavender, granular masses interstitial; to large spodumene laths, and often intergrown with microlite, at the Harding pegmatite, Dixon, Taos Co., NM; at the Brown Derby mine, Gunnison Co., CO; abundant in pegmatites of the Black Hills, SD; $2M_2$ with $2M_1$ lithian muscovite at the Stewart mine, Pala, and 1M at Himalaya mine, Mesa Grande, at the Little Three mine in the Ramona district, and other pegmatites of San Diego Co., CA; in the Tanco pegmatite, Bernic Lake, NB, Canada. 1M and $2M_2$ at the Varuträsk pegmatite, and $3M_2$ at *Skuleboda, Sweden*; 3T with lithian muscovite at Kimito, Finland; in pegmatites of the Isle of Elba, Italy; in large sheets at Penig, Saxony, Germany; massive material at Rožna, Moravia, and $2M_1$ near *Bisupice, Czech Republic*; at Alabashchka, near Mursinka, Ural Mts., Russia; at Minagi, Okayama Pref., Honshu, Japan; 3T at *Londonderry pegmatite, Coolgardie, WA, Australia*; abundant, often as large sheets at Antsomgombato, south of Betafo, and at Ampangabé, near Miandrarivo, and also at Maharitra on Mt. Bity, and elsewhere in Madagascar; in short prismatic, somewhat rounded crystals(!) at the Xanda mine, Virgem da Lapa, and in botryoidal form at Itinga, MG, Brazil. Uses: An important source of lithium compounds, and used directly in the manufacture of glass and enamels, thereby increasing strength and lowering the coefficient of expansion. AR *DHZI 3:85(1962)*, *AM 63:203(1978)*, *MSARM 13:579(1984)*, *MR 20:109(1989)*.

71.2.2b.8   Polylithionite-1M $KLi_2Al[Si_4O_{10}](F,OH)_2$

Named in 1884 by Lorenzen from the Greek for *many* and its high lithium content, the highest for any mineral known at the time. Mica group, biotite subgroup. MON $C2/m$. $a = 5.186$, $b = 8.968$, $c = 10.029$, $\beta = 100.40°$, $Z = 2$, $D = 2.823$. *21-952*(syn): 4.93_9 3.59_{10} 3.31_{10} 3.29_9 3.07_{10} 2.87_7 2.58_7 1.974_9. Colorless, pink, purple, amber; vitreous to pearly luster; transparent to translucent. Cleavage {001}, perfect. H = $3-3\frac{1}{2}$. G = 2.85. Biaxial (−); 2V small–30°; $N_x = 1.524-1.537$, $N_y = 1.543-1.563$, $N_z = 1.545-1.566$, these indices overlapping the common range for lepidolite. Bright yellow-green fluorescence in UV distinguishes it from taeniolite, with which it may occur (Mt. St.-Hilaire). Found at Point of Rocks, near Springer, NM. Relatively abundant at Mt. St.-Hilaire, QUE, Canada, in pegmatite pipes and veins as thin and fairly loose lamellae or as pseudohexagonal books or plates of up to 4 cm in diameter(!), and rarely in marble xenoliths. Found in the syenite pegmatites of *Kangerdluarsuk, Ilimaussaq, Greenland*, as tablets up to 10 cm in diameter(!); in the Oslo region, Norway; and the Kola Penin., Russia. AR *AM 53:1490(1968)*, *MR 21:330(1990)*.

71.2.2b.9   Taeniolite-1M,2M₁ $KLiMg_2[Si_4O_{10}]F_2$

Named in 1900 by Flink from the Greek for *band* or *strip*, in allusion to the tabular habit of the crystals. Mica group, biotite subgroup. 1M: MON $C2/m$.

3.09_3 2.59_8 2.56_8. **2M$_1$**: MON $C2/c$. $a = 5.199$, $b = 9.026$, $c = 19.969$, $\beta = 95.41°$, $Z = 4$, $D = 2.87$. *24-594:* 4.48_5 3.73_5 3.20_6 2.98_6 2.59_{10} 2.56_{10} 2.38_8 1.504_{10}. **3M$_2$**: MON $C2$. $a = 5.239$, $b = 9.070$, $c = 29.886$, $\beta = 92.58°$, $Z = 6$, $D = 3.11$. *PD:* 9.97_8 4.49_6 3.35_7 3.31_7 3.17_3 2.60_{10} 1.989_3 1.513_6. Structure: 1M, 2M$_2$, *AM 66:1221(1981)*; 3T, *AM 63:332(1978)*; 2M$_1$ *CCM 29:81(1981)*; 3M$_2$, *AM 72:1163(1987)*. Lithium often preferentially occupies the smaller M1 octahedral sites. Vacancies may exist in both M1 and M2 sites. Lithian muscovites are dioctahedral or intermediate between di- and trioctahedral structures, usually of the 2M$_1$ polytype. **Habit:** Pseudohexagonal or hexagonal, tabular {001} to short prismatic [001] crystals, the terminal faces often rounded and the [001] zone faces commonly rough or etched; also as concentric, curved plates. Most commonly as fine- to coarse-grained masses of interlocking scales and plates. Zonal or epitactic growths with muscovite. Twinning common according to the mica law. **Physical properties:** Gray, blue-gray, violet, lavender, violet-pink, also light yellow to brownish; vitreous to pearly luster on cleavage, often dull on [001] faces; transparent to translucent. Cleavage {001}, perfect, yielding flexible and elastic sheets. H = $2\frac{1}{2}$-$3\frac{1}{2}$. G = 2.8–2.9. **Optics:** Lepidolites are biaxial (−) or uniaxial (−) with $2V = 0$–$58°$; indices are commonly in the range $N_x = 1.529$–1.542, $N_y = 1.548$–1.567, $N_z = 1.553$–1.570, the lower part of the range overlapping the values for polylithionite; higher indices to $N_x = 1.548$, $N_y = 1.585$, $N_z = 1.587$ are also reported with increasing Fe or Mn content; $r > v$; $Y = b$, $Z \wedge a = 0$–$7°$. Weakly pleochroic. $X < Y = Z$; pinkish violet for Mn- and Fe^{3+}-bearing examples, and yellowish with Fe^{2+}-bearing examples. **Chemistry:** The majority of lepidolites represent a solid solution of polylithionite, $K(Li_2Al)[Si_4O_{10}](F,OH)_2$, and a hypothetical end member, trilithionite represented by $K(Li_{1.5}Al_{1.5})[Si_3AlO_{10}](F,OH)_2$. There appears to be complete solid solution with muscovite, so that only part of the larger M2 sites may be occupied; the transition between true lepidolite of 2M$_2$ type and lithian muscovite of 2M$_1$ type occurs at about 3.3 wt % Li$_2$O. At least partial solid solution exists with the magnesian, ferroan, and manganoan lithium micas, taeniolite, zinnwaldite, and masutomilite, and with ephesite, the sodium analog of lithian muscovite. In the absence of Fe^{2+}, the increasing Mn : Fe^{3+} ratio results in the deeper purple colors, whereas ferrous iron leads to yellowish or brownish colors. FeO and Fe_2O_3 may each attain 1.5 wt %, MgO 0.5 wt %, and MnO 1.0 wt %; titanium is present only in trace amounts. Potassium is always the dominant large alkali cation, but Na_2O, Rb_2O, and Cs_2O have been reported up to ≈ 1.0, ≈ 3.8, and ≈ 2.0 wt %, respectively. Fluorine is markedly dominant, and only rarely is it equal to or less than the hydroxyl content. **Occurrence:** Occurs almost exclusively in granitic pegmatite, where it is associated with other lithium minerals, elbaite, topaz, beryl, quartz, etc. It forms by metasomatic replacement of biotite and/or muscovite, and often forms granular to well-crystallized overgrowths on the latter. Rose muscovite is often later than associated lepidolite, on which it may form overgrowths. Also reported from granite, aplite, and high-temperature hydrothermal veins that are often tin-bearing. **Localities:** Abundant in the lithian pegmatites of

$a = 5.231$, $b = 9.065$, $c = 10.140$, $\beta = 99.86°$, $Z = 2$, $D = 2.84$. *31-1045*(syn): 9.95_9 4.98_4 4.51_3 4.48_3 3.33_{10} 3.11_3 2.40_4 1.995_3. 2M$_1$: MON $C2/c$. $a \approx 5.2$, $b \approx 9.1$, $c \approx 20$, $\beta \approx 95°$, $Z = 4$. *12-236*(syn): 10.0_8 5.01_8 4.52_8 4.48_6 3.61_8 3.11_8 2.60_{10} 2.40_8. Structure: *ZK 146:73(1977)*. The 1M polytype, for which there are detailed structure data, appears to be the most abundant, and a 3T polytype is also said to occur. Greenish brown, colorless, silvery; vitreous luster, pearly on cleavage; transparent to translucent. Cleavage {001}, perfect; flexible, somewhat elastic. $H = 2\frac{1}{2}$–$3\frac{1}{2}$. $G = 2.82$–2.90. Biaxial $(-)$; $2V = 2$–$5°$; $N_x = 1.522$–1.540, $N_y \approx N_z = 1.553$–1.570; $Z = b$, X perpendicular to cleavage; $r > v$, also $r < v$. Weakly pleochroic; Z, pale yellow; X, colorless. Also uniaxial $(-)$. Fusibility 3. May be regarded as a lithian phlogopite or magnesian lepidolite. Higher indices and darker color due largely to Fe^{2+} substitution for Mg. Occurs in nepheline–syenite pegmatites as tabular, pseudohexagonal crystals up to 5 cm in diameter and 0.5 cm thick. Found at Magnet Cove, Hot Spring Co., AR, and at Coyote Peak, Humboldt Co., CA. Occurs as thin lamellae and also as tapering pseudohexagonal prisms in breccia pipes and marble xenoliths at Mt. St.-Hilaire, PQ, Canada; at *Narsarsuk, Greenland*(!); and in the Lovozero massif, Kola Penin., Russia. AR *AM 47:1049(1962)*, *MR 21:342(1990)*.

71.2.2b.10 Zinnwaldite-1M,2M$_1$,3T   KLiFe^{2+}Al[AlSi$_3$O$_{10}$](F,OH)$_2$

Named in 1845 by Haidinger for the former name of the locality. Mica group, biotite subgroup. 1M: MON $C2/m$. $a = 5.296$, $b = 9.140$, $c = 10.096$, $\beta = 100.83°$, $Z = 2$, $D = 2.99$. *42-604*: 9.90_{10} 3.35_6 3.30_6 3.09_5 2.59_6 1.655_5 1.640_5 1.578_5; also *42-1399*. Structure: *AM 62:1158(1977)*. Ordering of cations may lead to a decrease in symmetry to Cm. Zonal growth of 1M and 3T polytypes has been reported. Occurs as pseudohexagonal tabular {001} or short prismatic [001] crystals, often in clusters or rosettes; also as disseminated scales. Twinning according to the mica law common. Gray-brown, yellowish brown, less commonly green or pale violet; vitreous to pearly luster; transparent to translucent. Cleavage {001}, perfect yielding elastic, flexible sheets. $H = 2\frac{1}{2}$–4. $G = 2.90$–3.02. Biaxial $(-)$ $2V = 0$–$40°$; or uniaxial $(-)$; $N_x = 1.535$–1.558, $N_y = 1.570$–1.589, $N_z = 1.572$–1.590; $r > v$, weak; for 1M type, $Y = b$, $Z \wedge a = 0$–$2°$. Weakly pleochroic; $X < Y < Z$; colorless to yellow-brown or gray-brown. The octahedral site occupancy may be represented as $[Li_{0.5-1.5}Fe^{2+}_{1.5-0.5}(Al, Fe^{3+})]$ with commensurate substitution in tetrahedral sites to maintain charge balance. As with lepidolite, the octahedral sites may not be fully occupied. In addition, significant Mn may be present (up to ≈ 2.0 wt %), as well as minor Ti. Potassium dominates the interlayer cation sites, and fluorine usually predominates over the hydroxyl ion. Occurs in granite pegmatites and high-temperature hydrothermal tin-bearing veins. It is associated with other lithium minerals, cassiterite, tourmaline, fluorite, topaz, etc. Found at Amelia Court House, VA; in growth-zoned crystals at Wigwam Creek and in other localities, in the Pikes Peak batholith, CO; the

White–Picacho district, Maricopa Co., AZ; with masutomilite in miarolitic cavities in the Sawtooth batholith, ID; the York district, Seward Penin., AK; in the Mourne Mts., Ireland, and in the tin deposits of Cornwall, England; at Altenburg, Saxony, Germany, and at *Cínovec (Zinnwald), Krusné Hory (Erzgebirge), Czech Republic*; at Novara, Baveno, Italy; in Japan at Kurobera, Yamanshi Pref., Naegi, Gifu Pref., Tanokamiyama, Shiga Pref., and the Takazawa mine, Okayama Pref. Also at Laouni, Algeria, and at Virgem da Lapa, MG, Brazil. AR *ZK 97:519(1937), MJJ 15:73(1990)*.

71.2.2b.11   Norrishite-1M $KLiMn_2^{3+}[Si_4O_{10}]O_2$

Named in 1989 by Eggleton and Ashley for Keith Norrish of the Division of Soils, Commonwealth Scientific and Industrial Organization (Australia), in recognition of his contribution to layer-silicate research. Mica group, biotite subgroup. MON $C2/m$. $a = 5.289$, $b = 8.914$, $c = 10.062$, $\beta = 98.22°$, $Z = 2$, $D = 3.21$. PD: 10.01_{10} 4.46_5 3.57_6 3.33_8 3.16_7 2.62_5 2.55_5 2.37_6. Structure: *AM 76:266(1991)*. Black to dark brown, opaque to translucent. Cleavage {001}, perfect; {010}, {100}, poor. H = $2\frac{1}{2}$. G = 3.25. Biaxial (+); 2V = 74°; ($2V_{blue} = 71°$, $2V_{red} = 75°$) for NaD: $N_x = 1.636$, $N_y = 1.687$, $N_z = 1.785$; optic axis dispersion strong with r > v. Pleochroic; X, yellow; Y, lime- to olive-green; Z, yellow-brown; color zoning is evident for Y. The composition is close to the ideal; the water content is the lowest for all analyzed micas and is consistent with the high octahedral layer charge. Occurs as millimeter-size crystals in rock-forming amounts on the *dumps of the abandoned Hoskins mine, 3 km W of Grenfell, NSW, Australia*, where it is associated with manganoan alkali clinopyroxene, pectolite–serandite, braunite, Ca and Ba carbonates, albite, K-feldspar, quartz, and barite. AR *AM 74:1360(1989)*.

71.2.2b.12   Masutomilite-1M, 2M₁ $K(Li,Al,Mn^{2+})_3[(Si,Al)_4O_{10}](F,OH)_2$

Named in 1976 by Harada et al. for Kazunosuke Masutomi, Japanese pharmacist and prominent amateur mineralogist and mineral collector. Mica group, biotite subgroup. 1M: MON $C2/m$. $a = 5.262$, $b = 9.102$, $c = 10.094$, $\beta = 100.83°$, $Z = 2$, $D = 2.94$. 29-822: 10.10_7 4.99_3 3.64_4 3.35_7 3.32_{10} 3.09_6 2.90_3 1.989_5. Structure: *MJJ 13:13(1986)*. Pale purplish pink, vitreous to pearly luster, transparent to translucent. Cleavage {001}, perfect. H = $2\frac{1}{2}$. G = 2.94. Biaxial (−); 2V = 29–31°; $N_x = 1.534$, $N_y = 1.568$, $N_z = 1.570$; $Y = b$, $Z \wedge a = 3°$; r > v, weak. Pleochroic; Y > Z = X; pale purple to colorless. Found with zinnwaldite in miarolitic cavities in the Sawtooth batholith, ID. Reported from a Li-pegmatite in western Moravia, Czech Republic. Occurs as druses and pseudohexagonal crystals up to 3 cm in diameter and 1 cm thick, associated with topaz, schorl, albite, and quartz in the *Tanakamiyama district, Shiga Pref.*, and at Tawara, Gifu Pref., *Honshu, Japan*. AR *MJJ 8:95(1976), AM 63:594(1977)*.

71.2.2b.13 Sodium Phlogopite $NaMg_3[AlSi_3O_{10}](OH)_2$

Named in 1980 by Schreyer et al. from the composition and relationship to phlogopite. Mica group, biotite subgroup. An inadequately described species. MON Space group unknown, probably $C2/m$. No lattice constants or X-ray diffraction patterns are documented. Silvery. Cleavage {001}, perfect. The average composition of 10 microprobe analyses is very close to the ideal. Occurs as minute flakes, rimmed by normal phlogopite in a massive, sugary dolomite with albite, talc, mixed-layer talc–chlorite, pyrite, and rounded tourmaline grains in a metamorphic evaporite sequence at *Derag, Tell Atlas, Algeria*. AR *CMP 74:223(1980)*.

71.2.2b.14 Wonesite-1Md $(Na,K)_{0.5}\square_{0.5}(Mg,Fe^{2+})_{2.5}Al_{0.5}[AlSi_3O_{10}](OH,F)_2$

Named in 1981 by Spear et al. for David R. Wones (1932–1984), Professor of Geology, Virginia Polytechnique Institute. Mica group, biotite subgroup. Also referred to as sodium phlogopite, but distinct from it in having a deficiency of interlayer alkali ions. MON $C2/m$. $a = 5.312$, $b = 9.163$, $c = 9.825$, $\beta = 103.18°$, $Z = 2$, $D = 2.875$. Partial structure data support the partial occupancy of the interlayer cation sites. Irregular cleavage plates; twinned on (001) with [310] twin axis; (001) epitactic intergrowths with phlogopite and talc. Brown, vitreous luster, transparent to translucent. Cleavage {001}, perfect. Biaxial (−) $2V = 0–5°$; or uniaxial (−); $N_x = 1.544$, $N_y = N_z = 1.608$. Pleochroic; X, pale brown; Y,Z, dark brown. Closely resembles phlogopite in thin section, but distinguished in thin section by slightly lighter color where the two occur together. Occurs as a rock-forming component of metamorphosed Ordovician Post Pond volcanics in the *southwestern corner of the Mt. Cube quadrangle, VT*. AR *AM 66:100(1981)*.

71.2.2b.15 Preiswerkite-1Md, 2M₁ $NaMg_2Al[Al_2Si_2O_{10}](OH)_2$

Named in 1980 by Keusem and Peters for H. Preiswerk (1876–1940), Basel, Switzerland. Mica group, biotite subgroup. 1Md: MON $C2/m$. $a = 5.227$, $b = 9.049$, $c = 9.804$, $\beta = 100.23°$, $Z = 2$. *42:605*: 9.65_8 4.50_3 3.22_9 2.57_{10} 2.46_3 2.12_4 1.508_8 1.488_4. 2M₁: MON $C2/c$. $a = 5.22$, $b = 9.05$, $c = 19.42$, $\beta = 95.17°$, $Z = 4$, $D = 2.94$. *33-1259*: 9.64_5 4.52_6 3.22_5 2.78_5 2.57_{10} 2.46_7 2.15_6 1.508_{10}. Structure: *IMA 355(1990)*. Pale green, colorless; vitreous luster; transparent. Cleavage {001}, perfect. $H = 2\frac{1}{2}$. $G = 2.96$. Biaxial (−); $2V = 5–7°$; $N_x = 1.560$, $N_y = 1.614$, $N_z = 1.615$. Found as platy grains up to 1 mm in nodular aggregates of Al-pargasite and zoisite in a rodingite dike in the *Geisspfad ultramafic complex, Binntal, Valais, Switzerland*. AR *AM 65:1134(1980), MSARM 13:582(1984)*.

71.2.2b.16 Ephesite-2M₁, 1M $NaLiAl_2[Al_2Si_2O_{10}](OH)_2$

Named in 1851 by Smith for the locality. Mica group, biotite subgroup. MON $C2/c$. $a = 5.120$, $b = 8.853$, $c = 19.303$, $\beta = 95.08°$, $Z = 4$, $D = 2.97$. PD: 9.63_6 4.40_8 3.21_6 2.54_{10b} 2.41_8 2.09_5 1.605_4 1.479_{10}; also *19-1181*; *35-423*(syn): 9.61_8

4.38_7 3.26_4 3.20_{10} 3.14_4 2.52_9 1.479_7 1.473_5. Structure: *NJMM 275(1987)*. The $2M_1$ polytype appears to be the more common, but most recent structure studies assign it to space group P1. $a = 5.123$, $b = 8.872$, $c = 19.307$, $\alpha = 89.97°$, $\beta = 95.15°$, $\gamma = 89.96°$, which would properly be called a 2A polytype. Forms indistinct crystals up to 13 mm in diameter and 10 mm long, and as flakes up to 5 mm in diameter; often twinned by rotation about [310] or [3̄10]. Pink to pinkish brown; vitreous luster, pearly on perfect {001} cleavage; translucent. Brittle. $H = 3\frac{1}{2}-4\frac{1}{2}$. $G = 2.84$. Biaxial $(-)$; $2V = 18-28°$; $N_x = 1.592-1.595$, $N_y = 1.624-1.625$, $N_z = 1.625-1.627$; $Z = b$, $Y \wedge a = 7°$; $r < v$. Although the brittleness and relatively high hardness suggest that the species should be grouped with the brittle micas, unlike them, the triple-layer charge is only 1^- and the alkaline earth content is low. Found in the Ilimaussaq complex, Greenland; in the emery deposit at *Gumuchdagh, Ephesus, near Izmir, Turkey*; with manganoan diaspore, braunite, and bixbyite in the Postmasburg district, South Africa. AR *AM 52:1689(1967), MSARM 13:581(1984), CMP 85:74(1984).*

71.2.2c.1   Margarite-$2M_1$ $CaAl_2[Al_2Si_2O_{10}](OH)_2$

Named prior to 1823 by Fuchs, from the Greek for *pearl*, in allusion to the pearly luster of its platy aggregates. Mica group, margarite subgroup. MON $C2/c$. $a = 5.108$, $b = 8.844$, $c = 19.156$, $\beta = 95.48°$, $Z = 4$, $D = 3.069$. *18-276:* 3.18_{10} 3.12_1 2.52_2 2.51_2 2.41_1 1.908_3 1.903_2 1.466_2. Structure: *ZK 165:295(1983)*. Dioctahedral, the Ca analog of muscovite, but with some buckling of the basal oxygens of the sheets due to anion repulsion where the smaller interlayer cations bring oxygens of adjacent sheets into proximity. Rarely as tabular {001} pseudohexagonal crystals; more commonly as random platy aggregates or scaly masses. Twinning according to the mica law common. Light pink to grayish pink, pale yellow, pale green; pearly luster on cleavage; transparent to translucent. Cleavage {001}, perfect, yielding inflexible plates; brittle. $H = 3\frac{1}{2}-4\frac{1}{2}$ on {001}, $H = 6$ perpendicular to {001}. $G = 3.0-3.1$. Biaxial $(-)$; $2V = 40-67°$; $N_x = 1.630-1.638$, $N_y = 1.642-1.648$, $N_z = 1.644-1.650$; $Z = b$, $Y \wedge a = 6-8°$; $r < v$; pleochroism very weak to absent. Fusible with great difficulty. Only partially decomposed by boiling acids. Most margarite is close to the end-member composition. Ca may be replaced by minor Ba, Sr, and K, and by significant Na. In the latter case the charge balance may be restored by replacing some tetrahedral Al by Si and thus forming a solid solution with paragonite, or by excess $(OH)^-$ as suggested by some published analyses. Fluorine is rarely present. Typically occurs in metamorphic emery (corundum) deposits; also in chlorite and mica schists with staurolite and schorl. Noteworthy localities are at the Emery mines, Chester, Hampden Co., MA, with corundum, magnetite, and diaspore; at Corundum Hill, Chester Village Green, and Delaware Co., PA; chromian at Line pit, State Line district, MD; near Buck Creek, Clay Co., NC; with corundum near Meadow Valley, Plumas Co., CA. Found at *Zillerthal, Tyrol, Austria*; sodian margarite at a pegmatite–amphibolite intrusion boundary at Mt. Yatyrgvata, northern Caucasus, Russia; at the Shin-Kiura mine, Oita Pref., Japan; embedded in corundum at Gibraltar, WA, Australia. AR *DHZI 3:95(1962), AM 63:186(1978)*.

71.2.2c.2 Clintonite-1M, 2M$_1$ CaMg$_2$Al[Al$_3$SiO$_{10}$](OH)$_2$

Named in 1843 by Mather for DeWitt Clinton (1769–1828), American lawyer and statesman who had an interest in the geological sciences. Mica group, margarite subgroup. Synonymous with xanthophyllite, although earlier use applied this term to material with OAP parallel to {010} (1M), and clintonite was reserved for that with OAP perpendicular to {010} (2M$_1$). 1M: MON $C2/m$. $a = 5.204$, $b = 9.013$, $c = 9.814$, $\beta = 100.26°$, $Z = 2$, $D = 3.12$. 20-321: 9.68_5 3.21_7 2.56_{10} 2.45_5 2.37_4 2.11_7 1.505_6 1.485_5. Structure: *AM 73:365(1988)*. A trioctahedral mica; as with margarite there is considerable distortion of the layer units. Crystals tabular {001} and pseudohexagonal; lamellar, in foliated or radiating, massive aggregates. Twinning according to the mica law. Colorless, pale yellow to greenish yellow, orange to reddish brown; uncolored streak or pale tan for darker varieties; vitreous to pearly luster, sometimes submetallic; transparent to translucent. Cleavage {001}, perfect, yielding inflexible plates; brittle. $H = 3\frac{1}{2}$ parallel to {001}, $H = 6$ perpendicular to {001}. $G = 3.0$–3.1. Biaxial $(-)$; $2V = 0$–$32°$; $N_x = 1.643$–1.649, $N_y = 1.655$–1.662, $N_z = 1.655$–1.663; $Y = b$ for 1M, $Z = b$ for 2M$_1$; $r < v$ weak. Pleochroism absent to weak, $X < Y = Z$, colorless to green or brown. Most clintonites have slightly less Si and more Al than the formula above, and thus represent a high degree of substitution of Al in tetrahedral sites. Aluminum is also present in about one-fourth to one-third of the octahedral sites. Fe^{2+} and Fe^{3+} also substitute in octahedral sites. Sodium may replace calcium, and fluorine is usually absent. Clintonite occurs with talc in chlorite schists and in metasomatized limestones with spinel, grossular, vesuvianite, diopside, and minerals of the humite group. Found at *Amity, Orange Co., NY*; in metamorphosed diorite at the Christmas mine, Gila Co., AZ; with blue calcite and monticellite at Crestmore, Riverside Co., CA; at Prince of Wales Is., AK. Found at Pargas, Finland; on Mt. Monzoni, Val di Fassa, Trentino, and Vacca Lake, Adamello, Italy; in the Shiskimskaya Mts., near Zlatoust, and near Achmatovsk, both in southern Ural Mts., Russia. AR *DHZI 3:99(1962), AM 71:1194(1986)*.

71.2.2c.3 Bityite-2M$_1$ CaLiAl$_2$[BeAlSi$_2$O$_{10}$](OH)$_2$

Named in 1908 by Lacroix for the type locality. Mica group, margarite subgroup. MON $C2/c$. $a = 5.058$, $b = 8.763$, $c = 19.111$, $\beta = 95.39°$, $Z = 4$, $D = 3.05$. 11-400: 4.29_6 3.14_8 2.48_{10} 2.36_6 2.04_9 1.878_7 1.570_6 1.450_{10}. Structure: *AM 68:130(1983)*. Colorless, white, yellow, brownish; vitreous luster; transparent. Cleavage {001}, perfect. $H = 5\frac{1}{2}$. $G = 3.02$–3.07. Biaxial $(-)$; $2V = 35$–$52°$; $N_x = 1.643$–1.651, $N_y = 1.652$–1.659, $N_z = 1.654$–1.661. X perpendicular to cleavage; cleavage plates commonly divided into six sectors, with lamellar twinning visible in each sector. Insoluble in acids. Easily fusible.

Occurs sparingly in lithium pegmatite as crusts and small scales sometimes with hexagonal outline. Found at Strickland quarry, Portland, CT; at the Foote mine, Kings Mt., Cleveland Co., NC, and the Harding pegmatite, Taos Co., NM; with albite, beryl, bavenite, columbite, and cassiterite at Londonderry, WA, Australia; at *Maharitra, Mt. Bity, Madagascar*, with pink tourmaline, albite, and lepidolite. AR *DT: 657(1932), ZK 107:325(1956)*.

71.2.2c.4 Anandite-2M$_1$, 2O   (Ba,K)(Fe^{2+},Mg,Fe^{3+})$_3$[(Si,Fe^{3+})$_4$O$_{10}$](OH,S)$_2$

Named (2M$_1$) in 1967 by Pattriaratchi et al. for Ananda Kentish Coomaraswamy (1877–1947), first director of the Mineral Survey of Ceylon (Sri Lanka). Polytype 2O described in 1972 by Giuseppetti and Tadini. Mica group, margarite subgroup. 2M$_1$: MON $C2/c$. $a = 5.412$, $b = 9.434$, $c = 19.953$, $\beta = 94.87°$, $Z = 4$, $D = 3.97$. *19-78:* 9.92$_6$ 4.99$_9$ 3.32$_{10}$ 2.72$_5$ 2.68$_4$ 2.49$_8$ 1.991$_3$ 1.660$_3$. 2O: ORTH *Pnmn*. $a = 5.439$, $b = 9.509$, $c = 19.878$, $Z = 4$, $D = 4.22$. *38-423:* 3.43$_4$ 3.31$_3$ 3.04$_4$ 2.71$_{10}$ 2.64$_4$ 2.11$_4$ 1.607$_4$ 1.357$_3$. Structure: *AM 70:1298(1985)*. Detailed structural data are available for the 2O polytype, the more common of the two. The usual *Ccmm* symmetry of the standard 2O polytype is reduced to *Pnmn*, due to the ordering of Fe^{3+} and Si in tetrahedral sites, Fe^{2+} and Mg in octahedral sites, and the (OH) and S anions. The 1M polytype has also been reported from the type locality. Black, nearly opaque. Cleavage {001}, perfect. H = 3–4. G = 3.94. Biaxial (+?); $N_y = 1.855$, $N_z > 1.88$; $Y = b$, $Z \wedge a = 12°$. Pleochroic; Y, green; Z, brown. Sulfur occupies about one-half of the sites normally occupied by (OH), and significant chlorine and minor fluorine are also present; Fe^{3+} occupies approximately one-third of the tetrahedral positions as well as $\approx 10\%$ of the octahedral sites. Tentatively identified in the Rush Creek–Big Creek area, Fresno Co., CA. Both polytypes are found as 1- to 5-cm-thick bands and lenses in iron ore at the *Wilagedera prospect, North Western Prov., Sri Lanka*, where it is associated with magnetite, chalcopyrite, pyrite, and pyrrhotite. AR *MM 36:1(1967), TMPM 18:169(1972)*.

71.2.2c.5 Kinoshitalite-1M   (Ba,K)(Mg,Mn^{2+},Al)$_3$[Al$_2$Si$_2$O$_{10}$](OH)$_2$

Named in 1973 by Yoshii et al. for Kameki Kinoshita (1896–1974), Japanese geologist at Kyushu University and investigator of ore deposits of Japan. Mica group, margarite subgroup. MON $C2/m$. $a = 5.345$, $b = 9.250$, $c = 10.256$, $\beta = 99.99°$, $Z = 2$, $D = 3.33$. *29-180:* 10.1$_4$ 5.05$_5$ 3.37$_{10}$ 3.16$_1$ 2.93$_1$ 2.52$_5$ 2.02$_5$ 1.684$_1$. Structure: *MJJ 12:1(1984)*. The structure is of 1M polytype. Yellow-brown, vitreous luster, translucent. Cleavage {001}, perfect. H = 2½–3. G = 3.30. Biaxial (−); 2V = 23°; $N_x = 1.619$, $N_y = 1.633$, $N_z = 1.635$. Weakly pleochroic; X, very light yellow; Y,Z, yellow. Type material contains (wt %): BaO, 17.8; K$_2$O, 3.3; MnO, 7.4; Mn$_2$O$_3$, 3.2. Reported from Trumbull Peak, Mariposa Co., CA, and at Netra, Balghat district, Madhya Pradesh, India. Found at Hokkejino, Kyoto Pref., Honshu, and in the *Misago orebody, Noda-Tamagawa mine, Iwate Pref., Honshu, Japan*, where it occurs as < 1-mm scales associated with celsian, quartz, spessartine, rhodonite, chalcopyrite, and pyrrhotite in hausmannite–tephroite ore. AR *AM* 60:486(1975), 74:200(1989).

71.2.2d.1 Hydrobiotite [K(Mg,Fe)$_3$Si$_3$AlO$_{10}$(OH)$_2$/{Mg$_{0.35\pm}$(H$_3$O$_{16,12,8,3,0}$)}(Mg,Fe)$_3$Si$_3$AlO$_{10}$(OH)$_2$]

Named in 1882 by Schrauf for its relationship to biotite; redefined in 1980. Mica group. Synonyms: hydromica, jefferisite, vermiculite (71.2.2d.3). ORTH Space group unknown. $a = 5.2$, $b = 9.0$, $c = 23.3$–25.5. Also given as HEX $Z = 1?$. *PD*(oriented flake): 24.3$_1$ 12.23$_6$ 8.27$_{<1}$ 4.91$_3$ 3.49$_5$ 3.07$_{1.5}$ 2.73$_2$ 2.04$_2$; *PD*(oriented powder): 23.6$_2$ 12.26$_{10}$ 8.18$_{<1}$ 4.88$_{1.5}$ 3.50$_5$ 3.05$_2$ 2.73$_1$ 2.04$_{1.5}$. *AM 68:420(1983)*. Hydrobiotite is a regularly interstratified 24- to 25-Å phyllosilicate with biotite and trioctahedral vermiculite layers. X-ray diffraction data nearly nonexistent except for basal reflections. **Habit:** Generally fine-grained, then earthy to indurated, sometimes 2–5 μm, but can be to several centimeters. **Physical properties:** Golden yellow to golden brown, rarely brownish black; white to tan streak, sometimes stained; dull to waxy luster, also pearly to bronzy on cleavages. Cleavage {001}, perfect; flexible, but inelastic, almost brittle. Partly decomposed by strong mineral acids. H = 1–2. **Chemistry:** The chemical distinction between hydrobiotite and biotite or vermiculite resides in the intermediate water contents and intermediate K$_2$O and Fe$_2$O$_3$ values compared with pure biotite or pure vermiculite. The mica layer is typically ferruginous phlogopite. *AM 68:420(1983)*. The vermiculite layer is probably hypersilicic, as with vermiculite proper. CEC of hydrobiotite is about 50–75 meq per 100 g, ammonium is absorbed from soils. *MA 78:4853(1977)*. The anion-exchange capacity should be about 2 meq per 100 g. Interlayer charge deficiency could be slightly compensated by conversion of OH$^-$ to O^{2-}. Many of the data on "hydrobiotite" were collected on materials without demonstrating their 1:1 interstratification requirement. **Optics:** Biaxial (–); N$_y$ = 1.560–1.562, N$_z$ = 1.565–1.567; 2V ≈ 10°; birefringence = 0.045–0.055; yellow-brown, brown, greenish brown, pleochroic in yellows, browns, or greens. Optical data on hydrobiotite are insufficient. **Occurrence:** Found in a wide variety of sedimentary environments and frequently coexisting with a variety of clay minerals, including biotite, vermiculite, smectites, chlorites, and so on. Hydrobiotite is frequently an important-to-exclusive component of commercial "vermiculite." Hydrobiotite has been synthesized from vermiculite. *CCM 41:580(1993)*. **Tests:** Hydrobiotite d$_{001}$ collapses to 10 Å when heated to 500° with disappearance of 14- and 7-Å peaks. The d$_{001}$ of hydrobiotite should not change more than negligibly upon glycolation, but smectite can be randomly present, giving false values. K-intersalination collapses d$_{001}$ to 10 Å with loss of 14- and 7-Å peaks. The original locality does not contain a predominance of hydrobiotite as redefined, but contains mostly vermiculite and biotite mixed with a small quantity of hydrobiotite. **Localities:** Well-studied hydrobiotite localities include: Brintons quarry, West Chester, Chester Co., PA; Enoree, Spartanburg Co., SC; Rainy Creek deposit, Libby, Lincoln Co., MT; Chanon granite, Montebras, Creuse, France; Fosso Pisciarello and Fosso di Monte Acuto, Monti Ernici, Lazio, Italy; Haugabreen and Austerdalsbreen regions, Norway; Schoeninger Mt., Stupná near Křemža, Czech Republic; northwestern and northeastern Transvaal, South Africa. VK, EF

71.2.2d.2 Illite $[K,(H_3O^+)]Al_2[Si_3AlO_{10}](OH)_2$

Named in 1937 by Grim et al. for the locality. Synonyms: hydromuscovite, hydromica, gümbelite. MON $C2/m$, $C2/c$, or Cc. $a = 5.19$–5.225, $b = 8.950$–9.020, $c = 9.95$–10.447, $\beta = 94.87$–$95.18°$, $(101.47$–$101.68°)$, $Z = 2, 4$, $D = 2.82$–2.61. 26-911($2M_1$): 3.34_{10} 10.00_9 2.005_5 5.02_5 2.99_2 4.48_2 3.20_1 4.44_1; 29-1496($1M$): 10.7_4 5.0_3 4.43_{10} 3.66_4 3.33_4 3.06_4 2.56_9 1.50_4. Illite is characterized as a 10-Å phyllosilicate and is generally a dioctahedral degradation product of muscovite, while trioctahedral "illites" are usually hydrobiotites; extensive alteration of illite can produce a dioctahedral vermiculite (HIV-vermiculite, 71.2.2d.3). *SSAJ 52:1486,1808(1988)*, *CCM 40:32(1992)*. The illite structure is essentially the same as muscovite: two infinite sheets of pseudohexagonal rings of silicon–aluminum tetrahedra with unpolymerized apices directed toward and forming the coordination of a central set of octahedral sites occupied by aluminum, but some structural Mg has been suggested to be helpful in the generation of illite. The 2:1 structure has a net negative charge that is compensated by interlayer cations—originally potassium, but subsequently partially replaced by H_3O^+. The interlayer potassium prevents significant intercalation by water or hydrated cations; however, intercalation by organic liquids has been accomplished. *AM 54:858(1969)*. Revised polytype determination methods exist. *CCM 41:45 389(1993)*. The most common polytype is 1Md with $2M_1$ and $2M_2$ very much less common and 3T rare. Layer stacking is disordered and reflections are broadened, but basal reflections are strong to distinct. Illite crystallinity decreases with lower K_2O. Frequently interstratified with smectite, chlorite, and others. **Habit:** Generally fine-grained, sometimes 2–5 µm, earthy to indurated. **Physical properties:** Gray-white, silvery-gray, gray-green, sometimes stained; white streak; dull to waxy luster. Cleavage {001}, perfect; flexible, but inelastic. H = 1–2. G = 2.79–2.8. Decomposed by strong mineral acids. **Chemistry:** Illite is $X_zY_3[(Si_{4-z}Al_z)O_{10}](OH,O)_2$; z is generally 0.5–0.7 and therefore might be hypersilicic. X, the interlayer site, is generally partly occupied by K and H_3O^+ : K_2O is generally in the range 6–8 wt % and therefore usually distinctly lower than normal muscovite, and very low values are suggestive of interstratification by nonillitic layers, while illitic muscovite with 11.00 wt % K_2O has been reported from a Texas shale. Y is virtually always occupied by Al, but Mg, Fe^{2+}, and Fe^{3+} have been reported. The cation-exchange capacity of illite is about 10–40 meq per 100 g, while the anion-exchange capacity is generally negligible. Cation ion-exchange values higher than 15 might indicate the presence of nonillite layers. Interlayer charge deficiency can be slightly compensated by conversion of OH^- to O^{2-}. A series probably extends, by Na substitution, to brammalite. Minor substitutions* include (wt %): TiO_2, 1.92 (Grès à Voltzia); Fe_2O_3, 4.99 (Fithian); FeO, 2.15 (Fithian); MnO, 1.58 (Grès à Voltzia); MgO, 4.2 (Grès à Voltzia); CaO, 3.0 (Grès à Voltzia); Cr_2O_3, 14.6 (Takova); $(NH_4)_2O$, 4.90 (calculated, Llewellyn). Trace elements found (in ppm, Grès à Voltzia) include: Sr, 198; Ba, 487; V, 178; Ni, 71; Co, 13; Cr, 126; B, 251; Zn, 6; Ga, 39; Cu, 87; Pb, 45; Li, 93; Rb, 221. Can be found naturally intercalated by small amounts of "humic" molecules. (*Many illite analyses represent whole-rock or whole-sediment compositions. Due to concerns of sample species purity, analyses of illites must be made on well-characterized materials. Many high-Fe_2O_3 or high-MgO samples could contain hydrobiotite, etc.). **Optics**: Biaxial (−); $N_x \approx$ perpendicular to (001), most have $N_x = 1.535–1.572$, $N_y = 1.555–1.600$, $N_z = 1.565–1.605$; 2V generally $< 10°$ rangingto25°;birefringence $= 0.025–0.037$;colorless,nonpleochroic.Dehydrated illite has slightly higher RI. **Occurrence**: Found in a wide variety of environments and frequently coexisting with a wide variety of minerals, including other micas, smectites, kaolinites, chlorites, vermiculite, and so on. Illite appears to have been formed by weathering or hydrothermal alteration of muscovite–phengite, but some illite is authigenic or could be derived from alteration of K-feldspars or recrystallization of smectites. Common in sediments, underclays, marls, shales, some slates. Illite from oil well drill cuttings increases in proportion with depth, apparently at the expense of kaolinite, and so on. Interstratified illite–smectite converts to illite at depth. *CCM 41:26,119,134(1993)*. Low-grade metamorphism can convert illite to true muscovite. Observed altered to smectite or dioctahedral HIV–vermiculite. **Tests**: Illite does not change d values when heated to 500°; basal spacings can be broad with widths $= 0.6–1° \, 2\theta$, d_{100} is asymmetrical with a low-angle "tail"; a slight basal peak shift is detected with glycol solvation. Sometimes confused with muscovite, sericite, glauconite, phengite, and others. **Localities**: Abundant worldwide; only selected localities with well-characterized materials are given here. Found at State College, Centre Co., Llewellyn and Pottsville formations, PA; *Maquoketa shale, Gilead, Calhoun Co.*, Salt Fork Creek, Fithian, Vermilion Co., Illinois Clay Products mine, Goose Lake area, Grundy Co., IL; Ouachita Mts., OK; Point Chevrecil and Atchafalaya Bay, LA; Marblehead, Fond du Lac Co., WI; Glacier National Park, MT; St. Austell, Cornwall, England; Ballater, Aberdeenshire, Scotland; Ogofau, Wales; Grès à Voltzia sandstone, Vosges Mts., France; Göschwitz, Thuringia, Germany; Saraspatak, Hungary; *Takova, Yugoslavia*; Candeias formation, Bahia, Brazil; Eureka mine, Cordoba Prov. and Barker, Buenos Aires Prov., Argentina; Nile delta, Egypt; Al-Habbaniya Lake, Iraq; Dhauladhar Mts., Siwaliks, India; Ichinomiya, Hyogo Pref., Japan; Tengchong, Yunnan Prov., China; Bidor, Perak, Indonesia. VK, EF

71.2.2d.3 Vermiculite [Mg$_{0.35\pm}$(H$_2$O)$_{16,12,8,3,0}$](Mg,Fe^{3+})$_3$ [Si$_3$(Al,Fe^{3+})O$_{10}$](OH)$_2$

Named in 1824 by Webb for *vermiculor*, in allusion to the swelling property of the mineral: "I breed or produce worms." Synonyms: hydromica, jefferisite, hydrobiotite (71.2.2d.1). MON $C2/m$, $C2/c$, $C2$, or Cc. $a = 5.326$–5.358, $b = 9.18$–9.28, c (for Mg hydrates) $= 20.6$, 14.81, 14.36, 13.82, 11.59, 9.02, $\beta = 96.56$–$97.12°$; for 2M vermiculite: $a = 5.24$, $b = 9.17$, $c = 28.60$, $\beta = 94.6°$, $Z = 2$, $D = 2.26$. *16-613*(2M nat): 14.2_{10} 4.57_6 2.85_3 2.61_5 2.57_5 2.52_5 2.36–2.38_4 1.53_7. Vermiculite is virtually always a trioctahedrally occupied 2:1 mica structure with interlayer water and generally a divalent, cation-exchangeable interlayer ion, usually Mg, rarely Ca. The interlayer cation is increasingly shown to be Al, which is nonexchangeable. Due to the inadequacy of the definition of vermiculite to accommodate the nonexchangeable variety, the term *HIV-vermiculite* (hydroxy-interlayer vermiculite) is sometimes used. *CCM 40:335(1992), SSAJ 52:1486,1808(1988)*. The two-layer hydrate provides octahedral coordination for the interlayer cation. Fine-grained dioctahedral "vermiculite" is known. Vermiculites have a 2:1 layer charge $= -0.6$ to -1.5. The distinction between vermiculite and smectite has been placed at a layer charge of -0.6, but the significance of layer charges between -0.5 and -0.7 is not obvious. *CCM 36:184(1988)*. Vermiculite stacking sequences are usually, at least partially, disordered. Additionally, interlayer cations are less regularly ordered than in the presumed parent mica; interlayer cations are fewer than the available sites and are frequently situated asymmetrically in the clefts of opposing bases of tetrahedra (usually occupied by Al), with repulsion effects dictating that occupied sites be interspersed with vacancies. Interlayer cations do not appear to migrate upon dehydration, although several interlayer cation sites exist. Domain-ordered Mg-hydrate vermiculite has been recognized, while some linear regularity of interlayer Ca has been suggested. Intercalated organic molecules show ordering. *AM 55:1550(1970); CCM 32:223(1984), 40:240(1992)*. Parent mica polytypes are usually 1M, rarely 2M$_1$, 2M$_2$, or 3T, with interlayer cations of the vermiculite partly controlling the derivative polytype or stacking mode. *AM 69:237(1984)*. Mössbauer spectroscopy indicated Phalaborwa vermiculite has 70% of its total Fe tetrahedral: ^{IV}Fe$^{3+}_{1.24}$ vs. ^{VI}Fe$^{3+}_{0.39}$ + ^{VI}Fe$^{2+}_{0.08}$. *CCM 39:467(1991)*. Partial interlayer water is easily lost or gained due to low or high relative humidity, respectively. Rapid intense heating (300°) of hydrated vermiculite generates interlayer steam, resulting in enormous volume expansion of the aggregate sample; structural Fe^{3+} sometimes partly reduced by dithionite–citrate–bicarbonate treatment; and dry grinding of vermiculite induces disorder of stacking. Vermiculite d$_{001}$ is usually sensitive to external relative humidity and thermal conditions (see HIV-vermiculite), but the water exchangeability is also inversely proportional to grain size (d$_{001}$ = 11.59 Å dehydrates very rapidly in dry air). When in a water-saturated medium, d$_{001\ max}$ = 14.81 Å (indicating 16–13 interlayer H$_2$O), while vermiculite in a dry medium is considered fully hydrated (12–9 H$_2$O) when d$_{001}$ = 14.36 Å . At lower relative humidity and/or higher temperature, vermiculite can dehydrate to $d_{001} = 13.82$ Å (8–9 H_2O) with an ordered double water molecule interlayer sheet, then to $d_{001} = 11.59$ Å (8–3 H_2O), indicating a single water molecule layer. Near dehydration (less than 3 H_2O), alternate layers contain single water molecule layers and water-free layers yielding $d_{001} = 20.6$ Å before becoming exclusively dehydrated with $d_{001} = 9.02$ Å. The actual interlayer region occupied by water is 4.98 Å. The double hydrate interlayer superficially resembles the interlayer of chlorites, but with only one-sixth, or less, of the cation sites and two-thirds of the anion sites occupied. Hydrogen bonds link water molecules to silicate ring bases. During dehydration, b changes in stepwise fashion as d_{001} collapse occurs. Cation-exchanged vermiculites show spacings related to the kind or absence of interlayer hydration associated with the exchanged cation: Mg, 14.33–14.39 Å; Ca, 15.07 Å; Sr, 15.0 Å; Ba, 12.3–12.56 Å; Li, 12.2–12.56 Å; Na, 14.8–12.56 Å; K, 10.6–10.42 Å; NH_4, 10.8–11.24 Å; Rb, 11.24 Å; Cs, 11.97 Å. *AM 33:655(1948)*. TGA data: e.g., *CM 14:399(1976)*. Vermiculites which show a resistance to d_{100} collapse are usually attributed to a propping effect of hydroxy-interlayer polymers (HIV; $[Al_{13}O_4(OH)_{24}(H_2O)_{12}]^{7+}$). Vermiculite sheets regularly interstratified with other structures are very common and have been treated as separate materials (e.g., vermiculite–biotite = hydrobiotite; vermiculite–smectite = HLC corrensite; vermiculite–chlorite = LLC corrensite). **Physical properties:** Generally foliated, crystals rare. Flaky (many centimeters) to fine-grained earthy, sometimes 2–5 μm. Gray-white, golden brown, green to olive-green to brownish black; white streak; oily to earthy luster, frequently bronzy. H = 1–2. G = 2.28–2.77. Cleavage {001}, perfect; flexible, but generally inelastic. **Chemistry:** Vermiculites can be chemically indistinguishable from some smectites or can range to hydrobiotite. A general formula for trioctahedral vermiculite is $X_{1-z}Y_3[(Si_{4-z}[Al, Fe^{3+}]_z)O_{10}](OH)_2$. z is generally 0.5–0.7. X, the interlayer site, is generally occupied by Mg plus a double (16–9 H_2O) or single water layer (8–3 H_2O). Y can be occupied by Mg, Fe^{2+}, Fe^{3+}, Al, and so on. The cation-exchange capacity of vermiculite is about 100–150, to 260 meq per 100 g, while the anion-exchange capacity is about 4 meq. While much vermiculite contains dominant interlayer Mg, the interlayer composition can vary due to exchange similar to smectites. Ferric iron is almost invariably present, several wt % to 19.22 (Ukraine), while FeO is generally small, generally < 1.0 to 8.61 wt % (Westcliffe). Leaching of original interlayer K^+ in micas is usually compensated by interlayer Mg and is additionally compensated by low-temperature conversion of ferrous to ferric iron and/or conversion of hydroxyl to O^{2-}. CaO can replace interlayer MgO, generally < 1 wt %, but to 2.70 wt % (Saskatchewan). Additional substitutions (wt %) include TiO_2, 2.23 (Saskatchewan); NiO, 11.25 (Webster); MnO, 0.22 (Maaninka); Na_2O, 0.39 (Röhrenhof); K_2O, 0.43 (Westcliffe); F, 0.05 (Röhrenhof). Vermiculites with natural 0.45 interlayer Ba (Llano) and 0.30 Ca (Malawi) have been observed. **Optics:** RI of vermiculite increases rapidly with increasing iron contents. Biaxial to pseudo-uniaxial, X approximately perpendicular to (001), sometimes pleochroic, green-brown. RI has been related to total iron

(DHZ). Uniaxial $(-)$; $N_O = 1.540$–1.648, $N_E = 1.517$–1.622; birefringence $= 0.019$–0.026 (rarely to 0.047). Biaxial $(-)$; $N_x = 1.520$–1.622, $N_y = 1.530$–1.648, $N_z = 1.530$–1.651; $2V = 0$–$8°$, up to $18°$; birefringence $= 0.020$–0.029 (rarely to 0.047) ($Fe_2O_3 = 4.2$–19.2 wt %). **Occurrence:** Found in a wide variety of environments and frequently coexisting with a wide variety of minerals. Potassium fixation by vermiculite in soils is common. Frequently found as alteration fringes on phlogopite or hydrobiotite. Vermiculite appears to have usually formed by weathering or hydrothermal alteration of iron-bearing phlogopite or annite at temperatures generally $300°$ or below and frequently at surface conditions, usually at acid pH. HIV-vermiculite appears to form directly from a muscovite precursor. *CCM 40:32(1992)*. Observed to form from amphibole, rarely clinochrysotile. Mg has been suggested to diffuse from octahedra in the 2:1 layer into the interlayer. *JG 65:603(1957)*. Also found in a contact zone near felsic intrusives, including pegmatites. Parent rocks include basic and ultrabasic rocks, including kimberlites, also gneisses, schists, or Mg skarns. In some carbonatites, meta-limestone, and others. Uncommon in marine sediments due to frequent K^+ in pore fluids. Observed to alter to chlorite or clinochrysotile. *CCM 7:135(1960), 18:213(1970)*. **Tests:** Stepwise dehydration 14.81–9.02 Å characteristic; d_{100} more intense than lower orders (unlike chlorite or HIV-vermiculite); $d = 7$ Å remains on heating to $110°$; vermiculite is not permanently affected by heating until about $550°$; intercalation with glycol generally does not expand Mg-rich vermiculite beyond 14.5 Å; generally contracts, almost irreversibly, to $d = 10^+$ Å with K^+ intercalation. HIV-vermiculite 14-Å peaks broaden and shift to ~ 12 Å upon heating. Intercalation of n-alkylammonium ions has been used to estimate interlayer charge. *CCM 40:240(1992)*. **Localities:** Only localities with well-characterized and/or abundant amounts of vermiculite are mentioned here. Found at *Millbury, Worcester Co., MA*; Kent, CT; Brintons quarry, West Chester, Chester Co., PA; Bare Hills and Pilot, MD; Culsagee mine, Macon Co., Webster, NC; Tigerville, Greenville Co., coastal plain sediments, FL; Magnet Cove, AR; Texas magnesite quarry, Llano Co., TX; Westcliffe, Custer Co., CO; Rainy Creek mine, Libby, Lincoln Co., MT; Gold Butte, Virgin Mts., Clark Co., NV; Santa Cruz mine, Pinal Co., AZ; Stanleyville, Lanark Co., ONT, gray luvisol profile, SAS, Canada; Cerro de Pedregosa, Coahuila, Mexico; Prayssac, Aveyron, France; La Garrenchosa, Santa Olalla, Huelva; Benahavis and Ojen massif, Serrania de Ronda, Malaga; Ronquillo and Real de la Jara, Seville, Spain; Maaninka, Posio, Finland; Junosuando, Pajala, Sweden; Röhrenhof, Fichtberg, Bavaria, Kropfmühl bei Passau, Germany; Nova Ves, Czech Republic; Batn El-Ghoul, Jordan; Kovdorsk, Murmansk Oblast, Russia; Paulistana, Piaui; Catalao Deposit, Goias; and Pernambuco, Brazil; Wadi as Shati' deposit, Fazzan, Libya; Kapirikamodzi, Malawi; Kibara Formation, Shaba, Zaire; Palabora mine, Phalaborwa and Loolekop, Transvaal, South Africa; Pauni mine, Pauni, Maharashtra; and Sevvattur, Tiruppattur Taluk, Tamil Nadu, India; Matsusaka, Japan; Mud Tank deposit, Strangways Mts., Australia. VK, EF *CCM 35:203(1987), 39:174(1991), 36:481(1988); AM 60:175(1975)*.

71.2.2d.4 Brammallite [Na,(H$_3$O$^+$)]Al$_2$[Si$_3$AlO$_{10}$](OH)$_2$

Named in 1943 by Bannister for Alfred Brammall (b.1879), British geologist and mineralogist. Synonyms: sodium hydromica, sodium illite, hydroparagonite. MON $C2/c$. $a = 5.12$–5.2, $b = 8.91$–9.0, $c = 19.2$–19.26, $\beta = 95.83°$, $Z = 4$, $D = 2.83$–2.88. 27-0020(2M$_1$): 9.77$_{10}$ 4.78$_8$ 4.41$_9$ 3.17$_{10}$ 2.54$_9$ 2.41$_6$ 2.11$_6$ 1.485$_{10}$. Brammallite is believed to be a degradation product of paragonite. The structure has not been determined but is believed to be essentially the same as that of paragonite: two infinite sheets of pseudohexagonal Si/Al rings with the unpolymerized apices directed toward and forming the coordination of a central set of octahedral sites occupied by Al. This 2:1 structure has a net negative charge which is compensated by interlayer cations—mostly Na, but partially replaced by H$_3$O$^+$. Interlayer Na prevents significant intercalation by water or hydrated cations, but brammallite should behave similarly to illite. Generally fine-grained to visibly fibrous, sometimes 2–5 μm, earthy to indurated, sometimes unctuous. White to pale greenish white, white streak, dull to waxy luster. Cleavage {001}, perfect; flexible. H = 1–2. G = 2.69. Brammallites are generally intermediate in chemical composition (wt %): Na$_2$O, 4.66–5.85; K$_2$O, 2.58–2.70 or less. Significant substitutions (wt %) are few: Fe$_2$O$_3$, 2.17 (Missouri); P$_2$O$_5$, 0.45 (Kazakhstan). Minor elements detected include: Mn, Ni, Co, Ti, V, Zr, Cu, Pb, and Ga (all < 0.001 wt %, Kazakhstan). Data for brammallite are few. CEC probably similar to illite and interlayer charge deficiency could be compensated by conversion of OH$^-$ to O^{2-}. A series probably extends, by K substitution, to illite. A mineral attributed to brammallite (Pilot Knob) is given as (Na$_{0.66}$, K$_{0.23}$)(Al$_{1.04}$, Fe$^{3+}_{0.01}$)[(Si$_{3.07}$Al$_{0.94}$)O$_{10}$(OH)$_{2.51}$. *AM 20:384(1935)*. DTA (Pilot Knob) shows 0.3–0.4 wt % H$_2$O loss at 100–120° and 0.6–0.7 wt % loss by ∼ 375°; similar results found at different temperature for Kazakhstan brammallite. Biaxial (−); N$_x$ = 1.561–1.567, N$_y$ = 1.58, N$_z$ = 1.579–1.585; 2V large; birefringence = 0.018°; nonpleochroic, colorless. Found in crevices in shale or weathered conglomerate (Missouri). Brammallite could have been formed by weathering or hydrothermal alteration of paragonite. Occurs at Pittsburgh coal seam, Pursglove, Monongalia Co., WV; Pilot Knob, Ironton, MO; *Llandebie, Dyfed, Wales*; Ordovician mudstone, northern Kazakhstan. VK, EF *MM 26:304(1943), DANS-ESS 208:157(1973)*.

71.2.3.1 Cuprorivaite CaCu[Si$_4$O$_{10}$]

Named in 1938 by Minguzzi for the composition and for Carol Riva (1872–1902), mineralogist. Equivalent to the synthetic pigment Egyptian blue. TET $P4/ncc$. (Syn): $a = 7.30$, $c = 15.12$, $Z = 4$, $D = 3.09$. 12-512(syn): 7.63$_4$ 3.78$_9$ 3.36$_8$ 3.29$_{10}$ 3.19$_5$ 3.00$_9$ 2.27$_5$ 1.831$_6$. Structure: *AC 12:733(1959)*. Silicate sheets lie parallel to {001}. Synthetic material habit is tabular {001} modified by {110}, and {102}; only {001} observed, but rarely, on natural material. Blue, vitreous luster, translucent. Cleavage {001}, perfect; brittle. H = 5. G = 2.87–3.08. Uniaxial (−); N$_O$ = 1.627–1.633, N$_E$ = 1.590. Pleochroic; O, blue; E, pale rose to pale yellow. Insoluble in HCl. Natural material slightly hydrated in part. Found at Summit Rock, Douglas Co., OR; at Wheal Edward, Cornwall England; as minute, crude crystals with quartz at Mt. Vesuvius, Campania, Italy; the Eifel district, Germany; at Messina, Transvaal, South Africa. AR *AM 47:409(1962)*.

71.2.3.2 Gillespite BaFeSi$_4$O$_{10}$

Named in 1922 by W. T. Schaller for Frank Gillespie, who found the first specimen near his mining claim in Alaska. At high pressure (18–45 kbar) transforms to an orthorhombic phase. See cuprorivaite (71.2.3.1), the Ca–Cu analog. TET $P4/ncc$. $a = 7.5164$–7.522, $c = 16.0768$–16.082, $Z = 4$, $D = 3.40$. *37-472*(nat): 8.06$_{10}$ 4.44$_4$ 4.02$_4$ 3.55$_4$ 3.41$_6$ 3.21$_8$ 2.75$_3$ 2.39$_5$. A layer silicate with four-membered rings of SiO$_4^{4-}$ tetrahedra forming the basic building blocks. *AM 59:1166(1974)*. Micaceous scales and in aggregates; aggregates to 6 cm at Big Creek, CA. Red, rose; vitreous luster. Cleavage {001}, good; {100}, poor; {110}, indistinct. H = 4. G = 3.33–3.40. Spectral data: *JCP 47:4240(1967)*. Readily decomposed by HCl. Traces of Mn + Ti. Uniaxial (−); N$_O$ = 1.619–1.621, N$_E$ = 1.617–1.619. O, colorless; E, red. Synthetic has N$_O$ = 1.620 and N$_E$ = 1.617. Associated with sanbornite and other Ba-silicates at five localities in California and elsewhere. In contact zones of granitic intrusives, and metasomatic rocks. From a moraine in Alaska. Occurs at Trumbull Peak, Incline, Mariposa Co., CA; Rush Creek, Fresno Co., CA; crystals to 6 cm or more at Big Creek, Fresno Co., CA; *Dry Delta, Alaska Range, Alaska (100 miles SE of Fairbanks)*; Ross R., Yukon, and Gunn claim, Itsi Mtns, deposit, YT, Canada; La Madrelena mine, Tres Pozos, and Rosario, Baja California Norte, Mexico. EF *AM 69:358(1984)*, *MIN 4(2):454(1992)*.

71.2.3.3 Effenbergerite BaCu[Si$_4$O$_{10}$]

Described in 1994 by Giester and Rieck and named for Herta S. Effenberger, (b. 1954), mineralogist, University of Vienna, Austria. See cuprorivaite (71.2.3.1), CaCu[Si$_4$O$_{10}$], and gillespite (71.2.3.2), BaFe[Si$_4$O$_{10}$]. TET $P4/ncc$. $a = 7.422$, $c = 16.133$, $Z = 4$, $D = 3.52$. *PD*: 8.06$_{10}$ 4.03$_4$ 3.54$_3$ 3.20$_4$ 2.69$_2$ 2.39$_4$ 2.02$_3$ 1.95$_2$. The structure consists of silicate sheets [Si$_8$O$_{20}$]$^{8-}$ parallel to (001) formed by corner linkage of silicate four-membered rings. The copper(II) atom is nearly planar four-coordinated; the barium atom has a distorted cube-like environment of oxygen atoms. *MM 58:663(1994)*. The mineral occurs as transparent blue ditetragonal dipyramidal platelets with a perfect cleavage parallel to {001} in sizes up to 8.0 mm × 8.0 mm × 0.1 mm; one very poor cleavage parallel to {110}. Forms: {001}, {100}, {110}, {102}. No twinning observed. Pale blue streak, vitreous luster on cleavage planes, resinous on crystal faces, and subvitreous on fracture surfaces. Subconchoidal fracture, brittle. H = 4–5. G = 3.57. Nonfluorescent. Insoluble in all common acids except HF. IR absorption spectrum given contains minor amounts of Al$_2$O$_3$: 0.4–0.7 wt %. Uniaxial (−); N$_O$ = 1.633, N$_E$ = 1.593, r > v, weak. Strongly pleochroic; O, intense blue; E, colorless. A hydrothermal mineral, effenbergerite occurs in the central-eastern orebody of the *Wessels mine in the Kalahari manganese field, northwestern Cape Prov., South Africa*. A matrix of braunite, sugilite, and hausmannite is cut by pectolite veinlets containing effenbergerite. Associated minerals are native copper, calcite, quartz, clinozoisite and two unknown species. EF

71.2.4.1 Ferrisurite ideally $(Pb,Ca)_{2-3}(CO_3)_{1.5-2}(OH,F)_{0.5-1}$ $[(Fe,Al)_2Si_4O_{10}(OH)_2] \cdot nH_2O$

Named in 1992 by Kampf et al. for the composition and as an analogue of surite. Compare: surite. MON (pseudo-ORTH). $P2$ or $P2_1/m$. $a = 5.241$Å, $b = 9.076$Å, $c = 16.23$Å, $\beta = 90.03°$, $Z = 2$, $D = 3.89$ (nat.) 16.1_4 $5.40_{2.5}$ 4.53_{10} $3.73_{3.5}$ 3.24_9 2.61_8 2.27_5 $1.71_{2.5}$ (AM 77:1107(1992). Structure: consists of smectite layers between which are cerussite- and hydrocerussite-like regions of variable composition (interlayer). Forest green, transparent; greenish yellow to olive green; fibrous, {010} is prominent; occurs in quartz and cerussite, crystals to 2 mm long on a, 0.001 mm on $b \times 0.04$ on c; no twinning observed, fibers are very flexible; silky; $H = 2-2\frac{1}{2}$; $G = 4.01$; IR data = (AM 77:1107(1992)); no fluorescence observed in SW or LW; effervesces in cold 1:1 HCL leaving a gelatinous residue, Analysis (wt. %) Na_2O 0.3, MgO 0.2, CaO 3.4, FeO 0.8, BaO 0.1, PbO 42.7, Al_2O_3 3.2, Fe_2O_3 10.5, SiO_2 26.6, F 0.8, CO_2 8.2, H_2O 3.5, = 100.3. Biaxial (+); $2V = 76°_{calc}$; $N_x = 1.757$, $N_y = 1.763$, $N_z = 1.773$; X = yellow, Y = brown, Z = light green; XYZ = cba. Occurs in an oxidized lead deposit in contact metamorphosed and faulted shaly limestone. Associated with galena, pyrite, chalcopyrite, covellite, chalcocite, quartz, calcite, hematite, cerussite, mimetite, wulfenite and malachite. Early stage oxidation mineral. Occurs at the *Shirley Ann claim, Ubehebe district, Inyo Co., CA*. EF, VK.

71.2.4.2 Surite $(Pb,Ca)_{2-3}(CO_3)_{1.5-2}(OH,F)_{0.5-1}[(Al,Fe)_2(Si,Al)_4O_{10}OH_2)] \cdot nH_2O$

Named in 1978 by Hayase et al. for the locality. Compare: ferrisurite. MON $P2_1$ or $P2_1/m$. $a = 5.241$Å, $b = 8.95$, $c = 16.20$, $\beta = 90.0$, $Z = 2$; $D = 3.99$; PD (no ICDD card data), (nat.): 16.2_7 5.4_7 4.48_2 4.05_{10} 3.24_6 2.70 2.59_2 2.31_5 (AM 77:1107(1992). Cell reported in 1978 (AM 63:1175(1978) was: $a = 5.22$, $b = 8.97$, $c = 16.3$, $\beta = 96.1°$. Structure: heating to below 550°C causes little change, at 550°C all X-ray peaks are weakened without changes in spacings. At 650°C all peaks disappear. White to pale green; streak white; tabular, lath-shaped; compact aggregates; twinning-not stated; 1 cleavage parallel to elongation {001}; $H = 2-3$; $G = 4.0$; IR, DTA and TGA data (AM 63:1175(1978); effervesces in 0.75% HCl at 25°C, leaves gel-like residue. Na_2O 0.77, MgO 1.29, CaO 4.75, CuO 0.07, PbO 45.3, Al_2O_3 11.27, Fe_2O_3 0.41, SiO_2 23.58, CO_2 9.45, H_2O 3.72, FeO –, BaO –, F –, Σ100.63. HCl treatment removes PbO & CO_2, leaving *smectite* component. Equivalent to a 2:1 dioctahedral smectite interlayered with $PbCO_3$. Biaxial (+); 2V – not determined, $N_x = 1.693$, $N_z = 1.738$; birefringence – 0.045–0.047; parallel extinction. Occurs at the

Cruz del Sur mine, Rio Negro Prov., Argentina; in oxidation zone of Pb–Zn–Cu deposits (fissure-filling veins). Associated with common secondary ore minerals. EF

71.2.5.1 Macaulayite $Fe^{3+}_{24}Si_4O_{43}(OH)_2$

Named by Wilson et al. in 1984 for the Macaulay Institute for Soil Research. MON [C] $a = 5.038$Å, $b = 8.726$Å, $c = 36.34$Å, $\beta = 92°$, $Z = 2$, $D = 4.41$–4.52; PD 42-596 (nat.): 36.6_{10} 18.2_{10} $4.38_{1.5}$ $3.70_{2.5}$ $2.72_{3.5}$ 2.53_{10} 2.21_2 $1.46_{3.5}$. IR response has been studied. Contracts to 34Å on heating to 300°C and expands to 40–41Å with glycerol or ethylene glycol. "The ideal formula is consistent with an arrangement consisting of twelve Fe octahedral sheets bounded by two Si tetrahedral sheets, corresponding essentially to a double hematite unit terminated on both sides by silicate sheets. Adjacent basal oxygen surfaces of these sheets are only weakly bonded, allowing structural expansion and the introduction of a layer of water molecules in the interlamellar space." (MM 48:127(1984). Bright red, earthy (2 micron subangular platelets), decomposed by HF. $N > 1.734$, birefringent?, {001} cleavage probable, pleochroism uncertain, pale red to yellow. Chemistry of type material (anhydrous totals), in wt. %: FeO 84.67; Al_2O_3 4.01; SiO_2 11.32; H_2O(TGA) 7.4 (apportioned 2.5 absorbed, 3.7 edge hydroxyl and interstitial water, and 1.2 structural hydroxyl). CEC = < 6 meg/100g. Found in red patches intimately associated with kaolinite and illite derived from deeply weathered granite. *Bennachie car park, 7 km NW of Inverurie, Aberdeenshire, Scotland, UK.* EF, VK, VK, CM 16:261(1981) MM 48:127(1984) Minerali 4(2):607(1992).

SMECTITE GROUP

The smectite group minerals are phyllosilicate "clay" minerals corresponding to the general formula

$$AM_2(Si,Al)_4O_{10}(OH)_2 \cdot nH_2O$$

where

A = interlayer cation = Na, Ca, K, Li
M = octahedral t-o-t site = Al, Mg, Fe^{3+}, Cr^{3+}, Zn, Cu^{2+}

As with all phyllosilicates, the basic structural element is a *t-o-t layer* (tetrahedral–octahedral–tetrahedral layer, also called a 2:1 layer). The t-o-t layer is made up of two infinite sheets of $(Si, Al)O_4$ tetrahedra, linked together in nearly hexagonal rings, with the apical oxygens of one sheet facing the apical oxygens of the other sheet, thus creating nearly close-packed octahedral sites between the sheets. Octahedral sites may be fully occupied by divalent cations (thus producing *trioctahedral layers*), or they may be 2/3 occupied by trivalent cations (which produces *dioctahedral layers*). Subgroup designations are based on this octahedral site occupancy.

71.3.1a.1 Beidellite

By substituting Al^{3+} for Si^{4+} in the t-o-t layer tetrahedra, the oxygen atoms of the surfaces of the layer can acquire a net negative charge. Charge may also arise from other causes: substitution in the octahedral sites (e.g., Mg ↔ Al). In the smectite group minerals, total layer charge below -0.25 is rare, -0.3 to -0.35 is normal, and above -0.6 is again rare.

In the smectite group structures, the interlayer space between two t-o-t layers is occupied by water molecules and the A cations. The number of water molecules depends on the circumstances, and the spacing between the layers can vary from about 9.6 Å in the completely dehydrated situation, to about 12.5 Å for a single water molecule layer, to 14.5–15.5 Å for a two-water-molecule layer (normal air-dried), up to $\approx$ 18 Å for three water molecules.

SMECTITE GROUP: DIOCTAHEDRAL SUBGROUP

Mineral	Formula	
Beidellite	$(Na,Ca_{0.5})_{0.33}Al_2[Si_{3.5}Al_{0.5}O_{10}(OH)_2] \cdot nH_2O$	71.3.1a.1
Montmorillonite	$(Na,Ca_{0.5})_{0.33}(Al,Mg)_2[(SiAl)_4O_{10}(OH)_2] \cdot nH_2O$	71.3.1a.2
Nontronite	$(Na,Ca_{0.5})_{0.33}(Fe^{3+},Al)_2[(SiAl)_4O_{10}(OH)_2] \cdot nH_2O$	71.3.1a.3
Volkonskoite	$(Ca_{0.5})_{0.33}(Cr^{3+},Mg,Fe^{3+})_2[Si_{3.5}Al_{0.5}O_{10}(OH)_2] \cdot 4H_2O$	71.3.1a.4
Swinefordite	$(Li,Ca_{0.5}Na)_{0.33}(Li,Al,Mg)_2[(SiAl)_4O_{10}(OH,F)_2] \cdot 2H_2O$	71.3.1a.5

SMECTITE GROUP: TRIOCTAHEDRAL SUBGROUP

Mineral	Formula	
Hectorite	$(Na,Ca_{0.5})_{0.33}(Mg_{2.67}Li_{0.33})[Si_4O_{10}(F,OH)_2] \cdot nH_2O$	71.3.1b.1
Pimelite	$(Ca_{0.5},Na)_{0.33}Ni_3[Si_4O_{10}(OH)_2] \cdot nH_2O$	71.3.1b.2
Saponite	$(Ca_{0.5},Na)_{0.33}(Mg,Fe^{2+})_3[Si_{3.5}Al_{0.5}O_{10}(OH)_2] \cdot 4H_2O$	71.3.1b.3
Sauconite	$(Na,Ca_{0.5})_{0.33}Zn_3[(Si_{3.5}Al_{0.5})O_{10}(OH)_2] \cdot 4H_2O$	71.3.1b.4
Stevensite	$(Ca_{0.5})_{0.33}(Mg,Fe^{2+})_3[Si_4O_{10}(OH)_2] \cdot 4H_2O$	71.3.1b.5
Sobotkite	$(Na,Ca_{0.5})_{0.33}(Mg_2Al)[Si_3AlO_{10}(OH)_2] \cdot 5H_2O$	71.3.1b.6
Yakhontovite	$(Ca,K)_{0.5}(Cu^{2+},Fe^{3+})_{2.37}[Si_4O_{10}(OH)_2] \cdot 3H_2O$	71.3.1b.7
Zincsilite	$Zn_3[Si_4O_{10}(OH)_2] \cdot 4H_2O$	71.3.1b.8

Note: Smectites constitute a major industrial mineral group, with consumption of millions of metric tons annually, principally montmorillonite and beidellite. Many uses take advantage of the thixotropic effects, where a stationary moist smectite is somewhat rigid but rapidly transforms to a low-viscosity fluid with vibration. Uses include drilling muds, oven cleaners, greases, lubricants, toothpaste, mold coating, mortars, plasters, caulking compounds, coatings to improve the pelletization of iron ores and animal feeds, fillers in paints, paper, and cosmetics, and as antistick coatings.

71.3.1a.1 Beidellite $Na_{0.50}Al_2(Si_{3.50}Al_{0.50})O_{10}(OH)_2 \cdot nH_2O$

Named in 1925 by Larsen and Wherry for the locality. Last redefined in *AM* 47:137(1962). Smectite group. Synonyms: montmorillonite, Ca-montmorillonite, Na-montmorillonite, bentonite. MON Probably $C2/m$ or ORTH, sometimes pseudo-HEX $a = 5.14$–5.18, $b = 8.93$–9.00, $c = 15.0$–15.5, $\beta = 99.54°$, $Z = 1$. 19-150(18 Å, glycerol): 17.6_{10} 4.42_{10} 3.95_{10} 3.54_{10} 2.57_8 2.50_{10} 2.36_8 1.50_{10}. Beidellite has the typical smectite structure with 15-Å spacing when divalent cations are dominant in the interlayer region. While montmorillonite has its layer charge resulting from octahedral substitutions ^{VI}Al ↔ Mg and is essentially tetrasilicic with minor ^{IV}Al, beidellite, and beidellite-like smectites have considerable ^{IV}Al. Beidellites have net layer charges dominantly arising from tetrahedral substitutions, while net layer charge in montmorillonite is the result of octahedral substitutions. The pattern of ^{IV}Al vs. ^{VI}Al source of net layer charge is paralleled by other smectites and intermediate substitutional schemes are normal. IR data: e.g., *AM 70:1004(1985)*. **Physical properties:** White to buff, pale yellow, greenish yellow, light greenish gray, can be stained red or brown; white streak; earthy to waxy luster. Cleavage {001}, perfect. H = 1–2. G = 2.06–2.3. Decomposed or gelatinized by common acids. Stained blue-green by benzidine. An "average" beidellite, as generally used in the literature, was $(Ca/2_{0.31}K_{0.04}Na_{0.01})(Al_{1.34-1.98}Fe^{3+}_{0.02-0.39}Fe^{2+}_{-0-0.02}Mg_{0.01-0.28})$ $(Si_{3.26-3.73}Al_{0.27-0.74})O_{10}(OH)_2 \cdot nH_2O$. *DS 15:55(1973)*. An "end-member" formula proposed [*USGSPP:205B* (1945)] was $Na_{0.33}Al_{2.22}(Si_3Al)O_{10}(OH)_2 \cdot nH_2O$, but the closest approximation to the end member observed by those authors was $Na_{0.33}Al_{2.17}(Si_{3.17}Al_{0.87})O_{10}(OH)_2 \cdot nH_2O$. The midpoint between beidellite and montmorillonite would be represented by $Na_{0.33}Al_{2.06}(Si_{3.5}Al_{0.50})O_{10}(OH)_2 \cdot nH_2O$ if only ^{IV}Al determined the net layer charge and which was kept constant at about −0.33. Beidellite is midway between montmorillonite and sobotkite. Due to the uncertainty of how beidellite was defined, significant erosion of the beidellite definition occurred, eventually with only slightly aluminian montmorillonites being called beidellite. Beidellite was redefined [*AM 47:137(1962)*] with a layer charge arising from tetrahedral substitutions and with exactly dioctahedral occupancy. An end-member formula was proposed: $Na_{0.45}K_{0.01}(Al_{1.98}Fe^{3+}_{0.02}Mg_{0.01})$ $(Si_{3.48}Al_{0.52})O_{9.98}(OH)_2 \cdot nH_2O$. The boundary between beidellite and montmorillonite was placed at the point where net layer charge had equal or greater contribution from the tetrahedral sheet than the octahedral sheet, but beidellite does not necessarily have a total layer charge higher than montmorillonite. *DS 15:55(1973)*. Additional Al behavior was noted: "Once the threshold value of 2.0–2.1 Al is exceeded, additional Al is divided between tetrahedral and octahedral sheets in the approximate proportion of 70% tetrahedral and 30% octahedral" and "When the total amount of Al in the structure is enough to fill more than two structural positions, the amount of octahedral Al decreases by approximately 0.25. These positions are filled by the larger Mg ... and Fe ... ions." **Tests:** Identical to montmorillonite (71.3.1a.2). **Optics:** Biaxial (−); N_x = 1.493–1.503, 1.530 (2.26 wt % Fe_2O_3), N_z = 1.526–1.533, 1.556 (2.26 wt % Fe_2O_3); 2V small, 10–15°; birefringence = 0.026–0.033; dehydration increases RI and birefringence; RI increases rapidly with increasing Fe_2O_3; positive elongation; generally nonpleochroic, but feruginous varieties can show red-brown pleochroism. **Chemistry:** The chemistry of beidellite is generally aluminous, but Al_2O_3 values of 16.07–31.30 wt % have given high enough tetrahedral Al occupancy to merit the name *beidellite*, but octahedral substitutions greatly affect the total Al. Silica has a smaller range, due to its appearing in only one coordination type:

44.02–50.72 wt %. A complete series extends to nontronite with increasing Fe_2O_3, but normally ranges from a fraction to several wt %. Other substitutions observed (wt %) include: Cr_2O_3, to 5.02 (Ural Mts.); FeO, 0–0.71, 7.83 (Gedmayshikh); MnO, 0–0.49; NiO, 3.83 (WI); TiO_2, 0–1.01; NiO, 0–0.30; MgO, 0.18–6.99; CaO, 0–2.76; K_2O, 0–1.04; Na_2O, 0–1.17; P_2O_5, 0–0.15 (2.16, Putnam); H_2O^-, to 16.66; H_2O^+, to 12.90. CEC = 115–130 meq per 100 g. **Occurrence:** Major component of some altered ash flows and bentonite layers. Occurs as an alteration product in hydrothermal ore deposits. Found in a variety of sedimentary environments. Also found in some granitic pegmatites as "pocket clay." Can be found with illite, kaolinite, and other clays, quartz, amorphous silica, detrital minerals, etc. **Localities:** Although it is not always clear how the identification was arrived at, the following have been seemingly reasonably called beidellite: Pontotoc, MS; Putnam, MO; Nashville, AR; South Bosque and Houston, TX; Dever Basin, Wagon Wheel Gap, Mineral Co.; *Esperanza mine, Beidell, Saguache Co., CO;* Black Jack mine, Carson district, Owyhee Co., ID; Candelaria, Mineral Co.; Fairview, Sanpete Co., UT; Carson district, NV; Embudo, NM; Himalaya dike system, Mesa Grande, San Diego Co.; Los Angeles, Los Angeles Co.; Castle Mt., Ivanpah, CA; Maniquipi, Mexico; Ancon, Canal Zone, Panama; Rothamstead, England; Gemmano, Italy; Unterrupsroth, Rhône, Germany; Takovaya and Cheget-Lakhran Valley, Ural Mts.; Gedmayshikh Valley, Russia; Velka Kopan, near Chust, Ukraine; Ropp, Nigeria; Akkar plain, Safita, Syria; Hathi-Ki-Dhani, India; Tottori, Japan. VK, EF *MIN 4(2):44(1992)*.

71.3.1a.2 Montmorillonite $(Na,Ca_{0.5})_{0.33}(Al,Mg)_2Si_4O_{10}(OH)_2 \cdot nH_2O$

Named in 1847 by Salvetat for the locality. Smectite group. Synonyms: Na-montmorillonite, Ca-montmorillonite, bentonite. May form random or regular interstratifications with other phyllosilicates (e.g., kaolinite, cookeite, illite, muscovite). MON (sometimes given as pseudo-HEX) $C2/m$. $a = 4.93$–5.21, $b = 8.94$–9.02, $c = 12.4, 15.0$–15.5, $\beta = 99.54°$, $Z = 1, 2$. *12-219*(18 Å, nat): 17.6_{10} 9.0_5 4.49_8 3.58_4 2.99_3 2.57_4 1.70_2 1.50_6; *13-135*(15 Å, nat): 15.0_{10} 5.01_6 4.50_8 3.77_2 3.02_6 1.70_3 1.50_5 1.49_5; *13-259*(14 Å, nat): 13.6_{10} 4.47_2 3.34_1 3.23_1 2.92_1 2.59 and $2.49_{0.5}$; *29-1498*(15 Å, nat): 13.6_{10} 5.16_1 $4.46_{6.5}$ 2.56_2 1.69_1 1.495_1; *29-1499*(21 Å, nat): 21.5_{10} 10.6_2 $4.45_{5.5}$ $2.56_{3.5}$ 1.69_1 $1.495_{2.5}$ 1.325_1. Montmorillonite has the typical smectite structure. It has been suggested that true montmorillonite has its layer charge resulting from octahedral substitutions, while beidellites are not necessarily high in Al compared to montmorillonite, but have significant layer charges arising from tetrahedral Al substitution. Expands to 17–18 Å with glycerol and collapses to ~ 10 Å with heating to 550°. **Physical properties:** Individual crystals are typically 1–3 µm and are platy or scaly, tabular on {001}. White to buff, pale yellow, light greenish gray, also pink, rose red to deep terra-cotta; earthy, also compact, dense and waxy. Cleavage {001}, perfect. H = 1–2. G = 2.06–2.3 (porous), but up to 2.7. Decomposed or gelatinized by common acids. Stained blue-green by benzidine. **Optics:** Biaxial (−); $X \sim c$, $Y = b$, $Z \sim a$; $N_x = 1.48$–1.503,

$N_y = 1.50–1.534$, $N_z = 1.50–1.534$; 2V small, 10–25°; birefringence = 0.018–0.031; positive elongation. Generally colorless, can be slightly pleochroic with increasing Fe_2O_3. **Chemistry:** A complete series extends to nontronite with increasing Fe_2O_3 and to beidellite with increasing ^{IV}Al. Chromian montmorillonite has been called volkonskoite, but true volkonskoite has $^{VI}Cr > {}^{VI}Al$. Chemical analyses have been reviewed, and the mean of 101 analyses. Weaver and Pollard [*DS 15:55(1973)*] gave $(Na_{0.01}Ca_{0.007}K_{0.004})$ $(Al_{1.492}Fe^{3+}0.187Fe^{2+}0.007Mg_{0.354})(Si_{3.837}Al_{0.158})O_{10}(OH)_2 \cdot nH_2O$. Two-thirds of the analyses chosen had $^{IV}Al < 0.2$, per four tetrahedral sites, and few had $^{IV}Al > 0.4$, with bimodal points at 0.10 and 0.30. Low-Mg montmorillonite (Wyoming-type; ~ 0.19 octahedral position) and high-Mg montmorillonite (Cheto-type; 0.55 octahedral position) have been distinguished [*AM 46:1329(1961)*] with no significant differences in tetrahedral occupancy, but with differing thermal behavior. The two types can be present in a single sample with the coarser size fraction more magnesian (Cheto type) than the finer fraction (Wyoming type), but observed MgO values do not show enough bimodal distribution to verify the high/low-Mg montmorillonite types as important features of the species unless the mixing of types within a sample is a common occurrence. Montmorillonites have also been subdivided into categories (Wyoming, Tatatilla and Chambers, and Otay types) based on their total layer charge and on their percentage contribution to net layer charge by tetrahedral substitutions. *CCM 17:115(1969)*. Mg may be helpful in the genesis of montmorillonite. *USGSPP:205B(1945)*. Frequently found with amorphous silica- and/or alumina-bearing material, can be impure and mixed with other clay minerals. Observed ranges of major elements (wt %) include: SiO_2, 50.10–65.00; Al_2O_3, 15.20–34.00; Fe_2O_3, 0.00–13.61; FeO, 0.00–1.61; MgO, 0.09–7.38; MnO, 0–0.03; CaO, 0.00–4.23; Na_2O, 0.00–3.74; K_2O, 0.00–2.80; TiO_2, 0.00–2.90; P_2O_5, 0–0.09; also (ppm): Li_2O, 4–88; F, 7100 (Chambers); H_2O^+, 5.21–13.75. CEC = 70–130 meq per 100 g. Interlayer cations to 0.465. Can alter to illite. **Occurrence:** A product of alkaline weathering conditions and poor drainage. Major component of altered ash flows and bentonite layers. Found in a wide variety of sedimentary environments. Also in replacement zones in some granite pegmatites, and in hydrothermal mineral deposits. Frequently noted with a mixture of clay minerals, also with detrital minerals, quartz, amorphous silica, and so on. **Localities:** Only selected occurrences can be listed here: Greenwood and Stoneham, Oxford Co., ME; Branchville, CT; Polkville, Aberdeen, and Armory, MS; Helms Park, Gonzales; Houston and San Saba, TX; Bayard and Santa Rita, Grant Co., NM; Pierre shale; Belle Fourche, Butte Co., SD; Bates Park, Casper; Clay Spur; J.C. Lane tract, Upton, Weston Co.; Newcastle, Crook Co., WY; Ely, NV; Chambers, Apache Co.; Cheto, AZ; Bearpaw shale, Garfield, Rosebud, and Horn Cos., MT; Plymouth, UT; Shoshone, Amargosa Valley; Otay, CA; Umiat, AK; Aina Haina, HI; Dorothy and Wayne, ALB; Pembina, MAN, Canada; Santa Rosa; Tatatilla, Vera Cruz, Mexico; Elliots, Antigua; Giants Causeway, County Antrim, Northern Ireland; Woburn, Bedfordshire; Red Hill, Surrey, England; *Montmorillon, Vienne*, Confolens, Charente Dept.; Plombières, Ludes, Montjavoult, and Virolet, *France*; Vallortigara-Posina, Schio, Venezia; Castigligoncello, Livorno; Cala Aqua mine, Ponza Is., Italy; Groschlattengrün, Fichtelgebirge, Bavaria, Germany; Cilly, Untersteiermark, Austria; Stritez, Bohemia, Czech Republic; Barataka, Transylvania, Romania; Selongin Daura, Russia; Malka River area, Caucasus Mts., Kazakhstan; Nahal Ayalon and Hatrurim formations, Israel; Emilia, Calingasta and Burrera mine, Jachal and El Retamito and Mario Don Fernando mine, Retimito, San Juan Prov.; Tala, Heras and Santa Elena, Potrerillos and San Gabriel, Mendoza Prov., Argentina; Taourirt, Morocco; Marnia, Fadli-Mostaganem, and Camp Berteaux, Algeria; Thies, Senegal; Youe, Chad; Iriba, Cameroon; Ankaratra, Malagasy; Reunion Is.; Itoigawa, Niigata Pref.; Usui and Hojun Mts., Gumma Pref.; Yakote and Hanaoka Mts., Akita Pref., Rokkaku, Yamagata Pref.; Tottori, Japan. VK, EF *MIN 4(2):15(1992)*.

71.3.1a.3   Nontronite $(Na,Ca_{0.5})_{0.33}(Fe^{3+},Al)_2(Si,Al)_4O_{10}$ $(OH)_2 \cdot nH_2O$

Named in 1827 by Berthier for the locality. Smectite group. Also originally called chloropal. MON $C2/m$. $a = 5.23$–5.264, $b = 9.06$–9.18, $c \sim 10, 14.80, 15.0$–15.5, $\beta = 90°, 99°$, $Z = 1, 2$, $D = 2.29$–2.36. *29-1497*(nat): 15.2_{10} $4.48_{5.5}$ 3.58_2 3.05_2 $2.564_{2.5}$ $2.560_{2.5}$ 1.51_1 1.34_1; *34-842*(nat): 14.6_{10} 4.53_{10} 3.67_2 3.01_3 2.60_5 1.52_8 1.31_3 1.27_3. Typical smectite structure with air-dried basal spacings 13.6–15.0 Å normal. Rarely naturally found with collapsed spacings ~ 10 Å. Ethylene glycol causes expansion to 16.9–17.3 Å. Collapses on heating (550°) 9.6–10.4 Å. K-intersalination generally causes some collapse from air-dried spacings to 11.9–13.1 Å. Mössbauer studies show some $^{IV}Fe^{3+}$. *AM 73:1346(1988)*. Nontronites can contain significant ^{IV}Al and parallel beidellite-like smectites: $Ca_{0.18}(Fe^{3+}1.94Mg_{0.10}Al_{0.06}Fe^{2+}0.05)(Si_{3.55}Al_{0.45})_{S4}$ $O_{10}(OH)_2 \cdot nH_2O$. *AM 31:294(1946)*. A pseudo-anhydrous nontronitelike mineral has been reported [Arkansas, *GSAPA 18(6):528(1986)*]. The Arkansas nontronite has a collapsed spacing (~ 10 Å), expands with glycol, but does not hydrate upon immersion in water, and has a near-end-member composition: one water molecule per formula is apparently ordered within the opening of the tetrahedral ring of the silicate layer, yet does not contribute to non-mica-like spacings, $(Ca, Mg, Mn)_{0.135}$ $(Fe^{3+}1.785Fe^{2+}0.185V^{5+}0.04)_{\Sigma 2.01}$ $(Si_{3.8}Fe^{3+}0.18Al_{0.02})_{\Sigma 4}O_{10}$ $(OH)_2 \cdot H_2O$. **Physical properties:** Pale to dark yellow-green to olive, also bright yellow-green, resinous to waxy luster with conchoidal fracture to earthy, sometimes fibrous. $H = 1$–2. $G = 2.06$–2.32. Cleavage $\{001\}$, perfect. Decomposed or gelatinized by common acids. **Optics:** Biaxial $(-)$; $Y = b$, $Z \sim c$, high RI material has $X \sim c$; $N_x = 1.545$–1.625, $N_y = 1.569$–1.650, $N_z = 1.570$–1.655; 2V moderate, 33–40°, ranges 25–70°; $r < v$; birefringence $= 0.035$–0.044 (0.010 low Fe); very strong increase in RI with increasing Fe^{3+}; positive elongation; usually pleochroic in yellowish $\wedge$ (001) to brownish greens parallel to (001); pleochroic. X, pale yellow; Y, olive-green; Z, yellow-green. **Chemistry:** Forms a series with montmorillonite by Fe^{3+} exchange with Al; most compositions yield < 1.0 wt % Al_2O_3, SiO_2,

40.8–48.8; Fe_2O_3, 20.4–32.4; FeO usually < 0.50; MgO, 0.61–2.1; CaO, 0.77–2.2; Na_2O, 0–0.29; K_2O, 0–0.24; TiO_2, 0.08–1.72; P_2O_5, 0–0.02; H_2O^+, 7.8–11.4; H_2O^-, 7.1–14.75; H_2O (total), 14.5–23.0; CEC = 60 meq per 100 g. Can replace olivine, chlorite, or pyroxene or be replaced by chlorite. **Occurrence:** Found with other clay minerals, including kaolinite, glauconite, chlorite, and vermiculite. Found in weathered basalts, ophiolites, basic and ultrabasic rocks, and fractures therein, also in basalt vesicles. Abundant in vertisols, also in submarine sediments, usually authigenic. As a sediment on sea mounts, also around midocean-ridge hydrothermal vents. As a coating in fractures in some granite pegmatites. **Localities:** Topsham, Sagadahoc Co., ME; Lehigh Mt., Mountainville; Bethlehem(!), PA; Chevy Chase, MD; North Garden, Albemarle Co., VA; Sugar Grove, Pendleton Co., WV; Sandy Ridge; Spruce Pine, Mitchell Co., NC; Wilson mine, Potash Sulphur Springs, AR; Green River Fm, Shirley Basin, WY; Bingham, UT; Santa Rita, NM; Miami, Gila Co., AZ; Woody, Kern Co.; Crestmore, Riverside Co., CA; Garfield(!) and Colfax, Whitman Co., Spokane, Manito, Excelsior, and Valleyford, Spokane Co., WA; lower Cook Inlet, AK; Loihi submarine volcano, HI; Peace River deposit, ALB, Canada; Pinares de Mayari deposit, Oriente Prov., Cuba; Concepcion del Oro, ZAC; Santa Eulalia, CHIH, Mexico; Smallacombe, Devon, England; Isle MaGee, County Antrim, Northern Ireland; *Perigueux Mn mine, St. Pardoux, Nontron, Dordogne, France*; Froland, Norway; Starbo, Sweden; Tirschenreuth and Passau, Bavaria; Meenser Steinberg, Göttingen, Hanover; St. Andreasberg, Zwickau, and Wolkenstein, Saxony; Menzenberg, Siebengebirge, Rhineland, Germany; Uzhorod, Ruthenia (formerly Unghwar); Sitno complex, Pukanec; Sternberg, Moravia, Czech Republic; Urkut deposit, also Szekes–Fejevar, Hungary; Starog Cikatova, Kosovo; Goles, Yugoslavia; eastern Rhodopes Mt., Bulgaria; Troodos ophiolite, Cyprus; Suakin and Atlantis II deeps, Red Sea; Stary-Krym, near Mariupol; Petrovsk, Krivoi-Rog region; Kerch, Ukraine; Balkan mine, Ural Mts., Russia; Baltatarak, Kazakhstan; Galapagos Is., Ecuador; Niquelandia, Goias; Xanda mine, Virgem da Lapa, MG; Campo Formo, Bahia, Brazil; Lake Malawi, Malawi; Lake Chad, Chad; Black Rock and Hotazel, Cape Prov., South Africa; Faratsiho and Behenjy, Malagasy; Shulan, Jilin Prov., China; Yamashiro-cho, Saga Pref., Japan; Mururoa Atoll, French Polynesia; Mariana Trench; Kasuga and other Pacific sea mounts; Burra Burra, SA, Australia; Waipiate, New Zealand; remotely sensed at *Viking* land site on Mars. VK, EF

71.3.1a.4 Volkonskoite $(Ca_{0.5})_{0.33}(Cr^{3+},Mg,Fe^{3+})_2(Si,Al)_4O_{10}(OH)_2 \cdot 4H_2O$

Named in 1830 by Volkov for A. Volkonsky, Russian nobleman. Smectite group. Synonyms: wolchonskoite, volchonskoite. MON Space group unknown. $a = 5.16$, $b = 8.94$, $c = 14.40$, $Z = 2$; also HEX $a = 5.172$, $c = 15.12$, $Z = 1$, $D = 2.29$. *42-619*(nat): 15.0_{10} $5.02_{<1}$ 4.49_5 $3.05_{<1}$ $2.56_{3.5}$ $2.51_{<1}$ 1.69_1 $1.50_{3.5}$. Volkonskoite is defined as a dioctahedral smectite with $Cr^{3+} \geq \sum[Al, Fe^{3+}, Mg, etc.]$. A structural formula of volkonskoite based on type material has been calculated: $(Ca_{0.11}Mg_{0.11}Fe^{2+}0.03K_{0.02})(Cr_{1.18}Mg_{0.78},$ $Fe^{3+}0.29Ca_{0.02})(Si_{3.50}Al_{0.51})O_{10}(OH)_2 \cdot 3.64H_2O$. Essentially the Cr^{3+} analog of beidellite. Volkonskoite expands on glycolation (typically to 16–17 Å). *CCM 35:139(1987)*. Deep to bright green, sometimes deep blue-green or blackish green; bluish green to green streak; resinous to earthy luster, sometimes fibrous. Conchoidal fracture. H = 2.5. G = 2.05–2.3. Cleavage {001}, perfect. Slowly soluble in water. Decomposed or gelatinized by hot HCl. Infusible. Mössbauer and TGA data: *CCM 35:139(1987)*. Biaxial (−); $N_x = 1.551$–1.569, $N_z = 1.564$–1.569; 2V typically 30–40°, but can be very low; birefringence = 0.004–0.013; sometimes slightly pleochroic in shades of green. Refractive indices are principally a function of the Cr content. Volkonskoite usually requires 15 wt % Cr_2O_3 to become the dominant octahedral cation in a smectite and thus satisfy nomenclature requirements, but the mineral from listed occurrences has 19.87–29.81 wt %. Due to the strong chromophoric effect of chromium, some bright green smectites containing relatively low amounts of chromium, and insufficient for volkonskoite, have been misidentified as volkonskoite. The only verified localities are listed below. Volkonskoites are usually magnesian 0.58–7.73 wt % MgO and Al_2O_3 is generally 4–5 wt %. An aberrant, high-Cr^{3+}, essentially dioctahedral smectite has been described where $^{VI}Mg \geq {}^{VI}Cr^{3+}$ [*CM 19:43(1984)*], but as the total trivalent elements exceed the total divalent elements, dioctahedral nomenclature should prevail. Iron is nearly completely Fe_2O_3 (90% or more) with a possible series to nontronite. Trace elements in volkonskoite include [*CCM 35:139(1987)*]: TiO_2, 0.015 wt %; also (ppm): Ag, 3; B, 70; Ba, 300; Be, 3; Cu, 300; Ga, 30; Li, 30; Mn, 500; Sc, 200; Sr, 70; V, 300; Y, 50; Zr, 100; Yb, 3 (Okhansk). Sometimes has organic impurities [see ref. by Gudoshnikov (1968) in *CCM 35:139(1987)*]. CEC (values not given), TGA, Mössbauer studies: *CCM 35:139(1987)*. Other substitutions (wt %) include: V_2O_5, 0.36 (Samosadka); MnO, 0.36 (Okhansk); P_2O_5, 0.11 (Samosadka); SO_3, 0.22 (Akkerman); NiO, 0.24 (Akkerman); CoO, 0.04 (Akkerman). The original volkonskoite was found in sedimentary red beds where it was the mineralizer of fossil plants. Well-characterized localities include: Nevrokop, Bulgaria; Akkerman area, northern Ural Mts., Uchtym and Udmurtia (Vyatka); Samosadka, Kama area; *Mount Efimyata, Efimyata, Okhansk region, middle Kama R. area, Perm Basin, Russia*; Daba area, Jordan. VK, EF *MIN 4(2):68(1992)*.

71.3.1a.5 Swinefordite $(Li,Ca_{0.5},Na)_{0.72}(Li,Al,Mg)_{2.66}(Si,Al)_4O_{10}(OH,F)_2 \cdot 2H_2O$

Named in 1975 by Tien et al. for Ada Swineford (1917–1993), clay mineralogist and professor of geology at Western Washington State College, Bellingham, WA. Smectite group. Probably ORTH or MON Space group unknown, probably $C2/m$. $a = 5.2$, $b = 9.0$, $c = 13.0$, $Z = 2$, $D = 2.21$. *29-809*(13 Å, nat): 12.96_{10} 6.27_4 4.53_9 3.09_8 2.62_6 1.71_6 1.51_9 1.30_6. Swinefordite is a trioctahedral 13-Å smectite. When fully hydrated, the mineral has "large" d values, which on dehydration shift to a 40-Å spacing and which continue to collapse on dehydration to 19.6 Å, then 16 Å, then 14.5 Å (at 60% relative humidity), and stably to 13 Å. The observed basal reflections of air-dried material are very much sharper than is usual for a smectite, although precession photographs indicate significant stacking disorder. The layer charge of swinefordite is probably near the upper limit allowed by a smectite. Swinefordite expands to 18 Å with glycerol and collapses to 9.8 Å on heating to 450°, but spontaneously rehydrates and expands to 11.3 Å within a few hours of returning to ambient temperature, while further spontaneous rehydration, to 13 Å, may take months. Heating to 575° postpones spontaneous rehydration expansion by 6 and 12 months, respectively. IR and DTA data: *AM 60:540(1975)*. Usually in leathery mats of fibrous, thin ribbons when dry. Gray-olive to greenish gray when dry, and greenish yellow to light olive when wet; white streak; also orange red scaly (China), tough when dry and greasy when wet, unctuous to gelatinous when wet and resembles a thick grease or petroleum jelly; glistening luster when wet and dull when dry. Cleavage {001}, perfect. H = 1. Biaxial (−); X ~ c, Y = a; $N_{x\ calc} = 1.492$, $N_y = 1.524$, $N_z = 1.526$; 2V = 29–45°; birefringence = 0.034; pleochroic. Y, colorless; Z, light yellow-brown. Swinefordite is richer in Li_2O than hectorite: 1.22 wt % (Hector) vs. 4.6 (unheated) or 5.66 wt % (heated to 150°) (Kings Mt.) $(Li_{0.41}Ca_{0.17}Na_{0.07}Mg_{0.01}K_{0.06})$ $(Li_{1.02}Al_{0.85}Mg_{0.66}Fe^{3+}0.10Fe^{2+}0.03)_{2.66}$ $(Si_{3.81}Al_{0.19})O_{10}[(OH)_{1.68}F_{0.32}] \cdot 2H_2O$. F 1.49 wt %. The $^{VI}Li:^{VI}Al \sim 1$ in swinefordite and the limit of Li contents is probably analogous to that of lepidolite: $(Li_{1.50}Al_{1.50})$. Type material contains (wt %): MgO, to 7.38; FeO, to 3.67; MnO, to 0.29. A high Al_2O_3 and low MgO (China) were indicated. *MA: 96(1995)*. CEC = 93–102 meq per 100 g. Found in greasy patches (wet) or thin mats or films (dry) lining fractures with fluorapatite and/or rare minerals in spodumene-bearing granite pegmatite. *Foote Quarry, Kings Mountain, Cleveland Co., NC*; Pegmatite No. 3, Keketuohai district, Xinjiang, China. VK, EF *MIN 4(2):51(1992), KY 14(1):1(1994)*.

71.3.1b.1 Sobotkite $(K,Ca_{0.5})_{0.33}(Mg_{0.66}Al_{0.33})_3(Si_3Al)O_{10}(OH)_2 \cdot$ $1-5H_2O$

Named in 1976 by Haranczyk and Prchazka for the locality. Smectite group. MON Space group unknown. b = 9.16–9.21, c = 14.5. PD: 14.5_7 4.48_9 2.61_6 2.51_6 2.41_5 2.35_5 1.527_6. Sobotkite is a trioctahedral smectite, but with ^{IV}Al equivalent to a mica. Part of the tetrahedral charge is balanced by corresponding ^{VI}Al or high interlayer cations. Structural formula of type material yields: $(Ca_{0.13}K_{0.015})(Mg_{1.91}Al_{0.95})(Si_{3.06}Al_{0.94})O_{10}(OH)_2 \cdot 5.18H_2O$, while sobotkite from Caslav has a similar formula: $(Ca_{0.33}Na_{0.08}K_{0.06})(Mg_{2.42}Fe^{2+}0.32Fe^{3+}0.21Al_{0.10}Mn_{0.03})(Si_{3.16}Al_{0.84})O_{10}(OH)_2 \cdot 1.97H_2O$ [*UC 1:177(1955)*], while sobotkite from Roseland gave $(K, Ca, Na)_{0.19}(Al_{1.84}Fe^{3+}0.31Mg_{0.02})(Si_{3.18}Al_{0.82})O_{10}(OH)_2 \cdot nH_2O)$ [*USGSPP:205-B(1945)*]. DTA data: *PMZ 22:3(1974)*. Sobotkite has a 14.5-Å spacing and "swells" with glycerol (data not given). Pale green, fine-grained, unctuous. H = 3. G = 2.31. Biaxial; $N_x = 1.513$, $N_{av} = 1.525$, $N_z = 1.536$. Analysis (wt %): SiO_2, 39.68–44.23;

Al_2O_3, 20.98–31.30; Fe_2O_3, 5.77; MgO, 0.18–16.72; NiO, 0.014; CaO, 0.45–1.58; K_2O, 0.15–0.95; Na_2O, 0.19 (Roseland); TiO2, 1.01 (Roseland); H_2O_{tot}, 16.02–20.71. Type material was found with chrysotile, nickeloan lizardite, and pimelite (willemseite) in serpentinite. Roseland, VA; *Mt. Sobotka, Godolow–Jordanow massif, Wiry, Lower Silesia, Poland*; Caslav, Czech Republic. VK, EF *MIN 4(2):83(1992)*.

71.3.1b.2 Saponite $(Ca_{0.5},Na)_{0.33}(Mg,Fe^{2+})_3(Si,Al)_4O_{10}(OH)_2 \cdot 4H_2O$

First recognized in 1758 by Axel Cronstedt and named in 1830 by Svanberg from the Greek word *sapo*, in allusion to its soapy feel. Smectite group. Synonym: griffithite. MON Space group unknown. $a = 5.232$–5.314, $b = 9.148$–9.211, $c = 12.4, 14.52$–15.56, Z = 2, D = 2.10. *6-2* (18 Å, glycerol): 18.8_{10} 9.1_5 4.55_5 3.61_5 3.01_4 2.61_6 1.74_4 1.54_4; *12-168*(17 Å, glycol, syn): 17.0_{10} 8.5_5 5.69_4 4.58_5 3.37_8 2.81_3 2.58_4 1.535_7; *13-86*(15 Å, nat): 14.2_{10} 4.96_4 4.57_5 3.67_8 2.09_3 1.84_3 1.53_9 1.32_4; *13-305*(15 Å, ferroan): 15.4_{10} 7.9_5 5.28_1 4.6_5 3.13_5 2.65 and 2.56_5 $1.54_{6.5}$ 1.33_3; *29-1491*(15 Å, nat): 15.5_{10} $7.73_{<1}$ $5.11_{<1}$ 4.57_1 $3.83_{<1}$ 3.07_1 $2.61_{<1}$ $1.53_{<1}$; *30-789*(15 Å, aluminian, also called sobotkite): 14.5_8 7.38_4 4.48_{10} 2.61_7 2.51_7 2.41_6 2.35_6 1.53_7. Saponite has the normal trioctahedral smectite structure and frequently has a 15-Å basal spacing with interlayer Ca or 12.5-Å spacing with interlayer Na. Saponite can have a net layer charge arising from tetrahedral Al substitutions and can grade from saponite to sobotkite [see montmorillonite (71.3.1a.2) and beidellite (71.3.1a.1)] but ^{IV}Al is frequently 0.25–0.50 atom per formula unit and distinguished from the essentially Al-deficient stevensite. Expands with ethylene glycol to 17–18 Å. DTA studies have been made [*CCM 32:147(1984)*] which show a shifting of basal spacings down to 12.5 Å, with an abrupt shift to 9.85 Å at 600°, indicating collapse due to dehydration, but the spacing spontaneously expands to 14 Å by rehydration on cooling. IR has also been studied. *CCM 32:147(1984)*. **Physical properties:** White, pale yellow to dark brown, gray-green, bluish green, light blue; can be stained orange to terra-cotta; earthy to resinous, rarely fibrous, unctuous when wet. H = 1 or less or up to 2. G = 2.10–2.30 for porous material, but can be 2.79–2.87. Cleavage {001}, perfect. Fusibility 4 or with difficulty. Decomposed or gelatinized by common acids. Can be visually similar to fine-grained chlorite in mid-ocean basalts. **Optics:** Biaxial (−); $N_x = 1.479$–1.53, $N_y = 1.486^*$ (very low iron), 1.510–1.565, $N_z = 1.50^*$ (low iron), 1.510–1.59; 2V usually 20–38°, but can be nearly 0°; birefringence = 0.020–0.044, 0.003 (high iron), N_x perpendicular to (001); positive elongation; high-iron varieties are pleochroic in brownish greens. Also uniaxial (−); $N_O = 1.54$–1.573, $N_E = 1.560$–1.593; birefringence = 0.020–0.037. **Chemistry:** Can be nickeloan forming a series with pimelite, also mistaken for volkonskoite when Cr-bearing. Chemical analyses (wt %) are variable: SiO_2, 42.78–53.88; Al_2O_3, 3.89–6.44; Fe_2O_3, generally < 1%; MgO, 22.96–31.61; MnO, 0.06; TiO_2, 0.37 (Irisugawa); FeO, 7.83 (Griffith Park); H_2O^+, 7.25 (Ballarat); H_2O^-, 10.76 (Ballarat). Interlayer cations can contain Mg and intersalination techniques are required to remove this Mg in order to have a rigorous understanding of its structural role. CEC = 76–109. **Occurrence:** Can be associated with other clays, quartz, chlorite, adularia, sepiolite, celadonite and other micas, calcite, chalcedony, zeolites, and so on. Saponite can be a major component of iddingsite alteration of olivine. *CCM 32:1(1984)*. Found lining cavities in basalt or as a pervasive alteration of olivine or plagioclase in submarine as well as continental basalts. Oxygen isotope analyses suggest that some saponite was derived from basalt while in equilibrium with seawater [*CMP 73:323(1980)*] and have demonstrated an authigenic reaction phlogopite → vermiculite → saponite. *CCM 33:214(1985)*. Found in lateritic profiles above, or as a hydrothermal alteration along fractures in basic or ultrabasic rocks. A component in some volcanic ash. **Localities:** Only selected saponite localities are listed here: Roaring Brook, New Haven, New Haven Co., CT; Sugar Grove, WV; Ahmeek and South Kearsarge mines and elsewhere, Keweenaw Penin., MI; Silver Bay; Pigeon Point to Fond du Lac, Lake Superior region, MN; Hills Pond peridotite, Woodson Co., KS; Italian and northern Italian Mts., Gunnison Co., CO; Ash Meadows; Trucker Canyon, Reno, NV; Milford, Beaver Co., UT; South Panamint Valley, Ballarat; Cahuenga Pass, Griffith Park, Los Angeles, Los Angeles Co., CA; Mt. St. Helens ash; Grand Rone Formation, WA; Georges Is., PEI, Canada; Faroe Is.; Orrock, Fife; Allt Ribhein, Fiskavaig Bay, Skye; Cathkin Hills, Carmunnock, Lanarkshire, Scotland; Lizard, Cornwall, England; Svärdsjö, Kopparberg, Dalarne, Sweden; Mormoiron, France; Sasbach, Kaiserstuhl, Germany; Kozakov, Czech Republic; Ksabi, Morocco; Krugersdorp, Transvaal, South Africa; Wakamatsu mine, Tari district, Tottori Pref.; Irisugawa, Gunma Pref., Japan. VK, EF

71.3.1b.3 Sauconite $(Ca_{0.5},Na)_{0.33}Zn_3(Si_{3.5}Al_{0.5})O_{10}(OH)_2 \cdot 4H_2O$

Named in 1875 by Frederick Genth for the locality. Smectite group. Synonym: tallow clay. MON Space group unknown. $a = 5.2$–5.35, $b = 9.1$–9.32, $c = 15.4$–15.8, $\beta = 90°$, $D = 2.43$–2.53. 8-243(15 Å, nat): 15.4_{10} 7.9_6 4.55_6 3.94_1 3.16_5 2.67_6 1.55_8 1.33_4; 8-444(17 Å, nat): 16.6_{10} 8.33_{10} 4.60_{10} $3.34_{7.5}$ $2.62_{7.5}$ 1.55_{10} 1.33_5 $1.01_{2.5}$; 8-445(15 Å, nat): 15.4_{10} 7.77_9 4.60_9 3.87_5 3.09_5 2.67_{10} 1.54_{10} $1.33_{7.5}$; 29-1500(15 Å, nat): 15.5_{10} 5.0_1 4.53_5 3.14_1 2.59_1 1.52_1. Sauconite has the normal trioctahedral smectite structure and shows 15.0- to 15.6-Å (air-dried) basal spacings and expands to 16.1–16.6 Å with ethylene glycol. *AM 36:795(1951)*. Untreated air-dried samples collapsed from 15.4 Å to 14.5 Å on heating to 327°. Sauconites usually have high ^{IV}Al and are analogous to beidellite or even sobotkite. Fine-grained, usually in dense claylike masses of small micaceous plates in laminated to compact masses. Ideally white, but usually stained yellow to brownish yellow, red-brown; luster is usually dull to greasy. Cleavage {001}, perfect. Like other smectites, soluble in water, decomposed by HCl. $H = 1\frac{1}{2}$–2. $G = 2.61$–2.71(for about 30 wt % ZnO); 2.76–2.80 (for about 40 wt % ZnO). DTA data: *AM 36:795(1951)*. May fluoresce bluish white in UV light. Biaxial (−); X perpendicular to (001),

$Y = b$; $N_x = 1.530, 1.550$–1.575, $N_y = 1.550, 1.59$–1.614, $N_z = 1.551, 1.592$–1.615; $2V = 0$–$20°$; birefringence $= 0.035$–0.042; positive elongation. Chemistry is variable (wt %): TiO_2, 0.24; CuO, to 0.13; MgO, to 2.4; Fe_2O_3, to 6.2; MnO, to 0.06; CaO, to 2.7; Na_2O, to 0.4; K_2O, to 0.65; Al_2O_3, 6.4–17.0; ZnO, 22.5–42.5; SiO_2, 29.6–38.7. Calculated formulas range from $(Ca_{0.13}Na_{0.09}(Zn_{2.40}Al_{0.22}Mg_{0.18}Fe^{3+}_{0.17})(Si_{3.47}Al_{0.53})O_{10}(OH)_2 \cdot nH_2O$ to $K_{0.33}(Zn_{1.48}Al_{0.74}Fe^{3+}0.40Mg_{0.14})Si_{3.01}Al_{0.99}O_{10}(OH)_2 \cdot nH_2O$. *AM 31:411 (1946)*. Frequently associated with hemimorphite. Found as small-to-large masses in rock or cavities in oxidized zones of zinc deposits, sometimes in considerable minable quantities. Also found with smithsonite, chrysocolla, and Fe-oxides. *Überroth mine, Friedensville, Saucon Valley, Lehigh Co., PA*; Liberty mine, Meekers Grove, Lafayette Co., and at Platteville, WI; Coon Hollow mine, Zinc, Boone Co., AR; Lower Waterloo, New Discovery, and Yankee Doodle mines, Leadville, Lake Co., CO; Defiance mine, Gleeson, Castle Dome, Globe–Miami district, and the 79 mine, Gila Co. and Magma mine, Superior, Pinal Co., AZ; Mohawk mine, San Bernardino Co., CA; Altenberge, Aachen, Germany; Moresnet, Liège, Belgium; Garperberg, Sweden; found in about 25 veins and dikes of pegmatite on Mts. Nepakh, Kuivchorr, Mannepakh, and others, Lovozero massif, Kola Penin., and central Tien-Shan, Russia; pink, Co-bearing (CoO = 4.15 wt %) from the Dashkesan Fe ore deposit, Azerbaijan. Well known from weathering zone of sulfide deposits of Kazakhstan; eastern Transbaikal, Siberia, and Yaroslav deposit, Primorye, Russia. VK, EF *MIN 4(2):96(1992)*.

71.3.1b.4 Hectorite $(Na,Ca_{0.5})_{0.33}(Mg_{2.67}Li_{0.33})Si_4O_{10}(F,OH)_2 \cdot H_2O$

Found in 1936 [*AM 21:238(1936)*] and named in 1941 by Strese and Hofmann for the type locality. Smectite group. MON $C2/m$, also indexed as pseudo-HEX: $a = 5.2$–5.274, $b = 9.16$–9.18, $c = 15.93$–16.13, $Z = 1$. *9-31* (15 Å, nat): 15.8_8 4.58_{10} 2.66_8 2.48_6 1.72_6 1.53_{10} 1.32_8 1.30_8; *25-1385* (15 Å, syn): 4.5_{10} 2.6_{10} 1.74_6 1.53_{10} 1.31_6 1.28_4 0.98_6 0.875_6. Hectorite is a trioctahedral 12.4-Å smectite and is the lithium-bearing equivalent of stevensite: $(Na_{0.28}Ca_{0.02}K_{0.01})Mg_{2.65}Li_{0.33}Al_{0.02})Si_4O_{10}(OH)_2 \cdot H_2O$. EPR data: *AM 70:996(1985)*. NMR data: *AM 72:935(1987), GCA 54:1655(1990)*. Swells with ethylene glycol and untreated material collapses on heating to 550°. Pilaring with $[Al_{13}O_4(OH)_{24}(H_2O)_{12}]^{7+}$ has been produced experimentally. *AM 41:598 (1993)*. Earthy aggregates, individual crystals, to several micrometers, are lathlike or brick-shaped, elongated parallel to a. White to buff, white streak, earthy or dull to waxy luster, unctuous when wet. Cleavage {001}, perfect; uneven fracture. H = 1–2. G = 2.34–2.5. Decomposed or gelatinized by common acids. IR, DTA, and other spectra data: *MIN 4(2):91(1992)*. Biaxial (−); $N_x = 1.485$–1.489, $N_y = 1.499$, $N_z = 1.510$–1.524; 2V small; birefringence = 0.021–0.031 (optical data are few). Heating raises the refractive indices: (100°): $N_x = 1.524$, $N_z = 1.553$; (200°): $N_x = 1.555$, $N_z = 1.584$. Analyses of natural material are few. Hectorite is virtually a nonaluminous smectite with

Al$_2$O$_3$ to 1.40 wt %. Laponite is a synthetic hectorite transformed from other smectites. Hectorite from Hector, CA, frequently has 0.3–0.57 Li in trioctahedral sites, while that from Romania varies from 0.33 to 0.45 but has no F. Some Li-bearing stevensites have been called hectorite. Fourier analysis of California material yielded 0.71 Li. *CR(D) 274(2):149(1972)*. Other substitutions include (wt %): MgO, 24.00–28.68; SiO$_2$, 49.3–59.76; Fe$_2$O$_3$, to 1.82; Li$_2$O, 0.36–1.35; Na$_2$O, 0.07–3.37; CaO, 0.1–1.80; K$_2$O, to 0.43; B$_2$O$_3$ to 0.035 (Hector); H$_2$O$^+$, 8.68; H$_2$O$^-$, 10.82; F, to 6.0 (Hector); Cl, to 0.31 (Hector). Some hectorite appears to contain very little F substituting for (OH). CEC = 40–115 meq per 100 g. Found as an alteration of clinoptilolite in bentonite in an alkaline lake environment and can be admixed with saponite; also by action of hot springs or relatively hot meteoric waters. *EG 53:22(1958)*. Usually contains minor fine-grained calcite. Also found in argilized skarn. Used in an enormous variety of industrial applications, particularly due to its high viscosity, absorbancy, and interesting rheology, with a correspondingly large research literature, sometimes as the synthetic equivalent called laponite (CEC = 71–75 meq per 100 g). Occurs at the Lyles mine, near Kirkland, Yavapai Co.; Date Creek basin, near Wickenburg, AZ; Disaster Peak, Montana Mts., Humboldt Co., NV; McDermitt caldera, NV–OR; 5 km S of *Hector, Mojave desert, San Bernardino Co., CA*; with calcite veins in acid tuffs at Puy Chalard near Verteson, Dept. Puy de Dôme, France; Ebro Basin and other Tertiary basins, Spain; Moldova Noua, Romania; Djebel–Gassul deposit, Central Atlas Mts., eastern Morocco; Balikesir borate deposit, Balikesir Prov., Turkey. VK, EF

71.3.1b.5   Pimelite   Ni$_3$Si$_4$O$_{10}$(OH)$_2$ · 4H$_2$O

Discovered in 1788 by Martin Klaproth and renamed in 1800 by Karsten from the Greek word for *fat*, in allusion to the mineral's appearance. Smectite group. Synonyms: röttisite, revdanskite, desaulsite. Apparently constitutes part of the mixtures called genthite, nickel–gymnite, garnierite, nepouite, and noumeite. HEX Space group unknown. $a = 5.256$, $c = 14.822$, $Z = 1$, $D = 2.68$. *43-664*(nat): 15.1$_{10}$ 4.56$_4$ 3.34$_2$ 2.66$_4$ 2.27$_2$ 1.72$_1$ 1.52$_8$. In 1938, pimelite from an unstated locality, but probably not holotype material, was determined to be a smectite. *CMA:360(1938)*. Additional studies have shown that a nickel-rich smectite exists at the type locality. *AM 51:279(1966)*. Earlier, pimelite was studied by powder diffraction film methods, but large spacings may have been unrecorded. *BSFM 59:163(1936)*. Garnierite from Riddle was found to consist of a mixture of smectitic pimelite, nickeloan deweylite, and nickeloan serpentine. *EG 44:13(1949)*. Generally low in Al with consequent low layer charge. Air-dried material (Brazil) 14.4 Å expands to 17.35 Å$_{001}$ with glycol, and collapses to 9.83 at 490°. Noumea smectitic pimelite is apparently a mixture that expands on glycolation from naturally collapsed 10.2 Å up to 15.3 Å, leaving a 9.6-Å residual. *AM 70:549(1985)*. A confusion has arisen, as a nickel-rich talc also has been observed in specimens from the type locality [*AM 64:615(1979)*] which has variously been called pimelite or kerolite. Smectitic pimelite has been synthesized (Macewan, 1961). It has been stated that a Ni-dominant smectite has never been found in nature, but these statements may have been made by workers studying Ni-dominant talc and not type specimens of pimelite. As röttisite, desaulsite, and revdanskite are also Ni-dominant smectites [AM 51:279(1966), but data not given], further systematic work is needed to establish the true nature of pimelite vs. kerolitic willemseite. An aberrant Ni-dominant smectite $(Ca_{0.01}K_{0.05})$ $(Ni_{1.47}Fe_{0.50}Mg_{0.48}Al_{0.17}Cr_{0.02})(Si_{3.92}Al_{0.08})O_{10}(OH)_2 \cdot nH_2O$ has been found at Niquelandia, Goias, Brazil. CCM 35:1(1987). The Ni-dominant smectite has high trivalent ion substitution (Fe^{3+}, Al, Cr) and so is dioctahedral. The Niquelandia clay sample contains Ni-rich and Ni-poor size fractions similar to Cheto- and Wyoming-type size fractions of montmorillonite. **Physical properties:** Bright green, apple-green, yellowish green, emerald-green; greenish white streak; usually waxy with conchoidal fracture, also fibrous. $H = 2-2\frac{1}{2}$. $G = 2.23-2.98$. Decomposed by acids. Optical absorption spectra: *AM 70:549(1985)*. **Optics:** Biaxial (−); $N_y = 1.592-1.615$; $2V = 30-80°$; birefringence $= 0.014-0.025$, but can appear isotropic due to fine grain size; pleochroic, pale green–colorless to light yellow green. **Chemistry:** Analyses are few; material from the type locality gave [AM 51:279(1966)] (wt %): SiO_2, 47.20; Al_2O_3, 0.22; Fe_2O_3, 0.20; MgO, 11.42; NiO, 27.66; CaO, 0.16; H_2O^+, 9.38; H_2O^-, 3.64. Type material is magnesian and pimelite could form a series with stevensite or saponite. Additional ranges of composition (wt %) include: NiO, 38.22 (Franklin); FeO, 12.15 (Revdansk); ZnO, 4.0 (Franklin); CaO, 0.70 (Franklin); As_2O_5, 4.77 (Franklin). **Occurrence:** Found in lateritic profiles above serpentinites or dunites. Frequently admixed with nickeloan serpentines or quartz. The color of chrysoprase has occasionally been attributed to pimelite inclusions. **Localities:** Reported pimelite occurrences that appear to be smectite include Wood's chrome mine, Texas, Lancaster Co., PA; Franklin, Sussex Co., NJ; Michipicoten, ID; Riddle, Douglas Co., OR; Wiley Well district, CA; Malaga, Spain; Saasthal, Walais, Switzerland; Röttis, Vogtland, Saxony; *Kosemütz, near Frankenstein, Upper Silesia, Poland*; Steinberg Mt., near Jordansmuhl, Silesia; Revdansk, Ural Mts., Russia; Sarykulbod and Karaganda, Kazakhstan; Jacuba mine, Niquelandia, Goias, Brazil; Marlborough Creek, Rockhampton district, Q, Australia; Noumea, and Poro, New Caledonia. VK, EF Karsten, *Mineralogische Tabellen* (1800), *MIN 4(1):248(1992)*.

71.3.1b.6 Stevensite $(Ca_{0.5},Na)_{0.33}(Mg,Fe^{2+})_3Si_4O_{10}(OH)_2 \cdot nH_2O$

Named in 1873 by A. R. Leeds for E. A. Stevens, founder of Stevens Institute of Technology, Hoboken, NJ. Smectite group. ORTH or MONO Space group unknown. $a = 5.26-5.4$, $b = 9.108-9.162$, $c = 15.3$, $Z = 2$, $D = 2.07$. 25-1498(15 Å, nat): 15.5_{10} 5.0_1 $4.53_{3.5}$ 3.10_1 2.54_1 $2.27_{<1}$ 1.87_1 $1.52_{1.5}$. Stevensite has the typical smectite structure and shows a 14-Å basal spacing. Expands to 15.1 Å, rarely to 16.9–18 Å, with ethylene glycol. Some "stevensites" show 24.5- to 26.5-Å spacings, suggesting they have a corrensite impurity. *AM*

44:342(1959). North Tyne stevensite might be naturally collapsed in part. *MM 32:218(1959)*. DTA and IR data: *AM 44:342(1959)*. Generally massive, individual crystals occur as platy aggregates to 2 μm. Pink, cream white, tan, buff, light brown; earthy or dull to resinous luster. Cleavage {001}, perfect; brittle when dry. Fusibility 3. H = 2.5. G = 2.15–2.565. Sometimes thermoluminescent. Decomposed by HCl. Stained by malachite-green/nitrobenzene deep emerald-green (natural) and very pale yellow (acid treated). *AM 38:973(1953)*. IR and DTA data: *MIN 4(2):85(1992)*. The chemical composition (wt %) is fairly constant: MgO, 23.7–27.9; CaO, to 2.2; MnO, 0.02–1.6; Al_2O_3, 0.1–1.0; Fe_2O_3, 0.3–1.75; SiO_2, 51.2–57.3; H_2O^+, 6.4–9.9; H_2O^-, 4.8–12.4; traces of Na_2O and K_2O. Isotropic; $N_{av} = 1.50$–1.511; birefringence = 0.0–0.02; also 2V and sign, not reported; $N_x = 1.515$(white), 1.552(brownish green), $N_z = 1.509$, 1.535(white), 1.571(brownish green), birefringence = 0.012, 0.009. Optical variations might be due to water contents, also similar to sepiolite [*MM 32:218(1959)*], but the X-ray diffraction data are identical for the two types. A structural formula of stevensite (North Tyne) is $(Ca_{0.5}, K, Na)_{0.273}$ $(Mg_{2.896}Fe^{2+}_{0.026}Mn^{2+}_{0.026}Fe^{3+}_{0.017})$ $Si_{3.925}Al_{0.075}O_{10}$ $(OH)_2 \cdot nH_2O$. Stevensite is usually very low in Al_2O_3 usually < 1.0 wt % and hence is nearly tetrasilicic, unlike saponite, but frequently has minor substitutions (wt %) including: TiO_2, 0.02 (Italy); Fe_2O_3, 1.29 (Italy); MnO, 0.75 (Jersey City); FeO, 1.33 (Scotland); Li_2O, 0.33 (Romania); H_2O^+, 9.9 (North Tyne); H_2O^-, 6.69 (Springfield). CEC = 36 meq per 100 g. Stevensite differs from saponite and hectorite by being low or deficient in Al_2O_3, Li_2O, and/or F; however, lithian stevensites have been described. *CCM 38:171(1990)*. Sometimes these lithian stevensites (with ~ 0.11–0.12 Li) have been called hectorite. Originally found as earthy replacements of pectolite in basalt cavities with quartz, calcite, datolite, and so on. A component of some deweylite. *MM 41:443(1977)*. Also as hydrothermal alteration of dunite, skarn, or in oolitic iron ore. Rarely as a component of volcanic ash. Well-characterized stevensite localities include: Willsboro, Essex Co., NY; Hartshorn quarry, Springfield, Essex Co.; railway tunnel, Jersey City; *railway tunnel, Bergen Hill, Hudson Co., and railway tunnel, Hoboken*; New Street Quarry, Paterson, *Passaic Co., NJ*; Cedar Hill quarry, Texas, Lancaster Co., PA; Mine Creek, Bakersville, Mitchell Co., NC; Mayville, WI; Green River formation, WY; Holbrook mine, Cochise Co., AZ; Whin Sill, Gunnerton, North Tyne, Northumberland, England; Corstorphine Hill, Edinburgh, Scotland; Puy Chalard, Puy de Dôme; Salinelles, France; pseudomorphous after periclase from Prócida Is., Italy; Burg Herborn, Germany; Moldova Noua, Romania; pseudomorphous after wollastonite from the Balkan deposit, southern Ural Mts., Russia; Ghassoul, Morocco; Ohiri mine, Yamagata Pref.; Akatani mine, Niigata Pref.; Obori mine, Yamagata Pref., and Kamaishi mine, Iwate Pref., Japan. EF *NJBM:556(1983)*. VK, EF.

71.3.1b.7 Yakhontovite $(Ca,K)_{0.5}(Cu^{2+},Fe^{3+})_{2.37}Si_4O_{10}(OH)_2 \cdot 3H_2O$

Named in 1986 by Postnikova et al. for Liia Konstantinovna Yakhontova (b.1926), Russian mineralogist, University of Moscow, a specialist in cobalt deposits. Smectite group. Probably MON Space group unknown. $a = 5.26$, $b = 9.108$, $c = 13.89$ ($c \times \sin \beta$), $Z = 1$, $D = 2.48$. *41-1447*(nat): 13.9_{10} 7.3_2 4.5_8 3.2_1 2.88_6. Yakhontovite is an unusual smectite. The dominant cation is copper, which should yield a trioctahedral occupancy, but the abundance of Fe^{3+} decreases the occupancy to 2.37. Also, copper occupies less than half of the octahedral positions: $(Ca_{0.20}K_{0.01})(Cu_{0.84}Fe^{3+}_{0.83}Mg_{0.67}Zn_{0.02}Al_{0.01})Si_4O_{10}(OH)_2 \cdot nH_2O$. As Yakhontovite is very nearly cuprian nontronite and as it was intimately associated with malachite and pseudomalachite, additional study should be directed at establishing all of the physical constants of the mineral. Yakhontovite d_{001} expands to 17.9 Å with glycerin and 17.6 Å with ethylene glycol. IR and DTA are similar to nontronite. *MZ 8(6):80(1986)*. Pistachio-green, dull luster. Conchoidal fracture. $H = 2$–3. Biaxial; $Z \wedge c = 12$–$14°$; $N_x = 1.530$–1.547, $N_z = 1.560$–1.570; positive elongation; pleochroic, dark to medium blue-green. Found as veinlets with malachite, pseudomalachite, chrysocolla, chalcopyrite, pyrrhotite, stannite, quartz, and so on. *Komsomolsk region, Far East Region, Russia.* VK, EF *MIN 4(2):83(1992)*.

71.3.1b.8 Zincsilite $Zn_3Si_4O_{10}(OH)_2 \cdot 4H_2O$

Named in 1960 by Smol'yaninova et al. for the chemical composition. Smectite group. MON Space group not certain. $b = 9.17$, $c = 15.3$; expands to 17.6 Å after glycolation, and collapses to 10 Å after heating to 400° for 10 minutes. *PD*(nat): 15.3_{10} 7.56_2 4.56_2 4.09_4 2.56_5 1.53_7; (glycolated): 17.6_{10} 8.50_4 4.56_4 4.05_7 2.57_4 1.53_6; (heated to 400°): 10.0_{10} 4.56_6 3.95_7 2.57_7 1.53_{10}. Zincsilite is indistinguishable from sauconite, except for optical properties, which may depend on Zn ↔ Al substitution. If tetrahedral Al were present, a layer charge would be developed and an interlayer cation would be necessary. Zincsilite would then represent an end member of the sauconites. Diagnostics are similar to dioctahedral smectite. Foliae to 2 mm. White to bluish white, pearly luster. Cleavage {001}, perfect. Decomposed by HCl. $H = 1\frac{1}{2}$–2. $G = 2.67$–2.71. Fusibility 5. IR and DTA data: *MIN 4(2):101(1992)*. Chemical analyses (wt %) for white and green varieties, respectively: CuO, 0.60, 3.07; MgO, 4.62, 1.08; CaO, 6.40, 2.00; MnO, 0.40, none; ZnO, 26.64, 35.00; Al_2O_3, 0.84, 0.70; Fe_2O_3, 2.16, 1.55; SiO_2, 47.6, 42.75; H_2O^+, 6.35, 8.50; H_2O^-, 4.35, 6.00. Biaxial (−); $Z \wedge c = 3°$, X approximately perpendicular to {001}, OAP parallel to (010); $N_x = 1.514$, $N_y = 1.559$, $N_z = 1.562$; $2V = 0$–$22°$. Heating to 400° produces pleochroic material in shades of brownish green to pale green to colorless. After heating to 400°, $N_x = 1.512$ and $N_z = 1.570$. Found as pseudomorphs after diopside with chrysocolla, fluorite, opal, and manganese oxides in an oxide zone of a galena–sphalerite–chalcopyrite skarn. *Batystau, central Kazakhstan*; Gifu Pref., Japan. VK, EF

CHLORITE GROUP

The chlorite group minerals are phyllosilicates corresponding to the general formula

$$A_{5-6}Z_4O_{10}(OH)_8$$

where

A = Al, Fe^{2+}, Fe^{3+}, Li, Mg, Mn, Ni (or a combination of two or more of these elements)
Z = Al, Si, Fe^{3+} (or a combination)

Allowing for substitution, the formula may be written somewhat more exactly as

$$(Mg, Fe^{2+}, Fe^{3+}, Al)_3(Al, Si)_4O_{10}(OH)_2(Mg, Fe^{2+}, Fe^{3+}, Al)_3(OH)_6$$

or

$$X_{2-3}Y_{2-3}[Si_{3\pm x}Al_{1\pm x}O_{10}(OH, O)_8]$$

The structure consists of an alternation of layers of two distinctly different types. The first type is called the *tetrahedral–octahedral–tetrahedral* (t-o-t) *layer* (in the literature this layer is also sometimes called the *2:1 silicate layer, talc layer*, or *pyrophyllite layer*). The second type of layer we call the interlayer (in the literature it has sometimes been called the *gibbsite layer* or *brucite layer*).

The t-o-t layer is itself made up of two infinite sheets of six-membered rings of SiO_4 and AlO_4 tetrahedra. The two infinite sheets are oriented toward one another, apical oxygens and hydroxyl groups of one sheet facing the apical oxygens and hydroxyl groups of the other sheet, so that between the two sheets there is an interior set of octahedral sites, the X sites, for cations. The exterior sides of the t-o-t layer are generally oxygen only. The usual ratio of Si to Al in the tetrahedra is 3:1. As Si^{4+} is tetravalent and Al^{3+} is trivalent, the substitution of an aluminum for a silicon frees an electron, and the t-o-t layer can thus acquire a net negative charge. The overall chemical formula of the t-o-t layer can be given as, for example in the case of magnesium, $Mg_3[(AlSi_3O_{10})(OH)_2]^-$.

The interlayer is also made up of two infinite sheets, this time sheets of close-packed hydroxyl groups, with the interstices between the layers providing octahedral sites, Y sites, for cations. The interlayer generally has a net positive charge, an example of the formula being $[Mg_2Al(OH)_6]^{1+}$.

The interlayer and the t-o-t layer are bound together by electrostatic and hydrogen-bonding forces. Normal to the layers, in the c direction the usual repeat distance is 14 Å; that is, 14 Å is the usual thickness of one t-o-t layer plus one interlayer. The a and b directions of the t-o-t layer may be oriented relative to the a and b directions of the interlayer in 12 different ways, resulting in a series of 12 unique polytypes discussed in *AM 46:819(1962), CCM 37:193(1989), MJJ 4:11(1963), REVM 19, MSM 1(5):86(1980)*.

71.3.1b.8 Zincsilite

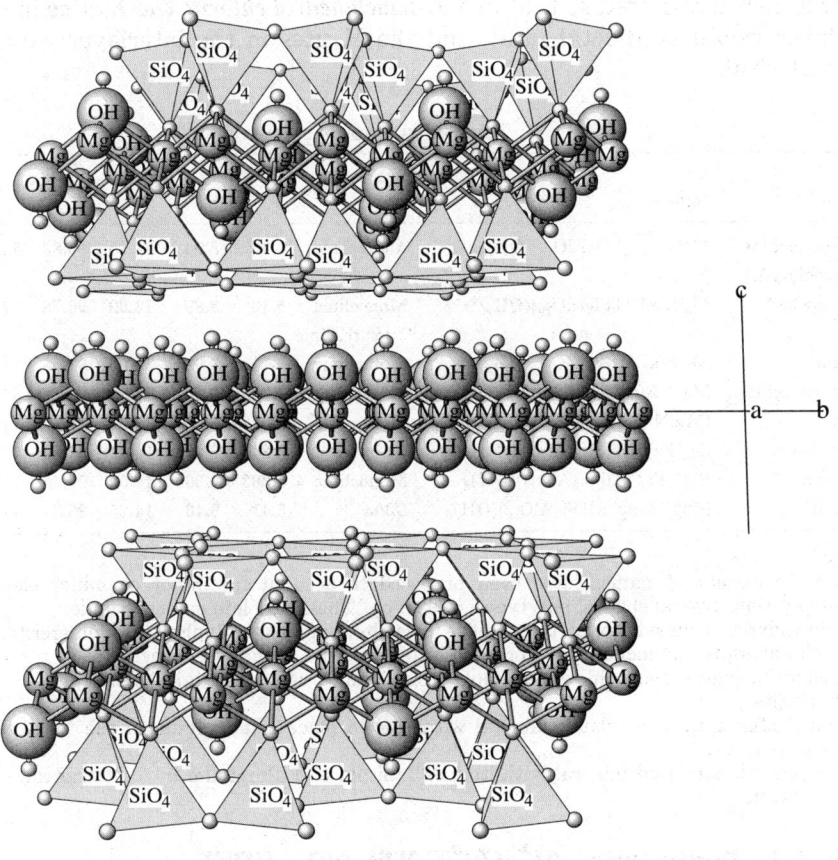

CHLORITE: $Mg_3Si_4O_{10}(OH)_2 \cdot Mg_3(OH)_6$
This is a double sandwich structure, with talc layers alternating with $Mg(OH)_2$ (brucite) layers.

The octahedral cation sites in both types of layers may be occupied by divalent or trivalent cations. When the cation is divalent, three of three adjacent cation sites may all be occupied, and this is electrostatically neutralized by $6OH^-$ ions (or by six O's from SiO_4 tetrahedra), leading to a group of zero net charge consisting of $[Mg_3^{2+}(OH^-)_6]^0$. A layer made up of such groups is referred to as *trioctahedral*, and each oxygen is surrounded by three divalent cations. Another contrasting situation exists when the cation is trivalent, such as Al^{3+}. The equivalent grouping consists of two trivalent cations plus an empty site: $[Al_2^{3+}\square(OH^-)_6]^0$. Sheets made up of this kind of group that involves vacancies are called *dioctahedral* since each oxygen is surrounded by only two octahedrally coordinated cations. With the chlorites the octahedral site occupancies are such that one can have two dioctahedral layers, two trioctahedral layers, or one of each. In the nomenclature of the group, the t-o-t layer is mentioned first, so that in a *di-trioctahedral* chlorite the X sites in the t-o-t layer would be dioctahedral, and the Y sites in the interlayer would be trioctahedral.

CHLORITE GROUP

Mineral	Formula	Space Group	a	b	c	β	
Di-di-donbassite (Ia polytype)	$Al_2^{3+}[Al_{2.33}^{3+}][Si_3AlO_{10}](OH)_8$	Monoclinic	5.16	8.94	14.15	93.83	71.4.1.1
Di-tri-cookeite	$Al_2[LiAl_2^{3+}][Si_3AlO_{10}](OH)_8$	Monoclinic or triclinic	5.13	8.89	14.09	96.78	71.4.1.2
Sudoite	$Al_2[Mg_2Al][Si_3AlO_{10}](OH)_8$	Monoclinic	5.21	9.02	14.00	97.03	71.4.1.3
Tri-tri-clinochlore	$Mg_3^{2+}[Mg_2^{2+}Al][Si_3AlO_{10}](OH)_8$	$C2/m$	5.306	9.203	14.18	96.83	71.4.1.4
Nimite	$(Mg,Ni)_3[Ni_2Al][Si_3AlO_{10}](OH)_8$	$C2/m$	5.320	9.214	14.302	97.10	71.4.1.5
Baileychlore	$Zn_3[Fe_2^{2+}Al][Si_3AlO_{10}](OH)_8$	$C1$ or $C\bar{1}$	5.346	9.257	14.402	97.12	71.4.1.6
Chamosite	$Fe_3^{2+}[Fe_2^{2+}Al][Si_3AlO_{10}](OH)_8$	Monoclinic	5.363	9.306	14.00	97.43	71.4.1.7
Pennantite	$Mn_3^{2+}[Mn_2^{2+}Al][Si_3AlO_{10}](OH)_8$	$C2/m$	5.41	9.40	14.25	96.8	71.4.1.8

Notes:
1. A wide variety of names have been proposed to account for major or minor elemental composition, crystal shapes, polytypes, and so on. These obsolete names include: aphrosiderite, brunsvigite, corundophilite, delessite, diabantite, daphnite, kämmererite, leuchtenbergite, orthochamosite, orthochlorite, mennine, menninite, prochlorite, pseudothuringite, pycnochlorite, ripidolite, sheridanite, strigovite, talc–chlorite, and thuringite.
2. Franklinfurnacite is a related mineral with zinc in place of aluminum in the tetrahedra of the t-o-t layer.
3. Gonyerite is a related mineral with iron Fe^{3+} in place of aluminum in the tetrahedra of the t-o-t layer.

71.4.1.1 Donbassite $Al_2^{3+}[Al_{2.33}^{3+}][Si_3AlO_{10}](OH)_8$

Named in 1940 by Lazarenko for the location. Chlorite group. Not currently approved by the IMA-CNMMN. Originally called α-chloritite based on chemistry in 1906, but without physical data. MON $C2/m$, Cc, or $C2$. $a = 5.16$–5.174, $b = 8.94$–8.956, $c = 14.15$–14.26, $\beta = 93.83$–$97.83°$ (two-layer polytype), Z = 2, D = 2.66. *45-1375*(calc pattern): 14.1_6 7.06_7 4.71_{10} 4.48_5 3.98_9 3.53_6 2.52_5 2.33_{10}. *CCM 37:193(1989)*, *SPC 17:456(1972)*. Structure generally Ia polytype. Ordering of Al into a specific tetrahedron has been observed with a partly vacant site above and below the Al ordered tetrahedron. *SPC 17:456(1972)*. Generally fine-grained earthy, also flaky aggregates. White to slightly greenish or yellowish white, pearly luster. Cleavage {001}, perfect. H = 2–$2\frac{1}{2}$. G = 2.63–2.64. Slightly decomposed by HCl and H_2SO_4, unaffected by HNO_3. Flexible lamellae. Frequently near end member, but can grade to aluminian sudoite or cookeite. Total octahedral occupancy generally varies from 4.27 to 4.49, with $^{IV}Al = 0.6$–1.3. Minor substitutions (wt %) include: FeO, 0.54 (Novaya Zemlya); Fe_2O_3, 3.34 (Kesselberg); MgO, 1.75 (Utrennyaia); MnO, 0.08 (Spain); CaO, 2.03 (Utrennyaia); TiO_2, 1.2 (St-Paul-de-Fenouillet); Na_2O, 1.08 (Zhuravaka); K_2O, 3.95 (Kesselberg); Li_2O, 1.43 (Itaya). Biaxial (+); $Z \wedge c = 0°$; $N_x = 1.560$–1.578, $N_y = 1.566$–1.582,

$N_z = 1.572$–1.596; $2V = 48$–$55°$; birefringence $= 0.006$–0.018; $r > v$; negative elongation; OAP parallel to (010). Probably a common mineral in high alumina, sedimentary, or altered wall or breccia-zone rocks of hydrothermal ore deposits (with hematite or arsenopyrite, etc.), fluorite veins, soils, sediments, and slickensides in coal. Frequently associated with kaolinite, muscovite or illite, pyrophyllite, interstratified phyllosilicates, and other minerals. Also occurs in granitic pegmatites. Observed as a hydrothermal alteration product of andalusite and spodumene. Major localities include: Alberni soils, BC, Canada; St-Paul-de-Fenouillet, eastern Pyrenees Mts., France; Cornberger sandstone, Hesse; Kesselberg, Schwartzwald, Germany; Zhuravaka and Nikitovsk ore field, and *Utrennyaia shaft and Uralskaya vein, Nagolyn area, Donbass region, Ukraine*; Karagand basin, Kazakhstan; Berezovskoye and Novaya Zemlya, Ural Mts., Russia; Kamp d'Atondo, Zaire; Namivu, Alto Ligonha, Mozambique; Itaya, Okayama Pref., Japan; Szabo Bluff, Scott Glacier, Antarctica. VK, EF *MS 21:40(1967)*, *MM 52:396(1988)*, *MIN 4(2):156(1992)*.

71.4.1.2 Cookeite $Al_2[LiAl_2^{3+}][Si_3AlO_{10}](OH)_8$

Named in 1866 by Brush for Josiah B. Cooke, Jr. (1827–1894), American chemist and mineralogist, Harvard University. Chlorite group. MON $C2/c$ or Cc. $a = 5.13$–5.17, $b = 8.89$–8.95, $c = 14.09$–14.35, $\beta = 96.78$–$98.75°$, $Z = 2$, $D = 2.68$–2.70. Rarely TRIC $C\bar{1}$. *AM 60:1041(1975)*. $a = 5.14$, $b = 8.90$, $c = 14.15$, $\alpha = 90.58°$, $\beta = 96.20°$, $\gamma = 90°$, $Z = 2$, $D = 2.65$. 38-416($1M_{Ia}$): 14.1_4 7.05_3 4.71_{10} 4.45_3 3.53_6 2.51_4 2.32_6 $1.96_{3.5}$; Ia(s): $14.1_{7.5}$ 7.05_6 4.68_{10} 4.44_3 3.51_6 2.50_5 2.32_9 1.49_5; Ia(r): 14.1_9 4.70_6 3.52_6 $2.81_{4.5}$ 2.56_6 2.505_8 2.32_{10} $1.49_{7.5}$; IIb: 14.1_8 4.70_8 $4.45_{6.5}$ 2.47_{10} 2.37_8 $1.96_{7.5}$ 1.49_9. *CCM 37:193(1989)* and *45-1376, 45-1377, 45-1476*. Polytype is usually Ia, but IIb has been reported. *LIT 4:7(1971)*. Ordering of tetrahedral Si and Al has been suggested. Thin tabular crystals usually in radial rosettes; twinning sometimes present, around [310], with {001} composition plane. White or pale yellow or pale pink, frequently pale green, occasionally orange, rarely deep rose to purple; white streak; oily to subvitreous luster. Cleavage {001}, perfect; flexible, inelastic. $H = 2\frac{1}{2}$–$3\frac{1}{2}$. $G = 2.58$–2.69. Insoluble in common acids, soluble in HF. Cookeites generally have 4.9–5.35 octahedral sites occupied and can grade to lithian donbassite. Si is nearly constant at 3.0 Si atoms per four tetrahedral positions. Be and B may substitute for tetrahedral Al. The main octahedral substitution is $3(Li, alkali)^+/Al^{3+}$. Total Li varies from 0.75 to 1.50 atoms per six octahedral positions. High-Al_2O_3 and lower-Li_2O cookeites could grade to donbassite, or with concomitant higher MgO, grade toward sudoite. Numerous minor substitutions (wt %) include: TiO_2, 0.13 (Auburn); Fe_2O_3, 1.72 (Allakurtti); B_2O_3, 1.66 (Radkovice); FeO, 1.69 (Allakurtti); MnO, 0.86 (Vostoknoye); CaO, 1.63 (B.C.); MgO, 1.32 (Lipovka); BeO, 1.06 (Muiâne); Na_2O, 1.12 (Lipovka); K_2O, 0.67 (Allakurtti); P_2O_5, 0.11 (Varuträsk); F, 0.46 (Hebron); B_2O_3, 0.38 (Muiâne). Biaxial (+); $Y = b$, $X \wedge a = 0$–$3°$, $Z \wedge a = 90$–$87°$, $Z \approx c$; $N_x = 1.572$–1.585, $N_y = 1.579$–1.590,

$N_z = 1.589–1.607$; $2V = 0–90°$, usually $35–60°$; $r < v$; birefringence = 0.014–0.031; negative elongation; six twin sectors common and sometimes about a uniaxial core. May be pleochroic. $X = Y > Z$; X,Y, pale green to pink; Z, colorless to pale yellow. Most noted from late-stage mineralization in gem-pocket-bearing granite pegmatites associated with cleavelandite, elbaite, lepidolite, spodumene, and so on; also from some relatively simple quartz crystal-lined veins; increasingly recognized from sedimentary environments with kaolinite, diaspore, boehmite, illite, sandstones, bauxites (sometimes with fine-grained pyrophyllite), and others. Synthesized in the range 260–480°, 1–14 kbar. *CMP 108:72(1991)*. Sometimes found among the corrosion products and as pseudomorphs of pegmatitic spodumene, petalite, and/or elbaite. Also noted as an alteration of kaolinite and diaspore, by Li-bearing solutions, releasing quartz. Notable localities include: Pulsifer mine, Auburn, Androscoggin Co., Bennett mine(!), Buckfield, Waisanen–Tamminen mine, Greenwood; *Mt. Rubellite mine, Hebron, and Mt. Mica mine(!), Paris, Oxford Co., ME*; Gillette mine, Haddam Neck, Middlesex Co., CT; Jeffrey mine, northern Little Rock, Pulaski Co., Stand on Your Head No. 2 mine(!), Saline Co., AR; Ozark Uplift, MO; Rib Mt., Marathon Co., WI; Tin Mountain mine, Custer, Custer Co., SD; Pala district, Himalaya mine, Mesa Grande, San Diego Co., CA; Wait-A-Bit Creek, BC, Canada; Ogofau mine, Carmarthenshire, Wales; northern Pennine ore field, England; Castelnau-de-Brassac, Tarn, Vanoise, Savoie, Pyrenees Mts., Haute Garrone, France; Batallero and Formigal mines, Tena Valley, Huesca, Spain; Richelsdorf Mts., Hesse, Germany; Dobrá Vodá and Radkovice, Czech Republic; Varuträsk, Västerbotten, Sweden; triclinic from Djalair bauxite deposit, Middle Asia, Allakurtti, Murmansk Oblast, Lipovka, Ural Mts., Vostoknoye, Siberia, and Kalbinski Mts., Ural Mts., Russia; Golconda mine(!), Governador Valadares, Lavra de Genipapo, Araçuai, MG, Brazil; Carcota mine, Bolivia; Muiâne pegmatite, Alto Ligonha district, Zambezia, Mozambique; Manono, Kivu, Zaire; Padar, Kashmir, Pakistan; Back Creek deposit, Pambula, NSW, Londonderry, WA, Australia. VK, EF *CM 10:636(1970), AM 60:1041(1975)*.

71.4.1.3 Sudoite $Al_2[Mg_2Al][Si_3AlO_{10}](OH)_8$

Named in 1962 by von Engelhardt, Müler, and Kromer for T. Sudo (b.1911), Japanese mineralogist. Chlorite group. MON $C2/m$. $a = 5.21–5.238$, $b = 9.02–9.080$, $c = 14.00–14.29$, $\beta = 97.03–97.45°$. Or $C2/c$. $a = 5.225$, $b = 9.058$, $c = 14.19$, $\beta = 93.67°$, $Z = 2$, $D = 2.65$. May also be $C\bar{1}$. $19-751(1M_{IIb}, \text{nat})$: 14.2_8 7.13_7 4.74_8 $4.52_{8.5}$ $3.55_{6.5}$ 2.50_{10} 2.41_5 1.51_6. Ia and IIb polytypes known. A di- or trioctahedral chlorite. IIb polytype: A characteristic feature of the one-level layers is the ordered distribution of the cations, in which two of the three independent positions are almost completely occupied by Mg, with one by Al. The interatomic distances show that the Al replaces the Si in an ordered fashion. Generally fine-grained earthy, sometimes 3–5 µm, may form small plates (1 mm) in rosettes, or as scaly films. White to pale green, white streak, pearly luster. Cleavage {001}, perfect. H = $2–2\frac{1}{2}$. G = 2.67–2.70. Sometimes slowly decomposed by acids. Octahedral occupancy generally 4.5–5.1 per 6.0 sites, while $^{IV}Al = 0.4$–1.1 with $^{VI}Al = 2.5$–3.4 and $Mg = 1.2$–2.5. Can grade to magnesian donbassite, within a small distance. Intermediate compositions can grade to magnesian cookeite or aluminian clinochlore. Excess interlayer Mg commonly found. FeO, 6.27 wt % (Monticiano–Rocastrada) ("impure" specimens with Fe > Mg, Geco mine); Fe_2O_3, 7.02 (Sivagli, Yakutia); TiO_2, 0.34 (Rudnii Altai); MnO, 0.31 (Ardennes); CaO, 1.80 (Novo Zolotushinskoye); Na_2O, 1.41; K_2O, 1.52 (Furotobe); MnO, 0.28 (Ottré); Cr_2O_3, 0.01 (Cigar Lake); BaO, 0.03 (Cigar Lake); Cu, 0.45 (Svagli, Yakutia); ZnO, 0.09 (Regné); NiO, 0.09 (Ottré); F, 0.14 (Cigar Lake); Li_2O, 1.60 (Echassière), and could grade to cookeite. Biaxial (+); $Z \wedge c = 7°$, Z perpendicular to (001); $N_x = 1.563$–1.583, $N_y = 1.569$–1.589, $N_z = 1.581$–1.601; $2V = 50$–$78°$; birefringence = 0.018; negative elongation. Biaxial (−); $Z \wedge c \approx 90°$; $N_x = 1.569$, $N_y = 1.579$, $N_z = 1.583$; $2V = 78°$; birefringence = 0.014; r < v; negative elongation. Found in a wide variety of environments and frequently coexisting with a wide variety of minerals; can be found with and confused with pyrophyllite. Also sometimes associated with Mg-chlorites, kaolinite, dickite, nacrite, muscovite, illite, chloritoid, böhmite, and so on; often associated with metallic ore deposits, also in marine sediments, soils, low-grade metamorphic rocks, quartz veins, late-stage granite pegmatite alterations, hydrothermally altered sedimentary and igneous rocks, bentonites, diagenetic rocks. Can coexist with trioctahedral chlorites and may be a very common mineral. Observed to alter from andalusite. Only selected localities are given here: Chesapeake Bay, MD; Tracy mine, Negaunee, Marquette Co., MI; Colorado Plateau, CO; Geco mine, Manitouwadge, ONT; Cigar Lake Deposit, SAS, Canada; Anhee, Ottré and Regné, Ardennes, Belgium; Echassières, Massif Central, France; Wippra Zone, Harz Mts., *Knollenmergel-Keuper, Lützelbach bei Plochingen, Württemburg, Germany*; Kaiserbach Valley, southern Pfalz, Germany; Monticiano–Roccastrada area, Apennine Mts., Italy; Srnetica Mt., Bosnia; Crete Is., Greece; Dneprovsk–Donetsk, Ukraine; Beresovsk, Ural Mts., Russia; Madneuli, Caucasus Mts., Novo Zolotushinskoye formation, and Orlovskoye, Rudnii Altai, Russia; Namivu, Mozambique; Kamikita mine, Honko deposit, Aomori Pref. and Hanaoka, Kosaka, Kamikita, and Furutobe mines and Niida mine, Odate, Akita Pref., and Shinyo, Japan; Welkom Goldfield, Witwatersrand Basin, South Africa; Chalmers, Q, Australia; Szabo Bluff, Scott Glacier, Antarctica. EF *JSP 27:355(1957)*, *AC 21(7):A172(1966)*, *AM 52:673(1967)*, *SPC 18:50(1973)*, *CM 16:365(1978)*, *CMP 86:409(1984)*, *MIN 4(2):161(1992)*.

71.4.1.4 Clinochlore $Mg_3^{2+}[Mg_2^{2+}Al][Si_3AlO_{10}](OH)_8$

Named in 1851 by Blake in allusion to its symmetry and relationship to chlorite. Chlorite group. MON $C2/m$. $Z = 4$. TRIC $P1$. $a = 5.306$–5.37, $b = 9.203$–9.40, $c = 14.18$–14.40, $\beta = 96.83$–$99.80°$. Polytypes Ia, Ib, IIb, rarely IIA; $Z = 2$, $D = 2.47$–2.88. *16-362*(ferroan clinochlore, $1M_{Ia}$); *16-351*(ferroan clinochlore, $1M_{Ib}$); *12-242*($1M_{IIb}$); *24-506*($1M_{IIb}$); *7-160*(chromian $1M_{IIb}$); 7-78(ferrian $1M_{IIb}$); 29-701(ferroan $1M_{IIb}$); 20-671(chromian clinochlore, IIA); 45-1321(manganoan $1M_{IIb}$). Crystals usually only in vein cavities or fracture linings, tabular pseudohexagonal to blocky pseudorhombohedral, almost always horizontally grooved and almost always forming arcuate vermiform clusters. Generally green to very dark green or greenish black, rarely grading to white (low iron, high aluminum), rose to purple (> 2 wt % Cr_2O_3), or pinkish orange (Mn). H = 2–3. G = 2.30–2.85 (manganoan). Sector twinning common. Also twinned on {001} by mica and penninite twin laws. Cleavage {001}, perfect; flexible, but inelastic. Usually decomposed by acids, especially H_2SO_4. Clinochlore forms a continuous series with chamosite and almost always has a small ferric iron content: Fe_2O_3 = 8.42 wt % (Chester Co.). Additional substitutions (wt %) include: Cr_2O_3, 8.31 (Gümüshane), 13.46 (northern Sweden); TiO_2, 0.85 (Miask, Urals); MnO, 20.1 (Falotta, Graubunden, Switzerland); CaO, 1.24 (Deer Creek); Na_2O, 0.53 (Chester Co.); NiO, 7.60 (Ukraine). Biaxial (− or +), frequently uniaxial; N_x = 1.555–1.670, N_y = 1.556–1.685, N_z = 1.558–1.685; birefringence = 0.003–0.014, noticeable dispersion, frequently with anomalous blue interference colors, RI increases with increasing FeO or MnO and rises rapidly with increasing Fe_2O_3. Uniaxial (+); N_O = 1.562–1.587, 1.618, N_E = 1.576–1.594, 1.621; birefringence = 0.003–0.014; negative elongation; low RI colorless, greenish gray to pleochroic. O, colorless; E, green. Sometimes 2V = 0–5°, usually resistant to H_2SO_4 decomposition. Uniaxial (−); N_O − 1.579–1.619, N_E = 1.577–1.605; birefringence = 0.002–0.014; positive elongation; sometimes 2V = 0–5°; pleochroic. O, green; E, pale green to colorless. Anomalous blue interference colors. Biaxial (+); Z ≈ c; N_x = 1.553–1.599, N_y = 1.554–1.601, N_z = 1.558–1.609; 2V usually small 0–15°, but can be to 22°, and rarely to 50°; r < v; birefringence = 0.005–0.010; negative elongation. Insoluble in H_2SO_4. Biaxial (−); X = c; N_x = 1.576–1.605, N_y = 1.578–1.619, N_z = 1.578–1.619; 2V = 0°; r > v; birefringence = 0.003–0.014; positive elongation; anomalous blue interference colors. Pleochroic; X, colorless to pale green; Y,Z, green. Decomposed by H_2SO_4. RI increases with FeO and more rapidly increases with Fe_2O_3. Frequently, a product of contact or low-grade regional metamorphism. Found in fissure veins, soils, sediments, and amygdules in volcanic rocks. Frequently associated with feldspars, quartz, biotite, garnet, calcite group, dolomite group, magnetite, chromite, talc, serpentine, chloritoid, rutile, ilmenite, titanite, common sulfides, zircon, zeolites; rarely with corundum, diaspore, margarite, or spinel. Can be slightly to completely oxidized or replaced by other minerals. Found as a replacement of a variety of species, including garnet, pyroxene, amphibole, and so on. Found at Belvidere mine(!), Lowell, Orleans Co., Vermont; Acushnet, Bristol Co., MA; Tilly Foster mine(!), Brewster, Putnam Co., NY; Unionville, *Brinton mine, West Chester*, Chester Co., Cedar Hill mine(!), Texas, Wood's mine(!), Lancaster Co., PA; Culsagee mine, Macon Co., NC; Deer Creek, WY; Miles City, Custer Co., MT; Challis Custer Co., ID; Calaveras and San Benito Cos., and Sierra Nevada Mts., CA; Black Lake mine, Coleraine, Jeffrey mine(!), Asbestos, PQ, Canada; Banffshire, Scotland; Harstig mine, Påjsberg, Värmland,

Sweden; St. Christophe, Isére, France; St. Gotthard(!), Rimpfischwäge, Zermatt(!), Vallais, Switzerland; Marienberg, Saxony, Germany; Zillertal, Burgumer Alp, Pfisch, Tirol, Austria; Traversella(!), Ala, Aosta, Ossola, and Fassa valleys, Piedmonte, Italy; Frankenstein, Silesia, Poland; Kop Chrom mine(!) and elsewhere, Kop Doglari, Askale, near Erzurum, Erzincan, and Gümüshane, Anatolia, Turkey; Saranovskii mine(!), Saranui, Bazhenovsk mine(!), Asbest, Akmatovsk(!), Zlatoust(!), Syssertsk, Karkadansk, and Nasiamsk, Ural Mts., Korshunovsk deposit, Siberia, Russia; [Ushkatin] Atasui deposits, central Kazakhstan; Pedra Preta mine, Brumado, Bahia, Brazil; Kumanohata mine, Shiga Pref., Japan; Nelson, New Zealand.SPC 18:50(1973), VK, EF *MIN 4(2):179(1992)*.

71.4.1.5 Nimite (Mg,Ni)$_3$[Ni$_2$Al][Si$_3$AlO$_{10}$](OH)$_8$

Named in 1968 by De Waal for the acronym of the National Institute for Metallurgy, South Africa. Chlorite group. MON $C2/m$. $a = 5.320$–5.346, $b = 9.214$–9.268, $c = 14.302, 28.649$, $\beta = 97.10°$, $Z = 2$ (IIb) or 4 (Ia), $D = 3.20$–3.21. *22-712*(1M$_{IIb}$, nat): 14.2$_3$ 7.10$_{10}$ 4.74$_2$ 4.60$_{<1}$ 3.55$_5$ 2.84$_1$ 2.58$_{<1}$ 2.00$_{<1}$. *38-424*(1M$_{Ia}$, nat): 7.11$_{10}$ 4.74$_{5.5}$ 3.56$_6$ 2.65$_4$ 2.55$_3$ 2.39$_5$ 2.01$_3$ 1.54$_5$. Nimite is mostly composed of the IIb polytype. Fine-grained. Yellowish green, white streak, oily to vitreous luster. Cleavage {001}, perfect. H = 3. G = 3.12–3.19. IR data: *MIN 4(2):223(1992)*. Type material is magnesian: MgO = 10.13. Minor substitutions (wt %) include: Fe$_2$O$_3$, 4.35; Cr$_2$O$_3$, <0.01; FeO, 2.78; MnO, 0.06; CaO, 0.38; CoO, 0.38. Biaxial $(-)$; $N_x = 1.637$, $N_y = N_z = 1.647$; 2V = 15°; birefringence = 0.010. Weakly pleochroic; X, yellow-green; Y,Z, apple-green. Found in nickeliferous serpentinite with bonaccordite, bunsenite, gaspeite, goethite, liebenbergite, millerite, opal, reevesite, trevorite, violarite, and/or willemseite. *3+ km W of Scotia Talc mine, Bon Accord area, Barberton district, Transvaal, South Africa*; Woodline Well, WA, Australia. VK, EF *AM 55:18(1970)*, *MSM 5:94(1980)*.

71.4.1.6 Baileychlore Zn$_3$[Fe$_2^{2+}$Al][Si$_3$AlO$_{10}$](OH)$_8$

Named in 1988 by Rule and Radke for Sturges W. Bailey (b.1919), mineralogist, University of Wisconsin. Chlorite group. TRIC $C1$ or $C\bar{1}$. $a = 5.346$, $b = 9.257$, $c = 14.402$, $\alpha = 90°$, $\beta = 97.12°$, $\gamma = 90°$, $Z = 2$, $D = 3.20$. *AM 73:135(1988)*. *42-1335* gives MON $C2/m$. $a = 5.344$, $b = 9.249$, $c = 14.392$, $\beta = 97.06°$. *42-1335:* 14.3$_9$ 7.14$_{10}$ 4.60$_3$ 3.573$_4$ 2.66$_5$ 2.450$_4$ 1.542$_6$ 1.508$_3$. Type material contains 0.36 vacancy out of six octahedral sites and is the Ib polytype with broadened *hkl* reflections. A one-dimensional electron-density analysis and an asymmetry value of +1.29 suggest a composition ^{VI}(Zn$_{2.50}$Al$_{0.14}$ vacancy$_{0.36}$)$^{0.58-}$ for the 2:1 layer and ^{VI}(Fe^{2+}$_{1.20}$Al$_{1.03}$Mg$_{0.76}$Mn^{2+}$_{0.01}$)$^{1.03+}$ for the interlayer sheet. *AM 73:135(1988)*. Dark green fibrous veinlets. G = 3.18. Cleavage {001}, perfect. DTA-TGA. Type material is a ferroan baileychlore with FeO to 14.5 wt %. Additional substitutions (wt %) include: MgO, to 4.7; MnO, to 0.4; CaO, to 1.0. Biaxial $(+)$; $N_x = 1.582$, $N_{av} = 1.59$, $N_z = 1.614$; weakly pleochroic green–yellow green. Type material is optically zoned green to yellow-green along the fiber length. Found as selvages (to 1 mm) lining calcite veins in mineralized collapse karst–breccia, composed of altered andesite, garnet–vesuvianite skarn, sandstone, felite porphyry, and chert. Associated with chalcopyrite, zincian clinochlore, copper, cuprite, goethite, hematite, and malachite. *Red Dome deposit, Mungana, 15 km WNW of Chillagoe, Lynd Co., Q, Australia.* VK, EF

71.4.1.7 Chamosite $Fe_3^{2+}[Fe_2^{2+}Al][Si_3AlO_{10}](OH)_8$

Named in 1820 by Berthier for the locality. Chlorite group. Synonyms: daphnite, delessite, brunsvigite. MON $C2/m$, $C2$, and ORTH. $a = 5.363$–5.40, $b = 9.286$–9.375, $c = 14.00$–14.222, $\beta = 90°, 97.43$–$97.88°$, $99°$; $Z = 2$, $D = 3.10$–3.13. *21-1227*($1M_{IIb}$): 14.1_7 7.05_{10} 3.52_{10} 2.60_9 2.55_7 2.45_7 2.39_8 1.55_9; *13-29*($1O_{Ib}$): 14.1_4 7.05_{10} 4.71_4 3.53_8 2.84_3 2.52_5 2.15_4 1.56_4. Crystals tabular pseudohexagonal to blocky pseudorhombohedral. Green to greenish black, oily to vitreous luster. H = $2\frac{1}{2}$–3. G = 2.95–3.30. Cleavage {001}, perfect; foliated, flexible but inelastic. Decomposed by hot HCl. Grades toward clinochlore with increasing MgO and toward pennantite with increasing MnO; almost always contains ferric iron (wt %): Fe_2O_3, 18.97 (Messina); 8.70 (Schmiedefeld). Additional substitutions (wt %) include: TiO_2, 0.35 (NZ); MnO, 21.14 [Ushkatin] Atasui deposits, Kazakhstan; ZnO, 0.16 [Ushkatin]; CaO, 3.34 (Frodingham); Na_2O, 1.1 (Penzance); K_2O, 0.50 (Mourne). Displays a wide variety of optical behavior. Uniaxial (−); $N_O = 1.637$–1.643, $N_E = 1.615$–1.649; birefringence = 0.006–0.022; positive elongation. O, olive-green; E, pale yellow to brownish yellow. Biaxial (+); $Z \approx c$; $N_x = 1.620$–1.630, $N_y = 1.620$–1.630, $N_z = 1.625$–1.635; $2V = 0$–$5°$; $r > v$; birefringence = 0.005; negative elongation; pleochroic. X,Y, deep green; Z, greenish brown. Not decomposed by acids. Biaxial (−); $X = c$; $N_x = 1.643$–1.670, $N_y = 1.649$–1.685, $N_z = 1.649$–1.685; $2V = 0$–$6°$; birefringence = 0.006–0.015; positive elongation. X, pale yellow; Y,Z, olive-green. Decomposed by acids. Biaxial (−); $N_x = 1.615$, $N_y = 1.637$, $N_z = 1.638$; $2V = 15°$; birefringence = 0.023, positive elongation. Pleochroic; X, olive; Y,Z, brownish yellow. Decomposed by acids. RI increases with increasing FeO and increases rapidly with Fe_2O_3. Formed in low-grade regional and contact-metamorphosed rocks, granite pegmatites, also forms as a hydrothermal alteration of wall rock around sulfide deposits; amygdules in volcanic rocks. Abundant in soils and marine clays. Observed in coal deposits. Found variously associated with quartz, feldspars, garnet, muscovite, biotite, amphiboles, pyroxenes, dolomite group, chromite, talc, serpentine, chloritoid, ilmenite, titanite, common sulfides, zircon, zeolites, siderite, calcite, tourmaline, magnetite, hematite, rutile, and others. Observed as a replacement of a wide variety of silicates, including garnet. Alters by recrystallization and reaction with coexisting iron minerals, including siderite, to a fine-grained, more iron-rich chamosite. Also alters to limonite. Excellent quality or analyzed material found at: Rangeley formation, Byron, Oxford Co., Maple Mt., Aroostook Co., ME; State Line quarry, York Co., French Creek, Chester Co., PA; Harper's Ferry, Jefferson Co., WV; Hot Springs, Garland Co., AR; Michigamme, Marquette Co., MI; Climax, Lake Co., CO; Liskeard, Tolgus mine, Redruth, Coradon mine, St. Clear, and Penzance, Cornwall, Wickwar, Gloucestershire, Frodingham, Lincolnshire, England; Silent Valley quarry, Mourne Mts., Northern Ireland; Bas Vallon, Brittany, St. Christophe, Isére, France; Windgälle, Uri, *Chamoson, Valais, Switzerland*; Schmiedefeld, Saalfeld, Thuringen, Büchenberg, Harz, Germany; Grängesberg, Sweden; Nučic, Chrustenic, and Kank, Čechy, Czech Republic; Strzegom (Streigau), Wroclaw, Poland; [Ushkatin] Atasui deposits, Kazakhstan; Messina, Transvaal, South Africa; Randalls and Kalgoorlie, WA, Australia; Tateyama Hot Springs, Jodojama, Toyama Pref., Kumanohata mine, Shiga Pref., Japan; Great Is., New Zealand. VK, EF *MM 25:441(1939), 30:277(1954); MIN 4(2):207(1992).*

71.4.1.8 Pennantite $Mn_3^{2+}[Mn_2^{2+}Al][Si_3AlO_{10}](OH)_8$

Named in 1946 by Campbell et al. for Thomas Pennant (1726–1798), Welsh naturalist. Chlorite group. MON $C2/m$. $a = 5.41$–5.51, $b = 9.40$–9.54, $c = 14.25$–14.40, $\beta = 96.8$–$97.3°$, $Z = 2$, $D = 3.18$. 42-594($1M_{IIb}$, nat): 7.12_{10} 3.56_9 2.65_7 2.60_5 2.43_5 2.30_5 2.03_6 1.58_{10}; 29-884($1M_{Ia}$, nat): 14.3_4 7.1_{10} 4.75_3 3.57_8 2.70_4 2.43_8 2.03_4 1.68_3. Known as Ia ("grovesite") and IIb polytypes. Generally fine-grained or closely packed small rosettes, rarely platy. Orange-brown, blackish brown to dark red; waxy to oily luster. $H = 2$–$2\frac{1}{2}$. $G = 3.04$–3.15. Cleavage $\{001\}$, perfect. Grades toward magnesian clinochlore. Substitutions observed (wt %) include: Fe_2O_3, 6.67 (Ushkatin); MgO, 3.4 (Franklin); BaO, 1.33 (Benalt); ZnO, 15.9 (Franklin); minor Pb, Na, K, and Ge detected (Ushkatin). Uniaxial $(-)$; $N_O = 1.667$–1.673, $N_E = 1.658$–1.664; birefringence $= 0.009$; positive elongation; pleochroic. O, dark brown; E, perpendicular to $\{001\}$, red-brown. For Ia: biaxial $(-)$; X perpendicular to $\{001\}$; $N_x = 1.646$–1.664, $N_y = 1.661$–1.673; $2V = 0$–$1°$; $r \ll v$; birefringence $=0.009$–0.029; positive elongation. Pleochroic; X, red-brown; Y,Z, dark brown. Found in a variety of metamorphic or hydrothermally altered manganese- and zinc-bearing assemblages. Found associated with alleghanyite, analcime, banalsite, barite, calcite, kellyite, kutnahorite, paragonite, prehnite, rhodochrosite, and/or willemite. Mine Hill, Franklin, NJ; Bald Knob, Sparta, Alleghany Co., NC; *Ty Canol incline, Benalt mine, Rhiw, Carnarvonshire, Lleyn Peninsula, Gwynedd, Wales*; French Alps (Dept. Savoie); [Ushkatin] Atasui deposits, central Kazakhstan. VK, EF *MM 27:217(1946), 30:645(1954); CM 21:545(1983); MIN 4(2):229(1992).*

REGULARLY INTERSTRATIFIED PHYLLOSILICATE GROUP

The great abundance of clay minerals with superlattice (long period basal) reflections has led to the recognition of phyllosilicates composed of a mixture of layered structures. Although clay minerals are generally too small to be subjected to single-crystal structural analysis techniques, one-dimensional Fourier calculations have generally substantiated the belief that a portion of the clay minerals consist of regular, 1:1 interstratifications of mica-like, talc/pyrophyllite-like, chlorite-like, vermiculite-like, serpentine-like or smectite-like layers: ". . . non-random interstratification involving a proportion of two types of layers different from 1:1 is less easily recognized, because it gives rise to a sequence of $(00l)$ reflections that is non-integral and is often mistakenly assumed to be random." *(CCM 18:25(1970))*. Trioctahedral serpentine-like layers have not been observed in regularly interstratified phyllosilicates, but dioctahedral kaolinite-like layers have not been well characterized. One dimensional Fourier methods of structure calculations have been summarized *(CCM 18:25(1970))*, Int. Clay Conf., Denver 1985:33). HRTEM and NMR studies are numerous enough that regular interstratification of phyllosilicates is verified, independent of XRD studies. These individual layers can be dioctahedral or trioctahedral, but are generally of the same type within a particular sample. (As the individual layers are not a true fragment of the mineral they represent, the suffix "like" is added to the layer name to clarify the use of smectite-like vs. smectite, for instance. Dioctahedral smectite is preferable to "montmorillonite" unless it was defined as a necessary component of a particular 1:1 interstratified mineral). Recent NMR analysis of the rectorite structure has broad implications as to the actual character of the "species-like" layers. NMR study has suggested that similar Al ↔ Si substitutions occur in the tetrahedral sheets facing the interlayer cations, rather than being on either side of the octahedral sites within a layer. Therefore, juxtaposed tetrahedral sheets might be both mica-like (Si_3Al) or both margarite-like (Si_2Al_2) about the fixed interlayer cation and both smectite-like ($Si_{4-x}Al_x$) tetrahedral sheets facing the exchangeable cation. The interstratifications could therefore be considered as hybrid layers where each and every layer was composed of smectite-like and mica-like or margarite-like tetrahedral sheets with alternation of "up-down" hybrid layer orientations *(AM 80:247(1995))*. Some regularly interstratified minerals appear more likely to be the result of constructive changes (e.g. LLC corrensite), while some may be more likely to be degradation products (hydrobiotite). Conjecture has been directed at interstratified phyllosilicates concerning their stability and whether or not they would spontaneously transform, through time, into inter-stratified phyllosilicates. Many interstratified clays are not 1:1 regular structures, but have random stacking and/or greater proportions of one layer compared to another. The regular interstratifications yield superlattice reflections equivalent to the sum of the thickness of the component layers, while semi-randomly or disproportionately (i.e. non-1:1) interstratified structures do not coherently diffract X-rays and so yield spacings only of their individual layers. Ordered interstratifications can be grouped in packets within otherwise non-interstratified or randomly interstratisfied clay particles. It is frequently possible that recognition of "regular" interstratifications to 2:1 is possible, but with diminished intensities and increasingly non-integral spacings. High ratios of interlayers do not provide sufficient intensities to be observed. Interstratified minerals are variously reported as hexagonal or orthorhombic, but few data have been presented to support these high symmetry categorizations. Not all interstratified layers have similar elemental ratios. For example, the chemical variability of chlorite-like layers appears greater than their co-interstratified smectite-like layers *(CCM 32:391(1984))*. Major problems of studies with regularly interstratified phyllosilicates have been attempts to rationalize new kinds of interstratifications or chemical variations with pre-existing classifications instead of recognizing a new mineral species.

Definitions. The AIPEA (Associated Internationale Pour l'Etude des Argiles) has adopted a set of criteria by which they will approve newly named 1:1 regular interstratifications *(CM 17:243(1982), CCM 30:76(1982))*. The most important criterion includes: "To merit a name, an interstratification of two layer types A and B should have sufficient regularity of alternation to give a well-defined series of at least ten 00- summation spacings $d^{AB} = d^A + d^B$, for which the suborders are integral and the even and odd suborders have closely similar diffraction breadths. If any odd 00- suborders are absent, calculations must be given to show that their intensities are too small to be observed. The coefficient of variation (CV) of the d(00-) values should be less than 0.75 to demonstrate adequate regularity of alternation." Additional details are considered. The limits of interstratification proportions have not been rigorously defined. Use of a species name suggests 1:1 proportions, but almost all natural samples deviate from this ratio slightly to significantly and still show characteristic long period spacings.

Types of Interstratification. The thickness of the various layers are constant: serpentine-like ≈ 7.3Å, talc/pyrophyllite-like $\approx 9.1 - 9.4$Å, mica-like ≈ 10Å, vermiculite-like $\approx 12 - 14$Å, chlorite-like ≈ 14Å, smectite-like $\approx 15-17$Å. Therefore, hydrobiotite, 1:1 mica-like/vermiculite-like interstratifications should have a spacing 10Å+14Å= 24Å+/−. Observed basal spacings alone are not unambiguous, especially when a given layer can be variously hydrated. Various diagnostic tests have been devised to verify the swelling or collapsing characteristics of the various layers involved, including intersalination, intercalation by organic liquids, or dehydration by heat. Computer programs have been devised to analyse interstratified phyllosilicate data *(CM 28:445(1993))*. A number of observed regular interstratifications have not been proposed for species status for different reasons, including high CV, ambiguous data, etc. Multiple interstratifications have been proposed, such as tarasovite (mica-like/rectorite-like, chlorite-like/corrensite-like, etc., but these materials are not 1:1 regular interstratifications and the CV is usually unfavorable.

Regularly interstratified phyllosilicates assemblages and localities which have been extensively studied (e.g. rectorite and corrensite) show a tendency to have shorter period phyllosilicate assemblages with increased age or with subjection to increased temperature exposure. For example illite-like/smectite-like interstratifications disappear at depth. Also basin-wide patterns show corrensite distributed between non-interstratified smectite-bearing and chlorite-bearing zones. Some interstratifications are not predicted. For example, vermiculite is a trioctahedral mineral by definition, and a report of a dioctahedral "vermiculite-like" layer would be a misnomer. Table xx is a summary of the 1:1 regularly interstratified phylosillicates.

Kaolinite-like/smectite-like interstratifications have been occasionally reported *(CM 26:343(1991)*; Murray et al., *Kaolin Genesis and Utilization*, 1993). The interstratification is generally poorly crystalline, yielding too few peaks to allow characterization or is randomly interstratified, yielding nonintegral basal reflections.

TABLE 1. 1:1 REGULARLY INTERSTRATIFIED PHYLLOSILICATES

	mica-like	talc/pyrophyllite-like	smectite-like	vermiculite-like	serpentine-like	layer types
pyrophyllite	–	–	–	‡	–	di-, dioctahedral
talc	–	–	aliettite	–	–	tri-, trioctahedral
chlorite	*	–	Al-tosudite	‡	–	di-,di-, dioctahedral
	saliotite	lunijianlaite	Mg-tosudite	‡	–	di-, tri-, dioctahedral
			Li-tosudite			"
	–	kulkeite	HLC corrensite	LLC corrensite	dozyite	tri-,tri-, trioctahedral
vermiculite	–	–	–			di-, dioctahedral
	hydrobiotite	**	–			tri-, trioctahedral
smectite	Na-rectorite					di-, dioctahedral
	K-rectorite					"
	–			di-, trioctahedral		
	–			tri-, trioctahedral		

* indicate that a material of this category has been reported without a specific proposal being made as to its nomenclature. *NJGPA 1988: 176 **sometimes has been called rectorite, but this interstratification is unknown.
‡ = not predicted [The octahedral site occupancies are designated for all layers. Whichever layer sequence is listed first, that layer has its site occupancy listed first. For chlorite-like interstratifications, the first designator is for the silicate layer, while the second designator is for the interlayer: di-, trioctahedral therefore signifies a dioctahedral silicate layer with a trioctahedral interlayer.]

71.4.2.1 Dozyite [$(Mg,Fe^{2+})_3SiAlO_5(OH)_4 \cdot (Mg,Fe^{2+})_3[(Mg,Fe^{2+},Fe^{3+})_2Al](Si,Al)_4O_{10}(OH)_8$]

Named in 1995 by Bailey et al. for Jean Jacques Dozy, Dutch geologist. MON Cm. $a = 5.323$, $b = 9.215$, $c = 21.45$, $\beta = 94.44°$, also 90° (Pennsylvania), $Z = 2$. Oriented basal reflections noted $d_{(001)}$, $21.4_{1.4 \text{ calc}}$; $d_{(002)}$, 10.7_1; $d_{(003)}$, 7.12_3; $d_{(004)}$, 5.31_2; $d_{(005)}$, 4.30_2; $d_{(006)}$, 3.56_{10}; $d_{(007)}$, 3.06_2; $d_{(00.15)}$, 1.426_2. AM 80:65(1995). Powder diffraction results: 7.1_{10}, $d_{(003)}$; 4.61_3, $d_{(02.11)}$; 3.56_2, $d_{(006)}$; 2.61_8, $d_{(201,13-2)}$; 2.56_5, $d_{(132,20-3)}$; 2.43_4, $d_{(203,13-4)}$; 1.536_6, $d_{(060,33-1)}$; 1.501_7, $d_{(332,063,33-4)}$. AM 80:65(1995). Indonesian dozyite is a Iaa-2,4,6 polytype. Dozyite is a regularly interstratified 21-Å phyllosilicate with alternating trioctahedral serpentine-like and trioctahedral chlorite-like layers, but stacking errors were detected and intergrowths with amesite noted. It is the first 1:1 layer with 2:1 layer phyllosilicate interstratification. HRTEM imaging confirmed the interstratification. CMP 117:137(1994).

Random serpentine-like–chlorite-like interstratifications have frequently been observed. *AM 80:65(1995)*. Does not swell with organic fluids, does not collapse on heating to 400°: "Detection of interstratification thus depends on a detailed comparison of observed and calculated peak intensities and widths." *AM 80:65(1995)*. Found as white to colorless foliae (to 2 mm) (Indonesia); also observed as pale purple or mauve fibers (chromian) (Pennsylvania); white streak; pearly luster. H = 2.5. G = 2.66. Cleavage (001), perfect. Biaxial (+); 2V < 5°; pseudouniaxial; OPA parallel to (010), Y = b, Z $\wedge$ c = 4°; N_x = N_y = 1.575, N_z = 1.581; weak dispersion. X,Y, colorless; Z, pale tan in thick crystals. Dozyite is difficult to distinguish optically from randomly interstratified serpentine-like–chlorite-like layers associated with it (Indonesia). Dozyite has the following formula: $(Mg_{2.00}Al_{0.85}Fe^{3+}0.15)^{+1.00}$ 3.00 $(Si_{1.00}Al_{1.00})^{-1.00}$ $2.00O_5(OH)_4$ $(Mg_{5.29}Al_{0.70}Fe^{3+}_{0.04})^{+0.80}$ $6.03(Si_{3.18}Al_{0.82})^{-0.82}$ $4.00O_{10}(OH)_8$ and is thus an interstratified amesite-like–clinochlore-like mineral. Chemical analyses (Indonesia, wt %) gave: SiO_2, 29.69; Al_2O_3, 20.29; MgO, 43.74; FeO, 1.63; CaO, 0.04; Na_2O, 0.07; K_2O, ~ 0; Li_2O, ~ 0; B_2O_3, ~ 0; BeO, ~ 0; Cl, 0.18; H_2O, 12.20. Low Al in dozyite from Pennsylvania rules out the likelihood of its having amesite as its serpentine component. An iron-rich dozyite consisting of berthierine-like layer–chamosite-like layer interstratification has been described. *GSAPA: A-146(1993)*. Found in Si, Fe, and Al metasomatized and hydrothermally altered limestone resulting in a "skarn." Associated minerals include lizardite, chrysotile pseudomorphic after forsterite, antigorite, amesite, clinochlore (IIbb, etc.), magnesite, dolomite, calcite, anhydrite, chalcopyrite, etc. (Indonesia). Also found with chromite, chromian clinochlore (IIbb), and chromian lizardite (Pennsylvania). *Wood's chrome mine, Lancaster Co., Pennsylvania*; Archean granulite–facies ironstone, eastern border of Beartooth Mts., MT; *drill core in skarn near Ertsberg East mine, 4,200-m elevation, Carstenz Mts., central Irian Jaya, Indonesia*. VK, EF

71.4.2.2 Lunijianlaite [$Al_2(LiAl_2)Si_3AlO_{10}(OH)_8/Al_2Si_4O_{10}(OH)_2$]= $LiAl_6(Si_7Al)O_{20}(OH)_{10}$

Named in 1990 by Kong et al. for "chlorite alternating (interstratified) with pyrophyllite." Probably MON Space group unknown. a = 5.09, b = 8.97, $c \sin \beta$ = 23.397. *AMS 10:289(1990)*. On hexagonal axes: a = 5.085, c = 23.367, D = 2.75–2.80. PD(nat): 14.27_2 11.79_2 7.80_2 4.70_{10} 3.90_1 3.54_5 3.34_5 2.92_4 2.83_2; *45-1397*: 23.5_1 11.79_2 7.80_2 4.70_{10} 3.90_1 3.34_5 2.92_4 1.95_1. Lunijianlaite is an interstratified cookeite-like–pyrophyllite-like interstratification with $\approx$ 23–24 Å basal spacing. Does not expand with glycol and does not collapse on heating to 500°. Colorless to white radial aggregates to 2 mm, pearly to vitreous luster, transparent. H = 2. G = 2.75. Cleavage (001), perfect. Biaxial (−); N_x = 1.576, N_y = 1.583, N_z = 1.587; 2V = 61°; birefringence = 0.011; extinction can be undulatory, extinction with cleavage can be parallel to 2–6°. Chemical analysis (wt %) of type material gave: SiO_2,

41.61; Al_2O_3, 44.80; Fe_2O_3, 0.60; Li_2O, 1.57; Na_2O, 0.063; K_2O, 0.012; H_2O^+, 11.296. Occurs with blue corundum aggregates with pyrophyllite in hydrothermally altered rhyolite. Can be in contact with diaspore and intergrown with minor cookeite. Additional associated species include chlorite, illite, halloysite, svanbergite, zeolites, and specular hematite. *Qingtian, Zhejiang Prov., China.* VK, EF

71.4.2.3 Saliotite $Al_2(LiAl_2)Si_3AlO_{10}(OH)_8/(Na,K)Al_2Si_3AlO_{10}(OH)_2$

Named in 1994 by Goffé for Pierre Saliot (b.1941), geologist at l'Ecole Normale Supérieure de Paris. MON $C2/m$. $a = 5.158$, $b = 8.914$, $c = 23.83$, $\beta = 94.23°$, 1M polytype, $Z = 4$, $D = 2.75$. PD(random orientation): 11.9_7 4.75_5 4.46_9 4.32_9 2.55_{10} 2.48_7 1.62_5 1.49_9; (oriented): 23.8_6 11.9_4 7.9_3 4.75_8 3.40_{10} 2.97_5. An ordered (1:1) 23.5- to 24-Å interstratification of cookeite (di-trioctahedral chlorite, 14 Å) and paragonite (dioctahedral mica, 9.5 Å). *EJM 6:897(1994)*. Should not expand with glycol or collapse with heating $> 500°$. Occurs as deformed lamellae 0.25–1 mm long and 0.05–0.1 mm thick. Sometimes in 0.5-mm-diameter rosettes included in quartz. White to colorless, nonfluorescent. Cleavage {001}, perfect. Low hardness and tenacity. Biaxial $(-)$; $N_x = 1.58-1.59$, $N_y = 1.58-1.59$, $N_z = 1.59-1.60$; $2V = 30-50°$; birefringence $= 0.007$. Traces of FeO (0.4 wt %), CaO (0.3 wt %), K_2O (0.35 wt %) present. Saliotite occurs in schists as a high-pressure, low-temperature metamorphic rock-forming mineral. Associated with pyrophyllite, paragonite, cookeite, and calcite. Found in the Alpujarrides nappes in the Sierra Alhamilla (Betic chain; Andalusia), Spain. An illite-like–sudoite-like interstratification "saliotite" was found at the Hanaoka mine, Kuroko, Akita Pref., Japan [*CS 4:45(1972)*]. VK, EF

71.4.2.4 Tosudite Chlorite-like–smectite-like layers; three varieties have been distinguished:
Al-tosudite $[Al_2(Al_{2.33})Si_3AlO_{10}(OH)_8/(Na,Ca)_{0.33}Al_2(Si,Al)_4O_{10}(OH)_2 \cdot nH_2O]$, Mg-tosudite $[Al_2(Mg_2Al)Si_3AlO_{10}(OH)_8/(Na,Ca)_{0.33}Al_2(Si,Al)_4O_{10}(OH)_2 \cdot nH_2O]$, and Li-tosudite $[Al_2(LiAl_2)Si_3AlO_{10}(OH)_8/(Na,Ca)_{0.33}Al_2(Si,Al)_4O_{10}(OH)_2 \cdot nH_2O]$

Named in 1963 by Frank-Kamenetsky et al. for Toshio Sudo (b.1911), Japanese mineralogist and crystallographer, University of Tokyo. Interstratified phyllosilicate group. Synonym: alushtite. HEX (given on *12-231* and *22-956*, varieties not stated) but probably ORTH, HEX cell: $a = 5.125-5.214$, $c = 29.78-29.80$, $Z = 1$. ORTH cell: $a = 5.0$, $b = 8.9$, $c = 29.8$, $Z = 2$. Some may be MON $C2$ [*MIN 4(2):271(1992)*], with $a = 5.15-5.21$, $b = 8.94-9.03$, $c = 24.2-25.1$, $\beta = 94°$ (desiccated material). D (for ORTH cell)$= 2.46$. *12-231*(nat): 30.0_6 $15.0_{<9}$ 4.97_8 4.53_{10} 3.29_3 2.60_3 2.54_5 $1.51_{<5}$; *22-956*(nat): 30.4_{10} 15.2_7 10.0_1 5.01_2 4.48_3 3.30_1 2.56_1 1.49_1. As currently defined, tosudite consists of three varieties. The original material (Al-tosudite) is an interstratified species that is di-,dioctahedral chlorite-like (donbassite-like) with dioctahedral smectite-like layers and is aluminum-rich, but also has significant, although minor, Mg.

71.4.2.4 Tosudite

Later discoveries include: di-, trioctahedral chloritelike "sudoite-like" with dioctahedral smectite-like layering (Mg-tosudite), and di-,trioctahedral chloritelike "cookeite-like" with smectite-like layering (Li-tosudite). Tosudites are 29.5- to 30.5-Å phyllosilicates with an X-ray powder pattern very similar to that of corrensite, tri-,trioctahedral chlorite-like with trioctahedral smectite-like layers, but with differences in peak intensities. Originally erroneously stated to be a pyrophyllite-like–hydroargillite-like interstratification. Octahedral occupancy for chlorites and chlorite smectites that do not contain lithium can be determined roughly from the position of the 060 reflection. A completely dioctahedral structure will give an 060 around 1.49 Å, a trioctahedral 060 is at about 1.54 Å, and a mixed di-,trioctahedral structure will give a peak about 1.51 Å (Bailey, 1975). Li-tosudite and cookeite, however, may give a dioctahedral 060 spacing around 1.49 Å, even though the hydroxide interlayer may be trioctahedral. The problem lies in the small b parameter of an (Al,Li)-hydroxide sheet. A chemical analysis is therefore required for identification. *CCM 26:327(1978)*. **Tests:** Basal spacing expands to about 31.1–32.1 Å with ethylene glycol. Collapses on heating to 500° to 23.3–24.0 Å. After heating to 700°, collapses to 23.7 Å. Can be rehydrated to 29.5 Å after heating to 500°. Observed to collapse from 30.5 Å to 26.8 Å with K-intersalination. **Physical properties:** White, pale yellow, pale green, pale blue, or pale pink; white streak; earthy to waxy luster. Cleavage {001}, {010}, very good; {100}, poor; conchoidal fracture in masses. Fine-grained masses, often pulverulent. H ~ 1. G = 2.42–2.50. IR data: *CCM 24:142(1976), 36:39(1988)*. Thermal data: *CCM 23:337(1975)*. Not soluble in common acids or bases or only poorly soluble; soluble in HF. **Optics:** Biaxial (−) with moderate 2V (58–68°); N_x = 1.542–1.564, N_y about 1.57 (few data), N_z = 1.548–1.574. Rarely uniaxial (−). Variable optical orientations observed. Tosudites are always Al-rich (33–37 wt % Al_2O_3) with SiO_2 39–47 wt %. May contain (wt %): TiO_2, 0.74 (Kamikita); Fe_2O_3, 1.82 (Tooho); FeO, 2.77 (Huy); Cr_2O_3, 1–3; MgO, to 8.2 (Mg-tosudite); CaO, to 2.56 (Hokuno); P_2O_5, 0.08 (Kamikita); S, 0.69 (Kamikita); Na_2O, 0.58 (Hokuno); K_2O, 1.40 (Takatama); Li_2O, to 1.7 (Li-tosudite), but Al-tosudite frequently has some Li. H_2O (total) 16–17 wt %. **Occurrence:** A hydrothermal alteration product of tuffs, andesites, sediments, and so on, and can occur with other clay minerals. Also occurs in altered granites, in granite pegmatites as late-stage cavity fillings and pseudomorphs after tourmaline or beryl. Occurs in some gold deposits as well. Also in conglomerate at two localities. May alter rarely to dickite. **Localities:** Occurs in more than 40 worldwide localities, only some of which are mentioned below. General tosudite localities include: Schasse, Hazle Twp., Luzerne Co., and at Delano, Schuylkill Co., PA; Geneva–Davis mine, Gogebic district, MI; Lisbon Valley, UT; Supai Group, Grand Canyon, AZ; Sespe formation, Santa Barbara, CA; Echassieres, Massif Central, and Montebras, Creuse, France; Ehrenfriedersdorf, Germany; Cr-bearing from Takovo, Yugoslavia; in Sakalinsk deposit, northern Caucasus, and at Borkut, Transcarpathians, Russia; Nikitovsk and Druzhkovsk–Konstantinovsk ore fields in the Donbass, Ukraine; Belogorsk Au deposit, Primorye, Russia; Dernberg pegmatite, Karibib, Namibia; Agades, Nigeria; in more than 20 "Kuroko" and "roseki" deposits in Japan: Makurazaki area, Kagoshima Pref., Japan; Bubsoo mine, Cheongsong-eup, Korea. **Li-tosudite:** Himalaya pegmatite–aplite dike system, Mesa Grande district, and Stewart mine, Pala district, San Diego Co., CA; *Huy, Belgium*; Tooho mine, Aichi Pref.; Hokuno mine, Gifu Pref., Japan. **Al-tosudite:** *Alushta, Crimea, Russia*; Takatama mine, Fukushima Pref., Uebi, Ehime Pref.; Uku and Kurata mines, Yamaguchi Pref.; Hanaoka mine, Akita Pref.; Izushi, Hyogo Pref.; Igashima mine, Niigata Pref.; Hiraki mine, Hyogo Pref.; Hokuno mine, Gifu Pref.; Amakusa, Kumamoto Pref.; Niida, Akita Pref.; Iwami mine, Shimane Pref., Japan. **Mg-tosudite:** *Kamakita mine, Aomori Pref., Japan.* VK, EF

71.4.2.5 Corrensite $(Mg,Fe^{2+},Fe^{3+})_3[(Mg,Fe^{2+},Fe^{3+})_2Al]$ $[(Si_3AlO_{10}(OH)_8]/(Ca,Na,Mg)_{0.35\pm},(Mg,Fe^{2+},Fe^{3+})_3(Si,Al)_4O_{10}(OH)_2$ (HLC corrensite) or $(Mg,Fe^{2+},Fe^{3+})_3[(Mg,Fe^{2+},Fe^{3+})_2Al][Si_3AlO_{10}(OH)_8]/[(Mg)_{0.35\pm},(H_2O)_{16,12,8,3,0}](Mg,Fe^{3+})_3[Si_3AlO_{10}(OH)_2]$ (LLC corrensite)

Originally recognized as long ago as 1938 [*AM 23:851(1938)*], but not named until 1954 by Lippmann for Carl Wilhelm Correns (1893–1980), German mineralogist and director of the Sedimentary Petrography Institute, University of Göttingen; redefined in 1980. Synonym: "swelling chlorite." ORTH Space group unknown. $a = 5.337$, $b = 9.215$–9.26, $c = 23.2$–$24.2, 28.5$–$29, 31$–32; or HEX P. $a = 5.337$, $c = 23.2$–$24.2, 28.5$–$29, 31$–32, $Z = 1$. ICDD *19-764* and *13-465* are older data. *31-794*(LLC corrensite): 29.0_3 14.0_{10} 7.83_3 7.08_6 4.72_3 4.62_3 3.53_6 2.57_3; *42-620*(glycerol-treated): 31.6_{10} 15.8_7 7.99_2 5.33_1 4.59_1 3.56_2 2.91_1. HEX P. $a = 5.337$, $c = 28.56$. Corrensite is a 1:1 regularly interstratified tri-,trioctahedral chlorite-like–trioctahedral smectite-like layers (high layer charge corrensite, HLC) or tri-,trioctahedral chlorite-like–(trioctahedral) vermiculite-like (low-layer-charge corrensite, LLC) 29- to 31-Å interstratified phyllosilicate. (Di-, dioctahedral chloritelike–dioctahedral smectite-like layerings are Al-tosudite, while di-,trioctahedral chlorite-like–dioctahedral smectite-like layerings are Mg-tosudite or Li-tosudite.) Additional nomenclature could be considered. *AM 67-394(1982)*. Structure not precisely determined, but surmised from oriented XRD and one-dimensional Fourier calculations. *CCM 35:150(1987)*. Site occupancies (octahedral, tetrahedral, and interlayer) have been calculated. *USGSOF:84-165 (1984)*, *CMP 105:123(1990)*. HRTEM study of HLC corrensite shows regular interstratification. *CCM 34:155(1986), CMP 105:123(1990)*. The interlayer cations in both high and low layer charge corrensite are exchangeable and therefore do not particularly distinguish the two types of corrensite. Fully hydrated corrensite can have $d_{100} = 31$–32 Å, while air-dried material is generally 28.5–29.2 Å. Cation-exchange treatment can result in 58-Å spacings. *CCM 33:128(1985)*. DTA studies of common and systematic intersalination of low-layer-charge corrensite shows double water layers with Li, Ca, Sr, and Ba and single layers with Na and absence of water layers with K. *AM*

71.4.2.5 Corrensite

49:556(1964). **Tests:** Untreated corrensite usually has $d_{100} = 28$–31 Å, which usually collapses to 21–28 Å, usually ≈ 24 Å, when heated to 500°; original spacings can be 24–28 Å, however. Intercalation with glycerol or ethylene glycol can increase d_{100} to 31–32 Å (HLC corrensite), while LLC corrensite should not expand. LLC corrensite can collapse from 28–30 Å to 24 Å on K-salination. HLC corrensite has relatively strong (003) while glycolated LLC corrensite has virtually undetectable (003). IR data: *CCM 33:458(1985), 35:150(1987)*. HLC corrensite is very common compared to true LLC corrensite. Many identifications are probably in error as to type of corrensite. **Physical properties:** Generally fine-grained to fibrous, typically 2–5 μm. Brown to light golden brown, gray-white, silvery gray, pale gray-green to yellow-green, sometimes stained; white streak; earthy to indurated, dull to waxy luster. Cleavage {001}, perfect; flexible, but inelastic. H = 1–2. Partly decomposed by strong HCl. **Chemistry:** Chemistry of corrensite is variable. *CCM 32:391(1984)*. Observed ranges in composition (wt %) are: HLC corrensite SiO_2, 31–38; Al_2O_3, 12–24 (high Al values might reflect sample purity or presence of tosudite); Fe_2O_3, to 13; FeO, to 21.85; MgO, 12–32; TiO_2, to 0.81; MnO, to 0.38; CaO, 0.18–3.49; Na_2O, 0.10–1.65; K_2O, 0.08–1.20; H_2O(tot.), to 15. Sedimentary corrensites are richer in silica, and hydrothermal corrensites are richer in Mg and Al. CEC of HLC corrensite is about 40–55 meq per 100 g, but data are few. **Optics:** Biaxial (−); $N_x = 1.560$–1.585, $N_y \approx N_z = 1.582$–1.612; $2V = 0$–$10°$; birefringence = 0.010 up to 0.021–0.027; also $N_{av} = 1.59$, birefringence = 0.000, approximately parallel to extinction with (001); positive elongation; N_x approximately perpendicular to {001}; pleochroic. X, yellowish brown; Y,Z, brownish green. RI and pleochroism increase with Fe (≈ 0.0015/at % Fe/Fe + Mg). RI between tri-trioctahedral chlorite and trioctahedral smectite. Corrensite (unspecified) birefringence = 0.01–0.03 similar to vermiculites (cf. random interstratifications < 0.01 and trioctahedral smectite > 0.03). Corrensites can have first-order pale yellow to orange yellow in thin section. **Occurrence:** Found in a variety of sedimentary environments, especially pelitic sediments, and frequently coexist with a wide variety of clay minerals almost universally, including illite, frequently found with chlorite, also found with smectite, micas, talc, and so on, as well as gypsum, dolomite, or zeolites. Corrensite not reported with kaolinite, but tosudites are. Both HLC and LLC corrensite have been found in the same sample area. Frequently found in evaporite sequences. Basin-wide zonation of clays observed with dioctahedral clays on basin margins, corrensite intermediate zone, and a chlorite zone at the basin center. *CCM 41:240(1993)*. Found as vesicle linings in volcanic rocks, also replacing volcanic glass in ash deposits. Noted as a hydrothermal alteration of phlogopite in marble [*CM 26:957(1988)*], also noted with talc, anthophyllite, antigorite, zeolites, and/or clinochrysotile. Corrensite has been suggested to form from the chloritization of smectite and from the hydrothermal alteration of phlogopite or as a hydrothermal alteration of amphibole in dolomite. Direct formation of corrensite from chlorite is debated. *CM 26:962(1988), CCM 32:391(1984)*. Corrensite can form diagenetically at 90–100°, ranging to 200° at low pressures [*Bull. Centre Rech. Pau, Soc. Nat. Pétroles d'Aquitaine* 7:543(1973), CS 6:103(1984)] or deuterically. **Localities:** Only selected localities are given here: Hartford Basin, MA–CT; West Rock Sill, New Haven, New Haven Co., CT; Hamilton, Adams Co., Hamburg Klippe, PA; Jordan No. 5 mine, Murphy, Durham, Durham Co., NC; Florida Keys, FL; Cumberland Plateau, TN; Valmeyeran carbonate, IL; Cary mine, Wellington formation, Lyons Co., KS; Brazer limestone, Juniper Canyon, Moffat Co., CO; DeMoines shale, OK; Ordovician limestone and Guadaloupe formation, TX; Salado formation, Delaware Basin, NM; Goldfield, NV; Gilson Dome, Paradox Basin and Hermosa formation, San Juan Co., UT; Dearborn River locality(!), MT; Point Sal, CA; Sherbot Lake and Carleton, Frontenac, Lanark, Leeds, and Renfrew Cos., ON, Shawinigan and Gaspé Peninsula, PQ, Canada; Iberian Mts., Saragossa Prov., Spain; Veitsivaara, Finland; Thones syncline, Haute Savoie, and Sancerre-Couy drill hole, Paris Basin, France; Keuper Marl, England; Caithness and Orkney, Scotland; *Zaiserweiher bei Maulbronn, Baden–Württemberg*, and *Hüntstollen* and Coburg, Keuper Basin, *Lower Saxony, Germany*; Taro and Ceno valleys, Italy; Zerevta asbestos deposits, Ural Mts., and Bugety-Say, Russia; abyssal basalt near St. Paul's Rocks; Cassiope, Parana, and Sergipe–Alagoas basins, Brazil; Sirban limestone, Riasi, Jammu, and Kashmir states, India; eastern Otago, NZ; Green Tuff formation, Yamanaka area, Ishikawa Pref.; Ohyu caldera, Aikita Pref.; Tsunemi, Fukuoka Pref.; Yamakata area, Ibaraki Pref.; Niigata oil field, and Matsukawa area, Japan. VK, EF *MIN 4(2):281(1992)*.

71.4.2.6 Kulkeite $(Mg,Fe^{2+},Fe^{3+})_3[(Mg,Fe^{2+},Fe^{3+})_2Al]Si_3AlO_{10}(OH)_8/(Mg,Fe^{2+})_3Si_4O_{10}(OH)_2$

Named in 1980 by Abraham et al. for Holger Kulke, geologist, Essen, Germany. MON Space group unknown. $a = 5.319$, $b = 9.195$, $c = 23.897$, $\beta = 97°$, $Z = 2$; also HEX $a = 5.304$, $c = 23.70$, $D = 2.70$. *42-584:* 11.9_8 7.9_{10} 4.74_6 3.38_8 2.96_5 2.55_6 2.46_5 1.53_{10}. Kulkeite is a regularly interstratified 23.5- to 23.7-Å trioctahedral chlorite-like–(trioctahedral) talc-like layer phyllosilicate. Unaffected by ethylene glycol or heating to 500° for 1 hour. Diagnostic peaks at $2\theta = 7.4°$, $11.2°$, $26.4°$, while peak at $3.7°$ is weak. The structure was verified by TEM and indicated numerous defects. *CMP 80:103(1982)*. The talc-like layer in kulkeite appears to be sodian while the talc-like layer in the chlorite does not appear to contain alkalis and considered with ^{IV}Al, the two talc-like layers are compositionally different. *CMP 80:103(1982)*. Kulkeite might not be a stable phase. Occurs as several millimeter-size porphyroblasts in dolomite or talc and usually intimately intergrown with clinochlore as a thin selvage between talc and kulkeite; platy to subhedral crystals. Colorless, white streak, pearly luster, transparent. Cleavage (001), perfect. $H \approx 2$. Biaxial $(-)$; $Z = b$, $Y \approx a$, OAP perpendicular to (010); $N_x = 1.552$, $N_y = 1.5605$, $N_z = 1.5610$; $2V_{obs} = 24°$; $r > v$; birefringence $= 0.009$; positive elongation. RI not directly intermediate between pure components. Distinguishable from talc by having lower birefringence and from chlorite by sign of elongation. Chemical analysis (wt %) of type material gave: SiO_2, 41.44, 40.53; Al_2O_3, 12.42, 12.64; MgO, 33.05, 33.19; CaO, 0.07, 0.06; Na_2O, 0.94, 1.20; K_2O, 0.20, 0.07; Li_2O, 0.09; H_2O_{calc}, 9.39, 9.24; total$_{determined}$ 88.12, 87.69. Associated with dolomite, albite, sodian–aluminian talc, quartz, dravite–uvite, clinochlore, phlogopite, sodium phlogopite, rutile, and pyrite in a meta-evaporite sequence probably subjected to less than 400°. *El Mourdur Hill, near Derrag, 35 km W of Ksar El Boukhari, Tell Atlas Mountains, Algeria.* VK, EF *FM 58:4(1980), MIN 4(2):250(1992).*

71.4.2.7 Rectorite $(Na,K)Al_2Si_3AlO_{10}(OH)_2/(Na,K,Ca)_{0.33}Al_2(Si,Al)_4O_{10}(OH)_2 \cdot nH_2O$

Named in 1891 by Brackett and Williams for E. W. Rector (b.1849), legislator and lawyer of Hot Springs, AR. Synonyms: Na-rectorite; K-rectorite; Ca-rectorite; illite–smectite (I/S); allevardite. Redefined by Brown and Weir in 1963 [also *CCM 30:76(1982)*] as paragonite-like–montmorillonite-like layer (= Na-rectorite), increasingly recognized as K-rectorite (illite-like or muscovite-like–dioctahedral smectite-like layers), also glauconite-like–dioctahedral smectite-like; earlier commonly thought to be a pyrophyllite-like–vermiculite-like interstratification. MON Space group unknown. $a = 5.12$–5.13, $b = 8.88$–8.92, $c = 23.77$–26.4, $\beta = 96.3°$, 99°; after heating to 600°, $d_{(001)} = 19.3$ Å; after ethylene glycol treatment, $d_{(001)} = 26.8$ Å; $Z = 2$, $D = 2.34$–2.39. *29-1495*(K-rectorite): 26.4_3 12.2_4 4.99_3 4.45_{10} 3.15_3 2.55_7 2.44_1 1.492_4; *25-0781*(Na-rectorite): 23.8_{10} $11.9_{6.5}$ 4.75_1 $3.96_{<1}$ 3.40_2 $2.54_{<1}$ 2.91_2 1.97_1. The structure of rectorite features a mica-like sheet that has tetrahedral aluminum and a strongly bonded interlayer cation alternating with a smectite-like sheet, which may or may not have tetrahedral aluminum, and does have an exchangeable interlayer cation with interlayer water. NMR study has suggested that similar Al ↔ Si substitutions occur in the tetrahedral sheets facing the interlayer cations, rather than on either side of the octahedral sites within a layer. Therefore, juxtaposed tetrahedral sheets might be both mica-like (Si_3Al) or both margarite-like (Si_2Al_2) about the fixed interlayer cation and both smectite-like ($Si_{4-x}Al_x$) tetrahedral sheets facing the exchangeable cation. The interstratifications could therefore be considered as hybrid layers where each and every layer was composed of smectite-like *and* mica-like or margarite-like tetrahedral sheets with alternation of "up-down" hybrid layer orientations. *AM 80:247(1995).* Interstratification model B: *CS 4:191(1974).* Mica layers of rectorite have been compared to muscovite, paragonite, and margarite spacings. Rectorite is a ~ 24- to-27-Å (usually ~ 25 Å) interstratified phyllosilicate which expands by intercalation (ethylene glycol = 26.4–28.8 Å, 29.5 Å [*CCM 28:241(1980), CM 16:91(1981)*] with low spacings 23.7–24.84 Å for K-rectorite); glycerol = 27.2–27.6 Å, also 28.8 Å, with smaller spacings, 23.48–23.72 Å for K-rectorite) and collapses to 19.2–19.8 Å on heating to 550°. Na-saturated rectorite can show $d_{001} = 22.0$–22.5 Å, Mg-saturated rectorite can show 24.6–25 Å, Ca-saturated rectorite can show 24.65–24.86 Å, and K-saturated can be 21.83–22.10 Å. *CM 7:63(1967), 16:91(1981).* Rectorite has been rehydrated after being heated > 500°. *CCM 40:92(1992)*. HRTEM verification of mica-like–smectite-like layer interstratification has been published [*CCM 34:155(1986)*] and shows dehydrated spacings. Illite-like–smectite-like interstratifications can be found outside the 1:1 rectorite definition, with loss of the 24- to 27-Å peak with increasing randomness of the interstratification. *CCM 28:401(1980)*. Mixtures of illite and smectite and randomly interstratified illite-like–smectite-like layers do not develop long period spacings, but near 1:1 interstratifications do yield diffractograms which appear to be mixtures of 27- and 17-Å layers (> 50% smectite) or 27- and 10-Å layers (< 50% smectite). Proportion diagnostics of illite to smectite layers have been presented. *AM 50:990(1965)*; *CCM 28:401(1980)*. Intermediate angle peak shifts have been used to determine the proportion of smectite layers in an "ordered" illite-like–smectite-like interstratification as well as nonintegral basal reflections that might split into two peaks when glycerated, indicating randomness in the interstratification. *CM 6:261(1966)*. "Only the relatively weak 002/003 and 002 serve to distinguish glauconite–smectite from illite–montmorillonite diffraction patterns." Many supposed "rectorites" do not show long-period X-ray diffraction peaks and these identifications can be questioned. Non-1:1 illite–smectite interstratifications are among the most abundant of the clay minerals observed, and the proportion of illite/smectite layers is open to calculation. *CCM 18:25(1970), 28:401(1980)*. Rectorite "ordering," presumably in packets, has been invoked for samples with as little as 20% illite/smectite layer proportions. "Kalkberg ordering" (ISII order) can show 47° peaks. DTA data: *AM 51:1035(1966), CM 16:91(1981)*. IR data: *CCM 19:61(1971), 32:154(1984)*. **Physical properties:** White to tan, sometimes pale green or brown, earthy; also thick leathery mats resembling "mountain leather," formed by aggregates of microscopic crystals; waxy luster (in mat form), infusible, unctuous, plastic and greasy when wet. Cleavage {001}, perfect; flexible and elastic (some sources say not elastic); heated (> 500°) material is brittle. Decomposed by hot HCl. H < 1. G about 2.2. **Optics:** Biaxial (−); $N_x = 1.519$, $N_y = 1.550$, $N_z = 1.559$; 2V = 5–20° (Arkansas), range 0–39°, also 49–63° [*USGSB:1627 1984*]; r > v; birefringence = 0.040; cleavage traces can show nearly parallel extinction. Optical data are few. **Chemistry:** The chemistry of rectorites is variable, but truly monomineralic samples are probably very rare. Interlayer water has been modeled with one-, two-, and three-layer hydrates. *AM 41:91(1956)*. CEC = 53–57 meq per 100 g (Pakistan and Makurazaki) to 62 meq per 100 g (Surges Bay). Shinzan rectorites had exchangeable Na and fixed K. *CCM 31:401(1983)*. Studies have suggested that the smectite-like component was beidellite-like. *AM 51:1035(1966), CCM 32:154(1984)*. Low- Na + K rectorites might seem chemically more like pyrophyllite-like–dioctahedral smectite-like or pyrophyllite-like–dioctahedral "vermiculite-like" layers and 22-Å "rectorite" has been reported. Due to the wide variety of possible mica-like–smectite-like interstratifications allowed by AIPEA definitions, chemical analyses of "rectorite" can be difficult to interpret without full X-ray diffraction characterization, including intercalation, even with one-dimensional Fourier modeling. Chemical analyses

71.4.2.7 Rectorite

of rectorites are few and some analyses (wt %) have been calculated to "anhydrous" totals: SiO_2, 44.62–55.2 (generally 51–54); Al_2O_3, 23.95–41.00; Fe_2O_3, 0.65–2.83; FeO, 0.16–0.97; MnO, 0.03–0.35; MgO, 0.07–1.46; TiO_2, 0.12–1.81; CaO, 0.33–1.61 (3.99, Ca-rectorite, Sano); K_2O, 0.10+ (K-rectorite, to 4.94); Na_2O, 0.10+ (Na-rectorite, to 5.31); SrO, 0.13 (Pakistan); Li_2O, 0.05 (Sano); P_2O_5, 0.15; total H_2O, 6.04–15.76 (not frequently determined). Rectorite has been synthesized from gels and smectite starting materials [*CCM 26:327(1978)*], as well as from sericite [*CCM 26:209(1978)*] and kaolinite [*PICC:303(1973)*]. **Occurrence:** Rectorite is one of the most common interstratified clay minerals and is found in a wide variety of sedimentary environments associated with quartz, carbonates, chlorite, and a wide variety of detrital minerals; usually found with mixtures of clays, including sericite, illite, montmorillonite, pyrophyllite, corrensite, tosudite, and so on. [Tosudite and other minerals might be present in some rectorite with a persistent 24-Å peak after heating to 550°. *CCM 28:241(1980)*.] Rectorite appears to be formed through diagenesis through the reaction: smectite $+K^+ + Al^{3+} \leftrightarrow$ illite $+ Si^{4+}$ over the temperature range ±100–175°. Temperature also seems to increase the order of interstratified illite-like–smectite-like layers. Noted in shales. Observed to replace volcanic glass shards. Rectorite has been frequently suggested to have altered to uninterstratified illite and sometimes to pyrophyllite. Found in a few bentonites. Also as a component in coal seams. Found in hydrothermal veins with quartz crystals and cookeite. As replacements of plagioclase in dacite, tuffs, and so on. Found in some Roseki clays in Japan. Variously reported in drill core samples in the eastern United States from the Cincinnati Arch through the Atlantic and Gulf of Mexico coastal plains, but much of the material is composed of non-1:1 interstratified mica-like–smectite-like layers. **Localities:** Selected localities include: *Blue Mt. district, Marble, Garland Co.*, Jeffrey quarry(!), northern Little Rock, Pulaski Co., AR; Cañon City, Fremont Co., and Buckeye Gulch, near Leadville, Lake Co., CO; Manning Canyon shale, UT; Tulameen, BC, Canada; Cala Abajo deposit, Puerto Rico; Woodbury; Alston block and Northumberland trough, England; Abersoch and Nanhoron, Llyn; Nantlle and Tremadog, Snowdonia, Wales; Allevard(!), Isère, Sibert, Rhône, France; Kinnekulle, Sweden; Király Hill, Mád, Hungary; Rhodope Mts., Bulgaria; Troodos ophiolite, Cyprus; Cufra, Libya; Berghersdorp, Cape Prov., Beatrix mine, South Africa; Muluti, Transkei; Donets Basin, Ukraine; Dagestan; Samara bend of Volga R., Russia; Brunjkaki, Fort Sandeman district, Baluchistan, Pakistan; Dafang Co., Guizhou Prov.; Longzhong, Hubei Prov., China; Ohira mine, Mitsuisi, Okayama Pref.; Makurazaki, Kawanabe, and Iwato, Kagoshima Pref.; Goto mine, Goto Is., Nagasaki Pref.; Sano mine, Kakuma, also Honami and Yonago mines, Nagano Pref.; Takatama mine, Fukushima Pref.; Shinzan, Aikita Pref.; Tooho mine, Aichi Pref., Japan; Maitland, NSW; Surges Bay, TAS, Australia; Wairaki geothermal field, New Zealand. VK, EF

71.4.2.8 Allietite $(Mg,Fe^{2+})_3Si_4O_{10}(OH)_2(Ca,Na)_{0.2-0.3}(Mg,Fe^{2+})_3$ $(Si,Al)_4O_{10}(OH)_2 \cdot 4H_2O$

Named in 1969 by Veniale and van der Marel for Andrea Alietti (b.1923), clay mineralogist, University of Modena, Italy. HEX Space group unknown. $a = 5.216$, $c = 24.6$, $Z = 1$; *07-0357:* 24.0_2 $12.0–11.2_{10}$ 4.54_x $3.50–3.00_6$ 2.62_9 1.725_4 1.520_9 1.314_7; oriented basal reflections noted (Greece, natural and *glycolated): $d_{(001)}$, 24.9,*27.1; $d_{(002)}$, 12.3,*13.4; $d_{(003)}$, 8.27,*8.90; $d_{(005)}$, 4.92,*5.32; $d_{(007)}$, 3.54; $d_{(008)}$, *3.32; $d_{(004,006)}$, absent. *CR 314(II):483(1992).* Allietite is a regularly interstratified 24- to 25-Å phyllosilicate with alternating talc-like and trioctahedral smectite-like layers, but significant deviation from 1:1 interstratification is the norm. X-ray diffraction data virtually nonexistent except for basal reflections, for which structure factors have been calculated. *CCM 28:388(1980).* Diagnostics: d_{001} untreated 24.8 Å; glycerol solvated 27.1 Å; natural, heated to 600° d_{001} collapses to 9.34 Å (Monte Chiaro), also heated to 490° d_{001} collapses to 9.6 Å (Greece). DTA data: *PICC 233(1969), CCM 28:388(1980), CR 314(II):483(1992).* IR data: *CCM 28:388(1980), PICC 1:233(1969).* Also described as a kerolite-like–stevensite-like or disordered talc-like–Mg-trioctahedral smectite-like interstratification. *CCM 30:321(1982).* Found as the clay-size fraction of rocks or sediments, also as "small, soft green plates which can be easily separated under the microscope." *CCM 28:388(1980).* Buff to gray, to dark green, depending on the host's coloration. Cleavage (001), perfect. Other physical data not reported. IR data: *MIN 4(2):248(1992).* $N = 1.556–1.567$; other optical properties not reported. Allietite is difficult to distinguish chemically from pure noninterstratified trioctahedral smectite. A calculated structural formula (Greece) gave: $(Mg_{0.15}K_{0.02}, Na_{0.02}, Ca_{0.02})$ $(Mg_{2.62}, Fe^{2+}0.38)$ $[Si_{3.62}Al_{0.23}$ Fe^{2+} $0.15O_{10}(OH)_2]$. "Allietite" from Amargosa desert region can have saponitelike or stevensitelike layers as the smectite component. Monte Chiaro material gave (wt %): Na_2O, 1.15; K_2O, 0.01; CaO, 1.02; MgO, 27.74; FeO, 1.71; MnO, 0.03; Al_2O_3, 2.95; Fe_2O_3, 3.90; TiO_2, 0.06; SiO_2, 55.43; H_2O^+, 6.00 (Amargosa desert region "anhydrous" analyses gave: Na_2O, 0.31–3.86, 0.31* (*for 50% expandable layers, i.e., 1:1 interstratification); K_2O, 1.00–1.68, 1.27*; Li_2O, 0.27–0.60, 0.32*; CaO, 0.01–3.22, 3.22*; MgO, 27.47–31.46, 31.46*; Al_2O_3, 1.44–3.76, 2.66*; Fe as Fe_2O_3, 0.69–1.76, 1.05*; TiO_2, 0.05–0.08, 0.08*; SiO_2, 59.64–63.15, 59.64*). CEC (Nevada) = 46–67, 53* meq per 100 g. Probably formed at low temperatures, 105–107°. *CR 314(II):483(1992).* Found with talc in serpentinites, dolomites; as a clay-size fraction in metasediments with saponite, illite–smectite, illite, HLC corrensite, orthochrysotile, and so on; also in ancient lake sediments with authigenic dolomite or calcite, sepiolite, or other minerals. Amargosa desert, Nye Co., NV; Hectorite–Whiting pit, Inyo Co., CA; *Monte Chiaro* and other localities, *Taro Valley, Parma Prov.,* Ferriere, Nure Valley, Piacenza Prov., Frassinoro, Modena, *Italy;* Beotian flysch, eastern Parnassus Mts., Hellenides, Greece; Kinsasha region, Shaba Prov., Zaire; Obori mine, Yamagata Pref., Japan. VK, EF *PICC 1:233(1969), MIN 4(2):248(1992).*

71.4.3.1 Franklinfurnaceite $Ca_2[Mn^{2+}Fe^{3+}][Mn_2^{2+}Mn^{3+}]$ $[Zn_2Si_2O_{10}](OH)_8$

Named by Dunn et al. in 1987 for an old name for the locality. Chlorite group. MON $C2$. $a = 5.483$, $b = 9.39$, $c = 14.51$, $\beta = 97.05°$, $Z = 2$, $D = 3.74$. 41-1412(nat): 14.4_5 3.60_4 3.35_5 3.20_4 2.71_8 2.60_7 2.305_{10} 1.58_4. Franklinfurnaceite is the only chlorite-related mineral with the IIa polytype. The interlayer has vacant sites over the Ca site. The overall structure is essentially that of a chlorite, but with unusual chemistry. The octahedral sheet of the 2:1 layer is trioctahedrally occupied with a pattern of six-membered Mn^{2+} (with minor Mg and Zn) octahedral "rings" with a centrally occupied Mn^{3+} octahedron. The tetrahedral sheets are nearly triangular, true six-membered rings with alternating large Zn and small Si tetrahedra. Outer oxygens of the tetrahedral sheets and hydroxyls of the interlayer sheet form a regular sixfold coordination of a Ca site with these sites above and below an interlayer octahedral vacancy. The dioctahedral interlayer consists of six-membered "rings" of alternatingly small Fe^{3+} (with minor Al) and large Mn^{2+} (with minor Mg, Zn) octahedra. Dark brown to black, opaque to subtranslucent crystals (to several millimeters) tabular triangular to pseudohexagonal, also plumose and in rosettes; brown streak; waxy to subvitreous luster. $H = 3$. $G = 3.66$. Cleavage {001}, perfect; brittle. Minor substitutions observed include MgO to 4.3 wt % and Al_2O_3 to 1.3 wt %. Biaxial $(-)$; $Z = b$, $Y \wedge c = 29°$; $N_x = 1.792$, $N_y = 1.798$, $N_z = 1.802$; $2V = 79°$; $r < v$; birefringence $= 0.010$; $X \gg Z > Y$; pleochroic. X, very dark brown; Z, deep brown; Y, brown. Usually found associated with hodgkinsonite, including clinohedrite, franklinite, hetaerolite, or willemite. *(Franklin mine) Franklin, Sussex Co., NJ.* VK, EF *AM 72:812(1987), 73:876(1988).*

71.4.4.1 Gonyerite $Mn_3^{2+}[Mn_2^{2+}Fe^{3+}][(Si,Fe^{3+})_4O_{10}](OH,O)_8$

Named in 1955 by Frondel for Forest A. Gonyer (1899–1971), chemist-analyst and mineralogist, Harvard University. Chlorite group. ORTH (pseudo-HEX) Space group unknown. $a = 5.47$, $b = 9.46$, $c = 28.8$, $Z = 4$, $D = 3.03$. 10-378(nat): 14.6_3 7.23_{10} 4.79_5 3.61_8 2.88_2 2.70_3 1.63_3 1.57_5. Originally described as a trioctahedral chlorite; however, X-ray powder data of the mineral show that it is not a true chlorite. *CM 13:179(1975).* Gonyerite is a modulated phyllosilicate structure commonly interleaved with 14-Å chlorite units. *REVM 19:707(1988).* Found in radially splayed plates. Dark brown to brownish black, or bright green; brown streak; waxy to subvitreous luster. Cleavage {001}, perfect; flexible, but inelastic. $H = 2\frac{1}{2}$. $G = 3.01$. Quickly soluble in concentrated HCl and insoluble in H_2SO_4. Gonyerite shows several unusual aspects involving charge balance, but about 5.9 octahedral sites are occupied. Type material is a magnesian gonyerite (MgO $= 12.22$). Gonyerite is a manganese-rich mineral; it is essentially aluminum deficient. A small quantity of ferric iron substitutes for ^{IV}Si. *AM 40:1090(1955).* Oxygen substitutes for a small amount of hydroxyl: $(Mn^{2+},Mg)_3[(Mn^{2+},Mg)_2Fe^{3+}][Si_{4-x}(Fe^{3+},Al)_x O_{10}](OH_{8-y}O_y)$. Minor substitutions (wt %) from the type locality include:

CaO, 0.07; PbO, 0.56; ZnO, 0.42; Al_2O_3, 0.58; $Na_2O + K_2O$, 0.31. Biaxial $(-)$; $N_x = 1.646$, $N_y = 1.664$, $N_z = 1.664$; $2V = 0°$; birefringence $= 0.018$; pleochroic. X, dark brown; Y,Z, light brown. Or $N_O = 1.646$, $N_E = 1.664$, pleochroic. O, light brown; E, dark brown. Found in veinlets in manganese-rich skarn. Associated with alleghanyite, barite, bementite, berzeliite, garnet, jacobsite, and so on. Karnes mine near Brinnon, Jefferson Co., WA; *Långban, Värmland, Sweden*; fairly common in Wessels mine, Kalahari manganese field, Cap Prov., South Africa, as thin brown bands in siliceous areas, and as small bright green crystals on hematite. VK, EF *MIN 4(2):234(1992)*.

71.5.1.1 Kegelite $Pb_8Al_4Si_8O_{20}(SO_4)_2(CO_3)_4(OH)_8$

Named in 1975 by Medenbach and Schmetzer for Friedrich Wilhelm Kegel, director of the Tsumeb mine from 1922 to 1938. MON $A2/m$, $A2$, or Am. $a = 21.04$, $b = 15.55$, $c = 8.986$, $\beta = 91.0°$, $Z = 3$, $D = 4.76$. PD: 21.0_{10} 7.01_5 4.49_3 3.82_8 3.74_4 3.01_5 2.59_9 2.34_5; also *29-790*. The suggested structure consists of a stacking of pyrophyllite-1A layers and leadhillite layers. White; pearly luster. Biaxial $(-)$; 2V small; N in plane of plates $= 1.81$. Found intergrown with hematite and mimetite in the deep oxidation zone at *Tsumeb, Namibia*. AR *AM 61:175(1976), 75:702(1990); MM 55:121(1991)*.

Class 72

Two-Dimensional Infinite Sheets with Other Than Six-Membered Rings

72.1.1.1 Ekanite $Ca_2(Th,U)Si_8O_{20}$

Named in 1961 by Anderson et al. for F. L. D. Ekanayake, who first found the mineral in Sri Lanka. See steacyite (63.1.1.1), iraqite-La (63.1.1.2). Metamict or TET $I422$. $a = 7.483$, $c = 14.893$, $Z = 2$, $D = 3.36$. *35-532*(Yukon, crystalline): 7.45_6 6.70_6 4.14_{10} 3.34_{10} 3.27_6 2.77_2 2.64_5 1.796_3; *PD*(Ceylon, 1250°C): 7.31_9 5.21_9 3.45_{10} 3.30_6 2.63_9 2.50_6 2.16_7 1.820_7; *PD*(E. Siberia, metamict, 800°C): 4.18_5 3.32_8 3.26_5 2.64_{10} 1.704_4 1.627_5 1.385_6 1.224_7. Structure: *CM 20:65(1982)*. Structure determined on naturally crystalline Yukon material indicates two-dimensionally infinite, puckered sheets, perpendicular to c, made up of four-membered rings. The unit cell is dimensionally equivalent to steacyite (63.1.1.1) and iraqite-La (63.1.1.2), that have four-membered double rings and a different space group. Tetragonal bipyramidal habit with {101} dominant and modified by {100}, {110}, and {001}; also anhedral, rounded masses in gem gravel. Yellow, green, red; white streak; vitreous luster; transparent to translucent. Cleavage {101}, distinct; irregular fracture; brittle. $H = 4\frac{1}{2}$. $G = 3.05–3.25$ (some reported physical properties may actually apply to steacyite). Uniaxial $(-)$, sometimes biaxial $(-)$; $2V = 10–15°$; $N_O = 1.570–1.597$, $N_E = 1.545–1.568$, Sri Lanka metamict gems 1.59–1.596. The composition is quite variable; Yukon and eastern Siberian material contains up to 11.5 wt % H_2O at least in part zeolitic; potassium is often present, calcium varies from 4–14 wt %, and uranium may replace up to 20% of the thorium. Found in syenite in the Tombstone Mts., YT, Canada; uranoan, in volcanic alkaline xenoliths at Pitigliano, Tuscany, Italy; the Murun massif, near Olkeminsk, Yakutia, Russia; in the *Ratnapura gem district, Eheliyagoda, Sri Lanka*. Sometimes faceted as a collector's gem (not to be worn due to high radioactivity), a handful of stones, one nearly 44 ct, have been reported. AR *NAT 190:997(1961)*, *MS 33:68(1979)*, *RSIMP 41:3(1986)*.

72.1.2.1 Thornasite $(Na,K)ThSi_{11}(O,F,OH)_{25} \cdot 8H_2O$

Named in 1987 by Ansell and Chao as a composite of element names and symbols: thorium, Na and Si. HEX-R $R32$. $a = 29.08$, $c = 17.30$, $Z = 18$, $D = 2.627$. *40-506:* 14.54_2 8.17_3 7.27_{10} 5.09_2 4.17_7 3.24_3 2.96_2 2.89_2. Structure undetermined; metamict in part. Colorless to pale green, vitreous to waxy luster. G = 2.62. Uniaxial (+); $N_O = 1.510$, $N_E = 1.512$. Bright green fluorescence in LWUV and SWUV. Found as anhedral grains, < 1 mm, embedded in brockite and associated with yofortierite, eudialyte, albite, serandite, etc., at the *DeMix quarry, Mt. St.-Hilaire, QUE, Canada*. AR *CM* 25:181(1987), *AM* 73:931(1988), *MR* 21:342(1990).

72.1.3.1 Prehnite $Ca_2Al_2Si_3O_{10}(OH)_2$

Named in 1789 by Werner for its discoverer, Hendrik von Prehn (1733–1785), Dutch military officer. ORTH *P2cm*. $a = 4.65$, $b = 5.49$, $c = 18.52$, $Z = 2$, $D = 2.91$. 29-290: 3.54_4 3.48_9 3.31_6 3.28_4 3.08_{10} 2.81_3 2.56_7 2.36_4. Structure: *EJM* 2:731(1990). The structure consists of sheets parallel to {001} consisting of four-membered rings whose tetrahedra share all oxygens, linked to tetrahedra that share only two oxygens with other tetrahedra. $Al^{[4]}$ is disordered with respect to Si in the ring-forming tetrahedra. The silicate sheets are interlayered with isolated AlO_6 octahedra and Ca polyhedra. **Habit**: Uncommon as individual crystals, and then short prismatic [001] to tabular {001}, often with curved faces; more commonly as fanlike, globular, spherical, or stalactitic forms of coarse to very fine crystals; in columnar, polycrystalline groups with a "bowtie" or "hourglass" morphology; granular to compact and, less commonly, fibrous. Lamellar twinning, probably of more than one type, is common. **Physical properties**: Pale to medium green, pinkish tan to tan, pale yellow, gray, white; vitreous luster to somewhat pearly {001}; translucent to nearly transparent. Cleavage {001}, good; {110}, poor; brittle with uneven fracture. $H = 6-6\frac{1}{2}$. $G = 2.90-2.95$. **Optics**: Biaxial (+); $2V = 65-69°$; $N_x = 1.611-1.632$, $N_y = 1.615-1.642$, $N_z = 1.632-1.655$; $XYZ = abc$; $r > v$; birefringence = 0.022–0.035; the refractive indices generally increasing with Fe content. Optical anomalies are common, with wavy or incomplete extinction, anomalous interference figures, and occasional crossed dispersion. These properties probably due to complex twinning or intergrowths of lamellae both parallel to and perpendicular to {001}. **Chemistry**: Little variation in composition is noted; FeO and Fe_2O_3 may total as much as 1.5 wt %; Mg and Mn are minor, as are K and Na. H_2O is lost in two endothermic steps at 787° and 868°C. Slowly gelatinized by HCl. **Occurrence**: Found as a late-forming mineral in veins and cavities in mafic volcanics, where it is associated with pectolite, quartz, and zeolite species; less commonly in veins in granite, monzonite, granite gneiss, etc; as pseudomorphs after laumontite, axinite, and clinozoisite. Found in contact-metamorphosed impure limestones, and in rocks subjected to calcium metasomatism; stable in the zeolite and greenschist facies. Shown to be a product of incipient metamorphism in schistose greywackes. Prehnite-bearing rocks may be of regional extent in a zone characterized by pumpellyite and/or prehnite. **Localities**: Common as light green botryoidal masses and encrustations(!) in Triassic basalts of the Newark series as at Hampden quarry, West Springfield, Hampden Co., MA; the Roncari quarry, East Granby, Hartford Co., CT; Bergen Hill, Hudson Co., and Paterson, Passaic Co., NJ; and the Fairfax quarry, Centreville, Fairfax Co., VA. At

Cornog, PA; with copper in the basalts of the Upper Peninsula, MI; at Bisbee, AZ; pinkish-tan to brown material in zoned pegmatite in quartz monzonite cutting magnesian limestone at Crestmore, Riverside Co., CA. Pale tan, somewhat curved crystals of several centimeters(!) at the Jeffrey quarry, Asbestos, QUE; colorless crystals in a fissure in peridotite on the east side of Bonaparte R., Ashcroft, BC, Canada. In the Kilpatrick Hills, Dunbarton, the Campsie Hills, Stirling, Scotland; at Bourg d'Oisans, Isère, France; at the Monte Redondo quarry, Portugal; at the Rauschermule quarry, Niederkirchen, and at Radauthal, Harzburg, Harz Mts., Germany; at Habachtal, Salzburg, Austria; at St. Gotthard, Tecino, Switzerland, and Val di Fassa, Italy; in pillow lava at Komiža Bay on Vis Island, Yugoslavia. With phlogopite at Emel'dzhak, southern Yakutia, Siberia, Russia; white fibrous prehnite in rodingite at Pastoki, Hindubagh, Pakistan; in numerous localities in the Deccan traps, as at Poona, and at the Bombay and Khandilavi quarries, Maharashtra, India. At numerous localities in Japan, as at Tokura, Okuzure, and Naoki, Shizuoka Pref., and Hiradani, Shiga Pref. In cavities in dolerite at the Emu quarry, Prospect Hill, near Sydney, NSW, Australia. Found at *Cradock, Cape Prov., South Africa*; well-crystallized(!) on the western slope of Tafelkop, Gobabusberg (also given as Brandberg), and as spherules to 1 cm about 70 km SE of Marienthal, Namibia; from North Victoria Land, Antarctica. Uses: Finds occasional use as a cabochon material and some chatoyant stones are known; faceted collector stones have been produced from Australian and Scottish material, and possibly from other sources. AR *D6:530(1892)*, *DHZI 3:263(1963)*, *MR 14:224(1983)*.

72.1.4.1 Amstallite $CaAl[(Si,Al)_4O_8](OH)_4 \cdot (H_2O,Cl)$

Named in 1987 by Quint for the locality. MON $C2/c$. $a = 18.83$, $b = 11.52$, $c = 5.19$, $\beta = 100.86$, $Z = 4$, $D = 2.38$. *41-1420:* 9.75_{10} 5.43_7 4.71_6 4.07_4 3.82_9 3.60_{10} 3.18_6 2.63_4. Structure: *NJMM: 253(1987)*. Structurally related to but distinct from prehnite; sheets lying parallel to {100} contain four-membered rings and are three tetrahedrons thick; tetrahedral Al is fully disordered with respect to Si. Al octahedra and Ca)[7] polyhedra form edge-sharing ribbons parallel to [001]. Prismatic or acicular unterminated crystals, striated and elongated [001] up to 1 cm long and 0.5 mm in diameter, with a rhombic to hexagonal cross section displaying {100} and/or {110}. Colorless, vitreous luster, transparent to translucent. Cleavage (100), good; conchoidal fracture. $H \approx 4$. $G = 2.90$. Biaxial (+); $2V_{obs} = 57°$; $N_x = 1.533$, $N_y = 1.534$, $N_z = 1.538$; $Y \wedge c \approx +10°$, $Z = b$; $r < v$. $Al^{[4]}$:Si $\approx 0.8:3.2$ and (H_2O):Cl $\approx 4:1$. Occurs in pegmatoid schlieren of low-pressure, low-temperature origin in a graphite deposit, and is associated with apatite, rutile, siderite, albite, laumontite, calcite, and rarely vivianite at the *Amstall quarry, Amstall, Austria*. AR *AM 73:1492(1988)*.

72.1.5.1 Eakerite $Ca_2SnAl_2Si_6O_{18}(OH)_2 \cdot 2H_2O$

Named in 1970 for Mr Jack Eaker (1930–   ) of Kings Mtn., NC who discovered the mineral (MR 1:92(1970). MON $P2_1/a$. $a = 15.829$Å, $b = 7.21$Å, $c = 7.438$Å, $\beta = 101.34°$, $Z = 2$, $D = 2.93$; 24-0218 (nat.) 7.31_8 6.905_5 5.94_5 5.29_9 4.81_{10} 4.00_5 3.88_5 3.40_4 3.35_6 3.30_5 3.02_8. The structure consists of crankshaft-like chains, similar to those in feldspars, of composition $AlSi_3O_9(OH)$, which are cross-linked to form a kinked sheet. Al is ordered, and the OH is bonded to it. Ca + Sn ions lie in sheets between the kinked alumino-silicate network. The Ca ions are coordinated by 4O, 2OH and $2H_2O$ in a square antiprism; these are edge-linked into chains which run across the alumino-silicate chains and are cross-linked by Sn octahedra. The OH and H_2O are bonded to Ca and by H bonds, and to other O; this strong bonding prevented their being distinguished by TGA. (AM 61:956(1976)). Colorless to milky white; prismatic, [001] prism zone is striated; elongated on c, with {001}, {201}, {$\bar{2}$01}, {210}, {410}, {11}, and {100}; to 5 mm long; no cleavage; conchoidal fracture; $H = 5\frac{1}{2}$; $G = 2.83$–2.93; TGA shows no loss of wt. below 350°C, 3% to 725°C, 6.5° to 775°C, 10.5% to 1175°C; mineral fuses at $\approx 1075°C$, not dissolved by acids or bases. SiO_2 46.75, Al_2O_3, 14.07, SnO_2 18.59, CaO 14.2, H_2O 6.70, total = 100.3. Biaxial (+); $2V = 35°$, $N_x = 1.584$, $N_y = 1.586$, $N_z = 1.660$; colorless; $X = b$, $Y \wedge c = +23\frac{1}{2}°$; disper. = $r > v$ marked. Occurs as a hydrothermal mineral in seams in spodumene pegmatite. Associated with quartz, albite and bavenite. Also occurs as an exsolution product from garnet (Russia). Occurs at *Kings Mountain, NC*; Kitelskoye skarn-tin deposit, north of Lake Laguda, Russia, probably also at the Staroye Ora deeposit, Pitkyaranta, Finland. EF ZVMO 101(5):286 (1972), DAN SSSR ESS 250:151(1982) Minerali 4(2):508(1992).

72.2.1.1 Dalyite $(K,Na)^2ZrSi_6O_{15}$

Named in 1952 by Van Tassel for Reginald Adsworth Daly (1871–1957), American geologist, Harvard University. See davanite (72.2.1.2). TRIC $P\bar{1}$. $a = 7.371$, $b = 7.730$, $c = 6.912$, $\alpha = 106.23°$, $\beta = 111.45°$, $\gamma = 100.0°$ (for $K_{1.7}Na_{0.3}$), $Z = 1$, $D = 2.798$. 23-1376(syn): 5.90_5 4.17_7 3.59_8 3.34_{10} 3.08_9 2.85_4 2.62_{10} 2.13_5. Structure: *ZK 121:349(1965)*. The silicate unit is a sheet composed of four-, six-, and eight-tetrahedron rings. Isostructural with davanite. Short prismatic crystals, 0.5 mm, displaying {$\bar{1}$01}, {100}, {1$\bar{1}$0}, {110}, and {$\bar{1}$11}. Twinned with composition plane (100). Colorless, vitreous luster, transparent to translucent. Cleavage {101}, {010}, good. $H = 7.5$. $G = 2.84$. Biaxial (–); $2V = 72°$; $N_x = 1.575$, $N_y = 1.590$, $N_z = 1.594$; $X \wedge c = 7°$; $r > v$, weak. Gelatinizes with acids. Fusibility 3. Limited Na substitution for K; Ti may substitute for Zr. Found as an accessory mineral (0.2%) in an alkali granite with aegirine, arfvedsonite, and aenigmatite at *Green Mountain, Ascension;* at Dalsfjorden, Sunnifjord, and associated with pyrophanite, elpidite, and kupletskite at Gjerdingen, Oslo area, Norway; in lamproite at Cancarix, Albacete, Spain; in the Murun massif and elsewhere in the Aldan shield, Yakutia, Russia; in the Straumsvola complex, Dronning Maud Land, Antarctica. AR *MM 29:850(1952), 47:93(1983)*.

72.2.1.2 Davanite K₂TiSi₆O₁₅

Named in 1984 by Lazebnick et al. for the locality. See dalyite (72.2.1.1). TRIC $P\bar{1}$. $a = 7.272$, $b = 7.480$, $c = 6.910$, $\alpha = 105.55°$, $\beta = 112.82°$, $\gamma = 99.42°$, $Z = 1$, $D = 2.725$. *40-483*: 5.87_5 4.10_7 3.49_8 3.35_8 3.18_7 3.02_{10} 2.79_5 2.62_5. Isostructural with dalyite. Colorless, vitreous luster, transparent to translucent. Subconchoidal fracture. $H = 5$. $G = 2.76$. Biaxial $(-)$; $2V_{calc} = 53°$; $N_x = 1.660$, $N_y = 1.684$, $N_z = 1.690$; also reported, biaxial $(+)$?; $N_x = 1.623$, $N_z = 1.668$. Minor Fe may be present. In lamproite from Smokey Butte, Garfield Co., MT; as pseudohexagonal grains to 5 mm diameter, with aegirine, pectolite, and titanite in the *Murun alkaline massif, near Davan Spring, Yakutia, Russia*, and possibly in platinum metals in the Inagli pluton, central Aldan. AR *AM 7:214(1985), 71:1473(1986), 72:1014(1987); DANS 309:690(1989)*.

72.2.1.3 Sazhinite-(Ce) Na₂Ce[Si₆O₁₄(OH)] · nH₂O (n ≥ 1.5)

Named in 1974 by Eskova et al. for Nikolai Petrovich Sazhin (1898–1969), founder of the Russian rare earth industry. See dalyite (72.2.1.1). ORTH *Pmm2*. $a = 7.50$, $b = 15.62$, $c = 7.35$, $Z = 2$, $D = 2.80$. *30-1176*(nat): 7.25_4 5.23_6 3.37_8 3.30_4 3.23_{10} 2.55_4 2.00_4 1.95_3. Structure: *SPC 25:419(1980)*. Structure is based on infinite corrugated networks formed by four-, six-, and eight-membered rings of Si tetrahedra. Tabular, platy prisms to 5 mm × 5 mm × 1 mm, also as irregular grains, and as dense fine-grained aggregates; striated prisms to 1.7 cm × 1 cm × 6 mm from Mt. St.-Hilaire. White to pale yellowish gray, cream; white streak; vitreous to pearly luster. Cleavage {100}, perfect; {010}, {001}, very good. $H = 2\frac{1}{2}$. $G = 2.61$. IR and DTA data: *ZVMO 103:338(1974)*. Probe data have shown La-dominant zones in crystals from Mt. St.-Hilaire. The La analog of sazhinite-(Ce) occurs in the 10-m-thick St.-Amable trachyphonolite sill, Varennes, Quebec. Analysis (wt %), Mount Karnasurt: SiO_2, 46.28; TiO_2, 1.06; Nb_2O_5, 0.65; Al_2O_3, 0.80; CaO, 0.50; Na_2O, 11.20; K_2O, 1.21; ThO_2, 1.30; Lu_2O_3, 21.15; P_2O_5, 1.05; H_2O^+, 9.58; H_2O^-, 4.46; Fe_2O_3, 0.26; MnO, 0.06. Ce > La = 54%/21%. Biaxial $(+)$; XYZ = *cba*; $N_x = 1.525$, $N_y = 1.528$, $N_z = 1.544$; $2V = 47°$. Occurs in cavities in sodalite xenoliths with vuonnemite, serandite, ussingite, sodalite, aegirine, and so on, at Mt. St.-Hilaire, QUE, Canada. Found in the central zone of the Jubilee pegmatite with natrolite, neptunite, and steenstrupine at *Mt. Karnasurt, Lovozero massif, Kola Penin., Russia*. EF

72.3.1.1 Fluorapophyllite KCa₄[Si₈O₂₀]F · 8H₂O

72.3.1.2 Hydroxyapophyllite KCa₄[Si₈O₂₀](OH) · 8H₂O

The term *apophyllite* from the Greek *away from* and *leaf*, in allusion to its tendency to exfoliate on heating, was probably first used by Haüy in 1800.

72.3.1.2 Hydroxyapophyllite

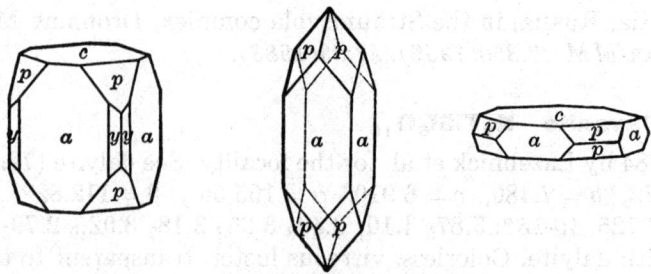

FLUORAPOPHYLLITE
Apophyllite, common habits. Forms: a {100}; c {001}; p {111}; y {310}.

The names of the chemically differentiated end members were coined in 1978 by Dunn and Wilson. The nonspecific term is retained for members of the series whose chemical composition has not been established. An orthorhombic polymorph of fluorapophyllite has been proposed. TET $P4/mnc$. (Fluor-): $a = 8.969$, $c = 15.706$, Z = 2, D = 2.347; (hydroxy-): $a = 8.985$, $c = 15.875$, Z = 2, D = 2.322. *19-82:* 7.81_{10} 4.51_4 3.90_5 3.57_6 3.17_4 2.95_5 2.49_9 2.10_3; *30-920*(calc): 7.81_{10} 4.55_4 3.89_4 3.58_6 3.17_4 2.99_5 2.50_6 2.49_5. Structure: *AC(B) 43:517(1987)*. The lattice dimensions and axial ratio increase linearly with increasing (OH) for F substitution. The silicate element is a two-dimensionally infinite sheet, perpendicular to [001], made up of four-tetrahedron rings with fourfold symmetry linked to one another to form intervening eight-tetrahedron rings with twofold symmetry; apical oxygens of alternate four-membered rings point in opposite directions. Calcium ions link the sheet via two oxygens each, as well as one (F,OH) and two waters. Potassium is eight-coordinated to water molecules. Crystals equant, to prismatic [001], to tabular {001}. The dominant forms are {100}, {001}, and {111} in the traditional morphological setting. Faces of prism zone are commonly striated parallel to [001]. The X-ray setting is rotated 45°; the transformation matrix, morphology to X-ray, is $1\bar{1}0/110/002$. Less commonly massive or lamellar aggregates. Colorless, gray, cream, pale yellow, pale to bright green, pink to reddish; vitreous luster, pearly on cleavage; transparent to translucent. Cleavage {001}, highly perfect; {110}$_{morph}$, imperfect; uneven fracture; brittle. H = $4\frac{1}{2}$–5. G = 2.33–2.37. Uniaxial (+); $N_O = 1.534$–1.542, $N_E = 1.536$–1.543, the indices increasing with increased (OH); strong dispersion; higher-index material may be isotropic or uniaxial (−); occasionally in biaxial sectors, (+) for red, (−) for blue with crossed axial planes. Soluble in acids leaving a silica residue. May undergo natural leaching to yield material of lower refractive index. The green color of some apophyllites has been attributed to vanadium. Found chiefly as a late mineral in amygdules and druses in basalts, where it is associated with pectolite, calcite, and zeolites. Less frequently found in cavities in granites and syenites, in metamorphic rocks and limestones and calc-silicate rocks, sometimes as an alteration of wollastonite, and as late product in hydrothermal ore deposits. Members of this series are found worldwide, but most have not been differentiated by chemical analysis. Fluorapophyllite crystals of excellent quality have been found in the traprocks at Bergen Hill, West Paterson, Prospect Park, and others in the vicinity of Paterson, NJ; high-hydroxyl material in large crystals at Cape Blomidon, NS, Canada, and as pale red crystals at Guanajuato, GTO, Mexico; as colorless crystals at St. Andreasberg, Harz, Germany. Light to deep green prismatic crystals(!) are found at several quarries near Khadakvasla, Pashan Hills, near Poona, and at several occurrences at Nasik and Ahmed Nagur, Maharashtra, India. Hydroxyapophyllite is found as thick druses of white crystals on chalcopyrite–pyrrhotite ore at the *Ore Knob mine, Jefferson, Ashe Co., NC*; as opaque, chalky crystals at Bergen Hill and Prospect Park, NJ; with fluorapophyllite at Mt. St.-Hilaire, QUE, Canada; at Guanajuato, GTO, Mexico, where it is more abundant than the fluorine end member. Found at Storr Mt., Isle of Skye, Scotland; at the Giken mine, Sulitjela, Nordland, Norway; at Khandivali and the Parvati Hills, near Bombay, Maharashtra, India; and at Kimberley, Transvaal, South Africa. Apophyllite (undifferentiated) is found at the Upper New Street quarry, Paterson, NJ; white, tabular crystals at Cornwall, Lebanon Co., PA; tabular, white to pale tan tabular crystals and rosettes(!) found at Centreville, Fairfax Co., VA, are a physical mixture of the two end members; at the Clarke mine, Keweenaw Co., MI. Found at the Jeffrey mine, Asbestos, QUE, Canada; in basalts at numerous localities in Iceland and at several localities in the Faeroe Is., Denmark. Found at numerous localities in Maharashtra, India, in excellent colorless, pale green, and white crystals, particularly Aurangabad, Malegaon, Nasik(!), Poona, Bombay, and the Jewel Tunnel; at the Pahua R., South Island, New Zealand; at the Wessels mine, Cape Prov., South Africa. Colorless to light green tabular crystals to 20 cm are found at Bento Gonsalves, RS, Brazil. Numerous other localities are known. An allegedly orthorhombic modification is reported from the Christmas mine, Gila Co., AZ. Colorless and green material from India has been faceted as a collector gem. AR *MR 9:95(1978); AM 63:196(1978), 77:1119(1992); MM 54:567(1990).*

72.3.1.3 Natroapophyllite $NaCa_4[Si_8O_{20}]F \cdot 8H_2O$

Named in 1981 by Matsueda et al. for the composition and structural relation to apophyllite. ORTH *Pnnm* or *Pnn*2. $a = 8.875$, $b = 8.881$, $c = 15.79$, $Z = 2$, $D = 2.30$. *41-1371*(potassian): 7.9_2 7.79_4 4.54_4 3.97_7 3.57_3 2.98_{10} 2.49_3 2.48_3. Distortion of the tetragonal apophyllite structure due to substitution of Na for K. Colorless, brownish yellow to yellow-brown; vitreous luster, pearly on cleavage; transparent to translucent. Cleavage {001}, perfect; {110}, poor. $H = 4–5$. $G = 2.5$. Biaxial (+); $2V = 32°$; $N_x = 1.536$, $N_y = 1.538$, $N_z = 1.544$. Significant potassium may replace sodium. Dissolved by HCl with separation of silica. Potassian material is found at Rancheria Palo Verde, SLP, Mexico; apophyllite from Boliden, Sweden, is, at least in part, natroapophyllite; as pyramid-truncated, prismatic crystals alternately zoned with fluorapophyllite and associated with zeophyllite, cuspidine, and calcite in skarn at the *Sampo mine, Okayama, Honshu, Japan*. AR *DHZI 3:260(1962), AM 66:410(1981).*

72.3.1.4 Carletonite $KNa_4Ca_4[Si_8O_{18}](CO_3)_4(OH,F)\cdot H_2O$

Named in 1971 by Chao for Carleton University, Ottawa, Canada, where it was first recognized. TET $P4/mbm$. $a = 13.178$, $c = 16.695$, $Z = 4$, $D = 2.426$. 25-628: 16.7_4 8.35_{10} 4.82_4 4.17_{10} 4.05_5 3.34_4 2.90_9 2.38_6. The silicate structural element consists of two apophyllite silicate layers condensed by sharing apical oxygens to form a layer of double thickness lying parallel to {001}; potassium occupies the voids in the layer. Blue to light pink, colorless in thin section or small grains; vitreous luster, pearly on cleavage. Cleavage {001}, perfect; {110}, distinct; conchoidal fracture. $H = 4$–$4\frac{1}{2}$. $G = 2.45$. Uniaxial (−); $N_O = 1.521$, $N_E = 1.517$ for Na(D); blue crystals weakly pleochroic; O, very pale blue; E, very pale pinkish brown. IR spectrum much more complex than for apophyllites due to lower site symmetries. No significant compositional difference noted for the several colors. Occurs in marble xenoliths in nepheline syenite as square prisms, {100} and {001} dominant, sometimes with color zoning or color phantoms, at *Mt. St.-Hilaire, QUE, Canada*, where crystals up to 6 cm long have been found. A few small gems (< 1 ct) have been cut for collectors. AR *AM 56:1855(1971), 57:763(1972); MR 21:301(1990)*.

TOBERMORITE GROUP

The tobermorite group minerals are hydrated Ca–Si, Ca–Al–Si, or Ca–B–Si minerals that are structurally heterogeneous but nevertheless appear to be somewhat related. All the minerals are rare, and for some complete structural information is lacking. A recent study and structural models are found in *ZVMO 124:36(1995)*.

TOBERMORITE GROUP

Mineral	Formula	Space Group	a	b	c	α	β	γ	
Tobermorite	$Ca_5Si_6O_{16}(OH)_2\cdot 4H_2O$	$C222_1$	11.233	7.372	22.56				72.3.2.1
Clinotobermorite	$Ca_5Si_6(O,OH)_{18}\cdot 5H_2O$	Cc or C2/c	11.331	7.353	22.67		96.59		72.3.2.2
Plombierite	$Ca_5Si_6O_{16}(OH)_2\cdot 6H_2O$	Imm2	11.250	7.344	27.99				72.3.2.3
Riversideite	$Ca_5Si_6O_{16}(OH)_2\cdot 2H_2O$	P***	5.571	3.641	18.79				72.3.2.4
Okenite	$Ca_3[Si_6O_{15}]\cdot 6H_2O$	P$\bar{1}$ or P1	9.69	7.28	22.0	87.5	100.1	110.9	72.3.2.5
Tacharanite	$Ca_{12}Al_2Si_{18}O_{33}(OH)_{36}$	A***	17.07	3.65	27.90		114.1		72.3.2.6
Nekoite	$Ca_3Si_6O_{15}\cdot 7H_2O$	P1	7.588	9.793	7.339	111.77	103.50	86.53	72.3.2.7
Oyelite	$Ca_{10}B_2Si_8O_{29}\cdot 12H_2O$	B***	11.23	7.24	20.42				72.3.2.8

72.3.2.1 Tobermorite

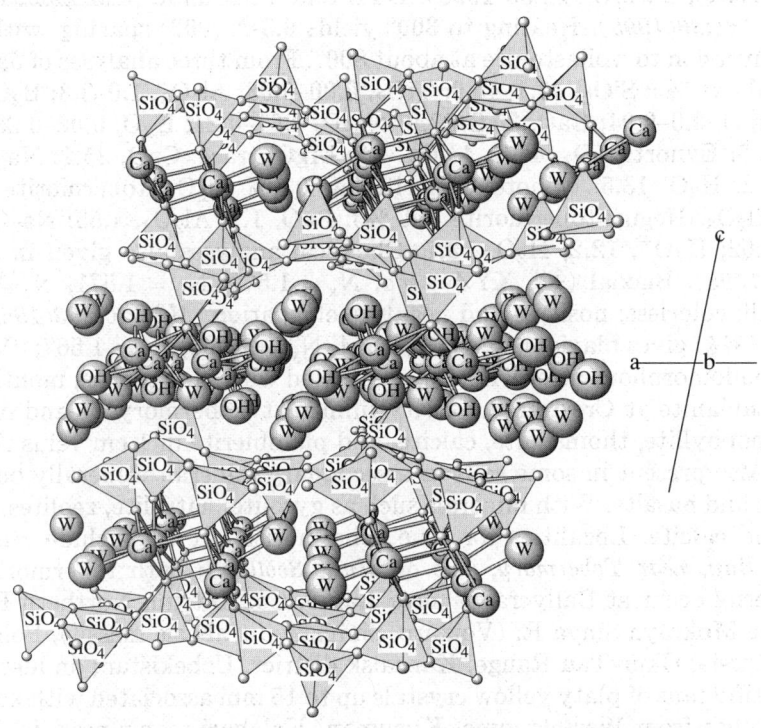

OKENITE: $Ca_{10}Si_{18}O_{46} \cdot 18H_2O$

This has two different types of silicate layer: (tobermorite group) a true sheet with five- and eight-membered rings, and a layer containing double chains with four- and six-membered rings. Ca atoms link the sheets, and water molecules occupy cavities in the larger rings. Other members of the tobermorite group have different layer structures.

72.3.2.1 Tobermorite $Ca_5Si_6O_{16}(OH)_2 \cdot 4H_2O$

Named in 1880 by M. F. Heddle for the locality. Tobermorite group. See plombierite (72.3.2.3) and riversideite (72.3.2.4). Polymorph: clinotobermorite. ORTH $C222_1$. $a = 11.233$–11.255, $b = 7.372$–7.375, $c = 22.56$–22.760, Z = 4, D = 2.58 (syn). 45-1480(nat): *JJMPE 84:374(1989)* (Fuka): 11.3_{10} 5.45_4 3.51_3 3.08_9 2.97_7 2.81_{7b} 2.00_2 1.84_{2b}; 19-1364(syn): 11.3_8 $5.48_{2.5}$ 3.53_2 3.08_{10} $2.98_{6.5}$ 2.82_4 2.00_2 1.84_4. Structure: *NAT 177:390(1956)*, *ZK 154:189(1981)*. At 140°, 11.3-Å tobermorite has double chains of SiO_4 tetrahedra. Partial dehydration of plombierite at 100° yields 11.3-Å tobermorite with single chains of SiO_4 tetrahedra. Porcelaneous, massive, also finely fibrous bundles, rosettes or sheaves, granular or {001} plates, acicular parallel to [010], platy crystals to 3 mm across. Colorless–white, pinkish white; white streak; vitreous–silky luster. Cleavage {001}, {100}, distinct. H = 3. G = 2.40–2.48 (Fuka). Material from Crestmore is sometimes weakly fluorescent white to yellow in SWUV and LWUV. IR data: *MM 42:229(1978), 56:353(1992); ZVMO 124:36(1995)*. DTA and TGA data: *MM 42:229(1978), ZVMO 124:36(1995)*. Heating to 300° yields 9.3-Å (002) spacing, with eventual conversion to wollastonite at about 800°. From three analyses of Japanese material (wt %): SiO_2, 45.1–49.0; TiO_2, 0.00–0.03; Al_2O_3, 1.0–3.3; B_2O_3, 0.0–0.23; MgO, 0.0–0.04; CaO, 35.2–37.9; Na_2O, 0.01–0.24; K_2O, 0.02–0.23; H_2O, 11.5–15.5; Eynort: SiO_2, 46.2; Al_2O_3, 4.3; MgO, trace; CaO, 35.2; Na_2O, 0.6; K_2O, 0.3; H_2O, 13.5. Clinotobermorite has 0.4% Al_2O_3, tobermorite has 1–>3% Al_2O_3. Heguri tobermorite (wt %): BaO, 1.7; Al_2O_3, 4.55; Na_2O, 0.45; K_2O, 0.52; H_2O^+, 12.2; H_2O^-, 3.7. Chemical analyses also given in *ZVMO 124:36(1995)*. Biaxial (+); XYZ = cba; $N_x = 1.570$, $N_y = 1.571$, $N_z = 1.575$; 2V small; colorless; positive and negative elongation. *MM 30:293(1954)*. *DA 25:613(1974)* gives biaxial (−); $N_x = 1.554$, $N_y = 1.566$, $N_z = 1.567$; 2V = 26–33°. Pseudomorphous after wilkeite, associated with blue calcite, monticellite, and vesuvianite at Crestmore. Cavity fillings at Tobermory, Island of Mull. With apophyllite, thomsonite, calcite, and plombierite in 1-cm veins that cut skarn. Also present in some soils as a secondary mineral. Generally occurs in gabbros and basalts. With minerals such as gyrolite, natrolite, zeolites, ettringite, and calcite. Localities for true tobermorite (11 Å) include *Pier, and Bloody Bay, near Tobermory, Isle of Mull, Scotland*; both tobermorite and plombierite occur at Ballycraigy near Larne, Co. Antrim, Northern Ireland; from the Mokraya Sinya R. (Voikaro–Sininsky ultrabasic massif), Polar Ural Mts., Russia; Okur-Tau Range, Kansaisk district, Uzbekistan; on inesite and as beautiful fans of platy yellow crystals up to 15 mm associated with xonotlite and datolite from Wessels mine, Kuruman, Kalahari manganese field, Cape Prov., South Africa; Ba-bearing tobermorite occurs at Heguri, Chiba Pref., Japan, and tobermorite, plombierite, and clinotobermorite occur at Fuka, Kawakami Twp., Okayama Pref., Japan. Other localities for "tobermorite," not clearly specified, from the literature include: Durham, NC; Bingham, UT; Goldfield, NV; Concepcion de Oro and the Noche Buena mine, Mazapil, ZAC, Mexico; Portree, Scotland; Loch Eynort, Skye, Scotland; Castle Hill, Kilbirnie, Ayrshire, Scotland; Höwenegg, Hegau, Germany; Arensberg near Zilsdorf, Eifel, Germany; Zeilberg quarry, Maroldsweisach, Franconia, Bavaria, Germany; Mont Alto di Castro (Viterbo), Italy; Pia de la Stua, and Monte Biaena, Trento, Italy; Klöch, southeastern Styria, Austria; Bazhenovskoye, Russia; Hatrurim Fm., Nahal Ayalon, Israel; Taiwan. EF *MIN 3(3):322(1981), ZK 154:268(1981), MM 56:353(1992), ZVMO 124:36(1995)*.

72.3.2.2 Clinotobermorite $Ca_5Si_6(O,OH)_{18} \cdot 5H_2O$

Described in 1989 by Henmi and Kasache and named in 1992 for the locality. Tobermorite group. The monoclinic polymorph of tobermorite. MON *Cc* or *C2/c*. $a = 11.331$, $b = 7.353$, $c = 22.67$, $\beta = 96.59°$, $Z = 2$, $D = 2.69$. 45-1479(nat): 11.25_{10} 3.74_4 3.30_5 3.07_4 3.03_6 3.01_4 2.81_4 2.79_6. *JJMPE 84:374(1989)*. Clinotobermorite is the low-temperature form of tobermorite. Upon heating at 300°, the (002) spacing is reduced from 11.3 Å to 9.3 Å.

Crystals are platy {001}, acicular [010], or tabular. Microtwinning is always observed on {011}, along with stacking disorder on {001}. Colorless–white, tabular crystals to 5 mm wide and acicular crystals to 2 mm long, white streak, vitreous luster. Cleavage {001}, perfect; {100}, poor. H = 4.5. G = 2.51–2.58. Thermal data: *JJMPE 84:374(1989)*, *MM 56:353(1992)*. Trace to minor amounts of Al, B, Mg, K, F and traces of Ti, Fe, Na, and Mn. No 2V measured because of microtwinning. $2V_{calc} = 90°$. $N_x = 1.575$, $N_y = 1.580$, $N_z = 1.585$, colorless. A vein-forming mineral in gehlenite–spurrite skarns. In hybrid rocks and in retrograde garnet-bearing rocks. Associated with tobermorite, calcite, apophyllite, plombierite, and oyelite. *Fuka, Bitchucho, Okayama Pref., Japan.* EF

72.3.2.3 Plombierite $Ca_5Si_6O_{16}(OH)_2 \cdot 6H_2O$

Described in 1858 by Daubree and named for the locality; redefined in 1954 by McConnell. Tobermorite group. Formerly called 14-Å tobermorite, redefined on material from Ballycraigy, Larne, County Antrim, Northern Ireland. Deposited from thermal waters on brick and mortar of an old Roman aqueduct. See riversideite (72.3.2.4) and tobermorite (72.3.2.1). ORTH $Imm2$. $a = 11.250$, $b = 7.344$, $c = 27.99$, $Z = 4$, $D = 2.31$. 29-331: 14.0_{10} $5.50_{2.5}$ $3.08_{6.5}$ $3.00_{4.5}$ 2.81_3 2.80_1 2.72_1 $1.835_{3.5}$. Also amorphous. Gelatinous masses (that harden in air) may be powdery, massive and compact to 5 cm, also as white fibers, radial aggregates, or lenses; sometimes elongated on b. Pinkish white, white; white streak; silky luster. Cleavage {001}, excellent; {100}, very good; brittle; masses have conchoidal fracture. $H = 2\frac{1}{2}$–4; G = 2.02–2.25 with some as high as 2.42. Material from Crestmore fluoresces weak pink-white in SWUV and weak to medium yellow-white in LWUV. IR data: *ZVMO 124:36(1995)*. Progressive heating (at 90°C) produces tobermorite and then riversideite (at 300°C). Essentially a hydrated calcium silicate that sometimes contains traces of K, Mg, Fe, and Al. Biaxial (+); 2V = 15°; variable optics are reported in the literature. $N_x = 1.570$, $N_y = 1.571$, $N_z = 1.575$, and Ny = 1.552(Israel) and 1.550(syn); length-slow or length-fast, OAP is perpendicular to length, $X = c$. Occurs in skarns and as deposits from thermal waters. Occurs at Bingham, Utah; in skarn at Crestmore, Riverside Co., CA; in skarn at Ballycraigy and Scawt Hill, near Larne, and at Carneal, County Antrim, Northern Ireland; at *Plombières, Vosges, France*; Klöch, Styria, Austria; in the Ozersk gabbro massif, West Baikal area, Siberia, Russia; Bazhenovskoye deposit, Russia; in the "mottled zone" in the Hatrurim area and Beersheba valley, northern Negev, Israel; in association with tobermoreite and oyelite in skarn that is mostly gehlenite and spurrite; Arimao Norte, Cuba; some parts of the skarns were retrogressively altered and cut by numerous veins of tobermorite, oyelite, bultfonteinite, scawtite, xonotlite, afwillite, and jennite at Fuka, Bitchu, Okayama Pref., Japan. EF *MIN 3(3):327(1981)*; *MM 56:353(1992)*; *ZVMO 123:111(1992), 124:36(1995)*.

72.3.2.4 Riversideite $Ca_5Si_6O_{16}(OH)_2 \cdot 2H_2O$

Named in 1917 by Eakle for the locality. Tobermorite group. ORTH P***. (*29-329*): $a = 5.571$, $b = 3.641$, $c = 18.79$, $Z = 1$; true cell is C centered: $2a = 11.142$, $2b = 7.282$, $c = 18.79$, $Z = 4$, $D = 2.87$. *29-329*(nat, heated plombierite from Crestmore): 9.4_2 4.82_1 3.59_2 3.15_2 3.01_{10} 2.78_2 2.74_1 1.82_1. Structure: *DAN SSSR 123:163(1958)*. White, white streak, fibrous precipitate, also massive length to 1 cm elongate on b. Cleavage {001}, excellent; {100}, less so. G given as 2.38 and 2.6–2.7. Material from Crestmore fluoresces a weak white in LWUV and SWUV. Biaxial (+); $XYZ = cba$; $N_x = 1.600$, $N_y = 1.601$, $N_z = 1.605$; $2V_{calc} = 53°$; length-fast or length-slow. Occurs as white fibrous precipitate from a gel, associated with tobermorite and wilkeite at *Crestmore, Riverside Co., CA*; also occurs at Ballycraigy, Larne, County Antrim, Northern Ireland; in xenoliths in basalt at Maroldsweisach, northern Bavaria, Germany; in the Hatrurim Fm., Israel. EF *MM 30:155(1953) 30:293(1954) 42:229(1978)*, *MIN 3(3):328(1981)*, *DA 46:69(1995)*.

72.3.2.5 Okenite $Ca_3[Si_6O_{15}] \cdot 6H_2O$

Named in 1828 by Von Kobell for Lorenz Oken (1779–1851), naturalist, Munich, Germany. Tobermorite group. TRIC $P\bar{1}$ or $P1$. $a = 9.69$–9.80, $b = 7.28$–7.29, $c = 22.0$, $\alpha = 87.5$–$92.7°$, $\beta = 100.1$–$102.0°$, $\gamma = 110.9$–$112.6°$, $Z = 9$. *AM 68:614(1983)* and *33-305*. The true cell (superstructure) has: (for $P\bar{1}$) $a = 9.81$, $b = 14.56$, $c = 22.02$, $\alpha = 87.3°$, $\beta = 106.0°$, $\gamma = 112.8°$; maintaining the same orientation as the subcell requires a C-centered cell ($C\bar{1}$): $a = 19.38$, $b = 14.56$, $c = 22.02$, $\alpha = 97.7°$, $\beta = 100.1°$, $\gamma = 110.9°$; $Z = 1$; another cell from a structure solution is: $a = 18.50$, $b = 7.29$, $c = 22.45$, $\alpha = 87.7°$, $\beta = 102.6°$, $\gamma = 91.5°$; $Z = 3$ [*ZK 159:37(1982)*], $D = 2.24$. *33-305*(nat): 21.0_{10} 8.84_8 3.11_6 3.00_6 2.95_6 2.92_6 2.80_5 1.82_7. A SOCOS layer is formed of infinite tetrahedral sheets S characterized by five- and eight-membered rings, double tetrahedral chains C, and double octahedral chains O. First example of a tetrahedral chain and sheet silicate. $R = 13\%$. *AM 68:614(1983)*. Structure solution in *ZK 159:37(1982)* yielded $R = 8\%$. Fibrous, bladed or lath-like crystals, elongate on [010], also radial aggregates of curved crystals, "balls" to 4 cm in diameter; repeated lamellar twinning across the cleavage plane {001} – $_{011}$[010] + $_{011}${001}. Colorless, white, pale yellow, or blue; white streak; pearly luster. Cleavage {001}, perfect; flexible fibers. $H = 4\frac{1}{2}$–5. $G = 2.2$–2.4. IR data: *MMS II 6:199(1966)*. DTA and TGA data: *MM 43:677(1980)*. Gelatinizes with acids; okenite transforms to β-$CaSiO_3$ (parawollastonite) on heating. The presence of tetrahedral ribbons in its structure instead of the octahedral layers in gyrolite appears to be a determining factor for its stability. May have trace to minor Na, K, Mg, Al, Fe. Na_2O to 1.2%. Biaxial (–); $Z \approx c$; $N_x = 1.512$–1.532, $N_y = 1.514$–1.535, $N_z = 1.515$–1.542; $2V = 60°$; length-fast. In cavities in basalt or related eruptive rocks. In limestone at Crestmore. Often associated with zeolites, apophyllite, calcite, prehnite, quartz and others. In carbonatites. Occurs at the Fairfax Quarry, Centreville, Fairfax Co., VA; Crestmore, Riverside Co.,

CA; Skookumchuck Dam, near Bucoda, Thurston Co., WA; Nova Scotia, and Jeffrey mine, Asbestos, QUE, Canada; Bordö, Faroe Is. (Denmark); *Disko Is., Greenland*; Kutdlisat, Ritenbenk district, Greenland; Scawt Hill, County Antrim, Northern Ireland; Isle of Mull and on Morven Peak, Grampian Highlands, Scotland; Branburg, Germany; Brèzno nad Labem, and Ústí nad Labem-Krásné Brèzno, Bohemia, Czech Republic; in carbonatite at Vuorijarvi, Kola Penin., Russia; Kerch and Taman Penin., Crimea, Ukraine; Aleshinsk deposit, Turgai, Russia; southern Mugodzhary, Kazakhstan; Poona and Deccan traps, Syhadree Mts., Bombay, MAH, Naskik, and Kolhapur district, MAH, India; Rio Putagan, Chile. EF *MM 31:5(1956) MIN 3(3):330(1981)*.

72.3.2.6 Tacharanite   Formula given variously as: $Ca_{12}AL_2Si_{18}O_{33}(OH)_{36}$; $Ca_{12}Al_2H_6Si_{18}O_{54} \cdot 15H_2O$; sodium-bearing from Japan: $NaCa_{12}AL_2H_6ALSi_{17}O_{54} \cdot 12H_2O$

Named in 1961 by Sweet et al. from the Gaelic *tacharan*, a changeling, because on standing in air the mineral breaks down to form other compounds (note that later investigators did not find this to be the case). Tobermorite group. MON A centered, pseudo-cell: $a = 17.05$–17.11, $b = 3.65$, $c = 27.8$–28.04, $\beta = 114.1°$, $Z = 1$, $D = 2.28$. A cell with b doubled has been reported [*BNSMTC 9:85(1983)*]: $a = 17.07$, $b = 7.276$, $c = 27.96$, $\beta = 114.0°$, $Z = 2$. 29-287(nat): 12.7_{10} 5.19_7 3.05_9 2.89_7 2.85_7 2.77_8 2.43_7 1.82_8. Unknown structure. Fine acicular, massive, porcelaneous masses and crusts, elongate laths, b parallel to fiber length. White, porcelaneous, cryptocrystalline; white streak. Conchoidal fracture. $H \approx 5$. $G = 2.21$–2.36. IR data: *MM 40:113(1975)*, very similar to tobermorite. DTA data: *MM 42:383(1978)*. TGA data: *MM 40:113(1975)*, very similar to tobermorite. Can heat to 700° before changes occur in the powder pattern; production of wollastonite at 900–1000°. Analysis (wt %): SiO_2, 48.4; Al_2O_3, 5.0; CaO, 30.4; Na_2O, 0.25; K_2O, 0.8; H_2O, 14.9; Japanese material: Fe_2O_3, 0.1; MgO, 0.6; Na_2O, 1.26; K_2O, 0.14; H_2O^+, 12.50; H_2O^-, 1.76. Substitution of (Na, Al) for (□, Si) occurs. Isotropic to biaxial (+) with large 2V; $N = 1.537$ (Portree cryptocrystalline), 1.530 (Carneal), 1.54 (Bramberg, variable crystallinity); 1.530 parallel to fiber direction and 1.518 perpendicular to fiber direction. For Tasmanian material, $N = 1.535$ parallel to the fiber direction and 1.525 perpendicular to the fiber direction. Colorless. Positive elongation of fibers. Occurs in vesicles of an olivine dolerite with zeolites and other hydrated Ca-silicates: tobermorite, xonotlite, gyrolite, and others. In Italy, in fractures in metagabbros with zeolites; in vesicular olivine nephelinite in Tasmania; in glassy pillow lava, Japan. A low-temperature (< 250°) mineral. Alteration: tacharanite altered to tobermorite, gyrolite, and xonotlite [*MM 32:745(1961)*], but no alteration was reported later [*MM 40:113(1975)*]. Alteration to tobermorite ± saponite also reported. Doe Hill, Highland Co., VA; Ritter Hot Springs, Oregon; *Portree, Isle of Skye, Scotland*; also Isle of Mull, and at the Binbill quarry, Cairnie, Huntley, Aberdeenshire, Scotland; Carneal, County Antrim, Northern Ireland; at Espalion, Aveyron,

France; Bramburg, near Göttingen, Lower Saxony, and at Arensberg, near Zilsdorf, Eifel district, Germany; Gruppo di Voltri, and along the Olbirella, near Tiglieto, Ligurian Alps, Italy; Scottsdale, Marrawah-Redpa, Brittons Swamp, and several other areas in Tasmania, Australia; Noaki, Yaizu city, Shizuoka Pref., Japan. EF *MM 42:383(1978), 51:467(1987); MIN 3(3):329(1981)*.

72.3.2.7 Nekoite $Ca_3Si_6O_{15} \cdot 7H_2O$

Named in 1956 by Gard and Taylor from *okenite* by reversing the first four letters; the minerals are very similar. Tobermorite group. See okenite (72.3.2.5). TRIC $P1$. $a = 7.588$, $b = 9.793$, $c = 7.339$, $\alpha = 111.77°$, $\beta = 103.50°$, $\gamma = 86.53°$, $Z = 1$, $D = 2.21$. *31-303*(nat): 9.1_{10} 6.64_7 3.38_6 3.32_9 3.30_6 2.83_6 2.80_6 1.83_8. *PM 48:5(1979)*: 9.09_5 7.38_8 3.69_4 3.45_6 3.38_{10} 2.83_6 2.82_6 2.19_7. The structure consists of tetrahedral sheets interlayered with octahedral chains. The tetrahedral Si_6O_{15} sheets are obtained by interconnecting a "dreierkette" so as to form alternating bands of five- and eight-membered rings. Ca-octahedral chains, which cross-link the tetrahedral sheets, are formed by a double strand of octahedra, of which three out of four are occupied and the fourth is empty. $R = 6.8\%$. *AM 65:1270(1980)*. Radial–fibrous nodular aggregates of fine acicular prismatic crystals to 5 mm long (Brazil); elongated parallel to c, polysynthetic twinning on {001} and {100}. Colorless, white streak, vitreous luster. Cleavage {001}, perfect. H = 5 or less. G = 2.21–2.24. Material from Crestmore fluoresces weak to medium white-yellow in SWUV and medium white-yellow in LWUV. IR, DTA, and TGA data: *PM 48:5(1979)*. Biaxial (+); Z perpendicular to (001), X $\wedge$ $a = 26°$; $N_y = 1.535$; $2V = 70°$; birefringence is very low; length-slow. Occurs at the Iron Cap mine, Landsman's Camp, Arivaipa District, Graham Co., AZ; in a contact metamorphosed limestone deposit, assoc. with apophyllite at *Crestmore quarry, Crestmore, Riverside Co., CA*; in cavities in basaltic rock at Caxias do Sul, Rio Grande do Sul, Brazil, along with heulandite and apophyllite. EF *MIN 3(3):335(1981)*.

72.3.2.8 Oyelite $Ca_{10}B_2Si_8O_{29} \cdot 12H_2O$

Named in 1984 by Kusachi et al. for Jiro Oye (1900–1968), professor of mineralogy, Okayama University, Japan. Tobermorite group. Previously reported and described as 10-Å tobermorite. ORTH Space group B***. $a = 11.23–11.25$, $b = 7.24–7.252$, $c = 20.42–20.46$, $Z = 2$, $D = 2.64$. *41-1386*(nat): 10.29_9 4.92_2 3.78_2 3.06_5 2.92_{10} 2.81_4 2.56_2 2.04_2. *JJMPE 81:138(1986)*. Acicular crystals 1–3 mm long, sprays and coatings. White, vitreous luster. H = 5. G = 2.62. IR, DTA, and TGA data: *JJMPE 81:138(1986)*. Analysis (wt %): SiO_2, 35.3; B_2O_3, 4.8; Al_2O_3, 0.3; CaO, 41.2; H_2O^+, 16.7; H_2O^-, 0.7; CO_2, 0.4. Biaxial (+); $N_x = 1.602$, $N_y = 1.606$, $N_z = 1.613$; $2V_{calc} = 74°$; colorless. Occurs at Crestmore Quarry, Riverside Co., CA, in association with bultfonteinite, scawtite, xonotlite, and calcite. In a vein cutting a spurrite zone of a spurrite–gehlenite skarn at *Fuka, Bitchu, Okayama Pref., Japan*. Also occurs at Suisho-dani, Ise

City, Mie Pref., Japan, where it is in a serpentinite body, in rodingitic rock derived from a gabbro pegmatite. Occurs associated with bultfonteinite at the N'chwaning II mine, Kalahari manganese field, Cape Prov., South Africa. EF *JJMPE 79:267(1984), MR 22:279(1991)*.

PYROSMALITE GROUP

The pyrosmalite group consists of several isostructural, polytypic, and derivative species conforming to the formula

$$R_8^{2+}[Si_6O_{15}](OH,Cl)_{10}$$

where

$R = Mn, Fe^{2+}, Mg$

The structure is characterized by a silicate layer built up of six-tetrahedron rings, each of which is joined by two tetrahedra to three adjacent rings whose apical oxygens point opposite to that of the central ring. The resultant layers, about 7.15 Å thick, have one 12-tetrahedron ring surrounding a large void, for every two six-tetrahedron rings.

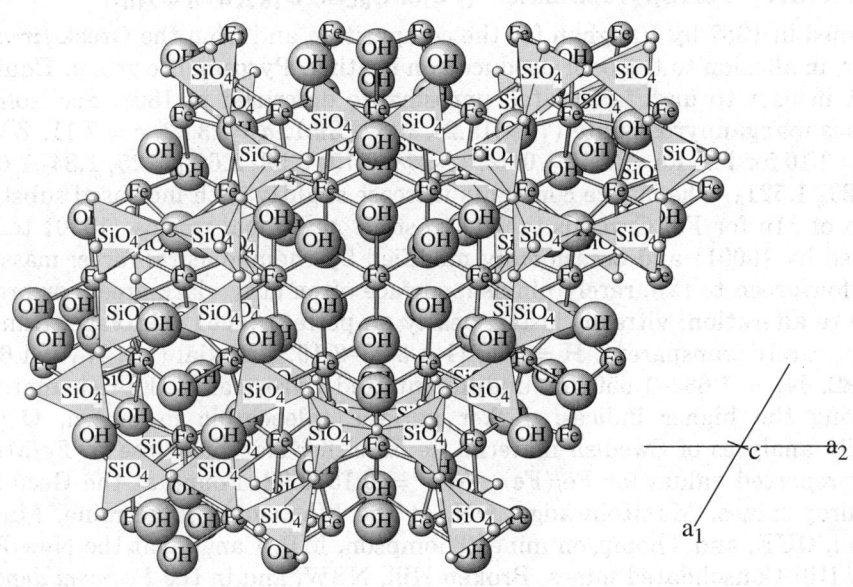

PYROSMALITE: $(MN,Fe)_{16}(Si_{12}O_{30})(OH,Cl)_{20}$
Perspective view approximately along *c*.

The several species differ from one another on the basis of the cation (Mn^{2+} or Fe^{2+}), the stacking sequence of the layers, the ordering of chlorine and hydroxyl, and the occupancy of silicate-layer voids by [$As^{3+}O_2(OH)$]. The stacking sequence appears not to be entirely independent of ordering and void occupancy, so that similar compositions of different symmetry may not be polytypes in *sensu strictu*. AM 66:1054(1981); CM 21:1(1983), 21:7(1983).

PYROSMALITE GROUP

Mineral	Formula	Space Group	a	b	c	β	
Ferropyrosmalite	$(Fe,Mn)_8[Si_6O_{15}](OH,Cl)_{10}$	$P\bar{3}m1$	13.33		7.11		72.4.1a.1
Manganpyrosmalite	$(Mn,Fe)_8[Si_6O_{15}](OH,Cl)_{10}$	$P\bar{3}m1$	13.442		7.165		72.4.1a.2
Schallerite	$Mn_{16}[As^{3+}O_2OH]_3[Si_{12}O_{30}](OH)_{14}$	$P\bar{3}m1(?)$	13.40		14.28		72.4.1a.3
Friedelite	$Mn_8[Si_6O_{15}](OH,Cl)_{10}$	$C2/m$	23.33	13.396	7.447	105.08	72.4.1b.1
Mcgillite	$Mn_8[Si_6O_{15}](OH)_8Cl_2$	$C2/m$	23.312	13.459	7.423	105.17	72.4.1b.2
Nelenite	$(Mn,Fe)_{16}[As^{3+}O_2OH]_3[Si_{12}O]_{36}(OH)_{17}$	$C2/m$	23.240	13.418	7.832	105.21	72.4.1b.3

72.4.1a.1 Ferropyrosmalite $(Fe,Mn)_8[Si_6O_{15}](OH,Cl)_{10}$

Named in 1987 by Vaughan for the composition and from the Greek *fire* and *odor*, in allusion to the odor produced on heating. Pyrosmalite group. Equivalent in part to undifferentiated pyrosmalite described in 1808. See isomorphous manganpyrosmalite (72.4.1a.2). HEX $P\bar{3}m1$. $a = 13.33$, $c = 7.11$, $Z = 2$, $D = 3.10$ for $Fe/(Fe + Mn) = 0.92$. *12-268*: 7.13_8 3.58_5 2.69_{10} 2.25_7 1.84_5 1.672_5 1.629_4 1.521_5. The lattice constants increase slightly with increased substitution of Mn for Fe. Crystals tablets to stout hexagonal prisms {$10\bar{1}0$} terminated by {0001} and occasionally modified by rhombohedral faces; massive. Yellow-green to tan, rarely pinkish; surface often gray, dark green, or brown due to alteration; vitreous luster, pearly on perfect {0001} cleavage; translucent, rarely transparent. $H = 4-5$. $G = 3.06-3.19$. Uniaxial $(-)$; $N_O = 1.664-1.682$, $N_E = 1.634-1.650$ for undifferentiated pyrosmalite, Fe-rich material having the higher indices; darker varieties pleochroic in brown, $O > E$. Early analyses of Swedish material lie on either side but close to $Fe/Mn = 1:1$; reported values for $Fe/(Fe + Mn) = 0.15-0.92$. Found at the Geco and Willroy mines, Manitouwadge, ON, at the Mattagami Lake mine, Mattagami, QUE, and Thompson mine, Thompson, MB, Canada; at the New Broken Hill Consolidated mines, Broken Hill, NSW, and in the *Pegmont deposit, Q, Australia*. Undifferentiated pyrosmalite: in transparent, pink microcrystals on silver at the El Bonanza mine, Port Radium, NWT, Canada; at Naut Francon, Wales, and the Treburland mine, Altarnum, Cornwall, England; in the iron ores of Dannemora, Uppsala, and *Nordmark, Värmland, Sweden(!)*, near midway toward Mn end member. See also manganpyrosmalite (72.4.1a.2). AR *D6:465(1892)*; *MM 50:527(1986)*, *51:174(1987)*; *CM 31:695(1993)*.

72.4.1a.2 Manganpyrosmalite $(Mn,Fe)_8[Si_6O_{15}](OH,Cl)_{10}$

Named in 1953 by Frondel and Bauer for the composition and from the Greek for *fire* and *odor*, in allusion to the odor produced on heating. Pyrosmalite group. Equivalent in part to undifferentiated pyrosmalite described in 1808. See isomorphous ferropyrosmalite (72.4.1a.1). HEX $P\bar{3}m1$. $a = 13.442$, $c = 7.165$, $Z = 2$, $D = 3.03$ for Fe/(Fe + Mn) = 0.18. *12-249*: 7.16_{10} 3.58_8 3.42_4 2.68_9 2.25_7 1.843_4 1.672_5 1.523_5. Structure: *CM 21:1(1983)*. The lattice constants increase slightly with increased substitution of Mn for Fe. Brown, translucent. Cleavage {0001}, perfect. H = $4\frac{1}{2}$. G = 3.13. Uniaxial $(-)$; $N_O = 1.669$, $N_E = 1.631$. Weakly pleochroic; O > E; light brown. Indices increase with increasing Fe substitution. Complete solid solution with ferropyrosmalite. Found at *Sterling Hill, Sussex Co., NJ*, with friedelite, bementite, and willemite in veinlets cutting franklinite ore. In Värmland and Uppsala, Sweden(!) (Mn/Fe near 1:1); in the Dzhumart and Ushkatyn deposits, Kazakhstan; Primorye Kray, Russia; Wafangzi, Liaoning, China; at the Kyurazawa mine, Tochigi Pref., Honshu, Japan; large composite crystals at Broken Hill mine, NSW, Australia; Lar mine, Tierra Amarilla, Chile. See also ferropyrosmalite AR *D6:465(1892)*, *AM 38:755(1953)*, *CM 31:695(1993)*.

72.4.1a.3 Schallerite $Mn_{16}[As^{3+}O_2OH]_3[Si_{12}O_{30}](OH)_{14}$

Named in 1925 by Gage et al. in honor of Waldemar Theodore Schaller (1882–1967), mineralogist, U.S. Geological Survey. Pyrosmalite group. HEX Space group unknown, probably $P3m1(?)$. $a = 13.40$, $c = 14.28$, $Z = 2$, $D = 3.45$. *12-253*: 3.55_4 2.83_3 2.67_6 2.47_5 2.02_5 1.975_4 1.688_{10} 1.511_6. $[As^{3+}O_2OH]^{2-}$ probably occupy the large voids in the 12-membered rings of the silicate sheets. Dimorphous with nelenite. Light brown to reddish brown; waxy luster, pearly on {0001} cleavage; translucent. H ≈ 5. G = 3.37. Uniaxial $(-)$; $N_O = 1.704$, $N_E = 1.679$. Ideally, As_2O_3 should be 12.87 wt %, but samples with as little as 1.2–6.9 wt % are known, suggesting a complex relationship to pyrosmalite and friedelite. Fe substitution for Mn appears to be limited. Found massive and as small (to 2 mm) hemimorphic pyramidal crystals with calcite in fractures perpendicular to banded willemite–franklinite ore at the 700-ft level at the *Franklin mine, Sussex Co., NJ*, and as spherical and rectangular aggregates in massive rhodonite; also in the Ködnitz Valley, Tyrol, Austria. AR *USGSPP 180:90(1935)*, *AM 66:1054(1981)*.

72.4.1b.1 Friedelite $Mn_8[Si_6O_{15}](OH,Cl)_{10}$

Named in 1876 by Bertrand for Charles Friedel (1832–1899), French chemist and mineralogist. Pyrosmalite group. The name once applied to the manganese end member of pyrosmalite. MON $C2/m$. $a = 23.33$, $b = 13.396$, $c = 7.447$, $\beta = 105.08°$, $Z = 4$, $D = 3.06$. *35-572*: 7.17_9 3.60_7 2.88_6 2.56_{10} 2.41_3 2.12_4

1.731_3 1.676_6. May be described as HEX-R with a supercell, $a = 13.40$, $c = 21.57 \approx (3 \times 7.15)$. Structure: *CM 21:7(1983)*. Crystals tabular, pseudo-hexagonal, and often hemimorphic; rarely slender needles; fibrous and lamellar aggregates; massive. Pale pink to dark red, brown to red-brown, rarely yellow; vitreous to pearly luster; transparent to translucent. Cleavage {001}, perfect; brittle. H = 4–5. G = 3.04–3.14. Uniaxial (−); $N_O = 1.654$–1.664, $N_E = 1.625$–1.629; also biaxial (−); 2V very small; $N_y = 1.623$, $N_y = 1.650$, $N_z = 1.651$; positive elongation to parallel cleavage. Pleochroic; X, colorless; Y,Z, greenish yellow. Decomposed by HCl; fusibility 4. The chlorine content appears to be Cl/OH $\leq$ 1:9; Fe^{2+} is low, as is Mg. Associated with franklinite and willemite at Franklin and Sterling Hill, Sussex Co., NJ; at the American tunnel, San Juan Co., CO. At *Adervielle, Louron Valley, Haute Pyrénées, France*; the Sjö mines, near Örebro, and the Harstig mine, Pajsberg, Sweden. At Broken Hill, NSW, Australia; Kalahari manganese field, Cape Prov., South Africa. AR *D6:465(1892)*; *AM 38:755(1953)*, *66:1054(1981)*.

72.4.1b.2 Mcgillite $Mn_8[Si_6O_{15}](OH)_8Cl_2$

Named in 1980 by Donnay et al. for McGill University, Montreal, Canada. Pyrosmalite group. MON $C2/m$. $a = 23.312$, $b = 13.459$, $c = 7.423$, $\beta = 105.17°$, Z = 4, D = 3.08, *33-891*: 11.7_2 7.16_7 3.57_4 2.89_6 2.56_{10} 2.41_2 2.12_4 1.683_4. Also regarded as HEX-R $R\bar{3}m$ with a supercell of $a = 13.459$, $c = 85.97 \approx (12 \times 7.16)$. An apparent polytype of friedelite but not in *sensu strictu*, the stacking sequence possibly may be compositionally dependent with regard to ordered site occupancy by Cl. Structure: *CM 21:7(1983)*. Possibly twinned on a micro scale by 120° about [110]. Light to dark pink, vitreous to pearly luster, translucent. Cleavage {001}, perfect. H $\approx$ 4. G = 2.98. Uniaxial (−); $N_O = 1.667$–1.670, $N_E = 1.640$–1.643. Decomposed by HCl. The chlorine content is close to that of the ideal formula. Found associated with sphalerite and minor boulangerite, jamesonite, galena, and quartz at the *Sullivan mine, Kimberley, BC, Canada*, and the Kyurazawa mine, Tochigi Pref., Honshu, Japan. AR *CM 18:31(1980)*.

72.4.1b.3 Nelenite $(Mn,Fe)_{16}[As^{3+}O_2OH]_3[Si_{12}O_{30}](OH)_{17}$

Named in 1984 by Dunn and Peacor for Joseph Aloysius Nelen (b.1923), Dutch–American chemist, Smithsonian Institution, Washington, DC. Pyrosmalite group. Equivalent to the name used earlier: ferroschallerite. MON $C2/m$. $a = 23.240$, $b = 13.418$, $c = 7.832$, $\beta = 105.21°$, Z = 2, D = 3.45; an HEX-R supercell has $a = 13.42$, $c = 85.5$. *38-355*: 7.10_4 3.55_6 2.88_7 2.55_{10} 2.40_4 2.10_4 1.723_5 1.677_6. Dimorphous (polytypic) with schallerite (72.4.1a.3). Light to medium brown, vitreous to resinous luster. Cleavage {001}, perfect. H $\approx$ 5. G = 3.46. Uniaxial (−); $N_O = 1.718$, $N_E = 1.700$; occasionally biaxial (−) with 2V very small. With actinolite, tirodite, albite, garnet, willemite, rhodonite, apatite, lennilenapeite, and stilpnomelane from the *Trotter shaft, Franklin, Sussex Co., NJ*. AR *MM 48:271(1984)*.

RHODESITE GROUP

The rhodesite group comprises silicates of composition near

$$R_4[Si_8O_{19}] \cdot nH_2O \quad \text{or a multiple thereof}$$

where

R_4 = H, K, Na, Ca, Ba, Y

and where there is possible substitution of $(OH)^-$ for O^{2-}. The structures of rhodesite and macdonaldite have been determined and are characterized by corrugated sheets having, in rhodesite, the composition $[Si_8O_{19}]^{-6}$. The assigned axial orientation is not self-consistent, but all have an edge of 23–25 Å.

RHODESITE GROUP

Mineral	Formula	Space Group	a	b	c	β	
Rhodesite	H(K,Na)Ca$_2$[Si$_8$O$_{19}$] · 5H$_2$O	Pmam	23.416	6.555	7.050		72.5.1.1
Monteregianite-(Y)	K$_2$Na$_4$Y$_2$[Si$_{16}$O$_{38}$] · 10H$_2$O	P2$_1$/n	9.512	23.956	9.617	93.85	72.5.1.2
Macdonaldite	BaCa$_4$[Si$_{16}$O$_{36}$](OH)$_2$ · 10H$_2$O	Cmcm	14.081	13.109	23.560		72.5.1.3
Delhayelite	(Na,K)$_{10}$Ca$_5$Al$_6$[Si$_{32}$O$_{80}$](Cl$_2$,F$_2$,SO$_4$)$_3$ · 18H$_2$O	Pnmm	13.06	24.65	7.04		72.5.1.4
Hydrodelhayelite	KCa$_2$[AlSi$_7$O$_{17}$](OH)$_2$ · 6H$_2$O	Pnm2$_1$	6.648	23.846	7.073		75.5.1.5

72.5.1.1 Rhodesite H(K,Na)Ca$_2$[Si$_8$O$_{19}$] · 5H$_2$O

Named in 1957 by Gard et al. for Cecil John Rhodes (1853–1902), British financier and colonial statesman who founded the DeBeers Mining Company. Rhodesite group. ORTH *Pmam*. $a = 23.416$, $b = 6.555$, $c = 7.050$, $Z = 2$, $D = 2.41$. *22-1253*: 6.55$_{10}$ 6.30$_3$ 5.90$_4$ 5.03$_3$ 4.39$_5$ 3.38$_2$ 2.86$_3$ 2.76$_3$. Structure: *ZK 199:25(1992)*. Matted silky [001] fibers and rosettes up to 2 mm in diameter. White, silky luster. Cleavage {100}, good. H = 4. G = 2.36. Biaxial (+); 2V$_{obs}$ small, 2V$_{calc}$ = 68–71°. N$_x$ = 1.501–1.504, N$_y$ = 1.505–1.508, N$_z$ = 1.513–1.518; XYZ = *bac*. Found with magadiite and montmorillonite in altered silicic lava 35 miles NW of Redding, Trinity Co., CA; with natrolite at Zeilberg, Unterfranken, Germany; San Venanzo, Rieti, Italy; with mountainite at *Bultfontein mine, Kimberley, South Africa*. AR *MM 31:607(1952)*.

72.5.1.2 Monteregianite-Y K$_2$Na$_4$Y$_2$[Si$_{16}$O$_{38}$] · 10H$_2$O

Named in 1978 by Chao for the Monteregian Hills, which include the type locality. Rhodesite group. Synonym: monteregianite. MON (pseudo-ORTH).

$P2_1/n$. $a = 9.512$, $b = 23.956$, $c = 9.617$, $\beta = 93.85°$, $Z = 2$, $D = 2.41$. *31-1087:* 12.0_{10} 7.03_{10} 6.55_4 6.02_5 4.42_{10} 3.41_5 3.03_5 2.87_8. Radiating clusters of acicular [100] crystals, mostly 1–5 mm long, but known up to 25 mm, displaying {100}, {010}, {001}, and occasionally {101}; also parallel groups of laths, elongated [100] and flattened on {010}. All indices are in the original morphological setting, the transformation matrix, morphological to structure being /10$\bar{1}$/010/101/. Colorless, white, gray, rarely mauve or green; vitreous luster when transparent to silky or greasy; transparent to opaque. Cleavage {010}, perfect; {001}, very good; {100}, fair. H = $3\frac{1}{2}$. G = 2.42. Biaxial (+); 2V = 87°; $N_x = 1.510$, $N_y = 1.513$, $N_z = 1.517$; $XYZ = cab$ in morphological setting. Bright green fluorescence in SWUV. Widespread in small amounts at *Mt. St.-Hilaire, QUE, Canada*, most commonly in igneous breccias and marble xenoliths, rarely in carbonate-rich miarolitic cavities, and very rarely in pegmatites. AR *CM 16:561(1978)*, *MR 21:323(1992)*.

72.5.1.3 Macdonaldite $BaCa_4[Si_{16}O_{36}(OH)_2] \cdot 10H_2O$

Named in 1965 by Alfors et al. for Gordon Andrew Macdonald (1911–1978), American volcanologist, University of Hawaii. Rhodesite group. ORTH *Cmcm*. $a = 14.081$, $b = 13.109$, $c = 23.560$, $Z = 4$, $D = 2.36$. *18-163:* 8.90_4 6.50_{10} 6.30_5 5.90_3 4.36_8 3.36_5 3.02_4 2.74_4. Structure: *SR 33A:489(1968)*. As acicular [100] crystals, 1–6 mm in length, displaying the three pinacoids. Colorless to white, satiny to vitreous luster. Cleavage {010}, perfect; {001}, good; {100}, poor. H = $3\frac{1}{2}$–4. G = 2.27. Biaxial (+, −); 2V ≈ 90°; $N_x = 1.518$, $N_y = 1.524$–1.525, $N_z = 1.530$; $XYZ = cba$. Insoluble in cold acids but decomposed by boiling HCl. Fusibility 5.5. Found in sanbornite-bearing metamorphic rocks near a granodiorite contact, exposed in a narrow zone $2\frac{1}{2}$ miles long *near Big Creek, eastern Fresno Co., CA.* AR *AM 50:314(1965)*.

72.5.1.4 Delhayelite $(Na, K)_{10}Ca_5Al_6[Si_{32}O_{80}](Cl_2,F_2,SO_4)_3 \cdot 18H_2O$

Named in 1959 by Sahama and Hytonen for Fernand Delhaye (1880–1946), Belgian geologist. Rhodesite group. ORTH *Pnmm*. $a = 13.06$, $b = 24.65$, $c = 7.04$, $Z = 1$, $D = 2.48$. *12-286:* 12.3_3 6.16_2 3.48_1 3.08_{10} 2.96_1 2.79_1 1.760_1 1.541_1. Colorless to light gray-green. Cleavage {010}, distinct. H ≈ 4. G = 2.517–2.615(impure). Biaxial (−); 2V = 83°; $N_y = 1.532$. Fluoresces orange in UV. Found at Mt. St.-Hilaire, QUE, Canada; at Yukspor Mt. and Mt Rasvumchorr, Khibiny massif, Kola Penin., Russia; and as platy crystals in kalsilite–melilite–nepheline lava at *Mt. Shaheru, Kivu, Zaire*. AR *MM 32:6(1959)*, *MR 10:108(1979)*.

72.5.1.5 Hydrodelhayelite $KCa_2[AlSi_7O_{17}(OH)]_2 \cdot 6H_2O$

Named in 1979 by Dorfman and Chiragov for its composition and relationship to delhayelite. Rhodesite group. ORTH $Pnm2_1$. $a = 6.648$, $b = 23.846$, $c = 7.073$, $Z = 2$, $D = 2.24$. *41-611:* 6.79_4 4.89_3 4.15_3 3.32_4 3.07_8 2.92_{10} 2.80_6 2.67_2. Grayish-white, vitreous luster, translucent. Cleavage {010}, perfect;

{100}, {001}, less so. H = 4. G = 2.17. Biaxial; $N_x = 1.503$, $N_z = 1.518$. Found as a supergene alteration product of delhayelite at *Mt. Rasvumchorr, Khibiny massif, Kola Penin., Russia*. AR *AM 72:1024(1987)*.

72.5.2.1 Leucosphenite $Na_4BaTi_2O_2[B_2Si_{10}O_{28}]$

Named in 1901 by Flink from the Greek for *white* and *wedge*, in allusion to its color and morphology. MON $C2/m$. $a = 9.814$, $b = 16.851$, $c = 7.210$, $\beta = 93.35°$, $Z = 2$, $D = 3.090$. *25-784:* 8.45_9 4.22_{10} 3.57_3 3.37_7 3.31_5 3.25_3 3.19_3 2.89_6. Structure: *SPD 26:372(1981)*. The structural unit consists of four-tetrahedron rings of two types. Two rings, $[Si_4O_{10}]$, share their apical oxygens with a ring, $[B_2Si_2O_8]$, in which B and Si are disordered; these are linked to form a layer, $[B_2Si_{10}O_{28}]^{14-}$. The individual rings lie perpendicular to the layer, which is parallel to {001}. Crystals prismatic [100] with wedge-shaped terminations; {001}, {010}, {130} dominant and lesser {101} and {110}. Also tabular to equant with the same dominant forms, yielding short, striated, pseudohexagonal prisms. Twinning with {001} composition plane common. Colorless, lemon yellow to brown, also pale blue; vitreous luster; transparent to translucent. Cleavage {010}, perfect; {001}, fair. $H = 6-6\frac{1}{2}$. $G = 3.05-3.09$. Biaxial (+); $2V \approx 76°$; $N_x = 1.643-1.649$, $N_y = 1.657-1.664$, $N_z = 1.681-1.691$; $X \approx a$, $Y \wedge c = -3°$, $Z = b$; $r > v$, weak. Fluoresces yellowish white in SWUV. Insoluble in acids. Calcium may replace sodium with concomitant additional boron replacement of silicon. Found with shortite and analcime in crystals in the Tertiary Green River formation, UK and WY; as blocky, prismatic [001] crystals(!), less commonly long [100] prisms, at Mt. St.-Hilaire, QUE, Canada. Occurs as [100] prisms at *Narsarsuk, Julianehaab district, Greenland*; in the Alaiski massif, Turkestan, and in the Inaglia alkaline massif, Yakutia, Russia. AR *AM 57:1801(1972)*, *CM 11:851(1972)*, *MR 21:319(1990)*.

72.5.2.2 Nordite-(Ce) $(Ce,La,Ca)(Sr,Ca)Na_2(Na,Mn)(Zn,Mg)Si_6O_{17}$

Named in 1958 by Semenov and Barinsky by analogy to the La-dominant analogue, nordite-(La). ORTH. *Pcca*. No cell data available. $Z = 4$. No powder data available. A metamict mineral. Recrystallized at 800°C for 10 hours. Structure not determined. Light brown, dark brown, nearly black, white; streak light brown; to 1 cm long, lamellar parallel to b; sometimes radial, fibrous aggregates of curved and lamellar crystals to 2.0 cm diameter; vitreous, greasy on fracture; cleavage = {010}; $H = 5$; $G = 3.43-3.49$; decomposes in acids. Biaxial ($-$); $2V = 30°$ or less, $N_\alpha = 1.619-1.621$, $N_\beta = 1.63-1.64$, $N_\gamma = 1.642-1.655$. Occurs associated with lueshite at Mont Saint-Hilaire, Quebec; on the *Chinglusuai River and Motchusuai River, Lovozero Massif, Kola Peninsula, Russia*; in sodalite syenite pegmatite. Also occurs in vuggy natrolite in the central zone of the Jubilee pegmatite and other locations on Mount Karnasurt, Lovozero Massif. Unspecified locality in the Khibiny Massif. Alteration: weathers readily to ochers of hydrous REE minerals and ancylite-(Ce). AM 55:1167(1970) Minerali 3(3):336(1981) New data on the mineralogy of rare elements of the Kola Peninsula, Apatity, p. 60–65(1991).

72.5.2.3 Nordite-(La) (La,Ca)(Sr,Ca)Na$_2$(Na,Mn)(Zn,Mg)Si$_6$O$_{17}$

Named in 1941 by Gerasimovsky from the word for North because of its northern origin in the Lovozero Massif, Kola Pen, Russia. Compare: Nordite-(Ce). ORTH. Pcca a 14.468 b 5.203 c 19.88 *(DAN, VMU 4:97(1992))* Z = 4 D = 3.51 a 14.369 b 5.180 c 19.766 (ICDD 41-584); *PD 41-584* (nat.): 7.2$_{10}$ 4.69$_5$ 4.20$_8$ 3.46$_7$ 3.31$_7$ 2.95$_6$ 2.86$_8$ 2.78$_5$. Structure: *(AM 55:1167(1970)* and *DAN VMU 4:97(1992))*. Light brown, dark brown to black; H 5-6; G 3.43-3.48. SiO$_2$ 45.5, Fe$_2$O$_3$ 1.84, Ce$_2$O$_3$ 8.77, La earths 10.48, Y earths 0.95, MnO 6.04, SrO 7.40, CaO 4.46, MgO 2.00, Na$_2$O 11.70, K$_2$O 0.08, Cl trace, total 99.27. Another REE analysis La$_2$O$_3$ 8.55, Ce$_2$O$_3$ 8.1, Pr$_2$O$_3$ 1.6, Nd$_2$O$_3$ 1.85. Biax. (−), 2V = 31°; N$_\alpha$1.619 N$_\beta$1.630-1.640 N$_\gamma$1.642; XYZ = abc. Sometimes uniaxial due to square distribution of heavy atoms on the yz plane. Occurs in Ne-syenite near the *Chinglusuai River, Lovozero Massif, Kola Peninsula, Russia*. Also at several localities in the Khibiny Massif, Kola Peninsula, Russia. EF *MIN 3(3):336(1981)*.

72.5.2.4 Semenovite-(Ce) (Ca,Ce,La,Na)$_{10-12}$ (Fe^{+2}, Mn)(Si,Be)$_{20}$(O, OH, F)$_{48}$ or Ce$_2$Na$_{0-2}$ (CSRE Mins. 1993) III

Named in 1972 by Petersen and Ronsbo for Yevgeny Ivanovich Semenov (1951–) Russian mineralogist, IMGRE, Moscow. ORTH *Pmmn*. a = 13.879Å, b = 13.835, c = 9.942, Z = 2, D = 3.17; *PD* 25-699 (nat.): 8.08$_7$ 6.95$_4$ 5.25$_4$ 3.28$_{10}$ 3.18$_4$ 3.10$_4$ 2.84$_{10}$ 2.73$_{10}$. Structure: R = 5% {110} twinning, always present, is responsible for pseudotetragonal symmetry. Structure related to miliphanite, leucophanite, gadolinite, nordite and hellandite. Aminoffite shows little similarity with regards to crystal structure *(AM 64:202(1979)*. Fresh – transparent brown, rarely orange-brown altered – dull gray and reddish brown to gray with brown hue, also colorless; streak white to pale brown; pseudotetragonal, bipyramidal crystals 0.1–1 mm usual, largest are 1 cm; twinning complex, interpenetrant, always twinned, TP = (110), twin after {110}, with composition plane = (120); vitreous; cleavage = none; fracture = uneven; H = 3$\frac{1}{2}$-4; G = 3.14; TGA data (LIT 5:163(1972). The chemistry is complex. Biaxial (−); 2V = 0 to 55°, N$_\alpha$ = 1.595, N$_\beta$ = 1.614, N$_\gamma$ = 1.614; OAP parallel to {010}, X parallel to [001]. Occurs on the Taseq slope, in the *Ilimaussaq complex, South Greenland*. Occurs in border zone of the only epididymite-eudidymite-bearing albitite, in cavities and fractures.

72.5.3.1 Kvanefjeldite Na$_4$(Ca,Mn)[Si$_3$O$_7$(OH)]$_2$

Named in 1984 by Petersen et al. for the locality. ORTH *Pcab*. a = 10.213, b = 15.878, c = 9.058, Z = 4, D = 2.53. *37-433:* 4.45$_6$ 4.36$_7$ 3.88$_7$ 3.67$_4$ 3.39$_6$ 3.31$_5$ 3.11$_{10}$ 2.58$_5$. Structure: *NJMM:505(1983)*. The structure is built of highly corrugated layers of composition [Si$_3$O$_7$(OH)]$^{3-}$ containing eight-membered rings. Violet-pink; vitreous luster, pearly on {010}. Cleavage {010}, good; {101}, imperfect. H = $5\frac{1}{2}$–6. G = 2.55. Biaxial (+); 2V = 0–9°; N_x = 1.522, N_y = 1.522, N_z = 1.543. Occurs as tabular {010} anhedral grains in veinlets and streaks in arfvedsonite–nepheline syenite with villiauminite and analcime in the *Ilimaussaq complex, Kvanefjeld Plateau, Greenland.* AR *CM 22:465(1984)*.

72.5.4.1 Armstrongite $CaZrSi_6O_{15} \cdot 3H_2O$

Named in 1973 for Neil Alden Armstrong (b.1930), American astronaut, first human being on the moon's surface, *Apollo 11* Lunar Mission. Not a lunar mineral. May form a series with elpidite, $Na_2ZrSi_6O_{15} \cdot 3H_2O$. MON $I2/m$, $I2$, or Im. a = 13.599, b = 14.114, c = 7.833, β = 103.41°, Z = 4, D = 2.70–2.71. *40-490:* 7.37_4 7.09_5 6.61_9 4.58_4 4.28_{10} 3.81_7 3.05_9 3.00_3. Subhedral crystals are 2–7 mm long and 2 mm wide (Strange Lake), with poikilitic crystals to 2 cm and aggregates to 50 cm × 50 cm from Mongolia; bladed habit with {100}, {010}, and {001} prominent; most crystals are twinned polysynthetically (patch twinned) by rotation about [100]. Colorless to brown (with iron), white streak, vitreous luster. Cleavage {001}, perfect; {100}, average; very brittle. H = $4\frac{1}{2}$. G = 2.56–2.59. DTA data on Mongolian material: *DANS 209:1185(1973)*. Dull blue cathodoluminescence. Can heat to 500° and remove the H_2O. If rehydrated, the H_2O is absorbed. Not attacked by HCl or HNO_3. Decomposed by HF. Mongolian material has 1.31 wt % Fe_2O_3 and 0.60 wt % Al_2O_3. Trace and minor elements in Strange Lake material are 0.4% Al_2O_3, 0.6% SnO_2, and 1.0% HfO_2 with a maximum content of 0.27% Na_2O. Biaxial (−); Z = b, X ∧ c = +4° in obtuse β; N_x = 1.563–1.567, N_y = 1.569–1.576, N_z = 1.573–1.577; 2V = 39°; colorless; r < v, weak. Intimately associated with gittinsite at Strange Lake; crystals are veined by turbid alteration minerals, predominantly gittinsite. In schlieren of alkali granite pegmatite at contact of arfvedsonite granite with xenoliths of acid volcanic rocks. Strange Lake alkalic complex, QUE–LAB, Canada; *Khan-Bogdin massif, Mongolia.* EF *DANS 209:1185(1973), PD 2:2(1987), CM 30:191(1992)*.

72.5.4.2 Elpidite $Na_2ZrSi_6O_{15} \cdot 3H_2O$

Named in 1894 by Lindström and Nordienskiöld from the Greek for *hope*, because there was expectancy of finding other interesting minerals from the same locality. See armstrongite (72.5.4.1), $CaZrSi_6O_{15} \cdot 3H_2O$, with which it may form a series. "Titanoelpidite" is labuntsovite. ORTH $Pbm2$. (Syn): a = 7.297, b = 14.302, c = 7.032; (nat): a = 7.111–7.31, b = 14.68–14.70, c = 7.13, Z = 2, D = 2.61–2.65. From structure refinement, cell also given as: a = 7.14, b = 14.68, c = 14.65, with Z = 4. *29-1294*(syn): 7.08_{10} 6.52_5 5.11_8 4.77_4 4.13_5 3.24_{10} 3.10_9 2.94_4. Structure: *K 25:620(1980), AM 58:106(1973)*. Usually massive, fine fibrous or columnar, with good prismatic cleavage. Dominant forms: {110}, {120}, {010}, {011}, {001}, and {100}. Elongated on c, rhombic. Polysynthetic twinning present, striated parallel to c. Fine crystals 1–3 cm, but some to 10 cm × 1 cm from Mt. St.-Hilaire. Crystals to 30 cm from Kazakhstan. Many colors: colorless, white, yellowish, tan, greenish gray, brick-red to red-brown; white streak; vitreous–greasy, sometimes silky luster. Two good cleavages parallel to c; {110}, good; splintery to uneven fracture; brittle. H = $6\frac{1}{2}$–7. G = 2.5–2.8. IR and thermal data: *MIN 3(3):317(1982)*. Sometimes has a dull blue cathodoluminescence. Decomposes only in HF. At Strange Lake, may have as much as 3.2 wt % CaO, indicating 33 mol % of the armstrongite end member. Mongolian material has to 3.5 wt % CaO. K_2O to 0.2 wt %. Dalyite is the anhydrous K analog. Burpala massif sample has 4.7 wt % HfO_2. Mt. Karnasurt, Lovozero, massif sample has 3.7 wt % TiO_2. Biaxial (+) in nearly all cases, but rarely (−); OAP parallel to (010), Z = a, X = c; N_x = 1.54–1.563, N_y = 1.568–1.570, N_z = 1.572–1.578; 2V ranges from 89° (−) to 71° (+); r > v, distinct, to r < v, distinct; negative elongation; pleochroic, colorless to yellow. In alkaline granites, aplite dikes, and pegmatites. Sometimes a rock-forming mineral. Also in nepheline syenites and their cogenetic pegmatites. Sometimes forms as a pseudomorph after eudialyte and calcite. Found as pseudomorphs after serandite at Mt. St.-Hilaire. May also form epitactic intergrowths with labuntsovite. Found in the Bearpaw Mountains, MT; authigenic from the Green River formation, UT; Golden Horn batholith, Okanogan Co., WA; Strange Lake complex, QUE-LAB; best material known is from the Desourdy quarry, Mt. St.-Hilaire, and St. Amable phonolite sill, Varennes, QUE, Canada; *Igalik, Harsarsuk, southern Greenland*; Kangerdlugsuak, Greenland; Lake Gjerdingen, Normarka, 30 km N of Oslo, Norway; Yumegot Pass, Mt. Yukspor, Khibiny massif, and Mt. Alluaiev, Mt. Allnayo (to 6 cm long), Mt. Karnasurt, and several other locations in the Lovozero massif, Kola Penin., Russia; to 30 cm in aegirine–quartz–feldspar pegmatites at Tarbagatai, eastern Kazakhstan, and Ulan-Erge, southeastern Tuva region, Russia; Maygunda R., Burpala massif, northern Pribaikal, Russia; Khan-Bogdin massif, Mongolia. EF *CM 9:286(1967), 30:191(1992); BM 97:433(1974)*.

72.5.5.1 Varennesite Na$_8$Mn$_2$Si$_{10}$O$_{25}$(OH,Cl)$_2$ · 12H$_2$O

Described in 1995 by Grice and Gault, and named for the type locality. ORTH. *Cmcm* a = 13.447Å, b = 15.022Å, c = 17.601Å, Z = 4, D = 2.32; *PD* 10.05$_{10}$ 8.82$_5$ 5.025$_2$ 4.14$_1$ 3.81$_2$ 3.66$_1$ 3.15$_1$ 2.72$_5$. The structure consists of alternating, undulating layers of silicate tetrahedra and Na and Mn octahedra. The structure has similarities to the monophyllosilicate structures of manganpyrosmalite, bementite, and apophyllite, but is unique in topology. Contains layers of unbranched tetrahedra of the type $[Si_2O_5]^{2-}$. Pale brownish yellow to orange, white streak, tabular crystals to 4 mm in length, showing the pinacoids {100}, and {010} and a set of {101} prism faces; the basal pinacoid {001} is also present. No fluorescence, G = 2.31, H = 4, brittle, good {010} cleavage and conchoidal fracture. Biaxial (+), (Na$_D$): N_α = 1.532, N_β = 1.540, N_γ = 1.550; 2V$_{meas.}$ = 89°, 2V$_{calc.}$ = 84°; dispersion is weak, with r < v. Pleochroic; Z > Y > X; Z = yellow green; X = very pale yellow. X parallel to c, Y parallel to a, Z parallel to b. Na$_2$O 19.25 wt. %, CaO 0.11, K$_2$O 0.21, SrO 0.15, MnO 6.50, FeO 3.25, TiO$_2$ 2.56, MgO 0.03, SiO$_2$ 48.26, Al$_2$O$_3$ 0.48, SO$_3$ 1.02, Cl 2.20, H$_2$O

18.47 (calc. from stoichiometry), total 102.00. Occurs in a peralkaline sill called the Saint-Amable sill, in the *Demix-Varennes quarry, between Varennes and Saint-Amable, Varennes Twp., Vercheres Co., Quebec, Canada*. A primary, late-stage mineral most common associated with eudialyte, aegirine, natrolite, serandite, mangan-neptunite, microcline, albite, and zakharovite. Other associated species include astrophyllite, villiaumite, makatite and vuonnemite. *CM 33:1073(1995)*.

72.6.1.1 Petalite LiAlSi$_4$O$_{10}$

Named in 1800 by d'Andrea from the Greek for *leaf*, in allusion to its perfect cleavage. Lithium was first discovered in this mineral. MON $P2/a$. $a = 11.754$, $b = 5.139$, $c = 7.630$, $\beta = 113.01°$, $Z = 2$, $D = 2.398$. *35-463*(calc, 1M polytype): 3.73_{10} 3.72_{10} 3.67_7 3.65_7 3.51_3 2.57_1 2.06_1 1.934_2. Structure: ZK 197:27(1991). The silicate structural element is a layer of tetrahedra, [Si$_4$O$_{10}$]$^{4-}$, lying parallel to {001}; the layers are linked by AlO$_4$ tetrahedra to form a three-dimensional framework, leading to its sometimes being regarded as a tektosilicate. A 2M polytype having a nearly identical X-ray powder pattern is known. Euhedral crystals are rare, usually small with {001}, {201}, {010} dominant; usually massive, as large, cleavable, blocky segregations. Polysynthetic twinning on {001}. Colorless to grayish white, less commonly light pink or light green; vitreous luster, on cleavage sometimes pearly; transparent to translucent. Cleavage {001}, perfect; {201}, distinct. H = 6–6$\frac{1}{2}$. G = 2.3–2.5. Biaxial (+); 2V = 83°; N$_x$ = 1.504–1.507, N$_y$ = 1.510–1.513, N$_z$ = 1.516–1.523; Z = b, X $\wedge$ a (cleavage trace) = 2–8°; r > v weak. Composition is usually close to ideal with but minor Na or K substitution for Li; Fe^{3+} may replace Al in very small amount. Insoluble in acids. Alters to montmorillonite. Characteristically found in granite pegmatites associated with albite (cleavelandite), quartz, and lepidolite. Widespread worldwide; notable localities are: at Peru, Oxford Co., ME; at Bolton, Worcester Co., MA; at Pala and Rincon, San Diego Co., CA; in the Tanco pegmatite, Bernic lake, MB, Canada. Found in the Varuträsk pegmatite and at *Utö, Stockholm, Sweden*; in the pegmatites of San Piero in Campo, Elba, Italy; and at a number of localities in the former USSR; at Nagatareyama, Fukuoka Pref., Japan. At the Londonderry quarry, Coolgardie, WA, Australia. In the Karibib district, Namibia, as massive, pink material; at Bikita, Zimbabwe; at Araçuai and other localities, MG, Brazil. A source of lithium. Brazilian material has been faceted into collector gems up to 50 ct. AR *D6:311(1892), VLA 2:35(1966), TMPM 27:129(1980)*.

Class 73

Condensed Tetrahedral Sheets

73.1.1.1 Naujakasite $(Na,K)_6(Fe,Mn)[Al_4Si_8O_{26}]$

Named in 1933 (Bøggild) for the type locality. MON $C2/m$, Cm or $C2$. $a = 15.039$, $b = 7.991$, $c = 10.487$, $\beta = 113.7°$, $Z = 2$, $D = 2.661$. *20-1113*: 3.99_{10} 3.69_6 3.56_7 3.44_6 3.06_6 2.79_6 2.61_6 2.26_7. Structure: *BGGU 116:11(1975)*. The structure has sheets of double-layer tetrahedra, each with 4- and 6-membered rings, lying perpendicular to c; aluminum shares two of the three tetrahedral sites with silicon. Crystals are tabular {001} and modified by {110} and {$\bar{1}$11}. Silver-white to grey, greenish when fresh; vitreous luster, pearly on {001}. Cleavage {001} perfect; {$\bar{4}$01} and {010} distinct; brittle. Biaxial $(-)$; $2V_{calc} = 68°$; $2V = 52–71°$ $(2V_{calc} = 68°)$; $N_x = 1.536–1.538$, $N_y = 1.548–1.551$, $N_z = 1.555–1.557$. Found as an aggregate of micaceous plates, < 1 mm thick and 1–5 mm in diameter, intimately associated with arfvedsonite at *Naujakasik, Tunugdliarfik Fjord, Greenland*. AR *AM 20:138(1935), 53:1780(1968)*.

73.1.2.1 Latiumite $(Ca,K)_8(Al,Mg,Fe)[(Si,Al)_{10}O_{25}](SO_4)$

Named in 1953 by Tilley and Henry for the locality. See tuscanite (73.1.2.2). MON $P2_1$. $a = 12.06$, $b = 5.08$, $c = 10.81$, $\beta = 106°$, $Z = 2$, $D = 2.90$. *8-174*: 4.60_5 4.50_5 3.83_5 3.38_5 3.28_5 3.06_9 2.96_8 2.86_{10}; *25-1212*(calc): 3.82_4 3.29_3 3.08_6 2.95_5 2.90_4 2.86_6 2.84_{10} 2.54_5. Structure: *AM 58:466(1973)*. The principal structural element is a double layer of ordered Si and Al tetrahedra lying parallel to the {100} cleavage, and joined to one another by the cations and sulfate groups. Each half of the double layer is made up of six- and eight-membered rings and the two halves are related by glide symmetry. Essentially dimorphous with tuscanite. Elongated tabular crystals; massive. Simple and multiple twinning on {100} common. Colorless to white, vitreous luster, transparent to translucent. Cleavage {100}, perfect. H = $5\frac{1}{2}$–6. G = 2.93. Biaxial $(+,-)$; $2V = 83°–90°$; $N_x = 1.600–1.603$, $N_y = 1.606–1.609$, $N_z = 1.614–1.615$; $Z = b$, $X \wedge c = 16–28°$; $r > v$, distinct; also biaxial $(-)$; $2V = 40°$; $N_x = 1.582$, $N_y = 1.590$, $N_z = 1.591$; extinction commonly mottled. Occurs with hedenbergite, Ca-garnet, melilite, leucite, and kaliophilite in two ejecta blocks in the *Alban Hills, Albano, Lazio (Latium), Italy*; also at Pitigliano, northern Grosseto, Tuscany, Italy. AR *MM 30:39(1953)*.

73.1.2.2 Tuscanite $K(Ca,Na)_6(Si,Al)_{10}O_{22}(SO_4,CO_3,(OH)_2) \cdot H_2O$

Named in 1977 by Orlandi et al. for the locality. See latiumite (73.1.2.1). MON $P2_1/a$. $a = 24.04$, $b = 5.11$, $c = 10.89$, $\beta = 106.96°$, $Z = 2$, $D = 2.77$. PD: 11.51_{10} 5.75_2 3.83_2 3.07_5 3.00_1 2.96_2 2.87_{10} 2.55_1. Structure: *AM 62:1114(1977)*. The structure is similar to that of latiumite except that the two halves of the tetrahedral double sheet are related by an inversion. Essentially dimorphous with latiumite. Colorless, vitreous to pearly luster, transparent. Cleavage {100}, distinct. H = $5\frac{1}{2}$–6. G = 2.83. Biaxial (−); 2V = 40°; $N_x = 1.581$, $N_y = 1.590$, $N_z = 1.591$. Found in ejecta in a pumice deposit near the *Latera caldera, Pitigliano, Tuscany, Italy*, associated with vesuvianite, grossular, andradite, pyroxene, wollastonite, anorthite, brandisite, and latiumite. AR *AM 62:1110(1977)*.

73.1.3.1 Fedorite $(K,Na)_{2.5}(Ca,Na)_7Si_{16}O_{38}(OH,F)_2 \cdot H_2O$

Named in 1965 by Kukharenko et al. for E. S. Fedorov (1853–1919), noted Russian mineralogist. TRIC $P\bar{1}$. $a = 9.650$, $b = 16.706$, $c = 13.153$, $\alpha = 93.42°$, $\beta = 114.92°$, $\gamma = 90.0°$, $Z = 2$, $D = 2.46$. *19-466*: 11.7_8 6.00_8 4.21_7 4.00_6 3.13_8 2.97_9 2.93_{10} 1.826_9. Structure: *SPC 33:763(1988)*. The structure, sparsely described, consists of double tetrahedral layers made up of two types of six-membered rings. Colorless to pink, vitreous luster, transparent to translucent. Cleavage {001}, perfect. G = 2.43. Biaxial (−); 2V = 32°; $N_x = 1.522$, $N_y = 1.530$, $N_z = 1.531$; XY ≈ *cb*; r < v. Insoluble in acids. Easily fusible. The formula above is based on the structure determination; composition also given as $KNa_4Ca_4(Si,Al)_{16}O_{36}(OH)_4 \cdot H_2O$. Found as pseudohexagonal, tabular crystals resembling muscovite at *Tur'yii Peninsula, Kola Penin., Russia*, and in the Murun massif, Chana R. area, Yakutia. AR *AM 52:561(1967)*.

73.1.4.1 Zeophyllite $Ca_{13}Si_{10}O_{28}(OH)_2F_8 \cdot 6H_2O$

Named in 1902 by Pelikan from the Greek for *to boil* and *leaf*, in allusion to the hemispherical, micaceous aggregates. TRIC Space group unknown. $a = 9.34$, $b = 9.34$, $c = 13.2$, $\alpha = 90°$, $\beta = 110°$, $\gamma = 120°$; pseudo-HEX-R $R\bar{3}$. $a = 9.36$, $c = 36.48$; $Z = 3$, $D = 2.79$. *35-715*: 12.3_{10} 6.13_5 3.71_5 3.20_4 3.05_8 2.55_4 2.34_5 1.770_7. Structure: *AC(B) 28:2726(1968)*. The silicate element consists of a pair of centrosymmetrically related sheets containing 12-tetrahedron rings, the tetrahedra sharing two or three oxygens. The pair of sheets are joined to one another by a sheet of six-coordinated Ca^{2+}, and the voids within these rings are occupied by three Ca^{2+}. Water lies between these t-o-t composite layers which are perpendicular to *c*. Platy crystals, in radiating spherical aggregates. White; translucent; tips of crystals often opaque. Cleavage {0001}, perfect. H = 3. G = 2.8. Uniaxial (−); $N_O = 1.565$, $N_E = 1.560$; or biaxial (−); 2V small; $N_x = 1.560$, $N_y = 1.565$, $N_z = 1.566$; uniaxial and biaxial segments. Gelatinizes in acids and exfoliates on heating. The formula above is based on the structure determination; the original formula given as $Ca_4Si_3O_8(OH,F)_4 \cdot 2H_2O$. Found in the Salmon River district, Idaho Co., ID; at Mont St. Hilaire, PQ, Canada; in the Oslo region, Norway; from

Göbnehall Yuglingarum, Skåne, Sweden; from the Lacher See volcanics, Rhineland, Germany; with apophyllite and zeolites in basalt at Březno and Lbem (*Gross-Priesen*), *Cechy (Bohemia)*, *Czech Republic*, and at Radicin and at Teplice; in recent lavas of Mte. Somma, Campania, Italy. AR *D6(Supp) 2:113(1909)*, *MM 47:397(1983)*.

73.1.5.1   Kanemite $NaHSi_2O_4(OH)_2 \cdot 2H_2O$ or $NaH[Si_2O_5] \cdot 3H_2O$

Named in 1972 by Johan and Maglione for the locality. ORTH *Pnmb*. $a = 7.282$, $b = 20.507$, $c = 4.956$, $Z = 4$, $D = 1.933$. *25-1309:* 10.3_{10} 4.01_{10} 3.44_9 3.16_7 3.09_7 2.48_8 2.39_6 2.00_5. Said to have a double tetrahedral layer structure. Colorless to white, translucent. Cleavage {010}, perfect; {100}, good. $H = 4$. $G = 1.93$. Biaxial (−); $2V = 46°$; $N_x = 1.451$, $N_y = 1.470$, $N_z = 1.478$; $XYZ = bac$; $r > v$. Decomposed by acids. Large dehydration endotherms at 160° and 220°. Found as spherulitic aggregates, 1–2 mm in diameter, associated with gaylussite in trona at *Lake Chad, Kanem region, Adaija, Chad*, and Lake Bogoria, Kenya. AR *AM 59:210(1974)*, *62:763(1977)*.

73.1.6.1   Vertumnite $Ca_8Al_4(Al_4Si_5)O_{12}(OH)_{36} \cdot 10H_2O$

Named in 1977 by Passaglia and Galli for the Etruscan god Vertumnus. MON (pseudo-HEX) $P2_1/m$. $a = 5.744$, $b = 5.766$, $c = 25.12$, $\beta = 119.72°$, $Z = 1$, $D = 2.15$. *29-291:* 12.6_7 6.28_6 4.28_2 4.19_{10} 2.87_2 2.49_1 2.44_1 2.08_1; this pattern as well as the bulk chemistry bears a strong resemblance to that of strätlingite (77.2.5.1) and may be regarded as a polytype thereof. Structure: *TMPM 25:33(1977)*. The structure is described as having a double tetrahedral layer. Occurs as pseudohexagonal tablets {001} and bounded by {100} and {010}. Twinned but twin law undefined. Colorless, vitreous luster, transparent. Conchoidal fracture, very brittle. $H = 5$. $G = 2.15$. Biaxial (+); $2V = 62°$; $N_x = 1.531$, $N_y = 1.535$, $N_z = 1.541$. Found in a "geode" in phonolite at *Campomorto, Montalto di Castro, Viterbo, Italy*. AR *TMPM 24:57(1977)*, *AM 62:1061(1977)*, *EJM 2:841(1990)*.

73.2.1.1   Jagoite $(Pb,Na)_3(Fe^{3+},Mg)Si_3O_{10}(Cl,OH)$

Named in 1957 by Blix et al. for John Bernard Jago Trelawny (b.1909), American lawyer and mineral collector of Palo Alto, CA. HEX $P\bar{6}2c$. $a = 8.528$, $c = 33.33$, $Z = 6$, $D = 5.34$. *36-417:* 6.16_5 5.56_7 4.13_6 3.38_{10} 3.07_5 2.98_8 2.78_9 2.05_5. Structure: *AM 66:852(1981)*. Structurally related to naujakasite, with alternating double and single tetrahedral sheets perpendicular to c, the latter being discontinuous with the missing tetrahedron replaced by an Fe^{3+} octahedron. Yellow-green, vitreous luster. Cleavage {0001}, perfect. $H = 3$. $G = 5.43$. Uniaxial (−); $N \approx 2.0$; birefringence = 0.025. The formula given is considerably simplified; Pb is, in part, four-coordinated, and in its larger coordination sites is replaced by both Na and Ca; Fe^{3+} is six-coordinated but also four-coordinated; and Be may also be present in tetrahedral sites. Occurs as fine-grained, micaceous aggregates, commonly surrounded by black melanotekite, in hematite ore at the *Canberra stope, Långban, Sweden.* AR *AM 43:387(1958).*

73.2.2a.1   Reyerite $(Na,K)_2Ca_{14}Al_2Si_{22}O_{58}(OH)_8 \cdot 6H_2O$

Named in 1907 by Cornu and Himmelbauer for the Austrian geologist Eduard Reyer (1849–1914), but the first samples were collected by Giesecke in 1811. See minehillite (73.2.2b.1) and truscottite (73.2.2a.2). HEX $P\bar{3}$ or $P3$. $a = 9.764$, $c = 19.07$, $Z = 1$, $D = 2.59$. *29-1039:* 19.0_4 4.23_6 3.51_7 3.15_{10} 3.03_4 2.85_9 2.65_6 1.84_6, also *17-760*. Structure: *MM 52:247(1988)*. The structure consists of alternating single, $[Si_8O_{20}]^{8-}$, and double, $[Si_{14}Al_2O_{58}]^{14-}$, tetrahedral sheets having six-membered rings and lying parallel to $\{0001\}$, with intervening layers of edge-sharing calcium octahedra. Al is ordered in one of the connecting tetrahedra of the double layer. Massive or as micaceous, spherical aggregates; sometimes tabular $\{0001\}$ with $\{10\bar{1}0\}$. Colorless, white; vitreous luster pearly on cleavage; transparent to translucent. Cleavage $\{0001\}$, perfect; brittle, uneven fracture. $H = 3–4$. $G = 2.54–2.578$. Uniaxial $(-)$; $N_O = 1.563–1.568$, $N_E = 1.558–1.563$. Found in chlorite-containing amygdules in diabase at the Rawlins quarry, Brunswick Co., VA; in a Triassic sill near Durham, NC; in tuff at *Niaqornat, Nuussuaq, Greenland*; and in basalt at 'S Airde Beinn, Mull, and at Alet Coir'a'Ghohbainn, Skye, Scotland. AR *D6(Supp) 2:88(1909)*, *AM 58:517(1973)*.

73.2.2a.2   Truscottite $(Ca,Mn^{2+})_{14}Si_{24}O_{58}(OH)_8 \cdot 2H_2O$

Named in 1914 by Höving for Samuel John Truscott (1870–1950), English mining geologist. See minehillite (73.2.2b.1) and reyerite (73.2.2a.1). HEX $P\bar{3}$ or $P3$. $a = 9.731$, $c = 18.836$, $Z = 1$, $D = 2.51$. *29-382:* 18.8_6 4.21_7 3.85_3 3.50_5 3.14_{10} 2.84_8 2.64_6 1.839_7. Isostructural with reyerite. Massive or as micaceous, spherical aggregates. White, pearly luster on cleavage, translucent. Cleavage $\{0001\}$, perfect. $G = 2.35$. Uniaxial $(-)$ or biaxial $(-)$; 2V very small; $N_O = 1.552$, $N_E = 1.530$. Found in a drill hole in the Lower Geyser Basin, Yellowstone National Park, WY; in the Hatrurim Formation, Israel; at the *Leong Donok gold mine, Benkulen, Sumatra, Indonesia*; and at the Toi mine, Tagata-gun, Shizuoka Pref., Japan. AR *MM 30:450(1954)*, *43:333(1979)*; *AM 44:470(1959)*.

73.2.2b.1   Minehillite $(K,Na)_{2-3}Ca_{28}Zn_4Al_4Si_{40}O_{112}(OH)_{16}$

Named in 1984 by Dunn et al. for Mine Hill, the hill on which the Franklin deposit cropped out. See reyerite (73.2.2a.1) and truscottite (73.2.2a.2). HEX $P6_3/mmc$, $P6_3mc$, or $P\bar{6}2c$. $a = 9.77$, $c = 33.01$, $Z = 1$, $D = 2.94$. *38-371:* 16.1_7 3.35_9 3.14_6 3.07_7 2.97_5 2.88_4 2.76_{10} 1.847_9. Dimensionally similar to reyerite, but space group has hexagonal symmetry. Colorless, white to grayish black; vitreous luster; transparent to translucent. Cleavage $\{0001\}$, perfect. $H \approx 4$. $G = 2.93$. Uniaxial $(-)$; $N_O = 1.607$, $N_E = 1.604$. Fluoresces a medium violet in UV. Found as aggregates of plates up to 5 mm in diameter, associated with diopside, calcite, grossular, vesuvianite, microcline, and wollastonite at the *Franklin mine, Franklin, Sussex Co., NJ.* AR *AM 69:1150(1984)*.

73.2.2b.2 Orlymanite $Ca_4Mn_3Si_8O_{20}(OH)_6 \cdot 2H_2O$

Named in 1990 by Peacor et al. for Orlando P. Lyman (1903–1986), founder of Lyman House Memorial Museum, Hilo, Hawaii. See gyrolite (73.2.2c.1), truscottite (73.2.2a.2), minehillite (73.2.2b.1), reyerite (73.2.2a.1), and fedorite (73.1.3.1). HEX-R $P\bar{3}$ or $P3$. $a = 9.60$, $c = 35.92$, $Z = 5$ [*AM 75:923(1990)*], $D = 2.93$. PD(nat): 7.15_7 4.18_7 3.60_{10} 3.13_8 2.54_4 2.47_4 1.84_9 1.62_2. Detailed structure not yet done; see *AM 75:923(1990)*. A possible link between the gyrolite family and conventional phyllosilicate minerals. Spheres to 2- to 3-mm, tightly packed rosettes, fibrils elongated on a rather than c. Dark brown, light brown streak, vitreous luster. Cleavage {001}, perfect. H = 4–5. G = 2.7–2.8. Analysis (wt %): SiO_2, 46.8; Al_2O_3, 0.2; Fe_2O_3, 0.8; MgO, 1.6; CaO, 21.6; MnO, 19.7; Na_2O, 0.6; H_2O, 8.68. Uniaxial $(-)$; $N_O = 1.598$, N_E not determined because of extreme fibrous shape. Pleochroic; O, pale brown; E′, dark brown. Occurs at *Wessels mine, Kalahari manganese field, Cape Prov., South Africa*; part of a vein assemblage 1.5–2.0 cm thick. Associated with inesite and hematite. One specimen known. EF

73.2.2c.1 Gyrolite $NaCa_{16}Si_{23}AlO_{60}(OH)_8 \cdot 14H_2O$

Named in 1851 by Anderson from the Greek for *round*, in allusion to its form as crystalline spherical concretions. Note that water content is from structure determination, may be more H_2O to 17 H_2O. See truscottite (73.2.2a.2), minehillite (73.2.2b.1), reyerite (73.2.2a.1), and orlymanite (73.2.2b.2). Converts to α-$CaSiO_3$, wollastonite, when heated (pseudowollastonite). TRIC $C\bar{1}$ or $P\bar{1}$ [*MM 52:377(1988)*]. $a = 9.74$, $b = 9.74$, $c = 22.40$ $\alpha = 95.7°$, $\beta = 91.5°$, $\gamma = 120.0°$, $Z = 1$, $D = 2.39$–2.46. *42-1452* [combination of Guinier and Debye–Scherrer data, $Ca_4(Si_6O_{15})(OH)_2 \cdot 3H_2O$, $Z = 4$]: 22.0_{10} 11.1_7 4.21_6 3.16_{10} $3.10_{8.5}$ 2.81_6 2.77_5 1.84_8; *9-449*(nat): 22.0_{10} 11.0_8 8.4_4 4.20_8 3.65_6 3.12_{10} 2.80_6 1.82_8; *12-217*(syn): 22.0_9 11.1_7 4.20_7, 3.72_5 3.16_{10} 3.10_9 2.83_7 1.83_6. Structure is built up of the structural units found in the crystal structure of reyerite, namely tetrahedral sheets S_1 and S_2 and octahedral sheets O. The tetrahedral and octahedral sheets are connected by corner-sharing to give rise to the complex layer which can be schematically described as $\bar{S}_2\bar{O}S_1OS_2$ where S_2 and $\bar{S}_2$ as well as O and $\bar{O}$ are symmetry-related units. Successive complex layers with composition $[Ca_{14}Si_{23}AlO_{60}(OH)_8]^{5-}$ are connected through an interlayer sheet made up of Ca and Na cations and H_2O molecules. There is appreciable disorder present in stacking of the structural layers. Disorder in stacking accounts for the high R factor (13.5%). *ZK 159:34(1982)*, *MM 52:377(1988)*. For additional details on the structure, see *AM 80:173(1995)*. Crystals to 5–6 mm, flakes to 3–4 cm, massive lamellar-radiating structure, concretionary. Colorless–white, rarely green, brown, gray; white streak; vitreous-pearly luster. Cleavage {0001}, perfect; uneven fracture; brittle. H = 3–4. G = 2.36–2.45. IR data: *AM 46:913(1961)*, *NBSM*

43:297(1960). DTA data: SIMP 39:695(1984). Decomposed by acids. The crystal chemical formula that accounts for most gyrolite samples is $Ca_{16}Si_{24}O_{60}(OH)_8 \cdot (14 + x)H_2O$ where $0 \leq x \leq 3$. Biaxial $(-)$; 2V small; sometimes uniaxial; $N_x = 1.537$, $N_y = 1.547$, $N_z = 1.549$ (perpendicular to perfect cleavage) and $N_O = 1.545$–1.551, $N_E = 1.534$–1.539. Occurs in basalt or diabase in amygdules. Secondary mineral; a stable phase under saturated steam conditions from 120 to 200°. Occurs in many worldwide localities; only the most notable are given here: Fort Point, San Francisco, New Almaden, Santa Clara Co., and Crestmore, Riverside Co., CA; Port George, Cape Blomidon, Black Rock, NS, Canada; Qarusait, Greenland; Faeroe Is.; *Portree, Skye*, and *s'Airde, Tobermory, Mull, Scotland*; Collinward, Belfast, County Antrim, Northern Ireland; Ortano, Elba, Monte Biaena, and Trento, Italy; Bramburg, Germany; Arsului and Korii valleys, Romania; Lower Tunguska R., Yakutia; Angara R., Siberia, Russia; Bombay and Pune (Poona), India (fine crystals); Ueda, Irakawa, and Sayama Lake, Japan; Mogy Guassu, São Paulo, Brazil; Mururoa, South Pacific. EF *SIMP 39:695(1984), MIN 4(2):311(1992).*

Class 74

Modulated Layers

74.1.1.1 Stilpnomelane $(K,Ca,Na)(Fe^{2+},Mg,Fe^{3+})_8(Si,Al)_{12}(O,OH)_{27} \cdot nH_2O$

Named in 1827 by Glocker from the Greek for *shining* and *black*, in allusion to the color and luster. See lennilenapeite (74.1.1.2), franklinphilite (74.1.1.3), and bannisterite (74.1.1.4). The name *ferristilpnomelane* has been applied to high-Fe^{3+} varieties, but such are called stilpnomelane by some, and *ferrostilpnomelane* is applied to Fe^{3+}-poor varieties. TRIC $P1$. $a = 21.995$, $b = 21.955$, $c = 17.619$, $\alpha = 124.83°$, $\beta = 95.96°$, $\gamma = 120.00°$, $Z = 6$, $D = 2.62$ also reported as MON. *25-174:* 12.1_{10} 4.04_8 3.03_7 2.72_4 2.57_{10} 2.35_7 1.585_6 1.571_5; also *15-48, 29-703*. Structure: *MM 38:693(1972)*. The structure is a modulated layer structure consisting of islands of 24 tetrahedra having roughly coplanar bases. These islands are joined to one another within a layer, and cross-linked to a second layer of oppositely facing islands by way of a double ring of approximately ditrigonal symmetry consisting of 12 tetrahedra; the resultant modulated layer is thus four tetrahedrons thick. The apical oxygens of the islands form the octahedral coordination of the smaller cations, and the larger cations lie between the facing basal oxygens of the islands within the modulated layer. Both tetrahedral and octahedral layers are warped, and successive modulated layers are shifted with respect to one another. Occurs as scales, foliated masses and coatings, and as fibrous druses, often spherulitic and with velvety surface. Golden brown, dark reddish brown to black, also dark green; pearly to vitreous luster, often submetallic and velvety on druses; translucent to nearly opaque. Cleavage {001}, perfect; {010}, imperfect; brittle. $H = 3\frac{1}{2}$. $G = 2.59$–2.96. Biaxial $(-)$; $2V \approx 0°$; $N_x = 1.543$–1.634, $N_y = N_z = 1.576$–1.745; X nearly perpendicular to {001}, $Y = b$; the indices increasing with increasing total Fe and increasing Fe^{3+}/Fe^{2+}. Strongly pleochroic with increasing Fe^{3+}; X, pale yellow to bright gold; Y,Z, deep green to very dark red-brown. Soluble in warm 1:1 H_2SO_4. Substitution of cations is highly complex; Ca and Na may replace K with Na frequently exceeding it. In octahedral sites, Fe^{3+} exceeds Fe^{2+} in many published analyses thus supporting the concept of two iron end members; the ratio of the two oxidation states is probably a function of oxidation-reduction conditions at the time of formation rather than the result of oxidation of earlier formed ferroan material. Manganese and aluminum may also enter octahedral sites, and vacancies exist. H^+ or hydronium ion as well as excess water may be present. Stilpnomelane occurs in iron-rich, low-grade regionally metamorphosed sediments and in associated ore veins with magnetite, goethite, pyrite, etc. It may also be an abundant constituent in chlorite schists. When occurring in small flakes it is easily dismissed as biotite, and many occurrences may thus be overlooked. Found in ores at the Sterling mine, Antwerp, Jefferson Co., NY, often in velvety druses; near Rockyhill, Somerset Co., NJ; manganoan at Bald Knob, NC; in ferruginous slate at Crystal Falls, MI; and in the Animikian iron formation of MN; at Rib Mt., Marathon Co., WI; and in glaucophane and chlorite schists at the San Juan Bautista mine, Santa Clara Co., CA. Occurs in vesicles in a granophyre inclusion in olive gabbro at Kangerdlugssuaq, Greenland; at Nordmark, Värmland, Sweden; *Obergrund, near Zuckmantel, Silesia, Poland Beroun na Morave, Beroun, Czech Republic;* in a polymetallic skarn deposit at Tetukhe in far-eastern Russia; in glaucophanic rocks at several localities in Japan; at the Great Cobar Mine, NSW, and the Hamersley Range, WA, Australia; ferroan and ferrian varieties in chlorite schists of Otago, New Zealand. AR *DHZI 3:103(1962), MM 42:361(1978).*

74.1.1.2 Lennilenapeite $K_{6-7}(Mg,Mn^{2+},Fe^{2+},Zn)_{48}(Si,Al)_{72}(O,OH)_{216} \cdot 16H_2O$

Named in 1984 by Dunn et al. for the Lenni Lenape Native American tribe, who first inhabited the type locality region. See stilpnomelane (74.1.1.1), franklinphilite (74.1.1.3), and bannisterite (74.1.1.4). TRIC (pseudo-HEX) Space group unknown. $a = 21.9$, $d_{001} = 12.18$, $Z = 1$. *34-403:* 12.11_{10} 4.07_2 3.04_2 2.73_3 2.58_4 2.37_3 1.593_3 1.578_3. Black; vitreous/resinous luster, translucent to opaque. Cleavage {001}, perfect; also imperfect perpendicular to {001}. $H \approx 3$. $G = 2.72$. Biaxial $(-)$, $2V = 0°$; pseudo-uniaxial; $N_x = 1.553$, $N_y = N_z = 1.594$. Strongly pleochroic; $Z = Y > X$; Z, dark brown; X, pale brown to colorless. Found as druses at *Franklin, Sussex Co., NJ,* associated with tirodite, willemite, and nelenite, or with sphalerite. AR *CM 22:259(1984), MSARM 19:675(1988).*

74.1.1.3 Franklinphilite $K_4(Mn^{2+},Mg,Fe^{3+},Zn)_{48}(Si,Al)_{72}(O,OH)_{216} \cdot 6H_2O$

Named in 1992 by Dunn et al. for the type locality and from the Greek *philos,* friend. See stilpnomelane (74.1.1.1), lennilenapeite (74.1.1.2), and bannisterite (74.1.1.4). TRIC $P1$ or $P\bar{1}$. A pseudo-HEX cell has $a = 22.08$, $d_{001}=12.19$, $Z = 1$; an ortho-HEX subcell has $a = 5.521$, $b = 9.560$, $c = 36.57$, $Z = \frac{3}{8}$, and $D = 2.66$. *PD:* 12.3_{10} 4.79_2 4.08_2 3.06_2 2.58_4 2.36_3 1.59_3 1.58_3. Dark brown to black, light brown streak, vitreous to slightly resinous luster, translucent to opaque. Cleavage {001}, imperfect; brittle. $H \approx 4$. $G = 2.6–2.8$ (impure). Biaxial $(-)$; $2V_{obs} = 10°$, $2V_{calc} = 0°$; $N_x = 1.545$, $N_y = 1.583$, $N_z = 1.583$; $X \wedge (001) = 6°$. Strongly pleochroic; $Z = Y > X$; dark brown to pale yellow. The manganese dominant analog of stilpnomelane and lennilenapeite, the holotype material having 56% of the end-member component; Na replaces

K in part, and minor Fe^{3+} may substitute in tetrahedral sites. Water determined by difference. Found as radial aggregates of platy crystals intimately associated with friedelite in a centimeter-wide vein cross-cutting a breccia of aegirine, calcite, chamosite and interlayered phyllosilicates, and in a second assemblage with nelenite, rhodonite, and tirodite, on the *Buckwheat dump, Franklin, Sussex Co., NJ*. AR *CM 22:259(1984), MR 23:465(1992)*.

74.1.1.4   Bannisterite $KCa(Mn^{2+},Fe^{2+},Mg,Zn)_{20}(Si,Al)_{32}O_{76}(OH)_{16} \cdot 4\text{--}12H_2O$

Named in 1936 by Foshag for Frederick Allen Bannister (b.1901), former keeper of minerals, British Museum (Natural History). See stilpnomelane (74.1.1.1), lennilenapeite (74.1.1.2), and franklinphilite (74.1.1.3). MON $A2/a$. $a = 22.27$, $b = 16.37$, $c = 22.67$, $\beta = 94.3°$, $Z = 4$, $D = 2.84$. *21-57:* 12.3_{10} 4.10_1 3.44_2 3.36_1 3.08_1 2.64_1 2.61_1 2.41_1. Structure: *CCM 40:129(1992)*. Prismatic, bladed crystals and aggregates to 20 cm; as anhedral, cleavage-bounded plates to 5 cm in diameter. Brown to nearly black, resinous luster, translucent to opaque. Cleavage {001}, perfect. $H = 4$. $G = 2.83\text{--}2.84$. Biaxial (−); 2V small to medium; $N_x = 1.544\text{--}1.574$, $N_y = 1.586\text{--}1.611$, $N_z = 1.589\text{--}1.612$; $r < v$, weak to moderate. Pleochroic; $X < Y = Z$; colorless to pale yellow or brown. Mn/Fe is ≈ 6:5 in Broken Hill material and ≈ 4:1 in Franklin material; Broken Hill material is reported as Zn-free and contains but minor Mg, whereas Franklin material contains 4.6 wt % ZnO and 3.1 wt % MgO; significant Na is present. Found in association with manganoan and zincian amphiboles, rhodonite, sphalerite, and calcite at Franklin, Sussex Co., NJ; at the *Bennalt mine, Rhiw, Lleyn Peninsula, Wales*; at Nyberget, Sweden; at the Tanohata mine, Iwate Pref., Honshu, and the Ananai mine, Kōci Pref., Shikoku, Japan; and with rhodonite, sphalerite, fluorite, and apophyllite at Broken Hill, NSW, Australia. AR *MM 36:893(1968), AM 66:1063(1981)*.

74.1.2.1   Ganophyllite $(K,Na)_2(Mn,Al,Mg)_8[(Si,Al)_{12}O_{29}](OH)_7 \cdot 8\text{--}9H_2O$

Named in 1890 by Hamberg from the Greek for *luster* and *leaf*, in allusion to its physical properties. See eggletonite (74.1.2.2). MON $A2/a$. $a = 16.60$, $b = 27.04$, $c = 50.34$, $\beta = 94.2°$, $Z = 24$, $D = 2.92$ (a slightly different formula, with approximately tripled subscripts, is given in the structure determination with $Z = 8$ and $D = 2.875$). *21-359:* 12.5_{10} 5.12_1 4.18_1 3.97_1 3.46_1 3.14_3 2.98_1 2.70_1. Structure: *MM 50:307(1986)*. The structure has modulated tetrahedral layers consisting of sixfold rings forming three-ring-wide strips. Brown to orange-red with paler streak, vitreous to greasy luster, transparent to translucent. Cleavage {001}, perfect. $H = 4\text{--}4\frac{1}{2}$. $G = 2.84\text{--}2.88$. Biaxial (−); 2V small; $N_x = 1.563\text{--}1.573$, $N_y = 1.593\text{--}1.611$, $N_z = 1.593\text{--}1.612$; $r < v$; positive elongation. Weakly pleochroic, $X > Y = Z$ ($X < Y = Z$ also reported), pale yellow to colorless. Gelatinizes with acids. Occurs as short prismatic to acicular crystals with rectangular cross section. Found at Franklin, Sussex Co., NJ; Alum Rock Park, Santa Clara Co., CA; with astrophyllite and labuntsovite which it closely resembles at Mt. St.-Hilaire, QUE, Canada; at the Sjö mine, near Grythyttan, and at the *Harstig mine, Pajsberg, Värmland, Sweden.* AR *USGSPP 180:114(1935), MM 36:893(1968), MR 21:311(1990).*

74.1.2.2 Eggletonite $Na_2Mn_8[Si_{11}AlO_{29}](OH)_7 \cdot 11H_2O$

Named in 1984 by Peacor et al. in honor of R. A. Eggleton, Australian National University. see ganophyllite (74.1.2.1). MON Ia or $I2/a$. $a = 5.554$, $b = 13.72$, $c = 25.00$, $\beta = 93.95°$, $Z = 2$, $D = 2.76$. *29-909:* 12.4_{10} 3.45_1 3.13_3 2.85_1 2.69_2 2.60_2 2.46_2 2.39_1. Undetermined structure, but probably isostructural or nearly so with ganophyllite. Occurs as radiating groups of acicular to prismatic, [100], crystals up to 1.5 mm long. Dominant forms are {110} and {001}. Twinned on (001). Dark golden brown with lighter streak, vitreous luster. Cleavage {001}, perfect; brittle. H = 3–4. G = 2.76. Biaxial (−); $2V = 9°$; $N_x = 1.566$, $N_y = 1.606$, $N_z = 1.606$. Found in miarolitic cavities in nepheline syenite at the 3M and *Big Rock quarry, Little Rock, Pulaski Co., AR.* AR *MM 48:93(1984).*

74.1.3.1 Parsettensite $(K,Na,Ca)(Mn,Al)_7Si_8O_{20}(OH)_8 \cdot 2H_2O(?)$

Named in 1923 by Jakob for the locality. ORTH or MON Space group unknown. $a = 39.1$, $b = 22.6$, $c = 12.6$, $\beta \approx 90°$, $Z = 1(?)$, $D = 2.58$. *25-8:* 12.1_{10} 4.2_6 3.84_5 3.70_5 2.79_8 2.65_{10} 1.634_8 1.617_6. The structure is a modulated 2:1 layer structure related to, but probably distinct from, stilpnomelane and ganophyllite. Copper-red, somewhat metallic luster, translucent. Cleavage {001}, perfect. H $\approx 1\frac{1}{2}(?)$. G = 2.59. Uniaxial (−) or biaxial (−); $2V = 0$–$8°$; $N_z = 1.546$, $N_y \approx N_z = 1.576$. Pleochroic; X, colorless; Y,Z, greenish yellow. Decomposed with hot acids. Fuses readily with intumescence. Found at the Foote mine, Kings Mt., Cleveland Co., NC; as somewhat micaceous masses in manganese deposits at *Val d'Err, Alp Parsettens, Sursass, Grischun, Switzerland.* Reported from the Molinella manganese mine, near Chiavari, Liguria, Italy; at the Noda-Tamagawa mine, Iwate Pref., Japan; at the N'Chwaning mine, Kuroman district, Cape Prov., South Africa. AR *SMPM 3:227(1923), MSARM 19:675(1985), AM 77:673(1990).*

74.1.4.1 Zussmanite $K(Fe^{2+},Mg,Mn)_{13}[AlSi_{17}O_{42}](OH)_{14}$

Named in 1964 by Agrell et al. for Jack Zussman (b.1924), crystallographer and emeritus professor of geology, Manchester University, England. HEX-R $R3$. $a = 11.66$, $c = 28.69$, $Z = 3$, $D = 3.14$. *19-1500:* 9.60_{10} 4.78_{10} 3.78_5 3.69_5 3.19_8 2.74_5 2.20_5 1.908_5. Structure: *MM 37:49(1969).* The tetrahedral layer consists of six-membered rings to which are joined three-membered rings akin to those of benitoite; these composite sheets are joined to a trioctahedral layer of the divalent cations. Pale green, vitreous luster, translucent. Cleavage {0001}, perfect. G = 3.146. Uniaxial (−); $N_O = 1.643$, $N_E = 1.623$. Weakly pleochroic; O, pale green; E, colorless. Found as tabular, {0001}, crystals in blueschist facies metamorphosed shale, ironstone, and impure limestone of the Franciscan formation at the *Laytonville quarry, Mendocino Co., CA*, and elsewhere in the Laytonville district. AR *AM 50:278(1965), MM 43:605(1980)*.

74.2.1.1 Wickenburgite $CaPb_3Al_2Si_{10}O_{27} \cdot 3H_2O$

Named in 1968 by Williams for the locality. HEX $P\bar{3}1c$. $a = 8.560$, $c = 20.190$, $Z = 2$, $D = 3.88$. *21-148:* 10.08_{10} 5.96_3 5.04_3 3.93_6 3.36_4 3.26_8 2.79_3 2.64_4. Structure: *CM 32:525(1994)*. The structure has layers of composition $[AlSi_5O_{15}]$, with ordered Al and Si tetrahedral sites, linked to layers of composition $[CaSi_5O_{17}]$, with octahedral Ca sites, to form thick slabs parallel to {0001}. Colorless to white, rarely pink; vitreous luster. Cleavage {0001}, indistinct. $H = 5$. $G = 3.85$. Uniaxial $(-)$; $N_O = 1.692$, $N_E = 1.648$. Fluoresces dull orange under SWUV. Occurs abundantly as 0.2- to 1.5-mm tabular crystals, associated with phoenicochroite, mimetite, cerussite, and willemite in oxidized zone of several prospects including the *Cramer claim, near Wickenburg, Maricopa Co., AZ*. AR *AM 53:1433(1968)*.

PALYGORSKITE–SEPIOLITE GROUP

The palygorskite–sepiolite group minerals are "clay" minerals that have typical phyllosilicate stoichiometry with the (Si,Al)/O ratio of 2:5. A general formula is

$$A_{4-5}(Si,Al)_8O_{20}(OH)_2 \cdot 5\text{–}9H_2O \quad \text{or} \quad A_{8-10}Si_{12}O_{30}(OH)_4 \cdot 12H_2O$$

where

A = Mg, Al, Na, Fe, Ni

and the structure is layered but different from that of the micas and with certain characteristics of the amphiboles.

In the micas the layers may be viewed as being made up of parallel, cross-linked amphibole ribbons, thus extending the structure infinitely in two dimensions; all apical oxygens point in the same direction. In palygorskite the layers are also built of cross-linked amphibole ribbons, but the apical oxygens of alternate ribbons point in opposite directions, resulting in a layer with a superimposed one-dimensional component. The layers, parallel to (001), are stacked so that apical oxygens of adjacent sheets point toward one another and are bonded to the remaining cations so as to form amphibole-like I-beam units. The linear component of these layers (parallel to c in the table below) has the characteristic $\approx$ 5-Å periodicity of the amphiboles. Hydroxyl ions are involved in the octahedral coordination of the I-beam cations. Water, both bonded and zeolitic, occupies the channels formed by the base-to-base facing portions of the layers. The structure of the sepiolite subgroup is similar except that the ribbons joined to form layers are three tetrahedra wide, as in the jimthompsonite ribbons.

Species in this group are fibrous or lath-like rather than lamellar, with fiber length rarely exceeding a few μm, and having a thickness or width of no more than a few hundred Å. Palygorskite and sepiolite commonly form matted aggregates of fibers to which terms such as "mountain leather", "mountain cork", "cardboard", "paper", "wool", etc. have been applied. The measured density is commonly less than the calculated value due to the highly porous nature of the fiber aggregates.

Adsorbed and zeolitic water is lost at $\approx 120°C$ (sepiolite) and $\approx 150°C$ (palygorskite). Palygorskite loses bound water in the range 450–500°C; in sepiolite the bound water is lost in two stages at $\approx 350°C$ and 500–550°C. About half of the bound water is lost at the first stage and the structure "folds" to close up the now unsupported channels.

Although relatively rare "clay" minerals, palygorskite and sepiolite have found considerable application due to their sorptive capabilities.

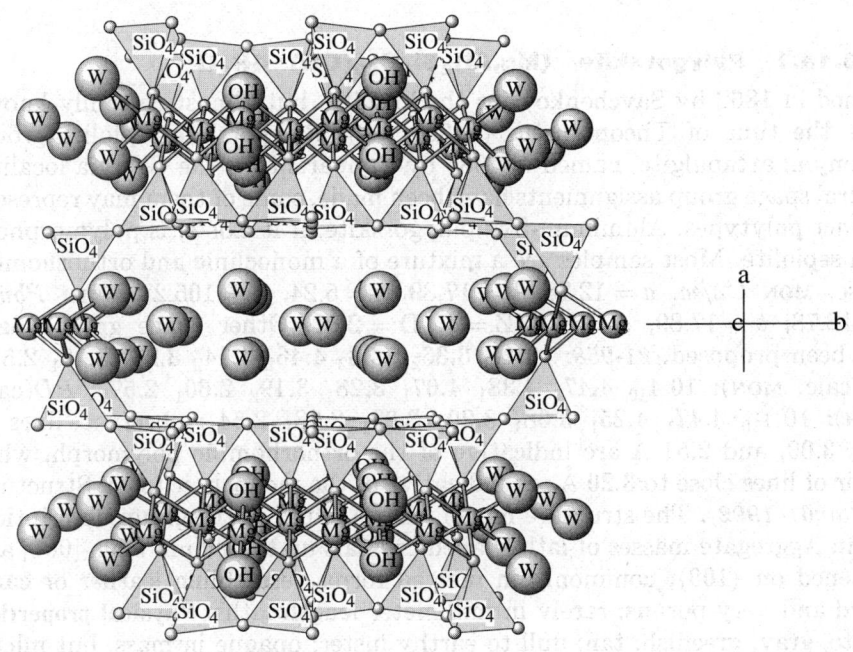

SEPIOLITE: $Mg_8Si_{12}O_{30}(OH)_4 \cdot nH_2O$

Perspective view along c of modulated layers formed of ribbons of composition $[Si_6O_{15}]$. Palygorskite is similar but the narrower ribbons have composition $[Si_4O_{10}]$.

	PALYGORSKITE SUBGROUP					
Mineral	Formula	Space Group	d_{100}	b	c	
Palygorskite	$(Mg,Al)_2Si_4O_{10}(OH) \cdot 4H_2O$	$C2/m$	12.33	17.89	5.24	74.3.1a.1
Tuperssuatsiaite	$Na_{1-2}Fe_3Si_8O_{20}(OH)_2 \cdot 5H_2O$	$C2/m$	13.30	18.000	4.828	74.3.1a.2
Yofortierite	$(Mn^{2+},Mg)_5Si_8O_{20}(OH)_2 \cdot 8-9H_2O$	$Pn(?)$	12.71	18.369	5.024	74.3.1a.3

	SEPIOLITE SUBGROUP					
Mineral	Formula	Space Group	a	b	c	
Sepiolite	$Mg_4Si_6O_{15}(OH)_2 \cdot 6H_2O$	$Pncn$	13.43	26.88	5.281	74.3.1b.1
Falcondoite	$(Ni,Mg)_4Si_6O_{15}(OH)_2 \cdot 6H_2O$	$Pncn$	13.5	26.9	5.24	74.3.1b.2
Laughlinite	$Na_4Mg_6Si_{12}O_{30}(OH)_4 \cdot nH_2O$	Ortho-rhombic(?)	?	?	?	74.3.1b.3

74.3.1a.1 Palygorskite $(Mg,Al)_2Si_4O_{10}(OH) \cdot 4H_2O$

Named in 1862 by Savchenkov for the locality, but almost certainly known since the time of Theophrastus ($\approx$ 300 B.C.). Palygorskite–sepiolite group. Synonym: attapulgite, named in 1935 by Lapparent for the Georgia locality. Several space group assignments have been made, some of them may represent distinct polytypes. Aluminum-free palygorskite, if it exists, is polymorphous with sepiolite. Most samples are a mixture of a monoclinic and orthorhombic phase. MON $C2/m$. $a = 12.78$, $b = 17.89$, $c = 5.24$, $\beta = 105.2°$. ORTH $Pbmn$. $a = 12.78$, $b = 17.89$, $c = 5.21$, $Z = 4$, $D = 2.31$. Other space groups have also been proposed. $21\text{-}958$: 10.4_{10} 6.36_2 5.38_2 4.46_2 4.14_2 3.23_1 3.18_1 2.51_1; PD(calc, MON): 10.4_{10} 4.47_1 4.38_1 4.07_1 3.28_1 3.19_2 2.60_1 2.57_1; PD(calc, ORTH): 10.4_{10} 4.47_1 4.25_1 3.68_1 3.20_2 3.09_1 2.68_1 2.54_1; observed lines at 4.25, 3.09, and 2.51 Å are indicative of the orthorhombic polymorph, while a pair of lines close to 3.20 Å are indicative of the monoclinic form. Structure: *CM 30:61(1992)*. The structure model will be found in the group description. **Habit:** Aggregate masses of lath-shaped crystals up to 300 μm long, [001] and flattened on {100}, commonly in matted forms resembling leather or cardboard and very porous; rarely in millimeter-length laths. **Physical properties:** White, gray, greenish, tan; dull to earthy luster; opaque in mass, but microscopically transparent. Cleavage {110}, good. H = 2–2½. G = 2.29–2.36 but often found lower due to high porosity. Biaxial (−); 2V = 61–36°; $N_x = 1.522–1.528$, $N_y = 1.530–1.546$, $N_z = 1.533–1.548$; the highest indices and smallest 2V for a sample with 6.6 wt % F_2O_3. Insoluble in acids. Infusible. Sometimes fluorescent, cream to bluish gray in SWUV, white in LWUV.

Chemistry: Aluminum substitution lies in the range 0.01–0.69 of the eight tetrahedral sites. Usually contains some iron and minor calcium; manganese is often present as well as minor sodium and potassium; fluorine may reach ≈ 5000 ppm. Its paucity in older than Tertiary sedimentary rocks suggests metastability. Occurrence: Found in hydrothermal veins; in basalts; in syenites and granites; and in granitic pegmatites. In deep-sea sediments and associated with active ridge and fracture zones in the Atlantic; in continental slope or margin deposits; in lacustrine sediments. Develops in soils of arid and semiarid regions. Formed as an alteration product in a diversity of rocks, including serpentinites. Localities: Found at the Lone Jack quarry, Rockbridge Co., VA; important commercial deposits at Attapulgus, Decatur Co., GA, and Quincy, Gadsden Co., FL; Sapillo, San Miguel Co., NM; at the New Cornelia mine, Ajo, AZ; in granite pegmatite at Mesa Grande, San Diego Co., and at the New Melones Lake Spillway, Calaveras Co., CA; Pend Oreille mine, Stevens Co., WA(!); Gustavus, Alexander Archipelago, AK; at Saucillo, CHIH, the Refugio mine, Concepcion del Oro, ZAC, and near Saculum village, northern Yucatan Penin., Mexico. In England at South Petherwin, Cornwall; Seaton, Devon; Endersby, Leicestershire; in Scotland at Cabrach, Aberdeenshire; and at Bakkasetter in the Shetland Is. At Torrejón, Cáceres, Spain; Schneeberg, Saxony, Germany; Idria, Venezia, Italy; Klesovo, Volyn, and Chekassk S of Kiev, Ukraine; *Palygorskaya, Ural Mts., Russia*; Jezda, Kazakhstan; Hyderabad, Andhara Pradesh, India; the Kuzuu district, Tochigi Pref., Japan; in the Flinders Range, Q, Australia; Taguenout-Hagueret, Algeria, and Tafrout, Morocco. Uses: Filtering and sorptive agent; as a thermal and acoustic insulator. Yucatan palygorskite, mixed with indigo, was used to form the highly prized "Maya blue" pigment in pre-Columbian times; the same material was used by local potters well into historic times. AR *AM 54:198(1969), SPC 16:183(1971), CM 28:329(1990).*

74.3.1a.2 Tuperssuatsiaite $Na_{1-2}Fe_3Si_8O_{20}(OH)_2 \cdot 5H_2O$

Named in 1985 by Karup-Møller and Petersen for the locality. Palygorskite–sepiolite group. MON $C2/m$ (material from Namibia reported as $C2/c$ or Cc). $a = 13.729$, $b = 18.000$, $c = 4.828$, $\beta = 104.28°$, $Z = 2$, $D = 2.468$. *38-372*(Greenland): 10.82_{10} 4.14_2 3.40_3 2.64_4 2.54_3 2.51_3 2.38_2 2.24_3. Fan-shaped aggregates to several centimeters in size, rosettes, or fibers, elongated on [001] and twinned on (001). Red-brown, vitreous luster, translucent. Cleavage {100}, good. Biaxial (+); 2V large; for Greenland material: $N_x \approx 1.54$, $N_y \approx 1.56$, $N_z \approx 1.58$; for Namibian material: $2V = 76.5°$; $N_x = 1.539$, $N_y = 1.560$, $N_z = 1.595$; $Y = b$, $Zc = 5-7°$. Pleochroic, $Z > Y > X$ dark red-brown to colorless. Greenland material contains $NaFe_3^{3+}$, whereas Namibian material approaches Na_2Fe_3, requiring part of the iron in the ferrous state. Manganese is present as a substituent for iron. A zinc-rich material is noted but not described further. Found in late, low-temperature veinlets associated with natrolite and aegirine in natrolite–albite rocks at *Tuperssuatsiait Bay,*

Ilimausaq, Greenland, and better crystallized material from the Aris phonolite, Windhoek, Namibia. AR NJMM:501(1984), 145(1992); AM 70:1332(1985).

74.3.1a.3 Yofortierite $(Mn^{2+},Mg)_5Si_8O_{20}(OH)_2 \cdot 8-9H_2O$

Named in 1975 by Perrault et al. in honor of Yves Oscar Fortier (b.1914), director of the Geological Survey of Canada, 1964–1972. Palygorskite–sepiolite group. MON Space group uncertain, probably Pn. $a = 12.759$, $b = 18.369$, $c = 5.024$, $\beta = 94.98°$, $Z = 2$, $D = 2.81$. 27-312: 10.5_{10} 4.41_2 3.68_1 3.30_9 2.62_3 2.53_1 2.51_2 1.564_1. Pink to violet, silky to dull luster. $H = 2\frac{1}{2}$. $G = 2.18$. Biaxial; $N_x = 1.530$, $N_z = 1.539$; $X \wedge$ fiber $= 8°$. Pleochroic; maximum absorption perpendicular to fiber. Found as radiating fibers up to 3 cm in length, associated with analcime, serandite, eudialyte, polylithionite, aegirine, microcline, and albite in pegmatite veins in nepheline syenite at the *DeMix quarry, Mt. St.-Hilaire, QUE, Canada*. AR CM 13:68(1975).

74.3.1b.1 Sepiolite $Mg_4Si_6O_{15}(OH)_2 \cdot 6H_2O$

Named in 1847 by Glöcker from the Greek for *cuttlefish*, whose "bone" is light and porous, as is the mineral, but the mineral was probably known nearly a century earlier. Palygorskite–sepiolite group. Synonym: meerschaum, German for *sea foam*, coined in 1788 by Werner. ORTH $Pncn$. $a = 13.43$, $b = 26.88$, $c = 5.281$, $Z = 4$, $D = 2.25$; alternate settings transpose a and c yielding $Pnan$. 13-595: 12.1_{10} 4.31_4 3.75_3 3.37_3 3.20_3 2.62_3 2.56_5 2.26_3. The structure model is presented in the group description. Usually as nodular masses of very fine, matted fibers and laths; aggregates commonly highly porous; rarely in laths to several millimeters long [001]. White, gray, and variously tinted in yellows, browns, and greens; dull, earthy luster; opaque in mass. Tough due to interlocking fibers. Cleavage {110}, good, but rarely discernible. $H = 2-2\frac{1}{2}$ but easily indented by pressure due to porous nature. $G = 2.0-2.2$ but often much lower in mass with dry pieces floating in water. Biaxial (−); $2V = 50-83°$; $N_x = 1.498-1.522$, $N_y = 1.507-1.553$, $N_z = 1.527-1.579$, the indices increasing with increased iron content; $Z = c$(elongation). *Meerschaum* is reported as isotropic with $N = 1.517$. Decomposed by HCl. Infusible. Compact varieties tend to adhere to tip of tongue when thus touched. Aluminum commonly replaces up to a few percent of the tetrahedral sites; Ca and alkalis are present only in small amount; Fe^{2+} and Fe^{3+} are commonly present with Fe_2O_3 up ≈ 10 wt % (Strunz has proposed the Fe-dominant ferrisepiolite); nickeloan material is known, and a series with falcondoite is likely; fluorine is commonly present in the range 0.5–1.3 wt %. Occurs in a wide variety of environments as described for palygorskite with which it is commonly associated. Forms as an authigenic mineral in lacustrine deposits; an important alteration product of several rock types particularly serpentinites and magnesite. Found in Chester and Delaware Cos., PA; as an authigenic mineral at Mound Lake, Lynn and Terry Cos., TX; at Sapillo, San Miguel Co., NM; at the Cole shaft and Southwest mine, Bisbee, Cochise Co., AZ; in Duchesne and Tooele Cos., UT; in small veins in calcite and as pellets in limestone at Crestmore, Riverside Co., CA; at the Cardiff mine, Wilberforce, ON, Canada. Important economic deposits occur at Vallecas–Vicalvaro, Madrid, Spain. The alleged type locality is at *Baldissero Canavese, Piemonte, Italy*; also in Moravia, Czech Republic; and at Thebes, Greece. Fine nodular masses (*meerschaum*) occur on the plains of Eskişehir, Turkey(!). Found at Liling, Hunan, China; in the Kuzuu district, Tochigi Pref., Japan; at Tintinara, SA, Australia (aluminian); as crystals(!) to several millimeters at Ampandradara, Madagascar; at Amboseli, Kenya; and in Morocco. Fine, compact meerschaum has been used extensively as a carving material, particularly in the manufacture of smoking pipes that were often executed in very ornate designs. Material from Spain is an important lightweight building stone and is also used in ceramic wares. AR *D6:680(1892), DT:679(1932), MSARM 19:631(1988)*.

74.3.1b.2   Falcondoite $(Ni,Mg)_4Si_6O_{15}(OH)_2 \cdot 6H_2O$

Named in 1976 by Springer for the contraction, Falcondo, of Falconbridge Dominica C. por A., the company mining at the type locality. Palygorskite–sepiolite group. The nickel analog of sepiolite. ORTH *Pncn*. $a = 13.5$, $b = 26.9$, $c = 5.24$, $Z = 4$, $D = 2.54$. *29-1433*: 12.2_{10} 3.33_3 3.19_2 2.62_3 2.53_3 2.44_3 2.39_2 2.26_2. Whitish green, translucent. Friable; cleavage described as "slightly schistose." $H = 2–3$. Biaxial; $N \approx 1.55$; positive elongation; extinction approximately parallel. The Ni/Mg ratio varies from 0.75 to 2.33 with an average of 1.38. DTA endotherms occur at 180° (zeolitic water), 360° and 540° (bound water), and 840° (hydroxyl). Occurs as tangled masses of 10-μm fibrous particles in 2-cm-wide fissures at a depth of 3 m below the surface in the *Loma Peguera laterite deposit at Bonao, Dominican Republic*. AR *CM 14:407(1976)*.

74.3.1b.3   Loughlinite $Na_2Mg_3[Si_6O_{15}](OH)_2 \cdot 7H_2O$

Named in 1960 by Fahey et al. for Gerald Francis Loughlin (1880–1946), chief geologist, U.S. Geological Survey. Palygorskite–sepiolite group. Probably ORTH Space group unknown. *13-310*: 12.8_{10} 7.60_3 4.80_5 4.45_{10} 3.79_{10} 3.65_{10} 2.90_7. White, silky to pearly luster. $G = 2.165$. Biaxial (+); $2V_{calc} = 60°$; $N_x = 1.500$, $N_y = 1.505$, $N_z = 1.525$. Decomposed by HCl; prolonged immersion in water produces sepiolite(?). Minor ferric and ferrous iron are present. Found as massive to fibrous fillings in veins up to 3 cm long and 1 cm thick, in dolomitic oil shale at three locations *18 to 20 miles W of Green River, Sweetwater Co., WY*. AR *AM 45:270(1960)*.

74.3.2.1   Chrysocolla $Cu_2H_2[Si_2O_5](OH)_4$

The name was first used by Theophrastus in 315 B.C., and derives from the Greek for *gold glue*, in allusion to its use, with other mineral materials as well, in the soldering of gold. The name often is used very broadly for any massive, globular, glassy, blue to green copper-bearing mineral whose specific identity is undetermined. Probably ORTH Space group unknown. $a = 5.7$, $b = 8.9$, $c = 6.7$. *27-188:* 17.9_8 7.9_6 4.45_2 4.07_6 2.90_8 2.56_7 1.602_4 1.486_{10}, weaker lines often diffuse; also *PD:* 4.39_{10} 2.80_5 2.60_4 2.42_1 2.32_1 1.600_3 1.482_8 1.321_2. **Structure:** *CR D271:1837(1970)*. For "well-crystallized" acicular material a structure is proposed that is related to that of palygorskite, with sheets made up of amphibolelike ribbons joined to one another with apical oxygens pointing in opposite directions in alternating ribbons. This superimposed linear component within the sheets accounts for the fibrous habit. **Habit:** As very fine acicular crystals to 5 mm in length in radiating clusters; fibrous; in botryoidal aggregates; cryptocrystalline. Often in glassy, opaline, or porcellaneous encrustations coating or replacing other phases. **Physical properties:** pale to deep blue, blue-green, green, brown to black due to impurities; pale blue to tan or gray streak; vitreous, waxy, porcellaneous, to earthy luster, translucent to opaque. Brittle, and in part, somewhat sectile. H = 2–4. G = 1.93–2.40. **Optics:** Biaxial (−); 2V small: $N_x = 1.575$, $N_y = 1.587$, $N_z = 1.600$; but optical properties highly variable. **Chemistry:** Al and Fe^{3+} may substitute for Cu^{2+} with Al_2O_3 reported up to 5.9 wt % and Fe_2O_3 up to 5.7 wt %; water content is variable. A formula written as $Cu_{2-x}(Al,Fe^{3+})_x H_{2-x}[Si_2O_5]\,(OH)_4 \cdot nH_2O$ takes the substitutions into account. Endothermic water loss is reported at 60° and 120°C, and approximately half of the water is lost by 200°C and the balance between 200° and 700°C. Many materials designated as chrysocolla are mixtures of such materials as several copper silicate species, amorphous material, cryptocrystalline quartz, and opaline silica. **Occurrence:** A ubiquitous phase in oxidized zones of copper deposits, often encrusting or replacing earlier secondary minerals, and commonly associated with malachite, azurite, tenorite, and clay minerals. **Localities:** Of worldwide occurrence. Some notable localities are: at the Chino pit, Santa Rita, Grant Co., and replacing hemimorphite and various copper minerals in the Hansonburg district, S of Bingham, Socorro Co., NM; in the Tintic district, UT; an important component of some large Cu orebodies in the Globe–Miami district, Gila Co., the Ray pit and San Manuel mine, Pinal Co., Morenci, Greenlee Co., and other deposits in AZ; at Cananea, SON, Mexico. Known from several copper deposits of Cornwall, England; at Lubietova (Libethen), Czech Republic; rich deposits from Nizhne Tagil, Ural Mts., Russia; "Eilat Stone" from Timna, southern Negev, Israel; at Broken Hill, Zambia; at Tsumeb, Namibia; at Lubumbashi, Likasi, Zaire; at the Pinnacles mine, Flinders range, SA, Broken Hill, NSW, and the Chillagoe district, Q, Australia. At Chuquicamata and Copiapó, Chile. **Uses:** A minor and incidental ore of copper from oxidized zones. A popular ornamental material fashioned in various decorative forms; chrysocolla, sometimes intergrown with malachite, azurite, and/or cuprite is often fashioned into cabochons ("Eilat Stone" in part). The ancient use as a component of soldering flux is no longer practiced. AR *D6:699(1892)*; AM *48:649(1963)*; AM *54:993(1969)*; IANS *6:29(1968)*.

74.3.3.1 Altisite Na₃K₆Ti₂[Al₂Si₈O₂₆]Cl₃

Named by Khomyakov et al. in 1994 for the principal high valence cations in the chemical formula: Al, Ti, Si. MON. $C2/m$, $a = 10.363$Å, $b = 16.310$Å, $c = 9.132$Å, $\beta = 105.34°$ at 20°C), $Z = 2$, $D = 2.67$; PD (ZVMO 123:82(1994) (nat.): 8.22_7 3.50_4 $3.16_{3.5}$, 3.05_{10}, 2.90_7, 2.835_8, 2.00_3, 1.97_3. The structure (R = 2.5%) consists of parallel (001) layers of Si and Al tetrahedra joined with Ti octahedra. Each infinite sheet of tetrahedra is grouped in 6 (two types) and 10-fold rings with an ordered distribution of Al and Si. In the structure, there are three types of channels filled with Na, K and Cl. Structure is related to that of lemoynite which has $2 \times c$ of altisite and $Z = 4$ (EJM 7:537(1995). Colorless, transparent, occurs as anhedral grains to 3 mm diameter, in the interstices of pectolite crystals; conchoidal fracture, vitreous luster, greater than that of nepheline, $H = 6$, $G = 2.64$. Not luminescent in UV light. Heated to 1000°C, then gives an X-ray pattern of KCl. Na₂O 9.3 wt. %, K₂O 20.7, BaO 1.0, Al₂O₃, 8.7, SiO₂ 40.3, ZrO₂ 0.1, TiO₂ 13.1, Nb₂O₅ 0.5, Cl 8.6, O for Cl 1.9, Σ100.4. Biaxial (+), $2V = 85°$ (calc. 86°); $N_\alpha = 1.601$, $N_\beta = 1.625$, $N_\gamma = 1.654$, weak dispersion, r < v. IR, DTA, TGA, and DTG data available (ZVMO 123:82(1994)). A hydrothermal mineral, formed late in the period of formation of ultra-agpaitic pegmatites. Found in core from a 470 m drill hole in pegmatitic rocks in the SW part of the Khibiny Massif in the region of the Olen Brook apatite deposit. Principal minerals are sodalite, nepheline, K-feldspar, and pectolite. Also associated with aegirine, shcherbakovite, tinaksite, Mg-astrophyllite, nefedovite, villiaumite, natrite, and rasvumite. EF.

74.3.3.2 Lemoynite (Na,K)₂CaZr₂[Si₁₀O₂₆]·5–6H₂O

Named in 1969 by Perrault et al. for Charles Lemoyne (1625–1685), prominent in the history of Canada. MON $C2/c$. $a = 10.384$, $b = 15.947$, $c = 18.601$, $\beta = 104.59°$, $Z = 4$, $D = 2.38$. 24-1072: 9.0_4 8.01_{10} 4.39_3 3.96_2 3.56_5 3.48_3 3.03_3 2.81_5. Structure: CM 14:132(1976). A pseudocell with c halved can be assigned to $C2/m$. The silicate structural element consists of double-width ribbons lying in (010) and parallel to [100]; these are condensed to form sheets parallel to (001) that, in turn, are linked by Zr octahedra to form an open framework. Prismatic to bladed crystals in spherical aggregates with the dominant forms {100}, {010}, {001}, and {011}; fine, almost fibrous. White to tan, rarely colorless, greenish-gray, yellowish white; dull luster but colorless crystals are vitreous. Cleavage {100}, {010}, perfect. $H = 4$. $G = 2.29$. Biaxial (+); $2V = 80°$; $N_x = 1.540$, $N_y = 1.553$, $N_z = 1.570$; $Y = b$, $Z \wedge a = 5°$; r < v, weak. As compact 3- to 8-mm spheres and spherical aggregates of bladed crystals in interstices of large microcline crystals in pegmatite at *Mt. St.-Hilaire, PQ, Canada*. AR CM 9:585(1969), MR 21:318(1990).

74.3.4.1 Sanbornite BaSi₂O₅

Named in 1932 for Frank B. Sanborn (1862–1936), mineralogist, California Department of Natural Resources. β-BaSi₂O₅, α-BaSi₂O₅ is the high-temperature form. ORTH *Pmnb*. (Syn): $a = 7.692$, $b = 13.525$, $c = 4.634$; *ZK 153:33(1980)*(nat): $a = 7.688$, $b = 13.523$, $c = 4.629$, $Z = 4$, $D = 3.774$. α-form, $B2/b$, has $a = 23.195$, $b = 13.613$, $c = 4.658$, $Z = 12$. *26-176*(syn): 6.77_4 3.97_9 3.42_5 3.34_7 3.10_{10} 2.73_4 2.71_4 2.28_4. Structure: *ZK 153:33(1980)*. Platy to 2–3 cm; polysynthetic twinning on (010). White, colorless; pearly luster. Cleavage (001), perfect; {010}, imperfect. H = 5. G = 3.70–3.77. May fluoresce cream, gray, or bluish in SWUV and LWUV light. Decomposed by acids. Biaxial (−); $Z \approx c$; $N_x = 1.597$, $N_y = 1.616$, $N_z = 1.624$; $2V = 66°$. Occurs as a major component of Ba-silicate assemblages at or near fringes of large granite intrusives. Associated with witherite and taramellite. Occurs at *Trumbull Peak, Incline, Mariposa Co., CA*; Rush Creek and Big Creek, Fresno Co., CA; Chicken Coop Canyon, Tulare Co., CA; Ross River, YT, Canada; La Madrelena mine, Tres Pozos, Baja Norte California, Mexico. EF

74.3.4.2   Krauskopfite $BaSi_2O_4(OH)_2 \cdot 2H_2O$

Named in 1965 by Alfors et al. for Konrad Bates Krauskopf (1910-). American geologist and geochemist, Stanford University. MON $P2_1/a$ $a = 8.460$-8.608 $b = 10.602$-10.622 $c = 7.831$-7.837 $\beta = 94.53$-$94.79°$ $Z = 4$ $D = 3.06$; *PD 18-192* (nat.): 6.7_3 6.36_5 5.34_5 3.94_3 3.84_{10} 3.66_3 3.01_4 2.88_3. Structure has been solved (Acad. Naz. dei Lincei 42:859(1967)). Colorless-white; streak white; equidimensional to short prismatic (on a), subhedral to anhedral grains to 6 mm; subvitreous, pearly on cleavage plates; cleavage = 2 perfect {010} and {001} at 90°, a third poor cleavage or fracture at a high angle to the other two; H 4; G 3.14; readily decomposed by cold acids (dilute) leaving a white to transparent silica residue. Not affected by weak bases. Traces of Ca, Mg and Fe present. Biax. (−), $2V = 88°$, $N_\alpha 1.574$, $N_\beta 1.587$, $N_\gamma 1.599$; colorless; $X = b$, $Y = 10\frac{1}{2}°$, fragments elongated on Y, thus can be either length fast or length slow. Dispersion = R > V distinct. Occurrence: A secondary mineral along with macdonaldite occuring in veinlets or lining cracks with Ba silicates in metasomatic rocks. Assoc. with opal and witherite, also replaces sanbornite. *Rush Creek, Fresno Co., CA*, Chickencoop Canayon, Tulare Co., CA. EF *AM 50:314(1965)*, *MIN 3(2):531(1981)*, *AM 69:358(1984)*.

74.3.5.1   Natrosilite $Na_2Si_2O_5$

Named in 1975 by Timoshenkov et al. from the composition: sodium (natrium) and silicon. MON $P2_1/a$. $a = 12.30$, $b = 4.88$, $c = 8.27$, $\beta = 104.23°$, $Z = 4$, $D = 2.51$; (syn): $a = 12.329$, $b = 4.848$, $c = 8.133$, $\beta = 104.24°$. *29-1261*(nat): 6.06_{10} 4.28_5 4.17_6 3.97_8 3.64_7 2.98_9 2.44_8 1.83_8; *23-529*(syn): 5.98_4 4.13_3 3.94_5 3.77_3 3.62_4 2.97_5 2.43_{10} 1.83_3. Structure: β-form of $Na_2Si_2O_5$. *AC(B)* *24:1077(1968)*. Pseudohexagonal thick tabular crystals to 6 cm × 6 cm × 4 cm; twins on {100} noted on synthetic material, but *not* in natural material. Colorless; silky luster on micaceous cleavage, also pearly; transparent—thin, translucent—thick. Cleavage {100}, micaceous; {001}, distinct; {011}, poor; even fracture. G = 2.48. IR data: *ZVMO 104:317(1975)*. Decomposed by water and weak acids, melts at $850 \pm 20°$ to a transparent yellowish glass. N = 1.505. Analysis (wt %): SiO_2, 66.03, 66.32; Na_2O, 33.96, 33.74; trace of K_2O. Biaxial $(-)$; X ≈ perpendicular to (100), Z = b, Y ∧ c at 484 nm = $-12°$, Y ∧ c at 645 nm = $1-2°$; $N_x = 1.507$, $N_y = 1.517$, $N_z = 1.521$; 2V = 63–64° (484 nm), 49–50° (645 nm); r < v; anomalous optical properties include undulatory extinction; anomalous blue interference colors. Occurs in pegmatite segregations of nepheline–syenites with microcline, analcime, natrolite, lomonosovite, ussingite, and vuonnemite. Occurs at Mt. St.-Hilaire, QUE, as white, elongated crystals to 0.5 mm, with good cleavage, associated with revdite in sodalite xenoliths; *Mt. Karnasurt, Lovozero massif, Kola Penin., Russia*; also in the Khibiny massif. EF

74.3.5.2 Makatite $Na_2Si_4O_8(OH)_2 \cdot 4H_2O$

Named in 1970 by Shepard et al. from the Masai *emakat*, soda, in allusion to the high sodium content. MON $P2_1/c$. $a = 7.388$, $b = 18.094$, $c = 9.523$, $\beta = 90.64°$, Z = 2, D = 2.05; originally reported as ORTH with a different cell. *23-703*: 9.04_5 8.42_3 5.09_{10} 4.22_3 3.42_4 3.13_3 2.99_6 2.88_4. Structure: *ZK 159:203(1982)*. The structure determined on synthetic material is made up of sheets, $[Si_2O_4(OH)]_n^{n+}$, parallel {010} and containing six-membered rings. Crystals acicular, 5–30 µm long and 0.05–2.0 µm wide; also in radiating aggregates and spherulites 0.05–0.3 mm in diameter; also as striated prisms of 3 mm length. White, dull vitreous luster, translucent. Cleaves into thin flakes presumably {010}. G = 2.072. Russian material is biaxial (+); 2V = 70°; $N_x = 1.472–1.475$, $N_y = 1.48$, $N_z = 1.487–1.490$; also reported as biaxial $(-)$. Often forms spongy pseudomorphs after bladed trona (Lake Magadi); possibly a pseudomorph after natrosilite (Kola). Found in sodalite xenoliths with vuonnemite at Mt. St.-Hilaire, QUE, Canada; at Howenegg, Hegau, Germany; at Mt. Alluaiv, Lovozero alkalic massif, Kola Penin., Russia; in cores of trona from *Lake Magadi, Kenya*; at the Aris mine, S of Windhoek, Namibia. AR *AM 55:358(1970)*, *DANS 255:181(1980)*.

74.3.5.3 Silinaite $NaLiSi_2O_5 \cdot 2H_2O$

Named in 1991 by Chao et al. for the composition: Si–Li–Na. MON $A2/n$. $a = 5.061$, $b = 8.334$, $c = 14.383$, $\beta = 96.67°$, Z = 4, D = 2.22–2.23. PD(nat): 7.14_{10} 4.24_8 4.14_{10} 4.02_8 2.85_{10} 2.70_5 1.61_4 1.58_4. Structure: *CM29:363(1991)*. Tabular crystals show {001}, fibrous clusters, poorly formed prismatic crystals, cloudy to earthy or powdery patches; crystals to 2 mm: twinning present, CP is parallel to (001), $[110]_{180°}$ = twin axis. Clear, colorless to white; white streak; vitreous to earthy luster; opaque to translucent. Cleavage {001}, perfect; {010}, good; {110}, distinct; conchoidal fracture; brittle. H = $4\frac{1}{2}$. G = 2.24. Dissolves slowly in concentrated HCl and HNO_3, more slowly in H_2SO_4, and leaves a gel residue. Analysis (wt %): SiO_2, 58.5; Al_2O_3, 0.0; CaO, 0.1–0.4; Na_2O, 15.0–13.7; Li_2O(calc), 7.25–7.28; H_2O(calc), 17.5–17.6. Biaxial (+); X = b, Y ∧ c = 16° in acute > β; Na(D): $N_x = 1.515$, $N_y = 1.516$, $N_z = 1.518$; 2V = 64°; strong, r > v, inclined; nonpleochroic. Occurs at the *Poudrette quarry, Mt. St.-Hilaire, QUE*, Canada in sodalite syenite xenoliths in nepheline syenite. Associated with many rare species, such as erdite, kogarkoite, tugtupite, and vitusite-(Ce). EF *CM 29:359(1991)*.

71.3.5.4 Searlesite $NaBSi_2O_5(OH)_2$

Named in 1914 for John W. Searles (1828–1897), an early California resident of the Searles Lake area, or possibly for his son (1865–1938). MON $P2_1$. *APK 194*, $a = 7.972$, $b = 7.049$, $c = 4.895$, $\beta = 93.93°$, $Z = 2$, $D = 2.46$. *AM 61:123(1976)*. $a = 7.981$, $b = 7.066$, $c = 4.905$, $\beta = 93.95°$, $Z = 2$, $D = 2.45$. *6-37*(nat): 8.01_{10} 4.31_3 4.06_5 3.54_3 3.48_4 3.24_4 3.21_4 2.92_3. Also *29-1181*(not in latest book). Structure: *AM 61:123(1976)*. Elongated on [001], striated, xls to 15 cm. White, colorless; white streak; vitreous luster; translucent. Cleavage (100), micaceous; (010), (001), average. $G = 2.46$–2.49. IR data: *ZSGD: 301(1977)*. DTA data: *USGSB: 1036-K(1957)*, *ZSGD:301(1977)*. Sometimes fluoresces orange in LWUV. Decomposed by HCl. Biaxial (−); $Z = b$, $X \wedge c = 30°$; $N_x = 1.515$–1.524, $N_y = 1.531$–1.533, $N_z = 1.535$; $2V = 55$–$73°$. Occurs in alkaline saline lakes, alkaline sedimentary rocks, and in alkalic rocks and their cogenetic pegmatites. Occurs in the Green River formation in Wyoming and Utah and at Coaldale, Esmeralda Co., NV; *Searles Lake, San Bernardino Co., CA*; also at China Lake, Owens Lake, and Lake Tecopa, Inyo Co., CA; Kramer borate deposit, Boron, CA; Nuka formation, Brooks Range, AK; Lopore, Bosnia; Kemna, Yugoslavia; Lovozero massif, Kola Penin., Russia; West Azzir, near Caspian Sea, Russia; Miyang depression, Henan Prov., China. EF

74.3.6.1 Revdite $Na_{16}[Si_4O_6(OH)_5]_2[Si_8O_{15}(OH)_6](OH)_{10} \cdot 28H_2O$

Named in 1980 by Khomyakov et al. for the locality, the nearby village of Revda. Compare natrosilite – revdite is the hydrous analogue. MON. $C2$. $a = 53.83$Å, $b = 9.972$, $c = 6.907$, $\beta = 96.78°$, $Z = 2$, $D = 1.93$; *PD 33-1279* (nat.): 13.4_9 4.46_{10} 3.79_4 3.34_7 3.12_2 2.88_4 2.50_5 2.23_3. Structure: main elements consist of bands of $[Si_4O_6(OH)_5]$, and the second type is a combination of two bands of the first type, displaced relative to one another (*SPC 37:632(1992)*. Colorless-white, transparent; acicular, spherical masses 1–2 cm diameter, irregular shaped; at Mont Saint-Hilaire, crystals are 1–2 mm long, to 2 mm at Lovozero; vitreous-pearly; cleavage = excellent, micaceous on (001) and one other nearly as good; $H = 2$; $G = 1.94$; IR and DTA data in *ZVMO 109:566(1980)*; dissolves slowly in H_2O at 20°C, giving an alkaline reaction to phenolphthalein; dissolves quickly in cold dilute acids. SiO_2, 45.21; Na_2O, 22.25; H_2O, 32.50; K_2O trace. Biaxial (−); $2V = 30$–$75°$, Khibiny: $N_\alpha = 1.460$, $N_\beta = 1.478$, $N_\gamma = 1.481$ and Lovozero: $N_\alpha = 1.469$, $N_\beta = 1.482$, $N_\gamma = 1.490$; dispersion = r > v. Occurs at Mont Saint-Hilaire in cavities and fractures in sodalite xenoliths; at Lovozero it occurs in ussingite veinlets cutting nepheline syenites. Mont Saint-Hilaire, Quebec, Canada; from deep workings on *Mount Karnasurt, Lovozero Massif, Kola Peninsula, Russia*; Mount Rasvumchorr, Khibiny Massif.

74.3.7.1 Cavansite $Ca(V^{+4}O)(Si_4O_{10} \cdot 4H_2O$

Named in 1973 for the composition Ca–V–Si. Polymorphous with pentagonite. ORTH. $Pcmn$ $a = 10.298$, $b = 13.99$, $c = 9.601$, $Z = 4$ (AM 58:412(1973)). D = 2.34, *25-182* (nat) 7.96_{10} 6.85_5 6.13_3 4.53_1 3.93_3 3.42_3 3.06_1 2.78_3 A layer silicate: contains zig-zag chains of $(SiO_3)_N$. Blue to greenish blue; elongate on c, prismatic aggregates with {110} & {101} dominant, rosettes to 5 mm from Oregon; India, crystals to >8 mm long; both prismatic & platelike xls, elongate parallel to b with {001}, {100}, {011} & {001}; vitreous; good cleavage parallel to b; H = 3–4; G = 2.21–2.31; difficulty soluble in acids; in HNO_3 – blue color bleaches out and turns olive brown; a fine white precipitate forms when heated in dil. H_2SO_4. Biaxial (+), $2V_z = 51$–$52°$; (India) $N_\alpha = 1.540$; $N_\beta = 1.545$; $N_\gamma = 1.550$; $(2V_{calc} = 90°)$; pleochroic from very light blue to deep blue. (Oregon) $N_\alpha = 1.542$, $N_\beta = 1.544$, $N_\gamma = 1.551$; N_x = colorless, N_y = blue, N_z = colorless. X = b, Y = b, Z = c. OAP perpendicular a, Bxa parallel to c. Dispersion r < v extreme. Occurs in basalts with zeolites, also in tuffs. Occurs in fractures in palagonitic tuff, association with zeolites and calcite at *Lake Owyhee State Park, Malheur Co., OR and Goble, Columbia Co., OR*; in tholeiitic basalt tuff breccia at Abanagur, near Pune (Poona), Maharashtra, India. !! (best material in the world). *AM 58:405(1973); MN 7:8(1991); Minerali 4(2):479(1992); AC B:1593(1975); LAP 14:39(1989).*

74.3.7.2 Pentagonite $Ca(V^{+4}O)Si_4O_{10} \cdot 4H_2O$

Named in 1973 by Staples et al. from the Greek words meaning 5 and angle. It occurs in prismatic crystals twinned to form fivelings with star-shaped cross sections. Polymorph: cavansite-dimorph. ORTH. $Ccm2_1$. $a = 10.298$Å, $b = 13.99$, $c = 8.891$, $Z = 4$, D = 2.34; *PD* 25-181 (nat.): 8.30_7, 6.07_{10} 4.45_3 3.92_{10} 3.76_{10} 3.50_4 2.64_3 2.57_4. Structure: A sheet structure. Both cavansite and pentagonite have silicate layer structures in which the layers are held together by VO^{2+} groups, and Ca^{2+} ions, but they differ in the way the SiO_4 tetrahedra link to form the layers (*AM 58:412(1973)*). Blue; streak light blue; prismatic, elongated parallel to c; twinning: yes, fivelings with star-shaped cross-section. Forms: {110}, {201} and {001}, twinned by reflection across the prism (110) which is the composition plane; vitreous; cleavage = 1 good, parallel to {010}; brittle; H = 3–4; G not done. Biaxial (−). 2V = $50°$; $N_\alpha = 1.533$, $N_\beta = 1.544$, $N_\gamma = 1.547$; pleochroic with X = colorless, Y = blue, Z = colorless; OAP perpendicular to a, Bxa perpendicular to b, X = b, Y = a, Z = c. Dispersion = r > v, strong. Occurs at *Lake Owyhee State Park, Malheur Co., Oregon*; in fractures in palagonitic tuff. Associated with analcime, stilbite, chabacite, thomsonite, heulandite, calcite and apophyllite. *AM 58:405(1973).*

Class 75

Si Tetrahedral Frameworks

75.1.1.1 Cristobalite SiO$_2$

Named in 1887 by vom Rath for the locality. The term *lussatite* has been used for a fibrous variety. Polymorphous with quartz, tridymite, coesite, and stishovite. The cristobalite structure proper has low-temperature (α) and high-temperature (β) configurations. β-Cristobalite is the stable form of SiO$_2$ from 1470°C to the melting point (1728°C). It exists as a metastable phase below 1470°C because of the sluggish nature of the reconstructive transition to tridymite. The transformation to quartz is further inhibited by the

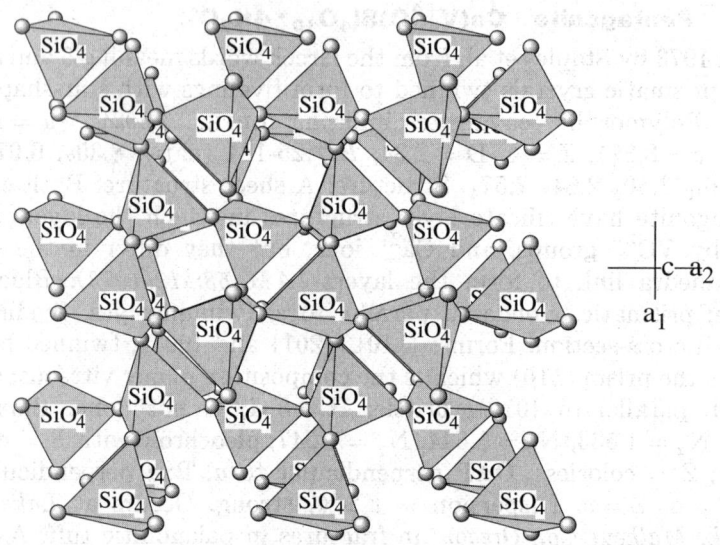

CRISTOBALITE
The structure of tetragonal a-cristobalite; Perspective along c.

75.1.1.1 Cristobalite

presence of foreign ions in the open framework of the cristobalite. The displacive transformation, β to α, takes place at 268°C for highly ordered, pure cristobalite but may be as low as 175°C where a high level of impurities exists. **Low(α)-cristobalite:** TET $P4_12_12$ or $P4_32_12$. $a = 4.969$, $c = 6.926$, $Z = 4$, $D = 2.331$. *39-1425*(syn): 4.04_{10} 3.14_1 2.84_1 2.49_1 2.47_1 2.12_1 1.929_1 1.612_1. **High(β)-cristobalite:** ISO $P2_13$ or $Fd3m$. $a = 7.16$ at 500°C, $Z = 8$, $D = 2.17$. $PD(500°C)$: 4.15_{10} 2.53_8 2.07_3 1.639_6 1.409_5 1.379_2 1.265_3 1.209_3. Structure: *ZK 201:125(1992)*. The structure is a three-dimensional framework of crosslinked, spiral chains of tetrahedra. In an idealized structure of β-cristobalite the silicon atoms form an array not unlike that of the carbons in diamond except that an oxygen lies at the midpoint between each pair of adjacent silicons resulting in a very open structure. Furthermore, the oxygens of β-cristobalite and β-tridymite, although not close packed, are related to one another in the manner of cubic and hexagonal close packing. **Habit:** Crystals, usually under 1 mm in diameter, are dominantly octahedral, rarely cubic; {110} and {331} have been reported, and all are paramorphs after high-cristobalite; the octahedra often skeletal and elongated on cell axes. β-Cristobalite occurs as simple (spinel type) or multiple twins on {100}. Oriented overgrowths on tridymite with cristobalite [110](111) parallel to tridymite [11$\bar{2}$0](0001). More commonly as spherulitic aggregates with radial-fibrous structure, often as the result of devitrification of natural and synthetic high-silica glass; microcrystalline aggregates; botryoidal and stalactitic forms; massive. **Physical properties:** Colorless to white, pale gray, yellowish, and brownish white; vitreous to dull luster; transparent to translucent. Conchoidal to splintery fracture, brittle. $H = 6\frac{1}{2}$. $G = 2.2$–2.33. Uniaxial ($-$); $N_O = 1.487$, $N_E = 1.484$ for α-cristobalite; inverted material is often so fine grained or complexly twinned as to appear isotropic. **Chemistry:** Substitution of aluminum for silicon and the very open structure permit the incorporation of alkali or alkaline earth ions in the voids of the framework. Published analyses all show less than 100% SiO_2, although the reported low value of 91% may be due to presence of impurities. The extreme case of substitution of half the Si sites by Al with half the voids being occupied by Na leads to the synthetic phase carnegieite that has corresponding high- and low-temperature modifications. The preparation of cristobalite from amorphous silica in the laboratory is facilitated by the presence of water and alkalis, although the latter are not necessary. **Occurrence:** Occurs as both α and β forms as a component of opal (75.2.1.1), the inversion of the latter apparently being inhibited by its being embedded in a colloidal matrix. Crystalline material occurs in siliceous volcanic rocks, apparently initially as the β form formed at relatively high temperatures but frequently as a metastable phase at lower temperatures (300°C). Characteristic in vesicles in rhyolite, andesite, and trachyte, where it is associated with tridymite, quartz, sanidine, pyroxene, fayalite, and magnetite, and also in basalts. Similarly, it occurs as thick-fibered, spherulitic aggregates in the groundmass of siliceous volcanics, and as large spherulites to several centimeters in diameter as devitrification products of obsidian, where it is often associated with tridymite and sanidine. Also found in thermally metamorphosed sandstones and sandstone xenoliths at the contact with basaltic and other basic rocks. Quartz grains in the fused walls of fulgurites may contain cristobalite. **Localities:** Of widespread occurrence throughout the world; see also under opal (75.2.1.1). Some noteworthy localities are: in the Tertiary lavas of the western United States, particularly in the San Juan Mts., CO, Yellowstone Park, WY, at Crater Lake, OR, and in the Columbia River basalts; as large spherulites in obsidian ("snowflake obsidian") at Glass Mt., Little Lake, and Sugarloaf Mt. near Coso Hot Springs, Inyo Co., CA. Found as octahedra to 4 mm associated with tridymite at *Cerro San Cristóbal, near Pachuca, HGO, Mexico*; with cassiterite in veinlets in rhyolite at the Santín mine, Santa Caterina, GTO, Mexico; as a component of a metamorphosed block in olivine–dolerite at Tievebulliagh, Northern Ireland; in trachyandesite from Puy de Clierzou, France, where it constitutes up to 10% of the rock; in basalts and other volcanics in the Mayen and Niedermendig region, Eifel district, Germany; in good crystals at Nezdenice, Čechy, Czech Republic; as twin crystals in rhyolite at Sárospatak, Hungary. In thermally metamorphosed sandstone at Arkhara, Transcaucasia, and in Quaternary volcanics of the Kamchatka Penin., Russia. As sharp octahedra and twinned octahedra perched on mordenite in basalt from Ellora, Hyderabad, India, where the temperature of formation is estimated at no more than 300°C. Found at numerous localities in Japan, some devitrification spherulites reaching up to 10 cm in diameter; at Tokatoka, New Zealand, where it makes up some 60% of a 10- to 15-cm-thick grossular–cristobalite contact zone between limestone and andesite. **Uses:** Snowflake obsidian is used as an ornamental stone. AR *D7 3:273(1962)*, *DHZI 4:177(1963)*, *ZK 138:274(1973)*.

75.1.2.1 Tridymite SiO$_2$

Named in 1868 by vom Rath from the Greek for *three* and *twin*, in allusion to the twinned crystals, which are often trillings. Polymorphous with quartz, cristobalite, coesite, lutecite, and stishovite. Tridymite occurs in a variety of polymorphs and polytypes. At 25°, α-tridymite (low form) is MON or rarely TRIC. With an increase in temperature, there are two ORTH (ORTH I and ORTH II, respectively) forms (now considered part of β-tridymite) and then finally two HEX forms (β-tridymite or high tridymite). All of the transformations between these structures are displacive (second order) but show variable degrees of hysteresis. The transitions from quartz to tridymite and from tridymite to cristobalite are reconstructive (first order). The latter transformation is topotactic, where {111} of cristobalite is parallel to {0001} of tridymite. Tridymite is the stable phase of SiO$_2$ between 867 and 1470° and its stability field is greatly reduced with increasing pressure. Due to the reconstructive nature of transformations to other polymorphs (i.e., quartz and cristobalite), tridymite tends to persist metastably upon rapid cooling. At 25°C, three coexisting polymorphs of tridymite (MC, PO-10, MX) have been found together along with lamellae of cristobalite. *AM 72:167(1987)*, *MM*

75.1.2.1 Tridymite

53:89(1989). The temperatures of transition for the different structures of tridymite are variable, chiefly because of substitution of variable amount of trace and minor elements such as Al, K, Na, Ti, and others. Transitions are reversible by slow heating or cooling. Microchemical data and the wide range of structural states occurring in tridymite crystals (and in cristobalite as well) even in the same rock sample indicate nonequilibrium during formation. Rapid growth rates also favor formation of tridymite. Terrestrial tridymites are generally triclinic or orthorhombic at 25° and sometimes monoclinic (e.g., Plumas Co., CA). Most synthetic and meteoritic tridymites are monoclinic. This is probably due to the amount of impurity elements present. Cell data for the most common structures are given below:

Name	Mc	TRIC PO-n	ORTH II OP	ORTH I OC	HEX	HEX HC
a	18.494	9.932	26.171	8.74	5.05	5.05
b	4.991	17.216	4.986	5.04		
c	23.758	81.864	8.916	8.24	8.25	8.27
β	105.79	all 90°	all 90°	all 90°	120°	120°
Space group	Cc	$F1$	$P2_12_12_1$	$C222_1$	$P6_322$	$P6_3/mmc$
Z	48	320	24	8	4	4
Temp. range	to about 105°C	to 70°C	105–180°C	180–350°C	350–465°C	465°C to transit. to β-cr.

Note: A slightly different sequence and temperatures of transition were observed upon heating and cooling a single crystal of synthetic monoclinic tridymite. *AM 63:1252(1978)*. Transformation of the Mc structure to an orthorhombic phase occurs at about 5 kbar. *AM 65:1283(1980)*. Additional details of the various phase transitions are given in *AM 79:606(1994)*. A review of the structural modifications of tridymite and the various cells for meteoritic and terrestrial tridymite are given in *RIM 29:1(1994)*. ICDD cards: *18-1170*(monoclinic, Cc): *42-1401*(orthorhombic, F-centered lattice); *18-1169*(hexagonal, $P6_3/mmc$); *14-260*(hexagonal, P-lattice). Monoclinic: 4.33_9 4.11_{10} 3.87_2 3.82_5 2.98_3 2.50_2 2.49_1 2.31_2; orthorhombic: 4.28_9 4.08_{10} 3.80_7 3.24_5 2.48_4 2.38_2 2.30_2 1.96_1; hexagonal (20H): 4.27_{10} 4.08_9 3.83_5 3.80_9 2.96_6 2.49_6 2.48_6 2.30_5. See also *AM 52:536(1967)* for powder data. D = 2.20–2.45. High tridymite is made up of layers of SiO_4 tetrahedra that are parallel to the {0001} plane. The mineral shows polytypism, and considerable variation occurs in the weak reflections. No evidence has been found for ordering of AL into particular tetrahedral sites. Natural low tridymite (orthorhombic, monoclinic, and triclinic) has up to 12 crystallographically distinct layers and the six-membered rings of tetrahedra which form these layers are all "oval," while meteoritic-synthetic low tridymite has only two crystallographically distinct layers and the layers contain one-third "oval" and two-thirds ditrigonal rings. High tridymite contains only ditrigonal rings. PO (pseudo-orthorhombic) tridymites are designated PO-n, where $n = 1, 1.5, 2, 5, 6, 10$, and 12 (multiple of 4.07 Å). Structural channels are present in tridymite running parallel to [001], which contain variable amounts of Na, K, and other cations. **Habit:** Crystals are generally flabelliform, thin to thick, tabular flattened on {0001}. Pseudohexagonal, as inversion pseudomorphs after β-tridymite. The pyramid {10$\bar{1}$3} is commonly present. Also as rosettes or in spherical and fan-shaped aggregates. Plates, often triplets. Twinning on {10$\bar{1}$6} is extremely common, contact and penetration; wedge-shaped, sheaflike, or as trillings. Twinning on {30$\bar{3}$4} is uncommon. Untwinned crystals are very rare. **Physical properties:** Generally white, or colorless; vitreous luster, pearly on {0001}. Cleavage {0001}, {10$\bar{1}$0}, imperfect; a parting parallel to c is sometimes observed; conchoidal fracture; brittle. H = 6.5–7. G = 2.26-2.33. Soluble in HF and boiling Na_2CO_3. Cathodoluminescent in shades of blue. *CLGM:49(1989)*. NMR spectroscopy: *NAT 303:223(1983)*. IR data: *MP 14:3(1983)*. Raman spectra: *AM 71:694(1986), 79:269(1994)*. **Chemistry:** May be pure but generally contains minor amounts of Na, K, Al, and Ti. Typical contents (wt %) are: Na, 0.1–0.7, Al, 0.2–0.7; K, 0.03–0.3; Ti, 0.03–0.1. Large cations substitute in the structural channels and charge compensation is accomplished by substitution of Al for Si (tetrahedral). **Optics:** High tridymite is uniaxial (+) with negative elongation. Low tridymite has a range of reported optical properties: (meteoritic and synthetic): biaxial (+); XZ = bc; $N_x = 1.468$–1.472, $N_y = 1.470$–1.473, $N_z = 1.474$–1.475; 2V = 40–75°; negative elongation; [natural (Transcarpathia)]: biaxial (+); $N_x = 1.482$, $N_y = 1.484$, $N_z = 1.486$; 2V = 86°; same orientation as given. Indices increase with an increase of Na, K, Al, and so on. Colorless, nonpleochroic. **Occurrence:** Chiefly found in magmatic rocks, especially felsic volcanic rocks. Occurs as phenocrysts to several centimeters across and as euhedral crystals in lithophysal cavities. Also occurs in meteorites. Occurs rarely as a constituent of xenoliths in lavas. Also a constituent of some opals and cherts. Often alters to quartz (i.e., quartz paramorphs after both cristobalite and tridymite). **Localities:** Because tridymite is such a common mineral, only some of the more notable occurrences are mentioned here: abundant in groundmass and as phenocrysts in andesite–rhyolite, San Juan Mountains, CO; in lithophysal rhyolite at Topaz Mountain, Juab Co., UT; in lithophysae of rhyolite and obsidian, Yellowstone Park, MT; Mt. Rainier, WA; Johnstown meteorite, Weld Co., CO; Drachenfels and Perlenhardt, Siebengebirge, Germany; Mts. Dore, Puy Capuchin, Puy de Dôme, France; Steinbach meteorite (Germany); Vesuvius, Mt. Aetna, Colli, and Euganei, Italy; crystals to 1 cm from Kremnička and Vechec, Czech Republic; *Cerro San Cristóbal, Pachuca, Mexico*; Mt. Pele, Martinique; crystals to 1.5 cm from Ishigayama, Japan; Niutoushan, southeastern China.   EF   DS *7(III):259(1962); AC 23:617(1967); AC(B) 32:2486(1976), 33:2615(1977), 34:391(1978); ZK 148:237(1978), 195:31(1991); SMPM 64:335(1984); EJM 5:605(1993); IMAMA 16:203.*

75.1.3.1 Quartz SiO_2

Known since antiquity, but the current name is apparently from the Saxon word *querkluftertz*, cross-vein ore, which became condensed to *querertz* and then to *quartz*. Clear, transparent, well-crystallized quartz is also referred to as *rock crystal* or derivatives thereof in various languages (e.g., Bergkristall). The Greeks and Romans used *cristallos* and *crystallus*, respectively, to refer to this type of quartz, and it is from these words that our word *crystal* is derived. Polymorphs include coesite, cristobalite, tridymite, and stishovite (not a tectosilicate). HEX-R (α-quartz) $P3_121$ (right-handed) or $P3_221$ (left-handed). HEX (β-quartz) $P6_422$ or $P6_222$, (α-quartz): $a_{synth} = 4.9133$, $c_{synth} = 5.4053$, $Z = 3$, $D = 2.66$. *33-1161*: 4.257_2 3.342_{10} 2.457_1 2.282_1 2.127_1 1.8179_1 1.5418_1, 1.3718_1. (β-quartz, at 575°): $a = 4.999$, $c = 5.457$, $Z = 3$. α-quartz is stable at low pressures and up to about 573°C, and β-quartz, also stable at low pressures, is stable from about 573°C to 870°C. It can exist metastably above 870°C. Enantiomorphic. High (β) quartz is made up of SiO_4 tetrahedra, with Si atoms at the center of each tetrahedron. Tetrahedra are grouped to form regular hexagonal and trigonal helices, and their heights (referred to their Si atoms) are expressed as fractions of the c repeat distance. The helices of Si—O—Si—O ⋯ atoms are either all right-handed or all left-handed in each of the enantiomorphous forms of quartz. Upon cooling, β-quartz undergoes a displacive (second order) transformation to the familiar α-form. The structure of α-quartz is similar to that of β-quartz but the SiO_4 tetrahedra are less regular and are rotated from their ideal positions. The α-to-β transition was formerly thought to be a simple second-order transformation. It now appears that α-quartz undergoes a first-order transformation to an incommensurately modulated phase that is stable over a 1.3° interval. The intermediate phase is an ordered mosaic of Dauphiné microtwins that occur as triangular prisms elongate along c. The prisms grow finer with increasing temperature. The subsequent transition is probably second order. *AM 76:1018(1991)*. **Habit**: Commonly anhedral, equant grains. When in crystals, commonly prismatic to short prismatic with the m$\{10\bar{1}0\}$ m faces horizontally striated; terminated commonly by the two rhombohedrons r$\{10\bar{1}1\}$ and z$\{01\bar{1}1\}$. When one rhombohedron predominates, it is in almost all cases, r. Often in double six-sided pyramids or quartzoids through equal development of r and z. When r is relatively large, the crystal then appears pseudocubic. Occasionally distorted, twisted or bent. Twinning is fairly common and the commonest twin laws observed in α-quartz are: Dauphiné, sometimes called electrical twinning because the electrical polarity of the piezoelectric a axes is reversed in the twinned parts (twin axis c, twin plane $\{10\bar{1}0\}$), in which two right- or left-handed crystals interpenetrate; Brazil, sometimes called optical twinning because it can be easily detected between crossed polaroids (twin plane $\{11\bar{2}0\}$), in which right- and left-handed crystals interpenetrate; $\{10\bar{1}1\}$ and more rarely $\{01\bar{1}1\}$ serve as the composition planes in Brazil twins formed during growth. Japanese (twin = composition plane $\{11\bar{2}2\}$), and combined Dauphiné-Brazil. All four possible types of Japan twinning are contact twins of two individuals, with the c axes inclined at 84°33'. Secondary Dauphiné twinning – Dauphiné twinning may be produced artificially in various ways: inversion from β-quartz when cooled through the α-β inversion point, by thermal shock produced by quenching from temperatures as low as 200°C, by the application of an intense electric field, by local pressure at room temperature or higher, by putting a thermal gradient in a cut plate as it cools through the inversion point, and by pure bending or torsion applied to cut plates at temperatures from 573° down to at least 300°. Dauphiné twinning disappears when quartz is heated over the 573° inversion temperature because the twin operation around the threefold axis of low quartz becomes an operation to identity in the sixfold axis of high-quartz. On cooling, Dauphiné twinning may or may not reappear. More than fifteen additional inclined-axis twin laws have been reported (*DS III:1962*). A number of twin laws of the {hoh̄l} type are known in β-quartz, including the Sardinian {101̄2}, Esterel {101̄1}, Belowda {303̄2}, Cornish {202̄1}, and doubtful {302̄1} laws. All can have geometrical equivalents in low-quartz. Of these only the Sella (Sardinian) {101̄2} and Reichenstein-Griesernthal {101̄1} laws have hitherto been described. The high-quartz Belowda (303̄2} law was recognized [*MM 21:368(1928)*] in a number of specimens from Belowda Beacon and Wheal Coates in Cornwall, England, and from the Esterel Mts., France. For β-quartz, twinning on {101̄1}, Esterel law, is very common; on {112̄2}, Verespatak law, less common; other twin laws are rare. So-called 'gwindels' are a growth phenomena where a series of quartz crystals joined with a prism face in common and parallel to one another have the c axis of each suceeding crystal in the stack slightly offset from the one under it. The result is the appearance that the composite crystal is a single crystal "twisted" about an a axis. They were first described by Kennegott in 1866 and are best known from the Swiss Alps, especially from many sites in the central Aar massif (e.g. the Grimsel region, Göschener Tal and Val Giuf). Other fine examples have been found in the region of Mont Blanc, France. **Physical properties:** Usually white or colorless. Also shades of gray, yellow to orange-yellow (citrine), purple and/or violet (amethyst), pink (rose quartz), and gray-brown, to brown, to nearly black (smoky quartz). May be colored by submicroscopic to visible inclusions of other minerals [e.g., chlorite (green), hematite (red-brown)]. Color-zoning commonly observed and may be patchy, irregular, or regular. Streak is white or may be faintly tinted in colored varieties. Luster is vitreous, sometimes greasy or waxy. Cleavage is rare on the positive rhombohedron (r) {101̄1} and negative rhombohedron (z) {011̄1}. Other rarely observed cleavages are on {101̄0}, {0001}, {112̄0}, {112̄1}, and {516̄1}. *AM 48:821(1963)*. Fracture is conchoidal, to a degree structurally controlled to produce flat conchoidal, uneven, splintery (massive varieties) and uneven, hackly or splintery (fibrous varieties). Brittle, cryptocrystalline varieties are tough. $H = 7$. $G = 2.65$–2.66. Gravity of chalcedonic quartz is less and is variable (2.59–2.63) because of porosity, admixed impurities, and a content of nonessential water. Pyroelectric and piezoelectric. Color centers in quartz are best studied by EPR and optical absorption spectroscopy EPR has proven to be the most informative

75.1.3.1 Quartz

method. Miscellaneous properties are discussed in *DS 7:116(1962)*. The hardness of β-quartz is also 7, and the specific gravity is about 2.35 (at 600°). Good $\{10\bar{1}1\}$ cleavage and a $\{10\bar{1}0\}$ parting is present in β-quartz. Fracture is conchoidal and the mineral is brittle. Crystals of β-quartz are always α-quartz paramorphs; typically dipyramidal; often rounded and rough. Enantiomorphous.

A list of the varieties of quartz in common use is as follows:

Crystalline varieties:
- Rock crystal—colorless, transparent.
- Amethyst—various shades of purple or violet; transparent to translucent.
- Citrine—pale yellow, yellow-brown, reddish brown. Transparent to translucent; often erroneously called smoky topaz.
- Smoky quartz—pale smoky brown to almost black. Transparent to nearly opaque; also called cairngorm or morion (black or very dark brown); often combined with amethyst.
- Rose quartz—pale pink to deep rose-red, occasionally with a purplish tinge; usually translucent, rarely clear; rarely in euhedral crystals; color may fade upon exposure to light or heat.
- Milky quartz—milky white to grayish white; translucent to nearly opaque; luster often greasy.
- Blue quartz—pale blue, grayish blue to lavender blue; translucent; somewhat chatoyant.

Crystalline varieties with inclusions:
- Rutilated quartz—contains distinct acicular rutile crystals in sprays or random orientation.
- Aventurine—spangled with glistening scales of micaceous minerals such as mica or hematite.
- Tiger-eye—finely fibrous, chatoyant; formed by complete or partial replacement of asbestos, especially crocidolite; yellow, yellow-brown, brown, reddish brown, bluish; also grayish green, green; cat's-eye quartz is closely related (shows opalescence).
- Ferruginous quartz—contains inclusions of goethite, hematite or other hydrous iron oxides; brown to red.

A variety of other minerals are found as inclusions in quartz crystals, including tourmaline, various amphiboles, stibnite, pyrite, scheelite, gold, copper, and epidote. Sagenitic quartz. May or may not produce asteriated quartz.

Cryptocrystalline varieties:
- Chalcedony—composed of microscopic fibers. Often mammillary, botryoidal, stalactitic, concretionary; as geodes; in veinlets; as pseudomorphs after other minerals; as a replacement of organic materials. Color white, grayish, blue, brown, black. Luster waxy. Transparent to translucent. Often fluorescent in various shades of green (due to incorporation of uranyl ion). Enhydros are nodules of chalcedony containing water. May be

intergrown with leuctecite (moganite). Some is optically (−), unlike true quartz. Structural water, as (OH), is present in Chalcedony *(AM 67:1248(1982))*.

- Agate—distinctly banded chalcedony with successive layers differing in color.
- Onyx—banded agate with layers in parallel straight lines; often used for cameos.
- Moss agate—gray, bluish, or milky transparent to translucent chalcedony containing dendritic inclusions.
- Sard—uniformly colored light brown to dark brown transparent chalcedony; sardonyx is banded sard.
- Carnelian—uniformly colored blood-red, flesh-red to reddish-brown translucent chalcedony.
- Chrysoprase—apple-green, Ni-bearing, translucent variety of chalcedony.
- Plasma—microgranular or microfibrous variety of quartz colored various shades of green.
- Prase—leek-green translucent variety of quartz.
- Heliotrope = bloodstone—greenish variety of chalcedony or plasma containing red spots of iron oxide.
- Flint—compact microcrystalline variety of quartz commonly found as nodules in chalk or marly limestone (e.g., White Cliffs of Dover).
- Chert—compact microcrystalline variety of quartz commonly found in bedded deposits or as lenses.
- Jasper—impure, opaque colored quartz; commonly red, also yellow, dark green, and grayish blue.
- Novaculite—compact microcrystalline variety of quartz of uniform texture frequently used for whetstones.
- Pseudomorphs of quartz, generally fine-grained, exist as replacements of minerals such as calcite, barite, fluorite, and siderite.
- Lechatelierite—naturally occurring fused quartz such as fulgarites from lightning strikes.

Various explanations have been given for the origin of the various colored varieties of quartz [e.g., *ANCH 12:283(1973)*]: (1) ions of transition metals or rare-earth metals that are incorporated in a state of atomic dispersion into the crystal structure, and cause selective absorption in the visible region of the spectrum because of their ligand field or electron transfer bands; (2) electronic defects, which are preferentially formed at structural faults (impurities, vacancies, or interstitial atoms) by ionizing radiation; by transfer of electrons, producing hole centers; (3) heterogeneous inclusions, which can cause selective absorption and scattering of light; in many cases these simply produce turbidity in the crystals. Color in SiO_2 has been summarized recently (RIM v.29, Ch13:433(1994). Iron can occupy both substitutional (S sites) and interstitial (I sites) sites in quartz. The substitutional site is the tetrahedral, oxygen-coordinated silicon site. At least two interstitial positions exist.

75.1.3.1 Quartz

Citrine—submicroscopic distribution of colloidal ferric hydroxide and oxides, as well as Fe^{3+} substituting for Si.

Rose quartz—Tyndall scattering by tiny oriented rutile needles and/or the presence of Ti^{3+} in channels and voids. Al^{3+} is usually also present. *MM 49:709(1985)*. Asterism in rose quartz is due to rutile needles. Massive rose quartz from pegmatites is colored by the above mentioned mechanisms. Transparent, single-crystals of pegmatitic rose quartz are colored by substitutional phosphorous (e.g. Sapucaia pegmatite, MG, Brazil).

Amethyst—color is produced by Fe^{3+} substituting for Si in tetrahedral coordination, and then the action of natural irradiation producing Fe^{4+}. The role of Fe-activated color centers was demonstrated conclusively. *PL 12:310(1964), S 144:289(1964), PCM 8:128(1982)*. Heating changes the color from blue-violet to gray-violet, to clear and sometimes to yellow (citrine). Heat-bleached material may be restored to its original color by X-radiation. Amethyst usually has both Fe and Al present in about equal amounts (Al is usually > Fe), to as much as 200 ppm. Heating of amethyst with ferrous iron present rather than ferric produces a transparent green quartz that has been called "praseolite" or "greened amethyst." The color of amethyst is usually unevenly distributed in natural crystals. It is usually concentrated under the major rhombohedral face r{1011} and often consists of a series of thin, dark bands which parallel the r and occasionally the z{0111} faces.

Smoky quartz—color is related to the presence of Al^{3+} in the tetrahedral site. If Fe^{3+} is present in greater concentration than Al^{3+}, irradiation of clear quartz produces smoky quartz at first but further irradiation further ionizes the iron and causes charge transfer between Fe^{4+} and a trapped hole on an oxygen atom of the Al^{3+} tetrahedron producing amethyst. *AM 70:1180(1985), 71:589(1986), 60:338(1975)*. Heating (to >225°) will decolorize smoky quartz.

Blue quartz—blue quartz phenocrysts such as those from the Llano rhyolite, Llano Co., TX, are colored by submicron-size inclusions of ilmenite which produce Rayleigh scattering. Other examples of blue quartz owe their color to submicron-size inclusions of rutile, tourmaline, or amphibole, and in rare cases to incorporation of Co. *ANCH 12:283(1973)*.

Ametrine is the trade name for unusual crystals of quartz that show amethyst and citrine sectors. *AM 71:1186(1986), GG 30:4(1994)*. Major rhombohedral r{10$\bar{1}$1} sectors are purple-bluish violet and the minor rhombohedral z{01$\bar{1}$1} sectors are orange-yellow. The r sectors are invariably Brazil-twinned with both right- and left-handed quartz. The orange-yellow sectors show no twinning. Orange-yellow sectors have 68–125 ppm Fe, while the purple sectors have 19–40 ppm Fe. Al contents are low; otherwise, the samples would be smoky. Citrine also has more water than the purple. Amethyst sectors are pleochroic. Most citrine available is heat-treated amethyst from Maraba or Rio Grande do Sul, Brazil and Uruguay. Heating citrine causes a loss of color (450–500°). The same is true for the amethyst (400°). The color of most amethyst fades upon exposure to sunlight. Some extremely dark Uruguayan amethyst does not seem to be affected by sunlight. Irradiation restores the color.

Various cathodoluminescent colors (CL) have been reported for quartz: red, blue, violet, gray, rust, and brown. *CLGM: 37-56(1989), MM 52:669(1988)*. The intensity of quartz CL is weak compared to that of feldspars and carbonates and many other minerals. The CL of authigenic quartz is dull or nonexistent. Quartz CL is apparently often controlled by defects and crystal damage effects rather than by activators. CL color may change with time under electron bombardment. "Long-lived" red, blue, and violet CL displayed by igneous and metamorphic quartz remains stable or changes only slightly during electron bombardment. "Short-lived" blue, bottle-green, and violet CL exists in hydrothermal quartz. Reddish to bluish-brown CL that increases in intensity during excitation by electrons occurs in quartz of igneous, metamorphic, hydrothermal, or diagenetic origin. Response to SWUV and LWUV light is usually nil. Cryptocrystalline varieties may fluoresce green due to the presence of uranyl ions. **Optics:** α-Quartz is uniaxial (+); refractive indices show little variation from $N_O = 1.544$ and $N_E = 1.553$. Accurate measurements of optical-grade quartz of known purity (Na light) gave $N_O = 1.544258$ and $N_E = 1.553380$ (at 18°). Quartz that has been shocked at high pressures, such as those from meteorite impact structures, has somewhat reduced refractive indices (as low as $N_O = 1.476$, $N_E = 1.481$) as well as one or more sets of characteristic deformation lamellae parallel to crystallographic directions in quartz with or without accompanying diaplectic glass ($N = 1.460$–1.466). In some metamorphic and igneous rocks quartz develops undulatory extinction due to strain, and in such quartz fine curvilinear lamellae (Böhm lamellae) may occur. These lamellae are not to be confused with true shock lamellae resulting from the sudden application of high pressures. Some quartz, particularly rose quartz, is distinctly biaxial, with a 2V value as high as 8 or 10°. The development of anomalous 2V generally results from strain and may be removed by heating or recrystallization. The color centers (Fe^{4+}) in amethyst are responsible for the biaxial character. The presence or absence of twinning can be determined only in thick/thin sections. Absorption is normally $N_E < N_O$ but $N_E > N_O$ has been observed. β-Quartz is uniaxial (+) with $N_O = 1.5329$ and $N_E = 1.5405$ (Na light at 580°). **Tests:** Insoluble in all acids except HF. Soluble in molten Na_2CO_3. **Chemistry:** Quartz is usually quite pure, but careful analytical studies have shown the presence of various trace elements in natural quartz. Ordinary clear and colorless quartz contains very few trace elements. Variable amounts (as much as 0.2% Al) of Al are present in all colored varieties of quartz and sometimes play a role in the development of color. Iron is usually very low in smoky quartz but is present to as much as 0.3% in rose quartz. Li, Mg, Ti, and Ca are other elements that have been detected in quartz. As much as 500 ppm Ti is present in rose quartz. Fluid inclusions are often present and result in the presence of water, halides, and other daughter products. **Occurrence:** β-Quartz occurs mainly in cavities or as phenocrysts in rhyolite and other acidic rocks. Next to the feldspars, quartz is the most abundant mineral

75.1.3.1 Quartz

in the Earth's crust. α-Quartz occurs in most geological environments and in most igneous, metamorphic, and sedimentary rocks. It also occurs in veins and in graphic granite. Quartz may be of multiple generations (e.g., authigenic overgrowths on detrital grains). **Localities:** Because quartz is one of the most common minerals, only those localities that yield unusual crystals or exceptionally fine specimens can be mentioned here. Brazil is the world's premier producer of the principal varieties of crystallized quartz (i.e. rock crystal, amethyst, citrine, rose, smoky, and rutilated), and is also the major producer of agate. The largest well-formed quartz crystal (20 ft long, 5 ft across each prism face, and weighing >44 tons) is from Manchao Felipe, Itapore, Goias, Brazil. Individual crystals weighing more than 800 lb were found in Alpine quartz veins in Switzerland. *Rock crystal and clear quartz*: Fine pseudo-cubic crystals occur at the Tamminen quarry, Greenwood, ME; "Herkimer diamonds" are found near Herkimer, Little Falls, Fonda, and Middleville, NY. The crystals occur in the Little Falls (Cambrian) Dolomite; they are highly perfect, transparent, usually equant, and range from 1 mm up to about 4cm, although larger ones are occasionally found. They are commonly single and usually show no point of attachment ("floaters"); associated minerals are normally lacking, except for occasional masses and included blebs of bitumen. Clear quartz occurs at Ellenville, Lake George, Diamond Point, and Diamond Isle, Warren Co., NY. Some of the most famous and prolific localities in the world for fine, clear, and beautifully shaped clear quartz crystal are near Hot Springs, Jessieville, and other areas in Garland Co., AR; also in Saline and Montgomery Counties, AR; Mt. Antero and Mt. White, Chaffee Co., CO; White Queen, Elizabeth R, and Tourmaline Queen mines, Pala district, Little Three mine, Ramona, and Himalaya dike system, Mesa Grande, San Diego Co., CA. "Lake County Diamonds" occur in rhyolite, Clear Lake region, Lake Co., CA; one of the most familiar occurrences to the public of generally colorless but rarely amethystine quartz are the linings of walls of hollow "coconut" concretions formed in an altered volcanic ash deposit in north-central Chihuahua about 120 km north of Chihuahua City; in England fine crystals come from Frizington, Cleator Moor, and Alston Moor, Cumberland; in Switzerland at the Zingenstock, Canton Bern and large crystals to 625 kg (1,400 pounds) from alpine clefts between Münster and Laax in Oberwallis (upper Valaise); in Austrak in the Pinzgau district, Salzburg; in the Stubach-tal in the Oberpinzbau in the Gross Glockner where the largest crystal found (in 1962) was 618 kg (1360 pounds); also in alpine veins at Gastein, Rauris, Kaprun, Zillertal, Gelbernbachtal, Habachtal, Untersulzbachtal, and Prägraten; from alpine veins in the Austrian Tyrol (e.g. Laperwitzkees); white, opaque exteriors on clear quartz at Suttrop in Westphalia, Germany; fine quality, limpid, equant crystals occur in the famous Carrara marble quarries, Tuscany, Italy; limpid crystals known as Paphos diamonds have been known since ancient times from the vicinity of Paphos, western tip of Cyprus; the gem-bearing pegmatites of the Nuristan district, Laghman Prov., Afghanistan often yield large quartz crystals in addition to the more famous spodumene, tourmaline, topaz and beryl crystals for which they are noted; "Kaysumah diamonds" occur at Al Quasumah, Saudia Arabia; at Sakangyi, Katha District, Burma, a 700 kg flawless crystal was recovered, from which a sphere nearly 33 cm in diameter and weighing more than 48 Kg was made, and is now at the Smithsonian Institution, Washington, D.C. For many centuries, Madagascar was the world's premier source of rock crystal, used almost exclusively in the arts; the crystals occur in cavernous veins in quartzite in the northern part of the island, and to a lesser extend from pegmatites, much of the latter being smoky. Occurs at a number of localities in Australia including: Stonyfell and Ashton near Adelaide; also in veins in Pre-Cambrian quartzite at the White Rock quarry at Horsenell Gulley, Magill, in the Giles Range, 30 km east of Adelaide; Newstead tin mines, New England, NSW; Maldon and Bendigo, northeast of Melbourne, VIC. The largest well-formed quartz crystal on record (6.1 m long, 1.5 m across each prism face, and weighing >44 tons) is from Manchao Felipe, Itapore, GO., Brazil. Two major districts for crystals, from veins in quartzite, are at Diamantina and Corinto, MG, and a further one at Cristalina, GO. Much of this quartz is used to make "lasca" (fragments of quartz crystals used as a raw material in the manufacturing of synthetic quartz crystals or for fusing); the production amounts to hundreds of tons/year. In addition, very large quantities of slightly less pure quartz are used for electric furnace production of silicon metal. Additional applications for clear crystals are for the electronics industry, and for optical purposes. Other Brazilian localities for fine crystals are Brumadao, BA (superb Japan law twins); the Morro Velho gold mine, Nova Lima, MG; Novo Horizonte (formerly Ibitiara) BA where fine gwindels have been found; and many of the pegmatites in eastern MG. In Chile from La Serena, 700 km north of Santiago and from Inca de Oro, 8 km north of Copiapo. *Smoky quartz:* from pegmatites in Oxford Co., and Androscoggin Co., ME; Moat Mountain, and Sugarloaf Mtn., NH; Lake George ring, Devils Head and other localities in the Pikes Peak batholith, CO; Sawtooth batholith, Salmon, Idaho; 19-mile district (Pipestone district), Jefferson Co., MT; Sierra Blanca, Otero Co., NM; Little Three mine, Ramona, San Diego Co., CA; Cairngorm Mtns., Banffshire, Scotland, and the Mourne Mountains, Ireland; with pink octahedcrons of fluorite from the Mont Blanc region, near Chamonix, France; the Swiss Alps, including St. Gotthard, Ticino (Grimsel region); Val Tavetsch, Grisons; Val Cavradi, Graubunden; Göschener Alp, Uri, and others such as Zinkenstock near Grimsel Pass and Tiefen Glacier in Uri which produced fine material in the last century. In Italy from Aosta, and other localities in Prov. Piemonte; from the Carrara marble quarries in Tuscany; from Poretta, Emilia; the island of Elba; classic localities in Austria include: Knappenwand, Salzburg; Saualpe, Carinthia; Schwemmhoisl quarry, Stiermark (Styria), and huge crystals from the Eiskögele, Salzburg; in Russia from the Ilmen Mtns., Urals; at Mursinka, Alabashka, Sherlova Gora, and other pegmatite fields; at Kingsgate mines, NSW, Australia; Alto Ligonha district, Mozambique; fine crystals from several localities in Madagascar. In Brazil many pegmatites produce fine smoky quartz, especially in MG: Arassuai, Itinga, Coronel Murta, Barra de Salinas, Galileia, São Geraldo do Baixio, Divino da Laranjeiras, Conselheiro Pena; and in ES: Baixo Guandu, Mimosa do Sul, and Fundão. Also from Conselheiro Pena, Corrego do Urucum, Corrego Frio, Cruzeiro, Arussuai area, Campo do Boa, Lavra da Ilha, near Taquaral, the Morro Velho gold mine at Nova Lima; Korea). *Rose quartz:* Discrete crystals of rose quartz are rare but have been found at a number of world localities: in ME from Dunton quarry (rough crystals to 1 cm), Nevel Quarry and Rose Quartz Crystal area, Newry, Oxford Co., also from the Red Hill quarry, Rumford; massive rose quartz (many tons) occurs at the Bumpus quarry and Scribner's Ledge, near Albany; also at the Whispering Pines quarry, Paris, Oxford Co.; rose quartz masses to 25 cm suitable for gem purposes and of the deepest rose red found anywhere at the Hibbs quarry, Hebron, Oxford Co. (including Antsirabé, Madagascar; the Sapucaia pegmatite, Conselheiro Peña, Lavra da Ilha on the Rio Jequitonhonha near Itinga, Taquaral, Pitorra quarry, Linopolis and several localities in Galileia; all in Minas Gerais, Brazil) in NH at the Charles Davis quarry, North Groton; also at Acworth; Williamsburg, MA; in pegmatite at La Grange, Troup Co., GA; the Scott Rose Quartz Quarry, French Creek, Custer Co., SD is one of the largest deposits of rose quartz in North America (reserves have been estimated at 600,000 tons); Rabenstein, near Zwiesel, at Hühnerkobel near Bodenmais, and at Schwarzenberg, in Bavaria, Germany; top quality rose quartz with saturated color occurs in beryl-bearing pegmatites near Sahanivotry, and Samiresy, Madagascar; in large amounts from Rössing, Swakopmund, Namibia; reported from Warangal, Hyderabad, and Chindwara, India. In Brazil, large amounts are produced at Alto Feio, Rio Grande do Norte, PB; Riberao do Largo, 45 km from Vitoria da Conquista, and Macarani in BA. Brazil has at least four major localities for crystalized rose quartz, the least of which is superior to any other known locality worldwide. All are in MG: Sapucaia do Norte, Municipo Divino de Laranjeiras; Piabanha and Pitorra, Mun. do Galileia; Lavra de Ilha (island in the Jequitinhonha River), Mun. do Itinga; and Coronel Murta. Other localities such as Taquaral and Sapucaia have also produced fine material. Superb, deep-colored massive material is also found at Joaima, MG (Jour. of Gemmology 22:273(1991)).

Amethyst: In ME very fine material is found in pegmatite at the Intergalactic pit, Deer Hill, Stow, Oxford Co., ME; very fine material occurs at the Warren farm, Pleasant Mountain, Denmark, Plumbago quarry, Sweden, Oxford Co., ME; Green's ledge, Milan Twp., and Diamond ledge, Stark Twp., Coos Co., NH; Hopkinton, RI; from PA at Media, Upper Providence Twp., and other localities in Delaware Co., and also from Chester Co.; in Va. from Amherst and Nelson Cos.; in NC at localities in Alexander, Iredell and Macon Cos. and at Iron Station, Lincoln County; Keweenaw peninsula, MI; Pohndorf pegmatite, 19-mile district, Jefferson Co., MT; Yellowstone Park, WY; in Colorado, at the Crystal Hill mine near La Garita, Saguache Co; massive amethyst occurs in the Creede district, Mineral Co., also at Red Feather Lakes area (Rainbow Lode mine or Pennoyer amethyst mine), Larimer Co.; Amethyst Queen claim, Unaweep Canyon, south of Grand Junction, in Mesa Co. Amethyst crystals line vugs in quartz veins and other voids in a brecciated zone in the Four Peaks area, Mazatzal Mountains, Maricopa Co., AZ; about

45 mi. NE of Phoenix (found about 1900); the Four Peaks amethyst is unique because of having well-developed "c" pinacoid faces; also heat-treatment turns the crystals green rather than citrine colored; amethyst and smoky quartz occurs at Peterson Mtn., Hallelujah Junction, NV–CA border. Some of the most spectacular and valuable amethyst specimens (sharp, lustrous scepters to 5-6 cm long) ever found in North America have come from Denny Mountain, King County, SE of Seattle, WA; in Canada at Thunder Bay, Ontario one of the largest amethyst occurrences in the world (associated minerals include fluorite, calcite, barite and minor amounts of the common sulfide minerals). Amethyst at Thunder Bay was mentioned as far back as 1846. Amethyst also occurs at the Dzuba mine at pass Lake-Elbow Lake, Ontario; Bay of Fundy, Nova Scotia; in Mexico at Mina la Veronica, Piedra Parada, Las Vigas, Ver.; Guanajuato, Gto.; royal purple crystals to 40 cm long at Amatitlan, Guerro; unlike most amethyst occurrences, the color is usually concentrated near the base, and the crystals often show a Dauphiné habit; deposits are found in France in the Auvergne, around Puy-de-Dome, near Vernet-la-Varenne, Condat, Chateauneauf, Champagnat, and Escout, and at Bansat, near Lamontgie; Teiser Kugele (geodes) occur in the Italian Tyrol near Teis (Tiso) and at Zillertal, in the Austrial Tyrol; at Schemnitz, Slovakia; and at Pokura near Nagy-Ag in Transylvania, Romania. Porkura, Hungary; ocurrences in Russia include: in pegmatites at Mursinka, middle Ural Mtns., Chassawarka in the southern Polar urals; Sanarka River area, southern Urals; Nerchinsk (now Chita) district, Transbaikal region and in Yakutia; at the Onyang mine, Onyang-Myon, Ulson-Gun, Korea; Madras, India; also from Sri Lanka. In Madagascar (particularly in pegmatites) at such localities as Vatomandry, Mahasolo, Broiziny, Tsaratanana, Soavinandriana, and Ambatofinahdrahana; large geodes (to two feet in diameter) lined with fine dark crystals that can be "burnt" to a fine "Madeira" citrine color occur on the Skeleton Coast, 65 km from Cape Fria, Namibia; spectacular material with clear, smoky and amethyst zones occur at Brandberg, about 160 km north of Swakopmund in NW Namibia; the most important amethyst deposit in Africia was discovered in 1956 in southern Zambia in the Kalomo district where it occurs in hydrothermal veins in granulite gneisses. The amethyst occurs in large crystals up to 1 foot long with zones of dark color from which stones called "Siberian" can be cut; amethyst localities in Australia include: Corona, near Broken Hill, NSW; Culladilla, near Mt. Isa, Q; the area of Beechworth in northeast VIC, 40 km southwest of Wodonga; the most important amethyst mine in Australia is at Wyloo Station in WA, about 985 miles north of Perth; In RS, Brazil, amethys occurs around Ametista do Sul (formerly Planalto); Iraí and Frederico Westphalen are nearby districts that produce smaller quantities, as well as from the San Eugenio district, Dept. Artigas, Uruguay. The amethyst occurs in geodes which can measure up to 3 m in length and weigh 1800 kgs. Individual crystals are up to 10 cm in length and the druses may be thinly studded with white to yellow calcite crystals. This is by far the world's largest source of amethyst geodes, both in quantity and quality; the total production in 1994 was in excess of 500

75.1.3.1 Quartz

tons. Amethyst is also found as veins in quartzites and arenites at Maraba and Pau d'Arco, PA; at the former, very large crystals up to 40 × 40 × 80 cm have been produced; the latter produces the best cutting-grade amethyst in Brazil. Other Brazilian localities are Brejinho das Ametisitas and Jacobina, Bahia; Santa Quiteria, CE; and as late-stage crystallization products around the quartz cores of pegmatites at Ataleia and Fernandes Tourinho, MG; and Fundão and Mimoso do Sul, ES where fine bicolor "scepters" over milky quartz are found. Amethyst–citrine combination (ametrine) occurs along faults in dolomitic limestone at the Anahí mine, in the Irma Bush Province of Santa Corazon Canton in the Dept. of Santa Cruz, eastern Bolivia, near the Brazilian border. Since 1989 more than 100 metric tons of rough crystals have been produced; most crystals are highly etched (G&G 30:4(1994)). Occurs in Brazil as geodes and cavity fillings in intermediate to acid flows in flood basalt provinces e.g., Parana Basin, RGS; at Brejinho das Ametistas, southwestern BA; also occurs as veins in quartzites and quartz arenites e.g. Maraba, near Alto Bonita, PB; Jacobina, Vitoria da Conquista, BA, and as a late-stage crystallization product around the quartz cores of pegmatites, e.g., MG and ES. *Citrine:* Natural citrine is rarer than amethyst and smoky quartz. It occurs chiefly at localities for amethyst, and principally as zones of citrine in amethyst: Isle of Arran, Scotland; Salamanca Prov., Spain; Dauphiné, France; Mursinka, Urals, Russia; Hyderabad, India; Madagascar; Anahí mine, Bolivia; excellent natural dark orange-brown citrine crystals were found at the Morro do Cristal, Campo Belo, MG Brazil, and at the Morro Redondo mine, Coronel Murta, large quantities of excellent dark citrine have been found since 1992, mostly broken pieces rather than well-formed crystals. *Blue quartz:* blue quartz is found in the Milford granite, Rhode Island; Cornog, PA; Nelson, Amherst, and Stafford Counties, VA; Llano, Llano Co., TX; and other world-wide localities. *Japan law twins:* rare, but fine-quality twins occur in the Hot Springs district, AR; at Capitan Mountain, Otero Co., NM; King County, WA; in AZ at Hamburg Cochise Co., and at the Holland mine, Washington Camp, Santa Cruz Co., which has produced thousands of twins, some of which may reach the size of 30 cm from tip to tip; also from the Cooper and Green Monster mines on Prince of Wales Island, AK; In France, fine twinned crystals from the La Gardette mine, near Bourg d'Oisans, Isère (Dauphiné), was the first locality noted for Japan-law twins which were known as La Gardette twins in 19th century literature; from Val Bedretto, Ticino Canton, Switzerland; in cavities in quartz porphyry at Saubach, Saxony; a Japan twin 26 cm high and 6 kg in weight was found at Gudzhivass in the Pamir Mountains, Tadzhikistan. The most famous locality for these twins is at Otomizaka, Yamanashi Prefecture, where beautiful twins up to half a meter across occur in granite pegmatite and quartz pegmatite. Also at Narushima Island, Shimo-goto, Nagasaki Pref.; from the Siglo XX mine at Llallagua, Dept. Oruro, Bolivia, in sulfide veins; some quite large twins and often showing virtually no re-entrant angle are found at Brumado, Bahia State, Brazil. A perfect quartz twin on $\{30\bar{3}2\}$ was found at the Ariranha mine, Teofilo Otoni, MG, Brazil. *Quartz with inclusions:* The finest locality for rutilated quartz is Novo Horizonte (formerly called Ibitiara), BA, Brazil. Here the rutile is a bright golden yellow color, sometimes radiating outward epitaxially in a six-rayed star from a crystal of hematite embedded in the quartz. Most rutile in quartz is brownish-red to red, and the gold color from this locality is due to an abnormally low iron content (around 0.5 wt. %); a more normal content would be at least double that. The needles may be curved, suggesting that the rutile was in place before the quartz began to crystallize. Another prolific producer of rutilated quartz is Oliveira dos Brejinhos, BA. Other minerals included in Brazilian quartz are anatase, schorl, various amphiboles, pyrite, and others. Anatase in quartz crystals is found at Diamantina and also at Novo Horizonte, BA; it can be caramel to brown colored, sometimes with delicate thin rutile, a dramatic combination. Well-formed, bright pyrite crystals in clear quartz crystals come from Serra do Cabral, MG; the crystals are usually pyritohedrons, and reach a size of 2 cm. Quartz with schorl needles is abundant at Jaguaraçū, MG and was mined there for decorative uses; minasgeraisite and milarite were found in the same deposit. *Quartz phantoms:* Blue "phantoms" quartz, from Morro do Serrote, Regisitro, SP is caused by a light coating of fine-grained riebeckite growing on the crystal termination at various stages of growth. White (etched or frosted surface), grey (sandy inclusions), green (chlorite), or reddish-brown (ferruginous matter) phantoms are found at Diamantina, also at Serra do Cipo, MG. The phantoms may be repeated several times, revealing successive stages in the growth of the crystal. Fantastic emerald-green phantoms come from Pakistan. *Shocked quartz:* Shocked quartz from the Cretaceous-Tertiary boundary occurs at numerous world-wide sites (e.g. Brownie-Butte, MT; Haiti, Stevns Klint, Denmark; Gubbio, and Pontedazzo, Italy; Caravaca, Spain; Woodside Creek, NZ ((S 224:867(1984), S 236:705(1987)). Such quartz also exists in the fallout ejecta and in impactite materials from meteorite impact structures (e.g. Ries crater, Germany). *Cryptocrystalline varieties:* Novaculite (used for sharpening stones), in the form of metaquartzite, occurs in the Novaculite mountains, Hot Springs Co., AR. Occurrences of carnelian, agate, chalcedony, jasper, and other cryptocrystalline forms of quartz are exceedingly numerous world-wide. Chrysocolla-stained chalcedony occurs in the Globe district, Gila Co., AZ and is especially prized because of its vibrant blue color. "Sweetwater agates" characterized by fine black dendrites occur scattered over the surface in an area north of the Sweetwater River in Fremont Co., WY. "Montana agate" occurs in gravel bars and river terraces along the course of the Yellowstone River in the central and eastern part of the state. The original source has never been located. Fine quality blue chalcedony comes from near Mojave, Mojave Co., CA; The Priday Ranch of the Richardson Recreational Ranch, near Prineville, central Oregon, is one of the best known localities for "thundereggs" (agate fillings of lithophysae in air-fall welded rhyolitic tuffs, or vitrophyres). Fine fire agate occurs at various localities in San Luis Potosi and north of Guadalajara, Jal., Mexico. Unusual chalcedony stalactites are found near Camborne, Cornwall, England. Colorful banded agate occurs near Fofarshire and Perthshire, Scotland; stalactites from amygdaloidal cavities are found in the Faeroe Islands. Fine carnelians and agates are found in India. Agate and carnelian were formerly found at and around Idar-Oberstein, Rhineland, Germany; the cutting and polishing of these stones started at least 700 years ago and resulted in the founding of this large community almost exclusively devoted to the lapidary arts. A famous locality for jasper for lapidary purposes is at Orsk, Ural Mountains, Russia; so-called Botswana agate, which is beautifully banded comes from unspecified localitites in Botswana; fine ($2.00/gram wholesale) blue chalcedony, known since the 1930's, comes from Namibia; important agate and jasper deposits are associated with the Deccan basalt flows in India; chalcedony and agate has been known for hundreds of years from near San'a, Yemen; fine agate and chalcedony occurs at Agate Creek, on the north slope of the Gregory Range, in northern Q, 350 miles inland from Cairns, Australia, "mookite", a colorful type of jasper occurs at Mooka Station, WA; in Brazil, enormously prolific agate deposits are found in the Parana Basin, RS, particularly on the outsksirts of Salto Grande on the banks of the Jacui River. The basalt flows from which these chalcedony and agate nodules are derived extend for 1200 km, from SP to PR, SC and RS states and are up to 1000 m thick. Most of the agate/chalcedony comes from the Salto do Jacui district, RS. In 1994, the production totaled about 2000 mt.; this is the world's main agate producer. In addition, fine jasper and bloodstone are produced in the district. Agate and carnelian are found in the provinces of Corrientes and Entre Rios, Argentina; Fine chrysoprase occurred at Exeter, Porterville, Visalia and Lindsay, Tulare Co., CA; deposits of fine chrysoprase occur near Frankenstein, Silesia, Poland, but they are now nearly exhausted. What many consider to be the world's best chrysoprase occurs at Marlbrough, a small town some 100 km northwest of Rockhampton, Queensland, Australia. "Tigereye", a silica replacement of "crocidolite" asbestos, occurs in several world-famous localities including: the area of Griquatown in the Doornbergen (Asbestos Mountains) north of the Orange River, South Africa, is the most important occurrence of yellow tigereye and hawkeye (colored blue due to the less altered crocidolite); mostly yellow-brown tigereye also occurs in iron ore deposits, near Wittenoom Gorge, WA, Australia. Petrifactions (replacements of organic material) of quartz are also numerous world-wide. Petrified wood replaced by silica is abundant. More famous localities include: perhaps the world's best known petrified forest is preserved in the Arizona Petrified Forest National Monument which is located northeast of Holbrook and straddles Navajo and Apache Counties, AZ; fossil dinosaur bone from numerous Mesozoic deposits in the western USA is replaced by silica; petrified forests are preserved in Yellowstone Nat. Park, Wyoming; at Hilbersdorf, a suburb of Chemnitz, Germany; another petrified forest occurs on the island of Mytilene in the northeastern part of the Aegean Sea, Greece; a petrified forest is preserved in the Cerro Cuadrado Petrified Forest, Neuquen Prov., Patagonia, Argentina. **Uses:** Quartz in its many forms, is one of the most important of all mineral commodities on a world-wide basis, and is exceeded in tonnage and value by only a few others such as iron ore, coal, or limestone. The principal uses in the U.S.

are, in decreasing order of value, the glass industry; foundry sand; abrasive sand; and hydraulic fracturing sand (used in the petroleum industry), with a total value of about $400,000,000. The manufacturing of silicon metal, ferrosilicon, and silicon alloys also consumes about a million tons/year of high purity quartz. This type of quartz, usually >99.6% SiO_2, is found in the quartz cores of pegmatites, pure quartzites, and other related geological formations. Even larger quantities of less pure quartz, >90% SiO_2, is used as a flux in making steel and in other metallurgical processes.

The first synthetic crystals of quartz of appreciable size and quality were produced in 1905. Crystals are grown hydrothermally, using as the raw material "lasca" (fragments of highly pure quartz crystals). These artificial crystals are preferred because they are free from twinning, to which most natural crystals are subject; twinning renders the crystals unsuitable for piezoelectric applications. The piezoelectric property of quartz has resulted in its use for oscillator plates for controlling and stabilizing the frequency of electronic oscillator circuits, e.g. for radio transmission or reception, or for time keeping (mostly satisfied by synthetic material).

Of all of the minerals used as gems, none compares with quartz in its diversity of occurrence and in abundance of varieties. There are two groups: coarsely crystalline varieties (which many refer to as quartz) and fine-grained, microcrystalline to crytocrystalline varieties (chalcedony, agate, etc.). EF *D6:183(1892)*, *DIII(1962)*, *MIN II(2):143(1965)*.

75.1.4.1 Coesite SiO_2

Named in 1954 by Sosman for Loring Coes, Jr. (1915–1973), American chemist, Norton Company, who first synthesized the phase. Polymorphous with quartz; tridymite, cristobalite, and stishovite. MON pseudo-HEX $C2/c$. $a = 7.135$, $b = 12.372$, $c = 7.174$, $\beta = 120.36°$, $Z = 16$, $D = 2.92$ (also reported as dimensionally hexagonal with $a = c$ and $\beta = 120°$). *14-654:* 3.44_7 3.10_{10} 2.77_5 2.71_7 2.30_5 2.19_5 2.04_7 1.717_7. Structure: *ZK 167:287(1984)*. The structure is dominated by four-tetrahedron rings linked into a three-dimensional tetrahedral framework, distinct from other tetrahedral silica structures but related to the framework of feldspars. Irregular grains 5–50 μm in diameter. Synthetic material forms minute, lathlike {010}, dominant crystals with a morphology reminiscent of gypsum. Twinning on {001} and {021}. Colorless to white, vitreous luster, transparent to translucent. Subconchoidal fracture. $H = 7$–8. $G = 3.01$. Biaxial (+); $2V = 54$–$64°$; $N_x = 1.594$, $N_y = 1.595$, $N_z = 1.599$; $X = b$; $r < v$, weak; in aggregate may appear isotropic with $N \approx 1.595$. Chemical analyses indicate >99 wt% SiO_2. Formed at elevated temperature (500–800°C and ≈ 30–35 kbar). Pseudomorphs of quartz after coesite are commonly encountered. Coesite forms in high-pressure, high-temperature environments such as shocked materials of impact craters, crustal eclogite facies, and possibly high-grade blueschist facies metamorphics. Found in fossil impact craters at Sinking Springs, OH, and at Kentland, Newton Co., IN; as a fine-grained, nearly isotropic groundmass in shocked Permian Coco-

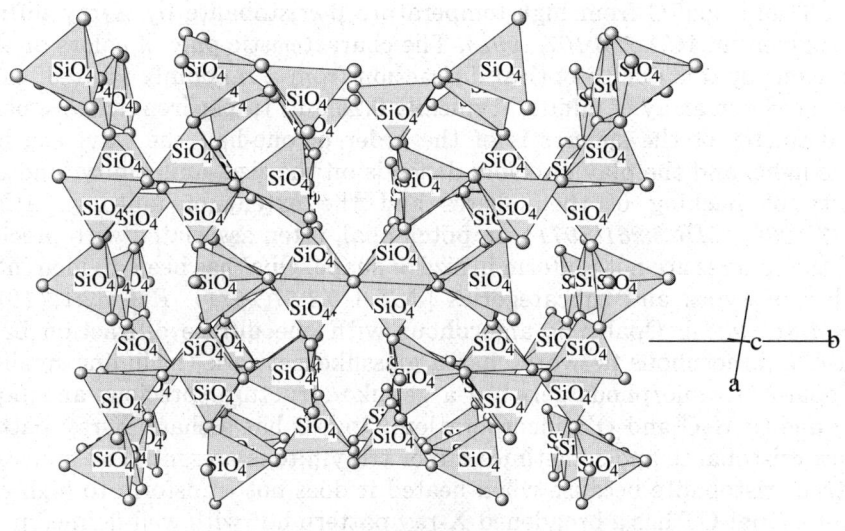

COESITE
Coesite: Perspective view approximately along c.

nino sandstone at *Meteor Crater, Coconino Co., AZ*; as inclusions in clinopyroxenes at Grytting, Norway; in stressed granite at the Rieskessel meteorite crater, Bavaria, Germany; in eclogite in the western Alps, Italy; at the Wabar meteorite crater, Al Hadida, Rub'-al-Khali, Saudi Arabia. Found as inclusions in zircon in diamond-bearing gneiss of the Kokchatov massif, Kazakhstan, and in diamond from the Anabar alluvial deposits, Yakutia, Russia; in eclogite near Xinxian, Henan, and Dhonghai Co., Jiangsu, China; as inclusions in omphacite in an impure marble at the northeastern edge of the Gourma Basin, Mali; and in the Roberts Victor kimberlite, Cape Prov., South Africa. AR *D73:310(1962), CMP 86:107(1984), MM 54:579(1990)*.

75.2.1.1 Opal $SiO_2 \cdot nH_2O$

Known as a gemstone since antiquity. The oldest known source of the word *opal* is from the Sanskrit *upala*, stone or precious stone. The Romans used the word *opalus* and the word *opal* is clearly derived from this. Amorphous (some opal) or more often disordered crystalline, for most common opal, possessing the structure (as disordered interlayers) of low-temperature cristobalite and/or low-temperature tridymite. Opals consist of structural units of amorphous SiO_2 and crystalline cristobalite (and/or tridymite). Gem opals are the least crystalline type (opal-AG) of opal. Precious opal (especially that from sedimentary deposits) is a mixture of amorphous and crystalline silica (low tridymite), with the former being dominant. The extent of crystallinity varies between samples from different localities. *38-448*: 4.08_{10} 3.14_1 2.86_1 2.51_3 $2.13_{0.5}$ $2.03_{0.5}$ $1.94_{0.5}$ $1.88_{0.5}$. ICDD card has schematic patterns for the three main types of opal: CT (cristobalite-tridymite), C (cristobalite), and A (amorphous). Information regarding the distinction between well-ordered opal-CT and opal-C from high-temperature β-cristobalite by X-ray diffraction is given in *ACA 286:107(1994)*. The characteristic play of colors present in precious opal is due to optical diffraction from a randomly faulted, close-packed ordered array of minute (typically 0.25 μm) transparent silica spheres. The diameter of the spheres is of the order of one-half the wavelength of visible light, and the play of colors depends on the size, uniformity, and uniformity of packing of the spheres and the extent of faulting. *AC(A) 24:427(1968)*, *MR 2:261(1971)*. In potch opal, often associated with precious opal, the spheres are not uniform in size or shape. Opal has been divided into a number of types and subcategories [*JGSA 18:57(1971)*, *FM 52:17(1974)*, *NJBM:81(1990)*]: Opal-A is amorphous with one diffuse diffraction band. Opal-AN (amorphous network) has a glasslike network (including hyalite), and opal-AG (amorphous gel) has a gel-like structure (precious and potch opal) due to H_2O and OH incorporation. Opal-C has a sharp X-ray pattern for low cristobalite and sometimes minor tridymite is present. The mineral is not true cristobalite because when heated it does not transform to high cristobalite. Opal-CT has a broadened X-ray pattern but with well-defined peaks for low cristobalite. Variable degrees of stacking disorder are present, leading to maxima attributable to low tridymite. Opal-CT typically occurs as lepispheres and also as fibers. Opal-CT includes common opal and some precious opal. Opal-CT has been further subdivided by some investigators, and a simple classification seems preferable: opal-CT_M, referred to as massy opal, shows no texture resolved with a polarizing microscope. Opal-CT_P is platy in habit. Opal-CT_L (L = lussatite) is fibrous with positive elongation, fiber axis [101] using indices of tetragonal low cristobalite. It can vary from opal-C_L to opal-CT. Opal-CT has also been interpreted as showing long-range ordering of oxygen atoms followed by short-range ordering of Si atoms. *AM 72:1195(1987)*. Opal-CT has a ^{29}Si MAS NMR spectrum much more like that of amorphous SiO_2 than that of cristobalite or tridymite. In volcanic rocks, one gets opal-AN or opal-CT_L. Opal-C occurs as horizontal layers on the bottom part of agate nodules and geodes of the Uruguay type, associated with chalcedony. Opal-C_M and opal-CT_M are massive in texture. All opal-CT will have $\bar{N}$ = 1.43–1.46. Opal C_L and opal-CT_L are birefringent. Massive and platy varieties are isotropic or nearly so. Some opal in fossil wood is all high tridymite (± quartz) or more rarely low cristobalite, along with variable water content. Chalcedony, on the other hand, is a fine-grained, fibrous variety of quartz with little or no water content. It has a negative optical character of fiber elongation (length-fast). The fiber axis is one of the trigonal *a* axes. The quartz fibers are composed of submicroscopic {10$\bar{1}$1} lamellae, polysynthetically twinned left- and right-handed according to the Brazil law. It may have 1–2 wt % H_2O. The water is present in open porosity on crystal surfaces and in closed micropores on inner surfaces. Water is also present in regions of accumulated defects. Quartzine is a length-slow chalcedony and has positive elongation, with the fiber axis being *c*. Opal-C may occur as horizontal layers in the lower part of agate nodules and geodes of the Uruguay type, associated with chalcedony. It can also occur in cavities in volcanic rocks, possibly formed by later recrystallization by further heating. Opal-CT is widespread and may be found as accretions in both volcanic and sedimentary rocks. It is also common in some earthy materials such as some deep-sea sediments and some forms of opal claystone, tripoli, and opoka (Russian). Opal-A is widespread in deep-sea deposits as the remains of diatoms and radiolaria, and on land as large deposits such as diatomite. It also occurs as potch opal, with which may be associated gem-quality material. Both opal-A and opal-CT may be found replacing wood, with remarkably detailed preservation of the cellular structures. Opal occurs as coatings, stalactites, nodules, concretions, crusts, pseudomorphs, after other minerals, such as ikaite ("opal pineapples"), gypsum, calcite, aragonite, apatite, siderite, and others. It occurs as botryoidal, reniform and globular masses, and various earthy aggregates or massive material. Numerous names have been given to opaline materials of differing morphology: geyserite and siliceous sinter—opal associated with volcanic activity and hot springs, forming terraces, basins, and cones; (diatomite) diatomaceous earth—earthy to soft deposits made up of the remains of tiny organisms such as diatoms, tripolite—a variety of diatomaceous earth, usually opal-A; opoka, opal claystone—earthy material often showing little evidence of biogenic remains; commonly opal-CT; fire opal—vitreous, transparent to semitransparent with yellow, orange, red, or brown body color, with or without play of color, sometimes cut as a gemstone; precious opal (= noble opal)—opal showing play of color caused by optical diffraction from the regular packing of silica spheres, with white and black subtypes being distinguished based on the body color; girasol opal—transparent opal showing some play of color, relatively transparent, with a rather uniform bluish or reddish floating or wavy type of internal light; common—a general term for any type of vitreous opal showing no play of color and is not associated with precious opal; hyalite (= water opal)—colorless and transparent opal that looks like congealed droplets of glass. Much of this type of opal is believed to have been deposited from an aqueous vapor phase ($> 100°$). Hyalite is thus genetically distinct from most opal. Regular opal is deposited from aqueous solutions or as a colloid (gel); moss—common opal patterned with dendritic growths of metal oxides; milk—translucent to opaque common opal with a milk white, pale bluish white, or greenish white color; wood opal—wood replaced by opal, with or without preservation of the woody structure; plant opal—opal particles formed in many types of plant growth, especially in grasses and eucalypts; cacholong—chalky white and completely opaque opal; hydrophane—a type of cacholong that has an affinity for water. Turns from opaque to highly transparent when placed in water; tabasheer—a milky white opaline silica deposited in the joints of bamboo plants; potch opal—a term applied to nonprecious opal, usually gray to black in color, associated with gem material, especially in the Australian opal fields. Other terms, such as cherry opal, pinfire opal, harlequin or mosaic opal, contra-luz opal, boulder opal, and jelly or water opal, are used to further describe precious opal. **Physical properties**: Opal is generally white or colorless but may also be just about any other color: yellow, red, orange, green, brown, blue, black, and others. They also show multicolored opalescence or pearly luster in the same sample. Opal ranges from perfectly transparent to nearly opaque. Luster ranges from vitreous, to dull, waxy, or greasy. Cleavage is absent. Fracture is conchoidal, uneven, irregular to splintery. Brittle to very brittle. H = $5\frac{1}{2}$–$6\frac{1}{2}$. G = 1.9–2.3 (depends partly on water content). Often luminescent and colors include white, yellow, yellow-green, green, or light blue. Fluorescent yellow to yellow-green due to presence of uranyl ions, especially hyalite opal. Black opal is inert to LWUV and SWUV. White opal may phosphoresce. Fire opal may be inert or may fluoresce a moderate greenish brown to LWUV and SWUV, and may also phosphoresce. IR data: *FM 52:17(1974)*. DTA and TGA data (precious opal): *MM 35:429(1965)*. Thermoluminescence data on synthetic and natural opals: *AM 63:737(1978)*. XAS studies: *AM 79:622(1994)*. Opal is both heat- and humidity-sensitive. Soluble in HF and KOH. Sometimes attacked by HCl. **Chemistry:** The presence of >1 wt % H_2O in opal is considered essential for the mineral. *JGSA 18:57(1971)*. Most opal has 3–9 wt % H_2O but it may have as much as 15–20 wt %, there being no clear distinction between opal-A and opal-CT on this basis. Opal-AN, however, usually contains around 3 wt % H_2O, while opal-AG contains 5–8 wt %, but sometimes contains much less. The water is present as (1) SiOH and as (2) H_2O (molecular). Opal-CT contains (1) 0.2–0.4 wt % and (2) widely varying amounts. Two types of molecular water are present: single, almost non-hydrogen-bonded molecules in small 0.35-nm cages of the SiO_2 matrix, and liquidlike, H-bonded water as films on inner surfaces. For additional details, see *FM 52:17(1974)* and for opal-C and chalcedony, *PCM 12:300(1985)*. At least five types of water occur in opal-A (Monterey diatomite). *AM 67:510(1982)*. Some opal contains trace to minor amounts of Fe, Al, Mg, Ca, alkalis, and sometimes Ni (prase opal). A wide variety of rare to common elements occur in variable amounts. As Fe can be precipitated at a low pH, it is sometimes present in significant amounts (up to 23% Fe_2O_3 reported). Unlike for cristobalite and tridymite, opal with high amounts of cristobalite is "dirty" (e.g., 0.25–3.5 wt % Al) and opal with tridymite is "clean" with 0.008–1.5 wt % Al. *AM 58:717(1973)*. **Optics:** Generally isotropic or with weak birefringence. Wood opal has a range of refractive indices between 1.410 and 1.453. The higher the water content, the lower the indices of refraction. All opal-CT has refractive indices between 1.43 and 1.46, with most being <1.44. Opal-AG and opal-AN usually have refractive indices between 1.45 and 1.46. **Occurrence:** Most opal is formed in a variety of low-temperature environments, generally in near-surface deposits. Precious opal and the associated potch (opal-AG) are deposited by colloidal processes in an aqueous environment. Geyserite (siliceous sinter) may be formed by deposition (in hot springs) from hot-water solutions containing 100-200 ppm SiO_2 in true solution. Opal-AG can be stable for long periods; unaltered sponge spicules, for example, have been found in Eocene opal claystone. Opal-AN in the form of hyalite is found mainly in volcanic rocks, at least in part having been formed by vapor-phase deposition. It occurs in sedimentary (e.g., siliceous shales) and diagenetic environments. It sometimes replaces organic matter and occurs as a fossil. Common in volcanic rocks, filling cavities and

75.2.1.1 Opal

fractures. The main constituent of diatomite. Most is deposited by Si-bearing waters at low temperatures (<100°) that evaporate. SiO_2 concentrations may be no greater than 100-200 ppm. The diagenetic alteration process in sediments generally involves opal-A → opal-CT (porcellanite) → chalcedony or cryptocrystalline quartz (chert). Under conditions of unrestricted growth, opal-CT usually forms spherical aggregates (lepispheres) consisting of opal-CT blades intergrown according to twinning laws of tridymite [i.e., $(30\bar{3}4)$ and $(10\bar{1}6)$]. Precious opal in Australia is chiefly in sedimentary host rocks, in the Andamooka, Coober Pedy, and Lightning Ridge fields. Elsewhere in the world, precious opal mostly occurs with volcanic rocks (e.g., Mexico). **Localities**: Opal, of varying types and quality, occurs worldwide. Only localities that have produced or are producing fine-quality, noteworthy, or otherwise unusual opal are mentioned here. Hyalite is present in Mitchell Co., NC; fine-quality precious opal in sandstone has recently been found at Cedar Grove, Houlton Parish, and near the Toledo Bend Reservoir, Sabine Parish, LA; opaline sinters occur in hot springs in Yellowstone National Park, WY; good-quality common opal and hyalite occur at Mt. Blackmore and Mt. Hyalite, Gallatin Co., MT; opaline sinters also occur in hot springs at Steamboat Springs, NV. Perhaps the most famous U.S. locality for precious opal is that of Virgin Valley, Humboldt Co., NV. The Rainbow Ridge, Royal Peacock, and Bonanza mines have produced fine black opals, including the 3.2-kg Hodson opal, the 3-kg Bonanza opal, and the 2402-ct Roebling black opal, all of which are opalized limb sections. One 130-lb log estimated to be > 30% precious opal was found at the Northern Lights mine. Most material from Virgin Valley is unstable and tends to crack and dehydrate. However, the Roebling opal has shown no degradation in more than 75 years. Milk opal occurs east of Spencer, ID. Hyalite is present in the Columbia R. basalts in the northwestern United States, and hyalite and other varieties of opal occur at Opal Butte, Morow Co., OR. Fine wood-opal occurs along Clover Creek, Bliss, Lincoln Co., ID, in the form of oak logs with well-preserved textures. Excellent wood opal also occurs in the Latah formation, in south-central Washington. Opal-CT occurs in the porcellanites of the Monterey Shale in California. Diatomites contain opal-A. Some fire opal occurs at the Okanagan opal mine, near Vernon, BC, Canada. Some of the world's best fire-opal and cherry-opal occurs near San Juan del Rio, QUE, Mexico. Good material also occurs near Magdalena, JAL, and Zimapan, HID, Mexico. Several areas in Honduras are producing good-quality opals. Before discovery of the Australian opal fields, the best known localities for precious opals were near Cervenica and Dubnik, now in Slovakia (formerly Hungary) on Mts. Simonka and Libanka in the Carpathian Mts. These localities have been known since the time of the Romans. Mining was discontinued in 1923, due to the competition from Australia. Fine hyalite occurs at Waltsch, Bohemia, and Puy de Dôme, France. Fire opal occurs at Aleksevskoye in Kazakhstan. Fine hyalite occurs at Tateyama and Sankyo, Japan. Other notable occurrences of opaline sinters are in Iceland and at Wairaki, New Zealand. A relatively little known but potentially significant brown fire opal occurs in Ethiopia. There are records of precious opal being found in and exported from India in ancient times, but the localities are not now known. Brazilian agate consists of chalcedony and opal-C. The world's finest precious opals occur in Australia: Mintabie, Coober Pedy, and Andamooka fields, South Australia; White Cliffs, and Lightning Ridge fields in New South Wales. The White Cliffs fields produce opal "pineapples" (to 700 g, 4.5 in. × 3.5 in.), which are pseudomorphs after ikaite(?). Lesser production comes from areas in Queensland, the best of which are the Opalton, Eromanga, Quilpie, and Yowah areas. In most areas the opal occurs in sandy clays of Cretaceous age. Precious opal also occurs in altered volcanic rocks in several areas in eastern NSW and Q but has never been commercial. Some 90% of the world's production of precious opal comes from the Australian opal fields, where it occurs as thin bands, nodules ("nobbies") or in ironstone concretions. Since 1960, the estimated total value of opal mined in Australia is about 2.2 billion (1988 Australian dollars). Uses: The chief use of opal, since antiquity, is as a gemstone. Usually cut *en cabachon*, but thin layers may also be used as doublets and triplets; larger pieces may be carved. Australia is by far the world's largest producer. For the past 20 years it has produced just about all of the world's precious opal. Coober Pedy is the single largest gem opal producer in the world. Mintabie is now ranked first. Australia currently accounts for 95% of the world's production of opal. The only other significant commercial sources at present are Mexico, where it occurs in volcanic rocks, and in Piaui State, Brazil, where although associated with volcanic rocks, it may have been formed by sedimentary processes; the opal is similar to that found in some of the Australian localities. Opal is also used in ceramics, chemical industries, and in insulating and acid-proof materials. Large quantities of opaline silica such as diatomite are used as filter insulating materials, as absorbents, and in refractory manufacture. More compact materials such as opal claystones are calcined for use as a grog for refractories and as pet litter. Small amounts of opal in aggregate can cause damage to concrete structures by reaction with alkali elements in the cement. Early examples of this were concrete roads in California and in the Parker Dam. Opal simulants (synthetics) are now known, such as Slocum stone, a glass product. The most convincing simulants are opalite and opal essence, a plastic material produced in Japan. Synthetic opal has been available commercially since the 1970s [*JG 14:215(1975)*] and was first marketed by Pierre Gilson, Saint Sulpice, Switzerland. Black opal and crystal opal types are sold by Inamori in Japan. EF *GSNA I:105(1959), II:49(1976); D7 III:287(1962); TOB(1961); LJ 47:36(1993)*.

75.2.2.1 Melanophlogite SiO_2 + (organic compounds)

Named in 1876 by Lasaulx from the Greek for *black* and *to be burned*, in allusion to the mineral blackening when heated in the blowpipe flame. TET (pseudo-ISO) $P4_2/nbc$. $a = 26.79$, $c = 13.395$ (pseudoisometric subcell with $a = 13.395$), $Z = 4$ ($Z = 128$ for SiO_2), $D = 2.06$. Structure: *SCI 148:232(1965)*. A three-dimensional, pentagonal dodecahedral framework of silica tetrahedra is proposed, the organic components within the voids being essential to maintaining the open framework. Crystals rarely as simple pseudocubes up to 4 mm on edge; more commonly as complex, rounded intergrowths forming crusts. Colorless, pale yellow to reddish brown; turbid white when altered; white streak; transparent to translucent. Brittle. H = $6\frac{1}{2}$–7. G = 2.005–2.052. Isotropic; N = 1.423–1.467, but occasionally weakly birefringent. The organic components are somewhat variable, 7–12 wt %. A formula based on Čechy material is given as $Si_{46}O_{92} \cdot C_2H_{17}O_5$; Sicilian material often contains significant sulfur. Insoluble in acids; blackens on heating. Found at Mount Hamilton, Santa Clara Co., CA; at Fortunillo, Livorno, Tuscany, and associated with sulfur at Caltanissett and at the *Giona mine, Racalmuto (Agrigento), Sicily, Italy*; as fine crystals on lussatite (fibrous cristobalite) in a vein cavity in a pyrite–rhodochrosite deposit at Chvaletice, Čechy (Bohemia), Czech Republic. AR *D7 3:283(1962); AM 48:854(1963), 57:779(1972).*

75.2.3.1 Lutecite SiO_2

Originally described in 1892 by Michel-Lévy and Munier-Chalmas. Synonyms: lutecine and moganite. The CNMMN IMA has not formally approved a name for this species. The name *moganite* was assigned in 1984, but *lutecite* has priority. Compare chalcedony (75.1.3.1). Polymorphous with the other forms of SiO_2: quartz, tridymite, cristobalite, and coesite. MON with many crystallites showing a TRIC superstructure. $I12/a1 = C2/c$. $a = 8.758$, $b = 4.876$, $c = 10.715$, $\beta = 90.09°$, Z = 12, D = 2.617. *38-360:* 4.45_1 3.39_5 3.33_{10} 3.11_1 2.29_1 1.83_1 1.38_1 1.37_1; also *DAN SSSR 320(2):428(1991)*: 4.26_5 3.33_{10} 2.44_4 2.27_5 1.80_8 1.54_4 1.37_6 1.18_5 [also *DAN ESS 320:169(1993)*]. Triclinic superstructure has: $a = 10.024$, $b = 9.752$, $c = 10.715$, $\alpha = 90.00°$, $\beta = 90.07°$, $\gamma = 119.11°$. The structural relationship to quartz is: $2a + b$, b, $2c$ (for a,b,c of quartz). The structure consists of corner-linked SiO_4 tetrahedra with six-, eight-, and four-membered rings. The tetrahedra are only slightly distorted. The structure can be thought of as consisting of alternating layers resembling (10$\bar{1}$1) slices through right- and left-handed quartz, which form a three-dimensional network of corner-sharing SiO_4 tetrahedra. The structural principle is thus a periodic twinning according to the Brazil law on a cell-dimension scale. **Habit:** The concentration of lutecite is dependent on depositional environment. Formation of the metastable polymorph is promoted by the substitution of $[Fe^{3+} + Na^+]$ for Si^{4+}. *IMAMA 16:170(1994)*. Occurs as microcrystalline fibrous silica fillings of cavities, fissures, and cooling cracks in ignimbrite flows (Gran Canaria); as weirdly shaped crusts, pockets, amygdules, and veinlets, in vesicles, in fractures, and in voids at intersection of joints in basalt (Gobi desert, Mongolia). Some pure lutecite occurs in masses 40 cm or more across. Often exhibit a typical and distinctive morphology: regular intergrowth of pseudohexagonal, strongly flattened individual crystals. Individual crystals have {10$\bar{1}$0} (of quartz), = {110} (monoclinic cell) dominant, lathlike, fibrous, and radiating plates. They are approximately 100

Å thick. Also in much agate and chalcedony throughout the world. *SCI 255:441(1992)*. Indistinguishable from chalcedony in hand specimens. **Physical properties:** Translucent gray to chalky white, cream. G = 2.52–2.58 (because of water that is present). NIRS and IR data: *EJM 4:693(1992)*. Raman spectra [*AM 79:269(1994)*] confirms that lutecite is a structurally distinct polymorph of SiO_2. **Optics:** Optically length-slow (rather than length-fast as for quartz), N = 1.524–1.531. Inclined extinction angle of 29–45°. **Chemistry:** Contains variable amounts of water (1.5–3.0 wt %) that is not a constituent of the structure. Also may contain traces of Mg, Mn, Ca, K, Na, and others. **Occurrence:** Occurs as veins, fillings, and crusts in rhyolitic ignimbrite flows, in Magadi-type chert (evaporite-type regimes), most microcrystalline quartz varieties contain some lutecite (agate, chert, flint, chalcedony, etc.). Formed at low temperature and pressure. Coatings on nodules may be >80% pure lutecite. **Localities:** *Gran Canaria, Canary Islands*; Lake Magadi, Kenya; Arts Bogdo range, NW Gobi desert, Mongolia. Fairly abundant and occurs worldwide. Most work has been done on material from Gran Canaria and Mongolia. EF *NJBA 149:325(1984), 163:19(1991)*.

75.2.4.1 Silhydrite $Si_3O_6 \cdot H_2O$

Named from the composition: a silica hydrate. ORTH *P* lattice. $a = 14.519$, $b = 18.80$, $c = 15.938$, Z = 28 (on three Si's) or 32 (on six O's), D = 2.116. *25-1332*(nat): 14.5_{10} 7.31_2 7.07_2 5.31_2 3.63_3 3.54_3 3.42_8 3.14_4. White; in masses, single pieces to 3–4 cm in longest dimension, but most are <5 mm across. Crystals are <4 µm long; earthy. Rough and subconchoidal fracture. H ≈ 1. G = 2.141. DTA and TGA data: *AM 57:1053(1972)*. Treatment with Na_2CO_3 produces a magadiitelike material; however, magadiite leached with dilute HCl yields a silica hydrate that differs from silhydrite. Analysis (wt %): SiO_2, 88.9; H_2O^+, 3.13; H_2O^-, 6.65; traces of Al, Fe, Mg, Ca, Na, and K. Biaxial; sign unknown; $N_{avg} = 1.466$. Occurs in a magadiite deposit as crystals remaining after Na has been leached from magadiite by spring water. Occurs east of *Trinity Center, Trinity Co., CA.* EF

75.3.1.1 Virgilite $Li_x[Al_xSi_{3-x}O_6]$

Named in 1978 by French et al. for Virgil Everett Barnes (b.1903), University of Texas, who pioneered the study of tektites and other natural glasses. Dimorphous with spodumene. HEX $P6_222$ or $P6_422$. $a = 5.132$, $c = 5.454$, Z = 1, D = 2.399. *31-707*: 4.44_1 3.44_{10} 2.57_1 2.24_2 1.870_1 1.605_1 1.429_1 1.408_1. Structure: *ZK 127:327(1968)*. The structure is a stuffed derivative of β-quartz, with Al–Si substitution disordered. Colorless, vitreous luster, transparent. H = $5\frac{1}{2}$–6. Uniaxial (−); $N_{av} = 1.520$; birefringence = 0.005. Al substitutes for Si such that $x = 0.5$–1.0. Found as hexagonal prisms and bipyramids, up to 50 µm, and as fibers on quartz associated with K-feldspar, biotite, and spinel in a stream cobble of volcanic glass at *Macusani, Peru.* AR *AM 63:461(1978)*.

Class 76

Al–Si Framework

FELDSPAR GROUP

The feldspar group minerals are tectosilicates with similar framework configuration, whose compositions lie between

$$R^{1+}[AlSi_3O_8] \quad \text{and} \quad R^{2+}[Al_2Si_2O_8]$$

where

R^{1+} = Na, K, or occasionally (NH_4)
R^{2+} = Ca, but also Ba and Sr

and where aluminum is sometimes replaced by boron (reedmergnerite).

Most naturally occurring feldspars are made up of three principal chemical components:

$$Or_x Ab_y An_{100-x-y}$$

where

Or = orthoclase = $KAlSi_3O_8$
Ab = albite = $NaAlSi_3O_8$
An = anorthoclase = $CaAl_2Si_2O_8$

The *alkali feldspars* are composed primarily of Or and Ab, while the *plagioclase series* feldspars are those containing primarily Ab and An. These two feldspar subgroups collectively constitute the dominant minerals of the upper lithosphere, being almost ubiquitous in igneous rocks, as well as important components of many metamorphic rocks, and also exist, although far less commonly, in sedimentary environments. Within these subgroups varying degrees of substitution and Al–Si ordering are found. Albite is common to the two subgroups but is usually included as a plagioclase, there being a discontinuity in the alkali feldspar series at low temperature.

The structural framework may be described in terms of its component structural elements. The fundamental building block is a ring of four tetrahedra. These four-membered rings are linked through common oxygens to form a chain, commonly described as a "double crankshaft" lying parallel to [100] with the rings alternately nearly perpendicular to [010] and somewhat inclined to [100]. The chains are cross-linked in a plane parallel to {20$\bar{1}$} to

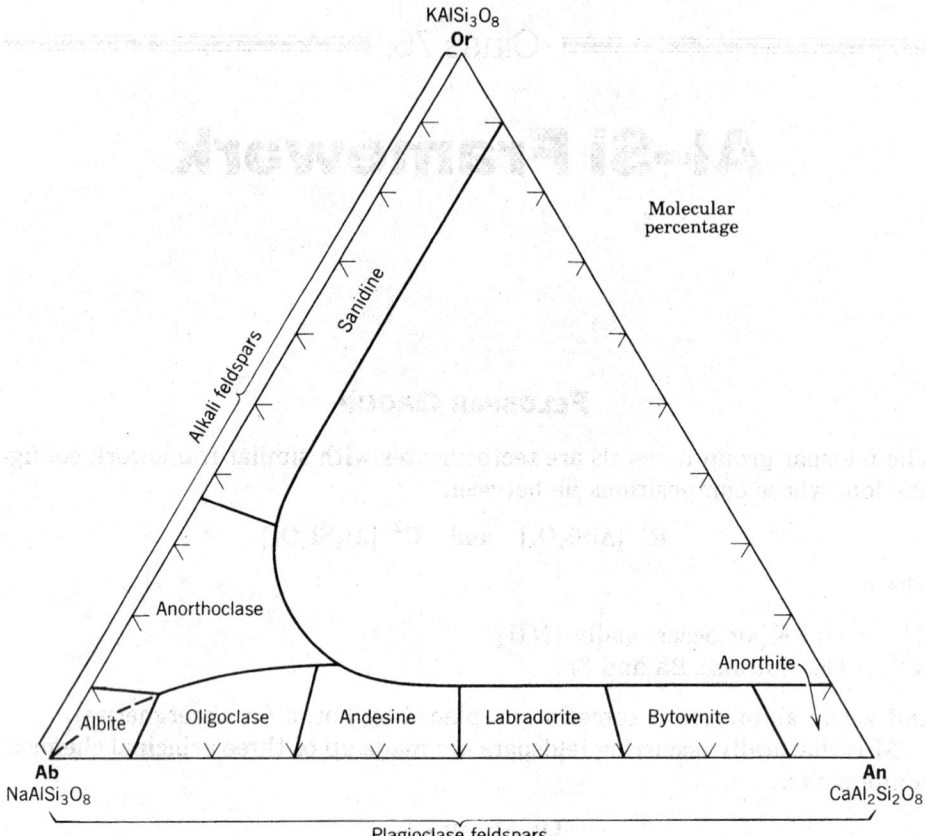

Nomenclature for the plagioclase feldspar series and the high-temperature alkali feldspars. (after Deer, Howie, and Zussman, 1963, *Rock Forming Minerals*, v. 4. Longman Group Ltd., p. 2.)

form puckered sheets. All feldspars display two cleavages, {001} perfect and {010} good, which represent severing of bonds that cross-link the double-crankshaft chains. The chain and sheet elements make up the framework, which has large "cages" accommodating the cations.

The conformation of apical oxygens within the rings comprising the puckered sheets is consistent in all feldspars; however, different conformations appear in similar sheets of some feldspar polymorphs and zeolites.

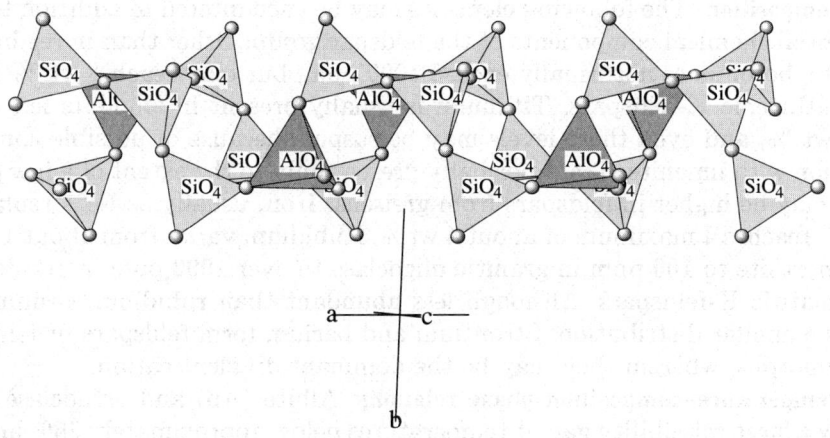

Feldspar Group
Partial structure of low-albite; the feldspar "crankshaft" extending parallel to a and made up of four-membered tetrahedral rings.

TRUE FELDSPARS: ALKALI FELDSPAR SUBGROUP

Mineral	Formula	Space Group	a	b	c	α	β	γ	
Orthoclase	$KAlSi_3O_8$	$C2/m$	8.56	12.94	7.21		116.3		76.1.1.1
Sanidine	$(K,Na)AlSi_3O_8$	$C2/m$	8.60	13.04	7.18		116.0		76.1.1.2
Hyalophane	$(K,Ba)(Al,Si)_4O_8$	$C2/m$	8.56	13.04	7.20		115.7		76.1.1.3
Celsian	$BaAl_2Si_2O_8$	$I2_1/c$	8.62	13.08	14.41		115.1		76.1.1.4
Microcline	$KAlSi_3O_8$	$C\bar{1}$	8.53	12.95	7.19	90.4	116.0	87.80	76.1.1.5
Anorthoclase	$(Na,K)AlSi_3O_8$	$C\bar{1}$	8.252	12.936	7.139	92.11	116.31	90.22	76.1.1.6
Buddingtonite	$(NH_4)AlSi_3O_8 \cdot 0.5H_2O$	$P2_1$ or $P2_1/m$	8.571	13.032	7.187		112.7		76.1.2.1

TRUE FELDSPARS: PLAGIOCLASE SUBGROUP

Mineral	Formula	Space Group	a	b	c	α	β	γ	
Albite (low)	$NaAlSi_3O_8$	$C\bar{1}$	8.14	12.79	7.16	94.3	116.6	87.7	76.1.3.1
Oligoclase (low)		$C\bar{1}$	8.15	12.82	7.14	94.0	116.5	88.6	76.1.3.2
Andesine		$C\bar{1}$ or $P\bar{1}$	8.15	12.83	14.21	93.6	116.2	89.7	76.1.3.3
Labradorite		$C\bar{1}$	8.18	12.87	7.10	93.46	116.09	90.5	76.1.3.4
Bytownite		$P\bar{1}$ or $I\bar{1}$	8.19	12.88	14.20	93.4	116.0	90.9	76.1.3.5
Anorthite	$CaAl_2Si_2O_8$	$P\bar{1}$	8.18	12.87	14.18	93.2	115.9	91.1	76.1.3.6
Reedmergnerite	$NaBSi_3O_8$	$C\bar{1}$	7.839	12.373	6.808	93.32	116.38	92.01	76.1.4.1

Order–disorder. Aluminum and silicon may be completely disordered to fully ordered in the tetrahedral sites, this being one cause of the complexity of polymorphic subspecies possible for a feldspar of a given composition.

Composition. The following elements may be encountered in addition to the essential chemical components of the feldspar group. Other than in reedmergnerite, boron does not usually exceed 1000 ppm, but may reach 1 wt % B_2O_3 in authigenic K-feldspars. Titanium is usually present in amounts less than 0.1 wt %, and even these levels may be suspect because of possible contamination with ilmenite. Tin is normally present only to the extent of a few ppm, but may be higher in feldspars from greisens. Iron, usually as Fe^{3+} replacing Al^{3+}, reaches a maximum of about 4 wt%. Rubidium varies from about 1 ppm in anorthite to 100 ppm in granitic oligoclase to over 1000 ppm in granitic or pegmatitic K-feldspars. Although less abundant than rubidium, cesium follows a similar distribution. Strontium and barium form feldspars or feldspar polymorphs, wherein they may be the dominant divalent cation.

Temperature–composition phase relations. Albite (Ab) and orthoclase (Or) show a large miscibility gap at temperatures below approximately 650° and an extensive region of solid solution above that temperature. The rate of cooling can be the essential factor in determining the phase preserved in the cooled

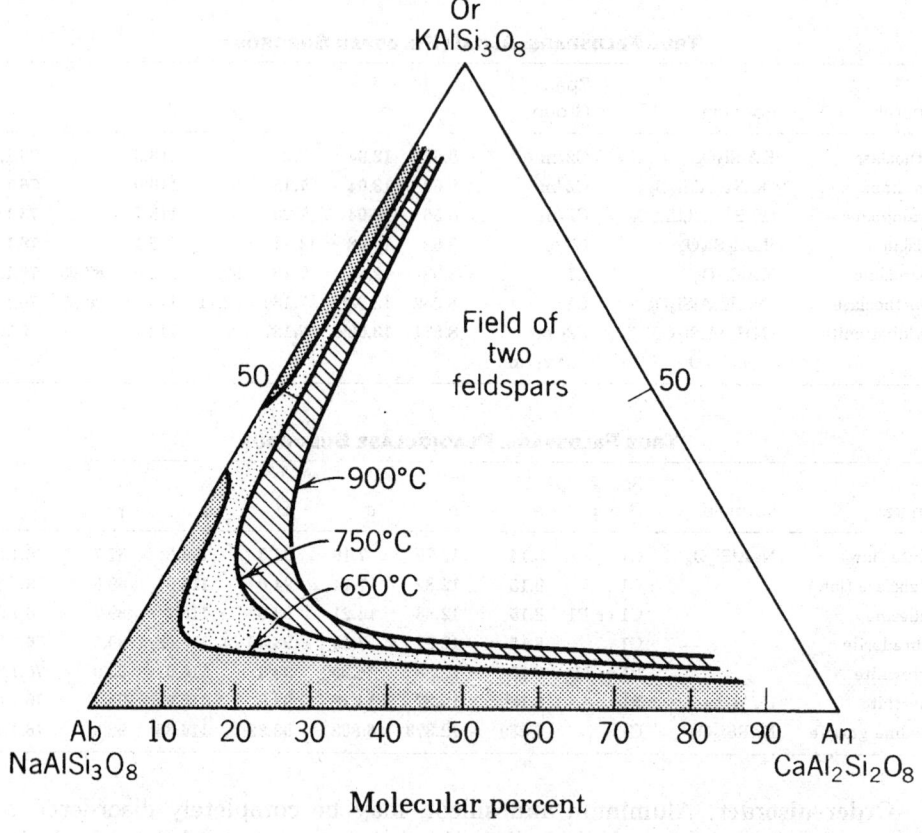

Experimentally determined extent of solid solution in the system Or-Ab-An at $P_{H20} = 1$ kbar (After Ribbe, *Reviews in Mineralogy*, v. 2, Mineralogical Society of America, 1975)

product. Sanidine, for example, can be quenched into the sanidine structure, or when cooled more slowly can result in either the monoclinic orthoclase or the triclinic microcline phase.

Exsolution. Many feldspars undergo exsolution resulting from decreased solubility at lower temperatures and order–disorder transformations. This may result in a variety of intergrowth textures and optical phenomena (schiller, adularescence, labradorescence, etc., as discussed under the individual species). The exsolved phase may occur as stringers, rods, blebs, interlocking masses, or other forms, and the size of its particles tends to increase with lower temperature of formation and time. However, the single phase domains are often beyond resolution by optical methods or by powder X-ray diffraction. (In such cases the concept of "average structure" may sometimes be introduced.) Exsolution commonly results in perthite (exsolved Na-rich feldspar in K-spar host) and antiperthite (exsolved K-spar in Na-rich feldspar host). Perthite (named for Perth, ONT, Canada) may thus occur as "sub-X-ray-perthite," "X-ray-perthite," cryptoperthite," and "microperthite," with sizes from < 15 Å to 100 µm, to "perthite," with exsolved regions exceeding 100 µm.

Twinning. Twinning in feldspars is very common and may arise in primary crystal growth, through deformation, or during thermal transformation. There are numerous twin laws, divided into (1) *normal twins*, with the twin axis normal to the composition plane that is a possible face; (2) *parallel twins*, with the twin axis being a zone axis and the composition plane including the twin axis; and (3) *combined twins*, involving a combination of a normal and a parallel twin law. Repeated (polysynthetic) twinning is common for some laws, and some laws apply only to triclinic phases. The most commonly observed twin laws are indicated in the table.

COMMON FELDSPAR TWIN LAWS

Name	Twin Axis	Composition	
Albite	$\perp$(010)	(010)	Triclinic only; repeated
Baveno (R,L)	$\perp$(021),(0$\bar{2}$1)	(021),(0$\bar{2}$1)	Contact; simple; rare in plagioclase
Carlsbad	[001]	(010), (hk0)	Contact and penetration; simple
Manebach	$\perp$(001)	(001)	Contact; simple
Pericline	[010]	($h0l$)	Triclinic only; repeated
Albite-Carlsbad		(010)	Triclinic only; repeated

In triclinic system pericline twins, the composition plane lies in zone [010], but its orientation is a function of the geometry of the cell; this composition plane is referred to as the "rhombic section." The orientation of the rhombic section relative to cleavage planes is useful in identification of the feldspar.

Notes:
1. The group name, feldspar, derives from the Swedish *fält spat*, through the German *feld spath*, meaning field spar, *spar* being a term used for a nonmetallic, often cleavable, lustrous stone. The usage is of mid-eighteenth-century origin, often attributed to Wallerius and/or Cronstedt, but probably first used in 1742 by Daniel Tillis.
2. The nomenclature of the feldspar minerals is highly complex, involving elements of composition, structure, order–disorder, exsolution, texture, and optical phenomena. An extensive discussion is presented in *JVS Ch 9*.
3. Polymorphism is common, and many of the true feldspars have structurally and chemically related polymorphic phases. These are classified in the Dana system after the true feldspars.

References: DHZI 4:1(1963), MSASC 2:R1(1975).

76.1.1.1 Orthoclase KAlSi$_3$O$_8$

Named in 1823 by Breithaupt from the Greek for *straight fracture*, in allusion to the perpendicular cleavages. Feldspar group. The varieties valencianite from the Valenciana mine, GTO, Mexico, and adularia from the Adular Mts., Switzerland, an erroneous placement of the St. Gotthard locality. Moonstone is sometimes included as an orthoclase microperthite but is more likely sanidine or anorthoclase. The term *K-spar* or *K-feldspar* for undifferentiated K-rich feldspars includes orthoclase. Dimorphous with microcline. MON $C2/m$. $a \approx 8.56$, $b \approx 12.94$, $c \approx 7.21$. $\beta \approx 116.3°$, $Z = 4$, $D = 2.54$. *31-966*: 4.22_7 3.77_8 3.45_5 3.31_{10} 3.29_6 3.24_7 2.99_5 2.90_3. Fully ordered orthoclase is considered by some to represent twinning of a triclinic cell on a submicroscopic scale. Adularia is reported to be triclinic in part, making it more appropriately a microcline, but some samples may have domains of the two symmetries. **Habit:** The habit is usually short prismatic, [100], with dominant {001}, {010}, {110}, and {20$\bar{1}$} or short prismatic [001] and somewhat tabular {010} with similar forms. The habit of adularia is commonly prismatic, with {110} dominant and modified by {001} and {10$\bar{1}$}. Massive, coarsely crystalline to granular; compact to cryptocrystalline. Usually anhedral or subhedral rock-bound grains, with euhedral crystals being found as phenocrysts in volcanic rocks and in pegmatites and fissure fillings. Twins are common on the *Carlsbad law*, the *Baveno Law*, and the *Manebach law*. **Physical properties:** Colorless to white, cream, pale yellow, pink, to brownish red; streak uncolored to very pale in dark varieties; vitreous luster, somewhat pearly on {001}; transparent to translucent; a pale blue to white Schiller effect, moonstone, is evident in some orthoclase microperthites. Cleavage {001}, perfect; {010}, less so; parting may occur on {110}, {100}, and {20$\bar{1}$}; subconchoidal to uneven fracture; brittle. $H = 6-6\frac{1}{2}$. $G = 2.55-2.63$. **Optics:** Biaxial $(-)$; $2V = 12-84°$; $N_x = 1.518-1.528$, $N_y = 1.521-1.533$, $N_z = 1.524-1.537$ increasing with Ab content; $X \wedge a = 3-14°$, OAP perpendicular to {010} and $\approx$ parallel to {001}; $r > v$; for $Or_{85}Ab_{14}An_1$, $2V = 46°$, $N_x = 1.519$, $N_y = 1.524$,

76.1.1.1 Orthoclase

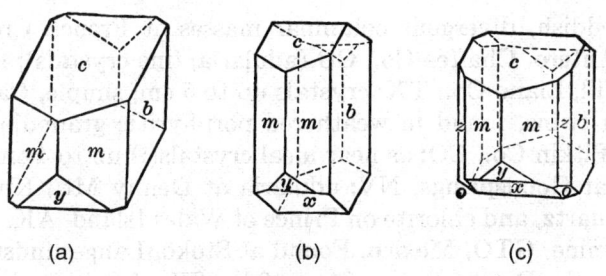

Typical habits of orthoclase. a) Short prismatic [100]. b) Prismatic [001]. c) Variety aularia. Forms: b, {010}; c, {001}; m, {110}; x, {$\bar{1}$01}, y, {$\bar{2}$01}.

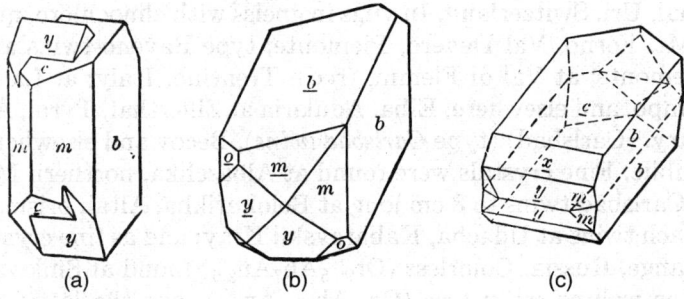

Orthoclase twin laws. a) Carlsbad penetration twin. b) Baveno twin. c) Manebach twin.

$N_z = 1.525$. **Chemistry:** Orthoclase has been reported with compositions from $Or_{>90-25}$, the balance being Ab and usually less than 4 mol % An, but some examples reported may in fact be sanidine or anorthoclase, the high-sodium examples having fine-scale perthitic textures. Adularia is always high in Or (>80). Silicon appears to be consistently shy of the ideal 3:1 Si:Al ratio. TiO_2 is generally low (0.1 wt %), and Fe_2O_3 is generally < 1 wt %, but may reach 3 wt % (Madagascar). BaO and SrO may be present in small amounts, but high-Ba analogs exist [see hyalophane (76.1.1.3)]. Alkali metal content, other than Na and K, is low. **Alteration:** Common alteration products are muscovite and clay minerals, particularly kaolin that may form pseudomorphs. Talc, chlorite, leucite, cassiterite, and calcite are known to form pseudomorphs after orthoclase. **Occurrence:** Typical of acidic and alkaline igneous rocks of plutonic and volcanic origin and in granitic pegmatites. In alpine-cleft environments (adularia) and in hydrothermal veins. Found in high-grade metamorphic environments; as porphyroblasts in contact zones. Found as weathered grains in arkosic sediments, often with authigenic overgrowths. **Localities:** Orthoclase is of common worldwide occurrence, and but a sampling of noteworthy localities is given. Found at Acworth, NH; as large crystals coated with clinochlore(!) at Acushnet, Bristol Co., MA; large crystals at Barre, Worcester Co., MA, and throughout the pegmatites of New

England; as reddish, divergent columnar masses at French Creek, Chester Co., PA; Mt. Antero, Chaffee Co., CO (adularia, fine crystals); large crystals at Barringer Hill, Llano Co., TX; crystals up to 5 cm, simple, Carlsbad twins and Manebach twins, found in weathered porphyritic granodiorite at West Maroon Pass, Pitkin Co., CO; as near ideal crystals(!) up to 4 cm long at Oro Grande, NM; at Goodsprings, NV; adularia at Denny Mt., King Co., WA; with epidote, quartz, and chlorite on Prince of Wales Island, AK. Valencianite at Valenciana mine, GTO, Mexico. Found at Stokö, Langesundsfjord, also as Baveno twins in the Drammen granite ≈ 40 km SW of Oslo, and at Fredricksvärn, near Larvik, Vestfold, Norway; at St. Agnes and elsewhere in Cornwall, England, often as Carlsbad twins sometimes replaced by cassiterite; adularia at Bourg d'Oisans, Dauphiné, France. Excellent crystals of adularia(!) at Val Tavetsch, Disentis, and elsewhere, Grisons, at St. Gotthard, Tessin, and at Unseren Thal, Uri, Switzerland. In vugs in gneiss with clinochlore, quartz, and sphene at Mt. Forno, Val Devero, Piemonte; type Baveno twins at Baveno, Novara, Piemonte; at Val di Flemm, (red), Trentino, Italy; at La Colta, St. Piero in Campo, and elsewhere, Elba. Adularia at Zillerthal, Tyrol, Austria; at Karlovy Vary, (Carlsbad) (type *Carlsbad twins*), Becov and elsewhere, Cechy, Czech Republic. Fine crystals were found at Albaschka, northern Mursinska, Ural Mts.; Carlsbad twins to 8 cm long at Belokurikha, Altai; prisms to 18 cm and Manebach twins at Udacha, Kabarovskii Kray; and as fine crystals in the Stanovoi Range, Russia. Colorless ($Or_{90.5}Ab_7An_{2.5}$) found at Sinkwa, Mogok, Myanmar, as well as *moonstone* ($Or_{20}Ab_{73.5}An_{6.5}$), but the latter may be a sanidine. Adularia found at Su Gu Myun, Su An Gun, and Hwang Hae Do, North Korea, and as white, 2-cm crystals heavily encrusting 20-cm quartz crystals at Nae Gun Gang Myun, Jun Yang Gun, Gang Weon Do, South Korea; adularia as drusy crystals in Alpine-type cavities in quartz veins in schist at Kukutzu, Taiwan; Baveno twins at Tanokamiyama, Shiga Pref., Japan; as fine, large, transparent yellow (2.56 wt % Fe_2O_3) crystals at Itrongay and elsewhere in northern Betroka, Madagascar. Uses: Orthoclase, together with other K-rich feldspars, is an important component of ceramic bodies. With other feldspars is used as a mild abrasive in cleaning compounds. Transparent yellow material from Madagascar has been fashioned into quite large faceted gems. If truly orthoclases, moonstone varieties are fashioned into cabochon and carved gems. AR *DHZI 4:1(1963), MSASC 2:R-1,ST-23,Sm-18(1975), MR 8:363(1977)*.

76.1.1.2 Sanidine (K,Na)AlSi$_3$O$_8$

Named in 1808 by Nose from the Greek *sanis*, tablet, in allusion to the crystal habit. Feldspar group. An earlier used term is *soda-orthoclase*. Much moonstone is cryptoperthitic sanidine. High- and low-sanidine modifications exist. High-sanidine forms a series with high-albite. MON $C2/m$. $a = 8.60$, $b = 13.04$, $c = 7.18$, $\beta = 116.0°$, $Z = 4$, $D = 2.54$. *19-1227*($Or_{54}Ab_{38}An_8$): 4.16_7 3.76_8 3.45_5 3.27_7 3.26_{10} 3.25_8 3.22_9 2.98_4; also *25-618*(disordered): 4.24_6 3.79_6 3.46_3 3.33_{10} 3.28_6 3.23_5 3.00_3 2.54_4. Structure: *CM 29:543(1991)*. Of all the feldspars, the structure of high sanidine is the best understood, and displays the double-crankshaft structural unit described earlier. High sanidine is fully disordered with respect to both Al/Si and Na/K. **Habit:** Crystals usually short prismatic, somewhat flattened on {010}, the prominent forms being {010}, {110}, {001}, and {20$\bar{1}$}; {010} sections may attain 50 cm. Also anhedral grains and cleavable masses. Twinning on Carlsbad law is common; Manebach and Baveno law twins, less so. **Physical properties:** Colorless, white, gray, other pale tints; white to light blue schiller is often present (moonstone); vitreous luster, somewhat pearly on cleavage; transparent to translucent. Cleavage {001}, perfect; {010}, good; {001} parting; conchoidal to uneven fracture; brittle. H = 6. G = 2.56–2.62. High: Biax(−); $2V = 0 - 40°$; $Y = b\, X \wedge a \approx 5°$; r>v weak. Low: Biax (−); $2V = 0 - 40°$; $Z = b\, X \wedge a = 5 - 9°$: r<w weak. High and low: $N_X = 1.518–1.527$ $N_Y = 1.523–1.532$ $N_Z = 1.524–1.534$; slightly higher indices are found in barian sanidines. **Chemistry:** Most published analyses lie in the range $Or_{40-80}Ab_{60-20}An_{0-10}$. Minor Ba and Sr may be present. Fe and Ti are each present at < 0.1 atom per 32 oxygens. Exsolution lamellae are common and not restricted to low-temperature forms. **Alteration:** Commonly alters to sericite and clay minerals. **Occurrence:** Sanidine is typical of silicic volcanic rocks such as rhyolite, phonolite, and trachyte, and in shallow intrusives. It is commonly cryptoperthitic. In highly silicic volcanics such as obsidian and rhyolite it may occur in spherulites intergrown with cristobalite and as acicular clusters. In metamorphic rocks formed at high temperature and low pressure (sanidinite facies). Less common than orthoclase and microcline in clastic sediments. **Localities:** Widespread worldwide; noteworthy localities are: Thomas Range, UT; in the Mitchell Mesa rhyolite, TX ($Or_{41.5}Ab_{57.7}An_{0.8}$); at Rabb Canyon, Grant Co., NM; as masses in trachybasalt at Black Point, near Huntingdon Lake, CA; Summit Rock, Klamath Co., OR; in Canada at Bernic Lake, MB, and Mt. St.-Hilaire, PQ. In blocky, somewhat tabular crystals(!) at Mina de la Pilio, La Cruz, CHIH, Mexico (cryptoperthitic moonstone, $Or_{38-40}Ab_{59-62}An_{0-1}$), the composition range overlaps that of anorthoclase, and that term has been applied to its descriptions. Mt. Dore, Auvergne, and Puys de Dôme, France; Mte. Somma, Campania, in the Alban Hills, near Rome, and at Mt. Cimino, near Viterbo, Lazio, Italy ($Or_{80}Ab_{17}An_3$); at Mendig, Mayen, and elsewhere, near Laacher See, Eifel, Germany ($Or_{78}Ab_{22}$); Mt. Somlyód, near Végardó, Slovakia (twinned, $Or_{63}Ab_{35.5}An_{1.5}$); phenocrysts of zoned, barian sanidine in dacite at Zvečan, Yugoslavia. *Moonstone* is found at a number of localities in southern India and Sri Lanka. Meisen Gun, Kanchin Prov., North Korea (iodiomorphic, colorless, moonstone). **Uses:** May find use in porcelain bodies. The moonstone variety is cut into cabochon gems. AR *DHZI 4:6,40,56(1963)*, *AM 63:1264(1978)*, *MR 26:222(1995)*.

76.1.1.3 Hyalophane (K,Ba)(Al,Si)$_4$O$_8$

Named in 1855 by von Walterhausen from Greek for *glass* and *to appear*, in allusion to its transparent crystals. Feldspar group. MON $C2/m$. $a = 8.56$,

$b = 13.04$, $c = 7.20$, $\beta = 115.7°$, $Z = 4$, $D = 2.88$. *19-2:* 3.78_5 3.46_5 3.31_9 3.24_{10} 3.00_7 2.91_4 2.57_5 2.17_4. Structure: *AC(B) 33:3073(1977)*. Habit often prismatic [001] with dominant {110}, similar to adularia, or equant to short prismatic as with ordinary orthoclase. Simple Carlsbad, Manebach, and Baveno law twins; also composite and multiple twins. Colorless, white, pale yellow, pale to deep pink; vitreous luster; transparent to translucent. Cleavage {001}, perfect; {010}, good; conchoidal fracture; brittle. $H = 6-6\frac{1}{2}$. $G = 2.58-2.82$. Biaxial $(-)$; $2V = 70-85°$, decreasing with increasing Ba content; $N_x = 1.522-1.542$, $N_y = 1.526-1.547$, $N_z = 1.528-1.551$; $X \wedge a = 0-19°$, $Z = b$; $r > v$, weak. Intermediate in composition between orthoclase and celsian, but with discontinuity in the series; sodium may be present to Ab_{25}, but only minor amounts of calcium are found. The celsian content is generally < 50 mol %. Typically in metamorphosed manganese deposits with Mn-epidote, rhodonite, spessartine, and so on. Found at Franklin, Sussex Co., NJ; high-Na (barium–sanidine) phenocrysts in analcite phonolite at Highwood Mts., MT; at Jakobsberg, Värmland, and the Sjö mines, near Örebro, Sweden; in dolomite marble with barite at the *Lengenbach quarry, Binn, Valais, Switzerland*; as large, complex twins(!) at Zogradski Creek, near Busovaca, Bosnia and Herzegovina; at Slyudyanka, Transbaikalia, Russia; at the Kaso mine, Tochigi Pref., Japan; at Broken Hill, NSW, Australia, and Pahau River, South Island, New Zealand; at Otjosondu, Namibia. AR *DHZI 4:166(1963)*, *MR 16:306(1985)*.

76.1.1.4 Celsian $BaAl_2Si_2O_8$

Named in 1895 by Sjogren for Anders Celsius (1701–1744), Swedish astronomer and naturalist. Feldspar group. Kasoite, from the locality, has been applied to K-rich material. Dimorphous with paracelsian (76.1.5.1). MON $I2/c$. $a = 8.62$, $b = 13.08$, $c = 14.41$, $\beta = 115.1°$, $Z = 8$, $D = 3.389$. *38-1450*(syn): 6.52_6 3.56_5 3.47_7 3.36_{10} 3.28_5 3.03_6 2.61_4 2.59_6; also *21-812, 19-9*. Structure: *AM 61:414(1976)*. The structure is similar to orthoclase but with a doubled c axis. Crystals are usually short prismatic, often similar to adularia in habit, with prominent {110} prism faces; also massive, cleavable. Twinning according to Carlsbad, Manebach, and Baveno laws common. Colorless, white, yellow; vitreous luster; transparent to translucent. Cleavage {001}, perfect; {010}, good; {110}, poor; uneven fracture; brittle. $H = 6-6\frac{1}{2}$. $G = 3.10-3.45$. Biaxial $(+,-)$; $2V_z = 83-92°$; $N_x = 1.579-1.587$, $N_y = 1.583-1.593$, $N_z = 1.588-1.600$; birefringence $= 0.009-0.013$; the highest indices being for the synthetic end member; $X \wedge a = 3-5°$, $Y = b$; . Occurs in amphibolite-grade metamorphic rocks, rich in Mn and Ba. Found in quartzite near Rush Creek and Big Creek, Fresno Co., CA, with quartz, diopside, witherite, and sanbornite; with sanbornite and gillespite near Incline, Mariposa Co., and at the Kalkar quarry, Santa Cruz, Santa Cruz Co., CA; in the Alaska Range, AK; at *Jakobsberg, Värmland, Sweden*; with paracelsian at the Benallt Mn mine, Rhiw, Lleyn Penin., Caernarvonshire, England; at Burutlas, W of Lake Balkhash, Kazakhstan; at the Kaso mine, Tochigi Pref., Japan; "The Piggery," Broken Hill, NSW, Australia; Otjosondu, Namibia, in Mn deposit. AR *DHZI 4:166(1963)*.

76.1.1.5 Microcline $KAlSi_3O_8$

Named in 1830 by Breithaupt from the Greek for *small angle* or *inclination*, in allusion to the slightly oblique intersection of the cleavage planes. Amazonite is named for the Amazon River. Feldspar group. Dimorphous with orthoclase (76.1.1.1). TRIC $C\bar{1}$. $a = 8.53$, $b = 12.95$, $c = 7.19$, $\alpha = 90.4°$, $\beta = 116.0°$, $\gamma = 87.80°$, $Z = 4$, $D = 2.59$. *19-926*(fully ordered, $Or_{96.5}Ab_{3.5}$): 4.22_{10} 3.70_4 3.49_3 3.37_4 3.29_6 3.26_8 3.25_8 3.24_4; *19-932*(intermediate, $Or_{84.4}Ab_{15.0}An_{0.6}$): 4.22_5 3.80_2 3.74_1 3.48_2 3.31_1 3.29_5 3.24_{10} 2.98_1; also *22-675*, *22-687*. Microcline lattice constants vary from nearly monoclinic to a maximum obliquity for "maximum microcline." The structure is similar to that of low albite. Structure: *BM 107:401(1984)*. **Habit**: Commonly as short prismatic, [001] or [100], crystals with dominant forms {001}, {010}, {110}, {1$\bar{1}$0}, plus minor modifying faces. Crystals from granite pegmatites may attain lengths measured in tens of meters and weights of several thousand metric tons. Twinning on the Carlsbad, Manebach, and Baveno laws are common; twinning on the albite and pericline laws are commonly observed in thin section. Perthitic textures formed by exsolution of albite are very common and distinctive for the species. Epitactic albite may be found, particularly on {001}. Cleavable masses, granular. **Physical properties**: Colorless, white, cream to pale yellow, salmon-pink to red, bright green to blue-green, generally colorless in thin section; vitreous luster, somewhat pearly on dominant cleavage; transparent to opaque. Cleavage {001}, perfect; {010}, good; parting may develop on {100}, {110}, {1$\bar{1}$0}, and {$\bar{2}$01}; conchoidal to uneven fracture; brittle. $H = 6-6\frac{1}{2}$. $G = 2.54-2.57$. **Optics**: Biaxial (−); $2V = 65-85°$, but may be as low as $35°$ ($2V_x = 103°$ has been reported but this may be due to exsolved albite lamellae); $N_x = 1.518$, $N_y = 1.522$, $N_z = 1.525$ for fully ordered microcline, sodic cryptoperthites have indices intermediate to those of low albite; $X \wedge a \approx Y \wedge c \approx Z \wedge b \approx 18°$; $r > v$, weak. On {001} in thin section, "Scotch plaid" albite–pericline twinning is usually coarser than in {100} sections of anorthoclase. **Chemistry**: True, fully ordered microcline is near the

Microline; "scotch-plaid" appearance in thin section due to combined albite and pericline twinning.

ideal composition, but most examples are actually perthitic on micro- to megascopic scale, and the bulk composition may reach $Or_{35}Ab_{65}$, although such Na-rich examples may indeed be antiperthites. The An content rarely reaches 4 mol %. Phosphorus content in feldspars of complex pegmatites may be as high as 1.2 wt %; in peraluminous feldspars Al^{3+} and P^{5+} may replace $2Si^{4+}$. Rb_2O may be present (3+ wt %) in the microcline of complex pegmatites, as may be Li_2O and Cs_2O but to a significantly lesser extent. Total iron oxides may reach ≈ 1 wt %, but only trace amounts of Ti are found. The green color of amazonite, which is lost at temperatures in excess of 250°C, is not fully understood; it has been attributed to the presence of Pb^{2+}, it is not uncommon in rubidian examples, but may be due to color centers caused by fluorine substitution for oxygen. **Occurrence:** The common feldspar of plutonic, felsic rocks such as granite, granodiorite, syenite, and in simple and complex granite pegmatites; in metamorphic rocks of greenschist and amphibolite facies; as a detrital component in sedimentary rocks wherein it may have authigenic overgrowths. Commonly associated with quartz, sodic plagioclase, muscovite, biotite, and hornblendic amphiboles. **Localities:** A widespread mineral, but a few noteworthy localities are given. Government pit, Carroll Co., NH; at the Gillette quarry, Haddam, Middlesex Co., CT, and elsewhere in pegmatites of New England; at the Buckwheat open pit, Franklin, Sussex Co., NJ; Poorhouse quarry, Chester Co., PA, as white crystals, low Na, on dolomite; Amelia C.T., Amelia Co., VA, amazonite; at Magnet Cove, Hot Springs Co., AR, having nearly ideal composition; amazonite at Rib Mt., Marathon Co., WI; in pegmatites of the Black Hills, SD. Fine specimens at numerous localities in Colorado, as on Mt. Antero, Chaffee Co.; amazonite(!) in pegmatites of the Pikes Peak granite in El Paso Co. and Teller Co. with large (to 25 cm) crystals, and Baveno and Manebach twins, with thin albite overgrowths(!) at the Ten-Percenter mine; large crystals with thick albite "caps" at Lake George, at Stevens Ranch, and at the Yucca Hill claim, Park Co., CO. In the Harding pegmatite, near Dixon, NM, as large cleavable masses often in pinkish tones due to partial replacement by manganoan muscovite. At the Blue Chihuahua, Hercules, and Himalaya mines, San Diego Co., CA. In Canada at Davis Hill, Bancroft, (antiperthite); Hybla; Smart mine, Eganville; Perth, Lanark Co., (perthite); Kuel Lake, Renfrew Co.; and Tory Hill, ONT; Yates mine, Pontiac Co., and at Mt. Saint Hilaire as euhedral crystals to 30 cm and Baveno twins to 15 cm, QUE. White to tan (8 cm) crystals with scheelite, dravite, and fluorapatite at the Guadalupeña mine, Santa Cruz, SON, Mexico. Carlsbad and Manebach twins at Narsarsuk, and *amazonite* (10 cm) at Torsukattak Fjord, Greenland. Found as cleavable masses in zircon syenite at *Fredriksvärn, Norway*; maximum microcline in pegmatite at Norra Kärr, and Rb microcline perthite in pegmatite at Varuträsk, Sweden; at Vigo, Galicia, Spain; at Baveno, Lago Maggiore, Piemonte, Italy; amazonite(!) in the Ilmen Mts., Ural Mts., and as large crystals at Naryn-Kunta, Irkutsk, Russia; in the pagmatites of the Mawi and Laghman districts, Afghanistan; 6- to 8-cm white crystals at Wae Gum Gang Myun, Go Sung Gun, Gang Weon Prov., South Korea; in the Ishikawa district, Fukushima Pref., Tadachi, Nagano Pref., and elsewhere in Japan; at Broken Hill, NSW, Australia; at Ambositra and Anjanbonoina, Madagascar. Widespread in the pegmatite of MG, Brazil, as at Fazenda do Bananal, Fazenda do Funil (amazonite), Taquaral district, and Virgem da Lapa. Uses: Used, as are other feldspars, in ceramics and mild abrasives. The amazonite variety is a popular ornamental stone. AR *DHZI 4:6,38(1963)*, *AM 69:440(1989)*, *CM 28:771(1990)*.

76.1.1.6 Anorthoclase (Na,K)AlSi$_3$O$_8$

Named in 1885 by Rosenbusch from the Greek for *not straight fracture* or *not orthoclase*, in allusion to the lower symmetry than that of orthoclase. Feldspar group. Much material displays a bluish schiller and may thus be included under the term *moonstone*. The name is often applied to multiphase material found in slightly alkalic lavas. The composition range slightly overlaps that of Na-rich sanidine, but true polymorphism may not exist as the overlap region may be a multiphase one. TRIC $C\bar{1}$. $a = 8.252$, $b = 12.936$, $c = 7.139$, $\alpha = 92.11°$, $\beta = 116.31°$; $\gamma = 90.22°$, $Z = 4$, $D = 2.60$. *9-478*(Or$_{32}$Ab$_{66}$An$_2$) 6.49$_1$ 6.42$_1$ 4.11$_2$ 3.77$_1$ 3.73$_1$ 3.24$_9$ 3.21$_{10}$ 2.16$_2$. Structure: *AM 67:975(1982)*. Habit: Crystals commonly prismatic [001] with dominant {110} and {1$\bar{1}$0} resembling somewhat those of adularia, or tabular {010}; massive, granular. Twinning occurs according to the Carlsbad, Manebach, and Baveno laws; albite and pericline law twins also occur, as is the case with microcline, but the pericline composition plane is different, and the reticulated pattern formed by these two laws is seen in thin section on {100}. Physical properties: Colorless, white, light to dark gray, cream, pale yellow, pink, green; may display a pale to deep blue, greenish, or gold Schiller; vitreous luster to somewhat pearly on cleavage; transparent to translucent. Cleavage {001}, perfect; {010}, good; parting may develop on {100}, {110}, {1$\bar{1}$0}, and {$\bar{2}$01}; conchoidal to uneven fracture; brittle. $H = 6-6\frac{1}{2}$. $G = 2.57-2.60$. Biaxial $(-)$; $2V = 0-55°$; $N_x = 1.519-1.529$, $N_y = 1.524-1.534$, $N_z = 1.527-1.536$, the indices increasing with Na content; $X \wedge a \approx 10°$, $Y \wedge c \approx 20°$, $Z \wedge b = 5°$; $r > v$ weak. On {100} in thin section, "Scotch plaid" albite–pericline twinning is usually finer than in {001} sections of microcline. Chemistry: The usual composition range of anorthoclase is Or$_{10-40}$Ab$_{60-90}$An$_{0-20}$, overlapping to some extent that of high-Na sanidine but distinguished therefrom by its lower symmetry. Total Fe may reach 0.2 atom per 32 oxygens. Substitution of Ti is low, as is that for Sr and Ba (<1 wt %). Exsolution lamellae of high-K and high-Na phases is common, yielding cryptoperthite intergrowths frequently displaying schiller effects. Occurrence: Found in sodian volcanic and hypabyssal rocks formed at elevated temperature. As phenocrysts intergrown with oligoclase in alkalic syenite (larvikite) in association with minor augite, titanian augite, apatite, ilmenite, etc. Localities: Found at Cripple Creek, Teller Co., CO; as

Manebach and Baveno twins at Obsidian Cliff, Yellowstone National Park, WY. A major component of larvikite at Larvik, near Oslo, Norway (blue schiller due to exsolution to oligoclase and orthoclase); phenocrysts ($Or_{17-22}Ab_{76-81}An_{1-2}$) in volcanic neck Brownhills, Fife, Scotland; Caldeira Grande, Azores ($Or_{30-36}Ab_{63-68}An_{0.1-1.7}$); on Ustica Island and in andesite of *Pantelleria Island, Sicily, Italy*. Found at Ogaya, Toyama Pref., Japan; in Australia, as massive white crystals ($Or_{0.7}Ab_{98}An_2$) at Tingha, NSW, and as phenocrysts ($Or_{19}Ab_{72}An_9$) at Mt. Anakie and ($Or_{21-22}Ab_{75-77}An_{2-3}$) at Camperdown, VIC; as large xenocrysts ($Or_{12-14}Ab_{83-84}An_{2-4}$) in the Deborah volcanic sequence Kakanui, New Zealand; loose cryptoperthitic crystals ($Or_{17-22}Ab_{63-72}An_{5-20}$) at Mt. Erebus crater, Antarctica. Rhomb-shaped cryptoperthitic phenocrysts ($Or_{19-22}Ab_{70-76}An_{4-8}$) in soda-trachyte at Ropp, Nigeria; at Mt. Kenya, Kenya, both lacking and displaying Schiller; and at Mt. Kilimanjaro, Tanzania. Uses: Material displaying Schiller may be cut en cabochon as a gemstone; coarse-textured larvikite containing Schillerized phenocrysts of anorthoclase is used as an ornamental building or facing stone. AR *DHZI 4:6,42,57(1963)*, *MM 33:949(1964)*.

76.1.2.1 Buddingtonite $(NH_4)AlSi_3O_8 \cdot 0.5H_2O$

Named in 1964 by Erd et al. for Arthur Francis Buddington (1890–1980), American petrologist, Princeton University. Feldspar group. MON $P2_1$ or $P2_1/m$. $a = 8.571$, $b = 13.032$, $c = 7.187$, $\beta = 112.7°$, $Z = 4$, $D = 2.38$. *17-517*: 6.52_{10} 5.91_4 4.33_7 3.81_{10} 3.38_7 3.26_6 3.23_7 3.01_4. Crystals similar to orthoclase in habit and of small size; more commonly as anhedral grains and cryptocrystalline. Colorless, vitreous luster, transparent. Cleavage {001}, good; {010}, distinct; subconchoidal fracture; brittle. $H = 5\frac{1}{2}$. $G = 2.32$. Biaxial (+); $2V_{calc} = 60°$; $N_x = 1.530$, $N_y = 1.531$, $N_z = 1.534$; $X' \wedge a = 0°$ on {001} and $4°$ on {010}, $Y \wedge c = 19°$. Found in compact masses pseudomorphous after plagioclase and as minute crystals lining cavities in Quaternary rocks altered by ammonia-bearing waters below the water table at the *Sulphur Bank mercury mine, Lake Co., CA*; also, widespread in the Meade Peak member of the Phosphoria formation, ID; at Tōshichi spa, Iwate Pref., Japan; and in the Condor oil shale, Q, Australia. AR *AM 49:831(1964)*, *MA 35:425(1984)*.

76.1.3.1	Albite	$NaAlSi_3O_8$	An_0–An_{10}
76.1.3.2	Oligoclase		An_{10}–An_{30}
76.1.3.3	Andesine		An_{30}–An_{50}
76.1.3.4	Labradorite		An_{50}–An_{70}
76.1.3.5	Bytownite		An_{70}–An_{90}
76.1.3.6	Anorthite	$CaAl_2Si_2O_8$	An_{90}–An_{100}

The term *plagioclase* is from the Greek *plagios*, oblique and fracture, in allusion to the slightly oblique intersection of the cleavage planes. Albite was named in 1815 by Gahn and Berzelius from the Latin *albus*, white, in allusion

76.1.3.6 Anorthite

to its usual color; the variety pericline from the Greek for *sloping on all sides*; the variety cleavelandite was named for Parker Cleaveland (1780-1858), American mineralogist, Bowdoin College, Maine; peristerite, which may encompass other species, is from the Greek *peristera*, pigeon, in allusion to the play of colors as on the neck feathers of a pigeon; oligoclase was named in 1826 by Breithaupt from the Greek for *little fracture*, again in reference to the small obliquity of the cleavage planes. Andesine was named in 1841 by Abich for the Andes Mountains; labradorite was named in 1780 by Werner for the locality; bytownite was named in 1835 by Thomson for Bytown, now Ottawa, Canada; anorthite was named in 1823 by Rose from the Greek *anorthos*, not straight, in allusion to the triclinic symmetry. Feldspar group. The compositional boundaries for names within the series are arbitrary, although there may be structurally valid boundaries that are not in use. Anorthite is polymorphous with dmisteinbergite (76.1.7.1) and svyatoslavite (76.1.6.4). Low- and high-temperature forms in the series are a function of the degree of ordering in the structure, and generally correspond to plutonic and volcanic occurrences, respectively. Structure: *ZK 133:43(1971), 169:35(1984); AM 65:81(1980), 75:135(1990); AC(B) 44:344(1988); CMP 104:471(1990)*. The structures of the series end members have been determined with reasonable accuracy, but those of intermediate members of the series are not fully understood, commonly involve exsolution phenomena, and often only "average structures" are reported. The space group is either $C\bar{1}$ or $P\bar{1}$, the latter with a doubled c axis; additionally, an I-cell with doubled c axis occurs in bytownites. The trend in lattice constants with increasing anorthite content is for a and b to increase, and for c or $c/2$ to decrease slightly. **Albite: Low-** TRIC $C\bar{1}$. $a = 8.14$, $b = 12.79$, $c = 7.16$, $\alpha = 94.3°$, $\beta = 116.6°$, $\gamma = 87.7°$, Z = 4, D = 2.62. *9-466*: 6.39_2 4.03_2 3.78_3 3.68_2 3.66_2 3.51_1 3.20_{10} 2.93_2; also *20-554, 20-572, 41-1480*. **High-** TRIC $C\bar{1}$. $a = 8.16$, $b = 12.87$, $c = 7.11$, $\alpha = 93.5°$, $\beta = 116.4°$, $\gamma = 90.3°$, Z = 4, D = 2.62. *20-528, 41-1481*. **Oligoclase: Low-** TRIC $C\bar{1}$. $a = 8.15$, $b = 12.82$, $c = 7.14$, $\alpha = 94.0°$, $\beta = 116.5°$, $\gamma = 88.6°$, Z = 4, D = 2.65. *PD:* 4.03_8 3.76_7 3.69_5 3.66_5 3.18_{10} 2.98_5 2.93_7 2.84_5. **High-** TRIC $C\bar{1}$. $a = 8.16$, $b = 12.88$, $c = 7.11$, $\alpha = 93.4°$, $\beta = 116.3°$, $\gamma = 90.3°$, Z = 4, D = 2.65. **Andesine:** TRIC $P\bar{1}$ (also $C\bar{1}$). $a = 8.15$, $b = 12.83$, $c = 14.21$, $\alpha = 93.6°$, $\beta = 116.2°$, $\gamma = 89.7°$, Z = 8, D = 2.68. *PD*(low): 4.04_8 3.76_7 3.65_7 3.21_{10} 3.18_9 3.14_7 2.93_7 2.53_7. **Labradorite:** TRIC $C\bar{1}$. $a = 8.18$, $b = 12.87$, $c = 7.10$, $\alpha = 93.46°$, $\beta = 116.09°$, $\gamma = 90.5°$, Z = 4, D = 2.70. *9-465*: 4.04_8 3.75_8 3.64_7 3.29_8 3.20_{10} 3.18_9 3.14_7 2.95_7. **Bytownite:** TRIC $P\bar{1}$ (also $I\bar{1}$). $a = 8.19$, $b = 12.88$, $c = 14.20$, $\alpha = 93.4°$, $\beta = 116.0°$, $\gamma = 90.9°$, Z = 8, D = 2.73. *PD:* 4.03_8 3.75_8 3.62_7 3.36_5 3.20_{10} 3.17_8 2.95_7 2.52_7. **Anorthite:** TRIC $P\bar{1}$. $a = 8.18$, $b = 12.87$, $c = 14.18$, $\alpha = 93.2°$, $\beta = 115.9°$, $\gamma = 91.1°$, Z = 8, D = 2.76. *20-20*: 4.04_5 3.78_2 3.62_3 3.26_5 3.21_6 3.19_{10} 3.18_9 3.12_4; also *41-1486*. **Habit:** Albite and anorthite occur as well-crystallized examples more commonly than the intermediate members of the series. Albite commonly forms equant to {010} tabular crystals with {010}, {110}, {1$\bar{1}$0}, and {001} dominant, the tabular or lamellar habit being pronounced in cleavelandite. The variety pericline is elongated on [010] with {001}, {$\bar{1}$01}, and {010} domi-

Striations appearing on {001} cleavage plane of polysynthetic twinning on the albite law.

nant and commonly twinned on the pericline law. Well-crystallized intermediate plagioclases have similar habit, but are more commonly euhedral to anhedral, rock-bound grains. Anorthite is commonly short prismatic, elongated on either [001] with {110} and {1$\bar{1}$0} dominant or elongated on [010] with {001}, {$\bar{1}$01} (equivalent to {$\bar{2}$01} in the older morphological setting), and {010} dominant. Commonly polysynthetically twinned on the albite law, with the twin lamellae appearing as striations on the {001} cleavage due to their slightly different orientations. Pericline and Carlsbad law twins are also common, and combined twins are commonly seen in thin section. **Physical properties:** Colorless, white to gray in mass, also yellow, pink, brown, black, pale blue, pale green; vitreous luster; transparent to translucent. Cleavage {001}, perfect, {010}, good, the two intersecting at 93–94°; {110}, imperfect; conchoidal to uneven fracture; brittle. H = 6–6$\frac{1}{2}$. G increases with Ca content [2.60–2.65(albite), 2.63–2.66(oligoclase), 2.66–2.68(andesine), 2.68–2.74(labradorite), 2.72–2.75(bytownite), 2.74–2.76 (anorthite)]. **Optics:** Plagioclases are biaxial (+ or −); $2V_z$ large to moderate; for low-plagioclase $2V_z$ rises from $\approx 88°$ at An_0 to $\approx 95°$ at $An_{\approx 25}$, then decreases to $\approx 88°$ at $An_{\approx 50}$ and rises to $\approx 104°$ at An_{100}; for high-plagioclase $2V_z$ varies from $\approx 132°$ for An_0 to a minimum of 76° at $An_{\approx 60}$ and rises again to $\approx 103°$ at An_{100}. The refractive indices of low- and high-plagioclase are nearly equal, the significant difference appearing in the albite–oligoclase range and in the bytownite range lie between 1.528 and 1.589; birefringence = 0.007–0.014 being least at An_{40-45} and greatest at An_{100}; the dispersion is weak, usually r < v. Extinction angles may be measured relative to either of the two cleavages or to twin composition planes; most useful for purposes of identification is the maximum extinction, $X' \wedge (010) = 0$–57°, measured in sections cut perpendicular to (010) (the Michel–Levy method), or the pair of extinction angles observed in similar sections for combined albite–Carlsbad twins. **Chemistry:** The plagioclase series was one of the first mineralogical solid-solution series to be described, and was initially represented as an ideal, simple, two-component series. In fact, the series is very complex with space group changes, order–disorder phenomena,

76.1.3.6 Anorthite

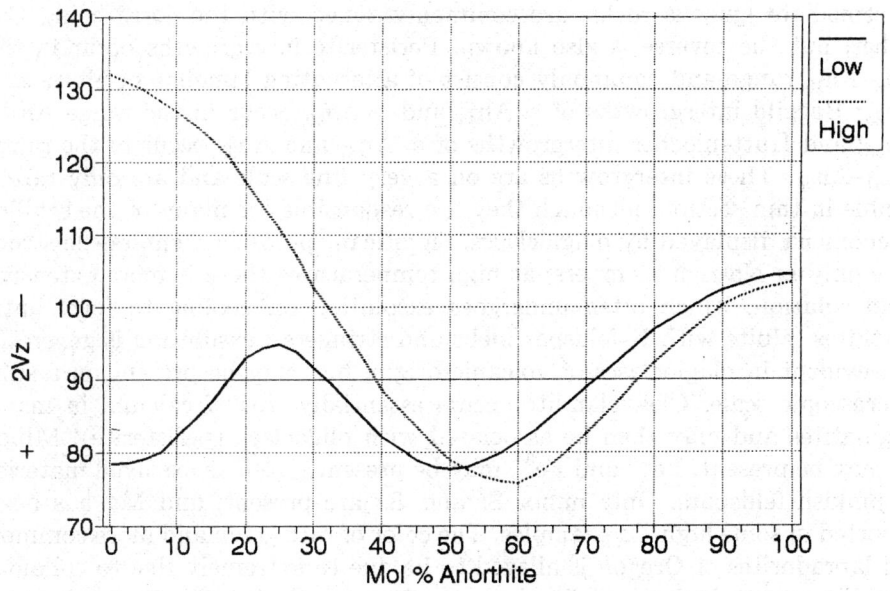

Approximate value of 2Vz for low and high-plagioclase

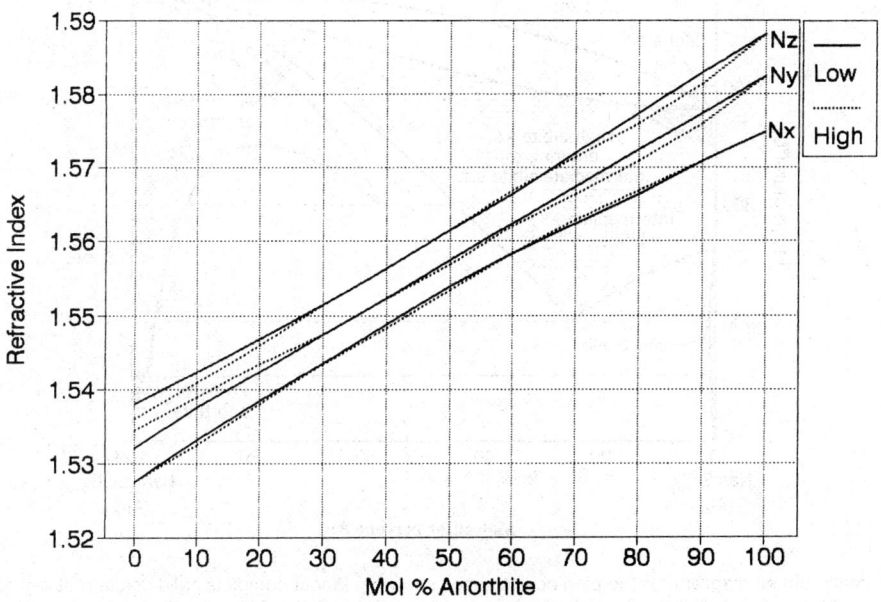

Approximate value of low and high-plagioclase

and exsolution phenomena. Plagioclases, particularly those of volcanic and intermediate igneous rocks, are commonly zoned with the cores being Ca-richer, but the reverse is also known. Peristerite intergrowths occur in the An_2–An_{16} range and commonly consist of alternating lamellae of albite and An_{25}; Bøggild intergrowths of $\approx An_{45}$ and $\approx An_{60}$ occur in the range An_{47}–An_{58}; and Huttenlocher intergrowths of $\approx An_{67}$ and An_{95} occur in the range An_{67}–An_{90}. These intergrowths are on a very fine scale and are only rarely visible in thin section, although they are responsible for many of the schiller phenomena displayed by plagioclases. Significant potassium enters the structure only in Na-rich members; at high temperatures there is more extensive solid solution, which often undergoes exsolution on cooling to form antiperthites (albite with K-feldspar blebs and stringers); exsolution is generally not evident in plagioclases of volcanic origin, but may be present on a submicroscopic scale. Cleavelandite occurs essentially free of calcium in many pegmatites and may then be associated with oligoclase (peristerite). Minor Ti may be present; Fe^{2+} and Fe^{3+} may be present, often as exsolved material in pinkish feldspars. Only minor Sr and Ba are present, and Mg has been reported in some high-Ca examples. The color of rare green and more common red labradorites of Oregon is alleged to be due to extremely fine to colloidal metallic copper inclusions. Plagioclases alter to sericite (fine-grained white mica), saussurite (sericite, epidote, albite), scapolite, prehnite, pyrophyllite,

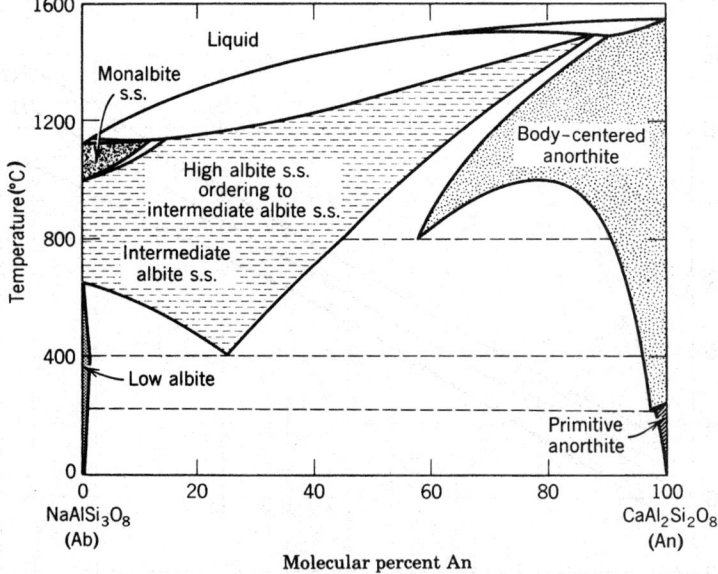

Schematic phase diagram for the plagiocalse series showing almost complete solid solution at elevated temperatures and miscibilkity gaps at lower temperatures. (After Smith & Brown, *Feldspar Minerals*, v. 1, Springer Verlag, 1988).

and various clay minerals. **Occurrence:** Plagioclases are widespread components of almost all types of igneous and metamorphic rocks, and in some sediments. Albite and oligoclase are common in acid igneous rocks, such as granite, granodiorite, rhyolite, syenite, and granite and syenite pegmatites, with cleavelandite being common in complex granite pegmatites. Intermediate igneous rocks are characterized by andesine; gabbros and basalts commonly contain labradorite and, less commonly, bytownite, either or both being the dominant constituents of anorthosites. Anorthite is much less common and is usually restricted to unusual mafic and ultramafic environments. In metamorphic rocks, plagioclase compositions are usually $<An_{50}$, the calcium content usually being higher with higher-grade metamorphism. Anorthite is sometimes found in metamorphosed carbonate rocks. In sedimentary rocks plagioclase is usually restricted to detrital grains, but authigenic albites may form by diagenesis of the detrital plagioclase. **Localities:** With the exception of anorthite, members of the plagioclase series are found worldwide, and but a few localities can be noted. Albite is particularly noteworthy for well-crystallized specimens. *Albite:* Found at Granville, Washington Co., NY (transparent); at the Gillette Quarry and others, Haddam Neck, CT; Topsham, MA; Amelia C. T., VA (cleavelandite); at various localities in the Pikes Peak granite, CO (cleavelandite, and as caps on green microcline); in the Green River formation, WY, CO, UT; Tip Top Mine, Custer, SD; in the Harding pegmatite, northern Dixon, Taos Co., NM (cleavable, massive cleavelandite with rose muscovite); in San Diego Co., CA (cleavelandite) (Cota, Hercules, Himalaya, Little Three, Mack, Stewart, Tourmaline King, Vandenberg, and White Queen mines). Albite antiperthite at Davis Hill and Silver Crater, northern Bancroft, ON; northern Sherbrooke (alpine cleft environment:albite and Carlsbad twins); at Mt. St.-Hilaire, QUE, Canada. Crystals to 4 cm at Ojos Negros, Ensenada, BCN, Mexico. Large crystals at Skarvebergbukten, Seeland, Norway; at Meylan, northern Grenoble, Isère, and Col du Bonhomme, near Madone, Savoie, France (in granular limestone); pericline in many of the same localities as adularia in Switzerland; near Mursinsk, Ural Mts. (cleavelandite with topaz), also near Miask, Ilmen Mts., and (iridescent) along the Bezymyannaya R., Slyudyanka, Irkutsk, Russia; cleavelandite in pegmatites of the Lagham and Jegdalek areas, Afghanistan. Cleavelandite is abundant in pegmatites of MG, Brazil, as at Campo do Boaitatiaia mine, Palmital, Corrego Frio mine, Cruzeiro mine, Galileia, Xanda mine, Jonas mine(!), the Pederneira mine, São José da Safira, and at the Morro Velho mine. *Oligoclase:* Found at the Palermo No. 1 mine, North Groton, NH; at Chester Emery mines, Hampton Co., MA; Macomb, St. Lawrence Co., NY; Mineral Hill, Delaware Co., PA (sunstone); at the Hawk Mine and others, Bakersville, Mitchell Co., NC ($Ab_{72.7}An_{24.8}Or_{2.5}$, gemmy fragments); in Scandinavia at Tredestvard, Kragero, and Arendal, Norway; at Kimito, Finland; at *Danviken, near Stockholm, Sweden*; and at the Monte Redondo quarry, 155 km N of Lisbon, Portugal. *Andesine:* Found at St. Raphael, Mt. Esterel, near Cannes (Carlsbad and Baveno twins), and northern Chagney, Haute-Saône, France; at Zabkowicz, Poland

(Frankenstein, Silesia); in Japan at Hishishiota, Shinano Pref., at Kaneda, Miyagi Pref., at Kuzuhara, Toyama Pref., and elsewhere; at *Marmato, Cauca, Colombia*, and elsewhere in the Andes Mts. of South America. *Labradorite:* An important component of anorthosites of the Adirondack Mts, NY, and northern MN; gemmy yellow, red, green material is mined in Lake Co., OR (in part bytownite), and in crystals N of Plush. In Canada at Taher Is., northern Nain, Labrador ($Ab_{47}An_{52}Or_{0.8}$); *Paul Is., Labrador, NF; and elsewhere;* at Santo Domingo, Puerto Peñasco, SON, Mexico (anhedral to 6 cm in Pinacate volcanic field). Found on Surtsey Is., S of Iceland; at Langesundfjord, Norway; at Lappeenranta, Finland; and at Etna, Sicily, and Mt. Vesuvius, Campania, Italy. Beautiful, nearly transparent material with a delicate labradorescence is mined in southern India. *Bytownite:* Found at Phoenixville, Chester Co., and Cornwall, Lebanon Co., PA; a major component of anorthosites and gabbros as at Crystal Bay and Duluth area, MN ($Ab_{70-75}An_{30-25}$), and in the Stillwater Complex, MT; gemmy material at Lakeview, Lake Co., OR (see also *labradorite*); in anorthosite in Foleyet Co., and at *Ottawa (Bytown), ONT, Canada;* E of Nueva Casas Grandes, CHIH, Mexico (8 cm, light yellow, transparent gems to 80 ct). Found at Fiskenesset, Greenland; Rhum Is., Scotland; and in the Bushveld Complex, South Africa. *Anorthite:* Found at Grass Valley, Nevada Co., CA, and Great Sitkin Is., AK; on Amitok Is., Labrador, NF, Canada. Found at Tunaberg, Sweden; in older lavas of *Mte., Somma, Campania, Italy;* northern Bogoslovsk and Barsovka, Ural Mts., Russia. In Japan at Miyakejima, Tokyo Pref. (pseudomonoclinic to 1.5, $An_{97}Ab_3$); at Otaru-shi (1.5 cm in andesite tuff) and Tarumae volcano, Hokkaido (tabular {001} and asymmetric); at Za volcano, Yamagata Pref., (2 cm in andesite breccia); at Hakone, Kanagawa Pref. Japan and elsewhere. Uses: Sodian plagioclases find such applications as in ceramics and cleaning compounds, much as for orthoclase. Various intermediate-plagioclase-rich rocks are utilized as building and ornamental stones, especially those displaying a prominent labradorescence, and the latter are also cut en cabochon and marketed under the names "fabulite" and "rainbow moonstone." Albite and oligoclase displaying adularescence and/or chatoyancy make up part of the moonstone gem feldspars; oligoclase with distinct orange-red schiller is known as sunstone and is cut en cabochon; transparent labradorite and bytownite are cut into faceted gems, as are the green and red varieties (Oregon), the latter also being designated as sunstone. AR *MM 30:306(1954), AM 43:1179(1958), DHZI 4:94(1963), BM 107:467(1984), FELD 1:14(1988)*.

76.1.4.1 Reedmergnerite $NaBSi_3O_8$

Named in 1960 by Milton et al. for Frank S. Reed (b.1894) and John L. Mergner (b.1894), laboratory technicians, U.S. Geological Survey. Feldspar group. TRIC $C\bar{1}$. $a = 7.839$, $b = 12.373$, $c = 6.808$, $\alpha = 93.32°$, $\beta = 116.38°$, $\gamma = 92.01°$, $Z = 4$, $D = 2.77$. *18-1201:* 6.08_5 3.88_7 3.56_9 3.23_9 3.08_9 3.04_{10} 2.96_5 2.84_6; also *PD:* 3.87_{10} 3.65_3 3.23_2 3.14_3 3.08_7 3.03_2 2.84_4 2.38_2. Structure: *AM 77:76(1992)*. Isostructural with low-albite, ordered with boron ideally in the T1 site. Order–disorder transformation for synthetic boron–albite occurs at 1 kbar H_2O pressure in the range 500–550°C. Isolated, short prisms up to 2 mm long, with wedge-shaped terminations; untwinned. Colorless, vitreous luster, transparent. Cleavage {001}, perfect; brittle. H = 6–$6\frac{1}{2}$. G = 2.78. Biaxial (−); 2V = 80°; N_x = 1.554–1.558, N_y = 1.565, N_z = 1.573–1.573; An authigenic mineral in black oil shale and brown dolomitic rock at the *Joseph Smith No. 1 well, Duchesne Co., UT*, and elsewhere in the Greenriver formation in CO, UT, and WY. Also found in peralkaline pegmatite, Dar-i-Pioz, Tien Shan, southern Tajikistan. AR *AM 45:188(1960), DANT 10:51(1967), EJM 5:971(1993)*.

76.1.5.1 Paracelsian $BaAl_2Si_2O_8$

Named in 1905 by Tacconi from the Greek *para*, akin to, and for its compositional identity to celsian. See slawsonite (76.1.5.2). Again described in 1942 by Spencer, the earlier description having been interpreted erroneously as being identical to celsian. Dimorphous with celsian (76.1.1.4). MON $P2_1/a$. $a = 9.065$, $b = 9.568$, $c = 8.578$, $\beta = 90.01°$, Z = 4, D = 3.351. *10-352*: 4.00_{10} 3.80_7 3.59_4 2.99_5 2.73_5 2.56_5 2.37_4 2.19_4. Structure: *AC 6:613(1953)*. The structure is pseudo-ORTH *Pnam*. The tetrahedral framework is similar to that of the feldspars but with a different cross-linkage of the double-crankshaft chains. Aluminum and silicon appear to be disordered or only partially ordered. The structure is closely related to that of danburite, which has symmetry *Pnam*, but in which the tetrahedral framework is ordered with the presence of distinct Si_2O_7 and B_2O_7 dimers. Isostructural with slawsonite. Prismatic crystals often well formed and resembling topaz; granular massive. Multiple, complex twinning. Colorless, white, pale yellow; vitreous luster. Cleavage {110}, poor; uneven fracture; brittle. H = $5\frac{1}{2}$–6. G = 3.29–3.32. Biaxial (−); 2V = 52.5°; N_x = 1.570, N_y = 1.582, N_z = 1.587. Found with celsian in a band in shale and sandstone associated with manganese ore at the Benallt mine, Rhiw, Lleyn Penin., Caernarvonshire, Wales; and as pale yellow granules in schist at *Candoglia, Piemonte, Italy*. AR *MM 26:231(1942), DHZI 4:172(1963), AM 70:969(1985)*.

76.1.5.2 Slawsonite $(Sr,Ca)Al_2Si_2O_8$

Named in 1977 by Griffen et al. for Chester Baker Slawson (1898–1964), mineralogist and professor, University of Michigan. See paracelsian (76.1.5.1). MON $P2_1/a$. $a = 8.888$, $b = 9.344$, $c = 8.326$, $\beta = 90.33°$, Z = 4, D = 3.08. *37-462*(Oregon): 3.93_8 3.71_8 3.22_{10} 2.94_4 2.92_5 2.68_4 2.67_5 2.55_4. Structure: *AM 62:31(1977)*. Isostructural with paracelsian. Aluminum and silicon are reported as being ordered; the difference in intensity distribution of the two samples may be due to different degrees of ordering. Fibrous to tabulate prisms. Colorless, gray; vitreous luster; transparent to translucent. Cleavage {001}, good; {110}, fair–good. H =$5\frac{1}{2}$–$6\frac{1}{2}$. G = 3.0. Biaxial (−); Oregon material: 2V = 82°; N_x = 1.573, N_y = 1.581, N_z = 1.585; Y $\wedge$ c = 11°; r > v, weak; Japan material: 2V = 55°; N_x = 1.570, N_y = 1.582, N_z = 1.586;

$Y \wedge c = 5°$; $r < v$, moderate. Japan material fluoresces pink to purplish red in SWUV. Up to 1.5 wt % BaO. In metamorphosed limestone of the *Martin Bridge formation, Wallowa Co., OR;* associated with celsian, cymrite, grossular, diopside, and others in veinlets cutting a rhodingite xenolith in chloritized serpentine at Sarusaka, Kagami, Kochi Pref., Japan. AR *AM 72:225(1987)*.

76.1.6.1 Banalsite $BaNa_2Al_4Si_4O_{16}$

Named in 1944 by Campbell Smith et al. from a contraction of the component cation symbols: Ba, Na, Al, and Si. ORTH *Ibam*. $a = 8.496$, $b = 9.983$, $c = 16.755$, $Z = 4$, $D = 3.045$. *23-651:* 8.50_8 5.20_9 4.20_6 3.77_7 3.53_{10} 3.21_8 2.90_8 2.09_8; also *38-401:* 6.48_6 3.52_8 3.24_{10} 3.10_4 3.02_5 2.91_8 2.09_6 2.08_6, the latter for a sample from Japan with $Ba_{0.81}Sr_{0.25}Na_{1.98}$. Structure: *MJ 7:262(1973)*. The structure is a framework of linked four-tetrahedron rings whose apical oxygens point alternately up and down, whereas in the feldspars the apical oxygen sequence is up–up–down–down. Within the framework the tetrahedra form layers parallel to (001). Al and Si are disordered in the tetrahedral sites; Ba and Na are 10- and six-coordinated, respectively. Isostructural with stronalsite. White, vitreous luster, translucent to transparent. Cleavage {110}, {001}, good. $H = 6$. $G = 3.065$. Biaxial (+); $2V = 41°$; $N_x = 1.5695$, $N_y = 1.5710$, $N_z = 1.5775$; Na(D): $XYZ = cab$, OAP parallel to {100}. Strontium substitutes for barium, and a solid solution with stronalsite is likely. Found in a vein in manganese ore at the *Benallt mine, Rhiw, Lleyn Penin., Caernarvonshire, Wales*; at Långban, Värmland, Sweden; and at the Shiromaru mine, Tokyo Pref., Honshu, Japan. AR *NAT 154:336(1944), MR 1:163(1970)*.

76.1.6.2 Stronalsite $SrNa_2[Al_4Si_4O_{16}]$

Named in 1987 by Hori et al. from a contraction of the names and symbols of the component cations: strontium, Na, Al, and Si. ORTH *Ibam* or *Iba*2. $a = 8.415$, $b = 9.901$, $c = 16.729$, $Z = 4$, $D = 2.943$. *40-1451:* 3.77_4 3.50_8 3.18_5 3.20_{10} 3.07_4 2.99_4 2.88_7 2.07_5. Isostructural with banalsite. White, colorless; vitreous luster; transparent to translucent. Cleavage not given but presumably like that of banalsite. $H = 6\frac{1}{2}$. $G = 2.95$. Biaxial (+); $2V = 32°$; $N_x = 1.563$, $N_{y\,calc} = 1.564$, $N_z = 1.574$. Minor barium is present. Found with jadeite in serpentinite at Mt. Ohsa, Okayama Pref., and with pectolite and slawsonite in veinlets in basic tuff xenoliths at *Rendai, Kochi City, Japan*. AR *MJ 13:368(1987), AM 72:226(1987)*.

76.1.6.3 Lisetite $CaNa_2Al_4Si_4O_{16}$

Named in 1986 by Smith et al. for the locality. ORTH $Pbc2_1$. $a = 8.260$, $b = 17.086$, $c = 9.654$, $Z = 4$, $D = 2.73$. *PD:* 4.16_4 3.53_3 3.47_4 3.20_{10} 3.16_2 3.10_2 3.01_1 2.95_3. Structure: *AM 71:1372(1986)*. The tetrahedral framework, with disordered Al and Si, is the same as that of banalsite, but cation occupancy and coordination is different (Na in two six-coordinated sites; Ca in seven-coordination). Dimensionally, $a_{lis} = c/2_{ban}$, $b_{lis} = 2a_{ban}$, and $c_{lis} = b_{ban}$.

Colorless, transparent. No cleavage noted. Refractive indices and birefringence similar to that of plagioclase, from which it is distinguished by micrometer-size inclusions and a corona texture. Chemically it is midway between anorthite and pure sodium nepheline. Found as anhedral, untwinned, 100-μm grains in a retrograde metamorphic environment in the *Liset eclogite pod, Selje district, Western Gneiss region, Norway*. AR

76.1.6.4 Svyatoslavite $CaAl_2Si_2O_8$

Named in 1989 by Chesnokov et al. for Svyatoslav Nestorovich Ivanov (b.1911), Russian geologist. Polymorphous with anorthite (76.1.3.6) and dmisteinbergite (76.1.7.1). ORTH $P2_12_12$ or $P2_12_12_1$. $a = 8.232$, $b = 8.606$, $c = 4.852$, $Z = 2$, $D = 2.687$. PD: 4.16_8 3.75_6 3.22_{10} 2.94_6 2.71_7 2.09_8 1.967_7 1.670_7. Undetermined structure, but having some dimensional correspondence to banalsite and lisetite; possibly $a_{svy} = c/2_{ban}$, $b_{svy} = a_{ban}$, $c_{svy} = b/2_{ban}$. Dimensionally equivalent orthorhombic and monoclinic anorthite compositions have been synthesized. As prismatic [100] crystals with dominant {011} and {100}. Colorless, white; vitreous luster; transparent to translucent. Cleavage {100}, poor. H ≈ 6. G = 2.695. Biaxial (−); $2V_{calc} = 37°$; $N_x = 1.552$, $N_y = 1.578$, $N_z = 1.581$; $XYZ = cba$. Weak yellow fluorescence in SWUV. Formed in a burning coal dump as a sublimate (700–900°C) on fracture walls in coal at *Kopeysk, Chelyabinsk coal basin, southern Ural Mts., Russia*, where it is associated with anorthite, troilite, cohenite, fayalite, titanite, and graphite. AR *ZVMO 118(2):111(1989), AM 76:300(1991)*.

76.1.7.1 Dmisteinbergite $CaAl_2Si_2O_8$

Named in 1990 by Chesnokov et al. for Dmitri Sergeievich Steinberg (1910–1992), Russian petrographer, Institute of Geology, Ekaterinburg. Feldspar group. Polymorphous with anorthite (76.1.3.6) and svyatoslavite (76.1.6.4). HEX $P6/mmm$. $a = 5.122$, $c = 14.781$, $Z = 2$, $D = 2.747$. PD: 4.48_6 3.83_6 3.73_{10} 2.85_7 2.57_8 2.14_3 1.847_7 1.481_4. Structure: *AC 12:465(1959)*. The structure has been determined on synthetic material. Colorless, vitreous luster. Cleavage {0001}, perfect. H = 6. G = 2.73. Uniaxial (+); $N_O = 1.575$, $N_E = 1.580$. Occurs as hexagonal tablets with well-developed {0001} and poorly developed {10$\bar{1}$0} on fractures in blocks from burned coal dumps from the *Chelyabinsk coal basin, southern Ural Mts., Russia*, where it is associated with cordierite, mullite, anorthite, wollastonite, tridymite, graphite, iron carbides, and fayalite. AR *ZVMO 119:43(1990), AM 77:446(1992)*.

NEPHELINE GROUP

The nepheline group minerals are tectosilicates of high aluminum content with structures that are derivatives of the β-tridymite silica structure. The nepheline minerals are, however, not isostructural. The general formula is

$$R(Al,Si,B)O_4$$

where

R = K, Na, Ca, or □

The formula may be restated in terms of the end members

$$Ne_xKs_yAn_zQ_{1-x-y-z}$$

where

Ne = soda nepheline = $NaAlSiO_4$
Ks = kalsilite = $KAlSiO_4$
An = anorthite = $CaAl_2Si_2O_8$
Q = silica
x,y,z = fractional percentages

The structures are layered, and the idealized structure of a single layer of each of the phases is illustrated along with the β-tridymite structure for comparison. Ignoring the orientation of the apical oxygens, the hexagonal six-

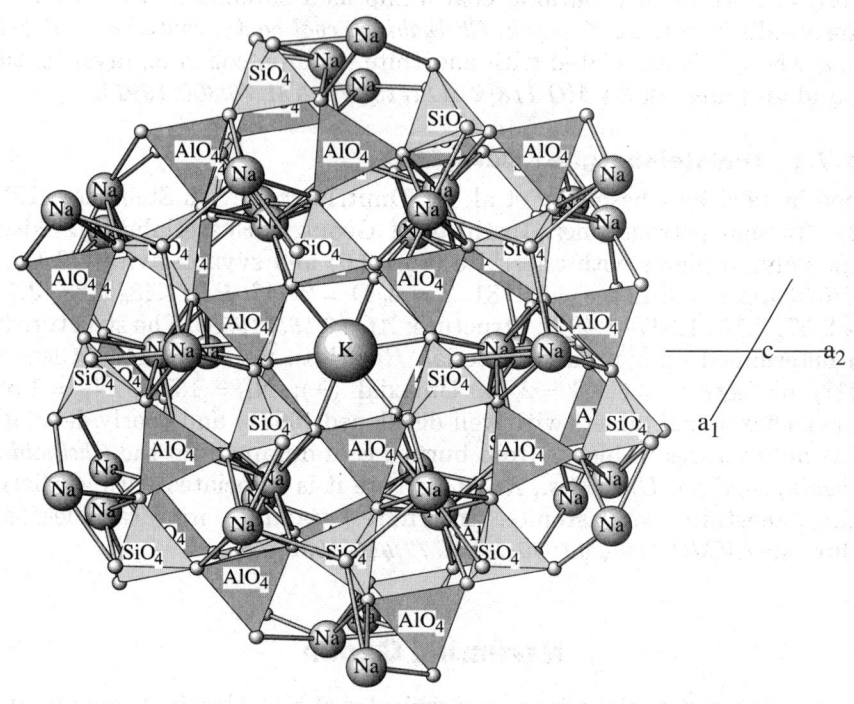

NEPHELINE: $KNa_3Al_4Si_4O_{16}$
Perspective view approximately along c of the hexagonal structure.

tetrahedron rings of tridymite are distorted to lower-symmetry ditrigonal and/or oval rings, alkali cations occupying the various voids. The resultant different ratio of ring types and accompanying voids is directly related to the limited solid solution among the several phases involved. Not all the voids need be occupied. Those that are may contain K, Na, and Ca. The tetrahedral site occupancy may have varying degrees of ordering, leading to polymorphs that may or may not exist in natural systems. *DHZI 4:231(1963)*, *FF 444(1984)*, *AM 77:19(1992)*.

NEPHELINE GROUP

Mineral	Formula	Space Group	a	c	Z	
Kalsilite	$K(AlSiO_4)$	$P6_3$	5.161	8.693	2	76.2.1.1
Nepheline	$(Na,K)(AlSiO_4)$	$P6_3$	9.989	8.380	8	76.2.1.2
Trikalsilite	$K_2Na(AlSiO_4)_3$	$P6_3$	15.339	8.501	6	76.2.1.3
Panunzite	$(K,Na)(AlSiO_4)$	$P6_3$	20.513	8.553	32	76.2.1.4
Kaliophilite	$K(AlSiO_4)$	$P6_3$	27.011	8.557	54	76.2.1.5
Yoshiokaite	$(Ca,\square)(Al,Si)O_4$	$P\bar{3}$	9.939	8.245	8	76.2.1.6

76.2.1.1 Kalsilite K(AlSiO₄)

Named in 1942 by Bannister for the symbols of the constituent elements K, Al, and Si and from the Greek *lithos*, stone. Nepheline group. Dimorphous with kaliophilite. HEX $P6_3$ or $P6_322$. $a = 5.161$, $c = 8.693$, $Z = 2$, $D = 2.619$. *11-579*(syn): 4.35_1 3.97_4 3.12_{10} 2.58_5 2.47_1 2.43_1 2.22_1 2.18_2. Structure: *MM 35:588(1965)*. Colorless, white, gray; vitreous to greasy luster; transparent to translucent. Cleavage {10$\bar{1}$0}, {0001}, poor; subconchoidal fracture; brittle. H = 6. G = 2.59–2.63. Uniaxial (−); $N_O = 1.538–1.543$, $N_E = 1.532–1.537$, reported slightly higher indices may be due to Na and/or Fe^{3+} substitution. Fusibility 3. Gelatinizes with acids. Excess silica is commonly present in the range $Q_{0.5-2.0}$, and soda nepheline up to Ne_{10}; calcium is present in very minor amount. Found as embedded grains in the groundmass of lavas and as phenocrysts of nepheline–kalsilite with a perthitic texture. Found in the Colima graben 50 km SSW of Guadalajara, JAL, Mexico; intergrown with K-feldspar the ultramafic Batbjerg intrusion, eastern Greenland; in lavas at San Venanzo, Umbria, Italy; in syenite in the Yakshinsk pluton, Russia; at *Kyambogo crater, Bunyaruguru field, Uganda*; as nepheline–kalsilite phenocrysts at Baruta crater, Nyiragongo area, Zaire. AR *AJS 255:282(1957)*, *DHZI 4:231(1963)*, *AM 77:19(1992)*.

76.2.1.2 Nepheline (Na,K)(AlSiO₄)

Named in 1800 by Haüy from the Greek for *cloud*, in allusion to its becoming cloudy or milky when immersed in strong acids. Nepheline group. For polymorphs see the nepheline group discussion. HEX $P6_3$. $a = 9.989$, $c = 8.380$, $Z = 8$, $D = 2.63$. *35-424*(syn): 4.32_4 4.17_8 3.83_{10} 3.27_7 3.00_{10} 2.88_4 2.57_5

2.34_4; *12-198*(syn, potassian): 4.25_8 4.02_7 3.35_6 3.07_{10} 2.56_6 2.39_3 2.36_4 2.13_5, the unit cell for this sample being $a = 10.237$, $c = 8.508$. Structure: *BM 107:499(1984)*. Many nephelines show satellite reflections in single-crystal diffraction patterns, indicating that the true unit cell is larger than reported here; these have been interpreted as indicating an ordering of Al and Si in tetrahedral sites. **Habit**: Commonly as hexagonal prisms $\{10\bar{1}0\}$ terminated by the pinacoid $\{0001\}$ and pyramids $\{10\bar{1}1\}$, but often with other prism and pyramid forms; faces, especially on large crystals, often rough and pitted; as rock-bound euhedral, often tabular, to anhedral grains; massive. Twinning on $\{10\bar{1}0\}$, $\{11\bar{2}2\}$, and $\{33\bar{6}5\}$. **Physical properties**: White, often with yellow or greenish tint, gray, red-brown, the color often due to microscopic inclusions; vitreous to greasy luster; transparent to nearly opaque. Cleavage $\{10\bar{1}0\}$, distinct; $\{0001\}$, imperfect; subconchoidal fracture; brittle. $H = 5\frac{1}{2}$–6. $G = 2.55$–2.67. **Optics**: Uniaxial ($-$); $N_O = 1.529$–1.546, $N_E = 1.526$–1.542; birefringence $= 0.003$–0.004. Substitution of K for Na increases the indices, as does excess silica. For synthetic material, introduction of Ca tends to reduce birefringence, but this is not borne out by natural examples; excess silica accompanied by cation vacancies reduces the indices. **Chemistry**: Pure sodium nepheline, $NaAlSiO_4$, has not been found in nature but has been prepared in the laboratory. At $1254°C$ it inverts to an isometric dimorph, carnegieite, which is also unknown in natural systems. Published analyses indicate a range of Ne/Ks from $\approx 3.5:0.5$ to $\approx 3:1$; a few reported lower ratios may actually represent trikalsilite. Although experimental data show that the nepheline structure may accommodate up to 35 mol % anorthite, calcium substitution in natural nepheline is no more than 0.03 Ca per formula unit, and usually much less. The Si/Al ratio is usually > 1; the excess silica can be represented as quartz and is accompanied by cation vacancies. On the basis of these observations the ideal unit cell formula of nepheline can be represented by $Na_6K_2[Al_8Si_8]O_{32}$, and that for the encountered substitutions by $Na_xK_yCa_z\square_{8-x-y-z}[Al_{x+y+2z}Si_{16-x-y-2z}O_{32}]$. Excess silica lies between $Q_{0.5-12.0}$. Nepheline–kalsilite "perthites" may arise from exsolution of phases formed at elevated temperatures, possibly "tetrakalsilite" (panunzite). **Occurrence**: A characteristic phase in alkaline igneous plutonic, hypabyssal, and volcanic rocks of quite varied bulk composition. It occurs in nepheline syenites and related rocks, where it is associated with aegirine–augite, aegirine, and alkali amphiboles, through basic and ultrabasic alkaline rocks, where it may occur with forsterite, spinel, and perovskite. Where crystallized early it forms euhedral crystals, but it may also form poikilitic masses of later crystallization. It forms textural intergrowths with K-feldspar and less often with plagioclase. The earlier formed nepheline usually contains less excess silica than that formed later. In volcanic rocks, the composition range of nephelines is broader than in those of plutonic origin. Nepheline may be of metasomatic origin, the "nephelinization" occurring through the actions of fluids from alkaline intrusives on country rock of varying composition, as in the Haliburton–Bancroft area, ON, Canada, and at Alnö, Sweden. Nepheline-bearing rocks may also form at the contact of Ca-rich sediments and basic intrusives. **Localities:** The following noteworthy and well-documented localities illustrate the widespread distribution. Found at Litchfield, Kennebec Co., ME; in the Magnet Cove area, Hot Springs, Co., AR; in the Highwood, Bearpaw, and Judith Mts., MT. Of numerous occurrences in Canada, that at Davis Hill, Dungannon Twp., Hastings Co., ON, is particularly noteworthy for large, though crude, 20-cm-diameter crystals(!) with biotite and albite–antiperthite; also at Mt. St.-Hilaire, QUE, sometimes as good, simple tabular crystals(!) to 3.5 cm; at the Lomo del Toro mine, Zimapán, HDG, Mexico. Abundant in the syenitic complexes of the Julianehaab district, Greenland; at Bratholmen, Langesundfjord, Norway; at Alnö, Sweden; with cancrinite at Iivaara, Finland; nepheline–syenite at Assynt, Sutherland, Scotland; at Scawt Hill, Northern Ireland; in leucite–nepheline diabase at Meiches, Hessen, and in the Laacher See district, Eifel, Germany; at Capo di Bove, Rome, and at Mte. Somma, Campania, Italy; at Ditro, Romania; near Miask, Ilmen Mts., Russia; in jacupirangite at Chaunggyi, and in ijolitic nepheline–syenite at Sinkwa, Moguk district, Myanmar; in Ul Len Co., Gyong Sang North Prov., and elsewhere in South Korea; at Nagahama, Shimane Pref., Japan; in phonolite at Abbot's Hill, Dunedin, New Zealand; nephelinite at Etinde, Cameroons; at Lake Kivu and Mt. Nyiragongo, Zaire; in phonolite at Nairobi, Kenya; in foyaite at Pilansberg and the Vaal River, Transvaal, South Africa; with analcime as pseudomorphs after leucite at Rio das Ostras, RJ, Brazil. **Use:** A component in ceramic bodies. AR *AJS 255:282(1957)*; *DHZI 4:231(1963)*; *MR20:439(1989), 21:235(1990)*.

76.2.1.3 Trikalsilite $K_2Na(AlSiO_4)_3$

Named in 1957 by Sahama and Smith for its similarity to kalsilite and the tripling of the empirical formula of the tetrahedral framework to account for the potassium/sodium ratio. Nepheline group. HEX $P6_3$. $a = 15.339$, $c = 8.501$, $Z = 6$, $D = 2.636$. *12-197*(syn): 4.27_3 3.93_3 3.38_2 3.08_{10} 3.05_9 2.66_2 2.56_4 2.41_3. Structure: *NJMM:559(1988)*. Colorless, vitreous luster, transparent. Subconchoidal fracture, brittle. H = 6. Occurs in parallel intergrowth with nepheline in the *Kabfumu lavas, North Kivu, Zaire*. AR *AM 42:286(1957)*, *JG 65:515(1957)*.

76.2.1.4 Panunzite $(K,Na)(AlSiO_4)$

Named in 1988 by Benedetti et al. for Achille Panunzi, University of Naples. Nepheline group. Possible synonym: tetrakalsilite. HEX $P6_3$. $a = 20.513$, $c = 8.553$, $Z = 32$, $D = 2.62$. *31-1081:* 4.28_2 3.93_7 3.07_{10} 2.91_3 2.56_4 2.38_2 2.27_2 2.18_2. X-ray powder data show sufficient difference from synthetic tetrakalsilite to make equivalence of the two uncertain. Structure: *NJMM: 322(1985)*. Prismatic [0001]. Colorless, vitreous luster, transparent. H = $5\frac{1}{2}$. G = 2.59. Uniaxial ($-$); $N_O = 1.540$, $N_E = 1.535$. Distinguished from associated nepheline by its greater transparency. Occurs as short, up to 4 mm, hexagonal prisms with nepheline, green augite, and biotite in ejected, pyroxene-rich blocks associated with the latest volcanic activity at *Mte. Somma–Vesuvius, Campania, Italy.* AR *AM 42:286(1957), 64:658(1979), 73:420(1988).*

76.2.1.5 Kaliophilite K(AlSiO$_4$)

Named in 1886 by Mierisch from the Latin *kalium*, potassium, and the Greek *philos*, friend, in allusion to the prominence of potassium. Nepheline group. Dimorphous with kalsilite. HEX $P6_3$ or $P6_322$. $a = 27.011$, $c = 8.557$, $Z = 54$, $D = 2.65$. *11-313*: 4.26_2 3.91_1 3.76_1 3.50_1 3.09_{10} 3.02_1 2.81_1 2.59_3. Structure: *EJM 4:1209(1992)*. Colorless to white, vitreous to silky luster, transparent to translucent. Cleavage {10$\bar{1}$0}, distinct. H = 6. D = 2.49–2.67. Uniaxial (−); $N_O = 1.530$–1.532, $N_E = 1.526$–1.527. Fusibility $3\frac{1}{2}$. Gelatinized by acids. Occurs in volcanic ejecta of biotite pyroxenite and augite–melilite–calcite rock at *Mte. Somma, Campania, Italy,* and in ejecta with leucite and haüyne at Albano, Lazio, Italy. AR *D6:427(1892), AJS 255:282(1957).*

76.2.1.6 Yoshiokaite Ca$_{1-x/2}$□$_{x/2}$(Al$_{2-x}$Si$_x$O$_4$)

Named in 1990 by Vaniman and Bish for T. Yoshioka (1935–1983), who synthesized the phase. Nepheline group. HEX $P\bar{3}$, pseudo-$P\bar{3}c1$ due to twinning. $a = 9.939$, $c = 8.245$, $Z = 8$, $D = 2.74$. PD: 8.54_5 4.94_3 3.71_1 3.02_2 2.97_{10} 2.86_2 2.30_5 2.06_2. Structure: *AM 75:1186(1990)*. The unit cell and structure are similar to those of nepheline except that the distortion of the parent tridymite structure has opposite senses of rotation in successive layers. Colorless, vitreous luster. Cleavage(?) {100}, poor. Uniaxial or biaxial with 2V < 5° due to strain; $N_{av} = 1.61$; birefringence ≈ 0.02. The value of x in the formula as given is 2.0–5.3. A lunar mineral occurring as a metastable phase due to devitrification in glassy, shocked grains in *Apollo 16* and *17*, and *Luna 24* soil samples, and in *Apollo 14* regolith, Luna. AR *AM 75:676(1990).*

76.2.2.1 Leucite K(AlSi$_2$O$_6$)

Named in 1791 by Werner from the Greek *leukos*, white, in allusion to its usual color. See ammonioleucite (76.2.2.2). *Low leucite* and *high leucite* refer to the low (tetragonal) and high (isometric/tetragonal) structures, respectively. Low: TET $I4_1/a$. $a = 13.09$, $c = 13.75$, $Z = 16$, $D = 2.461$. *15-47*: 5.54_5 5.39_8 3.44_9 3.27_{10} 2.92_7 2.84_7 2.81_6 2.37_7. High: ISO $Ia3d$ or TET $I4_1/acd$. $a = 13.4$, $Z = 16$. *31-967*(syn): 5.48_6 4.75_1 3.36_{10} 3.00_1 2.86_3 2.63_1 2.37_1 1.706_1. Structure: *AM 61:108(1976)*. The tetrahedral framework in both high- and low-temperature forms is made up of rings of six and four tetrahedra; the four- and six-membered rings are normal to the four- and threefold symmetry axes, respectively, of the high form. In the low form the four-membered rings, somewhat distorted, are normal to the c and a axes. The structure is closely related to the hydrous minerals analcime and pollucite. The transition from low to high form is gradual, but appears to be complete at $\approx 625°$. Euhedral crystals with trapezohedron {112} dominant in the high form, becoming {211}, {121}, and {112} on inversion to tetragonal low form; {110} may also be present. Upon inversion, multiple twinning is common on {110} and {101}, leading to a retention of the pseudoisometric morphology. Crystals may attain a diameter of 8–9 cm. As disseminated grains, granular and massive. White, colorless, gray; vitreous luster; transparent to translucent. Cleavage {110}, very poor; brittle. H = $5\frac{1}{2}$–6. G = 2.45–2.50. Uniaxial (+); sometimes anomalously biaxial (+); 2V very small; N_O = 1.508, N_E = 1.509. **Chemistry:** Most leucite is close to its ideal composition, with Al/Si near 1:2. TiO_2 is usually low, but reaches ≈ 1.4 wt % in material from Lake Kariya, Uganda. Na_2O may reach ≈ 2 wt %, but is usually lower; the presence of excess albite in the system is unfavorable to the formation of leucite. As cooling reaches 1020°C, leucite may react with liquid to form a mixture of nepheline and K-feldspar (pseudoleucite). Minor CaO may also be present, as well as FeO and MgO. Leucite is decomposed by HCl. **Occurrence:** Found as phenocrysts and in the matrix of K-rich mafic and ultramafic volcanics and hypabyssal rocks. Often associated with K-feldspar, analcime, nepheline, kalsilite, natrolite, and others. **Localities:** Of worldwide distribution; some important occurrences are: at Magnet Cove, Hot Springs Co., AR; as gray to white phenocrysts in volcanics from the Leucite Hills, Sweetwater Co., WY; in the Bearpaw Mts., Hill Co., MT; in the Sierra Nevada Range, CA. In Germany at the Laacher See, Eifel District; at Mte. Somma, Roccamonfina, and *Mt. Vesuvius, Campania, Italy,* and on Vulcano on Lipari Is. In Uganda, at Ruwenzori and Kariya; Mt. Kilimanjaro, Tanzania; and from Mt. Nyiragongo, Kivu, Zaire. AR *DHZI 4:276(1963)*, *ZK 127:213(1968)*, *PCM 16:298(1988)*.

76.2.2.2   Ammonioleucite $(NH_4,K)(AlSi_2O_6)$

Named in 1986 by Hori et al. from the composition and relationship to leucite. See leucite (76.2.2.1). TET (pseudo-ISO) $I4_1/a$. a = 13.214, c = 13.713, Z = 16, D = 2.24. *40-474:* 5.53_5 5.43_{10} 3.54_1 3.43_4 3.30_8 2.96_2 2.86_2 2.84_1. Isostructural with leucite. White, vitreous to resinous luster, translucent. G = 2.29. Uniaxial (+); N = 1.518; birefringence very low. The electron microprobe–based analysis shows a deficiency of Al compared to the ideal formula, and a concomitant deficiency in ammonium and potassium [$(NH_4)_{0.68}K_{0.19}$]. Found with dolomite and analcime in veinlets and cavities in hydrothermally altered schist as powdery replacements of analcime at *Fujioka, Tatarazawa, Sanbagawa, Gumma Pref, Japan.* AR *AM 71:1022(1986)*.

SODALITE GROUP

Species of the sodalite group proper have aluminum–silicon tetrahedral framework structures and a formula corresponding to

$Na_6(Na,Ca)_2(Al_6Si_6O_{24})X_{1-2} \cdot nH_2O$ or the corresponding halved formula

where

X = Cl, OH, SO_4, S
n = 0 or 1

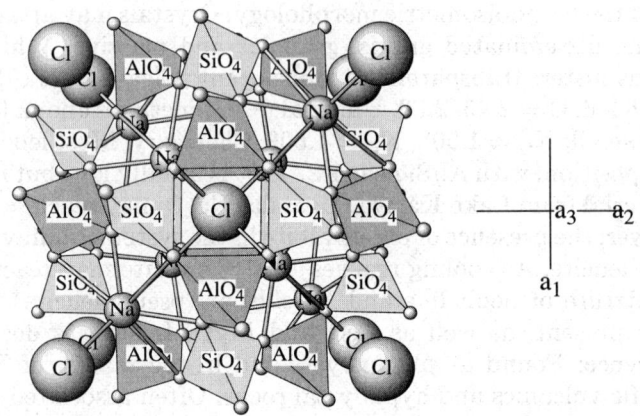

SODALITE: $Na_4(Al_3Si_3O_{12})Cl$
Perspective view along a.

In the structure, 24 tetrahedra (alternately occupied by Al and Si in a fully ordered scheme), and all shared by adjacent unit cells, make up a series of linked cages having cubo-octahedral configuration. The large cations, predominantly Na and Ca, and the anions Cl, $(SO_4)^{2-}$, $(OH)^{-1}$, S^{2-} and to a lesser extent, water, are located within the large voids of the cage. The exact position of these larger anionic groups is dependent on the particular combination encountered. The $P\bar{4}3n$ space group may represent an averaged structure, particularly for intermediate phases in which the substitutions are not wholly random or fully ordered and thus leading to domains of lower symmetry.

SODALITE MINERALS

Mineral	Formula	Space Group	a	
Sodalite	$Na_4(Al_3Si_3O_{12})Cl$	$P\bar{4}3n$	8.882	76.2.3.1
Nosean	$Na_8(Al_6Si_6O_{24})(SO_4) \cdot H_2O$	$P\bar{4}3n$	9.084	76.2.3.2
Haüyne	$Na_3Ca(Al_3Si_3O_{12})(SO_4)$	$P\bar{4}3n$	9.116	76.2.3.3
Lazurite	$Na_3Ca(Al_3Si_3O_{12})S$	$P\bar{4}3n$	9.105	76.2.3.4

In addition to the sodalite group, there are four related minerals with structures based on the fundamental sodalite framework.

SODALITE-RELATED MINERALS

Mineral	Formula	Space Group	a	b	c	
Bicchulite	$Ca_2(Al_2SiO_6)(OH)_2$	$I\bar{4}3m$	8.825			76.2.3.5
Kamaishilite	$Ca_2(Al_2SiO_6)(OH)_2$	$I4(?)$	8.850		8.770	76.2.3.6
Tugtupite	$Na_4(BeAlSi_4O_{12})Cl$	$I\bar{4}$	8.640		8.873	76.2.3.7
Tsargorotsevite	$Na(CH_3)_4(Si_{2.5}Al_{0.5}O_6)_2$	$I222$	8.984	8.937	8.927	76.2.3.8

Solid solution between sodalite and nosean, haüyne and lazurite, appears to be complete at elevated temperatures (approximately 1000°C and higher for increased pressure) but is limited at lower temperatures.

76.2.3.1 Sodalite $Na_4(Al_3Si_3O_{12})Cl$

The name was coined by Thomson in 1811 in allusion to the high sodium content. Sodalite group. The varietal name, hackmanite, honors Victor Axel Hackman (1866–1941), Finnish petrologist. ISO $P\bar{4}3n$. $a = 8.882$, $Z = 2$, $D = 2.30$. 37-476: 6.28_4 4.44_1 3.62_{10} 2.81_1 2.56_2 2.37_2 2.09_2 1.570_1. Structure: *AC(B) 40:6(1984)*. **Habit:** Dodecahedra, sometimes elongated, cubo-dodecahedra, octahedra, well formed and sharp to crude, from microcrystals to 10 cm in diameter; also as embedded grains and massive, sometimes crudely cleavable. Twinned on {111}, often as penetration twins and complex multiple twins. **Physical properties:** Gray, white, greenish or yellowish white, light to medium blue or violet blue, pink; white or very pale blue streak; transparent to translucent. Cleavage {110}, poor to distinct; uneven to conchoidal fracture; brittle. $H = 5\frac{1}{2}$–6. $G = 2.14$–2.30. **Optics:** Isotropic; $N = 1.483$–1.487. The color of yellow, pink, and violet varieties (hackmanite) tends to fade on exposure to visible light, but returns after storage in the dark for a few weeks or exposure to UV or X-radiation; these also display an intense orange to orange-red fluorescence in LWUV and SWUV, often with a short-lived yellowish-white phosphorescence. Deep violet and blue varieties appear to be color stable. Fusibility $3\frac{1}{2}$–4; gelatinized by HCl. Massive blue material is usually distinguished from lazurite by the absence of associated pyrite and its almost white streak. **Chemistry:** Substitutions for components of the ideal formula are very limited, but may strongly influence color. Al and Si are always close to a 1:1 ratio, with only minor Fe^{3+} substitution for aluminum; minor substitution of K^+ and Ca^{2+} is also possible. Other than the related tugtupite, sodalite is the only member of the group in which chlorine is important, but $(OH)^-$ may substitute for it and S^{2-} is almost invariably present in small amounts. The color and fluorescence phenomena have been attributed, in part, to the presence of sulfide or sodium polysulfide clusters. Complete solid solution between sodalite and the sulfate- or sulfide-rich members of the group exists only at elevated temperatures (>1000°C). **Occurrence:** Most frequently encountered

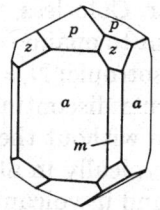

Scapolite, Monte Somma, Italy. Forms: a, {100}; m, {110}; p, {101}; z, {131}.

in nepheline syenites and associated pegmatitic rocks, and in volcanic ejecta. Less commonly it may be found in metasomatized calcareous rocks at the contact with alkaline igneous bodies. Localities: Only a sampling of noteworthy or well-documented occurrences is given. Massive blue sodalite with distinct cleavage associated with zircon and cancrinite at Litchfield and West Gardiner, ME; at Red Hill, NH; violet-blue in syenite with orthoclase, biotite, and zircon at Salem, Essex Co., MA; blue and white (hackmanite) at Magnet Cove, Hot Springs Co., AR; at Cripple Creek, Teller Co., CO. Sodalite, including hackmanite, is abundant at Mt. St.-Hilaire, QUE, sometimes in exceptional crystals(!), transparent, colorless to deep pink and pale violet, and up to several centimeters in diameter; hackmanite at the Davis quarry, east of Bancroft, and blue massive(!) at the Princess quarry, 4 km E of Bancroft, ON; in the Ice River complex, Ottertail Range, BC, Canada. Found on both sides of the *Kangerdluarssuk and Tunugdliarfik fjords, Greenland*; in the Langesundfjord area, Norway; in volcanic ejecta as white translucent dodecahedra with haüyne and sanidine, and rarely, as green cubododecahedra with vesuvianite and nepheline in limestone at Mte. Somma, and with sanidine in sodalite–trachyte at Scarrupata, Ischia, Campania, Italy; with nepheline and analcime at Val di Noto, Sicily; in the Laacher See area, Eifel, Germany; at several localities in the Kola Penin., and at Miask, Ilmen Mts., Russia. Blue and white (hackmanite) sodalite found at Kishengar, Rajputana, India; blue and white massive in Gil Ju Co., Ham Gyong North Prov., North Korea; dark blue in dike in nepheline syenite at Cerro Sapo, Cochabamba, and, of uncertain origin, in the ruins at Tiahuanaco, Bolivia. Uses: Massive blue material has been used as an ornamental stone and a carving material. Transparent Mt. St.-Hilaire material has been faceted into collector gems up to 15+ ct. AR *D6:429(1892), DHZI 4:289(1963), CM 21:549(1983), MR 21:339(1990), ZK 199:75(1992).*

76.2.3.2 Nosean Na$_8$(Al$_6$Si$_6$O$_{24}$)(SO$_4$)·H$_2$O

Named in 1815 by Klaproth for Karl Wilhelm Nose (1753?–1835), mineralogist, Brunswick, Germany. Sodalite group. Synonym: noselite. ISO $P\bar{4}3m$. $a = 9.084$, Z = 1, D = 2.21. *17-538*(syn): 9.09$_6$ 6.45$_7$ 3.71$_{10}$ 2.87$_5$ 2.63$_7$ 2.27$_2$ 2.14$_4$ 1.781$_3$. Structure: *CM 27:165(1989)*. The centrosymmetric space group is for an averaged structure made up of $P23$ domains occupied by Na$_4$(SO$_4$) and Na$_4$(H$_2$O); Al and Si are fully ordered. Dodecahedral crystals; twinned on {111}; usually massive, granular. Colorless, white, gray, brown, blue; vitreous luster; transparent to translucent to opaque due to inclusions. Cleavage {110}, poor. H = $5\frac{1}{2}$–6. G = 2.3–2.4. Isotropic; N = 1.470–1.495. Gelatinizes in acids without liberation of H$_2$S. Forms discontinuous solid solution with haüyne. The formula is commonly given without the H$_2$O, but it is almost invariably present in natural examples. Typically in alkali-rich, silica-deficient volcanic rocks, particularly phonolites, and in volcanic bombs. Found in phonolite and tinguaite in Lawrence Co., SD, in the Cripple Creek area, Teller Co., CO, and in the La Sal Mts., UT. Occurs as phenocrysts in phonolite in the Canary and

Cape Verde Is.; at Wolf Rock, Cornwall, England; at Cantal, France; from the Schellkopf quarry, Laacher See, Eifel, Germany; at Alban Hills, S of Rome, Italy; and elsewhere. It has been reported in syenites from northern China. AR *DHZI 4:289(1963)*, *AM 74:394(1989)*.

76.2.3.3 Haüyne $Na_3Ca(Al_3Si_3O_{12})(SO_4)$

Named in 1807 by Brunn-Neergard for René Just Haüy (1743–1822), French pioneer in crystallography. Sodalite group. ISO $P\bar{4}3n$. $a = 9.116$, $Z = 2$, $D = 2.41$. *37-474*: 6.47_2 3.72_{10} 2.87_1 2.62_2 2.43_1 2.14_1 1.936_1 1.781_1; also *20-1087, 25-802*. Structure: *CM 29:123(1991)*. Al and Si in the tetrahedral framework are fully ordered. The centrosymmetric space group is for an averaged structure made up of domains with $P23$ symmetry populated with either $Na_3Ca(SO_4)$ or $K_2Ca(OH)$, significant K being present in almost every case. Dodecahedral or octahedral; commonly in rounded grains. Twinning, contact, penetration, and polysynthetic, on {111} common. Blue, white, gray, yellow, green, pink; pale blue to white streak; vitreous to greasy luster; transparent to translucent. Cleavage {110}, distinct; uneven to conchoidal fracture; brittle. $H = 5\frac{1}{2}$–6. $G = 2.44$–2.50. Isotropic; $N = 1.496$–1.505. Fusibility $4\frac{1}{2}$. Gelatinizes with acids. As much as 5 wt % K_2O may be present; H_2O is also present (although not as much as in nosean), as well as Cl. Forms a discontinuous solid solution with nosean and with sodalite. Most commonly found in phonolites and related rocks, and less commonly in metamorphic rocks (marble). Haüyne of metamorphic origin at the Edwards mine, Edwards, St. Lawrence Co., NY; in nepheline alnöite with melilite, phlogopite, and apatite at Winnett, Petroleum Co., MT; in Lawrence Co., SD; the Cripple Creek district, Teller Co., CO. Also found at Aruca, Canary Is.; in the Auvergne, France; variously colored at Mte. Vulture, SW of Melfi, in the Alban Hills, Lazio and at *Mte. Somma, Campania, Italy*; at Niedermendig and Laacher See, Eifel, Germany; at Malaya Bistraya, Siberia, Russia; near Nanjing, Jiangsu Prov., China; and at Jebel Tourguejid, Morocco. AR *D6:431(1892)*, *CM 27:173(1989)*.

76.2.3.4 Lazurite $Na_3Ca(Al_3Si_3O_{12})S$

Named in 1890 by Brøgger from the Persian *lazhward*, blue, from the earlier-used lapis lazuli, which owes its color to lazurite but may incorporate several other phases. Sodalite group. The name is easily confused with the blue phosphate species lazulite of similar etymology. Equivalent to the *sapphiros* of Pliny (77a.d.). Ultramarine, the blue pigment now produced synthetically, was originally applied to powdered lazurite. ISO $P\bar{4}3n$. $a = 9.105$, $Z = 2$, $D = 2.39$. *17-749*: 6.43_4 3.71_{10} 2.87_5 2.62_8 2.27_3 2.14_4 1.782_3 1.606_2; Also *41-1392* for TRIC $P1$. $a = 9.09$, $b = 12.86$, $c = 25.71$, $\alpha = \beta = \gamma = 90°$; and *41-1393* for MON $P2/a$. $a = 36.36$, $b = 51.40$, $c = 51.40$, $\beta = 90°$; the four strongest lines of these patterns are identical, and the cell edges are either direct multiples of a of the $P\bar{4}3n$ cell or multiples of $2\frac{1}{2}a$ thereof. Structure: *AC(C) 41:827(1985)*. When well-crystallized form dodecahedra or cubododecahedra up to 5 cm;

more commonly compact, massive incorporating more or less of other minerals, primarily pyrite and calcite. Multiple contact or penetration twins on {111}, common but generally not apparent. Deep blue to lighter shades, violet blue, greenish blue, colorless; bright blue streak; dull to vitreous luster. Cleavage {110}, distinct; uneven fracture; brittle. H = 5–5$\frac{1}{2}$. G = 2.38–2.45. Isotropic; N = 1.500–1.522. Fusibility 3. Gelatinizes with acids with liberation of H_2S. As in nosean, Ca substitution is limited. Complete solid solution with nosean may exist, but is very limited with regard to sodalite. The low-symmetry, large-volume cells proposed for triclinic and monoclinic lazurites may reflect a nonrandom occupancy of anions in the framework cages. Occurs in metamorphosed limestones and dolomitic limestones and at pegmatite–limestone contacts. It has been reported from the Green River formation, CO, UT, and WY; at Ontario Peak, San Bernardino Co, CA; in ejected masses at Mte. Somma and Latium, Italy. At Lyadzhuar–Darinsk in the Pamir Mts., Tajikistan; and in dolomitic marble on the Slyudyanka, Malaya Bystraya, and Talaya Rivers, Sayan Mts., near Lake Baikal, Irkutsk, Russia. Massive lapis lazuli(!) has been mined since antiquity at *Sar-e-Sang, on the Kowkcheh R., Badakshan, Afghanistan*, where it occurs in white calcite marble, often enclosing pyrite, and sometimes in octahedral and cubo-octahedral crystals(!) to 5 cm in diameter; lapis lazuli of lighter color and less pyrite than Afghan material, associated with granite, at the headwaters of the Cazadero and Vias rivers, near Ovalle, Coquimbo, Chile. Long used as an ornamental stone in cabochons and small to large carvings, particularly the deep blue Afghan material. Crushed, it forms the artist's pigment, ultramarine, but now largely replaced by synthetic ultramarines of several blue to violet hues. Material of lesser or less uniform color has been dyed to enhance its appeal. AR *D6:432(1892)*, *DHZI 4:289(1963)*, *BM 89:333(1966)*, *ZVMO 119:110(1990)*, *PD 8:202(1993)*.

76.2.3.5 Bicchulite $Ca_2(Al_2SiO_6)(OH)_2$

Named in 1973 by Henmi et al. for the Japanese locality. Sodalite group. Dimorphous with kamaishilite. ISO $I\bar{4}3m$. $a = 8.825$, $Z = 4$, D = 2.824. PD: 3.60_9 3.04_4 2.96_4 2.79_{10} 2.75_9 2.60_5 2.46_4 1.559_5; also *27-66*(syn). Structure: *ZK 146:35(1977)*. Colorless, vitreous to dull luster, translucent. Cleavage not determined. $G_{syn} = 2.75$. Isotropic; N = 1.625. Found in skarns at *Carneal, County Antrim, Northern Ireland*, and at *Tuka, Bicchu, Okayama Pref., Japan* and at the Alkagane mine, Iwate pref. AR *MJJ 7:243(1973)*, *AM 59:1330(1974)*.

76.2.3.6 Kamaishilite $Ca_2(Al_2SiO_6)(OH)_2$

Named in 1981 by Uchida and Iiyama for the locality. Sodalite group. Dimorphous with bicchulite. TET *I*-lattice. $a = 8.850$, $c = 8.770$, $Z = 4$, D = 2.83. *35-673*: 3.61_{10} 2.80_8 2.78_4 2.55_4 2.36_2 2.35_2 2.09_3 2.08_2. Colorless, transparent. Uniaxial but nearly isotropic with N = 1.629 and no interference figure observed. Occurs as 0.1-mm grains associated with vesuvianite, garnet, perovskite, calcite, magnetite, and chalcopyrite in vesuvianite skarn at the Kamaishi mine, Kamaishi, Iwate Pref., Honshu, Japan. AR AM $67:855(1982)$.

76.2.3.7 Tugtupite $Na_4(BeAlSi_4O_{12})Cl$

Named in 1962 by Sørensen for the locality, but discovered in 1957. Sodalite group. TET $I\bar{4}$. $a = 8.640$, $c = 8.873$, $Z = 2$, $D = 2.34$. $38\text{-}472$: 6.19_4 6.11_1 4.43_1 3.59_4 3.54_{10} 2.52_2 2.06_1 2.04_1. Structure: CM $29:385(1991)$. The structural framework is intermediate between that of sodalite and helvite. Tugtupite involves an ordered substitution of BeAl for AlAl, thus reducing the symmetry. Most commonly as massive, granular material; as small globular aggregates; rarely in millimeter-size crystals sometimes in complex penetration trillings of pseudocubic habit with twin plane {101}, and also as pseudotrigonal contact twins on {10$\bar{1}$} and {01$\bar{1}$}. White, light pink, rose red; vitreous to greasy luster; transparent to translucent. Cleavage bipyramidal distinct; conchoidal fracture, brittle. H ≈ 4. G = 2.3. Uniaxial (+) sometimes biaxial (+); $2V < 10°$; $N_O = 1.496$, $N_E = 1.502$. The color tends to deepen on exposure to light, and fade when stored in darkness. Intensely fluorescent, dark crimson in SWUV and bright to pale crimson in LWUV. Found rarely as millimeter-size globular aggregates and crude pseudo-octahedra at Mt. St.-Hilaire, QUE, Canada; in hydrothermal veins associated with albite, analcime, aegirine, neptunite, and pyrochlore at several sites in the Ilimaussak Complex, as at Kangerdluarssuk, Kvanefjeld, and in the cliffs of Tugtup agatakôrfia, Tunugdliarfik, Greenland; also, but very sparsely, at Mt. Pengisschorr and Mt. Punkaruiv, Lovozero massif, Kola Penin., Russia. As a lapidary material it has produced cabochon gems, and small (<2.0 ct) faceted gems, principally for the collector. AR AM $48:1178(1963)$; MR $21:343(1990)$, $24:26(1993)$.

76.2.3.8 Tsargorodtsevite $N(CH_3)_4[Si_2(Si_{0.5}Al_{0.5})O_6]_2$

Named in 1993 by Pautov et al. for Sergei Vasilievich Tsargorodtsev (1953–1986), mineral collector. ORTH (pseudocubic). $I222$. $a = 8.984$, $b = 8.937$, $c = 8.927$, $Z = 2$, $D = 2.01$. PD(nat): 6.33_6 4.50_1 4.46_8 3.66_{10} 3.16_1 2.83_1 2.59_2 2.23_1. Powder pattern is very close to ammonian sodalite. Structure: $R = 4.7\%$, first mineral known with isolated polyatomic organic groups: $[N(CH_3)_4]$. A tetramethyl ammoniated aluminosilicate. Transforms to TET $I422$, at 870°, $a = 8.908$, $c = 8.925$, and then to cubic, $I432$, at 970°, $a = 8.817$. DAN $SSSR$ $317:884(1991)$, DAN RAN $332:309(1993)$. Pseudoisometric, pseudocubic habit with {100}, {001}, {101}, {011}, {110}, {010}; {100}, {001}, and {010} faces are dull, but others are lustrous; crystals to 10 mm across; sector twinning present. Colorless, very pale yellowish to opaque white; vitreous luster. No cleavage observed, conchoidal fracture, brittle. H ≈ 6. G = 2.04. IR data resemble those for sodalite. $ZVMO$ $122:128(1993)$. Raman spectral data and DTA and TGA data also given. No response to UV. Insoluble in boiling H_2SO_4, HNO_3, or HF. Partly dissolves with prolonged boiling in H_3PO_4. Analysis (%): Si, 31.0; Al, 6.45, N, 3.2; O, 45.6; need ≈ 2.8% H + 11.1% C as well. Biaxial (−); ($2V = 76°$; $N_x = 1.528$, $N_z = 1.531$; colorless; birefringence = 0.002; r < v, weak. Sector zoning is present in crystals, details given in *ZVMO 122:128(1993)*. Occurs in friable material filling tectonic fractures in muscovite–chlorite schist which occurs in the *Man'-Khambo Mts., Khanty–Mansiysk region, northern Ural Mts., Russia*. Description was done from specimens in the museum of the Ilmen State Preserve. Recorded as sodalite. On a sample of mica schist. Also on another specimen in the state museum in Yekaterinburg. Fluid inclusion homogenization temperature of 185°. *Man-Khambo Range, Polar Urals, Khanti-Mansi Autonomous Area*. EF MK *1:28(1994)*.

76.2.4.1 Helvite $Mn_4[Be_3Si_3O_{12}]S$

Named in 1817 by Werner from the Greek *helios*, sun, in allusion to its color. See danalite (76.2.4.2) and genthelvite (76.2.4.3). ISO $P\bar{4}3n$. $a = 8.291$, $Z = 2$, $D = 3.23$. *29-217*: 3.38_{10} 2.62_6 2.22_7 1.955_8 1.692_6 1.510_5 1.466_5 1.422_5. Structure: *AM 70:186(1985)*. The structural framework is similar to that of sodalite; Be and Si are completely ordered. The cell-edge length is related to the larger cations being greatest for Mn, intermediate for Fe, and least for Zn. The habit is markedly tetrahedral with {111} dominant, and modified by {1$\bar{1}$1}, {101}, and {2$\bar{1}$1}, up to several centimeters in diameter; less commonly pseudo-octahedral; also in rounded aggregates. Yellow, yellow-green, red-brown to brown, gray, darkens with weathering; vitreous to resinous luster; transparent to translucent. Cleavage {111}, distinct; conchoidal to uneven fracture; brittle. H = 6. G = 3.17–3.37. Isotropic; N = 1.728–1.747. Strong, deep red fluorescence (Mt. St-Hilaire) in LWUV and SWUV. Gelatinizes with acids. Forms complete solid solution with danalite and genthelvite. Found in granite and nepheline–syenite pegmatites, magnetite–fluorite skarns, and in hydrothermal environments. Found at Cumberland, RI; at Amelia, VA; at the Lexington and East Moulton mines, Butte, MT; at the Sunnyside mine, Silverton district, San Juan Co., CO; in skarn of banded fluorite, magnetite, and garnet at Iron Mt., Cuchillo Range, Sierra Co., NM; in tactite in the Dragoon Mts., Cochise Co., AZ; at the Miller mine, Beaver Co., UT; in the sawtooth batholith, ID, sometimes enclosed in topaz; and at Pala and Rincon, San Diego Co., CA. Rarely as small, sharp tetrahedra at Mt. St.-Hilaire, QUE, Canada; as sharp tan–yellow crystals to 4 mm with rhodochrosite and ilvaite at the Potosi mine, Santa Eulalia, CHIH, Mexico. At Langesundfjord and elsewhere in Norway; with magnetite and fluorite at Lupikko, Finland; at Breitenbrunn and *Schwarzenberg, Saxony, Germany*; at Kapnik, Hungary; as large spherical masses with topaz, phenakite, and monazite in the Ilmen Mts., near Miask, Russia. Abundant near Shantou, Guangdong, China, in quartz veins carrying nearly pure helvite–beryl intergrowth; massive at Saiya, Ladimi Bouchi, Nigeria; as sharp, lustrous tetrahedra to 12 cm(!) at the

Chingolo mine, Cosquin, Cordoba, Argentina. AR *D6:434(1892)*, *MM 40:627(1976)*.

76.2.4.2 Danalite $Fe_4[Be_3Si_3O_{12}]S$

Named in 1866 by Cook for James Dwight Dana (1813–1895), American mineralogist, Yale University. See helvite (76.2.4.1) and genthelvite (76.2.4.3). ISO $P\bar{4}3n$. $a = 8.218$–8.232, $Z = 2$, $D = 3.40$. *11-491:* 3.68_4 3.35_{10} 2.19_5 1.932_7 1.451_4 1.118_5 1.043_5 1.012_6. Structure: *AM 70:186(1985)*. Isostructural with helvite and genthelvite. The habit is commonly pseudo-octahedral or dodecahedral; also in rounded aggregates. Yellow, pink, red-brown to brown, gray; darkens with weathering; vitreous to resinous or greasy luster; transparent to translucent. Cleavage {111}, trace; conchoidal to uneven fracture; brittle. H = 6. G = 3.31–3.46. Isotropic; N = 1.753–1.771. Gelatinizes with acids. Forms complete solid solution with helvite and genthelvite. Found in granite pegmatites, gneiss, and contact-metasomatic and hydrothermal environments. Found at Bartlett, Carroll Co., NH; in the *Rockport granite, Cape Ann, Essex Co., MA*; at Grants Mills, RI; in the Pikes Peak district and St. Peter's Dome, El Paso Co., CO; and near Jerome, Yavapaii Co., AZ. Found at Needlepoint Mt., McDame area, BC, Canada; near Redruth, St. Just, Falmouth, and Lanivet, Cornwall, England; Sweden; Russia; at the Mihara fluorite mine, Hiroshima Pref., Honshu, Japan; and at Coolgardie, WA, Australia. AR *D6:435(1892)*, *CM 6:68(1957)*.

76.2.4.3 Genthelvite $Zn_4[Be_3Si_3O_{12}]S$

Named in 1944 by Glass et al. for its relationship to helvite and for Fredrick August Ludwig Karl Wilhelm Genth (1820–1893), German–American chemist and mineralogist. See danalite (76.2.4.2) and helvite (76.2.4.1). ISO $P\bar{4}3n$. $a = 8.149$, $Z = 2$, $D = 3.64$. *38-467:* 3.32_{10} 2.57_6 2.17_7 1.916_8 1.657_6 1.483_5 1.435_5 1.393_5. Isostructural with helvite and danalite. Crystals commonly tetrahedral, up to 5 cm in diameter; as large rounded masses. Colorless, white, yellow, green, pink to red; darkens to brown to black on weathering; vitreous luster; transparent to translucent. Uneven fracture, brittle. H = 6–6$\frac{1}{2}$. G = 3.62. Isotropic; N =1.742–1.745. Strong green fluorescence in LWUV and SWUV; occasionally, short-lived phosphorescence. May form complete solid solution with helvite and danalite. Gelatinizes with acids. Found in granite and syenite pegmatites. Found at *Rockport, MA; Cheyenne Canyon, near Cookstove Mt., El Paso Co., CO*, and several nearby localities; at Iron Mt., Cuchillo Range, Sierra Co., NM. In excellent crystals associated with analcime at Mt. St.-Hilaire, QUE, Canada(!); in the Cairngorm Mts., Scotland; the Lovozero massif, Kola Penin., Russia; in the Air Mts., Niger; and in columbite-bearing granites of the Jos–Bukuru complex, Nigeria. AR *AM 29:163(1944), 73:1384(1988); VLA 2:119(1966)*.

CANCRINITE GROUP

The cancrinite group minerals are feldspathoids, tectosilicates characterized by an aluminosilicate framework. The general formula is

$$A_{6-9}(Si,Al)_{12}O_{24}B_{1-4} \cdot nH_2O$$

where

A = Na, K, Ca
B = SO_4, CO_3, OH, Cl

and a structure based on a common framework of sixfold rings of tetrahedra.

The structure of the cancrinite minerals is based on a sixfold ring of tetrahedra: each ring is linked to three rings in the preceding layer and to three other rings in the succeeding one, resulting in a three-dimensional framework. The layers succeed each other along c according to the stacking sequence ABAB. Weak and diffuse reflections exist, indicating that superstructures exist with values of c equal to $5c$, $8c$, $11c$, $16c$, and $21c$, where $c = 5.13$ Å.

The aluminosilicate framework is constant for the various members of the group, and they differ only in the ordering of the cations outside the tetrahedral framework, which leads to the appearance of satellite reflections.

Most of the cancrinite minerals have Na as the dominant A-site cation, except for liottite, which has Ca dominant. A potassium-dominant vishnevite also exists. The Si/Al ratio is ideally 1:1 and is usually this value, but may be somewhat greater and is rarely less.

CANCRINITE GROUP

Mineral	Formula	Space Group	a	c	
Afghanite	$(Na,Ca,K)_8(Si,Al)_{12}O_{24}(SO_4,Cl,CO_3)_3 \cdot H_2O$	P31c	12.770	21.35	76.2.5.1
Bystrite	$Ca(Na,K)_7Si_6Al_6O_{24}(S^{2-})_{1.5} \cdot H_2O$	P31c	12.855	10.700	76.2.5.2
Cancrinite	$Na_6Ca_2[Al_6Si_6O_{24}](CO_3)_2 \cdot 2H_2O$	$P6_3$	12.75	5.15	76.2.5.3
Cancrisilite	$Na,\square)_8(Si,AL)_{12}O_{24}(CO_3)_{1-2} \cdot 3H_2O$	$P6_3mc$	12.575	5.105	76.2.5.4
Davyne	$Na_4K_2Ca_2(Si_6Al_6O_{24})(Cl,SO_4,CO_3)_{2-3}$	$P6_3/m$	12.8	5.37	76.2.5.5
Franzinite	$(Na,Ca)_7(Si,Al)_{12}O_{24}(SO_4,CO_3,OH,Cl)_3 \cdot H_2O$	$P\bar{3}m1$ or P321	12.88	26.58	76.2.5.6
Giuseppettite	$(Na,K,Ca)_{7-8}(Si,Al)_{12}O_{24}(SO_4,Cl)_{1-2}$	$P6_3mc$ or $P\bar{6}2c$ or $P6_3/mmc$	12.850	42.22	76.2.5.7
Hydroxycancrinite	$Na_8[Al_6Si_6O_{24}](OH)_2 \cdot 2H_2O$	P3	12.7	5.18	76.2.5.8
Liottite	$(Ca,Na,K,Fe)_8(Si,Al)_{12}O_{24}[(SO_4),(CO_3),(Cl,OH)]_4 \cdot H_2O$	$P\bar{6}$	12.870	16.096	76.2.5.9

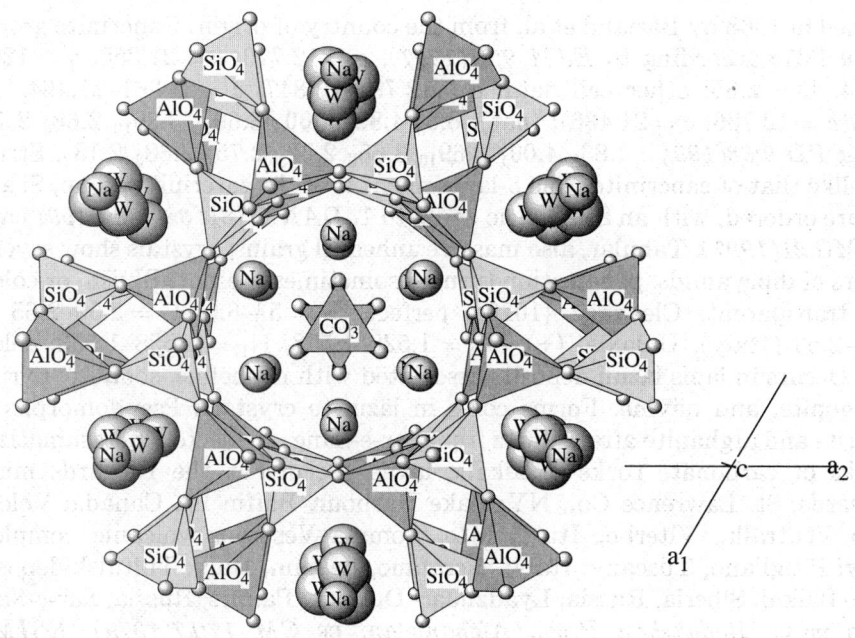

CANCRINITE: Na$_8$Al$_6$Si$_6$O$_{24}$ · 2CO$_3$ · 5H$_2$O

A framework with strictly alternating Al and Si tetrahedra forms 6- and 12-membered rings. Na$_1$ atoms at the center of the six-membered rings are coordinated by disordered oxygen, and Na$_2$ atoms in the 12-membered rings are coordianted partially by disordered Co$_3$ groups—less than half of the oxygen sites in the interring cavities are occupied.

CANCRINITE GROUP

Mineral	Formula	Space Group	a	c	
Microsommite	(Na,Ca,K)$_{7-8}$(Si,Al)$_{12}$O$_{24}$(Cl$_{2.5}$, SO$_4$, CO$_3$)$_{2-3}$	P6$_3$22 or P6$_3$	22.15	5.28	76.2.5.10
Pitiglianoite	Na$_6$K$_2$Si$_6$Al$_6$O$_{24}$(SO$_4$) · 2H$_2$O	P6$_3$	22.121	5.211	76.2.5.11
Quadridavyne	(Na,K)$_6$Ca$_2$[Al$_6$Si$_6$O$_{24}$]Cl$_4$	P6$_3$/m	25.771	5.371	76.2.5.12
Sacrofanite	(Na,Ca,K)$_9$Si$_6$Al$_6$O$_{24}$(OH,SO$_4$, CO$_3$Cl)$_4$ · nH$_2$O	P6$_3$/mmc	12.865	74.240	76.2.5.13
Tounkite	(Na,Ca,K)$_8$(Al$_6$Si$_6$O$_{24}$)(SO$_4$)$_2$Cl H$_2$O	P6$_2$22	12.843	32.239	76.2.5.14
Vishnevite	(Na,Ca,K)$_6$(Si$_6$Al$_6$O$_{24}$)(SO$_4$,CO$_3$, Cl)$_{2-4}$ · 2H$_2$O	P6$_3$	12.70	5.18	76.2.5.15

Note: Wenkite (76.2.5.16) is a related structure.

76.2.5.1 Afghanite (Na,Ca,K)$_8$(Si,Al)$_{12}$O$_{24}$(SO$_4$,Cl,CO$_3$)$_3 \cdot$ H$_2$O

Named in 1968 by Bariand et al. from the country of origin. Cancrinite group. HEX-R $P31c$ according to *EJM 9:21(1997)*. $a = 12.770$, $c = 21.350$, $\gamma = 120°$, $Z = 4$, $D = 2.65$; other cell data: $a = 12.796$–12.847, $c = 21.361$–21.464. *20-1086*($a = 12.786$, $c = 21.436$); (nat): 6.0$_5$ 4.82$_8$ 4.00$_6$ 3.69$_{10}$ 3.30$_{10}$ 2.68$_6$ 2.46$_5$ 2.13$_6$; *PD 9:68(1994)*: 4.83$_3$ 4.00$_1$ 3.69$_{10}$ 3.65$_6$ 2.77$_1$ 2.75$_1$ 2.68$_3$ 2.13$_2$. Structure like that of cancrinite. The 8-layer member of the carcrinite group, Si and AL are ordered, with an Si/Al ratio equal to 1. *DAN SSSR 320:882(1991)* and *L-JMG:21(1997)*. Tabular, also massive anhedral grains; crystals show several orders of dipryamids, penetration turning sometimes present. Bluish, or colorless transparent. Cleavage {10$\bar{1}$0}, perfect. H = 5$\frac{1}{2}$–6.0. G = 2.55–2.65 or 2.44–2.53 (Italy). Uniaxial (+); N$_O$ = 1.522–1.528, N$_E$ = 1.528–1.533; colorless. Occurs in lapis lazuli deposits associated with nepheline, sodalite, pyrite, phlogopite, and olivine. Forms cores in lazurite crystals. Pseudomorphs of lazurite and afghanite after quartz from Sar-e-Sang. In ejected metasomatized blocks of carbonate rocks in alkalic lavas. Occurs at the Edwards mine, Edwards, St. Lawrence Co., NY; Lake Harbour, Baffin Is., Canada; Volcan Vico Ventralla, Viterbo, Italy; Mte. Somma–Vesuvius volcanic complex, Italy; Pitigliano, Tuscany, Italy; Sacrafano, Latium, Italy; Tultuisk deposit, Lake Baikal, Siberia, Russia; Lyadzhuar–Darinsk, Pamirs, Russia; *Sar-e-Sang lapis mine, Badakshan Prov., Afghanistan*. EF *CM 17:47(1979)*, *RSIMP 35:713(1980)*.

76.2.5.2 Bystrite Ca(Na,K)$_7$Si$_6$Al$_6$O$_{24}$(S^{2-})$_{1.5} \cdot$ H$_2$O

Named in 1991 by Sapozhnikov et al. for the locality. Cancrinite group. See the other members of the group. HEX Originally stated to be $P6_3/mmc$ or $P\bar{6}2c$ or $P6_3mc$ (all from powder data); however, data from the structure refinement indicates HEX-R $P31c$. $a = 12.855$, $c = 10.700$, $Z = 2$, $D = 2.45$. *45-1373*(nat): 4.82$_7$ 3.92$_8$ 3.72$_{10}$ 3.38$_9$ 2.78$_3$ 2.69$_4$ 2.68$_7$ 2.47$_4$. Structure: R = 8.7%. *DAN SSSR 319:873(1991)*. Prismatic to flat plates to 5 mm long; {10$\bar{1}$0} dominant; no twinning. Yellow, transparent; white streak; vitreous luster; after heating at 800°, becomes dark green. Cleavage {10$\bar{1}$0}, good; weak pinacoidal cleavage or parting. H = 5–5$\frac{1}{2}$. G = 2.43. Nonfluorescent. Readily soluble in dilute acids, producing H$_2$S. Analysis (wt %): SiO$_2$, 33.4; Al$_2$O$_3$, 26.7; CaO, 5.10; Na$_2$O, 14.4; K$_2$O, 7.6; S, 12.0; SO$_3$, 0.50; CO$_2$, 0.32; Cl, 0.25; H$_2$O, 1.62. Uniaxial (+); N$_O$ = 1.584, N$_E$ = 1.660; birefringence = 0.076; negative elongation; O > E. O, dense yellow; E, nearly colorless. Found at the *Malo-Bystrinskoye lazurite deposit, western Pribaikal, Siberia, Russia*; in close association with lazurite in lazurite-bearing metasomatite—two generations, the second with lazurite, diopside, and calcite. It sometimes replaces lazurite. EF *ZVMO 120:97(1991)*.

76.2.5.3 Cancrinite Na$_6$Ca$_2$[Al$_6$Si$_6$O$_{24}$](CO$_3$)$_2 \cdot$ 2H$_2$O

Named by Rose in 1839 for Georg Cancrin (1774–1845), German–Russian, minister of finance of Russia. Cancrinite group. See vishnevite (76.2.5.15), microsommite (76.2.5.10), davyne (76.2.5.5), and the other members of the group. HEX-R $P6_3$. $a = 12.590$–12.75, $c = 5.117$–5.193, $Z = 1$, $D = 2.38$–2.56. (*25-776* nat): 10.92_4 4.64_9 3.64_7 3.21_{10} 2.73_4 2.10_7 1.49_5 1.45_5 (superstructure present, $c = 11c$); *34-176:* 6.30_5 4.63_9 3.63_5 3.21_{10} 2.73_3 2.60_2 2.56_3 2.41_2. AlO_4 and SiO_4 tetrahedra. The Al and Si atoms are completely ordered. Tetrahedra are corner-linked to form a framework of a chain of cages parallel to the Z axis. The cages are bounded at the top and bottom by parallel six-membered rings consisting of alternating AlO_4 and SiO_4 tetrahedra. Basic cancrinite (hydroxyl) is stoichiometric with respect to the interframework Na cations; in particular, the Na_2 site is fully occupied by Na atoms. Therefore, the Na atoms on the Na_2 site do not contribute to the satellite reflections, which arise from ordering of the (OH) and H_2O groups in the channels. The OH and H_2O atoms are in positions similar to those of the oxygen atoms of the CO_3 group in cancrinite. Another superstructure has $8 \times c$ axis. *CM 30:49(1992)*. In regular CO_3-rich cancrinite, superstructures are explained by ordering of cations, anions, and vacancies that occur in the channels. Fine-grained and commonly massive, also {0001} fibrous columnar masses and prisms with {10$\bar{1}$0} and {11$\bar{2}$0}, {10$\bar{1}$1} pyramid, {11$\bar{1}$0} elongated parallel to [0001]; crystals to several centimeters across. Lamellar twinning rare. Pale to dark yellow and pale orange, gray, green, small crystals are pale violet, pink to purple; sometimes colorless or white and also light blue to grayish blue; white streak; vitreous to subvitreous luster. Cleavage {10$\bar{1}$0}, perfect; {001}, poor; uneven fracture; brittle. $H = 5$–6. $G = 2.40$–2.45. Nonfluorescent. IR, DTA, TG, TGA data: *ZVMO 118:78(1989)*. Gelatinizes with acids. Heating to 900° does not appreciably change the cell dimensions. Uniaxial $(-)$; $N_O = 1.498$–1.528, $N_E = 1.492$–1.503. Has been synthesized many times [e.g., *26:822(1981)*, *SPC 26:33(1981)*]. Occurs as a primary mineral in alkaline rocks (plutonic and volcanic), in sodalite syenite at Mt. St.-Hilaire. Often in nepheline syenites. Also in skarn-zone silicate lavas and alkalic pegmatites. Only localities for fine specimens of cancrinite are given here: occurs as an alteration product of nepheline, in masses to 10 cm associated with sodalite at Litchfield, Kennebec Co., ME; York R., Bancroft, and Dungannon Twp., ONT, Canada; Mt. St.-Hilaire, QUE, Canada; Laacher See, Eifel, Germany; Val Malenco, Italy; Mt. Alluaiev and Mt. Karnasurt, Lovozero massif, Kola Penin., Russia. EF *NJBM 3:97(1990)*, *CM 20:239(1982) 28:341(1990)*.

76.2.5.4   Cancrisilite $Na_7[Al_5Si_7O_{24}]CO_3 \cdot 3H_2O$

Named in 1991 by Khomyakov et al. for the composition: silica-rich cancrinite. Cancrinite group. Formerly called carbonate–vishnevite. HEX $P6_3mc$. $a = 12.575$, $c = 5.105$, $Z = 1$, $D = 2.39$. PD(nat) (*ZVMO 120:80(1991)*): 6.30_7 4.61_5 3.65_9 3.22_{10} 2.72_5 2.60_2 2.40_2 2.10_2. Structure is as for other members of the cancrinite group. Symmetry is higher than that of cancrinite ($P6_3$). In cancrisilite, the 12 formula atoms of Si and Al are distributed as 7.2 Si and 4.8 Al, so that Si/Al = 1.5. Xenomorphic grains, anhedral, euhedral crystals to 1–3 mm across and aggregates to 10–15 mm. Lilac colored, clear to turbid; white streak; vitreous luster. Conchoidal fracture, brittle. H = 5. G = 2.40. IR data: *ZVMO 120:80(1991)*. In UV light, fluoresces bright yellow, characteristic of feldspathoids containing S^{2-} or SO_4^{2-} ions. Dissolves readily with effervescence of CO_2 at 25° in 10% HCl, HNO_3, and H_2SO_4. Analysis (%): Na_2O, 21; K_2O, 0.3; CaO, 0.6; MnO, 0.1; Fe_2O_3, 0.35; Al_2O_3, 24.5; SiO_2, 43.4; CO_2, 4.4; SO_3, 0.2; H_2O, 5.5. Uniaxial (−); $N_O = 1.509$, $N_E = 1.490$; colorless. Occurs in ultra-agpaitic pegmatites on *Mt. Alluaiev and along the Chinglusuai R., Lovozero massif, Kola Penin., Russia.* EF *RSIMP 35:713(1979)*.

76.2.5.5 Davyne $Na_4K_2Ca_2(Si_6Al_6O_{24})(Cl,SO_4,CO_3)_{2-3}$ (ideal formula)

Named in 1825 by Monticelli and Covelli for Sir Humphry Davy (1778–1829), English chemist. Cancrinite group. HEX $P6_3/m$ [*NJBM 3:97(1990)*] and $P6_3$ [*CM 28:341(1990)*]. $a = 12.705$–12.854, $c = 5.357$–5.368, $Z = 1$, $D = 2.43$. *20-379*(nat): 4.80_{10} 3.67_{10} 3.28_{10} 2.76_5 2.67_6 2.48_5 2.12_6. Structure details given in two papers: Ca–Cl–Ca–Cl chains along the threefold axes; Ca cations in center of sixfold rings of tetrahedra (references as above). Na and K together with sulfate groups and chlorine anions are located in the large channels along the sixfold axis. Ordering is present. Colorless, white; vitreous to pearly luster. Cleavage {10$\bar{1}$0}, {0001}, excellent. Conchoidal to uneven fracture, brittle. H = $5\frac{1}{2}$. G = 2.4. Gelatinizes with acids. Summary of analyses given in *NJBM 3:97(1990)* (wt %): SiO_2, ≈ 32; Al_2O_3, ≈28; Na_2O, ≈10; K_2O, ≈7; CaO, ≈14; SO_3 ≈4; Cl, ≈7; CO_3, ≈7; O for Cl, ≈2%. Uniaxial (+); $N_O = 1.518$–1.519, $N_E = 1.520$–1.521. Also isotropic and uniaxial (−). Occurs in volcanic ejecta and blocks from Mte. Somma–Vesuvius. Associated with leucite, orthoclase, garnet, vesuvianite, scapolite, and others. *Mte. Somma and Mt. Vesuvius, Napoli, Campania, Italy. Ptigliano, Tuscany, Italy.* EF *AM 60:435(1975)*.

76.2.5.6 Franzinite $(Na,Ca)_7(Si,Al)_{12}O_{24}(SO_4, CO_3, OH, Cl)_3 \cdot H_2O$

Named in 1977 by Merlino and Orlandi for Marco Franzini (b.1938), Italian mineralogist, University of Pisa. Cancrinite group. HEX $P\bar{3}m1$ or $P321$. $a = 12.861$–12.906, $c = 26.45$–26.62, $Z = 5$, $D = 2.52$. *30-1170*(nat): 3.81_4 3.72_{10} $3.59_{4.5}$ 3.56_4 3.30_2 3.05_2 2.66_2 2.15_3. Squat prisms to 1 cm in diameter. Pearly white to colorless. Cleavage {0001}, excellent. H = 5. G = 2.46–2.52. Analysis (wt %): SiO_2, 32.44; Al_2O_3, 25.21; CaO, 12.08; Na_2O, 11.50; K_2O, 4.24; SO_3, 10.65; CO_2, 1.54; H_2O, 1.88; minor components: Fe_2O_3, 0.04; MgO, 0.14; Cl, 0.36. Uniaxial (+); $N_O = 1.504$–1.510, $N_E = 1.506$–1.512. Occurs in a pumice deposit with afghanite and liottite in ejected blocks. Associated with diopside and vesuvianite. *Pitigliano, Tuscany*; Ariccia, Tuscany; Sacrofano, Latium, *Italy*. EF *NJBM 4:163(1977), RSIMP 35:713(1979)*.

76.2.5.7 Giuseppettite $(Na,K,Ca)_{7-8}(Si,Al)_{12}O_{24}(SO_4,Cl)_{1-2}$

Named in 1981 by Mazzi and Tadini for Guiseppe Guiseppetti, the University of Pavia. Cancrinite group. HEX $P6_3mc$, $P\bar{6}2c$, or $P6_3/mmc$. $a = 12.850$,

$c = 42.22$, $Z = 8$, $D = 2.37$. $35\text{-}479$(nat): 6.42_6 $4.32_{5.5}$ 3.71_{10} 3.45_8 3.28_5 3.13_7 2.64_6 $2.14_{6.5}$. Detailed structure not yet done. Si/Al ratio is 0.99 and is a strong indication for an ordered distribution of Si and Al in the octahedral framework. Anhedral, allomorphic masses, no crystals. Light violet-blue. No cleavage. $H = 6\text{-}7$. $G = 2.35$. IR data: confirmed no H_2O, OH, or CO_3. *NJBM 3:103(1981)*. Analysis (wt %): SiO_2, 33.25; Al_2O_3, 28.56; Fe_2O_3, 0.03; CaO, 4.85; Ha_2O, 14.37; K_2O, 8.00; SO_3, 9.92; Cl, 0.78. Uniaxial (+); Na(D): $N_O = 1.491$, $N_E = 1.507$. Occurs in fragmented, pale violet-blue veinlets a few millimeters thick in an ejected sanidinite block at *Sacrofano, Biachella Valley, Latium, Italy*. EF

76.2.5.8 Hydroxycancrinite $Na_8[Al_6Si_6O_{24}](OHCO_3)_2 \cdot 2H_2O$

Named in 1993 by Khomyakov et al. for the composition and cancrinite. Cancrinite group. See the other members of the group. HEX-R $P3$. $a = 12.664\text{-}12.740$, $c = 5.159\text{-}5.187$, $Z = 1$, $D = 2.26$. PD(nat): $6.43_{2.5}$ 4.70_6 4.17_2 3.68_7 3.26_{10} 2.76_5 2.59_2 2.43_3. Structure: *ZVMO 121:100(1993)*. R = 3.9%. Massive aggregates 10–15 mm. Blue, light blue; white streak; vitreous luster. Cleavage {$10\bar{1}0$}, excellent; steplike fracture; brittle. $H = 6$. $G = 2.32$. IR, DTA, TGA data: *ZVMO 121:100(1993)*. Lost 8.4 wt % to 1000°. Nonfluorescent under UV. Dissolves easily at room temperature in HCl, HNO_3, and H_2SO_4 with slight effervescence. Analysis (wt %): Na_2O, 23.4; K_2O, 0.45; CaO, 0.92; MgO, 0.11; MnO, 0.03; Fe_2O_3, 0.18; Al_2O_3, 31.15; SiO_2, 36.32; CO_2, 1.59; H_2O, 5.41. Uniaxial (+); $N_O = 1.494$, $N_E = 1.501\text{-}1.503$; colorless. Occurs in the northern part (Mt. Karnasurt) of the *Lovozero massif, Kola Penin., Russia*; in pegmatoid veins (1–5 cm wide), in ultra-agpaitic rocks (urtite–lujavrite–foyaite), associated with other alkalic minerals such as Natrolite, Steenstrupine, Vuonnemite, Epistolite, and others. EF *AC(B) 38:893(1982), MZ 6:50(1984), SPC 36:325(1991)*.

76.2.5.9 Liottite $(Ca,Na,K,Fe)_8(Si,Al)_{12}O_{24}[(SO_4),(CO_3), (Cl,OH)]_4 \cdot H_2O$

Named in 1977 by Merlino and Orlandi for Luciano Liotti (b.1932), Italian mineral collector, who supplied the original specimen from Tuscany. Cancrinite group. HEX $P\bar{6}$. $a = 12.870$, $c = 16.096$, $Z = 3$ [*CM 34:1021(1996)*], $D = 2.61$ also $a = 12.8575$ and $c = 16.0905$; $29\text{-}1187$(nat): $4.84_{3.5}$ 3.72_{10} $3.32_{7.5}$ 2.79_1 2.69_1 2.47_1 $2.14_{2.5}$ 1.80_1. Improved powder data in *JPD 10:13(1995)*. The 6-layer member of the cancrinite group. It has a perfectly (Si, AL)-ordered framework. Transparent, colorless crystals to 1 cm in diameter, flattened hexagonal prisms with dominant {0001}, {$10\bar{1}0$}, {$11\bar{1}0$}. $H = 5$. $G = 2.56$. Analysis (wt %): SiO_2, 30.51; Al_2O_3, 24.92; Fe_2O_3, 0.36; CaO, 16.71; Na_2O, 7.97; K_2O, 4.98; SO_3, 8.68; CO_2, 2.1; Cl, 2.57; H_2O, 1.8. Uniaxial (+); $N_O = 1.530$, $N_E = 1.528$. Occurs at *Pitigliano, Tuscany, Italy*; in ejecta blocks in pumice. The blocks are a product of syntexis between carbonate rocks and trachytic magma associated minerals include melilite, latiumite, clintonite, anorthite, vesuviantite, grossular and andradite. EF *AM 62:321(1977)*.

76.2.5.10 Microsommite (Na,Ca,K)$_{7-8}$(Si,Al)$_{12}$O$_{28}$(Cl$_{2.5}$,SO$_4$, CO$_3$)$_{2-3}$

Named in 1872 by Scacchi from the Greek for *small*, in allusion to the minute prismatic crystals, and for the locality. Cancrinite group. See the other members of the group. HEX $P6_322$. $a = 22.08$–22.156, $c = 5.237$–5.337, $Z = 3$, $D = 2.34$. *20-743*(nat): 4.81_{10} 3.69_{10} 3.29_{10} 2.765_6 2.67_8 2.66_8 2.455_6 2.13_6. Detailed structure not yet done. Hexagonal prisms; crystals to 14 cm long and 2 cm wide from St. John's Island (may be davyne). Colorless. {10$\bar{1}$0} perfect and imperfect {0001} cleavage. Vitreous, silky to brilliant luster on {0001}, $G = 2.37$–2.45. Analysis (wt %): SiO$_2$, 31; Al$_2$O$_3$, 28; CaO, 13.4; K$_2$O, 6.1; SO$_3$, 8.7; Cl, 8. Six wet analyses given in *BM 91:34(1968)*. St. John's Island complete analysis (wt %): Na$_2$O, 14.9; CaO, 10.4; K$_2$O, 0.17; SO$_3$, 6.60; Cl, 5.70; Al$_2$O$_3$, 29.4; SiO$_2$, 32.50; O for Cl, 1.36. This may be davyne because single-crystal work was not done to determine the cell dimensions. Uniaxial (+); N$_O$ = 1.521, N$_E$ = 1.529; positive elongation. Occurs from ejecta and leucite lavas. Also in fractures in metabasalt. *Mte. Somma, Vesuvius, Naples, Catania, Italy*; Pitigliano, southern Tuscany, Italy; possibly at St. John's Island, Red Sea. St. John's Is. material is extensively altered to a mixture of spinel, anorthite, calcite, corundum and amesite. EF *RSIMP 35:713(1979), NJBM 8:345(1980), CM 28:341(1990)*.

76.2.5.11 Pitiglianoite Na$_6$K$_2$Si$_6$Al$_6$O$_{24}$(SO$_4$)·2H$_2$O

Named in 1991 by Merlino et al. for the locality. Cancrinite group. Discovered in 1979, and called microsommite. *RSIMP 35:713(1979)*. Structurally very similar; however, the chemical composition is different. Na$^+$ + H$_2$O (pitiglianoite), Ca^{2+} + Cl$^-$ (microsommite). HEX $P6_3$. $a = 22.121$, $c = 5.211$, $Z = 3$, $D = 2.394$. *PD*(nat): $6.39_{str.}$ $4.77_{v.str.}$ $3.69_{med.}$ $3.27_{v.str.}$ $2.77_{med.}$ $2.65_{med.}$ (8 lines not given). *AM 76:2003(1991)*. A three-dimensional framework of alternating [SiO$_4$] and [AlO$_4$] tetrahedra with alkali cations, H$_2$O molecules, and sulfate groups in the channels and cages. Detailed structure not yet done. Hexagonal prisms to 4 mm long, 1 mm in diameter. Elongate parallel to [001]. Dominant forms are {0001} and {10$\bar{1}$0}; twinning present. Colorless, transparent; white streak; vitreous luster. Subconchoidal fracture, brittle. H ≈ 5. $G = 2.37$. Analysis (wt %): SiO$_2$, 35.0; Al$_2$O$_3$, 29.05; CaO, 0.07; Na$_2$O, 17.10; K$_2$O, 9.4; SO$_3$, 7.58; Cl, 0.01; H$_2$O, 3.46* (assuming two molecules of H$_2$O). Uniaxial (−); Na(D): N$_O$ = 1.508, N$_E$ = 1.506. Occurs at *Casa Collina, Ptigliano, Tuscany, Italy*, in a pumice quarry, in ejected blocks showing metasomatic effects. Associated with apatite, diopside, and grossular. A product of reaction between trachitic magma and carbonate and sulfate rocks of the volcanic vent. EF

76.2.5.12 Quadridavyne $(Na,K)_6Ca_2(Al_6Si_6O_{24})Cl_4$

Named in 1994 by Bonaccorsi et al. from its relationship to and quadruple cell volume of davyne. Cancrinite group. HEX $P6_3/m$. $a = 25.771$, $c = 5.371$, $Z = 4$, $D = 2.354$. PD: 4.85_8 3.71_{vs} 3.31_{vs} 2.79_s 2.68_m 2.47_m 2.15_m 1.80_m. The structure is topologically equivalent to that of davyne, the only difference being the long-range ordering of the alkali metals, and the chlorine in open channels parallel to [0001] resulting in the doubling of the a parameter. As hexagonal prisms, $\{10\bar{1}0\}$, to 2 mm in length terminated by $\{0001\}$; twinning on $\{10\bar{1}0\}$ common. Colorless, vitreous luster, transparent. Cleavage $\{0001\}$, perfect; $\{11\bar{2}0\}$, distinct; brittle. H ≈ 5. G = 2.335. Analysis (wt. %): SiO_2 33.1; Al_2O_3 27.6; CaO 11.35; Na_2O 11.21; K_2O 5.93; SO_3 1.08; CL 12.13. Uniaxial (+); $N_O = 1.529$, $N_E = 1.532$. In altered ash of the 1906 eruption of Vesuvius at *Ottaviano, near Naples, Campania, Italy*. EF, AR *EJM 6:481(1994)*.

76.2.5.13 Sacrofanite $(Na,Ca,K)_9Si_6Al_6O_{24}(OH,SO_4,CO_3,Cl)_4 \cdot nH_2O$

Named in 1980 by Burragato et al. for the locality. Cancrinite group. See the remainder of the cancrinite group. HEX Probably $P6_3/mmc$. $a = 12.865$, $c = 74.240$, $Z = 14$, $D = 2.446$. Also $a = 12.8945$ and $c = 74.213$. *35-653*(nat): 11.1_1 $3.74_{2.5}$ 3.73_{10} 3.69_3 3.48_4 2.90_1 2.65_3 2.15_2. improved powder data *JPD 10:13{1995}*). Detailed structure not yet done. Flattened hexagonal prisms to 2 cm. Colorless, pearly luster, transparent. Cleavage $\{0001\}$, very perfect; $\{0\bar{1}10\}$, perfect probably $\{10\bar{1}0\}$. H = $5\frac{1}{2}$–6. G = 2.42. IR data: *NJBA 140:102(1980)*. Analysis (wt %): SiO_2, 33.06; Al_2O_3, 24.94; Fe_2O_3, 0.35; CaO, 8.76; Na_2O, 16.50; K_2O, 5.56; H_2O, 2.47; CO_2, 1.00; SO_3, 7.77; Cl, 0.59. Uniaxial (−); $N_O = 1.505$, $N_E = 1.486$. Occurs on *Mt. Sabatini, Sacrafano volcano area, Latium, Italy*; in a cavity of an ejecta block associated with sanidine, andradite, fassaite, pyroxene, leucite, and hauyne. EF

76.2.5.14 Tounkite $(Na,Ca,K)_8(Al_6Si_6O_{24})(SO_4)_2Cl \cdot H_2O$

Named in 1992 by Ivanov et al. for the locality. Cancrinite group. See the other members of the group. HEX $P6_222$. $a = 12.843$, $c = 32.239$, $Z = 6$, $D = 2.60$. PD(nat): 4.84_4 3.71_{10} 3.31_8 3.04_2 2.99_2 2.69_3 2.47_2 2.14_3. Detailed structure not yet done. Columnar crystals to 1 cm at Tultui. Bottle green with yellowish or dark blue shade (tint), white streak, vitreous luster. Cleavage $(10\bar{1}0)$, medium. H = 5–$5\frac{1}{2}$. G = 2.557. Mineral is readily soluble in dilute common acids, at the same time losing color. Analysis (wt %): SiO_2, 30.7; Al_2O_3, 25.4; CaO, 10.8; Na_2O, 10.25; K_2, 6.5; SO_3, 13.3; Cl, 2.90; H_2O, 0.77. Uniaxial (+); $N_O = 1.528$, $N_E = 1.543$; colorless or pleochroic in yellow-green to colorless, E > O; birefringence = 0.15. Associated with lazurite metasomatites. At Malo-Bystri it is associated with diopside–lazurite rocks. At Tultui it is at contact with lazurite calciphyres, associated with calcite, diopside, pyrite, and apatite. *Malo-Bystri and Tultui lazurite deposits, Tounka Valley, southern Pribaikal, Russia*. EF *ZVMO 121:92(1992)*.

76.2.5.15 Vishnevite (Na,Ca,K)$_6$(Si$_6$Al$_6$O$_{24}$)(SO$_4$,CO$_3$,Cl)$_{2-4}$ · 2H$_2$O or (Na,K)$_8$(Si$_6$Al$_6$O$_{24}$)(SO$_4$) · 2H$_2$O (ideal)

Named in 1931 by Belyankin for the locality. Cancrinite group. Synonym: sulfatic cancrinite. See the other members of the group. HEX $P6_3$. $a = 12.685$–12.789, $c = 5.179$–5.236, $Z = 1$, $D = 2.40$; for high-K vishnevite: $a = 12.839$, $c = 5.272$. *25-1499* and *25-1500* (drastically overexposed film, 7 lines = 100 Int.). *25-1499*(nat): 3.20_5 2.96_9 1.97_{10} 1.61_8 1.49_6 1.44_6 1.40_6 1.25_8; *ZVMO 118:78(1989)*: $6.4_{2.5}$ 4.74_5 4.19_2 3.70_9 3.27_{10} $2.77_{6.5}$ 2.65_3 2.13_3; *ZVMO 119:72(1989)*(K-rich vishnevite): 6.42_6 4.79_9 3.72_8 3.29_{10} 2.68_7 2.14_7 1.34_6. The observed superstructure in vishnevite is attributed to ordering of the SO$_4$ group. A unit subcell in vishnevite contains two symmetry-unrelated orientations of the SO$_4$ group (apical oxygen atom pointing either up or down). Equal occurrences of each SO$_4$ orientation would give a random pattern and no superstructure. Ordering of the SO$_4$ groups together with a sympathetic ordering of the cations (Na, K, and Ca) on the Na$_2$ site within the channels give rise to the observed satellite reflections in vishnevite. The superstructure features a doubling of c. *CM 22:333(1984)*, *30:49(1992)*. Crystal structure of high-K vishnevite: *SPC 34:37(1989)*. Cell parameters for vishnevite decrease upon heating to 1100°. Crystals prismatic to tabular, usually massive; lamellar twinning, rare. Blue, white, yellow, orange, pink, colorless; white streak; vitreous to pearly luster. Cleavage {10$\bar{1}$0}, perfect; {0001}, poor; uneven fracture; brittle. H =5–6; G = 2.46–2.48. IR, TG, DTA, TGA data: *ZVMO 118:78(1989)*. High-K vishnevite has G = 2.39. Analysis (wt %): SiO$_2$, 33.38; Al$_2$O$_3$, 28.84; Fe$_2$O$_3$, 0.14; MgO, 0.31; CaO, 2.77; Na$_2$O, 20.56; K$_2$O, 1.29; SO$_3$, 6.36; CO$_2$, 1.10; Cl, 0.02; H$_2$O$^+$, 5.25; H$_2$O$^-$, 0.22. High K-containing vishnevite from Synnyrsky massif, northern Pribaikal, has 12.7 wt % K$_2$O. Uniaxial $(-)$; N$_O$ = 1.501, N$_E$ = 1.495; sometimes biaxial with 2V = 5–10°. For high-K vishnevite, N$_O$ = 1.491, N$_E$ = 1.490. Occurs as a primary mineral in certain alkalic rocks and as an alteration product of nepheline. Occurs at Beaver Creek, Iron Hill, Gunnison Co., CO; Allt a' Mhuillin, Loch Borolan, Assynt, Scotland; Alnö Is., Sweden; Iivaara, Kuusamo, Finland; *Kurochkin Valley, Vishnevye Mts., Ural Mts.*, Urusai, and Chekindy, Altai Ridge, *Russia*; high-K material from Synnyr massif, northern Pribaikal, Russia; Oldoinyo Lengai volcano, Tanzaniá. EF *CM 22:333(1984)*, *30:49(1992)*.

76.2.5.16 Wenkite Ba$_4$Ca$_6$(Si,Al)$_{20}$O$_{39}$(OH)$_2$(SO$_4$)$_3$ · nH$_2$O (?)

Named in 1962 by Papageorgakis for Eduard Wenk (b.1907), Swiss mineralogist and petrologist, University of Basel. Cancrinite group. HEX $P31m$. $a = 13.512$–13.528, $c = 7.462$–7.471, $Z = 1$, $D = 3.23$. *19-1418*(nat): 3.54_7 3.46_{10} 3.38_8 3.25_7 3.15_6 2.70_6 2.69_9 2.23_8; also *27-31*. An interrupted framework silicate with six- and eight-fold Al/Si tetrahedral ring channels linked by lone Al/Si tetrahedra that have either $+\bar{c}$ or $-\bar{c}$ orientations. *AC(B) 30:1262(1974)* has (Ba,K)$_4$(Ca,Na)$_6$(Si,Al)$_{20}$O$_{41}$(OH)$_2$(SO$_4$)$_3$ · H$_2$O. Prismatic to 5 cm long and 1 cm wide, dominant forms: {0001}, {10$\bar{1}$1}, {10$\bar{1}$0}.

76.3.1.2 Meionite

Light gray; vitreous to pearly luster, greasy on fractures. Prismatic cleavage is very poor, parting is pinacoidal, brittle. H = 6. G = 3.13–3.19. IR data: *SMPM 42:269(1962)*. DTA data: *ZVMO 100:492(1971)*. Nonluminescent. Analysis (wt %): SiO_2, 31.0; Al_2O_3, 19.8; CaO, 10.0; SrO, 0.8; BaO, 27.6; Na_2O, 0.6; K_2O, 2.3; SO_3, 6.7; Cl, trace. Unaxial (−); Na(D): N_O = 1.595–1.604, N_E = 1.589–1.594; birefringence = 0.006–0.008, colorless; sometimes anomalously biaxial, 2V to 10°. Occurs between barite layers and calc–silicate rock as a product of strong metamorphism of marbles. In a barite–polymetal deposit in Kazakhstan. *Cava Mergozzoni, Candoglia, Val d'Ossola, Piedmonte, Italy*; Karagaila deposit, Kazakhstan. EF *ZK 137:113(1973)*, *AM 59:1135(1974)*.

76.3.1.1 Marialite Na₄[Al₃Si₉O₂₄]Cl

76.3.1.2 Meionite Ca₄[Al₆Si₆O₂₄](CO₃, SO₄)

The name *scapolite*, now applied to the series as a whole, was introduced in 1800 by d'Andrade and derives from the Greek *skapos*, shaft, in allusion to the prismatic habit; marialite (Ma) was named in 1866 by vom Rath in honor of his wife, Maria Rosa vom Rath (1830–1888); meionite (Me), named in 1801

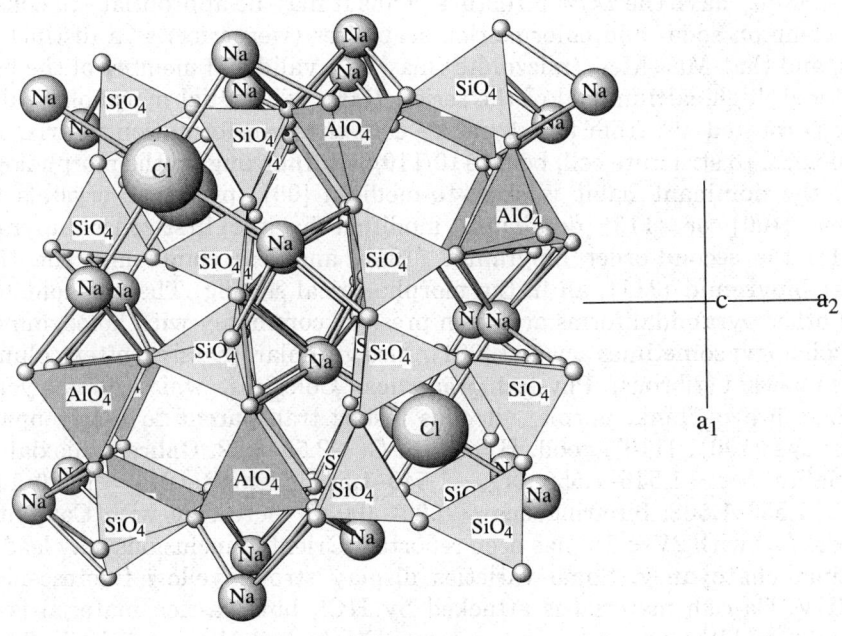

MARIALITE
Perspectivie view along *c*.

by Haüy, is from the Greek *meion*, less, in allusion to its being distinguished from vesuvianite by its less steep pyramidal faces. Wernerite, applied to the "common" scapolite, and honoring Abraham Gottlob Werner, was also named in 1800 by d'Andrade; mizzonite is from the Greek *meizon*, greater, in reference to its greater c axis relative length than for other members of the series; dipyre, a synonym for Na-rich intermediate members of the series but also used for some mizzonite, is from the Greek for *twice* and *fire*, in allusion to the two effects of heat, fusion and phosphorescence. The names *wernerite*, *dipyre*, and *mizzonite* have generally been discarded but are retained here because of evidence of space group variation within the series. **Marialite:** TET $P4_2/n$. $a = 12.059$, $c = 7.587$, $Z = 2$, $D = 2.543$. **Meionite** ($Ma_{16}Me_{84}$): TET $I4/m$. $a = 12.179$, $c = 7.571$, $Z = 2$, $D = 2.76$. *31-1279*($Ma_{65}Me_{35}$): 3.82_5 3.56_2 3.47_{10} 3.06_7 3.02_3 2.73_1 2.69_3 1.914_1; *29-1036*($Ma_{48}Me_{52}$): 6.04_2 3.82_6 3.46_{10} 3.07_7 3.03_6 2.70_3 2.69_3 1.912_3. Structure: *NJMA 149:309(1984)*. The structure is made up of four-tetrahedron rings with alternating up–down apical oxygens linked into columns parallel to [001]; the columns are cross-linked by additional four-tetrahedron rings to form the framework of disordered Al and Si tetrahedra. The large anions occupy a central location between the cross-linking rings and are surrounded by coplanar sodium and calcium. The proposed P and I structures differ in the manner of linkage of columns to rings. Scapolites in the range Me_0–Me_{75} crystallize in $P4_2/n$ except that pure end-member marialite is reported to crystallize in $I4/m$. Scapolites in the range Me_{75}–Me_{100} have the $I4/m$ structure. Thus it may be appropriate to consider the common soda- and chlorine-rich scapolites (wernerite) as a distinct species, and that $Ma_{25}Me_{75}$ (mizzonite) may be a valid end member of the body-centered, high-calcium end of the series. **Habit:** The usual morphological setting is rotated 45° from the structure cell, the transformation matrix, morphological to structure cell, being $1\bar{1}0/110/002$. Indexing on the morphological cell, the dominant habit is short-to-medium [001] prismatic crystals with either {100} or {110} dominant, modified by the first-order bipyramid {111}, the second-order bipyramid {101}, and less commonly, the third-order bipyramid {311}, all in the morphological setting. The pinacoid {001} and other pyramidal forms are often present, commonly with holosymmetric morphology; sometimes acicular [001]. Also granular, massive, often columnar to somewhat fibrous. **Physical properties:** Colorless, white, gray, yellow, orange, brown, pink, purple; vitreous luster; transparent to nearly opaque. Cleavage {100}, {110}, good. H = $5\frac{1}{2}$–6. G = 2.55–2.72. **Optics:** Uniaxial (−); marialite: $N_O = 1.546$–1.550, $N_E = 1.540$–1.541; meionite: $N_O = 1.590$–1.600, $N_E = 1.556$–1.562; birefringence = 0.004–0.037, increasing with Ca content; biaxial (−) with $2V < 10°$ has been reported. Oriented inclusions may lead to a distinct chatoyancy. Some varieties display strong yellow fluorescence in LWUV. Ca-rich material is attacked by HCl, but Na-rich material is not. **Chemistry:** Although end-member compositions have been synthesized, they have not been reported in nature, and examples within 80 mol % of either end member are very rare. The principal cation substitution is the Na–Ca couple accompanied by commensurate Si–Al substitution in tetrahedral sites. Minor potassium may substitute for sodium. Other cations, such as Mg, Fe^{2+}, and Fe^{3+}, are not significant. Anionic substitution, however, is significant, with the $Cl-CO_3$ couple generally following the Na–Ca substitution. Fluorine is a common substituent for Cl, and SO_4 may substitute for CO_3; hypothetical sulfate-marialite and sulfate-meionite have been proposed. **Alteration:** Commonly becomes cloudy and subsequently alters to a mixture of calcite, plagioclase, and sericitic mica. Alteration to aggregates of epidote, plagioclase, and vesuvianite has been reported, as well as alteration to harmotome and clay minerals as a result of barium metasomatism. **Occurrence:** Scapolites are usually confined to metamorphic and metasomatic environments. They form in regionally metamorphosed rocks such as marble, calcareous gneiss, and greenschists, most commonly in the amphibolite facies but spanning a wide range, perhaps from the zeolite to eclogite facies. In marbles they are often associated with pyroxene (augite), apatite species, sphene, phlogopite, and zircon. Large crystal size is common in the marble environments. Scapolites also form in metasomatic environments (scapolitization) at intrusive contacts, often those of feldspathic pegmatites and gabbros or pyroxenites. They may form at the expense of calcic plagioclase or in place of it in systems high in calcium and CO_2 partial pressure. Chlorine and carbon dioxide may derive from components of original sedimentary rocks. Less commonly, they may form by direct crystallization from pegmatitic fluids. **Localities:** Found as large crystals at Bolton and Boxborough, and massive at Westfield, MA; massive at Marlborough, VT; white, massive, somewhat fibrous at Monroe, CT; as good crystals near Ticonderoga, Essex Co., and Cl-rich, SO_4-poor in the contact zones of marble in the Amity, Orange Co., NY, area, where crystals(!) may reach up to 25 cm in length; similarly at Franklin and Ogdensburg, Sussex Co., NJ; at the Elizabeth mine, French Creek, Chester Co., with magnetite and pyrite in cavities in corroded grossular, and in scapolite–metagabbros ($Me_{\approx 63}$) interbanded with scapolite-free metagabbro in Bucks Co., PA; of primary crystallization in pyroxene aplite of the Phillipsburg Quadrangle, MT; from the Holbrook Shaft, Bisbee, Cochise Co., AZ. In Canada it is common throughout the Grenville Province, often with phlogopite and fluorapatite, and occurring as large (>10 cm) crystals(!), as at Bobs Lake mine, Bedford Twp., Frontenac Co., at Highland Grove, Haliburton Co., at the Craigmont mine, Raglan Twp., Renfrew Co., and mauve-colored at Gooderham, Glamorgan Twp., ONT; and Bear Lake and Leslie Lake, Pontiac Co., QUE. Found with augite and zircon in coarse marble at La Panchita mine, Ayoquezco, OAX, Mexico. Found at numerous localities in Scandinavia, as at Ødergården, Norway, the metasomatic deposits of the Kiruna region, northern Sweden, and marialite, $Me_{\approx 14}$, on Pezh Is., Karelia, Finland. Found at Wehrer Kessel, Laacher See district, Eifel, Germany; $Ma_{\approx 80}$ at *Pianura, Napoli*, and $Me_{\approx 90}$ at *Mte. Somma–Vesuvius, Campania, Italy*, in "geodes" in limestone; at Zdár Mt., Ruda, Moravia, Cecy, Czech Republic; in iron deposits of the Ural Mts., with Cl-rich amphibole in Tuva, K-rich scapolite from southern Yakutia, and associated with phlogopite at Slyudyanka, Lake Baikal, southern Irkutsk, Russia. Gemmy, chatoyant in part, reddish purple, Na-rich "scapolite" reported from western Xinjiang Prov., China. Gem material from the Ratnapura stone tract, Sri Lanka; pink and yellow, chatoyant, gemmy material in the Mogok stone tract, Myanmar; $Me_{\approx 57}$ at Milendella, Mt. Lofty Ranges, SA, Australia. In Africa, $Me_{\approx 55}$ at Tsarasaotra, Madagascar; yellow, orange, white, and violet transparent, frequently etched, terminated crystals(!) in the Umba Valley and nearby tracts near Arusha, Tanzania; yellow and pink, gemmy crystals from pegmatite east of Entre Rios, Mozambique; $Me_{\approx 65}$ in garnet–hornblende–pyroxene–scapolite gneiss at Shai Hills, Ghana. Yellow, gemmy scapolite found in ES, Brazil; at the Llanca mine, Coquimbo, Chile. Uses: Transparent material of various colors is fairly plentiful and is cut into faceted gems. Translucent chatoyant material produces attractive cat's-eye cabochons. AR *D6:466(1892)*, *DHZI 4:321(1963)*, *AC(B) 29:1272(1973)*, *MR 13:78(1982)*, *CGEM 1:12(1992)*, *ZVMO 123:90(1993)*.

76.3.2.1 Sarcolite $NaCa_6Al_4Si_6O_{24}F$ (?)

Named in 1807 by Thompson from the Greek for *flesh* and *stone*, in allusion to its pink-red color. TET $I4/m$. $a = 12.343$, $c = 15.463$, $Z = 4$, $D = 2.93$, the formula for the determined structure being given as $Na(Na,K,Fe,Mg)_{<1}Ca_6[Al_4Si_6O_{23}](OH,H_2O)_{<2}(SiO_4,SO_4,CO_3,Cl)$. *42-1367*: 4.85_5 3.91_3 3.35_8 3.28_5 2.95_9 2.86_8 2.76_{10}. Structure: *TMPM 24:1(1977)*. Although morphologically similar to the scapolites, and having a dimensionally related unit cell ($a_{sar} \approx a_{scp}$ and $c_{sar} \approx 2c_{scp}$), the structure is unique in being composed of eight-membered rings of alternating Al and Si tetrahedra that are cross-linked to form a framework, and further joined to Si_2O_7 dimers. The structural cell is rotated 45° from the older morphological cell, the transformation matrix, morphological to structural, being $1\bar{1}0/110/002$. The habit, in morphological setting, is as equant to short prismatic [001] with {100} dominant crystals, and highly modified by other prisms, bipyramids of all three orders and the basal pinacoid. Simpler crystals often resembling cubo-octahedrons. Also as irregular masses. Pink, red, pinkish white; vitreous luster; transparent to translucent. $H = 6$. $G = 2.545$. Uniaxial (+); $N_O = 1.604$, $N_E = 1.615$; but also reported as biaxial (+); 2V small; $N_x = N_y = 1.640$, $N_z = 1.657$; r > v, strong. Gelatinized by acids. Fusibility 3(?). Found in volcanic ejecta at *Mte. Somma, Campania, Italy*. AR *MM 48:107(1984)*, *CM 25:731(1987)*.

76.3.3.1 Ussingite $Na_2[AlSi_3O_8](OH)$

Named in 1915 (Bøggild) in honor of Prof. Niels Viggo Ussing (1864–1911), Copenhagen, Denmark. TRIC $P\bar{1}$. $a = 7.256$, $b = 7.686$, $c = 8.683$, $\alpha = 90.75$, $\beta = 99.75$, $\gamma = 122.48$, $Z = 2$, $D = 2.51$. *14-426* 6.88_7 6.35_9 4.92_7 4.18_7 3.84_7 3.47_7 2.95_{10} 2.69_{10}. Structure: *AM 59:335(1974)*. The structure is a unique alumino-silicate framework with ordered Al. Occurs as subparallel tabular to blocky crystals with {100}, {010}, and {001} dominant; more commonly as

76.3.3.1 Ussingite

fine-grained masses and aggregates. Twinning on (010) common. Pale reddish violet; vitreous luster; transparent to translucent. Cleavage {001} and {110} perfect. H = 6–7. G = 2.48. Biaxial (+); 2V = 32–39°; $N_x = 1.504$, $N_y = 1.508–1.509$, $N_z = 1.543–1.545$; Z ∧ $b = 3°$; pleochroic. Easily fusible. Gelatinizes with acids. Distinguished from similar appearing sodalite (St. Hilaire) by its lack of fluorescence in UV. Found in syenite pegmatites. In good crystals, 5 mm, and masses in sodalite xenoliths at Mont. St Hilaire, QUE, Canada; at *Kangerdluarsuk, Ilimaussaq, Greenland*; in the Khibiny and Lovozero massifs, Kola Peninsula, Russia. *DT 551(1932)*, MR 21:284(1990).

= Class 77 =

Zeolites

ZEOLITE GROUP

The zeolites are a very large group of minerals. They may be defined as crystalline, hydrated aluminosilicates of alkali and alkaline earth cations having infinite, three-dimensional structures. They are further characterized by an ability to lose and gain water reversibly and to exchange constituent cations without major changes in structure. *MASCN:4(1977)*. The general formula is

$$(Na,Ca,Ba)_{1-2}(Al,Si)_5O_{10} \cdot nH_2O$$

The structure is based on an arrangement of SiO_4 and AlO_4 tetrahedral groups positioned to create an open, three-dimensional framework of channels. In all zeolites, if a tetrahedral site forms after an alkali ion occupies the channel, an Al^{3+} ion will preferentially occupy the tetrahedral site. If a tetrahedral site forms before an alkali ion occupies the channel, an Si^{4+} ion will occupy the tetrahedral site. If the growth steps are inclined to the mirror or glide plane, the two symmetry-related sites will not be equivalent, and thus ordering will occur and the space group symmetry will be reduced; therefore the mirror plane changes into a twin plane. Indeed, during crystal growth various degrees of Al–Si ordering are produced by electron charge balance between alkali or alkaline cations and the Si–Al atoms on the crystal surface. Optical phenomena inconsistent with the apparent external or even the x-ray symmetry suggest the presence of complicated optical sectors with various symmetries, and no truly single crystals are produced during growth.

Because of the variety and complexity, commercial importance, practical utility, and beauty of the zeolite structures, the field has been very well studied and published reports include: *NZ:18(1985)*, *ZW (1992)*, *AM 77:685(1992)*, *EJM 7:501(1995)*, and many others.

77.1.1.1   Analcime   Na(AlSi$_2$)O$_6$ · H$_2$O

Named in 1801 by Haüy from the Greek for *weak*, in alluison to its weak electrical property when heated or rubbed. Zeolite group. See pollucite (77.1.1.2) and wairakite (77.1.1.3). ISO, TET, ORTH, MON, and TRIC; reported space groups include: $Ia3d$, $I4_1/acd$ or $I4_1a$, $Ibca$, $I2/a$, and $P1$. Monoclinic cell: $a = 13.72$–13.73, $b = 13.71$–13.73, $c = 13.69$–13.74, $\beta = 90.0$–$90.5°$,

77.1.1.1 Analcime

$Z = 16$, $D = 2.27$. Orthorhombic cell: $a = 13.720$–13.733, $b = 13.715$–13.729, $c = 13.709$–13.740. *Z 8:247(1988)*. *41-1478*(ISO): 5.59_6 4.84_1 3.43_{10} 2.92_4 2.69_1 2.50_1 2.22_1 1.74_2; also *42-1378*(magnesian), *19-1180*(ORTH). Analcime has a wide variation in Ca, Na, Cs, and Mg, along with order–disorder of Si–Al in the framework, which causes it to be classified in several crystal systems. The Al–Si framework is the same as for pollucite and wairakite. The framework is composed of four-ring chains wrapped around square prisms that are interconnected to form cages. Cubic analcime has disordered Al and Si in the framework. Ordering of Al and Si in the tetrahedra and related Na–Ca occupancy results in lower symmetries. *NZ:321(1985)*, *ZK 184:63(1988)*. Neutron-diffraction study: *ZK 135:240(1972)*. **Habit:** Trapezohedral {211} common with small cube {100} faces. Rare: {110}, {112}, {233}, {345}, {210}, and {421}. Noncubic analcime is polysynthetically twinned. Pseudo-merohedral twins with two sets of twin lamellae at 90° are common (contact plane {101}?). Crystals may be 25 cm across or more (e.g., Mt. St.-Hilaire). **Physical properties:** Generally white or colorless, but may be yellow, pink, orange, red, green, blue, brown, or black; white streak; vitreous luster. Cleavage {100}, poor; subconchoidal fracture; brittle. $H = 5$–$5\frac{1}{2}$. $G = 2.22$–2.63. Soluble in HCl. IR and Raman data: *PCM 7:96(1981)*. Thermal data: *NZ(1985)*. **Chemistry:** Si/Al ratio ranges from 1.5 to 3. T_{Si} varies widely 0.60–0.73. The terms *low-silica analcime* and *high-silica analcime* may be used where T_{Si} is significantly less or greater than the common value of approximately 0.67. Series between analcime and pollucite and wairakite are well established; most are Na-dominant. Mg-rich analcime (23 at .7%) has been found. **Optics:** $N = 1.479$–1.493. Isotropic or biaxial with a small $(-)$ 2V. Birefringence to 0.001. **Occurrence:** In a variety of geologic environments: continental and marine sedimentary rocks (deep-sea sediments, saline alkaline lakes), altered and unaltered pegmatites, veins and miarolitic cavities. Occurs in igneous rocks as phenocrysts, but there is still controversy as to their primary or secondary (formed from leucite) origin. *AM 78:225,230(1993)*. Some may be primary in basaltic magmas. **Localities:** Many worldwide localities. Display-quality specimens are known from the following: Watchung basalts at Paterson and Bergen Hill, NJ, particularly the Upper and Lower New Street quarries and the Prospect Park quarry; also at Chimney Rock and Bound Brook quarries, Somerset Co., NJ; Cornwall, Lebanon Co., PA; Copper Falls mine, Eagle Harbor, Keweenaw Co., MI; twinned crystals from the Phoenix mine, Houghton, MI; North and South Table Mts., Golden, CO; Skookumchuck Dam, near Tenino and Bucoda, Thurston Co., and 3-cm crystals at Kalama, Cowlitz Co., WA; fine, transparent crystals to 6 cm at the Dek and Two Hug quarries, Price Creek, Kings Valley, Benton Co., OR; also at the Springfield Butte quarry, Springfield, Lane Co., OR; Mt. St.-Hilaire, QUE (to 25 cm across), Wasson's Bluff and Cape Blomidon, Bay of Fundy, NS, and deep red phenocrysts at Crowsnest Pass, Coleman, ALB, Canada; Dumbarton, and Salisbury Crags, Edinburgh, Scotland; to 4.5 cm at the Dean quarry, Keverne, Lizard Penin., Cornwall, England; in the Faroe Is. at Eysteroy, Mykines, and Stremoy Is.; *Cyclope Is., Catania, Sicily, Italy*; pinkish-white crystals to 8 cm with apophyllite at Val di Fassa, Trento, Italy; white well-formed crystals to 18 cm at Tura and along the Nidym R. area, Tunguska R. region, Yakutia, Siberia, Russia; and crystals to 4 cm from along the Amudikha R., Tunguska R. region, near Mirny; 2-cm crystals with gmelinite at Cairns Bay, Flinders, Victoria, and Cape Grim, TAS, Australia; with natrolite at Pahau River, Culverden, North Canterbury, New Zealand. EF *INDM 6:127(1965); AM 63:448(1978), 66:403(1981), 76:189(1991); PT 1:1-22(1979); MR 21:284(1990); ZW (1992).*

77.1.1.2 Pollucite $(Cs > Na > Rb > Li)_xAl_xSi_{3-x}O_6 \cdot nH_2O$

Described by Breithaupt in 1846, and named for Pollux, one of the twin brothers (Gemini) of Greek and Roman mythology, when two new minerals were found together that resembled one another. Zeolite group. Castor (= castorite), the associated mineral, is now known as petalite. Isostructural with analcime, wairakite, leucite, and ammonioleucite. Most pollucite is cubic and is isotropic but is not completely disordered. Single-phase primary pollucite reequilibrates with decreasing temperature. Na-Si-enriched and Cs-Al-enriched domains (usually less than 200 µm) form. Quartz may exsolve, leading to Si-poor but cation-Al-rich pollucite. Forms a complete solid-solution series with analcime. ISO $Ia3d$ (nat) and TET $I4_1/a$ (syn anhydrous equivalents of $MAlSi_2O_6$, where M = K,Rb,Cs). Isometric: $a = 13.64$–13.71, $Z = 16$, $D = 3.25$ (for pure end member). 25-194(nat): 3.65_3 3.42_{10} $2.91_{4.5}$ 2.42_2 2.22_2 1.86_2 1.74_2 1.71_1; 29-407(syn): 5.58_1 3.66_5 3.42_{10} $2.91_{4.5}$ 2.42_3 2.20_2 1.86_2 1.74_2. Significant changes in peak intensities occur in X-ray diffraction patterns between pollucite and analcime. *CM 12:334(1974)*. Pollucite has the same aluminosilicate framework as analcime and wairakite. Four-ring chains are wrapped around square prisms and interconnected to form cages that are occupied by Cs. Large voids in the framework exist that are occupied by water molecules and Cs. Na is located between water molecules, in the same position as in analcime, but in randomly distributed clusters. Si/Al varies from about 2.1 to 2.6. High-Cs-containing samples have Si/Al approaching 2.0. Complete disorder (i.e., totally cubic) was not observed even in pollucite. *CM 32:69(1994)*. **Physical properties:** Commonly massive or as rounded, corroded crystals. Sharp, euhedral crystals are rare. Forms {100}, {211}, common; {110}, {210}, rare. Crystals to 10 cm or more in diameter. Colorless, white, gray, pink, blue, or violet; white streak; vitreous to greasy luster. No cleavage, conchoidal to uneven fracture. $H = 6\frac{1}{2}$–7. $G = 2.67$–3.03. NMR and MAS study of Al–Si order in pollucite–analcime series: *CM 32:69(1994)*. X-ray luminescence spectra: MS *36:116(1982)*. Dissolves with difficulty in hot HCl, but is readily soluble in HF. **Chemistry:** As the Na content increases, so does the H_2O content. Traces of K, Li, Mg, Fe, and Mn may be present. Ca is found in trace amounts only in sodian pollucite. Rb_2O to 3.0 wt %, Li_2O to 0.4 wt %, P_2O_5 to 1.5 wt % ($Al^{3+} + P^{5+} = 2Si^{4+}$). Specimens from $Poll_{50}$ to $Poll_{100}$ have been found. There is no solid solution toward wairakite or leucite. CRK $= 100(Cs + Rb + K)/\Sigma$ cations $= 69$ to 89, with a mean of 81 for pollucite from about 70% of known localities. Si/Al ratios for primary pollucite are usually 2.31–2.57, mean 2.48. **Optics**: Isotropic or very weakly anisotropic; N = 1.508–1.527. One can determine the Cs content from the refractive index. **Occurrence**: Restricted to highly fractionated rare-element, complex Li-Cs-bearing, evolved, granitic pegmatites. Forms at 300–400° and 2–4 kbar. Associated with quartz, feldspars, spodumene, lepidolite, elbaite, petalite, eucryptite, and apatite. **Alteration**: Quartz pseudomorphs to 23 cm across have been found. Also alters to quartz, analcime, adularia, lepidolite, spodumene, and clays. **Localities**: Approximately 100 worldwide localities are known. Only noteworthy localities are given here: gem-quality material from Dudley's Ledge, Bennett quarry at Buckfield, Oxford Co., ME; also Mt. Mica, Mt. Marie(!), Paris; Tamminen quarry, Greenwood; Dunton gem mine at Newry and other locations in Oxford Co., ME; Portland and Haddam, Middlesex Co., CT; Tin Mountain mine, Custer Co., SD; Himalaya dike system, Mesa Grande district, San Diego Co., CA; major reserves (350,000 tons) at the Tanco pegmatite, Bernic Lake, MAN, Canada, where crystals are 1–2 m across and bodies and pods are as much as 180 m × 75 m × 12 m(!); sharp, colorless crystals to 2.5 cm across from the La Speranza dike, *San Piero di Campo, Elba, Italy*; additional localities on Elba exist; sodian pollucite from the Varutrask pegmatite, near Skellefteå, Västerbotten, Sweden; Luolamaki, Viitaniemi, Haapaluoma, and Oriselka pegmatites in Finland; in pegmatites from Shmakat, Alingar, Sorkh Rud, Darya-i-Pech, and Kurgai pegmatite fields, Afghanistan; also from Nilaw-Kulam and Parown fields, Nuristan, Laghman Prov., and crystals to 60 cm across at the Paprock prospect, Kamdesh, Nuristan, Kunar Prov., Afghanistan. Crystals to 8 cm from Shingus, Gilgit Division, NWF, Pakistan; Bikita, Zimbabwe; Karibib, Namibia; Taquaral district, Itinga, MG, Brazil. **Uses**: Is used as a minor gemstone and as a source for Cs. EF *ZK 129:280(1969), MS 29:28(1975), NZ(1985), ZW(1992), IMA (Pisa, Italy):125(1994).*

77.1.1.3 Wairakite $(Ca,Na_2)[Al_2Si_4O_{12}] \cdot 2H_2O$

Named in 1955 by Steiner for the locality. Zeolite group. See analcime (77.1.1.1), the Na end of the series. MON, ISO, TET, ORTH, and TRIC $I*/a$, $I2/a$, $Ia3d$, $I4_1/acd$, and $Ibca$. $a = 13.69$–13.70, $b = 13.64$–13.58, $c = 13.56$–13.58, $\beta = 90.0$–$90.5°$, Z = 8, D = 2.27–2.28; *42-1451*: $a = 13.695$, $b = 13.638$, $c = 13.550$, $\beta = 90.33°$. PD *42-1451*(nat, monoclinic): 6.83_3 5.56_8 3.42_4 3.41_5 3.39_{10} 2.92_4 2.91_3 2.89_3; *11-156*(syn, cubic): 6.81_4 5.57_9 4.83_4 3.41_{10} 2.90_8 2.675_4 2.49_5 1.73_4. Aluminosilicate framework is the same as for analcime and pollucite. Isometric wairakite has a disordered Si and Al framework, while those with TET, ORTH, MON, and TRIC symmetry are partially or fully ordered. The Al is preferentially located in two tetrahedra sites near the Ca. Ca is located in one site with two sites empty or containing Na. Commonly forms trapezohedra with {211}, rarely modified by {100} and the dodecahedron {110}; most crystals <1 mm in diameter, some to 15 mm in diameter; lamellar twinning on {110} is commonly seen in monoclinic varieties. Colorless or white, white streak, vitreous to dull luster. Cleavage {100}, distinct; conchoidal fracture; brittle. H = $5\frac{1}{2}$–6. G = 2.26. A complete Ca–Na series between wairakite and analcime exists. A chemical series between warakite and pollucite has not been found. Analysis (wt %), Japan (Toi): SiO_2, 54.5; Al_2O_3, 23.9; Fe_2O_3, 0.1; MgO, 0.1; CaO, 12.4; Na_2O, 0.25; K_2O, 0.34; H_2O, 8.51. Biaxial (+) and (−); XYZ ≈ bac; N_x = 1.498–1.500, N_z = 1.502; 2V = 70–105°; birefringence = 0.00–anisotropic. Has been synthesized from stilbite and heulandite at low pressure and 300°. *PJA 42(8):925(1966)*. Occurs in geothermal wells and metamorphosed volcanics. 60–300°, 72–1600 m. Occurs in the deepest and hottest geothermal wells in Iceland. Occurs at a number of localities; among them: Spanish Peaks, CO; Lower Geyser Basin, Yellowstone National Park, WY; The Geysers, Sonoma Co., and near Rosamond, Kern Co., CA; Mt. Rainier National Park, Pierce Co., WA; St. Thomas and St. Croix, U.S. Virgin Is.; Buttle Lake area, Vancouver Island, BC, Canada; western and northern Iceland; *Wairakei geothermal field, North Island, New Zealand;* also in the Ngatamariki geothermal field, NE of Wairakei; Tui mine, Te Aroha, Auckland; a number of localities in Japan, in geothermal areas and past geothermal areas, including Nishi-Iburi district, Hokkaido; Seigoshi mine, Toi, and at Kawazu, Shizuoka Pref.; Onikobe, Miyagi Pref., Koriyama, Fukushima Pref., Yugami district, Fukui Pref., and Hikihara, Hyogo Pref. EF *AM 57:924(1972), 64:993(1979), 65:1212(1980); NZ (1985), ZW (1992)*.

77.1.1.4 Laumontite $CaAl_2Si_4O_{12} \cdot 4H_2O$

Initially described in 1801 as "zeolithe efflorescente" by Haüy, but the mineral was considered as an independent species by Werner who named it lomonite for François Pierre Nicolas Gillet de Laumont (1747–1834), who collected the sample studied by Haüy. Haüy (1809) changed the spelling to "laumonite"; finally, the name *laumontite* was proposed by Leonhard (1821) and is now currently used. *NZ (1985)*. *Leonhardite* is a now discredited term that is still used occasionally in the literature for a partially dehydrated laumontite. The transition *laumontite* to *leonhardite* is normally reversible. Zeolite group. MON $C2/m$. For fully hydrated material with 16–18H_2O for Z = 1: a = 14.82–14.90, b = 13.10–13.19, c = 7.54–7.565, β = 110.2–110.7°, Z = 4, D = 2.12. For leonhardite with 14H_2O: a = 14.77, b = 13.09, c = 7.58, β = 112.0°; for 12H_2O: a = 14.71, b = 13.09, c = 7.46, β = 112.1°. *45-1325*(nat): 9.48_{10} 6.87_4 4.51_1 4.16_2 3.68_2 3.51_5 3.28_1 3.04_1; *26-1047*(syn): 9.5_6 6.84_4 4.16_{10} $3.66_{3.5}$ 3.51_8 3.36_3 $3.27_{5.5}$ 3.20_3. Fully deuterated (18D_2O) laumontite has a = 14.863, b = 13.169, c = 7.537, β = 110.18°. *EJM 5:851(1993)*. The aluminosilicate framework is composed of singly connected four-ring chains in TTTT configuration, linked by Al–O tetrahedra. Si–Al distribution is perfectly ordered. Ca is present in the larger channel (parallel to c) connected to H_2O molecules and four framework oxygens of the Al-bearing tetrahedra. Four water sites are present in the structure: two are coordinated to Ca, the other two act as bridges between the Ca-coordinated waters through hydrogen bonding. All water sites are disordered to some degree and are essentially fully occupied in fully hydrated laumontite. Crystal structure determinations for fully hydrated laumontite: Z 13:249(1993), NJBM: 385(1992). Additional structure determination: SPC 30:624(1985). The number of water molecules in laumontite changes from $12H_2O$ at 0–5% relative humidity, to $14H_2O$ at 10–70%, and to $16–18H_2O$ at 80–100%. **Habit:** Often occurs as simple prisms, elongate on c. Square cross section, equant crystals are less common. Crystals to 15 cm long, shorter crystals to 5 cm long are 1–2.5 cm wide. Common forms include {110}, {$\bar{2}$01}, {100}, {010}. Twinning is present on {100} to form "swallowtail" reentrant twins. **Physical properties:** White, sometimes stained by Fe or Mn oxides; colorless, pink, or white when hydrated; gray, pink, yellow when partially dehydrated; white streak; vitreous luster, pearly on cleavages, and chalky (when dehydrated). Cleavage {010}, {110}, perfect; {100}, imperfect; uneven fracture. H = 3–4. G = 2.20–2.41. Laumontite sometimes fluoresces white to cream or yellow in LWUV and SWUV light. May be piezoelectric. **Chemistry:** Always very Ca-rich. Some Na (to 2.6 wt % Na_2O) and K (to 1.75 wt % K_2O). So-called "primary leonhardite" has 4.36 wt % K_2O and 2.58 wt % Na_2O), and this chemistry may be the cause, being unable to get a fully hydrated phase. Dehydrated or partly dehydrated crystals crumble readily. **Optics:** Biaxial (−); $Z \wedge c = 8–32°$, $X \wedge a = 10–26°$, $Y = b$; $N_x = 1.501–1.514$, $N_y = 1.508–1.522$, $N_z = 1.512–1.525$; 2V = 26–82° (most are 30–50°; r < v, distinct, weakly inclined. **Occurrence:** Found in a wide range of geologic conditions. Occurs in the weathering environment, to zeolite-grade (facies) metamorphic rocks. Occurs in hydrothermal ore deposits, volcanics, pegmatites, intrusive rocks, and hot springs. Rarely in deep-sea sediments. May be an authigenic cement in sandstones. **Localities:** Occurs in many worldwide localities, and only selected localities are given here: Paterson, Passaic Co., NJ; Centerville, Fairfax Co., VA; crystals to 15 cm long and 15 mm wide from the Pine Creek tungsten mine, near Bishop, Inyo Co., CA; to 1 cm from the Himalaya pegmatite–aplite dike system, Mesa Grande District, San Diego Co., CA; pink crystals to 6 cm long and 3 cm wide from the Bear Creek quarry, Drain, Douglas Co., OR; prisms to 4–10 cm long from Tum Tum Mountain, Clark Co., WA; *Bleigrube (lead mine), Huelgoët, Finistere, Brittany, France*; to 4 cm long from the Gall quarry, Frassgraben, Koralpe, Carinthia, Austria; colorless to white prisms to 5 cm long (common) to >20 cm long and 3 cm wide are found at the "Bombay quarry" near Khandivali, Bombay district, MAH, India. Also at Nasik and Pune (Poona). Crystals to 4–6 cm long from Alamos, SON, Mexico. EF *NZ (1985), CS 8:79(1991), ZW(1992).*

77.1.1.5 Hsianghualite $Ca_3Li_2Be_3(SiO_4)_3F_2$

Named in 1958 by Huang et al. for the locality, whose name means *fragrant flower*. ISO $I2_13$ (originally reported as $I4_132$). a = 12.879, Z = 8, D = 2.965. *36-412:* 3.44_6 3.22_4 2.75_{10} 2.35_4 2.21_{10} 2.09_9 1.753_7 1.691_5. **Structure:** *DANS 316:624(1991).* The structure is reported to be analogous to that of analcime, with Be and Si in tetrahedral coordination forming a three-dimensional framework, but no reference is made as to the possible ordering of these two elements. The observed and calculated densities are high for a phase of such structure and composition. White, vitreous luster, transparent to translucent. $H = 6\frac{1}{2}$. $G = 2.90$, also 2.97–3.00. Isotropic; $N = 1.613$. Minor iron, magnesium, and sodium are present. The interpretation of the chemical analysis is subject to some question. Occurs as dodecahedral and trapezohedral crystals, and as granular masses in phlogopite veins in a light-colored band in green- and white-banded, metamorphosed Devonian limestone at *Hsianghualing, in the Nanling Range, Linwu Co., Hunan Prov., China*, associated with fluorite, liberite, chrysoberyl, taaffeite, and nigerite. AR *AM 44:1327(1959), 46:244(1961)*.

77.1.2.1 Chabazite-Ca (Ca,Na$_2$,K$_2$,Sr,Mg)[Al$_2$Si$_4$O$_{12}$ · 6H$_2$O

Named in 1792 by Bosc d'Antic from the Greek *chabazios* or *chalazios*, an ancient name of a stone. The name of the last of the 20 stones celebrated for their virtues and mentioned in a poem ascribed to Orpheus. Zeolite group. Chabazite-(Na) and chabazite-K also exist and are approved by the CNMMN IMA. Chabazite-Na includes "herschelite." Synonyms include "phacolite," a bean or lens-shaped variety, which is repeatedly twinned about {0001}. Often intergrown with gmelinite. TRIC [pseudo-HEX (TRIG, $R\bar{3}m$) $P\bar{1}$. $a = 9.41$, $b = 9.42$, $c = 9.42$, $\alpha = 94.18°$, $\beta = 94.27°$, $\gamma = 94.35°$. On a hexagonal cell, dimensions range from $a = 13.71$–13.86 and $c = 14.83$–15.42. $Z = 6$, $D = 2.05$–2.16. *34-137*(rhombohedral axes): 9.34$_5$ 5.55$_3$ 5.00$_3$ 4.32$_{10}$ 3.87$_2$ 3.58$_4$ 2.93$_9$ 2.88$_{4.5}$. Patterns for three different chabazites are given in *NZ:333(1985)*. Significant variations in intensities exist. (Rhombohedral axes): strontian chabazite, *45-1427*: 9.43$_{10}$ 5.56$_5$ 5.12$_5$ 4.33$_8$ 3.97$_3$ 3.64$_5$ 2.93$_{10}$ 2.53$_5$. Al–Si tetrahedra are linked to form parallel double six-membered rings stacked in three different positions (A,B,C) in the repeat arrangement AABBCCAABBCC ···, forming large cages. The maximum symmetry compatible with this framework is $R\bar{3}m$, the highest symmetry which has been attributed to chabazite. Si and Al in the tetrahedra are disordered, but some areas show regular (1:1) ordering. Complete order of Al and Si tetrahedra is found in the variety [*ZW (1992)*] or species (according to the IMA at present) willhendersonite. If the amount of ordering in a mineral is not a basis for a separate species, the name *willhendersonite* should be discontinued. The structure of a K-exchanged chabazite ($R\bar{3}m$) was reported. *Z 3:205(1983)*. **Habit:** Basic crystals are distorted cubes or pseudorhombohedrons that are actually composed of six triclinic twins. Crystals may also be prismatic parallel to [001]$_{\text{hex}}$, tabular on {001}$_{\text{hex}}$. Twinning is common, often along several different planes or axes. Contact twins on the dominant rhombohedral face {10$\bar{1}$1} are relatively rare. Penetration twins are very common. Twinning around c results in bean- or lens-shaped crystals, which are said to have a "phacolitic" habit. Detailed morphological and twin drawings are given in *ZW (1992)*. **Physical properties:** Colorless, white, cream, yellow, pink, red, green, or brown; white streak; vitreous luster. Cleavage is parallel to the pseudorhom-

77.1.2.1 Chabazite-Ca

bohedron. Brittle. H = 3–5. G = 1.97–2.20. Thermal and optical spectral data: NZ (1985). Decomposed by HCl with separation of silica gel. Mostly nonfluorescent, but sometimes is fluorescent green under SWUV (probably uranyl ions). Fluorescent material occurs at Bancroft, ONT, and the New Street quarry, West Paterson, NJ. A complete series exists between Na and Ca members, but most chabazites are Ca-dominant. K-dominant and ordered (willhendersonite) material is also known. K-dominant chabazite may exist [NZ:191(1985)]. The old variety herschelite was defined as Na-dominant. Sr (to >8 wt % SrO) and Mg (to >0.5 wt % MgO) may be present. Si/Al ranges from about 1.5 to 4.1. T_{Si} varies from 0.6 to 0.8. **Optics:** Optical properties are variable. May be uniaxial (−) or (+), but very often are biaxial (+). As expected, high Na-chabazite has lower indices than Ca-rich chabazite, and Sr-bearing chabazite has the highest reported indices. Range: N_O = 1.461–1.517, N_E = 1.460–1.514; for biaxial chabazite: N_x = 1.485, N_y = 1.485, N_z = 1.486; 2V (+) = 67–75. **Occurrence:** Occurs in volcanic rocks, altered volcanic ash deposits in both fresh and saline lakes, pegmatites, and granitic and metamorphic rocks. **Localities:** Exceptional display-quality material occurs: crystals to 4.5 cm in Triassic basalts at Paterson, Passaic Co., NJ, particularly the Upper and Lower New Street quarries and the Prospect Park quarry; also Laurel Hill, Secacus, NJ; numerous localities in Oregon: the best (twinned phacolite crystals to 5 cm) is the Springfield Butte quarry, Springfield, Lane Co., OR; Oxbow Dam on the Snake R., near Copperfield, Baker Co., OR; rhombohedra to 3 cm at Goble, Columbia Co., and Burnt Cabin Creek, Wheeler Co.; also Two Hug quarry, Price Creek, Kings Valley, Benton Co.; salmon-orange crystals to 2.5 cm in diameter at Parrsboro, and Wasson's Bluff, Bay of Fundy, NS, Canada; in the Faroe Is.: colorless to white crystals to 4 cm across from Vidoy, Eysteroy, and Sandoy Is.; colorless, twinned phacolite crystals to 3 cm from the Craigahulliar quarry, Portrush, Northern Ireland; (some deep amber crystals to 2 cm fluoresce cream under LWUV); Kilmalcolm, Renfrewshire, Scotland; colorless rhombohedra to 2 cm across occur at *Idar Oberstein, Germany*; strontian (8% SrO) material from Alba, Ardeche, France, and at Kaiserstuhl, Baden, Germany; 2-cm crystals from Vogelsberg, Hessen, Germany; deep red crystals to 2 cm from Striegau, Poland; Rubendorfel, Czech Republic; twinned phacolite crystals to 8 cm at Csodi Mt., Dunabogdany, near Budapest, Hungary; white, colorless, or pink crystals to 4 cm across from the No. 10 Bombay quarry, Khandivali, MAH, India; also from Panvil; colorless, twinned phacolite crystals to 3 cm across from Fairy Mount, Kyogle, NSW, Australia. Sodium-rich crystals from the Bundoora quarry, Whittesea and Clifton Hill, Melbourne, VIC, Australia; also Sutherlands Bluff, Phillip Is., VIC, and twinned phacolite crystals from Gads Hill, Leina, TAS, Australia. EF *AM 55:1278(1970), LIT 14:17(1981), NJBM:461(1983), BM 111:671(1988), ZW (1992).*

77.1.2.2 Herschelite (Na,Ca,K)AlSi$_2$O$_6$ · 3H$_2$O

Described by Levy in 1825 and named for John Frederick William Herschel (1792–1871), British astronomer. Zeolite group. The status of this zeolite is now considered questionable. It is the Na analog of chabazite. HEX-R R$\bar{3}$m. $a = 13.799$, $c = 15.102$, $Z = 12$, $D = 2.05$; actually triclinic, P$\bar{1}$ (based on optics and detailed X-ray diffraction studies). 19-1178(nat): 9.4$_5$ 6.89$_2$ 5.03$_4$ 4.32$_{6.5}$ 3.88$_{2.5}$ 3.60$_2$ 2.93$_{10}$ 2.90$_3$. Structure is the same as that of chabazite (77.1.2.1). Heating to only 300° produces structural changes (i.e., lines weaken and disappear). Ca–Sr–K-rich material is most unstable (chabazite). Clean, Ca-rich material is very resistant to heating, and is stable to 650°. It appears that the rate of crystal growth governs whether herschelite or chabazite results. If growth is slow, Na can be excluded, and one gets chabazite; if fast, Na is incorporated within the chabazite structure, and herschelite results. Some K-dominant samples exist (e.g., Vallerano quarry, Rome, Italy) and are willhendersonite. Hexagonal plates that are either flat or biconvex, with prominent {10$\bar{1}$0, {0001}, and {10$\bar{1}$1}; the predominant prism and c-pinacoid give herschelite a distinctive morphology. Also occurs as aggregates of plates, composed of stacked individuals; rarely occurs as minute spherules. Gmelinite at Boron, CA, is epitaxially overgrown by herschelite, attached by respective (0001) faces. All herschelite is twinned, and several types of twinning exist. {0001} sectors show rotation twins around the c axis. {10$\bar{1}$0} sectors are polysynthetic twins of platy crystals parallel to (0001). Generally colorless or white, white streak, vitreous luster. Cleavage {10$\bar{1}$1}, distinct; uneven fracture; brittle. H = 4–5. G = 2.08–2.16; most are 2.0–2.08. Gelatinizes readily in common acids. There is a range of Na and Ca contents between the two end members; K$_2$O is variable, from 0.1 to 3.2 wt %. Uniaxial (+) or (−); N$_O$ = 1.471–1.479, N$_E$ = 1.470–1.481; $\bar{N}$ = 1.472, and some are even lower: 1.460–1.462. Chabzite has $\bar{N}$ = 1.485. Birefringence is 0.003–0.004. A low-temperature zeolite found in basalts through rhyolites (basic to acid rocks). Si-rich in rhyolite glass and tuff and in lacustrine saline lake environments, and Si-poor in mafic rocks. Associated with other zeolites such as chabazite, phillipsite, analcime, clinoptilolite, and stilbite. Occurs in many worldwide localities and only selected localities are given here: near Bowie, Cochise Co., and Horseshoe Dam area, Maricopa Co., AZ; Saddleback basalt, Boron, Inyo Co., CA; Fossil Canyon, San Bernardino Co., CA; Vaalvo, and Dalsnipa, on Sandoy, Faeroe Is.; Ilimaussaq intrusion, southern Greenland; flanks of Mt. Aetna, *Aci Castello, Sicily, Italy*; also at Aci Reale and Aci Trezza, Sicily, Italy; Cyclopean Is., Italy; at Richmond, Melbourne, VIC, Australia; Tokatoka district, about 150 km N of Auckland, New Zealand; Hayata, Saga Pref., Japan. It should be noted at this point that the material from near Bowie, Cochise Co., AZ, and Fossil Canyon, CA, is Na-rich but the habit is not that of herschelite. The term *herschelite* should be abandoned, as it is just a morphological variety of chabazite. EF *AM 51:909(1966), 55:1278(1970), 74:1337(1989); NZ(1985); CCM 36:131(1988); ZW (1992).*

77.1.2.3 Willhendersonite $KCa[Al_3Si_3O_{12}] \cdot 5H_2O$

Named in 1984 by Peacor et al. for research chemist and mineral collector William A. Henderson, Stamford, CT. Zeolite group. See chabazite (77.1.2.1), isostructural, $(Ca,Na_2)[Al_2Si_4O_{12}] \cdot 6H_2O$. Ordered chabazite. TRIC $P\bar{1}$. $a = 9.206-9.23$, $b = 9.21-9.216$, $c = 9.500-9.52$, $\alpha = 92.34-92.7°$, $\beta = 92.4-92.70°$, $\gamma = 90-90.12°$, $Z = 2$, $D = 2.17-2.20$. $35\text{-}0643$(nat): 9.16_{10} 5.18_3 4.09_4 3.93_2 3.82_2 3.71_3 2.91_6 2.80_5. Isostructural with chabazite, but with complete Si–Al ordering. The tectosilicate (SiO_4 and AlO_4) tetrahedral frameworks of both minerals are topologically similar and consist, in part, of double six-membered rings of tetrahedra, although the coordinations of Ca are very different. ZK 159:125(1982), NJBM 547(1984). "Trellis-like" twinned aggregates (San Venanzo), forms {001}, {100}, and {010} and tabular crystals, flattened on {001} (Ettringer Bellerberg, Mayen, Germany); Mayen crystals are 60 μm × 30 μm × 10 μm; twinning by rotation about [111], additional complex twinning of two types is also present; Mayen, Germany, material is untwinned. Colorless, white streak, vitreous luster, transparent. Cleavage parallel to {100}, {010}, and {001}, perfect; these three cleavages are equivalent to the rhombohedral cleavage of chabazite; brittle. $H = 3$. $G = 2.10-2.18$. Nonfluorescent. Analysis (wt %): K_2O, 6.8–8.0; Na_2O, trace; CaO, 10.7–11.6; Al_2O_3, 28.1–30.1; SiO_2, 34.8–35.5; H_2O, 16.0–18.4. The most common ratio of Si/Al in willhendersonite is 1:1, while it is 2:1 in chabazites. Biaxial (+); $2V = 87°$; Mayen: $N_x = 1.503$, $N_y = 1.511$ (calc.) for $2V = 80°$ (+), $N_z = 1.517$; for Terni crystals, optical properties not determined because of complex polysynthetic twinning, small size, and so on. Formed by low-temperature hydrothermal alteration. Occurs in cavities (Terni, Italy) of quaternary lavas and basic potassic lavas. Associated with phillipsite and other minerals. At Mayen, crystals are in cavities in a limestone xenolith in basalt. Associated with other zeolites. *San Venanzo quarry, Terni, Umbria, Italy*; Ettringer Bellerberg volcano, Mayen, Eifel, Germany; Stradnerkogel, Wilhelmsdorf, Styria, Austria. EF AM 55:1278(1970), 69:186(1984); NJBM:547(1984); NZ (1985); ZW (1992).

77.1.2.4 Offretite $(Ca,K_2,Mg)_{2.5}[Al_5Si_{13}O_{36}] \cdot 15H_2O$

Named in 1890 by Gonnard for Albert Jules Joseph Offret (b.1857), professor, Lyon, France. Zeolite group. HEX $P\bar{6}m2$. $a = 13.29-13.32$, $c = 7.56-7.61$, $Z = 1$, $D = 1.91-2.06$. $22\text{-}803$: 11.5_{10} 6.64_2 $5.76_{3.5}$ 4.35_6 $3.84_{4.5}$ 3.32_2 $2.88_{6.5}$ 2.51_2; $25\text{-}1186$(nat, calc pattern): 11.5_{10} $7.58_{1.5}$ 6.65_1 5.76_1 $4.35_{1.5}$ 3.77_2 3.60_1 2.86_2. Structure done on $(K_{1.1}Ca_{1.1}Mg_{0.7})_{\Sigma 2.9}[Si_{12.8}Al_{5.6}O_{36}] \cdot 15.2H_2O$. Aluminosilicate framework is based on double six-membered rings and single six-membered rings in two positions (A and B), linked in a regularly repeating stacking sequence AAB. The exchangeable cations are in three positions: K in cancrinite cages, Ca in double six-membered rings in wide channels, and Mg in gmelinite-type cavities. Similarity of structural units of offretite and erionite with levyne account for their epitaxial overgrowth on levyne. AC(B) 28:825(1972). Usually simple hexagonal prisms with {10$\bar{1}$0} terminated by

{0001} pinacoid, bundles of parallel crystals and radiating groups, sometimes barrel-shaped, offretite is commonly epitaxial on levyne or on erionite. Colorless, white, yellow, golden; white streak; vitreous, silky luster. H = 4. G = 2.10–2.13. Pure synthetic offretite, phase "O," has been produced by Aiello and Barrer (1970). *ZW (1992)*. Both Ca- and Mg-dominant varieties are known; Ca + Mg > K + Na. Ca is commonly more abundant than K. One example of Mg-dominant offretite is known (Sasbach, Germany). Material from Philip Is., and Flinders, VIC, Australia, also have Ca + Mg + Ba + Sr << K + Na. Uniaxial (−); $N_O = 1.489$–1.493, $N_E = 1.486$–1.491; negative elongation, elongated parallel to *c*. Intergrown with erionite, most offretite crystals should be considered an intergrowth of both offretite and erionite. *Offretite* = Ca + Mg > K + Na, Si/Al = 1.99–2.80, uniaxial (−), RI > 1.485, simple X-ray diffraction pattern; and *Erionite* = K + Na > Ca + Mg, Si/Al = 2.85–3.60, uniaxial (+), RI < 1.485, complex X-ray diffraction pattern. Occurs in veins and fractures in basaltic rocks. Not found in Si-rich volcanics, pegmatites, or altered volcanic ash tuffs (where erionite is abundant). May be intergrown with other zeolites. Of hydrothermal origin. Display specimens are rare, principally micromounts. Crystals to several millimeters have been found. From Queen Creek, AZ, CA, CO, NJ, Beech Creek, OR, and WA; 2–3-mm long needles on plates of levyne from Westwold, BC, Canada; several localities in County Antrim, Northern Ireland; barrel-shaped crystals to 0.3 mm long from *Mt. Semiol, Montbrison, Loire, France*; chabazite–offretite epitaxial overgrowths from Passo Forcel Rosso, Adamello, Brescia, Italy; Zezice, Czech Republic; on plates of levyne from Hunter Valley, NSW, Australia; Merriwa district, also Flinders, Jindivick, and Philip Is., VIC, Australia. EF *NZ (1985), EJM 6:397(1994)*.

77.1.2.5 Erionite $(K_2Na_2Ca,Mg)_2[Al_4Si_{14}O_{36}] \cdot 15H_2O$

Named in 1898 by A. S. Eakle from the Greek for *wool*, in allusion to its white, fibrous, wool-like appearance. Zeolite group. HEX $P6_3/mmc$. $a = 13.186$–13.34, $c = 15.041$–15.13, Z = 2, D = 2.09–2.13. 39-1379(nat): 11.56_8 4.35_8 3.84_6 3.77_9 3.59_6 2.88_8 2.85_{10} $2.83_{6.5}$. Aluminosilicate framework is based on double six-membered rings and single six-membered rings in three positions (A, B, and C) linked in a regularly repeating stacking sequence (AABAAC). Structures of erionite, offretite, and levyne all have hexagonal layers A, B, and C. Errors in the stacking sequences for erionite and offretite (AAB) produce an intergrowth of these two species. Exchangeable cations are in three positions: K in cancrinite cages, double six-membered rings are empty or contain some Ca, and a large erionite cage containing Ca, Na, and Mg in several sites. {1000}, {0001} and rare {10$\bar{1}$2}, may be wool-like masses, for which named, frequently striated parallel to *c*, sometimes hexagonal pyramid developed, simple hexagonal prisms, terminated by {0001}, may be epitaxial on levyne, or vice versa, commonly intergrown with offretite; generally small, <3 mm long, some hairs to 15 mm long, radiating aggregates to 8 mm in diameter; no twinning. Colorless, white, green, gray, orange; white streak; vitreous, silky luster. Cleavage is parallel to c axis; fracture is splintery. H = $3\frac{1}{2}$–4. G = 2.02–2.13. K-, Mg-, and Ca-dominant members exist. Commonly K + Na > Ca + Mg, although some varieties have considerable Ca. Cape Lookout, Tillamook Co., Oregon: Ca-rich. Mg-rich from Sasbach, Kaiserstuhl, Germany, also Ca-dominant from Montresta, Nuovo, Sardinia, Italy, and Nidym R., Siberia, Russia. Ts$_i$ ranges from 0.71–0.79. Uniaxial (+); N_O = 1.455–1.483, N_E = 1.457–1.485; generally positive elongation, but may show negative elongation as well (e.g., Faedo, Italy); often intergrown with offretite; erionite: K + Na > Ca + Mg, Si/Al = 2.78–3.60, uniaxial (+), RI < 1.485, complex X-ray diffraction pattern and offretite: Ca + Mg > K + Na, Si/Al 1.99–2.80, uniaxial (−), RI > 1.485; simple X-ray diffraction pattern; has been synthesized, with intergrowths of offretite. Occurs in volcanic and sedimentary rocks. Common in altered silicic tuffs, especially in saline lacustrine deposits. Hydrothermal erionite in cavities in volcanic rocks. Known from more than 40 localities; only noteworthy localities are mentioned here and only for pure or relatively pure material. Occurs at several areas in Arizona; *Durkee opal mine, Swayze Creek, near Durkee, Baker Co., OR*; best specimens come from Cape Lookout, Tillamook Co., and Agate Beach, Lincoln Co., OR; Lake Tecopa, Inyo Co., CA; Rock Island Dam, WA; excellent specimens without any intergrown offretite at Chase Creek, N of Falkland, BC, Canada; aggregates to 15 mm on the Nidym R., Siberia, Russia. Other localities include: Eastgate, Churchill Co., Reese R., Lander Co., Jersey Valley, Pershing Co., and Pine Valley, Eureka Co., NV; Hvalstod and Berufjord, Iceland; Parkgate quarry, Templepatrick, County Antrim, Northern Ireland; Mazi, Niigata Pref., Japan; Moeraki, Otago, South Island, New Zealand. EF *MM 32:261(1959); AM 54:875(1969), 61:853(1976); MGNZ (1977); AC(B) 33:3265(1977); NZ (1985); ZW (1992); NJBM:185(1995)*.

77.1.2.6 Gmelinite (Na$_2$,Ca)Al$_2$Si$_4$O$_{12}$ · 6H$_2$O

Described by Brewster in 1825 and named for Christian Gottlob Gmelin (1792–1860), professor of chemistry and mineralogist, Tübingen, Germany. Zeolite group. HEX $P6_3/mmc$. Regular Na–Ca-dominant gmelinite: a = 13.737–13.805, c = 9.974–10.077; a = 13.621–13.696, c = 10.203–10.254; K-dominant gmelinite: Z = 4, D = 2.07. *38-435*(nat): 11.9$_6$ 7.68$_3$ 5.02$_3$ 4.11$_{10}$ 3.28$_4$ 2.98$_{5.5}$ 2.86$_4$ 2.69$_4$; *NJBM:504(1990)* (K-rich gmelinite): 11.9$_3$ 5.14$_5$ 4.11$_{10}$ 3.29$_4$ 2.96$_8$ 2.85$_6$ 2.72$_8$ 2.59$_2$. The aluminosilicate framework is composed of tetrahedra linked to form parallel double six-membered rings stacked in two different positions (A and B) in the repeating arrangement AAB-BAABB. The framework has no Al–Si order although Na-rich gmelinite has better order and is more stable than gmelinite containing dominant Ca. Two cation sites, C1 and C2, contain either Na or Ca. In K-dominant gmelinite, C2 is nearly empty (3% occupancy). Stretching along c occurs because of K substitution. Structure of Na- and Ca-rich gmelinite: *NJBM:145(1982)*; K-rich gmelinite: *SPC 29:256(1984), NJBM:504(1990)*. Has been synthesized at temperatures less than 95°. *CR 272:ser. D:181(1971)*. Crystals intergrown with

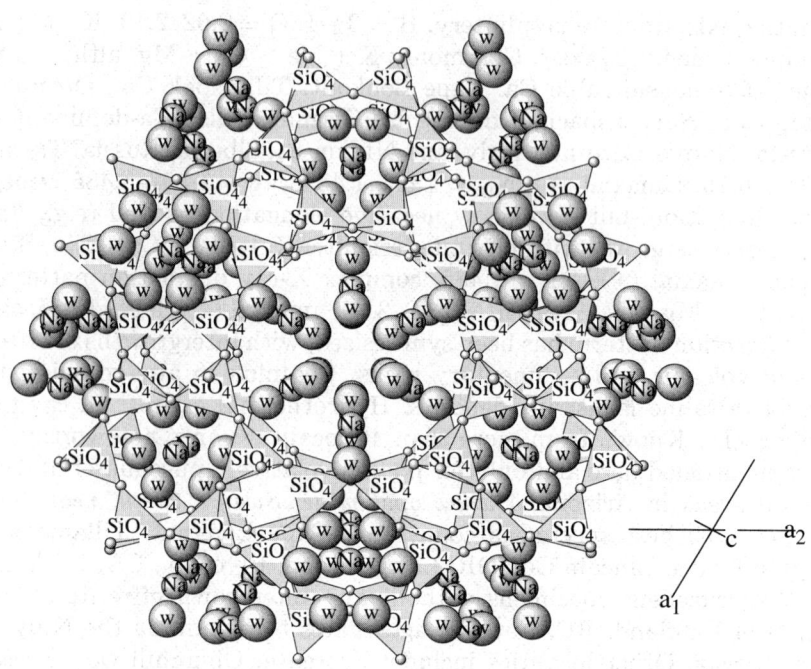

Gmelinite

chabazite and may reach 4 cm across. Hexagonal plates, or short prisms, showing hexagonal dipyramids, pyramids, and basal pinacoid: $\{10\bar{1}0\}$, $\{10\bar{1}1\}$, $\{0001\}$ dominant. May also be tabular or rhombohedral in habit. Striated parallel to (0001) or less commonly parallel to [0001}. Twinning is rare, on (1011); interpenetrant twins consist of four individuals; three are at 90° to the other and at 60° to each other. Colorless, white, pale yellow, lemon-yellow, greenish white, orange, pale green, pink, salmon, red, brown, and gray(!); white streak; dull to vitreous luster. Cleavage $\{10\bar{1}0\}$, good; $\{0001\}$ parting; uneven to conchoidal fracture; brittle. H = $4\frac{1}{2}$. G = 2.04–2.17. Piezoelectric. Thermal data: *NJBM:310(1978)*, *NZ(1985)*; for K-rich: *NJBM:504(1990)*. Soluble in cold 10% HCl, and gelatinizes with a residue. Na-rich variety occurs at Boron, CA; is also richer in Si than other gmelinites. The mineral is generally Na-dominant; however, considerable Ca, K (may be dominant), and Sr may be present (to 0.35 atom). Mg is very low and Ba is absent. Usually uniaxial (−) or (+); may be anomalously biaxial (−) or (+) with a small 2V; for Na–Ca gmelinite: N_O = 1.476–1.494, N_E = 1.474–1.480; birefringence = 0.002–0.008; for K-dominant gmelinite: N_O = 1.470, N_E = 1.474; colorless. Generally occurs in Si-poor volcanic rocks, marine basalts, and breccias. Also in Na-rich pegmatites in alkalic rocks. No sedimentary gmelinite has been found. A product of diagenetic reactions, hydrothermal activity, or deuteric alteration of basalt. Occurs in 55–75° geothermal wells. K-dominant samples have probably undergone cation exchange after deposition. Associated with other zeolites, quartz, aragonite, calcite, and so on. A widespread zeolite but occurring in small amounts. Localities for well-characterized material include: Bergen Hill, Hudson Co., and Great Notch, Paterson, and Prospect Park, Passaic Co., NJ; Na- and Si-richest gmelinite from Boron, San Bernardino Co., CA; Springfield, Lane Co., OR; Mt. St.-Hilaire, QUE, and at Pinnacle Rock, Five Is., and Two Is., Bay of Fundy, NS, Canada; Ilimaussaq intrusion, southern Greenland; *Little Deer Park, Glenarm, and elsewhere in County Antrim, Northern Ireland; Isle of Skye, Scotland; Montecchio Maggiore, Vicenza, Italy*; K-dominant from Cava Fortelongo, Fara Vicentina, Vicenza, Italy, and from Mt. Alluiav, Lovozero massif, Kola Penin., Russia; Bekiady, Madagascar; exceptional display specimens, to 4 cm in diameter, from the Mornington Penin., between Flinders and Cape Schank, VIC, Australia. EF *ZW (1992)*.

77.1.2.7 Faujasite $(Na_2,Ca,Mg)_{3.5}[Al_7Si_{17}O_{48}] \cdot 32H_2O$, or $(Na_2,Ca)Al_2Si_4O_{12} \cdot 8H_2O$

Named in 1842 by Damour for Barthélemy Faujas de Saint Fond (1741–1819), French geologist and writer on origin of volcanoes. Zeolite group. ISO $Fd3m$. $a = 24.64$–24.78, $Z = 8$, $D = 1.94$. *39-1380*(nat): 14.28_{10} 5.67_8 4.75_4 4.37_5 $3.76_{9.5}$ 3.30_7 2.91_4 2.85_8. Has the most open structure of all the zeolites.

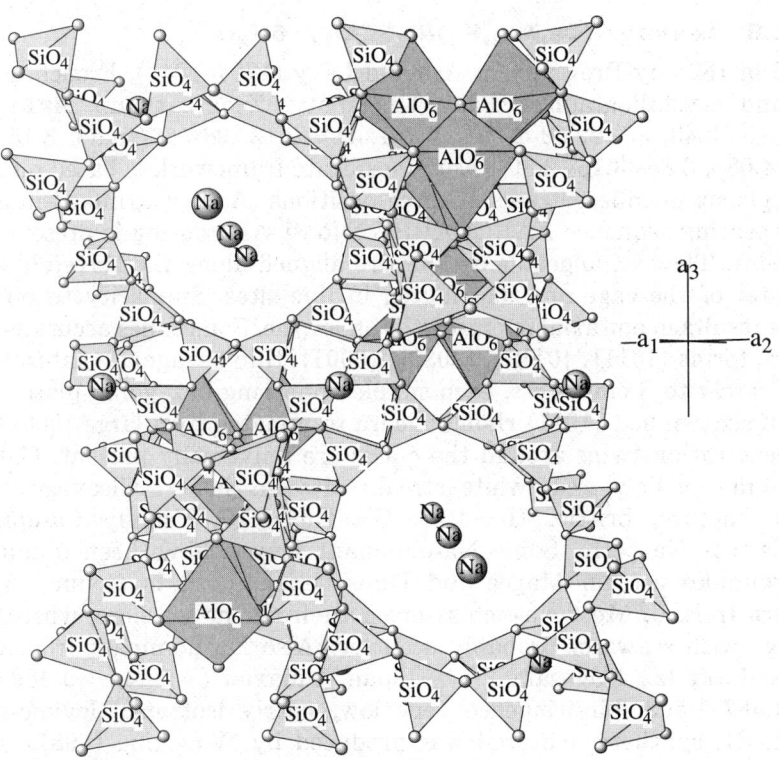

FAUJASITE

Structure: *NJBM:193(1958)*, *AM 49:697(1964)*. {111}, rare {100}, and {556}, aggregates to 10 mm across; small, single octahedra 1–4 mm in diameter common; contact and penetration twinning on spinel law {111} is common. Colorless–white or pale brown, white streak, vitreous luster, transparent to translucent. Cleavage {111], perfect; conchoidal to uneven fracture; brittle. H = $4\frac{1}{2}$–5. G = 1.92–1.93. NMR studies of Al, Si distribution: *ACS 104:4859(1982)*, *JPC 86:3489(1982)*. Na and Ca may both be dominant, but Na is dominant at most locations. Mg and K are present in most specimens. Isotropic; N = 1.47–1.48. Usually occurs in basaltic volcanics; metapyroxenite; also alteration product of volcanic tuff. Found at Cima Dome, near Valley Wells, San Bernardino Co., CA; at several localities on Oahu, Hawaii; Davis Hill and Khartoum, ONT; Hincks Bridge, Laurel, and Notre-Dame-de-la-Salette, and the Daisy Mica mine, Ottawa Co., QUE, Canada; *Sasbach, Kaiserstuhl, Baden, Germany* (principal source since 1842) and other localities in Baden–Württemburg; Pflasterkaute, near Eisenach, Thuringia; at Annerod, near Giessen, and at a number of localities around the Vogelsberg volcano, Hesse, Germany; Aci Reale, and Aci Castello, Catania, Sicily; several other worldwide localities are known. Synthetic counterparts are used for sorbents and catalysts. EF *AM 67:794(1982)*, *NZ (1985)*, *ZW (1992)*.

77.1.2.8 Levyne (Ca,Na$_2$,K$_2$)Al$_2$Si$_4$O$_{12}$ · 6H$_2$O

Named in 1825 by Brewster for Armand Lévy (1794–1841), French mineralogist and crystallographer, Paris University. Zeolite group. HEX-R $R\bar{3}m$. a = 13.34–13.43, c = 22.69–23.01, Z = 1, D = 2.12. *26-1381*(nat): 8.15$_2$ 6.69$_2$ 4.25$_{1.5}$ 4.08$_{10}$ 3.85$_2$ 3.16$_2$ 2.80$_6$. Aluminosilicate framework is based on double and single six-membered rings in three positions (A, B, C), linked in a regularly repeating sequence AABCCABBC. No Si–Al ordering is present in the tetrahedra. The exchangeable cations are aligned along the threefold axis at the center of the cage and distributed in five sites. Similarity to offretite–erionite results in epitaxial overgrowths on levyne. Sometimes occurs as rhombohedra, forms {10$\bar{1}$1}, {01$\bar{1}$2}, {30$\bar{3}$2}, {0001}; thin hexagonal plates 2–5 mm across, rarely to 3 cm across, each simple-appearing hexagonal plate is composed of six twinned {10$\bar{1}$1} rhombohedra terminated by a large {0001} pinacoid; penetration twins around the c axis are universally present. Colorless, white, gray, yellow, red; white streak; vitreous luster. Cleavage {10$\bar{1}$1}, uneven fracture, brittle. H = 4–$4\frac{1}{2}$. G = 2.09–2.16. Usually Ca-dominant with Ca >> Na > K. Some Na-dominant levyne have been found. Two such examples are Is. Magee and Dunseverick, both in County Antrim, Northern Ireland. Most Na-rich samples are from the seashore, where cation exchange with seawater probably occurred. Also Na dominant from Chojabaru, and Iki Is., Nagasaki Pref., Japan. Uniaxial (−); N$_O$ = 1.489–1.510, N$_E$ = 1.487–1.502; birefringence very low, nearly isotropic; levyne-na has N$_O$ = 1.481, synthetic material was produced by Wirsching (1981). Occurrences are most common in low-silica volcanic rocks (e.g., olivine basalt and rarely in andesite). Often with other zeolites, especially offretite and erionite.

Occurs at numerous worldwide localities; only localities for fine-quality and/or unusual specimens are given here. Best display specimens are from Burnt Cabin Creek, Spray, Wheeler Co., OR (to 12-mm-diameter colorless hexagonal plates); Beech Creek, Grant Co., OR(!) (to 1 cm); largest crystals (to 2 cm) are from Goble, Columbia Co., OR; twinned crystals without {0001} are found at Elk Mountain, WA, and Green Peter Dam, OR; hexagonal plates to 2 cm from near Berufjordur, Dyrafjord, Onundarfjord, and Mjöadalsa, Iceland; *Dalsnipa, Sandoy Is., Faroe Is.*; crystals to 3 cm are found at Rossafelli and Nesvik on Stremoy Is., Faroe Is. Other localities include North and South Table Mts., Golden, Jefferson Co., CO; Is. Magee and the Parkgate quarry, Templepatrick, County Antrim, Northern Ireland; Matgilligan and Dungiven, County Londonderry, Ireland; Cinchwad, Pune district, MAH, India; Flinders and Victoria Is., VIC, Australia. EF *AM 59:837(1974), 61:853(1976); TMPM 22:117(1975); NZ (1985); ZW (1992).*

77.1.3.1 Gismondine $Ca_2[Al_4Si_4O_{16}] \cdot 9H_2O$

Named in 1817 by Leonhard for Carlo Giuseppe Gismondi (1762–1824), Italian mineralogist, who previously discovered a similar mineral, zeagonite, a mixture of gismondine and phillipsite. Zeolite group. The oldest and best established member of the gismondine group [see amicite (77.1.3.2), garronite (77.1.3.3), and gobbinsite (77.1.3.4)]. Has been divided by chemistry and amount of disorder in the framework. MON (pseudo-TET) $P2_1/c$. $a = 10.021$, $b = 10.637$, $c = 9.836$, $\beta = 92.5°$, $Z = 2$, $D = 2.22$–2.28. *39-1373*(nat): 7.3_6 4.91_5 4.27_{10} 4.21_5 3.19_9 2.74_8 2.71_6 2.69_8 2.66_7; *20-452*: 7.28_2 4.91_2 $4.27_{3.5}$ 3.34_{10} 3.19_2 3.13_1 2.70_2 1.82_2; also *21-840* (preferred orientation may affect intensities). Single four-membered rings stacked in two-dimensional arrays of double-crankshaft chains parallel to [100] and [010]. Gismondine has perfect Si–Al ordering in the framework with alternation of Si and Al at the centers of the tetrahedra. Ca is in a fourfold position bound by two framework oxygens on one side and H_2O molecules. Gismondine is the ordered equivalent of garronite, the Ca equivalent of amicite, and the ordered Ca equivalent of gobbinsite. Appear to be pseudotetragonal dipyramids with {111}; in reality, they are interpenetration twins of two individuals with four faces of the form {$\bar{2}32$} and four faces of its twin, with twinning perpendicular to {100}, with twin plane approximately parallel to (100) and (001); commonly intergrown with phillipsite (epitaxial); aggregates to >15 mm across, individual crystals to >6 mm across; also in radiating spherulitic aggregates; rare twinning on {110} around a or b. Colorless–white, very pale pink; white streak; vitreous luster. Two poor cleavages parallel to form {$\bar{2}32$}, conchoidal fracture. $H = 4.5$. $G = 2.12$–2.28. IR data: *CMG 26:143(1981)*. Nonluminescent. Piezoelectric. Has a narrow range of composition, small amounts of Na, K, or Sr with no Ba, Fe, or Mg. Tsi ranges from 0.50–0.60 Biaxial $(-)$; X perpendicular to [010], $Y = b$, $Z \wedge c = 42°$; $N_x = 1.512$–1.538, $N_y = 1.516$–1.544, $N_z = 1.523$–1.549; $2V = 60$–$90°$, but as low as $15°$; length-fast; $r < v$, weak. Has been synthesized. *CCM 29:171(1981)*. Occurs in olivine basalt (most common), less commonly in nepheline basalt, and palagonite tuffs, schist, or pegmatites, also in metapyroxenite. Best localities for display-quality specimens: Italy, Germany, Oregon, Northern Ireland, and New Zealand. Also found in Australia, Austria, Canada, Canary Islands, Czech Republic, Faroe Is., France, Hungary, Iceland, Japan, Mexico, Poland, Russia, Senegal, South Africa, and Switzerland. Type locality is *Capo di Bove, Lazio, Italy*. Selected detailed localities include: Round Top volcano, Oahu, and Alexander Dam, Kaui, HI; Reydarfjord, Faskrudsfjord, and Fagridalur, Iceland; Ballyclare, County Antrim, Northern Ireland; Montalto di Castro, Viterbo, Italy; Gorner glacier, near Zermatt, Switzerland; in Germany from the Eifel district (e.g., Schellkopf near Brenk); Schlauroth, Görlitz, Saxony; Hohenberg near Bühne, Westphalia; Frauenberg, Fulda, Hesse; Arensberg, Zilsdorf; Zalezly, Czech Republic; Concepcion del Oro, ZAC, Mexico. EF *MM 33:187(1962), 43:841(1980); BM 107:805(1984); NZ (1985); Z6:361(1986); ZW (1992)*.

77.1.3.2   Amicite $K_2Na_2(Al_4Si_4)O_{16} \cdot 5H_2O$

Named in 1979 for Giovan Battista Amici (1786–1863), physicist, optician, and inventor of the Amici lens. Zeolite group. MON $I2$. $a = 10.23$–10.26, $b = 10.42$–10.44, $c = 9.88$–9.92, $\beta = 91.50$–$91.68°$, $Z = 2$, $D = 2.18$. *33-1273*: 7.30_6 5.11_4 4.22_9, 4.18_4 3.24_5 3.14_8 2.72_{10} 2.70_5. Characterized by untwisted double-crankshaft chains developed in two perpendicular directions, as in gismondine and the disordered phases gobbinsite and garronite, but with the symmetry and distribution of cations and water molecules being completely different. The Si–Al and Na–K distributions are perfectly ordered. Amicite can be considered as the ordered equivalent of gobbinsite. Na replaces Ca in gismondine sites and occupies another site as well. K occupies sites that are filled by H_2O in gismondine. *AC(B) 35:2866(1979)*. Rhombic prisms most common. {110} and {011} dominant, with {111}, {101}, and {010} uncommon. Lamellar twinning on {011} or {110}. Pyramidal crystals to 3 mm and rarely to 1 cm in length. Colorless, white streak, vitreous luster. No cleavage, conchoidal fracture. $H = 4\frac{1}{2}$. $G = 2.06$–2.23. Thermal data: *NZ (1985)*. Slowly decomposed by cold 10% HCl. Minor Ca and traces of Fe may be present. Biaxial $(-)$; $Y = b$, $Z \wedge c = 12°$; $N_x = 1.485$, $N_y = 1.490$, $N_z = 1.494$–1.498; $2V = 82°$. A low-temperature zeolite in veins cutting alkalic volcanic rocks (Hegau) and as veins several centimeters thick cutting alkalic pegmatites (Khibiny massif). Found at *Höwenegg, near Hegau, Eifel, Germany*, sometimes with merlinoite, in the S.M. Kirov mine, Mt. Kukisvumchorr, Khibiny massif, Kola Penin., Russia; also in cavities in basalt south of Ciudad Real, Ciudad Real Prov., Spain. EF *NJBM 11:481(1979), DANS-ESS 263:135(1982), NZ (1985); ZW (1992)*.

77.1.3.3   Garronite $Na_2Ca_5Al_{12}Si_{20}O_{64} \cdot 27H_2O$

Described in 1962 by Walker and named for the locality. Zeolite group. ORTH (pseudo-TET) $I\bar{4}m2$. $a = 9.871$–9.93, $b = 9.871$–9.893, $c = 10.292$–10.303, $Z = 0.5$, $D = 2.19$. *39-1374*(nat): $7.15_{7.5}$ 4.95_6 4.15_8 4.07_3 3.24_5 3.14_{10} 2.67_7

2.57_1 16-905(syn): 7.13_8 4.95_6 4.15_7 4.07_5 3.24_4 3.14_{10} 2.71_1 2.67_6. The mineral has an aluminosilicate framework that is the same as that of gismondine but with disordered Si–Al distribution. Probably grew as a disordered phase that later underwent a partial ordering with four differently oriented domains. Deviations from the maximum framework symmetry can be explained in terms of cation and water arrangements in the zeolitic cavities. *AM 77:189(1992)*. Generally compact, massive, sometimes fibrous, radiating. Rare single crystals to 4 mm; prismatic, rarely dipyramidal. Forms: {101}, {011}, and {112}. Often covered by or associated with phillipsite and as epitaxial overgrowths on phillipsite. Internal twinning on {1$\bar{1}$0}, and further twinning on {011} leads to characteristic groups. Various colors, including pale yellow, orange-yellow, yellowish brown, colorless, white, and tan; white streak; dull to greasy–vitreous luster. Two good cleavages: {100}, {010}; series of concentric fractures perpendicular to the length of the fibers (elongate on c). $H = 4\frac{1}{2}$–5. $G = 2.13$–2.18. Always Ca dominant. Na and K contents vary considerably. Garronite can be considered the Ca equivalent of gobbinsite and may have been the precursor that was altered by Na–K-rich solutions to gobbinsite. Some garronite has very little Na (about 0.2 atom). Biaxial ($-$) and ($+$), $2V = 25$–$30°$; also uniaxial ($+$) and ($-$); $N_x = 1.500$–1.512, $N_y = 1.502$–1.511, $N_z = 1.512$–1.515; $N_O = 1.500$–1.515, $N_E = 1.502$–1.512; length slow and fast; parallel or inclined extinction; $Y = c$ (elongation). Most common in vesicles in basalts and diabases; some metamorphic rocks, and in sodalite xenoliths at Mt. St.-Hilaire. North and South Table Mts., Golden, Jefferson Co., CO; Neer Road pit, Goble, Columbia Co., OR; Pete's Point, Aneroid Lake, Joseph, Wallowa Co., OR; Capitol Peak, Thurston Co., WA; Davis Hill, ONT; Mt. St.-Hilaire, QUE, Canada; at numerous localities in Iceland, between Berufjord and Seydisfjord (e.g., Skessa, Reydarfjord); *Glenariff Valley, Garron Plateau, and many other localities, County Antrim, Northern Ireland*; Storr, Isle of Skye, Scotland; Fara Vicentina, and San Giorgio di Perlena, Vincenza, Italy; Gignat, Puy de Dôme, France; Höwenegg quarry, Hegau, Baden–Württemberg, Germany. EF *MM 33:173(1962), 47:567(1983); NZ (1985); NJBM(2):91(1994)*.

77.1.3.4 Gobbinsite $Na_4(Ca,Mg,K_2)Al_6Si_{10}O_{32} \cdot 12H_2O$

Named in 1982 by Nawaz and Malone for the Gobbins escarpment, near Hills Port, Island Magee, County Antrim, Northern Ireland. Zeolite group. See garronite (77.1.3.3), phillipsite (77.1.3.6), gismondine (77.1.3.1), amicite (77.1.3.2), and merlinoite (77.1.3.11). ORTH (pseudo-TET) $Pmn2_1$. $a = 10.10$–10.15, $b = 9.77$–9.80, $c = 10.10$–10.17, $Z = 1$, $D = 2.15$. 35-459(nat): 7.11_{10} 5.06_5 4.89_3 4.12_{10} 3.20_{10} 3.11_8 2.70_8 2.65_4. Tetragonal cell is $P4_22_12$, with $a = 10.115$ and $c = 9.766$. The framework (Al–Si) is the same as gismondine. It is composed of stacked two-dimensional arrays of double-crankshaft chains parallel to [100] and [010], with Na and Ca ions located in more than half of the eight-membered rings perpendicular to a, where K ions alternate in eight-membered rings perpendicular to b. Distribution of Al and Si in the framework is disordered. *ZK 171:281(1985), MM 58:615(1994)*. Rietveld refinement. Terminated crystals are very rare, pyramidal crystals have external morphology and twinning of phillipsite, also steep and spikelike to about 2 mm, elongated on c; masses to 2 cm in diameter; twinning like phillipsite, on $\{110\}$. White to tan, cream; white streak; dull to greasy (chalky) luster. No cleavage observed, brittle, poor parting perpendicular to length of fibers. H probably ≈ 4. G = 2.19. Always Na dominant. K varies from being $\approx$ Na, to only a trace. Ca is present in considerable amounts in some samples. Mg present in small amounts. Tsi ranges from 0.62–0.68 Uniaxial $(-)$; $N_O = 1.494$ perpendicular to length, $N_E = 1.489$ parallel to length; or biaxial, with small 2V; negative elongation, parallel extinction; c parallel to direction of elongation. Occurs at Mt. St.-Hilaire, found in small cavities in a eudialyte-rich pegmatite layer in nepheline syenite. Has overgrowth of phillipsite. Gobbinsite appeared to form as a replacement of garronite during a late Na–K-rich phase. In olivine basalt and sodalite syenite (Ireland). Occurs at the Poudrette quarry, Mt. St.-Hilaire, QUE, Canada; this was the first locality for well-crystallized specimens, and most have thin, epitaxial overgrowths of phillipsite. *Gobbins escarpment, near Hills Port, Island Magee, County Antrim, Northern Ireland*; K-rich gobbinsite occurs near Two-Mouth Cave, Portmuck, Island Magee, and at Dunseverick, near the Giant's Causeway, County Antrim, also at Maghera Morne, Larne, Northern Ireland; Aranga, North Island, New Zealand. This may be gismondine or an intergrowth of garronite–gobbinsite. EF *MM 46:365(1982), 47:567(1983); NZ (1985); ZK 171:281(1985); ZW (1992)*.

77.1.3.5 Harmotome $(Ba,K)_{1-2}(Si,Al)_8O_{16} \cdot 6H_2O$

Described in 1801 by Haüy and named from the Greek for *joint cutting*; defined as "that which divides on the joints," in allusion to the morphology of twinned crystals. Zeolite group. See phillipsite (77.1.3.6), the Ca-, Na-, or K-dominant analog. MON $P2_1/m$. $a = 9.86$–9.91, $b = 14.12$–14.17, $c = 8.67$–8.72, $\beta = 124.43$–$124.74°$, Z = 2, D = 2.45. *39-1377*(nat): 8.12_6 $7.16_{6.5}$ $6.39_{9.5}$ 4.08_7 3.24_6 3.17_7 3.13_{10} 2.67_6. The framework consists of layers of tetrahedra forming chains of doubly connected four-membered rings (double-crankshaft chains) on planes where two nearby chains have been alternately rotated 180° and mirrored between successive planes. Two sets of intersecting channels are present, one parallel to a and the other parallel to b. Partial Si–Al ordering exists as determined by X-ray diffraction and optics. Ordering of Ba causes deviations from orthorhombic to monoclinic symmetry, while Si–Al order causes deviation, up to 2°, from monoclinic to triclinic symmetry. The topological orthorhombic symmetry (*Cmcm*) that reduces to monoclinic ($P2_1/m$) from cation ordering and finally to triclinic ($P1$) because of Si–Al ordering. *AC(B) 30:2426(1974); AM 63:960(1978), 70:822(1985); EJM 2:861(1990)*. May be blocky, equant, or radiating. Pseudo-orthorhombic, twinned, prismatic crystals, elongate on a or b. Crystals to about 4 cm long. All crystals are repeatedly twinned on $\{001\}$, $\{021\}$, and $\{110\}$, with a multitude of twin types. Harmotome appears to nucleate as twins. The major twin types are:

Morvenite: fourling twin present in all crystals; Marburg; a cruciform twin; Perrier: a cruciform twin; Stempl: a multiple twin of Morvenite, Marburg, or Perrier types and twinned around two or three axes. Three double twins may further twin to simulate a pseudotetragonal prism or dodecahedron. Dominant forms include {010}, {001}, {110}, {100}, {520}, {410}. Colorless, white, gray, red, yellow, orange, brown; white streak; vitreous luster. Cleavage {010}, good; {100}, fair; uneven to conchoidal fracture; brittle. H = $4\frac{1}{2}$-5. G = 2.38–2.50. IR data: *MPOL 16:3(1985)*, *CMG 26:143(1981)*. Soluble in common acids. A continuous solid-solution series exists between harmotome and phillipsite. Harmotome commonly contains little Sr, K, and Ca. However, considerable and dominant amounts of K and Na have been found in some crystals. Tsi = 0.70–0.72 Biaxial (+); Z = b, X $\wedge$ a = 62–67°; N_x = 1.498–1.508, N_y = 1.503–1.509, N_z = 1.506–1.514; 2V = 75–85°; weak, crossed dispersion. For triclinic material: Z $\wedge$ b = 0–2°, X $\wedge$ c = −52 to −58°, Z = b, Y $\wedge$ c = 27–34° in the acute angle. A low-temperature zeolite. Most occurrences are in late-stage hydrothermal veins. Some are in volcanics (e.g., basalts, phonolites, and trachytes), due to hydrothermal alteration of Ba-containing feldspars or other Ba-containing minerals. Harmotome is rarely found in altered tuffs and deep-sea sediments. Associated with such minerals as other zeolites, quartz, calcite, barite, pyrite, galena, hyalophane, clay minerals, and leucite. Occurs in many worldwide localities, but only localities for display-quality specimens and unusual specimens are given here. Crystals to 2 cm from Sing Sing, near Ossining, Westchester Co., NY; Glen Riddle, Delaware Co., PA; Beaver mine, Thunder Bay, ONT, Canada; crystals to 2.5 cm from several mines, most from the Bellsgrove mine, Strontian, Argyllshire, Scotland; crystals to 2.5 cm across, elongated on b, from the Korsnäs lead mine, south of Vaasa, Finland; Sarrabus, Sardinia, Italy; *Andreasburg, Harz Mts. and at Idar–Oberstein, Rhineland–Palatinate*, and Silberberg, near Bodenmais, Bavaria, *Germany*; Pribram, Czech Republic; Kupferberg–Rudelstadt, Silesia, Poland; crystals to 3 cm from Bukan, Kotui R. basin, Taimir, Russia; Batopilas, CHIH, Mexico. EF *MM 36:444(1967)*, *NZ (1985)*, *ZW (1992)*.

77.1.3.6 Phillipsite $(K,Na,Ca,Ba)_{1-2}(Si,Al)_8O_{16} \cdot 6H_2O$

Described in 1825 by Levy and named for William Phillips (1775–1829), British mineralogist and founder of the Geological Society of London. Zeolite group. Forms a series with harmotome. Wellsite is a barian phillipsite. MON $P2_1/m$. a = 9.84–10.02, b = 13.85–14.32, c = 8.64–8.73, β = 124.3–125.07°, Z = 2, D = 2.17. *39-1375*(nat): 7.18_6 7.16_7 4.13_4 4.12_4 3.28_4 3.21_{10} 2.75_4 2.70_4; *CCM 37:243(1989)*(sedimentary): 7.10_8 5.35_2 5.01_3 4.97_5 3.25_4 3.17_{10} 2.73_2 2.68_3. Structure is like that of harmotome (77.1.3.5). X-ray diffraction shows no Si–Al ordering. However, optics show fine-scale ordering. Ordering occurs when cations bond to the framework oxygens at Al-populated sites. If the cation is completely surrounded by water molecules, the Si–Al structure is disordered. Ordering of the exchangeable cations causes deviations from orthorhombic to monoclinic symmetry, while ordering of Al and Si causes deviations (up to 2°) from monoclinic to triclinic symmetry. Topologically orthorhombic ($Cmcm$) → monoclinic ($P2_1/m$) → triclinic ($P1$). Occurs as pseudo-orthorhombic, twinned, prismatic crystals, commonly elongated on a. Also, blocky, equant crystals and radiating aggegates and spherulites are common. Common forms are {010}, {001}, {110}, {100}, {520}, {410}. Always forms complex penetration twins. Untwinned phillipsite has not been found. Four types of penetration twins: Morvenite, Marburg, Perrier with {021} twin plane, and Stempl. May be single- or double-penetration twins. Twin plane can be {001} and {110} as well. The Perrier and Marburg twins are also cruciform twins. Colorless, white, pink, red, light yellow; white streak; vitreous luster. Cleavage {010}, distinct; {100}, indistinct; uneven fracture; brittle. H = $4-4\frac{1}{2}$. G = 2.19–2.20. Gelatinizes with common acids. Synthesized: *CCM 29:171(1981)*. IR and Raman data: *CPMG 26:143(1981)*, *CRT 16:917(1981)*. Thermal data: *NZ (1985)*. Dominant cations may be K, Na, Ca, or Ba, but Ba may not exceed 50% of the total exchangeable cations. Sr-bearing phillipsite is known. Most phillipsite has Ca as the dominant cation, but others are often dominant. Sedimentary phillipsites are generally very siliceous and alkalic [$Si/(Al + Fe^{3+})$ = 3.08–3.37]. Tsi varies from 0.56–0.77, may be biaxial (+) or (−); N_x = 1.483–1.505, N_y = 1.484–1.511, N_z = 1.486–1.514; $2V_z$ = 60–90° or $2V_x$60–90°; sedimentary phillipsites have a much lower average N than for other phillipsites because of their high Si and alkali content; dispersion is r < v moderate in optically positive phillipsite and r > v in negative phillipsites. X = b, Z ∧ c = 19° (11–30°) in the acute angle, positive elongation for optically positive phillipsite. For triclinic phillipsite: Z ∧ b = 2°, X ∧ c = (−)52–58°, Y ∧ c = 27–34° in the acute β. Phillipsite is a common zeolite in volcanic rocks (e.g., in cavities in basalts); ore veins; is a common constituent in diagenetically altered rhyolitic vitric tuffs of Cenozoic saline, alkaline lake deposits; on the ocean floor. A low-temperature zeolite. May be associated with other zeolites, apophyllite, calcite, celadonite, and other minerals. Display-quality specimens of phillipsite are rare. A Na-dominant (11.0 wt% Na_2O) of phillipsite occurs at Boron, Mojave Co., CA; display-quality specimens: crystals to 2 cm long from Wall Creek, N of Monument, Grant Co., OR; also Mt. Vernon, Grant Co., Oregon; Cape Lookout, Tillamook Co., and Spray, Wheeler Co., and other localities in Oregon; Mt. St.-Hilaire, QUE, Canada; at a number of localities, including the Giant's Causeway, County Antrim, Northern Ireland; with natrolite at *Aci Castello, Sicily, Italy;* Capo di Bove and other localities around Rome, Lazio, Italy; Asbach, Westerwald, and at Annerod, near Giessen, Hesse, Germany; to 8 mm long from Sumeg, Hungary; on heulandite to 12 mm long from Quebrada Grande, Costa Rica; water-clear to orange crystals to 4 mm long from near Melbourne, and other localities in VIC and at Gads Hill, near Liena, TAS, Australia; Maze, Niigata Pref., Japan. EF *AM 49:1366(1964)*, *57:1125(1972)*, *70:822(1985)*; *AC(B) 30:2426(1974)*; *CCM 36:131(1988)*; *NZ (1985)*; *ZW (1992)*.

77.1.3.8 Perlialite K₉Na(Ca,Sr)[Al₁₂Si₂₄O₇₂] · 15H₂O

Named in 1984 by Menshikov for Lilliya Alekseevna Perekrest, instructor of mineralogy, Kirov Mining School. Zeolite group. HEX $P6/mmm$. $a = 18.490$–18.535, $c = 7.51$–7.535, $Z = 1$, $D = 2.15$. *38-0395*(nat), *ZVMO 113:607(1984)*: 16.0_{10} 6.0_7 4.62_{10} 3.94_7 3.20_9 2.67_8 1.54_8 1.31_{10}; Murun massif, *ZVMO 115:200(1986)*: 16.1_8 6.07_3 4.64_{10} 3.95_5 3.21_{10} 3.19_3 2.68_9 2.21_3. Not related structurally to any of the other natural zeolites but similar to synthetic zeolites L, K, and Ba-G-g. X-ray diffraction patterns resemble synthetic zeolites L and K. Fibrous, radiating needles 2–5 mm long or opal-like masses to 2 cm in diameter; curved, flexible, and elastic needles. White, white streak, pearly luster. $H = 4$–5. $G = 2.14$–2.18. IR, DTA, TGA, DTG data: *ZVMO 115:200(1986)*. Kola Penin. = $K_{9.07}Na_{0.91}Ca_{0.49}Sr_{0.46}Fe_{0.08}Mg_{0.05}$ [$Al_{11.35}Fe_{0.33}Si_{24.27}O_{71.12}$] · 15.2H₂O and Murun = $K_{8.48}Na_{1.04}Ca_{0.61}Ba_{0.37}$ $Mg_{0.17}Sr_{0.15}$ $Mn_{0.02}[Al_{12.64}Si_{23.47}O_{72}]$ · 18H₂O. Uniaxial (+); $N_O = 1.479$–1.483, $N_E = 1.488$–1.489; parallel extinction; positive elongation. Occurs as radiating fibrous aggregates in nepheline–microcline and sodalite–microcline pegmatites, associated with pectolite, kalsilite, aegirine, wadeite, and yuksporite. *Mt. Eveslogchorr and Mt. Yukspor, Khibiny massif, Kola Penin., Russia;* also at Chioiny, Kola Penin.; Murun massif, Transbaikalia, Russia. EF *ZW (1992)*.

77.1.3.9 Paulingite (K₂,Ca,Na₂Ba)₅[Al₁₀Si₃₂O₈₄] · 34–44H₂O or (Ca₂.₆K₂.₃Ba₁.₄Na₀.₈)(Al₁₁.₅Si₃₀.₆O₈₄) · 33H₂O [IMA 16th Meet.: 238(1994)]

Named in 1960 by Kamb and Oke for Linus Carl Pauling (1901–1994), professor of chemistry, California Institute of Technology, Pasadena, CA. Zeolite group. ISO $Im3m$. $a = 35.049$–35.114; $Z = 16$, $D = 2.10$–2.22. *39-1378*(nat): 8.28_6 6.89_7 4.78_8 3.58_6 3.35_7 3.26_9 3.13_6 3.08_{10}. Has been synthesized. Extremely complex aluminosilicate framework composed of large polyhedral cages consisting of double eight-membered rings; gmelinite cages, and truncated cubo-octahedra, where the gmelinite cage is completely surrounded by H₂O molecules. No Al–Si order was detected. The size of the unit cell is the largest for any inorganic compound. *SCI 154:1004(1966)*, *JSSC 46:265(1983)*. Small, 0.1–0.8 mm usual, dodecahedra {110} dominant, very small cube {100} and trapezohedral {211} modifications are present, a few crystals are 3–4 mm across; no twinning. Colorless, light yellow, orange, red; white streak; vitreous to greasy luster. No cleavage, conchoidal fracture. $H = 5$. $G = 2.085$–2.24. A K–Ca–Ba–Na-rich zeolite. K is commonly the dominant cation, with considerable variation in the amounts of the exchangeable cations Ca, Ba, Na. A Ca-dominant variety has been found. When large cations are present, $D = 2.24$ and water is 44H₂O, while the smaller cations produce a cell with $D = 2.08$, and 34H₂O. Isotropic; $N = 1.472$–1.484. Some lamellae seen optically may indicate the presence of twinning. Occurs in tholeiitic basalt and sandstone xenoliths in basalt (melilite–nephelinite–basalt). Associated with erionite and pyrite. Occurs at the *Rock Island Dam, on the Columbia R., Wenatchee, Douglas Co., WA*; Paulingite-Ca at Three Mile School, NW of Ritter, Grant Co., OR;

Riggins, Idaho Co., ID; Chase Creek, N of Falkland, and Twig Creek, Monte Hills, BC, Canada; on the northern coast of County Antrim, particularly at Port Ganny, Giant's Causeway, and Craigahulliar, Portrush, Northern Ireland; Höwenegg quarry, Hegau, and Ortenberg quarry, Vogelsberg, Hesse, Germany; Vinarice, near Kladno, Bohemia, Czech Republic. EF *AM 45:79(1960), 67:799(1982); NZ (1985); CMG 33:109(1988); ZW (1992); IMAMA 238(1994).*

77.1.3.10 Mazzite

Named in 1974 by Galli et al. for Fiorenzo Mazzi (b.1924), professor of mineralogy, University of Pavia, Italy. Zeolite group. HEX Probably $P6_3/mmc$ (other possible space groups are $P\bar{6}2c$ and $P6_3mc$). $a = 18.38$–18.39, $c = 7.64$–7.65, $Z = 1$, $D = 2.11$. *38-426*(nat): 9.20_6 6.02_5 4.73_5 $3.82_{9.5}$ 3.65_5 3.53_9 3.19_{10} 2.94_{10}. The Al–Si framework consists of doubly connected four-membered ring chains that form gmelinite-type cages, stacked to form columns parallel to c. Adjacent columns are shifted by one-half the c-axis length and cross-linked to form two different types of channels parallel to c; the large channels are composed of 12-membered rings surrounded by six cages, while smaller channels are formed by distorted eight-membered rings between adjacent pairs of cages. Exchangeable cations are found in three positions. The Si–Al in the framework is disordered. *RSIMP 31:599(1975)*. Radiating groups of crystals, simple hexagonal prisms, with {1000} and {0001}; needles to 15 mm known. Colorless, white streak, vitreous luster. Monte Semiol, France: $K_{2.5}Mg_{2.1}Ca_{1.4}Na_{0.3}[Al_{9.9}Si_{26.5}O_{72}] \cdot 28.03H_2O$. Uniaxial $(-)$; $N_O = 1.5062$, $N_E = 1.4990$; birefringence: looks similar to other hexagonal zeolites and mordenite; negative elongation. Occurs in basalt, associated with phillipsite, chabazite, and offretite. Of hydrothermal origin. Nearly pure Na member to 1.5 cm long from Boron, Kern Co., CA; at *Mte. Semiol, Montbrison, Loire, France,* with other zeolites, calcite, and siderite. EF *AC(B) 31:1603(1975), CMP 45:99(1974).*

77.1.3.11 Merlinoite (K,Na)$_5$(Ca,Ba)$_2$[(Al$_9$Si$_{23}$)O$_{64}$] · 22H$_2$O

Described in 1977 by Passaglia et al. and named for Stefano Merlino (b.1938), professor of crystallography, University of Pisa. Zeolite group. ORTH $Immm$. $a = 14.05$–14.116, $b = 14.10$–14.25, $c = 9.946$–9.99, $Z = 1$, $D = 2.19$–2.23. *29-0989*(nat): 7.12_9 7.08_9 5.36_4 4.48_4 3.26_4 3.24_4 3.23_4 3.18_{10}. Structure consists of a framework with doubly connected eight-ring chains. Double octagonal rings that deviate from regularity. Deviation from tetragonal symmetry is due to the different arrangement of two (K, Ba) sites that would otherwise be equivalent to the tetragonal space group. Four double-crankshaft chains around a tetrad. Topological symmetry $= I_4/mmm$. Al and Si are completely disordered. *NJBM:1(1979)*. Because of isostructural similarity in structure, epitaxial intergrowths of phillipsite and merlinoite are common. Crystals to 4 mm long, in radial aggregates; pseudotetragonal, with {001}, {101}, {011}, and {100}, prismatic to fibrous, elongate on c; 90° rotation twinning about c

and interpenetrant twinning. Colorless to white, white streak, vitreous–dull luster. Cleavage {010}, good; steplike fracture. G = 2.14–2.27. Thermal data: *NZ (1985)*. Na or K may be dominant in the A site, and Ba or Ca may be dominant in the B site. BaO to 15.9 and SrO to 7.3 wt % (Khibiny). Traces of iron (0.72 wt % Fe_2O_3) are present in Cupaello merlinoite. Biaxial (−); XYZ = *bca*; N_x = 1.494–1.499, N_y = 1.500, N_z = 1.501; 2V = 56°; r > v. A secondary mineral occurring in cavities in basalt; at contact between basalt and marine sediments; from nepheline melilitite and other silica-undersaturated volcanic rocks; in cataclastic pegmatitic rocks, in pyroxene-rich volcanic ejecta; in manganese nodules and as a diagenetic alteration product of mafic to intermediate composition volcanic ash falls. Searles Lake, San Bernardino Co., CA; Tjornes Penin., northern Iceland; *Cupaello, Santa Rufina, Rieti*, and Valle Biachella, Sacrofano, Rome, *Italy*; twinned crystals from a sandstone xenolith in basalt at Ortenberg quarry, Vogelsberg, Hessen, also at Höwenegg, Hegau, Germany; Vuonemiok R., Khibiny massif, Kola Penin., Russia; in manganese nodules on the south Indian Ridge, Indian Ocean. EF *NJBM:355(1977), 481(1979); MM 51:749(1987); DA 41:295(1990); ZW (1992)*.

77.1.3.12   Montesommaite $(K,Na)_9Al_9Si_{23}O_{64} \cdot 10H_2O$

Named in 1990 by Rouse et al. for the locality. Zeolite group. ORTH (pseudo-TET) *Fdd*2. a = 10.099, b = 10.099, c = 17.307, Z = 1, D = 2.30. PD(nat): $6.59_{7.5}$ 4.33_4 3.30_{10} 3.13_{10} 2.80_3 2.51_2 2.35_2 1.78_2. *AM 75:1415(1990)*. Related to gismondine group and to merlinoite. Very nearly tetragonal $I\bar{4}2d$ or $I4_1md$. It probably has disordered intergrowths in an otherwise ordered arrangement of Si and Al in the framework. K is coordinated with two H_2O molecules and six oxygen atoms in the center of eight-membered ring channels. *AM 75:1415(1990)*. Crystals to 2 mm long, dipyramidal, {001 + 00$\bar{1}$}, {*hkl*} + {*hk$\bar{l}$*}, {504}, {50$\bar{4}$}, {054}, {05$\bar{4}$}; to 0.1 mm in size; twinning not observed. Colorless, white streak, vitreous luster. No cleavage. G = 2.34. Nonluminescent. Analysis (wt %): K_2O, 16.7; Na_2O, 0.2; Al_2O_3, 19.8; SiO_2, 55.7; H_2O (7.6) by difference = 100%. Biaxial (−); X = *c*, Y, Z = *a* or *b*; N_x = 1.498, N_y = 1.506, N_z = 1.507; 2V = 35° ($2V_{calc}$ = 39°); no dispersion. Occurs in a 3-mm vesicle in scoria. Associated minerals are dolomite, calcite, chabazite, and natrolite. Additional specimens were later found. Occurs at *Pollena in the Mte. Somma–Vesuvius volcanic complex, Naples, Italy*. EF *ZW (1992)*.

77.1.4.1   Heulandite-Ca $(Ca,Na,K)_{2-3}Al_3(Al,Si)_2Si_{13}O_{36} \cdot 12H_2O$

Described by Brooke in 1822 and named for John Henry Heuland (1778–1856), a British mineral collector and dealer. Zeolite group. Heulandite-(Na) and heulandite-(K) also approved by the CNMMN IMA. Clinoptilolite is considered by some authors [e.g., *KL:793(1980), ZW (1992)*] to be a high-silica variety of heulandite. See clinoptilolite (77.1.4.2) for a discussion of the thermal stability of the two minerals and the differentiation between the two. Three methods (cation ratios, Si/Al ratio, and thermal behavior) have been used to differentiate the two minerals, but there is a great lack of consistency between the three methods. MON $C2/m$. $a = 17.498$–17.74, $b = 17.816$–18.025, $c = 7.396$–7.529, $\beta = 116.07$–$116.60°$; K-rich material (Denmark): $a = 7.405$, $b = 17.885$, $c = 15.821$, $\beta = 91.39°$ (on a pseudo-orthorhombic cell); $Z = 2$, $D = 2.16$. 41-1357(nat): 8.96_{10} 4.65_3 $3.98_{6.5}$ 3.90_4 3.82_2 2.99_3 2.97_9 2.81_2. There have been at least five structural studies on untreated material [summarized in AM 76:$1872(1991)$]. There are four metal sites. Al and Si in the framework have some ordering. Low-Si heulandite contains two additional sites containing water molecules that are not present in silica-rich heulandite. All of the exchangeable cations are in two sites within channels parallel to c. One site contains Ca, and the other site contains all of the other cations (Na, K, Mg, Sr). Neutron diffraction studies of the structure: $JSSC$ 54:$1(1984)$. Structure of K-exchanged heulandite: $AC(B)$ 39:$189(1983)$. Heulandite has been synthesized [e.g., $JCSSJ$ 30:$31(1990)$]. Synthetic heulandite tends to change morphologically from prismatic to platy habit with an increase of Ca. The thermal stability of heulandite increases gradually with an increase of the Na–K or Si contents. **Habit:** Crystals to 3 cm long relatively common, as much as 14 cm long (elongated on a). Commonly tabular parallel to {010}, and elongated on a, as well as widest at the center—thus the common "coffin-shaped" crystals. Also platy, trapezoidal crystals with {010} dominant. Some crystals may be elongated on b or c. May be granular to massive. Common forms (pseudo-orthorhombic cell): {100}, {010}, {011}, {101}, {$\bar{1}$01}, {110}. Twinning sometimes present with {100} as the twin and contact plane. **Physical properties:** Colorless, white, gray, yellow, pink, red, brown, green, and black; white streak; vitreous to pearly luster (on {010}). Uneven fracture, brittle. $H = 3\frac{1}{2}$–4. $G = 2.10$–2.29. Soluble in HCl. May show photoluminescence. CRT 18:$213(1983)$. **Chemistry:** The mineral has a wide range in exchangeable cations, and Al–Si in the framework. Most large heulandite crystals are Ca-dominant, with a relatively low Si content. There is a continuous series between heulandite and Ca-clinoptilolite with $R^{2+} + 2Al \leftrightarrow 2Si$ substitution, where monovalent cations (Na, K) are contained in the range one to two in a unit formula based on 72 oxygens. Clinoptilolite is usually dominant in Na or K and is Si-rich. However, a K-dominant heulandite from Denmark contained 14.5 wt % K_2O. MM 54:$91(1990)$. An Na-rich heulandite from Boron, CA, contains 8.2 wt % Na_2O. Sr and Mg are present in small amounts and rarely are the dominant cations (SrO to 8.7 wt % from Campegli, Italy). $NJBM$: $541(1982)$. Ba may be present as well (to 6.9 wt % BaO). CM 12:$188(1973)$. T_{Si} varies widely from 0.71 to 0.83. H_2O content varies over a wide range. Low Si: $(Ca,Na_2,K_2,Mg)_9[Al_9Si_{27}O_{72}] \cdot 24H_2O$; high Si: $(Na_2,K_2,Ca,Mg)_6[Al_6\ Si_{30}O_{72}] \cdot 20H_2O$ (= clinoptilolite). In general, the exchangeable cations have a profound effect on the dehydration behavior. It is more appropriate to speak of a continuum of H_2O-molecule binding energies than to refer to loosely bound or tightly bound zeolitic water. Material with univalent cations dominant dehydrate more easily than those with dominant divalent cations. **Optics:** Biaxial (+); $N_x = 1.476$–1.506, $N_y = 1.479$–1.510, $N_z = 1.479$–1.517; $2V = 10$–$48°$; OAP perpendicular to {010}. **Occurrence:** A low-temperature (< 60–200°) zeolite found in a wide variety of geological environments: volcanic rocks, metamorphic rocks, pegmatites, tuffs, and deep-sea sediments. Occurs in vesicles and cavities in basaltic and andesitic rocks; in weathered andesites and diabases; as a devitrification product of silicic volcanic glasses and tuffs. Associated with other zeolites, calcite, apophyllite, datolite, and feldspars. **Localities:** Occurs in many worldwide localities; only the most important in terms of quality, amount, and unusual character are given here. Best display specimens are from India, Russia, Iceland, Brazil, New South Wales, New Jersey, Washington, and Oregon. Many different colors and crystals to >5 cm long from Paterson, Passaic Co., NJ, particularly the Upper and Lower New Street quarries and the Prospect Park quarry; also from Bergen Hill, Hudson Co., NJ; 6.5-cm crystals from Lookout Point Dam, Dexter, Lane Co. and 3.5-cm crystals from Dorena Dam, Cottage Grove, Lane Co., OR; also from Rickreal Creek quarry, Dallas, Polk Co., Mitchell, Wheeler Co., and Succor Creek, Malheur Co., OR; pink crystals to 2.5 cm long from Skookumchuck Dam, near Tenino and Bucoda, Thurston Co., WA; also on South Fork of the Toutle R., Mt. St. Helens, Cowlitz Co., and Kosmos, Lewis Co., WA; exceptional crystals to 5 cm long from Minas Basin and Bay of Fundy, Nova Scotia; excellent crystals to 10 cm long from Teigarhorn, Berufjord, and at Fassarfell and Randafell, Iceland; deep red crystals from around *Glasgow, Dumbartonshire, Scotland*; K-dominant from Sangstrup Klint, Denmark; crystals to 7 cm from the Tunguska R. area, Krasnoyarsk Kray, Yakutia, Russia; in India, exceptional crystals to 12 cm long (blocky) from the Poona (Pashan Hills and Lonavala) and Nasik districts (Pathardy, Pandulena, and Sayad Pimpri quarries); pink to maroon crystals to 12 cm with quartz and apophyllite from Jalgaon and 4-cm crystals on mordenite from Aurangabad, MAH, India; reddish-orange crystals to 2 cm at Tambar Springs, near Gunnedah, NSW, Australia; crystals to 4 cm from the Rio das Antas railway tunnel; Bento Goncalves, Rio Grande do Sul, and brown crystals from Pato Branco, Parana, Brazil. EF *AM 57:1448(1972), 57:1463(1972); TMPM 31:259(1983); NZ (1985); CCM 36:131(1988); ZW (1992).*

77.1.4.2 Clinoptilolite $(K,Na,Ca)_{2-3}Al_3(Al,Si)_2Si_{13}O_{36} \cdot 12H_2O$

Named in 1932 by Schaller for its inclined optics, from the Greek for *wing* or *down*, in allusion to the light, downy nature of aggregates of the mineral, and for ptilolite (mordenite). Zeolite group. Some authors consider this mineral to be a high-silica variety of heulandite. *KL:793(1980), ZW (1992).* However, clinoptilolite is distinct from heulandite, particularly in thermal behavior. *AM 45:351(1960).* Upon heating to about 450°, the structure of heulandite is destroyed, whereas that of clinoptilolite persists to about 700°. At a temperature below 450°, heulandite is transformed to a phase called heulandite-B; the temperature needed to start the transformation was reported to be 230° to about 400°. The structure of heulandite, which is thus sensitive to temperature compared to clinoptilolite, can be stabilized by replacement of Ca by K.

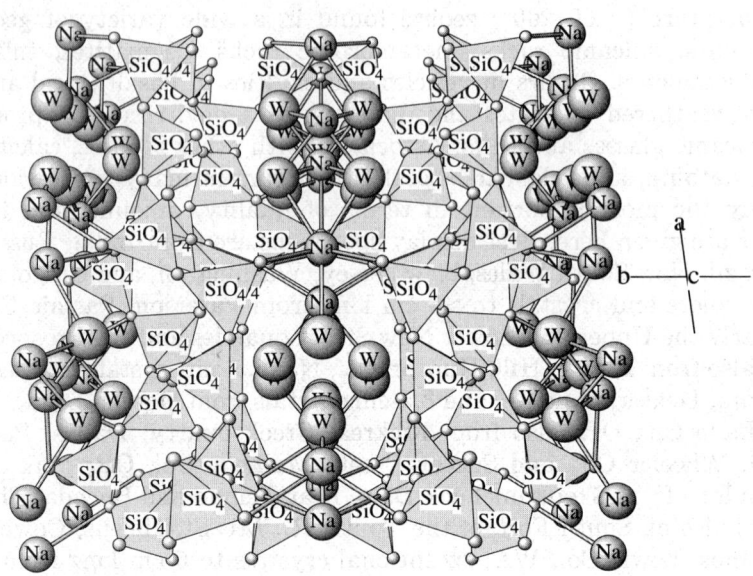

CLINOPTILOLITE: (Ca, Na$_2$)Al$_2$Si$_7$O$_{18}$ · 6H$_2$O (variable)
There are two types of large channels running down the c direction shown in this view. There are also channels in the a direction.

On the contrary, Ca-exchanged clinoptilolite behaves thermally similarly to heulandite. Clinoptilolite and heulandite have also been distinguished from one another on the basis of the ratio of divalent to monovalent cations: clinoptilolite has (Na + K) > 2Ca, and heulandite has (Na + K) < 2Ca. *AM 45:341(1960)*. An additional definition [*AM 57:1463(1972)*] has clinoptilolite with Si/Al > 4.0 and heulandite with Si/Al < 4.0. The Zeolite Subcommittee of the IMA will probably recommend that this name be retained because it is so entrenched in the literature and in use by industry. MON $C2/m$. $a = 17.622$–17.688, $b = 17.895$–17.941, $c = 7.399$–7.410, $\beta = 116.37$–$116.50°$, $Z = 2$, $D = 2.17$–2.29. 39-1383(nat): 8.95_{10} 4.65_2 3.98_6 3.96_6 3.91_5 3.42_2 3.00_2 2.97_5. PD for K-clinoptilolite [*EJM 2:819(1990)*] is somewhat different: 8.93_{10} 4.67_2 3.98_4 3.955_4 3.90_3 3.125_4 3.07_2 $2.80_{2.5}$. Structure solutions have been done on material from four localities: Agoura, CA; Alpe di Siusi, Italy; Kuruma Pass, Japan; and Richardson Ranch, OR. *AM 75:522(1990), 78:260(1993)*. The crystal structure of clinoptilolite at 350° is triclinic, probably $C\bar{1}$ rather than monoclinic. Cell dimensions are: $a = 17.698$, $b = 17.511$, $c = 7.392$, $\alpha = 90.21°$, $\beta = 116.76°$, $\gamma = 90.19°$. The potassium atom located at the M3 site at room temperature is still preserved at the same site, while those located at the cation sites other than M3 are missing. The occupancy of cations at the M3 site is even higher than that observed at room temperature. This occupancy and nearly even coordination by six framework oxygen atoms is responsible for the thermal stability of clinoptilolite. *AC(B) 39:189(1983), MJJ*

11:392(1983). **Habit:** Platy crystals to 1 cm; also tabular, elongated on a, flattened on b; commonly fine-grained and massive. Common forms include: {010}, {001}, {101}, and {20$\bar{1}$}. **Physical properties:** Colorless to white, vitreous to pearly luster, transparent to translucent. Cleavage {010}, perfect. $H = 3\frac{1}{2}$–4. $G = 2.10$–2.17. Readily decomposed by HCl. **Chemistry:** A wide range in chemistry has been reported for clinoptilolite. Sodium, potassium, or calcium may be the dominant extraframework cation. Sodium is the dominant cation in most sedimentary clinoptilolites. However, deep-sea clinoptilolites are predominantly potassic. K-dominant clinoptilolite also occurs in vitric tuffs and in alkali saline lake deposits. *AM 50:244(1965)*. The most Na-rich clinoptilolite reported is that from Boron, CA, with 7.9 wt % Na_2O. Clinoptilolite with as much as 9.6 wt % K_2O occurs in deep-sea sediments adjacent to the Japanese islands. Minor amounts of SrO (to 1.4 wt %), and trace to minor levels of Fe and Mg may be present. Ca-dominant clinoptilolite, such as that from Australia, contains as much as 5.0 wt % CaO. *NZ 93':15(1995)*. Clinoptilolite and heulandite may be found together in chemically zoned crystals, such as clinoptilolite cores and heulandite rims on single crystals from a deep bore hole off Japan. **Optics:** Biaxial $(-)$ or $(+)$; $Y = b$, $Z \wedge a = 30$–$45°$ (optically positive material); $X = b$, $Z \wedge a = 15°$ (optically negative material); $N_x = 1.476$–1.491, $N_y = 1.479$–1.493, $N_z = 1.479$–1.497; $2V = 31$–$48°$; birefringence $= 0.003$–0.006; $r > v$ (optically positive material), $r < v$, strong (for optically negative material). **Occurrence:** Occurs in a variety of geologic environments. Most commonly formed as a devitrification product of silicic volcanic glass from tuffs (air fall and ash flow). Also occurs in cavities in rhyolites and more-mafic rocks such as andesites and basalts. Often present in saline, alkaline lake sedimentary deposits derived from volcanic material. K-rich material is present in deep-sea sediments. *AM 57:1448(1972)*. As much as 5.0 wt % CaO occurs in clinoptilolite. 93:15(1995). Commonly associated with other zeolites, montmorillonite, hectorite, quartz, opal, calcite, and potassium feldspar. Other minerals that may be associated are halite, celadonite, gaylussite, and thenardite. **Localities:** Occurs in many worldwide localities, and only localities with fine crystals, large amounts, or special characteristics are given here. It is especially abundant, with some deposits being exploited commercially, in the western United States. Examples of large sedimentary deposits currently being mined include: near Fort La Clede, Sweetwater Co., WY; Cuchillo deposit, near Winston, Sierra Co., NM; near Wikieup, Mohave Co., AZ; near Oreana, Owyhee Co., ID; near Succor Creek, Malheur Co., OR; Owens Lake and China Lake, Inyo Co., CA; near Barstow and Hector, San Bernardino Co., CA; and near Tilden, McMullen Co., TX. Two large sedimentary provinces, Drummond and New England, occur in eastern Australia (NSW and central Q). In moderate to large crystals from Agate Beach, Lincoln Co., Succor Creek, Malheur Co., and at Richardson Ranch, near Bend, Jefferson Co., OR; Agoura, Los Angeles Co., CA; at Kamloops Lake, near Kamloops, BC, Canada; Kapfenberg, Styria, Austria; Vogelsberg, Hesse, Germany; Val di Fassa and Alpe di Siusi, Trentino–Alto Adige, Italy. **Uses:** An economically important zeolite because of its cation-exchange properties. Used for removal of radioactive Cs and Sr from low-level waste streams of nuclear installations and ammonium extraction from sewage and agricultural effluents. Also used as a soil conditioner, odor absorber, stock-feed additions, and heavy metal extraction and filtration. EF *AM 54:887(1969)*, *ZK 145:216(1977)*, *NZ (1985)*, *SSAJ 50:1618(1986)*, *CCM 36:131(1988)*, *ZW (1992)*.

77.1.4.3 Stilbite-(Ca) NaCa$_4$[Al$_8$Si$_{28}$O$_{72}$] · nH$_2$O to NaCa$_4$ [Al$_{10}$Si$_{26}$O$_{72}$] · nH$_2$O (n = 28-32)

Described in 1801 by Haüy and named from the Greek for *to shine*, in allusion to the pearly to vitreous luster. Zeolite group. Stilbite-(Na) also approved by the CNMMN IMA. See barrerite (77.1.4.5), an Na-rich, orthorhombic variety, and stellerite (77.1.4.4), a Ca-rich, orthorhombic variety. TRIC and MON $C1$ and $C2/m$. Orthorhombic cell: $a = 13.595$–13.69, $b = 18.197$–18.31, $c = 17.775$–17.86, $\beta = 90.00$–$90.91°$; monoclinic cell: $a = 13.595$–13.69, $b = 18.197$–18.31, $c = 11.265$–11.30, $\beta = 127.94$–$128.1°$; Z = 2, D = 2.20–2.23. *24-894*(nat), *26-584*(calc. pattern), *44-1479*(nat): 9.14$_{10}$ 4.67$_2$ 4.63$_2$ 4.06$_9$ 3.40$_2$ 3.20$_2$ 3.03$_5$ 2.78$_2$. Structure is related to heulandite. Stilbite, stellerite, and barrerite have the same tetrahedral framework, differing only in symmetry, due to minor differences from cation positioning. With an increase in the Na content, β increases from 90.0° to 90.91° (triclinic). There is a continuous series between stellerite and stilbite. Therefore, the change from orthorhombic to triclinic is gradual. Repulsion between Na and Ca pushes the Ca atoms from the mirror planes and forces the framework to rotate, thus reducing the symmetry to C1 (triclinic). Also, Al–Si ordering occurs, producing triclinic symmetry, not detected by X-ray diffraction but is detected by optics. There is also a continuous series between stilbite and barrerite. Recent work [*EJM 5:839(1993)*] has shown that the {001} growth sector is orthorhombic with respect to X-rays, whereas the {101} sector is monoclinic. Thus natural stilbite crystals consist of monoclinic and orthorhombic sectors rather than single monoclinic or orthorhombic crystals. **Habit:** Generally tabular, flattened on {010} and elongated on *a*. Crystals may be flat-topped or pointed. Aggregates may be sheaflike or in "bowties." Morphology is described in the monoclinic system: {010} broad, smaller, striated {001}; sometimes {$\bar{1}$01} termination. Twinning, which is ubiquitous on {001}, produces four equivalent {110} faces, due to twinning on (010), (001), (101), and (10$\bar{1}$). If indexed in the orthorhombic system: {010}, {001}, {111}, and flat {100}. The sector twinning is explained by growth ordering. **Physical properties:** Many colors, usually colorless or white, but may be yellow, brown, pink, salmon, red, green, blue, or black; white streak; vitreous luster, pearly on {010}. Cleavage {010}, perfect; uneven and brittle fracture. H = $3\frac{1}{2}$–4. G = 2.12–2.22. **Chemistry:** Shows a wide variation in exchangeable cations: Ca, Na, K. Sr, Ba, and Mg are essentially absent. Most stilbites are very rich in Ca, but some are Na- or K-dominant. The natural pure Na equivalent to stellerite has not been found. A continuous Ca–Na series between stellerite and barrerite is not possible because that compositional area is occupied by triclinic stilbite. T_{Si} ranges from 0.73 to 0.78. Biaxial $(-)$; X $\wedge$ $a = 5°$, Y $\wedge$ $b = 2$–$8°$, Z $\wedge$ $c = 5°$, elongate on a; $N_x = 1.479$–1.492, $N_y = 1.485$–1.500, $N_z = 1.489$–1.505, $2V = 22$–$79°$; inclined (wavy) extinction angle of 2–$9°$; $r < v$, weak. **Occurrence**: Occurs in many geologic environments: in basic igneous rocks, other volcanic rocks, granitic pegmatites, gneisses and schists, sedimentary tuffs, hot-spring deposits, and deep-sea sediments. **Localities**: A common zeolite; only noteworthy localities are mentioned here. Very fine crystals to 10 cm long from Paterson, Passaic Co., and Summit, Union Co., NJ; to 8 cm long at the Kibblehouse quarry, Perkiomenville, Montgomery Co., PA; at many localities in Oregon, but the best are Roadside quarry, Goble, Columbia Co., Rickreall Creek quarry, Dallas, Polk Co., Calapooya R., near Dollar, Linn Co., Roadside quarry, Goble, Columbia Co., Lookout Point Dam, Dexter, Lane Co., Mitchell, Wheeler Co., and Ritter, Grant Co.; fine bowties to 10 cm long in Washington at Skookumchuck Dam, near Tenino and Bucoda, Thurston Co.; Poison Creek, Skamania Co.; Mossyrock Dam, Lewis Co., South Fork of the Toutle R., near Mt. St. Helens, Cowlitz Co., and Kalama, Cowlitz Co.; on chabazite from Minas basin and Bay of Fundy, NS, Canada; bowties to 6 cm from the Teigarhorn in Berufjord, Iceland; Faroe Is.; four places are mentioned by Hauy as type localities: (1) Helgusta (Helgustadir) calcite mine, Reydarfjord, Iceland; (2) Andreasburg, Harz, Germany; (3) Dauphine Alps, France; and (4) Mt. Abildgaard, Arendal, Norway; numerous places in the United Kingdom, such as the Glasgow district, Dumbartonshire, Scotland, and the Isle of Skye; deep yellow-brown crystals from Striegau, Poland; golden-yellow crystals to 10 cm long associated with other zeolites occur in basalt along the Amudikha R., from the lower Tunguska R. area, near Mirny (several localities), Yakutia, Russia; gray, flat-topped crystals near Lake Baikal, Siberia, Russia; from numerous localities in Kazakhstan; localities in India within the Deccan Trap basalts are the most prolific in the world, most abundant at Nasik and Poona, decreasing toward the coast at Bombay, MAH; in the Nasik district, pink to salmon-orange bowties to 14 cm long occur at the Pandulena, Pathardy, and Syed Pimpri quarries; in the Pune (Poona) district, 12-cm-long white bowties and fanlike groups, up to 12 cm long, are found at Pashan Hills, Hadspar, and Lanavala; quarries near Jalgaon produce light pink crystals with quartz and colorless apophyllite, orange flat-topped crystals are found on mordenite at Aurangabad; to 5 cm long with 20-cm apophyllite crystals from Rio de Antas, Bento Gonsalves, Rio Grande do Sul, Brazil; also bowties from Volta Grande, in Santa Catarina; large orange crystals near Gunnedah, and Tanbar Springs, NSW, and Arkaroola, SA, Australia; Guanajuato, Mexico. EF *BM 101:368(1978)*, *AM 70:814(1985)*, *NZ (1985)*, *ZW (1992)*.

77.1.4.4 Stellerite $CaAl_2Si_7O_{18} \cdot 7H_2O$

Described by Morozewicz in 1909 and named for Georg Wilhelm Steller (1709–1746), German explorer and zoologist. Zeolite group. See barrerite

(77.1.4.5), the Na-dominant analog of stellerite. ORTH $Fmmm$. $a = 13.599$, $b = 18.222$, $c = 17.683$, $Z = 8$, $D = 2.12$. 25-124(nat): 9.03_{10} $4.66_{1.5}$ $4.06_{4.5}$ 3.40_1 3.18_1 $3.03_{2.5}$ 3.00_1 2.77_1. The X-ray pattern for barrerite is very similar to that of stellerite, and it may be difficult to tell them apart. The structure is like that of stilbite, with no Si–Al order, but a higher symmetry, $F2/m\ 2/m\ 2/m$, which coincides with the topological symmetry for the framework. In stilbite, electrostatic repulsive forces between extra-framework cations are strong enough to push the Ca atoms out of the mirror planes, so allowing only symmetry $C2/m$. In stellerite there is no Na present, and its site is partially occupied by water, so Ca can remain on the mirror planes. Lamellar to tabular crystals with dominant forms {010}, {100}, {001}, {111}. Commonly in spheres to 14 cm, of radiating crystals; also crystal aggregates. Ideally no twinning, but some are optically sector-twinned, indicating triclinic symmetry. Usually white, but may be pink, salmon-orange, brown, or red; white streak. Cleavage {010}, perfect. $H = 4\frac{1}{2}$. $G = 2.13$. Typical analysis (wt %): SiO_2, 59.15; Al_2O_3, 14.21; CaO, 7.45; Na_2O, 0.19; K_2O, 0.23; Fe_2O_3, 0.17; H_2O, 17.79; traces of other elements (Sr, Ba, Mg) present. Biaxial $(-)$; $XYZ = abc$; $N_x = 1.485$, $N_y = 1.496$, $N_z = 1.498$; $2V = 47°$. Occurs principally in veins and geodes in basalts and other basic volcanic rocks; on fracture surfaces as well. Also occurs in mica schist, skarns, gneisses, and graniodiorites. Occurs in geothermal fields. Associated with other zeolites, tridymite, prehnite, apophyllite, and others. An increasingly recognized zeolite that occurs in numerous worldwide localities; only selected localities are given here: various quarries in the Paterson area, Passaic Co., NJ; Fanwood quarry, Fanwood, Somerset Co., NJ; Ritter Hot Springs, Grant Co., OR; Villanova, Monte Leone, Sassari, Sardinia, Italy; Gelbe Birke, Furstenberge bei Schwartzenberg, Erzgebirge, Germany; Kongsberg, Norway; Szob, Malomvolgy, and Nadap, Hungary; orange stellerite from the Sarbayskaya quarry, near Rudny, Kusteni Oblast, northern Kazakhstan; on Copper Island in the Commander Islands (Kommandorskiye Ostrova), Bering Sea, Russia; Franca, São Paulo, Brazil; Chinchwad, NW of Pune (Poona), and in the Nasik district, MAH, India; Girrawaillie Creek, and Mitchells Creek, Garrawilla, NSW, Australia; also from near Gunnedah, NSW, and at Harcourt, Dookie, and Corop, VIC, Australia. EF *LIT 6:83(1973); BM 98:11(1975), 101:368(1978); NZ (1985); ZW (1992).*

77.1.4.5 Barrerite $Na_8(Al_8Si_{28}O_{72}) \cdot 26H_2O$

Named in 1975 by Passaglia and Pongiluppi for Richard Maling Barrer (b.1910), British chemist, Imperial College, London, who has contributed greatly to the investigation of zeolites and to the chemistry of molecular sieves. Zeolite group. *ZW (1992)* considers this to be an orthorhombic variety of stilbite. See stilbite (77.1.4.3) and stellerite (77.1.4.4). Stellerite is Ca-dominant. ORTH $Amma$. $a = 13.643$, $b = 18.200$, $c = 17.842$, $Z = 2$, $D = 2.11$. 29-1185(nat): 9.1_{10} 4.66_2 4.05_{10} 3.19_2 3.03_8 $3.00_{2.5}$ 2.77_2 1.82_2. Structure same as for stellerite (77.1.4.4). (010) lamellae to 5 cm diameter, (010) platy crystals with {001} and {111} are rare. White to slightly pink, vitreous luster, transparent to translucent. Cleavage {010}, perfect. G = 2.13. Analysis (wt %): SiO_2, 58.8; Al_2O_3, 14.75; Fe_2O_3, 0.04; MgO, 0.24; CaO, 1.66; Na_2O, 5.97; K_2O, 1.76; H_2O, 16.40. Biaxial (−); XYZ = abc; $N_x = 1.479$, $N_y = 1.485$, $N_z = 1.489$; $2V_{calc} = 78°$. Occurs in a monomineralic vein in lava (andesitic to rhyolitic). Seams in altered calc–silicate rock. Associated with heulandite. Occurs at Tosenvegen, Vefsn, Nordland, Norway; *Capo Pula, Nora, Sardinia, Italy.* EF *LIT 7:69(1974), MM 40:208(1975), BM 98:331(1975), NZ (1985)*.

77.1.5.1 Natrolite $Na_2Al_2Si_3O_{10} \cdot 2H_2O$

Described by Klaproth in 1803 and named from the Greek *natrium*, soda, in allusion to the Na content of the mineral. Zeolite group. Dimorphous with tetranatrolite. ORTH *Fdd2*. $a = 18.27–18.354$, $b = 18.587–18.67$, $c = 6.56–6.608$, Z = 8, D = 2.25. A study of 38 natrolites [*CE 49:297(1989)*] showed a range of: $a = 18.26–18.48$, $b = 18.52–18.70$, $c = 6.35–6.66$. For monoclinic natrolite, a to 19.00, b to 19.04, c to 6.60, β to 90.52°. *39-226*(calc. pattern), *45-1413*(nat): 6.54_5 5.89_{10} 4.40_4 4.36_4 3.20_3 $2.94_{3.5}$ 2.87_5 2.85_5. Earlier cards (*19-1185* and *20-759*) show preferred orientation effects. The aluminosilicate framework is the same as that of mesolite and scolecite. Long chains of tetrahedra are rotated by 24°. Cations are in four sites in the channels, parallel to c. The channels are occupied by two Na and two H_2O molecules. Si and Al in an ideal natrolite framework are highly ordered into three independent tetrahe-

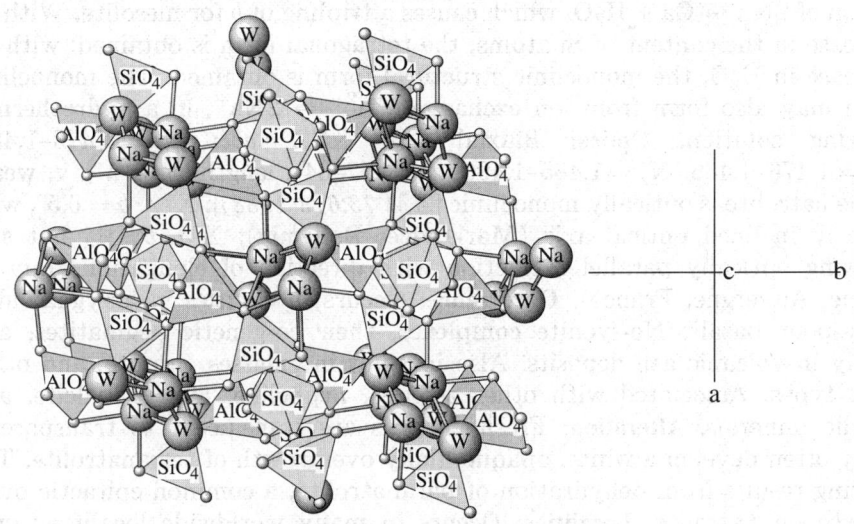

NATROLITE: $Na_2Al_2Si_3O_{10} \cdot 2H_2O$

Al tetrahedra green; Na atoms magenta; H_2O cyan. This is a view down the large zeolitic channels, which contain Na and H_2O.

dral sites. There appears to be a complete series between fully ordered natrolite and completely disordered tetranatrolite [i.e., Na-rich gonnardite according to *ZW (1992)*]. Tetranatrolite has from 17.5 to 0 Å (>50% disorder); $b - a = 0.35$ Å (fully ordered natrolite), $b - a = 17.5$ Å (disordered natrolite, with 50% disorder). Truly completely ordered natrolite has not been found thus far and should be referred to as strongly ordered rather than fully ordered. As much as 30% disorder (30% of the Al site is occupied by Si and about 20% of the Si sites are occupied by Al) in Si and Al has been found for some natrolite. *AC(B) 37:781(1981)*, *EJM 2:799(1990)*. **Habit:** Most commonly occurs as long, slender needles, striated parallel to elongation, prismatic (pseudotetragonal) on *c*; crystals to more than 1 m long; in stellate or jackstraw groups, radial aggregates, and compact masses. Frequently epitaxially intergrown with mesolite, tetranatrolite, gonnardite, paranatrolite, thomsonite, and rarely scolecite. Sometimes flattened, with {111} dominant. Common forms are {110} and {111}, with complex terminations. Also {100}, {010}, {101}, {331}, and {551}. Twinning is common, on {110}, {011}, and {031}. **Physical properties:** Pale pink, colorless, white, gray, red, yellow, green, or brown; white streak; vitreous to silky luster but may be greasy or dull. Cleavage {110}, perfect; {010} parting; uneven fracture; brittle. H = 5–5½. G = 2.20–2.27. IR data: *CMG 25:239(1980)*. Natrolite dehydrates to "metanatrolite," monoclinic, $a = 16.0$, $b = 16.73$, $c = 6.40$, $\beta = 90°$. *NJBM:135(1983)*. Gelatinizes with common acids. Pyroelectric and piezoelectric. Sometimes fluorescent orange to yellow in LWUV and SWUV. **Chemistry:** A very Na-rich zeolite. Contains very small amounts of Ca and K, traces of Mg, Sr, Ba. Chemically, natrolite differs from mesolite by an ordered substitution of 2Na ↔ Ca + H_2O, which causes a tripling of *b* for mesolite. With an increase in the content of Si atoms, the tetragonal form is obtained; with an increase in H_2O, the monoclinic structural form is obtained. The monoclinic form may also form from ion exchange, Ca^{2+} ↔ $2Na^+$, in a hydrothermal altering solution. **Optics:** Biaxial (+); XYZ = *abc*; N_x = 1.473–1.483, N_y = 1.476–1.486, N_z = 1.485–1.496; 2V = 57–64°; length-slow, r < v, weak. Some natrolite is optically monoclinic [*AM 73:613(1988)*]: X ∧ b = 0.5°, with slightly inclined optical axis (Markovice, Bohemia); X ∧ b = 1° but still showing optically parallel extinction with direction of elongation (Puy de Dôme, Auvergne, France). **Occurrence:** Occurs in cavities in amygdaloidal, silica-poor basalt, Ne-syenite complexes, their co-genetic pegmatites, and rarely in volcanic ash deposits. Also in veins in gneisses, granite, and other rock types. Associated with other zeolites, nepheline, quartz, calcite, and alkalic minerals. **Alteration:** Fresh crystals are translucent to transparent. They often develop a white, opaque, flaky overgrowth of tetranatrolite. The coating results from dehydration of paranatrolite, a common epitactic overgrowth on natrolite. **Localities:** Occurs in many worldwide localities; only noteworthy localities are given here: Bergen Hill, Hudson Co., and Prospect Park, Paterson, Passaic Co., NJ; truly orthorhombic (by optics) crystals to 18 cm long and 3 cm wide from the Chimney Rock quarry, Bound Brook, Passaic Co., NJ; to 4.5 cm long from the Dallas gem mine (Benitoite mine), and other nearby areas, San Benito Co., CA; new and old Springfield quarries, Springfield Butte, Lane Co., OR; to 4 cm long from Robertson quarry, near Dayton, Mason Co., WA; natrolite capped by mesolite from the Imnaha basalt, on the Snake R., 10 km W of Clarkston, WA; to 2.5 cm from Pinnacle Island, Bay of Fundy, NS; to 15 cm long and 2 cm wide from Mt. St.-Hilaire, and large crystals from the Johnston asbestos mine, near Thetford, QUE, Canada; to 25 cm long from the Ice River complex, near Golden, BC, Canada; coarse 2-cm needles from White Head, and to 4 cm at Larne, County Antrim, Northern Ireland; Dean quarry, St. Keverne, Lizard Penin., Cornwall, England; from the Langesundfjord and Tvedalen areas, Norway; at Puy de Marman, near Veyre, Puy de Dôme, Auvergne, France; *Hohentwiel, Hegau, Baden–Württemburg, Germany*; also from the Höwenegg quarry; gemmy, colorless crystals to 30 cm long and 13 cm in diameter in pegmatites at the Jubilee mine, Lovozero massif, Kola Penin., Russia; also from the Khibiny massif; exceptional groups of milky crystals 10 cm long forming groups 15 cm across with a thin natrolite overgrowth on mesolite and scolecite at Khandivali, Maharashtra state, India; at Jacupiranga, Sao Paulo, Brazil; crystals to 7 cm long and 5 mm diam. from Culverden, Canterbury, South Island, New Zealand; thin needles at Ardglen, NSW; Cairns Bay, Flinders, Victoria; groups of needles 2 cm across at Cape Grim, Tasmania, and on gonnardite at Bundoora, Victoria, Australia. *ZK 164:19(1983), 185:619(1988); NZ (1985); AM 77:685(1992).*

77.1.5.2 Tetranatrolite $Na_2Al_2Si_3O_{10} \cdot 2H_2O$ and $(Na_{1.5}Ca_{0.5})_2Al_2Si_3O_{10} \cdot 2H_2O$

Named in 1980 by Chen and Chao for the symmetry (tetragonal) and natrolite (same formula). Zeolite group. A tetragonal dimorph of natrolite. Ranite is a calcian variety; natrolite does not contain Ca. Tetranatrolite is interpreted as the result of partial dehydration of parantrolite. *EJM 7:501(1995).* TET $I\bar{4}2d$. $a = 13.02$–13.25, $c = 6.615$–6.704, $Z = 4$, $D = 2.22$–2.23. 33-1205(nat): 6.55_5 5.91_4 4.63_4 4.39_5 4.14_4 3.19_5 2.87_{10} 2.44_4; 42-1381(calcian): 6.62_6 5.90_9 4.68_6 4.41_8 3.21_8 3.11_6 2.96_7 2.90_{10}. Structure: *SPC 31:254(1986), ZK 189:191(1989).* Prismatic crystals and fine-grained coatings up to 1 mm, silky, fibrous, thick coatings on natrolite. White, rarely pale pink; white streak; silky to dull luster. No cleavage. H $\approx$ 5. G = 2.21–2.28. IR and TGA data: *CM 18:77(1980).* Dissolves readily in 1:1 HCl, slowly in 1:1 HNO_3 with gelatinous residues. Shows large deviations from ideal stoichiometry. Analysis (wt %), calcian: SiO_2, 39.36; Al_2O_3, 30.73; Fe_2O_3, 0.33; CaO, 6.34; SrO, 0.04; BaO, 0.02; Na_2O, 11.94; K_2O, 0.04; H_2O, 11.2; regular, Mt. St.-Hilaire: SiO_2, 46.9; Al_2O_3, 25.6; CaO, 1.48; Na_2O, 14.0; K_2O, 1.12; H_2O^+, 9.59; H_2O^-, 1.31; minor TiO_2 and FeO. The Si/Al ratio is ≥ 1.5 in tetranatrolite. Uniaxial (+); 589 nm; $N_O = 1.480$–1.481, $N_E = 1.493$–1.496. Occurs in all mineralogical environments at Mt. St.-Hilaire. Associated with natrolite. Forms coatings and crusts on natrolite. Also epitactic overgrowths. Also occurs in phonolite and basalt and in hydrothermal veins. Alteration: the end product of dehydration of paranatrolite, which forms epitactic overgrowths on natrolite. Point of Rocks Mesa, Springer, Colfax Co., NM; *Mt. St.-Hilaire*, and St.-Amable sill, Varennes, QUE, *Canada*; Tugtup Agtakorfia, Ilimaussaq, southern Greenland; Hills Port, Island Magee, County Antrim, Northern Ireland; Tvedalen, and Lamo, Langesundsfjord, Norway; Klöch, Austria; Schellkopf, Brenk, Eifel district, Germany; Marianberg, Bohemia, and Usti nad Labem, Czech Republic; Mt. Kukisvumchorr, Khibiny massif, and Lovozero intrusion, Kola Penin., Russia. EF *MGR 181:1(1969)*, NZ *(1985)*, *MM 52:207(1988)*, *CE 49:297(1989)*, *AM 77:685(1992)*.

77.1.5.3 Paranatrolite $Na_2Al_2Si_3O_{10} \cdot 3H_2O$

Named in 1980 by Chao for the mineral's epitaxial relationship with natrolite and similarity in chemical composition to natrolite. Zeolite group. Considered a doubtful species by the CNMMN IMA. Polymorphous with natrolite. MON (pseudo-ORTH) *Fmm2*, *Fm2m*, *F2mm*, *F222*, or *Fmnm* for the pseudo-orthorhombic cell. $a = 19.07$, $b = 19.13$, $c = 6.580$, $Z = 8$, $D = 2.20$; *42-1386:* $a = 18.93$, $b = 19.21$, $c = 6.589$. *35-458*(nat): 6.76_2 5.92_6 4.78_3 4.44_4 $3.26_{1.5}$ $3.12_{1.5}$ 2.94_{10} $2.65_{1.5}$; *42-1386*: 6.75_{10} 4.79_3 4.74_3 2.95_7 2.92_4 2.90_5 2.64_3 2.51_2 (O assigned because of poor agreement of cell to data). Crystals must be kept in H_2O from the time of removal. All tetranatrolite crystals are probably pseudomorphs after paranatrolite. Aluminosilicate framework is the same as natrolite, mesolite, scolecite, gonnardite, and tetranatrolite. It is an overhydrated monoclinic polymorph of natrolite that is unstable in air. Dehydrates irreversibly to tetranatrolite. *EJM 7:501(1995)*. Si and Al are disordered. Extra H_2O molecules warp the natrolite structure to monoclinic symmetry. Radial fibrous, sheaflike, parallel columnar aggregates; habit is the same as natrolite and tetranatrolite, {110} dominant, commonly terminated by {111} dipyramid or rarely {001}; most occurs as epitaxial overgrowth to 1 mm thick on natrolite; composite crystals to 5 cm, pure crystals to 2 mm long have been found, identical in appearance to natrolite. White, colorless, grayish, pale pink, yellow; white streak; greasy–vitreous luster. No cleavage, conchoidal fracture. H undetermined, ≈ 5. $G = 2.21$–2.29. Natrolite inverts to paranatrolite at >360°. *Acta Mont. 63:87(1983)*. Mt. St.-Hilaire: $(Na_{1.75}Ca_{0.10}K_{0.09})[Al_{1.95}Fe_{0.01}Si_{3.02}O_{10}] \cdot 2.98H_2O$; natrolite has $2H_2O$ rather than 3. Analysis (wt %), Khibiny massif: SiO_2, 40.07; Al_2O_3, 28.14; CaO, 0.30; Na_2O, 15.19; K_2O, 2.54; H_2O, 13.6. Biaxial (−) wet and (+) dry; 2V small (under 10°); $N_x = 1.479$, $N_y = 1.481$, $N_z = 1.493$; Khibiny massif: $N_x = 1.493$, $N_y = 1.499$, $N_z = 1.505$, 2V (±) 90°, undulating inclined extinction, to 21°. Occurs at Mt. St.-Hilaire in epitactic overgrowths on natrolite crystals and in miarolitic cavities, pegmatite veins, sodalite syenite, uncommon in igneous breccia, very rare in marble and sodalite xenoliths. In pegmatites and basic volcanics. On aegirine at Point of Rocks Mesa, near Springer, Colfax Co., NM; originally found at *Mt. St.-Hilaire, QUE, Canada*; small amounts from Gobbins Cliff, Hillsport, Is. Magee, County Antrim, Northern Ireland; small amounts from Aci Castello, Sicily, Italy; from Schellkopf, near Brenk, Eifel, also at Teichelberg, near Pechbruun, Bavaria, Germany; Mt. Kukisvumchorr,

Khibiny massif, and Mt. Alluaiv, Lovozero massif, Kola Penin., Russia. EF
CM 18:85(1980), NZ (1985), DAN-ESS 288:136(1986), CRT 23:467(1988),
ZW (1992).

77.1.5.4 Mesolite $Na_2Ca_2[Al_6Si_9O_{30}] \cdot 8H_2O$

Described in 1816 by Fuchs and Gehlen and named from the Greek for *middle* and *stone*, because it was intermediate in chemical composition between natrolite and scolecite. Zeolite group. ORTH $Fdd2$. $a = 18.39$–18.47, $b = 56.45$–56.84, $c = 6.53$–6.58, $Z = 8$, $D = 2.30$. *24-1064*(nat): 6.6_6 $5.9_{6.5}$ $4.72_{4.5}$ $4.41_{5.5}$ $4.37_{4.5}$ $2.94_{4.5}$ 2.89_{10} 2.87_7. Is isostructural with natrolite and scolecite. Structure has long Si–Al ordered chains, rotated by 24°, with Na–Ca cations and water molecules in four ion sites in the channels parallel to c. Mesolite can be visualized as sheets of natrolite-type channels (two Na and two H_2O) alternated with two sheets of scolecite-type channels (one Ca and three H_2O) that result in a b-axis dimension three times that of natrolite. Asymmetry with regard to the (100) plane causes mesolite to be monoclinic. Disorder in the structure (cation) causes it to be monoclinic. Disordered Si–Al varieties have not been recognized. *AC(C) 42:937(1986); AM 1988:613(1988), 73:613(1988)*. Best identified by optical methods because X-ray diffraction patterns of natrolite–scolecite–mesolite are very similar. Normally long, slender needles, elongate on c; may be in hairlike tufts and aggregates of fibers; divergent, radiating groups to 20 cm in diameter; less commonly as compact masses, fibrous stalactites, or massive, porcelaneous habit; epitaxial intergrowths with natrolite and scolecite in single needles is common; dominant forms: {110}, {111}, with less development of {100}, {010}, {101}, and {011}. Typically twinned on {010} or {100}, crystals consist of {111} sectoral twins without {110} growth sectors. Colorless, white, pink, red, yellowish, or green; white streak; vitreous–silky luster. Cleavage {110}, {$\bar{1}$10}, perfect; uneven fracture; brittle with compact masses being tough. $H = 5$. $G = 2.25$–2.26. Gelatinizes with common acids. May show a small pyroelectric effect and is also piezoelectric. The atomic percentages of Ca and Na are nearly equal. Sometimes Na is slightly greater and sometimes Ca is greater. Traces of K are present. Sr, Ba, and Mg are not present. The ordering of the cations makes mesolite a unique member intermediate between natrolite and scolecite. A continuous chemical series does not exist between natrolite (Na), mesolite (1:1), and scolecite (Ca). Biaxial (+); $X \wedge a = 10°$, $Y = c$, $Z = b$; $N_x = 1.504$–1.505, $N_y = 1.505$–1.506, $N_z = 1.505$–1.507; $2V = 80$–$87°$; nearly isotropic; length-slow, some are length-fast; parallel extinction with elongation direction; $r > v$, strong; 10° inclination extinction angle can only be observed looking down the end of a needle. Commonly found in basaltic rocks, but also in andesites, porphyries, and hydrothermal veins. Occurs in many worldwide localities; dominant localities having exceptional display specimens are given here: Table Mts., Golden, Jefferson Co., CO; to 4 cm long from Bear Creek quarry, Drain, Douglas Co., OR; also from Goble, Columbia Co., near Dollar, and on Shotgun Creek, Linn Co; Springfield,

Lane Co.; Ritter Hot Springs, Grant Co., OR; to 10 cm long from Skookumchuk Dam, near Tenino, Thurston Co., WA; western Bay of Fundy, Nova Scotia, Canada, including Gates Mt., North Mt., and Cape Blomiden; to 10 cm long from Naalsoy, Stremoy, the Faroe Is.; Neubauerberg, Bohemia, Czech Republic; to 10 cm long from the Pune (Poona) district, MAH, India; sprays to 20 cm from Ahmadnagar; also large crystals, 15% mesolite, 85% scolecite, from the No. 2 quarry at Khandivali, Bombay, India; to 10 cm long from Westhaven Nunatak, Darwin Glacier, Victoria Land, Antarctica. EF *NJBA 143:231(1982)*, *NZ (1985)*, *ZW (1992)*, *AM 77:685(1992)*.

77.1.5.5 Scolecite $CaAl_2Si_3O_{10} \cdot 3H_2O$

Described in 1813 by Gehlen and Fuchs and named from the Greek for *worm* because of scolecite's tendency to curl up in a wormlike fashion when heated in a blowpipe. Zeolite group. MON Cc. $a = 6.516–6.517$, $b = 18.948–18.956$, $c = 9.761–9.765$, $\beta = 108.86–108.98°$, $Z = 4$, $D = 2.275$. *41-1355*(nat): 6.63_{10} 5.87_6 4.74_7 4.62_5 2.90_4 2.89_5 2.88_5 2.86_4. Cell may also be expressed as: $a = 18.48–18.508$, $b = 18.891–18.96$, $c = 6.527–6.548$, $\beta = 90.64–90.75°$ with $Z = 8$ (pseudotetragonal). The structure of the aluminosilicate framework is the same as for natrolite and mesolite. Long, ordered chains, rotated 24°, with a small distortion caused by a slight rotation of the center of the silicon tetrahedron of the chains around the chain axis. One Ca cation and three H_2O are in four ion sites in the channels parallel to c. It is monoclinic with no indication of Si–Al disorder. Scolecite with a disordered Si–Al framework similar to gonnardite should be possible. Optical studies [*AM 73:613(1988)*] indicate that scolecite has some disorder on Al and Si positions because growth sectors are present with triclinic symmetry. There is also a symmetrical extinction angle of 3° from the b axis. Occurs commonly as thin, prismatic needles, frequently flattened on {010}, elongated on c, and striated parallel to c, and crystals appear to be pseudo-orthorhombic or pseudotetragonal. Crystals are as much as 35 cm long. Also occurs as radiating groups and fibrous masses. May be nodular and massive. Epitaxial intergrowths with mesolite are common. Natrolite does not directly form epitaxial overgrowths on scolecite. May have all three minerals developed in the same crystal. Crystals are commonly twinned on {100}, producing "v" or fishtail terminations. Rare twinning on {001} and {110} also occurs. Scolecite crystals consist of {110}, {010}, and {111} growth sectors with triclinic symmetry (by optics) and appear to be monoclinic by X-ray analysis. Common forms include {111}, {$\bar{1}$11}, {101}, {110}, {010}. Colorless, white, pink, salmon, red, or green; white streak; vitreous–silky luster. Cleavage {110}, {1$\bar{1}$0}, perfect; uneven fracture; brittle. H = $5–5\frac{1}{2}$. G = 2.24–2.31. Pyroelectric and piezoelectric. Sometimes fluorescent yellow to brown in SWUV and LWUV. IR data: *CMG 25:239(1980)*. Synthesized: *JG 68:41(1960)*, *CCM 29:171(1981)*. Soluble in common acids. Because of an ordered Si–Al framework, scolecite has a narrow range of composition. Only minor amounts of Na and traces of K substitute for Ca. There is an absence of Ba, Sr, Fe, and Mg. Scolecite is isostructural with mesolite and natrolite but does not form a continuous chemical series with them. There is not much deviation from ideal stoichiometry. Biaxial $(-)$; $N_x = 1.507–1.513$, $N_y = 1.516–1.520$, $N_z = 1.517–1.521$; $2V = 35–56°$; $r < v$, strong; negative elongation, inclined extinction. For triclinic material: $X \wedge c = 16–18°$, $Y \wedge a = 3°$ (also reported as -14 to $-17°$, $Z \wedge b = 3°$. For monoclinic optics, $Z = b$. A common zeolite, widely found with low-Si zeolites. Occurs in basalts, andesites, gneisses, and amphibolites, and in contact-metamorphic zones. Also in laccoliths and dikes of gabbroic to syenitic composition. A hydrothermal mineral. Associated with other zeolites, calcite, quartz, and prehnite. No type locality is known for this mineral. Many worldwide localities; only selected localities are given here: coarse prisms 4–15 mm wide and 9–14 cm long from the south fork of the Toutle R., near Mt. St. Helens, Cowlitz Co., WA; fine sprays to 8–10 cm long from Teigarhorn, Berufjord, Iceland; Ben More, Isle of Mull, and Talisker Bay, Isle of Skye, Scotland; from Val Giuf-Fallital area, Graubunden, Switzerland; most of the world's finest scolecite is from Nasik and Pune (Poona), also Bombay, MAH, India, where crystals are as much as 2 cm wide and 35 cm long; Charcas, San Luis Potosi, Mexico; fine, large crystals associated with other zeolites from Rio Pelotas and Bento Goncalves, Rio Grande do Sul, Brazil. EF *MM 38:72(1971), NJBA 143:231(1982), ZK 166:219(1984), P6IZC:842(1984), ZK 171:141(1985), NZ (1985), ZW (1992)*.

77.1.5.6 Edingtonite $(Ba,Ca)Al_2Si_3O_{10} \cdot 4H_2O$

Named in 1825 by Haidinger for James Edington (1787–1844), Glasgow, Scotland, who discovered the mineral. Zeolite group. Three structural varieties: tetragonal, orthorhombic, and triclinic (from optics). TET $P\bar{4}2_1m$. $a = 9.581$–9.585, $c = 6.524$–6.530. ORTH $P2_12_12_1$. $a = 9.537$–9.583, $b = 9.651$–9.680, $c = 6.509$–6.530, $Z = 2$, $D = 2.82$–2.83. 25-60: 6.51_{10} 5.37_6 $4.82.5$ 4.70_7 $3.59_{6.5}$ $3.57_{5.5}$ $2.76_{5.5}$ $2.74_{5.5}$; 25-61(nat): 6.51_8 5.38_6 4.79_3 4.69_5 3.58_{10} $2.74_{7.5}$ $2.59_{4.5}$ 2.28_3. Chains of SiO_4 and AlO_4 tetrahedra parallel to c. Each chain is linked through oxygen bridges to four neighboring chains, at the same level, to form a three-dimensional network. Seven of eight water positions are located in the zeolite channels (planes perpendicular to c). Ba atoms are surrounded by framework oxygen and H_2O molecules. Orthorhombic edingtonite has a framework with nearly perfect Al–Si order. Tetragonal varieties have a disordered Si–Al distribution. Triclinic material also exists. Material from the Ice River Complex, Canada, described as tetragonal [*NJBM 8:373(1984)*], has an orthorhombic core and a triclinic rim. Bipyramidal, pseudotetragonal, also prismatic, equant to elongate, elongate on c; common forms include {111}, {110}, {001}, {121}, {100}, {010}; 1–10 mm long with exceptional prisms as much as 10 cm long and 2.5 cm wide; penetration twinning (rotated 90° about

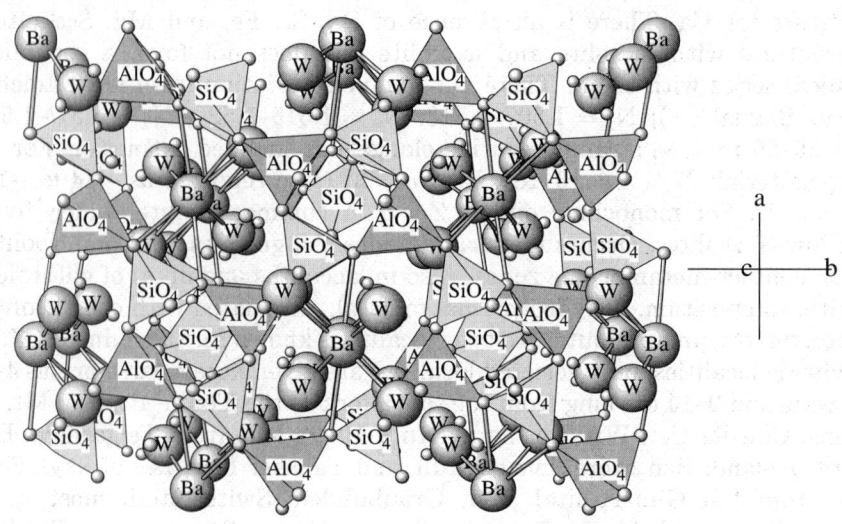

EDINGTONITE: BaAl$_2$Si$_3$O$_{10}$ · 4H$_2$O

This zeolite has eight- and four-membered rings and the framework is the same as that of natrolite—the only differences is in the channels where there is one Ba in place of two Na, but twice as many water molecules, and the orientation is quite different.

the c axis), and twins on {110} are rare. Doubly terminated crystals display hemimorphic morphology and rare forms. Colorless, white, gray, pink, brown, and yellow; white streak; vitreous luster. Cleavage {110}, perfect; uneven fracture. H = 4–5. G = 2.27–2.82. Pyroelectric and piezoelectric. IR data: *CMG 25:239(1980)*. DTA data: *NZ(1985)*. Gelatinizes with acids. Ba dominant. Varying smaller amounts of Na, K, Ca. A Ca-dominant variety was found in Russia. *DAN-ESS 234:170(1977)*. Biaxial (−); N$_x$ = 1.535–1.546, 2V = 10–63°, N$_y$ = 1.542–1.558, N$_z$ = 1.545–1.562; 2V increases with degree of Si–Al order; tetragonal material has 2V = 10–22°; orthorhombic and triclinic material has 2V = 50–63°; positive elongation; parallel or inclined extinction of 1–10° (for triclinic only); r < v, weak to moderate. For orthorhombic material: XYZ = *cba*. In hydrothermal ore veins, late hydrothermal stages of syenite complexes and volcanics. Carbonatites, alkaline igneous rocks. In dolerite in England. A relatively uncommon zeolite. In Franciscan greywacke at Ash Creek, Mendocino–Sonoma Co. line, CA; to 14 mm diameter and 4.5 cm long from Bathhurst, Gloucester Co., NB, very rare and in miarolitic cavities at Mt. St.-Hilaire, QUE, and zoned crystals (orthorhombic and triclinic) from Ice River, BC, Canada; *Old Kilpatrick, Dumbartonshire, Scotland*, and other localities in the Kilpatrick Hills; Squilver Quarry, Disgwylfa Hill, Shropshire, England; to 10 cm long and 3 cm wide from Böhlet, Västergötland, Sweden; Stare Ranski, eastern Bohemia, Czech Republic; Tulilukht Bay, Khibiny massif, Kola Penin., and Ca-dominant variety from

Castello, Aci Trezza, Osilo, and other localities on Sardinia; Arendahl, Tvedalen, and Brevik in the Langesund district, Norway; Schellkopf, near Brenk, Eifel district, and in the Höwenegg quarry, Hegau, Baden-Württenberg, and Arensberg, Zilsdorf, Eifel, Germany; Klöch, Styria, Austria; best display specimens are from Austria, Mt. St.-Hilaire, Germany, and Australia. Bundoora, near Melbourne, VIC, Australia; Maze, Niigata Pref., Japan. EF *MM 31:265(1956), NJBM 5:219(1986)*.

77.1.5.8 Cowlesite $Ca[Al_2Si_3O_{10}] \cdot 5\text{--}6H_2O$

Named in 1975 by Wise and Tschernich for John George Cowles (1907–1985), American amateur mineralogist and zeolite collector. Zeolite group. ORTH Space group unknown, probably $P222_1$. Cell data commonly given as $a = 11.27\text{--}11.29$, $b = 15.24\text{--}15.25$, $c = 12.61\text{--}12.68$, $Z = 6$, $D = 2.05$; however, new data from *MM 48:565(1984)* and *MM 56:575(1992)* indicate a cell with all axes doubled: $a = 23.3$, $b = 30.6$, $c = 25.0$, $Z = 52$. *29-286*(nat): 15.2_{10} $7.62_{1.5}$ 5.08_2 $3.81_{3.5}$ $3.75_{1.5}$ 3.05_2 $2.96_{3.5}$ $2.93_{2.5}$; *MM 56:575(1992)*: 15.2_{10} $7.64_{1.5}$ 3.83_1 3.06_1 2.98_1 2.94_1 $2.83_{0.5}$. Because of extreme thinness of single crystals, structure has not been determined. Thus cowlesite is the only natural zeolite whose crystal structure has not been described. Cell data derived from powder data. Structure now being attempted on synchotron data. Probably an ordered (Si,Al) distribution in the tetrahedral framework. Pointed, blade-shaped laths; broad {010}, extremely narrow {100}, terminated by {101}; 0.1 mm wide, only 2 μm thick, 0.2–2 mm long, ball-like aggregates are 2–3 mm across, some to 8 mm; twinning not observed. Colorless, gray, white, yellow; white streak; variable luster: vitreous, greasy, pearly. Cleavage (010), perfect. $H = 2$. $G = 2.05\text{--}2.14$. TGA data: *AM 60:951(1975)*. Very little variation from the ideal formula. Small amounts of Na present. Traces of K and Mg locally, also Sr, Ba, and Fe in trace levels. $Si/(Si + Al + Fe) = 0.603\text{--}0.622$. H_2O ranges from 30.2 to 32.8, ideal = 36. Biaxial $(-)$; $XYZ = abc$; $N_x = 1.505\text{--}1.513$, $N_y = 1.509\text{--}1.516$, $N_z = 1.509\text{--}1.518$; $2V = 30\text{--}53°$; positive elongation; colorless; no dispersion. Occurs only in low-silica volcanic rocks (e.g., olivine basalts). A low-silica zeolite from hydrothermal solutions. A low-temperature zeolite (70–110°C). Occurs 8 km S of Superior, Pinal Co., AZ; Table Mt., Jefferson Co., CO; *Neer Road, N of Goble, Columbia Co., OR*(!); also at Burnt Cabin Creek, Spray, Wheeler Co., OR(!); Beech Creek, Grant Co., OR(!); Capitol Peak, Thurston Co., WA; Monte Lake, BC, Canada; other localities in ONT and QUE, Canada; Dalsnipa, on Sandoy and Satan, on Streymoy, Faroe Is.; Mjoädalsa Canyon, Hvammur, Iceland; Dunseverick and Bally Clare quarry, County Antrim, Northern Ireland (10 localities total in Northern Ireland); Kingsburgh, Skye, Scotland; North and South Islands, New Zealand; Cairns Bay, Flinders, and Jindivick, VIC, Australia; Kuniga tunnel, Nishi-no-Shima, Dozen, Oki Is., Japan. EF *AM 60:951(1975), NZ (1985), ZW (1992)*.

pyrite deposits at Podolskoye, southern Ural Mts., Russia; in carbonatite at Jacupiranga, SW of São Paulo, SP, Brazil. EF *JCP 79:2356(1983); NJBM 8:373(1984), 541(1986); NZ (1985); AM 71:1510(1986); ZW (1992); MM 57:349(1993)*.

77.1.5.7 Gonnardite $(Na,Ca)_{12-15}(Al,Si)_{40}O_{80} \cdot nH_2O$

Described in 1896 by Lacroix and named for Ferdinand Gonnard, French mineralogist, who described gonnardite in 1871 as "mesole." Zeolite group. Described as ORTH and TET $I\bar{4}2d$. Orthorhombic axes: $a = 13.04$–13.45, $b = 13.04$–13.42, $c = 6.22$–6.67, $Z = 0.5$, $D = 2.25$–2.33. *AM 77:685(1992)* gives for a possible orthorhombic pseudocell: $a = 18.56$, $b = 18.64$, $c = 6.62$ ($Z = 1$). Gonnardite may well be a mixed structure, composed of slabs of natrolite and thomsonite. According to *AM 77:685(1992)*, the structure of gonnardite is still unknown. However, the crystal structures of four natural gonnardite samples were refined by full-profile Rietveld techniques. *MSF 79-82:845(1991)*. The Al–Si framework is the same as natrolite, mesolite, scolecite, tetranatrolite, and paranatrolite. Si and Al are disordered in both gonnardite and tetranatrolite, where Si/Al < 1.5 in gonnardite and ≥ 1.5 in tetranatrolite. *EJM 7:501(1995)*. Some of the water sites in the disordered natrolite structure of gonnardite are empty. A series, based on the disorder of the Si–Al in the framework, appears to exist between Na-rich gonnardite (i.e., tetranatrolite) and natrolite. Crystals are prismatic, bound by {110} and {111}, as well as {100} and {001}. Also occurs as radial hemispheres to 3 cm. Gonnardite and tetranatrolite form overgrowths on natrolite. Commonly found as zoned prisms or aggregates with thomsonite, natrolite, and paranatrolite. Colorless, white, yellow, pink to salmon-orange; white streak; vitreous to silky or dull luster. Cleavage not determined. $H = 4\frac{1}{2}$–5. $G = 2.21$–2.36. IR and thermal data: *JJMPE 77:78(1982), NZ (1985)*. Gonnardite and tetranatrolite have the same disordered Al–Si framework and differ only by the amount of Ca and Na present. *NJBM:219(1986)* showed the existence of a complete solid-solution series between tetranatrolite and gonnardite. Tetranatrolite is considered as Na-rich gonnardite by *ZW (1992)*. *MM 52:207(1988)* further showed the existence of a complete series but divided it into tetranatrolite–ranite–gonnardite, based on chemistry, optics, and X-ray diffraction patterns. Small amounts of Mg, Sr, Ba, and K may be present (0–2 wt % of each oxide). Uniaxial (−) and (+), also biaxial (−) or (+); $N_x = 1.480$–1.513, $N_z = 1.480$–1.515; 2V to 50°; Na-rich material is length-slow, Ca-bearing is length-fast or length-slow, with parallel extinction. Occurs in silica-poor volcanic rocks and pegmatites; rarely in skarn. Often intergrown with thomsonite, with cores of gonnardite and rims of thomsonite. Found in Crestmore, Riverside Co., CA; Table Mts., Golden, Jefferson Co., CO; Honolulu, Oahu, HI; Connecticut, New Mexico, Oregon, and Washington; Na-rich gonnardite (originating as paranatrolite) from Mt. St.-Hilaire, QUE, Canada; Hillsport, Is. Magee, County Antrim, Northern Ireland; *Chaux de Bergonne, Gignat, Puy de Dôme, France*; Capo di Bove, Lazio, near Rome; Aci

77.1.5.9 Thomsonite $(Ca,Sr)_2NaAl_5Si_5O_{20} \cdot 6H_2O$

Named in 1820 by Brooke for Thomas Thomson (1773–1852), professor of chemistry at Glasgow, Scotland, who analyzed the mineral. Zeolite group. ORTH (pseudo-TET); some is MON based on optics. $Pcnn$ (many weak hkl reflections suggest that the space group may be $Pcmn$ or $Pcn2$. $a = 13.00$–13.15, $b = 13.01$–13.09, $c = 13.20$–13.26; and at 13 K: $a = 13.0569$, $b = 13.1043$, $c = 13.2463$, Z = 4, D = 2.38 at 13 K. 35-498(nat): 6.57_7 5.90_6 4.62_{10} 3.50_7 3.19_7 2.94_{10} 2.86_{10} 2.67_{10}. Synthesized: CCM $29:171(1981)$. The aluminosilicate framework is long chains of tetrahedra rotated by 23° (similar to natrolite), but are cross-linked at two levels. Ideal thomsonite has an ordered framework. There is some Si–Al disorder in high-Si thomsonite. Degree of disorder is measured by the difference between the length of b vs. a. This is 5–10%. Two sets of channels parallel to c are present. The first set is fully occupied by equal amounts of Na and Ca, where the second channel is only partly occupied by (up to 50%) Ca or Sr. Four fully occupied H_2O sites are present. Two with two H_2O molecules (along with Ca and Sr) in the channels, and the other two sites with $2H_2O$ molecules between the channels. At 13 K, there is no cation ordering. Common forms are: {100}, {010}, {001}, {110}, has a wide range of habits: thin, rectangular, blades to blocky prismatic, rarely long, coarse, acicular prisms, elongated on c; twinning = {110} and {041} common. Colorless, white, pink, red, green, orange, yellow, blue; white streak; vitreous to pearly luster. Cleavage {100}, perfect; {010}, good; uneven to subconchoidal fracture; brittle. H = 5–5$\frac{1}{2}$. G = 2.25–2.44; IR data: CMG $25:239(1980)$. Gelatinizes with acids. Most is Ca-dominant, with considerable Na. Rarely, small amounts of Sr (to 10% of cations). Na-dominant thomsonite (Tyamyr, Russia) has been reported. Well ordered, Si/Al = 1.00. Si-rich, Si/Al = 1.18–1.3 has greatest disorder. High SrO ($\approx$ 6.6% SrO) in specimens from Honshu, Japan. Also 6.65% SrO in material from eclogite in western Norway. Biaxial (+); XYZ = bca, sometimes XYZ = acb; $N_x = 1.497$–1.530, $N_y = 1.513$–1.536, $N_z = 1.518$–1.544; 2V = 42–75°; r > v, distinct. Thomsonite is a common zeolite found in low-Si basalts and hypabyssal rocks. Less common in contact-metamorphic zones, pegmatites, and rarely in Si-rich tholeiitic basalt. Also in some alkalic igneous rocks. Not found in altered volcanic ash deposits. Occurs as an authigenic cement in some sandstones. Many worldwide localities, only notable localities are mentioned here. Scarce at Paterson, Passaic Co., NJ, but some exceptional aggregates to 8 cm occur; aggregates to 5 cm from Prospect Park, NJ; thomsonite is the most prominent zeolite at North and South Table Mts., Golden, Jefferson Co., CO; Jaquish Road and Neer Road, Goble, Columbia Co., Ritter Hot Springs, Grant Co., Springfield, Lane Co., and Drain, Douglas Co., OR; Stremoy, Nolsoy, and Essuroy Is., Faroe Is.; *Old Kilpatrick, Dumbartonshire, Scotland*; to 10 cm long and 3–10 cm wide from Boyleston quarry, Barrhead, Glasgow, Scotland; Mt. Monzoni, Val di Fassa, Trentino–Alto Adige, Italy; in the Eifel volcanics, Germany; also at Rossberg, near Darmstadt; Mettweiler, Rhein-Preussen, Germany; Eulenberg, Vinaricka Hill, Kladno, Bohemia, Usti nad

Labem, and Kadan, Czech Republic; Krasnoyarsk region, Yakutia, Russia; high SrO-containing (6.6 wt % SrO) from Honshu, Japan. EF *CM 16:487(1978); MM 44:231(1981); Z1:91(1981), 5:74(1985); NZ (1985); AC(C) 46:1370(1990); AM 77:685(1992).*

77.1.6.1 Mordenite $(Na_2,Ca,K_2)Al_2Si_{10}O_{24} \cdot 7H_2O$

Described in 1864 by How and named for the locality. Zeolite group. Old synonym: ptilolite. ORTH $Cmc2_1$. $a = 18.09–18.17$, $b = 20.40–20.58$, $c = 7.49–7.54$, $Z = 4$, $D = 2.09–2.13$. *6-239*(nat): 9.1_9 6.6_9 4.53_8 4.00_9 3.84_6 3.48_{10} 3.39_9 3.22_{10}; *29-1257*(nat): 9.06_{10} 4.53_3 4.00_7 $3.48_{4.5}$ $3.39_{3.5}$ 3.22_4 $3.20_{3.5}$ 13.6_2. The aluminosilicate framework is made up of complex chains of five-membered rings of SiO_4 tetrahedra and single AlO_4 tetrahedra cross-linked by four-membered rings that form twisted 12-membered rings around nearly cylindrical channels parallel to c. Si occupies 80–85% of the tetrahedra with Al in the remaining tetrahedra. Al–Si distribution is partly ordered. Cations are on two sites. Many other K and Na ions are not localized. Some mordenite needles have an inclined extinction of 3–5° and are monoclinic. *ZK 175:249(1986)*. Ion-exchanged mordenite is $Cmcm$. *CSJB 58:3035(1985)*. Slender needles elongated on c, and striated parallel elongation. Crystals to 5 cm long. Also radiating needles, cottony masses. Compact porcelaneous masses, and smooth radiating hemispheres. Forms: {100}, {010}, {110} in the prism zone; {101}, {001}, {111} terminal forms. Twinning present, parallel to c. Colorless, white, yellow, pink, orange, red; white streak; vitreous to silky luster, pearly on {010}. Cleavage {100}, perfect; {010}, distinct; brittle. $H = 4–5$. $G = 2.10–2.15$. Gelatinizes in common acids. Most mordernite is Na-dominant, but some is K-dominant or Ca-dominant. Mordenite derived from volcanic ash has the highest Si content. Biaxial (+) or (−); $XYZ = cab$; $N_x = 1.471–1.483$, $N_y = 1.475–1.485$, $N_z = 1.476–1.487$; $2V_z = 76–104°$; always negative elongation; length-fast, parallel to 3–5° inclined extinction, $r < v$, weak. Occurs in silica-rich (e.g., rhyolitic) volcanics and as a hydration product of volcanic glasses; volcanic ash beds, tuffs, and rarely in olivine–basalt. Of hydrothermal or sedimentary authigenic (mostly saline lake) origin. A widespread zeolite, and only notable occurrences are given here: Table Mts., Golden, Jefferson Co., CO; to 2 cm long from Wolf Creek Pass, Mineral Co., CO; Roadside quarry and Jaquish Road, Goble, Columbia Co., OR; Rickreall Creek quarry, Dallas, Polk Co., OR; to 5 cm long from South Point quarry, Jefferson Co., WA; also from near Stevenson, Skamania Co., WA; Crestmore, Riverside Co., CA; *Morden, King's Co., Nova Scotia, Canada*, and elsewhere along the Bay of Fundy; high-K-containing (5.3 wt % K_2O) from Mydzk, Volniya, Ukraine; to 5 cm long and spheres to 5 cm diameter at Chinchwad, Pune (Poona) district, MAH, India. Crystals in 5-cm groups occur at Kawazu, Shizuoka Pref., Japan. Only a few applications for natural mordenite, but synthetic material is widely used for adsorbers and catalysts. EF *CMP 50:65(1975), NZ (1985), ZW (1992)*.

77.1.6.2 Epistilbite (Ca,Na$_2$)[Al$_2$Si$_6$O$_{16}$] · 5H$_2$O

Named in 1826 by Rose from the Greek for *near* and *stilbite*, a mineral that is similar in many respects. Zeolite group. See goosecreekite (77.1.7.3), the dimorph. TRIC and MON $C1$, $C2$, Cm; $a = 8.92$–9.12, $b = 17.73$–17.77, $c = 10.21$–10.25, $\beta = 124.57$–$124.65°$, and $a = 9.083$, $b = 17.738$, $c = 10.209$, $\alpha 89.95°$, $\beta 124.58°$, $\alpha 90.00°$, $Z = 3$, $D = 2.27$–2.31. *39-1381*(nat): 8.9_{10} 6.9_3 $4.92_{5.5}$ 3.92_2 3.87_7 3.45_9 3.21_8 2.92_5. Aluminosilicate framework contains chains of four-membered rings joining sheets parallel to (010). The cation sites, coordinating three oxygens of the four-membered rings plus six H$_2$O molecules lay on the mirror plane. Si and Al in the framework are highly ordered, although streaking of single-crystal photos indicates some disorder. Si–Al ordering differs in each sector of the crystal. Structure has two tetrahedral modifications (R = 1.82%, 2.18%), resulting in $C2$ symmetry. Atomic ordering that is apparent in optical analysis is too small to be detected by X-ray diffraction. *ZK 173:257(1985)*, *AM 73:1434(1988)*. {110} is dominant, also {001}, {011}, {010}, {013}, {$\bar{1}$01}, prismatic, commonly elongated on c, with a diamond-shaped cross section; frequently 3–10 mm long, exceptional aggregates to 3 cm long; twinning = rarely on {110} to form penetration twins; all are twinned on {100} to form a pseudo-orthorhombic crystal. Two kinds of sectoral twins are observed; one is that of structure with symmetry $C2/m$, and the other is due to (Al,Si) order. Colorless, white, orange fresh or red; white streak; vitreous luster. Cleavage {010}, very good; uneven fracture; brittle. $H = 4$–$4\frac{1}{2}$. $G = 2.22$–2.28. Piezoelectric. Very little variation in exchangeable cations. Always Ca dominant. Moderate Na, K is scarce or absent. Mg, Ba, and Sr are generally not present. Ratio Ca/(Na + K) = 9 to 2. Biaxial (−); $Y = b$, $Y \wedge b = 2°$, $Z \wedge c = 5$–$10°$, $X \wedge a = -24°$, also $Z \wedge c = -10°$, and $X \wedge a = 11°$; $N_x = 1.485$–1.505, $N_y = 1.497$–1.515, $N_z = 1.497$–1.519; $2V = 44$–$46°$; colorless; length-slow; r < v, strong. Has been synthesized. Occurs in geothermal wells in basalt in Iceland (80–160°). Generally found in silica-rich tholeiitic basalt, olivine basalt, also in gneisses. No diagenetic epistilbite has been reported. A relatively rare zeolite, but it has been reported from more than 25 localities. Localities for fine-quality material include 1 to 2.5-cm crystals from Riffe Lake, Kosmos, Lewis Co., WA; crystals from 1 to 2.8 cm long associated with quartz and high-Si zeolites from *Berufjord, Iceland*; Fonte del Prete, San Piero in Campo, Elba, Italy; "Bombay quarry" at Khandivali and near Nasik, Bombay, MAH, India; also from the Pune district, India. Lesser-quality material occurs at Goble, Columbia Co., OR; Giebelsbach, Fiesch, Valais, Switzerland; Kuroiwa, Niigata Pref., and Yugawara, Kanagawa Pref., Japan. EF *AM 59:1055(1974)*, *GNZ (1977)*, *NZ (1985)*, *ZW (1992)*.

77.1.6.3 Maricopaite Pb$_7$Ca$_2$Al$_{12}$Si$_{36}$(O,OH)$_{100}$ · 32(H$_2$O,OH)

Named in 1988 by Peacor for the locality. Zeolite group. See mordenite (77.1.6.1). ORTH $Cm2m$, $Cmmm$, $C2mm$, $Cmm2$, or $C222$. $a = 19.65$–19.713, $b = 19.385$–19.40, $c = 7.522$–7.531, $Z = 1$. $D = 2.96$. PD(nat): 13.7_{100} 9.86_{40}

4.79_{20} 4.35_{30} 3.357_{40} 3.216_{50} 2.978_{40} 2.845_{40}. *CM 26:309(1988)*. Structure is similar to mordenite and other zeolites. Based on an interrupted mordenite-like framework. Maricopaite is not a true tetrasilicate because of the interrupted T-O-T linkages; one of a small group of natural and synthetic compounds having interrupted octrahedral framework structures. *AM 79:175(1994)*. Pb atoms form $Pb_4(O,OH)_4$ clusters with Pb_4 tetrahedra within channels. Acicular, sprays of crystals, elongated on [001]; radial sprays to 1 mm in diameter. Translucent white, white streak, silky to vitreous luster. Cleavage {010}, imperfect. Soft and friable. G = 2.94. IR and TGA. Analysis (wt %): SiO_2, 42.0; Al_2O_3, 11.4; CaO, 2.4; PbO, 30.8; H_2O, 11.0. Biaxial (−); $N_x = 1.563$, $N_y = 1.582$, $N_{z\,calc} = 1.592$; 2V = 70°; XYZ = *acb*; r > v, strong. Occurs with mimetite on quartz matrix. A secondary mineral in a single calcite–fluorite vein at the *Moon Anchor mine, Tonopah, Maricopa Co., AZ*. EF.

77.1.6.4 Dachiardite $(Ca,Na_2K_2)_5Al_{10}Si_{38}O_{96} \cdot 25H_2O$

Described in 1906 by D'Achiardi and named for Antonio D'Achiardi (1839–1902), an Italian mineralogist. The mineral was investigated by his son. A second occurrence of the mineral was not reported until 1964, from Japan; in 1975, the Na-dominant analog was found in Italy. Zeolite group. Svetlozarite is twinned and highly faulted dachiardite. MON $C2/m$. $a = 18.625$–18.69, $b = 7.489$–7.52, $c = 10.239$–10.30, $\beta = 107.58$–108.35°, Z = 0.5 (for 96 O), D = 2.14–2.17. *18-467*(nat): 8.90_5 6.91_5 4.97_5 4.88_5 3.93_5 3.80_5 3.45_{10} 3.20_{10}. The structure consists of sheets of six-membered rings (Al-free) connected with four-membered rings (which are Al-rich), while the exchangeable cations are distributed between the six-ring sheets. Dachiardite from Elba and Hokiya-dake shows (Al,Si) order, while bladed dachiardite from other localities (most) shows irregular {001} twinning and {100} stacking faults that indicate disorder. Sharp patterns for samples from Elba, Hokiya-dake, and Yellowstone Park. Most others are diffuse or show streaks indicating some disorder. *ZK 166:63(1983)*. Domain structure was shown to exist in dachiardite. *ZK 185:620(1988)*. More recently, structural studies of dachiardite show the coexistence of domains of "modified" dachiardite and "normal" dachiardite. *EJM 2:187(1990)*. Two acentric framework configurations are present with the same frequency, so that the statistical symmetry $C2/m$ is maintained. Crystals are fibers or blades, to 5 mm long; also radiating aggregates; Cs-bearing material from Elba forms cyclic-twinned eightlings; twinned on {110} at nearly 45°. Blades are elongated on *b* or *c*. Twinning on {001} is also common, rarely on {100}. Simple, untwinned crystals are also found with the "eightlings." Repeated parallel twins on {001} with {100} as the twin axis. Also, polysynthetic twinning on [100] twin planes. Colorless, white, pink, or orange-red; white streak; vitreous to silky luster. Cleavage {100}, {001}, perfect; conchoidal to uneven fracture; brittle. H = 4–4½. G = 2.14–2.21. Decomposed by common acids. Calcium is the dominant exchangeable cation in dachiardite, but Na may be dominant (sodium–dachiardite) as well as K (Yellowstone Park). A high-silica zeolite. Cs_2O-bearing dachiardite from

Elba has about 1 wt % Cs_2O and is nonfibrous in habit. It is bladed and has an ordered Al–Si framework. Na-rich occurrences (sodium–dachiardite) are about 2× those of Ca-rich dachiardite. Si/Al + Fe = 3.6–5.7, $R^{2+}/(R^+ + R^{2+})$ = 0.98–0.01 for the complete series dachiardite–Na-dachiardite. Biaxial (+); X = b, Z $\wedge$ c = –32 to –38°; N_x = 1.488–1.494, N_y = 1.490–1.496, N_z = 1.494–1.499; 2V = 58–73°; r > v, moderate to strong. Si/(Al + Fe) = 3.66–3.81 and $R^{2+}/(R^+ + R^{2+})$ = 0.53–0.84. *CM 25:475(1987)*. A hydrothermal zeolite that generally occurs in silica-rich environments; in late stages of pegmatites and Si-rich volcanoes, but also occurs in basalts. Yellowstone Park drill holes show temperatures of 100–200°. Also in hydrothermally altered pumiceous tuffaceous sediments. May be associated with mordenite, ferrierite, and heulandite. Only selected localities are given here: Hassayampa, AZ; Ca-richest dachiardite from Lower Geyser Basin, Yellowstone National Park, WY; Agoura Hills, Los Angeles Co., Ca; Yuquina Head, Lincoln Co., OR; Cape Lookout, Tillamook Co., OR; Altoona, Wahkiakum Co., WA; *Filone della Speranza, Mte. Capanne, San Piero di Campo, Elba, Italy*; Alpe di Siusi, Trentino–Alto Adige, Italy; Zeilberg quarry, Maroldsweisach, Franconia, Bavaria; Zvezdel, Rhodope Mts., Bulgaria; Onoyama mine, Kagoshima Pref., Hatsuneura, Ogasawara Is., Hokiya-dake, Nagano Pref., Japan. EF *MM 43:548(1979), MJJ 10:371(1981), NZ(1985), JCSSJ 30:45(1990), ZW (1992)*.

77.1.6.5 Sodium Dachiardite $(Na_2,Ca,K_2)_{4-5}Al_8Si_{40}O_{96} \cdot 26H_2O$

Described in 1977 by Yoshimura and Wakabayashi and named as the Na-dominant analog of dachiardite. Zeolite group. MON $C2/m$. a = 18.639–18.666, b = 7.506–7.512, c = 10.277–10.299, β = 108.35–108.68°, Z = 0.5 (for 96 oxygens), D = 2.14–2.17. *30-1149*(nat): 9.77_2 $8.66_{7.5}$ 4.88_9 $3.96_{2.5}$ 3.79_4 3.45_{10} 2.97_3 $2.86_{2.5}$. See dachiardite (77.1.6.4) for a description of the structure. Divergent bundles of bladed and acicular crystals to 1.5 cm long, twinning as for dachiardite. Colorless, white, reddish orange; white streak; vitreous to greasy luster, silky on cleavage surfaces. Two perfect cleavages parallel to [010], uneven fracture, brittle. H = $4-4\frac{1}{2}$. G = 2.14–2.17. Na is the dominant exchangeable cation. $R^{2+}/(R^+ + R^{2+})$ = 0.00–0.20, Si/(Al + Fe^{3+}) ≥ 4.46. The K-dominant equivalent of dachiardite has not yet been named. Material from near the Belogorskoye gold mine in the northern Sikote-Alin Range, Russia, has K dominant. *ZVMO 105:449(1976)*. Some dachiardite from Yellowstone Park is also K-dominant. Biaxial (–); X = b, Z $\wedge$ c = –8 to –20°, Y ≈ a; N_x = 1.471–1.480, N_y = 1.475–1.481, N_z = 1.476–1.484; 2V = 52–80°; r < v, moderate. Distinct from dachiardite, which is optically (+), sodium dachiardite has higher indices of refraction and an opposite sense of dispersion. Occurs in veins and cavities in basic to acid volcanic rocks, and alkalic rocks (e.g., silicocarbonatite). A hydrothermal mineral. Associated with other zeolites such as mordenite, analcime, heulandite, ferrierite, and clinoptilolite, as well as carbonates and quartz. Localities include: Yaquina Head, near Agate Beach, Lincoln Co., OR; strongly Na–Ca zoned crystals from Altoona, Wahkiakum Co., WA; Francon quarry, St. Michel, Montreal Is., QUE, Canada;

Alpe di Siusi and Val di Fassa, Trentino–Alto Adige, Italy; Tanzenberg, Kapfenberg, Styria, Austria; Zielberg, near Maroldsweisach, Bavaria, Germany; *Yanagi-Shiden, Tsugawa district, Niigata Pref., Japan*; Chichijima, Ogasawara (Bonin) Is., Japan. EF *CMP 49:63(1975), SRNU E4:49(1977), NZ (1985), JCSSJ 30:45(1990), ZW (1992)*.

77.1.6.6   Ferrierite $(Na_2,K_2,Mg,Ca)_{3-5}[Al_{5-7.5}Si_{27.5-31}O_{72}] \cdot 18H_2O$

Named in 1918 by Graham for Walter Frederick Ferrier (1865–1950), Canadian geologist and mining engineer. Zeolite group. Found in 1918 and then note reported again until 1967! The monoclinic polymorph was first reported in 1976. *AM 61:60(1976)*. Orthorhombic and monoclinic polymorphs. The monoclinic polymorph is rarer than the orthorhombic form. Three to five exchangeable cations (1^+ and 2^+) per 72 oxygen atoms. The univalent cations range from 21 to 85%; the Si–Al framework varies from ($Al_{7.5}Si_{27.5}O_{72}$ to ($Al_5Si_{31}O_{72}$). The a cell dimension varies linearly with SiO_2 content: high SiO_2, smaller a; low SiO_2, larger a. ORTH *Immm*. $a = 18.90$–19.45, $b = 14.12$–14.28, $c = 7.43$–7.53, $Z = 1$, $D = 2.10$. *39-1382*(nat): 9.60_{10} 5.84_2 3.97_4 3.89_1 3.71_3 3.56_1 3.53_3 3.49_2. MON $P2_1/m$. $a = 18.89$, $b = 14.18$, $c = 7.47$, $\beta = 90.0°$, $Z = 1$; no PD for the monoclinic polymorph. A Si-rich zeolite characterized by $(Si,Al)O_4$ tetrahedra forming five-membered rings linked in complex chains parallel to c. Between the chains in each unit cell are two large cation cages, which contain hydrated cations, commonly Mg, and two channels containing the loosely bound and highly disordered remaining cations and H_2O molecules. Orthorhombic structure; *AC 21:983(1966), ZK 178:249(1987)*; monoclinic structure: *AM 70:619(1985)*. Structure of both polymorphs is the same but with slight deviations in the monoclinic form. The amounts of Mg and Na appear to control the structure. Orthorhombic crystals are elongated on c, platy on {100}, commonly {100} dominant, with small {010} and {110}, terminated by {101}, rare {010} dominant variety has small {100} and {101}; usually 5–10 mm long, rarely to 8 cm long. Colorless, white, pink, orange, and red (due to Fe^{3+}); white streak; vitreous–silky luster. Cleavage {100}, perfect; {001}, imperfect; uneven fracture. $H = 3$–$3\frac{1}{2}$. $G = 2.06$–2.23. Specimens of the mineral from Agoura, CA, fluoresce and phosphoresce a moderately strong blue-white under SWUV and LWUV. They phosphoresce for about 10 seconds. IR data (orthorhombic): *NZ (1985)*. Monoclinic crystals are thin, diamond-shaped, tablets with {010} dominant, showing {201}, {100}, and {010}. Highest BaO (2.5%) and SrO (0.4) in Silver Mt. ferrierite. May have Mg, Na, or K dominant. A high Si-zeolite. Biaxial (+) and (−); XYZ = abc and material from Tapu has XYZ = bca; $N_x = 1.473$–1.489, $N_y = 1.474$–1.489, $N_z = 1.477$–1.492, and Tapu (Ba-Mg rich) $N_x = 1.487$, $N_y = 1.489$; 2V ranges from 50° about Z to 55° about X; positive elongation. Monoclinic material generally has $N_x = 1.473$–1.478, $N_y = 1.474$–1.479, and $N_z = 1.477$–1.482. Occurs in cavity fillings in basaltic rocks and as a diagenetic mineral in rhyolitic tuffaceous sediments. Also in metamorphic rocks. In rhyolitic pyroclastic rocks as a result of hydration reactions. Occurs at more than 30 worldwide localities; only selected localities are mentioned here. Minable amounts of K-dominant ferrierite in tuffaceous lake sediments at Lovelock, Pershing Co. NV; both orthorhombic and monoclinic polymorphs occur in vesicles and open spaces in brecciated basalt at *Altoona, Wahkiakum Co., WA* (for monoclinic polymorph); Na-dominant and both forms probably occur and are fluorescent and phosphorescent from the Canwood Mall site at Agoura in the Agoura Hills, CA; Leavitt Park, Sonora Pass, Tuolumne Co., CA; good display specimens from *Kamloops Lake* (orthorhombic polymorph), Monte Lake, and Pinaus Lake, near Westwold, BC, Canada; the Kamloops Lake locality is still the best and most prolific locality for ferrierite. Monoclinic polymorph and Mg-rich from Albero Basso and Val Timonchio, Vicenza, Italy. Ba–Mg-rich from Tapu, Thames, Coromandel Penin., North Island, New Zealand. Occurs at several localities in Japan. EF *NZ (1985), MM 50:63(1986), ZW (1992), AM 77:314(1992).*

77.1.6.7 Boggsite $(Ca,Na,K,Mg)_{11}(Fe,Al,Si)_{96}O_{192} \cdot 7OH_2O$

Described in 1990 and named for Robert Maxwell Boggs (b.1918) and Russell Calvin Boggs (b.1952), father and son, who discovered the mineral. Zeolite group. ORTH *Imma.* $a = 20.236–20.25$, $b = 23.798–23.82$, $c = 12.78–12.789$; also $a = 20.21$, $b = 23.77$, $c = 12.80$; $Z = 1$, $D = 1.994$. *AM 75:1200(1990).* PD: 11.80_3 11.30_{10} 10.2_3 4.43_7 3.86_8 3.63_3 3.37_{10}; $42-1379$(calc. pattern). A high-silica zeolite with the first reported three-dimensional channel system bounded by both 12- and 10-membered rings of tetrahedra. *AM 75:501(1990).* Chisel-shaped blades elongated on b. Radiating hemispheres 0.5–2 mm in diameter. Individuals crystals to 0.5 mm long and 0.2 mm wide. No twinning observed. Prominent forms are {102}, {010}, and {011}. {001}, {010}, and {100} are poorly developed. Morphology is similar to thomsonite. Colorless–white, white streak, vitreous to dull luster. No cleavage, conchoidal fracture, brittle. $H = 3\frac{1}{2}$. $G = 1.98$. White zones fluoresce a weak blue-white in LWUV and SWUV. Highly resistant to acids. $6\,N$ HCl for 1 hour leaves it unaffected. $Si/(Al + Fe) = 4.2$. The ratio of monovalent/divalent extra-framework cations ranges from 0.39 (Goble) to 1.03 (Mt. Adamson). Biaxial $(-)$; $XYZ = cab$; $N_x = 1.480$, $N_y = 1.481$, $N_z = 1.487$; $2V_x = 25°$; nonpleochroic; length-slow, parallel extinction; $r > v$, strong. Some material has a lower average refractive index (1.46–1.47). In one area (3 mm × 2 mm × 1 m) of a basalt flow, associated with tschernichite. Very early stage of zeolite growth. Other minerals include chalcedony, heulandite, apophyllite, levyne, chabazite, calcite, aragonite, and opal. *Neer Road pit, Goble, Columbia Co., OR;* Mt. Adamson, northern Victoria Land, Antarctica. EF *ZW (1992), EJM 7:1029(1995).*

77.1.7.1 Brewsterite $(Sr,Ba,Ca)Na,Al_2Si_6O \cdot 5H_2O$

Discovered and named by H. J. Brooke in 1822 for David Brewster (1781–1868), Scottish physicist, who studied the optical properties of crystals. Zeolite group. See Ba analog of brewsterite (77.1.7.5). TRIC *P*1 (pseudomonoclinic) or MON $P2_1/m$. $a = 6.75–6.82$, $b = 17.46–17.51$, $c = 7.73–7.76$, $\beta = 94.27–$

94.47°, Z = 2, D = 2.38. *41-1356*: 8.78_3 4.66_7 3.27_4 3.11_2 2.92_{10} 2.73_2 2.72_2 2.19_4; also 15-0582(nat). Ordering of Al and Si in the framework causes a slight distortion of only 0.5° from the *c* axis, producing a triclinic framework. A sheet-type structure composed of chains of parallel four-membered rings. *AC 17:857(1964)*, *AC(C) 41:492(1985)*, *AM 72:645(1987)*, *AC(B) 33:2907(1977)*. Common forms are: {010}, {001}, {100}, {011}, and {610}; tabular on (010) striated parallel to *a*, often flattened along *c*, blocky to prismatic, pseudomonoclinic, elongated on *a*; large crystals are 1–2 mm in diameter, 4 mm long common; microscopic twinning frequent, and parallel to the dominant {010} cleavage; mineral is piezoelectric (i.e., noncentric), tricinicity of brewsterite is too slight to be detected by XRD or neutron diffraction. Colorless, white, pink, light yellow, and brown; white streak; vitreous luster, pearly on (010). Cleavage {010}, perfect; {001}, poor; uneven fracture; brittle. H = $4\frac{1}{2}$–$5\frac{1}{2}$). G = 2.32–2.45. Si and Al contents varied: Strontian (Scotland) = $Sr_{0.58}Ba_{0.30}Ca_{0.14}[Al_{1.99}Si_{6.04}O_{16}] \cdot 4.9H_2O$. Yellow Lake = $Sr_{0.85}Na_{0.24}K_{0.02}Ca_{0.02}[Al_{2.17}Si_{5.84}O_{16}] \cdot xH_2O$. Burpala massif = $Sr_{0.75}Ba_{0.13}Na_{0.04}K_{0.02}$ $[Al_{2.06}Si_{5.94}O_{16}] \cdot 5.04H_2O$. Biaxial (+); X ∧ *a* = 22–28° in obtuse β, Y ∧ *c* = 0.5°, Z = *b*; N_x = 1.506–1.510, N_y = 1.510–1.512, N_z = 1.522–1.523; 2V = 55–65°; r > v; positive elongation. See also *MM 52:416(1988)* for a discussion of morphology and optics of brewsterite; not tied to X-ray orientation. A hydrothermal mineral; occurs in vugs of massive volcanic rocks. Also in hydrothermal veins. Occurs at Yellow Lake, Ollala, BC, Canada; in the Whitesmith, Middleshop, and Bellsgrove mines at *Strontian, Argyleshire, Scotland*; St. Cristophes, Bourg d'Oisans, Val d'Isere, France; at several localities in the high Pyrenees (e.g., in the Riou Maou Mountains between Gararie and Luz, and others). EF *MGNZ (1977)*, *NZ (1985)*, *ZW (1992)*, *MM 54:654(1990)*.

77.1.7.2 Yugawaralite $CaAl_2Si_6O_{16} \cdot 4H_2O$

Described in 1952 by Sakurai and Hayashi and named for the locality. Zeolite group. MON *Pc* (by X-ray analysis), TRIC *P*1 (by optics). *a* = 6.70–6.730, *b* = 13.95–14.008, *c* = 10.03–10.54, β = 111.07–111.50°, Z = 2, D = 2.22. *39-1372*(nat): 7.01_3 $5.82_{5.5}$ 4.67_{10} 4.30_3 3.24_4 3.05_9 2.94_2 2.72_2. This and other zeolites showing metastable crystal growth, may show noncentrosymmetry, but be centrosymmetric by X-ray analysis. The electrostatic charge is balanced on the growth surface (two-dimensional) of the crystal rather than the three-dimensional structure. The Al–Si framework consists of singly connected four-membered rings, linked to form five- and eight-membered rings, with a single Ca cation in the main channel that runs parallel to *a*, and is coordinated by four framework oxygens on one side and four H_2O molecules on the other side. Two of the water sites are fully occupied, whereas the other two have an occupancy of <20%. Perfect Al–Si order is found for Iceland samples, and only partial ordering is found for Yugawara, Japan, material. *MM 51:615(1987)* correlated optical symmetry to Si–Al ordering and found three possible variations: (1) crystals are monoclinic, (2) crystals are triclinic, and (3) crystals are monoclinic and triclinic (sectors alternate). Low-temperature hydrothermal yugawaralite is monoclinic–triclinic because unstable atom positions are frozen in place during growth. At higher temperature, atoms can reach a stable position and homogeneous structure. Thin, tabular, lathlike blades, elongate on c; broad {010} pinacoid, {001}, {010}, {100}, {011}, {120}, {110}, and {$\bar{1}$11} all common; usually less than 10 mm long, exceptional crystals to 6 cm long. Colorless, white, pink; white streak; vitreous luster, pearly on {010}. Cleavage {$\bar{1}$04}, {001}, distinct; {$\bar{1}$01}, imperfect; {010}, parting; conchoidal fracture; brittle. $H = 4\frac{1}{2} - 5\frac{1}{2}$. $G = 2.19-2.25$. IR, DTA, TGA data: *AM 56:1699(1971)*, *MJJ 8:456(1977)*, *NZ (1985)*. Insoluble or nearly so in hot concentrated HCL, HNO_3, H_2SO_4, aqua regia, and water. Only slightly attacked by concentrated hot KOH. Soluble in HF. Decomposes readily in KOH and Na_2CO_3 fusions. Is piezoelectric and pyroelectric (i.e., no center of symmetry). Nonmagnetic and nonfluorescent. Always Ca-rich, with trace to minor amounts of Na, K, Mg, Fe, and Sr. Biaxial (−) and (+), has anomalous optical properties; type material has: $N_x = 1.496$, $N_y = 1.497$, $N_z = 1.504$; $2V = 78°(+)$; $r > v$, weak; other samples of Yugawara material have: $N_x = 1.493$, $N_y = 1.499$, $N_z = 1.503$; $2V = 72°(-)$. Dispersion $r < v$ distinct (horizontal dispersion). Total range of indices is: $N_x = 1.492-1.500$, $N_y = 1.497-1.505$, $N_z = 1.501-1.512$; $2V_x = 65-82°$; $Y \wedge c = -6$ to $-14°$; $X \wedge a = -9$ to $+7°$; $Z = b$ or $Z \wedge b = 1°$. Occurs in natural geothermal systems (active or past). Hosts can be rhyolite, dacite, or basalt. Associated with other zeolites, such as wairakite and calcite. In drill core from Lower Geyser Basin, Yellowstone National Park, WY; Chena Hot Springs, Chena R., E of Fairbanks, AK; Yellow Lake, near Olalla, BC, Canada; at least three localities in Iceland, two are Heinabergsjökull and Hvalsgod, Hvalfjord; at least two localities on Sardinia: Osilo, Sassari, Sardinia, Italy; Khandivali quarry, near Bombay, MAH, India; Takitimu Mountains, South Island, New Zealand's Nukabira district; central Hokkaido, Japan; Shimoda district, and in the Toi and Seikoshi mines, Shizuoka Pref., Japan; *Yugawara Hot Springs, Kanagawa Pref., Honshu, Japan*; also at Onikobe, Miyagi Pref., in the Tanzawa Mts. in Shizuoka Pref., also at Kakkonda Hot Springs, and Takinoue, Iwate Pref. EF *AC(B) 25:1183(1969)*, *ZK 130:88(1969)*, *NZ (1985)*, *ZK 174:265(1986)*, *ZW (1992)*.

77.1.7.3 Goosecreekite $Ca[Al_2Si_6O_{16}] \cdot 5H_2O$

Named in 1980 by Dunn et al. for the locality. Zeolite group. MON (pseudo-ORTH) $P2_1$. $a = 7.401-7.52$, $b = 17.439-17.56$, $c = 7.293-7.35$, $\beta = 105.73°$, $Z = 2$, $D = 2.23$. *35-469*(nat): 7.19_5 5.59_5 4.91_5 4.53_{10} $3.53_{2.5}$ 3.35_4 $3.28_{2.5}$ $3.07_{2.5}$. Al–Si framework has (010) layers that resemble those of brewsterite but are cross-linked by Si tetrahedra (sharing two vertices with each adjacent layer) to form a three-dimensional framework. Si and Al are ordered in the framework, with only a few sites showing an exchange of Al and Si. Details given in *AM 71:1494(1986)*. Crystals equant or elongated on b, similar to epistilbite, distinguished by the presence of highly curved and twisted faces, {100}, {010}, {001}; crystals to 4 cm long, aggregates from India to 6 cm long,

10 cm in diameter; no twinning shown. Colorless, milky white; white streak; vitreous–pearly luster. Cleavage {010}, perfect. H = $4\frac{1}{2}$. G = 2.16–2.23. Insoluble in HCl or HNO_3. Only material from VA has been analyzed. No other exchangeable cations besides Ca. Uniaxial (−); Y parallel to b, $c \wedge Z = 46°$, X and Z lie in the cleavage plane; $N_x = 1.495$, $N_y = 1.498$, $N_z = 1.502$; 2V = 82°. Occurs in basalt–diabase and granitic pegmatites, associated with minerals such as actinolite, chlorite, epidote, quartz, prehnite, stilbite, albite, and apophyllite. *Goose Creek quarry, Leesburg, Loudon Co., VA;* Jensen quarry, Riverside, Riverside Co., CA; Oberbaumühle quarry, Windischeschenbach, Bavaria, Germany; Pandulena quarry, Nasik, MAH, India (crystals to more than 4 cm long and hemispheres to 8 cm in diameter); also from several other nearby localities. EF *CM 18:323(1980), NZ (1985), ZW (1992).*

77.1.7.4 Roggianite $Ca_2[Be(OH)_2Al_2Si_4O_{13}] \cdot 2.5H_2O$

Named in 1969 by Passaglia for Aldo G. Roggiani, teacher of natural sciences, who found the mineral. Zeolite group. Previously called ginzburgite. TET $I4/mcm$. $a = 18.366$–18.377, $c = 9.164$–9.187, Z = 1, D = 2.12. *39-366 [MZ 8(4):85(1986)]* (nat): 12.96_{10} $9.16_{4.5}$ 6.12_1 5.80_1 3.60_1 3.41_4 3.20_2 3.15_2; *MM 52:201(1988)*: 12.99_{10} 9.18_5 6.12_4 5.81_2 3.60_4 3.41_7 3.20_3 3.15_4. R = 6%. Characterized by an interrupted (Si,Al) framework with some OH vertices unshared by two tetrahedra; Ca cations and H_2O molecules lie in cavities where the framework is interrupted and in the largest channels, respectively. Be in Al tetrahedra. *NJBM: 307(1991)*. Elongated prisms with {100} {010}, {110}; long laths, 70–600 nm wide and 40–120 nm thick. Whitish yellow, colorless; white streak. Cleavage {100}, perfect. G = 2.02 (certainly too low) and 2.30. *MZ 8(4):85(1986)*. DTA data: *CLM 8:107(1969)*. TGA data: *MM 52:201(1988)*. Breaks down at > 850° to a feldspar. Analysis (wt %): SiO_2, 41.0; Al_2O_3, 18.7; CaO, 19.6; SrO, 0.02; Na_2O, 0.29; K_2O, 0.64 (Σ 80.34; Alpe Rosso, Italy) + $18.1H_2O$ and 3.2BeO; Pizzo Marcio: SiO_2, 44.1; Al_2O_3, 18.5; Fe_2O_3, 0.05; CaO, 20.0; SrO, 0.04; BaO, 0.09; K_2O, 0.02; H_2O, 17.2 (by difference). Roggianite is the first and, up to now, only zeolite with many tetrahedral sites occupied by Be. Uniaxial (+) and (−); $N_O = 1.527$, $N_E = 1.535$ (+) and $N_O = 1.526$, $N_E = 1.519$ (−) [*MZ 8(4):85(1986)*]; elongation = c. Occurs in coating fractures in a sodium feldspar dike cutting gneiss at *Alpe Rosso, Val Vigezzo, Novara Prov., Italy*. Also in another albitite dike 1 km away at Pizzo Marcio. Hydrothermal origin. Also occurs in desilicated pegmatites in the Ural Mts., Russia. EF *PM 48:15(1979), P5ICZ: 205(1980), NZ (1985).*

77.1.7.5 Ba Analog of Brewsterite   Ideally, $BaAl_2Si_6O_{16} \cdot 5H_2O$

Described in 1993 by Cabella et al. [*EJM 5:353(1993)*] and subsequently in the same year by Robinson and Grice [*CM 31:687(1993)*], but not named. Zeolite group. See brewsterite (77.1.7.1). MON $P2_1/m$. New York: $a = 6.780$, $b = 17.599$, $c = 7.733$, β = 94.47°, Z = 2; Italy: $a = 6.790$, $b = 17.581$, $c = 7.735$, β = 94.50(3)°, D = 2.52. *PD*(nat): 2.93_{10} 4.68_8 6.31_8 3.27_6 5.07_3

4.53_3 2.02_3 1.66_3. R = 5.1%. Structure like brewsterite (77.1.7.1). Platy aggregates of flattened pinacoidal crystals to 1 mm across (NY) and subparallel, tabular crystals to 3 mm and (Italy) colorless to pale pink prismatic crystals to 0.2 mm long, also radiating and parallel aggregates; prismatic to tabular on {010}, showing forms {100}, {010}, and {001}; twinning on {010}. Pale yellow to colorless, transparent. Cleavage {010}, perfect; conchoidal fracture; brittle. H = 4. G = 2.50. IR data: FTIR spectrum is like that of brewsterite; NY material fluoresces bright yellow-green in SWUV. New York material is zoned with respect to Ba and Sr. Analysis (wt %), most Ba-rich zone: BaO, 19.1; SrO, 1.99; CaO, 0.06; K_2O, 0.08; Na_2O, 0.06; Al_2O_3, 15.49; SiO_2, 52.48; Σ = 89.28 with empirical formula of: $(Ba_{0.85}Sr_{0.13}Ca_{0.01}K_{0.01})_{\Sigma 1.01}(Al_{2.07}Si_{5.95})$ O_{16} · $5H_2O$. Italian material has: $(Ba_{1.80}Mg_{0.08}Na_{0.07}K_{0.04}Ca_{0.03}Sr_{0.02})_{\Sigma 2.04}$ $(Si_{11.78}Al_{4.29}Fe_{0.02})_{\Sigma 12.09}O_{32}$ · $9.82H_2O$. Biaxial (+); Na(D) (New York): $N_x = 1.513$, $N_y = 1.517$, $N_z = 1.527$; nonpleochroic, 2V = 57°, $2V_{calc} = 64°$; (Italy): $N_x = 1.514$, $N_y = 1.516$, $N_z = 1.528$; 2V = 45°, $2V_{calc} = 44°$, X ∧ c = 36° (β obtuse) and Z parallel to b; no dispersion. A late-stage hydrothermal mineral that occurs in New York in cavities in massive prehnite at a wollastonite mine. Also associated with quartz, diopside, calcite, and microcline. In Italy, occurs in fractures in radiolarian metacherts and schistose metapelite. *Gouverneur Talc Co. No. 4 wollastonite quarry (Valentine property), Harrisville, Diana, Lewis Co., NY; Cerchiara mine, Faggiona, La Spezia, eastern Liguria, Italy.* EF

77.1.8.1 Bellbergite $(K,Ba,Sr)_2Sr_2Ca_2(Ca,Na)_4Al_{18}Si_{18}O_{72}$ · $30H_2O$

Named in 1993 by Rüdinger et al. for the locality. Zeolite group. HEX Space group unknown, possibly $P6_3/mmc$, $P\bar{6}2c$, or $P6_3mc$. a = 13.244, c = 15.988, Z = 1, D = 2.19. *45-1482*(nat): 9.25_4 6.58_8 3.80_{10} 2.95_7 2.70_5 2.50_5 2.21_7 1.83_5. *MP 48:147(1993)*. Zeolite structure type EAB. Not completely ordered. Six-fold rings of TO_4-tetrahedra form layers perpendicular to c. The stacking sequence of the layers is ABBACC. *MP 48:147(1993)*. Dipyramids to 0.3 mm, elongate on [0001], {10$\bar{1}$2} dipyramid; no twinning. Colorless-white, white streak, vitreous luster. No cleavage observed, conchoidal fracture. H = 5. G = 2.20. Analysis (wt %): SiO_2, 34.4; Al_2O_3, 27.9; BaO, 1.22; Na_2O, 0.70; K_2O, 1.95; SrO, 7.63; CaO, 9.31 (Σ 83.12). Uniaxial (−); $N_O = 1.522$, $N_E = 1.507$; colorless, nonpleochroic. Occurs in Ca-rich xenoliths in leucite tephrite lava. In thin marginal contact zones. *Bellberg volcano, Mayen, Laacher See area, Eifel, Germany.* EF

77.1.8.2 Tschernichite $Ca[Al_2Si_6O_{16}]$ · $8H_2O$ to $(Ca,Na,K,Mg)Al_{1.67}Si_{6.33}O_{16}$ · $4H_2O$

Named in 1991 by Smith et al. for Rudy Warren Tschernich (b.1945), mineralogist and zeolite specialist, who found the mineral in 1971. Zeolite group. TET [or pseudo-TET (some is MON from optics)]. $P4/mmm$ (most probable). a = 12.880, c = 25.020, Z = 8, D = 2.12. PD(nat): 12.5_1 11.6_3 4.03_{10} 3.57_1 3.16_2 3.06_1 2.73_1 2.11_2. *AM 78:882(1993)*. The aluminosilicate framework is the natural analog of synthetic zeolite beta. The distribution of Al and Si in tschernichite is disordered in two different arrangements, in approximately equal amounts, that correspond to the ordering in the polymorphs A + B of zeolite β. An intergrowth of an enantiomorphic pair having the tetragonal space groups $P4_122$ and $P4_322$, with $a = 12.447$ and $c = 26.56$ (polymorph A) and a triclinic polymorph (B) with space group $P\bar{1}$, with $a \approx b = 12.4$, $c = 14.5$, $\alpha \approx \beta = 73°$, and $\gamma = 90°$. *JCS 6:363(1991)*. A tschernichite-type mineral has been reported from Antarctica and the mineral shows significant differences in the X-ray diffraction pattern from Goble tschernichite, mainly in the low 2θ region, suggesting a different ratio of the two frameworks found in the Goble tschernichite and/or the highly probable presence of at least one other distinct beta-type framework in the Antarctic tschernichite. Generally elongate on c, steep, lustrous, tetragonal dipyramids, radiating to blocky or stout; forms: {302}, {502}, {101}, {001}, sometimes terminated with {001}, 0.1- to 1-mm hemispheres to drusy linings 1–2 mm thick, largest crystal was 1 cm long, single crystals from <0.1 mm to 1 cm, most are <3 mm in length; twinning on {302}, {101}, {304}, multiple twins on {101}. White–colorless, white streak, vitreous luster, transparent–translucent. No cleavage, no parting, conchoidal fracture, brittle. H – $4\frac{1}{2}$. G = 2.02. Sometimes fluorescent, a very pale yellow under LWUV and SWUV. Analysis (wt %): SiO_2, 54.1, 65.8; Al_2O_3, 15.4, 14.4; CaO, 8.3, 6.6; FeO, 0.3, N.D.; MgO, 0.5, N.D.; Na_2O, 0.2, N.D.; H_2O, 22.7, 13.15. $Ca_{0.97}Na_{0.05}Mg_{0.08}Al_2Si_6O_{16} \cdot 8H_2O$ (large crystal) and $Ca_{0.73}Na_{0.11}K_{0.02}Al_{1.69}Si_{6.33}O_{16} \cdot 4H_2O$ (druse). Uniaxial (−), also biaxial (−); $N_O = 1.484$, $N_E = 1.483$; 2V = 0–35°; length-fast, and commonly twinned. First zeolite to form in its paragenesis. Occurs in vesicular olivine basalt only in an area 2 m × 3 m and 1 m thick. Associated with other zeolites. May be associated with boggsite. A drusy coating. *Neer Road pit, Goble Creek, N of Goble, Columbia Co., OR*; Santiago de Puriscal, SW of San Jose, Costa Rica; Mt. Adamson, northern Victoria Land, Antarctica. EF *ZW (1992)*, *MP VII6:4(1992)*, *EJM 7:1029(1995)*.

77.2.1.1 Bikitaite $LiAlSi_2O_6 \cdot H_2O$

Named in 1957 by Hurlbut for the locality. Zeolite group. TRIC $P1$ (pseudo-MON). $a = 8.607$, $b = 4.954$, $c = 7.597$, $\alpha = 89.90°$, $\beta = 114.44°$, $\gamma = 89.99°$ [Z *9:303(1989)*], Z = 2, D = 2.28. *14-168:* 7.87_8 6.93_5 4.37_4 4.20_9 3.46_{10} 3.37_{10} 3.28_4 2.48_9. Originally stated to be monoclinic but recent work (both X-ray and neutron diffraction) has shown the mineral to be triclinic. Complete Al–Si order results in triclinic symmetry. If only partial Al–Si order is shown to exist, the space group would be $P2_1$. Al–Si framework consists of five-membered rings linked by Al tetrahedra. The framework contains channels parallel to [010] which are occupied by tetrahedrally coordinated Li atoms and water molecules. Several millimeters to 6 cm long, pseudo-orthorhombic prisms flattened on {001}, {100}, or {$\bar{1}$01}, flattened blades are elongate along b, and highly striated parallel to b. Colorless to cream colored, vitreous luster. Cleavage {100}, perfect; {001}, good; conchoidal fracture; brittle. H = 6. G = 2.29–

2.34. DTA, TGA data: *AM 59:71(1974)*. Biaxial (−); X ∧ $c = 28°$ in the obtuse angle, $Z = b$; $N_x = 1.509$–1.510, $N_y = 1.520$–1.521, $N_z = 1.522$–1.523; $2V = 45°$; positive elongation. The only lithium-bearing zeolite known at this time. It is like spodumene with the addition of water. Occurs in Li-rich pegmatites and forms prior to stilbite. Foote mine, Kings Mt., Cleveland Co., NC; *Bikita mine, Zimbabwe*. Also occurs at the El Hayat mine and Nolan mine at Bikita. EF *AM 53:1202(1968), NZ (1985), NJBM 6:241(1986), ZW (1992)*.

77.2.2.1 Lithosite $K_6Al_4Si_8O_{25} \cdot H_2O$

Named in 1983 by Khomyakov et al. from the Greek *lithos*, stone, because it contains the most abundant components of Earth's crust. Zeolite group. MON (pseudo-ORTH) $P2_1$. $a = 15.197$, $b = 10.233$, $c = 8.435$, $\beta = 90.21°$, $Z = 2$, $D = 2.54$. *37-457*(nat): 3.46_9 3.26_9 3.16_9 3.07_{10} 2.82_8 2.10_9 2.05_5 1.99_5. Structure: *DAN SSSR 291(6):1370(1986)*. Irregular grains to 3 mm across. Colorless, clear; white streak; vitreous luster. Conchoidal fracture. $H = 5\frac{1}{2}$. $G = 2.51$. IR data indicate presence of H_2O; under X-rays acquires a bright rose color which persists for at least 8 months, and then is pleochroic. X, Y, colorless; Z, bright rose. Easily decomposed by cold 10% HCl. Analysis (wt %): SiO_2, 50.0, 49.6; Al_2O_3, 20.7, 20.4; K_2O, 28.4, 28.0; H_2O LOI, 2.34. Biaxial (+); $Z = b$, Y near a, X near c; $N_x = 1.510$, $N_y = 1.513$, $N_z = 1.527$; $2V = 47°$. XY, nearly colorless; Z, pale rose. Occurs in veins of ultra-agpaitic pegmatite cutting nepheline syenite. In cavernous parts of drill cores. Associated with other alkalic minerals. *Vuonemiok R., southern Khibiny massif, Kola Penin., Russia*. EF *ZVMO 112:218(1983), NZ (1985), ZW (1992)*.

77.2.3.1 Mountainite $KNa_3Ca_2Si_8O_{20} \cdot 8H_2O$ or $(Ca,Na_2K_2)_2Si_4O_{10} \cdot 3H_2O$

Described in 1957 by Gard et al. and named for Edgar Donald Mountain (1901–1985), British mineralogist and geologist, Rhodes University, Grahamstown, South Africa. MON $P2_1/a$. $a = 13.51$–13.6, $b = 13.10$–13.2, $c = 13.51$–13.6, $\beta = 104°$, $Z = 4$ or 8 (depending on formula), $D = 2.37$–2.49. *25-676:* 13.1_8 6.69_4 4.67_8 4.18_6 2.94_{10} 2.80_7 1.97_7 1.72_6. Structure has not yet been solved in detail. The true b might be 6.55 Å rather than 13.1 Å. White, matted silky fibers in rosettes up to 2 mm, flattened prisms to 2 mm long (Lovozero), elongated on b; segregations to 2 cm; white streak; silky to vitreous luster. Cleavage (001), good. $H = 3$. $G = 2.36$–2.38. IR data: *DAN-ESS 210:837(1974)*. DTA data: *MM 31:611(1957)*. Analysis (wt %), Lovozero: SiO_2, 51.6; Na_2O, 10.15; K_2O, 4.43; CaO, 10.1; Al_2O_3, 2.3; H_2O, 17.8; South Africa: SiO_2, 58.5; CaO, 13.4; MgO, 0.2; Na_2O, 7.9; K_2O, 6.0; H_2O, 13.4. Biaxial (+); $Y = b$, $Z \wedge c = 18°$; $N_x = 1.500$–1.509, $N_y = 1.505$–1.513, $N_z = 1.513$–1.519; 2V medium to large. Occurs with rhodesite in serpentinized kimberlite; also as a late hydrothermal mineral in natrolite and aegerine in alkaline pegmatite, and with zeolites in vugs of a melilite–nephelinite. May form pseudomorphs after eudialyte and lomonosovite. Occurs in volcanics at Höwenegg, Hegau, Germany; in the Jubilee pegmatite body on Mt. Karnasurt, Lovozero massif, Kola Penin., Russia; and in the *Bultfontein kimberlite pipe, Kimberley, South Africa*. EF *DA 25:613(1974)*, *DAN-ESS 260:134(1983)*.

77.2.4.1 Partheite $Ca_2Al_4Si_4O_{15}(OH)_2 \cdot 4H_2O$

Named in 1979 by Sarp et al. for Erwin Parthé (b.1928), crystallographer at the University of Geneva. Zeolite group. See gismondine (77.1.3.1), a similar cell. Dimorph of lawsonite. MON $C2/c$. *SMPM 59:5(1979)*. $a = 21.59$, $b = 8.78$, $c = 9.31$, $\beta = 91.47°$, $Z = 8$, $D = 2.37$ and $a = 21.555$, $b = 8.761$, $c = 9.304$, $\beta = 91.55°$. *36-378[SMPM 59:5(1979)]*(nat): 10.79_{10} 8.12_8 6.10_7 3.74_5 3.60_4 3.19_4 3.05_3 2.90_3. Structure: *AC(A) 40:247(1984)*, *ZK 169:165(1984)*. Radial, fibrous crystals to 0.1–0.3 mm; no twinning observed. Colorless to white, colorless to white streak, vitreous luster. Cleavage {100}, {110}, distinct. $H = $ ca.4. $G = 2.39$–2.45. IR data: *ZVMO 111:209(1984)*, *SMPM 65:129(1985)*. Traces of Na_2O to 0.3% and $K_2O = 0.2\%$. Biaxial (+); $Y = b$, 23–30° extinction angle between X and c; $N_x = 1.547$–1.550, $N_y = 1.549$–1.552, $N_z = 1.559$–1.565; $2V = 48°$; $r > v$, moderate; negative elongation. Occurs about 7 km SE of the village of Doganbaba in the *Taurus Mts., Burdur Prov., southwestern Turkey*, in an ophiolite zone, in rodingitic rocks with prehnite, thomsonite, and augite. In a gabbro pegmatite at Denezhkin Kamen, Ural Mts., Russia. EF

77.2.5.1 Strätlingite $Ca_2Al_2SiO_2(OH)_{10} \cdot 2.25H_2O$

Named in 1976 by Hentschel and Kuzel for W. Strätling, who synthesized the compound in 1938. HEX-R $R\bar{3}m$. $a = 5.745$, $c = 37.77$, $Z = 3$, $D = 1.98$. *29-285*: 12.5_{10} 6.20_7 4.90_3 4.16_{10} 2.87_7 2.61_4 2.49_4 2.12_4; this pattern as well as the bulk chemistry bears a strong resemblance to that of vertumnite (73.1.6.1). Structure: *EJM 2:841(1990)*. The structure is reported to be related to the zeolite chabazite. Colorless to pale green, translucent. Cleavage {0001}, perfect; brittle. $G = 1.9$. Uniaxial (−); $N_O = 1.534$. Occurs as hexagonal plates < 1 mm in diameter, with nepheline, melilite, thomsonite, gismondine, and ettringite in a limestone inclusion in basalt at *Bellerberg Volcano, 2 km N of Mayen, Eifel, Germany*, and with tobermorite, ettringite, and vertumnite at Campomorto, Montalto di Castro, Lazio, Italy. AR *NJMM 326(1976)*, *AM 62:395(1977)*.

77.2.6.1 Gaultite $Na_4Zn_2Si_7O_{18} \cdot 5H_2O$

Named in 1994 for Robert Allen Gault (b.1943), Canadian Museum of Nature. ORTH $F2dd$. $a = 10.211$, $b = 39.88$, $c = 10.304$, $Z = 8$, $D = 2.52$. *PD*(nat): 6.35_{10} 5.36_2 4.96_3 3.24_6 3.17_4 3.14_4 3.00_2 2.82_3. Structure shows no chemical disorder. There are five tetrahedral sites, four with Si, one with Zn. Structure is similar to that of lovdarite. A zeolitelike mineral, because Zn is a framework cation. Euhedral, equant crystals to 0.5 mm in diameter, predominant forms are {010}, {110}, {$\bar{1}$10}, {021}, {151} + {$\bar{1}$51}. Colorless–pale mauve, white streak, vitreous luster. Cleavage {101} + {010}, perfect; {021}, poor; conchoidal fracture on some surfaces. $H = 6$. $G = 2.52$. IR data are similar to stilbite. *CM 32:855(1994)*. Bright apple-green fluorescence

77.2.8.1 Verplanckite

under SWUV. Does not readily dissolve in 1:1 HCl, but it is etched after prolonged exposure. Analysis (wt %): Na_2O, 16.6; ZnO, 19.2; SiO_2, 52.6; H_2O calc., 11.3. Biaxial (+); XYZ parallel to acb; Na light: $2V = 61°$, $2V_{calc} = 60°$; $N_x = 1.520$, $N_y = 1.521$, $N_z = 1.524$; nonpleochroic; r > v, weak to moderate for all directions except B×a. Occurs in a sodalite inclusion within Ne-sodalite syenite. Many associated alkaline minerals. *Poudrette quarry, Mt. St.-Hilaire, Rouville Co., QUE, Canada.* EF

77.2.7.1 Kalborsite $K_6Al_4Si_6O_{20}(OH)_4Cl$

Named in 1980 by Khomyakov et al. for the composition. TET $P42\bar{1}c$. $a = 9.851$, $c = 13.060$, $Z = 2$, $D = 2.48$ (*DAN SSSR ESS 252:131(1982)*); *PD 33-0999* (nat.): 3.44_8 3.26_5 3.08_{10} 2.94_8 2.79_9 2.24_5 2.08_5 1.17_5. The structure is a three-dimensional backbone of SiO_4 and AlO_4 tetrahedra. Channels are along c in which K atoms are seperated from each other by Cl atoms and isolated $B(OH)_4$ tetrahedra (*DAN SSSR 252:611(1980)*). Faint pinkish brown, transparent; 1–2 mm grains; vitreous luster, pearly on (110) cleavage planes; perfect {110} cleavage; $H = 6$; $G = 2.5$; IR data (*DAN SSSR ESS 252:131(1982)*); does not decompose in cold H_2O or 10% HCl. When heated to 600°C, the mineral becomes semitransparent and the R.I. decreases but XRD pattern is about the same. Uniaxial (+); $N_o \approx N_c = 1.525$; colorless, birefringence = < 0.001. Occurs on *Mount Rasvumchorr, Khibiny Massif, Kola Peninsula, Russia* in veins of rischorrite pegmatite associated with pectolite which forms 2–5 mm thick rims around circular segregations of lomonsovite. May have formed as a result of reaction of eudialyte and zirsinalite with hydrothermal solutions. In a lovozerite-rich pegmatite.

77.2.8.1 Verplanckite $Ba_{12}(Mn^{2+}Ti,Fe^{2+})_6(OH,O)_2[Si_4O_{12}]_3Cl_9(OH,H_2O)_7$

Named in 1965 by Alfors et al. for William E. VerPlanck (1916–1963) American geologist, California Division of Mines & Geology. HEX $P6/mmm$. $a = 16.398$Å, $c = 7.200$, $Z = 1$, $D = 3.46$ or 3.33; *PD 18-175* (nat.): 13.8_7 5.39_5 3.95_{10} 3.58_3 2.97_7 2.86_3 2.74_7 2.65_3. Structure: *AM 58:1103(1973)* and *AC 29:2019(1973)*. $R = 10\%$. Brownish orange, brownish yellow; streak pale orange; elongate parallel to c, radial masses of prismatic crystals to 3 mm long, also as disseminated grains in sanbornite-qtz rock, dominant forms are {10$\bar{1}$0} and {11$\bar{2}$0}; vitreous; cleavage = 1 good on (11$\bar{2}$0) plus a poor cleavage or fracture parallel to (0001); $H = 2\frac{1}{2}$–3; $G = 3.52$; IR data (*MZ 1:3(1979)*); no fluorescence; completely but slowly soluble in dil. HCl, quickly bleached and the white or transparent fragments are dissolved. Reacts the same in other acids, but the bleached fragments do not dissolve completely; no change in NaOH. Uniaxial (−); $N_o = 1.683$, $N_c = 1.672$; pleochroic with N_o = orange-yellow, and N_c = colorless; length fast. Occurs at or near the fringes of large granitic intrusives – metasomatic rocks; in metasediments surrounded by late Jurassic biotite-quartz diorite. *Esquire no. 7 mine, Big Creek and Rush Creek, Fresno Co., CA. Minerali 3(2):41(1981); AM 69:358(1984).*

Class 78

Unclassified Silicates

78.1.1.1 Cebollite $Ca_5Al_2(SiO_4)_3(OH)_4$

Named in 1914 by Larsen and Schaller for the locality. Probably ORTH Space group unknown. PD: 3.29_8 3.08_9 2.90_{10} 2.73_5 2.58_5 2.28_6 1.753_7 1.464_6. Structural data, composition, and classification uncertain. Fibrous masses. Colorless, white to reddish brown; vitreous luster; translucent. H = 5. G = 2.96. Biaxial (+); 2V = 58°; $N_x = 1.595$, $N_y = 1.60$, $N_z = 1.628$; parallel extinction, length-slow. Formed by deuteric alteration of melilite, with natrolite as an alteration of plagioclase-rich xenoliths and as a late-stage primary mineral in kimberlites. Found at *Iron Hill, Cebolla Creek, Gunnison Co., CO*; at Scawt Hill, County Antrim, Northern Ireland; in kimberlites at Letseng-La-Terai, Lesotho, and at the DeBeers mine, Kimberley, Cape Prov., South Africa. AR DT:607(1932), MM 43:583(1980).

78.1.2.1 Chantalite $CaAl_2(SiO_4)(OH)_4$

Named in 1977 by Sarp et al. for Chantal Sarp (b.1944), Swiss wife of its discoverer, Halil Sarp. TET $I4_1/a$. $a = 4.952$, $c = 23.275$, Z = 4, D = 2.97. 29-1410: 5.81_2 4.83_4 4.17_7 3.35_6 2.60_{10} 2.24_5 2.20_3 1.453_6. Structure: ZK 150:53(1950). Colorless to white, vitreous luster. G = 2.8–2.9. Uniaxial (−); $N_O = 1.653$, $N_E = 1.642$. Found as anhedral, 0.1- to 0.3-mm grains with vuagnatite, prehnite, hydrogrossular, chlorite, and calcite in rodingitic dikes of the ophiolite zone at *Doganbaba, Burdur, Taurus Mts., southwestern Turkey*. AR SMPM 57:149(1977), MA 78:3469(1978).

78.1.3.1 Cymrite $BaAl_2Si_2O_8 \cdot H_2O$

Described and named in 1949 for Cymru, the old Welsh name for Wales. MON (pseudo-HEX) $P2_1$. $a = 5.33$–5.35, $b = 36.6$–37.0, $c = 7.67$–7.71, $\beta = 90.0°$, Z = 8, D = 3.45. CA 97:58727(1982), K 20:280(1975). PD(nat, PDF 17-507): 7.71_5 3.96_9 2.96_{10} 2.67_7 2.24_4 2.21_4 1.925_3 1.85_4; PD(syn): 3.96_{10} 2.95_4 2.67_7 2.31_2 2.24_3 2.19_2 1.92_2 1.59_2. Platy, with {001} dominant; {010}, {100}, {140}, and {120} also present. Often twinned. Structure: K 20:280(1975), CA 97:58727(1982). As much as 2 mm across. Usually white, but may be colorless, green, or brown; satiny–vitreous luster. Cleavage {001}, good; perpendicular to {001}, fair; irregular fracture. H = 2–3. G = 3.41. IR data: MM 46:63(1982); AM 49:158(1964). Slightly soluble in HCl and other acids.

Usually fairly pure with traces of Ti, Fe, Na, K present. Minor to trace amounts of Ca substitute for Ba. Uniaxial or biaxial $(-)$; $2V = 0$–$5°$; $N_x = 1.611$, $N_y = 1.619$, $N_z = 1.621$, $N_O = 1.619$–1.622, $N_E = 1.611$–1.616; positive elongation; colorless. Most occurrences of this mineral are in bedded manganese ore deposits that have undergone low- and medium-grade (amphibolite facies) metamorphic rocks, often as part of Ba–Mn mineralization and associated with other Ba silicates and Mn minerals. It is stable at high pressures. Alters by dehydration to celsian. Celsian + H_2O = cymrite. Occurs at Pacheco Pass, San Benito Co., CA; Bonanza Creek, Brooks Range, and Ruby Creek, Bornite, AK; Navarana Fjord, Greenland; Aberfeldy, Scotland; *Bennalt mine, Rhiw, Carnarvonshire, Scotland*; Ireland; Arrens, Långban, Värmland, Sweden; High Pyrenees, France; La Zarza, Huelva, Spain; Blazna Valley, Romania; Andros Island, Greece; Pribaikal, and Saureyskoye deposit, Polar Ural Mts., Russia; Shiromaru mine, Tokyo, and Omiya mine, Saitama Pref.; Sarusaka, Kagami, and Niro mines, Tosayamada, Kochi Pref., Shikoku, Japan; Kalahari manganese field, N of Shishen, Cape Prov., South Africa; northwestern Nelson, South Island, New Zealand. EF *MM 28:676(1949)*.

78.1.4.1 Dellaite $Ca_6Si_3O_{11}(OH)_2$

Named in 1965 by Agrell for Della M. Roy (b.1926), Pennsylvania State University geochemist, who synthesized the phase. TRIC $P\bar{1}$. $a = 6.815$, $b = 6.934$, $c = 12.882$, $\alpha = 90.67°$, $\beta = 97.70°$, $\gamma = 98.14°$, $Z = 2$, $D = 2.97$. *29-376*(syn): 6.90_2 3.44_7 3.35_2 3.26_2 3.07_5 2.99_4 2.82_3 2.29_{10}. The natural phase is not well characterized. Structure: *AM 43:1009(1958)*. Found as minute elongated [100] grains twinned on {010}. Colorless, vitreous luster, transparent. Biaxial $(-)$; $2V < 30°$: $N_x = 1.650$, $N_y = 1.661$, $N_z = 1.664$; $Z' \wedge [100] = 20°$; also $2V = 65°$: $N_x = 1.650$, $N_y = 1.657$, $N_z = 1.660$. Found in late-stage veins cutting a metamorphic assemblage at *Kilchoan, Ardnamurchan, Scotland*. AR *MM 34:1(1965)*.

78.1.5.1 Jinshajiangite $(Na,K)_5(Ba,Ca)_4(Fe,Mn)_{15}(Ti,Fe^{3+}Nb,Zr)_8$ $(SiO_4)_{15}(F,O,OH)_{10}$

Named in 1982 for the locality. According to *CM 29:355(1991)*, the mineral is the Fe-dominant analog of perraultite (78.1.5.2). A recalculation of the formula yields $Na_2KBaCa(Fe,Mn)_8Ti_4Si_8O_{32}(OH,F,H_2O,OH)_6$. $Mn^{2+} \leftrightarrow Fe^{2+}$, and $2Nb \leftrightarrow 2Ti + Ca$ are the main substitutions. MON $C2/m$, Cm, or $C2$. $a = 10.732$, $b = 13.847$, $c = 20.817$, $\beta = 95.05°$, $Z = 2$, $D = 3.56$. *35-722*(nat): 10.2_7 4.4_4 3.44_{10} 3.15_8 2.85_7 2.63_7 2.57_8 1.72_5. Thin tabular prismatic crystals to 2×20 mm, elongation [001]. Blackish red, brownish red, or golden red; light yellow streak; vitreous luster. Cleavage {010}, {100}, perfect. H = 430 kg/mm². $G = 3.61$. Infrared, Mossbauer, DTA, and TGA data: *G(C) 1:458(1982)*. Analysis (wt %): SiO_2, 27.10; TiO_2, 15.90; $(Zr,Hf)O_2$, 0.70; Fe_2O_3, 1.64; Nb_2O_5, 1.03; BaO, 9.80; FeO, 19.07; MnO, 12.93; CaO, 2.94; K_2O, 2.30; Na_2O, 3.15; F, 2.65; minor: Al_2O_3, 0.36; Ln_2O_3, 0.30; Ta_2O_5, 0.07; SrO, 0.08; MgO, 0.28; H_2O^+, 0.33; H_2O^-, 0.36. Biaxial $(+)$; $c \wedge X$

$= 13°$; $N_x = 1.729$, $N_y = 1.802$, $N_z = 1.852$; $2V_{rep} = 72°$, $2V_{calc} = (-)76°$; $r < v$; pleochroic. X, light golden yellow; Y, brownish yellow; Z, brownish red. Occurs in an albite dike, associated with arfvedsonite and other alkalic minerals at a locality on the *Jinshajiang R., Sichuan Prov., China.* EF

78.1.5.2 Perraultite $Na_2KBaMn_8(Ti,Nb)_4Si_8O_{32}(OH)_5 \cdot 2H_2O$

Named in 1991 by Chao for Guy Perrault (b.1937) of the Ecole Polytechnique, Montreal. See jinshajiangite (78.1.5.1); these two minerals are probably analogs. MON $C2/m$, Cm, or $C2$. $a = 10.820$, $b = 13.843$, $c = 20.93$, $\beta = 95.09°$, $Z = 4$ [*CM 29:355(1991)*], $D = 3.81$. *45-1346*(nat): 10.4_4 3.47_{10} 3.19_2 2.87_1 2.79_1 2.61_4 2.08_2 1.73_1. Bound by $\{001\}$, $\{010\}$, $\{100\}$, and possibly $\{101\}$; elongate [100], flattened on $\{010\}$, prismatic to tabular, wedge-shaped terminations; to 1 mm; "swallowtail" contact twins common, twin plane and contact plane = $\{001\}$; has a pronounced subcell with $A2/m$, $A2$, or Am. Has $a' = a/2$, $b' = b/2$, $c' = c$, $\beta' = \beta$, $Z = 1$. Orange-brown; vitreous, slightly waxy luster; transparent to translucent. Cleavage $\{001\}$, perfect; uneven to irregular fracture; very brittle. $H = 4$. $G = 3.71$. IR and TGA data: *CM 29:355(1991)*. Analysis (wt %): Na_2O, 3.52; K_2O, 2.68; MgO, 0.06; MnO, 31.14; FeO, 1.12; BaO, 8.88; Al_2O_3, 0.03; SiO_2, 27.32; TiO_2, 9.44; ZrO_2, 0.12; Nb_2O_5, 13.35; H_2O, 3.49; F, 0.84; O = F, 0.35. Biaxial $(-)$; $X = b$, $Z \wedge c = 9.6°$ in acute angle β, $Y \wedge a = 19°$ in obtuse angle β; Na(D): $N_x = 1.785$, $N_y = 1.81$, $N_z = 1.82$; $2V = 66°$; $r \ll v$, strong; pleochroic. X,Y, light yellow; Z, dark orange-brown. Occurs at the *DeMix quarry, Mt. St.-Hilaire, QUE, Canada*; in pegmatite dikes with microcline, aegirine, albite, astrophyllite, analcime, rhodochrosite, catapleiite, natrolite, tetranatrolite, ancylite-(Ce), pyrochlore, and kupletskite. EF *MR 21:284(1990)*.

78.1.6.1 Kurumsakite $(Zn,Ni,Cu)_8Al_8V_2^{5+}Si_5O_{35} \cdot 27H_2O$

Named in 1954 by Ankinovich for the locality. An inadequately described mineral. Probably ORTH Space group unknown. *29-571*(nat): $4.91_{2.5}$ $3.91_{7.5}$ 2.61_5 2.42_1 2.28_4 1.99_1 1.53_{10} $1.42_{2.5}$. *IANKSG 19:116(1954)*. Greenish to bright yellow, radiating fibrous to felted masses; vitreous–silky luster. $G = 4.03$. Biaxial $(+)$; $N_x = 1.616$, $N_y = 1.617$, $N_z = 1.622–1.623$; $2V = 35°$; birefringence = 0.006; parallel extinction; positive elongation. Analysis of type material [*AM 42:583(1957), MA 13:207(1957)*] (wt %) SiO_2, 13.82; Al_2O_3, 20.51; Fe_2O_3, 2.15; V_2O_5, 8.50; ZnO, 17.55; CuO, 3.05; NiO, 7.33; MgO, 0.92; CaO, 1.24; H_2O, 23.25; SO_3, 1.15. Found in fractures in bituminous schists. *Kurumsak, near Dzhambul, Karatau Mts., Kazakhstan.* EF

78.1.7.1 Plumbotsumite $Pb_5Si_4O_8(OH)_{10}$

Named in 1982 for the composition and the locality. ORTH $C222_1$. $a = 15.875$, $b = 9.261$, $c = 29.364$, $Z = 10$, $D = 5.56$. *35-0858*: 14.7_7 7.99_6 4.63_5 3.65_{10} 3.12_8 2.47_7 1.633_6 1.476_6. Thick tabular $\{001\}$; irregular skeletal grains up to 1 mm $\times$ 1 mm $\times$ 0.5 mm. Twinned. Cleavage $\{001\}$, perfect. Colorless, adamantine luster, transparent. $H \approx 2$. $G = 5.6$. Biaxial $(-)$; $2V = 32°$; $N_x = 1.922$,

$N_y = 1.933$, $N_z = 1.938$; $XYZ = cba$; $r > v$. On alamosite and encrusted by melanotekite at *Tsumeb mine, Namibia*. AR *CE 41:1(1982)*; *AM 67:1075(1982)*.

78.1.8.1 Poldervaartite $Ca(Ca_{0.5}Mn_{0.5})[SiO_3(OH)](OH)$

Named in 1993 by Dai et al. for Arie Poldervaart (1918–1964), Professor of Petrology, Columbia University, New York City. ORTH *Pbca*. $a = 9.398$, $b = 9.139$, $c = 10.535$, $Z = 8$, $D = 2.90$. PD: 4.18_5 3.23_{10} 2.85_4 2.79_4 2.62_3 2.58_3 2.39_4 2.04_3. Structurally akin to $Ca_2(SiO_3OH)(OH)$ found in cement materials. Wheat-sheaf groups to euhedral prismatic crystals with {110}, {100}, {010}, and {001}. Milky white rims with colorless cores; translucent–transparent (cores); subvitreous luster, vitreous on fracture. Very brittle. $H = 5$. $G = 2.91$. Deep red fluorescence in SWUV. Biaxial (+); $2V = 65°$; $N_x = 1.634$, $N_y = 1.640$, $N_z = 1.656$; $XYZ = bac$; $r < v$, weak. Weakly pleochroic, $Z > Y > X$; colorless to bluish gray. Found in association with henritermierite and calcite, and coated with bultfonteinite and hematite in a pocket in massive braunite–hausmannite at the *Wessels mine, Kalahari manganese field, NW of Kuroman Hill, Cape Prov., South Africa*. AR *AM 78:1082(1993)*.

78.1.9.1 Stringhamite $CaCuSiO_4 \cdot H_2O$

Named in 1976 by Hindman for Bronson F. Stringham (1907–1968), University of Utah. MON $P2_1/c$. $a = 5.030$, $b = 16.135$, $c = 5.343$, $\beta = 102.96°$, $Z = 4$, $D = 3.359$. *29-318*: 8.05_1 4.88_2 4.69_2 3.93_3 3.24_4 2.77_{10} 2.52_7 1.614_2. Structure: *TMPM 34:15(1985)*. Flattened, pseudorhombohedra (0.1 mm) with {011}, {101} dominant, {010}, and {$\bar{1}$11}. Deep azure blue, light blue streak, vitreous luster, translucent to transparent. Cleavage {001}, good; {100}, imperfect. Biaxial (+); $2V = 80°$; $N_x = 1.709$, $N_y = 1.717$, $N_z = 1.729$. $X = b$, $Y \wedge c = 2.5°$. Pleochroic in blues; $Z > Y > X$. Occurs with other secondary and primary copper minerals in diopside–magnetite skarn at *Bawana mine, near Milford, Beaver Co., UT*; at the Christmas mine, Hayden, Gila Co., and in Pinal Co., AZ; at the Crestmore quarry, Riverside, CA; and the El Bronce mine, Tierra Amarilla, Chile. AR *AM 61:189(1976)*.

78.1.10.1 Vistepite $Mn_5SnB_2Si_5O_{20}$

Named in 1992 by Pautov et al. for Viktor Ivanovich Stepanov (1924–1988), Russian mineralogist and mineral collector. MON $P2/m$. $a = 28.77$, $b = 7.01$, $c = 13.72$, $\beta = 96.6°$, $Z = 7$, $D = 3.70$. PD(nat): 6.82_4 3.41_8 3.22_8 2.83_{10} 2.81_{10} 2.24_7 1.75_6 1.70_5. Structure not done. Radial aggregates to 15 mm, elongated on b; polysynthetic twinning parallel to elongation. Orange-yellow, vitreous luster, transparent. Three perfect cleavages, brittle. $H = 4\frac{1}{2}$. $G = 3.67$. IR and DTA data: *ZVMO 121:107(1992)*. Nonfluorescent. Traces of Al, Ca, Fe as well. Biaxial (−); $N_x = 1.696$, $N_y = 1.711$, $N_z = 1.715$; $2V = 57°$; nonpleochroic; OAP perpendicular to length; $r > v$, strong. Occurs in a rhodonite (plus quartz and galena) in biotite–quartz-bearing hornfels at the exocontact of the Inylchek Sn-bearing granite massif, at the *northern end of the Inylchek Range, Kirgizia.* EF

78.2.1.1 Ganomalite $Ca_5Pb_9MnSi_9O_{33}$

Named in 1876 by Nordenskiöld from the Greek *ganoma*, luster, in allusion to its vitreous to resinous appearance. HEX $P3$. $a = 9.82$, $c = 10.13$, $Z = 1$, $D = 5.69$. *25-150:* 4.92_7 4.43_7 3.53_9 3.38_8 3.06_{10} 2.71_7 1.985_7 1.797_7. The empirical $[Si_9O_{33}]$ silicate unit can be rendered as $[(SiO_4)_3(Si_2O_7)_3]$. Originally described as tetragonal. If the space group is given correctly, Mn must play the same role as Ca. Simple tabular crystals bounded by the prism {1010}; usually granular, massive or in curved plates. Colorless, white, gray; resinous to vitreous luster, somewhat pearly on plates; transparent to translucent. Cleavage {1010}, perfect; {0001}, less so. $H = 3$. $G = 5.74$. Uniaxial (+); $N_O = 1.910$, $N_E = 1.945$; or biaxial (+); 2V small. Easily fusible; gelatinized by HNO_3. Previously interpreted as the hydroxyl analog of nasonite. Found in tabular, colorless crystals at Franklin, Sussex Co., NJ; crystals of uncertain habit and massive at the *Långban mine, Värmland*, and at the Jakobsberg mine, N of Nordmarken, *Sweden*. AR *D6:422(1892), MR 10:47(1979), MM 49:579(1985)*.

78.3.1.1 Mathewrogersite $Pb_7(Fe^{2+},Cu^{2+})Al_3GeSi_{12}O_{36}(OH,H_2O)_6$

Named in 1986 by Keller and Dunn in honor of Mathew Rogers, first European prospector at Tsumeb, Namibia. HEX-R $R3$, $R\bar{3}$, $R32$, $R3/m$, or $R\bar{3}2/m$. $a = 8.457$, $c = 45.970$, $Z = 3$, $D = 4.68$. *40-476:* 15.3_7 7.68_6 4.08_5 3.26_{10} 2.86_5 2.77_6 2.03_7 1.762_6. Colorless, white, to pale greenish yellow; pearly luster; translucent. $H = 2$. $G = 4.7$. Uniaxial (−); $N_O = 1.810$, $N_E = 1.745$. The composition is close to that of the formula. Found at *Tsumeb, Namibia*, associated with queitite, alamosite, melanotekite, kegelite, schaurteite, etc. in corroded Pb–Zn ores of the lower oxidation zone. AR *AM 72:1025(1987)*.

78.3.2.1 Shafranovskite $(Na,K)_6(Mn,Fe)_3Si_9O_{24} \cdot 6H_2O$

Named in 1982 by Khomyakov et al. for Ilarion Ilarionovich Shafranovskii (b.1907), professor of mineralogy and crystallography, Leningrad Mining Institute. HEX $P3m1$ or $P31m$. $a = 14.58$, $c = 21.01$, $Z = 6$, $D = 2.78$. *37-453:* 10.8_3 10.5_{10} 3.60_4 3.51_7 3.16_3 2.98_5 2.79_6 2.67_3. Dark green, olive-green, yellow-green; vitreous luster. $H = 2$–3 (for grain aggregate). Uniaxial (−); $N_O = 1.587$, $N_E = 1.570$. Decomposed by dilute HCl; reacts with H_2O to give alkaline solution. Found as granular aggregates associated with thermonatrite, nacaphite, olympite, sidorenkite, and rasvumite in alkalic pegmatite at several localities in the Lovozero massif, and at *Mt. Rasvumchorr, Khibiny massif, Kola Penin., Russia*. AR *ZVMO 111:475(1982), AM 68:644(1983)*.

78.3.3.1 Zakharovite $Na_4Mn_5^{2+}Si_{10}O_{24}(OH)_6 \cdot 6H_2O$

Named in 1982 by Khomyakov et al. for E. E. Zakharov (1902–1980), director of the Moscow Geological Exploration Institute. HEX $P31m$. $a = 14.58$,

$c = 37.71$, $Z = 9$, $D = 2.67$. 39-405: 12.6_{10} 3.99_1 3.43_1 3.16_4 3.06_1 2.82_1 2.63_1 2.39_1. Yellowish to bright yellow, pearly to waxy luster. Cleavage {0001}, perfect; conchoidal fracture. $H = 2$. $G = 2.58$, 2.64. Uniaxial $(-)$; $N_O = 1.565$, $N_E = 1.535$. Found as platy aggregates with ussingite in veinlets cutting foyaite at *Mt. Karnarsut, Lovozero massif, and in the Yukspor and Koashva Mts., Khibiny massif, Kola Penin., Russia*, the several localities reported simultaneously. AR *ZVMO 111:491(1982)*, *AM 68:1040(1983)*.

78.4.1.1 Apachite $Cu_9Si_{10}O_{29} \cdot H_2O$

Named in 1980 by Cesbron and Williams for the Apache Indians, whose reservation is near the locality. Probably MON Space group unknown. $a = 12.89$, $b = 6.055$, $c = 19.11$, $\beta = 90.42°$, $Z = 2$, $D = 3.27$. 35-467: 12.9_{10} 10.6_4 9.56_4 7.66_5 4.49_4 4.17_4 3.17_7 2.89_4. Blue, translucent. $H = 2$. $G = 2.80$ (in poor agreement with D). Biaxial $(-)$; $2V \approx 15°$; $N_x = 1.610$, $N_y \approx N_z = 1.650$; Y approximately parallel to elongation. Spherules of radiating fibers associated with gilalite, kinoite, and apophyllite in a garnet diopside tactite zone at the *Christmas mine, Gila Co., AZ*. AR *MM 43:639(1980)*, *AM 65:1065(1980)*.

78.4.2.1 Ashburtonite $HPb_4Cu_4Si_4O_{12}(HCO_3)_4(OH)_4Cl$

Named in 1991 for the locality. TET $I4/m$. $a = 14.234$, $c = 6.103$, $Z = 2$, $D = 4.69$. *PD*: 10.2_{10} 5.64_7 4.50_{10} 3.33_{10} 3.01_9 2.81_3 2.61_5 2.01_3. Prismatic, elongated on [001] with {110}, {100}, {001}, and {301} present. 30×400 µm, in aggregates to 20 mm in diameter. Blue, pale blue streak, vitreous luster, transparent. No cleavage, conchoidal fracture, very brittle. $G > 4.09$. IR data: *AM 76:1701(1991)*. The mineral reacts in index liquids > 1.785, leaving an opaque residue. Decomposes rapidly in concentrated HCl, slowly in 0.2 M HCl, 1:1 HNO_3, and concentrated H_2SO_4. Uniaxial $(+)$; $N_O = 1.786$, $N_E = 1.800$ (sodium light). In a secondary Pb–Cu mineral assemblage in a weathered shear zone that cuts a series of shales and greywackes. *Anticline prospect, Ashburton Downs, WA, Australia*. EF *AM 76:1701 (1991)*.

78.4.3.1 Fukalite $Ca_4Si_2O_6CO_3(OH,F)_2$

Named in 1977 for the locality. ORTH Space group unknown. $a = 5.48$, $b = 3.78$, $c = 23.42$, $Z = 2$, $D = 2.78$. 29-308(nat): 5.86_3 3.90_2 3.60_2 3.08_9 2.93_7 2.85_{10} 2.34_3 1.76_3. Crystals to 0.2 mm long. Light brown, white, translucent. Analysis (wt %): SiO_2, 20.1; Al_2O_3, 0.55; CaO, 54.4; H_2O^+, 4.4; CO_2, 10.3; minor Fe_2O_3, 0.1; MgO, 0.14; Na_2O, 0.17; K_2O, 0.01; H_2O^-, 0.23; P_2O_5, 0.01; F, 0.32. H of aggregate is ≈ 4. $G = 2.77$. Biaxial $(\pm)$; $N_x = 1.595$, $N_y = 1.605$, $N_z = 1.626$; colorless, $2V = 90°$; negative elongation. Occurs as an alteration product of spurrite, and is associated with calcite $\pm$ xonotlite. In spurrite–gehlenite skarns. *Bicchu, Fuka, Okayama Pref., Mihara, Okayama Pref., and Kushiro, Hiroshima Pref., Japan*; *Cerro Mazahua, Zitacuaro, MICH, Mexico*. EF *MJJ 8:374(1977)*.

78.4.4.1 Gilalite $Cu_5Si_6O_{17} \cdot 7H_2O$

Named in 1980 by Cesbron and Williams for the locality. MON $a = 13.38$, $b = 19.16$, $c = 9.026$, $\beta = 90°$, $Z = 4$, $D = 2.54$. *35-468:* 13.4_{10} 11.0_3 7.79_5 6.68_3 4.79_5 3.90_4 3.32_3 3.15_3. Green to blue-green, translucent. $H = 2$. $G = 2.82$ (in poor agreement with D). Biaxial (−); 2V small; $N_x = 1.560$, $N_y \approx N_z = 1.635$. Found abundantly as spherules of radiating fibers (0.2 mm) with apachite, kinoite, and apophyllite in a garnet–diopside tactite zone at the *Christmas mine, Gila Co., AZ*. AR *MM 43:639(1980)*.

78.4.5.1 Grumantite $NaHSi_2O_5 \cdot H_2O$

Named in 1987 by Khomyakov et al. from the old Russian name for the Spitzbergen Archipelago, Arctic Ocean. ORTH *Fdd2*. $a = 15.979$, $b = 18.25$, $c = 7.16$, $Z = 16$, $D = 2.264$. *41-1331:* 6.20_5 6.05_5 4.46_5 3.51_{10} 3.35_2 3.09_5 3.01_{10} 2.23_2. The poorly described structure is stated to have four-tetrahedron spiral chains parallel to [001], suggesting that the formula be rendered as $Na_2[Si_4O_6](OH)_6$. White, silky to vitreous luster, translucent. Cleavage {110}, perfect. $H = 4\frac{1}{2}$. $G = 2.21$. Biaxial (+); $2V = 85°$; $N_x = 1.494$, $N_y = 1.507$, $N_z = 1.523$; $XYZ = acb$; $r > v$, weak. Weak, blue-white fluorescence in UV. Found as fine-grained compact to porous veins with ussingite margins, and as 2- to 5-mm irregular crystals replacing makatite in veins containing kazakovite, tisinalite, nordite, and sodalite at *Mt. Alluaiev, Lovozero massif, Kola Penin., Russia*. AR *ZVMO 116:244(1987)*, *AM 73:440(1988)*, *ZK 185:612(1988)*.

78.5.1.1 Ajoite $(K,Na)Cu_7AlSi_9O_{24}[OH]_6 \cdot 3H_2O$

Named in 1958 for the locality. TRIC $P\bar{1}$ or $P1$. *AM 66:201(1981)*. $a = 13.637$, $b = 14.507$, $c = 13.620$, $\alpha = 107.45°$, $\beta = 105.45°$, $\gamma = 110.57°$, $Z = 3$. Also reported as MON $P2_1/m$ or $P2_1$. $a = 15.218$, $b = 24.712$, $c = 13.632$, $\beta = 92.91°$ [*MJJ 8:234(1976)*], but $Z = 7$, which is unlikely; $D = 2.951$. *35-0477*(nat): 12.3_{10} 4.08_1 3.38_1 3.06_1 2.52_1 2.46_1 2.26_1. Laths, plates or massive; bladed 0.01 mm × 0.1 mm × 0.4 mm elongated on c, flattened on (010); dominant form is {010}; {1$\bar{1}$0} and {100} much less prominent. Bluish green, translucent. Cleavage {010} for triclinic cell. $G = 2.96$. TGA data: *AM 66:201(1981)*. Readily decomposed by acids such as HCl and HNO_3, leaving a coherent white mass of hydrated silica; no reaction in NH_4OH. Contains minor MnO, FeO, CaO. Biaxial (+); $X = b$, $Z \wedge c = 15°$; $N_x = 1.550$(Na light), 1.565, $N_y = 1.583$(Na light), 1.590, $N_z = 1.641$(Na light), 1.650; $2V = 68–80°$. X, very light bluish green; Y,Z, brilliant bluish green. Occurs in massive fracture coatings, vein fillings, and in vugs, associated with shattuckite, quartz, sericite, pyrite, and others. Alteration: may form from shattuckite and also is replaced by shattuckite. Occurs at the *New Cornelia mine, Ajo, Pima Co., AZ*; fine specimens also as inclusions in quartz crystals in the Messina mining district, South Africa. EF *AM 43:1107(1958)*; *MR 14:283(1983), 22:187(1991)*.

78.5.2.1 Bementite $Mn_8^{2+}Si_6O_{15}(OH)_{10}$ (?)

Named in 1887 by Koenig for Clarence Sweet Bement (1843–1923), manufacturer of machine tools and a collector of minerals, Philadelphia, PA. Possibly related to and confused with caryopilite. MON Space group unknown. $a = 7.5$, $b = 9.8$, $c = 5.65$, $Z = 1$ (?), but also reported as ORTH $P2_12_12_1$. $a = 14.5$, $b = 17.5$, $c = 29.1$, $Z = 18$ (?), $D = 3.06$. *25-546:* 7.25_9 3.66_{10} 3.58_9 3.43_1 3.30_5 2.44_1 2.11_1 2.10_1. Structure not determined but suggested to be related to serpentine or friedelite. (Calculated Z for an assumed $G \approx 3.0$ and using any of the various formulations, does not agree well with the values reported above. AR). Fibrous to lamellar, in radiating or stellate masses; granular, massive. Gray, pale yellow to brown, darkening on exposure to light; waxy luster, pearly on cleavage; translucent. Cleavage {001}, perfect; {100}, {010}, less so. $H = 4$–6 (?), also reported as soft. $G = 2.9$–3.1. Uniaxial $(-)$ or biaxial $(-)$; $2V \approx 0°$; $N_x = 1.602, 1.624$, $N_y = N_z = 1.632, 1.650$; $X = c$. Weakly pleochroic. X, colorless; Y,Z, pale yellow. The composition has been variously formulated as $Mn_8Si_7O_{17}(OH)_{10}$, $Mn_8Si_6O_{15}(OH)_{10}$, and $Mn_5Si_4O_{10}(OH)_6$. Minor Mg, Zn, and Fe^{2+} are present. Found at *Franklin, Sussex Co., NJ*; at various localities in CA; the Olympic Mts., WA; at Pajsberg, Värmland, Sweden (caryopilite?); and at Broken Hill, NSW, Australia. AR *D6:704(1892)*, *DT:685(1932)*, *AM 56:416(1971)*.

78.5.3.1 Burckhardtite $Pb_2(Fe^{3+},Mn^{3+})Te^{4+}(AlSi_3O_{10})O_2(OH)_2 \cdot H_2O$

Named in 1979 by Gaines et al. for Carlos (Carl) Burckhardt (1869–1935), Swiss geologist working in Mexico. MON (pseudo-HEX) Cm or $C2/m$. $a = 5.21$, $b = 9.04$, $c = 12.85$, $\beta = 90°$, $Z = 2$, $D = 4.96$. *33-0730:* 12.8_9 3.78_9 3.11_{10} 2.60_7 2.26_3 2.24_3 1.840_6 1.590_5. Pseudohexagonal tabular {001} crystals. Violet-red to pale pink, pale red streak, adamantine/pearly luster. Cleavage {001}, perfect. $H \approx 2$. $G = 3.2$ for impure sample. Biaxial $(-)$; $2V \approx 0°$; $N_x = 1.82$, $N_y \approx N_z = 1.85$; $X = c$ Pleochroic; X, pale magenta; Y,Z, carmine. A secondary mineral intergrown with dickite and often with cores of the latter. *Moctezuma mine, Moctezuma, SON, Mexico*, often accompanied by moctezumite. AR *AM 64:355(1979)*.

78.5.4.1 Canasite $K_3Na_2Ca_4(Na,Ca)_2[Si_{12}O_{30}]F(OH)_3$ (mon) and $K_3NaCa(Na,Ca)_6[Si_{12}O_{30}]F_3(OH) \cdot H_2O$ (tric)

Described in 1959 by Dorfman et al. and named for the composition: Ca–Na–Si. MON $C2/m$. $a = 18.836$, $b = 7.244$, $c = 12.636$, $\beta = 111.76°$, $Z = 2$, $D = 2.67$. TRIC $P1$. $a = 10.094$, $b = 12.691$, $c = 7.240$, $\alpha = 90.0°$, $\beta = 111.02°$, $\gamma = 110.2°$, $Z = 1$, $D = 2.69$. 13-553(mon): 4.81_6 4.69_7 4.27_4 4.20_5 3.08_{10} 2.91_8 2.36_6 1.64_8; *45-1398* and *MZ 14:71(1992)* (tric): 8.74_4 4.70_5 4.21_4 3.08_2 $3.01_{2.5}$ 2.92_{10} 2.35_3 2.31_2. Additional powder data in *ZVMO 123:104(1994)*. Structure done three times on Khibiny material. A threefold increase in F in triclinic canasite is accompanied by changes in ordering of Ca and Na. *MZ 14:71(1992)*. Subhedral to anhedral grains to 3 cm or more across. Polysynthetic twinning; twin plane is at an angle of 8° to (100). Greenish yellow, gray, lilac-gray, greenish brown; triclinic material is lilac-gray, blue-gray, rarely greenish; white streak; vitreous luster. Cleavage (mon) (001), excellent; (100), perfect; (tric) (010), excellent; (100), perfect; 118° between cleavages. Splintery fracture, brittle. G = 2.68–2.71. IR and thermal data: *ZVMO 123:104(1994)*. Variations in chemistry cause a variation in symmetry. Contains minor Al_2O_3 (to 0.6 wt %), Fe_2O_3 (to 1.1 wt %), MnO (to 0.5 wt %), and MgO (to 0.4 wt %). Biaxial (−) or (+); (mon): Y = b; N_x = 1.533–1.534, N_y = 1.538–1.539, N_z = 1.541–1.543; $2V_x$ = 58°; (tric): X ∧ a = 34°, Y = c, Z ∧ b = 5°; N_x = 1.536, N_y = 1.539, N_z = 1.542; $2V_z$ = 65–76°, colorless. Occurs in alkalic pegmatites with fenaksite and lamprophyllite (Khibiny massif), in the Lilac Stone deposit of charoitite rocks, associated with tinaksite, charoite, and aegirine, in the alkalic Murun massif (Yakutia). Several generations of canasite occur. *Khibiny massif, Kola Penin., Russia*; Murun massif, Chary R., southern Yakutia (now Sakha Republic). EF *TMMAN 9:158(1959), MIN 3(3):376(1981), L&L:32–36(1981)*.

78.5.5.1 Erlianite $(Fe^{2+},Mg)_4(Fe^{3+},V^{3+})_2[Si_6O_{15}](O,OH)_8$

Named in 1986 by Feng and Yang for the locality. ORTH $Pm2_1n$ or $Pmmn$. a = 23.2, b = 9.2, c = 13.2, Z = 6, D = 3.11. PD: 11.5_{10} 3.05_5 2.89_6 2.61_6 2.52_5 2.42_3 2.31_3 1.56_5. Black, brown streak, silky luster, nearly opaque. Cleavage {001}, {100}, perfect. H = $3\frac{1}{2}$–4. G = 3.11. Biaxial (−); 2V = 56–59°; N_x = 1.667, N_y = 1.674, N_z = 1.679. Occurs as 1- to 2-cm lathlike aggregates of fibrous to platy crystals associated with quartz, magnetite, siderite, albite, stilpnomelane, minnesotaite, and deerite at the *Harhada iron deposit, near the Jining–Erlian railway, Inner Mongolia, China*. AR *MM 50:285(1986)*.

78.5.6.1 Karpinskite $(Mg,Ni)_2Si_2O_5(OH)_2$ (?)

Named in 1956 by Rukavishnikova for Alexander Petrovich Karpinsky (1846–1936), Russian geologist and president of the Russian Academy of Sciences. An inadequately described mineral. Probably MON Space group unknown. *42-1313*(nat): 11.0_{10} 7.71_4 4.76_4 3.75_4 3.32_2 2.72_2 2.55_3 1.555_4. *KV 2:124(1956)*. Apparently an 11-Å phyllosilicate. DTA has been reported. *AM 42:584(1957)*. Fine-grained veinlets to massive, microscopic plates or prisms (to 0.8 mm long). Greenish blue or blue to white, dull to greasy luster. Trace cleavage parallel to c. H = $2\frac{1}{2}$–3. G = 2.53–2.63. Visually similar to nepouite, dyed deep blue by benzadine. Biaxial (−); Z ∧ c =12°, 0–12° (second data set for nickeloan material with NiO = 28.07 wt%; N_x = 1.553, 1.570, N_z = 1.569, 1.594; birefringence = 0.016, 0.024; nonpleochroic; positive elongation. Analysis of type material (wt %): SiO_2, 47.55; Al_2O_3, 0.48; MgO, 17.56; NiO, 21.12 (also 28.07); CaO, 0.80; CuO, 0.01; H_2O^-, 3.50; H_2O^+, 6.50; LOI, 2.30. Found in veinlets in kerolitized serpentinite from the *Nizhne Tagilsk massif, Ural Mts., Russia*.

78.5.7.1 Molybdophyllite $Pb_9Mg_9Si_9O_{24}(OH)_{24}$(14 Å) or $Pb_3Mg_2Si_2O_7(OH)_4$ (18 Å)

Named in 1901 by Flink from the Greek for *lead* and *leaf*, in allusion to its composition and foliated habit. There are two structural forms of the mineral: 14 Å and 18 Å. HEX $P6_322$. (14 Å): $a = 9.371$, $c = 27.44$, $Z = 2$, $D = 4.98$. (18 Å): $a = 9.34$. $c = 36.79$, $Z = 10$, $D = 5.41$. *42-1384*(14 Å, nat): 14.2_8 8.16_4 4.69_4 3.07_6 2.70_7 2.68_7 2.41_7 1.77_{10}; *42-1385* (18 Å): 18.6_{10} 4.32_6 4.18_6 4.05_6 3.29_6 3.07_7 2.70_9 2.34_6. No structural information available. Irregular foliated masses. Light green, pale yellow to colorless; white streak; pearly luster; vitreous luster; transparent. Perfect basal cleavage. $H = 3–4$. $G = 4.72$. Analysis (wt %), 14 Å: SiO_2, 18.25; Al_2O_3, 0.32; MgO, 11.94; MnO, 0.18; CaO, 0.01; PbO, 59.47; BaO, 0.25; 18 Å: SiO_2, 12.4; Al_2O_3, 0.35; MgO, 7.98; MnO, 0.03; CaO, 0.01; PbO, 70.9; BaO, 0.13; Cl, 0.07; F, 0.05. Uniaxial $(-)$; $N_O = 1.815$, $N_E = 1.761$. Occurs with calcite and hausmannite at *Långban, Filipstad, Värmland, Sweden*. G. Charalampides, Ph.D. thesis, Dept. Geology, Stockholm Univ., Sweden, 1988. EF

78.5.8.1 Raite $(Na,Ca)_4(Mn,Ti,Fe)_3Si_8(O,OH)_{24} \cdot 9H_2O$

Named in 1973 by Me'kov et al. for the participants in the 1969–1990 voyage of the papyrus ship *RA* under the leadership of Thor Hyerdahl. ORTH $C222$. $a = 30.6$, $b = 5.31$, $c = 18.20$, $Z = 4$, $D = 2.32$. *25-1318*: 11.4_{10} 4.50_8 2.94_{10} 2.65_{10} 2.48_6 1.640_6 1.575_6 1.514_6. Red, reddish brown, brown, gold. Cleavage $\{100\}$, $\{010\}$, $\{001\}$, perfect. $H = 3$. $G = 2.39$. Biaxial $(+)$; $2V = 53°$; $N_x = 1.540$, $N_y = 1.542$, $N_z = 1.550$; $Z = c$. Pleochroic; $Z > Y > X$; golden brown to colorless. Occurs as radiating fiber aggregates. Rare in pegmatite with rhodochrosite, serandite, nenadkevichite, natrolite, etc., and in sodalite xenoliths with villiauminite, eudialite, serandite, ussingite, steenstrupine, vuonnemite, and lovozerite at Mt. St.-Hilaire, QUE, Canada. Found in alkali syenite pegmatite of the Lovozero massif with zorite, aegerine, mountainite, and natrolite at *Mt. Karnarsut, Kola Penin., Russia*. AR *ZVMO 102:54(1973)*, *ZM 58:1113(1973)*, *MR 10:99(1979)*.

78.5.9.1 Tungusite $Ca_4Fe_2Si_6O_{15}(OH)_6$

Named in 1966 by Kudryasheva for the locality. An inadequately described mineral. System and space group unknown. *19-231*: 4.17_8 3.58_6 3.12_8 3.01_8 2.79_2 1.818_{10} 1.605_3 1.570_3. Yellow-green to grass-green, somewhat pearly luster. Separable into very fine, elastic platelets. $H \approx 2$. $G = 2.59$. Uniaxial $(-)$ or biaxial $(-)$; $2V = 0°$; $N_O(?) = 1.586$. Found as plates up to 0.5 cm, forming crusts of radiating structure, associated with analcime, apophyllite, gyrolite, "zeolites," calcite, and quartz on the walls of amygdules in lava from the *right bank of the Tunguska R., 2 km above Tura, Siberia, Russia*. AR *AM 52:927(1967)*.

78.6.1.1 Hyalotekite

Named in 1877 by Nordenskiold from the Greek for *glass* and *to melt*, in allusion to its easy fusion to a clear glass. TRIC $I\bar{1}$. $a = 11.310$, $b = 10.995$, $c = 10.317$, $\alpha = 90.43°$, $\beta = 90.02°$, $\gamma = 90.16°$, $Z = 2$, $D = 3.829$. *19-572*: 7.70_6 4.32_5 3.81_7 3.53_8 3.45_{10} 2.94_8 2.30_6 2.14_6. Structure: *AM 67:1012(1982)*. The structure is an incomplete tetrahedral framework related to layer-structured gillespite. White to gray, vitreous luster, translucent. Two cleavages at $\approx 90°$, brittle. H = 5 – $5\frac{1}{2}$. G = 3.80–3.82. Biaxial (+); 2V = 15°; $N_x = 1.963$, $N_y = 1.963$, $N_z = 1.966$; OAP perpendicular to cleavage zone axis; r < v, strong, . Easily fusible. Insoluble in acids. Found as coarsely crystalline masses associated with barylite at *Långban mine, near Filipstad, Värmland, Sweden*, and in the Dara-i-Pioz massif, Alai Range, Tajikistan. *D6 422(1892), MM 58:285(1994)*.

78.6.2.1 Lovdarite $K_2Na_6[Be_4Si_{14}O_{36}] \cdot 9H_2O$

Named in 1973 by Men'shikov et al. from the Russian for *gift of Lovozero*. ORTH *Pma2* (also reported as $P2_12_12_1$, and as $P2_122$ with doubled *a* axis). $a = 39.576$, $b = 6.931$, $c = 7.153$, $Z = 2$, $D = 2.34$. *12-1305*: 6.56_6 4.96_9 3.52_4 3.29_{10} 3.14_{10} 2.48_6 2.29_6 1.785_5. Structure: *EJM 2:809(1990)*. The structure is a three-dimensional framework of Be and Si tetrahedra with the beryllium ordered; the framework is unusual in that it contains 5-tetrahedron components consisting of 3-membered rings sharing a common tetrahedron to form a 5-tetrahedron element. Colorless, white to yellowish; vitreous luster; translucent. Cleavage {100}, {010}, {001}, distinct; {110}, poor. H = 5–6. G = 2.33. Biaxial (+,–); 2V ≈ 90°; $N_x = 1.513$, $N_y = 1.515$–1.516, $N_z = 1.518$. Readily fusible. Insoluble in acids. Occurs as a late hydrothermal phase replacing chkalovite and, rarely, as 1- to 2-mm prisms lining cavities in alkalic pegmatite at *Mt. Karnarsut, Lovozero massif, Kola Penin., Russia*. AR *DANES 213:130(1973)*, *AM 68:474(1983)*.

78.6.3.1 Pellyite $Ba_2Ca(Fe,Mg)_2 \cdot Si_6O_{17}$

Named in 1972 by Montgomery et al. for the locality. ORTH *Cmcm*. $a = 15.677$, $b = 7.151$, $c = 14.209$, $Z = 4$, $D = 3.62$. *25-95*(nat): 3.83_5 3.46_6 3.43_{10} 3.19_7 3.17_5 3.15_5 2.31_6 2.11_5. Fe^{2+} (and minor substituting Mg, Zn, and Mn) occur in tetrahedral coordination. Ca is octahedrally coordinated. A nearly continuous framework of Si and Fe tetrahedra with Ba occupying a large void of 10-fold coordination within the tetrahedral framework. *AM 61:67(1976)*. Average grain size is 2 mm, massive. Colorless, light yellow; white streak; vitreous luster. Cleavage poor prismatic; conchoidal fracture. H = 6. G = 3.51. Decomposes slowly in dilute HCl, leaving a white residue. Probe analysis (wt %): BaO, 36.00; CaO, 6.02; FeO, 12.18; MgO, 0.98; ZnO, 1.20; MnO, 1.54; Al_2O_3, 0.28; SiO_2, 40.86; wet analysis: SiO_2, 40.5; BaO, 34.16; CaO, 6.25; FeO, 12.46; MnO, 0.57; MgO, 1.46; ZnO, 1.05; Al_2O_3, 3.53. Biaxial (+); Na(D): $N_x = 1.643$, $N_y = 1.645$, $N_z = 1.649$; 2V = 47°; rhombic dispersion, r > v, very strong. Occurs in a contact-metasomatic deposit adjacent to a porphyritic quartz monzonite stock. With other Ba-silicates, sanbornite, quartz, and so on, in metamorphic rocks. Big Creek, Fresno Co., and Trumbull Peak area, Mariposa Co., CA; headwaters of the *Pelly R. and Ross R., YT, Canada*; La Madrelena mine, Tres Pozos, northern Baja California, Mexico. EF *CM 11:444(1972), AM 69:358(1984)*.

78.6.4.1   Tiettaite $(Na,K)_{17}Fe^{3+}TiSi_{16}O_{29}(OH)_{30} \cdot 2H_2O$ (Koashva) and $(Na,K,Ca)_{20}Fe^{3+}(Ti,Fe^{3+})_2Si_{16}O_{32}(OH)_{28.5} \cdot 8H_2O$ (Rasvumchorr)

Named in 1993 by Khomyakov et al. from the Finnish *tietta*, science or knowledge, for the first science station in Khibiny. ORTH $Cmcm$, $Cmc2_1$, or $C2cm$. $a = 29.77$, $b = 11.03$, $c = 17.111$, $Z = 4$, $D = 2.39$. PD(nat): 14.79_5 10.38_{10} 4.96_6 4.52_8 3.70_6 3.41_5 3.22_7 3.10_8. In aggregate grayish white, sometimes dark gray; masses (spheres) to 1 cm, grains 0.3–0.5 mm; rounded, friable aggregates to 1 cm, containing acicular hairs elongate [001] and flattened on (100); vitreous to silky luster. Cleavage {100}, {010}, excellent; steplike fracture. $H = 3$. $G = 2.42$ for both localities. IR, DTA, and TGA data: *ZVMO 122:121(1993)*. Decomposes between 200 and 300°. Tarnished with prolonged exposure in air; soluble in weak HCl and HNO_3. 15.4% weight loss to 1000°. H_2O analysis gave 14.73%. Biaxial (−); XYZ = acb; (Koashva): $N_x = 1.532$, $N_y = 1.548$, $N_z = 1.559$; (Rasvumchorr): $N_x = 1.530$, $N_y = 1.549$, $N_z = 1.559$; $2V = 79°$; colorless; $r < v$, moderate. Occurs on *Mt. Rasvumchorr and Mt. Koashva, Khibiny massif, Kola Penin., Russia*; in ultra-agpaitic pegmatites associated with other alkalic minerals. EF

78.7.1.1   Bostwickite $CaMn_6^{3+}Si_3O_{16} \cdot 7H_2O$

Named in 1983 by Dunn and Leavens for Richard C. Bostwick (b.1943), technician with SPEX Industries, and collector and compiler of data on Franklin and Sterling Hill, NJ mineralogy, particularly fluorescent species. A corresponding but impure phase has been known since 1874. ORTH *35-616*: 11.30_{10} 3.55_3 2.90_3 2.57_4 2.48_2 2.26_2 2.24_2 1.470_2. Dark red, vitreous to submetallic luster. $H = 1$. $G = 2.93$. Biaxial (−); $2V = 25°$; $N_x = 1.775$, $N_y = 1.798$, $N_z = 1.800$; $r < v$, strong. Strongly pleochroic; X,Y, red-brown; Z, yellow-brown. Found as divergent sprays of bladed, prismatic crystals and as hemispherules up to 3 mm in diameter, associated with franklinite, calcite, fluorite, and minor willemite at the *Franklin mine, Franklin, Sussex Co., NJ*. AR *MM 47:387(1983)*.

78.7.2.1   Creaseyite $Cu_2Pb_2(Fe,Al)_2Si_5O_{17} \cdot 6H_2O$

Named in 1975 by Williams and Bideaux for S. C. Creasey (b.1917) in recognition of his studies of the Mammoth–St. Anthony mine. ORTH mmm. $a = 12.483$, $b = 21.395$, $c = 7.283$, $Z = 4$, $D = 4.01$. *29-566*: 10.7_{10} 6.02_5 5.35_4 4.07_5 3.56_5 3.01_5 2.98_5 2.70_5. Fibers [001] up to 0.5 mm; also spherules and matted aggregates. Yellow-green, pale green streak. $H = 2\frac{1}{2}$. $G = 4.1$. Biaxial (+); $2V = 69°$; $N_x = 1.737$, $N_y = 1.747$, $N_z = 1.768$; XYZ = abc. Pleochroic; Z = X > Y in yellow-greens. Found in oxidized zones with wulfenite, willemite, mimetite, and other copper silicates at *Mammoth mine, Tiger, Pinal Co., AZ*, and at Wickenburg, Maricopa Co., with fluorite and ajoite. Mostly replaced by chrysocolla at Caborca, SON, Mexico. AR *MM 49:227(1975)*.

78.7.3.1 Ertixiite Na$_2$Si$_4$O$_9$

Named in 1985 for the locality. ISO *Pa3*. $a = 5.975$, $Z = 1$, $D = 2.35$. *39-382*(nat): 3.44_2 2.99_2 2.67_2 2.19_2 2.00_8 1.80_{10} 1.60_2 1.49_2. Structure not yet done, no single-crystal work done. *CJG 4(2):192(1985)*. Granular, 0.1–0.5 mm. Colorless, white streak, vitreous luster, transparent. No cleavage, poor subconchoidal fracture. $H = 5.8$–6.5. $G = 2.35$. IR data: bands at 1050 and 475 cm^{-1} due to (SiO$_4$)$^{4-}$ vibrations and a band at 775 cm^{-1} resulting from Al–O–Al vibrations. Na loss occurred under the electron beam. (Na$_{1.75}$Ca$_{0.15}$)$_{1.90}$(Al$_{0.09}$Si$_{3.92}$)$_{4.01}$O$_9$. Analysis (wt %): Na$_2$O, 17.97; CaO, 2.82; SiO$_2$, 77.86; Al$_2$O$_3$, 1.45; FeO, 0.05. Isotropic; $N = 1.502$; colorless. Occurs in miarolytic cavities in a complex Ta–Nb–Be-bearing granitic pegmatite. Associated with topaz, albite, muscovite, quartz, apatite, and garnet. *Altai No. 3 pegmatite, Ertixi R., Fuyun Co., Xinjiang Autonomous Region, China.* EF

78.7.4.1 Ilimaussite-(Ce) Ba$_2$Na$_4$(Ce,La)Fe^{3+}Nb$_2$Si$_8$O$_{28}$ · 5H$_2$O

Named in 1968 for the locality. HEX Space group and unit cell data not certain. *DANS-ESS 182:139(1968)*. Superstructure appears to be present with $a = 10.72$ and $c = 60.54$. If C centered, possible space groups are $C6/mcm$, $C6cm$, and $C62c$. If P cell, possible space groups are $P6_3mmc$, $P6_3mc$, or $P62c$. $a = 10.72$–10.80, $c = 20.31$ or 60.54, $Z = 1$ or 3 [*MGR 181:3(1968)*], $D = 3.70$. *21-399*: 3.25_6 3.12_5 2.98_4 2.67_{10} 2.50_2 2.24_3 2.02_2 1.82_2. Structure not yet determined. Polysynthetic twinning sometimes present; forms lamellar and flaky {0001} aggregates to 15 × 10 × 3 mm. Brownish yellow, vitreous to resinous luster. Conchoidal fracture, very brittle. $H = 4$. $G = 3.6$–3.7. IR data: *DANS-ESS 182:139(1968), GZ 36:54(1976)*. Ilimaussaq material contains 3.80% K$_2$O and 1.64 wt % TiO$_2$. Khibiny material contains 6.3 wt % TiO$_2$. Substitution: Ba + Ti ↔ (K,Na) + Nb. Uniaxial (+); $N_O = 1.686$–1.689, $N_E = 1.695$–1.698; negative elongation; pleochroic. O, pale yellow; E, bright yellow. Occurs in a hydrothermal ussingite–analcine vein cutting a nepheline–sodalite syenite, *Nakalaq, Ilimaussaq massif, southern Greenland*. Associated minerals are chkalovite and epistolite. Alters readily, becoming dull and nontransparent. Also occurs in the Material'naya adit, on Mt. Yukspor, Khibiny massif, Kola Penin., Russia, in a cavity in nepheline–aegirine–feldspar pegmatite. EF

78.7.5.1 Ilmajokite (Na,Ce,La,Ba)$_2$TiSi$_3$O$_5$(OH)$_{10}$ · nH$_2$O (?)

Named in 1972 by Bussen et al. for the locality. MON $C2/c$ or Cc. $a = 39.80$, $c = 29.83$, $\beta = 96.63°$, $G = 2.20$. *25-0783*(nat): 11.5_{10} 10.9_7 10.2_9 4.3_{10} 3.7_7 3.1_9 2.48_7 2.44_{10}. Hydrous salt of metasilicic acid of the type MRO$_3$. 3M^{2+} ↔ Na$_2$Ti. Prismatic and platy; granular deposits, crusts, and aggregates of crystals to 2 mm long. Very unstable and decomposes in both damp and dry atmospheres. It dries and cracks in air, then becomes turbid and disintegrates. Bright yellow, transparent, vitreous luster. Perfect cleavage on the rhombic prism and pinacoid, intersecting at 72°; brittle. H = 1. G = 2.20. IR, DTA, and TGA data: *ZVMO 101:75(1972)*. Contains OH + H_2O; in H_2O at 20° it increases in volume and disintegrates into quartzine fibers with $N_x = 1.541$ and $N_z = 1.543$; in 1:1 HCl at 20° it effervesces and dissolves; decomposed by cold water or by acids, it fuses readily to a white enamel. Biaxial (+); $N_x = 1.573$, $N_y = 1.576$, $N_z = 1.579$; 2V = 90°. Symmetrical extinction with respect to cleavage. Occurs in cavities in the central (natrolitic) zone of the Jubilee stratified pegmatite and along the border layers of lujavrite and ijolite (foyaite) on Mt. Karnarsut in the Lovozero alkaline massif, Kola Penin., Russia. In one of the other mines of Mt. Karnarsut, ilmajokite forms granular deposits on the walls of cavities in white natrolite apophyses, associated with aegirine, cleiophane, halite, and mountainite. EF *MU (1990)*.

78.7.6.1 Juanite $Ca_{10}Mg_4Al_2Si_{11}O_{39} \cdot 4H_2O$ (?)

Named in 1932 by Larsen and Goranson for the San Juan Mountains, Colorado. An inadequately described species. Probably ORTH Space group unknown. *29-335:* 3.58_5 3.41_5 3.40_5 3.27_{10} 2.98_{10} 290_9 1.97_8 1.93_{10}. White to pistachio-green to brown, translucent. G = 3.01. Biaxial (+); 2V = 50°; $N_x = 1.640$, $N_y = 1.642$, $N_z = 1.647$; positive elongation. Decomposed by acids. Formula given for Kuznetzk material is $Ca_9(Mg,Fe)_3(Si,Al)_{12}(O,OH)_{36} \cdot 7H_2O$, and includes 5.18 wt % Fe_2O_3. A fibrous alteration product of melilite found at *Iron Hill, Powderhorn–Cebolla district, Gunnison Co., CO*; in the Kiya–Shatyrsk massif, Kuznetsk Ala-Tau, Kazakhstan. AR *AM 17:343(1932)*, *GG 12:62(1971)*.

78.7.7.1 Kenyaite $Na_2Si_{22}P_{41}(OH)_8 \cdot 6H_2O$

Named in 1967 by Eugster for the country of origin. MON *P*. a = 7.79, b = 19.72, c = 6.91, β = 95.9°, Z = 1, D = 2.46; *20-1157*(nat): 19.68_{10} 9.93_5 4.97_4 4.69_3 3.53_2 3.43_9 3.32_5 3.20_6; *AM 68:818(1983)*(syn, air-dried): 19.8_{10} 9.88_3 4.89_1 4.24_1 3.79_1 3.43_8 3.19_8 2.93_2. X-ray patterns for both natural and synthetic kenyaite can be sensitively changed by variations in the amount of interlayer water, and partial or complete exchange of the interlayer Na by protons if exposed to H_2O. Nodular concretions, sometimes with cores of chert. White, white streak. G ≈ 3.18. DTA data: *AM 53:2061(1968)*. Material from New Mexico fluoresces very pale yellow to white in SWUV. Biaxial (+), N = 1.48, small 2V. Occurs in saline playa evaporite alkaline lake deposits. Formed by leaching of magadiite, and from saline brines. Point of Rocks Mesa, Colfax Co., NM; Alkali Lake, Oregon; Trinity Co., CA; *Lake Magadi, Kenya, and vicinity (e.g., High Natron Beds)*; Lake Tchad, Africa; Lake Natron, Tanzania; Kaffi Basin, Niger. EF *CMP 22:1(1969)*, *MIN 4(2):437(1992)*.

78.7.8.1 Kuliokite-(Y) (Y,REE)$_4$Al(SiO$_4$)$_2$(OH)$_2$F$_5$

Named in 1984 by Voloshin et al. for the locality. TRIC P1. $a = 8.606$, $b = 8.672$, $c = 4.317$, $\alpha = 102.79°$, $\beta = 97.94°$, $\gamma = 116.66°$, $Z = 1$, $D = 4.24$–4.28. 40-467(nat): 3.71_9 3.49_9 3.31_6 2.79_{10} 2.46_8 2.14_7 1.86_6 1.70_8. Structure: *SPD 31:601(1986)*. Crystals to 0.5 mm in maximum dimension; crystals from Norway are intimately and complexly twinned, centimeter-size crystals from Norway, largest to 3 cm, pseudomonoclinic. Norwegian material recognized as a new species in 1961, but twinning prevented description. Colorless, transparent (Norway: light pink); white streak; vitreous to greasy luster. One good cleavage parallel or nearly parallel to the OAP, two other more or less perfect cleavages or partings present. $H = 4$–5. $G = 4.2$–4.3. IR data: *MZ 8:94(1984)*. Yellow-green cathodoluminescence. Analysis (wt %): Y$_2$O$_3$, 56.4; Yb$_2$O$_3$, 2.71; Er$_2$O$_3$, 2.24; Dy$_2$O$_3$, 1.38; Lu$_2$O$_3$, 0.11; Gd$_2$O$_3$, 0.39; Tm$_2$O$_3$, 0.10; Ho$_2$O$_3$, 0.19; Al$_2$O$_3$, 7.36; SiO$_2$, 17.75; F, 13.45; H$_2$O, 3.58. Chemistry of Norwegian material is similar. Biaxial $(-)$; X $\wedge$ $a = 7°$, Y $\wedge$ $b = 28°$, Z $= c$; Na(D): $N_x = 1.656$, $N_y = 1.700$, $N_z = 1.703$; $2V = 10$–$19°$; $r > v$. Occurs in fissure fillings in pegmatites, also inclusions in fluorite. Associated with thalenite-(Y), xenotime-(Y), kainosite-(Y), and bastnaesite. In amazonite-bearing pegmatite at Hoydalen, Tordal, Norway; the mineral alters to kamphaugite-(Y) and other carbonate-bearing species [e.g., kainosite-(Y) and tengerite-Y)]; *unspecified pegmatite near the Kuliok R., Kola Penin., Russia*. EF *EJM 5:691(1993)*.

78.7.9.1 Laplandite-(Ce) (Na,K,Ca)$_4$(Ce,Th)(Ti,Mg,Al,Nb)PSi$_7$O$_{22}$ · 5H$_2$O

Named in 1974 by Eskova et al. for the locality, the area called Lapland in the northern part of Fennoscandia at Mt. Karnasurt. ORTH Pmmm. $a = 7.27$, $b = 14.38$, $c = 22.25$, $Z = 4$, $D = 2.71$. 27-673(nat): 7.27_{10} 6.47_7 3.76_{10} 3.36_8 3.35_{10} 3.03_{10} 2.80_8 2.01_7. Structure not done because of lack of suitable single crystals. Radiating fibrous groups (to 1 cm in diameter) of crystals 0.5–1 mm in diameter, also fan-shaped platy aggregates ≈ 0.01 mm thick, crystals flattened on c. Light gray to yellowish, rarely bluish; vitreous or silky to greasy luster. Splintery fracture. $H = 2$–3. $G = 2.83$. IR and thermal data: *ZVMO 103:571(1974)*. Analysis (wt %): SiO$_2$, 40.9; P$_2$O$_5$, 9.6; TiO$_2$, 4.1; Nb$_2$O$_5$, 1.9; Al$_2$O$_3$, 0.64; MgO, 1.02; MnO, 0.20; CaO, 0.56; ThO$_2$, 1.32; REE$_2$O$_3$, 16.8; Na$_2$O, 9.81; K$_2$O, 1.88; H$_2$O$^+$, 8.96; H$_2$O$^-$, 0.37, for yellowish sample; bluish sample has 4.15 wt %. A poorly studied P-free analog of laplandite-(Ce) also occurs in the Jubilee pegmatite. Biaxial $(-)$; X $= c$; $N_x = 1.568$, $N_y = 1.584$, $N_z = 1.585$; $2V_{calc} = 28°$; colorless. Laplandite may form pseudomorphs after steenstrupine, along with vitusite, belovite, and others. Occurs on *Mt. Karnarsut, Lovozero massif, Kola Penin., Russia*; in the central natrolite zone of an alkalic pegmatite (Jubilee pegmatite), associated with belovite, nordite, serandite, thermonatrite, steenstrupine, leucosphenite, sphalerite, ilmajokite, raite, and zorite. EF

78.7.10.1 Leifite $Na_6Be_2Al_2Si_{16}O_{39}(OH)_2 \cdot 1.5H_2O$

Named in 1915 by Bøggild for the tenth-century Norse mariner and explorer Leif Ericson. HEX $P3m1$. $a = 14.352$, $c = 4.852$, $Z = 1$, $D_c = 2.59$. 27-1: 12.4_2 4.70_4 4.52_2 4.14_2 3.59_2 3.38_7 3.15_{10} 2.46_2. Hexagonal prisms {1010} terminated by {11$\bar{2}$1}, {0001}, fine needles to fibrous, 1 mm to 5 cm long; radiating aggregates and rosettes. Penetration trillings with {10$\bar{1}$0} as twin plane (Ilimaussaq). White, colorless to pale violet; vitreous luster; translucent. Prismatic cleavage {10$\bar{1}$0}, distinct; brittle. H = 6. G = 2.57. Uniaxial (+); $N_O = 1.516$–1.518, $N_E = 1.520$–1.521. In cavities in alkali–pegmatite veins. At Mt. St-Hilaire, QUE, Canada, in crystals(!) up to 5 cm long and as disk-shaped aggregates and radially fibrous spheres with albite, natrolite, serandite, and catapleiite; at Ilimaussaq and *Narssarssuk, Greenland*, with microcline, calcite, zinnwaldite, and acmite; Lovozero massif, Kola Penin., Russia. AR *DT:535(1932)*, *BGSD 20:134(1970)*, *AC(B) 30:396(1974)*, *MR 24:45(1993)*, *NJMM 2:83(1994)*.

78.7.11.1 Luddenite $Cu_2Pb_2Si_5O_{14} \cdot 14H_2O$

Named in 1982 by Williams for Raymond W. Ludden, chief geologist for western exploration, Phelps-Dodge Corporation. MON Space group unknown. $a = 7.85$, $b = 20.06$, $c = 14.72$, $\beta = 90.78°$ (from powder data), $Z = 6$ (originally reported as 4), D = 4.98. 35-560: 7.36_{10} 5.22_7 4.23_5 3.52_5 3.41_5 3.17_{10} 2.92_8 2.89_5. Green, translucent. Cleavage in one unspecified direction. H = 4. G = 4.45. Biaxial; $2V = 40°$; $N_x = 1.852$, $N_z = 1.867$. Pleochroic; Y,X, yellow-green; Z, emerald-green. Soluble with difficulty in acids. Found as rosettes and fan-shaped aggregates with galena, chalcopyrite, fluorite, alamosite, melenotekite, and hyalotekite in a Pb–Cu–Ag prospect *near Artillery Peak, Mohave Co., AZ*. AR *MM 46:363(1982)*.

78.7.12.1 Magadiite $NaSi_7O_{13}(OH)_3 \cdot 4H_2O$

Named in 1967 by Eugster for the locality. MON P. (nat): $a = 7.25$, $b = 7.25$, $c = 15.69$, $\beta = 96.8°$, $Z - 2$; (syn): $a = 7.30$, $b = 7.28$, $c = 15.71$, $\beta = 96.4°$ for $Z = 1$ ($Na_2Si_{14}O_{29} \cdot 10H_2O$), D = 2.17–2.25. 24-698(nat): 15.6_{10} 5.19_2 5.01_1 4.47_2 3.54_2 3.43_8 3.30_4 3.15_6; 42-1350(syn): 15.46_8 5.18_2 3.64_2 3.56_2 3.45_{10} 3.31_7 3.15_{10} 1.82_2. Dehydrated form is stable to 500°C and can be completely rehydrated; 500–700° produces quartz with or without cristobalite; >700° produces tridymite. Structure not determined. Thin rectangular plates ≈ 1 × 1 μm, a white powder; when put in vacuum, the (001) spacing decreases to 13.5 Å but recovers the original spacing when exposed again to H_2O vapor. White, white streak, usually earthy in texture, dull to vitreous luster. IR and NMR data: *AM 54:1034(1969)*, *ZAAC 597:183(1991)*. DTA and TGA data: *AM 53:2061(1968)*, *60:642(1975)*. In dilute acids at 20°, a hydrous form of silica results; composition near $6SiO_2 \cdot H_2O$. May contain traces of K, Al, Fe, Mg, and Ca in addition to Na, Si, and water. Biaxial; N ≈ 1.48; positive elongation. At Mt. St.-Hilaire, occurs in cavities in marble xenoliths. Usually occurs in alkaline lakes, particularly those of the sodium carbonate–bicarbonate variety. Brines with high pH and silica content. Magadiite has not been reported from lacustrine deposits older than late Pleistocene. Occurs as white puttylike laminae in silts. In veins at Alkali Lake, Oregon. Evaporites. Alteration to chert with loss of water and alkalis. Trinity R. area, Trinity Co., CA; Alkali Lake, Lake Co., OR; Mt. St.-Hilaire, QUE, Canada; the Ca analog of magadiite occurs in cavities to several centimeters across in the St.-Amable 10-m-thick sill of trachytic phonolite at Varennes, QUE. Associated with other rare alkalic minerals [*RMS:8–9(1993)*]; *Lake Magadi*, Olduvai Gorge, and Kanem, Lake Chad, *Kenya*; El Fraile Municipio de Huajuapan de Leon, OX, Mexico. EF *CMPM 22:1(1969)*.

78.7.13.1 Maufite $(Mg,Ni)Al_4Si_3O_{13} \cdot 4H_2O$ (?)

Very poorly described mineral. *TGSSA 32:103(1930)*, *AM 15:275(1930)*.

78.7.14.1 Spadaite $MgSiO_3 \cdot 2H_2O$ or $MgSiO_2(OH)_2 \cdot H_2O$ (?)

Named in 1843 by von Kobell for Lavino Spada (de Medici)(1801–1863), Italian politician and amateur mineralogist. Probably ORTH Space group unknown. Dense massive to microscopic platy or columnar. Cream white to pink, pearly to greasy luster. Conchoidal fracture when dense, sometimes splintery. Decomposed by HCl. H = 2.5. G = 2.2–2.32. Fusible at 5 to a white enamel, decomposed by HCl with separation of sandy silica. Superficially resembles a light-colored serpentine. Biaxial (+); $N_x = 1.521$, $N_y = 1.525$–1.53, $N_z = 1.545$; $2V = 37$–$61°$; birefringence = 0.024; negative and positive elongation; parallel extinction to slightly inclined; colorless and nonpleochroic in thin section. Type material and Utah [dehydrated and corrected, *AM 16:231(1931)*] spadaite gave (wt %), respectively: SiO_2, 56.00, 53.96; Al_2O_3, 0.66, —; FeO, 0.66, 0.35; MgO, 30.67, 32.08; H_2O, 11.34, 13.64. Loses half its H_2O [Utah, *AM 16:231(1931)*] at 110°. Found replacing tactite minerals, and associated with quartz, chalcopyrite, bornite, garnet, diopside, calcite, wollastonite, and gold. Also in dolerite–diabase and leucite-bearing lava. Spadaite is observed to replace wollastonite. Cane Springs, Alvarado, and Midas mines, Gold Hill, Tooele Co., UT; Sasbach, Kaiserstuhl, Baden-Württemberg, Germany; *Capo di Bove, near Rome, Lazio, Italy*. EF

78.7.15.1 Terskite $Na_4ZrSi_6O_{15}(OH)_2 \cdot H_2O$

Named in 1983 by Khomyakov et al. for Terskii berreg, the southern shore of the Kola Peninsula. ORTH $Pnc2$. $a = 14.195$, $b = 14.750$, $c = 7.511$, $Z = 4$ [*DAN SSSR 316:645(1991)*], D = 2.74. 35-712(nat): 4.09_6 3.32_{10} 3.30_{10} 3.26_{10} 3.19_8 3.13_6 2.62_7 2.56_6. Structure: *DAN SSSR 316:645(1991)*. White, pink, and rarely light cream to light violet-lilac to dark brown; at Mt. St.-Hilaire, powdery to porcelaneous; in Russia, plates, 1–3.5 mm, pseudotetragonal, vitreous to porcelaneous. H ≈ 5. G = 2.71. IR, DTA, TGA data: *ZVMO 112:226(1983)*. Weak yellowish-white fluorescence under SWUV and LWUV, which shows bright green photoluminescence. Insoluble in 10% HCl or HNO_3. Analysis (wt %): SiO_2, 56.3; ZrO_2, 17.8; Na_2O, 19.0; H_2O, 6.37.

Biaxial $(-)$; $N_x = 1.576$, $N_y = 1.582$, $N_z = 1.584$; $2V = 53°$; $r > v$, weak. Occurs at Mt. St.-Hilaire in sodalite xenoliths. Occurs in syenites at Mt. Alluaiev and in the Jubilee pegmatite on Mt. Karnarsut. In pegmatite veins in syenites. Associated with alkalic minerals. Alteration: hollow pseudomorphs replacing lovozerite. Crude, twinned dodecahedra. Replaces eudialyte. Mt. St.-Hilaire and St.-Amable sill, Varennes, QUE, Canada; Ilimaussaq, Greenland; Mt. Alluaiev and Mt. Karnarsut, Lovozero massif, Kola Penin., Russia. EF

78.7.16.1 Tranquillityite $Fe_8^{2+}Ti_3(Zr,Y)_2Si_3O_{24}$

Named in 1971 for the locality. HEX Primitive lattice. $a = 11.69$, $c = 22.25$, $Z = 3$ or 6, $D = 4.70$. Alternative cell data given in GCA Suppl., Proc. 8th Lunar Sci. Conf., 2, 1831(1977): rhombohedral cell with $a_{rh} = 4.743$ and $\alpha_{rh} = 88.57°$. A number of weak reflections could be indexed with $a_{rh} = 9.486$ (2×) and $D = 4.83$. If the original hexagonal cell is transformed to rhombohedral coordinates, the results are $a_{rh} = 10.83$ and $\alpha_{rh} = 71.31°$, which do not agree well with the subsequently proposed rhombohedral cell. Probably, two different minerals are involved. 26-1143(nat): 4.04_5 3.34_4 3.23_{10} 3.18_4 3.13_4 2.92_4 2.16_6 1.78_7. Nearly opaque, dark red, black. Thin laths, from a few microns to 65×15 μm. Reported as being metamict. GCA Suppl., Proc. 8th Lunar Sci. Conf., 2, 1831(1977). Analysis (wt %): SiO_2, 14.0; TiO_2, 19.5; Al_2O_3, 1.1; FeO, 42.5; CaO, 1.3; ZrO_2, 17.2; Y_2O_3, 2.8; traces of Gd_2O_3, Cr_2O_3, MnO, HfO_2, Nb_2O_5, Pr_2O_3, Nd_2O_3. Biaxial; $N = 2.12$ (calculated from r values); $2V = 40°$ (Proc. Apollo 11 Lunar Science Conference, GCA Suppl. 1(1):315(1970)); reflectance at 546 nm = 12.9%. Occurs in lunar basalts, associated with interstitial phases such as troilite, pyroxferroite, cristobalite, and alkali feldspar. Found at Tranquility Base, Moon, and other sites with mare and non-mare basalts. AR Proc. Second Lunar Sci. Conf., GCA Suppl. 2(1):39(1971), NAT 236:215(1972).

78.7.17.1 Umbozerite $Na_3Sr_4ThSi_8(O,OH)_{24}$

Named in 1974 by Es'kova et al. for the locality. Amorphous. 26-1384(heated to 1100°): 3.38_6 3.29_{10} 2.98_5 2.82_5 2.09_4 2.00_6 1.70_{18} 1.290_4. Green to greenish brown, vitreous luster. Conchoidal fracture. $H \approx 5$. $G = 3.60$. Isotropic; $N = 1.640$; metamict. Occurs as crude crystals of apparent tetragonal morphology up to 3 mm and as irregular masses associated with sphalerite, belovite, schizolite, and niobium minerals of the lomonosovite group in ussingite veinlets cutting alkalic rocks at Umbozero (Lake Umba), Kola Penin., Russia. AR AM 60:341(1975).

78.7.18.1 Vyuntspakhkite-(Y) $(Y,Yb,Er)_4Al_2AlSi_5O_{18}(OH)_5$

Named in 1983 by Voloshin et al. for the locality. MON $P2_1/a$. $a = 5.830$, $b = 14.763$, $c = 6.221$, $\beta = 123.05°$, $Z = 1$, $D = 4.04$. 35-0708(nat): 7.4_6 4.98_6 4.92_6 3.47_{10} 2.95_5 2.86_6 2.60_8 2.27_5. Structure: SPC 29:141(1984). Slender prismatic crystals 0.5–0.7 mm long and 0.05–0.2 mm wide; vacancies exist in the two Al positions because of the short distances between the two sites (1.68 Å), cannot be occupied simultaneously. Colorless, white streak, adamantine luster, transparent. No cleavage, brittle. H = 6–7. G = 4.02. IR data: *SPC 29:141(1984)*. Nonfluorescent, very pale yellow-green cathodoluminescence; not soluble in HCl. Average of four probe analyses (wt %): Y_2O_3, 24.18; Yb_2O_3, 15.71; Er_2O_3, 5.74; Dy_2O_3, 3.02; Lu_2O_3, 2.22; Tm_2O_3, 1.57; Al_2O_3, 13.14; SiO_2, 28.69; H_2O, 4.29; Gd_2O_3, ≈ 0.4, Ho_2O_3, ≈ 0.4, Tb_2O_3, ≈ 0.3. Biaxial (+); Z ∧ a = 68°, X ∧ c = 40°, OAP = (010); N_x = 1.680, N_y = 1.692–1.695, N_z = 1.720–1.722. Occurs on *Mt. Vyuntspakh, Kola Penin., Russia*, from amazonite pegmatites. In cavities within fluorite in amazonite pegmatite. Varieties rich in Y give faint yellow-green luminescence in cathode ray. Two generations; chemistry varies widely: Y_2O_3, 17.8–34.7; associated with fluorite, xenotime, and bastnaesite. EF *MZ 5:89(1983)*.

78.7.19.1 Wawayandaite $Ca_{12}Be_{18}Mn_4B_2Si_{12}O_{46}(OH,Cl)_{30}$

Named in 1990 by Dunn et al. from the Lenni Lenape Native American *wawayanda*, many or several windings, in allusion to the grossly curved and winding habit of its crystals. MON Pc or $P2/c$. a = 15.59, b = 4.87, c = 18.69, β = 101.84°, Z = 1, D = 2.98. *PD:* 15.1_9 9.20_3 3.16_{10} 3.00_7 2.72_6 2.63_7 2.58_3 2.25_5. Usually as very thin plates, curved and warped in all dimensions. Rarely as minute (0.1 mm) sharp and thin crystals tabular {100} and twinned on (100); other forms are not discernible. Colorless and transparent; fine-grained aggregates white and opaque; pearly luster is obvious on curved crystals, but less so on aggregates. Cleavage {100}, perfect. Flexible. G ≈ 3.0. Biaxial (−); 2V = 85.2°; $N_{x\ calc}$ = 1.619, N_y = 1.631, N_z = 1.641 for Na(D); X ∧ a = 11.5° in obtuse β, YZ = bc; r < v strong. Observed on a museum specimen from the *Franklin mine, Franklin, Sussex Co., NJ*, associated with superb willemite crystals and with calcite, hodgkinsonite, cahnite, and friedelite. AR *AM 75:405(1990)*.

Appendix

NEW MINERALS FOR 1996

	Mineral Name	Formula	Reference
1.	Androsite-(La)	$(Mn,Ca)(REE)AlMn^{3+}Mn^{2+}(SiO_4)(Si_2O_7)O(OH)$	AM 81:735(1996)
2.	Babkinite	$Pb_2Bi_2(S,Se)_3$	DAN 346(5):656(1996), and AM 81:1513(1996)
3.	Baksanite	$Bi_6(Te_2S_3)$	DAN 347(6):787(1996)
4.	Bechereite	$(Zn,Cu)_6Zn_2(OH)_{13}[(S,Si)(O,OH)_4]_2$	AM 81:244(1996)
5.	Belovite-(La)	$Sr_3Na(La,Ce)[PO_4]_3(F,OH)$	PRMS 125(3):101(1996)
6.	Benavite	$HSrFe^{3+}_3(PO_4)_2(OH)_6$	Chemie der Erde 56:171(1996)
7.	Calcioaravaipaite	$PbCa_2Al(F,OH)_9$	MR 27:293(1996)
8.	Christelite	$Zn_3Cu_2(SO_4)_2(OH)_6 \cdot 4H_2O$	NJBM 188(1996)
9.	Clerite	$MnSb_2S_4$	PRMS 125(3):95(1996)
10.	Clinoatacamite	$Cu_2(OH)_3Cl$	CM 34:61(1996) and CM 34:73(1996)
11.	Deloneite-(Ce)	$NaCa_2SrCe(PO_4)_3F$	PRMS 125(5):83(1996)
12.	Dusmatovite	$K(K,Na)(Mn^{2+},Y,Zr)_2(Zn,Li)_3Si_{12}O_{30}$	Vestnik Moscow U. Sreies 4, Geology Z, pp54–60 and DAN344 (5)1607–610(1996)
13.	Fettelite	$Ag_{24}HgAs_5S_{20}$	NJBM 313(1996)
14.	Fianelite	$Mn_2 V(V,As)O_7 \cdot 2H_2O$	AM 81:1270(1996)
15.	Fluor-cannilloite	$CaCa_2(Mg_4Al)(Si_5Al_3)O_{22}F_2$	AM 81:995(1996)
16.	Fluor-ferro-leakeite	$NaNa_2(Fe^{2+}_2Fe^{3+}_2Li)Si_8O_{22}F_2$	AM 81:226(1996)
17.	Frankamenite	$K_3Na_3Ca_5(Si_{12}O_{30})(F_3OH) \cdot H_2O$	crystal structure MM 60:897(1996) and PRMS 125(2):106(1996)
18.	Gallobeudantite	$PbGa_3[(AsO_4),(SO_4)]_2(OH)_6$	CM 34:1305(1996)
19.	Gottardiite	$Na_{2.5}K_{0.2}Mg_{3.1}Ca_{4.8}Al_{18.8}Si_{117.2}O_{272} \cdot 93H_2O$	EJM 8:687(1996) and EJM 8:69(1996)
20.	Hanawaltite	$Hg^{1+}Hg^{2+}[Cl,(OH)]_2O_3$	Powder Diffraction 11:45(1996)
21.	Hyttsjoite	$Pb_{18}Ba_2Ca_5Mn^{2+}_2Fe^{3+}_2Si_{30}O_{90}Cl \cdot 6H_2O$	AM 81:743(1996)
22.	Jáchymovite	$(UO_2)_8(SO_4)(OH)_{14} \cdot 13H_2O$	NJBA 170:155(1996)
23.	Jensenite	$Cu_3Te^{6+}O_6 \cdot 2H_2O$	CM 34:49(1996) and CM 34:55(1996)
24.	Krasnovite	$Ba(Al,Mg)(PO_4,CO_3)(OH)_2 \cdot H_2O$	PRMS 125(3):110(1996)
25.	Kukharenkoite-(Ce)	$Ba_2Ce(CO_3)_3F$	EJM 8:1327(1996)
26.	Malinite	$Cu(Pt,Ir)_2S_4$	MP 56:25(1996) and Acta Geologica Sinica 70(4):309(1996)
27.	Mahnertite	$(Na,Ca)Cu_3(AsO_4)_2Cl \cdot 5H_2O$	Archives de Science Geneve 49(2):119(1996)
28.	Leisingite	$Cu(Mg,Cu,Fe,Zn)_2Te^{6+}O_6 \cdot 6H_2O$	MM 60:653(1996)
29.	Medenbachite	$Bi_2Fe(Cu,Fe)(O,OH)_2(OH)_2(AsO_4)_2$	AM 81:505(1996)
30.	Meurigite	$KFe^{3+}_7(PO_4)_5(OH)_7 \cdot 8H_2O$	MM 60:787(1996)
31.	Natroxalate	$Na_2C_2O_4$	PRMS 125(1):126(1996)
32.	Nežilovite	$PbZn_2(Mn^{4+},Ti^{4+})_2Fe_8O_{19}$	CM 34:1287(1996)
33.	Noelbensonite	$BaMn^{3+}_2[Si_2O_7](OH)_2 \cdot H_2O$	MM 60:369(1996)
34.	Oulankaite	$(Pd,Pt)_5(Cu,Fe)_4SnTe_2S_2$	EJM 8:311(1996)
35.	Penobsquisite	$Ca_2FeCl[B_9O_{13}(OH)_6] \cdot 4H_2O$	CM 34:657(1996)

36.	Piretite	$Ca(UO_2)_3(SeO_3)_2(OH)_4 \cdot 4H_2O$	CM 34:1317(1996)
37.	Pyatenkoite-(Y)	$Na_5(Y,Dy,Gd)TiSi_6O_{18} \cdot 6H_2O$	PRMS 125(4):72(1996)
38.	Rosiaite	$PbSb_2O_6$	EJM 8:487(1996)
39.	Shkatulkalite	$Na_{10}(Mn,Ca,Sr)Ti_3Nb_3(Si_2O_7)_6(OH)_2F \cdot 12H_2O$	PRMS 125(1):120(1996)
40.	Sigismundite	$(Ba,K,Pb)Na_3(Ca,Sr)(Fe,Mg,Mn)_{14}Al(OH)_2(PO_4)_{12}$	CM 34:827(1996)
41.	Smrkovecite	$Bi_2O(OH)(PO_4)$	NJBM 97(1996)
42.	Vianeite	$(Fe,Pb)_4O_8O$	EJM 8:93(1996)
43.	Wesselsite	$SrCu[Si_4O_{10}]$	MM 60:795(1996)
44.	Zajacite-(Ce)	$Na[REE_xCa_{1-x}][REE_yCa_{1-y}]F_6$ where x is not equal to y	CM 34:1299(1966) NJBM 103(1996)

Index of Mineral Names in Numerical Order

1.1.1.1	Gold 2		1.1.22.1	Perryite 16
1.1.1.2	Silver 3		1.1.23.1	Suessite 16
1.1.1.3	Copper 4		1.1.23.2	Gupeiite 16
1.1.1.4	Lead 5		1.1.23.3	Xifengite 16
1.1.1.5	Aluminum 5		1.2.1.1	Platinum 17
1.1.2.1	Auricupride 5		1.2.1.2	Iridium 17
1.1.2.2	Tetraauricupride 6		1.2.1.3	Rhodium 18
1.1.2.3	Yuanjiangite 6		1.2.1.4	Palladium 18
1.1.3.1	Maldonite 6		1.2.2.1	Osmium 19
1.1.4.1	Anyuiite 6		1.2.2.2	Ruthenium 19
1.1.5.1	Zinc 6		1.2.4.1	Tetraferroplatinum 21
1.1.5.2	Cadmium 7		1.2.4.2	Tulameenite 22
1.1.6.1	Danbaite 7		1.2.4.3	Ferronickelplatinum 22
1.1.6.2	Zhanghengite 7		1.2.4.4	Potarite 22
1.1.7.1	Mercury 7		1.2.5.1	Isoferroplatinum 24
1.1.8.1	Moschallandsbergite 7		1.2.5.2	Rustenbergite 24
1.1.8.2	Schachnerite 8		1.2.5.3	Atokite 24
1.1.8.3	Paraschachnerite 8		1.2.5.4	Zvyagintsevite 24
1.1.8.4	Luanheite 8		1.2.5.5	Chengdeite 24
1.1.8.5	Eugenite 8		1.2.6.1	Paolovite 25
1.1.8.6	Weishanite 8		1.2.7.1	Taimyrite 25
1.1.9.1	Kolymite 8		1.2.8.1	Cabriite 25
1.1.9.2	Belendorffite 9		1.2.9.1	Stannopalladinite 25
1.1.10.1	Leadamalgam 9		1.2.10.1	Niggliite 25
1.1.11.1	Kamacite 10		1.2.11.1	Plumbopalladinite 26
1.1.11.2	Taenite 10		1.3.1.1	Arsenic 26
1.1.11.3	Tetrataenite 10		1.3.1.2	Antimony 27
1.1.11.4	Awaruite 12		1.3.1.3	Stibarsen 28
1.1.11.5	Nickel 12		1.3.1.4	Bismuth 28
1.1.11.6	Wairauite 12		1.3.1.5	Stistaite 29
1.1.12.1	Chromium 12		1.3.2.1	Arsenolamprite 29
1.1.12.2	Chromferide 13		1.3.3.1	Paradocrasite 29
1.1.12.3	Ferchromide 13		1.3.4.1	Selenium 30
1.1.13.1	Tin 13		1.3.4.2	Tellurium 30
1.1.14.1	Indium 13		1.3.5.1	Sulfur 30
1.1.15.1	Cupalite 13		1.3.5.2	Rosickyite 31
1.1.15.2	Khatyrkite 13		1.3.6.1	Diamond 32
1.1.16.1	Cohenite 14		1.3.6.2	Graphite 33
1.1.16.2	Haxonite 14		1.3.6.3	Lonsdaleite 35
1.1.17.1	Tongbaite 14		1.3.6.4	Chaoite 35
1.1.18.1	Siderazot 14		2.1.1.1	Algodonite 37
1.1.18.2	Roaldite 14		2.1.2.1	Horsfordite 37
1.1.19.1	Osbornite 15		2.1.3.1	Telargpalite 37
1.1.19.2	Khamrabaevite 15		2.1.4.1	Duranusite 37
1.1.20.1	Carlsbergite 15		2.1.5.1	Bezsmertnovite 38
1.1.21.1	Barringerite 15		2.1.6.1	Bilibinskite 38
1.1.21.2	Schreibersite 15		2.2.1.1	Dyscrasite 38

1724 Index of Mineral Names in Numerical Order

2.2.1.2	Allargentum 38	2.6.2.7	Rucklidgeite 58
2.2.2.1	Domeykite 39	2.6.3.1	Kochkarite 59
2.2.2.2	Kutinaite 39	2.6.3.2	Aleksite 59
2.2.2.3	Dienerite 39	2.6.3.3	Hedleyite 59
2.2.3.1	Bogdanovite 39	2.6.4.1	Genkinite 59
2.2.4.1	Atheneite 39	2.6.5.1	Temagamite 60
2.2.5.1	Vincentite 40	2.6.6.1	Donharrisite 60
2.2.6.1	Keithconnite 40	2.7.1.4	Shadlunite 62
2.3.1.1	Koutekite 40	2.7.1.5	Manganese Shadlunite 62
2.3.2.1	Orcelite 40	2.7.1.6	Geffroyite 63
2.3.3.1	Stibiopalladinite 41	2.7.2.1	Mackinawite 63
2.3.3.2	Palarstanide 41	2.7.3.1	Yarrowite 63
2.3.4.1	Parkerite 41	2.7.4.1	Godlevskite 63
2.3.5.1	Shandite 41	2.7.5.1	Kharaelakhite 64
2.3.5.2	Rhodplumsite 41	2.8.1.1	Galena 64
2.3.6.1	Vozhminite 42	2.8.1.2	Clausthalite 65
2.4.1.1	Acanthite 42	2.8.1.3	Altaite 65
2.4.1.2	Naumannite 42	2.8.1.7	Borovskite 67
2.4.1.3	Aguilarite 42	2.8.1.8	Crerarite 67
2.4.2.1	Hessite 43	2.8.2.1	Sphalerite 68
2.4.4.1	Jalpaite 44	2.8.2.2	Stilleite 69
2.4.5.1	Mckinstryite 44	2.8.2.3	Metacinnabar 69
2.4.6.1	Stromeyerite 45	2.8.2.4	Tiemannite 69
2.4.6.2	Eucairite 45	2.8.2.5	Coloradoite 70
2.4.7.1	Chalcocite 46	2.8.2.6	Hawleyite 70
2.4.7.2	Djurleite 47	2.8.3.1	Polhemusite 70
2.4.7.3	Digenite 47	2.8.4.1	Xingzhongite 70
2.4.7.4	Roxbyite 48	2.8.5.1	Cooperite 71
2.4.7.5	Anilite 48	2.8.5.2	Vysotskite 71
2.4.7.7	Spionkopite 49	2.8.5.3	Braggite 71
2.4.8.1	Weissite 49	2.8.6.1	Polarite 71
2.4.9.1	Bellidoite 49	2.8.7.1	Wurtzite 72
2.4.10.1	Berzelianite 49	2.8.7.2	Greenockite 73
2.4.11.1	Cuprostibite 50	2.8.7.3	Cadmoselite 73
2.4.12.1	Crookesite 50	2.8.8.1	Hypercinnabar 73
2.4.12.2	Sabatierite 50	2.8.9.1	Troilite 74
2.4.13.1	Carlinite 50	2.8.10.1	Pyrrhotite 74
2.4.14.1	Palladoarsenide 51	2.8.10.2	Smythite 75
2.4.15.1	Palladobismutharsenide 51	2.8.11.1	Nickeline 76
2.4.16.1	Majakite 51	2.8.11.2	Breithauptite 77
2.4.17.1	Petrovskaite 51	2.8.11.3	Sederholmite 77
2.4.18.1	Novakite 51	2.8.11.4	Hexatestibiopanickelite 77
2.5.1.1	Umangite 52	2.8.11.9	Langisite 79
2.5.2.1	Bornite 52	2.8.11.10	Freboldite 79
2.5.3.1	Heazlewoodite 53	2.8.12.1	Covellite 79
2.5.4.1	Oregonite 53	2.8.12.2	Klockmannite 79
2.5.5.1	Thalcusite 54	2.8.13.1	Vulcanite 80
2.5.5.2	Bukovite 54	2.8.14.1	Cinnabar 80
2.5.5.3	Murunskite 54	2.8.15.1	Matraite 80
2.5.6.1	Argyrodite 54	2.8.16.1	Millerite 81
2.5.6.2	Canfieldite 55	2.8.16.2	Mäkinenite 81
2.5.7.1	Daomanite 55	2.8.17.1	Ruthenarsenite 81
2.5.8.1	Imiterite 55	2.8.17.2	Cherepanovite 81
2.5.9.1	Chvilevaite 55	2.8.20.1	Tsumoite 82
2.6.1.1	Dimorphite 56	2.8.20.2	Sulphotsumoite 83
2.6.2.1	Joséite 56	2.8.20.3	Nevskite 83
2.6.2.2	Joséite-B 57	2.8.20.4	Ingodite 84
2.6.2.3	Ikunolite 58	2.8.21.1	Platynite 84
2.6.2.4	Laitakarite 58	2.8.22.1	Realgar 84
2.6.2.5	Pilsenite 58	2.8.22.2	Pararealgar 85
2.6.2.6	Poubaite 58	2.8.22.3	Uzonite 85

Index of Mineral Names in Numerical Order

2.8.22.4	Alacranite 85		2.10.2.1	Wilkmanite 103
2.9.1.1	Chalcopyrite 87		2.10.2.2	Brezinaite 103
2.9.1.2	Eskebornite 88		2.10.2.3	Heideite 103
2.9.1.3	Gallite 88		2.10.3.1	Rhodostannite 103
2.9.1.4	Roquesite 88		2.10.3.2	Toyohaite 103
2.9.1.5	Lenaite 89		2.10.4.1	Konderite 104
2.9.2.1	Stannite 89		2.10.4.2	Inaglyite 104
2.9.2.2	Cernýite 90		2.11.1.1	Orpiment 104
2.9.2.3	Briartite 90		2.11.2.2	Antimonselite 107
2.9.2.4	Kuramite 90		2.11.2.3	Bismuthinite 107
2.9.2.5	Sakuraiite 90		2.11.2.4	Guanajuatite 108
2.9.2.6	Hocartite 91		2.11.3.1	Metastibnite 108
2.9.2.7	Pirquitasite 91		2.11.4.1	Wakabayashilite 108
2.9.2.8	Velikite 91		2.11.5.1	Pääkkönenite 108
2.9.2.9	Kesterite 91		2.11.6.1	Laphamite 109
2.9.2.10	Ferrokesterite 91		2.11.7.1	Tetradymite 110
2.9.3.1	Mawsonite 92		2.11.7.2	Tellurobismuthite 110
2.9.3.2	Chatkalite 92		2.11.7.3	Tellurantimony 111
2.9.3.3	Stannoidite 92		2.11.7.4	Paraguanajuatite 111
2.9.4.1	Renierite 92		2.11.7.5	Kawazulite 111
2.9.4.2	Germanite 93		2.11.7.6	Skippenite 111
2.9.5.1	Morozeviczite 93		2.11.8.1	Montbrayite 111
2.9.5.2	Polkovicite 93		2.11.9.1	Ottemannite 112
2.9.6.1	Hemusite 93		2.11.10.1	Nagyagite 112
2.9.6.2	Kiddcreekite 93		2.11.11.1	Buckhornite 112
2.9.7.1	Putoranite 94		2.11.12.1	Bowieite 112
2.9.8.1	Talnakhite 94		2.11.12.2	Kashinite 113
2.9.8.2	Haycockite 94		2.12.1.1	Pyrite 114
2.9.8.3	Mooihoekite 94		2.12.1.2	Vaesite 115
2.9.9.1	Raguinite 94		2.12.1.3	Cattierite 115
2.9.10.1	Teallite 95		2.12.1.4	Penroseite 115
2.9.11.1	Rasvumite 95		2.12.1.5	Trogtalite 116
2.9.12.1	Sternbergite 95		2.12.1.6	Villamaninite 116
2.9.12.2	Picotpaulite 95		2.12.1.7	Fukuchilite 116
2.9.13.1	Cubanite 96		2.12.1.8	Krutaite 116
2.9.13.2	Argentopyrite 96		2.12.1.9	Hauerite 117
2.9.13.3	Isocubanite 96		2.12.1.10	Laurite 117
2.9.14.1	Idaite 96		2.12.1.11	Aurostibite 117
2.9.15.1	Nukundamite 97		2.12.1.12	Krutovite 117
2.9.16.1	Mohite 97		2.12.1.13	Sperrylite 118
2.9.17.1	Caswellsilverite 97		2.12.1.14	Geversite 118
2.9.17.2	Schöllhornite 97		2.12.1.15	Insizwaite 118
2.9.18.1	Petrukite 97		2.12.1.16	Erlichmanite 118
2.9.19.1	Bartonite 98		2.12.1.17	Dzarkenite 119
2.10.1.1	Linnaeite 99		2.12.1.18	Gaotaiite 119
2.10.1.2	Carrollite 99		2.12.1.19	Mayingite 119
2.10.1.3	Fletcherite 100		2.12.2.1	Marcasite 120
2.10.1.4	Tyrrellite 100		2.12.2.2	Ferroselite 121
2.10.1.5	Bornhardtite 100		2.12.2.3	Frohbergite 121
2.10.1.6	Siegenite 100		2.12.2.4	Hastite 122
2.10.1.7	Polydymite 100		2.12.2.5	Mattagamite 122
2.10.1.8	Violarite 101		2.12.2.6	Kullerudite 122
2.10.1.9	Trüstedtite 101		2.12.2.7	Omeiite 122
2.10.1.10	Greigite 101		2.12.2.8	Anduoite 122
2.10.1.11	Daubréelite 101		2.12.2.9	Löllingite 122
2.10.1.12	Indite 102		2.12.2.10	Seinäjokite 123
2.10.1.13	Kalininite 102		2.12.2.11	Safflorite 123
2.10.1.14	Florensovite 102		2.12.2.12	Rammelsbergite 123
2.10.1.15	Cuproiridisite 102		2.12.2.13	Nisbite 124
2.10.1.16	Cuprorhodisite 102		2.12.3.1	Cobaltite 125
2.10.1.17	Malanite 103		2.12.3.2	Gersdorffite 125

2.12.3.3	Ullmannite 125		2.15.3.1	Kolarite 143
2.12.3.4	Willyamite 126		2.15.4.1	Radhakrishnaite 143
2.12.3.5	Tolovkite 126		2.15.5.1	Perroudite 143
2.12.3.6	Platarsite 126		2.15.6.1	Capgarronite 144
2.12.3.7	Irarsite 126		2.16.1.1	Mertieite-I 144
2.12.3.8	Hollingworthite 127		2.16.2.1	Isomertieite 144
2.12.3.9	Jolliffeite 127		2.16.3.1	Mertieite-II 144
2.12.3.10	Padmaite 127		2.16.4.1	Stillwaterite 144
2.12.3.11	Michenerite 127		2.16.5.1	Arsenopalladinite 145
2.12.3.12	Maslovite 127		2.16.6.1	Telluropalladinite 145
2.12.3.13	Testibiopalladite 128		2.16.7.1	Balkanite 145
2.12.4.1	Arsenopyrite 129		2.16.7.2	Danielsite 145
2.12.4.2	Gudmundite 129		2.16.8.1	Betekhtinite 145
2.12.4.3	Osarsite 130		2.16.9.1	Larosite 146
2.12.4.4	Ruarsite 130		2.16.10.1	Sopcheite 146
2.12.4.5	Iridarsenite 130		2.16.11.1	Henryite 146
2.12.4.6	Clinosafflorite 130		2.16.12.1	Furutobeite 146
2.12.5.1	Pararammelsbergite 131		2.16.13.1	Stützite 146
2.12.5.2	Paxite 131		2.16.14.1	Gortdrumite 147
2.12.6.1	Glaucodot 131		2.16.15.1	Rickardite 147
2.12.6.2	Alloclasite 131		2.16.15.2	Oosterboschite 147
2.12.7.1	Costibite 132		2.16.16.1	Maucherite 147
2.12.7.2	Paracostibite 132		2.16.17.1	Athabascaite 148
2.12.8.1	Lautite 132		2.16.18.1	Penzhinite 148
2.12.9.1	Bambollaite 132		2.16.19.1	Palladseite 148
2.12.10.1	Molybdenite 132		2.16.20.1	Cameronite 148
2.12.10.2	Drysdallite 134		2.16.21.1	Patronite 148
2.12.10.3	Tungstenite 134		2.16.22.1	Vasilite 149
2.12.11.1	Jordisite 134		2.16.23.1	Luberoite 149
2.12.12.1	Jeromite 134		3.1.1.1	Colusite 150
2.12.13.1	Krennerite 134		3.1.1.2	Germanocolusite 151
2.12.13.2	Calaverite 135		3.1.1.3	Nekrasovite 151
2.12.13.3	Sylvanite 135		3.1.1.4	Stibiocolusite 151
2.12.13.4	Kostovite 136		3.1.2.1	Vinciennite 151
2.12.14.1	Melonite 137		3.1.3.1	Levyclaudite 151
2.12.14.2	Kitkaite 137		3.1.4.1	Cylindrite 152
2.12.14.3	Moncheite 137		3.1.4.2	Franckeite 152
2.12.14.4	Merenskyite 138		3.1.4.3	Incaite 152
2.12.14.5	Berndtite 138		3.1.4.4	Potosiite 152
2.12.14.6	Shuangfengite 138		3.1.5.1	Miharaite 153
2.12.15.1	Froodite 138		3.1.6.1	Billingsleyite 153
2.12.15.2	Urvantsevite 139		3.1.7.1	Arsenpolybasite 153
2.12.16.1	Borishanskiite 139		3.1.7.2	Polybasite 153
2.12.17.1	Skutterudite 139		3.1.8.1	Pearceite 153
2.12.17.2	Nickel-Skutterudite 139		3.1.8.2	Antimonpearceite 154
2.12.17.3	Kieftite 140		3.1.9.1	Petrovicite 154
2.13.1.1	Kermesite 140		3.1.10.1	Benleonardite 154
2.13.2.1	Sarabauite 140		3.1.11.1	Aschamalmite 154
2.13.3.1	Cetineite 140		3.1.12.1	Tsnigriite 155
2.14.1.1	Valleriite 141		3.2.1.1	Enargite 155
2.14.2.1	Tochilinite 141		3.2.2.1	Luzonite 156
2.14.3.1	Haapalaite 141		3.2.2.2	Famatinite 156
2.14.3.2	Yushkinite 141		3.2.2.3	Permingeatite 156
2.14.4.1	Vyalsovite 142		3.2.3.1	Sulvanite 157
2.14.5.1	Erdite 142		3.2.3.2	Arsenosulvanite 157
2.14.6.1	Coyoteite 142		3.2.4.1	Stephanite 157
2.14.7.1	Orickite 142		3.2.4.2	Selenostephanite 157
2.15.1.1	Ardaite 142		3.2.5.1	Hauchecornite 158
2.15.2.1	Djerfisherite 143		3.2.5.2	Bismutohauchecornite 159
2.15.2.2	Thalfenisite 143		3.2.5.3	Tellurohauchecornite 159
2.15.2.3	Owensite 143		3.2.5.4	Arsenohauchecornite 159

Index of Mineral Names in Numerical Order

3.2.5.5	Tučekite 159	3.5.1.1	Neyite 178
3.2.6.1	Arcubisite 159	3.5.2.1	Boulangerite 178
3.2.7.1	Chalcothallite 160	3.5.2.2	Falkmanite 179
3.2.8.1	Chaméanite 160	3.5.3.1	Sterryite 179
3.2.9.1	Fangite 160	3.5.4.1	Diaphorite 179
3.3.1.1	Jordanite 160	3.5.5.1	Schirmerite 179
3.3.1.2	Geocronite 160	3.5.6.1	Ourayite 180
3.3.2.1	Gratonite 161	3.5.7.1	Madocite 180
3.3.3.1	Heyrovskyite 161	3.5.8.1	Wallisite 180
3.3.4.1	Nuffieldite 161	3.5.8.2	Hatchite 180
3.3.5.1	Meneghinite 162	3.5.9.1	Cosalite 180
3.3.6.1	Tetrahedrite 162	3.5.9.2	Veenite 181
3.3.6.2	Tennantite 164	3.5.9.3	Dufrénoysite 181
3.3.6.3	Freibergite 164	3.5.10.1	Owyheeite 181
3.3.6.4	Hakite 164	3.5.11.1	Imhofite 181
3.3.6.5	Giraudite 164	3.5.12.1	Izoklakeite 182
3.3.6.6	Goldfieldite 165	3.5.12.2	Giessenite 182
3.3.6.7	Argentotennantite 165	3.5.13.1	Jaskólskiite 182
3.3.7.1	Lengenbachite 165	3.5.14.1	Zoubekite 182
3.4.1.1	Proustite 165	3.5.15.1	Watanabeite 182
3.4.1.2	Pyrargyrite 166	3.6.1.1	Proudite 183
3.4.2.1	Xanthoconite 166	3.6.2.1	Eskimoite 183
3.4.2.2	Pyrostilpnite 167	3.6..3.1	Treasurite 183
3.4.3.1	Seligmanite 167	3.6.4.1	Playfairite 183
3.4.3.2	Bournonite 167	3.6.5.1	Cannizzarite 183
3.4.3.3	SoucŬekite 168	3.6.5.2	Wittite 184
3.4.4.1	Lapieite 168	3.6.6.1	Launayite 184
3.4.4.2	Mückeite 168	3.6.7.1	Jamesonite 184
3.4.5.1	Aikinite 169	3.6.7.2	Benavidesite 184
3.4.5.2	Krupkaite 170	3.6.8.1	Dadsonite 185
3.4.5.3	Gladite 170	3.6.9.1	Parajamesonite 185
3.4.5.4	Hammarite 170	3.6.10.1	Eclarite 185
3.4.5.5	Friedrichite 170	3.6.11.1	Vikingite 185
3.4.5.6	Pekoite 170	3.6.12.1	Sorbyite 185
3.4.5.7	Lindströmite 171	3.6.13.1	Baumhauerite 186
3.4.6.1	Marrite 171	3.6.14.1	Sinnerite 186
3.4.6.2	Freieslebenite 171	3.6.15.1	Berryite 186
3.4.7.2	Mgriite 171	3.6.16.1	Robinsonite 186
3.4.8.1	Wittichenite 172	3.6.17.1	Liveingite 187
3.4.8.2	Skinnerite 172	3.6.18.1	Weibullite 187
3.4.9.1	Ellisite 172	3.6.19.1	Kobellite 187
3.4.10.1	Christite 172	3.6.19.2	Tintinaite 187
3.4.10.2	Laffittite 173	3.6.20.1	Fülöppite 188
3.4.11.1	Routhierite 173	3.6.20.2	Plagionite 188
3.4.11.2	Stalderite 173	3.6.20.3	Heteromorphite 189
3.4.12.1	Samsonite 173	3.6.20.4	Semseyite 189
3.4.13.1	Nowackiite 173	3.6.20.5	Rayite 190
3.4.13.2	Aktashite 174	3.7.1.1	Matildite 190
3.4.13.3	Gruzdevite 174	3.7.1.2	Bohdanowiczite 190
3.4.14.1	Galkhaite 174	3.7.1.3	Volynskite 190
3.4.15.1	Lillianite 175	3.7.2.1	Trechmannite 191
3.4.15.2	Bursaite 176	3.7.3.1	Smithite 191
3.4.15.3	Gustavite 176	3.7.3.2	Miargyrite 191
3.4.15.4	Andorite 176	3.7.4.1	Aramayoite 191
3.4.15.5	Uchucchacuaite 177	3.7.5.1	Chalcostibite 192
3.4.15.6	Ramdohrite 177	3.7.5.2	Emplectite 192
3.4.15.7	Roshchinite 177	3.7.6.1	Lorandite 192
3.4.15.8	Fizelyite 177	3.7.7.1	Weissbergite 193
3.4.16.1	Xilingolite 178	3.7.8.1	Sartorite 193
3.4.17.1	Kirkiite 178	3.7.8.2	Guettardite 193
3.4.18.1	Erniggliite 178	3.7.8.3	Twinnite 193

3.7.9.1	Galenobismutite 194		4.3.3.3	Loparite-(Ce) 224
3.7.9.2	Sakharovaite 194		4.3.3.4	Lueshite 224
3.7.9.3	Berthierite 194		4.3.3.5	Tausonite 224
3.7.9.4	Garavellite 195		4.3.4.1	Natroniobite 225
3.7.10.1	Simonite 195		4.3.5.1	Ilmenite 226
3.7.10.2	Edenharterite 195		4.3.5.2	Geikielite 227
3.7.11.1	Livingstonite 195		4.3.5.3	Pyrophanite 228
3.7.12.1	Rathite 195		4.3.5.4	Ecandrewsite 228
3.7.13.1	Nordströmite 196		4.3.5.5	Melanostibite 228
3.7.14.1	Junoite 196		4.3.5.6	Briziite 229
3.7.15.1	Vrbaite 196		4.3.6.1	Macedonite 229
3.7.16.1	Rohaite 196		4.3.7.1	Maghemite 229
3.7.17.1	Dervillite 196		4.3.7.2	Bixbyite 230
3.7.18.1	Watkinsonite 197		4.3.8.1	Avicennite 231
3.8.1.1	Zinkenite 197		4.3.9.1	Arsenolite 231
3.8.2.1	Cuprobismutite 197		4.3.9.2	Sénarmontite 231
3.8.3.1	Tvalchrelidzeite 197		4.3.10.1	Claudetite 232
3.8.4.1	Hodrushite 198		4.3.10.2	Bismite 232
3.8.4.2	Paderaite 198		4.3.11.1	Valentinite 233
3.8.6.1	Hutchinsonite 198		4.3.12.1	Sillénite 233
3.8.7.1	Pierrotite 198		4.3.13.1	Sphaerobismoite 233
3.8.8.1	Gerstleyite 198		4.4.1.1	Rutile 235
3.8.9.1	Gillulyite 199		4.4.1.2	Ilmenorutile 237
3.8.10.1	Pavonite 199		4.4.1.3	Strüverite 237
3.8.10.2	Makovickyite 200		4.4.1.4	Pyrolusite 238
3.8.10.3	Benjaminite 200		4.4.1.5	Cassiterite 239
3.8.10.4	Mummeite 201		4.4.1.6	Plattnerite 242
3.8.10.5	Borodaevite 201		4.4.1.7	Argutite 243
3.8.10.6	Cupropavonite 201		4.4.1.8	Squawcreekite 243
3.8.11.1	Ustarasite 201		4.4.1.9	Stishovite 243
3.8.12.1	Chabournéite 201		4.4.2.1	Varlamoffite 244
3.8.13.1	Rebulite 202		4.4.3.1	Downeyite 244
3.8.13.2	Jankovićite 202		4.4.3.2	Paratellurite 244
3.8.14.1	Parapierrotite 202		4.4.4.1	Anatase 245
3.8.14.2	Bernardite 202		4.4.5.1	Brookite 246
3.8.15.1	Vaughanite 202		4.4.6.1	Tellurite 247
3.9.1.1	Criddleite 203		4.4.6.2	Scrutinyite 247
4.1.1.1	Cuprite 204		4.4.7.1	Ramsdellite 247
4.1.2.1	Ice 205		4.4.8.1	Nsutite 248
4.2.1.1	Periclase 209		4.4.9.1	Vernadite 248
4.2.1.2	Bunsenite 209		4.4.10.1	Akhtenskite 249
4.2.1.3	Manganosite 209		4.4.11.1	Paramontroseite 249
4.2.1.4	Monteponite 210		4.4.12.1	Cerianite-(Ce) 249
4.2.1.5	Lime 210		4.4.13.1	Tazheranite 250
4.2.1.6	Wüstite 210		4.4.13.2	Srilankite 250
4.2.1.7	Hongquiite 210		4.4.14.1	Baddeleyite 250
4.2.2.1	Zincite 211		4.4.15.1	Tugarinovite 251
4.2.2.2	Bromellite 211		4.4.16.1	Cervantite 251
4.2.3.1	Tenorite 211		4.4.17.1	Lenoblite 251
4.2.4.1	Litharge 212		4.5.1.1	Molybdite 251
4.2.5.1	Romarchite 212		4.5.2.1	Tungstite 252
4.2.6.1	Montroydite 212		4.5.3.1	Meymacite 252
4.2.7.1	Massicot 213		4.5.4.1	Alumotungstite 252
4.3.1.1	Corundum 214		4.5.5.1	Ferritungstite 252
4.3.1.2	Hematite 217		4.5.6.1	Sidwillite 253
4.3.1.3	Eskolaite 220		4.6.1.1	Shcherbinaite 253
4.3.1.4	Karelianite 220		4.6.2.1	Navajoite 253
4.3.2.1	Akdalaite 221		4.6.3.1	Ilsemannite 253
4.3.2.2	Ferrihydrite 221		4.6.4.1	Paramelaconite 254
4.3.3.1	Perovskite 223		4.6.5.1	Murdochite 254
4.3.3.2	Latrappite 223		4.6.6.1	Tantite 254

7.2.8.1	Minium 306		7.11.8.1	Nigerite 6H 329
7.2.9.1	Chrysoberyl 307		7.11.8.2	Nigerite 24R 329
7.2.10.1	Marokite 309		7.11.9.1	Pengzhizhongite-6H 330
7.2.11.1	Taaffeite 309		7.11.9.2	Pengzhizhongite-24R 330
7.2.12.1	Musgravite 309		7.11.10.1	Kleberite 330
7.2.12.2	Pehrmanite 309		7.11.11.1	Mongshanite 330
7.2.13.1	Filipstadite 310		7.11.12.1	Diaoyudaoite 330
7.2.14.1	Yafsoanite 310		7.11.13.1	Chiluite 330
7.3.1.1	Welinite 310		7.11.14.1	Rilandite 331
7.3.1.2	Franciscanite 311		7.11.15.1	Lindqvistite 331
7.3.1.3	Örebroite 311		8.1.1.1	Fergusonite-(Y) 332
7.3.2.1	Grossite 311		8.1.1.2	Formanite-Y 332
7.4.1.1	Hibonite 311		8.1.1.3	Fergusonite-(Ce) 333
7.4.1.2	Yimengite 312		8.1.1.4	Fergusonite-(Nd) 333
7.4.1.3	Hawthorneite 312		8.1.2.1	β-Fergusonite-(Y) 333
7.4.2.1	Magnetoplumbite 312		8.1.2.2	β-Fergusonite-(Ce) 333
7.5.1.1	Braunite I 312		8.1.2.3	β-Fergusonite-(Nd) 333
7.5.1.2	Neltnerite 314		8.1.3.1	Yttrotantalite-(Y) 334
7.5.1.3	Braunite II 314		8.1.3.2	Yttrocolumbite-(Y) 334
7.5.1.4	Abswurmbachite 314		8.1.4.1	Ishikawaite 334
7.5.2.1	Painite 314		8.1.5.1	Loranskite-(Y) 334
7.5.3.1	Birnessite 315		8.1.6.1	Stibiocolumbite 334
7.5.4.1	Zenzénite 315		8.1.6.2	Stibiotantalite 335
7.6.1.1	Cesárolite 315		8.1.6.3	Bismutotantalite 335
7.6.2.1	Bartelkeite 316		8.1.6.4	Bismutocolumbite 335
7.6.3.1	Kamiokite 316		8.1.7.1	Alumotantite 336
7.7.1.1	Pseudobrookite 316		8.1.8.1	Wodginite 337
7.7.1.2	Armalcolite 317		8.1.8.2	Ferrowodginite 338
7.7.2.1	Berdesinskiite 317		8.1.8.3	Titanowodginite 338
7.8.1.1	Todorokite 317		8.1.8.4	Lithiowodginite 338
7.8.1.2	Woodruffite 319		8.1.9.1	Liandratite 339
7.8.2.1	Chalcophanite 319		8.1.9.2	Petscheckite 339
7.8.2.2	Aurorite 319		8.1.10.1	Ixiolite 339
7.8.2.3	Jianshuiite 320		8.1.11.1	Samarskite-(Y) 340
7.8.2.4	Ernienickelite 320		8.1.12.1	Qitianlingite 340
7.8.3.1	Lazarenkoite 320		8.2.1.1	Pyrochlore 343
7.9.1.1	Hollandite 321		8.2.1.2	Kalipyrochlore 344
7.9.1.2	Cryptomelane 321		8.2.1.3	Bariopyrochlore 344
7.9.1.3	Manjiroite 322		8.2.1.4	Yttropyrochlore-(Y) 344
7.9.1.4	Coronadite 323		8.2.1.5	Ceriopyrochlore-(Ce) 344
7.9.2.1	Romanechite 323		8.2.1.6	Plumbopyrochlore 345
7.9.3.1	Freudenbergite 324		8.2.1.7	Uranpyrochlore 345
7.9.4.1	Priderite 324		8.2.1.8	Strontiopyrochlore 345
7.9.4.2	Ankangite 324		8.2.2.1	Microlite 345
7.9.5.1	Mannardite 324		8.2.2.2	Bariomicrolite 346
7.9.5.2	Redledgeite 325		8.2.2.3	Plumbomicrolite 346
7.10.1.1	Ranciéite 325		8.2.2.4	Uranmicrolite 346
7.10.1.2	Takanelite 325		8.2.2.5	Bismutomicrolite 347
7.10.2.1	Karibibite 325		8.2.2.6	Stannomicrolite 347
7.10.3.1	Otjisumeite 326		8.2.2.7	Stibiomicrolite 347
7.11.1.1	Muskoxite 326		8.2.3.1	Betafite 347
7.11.2.1	Brownmillerite 326		8.2.3.2	Yttrobetafite-(Y) 348
7.11.2.2	Srebrodol'skite 327		8.2.3.3	Plumbobetafite 348
7.11.3.1	Mayenite 327		8.2.3.4	Stibiobetafite 348
7.11.4.1	Hematophanite 327		8.2.3.5	Calciobetafite 348
7.11.5.1	Plumboferrite 327		8.2.4.1	Cesstibtantite 349
7.11.6.1	Carboirite 328		8.2.4.2	Natrobistantite 349
7.11.7.1	Högbomite 4H, 5H, 6H, 15H 328		8.2.5.1	Zirkelite 349
			8.2.5.2	Zirconolite-3O 349
7.11.7.2	Högbomite 8H 328		8.2.5.3	Zirconolite-2M 350
7.11.7.3	Högbomite 15R, 18R, 24R 328		8.2.5.4	Zirconolite-3T 350

4.6.7.1	Nolanite 254	6.2.13.1	Cianciulliite 280
5.1.1.1	Uraninite 255	6.3.1.1	Gibbsite 280
5.1.1.2	Thorianite 256	6.3.2.1	Bayerite 281
5.2.1.1	Metaschoepite 257	6.3.3.1	Nordstrandite 281
5.2.1.2	Paraschoepite 257	6.3.4.1	Doyleite 282
5.2.1.3	Schoepite 257	6.3.5.1	Dzhalindite 282
5.2.2.1	Masuyite 258	6.3.5.2	Söhngeite 283
5.3.1.1	Studtite 258	6.3.5.3	Bernalite 283
5.3.1.2	Metastudtite 259	6.3.6.1	Wickmanite 284
5.3.2.1	Vandenbrandeite 259	6.3.6.2	Schoenfliesite 285
5.4.1.1	Clarkeite 259	6.3.6.3	Natanite 285
5.4.2.1	Calciouranoite 259	6.3.6.4	Vismirnovite 285
5.4.2.2	Bauranoite 260	6.3.6.5	Burtite 285
5.4.3.1	Metacalciouranoite 260	6.3.6.6	Mushistonite 286
5.4.3.2	Wölsendorfite 260	6.3.7.1	Stottite 286
5.5.1.1	Agrinierite 260	6.3.7.2	Tetrawickmanite 286
5.5.2.1	Rameauite 260	6.3.7.3	Jeanbandyite 286
5.5.3.1	Protasite 261	6.3.7.4	Mopungite 287
5.6.1.1	Ianthinite 261	6.3.8.1	Jamborite 287
5.7.1.1	Compreignacite 261	6.3.9.1	Bottinoite 287
5.7.1.2	Becquerelite 261	6.4.1.1	Lithiophorite 288
5.7.1.3	Billietite 262	6.4.1.2	Quenselite 288
5.8.1.1	Vandendriesscheite 262	6.4.2.1	Hydroromarchite 288
5.8.1.2	Metavandendriesscheite 262	6.4.3.1	Häggite 289
5.9.1.1	Uranosphaerite 263	6.4.4.1	Hydrocalumite 289
5.9.2.1	Fourmarierite 263	6.4.5.1	Iowaite 289
5.9.3.1	Curite 263	6.4.6.1	Meixnerite 289
5.9.4.1	Mourite 264	6.4.7.1	Janggunite 290
5.9.5.1	Richetite 264	6.4.8.1	Kimrobinsonite 290
5.9.6.1	Sayrite 264	6.4.9.1	Asbolane 290
6.1.1.1	Diaspore 266	6.4.10.1	Schwertmannite 290
6.1.1.2	Goethite 268	7.1.1.1	Delafossite 292
6.1.1.3	Groutite 269	7.1.1.2	Mcconnellite 292
6.1.1.4	Montroseite 269	7.1.2.1	Crednerite 292
6.1.1.5	Bracewellite 269	7.2.1.1	Spinel 295
6.1.2.1	Boehmite 270	7.2.1.2	Galaxite 297
6.1.2.2	Lepidocrocite 270	7.2.1.3	Hercynite 297
6.1.2.3	Guyanaite 271	7.2.1.4	Gahnite 297
6.1.3.1	Manganite 271	7.2.2.1	Magnesioferrite 298
6.1.4.1	Heterogenite-2H 272	7.2.2.2	Jacobsite 298
6.1.4.2	Heterogenite-3R 272	7.2.2.3	Magnetite 299
6.1.4.3	Feitknechtite 272	7.2.2.4	Franklinite 300
6.1.4.4	Feroxyhyte 273	7.2.2.5	Trevorite 300
6.1.5.1	Grimaldiite 273	7.2.2.6	Cuprospinel 301
6.1.6.1	Akaganéite 273	7.2.2.7	Brunogeierite 301
6.2.1.1	Brucite 275	7.2.3.1	Magnesiochromite 301
6.2.1.2	Amakinite 275	7.2.3.2	Manganochromite 301
6.2.1.3	Pyrochroite 276	7.2.3.3	Chromite 302
6.2.1.4	Portlandite 276	7.2.3.4	Nichromite 302
6.2.1.5	Theophrastite 276	7.2.3.5	Cochromite 303
6.2.2.1	Behoite 277	7.2.3.6	Zincochromite 303
6.2.3.1	Clinobehoite 277	7.2.4.1	Vuorelainenite 303
6.2.4.1	Spertiniite 277	7.2.4.2	Coulsonite 303
6.2.5.1	Calumetite 277	7.2.4.3	Magnesiocoulsonite 303
6.2.6.1	Anthonyite 278	7.2.5.1	Qandilite 304
6.2.7.1	Duttonite 278	7.2.5.2	Ulvöspinel 304
6.2.8.1	Tivanite 278	7.2.6.1	Kusachiite 304
6.2.9.1	Sweetite 278	7.2.6.2	Iwakiite 305
6.2.10.1	Wülfingite 279	7.2.7.1	Hausmannite 305
6.2.11.1	Ashoverite 279	7.2.7.2	Hetaerolite 305
6.2.12.1	Paraotwayite 279	7.2.7.3	Hydrohetaerolite 306

Index of Mineral Names in Numerical Order

8.2.5.5	Zirconolite 350	8.7.10.1	Zimbabweite 372
8.2.6.1	Scheteligite 350	8.7.11.1	Cesplumtantite 372
8.2.7.1	Orthobrannerite 351	8.7.12.1	Koragoite 372
8.2.8.1	Parabariomicrolite 351	9.1.1.1	Halite 373
8.3.1.1	Ferrotapiolite 351	9.1.1.2	Sylvite 375
8.3.1.2	Manganotapiolite 352	9.1.1.3	Villiaumite 375
8.3.2.1	Ferrotantalite 354	9.1.1.4	Carobbiite 376
8.3.2.2	Ferrocolumbite 354	9.1.1.5	Griceite 376
8.3.2.3	Manganotantalite 355	9.1.2.1	Hydrohalite 376
8.3.2.4	Manganocolumbite 355	9.1.3.1	Sal Ammoniac 377
8.3.2.5	Magnocolumbite 356	9.1.4.1	Chlorargyrite 377
8.3.3.1	Fersmite 356	9.1.4.2	Bromargyrite 378
8.3.4.1	Brannerite 356	9.1.5.1	Iodargyrite 378
8.3.4.2	Thorutite 357	9.1.6.1	Tocornalite 378
8.3.5.1	Lucasite-(Ce) 357	9.1.7.1	Nantokite 379
8.3.6.1	Aeschynite-(Ce) 357	9.1.7.2	Miersite 379
8.3.6.2	Niobo-aeschynite-(Ce) 358	9.1.7.3	Marshite 379
8.3.6.3	Aeschynite-(Y) 358	9.1.8.1	Calomel 379
8.3.6.4	Tantalaeschynite-(Y) #358	9.1.8.2	Kuzminite 380
8.3.6.5	Aeschynite-(Nd) 359	9.1.8.3	Moschelite 380
8.3.7.1	Vigezzite 359	9.2.1.1	Fluorite 380
8.3.7.2	Rynersonite 359	9.2.1.3	Tveitite-(Y) 384
8.3.8.1	Polycrase-(Y) 359	9.2.2.1	Sellaite 384
8.3.8.2	Euxenite-(Y) 360	9.2.3.1	Lawrencite 384
8.3.8.3	Tanteuxenite-(Y) 360	9.2.3.2	Scacchite 384
8.3.8.4	Yttrocrasite-(Y) 361	9.2.3.3	Chloromagnesite 385
8.3.8.5	Uranopolycrase 361	9.2.4.1	Rokühnite 385
8.3.9.1	Kassite 361	9.2.5.1	Sinjarite 385
8.3.10.1	Thoreaulite 361	9.2.6.1	Antarcticite 385
8.3.10.2	Foordite 362	9.2.7.1	Cotunnite 386
8.3.11.1	Changbaiite 362	9.2.7.2	Hydrophilite 386
8.3.12.1	Kobeite-(Y) 362	9.2.8.1	Eriochalcite 386
8.4.1.1	Schreyerite 362	9.2.8.2	Tolbachite 387
8.4.1.2	Olkhonskite 363	9.2.9.1	Bischofite 387
8.4.1.3	Kyzylkumite 363	9.2.9.2	Nickelbischofite 387
8.4.2.1	Pseudorutile 363	9.2.10.1	Laurelite 387
8.5.1.1	Landauite 365	9.2.11.1	Matlockite 388
8.5.1.2	Loveringite 365	9.2.11.2	Rorisite 388
8.5.1.3	Crichtonite 366	9.3.1.1	Molysite 388
8.5.1.4	Senaite 366	9.3.2.1	Hydromolysite 388
8.5.1.5	Davidite-(La) 366	9.3.3.1	Chloraluminite 389
8.5.1.6	Davidite-(Ce) 366	9.3.4.1	Fluocerite-(Ce) 389
8.5.1.7	Mathiasite 367	9.3.4.2	Fluocerite-(La) 389
8.5.1.8	Lindsleyite 367	9.3.5.1	Gananite 389
8.6.1.1	Franconite 367	9.3.6.1	Rosenbergite 390
8.6.1.2	Hochelagaite 368	10.1.1.1	Atacamite 391
8.6.2.1	Calciotantite 368	10.1.1.2	Hibbingite 391
8.6.2.2	Natrotantite 368	10.1.2.1	Paratacamite 392
8.6.2.3	Irtyshite 368	10.1.3.1	Botallackite 392
8.7.1.1	Murataite 369	10.1.4.1	Kempite 392
8.7.2.1	Uhligite 369	10.1.5.1	Korshunovskite 392
8.7.3.1	Cafetite 369	10.2.1.1	Zavaritskite 393
8.7.4.1	Calzirtite 369	10.2.1.2	Bismoclite 393
8.7.5.1	Simpsonite 370	10.2.1.3	Daubréeite 393
8.7.6.1	Rankamaite 370	10.2.2.1	Laurionite 394
8.7.6.2	Sosedkoite} 370	10.2.3.1	Paralaurionite 394
8.7.7.1	Lithiotantite 371	10.2.4.1	Blixite 394
8.7.8.1	Belyankinite 371	10.2.5.1	Perite 394
8.7.8.2	Manganbelyankinite 371	10.2.5.2	Nadorite 395
8.7.8.3	Gerasimovskite 371	10.2.6.1	Thorikosite 395
8.7.9.1	Jeppeite 372	10.2.7.1	Asisite 395

10.2.8.1	Zharchikhite 395		11.6.3.1	Colquiriite 413
10.2.9.1	Pinalite 396		11.6.4.1	Cryolithionite 413
10.3.1.1	Mendipite 396		11.6.5.1	Pachnolite 413
10.3.2.1	Fiedlerite 396		11.6.6.1	Thomsenolite 414
10.3.3.1	Corderoite 396		11.6.7.1	Carlhintzeite 414
10.3.4.1	Lavrentievite 397		11.6.8.1	Gearksutite 414
10.3.4.2	Arzakite 397		11.6.9.1	Prosopite 414
10.3.5.1	Grechishchevite 397		11.6.10.1	Jarlite 415
10.3.6.1	Radtkeite 397		11.6.10.2	Calcjarlite 415
10.4.1.1	Penfieldite 398		11.6.11.1	Chiolite 415
10.4.2.1	Terlinguaite 398		11.6.12.1	Ralstonite 416
10.4.3.1	Kleinite 398		11.6.13.1	Weberite 416
10.4.4.1	Melanothallite 398		11.6.14.1	Usovite 416
10.5.1.1	Cadwaladerite 399		11.6.15.1	Yaroslavite 416
10.5.2.1	Poyarkovite 399		11.6.16.1	Tikhonenkovite 417
10.5.3.1	Pinchite 399		11.6.17.1	Acuminite 417
10.5.4.1	Eglestonite 399		11.6.18.1	Artroeite 417
10.5.4.2	Kadyrelite 400		11.6.19.1	Aravaipaite 417
10.5.5.1	Comancheite 400		11.6.20.1	Bøgvadite 418
10.5.6.1	Claringbullite 400		11.6.21.1	Karasugite 418
10.5.7.1	Onoratoite 401		12.1.1.1	Stenonite 419
10.5.8.1	Simonkolleite 401		12.1.2.1	Grandreefite 419
10.5.9.1	Abhurite 401		12.1.3.1	Pseudograndreefite 419
10.5.10.1	Damaraite 401		12.1.4.1	Creedite 420
10.5.11.1	Parkinsonite 402		12.1.5.1	Chukhrovite-(Y) 420
10.5.12.1	Lorettoite 402		12.1.5.2	Chukhrovite-(Ce) 420
10.5.13.1	Hanawaltite 402		12.1.6.1	Boggildite 420
10.6.1.1	Diaboleite 402		12.1.7.1	Barstowite 421
10.6.2.1	Koenenite 403		12.1.8.1	Arzrunite 421
10.6.3.1	Yedlinite 403		13.1.1.1	Nahcolite 422
10.6.4.1	Chloroxiphite 403		13.1.2.1	Kalicinite 423
10.6.5.1	Zirklerite 404		13.1.3.1	Teschemacherite 423
10.6.6.1	Boleite 404		13.1.4.1	Trona 423
10.6.7.1	Cumengéite 404		13.1.5.1	Nesquehonite 424
10.6.8.1	Pseudoboleite 404		13.1.6.1	Wegscheiderite 424
10.6.9.1	Bideauxite 405		13.1.7.1	Sergeevite 424
10.6.10.1	Chlormagaluminite 405		13.1.8.1	Barentsite 425
10.6.11.1	Kelyanite 405		14.1.1.1	Calcite 428
10.6.12.1	Ponomarevite 406		14.1.1.2	Magnesite 434
11.1.1.1	Neighborite 407		14.1.1.3	Siderite 435
11.1.2.1	Carnallite 407		14.1.1.4	Rhodochrosite 436
11.1.3.1	Chlorocalcite 408		14.1.1.5	Sphaerocobaltite 438
11.2.1.1	Pseudocotunnite 408		14.1.1.6	Smithsonite 438
11.2.2.1	Avogadrite 408		14.1.1.7	Otavite 439
11.2.3.1	Ferruccite 409		14.1.1.8	Gaspéite 439
11.2.4.1	Barberiite 409		14.1.2.1	Vaterite 440
11.3.1.1	Douglasite 409		14.1.2.2	Gregoryite 440
11.3.2.1	Mitscherlichite 409		14.1.3.2	Witherite 444
11.4.1.1	Erythrosiderite 409		14.1.3.3	Strontianite 445
11.4.1.2	Kremersite 410		14.1.3.4	Cerussite 446
11.5.1.1.	Hieratite 410		14.1.4.1	Rutherfordine 447
11.5.1.2	Cryptohalite 410		14.1.5.1	Widenmannite 447
11.5.2.1	Malladrite 410		14.2.1.1	Dolomite 450
11.5.2.2	Bararite 411		14.2.1.2	Ankerite 451
11.5.3.1	Rinneite 411		14.2.1.3	Kutnohorite 452
11.5.4.1	Chlormanganokalite 411		14.2.1.4	Minrecordite 452
11.5.5.1	Tachyhydrite 411		14.2.2.1	Norsethite 452
11.5.6.1	Gagarinite-(Y) 411		14.2.2.2	Paralstonite 453
11.5.7.1	Zajacite-(Ce) 412		14.2.2.3	Olekminskite 453
11.6.1.1	Cryolite 412		14.2.3.1	Benstonite 453
11.6.2.1	Elpasolite 413		14.2.4.1	Ewaldite 453

Index of Mineral Names in Numerical Order

14.2.5.1	Alstonite	454
14.2.6.1	Barytocalcite	454
14.3.1.1	Bütschliite	455
14.3.2.1	Eitelite	455
14.3.3.1	Fairchildite	455
14.3.3.2	Zemkorite	456
14.3.4.1	Nyerereite	456
14.3.5.1	Natrofairchildite	456
14.4.1.1	Shortite	456
14.4.2.1	Sahamalite-(Ce)	457
14.4.3.1	Huntite	457
14.4.4.1	Burbankite	459
14.4.4.2	Khanneshite	459
14.4.4.3	Calcioburbankite	460
14.4.5.1	Remondite-(Ce)	460
14.4.5.2	Petersenite-(Ce)	460
14.4.6.1	Carbocernaite	461
15.1.1.1	Thermonatrite	462
15.1.2.1	Natron	462
15.1.3.1	Monohydrocalcite	463
15.1.4.1	Ikaite	463
15.1.5.1	Barringtonite	463
15.1.6.1	Lansfordite	464
15.1.7.1	Hellyerite	464
15.1.8.1	Joliotite	464
15.2.1.1	Pirssonite	465
15.2.2.1	Gaylussite	465
15.2.3.1	Chalconatronite	465
15.2.4.1	Baylissite	466
15.2.5.1	Andersonite	466
15.2.6.1	Grimselite	467
15.3.1.1	Zellerite	467
15.3.1.2	Metazellerite	467
15.3.2.1	Liebigite	468
15.3.3.1	Bayleyite	468
15.3.3.2	Swartzite	469
15.3.4.1	Donnayite-(Y)	469
15.3.4.2	Mckelveyite-(Y)	469
15.3.4.3	Mckelveyite-(Nd)	470
15.3.4.4	Weloganite	470
15.3.5.1	Voglite	471
15.3.6.1	Fontanite	471
15.4.1.1	Calkinsite-(Ce)	471
15.4.2.1	Lanthanite-(La)	472
15.4.2.2	Lanthanite-(Nd)	472
15.4.2.3	Lanthanite-(Ce)	472
15.4.3.1	Tengerite-(Y)	473
15.4.4.1	Lokkaite-(Y)	473
15.4.5.1	Kimuraite-(Y)	473
15.4.6.1	Tuliokite	474
15.4.7.1	Shomiokite-(Y)	474
16a.1.1.1	Bastnäsite-(Ce)	476
16a.1.1.2	Bastnäsite-(La)	478
16a.1.1.3	Bastnäsite-(Y)	478
16a.1.2.1	Hydroxylbastnäsite-(Ce)	478
16a.1.2.2	Hydroxylbastnäsite-(La)	478
16a.1.2.3	Hydroxylbastnäsite-(Nd)	479
16a.1.3.1	Synchysite-(Ce)	479
16a.1.3.2	Synchysite-(Y)	479
16a.1.3.3	Synchysite-(Nd)	480
16a.1.4.1	Huanghoite-(Ce)	480
16a.1.5.1	Parisite-(Ce)	480
16a.1.5.2	Parisite-(Nd)	481
16a.1.6.1	Röntgenite-(Ce)	481
16a.1.7.1	Cordylite-(Ce)	481
16a.1.8.1	Zhonghuacerite-(Ce)	482
16a.1.9.1	Cebaite-(Ce)	482
16a.2.1.1	Azurite	482
16a.2.2.1	Hydrocerussite	485
16a.2.3.1	Beyerite	486
16a.3.1.1	Rosasite	487
16a.3.1.2	Glaukospherite	487
16a.3.1.3	Kolwezite	488
16a.3.1.4	Zincrosasite	488
16a.3.1.5	Mcguinnessite	488
16a.3.2.1	Malachite	488
16a.3.2.2	Nullaginite	490
16a.3.2.3	Pokrovskite	490
16a.3.3.1	Georgeite	490
16a.3.4.1	Phosgenite	491
16a.3.5.1	Bismutite	491
16a.3.6.1	Brenkite	492
16a.3.7.1.	Kettnerite	492
16a.3.8.1	Dawsonite	492
16a.3.9.1	Northupite	493
16a.4.1.1	Hydrozincite	493
16a.4.2.1	Aurichalcite	494
16a.5.1.1	Plumbonacrite	494
16a.5.2.1	Tunisite	494
16a.5.3.1	Loseyite	495
16a.5.3.2	Sclarite	495
16a.5.4.1	Sabinaite	495
16a.5.5.1	Rouvilleite	496
16b.1.1.1	Ancylite-(Ce)	497
16b.1.1.2	Calcioancylite-(Ce)	498
16b.1.1.3	Calcioancylite-(Nd)	498
16b.1.1.4	Gysinite-(Nd)	498
16b.1.2.1	Thorbastnäsites	499
16b.1.3.1	Indigirite	499
16b.1.4.1	Schuilingite-(Nd)	499
16b.1.5.1	Shabaite-(Nd)	499
16b.1.6.1	Astrocyanite-(Ce)	500
16b.1.7.1	Kamphaugite-(Y)	500
16b.2.1.1	Dundasite	501
16b.2.1.2	Dresserite	501
16b.2.1.3	Strontiodresserite	501
16b.2.2.1	Hydrodresserite	501
16b.2.3.1	Alumohydrocalcite	502
16b.2.3.2	Para-alumohydrocalcite	502
16b.2.4.1	Bijvoetite-(Y)	502
16b.3.1.1	Artinite	503
16b.3.2.1	Otwayite	503
16b.3.3.1	Kamotoite-(Y)	503
16b.4.1.1	Zaratite	504
16b.4.2.1	Defernite	504
16b.4.2.2	Holdawayite	505
16b.4.3.1	Claraite	505
16b.4.4.1	Peterbaylissite	505
16b.5.1.1	Callaghanite	506
16b.5.2.1	Urancalcarite	506
16b.5.3.1	Montroyalite	506
16b.6.1.1	Manasseite	508

16b.6.1.2	Barbertonite 509		21.1.2.1	Brüggenite 531	
16b.6.1.3	Sjögrenite 509		21.1.3.1	Bellingerite 531	
16b.6.2.1	Hydrotalcite 509		22.1.1.1	Salesite 532	
16b.6.2.2	Stichtite 510		22.1.2.1	Seeligerite 532	
16b.6.2.3	Pyroaurite 510		22.1.3.1	Schwartzembergite	532
16b.6.2.4	Desautelsite 511		23.1.1.1	Dietzeite 533	
16b.6.3.1	Reevesite 511		23.1.2.1	Fuenzalidaite 533	
16b.6.3.2	Takovite 511		23.1.2.2	Carlosruizite 533	
16b.6.3.3	Comblainite 512		24.1.1.1	Sinhalite 534	
16b.6.4.1	Quintinite-2H 512		24.1.2.1	Behierite 534	
16b.6.4.2	Quintinite-3T 512		24.2.1.1	Ludwigite 535	
16b.6.4.3	Caresite 513		24.2.1.2	Vonsenite 536	
16b.6.4.4	Charmarite-2H 513		24.2.1.3	Azoproite 536	
16b.7.1.1	Hydromagnesite 513		24.2.1.4	Bonaccordite 536	
16b.7.1.2	Widgiemoolthaite 514		24.2.1.5	Chestermanite 537	
16b.7.2.1	Dypingite 514		24.2.1.6	Fredrikssonite 537	
16b.7.2.2	Giorgiosite 514		24.2.2.1	Warwickite 537	
16b.7.3.1	Rabbittite 515		24.2.2.2	Yuanfulite 537	
16b.7.4.1	Wyartite 515		24.2.3.1	Hulsite 538	
16b.7.5.1	Brugnatellite 515		24.2.3.2	Magnesiohulsite 538	
16b.7.6.1	Coalingite 516		24.2.4.1	Pinakiolite 538	
16b.7.7.1	Carbonate-Cyanotrichite	516	24.2.5.1	Orthopinakiolite 538	
16b.7.8.1	Scarbroite 516		24.2.6.1	Takéuchiite 538	
16b.7.9.1	Hydroscarbroite 517		24.2.7.1	Blatterite 538	
16b.7.10.1	Sharpite 517		24.3.1.1	Sassolite 539	
16b.7.11.1	Albrechtschraufite 517		24.3.2.1	Kotoite 539	
16b.7.12.1	Kambaldaite 517		24.3.2.2	Jimboite 539	
16b.7.13.1	Roubaultite 518		24.3.3.1	Nordenskiöldine 539	
16b.7.14.1	Znucalite 518		24.3.3.2	Tusionite 540	
16b.7.15.1	Szymanskiite 518		24.3.4.1	Peprossiite-(Ce) 540	
17.1.1.1	Tychite 519		24.3.5.1	Takedaite 540	
17.1.1.2	Ferrotychite 519		24.4.1.1	Suanite 540	
17.1.1.3	Manganotychite 519		24.4.2.1	Kurchatovite 540	
17.1.2.1	Leadhillite 520		24.4.3.1	Clinokurchatovite 541	
17.1.3.1	Susannite 520		24.5.1.1	Metaborite 541	
17.1.4.1	Macphersonite 521		24.5.2.1	Calciborite 541	
17.1.5.1	Schröckingerite 521		24.5.3.1	Johachidolite 541	
17.1.6.1	Nasledovite 521		24.5.4.1	Diomignite 541	
17.1.7.1	Motukoreaite 522		25.1.1.1	Hambergite 542	
17.1.8.1	Canavesite 522		25.1.2.1	Fluoborite 542	
17.1.9.1	Harkerite 522		25.1.3.1	Frolovite 542	
17.1.10.1	Daqingshanite-(Ce) 523		25.1.4.1	Teepleite 543	
17.1.11.1	Tundrite-(Ce) 523		25.1.4.2	Bandylite 543	
17.1.11.2	Tundrite-(Nd) 523		25.1.5.1	Vimsite 543	
17.1.12.1	Lepersonnite-(Gd) 524		25.1.6.1	Olshanskyite 543	
17.1.13.1	Qilianshanite 524		25.2.1.1	Sussexite 544	
17.1.14.1	Mineevite-(Y) 524		25.2.1.2	Szaibelyite 544	
17.1.15	Brianyoungite 525		25.2.2.1	Sibirskite 544	
18.1.1.1	Nitratine 526		25.2.3.1	Pinnoite 544	
18.1.2.1	Niter 526		25.3.1.1	Ameghinite 545	
18.2.1.1	Nitrobarite 527		25.3.2.1	Solongoite 545	
18.2.2.1	Nitrocalcite 527		25.3.3.1	Fabianite 545	
18.2.3.1	Nitromagnesite 527		25.3.4.1	Uralborite 545	
19.1.1.1	Gerhardtite 528		25.3.5.1	Howlite 545	
19.1.2.1	Buttgenbachite 528		25.4.1.1	Roweite 546	
19.1.3.1	Sveite 528		25.4.1.2	Federovskite 546	
19.1.4.1	Mbobomkulite 528		25.5.1.1	Priceite 546	
19.1.4.2	Hydrombobomkulite 529		25.6.1.1	Boracite 548	
19.1.5.1	Likasite 529		25.6.1.2	Ericaite 548	
20.1.1.1	Darapskite 530		25.6.1.3	Chambersite 548	
21.1.1.1	Lautarite 531		25.6.2.1	Congolite 548	

Index of Mineral Names in Numerical Order

25.6.2.2	Trembathite 549	27.1.1.1	Carboborite 565
25.6.3.1	Strontioborite 549	27.1.2.1	Gaudefroyite 565
25.7.1.1	Preobrazhenskite 549	27.1.3.1	Borcarite 565
25.8.1.1	Jeremejevite 549	27.1.4.1	Sakhaite 565
25.8.2.1	Rhodizite 549	27.1.5.1	Sulfoborite 566
26.1.1.1	Berborite 551	27.1.6.1	Teruggite 566
26.1.2.1	Wightmanite 551	27.1.7.1	Garrelsite 566
26.1.3.1	Shabynite 551	27.1.8.1	Iquiqueite 566
26.1.4.1	Hexahydroborite 551	27.1.9.1	Moydite-(Y) 566
26.1.5.1	Henmilite 552	27.1.10.1	Wiserite 567
26.2.1.1	Pentahydroborite 552	28.1.1.1	Mercallite 568
26.3.1.1	Inyoite 552	28.1.2.1	Misenite 568
26.3.1.2	Inderborite 552	28.1.3.1	Letovicite 568
26.3.1.3	Inderite 553	28.2.1.1	Mascagnite 568
26.3.2.1	Meyerhofferite 553	28.2.1.2	Arcanite 569
26.3.3.1	Kurnakovite 553	28.2.2.1	Aphthitalite 569
26.3.4.1	Hydrochlorborite 553	28.2.3.1	Thenardite 570
26.3.5.1	Colemanite 553	28.2.4.1	Gianellaite 570
26.3.6.1	Hydroboracite 554	28.3.1.1	Barite 572
26.3.7.1	Nifontovite 554	28.3.1.2	Celestine 574
26.4.1.1	Borax 554	28.3.1.3	Anglesite 575
26.4.2.1	Tincalconite 555	28.3.2.1	Anhydrite 576
26.4.3.1	Hungchaoite 555	28.3.3.1	Chalcocyanite 577
26.4.4.1	Halurgite 555	28.3.4.1	Yavapaiite 578
26.4.5.1	Kernite 556	28.3.5.1	Sabieite 578
26.5.1.1	Sborgite 556	28.3.5.2	Godovikovite 578
26.5.2.1	Santite 556	28.4.1.1	Vanthoffite 578
26.5.3.1	Ammonioborite 556	28.4.2.1	Glauberite 579
26.5.4.1	Larderellite 556	28.4.3.1	Palmierite 580
26.5.5.1	Ezcurrite 557	28.4.3.2	Kalistrontite 580
26.5.6.1	Nasinite 557	28.4.3.3	Mereiterite 580
26.5.7.1	Biringuccite 557	28.4.4.1	Langbeinite 581
26.5.8.1	Gowerite 557	28.4.4.2	Manganolangbeinite 581
26.5.9.1	Veatchite 558	28.4.4.3	Efremovite 581
26.5.10.1	Veatchite-A 558	28.4.5.1	Millosevichite 581
26.5.11.1	Ulexite 558	28.4.5.2	Mikasaite 582
26.5.12.1	Probertite 558	28.4.6.1	Klyuchevskite 582
26.5.12.2	Tuzlaite 559	28.4.6.2	Alumoklyuchevskite 582
26.5.13.1	Kaliborite 559	29.1.1.1	Rhomboclase 583
26.5.14.1	Hilgardite 559	29.1.2.1	Matteuccite 583
26.5.15.1	Tyretskite 560	29.1.3.1	Monsmedite 583
26.5.16.1	Volkovskite 560	29.2.1.1	Lecontite 584
26.5.17.1	Pringleite 560	29.2.2.1	Mirabilite 584
26.5.17.2	Ruitenbergite 560	29.3.1.1	Syngenite 584
26.6.1.1	Rivadavite 560	29.3.1.2	Koktaite 585
26.6.2.1	McAllisterite 561	29.3.2.1	Kröhnkite 585
26.6.3.1	Admontite 561	29.3.3.1	Blödite 585
26.6.4.1	Aksaite 561	29.3.3.2	Nickelblödite 586
26.6.5.1	Aristarainite 561	29.3.3.3	Leonite 586
26.6.6.1	Nobleite 562	29.3.4.1	Wattevillite 586
26.6.6.2	Tunellite 562	29.3.5.1	Konyaite 586
26.6.7.1	Ginorite 562	29.3.6.1	Picromerite 588
26.6.7.2	Strontioginorite 562	29.3.6.2	Cyanochroite 588
26.7.1.1	Korzhinskite 563	29.3.7.1	Mohrite 588
26.7.2.1	Chelkarite 563	29.3.7.2	Boussingaultite 588
26.7.3.1	Ekaterinite 563	29.3.7.3	Nickel-Boussingaultite 589
26.7.4.1	Satimolite 563	29.3.8.1	Mosesite 589
26.7.5.1	Tertschite 563	29.4.1.1	Hydroglauberite 589
26.7.6.1	Braitschite-(Ce) 563	29.4.2.1	Eugsterite 590
26.7.7.1	Wardsmithite 564	29.4.3.1	Löweite 590
26.7.8.1	Studenitsite 564	29.4.4.1	Ferrinatrite 590

29.4.5.1	Polyhalite 590	29.8.4.1	Paracoquimbite 619
29.4.5.2	Leightonite 591	29.8.5.1	Quenstedtite 619
29.4.6.1	Metavoltine 591	29.8.6.1	Alunogen 620
29.4.7.1	Görgeyite 591	29.8.7.1	Meta-alunogen 620
29.5.1.1	Krausite 591	29.9.1.1	Voltaite 620
29.5.2.1	Goldichite 592	29.9.1.2	Zincovoltaite 621
29.5.3.1	Tamarugite 592	29.9.2.1	Zircosulfate 621
29.5.3.2	Amarillite 592	29.9.3.1	Bazhenovite 621
29.5.4.1	Mendozite 593	30.1.1.1	Sundiusite 623
29.5.4.2	Kalinite 593	30.1.2.1	Elyite 623
29.5.5.1	Potassium alum 593	30.1.3.1	Brochantite 623
29.5.5.2	Sodium alum 593	30.1.5.1	Klebelsbergite 624
29.5.5.3	Tschermigite 594	30.1.6.1	Kogarkoite 624
29.5.5.4	Lonecreekite 594	30.1.7.1	Sulphohalite 624
29.6.1.1	Bassanite 594	30.1.8.1	Galeite 625
29.6.2.1	Kieserite 596	30.1.9.1	Schairerite 625
29.6.2.2	Szomolnokite 596	30.1.10.1	D'Ansite 625
29.6.2.3	Szmikite 596	30.1.11.1	Chlorothionite 625
29.6.2.4	Poitevinite 597	30.1.12.1	Antlerite 626
29.6.2.5	Gunningite 597	30.1.13.1	Schuetteite 626
29.6.2.6	Dwornikite 597	30.1.14.1	Mammothite 626
29.6.3.1	Gypsum 598	30.1.15.1	Ye'elimite 627
29.6.6.1	Rozenite 603	30.1.16.1	Chenite 627
29.6.6.2	Starkeyite 603	30.1.17.1	Nabokoite 627
29.6.6.3	Ilesite 603	30.1.17.2	Atlasovite 627
29.6.6.4	Aplowite 603	30.1.18.1	Coquandite 628
29.6.6.5	Boyleite 604	30.2.1.1	Lanarkite 628
29.6.7.1	Chalcanthite 605	30.2.2.1	Dolerophanite 628
29.6.7.2	Siderotil 606	30.2.3.1	Linarite 629
29.6.7.3	Pentahydrite 606	30.2.4.1	Alunite 631
29.6.7.4	Jokokuite 606	30.2.4.3	Schlossmacherite 632
29.6.8.1	Hexahydrite 608	30.2.4.4	Osarizawaite 632
29.6.8.2	Bianchite 608	30.2.4.5	Minamiite 633
29.6.8.3	Ferrohexahydrite 608	30.2.4.6	Ammonioalunite 633
29.6.8.4	Nickel-Hexahydrite 609	30.2.4.7	Walthierite 633
29.6.8.5	Moorhouseite 609	30.2.4.8	Huangite 633
29.6.8.6	Chvaleticeite 609	30.2.5.1	Jarosite 634
29.6.9.1	Retgersite 609	30.2.5.4	Ammoniojarosite 635
29.6.10.1	Melanterite 611	30.2.5.5	Argentojarosite 635
29.6.10.2	Boothite 611	30.2.5.6	Plumbojarosite 635
29.6.10.3	Zinc-melanterite 612	30.2.5.7	Beaverite 636
29.6.10.4	Bieberite 612	30.2.5.8	Dorallcharite 636
29.6.10.5	Mallardite 612	30.2.6.1	Itoite 636
29.6.11.1	Epsomite 612	30.2.7.1	Piypite 636
29.6.11.2	Goslarite 613	30.2.8.1	Kamchatkite 637
29.6.11.3	Morenosite 613	30.2.9.1	Cannonite 637
29.6.12.1	Minasragrite 614	30.3.1.1	Euchlorine 637
29.6.13.1	Stanleyite 614	30.3.2.1	Caracolite 638
29.7.1.1	Ransomite 614	30.3.3.1	Cesanite 638
29.7.2.1	Römerite 615	30.3.4.1	Fedotovite 638
29.7.2.2	Lishizhenite 615	30.4.1.1	Klyuchevskite 638
29.7.3.1	Pickeringite 616	31.1.1.1	Connellite 639
29.7.3.2	Halotrichite 617	31.1.2.1	Shigaite 639
29.7.3.3	Apjohnite 617	31.1.3.1	Mooreite 639
29.7.3.4	Dietrichite 617	31.1.4.1	Torreyite 640
29.7.3.5	Bilinite 618	31.1.4.2	Lawsonbauerite 640
29.7.3.6	Redingtonite 618	31.1.5.1	Spangolite 640
29.7.3.7	Wupatkiite 618	31.1.6.1	Schulenbergite 641
29.8.1.1	Lausenite 618	31.2.1.1	Cyanotrichite 641
29.8.2.1	Kornelite 619	31.2.2.1	Woodwardite 641
29.8.3.1	Coquimbite 619	31.2.3.1	Zincaluminite 641

Index of Mineral Names in Numerical Order

31.2.4.1	Guarinoite 641		31.10.4.1	Zippeite 660
31.2.5.1	Theresemagnanite 642		31.10.4.2	Sodium Zippeite 661
31.2.6.1	Uranopilite 642		31.10.4.3	Magnesium Zippeite 661
31.2.7.1	Meta-Uranopilite 642		31.10.4.4	Nickel Zippeite 661
31.3.1.1	Chalcoalumite 642		31.10.4.5	Zinc Zippeite 661
31.3.1.2	Nickelalumite 643		31.10.4.6	Cobalt Zippeite 661
31.3.2.1	Wermlandite 643		31.10.5.1	Copiapite 663
31.4.1.1	Posnjakite 643		31.10.5.2	Magnesiocopiapite 663
31.4.2.1	Wroewolfeite 644		31.10.5.3	Cuprocopiapite 664
31.4.3.1	Langite 644		31.10.5.4	Ferricopiapite 664
31.4.4.1	Felsöbanyaite 644		31.10.5.5	Calciocopiapite 664
31.4.5.1	Basaluminite 644		31.10.5.6	Zincocopiapite 664
31.4.6.1	Hydrobasaluminite 645		31.10.5.7	Aluminocopiapite 664
31.4.7.1	Namuwite 645		31.10.6.1	Honessite 665
31.4.8.1	Glaucocerinite 645		31.10.7.1	Hydrohonessite 665
31.4.9.1	Ramsbeckite 645		31.10.8.1	Clairite 665
31.5.1.1	Vonbezingite 646		31.10.9.1	Caminite 665
31.6.1.1	Devilline 646		32.1.1.1	Burkeite 666
31.6.1.2	Lautenthalite 646		32.1.2.1	Hectorfloresite 666
31.6.2.1	Serpierite 647		32.1.3.1	Olsacherite 666
31.6.3.1	Ktenasite 647		32.2.1.1	Rapidcreekite 667
31.6.4.1	Peretaite 647		32.2.2.1	Humberstonite 667
31.6.6.1	Campigliaite 647		32.2.3.1	Ungemachite 667
31.6.7.1	Orthoserpierite 648		32.3.1.1	Hanksite 667
31.6.8.1	Rabejacite 648		32.3.2.1	Caledonite 668
31.7.1.1	Kainite 648		32.3.3.1	Wherryite 668
31.7.2.1	Uklonskovite 649		32.3.4.1	Hauckite 668
31.7.3.1	Clinoungemachite 649		32.3.5.1	Heidornite 669
31.7.4.1	Aluminite 649		32.4.1.1	Nakauriite 669
31.7.5.1	Meta-Aluminite 649		32.4.2.1	Tatarskite 669
31.7.6.1	Despujolsite 650		32.4.3.1	Mountkeithite 669
31.7.6.2	Schaurteite 650		32.4.4.1	Charlesite 670
31.7.6.3	Fleischerite 650		32.4.4.2	Sturmanite 670
31.8.1.1	Natrochalcite 650		32.4.4.3	Jouravskite 670
31.8.2.1	Johannite 650		32.4.4.4	Thaumasite 671
31.8.3.1	Sideronatrite 651		32.4.5.1	Hotsonite 671
31.8.4.1	Metasideronatrite 651		32.4.6.1	Chessexite 671
31.9.1.1	Butlerite 651		33.1.1.1	Schmiederite 672
31.9.2.1	Parabutlerite 652		33.1.2.1	Xocomecatlite 672
31.9.3.1	Amarantite 652		33.1.3.1	Khinite 672
31.9.4.1	Hohmannite 652		33.1.4.1	Frankhawthorneite 673
31.9.5.1	Metahohmannite 653		33.2.1.1	Kuranakhite 673
31.9.6.1	Botryogen 653		33.2.2.1	Montanite 673
31.9.6.2	Zincobotryogen 653		33.2.3.1	Cuzticite 673
31.9.6.3	Xitieshanite 654		33.2.4.1	Parakhinite 674
31.9.7.1	Guildite 654		33.2.5.1	Mcalpineite 674
31.9.7.2	Chaidamuite 654		33.2.6.1	Jensenite 674
31.9.8.1	Aubertite 654		33.2.7.1	Cesbronite 674
31.9.8.2	Svyazhinite 655		33.3.1.1	Schieffelinite 675
31.9.8.3	Magnesioaubertite 655		33.3.2.1	Tlalocite 675
31.9.9.1	Wilcoxite 655		33.3.3.1	Girdite 675
31.9.10.1	Jurbanite 655		33.3.5.1	Cheremnykhite 676
31.9.11.1	Khademite 656		33.3.5.2	Kuksite 676
31.9.11.2	Rostite 656		33.3.5.3	Dugganite 676
31.9.12.1	Fibroferrite 656		34.1.1.1	Molybdomenite 677
31.9.13.1	Slavikite 657		34.1.1.2	Scotlandite 677
31.9.14.1	Lannonite 657		34.1.2.1	Fairbankite 677
31.10.1.1	Carrboydite 657		34.1.3.1	Balyakinite 678
31.10.2.1	Ettringite 659		34.1.4.1	Plumbotellurite 678
31.10.2.2	Bentorite 659		34.1.5.1	Moctezumite 678
31.10.3.1	Zaherite 659		34.1.6.1	Schmitterite 678

34.1.7.1	Magnolite 679		37.1.4.2	Archerite 698
34.1.8.1	Smirnite 679		37.1.5.1	Nahpoite 699
34.2.1.1	Graemite 679		37.1.6.1	Lotharmeyerite 699
34.2.2.1	Chalcomenite 680		37.1.6.2	Ferrilotharmeyerite 699
34.2.2.2	Teineite 680		38.1.1.1	Triphylite 700
34.2.3.1	Clinochalcomenite 680		38.1.1.2	Lithiophilite 701
34.2.3.2	Cobaltomenite 680		38.1.1.3	Natrophilite 701
34.2.3.3	Ahlfeldite 681		38.1.2.1	Maricite 702
34.2.4.1	Choloalite 681		38.1.2.2	Buchwaldite 702
34.2.5.1	Hannebachite 681		38.1.3.1	Olgite 702
34.2.5.2	Gravegliaite 682		38.1.4.1	Ferrisicklerite 703
34.3.1.1	Cliffordite 682		38.1.4.2	Sicklerite 703
34.3.2.1	Zemannite 682		38.1.5.1	Beryllonite 703
34.3.2.2	Kinichilite 682		38.1.6.1	Panethite 704
34.3.2.3	Keystoneite 683		38.1.7.1	Brianite 704
34.3.3.1	Emmonsite 683		38.1.8.1	Vitusite-Ce 704
34.3.4.1	Mandarinoite 683		38.2.1.1	Berzeliite 705
34.4.1.1	Denningite 684		38.2.1.2	Manganberzeliite 705
34.4.2.1	Rajite 684		38.2.1.3	Palenzonaite 705
34.5.1.1	Spiroffite 684		38.2.2.1	Caryinite 708
34.5.2.1	Winstanleyite 685		38.2.2.2	Arseniopleite 708
34.5.3.1	Carlfriesite 685		38.2.3.1	Ferrohagendorfite 708
34.5.4.1	Pingguite 685		38.2.3.2	Hagendorfite 708
34.6.1.1	Mackayite 685		38.2.3.3	Varulite 709
34.6.2.1	Rodalquilarite 686		38.2.3.4	Maghagendorfite 709
34.6.3.1	Quetzalcoatlite 686		38.2.3.5	Ferroalluaudite 709
34.6.4.1	Chekhovichite 686		38.2.3.6	Alluaudite 709
34.6.5.1	Sophiite 686		38.2.3.7	O'Danielite 710
34.6.6.1	Francisite 687		38.2.3.8	Johillerite 710
34.7.1.1	Sonoraite 687		38.2.3.9	Nickenichite 710
34.7.2.1	Cesbronite 687		38.2.4.1	Ferrowyllieite 711
34.7.3.1	Guilleminite 688		38.2.4.2	Wyllieite 711
34.7.4.1	Marthozite 688		38.2.4.3	Rosemaryite 711
34.7.5.1	Derriksite 688		38.2.4.4	Qingheiite 712
34.7.6.1	Demesmaekerite 688		38.2.4.5	Bobfergusonite 712
34.7.7.1	Haynesite 689		38.2.5.1	Fillowite 712
34.8.1.1	Poughite 689		38.2.5.2	Johnsomervilleite 713
34.8.2.1	Tlapallite 689		38.2.5.3	Chladniite 713
34.8.3.1	Oboyerite 690		38.3.1.1	Sarcopside 713
34.8.4.1	Mroseite 690		38.3.1.2	Farringtonite 714
34.8.5.1	Eztlite 690		38.3.2.1	Xanthiosite 714
34.8.6.1	Yecoraite 691		38.3.3.1	Graftonite 714
34.8.7.1	Orschallite 691		38.3.3.2	Beusite 715
35.1.1.1	Tarapacaite 692		38.3.4.1	Whitlockite 715
35.1.2.1	Phoenicochroite 692		38.3.4.2	Strontiowhitlockite 716
35.2.1.1	Lopezite 692		38.3.4.3	Merrillite-(Ca) 717
35.3.1.1	Crocoite 693		38.3.5.1	Stanfieldite 717
35.3.2.1	Chromatite 693		38.3.6.1	Hurlbutite 717
35.3.3.1	Hashemite 693		38.3.7.1	Stranskiite 718
35.4.1.1	Santanaite 694		38.3.8.1	Keyite 718
35.4.2.1	Wattersite 694		38.3.9.1	Lammerite 718
35.4.3.1	Deanesmithite 694		38.3.10.1	Mcbirneyite 719
35.4.4.1	Edoylerite 695		38.4.1.1	Heterosite 719
36.1.1.1	Iranite 696		38.4.1.2	Purpurite 719
36.1.1.2	Hemihedrite 696		38.4.2.1	Berlinite 720
36.1.2.1	Macquartite 696		38.4.3.1	Monazite-(Ce) 722
37.1.1.1	Monetite 697		38.4.3.2	Monazite-(La) 724
37.1.1.2	Weilite 697		38.4.3.3	Cheralite 724
37.1.2.1	Schultenite 698		38.4.3.4	Brabantite 724
37.1.3.1	Phosphammite 698		38.4.3.5	Monazite-(Nd) 724
37.1.4.1	Biphosphammite 698		38.4.3.6	Gasparite-(Ce) 725

Index of Mineral Names in Numerical Order

38.4.4.1	Rooseveltite 725		40.1.3.2	Nabaphite 748
38.4.5.1	Tetrarooseveltite 725		40.1.4.1	Pottsite 749
38.4.6.1	Pucherite 725		40.1.5.1	Mahlmoodite 749
38.4.7.1	Clinobisvanite 726		40.2.1.1	Anapaite 749
38.4.8.1	Dreyerite 726		40.2.2.1	Fairfieldite 751
38.4.9.1	Ximengite 727		40.2.2.2	Messelite 752
38.4.10.1	Lithiophosphate 727		40.2.2.3	Collinsite 752
38.4.10.2	Olympite 727		40.2.2.4	Cassidyite 752
38.4.10.3	Nalipoite 728		40.2.2.5	Talmessite 753
38.4.9.1	Xenotime-(Y) 729		40.2.2.6	Gaitite 753
38.4.11.2	Chernovite-(Y) 730		40.2.2.7	β-Roselite 753
38.4.11.3	Wakefieldite-(Y) 731		40.2.2.8	Parabrandtite 753
38.4.11.4	Wakefieldite-(Ce) 731		40.2.3.1	Roselite 754
38.4.12.1	Kosnarite 731		40.2.3.2	Brandtite 755
38.5.1.1	Chursinite 731		40.2.3.3	Zincroselite 755
38.5.2.1	Lyonsite 732		40.2.3.4	Wendwilsonite 755
38.5.3.1	Howardevansite 732		40.2.4.1	Prosperite 756
38.5.4.1	Ludlockite 732		40.2.5.1	Parascholzite 756
38.5.5.1	Chervetite 733		40.2.6.1	Scholzite 756
38.5.5.2	Ziesite 733		40.2.7.1	Phosphophyllite 757
38.5.6.1	Blossite 733		40.2.8.1	Brackebuschite 758
38.5.7.1	Simferite 734		40.2.8.2	Arsenbrackebuschite 759
38.5.9.1	Namibite 734		40.2.9.1	Tsumcorite 760
38.5.10.1	Aerugite 734		40.2.9.2	Helmutwinklerite 760
39.1.1.1	Brushite 735		40.2.9.3	Thometzekite 761
39.1.1.2	Pharmacolite 735		40.2.9.4	Mawbyite 761
39.1.2.1	Fluckite 736		40.2.10.1	Wicksite 761
39.1.3.1	Krautite 736		40.2.11.1	Grischunite 762
39.1.4.1	Koritnigite 736		40.2a.1.1	Autunite 766
39.1.4.2	Cobaltkoritnigite 737		40.2a.1.2	Meta-Autunite 766
39.1.5.1	Haidingerite 737		40.2a.1.3	Pseudoautunite 767
39.1.6.1	Newberyite 737		40.2a.2.1	Uranospinite 767
39.1.7.1	Brassite 738		40.2a.2.2	Meta-uranospinite 768
39.1.8.1	Dorfmanite 738		40.2a.3.1	Uranocircite 768
39.1.9.1	Rösslerite 738		40.2a.3.2	Meta-Uranocircite 768
39.1.9.2	Phosphorrösslerite 739		40.2a.4.1	Heinrichite 769
39.2.1.1	Huréaulite 739		40.2a.4.2	Meta-Heinrichite 769
39.2.1.2	Sainfeldite 740		40.2a.5.1	Sodium Autunite 769
39.2.1.3	Villyaellenite 740		40.2a.6.1	Sodium Uranospinite 770
39.2.2.1	Vladimirite 740		40.2a.7.1	Uramphite 770
39.2.2.2	Guerinite 741		40.2a.8.1	Meta-Ankoleite 770
39.2.3.1	Ferrarisite 741		40.2a.9.1	Abernathyite 771
39.2.4.1	Picropharmacolite 741		40.2a.10.1	Novacekite 771
39.2.5.1	Irhtemite 742		40.2a.10.2	Meta-Novacekite 771
39.2.6.1	Chudobaite 742		40.2a.11.1	Saleeite 772
39.2.6.2	Geigerite 742		40.2a.12.1	Seelite 772
39.2.7.1	Lindackerite 742		40.2a.13.1	Torbernite 773
39.3.1.1	Stercorite 743		40.2a.13.2	Meta-Torbernite 773
39.3.2.1	Schertelite 743		40.2a.14.1	Zeunerite 774
39.3.3.1	Mundrabillaite 743		40.2a.14.2	Meta-Zeunerite 774
39.3.4.1	Swaknoite 744		40.2a.15.1	Kahlerite 775
39.3.5.1	Hannayite 744		40.2a.15.2	Meta-Kahlerite 775
39.3.5.2	Francoanellite 744		40.2a.16.1	Bassetite 775
39.3.6.1	Taranakite 744		40.2a.16.2	Lehnerite 776
39.3.7.1	McNearite 745		40.2a.17.1	Meta-Kirchheimerite 776
39.3.8.1	Machatschkiite 745		40.2a.18.1	Meta-Lodevite 776
39.3.9.1	Kaatialaite 745		40.2a.19.1	Chernikovite 777
40.1.1.1	Struvite 747		40.2a.20.1	Trögerite 777
40.1.2.1	Dittmarite 748		40.2a.21.1	Przhevalskite 778
40.1.2.2	Niahite 748		40.2a.22.1	Uranospathite 778
40.1.3.1	Nastrophite 748		40.2a.23.1	Arsenuranospathite 778

40.2a.24.1	Sabugalite 779		40.4.8.2	Lermontovite 808	
40.2a.25.1	Fritzscheite 781		40.4.8.3	Vyacheslavite 809	
40.2a.26.1	Tyuyamunite 781		40.5.1.1	Fransoletite 809	
40.2a.26.2	Meta-Tyuyamunite 781		40.5.2.1	Parafransoleite 809	
40.2a.27.1	Francevillite 782		40.5.3.1	Faheyite 810	
40.2a.27.2	Curienite 782		40.5.4.1	Gainesite 810	
40.2a.28.1	Carnotite 782		40.5.4.2	Mccrillisite 810	
40.2a.28.2	Margaritasite 783		40.5.4.3	Selwynite 811	
40.2a.29.1	Strelkinite 783		40.5.5.1	Wycheproofite 811	
40.2a.30.1	Vanuranylite 784		40.5.6.1	Ehrleite 811	
40.2a.31.1	Parsonsite 784		40.5.7.1	Pahasapaite 812	
40.2a.32.1	Hallimondite 784		40.5.8.1	Smolianinovite 812	
40.2a.33.1	Ulrichite 785		40.5.8.2	Fahleite 812	
40.2a.34.1	Triangulite 785		40.5.9.1	Walpurgite 813	
40.3.1.1	Warikahnite 785		40.5.9.2	Orthowalpurgite 813	
40.3.2.1	Phosphoferrite 787		40.5.10.1	Walentaite 813	
40.3.2.2	Kryzhanovskite 787		40.5.11.1	Canaphite 814	
40.3.2.3	Reddingite 788		40.5.12.1	Geminite 814	
40.3.2.4	Landesite 788		40.5.13.1	Kolovratite 814	
40.3.2.5	Garyansellite 788		41.1.1.1	Chlorophoenicite 815	
40.3.3.1	Parahopeite 789		41.1.1.2	Magnesium-chlorophoenicite	815
40.3.4.1	Hopeite 789		41.1.1.3	Jarosewichite 815	
40.3.5.1	Ludlamite 790		41.1.2.1	Theisite 816	
40.3.5.2	Metaswitzerite 790		41.1.3.1	Georgiadesite 816	
40.3.5.3	Sterlinghillite 790		41.1.4.1	Sahlinite 816	
40.3.5.4	Switzerite 791		41.1.4.2	Kombatite 817	
40.3.6.1	Vivianite 793		41.2.1.1	Allactite 817	
40.3.6.2	Baricite 794		41.2.2.1	Heyite 817	
40.3.6.3	Erythrite 794		41.3.1.1	Clinoclase 818	
40.3.6.4	Annabergite 795		41.3.2.1	Cornetite 818	
40.3.6.5	Köttigite 795		41.3.3.1	Flinkite 818	
40.3.6.6	Parasymplesite 796		41.3.4.1	Retzian-(Ce) 819	
40.3.6.7	Hörnesite 796		41.3.4.2	Retzian-(Nd) 819	
40.3.6.8	Arupite 796		41.3.4.3	Retzian-(La) 819	
40.3.7.1	Bobierrite 797		41.3.5.1	Viitaniemiite 820	
40.3.7.2	Manganese-Hörnesite 797		41.3.6.1	Nacaphite 820	
40.3.8.1	Symplesite 797		41.3.7.1	Gerdtremmelite 820	
40.3.8.2	Metaköttigite 798		41.3.8.1	Paulkellerite 820	
40.3.9.1	Metavivianite 798		41.4.1.1	Arsenoclasite 821	
40.3.10.1	Volborthite 798		41.4.1.2	Gatehouseite 821	
40.3.11.1	Rauenthalite 799		41.4.2.1	Cornubite 821	
40.3.12.1	Phaunouxite 799		41.4.2.2	Cornwallite 822	
40.3.13.1	Ferrazite 799		41.4.3.1	Pseudomalachite 822	
40.4.1.1	Variscite 800		41.4.3.2	Reichenbachite 823	
40.4.1.2	Strengite 801		41.4.3.3	Ludjibaite 823	
40.4.1.3	Scorodite 802		41.4.4.1	Turanite 823	
40.4.1.4	Mansfieldite 802		41.4.5.1	Stoiberite 823	
40.4.1.5	Yanomamite 803		41.4.6.1	Nealite 824	
40.4.2.1	Koninckite 803		41.4.7.1	Tancoite 824	
40.4.3.1	Metavariscite 803		41.4.8.1	Reppiaite 824	
40.4.3.2	Phosphosiderite 804		41.5.1.1	Adelite 826	
40.4.3.3	Kolbeckite 804		41.5.1.2	Conichalcite 826	
40.4.4.1	Kankite 804		41.5.1.3	Austinite 827	
40.4.5.1	Steigerite 805		41.5.1.4	Duftite-β 827	
40.4.6.1	Churchite-(Y) 805		41.5.1.5	Gabrielsonite 828	
40.4.7.1	Rhabdophane-(Ce) 807		41.5.1.6	Tangeite 828	
40.4.7.2	Rhabdophane-(La) 807		41.5.1.7	Nickelaustinite 828	
40.4.7.4	Grayite 808		41.5.1.8	Cobaltaustinite 829	
40.4.7.5	Brockite 808		41.5.1.9	Arsendescloizite 829	
40.4.7.6	Tristramite 808		41.5.2.1	Descloizite 829	
40.4.8.1	Ningyoite 808		41.5.2.2	Mottramite 830	

Index of Mineral Names in Numerical Order

41.5.2.3	Pyrobelonite 830		41.8.1.3	Hydroxylapatite 861
41.5.2.4	Cechite 831		41.8.1.4	Carbonate-Fluorapatite 862
41.5.2.5	Duftite-α 831		41.8.1.6	Belovite 862
41.5.3.1	Babefphite 831		41.8.1.7	Strontium-Apatite 863
41.5.4.1	Herderite 832		41.8.2.1	Hedyphane 863
41.5.4.2	Hydroxylherderite 832		41.8.3.1	Svabite 864
41.5.4.3	Väyrynenite 833		41.8.3.2	Turneaureite 864
41.5.4.4	Bergslagite 833		41.8.3.3	Johnbaumite 865
41.5.5.1	Lacroixite 834		41.8.3.4	Fermorite 865
41.5.5.2	Durangite 834		41.8.4.1	Pyromorphite 865
41.5.5.3	Maxwellite 835		41.8.4.2	Mimetite 866
41.5.6.1	Tilasite 835		41.8.4.3	Vanadinite 867
41.5.6.2	Isokite 835		41.8.5.1	Morelandite 868
41.5.6.3	Panasqueiraite 836		41.8.5.2	Alforsite 868
41.5.7.1	Brazilianite 836		41.8.6.1	Clinomimetite 868
41.5.8.1	Amblygonite 836		41.9.1.1	Kulanite 870
41.5.8.2	Montebrasite 837		41.9.1.2	Penikisite 870
41.5.8.3	Natromontebrasite 838		41.9.1.3	Bjarebyite 870
41.5.9.1	Tavorite 838		41.9.1.4	Perloffite 871
41.5.10.1	Dussertite 839		41.9.2.1	Rockbridgeite 871
41.5.10.2	Florencite-(Ce) 839		41.9.2.2	Frondelite 872
41.5.10.3	Florencite-(La) 840		41.9.3.1	Griphite 872
41.5.10.4	Florencite-(Nd) 840		41.10.1.1	Lazulite 873
41.5.11.1	Arsenoflorencite-(Ce) 840		41.10.1.2	Scorzalite 874
41.5.11.2	Arsenoflorencite-(Nd) 840		41.10.1.3	Hentschelite 875
41.5.11.3	Arsenoflorencite-(La) 841		41.10.1.4	Barbosalite 875
41.5.12.1	Waylandite 841		41.10.2.1	Lipscombite 875
41.5.12.2	Eylettersite 841		41.10.3.1	Jagowerite 876
41.5.12.3	Zairite 841		41.10.4.1	Goedkenite 876
41.5.13.1	Vesignieite 842		41.10.4.2	Bearthite 876
41.5.14.1	Bayldonite 842		41.10.4.3	Gamagarite 877
41.5.15.1	Curetonite 842		41.10.5.1	Melonjosephite 877
41.5.16.1	Thadeuite 843		41.10.6.1	Carminite 877
41.5.17.1	Leningradite 843		41.10.7.1	Mounanaite 878
41.5.18.1	Arctite 843		41.10.8.1	Samuelsonite 878
41.6.1.1	Zweiselite 845		41.10.9.1	Preisingerite 878
41.6.1.2	Triplite 845		41.10.9.2	Schumacherite 879
41.6.1.3	Magniotriplite 846		41.10.10.1	Petitjohnite 879
41.6.2.1	Wagnerite 846		41.10.11.1	Drugmanite 880
41.6.3.1	Wolfeite 847		41.11.1.1	Trolleite 880
41.6.3.2	Triploidite 847		41.11.2.1	Jamesite 880
41.6.3.3	Sarkinite 848		41.11.3.1	Fingerite 880
41.6.4.1	Satterlyite 848		41.11.4.1	Nefedovite 881
41.6.4.2	Holtedahlite 848		41.11.5.1	Atelestite 881
41.6.5.1	Althausite 849		42.1.1.1	Liskeardite 882
41.6.6.1	Olivenite 849		42.1.2.1	Evansite 882
41.6.6.2	Libethenite 849		42.2.1.1	Liroconite 882
41.6.6.3	Adamite 850		42.2.2.1	Ranunculite 883
41.6.6.4	Eveite 850		42.2.3.1	Veszelyite 883
41.6.7.1	Tarbuttite 851		42.2.4.1	Kipushite 883
41.6.7.2	Paradamite 851		42.2.4.2	Philipsburgite 884
41.6.8.1	Augelite 851		42.3.1.1	Rusakovite 884
41.6.9.1	Arsenobismite 852		42.4.1.1	Akrochordite 884
41.6.10.1	Angelellite 852		42.4.2.1	Morinite 885
41.6.11.1	Spodiosite 852		42.4.3.1	Tyrolite 885
41.7.1.1	Palermoite 853		42.4.4.1	Clinotyrolite 885
41.7.1.2	Bertossaite 853		42.4.5.1	Dumontite 886
41.7.2.1	Arrojadite 853		42.4.5.2	Hügelite 886
41.7.2.2	Dickinsonite 854		42.4.5.3	Bergenite 886
41.8.1.1	Fluorapatite 858		42.4.6.1	Phuralumite 886
41.8.1.2	Chlorapatite 860		42.4.7.1	Phurcalite 887

42.4.8.1	Phosphuranylite 887	42.7.14.2	Metavanmeerscheite 907
42.4.9.1	Arsenuranylite 887	42.7.15.1	Richelsdorfite 908
42.4.9.2	Renardite 888	42.7.16.1	Camgasite 908
42.4.9.3	Kivuite 888	42.8.1.1	Pharmacosiderite 908
42.4.10.1	Glucine 888	42.8.1.2	Alumopharmacosiderite 909
42.4.11.1	Asselbornite 888	42.8.1.3	Barium Pharmacosiderite 909
42.4.12.1	Wallkilldellite 889	42.8.1.4	Barium-alumopharmacosiderite 910
42.4.13.1	Althupite 889		
42.4.14.1	Françoisite-(Nd) 889	42.8.1.5	Sodium Pharmacosiderite 910
42.4.15.1	Mrazekite 889	42.8.2.1	Kidwellite 910
42.5.1.1	Mixite 890	42.8.3.1	Schoonerite 910
42.5.1.2	Agardite-(Y) 890	42.8.4.1	Mitridatite 911
42.5.1.3	Goudeyite 891	42.8.4.2	Robertsite 911
42.5.2.1	Petersite-(Y) 891	42.8.4.3	Arseniosiderite 912
42.5.3.1	Zapatalite 891	42.8.5.1	Kolfanite 912
42.6.1.1	Moraesite 891	42.8.5.2	Pararobertsite 912
42.6.1.2	Bearsite 892	42.8.6.1	Yukonite 913
42.6.2.1	Isoclasite 892	42.8.7.1	Mapimite 913
42.6.3.1	Euchroite 892	42.8.8.1	Shubnikovite 913
42.6.4.1	Legrandite 893	42.9.1.1	Burangaite 913
42.6.4.2	Spencerite 893	42.9.1.2	Dufrenite 914
42.6.5.1	Strashimirite 893	42.9.1.3	Natrodufrenite 914
42.6.5.2	Arhbarite 894	42.9.2.1	Souzalite 914
42.6.6.1	Delvauxite 894	42.9.2.2	Gormanite 915
42.6.7.1	Senegalite 894	42.9.3.1	Turquoise 917
42.6.7.2	Bulachite 895	42.9.3.2	Coeruleolactite 918
42.6.8.1	Bolivarite 895	42.9.3.3	Faustite 919
42.6.9.1	Fluellite 895	42.9.3.4	Chalcosiderite 919
42.6.10.1	Sengierite 895	42.9.3.5	Aheylite 919
42.6.11.1	Tiptopite 896	42.9.3.6	Planerite 920
42.6.12.1	Yingjiangite 896	42.9.4.1	Sampleite 920
42.7.1.1	Childrenite 896	42.9.4.2	Lavendulan 920
42.7.1.2	Eosphorite 897	42.9.4.3	Zdenekite 921
42.7.1.3.	Ernstite 897	42.9.5.1	Duhamelite 921
42.7.1.4	Sinkankasite 898	42.9.6.1	Santafeite 921
42.7.2.1	Foggite 898	42.9.7.1	Ogdensburgite 922
42.7.3.1	Crandallite 899	42.9.8.1	Dewindtite 922
42.7.3.2	Gorceixite 900	42.10.1.1	Ferrisymplesite 922
42.7.3.3	Goyazite 900	42.10.2.1	Wavellite 922
42.7.3.4	Lusungite 900	42.10.3.1	Kingite 923
42.7.3.5	Plumbogummite 901	42.10.4.1	Gutsevichite 924
42.7.3.6	Kintoreite 901	42.11.1.1	Overite 925
42.7.4.1	Arsenocrandallite 901	42.11.1.2	Segelerite 925
42.7.4.2	Philipsbornite 902	42.11.1.3	Manganosegelerite 926
42.7.4.3	Arsenogoyazite 902	42.11.1.4	Lun'okite 926
42.7.4.4	Segnitite 902	42.11.1.5	Wilhelmvierlingite 926
42.7.5.1	Nissonite 902	42.11.1.6	Kaluginite 927
42.7.6.1	Uralolite 903	42.11.2.1	Jahnsite-(CaMnMg) 929
42.7.6.2	Weinebenite 903	42.11.3.1	Whiteite-(CaFeMg) 930
42.7.7.1	Roscherite-A 904	42.11.3.4	Rittmannite 931
42.7.7.3	Zanazziite 904	42.11.4.1	Keckite 931
42.7.8.1	Cyrilovite 905	42.11.5.1	Minyulite 932
42.7.8.2	Wardite 905	42.11.6.1	Leucophosphite 932
42.7.9.1	Millisite 905	42.11.6.2	Tinsleyite 932
42.7.10.1	Chenevixite 906	42.11.6.3	Spheniscidite 933
42.7.10.2	Luetheite 906	42.11.7.1	Giniite 933
42.7.11.1	Olmsteadite 906	42.11.8.1	Montgomeryite 935
42.7.11.2	Johnwalkite 907	42.11.8.2	Kingsmountite 935
42.7.12.1	Upalite 907	42.11.8.3.	Calcioferrite 936
42.7.13.1	Mundite 907	42.11.8.4	Zodacite 936
42.7.14.1	Vanmeersscheite 907	42.11.9.1	Strunzite 938

Index of Mineral Names in Numerical Order

42.11.9.2	Ferrostrunzite 939		43.4.1.2	Corkite 961
42.11.9.3	Ferristrunzite 939		43.4.1.3	Hidalgoite 961
42.11.10.1	Laueite 940		43.4.1.4	Orpheite 961
42.11.10.2	Stewartite 940		43.4.1.5	Hinsdalite 962
42.11.10.3	Pseudolaueite 941		43.4.1.6	Svanbergite 962
42.11.10.4	Ushkovite 941		43.4.1.7	Kemmlitzite 962
42.11.11.1	Metavauxite 941		43.4.1.8	Woodhouseite 962
42.11.12.1	Gatumbaite 942		43.4.1.9	Weilerite 962
42.11.12.2	Kleemanite 942		43.4.2.1	Tsumebite 963
42.11.13.1	Vanuralite 942		43.4.2.2	Arsentsumebite 963
42.11.13.2	Metavanuralite 942		43.4.3.1	Vauquelinite 963
42.11.13.3	Threadgoldite 942		43.4.3.2	Fornacite 963
42.11.14.1	Vauxite 943		43.4.3.3	Molybdofornacite 964
42.11.14.2	Paravauxite 943		43.4.4.1	Cahnite 964
42.11.14.3	Sigloite 943		43.4.5.1	Seamanite 964
42.11.14.4	Gordonite 944		43.4.6.1	Hematolite 964
42.11.14.5	Mangangordonite 944		43.4.7.1	Holdenite 964
42.11.15.1	Xanthoxenite 944		43.4.8.1	Kolicite 965
42.11.16.1	Beraunite 945		43.4.9.1	McGovernite 965
42.11.17.1	Bermanite 945		43.4.10.1	Kraisslite 965
42.11.18.1	Whitmoreite 947		43.4.11.1	Heneuite 965
42.11.18.2	Arthurite 947		43.4.12.1	Cervandonite-(Ce) 965
42.11.18.3	Ojuelaite 948		43.4.13.1	Attakolite 966
42.11.18.4	Earlshannonite 948		43.5.1.1	Sarmientite 966
42.11.19.1	Sincosite 948		43.5.1.2	Bukovskyite 966
42.11.20.1	Furongite 949		43.5.1.3	Sanjuanite 966
42.11.21.1	Paulkerrite 949		43.5.2.1	Diadochite 966
42.11.21.2	Mantiennéite 949		43.5.3.1	Pitticite 967
42.11.21.3	Matveevite 950		43.5.3.2	Zykaite 967
42.11.21.4	Benyacarite 950		43.5.4.1	Sasaite 967
42.11.22.1	Mcauslanite 950		43.5.5.1	Coconinoite 967
42.11.23.1	Vochtenite 950		43.5.6.1	Xiangjiangite 967
42.12.1.1	Vantasselite 951		43.5.7.1	Kribergite 968
42.12.2.1	Vashegyite 951		43.5.8.1	Kovdorskite 968
42.12.3.1	Moreauite 951		43.5.9.1	Viséite 968
42.12.4.1	Tinticite 952		43.5.10.1	Perhamite 968
42.12.5.1	Tooeleite 952		43.5.11.1	Lünebergite 968
42.13.1.1	Aldermanite 952		43.5.12.1	Synadelphite 968
42.13.2.1	Richellite 953		43.5.13.1	Parnauite 969
42.13.3.1	Matulaite 953		43.5.14.1	Chalcophyllite 969
42.13.4.1	Jungite 953		43.5.15.1	Peisleyite 969
42.13.5.1	Cacoxenite 953		43.5.16.1	Voggite 969
42.13.6.1	Ceruleite 954		43.5.17.1	Girvasite 970
42.13.7.1	Rosiérésite 954		43.5.18.1	Hotsonite 970
42.13.8.1	Englishite 954		43.19.1	Quadruphite 970
42.13.9.1	Natrophosphate 954		43.20.1	Polyphite 970
42.13.10.1	Phosphofibrite 955		44.1.1.1	Stibiconite 971
42.13.11.1	Kamitugaite 955		44.1.1.2	Bindheimite 971
42.13.12.1	Sieleckiite 955		44.1.1.3	Roméite 971
42.13.13.1	Alvanite 956		44.1.1.4	Monimolite 972
43.1.1.1	Ardealite 957		44.1.1.5	Stetefeldtite 972
43.2.1.1	Bradleyite 957		44.1.1.6	Bismutostibiconite 972
43.2.1.2	Sidorenkite 958		44.1.1.7	Partzite 972
43.2.1.3	Bonshtedtite 958		44.1.2.1	Ingersonite 972
43.2.1.4	Crawfordite 959		44.2.1.1	Byströmite 973
43.2.2.1	Gartrellite 959		44.2.1.2	Ordoñezite 973
43.3.1.1	Schoderite 959		44.2.1.3	Tripuhyite 973
43.3.1.2	Metaschoderite 959		44.3.1.1	Swedenborgite 973
43.3.2.1	Embreyite 959		44.3.2.1	Manganostibite 973
43.3.2.2	Cassedanneite 960		44.3.3.1	Parwelite 974
43.4.1.1	Beudantite 961		44.3.4.1	Långbanite 974

44.3.5.1	Katoptrite 974		47.3.3.1	Schubnelite 989	
44.3.6.1	Yeatmanite 974		47.3.4.1	Fervanite 990	
44.3.7.1	Bahianite 974		47.3.5.1	Bannermanite 990	
44.3.8.1	Shakhovite 975		47.3.6.1	Melanovanadite 990	
44.3.9.1	Cyanophyllite 975		47.4.1.1	Satpaevite 990	
44.3.10.1	Cualstibite 975		47.4.2.1	Vanalite 990	
44.3.11.1	Camerolaite 975		47.4.3.1	Pintadoite 991	
44.3.12.1	Brizziite 975		47.4.4.1	Rauvite 991	
45.1.1.1	Reinerite 977		47.4.5.1	Uvanite 991	
45.1.2.1	Trigonite 977		47.4.6.1	Vanoxite 991	
45.1.2.2	Rouseite 977		47.4.7.1	Doloresite 991	
45.1.3.1	Asbecasite 977		47.4.8.1	Simplotite 991	
45.1.4.1	Cafarsite 978		48.1.1.1	Hübnerite 993	
45.1.5.1	Trippkeite 978		48.1.1.2	Ferberite 994	
45.1.6.1	Schafarzikite 978		48.1.1.3	Sanmartinite 996	
45.1.7.1	Leiteite 978		48.1.2.1	Scheelite 996	
45.1.8.1	Paulmooreite 978		48.1.2.2	Powellite 998	
45.1.9.1	Apuanite 979		48.1.3.1	Wulfenite 999	
45.1.10.1	Versiliaite 979		48.1.3.2	Stolzite 1000	
45.1.11.1	Stibivanite-2M 979		48.1.4.1	Raspite 1001	
45.1.11.2	Stibivanite-20 979		48.2.1.1	Russellite 1001	
45.1.12.1	Schneiderhöhnite 979		48.2.2.1	Koechlinite 1001	
45.1.13.1	Fetiasite 979		48.2.3.1	Tungstibite 1002	
45.1.14.1	Ludlockite 980		48.3.1.1	Lindgrenite 1002	
46.1.1.1	Finnemanite 981		48.3.2.1	Cuprotungstite 1002	
46.1.2.1	Nanlingite 981		48.3.3.1	Anthoinite 1002	
46.1.3.1	Magnussonite 981		48.3.4.1	Deloryite 1003	
46.1.4.1	Stenhuggarite 981		48.3.5.1	Szenicsite 1003	
46.1.5.1	Freedite 982		48.4.1.1	Sedovite 1003	
46.1.6.1	Nealite 982		48.4.2.1	Jixianite 1004	
46.2.1.1	Ecdemite 982		49.1.1.1	Hydrotungstite 1005	
46.2.2.1	Heliophyllite 982		49.1.2.1	Moluranite 1005	
46.2.3.1	Tomichite 982		49.2.1.1	Ferrimolybdite 1005	
46.2.3.2	Derbylite 983		49.2.2.1	Umohoite 1006	
46.2.4.1	Hemloite 983		49.2.3.1.	Iriginite 1006	
46.2.5.1	Gebhardite 983		49.2.4.1	Phyllotungstite 1006	
46.2.6.1	Manganarsite 983		49.3.1.1	Calcurmolite 1007	
46.2.7.1	Armangite 983		49.3.1.2	Tengchongite 1007	
46.2.8.1	Dixenite 983		49.3.2.1	Cousinite 1007	
46.2.9.1	Seelite 984		49.3.3.1	Cerotungstite-(Ce) 1007	
47.1.1.1	Rossite 985		49.3.3.2	Yttrotungstite-(Y) 1007	
47.1.1.2	Metarossite 985		49.3.4.1	Mpororoite 1008	
47.1.2.1	Delrioite 985		49.3.5.1	Uranotungstite 1008	
47.1.2.2	Metadelrioite 985		49.3.6.1	Rankachite 1008	
47.1.3.1	Munirite 986		49.4.1.1	Betpakdalite 1008	
47.1.3.2	Metamunirite 986		49.4.2.1	Obradovicite 1009	
47.2.1.1	Pascoite 986		49.4.3.1	Melkovite 1009	
47.2.2.1	Hummerite 986		49.4.4.1	Sodium Betpakdalite 1009	
47.2.3.1	Huemulite 987		49.4.5.1	Mendozavilite 1010	
47.2.4.1	Sherwoodite 987		49.4.6.1	Paramendozavilite 1010	
47.3.1.1	Hewettite 987		49.4.7.1	Paraniite-(Y) 1010	
47.3.1.2	Metahewettite 987		50.1.1.1	Whewellite 1011	
47.3.1.3	Barnesite 988		50.1.2.1	Weddellite 1011	
47.3.1.4	Hendersonite 988		50.1.3.1	Humboldtine 1011	
47.3.1.5	Grantsite 988		50.1.3.2	Glushinskite 1012	
47.3.2.1	Straczekite 988		50.1.4.1	Minguzzite 1012	
47.3.2.2	Corvusite 988		50.1.5.1	Oxammite 1012	
47.3.2.3	Fernandinite 989		50.1.6.1	Moolooite 1013	
47.3.2.4	Bariandite 989		50.1.7.1	Stepanovite 1013	
47.3.2.5	Bokite 989		50.1.7.2	Zhemchuzhnikovite 1013	
47.3.2.6	Kazakhstanite 989		50.1.8.1	Wheatleyite 1013	

Index of Mineral Names in Numerical Order

50.2.1.1	Mellite 1013		51.5.2.2	Hafnon 1056
50.2.2.1	Earlandite 1014		51.5.2.3	Thorite 1057
50.2.3.1	Julienite 1014		51.5.2.4	Coffinite 1058
50.2.4.1	Calclacite 1014		51.5.2.5	Thorogummite 1058
50.2.5.1	Kafehydrocyanite 1014		51.5.3.1	Huttonite 1059
50.3.1.1	Kratochvilite 1015		51.5.3.2	Tombarthite-(Y) 1060
50.3.2.1	Ravatite 1015		51.5.4.1	Eulytite 1060
50.3.3.1	Simonellite 1015		52.1.1.1	Beryllite 1061
50.3.4.1	Fichtelite 1015		52.2.1.1	Euclase 1061
50.3.5.1	Dinite 1015		52.2.1.2	Clinohedrite 1062
50.3.6.1	Evenkite 1016		52.2.1.3.	Hodgkinsonite 1063
50.3.7.1	Karpatite 1016		52.2.1.4	Gerstmannite 1063
50.3.8.1	Idrialite 1016		52.2.2a.1	Sillimanite 1064
50.4.1.1	Refikite 1017		52.2.2a.2	Mullite 1065
50.4.2.1	Hoelite 1017		52.2.2b.1	Andalusite 1066
50.4.3.1	Flagstaffite 1017		52.2.2b.2	Kanonaite 1068
50.4.4.1	Uricite 1017		52.2.2b.3	Yoderite 1068
50.4.5.1	Guanine 1017		52.2.2c.1	Kyanite 1069
50.4.6.1	Urea 1018		52.2.3.1	Staurolite 1071
50.4.7.1	Acetamide 1018		52.3.1.1	Topaz 1073
50.4.8.1	Kladnoite 1018		52.3.2a.1	Norbergite 1078
50.4.9.1	Abelsonite 1018		52.3.2b.1	Alleghanyite 1079
51.1.1.1	Phenakite 1021		52.3.2b.2	Chondrodite 1080
51.1.1.2	Willemite 1022		52.3.2b.3	Reinhardbraunsite 1081
51.1.1.3	Eucryptite 1022		52.3.2b.4	Ribbeite 1081
51.1.2.1	Liberite 1023		52.3.2c.1	Humite 1082
51.2.1.1	Trimerite 1023		52.3.2c.2	Leucophoenicite 1083
51.2.1.2	Larsenite 1023		52.3.2c.3	Manganhumite 1083
51.2.1.3	Esperite 1024		52.3.2d.1	Clinohumite 1084
51.2.2.1	Sverigeite 1024		52.3.2d.2	Jerrygibbsite 1085
51.3.1.1	Fayalite 1025		52.3.2d.3	Sonolite 1085
51.3.1.2	Forsterite 1026		52.3.3.1	Chloritoid-A,M 1088
51.3.1.3	Liebenbergite 1029		52.3.3.2	Magnesiochloritoid-A,M 1089
51.3.1.4	Tephroite 1029		52.3.3.3	Ottrelite-A,M 1089
51.3.1.5	Laihunite 1031		52.3.3.4	Carboirite 1090
51.3.2.1	Monticellite 1032		52.4.1.1	Örebroite 1090
51.3.2.2	Kirschsteinite 1033		52.4.1.2	Welinite 1090
51.3.2.3	Glauchroite 1034		52.4.1.3	Kittatinnyite 1091
51.3.3.1	Ringwoodite 1034		52.4.1.4	Franciscanite 1091
51.3.4.1	Wadsleyite 1035		52.4.2.1	Mozartite 1091
51.4.1.1	Bredigite 1035		52.4.2.2	Vuagnatite 1092
51.4.2.1	Merwinite 1036		52.4.3.1	Titanite 1092
51.4.3a.1	Pyrope 1038		52.4.3.2	Malayaite 1095
51.4.3a.2	Almandine 1039		52.4.3.3	Vanadomalayaite 1095
51.4.3a.3	Spessartine 1041		52.4.4.1	Natisite 1096
51.4.3a.4	Knorringite 1043		52.4.4.2	Paranatisite 1096
51.4.3a.5	Majorite 1043		52.4.4.3	Yftisite-(Y) 1096
51.4.3a.6	Calderite 1044		52.4.5.1	Törnebohmite-(Ce) 1096
51.4.3b.1	Andradite 1044		52.4.5.2	Törnebohmite-(La) 1096
51.4.3b.2	Grossular 1046		52.4.6.1	Cerite-(Ce) 1097
51.4.3b.3	Uvarovite 1048		52.4.7.1	Afwillite 1098
51.4.3b.4	Goldmanite 1049		52.4.7.2	Bultfonteinite 1098
51.4.3c.1	Schorlomite 1050		52.4.7.3	Hatrurite 1099
51.4.3c.2	Kimzeyite 1050		52.4.7.4	Jasmundite 1099
51.4.3c.3	Morimotoite 1051		52.4.8.1	Trimounsite-(Y) 1099
51.4.3d.1	Hibschite 1051		52.4.9.1	Britholite-(Ce) 1100
51.4.3d.2	Katoite 1052		52.4.9.2	Britholite-(Y) 1101
51.4.4.1	Henritermierite 1052		52.4.9.3	Fluorellestadite 1102
51.4.5.1	Wadalite 1053		52.4.9.4	Hydroxylellestadite 1102
51.5.1.1	Larnite 1053		52.4.9.5	Chlorellestadite 1103
51.5.2.1	Zircon 1054		52.4.9.6	Mattheddleite 1103

52.4.9.7	Karnasurtite-(Ce) 1104		55.4.1.1	Åkermanite 1143
52.4.9.8	Fluorbritholite-(Ce) 1104		55.4.1.2	Gehlenite 1143
52.4.10.1	Ellenbergerite 1104		55.4.1.3	Melilite 1144
52.4.11.1	Sitinakite 1105		55.4.2.1	Fresnoite 1145
53.1.1.1	Spurrite 1106		55.4.2.2	Hardystonite 1145
53.1.2.1	Paraspurrite 1106		55.4.2.3	Jeffreyite 1145
53.1.3.1	Iimoriite-(Y) 1107		55.4.2.4	Leucophanite 1146
53.2.1.1	Saryarkite-Y 1107		55.4.2.5	Meliphanite 1146
53.2.2.1	Nagelschmidtite 1108		55.4.2.6	Gugiaite 1147
53.3.1.1	Kasolite 1110		55.5.1.1	Edgarbaileyite 1147
53.3.1.2	Uranophane 1111		56.1.1.1	Bertrandite 1149
53.3.1.3	Sklodowskite 1111		56.1.2.1	Hemimorphite 1150
53.3.1.4	Cuprosklodowskite 1112		56.2.1.1	Junitoite 1152
53.3.1.5	Boltwoodite 1112		56.2.2.1	Ferroaxinite 1154
53.3.1.6	Sodium-Boltwoodite 1113		56.2.2.2	Magnesioaxinite 1155
53.3.1.7	Oursinite 1113		56.2.2.3	Manganaxinite 1156
53.3.1.8	Swamboite 1113		56.2.2.4	Tinzenite 1156
53.3.1.9	Beta-uranophane 1113		56.2.3.1	Lawsonite 1157
53.3.2.1	Weeksite 1114		56.2.3.2	Hennomartinite 1158
53.3.2.2	Haiweeite 1114		56.2.3.3	Ilvaite 1158
53.3.2.3	Metahaiweeite 1115		56.2.4.1	Baghdadite 1161
53.3.3.1	Soddyite 1115		56.2.4.2	Burpalite 1161
53.3.3.2	Uranosilite 1115		56.2.4.3	Cuspidine 1161
54.1.1.1	Grandidierite 1116		56.2.4.4	Låvenite 1162
54.1.2.1	Dumortierite 1117		56.2.4.5	Wöhlerite 1163
54.1.2.2	Magnesiodumortierite 1118		56.2.4.6	Niocalite 1164
54.1.3.1	Holtite 1119		56.2.4.7	Hiortdahlite 1164
54.2.1a.1	Datolite 1121		56.2.4.8	Rosenbuschite 1165
54.2.1a.2	Hingganite-(Ce) 1122		56.2.4.9	Hainite 1166
54.2.1a.3	Hingganite-(Y) 1123		56.2.4.10	Janhaugite 1166
54.2.1a.4	Hingganite-(Yb) 1123		56.2.4.11	Jennite 1166
54.2.1b.1	Bakerite 1123		56.2.4.12	Komarovite 1167
54.2.1b.2	Gadolinite-(Ce) 1124		56.2.4.13	Na-Komarovite 1167
54.2.1b.3	Gadolinite-(Y) 1124		56.2.4.14	Suolunite 1168
54.2.1b.4	Calciogadolinite 1126		56.2.4.15	Mongolite 1168
54.2.1b.5	Homilite 1126		56.2.5.1	Mosandrite 1168
54.2.1b.6	Minasgeraisite-(Y) 1126		56.2.5.2	Nacareniobsite-(Ce) 1169
54.2.2.1	Hellandite-(Y) 1127		56.2.5.3	Fersmanite 1170
54.2.2.2	Tadzhikite-(Y) 1128		56.2.5.4	Götzenite 1170
54.2.3.1	Cappelenite- (Y) 1128		56.2.6a.1	Seidozerite 1171
54.2.3.2	Stillwellite-(Ce) 1129		56.2.6b.1	Bafertisite 1171
54.2.4.1	Garrelsite 1129		56.2.6b.2	Hejtmanite 1172
54.2.4.3	Okanoganite-(Y) 1130		56.2.6c.1	Barytolamprophyllite 1174
54.2.5.1	Tritomite-(Ce) 1130		56.2.6c.2	Delindeite 1174
54.2.5.2	Tritomite-(Y) 1131		56.2.6c.3	Ericssonite 1175
54.2.5.3	Melanocerite-(ce) 1131		56.2.6c.4	Lamprophyllite 1175
54.3.1.1.	Vicanite-(Ce) 1132		56.2.6c.5	Orthoericssonite 1175
55.1.1.1	Barylite 1133		56.2.6c.6	Andremeyerite 1176
55.2.1a.1	Gittinsite 1134		56.2.7.1	Epistolite 1176
55.2.1a.2	Keiviite-(Y) 1135		56.2.7.2	Murmanite 1177
55.2.1a.3	Keiviite-(Yb) 1136		56.2.8.1	Chevkinite-(Ce) 1177
55.2.1a.4	Thortveitite 1136		56.2.8.2	Strontiochevkinite 1179
55.2.1b.1	Rowlandite-(Y) 1137		56.2.8.3	Perrierite-(Ce) 1179
55.2.1b.2	Thalénite-(Y) 1137		56.2.8.4	Orthochevkinite 1180
55.2.1b.3	Yttrialite-(Y) 1138		56.2.9.1	Tilleyite 1180
55.2.2.1	Keldyshite 1139		56.2.9.2	Killalaite 1181
55.2.2.2	Khibinskite 1139		56.2.9.3	Foshallasite 1181
55.2.2.3	Parakeldyshite 1140		56.2.10.1	Melanotekite 1182
55.2.3.1	Barysilite 1140		56.2.10.2	Kentrolite 1182
55.3.1.1	Rankinite 1140		56.2.11.1	Nasonite 1183
55.3.2.1	Jaffeite 1141		56.3.1.1	Danburite 1183

Index of Mineral Names in Numerical Order

56.3.2.1	Werdingite 1184		59.2.2.2	Calcium Catapleiite 1227
56.4.1.1	Lomonosovite 1184		59.2.2.3	Gaidonnayite 1227
56.4.1.2	Polyphite 1185		59.2.2.4	Georgechaoite 1228
56.4.1.3	Quadruphite 1186		59.2.2.5	Loudounite 1229
56.4.1.4	Sobolevite 1186		59.2.3.1	Hilairite 1229
56.4.2.1	Bornemanite 1187		59.2.3.2	Calciohilairite 1229
56.4.3.1	Vuonnemite 1187		59.2.3.3	Sazykinaite-(Y) 1230
57.1.1.1	Aminoffite 1188		59.2.4.1	Komkovite 1230
57.1.1.2	Harstigite 1188		60.1.1a.1	Joaquinite-(Ce) 1232
57.1.2.1	Kinoite 1188		60.1.1a.2	Strontiojoaquinite 1232
57.1.3.1	Trabzonite 1189		60.1.1b.1	Orthojoaquinite-(Ce) 1232
57.1.4.1	Rosenhahnite 1189		60.1.1b.2	Strontio-orthojoaquinite 1232
57.2.1.1	Tiragalloite 1190		60.1.1b.3	Bario-orthojoaquinite 1233
57.2.2.1	Ruizite 1190		60.1.1b.4	Byelorussite-(Ce) 1233
57.2.3.1	Akatoreite 1191		60.1.2.1	Baotite 1233
57.3.1.1	Zunyite 1191		60.1.3.1	Nenadkevichite 1234
57.4.1.1	Medaite 1192		60.1.3.2	Labuntsovite 1235
58.1.1.1	Kornerupine 1193		60.1.4.1	Papagoite 1236
58.1.2.1	Davreuxite 1194		60.2.1.1	Kainosite-(Y) 1236
58.1.3.1	Queitite 1194		61.1.1.1	Beryl 1239
58.2.1a.1	Allanite-(Ce) 1196		61.1.1.2	Bazzite 1244
58.2.1a.2	Allanite-(La) 1197		61.1.1.3	Indialite 1245
58.2.1a.3	Allanite-(Y) 1198		61.1.2a.1	Lovozerite 1247
58.2.1a.4	Clinozoisite 1198		61.1.2a.2	Kazakovite 1247
58.2.1a.5	Dissakisite-(Ce) 1199		61.1.2a.3	Tisinalite 1248
58.2.1a.6	Dollaseite-(Ce) 1199		61.1.2a.4	Zirsinalite 1248
58.2.1a.7	Epidote 1200		61.1.2a.5	Combeite 1248
58.2.1a.8	Hancockite 1201		61.1.2b.1	Imandrite 1249
58.2.1a.9	Khristovite-(Ce) 1201		61.1.2b.2	Koashvite 1249
58.2.1a.10	Mukhinite 1202		61.1.2b.3	Petarasite 1249
58.2.1a.11	Piemontite 1202		61.1.3.1	Dioptase 1249
58.2.1a.12	Strontiopiemontite 1203		61.1.4.1	Baratovite 1250
58.2.1b.1	Zoisite 1203		61.1.4.2	Katayamalite 1250
58.2.2.1	Julgoldite-(Fe^{2+}) 1207		61.1.5.1	Odintsovite 1251
58.2.2.2	Julgoldite-(Fe^{3+}) 1208		61.2.1.1	Cordierite 1251
58.2.2.3	Okhotskite-(Mg) 1208		61.2.1.2	Sekaninaite 1253
58.2.2.5	Pumpellyite-(Fe^{2+}) 1209		61.2.2.1	Jonesite 1254
58.2.2.6	Pumpellyite-(Fe^{3+}) 1209		61.2.3.1	Lourenswalsite 1254
58.2.2.7	Pumpellyite-(Mg) 1210		61.3.1.1	Foitite 1257
58.2.2.8	Pumpellyite-(Mn^{2+}) 1211		61.3.1.2	Liddicoatite 1257
58.2.2.9	Shuiskite-(Mg) 1211		61.3.1.3	Uvite 1258
58.2.3.1	Sursassite 1212		61.3.1.4	Feruvite 1259
58.2.3.2	Macfallite 1212		61.3.1.5	Buergerite 1259
58.2.4.1	Vesuvianite 1213		61.3.1.6	Povondraite 1260
58.2.5.1	Rustumite 1215		61.3.1.7	Olenite 1260
58.2.6.1	Yoshimuraite 1216		61.3.1.8	Elbaite 1261
58.2.6.2	Innelite 1216		61.3.1.9	Dravite 1262
58.2.7.1	Samfowlerite 1217		61.3.1.10	Schorl 1263
58.3.1.1	Ardennite 1217		61.3.1.11	Chromdravite 1264
58.3.1.2	Orientite 1218		61.4.1.1	Abenakiite-(Ce) 1264
58.3.2.1	Kilchoanite 1219		61.4.2.1	Steenstrupine-(Ce) 1265
59.1.1.1	Bazirite 1222		61.4.2.2	Thorosteenstrupine 1265
59.1.1.2	Benitoite 1222		62.1.1.1	Muirite 1267
59.1.1.3	Pabstite 1223		62.1.2.1	Megacyclite 1267
59.1.1.4	Wadeite 1223		63.1.1.1	Steacyite 1268
59.1.2.1	Walstromite 1224		63.1.1.2	Iraqite-La 1269
59.1.2.2	Margarosanite 1224		63.2.1a.1	Brannockite 1271
59.2.1.1	Umbite 1224		63.2.1a.2	Chayesite 1272
59.2.1.2	Paraumbite 1225		63.2.1a.3	Darapiosite 1272
59.2.1.3	Kostylevite 1225		63.2.1a.4	Eifelite 1272
59.2.2.1	Catapleiite 1226		63.2.1a.5	Merrihueite 1273

63.2.1a.6	Osumilite 1273		65.2.4.1	Sorensenite 1322
63.2.1a.8	Poudretteite 1275		65.3.1.1	Haradaite 1322
63.2.1a.9	Sugilite 1275		65.3.1.2	Suzukiite 1323
63.2.1a.10	Yagiite 1276		65.3.2.1	Balangeroite 1323
63.2.1a.11	Dusmatovite 1276		65.3.2.2a	Gageite-2M 1323
63.2.1a.12	Milarite 1277		65.3.3.1	Taikanite 1324
63.2.1a.13	Sogdianite 1278		65.3.4.1	Batisite 1324
63.2.1a.14	Roedderite 1278		65.3.4.2	Shcherbakovite 1325
63.2.1b.1	Armenite 1279		65.3.4.3	Ohmilite 1325
64.1.1.1	Eudialyte 1281		65.4.1.1	Rhodonite 1327
64.1.1.2	Alluaivite 1282		65.4.1.2	Babingtonite 1328
64.2.1.1	Scawtite 1283		65.4.1.3	Manganbabingtonite 1329
64.2.2.1	Roeblingite 1284		65.4.1.4	Nambulite 1329
64.3.1.1	Taramellite 1284		65.4.1.5	Natronambulite 1330
64.3.1.2	Titantaramellite 1285		65.4.1.6	Marsturite 1330
64.3.1.3	Nagashimalite 1285		65.4.1.7	Lithiomarsturite 1331
64.3.2.1	Strakhovite 1286		65.4.2.1	Santaclaraite 1331
64.4.1.1	Tienshanite 1286		65.5.1.1	Stokesite 1331
64.5.1.1	Traskite 1286		65.5.2.1	Chkalovite 1332
65.1.1.1	Clinoenstatite 1291		65.6.1.1	Pyroxmangite 1332
65.1.1.2	Clinoferrosilite 1292		65.6.1.2	Pyroxferroite 1333
65.1.1.3	Kanoite 1292		65.7.1.1	Alamosite 1333
65.1.1.4	Pigeonite 1292		66.1.1.1	Magnesiocummingtonite 1339
65.1.2.1	Enstatite 1294		66.1.1.4	Tirodite 1342
65.1.2.2	Ferrosilite 1295		66.1.1.6	Magnesioclinoholmquistite 1343
65.1.2.3	Donpeacorite 1296		66.1.2.1	Magnesioanthophyllite 1343
65.1.3a.1	Diopside 1296		66.1.2.5	Magnesiogedrite 1345
65.1.3a.2	Hedenbergite 1298		66.1.2.8	Sodium-gedrite 1345
65.1.3a.3	Augite 1299		66.1.2.9	Magnesioholmquistite 1346
65.1.3a.4	Johannsenite 1301		66.1.3a.1	Tremolite 1346
65.1.3a.5	Petedunnite 1302		66.1.3a.4	(Alumino)magnesio-
65.1.3a.6	Esseneite 1302			hornblende 1350
65.1.3b.1	Omphacite 1302		66.1.3a.6	(Alumino)tschermakite 1352
65.1.3b.2	Aegirine-augite 1303		66.1.3a.10	Edenite 1353
65.1.3c.1	Jadeite 1304		66.1.3a.12	Pargasite 1354
65.1.3c.2	Aegirine 1305		66.1.3a.16	Magnesiosadanagaite 1356
65.1.3c.3	Namansilite 1306		66.1.3a.18	Kaersutite 1356
65.1.3c.4	Kosmochlor 1307		66.1.3b.1	(Alumino)winchite 1357
65.1.3c.5	Natalyite 1307		66.1.3b.5	(Alumino)barroisite 1358
65.1.3c.6	Jervisite 1307		66.1.3b.9	Richterite 1358
65.1.4.1	Spodumene 1308		66.1.3b.11	Magnesio(alumino)kato-
65.1.5.1	Carpholite 1311			phorite 1359
65.1.5.2	Ferrocarpholite 1311		66.1.3b.15	Magnesio(alumino)taramite 1360
65.1.5.3	Magnesiocarpholite 1311		66.1.3c.1	Glaucophane 1361
65.1.5.4	Balipholite 1311		66.1.3c.6	Eckermannite 1364
65.1.6.1	Lorenzenite 1312		66.1.3c.10	Kozulite 1365
65.1.6.2	Kukisvumite 1312		66.1.3c.11	Nyböite 1366
65.1.6.3	Lintisite 1313		66.1.3c.12	Leakeite 1366
65.1.7.1	Shattuckite 1313		66.1.3c.13	Kornite 1366
65.1.8.1	Nchwaningite 1313		66.1.3c.14	Ungarettiite 1367
65.2.1.1a	Wollastonite-1A 1315		66.1.3c.15	Fluor-ferro-leakeite 1367
65.2.1.1b	Wollastonite-2M 1317		66.1.4.1	Joesmithite 1367
65.2.1.1c	Wollastonite-3A,4A,5A,7A 1317		66.1.5.1	Ershovite 1368
65.2.1.2	Bustamite 1318		66.2.1.1	Plancheite 1368
65.2.1.3	Ferrobustamite 1318		66.2.2.1	Vlasovite 1369
65.2.1.4a	Pectolite-1A 1319		66.3.1.1	Xonotlite 1369
65.2.1.5	Serandite 1320		66.3.1.2	Zorite 1370
65.2.1.6	Cascandite 1320		66.3.1.3	Euddidymite 1370
65.2.1.7	Denisovite 1321		66.3.1.4	Epididymite 1371
65.2.2.1	Foshagite 1321		66.3.1.5	Yuksporite 1371
65.2.3.1	Hillebrandite 1321		66.3.3.1	Inesite 1372

Index of Mineral Names in Numerical Order

66.3.4.1	Tuhualite 1374		71.1.2c.4	Fraipontite 1425
66.3.4.2	Zektzerite 1374		71.1.2c.5	Kellyite 1426
66.3.4.3	Emeleusite 1375		71.1.2c.6	Manandonite 1426
67.1.1.1	Jimthompsonite 1376		71.1.2c.7	Cronstedtite 1427
67.1.1.2	Clinojimthompsonite 1376		71.1.2d.1	Clinochrysotile 1428
67.1.2.1	Carlosturanite 1376		71.1.2d.4	Pecoraite 1430
67.2.1.1	Tinaksite 1377		71.1.3.1	Bismutoferrite 1431
67.2.1.2	Tokkoite 1377		71.1.3.2	Chapmanite 1431
68.1.1.1	Chesterite 1379		71.1.5.1	Allophane 1432
68.1.2.1	Vinogradovite 1379		71.1.5.2	Hisingerite 1434
68.1.3.1	Aërinite 1380		71.1.5.3	Imogolite 1435
69.1.1.1	Astrophyllite 1382		71.1.5.4	Neotocite 1436
69.1.1.2	Kupletskite 1383		71.2.1.1	Pyrophyllite-1A, 2M 1439
69.1.1.3	Cesium Kupletskite 1383		71.2.1.2	Ferripyrophyllite 1441
69.1.1.4	Niobophyllite 1383		71.2.1.3	Talc-1A 1441
69.1.1.5	Zircophyllite 1384		71.2.1.4	Willemseite 1443
69.1.1.6	Hydroastrophyllite 1384		71.2.1.5	Minnesotaite 1443
69.1.1.7	Magnesium Astrophyllite 1384		71.2.2a.1	Muscovite-$2M_1$, 1M, 3T, $2M_2$ 1448
69.2.1a.1	Aenigmatite 1386			
69.2.1a.2	Dorrite 1387		71.2.2a.2	Paragonite-$2M_1$, 3T, 1M 1452
69.2.1a.3	Høgtuvaite 1387		71.2.2a.3	Chernykhite-$2M_1$ 1452
69.2.1a.4	Krinovite 1388		71.2.2a.4	Roscoelite-1M 1453
69.2.1a.5	Rhönite 1388		71.2.2a.5	Glauconite-1M 1453
69.2.1a.6	Serendibite 1389		71.2.2a.6	Celadonite-1M 1454
69.2.1a.7	Welshite 1390		71.2.2a.7	Tobelite-1M 1455
69.2.1a.8	Wilkinsonite 1391		71.2.2a.8	Nanpingite-$2M_1$ 1455
69.2.1b.1	Sapphirine 1391		71.2.2a.9	Boromuscovite-1M, $2M_1$ 1456
69.2.1b.2	Surinamite 1393		71.2.2a.10	Montdorite-1M, 3T 1456
69.2.1c.1	Magbasite 1393		71.2.2b.1	Phlogopite-1M, 3T, $2M_1$ 1456
69.2.2.1	Saneroite 1393		71.2.2b.2	Biotite-1M, 3T, $2M_1$ 1458
69.2.3.1	Howieite 1394		71.2.2b.3	Annite-1M 1461
69.2.3.2	Taneyamalite 1394		71.2.2b.4	Ferriannite-1M 1461
69.2.3.3	Deerite 1395		71.2.2b.5	Siderophyllite-1M 1461
69.2.4.1	Liebauite 1395		71.2.2b.6	Hendricksite-1M 1462
69.2.5.1	Johninnesite 1395		71.2.2b.7	Lepidolite 1462
70.1.1.1	Litidionite 1397		71.2.2b.8	Polylithionite-1M 1464
70.1.1.2	Fenaksite 1397		71.2.2b.9	Taeniolite-1M,$2M_1$ 1464
70.1.1.3	Manaksite 1398		71.2.2b.10	Zinnwaldite-1M,$2M_1$,3T 1465
70.1.1.4	Agrellite 1398		71.2.2b.11	Norrishite-1M 1466
70.1.2.1	Narsarsukite 1398		71.2.2b.12	Masutomilite-1M, $2M_1$ 1466
70.1.2.2	Caysichite-(Y) 1399		71.2.2b.13	Sodium Phlogopite 1467
70.1.2.3	Charoite 1399		71.2.2b.14	Wonesite-1Md 1467
70.2.1.1	Miserite 1400		71.2.2b.15	Preiswerkite-1Md, $2M_1$ 1467
70.2.2.1	Penkvilksite 1400		71.2.2b.16	Ephesite-$2M_1$, 1M 1467
70.3.1.1	Ashcroftine-(Y) 1401		71.2.2c.1	Margarite-$2M_1$ 1468
70.4.1.1	Neptunite 1401		71.2.2c.2	Clintonite-1M, $2M_1$ 1469
70.4.1.2	Manganneptunite 1402		71.2.2c.3	Bityite-$2M_1$ 1469
70.5.1.1	Chiavennite 1402		71.2.2c.4	Anandite-$2M_1$, 2O 1470
70.5.2.1	Tvedalite 1403		71.2.2c.5	Kinoshitalite-1M 1470
70.5.3.1	Bavenite 1403		71.2.2d.1	Hydrobiotite 1471
71.1.1.1	Dickite 1407		71.2.2d.2	Illite 1472
71.1.1.4	Halloysite 1411		71.2.2d.3	Vermiculite 1474
71.1.1.5	Odinite 1414		71.2.2d.4	Brammallite 1477
71.1.2a.1	Antigorite 1415		71.2.3.1	Cuprorivaite 1477
71.1.2b.1	Caryopilite 1417		71.2.3.2	Gillespite 1478
71.1.2b.2	Lizardite 1418		71.2.3.3	Effenbergerite 1478
71.1.2b.3	Nepouite 1420		71.2.4.1	Ferrisurite 1479
71.1.2b.4	Greenalite 1421		71.2.4.2	Surite 1479
71.1.2c.1	Amesite 1422		71.2.5.1	Macaulayite 1480
71.1.2c.2	Berthierine 1423		71.3.1a.1	Beidellite 1481
71.1.2c.3	Brindleyite 1424		71.3.1a.2	Montmorillonite 1483

71.3.1a.3	Nontronite 1485		72.5.2.2	Nordite-(Ce) 1541
71.3.1a.4	Volkonskoite 1486		72.5.2.3	Nordite-(La) 1542
71.3.1a.5	Swinefordite 1487		72.5.2.4	Semenovite-(Ce) 1542
71.3.1b.1	Sobotkite 1488		72.5.3.1	Kvanefjeldite 1542
71.3.1b.2	Saponite 1489		72.5.4.1	Armstrongite 1543
71.3.1b.3	Sauconite 1490		72.5.4.2	Elpidite 1543
71.3.1b.4	Hectorite 1491		72.5.5.1	Varennesite 1544
71.3.1b.5	Pimelite 1492		72.6.1.1	Petalite 1545
71.3.1b.6	Stevensite 1493		73.1.1.1	Naujakasite 1546
71.3.1b.7	Yakhontovite 1495		73.1.2.1	Latiumite 1546
71.3.1b.8	Zincsilite 1495		73.1.2.2	Tuscanite 1547
71.4.1.1	Donbassite 1498		73.1.3.1	Fedorite 1547
71.4.1.2	Cookeite 1499		73.1.4.1	Zeophyllite 1547
71.4.1.3	Sudoite 1500		73.1.5.1	Kanemite 1548
71.4.1.4	Clinochlore 1501		73.1.6.1	Vertumnite 1548
71.4.1.5	Nimite 1503		73.2.1.1	Jagoite 1548
71.4.1.6	Baileychlore 1503		73.2.2a.1	Reyerite 1549
71.4.1.7	Chamosite 1504		73.2.2a.2	Truscottite 1549
71.4.1.8	Pennantite 1505		73.2.2b.1	Minehillite 1549
71.4.2.1	Dozyite 1508		73.2.2b.2	Orlymanite 1550
71.4.2.2	Lunijianlaite 1509		73.2.2c.1	Gyrolite 1550
71.4.2.3	Saliotite 1510		74.1.1.1	Stilpnomelane 1552
71.4.2.4	Tosudite 1510		74.1.1.2	Lennilenapeite 1553
71.4.2.5	Corrensite 1512		74.1.1.3	Franklinphilite 1553
71.4.2.6	Kulkeite 1514		74.1.1.4	Bannisterite 1554
71.4.2.7	Rectorite 1515		74.1.2.1	Ganophyllite 1554
71.4.2.8	Allietite 1518		74.1.2.2	Eggletonite 1555
71.4.3.1	Franklinfurnaceite 1519		74.1.3.1	Parsettensite 1555
71.4.4.1	Gonyerite 1519		74.1.4.1	Zussmanite 1555
71.5.1.1	Kegelite 1520		74.2.1.1	Wickenburgite 1556
72.1.1.1	Ekanite 1521		74.3.1a.1	Palygorskite 1558
72.1.2.1	Thornasite 1521		74.3.1a.2	Tuperssuatsiaite 1559
72.1.3.1	Prehnite 1522		74.3.1a.3	Yofortierite 1560
72.1.4.1	Amstallite 1523		74.3.1b.1	Sepiolite 1560
72.1.5.1	Eakerite 1524		74.3.1b.2	Falcondoite 1561
72.2.1.1	Dalyite 1524		74.3.1b.3	Loughlinite 1561
72.2.1.2	Davanite 1525		74.3.2.1	Chrysocolla 1561
72.2.1.3	Sazhinite-(Ce) 1525		74.3.3.1	Altisite 1563
72.3.1.1	Fluorapophyllite 1525		74.3.3.2	Lemoynite 1563
72.3.1.3	Natroapophyllite 1527		74.3.4.1	Sanbornite 1563
72.3.1.4	Carletonite 1528		74.3.4.2	Krauskopfite 1564
72.3.2.1	Tobermorite 1529		74.3.5.1	Natrosilite 1564
72.3.2.2	Clinotobermorite 1530		74.3.5.2	Makatite 1565
72.3.2.3	Plombierite 1531		74.3.5.3	Silinaite 1565
72.3.2.4	Riversideite 1532		71.3.5.4	Searlesite 1566
72.3.2.5	Okenite 1532		74.3.6.1	Revdite 1566
72.3.2.6	Tacharanite 1533		74.3.7.1	Cavansite 1567
72.3.2.7	Nekoite 1534		74.3.7.2	Pentagonite 1567
72.3.2.8	Oyelite 1534		75.1.1.1	Cristobalite 1568
72.4.1a.1	Ferropyrosmalite 1536		75.1.2.1	Tridymite 1570
72.4.1a.2	Manganpyrosmalite 1537		75.1.3.1	Quartz 1573
72.4.1a.3	Schallerite 1537		75.1.4.1	Coesite 1586
72.4.1b.1	Friedelite 1537		75.2.1.1	Opal 1587
72.4.1b.2	Mcgillite 1538		75.2.2.1	Melanophlogite 1592
72.4.1b.3	Nelenite 1538		75.2.3.1	Lutecite 1593
72.5.1.1	Rhodesite 1539		75.2.4.1	Silhydrite 1594
72.5.1.2	Monteregianite-Y 1539		75.3.1.1	Virgilite 1594
72.5.1.3	Macdonaldite 1540		76.1.1.1	Orthoclase 1600
72.5.1.4	Delhayelite 1540		76.1.1.2	Sanidine 1602
72.5.1.5	Hydrodelhayelite 1540		76.1.1.3	Hyalophane 1603
72.5.2.1	Leucosphenite 1541		76.1.1.4	Celsian 1604

Index of Mineral Names in Numerical Order

76.1.1.5	Microcline 1605		77.1.2.7	Faujasite 1659
76.1.1.6	Anorthoclase 1607		77.1.2.8	Levyne 1660
76.1.2.1	Buddingtonite 1608		77.1.3.1	Gismondine 1661
76.1.3.1	Albite 1608		77.1.3.2	Amicite 1662
76.1.4.1	Reedmergnerite 1614		77.1.3.3	Garronite 1662
76.1.5.1	Paracelsian 1615		77.1.3.4	Gobbinsite 1663
76.1.5.2	Slawsonite 1615		77.1.3.5	Harmotome 1664
76.1.6.1	Banalsite 1616		77.1.3.6	Phillipsite 1665
76.1.6.2	Stronalsite 1616		77.1.3.8	Perlialite 1667
76.1.6.3	Lisetite 1616		77.1.3.9	Paulingite 1667
76.1.6.4	Svyatoslavite 1617		77.1.3.10	Mazzite 1668
76.1.7.1	Dmisteinbergite 1617		77.1.3.11	Merlinoite 1668
76.2.1.1	Kalsilite 1619		77.1.3.12	Montesommaite 1669
76.2.1.2	Nepheline 1619		77.1.4.1	Heulandite-Ca 1669
76.2.1.3	Trikalsilite 1621		77.1.4.2	Clinoptilolite 1671
76.2.1.4	Panunzite 1621		77.1.4.3	Stilbite-(Ca) 1674
76.2.1.5	Kaliophilite 1622		77.1.4.4	Stellerite 1675
76.2.1.6	Yoshiokaite 1622		77.1.4.5	Barrerite 1676
76.2.2.1	Leucite 1622		77.1.5.1	Natrolite 1677
76.2.2.2	Ammonioleucite 1623		77.1.5.2	Tetranatrolite 1679
76.2.3.1	Sodalite 1625		77.1.5.3	Paranatrolite 1680
76.2.3.2	Nosean 1626		77.1.5.4	Mesolite 1681
76.2.3.3	Haüyne 1627		77.1.5.5	Scolecite 1682
76.2.3.4	Lazurite 1627		77.1.5.6	Edingtonite 1683
76.2.3.5	Bicchulite 1628		77.1.5.7	Gonnardite 1685
76.2.3.6	Kamaishilite 1628		77.1.5.8	Cowlesite 1686
76.2.3.7	Tugtupite 1629		77.1.5.9	Thomsonite 1687
76.2.3.8	Tsargorodtsevite 1629		77.1.6.1	Mordenite 1688
76.2.4.1	Helvite 1630		77.1.6.2	Epistilbite 1689
76.2.4.2	Danalite 1631		77.1.6.3	Maricopaite 1689
76.2.4.3	Genthelvite 1631		77.1.6.4	Dachiardite 1690
76.2.5.1	Afghanite 1634		77.1.6.5	Sodium Dachiardite 1691
76.2.5.2	Bystrite 1634		77.1.6.6	Ferrierite 1692
76.2.5.3	Cancrinite 1634		77.1.6.7	Boggsite 1693
76.2.5.4	Cancrisilite 1635		77.1.7.1	Brewsterite 1693
76.2.5.5	Davyne 1636		77.1.7.2	Yugawaralite 1694
76.2.5.6	Franzinite 1636		77.1.7.3	Goosecreekite 1695
76.2.5.7	Giuseppettite 1636		77.1.7.4	Roggianite 1696
76.2.5.8	Hydroxycancrinite 1637		77.1.7.5	Ba Analog of Brewsterite 1696
76.2.5.9	Liottite 1637		77.1.8.1	Bellbergite 1697
76.2.5.10	Microsommite 1638		77.1.8.2	Tschernichite 1697
76.2.5.11	Pitiglianoite 1638		77.2.1.1	Bikitaite 1698
76.2.5.12	Quadridavyne 1639		77.2.2.1	Lithosite 1699
76.2.5.13	Sacrofanite 1639		77.2.3.1	Mountainite 1699
76.2.5.14	Tounkite 1639		77.2.4.1	Partheite 1700
76.2.5.15	Vishnevite 1640		77.2.5.1	Strätlingite 1700
76.2.5.16	Wenkite 1640		77.2.6.1	Gaultite 1700
76.3.1.1	Marialite 1641		77.2.7.1	Kalborsite 1701
76.3.2.1	Sarcolite 1644		77.2.8.1	Verplanckite 1701
76.3.3.1	Ussingite 1644		78.1.1.1	Cebollite 1702
77.1.1.1	Analcime 1646		78.1.2.1	Chantalite 1702
77.1.1.2	Pollucite 1648		78.1.3.1	Cymrite 1702
77.1.1.3	Wairakite 1649		78.1.4.1	Dellaite 1703
77.1.1.4	Laumontite 1650		78.1.5.1	Jinshajiangite 1703
77.1.1.5	Hsianghualite 1651		78.1.5.2	Perraultite 1704
77.1.2.1	Chabazite-Ca 1652		78.1.6.1	Kurumsakite 1704
77.1.2.2	Herschelite 1654		78.1.7.1	Plumbotsumite 1704
77.1.2.3	Willhendersonite 1655		78.1.8.1	Poldervaartite 1705
77.1.2.4	Offretite 1655		78.1.9.1	Stringhamite 1705
77.1.2.5	Erionite 1656		78.1.10.1	Vistepite 1705
77.1.2.6	Gmelinite 1657		78.2.1.1	Ganomalite 1706

78.3.1.1	Mathewrogersite 1706		78.6.4.1	Tiettaite 1713
78.3.2.1	Shafranovskite 1706		78.7.1.1	Bostwickite 1713
78.3.3.1	Zakharovite 1706		78.7.2.1	Creaseyite 1713
78.4.1.1	Apachite 1707		78.7.3.1	Ertixiite 1714
78.4.2.1	Ashburtonite 1707		78.7.4.1	Ilimaussite-(Ce) 1714
78.4.3.1	Fukalite 1707		78.7.5.1	Ilmajokite 1714
78.4.4.1	Gilalite 1708		78.7.6.1	Juanite 1715
78.4.5.1	Grumantite 1708		78.7.7.1	Kenyaite 1715
78.5.1.1	Ajoite 1708		78.7.8.1	Kuliokite-(Y) 1716
78.5.2.1	Bementite 1709		78.7.9.1	Laplandite-(Ce) 1716
78.5.3.1	Burckhardtite 1709		78.7.10.1	Leifite 1717
78.5.4.1	Canasite 1709		78.7.11.1	Luddenite 1717
78.5.5.1	Erlianite 1710		78.7.12.1	Magadiite 1717
78.5.6.1	Karpinskite 1710		78.7.13.1	Maufite 1718
78.5.7.1	Molybdophyllite 1711		78.7.14.1	Spadaite 1718
78.5.8.1	Raite 1711		78.7.15.1	Terskite 1718
78.5.9.1	Tungusite 1711		78.7.16.1	Tranquillityite 1719
78.6.1.1	Hyalotekite 1712		78.7.17.1	Umbozerite 1719
78.6.2.1	Lovdarite 1712		78.7.18.1	Vyuntspakhkite-(Y) 1719
78.6.3.1	Pellyite 1712		78.7.19.1	Wawayandaite 1720

Index of Mineral Names in Alphabetical Order

β-Fergusonite-(Ce) 8.1.2.2   p.333
β-Fergusonite-(Nd) 8.1.2.3   p.333
β-Fergusonite-(Y)   8.1.2.1   p.333
β-Roselite 40.2.2.7 p.753
Abelsonite 50.4.9.1 p.1018
Abenakiite-(Ce) 61.4.1.1 p.1264
Abernathyite 40.2a.9.1 p.771
Abhurite 10.5.9.1 p.401
Abswurmbachite 7.5.1.4 p.314
Acanthite 2.4.1.1 p.42
Acetamide 50.4.7.1 p.1018
Acuminite 11.6.17.1 p.417
Adamite 41.6.6.3 p.850
Adelite 41.5.1.1 p.826
Admontite 26.6.3.1 p.561
Aegirine-augite 65.1.3b.2 p.1303
Aegirine 65.1.3c.2 p.1305
Aenigmatite 69.2.1a.1 p.1386
Aerugite 38.5.10.1 p.734
Aeschynite-(Ce) 8.3.6.1 p.357
Aeschynite-(Nd) 8.3.6.5 p.359
Aeschynite-(Y) 8.3.6.3 p.358
Aërinite 68.1.3.1 p.1380
Afghanite 76.2.5.1 p.1634
Afwillite 52.4.7.1 p.1098
Agardite-(Y) 42.5.1.2 p.890
Agrellite 70.1.1.4 p.1398
Agrinierite 5.5.1.1 p.260
Aguilarite 2.4.1.3 p.42
Aheylite 42.9.3.5 p.919
Ahlfeldite 34.2.3.3 p.681
Aikinite 3.4.5.1 p.169
Ajoite 78.5.1.1 p.1708
Akaganéite 6.1.6.1 p.273
Akatoreite 57.2.3.1 p.1191
Akdalaite 4.3.2.1 p.221
Akhtenskite 4.4.10.1 p.249
Akrochordite 42.4.1.1 p.884
Aksaite 26.6.4.1 p.561
Aktashite 3.4.13.2 p.174
Alacranite 2.8.22.4 p.85
Alamosite 65.7.1.1 p.1333
Albite 76.1.3.1 p.1608
Albrechtschraufite 16b.7.11.1 p.517
Aldermanite 42.13.1.1 p.952
Aleksite 2.6.3.2 p.59
Alforsite 41.8.5.2 p.868

Algodonite 2.1.1.1 p.37
Allactite 41.2.1.1 p.817
Allanite-(Ce) 58.2.1a.1 p.1196
Allanite-(La) 58.2.1a.2 p.1197
Allanite-(Y) 58.2.1a.3 p.1198
Allargentum 2.2.1.2 p.38
Alleghanyite 52.3.2b.1 p.1079
Allietite 71.4.2.8 p.1518
Alloclasite 2.12.6.2 p.131
Allophane 71.1.5.1 p.1432
Alluaivite 64.1.1.2 p.1282
Alluaudite 38.2.3.6 p.709
Almandine 51.4.3a.2 p.1039
Alstonite 14.2.5.1 p.454
Altaite 2.8.1.3 p.65
Althausite 41.6.5.1 p.849
Althupite 42.4.13.1 p.889
Altisite 74.3.3.1 p.1563
Aluminite 31.7.4.1 p.649
Aluminocopiapite 31.10.5.7 p.664
(Alumino)barroisite 66.1.3b.5 p.1358
(Alumino)magnesiohornblende 66.1.3a.4 p.1350
(Alumino)tschermakite 66.1.3a.6 p.1352
(Alumino)winchite 66.1.3b.1 p.1357
Aluminum 1.1.1.5 p.5
Alumohydrocalcite 16b.2.3.1 p.502
Alumoklyuchevskite 28.4.6.2 p.582
Alumopharmacosiderite 42.8.1.2 p.909
Alumotantite 8.1.7.1 p.336
Alumotungstite 4.5.4.1 p.252
Alunite 30.2.4.1 p.631
Alunogen 29.8.6.1 p.620
Alvanite 42.13.13.1 p.956
Amakinite 6.2.1.2 p.275
Amarantite 31.9.3.1 p.652
Amarillite 29.5.3.2 p.592
Amblygonite 41.5.8.1 p.836
Ameghinite 25.3.1.1 p.545
Amesite 71.1.2c.1 p.1422
Amicite 77.1.3.2 p.1662
Aminoffite 57.1.1.1 p.1188
Ammonioalunite 30.2.4.6 p.633
Ammonioborite 26.5.3.1 p.556
Ammoniojarosite 30.2.5.4 p.635
Ammonioleucite 76.2.2.2 p.1623
Amstallite 72.1.4.1 p.1523

1753

Analcime 77.1.1.1 p.1646
Anandite-2M_1, 2O 71.2.2c.4 p.1470
Anapaite 40.2.1.1 p.749
Anatase 4.4.4.1 p.245
Ancylite-(Ce) 16b.1.1.1 p.497
Andalusite 52.2.2b.1 p.1066
Andersonite 15.2.5.1 p.466
Andorite 3.4.15.4 p.176
Andradite 51.4.3b.1 p.1044
Andremeyerite 56.2.6c.6 p.1176
Anduoite 2.12.2.8 p.122
Angelellite 41.6.10.1 p.852
Anglesite 28.3.1.3 p.575
Anhydrite 28.3.2.1 p.576
Anilite 2.4.7.5 p.48
Ankangite 7.9.4.2 p.324
Ankerite 14.2.1.2 p.451
Annabergite 40.3.6.4 p.795
Annite-1M 71.2.2b.3 p.1461
Anorthoclase 76.1.1.6 p.1607
Antarcticite 9.2.6.1 p.385
Anthoinite 48.3.3.1 p.1002
Anthonyite 6.2.6.1 p.278
Antigorite 71.1.2a.1 p.1415
Antimonpearceite 3.1.8.2 p.154
Antimonselite 2.11.2.2 p.107
Antimony 1.3.1.2 p.27
Antlerite 30.1.12.1 p.626
Anyuiite 1.1.4.1 p.6
Apachite 78.4.1.1 p.1707
Aphthitalite 28.2.2.1 p.569
Apjohnite 29.7.3.3 p.617
Aplowite 29.6.6.4 p.603
Apuanite 45.1.9.1 p.979
Aramayoite 3.7.4.1 p.191
Aravaipaite 11.6.19.1 p.417
Arcanite 28.2.1.2 p.569
Archerite 37.1.4.2 p.698
Arctite 41.5.18.1 p.843
Arcubisite 3.2.6.1 p.159
Ardaite 2.15.1.1 p.142
Ardealite 43.1.1.1 p.957
Ardennite 58.3.1.1 p.1217
Argentojarosite 30.2.5.5 p.635
Argentopyrite 2.9.13.2 p.96
Argentotennantite 3.3.6.7 p.165
Argutite 4.4.1.7 p.243
Argyrodite 2.5.6.1 p.54
Arhbarite 42.6.5.2 p.894
Aristarainite 26.6.5.1 p.561
Armalcolite 7.7.1.2 p.317
Armangite 46.2.7.1 p.983
Armenite 63.2.1b.1 p.1279
Armstrongite 72.5.4.1 p.1543
Arrojadite 41.7.2.1 p.853
Arsenbrackebuschite 40.2.8.2 p.759
Arsendescloizite 41.5.1.9 p.829
Arsenic 1.3.1.1 p.26
Arseniopleite 38.2.2.2 p.708
Arseniosiderite 42.8.4.3 p.912
Arsenobismite 41.6.9.1 p.852

Arsenoclasite 41.4.1.1 p.821
Arsenocrandallite 42.7.4.1 p.901
Arsenoflorencite-(Ce) 41.5.11.1 p.840
Arsenoflorencite-(La) 41.5.11.3 p.841
Arsenoflorencite-(Nd) 41.5.11.2 p.840
Arsenogoyazite 42.7.4.3 p.902
Arsenohauchecornite 3.2.5.4 p.159
Arsenolamprite 1.3.2.1 p.29
Arsenolite 4.3.9.1 p.231
Arsenopalladinite 2.16.5.1 p.145
Arsenopyrite 2.12.4.1 p.129
Arsenosulvanite 3.2.3.2 p.157
Arsenpolybasite 3.1.7.1 p.153
Arsentsumebite 43.4.2.2 p.963
Arsenuranospathite 40.2a.23.1 p.778
Arsenuranylite 42.4.9.1 p.887
Arthurite 42.11.18.2 p.947
Artinite 16b.3.1.1 p.503
Artroeite 11.6.18.1 p.417
Arupite 40.3.6.8 p.796
Arzakite 10.3.4.2 p.397
Arzrunite 12.1.8.1 p.421
Asbecasite 45.1.3.1 p.977
Asbolane 6.4.9.1 p.290
Aschamalmite 3.1.11.1 p.154
Ashburtonite 78.4.2.1 p.1707
Ashcroftine-(Y) 70.3.1.1 p.1401
Ashoverite 6.2.11.1 p.279
Asisite 10.2.7.1 p.395
Asselbornite 42.4.11.1 p.888
Astrocyanite-(Ce) 16b.1.6.1 p.500
Astrophyllite 69.1.1.1 p.1382
Atacamite 10.1.1.1 p.391
Atelestite 41.11.5.1 p.881
Athabascaite 2.16.17.1 p.148
Atheneite 2.2.4.1 p.39
Atlasovite 30.1.17.2 p.627
Atokite 1.2.5.3 p.24
Attakolite 43.4.13.1 p.966
Aubertite 31.9.8.1 p.654
Augelite 41.6.8.1 p.851
Augite 65.1.3a.3 p.1299
Aurichalcite 16a.4.2.1 p.494
Auricupride 1.1.2.1 p.5
Aurorite 7.8.2.2 p.319
Aurostibite 2.12.1.11 p.117
Austinite 41.5.1.3 p.827
Autunite 40.2a.1.1 p.766
Avicennite 4.3.8.1 p.231
Avogadrite 11.2.2.1 p.408
Awaruite 1.1.11.4 p.12
Azoproite 24.2.1.3 p.536
Azurite 16a.2.1.1 p.482
Ba Analog of Brewsterite 77.1.7.5 p.1696
Babefphite 41.5.3.1 p.831
Babingtonite 65.4.1.2 p.1328
Baddeleyite 4.4.14.1 p.250
Bafertisite 56.2.6b.1 p.1171
Baghdadite 56.2.4.1 p.1161
Bahianite 44.3.7.1 p.974
Baileychlore 71.4.1.6 p.1503

Index of Mineral Names in Alphabetical Order

Bakerite 54.2.1b.1 p.1123
Balangeroite 65.3.2.1 p.1323
Balipholite 65.1.5.4 p.1311
Balkanite 2.16.7.1 p.145
Balyakinite 34.1.3.1 p.678
Bambollaite 2.12.9.1 p.132
Banalsite 76.1.6.1 p.1616
Bandylite 25.1.4.2 p.543
Bannermanite 47.3.5.1 p.990
Bannisterite 74.1.1.4 p.1554
Baotite 60.1.2.1 p.1233
Bararite 11.5.2.2 p.411
Baratovite 61.1.4.1 p.1250
Barberiite 11.2.4.1 p.409
Barbertonite 16b.6.1.2 p.509
Barbosalite 41.10.1.4 p.875
Barentsite 13.1.8.1 p.425
Bariandite 47.3.2.4 p.989
Baricite 40.3.6.2 p.794
Bario-orthojoaquinite 60.1.1b.3 p.1233
Bariomicrolite 8.2.2.2 p.346
Bariopyrochlore 8.2.1.3 p.344
Barite 28.3.1.1 p.572
Barium Pharmacosiderite 42.8.1.3 p.909
Barium-alumopharmacosiderite 42.8.1.4 p.910
Barnesite 47.3.1.3 p.988
Barrerite 77.1.4.5 p.1676
Barringerite 1.1.21.1 p.15
Barringtonite 15.1.5.1 p.463
Barstowite 12.1.7.1 p.421
Bartelkeite 7.6.2.1 p.316
Bartonite 2.9.19.1 p.98
Barylite 55.1.1.1 p.1133
Barysilite 55.2.3.1 p.1140
Barytocalcite 14.2.6.1 p.454
Barytolamprophyllite 56.2.6c.1 p.1174
Basaluminite 31.4.5.1 p.644
Bassanite 29.6.1.1 p.594
Bassetite 40.2a.16.1 p.775
Bastnäsite-(Ce) 16a.1.1.1 p.476
Bastnäsite-(La) 16a.1.1.2 p.478
Bastnäsite-(Y) 16a.1.1.3 p.478
Batisite 65.3.4.1 p.1324
Baumhauerite 3.6.13.1 p.186
Bauranoite 5.4.2.2 p.260
Bavenite 70.5.3.1 p.1403
Bayerite 6.3.2.1 p.281
Bayldonite 41.5.14.1 p.842
Bayleyite 15.3.3.1 p.468
Baylissite 15.2.4.1 p.466
Bazhenovite 29.9.3.1 p.621
Bazirite 59.1.1.1 p.1222
Bazzite 61.1.1.2 p.1244
Bearsite 42.6.1.2 p.892
Bearthite 41.10.4.2 p.876
Beaverite 30.2.5.7 p.636
Becquerelite 5.7.1.2 p.261
Behierite 24.1.2.1 p.534
Behoite 6.2.2.1 p.277
Beidellite 71.3.1a.1 p.1481

Belendorffite 1.1.9.2 p.9
Bellbergite 77.1.8.1 p.1697
Bellidoite 2.4.9.1 p.49
Bellingerite 21.1.3.1 p.531
Belovite 41.8.1.6 p.862
Belyankinite 8.7.8.1 p.371
Bementite 78.5.2.1 p.1709
Benavidesite 3.6.7.2 p.184
Benitoite 59.1.1.2 p.1222
Benjaminite 3.8.10.3 p.200
Benleonardite 3.1.10.1 p.154
Benstonite 14.2.3.1 p.453
Bentorite 31.10.2.2 p.659
Benyacarite 42.11.21.4 p.950
Beraunite 42.11.16.1 p.945
Berborite 26.1.1.1 p.551
Berdesinskiite 7.7.2.1 p.317
Bergenite 42.4.5.3 p.886
Bergslagite 41.5.4.4 p.833
Berlinite 38.4.2.1 p.720
Bermanite 42.11.17.1 p.945
Bernalite 6.3.5.3 p.283
Bernardite 3.8.14.2 p.202
Berndtite 2.12.14.5 p.138
Berryite 3.6.15.1 p.186
Berthierine 71.1.2c.2 p.1423
Berthierite 3.7.9.3 p.194
Bertossaite 41.7.1.2 p.853
Bertrandite 56.1.1.1 p.1149
Beryllite 52.1.1.1 p.1061
Beryllonite 38.1.5.1 p.703
Beryl 61.1.1.1 p.1239
Berzelianite 2.4.10.1 p.49
Berzeliite 38.2.1.1 p.705
Beta-uranophane 53.3.1.9 p.1113
Betafite 8.2.3.1 p.347
Betekhtinite 2.16.8.1 p.145
Betpakdalite 49.4.1.1 p.1008
Beudantite 43.4.1.1 p.961
Beusite 38.3.3.2 p.715
Beyerite 16a.2.3.1 p.486
Bezsmertnovite 2.1.5.1 p.38
Bianchite 29.6.8.2 p.608
Bicchulite 76.2.3.5 p.1628
Bideauxite 10.6.9.1 p.405
Bieberite 29.6.10.4 p.612
Bijvoetite-(Y) 16b.2.4.1 p.502
Bikitaite 77.2.1.1 p.1698
Bilibinskite 2.1.6.1 p.38
Bilinite 29.7.3.5 p.618
Billietite 5.7.1.3 p.262
Billingsleyite 3.1.6.1 p.153
Bindheimite 44.1.1.2 p.971
Biotite-1M, 3T, 2M$_1$ 71.2.2b.2 p.1458
Biphosphammite 37.1.4.1 p.698
Biringuccite 26.5.7.1 p.557
Birnessite 7.5.3.1 p.315
Bischofite 9.2.9.1 p.387
Bismite 4.3.10.2 p.232
Bismoclite 10.2.1.2 p.393
Bismuthinite 2.11.2.3 p.107

Bismuth 1.3.1.4 p.28
Bismutite 16a.3.5.1 p.491
Bismutocolumbite 8.1.6.4 p.335
Bismutoferrite 71.1.3.1 p.1431
Bismutohauchecornite 3.2.5.2 p.159
Bismutomicrolite 8.2.2.5 p.347
Bismutostibiconite 44.1.1.6 p.972
Bismutotantalite 8.1.6.3 p.335
Bityite-2M_1 71.2.2c.3 p.1469
Bixbyite 4.3.7.2 p.230
Bjarebyite 41.9.1.3 p.870
Blatterite 24.2.7.1 p.538
Blixite 10.2.4.1 p.394
Blossite 38.5.6.1 p.733
Blödite 29.3.3.1 p.585
Bobfergusonite 38.2.4.5 p.712
Bobierrite 40.3.7.1 p.797
Boehmite 6.1.2.1 p.270
Bogdanovite 2.2.3.1 p.39
Boggildite 12.1.6.1 p.420
Boggsite 77.1.6.7 p.1693
Bohdanowiczite 3.7.1.2 p.190
Bokite 47.3.2.5 p.989
Boleite 10.6.6.1 p.404
Bolivarite 42.6.8.1 p.895
Boltwoodite 53.3.1.5 p.1112
Bonaccordite 24.2.1.4 p.536
Bonshtedtite 43.2.1.3 p.958
Boothite 29.6.10.2 p.611
Boracite 25.6.1.1 p.548
Borax 26.4.1.1 p.554
Borcarite 27.1.3.1 p.565
Borishanskiite 2.12.16.1 p.139
Bornemanite 56.4.2.1 p.1187
Bornhardtite 2.10.1.5 p.100
Bornite 2.5.2.1 p.52
Borodaevite 3.8.10.5 p.201
Boromuscovite-1M, 2M_1 71.2.2a.9 p.1456
Borovskite 2.8.1.7 p.67
Bostwickite 78.7.1.1 p.1713
Botallackite 10.1.3.1 p.392
Botryogen 31.9.6.1 p.653
Bottinoite 6.3.9.1 p.287
Boulangerite 3.5.2.1 p.178
Bournonite 3.4.3.2 p.167
Boussingaultite 29.3.7.2 p.588
Bowieite 2.11.12.1 p.112
Boyleite 29.6.6.5 p.604
Brabantite 38.4.3.4 p.724
Bracewellite 6.1.1.5 p.269
Brackebuschite 40.2.8.1 p.758
Bradleyite 43.2.1.1 p.957
Braggite 2.8.5.3 p.71
Braitschite-(Ce) 26.7.6.1 p.563
Brammallite 71.2.2d.4 p.1477
Brandtite 40.2.3.2 p.755
Brannerite 8.3.4.1 p.356
Brannockite 63.2.1a.1 p.1271
Brassite 39.1.7.1 p.738
Braunite II 7.5.1.3 p.314
Braunite I 7.5.1.1 p.312

Brazilianite 41.5.7.1 p.836
Bredigite 51.4.1.1 p.1035
Breithauptite 2.8.11.2 p.77
Brenkite 16a.3.6.1 p.492
Brewsterite 77.1.7.1 p.1693
Brezinaite 2.10.2.2 p.103
Brianite 38.1.7.1 p.704
Brianyoungite 17.1.15 p.525
Briartite 2.9.2.3 p.90
Brindleyite 71.1.2c.3 p.1424
Britholite-(Ce) 52.4.9.1 p.1100
Britholite-(Y) 52.4.9.2 p.1101
Briziite 4.3.5.6 p.229
Brizziite 44.3.12.1 p.975
Brochantite 30.1.3.1 p.623
Brockite 40.4.7.5 p.808
Bromargyrite 9.1.4.2 p.378
Bromellite 4.2.2.2 p.211
Brookite 4.4.5.1 p.246
Brownmillerite 7.11.2.1 p.326
Brucite 6.2.1.1 p.275
Brugnatellite 16b.7.5.1 p.515
Brunogeierite 7.2.2.7 p.301
Brushite 39.1.1.1 p.735
Brüggenite 21.1.2.1 p.531
Buchwaldite 38.1.2.2 p.702
Buckhornite 2.11.11.1 p.112
Buddingtonite 76.1.2.1 p.1608
Buergerite 61.3.1.5 p.1259
Bukovite 2.5.5.2 p.54
Bukovskyite 43.5.1.2 p.966
Bulachite 42.6.7.2 p.895
Bultfonteinite 52.4.7.2 p.1098
Bunsenite 4.2.1.2 p.209
Burangaite 42.9.1.1 p.913
Burbankite 14.4.4.1 p.459
Burckhardtite 78.5.3.1 p.1709
Burkeite 32.1.1.1 p.666
Burpalite 56.2.4.2 p.1161
Bursaite 3.4.15.2 p.176
Burtite 6.3.6.5 p.285
Bustamite 65.2.1.2 p.1318
Butlerite 31.9.1.1 p.651
Buttgenbachite 19.1.2.1 p.528
Byelorussite-(Ce) 60.1.1b.4 p.1233
Bystrite 76.2.5.2 p.1634
Byströmite 44.2.1.1 p.973
Bütschliite 14.3.1.1 p.455
Bøgvadite 11.6.20.1 p.418
Cabriite 1.2.8.1 p.25
Cacoxenite 42.13.5.1 p.953
Cadmium 1.1.5.2 p.7
Cadmoselite 2.8.7.3 p.73
Cadwaladerite 10.5.1.1 p.399
Cafarsite 45.1.4.1 p.978
Cafetite 8.7.3.1 p.369
Cahnite 43.4.4.1 p.964
Calaverite 2.12.13.2 p.135
Calciborite 24.5.2.1 p.541
Calcioancylite-(Ce) 16b.1.1.2 p.498
Calcioancylite-(Nd) 16b.1.1.3 p.498

Index of Mineral Names in Alphabetical Order

Calciobetafite 8.2.3.5 p.348
Calcioburbankite 14.4.4.3 p.460
Calciocopiapite 31.10.5.5 p.664
Calcioferrite 42.11.8.3. p.936
Calciogadolinite 54.2.1b.4 p.1126
Calciohilairite 59.2.3.2 p.1229
Calciotantite 8.6.2.1 p.368
Calciouranoite 5.4.2.1 p.259
Calcite 14.1.1.1 p.428
Calcium Catapleiite 59.2.2.2 p.1227
Calcjarlite 11.6.10.2 p.415
Calclacite 50.2.4.1 p.1014
Calcurmolite 49.3.1.1 p.1007
Calderite 51.4.3a.6 p.1044
Caledonite 32.3.2.1 p.668
Calkinsite-(Ce) 15.4.1.1 p.471
Callaghanite 16b.5.1.1 p.506
Calomel 9.1.8.1 p.379
Calumetite 6.2.5.1 p.277
Calzirtite 8.7.4.1 p.369
Camerolaite 44.3.11.1 p.975
Cameronite 2.16.20.1 p.148
Camgasite 42.7.16.1 p.908
Caminite 31.10.9.1 p.665
Campigliaite 31.6.6.1 p.647
Canaphite 40.5.11.1 p.814
Canasite 78.5.4.1 p.1709
Canavesite 17.1.8.1 p.522
Cancrinite 76.2.5.3 p.1634
Cancrisilite 76.2.5.4 p.1635
Canfieldite 2.5.6.2 p.55
Cannizzarite 3.6.5.1 p.183
Cannonite 30.2.9.1 p.637
Capgarronite 2.15.6.1 p.144
Cappelenite- (Y) 54.2.3.1 p.1128
Caracolite 30.3.2.1 p.638
Carboborite 27.1.1.1 p.565
Carbocernaite 14.4.6.1 p.461
Carboirite 52.3.3.4 p.1090
Carboirite 7.11.6.1 p.328
Carbonate-Cyanotrichite 16b.7.7.1 p.516
Carbonate-Fluorapatite 41.8.1.4 p.862
Caresite 16b.6.4.3 p.513
Carletonite 72.3.1.4 p.1528
Carlfriesite 34.5.3.1 p.685
Carlhintzeite 11.6.7.1 p.414
Carlinite 2.4.13.1 p.50
Carlosruizite 23.1.2.2 p.533
Carlosturanite 67.1.2.1 p.1376
Carlsbergite 1.1.20.1 p.15
Carminite 41.10.6.1 p.877
Carnallite 11.1.2.1 p.407
Carnotite 40.2a.28.1 p.782
Carobbiite 9.1.1.4 p.376
Carpholite 65.1.5.1 p.1311
Carrboydite 31.10.1.1 p.657
Carrollite 2.10.1.2 p.99
Caryinite 38.2.2.1 p.708
Caryopilite 71.1.2b.1 p.1417
Cascandite 65.2.1.6 p.1320
Cassedanneite 43.3.2.2 p.960

Cassidyite 40.2.2.4 p.752
Cassiterite 4.4.1.5 p.239
Caswellsilverite 2.9.17.1 p.97
Catapleiite 59.2.2.1 p.1226
Cattierite 2.12.1.3 p.115
Cavansite 74.3.7.1 p.1567
Caysichite-(Y) 70.1.2.2 p.1399
Cebaite-(Ce) 16a.1.9.1 p.482
Cebollite 78.1.1.1 p.1702
Cechite 41.5.2.4 p.831
Celadonite-1M 71.2.2a.6 p.1454
Celestine 28.3.1.2 p.574
Celsian 76.1.1.4 p.1604
Cerianite-(Ce) 4.4.12.1 p.249
Ceriopyrochlore-(Ce) 8.2.1.5 p.344
Cerite-(Ce) 52.4.6.1 p.1097
Cerotungstite-(Ce) 49.3.3.1 p.1007
Ceruleite 42.13.6.1 p.954
Cerussite 14.1.3.4 p.446
Cervandonite-(Ce) 43.4.12.1 p.965
Cervantite 4.4.16.1 p.251
Cesanite 30.3.3.1 p.638
Cesárolite 7.6.1.1 p.315
Cesbronite 33.2.7.1 p.674
Cesbronite 34.7.2.1 p.687
Cesium Kupletskite 69.1.1.3 p.1383
Cesplumtantite 8.7.11.1 p.372
Cesstibtantite 8.2.4.1 p.349
Cetineite 2.13.3.1 p.140
Chabazite-Ca 77.1.2.1 p.1652
Chabournéite 3.8.12.1 p.201
Chaidamuite 31.9.7.2 p.654
Chalcanthite 29.6.7.1 p.605
Chalcoalumite 31.3.1.1 p.642
Chalcocite 2.4.7.1 p.46
Chalcocyanite 28.3.3.1 p.577
Chalcomenite 34.2.2.1 p.680
Chalconatronite 15.2.3.1 p.465
Chalcophanite 7.8.2.1 p.319
Chalcophyllite 43.5.14.1 p.969
Chalcopyrite 2.9.1.1 p.87
Chalcosiderite 42.9.3.4 p.919
Chalcostibite 3.7.5.1 p.192
Chalcothallite 3.2.7.1 p.160
Chambersite 25.6.1.3 p.548
Chamosite 71.4.1.7 p.1504
Chaméanite 3.2.8.1 p.160
Changbaiite 8.3.11.1 p.362
Chantalite 78.1.2.1 p.1702
Chaoite 1.3.6.4 p.35
Chapmanite 71.1.3.2 p.1431
Charlesite 32.4.4.1 p.670
Charmarite-2H 16b.6.4.4 p.513
Charoite 70.1.2.3 p.1399
Chatkalite 2.9.3.2 p.92
Chayesite 63.2.1a.2 p.1272
Chekhovichite 34.6.4.1 p.686
Chelkarite 26.7.2.1 p.563
Chenevixite 42.7.10.1 p.906
Chengdeite 1.2.5.5 p.24
Chenite 30.1.16.1 p.627

Cheralite 38.4.3.3 p.724
Cheremnykhite 33.3.5.1 p.676
Cherepanovite 2.8.17.2 p.81
Chernikovite 40.2a.19.1 p.777
Chernovite-(Y) 38.4.11.2 p.730
Chernykhite-2M$_1$ 71.2.2a.3 p.1452
Chervetite 38.5.5.1 p.733
Chessexite 32.4.6.1 p.671
Chesterite 68.1.1.1 p.1379
Chestermanite 24.2.1.5 p.537
Chevkinite-(Ce) 56.2.8.1 p.1177
Chiavennite 70.5.1.1 p.1402
Childrenite 42.7.1.1 p.896
Chiluite 7.11.13.1 p.330
Chiolite 11.6.11.1 p.415
Chkalovite 65.5.2.1 p.1332
Chladniite 38.2.5.3 p.713
Chloraluminite 9.3.3.1 p.389
Chlorapatite 41.8.1.2 p.860
Chlorargyrite 9.1.4.1 p.377
Chlorellestadite 52.4.9.5 p.1103
Chloritoid-A,M 52.3.3.1 p.1088
Chlormagaluminite 10.6.10.1 p.405
Chlormanganokalite 11.5.4.1 p.411
Chlorocalcite 11.1.3.1 p.408
Chloromagnesite 9.2.3.3 p.385
Chlorophoenicite 41.1.1.1 p.815
Chlorothionite 30.1.11.1 p.625
Chloroxiphite 10.6.4.1 p.403
Choloalite 34.2.4.1 p.681
Chondrodite 52.3.2b.2 p.1080
Christite 3.4.10.1 p.172
Chromatite 35.3.2.1 p.693
Chromdravite 61.3.1.11 p.1264
Chromferide 1.1.12.2 p.13
Chromite 7.2.3.3 p.302
Chromium 1.1.12.1 p.12
Chrysoberyl 7.2.9.1 p.307
Chrysocolla 74.3.2.1 p.1561
Chudobaite 39.2.6.1 p.742
Chukhrovite-(Ce) 12.1.5.2 p.420
Chukhrovite-(Y) 12.1.5.1 p.420
Churchite-(Y) 40.4.6.1 p.805
Chursinite 38.5.1.1 p.731
Chvaleticeite 29.6.8.6 p.609
Chvilevaite 2.5.9.1 p.55
Cianciulliite 6.2.13.1 p.280
Cinnabar 2.8.14.1 p.80
Clairite 31.10.8.1 p.665
Claraite 16b.4.3.1 p.505
Claringbullite 10.5.6.1 p.400
Clarkeite 5.4.1.1 p.259
Claudetite 4.3.10.1 p.232
Clausthalite 2.8.1.2 p.65
Cliffordite 34.3.1.1 p.682
Clinobehoite 6.2.3.1 p.277
Clinobisvanite 38.4.7.1 p.726
Clinochalcomenite 34.2.3.1 p.680
Clinochlore 71.4.1.4 p.1501
Clinochrysotile 71.1.2d.1 p.1428
Clinoclase 41.3.1.1 p.818

Clinoenstatite 65.1.1.1 p.1291
Clinoferrosilite 65.1.1.2 p.1292
Clinohedrite 52.2.1.2 p.1062
Clinohumite 52.3.2d.1 p.1084
Clinojimthompsonite 67.1.1.2 p.1376
Clinokurchatovite 24.4.3.1 p.541
Clinomimetite 41.8.6.1 p.868
Clinoptilolite 77.1.4.2 p.1671
Clinosafflorite 2.12.4.6 p.130
Clinotobermorite 72.3.2.2 p.1530
Clinotyrolite 42.4.4.1 p.885
Clinoungemachite 31.7.3.1 p.649
Clinozoisite 58.2.1a.4 p.1198
Clintonite-1M, 2M$_1$ 71.2.2c.2 p.1469
Coalingite 16b.7.6.1 p.516
Cobalt Zippeite 31.10.4.6 p.661
Cobaltaustinite 41.5.1.8 p.829
Cobaltite 2.12.3.1 p.125
Cobaltkoritnigite 39.1.4.2 p.737
Cobaltomenite 34.2.3.2 p.680
Cochromite 7.2.3.5 p.303
Coconinoite 43.5.5.1 p.967
Coeruleolactite 42.9.3.2 p.918
Coesite 75.1.4.1 p.1586
Coffinite 51.5.2.4 p.1058
Cohenite 1.1.16.1 p.14
Colemanite 26.3.5.1 p.553
Collinsite 40.2.2.3 p.752
Coloradoite 2.8.2.5 p.70
Colquiriite 11.6.3.1 p.413
Colusite 3.1.1.1 p.150
Comancheite 10.5.5.1 p.400
Combeite 61.1.2a.5 p.1248
Comblainite 16b.6.3.3 p.512
Compreignacite 5.7.1.1 p.261
Congolite 25.6.2.1 p.548
Conichalcite 41.5.1.2 p.826
Connellite 31.1.1.1 p.639
Cookeite 71.4.1.2 p.1499
Cooperite 2.8.5.1 p.71
Copiapite 31.10.5.1 p.663
Copper 1.1.1.3 p.4
Coquandite 30.1.18.1 p.628
Coquimbite 29.8.3.1 p.619
Corderoite 10.3.3.1 p.396
Cordierite 61.2.1.1 p.1251
Cordylite-(Ce) 16a.1.7.1 p.481
Corkite 43.4.1.2 p.961
Cornetite 41.3.2.1 p.818
Cornubite 41.4.2.1 p.821
Cornwallite 41.4.2.2 p.822
Coronadite 7.9.1.4 p.323
Corrensite 71.4.2.5 p.1512
Corundum 4.3.1.1 p.214
Corvusite 47.3.2.2 p.988
Cosalite 3.5.9.1 p.180
Costibite 2.12.7.1 p.132
Cotunnite 9.2.7.1 p.386
Coulsonite 7.2.4.2 p.303
Cousinite 49.3.2.1 p.1007
Covellite 2.8.12.1 p.79

Cowlesite 77.1.5.8 p.1686
Coyoteite 2.14.6.1 p.142
Crandallite 42.7.3.1 p.899
Crawfordite 43.2.1.4 p.959
Creaseyite 78.7.2.1 p.1713
Crednerite 7.1.2.1 p.292
Creedite 12.1.4.1 p.420
Crerarite 2.8.1.8 p.67
Crichtonite 8.5.1.3 p.366
Criddleite 3.9.1.1 p.203
Cristobalite 75.1.1.1 p.1568
Crocoite 35.3.1.1 p.693
Cronstedtite 71.1.2c.7 p.1427
Crookesite 2.4.12.1 p.50
Cryolite 11.6.1.1 p.412
Cryolithionite 11.6.4.1 p.413
Cryptohalite 11.5.1.2 p.410
Cryptomelane 7.9.1.2 p.321
Cualstibite 44.3.10.1 p.975
Cubanite 2.9.13.1 p.96
Cumengéite 10.6.7.1 p.404
Cupalite 1.1.15.1 p.13
Cuprite 4.1.1.1 p.204
Cuprobismutite 3.8.2.1 p.197
Cuprocopiapite 31.10.5.3 p.664
Cuproiridisite 2.10.1.15 p.102
Cupropavonite 3.8.10.6 p.201
Cuprorhodisite 2.10.1.16 p.102
Cuprorivaite 71.2.3.1 p.1477
Cuprosklodowskite 53.3.1.4 p.1112
Cuprospinel 7.2.2.6 p.301
Cuprostibite 2.4.11.1 p.50
Cuprotungstite 48.3.2.1 p.1002
Curetonite 41.5.15.1 p.842
Curienite 40.2a.27.2 p.782
Curite 5.9.3.1 p.263
Cuspidine 56.2.4.3 p.1161
Cuzticite 33.2.3.1 p.673
Cyanochroite 29.3.6.2 p.588
Cyanophyllite 44.3.9.1 p.975
Cyanotrichite 31.2.1.1 p.641
Cylindrite 3.1.4.1 p.152
Cymrite 78.1.3.1 p.1702
Cyrilovite 42.7.8.1 p.905
Cernýite 2.9.2.2 p.90
D'Ansite 30.1.10.1 p.625
Dachiardite 77.1.6.4 p.1690
Dadsonite 3.6.8.1 p.185
Dalyite 72.2.1.1 p.1524
Damaraite 10.5.10.1 p.401
Danalite 76.2.4.2 p.1631
Danbaite 1.1.6.1 p.7
Danburite 56.3.1.1 p.1183
Danielsite 2.16.7.2 p.145
Daomanite 2.5.7.1 p.55
Daqingshanite-(Ce) 17.1.10.1 p.523
Darapiosite 63.2.1a.3 p.1272
Darapskite 20.1.1.1 p.530
Datolite 54.2.1a.1 p.1121
Daubréeite 10.2.1.3 p.393
Daubréelite 2.10.1.11 p.101

Davanite 72.2.1.2 p.1525
Davidite-(Ce) 8.5.1.6 p.366
Davidite-(La) 8.5.1.5 p.366
Davreuxite 58.1.2.1 p.1194
Davyne 76.2.5.5 p.1636
Dawsonite 16a.3.8.1 p.492
Deanesmithite 35.4.3.1 p.694
Deerite 69.2.3.3 p.1395
Defernite 16b.4.2.1 p.504
Delafossite 7.1.1.1 p.292
Delhayelite 72.5.1.4 p.1540
Delindeite 56.2.6c.2 p.1174
Dellaite 78.1.4.1 p.1703
Deloryite 48.3.4.1 p.1003
Delrioite 47.1.2.1 p.985
Delvauxite 42.6.6.1 p.894
Demesmaekerite 34.7.6.1 p.688
Denisovite 65.2.1.7 p.1321
Denningite 34.4.1.1 p.684
Derbylite 46.2.3.2 p.983
Derriksite 34.7.5.1 p.688
Dervillite 3.7.17.1 p.196
Desautelsite 16b.6.2.4 p.511
Descloizite 41.5.2.1 p.829
Despujolsite 31.7.6.1 p.650
Devilline 31.6.1.1 p.646
Dewindtite 42.9.8.1 p.922
Diaboleite 10.6.1.1 p.402
Diadochite 43.5.2.1 p.966
Diamond 1.3.6.1 p.32
Diaoyudaoite 7.11.12.1 p.330
Diaphorite 3.5.4.1 p.179
Diaspore 6.1.1.1 p.266
Dickinsonite 41.7.2.2 p.854
Dickite 71.1.1.1 p.1407
Dienerite 2.2.2.3 p.39
Dietrichite 29.7.3.4 p.617
Dietzeite 23.1.1.1 p.533
Digenite 2.4.7.3 p.47
Dimorphite 2.6.1.1 p.56
Dinite 50.3.5.1 p.1015
Diomignite 24.5.4.1 p.541
Diopside 65.1.3a.1 p.1296
Dioptase 61.1.3.1 p.1249
Dissakisite-(Ce) 58.2.1a.5 p.1199
Dittmarite 40.1.2.1 p.748
Dixenite 46.2.8.1 p.983
Djerfisherite 2.15.2.1 p.143
Djurleite 2.4.7.2 p.47
Dmisteinbergite 76.1.7.1 p.1617
Dolerophanite 30.2.2.1 p.628
Dollaseite-(Ce) 58.2.1a.6 p.1199
Dolomite 14.2.1.1 p.450
Doloresite 47.4.7.1 p.991
Domeykite 2.2.2.1 p.39
Donbassite 71.4.1.1 p.1498
Donharrisite 2.6.6.1 p.60
Donnayite-(Y) 15.3.4.1 p.469
Donpeacorite 65.1.2.3 p.1296
Dorallcharite 30.2.5.8 p.636
Dorfmanite 39.1.8.1 p.738

Dorrite 69.2.1a.2 p.1387
Douglasite 11.3.1.1 p.409
Downeyite 4.4.3.1 p.244
Doyleite 6.3.4.1 p.282
Dozyite 71.4.2.1 p.1508
Dravite 61.3.1.9 p.1262
Dresserite 16b.2.1.2 p.501
Dreyerite 38.4.8.1 p.726
Drugmanite 41.10.11.1 p.880
Drysdallite 2.12.10.2 p.134
Dufrenite 42.9.1.2 p.914
Dufrénoysite 3.5.9.3 p.181
Duftite-α 41.5.2.5 p.831
Duftite-b 41.5.1.4 p.827
Dugganite 33.3.5.3 p.676
Duhamelite 42.9.5.1 p.921
Dumontite 42.4.5.1 p.886
Dumortierite 54.1.2.1 p.1117
Dundasite 16b.2.1.1 p.501
Durangite 41.5.5.2 p.834
Duranusite 2.1.4.1 p.37
Dusmatovite 63.2.1a.11 p.1276
Dussertite 41.5.10.1 p.839
Duttonite 6.2.7.1 p.278
Dwornikite 29.6.2.6 p.597
Dypingite 16b.7.2.1 p.514
Dyscrasite 2.2.1.1 p.38
Dzarkenite 2.12.1.17 p.119
Dzhalindite 6.3.5.1 p.282
Eakerite 72.1.5.1 p.1524
Earlandite 50.2.2.1 p.1014
Earlshannonite 42.11.18.4 p.948
Ecandrewsite 4.3.5.4 p.228
Ecdemite 46.2.1.1 p.982
Eckermannite 66.1.3c.6 p.1364
Eclarite 3.6.10.1 p.185
Edenharterite 3.7.10.2 p.195
Edenite 66.1.3a.10 p.1353
Edgarbaileyite 55.5.1.1 p.1147
Edingtonite 77.1.5.6 p.1683
Edoylerite 35.4.4.1 p.695
Effenbergerite 71.2.3.3 p.1478
Efremovite 28.4.4.3 p.581
Eggletonite 74.1.2.2 p.1555
Eglestonite 10.5.4.1 p.399
Ehrleite 40.5.6.1 p.811
Eifelite 63.2.1a.4 p.1272
Eitelite 14.3.2.1 p.455
Ekanite 72.1.1.1 p.1521
Ekaterinite 26.7.3.1 p.563
Elbaite 61.3.1.8 p.1261
Ellenbergerite 52.4.10.1 p.1104
Ellisite 3.4.9.1 p.172
Elpasolite 11.6.2.1 p.413
Elpidite 72.5.4.2 p.1543
Elyite 30.1.2.1 p.623
Embreyite 43.3.2.1 p.959
Emeleusite 66.3.4.3 p.1375
Emmonsite 34.3.3.1 p.683
Emplectite 3.7.5.2 p.192
Enargite 3.2.1.1 p.155

Englishite 42.13.8.1 p.954
Enstatite 65.1.2.1 p.1294
Eosphorite 42.7.1.2 p.897
Ephesite-$2M_1$, $1M$ 71.2.2b.16 p.1467
Epididymite 66.3.1.4 p.1371
Epidote 58.2.1a.7 p.1200
Epistilbite 77.1.6.2 p.1689
Epistolite 56.2.7.1 p.1176
Epsomite 29.6.11.1 p.612
Erdite 2.14.5.1 p.142
Ericaite 25.6.1.2 p.548
Ericssonite 56.2.6c.3 p.1175
Eriochalcite 9.2.8.1 p.386
Erionite 77.1.2.5 p.1656
Erlianite 78.5.5.1 p.1710
Erlichmanite 2.12.1.16 p.118
Ernienickelite 7.8.2.4 p.320
Erniggliite 3.4.18.1 p.178
Ernstite 42.7.1.3. p.897
Ershovite 66.1.5.1 p.1368
Ertixiite 78.7.3.1 p.1714
Erythrite 40.3.6.3 p.794
Erythrosiderite 11.4.1.1 p.409
Eskebornite 2.9.1.2 p.88
Eskimoite 3.6.2.1 p.183
Eskolaite 4.3.1.3 p.220
Esperite 51.2.1.3 p.1024
Esseneite 65.1.3a.6 p.1302
Ettringite 31.10.2.1 p.659
Eucairite 2.4.6.2 p.45
Euchlorine 30.3.1.1 p.637
Euchroite 42.6.3.1 p.892
Euclase 52.2.1.1 p.1061
Eucryptite 51.1.1.3 p.1022
Eudialyte 64.1.1.1 p.1281
Eudidymite 66.3.1.3 p.1370
Eugenite 1.1.8.5 p.8
Eugsterite 29.4.2.1 p.590
Eulytite 51.5.4.1 p.1060
Euxenite-(Y) 8.3.8.2 p.360
Evansite 42.1.2.1 p.882
Eveite 41.6.6.4 p.850
Evenkite 50.3.6.1 p.1016
Ewaldite 14.2.4.1 p.453
Eylettersite 41.5.12.2 p.841
Ezcurrite 26.5.5.1 p.557
Eztlite 34.8.5.1 p.690
Fabianite 25.3.3.1 p.545
Faheyite 40.5.3.1 p.810
Fahleite 40.5.8.2 p.812
Fairbankite 34.1.2.1 p.677
Fairchildite 14.3.3.1 p.455
Fairfieldite 40.2.2.1 p.751
Falcondoite 74.3.1b.2 p.1561
Falkmanite 3.5.2.2 p.179
Famatinite 3.2.2.2 p.156
Fangite 3.2.9.1 p.160
Farringtonite 38.3.1.2 p.714
Faujasite 77.1.2.7 p.1659
Faustite 42.9.3.3 p.919
Fayalite 51.3.1.1 p.1025

Index of Mineral Names in Alphabetical Order

Federovskite 25.4.1.2 p.546
Fedorite 73.1.3.1 p.1547
Fedotovite 30.3.4.1 p.638
Feitknechtite 6.1.4.3 p.272
Felsöbanyaite 31.4.4.1 p.644
Fenaksite 70.1.1.2 p.1397
Ferberite 48.1.1.2 p.994
Ferchromide 1.1.12.3 p.13
Fergusonite-(Ce) 8.1.1.3 p.333
Fergusonite-(Nd) 8.1.1.4 p.333
Fergusonite-(Y) 8.1.1.1 p.332
Fermorite 41.8.3.4 p.865
Fernandinite 47.3.2.3 p.989
Feroxyhyte 6.1.4.4 p.273
Ferrarisite 39.2.3.1 p.741
Ferrazite 40.3.13.1 p.799
Ferriannite-1M 71.2.2b.4 p.1461
Ferricopiapite 31.10.5.4 p.664
Ferrierite 77.1.6.6 p.1692
Ferrihydrite 4.3.2.2 p.221
Ferrilotharmeyerite 37.1.6.2 p.699
Ferrimolybdite 49.2.1.1 p.1005
Ferrinatrite 29.4.4.1 p.590
Ferripyrophyllite 71.2.1.2 p.1441
Ferrisicklerite 38.1.4.1 p.703
Ferristrunzite 42.11.9.3 p.939
Ferrisurite 71.2.4.1 p.1479
Ferrisymplesite 42.10.1.1 p.922
Ferritungstite 4.5.5.1 p.252
Ferroalluaudite 38.2.3.5 p.709
Ferroaxinite 56.2.2.1 p.1154
Ferrobustamite 65.2.1.3 p.1318
Ferrocarpholite 65.1.5.2 p.1311
Ferrocolumbite 8.3.2.2 p.354
Ferrohagendorfite 38.2.3.1 p.708
Ferrohexahydrite 29.6.8.3 p.608
Ferrokesterite 2.9.2.10 p.91
Ferronickelplatinum 1.2.4.3 p.22
Ferropyrosmalite 72.4.1a.1 p.1536
Ferroselite 2.12.2.2 p.121
Ferrosilite 65.1.2.2 p.1295
Ferrostrunzite 42.11.9.2 p.939
Ferrotantalite 8.3.2.1 p.354
Ferrotapiolite 8.3.1.1 p.351
Ferrotychite 17.1.1.2 p.519
Ferrowodginite 8.1.8.2 p.338
Ferrowyllieite 38.2.4.1 p.711
Ferruccite 11.2.3.1 p.409
Fersmanite 56.2.5.3 p.1170
Fersmite 8.3.3.1 p.356
Feruvite 61.3.1.4 p.1259
Fervanite 47.3.4.1 p.990
Fetiasite 45.1.13.1 p.979
Fibroferrite 31.9.12.1 p.656
Fichtelite 50.3.4.1 p.1015
Fiedlerite 10.3.2.1 p.396
Filipstadite 7.2.13.1 p.310
Fillowite 38.2.5.1 p.712
Fingerite 41.11.3.1 p.880
Finnemanite 46.1.1.1 p.981
Fizelyite 3.4.15.8 p.177

Flagstaffite 50.4.3.1 p.1017
Fleischerite 31.7.6.3 p.650
Fletcherite 2.10.1.3 p.100
Flinkite 41.3.3.1 p.818
Florencite-(Ce) 41.5.10.2 p.839
Florencite-(La) 41.5.10.3 p.840
Florencite-(Nd) 41.5.10.4 p.840
Florensovite 2.10.1.14 p.102
Fluckite 39.1.2.1 p.736
Fluellite 42.6.9.1 p.895
Fluoborite 25.1.2.1 p.542
Fluocerite-(Ce) 9.3.4.1 p.389
Fluocerite-(La) 9.3.4.2 p.389
Fluor-ferro-leakeite 66.1.3c.15 p.1367
Fluorapatite 41.8.1.1 p.858
Fluorapophyllite 72.3.1.1 p.1525
Fluorbritholite-(Ce) 52.4.9.8 p.1104
Fluorellestadite 52.4.9.3 p.1102
Fluorite 9.2.1.1 p.380
Foggite 42.7.2.1 p.898
Foitite 61.3.1.1 p.1257
Fontanite 15.3.6.1 p.471
Foordite 8.3.10.2 p.362
Formanite-Y 8.1.1.2 p.332
Fornacite 43.4.3.2 p.963
Forsterite 51.3.1.2 p.1026
Foshagite 65.2.2.1 p.1321
Foshallasite 56.2.9.3 p.1181
Fourmarierite 5.9.2.1 p.263
Fraipontite 71.1.2c.4 p.1425
Francevillite 40.2a.27.1 p.782
Franciscanite 52.4.1.4 p.1091
Franciscanite 7.3.1.2 p.311
Francisite 34.6.6.1 p.687
Franckeite 3.1.4.2 p.152
Francoanellite 39.3.5.2 p.744
Franconite 8.6.1.1 p.367
Frankhawthorneite 33.1.4.1 p.673
Franklinfurnaceite 71.4.3.1 p.1519
Franklinite 7.2.2.4 p.300
Franklinphilite 74.1.1.3 p.1553
Fransoletite 40.5.1.1 p.809
Franzinite 76.2.5.6 p.1636
Françoisite-(Nd) 42.4.14.1 p.889
Freboldite 2.8.11.10 p.79
Fredrikssonite 24.2.1.6 p.537
Freedite 46.1.5.1 p.982
Freibergite 3.3.6.3 p.164
Freieslebenite 3.4.6.2 p.171
Fresnoite 55.4.2.1 p.1145
Freudenbergite 7.9.3.1 p.324
Friedelite 72.4.1b.1 p.1537
Friedrichite 3.4.5.5 p.170
Fritzscheite 40.2a.25.1 p.781
Frohbergite 2.12.2.3 p.121
Frolovite 25.1.3.1 p.542
Frondelite 41.9.2.2 p.872
Froodite 2.12.15.1 p.138
Fuenzalidaite 23.1.2.1 p.533
Fukalite 78.4.3.1 p.1707
Fukuchilite 2.12.1.7 p.116

Furongite 42.11.20.1 p.949
Furutobeite 2.16.12.1 p.146
Fülöppite 3.6.20.1 p.188
Gabrielsonite 41.5.1.5 p.828
Gadolinite-(Ce) 54.2.1b.2 p.1124
Gadolinite-(Y) 54.2.1b.3 p.1124
Gagarinite-(Y) 11.5.6.1 p.411
Gageite-2M 65.3.2.2a p.1323
Gahnite 7.2.1.4 p.297
Gaidonnayite 59.2.2.3 p.1227
Gainesite 40.5.4.1 p.810
Gaitite 40.2.2.6 p.753
Galaxite 7.2.1.2 p.297
Galeite 30.1.8.1 p.625
Galena 2.8.1.1 p.64
Galenobismutite 3.7.9.1 p.194
Galkhaite 3.4.14.1 p.174
Gallite 2.9.1.3 p.88
Gamagarite 41.10.4.3 p.877
Gananite 9.3.5.1 p.389
Ganomalite 78.2.1.1 p.1706
Ganophyllite 74.1.2.1 p.1554
Gaotaiite 2.12.1.18 p.119
Garavellite 3.7.9.4 p.195
Garrelsite 27.1.7.1 p.566
Garrelsite 54.2.4.1 p.1129
Garronite 77.1.3.3 p.1662
Gartrellite 43.2.2.1 p.959
Garyansellite 40.3.2.5 p.788
Gasparite-(Ce) 38.4.3.6 p.725
Gaspéite 14.1.1.8 p.439
Gatehouseite 41.4.1.2 p.821
Gatumbaite 42.11.12.1 p.942
Gaudefroyite 27.1.2.1 p.565
Gaultite 77.2.6.1 p.1700
Gaylussite 15.2.2.1 p.465
Gearksutite 11.6.8.1 p.414
Gebhardite 46.2.5.1 p.983
Geffroyite 2.7.1.6 p.63
Gehlenite 55.4.1.2 p.1143
Geigerite 39.2.6.2 p.742
Geikielite 4.3.5.2 p.227
Geminite 40.5.12.1 p.814
Genkinite 2.6.4.1 p.59
Genthelvite 76.2.4.3 p.1631
Geocronite 3.3.1.2 p.160
Georgechaoite 59.2.2.4 p.1228
Georgeite 16a.3.3.1 p.490
Georgiadesite 41.1.3.1 p.816
Gerasimovskite 8.7.8.3 p.371
Gerdtremmelite 41.3.7.1 p.820
Gerhardtite 19.1.1.1 p.528
Germanite 2.9.4.2 p.93
Germanocolusite 3.1.1.2 p.151
Gersdorffite 2.12.3.2 p.125
Gerstleyite 3.8.8.1 p.198
Gerstmannite 52.2.1.4 p.1063
Geversite 2.12.1.14 p.118
Gianellaite 28.2.4.1 p.570
Gibbsite 6.3.1.1 p.280
Giessenite 3.5.12.2 p.182

Gilalite 78.4.4.1 p.1708
Gillespite 71.2.3.2 p.1478
Gillulyite 3.8.9.1 p.199
Giniite 42.11.7.1 p.933
Ginorite 26.6.7.1 p.562
Giorgiosite 16b.7.2.2 p.514
Giraudite 3.3.6.5 p.164
Girdite 33.3.3.1 p.675
Girvasite 43.5.17.1 p.970
Gismondine 77.1.3.1 p.1661
Gittinsite 55.2.1a.1 p.1134
Giuseppettite 76.2.5.7 p.1636
Gladite 3.4.5.3 p.170
Glauberite 28.4.2.1 p.579
Glauchchroite 51.3.2.3 p.1034
Glaucocerinite 31.4.8.1 p.645
Glaucodot 2.12.6.1 p.131
Glauconite-1M 71.2.2a.5 p.1453
Glaucophane 66.1.3c.1 p.1361
Glaukospherite 16a.3.1.2 p.487
Glucine 42.4.10.1 p.888
Glushinskite 50.1.3.2 p.1012
Gmelinite 77.1.2.6 p.1657
Gobbinsite 77.1.3.4 p.1663
Godlevskite 2.7.4.1 p.63
Godovikovite 28.3.5.2 p.578
Goedkenite 41.10.4.1 p.876
Goethite 6.1.1.2 p.268
Goldfieldite 3.3.6.6 p.165
Goldichite 29.5.2.1 p.592
Goldmanite 51.4.3b.4 p.1049
Gold 1.1.1.1 p.2
Gonnardite 77.1.5.7 p.1685
Gonyerite 71.4.4.1 p.1519
Goosecreekite 77.1.7.3 p.1695
Gorceixite 42.7.3.2 p.900
Gordonite 42.11.14.4 p.944
Gormanite 42.9.2.2 p.915
Gortdrumite 2.16.14.1 p.147
Goslarite 29.6.11.2 p.613
Goudeyite 42.5.1.3 p.891
Gowerite 26.5.8.1 p.557
Goyazite 42.7.3.3 p.900
Graemite 34.2.1.1 p.679
Graftonite 38.3.3.1 p.714
Grandidierite 54.1.1.1 p.1116
Grandreefite 12.1.2.1 p.419
Grantsite 47.3.1.5 p.988
Graphite 1.3.6.2 p.33
Gratonite 3.3.2.1 p.161
Gravegliaite 34.2.5.2 p.682
Grayite 40.4.7.4 p.808
Grechishchevite 10.3.5.1 p.397
Greenalite 71.1.2b.4 p.1421
Greenockite 2.8.7.2 p.73
Gregoryite 14.1.2.2 p.440
Greigite 2.10.1.10 p.101
Griceite 9.1.1.5 p.376
Grimaldiite 6.1.5.1 p.273
Grimselite 15.2.6.1 p.467
Griphite 41.9.3.1 p.872

Index of Mineral Names in Alphabetical Order

Grischunite 40.2.11.1 p.762
Grossite 7.3.2.1 p.311
Grossular 51.4.3b.2 p.1046
Groutite 6.1.1.3 p.269
Grumantite 78.4.5.1 p.1708
Gruzdevite 3.4.13.3 p.174
Guanajuatite 2.11.2.4 p.108
Guanine 50.4.5.1 p.1017
Guarinoite 31.2.4.1 p.641
Gudmundite 2.12.4.2 p.129
Guerinite 39.2.2.2 p.741
Guettardite 3.7.8.2 p.193
Gugiaite 55.4.2.6 p.1147
Guildite 31.9.7.1 p.654
Guilleminite 34.7.3.1 p.688
Gunningite 29.6.2.5 p.597
Gupeiite 1.1.23.2 p.16
Gustavite 3.4.15.3 p.176
Gutsevichite 42.10.4.1 p.924
Guyanaite 6.1.2.3 p.271
Gypsum 29.6.3.1 p.598
Gyrolite 73.2.2c.1 p.1550
Gysinite-(Nd) 16b.1.1.4 p.498
Görgeyite 29.4.7.1 p.591
Götzenite 56.2.5.4 p.1170
Haapalaite 2.14.3.1 p.141
Hafnon 51.5.2.2 p.1056
Hagendorfite 38.2.3.2 p.708
Haidingerite 39.1.5.1 p.737
Hainite 56.2.4.9 p.1166
Haiweeite 53.3.2.2 p.1114
Hakite 3.3.6.4 p.164
Halite 9.1.1.1 p.373
Hallimondite 40.2a.32.1 p.784
Halloysite 71.1.1.4 p.1411
Halotrichite 29.7.3.2 p.617
Halurgite 26.4.4.1 p.555
Hambergite 25.1.1.1 p.542
Hammarite 3.4.5.4 p.170
Hanawaltite 10.5.13.1 p.402
Hancockite 58.2.1a.8 p.1201
Hanksite 32.3.1.1 p.667
Hannayite 39.3.5.1 p.744
Hannebachite 34.2.5.1 p.681
Haradaite 65.3.1.1 p.1322
Hardystonite 55.4.2.2 p.1145
Harkerite 17.1.9.1 p.522
Harmotome 77.1.3.5 p.1664
Harstigite 57.1.1.2 p.1188
Hashemite 35.3.3.1 p.693
Hastite 2.12.2.4 p.122
Hatchite 3.5.8.2 p.180
Hatrurite 52.4.7.3 p.1099
Hauchecornite 3.2.5.1 p.158
Hauckite 32.3.4.1 p.668
Hauerite 2.12.1.9 p.117
Hausmannite 7.2.7.1 p.305
Hawleyite 2.8.2.6 p.70
Hawthorneite 7.4.1.3 p.312
Haxonite 1.1.16.2 p.14
Haycockite 2.9.8.2 p.94

Haynesite 34.7.7.1 p.689
Haüyne 76.2.3.3 p.1627
Heazlewoodite 2.5.3.1 p.53
Hectorfloresite 32.1.2.1 p.666
Hectorite 71.3.1b.4 p.1491
Hedenbergite 65.1.3a.2 p.1298
Hedleyite 2.6.3.3 p.59
Hedyphane 41.8.2.1 p.863
Heideite 2.10.2.3 p.103
Heidornite 32.3.5.1 p.669
Heinrichite 40.2a.4.1 p.769
Hejtmanite 56.2.6b.2 p.1172
Heliophyllite 46.2.2.1 p.982
Hellandite-(Y) 54.2.2.1 p.1127
Hellyerite 15.1.7.1 p.464
Helmutwinklerite 40.2.9.2 p.760
Helvite 76.2.4.1 p.1630
Hematite 4.3.1.2 p.217
Hematolite 43.4.6.1 p.964
Hematophanite 7.11.4.1 p.327
Hemihedrite 36.1.1.2 p.696
Hemimorphite 56.1.2.1 p.1150
Hemloite 46.2.4.1 p.983
Hemusite 2.9.6.1 p.93
Hendersonite 47.3.1.4 p.988
Hendricksite-1M 71.2.2b.6 p.1462
Heneuite 43.4.11.1 p.965
Henmilite 26.1.5.1 p.552
Hennomartinite 56.2.3.2 p.1158
Henritermierite 51.4.4.1 p.1052
Henryite 2.16.11.1 p.146
Hentschelite 41.10.1.3 p.875
Hercynite 7.2.1.3 p.297
Herderite 41.5.4.1 p.832
Herschelite 77.1.2.2 p.1654
Hessite 2.4.2.1 p.43
Hetaerolite 7.2.7.2 p.305
Heterogenite-2H 6.1.4.1 p.272
Heterogenite-3R 6.1.4.2 p.272
Heteromorphite 3.6.20.3 p.189
Heterosite 38.4.1.1 p.719
Heulandite-Ca 77.1.4.1 p.1669
Hewettite 47.3.1.1 p.987
Hexahydrite 29.6.8.1 p.608
Hexahydroborite 26.1.4.1 p.551
Hexatestibiopanickelite 2.8.11.4 p.77
Heyite 41.2.2.1 p.817
Heyrovskyite 3.3.3.1 p.161
Hibbingite 10.1.1.2 p.391
Hibonite 7.4.1.1 p.311
Hibschite 51.4.3d.1 p.1051
Hidalgoite 43.4.1.3 p.961
Hieratite 11.5.1.1. p.410
Hilairite 59.2.3.1 p.1229
Hilgardite 26.5.14.1 p.559
Hillebrandite 65.2.3.1 p.1321
Hingganite-(Ce) 54.2.1a.2 p.1122
Hingganite-(Y) 54.2.1a.3 p.1123
Hingganite-(Yb) 54.2.1a.4 p.1123
Hinsdalite 43.4.1.5 p.962
Hiortdahlite 56.2.4.7 p.1164

Hisingerite 71.1.5.2 p.1434
Hocartite 2.9.2.6 p.91
Hochelagaite 8.6.1.2 p.368
Hodgkinsonite 52.2.1.3. p.1063
Hodrushite 3.8.4.1 p.198
Hoelite 50.4.2.1 p.1017
Hohmannite 31.9.4.1 p.652
Holdawayite 16b.4.2.2 p.505
Holdenite 43.4.7.1 p.964
Hollandite 7.9.1.1 p.321
Hollingworthite 2.12.3.8 p.127
Holtedahlite 41.6.4.2 p.848
Holtite 54.1.3.1 p.1119
Homilite 54.2.1b.5 p.1126
Honessite 31.10.6.1 p.665
Hongquiite 4.2.1.7 p.210
Hopeite 40.3.4.1 p.789
Horsfordite 2.1.2.1 p.37
Hotsonite 32.4.5.1 p.671
Hotsonite 43.5.18.1 p.970
Howardevansite 38.5.3.1 p.732
Howieite 69.2.3.1 p.1394
Howlite 25.3.5.1 p.545
Hsianghualite 77.1.1.5 p.1651
Huanghoite-(Ce) 16a.1.4.1 p.480
Huangite 30.2.4.8 p.633
Huemulite 47.2.3.1 p.987
Hulsite 24.2.3.1 p.538
Humberstonite 32.2.2.1 p.667
Humboldtine 50.1.3.1 p.1011
Humite 52.3.2c.1 p.1082
Hummerite 47.2.2.1 p.986
Hungchaoite 26.4.3.1 p.555
Huntite 14.4.3.1 p.457
Hurlbutite 38.3.6.1 p.717
Huréaulite 39.2.1.1 p.739
Hutchinsonite 3.8.6.1 p.198
Huttonite 51.5.3.1 p.1059
Hyalophane 76.1.1.3 p.1603
Hyalotekite 78.6.1.1 p.1712
Hydroastrophyllite 69.1.1.6 p.1384
Hydrobasaluminite 31.4.6.1 p.645
Hydrobiotite 71.2.2d.1 p.1471
Hydroboracite 26.3.6.1 p.554
Hydrocalumite 6.4.4.1 p.289
Hydrocerussite 16a.2.2.1 p.485
Hydrochlorborite 26.3.4.1 p.553
Hydrodelhayelite 72.5.1.5 p.1540
Hydrodresserite 16b.2.2.1 p.501
Hydroglauberite 29.4.1.1 p.589
Hydrohalite 9.1.2.1 p.376
Hydrohetaerolite 7.2.7.3 p.306
Hydrohonessite 31.10.7.1 p.665
Hydromagnesite 16b.7.1.1 p.513
Hydrombobomkulite 19.1.4.2 p.529
Hydromolysite 9.3.2.1 p.388
Hydrophilite 9.2.7.2 p.386
Hydroromarchite 6.4.2.1 p.288
Hydroscarbroite 16b.7.9.1 p.517
Hydrotalcite 16b.6.2.1 p.509
Hydrotungstite 49.1.1.1 p.1005

Hydroxycancrinite 76.2.5.8 p.1637
Hydroxylapatite 41.8.1.3 p.861
Hydroxylbastnäsite-(Ce) 16a.1.2.1 p.478
Hydroxylbastnäsite-(La) 16a.1.2.2 p.478
Hydroxylbastnäsite-(Nd) 16a.1.2.3 p.479
Hydroxylellestadite 52.4.9.4 p.1102
Hydroxylherderite 41.5.4.2 p.832
Hydrozincite 16a.4.1.1 p.493
Hypercinnabar 2.8.8.1 p.73
Häggite 6.4.3.1 p.289
Högbomite 15R, 18R, 24R 7.11.7.3 p.328
Högbomite 4H, 5H, 6H, 15H 7.11.7.1 p.328
Högbomite 8H 7.11.7.2 p.328
Hörnesite 40.3.6.7 p.796
Hübnerite 48.1.1.1 p.993
Hügelite 42.4.5.2 p.886
Høgtuvaite 69.2.1a.3 p.1387
Ianthinite 5.6.1.1 p.261
Ice 4.1.2.1 p.205
Idaite 2.9.14.1 p.96
Idrialite 50.3.8.1 p.1016
Iimoriite-(Y) 53.1.3.1 p.1107
Ikaite 15.1.4.1 p.463
Ikunolite 2.6.2.3 p.58
Ilesite 29.6.6.3 p.603
Ilimaussite-(Ce) 78.7.4.1 p.1714
Illite 71.2.2d.2 p.1472
Ilmajokite 78.7.5.1 p.1714
Ilmenite 4.3.5.1 p.226
Ilmenorutile 4.4.1.2 p.237
Ilsemannite 4.6.3.1 p.253
Ilvaite 56.2.3.3 p.1158
Imandrite 61.1.2b.1 p.1249
Imhofite 3.5.11.1 p.181
Imiterite 2.5.8.1 p.55
Imogolite 71.1.5.3 p.1435
Inaglyite 2.10.4.2 p.104
Incaite 3.1.4.3 p.152
Inderborite 26.3.1.2 p.552
Inderite 26.3.1.3 p.553
Indialite 61.1.1.3 p.1245
Indigirite 16b.1.3.1 p.499
Indite 2.10.1.12 p.102
Indium 1.1.14.1 p.13
Inesite 66.3.3.1 p.1372
Ingersonite 44.1.2.1 p.972
Ingodite 2.8.20.4 p.84
Innelite 58.2.6.2 p.1216
Insizwaite 2.12.1.15 p.118
Inyoite 26.3.1.1 p.552
Iodargyrite 9.1.5.1 p.378
Iowaite 6.4.5.1 p.289
Iquiqueite 27.1.8.1 p.566
Iranite 36.1.1.1 p.696
Iraqite-La 63.1.1.2 p.1269
Irarsite 2.12.3.7 p.126
Irhtemite 39.2.5.1 p.742
Iridarsenite 2.12.4.5 p.130
Iridium 1.2.1.2 p.17
Iriginite 49.2.3.1. p.1006

Index of Mineral Names in Alphabetical Order

Irtyshite 8.6.2.3 p.368
Ishikawaite 8.1.4.1 p.334
Isoclasite 42.6.2.1 p.892
Isocubanite 2.9.13.3 p.96
Isoferroplatinum 1.2.5.1 p.24
Isokite 41.5.6.2 p.835
Isomertieite 2.16.2.1 p.144
Itoite 30.2.6.1 p.636
Iwakiite 7.2.6.2 p.305
Ixiolite 8.1.10.1 p.339
Izoklakeite 3.5.12.1 p.182
Jacobsite 7.2.2.2 p.298
Jadeite 65.1.3c.1 p.1304
Jaffeite 55.3.2.1 p.1141
Jagoite 73.2.1.1 p.1548
Jagowerite 41.10.3.1 p.876
Jahnsite-(CaMnMg) 42.11.2.1 p.929
Jalpaite 2.4.4.1 p.44
Jamborite 6.3.8.1 p.287
Jamesite 41.11.2.1 p.880
Jamesonite 3.6.7.1 p.184
Janggunite 6.4.7.1 p.290
Janhaugite 56.2.4.10 p.1166
Jankovićite 3.8.13.2 p.202
Jarlite 11.6.10.1 p.415
Jarosewichite 41.1.1.3 p.815
Jarosite 30.2.5.1 p.634
Jaskólskiite 3.5.13.1 p.182
Jasmundite 52.4.7.4 p.1099
Jeanbandyite 6.3.7.3 p.286
Jeffreyite 55.4.2.3 p.1145
Jennite 56.2.4.11 p.1166
Jensenite 33.2.6.1 p.674
Jeppeite 8.7.9.1 p.372
Jeremejevite 25.8.1.1 p.549
Jeromite 2.12.12.1 p.134
Jerrygibbsite 52.3.2d.2 p.1085
Jervisite 65.1.3c.6 p.1307
Jianshuiite 7.8.2.3 p.320
Jimboite 24.3.2.2 p.539
Jimthompsonite 67.1.1.1 p.1376
Jinshajiangite 78.1.5.1 p.1703
Jixianite 48.4.2.1 p.1004
Joaquinite-(Ce) 60.1.1a.1 p.1232
Joesmithite 66.1.4.1 p.1367
Johachidolite 24.5.3.1 p.541
Johannite 31.8.2.1 p.650
Johannsenite 65.1.3a.4 p.1301
Johillerite 38.2.3.8 p.710
Johnbaumite 41.8.3.3 p.865
Johninnesite 69.2.5.1 p.1395
Johnsomervilleite 38.2.5.2 p.713
Johnwalkite 42.7.11.2 p.907
Jokokuite 29.6.7.4 p.606
Joliotite 15.1.8.1 p.464
Jolliffeite 2.12.3.9 p.127
Jonesite 61.2.2.1 p.1254
Jordanite 3.3.1.1 p.160
Jordisite 2.12.11.1 p.134
Joséite-B 2.6.2.2 p.57
Joséite 2.6.2.1 p.56

Jouravskite 32.4.4.3 p.670
Juanite 78.7.6.1 p.1715
Julgoldite-(Fe^{2+}) 58.2.2.1 p.1207
Julgoldite-(Fe^{3+}) 58.2.2.2 p.1208
Julienite 50.2.3.1 p.1014
Jungite 42.13.4.1 p.953
Junitoite 56.2.1.1 p.1152
Junoite 3.7.14.1 p.196
Jurbanite 31.9.10.1 p.655
Kaatialaite 39.3.9.1 p.745
Kadyrelite 10.5.4.2 p.400
Kaersutite 66.1.3a.18 p.1356
Kafehydrocyanite 50.2.5.1 p.1014
Kahlerite 40.2a.15.1 p.775
Kainite 31.7.1.1 p.648
Kainosite-(Y) 60.2.1.1 p.1236
Kalborsite 77.2.7.1 p.1701
Kaliborite 26.5.13.1 p.559
Kalicinite 13.1.2.1 p.423
Kalininite 2.10.1.13 p.102
Kalinite 29.5.4.2 p.593
Kaliophilite 76.2.1.5 p.1622
Kalipyrochlore 8.2.1.2 p.344
Kalistrontite 28.4.3.2 p.580
Kalsilite 76.2.1.1 p.1619
Kaluginite 42.11.1.6 p.927
Kamacite 1.1.11.1 p.10
Kamaishilite 76.2.3.6 p.1628
Kambaldaite 16b.7.12.1 p.517
Kamchatkite 30.2.8.1 p.637
Kamiokite 7.6.3.1 p.316
Kamitugaite 42.13.11.1 p.955
Kamotoite-(Y) 16b.3.3.1 p.503
Kamphaugite-(Y) 16b.1.7.1 p.500
Kanemite 73.1.5.1 p.1548
Kankite 40.4.4.1 p.804
Kanoite 65.1.1.3 p.1292
Kanonaite 52.2.2b.2 p.1068
Karasugite 11.6.21.1 p.418
Karelianite 4.3.1.4 p.220
Karibibite 7.10.2.1 p.325
Karnasurtite-(Ce) 52.4.9.7 p.1104
Karpatite 50.3.7.1 p.1016
Karpinskite 78.5.6.1 p.1710
Kashinite 2.11.12.2 p.113
Kasolite 53.3.1.1 p.1110
Kassite 8.3.9.1 p.361
Katayamalite 61.1.4.2 p.1250
Katoite 51.4.3d.2 p.1052
Katoptrite 44.3.5.1 p.974
Kawazulite 2.11.7.5 p.111
Kazakhstanite 47.3.2.6 p.989
Kazakovite 61.1.2a.2 p.1247
Keckite 42.11.4.1 p.931
Kegelite 71.5.1.1 p.1520
Keithconnite 2.2.6.1 p.40
Keiviite-(Y) 55.2.1a.2 p.1135
Keiviite-(Yb) 55.2.1a.3 p.1136
Keldyshite 55.2.2.1 p.1139
Kellyite 71.1.2c.5 p.1426
Kelyanite 10.6.11.1 p.405

Kemmlitzite 43.4.1.7 p.962
Kempite 10.1.4.1 p.392
Kentrolite 56.2.10.2 p.1182
Kenyaite 78.7.7.1 p.1715
Kermesite 2.13.1.1 p.140
Kernite 26.4.5.1 p.556
Kesterite 2.9.2.9 p.91
Kettnerite 16a.3.7.1. p.492
Keyite 38.3.8.1 p.718
Keystoneite 34.3.2.3 p.683
Khademite 31.9.11.1 p.656
Khamrabaevite 1.1.19.2 p.15
Khanneshite 14.4.4.2 p.459
Kharaelakhite 2.7.5.1 p.64
Khatyrkite 1.1.15.2 p.13
Khibinskite 55.2.2.2 p.1139
Khinite 33.1.3.1 p.672
Khristovite-(Ce) 58.2.1a.9 p.1201
Kiddcreekite 2.9.6.2 p.93
Kidwellite 42.8.2.1 p.910
Kieftite 2.12.17.3 p.140
Kieserite 29.6.2.1 p.596
Kilchoanite 58.3.2.1 p.1219
Killalaite 56.2.9.2 p.1181
Kimrobinsonite 6.4.8.1 p.290
Kimuraite-(Y) 15.4.5.1 p.473
Kimzeyite 51.4.3c.2 p.1050
Kingite 42.10.3.1 p.923
Kingsmountite 42.11.8.2 p.935
Kinichilite 34.3.2.2 p.682
Kinoite 57.1.2.1 p.1188
Kinoshitalite-1M 71.2.2c.5 p.1470
Kintoreite 42.7.3.6 p.901
Kipushite 42.2.4.1 p.883
Kirkiite 3.4.17.1 p.178
Kirschsteinite 51.3.2.2 p.1033
Kitkaite 2.12.14.2 p.137
Kittatinnyite 52.4.1.3 p.1091
Kivuite 42.4.9.3 p.888
Kladnoite 50.4.8.1 p.1018
Klebelsbergite 30.1.5.1 p.624
Kleberite 7.11.10.1 p.330
Kleemanite 42.11.12.2 p.942
Kleinite 10.4.3.1 p.398
Klockmannite 2.8.12.2 p.79
Klyuchevskite 28.4.6.1 p.582
Klyuchevskite 30.4.1.1 p.638
Knorringite 51.4.3a.4 p.1043
Koashvite 61.1.2b.2 p.1249
Kobeite-(Y) 8.3.12.1 p.362
Kobellite 3.6.19.1 p.187
Kochkarite 2.6.3.1 p.59
Koechlinite 48.2.2.1 p.1001
Koenenite 10.6.2.1 p.403
Kogarkoite 30.1.6.1 p.624
Koktaite 29.3.1.2 p.585
Kolarite 2.15.3.1 p.143
Kolbeckite 40.4.3.3 p.804
Kolfanite 42.8.5.1 p.912
Kolicite 43.4.8.1 p.965
Kolovratite 40.5.13.1 p.814

Kolwezite 16a.3.1.3 p.488
Kolymite 1.1.9.1 p.8
Komarovite 56.2.4.12 p.1167
Kombatite 41.1.4.2 p.817
Komkovite 59.2.4.1 p.1230
Konderite 2.10.4.1 p.104
Koninckite 40.4.2.1 p.803
Konyaite 29.3.5.1 p.586
Koragoite 8.7.12.1 p.372
Koritnigite 39.1.4.1 p.736
Kornelite 29.8.2.1 p.619
Kornerupine 58.1.1.1 p.1193
Kornite 66.1.3c.13 p.1366
Korshunovskite 10.1.5.1 p.392
Korzhinskite 26.7.1.1 p.563
Kosmochlor 65.1.3c.4 p.1307
Kosnarite 38.4.12.1 p.731
Kostovite 2.12.13.4 p.136
Kostylevite 59.2.1.3 p.1225
Kotoite 24.3.2.1 p.539
Köttigite 40.3.6.5 p.795
Koutekite 2.3.1.1 p.40
Kovdorskite 43.5.8.1 p.968
Kozulite 66.1.3c.10 p.1365
Kraisslite 43.4.10.1 p.965
Kratochvilite 50.3.1.1 p.1015
Krausite 29.5.1.1 p.591
Krauskopfite 74.3.4.2 p.1564
Krautite 39.1.3.1 p.736
Kremersite 11.4.1.2 p.410
Krennerite 2.12.13.1 p.134
Kribergite 43.5.7.1 p.968
Krinovite 69.2.1a.4 p.1388
Kröhnkite 29.3.2.1 p.585
Krupkaite 3.4.5.2 p.170
Krutaite 2.12.1.8 p.116
Krutovite 2.12.1.12 p.117
Kryzhanovskite 40.3.2.2 p.787
Ktenasite 31.6.3.1 p.647
Kukisvumite 65.1.6.2 p.1312
Kuksite 33.3.5.2 p.676
Kulanite 41.9.1.1 p.870
Kuliokite-(Y) 78.7.8.1 p.1716
Kulkeite 71.4.2.6 p.1514
Kullerudite 2.12.2.6 p.122
Kupletskite 69.1.1.2 p.1383
Kuramite 2.9.2.4 p.90
Kuranakhite 33.2.1.1 p.673
Kurchatovite 24.4.2.1 p.540
Kurnakovite 26.3.3.1 p.553
Kurumsakite 78.1.6.1 p.1704
Kusachiite 7.2.6.1 p.304
Kutinaite 2.2.2.2 p.39
Kutnohorite 14.2.1.3 p.452
Kuzminite 9.1.8.2 p.380
Kvanefjeldite 72.5.3.1 p.1542
Kyanite 52.2.2c.1 p.1069
Kyzylkumite 8.4.1.3 p.363
Labuntsovite 60.1.3.2 p.1235
Lacroixite 41.5.5.1 p.834
Laffittite 3.4.10.2 p.173

Index of Mineral Names in Alphabetical Order

Laihunite   51.3.1.5   p.1031
Laitakarite   2.6.2.4   p.58
Lammerite   38.3.9.1   p.718
Lamprophyllite   56.2.6c.4   p.1175
Lanarkite   30.2.1.1   p.628
Landauite   8.5.1.1   p.365
Landesite   40.3.2.4   p.788
Langbeinite   28.4.4.1   p.581
Langisite   2.8.11.9   p.79
Langite   31.4.3.1   p.644
Lannonite   31.9.14.1   p.657
Lansfordite   15.1.6.1   p.464
Lanthanite-(Ce)   15.4.2.3   p.472
Lanthanite-(La)   15.4.2.1   p.472
Lanthanite-(Nd)   15.4.2.2   p.472
Laphamite   2.11.6.1   p.109
Lapieite   3.4.4.1   p.168
Laplandite-(Ce)   78.7.9.1   p.1716
Larderellite   26.5.4.1   p.556
Larnite   51.5.1.1   p.1053
Larosite   2.16.9.1   p.146
Larsenite   51.2.1.2   p.1023
Latiumite   73.1.2.1   p.1546
Latrappite   4.3.3.2   p.223
Laueite   42.11.10.1   p.940
Laumontite   77.1.1.4   p.1650
Launayite   3.6.6.1   p.184
Laurelite   9.2.10.1   p.387
Laurionite   10.2.2.1   p.394
Laurite   2.12.1.10   p.117
Lausenite   29.8.1.1   p.618
Lautarite   21.1.1.1   p.531
Lautenthalite   31.6.1.2   p.646
Lautite   2.12.8.1   p.132
Lavendulan   42.9.4.2   p.920
Lavrentievite   10.3.4.1   p.397
Lawrencite   9.2.3.1   p.384
Lawsonbauerite   31.1.4.2   p.640
Lawsonite   56.2.3.1   p.1157
Lazarenkoite   7.8.3.1   p.320
Lazulite   41.10.1.1   p.873
Lazurite   76.2.3.4   p.1627
Leadamalgam   1.1.10.1   p.9
Leadhillite   17.1.2.1   p.520
Lead   1.1.1.4   p.5
Leakeite   66.1.3c.12   p.1366
Lecontite   29.2.1.1   p.584
Legrandite   42.6.4.1   p.893
Lehnerite   40.2a.16.2   p.776
Leifite   78.7.10.1   p.1717
Leightonite   29.4.5.2   p.591
Leiteite   45.1.7.1   p.978
Lemoynite   74.3.3.2   p.1563
Lenaite   2.9.1.5   p.89
Lengenbachite   3.3.7.1   p.165
Leningradite   41.5.17.1   p.843
Lennilenapeite   74.1.1.2   p.1553
Lenoblite   4.4.17.1   p.251
Leonite   29.3.3.3   p.586
Lepersonnite-(Gd)   17.1.12.1   p.524
Lepidocrocite   6.1.2.2   p.270

Lepidolite   71.2.2b.7   p.1462
Lermontovite   40.4.8.2   p.808
Letovicite   28.1.3.1   p.568
Leucite   76.2.2.1   p.1622
Leucophanite   55.4.2.4   p.1146
Leucophoenicite   52.3.2c.2   p.1083
Leucophosphite   42.11.6.1   p.932
Leucosphenite   72.5.2.1   p.1541
Levyclaudite   3.1.3.1   p.151
Levyne   77.1.2.8   p.1660
Liandratite   8.1.9.1   p.339
Liberite   51.1.2.1   p.1023
Libethenite   41.6.6.2   p.849
Liddicoatite   61.3.1.2   p.1257
Liebauite   69.2.4.1   p.1395
Liebenbergite   51.3.1.3   p.1029
Liebigite   15.3.2.1   p.468
Likasite   19.1.5.1   p.529
Lillianite   3.4.15.1   p.175
Lime   4.2.1.5   p.210
Linarite   30.2.3.1   p.629
Lindackerite   39.2.7.1   p.742
Lindgrenite   48.3.1.1   p.1002
Lindqvistite   7.11.15.1   p.331
Lindsleyite   8.5.1.8   p.367
Lindströmite   3.4.5.7   p.171
Linnaeite   2.10.1.1   p.99
Lintisite   65.1.6.3   p.1313
Liottite   76.2.5.9   p.1637
Lipscombite   41.10.2.1   p.875
Liroconite   42.2.1.1   p.882
Lisetite   76.1.6.3   p.1616
Lishizhenite   29.7.2.2   p.615
Liskeardite   42.1.1.1   p.882
Litharge   4.2.4.1   p.212
Lithiomarsturite   65.4.1.7   p.1331
Lithiophilite   38.1.1.2   p.701
Lithiophorite   6.4.1.1   p.288
Lithiophosphate   38.4.10.1   p.727
Lithiotantite   8.7.7.1   p.371
Lithiowodginite   8.1.8.4   p.338
Lithosite   77.2.2.1   p.1699
Litidionite   70.1.1.1   p.1397
Liveingite   3.6.17.1   p.187
Livingstonite   3.7.11.1   p.195
Lizardite   71.1.2b.2   p.1418
Lokkaite-(Y)   15.4.4.1   p.473
Lomonosovite   56.4.1.1   p.1184
Lonecreekite   29.5.5.4   p.594
Lonsdaleite   1.3.6.3   p.35
Loparite-(Ce)   4.3.3.3   p.224
Lopezite   35.2.1.1   p.692
Lorandite   3.7.6.1   p.192
Loranskite-(Y)   8.1.5.1   p.334
Lorenzenite   65.1.6.1   p.1312
Lorettoite   10.5.12.1   p.402
Loseyite   16a.5.3.1   p.495
Lotharmeyerite   37.1.6.1   p.699
Loudounite   59.2.2.5   p.1229
Loughlinite   74.3.1b.3   p.1561
Lourenswalsite   61.2.3.1   p.1254

Lovdarite 78.6.2.1 p.1712
Loveringite 8.5.1.2 p.365
Lovozerite 61.1.2a.1 p.1247
Luanheite 1.1.8.4 p.8
Luberoite 2.16.23.1 p.149
Lucasite-(Ce) 8.3.5.1 p.357
Luddenite 78.7.11.1 p.1717
Ludjibaite 41.4.3.3 p.823
Ludlamite 40.3.5.1 p.790
Ludlockite 38.5.4.1 p.732
Ludlockite 45.1.14.1 p.980
Ludwigite 24.2.1.1 p.535
Lueshite 4.3.3.4 p.224
Luetheite 42.7.10.2 p.906
Lun'okite 42.11.1.4 p.926
Lunijianlaite 71.4.2.2 p.1509
Lusungite 42.7.3.4 p.900
Lutecite 75.2.3.1 p.1593
Luzonite 3.2.2.1 p.156
Lyonsite 38.5.2.1 p.732
Löllingite 2.12.2.9 p.122
Löweite 29.4.3.1 p.590
Lünebergite 43.5.11.1 p.968
Långbanite 44.3.4.1 p.974
Låvenite 56.2.4.4 p.1162
Macaulayite 71.2.5.1 p.1480
Macdonaldite 72.5.1.3 p.1540
Macedonite 4.3.6.1 p.229
Macfallite 58.2.3.2 p.1212
Machatschkiite 39.3.8.1 p.745
Mackayite 34.6.1.1 p.685
Mackinawite 2.7.2.1 p.63
Macphersonite 17.1.4.1 p.521
Macquartite 36.1.2.1 p.696
Madocite 3.5.7.1 p.180
Magadiite 78.7.12.1 p.1717
Magbasite 69.2.1c.1 p.1393
Maghagendorfite 38.2.3.4 p.709
Maghemite 4.3.7.1 p.229
Magnesio(alumino)katophorite 66.1.3b.11 p.1359
Magnesio(alumino)taramite 66.1.3b.15 p.1360
Magnesioanthophyllite 66.1.2.1 p.1343
Magnesioaubertite 31.9.8.3 p.655
Magnesioaxinite 56.2.2.2 p.1155
Magnesiocarpholite 65.1.5.3 p.1311
Magnesiochloritoid-A,M 52.3.3.2 p.1089
Magnesiochromite 7.2.3.1 p.301
Magnesioclinoholmquistite 66.1.1.6 p.1343
Magnesiocopiapite 31.10.5.2 p.663
Magnesiocoulsonite 7.2.4.3 p.303
Magnesiocummingtonite 66.1.1.1 p.1339
Magnesiodumortierite 54.1.2.2 p.1118
Magnesioferrite 7.2.2.1 p.298
Magnesiogedrite 66.1.2.5 p.1345
Magnesioholmquistite 66.1.2.9 p.1346
Magnesiohulsite 24.2.3.2 p.538
Magnesiosadanagaite 66.1.3a.16 p.1356
Magnesite 14.1.1.2 p.434
Magnesium Astrophyllite 69.1.1.7 p.1384

Magnesium Zippeite 31.10.4.3 p.661
Magnesium-chlorophoenicite 41.1.1.2 p.815
Magnetite 7.2.2.3 p.299
Magnetoplumbite 7.4.2.1 p.312
Magniotriplite 41.6.1.3 p.846
Magnocolumbite 8.3.2.5 p.356
Magnolite 34.1.7.1 p.679
Magnussonite 46.1.3.1 p.981
Mahlmoodite 40.1.5.1 p.749
Majakite 2.4.16.1 p.51
Majorite 51.4.3a.5 p.1043
Makatite 74.3.5.2 p.1565
Makovickyite 3.8.10.2 p.200
Malachite 16a.3.2.1 p.488
Malanite 2.10.1.17 p.103
Malayaite 52.4.3.2 p.1095
Maldonite 1.1.3.1 p.6
Malladrite 11.5.2.1 p.410
Mallardite 29.6.10.5 p.612
Mammothite 30.1.14.1 p.626
Manaksite 70.1.1.3 p.1398
Manandonite 71.1.2c.6 p.1426
Manasseite 16b.6.1.1 p.508
Mandarinoite 34.3.4.1 p.683
Manganarsite 46.2.6.1 p.983
Manganaxinite 56.2.2.3 p.1156
Manganbabingtonite 65.4.1.3 p.1329
Manganbelyankinite 8.7.8.2 p.371
Manganberzeliite 38.2.1.2 p.705
Manganese Shadlunite 2.7.1.5 p.62
Manganese-Hörnesite 40.3.7.2 p.797
Mangangordonite 42.11.14.5 p.944
Manganhumite 52.3.2c.3 p.1083
Manganite 6.1.3.1 p.271
Manganneptunite 70.4.1.2 p.1402
Manganochromite 7.2.3.2 p.301
Manganocolumbite 8.3.2.4 p.355
Manganolangbeinite 28.4.4.2 p.581
Manganosegelerite 42.11.1.3 p.926
Manganosite 4.2.1.3 p.209
Manganostibite 44.3.2.1 p.973
Manganotantalite 8.3.2.3 p.355
Manganotapiolite 8.3.1.2 p.352
Manganotychite 17.1.1.3 p.519
Manganpyrosmalite 72.4.1a.2 p.1537
Manjiroite 7.9.1.3 p.322
Mannardite 7.9.5.1 p.324
Mansfieldite 40.4.1.4 p.802
Mantiennéite 42.11.21.2 p.949
Mapimite 42.8.7.1 p.913
Marcasite 2.12.2.1 p.120
Margaritasite 40.2a.28.2 p.783
Margarite-2M$_1$ 71.2.2c.1 p.1468
Margarosanite 59.1.2.2 p.1224
Marialite 76.3.1.1 p.1641
Maricite 38.1.2.1 p.702
Maricopeite 77.1.6.3 p.1689
Marokite 7.2.10.1 p.309
Marrite 3.4.6.1 p.171
Marshite 9.1.7.3 p.379
Marsturite 65.4.1.6 p.1330

Index of Mineral Names in Alphabetical Order

Marthozite 34.7.4.1 p.688
Mascagnite 28.2.1.1 p.568
Maslovite 2.12.3.12 p.127
Massicot 4.2.7.1 p.213
Masutomilite-1M, 2M$_1$ 71.2.2b.12 p.1466
Masuyite 5.2.2.1 p.258
Mathewrogersite 78.3.1.1 p.1706
Mathiasite 8.5.1.7 p.367
Matildite 3.7.1.1 p.190
Matlockite 9.2.11.1 p.388
Matraite 2.8.15.1 p.80
Mattagamite 2.12.2.5 p.122
Matteuccite 29.1.2.1 p.583
Mattheddleite 52.4.9.6 p.1103
Matulaite 42.13.3.1 p.953
Matveevite 42.11.21.3 p.950
Maucherite 2.16.16.1 p.147
Maufite 78.7.13.1 p.1718
Mawbyite 40.2.9.4 p.761
Mawsonite 2.9.3.1 p.92
Maxwellite 41.5.5.3 p.835
Mayenite 7.11.3.1 p.327
Mayingite 2.12.1.19 p.119
Mazzite 77.1.3.10 p.1668
Mbobomkulite 19.1.4.1 p.528
McAllisterite 26.6.2.1 p.561
Mcalpineite 33.2.5.1 p.674
Mcauslanite 42.11.22.1 p.950
Mcbirneyite 38.3.10.1 p.719
Mcconnellite 7.1.1.2 p.292
Mccrillisite 40.5.4.2 p.810
Mcgillite 72.4.1b.2 p.1538
McGovernite 43.4.9.1 p.965
Mcguinnessite 16a.3.1.5 p.488
Mckelveyite-(Nd) 15.3.4.3 p.470
Mckelveyite-(Y) 15.3.4.2 p.469
Mckinstryite 2.4.5.1 p.44
McNearite 39.3.7.1 p.745
Medaite 57.4.1.1 p.1192
Megacyclite 62.1.2.1 p.1267
Meixnerite 6.4.6.1 p.289
Melanocerite-(ce) 54.2.5.3 p.1131
Melanophlogite 75.2.2.1 p.1592
Melanostibite 4.3.5.5 p.228
Melanotekite 56.2.10.1 p.1182
Melanothallite 10.4.4.1 p.398
Melanovanadite 47.3.6.1 p.990
Melanterite 29.6.10.1 p.611
Melilite 55.4.1.3 p.1144
Meliphanite 55.4.2.5 p.1146
Melkovite 49.4.3.1 p.1009
Mellite 50.2.1.1 p.1013
Melonite 2.12.14.1 p.137
Melonjosephite 41.10.5.1 p.877
Mendipite 10.3.1.1 p.396
Mendozavilite 49.4.5.1 p.1010
Mendozite 29.5.4.1 p.593
Meneghinite 3.3.5.1 p.162
Mercallite 28.1.1.1 p.568
Mercury 1.1.7.1 p.7
Mereiterite 28.4.3.3 p.580

Merenskyite 2.12.14.4 p.138
Merlinoite 77.1.3.11 p.1668
Merrihueite 63.2.1a.5 p.1273
Merrillite-(Ca) 38.3.4.3 p.717
Mertieite-II 2.16.3.1 p.144
Mertieite-I 2.16.1.1 p.144
Merwinite 51.4.2.1 p.1036
Mesolite 77.1.5.4 p.1681
Messelite 40.2.2.2 p.752
Meta-Aluminite 31.7.5.1 p.649
Meta-alunogen 29.8.7.1 p.620
Meta-Ankoleite 40.2a.8.1 p.770
Meta-Autunite 40.2a.1.2 p.766
Meta-Heinrichite 40.2a.4.2 p.769
Meta-Kahlerite 40.2a.15.2 p.775
Meta-Kirchheimerite 40.2a.17.1 p.776
Meta-Lodevite 40.2a.18.1 p.776
Meta-Novacekite 40.2a.10.2 p.771
Meta-Torbernite 40.2a.13.2 p.773
Meta-Tyuyamunite 40.2a.26.2 p.781
Meta-Uranocircite 40.2a.3.2 p.768
Meta-Uranopilite 31.2.7.1 p.642
Meta-uranospinite 40.2a.2.2 p.768
Meta-Zeunerite 40.2a.14.2 p.774
Metaborite 24.5.1.1 p.541
Metacalciouranoite 5.4.3.1 p.260
Metacinnabar 2.8.2.3 p.69
Metadelrioite 47.1.2.2 p.985
Metahaiweeite 53.3.2.3 p.1115
Metahewettite 47.3.1.2 p.987
Metahohmannite 31.9.5.1 p.653
Metaköttigite 40.3.8.2 p.798
Metamunirite 47.1.3.2 p.986
Metarossite 47.1.1.2 p.985
Metaschoderite 43.3.1.2 p.959
Metaschoepite 5.2.1.1 p.257
Metasideronatrite 31.8.4.1 p.651
Metastibnite 2.11.3.1 p.108
Metastudtite 5.3.1.2 p.259
Metaswitzerite 40.3.5.2 p.790
Metavandendriesscheite 5.8.1.2 p.262
Metavanmeerscheite 42.7.14.2 p.907
Metavanuralite 42.11.13.2 p.942
Metavariscite 40.4.3.1 p.803
Metavauxite 42.11.11.1 p.941
Metavivianite 40.3.9.1 p.798
Metavoltine 29.4.6.1 p.591
Metazellerite 15.3.1.2 p.467
Meyerhofferite 26.3.2.1 p.553
Meymacite 4.5.3.1 p.252
Mgriite 3.4.7.2 p.171
Miargyrite 3.7.3.2 p.191
Michenerite 2.12.3.11 p.127
Microcline 76.1.1.5 p.1605
Microlite 8.2.2.1 p.345
Microsommite 76.2.5.10 p.1638
Miersite 9.1.7.2 p.379
Miharaite 3.1.5.1 p.153
Mikasaite 28.4.5.2 p.582
Milarite 63.2.1a.12 p.1277
Millerite 2.8.16.1 p.81

Millisite 42.7.9.1   p.905
Millosevichite 28.4.5.1   p.581
Mimetite 41.8.4.2   p.866
Minamiite 30.2.4.5   p.633
Minasgeraisite-(Y) 54.2.1b.6   p.1126
Minasragrite 29.6.12.1   p.614
Mineevite-(Y) 17.1.14.1   p.524
Minehillite 73.2.2b.1   p.1549
Minguzzite 50.1.4.1   p.1012
Minium 7.2.8.1   p.306
Minnesotaite 71.2.1.5   p.1443
Minrecordite 14.2.1.4   p.452
Minyulite 42.11.5.1   p.932
Mirabilite 29.2.2.1   p.584
Misenite 28.1.2.1   p.568
Miserite 70.2.1.1   p.1400
Mitridatite 42.8.4.1   p.911
Mitscherlichite 11.3.2.1   p.409
Mixite 42.5.1.1   p.890
Moctezumite 34.1.5.1   p.678
Mohite 2.9.16.1   p.97
Mohrite 29.3.7.1   p.588
Moluranite 49.1.2.1   p.1005
Molybdenite 2.12.10.1   p.132
Molybdite 4.5.1.1   p.251
Molybdofornacite 43.4.3.3   p.964
Molybdomenite 34.1.1.1   p.677
Molybdophyllite 78.5.7.1   p.1711
Molysite 9.3.1.1   p.388
Monazite-(Ce) 38.4.3.1   p.722
Monazite-(La) 38.4.3.2   p.724
Monazite-(Nd) 38.4.3.5   p.724
Moncheite 2.12.14.3   p.137
Monetite 37.1.1.1   p.697
Mongolite 56.2.4.15   p.1168
Mongshanite 7.11.11.1   p.330
Monimolite 44.1.1.4   p.972
Monohydrocalcite 15.1.3.1   p.463
Monsmedite 29.1.3.1   p.583
Montanite 33.2.2.1   p.673
Montbrayite 2.11.8.1   p.111
Montdorite-1M, 3T 71.2.2a.10   p.1456
Montebrasite 41.5.8.2   p.837
Monteponite 4.2.1.4   p.210
Monteregianite-Y 72.5.1.2   p.1539
Montesommaite 77.1.3.12   p.1669
Montgomeryite 42.11.8.1   p.935
Monticellite 51.3.2.1   p.1032
Montmorillonite 71.3.1a.2   p.1483
Montroseite 6.1.1.4   p.269
Montroyalite 16b.5.3.1   p.506
Montroydite 4.2.6.1   p.212
Mooihoekite 2.9.8.3   p.94
Moolooite 50.1.6.1   p.1013
Mooreite 31.1.3.1   p.639
Moorhouseite 29.6.8.5   p.609
Mopungite 6.3.7.4   p.287
Moraesite 42.6.1.1   p.891
Mordenite 77.1.6.1   p.1688
Moreauite 42.12.3.1   p.951
Morelandite 41.8.5.1   p.868

Morenosite 29.6.11.3   p.613
Morimotoite 51.4.3c.3   p.1051
Morinite 42.4.2.1   p.885
Morozeviczite 2.9.5.1   p.93
Mosandrite 56.2.5.1   p.1168
Moschallandsbergite 1.1.8.1   p.7
Moschelite 9.1.8.3   p.380
Mosesite 29.3.8.1   p.589
Mottramite 41.5.2.2   p.830
Motukoreaite 17.1.7.1   p.522
Mounanaite 41.10.7.1   p.878
Mountainite 77.2.3.1   p.1699
Mountkeithite 32.4.3.1   p.669
Mourite 5.9.4.1   p.264
Moydite-(Y) 27.1.9.1   p.566
Mozartite 52.4.2.1   p.1091
Mpororoite 49.3.4.1   p.1008
Mrazekite 42.4.15.1   p.889
Mroseite 34.8.4.1   p.690
Muirite 62.1.1.1   p.1267
Mukhinite 58.2.1a.10   p.1202
Mullite 52.2.2a.2   p.1065
Mummeite 3.8.10.4   p.201
Mundite 42.7.13.1   p.907
Mundrabillaite 39.3.3.1   p.743
Munirite 47.1.3.1   p.986
Murataite 8.7.1.1   p.369
Murdochite 4.6.5.1   p.254
Murmanite 56.2.7.2   p.1177
Murunskite 2.5.5.3   p.54
Muscovite-2M$_1$, 1M, 3T, 2M$_2$ 71.2.2a.1   p.1448
Musgravite 7.2.12.1   p.309
Mushistonite 6.3.6.6   p.286
Muskoxite 7.11.1.1   p.326
Mäkinenite 2.8.16.2   p.81
Mückeite 3.4.4.2   p.168
Na-Komarovite 56.2.4.13   p.1167
Nabaphite 40.1.3.2   p.748
Nabokoite 30.1.17.1   p.627
Nacaphite 41.3.6.1   p.820
Nacareniobsite-(Ce) 56.2.5.2   p.1169
Nadorite 10.2.5.2   p.395
Nagashimalite 64.3.1.3   p.1285
Nagelschmidtite 53.2.2.1   p.1108
Nagyagite 2.11.10.1   p.112
Nahcolite 13.1.1.1   p.422
Nahpoite 37.1.5.1   p.699
Nakauriite 32.4.1.1   p.669
Nalipoite 38.4.10.3   p.728
Namansilite 65.1.3c.3   p.1306
Nambulite 65.4.1.4   p.1329
Namibite 38.5.9.1   p.734
Namuwite 31.4.7.1   p.645
Nanlingite 46.1.2.1   p.981
Nanpingite-2M$_1$ 71.2.2a.8   p.1455
Nantokite 9.1.7.1   p.379
Narsarsukite 70.1.2.1   p.1398
Nasinite 26.5.6.1   p.557
Nasledovite 17.1.6.1   p.521
Nasonite 56.2.11.1   p.1183

Nastrophite 40.1.3.1   p.748
Natalyite 65.1.3c.5   p.1307
Natanite 6.3.6.3   p.285
Natisite 52.4.4.1   p.1096
Natroapophyllite 72.3.1.3   p.1527
Natrobistantite 8.2.4.2   p.349
Natrochalcite 31.8.1.1   p.650
Natrodufrenite 42.9.1.3   p.914
Natrofairchildite 14.3.5.1   p.456
Natrolite 77.1.5.1   p.1677
Natromontebrasite 41.5.8.3   p.838
Natronambulite 65.4.1.5   p.1330
Natroniobite 4.3.4.1   p.225
Natron 15.1.2.1   p.462
Natrophilite 38.1.1.3   p.701
Natrophosphate 42.13.9.1   p.954
Natrosilite 74.3.5.1   p.1564
Natrotantite 8.6.2.2   p.368
Naujakasite 73.1.1.1   p.1546
Naumannite 2.4.1.2   p.42
Navajoite 4.6.2.1   p.253
Nchwaningite 65.1.8.1   p.1313
Nealite 41.4.6.1   p.824
Nealite 46.1.6.1   p.982
Nefedovite 41.11.4.1   p.881
Neighborite 11.1.1.1   p.407
Nekoite 72.3.2.7   p.1534
Nekrasovite 3.1.1.3   p.151
Nelenite 72.4.1b.3   p.1538
Neltnerite 7.5.1.2   p.314
Nenadkevichite 60.1.3.1   p.1234
Neotocite 71.1.5.4   p.1436
Nepheline 76.2.1.2   p.1619
Nepouite 71.1.2b.3   p.1420
Neptunite 70.4.1.1   p.1401
Nesquehonite 13.1.5.1   p.424
Nevskite 2.8.20.3   p.83
Newberyite 39.1.6.1   p.737
Neyite 3.5.1.1   p.178
Niahite 40.1.2.2   p.748
Nichromite 7.2.3.4   p.302
Nickel Zippeite 31.10.4.4   p.661
Nickel-Boussingaultite 29.3.7.3   p.589
Nickel-Hexahydrite 29.6.8.4   p.609
Nickel-Skutterudite 2.12.17.2   p.139
Nickelalumite 31.3.1.2   p.643
Nickelaustinite 41.5.1.7   p.828
Nickelbischofite 9.2.9.2   p.387
Nickelblödite 29.3.3.2   p.586
Nickeline 2.8.11.1   p.76
Nickel 1.1.11.5   p.12
Nickenichite 38.2.3.9   p.710
Nifontovite 26.3.7.1   p.554
Nigerite 24R 7.11.8.2   p.329
Nigerite 6H 7.11.8.1   p.329
Niggliite 1.2.10.1   p.25
Nimite 71.4.1.5   p.1503
Ningyoite 40.4.8.1   p.808
Niobo-aeschynite-(Ce) 8.3.6.2   p.358
Niobophyllite 69.1.1.4   p.1383
Niocalite 56.2.4.6   p.1164

Nisbite 2.12.2.13   p.124
Nissonite 42.7.5.1   p.902
Niter 18.1.2.1   p.526
Nitratine 18.1.1.1   p.526
Nitrobarite 18.2.1.1   p.527
Nitrocalcite 18.2.2.1   p.527
Nitromagnesite 18.2.3.1   p.527
Nobleite 26.6.6.1   p.562
Nolanite 4.6.7.1   p.254
Nontronite 71.3.1a.3   p.1485
Norbergite 52.3.2a.1   p.1078
Nordenskiöldine 24.3.3.1   p.539
Nordite-(Ce) 72.5.2.2   p.1541
Nordite-(La) 72.5.2.3   p.1542
Nordstrandite 6.3.3.1   p.281
Nordströmite 3.7.13.1   p.196
Norrishite-1M 71.2.2b.11   p.1466
Norsethite 14.2.2.1   p.452
Northupite 16a.3.9.1   p.493
Nosean 76.2.3.2   p.1626
Novacekite 40.2a.10.1   p.771
Novakite 2.4.18.1   p.51
Nowackiite 3.4.13.1   p.173
Nsutite 4.4.8.1   p.248
Nuffieldite 3.3.4.1   p.161
Nukundamite 2.9.15.1   p.97
Nullaginite 16a.3.2.2   p.490
Nyböite 66.1.3c.11   p.1366
Nyerereite 14.3.4.1   p.456
O'Danielite 38.2.3.7   p.710
Oboyerite 34.8.3.1   p.690
Obradovicite 49.4.2.1   p.1009
Odinite 71.1.1.5   p.1414
Odintsovite 61.1.5.1   p.1251
Offretite 77.1.2.4   p.1655
Ogdensburgite 42.9.7.1   p.922
Ohmilite 65.3.4.3   p.1325
Ojuelaite 42.11.18.3   p.948
Okanoganite-(Y) 54.2.4.3   p.1130
Okenite 72.3.2.5   p.1532
Okhotskite-(Mg) 58.2.2.3   p.1208
Olekminskite 14.2.2.3   p.453
Olenite 61.3.1.7   p.1260
Olgite 38.1.3.1   p.702
Olivenite 41.6.6.1   p.849
Olkhonskite 8.4.1.2   p.363
Olmsteadite 42.7.11.1   p.906
Olsacherite 32.1.3.1   p.666
Olshanskyite 25.1.6.1   p.543
Olympite 38.4.10.2   p.727
Omeiite 2.12.2.7   p.122
Omphacite 65.1.3b.1   p.1302
Onoratoite 10.5.7.1   p.401
Oosterboschite 2.16.15.2   p.147
Opal 75.2.1.1   p.1587
Orcelite 2.3.2.1   p.40
Ordoñezite 44.2.1.2   p.973
Oregonite 2.5.4.1   p.53
Orickite 2.14.7.1   p.142
Orientite 58.3.1.2   p.1218
Orlymanite 73.2.2b.2   p.1550

Orpheite 43.4.1.4 p.961
Orpiment 2.11.1.1 p.104
Orschallite 34.8.7.1 p.691
Orthobrannerite 8.2.7.1 p.351
Orthochevkinite 56.2.8.4 p.1180
Orthoclase 76.1.1.1 p.1600
Orthoericssonite 56.2.6c.5 p.1175
Orthojoaquinite-(Ce) 60.1.1b.1 p.1232
Orthopinakiolite 24.2.5.1 p.538
Orthoserpierite 31.6.7.1 p.648
Orthowalpurgite 40.5.9.2 p.813
Osarizawaite 30.2.4.4 p.632
Osarsite 2.12.4.3 p.130
Osbornite 1.1.19.1 p.15
Osmium 1.2.2.1 p.19
Osumilite 63.2.1a.6 p.1273
Otavite 14.1.1.7 p.439
Otjisumeite 7.10.3.1 p.326
Ottemannite 2.11.9.1 p.112
Ottrelite-A,M 52.3.3.3 p.1089
Otwayite 16b.3.2.1 p.503
Ourayite 3.5.6.1 p.180
Oursinite 53.3.1.7 p.1113
Overite 42.11.1.1 p.925
Owensite 2.15.2.3 p.143
Owyheeite 3.5.10.1 p.181
Oxammite 50.1.5.1 p.1012
Oyelite 72.3.2.8 p.1534
Örebroite 7.3.1.3 p.311
Pabstite 59.1.1.3 p.1223
Pachnolite 11.6.5.1 p.413
Paderaite 3.8.4.2 p.198
Padmaite 2.12.3.10 p.127
Pahasapaite 40.5.7.1 p.812
Painite 7.5.2.1 p.314
Palarstanide 2.3.3.2 p.41
Palenzonaite 38.2.1.3 p.705
Palermoite 41.7.1.1 p.853
Palladium 1.2.1.4 p.18
Palladoarsenide 2.4.14.1 p.51
Palladobismutharsenide 2.4.15.1 p.51
Palladseite 2.16.19.1 p.148
Palmierite 28.4.3.1 p.580
Palygorskite 74.3.1a.1 p.1558
Panasqueiraite 41.5.6.3 p.836
Panethite 38.1.6.1 p.704
Panunzite 76.2.1.4 p.1621
Paolovite 1.2.6.1 p.25
Papagoite 60.1.4.1 p.1236
Para-alumohydrocalcite 16b.2.3.2 p.502
Parabariomicrolite 8.2.8.1 p.351
Parabrandtite 40.2.2.8 p.753
Parabutlerite 31.9.2.1 p.652
Paracelsian 76.1.5.1 p.1615
Paracoquimbite 29.8.4.1 p.619
Paracostibite 2.12.7.2 p.132
Paradamite 41.6.7.2 p.851
Paradocrasite 1.3.3.1 p.29
Parafransoleite 40.5.2.1 p.809
Paragonite-$2M_1$, 3T, 1M 71.2.2a.2 p.1452
Paraguanajuatite 2.11.7.4 p.111

Parahopeite 40.3.3.1 p.789
Parajamesonite 3.6.9.1 p.185
Parakeldyshite 55.2.2.3 p.1140
Parakhinite 33.2.4.1 p.674
Paralaurionite 10.2.3.1 p.394
Paralstonite 14.2.2.2 p.453
Paramelaconite 4.6.4.1 p.254
Paramendozavilite 49.4.6.1 p.1010
Paramontroseite 4.4.11.1 p.249
Paranatisite 52.4.4.2 p.1096
Paranatrolite 77.1.5.3 p.1680
Paraniite-(Y) 49.4.7.1 p.1010
Paraotwayite 6.2.12.1 p.279
Parapierrotite 3.8.14.1 p.202
Pararammelsbergite 2.12.5.1 p.131
Pararealgar 2.8.22.2 p.85
Pararobertsite 42.8.5.2 p.912
Paraschachnerite 1.1.8.3 p.8
Paraschoepite 5.2.1.2 p.257
Parascholzite 40.2.5.1 p.756
Paraspurrite 53.1.2.1 p.1106
Parasymplesite 40.3.6.6 p.796
Paratacamite 10.1.2.1 p.392
Paratellurite 4.4.3.2 p.244
Paraumbite 59.2.1.2 p.1225
Paravauxite 42.11.14.2 p.943
Pargasite 66.1.3a.12 p.1354
Parisite-(Ce) 16a.1.5.1 p.480
Parisite-(Nd) 16a.1.5.2 p.481
Parkerite 2.3.4.1 p.41
Parkinsonite 10.5.11.1 p.402
Parnauite 43.5.13.1 p.969
Parsettensite 74.1.3.1 p.1555
Parsonsite 40.2a.31.1 p.784
Partheite 77.2.4.1 p.1700
Partzite 44.1.1.7 p.972
Parwelite 44.3.3.1 p.974
Pascoite 47.2.1.1 p.986
Patronite 2.16.21.1 p.148
Paulingite 77.1.3.9 p.1667
Paulkellerite 41.3.8.1 p.820
Paulkerrite 42.11.21.1 p.949
Paulmooreite 45.1.8.1 p.978
Pavonite 3.8.10.1 p.199
Paxite 2.12.5.2 p.131
Pearceite 3.1.8.1 p.153
Pecoraite 71.1.2d.4 p.1430
Pectolite-1A 65.2.1.4a p.1319
Pehrmanite 7.2.12.2 p.309
Peisleyite 43.5.15.1 p.969
Pekoite 3.4.5.6 p.170
Pellyite 78.6.3.1 p.1712
Penfieldite 10.4.1.1 p.398
Pengzhizhongite-24R 7.11.9.2 p.330
Pengzhizhongite-6H 7.11.9.1 p.330
Penikisite 41.9.1.2 p.870
Penkvilksite 70.2.2.1 p.1400
Pennantite 71.4.1.8 p.1505
Penroseite 2.12.1.4 p.115
Pentagonite 74.3.7.2 p.1567
Pentahydrite 29.6.7.3 p.606

Index of Mineral Names in Alphabetical Order

Pentahydroborite 26.2.1.1 p.552
Penzhinite 2.16.18.1 p.148
Peprossiite-(Ce) 24.3.4.1 p.540
Peretaite 31.6.4.1 p.647
Perhamite 43.5.10.1 p.968
Periclase 4.2.1.1 p.209
Perite 10.2.5.1 p.394
Perlialite 77.1.3.8 p.1667
Perloffite 41.9.1.4 p.871
Permingeatite 3.2.2.3 p.156
Perovskite 4.3.3.1 p.223
Perraultite 78.1.5.2 p.1704
Perrierite-(Ce) 56.2.8.3 p.1179
Perroudite 2.15.5.1 p.143
Perryite 1.1.22.1 p.16
Petalite 72.6.1.1 p.1545
Petarasite 61.1.2b.3 p.1249
Petedunnite 65.1.3a.5 p.1302
Peterbaylissite 16b.4.4.1 p.505
Petersenite-(Ce) 14.4.5.2 p.460
Petersite-(Y) 42.5.2.1 p.891
Petitjohnite 41.10.10.1 p.879
Petrovicite 3.1.9.1 p.154
Petrovskaite 2.4.17.1 p.51
Petrukite 2.9.18.1 p.97
Petscheckite 8.1.9.2 p.339
Pharmacolite 39.1.1.2 p.735
Pharmacosiderite 42.8.1.1 p.908
Phaunouxite 40.3.12.1 p.799
Phenakite 51.1.1.1 p.1021
Philipsbornite 42.7.4.2 p.902
Philipsburgite 42.2.4.2 p.884
Phillipsite 77.1.3.6 p.1665
Phlogopite-1M, 3T, $2M_1$ 71.2.2b.1 p.1456
Phoenicochroite 35.1.2.1 p.692
Phosgenite 16a.3.4.1 p.491
Phosphammite 37.1.3.1 p.698
Phosphoferrite 40.3.2.1 p.787
Phosphofibrite 42.13.10.1 p.955
Phosphophyllite 40.2.7.1 p.757
Phosphorrösslerite 39.1.9.2 p.739
Phosphosiderite 40.4.3.2 p.804
Phosphuranylite 42.4.8.1 p.887
Phuralumite 42.4.6.1 p.886
Phurcalite 42.4.7.1 p.887
Phyllotungstite 49.2.4.1 p.1006
Pickeringite 29.7.3.1 p.616
Picotpaulite 2.9.12.2 p.95
Picromerite 29.3.6.1 p.588
Picropharmacolite 39.2.4.1 p.741
Piemontite 58.2.1a.11 p.1202
Pierrotite 3.8.7.1 p.198
Pigeonite 65.1.1.4 p.1292
Pilsenite 2.6.2.5 p.58
Pimelite 71.3.1b.5 p.1492
Pinakiolite 24.2.4.1 p.538
Pinalite 10.2.9.1 p.396
Pinchite 10.5.3.1 p.399
Pingguite 34.5.4.1 p.685
Pinnoite 25.2.3.1 p.544
Pintadoite 47.4.3.1 p.991

Pirquitasite 2.9.2.7 p.91
Pirssonite 15.2.1.1 p.465
Pitiglianoite 76.2.5.11 p.1638
Pitticite 43.5.3.1 p.967
Piypite 30.2.7.1 p.636
Plagionite 3.6.20.2 p.188
Plancheite 66.2.1.1 p.1368
Planerite 42.9.3.6 p.920
Platarsite 2.12.3.6 p.126
Platinum 1.2.1.1 p.17
Plattnerite 4.4.1.6 p.242
Platynite 2.8.21.1 p.84
Playfairite 3.6.4.1 p.183
Plombierite 72.3.2.3 p.1531
Plumbobetafite 8.2.3.3 p.348
Plumboferrite 7.11.5.1 p.327
Plumbogummite 42.7.3.5 p.901
Plumbojarosite 30.2.5.6 p.635
Plumbomicrolite 8.2.2.3 p.346
Plumbonacrite 16a.5.1.1 p.494
Plumbopalladinite 1.2.11.1 p.26
Plumbopyrochlore 8.2.1.6 p.345
Plumbotellurite 34.1.4.1 p.678
Plumbotsumite 78.1.7.1 p.1704
Poitevinite 29.6.2.4 p.597
Pokrovskite 16a.3.2.3 p.490
Polarite 2.8.6.1 p.71
Poldervaartite 78.1.8.1 p.1705
Polhemusite 2.8.3.1 p.70
Polkovicite 2.9.5.2 p.93
Pollucite 77.1.1.2 p.1648
Polybasite 3.1.7.2 p.153
Polycrase-(Y) 8.3.8.1 p.359
Polydymite 2.10.1.7 p.100
Polyhalite 29.4.5.1 p.590
Polylithionite-1M 71.2.2b.8 p.1464
Polyphite 43.20.1 p.970
Polyphite 56.4.1.2 p.1185
Ponomarevite 10.6.12.1 p.406
Portlandite 6.2.1.4 p.276
Posnjakite 31.4.1.1 p.643
Potarite 1.2.4.4 p.22
Potassium alum 29.5.5.1 p.593
Potosiite 3.1.4.4 p.152
Pottsite 40.1.4.1 p.749
Poubaite 2.6.2.6 p.58
Poudretteite 63.2.1a.8 p.1275
Poughite 34.8.1.1 p.689
Povondraite 61.3.1.6 p.1260
Powellite 48.1.2.2 p.998
Poyarkovite 10.5.2.1 p.399
Prehnite 72.1.3.1 p.1522
Preisingerite 41.10.9.1 p.878
Preiswerkite-1Md, $2M_1$ 71.2.2b.15 p.1467
Preobrazhenskite 25.7.1.1 p.549
Priceite 25.5.1.1 p.546
Priderite 7.9.4.1 p.324
Pringleite 26.5.17.1 p.560
Probertite 26.5.12.1 p.558
Prosopite 11.6.9.1 p.414
Prosperite 40.2.4.1 p.756

Protasite 5.5.3.1 p.261
Proudite 3.6.1.1 p.183
Proustite 3.4.1.1 p.165
Przhevalskite 40.2a.21.1 p.778
Pseudoautunite 40.2a.1.3 p.767
Pseudoboleite 10.6.8.1 p.404
Pseudobrookite 7.7.1.1 p.316
Pseudocotunnite 11.2.1.1 p.408
Pseudograndreefite 12.1.3.1 p.419
Pseudolaueite 42.11.10.3 p.941
Pseudomalachite 41.4.3.1 p.822
Pseudorutile 8.4.2.1 p.363
Pucherite 38.4.6.1 p.725
Pumpellyite-(Fe^{2+}) 58.2.2.5 p.1209
Pumpellyite-(Fe^{3+}) 58.2.2.6 p.1209
Pumpellyite-(Mg) 58.2.2.7 p.1210
Pumpellyite-(Mn^{2+}) 58.2.2.8 p.1211
Purpurite 38.4.1.2 p.719
Putoranite 2.9.7.1 p.94
Pyrargyrite 3.4.1.2 p.166
Pyrite 2.12.1.1 p.114
Pyroaurite 16b.6.2.3 p.510
Pyrobelonite 41.5.2.3 p.830
Pyrochlore 8.2.1.1 p.343
Pyrochroite 6.2.1.3 p.276
Pyrolusite 4.4.1.4 p.238
Pyromorphite 41.8.4.1 p.865
Pyrope 51.4.3a.1 p.1038
Pyrophanite 4.3.5.3 p.228
Pyrophyllite-1A, 2M 71.2.1.1 p.1439
Pyrostilpnite 3.4.2.2 p.167
Pyroxferroite 65.6.1.2 p.1333
Pyroxmangite 65.6.1.1 p.1332
Pyrrhotite 2.8.10.1 p.74
Pääkkönenite 2.11.5.1 p.108
Qandilite 7.2.5.1 p.304
Qilianshanite 17.1.13.1 p.524
Qingheiite 38.2.4.4 p.712
Qitianlingite 8.1.12.1 p.340
Quadridavyne 76.2.5.12 p.1639
Quadruphite 43.19.1 p.970
Quadruphite 56.4.1.3 p.1186
Quartz 75.1.3.1 p.1573
Queitite 58.1.3.1 p.1194
Quenselite 6.4.1.2 p.288
Quenstedtite 29.8.5.1 p.619
Quetzalcoatlite 34.6.3.1 p.686
Quintinite-2H 16b.6.4.1 p.512
Quintinite-3T 16b.6.4.2 p.512
Rabbittite 16b.7.3.1 p.515
Rabejacite 31.6.8.1 p.648
Radhakrishnaite 2.15.4.1 p.143
Radtkeite 10.3.6.1 p.397
Raguinite 2.9.9.1 p.94
Raite 78.5.8.1 p.1711
Rajite 34.4.2.1 p.684
Ralstonite 11.6.12.1 p.416
Ramdohrite 3.4.15.6 p.177
Rameauite 5.5.2.1 p.260
Rammelsbergite 2.12.2.12 p.123
Ramsbeckite 31.4.9.1 p.645

Ramsdellite 4.4.7.1 p.247
Ranciéite 7.10.1.1 p.325
Rankachite 49.3.6.1 p.1008
Rankamaite 8.7.6.1 p.370
Rankinite 55.3.1.1 p.1140
Ransomite 29.7.1.1 p.614
Ranunculite 42.2.2.1 p.883
Rapidcreekite 32.2.1.1 p.667
Raspite 48.1.4.1 p.1001
Rasvumite 2.9.11.1 p.95
Rathite 3.7.12.1 p.195
Rauenthalite 40.3.11.1 p.799
Rauvite 47.4.4.1 p.991
Ravatite 50.3.2.1 p.1015
Rayite 3.6.20.5 p.190
Realgar 2.8.22.1 p.84
Rebulite 3.8.13.1 p.202
Rectorite 71.4.2.7 p.1515
Reddingite 40.3.2.3 p.788
Redingtonite 29.7.3.6 p.618
Redledgeite 7.9.5.2 p.325
Reedmergnerite 76.1.4.1 p.1614
Reevesite 16b.6.3.1 p.511
Refikite 50.4.1.1 p.1017
Reichenbachite 41.4.3.2 p.823
Reinerite 45.1.1.1 p.977
Reinhardbraunsite 52.3.2b.3 p.1081
Remondite-(Ce) 14.4.5.1 p.460
Renardite 42.4.9.2 p.888
Renierite 2.9.4.1 p.92
Reppiaite 41.4.8.1 p.824
Retgersite 29.6.9.1 p.609
Retzian-(Ce) 41.3.4.1 p.819
Retzian-(La) 41.3.4.3 p.819
Retzian-(Nd) 41.3.4.2 p.819
Revdite 74.3.6.1 p.1566
Reyerite 73.2.2a.1 p.1549
Rhabdophane-(Ce) 40.4.7.1 p.807
Rhabdophane-(La) 40.4.7.2 p.807
Rhodesite 72.5.1.1 p.1539
Rhodium 1.2.1.3 p.18
Rhodizite 25.8.2.1 p.549
Rhodochrosite 14.1.1.4 p.436
Rhodonite 65.4.1.1 p.1327
Rhodostannite 2.10.3.1 p.103
Rhodplumsite 2.3.5.2 p.41
Rhomboclase 29.1.1.1 p.583
Rhönite 69.2.1a.5 p.1388
Ribbeite 52.3.2b.4 p.1081
Richellite 42.13.2.1 p.953
Richelsdorfite 42.7.15.1 p.908
Richetite 5.9.5.1 p.264
Richterite 66.1.3b.9 p.1358
Rickardite 2.16.15.1 p.147
Rilandite 7.11.14.1 p.331
Ringwoodite 51.3.3.1 p.1034
Rinneite 11.5.3.1 p.411
Rittmannite 42.11.3.4 p.931
Rivadavite 26.6.1.1 p.560
Riversideite 72.3.2.4 p.1532
Roaldite 1.1.18.2 p.14

Index of Mineral Names in Alphabetical Order

Robertsite 42.8.4.2 p.911
Robinsonite 3.6.16.1 p.186
Rockbridgeite 41.9.2.1 p.871
Rodalquilarite 34.6.2.1 p.686
Roeblingite 64.2.2.1 p.1284
Roedderite 63.2.1a.14 p.1278
Roggianite 77.1.7.4 p.1696
Rohaite 3.7.16.1 p.196
Rokühnite 9.2.4.1 p.385
Romanechite 7.9.2.1 p.323
Romarchite 4.2.5.1 p.212
Roméite 44.1.1.3 p.971
Rooseveltite 38.4.4.1 p.725
Roquesite 2.9.1.4 p.88
Rorisite 9.2.11.2 p.388
Rosasite 16a.3.1.1 p.487
Roscherite-A 42.7.7.1 p.904
Roscoelite-1M 71.2.2a.4 p.1453
Roselite 40.2.3.1 p.754
Rosemaryite 38.2.4.3 p.711
Rosenbergite 9.3.6.1 p.390
Rosenbuschite 56.2.4.8 p.1165
Rosenhahnite 57.1.4.1 p.1189
Roshchinite 3.4.15.7 p.177
Rosickyite 1.3.5.2 p.31
Rosiérésite 42.13.7.1 p.954
Rossite 47.1.1.1 p.985
Rostite 31.9.11.2 p.656
Roubaultite 16b.7.13.1 p.518
Rouseite 45.1.2.2 p.977
Routhierite 3.4.11.1 p.173
Rouvilleite 16a.5.5.1 p.496
Roweite 25.4.1.1 p.546
Rowlandite-(Y) 55.2.1b.1 p.1137
Roxbyite 2.4.7.4 p.48
Rozenite 29.6.6.1 p.603
Ruarsite 2.12.4.4 p.130
Rucklidgeite 2.6.2.7 p.58
Ruitenbergite 26.5.17.2 p.560
Ruizite 57.2.2.1 p.1190
Rusakovite 42.3.1.1 p.884
Russellite 48.2.1.1 p.1001
Rustenbergite 1.2.5.2 p.24
Rustumite 58.2.5.1 p.1215
Ruthenarsenite 2.8.17.1 p.81
Ruthenium 1.2.2.2 p.19
Rutherfordine 14.1.4.1 p.447
Rutile 4.4.1.1 p.235
Rynersonite 8.3.7.2 p.359
Römerite 29.7.2.1 p.615
Röntgenite-(Ce) 16a.1.6.1 p.481
Rösslerite 39.1.9.1 p.738
Sabatierite 2.4.12.2 p.50
Sabieite 28.3.5.1 p.578
Sabinaite 16a.5.4.1 p.495
Sabugalite 40.2a.24.1 p.779
Sacrofanite 76.2.5.13 p.1639
Safflorite 2.12.2.11 p.123
Sahamalite-(Ce) 14.4.2.1 p.457
Sahlinite 41.1.4.1 p.816
Sainfeldite 39.2.1.2 p.740

Sakhaite 27.1.4.1 p.565
Sakharovaite 3.7.9.2 p.194
Sakuraiite 2.9.2.5 p.90
Sal Ammoniac 9.1.3.1 p.377
Saleeite 40.2a.11.1 p.772
Salesite 22.1.1.1 p.532
Saliotite 71.4.2.3 p.1510
Samarskite-(Y) 8.1.11.1 p.340
Samfowlerite 58.2.7.1 p.1217
Sampleite 42.9.4.1 p.920
Samsonite 3.4.12.1 p.173
Samuelsonite 41.10.8.1 p.878
Sanbornite 74.3.4.1 p.1563
Saneroite 69.2.2.1 p.1393
Sanidine 76.1.1.2 p.1602
Sanjuanite 43.5.1.3 p.966
Sanmartinite 48.1.1.3 p.996
Santaclaraite 65.4.2.1 p.1331
Santafeite 42.9.6.1 p.921
Santanaite 35.4.1.1 p.694
Santite 26.5.2.1 p.556
Saponite 71.3.1b.2 p.1489
Sapphirine 69.2.1b.1 p.1391
Sarabauite 2.13.2.1 p.140
Sarcolite 76.3.2.1 p.1644
Sarcopside 38.3.1.1 p.713
Sarkinite 41.6.3.3 p.848
Sarmientite 43.5.1.1 p.966
Sartorite 3.7.8.1 p.193
Saryarkite-Y 53.2.1.1 p.1107
Sasaite 43.5.4.1 p.967
Sassolite 24.3.1.1 p.539
Satimolite 26.7.4.1 p.563
Satpaevite 47.4.1.1 p.990
Satterlyite 41.6.4.1 p.848
Sauconite 71.3.1b.3 p.1490
Sayrite 5.9.6.1 p.264
Sazhinite-(Ce) 72.2.1.3 p.1525
Sazykinaite-(Y) 59.2.3.3 p.1230
Sborgite 26.5.1.1 p.556
Scacchite 9.2.3.2 p.384
Scarbroite 16b.7.8.1 p.516
Scawtite 64.2.1.1 p.1283
Schachnerite 1.1.8.2 p.8
Schafarzikite 45.1.6.1 p.978
Schairerite 30.1.9.1 p.625
Schallerite 72.4.1a.3 p.1537
Schaurteite 31.7.6.2 p.650
Scheelite 48.1.2.1 p.996
Schertelite 39.3.2.1 p.743
Scheteligite 8.2.6.1 p.350
Schieffelinite 33.3.1.1 p.675
Schirmerite 3.5.5.1 p.179
Schlossmacherite 30.2.4.3 p.632
Schmiederite 33.1.1.1 p.672
Schmitterite 34.1.6.1 p.678
Schneiderhöhnite 45.1.12.1 p.979
Schoderite 43.3.1.1 p.959
Schoenfliesite 6.3.6.2 p.285
Schoepite 5.2.1.3 p.257
Scholzite 40.2.6.1 p.756

Schoonerite 42.8.3.1 p.910
Schorlomite 51.4.3c.1 p.1050
Schorl 61.3.1.10 p.1263
Schreibersite 1.1.21.2 p.15
Schreyerite 8.4.1.1 p.362
Schröckingerite 17.1.5.1 p.521
Schubnelite 47.3.3.1 p.989
Schuetteite 30.1.13.1 p.626
Schuilingite-(Nd) 16b.1.4.1 p.499
Schulenbergite 31.1.6.1 p.641
Schultenite 37.1.2.1 p.698
Schumacherite 41.10.9.2 p.879
Schwartzembergite 22.1.3.1 p.532
Schwertmannite 6.4.10.1 p.290
Schöllhornite 2.9.17.2 p.97
Sclarite 16a.5.3.2 p.495
Scolecite 77.1.5.5 p.1682
Scorodite 40.4.1.3 p.802
Scorzalite 41.10.1.2 p.874
Scotlandite 34.1.1.2 p.677
Scrutinyite 4.4.6.2 p.247
Seamanite 43.4.5.1 p.964
Searlesite 71.3.5.4 p.1566
Sederholmite 2.8.11.3 p.77
Sedovite 48.4.1.1 p.1003
Seeligerite 22.1.2.1 p.532
Seelite 40.2a.12.1 p.772
Seelite 46.2.9.1 p.984
Segelerite 42.11.1.2 p.925
Segnitite 42.7.4.4 p.902
Seidozerite 56.2.6a.1 p.1171
Seinäjokite 2.12.2.10 p.123
Sekaninaite 61.2.1.2 p.1253
Selenium 1.3.4.1 p.30
Selenostephanite 3.2.4.2 p.157
Seligmanite 3.4.3.1 p.167
Sellaite 9.2.2.1 p.384
Selwynite 40.5.4.3 p.811
Semenovite-(Ce) 72.5.2.4 p.1542
Semseyite 3.6.20.4 p.189
Senaite 8.5.1.4 p.366
Senegalite 42.6.7.1 p.894
Sengierite 42.6.10.1 p.895
Sepiolite 74.3.1b.1 p.1560
Serandite 65.2.1.5 p.1320
Serendibite 69.2.1a.6 p.1389
Sergeevite 13.1.7.1 p.424
Serpierite 31.6.2.1 p.647
Shabaite-(Nd) 16b.1.5.1 p.499
Shabynite 26.1.3.1 p.551
Shadlunite 2.7.1.4 p.62
Shafranovskite 78.3.2.1 p.1706
Shakhovite 44.3.8.1 p.975
Shandite 2.3.5.1 p.41
Sharpite 16b.7.10.1 p.517
Shattuckite 65.1.7.1 p.1313
Shcherbakovite 65.3.4.2 p.1325
Shcherbinaite 4.6.1.1 p.253
Sherwoodite 47.2.4.1 p.987
Shigaite 31.1.2.1 p.639
Shomiokite-(Y) 15.4.7.1 p.474

Shortite 14.4.1.1 p.456
Shuangfengite 2.12.14.6 p.138
Shubnikovite 42.8.8.1 p.913
Shuiskite-(Mg) 58.2.2.9 p.1211
Sibirskite 25.2.2.1 p.544
Sicklerite 38.1.4.2 p.703
Siderazot 1.1.18.1 p.14
Siderite 14.1.1.3 p.435
Sideronatrite 31.8.3.1 p.651
Siderophyllite-1M 71.2.2b.5 p.1461
Siderotil 29.6.7.2 p.606
Sidorenkite 43.2.1.2 p.958
Sidwillite 4.5.6.1 p.253
Siegenite 2.10.1.6 p.100
Sieleckiite 42.13.12.1 p.955
Sigloite 42.11.14.3 p.943
Silhydrite 75.2.4.1 p.1594
Silinaite 74.3.5.3 p.1565
Sillimanite 52.2.2a.1 p.1064
Sillénite 4.3.12.1 p.233
Silver 1.1.1.2 p.3
Simferite 38.5.7.1 p.734
Simonellite 50.3.3.1 p.1015
Simonite 3.7.10.1 p.195
Simonkolleite 10.5.8.1 p.401
Simplotite 47.4.8.1 p.991
Simpsonite 8.7.5.1 p.370
Sincosite 42.11.19.1 p.948
Sinhalite 24.1.1.1 p.534
Sinjarite 9.2.5.1 p.385
Sinkankasite 42.7.1.4 p.898
Sinnerite 3.6.14.1 p.186
Sitinakite 52.4.11.1 p.1105
Sjögrenite 16b.6.1.3 p.509
Skinnerite 3.4.8.2 p.172
Skippenite 2.11.7.6 p.111
Sklodowskite 53.3.1.3 p.1111
Skutterudite 2.12.17.1 p.139
Slavikite 31.9.13.1 p.657
Slawsonite 76.1.5.2 p.1615
Smirnite 34.1.8.1 p.679
Smithite 3.7.3.1 p.191
Smithsonite 14.1.1.6 p.438
Smolianinovite 40.5.8.1 p.812
Smythite 2.8.10.2 p.75
Sobolevite 56.4.1.4 p.1186
Sobotkite 71.3.1b.1 p.1488
Sodalite 76.2.3.1 p.1625
Soddyite 53.3.3.1 p.1115
Sodium alum 29.5.5.2 p.593
Sodium Autunite 40.2a.5.1 p.769
Sodium Betpakdalite 49.4.4.1 p.1009
Sodium Dachiardite 77.1.6.5 p.1691
Sodium Pharmacosiderite 42.8.1.5 p.910
Sodium Phlogopite 71.2.2b.13 p.1467
Sodium Uranospinite 40.2a.6.1 p.770
Sodium Zippeite 31.10.4.2 p.661
Sodium-Boltwoodite 53.3.1.6 p.1113
Sodium-gedrite 66.1.2.8 p.1345
Sogdianite 63.2.1a.13 p.1278
Solongoite 25.3.2.1 p.545

Index of Mineral Names in Alphabetical Order

Sonolite 52.3.2d.3 p.1085
Sonoraite 34.7.1.1 p.687
Sopcheite 2.16.10.1 p.146
Sophiite 34.6.5.1 p.686
Sorbyite 3.6.12.1 p.185
Sorensenite 65.2.4.1 p.1322
Sosedkoite} 8.7.6.2 p.370
SoucÜekite 3.4.3.3 p.168
Souzalite 42.9.2.1 p.914
Spadaite 78.7.14.1 p.1718
Spangolite 31.1.5.1 p.640
Spencerite 42.6.4.2 p.893
Sperrylite 2.12.1.13 p.118
Spertiniite 6.2.4.1 p.277
Spessartine 51.4.3a.3 p.1041
Sphaerobismoite 4.3.13.1 p.233
Sphaerocobaltite 14.1.1.5 p.438
Sphalerite 2.8.2.1 p.68
Spheniscidite 42.11.6.3 p.933
Spinel 7.2.1.1 p.295
Spionkopite 2.4.7.7 p.49
Spiroffite 34.5.1.1 p.684
Spodiosite 41.6.11.1 p.852
Spodumene 65.1.4.1 p.1308
Spurrite 53.1.1.1 p.1106
Squawcreekite 4.4.1.8 p.243
Srebrodol'skite 7.11.2.2 p.327
Srilankite 4.4.13.2 p.250
Stalderite 3.4.11.2 p.173
Stanfieldite 38.3.5.1 p.717
Stanleyite 29.6.13.1 p.614
Stannite 2.9.2.1 p.89
Stannoidite 2.9.3.3 p.92
Stannomicrolite 8.2.2.6 p.347
Stannopalladinite 1.2.9.1 p.25
Starkeyite 29.6.6.2 p.603
Staurolite 52.2.3.1 p.1071
Steacyite 63.1.1.1 p.1269
Steenstrupine-(Ce) 61.4.2.1 p.1265
Steigerite 40.4.5.1 p.805
Stellerite 77.1.4.4 p.1675
Stenhuggarite 46.1.4.1 p.981
Stenonite 12.1.1.1 p.419
Stepanovite 50.1.7.1 p.1013
Stephanite 3.2.4.1 p.157
Stercorite 39.3.1.1 p.743
Sterlinghillite 40.3.5.3 p.790
Sternbergite 2.9.12.1 p.95
Sterryite 3.5.3.1 p.179
Stetefeldite 44.1.1.5 p.972
Stevensite 71.3.1b.6 p.1493
Stewartite 42.11.10.2 p.940
Stibarsen 1.3.1.3 p.28
Stibiconite 44.1.1.1 p.971
Stibiobetafite 8.2.3.4 p.348
Stibiocolumbite 8.1.6.1 p.334
Stibiocolusite 3.1.1.4 p.151
Stibiomicrolite 8.2.2.7 p.347
Stibiopalladinite 2.3.3.1 p.41
Stibiotantalite 8.1.6.2 p.335
Stibivanite-2O 45.1.11.2 p.979

Stibivanite-2M 45.1.11.1 p.979
Stichtite 16b.6.2.2 p.510
Stilbite-(Ca) 77.1.4.3 p.1674
Stilleite 2.8.2.2 p.69
Stillwaterite 2.16.4.1 p.144
Stillwellite-(Ce) 54.2.3.2 p.1129
Stilpnomelane 74.1.1.1 p.1552
Stishovite 4.4.1.9 p.243
Stistaite 1.3.1.5 p.29
Stoiberite 41.4.5.1 p.823
Stokesite 65.5.1.1 p.1331
Stolzite 48.1.3.2 p.1000
Stottite 6.3.7.1 p.286
Straczekite 47.3.2.1 p.988
Strakhovite 64.3.2.1 p.1286
Stranskiite 38.3.7.1 p.718
Strashimirite 42.6.5.1 p.893
Strelkinite 40.2a.29.1 p.783
Strengite 40.4.1.2 p.801
Stringhamite 78.1.9.1 p.1705
Stromeyerite 2.4.6.1 p.45
Stronalsite 76.1.6.2 p.1616
Strontianite 14.1.3.3 p.445
Strontio-orthojoaquinite 60.1.1b.2 p.1232
Strontioborite 25.6.3.1 p.549
Strontiochevkinite 56.2.8.2 p.1179
Strontiodresserite 16b.2.1.3 p.501
Strontioginorite 26.6.7.2 p.562
Strontiojoaquinite 60.1.1a.2 p.1232
Strontiopiemontite 58.2.1a.12 p.1203
Strontiopyrochlore 8.2.1.8 p.345
Strontiowhitlockite 38.3.4.2 p.716
Strontium-Apatite 41.8.1.7 p.863
Strunzite 42.11.9.1 p.938
Struvite 40.1.1.1 p.747
Strätlingite 77.2.5.1 p.1700
Strüverite 4.4.1.3 p.237
Studenitsite 26.7.8.1 p.564
Studtite 5.3.1.1 p.258
Sturmanite 32.4.4.2 p.670
Stützite 2.16.13.1 p.146
Suanite 24.4.1.1 p.540
Sudoite 71.4.1.3 p.1500
Suessite 1.1.23.1 p.16
Sugilite 63.2.1a.9 p.1275
Sulfoborite 27.1.5.1 p.566
Sulfur 1.3.5.1 p.30
Sulphohalite 30.1.7.1 p.624
Sulphotsumoite 2.8.20.2 p.83
Sulvanite 3.2.3.1 p.157
Sundiusite 30.1.1.1 p.623
Suolunite 56.2.4.14 p.1168
Surinamite 69.2.1b.2 p.1393
Surite 71.2.4.2 p.1479
Sursassite 58.2.3.1 p.1212
Susannite 17.1.3.1 p.520
Sussexite 25.2.1.1 p.544
Suzukiite 65.3.1.2 p.1323
Svabite 41.8.3.1 p.864
Svanbergite 43.4.1.6 p.962
Sveite 19.1.3.1 p.528

Sverigeite 51.2.2.1   p.1024
Svyatoslavite 76.1.6.4   p.1617
Svyazhinite 31.9.8.2   p.655
Swaknoite 39.3.4.1   p.744
Swamboite 53.3.1.8   p.1113
Swartzite 15.3.3.2   p.469
Swedenborgite 44.3.1.1   p.973
Sweetite 6.2.9.1   p.278
Swinefordite 71.3.1a.5   p.1487
Switzerite 40.3.5.4   p.791
Sylvanite 2.12.13.3   p.135
Sylvite 9.1.1.2   p.375
Symplesite 40.3.8.1   p.797
Synadelphite 43.5.12.1   p.968
Synchysite-(Ce) 16a.1.3.1   p.479
Synchysite-(Nd) 16a.1.3.3   p.480
Synchysite-(Y) 16a.1.3.2   p.479
Syngenite 29.3.1.1   p.584
Szaibelyite 25.2.1.2   p.544
Szenicsite 48.3.5.1   p.1003
Szmikite 29.6.2.3   p.596
Szomolnokite 29.6.2.2   p.596
Szymanskiite 16b.7.15.1   p.518
Sénarmontite 4.3.9.2   p.231
Söhngeite 6.3.5.2   p.283
Taaffeite 7.2.11.1   p.309
Tacharanite 72.3.2.6   p.1533
Tachyhydrite 11.5.5.1   p.411
Tadzhikite-(Y) 54.2.2.2   p.1128
Taeniolite-1M,2M$_1$ 71.2.2b.9   p.1464
Taenite 1.1.11.2   p.10
Taikanite 65.3.3.1   p.1324
Taimyrite 1.2.7.1   p.25
Takanelite 7.10.1.2   p.325
Takedaite 24.3.5.1   p.540
Takovite 16b.6.3.2   p.511
Takéuchiite 24.2.6.1   p.538
Talc-1A 71.2.1.3   p.1441
Talmessite 40.2.2.5   p.753
Talnakhite 2.9.8.1   p.94
Tamarugite 29.5.3.1   p.592
Tancoite 41.4.7.1   p.824
Taneyamalite 69.2.3.2   p.1394
Tangeite 41.5.1.6   p.828
Tantalaeschynite-(Y) 8.3.6.4   p.
Tanteuxenite-(Y) 8.3.8.3   p.360
Tantite 4.6.6.1   p.254
Taramellite 64.3.1.1   p.1284
Taranakite 39.3.6.1   p.744
Tarapacaite 35.1.1.1   p.692
Tarbuttite 41.6.7.1   p.851
Tatarskite 32.4.2.1   p.669
Tausonite 4.3.3.5   p.224
Tavorite 41.5.9.1   p.838
Tazheranite 4.4.13.1   p.250
Teallite 2.9.10.1   p.95
Teepleite 25.1.4.1   p.543
Teineite 34.2.2.2   p.680
Telargpalite 2.1.3.1   p.37
Tellurantimony 2.11.7.3   p.111
Tellurite 4.4.6.1   p.247

Tellurium 1.3.4.2   p.30
Tellurobismuthite 2.11.7.2   p.110
Tellurohauchecornite 3.2.5.3   p.159
Telluropalladinite 2.16.6.1   p.145
Temagamite 2.6.5.1   p.60
Tengchongite 49.3.1.2   p.1007
Tengerite-(Y) 15.4.3.1   p.473
Tennantite 3.3.6.2   p.164
Tenorite 4.2.3.1   p.211
Tephroite 51.3.1.4   p.1029
Terlinguaite 10.4.2.1   p.398
Terskite 78.7.15.1   p.1718
Tertschite 26.7.5.1   p.563
Teruggite 27.1.6.1   p.566
Teschemacherite 13.1.3.1   p.423
Testibiopalladite 2.12.3.13   p.128
Tetraauricupride 1.1.2.2   p.6
Tetradymite 2.11.7.1   p.110
Tetraferroplatinum 1.2.4.1   p.21
Tetrahedrite 3.3.6.1   p.162
Tetranatrolite 77.1.5.2   p.1679
Tetrarooseveltite 38.4.5.1   p.725
Tetrataenite 1.1.11.3   p.10
Tetrawickmanite 6.3.7.2   p.286
Thadeuite 41.5.16.1   p.843
Thalcusite 2.5.5.1   p.54
Thalfenisite 2.15.2.2   p.143
Thalénite-(Y) 55.2.1b.2   p.1137
Thaumasite 32.4.4.4   p.671
Theisite 41.1.2.1   p.816
Thenardite 28.2.3.1   p.570
Theophrastite 6.2.1.5   p.276
Theresemagnanite 31.2.5.1   p.642
Thermonatrite 15.1.1.1   p.462
Thometzekite 40.2.9.3   p.761
Thomsenolite 11.6.6.1   p.414
Thomsonite 77.1.5.9   p.1687
Thorbastnäsite 16b.1.2.1   p.499
Thoreaulite 8.3.10.1   p.361
Thorianite 5.1.1.2   p.256
Thorikosite 10.2.6.1   p.395
Thorite 51.5.2.3   p.1057
Thornasite 72.1.2.1   p.1521
Thorogummite 51.5.2.5   p.1058
Thorosteenstrupine 61.4.2.2   p.1265
Thortveitite 55.2.1a.4   p.1136
Thorutite 8.3.4.2   p.357
Threadgoldite 42.11.13.3   p.942
Tiemannite 2.8.2.4   p.69
Tienshanite 64.4.1.1   p.1286
Tiettaite 78.6.4.1   p.1713
Tikhonenkovite 11.6.16.1   p.417
Tilasite 41.5.6.1   p.835
Tilleyite 56.2.9.1   p.1180
Tinaksite 67.2.1.1   p.1377
Tincalconite 26.4.2.1   p.555
Tinsleyite 42.11.6.2   p.932
Tinticite 42.12.4.1   p.952
Tintinaite 3.6.19.2   p.187
Tinzenite 56.2.2.4   p.1156
Tin 1.1.13.1   p.13

Index of Mineral Names in Alphabetical Order

Tiptopite   42.6.11.1   p.896
Tiragalloite   57.2.1.1   p.1190
Tirodite   66.1.1.4   p.1342
Tisinalite   61.1.2a.3   p.1248
Titanite   52.4.3.1   p.1092
Titanowodginite   8.1.8.3   p.338
Titantaramellite   64.3.1.2   p.1285
Tivanite   6.2.8.1   p.278
Tlalocite   33.3.2.1   p.675
Tlapallite   34.8.2.1   p.689
Tobelite-1M   71.2.2a.7   p.1455
Tobermorite   72.3.2.1   p.1529
Tochilinite   2.14.2.1   p.141
Tocornalite   9.1.6.1   p.378
Todorokite   7.8.1.1   p.317
Tokkoite   67.2.1.2   p.1377
Tolbachite   9.2.8.2   p.387
Tolovkite   2.12.3.5   p.126
Tombarthite-(Y)   51.5.3.2   p.1060
Tomichite   46.2.3.1   p.982
Tongbaite   1.1.17.1   p.14
Tooeleite   42.12.5.1   p.952
Topaz   52.3.1.1   p.1073
Torbernite   40.2a.13.1   p.773
Torreyite   31.1.4.1   p.640
Tosudite   71.4.2.4   p.1510
Tounkite   76.2.5.14   p.1639
Toyohaite   2.10.3.2   p.103
Trabzonite   57.1.3.1   p.1189
Tranquillityite   78.7.16.1   p.1719
Traskite   64.5.1.1   p.1286
Treasurite   3.6..3.1   p.183
Trechmannite   3.7.2.1   p.191
Trembathite   25.6.2.2   p.549
Tremolite   66.1.3a.1   p.1346
Trevorite   7.2.2.5   p.300
Triangulite   40.2a.34.1   p.785
Tridymite   75.1.2.1   p.1570
Trigonite   45.1.2.1   p.977
Trikalsilite   76.2.1.3   p.1621
Trimerite   51.2.1.1   p.1023
Trimounsite-(Y)   52.4.8.1   p.1099
Triphylite   38.1.1.1   p.700
Triplite   41.6.1.2   p.845
Triploidite   41.6.3.2   p.847
Trippkeite   45.1.5.1   p.978
Tripuhyite   44.2.1.3   p.973
Tristramite   40.4.7.6   p.808
Tritomite-(Ce)   54.2.5.1   p.1130
Tritomite-(Y)   54.2.5.2   p.1131
Trogtalite   2.12.1.5   p.116
Troilite   2.8.9.1   p.74
Trolleite   41.11.1.1   p.880
Trona   13.1.4.1   p.423
Truscottite   73.2.2a.2   p.1549
Trögerite   40.2a.20.1   p.777
Trüstedtite   2.10.1.9   p.101
Tsargorodtsevite   76.2.3.8   p.1629
Tschermigite   29.5.5.3   p.594
Tschernichite   77.1.8.2   p.1697
Tsnigriite   3.1.12.1   p.155

Tsumcorite   40.2.9.1   p.760
Tsumebite   43.4.2.1   p.963
Tsumoite   2.8.20.1   p.82
Tučekite   3.2.5.5   p.159
Tugarinovite   4.4.15.1   p.251
Tugtupite   76.2.3.7   p.1629
Tuhualite   66.3.4.1   p.1374
Tulameenite   1.2.4.2   p.22
Tuliokite   15.4.6.1   p.474
Tundrite-(Ce)   17.1.11.1   p.523
Tundrite-(Nd)   17.1.11.2   p.523
Tunellite   26.6.6.2   p.562
Tungstenite   2.12.10.3   p.134
Tungstibite   48.2.3.1   p.1002
Tungstite   4.5.2.1   p.252
Tungusite   78.5.9.1   p.1711
Tunisite   16a.5.2.1   p.494
Tuperssuatsiaite   74.3.1a.2   p.1559
Turanite   41.4.4.1   p.823
Turneaureite   41.8.3.2   p.864
Turquoise   42.9.3.1   p.917
Tuscanite   73.1.2.2   p.1547
Tusionite   24.3.3.2   p.540
Tuzlaite   26.5.12.2   p.559
Tvalchrelidzeite   3.8.3.1   p.197
Tvedalite   70.5.2.1   p.1403
Tveitite-(Y)   9.2.1.3   p.384
Twinnite   3.7.8.3   p.193
Tychite   17.1.1.1   p.519
Tyretskite   26.5.15.1   p.560
Tyrolite   42.4.3.1   p.885
Tyrrellite   2.10.1.4   p.100
Tyuyamunite   40.2a.26.1   p.781
Törnebohmite-(Ce)   52.4.5.1   p.1096
Törnebohmite-(La)   52.4.5.2   p.1096
Uchucchacuaite   3.4.15.5   p.177
Uhligite   8.7.2.1   p.369
Uklonskovite   31.7.2.1   p.649
Ulexite   26.5.11.1   p.558
Ullmannite   2.12.3.3   p.125
Ulrichite   40.2a.33.1   p.785
Ulvöspinel   7.2.5.2   p.304
Umangite   2.5.1.1   p.52
Umbite   59.2.1.1   p.1224
Umbozerite   78.7.17.1   p.1719
Umohoite   49.2.2.1   p.1006
Ungarettiite   66.1.3c.14   p.1367
Ungemachite   32.2.3.1   p.667
Upalite   42.7.12.1   p.907
Uralborite   25.3.4.1   p.545
Uralolite   42.7.6.1   p.903
Uramphite   40.2a.7.1   p.770
Urancalcarite   16b.5.2.1   p.506
Uraninite   5.1.1.1   p.255
Uranmicrolite   8.2.2.4   p.346
Uranocircite   40.2a.3.1   p.768
Uranophane   53.3.1.2   p.1111
Uranopilite   31.2.6.1   p.642
Uranopolycrase   8.3.8.5   p.361
Uranosilite   53.3.3.2   p.1115
Uranospathite   40.2a.22.1   p.778

Uranosphaerite 5.9.1.1 p.263
Uranospinite 40.2a.2.1 p.767
Uranotungstite 49.3.5.1 p.1008
Uranpyrochlore 8.2.1.7 p.345
Urea 50.4.6.1 p.1018
Uricite 50.4.4.1 p.1017
Urvantsevite 2.12.15.2 p.139
Ushkovite 42.11.10.4 p.941
Usovite 11.6.14.1 p.416
Ussingite 76.3.3.1 p.1644
Ustarasite 3.8.11.1 p.201
Uvanite 47.4.5.1 p.991
Uvarovite 51.4.3b.3 p.1048
Uvite 61.3.1.3 p.1258
Uzonite 2.8.22.3 p.85
Vaesite 2.12.1.2 p.115
Valentinite 4.3.11.1 p.233
Valleriite 2.14.1.1 p.141
Vanadinite 41.8.4.3 p.867
Vanadomalayaite 52.4.3.3 p.1095
Vanalite 47.4.2.1 p.990
Vandenbrandeite 5.3.2.1 p.259
Vandendriesscheite 5.8.1.1 p.262
Vanmeersscheite 42.7.14.1 p.907
Vanoxite 47.4.6.1 p.991
Vantasselite 42.12.1.1 p.951
Vanthoffite 28.4.1.1 p.578
Vanuralite 42.11.13.1 p.942
Vanuranylite 40.2a.30.1 p.784
Varennesite 72.5.5.1 p.1544
Variscite 40.4.1.1 p.800
Varlamoffite 4.4.2.1 p.244
Varulite 38.2.3.3 p.709
Vashegyite 42.12.2.1 p.951
Vasilite 2.16.22.1 p.149
Vaterite 14.1.2.1 p.440
Vaughanite 3.8.15.1 p.202
Vauquelinite 43.4.3.1 p.963
Vauxite 42.11.14.1 p.943
Veatchite-A 26.5.10.1 p.558
Veatchite 26.5.9.1 p.558
Veenite 3.5.9.2 p.181
Velikite 2.9.2.8 p.91
Vermiculite 71.2.2d.3 p.1474
Vernadite 4.4.9.1 p.248
Verplanckite 77.2.8.1 p.1701
Versiliaite 45.1.10.1 p.979
Vertumnite 73.1.6.1 p.1548
Vesignieite 41.5.13.1 p.842
Vesuvianite 58.2.4.1 p.1213
Veszelyite 42.2.3.1 p.883
Vicanite-(Ce) 54.3.1.1. p.1132
Vigezzite 8.3.7.1 p.359
Viitaniemiite 41.3.5.1 p.820
Vikingite 3.6.11.1 p.185
Villamaninite 2.12.1.6 p.116
Villiaumite 9.1.1.3 p.375
Villyaellenite 39.2.1.3 p.740
Vimsite 25.1.5.1 p.543
Vincentite 2.2.5.1 p.40
Vinciennite 3.1.2.1 p.151

Vinogradovite 68.1.2.1 p.1379
Violarite 2.10.1.8 p.101
Virgilite 75.3.1.1 p.1594
Vishnevite 76.2.5.15 p.1640
Vismirnovite 6.3.6.4 p.285
Vistepite 78.1.10.1 p.1705
Viséite 43.5.9.1 p.968
Vitusite-Ce 38.1.8.1 p.704
Vivianite 40.3.6.1 p.793
Vladimirite 39.2.2.1 p.740
Vlasovite 66.2.2.1 p.1369
Vochtenite 42.11.23.1 p.950
Voggite 43.5.16.1 p.969
Voglite 15.3.5.1 p.471
Volborthite 40.3.10.1 p.798
Volkonskoite 71.3.1a.4 p.1486
Volkovskite 26.5.16.1 p.560
Voltaite 29.9.1.1 p.620
Volynskite 3.7.1.3 p.190
Vonbezingite 31.5.1.1 p.646
Vonsenite 24.2.1.2 p.536
Vozhminite 2.3.6.1 p.42
Vrbaite 3.7.15.1 p.196
Vuagnatite 52.4.2.2 p.1092
Vulcanite 2.8.13.1 p.80
Vuonnemite 56.4.3.1 p.1187
Vuorelainenite 7.2.4.1 p.303
Vyacheslavite 40.4.8.3 p.809
Vyalsovite 2.14.4.1 p.142
Vysotskite 2.8.5.2 p.71
Vyuntspakhkite-(Y) 78.7.18.1 p.1719
Väyrynenite 41.5.4.3 p.833
Wadalite 51.4.5.1 p.1053
Wadeite 59.1.1.4 p.1223
Wadsleyite 51.3.4.1 p.1035
Wagnerite 41.6.2.1 p.846
Wairakite 77.1.1.3 p.1649
Wairauite 1.1.11.6 p.12
Wakabayashilite 2.11.4.1 p.108
Wakefieldite-(Ce) 38.4.11.4 p.731
Wakefieldite-(Y) 38.4.11.3 p.731
Walentaite 40.5.10.1 p.813
Wallisite 3.5.8.1 p.180
Wallkilldellite 42.4.12.1 p.889
Walpurgite 40.5.9.1 p.813
Walstromite 59.1.2.1 p.1224
Walthierite 30.2.4.7 p.633
Wardite 42.7.8.2 p.905
Wardsmithite 26.7.7.1 p.564
Warikahnite 40.3.1.1 p.785
Warwickite 24.2.2.1 p.537
Watanabeite 3.5.15.1 p.182
Watkinsonite 3.7.18.1 p.197
Wattersite 35.4.2.1 p.694
Wattevillite 29.3.4.1 p.586
Wavellite 42.10.2.1 p.922
Wawayandaite 78.7.19.1 p.1720
Waylandite 41.5.12.1 p.841
Weberite 11.6.13.1 p.416
Weddellite 50.1.2.1 p.1011
Weeksite 53.3.2.1 p.1114

Index of Mineral Names in Alphabetical Order

Wegscheiderite 13.1.6.1   p.424
Weibullite 3.6.18.1   p.187
Weilerite 43.4.1.9   p.962
Weilite 37.1.1.2   p.697
Weinebenite 42.7.6.2   p.903
Weishanite 1.1.8.6   p.8
Weissbergite 3.7.7.1   p.193
Weissite 2.4.8.1   p.49
Welinite 52.4.1.2   p.1090
Welinite 7.3.1.1   p.310
Weloganite 15.3.4.4   p.470
Welshite 69.2.1a.7   p.1390
Wendwilsonite 40.2.3.4   p.755
Wenkite 76.2.5.16   p.1640
Werdingite 56.3.2.1   p.1184
Wermlandite 31.3.2.1   p.643
Wheatleyite 50.1.8.1   p.1013
Wherryite 32.3.3.1   p.668
Whewellite 50.1.1.1   p.1011
Whiteite-(CaFeMg) 42.11.3.1   p.930
Whitlockite 38.3.4.1   p.715
Whitmoreite 42.11.18.1   p.947
Wickenburgite 74.2.1.1   p.1556
Wickmanite 6.3.6.1   p.284
Wicksite 40.2.10.1   p.761
Widenmannite 14.1.5.1   p.447
Widgiemoolthaite 16b.7.1.2   p.514
Wightmanite 26.1.2.1   p.551
Wilcoxite 31.9.9.1   p.655
Wilhelmvierlingite 42.11.1.5   p.926
Wilkinsonite 69.2.1a.8   p.1391
Wilkmanite 2.10.2.1   p.103
Willemite 51.1.1.2   p.1022
Willemseite 71.2.1.4   p.1443
Willhendersonite 77.1.2.3   p.1655
Willyamite 2.12.3.4   p.126
Winstanleyite 34.5.2.1   p.685
Wiserite 27.1.10.1   p.567
Witherite 14.1.3.2   p.444
Wittichenite 3.4.8.1   p.172
Wittite 3.6.5.2   p.184
Wodginite 8.1.8.1   p.337
Wolfeite 41.6.3.1   p.847
Wollastonite-1A 65.2.1.1a   p.1315
Wollastonite-2M 65.2.1.1b   p.1317
Wollastonite-3A,4A,5A,7A 65.2.1.1c   p.1317
Wonesite-1Md 71.2.2b.14   p.1467
Woodhouseite 43.4.1.8   p.962
Woodruffite 7.8.1.2   p.319
Woodwardite 31.2.2.1   p.641
Wroewolfeite 31.4.2.1   p.644
Wulfenite 48.1.3.1   p.999
Wupatkiite 29.7.3.7   p.618
Wurtzite 2.8.7.1   p.72
Wyartite 16b.7.4.1   p.515
Wycheproofite 40.5.5.1   p.811
Wyllieite 38.2.4.2   p.711
Wöhlerite 56.2.4.5   p.1163
Wölsendorfite 5.4.3.2   p.260
Wülfingite 6.2.10.1   p.279
Wüstite 4.2.1.6   p.210

Xanthiosite 38.3.2.1   p.714
Xanthoconite 3.4.2.1   p.166
Xanthoxenite 42.11.15.1   p.944
Xenotime-(Y) 38.4.9.1   p.729
Xiangjiangite 43.5.6.1   p.967
Xifengite 1.1.23.3   p.16
Xilingolite 3.4.16.1   p.178
Ximengite 38.4.9.1   p.727
Xingzhongite 2.8.4.1   p.70
Xitieshanite 31.9.6.3   p.654
Xocomecatlite 33.1.2.1   p.672
Xonotlite 66.3.1.1   p.1369
Yafsoanite 7.2.14.1   p.310
Yagiite 63.2.1a.10   p.1276
Yakhontovite 71.3.1b.7   p.1495
Yanomamite 40.4.1.5   p.803
Yaroslavite 11.6.15.1   p.416
Yarrowite 2.7.3.1   p.63
Yavapaiite 28.3.4.1   p.578
Ye'elimite 30.1.15.1   p.627
Yeatmanite 44.3.6.1   p.974
Yecoraite 34.8.6.1   p.691
Yedlinite 10.6.3.1   p.403
Yftisite-(Y) 52.4.4.3   p.1096
Yimengite 7.4.1.2   p.312
Yingjiangite 42.6.12.1   p.896
Yoderite 52.2.2b.3   p.1068
Yofortierite 74.3.1a.3   p.1560
Yoshimuraite 58.2.6.1   p.1216
Yoshiokaite 76.2.1.6   p.1622
Yttrialite-(Y) 55.2.1b.3   p.1138
Yttrobetafite-(Y) 8.2.3.2   p.348
Yttrocolumbite-(Y) 8.1.3.2   p.334
Yttrocrasite-(Y) 8.3.8.4   p.361
Yttropyrochlore-(Y) 8.2.1.4   p.344
Yttrotantalite-(Y) 8.1.3.1   p.334
Yttrotungstite-(Y) 49.3.3.2   p.1007
Yuanfulite 24.2.2.2   p.537
Yuanjiangite 1.1.2.3   p.6
Yugawaralite 77.1.7.2   p.1694
Yukonite 42.8.6.1   p.913
Yuksporite 66.3.1.5   p.1371
Yushkinite 2.14.3.2   p.141
Zaherite 31.10.3.1   p.659
Zairite 41.5.12.3   p.841
Zajacite-(Ce) 11.5.7.1   p.412
Zakharovite 78.3.3.1   p.1706
Zanazziite 42.7.7.3   p.904
Zapatalite 42.5.3.1   p.891
Zaratite 16b.4.1.1   p.504
Zavaritskite 10.2.1.1   p.393
Zdenekite 42.9.4.3   p.921
Zektzerite 66.3.4.2   p.1374
Zellerite 15.3.1.1   p.467
Zemannite 34.3.2.1   p.682
Zemkorite 14.3.3.2   p.456
Zenzénite 7.5.4.1   p.315
Zeophyllite 73.1.4.1   p.1547
Zeunerite 40.2a.14.1   p.774
Zhanghengite 1.1.6.2   p.7
Zharchikhite 10.2.8.1   p.395

Zhemchuzhnikovite 50.1.7.2 p.1013
Zhonghuacerite-(Ce) 16a.1.8.1 p.482
Ziesite 38.5.5.2 p.733
Zimbabweite 8.7.10.1 p.372
Zinc Zippeite 31.10.4.5 p.661
Zinc-melanterite 29.6.10.3 p.612
Zincaluminite 31.2.3.1 p.641
Zincite 4.2.2.1 p.211
Zincobotryogen 31.9.6.2 p.653
Zincochromite 7.2.3.6 p.303
Zincocopiapite 31.10.5.6 p.664
Zincovoltaite 29.9.1.2 p.621
Zincrosasite 16a.3.1.4 p.488
Zincroselite 40.2.3.3 p.755
Zincsilite 71.3.1b.8 p.1495
Zinc 1.1.5.1 p.6
Zinkenite 3.8.1.1 p.197
Zinnwaldite-1M,2M$_1$,3T 71.2.2b.10 p.1465
Zippeite 31.10.4.1 p.660
Zirconolite-2M 8.2.5.3 p.350
Zirconolite-3O 8.2.5.2 p.349

Zirconolite-3T 8.2.5.4 p.350
Zirconolite 8.2.5.5 p.350
Zircon 51.5.2.1 p.1054
Zircophyllite 69.1.1.5 p.1384
Zircosulfate 29.9.2.1 p.621
Zirkelite 8.2.5.1 p.349
Zirklerite 10.6.5.1 p.404
Zirsinalite 61.1.2a.4 p.1248
Znucalite 16b.7.14.1 p.518
Zodacite 42.11.8.4 p.936
Zoisite 58.2.1b.1 p.1203
Zorite 66.3.1.2 p.1370
Zoubekite 3.5.14.1 p.182
Zunyite 57.3.1.1 p.1191
Zussmanite 74.1.4.1 p.1555
Zvyagintsevite 1.2.5.4 p.24
Zweiselite 41.6.1.1 p.845
Zykaite 43.5.3.2 p.967
Åkermanite 55.4.1.1 p.1143
Örebroite 52.4.1.1 p.1090

General Index

A

Abelsonite (nickel porphyrin), *1018-1019*
Abenakiite-(Ce), *1264-1265*
Abernathyite, 765, *771*
Abhurite, *401*
Abswurmbachite, *314*
Abukumalite (**britholite-(Y)**), *1101-1102*
Acanthite, *42*, 43
Acetamide (ethanamide), *1018*
Acmite (**aegirine**), 1291, 1303, *1305-1306*, 1401
Actinolite, 1338, *1345*
Acuminite, *417*
Adamite, *850*
Adelite, 825, *826*
Adelite group, 825, 826-831, 1091, 1092. See also Descloizite subgroup
Admontite, *561*
Adularia, 1600, 1604
Aegirine (acmite), 1291, 1303, *1305-1306*, 1401
Aegirine-augite, 1291, *1303-1304*
Aenigmatite, 1385, *1386-1387*
Aenigmatite group, 1384-1391, 1393
Aërinite, *1380*
Aerugite, *734*
Aeschynite-(Ce), *357-358*
Aeschynite-(Nd), *359*
Aeschynite-(Y), *358*
Afghanite, 1632, *1634*
Afwillite, *1098*
Agalmatolite, 1439
Agardite-(Ce), 890
Agardite-(La), 890
Agardite-(Y), *890*, 891
Agrellite, *1398*
Agrinierite, *260*
Aguilarite, *42-43*
Aheylite, 917, *919-920*
Ahlfeldite, *681*
Aikinite, *169*
Aikinite group, 168-171
Ajoite, *1708*
Akaganéite, 268, 270, *273-274*
Akatoreite, *1191*
Akdalaite, *221*
Åkermanite, 1142, *1143*, 1144
Akhtenskite, 238, 247, *249*

Akrochordite, *884-885*
Aksaite, *561*
Aktashite, *174*
Alabandite, 64, *66*, 440
Alacranite, *85-86*
Alamosite, *1333-1334*
Albite, 1022, 1597, 1599, *1608*
Albrechtschraufite, *517*
Aldermanite, *952*
Aleksite, *59*
Alforsite, 857, *868*
Algodonite, 4, *37*, 39
Alietite, *1518*
Alkali feldspar subgroup, 1597-1608. See also Feldspar group
Allactite, *817*
Allanate, 1200
Allanite-(Ce) (orthite), *1196-1197*
Allanite-(La), 1196, *1197*
Allanite-(Y), 1196, *1198*
Allargentum, *38*
Alleghanyite, *1079-1080*, 1081
Alloclasite, *131-132*
Allophane group, 1432-1437
Allophane (protoallophane), *1432-1434*
Alluaivite, *1282-1283*
Alluaudite group, 706-712. See also Wyllieite subgroup
Alluaudite (hühnerkobelite), 706, 707, *709-710*
Almandine, 1037, *1039-1041*
Alstonite, 444, 453, *454*
Altaite, 64, *65-66*
Althausite, *849*
Althupite, *889*
Altisite, *1563*
Aluminilite (**alunite**), 630, *631-632*, 899
Aluminite, *649*
(Alumino)barroisite, 1338, *1358*
Aluminocopiaite, 663, *664-665*
(Alumino)ferrohornblende, 1338, *1350-1352*
Aluminohydrocalcite, 506
(Alumino)katophorite, 1338, *1359*
(Alumino)magnesiohornblende, 1338, *1350*
(Alumino)taramite, 1338, *1360*
(Alumino)tschermakite, 1338, *1352*
(Alumino)winchite, 1338, *1357*
Aluminum, 2, *5*

1783

Alumohydrocalcite, 501, *502*
Alumoklyuchevskite, *582*
Alumopharmacosiderite, 908, *909*
Alumotantite, *336*
Alumotungstite, *252*
Alunite (aluminilite), 630, *631-632*, 899
Alunite group, 629-636
Alunogen, *620*
Alushtite (tosudite), *1510-1512*
Alvanite, *956*
Amakinite, *275-276*
Amarantite, *652*
Amarillite, *592*
Amblygonite, *836-837*
Ameghinite, *545*
Amesite, 221, 1406, *1422-1423*, 1426
Amianthus (parachrysotile), 1406, *1428-1430*
Amicite, *1662*, 1663
Aminoffite, *1188*
Ammonia alum (tschermigite), *593*
Ammonian arcanite (arcanite), *569*
Ammonioalunite, 631, *633*
Ammonioborite, *556*
Ammoniojarosite, 631, *635*
Ammonioleucite, 1622, *1623*, 1648
Amosite (grunerite), 1337, *1339-1342*
Amphibole group, 1335-1367. See also Anthophyllite subgroup; Cummingtonite subgroup; Glaucophane subgroup; Richterite subgroup; Tremolite subgroup
Amstallite, *1523*
Analcime, 224, *1646-1648*, 1649, 1651
Anandite, 1447, *1470*
Anapaite, *749*
Anatase (octahedrite), 235, *245-246*
Anauxite, 1407
Ancylite-(Ce), 496, *497*
Ancylite group, 496-499
Andalusite, 1064, *1066-1068*, 1069, 1116
Andalusite group, 1063-1073
Andersonite, *466*
Andesine, 1597, *1608*
Andorite, 175, *176-177*
Andradite, 221, 1037, 1039, 1043, *1044-1046*, 1049, 1050
Andremeyerite, *1176*
Anduoite, 120, *122*
Angelellite, *852*
Anglesite, 572, *575-576*
Anhydrite, *576-577*
Anilite, 46, *48*
Ankangite, *324*
Ankerite, 449, *451*
Annabergite, 792, *795*
Annite, 1447, 1458, *1461*
Anorthite, 1597, *1608-1614*, 1617
Anorthoclase, 1597, 1600, *1607-1608*
Anorthosite, 227
Antarcticite, *385-386*
Anthoinite, *1002-1003*
Anthonyite, *278*

Anthophyllite, 1337, *1343*
Anthophyllite subgroup, 1337, 1343-1346. See also Amphibole group
Anthraquinone (hoelite), *1017*
Antigorite (williamsite, bowenite, jenkinsite, picrosmine, picrolite), 1406, *1415-1417*
Antimonpearceite, *154*
Antimonselite, 106, *107*
Antimony, 26, *27-28*, 29, 228
Antlerite, 5, *626*
Anyuiite, *6*
Apachite, *1707*
Aparite group, 1100
Apatite, 224, 638, 704, 715, 727, 918, 981
Apatite group, 854-868, 1100-1103, 1104
Aphthitalite (glaserite), *569-570*, 704
Apjohnite, 616, *617*
Aplowite, 602, *603*
α-Apodumene (spodumene), 1291, *1308-1309*, 1594
Apuanite, *979*
Aquacreptite, 1428
Aragonite, 5, 426, 428, 440, 441, *442-444*, 644, 692
Aragonite group, 440-447
Aramayoite, *191-192*
Aravaipaite, *417-418*
Arcanite (taylorite, ammonian arcanite), *569*
Archerite, *698-699*
Arctite, *843*
Arcubisite, *159*
Ardaite, *142*
Ardealite, *957*
Ardennite, *1217-1218*
Ardennite group, 1212
Arfvedsonite, 1339, 1359, *1364-1365*
Argentojarosite, 631, *635*
Argentopentlandite, 60, *62*
Argentopyrite, *96*
Argentotennantite, 162, *165*
Argutite, 235, *243*
Argyrodite, *54*, 55
Arhbarite, *894*
Aristarainite, *561*
Armalcolite, *317-318*
Armangite, *983*
Armenite (calciocelsian), 1271, *1279-1280*
Armstrongite, *1543*
Arrojadite, *853-854*
Arsenbrackebushite, 758, *759*
Arsendescloizite, 826, *829*
Arsenic, *26-27*, 29
Arsenic group, 26-29
Arseniopleite, 706, *708*
Arseniosiderite, 911, *912*
Arsenite, 868
Arsenobismite, *852*
Arsenoclasite, *821*
Arsenocrandallite, 899, *901-902*
Arsenoflorencite-(Ce), 839, *840*

Arsenoflorencite group. *See* Florencite/
   arsenoflorencite group
Arsenoflorencite-(La), 839, *841*
Arsenoflorencite-(Nd), 839, *840*, 841
Arsenogoyazite, 840, 899, *902*
Arsenohauchecornite, 158, *159*
Arsenolamprite, 26, *29*
Arsenolite, *231*, 232
Arsenopalladinite, *145*
Arsenopyrite, 128, *129*, 803
Arsenosulvanite, *157*
Arsenpolybasite, *153*
Arsentsumebite, 758, *963*
Arsenuranospathite, 764, *778*
Arsenuranylite, *887*
Arthurite, 947, *947-948*
Artinite, *503*
Artroeite, *417*
Arupite, 792, *796-797*
Arzakite, *397*
Arzrunite, *421*
Asbecasite, *977*
Asbolane (cobaltian wad), *290*
Aschamalmite, *154-155*
Ashburtonite, *1707*
Ashcroftine-(Y), *1401*
Ashoverite, 278, *279*
Asisite, *395*
Asselbornite, *888*
Astrocyanite-(Ce), *500*
Astrophyllite, 1381, *1382-1383*
Astrophyllite group, 1381-1384
Atacamite, 5, *391*, 392
Atelestite, *881*
Athabascaite, *148*
Atheneite, *39-40*
Atlasovite, *627-628*
Atokite, 23, *24*
Attakolite, *966*
Attapulgite (**palygorskite**), *1558-1559*, 1562
Aubertite, *654-655*, 655
Augelite, *851-852*
Augite, 1291, *1299-1301*, 1303
Aurichalcite, *494*, 503
Auricupride, *5-6*
Aurorite, *319*
Aurostibite, 114, *117*
Austinite, 826, *827*, 828, 829
Autunite, 763, 764, *766*, 769, 773
Autunite group, 762-779
Avicennite, *231*
Avogadrite, *408*, 572
Awaruite (josephinite, souesite, bobrovskite),
   10, *12*
Axinite group, 1152-1157
Azoproite, 535, *536*
Azurite, 5, *482-485*, 488, 490

B

Ba analog of brewsterite, *1696-1697*
Babefphite, *831-832*

Babingtonite, 1326, *1328-1329*
Baddeleyite, *250*
Bafertisite, *1171-1172*
Baghdadite, 1160, *1161*, 1163
Bahianite, 220, *974-975*
Baileychlore, 1498, *1503-1504*
Bakerite, 1120, *1123-1124*
Balangeroite, *1323*
Balipholite, 1310, *1311-1312*
Balkanite, *145*
Baltimorite, 1428
Balyakinite, *678*
Bambollaite, *132*
Banalsite, *1616*
Bandylite, *543*
Bannermanite, *990*
Bannisterite, 1552, *1554*
Baotite, *1233-1234*
Bararite, 410, *411*
Baratovite, *1250*, 1251
Barberlite, *409*
Barbertonite, 507, *509*, 510
Barbosalite, 873, *875*
Barentsite, *425*
Bariandite, *989*
Baricite, 792, *794*
Bariomicrolite (rijkeboerite), 342, *346*
Bario-orthojoaquinite, 1231, *1233*
Bariopyrochlore (pandaite), 342, *344*
Barite (barytes), 4, 464, 571, *572-573*, 574,
   576, 603, 636
Barite group, 571-576, 636
Barium-alumopharmacosiderite, *910*
Barium pharmacosiderite, 908, *909-910*
Barium phosphuranylite, 886
Barnesite, *988*
Barrerite, 1675, *1676-1677*
Barringerite, *15*
Barringtonite, *463-464*
Barstowite, *421*
Bartelkeite, *316*
Bartonite, *98*
Barylite, *1133*
Barysilite, *1140*
Barytes (**barite**), 4, 464, 571, *572-573*, 574,
   576, 603, 636
Barytocalcite, 444, 453, *454*
Barytolamprophyllite, *1174*, 1175
Basaluminite, *644-645*
Bassanite (plaster of Paris), *593*, 599
Bassetite, 765, 769, 773, *775-776*
Bastite, 1418, 1428
Bastnäsite-(Ce), *476-477*, 478, 481
Bastnäsite group, 475, 476-479, 496, 499
Bastnäsite-(La), 476, *478*
Bastnäsite/synchysite/parasite groups. *See*
   Bastnäsite group; Parisite group;
   Synchysite group
Bastnäsite-(Y), 476, *478*
Batisite, *1324-1325*
Baumhauerite, *186*, 1191

Baumite, 1418
Bauranoite, *260*
Bavenite, *1403-1404*
Bayerite, 280, *281*, 282
Bayldonite, *842*
Bayleyite, *468-469*
Baylissite, *466*
Bazhenovite, *621-622*
Bazirite, 1221, *1222*
Bazzite, 1238, *1244*, 1277
Bearhite, 758
Bearsite, 891, *892*
Bearthite, *876-877*
Beaverite, 631, 632, *636*
Beckelite (**britholite-(Ce)**), *1100-1101*, 1104
Becquerelite, *261-262*
Behierite, *534*, 1053
Behoite, *277*
Beidellite (**Ca-montmorillonite, Na-montmorillonite, bentonite**), *1481-1483*
Belendorffite, 8, *9*
Bellbergite, *1697*
Bellidoite, *49*
Bellingerite, *531*
Belovite, 856, *862-863*
Belyankinite, *371*
Bementite, *1709*
Benavidesite, *184-185*
Benitoite, 1220, 1221, 1222, *1222-1223*
Benitoite group, 1220-1223. *See also* Catapleiite subgroup
Benjaminite, 199, *200*
Benleonardite, *154*
Benstonite, *453*
Bentonite (**montmorillonite, beidellite**), *1481-1483*, *1483-1485*
Bentorite, 658, *659*
Benyacarite, *950*
Beraunite, *945*
Berborite, *551*
Berdesinskiite, *318*
Bergenite, *886*
Bergslagite, *833*
Berlinite, *720*
Bermanite, *945-946*
Bernalite, *283*
Bernardite, *202*
Berndtite, 136, 137, *138*
Bernessite, 248
Berryite, *186*
Berthierine, 1406, 1414, *1423-1424*, 1425
Berthierite, *194*
Bertossaite, *853*
Bertrandite, *1149-1150*
Beryl, 1238, *1239-1244*, 1277
Beryl group, 1238-1245, 1251
Beryllite, *1061*
Beryllonite, *703-704*
Berzelianite, *49-50*
Berzeliite, *705*, 708
Betafite (**samiresite**), 342, *347-348*

Betekhtinite, *145-146*
Betpakdalite, *1008-1009*
Beudantite, 960, *961*
Beudantite group, 960-963
Beusite, *715*
Beyerite, *486*
Bezsmertnovite, *38*
Bianchite, *608*
Bicchulite, 1624, *1628*
Bideauxite, *405*
Bieberite, 610, *612*
Bijvoetite-(Y), *502-503*
Bikitaite, *1698-1699*
Bilibinskite, *38*
Bilinite, 616, *618*
Billietite, *262*
Billingsleyite, *153*
Bindheimite, *971*
Biotite, 224, 228, 1447, 1456, *1458-1460*, 1471
Biotite subgroup, 1447, 1456-1468. *See also* Mica group
Biphosphammite, *698*
Biringuccite, *557*
Birnessite, *315*
Bischofite, *387*
Bismite, 232, *233*
Bismoclite, *393*
Bismuth, 26, *28-29*
Bismuthinite, 29, 106, *107*, 491, 725
Bismutite, 486, *491-492*
Bismutocolumbite, *335-336*
Bismutoferrite, *1431*
Bismutohauchecornite, 158, *159*
Bismutomicrolite (**westgrenite**), 342, *347*, 349
Bismutostibiconite, *972*
Bismutotantalite, *335*, 336
Bityite, 1447, *1469-1470*
Bixbyite, *230-231*
Bjarebyite, 869, *870-871*
Bjarebyite group, 869-871
Blatterite, *538-539*
Blaubleibender Covellit (**yarrowite**), 63
Blixite, *394*
Blödite (**bloedite**), *585*, 586
Bloedite (**blödite**), *585*, 586
Blossite, *733*
Blue-remaining covellite (**yarrowite**), 63
Bobfergusonite, 707, *712*
Bobierrite, 792, *797*
Bobrovskite (**awaruite**), 10, *12*
Boehmite, *270*
Bogdanovite, *39*
Boggildite, *420*
Boggsite, *1693*
Bogvadite, *418*
Bohdanowiczite, *190*
Böhmite, 266
Bokite, *989*
Boleite, *404*
Bolivarite, 894, *895*
Boltwoodite (**nenadkevite**), 1109, *1112*, 1113

Bonaccordite, 535, *536*
Bonattite, *601*
Bonshtedtite, 957, *958-959*
Boothite, 610, *611*
Boracite, 547, *548*, 549
Boracite group, 547-549
Borax, *554-555*
Borcarite, *565*
Borickite, *894*
Borishanskiite, *139*
Bornemanite, *1187*
Bornhardtite, 98, *100*
Bornite, 4, 47, *52-53*, 92, 151
Borodaevite, 199, *201*
Boromuscovite, 1446, *1456*
Borovskite, *67*
Bostwickite, *1713*
Botallackite, 391, *392*
Botryogen, *653*, 653
Bottinoite, *287*
Boulangerite, *178-179*
Bournonite, *167-168*
Boussingaultite, 581, 587, *588-589*
Bowenite (antigorite), 1406, *1415-1417*
Bowieite, *112*
Boyleite, 602, *604*
Brabanite, 724
Brabantite (lingaitukuang), *724*
Bracewellite, 220, 266, *269-270*, 271, 273
Brackebushite, *758-759*, 877
Brackebushite group, 757-759, 876, 963, 964, 1097
Bradleyite, *957-958*, 959
Bradleyite group, 957-959
Braggite, 71
Braitschite-(Ce), *563-564*
Brammallite (sodium hydromica, sodium illite, hydroparagonite), *1477*
Brandite, 750, 753, *755*
Brannerite (orthobrannerite), 351, *356-357*
Brannockite, *1271-1272*
Brassite, *738*
Braunite I, *312-314*
Braunite II, *314*
Brazilianite, *836*
Bredigite, *1035*, 1052
Breithauptite, 76, *77*
Brenkite, *492*
Brewsterite, *1693-1694*, 1695, 1696-1697
Brezinaite, *103*
Brianite, *704*
Brianyoungite, *525*
Briartite, 89, *90*
Brindleyite (nimesite), 1406, *1424-1425*
Britholite-(Ce) (beckelite, lessingite, pravdite), *1100-1101*, 1104
Britholite subgroup, 1100-1104
Britholite-(Y) (abukumalite), *1101-1102*
Briziite, *229*
Brizziite, *975-976*
Brochantite, *623-624*

Brockite, 807, *808*
Bromargyrite, *378*
Bromellite, *211*
Brookite, 235, *246-247*
Brownmillerite, *326*
Brucite, 209, 274, *275*, 1216
Brucite group, 274-277
Brüggenite, *531*
Brugnatellite, *515-516*
Brunogeierite, 294, *301*
Brunsvigite (chamosite), 1423, 1498, *1504-1505*
Brushite, 598, 697, *735*
Buchwaldite, *702*
Buckhornite, *112*
Buddingtonite, 1597, *1608*
Buergerite, 1256, *1259-1260*
Bukovite, *54*
Bukovskyite, *966*
Bulachite, 894, *895*
Bultfonteinite, *1098-1099*
Bunsenite, 208, *209*
Burangaite, *913*, 914
Burbankite, 458, *459*, 460, 461
Burbankite group, 458-461
Burckhardtite, *1709*
Burkeite, *666*
Burpalite, 1160, *1161*, 1163
Bursaite, 175, *176*
Burtite, 284, *285-286*
Bustamite, 1314, *1318*
Butlerite, *651-652*
Bütschliite, *455*
Buttgenbachite, *528*, 639
Byelorussite-(Ce), 1231, *1233*
Byssolite, 1346
Bystrite, 1632, *1634*
Byströmite, *973*
Bytownite, 1597, *1608*

C

Cabriite, *25*
Cacoxenite, 944, *953-954*
Cadmium, *7*
Cadmium oxide (monteponite), 208, *210*
Cadmoselite, 72, *73*
Cadwaladerite, *399*
Cafarsite, *978*
Cafetite, *369*
Cahnite, *964*
Calamine (hemimorphite), *1150-1152*
Calaverite, *135*
Calciborite, *541*, 542
Calcioancylite-(Ce), 496, *498*
Calcioancylite-(Nd), 496, *498*
Calciobetafite, 342, *348*
Calcioburbankite, 459, *460*
Calciocelsian (armenite), 1271, *1279-1280*
Calciochondrodite (reinhardbraunsite), 1078, *1081*
Calciocopiaite, 663, *664*

Calcioferrite, 935, *936*
Calciogadolinite, 1120, *1126*
Calciohilarite, 1221*1229-1230*
Calcio-olivine, 1025
Calciotantite, *368*
Calciouraniote, *259*
Calciovolborthite (**tangeite**), 826, *828*
Calcite (**calcspar, kalkspar, kalkstein, doppelstein, Iceland spar, lublinite, satin spar, slate spar, cave pearls, stinckcalc, sand calcite**), 4, 5, 54, 211, 213, 220, 224, 228, 427, *428-434*, 436, 440, 442, 463, 495, 572
Calcite group, 426-440, 448
Calcium catapleiite, 1221, 1226, *1227*
Calcium-larsenite (**esperite**), *1024*
Calcjarlite, *415*
Calclacite, *1014*
Calcspar (**calcite**), 427, *428-434*
Calcurmolite, *1007*
Caldedonite, 213
Calderite, 1037, *1043-1044*
Caledonite, *668*
Calkinsite-(Ce), *471-472*
Callaghanite, *506*
Calomel, 213, *379-380*
Calumetite, *277-278*
Calzirtite, *369-370*
Camerolaite, *975*
Cameronite, *148*
Camgasite, *908*
Caminite, *665*
Ca-montmorillonite (**montmorillonite,beidellite**), *1481-1483, 1483-1485*
Campigliaite, *647-648*
Canaphite, *814*
Canasite, *1709-1710*
Canavesite, *522*
Canbyite (**hisingerite**), 1432, *1434-1435*
Cancrinite, 1632, *1634-1635*
Cancrinite group, 896, 1632-1641
Cancrisilite (**carbonate-vishnevite**), 1632, *1635-1636*
Canfieldite, 55
Cannizzarite, *183-184*
Cannonite, *637*
Capgarronite (**tocornalite**), *144*
Cappelenite-(Y), 1121, *1128-1129*
Caracolite, *638*
Carboborite, *565*
Carbocernaite, *461*
Carboirite, *328*, 1087, *1090*
Carbonate-cyanotrichite, *516*
Carbonate-fluorapatite, 856, *862*
Carbonate-hydroxylapatite, 856, *862*
Carbonate-vishnevite (**cancrisilite**), 1632, *1635-1636*
Carbonatite, 223, 224
Carborundum (synthetic moissanite), 35
Ca-rectorite (**rectorite**), *1515-1517*

Caresite, *513*
Carletonite, *1528*
Carlfriesite, *685*
Carlhintzeite, *414*
Carlinite, *50-51*
Carlosruizite, *533*
Carlosuranite, *1376-1377*
Carlsbergite, *15*
Carminite, 761, *877-878*
Carnallite, *407-408*
Carnotite, 780, *782-783*
Carnotite group, 779-784
Carobbiite, 373, *376*
Carpholite, 1310, *1311*
Carpholite group, 1309-1312
Carphosiderite (**hydronium jarosite**), 631, *634-635*
Carrboydite, *657*
Carrollite, 98, *99*
Caryinite, 706, *708*
Caryopilite, 1406, *1417*, 1709
Cascandite, 1314, *1320-1321*
Cassedanneite, *960*
Cassidyite, 751, *752*
Cassiterite, 212, 220, 235, *239-242*
Caswellsilverite, *97*
Cataphorite (**ferrikatophorite**), 1338, *1359-1360*
Catapleiite, 1221, *1226-1227*
Catapleiite subgroup, 1221, 1226-1230. *See also* Benitoite group
Catoptrite (**katoptrite**), *974*
Cattierite, 114, *115*
Cavansite, *1567*
Cave pearls (**calcite**), 427, *428-434*
Caysichite-(Y), *1399*
Cebaite-(Ce), *482*
Cebollite, *1702*
Celadonite, 1446, *1454-1455*
Celestine (**celestite**), 572, *574-575*
Celestite (**celestine**), 572, *574-575*
Celsian, 1597, *1604-1605*, 1615
Cementite (**cohenite**), *14*
Cenosite (**kainosite-(Y)**), *1236-1237*
Centrallasite (**gyrolite**), 1181
Cerargyrite (**chlorargyrite**), *377-378*
Cerianite, 257
Cerianite-(Ce), *249*
Ceriopyrochlore-(Ce) (**marignacite**), 342, *344-345*
Cerite-(Ce), 715, *1097*
Cernyite, 89, *90*
Cerotungstite-(Ce), *1007*
Ceruleite, *954*
Cerussite, 212, 442, *446-447*
Cervandonite-(Ce), *965*
Cervantite, *251*
Cervelleite, *43*
Cesanite, *638*, 858
Cesárolite, *315-316*
Cesbronite, *674-675, 687-688*

Cesium kupletskite, 1381, *1383*
Cesplumtantile, *372*
Cesstibtantite, 343, *349*
Cetineite, *140*
Chabazite-(Ca) (phacolite), *1652-1653*, 1655
Chabazite-(K), *1652-1653*
Chabazite-(Na), *1652-1653*
Chabournéite, *201*
Chaidamuite, *654*
Chalcanthite, 604, *605-606*
Chalcanthite group, 604-607
Chalcoalumite, *642-643*
Chalcocite, 4, 5, *46-47*, 718
Chalcocite group, 45-49
Chalcocyanite (chalcokyanite, hydrocyanite), *577-578*
Chalcokyanite (**chalcocyanite**), *577-578*
Chalcomenite, *680*
Chalconatronite, *465-466*
Chalcophanite, *319*
Chalcophyllite, *969*
Chalcopyrite, 47, 51, *87-88*, 220, 221
Chalcopyrite group, 86-89
Chalcosiderite, 917, *919*
Chalcostibite, *192*
Chalcothallite, *160*
Chalybite (siderite), 427, *435-436*
Chambersite, 547, *548*
Chaméanite, *160*, 171
Chamosite (daphnite, delessite, brunsvigite), 1423, 1498, *1504-1505*
Changbaiite, *362*
Chantalite, *1702*
Chaoite, 32, 33, *35*
Chapmanite, *1431-1432*
Charlesite, 658, *670*
Charmarite-2H, *513*
Charmarite-3T, *513*
Charoite, *1399-1400*
Chatkalite, *92*
Chayesite, 1271, *1272*, 1278
Chekhovichite, *686*
Chelkarite, *563*
Chenevixite, *906*
Chengdeite, 23, *24-25*
Chenite, *627*
Cheralite, 723, *724*, 1266
Cheremnykhite, *676*
Cherepanovite, *81*
Chernikovite (hydrogen-autunite), 765, *777*
Chernovite-(Y), 728, *730*
Chernykhite, 1446, *1452*
Chervetite, *733*
Cherysoberyl, *307-309*
Chessexite, *671*
Chesterite, *1379*
Chestermanite, 535, *537*
Chevkinite-(Ce), *1177-1178*, 1179
Chiastolite, 1066, 1067
Chiavennite, *1402-1403*
Childrenite, *896-897*

Chiluite, *330-331*
Chiolite, *415*
Chkalovite, *1332*
Chladniite, *713*
Chloraluminite, *389*
Chlorapatite, 856, 858, *860-861*, 864, 866
Chlorargyrite (cerargyrite, horn silver), *377-378*
Chlorellestadite, 1102, *1103*
Chlorite, 1497
Chlorite group, 1496-1505
Chloritoid-A, 1087, *1088-1089*
Chloritoid group, 1086-1090
Chloritoid-M, 1087, *1088-1089*
Chlormagaluminite (chlor-manasseite), *405*
Chlor-manasseite (**chlormagaluminite**), *405*
Chlormanganokalite, *411*
Chlorocalcite, *408*
Chloromagnesite, *385*
Chloromelanite, 1302, 1303
Chloropal (nontronite), 1481, *1485-1486*
Chlorophoenicite, *815*
Chlorothionite, *625*
Chloroxiphite, *403*
Choloalite, *681*
Chondrodite, 209, 227, 1077, 1078, *1080-1081*
Christite, *172*
Chromatite, *693*
Chromdravite, 1257, *1264*
Chromferide, *13*
Chromite, 294, *302*
Chromium, *12*
Chrompicotite (**magnesiochromite**), 294, *301*
Chromrutile (**redledgeite**), *325*
Chrysocolla, 212, *1561-1562*
Chrysolite (**forsterite**), 1025, *1026-1029*
Chrysotile, 1406, *1428-1430*
Chudobaite, *742*
Chukhrovite-(Ce), *420*
Chukhrovite-(Y), *420*
Churchite-(Nd), 598
Churchite-(Y) (weinschenkite), 598, *805-806*
Chursinite, *731-732*
Chvaleticeite, 608, *609*
Chvilevaite, *55*
Cianciulliite, *280*
Cinnabar, 7, 69, 73, *80*, 213, 306, 400
Cis-terpin hydrate (**flagstaffite**), *1017*
Clairite, *665*
Claraite, *505*
Claringbullite, *400*
Clarkeite, *259*
Claudetite, 231, *232*, 732
Clausthalite, 64, *65*
Cliffordite, *682*
Clinobehoite, *277*
Clinobisvanite, 725, *726*, 728n
Clinochalcomenite, *680*
Clinochlore, 1498, *1501-1503*
Clinochrysotile, 1406, *1428*
Clinoclase, *818*

Clinoenstatite (clinohypersthene), 1289, *1291-1292*
Clinoferrosilite (clinohypersthene), 1289, *1292*, 1295
Clinohedrite, *1062*
Clinoholmquistite, 1337, *1343*, 1346
Clinohumite, 1078, *1084-1085*
Clinohypersthene (**clinoenstatite**), 1289, *1291-1292*
Clinohypersthene (**clinoferrosilite**), 1289, *1292*, 1295
Clinojimthompsonite, *1376*
Clinokurchatovite, 540, *541*
Clinomimetite, 857, 866, *868*
Clinoptilolite, *1671-1674*
Clinopyroxene kanoite, 1296
Clinopyroxene subgroup, 1289, 1291-1294, 1296-1309. See also Pyroxene group
Clinosafflorite, 129, *130*
Clinotobermorite, 1528, 1529, *1530-1531*
Clinotyrolite, *885-886*
Clinoungemachite, *649*
Clinozoisite, 1196, *1198-1199*, 1202, 1204
Clintonite (xanthophyllite), 1447, *1469*
Coalingite, *516*
Cobalt, 4
Cobaltaustinite, 826, *829*
Cobaltian wad (**asbolane**), *290*
Cobaltite, 124, *125*, 737
Cobaltite group, 124-128
Cobaltkoritnigite, *737*
Cobaltocalcite (**sphaerocobaltite**), 427, *438*
Cobaltomenite, *680-681*
Cobalt pentlandite, 60, *62*
Cobalt zippeite, *661-662*
Cochromite, 294, *303*
Coconinoite, *967*
Coeruleolactite, 917, *918-919*
Coesite, 243, 1570, 1573, *1586-1587*, 1593
Coffinite, 1053, 1056, *1057-1058*
Cohenite (cementite), *14*
Colemanite, *553-554*
Collinsite, 751, *752*
Coloradoite, 67, *70*
Colquiriite, *413*
Columbite, 340, 353, 356
Columbite group, 352-356
Colusite, *150*, 151
Colusite group, 150-151
Comancheite, *400*
Combeite, 1246, *1248*
Comblainite, 508, *512*
Compreignacie, *261*
Congolite, 547, *548*
Conichalcite, *826-827*
Connellite, *639*
Cookeite, 1426, 1498, *1499-1500*
Cooperite, *71*
Copiaite, 662, *663*, 664
Copiaite group, 662-665

Copper, 1, 2, *4-5*, 11, 46, 47, 79, 87, 204, 205, 212, 465
Coquandite, *628*
Coquimbite, *619*
Corderoite, *396*, 397
Cordierite, 1238, 1245, 1253, 1279
α-Cordierite, 1238, 1245
Cordierite (iolite, dichroite), *1251-1253*
Cordylite-(Ce), *481-482*
Corkite, 960, *961*
Cornetite, *818*
Cornubite, *821-822*
Cornwallite, 821, *822*
Coronadite, 321, *323*
Coronene (**karpatite**), *1016*
Corrensite (swelling chlorite), *1512-1514*
Corundum, *214-217*
Corundum-hematite group, 213-221
Corvusite, *988-989*
Cosalite, *180-181*
Cosmochlore (**kosmochlor**), 1291, *1307*
Costibite, 126, *132*
Cotunnite, *386*, 408
Coulsonite, 221, 294, *303*
Cousinite, *1007*
Covellite, *79*
Cowlesite, *1686*
Coyoteite, *142*
Craigite, 207
Crandallite, *899*
Crandallite group, 821, 838, 839, 840, 898-902
Crawfordite, 957, *959*
Creaseyite, *1713-1714*
Crednerite, *292-293*
Creedite, *420*
Crerarite, *67*
Crichtonite, 365, *366*
Crichtonite group, 364-367
Criddleite, *203*
Cristobalite, 243, *1568-1570*, 1573, 1586, 1587, 1593
Crocoite, *693*
Cronstedtite, 1406, *1427-1428*
Crookesite, *50*
Crossite, 1339, *1361*
Cryolite, *412*
Cryolithionite, *413*
Cryptohalite, *410*, 411
Cryptomelane, 320, 321, *322*
Cryptomelane group, 321-323, 324
Cualstibite, *975*
Cubanite, *96*
Cumengéite (**cumengite**), *404*
Cumengite (**cumengéite**), *404*
Cummingtonite, 1337, *1339*, 1341, 1343
Cummingtonite subgroup, 1337, 1339-1343. See also Amphibole group
Cupalite, *13*
Cuprite, 5, *204-205*, 212
Cuprobismutite, *197*
Cuprocopiaite, 663, *664*

Cuproiridiste, 99, *102*
Cupropavonite, 199, *201*
Cuprorhodisite, 99, *102*
Cuprorivaite, *1477-1478*
Cuprosklodowskite (jachymovite, jachimovite), 1109, *1112*
Cuprospinel, 294, *301*
Cuprostibite, *50*
Cuprotungstite, *1002*
Curetonite, *842*
Curienite, 780, *782*
Curite, *263-264*
Curtisite (**idrialite**), *1016*
Cuspidine, 1160, *1161-1162*, 1163, 1164
Cuspidine-wöhlerite group, 1160-1168, 1181
Cuzticite, *673*
Cyanite (**kyanite**), 1064, 1066, 1067, *1069-1070*
Cyanochroite, 587, *588*
Cyanophyllite, *975*
Cyanotrichite, 516, *641*
Cylindrite, *152*
Cymrite, *1702-1703*
Cyrilovite, *905*
Cyrtolite (**zircon**), *1053-1056*

D

Dachiardite, *1690-1691*
Dadsonite, *185*
Dalyite, *1524-1525*
Damaraite, *401*
Danalite, 1630, *1631*
Danbaite, *7*
Danburite, *1183-1184*
Danielsite, *145*
Dannemorite, 1337, *1342-1343*
D'Ansite, *625*
Daomanite, *55*
Daphnite (**chamosite**), 1498, *1504-1505*
Daqingshanite-(Ce), *523*
Darapiosite, 1271, *1272*
Darapskite, *530*
Datolite, 1120, *1121-1122*
Datolite-gadolinite group, 1119-1130
Daubréeite, *393*
Daubréelite, 99, *101-102*
Davanite, 1524, *1525*
Davidite-(Ce), 365, *366-367*
Davidite-(La), 365, *366*
Davisonite, 727
Davreuxite, *1194*
Davyne, 1632, *1636*, 1639
Dawsonite, *492-493*
Deanesmithite, *694-695*
Deerite, *1395*
Defernite, *504-505*
Delafossite, *292*
Delessite (**chamosite**), 1498, *1504-1505*
Delhayelite, 1539, *1540*
Delindeite, *1174*
Dellaite, *1703*

Deloryite, *1003*
Delrioite, *985*, 986
Delta13-dihydro-d -pimaric acid (**refikite**), *1017*
Delvauxite, *894*
Demantoid, 1044
Demesmaekerite, *688-689*
Denisovite, 1314, *1321*
Denningite, *684*
Derbylite, *983*
Derriksite, *688*
Dervillite, *196*
Desaulsite (**pimelite**), 1481, *1492-1493*
Desautelsite, 508, *511*
Descloizite, 826, *829-830*
Descloizite subgroup, 825, 826, 829-831. See also Adelite group
Desourdyite (**steacyite**), *1269*, 1521
Despujolsite, *650*
Devilline, *646*
Deweylite, 1428
Dewindtite, *922*
Diaboleite, *402-403*
Diadochite, *966-967*
Diamond, 12, *32-33*
Diaoyudaoite, *330*
Diaphorite, *179*
Diasphore, 266-267
Diasphore group, 266-270
Diaspore, 270
Dichroite (**cordierite**), *1251-1253*
Dickinsonite, 701, 751, 853, *854*
Dickite, 1406, *1407*
Di-di-donbassite. See **Donbassite**
Dienerite, *39*
Dietrichite, 616, *617-618*
Dietzeite, *533*
Digenite, 46, *47-48*
1,1-Dimethyl-7-isopropyl-1,2,3,4-tetrahydrophenanthrene (**simonellite**), *1015*
Dimorphite, *56*
Dinite, *1015-1016*
Diomignite, *541*
Diopside, 221, 224, 228, 1291, *1296-1298*
Dioptase, *1249-1250*
Diorite, 227
Dissakisite-(Ce), 1196, *1199*
Disthene (**kyanite**), 1064, 1066, 1067, *1069-1070*
Di-tri-cookeite. See **Cookeite**
Dittmarite, *748*
Dixenite, *983-984*
Djalmaite (**uranmicrolite**), 342, *346-347*
Djerfisherite, *143*
Djurieite, 46, *47*
Dmisteinbergite, *1617*
Dolerophanite, *628-629*
Dollaseite-(Ce) (magnesium orthite), 1196, *1199*, 1202

Dolomite, 50, 431, 434, 437, 449, *450-451*, 454, 539, 540
Dolomite group, 448-452
Doloresite, *991*
Domeykite, *39*
Donathite, 302
Donbassite, *1498-1499*
Donharrisite, *60*
Donnayite-(Y), *469*, 470
Donpeacorite, 1289, 1292, *1296*
Doppelstein (**calcite**), 427, *428-434*
Dorallcharite, 631, *636*
Dorfmanite, *738*
Dorrite, 1386, *1387*
Douglasite, *409*
Doverite (**synchysite-(Y)**), 476, *479*
Downeyite, *244*
Doyleite, 280, 281, *282*
Dozyite, *1508-1509*
Drauite, 1255
Dravite, 1257, *1262-1263*, 1264
Dresserite, *501*
Dreyerite, 725, *726*, 728
Drugmanite, *880*
Drysdallite, *134*
Dufrenite, 913, *914*
Dufrénoysite, *181*
Duftite-α, 826, 827, *831*
Duftite-β, 826, *827-828*, 831
Dugganite, *676*
Duhamelite, *921*
Dumontite, *886*
Dumortierite, *1117-1118*, 1119
Dundasite, *501*
Dunite, 1028
Durangite, *834-835*
Duranusite, *37*
Dusmatovite, 1271, *1276-1277*
Dussertite, *839*
Duttonite, *278*
Dwornikite, 595, *597*
Dypingite, *514*
Dyscrasite, *38*
Dzarkenite, 114, *119*
Dzhalindite, *282*

E

Eakerite, *1524*
Earlandite, *1014*
Earlshannonite, 947, *948*
Eastonite, 1456
Ecandrewsite, 225, *228*
Ecdemite, *982*
Eckermannite, 1339, *1364*
Eclarite, *185*
Edenharterite, *195*
Edenite, 1338, *1353*
Edgarbaileyite, *1147-1148*
Edingtonite, *1683-1685*
Edoylerite, *695*
Effenbergerite, *1478-1479*

Efremovite, *581*
Eggletonite, 1554, *1555*
Eglestonite, 213, *399-400*
Ehrleite, *811-812*
Eifelite, 1271, *1272-1273*, 1278
Eitelite, *455*
Ekanite, 1265, *1521* . See also **Steacyite**
Ekaterinite, *563*
Elbaite, 1256, 1258, *1261-1262*
Ellenbergerite, *1104-1105*
Ellestadite, 832
Ellisite, *172*
Elpasolite, *413*
Elpidite, *1543-1544*
Elyite, *623*
Embreyite, *959-960*
Emeleusite, 1373, *1375*
Emmonsite, *683*
Emplectite, *192*
Empressite, *86*
Enargite, 47, *155-156*
Endellite, 1411
Endiopside, 1297
Englishite, *954*
Enstatite (**orthoenstatite**), 66, 1289, 1291, *1294-1295*
Eosphorite, 701, 896, *897*
Ephesite, 1447, *1467-1468*
Epididymite, *1371*
Epidote, 4, 1188, 1195, 1196, *1200-1201*
Epidote group, 1195-1205
Epistilbite, *1689*
Epistolite, *1176-1177*, 1187
Epsomite, 608, *612-613*, 614
Erdite, *142*
Ericaite, 547, *548*
Ericssonite, 1174, *1175*
Eriochalcite, *386*
Erionite, *1656-1657*
Erlianite, *1710*
Erlichmanite, 114, 117, *118*
Ernienickelite, *320-321*
Erniggliite, *178*
Ernstite, *897-898*
Ershovite, *1368*
Ertixiite, *1714*
Erythrite, 755, 792, *794-795*
Erythrosiderite, *409-410*
Eskebornite, 87, *88*
Eskimoite, *183*
Eskolaite, 214, *220*
Esperite (**calcium-larsenite**), 703, *1024*
Esseneite, 1291, *1302*
Essonite (**hessonite**), 1046, 1047
Ethanamide (**acetamide**), *1018*
Ettringite, 658, *659*, 670, 671
Ettringite group, 657-659, 670, 671
Eucairite, *45*
Euchlorine, *637*
Euchroite, *892*
Euclase, *1061-1062*

Eucolite, 1281
Eucryptite, 1020, *1022-1023*
Eudialyte, 1281, *1281-1282*
Eudidymite, *1370-1371*
Eugenite, *8*
Eugsterite, *590*
Eulite, 1295
Eulytine (eulytite), *1059-1060*
Eulytite (eulytine), *1059-1060*
Euxenite-(Y), *360*
Evansite, *882*
Eveite, *850-851*
Evenkite, *1016*
Ewaldite, *453-454*, 469
Eylettersite, 839, *841*
Ezcurrite, *557*
Eztlite, *690*

F

Fabianite, *545*
Faheyite, *810*, 812
Fahleite, *812*
Fairbankite, *677-678*
Fairchildite, *455*, 456
Fairfieldite, 701, *751*
Fairfieldite subgroup, 750, 751-754. *See also* Roselite subgroup
Falcondoite, 1558, *1561*
Falkmanite, *179*
Famatinite, *156*
Fangite, *160*
Farringtonite, *714*
Faujasite, *1659-1660*
Faujasit group, 812
Faustite, 917, *919*
Fayalite (knebelite), *1025-1026*, 1027
Federovskite, *546*
Fedorite, *1547*
Fedotovite, *638*
Feitknechtite, 269, 271, *272-273*
Feldspar group, 1595-1615. *See also* Alkali feldspar subgroup; Plagioclase subgroup
Felsöbanyaite, *644*
Fenaksite, *1397-1398*
Ferberite, 252, 993, *994-996*
Ferchromide, *13*
Fergusonite-(Ce), *333*
β-Fergusonite-(Ce), *333*, 334
Fergusonite-(Nd), *333*
β-Fergusonite-(Nd), *333-334*
Fergusonite-(Y), *332*, 334
β-Fergusonite-(Y), *333*
Fermorite, 856, *865*
Fernandinite, *989*
Feroxyhite, 249
Feroxyhyte, 268, 270, *273*
Ferrarisite, *741*
Ferrazite, *799*
Ferri-alluaudite (**ferroalluaudite**), 706, *709*
Ferriannite, 1447, 1458, *1461*
Ferribarroisite, 1338, *1358*

Ferricopiaite, 663, *664*
Ferrierite, *1692-1693*
Ferrihydrite, *221*
Ferrikatophorite (cataphorite), 1338, *1359-1360*
Ferrilotharmeyerite, *699*
Ferrimolybdite, *1005-1006*
Ferrinatrite, *590*
Ferripyrophyllite, 1439, *1441*
Ferrisicklerite, *703*, 715
Ferristilpnomelane, 1552
Ferristrunzite, 938, *939-940*
Ferrisurite, *1479*
Ferrisymplesite, 797, *922*
Ferritaramite, 1338, *1360-1361*
Ferritschermakite, 1338, *1352*
Ferritungstite, *252*
Ferriwinchite, 1338, *1357*
Ferroactinolite, 1338, *1345-1350*
Ferroalluaudite (ferri-alluaudite), 706, *709*
Ferro(alumino)barroisite, 1338, *1358*
Ferro(alumino)tschermakite, 1338, *1352*
Ferro(alumino)winchite, 1338, *1357*
Ferroanthophyllite, 1337, *1343*
Ferroaxinite, 1153, *1154-1155*
Ferrobustamite (iron-wollastonite, ferrowollastonite), 1314, *1318-1319*
Ferrocarpholite, 1310, *1311*
Ferroclinoholmquistite, 1337, *1343*
Ferrocolumbite, 353, *354-355*, 356
Ferroeckermannite, 1339, *1364*
Ferroedenite, 1338, *1353-1354*
Ferroferribarroisite, 1338, *1358*
Ferroferritschermakite, 1338, *1352-1353*
Ferroferriwinchite, 1338, *1357-1358*
Ferrogedrite, 1337, *1345*
Ferroglaucophane, 1339, *1361*
Ferrohagendorfite, 706, *708*, 709
Ferrohedenbergite, 1298
Ferrohexahydrite, *608*
Ferroholmquistite, 1337, *1346*
Ferrohornblende, 1359
Ferrohortonolite, 1027
Ferrohypersthene, 1295
Ferrokaersutite, 1338, *1356-1357*
Ferrokesterite, 89, *91*
Ferronickelplatinum, 21, *22*
Ferropargasite, 1338, *1354*
Ferropigeonite, 1292
Ferropyrosmalite, *1536-1537*
Ferrorichterite, 1338, *1358-1359*
Ferrosalite, 1298
Ferroschallerite (nelenite), 1536, 1537, *1538*
Ferroselite, 120, *121*
Ferrosilite (orthoferrosilite), 1289, 1292, *1295-1296*
Ferrostrunzite, 938, *939*
Ferrotantalite, 353, *354*
Ferrotapiolite (tapiolite), *351-352*
Ferrotychite, *519*
Ferrowodginite, 337, *338*

Ferrowollastonite (**ferrobustamite**), 1314, *1318-1319*
Ferrowyllieite, 707, *711*
Ferruccite, *409*
Fersmanite, *1170*
Fersmite, *356*
Feruvite, 1256, *1259*
Fervanite, *990*
Fetiasite, *979-980*
Fibroferrite, *656*
Fibrolite, 1064
Fichtelite, *1015*
Fiedlerite, *396*
Filipstadite, *310*
Fillowite, 701, *712-713*
Fingerite, *880-881*
Finnemanite, 857, 868, *981*
Fischesserite, *44*
Fizelyite, 175, *177-178*
Flagstaffite (*cis* -terpin hydrate), *1017*
Fleischerite, *650*
Fletcherite, 98, *100*
Flinkite, *818-819*
Florencite/arsenoflorencite group, 838-842, 898
Florencite-(Ce), *839-840*, 841
Florencite-(La), 839, *840*
Florencite-(Nd), 839, *840*
Florensovite, 99, *102*
Fluckite, *736*
Fluellite, *895*
Fluoborite, *542*
Fluocerite-(Ce) (tysonite, fluocerite), *389*
Fluocerite (**fluocerite-(Ce)**), *389*
Fluocerite (**fluocerite-(La)**), *389*
Fluocerite-(La) (fluocerite), *389*
Fluorapatite, 855, 856, *858-860*, 861
Fluorapophyllite, *1525*, 1526
Fluorbritholite-(Ce), *1104*
Fluorellestadite, *1102*, 1103
Fluorferroleakeite, 1339, *1367*
Fluorite (fluorspar), 4, 221, 255, *380-383*
Fluorspar (**fluorite**), 4, 221, 255, *380-383*
Foggite, *898*
Foitite, 1256, *1257*
Fontanite, *471*
Foordite, *362*
Forcherite, *894*
Formanite-(Y), *332-333*, 334
Fornacite, 758, *963*, 964
Forsterite (chrysolite, peridot), 1025, *1026-1029*
Foshagite, *1321*
Foshallasite, 1160, *1181*
Fosterite, 1034, 1140
Fourmarierite, *263*
Fraipontite, 1406, *1425-1426*
Francevillite, 733, 780, *782*
Franciscanite, *311*, 1090, *1091*
Francisite, *687*
Franckeite, *152*

Francoanellite, *744*, 745
Françoisite-(Nd), *889*
Franconite, *367*, 368
Frankdicksonite, *383-384*
Frankhawthorneite, *673*
Franklinfurnaceite, *1519*
Franklinite, 211, 294, *300*
Franklinphilite, 1552, *1553-1554*, 1554
Fransoletite, *809*
Franzinite, 1632, *1636*
Freboldite, 76, *79*
Fredrikssonite, 535, *537*
Freedite, *982*
Freibergite, 162, 163, *164*
Freieslebenite, *171*, 179
Fresnoite, 1143, *1146*
Fresnoite subgroup. *See* Melilite-fresnoite group
Freudenbergite, *324*
Friedelite, 1536, *1537-1538*
Friedrichite, 169, *170*
Fritzscheite, 780, *781*
Fritzschite, 776
Frohbergite, 120, *121*
Frolovite, *542-543*
Frondelite, *872*
Froodite, *138*, 139
Fuchsite, 1448
Fuenzalidaite, *533*
Fukalite, *1707*
Fukuchilite, 114, *116*
Fülöppite, *188*
Fülöppite group, 188-190
Furongite, *949*
Furutobeite, *146*

G

Gabrielsonite, 826, *828*
Gadolinite-(Ce), 1120, 1123, *1124*
Gadolinite group, 880. *See* Datolite-gadolinite group
Gadolinite-(Y), 1120, 1123, *1124-1126*
Gagarinite-(Y), *411-412*
Gageite-1A, *1323-1324*
Gageite-2M, *1323*
Gahnite, 228, 294, *297-298*
Gaidonnayite, 1221, 1226, *1227-1228*
Gainesite, *810*, 811
Gaitite, 751, *753*
Galaxite, 294, *297*
Galeite, *625*
Galena, *64-65*
Galena group, 64-66
Galenobismutite, *194*
Galkhaite, *174*
Gallite, 87, *88*
Gamagarite, 758, *877*
Gananite, *389*
Ganomalite, *1706*
Ganophyllite, 228, *1554-1555*
Gaotaiite, 114, *119*

General Index

Garavellite, *195*
Garnet, 228, 705, 872
Garnet group, 1036-1052
Garnierite, 1418, 1492
Garrelsite, *566*, 1121
Garronite, 1661, *1662-1663*, 1663
Gartrellite, *959*
Garyansellite, 787, *788-789*
Gasparite-(Ce), 721, *725*
Gaspéite, 427, *439-440*
Gatehouseite, *821*
Gatumbaite, *942*
Gaudefroyite, *565*
Gaultite, *1700-1701*
Gaylussite, *465*
Gearksutite, *414*
Gebhardite, *983*
Gedrite, 1337, *1345*
Geerite, 46, *49*
Geffroyite, 60, *63*
Gehlenite, 1142, *1143-1144*
Geigerite, *742*
Geikielite, 225, *227-228*
Gelbertrandite, 1149
Geminite, *814*
Genkinite, *59*
Genthelvite, 1630, *1631*
Genthite (**nepouite**), 1406, *1420-1421*, 1492
Geocronite, *160-161*
Georgechaoite, 1221, 1227, *1228*
Georgeite, *490-491*
Georgiadesite, *816*
Gerasimovskite, *371*
Gerdtremmelite, *820*
Gerhardtite, *528*
Germanite, *93*
Germanocolusite, 150, *151*
Gerrelsite, 1121, *1129-1130*
Gersdorffite, 124, *125*
Gerstleyite, *198*
Gerstmannite, *1063*
Getchellite, *105*
Geversite, 114, *118*
Gianellaite, *570*
Gibbsite, *280-281*, 282
Giessenite, *182*
Gilalite, *1708*
Gillespite, *1478*
Gillulyite, *199*
Giniite, *933*
Ginorite, *562*, 562
Ginzburgite (**roggianite**), *1696*
Giorgiosite, *514-515*
Gips (**gypsum**), *598-601*
Giraudite, 162, *164*
Girdite, *675*
Girvasite, *970*
Gismondine, *1661-1662*, 1663, 1700
Gittinsite, *1134-1135*
Giuseppettite, 1632, *1636-1637*
Gladite, 169, *170*

Glaserite (**aphthitalite**), *569-570*, 704
Glauberite, *579-580*, 589
Glauber's salt (**mirabilite**), 579, *584*
Glaucocerinite, *645*
Glaucochroite, 1025, *1034*
Glaucodot, *131*, 737
Glauconite, 1421, 1446, *1453-1454*
Glaucophane, 1339, *1361*
Glaucophane subgroup, 1339, 1361-1367. See *also* Amphibole group
Glaukospherite, *487*
Glucine, *888*
Glushinskite, *1012*
Gmelinite, *1657-1659*
Gobbinsite, 1661, *1663-1664*
Godlevskite, *63-64*
Godovikovite, *578*
Goedkenite, 758, *876*, 877
Goethite (**limonite**), 266, *268-269*, 270, 271, 273, 464, 634, 673, 900
Gold, 1, *2-3*, 11, 30
Goldfieldite, 162, *165*
Gold group, 1-9
Goldichite, *592*
Goldmanite, 1037, *1049*
Gonnardite, *1685-1686*
Gonyerite, *1519-1520*
Goosecreekite, 1689, *1695-1696*
Gorceixite, 840, 899, *900*
Gordonite, 899, 937, *944*
Görgeyite, *591*
Gormanite, *915*
Gortdrumite, *147*
Goslarite, *613*
Götzenite, 1166, *1170-1171*
Goudeyite, *891*
Gowerite, *557*
Goyazite, 899, *900*, 902
Graemite, *679*
Graftonite, *714-715*
Grandidierite, *1116-1117*, 1184
Grandite, 1039
Grandreefite, *419*
Grantsite, *988*
Graphite, 13, 32, *33-34*
Gratonite, *161*
Gravegliaite, *682*
Grayite, 807, *808*
Grechishchevite, *397*
Greenalite, 1406, 1417, *1421-1422*
Greenockite, 70, 72, *73*
Gregoryite, *440*
Greigite (**melnikovite**), 99, *101*
Griceite, 373, *376*
Griffithite (**saponite**), 1481, *1489-1490*
Grimaldiite, 220, 269, 271, 272
Grimaldite, *273*
Grimselite, *467*
Griphite, *872*
Grischunite, *762*
Grossite, *311*, *331*

Grossular, 1037, 1039, 1044, *1046-1048*, 1049, 1051, 1052
Groutite, 248, 266, *269*, 271, 272
Grovesite, 1505
Grumantite, *1708*
Grunerite (amosite), 1337, *1339-1342*
Grünlingite (ingodite), 82, *84*
Gruzdevite, *174*
Guanajuatite, 106, *108*
Guanine, *1017-1018*
Guarinoite, *641-642*
Gudmundite, *129-130*
Guerinite, *741*
Guettardite, *193*
Gugliaite, 1143, 1146, *1147*
Guildite, *654*
Guilleminite, *688*
Guitermanite, 1191
Gümbelite (illite), *1472-1474*
Gunningite, 595, *597*, 604
Gupeiite, *16*
Gustavite, 175, *176*
Gutsevichite, *924*
Guyanaite, 220, 269, *271*, 273
Gypsum (gips, selenite), 593, *598-601*, 651, 735
Gypsum group, 597-601, 735
Gyrolite (centrallasite), 1181, 1533, *1550-1551*
Gysinite-(Nd), 496, *498-499*

H

Haapalaite, *141*
Haarsalz (halotrichite), 616, *617*
Hackmanite, 1625
Hafnon, 1053, *1056*
Hagendorfite, 706, *708-709*
Häggite, *289*
Haidingerite, 697, 736, *737*
Hainite, 1160, *1166*, 1170
Haiweeite (ranquilite), *1114*, 1115
Hakite, 162, *164*
Halite group, 373-376
Halite (salt, rock salt), 13, *373-375*
Hallimondite, *784-785*
Halloysite, 1406, 1407, *1411-1414*
Halotrichite group, 615-618
Halotrichite (haarsalz), 616, *617*
Halurgite, *555*
Hambergite, *542*
Hammarite, 169, *170*
Hanawaltite, *402*
Hancockite, 1196, *1201*
Hanksite, *667-668*
Hannayite, *744*
Hannebachite, *681-682*
Hardystonite, 1142, 1143, *1145-1146*
Harkerite, *522-523*
Harmotome, *1664-1665*
Harstigite, *1188*
Hashemite, 572, *693-694*
Hastingsite, 1338, *1354-1355*
Hastite, 120, *122*
Hatchettolite (uranpyrochlore), 342, *345*
Hatchite, *180*
Hatrurite, *1099*
Hauchecornite, *158-159*
Hauchecornite group, 158-159
Hauckite, *668-669*
Hauerite, 114, *117*
Hausmannite, *305*
Haüyne, 1624, *1627*
Hawleyite, 67, *70*
Hawthorneite, *312*
Haxonite, *14*
Haycockite, *94*
Haynesite, *689*
Heazlewoodite, *53*
Hectorfloresite, *666*
Hectorite, 1481, *1491-1492*
Hedenbergite, 1291, *1298-1299*
Hedleyite, *59*
Hedyphane, 857, *863-864*
Heideite, *103*
Heidornite, *669*
Heinrichite, 764, *769*
Hejtmanite, *1172*
Heliophyllite, *982*
Hellandite-(Y), 1121, *1127-1128*
Hellyerite, *464*
Helmutwinklerite, *760-761*
Helmutwinklerite subgroup, 760-761
Helvite, *1630-1631*
Hematite, 17, 214, 505
Hematite (specularite, oligiste), 214, *217-220*
Hematolite, *964*
Hematophanite, *327*
Hemihedrite, *696*
Hemimorphite (calamine), *1150-1152*
Hemloite, *983*
Hemusite, *93*
Hendersonite, *988*
Hendricksite, 1447, *1462*
Heneuite, *965*
Henmilite, *552*
Hennomartinite, 1157, *1158*
Henritermierite, *1052*
Hentschelite, 873, *875*
Heradaite, *1322-1323*
Hercynite, 294, *297*
Herderite, *832*
Herschelite, 1652
Herschelite, *1654*
Heryite, *146*
Herzenbergite, *86*
Hessite, *43*
Hessonite (essonite), 1046, 1047
Hetaerolite, *305-306*
Heterogenite-2H, *272*
Heterogenite-3R, *272*
Heteromorphite, 188, *189*
Heterosite, 710, 715, *719*
Heulandite-(Ca), *1669-1671*

General Index

Heulandite-(K), 1669
Heulandite-(Na), 1669
Hewettite, 987
Hexagonite, 1346
Hexahydrite, 607, 608
Hexahydrite group, 607-609
Hexahydroborite, 551
Hexatestibiopanickelite, 76, 77
Heyite, 817
Heyrovskyite, 161
Hibbingite, 391-392
Hibonite, 311-312
Hibschite (hydrogrossular, plazolite), 1037, 1051
Hidalgoite, 960, 961
Hieratite, 410
Hilairite, 1221, 1229, 1230
Hilgardite, 559
Hillebrandite, 1321-1322
Hingganite-(Ce), 1120, 1122
Hingganite-(Yb), 1120, 1123
Hingganite-(Y) (yttroceberysite, yberisilite), 1120, 1122, 1123
Hinsdaleite, 960, 961, 962
Hiortdahlite (I and II), 1160, 1164-1165
Hisingerite (canbyite, scotiolite, sturtite), 1432, 1434-1435, 1436
Hocartite, 89, 91
Hochelagaite, 368
Hodgkinsonite, 1063
Hodrushite, 198
Hoelite (anthraquinone), 1017
Högbomite 8H, 328
Högbomite 4H,5H,6H,15H, 328
Högbomite 15R,18R,24R, 328-329
Hogtuvaite (makarochkinite), 1386, 1387-1388
Hohmannite, 652-653
Holdawayite, 505
Holdenite, 964
Hollandite, 321, 322
Hollingworthite, 124, 127
Holmquistite, 1337, 1343, 1346
Holtedahlite, 848
Holtite, 1117, 1119
Homilite, 1120, 1126
Honessite, 511, 665
Honguiite, 208, 210
Hongshiite, 20
Hopeite, 789-790
Hörnesite, 792, 796, 797
Horn silver (chlorargyrite), 377-378
Horsfordite, 37
Hortonolite, 1027
Hotsonite, 671, 970
Howardevansite, 732
Howieite, 1394
Howlite, 545-546, 1121
Hsianghualite, 1651-1652
Huanghoite-(Ce), 481, 482
Huangite, 631, 633-634
Hübnerite, 993-994, 995

Huemulite, 987
Hügelite, 886
Hühnerkobelite (alluaudite), 706, 707, 709-710
Hulsite, 538
Humberstonite, 667
Humboldtilite (melilite), 1142, 1144-1145
Humboldtine, 1011-1012
Humite, 1078, 1082, 1083
Humite group, 1076-1086
Hummerite, 986
Hungchaoite, 555
Huntite, 424, 434, 457
Huréaulite, 739-740
Hurlbutite, 717-718
Hutchinsonite, 198
Huttonite, 723, 1058-1059, 1104
Hyacinth (zircon), 1053-1056
Hyalophane, 1597, 1603-1604
Hyalosiderite, 1027
Hyalotekite, 1712
Hydrated carbonates, 462-474
Hydroastrophyllite, 1381, 1384
Hydrobasaluminite (winebergite), 645
Hydrobiotite (hydromica, jefferisite, vermiculite), 1471-1472, 1474-1476
Hydroboracite, 554
Hydrocalumite, 289
Hydrocerite, 1104
Hydrocerussite, 485-486, 494
Hydrochlorborite, 553
Hydrocyanite (chalcocyanite), 577-578
Hydrodelhayelite (monteregianite), 1539, 1540-1541
Hydrodresserite, 501-502
Hydrogen-autunite (chernikovite), 765, 777
Hydroglauberite, 589
Hydrogrossular (hibschite), 1037, 1051
Hydrohalite, 376
Hydrohetaerolite, 306
Hydrohonessite, 511, 665
Hydromagnesite, 513-514
Hydrombobomkulite, 529
Hydromica (illite), 1472-1474
Hydromica (vermiculite,hydrobiotite), 1471-1472, 1474-1476
Hydromolysite, 388-389
Hydromuscovite (illite), 1472-1474
Hydronium jarosite (carphosiderite), 631, 634-635
Hydroparagonite (brammallite), 1477
Hydrophilite, 386
Hydroromarchite, 212, 288
Hydroscarbroite, 516, 517
Hydrotalcite, 508, 509-510
Hydrotalcite subgroup. See Sjögrenite-hydrotalcite group
Hydrotungstite, 1005
Hydroxyapophyllite, 1525-1527
Hydroxycancrinite, 1632, 1637
Hydroxylapatite, 856, 858, 861, 865
Hydroxylbastnäsite-(Ce), 476, 478

Hydroxylbastnäsite-(La), 476, *478*
Hydroxylbastnäsite-(Nd), 476, *479*
Hydroxylellestadite, *1102-1103*
Hydroxylherderite, *832*, 833
Hydrozincite, *493-494*, 525
Hypercinnabar, *73-74*

I

Ianthinite, *261*
Ice, *205-208*
Iceland spar (calcite), 427, *428-434*
Idaite, *96-97*
Idocrase (vesuvianite), *1213-1215*
Idrialite (picene, curtisite), *1016*
Igdloite, 224
Iimoriite-(Y), *1107*
Ikaite, *463*
Ikunolite, 56, 57, *58*
Ilesite, 602, *603*
Ilimaussite-(Ce), *1714*
Illite (hydromuscovite, hydromica, gümbelite), *1472-1474*
Ilmajokite, *1714-1715*
Ilmenite, 225, *226-227*, 330, 363, 366
Ilmenite group, 225-229
Ilmenorutile, 235, *237*
Ilsemannite, *253*
Ilvaite (lievrite), *1158-1159*
Imandrite, 1247, *1249*
Imhofite, *181-182*
Imiterite, *55*
Imogolite, *1432-1436*, 1435
Inaglyite, *104*
Incaite, *152*
Inderborite, *552*
Inderite, *553*
Indialite, 1238, *1245*, 1251
Indigirite, *499*
Indite, 99, *102*
Indium, *13*, 102
Inesite, *1372*
Ingersonite, *972-973*
Ingodite (grünlingite), 82, *84*
Innelite, *1216*
Insizwaite, 114, *118*
Inyoite, *552*
Iodargyrite, *378*
Iolite (cordierite), *1251-1253*
Iowaite, *289*
Iquiqueite, *566*
Iranite, *696*
Iraqite-La (desourdyite, ekanite), *1269*, 1521
Irarsite, 124, *126-127*
Irhtemite, 741, *742*
Iridarsenite, 129, *130*
Iridium, *17-18*, 20
Iridosmine (osmium), 18, *19*
Iriginite, 1005, *1006*
Iron, 4, 11
Iron-nickel group, 9-12

Iron-wollastonite (ferrobustamite), 1314, *1318-1319*
Irtyshite, *368-369*
Ishikawaite, *334*
Isoclasite, *892*
Isocubanite, *96*
Isoferroplatinum, 23, *24*
Isoferroplatinum group, 23-25
Isokite, 834, *835-836*
Isomertieite, *144*
Itoite, *636*
Itolite, 572
Iwakiite (α-vredenbergite), 298, *305*
Ixiolite, *339-340*
Izoklakeite, *182*

J

Jachimovite cuprosklodowskite, 1109, *1112*
Jachymovite cuprosklodowskite, 1109, *1112*
Jacobsite, 294, *298*
Jadeite, 1291, *1304-1305*
Jaffeite, *1141*
Jagoite, *1548-1549*
Jagowerite, *876*
Jahnsite-(CaMnFe), 928, *929-930*
Jahnsite-(CaMnMg), 928, *929-930*
Jahnsite-(CaMnMn), 928, *929-930*
Jahnsite-(MnMnMn), 928, *929-930*
Jahnsite subgroup, 927-930, 931. *See also* Whiteite subgroup
Jalpaite, *44*
Jamborite, *287*
Jamesite, *880*
Jamesonite, *184*, 185
Janggunite, *290*
Janhaugite, 1160, *1166*
Jankovicite, *202*
Jarlite, *415*
Jarosewichite, *815-816*
Jarosite, 630, 631, *634-635*
Jaskólskiite, *182*
Jasmundite, *1099*
Jeanbandyite, 284, *286-287*
Jefferisite (vermiculite,hydrobiotite), *1471-1472*, *1474-1476*
Jeffersonite, 1298
Jeffreyite, 1143, *1146*, 1147
Jenkinsite (antigorite), 1406, *1415-1417*, 1428
Jennite, 1160, *1166-1167*
Jensenite, *674*
Jeppeite, *372*
Jeremejevite, *549*
Jeromite, *134*
Jerrygibbsite, 1078, *1085*
Jervisite, 1291, *1307*
Jezekite, *885*
Jianshuiite, *320*
Jimboite, *539*
Jimthompsonite, *1376*
Jinshajiangite, *1703-1704*
Jixianite, *1004*

General Index

Joaquinite-(Ce), 1231, *1232*
Joaquinite group, 1231-1237
Joesmithite, *1367-1368*
Johachidoliite, *541*
Johannite, *650-651*
Johannsenite, 1291, *1301-1302*
Johillerite, 706, *710*
Johnbaumite, 856, 864, *865*
Johninnesite, *1395-1396*
Johnsomervilleite, 712, *713*
Johnstrupite (mosandrite), *1168-1169*
Johnwalkite, *907*
Jokokuite, 604, *606-607*
Joliotite, *464*
Jolliffeite, 124, *127*
Jonesite, *1254*
Jordanite, *160*
Jordisite, *134*
Joséite, *56-57*
Joséite-B, 56, *57*
Joséite-C, 56
Joséite group, 56-59
Josephinite (awaruite), 10, *12*
Jouravskite, 658, *670*
Juanite, *1715*
Julgoldite-(Fe^{2+}), 1206, *1207*
Julgoldite-(Fe^{3+}), 1206, 1207, *1208*, 1209
Julienite (sodium cobalt thiocyanate), *1014*
Jungite, *953*
Junitoite, *1152*
Junoite, *196*
Jurbanite, *655-656*

K

Kaatialaite, *745-746*
Kadyrelite, *400*
Kaersutite, 1338, *1356*
Kafehydrocyanite, *1014*
Kahlerite, 764, *775*
Kainite, *648*
Kainosite-(Y) (cenosite), *1236-1237*
Kalborsite, *1701*
Kaliborite, *559*
Kalicinite, *423*
Kalininite, 99, *102*
Kalinite, *593*
Kaliophilite, 1619, *1622*
Kalipyrochlore, 342, *344*
Kalistrontite, *580*
Kalithompsonite, *1401*
Kalkspar (calcite), 427, *428-434*
Kalkstein (calcite), 427, *428-434*
Kalsilite, *1619*, 1622
Kaluginite, *927*
Kamacite, 9, *10*, 11, 16
Kamaishilite, 1624, *1628-1629*
Kambaldaite, *517*
Kamchatkite, *637*
Kamiokite, *316*
Kamitugaite, *955*
Kamotoite-(Y), *503-504*

Kamphaugite-(Y), *500*
Kanemite, *1548*
Kankite, *804-805*
Kanoite, 1289, *1292*, 1296
Kanonaite, 1064, 1067, *1068*
Kaolinite, 1067, 1406, *1407-1411*, 1412
Kaolinite-serpentine group, 1405-1431
Karasugite, *418*
Karelianite, 214, *220-221*
Karibibite, *325-326*
Karnasurtite-(Ce), *1104*
Karpatite (coronene, pendletonite), *1016*
Karpinskite, *1710*
Kashinite, *113*
Kasoite, 1604
Kasolite, 1109, *1110*
Kassite, *361*
Katayamalite, *1250-1251*
Katoite, 1037, *1051-1052*
Katoptrite (catoptrite), *974*
Kawazuite, 109, 110, *111*
Kazakhstanite, *989*
Kazakovite, 1246, *1247-1248*
Keckite, 928, *931*
Kegelite, *1520*
Keithconnite, *40*
Keiviite-(Y), 1134, *1135*
Keiviite-(Yb), 1134, *1135-1136*
Keldyshite, *1139*
Kellyite, 1406, *1426*
Kelyanite, *405*
Kemmlitzite, 960, *962*
Kempite, *392*
Kentrolite, *1182-1183*
Kenyaite, *1715*
Kerchite, *793*
Kermesite, *140*
Kernite, *556*
Kerolite, 1441, 1443
Kertschenite (metavivianite), 792, 793, *798*
Kesterite, 89, *91*
Kettnerite, *492*
Keyite, *718*
Keystoneite, *683*
Khademite, *656*
Khamrabaevite, *15*
Khanneshite, *459-460*
Kharaelakhite, *64*
Khatyrkite, *13-14*
Khibinskite, *1139*
Khinite, *672*, 674
Khristovite-(Ce), 1196, *1201-1202*
Kiddcreekite, *93*
Kidwellite, *910*
Kieftite, *140*
Kieserite, 595, *596*
Kieserite group, 595-597
Kilchoanite, *1219*
Killalaite, 1160, *1181*
Kimrobinsonite, *290*
Kimuraite-(Y), *473-474*, 500

Kimzeyite, 1037, *1050*
Kingite, *923*
Kingsmountite, *935-936*
Kinichilite, *682-683*
Kinoite, *1188-1189*
Kinoshitalite, 1447, *1470-1471*
Kintoreite, 899, *901*
Kipushite, *883-884*
Kirchsteinite, 1025, 1032, *1033*
Kirkiite, *178*
Kirovite, 611
Kitkaite, *137*
Kittatinnyite, 889, *1091*
Kivuite, *888*
Kladnoite (phthalimide), *1018*
Klebelsbergite, *624*
Kleberite, *330*
Kleemanite, *942*
Kleinite, *398*
Klockmannite, *79-80*
Klyuchevskite, *582*, *638*
Knebelite (fayalite), *1025-1026*
Knorringite, 1037, *1042-1043*
Koashvite, 1247, *1249*
Kobeite-(Y), *362*
Kobellite, *187*
Kochkarite, *59*
Koechlinite, *1001*
Koenenite, *403*
Kogarkoite, *624*
Koktaite, *585*
Kolarite, *143*
Kolbeckite, 800, *804*
Kolchoanite, 1141
Kolfanite, *912*
Kolicite, *965*
Kolovratite, *814*
Kolwezite, 486, 487, *488*
Kolymite, *8-9*
Komarovite, 1160, *1167*
Kombatite, 816, *817*
Komkovite, *1230*
Konderite, *104*
Koninckite, *803*
Konyaite, *586*
Koragoite, *372*
Koritnigite, *736-737*
Kornelite, *619*
Kornerupine, *1193-1194*
Kornite, 1339, *1366-1367*
Korshunovskite, *392-393*
Korzhinskite, *563*
Kosmochlor (cosmochlore, ureyite), 1291, *1307*
Kosnarite, *731*
Kostovite, *136*
Kostylevite, 1224, *1225-1226*
Kotoite, *539*
Köttigite, 792, *795*, 798, 913
Kotulskite, 76, *78*
Koutekite, *40*
Kovdorskite, *968*

Kozulite, 1339, *1365-1366*
Kraisslite, *965*
Kratochvilite, *1015*
Krausite, *591-592*
Krauskopfite, *1564*
Krautite, 735, *736*
K-rectorite (rectorite), *1515-1517*
Kremersite, *410*
Krennerite, *134-135*
Kribergite, *968*
Krinovite, 1386, *1388*
Kröhnkite, *585*
Krupkaite, 169, *170*
Krutaite, 114, *116*
Krutovite, 114, *117*, 123
Kryzhanovskite, *787-788*
Ktenasite, *647*
Kukisvumite, *1312*
Kuksite, *676*
Kulanite, *870*
Kuliokite-(Y), *1716*
Kulkeite, *1514-1515*
Kullerudite, 116, 120, *122*
Kupletskite, 1381, 1382, *1383*
Kuramite, 89, *90*
Kuranakhite, *673*
Kurchatovite, *540-541*
Kurnakovite, *553*
Kurumsakite, *1704*
Kusachiite, *304-305*
Kusuite (wakefieldite-(Ce)), 728, *731*
Kutinaite, *39*
Kutnohorite (kutnahorite), 437, 449, *452*
Kuzminite, *380*
Kvanefjeldite, *1542-1543*
Kyanite (cyanite, disthene), 1064, 1066, 1067, *1069-1070*
Kyzylkumite, *363*

L

Labradorite, 1597, *1608*
Labuntsovite, 1167, 1234, *1235-1236*, 1543
Lacroixite, *834*
Laffittite, *173*
Laihunite, 1025, *1031-1032*
Laitakarite, 56, *58*
Lammerite, *718-719*
Lamprophyllite, 1173, 1174, *1175*
Lamprophyllite group, 1173-1176
Lanarkite, *628*
Landauite, *365*
Landesite, 787, *788*
Långbanite, *974*
Langbeinite, *581*, 581
Langisite, 76, *79*
Langite, *644*
Lannaeite, 98, *99*
Lannaeite group, 98-103
Lannonite, *657*
Lansfordite, 424, *464*
Lanthanite-(Ce), *472-473*

Lanthanite-(La), *472*
Lanthanite-(Nd), *472*
Laphamite, *109*
Lapieite, *168*
Laplandite-(Ce), *1716*
Laponite, 1492
Lapparentite, 656
Larderellite, *556-557*
Larnite, *1052-1053*
Larosite, *146*
Larsenite, *1023-1024*
Latiumite, *1546*
Latrappite, 222, *223-224*
Laueite, 937, 938, *940*, 941
Laueite-paravauxite group, 936-941, 943-944.
   See also Strunzite subgroup
Laumontite (lomonite, leonhardite), *1650-1651*
Launayite, *184*
Laurelite, *387-388*
Laurionite, *394*
Laurite, 114, *117*, 118
Lausenite, *618*
Lautarite, *531*
Lautenthalite, *646-647*
Lautite, *132*
Lavendulan, *920-921*
Låvenite, 1160, *1162-1163*, 1164
Lavrentievite, 396, *397*
Lawonite, *1157*, 1158
Lawrencite, *384*
Lawsonbauerite, 639, *640*
Lazarenkoite, *321*
Lazulite, *873-874*, 880
Lazulite group, 873-875
Lazurite, 1624, *1627-1628*
Lead, 2, *5*, 13, 64, 65, 212
Leadamalgam, *9*
Leadhillite, 212, *520*, 521
Leakeite, 1339, *1366*
Lecontite, *584*
Legrandite, *893*
Lehnerite, 765, *776*
Leifite, *1717*
Leightonite, *591*
Leiteite, *978*
Lemoynite, *1563*
Lenaite, *89*
Lengenbachite, *165*
Leningradite, *843*
Lennilenapeite, 1552, *1553*, 1554
Lenoblite, *251*
Leonhardite (laumontite), *1650-1651*
Leonite, 580, *586*
Lepersonnite-(Gd), *524*
Lepidocrocite, 268, *270-271*, 273, *273*
Lepidolite, 1447, *1462-1464*
Lepidomelane, 1090, 1458
Lermontovite, *808-809*
Lessingite (britholite-(Ce)), *1100-1101*, 1104
Letovicite, *568*
Leucite, *1622-1623*, 1648

Leucophanite, 1143, *1146*, 1147
Leucophoenicite, 1078, *1083*
Leucophosphite, *932*
Leucosphenite, *1541*
Levyclaudite, *151*
Levyne, *1660-1661*
Liandratite, *339*
Liberite, *1023*
Libethenite, *849-850*
Liddicoatite, 1256, *1257-1258*
Liebauite, *1395*
Liebenbergite, 1025, *1029*
Liebigite, *468*
Lievrite (ilvaite), *1158-1159*
Likasite, *529*
Lillianite, *175-176*
Lillianite group, 174-178
Lime, 208, *210*, 428
Limonite (goethite), *268-269*, 270, 271, 273, 464, 634, 673, 900
Linarite, *629*, 672
Lindackerite, *742-743*
Lindgrenite, *1002*
Lindqvistite, *331*
Lindsleyite, 364, 365, *367*
Lindströmite, 169, *171*
Lingaitukuang (brabantite), *724*
Lintisite, *1313*
Liottite, 1632, *1637-1638*
Lipscombite, *875-876*
Liroconite, *882-883*
Lisetite, *1616-1617*
Lishizhenite, *615*
Liskeardite, *882*
Litharge, *212*, 213
Lithidionite (litidionite), *1397*
Lithiomarsturite, 1327, *1331*
Lithiophilite, 700, *701*, 703
Lithiophorite, *288*
Lithiophosphate, *727*
Lithiotantite, *371*
Lithiowodginite, 337, *338*
Lithosite, *1699*
Litidionite (lithidionite), *1397*
Liveingite (rathite-II), *187*
Livingstonite, *195*
Lizardite (orthoantigorite), 1406, *1418-1420*
Loeweite (löweite), *590*
Lokkaite-(Y), *473*, 500
Löllingite, 120, *122-123*
Lomonite (laumontite), *1650-1651*
Lomonosovite, 1177, *1184-1185*, 1186, 1187
Lonecreekite, *593*
Lonsdaleite, 32, 33, *35*
Loparite, 225
Loparite-(CE), 222, *224*
Lopezite, *692-693*
Lorandite, *192-193*
Loranskite-(Y), *334*
Lorenzenite, *1312*
Lorenzenite group, 1312-1313

Lorettoite, *402*
Loseyite, *495*
Lotharmeyerite, *699*
Loudounite, *1229*
Loughlinite, 1558, *1561*
Lourenswalsite, *1254*
Lovdarite, *1712*
Loveringite, *365*
Lovozerite, 1246, *1247*
Lovozerite group, 1245-1249
Löweite (loeweite), *590*
Luanheite, *8*
Luberoite, *149*
Lublinite (calcite), 427, *428-434*
Lucasite-(Ce), *357*
Luddenite, *1717*
Ludjibaite, 822, *823*
Ludlamite, 776, *790*
Ludlockite, *732*, *980*
Ludwigite, *535-536*
Ludwigite group, 534-537
Lueshite, 222, *224*
Luetheite, *906*
Lünebergite, *968*
Lunijianiaite, *1509-1510*
Lun'okite, *926*
Lussatite, 1568
Lusungite, 899, *900-901*
Lutecine (lutecite), *1593-1594*
Lutecite (lutecine, moganite), 243, 1570, *1593-1594*
Luzonite, *156*
Lyonsite, *732*

M

Macaulayite, *1480*
Macdonaldite, 1539, *1540*
Macedonite, *229*
Macfallite, *1212*
Machatschkiite, *745*
Mackayite, *685-686*
Mackinawite, *63*
Macphersonite, 520, *521*
Macquartite, *696*
Madocite, *180*
Magadiite, *1717-1718*
Magbasite, *1393*
Maghagendorfite, 706, *709*
Maghemite, *229-230*
Magnesio(alumino)katophorite, 1338, *1359*
Magnesio(alumino)taramite, 1338, *1360*
Magnesioanthophyllite, 1337, *1343*
Magnesioarfvedsonite, 1339, *1364*
Magnesioaubertite, *655*
Magnesioaxinite, 1154, *1155*
Magnesiocarpholite, 1310, *1311*
Magnesiochloritoid-A, 1087, *1089*
Magnesiochloritoid-M (sismondine), 1087, *1089*
Magnesiochromite (chrompicotite), 294, *301*, 505

Magnesioclinoholmquistite, 1337, *1343*
Magnesiocopiaite, *663*
Magnesiocoulsonite, 294, *303-304*
Magnesiocummingtonite, 1337, *1339*
Magnesiodumortierite, *1118-1119*
Magnesioferrikatophorite, 1338, *1359*
Magnesioferritaramite, 1338, *1360*
Magnesioferrite, 294, *298*
Magnesiogedrite, 1337, *1345*
Magnesiohastingsite, 1338, *1354*
Magnesioholmquistite, 1337, *1346*
Magnesiohulsite, *538*
Magnesioriebeckite, 1339, *1361*
Magnesiosadanagaite, 1338, *1356*
Magnesite, 427, 431, *434-435*, 436
Magnesium astrophyllite, 1381, *1384*
Magnesium-chlorophoenicite, *815*
Magnesium orthite (**dollaseite-(Ce)**), 1196, *1199*
Magnesium zippeite, 660, *661*
Magnetite, 215, 221, 294, *299*
Magnetoplumbite, *312*
Magniotriplite, 845, *846*
Magnocolumbite, 353, *356*
Magnolite, *679*
Magnussonite, *981*
Mahlmoodite, *749*
Majakite, *51*
Majorite, 1037, *1043*
Makarochkinite (**hogtuvaite**), 1386, *1387-1388*
Makatite, *1565*
Mäkinenite, *81*
Makovickyite, 199, *200*
Malachite, 5, 482-485, *488-490*, 917
Malacon (zircon), *1053-1056*
Malanite, 99, *103*
Malayaite, 1092, *1095*
Maldonite, *6*
Malladrite, *410*
Mallardite, 610, *612*
Mammothite, *626-627*
Manaksite, 1397, *1398*
Manandonite, 1406, *1426-1427*
Manasseite, 507, *508-509*
Mandarinoite, *683-684*
Manganarsite, *983*
Manganaxinite, 1154, *1156*
Manganbabingtonite, 1326, 1328, *1329*
Manganbelyankinite, *371*
Manganberzeliite, *705*
Manganese-hörnesite, 792, *797*
Manganese shadlunite, 60, *62-63*
Mangangordonite, 937, *944*
Manganhumite, 1078, *1083-1084*
Manganite, 269, *271-272*, 272
Manganneptunite, *1402*
Manganoan nsutite, 248
Manganochromite, 294, *301-302*
Manganocolumbite, 353, *355-356*
Manganolangbeinite, *581*
Manganophyllite, 211, 228

General Index

Manganophyllite, 1456
Manganosegelerite, *926*
Manganosite, 208, *209-210*
Manganostibite, *973-974*
Manganotantalite, 353, *355*
Manganotapiolite, *352*
Manganotychite, *519-520*
Manganpyrosmalite, 1536, *1537*
Manjiroite, 321, *322*
Mannardite, *324-325*
Mansfieldite, 800, *802-803*
Mantiennéite, *949*
Mapimite, *913*
Marcasite, 114, *120-121*, 129, 611
Marcasite group, 119-124
Margaritasite, 780, *783*
Margarite, 1447, *1468-1469*
Margarite subgroup, 1447, 1468-1471. *See also* Mica group
Margarosanite, *1224*
Marialite, *1641*
Maricite, *702*
Maricopaite, *1689-1690*
Marignacite (**ceriopyrochlore-(Ce)**), 342, *344-345*
Marmolite, 1418, 1428
Marokite, *309*
Marrite, *171*
Marshite, *379*
Marsturite, 1327, *1330-1331*
Marthozite, *688*
Mascagnite, *568-569*
Maslovite, 125, *127-128*
Massicot, 212, *213*
Masutomilite, 1447, *1466*
Masuyite, *258*
Mathewwrogersite, *1706*
Mathiasite, 365, *367*
Matildite, *190*
Matlockite, *388*
Matraite, 72, *80*
Mattagamite, 120, *122*
Matteuccite, *583*
Mattheddleite, *1103*
Matulaite, 951, *953*
Matveevite, *950*
Maucherite, *147-148*
Maufite, *1718*
Mawbyite, 760, *761*
Mawsonite, *92*
Maxwellite, 834, *835*
Mayenite, *327*
Mayingite, 114, *119*
Mazzite, *1668*
Mbobomkulite, *528*, 529
McAllisterite, *561*
Mcalpineite, *674*
Mcauslanite, *950*
Mcbirneyite, 718, *719*
Mcconnellite, 220, *292*
Mccrillisite, *810-811*

Mcgillite, 1536, *1538*
McGovernite, *965*
Mcguinnessite, 487, *488*
Mckelveyite-(Nd), *470*
Mckelveyite-(Y), *469-470*
Mckinstryite, *44-45*
McNearite, *745*
Medaite, *1192*
Meerschaum (**sepiolite**), 1557, 1558, *1560-1561*
Megacyclite, *1267-1268*
Meionite, *1641-1644*
Meixnerite, *289-290*
Melaconite (**tenorite**), *211-212*
Melanite, 1044, 1049
Melanoccrite, 1100
Melanocerite-(Ce), *1131*, 1137
Melanophlogite, *1592-1593*
Melanostibite, 225, *228-229*
Melanotekite, *1182*
Melanothallite, *398-399*
Melanovanadite, *990*
Melanterite, 608, 610, *611*, 612
Melanterite group, 610-612
Melilite-fresnoite group, 1142-1148
Melilite (humboldtilite, velardeñite), 1142, *1144-1145*
Meliphanite, 1143, *1146-1147*
Melkovite, *1009*
Mellite, *1013-1014*
Melnikovite (**greigite**), 99, *101*
Melonite, *137*
Melonite group, 136-138
Melonjosephite, *877*
Mendipite, *396*
Mendozavilite, *1010*
Mendozite, *593*, 593
Meneghinite, *162*
Mercallite, *568*
Mercury, 7, 80, 213, 378-379
Mereiterite, *580-581*, 586
Merenskyite, 137, *138*
Merlinoite, 1663, *1668-1669*
Merrihueite, 1271, *1273*, 1374
Merrillite, 715
Merrillite-(Ca), *717*
Merrillite-(Na), *717*
Merrillite-(Y), *717*
Mertieite-I, *144*
Mertieite-II, *144*
Merwinite, 704, *1035-1036*
Mesole, 1685
Mesolite, *1681-1682*
Messelite, 751, *752*
Meta-aluminite, *649*
Meta-alunogen, *620*
Meta-ankoleite, 765, *770*
Meta-autunite, 765, *766-767*, 948
Metaborite, *541*
Metacalciouraniote, *260*
Metacinnabar (onofrite (selenian), saukovite (cadmian)), 67, *69*, 73

Metadelriotie, *985-986*
Metahaiweeite, *1115*
Metahalloysite, 1411
Meta-heinrichite, 765, *769*
Metahewettite, *987*
Metahohmannite, *653*
Meta-jennite, 1167
Meta-kahlerite, 765, *775*
Meta-kingite, 923
Meta-kirscheimerite, 765, *776*
Metaköttigite, 792, *798*
Meta-lodevite, 765, *776-777*
Metamunirite, *986*
Meta-novacekite, 765, *771*
Metarossite, *985*
Meta-saleeite, 772
Metaschoderite, *959*
Metaschoepite, *257*
Metasideronatrite, *651*
Metastibnite, *108*
Metastrengite (phosphosiderite), 800, 801, 803, *804*
Metastudtite, *259*
Metaswitzerite, *790*, 791
Meta-torbernite, 765, *773-774*
Meta-tyuyamunite, 780, *781-782*
Meta-uranocircite, *768-769*, 774, 779
Meta-uranopilite, *642*
Meta-uranospinite, 765, 767, *768*
Metavandendriesscheite, *262-263*
Metavanmeersscheite, *907-908*
Metavanuralite, *942*
Metavariscite, 800, *803*
Metavauxite, 937, *941*, 943
Metavivianite (kertschenite, oxykertschenite), 792, 793, *798*
Metavoltine, *591*
Metaxite, 1428
Metazellerite, *467*
Meta-zeunerite, 765, 770, *774-775*
Meyerhofferite, *553*
Meymacite, *252*
Mgriite, 160, *171-172*
Miargyite, *191*
Mica group, 1444-1471
Michenerite, 125, *127*
Microcline, 1597, *1605-1607*
Microlite, 342, *345-346*
Microsommite, 1633, *1638*
Miersite, *379*
Miharaite, *153*
Mikasaite, *582*
Milarite, 1271, *1277-1278*
Milarite group, 1270-1280
Millerite, *81*
Millisite, *905-906*
Millosevichite, *581-582*
Mimetite, 857, 863, 864, 865, *866-867*, 868
Minamiite, 631, *633*
Minasgeraisite-(Y), 1120, *1126-1127*
Minasragrite, *614*

Mineevite-(Y), *524-525*
Minehillite, *1549-1550*
Minguzzite, *1012*
Minium, 213, *306*
Minnesotaite, 1439, *1443-1444*
Minrecordite, 449, *452*
Minyulite, *932*
Mirabilite (Glauber's salt), 579, *584*
Misenite, *568*
Miserite, *1400*
Mitridatite, *911*, 912
Mitscherlichite, *409*
Mixite, *890*
Mizzonite, 1642
Moctezumite, *678*
Modderite, *82*
Moganite (lutecite), *1593-1594*
Mohite, *97*
Mohrite (Mohr's salt), 587, *588*
Mohr's salt (mohrite), 587, *588*
Moissanite, *35*
Moluranite, *1005*
Molybdenite, *132-134*, 134
Molybdite, *251-252*
Molybdofornacite, 758, *964*
Molybdomenite, *677*, 680
Molybdophyllite, *1711*
Molysite, *388*
Monazite-(Ce), 721, *722-724*, 1058, 1059
Monazite group, 720-725
Monazite-(La), 721, *724*
Monazite-(Nd), 721, *724-725*
Moncheite, *137-138*
Monetite, *697*
Mongolite, 1160, *1168*
Mongshanite, *330*
Monimolite, *972*
Monohydrocalcite, *463*
Monsmedite, *583*
Montanite, *673*
Montbrayite, *111-112*
Montdorite, 1446, *1456*
Montebrasite, 727, 836, *837-838*
Monteponite (cadmium oxide), 208, *210*
Monteregianite-(Y), *1539-1540*
Montesommaite, *1669*
Montgomeryite, 934, *935*, 936
Montgomeryite group, 934-936
Monticellite, 1025, *1032-1033*
Montmorillonite (Ca-montmorillonite, Na-montmorillonite, bentonite), 1481, *1483-1485*
Montroseite, 249, 266, *269*
Montroyalite, *506*
Montroydite, *212-213*
Mooihoekite, *94*
Moolooite, *1013*
Mooreite, *639-640*
Moorhouseite, 608, *609*
Mopungite, 284, *287*
Moraesite, *891-892*

General Index

Mordenite, *1688*, 1689
Moreauite, *951-952*
Morelandite, 857, 864, *868*
Morenosite, *613-614*
Morimotoite, 1037, 1049, *1050-1051*
Morinite, *885*
Morozeviczite, *93*
Mosandrite (rinkite, rinkolite, johnstrupite), *1168-1169*
Moschallandsbergite, *7*
Moschelite, *380*
Moschellandsbergite, *213*
Mosesite, *589*
Mottramite, 826, *830*
Motukoreaite, *522*
Mounanaite, *878*
Mountainite, *1699-1700*
Mountkeithite, *669-670*
Mourite, *264*
Moydite-(Y), *566-567*
Mozartite, *1091-1092*
Mpororoite, *1008*
Mrazekite, *889-890*
Mroseite, *690*
Mückeite, *168*
Muirite, *1267*
Mukhinite, 1196, *1202*
Mullite, 1064, *1065-1066*, 1184
Mummeite, 199, *201*
Mundite, *907*
Mundrabillaite, *743-744*
Munirite, *986*
Murataite, *369*
Murdochite, *254*
Murmanite, 1176, *1177*, 1187
Murunskite, *54*
Muscovite, 221, 1439, 1446, *1448-1451*, 1452, 1456, 1468
Muscovite subgroup, 1446, 1448-1456. See also Mica group
Musgravite, *309*
Mushistonite, 284, *286*
Muskoxite, *326*
Muthmannite, *86*

N

Nabaphite, *748*
Nabokoite, *627*, 628
Nacaphite, *820*
Nacareniobsite-(Ce), *1169*
Nacrite, 1406, *1407-1411*
Nadorite, *395*
Nagashimalite, 1284, *1285-1286*
Nagelschmidtite, *1108*
Nagyagite, *112*
Nahcolite (sodium bicarbonate), *422-423*
Nahpoite, *699*
Nakauriite, *669*
Na-komarovite, 1160, *1167-1168*
Nalipoite, *728*
Namansilite, 1291, *1306*
Nambulite, 1326, *1329-1330*
Namibite, *734*
Na-montmorillonite (**montmorillonite,beidellite**), *1481-1483*, *1483-1485*
Namuwite, *645*
Nanlingite, *981*
Nanpingite, 1446, *1455*
Nantokite, *379*
Na-rectorite (**rectorite**), *1515-1517*
Narsarsukite, *1398-1399*
Nasinite, *557*
Nasledovite, *521-522*
Nasonite, *1183*
Nastrophite, *748*
Natalyite, 1291, *1307*
Natanite, 284*285*
Natisite, *1096*
Natrite, *448*
Natroalunite, 630, *631-632*
Natroapophyllite, *1527*
Natrobistantite, 343, *349*
Natrochalcite, *650*
Natrodufrenite, 913, *914*
Natrofairchildite, 455, *456*
Natrojarosite, 631, *634-635*
Natrolite, *1677-1679*, 1680, 1681, 1682
Natromontebrasite, 836, 837, *838*
Natron, *462-463*, 465
Natronambulite, 1327, 1329, *1330*
Natroniobite, *225*
Natrophilite, *701-702*
Natrophosphate, *954-955*
Natrosilite, *1564-1565*, 1566
Natrotantite, *368*
Naujakasite, *1546*
Naumannite, *42*
Navajoite, *253*
Nchwaningite, *1313-1314*
Nealite, *824*, *982*
Nefedovite, *881*
Neighborite, *407*
Nekoite, 1528, *1534*
Nekrasovite, 150, *151*
Nelenite (ferroschallerite), 1536, 1537, *1538*
NeltneriteI, *314*
Nenadkevichite, 1167, *1234-1235*
Nenadkevite (**boltwoodite**), 1109, *1112*, 1113
Neotocite (stratopeite, penwithite), 1435, *1436-1437*
Nepheline, 223, 1618, *1619-1621*
Nepheline group, 1617-1622
Nephrite, 1346
Nepouite (genthite), 1406, *1420-1421*, 1430, 1492
Neptunite, 224, *1401-1402*
Nesquehonite, *424*, 464
Nevskite, 82, *83-84*
Nevyanskite (**osmium**), 18, *19*
Newberyite, *737-738*
Neyite, *178*

Niahite, *748*
Niccolite (**nickeline**), *76-77*
Nichromite, 294, *302*
Ni-chrysotile (**pecoraite**), 1406, *1430-1431*
Nickel, 1, 4, 11, *12*
Nickelalumite, *643*
Nickelaustinite, 826, *828-829*
Nickelbischofite, *387*
Nickelblödite, *586*
Nickel-boussingaultite, 587, *589*
Nickel-cobalt, 4
Nickel-hexahydrite, 608, *609*
Nickeline group, 75-79
Nickeline (niccolite), *76-77*
Nickel porphyrin (**abelsonite**), *1018-1019*
Nickel-Skutterudite, *139*
Nickel zippeite, 660, *661*, 662
Nickenichite, 706, *710-711*
Nierite, *36*
Nifontovite, *554*
Nigerite 6H, *329*
Nigerite 24R, *329*, 330
Niggliite, *25-26*
Nimesite (**brindleyite**), 1406, *1424-1425*
Nimite, 1498, *1503*
Ningyoite, 807, *808*
Niningerite, 64, *66*
Niobian rutile, 235, 238
Niobo-aeschynite-(Ce), *358*
Niobophyllite, 1381, *1383-1384*
Niocalite, 1160, 1163, *1164*
Nisbite, 120, *124*
Nissonite, *902-903*
Niter (saltpeter), *526-527*
Nitratine, *526*
Nitrobarite, *527*
Nitrocalcite, *527*
Nitromagnesite, *527*
Nobleite, *562*
Nolanite, *254*
Nontronite (chloropal), 1481, *1485-1486*
Norbergite, *1078-1079*
Nordenskiöldine, *539-540*
Nordite-(Ce), *1541-1542*
Nordite-(La), *1542*
Nordstrandite, 280, *281-282*
Nordströmite, *196*
Norrishite, 1447, *1466*
Norsethite, *452*
Northupite, *493*
Nosean (noselite), 1624, *1626-1627*
Noselite (**nosean**), 1624, *1626-1627*
Novacekite, 764, *771*, 772
Novakite, *51-52*
Nowackiite, *173*
Nsutite, *248*
Nuffieldite, *161*
Nukundamite, *97*
Nullaginite, *490*
Nyböite, 1339, *1366*
Nyerereite, 440, *456*

O

Oboyerite, *690*
Obradovicite, *1009*
Obruchevite (**yttropyrochlore-(Y)**), 342, *344*
Octahedrite (**anatase**), *245-246*
O'Danielite, 706, *710*
Odinite (phyllite), 1406, *1414-1415*
Odintsovite, *1251*
Offretite, *1655-1656*
Ogdensburgite, *922*
Ohmilite, *1325*
Ojuelaite, 947, *948*
Okanoganite-(Y), 1121, *1130*
Okenite, 1528, *1532-1533*, 1534
Okhotskite-(Mg), 1206, *1208*
Okhotskite-(Mn^{2+}), 1206, *1208-1209*
Oldhamite, 64, *66*
Olekminskite, *453*
Olenite, 1256, *1260*
Olgite, *702*
Oligiste (**hematite**), 214, *217-220*
Oligoclase, 1597, *1608*
Olivenite, *849*
Olivine, 12, 307, 700, 701, 703, 713
Olivine group, 1024-1034, 1077
Olkhonskite, *363*
Olmsteadite, *906-907*
Olsacherite, 572, *666-667*
Olshanskyite, *543*
Olympite, *727*
Omeiite, 120, *122*
Omphacite, 1291, *1302-1303*
Onofrite (**selenian, metacinnabar**), 67, *69*, 73
Onoratoite, *401*
Oosterboschite, *147*
Opal, *1587-1592*
Orange bornite (**renierite**), *92-93*
Orcelite, *40*
Ordoñezite, *973*
Örebroite, *311*, *1090*, 1091
Oregonite, *53*
Orickite, *142*
Orientite, *1218-1219*
Orlymanite, *1550*
Orpheite, 960, *961*
Orpiment, *104-105*, 212
Orschallite, *691*
Orthite (**allanite-(Ce)**), *1196-1197*
Orthoantigorite (**lizardite**), 1406, *1418-1420*
Orthobrannerite (**brannerite**), *351*
Orthochevkinite, *1180*
Orthochrysotile, 1406, *1428*
Orthoclase, 1597, *1600-1602*, 1605
Orthoenstatite (**enstatite**), 1289, *1294-1295*
Orthoericssonite, 1174, *1175-1176*
Orthoferrosilite (**ferrosilite**), 1289, 1292, *1295-1296*
Orthojoaquinite-(Ce), 1231, *1232*
Orthojoaquinite subgroup, 1231, 1232-1233.
    See also Joaquinite group
Orthopinakiolite, 537, *538*

Orthopyroxene subgroup, 1289, 1294-1296.
  See also Pyroxene group
Orthorhombic låvenite (**burpalite**), 1160, *1161*
Orthoserpierite, 647, *648*
Orthowalpurgite, *813*
Osarizawaite, 630, *632*
Osarsite, 129, *130*
Osbornite, *15*
Osmiridium (**iridium**), *17-18*
Osmium group, 6, 7, 18-20
Osmium (**iridosmine, siserskite, nevyanskite**), 18, *19*
Osumilite-(Fe), 1271, *1273-1275*
Osumilite-(Mg), 1270, 1271, *1273-1275*
Otavite, 427, *439*
Otjisumeite, *326*
Ottemannite, *112*
Ottrelite-M, 1087, 1088, *1089-1090*
Otwayite, 279, *503*
Ourayite, *180*
Oursinite, 1109, *1113*
Overite, 924, *925*
Overite group, 924-927
Owensite, *143*
Owyheeite, *181*
Oxammite, *1012*
Oxykertschenite (**metavivianite**), 792, 793, *798*
Oyelite, 1528, *1534-1535*

P

Pääkkönenite, *108*
Pabstite, 1221, *1223*
Pachnolite, *413*, 414
Paderaite, *198*
Padmaite, 125, *127*
Pahasapaite, *812*
Painite, *314-315*
Palarstanide, *41*
Palenzonaite, *705-706*
Palermoite, *853*
Palladium, 17, *18*
Palladoarsenide, *51*
Palladobismutharsenide, *51*
Palladseite, *148*
Palmierite, *580*
Palygorskite (attapulgite), *1558-1559*, 1562
Palygorskite subgroup, 1556-1560. *See also* Sepiolite subgroup
Panasqueiraite, 834, *836*
Pandaite (**bariopyrochlore**), 342, *344*
Panethite, *704*
Panunzite, 1619, *1621-1622*
Paolovite, *25*
Papagoite, *1236*
Para-alumohydrocalcite, *502*
Parabariomicrolite, *351*
Parabrandtite, 751, *753-754*
Parabutlerite, 651, *652*
Paracelsian, 1604, *1615*
Parachrysotile (amianthus), 1406, *1428-1430*
Paracoquimbite, *619*

Paracostibite, 126, *132*
Paradamite, *851*
Paradocrasite, *29-30*
Parafransoletite, *809-810*
Paragonite, 1446, *1452*
Paraguanajuatite, 110, *111*
Parahilgardite, 559
Parahopeite, *789*
Parajamesonite, *185*
Parakeldyshite, *1139-1140*
Parakhinite, 672, *674*
Paralaurionite, *394*
Paralstonite, 444, *453*, 454
Paramelaconite, *253-254*
Paramendozavilite, *1010*
Paramontroseite, *249*
Paranatisite, *1096*
Paraniite-(Y), *1010*
Parantrolite, 1679, *1680-1681*
Paraotwayite, *279*
Parapierrotite, *202*
Pararammelsbergite, 123, *131*
Pararealgar, *85*
Pararobertsite, 911, *912-913*
Paraschachnerite, *8*
Paraschoepite, *257*
Parascholzite, *756*
Paraspurrite, *1106-1107*
Parasymplesite, 792, *796*, 797
Parasymplessite, 913
Paratacamite, 391, *392*
Paratellurite, *244-245*, 247
Paraumbite, *1225*
Paravauxite, 937, *943*
Paravauxite subgroup. *See* Laueite-paravauxite group; Strunzite subgroup
Parawollastonite (**wallastonite-2M**), 1314, 1315, *1317*
Pargasite, 1338, *1354*
Parisite-(Ce), 476, *480-481*, 499
Parisite group, 475, 476, 480-481
Parisite-(Nd), 476, *481*
Parkerite, *41*
Parkinsonite, *402*
Parnauite, *969*
Parsettensite, *1555*
Parsonsite, *784*
Partheite, 1157, *1700*
Partridgeite, 230
Partzite, *972*
Parwelite, *974*
Pascoite, *986*
Patronite, *148-149*
Paulingite, *1667-1668*
Paulkellerite, *820-821*
Paulkerrite, *949*, 950
Paulmooreite, *978*
Pavonite, *199-200*
Pavonite group, 199-201
Paxite, *131*
Pearceite, *153-154*

Pecoraite, 490
Pecoraite (ni-chrysotile), 1406, *1430-1431*
Pehrmanite, *309-310*
Peisleyite, *969*
Pekoite, 169, *170-171*
Pellyite, *1712-1713*
Pendletonite (**karpatite**), *1016*
Penfieldite, *398*
Pengzhizhongite-6H, *330*
Pengzhizhongite-24R, *330*
Penikisite, *870*
Penkvilksite, *1400-1401*
Pennantite, 1498, *1505*
Penroseite, 114, *116*
Pentagonite, *1567*
Pentahydrite, 604, *606*
Pentahydroborite, *552*
Pentlandite, 40, 60, *61-62*, 220
Pentlandite group, 60-63
Penwithite (**neotocite**), 1435, *1436-1437*
Penzhinite, *148*
Peprossiite-(Ce), *540*
Peretaite, *647*
Perhamite, *968*
Periclase, 208, *209*
Periclase group, 208-210
Peridot (**forsterite**), 1025, *1026-1029*
Peristerite, 1609
Perite, *394-395*
Perlialite, *1667*
Perloffite, *871*
Permingeatite, *156*
Perovskite, 222, *223*, 369
Perovskite group, 221-225
Perraultite, 1703, *1704*
Perrierite-(Ce), 1177, *1179-1180*
Perroudite, *143-144*
Perryite, *16*
Petalite, *1545*
Petarasite, 1247, *1249*
Petedunnite, 1291, *1302*
Peterbaylissite, *505-506*
Petersenite-(Ce), 459, *460-461*
Petersite-(Y), *891*
Petitjohnite, *879*
Petolite-1A, 1314, *1319*
Petolite-M2abc, *1319-1320*
Petrovicite, *154*
Petrovskaite, *51*
Petrukite, *97-98*
Petscheckite, *339*
Petzite, *44*
Phacolite (**chabazite-(Ca)**), *1652-1653*, 1655
Pharmacolite, 598, 697, *735-736*, 741
Pharmacosiderite, *908-909*
Phaunouxite, *799*
Phenacite, 211
Phenakite, 1020, *1021-1022*
Phenakite group, 1020-1023
Phenanthrene (**ravatite**), *1015*
Phengite, 1448

Philipsbornite, 899, *902*
Philipsburgite, 883, *884*
Phillipsite, 1661, 1663, 1664, *1665-1666*
Phlogopite, 1447, *1456-1458*
Phoenicochroite, *692*
Phosgenite, *491*
Phosphammite, *698*
Phosphoferrite, 786, *787*, 788
Phosphoferrite group, 786-789
Phosphofibrite, *955*
Phosphophyllite, *757*
Phosphorrösslerite, *739*
Phosphosiderite (metastrengite), 800, 801, 803, *804*
Phosphuranylite, 886, *887*, 907
Phthalimide (**kladnoite**), *1018*
Phuralumite, *886-887*
Phurcalite, *887*
Phyllite (**odinite**), 1406, *1414-1415*
Phyllotungstite, *1006*
Picene (**idrialite**), *1016*
Picite, *894*
Pickeringite, *616-617*
Picotpaulite, *95*
Picrolite (**antigorite**), 1406, *1415-1417*, 1418, 1428
Picromerite, 587, *588*
Picromerite group, 587-589
Picropharmacolite, *741*, 742
Picrosmine (**antigorite**), 1406, *1415-1417*
Piemontite, 1196, *1202-1203*, 1204
Pierrotite, *198*
Pigeonite, 1289, *1292-1294*
Pilsenite, 56, *58*
Pimelite (**röttisite, revdanskite, desaulsite**), 1443, 1481, *1492-1493*
Pinakiolite, 537, *538*
Pinalite, *396*
Pinchite, *399*
Pingguite, *685*
Pinnoite, *544-545*
Pintadoite, *991*
Pirquitasite, 89, *91*
Pirssonite, *465*
Pisanite, 611
Pitchblende (**uraninite**). See Uraninite
Pitiglianoite, 1633, *1638*
Pitticite, *967*
Piypite, *636-637*
Plagioclase subgroup, 1597-1600, 1608-1615
Plagionite, *188-189*
Plancheite, *1368-1369*
Planerite, 917, *920*
Plaster of Paris (**bassanite**), *593*, 599
Platarsite, 124, *126*
Platinian iridium (**iridium**), *17-18*
Platiniridium (**iridium**), *17-18*
Platinum, 1, 11, *17*
Platinum group, 17-18
Plattnerite, 235, *242*, 247
Platynite, *84*

Playfairite, *183*
Plazolite (hibschite), 1037, *1051*
Plombierite, 1528, 1529, *1531*
Plumbobetafite, 342, *348*
Plumboferrite, *327*
Plumbogummite, 899, *901*
Plumbojarosite, 631, *635-636*
Plumbomicrolite, 342, *346*
Plumbonacrite, *494*
Plumbopalladinite, *26*
Plumbopyrochlore, 342, *345*
Plumbotellurite, 677, *678*
Plumbotsumite, *1704-1705*
Poitevinite, 595, *597*
Pokrovskite, 488, *490*
Polarite, *71*
Poldervaartite, *1705*
Polhemusite, *70*
Polianite, 238
Polkovicite, *93*
Pollucite, 1646, 1647, *1648-1649*
Polybasite, *153*
Polycrase-(Y), *359-360*, 361, 362
Polydymite, 98, *100-101*
Polyhalite, 584, *590-591*
Polylithionite, 1447, 1462, *1464*
Polymignite (zirconolite), *350*
Polyphite, *970*, *1185-1186*
Ponomarevite, *406*
Portlandite, 275, *276*
Posnjakite, *643*
Potarite, 21, *22*
Potash alum (**potassium alum**), *593*
Potassium alum (potash alum), *593*
Potosiite, *152*
Pottsite, *749*
Poubaite, 56, *58*
Poudretteite, 1271, *1275*
Poughite, *689*
Povondraite, 1256, *1260*
Powellite, 997, *998-999*
Poyarkovite, *399*
Pravdite (britholite-(Ce)), *1100-1101*, 1104
Prehnite, *1522-1523*
Preisingerite, *878-879*
Preiswerkite, 1447, *1467*
Preobrazhenskite, *549*
Priceite, *546*
Priderite, *324*
Pringleite, *560*
Probertite, *558-559*
Prosopite, *414-415*
Prosperite, *756*
Protasite, *261*
Protoallophane (**allophane**), *1432-1434*
Protoenstatite, 1291
Proudite, *183*
Proustite, *165*, 166-167
Przhevalskite, *778*
Pseudoautunite, *767*
Pseudoboleite, *404-405*

Pseudobrookite, *316-317*
Pseudocotunnite, *408*
Pseudograndreefite, *419*
Pseudolaueite, 937, *941*
Pseudomalachite, 821, *822-823*
Pseudorutile, *363*
Pseudowollastonite, 1315, 1317, 1550
Psilomelane (romanechite), *323*
Pucherite, *725-726*, 728n
Pumpellyite-(Fe^{2+}), 1206, 1207, *1209*, 1217
Pumpellyite-(Fe^{3+}), 1206, *1209-1210*, 1217
Pumpellyite group, 1205-1211
Pumpellyite-(Mg), 1206, 1207, 1208, *1210*
Pumpellyite-(Mn^{2+}), 1207, 1208, *1211*
Purpurite, 703, 710, *719-720*
Putoranite, *94*
Pyrargyrite, *166*
Pyrite, 47, 96, 113, *114-115*, 120, 228, 632, 634
Pyrite group, 113-119
Pyroaurite, 508, 509, *510-511*, 516, 522
Pyrobelonite, 826, *830-831*
Pyrobelonite group, 1097
Pyrochlore, 341, 342, *343-344*
Pyrochlore group, 341-349
Pyrochroite, 275, *276*
Pyrolusite, 17, 235, *238-239*, 247
β-Pyrolusite, 238
Pyromorphite, 857, *865-866*
Pyrooxene, 228
Pyrope, *1037-1039*, 1042
Pyrophanite, 225, 228
Pyrophyllite-1A,2M, 1438, *1439-1441*
Pyrophyllite talc group, 1437-1444
Pyrosmalite, 1535
Pyrosmalite group, 1535-1538
Pyrostilpnite, 166, *167*
Pyroxene, 228
Pyroxene group, 1288-1309
Pyroxferroite, 1332, *1333*
Pyroxmangite, 1327, *1332-1333*
Pyrrhotite, 61, *74-75*, 220, 221, 227, 228

Q

Qandilite, 295, *304*
Qilianshanite, *524*
Qingheiite, 707, *712*
Qitianlinite, *340*
Quadridavyne, 1633, *1639*
Quadruphite, *970*, 1185, *1186*
Quartz, 4, 13, 17, 192, 220, 243, 720, 1570, *1573-1586*, 1593
Quatrandorite, *177*
Queitite, *1194*
Quenselite, *288*
Quenstedtite, *619-620*
Quetzalcoatlite, *686*
Quintinite-2H, *512*
Quintinite-3T, *512*

R

Rabbittite, 515
Rabejacite, 648
Radhakrishnaite, 143
Radtkeite, 397
Raguinite, 94
Raite, 1711
Rajite, 684
Ralstonite, 416
Ramdohrite, 175, 177
Rameauite, 260-261
Rammelsbergite, 120, 123-124, 131
Ramsayite, 1312
Ramsbeckite, 645-646
Ramsdellite, 238, 247-248
Ranciéite, 325
Ranite, 1679
Rankachite, 1008
Rankamaite, 370, 371
Rankinite, 1141, 1219
Ranquilite (haiweeite), 1114, 1115
Ransomite, 614
Ranunculite, 883
Rapidcreekite, 667
Raspite, 1000, 1001
Rasvumite, 95
Rathite, 195
Rathite-II (liveingite), 187
Rauenthalite, 799
Rauvite, 991
Ravatite (phenanthrene), 1015
Rayite, 188, 190
Realgar, 84-85, 212
Rebulite, 202
Rectorite (Na-rectorite, K-rectorite, Ca-rectorite), 1515-1517
Reddingite, 701, 787, 788
Redingtonite, 616, 618
Redledgeite (chromrutile), 325
Reedmergnerite, 1597, 1614-1615
Reevesite, 508, 511
Refikite (delta13-dihydro-d-pimaric acid), 1017
Regularly interstratified phyllosilicate group, 1505-1518
Reichenbachite, 822, 823
Reinerite, 977
Reinhardbraunsite (calciochondrodite), 1078, 1081
Remondite-(Ce), 459, 460, 461
Renardite, 886, 888
Renierite (orange bornite), 92-93
Reppiaite, 824
Retgersite, 609, 613
Retinalite, 1418, 1428
Retzian-(Ce), 819
Retzian-(La), 819
Retzian-(Nd), 819
Revdanskite (pimelite), 1481, 1492-1493
Revdite, 1566
Reyerite, 1549, 1550

Rhabdite (schreibersite), 15
Rhabdophane-(Ce), 806, 807
Rhabdophane group, 806-808
Rhabdophane-(La), 807
Rhabdophane-(Nd), 807
Rhenium, 18, 20
Rhodesite, 1539
Rhodesite group, 1539-1541
Rhodium, 17, 18
Rhodizite, 549-550
Rhodochrosite, 427, 436-438, 701, 751
Rhodolite, 1037
Rhodonite, 1326, 1327-1328, 1331
Rhodonite group, 1325-1331
Rhodostannite, 103
Rhodplumsite, 41-42
Rhomboclase, 583
Rhönite, 1386, 1387, 1388-1389
Ribbeite, 1078, 1079, 1081-1082
Richellite, 953
Richelsdorfite, 908
Richetite, 264
Richterite, 211, 1338, 1358
Richterite subgroup, 1338, 1357-1361. See also Amphibole group
Rickardite, 147
Riebeckite, 1339, 1361-1364
Rijkeboerite (bariomicrolite), 342, 346
Rilandite, 331
Ringwoodite, 1034-1035, 1140
Rinkite group, 1168
Rinkite (mosandrite), 1168-1169
Rinkite-seidozerite group, 1165
Rinkolite (mosandrite), 1168-1169
Rinneite, 411
Rittmannite, 929, 931
Rivadavite, 560-561
Riversideite, 1528, 1531, 1532
Roaldite, 14
Robertsite, 911, 912
Robinsonite, 186-187
Rockbridgeite, 871-872
Rock salt (halite), 373-375
Rodalquilarite, 686
Roeblingite, 1284
Roedderite, 1271, 1272, 1273, 1278-1279
Roggianite (ginzburgite), 1696
Rohaite, 196
Roküknite, 385
Romanechite (psilomelane), 323
Romarchite, 212, 288
Roméite, 971-972
Römerite, 615
Röntgenite-(Ce), 481
Rooseveltite, 721, 725
Roquesite, 87, 88-89
Rorisite, 388
Rosasite, 487
Rosasite group, 486-488
Roscherite-A, 904
Roscherite-M, 904

Roscoelite, 1446, *1453*
Roselite, 750, *754*, 755
β-Roselite, 751, *753*
Roselite subgroup, 750, 754-756. *See also*
 Fairfieldite subgroup
Rosemaryite, 707, *711-712*
Rosenbergite, *390*
Rosenbuschite, 1160, *1165*, 1166, 1170
Rosenhahnite, *1189-1190*
Roshchinite, 175, *177*
Rosickyite, 30, *31-32*
Rosiérésite, *954*
Rossite, *985*
Rösslerite (wapplerite), *738-739*
Rostite, 655, *656*
Röttisite (pimelite), 1481, *1492-1493*
Roubaultite, *518*
Rouseite, *977*
Routhierite, *173*
Rouvilleite, *496*
Roweite, *546*
Rowlandite-(Y), *1137-1138*
Roxbyite, 46, *48*
Rozenite, 602, *603*
Rozenite group, 602-604
Ruarsite, 129, *130*
Rucklidgeite, 56, *58-59*
Ruitenbergite, *560*
Ruizite, *1190-1191*
Rusakovite, *884*
Russellite, *1001*
Rustenbergite, 23, *24*
Rustumite, *1215*
Ruthenarsenite, *81*
Rutheniridosmine, 18, *20*
Ruthenium, 18, *19-20*
Rutherfordine, *447*
Rutile, 227, 234, *235-237*, 246, 363
Rutile group, 234-244
Rynersonite, *359*

S

Sabatierite, *50*
Sabieite, *578*
Sabinaite, *495-496*
Sabugalite, *779*
Sacrofanite, 1633, *1639*
Sadanagaite, 1338, *1356*
Safflorite, 120, *123*
Sahamalite, *457*
Sahlinite, *816-817*
Sainfeldite, *740*, 741
Sakhaite, 522, *565-566*
Sakharovaite, *194*
Sakuraiite, 89, *90*
Sal Ammoniac, *377*
Saleeite, 764, *772*
Salesite, *532*
Saliotite, *1510*
Salite, 1297
Salt (halite), *373-375*

Saltpeter (niter), *526-527*
Samarskite-(Y), *340*
Samfowlerite, *1217*
Samiresite (betafite), 342, *347-348*
Sampleite, *920*
Samsonite, *173*
Samuelsonite, *878*
Sanbornite, *1563-1564*
Sand calcite (calcite), 427, *428-434*
Sanderite, *601*
Saneroite, *1393-1394*
Sanidine, 1597, 1600, *1602-1603*
Sanjuanite, *966*
Sanmartinite, 995, *996*
Santaclaraite, *1331*
Santafeite, *921*
Santanaite, *694*
Santite, *556*
Saponite (griffithite), 1481, *1489-1490*
Sapphirine, *1391-1392*
Sarabauite, *140*
Sarcolite, *1644*
Sarcopside, *713-714*
Sarkinite, 845, *848*, 850
Sarmientite, *966*
Sartorite (scleroclase, skleroklas), *193*
Saryarkite (saryarkite-(Y)), *1107-1108*
Saryarkite-(Y) (saryarkite), *1107-1108*
Sasaite, *967*
Sassolite, *539*
Satimolite, *563*
Satin spar (calcite), 427, *428-434*
Satpaevite, *990*
Satterlyite, *848*
Sauconite (tallow clay), 1481, *1490-1491*, 1495
Saukovite (cadmian, metacinnabar), 67, *69*, 73
Sayrite, *264-265*
Sazhinite-(Ce), *1525*
Sazykinaite-(Y), *1230*
Sborgite, *556*
Scacchite, *384-385*
Scarbroite, *516-517*
Scawtite, *1283*
Schachnerite, *8*
Schafarzikite, *978*
Schairerite, *625*
Schallerite, 1536, *1537*, 1538
Schaurteite, *650*
Scheelite, 252, *996-998*, 1010
Schefferite, 1298
Schertelite, *743*
Scheteligite, *350-351*
Schieffelinite, *675*
Schirmerite, *179*
Schizolite, 1319
Schlossmacherite, 630, *632*
Schmiederite, *672*
Schmitterite, *678-679*
Schneiderhöhnite, *979*
Schoderite, *959*
Schoenfliesite, 284, *285*

Schoepite, 257-258
Schöllhornite, 97
Scholzite, 756-757
Schoonerite, 910-911
Schorl, 1257, 1263-1264
Schorlomite, 1037, 1044, 1049-1050
Schreibersite (rhabdite), 15
Schreyerite, 362-363
Schröckingerite, 521
Schubnelite, 989-990
Schuetteite, 626
Schuilingite-(Nd), 499
Schulenbergite, 641
Schultenite, 698
Schumacherite, 879
Schwartzembergite, 532
Schweizerite, 1428
Schwertmannite, 290-291
Sclarite, 495
Scleroclase (sartorite), 193
Scolecite, 1681, 1682-1683
Scorodite, 800, 801, 802, 803
Scorzalite, 873, 874-875
Scotiolite (hisingerite), 1432, 1434-1435
Scotlandite, 677
Scrutinyite, 242, 247
Seamanite, 964
Searlesite, 1566
Sederholmite, 76, 77
Sedovite, 1003-1004
Seeligerite, 532
Seelite, 764, 772, 984
Segelerite, 925-926
Segnitite, 899, 902
Seidozerite, 1171
Seinäjokite, 120, 123
Sekaninaite, 1238, 1245, 1253-1254
Selenite (gypsum), 598-601
Selenium, 30
Selenostephanite, 157
Seligmanite, 167
Sellaite, 384
Selwynite, 810, 811
Semenovite-(Ce), 1542
Semseyite, 188, 189
Senaite, 365, 366
Senandorite, 177
Sénarmontite, 231-232, 233
Senegalite, 894-895
Sengierite, 895-896
Sepiolite (meerschaum), 1557, 1558, 1560-1561
Sepiolite subgroup, 1556, 1558, 1560-1561. See also Palygorskite subgroup
Serandite, 1314, 1319, 1320
Serendibite, 1386, 1389-1390
Sergeevite, 424-425
Sericite, 1067, 1448
Serophite, 1418
Serpentine, 1415, 1418
Serpentine group. See Kaolinite-serpentine group

Serpentinite, 6, 13, 14, 464, 1323
Serpierite, 647, 648
Severginite (tinzenite), 1154, 1156-1157
Shabaite-(Nd), 499-500
Shabynite, 551
Shadlunite, 60, 62
Shafranovskite, 1706
Shakhovite, 975
Shandite, 41
Sharpite, 517
Shattuckite, 1313
Shcherbakovite, 1325
Shcherbinaite, 253
Sherwoodite, 987
Shigaite, 639
Shomiokite-(Y), 474
Shortite, 456-457
Shuangfengite, 138
Shubnikovite, 913
Shuiskite-(Mg), 1207, 1211
Sibirskite, 544
Sicklerite, 701, 703
Siderazot, 14
Siderite (chalybite), 427, 435-436, 436, 603
Sideronatrite, 590, 651
Siderophyllite, 1447, 1458, 1461-1462
Siderotil, 604, 606
Sidorenkite, 957, 958
Sidwillite, 253
Siegenite, 98, 100
Sieleckiite, 955
Sigloite, 937, 943-944
Silhydrite, 1594
Silicon, 35
Silinaite, 1565-1566
Sillénite, 233
Sillimanite, 1064-1065, 1066, 1069, 1184
Silver, 2, 3-4, 28, 65, 378-379
Silver mercury iodide (tocornalite), 378-379
Simferite, 734
Simonellite (1,1-dimethyl-7-isopropyl-1,2,3,4-tetrahydrophenanthrene), 1015
Simonite, 195
Simonkolleite, 401
Simplotite, 991-992
Simpsonite, 370
Sincosite, 948-949
Sinhalite, 534
Sinjarite, 385
Sinkankasite, 898
Sinnerite, 186
Sinoite, 36
Siserskite (osmium), 18, 19
Sismondine (magnesiochloritoid-M), 1087, 1089
Sitinakite, 1105
Sjögrenite, 507, 509, 510, 516, 522
Sjögrenite-hydrotalcite group, 507-512, 516
Skinnerite, 172
Skippenite, 110, 111
Skleroklas (sartorite), 193

Sklodowskite, 1109, *1111-1112*
Skutterudite (smaltite), *139*
Slate spar (calcite), 427, *428-434*
Slavíkite, *657*
Slawsonite, *1615-1616*
Smaltite (skutterudite), *139*
Smectite group, 1480-1495
Smirnite, *679*
Smithite, *191*
Smithsonite, 427, *438-439*
Smolianinovite, *812*
Smythite, *75*
Soapstone, 1441
Sobolevskite, 76, *78*
Sobotkite, 1481, *1488-1489*
Soda alum (sodium alum), *593*
Sodalite, 50, 1624, *1625-1626*, 1630
Sodalite group, 1623-1630
Soddyite, *1115*
Sodium alum (soda alum), *593*
Sodium-anthophyllite, 1337, *1343-1344*
Sodium autunite, 764, 766, *769-770*
Sodium betpakdalite, *1009*
Sodium bicarbonate (nahcolite), *422-423*
Sodium-boltwoodite, 1109, *1113*
Sodium cobalt thiocyanate (julienite), *1014*
Sodium dachiardite, *1691-1692*
Sodium-gedrite, 1337, *1345*
Sodium hydromica (brammallite), *1477*
Sodium illite (brammallite), *1477*
Sodium pharmacosiderite, 908, *910*
Sodium phlogopite, 1447, *1467*
Sodium uranospinite, 765, *770*
Sodium zippeite, 660, *661*
Sogdianite, 1271, *1278*, 1374
Söhngeite, *283*
Solongoite, *545*
Solonite, 1078, *1085-1086*
Sonoraite, *687*
Sopcheite, *146*
Sophiite, *686-687*
Sorbyite, *185-186*
Sorensenite, *1322*
Sosedkoite, *370-371*
Soucekite, *168*
Souesite (awaruite), 10, *12*
Souzalite, *914-915*
Sovolevite, 1185, *1186-1187*
Spadaite, *1718*
Spangolite, *640*
Specularite (hematite), 214, *217-220*
Spenceite (tritomite-(Y)), 1130, *1131*
Spencerite, *893*
Sperrylite, 114, *118*
Spertiniite, *277*
Spessartine, 1037, 1039, *1041-1042*, 1043
Sphaerobertrandite, 1149
Sphaerobismoite, *233-234*
Sphaerocobaltite (cobaltcalcite), 427, *438*
Sphalerite, 6, 55, 65, 67, *68-69*, 70, 72, 80, 243, 613, 803

Sphalerite group, 67-70
Sphene (titanite), *1092-1094*, 1095
Spheniscidite, 932, *933*
Spinel, 209, 220, 221, 227, 228, 293, 294, *295-297*
Spinel group, 293-304
Spionkopite, 46, *49*
Spiroffite, *684*
Spodiosite, *852*
Spodumene (α-apodumene, triphane), 1291, *1308-1309*, 1594
α-Spurrite (spurrite), *1106*, 1107
β-Spurrite (spurrite), *1106*, 1107
Spurrite (β-spurrite, α-spurrite), *1106*, 1107
Squawcreekite, 235, *243*
Srebrodol'skite, *327*
Srilankite, *250*
Stalderite, *173*
Stanfieldite, *717*
Stanleyite, *614*
Stannite, *89-90*, *92*, 103
Stannite group, 89-91
Stannoidite, 92
Stannomicrolite (sukulaite), 342, *347*
Stannopalladinite, *25*
Starkeyite, 602, *603*
Staurolite, 1064, *1071-1073*
Steacyite (desourdyite), *1269*, 1521. See also Ekanite
Steatite, 1441
Steenstrupine-(Ce), *1265*
Steigerite, *805*
Stellerite, *1675-1676*
Stenhuggarite, *981*
Stenonite, *419*
Stepanovite, *1013*
Stephanite, *157*
Stercorite, *743*
Sterlinghillite, *790-791*
Sternbergite, *95*
Sterryite, *179*
Stetefeldite, *972*
Stevensite, 1481, 1491, *1493-1494*
Stewartite, 937, *940-941*
Stibarsen, 26, *28*
Stibiconite, *971*
Stibiobetafite, 342, *348*
Stibiocolumbite, *334-335*
Stibiocolusite, 150, *151*
Stibiomicrolite, 342, *347*
Stibiopalladinite, *41*
Stibiotantalite, *335*
Stibium (stibnite), *106-107*
Stibivanite-2M, *979*
Stibivanite-2O, *979*
Stibnite group, 105-108
Stibnite (stibium), *106-107*, 140, 251
Stichtite, 508, 509, *510*
Stilbite-(Ca), *1674-1675*
Stilbite-(Na), 1674
Stilleite, 67, *69*

Stillwaterite, *144-145*
Stillwellite-(Ce), 1121, *1129*
Stilpnomelane, *1552-1553*, 1554, 1555
Stinkcalc (calcite), 427, *428-434*
Stishovite, 235, *243-244*, 1570, 1573, 1586
Stistaite, *29*
Stoiberite, *823-824*
Stokesite, *1331-1332*
Stolzite, 999, *1000-1001*
Stottite, 284, *286*
Straczekite, *988*
Strakhovite, *1286*
Stranskiite, *718*, 719
Strashimirite, *893-894*
Strätlingite, *1700*
Stratopeite (neotocite), 1435, *1436-1437*
Strelkinite, 780, *783*
Strengite, 800, *801-802*, 804
Stringhamite, *1705*
Stromeyerite, *45*
Stronalsite, *1616*
Strontianite, 441, 442, *445-446*, 959
Strontioborite, *549*
Strontiochevkinite, 1177, *1179*
Strontiodresserite, *501*
Strontioginorite, *562*
Strontiojoaquinite, 1231, *1232*
Strontio-orthojoaquinite, 1231, *1232-1233*
Strontiopiemontite, 1196, *1203*
Strontiopyrochlore, 342, *345*
Strontiowhitlockite, 715, *716*
Strontium-apatite, 856, 862, *863*
Strunzite, *938-939*
Strunzite subgroup, 938-940
Strüverite, 235, *237-238*
Struvite, *747-748*
Studenitsite, *564*
Studtite, *258*
Stumpflite, 76, *78*
Sturmanite, 658, *670*
Sturtite (hisingerite), 1432, *1434-1435*
Stützite, *146-147*
Suanite, *540*
Subugalite, 764
Sudburyite, 76, *78*
Sudoite, 1498, *1500-1501*
Suessite, *16*
Sugilite, 1271, *1275-1276*
Sukulaite (stannomicrolite), 342, *347*
Sulfatic cancrinite (vishnevite), 1633, *1640*
Sulfoborite, *566*
Sulfur, *30-31*
Sulphohalite, *624-625*
Sulphotsumoite, 82, *83*
Sulvanite, *157*
Sundiusite, *623*
Suolunite, 465, 1160, *1168*
Surinamite, *1393*
Surite, *1479-1480*
Sursassite, *1212*, 1217
Susannite, *520-521*

Sussexite, *544*
Suzukiite, 1322, *1323*
Svabite, 856, *864*, 865
Svanbergite, 960, *962*
Sveite, *528*
Sverigeite, *1024*
Svetlozarite, 1690
Svyatoslavite, *1617*
Svyazhinite, *655*
Swaknoite, 743, *744*
Swamboite, 1109, *1113*
Swartzite, *469*
Swedenborgite, 211, *973*
Sweetite, *278-279*
Swelling chlorite (corrensite), *1512-1514*
Swinefordite, 1481, *1487-1488*
Switzerite, 790, *791*
Sylvanite, *135*
Sylvite, 373, *375*
Symplesite, 792, 796, *797*, 798, 922
Synadelphite, *968-969*
Synchysite-(Ce), 476, *479*, 480, 481
Synchysite group, 475, 476, 479-480, 481-482, 496
Synchysite-(Nd), 476, *480*
Synchysite-(Y) (doverite), 476, *479*
Syngenite, *584-585*
Szaibelyite, *544*
Szenicsite, *1003*
Szmikite, 595, *596*
Szomolnokite, 595, *596*
Szymanskiite, *518*

T

Taaffeite, *309*
Tacharanite, 1528, *1533-1534*
Tachyhydrite, *411*
Tadzhikite-(Y), 1121, 1127, *1128*
Taeniolite, 1447, *1464-1465*
Taenite, 1, 9, *10*, 11
Taikanite, *1324*
Taimyrite, *25*
Takanelite, *325*
Takedaite, *540*
Takéuchiite, 537, *538*
Takovite, 508, *511-512*
Talc, 1438
Talc-1A, 1439, *1441-1443*
Talc group. See Pyrophyllite talc group
Tallow clay (sauconite), 1481, *1490-1491*, 1495
Talmessite, 751, *753*
Talnakhite, *94*
Tamarugite, *592*
Tancoite, *824*
Taneyamalite, *1394*
Tangeite (calciovolborthite), 826, *828*
Tantalaeschynite-(Y), *358-359*
Tantalian rutile, 235, 238
Tantalite, 338, 339
Tanteuxenite-(Y), *360-361*
Tantite, *254*

Tanzanite, 1204
Tapiolite (ferrotapiolite), *351-352*
Taramellite, *1284-1285*
Taranakite, *744-745*
Tarapacaite, *692*
Tarbuttite, *851*
Tatarskite, *669*
Tausonite, 222, *224-225*
Tavorite, *838*
Tawmawite, 1200
Taylorite (**arcanite**), *569*
Tazheranite, *249-250*
Teallite, *95*
Teepleite, *543*
Teineite, *680*
Telargpalite, *37*
Tellurantimony, 110, *111*
Tellurite, 244, *247*
Tellurium, *30*
Tellurobismuthite, *110-111*
Tellurohauchecornite, 158, *159*
Telluropalladinite, *145*
Temagamite, *60*
Tengchongite, *1007*
Tengerite-(Y), *473*, 500
Tennantite, 162, 163, *164*, 718
Tenorite (melaconite), *211-212*
Tephroite, 1025, *1029-1031*
Terlinguaite, 213, *398*
Terskite, *1718-1719*
Tertschite, *563*
Teruggite, *566*
Teschemacherite, *423*
Testibiopalladite, 125, *128*
Tetraauricupride, *6*
Tetradymite, *110*
Tetradymite group, 109-111
Tetraferroplatinum, *21-22*
Tetraferroplatinum group, 6, 21-22
Tetrahedite, *162-163*, 164
Tetrahedite group, 162-165
Tetramatrolite, 1677, *1679-1680*
Tetrarooseveltite, *725*
Tetrataenite, *10*, 12
Tetrawickmanite, 284, *286*
Thadeuite, *843*
Thalcusite, *54*
Thalénite-(Y), *1137*
Thalfenisite, *143*
Thaumasite, 658, 670, *671*
Theisite, *816*
Thenardite, *570*
Theophrasite, 275, *276-277*
Theresemagnanite, *642*
Thermonatrite, 448, *462*, 463, 820
Thometzekite, 760, *761*
Thomsenolite, 413, *414*
Thomsonite, *1687-1688*
Thorbastnäsite, *499*
Thoreaulite, *361-362*
Thorianite, 249, *256-257*

Thorikosite, *395*
Thorite, 499, 1053, *1056-1057*, 1058
Thornasite, *1521-1522*
Thorogummite, 1053, *1058*
Thorosteenstrupine, *1265-1266*
Thortveitite, 798, 1134, 1135, *1136*, 1137
Thortveitite group, 1133-1136, 1137
Thorutite, *357*
Threadgoldite, *942-943*
Thulite, 1204
Tiemannite, 67, *69-70*
Tienshanite, *1286*
Tiettaite, *1713*
Tikhonenkovite, *417*
Tilasite, 826, 834, *835*
Tilasite group, 833-836, 1092
Tilleyite, *1180-1181*
Tin, *13*, 29, 212
Tinaksite, *1377*
Tincalconite, 543, *555*, 556
Tinsleyite, *932-933*
Tinticite, *952*
Tintinaite, *187*
Tinzenite (severginite), 1154, *1156-1157*
Tiptopite, *896*
Tiragalloite, *1190*
Tirodite, 1337, *1342*
Tisinalite, 1246, *1248*
Titanite (sphene), *1092-1094*, 1095
Titan-låvenite, 1162
Titanoelpidite, 1235, 1543
Titanonenadkevichite, 1234
Titanowodginite, 337, *338*
Titantaramellite, 1284, *1285*
Tivanite, *278*
Tlalocite, *675*
Tlapallite, *689-690*
Tobelite, 1446, *1455*
Tobermorite, 1528, *1529-1530*, 1533
Tobermorite group, 1528-1535
Tochilinite, *141*
Tocornalite (**capgarronite**), *144*
Tocornalite (silver mercury iodide), *378-379*
Todorokite, 317, *318-319*
Tokkoite, *1377-1378*
Tolbachite, *387*
Tolovkite, 124, *126*
Tombarthite-(Y), *1059*
Tomichite, *982*
Tongbaite, *14*
Tooeleite, *952*
Topaz, *1073-1076*, 1183
Topazolite, 1044
Torbernite, 764, 766, *773*, 774
Törnebohmite-(Ce), *1096*
Törnebohmite-(La), *1096-1097*
Torreyite, 639, *640*
Tosalite, 1421
Tosudite (alushtite), *1510-1512*
Tounkite, 1633, *1639*
Tourmaline group, 1254-1262

Toyohaite, *103-104*
Trabzonite, *1189*
Tranquillityite, *1719*
Traskite, *1286-1287*
Treasurite, *183*
Trechmannite, *191*
Trembathite, 547, *549*
Tremolite, 1336, 1338, *1345*
Tremolite subgroup, 1338, 1346-1357. *See also*
  Amphibole group
Trevorite, 294, *300-301*
Triangulite, *785*
Tridymite, 243, *1570-1572*, 1573, 1586, 1587, 1593
Trigonite, *977*
Trikalsilite, 1619, *1621*
Trilithionite, 1462
Trimerite, *1023*
Trimounsite-(Y), *1099-1100*
Triphane (**spodumene**), 1291, *1308-1309*, 1594
Triphylite, *700-701*, 703, 714, 715
Triplite, 844, *845-846*, 847
Triplite-triploidite group, 844-848
Triploidite, 845, *847*
Trippkeite, *978*
Tripuhyite, *973*
Tristramite, 807, *808*
Tritomite-(Ce) (tritomite), *1130-1131*, 1131, 1137
Tritomite (**tritomite-(Ce)**), *1130-1131*, 1131
Tritomite-(Y) (spenceite), 1130, *1131*
Tri-tri-clinochlore. *See* Clinochlore
Trögerite, 764, 765, 774, *777*
Trogtalite, 114, *116*
Troilite, *74*, 101
Trolleite, *880*
Trona, *423-424*, 462
Troostite (**willemite**), 1020, *1022*
Truscottite, *1549*, 1550
Trüstedtite, 98, *101*
Tsargorotsevite, 1624, *1629-1630*
Tsavorite, 1046, 1047
Tschermigite (ammonia alum), *593*
Tschernichite, *1697-1698*
Tsnigriite, *155*
Tsumcorite, *760*, 761
Tsumebite, 758, 877, *963*
Tsumoite, *82-83*
Tsumoite group, 82-84
Tucekite, 158, *159*
Tugarinovite, *250-251*
Tugtupite, 1624, *1629*
Tuhualite, 1373, *1374*
Tuhualite group, 1373-1375
Tulameenite, 21, *22*
Tuliokite, *474*
Tundrite-(Ce), *523*
Tundrite-(Nd), *523-524*
Tunellite, *562*
Tungstenite, *134*
Tungstibite, *1002*

Tungstite, *252*
Tungusite, *1711*
Tunisite, *494-495*
Tuperssuatsiaite, 558, 1558, *1559-1560*
Turanite, *823*
Turneaureite, 856, *864-865*
Turquoise, 916, *917-918*, 919
Turquoise group, 915-920
Tuscanite, 1546, *1547*
Tusionite, *540*
Tuzlaite, *559*
Tvaichrelidzeite, *197*
Tvedalite, 1402, *1403*
Tveitite-(Y), *384*
Twinnite, *193-194*
Tychite, *519*
Tyretskite, 560
Tyrolite, *885*
Tyrrellite, 98, *100*
Tysonite (**fluocerite-(Ce)**), *389*
Tyuyamunite, 780, *781*, 784

U

Uchucchacuaite, 175, *177*
Uhligite, *369*
Uklonskovite, *649*
Ulexite, *558*
Ullmannite, 124, *125-126*
Ulrichite, *785*
Ulvospinel, 295, *304*
Umangite, *52*
Umbite, *1224-1225*
Umbozerite, *1719*
Umohoite, *1006*
Ungarettiite, 1339, *1367*
Ungemachite, 649, *667*
Upalite, *907*
Uralborite, *545*
Uralolite, *903*
Uramphite, 765, *770*
Urancalcarite, *506*
Uraninite, 4, 249, *255-256*, 500, 512, 651, 766
Uranmicrolite (djalmaite), 342, *346-347*
Uranocircite, 764, *768*
Uranophane group, 1108-1114
Uranophane-α (**uranophane**), 1109, *1111*, 1113
α-Uranophane (**uranophane**), 1109, *1111*, 1113
β-uranophane (β-uranotile, uranophane-β), 1109, *1113*
Uranophane (uranotile, α-uranophane, uranophane-α), 1109, *1111*, 1113
Uranopilite, *642*
Uranopolycrase, *361*
Uranosilite, *1115*
Uranospathite, 764, 773, *778*, 779
Uranospharite, *263*
Uranospinite, 764, *767*, 768, 770, 774
Uranotile (**uranophane**), 1109, *1111*, 1113
Uranotungstite, *1008*
Uranpyrochlore (hatchettolite), 342, *345*
Urea, *1018*

General Index

Ureyite (**kosmochlor**), 1291, *1307*
Uric acid (**uricite**), *1017*
Uricite (uric acid), *1017*
Urvantsevite, *139*
Ushkovite, 937, *941*
Usovite, *416*
Ussingite, *1644-1645*
Ustarasite, *201*
Uvanite, *991*
Uvarovite, 220, 1037, 1044, 1046, *1048*
Uvite, 1256, *1258-1259*
Uytenbogaardtite, *43*
Uzonite, *85*

V

Vaesite, 114, *115*
Valencianite, 1600
Valentinite, 231, *233*
Valleriite, *141*
Vanadinite, 213, 830, 857, *867*
Vanadomalayaite, *1095*
Vanalite, *990-991*
Vandenbrandeite, *259*
Vandendriesscheite, *262*
Vanmeersscheite, 888, *907*
Vanoxite, *991*
Vanquelinite, 758
Vantasselite, *951*
Vanthoffite, *578*
Vanuralite, 784, *942*
Vanuranylite, 780, *784*
Varennesite, *1544-1545*
Variamoffite, *244*
Variscite, *800-801*
Variscite group, 799-804
Varulite, 706, *709*
Vashegyite, *951*
Vasilite, *149*
Vaterite, *440*, 442
Vaughanite, *202*
Vauquelinite, *963*
Vauxite, *943*
Väyrynenite, *833*
Veatchite, *558*
Veatchite-A, *558*
Veenite, *181*
Velardeñite (**melilite**), 1142, *1144-1145*
Velikite, 89, *91*
Vermiculite (hydromica, jefferisite,hydrobiotite), *1471-1472, 1474-1476*
Vernadite, *248-249*
Verplanckite, *1701*
Versiliaite, *979*
Vertumnite, *1548*
Vesignieite, *842*
Vesuvianite, 221
Vesuvianite (idocrase), *1213-1215*
Veszelyite, *883*
Vicanite-(Ce), *1132*
Vigezzite, *359*

Viitaniemiite, *820*
Vikingite, *185*
Villamaninite, 114, *116*
Villiaumite, 373, *375-376*
Villyaellenite, *740*
Vimsite, *543*
Vincentite, *40*
Vinciennite, *151*
Vinogradovite, *1379-1380*
Violarite, 98, *101*
Virgilite, *1594*
Viséite, *968*
Vishnevite (sulfatic cancrinite), 1633, 1634, *1640*
Vismirnovite, 284, *285*
Vistepite, *1705-1706*
Vitusite-Ce, *704-705*
Vitusite-Nd, 704
Vivianite, 792, *793-794*, 796, 798, 917
Vivianite group, 791-798
Vladimirite, *740*
Vlasovite, *1369*
Vochtenite, *950-951*
Voggite, *969-970*
Voglite, *471*
Volborthite, *798-799*, 828
Volchonskoite (**volkonskoite**), 1481, *1486-1487*
Volkonskoite (wolchonskoite, volchonskoite), 1481, *1486-1487*
Volkovskite, *560*
Voltaite, 591, *620-621*
Volynskite, *190*
Vonbezingite, *646*
Vonsenite, 535, *536*
Vozhminite, *42*
Vrbaite, *196*
α-Vredenbergite (**iwakiite**), *305*
Vuagnatite, 1091, *1092*
Vulcanite, *80*
Vuonnemite, *1187*
Vuorelainenite, 294, *303*
Vyacheslavite, *809*
Vyalsovite, *142*
Vysotskite, *71*
Vyuntspakhkite-(Y), *1719-1720*

W

Wadalite, *1052*
Wadeite, 1221, *1223*
Wadsleyite, 1034, *1140*
Wagnerite, 845, *846-847*
Wairakite, 1646, 1647, 1648, *1649-1650*
Wairauite, *12*
Wakabayashilite, *108*
Wakefieldite-(Ce) (kusuite), 728, *731*
Wakefieldite-(Y), 728, *731*
Walentaite, *813*
Wallastonite-1A, 1314, *1315-1316*, 1317
Wallastonite-3A, 1314, *1317*
Wallastonite-4A, 1314, *1317*
Wallastonite-5A, 1314, *1317*

Wallastonite-7A, 1314, *1317*
Wallastonite group, 1314-1321
Wallastonite-2M (parawollastonite), 1314, 1315, *1317*
Wallisite, *180*
Wallkilldellite, *889*, 1091
Walpurgite, *813*
Walstromite, *1224*
Walthierite, 631, *633*
Wapplerite (rösslerite), *738-739*
Wardite, *905*
Wardsmithite, *564*
Warikahnite, *785*
Warwickite, *537*
Watanabeite, *182-183*
Watkinsonite, *197*
Wattersite, *694*
Wattevillite, *586*
Wavellite, *922-923*
Wawayandaite, *1720*
Waylandite, 839, *841*
Weberite, *416*
Weddellite, *1011*
Weeksite, *1114*
Wegscheiderite, *424*
Weibullite, *187*
Weilerite, 960, *962-963*
Weilite, *697*
Weinebenite, *903*
Weinschenkite (churchite-(Y)), *805-806*
Weishanite, *8*
Weissbergite, *193*
Weissite, *49*
Welinite, *310*, 1090-1091
Wellsite, 1665
Weloganite, 469, *470-471*
Welshite, 1386, *1390*
Wendwilsonite, 750, *755-756*
Wenkite, *1640-1641*
Werdingite, *1184*
Wermlandite, *643*
Westerveldite, *82*
Westgrenite (bismutomicrolite), 342, *347*, 349
Wheatleyite, *1013*
Wherryite, *668*
Whewellite, *1011*
Whiteite-(CaFeMg), 929, *930-931*
Whiteite-(CaMnMg), 929, *930-931*
Whiteite-(MnFeMg), 929, *930-931*
Whiteite subgroup, 927, 929, 930-931. *See also* Jahnsite subgroup
Whitlockite, *715-716*, 717
Whitmoreite, 946, *947*, 948
Whitmoreite group, 946-948
Whitneyite, 4
Wickenburgite, *1556*
Wickmanite, *284-285*, 286
Wickmanite group, 283-287
Wicksite, *761-762*
Widenmannite, *447-448*
Widgiemoolthaite, *514*

Wightmanite, *551*
Wilcoxite, *655*
Wilhelmvierlingite, *926-927*
Wilkeite, 209
Wilkinsonite, 1386, *1391*
Wilkmanite, *103*
Willemite (troostite), 211, 996, 1020, *1022*
Willemseite, 1439, *1443*
Willhendersonite, *1655*
Williamsite (antigorite), 1406, *1415-1417*
Willyamite, 124, *126*
Winebergite (hydrobasaluminite), *645*
Winstanleyite, *685*
Wiserite, *567*
Witherite, 441, 442, *444-445*, 572
Wittichenite, *172*
Wittite, *184*
Wodginite, *337-338*
Wodginite group, 336-338
Wöhlerite, 1160, *1163-1164*
Wolchonskoite (volkonskoite), 1481, *1486-1487*
Wolfeite, 845, *847*, 848
Wolframite, 20, 252, 340
Wolframite group, 993-996
Wollastonite, 1533, 1550
Wolsdendorfite, *260*
Wonesite, 1447, *1467*
Woodhouseite, 960, *962*
Woodruffite, *319*
Woodwardite, *641*
Wroewolfeite, *644*
Wulfenite, *999-1000*
Wülfingite, 278, *279*
Wupatkiite, 616, *618*
Wurtzite, *72-73*, 80
Wurtzite group, 72-73
Wüstite, 208, *210*
Wyartite, *515*
Wycheproofite, 810, *811*
Wyllieite, 707, *711*
Wyllieite subgroup, 706, 707, 711-712. *See also* Alluaudite group

X

Xanthiosite, *714*, 734
Xanthoconite, *166-167*
Xanthophyllite (clintonite), 1447, *1469*
Xanthoxenite, *944-945*
Xenotime group, 728-731
Xenotime-(Y), 728, *729-730*, 1053
Xiangjiangite, *967*
Xifengite, *16*
Xilingolite, 175, *178*
Ximengite, *727*
Xingzhongite, *70*
Xitieshanite, *654*
Xocomecatlite, *672*
Xonotlite, 1189, *1369-1370*, 1533

General Index

Y

Yafsoanite, *310*
Yagiite, 1271, *1276*
Yakhontovite, 1481, *1495*
Yamatoite, 1049
Yanomamite, 800, *803*
Yaroslavite, *416-417*
Yarrowite, *63*
Yavapaiite, *578*
Yberisilite (**hingganite-(Y)**), 1120, 1122, *1123*
Yeatmanite, *974*
Yecoraite, *691*
Yedlinite, *403*
Ye'elimite, *627*
Yftisite-(Y), *1096*
Yimengite, *312*
Yingjiangite, *896*
Yoderite, 1064, *1068-1069*
Yofortierite, 558, 1558, *1560*
Yoshimuraite, *1216*
Yoshiokaite, 1619, *1622*
Yttrialite-(Y), 1137, *1138*
Yttrobetafite-(Y), 342, *348*
Yttroceberysite (**hingganite-(Y)**), 1120, 1122, *1123*
Yttrocolumbite-(Y), *334*
Yttrocrasite-(Y), *361*
Yttropyrochlore-(Y) (obruchevite), 342, *344*
Yttrotantalite-(Y), *334*
Yttrotungstite-(Y), *1007-1008*
Yuanfulite, *537*
Yuanjiangite, *6*
Yugawaralite, *1694-1695*
Yukonite, *913*
Yuksporite, *1371-1372*
Yushkinite, *141-142*

Z

Zabuyelite, *448*
Zaherite, *659*
Zairite, 820, 839, *841-842*
Zajacite-(Ce), *412*
Zakharovite, *1706-1707*
Zanazziite, *904-905*
Zapatalite, *891*
Zaratite, 464, *504*
Zavaritskite, *393*
Zdenekite, *921*
Zeagonite, 1661
Zektzerite, 1373, *1374*
Zellerite, *467*
Zemannite, *682*, 683
Zemkorite, *456*
Zenzénite, *315*
Zeolite group, 4, 1646-1701
Zeophyllite, *1547-1548*
Zeunerite, 764, 773, *774*
Zhanghengite, *7*
Zharchikhite, *395*
Zhemchuzhnikovite, *1013*
Zhonghuacerite-(Ce), *482*
Ziesite, *733*
Zimbabweite, *372*
Zinc, *6-7*
Zincaluminite, *641*
Zincite, *211*
Zinc-melanterite, 610, *612*
Zincobotryogen, *653*
Zincochromite, 294, *303*
Zincocopiaite, 663, *664*
Zincovoltaite, *621*
Zincrosasite, 487, *488*
Zincroselite, 750, *755*
Zincsilite, 1481, *1495*
Zinc zippeite, 660, *661*
Zinkenite, *197*
Zinnwaldite, 1447, *1465-1466*
Zippeite, *660*
Zippeite group, 660-662
Zircon (cyrtolite, hyacinth, malacon), 729, *1053-1056*
Zircon group, 1053-1058
Zirconolite-3O, *349-350*
Zirconolite-2M (zirkelite), *350*
Zirconolite (polymignite), *350*
Zirconolite-3T (zirkelite), *350*
Zircophyllite, 1381, *1384*
Zircosulfate, *621*
Zirkelite, *349*
Zirkelite (zirconolite-2M), *350*
Zirkelite (zirconolite-3T), *350*
Zirklerite, *404*
Zirsinalite, 1246, *1248*
Znucalite, *518*
Zodacite, 935, *936*
Zoisite, 1196, 1198, *1203-1205*
Zorite, *1370*
Zoubekite, *182*
Zunyite, *1191-1192*
Zussmanite, *1555-1556*
Zvyagintsevite, 23, *24*
Zweiselite, *845*
Zykaite, *967*